Contents

ASM Handbook®

Formerly Tenth Edition, Metals Handbook

Volume 2
Properties and Selection: Nonferrous Alloys and Special-Purpose Materials

ASM INTERNATIONAL®
The Materials
Information Society

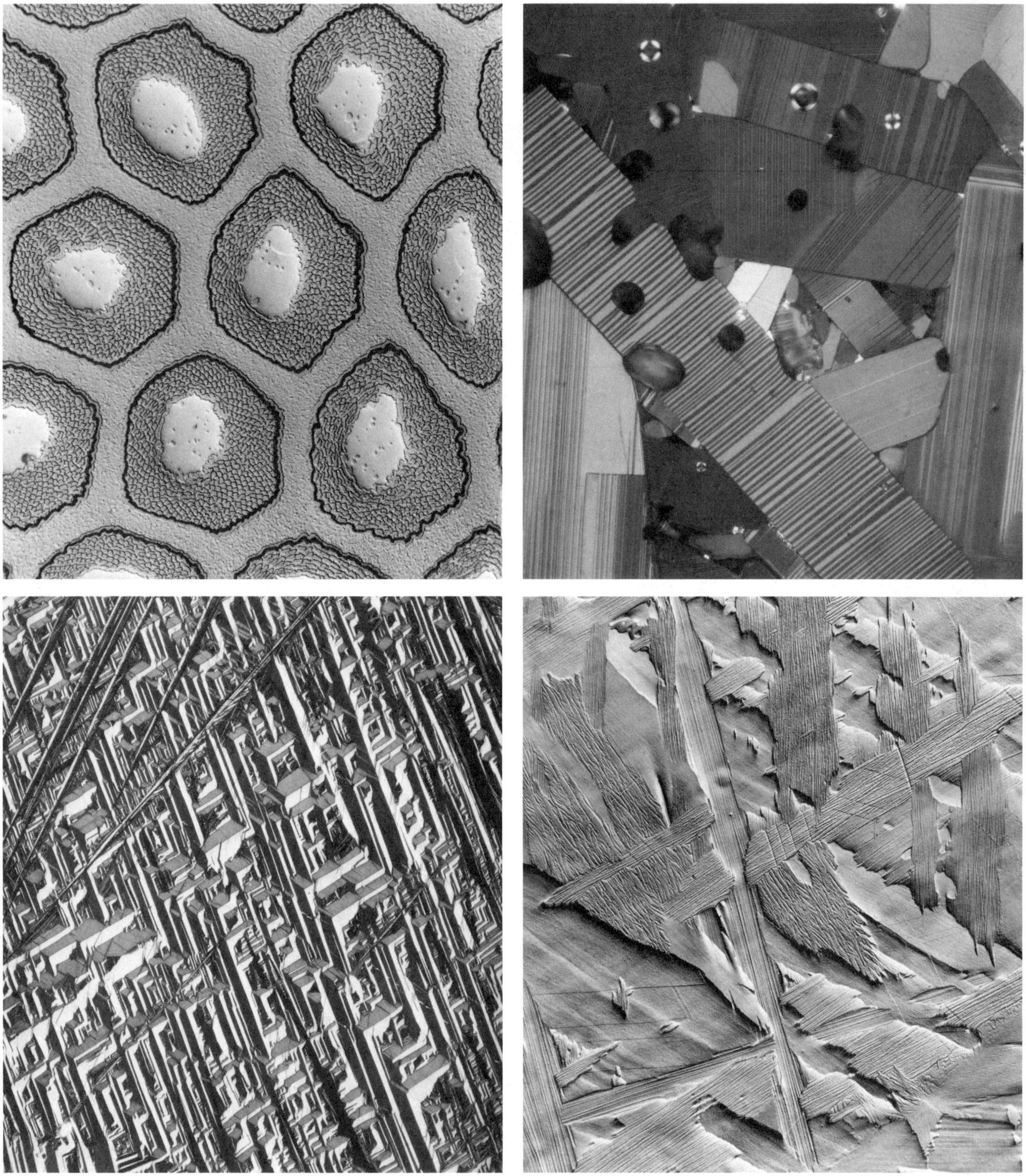

Examples of some of the many nonferrous alloys and special-purpose materials described in this Volume. Shown clockwise from the upper left-hand corner are: (1) a cross-section of a multifilament Nb_3Sn superconducting wire, 1000×; (2) a high-temperature ceramic $YBa_2Cu_3O_{7-x}$ superconductor, 600×; (3) beta martensite in a cast Cu-12Al alloy, 100×; and (4) alpha platelet colonies in a Zr-Hf plate, 400×. Courtesy of Paul E. Danielson, Teledyne Wah Chang Albany (micrographs 1 and 4) and George F. Vander Voort, Carpenter Technology Corporation (micrographs 2 and 3).

ASM Handbook®

Formerly Tenth Edition, Metals Handbook

Volume 2
Properties and Selection: Nonferrous Alloys and Special-Purpose Materials

Prepared under the direction of the
ASM International Handbook Committee

Joseph R. Davis, Manager of Handbook Development
Penelope Allen, Manager of Handbook Production
Steven R. Lampman, Technical Editor
Theodore B. Zorc, Technical Editor
Scott D. Henry, Assistant Editor
Janice L. Daquila, Assistant Editor
Alice W. Ronke, Assistant Editor
Janet Jakel, Word Processing Specialist
Karen Lynn O'Keefe, Word Processing Specialist

Robert L. Stedfeld, Director of Reference Publications

Editorial Assistance
Lois A. Abel
Robert T. Kiepura
Penelope Thomas
Heather F. Lampman
Nikki D. Wheaton

ASM INTERNATIONAL®
The Materials Information Society

First printing, October 1990
Second printing, January 1992
Third printing, November 1993
Fourth printing, October 1995
Fifth printing, January 1998
Sixth printing, December 2000

ASM Handbook is a collective effort involving thousands of technical specialists. It brings together in one book a wealth of information from world-wide sources to help scientists, engineers, and technicians solve current and long-range problems.

Great care is taken in the compilation and production of this Volume, but it should be made clear that no warranties, express or implied, are given in connection with the accuracy or completeness of this publication, and no responsibility can be taken for any claims that may arise.

Nothing contained in the ASM Handbook shall be construed as a grant of any right of manufacture, sale, use, or reproduction, in connection with any method, process, apparatus, product, composition, or system, whether or not covered by letters patent, copyright, or trademark, and nothing contained in the ASM Handbook shall be construed as a defense against any alleged infringement of letters patent, copyright, or trademark, or as a defense against liability for such infringement.

Comments, criticisms, and suggestions are invited, and should be forwarded to ASM International.

Library of Congress Cataloging-in-Publication Data

ASM International

Metals handbook.

Vol. 2: Prepared under the direction of
the ASM International Handbook Committee.
Includes bibliographies and indexes.
Contents: v. 2. Properties and selection—
nonferrous alloys and special-purpose materials.

1. Metals—Handbooks, manuals, etc.
I. ASM International. Handbook Committee.
TA459.M43 1990 620.1'6 90-115
ISBN 0-87170-378-5 (v. 2)
SAN 204-7586

Printed in the United States of America

Foreword to the Second Printing

With the second printing of this Volume, it takes its place in the new *ASM Handbook* series. The *ASM Handbook* was established to build upon the proud tradition of the *Metals Handbook* and to position the series to meet the needs of future engineers, researchers, technicians, and students. The *ASM Handbook* series will encompass volumes from both the 9th and 10th Editions of *Metals Handbook*—as well as new and revised Volumes as they are released—in order to establish one comprehensive set of reference materials. This will allow a much more flexible approach to updating information: technological advances will be the impetus behind Volume revisions or the addition of new Volumes to the series. The title of the new *ASM Handbook* series reflects the increasingly interrelated nature of materials technology and emphasizes the position of ASM International as the premier source of authoritative materials information.

The Editors

Foreword to the First Printing

Throughout the history of *Metals Handbook,* the amount of coverage accorded nonferrous alloys, special-purpose materials, and pure metals has steadily, if not dramatically, increased. That this trend has continued into the current 10th Edition is easily justified when one considers the significant developments that have occurred in the past decade. For example, metal-matrix composites, superconducting materials, and intermetallic alloys—materials described in detail in the present volume—were either laboratory curiosities or, in the case of high-temperature superconductors, not yet discovered when the 9th Edition Volume on this topic was published 10 years ago. Today, such materials are the focus of intensive research efforts and are considered commercially viable for a wide range of applications. In fact, the development of these new materials, combined with refinements and improvements in existing alloy systems, will ensure the competitive status of the metals industry for many years to come.

Publication of this Volume is also significant in that it marks the completion of a two-volume set on properties and selection of metals that serves as the foundation for the remainder of the 10th Edition. Exhaustive in scope, yet practical in approach, these companion volumes provide engineers with a reliable and authoritative reference that should prove a useful resource during critical materials selection decision-making.

On behalf of ASM International, we would like to extend our sincere thanks and appreciation to the authors, reviewers, and other contributors who so generously donated their time and efforts to this Handbook project. Thanks are also due to the ASM Handbook Committee for their guidance and unfailing support and to the Handbook editorial staff for their dedication and professionalism. This unique pool of talent is to be credited with continuing the tradition of quality long associated with *Metals Handbook*.

Klaus M. Zwilsky
President
ASM International

Edward L. Langer
Managing Director
ASM International

Policy on Units of Measure

By a resolution of its Board of Trustees, ASM International has adopted the practice of publishing data in both metric and customary U.S. units of measure. In preparing this Handbook, the editors have attempted to present data in metric units based primarily on Système International d'Unités (SI), with secondary mention of the corresponding values in customary U.S. units. The decision to use SI as the primary system of units was based on the aforementioned resolution of the Board of Trustees and the widespread use of metric units throughout the world.

For the most part, numerical engineering data in the text and in tables are presented in SI-based units with the customary U.S. equivalents in parentheses (text) or adjoining columns (tables). For example, pressure, stress, and strength are shown both in SI units, which are pascals (Pa) with a suitable prefix, and in customary U.S. units, which are pounds per square inch (psi). To save space, large values of psi have been converted to kips per square inch (ksi), where 1 ksi = 1000 psi. The metric tonne (kg $\times$ 10^3) has sometimes been shown in megagrams (Mg). Some strictly scientific data are presented in SI units only.

To clarify some illustrations, only one set of units is presented on artwork. References in the accompanying text to data in the illustrations are presented in both SI-based and customary U.S. units. On graphs and charts, grids corresponding to SI-based units appear along the left and bottom edges. Where appropriate, corresponding customary U.S. units appear along the top and right edges.

Data pertaining to a specification published by a specification-writing group may be given in only the units used in that specification or in dual units, depending on the nature of the data. For example, the typical yield strength of steel sheet made to a specification written in customary U.S. units would be presented in dual units, but the sheet thickness specified in that specification might be presented only in inches.

Data obtained according to standardized test methods for which the standard recommends a particular system of units are presented in the units of that system. Wherever feasible, equivalent units are also presented. Some statistical data may also be presented in only the original units used in the analysis.

Conversions and rounding have been done in accordance with ASTM Standard E 380, with attention given to the number of significant digits in the original data. For example, an annealing temperature of 1570 °F contains three significant digits. In this case, the equivalent temperature would be given as 855 °C; the exact conversion to 854.44 °C would not be appropriate. For an invariant physical phenomenon that occurs at a precise temperature (such as the melting of pure silver), it would be appropriate to report the temperature as 961.93 °C or 1763.5 °F. In some instances (especially in tables and data compilations), temperature values in °C and °F are alternatives rather than conversions.

The policy on units of measure in this Handbook contains several exceptions to strict conformance to ASTM E 380; in each instance, the exception has been made in an effort to improve the clarity of the Handbook. The most notable exception is the use of g/cm^3 rather than kg/m^3 as the unit of measure for density (mass per unit volume).

SI practice requires that only one virgule (diagonal) appear in units formed by combination of several basic units. Therefore, all of the units preceding the virgule are in the numerator and all units following the virgule are in the denominator of the expression; no parentheses are required to prevent ambiguity.

Preface

This is the second of two volumes in the *ASM Handbook* that present information on compositions, properties, selection, and applications of metals and alloys. In the first volume, irons, steels, and superalloys were described. In the present volume, nonferrous alloys, superconducting materials, pure metals, and materials developed for use in special applications are reviewed. In addition to being vastly expanded from the coverage offered in the 9th Edition, these companion volumes document some of the more important changes and developments that have taken place in materials science during the past decade—changes that undoubtedly will continue to impact materials engineering into the 21st century.

During the 1970s and '80s, the metals industry was forced to respond to the challenges brought about by rapid advancements in composite, plastic, and ceramic technology. During this time, the use of metals in a number of key industries declined. For example, Fig. 1 shows materials selection trends in the aircraft industries. As can be seen, the use of aluminum, titanium, and other structural materials is expected to level off during the 1990s, while polymer-matrix composites, carbon-carbon composites, and ceramic-matrix composites probably will continue to see increased application. However, this increasing competition has also spurred new alloy development that will ensure that metals will remain competitive in the aerospace industry. Some of these new or improved materials and methods include:

- Ingot metallurgy aluminum-lithium alloys for airframe components that have densities 7 to 12% lower and stiffnesses 15 to 20% higher than existing high-strength aluminum alloys
- High-strength aluminum P/M alloys made by rapid solidification or mechanical alloying
- Advances in processing of titanium alloys that have resulted in improved elevated-temperature performance
- The continuing development and research of metal-matrix composites and intermetallic alloys such as Ni_3Al, Fe_3Al, and Ti_3Al

These are but four of the many new developments in nonferrous metallurgy that are documented in Volume 2's 1300 pages.

Principal Sections

Volume 2 has been organized into five major sections:

- Specific Metals and Alloys
- Special-Purpose Materials
- Superconducting Materials
- Pure Metals
- Special Engineering Topics

A total of 62 articles are contained in these sections. Of these, 31 are completely new to the *ASM Handbook* series, 8 were completely rewritten, with the remaining revised and/or expanded. A summary of the content of the major sections is given in Table 1 and discussed below. Differences between the present volume and its *Metals Handbook,* 9th Edition predecessor are highlighted.

Specific Metals and Alloys are described in 36 articles. Extensive new data have been added to all major alloys groups. For example, more than 400 pages detail processing, properties, and applications of aluminum-base and copper-base alloys. Included are new articles on "Aluminum-Lithium Alloys," "High-Strength Aluminum P/M Alloys," "Copper P/M Products," and "Beryllium-Copper and Other Beryllium-Containing Alloys." When appropriate, separate articles describing wrought, cast, and P/M product forms for the same alloys system have been provided to assist in materials selection and comparison. Articles have also been added on technologically important, but less commonly used, metals and alloys

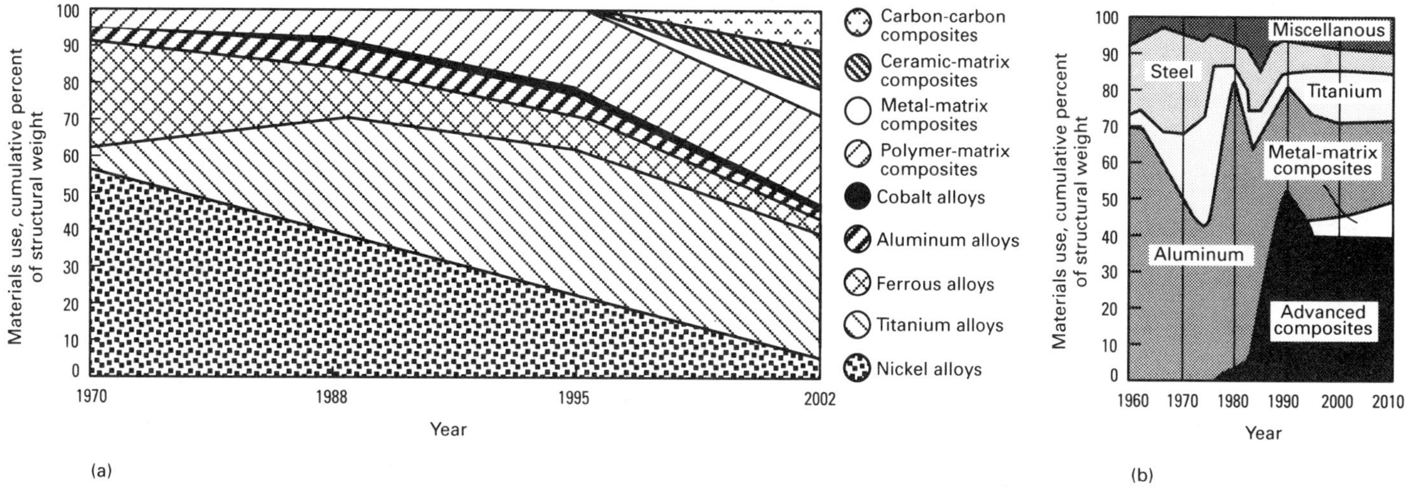

Fig. 1 Trends in materials usage for the aircraft industry. (a) Jet engine material usage. Source: Titanium Development Association and General Electric Company. (b) Airframe materials usage for naval aircraft. Source: Naval Air Development Center and Naval Air Systems Command

Table 1 Summary of contents for Volume 2, *ASM Handbook*

Section title	Number of articles	Pages	Figures(a)	Tables(b)	References
Specific Metals and Alloys	36	757	586	703	646
Special-Purpose Materials	15	265	292	142	694
Superconducting Materials	7	64	101	6	325
Pure Metals	2	111	156	230	622
Special Engineering Topics	2	67	26	21	384
Totals	**62**	**1 264**	**1 161**	**1 102**	**2 671**

(a) Total number of figure captions; some figures may include more than one illustration. (b) Does not include in-text tables or tables that are part of figures

such as beryllium, gallium and gallium arsenide (used in semiconductor devices), and rare earth metals.

Special-Purpose Materials. The 15 articles in this section, 7 of which are completely new, examine materials used for more demanding or specialized application. Alloys with outstanding magnetic and electrical properties (including rare earth magnets and metallic glasses), heat-resistant alloys, wear-resistant materials (cemented carbides, ceramics, cermets, synthetic diamond, and cubic boron nitride), alloys exhibiting unique physical characteristics (low-expansion alloys and shape memory alloys), and metal-matrix composites and advanced ordered intermetallics currently in use or under development for critical aerospace components are described.

Superconducting Materials. This is the first time that a significant body of information has been presented on superconducting materials in the *ASM Handbook*. This new section was carefully planned and structured to keep theory to a minimum and emphasize manufacture and applications of the materials used for superconductors. Following brief articles on the historical background and principles associated with superconductivity, the most widely used superconductors—niobium-titanium and A15 compounds (including Nb_3Sn)—are examined in detail. The remaining articles in the section discuss Chevrel-phase superconductors ($PbMo_6S_8$ and $SnMo_6S_8$), thin-film superconductors, and high-temperature oxide superconductors ($YBa_2Cu_3O_7$, $Bi_2Sr_2Ca_2Cu_3O_x$, and $Tl_2Ba_2Ca_2Cu_3O_x$).

Pure Metals are described in an extensive collection of data compilations that describe crystal structures, mass characteristics, as well as thermal, electrical/magnetic optical, nuclear, chemical, and mechanical properties for more than 80 elements. Also included is a review of methods used to prepare and characterize pure metals.

Special Engineering Topics. With environmental issues more important than ever, recycling behavior is becoming a key consideration for materials selection. The articles on recycling in Volume 2 cover a wide range of materials and topics—from the recycling of aluminum beverage cans to the reclaiming of precious metals from electronic scrap. Statistical information on scrap consumption and secondary recovery of metals supplements each contribution. A detailed review of the toxic effects of metals is also included in this section.

Acknowledgments

Volume 2 has proved to be one of the largest and most comprehensive volumes ever published in the 67-year history of the *ASM Handbook* (formerly *Metals Handbook*). The extensive data and breadth of information presented in this book were the result of the collective efforts of more than 400 authors, reviewers, and miscellaneous contributors. Their generous gifts of time, effort, and knowledge are greatly appreciated by ASM.

We are also indebted to the ASM Handbook Committee for their very active role in this project. Specifically, we would like to acknowledge the efforts of the following Committee members: Elwin L. Rooy, Aluminum Company of America, who organized and authored material on aluminum and aluminum alloys; William L. Mankins, Inco Alloys International, Inc., who coauthored the article "Nickel and Nickel Alloys"; Susan Housh, Dow Chemical U.S.A., who revised the articles on magnesium and magnesium alloys; Robert Barnhurst, Noranda Technology Centre, who prepared the article "Zinc and Zinc Alloys"; John B. Lambert, Fansteel Inc., who organized the committee that revised the material on refractory metals and alloys; Toni Grobstein, NASA Lewis Research Center, who contributed material on rhenium and metal-matrix composites containing tungsten fibers; and David V. Neff, Metaullic Systems, who organized the committee that prepared the article, "Recycling of Nonferrous Alloys."

Thanks to the spirit of cooperation and work ethic demonstrated by all of these individuals, a book of lasting value to the metals industry has been produced.

The Editors

Officers and Trustees of ASM International

Klaus M. Zwilsky
President and Trustee
National Materials Advisory Board
National Academy of Sciences

Stephen M. Copley
Vice President and Trustee
Illinois Institute of Technology

Richard K. Pitler
Immediate Past President and Trustee
Allegheny Ludlum Corporation
(retired)

Edward L. Langer
Secretary and Managing Director
ASM International

Robert D. Halverstadt
Treasurer
AIMe Associates

Trustees

John V. Andrews
Teledyne Allvac

Edward R. Burrell
Inco Alloys International, Inc.

H. Joseph Klein
Haynes International, Inc.
Kenneth F. Packer
Packer Engineering, Inc.
Hans Portisch
VDM Technologies Corporation
William E. Quist
Boeing Commercial Airplanes
John G. Simon
General Motors Corporation
Charles Yaker
Howmet Corporation
Daniel S. Zamborsky
Kennametal Inc.

Members of the ASM Handbook Committee (1990-1991)

Dennis D. Huffman
(Chairman 1986-; Member 1983-)
The Timken Company
Roger J. Austin (1984-)
Hydro-Lift
Roy G. Baggerly (1987-)
Kenworth Truck Company
Robert J. Barnhurst (1988-)
Noranda Technology Centre
Hans Borstell (1988-)
Grumman Aircraft Systems
Gordon Bourland (1988-)
LTV Aerospace and Defense Company
John F. Breedis (1989-)
Olin Corporation
Stephen J. Burden (1989-)
GTE Valenite
Craig V. Darragh (1989-)
The Timken Company
Gerald P. Fritzke (1988-)
Metallurgical Associates
J. Ernesto Indacochea (1987-)
University of Illinois at Chicago
John B. Lambert (1988-)
Fansteel Inc.
James C. Leslie (1988-)
Advanced Composites Products and Technology
Eli Levy (1987-)
The de Havilland Aircraft Company of Canada
William L. Mankins (1989-)
Inco Alloys International, Inc.
Arnold R. Marder (1987-)
Lehigh University
John E. Masters (1988-)
American Cyanamid Company
David V. Neff (1986-)
Metaullics Systems
David LeRoy Olson (1989-)
Colorado School of Mines
Dean E. Orr (1988-)
Orr Metallurgical Consulting Service, Inc.
Elwin L. Rooy (1989-)
Aluminum Company of America
Kenneth P. Young (1988-)
AMAX Research & Development

Previous Chairmen of the ASM Handbook Committee

R.S. Archer
(1940-1942) (Member, 1937-1942)
L.B. Case
(1931-1933) (Member, 1927-1933)
T.D. Cooper
(1984-1986) (Member, 1981-1986)
E.O. Dixon
(1952-1954) (Member, 1947-1955)
R.L. Dowdell
(1938-1939) (Member, 1935-1939)
J.P. Gill
(1937) (Member, 1934-1937)
J.D. Graham
(1966-1968) (Member, 1961-1970)
J.F. Harper
(1923-1926) (Member, 1923-1926)
C.H. Herty, Jr.
(1934-1936) (Member, 1930-1936)
J.B. Johnson
(1948-1951) (Member, 1944-1951)
L.J. Korb
(1983) (Member, 1978-1983)
R.W.E. Leiter
(1962-1963) (Member, 1955-1958, 1960-1964)
G.V. Luerssen
(1943-1947) (Member, 1942-1947)
G.N. Maniar
(1979-1980) (Member, 1974-1980)
J.L. McCall
(1982) (Member, 1977-1982)
W.J. Merten
(1927-1930) (Member, 1923-1933)
N.E. Promisel
(1955-1961) (Member, 1954-1963)
G.J. Shubat
(1973-1975) (Member, 1966-1975)
W.A. Stadtler
(1969-1972) (Member, 1962-1972)
R. Ward
(1976-1978) (Member, 1972-1978)
M.G.H. Wells
(1981) (Member, 1976-1981)
D.J. Wright
(1964-1965) (Member, 1959-1967)

Authors

J.H. Adams
Eagle-Picher Industries, Inc.
Mitchell Ammons
Martin Marietta Energy Systems
Howard S. Avery
Consulting Engineer
Robert J. Barnhurst
Noranda Technology Centre
John C. Bean
AT&T Bell Laboratories
B.J. Beaudry
Iowa State University
David F. Berry
SCM Metal Products, Inc.
William T. Black
Copper Development Association Inc.
Michael Bess
Certified Alloys, Inc.
R.J. Biermann
Harrison Alloys Inc.
Charles M. Blackmon
Naval Surface Warfare Center
Richard D. Blaugher
Intermagnetics General Corporation
Charles O. Bounds
Rhône-Poulenc
Jack W. Bray
Reynolds Metals Company
M.B. Brodsky
Argonne National Laboratory
Terrence K. Brog
Coors Ceramics Company
J. Capellen
Iowa State University
Paul J. Cascone
J.F. Jelenko & Company
J.E. Casteras
Alpha Metals, Inc.
Barrie Cayless
Alcan Rolled Products Company
M.W. Chase
National Institute of Standards and Technology
T.J. Clark
G.E. Superabrasives
Arthur Cohen
Copper Development Association Inc.
Barbara Cort
Los Alamos National Laboratory
W. Raymond Cribb
Brush Wellman Inc.
Paul Crook
Haynes International, Inc.
Donald Cunningham
Emerson Electric, Wiegand Division
Charles B. Daellenback
U.S. Bureau of Mines
Jack deBarbadillo
Inco Alloys International, Inc.
Gerald L. DePoorter
Colorado School of Mines
James D. Destefani
Bailey Controls Company
R.C. DeVries
G.E. Corporate Research & Development Center
Douglas Dietrich
Carpenter Technology Corporation
Lisa A. Dodson
Johnson Matthey, Inc.
R.E. Droegkamp
Fansteel Inc.
Paul S. Dunn
Los Alamos National Laboratory
Kenneth H. Eckelmeyer
Sandia National Laboratories
John L. Ellis
Consultant
Daniel Eylon
University of Dayton
J.A. Fahey
Bronx Community College
George Fielding
Harrison Alloys Inc.
J.W. Fiepke
Crucible Magnetics, Division of Crucible Materials Corporation
John Fischer
Inco Alloys International, Inc.
John V. Foltz
Naval Surface Warfare Center
Fred Foyle
Sandvik-Rhenium Alloys Corporation
Earl L. Frantz
Carpenter Technology Corporation
F.H. (Sam) Froes
University of Idaho
C.E. Fuerstenau
Lucas-Milhaupt, Inc.
Robert C. Gabler, Jr.
U.S. Bureau of Mines
Jeffrey Gardner
Texas Instruments, Inc.
Sam Gerardi
Fansteel Inc., Precision Sheet Metal Division
Claus G. Goetzel
Consultant & Lecturer
Robert A. Goyer
University of Western Ontario
Toni Grobstein
NASA Lewis Research Center
K.A. Gschneidner
Iowa State University
R.G. Haire
Oak Ridge National Laboratory
W.B. Hampshire
Tin Information Center
John C. Harkness
Brush Wellman Inc.
Darel E. Hodgson
Shape Memory Applications, Inc.
Susan Housh
Dow Chemical U.S.A.
J.L. Hunt
Kennametal Inc.
Richard S. James
Alcoa Technical Center
Walter Johnson
Michigan Technological University
William L. Johnson
California Institute of Technology
Bo Jönsson
Kanthal AB
Avery L. Kearney
Avery Kearney & Company
James R. Keiser
Oak Ridge National Laboratory
Kenneth E. Kihlstrom
Westmont College
Erhard Klar
SCM Metal Products, Inc.
James J. Klinzing
Johnson Matthey Inc.
C. Koch
North Carolina State University
Deborah A. Kramer
U.S. Bureau of Mines
T. Scott Kreilick
Hudson International Conductors
S. Lamb
Inco Alloys International, Inc.
John B. Lambert
Fansteel Inc.
S. Lampman
ASM International
D.C. Larbalestier
University of Wisconsin-Madison
Pat Lattari
Texas Instruments, Inc.

Luc LeLay
University of Wisconsin-Madison
H.M. Liaw
Motorola, Inc.
C.T. Liu
Oak Ridge National Laboratory
Thomas Lograsso
Iowa State University
W.L. Mankins
Inco Alloys International, Inc.
J.M. Marder
Brush Wellman Inc.
Barry Mikucki
Dow Chemical U.S.A.
L.F. Mondolfo
Consultant
Hugh Morrow
Cadmium Council, Inc.
Lester R. Morss
Argonne National Laboratory
Robert Mroczkowski
AMP Inc.
G.T. Murray
California Polytechnic State University
David V. Neff
Metaullics Systems
Jeremy R. Newman
TiTech International, Inc.
M. Nowak
Troy Chemical Corporation
John T. O'Reilly
The Doe Run Company
F.H. Perfect
Reading Alloys, Inc.
Donald W. Petrasek
NASA Lewis Research Center
C.W. Philp
Handy & Harman
Joseph R. Pickens
Martin Marietta Laboratories
Charles Pokrass
Brush Wellman Inc.
(formerly with Fansteel Inc.)
R. David Prengamen
RSR Corporation
John J. Rausch
Fansteel Inc.
Michael J. Readey
Coors Ceramic Company
William D. Riley
U.S. Bureau of Mines
A.M. Reti
Handy & Harman
A.R. Robertson
Engelhard Corporation
Peter Robinson
Olin Corporation
Elwin L. Rooy
Aluminum Company of America
(retired)
N.W. Rupp
National Institute of Standards and
Technology
M.J.H. Ruscoe
Sherritt Gordon Ltd.
A.T. Santhanam
Kennametal Inc.
James C. Schaeffer
JCS Consulting
Donald G. Schmidt
North Chicago Refiners and Smelters,
Division of R. Lavin & Sons, Inc.
Robert F. Schmidt
Colonial Metals
D.K. Schroder
Arizona State University
Yuan-Shou Shen
Engelhard Corporation
Michael Slovich
Garfield Alloys, Inc.
David B. Smathers
Teledyne Wah Chang Albany
J.F. Smith
Ames Laboratory
William D. Spiegelberg
Brush Wellman Inc.
Joseph Stephens
NASA Lewis Research Center
L.G. Stevens
Indium Corporation of America
Michael F. Stevens
Los Alamos National Laboratory
Archie Stevenson
Magnesium Elektron, Inc.
James O. Stiegler
Oak Ridge National Laboratory
A.J. Stonehouse
Brush Wellman Inc.
Michael Suisman
Suisman Titanium Corporation
John K. Thorne
TiTech International, Inc.
P. Tierney
Kennametal Inc.
Robert Titran
NASA Lewis Research Center
Louis Toth
Engelhard Corporation
Derek E. Tyler
Olin Corporation
J.H.L. Van Linden
Alcoa Technical Center
Carl Vass
Fansteel/Wellmon Dynamics
T.P. Wang
Thermo Electric Company, Inc.
William H. Warnes
Oregon State University
Leonard Wasserman
Suisman Titanium Corporation
R.M. Waterstrat
National Institute of Standards &
Technology
Robert A. Watson
Kanthal Corporation
R.T. Webster
Teledyne Wah Chang Albany
J.H. Westbrook
Sci-Tech Knowledge Systems
C.E.T. White
Indium Corporation of America
R.K. Williams
Oak Ridge National Laboratory
Keith R. Willson
Geneva College
G.M. Wityak
Handy & Harman
Anthony W. Worcester
The Doe Run Company
Ming H. Wu
Memry Corporation

Reviewers and Contributors

S.P. Abeln
EG&G Rocky Flats
Stanley Abkowitz
Dynamet Technology
D.J. Accinno
Engelhard Industries, Inc.
W. Acton
Axel Johnson Metals, Inc.
G. Adams
Cominco Metals
Roy E. Adams
TIMET
H.J. Albert
Engelhard Industries
(deceased)
John Allison
Ford Motor Company
Paul Amico
Handy & Harmon
L. Angers
Aluminum Company of America
R.H. Atkinson
Inco Alloys International, Inc.
(retired)
H.C. Aufderhaar
Union Carbide Corporation
Roger J. Austin
Hydro-Lift
R. Avery
Consultant to Nickel Development Institute
Denise M. Aylor
David W. Taylor Naval Ship Research and Development Center
Roy G. Baggerly
Kenworth Truck Company
A.T. Balcerzak
St. Joe Lead Company
T.A. Balliett
Carpenter Technology Corporation
William H. Balme
Degussa Metz Metallurgical Corporation
J.A. Bard
Matthey Bishop, Inc.
Robert J. Barnhurst
Noranda Technology Centre
E.S. Bartlett
Battelle Memorial Institute
Louis Baum
Remington Arms Company
J. Benford
Allegheny Ludlum Steel, Division of Allegheny Ludlum Corporation
R. Benn
Textron Lycoming
D. Bernier
Kester Solder
Michael Bess
Certified Alloys, Inc.
A.W. Blackwood
ASARCO Inc.
M. Bohlmann
Bohlmann TECHNET
G. Boiko
Billiton Witmetaal U.S.A.
Rodney R. Boyer
Boeing Commercial Airplane Company
Leonard Bozza
Engelhard Corporation
John F. Breedis
Olin Corporation
S. Brown
ASARCO Inc.
Stephen J. Burden
GTE Valenite
H.I. Burrier
The Timken Company
Alan T. Burns
S.K. Wellman Corp.
D. Burton
Perry Tool & Research
Donald W. Capone, II
Supercon, Inc.
S.C. Carapella, Jr.
ASARCO, Inc.
James F. Carney
Johnson Matthey, Inc.
F.E. Carter
Engelhard Industries, Inc.
Robert L. Caton
Carpenter Technology Corporation
L. Christodoulou
Martin Marietta Laboratories
Thomas M. Cichon
Arrow Pneumatics, Inc.
Byron Clow
International Magnesium Association
James Cohn
Sigmund Cohn Corporation
R. Cook
IBM Corporation
R.R. Corle
EG&G Rocky Flats
D.A. Corrigan
Handy & Harman
C.D. Coxe
Handy & Harman
(deceased)
M. Daeumling
IBM Research Laboratories
Paul E. Danielson
Teledyne Wah Chang Albany
J.H. DeVan
Oak Ridge National Laboratory
D. Diesburg
Climax Performance Materials
C. Di Martini
Alpha Metals Inc.
C. Dooley
U.S. Bureau of Mines
T. Duerig
Raychem Corporation
G. Dudder
Battelle Pacific Northwest Laboratories
Francois Duffaut
Imphy S.A.
B. Dunning
Consultant
W. Eberly
Consultant
C.E. Eckert
Alcoa Technical Center
T. Egami
University of Pennsylvania
A. Elshabini-Riad
Virginia Polytechnic Institute and State University
John Elwell
Phoenix Metallurgical Corporation
A. Epstein
Technical Materials, Inc.
S.G. Epstein
The Aluminum Association
S.C. Erickson
Dow Chemical U.S.A.
Daniel Eylon
University of Dayton
K. Faber
Northwestern University
L. Ferguson
Deformation Control Technology
D. Finnemore
Iowa State University
D.Y. Foster
Métalimphy Alloys Corporation
R. Frankena
Ingal International Gallium GmbH
Gerald P. Fritzke
Metallurgical Associates

T. Gambatese
S.K. Wellman Corp.
A. Geary
Nuclear Metals, Inc.
G. Geiger
North Star Steel Company
R. Gibson
Snap-On-Tool Corporation
G. Goller
Ligonier Powders, Inc.
J. Goodwill
Carnegie-Mellon Research Institute
F. Goodwin
International Lead Zinc Research Organization
Arnold Gottlieb
Harrison Alloys Inc.
T. Gray
Allegheny Ludlum Steel, Division of Allegheny Ludlum Corporation
R.B. Green
Radio Corporation of America
F. Greenwald
Arnold Engineering Company
C. Grimes
Teledyne Wah Chang Albany
A. Gunderson
Wright Patterson Air Force Base
B. Hanson
Hazen Research Institute, Inc.
Charles E. Harper, Jr.
Metallurgical & Environmental Testing Laboratories, Inc.
J. Hafner
Texas Instruments, Inc.
J.P. Hager
Colorado School of Mines
Robert Hard
Cabot Corporation
Douglas Hayduk
ASARCO Inc.
B. Heuer
Nooter Corporation
G.J. Hildeman
Aluminum Company of America
James E. Hillis
Dow Chemical U.S.A.
G.M. Hockaday
Titanium Development Association
Ernest W. Horvick
The Zinc Institute
G. Hsu
Reynolds Metals Company
E. Kent Hudson
Lake Engineering, Inc.
Dennis D. Huffman
The Timken Company
H.Y. Hunsicker
Aluminum Company of America
Mildred Hunt
The Chemists' Club Library
J. Ernesto Indacochea
University of Illinois at Chicago
E. Jenkins
Stellite Coatings
A. Johnson
TiNi Alloy Company
L. Johnson
G.E. Corporate Research & Development Center
Peter K. Johnson
Metal Powder Industries Federation
T. Johnson
Lanxide Corporation
J. Jolley
Precision Castparts Corporation
Willard E. Kemp
Fike Metal Products, Noble Alloy Valve Group
G. Kendall
Northrop Corporation
B. Kilbourn
Molycorp, Inc.
James J. Klinzing
Johnson Matthey, Inc.
G. Kneisel
Teledyne Wah Chang Albany
C.C. Koch
North Carolina State University
R.V. Kolarik
The Timken Company
R. Komanduri
Oklahoma State University
P. Koros
LTV Steel Company
K.S. Kumar
Martin Marietta Laboratories
Henry Kunzman
Eaton Corporation
John B. Lambert
Fansteel Inc.
D.C. Larbalestier
University of Wisconsin-Madison
T. Larek
IBM Corporation
J.A. Laverick
The Timken Company
J. Laughlin
Oregon Metallurgical Corporation
J. Lee
Spang & Company
M. Lee
General Electric
P. Lees
Technical Materials, Inc.
James C. Leslie
Advanced Composites Products & Technology
W.C. Leslie
University of Michigan (retired)
A. Levy
Lawrence Berkeley Laboratory
Eli Levy
The de Havilland Aircraft Company of Canada
Joseph Linteau
Climax Specialty Metals
Lloyd Lockwood
Dow Chemical U.S.A.
P. Loewenstein
Nuclear Metals, Inc. (retired/consultant)
G. London
Naval Air Development Center
Joseph B. Long
Tin Information Center
F. Luborsky
G.E. Corporate Research & Development Center
G. Ludtka
Martin Marietta Energy Systems
David Lundy
International Precious Metals Institute
Armand A. Lykens
Carpenter Technology Corporation
W. Stuart Lyman
Copper Development Association Inc.
C. MacKay
Microelectronics & Computer Technology Corporation
T. Mackey
Key Metals & Minerals Engineering Company
John H. Madaus
Callery Chemical Company
H. Makar
U.S. Bureau of Mines
W.L. Mankins
Inco Alloys International, Inc.
W. Marancik
Oxford Superconducting Technology
K. Marken
Battelle Memorial Institute
Daniel Marx
Materials Research Corporation
Lisa C. Martin
Lanxide Corporation
John E. Masters
American Cyanamid Company
Ian Masters
Sherrit Research Center
P. Matthews
U.S. Bronze Powders, Inc.
D.J. Maykuth
Battelle Memorial Institute
B. Maxwell
Nickel Development Institute
A.S. McDonald
Handy & Harman
A. McInturff
Fermi Accelerator Laboratory
K. McKee
Carboloy Inc.
W. Mihaichuk
Eastern Alloys
K. Minnick
Lukens Steel Company
J. Mitchell
Precision Castparts Corporation
J.D. Mitilineos
Sigmund Cohn Corporation
Melvin A. Mittnick
Textron Specialty Materials
J. Moll
Crucible Research
C.E. Mueller
Naval Surface Weapons Center
H. Muller
Brookhaven National Laboratory
Y. Murty
NGK Metals Corporation

S. Narasimhan
Hoeganaes Corporation
David V. Neff
Metaullics Systems
O. Edward Nelson
Oregon Metallurgical Corporation
Dale H. Nevison
Zinc Information Center, Ltd.
P. Noros
LTV Steel Company
R.S. Nycum
Consultant
B.F. Oliver
University of Tennessee
David L. Olson
Colorado School of Mines
Dean E. Orr
Orr Metallurgical Consulting Service, Inc.
R. Osman
Airco Specialty Gasses
Heinz H. Pariser
Heinz H. Pariser Alloy Metals & Steel Market Research
L. Pederson
Battelle Pacific Northwest Laboratory
D. Peterson
Iowa State University
R. Peterson
Reynolds Metals Company
C. Petzold
Exide Corporation
K. Pike
East Penn Manufacturing Company
W. Pollack
E.I. DuPont de Nemours & Company
P. Pollak
The Aluminum Association
A. Ponikvar
International Lead Zinc Research Organization
Paul Pontrelli
Joseph Oat Corporation
D. Pope
University of Pennsylvania
T. Porter
GA Avril Company
R. David Prengamen
RSR Corporation
B. Quigley
NASA Lewis Research Center
V. Ramachandran
ASARCO Inc.
U. Ranzi
IG Technologies, Inc.
H.T. Reeve
AT&T Bell Laboratories
H.F. Reid
American Welding Society
C. Revac
RMI Company
M.V. Rey
The Timken Company
F.W. Rickenbach
Titanium Development Association
W.C. Riley
Research Opportunities
P. Roberts
Nuclear Metals, Inc.
M. Robinson
SPS Technologies
T. Rogers
IMCO Recycling Inc.
Elwin L. Rooy
Aluminum Company of America (retired)
R. Roth
Howmet Corporation
Y. Sahai
Ohio State University
H. Sanderow
Management & Engineering Technologies
R. Scanlon
Lawrence Berkeley Laboratory
Robert D. Schelleng
Inco Alloys International, Inc.
J. Schemel
Sandvik Special Metals Corporation
S. Seagle
RMI Company
P. Seegopaul
Materials Research Corporation
J.E. Selle
Oak Ridge National Laboratory
Scott O. Shook
Dow Chemical U.S.A.
G.H. Sistare, Jr.
Handy & Harman (deceased)
Hendrick Slaats
Engelhard Corporation
Gerald R. Smith
U.S. Bureau of Mines
J.F. Smith
Lead Industries Association, Inc.
L.R. Smith
Ford Motor Company
R. Smith
Ametek
H. Clinton Snyder
Aluminum Company of America
Kathleen Soltow
Jet Engineering, Inc.
F. Spaepen
Harvard University
J.R. Spence
The Timken Company
C. Sponaugle
Haynes International, Inc.
H. Stadelmaier
North Carolina State University
M.D. Swintosky
The Timken Company
A. Taub
G.E. Corporate Research & Development Center
Peter J. Theisen
Eaton Corporation
R. Thorpe
AMP Inc.
C.D. Thurmond
AT&T Bell Laboratories
T. Tiegs
Oak Ridge National Laboratory
P.A. Tomblin
The de Havilland Aircraft Company of Canada
M. Topolski
Babcock & Wilcox
R.L. Trevison
Johnson Matthey Electronics
S. Trout
Molycorp, Inc.
W. Ullrich
Alcan Powders & Pigments, Division of Alcan Aluminum Corporation
George F. Vander Voort
Carpenter Technology Corporation
K. Vedula
Office of Naval Research
R.F. Vines
Inco Alloys International, Inc.
R. Volterra
Texas Instruments Metals & Controls Division
F. James Walnista
Wyman-Gordon Company
John Waltrip
Dow Chemical U.S.A.
William H. Warnes
Oregon State University
C. Wayman
University of Illinois
R.H. Weichsel
AB Consultants International Inc.
M. Wells
U.S. Army Material Technology Laboratory
E.M. Wise
Inco Alloys International, Inc.
Gerald J. Witter
Chugai USA, Inc.
D. Yates
Inco Alloys International, Inc.
J. Yerger
Aluminum Company of America
Stephen W.H. Yih
Consultant
Ernest M. Yost
Chemet Corporation
Leon Zollo
SPS Technologies
R.D. Zordan
Allison Gas Turbines
Edward D. Zysk
Engelhard Corporation (deceased)

Contents

Specific Metals and Alloys

Introduction to Aluminum and Aluminum Alloys

Elwin L. Rooy, Aluminum Company of America

ALUMINUM, the second most plentiful metallic element on earth, became an economic competitor in engineering applications as recently as the end of the 19th century. It was to become a metal for its time. The emergence of three important industrial developments would, by demanding material characteristics consistent with the unique qualities of aluminum and its alloys, greatly benefit growth in the production and use of the new metal.

When the electrolytic reduction of alumina (Al_2O_3) dissolved in molten cryolite was independently developed by Charles Hall in Ohio and Paul Heroult in France in 1886, the first internal-combustion-engine-powered vehicles were appearing, and aluminum would play a role as an automotive material of increasing engineering value. Electrification would require immense quantities of lightweight conductive metal for long-distance transmission and for construction of the towers needed to support the overhead network of cables which deliver electrical energy from sites of power generation. Within a few decades the Wright brothers gave birth to an entirely new industry which grew in partnership with the aluminum industry development of structurally reliable, strong, and fracture-resistant parts for airframes, engines, and ultimately, for missile bodies, fuel cells, and satellite components.

The aluminum industry's growth was not limited to these developments. The first commercial applications of aluminum were novelty items such as mirror frames, house numbers, and serving trays. Cooking utensils were also a major early market. In time, aluminum grew in diversity of applications to the extent that virtually every aspect of modern life would be directly or indirectly affected by its use.

Properties. Among the most striking characteristics of aluminum is its versatility. The range of physical and mechanical properties that can be developed—from refined high-purity aluminum (see the article "Properties of Pure Metals" in this Volume) to the most complex alloys—is remarkable. More than three hundred alloy compositions are commonly recognized, and many additional variations have been developed internationally and in supplier/consumer relationships. Compositions for both wrought and cast aluminum alloys are provided in the article "Alloy and Temper Designation Systems for Aluminum and Aluminum Alloys" that immediately follows.

The properties of aluminum that make this metal and its alloys the most economical and attractive for a wide variety of uses are appearance, light weight, fabricability, physical properties, mechanical properties, and corrosion resistance.

Aluminum has a density of only 2.7 g/cm^3, approximately one-third as much as steel (7.83 g/cm^3), copper (8.93 g/cm^3), or brass (8.53 g/cm^3). It can display excellent corrosion resistance in most environments, including atmosphere, water (including salt water), petrochemicals, and many chemical systems. The corrosion characteristics of aluminum are examined in detail in Volume 13, *Corrosion*, of the 9th Edition of *Metals Handbook*.

Aluminum surfaces can be highly reflective. Radiant energy, visible light, radiant heat, and electromagnetic waves are efficiently reflected, while anodized and dark anodized surfaces can be reflective or absorbent. The reflectance of polished aluminum, over a broad range of wave lengths, leads to its selection for a variety of decorative and functional uses.

Aluminum typically displays excellent electrical and thermal conductivity, but specific alloys have been developed with high degrees of electrical resistivity. These alloys are useful, for example, in high-torque electric motors. Aluminum is often selected for its electrical conductivity, which is nearly twice that of copper on an equivalent weight basis. The requirements of high conductivity and mechanical strength can be met by use of long-line, high-voltage, aluminum steel-cored reinforced transmission cable. The thermal conductivity of aluminum alloys, about 50 to 60% that of copper, is advantageous in heat exchangers, evaporators, electrically heated appliances and utensils, and automotive cylinder heads and radiators.

Aluminum is nonferromagnetic, a property of importance in the electrical and electronics industries. It is nonpyrophoric, which is important in applications involving inflammable or explosive-materials handling or exposure. Aluminum is also nontoxic and is routinely used in containers for foods and beverages. It has an attractive appearance in its natural finish, which can be soft and lustrous or bright and shiny. It can be virtually any color or texture.

Some aluminum alloys exceed structural steel in strength. However, pure aluminum and certain aluminum alloys are noted for extremely low strength and hardness.

Aluminum Production

All aluminum production is based on the Hall-Heroult process. Alumina refined from bauxite is dissolved in a cryolite bath with various fluoride salt additions made to control bath temperature, density, resistivity, and alumina solubility. An electrical current is then passed through the bath to electrolyze the dissolved alumina with oxygen forming at and reacting with the carbon anode, and aluminum collecting as a metal pad at the cathode. The separated metal is periodically removed by siphon or vacuum methods into crucibles, which are then transferred to casting facilities where remelt or fabricating ingots are produced.

The major impurities of smelted aluminum are iron and silicon, but zinc, gallium, titanium, and vanadium are typically present as minor contaminants. Internationally, minimum aluminum purity is the primary criterion for defining composition and value. In the United States, a convention for considering the relative concentrations of iron and silicon as the more important criteria has evolved. Reference to grades of unalloyed metal may therefore be by purity

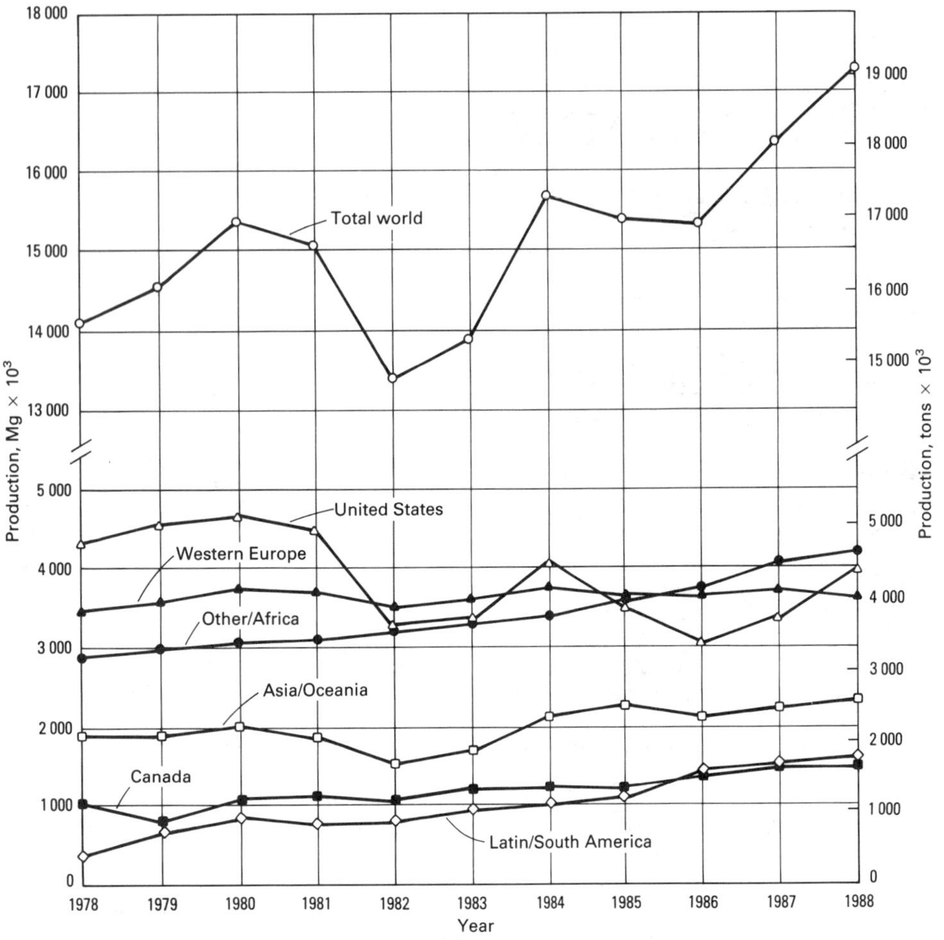

Fig. 1 Annual world production of primary aluminum. Source: Aluminum Association, Inc.

alone, for example, 99.70% aluminum, or by the method sanctioned by the Aluminum Association in which standardized P*xxx* grades have been established. In the latter case, the digits following the letter P refer to the maximum decimal percentages of silicon and iron, respectively. For example, P1020 is unalloyed smelter-produced metal containing no more than 0.10% Si and no more than 0.20% Fe. P0506 is a grade which contains no more than 0.05% Si and no more than 0.06% Fe. Common P grades range from P0202 to P1535, each of which incorporates additional impurity limits for control purposes.

Refining steps are available to attain much higher levels of purity. Purities of 99.99% are achieved through fractional crystallization or Hoopes cell operation. The latter process is a three-layer electrolytic process which employs molten salt of greater density than pure molten aluminum. Combinations of these purification techniques result in 99.999% purity for highly specialized applications.

Production Statistics. World production of primary aluminum totaled 17 304 thousand metric tonnes (17.304×10^6 Mg) in 1988 (Fig. 1). From 1978 to 1988, world production increased 22.5%, an annual growth rate of 1.6%. As shown in Fig. 2, the United States accounted for 22.8% of the world's production in 1988, while Europe accounted for 21.7%. The remaining 55.5% was produced by Asia (5.6%), Canada (8.9%), Latin/South America (8.8%), Oceania (7.8%), Africa (3.1%), and others (21.3%). The total U.S. supply in 1988 was 7 533 749 Mg in 1988, with primary production representing 54% of total supply, imports accounting for 20%, and secondary recovery representing 26% (Fig. 3). The source of secondary production is scrap in all forms, as well as the product of skim and dross processing. Primary and secondary production of aluminum are integrally related and complementary. Many wrought and cast compositions are constructed to reflect the impact of controlled element contamination that may accompany scrap consumption. A recent trend has been increased use of scrap in primary and integrated secondary fabricating facilities for various wrought products, including can sheet.

Aluminum Alloys

It is convenient to divide aluminum alloys into two major categories: casting compositions and wrought compositions. A further differentiation for each category is based on the primary mechanism of property development. Many alloys respond to thermal treatment based on phase solubilities. These treatments include solution heat treatment, quenching, and precipitation, or age, hardening. For either casting or wrought alloys, such alloys are described as heat treatable. A large number of other wrought compositions rely instead on work hardening through mechanical reduction, usually in combination with various annealing procedures for property development. These alloys are referred to as work hardening. Some casting alloys are essentially not heat treatable and are used only in as-cast or in thermally modified conditions unrelated to solution or precipitation effects.

Cast and wrought alloy nomenclatures have been developed. The Aluminum Association system is most widely recognized in the United States. Their alloy identification system employs different nomenclatures for wrought and cast alloys, but divides alloys into families for simplification (see the article "Alloy and Temper Designation Systems for Aluminum and Aluminum Alloys" in this Volume for details). For wrought alloys a four-digit system is used to produce a list of wrought composition families as follows:

- 1*xxx* Controlled unalloyed (pure) compositions
- 2*xxx* Alloys in which copper is the principal alloying element, though other elements, notably magnesium, may be specified
- 3*xxx* Alloys in which manganese is the principal alloying element
- 4*xxx* Alloys in which silicon is the principal alloying element
- 5*xxx* Alloys in which magnesium is the principal alloying element
- 6*xxx* Alloys in which magnesium and silicon are principal alloying elements
- 7*xxx* Alloys in which zinc is the principal alloying element, but other elements such as copper, magnesium, chromium, and zirconium may be specified
- 8*xxx* Alloys including tin and some lithium compositions characterizing miscellaneous compositions
- 9*xxx* Reserved for future use

Casting compositions are described by a three-digit system followed by a decimal value. The decimal .0 in all cases pertains to casting alloy limits. Decimals .1, and .2 concern ingot compositions, which after melting and processing should result in chemistries conforming to casting specification requirements. Alloy families for casting compositions are:

- 1*xx.x* Controlled unalloyed (pure) compositions, especially for rotor manufacture
- 2*xx.x* Alloys in which copper is the principal alloying element, but other alloying elements may be specified

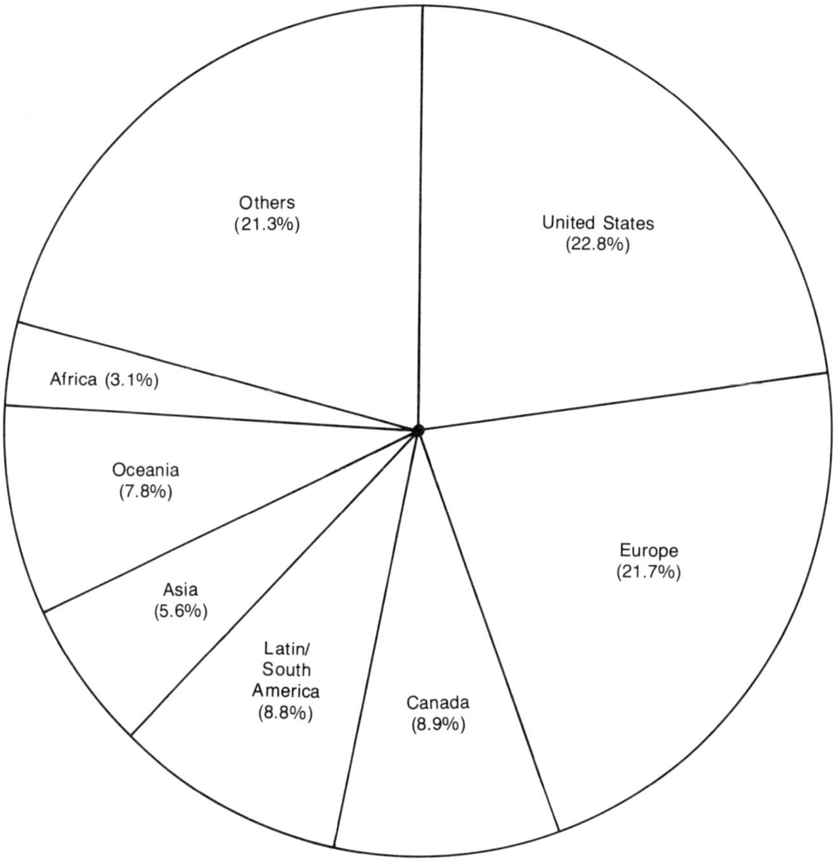

Fig. 2 Percentage distribution of world primary aluminum production in 1988. Source: Aluminum Association, Inc.

- 3*xx.x* Alloys in which silicon is the principal alloying element, but other alloying elements such as copper and magnesium are specified
- 4*xx.x* Alloys in which silicon is the principal alloying element
- 5*xx.x* Alloys in which magnesium is the principal alloying element
- 6*xx.x* Unused
- 7*xx.x* Alloys in which zinc is the principal alloying element, but other alloying elements such as copper and magnesium may be specified
- 8*xx.x* Alloys in which tin is the principal alloying element
- 9*xx.x* Unused

Manufactured Forms

Aluminum and its alloys may be cast or formed by virtually all known processes. Manufactured forms of aluminum and aluminum alloys can be broken down into two groups. Standardized products include sheet, plate, foil, rod, bar, wire, tube, pipe, and structural forms. Engineered products are those designed for specific applications and include extruded shapes, forgings, impacts, castings, stampings, powder metallurgy (P/M) parts, machined parts, and metal-matrix composites. A percentage distribution of major aluminum products is presented in Fig. 4. Properties and applications of the various aluminum product forms can be found in the articles "Aluminum Mill and Engineered Wrought Products" and "Aluminum Foundry Products" that follow.

Standardized Products

Flat-rolled products include plate (thickness equal to or greater than 6.25 mm, or 0.25 in.), sheet (thickness 0.15 mm through 6.25 mm, or 0.006 through 0.25 in.), and foil (thickness less than 0.15 mm, or 0.006 in.). These products are semifabricated to rectangular cross section by sequential reductions in the thickness of cast ingot by hot and cold rolling. Properties in work-hardened tempers are controlled by degree of cold reduction, partial or full annealing, and the use of stabilizing treatments. Plate, sheet, and foil produced in heat-treatable compositions may be solution heat treated, quenched, precipitation hardened, and thermally or mechanically stress relieved.

Sheet and foil may be rolled with textured surfaces. Sheet and plate rolled with specially prepared work rolls may be embossed to produce products such as tread plate. By roll forming, sheet in corrugated or other contoured configurations can be produced for such applications as roofing, siding, ducts, and gutters.

While the vast majority of flat-rolled products are produced by conventional rolling mills, continuous processes are now in use to convert molten alloy directly to reroll gages (Fig. 5). Strip casters employ counterrotating water-cooled cylinders or rolls to solidify and partially work coilable gage reroll stock in line. Slab casters of either twin-belt or moving block design cast stock typically 19 mm (0.75 in.) in thickness which is reduced in thickness by in-line hot reduction mill(s) to produce coilable reroll. Future developments based on technological and operational advances in continuous processes may be expected to globally affect industry expansions in flat-rolled product manufacture.

Wire, rod, and bar are produced from cast stock by extrusion, rolling, or combinations of these processes. Wire may be of any cross section in which distance between parallel faces or opposing surfaces is less than 9.4 mm (0.375 in.). Rod exceeds 9.4 mm (0.375 in.) in diameter and bar in square, rectangular, or regular hexagonal or octagonal cross section is greater than 9.4 mm (0.375 in.) between any parallel or opposing faces.

An increasingly large proportion of rod and wire production is derived from continuous processes in which molten alloy is cast in water-cooled wheel/mold-belt units to produce a continuous length of solidified bar which is rolled in line to approximately 9.4 to 12 mm (0.375 to 0.50 in.) diameter.

Engineered Products

Aluminum alloy castings are routinely produced by pressure-die, permanent-mold, green- and dry-sand, investment, and plaster casting. Shipment statistics are provided in Fig. 6. Process variations include vacuum, low-pressure, centrifugal, and pattern-related processes such as lost foam. Castings are produced by filling molds with molten aluminum and are used for products with intricate contours and hollow or cored areas. The choice of castings over other product forms is often based on net shape considerations. Reinforcing ribs, internal passageways, and complex design features, which would be costly to machine in a part made from a wrought product, can often be cast by appropriate pattern and mold or die design. Premium engineered castings display extreme integrity, close dimensional tolerances, and consistently controlled mechanical properties in the upper range of existing high-strength capabilities for selected alloys and tempers.

Extrusions are produced by forcing solid metal through aperture dies. Designs that are symmetrical around one axis are especially adaptable to production in extruded form. With current technology, it is also possible to extrude complex, mandrel-cored, and asymmetrical configurations. Precision extrusions display exceptional di-

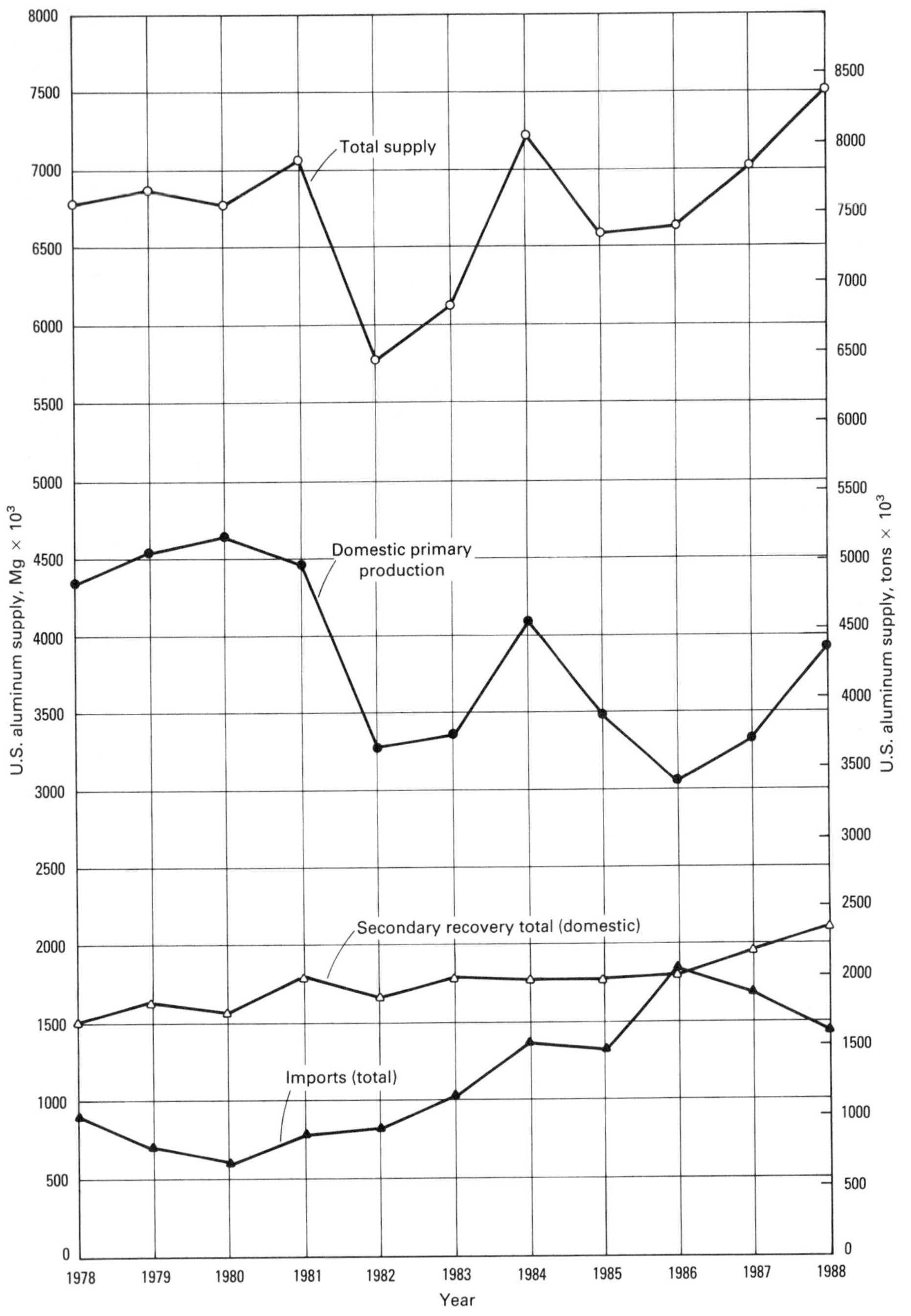

Fig. 3 U.S. aluminum production and supply statistics. Source: Aluminum Association, Inc.

mensional control and surface finish. Major dimensions usually require no machining; tolerance of the as-extruded product often permits completion of part manufacture with simple cutoff, drilling, broaching, or other minor machining operations. Extruded and extruded/drawn seamless tube competes with mechanically seamed and welded tube.

Forgings are produced by inducing plastic flow through the application of kinetic, mechanical, or hydraulic forces in either closed or open dies. Hand forgings are simple geometric shapes, formable between flat or modestly contoured open dies such as rectangles, cylinders (multiface rounds), disks (biscuits), or limited variations of these shapes. These forgings fill a frequent need in industry when only a limited number of pieces is required, or when prototype designs are to be proven.

Most aluminum forgings are produced in closed dies to produce parts with good surface finish, dimensional control, and exceptional soundness and properties. Precision forgings emphasize near net shape objectives, which incorporate reduced draft and more precise dimensional accuracy. Forgings are also available as rolled or mandrel-forged rings.

Impacts are formed in a confining die from a lubricated slug, usually cold, by a single-stroke application of force through a metal punch causing the metal to flow around the punch and/or through an opening in the punch or die. The process lends itself to high production rates with a precision part being produced to exacting quality and dimensional standards. Impacts are a combination of both cold extrusion and cold forging and, as such, combine advantages of each process.

There are three basic types of impact forming—reverse impacting, forward impacting, and a combination of the two—each of which may be used in aluminum fabrication. Reverse impacting is used to make shells with a forged base and extruded sidewalls. The slug is placed in a die cavity and struck by a punch, which forces the metal to flow back (upward) around the punch, through the opening between the punch and die, to form a simple shell. Forward impacting somewhat resembles conventional extrusion. Metal is forced through an orifice in the die by the action of a punch, causing the metal to flow in the direction of pressure application. Punch/die clearance limits flash formation. Forward impacting with a flat-face punch is used to form round, contoured, straight, and ribbed rods. With a stop-race punch, thin-walled parallel or tapered sidewall tubes with one or both ends open may be formed. In the combination method, the punch is smaller than an orificed die resulting in both reverse and forward metal flow.

Powder metallurgy (P/M) parts are formed by a variety of processes. For less demanding applications, metal powder is compressed in a shaped die to produce green compacts, and then the compacts are sintered (diffusion bonded) at elevated temperature under protective atmosphere. During sintering, the compacts consolidate and strengthen. The density of sintered compacts may be increased by re-pressing. When re-pressing is performed primarily to improve dimensional accuracy, it is termed "sizing;" when performed to alter configuration, it is termed "coining." Re-pressing may be followed by resintering, which relieves stresses induced by cold work and may further consolidate the structure. By pressing and sintering only, parts having densities of greater than 80% theoretical density can be produced. By re-pressing, with or without resintering, parts of 90% theoretical density or more can be produced. Additional information on conventionally pressed and sintered aluminum P/M products can be found in the Appendix to the article "High-Strength Aluminum P/M Alloys" in this Volume.

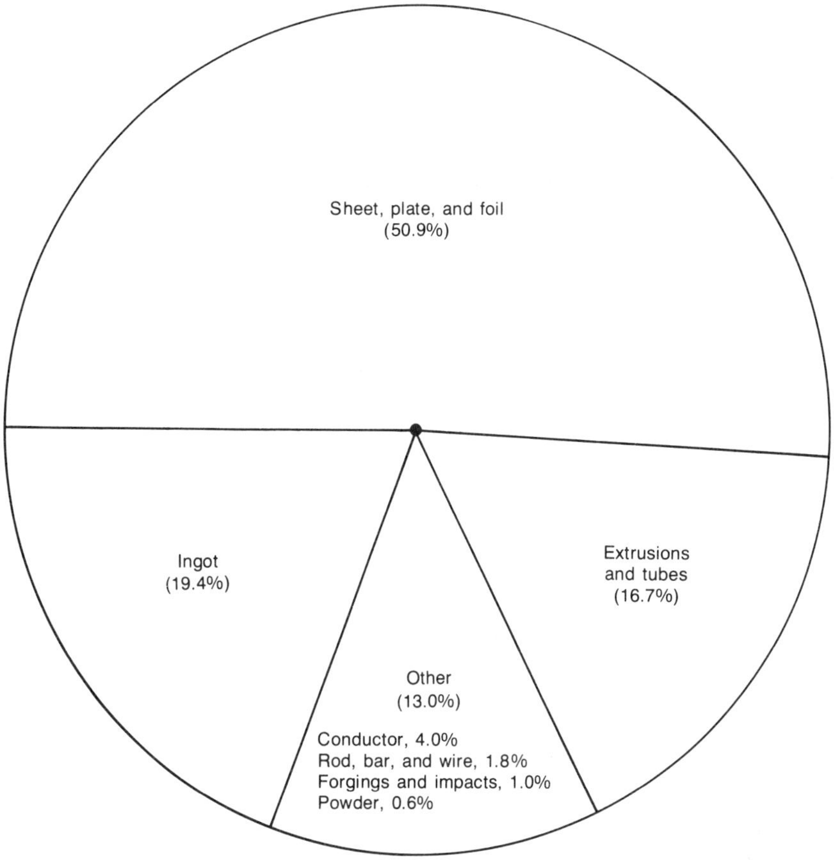

Fig. 4 Percentage distribution of major aluminum products in 1988. Source: Aluminum Association, Inc.

Fig. 5 Facility for producing aluminum sheet reroll directly from molten aluminum

For more demanding applications, such as aerospace parts or components requiring enhanced resistance to stress-corrosion cracking, rapidly solidified or mechanically attrited aluminum powders are consolidated by more advanced techniques that result in close to 100% of theoretical density. These consolidation methods include hot isostatic pressing, rapid omnidirectional compaction, ultra-high strain rate (dynamic) compaction, and spray deposition techniques. Using advanced P/M processing methods, alloys that cannot be produced through conventional ingot metallurgy methods are routinely manufactured. The aforementioned article "High-Strength Aluminum P/M Alloys" provides detailed information on advanced P/M processing.

Powder metallurgy parts may be competitive with forgings, castings, stampings, machined components, and fabricated assemblies. Certain metal products can be produced only by powder metallurgy; among these are oxide-dispersioned strengthened alloys and materials whose porosity (number distribution and size of pores) is controlled (filter elements and self-lubricating bearings).

Metal-matrix composites (MMCs) basically consist of a nonmetallic reinforcement incorporated into a metallic matrix. The combination of light weight, corrosion resistance, and useful mechanical properties, which has made aluminum alloys so popular, lends itself well to aluminum MMCs. The melting point of aluminum is high enough to satisfy many application requirements, yet is low enough to render composite processing reasonably convenient. Aluminum can also accommodate a variety of reinforcing agents. Reinforcements, characterized as either continuous or discontinuous fibers, typically constitute 20 vol% or more of the composite. The family of aluminum MMC reinforcements includes continuous boron; aluminum oxide; silicon carbide and graphite fibers; and various particles, short fibers, and whiskers. Figure 7 shows a variety of parts produced from aluminum MMCs. Information on the processing and properties of these materials can be found in the article "Metal-Matrix Composites" in this Volume.

Fabrication Characteristics

This section will briefly review important considerations in the machining, forming, forging, and joining of aluminum alloys. Additional information can be found in the articles "Aluminum Mill and Engineered Wrought Products" and "Aluminum Foundry Products" in this Volume and in articles found in other Handbooks that are referenced below.

Machinability of most aluminum alloys is excellent. Among the various wrought and cast aluminum alloys and among the tempers in which they are produced, there is considerable variation in machining characteristics, which may require special tooling or techniques (see the article "Machining of

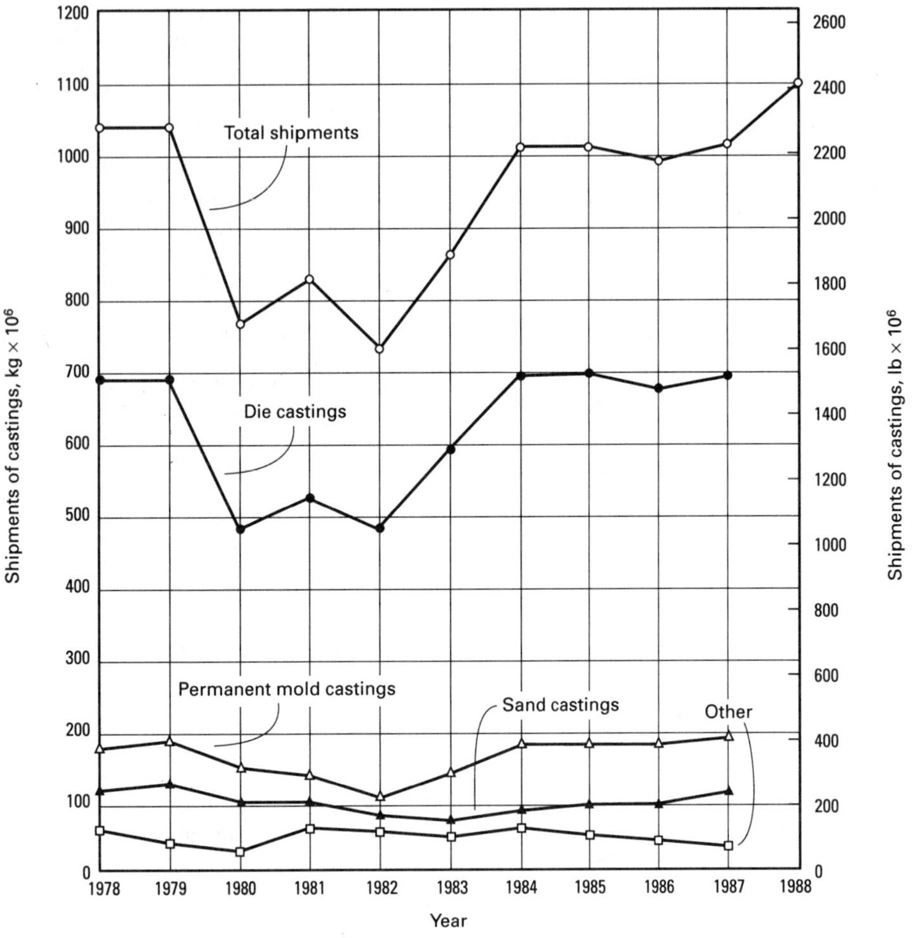

Fig. 6 U.S. casting shipments from 1978 through 1988. Source: Aluminum Association, Inc.

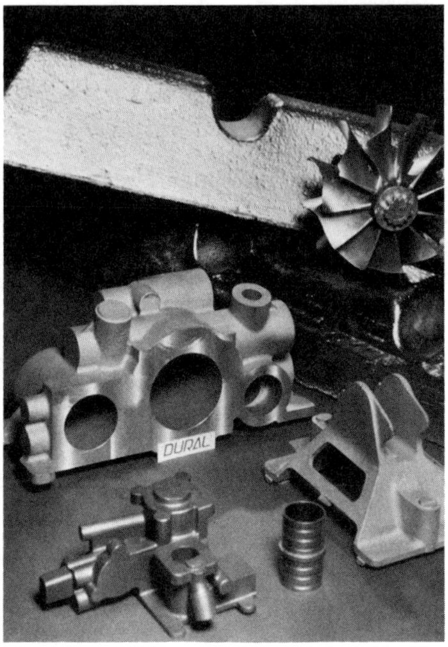

Fig. 7 Various parts made from aluminum MMCs. Courtesy of Alcan International

Aluminum and Aluminum Alloys'' in Volume 16, *Machining*, of the 9th Edition of *Metals Handbook*). Hardness and yield strength are variously used as approximations of machinability.

Chemical milling, the removal of metal by chemical attack in an alkaline or acid solution, is routine for specialized reductions in thickness. For complex large surface areas in which uniform metal removal is required, chemical milling is often the most economical method. The process is used extensively to etch preformed aerospace parts to obtain maximum strength-to-weight ratios. Integrally stiffened aluminum wing and fuselage sections are chemically milled to produce an optimum cross section and minimum skin thickness. Spars, stringers, floor beams, and frames are frequent applications as well. See the article ''Chemical Milling'' in Volume 16, *Machining*, of the 9th Edition of *Metals Handbook* for more information.

Formability is among the more important characteristics of aluminum and many of its alloys. Specific tensile and yield strengths, ductility, and respective rates of work hardening control differences in the amount of permissible deformation.

Ratings of comparable formability of the commercially available alloys in various tempers depend on the forming process, and are described in the article ''Forming of Aluminum Alloys'' in Volume 14, *Forming and Forging*, of the 9th Edition of *Metals Handbook*. Such ratings provide generally reliable comparisons of the working characteristics of metals, but serve as an approximate guide rather than as quantitative formability limits.

Choice of temper may depend on the severity and nature of forming operations. The annealed temper may be required for severe forming operations such as deep drawing, or for roll forming or bending on small radii. Usually, the strongest temper that can be formed consistently is selected. For less severe forming operations, intermediate tempers or even fully hardened conditions may be acceptable.

Heat-treatable alloys can be formed in applications for which a high strength-to-weight ratio is required. The annealed temper of these alloys is the most workable condition, but the effects of dimensional change and distortion caused by subsequent heat treatment for property development, and the straightening or other dimensional control steps that may be required, are important considerations. Alloys that are formed immediately following solution heat treatment and quench (T3, T4, or W temper) are nearly as formable as when annealed, and can be subsequently hardened by natural or artificial aging. Parts can be stored at low temperatures (approximately −30 to −35 °C, or −20 to −30 °F or lower) in the W temper for prolonged periods as a means of inhibiting natural aging and preserving an acceptable level of formability. Material that has been solution heat treated and quenched but not artificially aged (T3, T4, or W temper) is generally suitable only for mild forming operations such as bending, mild drawing, or moderate stretch forming if these operations cannot be performed immediately after quenching. Solution heat-treated and artificially aged (T6 temper) alloys are in general unsuitable for forming operations.

Forgeability. Aluminum alloys can be forged into a variety of shapes and types of forgings with a broad range of final part forging design criteria based on the intended application. Aluminum alloy forgings, particularly closed-die forgings, are usually produced to more highly refined final forging configurations than hot-forged carbon and/or alloy steels. For a given aluminum alloy forging shape, the pressure requirements in forging vary widely, depending primarily on the chemical composition of the alloy being forged, the forging process being employed, the forging strain rate, the type of forging being manufactured, the lubrication conditions, and the forging and die temperatures.

As a class of alloys, aluminum alloys are generally considered to be more difficult to forge than carbon steels and many alloy steels. Compared to the nickel/cobalt-base alloys and titanium alloys, however, aluminum alloys are considerably more forgeable, particularly in conventional forging process technology, in which dies are heated to 540 °C (1000 °F) or less. The factors influencing the forgeability of aluminum alloys as well as applicable forging methods are described in the article "Forging of Aluminum Alloys" in Volume 14, *Forming and Forging*, of the 9th Edition of *Metals Handbook*.

Joining. Aluminum can be joined by a wide variety of methods, including fusion and resistance welding, brazing, soldering, adhesive bonding, and mechanical methods such as riveting and bolting. Factors that affect the welding of aluminum include:

- Aluminum oxide coating
- Thermal conductivity
- Thermal expansion coefficient
- Melting characteristics
- Electrical conductivity

Aluminum oxide immediately forms on aluminum surfaces exposed to air. Before aluminum can be welded by fusion methods, the oxide layer must be removed mechanically by machining, filing, wire brushing, scraping, or chemical cleaning. If oxides are not removed, oxide fragments may be entrapped in the weld and will cause a reduction in ductility, a lack of fusion, and possibly weld cracking. During welding, the oxide must be prevented from re-forming by shielding the joint area with a nonoxidizing gas such as argon, helium, or hydrogen, or chemically by use of fluxes.

Thermal conductivity is the physical property that most affects weldability. The thermal conductivity of aluminum alloys is about one-half that of copper and four times that of low-carbon steel. This means that heat must be supplied four times as fast to aluminum alloys as to steel to raise the temperature locally by the same amount. However, the high thermal conductivity of aluminum alloys helps to solidify the molten weld pool of aluminum and, consequently, facilitates out-of-position welding.

The coefficient of linear thermal expansion, which is a measure of the change in length of a material with a change in its temperature, is another physical property of importance when considering weldability. The coefficient of linear thermal expansion for aluminum is twice that for steel. This means that extra care must be taken in welding aluminum to ensure that the joint space remains uniform. This may necessitate preliminary joining of the parts of the assembly by tack welding prior to the main welding operation.

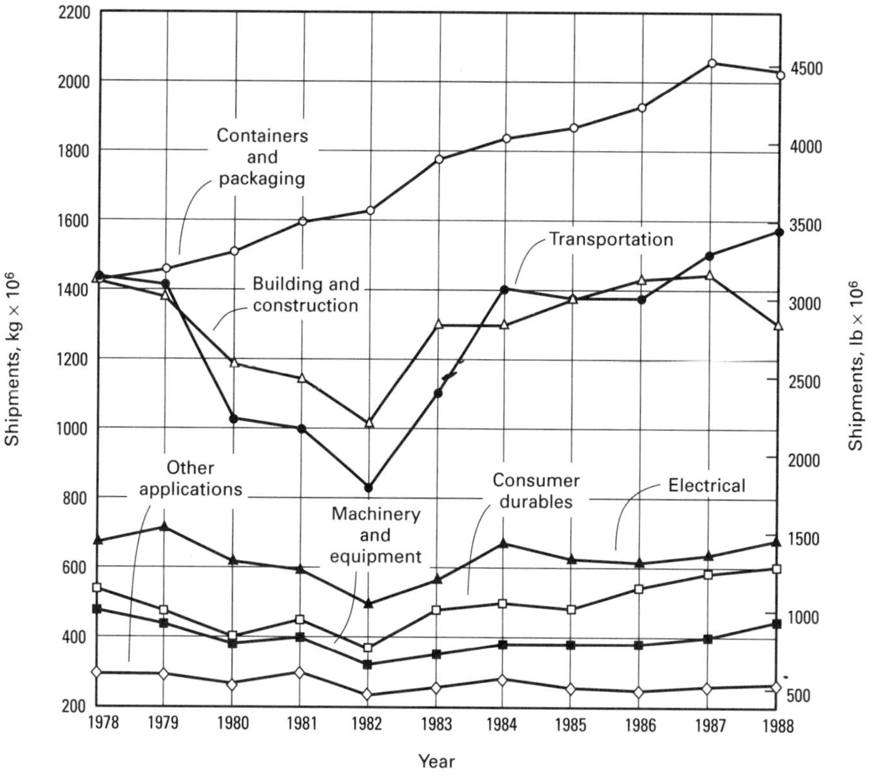

Fig. 8 U.S. net aluminum shipments by major market. Source: Aluminum Association, Inc.

The combination of high coefficient of thermal expansion and high thermal conductivity would cause considerable distortion of aluminum during welding were it not for the high welding speed possible.

Melt Characteristics. The melting ranges for aluminum alloys are considerably lower than those for copper or steel. Melting temperatures and the volumetric specific heats and heats of fusion of aluminum alloys determine that the amount of heat required to enter the welding temperature range is much lower for aluminum alloys.

Electrical conductivity has little influence on fusion welding but is a very important property for materials that are to be resistance welded. In resistance welding, resistance of the metal to the flow of welding current produces heat, which causes the portion of the metal through which the current flows to approach or reach its melting point. Aluminum has higher conductivity than steel, which means that much higher currents are required to produce the same heating effect. Consequently, resistance welding machines for aluminum must have much higher output capabilities than those normally used for steel, for welding comparable sections. More detailed information on welding of aluminum alloys as well as other joining methods can be found in Volume 6, *Welding, Brazing, and Soldering*, of the 9th Edition of *Metals Handbook* and in Volume 3, *Adhesives and Sealants*, of the *Engineered Materials Handbook*.

Product Classifications

In the United States the aluminum industry has identified its major markets as building and construction, transportation, consumer durables, electrical, machinery and equipment, containers and packaging, exports, and other end uses. As described below, each of these major markets comprises a wide range of end uses. Figure 8 provides data on annual U.S. shipments of aluminum by major markets. The percentage distribution of these markets is illustrated in Fig. 9.

Building and Construction Applications

Aluminum is used extensively in buildings of all kinds, bridges, towers, and storage tanks. Because structural steel shapes and plate are usually lower in initial cost, aluminum is used when engineering advantages, construction features, unique architectural designs, light weight, and/or corrosion resistance are considerations.

Static Structures. Design and fabrication of aluminum static structures differ little from practices used with steel. The modulus of elasticity of aluminum is one-third that of steel and requires special attention to compression members. However, it offers advantages under shock loads and in cases of minor misalignments. When properly designed, aluminum typically saves over 50% of the weight required by low-carbon steel in small structures; similar savings may be

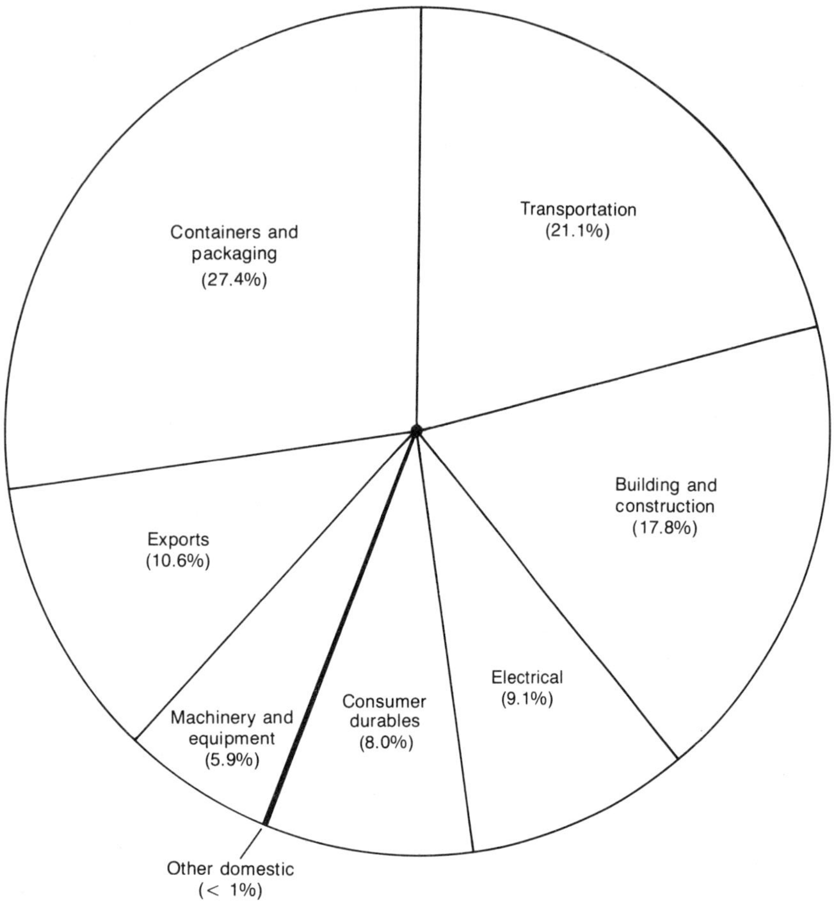

Fig. 9 Percentage distribution of net U.S. aluminum product shipments by major market. Source: Aluminum Association, Inc.

possible in long-span or movable bridges. Savings also result from low maintenance costs and in resistance to atmospheric or environmental corrosion.

Forming, shearing, sawing, punching, and drilling are readily accomplished on the same equipment used for fabricating structural steel. Since structural aluminum alloys owe their strength to properly controlled heat treatment, hot forming or other subsequent thermal operations are to be avoided. Special attention must be given to the strength requirements of welded areas because of the possibility of localized annealing effects.

Buildings. Corrugated or otherwise stiffened sheet products are used in roofing and siding for industrial and agricultural building construction. Ventilators, drainage slats, storage bins, window and door frames, and other components are additional applications for sheet, plate, castings, and extrusions.

Aluminum products such as roofing, flashing, gutters, and downspouts are used in homes, hospitals, schools, and commercial and office buildings. Exterior walls, curtain walls, and interior applications such as wiring, conduit, piping, ductwork, hardware, and railings utilize aluminum in many forms and finishes.

Aluminum is used in bridges and highway accessories such as bridge railings, highway guard rails, lighting standards, traffic control towers, traffic signs, and chain-link fences. Aluminum is also commonly used in bridge structures, especially in long-span or movable bascule and vertical-lift construction. Construction of portable military bridges and superhighway overpass bridges has increasingly relied on aluminum elements.

Scaffolding, ladders, electrical substation structures, and other utility structures utilize aluminum, chiefly in the form of structural and special extruded shapes. Cranes, conveyors, and heavy-duty handling systems incorporate significant amounts of aluminum. Water storage tanks are often constructed of aluminum alloys to improve resistance to corrosion and to provide attractive appearance.

Containers and Packaging

The food and drug industries use aluminum extensively because it is nontoxic, nonadsorptive, and splinter-proof. It also minimizes bacterial growth, forms colorless salts, and can be steam cleaned. Low volumetric specific heat results in economies when containers or conveyors must be moved in and out of heated or refrigerated areas. The nonsparking property of aluminum is valuable in flour mills and other plants subject to fire and explosion hazards. Corrosion resistance is important in shipping fragile merchandise, valuable chemicals, and cosmetics. Sealed aluminum containers designed for air, shipboard, rail, or truck shipments are used for chemicals not suited for bulk shipment.

Packaging has been one of the fastest-growing markets for aluminum. Products include household wrap, flexible packaging and food containers, bottle caps, collapsible tubes, and beverage and food cans. Aluminum foil works well in packaging and for pouches and wraps for foodstuffs and drugs, as well as for household uses.

Beverage cans have been the aluminum industry's greatest success story, and market penetrations by the food can are accelerating. Soft drinks, beer, coffee, snack foods, meat, and even wine are packaged in aluminum cans. Draft beer is shipped in alclad aluminum barrels. Aluminum is used extensively in collapsible tubes for toothpaste, ointments, food, and paints.

Transportation

Automotive. Both wrought and cast aluminum have found wide use in automobile construction (Table 1). Typical aluminum usage per unit of approximately 70 kg (150 lb) is expected to increase dramatically as average fuel economy mandates and emphasis on recycling continue. The most intensive use of aluminum in a passenger car approximates 295 kg (650 lb), defining the present target for further material substitutions. Aluminum sand, die, and permanent mold castings are critically important in engine construction; engine blocks, pistons, cylinder heads, intake manifolds, crankcases, carburetors, transmission housings, and rocker arms are proven components. Brake valves and brake calipers join innumerable other components in car design importance. Cast aluminum wheels continue to grow in popularity. Aluminum sheet is used for hoods, trunk decks, bright finish trim, air intakes, and bumpers. Extrusions and forgings are finding new and extensive uses. Forged aluminum alloy wheels are a premium option.

Trucks. Because of weight limitations and a desire to increase effective payloads, manufacturers have intensively employed aluminum in cab, trailer, and truck designs. Sheet alloys are used in truck cab bodies, and dead weight is also reduced using extruded stringers, frame rails, and cross members. Extruded or formed sheet bumpers and forged wheels are usual. Fuel tanks of aluminum offer weight reduction, corrosion resistance, and attractive appear-

Table 1 Trends in aluminum usage in the U.S. transportation industry

Usage	1988	1987	1986	1985	1984	1983	1982	1981	1980	1979	1978
Trucks & buses											
Ingot	235	205	200	200	212	183	151	148	136	216	199
Total mill products	271	265	236	219	254	174	122	162	151	281	250
Sheet	148	144	128	123	151	96	68	85	77	158	137
Plate	6	6	5	5	4	4	3	4	4	8	7
Foil	1	1	1	1	1	...	...	...	...	...	...
Rod and bar(a)	2	2	1	1	1	2	2	2	2	5	5
Extruded shapes	68	70	65	55	60	49	32	45	44	69	64
Extruded pipe and tube(b)	3	2	2	2	1	1	1	1	1	1	1
Drawn tube(b)	...	...	...	...	1	1	1	1	1	1	1
Bare wire	...	...	...	...	...	...	...	...	...	1	1
Forgings	42	39	33	31	34	20	14	23	21	37	33
Impacts	1	1	1	1	1	1	1	1	1	1	1
Total	**506**	**470**	**436**	**419**	**466**	**357**	**273**	**310**	**287**	**497**	**449**
Passenger cars											
Ingot	1253	1162	1101	1109	1108	917	662	781	746	1008	1003
Total mill products	493	468	444	438	434	362	249	324	300	493	568
Sheet	313	296	286	274	284	235	156	199	180	333	392
Plate	2	2	1	1	1	1	1	1	1	1	2
Foil	56	55	47	46	43	40	26	28	24	36	41
Rod and bar(a)	22	20	22	20	18	19	9	9	10	11	11
Extruded shapes	46	49	46	53	43	34	37	59	61	73	77
Extruded pipe and tube(b)	39	33	26	27	22	18	9	13	9	10	14
Drawn tube(b)	...	...	...	...	5	3	1	4	7	12	13
Bare wire	2	2	3	3	4	1	1	1	1	1	1
Forgings	11	9	11	12	12	10	8	8	6	14	14
Impacts	2	2	2	2	2	1	1	2	1	2	3
Total(c)	**1746**	**1630**	**1545**	**1547**	**1542**	**1279**	**911**	**1105**	**1046**	**1501**	**1571**
Trailers and semi-trailers											
Ingot	29	32	28	30	31	20	17	20	21	33	33
Total mill products	396	418	360	356	394	281	157	191	222	355	392
Sheet	153	167	143	140	158	124	62	78	88	143	165
Plate	9	9	10	11	9	7	4	7	9	14	15
Rod and bar(a)	1	1	1	1	1	1	1	1	1	2	2
Extruded shapes	229	237	203	201	222	147	88	103	122	194	208
Pipe and tube(b)	3	3	2	2	2	1	1	1	1	1	1
Bare wire	1	1	1	1	2	1	1	1	1	1	1
Forgings	...	...	...	...	...	...	...	...	...	...	...
Total	**425**	**450**	**388**	**386**	**425**	**301**	**174**	**211**	**243**	**388**	**425**

(a) Extruded rod and bar combined with rolled and continuous cast and rod bar. (b) Drawn tube combined with extruded pipe and tube. (c) Shipments to passenger cars cover new domestic automobile production, spare parts, accessories and after-market parts. Shipments for light trucks and vans are included in the trucks and buses classification. Source: Aluminum Association, Inc.

ance. Castings and forgings are used extensively in engines and suspension systems.

Truck trailers are designed for maximum payload and operating economy in consideration of legal weight requirements. Aluminum is used in frames, floors, roofs, cross sills, and shelving. Forged aluminum wheels are commonly used. Tanker and dump bodies are made from sheet and/or plate in riveted and welded assemblies.

Mobile homes and travel trailers usually are constructed of aluminum alloy sheet used bare or with mill-applied baked-enamel finish on wood, steel, or extruded aluminum alloy frames.

Bus manufacturers also are concerned with minimizing dead weight. Aluminum sheet, plate, and extrusions are used in body components and bumpers. Forged wheels are common. Engine and structural components in cast, forged, and extruded form are extensively used.

Bearings. Aluminum-tin alloys are used in medium and heavy-duty gasoline and diesel engines for connecting-rod and main bearings. Cast and wrought bearings may be composite with a steel backing and babbited or other plated overlay. Bearing alloys are further discussed in the article "Aluminum Foundry Products" in this Volume.

Railroad Cars. Aluminum is used in the construction of railroad hopper cars, box cars, refrigerator cars, and tank cars (Fig. 10). Aluminum is also used extensively in passenger rail cars, particularly those for mass transit systems.

Marine Applications. Aluminum is commonly used for a large variety of marine applications, including main strength members such as hulls and deckhouses, and other applications such as stack enclosures, hatch covers, windows, air ports, accommodation ladders, gangways, bulkheads, deck plate, ventilation equipment, lifesaving equipment, furniture, hardware, fuel tanks, and bright trim. In addition, ships are making extensive use of welded aluminum alloy plate in the large tanks used for transportation of liquefied gases.

The corrosion-resistant aluminum alloys in current use permit designs that save about 50% of the weight of similar designs in steel. Substantial savings of weight in deckhouses and topside equipment permit lighter supporting structures. The cumulative savings in weight improve the stability of the vessel and allow the beam to be decreased. For comparable speed, the lighter, narrower craft will require a smaller power plant and will burn less fuel. Consequently, 1 kg (2.2 lb) of weight saved by the use of lighter structures or equipment frequently leads to an overall decrease in displaced weight of 3 kg (6.5 lb). Aluminum also reduces maintenance resulting from corrosive or biological attack.

The relatively low modulus of elasticity for aluminum alloys offers advantages in structures erected on a steel hull. Flexure of the steel hull results in low stresses in an aluminum superstructure, as compared with the stresses induced in a similar steel superstructure. Consequently, continuous aluminum deckhouses may be built without expansion joints.

Casting alloys are used in outboard motor structural parts and housings subject to continuous or intermittent immersion, motor hoods, shrouds, and miscellaneous parts, including fittings and hardware. Ad-

Fig. 10 The intensive use of aluminum in all transportation systems minimizes dead weight and reduces operating and maintenance costs. Courtesy of Alcan International

ditional marine applications are in sonobuoys, navigation markers, rowboats, canoes, oars, and paddles.

Aerospace. Aluminum is used in virtually all segments of the aircraft, missile, and spacecraft industry (Fig. 11)—in airframes, engines, accessories, and tankage for liquid fuel and oxidizers. Aluminum is widely used because of its high strength-to-density ratio, corrosion resistance, and weight efficiency, especially in compressive designs.

Increased resistance to corrosion in salt water and other atmospheres is secured through the use of alclad alloys or anodic coatings. The exterior of aircraft exposed to salt water environment is usually fabricated from clad alloys. Anodized bare stock successfully resists corrosion when only occasional exposure to salt water is encountered. Corrosion resistance may be further enhanced by organic finishes or other protective coatings. Extensive reviews on the uses and corrosion properties of aluminum for aircraft and aerospace vehicles can be found in Volume 13, *Corrosion*, of the 9th Edition of *Metals Handbook* (see pages 1019 to 1106).

Fig. 11 Aluminum is used extensively in aircraft/aerospace vehicles such as the space shuttle shown in this figure.

Electrical Applications

Conductor Alloys. The use of aluminum predominates in most conductor applications. Aluminum of controlled composition is treated with trace additions of boron to remove titanium, vanadium, and zirconium, each of which increases resistivity. The use of aluminum rather than competing materials is based on a combination of low cost, high electrical conductivity, adequate mechanical strength, low specific gravity, and excellent resistance to corrosion.

The most common conductor alloy (1350) offers a minimum conductivity of 61.8% of the International Annealed Copper Standard (IACS) and from 55 to 124 MPa (8 to 18 ksi) minimum tensile strength, depending on size. When compared with IACS on a basis of mass instead of volume, minimum conductivity of hard drawn aluminum 1350 is 204.6%. Other alloys are used in bus bar, for service at slightly elevated temperatures, and in cable television installations.

Cable sheathing is achieved by extruding the sheath in final position and dimensions around the cable as it is fed through an axial orifice in the extrusion die. It can also be done by threading the cable through an oversized prefabricated tube and then squeezing the tube to final dimensions around the cable by tube reducers and draw dies.

Conductor accessories may be rolled, extruded, cast, or forged. Common forms of aluminum conductors are single wire and multiple wire (stranded, bunched, or rope layed). Each is used in overhead or other tensioned applications, as well as in nontensioned insulated applications.

Size for size, the direct current resistance of the most common aluminum conductor is from about 1.6 to 2.0 times IACS. For equivalent direct-current resistance, an aluminum wire that is two American Wire Gage sizes larger than copper wire must be used. Nevertheless, as a result of the lower specific gravity, the conductivity-based aluminum required weighs only about half as much as an equivalent copper conductor.

Aluminum conductors, steel reinforced (ACSR) consist of one or more layers of concentric-lay stranded aluminum wire around a high-strength galvanized or aluminized steel wire core, which itself may be a single wire or a group of concentric-lay strands. Electrical resistance is determined by the aluminum cross section, whereas tensile strength is determined on the composite with the steel core providing 55 to 60% of the total strength.

The ACSR construction is used for mechanical strength. Strength-to-weight ratio is usually about two times that of copper of equivalent direct-current resistance. Use of ACSR cables permits longer spans and fewer or shorter poles or towers.

Bus Bar Conductors. Commercial bus design in the United States utilizes four types of bus conductors: rectangular bar, solid round bar, tubular, and structural shapes.

Motors and Generators. Aluminum has long been used for cast rotor windings and structural parts. Rotor rings and cooling fans are pressure cast integrally with bars through slots of the laminated core in caged motor rotors.

Aluminum structural parts, such as stator frames and end shields, are often economically die cast. Their corrosion resistance may be necessary in specific environments—in motors for spinning natural and synthetic fiber, and in aircraft generators when light weight is equally important, for example.

Additional applications are field coils for direct-current machines, stator windings in motors, and transformer windings. Alloyed wire is used in extremely large turbogenerator field coils, where operating temperatures and centrifugal forces might otherwise result in creep failure.

Transformers. Aluminum windings have been extensively used in dry-type power transformers and have been adapted to secondary coil windings in magnetic-suspension type constant-current transformers. Their use decreases weight and permits the coil to float in electromagnetic suspension. In a closely associated application, aluminum is being used in concrete reactor devices that protect transformers from overloads.

Extruded shapes and punched sheet are used in radar antennas, extruded and roll-formed tubing in television antennas, rolled strips in coiled line traps; drawn or impact-extruded cans in condensers and shields, and vaporized high-purity coatings in cathode-ray tubes.

Examples of applications in which electrical properties other than magnetic are not dominant are chassis for electronic equipment, spun pressure receptacles for airborne equipment, etched name plates, and hardware such as bolts, screws, and nuts. In addition, finned shapes are used in electronic components to facilitate heat removal. Aluminum may be used as the cell base for the deposition of selenium in the manufacture of selenium rectifiers.

Lighting. Aluminum in incandescent and fluorescent lamp bases and other sheet alloys for sockets are established uses. Cast, stamped, and spun parts are used, often artistically, in table, floor, and other lighting fixtures. Aluminum reflector is common in fluorescent and other installed lighting systems.

Capacitors. Aluminum in the form of foil dominates all other metals in the construction of capacitor electrodes. Dry electrolytic and nonelectrolytic capacitors are the basic condenser types in extensive commercial use. Dry electrolytic capacitors usually employ parallel coiled or wrapped aluminum foil ribbons as electrodes. Paper saturated with an operative electrolyte, wrapped into the coil, mechanically separates the ribbons. In designs for intermittent use in alternating circuits, both electrodes are anodized in a hot boric acid electrolyte. The resulting thin anodic films constitute the dielectric element.

Only the anode foil is anodized in dry electrolytic assemblies intended for direct-current applications. Anodized electrodes are of high purity, whereas the nonanodized electrodes utilize foil ribbons of lower purity. Prior to anodizing the foil is usually, but not always, etched to increase effective surface area. Containers for dry electrolytic capacitors may be either drawn or impact extruded.

Ordinary clean foil ribbons serve as electrodes in commercial nonelectrolytic capacitors. Oil-impregnated paper separates the electrodes and adjacent coils of the wrap. Nonelectrolytic foil assemblies are packed in either aluminum alloy or steel cans.

Consumer Durables

Household Appliances. Light weight, excellent appearance, adaptability to all forms of fabrication, and low cost of fabrication are the reasons for the broad usage of aluminum in household electrical appliances. Light weight is an important characteristic in vacuum cleaners, electric irons, portable dishwashers, food processors, and blenders. Low fabricating costs depend on several properties, including adaptability to die casting and ease of finishing. Because of a naturally pleasing appearance and good corrosion resistance, expensive finishing is not necessary.

In addition to its other desirable characteristics aluminum's brazeability makes it useful for refrigerator and freezer evaporators. Tubing is placed on embossed sheet over strips of brazing alloy with a suitable flux. The assembly is then furnace brazed and the residual flux removed by successive washes in boiling water, nitric acid, and cold water. The result is an evaporator with high thermal conductivity and efficiency, good corrosion resistance, and low manufacturing cost.

With the exception of a few permanent mold parts, virtually all aluminum castings in electrical appliances are die cast. Cooking utensils may be cast, drawn, spun, or drawn and spun from aluminum. Handles are often joined to the utensil by riveting or spot welding. In some utensils, an aluminum exterior is bonded to a stainless steel interior; in others, the interior is coated with porcelain or Teflon. Silicone resin, Teflon, or other coatings enhance the utility of heated aluminum utensils. Many die castings in appliances are internal functional parts and are used without finish. Organic finishes are usually applied to external die-cast parts such as appliance housings.

Wrought forms fabricated principally from sheet, tube, and wire are used in approximately the same quantities as die castings. Wrought alloys are selected on the basis of corrosion resistance, anodizing characteristics, formability, or other engineering properties.

The natural colors some alloys assume after anodizing are extremely important for food-handling equipment. Applications include refrigerator vegetable/meat pans, ice cube trays, and wire shelves. In the production of wire shelves, full-hard wire is cold headed over extruded strips, which form the borders.

Furniture. Light weight, low maintenance, corrosion resistance, durability, and attractive appearance are the principal advantages of aluminum in furniture.

Chair bases, seat frames, and arm rests are cast, drawn or extruded tube (round, square, or rectangular), sheet, or bar. Frequently, these parts are formed in the annealed or partially heat-treated tempers, and are subsequently heat treated and aged. Designs are generally based on service requirements; however, styling often dictates overdesign or inefficient sections. Fabrication is conventional; joining is usually by welding or brazing. Various finishing procedures are used: mechanical, anodic, color anodized, anodized and dyed, enamel coated, or painted.

Tubular sections, usually round and frequently formed and welded from flat strip, are the most popular form of aluminum for lawn furniture. Conventional tube bending and mechanically fitted joints may be used. Finishing is usually by grinding and buffing and is frequently followed by clear lacquer coating.

Machinery and Equipment

Processing Equipment. In the petroleum industry, aluminum tops are used on steel storage tanks, exteriors are covered with aluminum pigmented paint, and aluminum pipelines are carriers of petroleum products. Aluminum is used extensively in the rubber industry because it resists all corrosion that occurs in rubber processing and is nonadhesive. Aluminum alloys are widely used in the manufacture of explosives because of their nonpyrophoric characteristics. Strong oxidants are processed, stored, and shipped in aluminum systems. Aluminum is especially compatible with sulfur, sulfuric acid, sulfides, and sulfates. In the nuclear energy industry, aluminum-jacketed fuel elements protect uranium from water corrosion, prevent the entry of reaction products into the cooling water, transfer heat efficiently from uranium to water, and contribute to minimizing parasitic capture of neutrons. Aluminum tanks are used to contain heavy water. The use of aluminum for each of the aforementioned industries is described in more detail in Volume 13, *Corrosion*, of the 9th Edition of *Metals Handbook*.

Textile Equipment. Aluminum is used extensively in textile machinery and equipment in the form of extrusions, tube, sheet, castings, and forgings. It is resistant to many corrosive agents encountered in textile mills and in the manufacture of yarns. A high strength-to-weight ratio reduces the inertia of high-speed machine parts. Permanent dimensional accuracy with light weight improves the dynamic balance of machine members running at high speeds, and reduces vibration. Painting is usually unnecessary. Spool beamheads and cores are usually permanent mold castings and extruded or welded tube, respectively.

Paper and Printing Industries. An interesting application of aluminum is found in returnable shipping cores. Cores may be reinforced with steel end-sleeves which also constitute wear-resistant drive elements. Processing or rewinding cores are fabricated of aluminum alloys. Fourdrinier or table rolls for papermaking machines are also of aluminum construction.

Curved aluminum sheet printing plates permit higher rotary-press speeds and minimize misregister by decreasing centrifugal force. Aluminum lithographic sheet offers exceptional reproduction in mechanical and electrograined finishes.

Coal Mine Machinery. The use of aluminum equipment in coal mines has increased in recent years. Applications include cars, tubs and skips, roof props, nonsparking tools, portable jacklegs, and shaking conveyors. Aluminum is resistant to the corrosive conditions associated with surface and deep mining. Aluminum is self cleaning and offers good resistance to abrasion, vibration, splitting, and tearing.

Portable Irrigation Pipe and Tools. Aluminum is extensively used in portable sprinkler and irrigation systems. Portable tools use large quantities of aluminum in electric and gas motors and motor housings. Precision cast housings and engine components, including pistons, are used for power drills, power saws, gasoline-driven chain saws, sanders, buffing machines, screwdrivers, grinders, power shears, hammers, various impact tools, and stationary bench tools. Aluminum alloy forgings are found in many of the same applications and in manual tools such as wrenches and pliers.

Jigs, Fixtures, and Patterns. Thick cast or rolled aluminum plates and bar, precisely machined to high finish and flatness, are used for tools and dies. Plate is suitable for hydropress form blocks, hydrostretch form dies, jigs, fixtures, and other tooling. Aluminum is used in the aircraft industry for drill jigs, as formers, stiffeners and stringers for large assembly jigs, router bases, and layout tables. Used in master tooling, cast aluminum eliminates warpage problems resulting from uneven expansion of the tool due to changes in ambient temperature. Large aluminum bars have been used to replace zinc alloys as a fixture base on spar mills with weight savings of two-thirds. Cast aluminum serves as matchplate in the foundry industry.

Instruments. On the basis of combinations of strength and dimensional stability, aluminum alloys are used in the manufacture of optical, telescopic, space guidance, and other precision instruments and devices. To assure dimensional accuracy and stability in manufacturing and assembling parts for such equipment, additional thermal stress-relief treatments are sometimes applied at stages of machining, or after welding or mechanical assembly.

Other Applications

Reflectors. Reflectivity of light is as high as 95% on especially prepared surfaces of high-purity aluminum. Aluminum is generally superior to other metals in its ability to reflect infrared or heat rays. It resists tarnish from sulfides, oxides, and atmospheric contaminants, and has three to ten times the useful life of silver for mirrors in searchlights, telescopes, and similar reflectors. Heat reflectivity may be as much as 98% for a highly polished surface. Performance is reduced only slightly as the metal weathers and loses its initial brilliance. When maximum reflectivity is desired, chemical or electrochemical brightening treatments are used; quick anodic treatment usually follows, sometimes finished by a coat of clear lacquer. Reflectors requiring less brightness may simply be buffed and lacquered. Etching in a mild caustic solution produces a diffuse finish, which may also be protected by clear lacquer, an anodic coating, or both.

Powders and Pastes. The addition of aluminum flakes to paint pigments exploits the intrinsic advantages of high reflectance, durability, low emissivity, and minimum moisture penetration. Other applications for powders and pastes include printing inks, pyrotechnics, floating soap, aerated concrete, thermite welding, and energy-enhancing fuel additives. Additional information can be found in Volume 7, *Powder Metallurgy*, of the 9th Edition of *Metals Handbook*.

Anode Materials. Highly electronegative aluminum alloys are routinely employed as sacrificial anodes, generally on steel structures or vessels such as pipelines, offshore construction, ships, and tank storage units. See the article "Cathodic Protection" in Volume 13, *Corrosion*, of the 9th Edition of *Metals Handbook* for additional information.

Alloy and Temper Designation Systems for Aluminum and Aluminum Alloys

R.B.C. Cayless, Alcan Rolled Products Company

SYSTEMS FOR DESIGNATING aluminum and aluminum alloys that incorporate the product form (wrought, casting, or foundry ingot), and its respective temper (with the exception of foundry ingots, which have no temper classification) are covered by American National Standards Institute (ANSI) standard H35.1. The Aluminum Association is the registrar under ANSI H35.1 with respect to the designation and composition of aluminum alloys and tempers registered in the United States.

Wrought Aluminum and Aluminum Alloy Designation System

A four-digit numerical designation system is used to identify wrought aluminum and aluminum alloys. As shown below, the first digit of the four-digit designation indicates the group:

Aluminum, ≥99.00%	1*xxx*
Aluminum alloys grouped by major alloying element(s):	
Copper	2*xxx*
Manganese	3*xxx*
Silicon	4*xxx*
Magnesium	5*xxx*
Magnesium and silicon	6*xxx*
Zinc	7*xxx*
Other elements	8*xxx*
Unused series	9*xxx*

For the 2*xxx* through 7*xxx* series, the alloy group is determined by the alloying element present in the greatest mean percentage. An exception is the 6*xxx* series alloys in which the proportions of magnesium and silicon available to form magnesium silicide (Mg_2Si) are predominant. Another exception is made in those cases in which the alloy qualifies as a modification of a previously registered alloy. If the greatest mean percentage is the same for more than one element, the choice of group is in order of group sequence: copper, manganese, silicon, magnesium, magnesium silicide, zinc, or others.

Aluminum. In the 1*xxx* group, the series 10*xx* is used to designate unalloyed compositions that have natural impurity limits. The last two of the four digits in the designation indicate the minimum aluminum percentage. These digits are the same as the two digits to the right of the decimal point in the minimum aluminum percentage when expressed to the nearest 0.01%. Designations having second digits other than zero (integers 1 through 9, assigned consecutively as needed) indicate special control of one or more individual impurities.

Aluminum Alloys. In the 2*xxx* through 8*xxx* alloy groups, the second digit in the designation indicates alloy modification. If the second digit is zero, it indicates the original alloy; integers 1 through 9, assigned consecutively, indicate modifications of the original alloy. Explicit rules have been established for determining whether a proposed composition is merely a modification of a previously registered alloy or if it is an entirely new alloy. The last two of the four digits in the 2*xxx* through 8*xxx* groups have no special significance, but serve only to identify the different aluminum alloys in the group.

Cast Aluminum and Aluminum Alloy Designation System

A system of four-digit numerical designations incorporating a decimal point is used to identify aluminum and aluminum alloys in the form of castings and foundry ingot. The first digit indicates the alloy group:

Aluminum, ≥99.00%	1*xx.x*
Aluminum alloys grouped by major alloying element(s):	
Copper	2*xx.x*
Silicon, with added copper and/or magnesium	3*xx.x*
Silicon	4*xx.x*
Magnesium	5*xx.x*
Zinc	7*xx.x*
Tin	8*xx.x*
Other elements	9*xx.x*
Unused series	6*xx.x*

For 2*xx.x* through 8*xx.x* alloys, the alloy group is determined by the alloying element present in the greatest mean percentage, except in cases in which the composition being registered qualifies as a modification of a previously registered alloy. If the greatest mean percentage is common to more than one alloying element, the alloy group is determined by the element that comes first in the sequence.

The second two digits identify the specific aluminum alloy or, for the aluminum (1*xx.x*) series, indicate purity. The last digit, which is separated from the others by a decimal point, indicates the product form, whether casting or ingot. A modification of an original alloy, or of the impurity limits for unalloyed aluminum, is indicated by a serial letter preceding the numerical designation. The serial letters are assigned in alphabetical sequence starting with A but omitting I, O, Q, and X, the X being reserved for experimental alloys. Explicit rules have been established for determining whether a proposed composition is a modification of an existing alloy or if it is a new alloy.

Aluminum Castings and Ingot. For the 1*xx.x* group, the second two of the four digits in the designation indicate the minimum aluminum percentage. These digits are the same as the two digits to the right of the decimal point in the minimum aluminum

Table 1 Alloying element and impurity specifications for ingots that will be remelted into sand, permanent mold, and die castings

	Composition, wt%			
	Casting			
Alloying element	Sand and permanent mold	Die	All	Ingot
Iron	≤0.15	...	...	Casting −0.03
	>0.15–0.25	...	...	Casting −0.05
	>0.25–0.6	...	...	Casting −0.10
	>0.6–1.0	...	...	Casting −0.2
	>1.0	...	...	Casting −0.3
	...	≤1.3	...	Casting −0.3
	...	>1.3	...	≤1.1
Magnesium	...	...	<0.50	Casting +0.05(a)
	...	...	≥0.50	Casting +0.1(a)
Zinc	...	>0.25 to 0.60	...	Casting −0.10
	...	>0.60	...	Casting −0.1

(a) Applicable only when the specified range for castings is >0.15% Mg. Source: Ref 1

percentage when expressed to the nearest 0.01%. The last digit indicates the product form: 1*xx*.0 indicates castings, and 1*xx*.1 indicates ingot.

Aluminum Alloy Castings and Ingot. For the 2*xx.x* through 9*xx.x* alloy groups, the second two of the four digits in the designation have no special significance but serve only to identify the different alloys in the group. The last digit, which is to the right of the decimal point, indicates the product form: *xxx*.0 indicates castings, and *xxx*.1 indicates ingot having limits for alloying elements the same as those for the alloy in the form of castings, except for those listed in Table 1.

Designations for Experimental Alloys

Experimental alloys also are designated in accordance with the systems for wrought and cast alloys, but they are indicated by the prefix X. The prefix is dropped when the alloy is no longer experimental. During development and before they are designated as experimental, new alloys may be identified by serial numbers assigned by their originators. Use of the serial number is discontinued when the ANSI H35.1 designation is assigned.

Cross-Referencing of Aluminum and Aluminum Alloy Products

Tables 2 and 3 cross-reference aluminum wrought and ingot/cast products according to composition, per Aluminum Association, Unified Numbering System (UNS), and International Organization for Standardization (ISO) standards.

Unified Numbering System. UNS numbers correlate many nationally used numbering systems currently administered by societies, trade associations, and individual users and producers of metals and alloys.

Aluminum Association International Alloy Designations. For wrought aluminum and aluminum alloys only, compositions may be registered with the Aluminum Association by a number of foreign organizations. These organizations are signatories of a Declaration of Accord on the Recommendation for an International Designation System for Wrought Aluminum and Wrought Aluminum Alloys. In addition to the United States, the countries represented by signatories are Argentina, Australia, Austria, Belgium, Brazil, Denmark, Finland, France, Germany, Italy, Japan, Netherlands, Norway, Spain, Sweden, Switzerland, South Africa, and the United Kingdom. The European Aluminum Association is also a signatory.

Under ANSI standard H35.1, wrought aluminum or aluminum alloys will be registered in decreasing priority as national variations, as modifications, or as a new four-digit number. A national variation that has composition limits very close but not identical to those registered by another country is identified by a serial letter following the numerical designation.

Castings and Foundry Alloys. There is no similar international accord for these aluminum or aluminum alloy products.

Foreign Alloy Designations. Historically, all major industrialized countries developed their own standard designations for aluminum and aluminum alloys. These are now being grouped under systems of the American National Standards Institute, the International Organization for Standardization, and the European Committee for Standardization.

The International Organization for Standardization has developed its own alphanumeric designation system for wrought aluminum and its alloys, based on the systems that have been used by certain European countries. The main addition element is distinguished by specifying the required content (middle of range) rounded off to the nearest 0.5:

5052=Al Mg2.5

5251=Al Mg2

If required, the secondary addition elements are distinguished by specifying the required content rounded off to the nearest 0.1, for two elements at most:

6181=Al Si1Mg0.8

The chemical symbols for addition elements should be limited to four:

7050=Al Zn6CuMgZr

If an alloy cannot otherwise be distinguished, a suffix in brackets is used:

6063=Al Mg0.7Si

6463=Al Mg0.7Si(B)

and international alloy registration

6063A=Al Mg0.7Si(A)

Note that suffixes (A), (B), and so on, should not be confused with suffixes of the Aluminum Association.

The proposed ISO chemical composition standard for aluminum and its alloys references Aluminum Association equivalents as well as its own identification system. A listing of these is given in Table 4.

European Committee for Standardization. This committee (Comité Européen de Normalisation, CEN) of European Common Market members has developed a composition standard based on the ISO standard, but is proposing new designations not included in that standard. Some of these new designations are already registered as German (Deutsche Industrial-Normen, DIN) standards. The proposed standard also references Aluminum Association equivalents.

Temper Designation System for Aluminum and Aluminum Alloys

The temper designation system used in the United States for aluminum and aluminum alloys is used for all product forms (both wrought and cast), with the exception of ingot. The system is based on the sequences of mechanical or thermal treatments, or both, used to produce the various tempers. The temper designation follows the alloy designation and is separated from it by a hyphen. Basic temper designations consist of individual capital letters. Major subdivisions of basic tempers, where required, are indicated by one or more digits following the letter. These digits designate specific sequences of treatments that produce specific combinations of characteristics in the product. Variations in treatment conditions within major subdivisions are identified by additional digits. The conditions during heat treatment (such as time, temperature, and quenching rate) used to produce a given temper in one alloy may differ from those employed to produce the same temper in another alloy.

Basic Temper Designations

Designations for the common tempers, and descriptions of the sequences of operations used to produce these tempers, are given in the following paragraphs.

Table 2 Composition of wrought unalloyed aluminum and wrought aluminum alloys

Grade designation: Aluminum Association	Grade designation: UNS No.	Grade designation: ISO No. R209	Composition, wt%: Si	Fe	Cu	Mn	Mg	Cr	Ni	Zn	Ga	V	Specified other elements	Ti	Unspecified other elements: Each	Unspecified other elements: Total	Al, minimum
1035	· · ·	· · ·	0.35	0.6	0.10	0.05	0.05	· · ·	· · ·	0.10	· · ·	0.05	· · ·	0.03	0.03	· · ·	99.35
1040	A91040	· · ·	0.30	0.50	0.10	0.05	0.05	· · ·	· · ·	0.10	· · ·	0.05	· · ·	0.03	0.03	· · ·	99.40
1045	A91045	· · ·	0.30	0.45	0.10	0.05	0.05	· · ·	· · ·	0.05	· · ·	0.05	· · ·	0.03	0.03	· · ·	99.45
1050	A91050	Al 99.5	0.25	0.40	0.05	0.05	0.05	· · ·	· · ·	0.05	· · ·	0.05	· · ·	0.03	0.03	· · ·	99.50
1060	A91060	Al 99.6	0.25	0.35	0.05	0.03	0.03	· · ·	· · ·	0.05	· · ·	0.05	· · ·	0.03	0.03	· · ·	99.60
1065	A91065	· · ·	0.25	0.30	0.05	0.03	0.03	· · ·	· · ·	0.05	· · ·	0.05	· · ·	0.03	0.03	· · ·	99.65
1070	A91070	Al 99.7	0.20	0.25	0.04	0.03	0.03	· · ·	· · ·	0.04	· · ·	0.05	· · ·	0.03	0.03	· · ·	99.70
1080	A91080	Al 99.8	0.15	0.15	0.03	0.02	0.02	· · ·	· · ·	0.03	0.03	0.05	· · ·	0.03	0.02	· · ·	99.80
1085	A91085	· · ·	0.10	0.12	0.03	0.02	0.02	· · ·	· · ·	0.03	0.03	0.05	· · ·	0.02	0.01	· · ·	99.85
1090	A91090	· · ·	0.07	0.07	0.02	0.01	0.01	· · ·	· · ·	0.03	0.03	0.05	· · ·	0.01	0.01	· · ·	99.90
1098	· · ·	· · ·	0.010	0.006	0.003	· · ·	· · ·	· · ·	· · ·	0.015	· · ·	· · ·	· · ·	0.003	0.003	· · ·	99.98
1100	A91100	Al 99.0 Cu	0.95 (Si + Fe)		0.05–0.20	0.05	· · ·	· · ·	· · ·	0.10	· · ·	· · ·	(a)	· · ·	0.05	0.15	99.00
1110	· · ·	· · ·	0.30	0.8	0.04	0.01	0.25	0.01	· · ·	· · ·	· · ·	· · ·	0.02 B, 0.03 (V + Ti)	· · ·	0.03	· · ·	99.10
1200	A91200	Al 99.0	1.00 (Si + Fe)		0.05	0.05	· · ·	· · ·	· · ·	0.10	· · ·	· · ·	· · ·	0.05	0.05	0.15	99.00
1120	· · ·	· · ·	0.10	0.40	0.05–0.35	0.01	0.20	0.01	· · ·	0.05	0.03	· · ·	0.05 B, 0.02 (V + Ti)	· · ·	0.03	0.10	99.20
1230	A91230	Al 99.3	0.70 (Si + Fe)		0.10	0.05	0.05	· · ·	· · ·	0.10	· · ·	0.05	· · ·	0.03	0.03	· · ·	99.30
1135	A91135	· · ·	0.60 (Si + Fe)		0.05–0.20	0.04	0.05	· · ·	· · ·	0.10	· · ·	0.05	· · ·	0.03	0.03	· · ·	99.35
1235	A91235	· · ·	0.65 (Si + Fe)		0.05	0.05	0.05	· · ·	· · ·	0.10	· · ·	0.05	· · ·	0.06	0.03	· · ·	99.35
1435	A91345	· · ·	0.15	0.30–0.50	0.02	0.05	0.05	· · ·	· · ·	0.10	· · ·	0.05	· · ·	0.03	0.03	· · ·	99.35
1145	A91145	· · ·	0.55 (Si + Fe)		0.05	0.05	0.05	· · ·	· · ·	0.05	· · ·	0.05	· · ·	0.03	0.03	· · ·	99.45
1345	A91345	· · ·	0.30	0.40	0.10	0.05	0.05	· · ·	· · ·	0.05	· · ·	0.05	· · ·	0.03	0.03	· · ·	99.45
1445	· · ·	· · ·	0.50 (Si + Fe)(b)		0.04(b)	· · ·	· · ·	· · ·	· · ·	· · ·	· · ·	· · ·	· · ·	· · ·	· · ·	0.05	99.45
1150	· · ·	· · ·	0.45 (Si + Fe)		0.05–0.20	0.05	0.05	· · ·	· · ·	0.05	· · ·	· · ·	· · ·	0.03	0.03	· · ·	99.50
1350	A91350	E-Al 99.5	0.10	0.40	0.05	0.01	· · ·	0.01	· · ·	0.05	0.03	· · ·	0.05 B, 0.02 (V + Ti)	· · ·	0.03	0.10	99.50
1260	A91260(c)	· · ·	0.40 (Si + Fe)		0.04	0.01	0.03	· · ·	· · ·	0.05	· · ·	0.05	(a)	0.03	0.03	· · ·	99.60
1170	A91170	· · ·	0.30 (Si + Fe)		0.03	0.03	0.02	0.03	· · ·	0.04	· · ·	0.05	· · ·	0.03	0.03	· · ·	99.70
1370	· · ·	E-Al 99.7	0.10	0.25	0.02	0.01	0.02	0.01	· · ·	0.04	0.03	· · ·	0.02 B, 0.02 (V + Ti)	· · ·	0.02	0.10	99.70
1175	A91175	· · ·	0.15 (Si + Fe)		0.10	0.02	0.02	· · ·	· · ·	0.04	0.03	0.05	· · ·	0.02	0.02	· · ·	99.75
1275	· · ·	· · ·	0.08	0.12	0.05–0.10	0.02	0.02	· · ·	· · ·	0.03	0.03	0.03	· · ·	0.02	0.01	· · ·	99.75
1180	A91180	· · ·	0.09	0.09	0.01	0.02	0.02	· · ·	· · ·	0.03	0.03	0.05	· · ·	0.02	0.02	· · ·	99.80
1185	A91185	· · ·	0.15 (Si + Fe)		0.01	0.02	0.02	· · ·	· · ·	0.03	0.03	0.05	· · ·	0.02	0.01	· · ·	99.85
1285	A91285	· · ·	0.08(d)	0.08(d)	0.02	0.01	0.01	· · ·	· · ·	0.03	0.03	0.05	· · ·	0.02	0.01	· · ·	99.85
1385	· · ·	· · ·	0.05	0.12	0.02	0.01	0.02	0.01	· · ·	0.03	0.03	· · ·	0.02 (V + Ti)(e)	· · ·	0.01	· · ·	99.85
1188	A91188	· · ·	0.06	0.06	0.005	0.01	0.01	· · ·	· · ·	0.03	0.03	0.05	(a)	0.01	0.01	· · ·	99.88
1190	· · ·	· · ·	0.05	0.07	0.01	0.01	0.01	0.01	· · ·	0.02	0.02	· · ·	0.01 (V + Ti)(f)	· · ·	0.01	· · ·	99.90
1193	A91193(c)	· · ·	0.04	0.04	0.006	0.01	0.01	· · ·	· · ·	0.03	0.03	0.05	· · ·	0.01	0.01	· · ·	99.93
1199	A91199	· · ·	0.006	0.006	0.006	0.002	0.006	· · ·	· · ·	0.006	0.005	0.005	· · ·	0.002	0.002	· · ·	99.99
2001	· · ·	· · ·	0.20	0.20	5.2–6.0	0.15–0.50	0.20–0.45	0.10	0.05	0.10	· · ·	· · ·	0.05 Zr(g)	0.20	0.05	0.15	rem
2002	· · ·	· · ·	0.35–0.8	0.30	1.5–2.5	0.20	0.50–1.0	0.20	· · ·	0.20	· · ·	· · ·	· · ·	0.20	0.05	0.15	rem
2003	· · ·	· · ·	0.30	0.30	4.0–5.0	0.30–0.8	0.02	· · ·	· · ·	0.10	· · ·	0.05–0.20	0.10–0.25 Zr(h)	0.15	0.05	0.15	rem
2004	· · ·	· · ·	0.20	0.20	5.5–6.5	0.10	0.50	· · ·	· · ·	0.10	· · ·	· · ·	0.30–0.50 Zr	0.05	0.05	0.15	rem
2005	· · ·	· · ·	0.8	0.7	3.5–5.0	1.0	0.20–1.0	0.10	0.20	0.50	· · ·	· · ·	0.20 Bi, 1.0–2.0 Pb	0.20	0.05	0.15	rem
2006	· · ·	· · ·	0.8–1.3	0.7	1.0–2.0	0.6–1.0	0.50–1.4	· · ·	0.20	0.20	· · ·	· · ·	· · ·	0.30	0.05	0.15	rem
2007	· · ·	· · ·	0.8	0.8	3.3–4.6	0.50–1.0	0.40–1.8	0.10	0.20	0.8	· · ·	· · ·	(i)	0.20	0.10	0.30	rem
2008	· · ·	· · ·	0.50–0.8	0.40	0.7–1.1	0.30	0.25–0.50	0.10	· · ·	0.25	· · ·	0.05	· · ·	0.10	0.05	0.15	rem
2011	A92011	AlCu6Bi Pb	0.40	0.7	5.0–6.0	· · ·	· · ·	· · ·	· · ·	0.30	· · ·	· · ·	(j)	· · ·	0.05	0.15	rem
2014	Al92014	AlCu4SiMg	0.50–1.2	0.7	3.9–5.0	0.40–1.2	0.20–0.8	0.10	· · ·	0.25	· · ·	· · ·	(k)	0.15	0.05	0.15	rem
2214	A92214	AlCu4SiMg	0.50–1.2	0.30	3.9–5.0	0.40–1.2	0.20–0.8	0.10	· · ·	0.25	· · ·	· · ·	(k)	0.15	0.05	0.15	rem
2017	A92017	AlCu4MgSi	0.20–0.8	0.7	3.5–4.5	0.40–1.0	0.40–0.8	0.10	· · ·	0.25	· · ·	· · ·	(k)	0.15	0.05	0.15	rem
2117	A92117	AlCu2.5Mg	0.20–0.8	0.7	3.5–4.5	0.40–1.0	0.40–1.0	0.10	· · ·	0.25	· · ·	· · ·	0.25 Zr + Ti	· · ·	0.05	0.15	rem
		AlCu2Mg	0.8	0.7	2.2–3.0	0.20	0.20–0.50	0.10	· · ·	0.25	· · ·	· · ·	· · ·	· · ·	0.05	0.15	rem
2018	A92018	· · ·	0.9	1.0	3.5–4.5	0.20	0.45–0.9	0.10	1.7–2.3	0.25	· · ·	· · ·	· · ·	· · ·	0.05	0.15	rem
2218	A92218	· · ·	0.9	1.0	3.5–4.5	0.20	1.2–1.8	0.10	1.7–2.3	0.25	· · ·	· · ·	· · ·	· · ·	0.05	0.15	rem
2618	A92618	· · ·	0.10–0.25	0.9–1.3	1.9–2.7	· · ·	1.3–1.8	· · ·	0.9–1.2	0.10	· · ·	· · ·	· · ·	0.04–0.10	0.05	0.15	rem
2219	A92219	AlCu6Mn	0.20	0.30	5.8–6.8	0.20–0.40	0.02	· · ·	· · ·	0.10	· · ·	0.05–0.15	0.10–0.25 Zr	0.02–0.10	0.05	0.15	rem
2319	A92319	· · ·	0.20	0.30	5.8–6.8	0.20–0.40	0.02	· · ·	· · ·	0.10	· · ·	0.05–0.15	0.10–0.25 Zr(a)	0.10–0.20	0.05	0.15	rem
2419	A92419	· · ·	0.15	0.18	5.8–6.8	0.20–0.40	0.02	· · ·	· · ·	0.10	· · ·	0.05–0.15	0.10–0.25 Zr	0.02–0.10	0.05	0.15	rem
2519	A92519	· · ·	0.25(l)	0.30(l)	5.3–6.4	0.10–0.50	0.05–0.40	· · ·	· · ·	0.10	· · ·	0.05–0.15	0.10–0.25 Zr	0.02–0.10	0.05	0.15	rem
2021	A92021(c)	· · ·	0.20	0.30	5.8–6.8	0.20–0.40	0.02	· · ·	· · ·	0.10	· · ·	0.05–0.15	0.10–0.25 Zr(m)	0.02–0.10	0.05	0.15	rem
2024	A92024	AlCu4Mg1	0.50	0.50	3.8–4.9	0.30–0.9	1.2–1.8	0.10	· · ·	0.25	· · ·	· · ·	(k)	0.15	0.05	0.15	rem
2124	A92124	· · ·	0.20	0.30	3.8–4.9	0.30–0.9	1.2–1.8	0.10	· · ·	0.25	· · ·	· · ·	(k)	0.15	0.05	0.15	rem

(continued)

(a) 0.0008 Be max for welding electrode and filler wire only. (b) (Si + Fe + Cu) = 0.50 max. (c) Obsolete. (d) 0.14 (Si + Fe) max. (e) 0.02 B max. (f) 0.01 B max. (g) 0.003 Pb max. (h) 0.05 to 0.20 Cd. (i) 0.20 Bi, 0.8 to 1.5 Pb, 0.20 Sn. (j) 0.20 to 0.6 Bi, 0.20 to 0.6 Pb. (k) A (Zr + Ti) limit of 0.20% maximum may be used for extruded and forged products when the supplier or producer and the purchaser have so agreed. (l) 0.40 (Si + Fe) max. (m) 0.05 to 0.20 Cd, 0.03 to 0.08 Sn. (n) 1.9 to 2.6 Li. (o) 1.7 to 2.3 Li. (p) 0.6 to 1.5 Bi, 0.05 Cd max. (q) 0.0008 Be max, 0.05 to 0.25 Zr. (r) 45 to 65% of Mg. (s) 0.40 to 0.7 Bi, 0.40 to 0.7 Pb. (t) 0.25 to 0.40 Ag. (u) 0.15 (Mn + Cr) min. (v) 0.08 to 0.20 Zr, 0.08 to 0.25 (Zr + Ti). (w) 0.20 (Ti + Zr) max. (x) 0.10 to 0.40 Co, 0.05 to 0.30 O. (y) A (Zr + Ti) limit of 0.25% maximum may be used for extruded and forged products when the supplier or producer and the purchaser have so agreed. (z) 0.20 to 0.50 O. (aa) 0.001 B max, 0.003 Cd max, 0.001 Co max, 0.008 Li max. (bb) 0.10 to 0.50 Bi, 0.10 to 0.25 Sn. (cc) 1.0 (Si + Fe) max. (dd) 0.02 to 0.08 Zr. (ee) 2.2 to 2.7 Li. (ff) 2.4 to 2.8 Li. (gg) 2.1 to 2.7 Li. (hh) 2.3 to 2.9 Li
Source: Ref 2, 3, 4

Table 2 (continued)

Grade designation: Aluminum Association	UNS No.	ISO No. R209	Composition, wt%: Si	Fe	Cu	Mn	Mg	Cr	Ni	Zn	Ga	V	Specified other elements	Ti	Unspecified other elements: Each	Total	Al, minimum
2224	A92224	· · ·	0.12	0.15	3.8–4.4	0.30–0.9	1.2–1.8	0.10	· · ·	0.25	· · ·	· · ·	· · ·	0.15	0.05	0.15	rem
2324	A92324	· · ·	0.10	0.12	3.8–4.4	0.30–0.9	1.2–1.8	0.10	· · ·	0.25	· · ·	· · ·	· · ·	0.15	0.05	0.15	rem
2025	A92025	· · ·	0.50–1.2	1.0	3.9–5.0	0.40–1.2	0.05	0.10	· · ·	0.25	· · ·	· · ·	· · ·	0.15	0.05	0.15	rem
2030	· · ·	AlCu4PbMg	0.8	0.7	3.3–4.5	0.20–1.0	0.50–1.3	0.10	· · ·	0.50	· · ·	· · ·	0.20 Bi, 0.8–1.5 Pb	0.20	0.10	0.30	rem
2031	· · ·	· · ·	0.50–1.3	0.6–1.2	1.8–2.8	0.50	0.6–1.2	· · ·	0.6–1.4	0.20	· · ·	· · ·	· · ·	0.20	0.05	0.15	rem
2034	· · ·	· · ·	0.10	0.12	4.2–4.8	0.8–1.3	1.3–1.9	0.05	· · ·	0.20	· · ·	· · ·	0.08–0.15 Zr	0.15	0.05	0.15	rem
2036	A92036	· · ·	0.50	0.50	2.2–3.0	0.10–0.40	0.30–0.6	0.10	· · ·	0.25	· · ·	· · ·	· · ·	0.15	0.05	0.15	rem
2037	A92037	· · ·	0.50	0.50	1.4–2.2	0.10–0.40	0.30–0.8	0.10	· · ·	0.25	· · ·	0.05	· · ·	0.15	0.05	0.15	rem
2038	A92038	· · ·	0.50–1.3	0.6	0.8–1.8	0.10–0.40	0.40–1.0	0.20	· · ·	0.50	0.05	0.05	· · ·	0.15	0.05	0.15	rem
2048	A92048	· · ·	0.15	0.20	2.8–3.8	0.20–0.6	1.2–1.8	· · ·	· · ·	0.25	· · ·	· · ·	· · ·	0.10	0.05	0.15	rem
2090	A92090	· · ·	0.10	0.12	2.4–3.0	0.05	0.25	0.05	· · ·	0.10	· · ·	· · ·	0.08–0.15 Zr(n)	0.15	0.05	0.15	rem
2091	· · ·	· · ·	0.20	0.30	1.8–2.5	0.10	1.1–1.9	0.10	· · ·	0.25	· · ·	· · ·	0.04–0.16 Zr(o)	0.10	0.05	0.15	rem
3002	A93002	· · ·	0.08	0.10	0.15	0.05–0.25	0.05–0.20	· · ·	· · ·	0.05	· · ·	0.05	· · ·	0.03	0.03	0.10	rem
3102	A93102	· · ·	0.40	0.7	0.10	0.05–0.40	· · ·	· · ·	· · ·	0.30	· · ·	· · ·	· · ·	0.10	0.05	0.15	rem
3003	A93003	AlMn1Cu	0.6	0.7	0.05–0.20	1.0–1.5	· · ·	· · ·	· · ·	0.10	· · ·	· · ·	· · ·	· · ·	0.05	0.15	rem
3103	· · ·	· · ·	0.50	0.7	0.10	0.9–1.5	0.30	0.10	· · ·	0.20	· · ·	· · ·	0.10 Zr + Ti	· · ·	0.05	0.15	rem
3203	· · ·	· · ·	0.6	0.7	0.05	1.0–1.5	· · ·	· · ·	· · ·	0.10	· · ·	· · ·	(a)	· · ·	0.05	0.15	rem
3303	A93303	AlMn1	0.6	0.7	0.05–0.20	1.0–1.5	· · ·	· · ·	· · ·	0.30	· · ·	· · ·	· · ·	· · ·	0.05	0.15	rem
3004	A93004	AlMn1Mg1	0.30	0.7	0.25	1.0–1.5	0.8–1.3	· · ·	· · ·	0.25	· · ·	· · ·	· · ·	· · ·	0.05	0.15	rem
3104	A93104	· · ·	0.6	0.8	0.05–0.25	0.8–1.4	0.8–1.3	· · ·	· · ·	0.25	0.05	0.05	· · ·	0.10	0.05	0.15	rem
3005	A93005	AlMn1Mg0.5	0.6	0.7	0.30	1.0–1.5	0.20–0.6	0.10	· · ·	0.25	· · ·	· · ·	· · ·	0.10	0.05	0.15	rem
3105	A93105	AlMn0.5Mg0.5	0.6	0.7	0.30	0.30–0.8	0.20–0.8	0.20	· · ·	0.40	· · ·	· · ·	· · ·	0.10	0.05	0.15	rem
3006	A93006	· · ·	0.50	0.7	0.10–0.30	0.50–0.8	0.30–0.6	0.20	· · ·	0.15–0.40	· · ·	· · ·	· · ·	0.10	0.05	0.15	rem
3007	A93007	· · ·	0.50	0.7	0.05–0.30	0.30–0.8	0.6	0.20	· · ·	0.40	· · ·	· · ·	· · ·	0.10	0.05	0.15	rem
3107	A93107	· · ·	0.6	0.7	0.05–0.15	0.40–0.9	· · ·	· · ·	· · ·	0.20	· · ·	· · ·	· · ·	0.10	0.05	0.15	rem
3207	· · ·	· · ·	0.30	0.45	0.10	0.40–0.8	0.10	· · ·	· · ·	0.10	· · ·	· · ·	· · ·	· · ·	0.05	0.10	rem
3307	· · ·	· · ·	0.6	0.8	0.30	0.50–0.9	0.30	· · ·	· · ·	0.25	· · ·	· · ·	· · ·	0.10	0.05	0.15	rem
3008	· · ·	· · ·	0.40	0.7	0.10	1.2–1.8	0.01	0.05	0.05	0.05	· · ·	· · ·	0.10–0.50 Zr	0.10	0.05	0.15	rem
3009	A93009	· · ·	1.0–1.8	0.7	0.10	1.2–1.8	0.10	0.05	0.05	0.05	· · ·	· · ·	0.10 Zr	0.10	0.05	0.15	rem
3010	A93010	· · ·	0.10	0.20	0.03	0.20–0.9	· · ·	0.05–0.40	· · ·	0.05	· · ·	0.05	· · ·	0.05	0.03	0.10	rem
3011	A93011	· · ·	0.40	0.7	0.05–0.20	0.8–1.2	· · ·	0.10–0.40	· · ·	0.10	· · ·	· · ·	0.10–0.30 Zr	0.10	0.05	0.15	rem
3012	· · ·	· · ·	0.6	0.7	0.10	0.50–1.1	0.10	0.20	· · ·	0.10	· · ·	· · ·	· · ·	0.10	0.05	0.15	rem
3013	· · ·	· · ·	0.6	1.0	0.50	0.9–1.4	0.20–0.6	· · ·	· · ·	0.50–1.0	· · ·	· · ·	· · ·	· · ·	0.05	0.15	rem
3014	· · ·	· · ·	0.6	1.0	0.50	1.0–1.5	0.10	· · ·	· · ·	0.50–1.0	· · ·	· · ·	· · ·	0.10	0.05	0.15	rem
3015	· · ·	· · ·	0.6	0.8	0.30	0.50–0.9	0.20–0.7	· · ·	· · ·	0.25	· · ·	· · ·	· · ·	0.10	0.05	0.15	rem
3016	· · ·	· · ·	0.6	0.8	0.30	0.50–0.9	0.50–0.8	· · ·	· · ·	0.25	· · ·	· · ·	· · ·	0.10	0.05	0.15	rem
4004	A94004	· · ·	9.0–10.5	0.8	0.25	0.10	1.0–2.0	· · ·	· · ·	0.20	· · ·	· · ·	· · ·	· · ·	0.05	0.15	rem
4104	A94104	· · ·	9.0–10.5	0.8	0.25	0.10	1.0–2.0	· · ·	· · ·	0.20	· · ·	· · ·	0.02–0.20 Bi	· · ·	0.05	0.15	rem
4006	· · ·	· · ·	0.8–1.2	0.50–0.8	0.05	0.03	0.01	0.20	· · ·	0.05	· · ·	· · ·	· · ·	· · ·	0.05	0.15	rem
4007	· · ·	· · ·	1.0–1.7	0.40–1.0	0.20	0.8–1.5	0.20	0.05–0.25	0.15–0.7	0.10	· · ·	· · ·	0.05 Co	0.10	0.05	0.15	rem
4008	A94008	· · ·	6.5–7.5	0.09	0.05	0.05	0.30–0.45	· · ·	· · ·	0.05	· · ·	· · ·	(a)	0.04–0.15	0.05	0.15	rem
4009	· · ·	· · ·	4.5–5.5	0.20	1.0–1.5	0.10	0.45–0.6	· · ·	· · ·	0.10	· · ·	· · ·	(a)	0.20	0.05	0.15	rem
4010	· · ·	· · ·	6.5–7.5	0.20	0.20	0.10	0.30–0.45	· · ·	· · ·	0.10	· · ·	· · ·	(a)	0.20	0.05	0.15	rem
4011	· · ·	· · ·	6.5–7.5	0.20	0.20	0.10	0.45–0.7	· · ·	· · ·	0.10	· · ·	· · ·	0.04–0.07 Be	0.04–0.20	0.05	0.15	rem
4013	· · ·	· · ·	3.5–4.5	0.35	0.05–0.20	0.03	0.05–0.20	· · ·	· · ·	0.05	· · ·	· · ·	(p)	0.02	0.05	0.15	rem
4032	A94032	· · ·	11.0–13.5	1.0	0.50–1.3	· · ·	0.8–1.3	0.10	0.50–1.3	0.25	· · ·	· · ·	· · ·	· · ·	0.05	0.15	rem
4043	A94043	AlSi5	4.5–6.0	0.8	0.30	0.05	0.05	· · ·	· · ·	0.10	· · ·	· · ·	(a)	0.20	0.05	0.15	rem
4343	A94343	· · ·	6.8–8.2	0.8	0.25	0.10	· · ·	· · ·	· · ·	0.20	· · ·	· · ·	· · ·	· · ·	0.05	0.15	rem
4543	A94543	· · ·	5.0–7.0	0.50	0.10	0.05	0.10–0.40	0.05	· · ·	0.10	· · ·	· · ·	· · ·	0.10	0.05	0.15	rem
4643	A94643	· · ·	3.6–4.6	0.8	0.10	0.05	0.10–0.30	· · ·	· · ·	0.10	· · ·	· · ·	(a)	0.15	0.05	0.15	rem
4044	A94044	· · ·	7.8–9.2	0.8	0.25	0.10	· · ·	· · ·	· · ·	0.20	· · ·	· · ·	· · ·	· · ·	0.05	0.15	rem
4045	A94045	· · ·	9.0–11.0	0.8	0.30	0.05	0.05	· · ·	· · ·	0.10	· · ·	· · ·	· · ·	0.20	0.05	0.15	rem
4145	A94145	· · ·	9.3–10.7	0.8	3.3–4.7	0.15	0.15	0.15	· · ·	0.20	· · ·	· · ·	(a)	· · ·	0.05	0.15	rem
4047	A94047	AlSi12	11.0–13.0	0.8	0.30	0.15	0.10	· · ·	· · ·	0.20	· · ·	· · ·	(a)	· · ·	0.05	0.15	rem
5005	A95005	AlMg1	0.30	0.7	0.20	0.20	0.50–1.1	0.10	· · ·	0.25	· · ·	· · ·	· · ·	· · ·	0.05	0.15	rem
5205	· · ·	AlMg1(B)	0.15	0.7	0.03–0.10	0.10	0.6–1.0	0.10	· · ·	0.05	· · ·	· · ·	· · ·	· · ·	0.05	0.15	rem
5006	A95006	· · ·	0.40	0.8	0.10	0.40–0.8	0.8–1.3	0.10	· · ·	0.25	· · ·	· · ·	· · ·	0.10	0.05	0.15	rem
5010	A95010	· · ·	0.40	0.7	0.25	0.10–0.30	0.20–0.6	0.15	· · ·	0.30	· · ·	· · ·	· · ·	0.10	0.05	0.15	rem
5013	· · ·	· · ·	0.20	0.25	0.03	0.30–0.50	3.2–3.8	0.03	0.03	0.10	· · ·	· · ·	0.05 Zr(g)	0.10	0.05	0.15	rem
5014	· · ·	· · ·	0.40	0.40	0.20	0.20–0.9	4.0–5.5	0.20	· · ·	0.7–1.5	· · ·	· · ·	· · ·	0.20	0.05	0.15	rem
5016	A95016	· · ·	0.25	0.6	0.20	0.40–0.7	1.4–1.9	0.10	· · ·	0.15	· · ·	· · ·	· · ·	0.05	0.05	0.15	rem
5017	· · ·	· · ·	0.40	0.7	0.18–0.28	0.6–0.8	1.9–2.2	· · ·	· · ·	· · ·	· · ·	· · ·	· · ·	0.09	0.05	0.15	rem
5040	A95040	· · ·	0.30	0.7	0.25	0.9–1.4	1.0–1.5	0.10–0.30	· · ·	0.25	· · ·	· · ·	· · ·	· · ·	0.05	0.15	rem
5042	A95042	· · ·	0.20	0.35	0.15	0.20–0.50	3.0–4.0	0.10	· · ·	0.25	· · ·	· · ·	· · ·	0.10	0.05	0.15	rem
5043	A95043	· · ·	0.40	0.7	0.05–0.35	0.7–1.2	0.7–1.3	0.05	· · ·	0.25	0.05	0.05	· · ·	0.10	0.05	0.15	rem
5049	· · ·	· · ·	0.40	0.50	0.10	0.50–1.1	1.6–2.5	0.30	· · ·	0.20	· · ·	· · ·	· · ·	0.10	0.05	0.15	rem
5050	A95050	AlMg1.5(C) AlMg1.5	0.40	0.7	0.20	0.10	1.1–1.8	0.10	· · ·	0.25	· · ·	· · ·	· · ·	· · ·	0.05	0.15	rem

(continued)

(a) 0.0008 Be max for welding electrode and filler wire only. (b) (Si + Fe + Cu) = 0.50 max. (c) Obsolete. (d) 0.14 (Si + Fe) max. (e) 0.02 B max. (f) 0.01 B max. (g) 0.003 Pb max. (h) 0.05 to 0.20 Cd. (i) 0.20 Bi, 0.8 to 1.5 Pb, 0.20 Sn. (j) 0.20 to 0.6 Bi, 0.20 to 0.6 Pb. (k) A (Zr + Ti) limit of 0.20% maximum may be used for extruded and forged products when the supplier or producer and the purchaser have so agreed. (l) 0.40 (Si + Fe) max. (m) 0.05 to 0.20 Cd, 0.03 to 0.08 Sn. (n) 1.9 to 2.6 Li. (o) 1.7 to 2.3 Li. (p) 0.6 to 1.5 Bi, 0.05 Cd max. (q) 0.0008 Be max, 0.05 to 0.25 Zr. (r) 45 to 65% of Mg. (s) 0.40 to 0.7 Bi, 0.40 to 0.7 Pb. (t) 0.25 to 0.40 Ag. (u) 0.15 (Mn + Cr) min. (v) 0.08 to 0.20 Zr, 0.08 to 0.25 (Zr + Ti). (w) 0.20 (Ti + Zr) max. (x) 0.10 to 0.40 Co, 0.05 to 0.30 O. (y) A (Zr + Ti) limit of 0.25% maximum may be used for extruded and forged products when the supplier or producer and the purchaser have so agreed. (z) 0.20 to 0.50 O. (aa) 0.001 B max, 0.003 Cd max, 0.001 Co max, 0.008 Li max. (bb) 0.10 to 0.50 Bi, 0.10 to 0.25 Sn. (cc) 1.0 (Si + Fe) max. (dd) 0.02 to 0.08 Zr. (ee) 2.2 to 2.7 Li. (ff) 2.4 to 2.8 Li. (gg) 2.1 to 2.7 Li. (hh) 2.3 to 2.9 Li

Source: Ref 2, 3, 4

Table 2 (continued)

Grade designation			Composition, wt%														
Aluminum Association	UNS No.	ISO No. R209	Si	Fe	Cu	Mn	Mg	Cr	Ni	Zn	Ga	V	Specified other elements	Ti	Unspecified other elements Each	Unspecified other elements Total	Al, minimum
5150	···	···	0.08	0.10	0.10	0.03	1.3–1.7	···	···	0.10	···	···	···	0.06	0.03	0.10	rem
5250	A95250	···	0.08	0.10	0.10	0.05–0.15	1.3–1.8	···	···	0.05	0.03	0.05	···	···	0.03	0.10	rem
5051	A95051	AlMg2	0.40	0.7	0.25	0.20	1.7–2.2	0.10	···	0.25	···	···	···	0.10	0.05	0.15	rem
5151	A95151	···	0.20	0.35	0.15	0.10	1.5–2.1	0.10	···	0.15	···	···	···	0.10	0.05	0.15	rem
5251	···	AlMg2	0.40	0.50	0.15	0.10–0.50	1.7–2.4	0.15	···	0.15	···	···	···	0.15	0.05	0.15	rem
5351	A95351	···	0.08	0.10	0.10	0.10	1.6–2.2	···	···	0.05	···	0.05	···	···	0.03	0.10	rem
5451	A95154	AlMg3.5	0.25	0.40	0.10	0.10	1.8–2.4	0.15–0.35	0.05	0.10	···	···	···	0.05	0.05	0.15	rem
5052	A95052	AlMg2.5	0.25	0.40	0.10	0.10	2.2–2.8	0.15–0.35	···	0.10	···	···	···	···	0.05	0.15	rem
5252	A95252	···	0.08	0.10	0.10	0.10	2.2–2.8	···	···	0.05	···	0.05	···	···	0.03	0.10	rem
5352	A95352	···	0.45 (Si + Fe)		0.10	0.10	2.2–2.8	0.10	···	0.10	···	···	···	0.10	0.05	0.15	rem
5552	A95652	···	0.04	0.05	0.10	0.10	2.2–2.8	···	···	0.05	···	0.05	···	···	0.03	0.10	rem
5652	A95652	···	0.40 (Si + Fe)		0.04	0.01	2.2–2.8	0.15–0.35	···	0.10	···	···	···	···	0.05	0.15	rem
5154	···	AlMg3.5	0.25	0.40	0.10	0.10	3.1–3.9	0.15–0.35	···	0.20	···	···	(a)	0.20	0.05	0.15	rem
5254	A95254	···	0.45 (Si + Fe)		0.05	0.01	3.1–3.9	0.15–0.35	···	0.20	···	···	···	0.05	0.05	0.15	rem
5454	A95454	AlMg3Mn	0.25	0.40	0.10	0.50–1.0	2.4–3.0	0.05–0.20	···	0.25	···	···	···	0.20	0.05	0.15	rem
5554	A95554	AlMg3Mn(A)	0.25	0.40	0.10	0.50–1.0	2.4–3.0	0.05–0.20	···	0.25	···	···	(a)	0.05–0.20	0.05	0.15	rem
5654	A95654	···	0.45 (Si + Fe)		0.05	0.01	3.1–3.9	0.15–0.35	···	0.20	···	···	(a)	0.05–0.15	0.05	0.15	rem
5754	A95754	AlMg3	0.40	0.40	0.10	0.50	2.6–3.6	0.30	···	0.20	···	···	0.10–0.6 (Mn + Cr)	0.15	0.05	0.15	rem
5854	···	···	0.45 (Si + Fe)		0.10	0.10–0.50	3.1–3.9	0.15–0.35	···	0.20	···	···	···	0.20	0.05	0.15	rem
5056	A95056	AlMg5 AlMg5Cr	0.30	0.40	0.10	0.05–0.20	4.5–5.6	0.05–0.20	···	0.10	···	···	···	···	0.05	0.15	rem
5356	A95356	AlMg5Cr(A)	0.25	0.40	0.10	0.05–0.20	4.5–5.5	0.05–0.20	···	0.10	···	···	(a)	0.06–0.20	0.05	0.15	rem
5456	A95456	AlMg5Mn1	0.25	0.40	0.10	0.50–1.0	4.7–5.5	0.05–0.20	···	0.25	···	···	···	0.20	0.05	0.15	rem
5556	A95556	···	0.25	0.40	0.10	0.50–1.0	4.7–5.5	0.05–0.20	···	0.25	···	···	(a)	0.05–0.20	0.05	0.15	rem
5357	A95357	···	0.12	0.17	0.20	0.15–0.45	0.8–1.2	···	···	0.05	···	···	···	···	0.05	0.15	rem
5457	A95457	···	0.08	0.10	0.20	0.15–0.45	0.8–1.2	···	···	0.05	···	0.05	···	···	0.03	0.10	rem
5557	A95557	···	0.10	0.12	0.15	0.10–0.40	0.40–0.8	···	···	···	···	0.05	···	···	0.03	0.10	rem
5657	A95657	···	0.08	0.10	0.10	0.03	0.6–1.0	···	···	0.05	0.03	0.05	···	···	0.02	0.05	rem
5280	···	···	0.35 (Si + Fe)		0.10	0.20–0.7	3.5–4.5	0.05–0.25	···	1.5–2.8	···	···	(q)	···	0.05	0.15	rem
5082	A95082	···	0.20	0.35	0.15	0.15	4.0–5.0	0.15	···	0.25	···	···	···	0.10	0.05	0.15	rem
5182	A95182	···	0.20	0.35	0.15	0.20–0.50	4.0–5.0	0.10	···	0.25	···	···	···	0.10	0.05	0.15	rem
5083	A95083	AlMg4.5Mn	0.40–0.7	0.40	0.10	0.40–0.10	4.0–4.9	0.05–0.25	···	0.25	···	···	···	0.15	0.05	0.15	rem
5183	A95183	AlMg4.5Mn	0.40–0.7(A)	0.40	0.10	0.50–1.0	4.3–5.2	0.05–0.25	···	0.25	···	···	(a)	0.15	0.05	0.15	rem
5283	···	···	0.30	0.30	0.03	0.50–1.0	4.5–5.1	0.05	0.03	0.10	···	···	0.05 Zr	0.03	0.05	0.15	rem
5086	A95086	AlMg4	0.40	0.50	0.10	0.20–0.7	3.5–4.5	0.05–0.25	···	0.25	···	···	···	0.15	0.05	0.15	rem
6101	A96101	E-AlMgSi	0.30–0.7	0.50	0.10	0.03	0.35–0.8	0.03	···	0.10	···	···	0.06 B	···	0.03	0.10	rem
6201	A96201	···	0.50–0.9	0.50	0.10	0.03	0.6–0.9	0.03	···	0.10	···	···	0.06 B	···	0.03	0.10	rem
6301	A96301	···	0.50–0.9	0.7	0.10	0.15	0.6–0.9	0.10	···	0.25	···	···	···	0.15	0.05	0.15	rem
6002	···	···	0.6–0.9	0.25	0.10–0.25	0.10–0.20	0.45–0.7	0.05	···	···	···	···	0.09–0.14 Zr	0.08	0.05	0.15	rem
6003	A96803	AlMg1Si	0.35–1.0	0.6	0.10	0.8	0.8–1.5	0.35	···	0.20	···	···	···	0.10	0.05	0.15	rem
6103	···	···	0.35–1.0	0.6	0.20–0.30	0.8	0.8–1.5	0.35	···	0.20	···	···	···	0.10	0.05	0.15	rem
6004	A96004	···	0.30–0.6	0.10–0.30	0.10	0.20–0.6	0.40–0.7	···	···	0.05	···	···	···	···	0.05	0.15	rem
6005	A96005	AlSiMg	0.6–0.9	0.35	0.10	0.10	0.40–0.6	0.10	···	0.10	···	···	···	0.10	0.05	0.15	rem
6105	A96105	···	0.6–1.0	0.35	0.10	0.10	0.45–0.8	0.10	···	0.10	···	···	···	0.10	0.05	0.15	rem
6205	A96205	···	0.6–0.9	0.7	0.20	0.05–0.15	0.40–0.6	0.05–0.15	···	0.25	···	···	0.05–0.15 Zr	0.15	0.05	0.15	rem
6006	A96006	···	0.20–0.6	0.35	0.15–0.30	0.15–0.20	0.45–0.9	0.10	···	0.10	···	···	···	0.10	0.05	0.15	rem
6106	···	···	0.30–0.6	0.35	0.25	0.05–0.20	0.40–0.8	0.20	···	0.10	···	···	···	···	0.05	0.10	rem
X6206	···	···	0.35–0.7	0.35	0.20–0.50	0.13–0.30	0.45–0.8	0.10	···	0.20	···	···	···	0.10	0.05	0.15	rem
6007	A96007	···	0.9–1.4	0.7	0.20	0.05–0.25	0.6–0.9	0.05–0.25	···	0.25	···	···	0.05–0.20 Zr	0.15	0.05	0.15	rem
6008	···	···	0.50–0.9	0.35	0.30	0.30	0.40–0.7	0.30	···	0.20	···	0.05–0.20	···	0.10	0.05	0.15	rem
6009	A96009	···	0.6–1.0	0.50	0.15–0.6	0.20–0.8	0.40–0.8	0.10	···	0.25	···	···	···	0.10	0.05	0.15	rem
6010	A96010	···	0.8–1.2	0.50	0.15–0.6	0.20–0.8	0.6–1.0	0.10	···	0.25	···	···	···	0.10	0.05	0.15	rem
6110	A96110	···	0.7–1.5	0.8	0.20–0.7	0.20–0.7	0.50–1.1	0.04–0.25	···	0.30	···	···	···	0.15	0.05	0.15	rem
6011	A96011	···	0.6–1.2	1.0	0.40–0.9	0.8	0.6–1.2	0.30	0.20	1.5	···	···	···	0.20	0.05	0.15	rem
6111	A96111	···	0.7–1.1	0.40	0.50–0.9	0.15–0.45	0.50–1.0	0.10	···	0.15	···	···	···	0.10	0.05	0.15	rem
6012	···	···	0.6–1.4	0.50	0.10	0.40–1.0	0.6–1.2	0.30	···	0.30	···	···	0.7 Bi, 0.40–2.0 Pb	0.20	0.05	0.15	rem
X6013	···	···	0.6–1.0	0.50	0.6–1.1	0.20–0.8	0.8–1.2	0.10	···	0.25	···	···	···	0.10	0.05	0.15	rem
6014	···	···	0.30–0.6	0.35	0.25	0.05–0.20	0.40–0.8	0.20	···	0.10	···	0.05–0.20	···	0.10	0.05	0.15	rem
6015	···	···	0.20–0.40	0.10–0.30	0.10–0.25	0.10	0.8–1.1	0.10	···	0.10	···	···	···	0.10	0.05	0.15	rem
6016	···	···	1.0–1.5	0.50	0.20	0.20	0.25–0.6	0.10	···	0.20	···	···	···	0.15	0.05	0.15	rem
6017	A96017	···	0.55–0.7	0.15–0.30	0.05–0.20	0.10	0.45–0.6	0.10	···	0.05	···	···	···	0.05	0.05	0.15	rem
6151	A96151	···	0.6–1.2	1.0	0.35	0.20	0.45–0.8	0.15–0.35	···	0.25	···	···	···	0.15	0.05	0.15	rem
6351	A96351	AlSi1Mg0.5Mn	0.7–1.3	0.50	0.10	0.40–0.8	0.40–0.8	···	···	0.20	···	···	···	0.20	0.05	0.15	rem
6951	A96951	···	0.20–0.50	0.8	0.15–0.40	0.10	0.40–0.8	···	···	0.20	···	···	···	···	0.05	0.15	rem
6053	A96053	···	(r)	0.35	0.10	···	1.1–1.4	0.15–0.35	···	0.10	···	···	···	···	0.05	0.15	rem
6253	A96253	···	(r)	0.50	0.10	···	1.0–1.5	0.04–0.35	···	1.6–2.4	···	···	···	···	0.05	0.15	rem

(continued)

(a) 0.0008 Be max for welding electrode and filler wire only. (b) (Si + Fe + Cu) = 0.50 max. (c) Obsolete. (d) 0.14 (Si + Fe) max. (e) 0.02 B max. (f) 0.01 B max. (g) 0.003 Pb max. (h) 0.05 to 0.20 Cd. (i) 0.20 Bi, 0.8 to 1.5 Pb, 0.20 Sn. (j) 0.20 to 0.6 Bi, 0.20 to 0.6 Pb. (k) A (Zr + Ti) limit of 0.20% maximum may be used for extruded and forged products when the supplier or producer and the purchaser have so agreed. (l) 0.40 (Si + Fe) max. (m) 0.05 to 0.20 Cd, 0.03 to 0.08 Sn. (n) 1.9 to 2.6 Li. (o) 1.7 to 2.3 Li. (p) 0.6 to 1.5 Bi, 0.05 Cd max. (q) 0.0008 Be max, 0.05 to 0.25 Zr. (r) 45 to 65% of Mg. (s) 0.40 to 0.7 Bi, 0.40 to 0.7 Pb. (t) 0.25 to 0.40 Ag. (u) 0.15 (Mn + Cr) min. (v) 0.08 to 0.20 Zr, 0.08 to 0.25 (Zr + Ti). (w) 0.20 (Ti + Zr) max. (x) 0.10 to 0.40 Co, 0.05 to 0.30 O. (y) A (Zr + Ti) limit of 0.25% maximum may be used for extruded and forged products when the supplier or producer and the purchaser have so agreed. (z) 0.20 to 0.50 O. (aa) 0.001 B max, 0.003 Cd max, 0.001 Co max, 0.008 Li max. (bb) 0.10 to 0.50 Bi, 0.10 to 0.25 Sn. (cc) 1.0 (Si + Fe) max. (dd) 0.02 to 0.08 Zr. (ee) 2.2 to 2.7 Li. (ff) 2.4 to 2.8 Li. (gg) 2.1 to 2.7 Li. (hh) 2.3 to 2.9 Li
Source: Ref 2, 3, 4

Table 2 (continued)

Grade designation: Aluminum Association	UNS No.	ISO No. R209	Composition, wt%: Si	Fe	Cu	Mn	Mg	Cr	Ni	Zn	Ga	V	Specified other elements	Ti	Unspecified other elements: Each	Total	Al, minimum
6060	A96060	AlMgSi	0.30–0.6	0.10–0.30	0.10	0.10	0.35–0.6	0.05	...	0.15	...	...	...	0.10	0.05	0.15	rem
6061	A96061	AlMg1SiCu	0.40–0.8	0.7	0.15–0.40	0.15	0.8–1.2	0.04–0.35	...	0.25	...	...	...	0.15	0.05	0.15	rem
6261	A96261	...	0.40–0.7	0.40	0.15–0.40	0.20–0.35	0.7–1.0	0.10	...	0.20	...	...	...	0.10	0.05	0.15	rem
6162	A96162	...	0.40–0.8	0.50	0.20	0.10	0.7–1.1	0.10	...	0.25	...	...	...	0.10	0.05	0.15	rem
6262	A96262	AlMg1SiPb	0.40–0.8	0.7	0.15–0.40	0.15	0.8–1.2	0.04–0.14	...	0.25	...	...	(s)	0.15	0.05	0.15	rem
6063	A96063	AlMg0.5Si	0.20–0.6	0.35	0.10	0.10	0.45–0.9	0.10	...	0.10	...	...	...	0.10	0.05	0.15	rem
6463	A96463	AlMg0.7Si	0.20–0.6	0.15	0.20	0.05	0.45–0.9	...	...	0.05	...	...	...	...	0.05	0.15	rem
6763	A96763	...	0.20–0.6	0.08	0.04–0.16	0.03	0.45–0.9	...	...	0.03	...	0.05	...	...	0.03	0.10	rem
6863	...	...	0.40–0.6	0.15	0.05–0.20	0.05	0.50–0.8	0.05	...	0.10	...	...	...	0.10	0.05	0.15	rem
6066	A96066	...	0.9–1.8	0.50	0.7–1.2	0.6–1.1	0.8–1.4	0.40	...	0.25	...	...	...	0.20	0.05	0.15	rem
6070	A96070	...	1.0–1.7	0.50	0.15–0.40	0.40–1.0	0.50–1.2	0.10	...	0.25	...	...	...	0.15	0.05	0.15	rem
6081	...	...	0.7–1.1	0.50	0.10	0.10–0.45	0.6–1.0	0.10	...	0.20	...	...	...	0.15	0.05	0.15	rem
6181	...	AlSiMg0.8	0.8–1.2	0.45	0.10	0.15	0.6–1.0	0.10	...	0.20	...	...	...	0.10	0.05	0.15	rem
6082	...	AlSi1MgMn	0.7–1.3	0.50	0.10	0.40–1.0	0.6–1.2	0.25	...	0.20	...	...	...	0.10	0.05	0.15	rem
7001	A97001	...	0.35	0.40	1.6–2.6	0.20	2.6–3.4	0.18–0.35	...	6.8–8.0	...	...	...	0.20	0.05	0.15	rem
7003	...	...	0.30	0.35	0.20	0.30	0.50–1.0	0.20	...	5.0–6.5	...	...	0.05–0.25 Zr	0.20	0.05	0.15	rem
7004	A97004	...	0.25	0.35	0.05	0.20–0.7	1.0–2.0	0.05	...	3.8–4.6	...	...	0.10–0.20 Zr	0.05	0.05	0.15	rem
7005	A97005	...	0.35	0.40	0.10	0.20–0.7	1.0–1.8	0.06–0.20	...	4.0–5.0	...	...	0.08–0.20 Zr	0.01–0.06	0.05	0.15	rem
7008	A97008	...	0.10	0.10	0.05	0.05	0.7–1.4	0.12–0.25	...	4.5–5.5	...	...	...	0.05	0.05	0.10	rem
7108	A97108	...	0.10	0.10	0.05	0.05	0.7–1.4	...	...	4.5–5.5	...	...	0.12–0.25 Zr	0.05	0.05	0.15	rem
7009	...	...	0.20	0.20	0.6–1.3	0.10	2.1–2.9	0.10–0.25	...	5.5–5.6	...	...	(t)	0.20	0.05	0.15	rem
7109	...	...	0.10	0.15	0.8–1.3	0.10	2.2–2.7	0.04–0.08	...	5.8–6.5	...	...	0.10–0.20 Zr (t)	0.10	0.05	0.15	rem
7010	...	AlZn6MgCu	0.12	0.15	1.5–2.0	0.10	2.1–2.6	0.05	0.05	5.7–6.7	...	...	0.10–0.16 Zr	0.06	0.05	0.15	rem
7011	A97011(c)	...	0.15	0.20	0.05	0.10–0.30	1.0–1.6	0.05–0.20	...	4.0–5.5	...	...	...	0.05	0.05	0.15	rem
7012	...	...	0.15	0.25	0.8–1.2	0.08–0.15	1.8–2.2	0.04	...	5.8–6.5	...	...	0.10–0.18 Zr	0.02–0.08	0.05	0.15	rem
7013	A97013	...	0.6	0.7	0.10	1.0–1.5	...	...	...	1.5–2.0	...	...	...	...	0.05	0.15	rem
7014	...	...	0.50	0.50	0.30–0.7	0.30–0.7	2.2–3.2	...	0.10	5.2–6.2	...	...	0.20 (Ti + Zr)	...	0.05	0.15	rem
7015	...	...	0.20	0.30	0.06–0.15	0.10	1.3–2.1	0.15	...	4.6–5.2	...	...	0.10–0.20 Zr	0.10	0.05	0.15	rem
7016	A97016	...	0.10	0.12	0.45–1.0	0.03	0.8–1.4	...	...	4.0–5.0	...	0.05	...	0.03	0.03	0.10	rem
7116	...	...	0.15	0.30	0.50–1.1	0.05	0.8–1.4	...	...	4.2–5.2	0.03	0.05	...	0.05	0.05	0.15	rem
7017	...	...	0.35	0.45	0.20	0.05–0.50	2.0–3.0	0.35	0.10	4.0–5.2	...	...	0.10–0.25 Zr (u)	0.15	0.05	0.15	rem
7018	...	...	0.35	0.45	0.20	0.15–0.50	0.7–1.5	0.20	0.10	4.5–5.5	...	...	0.10–0.25 Zr	0.15	0.05	0.15	rem
7019	...	...	0.35	0.45	0.20	0.15–0.50	1.5–2.5	0.20	0.10	3.5–4.5	...	...	0.10–0.25 Zr	0.15	0.05	0.15	rem
7020	...	AlZn4.5Mg1	0.35	0.40	0.20	0.05–0.50	1.0–1.4	0.10–0.35	...	4.0–5.0	...	...	(v)	...	0.05	0.15	rem
7021	A97021	...	0.25	0.40	0.25	0.10	1.2–1.8	0.05	...	5.0–6.0	...	...	0.08–0.18 Zr	0.10	0.05	0.15	rem
7022	...	...	0.50	0.50	0.50–1.0	0.10–0.40	2.6–3.7	0.10–0.30	...	4.3–5.2	...	...	0.20 (Ti + Zr)	...	0.05	0.15	rem
7023	...	...	0.50	0.50	0.50–1.0	0.10–0.6	2.0–3.0	0.05–0.35	...	4.0–6.0	...	...	...	0.10	0.05	0.15	rem
7024	...	...	0.30	0.40	0.10	0.10–0.6	0.50–1.0	0.05–0.35	...	3.0–5.0	...	...	...	0.10	0.05	0.15	rem
7025	...	...	0.30	0.40	0.10	0.10–0.6	0.8–1.5	0.05–0.35	...	3.0–5.0	...	...	...	0.10	0.05	0.15	rem
7026	...	...	0.08	0.12	0.6–0.9	0.05–0.20	1.5–1.9	...	...	4.6–5.2	...	...	0.09–0.14 Zr	0.05	0.03	0.10	rem
7027	...	...	0.25	0.40	0.10–0.30	0.10–0.40	0.7–1.1	...	...	3.5–4.5	...	...	0.05–0.30 Zr	0.10	0.05	0.15	rem
7028	...	...	0.35	0.50	0.10–0.30	0.15–0.6	1.5–2.3	0.20	...	4.5–5.2	...	...	0.08–0.25 (Zr + Ti)	0.05	0.05	0.15	rem
7029	A97029	...	0.10	0.12	0.50–0.9	0.03	1.3–2.0	...	...	4.2–5.2	...	0.05	...	0.05	0.03	0.10	rem
7129	A97129	...	0.15	0.30	0.50–0.9	0.10	1.3–2.0	0.10	...	4.2–5.2	0.03	0.05	...	0.05	0.05	0.15	rem
7229	...	...	0.06	0.08	0.50–0.9	0.03	1.3–2.0	...	...	4.2–5.2	...	0.05	...	0.05	0.03	0.10	rem
7030	...	...	0.20	0.30	0.20–0.40	0.05	1.0–1.5	0.04	...	4.8–5.9	0.03	...	0.03 Zr	0.03	0.05	0.15	rem
7039	A97039	...	0.30	0.40	0.10	0.10–0.40	2.3–3.3	0.15–0.25	...	3.5–4.5	...	...	...	0.10	0.05	0.15	rem
7046	A97046	...	0.20	0.40	0.25	0.30	1.0–1.6	0.20	...	6.6–7.6	...	...	0.10–0.18 Zr	0.06	0.05	0.15	rem
7146	A97146	...	0.20	0.40	...	...	1.0–1.6	...	...	6.6–7.6	...	...	0.10–0.18 Zr	0.06	0.05	0.15	rem
7049	A97049	...	0.25	0.35	1.2–1.9	0.20	2.0–2.9	0.10–0.22	...	7.2–8.2	...	...	...	0.10	0.05	0.15	rem
7149	A97149	...	0.15	0.20	1.2–1.9	0.20	2.0–2.9	0.10–0.22	...	7.2–8.2	...	...	...	0.10	0.05	0.15	rem
7050	A97050	AlZn6CuMgZr	0.12	0.15	2.0–2.6	0.10	1.9–2.6	0.04	...	5.7–6.7	...	...	0.08–0.15 Zr	0.06	0.05	0.15	rem
7150	A97150	...	0.12	0.15	1.9–2.5	0.10	2.0–2.7	0.04	...	5.9–6.9	...	...	0.08–0.15 Zr	0.06	0.05	0.15	rem
7051	...	...	0.35	0.45	0.15	0.10–0.45	1.7–2.5	0.05–0.25	...	3.0–4.0	...	...	...	0.15	0.05	0.15	rem
7060	...	...	0.15	0.20	1.8–2.6	0.20	1.3–2.1	0.15–0.25	...	6.1–7.5	...	...	0.003 Pb (w)	0.10	0.05	0.15	rem
X7064	...	...	0.12	0.15	1.8–2.4	...	1.9–2.9	0.06–0.25	...	6.8–8.0	...	...	0.10–0.50 Zr (x)	...	0.05	0.15	rem
7072	A97072	AlZn1	0.7 (Si + Fe)		0.10	0.10	0.10	...	...	0.8–1.3	...	...	...	...	0.05	0.15	rem
7472	A97472	...	0.25	0.6	0.05	0.05	0.9–1.5	...	...	1.3–1.9	...	...	...	...	0.05	0.15	rem
7075	A97075	AlZn5.5 MgCu	0.40	0.50	1.2–2.0	0.30	2.1–2.9	0.18–0.28	...	5.1–6.1	...	...	(y)	0.20	0.05	0.15	rem
7175	A97175	...	0.15	0.20	1.2–2.0	0.10	2.1–2.9	0.18–0.28	...	5.1–6.1	...	...	...	0.10	0.05	0.15	rem
7475	A97475	AlZn5.5MgCu(A)	0.10	0.12	1.2–1.9	0.06	1.9–2.6	0.18–0.25	...	5.2–6.2	...	...	...	0.06	0.05	0.15	rem
7076	A97076	...	0.40	0.6	0.30–1.0	0.30–0.8	1.2–2.0	...	...	7.0–8.0	...	...	...	0.20	0.05	0.15	rem

(continued)

(a) 0.0008 Be max for welding electrode and filler wire only. (b) (Si + Fe + Cu) = 0.50 max. (c) Obsolete. (d) 0.14 (Si + Fe) max. (e) 0.02 B max. (f) 0.01 B max. (g) 0.003 Pb max. (h) 0.05 to 0.20 Cd. (i) 0.20 Bi, 0.8 to 1.5 Pb, 0.20 Sn. (j) 0.20 to 0.6 Bi, 0.20 to 0.6 Pb. (k) A (Zr + Ti) limit of 0.20% maximum may be used for extruded and forged products when the supplier or producer and the purchaser have so agreed. (l) 0.40 (Si + Fe) max. (m) 0.05 to 0.20 Cd, 0.03 to 0.08 Sn. (n) 1.9 to 2.6 Li. (o) 1.7 to 2.3 Li. (p) 0.6 to 1.5 Bi, 0.05 Cd max. (q) 0.0008 Be max, 0.05 to 0.25 Zr. (r) 45 to 65% of Mg. (s) 0.40 to 0.7 Bi, 0.40 to 0.7 Pb. (t) 0.25 to 0.40 Ag. (u) 0.15 (Mn + Cr) min. (v) 0.08 to 0.20 Zr, 0.08 to 0.25 (Zr + Ti). (w) 0.20 (Ti + Zr) max. (x) 0.10 to 0.40 Co, 0.05 to 0.30 O. (y) A (Zr + Ti) limit of 0.25% maximum may be used for extruded and forged products when the supplier or producer and the purchaser have so agreed. (z) 0.20 to 0.50 O. (aa) 0.001 B max, 0.003 Cd max, 0.001 Co max, 0.008 Li max. (bb) 0.10 to 0.50 Bi, 0.10 to 0.25 Sn. (cc) 1.0 (Si + Fe) max. (dd) 0.02 to 0.08 Zr. (ee) 2.2 to 2.7 Li. (ff) 2.4 to 2.8 Li. (gg) 2.1 to 2.7 Li. (hh) 2.3 to 2.9 Li

Source: Ref 2, 3, 4

Table 2 (continued)

Grade designation: Aluminum Association	Grade designation: UNS No.	Grade designation: ISO No. R209	Composition, wt%: Si	Fe	Cu	Mn	Mg	Cr	Ni	Zn	Ga	V	Specified other elements	Ti	Unspecified other elements: Each	Unspecified other elements: Total	Al, minimum
7277	A97277	...	0.50	0.7	0.8–1.7	...	1.7–2.3	0.18–0.35	...	3.7–4.3	...	...	...	0.10	0.05	0.15	rem
7178	A97178	...	0.40	0.50	1.6–2.4	0.30	2.4–3.1	0.18–0.28	...	6.3–7.3	...	...	...	0.20	0.05	0.15	rem
7278	...	...	0.15	0.20	1.6–2.2	0.02	2.5–3.2	0.17–0.25	...	6.6–7.4	0.03	0.05	...	0.03	0.03	0.10	rem
7079	A97079	...	0.30	0.40	0.40–0.8	0.10–0.30	2.9–3.7	0.10–0.25	...	3.8–4.8	...	...	...	0.10	0.05	0.15	rem
7179	A97179	...	0.15	0.20	0.40–0.8	0.10–0.30	2.9–3.7	0.10–0.25	...	3.8–4.8	...	...	...	0.10	0.05	0.15	rem
7090	A97090	...	0.12	0.15	0.6–1.3	...	2.0–3.0	...	...	7.3–8.7	...	...	1.0–1.9 Co (z)	...	0.05	0.15	rem
7091	A97091	...	0.12	0.15	1.1–1.8	...	2.0–3.0	...	...	5.8–7.1	...	...	0.20–0.6 Co (z)	...	0.05	0.15	rem
8001	A98001	...	0.17	0.45–0.7	0.15	...	...	...	0.9–1.3	0.05	...	...	(aa)	...	0.05	0.15	rem
8004	...	...	0.15	0.15	0.03	0.02	0.02	...	...	0.03	...	...	...	0.30–0.7	0.02	0.15	rem
8005	...	...	0.20–0.50	0.40–0.8	0.05	...	0.05	...	...	0.05	...	...	...	...	0.05	0.15	rem
8006	A98006	...	0.40	1.2–2.0	0.30	0.30–1.0	0.10	...	...	0.10	...	...	...	...	0.05	0.15	rem
8007	A98007	...	0.40	1.2–2.0	0.10	0.30–1.0	0.10	...	...	0.8–1.8	...	...	...	...	0.05	0.15	rem
8008	...	...	0.6	0.9–1.6	0.20	0.50–1.0	...	...	...	0.10	...	...	...	...	0.05	0.15	rem
8010	...	...	0.40	0.35–0.7	0.10–0.30	0.10–0.8	0.10–0.50	0.20	...	0.40	...	...	...	0.10	0.05	0.15	rem
8011	A98011	...	0.50–0.9	0.6–1.0	0.10	0.20	0.05	0.05	...	0.10	...	...	...	0.08	0.05	0.15	rem
8111	A98111	...	0.30–1.1	0.40–1.0	0.10	0.10	0.05	0.05	...	0.10	...	...	...	0.08	0.05	0.15	rem
8112	A98112	...	1.0	1.0	0.40	0.6	0.7	0.20	...	1.0	...	...	...	0.20	0.05	0.15	rem
8014	A98014	...	0.30	1.2–1.6	0.20	0.20–0.6	0.10	...	...	0.10	...	...	...	0.10	0.05	0.15	rem
8017	A98017	...	0.10	0.55–0.8	0.10–0.20	...	0.01–0.05	...	...	0.05	...	...	0.04 B, 0.003 Li	...	0.03	0.10	rem
8020	A98020	...	0.10	0.10	0.005	0.005	...	...	...	0.005	...	0.05	(bb)	...	0.03	0.10	rem
8030	A98030	...	0.10	0.30–0.8	0.15–0.30	...	0.05	...	...	0.05	...	...	0.001–0.04 B	...	0.03	0.10	rem
8130	A98130	...	0.15 (cc)	0.40–1.0(cc)	0.05–0.15	...	...	...	...	0.10	...	...	...	...	0.03	0.10	rem
8040	A98040	...	1.0 (Si + Fe)		0.20	0.05	...	...	...	0.20	...	...	0.10–0.30 Zr	...	0.05	0.15	rem
8076	A98076	...	0.10	0.6–0.9	0.04	...	0.08–0.22	...	...	0.05	...	...	0.04 B	...	0.03	0.10	rem
8176	A98176	...	0.03–0.15	0.40–1.0	...	...	...	...	...	0.10	0.03	...	...	...	0.05	0.15	rem
8276	...	...	0.25	0.50–0.8	0.035	0.01	0.02	0.01	...	0.05	0.03	...	0.03 (V + Ti)(e)	...	0.03	0.10	rem
8077	A98077	...	0.10	0.10–0.40	0.05	...	0.10–0.30	...	...	0.05	...	...	0.05 B (dd)	...	0.03	0.10	rem
8177	A98177	...	0.10	0.25–0.45	0.04	...	0.04–0.12	...	...	0.05	...	...	0.04 B	...	0.03	0.10	rem
8079	A98079	...	0.05–0.30	0.7–1.3	0.05	...	...	...	...	0.10	...	...	...	...	0.05	0.15	rem
8280	A98280	...	1.0–2.0	0.7	0.7–1.3	0.10	...	...	0.20–0.7	0.05	...	...	5.5–7.0 Sn	0.10	0.05	0.15	rem
8081	A98081	...	0.7	0.7	0.7–1.3	0.10	...	...	...	0.05	...	...	18.0–22.0 Sn	0.10	0.05	0.15	rem
8090	...	...	0.20	0.30	1.0–1.6	0.10	0.6–1.3	0.10	...	0.25	...	...	0.04–0.16 Zr (ee)	0.10	0.05	0.15	rem
8091	...	...	0.30	0.50	1.6–2.2	0.10	0.50–1.2	0.10	...	0.25	...	...	0.08–0.16 Zr (ff)	0.10	0.05	0.15	rem
X8092	...	...	0.10	0.15	0.50–0.8	0.05	0.9–1.4	0.05	...	0.10	...	...	0.08–0.15 Zr (gg)	0.15	0.05	0.15	rem
X8192	...	...	0.10	0.15	0.40–0.7	0.05	0.9–1.4	0.05	...	0.10	...	...	0.08–0.15 Zr (hh)	0.15	0.05	0.15	rem

(a) 0.0008 Be max for welding electrode and filler wire only. (b) (Si + Fe + Cu) = 0.50 max. (c) Obsolete. (d) 0.14 (Si + Fe) max. (e) 0.02 B max. (f) 0.01 B max. (g) 0.003 Pb max. (h) 0.05 to 0.20 Cd. (i) 0.20 Bi, 0.8 to 1.5 Pb, 0.20 Sn. (j) 0.20 to 0.6 Bi, 0.20 to 0.6 Pb. (k) A (Zr + Ti) limit of 0.20% maximum may be used for extruded and forged products when the supplier or producer and the purchaser have so agreed. (l) 0.40 (Si + Fe) max. (m) 0.05 to 0.20 Cd, 0.03 to 0.08 Sn. (n) 1.9 to 2.6 Li. (o) 1.7 to 2.3 Li. (p) 0.6 to 1.5 Bi, 0.05 Cd max. (q) 0.0008 Be max, 0.05 to 0.25 Zr. (r) 45 to 65% of Mg. (s) 0.40 to 0.7 Bi, 0.40 to 0.7 Pb. (t) 0.25 to 0.40 Ag. (u) 0.15 (Mn + Cr) min. (v) 0.08 to 0.20 Zr, 0.08 to 0.25 (Zr + Ti). (w) 0.20 (Ti + Zr) max. (x) 0.10 to 0.40 Co, 0.05 to 0.30 O. (y) A (Zr + Ti) limit of 0.25% maximum may be used for extruded and forged products when the supplier or producer and the purchaser have so agreed. (z) 0.20 to 0.50 O. (aa) 0.001 B max, 0.003 Cd max, 0.001 Co max, 0.008 Li max. (bb) 0.10 to 0.50 Bi, 0.10 to 0.25 Sn. (cc) 1.0 (Si + Fe) max. (dd) 0.02 to 0.08 Zr. (ee) 2.2 to 2.7 Li. (ff) 2.4 to 2.8 Li. (gg) 2.1 to 2.7 Li. (hh) 2.3 to 2.9 Li
Source: Ref 2, 3, 4

F, As-Fabricated. This is applied to products shaped by cold working, hot working, or casting processes in which no special control over thermal conditions or strain hardening is employed. For wrought products, there are no mechanical property limits.

O, Annealed. O applies to wrought products that are annealed to obtain lowest-strength temper and to cast products that are annealed to improve ductility and dimensional stability. The O may be followed by a digit other than zero.

H, Strain-Hardened (Wrought Products Only). This indicates products that have been strengthened by strain hardening, with or without supplementary thermal treatment to produce some reduction in strength. The H is always followed by two or more digits, as discussed in the section "System for Strain-Hardened Products" in this article.

W, Solution Heat-Treated. This is an unstable temper applicable only to alloys whose strength naturally (spontaneously) changes at room temperature over a duration of months or even years after solution heat treatment. The designation is specific only when the period of natural aging is indicated (for example, W ½ h). See also the discussion of the T*x*51, T*x*52, and T*x*54 tempers in the section "System for Heat-Treatable Alloys" in this article.

T, Solution Heat-Treated. This applies to alloys whose strength is stable within a few weeks of solution heat treatment. The T is always followed by one or more digits, as discussed in the section "System for Heat-Treatable Alloys" in this article.

System for Strain-Hardened Products

Temper designations for wrought products that are strengthened by strain hardening consist of an H followed by two or more digits. The first digit following the H indicates the specific sequence of basic operations.

H1, Strain-Hardened Only. This applies to products that are strain hardened to obtain the desired strength without supplementary thermal treatment. The digit following the H1 indicates the degree of strain hardening.

H2, Strain-Hardened and Partially Annealed. This pertains to products that are strain-hardened more than the desired final amount and then reduced in strength to the desired level by partial annealing. The digit following the H2 indicates the degree of

Table 3 Composition of unalloyed and alloyed aluminum castings (*xxx*.0) and ingots (*xxx*.1 or *xxx*.2)

Grade designation				Composition, wt%												
Aluminum Association(a)	UNS No.	ISO(b)	Product(c)	Si	Fe	Cu	Mn	Mg	Cr	Ni	Zn	Sn	Ti	Unspecified other elements Each	Unspecified other elements Total	Al, min(d)
100.1	A01001	Al99.0	Ingot	0.15	0.6–0.8	0.10	(e)	...	(e)	...	0.05	...	(e)	0.03(e)	0.10	99.00
130.1	A01301	...	Ingot	(f)	(f)	0.10	(e)	...	(e)	...	0.05	...	(e)	0.03(e)	0.10	99.30
150.1	A01501	Al99.5	Ingot	(g)	(g)	0.05	(e)	...	(e)	...	0.05	...	(e)	0.03(e)	0.10	99.50
160.1	A01601	Al99.8	Ingot	0.10(g)	0.25(g)	...	(e)	...	(e)	...	0.05	...	(e)	0.03(e)	0.10	99.60
170.1	A01701	Al99.7	Ingot	(h)	(h)	...	(e)	...	(e)	...	0.05	...	(e)	0.03(e)	0.10	99.70
201.0	A02010	...	S	0.10	0.15	4.0–5.2	0.20–0.50	0.15–0.55	...	...	...	...	0.15–0.35	0.05(i)	0.10	rem
201.2	A02012	...	Ingot	0.10	0.10	4.0–5.2	0.20–0.50	0.20–0.55	...	...	...	...	0.15–0.35	0.05(i)	0.10	rem
A201.0	A12010	...	S	0.05	0.10	4.0–5.0	0.20–0.40	0.15–0.35	...	...	...	...	0.15–0.35	0.03(i)	0.10	rem
A201.1	A12011	...	Ingot	0.05	0.07	4.5–5.0	0.20–0.40	0.20–0.35	...	...	...	...	0.15–0.35	0.03(i)	0.10	rem
B201.0	A22010	...	S	0.05	0.05	4.5–5.0	0.20–0.50	0.25–0.35	...	...	...	...	0.15–0.35	0.05(j)	0.15	rem
203.0	A02030	...	S	0.30	0.50	4.5–5.5	0.20–0.30	0.10	...	1.3–1.7	0.10	...	0.15–0.25(k)	0.05(l)	0.20	rem
203.2	A02032	...	Ingot	0.20	0.35	4.8–5.2	0.20–0.30	0.10	...	1.3–1.7	0.10	...	0.15–0.25(k)	0.05(l)	0.20	rem
204.0	A02040	3522 AlCu4MgTi R164 AlCu4MgTi R2147 AlCu4MgTi	S, P	0.20	0.35	4.2–5.0	0.10	0.15–0.35	...	0.05	0.10	0.05	0.15–0.30	0.05	0.15	rem
204.2	A02042	...	Ingot	0.15	0.10–0.20	4.2–4.9	0.05	0.20–0.35	...	0.03	0.05	0.05	0.15–0.25	0.05	0.15	rem
206.0	A02060	...	S, P	0.10	0.15	4.2–5.0	0.20–0.50	0.15–0.35	...	0.05	0.10	0.05	0.15–0.30	0.05	0.15	rem
206.2	A02062	...	Ingot	0.10	0.10	4.2–5.0	0.20–0.50	0.20–0.35	...	0.03	0.05	0.05	0.15–0.25	0.05	0.15	rem
A206.0	A12060	...	S, P	0.05	0.10	4.2–5.0	0.20–0.50	0.15–0.35	...	0.05	0.10	0.05	0.15–0.30	0.05	0.15	rem
A206.2	A12062	...	Ingot	0.05	0.07	4.2–5.0	0.20–0.50	0.20–0.35	...	0.03	0.05	0.05	0.15–0.25	0.05	0.15	rem
208.0	A02080	...	S, P	2.5–3.5	1.2	3.5–4.5	0.50	0.10	...	0.35	1.0	...	0.25	...	0.50	rem
208.1	A02081	...	Ingot	2.5–3.5	0.9	3.5–4.5	0.50	0.10	...	0.35	1.0	...	0.25	...	0.50	rem
208.2	A02082	...	Ingot	2.5–3.5	0.8	3.5–4.5	0.30	0.03	...	...	0.20	...	0.20	...	0.30	rem
213.0	A02130	...	S, P	1.0–3.0	1.2	6.0–8.0	0.6	0.10	...	0.35	2.5	...	0.25	...	0.50	rem
213.1	A02131	...	Ingot	1.0–3.0	0.9	6.0–8.0	0.6	0.10	...	0.35	2.5	...	0.25	...	0.50	rem
222.0	A02220	...	S, P	2.0	1.5	9.2–10.7	0.50	0.15–0.35	...	0.50	0.8	...	0.25	...	0.35	rem
222.1	A02221	...	Ingot	2.0	1.2	9.2–10.7	0.50	0.20–0.35	...	0.50	0.8	...	0.25	...	0.35	rem
224.0	A02240	...	S, P	0.06	0.10	4.5–5.5	0.20–0.50	...	...	...	...	...	0.35	0.03(m)	0.10	rem
224.2	A02242	...	Ingot	0.02	0.04	4.5–5.5	0.20–0.50	...	...	...	...	...	0.25	0.03(m)	0.10	rem
240.0	A02400	...	S	0.50	0.50	7.0–9.0	0.30–0.7	5.5–6.5	...	0.30–0.7	0.10	...	0.20	0.05	0.15	rem
240.1	A02401	...	Ingot	0.50	0.40	7.0–9.0	0.30–0.7	5.6–6.5	...	0.30–0.7	0.10	...	0.20	0.05	0.15	rem
242.0	A02420	3522 AlCu4Ni2Mg2 R164 AlCu4Ni2Mg2	S, P	0.7	1.0	3.5–4.5	0.35	1.2–1.8	0.25	1.7–2.3	0.35	...	0.25	0.05	0.15	rem
242.1	A02421	...	Ingot	0.7	0.8	3.5–4.5	0.35	1.3–1.8	0.25	1.7–2.3	0.35	...	0.25	0.05	0.15	rem
242.2	A02422	...	Ingot	0.6	0.6	3.5–4.5	0.10	1.3–1.8	...	1.7–2.3	0.10	...	0.20	0.05	0.15	rem
A242.0	A12420	...	S	0.6	0.8	3.7–4.5	0.10	1.2–1.7	0.15–0.25	1.8–2.3	0.10	...	0.07–0.20	0.05	0.15	rem
A242.1	A12421	...	Ingot	0.6	0.6	3.7–4.5	0.10	1.3–1.7	0.15–0.25	1.8–2.3	0.10	...	0.07–0.20	0.05	0.15	rem
A242.2	A12422	...	Ingot	0.35	0.6	3.7–4.5	0.10	1.3–1.7	0.15–0.25	1.8–2.3	0.10	...	0.07–0.20	0.05	0.15	rem
243.0(a)	A02430	...	S	0.35	0.40	3.5–4.5	0.15–0.45	1.8–2.3	0.20–0.40	1.9–2.3	0.05	...	0.06–0.20	0.05(n)	0.15	rem
243.1	A02431	...	Ingot	0.35	0.30	3.5–4.5	0.15–0.45	1.9–2.3	0.20–0.40	1.9–2.3	0.05	...	0.06–0.20	0.05(n)	0.15	rem
295.0	A02950	...	S	0.7–1.5	1.0	4.0–5.0	0.35	0.03	...	...	0.35	...	0.25	0.05	0.15	rem
295.1	A02951	...	Ingot	0.7–1.5	0.8	4.0–5.0	0.35	0.03	...	...	0.35	...	0.25	0.05	0.15	rem
295.2	A02952	...	Ingot	0.7–1.2	0.8	4.0–5.0	0.30	0.03	...	...	0.30	...	0.20	0.05	0.15	rem
296.0	A02960	...	P	2.0–3.0	1.2	4.0–5.0	0.35	0.05	...	0.35	0.50	...	0.25	...	0.35	rem
296.1	A02961	...	Ingot	2.0–3.0	0.9	4.0–5.0	0.35	0.05	...	0.35	0.50	...	0.25	...	0.35	rem
296.2	A02962	...	Ingot	2.0–3.0	0.8	4.0–5.0	0.30	0.35	...	...	0.30	...	0.20	0.05	0.15	rem
305.0	A03050	...	S, P	4.5–5.5	0.6	1.0–1.5	0.50	0.10	0.25	...	0.35	...	0.25	0.05	0.15	rem
305.2	A03052	...	Ingot	4.5–5.5	0.14–0.25	1.0–1.5	0.05	...	...	...	0.05	...	0.20	0.05	0.15	rem
A305.0	A13050	...	S, P	4.5–5.5	0.20	1.0–1.5	0.10	0.10	...	...	0.10	...	0.20	0.05	0.15	rem
A305.1	A13051	...	Ingot	4.5–5.5	0.15	1.0–1.5	0.10	0.10	...	...	0.10	...	0.20	0.05	0.15	rem
A305.2	A13052	...	Ingot	4.5–5.5	0.13	1.0–1.5	0.05	...	...	...	0.05	...	0.20	0.05	0.15	rem
308.0	A03080	...	S, P	5.0–6.0	1.0	4.0–5.0	0.50	0.10	...	...	1.0	...	0.25	...	0.50	rem
308.1	A03081	...	Ingot	5.0–6.0	0.8	4.0–5.0	0.50	0.10	...	...	1.0	...	0.25	...	0.50	rem
308.2	A03082	...	Ingot	5.0–6.0	0.8	4.0–5.0	0.30	0.10	...	...	0.50	...	0.20	...	0.50	rem
319.0	A03190	3522 AlSi5Cu3 3522 AlSi5Cu3Mn 3522 AlSi6Cu4 3522 AlSi6Cu4Mn R164 AlSi5Cu3 R164 AlSi5Cu3Fe R164 AlSi6Cu4	S, P	5.5–6.5	1.0	3.0–4.0	0.50	0.10	...	0.35	1.0	...	0.25	...	0.50	rem
319.1	A03191	...	Ingot	5.5–6.5	0.8	3.0–4.0	0.50	0.10	...	0.35	1.0	...	0.25	...	0.50	rem
319.2	A03192	...	Ingot	5.5–6.5	0.6	3.0–4.0	0.10	0.10	...	0.10	0.10	...	0.20	...	0.20	rem
A319.0	A13190	3522 AlSi5Cu3 3522 AlSi5Cu3Mn 3522 AlSi6Cu4 3522 AlSi6Cu4Mn R164 AlSi5Cu3 R164 AlSi5Cu3Fe R164 AlSi6Cu4	S, P	5.5–6.5	1.0	3.0–4.0	0.50	0.10	...	0.35	3.0	...	0.25	...	0.50	rem

(continued)

(a) Serial letter prefix indicates modification: A, B, C, D, and F. (b) Per ISO standard No. R115 unless other standard (R164, R2147, or 3522) specified. (c) D, die casting; P, permanent mold; s, sand. Other products may pertain to the composition shown even though not listed. (d) The Al content for unalloyed aluminum by remelt is the difference between 100.00% and the sum of all other metallic elements present in amounts of 0.010% or more each, expressed to the second decimal before determining the sum. (e) (Mn + Cr + Ti + V) = 0.025% max. (f) Fe/Si ratio 2.5 min. (g) Fe/Si ratio 2.0 min. (h) Fe/Si ratio 1.5 min. (i) 0.40 to 1.0% Ag. (j) 0.50–1.0% Ag. (k) Ti + Zr = 0.50 max. (l) 0.20 to 0.30% Sb; 0.20 to 0.30% Co; 0.10 to 0.30% Zr. (m) 0.05–0.15% V; 0.10–0.25% Zr. (n) 0.06–0.20% V. (o) For Fe > 0.45%, Mn content shall not be less than one-half Fe content. (p) 0.04–0.07% Be. (q) 0.10–0.30% Be. (r) 0.15–0.30% Be. (s) A*xxx*.1 ingot is used to produce *xxx*.0 and A*xxx*.0 castings. (t) (Mn + Cr) = 0.8% max. (u) 0.25% Pb max. (v) 0.02–0.04% Be. (w) 0.08–0.15% V. (x) Used to coat steel. (y) Used with Zn to coat steel. (z) 0.10% Pb max. (aa) 0.003–0.007% Be; 0.005% B max. (bb) 0.003–0.007% Be; 0.002% B max
Source: Ref 3, 4, 5

Table 3 (continued)

Grade designation				Composition, wt%												
Aluminum Association(a)	UNS No.	ISO(b)	Product(c)	Si	Fe	Cu	Mn	Mg	Cr	Ni	Zn	Sn	Ti	Unspecified other elements Each	Unspecified other elements Total	Al, min(d)
A319.1	A13191	· · ·	Ingot	5.5–6.5	0.8	3.0–4.0	0.50	0.10	· · ·	0.35	3.0	· · ·	0.25	· · ·	0.50	rem
B319.0	A23190	· · ·	S, P	5.5–6.5	1.2	3.0–4.0	0.8	0.10–0.50	· · ·	0.50	1.0	· · ·	0.25	· · ·	0.50	rem
B319.1	A23191	· · ·	Ingot	5.5–6.5	0.9	3.0–4.0	0.8	0.15–0.50	· · ·	0.50	1.0	· · ·	0.25	· · ·	0.50	rem
320.0	A03200	· · ·	S, P	5.0–8.0	1.2	2.0–4.0	0.8	0.05–0.6	· · ·	0.35	3.0	· · ·	0.25	· · ·	0.50	rem
320.1	A03201	· · ·	Ingot	5.0–8.0	0.9	2.0–4.0	0.8	0.10–0.6	· · ·	0.35	3.0	· · ·	0.25	· · ·	0.50	rem
324.0	A03240	· · ·	P	7.0–8.0	1.2	0.40–0.6	0.50	0.40–0.7	· · ·	0.30	1.0	· · ·	0.20	0.15	0.20	rem
324.1	A03241	· · ·	Ingot	7.0–8.0	0.9	0.40–0.6	0.50	0.45–0.7	· · ·	0.30	1.0	· · ·	0.20	0.15	0.20	rem
324.2	A03242	· · ·	Ingot	7.0–8.0	0.6	0.40–0.6	0.10	0.45–0.7	· · ·	0.10	0.10	· · ·	0.20	0.05	0.15	rem
328.0	A03280	· · ·	S	7.5–8.5	1.0	1.0–2.0	0.20–0.6	0.20–0.6	0.35	0.25	1.5	· · ·	0.25	· · ·	0.50	rem
328.1	A03281	· · ·	Ingot	7.5–8.5	0.8	1.0–2.0	0.20–0.6	0.20–0.6	0.35	0.25	1.5	· · ·	0.25	· · ·	0.50	rem
332.0	A03320	· · ·	P	8.5–10.5	1.2	2.0–4.0	0.50	0.50–1.5	· · ·	0.50	1.0	· · ·	0.25	· · ·	0.50	rem
332.1	A03321	· · ·	Ingot	8.5–10.5	0.9	2.0–4.0	0.50	0.6–1.5	· · ·	0.50	1.0	· · ·	0.25	· · ·	0.50	rem
332.2	A03322	· · ·	Ingot	8.5–10.0	0.6	2.0–4.0	0.10	0.9–1.3	· · ·	0.10	0.10	· · ·	0.20	· · ·	0.30	rem
333.0	A03330	· · ·	P	8.0–10.0	1.0	3.0–4.0	0.50	0.05–0.50	· · ·	0.50	1.0	· · ·	0.25	· · ·	0.50	rem
333.1	A03331	· · ·	Ingot	8.0–10.0	0.8	3.0–4.0	0.50	0.10–0.50	· · ·	0.50	1.0	· · ·	0.25	· · ·	0.50	rem
A333.0	A13330	· · ·	P	8.0–10.0	1.0	3.0–4.0	0.50	0.05–0.50	· · ·	0.50	3.0	· · ·	0.25	· · ·	0.50	rem
A333.1	A13331	· · ·	Ingot	8.0–10.0	0.8	3.0–4.0	0.50	0.10–0.50	· · ·	0.50	3.0	· · ·	0.25	· · ·	0.50	rem
336.0	A03360	· · ·	P	11.0–13.0	1.2	0.50–1.5	0.35	0.7–1.3	· · ·	2.0–3.0	0.35	· · ·	0.25	0.05	· · ·	rem
336.1	A03361	· · ·	Ingot	11.0–13.0	0.9	0.50–1.5	0.35	0.8–1.3	· · ·	2.0–3.0	0.35	· · ·	0.25	0.05	· · ·	rem
336.2	A03362	· · ·	Ingot	11.0–13.0	0.9	0.50–1.5	0.10	0.9–1.3	· · ·	2.0–3.0	0.10	· · ·	0.20	0.05	0.15	rem
339.0	A03390	· · ·	P	11.0–13.0	1.2	1.5–3.0	0.50	0.50–1.5	· · ·	0.50–1.5	1.0	· · ·	0.25	· · ·	0.50	rem
339.1	· · ·	· · ·	Ingot	11.0–13.0	0.9	1.5–3.0	0.50	0.6–1.5	· · ·	0.50–1.5	1.0	· · ·	0.25	· · ·	0.50	rem
343.0	A03430	· · ·	D	6.7–7.7	1.2	0.50–0.9	0.50	0.10	0.10	· · ·	1.2–2.0	0.50	· · ·	0.10	0.35	rem
343.1	A03431	· · ·	Ingot	6.7–7.7	0.9	0.50–0.9	0.50	0.10	0.10	· · ·	1.2–1.9	0.50	· · ·	0.10	0.35	rem
354.0	A03540	· · ·	P	8.6–9.4	0.20	1.6–2.0	0.10	0.40–0.6	· · ·	· · ·	0.10	· · ·	0.20	0.05	0.15	rem
354.1	A03541	· · ·	Ingot	8.6–9.4	0.15	1.6–2.0	0.10	0.45–0.6	· · ·	· · ·	0.10	· · ·	0.20	0.05	0.15	rem
355.0	A03550	3522 AlSi5Cu1Mg R164 AlSi5Cu1	S, P	4.5–5.5	0.6(o)	1.0–1.5	0.50(o)	0.40–0.6	0.25	· · ·	0.35	· · ·	0.25	0.05	0.15	rem
355.1	A03551	· · ·	Ingot	4.5–5.5	0.50(o)	1.0–1.5	0.50(o)	0.45–0.6	0.25	· · ·	0.35	· · ·	0.25	0.05	0.15	rem
355.2	A03552	· · ·	Ingot	4.5–5.5	0.14–0.25	1.0–1.5	0.05	0.50–0.6	· · ·	· · ·	0.05	· · ·	0.20	0.05	0.15	rem
A355.0	A13550	· · ·	S, P	4.5–5.5	0.09	1.0–1.5	0.05	0.45–0.6	· · ·	· · ·	0.05	· · ·	0.04–0.20	0.05	0.15	rem
A355.2	A13552	· · ·	Ingot	4.5–5.5	0.06	1.0–1.5	0.03	0.50–0.6	· · ·	· · ·	0.03	· · ·	0.04–0.20	0.03	0.10	rem
C355.0	A33350	· · ·	S, P	4.5–5.5	0.20	1.0–1.5	0.10	0.40–0.6	· · ·	· · ·	0.10	· · ·	0.20	0.05	0.15	rem
C355.1	A33351	· · ·	Ingot	4.5–5.5	0.15	1.0–1.5	0.10	0.45–0.6	· · ·	· · ·	0.10	· · ·	0.20	0.05	0.15	rem
C355.2	A33352	· · ·	Ingot	4.5–5.5	0.13	1.0–1.5	0.05	0.50–0.6	· · ·	· · ·	0.05	· · ·	0.20	0.05	0.15	rem
356.0	A03560	3522 AlSi7Mg R2147 AlSi7Mg	S, P	6.5–7.5	0.6(o)	0.25	0.35(o)	0.20–0.45	· · ·	· · ·	0.35	· · ·	0.25	0.05	0.15	rem
356.1	A03561	· · ·	Ingot	6.5–7.5	0.50(o)	0.25	0.35(o)	0.25–0.45	· · ·	· · ·	0.35	· · ·	0.25	0.05	0.15	rem
356.2	A03562	· · ·	Ingot	6.5–7.5	0.13–0.25	0.10	0.05	0.30–0.45	· · ·	· · ·	0.05	· · ·	0.20	0.05	0.15	rem
A356.0	A13560	· · ·	S, P	6.5–7.5	0.20	0.20	0.10	0.25–0.45	· · ·	· · ·	0.10	· · ·	0.20	0.05	0.15	rem
A356.1	A13561	· · ·	Ingot	6.5–7.5	0.15	0.20	0.10	0.30–0.45	· · ·	· · ·	0.10	· · ·	0.20	0.05	0.15	rem
A356.2	A3562	· · ·	Ingot	6.5–7.5	0.12	0.10	0.05	0.30–0.45	· · ·	· · ·	0.05	· · ·	0.20	0.05	0.15	rem
B356.0	A23560	· · ·	S, P	6.5–7.5	0.09	0.05	0.05	0.25–0.45	· · ·	· · ·	0.05	· · ·	0.04–0.20	0.05	0.15	rem
B356.2	A23562	· · ·	Ingot	6.5–7.5	0.06	0.03	0.03	0.30–0.45	· · ·	· · ·	0.03	· · ·	0.04–0.20	0.03	0.10	rem
C356.0	A33560	· · ·	S, P	6.5–7.5	0.07	0.05	0.05	0.25–0.45	· · ·	· · ·	0.05	· · ·	0.04–0.20	0.05	0.15	rem
C356.2	A33562	· · ·	Ingot	6.5–7.5	0.04	0.03	0.03	0.30–0.45	· · ·	· · ·	0.03	· · ·	0.04–0.20	0.03	0.10	rem
F356.0	A63560	· · ·	S, P	6.5–7.5	0.20	0.20	0.10	0.17–0.25	· · ·	· · ·	0.10	· · ·	0.04–0.20	0.05	0.15	rem
F356.2	A63562	· · ·	Ingot	6.5–7.5	0.12	0.10	0.05	0.17–0.25	· · ·	· · ·	0.05	· · ·	0.04–0.20	0.05	0.15	rem
357.0	A03570	· · ·	S, P	6.5–7.5	0.15	0.05	0.03	0.45–0.6	· · ·	· · ·	0.05	· · ·	0.20	0.05	0.15	rem
357.1	A03571	· · ·	Ingot	6.5–7.5	0.12	0.05	0.03	0.45–0.6	· · ·	· · ·	0.05	· · ·	0.20	0.05	0.15	rem
A357.0	A13570	· · ·	S, P	6.5–7.5	0.20	0.20	0.10	0.40–0.7	· · ·	· · ·	0.10	· · ·	0.04–0.20	0.05(p)	0.15	rem
A357.2	A13572	· · ·	Ingot	6.5–7.5	0.12	0.10	0.05	0.45–0.7	· · ·	· · ·	0.05	· · ·	0.04–0.20	0.03(p)	0.10	rem
B357.0	· · ·	· · ·	S, P	6.5–7.5	0.09	0.05	0.05	0.40–0.6	· · ·	· · ·	0.05	· · ·	0.04–0.20	0.05	0.15	rem
B357.2	A23572	· · ·	Ingot	6.5–7.5	0.06	0.03	0.03	0.45–0.6	· · ·	· · ·	0.03	· · ·	0.04–0.20	0.03	0.10	rem
C357.0	· · ·	· · ·	S, P	6.5–7.5	0.09	0.05	0.05	0.45–0.7	· · ·	· · ·	0.05	· · ·	0.04–0.20	0.05(p)	0.15	rem
C357.2	· · ·	· · ·	Ingot	6.5–7.5	0.06	0.03	0.03	0.50–0.7	· · ·	· · ·	0.03	· · ·	0.04–0.20	0.03(p)	0.10	rem
D357.0	· · ·	· · ·	S	6.5–7.5	0.20	· · ·	0.10	0.55–0.6	· · ·	· · ·	· · ·	· · ·	0.10–0.20	0.05(p)	0.15	rem
358.0	A03580	· · ·	S, P	7.6–8.6	0.30	0.20	0.20	0.40–0.6	0.20	· · ·	0.20	· · ·	0.10–0.20	0.05(q)	0.15	rem
358.2	A03582	· · ·	Ingot	7.6–8.6	0.20	0.10	0.10	0.45–0.6	0.05	· · ·	0.10	· · ·	0.12–0.20	0.05(r)	0.15	rem
359.0	A03590	· · ·	S, P	8.5–9.5	0.20	0.20	0.10	0.50–0.7	· · ·	· · ·	0.10	· · ·	0.20	0.05	0.15	rem
359.2	A03592	· · ·	Ingot	8.5–9.5	0.12	0.10	0.10	0.55–0.7	· · ·	· · ·	0.10	· · ·	0.20	0.05	0.15	rem
360.0(s)	A03600(s)	3522 AlSi10Mg(s) R164 AlSi10Mg(s) R2147AlSi10Mg(s)	D	9.0–10.0	2.0	0.6	0.35	0.40–0.6	· · ·	0.50	0.50	0.15	· · ·	· · ·	0.25	rem
360.2	A03602	· · ·	Ingot	9.0–10.0	0.7–1.1	0.10	0.10	0.45–0.6	· · ·	0.10	0.10	0.10	· · ·	· · ·	0.20	rem
A360.0(s)	A13600	· · ·	D	9.0–10.0	1.3	0.6	0.35	0.40–0.6	· · ·	0.50	0.50	0.15	· · ·	· · ·	0.25	rem
A360.1(s)	A13601(s)	· · ·	Ingot	9.0–10.0	1.0	0.6	0.35	0.45–0.6	· · ·	0.50	0.40	0.15	· · ·	· · ·	0.25	rem
A360.2	A13602(s)	· · ·	Ingot	9.0–10.0	0.6	0.10	0.05	0.45–0.6	· · ·	· · ·	0.05	· · ·	· · ·	0.05	0.15	rem
361.0	A03610	· · ·	D	9.5–10.5	1.1	0.50	0.25	0.40–0.6	0.20–0.30	0.20–0.30	0.50	0.10	0.20	0.05	0.15	rem

(continued)

(a) Serial letter prefix indicates modification: A, B, C, D, and F. (b) Per ISO standard No. R115 unless other standard (R164, R2147, or 3522) specified. (c) D, die casting; P, permanent mold; s, sand. Other products may pertain to the composition shown even though not listed. (d) The Al content for unalloyed aluminum by remelt is the difference between 100.00% and the sum of all other metallic elements present in amounts of 0.010% or more each, expressed to the second decimal before determining the sum. (e) (Mn + Cr + Ti + V) = 0.025% max. (f) Fe/Si ratio 2.5 min. (g) Fe/Si ratio 2.0 min. (h) Fe/Si ratio 1.5 min. (i) 0.40 to 1.0% Ag. (j) 0.50–1.0% Ag. (k) Ti + Zr = 0.50 max. (l) 0.20 to 0.30% Sb; 0.20 to 0.30% Co; 0.10 to 0.30% Zr. (m) 0.05–0.15% V; 0.10–0.25% Zr. (n) 0.06–0.20% V. (o) For Fe > 0.45%, Mn content shall not be less than one-half Fe content. (p) 0.04–0.07% Be. (q) 0.10–0.30% Be. (r) 0.15–0.30% Be. (s) *Axxx*.1 ingot is used to produce *xxx*.0 and *Axxx*.0 castings. (t) (Mn + Cr) = 0.8% max. (u) 0.25% Pb max. (v) 0.02–0.04% Be. (w) 0.08–0.15% V. (x) Used to coat steel. (y) Used with Zn to coat steel. (z) 0.10% Pb max. (aa) 0.003–0.007% Be; 0.005% B max. (bb) 0.003–0.007% Be; 0.002% B max
Source: Ref 3, 4, 5

Table 3 (continued)

Grade designation: Aluminum Association(a)	UNS No.	ISO(b)	Product(c)	Composition, wt%: Si	Fe	Cu	Mn	Mg	Cr	Ni	Zn	Sn	Ti	Unspecified other elements: Each	Total	Al, min(d)
361.1	A03611	· · ·	Ingot	9.5–10.5	0.8	0.50	0.25	0.45–0.6	0.20–0.30	0.20–0.30	0.40	0.10	0.20	0.05	0.15	rem
363.0	A03630	· · ·	S, P	4.5–6.0	1.1	2.5–3.5	(t)	0.15–0.40	(t)	0.25	3.0–4.5	0.25	0.20	(u)	0.30	rem
363.1	A03631	· · ·	Ingot	4.5–6.0	0.8	2.5–3.5	(t)	0.20–0.40	(t)	0.25	3.0–4.5	0.25	0.20	(u)	0.30	rem
364.0	A03640	· · ·	D	7.5–9.5	1.5	0.20	0.10	0.20–0.40	0.25–0.50	0.15	0.15	0.15	· · ·	0.05(v)	0.15	rem
364.2	A03642	· · ·	Ingot	7.5–9.5	0.7–1.1	0.20	0.10	0.25–0.40	0.25–0.50	0.15	0.15	0.15	· · ·	0.05(v)	0.15	rem
369.0	A03690	· · ·	D	11.0–12.0	1.3	0.50	0.35	0.25–0.45	0.30–0.40	0.05	1.0	0.10	· · ·	0.05	0.15	rem
369.1	A03691	· · ·	Ingot	11.0–12.0	1.0	0.50	0.35	0.30–0.45	0.30–0.40	0.05	0.9	0.10	· · ·	0.05	0.15	rem
380.0(s)	A03800(s)	· · ·	D	7.5–9.5	2.0	3.0–4.0	0.50	0.10	· · ·	0.50	3.0	0.35	· · ·	· · ·	0.50	rem
380.2	A03802	· · ·	Ingot	7.5–9.5	0.7–1.1	3.0–4.0	0.10	0.10	· · ·	0.10	0.10	0.10	· · ·	· · ·	0.20	rem
A380.0(s)	A13800(s)	3522 AlSi8Cu3Fe R164 AlSi8Cu3Fe	D	7.5–9.5	1.3	3.0–4.0	0.50	0.10	· · ·	0.50	3.0	0.35	· · ·	· · ·	0.50	rem
A380.1(s)	A13801(s)	· · ·	Ingot	7.5–9.5	1.0	3.0–4.0	0.50	0.10	· · ·	0.50	2.9	0.35	· · ·	· · ·	0.50	rem
A380.2	A13802	· · ·	Ingot	7.5–9.5	0.6	3.0–4.0	0.10	0.10	· · ·	0.10	0.10	· · ·	· · ·	0.05	0.15	rem
B380.0	A23800	· · ·	D	7.5–9.5	1.3	3.0–4.0	0.50	0.10	· · ·	0.50	1.0	0.35	· · ·	· · ·	0.50	rem
B380.1	A28801	· · ·	Ingot	7.5–9.5	1.0	3.0–4.0	0.50	0.10	· · ·	0.50	0.9	0.35	· · ·	· · ·	0.50	rem
383.0	A03830	· · ·	D	9.5–11.5	1.3	2.0–3.0	0.50	0.10	· · ·	0.30	3.0	0.15	· · ·	· · ·	0.50	rem
383.1	A03831	· · ·	Ingot	9.5–11.5	1.0	2.0–3.0	0.50	0.10	· · ·	0.30	2.9	0.15	· · ·	· · ·	0.50	rem
383.2	A03832	· · ·	Ingot	9.5–11.5	0.6–1.0	2.0–3.0	0.10	0.10	· · ·	0.10	0.10	0.10	· · ·	· · ·	0.20	rem
384.0	A03840	· · ·	D	10.5–12.0	1.3	3.0–4.5	0.50	0.10	· · ·	0.50	3.0	0.35	· · ·	· · ·	0.50	rem
384.1	A03841	· · ·	Ingot	10.5–12.0	1.0	3.0–4.5	0.50	0.10	· · ·	0.50	2.9	0.35	· · ·	· · ·	0.50	rem
384.2	A03842	· · ·	Ingot	10.5–12.0	0.6–1.0	3.0–4.5	0.10	0.10	· · ·	0.10	0.10	0.10	· · ·	· · ·	0.20	rem
A384.0	A13840	· · ·	D	10.5–12.0	1.3	3.0–4.5	0.50	0.10	· · ·	0.50	1.0	0.35	· · ·	· · ·	0.50	rem
A384.1	A13841	· · ·	Ingot	10.5–12.0	1.0	3.0–4.5	0.50	0.10	· · ·	0.50	0.9	0.35	· · ·	· · ·	0.50	rem
385.0	A03850	· · ·	D	11.0–13.0	2.0	2.0–4.0	0.50	0.30	· · ·	0.50	3.0	0.30	· · ·	· · ·	0.50	rem
385.1	A03851	· · ·	Ingot	11.0–13.0	1.1	2.0–4.0	0.50	0.30	· · ·	0.50	2.9	0.30	· · ·	· · ·	0.50	rem
390.0	A03900	· · ·	D	16.0–18.0	1.3	4.0–5.0	0.10	0.45–0.65	· · ·	· · ·	0.10	· · ·	0.20	0.10	0.20	rem
390.2	A03902	· · ·	Ingot	16.0–18.0	0.6–1.0	4.0–5.0	0.10	0.50–0.65	· · ·	· · ·	0.10	· · ·	0.20	0.10	0.20	rem
A390.0	A13900	· · ·	S, P	16.0–18.0	0.50	4.0–5.0	0.10	0.45–0.65	· · ·	· · ·	0.10	· · ·	0.20	0.10	0.20	rem
A390.1	A13901	· · ·	Ingot	16.0–18.0	0.40	4.0–5.0	0.10	0.50–0.65	· · ·	· · ·	0.10	· · ·	0.20	0.10	0.20	rem
B390.0	A23900	· · ·	D	16.0–18.0	1.3	4.0–5.0	0.50	0.45–0.65	· · ·	0.10	1.5	· · ·	0.20	0.10	0.20	rem
B390.1	A23901	· · ·	Ingot	16.0–18.0	1.0	4.0–5.0	0.50	0.50–0.65	· · ·	0.10	1.4	· · ·	0.20	0.10	0.20	rem
392.0	A03920	· · ·	D	18.0–20.0	1.5	0.40–0.8	0.20–0.6	0.8–1.2	· · ·	0.50	0.50	0.30	0.20	0.15	0.50	rem
392.1	A03921	· · ·	Ingot	18.0–20.0	1.1	0.40–0.8	0.20–0.6	0.9–1.2	· · ·	0.50	0.40	0.30	0.20	0.15	0.50	rem
393.0	A03930	· · ·	S, P, D	21.0–23.0	1.3	0.7–1.1	0.10	0.7–1.3	· · ·	2.0–2.5	0.10	· · ·	0.10–0.20	0.05(w)	0.15	rem
393.1	A03931	· · ·	Ingot	21.0–23.0	1.0	0.7–1.1	0.10	0.8–1.3	· · ·	2.0–2.5	0.10	· · ·	0.10–0.20	0.05(w)	0.15	rem
393.2	A03932	· · ·	Ingot	21.0–23.0	0.8	0.7–1.1	0.10	0.8–1.3	· · ·	2.0–2.5	0.10	· · ·	0.10–0.20	0.05(w)	0.15	rem
408.2(x)	A04082(x)	· · ·	Ingot	8.5–9.5	0.6–1.3	0.10	0.10	· · ·	· · ·	· · ·	0.10	· · ·	· · ·	0.10	0.20	rem
409.2(x)	A04092(x)	· · ·	Ingot	9.0–10.0	0.6–1.3	0.10	0.10	· · ·	· · ·	· · ·	0.10	· · ·	· · ·	0.10	0.20	rem
411.2(x)	A04112(x)	· · ·	Ingot	10.0–12.0	0.6–1.3	0.20	0.10	· · ·	· · ·	· · ·	0.10	· · ·	· · ·	0.10	0.20	rem
413.0(s)	A04130(s)	3522 AlSi12CuFe(s) 3522 AlSi12 Fe(s) R164 AlSi12(s) R164 AlSi12Cu(s) R164 AlSi12CuFe(s) R164 AlSi12Fe(s) R2147 AlSi12(s)	D	11.0–13.0	2.0	1.0	0.35	0.10	· · ·	0.50	0.50	0.15	· · ·	· · ·	0.25	rem
413.2(s)	A04132(s)	· · ·	Ingot	11.0–13.0	0.7–1.1	0.10	0.10	0.07	· · ·	0.10	0.10	0.10	· · ·	· · ·	0.20	rem
A413.0(s)	A14130(s)	· · ·	D	11.0–13.0	1.3	1.0	0.35	0.10	· · ·	0.50	0.50	0.15	· · ·	· · ·	0.25	rem
A413.1(s)	A14131(s)	· · ·	Ingot	11.0–13.0	1.0	1.0	0.35	0.10	· · ·	0.50	0.40	0.15	· · ·	· · ·	0.25	rem
A413.2	A14132(s)	· · ·	Ingot	11.0–13.0	0.6	0.10	0.05	0.05	· · ·	0.05	0.05	0.05	· · ·	· · ·	0.10	rem
B413.0	A24130	· · ·	S, P	11.0–13.0	0.50	0.10	0.35	0.05	· · ·	0.05	0.10	· · ·	0.25	0.05	0.20	rem
B413.1	B24131	· · ·	Ingot	11.0–13.0	0.40	0.10	0.35	0.05	· · ·	0.05	0.10	· · ·	0.25	0.05	0.20	rem
435.2(y)	A04352(y)	· · ·	Ingot	3.3–3.9	0.40	0.05	0.05	0.05	· · ·	· · ·	0.10	· · ·	· · ·	0.05	0.20	rem
443.0	A04430	· · ·	S, P	4.5–6.0	0.8	0.6	0.50	0.05	0.25	· · ·	0.50	· · ·	0.25	· · ·	0.35	rem
443.1	A04431	· · ·	Ingot	4.5–6.0	0.6	0.6	0.50	0.05	0.25	· · ·	0.50	· · ·	0.25	· · ·	0.35	rem
443.2	A04432	· · ·	Ingot	4.5–6.0	0.6	0.10	0.10	0.05	· · ·	· · ·	0.10	· · ·	0.20	0.05	0.15	rem
A443.0	A14430	· · ·	S	4.5–6.0	0.8	0.30	0.50	0.05	0.25	· · ·	0.50	· · ·	0.25	· · ·	0.35	rem
A443.1	A14431	· · ·	Ingot	4.5–6.0	0.6	0.30	0.50	0.05	0.25	· · ·	0.50	· · ·	0.25	· · ·	0.35	rem
B443.0	A24430	3522 AlSi5 R164 AlSi5	S, P	4.5–6.0	0.8	0.15	0.35	0.05	· · ·	· · ·	0.35	· · ·	0.25	0.05	0.15	rem
B443.1	A24431	· · ·	Ingot	4.5–6.0	0.6	0.15	0.35	0.05	· · ·	· · ·	0.35	· · ·	0.25	0.05	0.15	rem
C443.0	A34430	R164 AlSi5Fe	D	4.5–6.0	2.0	0.6	0.35	0.10	· · ·	0.50	0.50	0.15	· · ·	· · ·	0.25	rem
C443.1	A34431	· · ·	Ingot	4.5–6.0	1.1	0.6	0.35	0.10	· · ·	0.50	0.40	0.15	· · ·	· · ·	0.25	rem
C443.2	A34432	· · ·	Ingot	4.5–6.0	0.7–1.1	0.10	0.10	0.05	· · ·	· · ·	0.10	· · ·	· · ·	0.05	0.15	rem
444.0	A04440	· · ·	S, P	6.5–7.5	0.6	0.25	0.35	0.10	· · ·	· · ·	0.35	· · ·	0.25	0.05	0.15	rem
444.2	A04442	· · ·	Ingot	6.5–7.5	0.13–0.25	0.10	0.05	0.05	· · ·	· · ·	0.05	· · ·	0.20	0.05	0.15	rem
A444.0	A14440	· · ·	P	6.5–7.5	0.20	0.10	0.10	0.05	· · ·	· · ·	0.10	· · ·	0.20	0.05	0.15	rem
A444.1	A14441	· · ·	Ingot	6.5–7.5	0.15	0.10	0.10	0.05	· · ·	· · ·	0.10	· · ·	0.20	0.05	0.15	rem
A444.2	A14442	· · ·	Ingot	6.5–7.5	0.12	0.05	0.05	0.05	· · ·	· · ·	0.05	· · ·	0.20	0.05	0.15	rem

(continued)

(a) Serial letter prefix indicates modification: A, B, C, D, and F. (b) Per ISO standard No. R115 unless other standard (R164, R2147, or 3522) specified. (c) D, die casting; P, permanent mold; s, sand. Other products may pertain to the composition shown even though not listed. (d) The Al content for unalloyed aluminum by remelt is the difference between 100.00% and the sum of all other metallic elements present in amounts of 0.010% or more each, expressed to the second decimal before determining the sum. (e) (Mn + Cr + Ti + V) = 0.025% max. (f) Fe/Si ratio 2.5 min. (g) Fe/Si ratio 2.0 min. (h) Fe/Si ratio 1.5 min. (i) 0.40 to 1.0% Ag. (j) 0.50–1.0% Ag. (k) Ti + Zr = 0.50 max. (l) 0.20 to 0.30% Sb; 0.20 to 0.30% Co; 0.10 to 0.30% Zr. (m) 0.05–0.15% V; 0.10–0.25% Zr. (n) 0.06–0.20% V. (o) For Fe > 0.45%, Mn content shall not be less than one-half Fe content. (p) 0.04–0.07% Be. (q) 0.10–0.30% Be. (r) 0.15–0.30% Be. (s) A*xxx*.1 ingot is used to produce *xxx*.0 and A*xxx*.0 castings. (t) (Mn + Cr) = 0.8% max. (u) 0.25% Pb max. (v) 0.02–0.04% Be. (w) 0.08–0.15% V. (x) Used to coat steel. (y) Used with Zn to coat steel. (z) 0.10% Pb max. (aa) 0.003–0.007% Be; 0.005% B max. (bb) 0.003–0.007% Be; 0.002% B max
Source: Ref 3, 4, 5

Table 3 (continued)

Grade designation				Composition, wt%												
														Unspecified other elements		
Aluminum Association(a)	UNS No.	ISO(b)	Product(c)	Si	Fe	Cu	Mn	Mg	Cr	Ni	Zn	Sn	Ti	Each	Total	Al, min(d)
445.2(x)	A04452(x)	...	Ingot	6.5–7.5	0.6–1.3	0.10	0.10	...	...	...	0.10	...	...	0.10	0.20	rem
511.0	A05110	...	S	0.30–0.7	0.50	0.15	0.35	3.5–4.5	...	...	0.15	...	0.25	0.05	0.15	rem
511.1	A05111	...	Ingot	0.30–0.7	0.40	0.15	0.35	3.6–4.5	...	...	0.15	...	0.25	0.05	0.15	rem
511.2	A05112	...	Ingot	0.30–0.7	0.30	0.10	0.10	3.6–4.5	...	...	0.10	...	0.20	0.05	0.15	rem
512.0	A05120	...	S	1.4–2.2	0.6	0.35	0.8	3.5–4.5	0.25	...	0.35	...	0.25	0.05	0.15	rem
512.2	A05122	...	Ingot	1.4–2.2	0.30	0.10	0.10	3.6–4.5	...	...	0.10	...	0.20	0.05	0.15	rem
513.0	A05130	...	P	0.30	0.40	0.10	0.30	3.5–4.5	...	...	1.4–2.2	...	0.20	0.05	0.15	rem
513.2	A05132	...	Ingot	0.30	0.30	0.10	0.10	3.6–4.5	...	...	1.4–2.2	...	0.20	0.05	0.15	rem
514.0	A05140	3522 AlMg3 R164 AlMg3; R2147AlMg3	S	0.35	0.50	0.15	0.35	3.5–4.5	...	...	0.15	...	0.25	0.05	0.15	rem
514.1	A05141	...	Ingot	0.35	0.40	0.15	0.35	3.6–4.5	...	...	0.15	...	0.25	0.05	0.15	rem
514.2	A05142	...	Ingot	0.30	0.30	0.10	0.10	3.6–4.5	...	...	0.10	...	0.20	0.05	0.15	rem
515.0	A05150	...	D	0.50–1.0	1.3	0.20	0.40–0.6	2.5–4.0	...	...	0.10	...	...	0.05	0.15	rem
515.2	A05152	...	Ingot	0.50–1.0	0.6–1.0	0.10	0.40–0.6	2.7–4.0	...	...	0.05	...	...	0.05	0.15	rem
516.0	A05160	...	D	0.30–1.5	0.35–1.0	0.30	0.15–0.40	2.5–4.5	...	0.25–0.40	0.20	0.10	0.10–0.20	0.05(z)	...	rem
516.1	A05161	...	Ingot	0.30–1.5	0.35–0.7	0.30	0.15–0.40	2.6–4.5	...	0.25–0.40	0.20	0.10	0.10–0.20	0.05(z)	...	rem
518.0	A05180	...	D	0.35	1.8	0.25	0.35	7.5–8.5	...	0.15	0.15	0.15	...	...	0.25	rem
518.1	A05181	...	Ingot	0.35	1.1	0.25	0.35	7.6–8.5	...	0.15	0.15	0.15	...	...	0.25	rem
518.2	A05182	...	Ingot	0.25	0.7	0.10	0.10	7.6–8.5	...	0.05	...	0.05	...	...	0.10	rem
520.0	A05200	3522 AlMg10 R164 AlMg10; R2147 AlMg10	S	0.25	0.30	0.25	0.15	9.5–10.6	...	...	0.15	...	0.25	0.05	0.15	rem
520.2	A05202	...	Ingot	0.15	0.20	0.20	0.10	9.6–10.6	...	...	0.10	...	0.20	0.05	0.15	rem
535.0	A05350	...	S	0.15	0.15	0.05	0.10–0.25	6.2–7.5	...	...	...	...	0.10–0.25	0.05(aa)	0.15	rem
535.2	A05352	...	Ingot	0.10	0.10	0.05	0.10–0.25	6.6–7.5	...	...	...	...	0.10–0.25	0.05(bb)	0.15	rem
A535.0	A15350	...	S	0.20	0.20	0.10	0.10–0.25	6.5–7.5	...	...	...	...	0.25	0.05	0.15	rem
A535.1	A15351	...	Ingot	0.20	0.15	0.10	0.10–0.25	6.6–7.5	...	...	...	...	0.25	0.05	0.15	rem
B535.0	A25350	...	S	0.15	0.15	0.10	0.05	6.5–7.5	...	...	...	...	0.10–0.25	0.05	0.15	rem
B535.2	A25352	...	Ingot	0.10	0.12	0.05	0.05	6.6–7.5	...	...	...	...	0.10–0.25	0.05	0.15	rem
705.0	A07050	...	S, P	0.20	0.8	0.20	0.40–0.6	1.4–1.8	0.20–0.40	...	2.7–3.3	...	0.25	0.05	0.15	rem
705.1	A07051	...	Ingot	0.20	0.6	0.20	0.40–0.6	1.5–1.8	0.20–0.40	...	2.7–3.3	...	0.25	0.05	0.15	rem
707.0	A07070	...	S, P	0.20	0.8	0.20	0.40–0.6	1.8–2.4	0.20–0.40	...	4.0–4.5	...	0.25	0.05	0.15	rem
707.1	A07071	...	Ingot	0.20	0.6	0.20	0.40–0.6	1.9–2.4	0.20–0.40	...	4.0–4.5	...	0.25	0.05	0.15	rem
710.0	A07100	...	S	0.15	0.50	0.35–0.65	0.05	0.6–0.8	...	...	6.0–7.0	...	0.25	0.05	0.15	rem
710.1	A07101	...	Ingot	0.15	0.40	0.35–0.65	0.05	0.65–0.8	...	...	6.0–7.0	...	0.25	0.05	0.15	rem
711.0	A07110	...	P	0.30	0.7–1.4	0.35–0.65	0.05	0.25–0.45	...	...	6.0–7.0	...	0.20	0.05	0.15	rem
711.1	A07111	...	Ingot	0.30	0.7–1.1	0.35–0.65	0.05	0.30–0.45	...	...	6.0–7.0	...	0.20	0.05	0.15	rem
712.0	A07120	...	S	0.30	0.50	0.25	0.10	0.50–0.65	0.40–0.6	...	5.0–6.5	...	0.15–0.25	0.05	0.20	rem
712.2	A07122	...	Ingot	0.15	0.40	0.25	0.10	0.50–0.65	0.40–0.6	...	5.0–6.5	...	0.15–0.25	0.05	0.20	rem
713.0	A07130	...	S, P	0.25	1.1	0.40–1.0	0.6	0.20–0.50	0.35	0.15	7.0–8.0	...	0.25	0.10	0.25	rem
713.1	A07131	...	Ingot	0.25	0.8	0.40–1.0	0.6	0.25–0.50	0.35	0.15	7.0–8.0	...	0.25	0.10	0.25	rem
771.0	A07710	...	S	0.15	0.15	0.10	0.10	0.8–1.0	0.06–0.20	...	6.5–7.5	...	0.10–0.20	0.05	0.15	rem
771.2	A07712	...	Ingot	0.10	0.10	0.10	0.10	0.85–1.0	0.06–0.20	...	6.5–7.5	...	0.10–0.20	0.05	0.15	rem
772.0	A07720	...	S	0.15	0.15	0.10	0.10	0.6–0.8	0.06–0.20	...	6.0–7.0	...	0.10–0.20	0.05	0.15	rem
772.2	A07722	...	Ingot	0.10	0.10	0.10	0.10	0.65–0.8	0.06–0.20	...	6.0–7.0	...	0.10–0.20	0.05	0.15	rem
850.0	A08500	...	S, P	0.7	0.7	0.7–1.3	0.10	0.10	...	0.7–1.3	...	5.5–7.0	0.20	...	0.30	rem
850.1	A08501	...	Ingot	0.7	0.50	0.7–1.3	0.10	0.10	...	0.7–1.3	...	5.5–7.0	0.20	...	0.30	rem
851.0	A08510	...	S, P	2.0–3.0	0.7	0.7–1.3	0.10	0.10	...	0.30–0.7	...	5.5–7.0	0.20	...	0.30	rem
851.1	A08511	...	Ingot	2.0–3.0	0.50	0.7–1.3	0.10	0.10	...	0.30–0.7	...	5.5–7.0	0.20	...	0.30	rem
852.0	A08520	...	S, P	0.40	0.7	1.7–2.3	0.10	0.6–0.9	...	0.9–1.5	...	5.5–7.0	0.20	...	0.30	rem
852.1	A08521	...	Ingot	0.40	0.50	1.7–2.3	0.10	0.7–0.9	...	0.9–1.5	...	5.5–7.0	0.20	...	0.30	rem
853.0	A08530	...	S, P	5.5–6.5	0.7	3.0–4.0	0.50	...	...	...	...	5.5–7.0	0.20	...	0.30	rem
853.2	A08532	...	Ingot	5.5–6.5	0.50	3.0–4.0	0.10	...	...	...	...	5.5–7.0	0.20	...	0.30	rem

(a) Serial letter prefix indicates modification: A, B, C, D, and F. (b) Per ISO standard No. R115 unless other standard (R164, R2147, or 3522) specified. (c) D, die casting; P, permanent mold; s, sand. Other products may pertain to the composition shown even though not listed. (d) The Al content for unalloyed aluminum by remelt is the difference between 100.00% and the sum of all other metallic elements present in amounts of 0.010% or more each, expressed to the second decimal before determining the sum. (e) (Mn + Cr + Ti + V) = 0.025% max. (f) Fe/Si ratio 2.5 min. (g) Fe/Si ratio 2.0 min. (h) Fe/Si ratio 1.5 min. (i) 0.40 to 1.0% Ag. (j) 0.50–1.0% Ag. (k) Ti + Zr = 0.50 max. (l) 0.20 to 0.30% Sb; 0.20 to 0.30% Co; 0.10 to 0.30% Zr. (m) 0.05–0.15% V; 0.10–0.25% Zr. (n) 0.06–0.20% V. (o) For Fe > 0.45%, Mn content shall not be less than one-half Fe content. (p) 0.04–0.07% Be. (q) 0.10–0.30% Be. (r) 0.15–0.30% Be. (s) A*xxx*.1 ingot is used to produce *xxx*.0 and A*xxx*.0 castings. (t) (Mn + Cr) = 0.8% max. (u) 0.25% Pb max. (v) 0.02–0.04% Be. (w) 0.08–0.15% V. (x) Used to coat steel. (y) Used with Zn to coat steel. (z) 0.10% Pb max. (aa) 0.003–0.007% Be; 0.005% B max. (bb) 0.003–0.007% Be; 0.002% B max
Source: Ref 3, 4, 5

strain hardening remaining after the product has been partially annealed.

H3, Strain-Hardened and Stabilized. This applies to products that are strain-hardened and whose mechanical properties are stabilized by a low-temperature thermal treatment or as a result of heat introduced during fabrication. Stabilization usually improves ductility. This designation applies only to those alloys that, unless stabilized, gradually age soften at room temperature. The digit following the H3 indicates the degree of strain hardening remaining after stabilization.

Additional Temper Designations. For alloys that age soften at room temperature, each H2*x* temper has the same minimum ultimate tensile strength as the H3*x* temper with the same second digit. For other alloys, each H2*x* temper has the same minimum ultimate tensile strength as the H1*x* with the same second digit, and slightly higher elongation.

The digit following the designations H1, H2, and H3, which indicates the degree of strain hardening, is a numeral from 1 through 9. Numeral 8 indicates tempers with ultimate tensile strength equivalent to that achieved by about 75% cold reduction (temperature during reduction not to exceed 50 °C, or 120 °F) following full annealing. Tempers between 0 (annealed) and 8 are designated by numerals 1 through 7. Material having an ultimate tensile strength ap-

Table 4 ISO equivalents of wrought Aluminum Association international alloy designations

Aluminum Association international designation	ISO designation	Aluminum Association international designation	ISO designation
1050A	Al 99.5	5086	Al Mg4
1060	Al 99.6	5154	Al Mg3.5
1070A	Al 99.7		
1080A	Al 99.8	5154A	Al Mg3.5(A)
1100	Al 99.0 Cu	5183	Al Mg4.5Mn0.7(A)
		5251	Al Mg2
1200	Al 99.0	5356	Al Mg5Cr(A)
1350	E-Al 99.5	5454	Al Mg3Mn
. . .	Al 99.3		
1370	E-Al 99.7	5456	Al Mg5Mn
2011	Al Cu6BiPb	5554	Al Mg3Mn(A)
		5754	Al Mg3
2014	Al Cu4SiMg	6005	Al SiMg
2014A	Al Cu4SiMg(A)	6005A	Al SiMg(A)
2017	Al Cu4MgSi		
2017A	Al Cu4MgSi(A)	6060	Al MgSi
2024	Al Cu4Mg1	6061	Al Mg1SiCu
		6063	Al Mg0.7Si
2030	Al Cu4PbMg	6063A	Al Mg0.7Si(A)
2117	Al Cu2.5Mg	6082	Al Si1MgMn
2219	Al Cu6Mn		
3003	Al Mn1Cu	6101	E-Al MgSi
3004	Al Mn1Mg1	6101A	E-Al MgSi(A)
		6181	Al Si1Mg0.8
3005	Al Mn1Mg0.5	6262	Al Mg1SiPb
3103	Al Mn1	6351	Al Si1Mg0.5Mn
3105	Al Mn0.5Mg0.5		
4043	Al Si5	7005	Al Zn4.5Mg1.5Mn
4043A	Al Si5(A)	7010	Al Zn6MgCu
		7020	Al Zn4.5Mg1
4047	Al Si12	7049A	Al Zn8MgCu
4047A	Al Si12(A)	7050	Al Zn6CuMgZr
5005	Al Mg1(B)		
5050	Al Mg1.5(C)	7075	Al Zn5.5MgCu
5052	Al Mg2.5	7178	Al Zn7MgCu
		7475	Al Zn5.5MgCu(A)
5056	Al Mg5Cr	. . .	Al Zn4Mg1.5Mn
5056A	Al Mg5	. . .	Al Zn6MgCuMn
5083	Al Mg4.5Mn0.7		

proximately midway between that of the 0 temper and the 8 temper is designated by the numeral 4, midway between the 0 and 4 tempers by the numeral 2, and midway between the 4 and 8 tempers by the numeral 6. Numeral 9 designates tempers whose minimum ultimate tensile strength exceeds that of the 8 temper by 10 MPa (2 ksi) or more. For two-digit H tempers whose second digits are odd, the standard limits for strength are the arithmetic mean of the standard limits for the adjacent two-digit H tempers whose second digits are even.

For alloys that cannot be sufficiently cold-reduced to establish an ultimate tensile strength applicable to the 8 temper (75% cold reduction after full annealing), the 6-temper tensile strength may be established by cold reduction of approximately 55% following full annealing, or the 4-temper tensile strength may be established by cold reduction of approximately 35% after full annealing.

When it is desirable to identify a variation of a two-digit H temper, a third digit (from 1 to 9) may be assigned. The third digit is used when the degree of control of temper or the mechanical properties are different from but close to those for the two-digit H temper designation to which it is added, or when some other characteristic is significantly affected. The minimum ultimate tensile strength of a three-digit H temper is at least as close to that of the corresponding two-digit H temper as it is to either of the adjacent two-digit H tempers. Products in H tempers whose mechanical properties are below those of H*x*1 tempers are assigned variations of H*x*1. Some three-digit H temper designations have already been assigned for wrought products in all alloys:

Hx11 applies to products that incur sufficient strain hardening after final annealing to fail to qualify as 0 temper, but not so much or so consistent an amount of strain hardening to qualify as H*x*1 temper.

H112 pertains to products that may acquire some strain hardening during working at elevated temperature and for which there are mechanical property limits.

Patterned or Embossed Sheet. Table 5 lists the three-digit H temper designations that have been assigned to patterned or embossed sheet.

System for Heat-Treatable Alloys

The temper designation system for wrought and cast products that are strengthened by heat treatment employs the W and T designations described in the section "Basic Temper Designations" in this article. The W designation denotes an unstable temper, whereas the T designation denotes a stable temper other than F, O, or H. The T is followed by a number from 1 to 10, each number indicating a specific sequence of basic treatments.

T1, Cooled From an Elevated-Temperature Shaping Process and Naturally Aged to a Substantially Stable Condition. This designation applies to products that are not cold worked after an elevated-temperature shaping process such as casting or extrusion and for which mechanical properties have been stabilized by room-temperature aging. It also applies to products are flattened or straightened after cooling from the shaping process, for which the effects of the cold work imparted by flattening or straightening are not accounted for in specified property limits.

T2, Cooled From an Elevated-Temperature Shaping Process, Cold Worked, and Naturally Aged to a Substantially Stable Condition. This variation refers to products that are cold worked specifically to improve strength after cooling from a hot-working process such as rolling or extrusion and for which mechanical properties have been stabilized by room-temperature aging. It also applies to products in which the effects of cold work, imparted by flattening or straightening, are accounted for in specified property limits.

T3, Solution Heat Treated, Cold Worked, and Naturally Aged to a Substantially Stable Condition. T3 applies to products that are cold worked specifically to improve strength after solution heat treatment and for which mechanical properties have been stabilized by room-temperature aging. It also applies to products in which the effects of cold work, imparted by flattening or straightening, are accounted for in specified property limits.

T4, Solution Heat Treated and Naturally Aged to a Substantially Stable Condition. This signifies products that are not cold worked after solution heat treatment and for which mechanical properties have been stabilized by room-temperature aging. If the products are flattened or straightened, the effects of the cold work imparted by flattening or straightening are not accounted for in specified property limits.

T5, Cooled From an Elevated-Temperature Shaping Process and Artificially Aged. T5 includes products that are not cold worked after an elevated-temperature shaping process such as casting or extrusion and for which mechanical properties have been substantially improved by precipitation heat treatment. If the products are flattened or straightened after cooling from the shaping process, the effects of the cold work imparted by flattening or straightening are not accounted for in specified property limits.

T6, Solution Heat Treated and Artificially Aged. This group encompasses products

Table 5 H temper designations for aluminum and aluminum alloy patterned or embossed sheet

Patterned or embossed sheet	Temper of sheet from which textured sheet was fabricated
H114	O
H124	H11
H224	H21
H324	H31
H134	H12
H234	H22
H334	H32
H144	H13
H244	H23
H344	H33
H154	H14
H254	H24
H354	H34
H164	H15
H264	H25
H364	H35
H174	H16
H274	H26
H374	H36
H184	H17
H284	H27
H384	H37
H194	H18
H294	H28
H394	H38
H195	H19
H295	H29
H395	H39

Source: Ref 1

that are not cold worked after solution heat treatment and for which mechanical properties or dimensional stability, or both, have been substantially improved by precipitation heat treatment. If the products are flattened or straightened, the effects of the cold work imparted by flattening or straightening are not accounted for in specified property limits.

T7, Solution Heat Treated and Overaged or Stabilized. T7 applies to wrought products that have been precipitation heat treated beyond the point of maximum strength to provide some special characteristic, such as enhanced resistance to stress-corrosion cracking or exfoliation corrosion. It applies to cast products that are artificially aged after solution heat treatment to provide dimensional and strength stability.

T8, Solution Heat Treated, Cold Worked, and Artificially Aged. This designation applies to products that are cold worked specifically to improve strength after solution heat treatment and for which mechanical properties or dimensional stability, or both, have been substantially improved by precipitation heat treatment. The effects of cold work, including any cold work imparted by flattening or straightening, are accounted for in specified property limits.

T9, Solution Heat Treated, Artificially Aged, and Cold Worked. This grouping is comprised of products that are cold worked specifically to improve strength after they have been precipitation heat treated.

T10, Cooled From an Elevated-Temperature Shaping Process, Cold Worked, and Artificially Aged. T10 identifies products that are cold worked specifically to improve strength after cooling from a hot-working process such as rolling or extrusion and for which mechanical properties have been substantially improved by precipitation heat treatment. The effects of cold work, including any cold work imparted by flattening or straightening, are accounted for in specified property limits.

Additional T Temper Variations. When it is desirable to identify a variation of one of the ten major T tempers described above, additional digits, the first of which cannot be zero, may be added to the designation.

Specific sets of additional digits have been assigned to stress-relieved wrought products:

Stress Relieved by Stretching, Compressing, or Combination of Stretching and Compressing. This designation applies to the following products when stretched to the indicated amounts after solution heat treatment or after cooling from an elevated-temperature shaping process:

Product form	Permanent set, %
Plate	1½–3
Rod, bar, shapes, and extruded tube	1–3
Drawn tube	½–3

- T*x*51 applies specifically to plate, to rolled or cold-finished rod and bar, to die or ring forgings, and to rolled rings. These products receive no further straightening after stretching
- T*x*510 applies to extruded rod, bar, shapes and tubing, and to drawn tubing. Products in this temper receive no further straightening after stretching
- T*x*511 refers to products that may receive minor straightening after stretching to comply with standard tolerances

This variation involves stress relief by compressing.

- T*x*52 applies to products that are stress relieved by compressing after solution heat treatment or after cooling from a hot-working process to produce a permanent set of 1 to 5%

The next designation is used for products that are stress relieved by combining stretching and compressing.

- T*x*54 applies to die forgings that are stress relieved by restriking cold in the finish die. (These same digits—and 51, 52, and 54—may be added to the designation W to indicate unstable solution-heat-treated and stress-relieved tempers)

Temper designations have been assigned to wrought products heat treated from the O or the F temper to demonstrate response to heat treatment:

- T42 means solution heat treated from the O or the F temper to demonstrate response to heat treatment and naturally aged to a substantially stable condition
- T62 means solution heat treated from the O or the F temper to demonstrate response to heat treatment and artificially aged

Temper designations T42 and T62 also may be applied to wrought products heat treated from any temper by the user when such heat treatment results in the mechanical properties applicable to these tempers.

System for Annealed Products

A digit following the "O" indicates a product in annealed condition having special characteristics. For example, for heat-treatable alloys, O1 indicates a product that has been heat treated at approximately the same time and temperature required for solution heat treatment and then air cooled to room temperature; this designation applies to products that are to be machined prior to solution heat treatment by the user. Mechanical property limits are not applicable.

Designation of Unregistered Tempers

The letter P has been assigned to denote H, T, and O temper variations that are negotiated between manufacturer and purchaser. The letter P follows the temper designation that most nearly pertains. The use of this type of designation includes situations where:

- The use of the temper is sufficiently limited to preclude its registration
- The test conditions are different from those required for registration with the Aluminum Association
- The mechanical property limits are not established on the same basis as required for registration with the Aluminum Association

Foreign Temper Designations

Unlike the agreement relating to wrought alloy designations, there is no Declaration of Accord for an international system of tempers to be registered with the Aluminum Association by foreign organizations. For the most part, the ANSI system is used, but because there is no international accord, reference to ANSI H35.1 properties and characteristics of aluminum alloy tempers registered with the Aluminum Association under ANSI 35.1 may not always reflect actual properties and characteristics associated with the particular alloy temper. In addition, temper designations may be created which are not registered with the Aluminum Association.

REFERENCES

1. "American National Standard Alloy and Temper Designation Systems for Aluminum," PP/2650/988/11, Aluminum Association, July 1988
2. "Registration Record of International Alloy Designations and Chemical Composition Limits for Wrought Aluminum and Wrought Aluminum Alloys," PP/2M/289/A1, Aluminum Association, Feb 1989
3. *Metals & Alloys in the Unified Numbering System*, 4th ed., Society of Automotive Engineers, 1986
4. J.G. Gensure and D.L. Potts, Ed., *International Metallic Materials Cross-Reference*, 3rd ed., Genium Publishing, 1983
5. "Registration Record of Aluminum Association Alloy Designations and Chemical Composition Limits for Aluminum Alloys in the Form of Casting and Ingot," Aluminum Association, Jan 1989

Aluminum Mill and Engineered Wrought Products

Jack W. Bray, Reynolds Metals Company

ALUMINUM mill products are those aluminum products that have been subjected to plastic deformation by hot- and cold-working mill processes (such as rolling, extruding, and drawing, either singly or in combination), so as to transform cast aluminum ingot into the desired product form. The microstructural changes associated with the working and with any accompanying thermal treatments are used to control certain properties and characteristics of the worked, or wrought, product or alloy.

Typical examples of mill products include plate or sheet (which is subsequently formed or machined into products such as aircraft or building components), household foil, and extruded shapes such as storm window frames. A vast difference in the mechanical and physical properties of aluminum mill products can be obtained through the control of the chemistry, processing, and thermal treatment.

Wrought Alloy Series

Aluminum alloys are commonly grouped into an alloy designation series, as described earlier in the article "Alloy and Temper Designation Systems for Aluminum and Aluminum Alloys" in this Volume. The general characteristics of the alloy groups are described below, and the comparative corrosion and fabrication characteristics and some typical applications of the commonly used grades or alloys in each group are presented in Table 1.

1*xxx* Series. Aluminum of 99.00% or higher purity has many applications, especially in the electrical and chemical fields. These grades of aluminum are characterized by excellent corrosion resistance, high thermal and electrical conductivities, low mechanical properties, and excellent workability. Moderate increases in strength may be obtained by strain hardening. Iron and silicon are the major impurities. Typical uses include chemical equipment, reflectors, heat exchangers, electrical conductors and capacitors, packaging foil, architectural applications, and decorative trim.

2*xxx* Series. Copper is the principal alloying element in 2*xxx* series alloys, often with magnesium as a secondary addition. These alloys require solution heat treatment to obtain optimum properties; in the solution heat-treated condition, mechanical properties are similar to, and sometimes exceed, those of low-carbon steel. In some instances, precipitation heat treatment (aging) is employed to further increase mechanical properties. This treatment increases yield strength, with attendant loss in elongation; its effect on tensile strength is not as great.

The alloys in the 2*xxx* series do not have as good corrosion resistance as most other aluminum alloys, and under certain conditions they may be subject to intergranular corrosion. Therefore, these alloys in the form of sheet usually are clad with a high-purity aluminum or with a magnesium-silicon alloy of the 6*xxx* series, which provides galvanic protection of the core material and thus greatly increases resistance to corrosion.

Alloys in the 2*xxx* series are particularly well suited for parts and structures requiring high strength-to-weight ratios and are commonly used to make truck and aircraft wheels, truck suspension parts, aircraft fuselage and wing skins, and structural parts and those parts requiring good strength at temperatures up to 150 °C (300 °F). Except for alloy 2219, these alloys have limited weldability, but some alloys in this series have superior machinability.

3*xxx* Series. Manganese is the major alloying element of 3*xxx* series alloys. These alloys generally are non-heat treatable but have about 20% more strength than 1*xxx* series alloys. Because only a limited percentage of manganese (up to about 1.5%) can be effectively added to aluminum, manganese is used as a major element in only a few alloys. However, three of them—3003, 3X04, and 3105—are widely used as general-purpose alloys for moderate-strength applications requiring good workability. These applications include beverage cans, cooking utensils, heat exchangers, storage tanks, awnings, furniture, highway signs, roofing, siding, and other architectural applications.

4*xxx* Series. The major alloying element in 4*xxx* series alloys is silicon, which can be added in sufficient quantities (up to 12%) to cause substantial lowering of the melting range without producing brittleness. For this reason, aluminum-silicon alloys are used in welding wire and as brazing alloys for joining aluminum, where a lower melting range than that of the base metal is required. Most alloys in this series are non-heat treatable, but when used in welding heat-treatable alloys, they will pick up some of the alloying constituents of the latter and so respond to heat treatment to a limited extent. The alloys containing appreciable amounts of silicon become dark gray to charcoal when anodic oxide finishes are applied and hence are in demand for architectural applications. Alloy 4032 has a low coefficient of thermal expansion and high wear resistance, and thus is well suited to production of forged engine pistons.

5*xxx* Series. The major alloying element in 5*xxx* series alloys is magnesium. When it is used as a major alloying element or with manganese, the result is a moderate-to-high-strength work-hardenable alloy. Magnesium is considerably more effective than manganese as a hardener, about 0.8% Mg being equal to 1.25% Mn, and it can be added in considerably higher quantities. Alloys in this series possess good welding characteristics and good resistance to corrosion in marine atmospheres. However, certain limitations should be placed on the amount of cold work and the safe operating temperatures permissible for the higher-magnesium alloys (over about 3.5% for operating temperatures above about 65 °C, or 150 °F) to avoid susceptibility to stress-corrosion cracking.

Uses include architectural, ornamental, and decorative trim; cans and can ends; household appliances; streetlight standards;

Table 1 Comparative corrosion and fabrication characteristics and typical applications of wrought aluminum alloys

Alloy temper	Resistance to corrosion: General(a)	Resistance to corrosion: Stress-corrosion cracking(b)	Workability (cold)(e)	Machinability(e)	Weldability(f): Gas	Weldability(f): Arc	Weldability(f): Resistance spot and seam	Brazeability(f)	Solderability(g)	Some typical applications of alloys
1050 0	A	A	A	E	A	A	B	A	A	Chemical equipment, railroad tank cars
H12	A	A	A	E	A	A	A	A	A	
H14	A	A	A	D	A	A	A	A	A	
H16	A	A	B	D	A	A	A	A	A	
H18	A	A	B	D	A	A	A	A	A	
1060 0	A	A	A	E	A	A	B	A	A	Chemical equipment, railroad tank cars
H12	A	A	A	E	A	A	A	A	A	
H14	A	A	A	D	A	A	A	A	A	
H16	A	A	B	D	A	A	A	A	A	
H18	A	A	B	D	A	A	A	A	A	
1100 0	A	A	A	E	A	A	B	A	A	Sheet-metal work, spun hollowware,
H12	A	A	A	E	A	A	A	A	A	fin stock
H14	A	A	A	D	A	A	A	A	A	
H16	A	A	B	D	A	A	A	A	A	
H18	A	A	C	D	A	A	A	A	A	
1145 0	A	A	A	E	A	A	B	A	A	Foil, fin stock
H12	A	A	A	E	A	A	A	A	A	
H14	A	A	A	D	A	A	A	A	A	
H16	A	A	B	D	A	A	A	A	A	
H18	A	A	B	D	A	A	A	A	A	
1199 0	A	A	A	E	A	A	B	A	A	Electrolytic capacitor foil, chemical
H12	A	A	A	E	A	A	A	A	A	equipment, railroad tank cars
H14	A	A	A	D	A	A	A	A	A	
H16	A	A	B	D	A	A	A	A	A	
H18	A	A	B	D	A	A	A	A	A	
1350 0	A	A	A	E	A	A	B	A	A	Electrical conductors
H12, H111	A	A	A	E	A	A	A	A	A	
H14, H24	A	A	A	D	A	A	A	A	A	
H16, H26	A	A	B	D	A	A	A	A	A	
H18	A	A	B	D	A	A	A	A	A	
2011 T3	D(c)	D	C	A	D	D	D	D	C	Screw-machine products
T4, T451	D(c)	D	B	A	D	D	D	D	C	
T8	D	B	D	A	D	D	D	D	C	
2014 0	...	...	...	D	D	D	B	D	C	Truck frames, aircraft structures
T3, T4, T451	D(c)	C	C	B	D	B	B	D	C	
T6, T651, T6510, T6511	D	C	D	B	D	B	B	D	C	
2024 0	...	...	...	D	D	D	D	D	C	Truck wheels, screw-machine
T4, T3, T351, T3510, T3511	D(c)	C	C	B	C	B	B	D	C	products, aircraft structures
T361	D(c)	C	D	B	D	C	B	D	C	
T6	D	B	C	B	D	C	B	D	C	
T861, T81, T851, T8510, T8511	D	B	D	B	D	C	B	D	C	
T72	...	...	...	B	...	...	...	...	...	
2036 T4	C	...	B	C	...	B	B	D	...	Auto-body panel sheet
2124 T851	D	B	D	B	D	C	B	D	C	Military supersonic aircraft
2218 T61	D	C	...	...	...	...	C	...	C	Jet engine impellers and rings
T72	D	C	...	B	D	C	B	D	C	
2219 0	...	...	...	...	D	A	B	D		Structural uses at high temperatures
T31, T351, T3510, T3511	D(c)	C	C	B	A	A	A	D	NA	(to 315 °C, or 600 °F) high-strength
T37	D(c)	C	D	B	A	A	A	D		weldments
T81, T851, T8510, T8511	D	B	D	B	A	A	A	D		
T87	D	B	D	B	A	A	A	D		
2618 T61	D	C	...	B	D	C	B	D	NA	Aircraft engines
3003 0	A	A	A	E	A	A	B	A	A	Cooking utensils, chemical equipment,
H12	A	A	A	E	A	A	A	A	A	pressure vessels, sheet-metal work,
H14	A	A	B	D	A	A	A	A	A	builder's hardware, storage tanks
H16	A	A	C	D	A	A	A	A	A	
H18	A	A	C	D	A	A	A	A	A	
H25	A	A	B	D	A	A	A	A	A	
3004 0	A	A	A	D	B	A	B	B	B	Sheet-metal work, storage tanks
H32	A	A	B	D	B	A	A	B	B	
H34	A	A	B	C	B	A	A	B	B	
H36	A	A	C	C	B	A	A	B	B	
H38	A	A	C	C	B	A	A	B	B	
3105 0	A	A	A	E	B	A	B	B	B	Residential siding, mobile homes,
H12	A	A	B	E	B	A	A	B	B	rain-carrying goods, sheet-metal
H14	A	A	B	D	B	A	A	B	B	work
H16	A	A	C	D	B	A	A	B	B	
H18	A	A	C	D	B	A	A	B	B	

(continued)

(a) Ratings A through E are relative ratings in decreasing order of merit, based on exposures to sodium chloride solution by intermittent spraying or immersion. Alloys with A and B ratings can be used in industrial and seacoast atmospheres without protection. Alloys with C, D, and E ratings generally should be protected at least on faying surfaces. (b) Stress-corrosion cracking ratings are based on service experience and on laboratory tests of specimens exposed to the 3.5% sodium chloride alternate immersion test. A = No known instance of failure in service or in laboratory tests. B = No known instance of failure in service; limited failures in laboratory tests of short transverse specimens. C = Service failures with sustained tension stress acting in short transverse direction relative to grain structure; limited failures in laboratory tests of long transverse specimens. D = Limited service failures with sustained longitudinal or long transverse stress. (c) In relatively thick sections the rating would be E. (d) This rating may be different for material held at elevated temperature for long periods. (e) Ratings A through D for workability (cold), and A through E for machinability, are relative ratings in decreasing order of merit. (f) Ratings A through D for weldability and brazeability are relative ratings defined as follows: A = Generally weldable by all commercial procedures and methods. B = Weldable with special techniques or for specific applications; requires preliminary trials or testing to develop welding procedure and weld performance. C = Limited weldability because of crack sensitivity or loss in resistance to corrosion and mechanical properties. D = No commonly used welding methods have been developed. (g) Ratings A through D and NA for solderability are relative ratings defined as follows: A = Excellent. B = Good. C = Fair. D = Poor. NA = Not applicable

Table 1 (continued)

Alloy temper	Resistance to corrosion: General(a)	Resistance to corrosion: Stress-corrosion cracking(b)	Workability (cold)(e)	Machinability(e)	Weldability(f): Gas	Weldability(f): Arc	Weldability(f): Resistance spot and seam	Brazeability(f)	Solderability(g)	Some typical applications of alloys
H25	A	A	B	D	B	A	A	B	B	
4032 T6	C	B	. . .	B	D	B	C	D	NA	Pistons
4043	B	A	NA	C	NA	NA	NA	NA	NA	Welding electrode
5005 0	A	A	A	E	A	A	B	B	B	Appliances, utensils, architectural,
H12	A	A	A	E	A	A	A	B	B	electrical conductors
H14	A	A	B	D	A	A	A	B	B	
H16	A	A	C	D	A	A	A	B	B	
H18	A	A	C	D	A	A	A	B	B	
H32	A	A	B	E	A	A	A	B	B	
H34	A	A	C	D	A	A	A	B	B	
H36	A	A	C	D	A	A	A	B	B	
H38	A	A		D	A	A	A	B	B	
5050 0	A	A	A	E	A	A	B	B	C	Builders' hardware, refrigerator trim,
H32	A	A	A	D	A	A	A	B	C	coiled tubes
H34	A	A	B	D	A	A	A	B	C	
H36	A	A	C	C	A	A	A	B	C	
H38	A	A	C	C	A	A	A	B	C	
5052 0	A	A	A	D	A	A	B	C	D	Sheet-metal work, hydraulic tube,
H32	A	A	B	D	A	A	A	C	D	appliances
H34	A	A	B	C	A	A	A	C	D	
H36	A	A	C	C	A	A	A	C	D	
H38	A	A	C	C	A	A	A	C	D	
5056 0	A(d)	B(d)	A	D	C	A	B	D	D	Cable sheathing, rivets for magnesium,
H111	A(d)	B(d)	A	D	C	A	A	D	D	screen wire, zippers
H12, H32	A(d)	B(d)	B	D	C	A	A	D	D	
H14, H34	A(d)	B(d)	B	C	C	A	A	D	D	
H18, H38	A(d)	C(d)	C	C	C	A	A	D	D	
H192	B(d)	D(d)	D	B	C	A	A	D	D	
H392	B(d)	D(d)	D	B	C	A	A	D	D	
5083 0	A(d)	A(d)	B	D	C	A	B	D	D	Unfired, welded pressure vessels,
H321, H116	A(d)	A(d)	C	D	C	A	A	D	D	marine, auto aircraft cryogenics, TV
H111	A(d)	B(d)	C	D	C	A	A	D	D	towers, drilling rigs, transportation
5086 0	A(d)	A(d)	A	D	C	A	B	D	D	equipment, missile components
H32, H116	A(d)	A(d)	B	D	C	A	A	D	D	
H34	A(d)	B(d)	B	C	C	A	A	D	D	
H36	A(d)	B(d)	C	C	C	A	A	D	D	
H38	A(d)	B(d)	C	C	C	A	A	D	D	
H111	A(d)	A(d)	B	D	C	A	A	D	D	
5154 0	A(d)	A(d)	A	D	C	A	B	D	D	Welded structures, storage tanks,
H32	A(d)	A(d)	B	D	C	A	A	D	D	pressure vessels, salt-water service
H34	A(d)	A(d)	B	C	C	A	A	D	D	
H36	A(d)	A(d)	C	C	C	A	A	D	D	
H38	A(d)	A(d)	C	C	C	A	A	D	D	
5182 0	A	A(d)	A	D	C	A	B	D	D	Automobile body sheet, can ends
H19	A	A(d)	D	B	C	A	A	D	D	
5252 H24	A	A	B	D	A	A	A	C	D	Automotive and appliance trim
H25	A	A	B	C	A	A	A	C	D	
H28	A	A	C	C	A	A	A	C	D	
5254 0	A(d)	A(d)	A	D	C	A	B	D	D	Hydrogen peroxide and chemical
H32	A(d)	A(d)	B	D	C	A	A	D	D	storage vessels
H34	A(d)	A(d)	B	C	C	A	A	D	D	
H36	A(d)	A(d)	C	C	C	A	A	D	D	
H38	A(d)	A(d)	C	C	C	A	A	D	D	
5356	A	A	NA	B	NA	NA	NA	NA	NA	Welding electrode
5454 0	A	A	A	D	C	A	B	D		Welded structures, pressure vessels,
H32	A	A	B	D	C	A	A	D		marine service
H34	A	A	B	C	C	A	A	D	NA	
H111	A	A	B	D	C	A	A	D		
5456 0	A(d)	B(d)	B	D	C	A	B	D		High-strength welded structures,
H111	A(d)	B(d)	C	D	C	A	A	D		storage tanks, pressure vessels,
H321, H115	A(d)	B(d)	C	D	C	A	A	D	NA	marine applications
5457 0	A	A	A	E	A	A	B	B	B	
5652 0	A	A	A	D	A	A	B	C	D	Hydrogen peroxide and chemical
H32	A	A	B	D	A	A	A	C	D	storage vessels
H34	A	A	B	C	A	A	A	C	D	
H36	A	A	C	C	A	A	A	C	D	
H38	A	A	C	C	A	A	A	C	D	
5657 H241	A	A	A	D	A	A	A	B		Anodized auto and appliance trim

(continued)

(a) Ratings A through E are relative ratings in decreasing order of merit, based on exposures to sodium chloride solution by intermittent spraying or immersion. Alloys with A and B ratings can be used in industrial and seacoast atmospheres without protection. Alloys with C, D, and E ratings generally should be protected at least on faying surfaces. (b) Stress-corrosion cracking ratings are based on service experience and on laboratory tests of specimens exposed to the 3.5% sodium chloride alternate immersion test. A = No known instance of failure in service or in laboratory tests. B = No known instance of failure in service; limited failures in laboratory tests of short transverse specimens. C = Service failures with sustained tension stress acting in short transverse direction relative to grain structure; limited failures in laboratory tests of long transverse specimens. D = Limited service failures with sustained longitudinal or long transverse stress. (c) In relatively thick sections the rating would be E. (d) This rating may be different for material held at elevated temperature for long periods. (e) Ratings A through D for workability (cold), and A through E for machinability, are relative ratings in decreasing order of merit. (f) Ratings A through D for weldability and brazeability are relative ratings defined as follows: A = Generally weldable by all commercial procedures and methods. B = Weldable with special techniques or for specific applications; requires preliminary trials or testing to develop welding procedure and weld performance. C = Limited weldability because of crack sensitivity or loss in resistance to corrosion and mechanical properties. D = No commonly used welding methods have been developed. (g) Ratings A through D and NA for solderability are relative ratings defined as follows: A = Excellent. B = Good. C = Fair. D = Poor. NA = Not applicable

Table 1 (continued)

Alloy temper	Resistance to corrosion: General(a)	Resistance to corrosion: Stress-corrosion cracking(b)	Workability (cold)(e)	Machinability(e)	Weldability(f): Gas	Weldability(f): Arc	Weldability(f): Resistance spot and seam	Brazeability(f)	Solderability(g)	Some typical applications of alloys
H25	A	A	B	D	A	A	A	B	NA	
H26	A	A	B	D	A	A	A	B		
H28	A	A	C	D	A	A	A	B		
6005 T5	B	A	C	C	A	A	A	A	NA	Heavy-duty structures requiring good corrosion-resistance applications, truck and marine, railroad cars, furniture, pipelines
6009 T4	A	A	A	C	A	A	A	A	B	Automobile body sheet
6010 T4	A	A	B	C	A	A	A	A	B	Automobile body sheet
6061 0	B	A	A	D	A	A	B	A	B	Heavy-duty structures requiring good corrosion resistance, truck and marine, railroad cars, furniture, pipelines
T4, T451, T4510, T4511	B	B	B	C	A	A	A	A	B	
T6, T651, T652, T6510, T6511	B	A	C	C	A	A	A	A	B	
6063 T1	A	A	B	D	A	A	A	A	B	Pipe railing, furniture, architectural extrusions
T4	A	A	B	D	A	A	A	A	B	
T5, T52	A	A	B	C	A	A	A	A	B	
T6	A	A	C	C	A	A	A	A	B	
T83, T831, T832	A	A	C	C	A	A	A	A	B	
6066 0	C	A	B	D	D	B	B	D		Forgings and extrusions for welded structures
T4, T4510, T4511	C	B	C	C	D	B	B	D	NA	
T6, T6510, T6511	C	B	C	B	D	B	B	D		
6070 T4, T4511	B	B	B	C	A	A	A	B	NA	Heavy-duty welded structures, pipelines
T6	B	B	C	C	A	A	A	B		
6101 T6, T63	A	A	C	C	A	A	A	A	NA	High-strength bus conductors
T61, T64	A	A	B	D	A	A	A	A		
6151 T6, T652	...	...	...	...	...	...	...	...	B	Moderate-strength, intricate forgings for machine and auto parts
6201 T81	A	A	...	C	A	A	A	A	NA	High-strength electric conductor wire
6262 T6, T651, T6510, T6511	B	A	C	B	A	A	A	A	NA	Screw-machine products
T9	B	A	D	B	A	A	A	A		
6351, T5, T6	B	A	C	C	A	A	A	A	B	Heavy-duty structures requiring good corrosion resistance, truck and tractor extrusions
6463 T1	A	A	B	D	A	A	A	A		Extruded architectural and trim sections
T5	A	A	B	C	A	A	A	A	NA	
T6	A	A	C	C	A	A	A	A		
7005 T53	B	B	C	A	B	B	B	B	B	Heavy-duty structures requiring good corrosion resistance, trucks, trailers, dump bodies
7049 T73, T7351, T7352	C	B	D	B	D	C	B	D	D	Aircraft and other structures
T76, T7651	C	B	D	B	D	C	B	D	D	
7050 T74, T7451, T7452	C	B	D	B	D	C	B	D	D	Aircraft and other structures
T76, T761	C	B	D	B	D	C	B	D	D	
7072	A	A	A	D	A	A	A	A	A	Fin stock, cladding alloy
7075 0	...	...	...	D	D	C	B	D	D	Aircraft and other structures
T6, T651, T652, T6510, T6511	C(c)	C	D	B	D	C	B	D	D	
T73, T7351	C	B	D	B	D	C	B	D	D	
7175, T74, T7452	C	B	D	B	D	C	B	D	D	Aircraft and other structures, forgings
7178 0	...	...	...	...	D	C	B	D	D	Aircraft and other structures
T6, T651, T6510, T6511	C(c)	C	D	B	D	C	B	D	D	
7475 T6, T651	C	C	D	B	D	C	B	D	D	Aircraft and other structures
T73, T7351, T7352	C	B	D	B	D	C	B	D	D	
T76, T7651	C	B	D	B	D	C	B	D	D	

(a) Ratings A through E are relative ratings in decreasing order of merit, based on exposures to sodium chloride solution by intermittent spraying or immersion. Alloys with A and B ratings can be used in industrial and seacoast atmospheres without protection. Alloys with C, D, and E ratings generally should be protected at least on faying surfaces. (b) Stress-corrosion cracking ratings are based on service experience and on laboratory tests of specimens exposed to the 3.5% sodium chloride alternate immersion test. A = No known instance of failure in service or in laboratory tests. B = No known instance of failure in service; limited failures in laboratory tests of short transverse specimens. C = Service failures with sustained tension stress acting in short transverse direction relative to grain structure; limited failures in laboratory tests of long transverse specimens. D = Limited service failures with sustained longitudinal or long transverse stress. (c) In relatively thick sections the rating would be E. (d) This rating may be different for material held at elevated temperature for long periods. (e) Ratings A through D for workability (cold), and A through E for machinability, are relative ratings in decreasing order of merit. (f) Ratings A through D for weldability and brazeability are relative ratings defined as follows: A = Generally weldable by all commercial procedures and methods. B = Weldable with special techniques or for specific applications; requires preliminary trials or testing to develop welding procedure and weld performance. C = Limited weldability because of crack sensitivity or loss in resistance to corrosion and mechanical properties. D = No commonly used welding methods have been developed. (g) Ratings A through D and NA for solderability are relative ratings defined as follows: A = Excellent. B = Good. C = Fair. D = Poor. NA = Not applicable

boats and ships, cryogenic tanks; crane parts; and automotive structures.

6*xxx* Series. Alloys in the 6*xxx* series contain silicon and magnesium approximately in the proportions required for formation of magnesium silicide (Mg_2Si), thus making them heat treatable. Although not as strong as most 2*xxx* and 7*xxx* alloys, 6*xxx* series alloys have good formability, weldability, machinability, and corrosion resistance, with medium strength. Alloys in this heat-treatable group may be formed in the T4 temper (solution heat treated but not precipitation heat treated) and strengthened after forming to full T6 properties by precipitation heat treatment. Uses include architectural applications, bicycle frames, transportation equipment, bridge railings, and welded structures.

7*xxx* Series. Zinc, in amounts of 1 to 8%, is the major alloying element in 7*xxx* series alloys, and when coupled with a smaller percentage of magnesium results in heat-treatable alloys of moderate to very high strength. Usually other elements, such as copper and chromium, are also added in small quantities. 7*xxx* series alloys are used in airframe structures, mobile equipment, and other highly stressed parts.

Higher strength 7*xxx* alloys exhibit reduced resistance to stress corrosion cracking and are often utilized in a slightly overaged temper to provide better combinations of strength, corrosion resistance, and fracture toughness.

Types of Mill Products

Commercial wrought aluminum products are divided basically into five major categories based on production methods as well as geometric configurations. These are:

- Flat-rolled products (sheet, plate, and foil)
- Rod, bar, and wire
- Tubular products
- Shapes
- Forgings

In the aluminum industry, rod, bar, wire, tubular products, and shapes are termed mill products, as they are in the steel industry, even though they often are produced by extrusion rather than by rolling. Aluminum forgings, although usually not considered mill products, are wrought products and are briefly reviewed in this section.

In addition to production method and product configuration, wrought aluminum products also may be classified into heat-treatable and non-heat-treatable alloys. Initial strength of non-heat-treatable (1*xxx*, 3*xxx*, 4*xxx*, and 5*xxx*) alloys depends on the hardening effects of elements such as manganese, silicon, iron, and magnesium, singly or in various combinations. Because these alloys are work hardenable, further strengthening is made possible by various degrees of cold working, denoted by the H series of tempers, as discussed earlier in this Volume in the article on temper designations of aluminum and aluminum alloys. Alloys containing appreciable amounts of magnesium when supplied in strain-hardened tempers usually are given a final elevated-temperature treatment, called stabilizing, to ensure stability of properties. Initial strength of heat-treatable (2*xxx*, 4*xxx*, 6*xxx*, 7*xxx*, and some 8*xxx*) alloys is enhanced by addition of alloying elements such as copper, magnesium, zinc, lithium, and silicon. Because these elements, singly or in various combinations, show increasing solid solubility in aluminum with increasing temperature, it is possible to subject them to thermal treatments that will impart pronounced strengthening.

Flat-rolled products include sheet, plate, and foil. They are manufactured by either hot or hot-and-cold rolling, are rectangular in cross section and form, and have uniform thickness.

Plate. In the United States, plate refers to a product whose thickness is greater than 0.250 in. (6.3 mm). Plate up to 8 in. (200 mm) thick is available in some alloys. It usually has either sheared or sawed edges. Plate can be cut into circles, rectangles, or odd-shape blanks. Plate of certain alloys—notably the high-strength 2*xxx* and 7*xxx* series alloys—also are available in Alclad form, which comprises an aluminum alloy core having on one or both sides a metallurgically bonded aluminum or aluminum alloy coating that is anodic to the core, thus electrolytically protecting the core against corrosion. Most often, the coating consists of a high-purity aluminum, a low magnesium-silicon alloy, or an alloy containing 1% Zn. Usually, coating thickness (one side) is from 2.5 to 5% of the total thickness. The most commonly used plate alloys are 2024, 2124, 2219, 7050, 7075, 7150, 7475, and 7178 for aircraft structures; 5083, 5086, and 5456 for marine, cryogenics, and pressure vessels; and 1100, 3003, 5052, and 6061 for general applications.

Sheet. In the United States, sheet is classified as a flat-rolled product with a thickness of 0.006 to 0.249 in. (0.15 to 0.63 mm). Sheet edges can be sheared, slit, or sawed. Sheet is supplied in flat form, in coils, or in pieces cut to length from coils. Current facilities permit production of a limited amount of extra-large sheet, for example, up to 200 in. (5 m) wide by 1000 in. (25 m) long. The term strip, as applied to narrow sheet, is not used in the U.S. aluminum industry. Aluminum sheet usually is available in several surface finishes such as mill finish, one-side bright finish, or two-side bright finish. It may also be supplied embossed, perforated, corrugated, painted, or otherwise surface treated; in some instances, it is edge conditioned. As with aluminum plate, sheet made of the heat-treatable alloys in which copper or zinc are the major alloying constituents, notably the high-strength 2*xxx* and 7*xxx* series alloys, also is available in Alclad form for increased corrosion resistance. In addition, special composites may be obtained such as Alclad non-heat-treatable alloys for extra corrosion protection, for brazing purposes, or for special surface finishes.

With a few exceptions, most alloys in the 1*xxx*, 2*xxx*, 3*xxx*, 5*xxx*, and 7*xxx* series are available in sheet form. Along with alloy 6061, they cover a wide range of applications from builders' hardware to transportation equipment and from appliances to aircraft structures.

Foil is a product with a thickness less than 0.006 in. (0.15 mm). Most foil is supplied in coils, although it is also available in rectangular form (sheets). One of the largest end uses of foil is household wrap. There is a wider variety of surface finishes for foil than for sheet. Foil often is treated chemically or mechanically to meet the needs of specific applications. Common foil alloys are limited to the higher-purity 1*xxx* series and 3003, 5052, 5056, 8111, and 8079 (Al-1.0Fe-0.15Si).

Bar, rod, and wire are all solid products that are extremely long in relation to their cross section. They differ from each other only in cross-sectional shape and in thickness or diameter. In the United States, when the cross section is round or nearly round and over ⅜ in. (10 mm) in diameter, it is called rod. It is called bar when the cross section is square, rectangular, or in the shape of a regular polygon and when at least one perpendicular distance between parallel faces (thickness) is over ⅜ in. (10 mm). Wire refers to a product, regardless of its cross-sectional shape, whose diameter or greatest perpendicular distance between parallel faces is less than ⅜ in. (10 mm).

Rod and bar can be produced by either hot rolling or hot extruding and brought to final dimensions with or without additional cold working. Wire usually is produced and sized by drawing through one or more dies, although roll flattening is also used. Alclad rod or wire for additional corrosion resistance is available only in certain alloys. Many aluminum alloys are available in bar, rod, and wire; among these alloys, 2011 and 6262 are specially designed for screw-machine products, 2117 and 6053 for rivets and fittings. Alloy 2024-T4 is a standard material for bolts and screws. Alloys 1350, 6101, and 6201 are extensively used as electrical conductors. Alloy 5056 is used for zippers and alclad 5056 for insect screen wire.

Tubular products include tube and pipe. They are hollow wrought products that are long in relation to their cross section and have uniform wall thickness except as affected by corner radii. Tube is round, elliptical, square, rectangular, or regular polygonal in cross section. When round tubular products are in standardized combinations of outside diameter and wall thickness, commonly designated by "Nominal Pipe Sizes" and "ANSI Schedule Numbers," they are classified as pipe.

Tube and pipe may be produced by using a hollow extrusion ingot, by piercing a solid extrusion ingot, or by extruding through a porthole die or a bridge die. They also may be made by forming and welding sheet. Tube may be brought to final dimensions by drawing through dies. Tube (both extruded and drawn) for general applications is available in such alloys as 1100, 2014, 2024, 3003, 5050, 5086, 6061, 6063, and 7075. For heat-exchanger tube, alloys 1060, 3003, alclad 3003, 5052, 5454, and 6061 are most widely used. Clad tube is available only in certain alloys and is clad only on one side (either inside or outside). Pipe is available only in alloys 3003, 6061, and 6063.

Shapes. A shape is a product that is long in relation to its cross-sectional dimensions and has a cross-sectional shape other than that of sheet, plate, rod, bar, wire, or tube. Most shapes are produced by extruding or by extruding plus cold finishing; shapes are now rarely produced by rolling because of economic disadvantages. Shapes may be solid, hollow (with one or more voids), or semihollow. The 6*xxx* series (Al-Mg-Si) alloys, because of their easy extrudability, are the most popular alloys for producing shapes. Some 2*xxx* and 7*xxx* series alloys are often used in applications requiring higher strength.

Standard structural shapes such as I beams, channels, and angles produced in alloy 6061 are made in different and fewer configurations than similar shapes made of steel; the patterns especially designed for aluminum offer better section properties and greater structural stability than the steel design by using the metal more efficiently. The dimensions, weights, and properties of the alloy 6061 standard structural shapes, along with other information needed by structural engineers and designers, are contained in the Aluminum Construction Manual, published by the Aluminum Association, Inc.

Most aluminum alloys can be obtained as precision extrusions with good as-extruded surfaces; major dimensions usually do not need to be machined because tolerances of the as-extruded product often permit manufacturers to complete the part with simple cutoff, drilling, or other minor operations.

In many instances, long aircraft structural elements involve large attachment fittings at one end. Such elements often are more economical to machine from stepped aluminum extrusions, with two or more cross sections in one piece, rather than from an extrusion having a uniform cross section large enough for the attachment fitting.

Aluminum Alloy Forgings. Aluminum alloys can be forged into a variety of shapes and types of forgings with a broad range of final part forging design criteria based on the intended application. As a class of alloys, however, aluminum alloys are generally considered to be more difficult to forge than carbon steels and many alloy steels. Compared to the nickel/cobalt-base alloys and titanium alloys, aluminum alloys are considerably more forgeable, particularly in conventional forging-process technology, in which dies are heated to 540 °C (1000 °F) or less.

Figure 1 illustrates the relative forgeability of ten aluminum alloys that constitute the bulk of aluminum alloy forging production. This arbitrary unit is principally based on the deformation per unit of energy absorbed in the range of forging temperatures typically employed for the alloys in question. Also considered in this index is the difficulty of achieving specific degrees of severity in deformation as well as the cracking tendency of the alloy under forging-process conditions. There are wrought aluminum alloys, such as 1100 and 3003, whose forgeability would be rated significantly above those presented; however, these alloys have limited application in forging because they cannot be strengthened by heat treatment.

The 15 aluminum alloys that are most commonly forged, as well as recommended temperature ranges, are listed in Table 2.

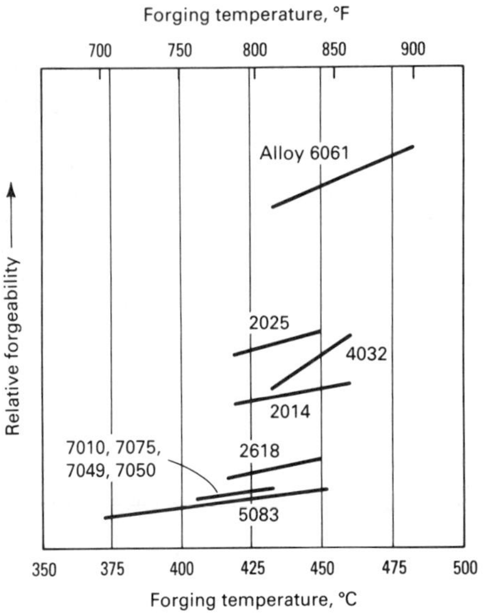

Fig. 1 Forgeability and forging temperatures of various aluminum alloys

All of these alloys are generally forged to the same severity, although some alloys may require more forging power and/or more forging operations than others. The forging temperature range for most alloys is relatively narrow (generally <55 °C, or 100 °F), and for no alloy is the range greater than 85 °C (155 °F). Obtaining and maintaining proper metal temperatures in the forging of aluminum alloys is critical to the success of the forging process. Die temperature and deformation rates play key roles in the actual forging temperature achieved.

Forging Methods. Aluminum alloys are produced by all of the current forging methods available, including open-die (or hand) forging, closed-die forging, upsetting, roll forging, orbital (rotary) forging, spin forging, mandrel forging, ring rolling, and extrusion. Selection of the optimal forging method for a given forging shape is based on the desired forged shape, the sophistication of the forged-shape design, and cost. In many cases, two or more forging methods are combined in order to achieve the desired forging shape and to obtain a thoroughly wrought structure. For example, open-die forging frequently precedes closed-die forging in order to prework the alloy (especially when cast ingot forging stock is being employed) and in order to preshape (or preform) the metal to conform to the subsequent closed dies and to conserve input metal.

Most aluminum alloy forgings are produced in closed dies. However, open-die forging is frequently used to produce small quantities of aluminum alloy forgings when the construction of expensive closed dies is not justified or when such quantities are needed during the prototype fabrication stages of a forging application. The quantity that warrants the use of closed dies varies

Table 2 Recommended forging temperature ranges for aluminum alloys

Aluminum alloy	Forging temperature range °C	°F
1100	315–405	600–760
2014	420–460	785–860
2025	420–450	785–840
2219	425–470	800–880
2618	410–455	770–850
3003	315–405	600–760
4032	415–460	780–860
5083	405–460	760–860
6061	430–480	810–900
7010	370–440	700–820
7039	380–440	720–820
7049	360–440	680–820
7050	360–440	680–820
7075	380–440	720–820
7079	405–455	760–850

considerably, depending on the size and shape of the forging and on the application for the part. However, open-die forging is by no means confined to small or prototype quantities, and in some cases, it may be the most cost-effective method of aluminum forging manufacture. For example, as many as 2000 pieces of biscuit forgings have been produced in open dies when closed dies did not provide sufficient economic benefits. Further information on the forging of aluminum alloys is given in *Forming and Forging*, Volume 14 of the 9th Edition of *Metals Handbook*.

Design of Shapes

Aluminum shapes can be produced in a virtually unlimited variety of cross-sectional designs that place the metal where needed to meet functional and appearance requirements. Full utilization of this capability of the extrusion process depends principally on the ingenuity of designers in creating new and useful configurations. The cross-sectional design of an extruded shape, however, can have an important influence on its producibility, production rate, cost of tooling, surface finish, and ultimate production cost. The optimum design of an extruded shape must take into account alloy thickness or thicknesses involved, and the size, type, and complexity of the shape. Therefore, the extruder should be consulted during design to ensure adequate dimensional control, satisfactory finish, and lowest cost while retaining the desired functional and appearance characteristics.

Classification of Shapes. The complexity of a shape producible as an extrusion is a function of metal-flow characteristics of the process and the means available to control flow. Control of metal flow places a few limitations on the design features of the cross section of an extruded shape that affect production rate, dimensional and surface quality, and costs. Extrusions are classified by shape complexity from an extrusion-production viewpoint into solid, hollow, and semihollow shapes. Each hollow shape—a shape with any part of its cross section completely enclosing a void—is further classified by increasing complexity as follows:

- Class 1: A hollow shape with a round void 25 mm (1 in.) or more in diameter and with its weight equally distributed on opposite sides of two or more equally spaced axes
- Class 2: Any hollow shape other than Class 1, not exceeding a 125 mm (5 in.) diam circle and having a single void of not less than 9.5 mm (0.375 in.) diam or 70 mm^2 (0.110 in.^{2}) area
- Class 3: Any hollow shape other than Class 1 or 2

A semihollow shape is a shape with any part of its cross section partly enclosing a void having the following ratios for the area of the void to the square of the width of the gap leading to the void:

Gap width: mm	Gap width: in.	Ratio
0.9–1.5	0.035–0.061	Over 2
1.6–3.1	0.062–0.124	Over 3
3.2–6.3	0.125–0.249	Over 4
6.4–12.6	0.250–0.499	Over 5
12.7 and greater	0.500 and greater	Over 6

Alloy Extrudability. Aluminum alloys differ in inherent extrudability. Alloy selection is important because it establishes the minimum thickness for a shape and has a basic effect on extrusion cost. In general, the higher the alloy content and the strength of an alloy, the more difficult it is to extrude and the lower its extrusion rate.

The relative extrudabilities, as measured by extrusion rate, for several of the more important commercial extrusion alloys are given below:

Alloy	Extrudability, % of rate for 6063
1350	160
1060	135
1100	135
3003	120
6063	100
6061	60
2011	35
5086	25
2014	20
5083	20
2024	15
7075	9
7178	8

Actual extrusion rate depends on pressure, temperature, and other requirements for the particular shape, as well as ingot quality.

Shape and Size Factors. The important shape factor of an extrusion is the ratio of its perimeter to its weight per unit length. For a single classification, increasing shape factor is a measure of increasing complexity. Designing for minimum shape factor promotes ease of extrusion.

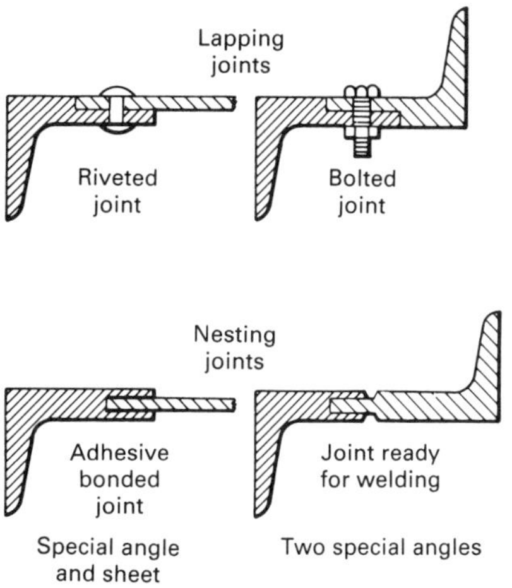

Fig. 2 Four examples of interconnecting extrusions that fit together or fit other products, and four examples of joining methods

The size of an extruded shape affects ease of extrusion and dimensional tolerances. As the circumscribing circle size (smallest diameter that completely encloses the shape) increases, extrusion becomes more difficult. In extrusion, the metal flows fastest at the center of the die face. With increasing circle size, the tendency for different metal flow increases, and it is more difficult to design and construct extrusion dies with compensating features that provide uniform metal-flow rates to all parts of the shape.

Ease of extrusion improves with increasing thickness; shapes of uniform thickness are most easily extruded. A shape whose cross section has elements of widely differing thicknesses increases the difficulty of extrusion. The thinner a flange on a shape, the less the length of flange that can be satisfactorily extruded. Thinner elements at the ends of long flanges are difficult to fill properly and make it hard to obtain desired dimensional control and finish. Although it is desirable to produce the thinnest shape feasible for an application, reducing thickness can cause an increase in cost of extrusion that more than offsets the savings in metal cost. Extruded shapes 1 mm (0.040 in.) thick and even less can be produced, depending on alloy, shape, size, and design. Manufacturing limits on minimum practical thickness of extruded shapes are given in Table 3.

Size and thickness relationships among the various elements of a shape can add to its complexity. Rod, bar, and regular shapes of uniform thickness are easily produced. For example, a bar 3.2 mm (0.125 in.) thick, a rod 25 mm (1 in.) in diameter, and an angle 19 by 25 mm (0.75 by 1 in.) in cross section, and 1.6 mm (0.0625 in.) thick are readily extruded, whereas extrusion of a 75 mm (3 in.) bar-type shape with a 3.2 mm (0.125 in.) flange is more difficult.

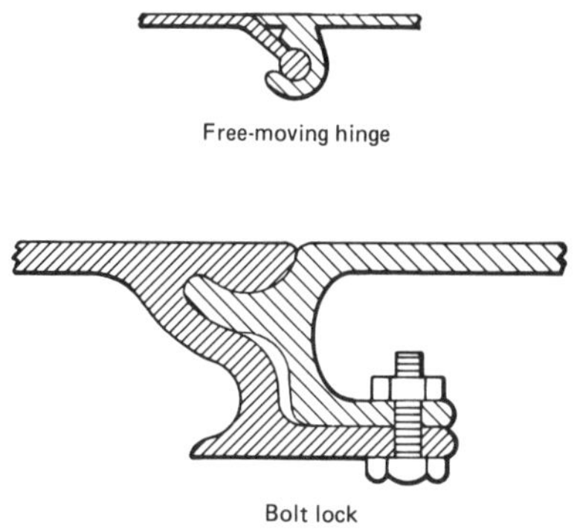

Fig. 3 Two examples of extrusions with nonpermanent interconnections

Semihollow and channel shapes require a tongue in the extrusion die, which must have adequate strength to resist the extrusion force. Channel shapes become increasingly difficult to produce as the depth-to-width ratio increases. Wide, thin shapes are difficult to produce and make it hard to control dimension. Channel-type shapes and wide, thin shapes may be fabricated if they are not excessively thin. Thin flanges or projections from a thicker element of the shape add to the complexity of an extruded design. On thinner elements at the extremities of high flanges, it is difficult to get adequate fill to obtain desired dimensions. The greater the difference in thickness of individual elements comprising a shape, the more difficult the shape is to produce. The effect of such thickness differences can be greatly diminished by blending one thickness into the other by tapered or radiused transitions. Sharp corners should be avoided wherever possible because they reduce maximum extrusion speed and are locations of stress concentrations in the die opening that can cause premature die failure. Fillet radii of at least 0.8 mm (0.031 in.) are desirable, but corners with radii of only 0.4 mm (0.015 in.) are feasible.

In general, the more unbalanced and unsymmetrical an extruded-shape cross section, the more difficult that shape is to produce. Despite this, production of grossly unbalanced and unsymmetrical shapes is the basis of the great growth that has occurred in the use of aluminum extrusions, and such designs account for the bulk of extruded shapes produced today.

Interconnecting Shapes. It is becoming increasingly common to include an interconnecting feature in the design of an extruded shape to facilitate its assembly to a similar shape or to another product. This feature can be a simple step to provide a smooth lapping joint, or a tongue and groove for a nesting joint (see Fig. 2). Such connections can be secured by any of the common joining methods. Of special interest when the joint is to be arc welded is the fact that lapping and nesting types of interconnections can be designed to provide

Table 3 Standard manufacturing limits (in inches) for aluminum extrusions

Diameter of circumscribing circle, in.	Minimum wall thickness, in.				
	1060, 1100, 3003	6063	6061	2014, 5086, 5454	2024, 2219, 5083, 7001, 7075, 7079, 7178
Solid and semihollow shapes, rod, and bar					
0.5–2	0.040	0.040	0.040	0.040	0.040
2–3	0.045	0.045	0.045	0.050	0.050
3–4	0.050	0.050	0.050	0.050	0.062
4–5	0.062	0.062	0.062	0.062	0.078
5–6	0.062	0.062	0.062	0.078	0.094
6–7	0.078	0.078	0.078	0.094	0.109
7–8	0.094	0.094	0.094	0.109	0.125
8–10	0.109	0.109	0.109	0.125	0.156
10–11	0.125	0.125	0.125	0.125	0.156
11–12	0.156	0.156	0.156	0.156	0.156
12–17	0.188	0.188	0.188	0.188	0.188
17–20	0.188	0.188	0.188	0.188	0.250
20–24	0.188	0.188	0.188	0.250	0.500
Class 1 hollow shapes(a)					
1.25–3	0.062	0.050	0.062	...	...
3–4	0.094	0.050	0.062	...	...
4–5	0.109	0.062	0.062	0.156	0.250
5–6	0.125	0.062	0.078	0.188	0.281
6–7	0.156	0.078	0.094	0.219	0.312
7–8	0.188	0.094	0.125	0.250	0.375
8–9	0.219	0.125	0.156	0.281	0.438
9–10	0.250	0.156	0.188	0.312	0.500
10–12.75	0.312	0.188	0.219	0.375	0.500
12.75–14	0.375	0.219	0.250	0.438	0.500
14–16	0.438	0.250	0.375	0.438	0.500
16–20.25	0.500	0.375	0.438	0.500	0.625
Class 2 and 3 hollow shapes(b)					
0.5–1	0.062	0.050	0.062	...	...
1–2	0.062	0.055	0.062	...	...
2–3	0.078	0.062	0.078	...	...
3–4	0.094	0.078	0.094	...	...
4–5	0.109	0.094	0.109	...	...
5–6	0.125	0.109	0.125	...	...
6–7	0.156	0.125	0.156	...	...
7–8	0.188	0.156	0.188	...	...
8–10	0.250	0.188	0.250	...	...

(a) Minimum inside diameter is one-half the circumscribing diameter, but never under 1 in. for alloys in first three columns or under 2 in. for alloys in last two columns. (b) Minimum hole size for all alloys is 0.110 sq. in. in area or 0.375 in. in diam.

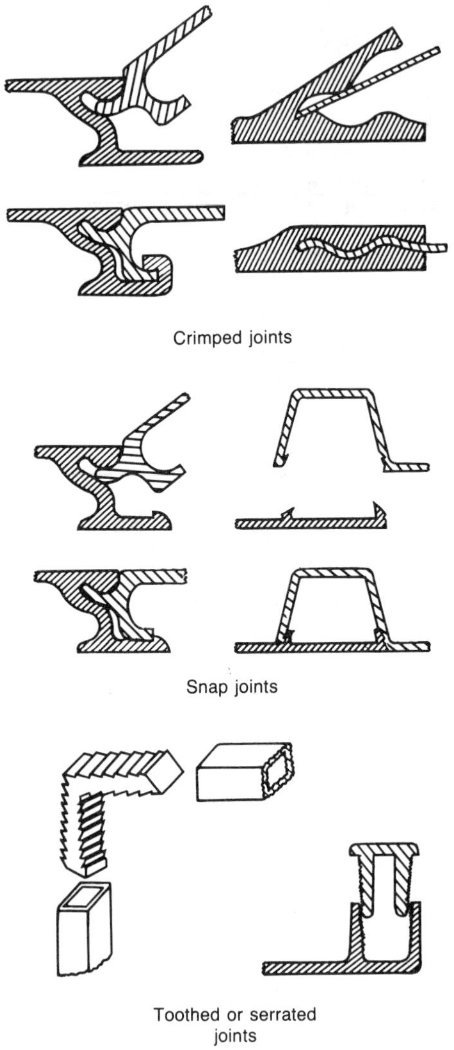

Fig. 4 Six examples of interconnecting extrusions that lock together or lock to other products

edge preparation and/or integral backing for the weld (see the sketch at bottom left in Fig. 2).

Interlocking joints can be designed to incorporate a free-moving hinge (see top sketch in Fig. 3) when one part is slid lengthwise into the mating portion of the next extrusion. Panel-type extrusions with hinge joints have found application in conveyor belts and roll-up doors.

A more common type of interlocking feature used in interconnecting extrusions is the nesting type that requires rotation of one part relative to the mating part for assembly (see bottom sketch in Fig. 3). Such joints can be held together by gravity or by mechanical devices. If a nonpermanent joint is desired, a bolt or other fastener can be used, as illustrated in the bottom sketch in Fig. 3.

When a permanent joint is desired, a snapping or crimping feature can be added to interlocking extrusions (see Fig. 4). Crimping also can be used to make a permanent joint between an interlocking extrusion and sheet (Fig. 4). Extrusions also can be provided with longitudinal teeth or serrations, which will permanently grip smooth surfaces as well as surfaces provided with mating teeth or serrations; this is illustrated in the sketch at the bottom of Fig. 4.

Applications for interconnecting extrusions include doors, wall, ceiling and floor panels, pallets, aircraft landing mats, highway signs, window frames, and large cylinders.

Physical Metallurgy

The principal concerns in the physical metallurgy of aluminum alloys include the effects of composition, mechanical working, and/or heat treatment on mechanical and physical properties. In terms of properties, strength improvement is a major objective in the design of aluminum alloys because the low strength of pure aluminum (about a 10 MPa , or 1.5 ksi, tensile yield strength in the annealed condition) limits its commercial usefulness. The two most common methods for increasing the strength of aluminum alloys are to:

- Disperse second-phase constituents or elements in solid solution and cold work the alloy (non-heat-treatable alloys)
- Dissolve the alloying elements into solid solution and precipitate them as coherent submicroscopic particles (heat-treatable or precipitation-hardening alloys)

The factors affecting these strengthening mechanisms and the fracture toughness and physical properties of aluminum alloys are discussed in the following portions of this section.

Phases in Aluminum Alloys

The elements that are most commonly present in commercial aluminum alloys to provide increased strength—particularly when coupled with strain hardening by cold working or with heat treatment, or both—are copper, magnesium, manganese, silicon, and zinc. These elements all have significant solid solubility in aluminum, and in all cases the solubility increases with increasing temperature (see Fig. 5).

Of all the elements, zinc has the greatest solid solubility in aluminum (a maximum of 66.4 at%). In addition to zinc, the solid

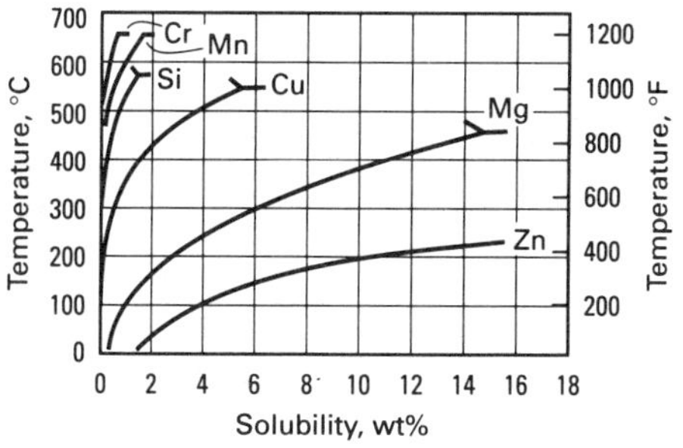

Fig. 5 Equilibrium binary solid solubility as a function of temperature for alloying elements most frequently added to aluminum

solubilities of silver, magnesium, and lithium are greater than 10 at% (in order of decreasing maximum solubility). Gallium, germanium, copper, and silicon (in decreasing order) have maximum solubilities of less than 10 but greater than 1 at%. All other elements are less soluble. With the one known exception of tin (which shows a retrograde solid solubility between the melting point of aluminum and the eutectic temperature, 228.3 °C, with a maximum of 0.10% at approximately 660 °C), the maximum solid solubility in aluminum alloys occurs at the eutectic, peritectic, or monotectic temperature. With decreasing temperature, the solubility limits decrease. This decrease from appreciable concentrations at elevated temperatures to relatively low concentrations at low temperatures is one fundamental characteristic that provides the basis for substantially increasing the hardness and strength of aluminum alloys by solution heat treatment and subsequent precipitation aging operations.

For those elements in concentrations below their solubility limits, the alloying elements are essentially in solid solution and constitute a single phase. However, no element is known to have complete miscibility with aluminum in the solid state. Among the commercial alloys, only the bright-finishing alloys such as 5657 and 5252, which contain 0.8 and 2.5% Mg (nominal), respectively, with very low limits on all impurities, may be regarded as nearly pure solid solutions.

Second-Phase Constituents. When the content of an alloying element exceeds the solid-solubility limit, the alloying element produces "second-phase" microstructural constituents that may consist of either the pure alloying ingredient or an intermetallic-compound phase. In the first group are silicon, tin, and beryllium. If the alloy is a ternary or higher-order alloy, however, silicon or tin may form intermetallic-compound phases. Most of the other alloying elements form such compounds with aluminum in binary alloys and more complex phases in ternary or higher-order alloys.

Manganese and chromium are included in the group of elements that form predominantly second-phase constituents, because in commercial alloys they have very low equilibrium solid solubilities. In the case of many compositions containing manganese, this is because iron and silicon are also present and form the quaternary-phase $Al_{12}(Fe,Mn)_3Si$. In alloys containing copper and manganese, the ternary-phase $Al_{20}Cu_2Mn_3$ is formed. Most of the alloys in which chromium is present also contain magnesium, so that during solid-state heating they form $Al_{12}Mg_2Cr$, which also has very low-equilibrium solid solubility. Smelter-grade primary metal, whether in ingot or wrought-product form, contains a small volume fraction of second-phase particles, chiefly iron-bearing phases—the metastable Al_6Fe, the stable Al_3Fe, which forms from Al_6Fe on solid-state heating, and $Al_{12}Fe_3Si$. Proportions of the binary and ternary phases depend on relative iron and silicon contents.

In quaternary systems, intermetallic phases of the respective binary and ternary systems are occasionally isomorphous, forming continuous series of solid solutions in equilibrium with aluminum solid solution. An important example is in the aluminum-copper-magnesium-zinc quaternary system where there are three such pairs: $CuMg_4Al_6$ + $Mg_3Zn_3Al_2$, Mg_2Zn_{11} + $Cu_6Mg_2Al_5$, and $MgZn_2$ + CuMgAl. The first pair have similar lattice parameters and form extensive mutual solid solution, the others less so. Neither $Cu_6Mg_2Al_5$ nor CuMgAl are equilibrium phases in aluminum-copper-magnesium, although both Mg_2Zn_{11} and $MgZn_2$ are equilibrium phases in aluminum-magnesium-zinc. Another instance is in the aluminum-iron-manganese-silicon quaternary system; here the stable phase $(FeMn)_3Si_2Al_{15}$ (body-centered cubic) can vary from $Mn_3Si_2Al_{15}$, a = 1.2652 nm (12.652 Å) to ~$(Mn_{0.1}Fe_{0.9})_3Si_2Al_{15}$, a = 1.2548 nm (12.548 Å). The stable phase of the closest composition in aluminum-iron-silicon is Fe_2SiAl_8 (hexagonal); the hexagonal-to-cubic transition is also accomplished by small additions of vanadium, chromium, molybdenum, and tungsten, and larger additions of copper (Ref 1). Such chemical stabilization effects, coupled with the metastability introduced by casting, frequently cause complex alloy structure.

Prediction of Intermetallic Phases in Aluminum Alloys. The wide variety of intermetallic phases in aluminum alloys, which occur because aluminum is highly electronegative and trivalent, has been the subject of considerable study (Ref 2-4). Details depend on ratios and total amounts of alloying elements present and require reference to the phase diagrams for prediction. It must be kept in mind, however, that metastable conditions frequently prevail that are characterized by the presence of phases that are not shown on the equilibrium diagrams. Transition metals, for example, exhibit frequent metastability, in which one phase introduced during fast solidification transforms in the solid state to another, for example, $FeAl_6 \rightarrow FeAl_3$, or a metastable variant precipitates from supersaturated solid solution such as $MnAl_{12}$.

Calculation of Phase Diagrams. Recently, considerable advances have been made in the thermodynamic evaluation of phase diagrams, particularly through the application of computer techniques (Ref 5). The available data and computational procedures have been systemized internationally since 1971 through the CALPHAD (Computer Coupling of Phase Diagrams and Thermochemistry) project (Ref 6). Application for multicomponent aluminum alloy phase diagram prediction has some inherent problems, particularly regarding the unexpected occurrence of ternary intermetallic phases, but is rapidly becoming an effective procedure. As of 1980, the following ternary aluminum-containing systems have been examined: Al-Fe-Ti, Al-Ga-Ge, Al-Ga-In, Al-Ge-Sn, Al-Li-Mg, and Al-Ni-Ti. As well as the 15 binaries required for these systems, phase diagrams of the following binary systems have also been examined: Al-Ca, Al-Ce, Al-Co, Al-Cr, Al-Cu, Al-Mn, Al-Mo, Al-Nb, Al-O, Al-P, and Al-Si (Ref 7). In principle, any ternary or quaternary combination of these binary systems can be analyzed.

Strengthening Mechanisms

The predominant objective in the design of aluminum alloys is to increase strength, hardness, and resistance to wear, creep, stress relaxation, or fatigue. Effects on these properties are specific to the different combinations of alloying elements, their alloy phase diagrams, and to the microstructures and substructures they form as a result of solidification, thermomechanical history, heat treatment, and/or cold working. These factors, to a large extent, depend on whether the alloy is a non-heat-treatable alloy or a heat-treatable (precipitation-strengthening) alloy.

Strength at elevated temperatures is improved mainly by solid-solution and second-phase hardening because at least for temperatures exceeding those of the precipitation-hardening range—230 °C (450 °F) and over—the precipitation reactions continue into the softening regime. For supersonic aircraft and space vehicle applications subject to aerodynamic heating, the heat-treatable alloys of the 2*xxx* group can be used for temperatures up to about 150 °C (300 °F).

Strengthening in non-heat-treatable alloys occurs from solid-solution formation, second-phase microstructural constituents, dispersoid precipitates, and/or strain hardening. Wrought alloys of this type are mainly those of the 3*xxx* and 5*xxx* groups containing magnesium, manganese, and/or chromium as well as the 1*xxx* aluminums

Table 4 Solid-solution effects on strength of principal solute elements in super-purity aluminum

Element	Difference in atomic radii, $r_x - r_{Al}$, %(a)	Strength/addition values(b): Yield strength/% addition(c) MPa/at%	ksi/at%	MPa/wt%	ksi/wt%	Tensile strength/% addition(d) MPa/at%	ksi/at%	MPa/wt%	ksi/wt%
Si	−3.8	9.3	1.35	9.2	1.33	40.0	5.8	39.6	5.75
Zn	−6.0	6.6	0.95	2.9	0.42	20.7	3.0	15.2	2.2
Cu	−10.7	16.2	2.35	13.8	2.0	88.3	12.8	43.1	6.25
Mn	−11.3	(e)	(e)	30.3	4.4	(e)	(e)	53.8	7.8
Mg	+11.8	17.2	2.5	18.6	2.7	51.0	7.4	50.3	7.3

(a) Listed in order of increasing percent difference in atomic radii. (b) Some property−percent addition relationships are nonlinear. Generally, the unit effects of smaller additions are greater. (c) Increase in yield strength (0.2% offset) for 1% (atomic or weight basis) alloy addition. (d) Increase in ultimate tensile strength for 1% (atomic or weight basis) alloy addition. (e) 1 at% of manganese is not soluble.

and some alloys of the 4*xxx* group that contain only silicon. Non-heat-treatable casting alloys are of the 4*xx.x* or 5*xx.x* groups, containing silicon or magnesium, respectively, and the 1*xx.x* aluminums.

Solid-Solution Strengthening. For those elements that form solid solutions, the strengthening effect when the element is in solution tends to increase with increasing difference in the atomic radii of the solvent (Al) and solute (alloying element) atoms. This factor is evident in data obtained from super-purity binary solid-solution alloys in the annealed state, presented in Table 4, but it is evident that other effects are involved, chief among which is an electronic bonding factor. The effects of multiple solutes in solid solution are somewhat less than additive and are nearly the same when one solute has a larger and the other a smaller atomic radius than that of aluminum as when both are either smaller or larger. Manganese in solid solution is highly effective in strengthening binary alloys. Its contribution to the strength of commercial alloys is less, because in these compositions, as a result of commercial mill fabricating operations, the manganese is largely precipitated.

The principal alloys that are strengthened by alloying elements in solid solution (often coupled with cold work) are those in the aluminum-magnesium (5*xxx*) series, ranging from 0.5 to 6 wt% Mg. These alloys often contain small additions of transition elements such as chromium or manganese, and less frequently zirconium to control the grain or subgrain structure and iron and silicon impurities that usually are present in the form of intermetallic particles. Figure 6 illustrates the effect of magnesium in solid solution on the yield strength and tensile elongation for most of the common aluminum-magnesium commercial alloys.

Strengthening From Second-Phase Constituents. Elements and combinations that form predominantly second-phase constituents with relatively low solid solubility include iron, nickel, titanium, manganese, and chromium, and combinations thereof. The presence of increasing volume fractions of the intermetallic-compound phases formed by these elements and the elemental silicon constituent formed by silicon during solidification or by precipitation in the solid state during postsolidification heating also increases strength and hardness. The rates of increase per unit weight of alloying element added are frequently similar to but usually lower than those resulting from solid solution. This "second-phase" hardening occurs even though the constituent particles are of sizes readily resolved by optical microscopy. These irregularly shaped particles form during solidification and occur mostly along grain boundaries and between dendrite arms.

Grain Refinement With Dispersed Precipitates. Manganese and/or chromium additions in wrought aluminum alloys allow the formation of complex precipitates that not only retard grain growth during ingot reheating but also assist in grain refinement during rolling. This method involves rapid solidification and cooling during the casting of ingots, so that a solid-solution state is formed with concentrations of manganese and/or chromium that greatly exceed their equilibrium solubility. During reheating of the as-cast ingot for wrought processing, this supersaturated metastable solid solution is designed to cause solid-state precipitation of complex phases. This precipitation does not cause appreciable hardening, nor is it intended that it should. Its purpose is to produce finely divided and dispersed particles that retard or inhibit recrystallization and grain growth in the alloy during subsequent heatings. The precipitate particles of $Al_{12}(Fe,Mn)_3Si$, $Al_{20}Cu_2Mn_3$, or $Al_{12}Mg_2Cr$ are incoherent with the matrix, and concurrent with their precipitation the original solid solution becomes less concentrated. These conditions do not provide appreciable precipitation hardening. Changes in electrical conductivity constitute an effective measure of the completeness of these precipitation reactions that occur in preheating.

The newer "in-line" or integrated processes that shorten the path from molten metal to wrought product, avoiding ingot preheating and reducing the overall time-temperature history, are changing this conventional or traditional picture. It seems very probable that in order to obtain the

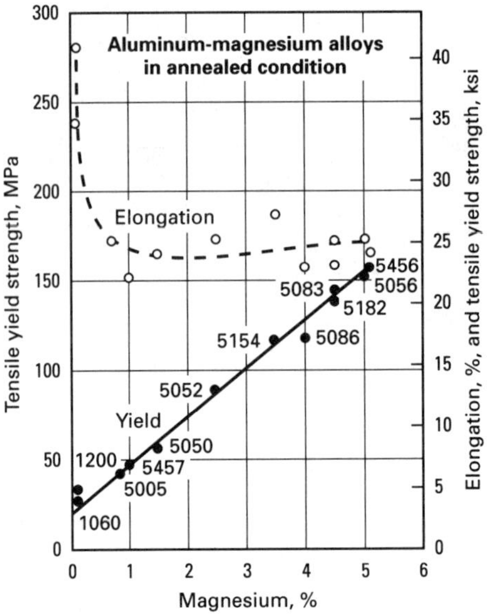

Fig. 6 Correlation between tensile yield, elongation, and magnesium content for some commercial aluminum alloys

best results from such processes, traditional alloy compositions should be adjusted taking into account the fact that larger proportions of these elements would be expected to remain in solid solution through such abbreviated and truncated thermomechanical operations. New capabilities may be obtained with currently standard alloys in some instances, but it would not be expected that a particular alloy would exhibit the same properties when produced by the two types of processes.

For alloys that are composed of both solid-solution and second-phase constituents and/or dispersoid precipitates, all of these components of microstructure contribute to strength, in a roughly additive manner. This is shown in Fig. 7 for Al-Mg-Mn alloys in the annealed condition.

Strain hardening by cold rolling, drawing, or stretching is a highly effective means of increasing the strength of non-heat-treatable alloys. Work- or strain-hardening curves for several typical non-heat-treatable commercial alloys (Fig. 8) illustrate the increases in strength that accompany increasing reduction by cold rolling of initially annealed temper sheet. This increase is obtained at the expense of ductility as measured by percent elongation in a tensile test and by reducing formability in operations such as bending and drawing. It is often advantageous to use material in a partially annealed (H2*x*) or stabilized (H3*x*) temper when bending, forming, or drawing is required, since material in these tempers has greater forming capability for the same strength levels than does strain-hardened only (H1*x*) material (see Table 5, for example).

All mill products can be supplied in the strain-hardened condition, although there are limitations on the amounts of strain that

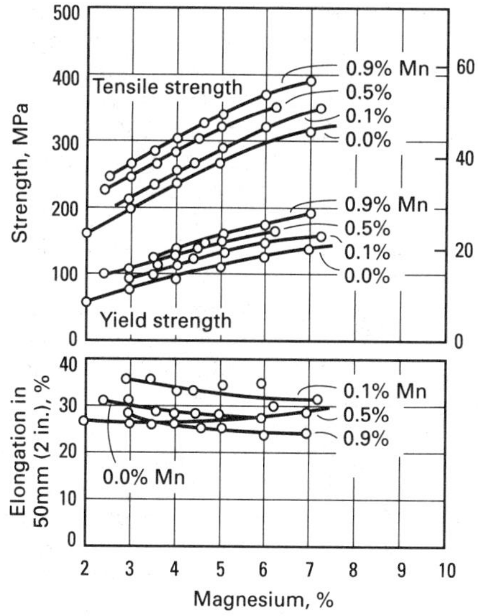

Fig. 7 Tensile properties in Al-Mg-Mn alloys in the form of annealed (O temper) plate 13 mm (0.5 in.) thick

can be applied to products such as die forgings and impacts. Even aluminum castings have been strengthened by cold pressing for certain applications. The heat-treatable alloys described below can also be subjected to strain hardening.

Heat-treatable (precipitation-hardening) aluminum alloys for wrought and cast products contain elements that decrease in solubility with decreasing temperature, and in concentrations that exceed their equilibrium solid solubility at room- and moderately higher temperatures. However, these features alone do not make an alloy capable of (precipitation hardening) during heat treatment. The strengths of most binary alloys containing magnesium, silicon, zinc, chromium, or manganese alone exhibit little change from thermal treatments regardless of whether the solute is completely in solid solution, partially precipitated, or substantially precipitated.

The mechanism of strengthening by age hardening involves the formation of coherent clusters of solute atoms (that is, the solute atoms have collected into a cluster but still have the same crystal structure as the solvent phase). This causes a great deal of strain because of mismatch in size between the solvent and solute atoms. The cluster stabilizes dislocations, because dislocations tend to reduce the strain, similar to the reduction in strain energy of a single solute atom by a dislocation. When dislocations are anchored or trapped by coherent solute clusters, the alloy is considerably strengthened and hardened.

However, if the precipitates are semicoherent (sharing a dislocation-containing interface with the matrix), incoherent (sharing a disordered interface, akin to a large-angle grain boundary, with the matrix), or are incapable of reducing strain behavior be-

Table 5 Tensile-property data illustrating typical relationships between strength and elongation for non-heat-treatable alloys in H1*x* versus H2*x* or H3*x* tempers

Alloy and temper	Tensile strength MPa	Tensile strength ksi	Yield strength MPa	Yield strength ksi	Elongation, %
3105–H14	172	25	152	22	5
3105–H25	179	26	159	23	8
3105–H16	193	28	172	25	4

cause they are too strong, a dislocation can circumvent the particles only by bowing into a roughly semicircular shape between them under the action of an applied shear stress. Consequently, the presence of the precipitate particles, and even more importantly the strain fields in the matrix surrounding the coherent particles, provide higher strength by obstructing and retarding the movement of dislocations. The characteristic that determines whether a precipitate phase is coherent or noncoherent is the closeness of match or degree of disregistry between the atomic spacings on the lattice of the matrix and on that of the precipitate.

Heat treatment for precipitation strengthening includes a solution heat treatment at a high temperature to maximize solubility, followed by rapid cooling or quenching to a low temperature to obtain a solid solution supersaturated with both solute elements and vacancies. Solution heat treatments are designed to maximize the solubility of elements that participate in subsequent aging treatments. They are most effective near the solidus or eutectic temperature, where maximum solubility exists and diffusion rates are rapid. However, care must be taken to avoid incipient melting of low-temperature eutectics and grain-boundary phases. Such melting results in quench cracks and loss in ductility. The maximum temperature may also be set with regard to grain growth, surface effects, and economy of operation. The minimum temperature should be above the solvus, or the desired properties derived from aging will not be realized. The optimum heat-treatment range may be quite small, with a margin of safety sometimes only ±5 K.

The high strength is produced by the finely dispersed precipitates that form during aging heat treatments (which may include either natural aging or artificial aging as described below). This final step must be accomplished not only below the equilibrium solvus temperature, but below a metastable miscibility gap called the Guinier-Preston (GP) zone solvus line. The supersaturation of vacancies allows diffusion, and thus zone formation, to occur much faster than expected from equilibrium diffusion coefficients. In the precipitation process, the saturated solid solution first develops solute clusters, which then become involved in the formation of transi-

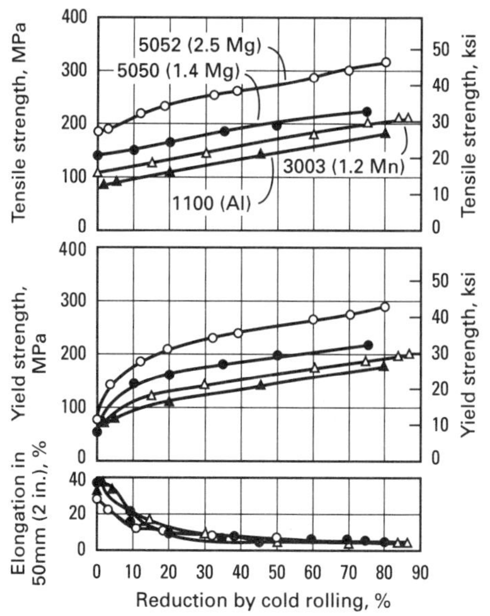

Fig. 8 Strain-hardening curves for aluminum (1100), and for Al-Mn (3003) and Al-Mg (5050 and 5052) alloys

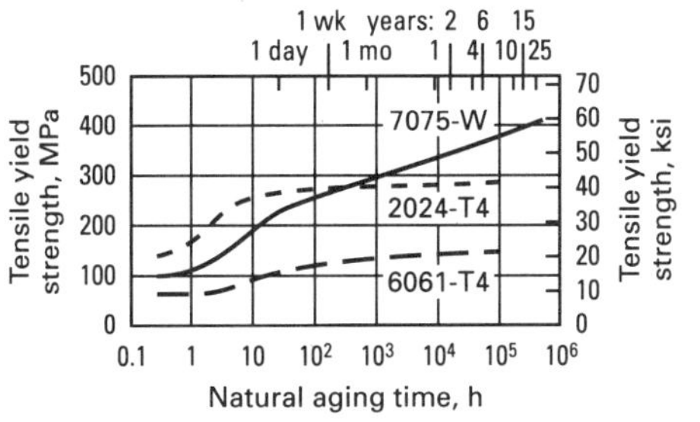

Fig. 9 Natural aging curves for three solution heat-treated wrought aluminum alloys

tional (nonequilibrium) precipitates. The final structure consists of equilibrium precipitates, which do not contribute to age hardening (precipitation strengthening).

Natural aging refers to the spontaneous formation of a G-P zone structure during exposure at room temperature. Solute atoms either cluster or segregate to selected atomic lattice planes, depending on the alloy system, to form the G-P zones, which are more resistant to movement of dislocations through the lattice, and hence are stronger. Curves showing the changes in tensile yield strength with time at room temperature (natural aging curves) for three wrought commercial heat-treatable alloys of different alloy systems are shown in Fig. 9. The magnitudes of increase in this property are considerably different for the three alloys, and the differences in rate of change with time are of practical importance. Because 7075 and similar alloys never become completely stable under these conditions, they are rarely used in the naturally aged temper. On the other hand, 2024 is widely used in this condition.

Of the binary alloys, aluminum-copper alloys exhibit natural aging after being solu-

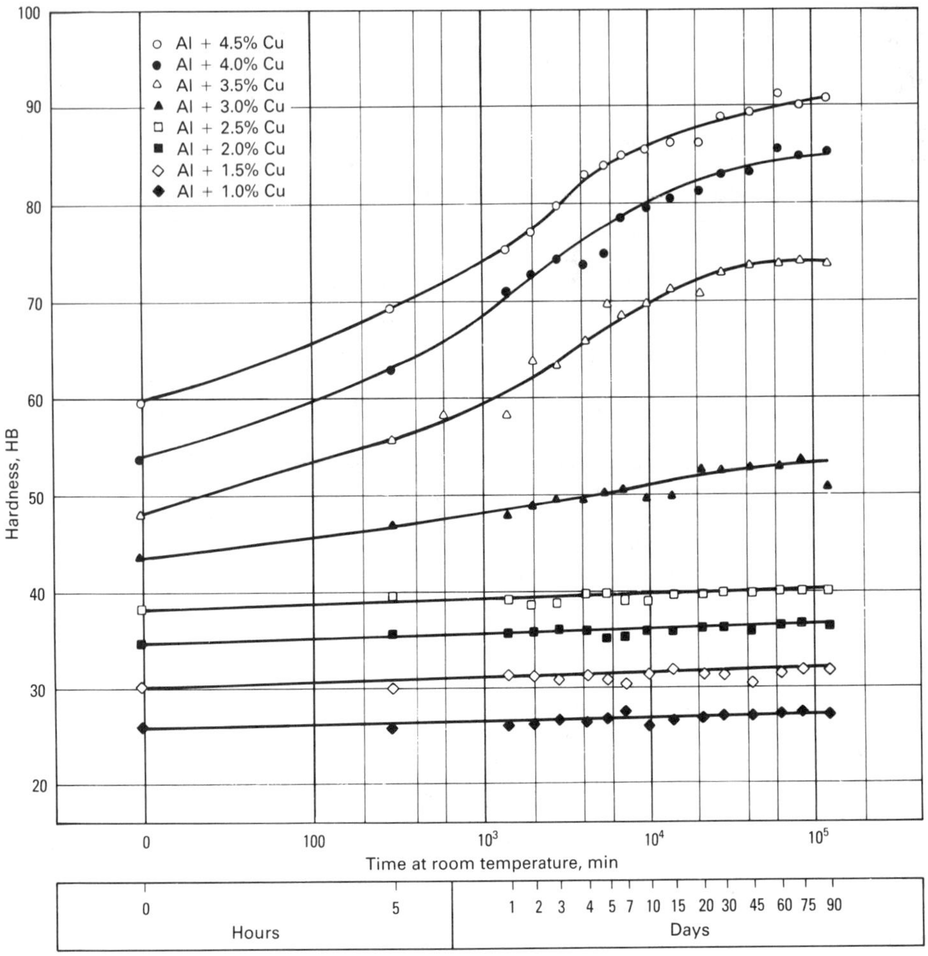

Fig. 10 Natural aging curves for binary Al-Cu alloys quenched in water at 100 °C (212 °F)

tion heat treated and quenched. The amounts by which strength and hardness increase become larger with time of natural aging and with the copper content of the alloy from about 3% to the limit of solid solubility (5.65%). Natural aging curves for slowly quenched, high-purity Al-Cu alloys with 1 to 4.5% Cu are shown in Fig. 10. The rates and amounts of the changes in strength and hardness can be increased by holding the alloys at moderately elevated temperatures (for alloys of all types, the useful range is about 120 to 230 °C, or 250 to 450 °F). This treatment is called precipitation heat treating or artificial aging. In the Al-Cu system, alloys with as little as 1% Cu, again slowly quenched, start to harden after about 20 days at a temperature of 150 °C, or 300 °F (see Fig. 11). The alloys of this system, having less than about 3% Cu, show little or no natural aging after low-cooling-rate quenching, which introduces little stress.

Artificial aging includes exposure at temperatures above room temperature so as to produce the transitional (metastable) forms of the equilibrium precipitate of a particular alloy system. These transitional precipitates remain coherent with the solid-solution matrix and thus contribute to precipitation strengthening. With further heating at temperatures that cause strengthening or at higher temperatures, the precipitate particles grow, but even more importantly convert to the equilibrium phases, which generally are not coherent. These changes soften the material, and carried further, produce the softest or annealed condition. Even at this stage, the precipitate particles are still too small to be clearly resolved by optical microscopy, although etching effects are readily observed—particularly in alloys containing copper.

Precipitation heat treatment or artificial aging curves for the Al-Mg-Si wrought alloy 6061 (which is widely used for structural shapes) are shown in Fig. 12. This is a typical family of curves showing the changes in tensile yield strength that accrue with increasing time at each of a series of temperatures. In all cases, the material had been given a solution heat treatment followed by a quench just prior to the start of the precipitation heat treatment. For detailed presentation of heat-treating operations, parameters, and practices, see *Heat Treating*, Volume 4 of the 9th Edition of *Metals Handbook*.

The commercial heat-treatable aluminum alloys are, with few exceptions, based on ternary or quaternary systems with respect to the solutes involved in developing strength by precipitation. The most prominent systems are: Al-Cu-Mg, Al-Cu-Si, and Al-Cu-Mg-Si, alloys of which are in the 2*xxx* and 2*xx.x* groups (wrought and casting alloys, respectively); Al-Mg-Si (6*xxx* wrought alloys); Al-Si-Mg, Al-Si-Cu, and Al-Si-Mg-Cu (3*xx.x* casting alloys); and Al-Zn-Mg and Al-Zn-Mg-Cu (7*xxx* wrought and 7*xx.x* casting alloys). In each case the solubility of the multiple-solute elements decreases with decreasing temperature.

These multiple alloying additions of both major solute elements and supplementary elements employed in commercial alloys are strictly functional and serve with different heat treatments to provide the many different combinations of properties—physical, mechanical, and electrochemical—that are required for different applications. Some alloys, particularly those for foundry production of castings, contain amounts of silicon far in excess of the amount that is soluble or needed for strengthening alone. The function here is chiefly to improve casting soundness and freedom from cracking, but the excess silicon also serves to increase wear resistance, as do other microstructural constituents formed by manganese, nickel, and iron. Parts made of such alloys are commonly used in gasoline and diesel engines (pistons, cylinder blocks, and so forth). The system of numerical nomenclature used to designate the alloys, and that for the strain-hardened and heat-treated tempers, is described in the article "Alloy and Temper Designation Systems for Aluminum and Aluminum Alloys" in this Volume.

Alloys containing the elements silver, lithium, and germanium are also capable of providing high strength with heat treatment, and in the case of lithium, both increased elastic modulus and lower density, which are highly advantageous—particularly for aerospace applications (see the article "Aluminum-Lithium Alloys" in this Volume). Commercial use of alloys containing these elements has been restricted either by cost or by difficulties encountered in producing them. Such alloys are used to some extent, however, and research is being directed toward overcoming their disadvantages.

In the case of alloys having copper as the principal alloying ingredient and no magnesium, strengthening by precipitation can be greatly increased by adding small fractional percentages of tin, cadmium, or indium, or combinations of these elements. Alloys based on these effects have been produced commercially but not in large volumes because of costly special practices and limitations required in processing, and in the case of cadmium, the need for special facilities to avoid health hazards from formation and release of cadmium vapor during alloying. Such alloys,

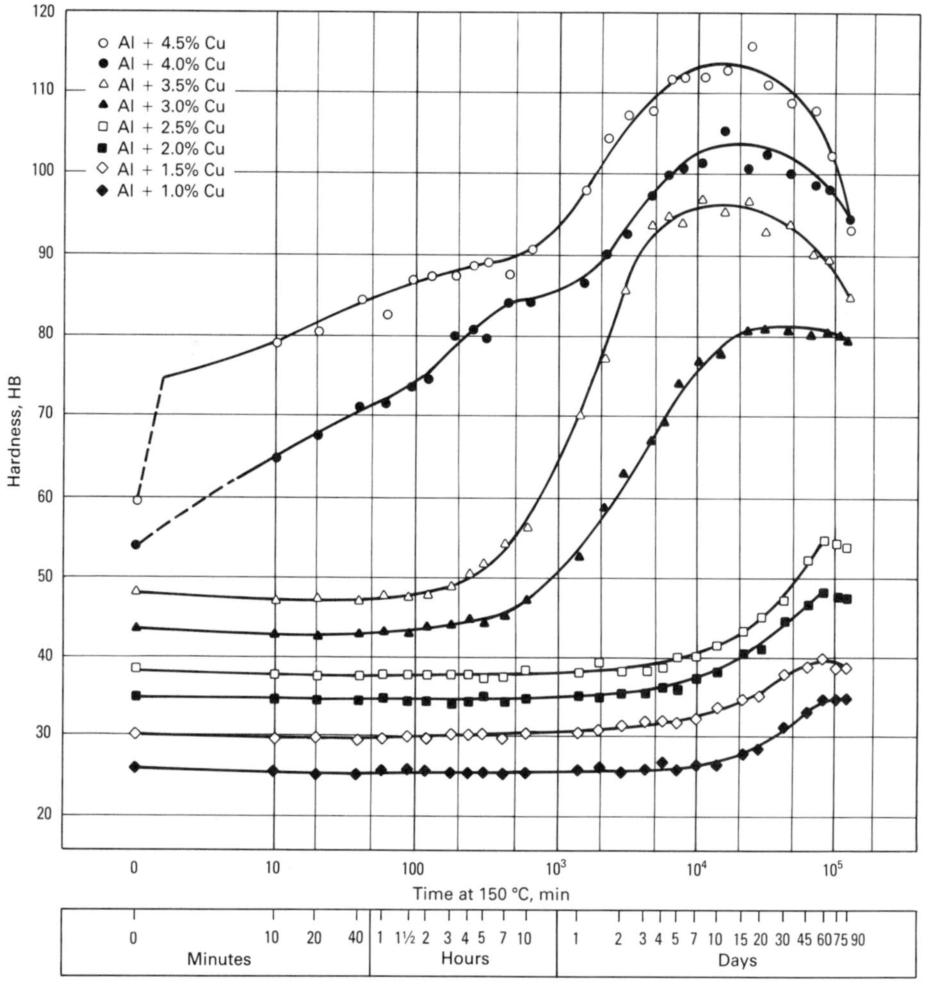

Fig. 11 Precipitation hardening curves for binary Al-Cu alloys quenched in water at 100 °C (212 °F) and aged at 150 °C (300 °F)

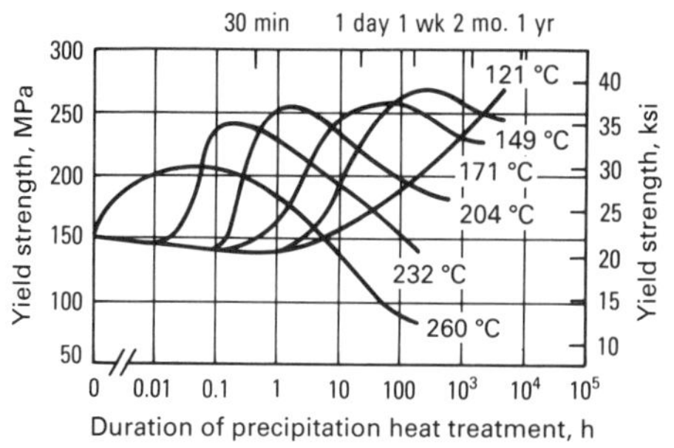

Fig. 12 Precipitation heat treatment or artificial aging curves for solution heat-treated aluminum alloy 6061

as well as those containing silver, lithium, or other particle-forming elements, may be used on a selective basis in the future.

Effects on Physical and Electrochemical Properties. The above description of the precipitation processes in commercial heat-treatable aluminum alloys (as well as the heat-treatable binary alloys, none of which is used commercially) affect not only mechanical properties but also physical properties (density and electrical and thermal conductivities) and electrochemical properties (solution potential). On the microstructural and submicroscopic scales, the electrochemical properties develop point-to-point nonuniformities that account for changes in corrosion resistance.

Measurements of changes in physical and electrochemical properties have played an important role in completely describing precipitation reactions and are very useful in analyzing or diagnosing whether heat-treatable products have been properly or improperly heat treated. Although they may be indicative of the strength levels of products, they cannot be relied upon to determine whether or not the product meets specified mechanical-property limits. Since elements in solid solution are always more harmful to electrical conductivity than the same elements combined with others as intermetallic compounds, thermal treatments are applied to ingots used for fabrication of electrical conductor parts. These thermal treatments are intended to precipitate as much as possible of the dissolved impurities. Iron is the principal element involved, and although the amount precipitated is only a few hundredths of a percent, the effect on electrical conductivity of the wire, cable, or other product made from the ingot is of considerable practical importance. These alloys may or may not be heat treatable with respect to mechanical properties. Electrical conductor alloys 6101 and 6201 are heat treatable. These alloys are used in tempers in which their strengthening precipitate, the transition form of Mg_2Si, is largely out of solid solution to optimize both strength and conductivity.

Metallurgical Factors of Other Mechanical Properties

Forming. The formability of a material is the extent to which it can be deformed in a particular process before the onset of failure. Aluminum sheet or aluminum shapes usually fail by localized necking or by ductile fracture. Necking is governed largely by bulk material properties such as work hardening and strain-rate hardening and depends critically on the strain path followed by the forming process. In dilute alloys, the extent of necking or limit strain is reduced by cold work, age hardening, gross defects, a large grain size, and the presence of alloying elements in solid solution. Ductile fracture occurs as a result of the nucleation and linking of microscopic voids at particles and the concentration of strain in narrow shear bands. Fracture usually occurs at larger strains than does localized necking and therefore is usually important only when necking is suppressed. Common examples where fracture is encountered are at small radius bends and at severe drawing, ironing, and stretching near notches or sheared edges.

Considerable advances have been made in developing alloys with good formability, but in general, an alloy cannot be optimized on this basis alone. The function of the formed part must also be considered and improvements in functional characteristics, such as strength and ease of machining, often tend to reduce the formability of the alloy.

The principal alloys that are strengthened by alloying elements in solid solution (often coupled with cold work) are those in the aluminum-magnesium (5*xxx*) series, ranging from 0.5 to 6 wt% Mg. Figure 6 illustrates the effect of magnesium in solid solution on the yield strength and tensile elongation for most of the common aluminum-magnesium commercial alloys. Note the large initial reduction in the tensile elongation with the addition of small amounts of magnesium.

The reductions in the forming limit produced by additions of magnesium and copper appear to be related to the tendency of the solute atoms to migrate to dislocations (strain age). This tends to increase work hardening at low strains, where dislocations are pinned by solute atoms, but produces a decrease in work hardening at large strains. Small amounts of magnesium or copper also

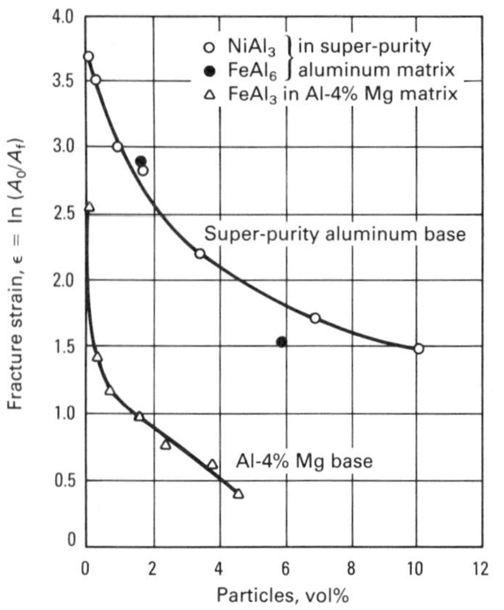

Fig. 13 Effect of volume percent fraction of micron-size intermetallic particles and composition of the matrix on the fracture strain of 5 mm (0.2 in.) diam tensile specimens. A_0 is initial cross-sectional area. A_f is area of fracture.

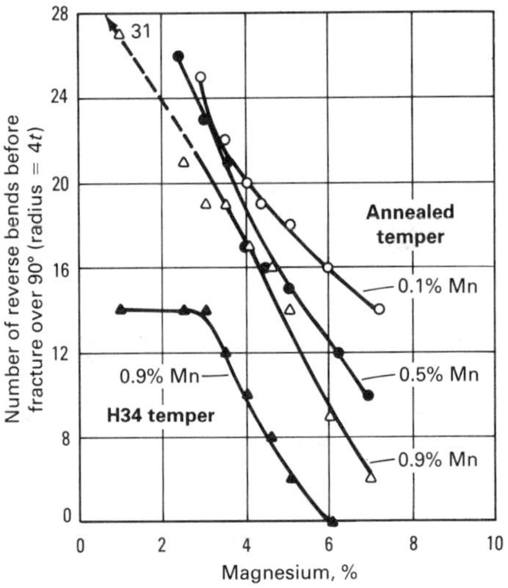

Fig. 14 Effect of magnesium and manganese on the formability of aluminum alloys in the annealed and H34 tempers; 1.6 mm (0.064 in.) thick sheet

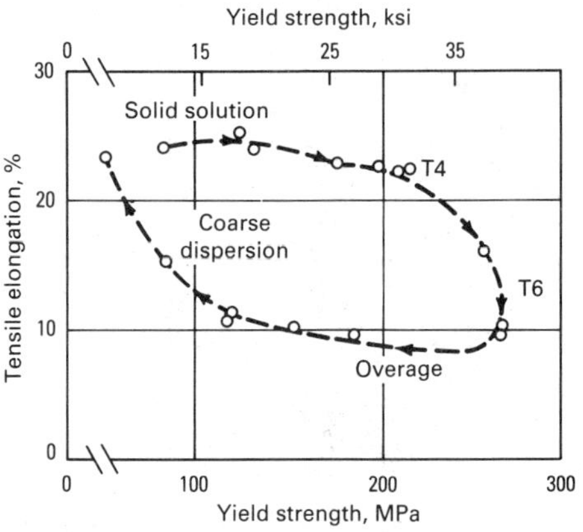

Fig. 15 Effect of precipitation on yield strength and elongation in alloy 2036

reduce the strain-rate hardening, which will reduce the amount of useful diffuse necking that occurs after the uniform elongation. Zinc in dilute alloys has little effect on work hardening or necking, and it does not cause strain aging.

Elements that have low solid solubilities at typical processing temperatures, such as iron, silicon, and manganese, are present in the form of second-phase particles and have little influence on either strain hardening or strain-rate hardening and thus a relatively minor influence on necking behavior. Second-phase particles do, however, have a large influence on fracture, as is shown in Fig. 13 and 14. In these examples an increase in the iron, nickel, or manganese content produces an increase in the number of microscopic particles that promote fracture. The addition of magnesium promotes an additional reduction in fracture strain because the higher flow stresses aid in the formation and growth of voids at the intermetallic particles. Magnesium in solid solution also promotes the localization of strain into shear bands, which concentrates the voids in a thin plane of highly localized strain.

Precipitation-strengthened alloys are usually formed in the naturally aged (T4) condition, or in the annealed (O) condition, but only very rarely in the peak strength (T6) condition where both the necking and fracture limits are low. In Fig. 15 the effect of a wide range of precipitate structures on some of the forming properties is illustrated for alloy 2036 (2.5% Cu-0.5% Mg). Curves similar in shape can be drawn for most of the precipitation-strengthened alloys in the 2*xxx* and 6*xxx* series. The properties in Fig. 15 were obtained from sheet tensile specimens first solution heat treated, then aged at temperatures ranging from room temperature to 350 °C (660 °F). This produced a full range of structures from solid solution (as-quenched) through T4 and T6 tempers to various degrees of overaging and precipitate agglomeration.

Fracture toughness and fatigue behavior, which are important characteristics of the high-strength aluminum alloys used in aerospace applications, are known to be influenced by the following three types of constituent particles:

Type	Size, μm	Size, mil	Typical examples
Constituent particles	2–50	0.08–2	Cu_2FeAl_7, $CuAl_2$, $FeAl_6$
Dispersoid particles	0.01–0.5	0.0004–0.02	$ZrAl_3$, $CrMg_2Al_{12}$
Strengthening precipitates	0.001–0.5	0.00004–0.02	Guinier-Preston zones

Consequently, the design of damage-tolerant aluminum alloys such as 7475, 7050, or 2124 has been primarily based upon the control of microstructure through composition and fabrication practice.

Effect of Second-Phase Constituents on Fracture Toughness. It is generally accepted that the fracture of brittle constituent particles leads to preferential paths for crack advance and reduced fracture toughness. Consequently, an often-used approach to improve the toughness of high-strength aluminum alloys has been the reduction of iron and silicon levels. The recent development of improved alloys such as 7475, 7050, and 2124 has hinged, in large part, upon the use of higher-purity base metal than 7075 or 2024. Figure 16 illustrates the influence of base metal purity on the fracture resistance of alloy 7475 sheet. The partially soluble constituents exert a similar effect on the fracture behavior of other high-strength alloys. Figure 17 shows the reduction in toughness experienced as the volume fraction of Al_2CuMg is increased in alloy 7050 plate.

For superior toughness, the amount of dispersoid-forming element should be held to the minimum required for control of grain structure, mechanical properties, or resistance to stress-corrosion cracking. Results for 7*xxx* alloy sheet (Fig. 18) show the marked decrease in unit propagation energy as chromium is increased. Substitution of other elements, such as zirconium or manganese, for chromium can also influence fracture toughness. However, the observed effects of the different dispersoids on fracture toughness can quite possibly be related to the particular toughness parameter chosen and the influence of the dispersoid on the grain structure of the wrought product.

The primary effect of hardening precipitates on fracture toughness of high-strength aluminum alloys is through the increase in yield strength and depends upon the particular working and heat-treatment practices

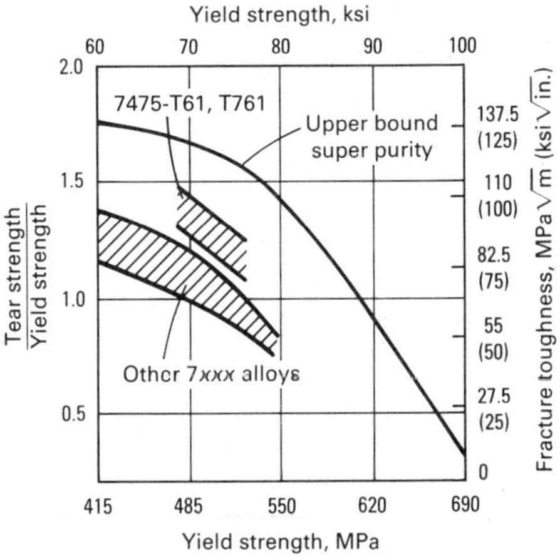

Fig. 16 Tear strength and yield strength ratio of alloy 7475 sheet

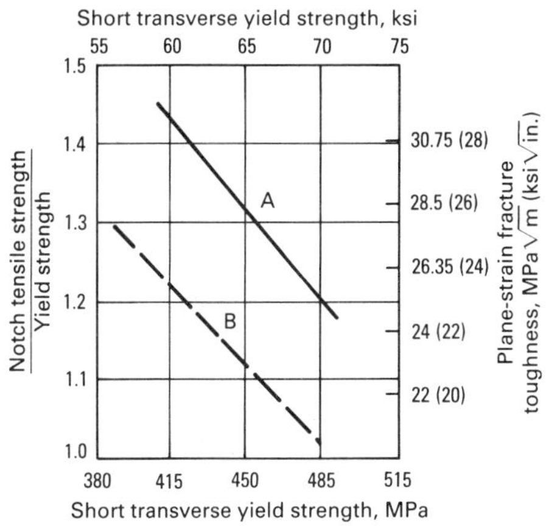

Fig. 17 Effects of amount of Al_2CuMg constituent on the toughness of 7050 plate

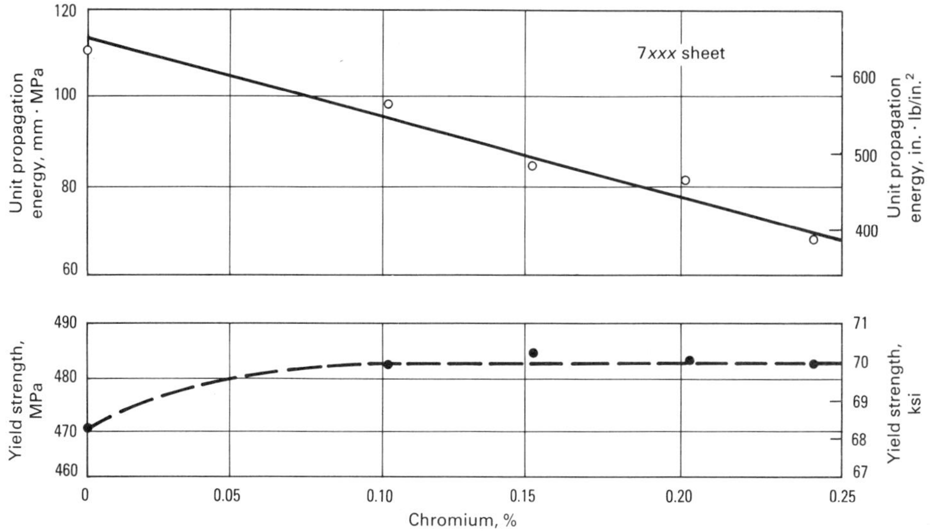

Fig. 18 Effect of chromium content on unit crack propagation energy and yield strength on a Zn-Mg-Cu aluminum alloy (5.5 Zn, 2.4 Mg, 1.4 Cu, 0.30 Fe, 0.08 Si, 0.03 Ti, 0.01 Mn)

applied to the wrought products. However, composition changes, particularly magnesium level, can produce significant effects on toughness of 7*xxx* alloys. These variations in composition do not alter the basic character of the hardening precipitates, but exert a subtle influence on the overall precipitate structure.

Effect of Second-Phase Constituents on Fatigue Behavior. Although the three types of constituent particles may influence fatigue behavior, the effect of constituent particles on fatigue behavior is highly dependent upon the type of fatigue test or the stress regime chosen for evaluation. Consequently, the design of aluminum alloys to resist failure by fatigue mechanisms has not proceeded to the same extent as for fracture toughness. In the case of large constituent particles, for example, reduced iron and silicon contents do not always result in improved fatigue resistance commensurate with the previously described improvements in fracture toughness. Increased purity level does not, for instance, produce any appreciable improvement in notched or smooth *S-N* fatigue strength.

In terms of fatigue crack growth (FCG) rates, no consistent differences have been observed for low- and high-purity 7*xxx* alloy variants at low to intermediate ΔK levels. However, at high stress-intensity ranges, FCG rates are notably reduced for low iron and silicon alloys. The reason for the observed improvement is undoubtedly related to the higher fracture toughness of high-purity metals. At high stress-intensity ranges, where crack growth per cycle (da/dN) values are large, localized fracture and void nucleation at constituent particles become the dominant FCG mechanism. For samples subjected to periodic spike overloads, low-purity alloys were shown to exhibit slower overall FCG rates than higher-purity materials. This effect was attributed to localized crack deviation induced by the insoluble constituents. Secondary cracks at these particles acted to lower crack tip stress-intensity values and to reduce measured FCG rates.

No clear-cut influence of dispersoid particles on the fatigue behavior of aluminum alloys has emerged. Two separate studies have concluded that dispersoid type has little effect on either FCG resistance or notched fatigue resistance of 7*xxx* alloys. The only expected effect of dispersoid type on fatigue performance should occur for high ΔK fatigue crack growth, where mechanisms similar to those for fracture toughness predominate.

Within a given alloy system, slight changes in composition that influence hardening precipitates have not been shown to influence the *S-N* fatigue resistance of aluminum alloys. However, significant differences have been observed in comparison of

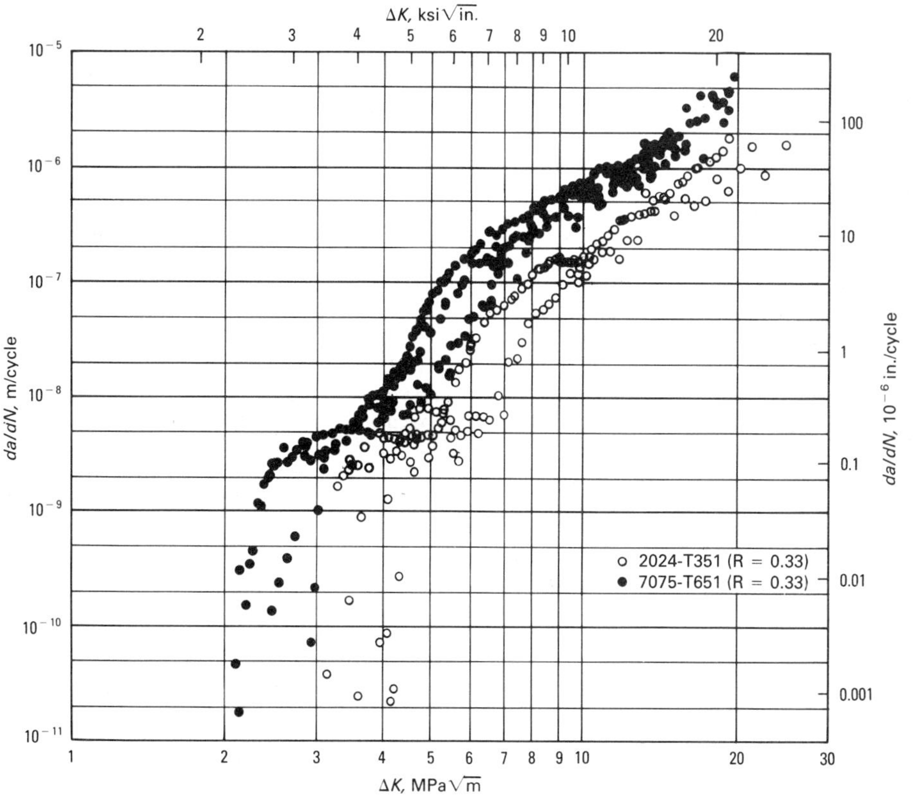

Fig. 19 Fatigue crack growth of 2024-T3 versus 7075-T6 plate over entire da/dN-ΔK range

alloys of different systems. For instance, 2024-T3 is known to outperform 7075-T6 at stresses where fatigue lives are short (~10^5 cycles). The superior fatigue performance of alloy 2024-T3 in the 10^5 cycle range has led most aircraft designers to specify it in preference to 7075-T6 in applications where tension-tension loads are predominant.

Alloy 2024-T3 shows a similar advantage over 7075-T6 and other 7*xxx* alloys in fatigue crack growth. The superior performance for 2024-T3 plate versus 7075-T6 extends over the entire da/dN-ΔK range, as shown in Fig. 19. Within the 7*xxx* alloy system, increasing copper content improves FCG performance in high humidity. This result was attributed to an increased resistance of the high-copper alloy to corrosion in the moist environment.

Fatigue designers are currently beginning to use increasingly complex "spectrum" FCG tests to predict the performance of materials in service. Early work in the area of spectrum fatigue showed that high-toughness alloy 7475-T76 performed better than either 2024-T3 or 7075-T6.

General Effects of Alloying

Although the predominant reason for alloying is to increase strength, alloying also has important effects on other characteristics of aluminum alloys. Some of these effects are discussed below. Effects of specific elements are discussed in the section "Specific Alloying Elements and Impurities" in this article.

Alloy Effects on Physical Properties. Most of the physical properties—density, melting-temperature range, heat content, coefficient of thermal expansion, and electrical and thermal conductivities—are changed by addition of one or more alloying elements. The rates of change in these properties with each incremental addition are specific for each element and depend, in many cases drastically, on whether a solid solution or a second phase is formed. In those cases in which the element or elements may be either dissolved or precipitated by heat treatment, certain of these properties, particularly density and conductivity, can be altered substantially by heat treatment. Density and conductivity of such alloys show relatively large differences from one temper to another.

Electrochemical properties and corrosion resistance are strongly affected by alloying elements that form either solid solutions, or additional phases, or both. For those systems exhibiting substantial changes in solid solubility with temperature, these properties may change markedly with heat-treated tempers, and although infrequently, even with room-temperature aging (for example, the stress-corrosion resistance of high-Mg 5*xxx* alloys in strain-hardened tempers). The strongest electrochemical effects are from copper or zinc in solid solution. Additions of copper in solid solution change the electrochemical solution potential in the cathodic direction at the rate of 0.047 V/wt% (0.112 V/at%), and additions of zinc change it in the anodic direction at the rate of 0.063 V/wt% (0.155 V/at%). These potentials are those measured in aqueous solution of 53 g NaCl + 3 g H_2O_2 per liter. Magnesium and silicon, which are the basis for the 4*xxx*, 5*xxx*, and 6*xxx* series wrought alloys and the 3*xx.x*, 4*xx.x*, and 5*xx.x* series casting alloys, and which are prominent in the compositions of many other alloys, have relatively mild effects on solution potential and are not detrimental to corrosion resistance.

Although aluminum is a thermodynamically reactive metal, it has excellent resistance to corrosion in most environments, which may be attributed to the passivity afforded by a protective film of aluminum oxide. This film is strongly bonded to the surface of the metal, and if damaged re-forms almost immediately. The continuity of the film is affected by the microstructure of the metal—in particular, by the presence and volume fraction of second-phase particles. Corrosion resistance is affected by this factor and by the solution-potential relationships between the second-phase particles or constituents and the solid-solution matrix in which they occur. In most environments, resistance to corrosion of unalloyed aluminum increases with increasing purity. The resistance of an alloy depends not only on the microstructural relationships involving the specific types, amounts, and distributions of the second-phase constituents but even more strongly on the nature of the solid solutions in which they are present. Copper reduces corrosion resistance despite the fact that when in solid solution it makes the alloy more cathodic (less active thermodynamically). This is explained by the fact that copper ions taken into solution in aqueous, corroding media replate on the aluminum alloy surface as minute particles of metallic copper, forming even more active corrosion couples because metallic copper is highly cathodic to the alloy. Manganese, which in solid solution changes solution potential in the cathodic direction as strongly as does copper, does not impair corrosion resistance of commercial alloys that contain it, because the amounts left in solid solution in commercial products, which undergo extensive solid-state heating in process, are very small, and the manganese does not replate from solution as does copper.

The differences in solution potential among alloys of different compositions are used to great advantage in the composite Alclad products. In these products, the structural component of the composite, usually a strong or heat-treatable alloy, is made the core of the product and is covered by a cladding alloy of a composition that not only is highly corrosion resistant but also has a solution potential that is anodic to that of the core. Analogous to the protection of the underlying steel afforded by zinc on the surfaces of galvanized steel products, the aluminum alloy core is protected electrolytically by the more-anodic cladding. The composition of the cladding material is designed specifically to protect the core alloy, so that, for those containing copper as the principal alloying ingredient (2*xxx* type), the more-anodic unalloyed aluminum (1*xxx* type) serves to protect the core electrolytically. In the case of the strong alloys containing zinc along with magnesium and copper (such as 7049, 7050, 7075, and 7178), an aluminum-zinc alloy (7072) or an aluminum-zinc-magnesium alloy (7008 or 7011) provides protection. The latter provides higher strength.

Impurity Effects. Although major differences in properties and characteristics are usually associated with alloying additions of one to several percent, many alloying elements produce highly significant effects when added in small fractions of 1% or when increased by such small amounts. With respect to mechanical properties, this is particularly true for combinations of certain elements. The interactions are quite complex, and a given element may be either highly beneficial or highly detrimental depending on the other elements involved and on the property or combination of properties needed.

The presence or absence of amounts on the order of one thousandth of one percent of certain impurities—sodium and calcium, for example—may make the difference between success or complete failure in fabricating high-magnesium 5*xxx* alloy ingots into useful wrought products. There are many other examples of equal practical importance. Impurity limits specified for commercial alloys reflect some of these effects, but producers of mill products must adhere to even more-restrictive limits in many cases to ensure good product recovery.

Silicon-Modifying Additions. Additions of similarly small percentages of both metallic and nonmetallic substances—sodium and phosphorus, for example—are used to enhance the mechanical and machining properties of silicon-containing casting alloys.

Grain-Refining Additions. Most alloys produced as "fabricating ingots" for fabricating wrought products, as well as those in the form of foundry ingot, have small additions of titanium or boron, or combinations of these two elements, in controlled proportions. The purpose of these additions is to control grain size and shape in the as-cast fabricating ingot or in castings produced from the foundry ingot. These grain-refining additions have little effect on changes in grain size that occur during or as a result of working or recrystallization. Welding filler alloys and casting alloys generally have higher contents of the grain-refining elements to ensure highest resistance to crack-

Table 6 Typical physical properties of aluminum alloys

Alloy	Average coefficient of thermal expansion(a) µm/m · °C	µin./in. · °F	Approximate melting range(b)(c) °C	°F	Temper	Thermal conductivity at 25 °C (77 °F) W/m · °C	Btu · in./ft² · h · °F	Electrical conductivity at 20 °C (68 °F), %IACS Equal volume	Equal weight	Electrical resistivity at 20 °C (68 °F) Ω · mm²/m	Ω · circ mil/ft
1060	23.6	13.1	645–655	1195–1215	0	234	1625	62	204	0.028	17
					H18	230	1600	61	201	0.028	17
1100	23.6	13.1	643–655	1190–1215	0	222	1540	59	194	0.030	18
					H18	218	1510	57	187	0.030	18
1350	23.75	13.2	645–655	1195–1215	All	234	1625	62	204	0.028	17
2011	22.9	12.7	540–643(d)	1005–1190(d)	T3	151	1050	39	123	0.045	27
					T8	172	1190	45	142	0.038	23
2014	23.0	12.8	507–638(e)	945–1180(e)	0	193	1340	50	159	0.035	21
					T4	134	930	34	108	0.0515	31
					T6	154	1070	40	127	0.043	26
2017	23.6	13.1	513–640(e)	955–1185(e)	0	193	1340	50	159	0.035	21
					T4	134	930	34	108	0.0515	31
2018	22.3	12.4	507–638(d)	945–1180(d)	T61	154	1070	40	127	0.043	26
2024	23.2	12.9	500–638(e)	935–1180(e)	0	193	1340	50	160	0.035	21
					T3, T4, T361	121	840	30	96	0.058	35
					T6, T81, T861	151	1050	38	122	0.045	27
2025	22.7	12.6	520–640(e)	970–1185(e)	T6	154	1070	40	128	0.043	26
2036	23.4	13.0	555–650(d)	1030–1200(d)	T4	159	1100	41	135	0.0415	25
2117	23.75	13.2	555–650(d)	1030–1200(d)	T4	154	1070	40	130	0.043	26
2124	22.9	12.7	500–638(e)	935–1180(e)	T851	152	1055	38	122	0.045	27
2218	22.3	12.4	505–635(e)	940–1175(e)	T72	154	1070	40	126	0.043	26
2219	22.3	12.4	543–643(e)	1010–1190(e)	0	172	1190	44	138	0.040	24
					T31, T37	112	780	28	88	0.0615	37
					T6, T81, T87	121	840	30	94	0.058	35
2618	22.3	12.4	550–638	1020–1180	T6	147	1020	37	120	0.0465	28
3003	23.2	12.9	643–655	1190–1210	0	193	1340	50	163	0.035	21
					H12	163	1130	42	137	0.0415	25
					H14	159	1100	41	134	0.0415	25
					H18	154	1070	40	130	0.043	26
3004	23.9	13.3	630–655	1165–1210	All	163	1130	42	137	0.0415	25
3105	23.6	13.1	635–655	1175–1210	All	172	1190	45	148	0.038	23
4032	19.4	10.8	532–570(e)	990–1060(e)	0	154	1070	40	132	0.043	26
					T6	138	960	35	116	0.050	30
4043	22.1	12.3	575–632	1065–1170	0	163	1130	42	140	0.0415	25
4045	21.05	11.7	575–600	1065–1110	All	172	1190	45	151	0.038	23
4343	21.6	12.0	577–613	1070–1135	All	180	1250	42	158	0.0415	25
5005	23.75	13.2	632–655	1170–1210	All	200	1390	52	172	0.033	20
5050	23.75	13.2	625–650	1155–1205	All	193	1340	50	165	0.035	21
5052	23.75	13.2	607–650	1125–1200	All	138	960	35	116	0.050	30
5056	24.1	13.4	568–638	1055–1180	0	117	810	29	98	0.060	36
					H38	108	750	27	91	0.063	38
5083	23.75	13.2	590–638	1095–1180	0	117	810	29	98	0.060	36
5086	23.75	13.2	585–640	1085–1185	All	125	870	31	104	0.055	33
5154	23.9	13.3	593–643	1100–1190	All	125	870	32	107	0.053	32
5252	23.75	13.2	607–650	1125–1200	All	138	960	35	116	0.050	30
5254	23.9	13.3	593–643	1100–1190	All	125	870	32	107	0.053	32
5356	24.1	13.4	570–635	1060–1175	0	117	810	29	98	0.060	36
5454	23.6	13.1	600–645	1115–1195	0	134	930	34	113	0.0515	31
					H38	134	930	34	113	0.0515	31
5456	23.9	13.3	568–638	1055–1180	0	117	810	29	98	0.060	36
5457	23.75	13.2	630–655	1165–1210	All	176	1220	46	153	0.038	23
5652	23.75	13.2	607–650	1125–1200	All	138	960	35	116	0.050	30
5657	23.75	13.2	638–657	1180–1215	All	205	1420	54	180	0.0315	19
6005	23.4	13.0	610–655(d)	1125–1210(d)	T1	180	1250	47	155	0.0365	22
					T5	190	1310	49	161	0.035	21
6053	23	12.8	575–650(d)	1070–1205(d)	0	172	1190	45	148	0.038	23
					T4	154	1070	40	132	0.043	26
					T6	163	1130	42	139	0.0415	25
6061	23.6	13.1	580–650(d)	1080–1205(d)	0	180	1250	47	155	0.0365	22
					T4	154	1070	40	132	0.043	26
					T6	167	1160	43	142	0.040	24
6063	23.4	13.0	615–655	1140–1210	0	218	1510	58	191	0.030	18
					T1	193	1340	50	165	0.035	21
					T5	209	1450	55	181	0.032	19
					T6, T83	200	1390	53	175	0.033	20
6066	23.2	12.9	565–645(e)	1045–1195(e)	0	154	1070	40	132	0.043	26
					T6	147	1020	37	122	0.0465	28
6070	...	...	565–650(e)	1050–1200(e)	T6	172	1190	44	145	0.040	24
6101	23.4	13.0	620–655	1150–1210	T6	218	1510	57	188	0.030	18
					T61	222	1540	59	194	0.030	18
					T63	218	1510	58	191	0.030	18
					T64	226	1570	60	198	0.028	17
					T65	218	1510	58	191	0.030	18

(continued)

(a) Coefficient from 20 to 100 °C (68 to 212 °F). (b) Melting ranges shown apply to wrought products of 6.35 mm (¼ in.) thickness or greater. (c) Based on typical composition of the indicated alloys. (d) Eutectic melting can be completely eliminated by homogenization. (e) Eutectic melting is not eliminated by homogenization. (f) Although not formerly registered, the literature and some specifications have used T736 as the designation for this temper. (g) Homogenization may raise eutectic melting temperature 10 to 20 °C (20 to 40 °F) but usually does not eliminate eutectic melting.

Table 6 (continued)

Alloy	Average coefficient of thermal expansion(a), μm/m · °C	Average coefficient of thermal expansion(a), μin./in. · °F	Approximate melting range(b)(c), °F	Approximate melting range(b)(c), °F	Temper	Thermal conductivity at 25 °C (77 °F), W/m · °C	Thermal conductivity at 25 °C (77 °F), Btu · in./ft² · h · °F	Electrical conductivity at 20 °C (68 °F), %IACS, Equal volume	Electrical conductivity at 20 °C (68 °F), %IACS, Equal weight	Electrical resistivity at 20 °C (68 °F), Ω · mm²/m	Electrical resistivity at 20 °C (68 °F), Ω · circ mil/ft
6105	23.4	13.0	600–650(d)	1110–1200(d)	T1	176	1220	46	151	0.038	23
					T5	193	1340	50	165	0.035	21
6151	23.2	12.9	590–650(d)	1090–1200(d)	0	205	1420	54	178	0.0315	19
					T4	163	1130	42	138	0.0415	25
					T6	172	1190	45	148	0.038	23
6201	23.4	13.0	607–655(d)	1125–1210(d)	T81	205	1420	54	180	0.0315	19
6253	...	...	600–650	1100–1205	...	...	...	...	...	...	...
6262	23.4	13.0	580–650(d)	1080–1205(d)	T9	172	1190	44	145	0.040	24
6351	23.4	13.0	555–650	1030–1200	T6	176	1220	46	151	0.038	23
6463	23.4	13.0	615–655	1140–1210	T1	193	1340	50	165	0.035	21
					T5	209	1450	55	181	0.0315	19
					T6	200	1390	53	175	0.033	20
6951	23.4	13.0	615–655	1140–1210	0	213	1480	56	186	0.0315	19
					T6	198	1370	52	172	0.033	20
7049	23.4	13.0	475–635	890–1175	T73	154	1070	40	132	0.043	26
7050	24.1	13.4	490–630	910–1165	T74(f)	157	1090	41	135	0.0415	25
7072	23.6	13.1	640–655	1185–1215	0	222	1540	59	193	0.030	18
7075	23.6	13.1	475–635(g)	890–1175(g)	T6	130	900	33	105	0.0515	31
7178	23.4	13.0	475–630(g)	890–1165(g)	T6	125	870	31	98	0.055	33
8017	23.6	13.1	645–655	1190–1215	H12, H22	...	...	59	193	0.030	18
					H212	...	...	61	200	0.028	17
8030	23.6	13.1	645–655	1190–1215	H221	230	1600	61	201	0.028	17
8176	23.6	13.1	645–655	1190–1215	H24	230	1600	61	201	0.028	17

(a) Coefficient from 20 to 100 °C (68 to 212 °F). (b) Melting ranges shown apply to wrought products of 6.35 mm (¼ in.) thickness or greater. (c) Based on typical composition of the indicated alloys. (d) Eutectic melting can be completely eliminated by homogenization. (e) Eutectic melting is not eliminated by homogenization. (f) Although not formerly registered, the literature and some specifications have used T736 as the designation for this temper. (g) Homogenization may raise eutectic melting temperature 10 to 20 °C (20 to 40 °F) but usually does not eliminate eutectic melting.

ing during solidification of welds and castings.

The elements that have relatively great and controlling effects on grain sizes and shapes produced by the mechanical working required to produce wrought products (their thermomechanical history) are manganese, chromium, and zirconium. Small amounts (fractional percentages) of these elements, singly or in combination, are included in the compositions of many alloys to control grain size and recrystallization behavior through fabrication and heat treatment. Such grain control has many purposes, which include ensuring good resistance to stress-corrosion cracking (SCC), high fracture toughness, and good forming characteristics. In specific alloys, these elements have highly significant supplementary beneficial effects on strength, resistance to fatigue, or strength at elevated temperatures. In order to fulfill their grain-control functions, these elements must be precipitated as finely distributed particles termed dispersoids. Their precipitation is accomplished primarily by the high-temperature, solid-state heating involved in ingot preheating.

Secondary Aluminum. Aluminum recovered from scrap (secondary aluminum) has been an important contributor to the total metal supply for many years. For some uses, secondary aluminum alloys may be treated to remove certain impurities or alloying elements. Chief among the alloying elements removed is magnesium, which is frequently present in greater amounts in secondary metal than in the alloys to be produced from it. Magnesium is usually removed by fluxing with chlorine gas or halide salts.

Specific Alloying Elements and Impurities

The important alloying elements and impurities are listed here alphabetically as a concise review of major effects. Some of the effects, particularly with respect to impurities, are not well documented and are specific to particular alloys or conditions.

Antimony is present in trace amounts (0.01 to 0.1 ppm) in primary commercial-grade aluminum. Antimony has a very small solid solubility in aluminum (<0.01%). It has been added to aluminum-magnesium alloys because it was claimed that by forming a protective film of antimony oxychloride, it enhances corrosion resistance in salt water. Some bearing alloys contain up to 4 to 6% Sb. Antimony can be used instead of bismuth to counteract hot cracking in aluminum-magnesium alloys.

Arsenic. The compound AsAl is a semiconductor. Arsenic is very toxic (as AsO_3) and must be controlled to very low limits where aluminum is used as foil for food packaging.

Beryllium is used in aluminum alloys containing magnesium to reduce oxidation at elevated temperatures. Up to 0.1% Be is used in aluminizing baths for steel to improve adhesion of the aluminum film and restrict the formation of the deleterious iron-aluminum complex. The mechanism of protection is attributed to beryllium diffusion to the surface and the formation of a protective layer.

Oxidation and discoloration of wrought aluminum-magnesium products are greatly reduced by small amounts of beryllium because of the diffusion of beryllium to the surface and the formation of an oxide of high-volume ratio. Beryllium does not affect the corrosion resistance of aluminum. Beryllium is generally held to <8 ppm in welding filler metal, and its content should be limited in wrought alloys that may be welded.

Beryllium poisoning is an allergic disease, a problem of individual hypersensitivity that is related to intensity and duration of exposure. Inhalation of dust containing beryllium compounds may lead to acute poisoning. Beryllium is not used in aluminum alloys that may contact food or beverages.

Bismuth. The low-melting-point metals such as bismuth, lead, tin, and cadmium are added to aluminum to make free-machining alloys. These elements have a restricted solubility in solid aluminum and form a soft, low-melting phase that promotes chip breaking and helps to lubricate the cutting tool. An advantage of bismuth is that its expansion on solidification compensates for the shrinkage of lead. A 1-to-1 lead-bismuth ratio is used in the aluminum-copper alloy, 2011, and in the aluminum-Mg_2Si alloy, 6262. Small additions of bismuth (20 to 200 ppm) can be added to aluminum-magnesium alloys to counteract the detrimental effect of sodium on hot cracking.

Boron is used in aluminum and its alloys as a grain refiner and to improve conductivity by precipitating vanadium, titanium, chromium, and molybdenum (all of which are harmful to electrical conductivity at

Table 7 Nominal densities of aluminum and aluminum alloys

Alloy	Density g/cm³	Density lb/in.³
1050	2.705	0.0975
1060	2.705	0.0975
1100	2.71	0.098
1145	2.700	0.0975
1175	2.700	0.0975
1200	2.70	0.098
1230	2.70	0.098
1235	2.705	0.0975
1345	2.705	0.0975
1350	2.705	0.0975
2011	2.83	0.102
2014	2.80	0.101
2017	2.79	0.101
2018	2.82	0.102
2024	2.78	0.101
2025	2.81	0.101
2036	2.75	0.100
2117	2.75	0.099
2124	2.78	0.100
2218	2.81	0.101
2219	2.84	0.103
2618	2.76	0.100
3003	2.73	0.099
3004	2.72	0.098
3005	2.73	0.098
3105	2.72	0.098
4032	2.68	0.097
4043	2.69	0.097
4045	2.67	0.096
4047	2.66	0.096
4145	2.74	0.099
4343	2.68	0.097
4643	2.69	0.097
5005	2.70	0.098
5050	2.69	0.097
5052	2.68	0.097
5056	2.64	0.095
5083	2.66	0.096
5086	2.66	0.096
5154	2.66	0.096
5183	2.66	0.096
5252	2.67	0.096
5254	2.66	0.096
5356	2.64	0.096
5454	2.69	0.097
5456	2.66	0.096
5457	2.69	0.097
5554	2.69	0.097
5556	2.66	0.096
5652	2.67	0.097
5654	2.66	0.096
5657	2.69	0.097
6003	2.70	0.097
6005	2.70	0.097
6053	2.69	0.097
6061	2.70	0.098
6063	2.70	0.097
6066	2.72	0.098
6070	2.71	0.098
6101	2.70	0.097
6105	2.69	0.097
6151	2.71	0.098
6162	2.70	0.097
6201	2.69	0.097
6262	2.72	0.098
6351	2.71	0.098
6463	2.69	0.097
6951	2.70	0.098
7005	2.78	0.100
7008	2.78	0.100
7049	2.84	0.103
7050	2.83	0.102
7072	2.72	0.098
7075	2.81	0.101
7178	2.83	0.102
8017	2.71	0.098
8030	2.71	0.098
8176	2.71	0.098
8177	2.70	0.098

their usual impurity level in commercial-grade aluminum). Boron can be used alone (at levels of 0.005 to 0.1%) as a grain refiner during solidification, but becomes more effective when used with an excess of titanium. Commercial grain refiners commonly contain titanium and boron in a 5-to-1 ratio. Boron has a high neutron capture cross section and is used in aluminum alloys for certain atomic energy applications, but its content has to be limited to very low levels in alloys used in reactor areas where this property is undesirable.

Cadmium is a relatively low-melting element that finds limited use in aluminum. Up to 0.3% Cd may be added to aluminum-copper alloys to accelerate the rate of age hardening, increase strength, and increase corrosion resistance. At levels of 0.005 to 0.5%, it has been used to reduce the time of aging of aluminum-zinc-magnesium alloys. It has been reported that traces of Cd lower the corrosion resistance of unalloyed aluminum. In excess of 0.1%, cadmium causes hot shortness in some alloys. Because of its high neutron absorption, cadmium has to be kept very low for atomic energy use. It has been used to confer free-cutting characteristics, particularly to aluminum-zinc-magnesium alloys; it was preferred to bismuth and lead because of its higher melting point. As little as 0.1% provides an improvement in machinability. Cadmium is used in bearing alloys along with silicon. The oral toxicity of cadmium compounds is high. In melting, casting, and fluxing operations cadmium oxide fume can present hazards.

Calcium has very low solubility in aluminum and forms the intermetallic $CaAl_4$. An interesting group of alloys containing about 5% Ca and 5% Zn have superplastic properties. Calcium combines with silicon to form $CaSi_2$, which is almost insoluble in aluminum and therefore will increase the conductivity of commercial-grade metal slightly. In aluminum-magnesium-silicon alloys, calcium will decrease age hardening. Its effect on aluminum-silicon alloys is to increase strength and decrease elongation, but it does not make these alloys heat treatable. At the 0.2% level, calcium alters the recrystallization characteristics of 3003. Very small amounts of calcium (10 ppm) increase the tendency of molten aluminum alloys to pick up hydrogen.

Carbon may occur infrequently as an impurity in aluminum in the form of oxycarbides and carbides, of which the most common is Al_4C_3, but carbide formation with other impurities such as titanium is possible. Al_4C_3 decomposes in the presence of water and water vapor, and this may lead to surface pitting. Normal metal transfer and fluxing operations usually reduce carbon to the ppm level.

Cerium, mostly in the form of mischmetal (rare earths with 50 to 60% Ce), has been added experimentally to casting alloys to increase fluidity and reduce die sticking. In alloys containing high iron (>0.7%), it is reported to transform acicular $FeAl_3$ into a nonacicular compound.

Chromium occurs as a minor impurity in commercial-purity aluminum (5 to 50 ppm). It has a large effect on electrical resistivity. Chromium is a common addition to many alloys of the aluminum-magnesium, aluminum-magnesium-silicon, and aluminum-magnesium-zinc groups, in which it is added in amounts generally not exceeding 0.35%. In excess of these limits, it tends to form very coarse constituents with other impurities or additions such as manganese, iron, and titanium. This limit is decreased as the content of transition metals increases. In casting alloys, excess chromium will produce a sludge by peritectic precipitation on holding.

Chromium has a slow diffusion rate and forms fine dispersed phases in wrought products. These dispersed phases inhibit nucleation and grain growth. Chromium is used to control grain structure, to prevent grain growth in aluminum-magnesium alloys, and to prevent recrystallization in aluminum-magnesium-silicon or aluminum-magnesium-zinc alloys during hot working or heat treatment. The fibrous structures that develop reduce stress corrosion susceptibility and/or improve toughness. Chromium in solid solution and as a finely dispersed phase increases the strength of alloys slightly. The main drawback of chromium in heat-treatable alloys is the increase in quench sensitivity when the hardening phase tends to precipitate on the preexisting chromium-phase particles. Chromium imparts a yellow color to the anodic film.

Cobalt is not a common addition to aluminum alloys. It has been added to some aluminum-silicon alloys containing iron, where it transforms the acicular β (aluminum-iron-silicon) into a more rounded aluminum-cobalt-iron phase, thus improving strength and elongation. Aluminum-zinc-magnesium-copper alloys containing 0.2 to 1.9% Co are produced by powder metallurgy.

Copper. Aluminum-copper alloys containing 2 to 10% Cu, generally with other additions, form important families of alloys. Both cast and wrought aluminum-copper alloys respond to solution heat treatment and subsequent aging with an increase in strength and hardness and a decrease in elongation. The strengthening is maximum between 4 and 6% Cu, depending upon the influence of other constituents present. The properties of aluminum-copper alloy sheet in a number of thermal conditions are assembled in Fig. 20. The aging characteristics of binary aluminum-copper alloys have been studied in greater detail than any other system, but there are actually very few commercial binary aluminum-copper alloys. Most commercial alloys contain other alloying elements.

Copper-Magnesium. The main benefit of adding magnesium to aluminum-copper alloys is the increased strength possible following solution heat treatment and quenching. In wrought material of certain alloys of this type, an increase in strength accompanied by high ductility occurs on aging at room temperature. On artificial aging, a further increase in strength, especially in yield strength, can be obtained, but at a substantial sacrifice in tensile elongation.

On both cast and wrought aluminum-copper alloys, as little as about 0.5% Mg is effective in changing aging characteristics. In wrought products, the effect of magnesium additions on strength can be maximized in artificially aged materials by cold working prior to aging (Fig. 21). In naturally aged materials, however, the benefit to strength from magnesium additions can decrease with cold working (Fig. 22). The effect of magnesium on the corrosion resistance of aluminum-copper alloys depends on the type of product and the thermal treatment.

Copper-Magnesium Plus Other Elements. The cast aluminum-copper-magnesium alloys containing iron are characterized by dimensional stability and improved bearing characteristics, as well as by high strength and hardness at elevated temperatures. However, in a wrought Al-4%Cu-0.5%Mg alloy, iron in concentrations as low as 0.5% lowers the tensile properties in the heat-treated condition, if the silicon content is less than that required to tie up the iron as the αFeSi constituent. In this event, the excess iron unites with copper to form the Cu_2FeAl_7 constituent, thereby reducing the amount of copper available for heat-treating effects. When sufficient silicon is present to combine with the iron, the properties are unaffected. Silicon also combines with magnesium to form Mg_2Si precipitate and contributes in the age-hardening process.

Silver substantially increases the strength of heat-treated and aged aluminum-copper-magnesium alloys. Nickel improves the strength and hardness of cast and wrought aluminum-copper-magnesium alloys at elevated temperatures. Addition of about 0.5% Ni lowers the tensile properties of the heat-treated, wrought Al-4%Cu-0.5%Mg alloy at room temperature.

The alloys containing manganese form the most important and versatile system of commercial high-strength wrought aluminum-copper-magnesium alloys. The substantial effect exerted by manganese on the tensile properties of aluminum-copper alloys containing 0.5% Mg is shown in Fig. 23. It is apparent that no one composition offers both maximum strength and ductility. In general, tensile strength increases with separate or simultaneous increases in magnesium and manganese, and the yield strength also increases, but to a lesser extent. Further increases in tensile and particularly yield strength occur on cold working after heat treatment. Additions of manganese and magnesium decrease the fabricating characteristics of the aluminum-copper alloys, and manganese also causes a loss in ductility; hence, the concentration of this element does not exceed about 1% in commercial alloys. Additions of cobalt, chromium, or molybdenum to the wrought Al-4%Cu-0.5%Mg type of alloy increase the tensile properties on heat treatment, but none offers a distinct advantage over manganese.

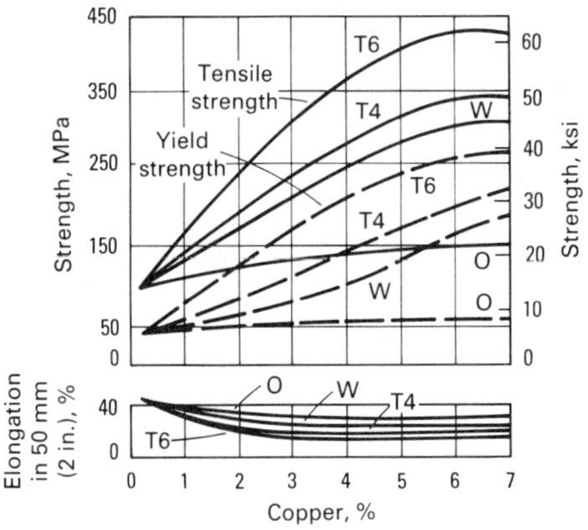

Fig. 20 Tensile properties of high-purity, wrought aluminum-copper alloys. Sheet specimen was 13 mm (0.5 in.) wide and 1.59 mm (0.0625 in.) thick. O, annealed; W, tested immediately after water quenching from a solution heat treatment; T4, as in W, but aged at room temperature; T6, as in T4, followed by precipitation treatment at elevated temperature

Alloys with lower copper content than the conventional 2024 and 2014 type alloys were necessary to provide the formability required by the automobile industry. Copper-magnesium alloys developed for this purpose are 2002, AU2G, and 2036 variations. These have acceptable formability, good spot weldability, reasonable fusion weldability, good corrosion resistance, and freedom from Lüder lines. The paint-baking cycle serves as a precipitation treatment to give final mechanical properties.

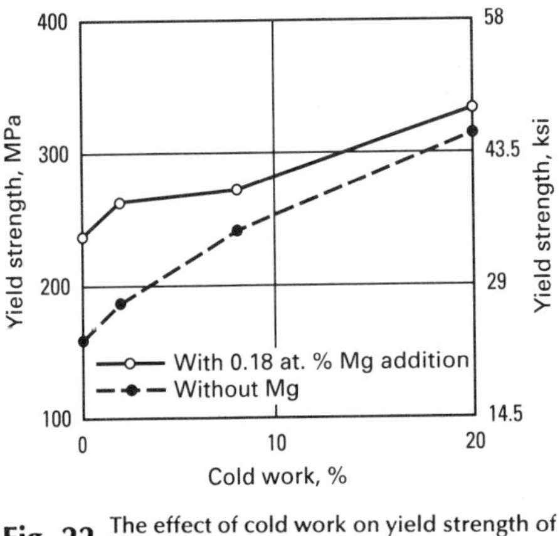

Fig. 22 The effect of cold work on yield strength of aluminum-copper alloy 2419 in naturally aged materials. Source: Ref 8

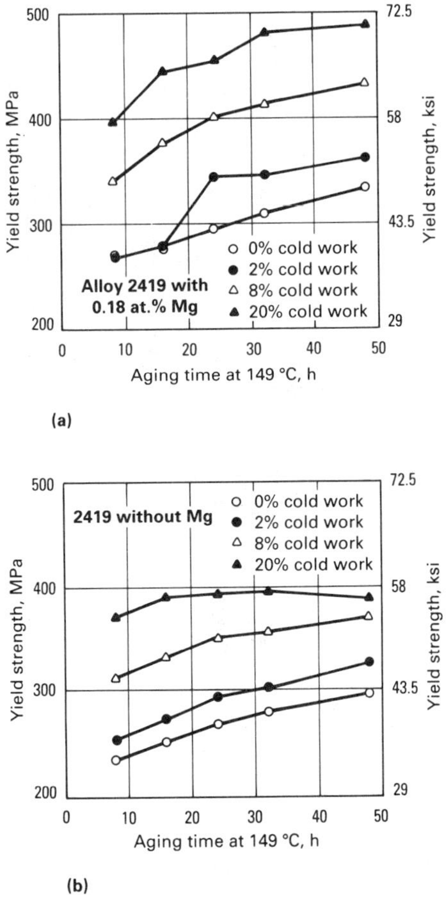

Fig. 21 Effect of cold work and Mg addition on alloy 2419. (a) The effect of cold work on the yield strength response to aging at 149 °C (300 °F) for the alloy with 0.18 at% Mg. (b) The effect of cold work on the yield strength response to aging at 149 °C (300 °F) for the alloy without Mg. Source: Ref 8

Copper and Minor Additions. In the wrought form, an alloy family of interest is the one containing small amounts of several metals known to raise the recrystallization temperature of aluminum and its alloys, specifically manganese, titanium, vanadium, or zirconium. An alloy of this nature retains its properties well at elevated temperatures, fabricates readily, and has good casting and welding characteristics. Figure

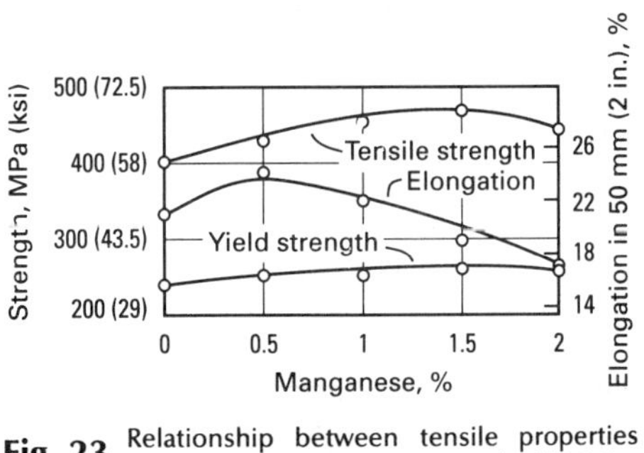

Fig. 23 Relationship between tensile properties and manganese content of Al-4%Cu-0.5%Mg alloy, heat treated at 525 °C (980 °F)

Table 8 Typical mechanical properties of various aluminum alloys

Alloy and temper	Ultimate tensile strength, MPa	Ultimate tensile strength, ksi	Tensile yield strength, MPa	Tensile yield strength, ksi	Elongation in 50 mm (2 in.), %: 1.6 mm (1/16 in.) thick specimen	Elongation in 50 mm (2 in.), %: 1.3 mm (1/2 in.) diam specimen	Hardness, HB(a)	Ultimate shearing strength, MPa	Ultimate shearing strength, ksi	Fatigue endurance limit(b), MPa	Fatigue endurance limit(b), ksi	Modulus of elasticity(c), GPa	Modulus of elasticity(c), 10^6 psi
1060-0	70	10	30	4	43	...	19	50	7	20	3	69	10.0
1060-H12	85	12	75	11	16	...	23	55	8	30	4	69	10.0
1060-H14	95	14	90	13	12	...	26	60	9	35	5	69	10.0
1060-H16	110	16	105	15	8	...	30	70	10	45	6.5	69	10.0
1060-H18	130	19	125	18	6	...	35	75	11	45	6.5	69	10.0
1100-0	90	13	35	5	35	45	23	60	9	35	5	69	10.0
1100-H12	110	16	105	15	12	25	28	70	10	40	6	69	10.0
1100-H14	125	18	115	17	9	20	32	75	11	50	7	69	10.0
1100-H16	145	21	140	20	6	17	38	85	12	60	9	69	10.0
1100-H18	165	24	150	22	5	15	44	90	13	60	9	69	10.0
1350-0	85	12	30	4	...	(d)	...	55	8	...	...	69	10.0
1350-H12	95	14	85	12	...	...	...	60	9	...	...	69	10.0
1350-H14	110	16	95	14	...	...	...	70	10	...	...	69	10.0
1350-H16	125	18	110	16	...	...	...	75	11	...	...	69	10.0
1350-H19	185	27	165	24	...	(e)	...	105	15	50	7	69	10.0
2011-T3	380	55	295	43	...	15	95	220	32	125	18	70	10.2
2011-T8	405	59	310	45	...	12	100	240	35	125	18	70	10.2
2014-0	185	27	95	14	...	18	45	125	18	90	13	73	10.6
2014-T4, T451	425	62	290	42	...	20	105	260	38	140	20	73	10.6
2014-T6, T651	485	70	415	60	...	13	135	290	42	125	18	73	10.6
Alclad 2014-0	175	25	70	10	21	...	...	125	18	...	...	72	10.5
Alclad 2014-T3	435	63	275	40	20	...	...	255	37	...	...	72	10.5
Alclad 2014-T4, T451	420	61	255	37	22	...	...	255	37	...	...	72	10.5
Alclad 2014-T6, T651	470	68	415	60	10	...	...	285	41	...	...	72	10.5
2017-0	180	26	70	10	...	22	45	125	18	90	13	72	10.5
2017-T4, T451	425	62	275	40	...	22	105	260	38	125	18	72	10.5
2018-T61	420	61	315	46	...	12	120	270	39	115	17	74	10.8
2024-0	185	27	75	11	20	22	47	125	18	90	13	73	10.6
2024-T3	485	70	345	50	18	...	120	285	41	140	20	73	10.6
2024-T4, T351	470	68	325	47	20	19	120	285	41	140	20	73	10.6
2024-T361(f)	495	72	395	57	13	...	130	290	42	125	18	73	10.6
Alclad 2024-0	180	26	75	11	20	...	...	125	18	...	...	73	10.6
Alclad 2024-T3	450	65	310	45	18	...	...	275	40	...	...	73	10.6
Alclad 2024-T4, T351	440	64	290	42	19	...	...	275	40	...	...	73	10.6
Alclad 2024-T361(f)	460	67	365	53	11	...	...	285	41	...	...	73	10.6
Alclad-2024-T81, T851	450	65	415	60	6	...	...	275	40	...	...	73	10.6
Alclad 2024-T861(f)	485	70	455	66	6	...	...	290	42	...	...	73	10.6
2025-T6	400	58	255	37	...	19	110	240	35	125	18	71	10.4
2036-T4	340	49	195	28	24	...	...	...	...	125(g)	18 (g)	71	10.3
2117-T4	295	43	165	24	...	27	70	195	28	95	14	71	10.3
2124-T851	485	70	440	64	...	8	...	...	...	...	...	73	10.6
2218-T72	330	48	255	37	...	11	95	205	30	...	...	74	10.8
2219-0	175	25	75	11	18	...	...	...	...	...	...	73	10.6
2219-T42	360	52	185	27	20	...	...	...	...	...	...	73	10.6
2219-T31, T351	360	52	250	36	17	...	...	...	...	...	...	73	10.6
2219-T37	395	57	315	46	11	...	...	...	...	...	...	73	10.6
2219-T62	415	60	290	42	10	...	...	...	...	105	15	73	10.6
2219-T81, T851	455	66	350	51	10	...	...	...	...	105	15	73	10.6
2219-T87	475	69	395	57	10	...	...	...	...	105	15	73	10.6
2618-T61	440	64	370	54	...	10	115	260	38	125	18	74	10.8
3003-0	110	16	40	6	30	40	28	75	11	50	7	69	10.0
3003-H12	130	19	125	18	10	20	35	85	12	55	8	69	10.0
3003-H14	150	22	145	21	8	16	40	95	14	60	9	69	10.0
3003-H16	180	26	170	25	5	14	47	105	15	70	10	69	10.0
3003-H18	200	29	185	27	4	10	55	110	16	70	10	69	10.0
Alclad 3003-0	110	16	40	6	30	40	...	75	11	...	...		
Alclad 3003-H12	130	19	125	18	10	20	...	85	12	...	...	69	10.0
Alclad 3003-H14	150	22	145	21	8	16	...	95	14	...	...	69	10.0
Alclad 3003-H16	180	26	170	25	5	14	...	105	15	...	...	69	10.0
Alclad 3003-H18	200	29	185	27	4	10	...	110	16	...	...	69	10.0
3004-0	180	26	70	10	20	25	45	110	16	95	14	69	10.0
3004-H32	215	31	170	25	10	17	52	115	17	105	15	69	10.0
3004-H34	240	35	200	29	9	12	63	125	18	105	15	69	10.0
3004-H36	260	38	230	33	5	9	70	140	20	110	16	69	10.0
3004-H38	285	41	250	36	5	6	77	145	21	110	16	69	10.0
Alclad 3004-0	180	26	70	10	20	25	...	110	16	...	...	69	10.0
Alclad 3004-H32	215	31	170	25	10	17	...	115	17	...	...	69	10.0
Alclad 3004-H34	240	35	200	29	9	12	...	125	18	...	...	69	10.0
Alclad 3004-H36	260	38	230	33	5	9	...	140	20	...	...	69	10.0
Alclad 3004-H38	285	41	250	36	5	6	...	145	21	...	...	69	10.0
3105-0	115	17	55	8	24	...	...	85	12	...	...	69	10.0

(continued)

(a) 500 kg load and 10 mm ball. (b) Based on 500 000 000 cycles of completely reversed stress using the R.R. Moore type of machine and specimen. (c) Average of tension and compression moduli. Compression modulus is about 2% greater than tension modulus. (d) 1350-0 wire will have an elongation of approximately 23% in 250 mm (10 in.). (e) 1350-H19 wire will have an elongation of approximately 1½% in 250 mm (10 in.). (f) Tempers T361 and T861 were formerly designated T36 and T86, respectively. (g) Based on 10^7 cycles using flexural type testing of sheet specimens. (h) T7451, although not previously registered, has appeared in literature and in some specifications as T73651.

Table 8 (continued)

Alloy and temper	Ultimate tensile strength		Tensile yield strength		Elongation in 50 mm (2 in.), %		Hardness, HB(a)	Ultimate shearing strength		Fatigue endurance limit(b)		Modulus of elasticity(c)	
	MPa	ksi	MPa	ksi	1.6 mm (1/16 in.) thick specimen	1.3 mm (1/2 in.) diam specimen		MPa	ksi	MPa	ksi	GPa	10^6 psi
3105-H12	150	22	130	19	7	...	...	95	14	...	...	69	10.0
3105-H14	170	25	150	22	5	...	...	105	15	...	...	69	10.0
3105-H16	195	28	170	25	4	...	...	110	16	...	...	69	10.0
3105-H18	215	31	195	28	3	...	...	115	17	...	...	69	10.0
3105-H25	180	26	160	23	8	...	...	105	15	...	...	69	10.0
4032-T6	380	55	315	46	...	9	120	260	38	110	16	79	11.4
5005-0	125	18	40	6	25	...	28	75	11	...	...	69	10.0
5005-H12	140	20	130	19	10	...	...	95	14	...	...	69	10.0
5005-H14	160	23	150	22	6	...	...	95	14	...	...	69	10.0
5005-H16	180	26	170	25	5	...	...	105	15	...	...	69	10.0
5005-H18	200	29	195	28	4	...	...	110	16	...	...	69	10.0
5005-H32	140	20	115	17	11	...	36	95	14	...	...	69	10.0
5005-H34	160	23	140	20	8	...	41	95	14	...	...	69	10.0
5005-H36	180	26	165	24	6	...	46	105	15	...	...	69	10.0
5005-H38	200	29	185	27	5	...	51	110	16	...	...	69	10.0
5050-0	145	21	55	8	24	...	36	105	15	85	12	69	10.0
5050-H32	170	25	145	21	9	...	46	115	17	90	13	69	10.0
5050-H34	195	28	165	24	8	...	53	125	18	90	13	69	10.0
5050-H36	205	30	180	26	7	...	58	130	19	95	14	69	10.0
5050-H38	220	32	200	29	6	...	63	140	20	95	14	69	10.0
5052-0	195	28	90	13	25	30	47	125	18	110	16	70	10.2
5052-H32	230	33	195	28	12	18	60	140	20	115	17	70	10.2
5052-H34	260	38	215	31	10	14	68	145	21	125	18	70	10.2
5052-H36	275	40	240	35	8	10	73	160	23	130	19	70	10.2
5052-H38	290	42	255	37	7	8	77	165	24	140	20	70	10.2
5056-0	290	42	150	22	...	35	65	180	26	140	20	71	10.3
5056-H18	435	63	405	59	...	10	105	235	34	150	22	71	10.3
5056-H38	415	60	345	50	...	15	100	220	32	150	22	71	10.3
5083-0	290	42	145	21	...	22	...	170	25	...	...	71	10.3
5083-H321, H116	315	46	230	33	...	16	...	...	...	160	23	71	10.3
5086-0	260	38	115	17	22	...	...	160	23	...	...	71	10.3
5086-H32, H116	290	42	205	30	12	...	...	...	...	...	...	71	10.3
5086-H34	325	47	255	37	10	...	...	185	27	...	...	71	10.3
5086-H112	270	39	130	19	14	...	...	...	...	...	...	71	10.3
5154-0	240	35	115	17	27	...	58	150	22	115	17	70	10.2
5154-H32	270	39	205	30	15	...	67	150	22	125	18	70	10.2
5154-H34	290	42	230	33	13	...	73	165	24	130	19	70	10.2
5154-H36	310	45	250	36	12	...	78	180	26	140	20	70	10.2
5154-H38	330	48	270	39	10	...	80	195	28	145	21	70	10.2
5154-H112	240	35	115	17	25	...	63	...	...	115	17	70	10.2
5252-H25	235	34	170	25	11	...	68	145	21	...	...	69	10.0
5252-H38, H28	285	41	240	35	5	...	75	160	23	...	...	69	10.0
5254-0	240	35	115	17	27	...	58	150	22	115	17	70	10.2
5254-H32	270	39	205	30	15	...	67	150	22	125	18	70	10.2
5254-H34	290	42	230	33	13	...	73	165	24	130	19	70	10.2
5254-H36	310	45	250	36	12	...	78	180	26	140	20	70	10.2
5254-H38	330	48	270	39	10	...	80	195	28	145	21	70	10.2
5254-H112	240	35	115	17	25	...	63	...	...	115	17	70	10.2
5454-0	250	36	115	17	22	...	62	160	23	...	...	70	10.2
5454-H32	275	40	205	30	10	...	73	165	24	...	...	70	10.2
5454-H34	305	44	240	35	10	...	81	180	26	...	...	70	10.2
5454-H111	260	38	180	26	14	...	70	160	23	...	...	70	10.2
5454-H112	250	36	125	18	18	...	62	160	23	...	...	70	10.2
5456-0	310	45	160	23	...	24	...	...	...	...	...	71	10.3
5456-H112	310	45	165	24	...	22	...	...	...	...	...	71	10.3
5456-H321, H116	350	51	255	37	...	16	90	205	30	...	...	71	10.3
5457-0	130	19	50	7	22	...	32	85	12	...	...	69	10.0
5457-H25	180	26	160	23	12	...	48	110	16	...	...	69	10.0
5457-H38, H28	205	30	185	27	6	...	55	125	18	...	...	69	10.0
5652-0	195	28	90	13	25	30	47	125	18	110	16	70	10.2
5652-H32	230	33	195	28	12	18	60	140	20	115	17	70	10.2
5652-H34	260	38	215	31	10	14	68	145	21	125	18	70	10.2
5652-H36	275	40	240	35	8	10	73	160	23	130	19	70	10.2
5652-H38	290	42	255	37	7	8	77	165	24	140	20	70	10.2
5657-H25	160	23	140	20	12	...	40	95	14	...	...	69	10.0
5657-H38, H28	195	28	165	24	7	...	50	105	15	...	...	69	10.0
6061-0	125	18	55	8	25	30	30	85	12	60	9	69	10.0
6061-T4, T451	240	35	145	21	22	25	65	165	24	95	14	69	10.0
6061-T6, T651	310	45	275	40	12	17	95	205	30	95	14	69	10.0
Alclad 6061-0	115	17	50	7	25	...	...	75	11	...	...	69	10.0
Alclad 6061-T4, T451	230	33	130	19	22	...	...	150	22	...	...	69	10.0

(continued)

(a) 500 kg load and 10 mm ball. (b) Based on 500 000 000 cycles of completely reversed stress using the R.R. Moore type of machine and specimen. (c) Average of tension and compression moduli. Compression modulus is about 2% greater than tension modulus. (d) 1350-0 wire will have an elongation of approximately 23% in 250 mm (10 in.). (e) 1350-H19 wire will have an elongation of approximately 1½% in 250 mm (10 in.). (f) Tempers T361 and T861 were formerly designated T36 and T86, respectively. (g) Based on 10^7 cycles using flexural type testing of sheet specimens. (h) T7451, although not previously registered, has appeared in literature and in some specifications as T73651.

Table 8 (continued)

Alloy and temper	Ultimate tensile strength, MPa	Ultimate tensile strength, ksi	Tensile yield strength, MPa	Tensile yield strength, ksi	Elongation in 50 mm (2 in.), %: 1.6 mm (1/16 in.) thick specimen	Elongation in 50 mm (2 in.), %: 1.3 mm (1/2 in.) diam specimen	Hardness, HB(a)	Ultimate shearing strength, MPa	Ultimate shearing strength, ksi	Fatigue endurance limit(b), MPa	Fatigue endurance limit(b), ksi	Modulus of elasticity(c), GPa	Modulus of elasticity(c), 10^6 psi
Alclad 6061-T6, T651	290	42	255	37	12	...	...	185	27	...	...	69	10.0
6063-0	90	13	50	7	...	...	25	70	10	55	8	69	10.0
6063-T1	150	22	90	13	20	...	42	95	14	60	9	69	10.0
6063-T4	170	25	90	13	22	...	...	...	...	...	...	69	10.0
6063-T5	185	27	145	21	12	...	60	115	17	70	10	69	10.0
6063-T6	240	35	215	31	12	...	73	150	22	70	10	69	10.0
6063-T83	255	37	240	35	9	...	82	150	22	...	...	69	10.0
6063-T831	205	30	185	27	10	...	70	125	18	...	...	69	10.0
6063-T832	290	42	270	39	12	...	95	185	27	...	...	69	10.0
6066-0	150	22	85	12	...	18	43	95	14	...	...	69	10.0
6066-T4, T451	360	52	205	30	...	18	90	200	29	...	...	69	10.0
6066-T6, T651	395	57	360	52	...	12	120	235	34	110	16	69	10.0
6070-T6	380	55	350	51	10	...	...	235	34	95	14	69	10.0
6101-H111	95	14	75	11	...	...	...	...	...	...	...	69	10.0
6101-T6	220	32	195	28	15	...	71	140	20	...	...	69	10.0
6351-T4	250	36	150	22	20	...	...	...	...	...	...	69	10.0
6351-T6	310	45	285	41	14	...	95	200	29	90	13	69	10.0
6463-T1	150	22	90	13	20	...	42	95	14	70	10	69	10.0
6463-T5	185	27	145	21	12	...	60	115	17	70	10	69	10.0
6463-T6	240	35	215	31	12	...	74	150	22	70	10	69	10.0
7049-T73	515	75	450	65	...	12	135	305	44	...	...	72	10.4
7049-T7352	515	75	435	63	...	11	135	295	43	...	...	72	10.4
7050-T73510, T73511	495	72	435	63	...	12	...	...	...	...	...	72	10.4
7050-T7451(h)	525	76	470	68	...	11	...	305	44	...	...	72	10.4
7050-T7651	550	80	490	71	...	11	...	325	47	...	...	72	10.4
7075-0	230	33	105	15	17	16	60	150	22	...	...	72	10.4
7075-T6, T651	570	83	505	73	11	11	150	330	48	160	23	72	10.4
Alclad 7075-0	220	32	95	14	17	...	...	150	22	...	...	72	10.4
Alclad 7075-T6, T651	525	76	460	67	11	...	...	315	46	...	...	72	10.4

(a) 500 kg load and 10 mm ball. (b) Based on 500 000 000 cycles of completely reversed stress using the R.R. Moore type of machine and specimen. (c) Average of tension and compression moduli. Compression modulus is about 2% greater than tension modulus. (d) 1350-0 wire will have an elongation of approximately 23% in 250 mm (10 in.). (e) 1350-H19 wire will have an elongation of approximately 1½% in 250 mm (10 in.). (f) Tempers T361 and T861 were formerly designated T36 and T86, respectively. (g) Based on 10^7 cycles using flexural type testing of sheet specimens. (h) T7451, although not previously registered, has appeared in literature and in some specifications as T73651.

24 illustrates the effect of 3 to 8% Cu on an alloy of Al-0.3%Mn-0.2%Zr-0.1%V at room temperature and after exposure at 315 °C (600 °F) for two different periods of time. The stability of the properties should be noted, as reflected in the small reduction in strength with time at this temperature.

Gallium is an impurity in aluminum and is usually present at levels of 0.001 to 0.02%. At these levels its effect on mechanical properties is quite small. At the 0.2% level, gallium has been found to affect the corrosion characteristics and the response to etching and brightening of some alloys. Liquid gallium metal penetrates very rapidly at aluminum grain boundaries and can produce complete grain separation. In sacrificial anodes, an addition of gallium (0.01 to 0.1%) keeps the anode from passivating.

Hydrogen has a higher solubility in the liquid state at the melting point than in the solid at the same temperature. Because of this, gas porosity can form during solidification. Hydrogen is produced by the reduction of water vapor in the atmosphere by aluminum and by the decomposition of hydrocarbons. Hydrogen pickup in both solid and liquid aluminum is enhanced by the presence of certain impurities, such as sulfur compounds, on the surface and in the atmosphere. Hydride-forming elements in the metal increase the pickup of hydrogen in the liquid. Other elements such as beryllium, copper, tin, and silicon decrease hydrogen pickup.

In addition to causing primary porosity in casting, hydrogen causes secondary porosity, blistering, and high-temperature deterioration (advanced internal gas precipitation) during heat treating. It probably plays a role in grain-boundary decohesion during SCC. Its level in melts is controlled by fluxing with hydrogen-free gases or by vacuum degassing.

Indium. Small amounts (0.05 to 0.2%) of indium have a marked influence on the age hardening of aluminum-copper alloys, particularly at low copper contents (2 to 3% Cu). In this respect, indium acts very much like cadmium in that it reduces room-tem-

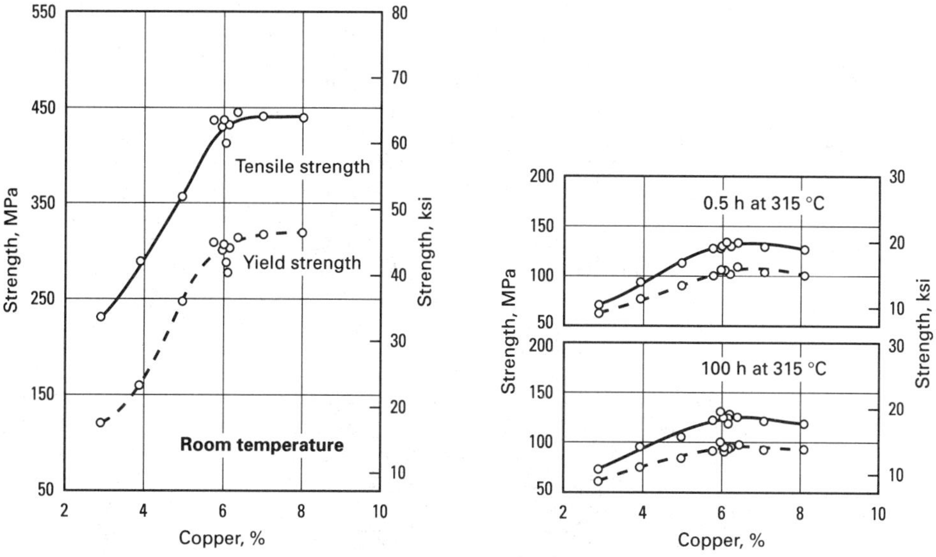

Fig. 24 Variation of tensile properties with copper content in Al-0.3%Mn-0.2%Zr-0.1%V alloy in the T6 temper

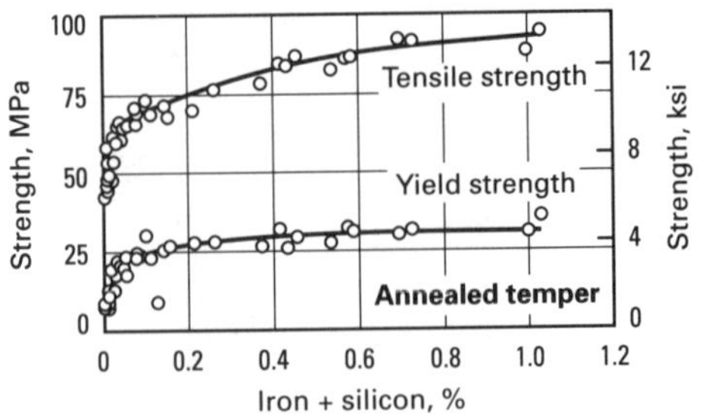

Fig. 25 Effect of iron plus silicon impurities on tensile strength and yield strength of aluminum

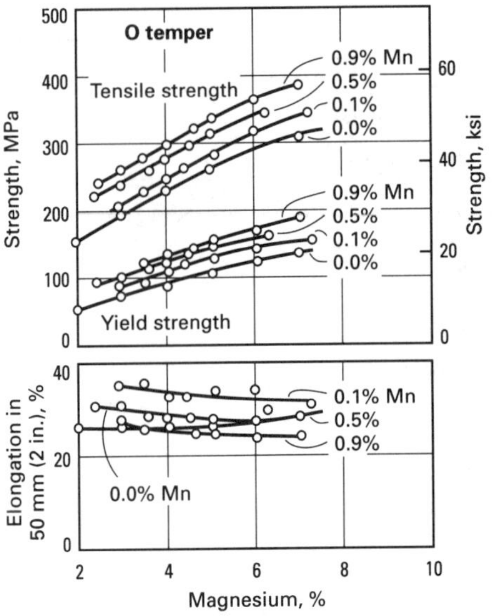

Fig. 26 Tensile properties of 13 mm (0.5 in.) aluminum-magnesium-manganese plate in O temper

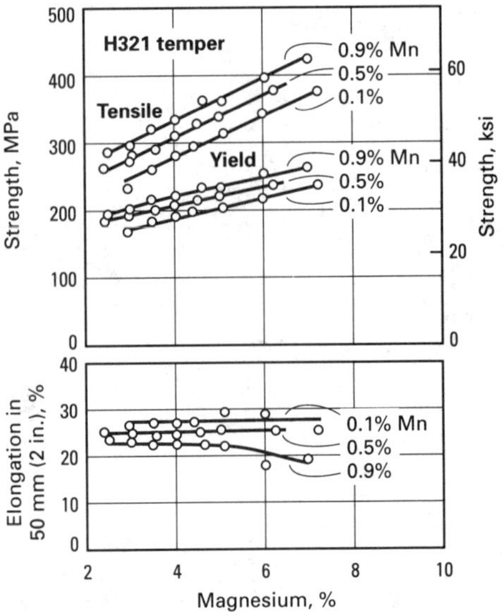

Fig. 27 Tensile properties of 13 mm (0.5 in.) aluminum-magnesium-manganese plate in H321 temper

perature aging but increases artificial aging. The addition of magnesium decreases the effect of indium. Small amounts of indium (0.03 to 0.5%) are claimed to be beneficial in aluminum-cadmium-bearing alloys.

Iron is the most common impurity found in aluminum. It has a high solubility in molten aluminum and is therefore easily dissolved at all molten stages of production. The solubility of iron in the solid state is very low (~0.04%) and therefore, most of the iron present in aluminum over this amount appears as an intermetallic second phase in combination with aluminum and often other elements. Because of its limited solubility, it is used in electrical conductors in which it provides a slight increase in strength (Fig. 25) and better creep characteristics at moderately elevated temperatures.

Iron reduces the grain size in wrought products. Alloys of iron and manganese near the ternary eutectic content, such as 8006, can have useful combinations of strength and ductility at room temperature and retain strength at elevated temperatures. The properties are due to the fine grain size that is stabilized by the finely dispersed iron-rich second phase. Iron is added to the aluminum-copper-nickel group of alloys to increase strength at elevated temperatures.

Lead. Normally present only as a trace element in commercial-purity aluminum, lead is added at about the 0.5% level with the same amount as bismuth in some alloys (2011 and 6262) to improve machinability. Additions of lead may be troublesome to the fabricator as it will tend to segregate during casting and cause hot shortness in aluminum-copper-magnesium alloys. Lead compounds are toxic.

Lithium. The impurity level of lithium is of the order of a few ppm, but at a level of less than 5 ppm it can promote the discoloration (blue corrosion) of aluminum foil under humid conditions. Traces of lithium greatly increase the oxidation rate of molten aluminum and alter the surface characteristics of wrought products. Binary aluminum-lithium alloys age harden but are not used commercially. Present interest is on the aluminum-copper-magnesium-lithium alloys, which can be heat treated to strengths comparable to present aircraft alloys (see the article "Aluminum-Lithium Alloys" in this Volume). In addition, the density is decreased and the modulus is increased. This type of alloy has a high volume fraction of coherent, ordered $LiAl_3$ precipitate. In addition to increasing the elastic modulus, the fatigue crack growth resistance is increased at intermediate levels of stress intensity.

Magnesium is the major alloying element in the 5*xxx* series of alloys. Its maximum solid solubility in aluminum is 17.4%, but the magnesium content in current wrought alloys does not exceed 5.5%. Magnesium precipitates preferentially at grain boundaries as a highly anodic phase (Mg_5Al_3 or Mg_5Al_8), which produces susceptibility to intergranular cracking and to stress corrosion. Wrought alloys containing up to 5% Mg properly fabricated are stable under normal usage. The addition of magnesium markedly increases the strength of aluminum without unduly decreasing the ductility. Corrosion resistance and weldability are good. In the annealed condition, magnesium alloys form Lüder lines during deformation.

Magnesium-Manganese. In wrought alloys, this system has high strength in the work-hardened condition, high resistance to corrosion, and good welding characteristics. Increasing amounts of either magnesium or manganese intensify the difficulty of fabrication and increase the tendency toward cracking during hot rolling, particularly if traces of sodium are present. The two main advantages of manganese additions are that the precipitation of the magnesium phase is more general throughout the structure, and that for a given increase in strength, manganese allows a lower magnesium content and ensures a greater degree of stability to the alloy.

The tensile properties of 13 mm (0.5 in.) plate at various magnesium and manganese concentrations are shown in Fig. 26 for the O temper and in Fig. 27 for a work-hardened temper. Increasing magnesium raises the tensile strength by about 35 MPa (5 ksi) for each 1% increment; manganese is about twice as effective as magnesium.

Magnesium-Silicide. Wrought alloys of the 6*xxx* group contain up to 1.5% each of magnesium and silicon in the approximate ratio to form Mg_2Si, that is, 1.73:1. The maximum solubility of Mg_2Si is 1.85%, and this decreases with temperature. Precipitation upon age hardening occurs by formation of Guinier-Preston zones and a very fine precipitate. Both confer an increase in strength to these alloys, though not as great as in the case of the 2*xxx* or the 7*xxx* alloys.

Al-Mg_2Si alloys can be divided into three groups. In the first group, the total amount of magnesium and silicon does not exceed 1.5%. These elements are in a nearly balanced ratio or with a slight excess of silicon. Typical of this group is 6063, widely used for extruded architectural sections. This easily extrudable alloy nominally contains 1.1% Mg_2Si. Its solution heat-treating temperature of just over 500 °C (930 °F) and its low quench sensitivity are such that this alloy does not need a separate solution treatment after extrusion but may be air quenched at the press and artificially aged to achieve moderate strength, good ductility, and excellent corrosion resistance.

The second group nominally contains 1.5% or more of magnesium + silicon and other additions such as 0.3% Cu, which increases strength in the T6 temper. Ele-

Table 9 Recommended minimum bend radii for 90° cold forming of sheet and plate

Alloy	Temper	0.4 mm (1/64 in.)	0.8 mm (1/32 in.)	1.6 mm (1/16 in.)	3.2 mm (1/8 in.)	4.8 mm (3/16 in.)	6.35 mm (1/4 in.)	9.5 mm (3/8 in.)	13 mm (1/2 in.)
		Radii(a)(b)(c)(d) for various thicknesses expressed in terms of thickness, t							
1100	0	0	0	0	0	½t	1t	1t	1½t
	H12	0	0	0	½t	1t	1t	1½t	2t
	H14	0	0	0	1t	1t	1½t	2t	2½t
	H16	0	½t	1t	1½t	...	...	...	...
	H18	1t	1t	1½t	2½t	...	...	...	...
2014	0	...	0	0	½t	1t	1t	2½t	4t
	T3	...	2½t	3t	4t	5t	5t	...	...
	T4	...	2½t	3t	4t	5t	5t	...	...
	T6	...	4t	4t	5t	6t	8t	...	...
2024	0	0	0	0	½t	1t	1t	2½t	4t
	T3	2½t	3t	4t	5t	5t	6t	...	...
	T361(d)	...	4t	5t	6t	6t	8t	8½t	9½t
	T4	2½t	3t	4t	5t	5t	6t	...	...
	T81	4½t	5½t	6t	7½t	8t	9t	...	...
	T861(d)	...	6t	7t	8½t	9½t	10t	11½t	11½t
2036	T4	...	1t	1t	...	...	...	...	...
3003	0	0	0	0	0	½t	1t	1t	1½t
	H12	0	0	0	½t	1t	1t	1½t	2t
	H14	0	0	0	1t	1t	1½t	2t	2½t
	H16	½t	1t	1t	1½t	...	...	...	...
	H18	1t	1½t	2t	2½t	...	...	...	...
3004	0	0	0	0	½t	1t	1t	1t	1½t
	H32	0	0	½t	1t	1t	1½t	1½t	2t
	H34	0	1t	1t	1½t	1½t	2½t	2½t	3t
	H36	1t	1t	1½t	2½t	...	...	...	...
	H38	1t	1½t	2½t	3t	...	...	...	...
3105	H25	½t	½t	½t	...	...	...	...	...
5005	0	0	0	0	0	½t	1t	1t	1½t
	H12	0	0	0	½t	1t	1t	1½t	2t
	H14	0	0	0	1t	1½t	1½t	2t	2½t
	H16	½t	1t	1t	1½t	...	...	...	...
	H18	1t	1½t	2t	2½t	...	...	...	...
	H32	0	0	0	½t	1t	1t	1½t	2t
	H34	0	0	0	1t	1½t	1½t	2t	2½t
	H36	½t	1t	1t	1½t	...	...	...	...
	H38	1t	1½t	2t	2½t	...	...	...	...
5050	0	0	0	0	½t	1t	1t	1½t	1½t
	H32	0	0	0	1t	1t	1½t	...	...
	H34	0	0	1t	1½t	1½t	2t	...	...
	H36	1t	1t	1½t	2t	...	...	...	...
	H38	1t	1½t	2½t	3t	...	...	...	...
5052	0	0	0	0	½t	1t	1t	1½t	1½t
	H32	0	0	1t	1½t	1½t	1½t	1½t	2t

Alloy	Temper	0.4 mm (1/64 in.)	0.8 mm (1/32 in.)	1.6 mm (1/16 in.)	3.2 mm (1/8 in.)	4.8 mm (3/16 in.)	6.35 mm (1/4 in.)	9.5 mm (3/8 in.)	13 mm (1/2 in.)
		Radii(a)(b)(c)(d) for various thicknesses expressed in terms of thickness, t							
5052	H34	0	1t	1½t	2t	2t	2½t	2½t	3t
	H36	1t	1t	1½t	2½t	...	...	...	...
	H38	1t	1½t	2½t	3t	...	...	...	...
5083	0	...	...	½t	1t	1t	1t	1½t	1½t
	H321	...	...	...	...	1½t	1½t	2t	2½t
5086	0	...	0	½t	1t	1t	1t	1½t	1½t
	H32	...	½t	1t	1½t	1½t	2t	2½t	3t
	H34	½t	1t	1½t	2t	2½t	3t	3½t	4t
	H36	1½t	2t	2½t	...	...	...	...	...
5154	...	...	0	½t	1t	1t	1t	1½t	1½t
	H32	...	½t	1t	1½t	1½t	2t	2½t	3½t
	H34	½t	1t	1½t	2t	2½t	3t	3½t	4t
	H36	1t	1½t	2t	3t	...	...	...	...
	H38	1½t	2½t	3t	4t	...	...	...	...
5252	H25	...	0	1t	...	...	...	...	...
	H28	...	1½t	2½t	...	...	...	...	...
5254	0	...	0	½t	1t	1t	1t	1½t	1½t
	H32	...	½t	1t	1½t	1½t	2t	2½t	3½t
	H34	½t	1t	1½t	2t	2½t	3t	3½t	4t
	H36	1t	1½t	2t	3t	...	...	...	...
	H38	1½t	2½t	3t	4t	...	...	...	...
5454	0	...	½t	1t	1t	1t	1½t	1½t	2t
	H32	...	½t	1t	2t	2t	2½t	3t	4t
	H34	...	1t	1½t	2t	2½t	3t	3½t	4t
5456	0	...	...	1t	1t	1½t	1½t	2t	2t
	H321	...	...	...	...	2t	2½t	3t	3½t
5457	0	...	0	0	...	...	...	...	...
5652	0	0	0	0	½t	1t	1t	1½t	1½t
	H32	0	0	1t	1½t	1½t	1½t	1½t	2t
	H34	0	1t	1½t	2t	2t	2½t	2½t	3t
	H36	1t	1t	1½t	2½t	...	...	...	...
	H38	1t	1½t	2½t	3t	...	...	...	...
5657	H25	...	0	0	...	...	...	...	...
	H28	...	1½t	2½t	...	...	...	...	...
6061	0	0	0	0	1t	1t	1t	1½t	2t
	T4	0	0	1t	1½t	2½t	3t	...	...
	T6	1t	1t	1½t	2½t	3t	3½t	...	...
7050	T7	...	...	...	...	...	8t	9t	9½t
7072	0	0	0	...	...	...	...	...	...
	H14	0	0	...	...	...	...	...	...
	H18	1t	1t	...	...	...	...	...	...
7075	0	0	0	1t	1t	1½t	2½t	3½t	4t
	T6	3t	4t	5t	6t	6t	8t	9t	9½t

(a) The radii listed are the minimum recommended for bending sheets and plates without fracturing in a standard press brake with air bend dies. Other types of bending operations may require larger radii or permit smaller radii. The minimum permissible radii will also vary with the design and condition of the tooling. (b) Alclad sheet in the heat-treatable alloys can be bent over slightly smaller radii than the corresponding tempers of the bare alloy. (c) Heat-treatable alloys can be formed over appreciably smaller radii immediately after solution heat treatment. (d) The H112 temper (applicable to non-heat-treatable alloys) is supplied in the as-fabricated condition without special property control but usually can be formed over radii applicable to the H14 (or H34) temper or smaller. (e) Tempers T361 and T861 formerly designated T36 and T86, respectively

ments such as manganese, chromium, and zirconium are used for controlling grain structure. Alloys of this group, such as the structural alloy 6061, achieve strengths about 70 MPa (10 ksi) higher than in the first group in the T6 temper. Alloys of the second group require a higher solution-treating temperature than the first and are quench sensitive. Therefore, they generally require a separate solution treatment followed by rapid quenching and artificial aging.

The third group contains an amount of Mg_2Si overlapping the first two but with a substantial excess silicon. An excess of 0.2% Si increases the strength of an alloy containing 0.8% Mg_2Si by about 70 MPa (10 ksi). Larger amounts of excess silicon are less beneficial. Excess magnesium, however, is of benefit only at low Mg_2Si contents because magnesium lowers the solubility of Mg_2Si. In excess silicon alloys, segregation of silicon to the boundary causes grain-boundary fracture in recrystallized structures. Additions of manganese, chromium, or zirconium counteract the effect of silicon by preventing recrystallization during heat treatment. Common alloys of this group are 6351 and the more recently introduced alloys 6009 and 6010. Additions of lead and bismuth to an alloy of this series (6262) improves machinability. This alloy has a better corrosion resistance than 2011, which also is used as a free-machining alloy.

Manganese is a common impurity in primary aluminum, in which its concentration normally ranges from 5 to 50 ppm. It decreases resistivity. Manganese increases strength either in solid solution or as a finely precipitated intermetallic phase. It has no adverse effect on corrosion resistance. Manganese has a very limited solid solubility in aluminum in the presence of normal impurities but remains in solution when chill cast so that most of the manganese added is substantially retained in solution, even in large ingots. As an addition, it is used to increase strength and to control the grain structure (Fig. 28). The effect of manganese is to increase the recrystallization temperature and to promote the formation of fibrous structure upon hot working. As a dispersed precipitate it is effective in slowing recovery and in preventing grain growth. The manganese precipitate increases the quench sensitivity of heat-treatable alloys.

Manganese is also used to correct the shape of acicular or of platelike iron constituents and to decrease their embrittling effect. Up to the 1.25% level, manganese is the main alloying addition of the 3*xxx* series of alloys, in which it is added alone or with magnesium. This series of alloys is used in large tonnages for beverage containers and general utility sheet. Even after high degrees of work hardening, these alloys are used to produce severely formed can bodies.

The combined content of manganese, iron, chromium, and other transition metals

Table 10 Minimum and typical room-temperature plane-strain fracture-toughness values for several high-strength aluminum alloys

Product form	Alloy and temper	Thickness, mm	Thickness, in.	Plane-strain fracture toughness (K_{Ic}): L-T direction(a), Minimum, MPa√m	L-T Minimum, ksi√in.	L-T Typical, MPa√m	L-T Typical, ksi√in.	T-L direction(b), Minimum, MPa√m	T-L Minimum, ksi√in.	T-L Typical, MPa√m	T-L Typical, ksi√in.	S-L direction(c), Minimum, MPa√m	S-L Minimum, ksi√in.	S-L Typical, MPa√m	S-L Typical, ksi√in.
Plate	7050-T7451	25.40–50.80	1.000–2.000	31.9	29.0	37	34	27.5	25.0	33	30	...	...	...	...
		50.83–76.20	2.001–3.000	29.7	27.0	36	33	26.4	24.0	32	29	23.1	21.0	28	25
		76.23–101.60	3.001–4.000	28.6	26.0	35	32	25.3	23.0	31	28	23.1	21.0	28	25
		101.63–127.00	4.001–5.000	27.5	25.0	32	29	24.2	22.0	29	26	23.1	21.0	28	25
		127.03–152.40	5.001–6.000	26.4	24.0	31	28	24.2	22.0	28	25	23.1	21.0	28	25
	7050-T7651	25.40–50.80	1.000–2.000	28.6	26.0	34	31	26.4	24.0	31	28	...	...	...	...
		50.83–76.20	2.001–3.000	26.4	24.0	...	...	25.3	23.0	...	...	22.0	20.0	26	24
	7475-T651			33.0	30.0	46	42	30.8	28.0	41	37	...	...	...	...
	7475-T7651			36.3	33.0	47	43	33.0	30.0	41	37	...	...	...	...
	7475-T7351			41.8	38.0	55	50	35.2	32.0	45	41	27.5	25.0	36	33
	7075-T651			...	...	29	26	...	...	25	23	...	...	20	18
	7075-T7651			...	...	30	27	...	...	24	22	...	...	20	18
	7075-T7351			...	...	32	30	...	...	29	26	...	...	20	18
	7079-T651			...	...	30	27	...	...	25	23	...	...	18	16
	2124-T851			26.4	24.0	32	29	22.0	20.0	26	24	19.8	18.0	26	24
	2024-T351			...	...	37	34	...	...	32	29	...	...	26	24
Die forgings	7050-T74, -T7452			27.5	25.0	38	35	20.9	19.0	32	29	20.9	19.0	29	26
	7175-T736, -T73652			29.7	27.0	38	35	23.1	21.0	34	31	23.1	21.0	31	28
	7075-T7352			...	...	32	29	...	...	30	27	...	...	29	26
Hand forgings	7050-T7452			29.7	27.0	36	33	18.7	17.0	28	25	17.6	16.0	29	26
	7075-T73, -T7352			...	...	42	38	...	...	28	25	...	...	28	25
	7175-T73652			33.0	30.0	40	36	27.5	25.0	30	27	23.1	21.0	28	25
	2024-T852			...	...	26	24	...	...	22	20	...	...	20	18
Extrusions	7050-T7651*x*			...	...	44	40	...	...	31	28	...	...	28	25
	7050-T7351*x*			...	...	...	...	...	...	...	...	...	...	...	...
	7075-T651*x*			...	...	34	31	...	...	22	20	...	...	20	18
	7075-T7351*x*			...	...	33	30	...	...	26	24	...	...	22	20
	7150-T7351*x*			24.2	22.0	31	28	...	...	...	...	...	...	...	...
	7175-T7351*x*			33.0	30.0	40	36	30.8	28.0	34	31	...	...	...	...

(a) L-T, crack plane and growth direction perpendicular to the rolling direction. (b) T-L, crack plane and growth direction parallel to the rolling direction. (c) S-L, short transverse fracture toughness

must be limited, otherwise large primary intermetallic crystals precipitate from the melt in the transfer system or in the ingot sump during casting. In alloys 3003 and 3004 the iron plus manganese content should be kept below about 2.0 and 1.7%, respectively, to prevent the formation of primary (Fe,Mn)Al_6 during casting.

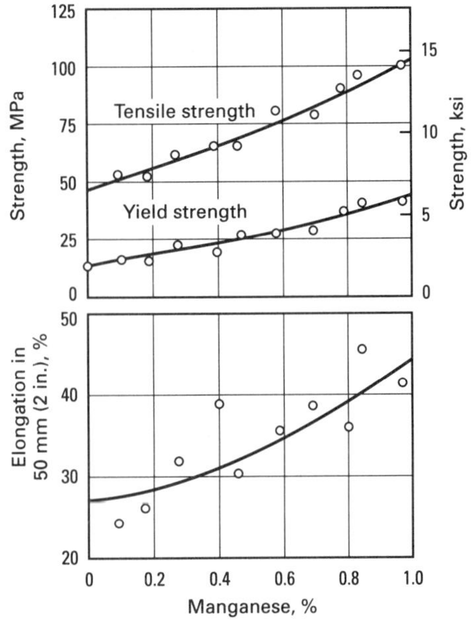

Fig. 28 Effect of manganese on tensile properties of wrought 99.95% Al, 1.6 mm (0.064 in.) thick specimens, quenched in cold water from 565 °C (1050 °F)

Mercury has been used at the level of 0.05% in sacrificial anodes used to protect steel structures. Other than for this use, mercury in aluminum or in contact with it as a metal or a salt will cause rapid corrosion of most aluminum alloys. The toxic properties of mercury must be kept in mind when adding it to aluminum alloys.

Molybdenum is a very low level (0.1 to 1.0 ppm) impurity in aluminum. It has been used at a concentration of 0.3% as a grain refiner, because the aluminum end of the equilibrium diagram is peritectic, and also as a modifier for the iron constituents, but it is not in current use for these purposes.

Nickel. The solid solubility of nickel in aluminum does not exceed 0.04%. Over this amount, it is present as an insoluble intermetallic, usually in combination with iron. Nickel (up to 2%) increases the strength of high-purity aluminum but reduces ductility. Binary aluminum-nickel alloys are no longer in use, but nickel is added to aluminum-copper and to aluminum-silicon alloys to improve hardness and strength at elevated temperatures and to reduce the coefficient of expansion. Nickel promotes pitting corrosion in dilute alloys such as 1100. It is limited in alloys for atomic reactor use, due to its high neutron absorption, but in other areas it is a desirable addition along with iron to improve corrosion resistance to high-pressure steam.

Niobium. As with other elements forming a peritectic reaction, niobium would be expected to have a grain refining effect on casting. It has been used for this purpose, but the effect is not marked.

Phosphorus is a minor impurity (1 to 10 ppm) in commercial-grade aluminum. Its solubility in molten aluminum is very low (~0.01% at 660 °C, or 1220 °F) and considerably smaller in the solid. Phosphorus is used as a modifier for hypereutectic aluminum-silicon alloys where aluminum-phosphide acts as nucleus for primary silicon, thus refining silicon and improving machinability. The aluminum-phosphorus compound reacts with water vapor to give phosphine (PH_3), but the level of phosphorus in aluminum is sufficiently low that this does not constitute a health hazard if adequate ventilation is used when machining phosphorus-nucleated castings. Phosphine can be a problem in furnace teardowns where phosphate-bonded refractories are used.

Silicon, after iron, is the highest impurity level in electrolytic commercial aluminum (0.01 to 0.15%). In wrought alloys, silicon is used with magnesium at levels up to 1.5% to produce Mg_2Si in the 6*xxx* series of heat-treatable alloys.

High-purity aluminum-silicon alloys are hot short up to 3% Si, the most critical range being 0.17 to 0.8% Si, but additions of silicon (0.5 to 4.0%) reduce the cracking tendency of aluminum-copper-magnesium

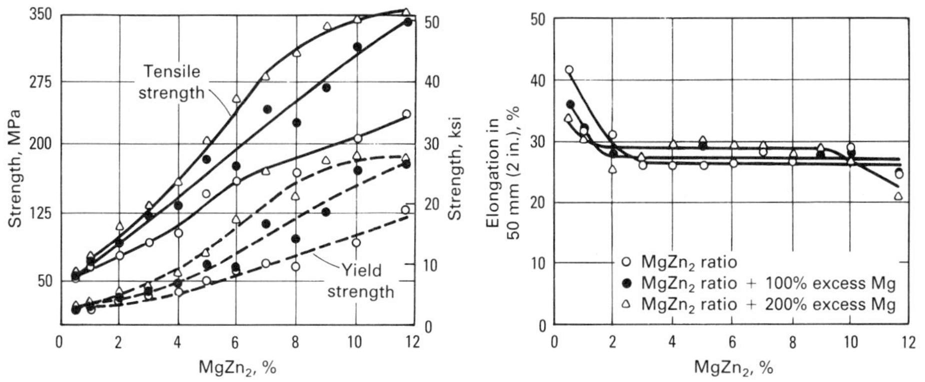

Fig. 29 Effect of $MgZn_2$ and $MgZn_2$ with excess magnesium on tensile properties of wrought 95% Al; 1.59 mm (0.0625 in.) specimens, quenched in cold water from 470 °C (875 °F)

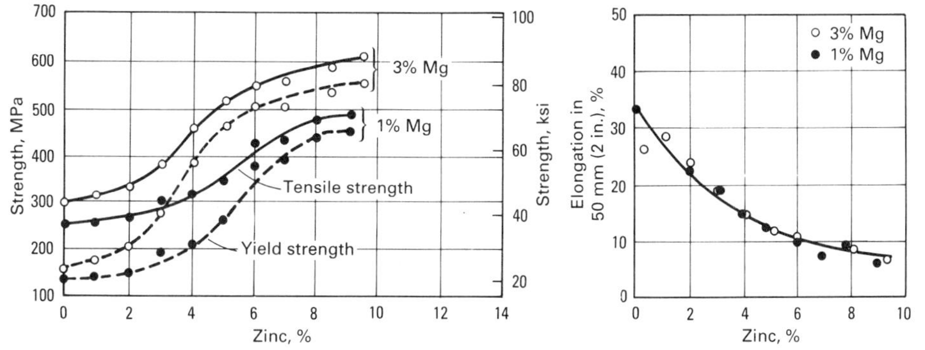

Fig. 30 Effect of zinc on aluminum alloy containing 1.5% Cu and 1 and 3% Mg; 1.6 mm (0.064 in.) thick sheet. Alloy with 1% Mg heat treated at 495 °C (920 °F); that with 3% Mg heat treated at 460 °C (860 °F). All specimens quenched in cold water, aged 12 h at 135 °C (275 °F)

alloys. Small amounts of magnesium added to any silicon-containing alloy will render it heat treatable, but the converse is not true as excess magnesium over that required to form Mg_2Si sharply reduces the solid solubility of this compound. Modification of the silicon can be achieved through the addition of sodium in eutectic and hypoeutectic alloys and by phosphorus in hypereutectic alloys. Up to 12% Si is added in wrought alloys used as cladding for brazing sheet. Alloys containing about 5% Si acquire a black color when anodized and are used for ornamental purposes.

Silver has an extremely high solid solubility in aluminum (up to 55%). Because of cost, no binary aluminum-silver alloys are in use, but small additions (0.1 to 0.6% Ag) are effective in improving the strength and stress-corrosion resistance of aluminum-zinc-magnesium alloys.

Strontium. Traces of strontium (0.01 to 0.1 ppm) are found in commercial-grade aluminum.

Sulfur. As much as 0.2 to 20 ppm sulfur are present in commercial-grade aluminum. It has been reported that sulfur can be used to modify both hypo- and hypereutectic aluminum-silicon alloys.

Tin is used as an alloying addition to aluminum—from concentrations of 0.03 to several percent in wrought alloys, to concentrations of about 25% in casting alloys. Small amounts of tin (0.05%) greatly increase the response of aluminum-copper alloys to artificial aging following a solution heat treatment. The result is an increase in strength and an improvement in corrosion resistance. Higher concentrations of tin cause hot cracking in aluminum-copper alloys. If small amounts of magnesium are present, the artificial aging characteristics are markedly reduced, probably because magnesium and tin form a noncoherent second phase.

The aluminum-tin bearing alloys, with additions of other metals such as copper, nickel, and silicon are used where bearings are required to withstand high speeds, loads, and temperatures. The copper, nickel, and silicon additions improve load-carrying capacity and wear resistance, and the soft tin phase provides antiscoring properties.

As little as 0.01% Sn in commercial-grade aluminum will cause surface darkening on annealing and increase the susceptibility to corrosion, which appears to be due to migration of tin to the surface. This effect may be reduced by small additions (0.2%) of copper. Aluminum-zinc alloys with small additions of tin are used as sacrificial anodes in salt water.

Titanium. Amounts of 10 to 100 ppm Ti are found in commercial-purity aluminum. Titanium depresses the electrical conductivity of aluminum, but its level can be reduced by the addition of boron to the melt to form insoluble TiB_2. Titanium is used primarily as a grain refiner of aluminum alloy castings and ingots. When used alone, the effect of titanium decreases with time of holding in the molten state and with repeated remelting. The grain-refining effect is enhanced if boron is present in the melt or if it is added as a master alloy containing boron largely combined as TiB_2. Titanium is a common addition to weld filler wire; it refines the weld structure and prevents weld cracking. It is usually added alone or with TiB_2 during the casting of sheet or extrusion ingots to refine the as-cast grain structure and to prevent cracking.

Vanadium. There is usually 10 to 200 ppm V in commercial-grade aluminum, and because it lowers conductivity, it generally is precipitated from electrical conductor alloys with boron. The aluminum end of the equilibrium diagram is peritectic, and therefore the intermetallic VAl_{11} would be expected to have a grain-refining effect upon solidification, but it is less efficient than titanium and zirconium. The recrystallization temperature is raised by vanadium.

Zinc. The aluminum-zinc alloys have been known for many years, but hot cracking of the casting alloys and the susceptibility to stress-corrosion cracking of the wrought alloys curtailed their use. Aluminum-zinc alloys containing other elements offer the highest combination of tensile properties in wrought aluminum alloys. Efforts to overcome the aforementioned limitations have been successful, and these aluminum-zinc base alloys are being used commercially to an increasing extent. The presence of zinc in aluminum increases its solution potential, hence its use in protective cladding (7072) and in sacrificial anodes.

Zinc-Magnesium. The addition of magnesium to the aluminum-zinc alloys develops the strength potential of this alloy system, especially in the range of 3 to 7.5% Zn. Magnesium and zinc form $MgZn_2$, which produces a far greater response to heat treatment than occurs in the binary aluminum-zinc system.

The strength of the wrought aluminum-zinc alloys also is substantially improved by the addition of magnesium. Increasing the $MgZn_2$ concentration from 0.5 to 12% in cold-water quenched 1.6 mm (0.062 in.) sheet continuously increases the tensile and yield strengths. The addition of magnesium in excess (100 and 200%) of that required to form $MgZn_2$ further increases tensile strength, as shown in Fig. 29.

On the negative side, increasing additions of both zinc and magnesium decrease the overall corrosion resistance of aluminum to the extent that close control over the microstructure, heat treatment, and composition are often necessary to maintain adequate

Table 11(a) Ultimate tensile strengths of various aluminum alloys at cryogenic and elevated temperatures

Alloy and temper	Ultimate tensile strength(a), MPa (ksi), at: −195 °C (−320 °F)	−80 °C (−112 °F)	0 °C (−18 °F)	24 °C (75 °F)	100 °C (212 °F)	150 °C (300 °F)	205 °C (400 °F)	260 °C (500 °F)	315 °C (600 °F)	370 °C (700 °F)
1100-0	172 (25)	103 (15)	97 (14)	90 (13)	70 (10)	55 (8)	40 (6)	28 (4)	20 (2.9)	14 (2.1)
1100-H14	207 (30)	138 (20)	130 (19)	125 (18)	110 (16)	97 (14)	70 (10)	28 (4)	20 (2.9)	14 (2.1)
1100-H18	235 (34)	180 (26)	172 (25)	165 (24)	145 (21)	125 (18)	40 (6)	28 (4)	20 (2.9)	14 (2.1)
2011-T3	. . .	. . .	. . .	380 (55)	325 (47)	193 (28)	110 (16)	45 (6.5)	21 (3.1)	16 (2.3)
2014-T6, T651	580 (84)	510 (74)	495 (72)	483 (70)	435 (63)	275 (40)	110 (16)	66 (9.5)	45 (6.5)	30 (4.3)
2017-T4, T451	550 (80)	448 (65)	440 (64)	427 (62)	393 (57)	275 (40)	110 (16)	62 (9)	40 (6)	30 (4.3)
2024-T3 (sheet)	585 (85)	503 (73)	495 (72)	483 (70)	455 (66)	380 (55)	185 (27)	75 (11)	52 (7.5)	35 (5)
2024-T4, T351 (plate)	580 (84)	490 (71)	475 (69)	470 (68)	435 (63)	310 (45)	180 (26)	75 (11)	52 (7.5)	35 (5)
2024-T6, T651	580 (84)	495 (72)	483 (70)	475 (69)	448 (65)	310 (45)	180 (26)	75 (11)	52 (7.5)	35 (5)
2024-T81, T851	585 (85)	510 (74)	503 (73)	483 (70)	455 (66)	380 (55)	185 (27)	75 (11)	52 (7.5)	35 (5)
2024-T861	635 (92)	558 (81)	538 (78)	517 (75)	483 (70)	372 (54)	145 (21)	75 (11)	52 (7.5)	35 (5)
2117-T4	385 (56)	310 (45)	303 (44)	295 (43)	248 (36)	207 (30)	110 (16)	52 (7.5)	32 (4.7)	20 (2.9)
2124-T851	593 (86)	525 (76)	503 (73)	483 (70)	455 (66)	372 (54)	185 (27)	75 (11)	52 (7.5)	38 (5.5)
2218-T61	495 (72)	420 (61)	407 (59)	407 (59)	385 (56)	283 (41)	152 (22)	70 (10)	38 (5.5)	28 (4)
2219-T62	503 (73)	435 (63)	415 (60)	400 (58)	372 (54)	310 (45)	235 (34)	185 (27)	70 (10)	30 (4.4)
2219-T81, T851	572 (83)	490 (71)	475 (69)	455 (66)	415 (60)	338 (49)	248 (36)	200 (29)	48 (7)	30 (4.4)
2618-T61	538 (78)	462 (67)	440 (64)	440 (64)	427 (62)	345 (50)	220 (32)	90 (13)	52 (7.5)	35 (5)
3003-0	228 (33)	138 (20)	117 (17)	110 (16)	90 (13)	75 (11)	59 (8.5)	40 (6)	28 (4)	19 (2.8)
3003-H14	240 (35)	165 (24)	152 (22)	152 (22)	145 (21)	125 (18)	97 (14)	52 (7.5)	28 (4)	19 (2.8)
3003-H18	283 (41)	220 (32)	207 (30)	200 (29)	180 (26)	160 (23)	97 (14)	52 (7.5)	28 (4)	19 (2.8)
3004-0	290 (42)	193 (28)	180 (26)	180 (26)	180 (26)	152 (22)	97 (14)	70 (10)	52 (7.5)	35 (5)
3004-H34	360 (52)	262 (38)	248 (36)	240 (35)	235 (34)	193 (28)	145 (21)	97 (14)	52 (7.5)	35 (5)
3004-H38	400 (58)	303 (44)	290 (42)	283 (41)	275 (40)	215 (31)	152 (22)	83 (12)	52 (7.5)	35 (5)
4032-T6	455 (66)	400 (58)	385 (56)	380 (55)	345 (50)	255 (37)	90 (13)	55 (8)	35 (5)	23 (3.4)
5050-0	255 (37)	152 (22)	145 (21)	145 (21)	145 (21)	130 (19)	97 (14)	62 (9)	40 (6)	27 (3.9)
5050-H34	303 (44)	207 (30)	193 (28)	193 (28)	193 (28)	172 (25)	97 (14)	62 (9)	40 (6)	27 (3.9)
5050-H38	317 (46)	235 (34)	220 (32)	220 (32)	215 (31)	185 (27)	97 (14)	62 (9)	40 (6)	27 (3.9)
5052-0	303 (44)	200 (29)	193 (28)	193 (28)	193 (28)	160 (23)	117 (17)	83 (12)	52 (7.5)	35 (5)
5052-H34	380 (55)	275 (40)	262 (38)	262 (38)	262 (38)	207 (30)	165 (24)	83 (12)	52 (7.5)	35 (5)
5052-H38	415 (60)	303 (44)	290 (42)	290 (42)	275 (40)	235 (34)	172 (25)	83 (12)	52 (7.5)	35 (5)
5083-0	407 (59)	295 (43)	290 (42)	290 (42)	275 (40)	215 (31)	152 (22)	117 (17)	75 (11)	40 (6)
5086-0	380 (55)	270 (39)	262 (38)	262 (38)	262 (38)	200 (29)	152 (22)	117 (17)	75 (11)	40 (6)
5154-0	360 (52)	248 (36)	240 (35)	240 (35)	240 (35)	200 (29)	152 (22)	117 (17)	75 (11)	40 (6)
5254-0	360 (52)	248 (36)	240 (35)	240 (35)	240 (35)	200 (29)	152 (22)	117 (17)	75 (11)	40 (6)
5454-0	372 (54)	255 (37)	248 (36)	248 (36)	248 (36)	200 (29)	152 (22)	117 (17)	75 (11)	40 (6)
5454-H32	407 (59)	290 (42)	283 (41)	275 (40)	270 (39)	220 (32)	172 (25)	117 (17)	75 (11)	40 (6)
5454-H34	435 (63)	317 (46)	303 (44)	303 (44)	295 (43)	235 (34)	180 (26)	117 (17)	75 (11)	40 (6)
5456-0	427 (62)	317 (46)	310 (45)	310 (45)	290 (42)	215 (31)	152 (22)	117 (17)	75 (11)	40 (6)
5652-0	303 (44)	200 (29)	193 (28)	193 (28)	193 (28)	160 (23)	117 (17)	83 (12)	52 (7.5)	35 (5)
5652-H34	380 (55)	275 (40)	262 (38)	262 (38)	262 (38)	207 (30)	165 (24)	83 (12)	52 (7.5)	35 (5)
5652-H38	415 (60)	303 (44)	290 (42)	290 (42)	275 (40)	235 (34)	172 (25)	83 (12)	52 (7.5)	35 (5)
6053-T6, T651	. . .	. . .	. . .	255 (37)	220 (32)	172 (25)	90 (13)	38 (5.5)	28 (4)	20 (2.9)
6061-T6, T651	415 (60)	338 (49)	325 (47)	310 (45)	290 (42)	235 (34)	130 (19)	52 (7.5)	32 (4.6)	21 (3)
6063-T1	235 (34)	180 (26)	165 (24)	152 (22)	152 (22)	145 (21)	62 (9)	31 (4.5)	22 (3.2)	16 (2.3)
6063-T5	255 (37)	200 (29)	193 (28)	185 (27)	165 (24)	138 (20)	62 (9)	31 (4.5)	22 (3.2)	16 (2.3)
6063-T6	325 (47)	262 (38)	248 (36)	240 (35)	215 (31)	145 (21)	62 (9)	31 (4.5)	22 (3.2)	16 (2.3)
6101-T6	295 (43)	248 (36)	235 (34)	220 (32)	193 (28)	145 (21)	70 (10)	33 (4.8)	21 (3)	17 (2.5)
6151-T6	393 (57)	345 (50)	338 (49)	330 (48)	295 (43)	193 (28)	97 (14)	45 (6.5)	35 (5)	28 (4)
6262-T651	415 (60)	338 (49)	325 (47)	310 (45)	290 (42)	235 (34)	. . .	. . .	. . .	. . .
6262-T9	510 (74)	427 (62)	415 (60)	400 (58)	365 (53)	262 (38)	103 (15)	59 (8.5)	32 (4.6)	21 (3)
7075-T6, T651	703 (102)	620 (90)	593 (86)	572 (83)	483 (70)	215 (31)	110 (16)	75 (11)	55 (8)	40 (6)
7075-T73, T7351	635 (92)	545 (79)	525 (76)	503 (73)	435 (63)	215 (31)	110 (16)	75 (11)	55 (8)	40 (6)
7178-T6, T651	730 (106)	648 (94)	627 (91)	607 (88)	503 (73)	215 (31)	103 (15)	75 (11)	59 (8.5)	45 (6.5)
7178-T76, T7651	730 (106)	627 (91)	607 (88)	572 (83)	475 (69)	215 (31)	103 (15)	75 (11)	59 (8.5)	45 (6.5)

(a) These data are based on a limited amount of testing and represent the lowest strength during 10 000 h of exposure at testing temperature under no load; stress applied at 34 MPa/min (5000 psi/min) to yield strength and then at strain rate of 0.05 mm/mm/min (0.05 in./in./min) to failure. Under some conditions of temperature and time, the application of heat will adversely affect certain other properties of some alloys.

resistance to stress corrosion and to exfoliatory attack. For example, depending upon the alloy, stress corrosion is controlled by some or all of the following:

- Overaging
- Cooling rate after solution treatment
- Maintaining a nonrecrystallized structure through the use of additions such as zirconium
- Copper or chromium additions (see zinc-magnesium-copper alloys)
- Adjusting the zinc-magnesium ratio closer to 3:1

Zinc-Magnesium-Copper. The addition of copper to the aluminum-zinc-magnesium system, together with small but important amounts of chromium and manganese, results in the highest-strength aluminum-base alloys commercially available. The properties of a representative group of these compositions, after one of several solution and aging treatments to which they respond, are shown in Fig. 30.

In this alloy system, zinc and magnesium control the aging process. The effect of copper is to increase the aging rate by increasing the degree of supersaturation and perhaps through nucleation of the $CuMgAl_2$ phase. Copper also increases quench sensitivity upon heat treatment. In general, copper reduces the resistance to general corrosion of aluminum-zinc-magnesium alloys, but increases the resistance to stress corrosion. The minor alloy additions, such as chromium and zirconium, have a marked effect on mechanical properties and corrosion resistance.

Zirconium additions in the range 0.1 to 0.3% are used to form a fine precipitate of intermetallic particles that inhibit recovery and recrystallization. An increasing number of alloys, particularly in the aluminum-zinc-magnesium family, use zirconium additions to increase the recrystallization temperature and to control the grain structure in wrought products. Zirconium additions leave this family of alloys less quench sen-

Table 11(b) Tensile yield strengths of various aluminum alloys at cryogenic and elevated temperatures

Alloy and temper	0.2% offset yield strength(a), MPa (ksi), at:									
	−195 °C (−320 °F)	−80 °C (−112 °F)	0 °C (−18 °F)	24 °C (75 °F)	100 °C (212 °F)	150 °C (300 °F)	205 °C (400 °F)	260 °C (500 °F)	315 °C (600 °F)	370 °C (700 °F)
1100-0	40 (6)	38 (5.5)	35 (5)	35 (5)	32 (4.6)	29 (4.2)	24 (3.5)	18 (2.6)	14 (2)	11 (1.6)
1100-H14	138 (20)	125 (18)	117 (17)	117 (17)	103 (15)	83 (12)	52 (7.5)	18 (2.6)	14 (2)	11 (1.6)
1100-H18	180 (26)	160 (23)	160 (23)	152 (22)	130 (19)	97 (14)	24 (3.5)	18 (2.6)	14 (2)	11 (1.6)
2011-T3	...	...	...	295 (43)	235 (34)	130 (19)	75 (11)	26 (3.8)	12 (1.8)	10 (1.4)
2014-T6, T651	495 (72)	448 (65)	427 (62)	415 (60)	393 (57)	240 (35)	90 (13)	52 (7.5)	35 (5)	24 (3.5)
2017-T4, T451	365 (53)	290 (42)	283 (41)	275 (40)	270 (39)	207 (30)	90 (13)	52 (7.5)	35 (5)	24 (3.5)
2024-T3 (sheet)	427 (62)	360 (52)	352 (51)	345 (50)	330 (48)	310 (45)	138 (20)	62 (9)	40 (6)	28 (4)
2024-T4, T351 (plate)	420 (61)	338 (49)	325 (47)	325 (47)	310 (45)	248 (36)	130 (19)	62 (9)	40 (6)	28 (4)
2024-T6, T651	470 (68)	407 (59)	400 (58)	393 (57)	372 (54)	248 (36)	130 (19)	62 (9)	40 (6)	28 (4)
2024-T81, T851	538 (78)	475 (69)	470 (68)	448 (65)	427 (62)	338 (49)	138 (20)	62 (9)	40 (6)	28 (4)
2024-T861	585 (85)	530 (77)	510 (74)	490 (71)	462 (67)	330 (48)	117 (17)	62 (9)	40 (6)	28 (4)
2117-T4	228 (33)	172 (25)	165 (24)	165 (24)	145 (21)	117 (17)	83 (12)	38 (5.5)	23 (3.3)	14 (2)
2124-T851	545 (79)	490 (71)	470 (68)	440 (64)	420 (61)	338 (49)	138 (20)	55 (8)	40 (6)	28 (4.1)
2218-T61	360 (52)	310 (45)	303 (44)	303 (44)	290 (42)	240 (35)	110 (16)	40 (6)	20 (3)	17 (2.5)
2219-T62	338 (49)	303 (44)	290 (42)	275 (40)	255 (37)	228 (33)	172 (25)	138 (20)	55 (8)	26 (3.7)
2219-T81, T851	420 (61)	372 (54)	360 (52)	345 (50)	325 (47)	275 (40)	200 (29)	160 (23)	40 (6)	26 (3.7)
2618-T61	420 (61)	380 (55)	372 (54)	372 (54)	372 (54)	303 (44)	180 (26)	62 (9)	31 (4.5)	24 (3.5)
3003-0	59 (8.5)	48 (7)	45 (6.5)	40 (6)	38 (5.5)	35 (5)	30 (4.3)	23 (3.4)	17 (2.4)	12 (1.8)
3003-H14	172 (25)	152 (22)	145 (21)	145 (21)	130 (19)	110 (16)	62 (9)	28 (4)	17 (2.4)	12 (1.8)
3003-H18	228 (33)	200 (29)	193 (28)	185 (27)	145 (21)	110 (16)	62 (9)	28 (4)	17 (2.4)	12 (1.8)
3004-0	90 (13)	75 (11)	70 (10)	70 (10)	70 (10)	70 (10)	66 (9.5)	52 (7.5)	35 (5)	20 (3)
3004-H34	235 (34)	207 (30)	200 (29)	200 (29)	200 (29)	172 (25)	103 (15)	52 (7.5)	35 (5)	20 (3)
3004-H38	295 (43)	262 (38)	248 (36)	248 (36)	248 (36)	185 (27)	103 (15)	52 (7.5)	35 (5)	20 (3)
4032-T6	330 (48)	317 (46)	317 (46)	317 (46)	303 (44)	228 (33)	62 (9)	38 (5.5)	22 (3.2)	14 (2)
5050-0	70 (10)	59 (8.5)	55 (8)	55 (8)	55 (8)	55 (8)	52 (7.5)	40 (6)	29 (4.2)	18 (2.6)
5050-H34	207 (30)	172 (25)	165 (24)	165 (24)	165 (24)	152 (22)	52 (7.5)	40 (6)	29 (4.2)	18 (2.6)
5050-H38	248 (36)	207 (30)	200 (29)	200 (29)	200 (29)	172 (25)	52 (7.5)	40 (6)	29 (4.2)	18 (2.6)
5052-0	110 (16)	90 (13)	90 (13)	90 (13)	90 (13)	90 (13)	75 (11)	52 (7.5)	38 (5.5)	21 (3.1)
5052-H34	248 (36)	220 (32)	215 (31)	215 (31)	215 (31)	185 (27)	103 (15)	52 (7.5)	38 (5.5)	21 (3.1)
5052-H38	303 (44)	262 (38)	255 (37)	255 (37)	248 (36)	193 (28)	103 (15)	52 (7.5)	38 (5.5)	21 (3.1)
5083-0	165 (24)	145 (21)	145 (21)	145 (21)	145 (21)	130 (19)	117 (17)	75 (11)	52 (7.5)	29 (4.2)
5086-0	130 (19)	117 (17)	117 (17)	117 (17)	117 (17)	110 (16)	103 (15)	75 (11)	52 (7.5)	29 (4.2)
5154-0	130 (19)	117 (17)	117 (17)	117 (17)	117 (17)	110 (16)	103 (15)	75 (11)	52 (7.5)	29 (4.2)
5254-0	130 (19)	117 (17)	117 (17)	117 (17)	117 (17)	110 (16)	103 (15)	75 (11)	52 (7.5)	29 (4.2)
5454-0	130 (19)	117 (17)	117 (17)	117 (17)	117 (17)	110 (16)	103 (15)	75 (11)	52 (7.5)	29 (4.2)
5454-H32	248 (36)	215 (31)	207 (30)	207 (30)	200 (29)	180 (26)	130 (19)	75 (11)	52 (7.5)	29 (4.2)
5454-H34	283 (41)	248 (36)	240 (35)	240 (35)	235 (34)	193 (28)	130 (19)	75 (11)	52 (7.5)	29 (4.2)
5456-0	180 (26)	160 (23)	160 (23)	160 (23)	152 (22)	138 (20)	117 (17)	75 (11)	52 (7.5)	29 (4.2)
5652-0	110 (16)	90 (13)	90 (13)	90 (13)	90 (13)	90 (13)	75 (11)	52 (7.5)	38 (5.5)	21 (3.1)
5652-H34	248 (36)	220 (32)	215 (31)	215 (31)	215 (31)	185 (27)	103 (15)	52 (7.5)	38 (5.5)	21 (3.1)
5652-H38	303 (44)	262 (38)	255 (37)	255 (37)	248 (36)	193 (28)	103 (15)	52 (7.5)	38 (5.5)	21 (3.1)
6053-T6, T651	...	...	...	220 (32)	193 (28)	165 (24)	83 (12)	28 (4)	19 (2.7)	14 (2)
6061-T6, T651	325 (47)	290 (42)	283 (41)	275 (40)	262 (38)	215 (31)	103 (15)	35 (5)	19 (2.7)	12 (1.8)
6063-T1	110 (16)	103 (15)	97 (14)	90 (13)	97 (14)	103 (15)	45 (6.5)	24 (3.5)	17 (2.5)	14 (2)
6063-T5	165 (24)	152 (22)	152 (22)	145 (21)	138 (20)	125 (18)	45 (6.5)	24 (3.5)	17 (2.5)	14 (2)
6063-T6	248 (36)	228 (33)	220 (32)	215 (31)	193 (28)	138 (20)	45 (6.5)	24 (3.5)	17 (2.5)	14 (2)
6101-T6	228 (33)	207 (30)	200 (29)	193 (28)	172 (25)	130 (19)	48 (7)	23 (3.3)	16 (2.3)	12 (1.8)
6151-T6	345 (50)	317 (46)	310 (45)	295 (43)	275 (40)	185 (27)	83 (12)	35 (5)	27 (3.9)	22 (3.2)
6262-T651	325 (47)	290 (42)	283 (41)	275 (40)	262 (38)	215 (31)	...	...	...	...
6262-T9	462 (67)	400 (58)	385 (56)	380 (55)	360 (52)	255 (37)	90 (13)	40 (6)	19 (2.7)	12 (1.8)
7075-T6, T651	635 (92)	545 (79)	517 (75)	503 (73)	448 (65)	185 (27)	90 (13)	62 (9)	45 (6.5)	32 (4.6)
7075-T73, T7351	495 (72)	462 (67)	448 (65)	435 (63)	400 (58)	185 (27)	90 (13)	62 (9)	45 (6.5)	32 (4.6)
7178-T6, T651	648 (94)	580 (84)	558 (81)	538 (78)	470 (68)	185 (27)	83 (12)	62 (9)	48 (7)	38 (5.5)
7178-T76, T7651	615 (89)	538 (78)	525 (76)	503 (73)	440 (64)	185 (27)	83 (12)	62 (9)	48 (7)	38 (5.5)

sitive than similar chromium additions. Higher levels of zirconium are employed in some superplastic alloys to retain the required fine substructure during elevated-temperature forming. Zirconium additions have been used to reduce the as-cast grain size, but its effect is less than that of titanium. In addition, zirconium tends to reduce the grain-refining effect of titanium plus boron additions so that it is necessary to use more titanium and boron to grain refine zirconium-containing alloys.

Properties of Wrought Aluminum Alloys

Property data on aluminum alloys are of two basic types:

- Typical property values
- Property limits

The data on wrought aluminum alloys presented in this section are primarily typical property values, although sources and tabular data for some mechanical property limits are also mentioned.

Typical values are considered nominal or representative values. Physical properties (Tables 6 and 7), for example, are median values determined in laboratory tests of representative commercial products. Typical mechanical properties (Table 8) are average or median values, near the peaks of distribution curves derived from routine quality-control tests of commercial products processed by standard mill procedures. The values listed are representative of products of moderate cross section or thickness, and are most useful for demonstrating relationships between alloys and tempers. These data are not intended to be used for critical design purposes. Static-strength values from tensile tests listed as typical do not represent the somewhat higher values (5 to 10% higher) obtained in tests (longitudinal direction) of extruded products of moderate section thickness nor do they represent the lower values expected in tests of very thick, heat-treated products.

Mechanical property limits are established on a statistical "A"-value basis, whereby 99% of the material is expected to conform at a confidence of 0.95. In most instances limits are based on a normally distributed database of a minimum of 100 tests from at least 10 different lots of material. Mechanical property limits are typically used for design or lot

Table 11(c) Elongation of various aluminum alloys at cryogenic and elevated temperatures

Alloy and temper	Elongation in 50 mm (2 in.), %, at:									
	−195 °C (−320 °F)	−80 °C (−112 °F)	0 °C (−18 °F)	24 °C (75 °F)	100 °C (212 °F)	150 °C (300 °F)	205 °C (400 °F)	260 °C (500 °F)	315 °C (600 °F)	370 °C (700 °F)
1100-0	50	43	40	40	45	55	65	75	80	85
1100-H14	45	24	20	20	20	23	26	75	80	85
1100-H18	30	16	15	15	15	20	65	75	80	85
2011-T3	...	...	...	15	16	25	35	45	90	125
2014-T6, T651	14	13	13	13	15	20	38	52	65	72
2017-T4, T451	28	24	23	22	18	15	35	45	65	70
2024-T3 (sheet)	18	17	17	17	16	11	23	55	75	100
2024-T4, T351 (plate)	19	19	19	19	19	17	27	55	75	100
2024-T6, T651	11	10	10	10	10	17	27	55	75	100
2024-T81, T851	8	7	7	7	8	11	23	55	75	100
2024-T861	5	5	5	5	6	11	28	55	75	100
2117-T4	30	29	28	27	16	20	35	55	80	110
2124-T851	9	8	8	9	9	13	28	60	75	100
2218-T61	15	14	13	13	15	17	30	70	85	100
2219-T62	16	13	12	12	14	17	20	21	40	75
2219-T81, T851	15	13	12	12	15	17	20	21	55	75
2618-T61	12	11	10	10	10	14	24	50	80	120
3003-0	46	42	41	40	43	47	60	65	70	70
3003-H14	30	18	16	16	16	16	20	60	70	70
3003-H18	23	11	10	10	10	11	18	60	70	70
3004-0	38	30	26	25	25	35	55	70	80	90
3004-H34	26	16	13	12	13	22	35	55	80	90
3004-H38	20	10	7	6	7	15	30	50	80	90
4032-T6	11	10	9	9	9	9	30	50	70	90
5050-0	...	...	...	...	...	...	...	...	...	...
5050-H34	...	...	...	...	...	...	...	...	...	...
5050-H38	...	...	...	...	...	...	...	...	...	...
5052-0	46	35	32	30	36	50	60	80	110	130
5052-H34	28	21	18	16	18	27	45	80	110	130
5052-H38	25	18	15	14	16	24	45	80	110	130
5083-0	36	30	27	25	36	50	60	80	110	130
5086-0	46	35	32	30	36	50	60	80	110	130
5154-0	46	35	32	30	36	50	60	80	110	130
5254-0	46	35	32	30	36	50	60	80	110	130
5454-0	39	30	27	25	31	50	60	80	110	130
5454-H32	32	23	20	18	20	37	45	80	110	130
5454-H34	30	21	18	16	18	32	45	80	110	130
5456-0	32	25	22	20	31	50	60	80	110	130
5652-0	46	35	32	30	30	50	60	80	110	130
5652-H34	28	21	18	16	18	27	45	80	110	130
5652-H38	25	18	15	14	16	24	45	80	110	130
6053-T6, T651	...	...	...	13	13	13	25	70	80	90
6061-T6, T651	22	18	17	17	18	20	28	60	85	95
6063-T1	44	36	34	33	18	20	40	75	80	105
6063-T5	28	24	23	22	18	20	40	75	80	105
6063-T6	24	20	19	18	15	20	40	75	80	105
6101-T6	24	20	19	19	20	20	40	80	100	105
6151-T6	20	17	17	17	17	20	30	50	43	35
6262-T651	22	18	17	17	18	20	...	...	...	...
6262-T9	14	10	10	10	10	14	34	48	85	95
7075-T6, T651	9	11	11	11	14	30	55	65	70	70
7075-T73, T7351	14	14	13	13	15	30	55	65	70	70
7178-T6, T651	5	8	9	11	14	40	70	76	80	80
7178-T76, T7651	10	10	10	11	17	40	70	76	80	80

Note: Same test conditions as those specified in the footnote of Table 11(a)

acceptance. The distinction between metric and English unit property limits can be important because of rounding from metric to English or vice versa.

Typical physical-property values (Table 6) are given only as a basis for comparing alloys and tempers and should not be specified as engineering requirements or used for design purposes. They are not guaranteed values, since in most cases they are averages for various sizes, product forms, and methods of manufacture and may not be exactly representative of any particular size or product. Density values for the annealed (O) temper are listed in Table 7.

Typical Mechanical Properties. Typical tensile strengths (ultimate and yield), tensile elongations, ultimate shear strengths, fatigue strengths (endurance limits), hardnesses, and elastic moduli are given in Table 8. As typical properties they are for comparative purposes and not design, as discussed previously. The table lists both heat-treatable and non-heat-treatable alloys, and in most cases the properties are averages for various sizes, product forms, and methods of manufacture.

Tensile property limits for various wrought aluminum alloys are given in the article "Properties of Wrought Aluminum and Aluminum Alloys" in this Volume. In addition, the current edition of *Aluminum Standards and Data*, published biennially by The Aluminum Association, provides tensile property limits for most alloy tempers and product forms.

Bend Properties. Recommended minimum 90° cold bend radii for sheet and plate are given in Table 9. Additional forming characteristics (Olsen ball, *n*, *r*, minimum bend radii painted sheet, bend radii bus bar) may be found in *Aluminum Standards and Data*, published by The Aluminum Association.

Classification of Alloys for Fracture Toughness. Fracture toughness is rarely, if ever, a design consideration in the 1000, 3000, 4000, 5000, and 6000 series alloys. The fracture toughness of these alloys is sufficiently high that thicknesses beyond those commonly produced would be required to obtain a valid test. Therefore, these alloys are excluded from further consideration in this article. Among the alloys for which fracture toughness is a meaningful design-related parameter, controlled-toughness high-strength alloys and conventional high-strength alloys merit discussion.

Controlled-toughness high-strength alloys were developed for their high fracture toughness and range in measured K_{Ic} values from about 20 $MPa\sqrt{m}$ (18 $ksi\sqrt{in.}$) upward. The alloys and tempers currently identified as controlled-toughness high-strength products include:

Alloy	Condition	Product form
2048	T8	Sheet and plate
2124	T3, T8	Sheet and plate
2419	T8	Sheet, plate, extrusions, and forgings
7049	T7	Plate, forgings, and extrusions
7050	T7	Sheet, plate, forgings, and extrusions
7150	T6	Sheet and plate
7175	T6, T7	Sheet, plate, forgings, and extrusions
7475	T6, T7	Sheet and plate

Typical applications include 2419-T851 used in the lower wing skins of the B-1 bomber and 7475-T7351 and 2124-T851 in the F-16 aircraft.

Conventional High-Strength Alloys. Although these alloys, tempers, and products are not used for fracture-critical components, fracture toughness can be a meaningful design parameter. Conventional aerospace alloys for which fracture toughness minimums may be useful in design include 2014, 2024, 2219, 7075, and 7079. These alloys have toughness levels that are inferior to those of their controlled-toughness counterparts. Consequently, toughness is not guaranteed.

Controlled-toughness alloys are often derivatives of conventional alloys. For example, 7475 alloy is a derivative of 7075 with maximum compositional limits on some elements that were found to decrease toughness.

Fracture toughness quality control and material procurement minimums are appropriate for controlled-toughness, high-strength alloys, tempers, and products, because checks on composition and tensile properties are inadequate assurances that the proper levels of toughness have been achieved. If the minimum specified fracture toughness value is not attained, the material is not acceptable.

Minimum and typical room-temperature plane-strain fracture toughness is listed for selected high-strength aluminum alloys in Table 10. The effect of alloying elements and microstructural constituents on fracture toughness is discussed in the section "Fracture Toughness and Fatigue Behavior" in this article.

Fatigue and Fatigue Crack Growth. Aluminum does not generally exhibit the sharply defined fatigue limit typically shown by low-carbon steel in *S-N* tests. For smooth

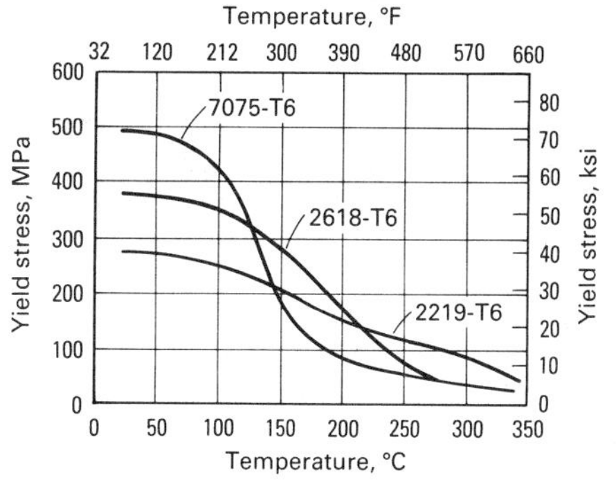

Fig. 31 Values of 0.2% yield stress of aluminum alloys after exposure for 1000 h at temperatures between 0 and 350 °C

or notched coupon tests, where lifetime is governed primarily by crack initiation, the fatigue resistance is expressed as a fatigue strength (stress) for a given number of cycles. Table 8 gives typical data on the fatigue strength of various aluminum alloys.

In tests where fatigue crack growth is of interest, the performance of aluminum is measured by recording the crack growth rate (da/dN) as a function of stress intensity range (ΔK). This type of FCG test is currently of prime importance for alloys used in aerospace applications. Fatigue crack growth can be influenced by alloy composition and microstructure, the presence of oxygen, temperature, load ratio (R), material thickness (or thickness in relation to plastic zone size), stress intensity range, and the processes used in preparing the alloys. It is recognized that the interactions among these variables complicate the proper interpretation and extrapolation of experimental data and introduce additional uncertainties with respect to damage-tolerant design and failure analysis.

Elevated-Temperature Properties. Tables 11(a), 11(b), and 11(c) list typical tensile properties of various aluminum alloys at elevated temperatures. The 7*xxx* series of age-hardenable alloys that are based on the Al-Zn-Mg-Cu system develop the highest room-temperature tensile properties of any aluminum alloys produced from conventionally cast ingots. However, the strength of these alloys declines rapidly if they are exposed to elevated temperatures (Fig. 31), due mainly to coarsening of the fine precipitates on which the alloys depend for their strength. Alloys of the 2*xxx* series such as 2014 and 2024 perform better above these temperatures but are not normally used for elevated-temperature applications.

Strength at temperatures above about 100 to 200 °C (200 to 400 °F) is improved mainly by solid-solution strengthening or second-phase hardening. Another approach to improve the elevated-temperature performance of aluminum alloys has been the use of rapid solidification technology to produce powders or foils containing high super-

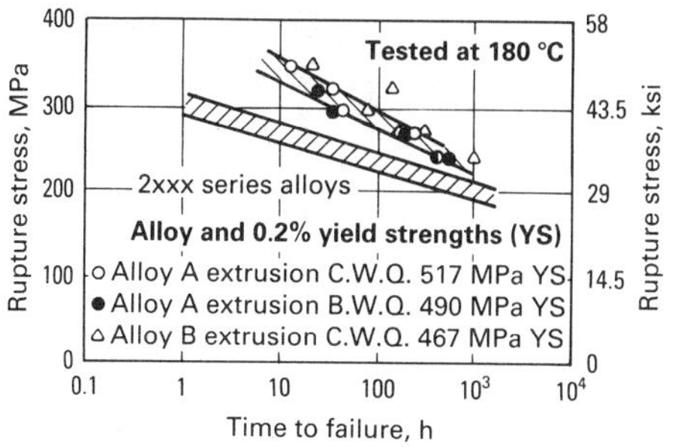

Fig. 32 Stress-rupture results for creep tests at 180 °C (355 °F) on aluminum alloys with silver additions compared with those for 2*xxx* series alloys. Alloy A: 6.3% Cu, 0.5% Mg, 0.5% Ag, 0.5% Mn, and 0.2% Zr. Alloy B: 6.0% Cu, 0.45% Mg, 0.5% Ag, 0.5% Mn, and 0.14% Zr. CWQ, cold-water quenched before aging; BWQ, boiling-water quenched before aging. Source: Ref 9

saturations of elements such as iron or chromium that diffuse slowly in solid aluminum. In this regard, several experimental materials are now available that have promising creep properties up to 350 °C (650 °F). An experimental Al-Cu-Mg alloy with silver additions has also resulted in improved creep properties (Fig. 32). Iron is also used to improve creep properties (see the heading "Iron" in this article).

Low-Temperature Properties. Aluminum alloys represent a very important class of structural metals for subzero-temperature applications and are used for structural parts for operation at temperatures as low as −270 °C (−450 °F). Below zero, most aluminum alloys show little change in properties; yield and tensile strengths may increase; elongation may decrease slightly; impact strength remains approximately constant. Consequently, aluminum is a useful material for many low-temperature applications; the chief deterrent is its relatively low elongation compared with certain austenitic ferrous alloys. This inhibiting factor affects principally industries that must work with public safety codes. A notable exception to this has been the approval, in the ASME unfired pressure vessel code, to use alloys 5083 and 5456 for pressure vessels within the range from −195 to 65 °C (−320 to 150 °F). With these alloys, tensile strength increases 30 to 40%, yield strength 5 to 10%, and elongation 60 to 100% between room temperature and −195 °C (−320 °F).

The wrought alloys most often considered for low-temperature service are alloys 1100, 2014, 2024, 2219, 3003, 5083, 5456, 6061, 7005, 7039, and 7075. Alloy 5083-O, which is the most widely used aluminum alloy for cryogenic applications, exhibits the following increases in tensile properties when cooled from room temperature to the boiling point of nitrogen (−195.8 °C, or −320.4 °F):

- About 40% in ultimate tensile strength
- About 10% in yield strength

Table 12 Fracture toughness of aluminum alloy plate

Alloy and condition	Room temperature yield strength, MPa	Room temperature yield strength, ksi	Specimen design	Orientation	Fracture toughness, K_{Ic} or $K_{Ic}(J)$ at: 24 °C (75 °F), MPa$\sqrt{m}$	24 °C (75 °F), ksi$\sqrt{in.}$	−196 °C (−320 °F), MPa$\sqrt{m}$	−196 °C (−320 °F), ksi$\sqrt{in.}$	−253 °C (−423 °F), MPa$\sqrt{m}$	−253 °C (−423 °F), ksi$\sqrt{in.}$	−269 °C (−452 °F), MPa$\sqrt{m}$	−269 °C (−452 °F), ksi$\sqrt{in.}$
2014-T651	432	62.7	Bend	T-L	23.2	21.2	28.5	26.1	...	...	...	...
2024-T851	444	64.4	Bend	T-L	22.3	20.3	24.4	22.2	...	...	...	...
2124-T851(a)	455	66.0	CT	T-L	26.9	24.5	32.0	29.1	...	...	...	...
	435	63.1	CT	L-T	29.2	26.6	35.0	31.9	...	...	...	...
	420	60.9	CT	S-L	22.7	20.7	24.3	22.1	...	...	...	...
2219-T87	382	55.4	Bend	T-S	39.9	36.3	46.5	42.4	52.5	48.0	...	...
			CT	T-S	28.8	26.2	34.5	31.4	37.2	34.0	...	...
	412	59.6	CT	T-L	30.8	28.1	38.9	32.7	...	...	...	...
5083-O	142	20.6	CT	T-L	27.0(b)	24.6(b)	43.4(b)	39.5(b)	...	...	48.0(b)	43.7(b)
6061-T651	289	41.9	Bend	T-L	29.1	26.5	41.6	37.9	...	...	...	...
7039-T6	381	55.3	Bend	T-L	32.3	29.4	33.5	30.5	...	...	...	...
7075-T651	536	77.7	Bend	T-L	22.5	20.5	27.6	25.1	...	...	...	...
7075-T7351	403	58.5	Bend	T-L	35.9	32.7	32.1	29.2	...	...	...	...
7075-T7351	392	56.8	Bend	T-L	31.0	28.2	30.9	28.1	...	...	...	...

(a) 2124 is similar to 2024 but with higher-purity base and special processing to improve fracture toughness. (b) $K_{Ic}(J)$. Source: Volume 3 of 9th Edition *Metals Handbook*

Table 13 Results of fatigue-life tests on aluminum alloys

Alloy and condition	Stressing mode	Stress ratio, R	K_t	Fatigue strength at 10^6 cycles, at: 24 °C (75 °F), MPa	24 °C (75 °F), ksi	−196 °C (−320 °F), MPa	−196 °C (−320 °F), ksi	−253 °C (−423 °F), MPa	−253 °C (−423 °F), ksi
2014-T6 sheet	Axial	−1.0	1	115	17	170	25	315	46
		+0.01	1	215	31	325	47	435	63
2014-T6 sheet, GTA welded, 2319 filler	Axial	−1.0	1	83	12	105	150	125	18
2219-T62 sheet	Axial	−1.0	1	130	19	15	22	255	37
			3.5	52	7.5	45	6.5	62	9
2219-T87 sheet	Axial	−1.0	1	150	22	115–170	17–25	275	40
			3.5	52	7.5	48	7	55	8
2219-T87 sheet, GTA welded, 2319 filler	Axial	−1.0	1	69	10	83	12	150	22
5083-H113 plate	Flex	−1.0	1	140	20.5	190	27.5	...	...
5083-H113 plate, GMA welded, 5183 filler	Flex	−1.0	1	90	13	130	18.8	...	...
6061-T6 sheet(a)	Flex	−1.0	1	160	23	220	32	235	34
6061-T6 sheet(b)	Flex	−1.0	1	165	24	230	33	230	33
7039-T6 sheet	Axial	−1.0	1	140	20	215	31	275	40
		+0.01	1	230	33	330	48	440	64
		−1.0	3.5	48	7	48	7	62	9
7075-T6 sheet	Axial	−1.0	1	96	14	145	21	250	36

(a) Surface finish, 150 μin. rms. (b) Surface finish, 20 μin. rms. Source: Volume 3 of 9th Edition *Metals Handbook*

- Sixty percent in elongation

Typical tensile properties of various aluminum alloys at cryogenic temperatures are given in Tables 11(a), 11(b), and 11(c).

Retention of toughness also is of major importance for equipment operating at low temperature. Aluminum alloys have no ductile-to-brittle transition; consequently, neither ASTM nor ASME specifications require low-temperature Charpy or Izod tests of aluminum alloys. Other tests, including notch-tensile and tear tests, assess the notch-tensile and tear toughness of aluminum alloys at low temperatures. The low-temperature characteristics of welds in the weldable aluminum alloys parallel those described above for unwelded material.

Fracture Toughness. Data on fracture toughness of several aluminum alloys at room and subzero temperatures are summarized in Table 12. Of the alloys listed in Table 12, 5083-O has substantially greater toughness than the others. Because this alloy is too tough for obtaining valid K_{Ic} data, the values shown for 5083-O were converted from J_{Ic} data. The fracture toughness of this alloy increases as exposure temperature decreases. Of the other alloys, which were all evaluated in various heat-treated conditions, 2219-T87 has the best combination of strength and fracture toughness, both at room temperature and at −196 °C (−320 °F), of all the alloys that can be readily welded.

Alloy 6061-T651 has good fracture toughness at room temperature and at −196 °C (−320 °F), but its yield strength is lower than that of alloy 2219-T87. Alloy 7039 also is weldable and has a good combination of strength and fracture toughness at room temperature and at −196 °C (−320 °F). Alloy 2124 is similar to 2024 but with a higher-purity base and special processing for improved fracture toughness. Tensile properties of 2124-T851 at subzero temperatures can be expected to be similar to those for 2024-T851.

Several other aluminum alloys, including 2214, 2419, 7050, and 7475, have been developed in order to obtain room-temperature fracture toughness superior to that of other 2000 and 7000 series alloys. Information on subzero properties of these alloys is limited, but it is expected that these alloys also would have improved fracture toughness at subzero temperatures as well as at room temperature.

Fatigue Strength. Results of axial and flexural fatigue tests at 10^6 cycles on aluminum alloy specimens at room temperature and at subzero temperatures are presented in Table 13. These data indicate that, for a fatigue life of 10^6 cycles, fatigue strength is higher at subzero temperatures than at room temperature for each alloy. This trend is not necessarily valid for tests at higher stress levels and shorter fatigue lives, but at 10^6 cycles results are consistent with the effect of subzero temperatures on tensile strength.

ACKNOWLEDGMENT

The information in this article is largely taken from four sources:

- Volume 2 of the 9th Edition of *Metals Handbook*
- "Introduction to Aluminum and Aluminum Alloys" in the *Metals Handbook Desk Edition* (1985)
- "Effects of Alloying Elements and Impurities on Properties" in *Aluminum: Proper-*

ties and Physical Metallurgy (ASM, 1984)
- *Aluminum Standards and Data 1988*, 9th Edition, Aluminum Association

REFERENCES

1. D. Munson, *J. Inst. Met.*, Vol 95, 1967, p 217-219
2. W. Hume-Rothery and G.V. Raynor, *The Structure of Metals and Alloys*, The Institute of Metals, 1962
3. W.B. Pearson, *Handbook of LatticeSpacings and Structures of Metals and Alloys*, Vol 2, Pergamon Press, 1967
4. L.F. Mondolfo, *Aluminum Alloys: Structure and Properties*, Butterworths, 1976
5. F.L. Kaufman and H. Bernstein, *Computer Calculation of Phase Diagrams*, Academic Press, 1970
6. CALPHAD (Computer Coupling of Phase Diagrams and Thermochemistry), L. Kaufman, Ed., Manlabs Inc.
7. CALPHAD, Pergamon Press, 1976-1980
8. R.K. Wyss and R.E. Sanders, Jr., "Microstructure-Property Relationship in a 2*xxx* Aluminum Alloy With Mg Addition," *Metall. Trans. A*, Vol 19A, 1988, p 2523-2530
9. I.J. Polmear and M.J. Couper, "Design and Development of an Experimental Wrought Aluminum Alloy for Use at Elevated Temperatures," *Metall. Trans. A*, Vol 19A, p 1027-1035

Properties of Wrought Aluminum and Aluminum Alloys

1050
99.5 Al min

Specifications

ASTM. B 491
UNS number. A91050
Foreign. Canada: CSA 9950. France: NF A5. United Kingdom: BS 1B. West Germany: DIN A199.5

Chemical Composition

Composition limits. 99.50 Al min, 0.25 Si max, 0.40 Fe max, 0.05 Cu max, 0.05 Mn max, 0.05 Mg max, 0.05 V max, 0.03 max other (each)

Applications

Typical uses. Extruded coiled tube for equipment and containers for food, chemical, and brewing industries; collapsible tubes; pyrotechnic powder

Mechanical Properties

Tensile properties. See Table 1.

Mass Characteristics

Density. 2.705 g/cm^3 (0.0977 lb/in.^{3}) at 20 °C (68 °F)

Thermal Properties

Liquidus temperature. 657 °C (1215 °F)
Solidus temperature. 646 °C (1195 °F)
Coefficient of thermal expansion. Linear:

Temperature range °C	°F	Average coefficient µm/m · K	µin./in. · °F
−50 to 20	−58 to 68	21.8	12.1
20 to 100	68 to 212	23.6	13.1
20 to 200	68 to 392	24.5	13.6
20 to 300	68 to 572	25.5	14.2

Volumetric: 68.1 × 10^{-6} m^3/m^3 · K (3.78 × 10^{-5} in.^{3}/in.^{3} · °F)
Specific heat. 900 J/kg · K (0.215 Btu/lb · °F) at 20 °C (68 °F)
Thermal conductivity. O temper, 231 W/m · K (133 Btu/ft · h · °F) at 20 °C (68 °F)

Electrical Properties

Electrical conductivity. Volumetric. O temper, 61.3% IACS at 20 °C (68 °F)
Electrical resistivity. O temper: 28.1 nΩ · m at 20 °C (68 °F); temperature coefficient, 0.1 nΩ · m per K at 20 °C (68 °F)

Table 1 Typical mechanical properties of 1050 aluminum

Temper	Tensile strength MPa	Tensile strength ksi	Yield strength MPa	Yield strength ksi	Elongation, %	Shear strength MPa	Shear strength ksi
O	76	11	28	4	39	62	9
H14	110	16	103	15	10	69	10
H16	131	19	124	18	8	76	11
H18	159	23	145	21	7	83	12

1060
99.60 Al min

Specifications

AMS. Sheet and plate: 4000
ASME. See Table 2.
ASTM. See Table 2.
SAE. J454
UNS number. A91060

Chemical Composition

Composition limits. 99.60 Al min, 0.25 Si max, 0.35 Fe max, 0.05 Cu max, 0.03 Mn max, 0.03 Mg max, 0.05 Zn max, 0.05 V max, 0.03 Ti max, 0.03 max other (each)

Applications

Typical uses. Applications requiring very good resistance to corrosion and good formability, but tolerate low strength. Chemical process equipment is typical.

Mechanical Properties

Tensile properties. See Tables 3 and 4.
Hardness. See Table 3.
Poisson's ratio. 0.33 at 20 °C (68 °F)
Elastic modulus. Tension, 69 GPa (10 × 10^6 psi)
Fatigue strength. See Table 3.

Mass Characteristics

Density. 2.705 g/cm^3 (0.0977 lb/in.^{3}) at 20 °C (68 °F)

Thermal Properties

Liquidus temperature. 657 °C (1215 °F)
Solidus temperature. 646 °C (1195 °F)
Coefficient of thermal expansion. Linear:

Temperature range °C	°F	Average coefficient µm/m · K	µin./in. · °F
−50 to 20	−58 to 68	21.8	12.1
20 to 100	68 to 212	23.6	13.1
20 to 200	68 to 392	24.5	13.6
20 to 300	68 to 572	25.5	14.1

Volumetric: 68 × 10^{-6} m^3/m^3 · K (3.8 × 10^{-5} in.^{3}/in.^{3} · °F)
Specific heat. 900 J/kg · K (0.215 Btu/lb · °F) at 20 °C (68 °F)
Thermal conductivity. 234 W/m · K (135 Btu/ft · h · °F) at 25 °C (77 °F)

Electrical Properties

Electrical conductivity. Volumetric at 20 °C (68 °F): O temper, 62% IACS: H18 temper, 61% IACS
Electrical resistivity. At 20 °C (68 °F): O temper, 27.8 nΩ · m; H18 temper, 28.3 nΩ · m. Temperature coefficient, O and H18 tempers, 0.1 nΩ · m per K at 20 °C (68 °F)
Electrolytic solution potential. −0.84 V versus 0.1 *N* calomel electrode in aqueous

Table 2 ASME and ASTM specifications for 1060 aluminum

Mill form and condition	Specification number ASME	ASTM
Sheet and plate	SB209	B 209
Wire, rod, and bar (rolled or cold finished)	· · ·	B 211
Wire, rod, bar, shapes, and tube (extruded)	SB221	B 221
Pipe (gas and oil transmission)	· · ·	B 345
Tube (condenser)	SB234	B 234
Tube (condenser with integral fins)	· · ·	B 404
Tube (drawn)	· · ·	B 483
Tube (drawn, seamless)	SB210	B 210
Tube (extruded, seamless)	SB241	B 241

Table 3 Typical mechanical properties of 1060 aluminum

Temper	Tensile strength, MPa	Tensile strength, ksi	Yield strength, MPa	Yield strength, ksi	Elongation(a), %	Hardness, HB(b)	Shear strength, MPa	Shear strength, ksi	Fatigue limit(c), MPa	Fatigue limit(c), ksi
O	69	10	28	4	43	19	48	7	21	3
H12	83	12	76	11	16	23	55	8	28	4
H14	97	14	90	13	12	26	62	9	34	5
H16	110	16	103	15	8	30	69	10	45	6.5
H18	131	19	124	18	6	35	76	11	45	6.5

(a) 1.6 mm (1/16 in.) thick specimens. (b) 500 kg load; 10 mm diam ball. (c) At 5 × 10[8] cycles; R.R. Moore type test

Table 4 Tensile-property limits for 1060 aluminum

Temper	Tensile strength Minimum, MPa	Tensile strength Minimum, ksi	Tensile strength Maximum, MPa	Tensile strength Maximum, ksi	Yield strength (min), MPa	Yield strength (min), ksi	Elongation (min), %(a)
Sheet and plate							
O	55	8.0	95	14.0	17	2.5	15–25
H12	75	11.0	110	16.0	62	9.0	6–12
H14	83	12.0	115	17.0	70	10.0	1–10
H18	110	16.0	...	...	83	12.0	1–4
H112							
0.250–0.499 in. thick	75	11.0	...	...	...	...	10
0.500–1.000 in. thick	70	10.0	...	...	...	...	20
1.001–3.000 in. thick	62	9.0	...	...	...	...	25
Drawn tube (0.010–0.500 in. wall thickness)							
O	58	8.5	...	...	17	2.5	...
H12	70	10.0	...	...	28	4.0	...
H14	83	12.0	...	...	70	10.0	...
H18	110	16.0	...	...	90	13.0	...
H112	58	8.5	...	...	17	2.5	...
Extruded tube							
O	58	8.5	95	14.0	17	2.5	...
H112	58	8.5	95(b)	14.0(b)	17	2.5	30(b)
Heat-exchanger tube (0.010–0.200 in. wall thickness)							
H14	83	12.0	...	...	70	10.0	...

(a) In 50 mm (2 in.) or 4*d*, where *d* is diameter of reduced section of tensile test specimen. Where a range of values appears in this column, specified minimum elongation varies with thickness of the mill product. (b) Applicable only to tube 25.4 to 114.3 mm (1.000 to 4.500 in.) diam by 1.27 to 4.29 mm (0.050 to 0.169 in.) wall thickness

Table 5 Standard specifications for 1100 aluminum

Mill form and condition	AMS	ASME	ASTM	Government
Sheet and plate	4001, 4003	SB209	B 209	QQ-A-250/1
Wire, rod, and bar (rolled or cold finished)	4102	...	B 211	QQ-A-225/1
Wire, rod, bar, shapes, and tube (extruded)	...	SB221	B 221	...
Tube (extruded, seamless)	...	SB241	B 241	...
Tube (extruded, coiled)	...	...	B 491	...
Tube (drawn)	...	...	B 483	...
Tube (drawn, seamless)	4062	...	B 210	WW-T-700/1
Tube (welded)	...	...	B 313, B 547	...
Rivet wire and rod	...	...	B 316	QQ-A-430
Spray gun wire	4180	...	...	MIL-W-6712
Forgings and forging stock	...	...	B 247	...
Welding rod and electrodes (bare)	...	...	...	QQ-R-566, MIL-E-16053
Impacts	...	...	...	MIL-A-12545
Foil	...	...	...	QQ-A-1876

Table 6 Typical room-temperature mechanical properties of 1100 aluminum

Temper	Tensile strength, MPa	Tensile strength, ksi	Yield strength, MPa	Yield strength, ksi	Elongation, % 1/16 in. thick specimens	Elongation, % 1/2 in. thick specimens	Hardness, HB(a)	Shear strength, MPa	Shear strength, ksi	Fatigue limit(b), MPa	Fatigue limit(b), ksi
O	90	13	34	5	35	45	23	62	9	34	5
H12	110	16	103	15	12	25	28	69	10	41	6
H14	124	18	117	17	9	20	32	76	11	48	7
H16	145	21	138	20	6	17	38	83	12	62	9
H18	165	24	152	22	5	15	44	90	13	62	9

(a) 500 kg load; 10 mm ball. (b) At 5 × 10[8] cycles; R.R. Moore type test

solution containing 53 g NaCl plus 3 g H_2O_2 per liter

Fabrication Characteristics

Annealing temperature. 345 °C (650 °F)

1100
99.00Al (min)-0.12Cu

Commercial Names

Common name. Aluminum

Specifications

AMS. See Table 5.
ASME. See Table 5.
ASTM. See Table 5.
SAE. J454
UNS number. A91100
Government. See Table 5.
Foreign. Canada: CSA 990C. France: NF A45. ISO: A199.0Cu

Chemical Composition

Composition limits. 99.00 Al min, 1.0 Si max + Fe, 0.05 to 0.20 Cu, 0.05 Mn max, 0.10 Zn max, 0.05 max other (each), 0.15 max others (total), 0.0008 Be max (welding electrode and filler wire only)

Applications

Typical uses. Applications requiring good formability and high resistance to corrosion where high strength is not necessary. Food and chemical handling and storage equipment, sheet metal work, drawn or spun hollowware, welded assemblies, heat exchangers, litho plate, nameplates, light reflectors

Mechanical Properties

Tensile properties. See Tables 6, 7, and 8.
Hardness. See Table 6.
Poisson's ratio. 0.33 at 20 °C (68 °F)
Elastic modulus. Tension, 69 GPa (10 × 10[6] psi); shear, 26 GPa (3.75 × 10[6] psi)

Mass Characteristics

Density. 2.71 g/cm³ (0.098 lb/in.³) at 20 °C (68 °F)

Thermal Properties

Liquidus temperature. 657 °C (1215 °F)
Solidus temperature. 643 °C (1190 °F)
Coefficient of thermal expansion. Linear:

Temperature range, °C	Temperature range, °F	Average coefficient, μm/m · K	Average coefficient, μin./in. · °F
−50 to 20	−58 to 68	21.8	12.1
20 to 100	68 to 212	23.6	13.1
20 to 200	68 to 392	24.5	13.6
20 to 300	68 to 572	25.5	14.1

Volumetric: 68 × 10^{-6} m³/m³ · K (3.8 × 10^{-5} in.³/in.³ · °F)
Specific heat. 904 J/kg · K (0.216 Btu/lb · °F) at 20 °C (68 °F)
Thermal conductivity. O temper, 222 W/m ·

Table 7 Tensile-property limits for 1100 aluminum

Temper	Tensile strength, Minimum, MPa	Tensile strength, Minimum, ksi	Tensile strength, Maximum, MPa	Tensile strength, Maximum, ksi	Yield strength (min), MPa	Yield strength (min), ksi	Elongation (min), %(a)
Sheet and plate							
O	75	11.0	105	15.5	25	3.5	15–28
H12	95	14.0	130	19.0	75	11.0	3–12
H14	110	16.0	145	21.0	95	14.0	1–10
H16	130	19.0	165	24.0	115	17.0	1–4
H18	150	22.0	...	...	...	...	1–4
H112							
0.250–0.499 in. thick	90	13.0	...	...	50	7.0	9
0.500–2.000 in. thick	83	12.0	...	...	35	5.0	14
2.001–3.000 in. thick	80	11.5	...	...	30	4.0	20
Wire, rod, and bar (rolled or cold finished)							
O	75	11.0	105	15.5	20	3.0	25
H112	75	11.0	...	...	20	3.0	...
H12(b)	95	14.0	...	...	...	...	...
H14(b)	110	16.0	...	...	...	...	...
H16(b)	130	19.0	...	...	...	...	...
H18(b)	150	22.0	...	...	...	...	...
Wire, rod, bar, and shapes (extruded)							
O	75	11.0	105	15.5	20	3.0	25
H112	75	11.0	...	...	20	3.0	...
Wire and rod (rivet and cold heading grade)							
O(c)	...	...	105	15.5	...	...	...
H14(c)	110	16.0	145	21.0	...	...	...
Drawn tube (0.014 to 0.500 in. wall thickness)							
O	...	...	105	15.5	...	...	...
H12	95	14.0	...	...	...	...	...
H14	110	16.0	...	...	...	...	...
H16	130	19.0	...	...	...	...	...
H18	150	22.0	...	...	...	...	...
Extruded tube							
O	75	11.0	105	15.5	20	3.0	25
H112	75	11.0	...	...	20	3.0	25

(a) In 50 mm (2 in.) or 4*d*, where *d* is diameter of reduced section of tensile test specimen. Where a range of values appears in this column, the specified minimum elongation varies with thickness of the mill product. (b) Nominal thickness up through 9.5 mm (0.374 in.). (c) Nominal diameter up through 25.4 mm (1.000 in.)

K (128 Btu/ft · h · °F); H18 temper, 218 W/m · K (126 Btu/ft · h · °F)

Electrical Properties

Electrical conductivity. Volumetric at 20 °C (68 °F): O temper, 59% IACS; H18 temper, 57% IACS
Electrical resistivity. At 20 °C (68 °F): O temper, 29.2 nΩ · m; H18 temper, 30.2 nΩ · m. Temperature coefficient at 20 °C (68 °F): O and H18 tempers, 0.1 nΩ · m per K
Electrolytic solution potential. All tempers, −0.83 V versus 0.1 *N* calomel electrode in aqueous solution containing 53 g NaCl plus 3 g H_2O_2 per liter at 25 °C (77 °F)

Optical Properties

Reflectance. Brightly polished or diffusely etched reflector: 86% for light from tungsten filament; 84% for light having a wavelength of 250 nm. See also Fig. 1.
Emittance. See Fig. 2.

Fabrication Characteristics

Annealing temperature. 343 °C (650 °F)

1145
99.45 Al min

Specifications

AMS. 4011
ASTM. B 373
Government. QQ-A-1876

Chemical Composition

Composition limits. 99.45 Al min, 0.55 Si max + Fe, 0.05 Cu max, 0.05 Mn max, 0.05 Mg max, 0.05 Zn max, 0.05 V max, 0.03 Ti max, 0.03 max other (each)

Applications

Typical uses. Foil for packaging, insulating, and heat exchangers

Mechanical Properties

Tensile properties. See Table 9.

Mass Characteristics

Density. 2.705 g/cm³ (0.0977 lb/in.³) at 20 °C (68 °F)

Thermal Properties

Liquidus temperature. 657 °C (1215 °F)
Solidus temperature. 646 °C (1195 °F)

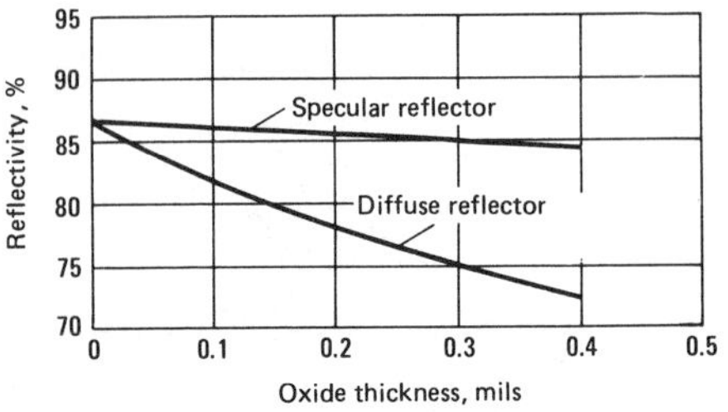

Fig. 1 Reflectivity of 1100 aluminum as a function of aluminum oxide coating thickness

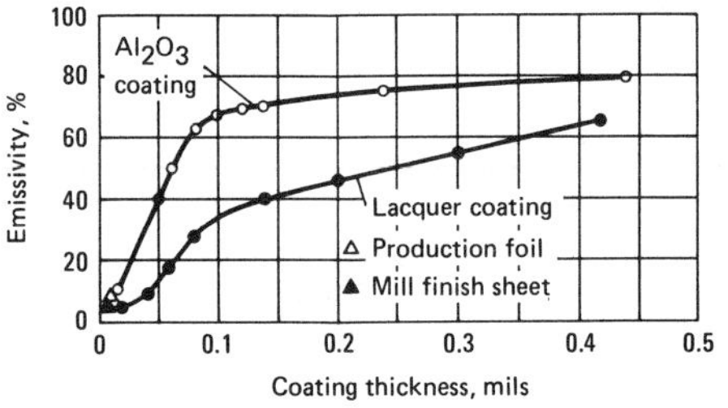

Fig. 2 Emissivity of 1100 aluminum foil as a function of coating thickness

Coefficient of thermal expansion. Linear:

Temperature range, °C	Temperature range, °F	Average coefficient, µm/m · K	Average coefficient, µin./in. · °F
−50 to 20	−58 to 68	21.8	12.1
20 to 100	68 to 212	23.6	13.1
20 to 200	68 to 392	24.5	13.6
20 to 300	68 to 572	25.5	14.1

Volumetric: 68 × 10^{-6} m³/m³ · K (3.8 × 10^{-5} in.³/in.³ · °F)
Specific heat. 904 J/kg · K (0.216 Btu/lb · °F) at 20 °C (68 °F)
Thermal conductivity. At 20 °C (68 °F): O temper, 230 W/m · K (133 Btu/ft · h · °F); H18 temper, 227 W/m · K (131 Btu/ft · h · °F)

Electrical Properties

Electrical conductivity. Volumetric at 20 °C (68 °F): O temper, 61% IACS; H18 temper, 60% IACS
Electrical resistivity. At 20 °C (68 °F): O temper, 28.3 nΩ · m; H18 temper, 28.7 nΩ · m. Temperature coefficient at 20 °C: O and H18 tempers, 0.1 nΩ · m per K

Optical Properties

Reflectance. 95 to 97% for λ = 0.3 to 10 µm
Emittance. 3 to 5% for λ = 9.3 µm at 20 °C (68 °F)

Fabrication Characteristics

Annealing temperature. 345 °C (650 °F)

1199
99.99 Al min

Commercial Names

Trade name. Super-purity aluminum, Raffinal

Table 8 Typical tensile properties of 1100 aluminum at various temperatures

Temperature °C	Temperature °F	Tensile strength MPa	Tensile strength ksi	Yield strength MPa	Yield strength ksi	Elongation, %
O temper						
−195	−320	170	25	41	6	50
−80	−112	105	15	38	5.5	43
−28	−18	97	14	34	5	40
24	75	90	13	34	5	40
100	212	69	10	32	4.6	45
149	300	55	8	29	4.2	55
204	400	41	6	24	3.5	65
260	500	28	4	18	2.6	75
316	600	20	2.9	14	2.0	80
371	700	14	2.1	11	1.6	85
H14 temper						
−196	−320	205	30	140	20	45
−80	−112	140	20	125	18	24
−28	−18	130	19	115	17	20
24	75	125	18	115	17	20
100	212	110	16	105	15	20
149	300	97	14	83	12	23
204	400	69	10	52	7.5	26
260	500	28	4	18	2.6	75
316	600	20	2.9	14	2.0	80
371	700	14	2.1	11	1.6	85
H18 temper						
−196	−320	235	34	180	26	30
−80	−112	180	26	160	23	16
−28	−18	170	25	160	23	16
24	75	165	24	150	22	15
100	212	145	21	130	19	15
149	300	125	18	97	14	20
204	400	41	6	24	3.5	65
260	500	28	4	18	2.6	75
316	600	20	2.9	14	2.0	80
371	700	14	2.1	11	1.6	85

Table 9 Tensile properties of 1145 aluminum foil

Temper	Tensile strength MPa	Tensile strength ksi	Yield strength MPa	Yield strength ksi	Elongation, %
Typical properties					
O	75	11	34	5	40
H18	145	21	117	17	5
Tensile strength limits(a)					
O	95 max	14 max	...	...	...
H19	140 min	20 min	...	...	...

(a) Unmounted foil 0.02 to 0.15 mm (0.0007 to 0.0059 in.) thick

Table 10 Typical tensile properties of 1199 aluminum

Reduction by cold rolling, %	Tensile strength MPa	Tensile strength ksi	Yield strength MPa	Yield strength ksi	Elongation, %
0 (annealed)	45	6.5	10	1.5	50
10	59	8.6	57	8.2	40
20	77	11.1	75	10.8	15
40	96	13.9	91	13.2	11
60	110	15.9	105	15.1	6
75	120	17.5	113	16.4	5

Common name. Super-purity aluminum, refined aluminum

Chemical Composition

Composition limits. 99.99 Al min, 0.006 Si max, 0.006 Fe max, 0.006 Cu max, 0.002 Mn max, 0.006 Mg max, 0.006 Zn max, 0.002 Ti max, 0.005 V max, 0.005 Ga max, 0.002 max other (each)
Consequence of exceeding impurity limits. See Fig. 3.

Applications

Typical uses. Electrolytic capacitor foil, vapor deposited coatings for optically reflecting surfaces

Mechanical Properties

Tensile properties. See Table 10 and Fig. 3.
Hardness. O temper, 15 HB; H18 temper, 27 HB. (500 kg load; 10 mm diam ball). See also Fig. 3.
Elastic modulus. Tension, 62 GPa (9.0 × 10^6 psi); shear, 25.0 GPa (3.62 × 10^6 psi)

Mass Characteristics

Density. 2.70 g/cm^3 (0.0975 lb/in.^{3}) at 20 °C (68 °F)

Thermal Properties

Melting point. 660 °C (1220 °F)
Coefficient of thermal expansion. Linear:

Temperature range °C	Temperature range °F	Average coefficient μm/m · K	Average coefficient μin./in. · °F
−50 to 20	−58 to 68	21.8	12.1
20 to 100	68 to 212	23.6	13.1
20 to 200	68 to 392	24.5	13.6
20 to 300	68 to 572	25.5	14.2

Specific heat. 900 J/kg · K at 25 °C (77 °F)
Heat of fusion. 390 kJ/kg · K
Thermal conductivity. O temper, 243 W/m · K (140 Btu/ft · h · °F) at 20 °C (68 °F)

Electrical Properties

Electrical conductivity. Volumetric, O temper: 64.5% IACS at 20 °C (68 °F)
Electrical resistivity. O temper: 26.7 nΩ · m at 20 °C (68 °F); temperature coefficient, O temper: 0.1 nΩ · m per K at 20 °C (68 °F)

Optical Properties

Reflectivity. 85 to 90% to visible light for an electrolytically brightened surface

1350
99.50 Al min

Commercial Names

Common name. Electrical conductor grade (EC)

Specifications

ASTM. Aluminum conductor steel reinforced B 232, B 401. Bus conductors: B 236. Communication wire: B 314. Rolled redraw rod: B 233. Round wire: B 230, B 609. Wire, rectangular and square: B 324. Round solid conductors: B 544. Stranded conductors: B 231, B 400
Foreign. France: NF A5/L. Spain: UNE AL99.5E. United Kingdom: BS1E. West Germany: DIN E-A199.5

Chemical Composition

Composition limits. 99.50 Al min, 0.10 Si max, 0.40 Fe max, 0.05 Cu max, 0.01 Mn max, 0.01 Cr max, 0.05 Zn max, 0.03 Ga max, 0.02 V max + Ti, 0.05 B max, 0.03 max other (each), 0.10 max others (total)
Consequence of exceeding impurity limits. Impurity elements in excess of limits degrade electrical conductivity.

Applications

Typical uses. Wire, stranded conductors, bus conductors, transformer strip

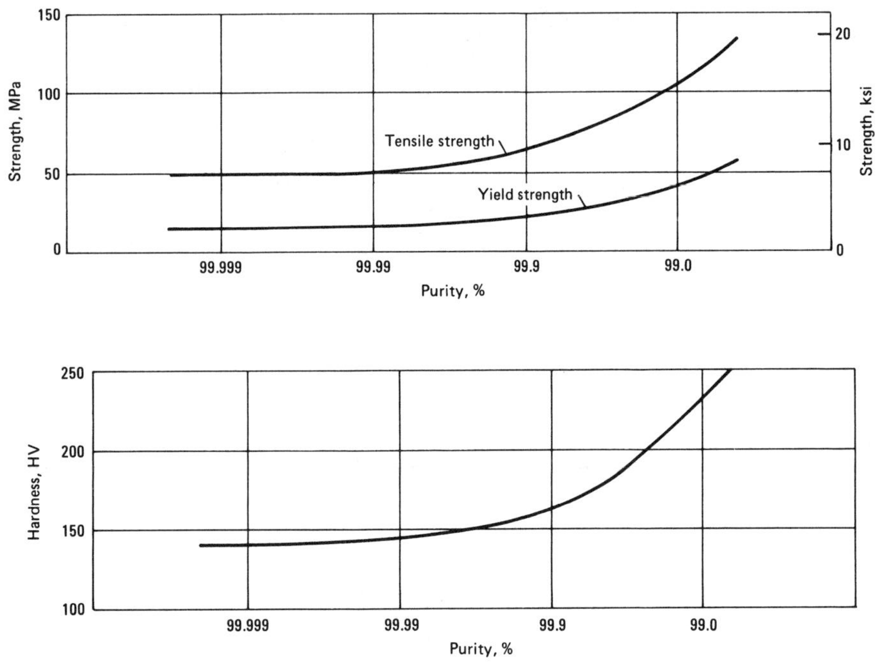

Fig. 3 Effect of purity on strength and hardness of unalloyed aluminum

Table 11 Typical mechanical properties of 1350 aluminum

Temper	Tensile strength MPa	Tensile strength ksi	Yield strength MPa	Yield strength ksi	Elongation(a), %	Shear strength MPa	Shear strength ksi
O	83	12	28	4	23	55	8
H12	97	14	83	12	...	62	9
H14	110	16	97	14	...	69	10
H16	124	18	110	16	...	76	11
H19	186	27	165	24	1.5	103	15

(a) In 250 mm (10 in.), value applicable to wire only

Table 12 Tensile-property limits for 1350 aluminum

Temper	Tensile strength Minimum MPa	Tensile strength Minimum ksi	Tensile strength Maximum MPa	Tensile strength Maximum ksi	Yield strength (min) MPa	Yield strength (min) ksi	Elongation (min), %(a)
Sheet and plate							
O	55	8.0	95	14.0	...	...	15–28
H12	83	12.0	115	17.0	...	...	3–12
H14	95	14.0	130	19.0	...	...	1–10
H16	110	16.0	145	21.0	...	...	1–4
H18	125	18.0	...	...	...	...	1–4
H112							
0.250–0.499 in.	75	11.0	...	...	...	...	10
0.500–1.000 in.	70	10.0	...	...	...	...	16
1.001–1.500 in.	62	9.0	...	...	...	...	22
Wire(b) and redraw rod(c)							
O	58	8.5	95	14.0	...	...	...
H12 and H22	83	12.0	115	17.0	...	...	...
H14 and H24	105	15.0	140	20.0	...		...
H16 and H26	115	17.0	150	22.0	...		...
Extrusions(d)							
H111	58	8.5	...	...	25	3.5	...
Rolled bar(e)							
H12	83	12.0	...	...	55	8.0	...
Sawed-plate bar							
H112							
0.125–0.499 in.	75	11.0	...	...	40	6.0	...
0.500–1.000 in.	70	10.0	...	...	28	4.0	...
1.001–1.500 in.	62	9.0	...	...	25	3.5	...

(a) In 50 mm (2 in.) or 4*d*, where *d* is diameter of reduced section of test specimen. Where a range of values appears in this column, specified minimum elongation varies with thickness of the mill product. (b) Up through 9.50 mm (0.374 in.) diam. (c) 9.52 mm (0.375 in.) diam. (d) Bar, rod, tubular products, and structural shapes. (e) 3 to 25 mm (0.125 to 1.0 in.) thick

Mechanical Properties

Tensile properties. Typical, see Table 11; property limits, see Tables 12 and 13.
Shear strength. See Table 11.
Poisson's ratio. 0.33 at 20 °C (68 °F)
Elastic modulus. Tension, 69 GPa (10×10^6 psi)
Fatigue strength. H19 temper, 48 MPa (7 ksi) at 5×10^8 cycles in an R.R. Moore type test

Mass Characteristics

Density. 2.705 g/cm^3 (0.0977 lb/in.^{3}) at 20 °C (68 °F)

Thermal Properties

Liquidus temperature. 657 °C (1215 °F)
Solidus temperature. 646 °C (1195 °F)
Coefficient of thermal expansion. Linear:

Temperature range °C	Temperature range °F	Average coefficient µm/m · K	Average coefficient µin./in. · °F
−50 to 20	−58 to 68	21.8	12.1
20 to 100	68 to 212	23.6	13.1
20 to 200	68 to 392	24.5	13.6
20 to 300	68 to 572	25.5	14.2

Volumetric: 68×10^{-6} m^3/m^3 · K (3.8×10^{-5} in.^{3}/in.^{3} · °F)
Specific heat. 900 J/kg · K (0.215 Btu/lb · °F) at 20 °C (68 °F)
Thermal conductivity. O temper, 234 W/m · K (135 Btu/ft · h · °F); H19 temper, 230 W/m · K (133 Btu/ft · h · °F)

Electrical Properties

Electrical conductivity. Volumetric, at 20 °C (68 °F). O temper, 61.8% IACS min; H1*x* tempers, 61.0% IACS min
Electrical resistivity. O temper, 27.9 nΩ · m max at 20 °C (68 °F); H1*x* tempers, 28.2 nΩ · m max at 20 °C. Temperature coefficient, all tempers: 0.1 nΩ · m per K at 20 °C
Electrolytic solution potential. −0.84 V versus 0.1 *N* calomel electrode in aqueous solution of 53 g NaCl plus 3 g H_2O_2 per liter at 25 °C (77 °F)

Fabrication Characteristics

Annealing temperature. 345 °C (650 °F)

2011
5.5Cu-0.4Pb-0.4Bi

Specifications

ASTM. Drawn, seamless tube: B 210. Rolled or cold finished wire, rod, and bar: B 211
SAE. J454
UNS number. A92011
Government. Rolled or cold finished wire, rod, and bar: QQ-A-225/3
Foreign. Canada: CSA CB60. France: NF A-U4Pb. United Kingdom: BS FC1. Germany: DINAL CuBiPb

Table 13 Tensile-property limits for 1350 aluminum wire, H19 temper

Wire diameter, in.	Minimum tensile strength: Individual(a) MPa	ksi	Average(b) MPa	ksi	Minimum elongation(c), %: Individual(a)	Average(b)
0.0105–0.0500	160	23.0	172	25.0	...	...
0.0501–0.0600	185	27.0	200	29.0	1.2	1.4
0.0601–0.0700	185	27.0	195	28.5	1.3	1.5
0.0701–0.0800	183	26.5	193	28.0	1.4	1.6
0.0801–0.0900	180	26.0	190	27.5	1.5	1.6
0.0901–0.1000	175	25.5	185	27.0	1.5	1.6
0.1001–0.1100	170	24.5	180	26.0	1.5	1.6
0.1101–0.1200	165	24.0	175	25.5	1.6	1.7
0.1201–0.1400	162	23.5	172	25.0	1.7	1.8
0.1401–0.1500	162	23.5	170	24.5	1.8	1.9
0.1501–0.1800	160	23.0	165	24.0	1.9	2.0
0.1801–0.2100	160	23.0	165	24.0	2.0	2.1
0.2101–0.2600	155	22.5	162	23.5	2.2	2.3

(a) Minimum value for any test in a given lot. (b) Minimum value for average of all tests for a given lot. (c) In 250 mm (10 in.)

Chemical Composition

Composition limits. 0.40 Si max, 0.7 Fe max, 5.0 to 6.0 Cu, 0.30 Zn max, 0.20 to 0.6 Pb, 0.05 max other (each), 0.15 others (total), bal Al

Applications

Typical uses. Wire, rod, and bar for screw machine products. Applications where good machinability and good strength are required

Mechanical Properties

Tensile properties. See Tables 14 and 15.
Compressive yield strength. Approximately equal to tensile yield strength
Hardness. See Table 14.
Poisson's ratio. 0.33 at 20 °C (68 °F)
Elastic modulus. Tension, 70 GPa (10.2 × 10^6 psi); shear, 26 GPa (3.8 × 10^6 psi)
Fatigue strength. At 5 × 10^8 cycles, R.R. Moore type test: T3 and T8 tempers, 124 MPa (18 ksi)

Mass Characteristics

Density. 2.82 g/cm^3 (0.102 lb/in.^{3}) at 20 °C (68 °F)

Thermal Properties

Liquidus temperature. 638 °C (1180 °F)
Solidus temperature. 541 °C (1005 °F)
Incipient melting temperature. 535 °C (995 °F)
Coefficient of thermal expansion. Linear:

Temperature range °C	°F	Average coefficient µm/m · K	µin./in. · °F
−50 to 20	−58 to 68	21.4	11.9
20 to 100	68 to 212	23.1	12.8
20 to 200	68 to 392	24.0	13.3
20 to 300	68 to 572	25.0	13.9

Volumetric: 67 × 10^{-6} m^3/m^3 · K (3.72 × 10^{-5} in.^{3}/in.^{3} · °F) at 20 °C (68 °F)
Specific heat. 864 J/kg · K (0.206 Btu/lb · °F) at 20 °C (68 °F)
Thermal conductivity. At 20 °C (68 °F): T3 and T4 tempers, 152 W/m · K (87.8 Btu/ft · h · °F); T8 temper, 173 W/m · K (99.9 Btu/ft · h · °F)

Electrical Properties

Electrical conductivity. Volumetric, at 20 °C (68 °F): T3 and T4 tempers, 39% IACS; T8 temper, 45% IACS
Electrical resistivity. At 20 °C (68 °F): T3 and T4 tempers, 44 nΩ · m; T8 temper, 38 nΩ · m; temperature coefficient. T3, T4, and T8 tempers, 0.1 nΩ · m per K at 20 °C (68 °F)
Electrolytic solution potential. At 25 °C (77 °F): −0.69 V (T3 and T4 tempers), −0.83 V (T8 temper) versus 0.1 *N* calomel electrode in an aqueous solution containing 53 g NaCl plus 3 g H_2O_2 per liter

Fabrication Characteristics

Annealing temperature. 413 °C (775 °F)
Solution temperature. 524 °C (975 °F)
Aging temperature. T8 temper, 160 °C (320 °F); 14 h at temperature

Table 14 Room-temperature mechanical properties of alloy 2011

Temper	Tensile strength MPa	ksi	Yield strength(a) MPa	ksi	Elongation, %	Hardness(b), HB	Shear strength MPa	ksi
Typical properties								
T3(c)	379	55	296	43	15(d)	95	221	32
T8(c)	407	59	310	45	12(d)	100	241	35
Property limits (minimum values for rolled or cold finished wire, rod, and bar)								
T3								
0.125 to 1.500 in. thick	310	45	260	38	10(a)(e)	...	...	...
1.501 to 2.000 in. thick	295	43	235	34	12(a)(e)	...	...	...
2.001 to 3.250 in. thick	290	42	205	30	14(a)(e)	...	...	...
T4, T451								
0.375 to 8.000 in. thick	275	40	125	18	16	...	...	...
T8								
0.125 to 3.250 in. thick	370	54	275	40	10	...	...	...

(a) Yield strength and elongation limits not applicable to wire less than 3.2 mm (0.125 in.) in thickness or diameter. (b) 500 kg (1100 lb) load; 10 mm diam ball. (c) Strengths and elongations generally unchanged or improved at low temperatures. (d) 13 mm (½ in.) diam specimen. (e) In 2 in. or 4*d*, where *d* is diameter of reduced section of tensile test specimen

Table 15 Typical tensile properties of alloy 2011-T3

Temperature °C	°F	Tensile strength MPa	ksi	Yield strength (0.2% offset) MPa	ksi	Elongation, %
24	75	379	55	296	43	15
100	212	324	47	234	34	16
149	300	193	28	131	19	25
204	400	110	16	76	11	35
260	500	45	6.5	26	3.8	45
316	600	21	3.1	12	1.8	90
371	700	16	2.3	10	1.4	125

(a) Lowest strength for exposures up to 10 000 h at temperature, no load; test loading applied at 35 MPa/min (6 ksi/min) to yield strength and then at strain rate of 5%/min to fracture

2014, Alclad 2014
4.4Cu-0.8Si-0.8Mn-0.5Mg

Specifications

AMS. See Table 16.
ASME. Rolled or cold finished wire, rod, and bar: SB211. Forgings: SB247
ASTM. See Table 16.
SAE. J454
UNS number. A92014
Government. See Table 16.
Foreign. Canada: CSA CS41N. France: NF A-U4SG. Germany: DIN AlCuSiMn. ISO: AlCu4SiMg. United Kingdom: BS H15

Chemical Composition

Composition limits of 2014. 3.9 to 5.0 Cu, 0.50 to 1.2 Si, 0.7 Fe max, 0.40 to 1.2 Mn, 0.20 to 0.8 Mg, 0.25 Zn max, 0.10 Cr max, 0.15 Ti max, 0.05 max other (each), 0.15 max others (total), bal Al
Composition limits of Alclad 2014. 6006 cladding—0.20 to 0.6 Si, 0.35 Fe max, 0.15 to 0.30 Cu, 0.05 to 0.20 Mn, 0.45 to 0.9 Mg, 0.10 Cr max, 0.10 Zn max, 0.10 Ti max, 0.05 max other (each), 0.15 max others (total), bal Al

Table 16 Standard specifications for alloy 2014

Mill form	Specification number AMS	ASTM	Government
Sheet and plate	4014	B 209	...
	4028	...	...
	4029	...	...
Rolled or cold finished wire, rod, and bar	4121	B 211	QQ-A-225/4
Extruded wire, rod, bar, shapes, and tube	4153	B 221	QQ-A-200/2
Extruded seamless tube	...	B 241	...
Drawn, seamless tube	...	B 210	...
Forgings	4133	B 247	QQ-A-367
	4134		MIL-A-22771
	4135		
Forging stock	4134	...	QQ-A-367
	4133	...	...
	4135	...	...
Impacts	...	...	MIL-A-12545
Sheet and plate (Alclad)	...	B 209	QQ-A-250/3

Table 17 Typical tensile properties of alloy 2014

Temper	Tensile strength MPa	ksi	Yield strength MPa	ksi	Elongation, %	Hardness, HB	Shear strength MPa	ksi	Fatigue strength MPa	ksi
Bare 2014										
O	186	27	97	14	18(a)	45	125	18	90	13
T4	427	62	290(b)	42(a)	20(a)	105	260	38	140	20
T6(c)	483	70	414	60	13(a)	135	240	42	125	18
Alclad 2014										
O	172	25	69	10	21(e)	...	125	18	...	...
T3(d)	434	63	276	40	20(e)	...	255	37	...	...
T4(d)	421	61	255	37	22(e)	...	255	37	...	...
T6(d)	469	68	414	60	10(e)	...	285	41	...	...

(a) Round bar 13 mm (½ in.) diam. (b) Die forgings have about 20% lower yield strength. (c) Extruded products more than 19 mm (¾ in.) thick have 15 to 20% higher strengths. (d) Sheet less than 1 mm (0.04 in.) thick has slightly lower strength. (e) Sheet 1.6 mm (1/16 in.) thick

Table 18 Typical tensile properties of alloy 2014-T6 or 2014-T651 at various temperatures

Lowest strength for exposures up to 10 000 h at temperature under no load; test loading applied at 35 MPa/min (5 ksi/min) to yield strength and then at strain rate of 5%/min to fracture

Temperature °C	°F	Tensile strength MPa	ksi	Yield strength(a) MPa	ksi	Elongation, %
−196	−320	579	84	496	72	14
−80	−112	510	74	448	65	13
−28	−18	496	72	427	62	13
24	75	483	70	414	60	13
100	212	439	63	393	57	15
149	300	276	40	241	35	20
204	400	110	16	90	13	38
260	500	66	9.5	52	7.5	52
316	600	45	6.5	34	5	65
371	700	30	4.3	24	3.5	72

(a) 0.2% offset

Applications

Typical uses. Heavy-duty forgings, plate, and extrusions for aircraft fittings, wheels, and major structural components, space booster tankage and structure, truck frame and suspension components. Applications requiring high strength and hardness including service at elevated temperatures

Mechanical Properties

Tensile properties. See Tables 17 to 19.
Compressive yield strength. Approximately the same as tensile yield strength
Hardness. O temper: 87 to 98 HRH; 45 HB. T4 temper: 65 to 73 HRB; 105 HB. T6 temper: 80 to 86 HRB; 135 HB. HB values obtained using 500 kg load and 10 mm diam ball
Poisson's ratio. 0.33 at 20 °C (68 °F)
Elastic modulus. Tension: 2014, 72.4 GPa (10.5 × 10^6 psi); Alclad 2014, 71.7 GPa (10.4 × 10^6 psi). Shear: 2014 and Alclad 2014, 28 GPa (4.0 × 10^6 psi). Compression: 2014, 73.8 GPa (10.7 × 10^6 psi); Alclad 2014, 73.1 GPa (10.6 × 10^6 psi)
Fatigue strength. O temper, 90 MPa (13 ksi); T4 temper, 140 MPa (20 ksi); T6 temper, 125 MPa (18 ksi); all at 5 × 10^8 cycles in an R.R. Moore type test

Mass Characteristics

Density. 2.80 g/cm³ (0.101 lb/in.³) at 20 °C (68 °F)

Thermal Properties

Liquidus temperature. 638 °C (1180 °F)
Solidus temperature. 507 °C (945 °F)
Coefficient of thermal expansion.

Linear:

Temperature range °C	°F	Average coefficient µm/m · K	µin./in. · °F
−50 to 20	−58 to 68	20.8	11.5
20 to 100	68 to 212	22.5	12.5
20 to 200	68 to 392	23.4	13.0
20 to 300	68 to 572	24.4	13.6

Volumetric: 65.1 × 10^{-6} m³/m³ · K (3.62 × 10^{-5} in.³/in.³ · °F)
Thermal conductivity. At 20 °C (68 °F): O temper, 192 W/m · K (111 Btu/ft · h · °F); T3, T4, T451 tempers, 134 W/m · K (77.4 Btu/ft · h · °F); T6, T651, T652 tempers, 155 W/m · K (89.5 Btu/ft · h · °F)

Electrical Properties

Electrical conductivity. At 20 °C (68 °F): O temper, 50% IACS, T3, T4, T451 tempers, 34% IACS; T6, T651, T652 tempers, 40% IACS
Electrical resistivity. At 20 °C (68 °F): O temper, 34 nΩ · m; T3, T4, T451 tempers, 51 nΩ · m; T6, T651, T652 tempers, 43 nΩ · m. Temperature coefficient: O, T3, T4, T451, T6, T651, T652 tempers, 0.1 nΩ · m per K at 20 °C (68 °F)
Electrolytic solution potential. At 25 °C (77 °F): −0.68 V (T3, T4, T451 tempers) or −0.78 V (T6, T651, T652 tempers) versus 0.1 *N* calomel electrode in an aqueous solution containing 53 g NaCl plus 3 g H_2O_2 per liter

Fabrication Characteristics

Annealing temperature. 413 °C (775 °F)
Solution temperature. 502 °C (935 °F)
Aging temperature. T6 temper. Sheet, plate, wire, rod, bar, shapes, and tube: 160 °C (320 °F) for 18 h at temperature. Forgings: 171 °C (340 °F) for 10 h at temperature

2017
4.0Cu-0.6Mg-0.7Mn-0.5Si

Specifications

ASTM. B 211 and B 316
SAE. J454
ANSI. H38.4 and H38.12
UNS number. A92017
Government. QQ-A-222/5, QQ-A-430, MIL-R-430
Foreign. France: A-U46. Germany: AlCuMg1 and 3.1325. Great Britain: L18 and 150A. Canada: CM41. Austria: AlCuMg1. ISO: AlCuMgSi

Chemical Composition

Composition limits. 0.20 to 0.80 Si, 0.7 max Fe, 3.5 to 4.5 Cu, 0.4 to 0.80 Mg, 0.40 to 1.0 Mn, 0.10 max Cr, 0.15 max Ti, 0.25 max Zn, 0.05 other (each), 0.15 others (total); bal Al

Applications

Typical uses. Alloy 2017, which was the first alloy developed in the Al-Cu-Mg series, is now in rather limited use, chiefly for

Table 19 Tensile-property limits for alloy 2014

Temper	Tensile strength, Minimum, MPa	Tensile strength, Minimum, ksi	Tensile strength, Maximum, MPa	Tensile strength, Maximum, ksi	Yield strength (min), MPa	Yield strength (min), ksi	Elongation(a), %
Flat products (bare)							
Sheet and plate, O							
0.020–0.499 in. thick	···	···	220	32	110 (max)	16 (max)	16
0.500–1.000 in. thick	···	···	220	32	···	···	10
Flat sheet, T3							
0.020–0.039 in. thick	405	59	···	···	240	35	15
0.040–0.249 in. thick	405	59	···	···	250	36	14
Coiled sheet, T4							
0.020–0.249 in. thick	405	59	···	···	240	35	14
Plate, T451(b)							
0.250–2.000 in. thick	400	58	···	···	250	36	14–12
2.001–3.000 in. thick	395	57	···	···	250	36	8
Sheet and plate, T42							
0.020–1.000 in. thick	400	58	···	···	235	34	14
Sheet, T6, T62							
0.020–0.039 in. thick	440	64	···	···	395	57	6
0.040–0.249 in. thick	455	66	···	···	400	58	7
Plate, T62, T651							
0.250–2.000 in. thick	460	67	···	···	405	59	7–4
2.001–2.500 in. thick	450	65	···	···	400	58	2
2.501–3.000 in. thick	435	63	···	···	395	57	2
3.001–4.000 in. thick	405	59	···	···	380	55	1
Flat products (Alclad)							
Sheet and plate, O							
0.020–0.499 in. thick	···	···	205	30	95 (max)	14 (max)	16
0.500–1.000 in. thick	···	···	220	32	···	···	10
Flat sheet, T3							
0.020–0.024 in. thick	370	54	···	···	230	33	14
0.025–0.039 in. thick	380	55	···	···	235	34	14
0.040–0.249 in. thick	395	57	···	···	240	35	15
Coiled sheet, T4							
0.020–0.024 in. thick	370	54	···	···	215	31	14
0.025–0.039 in. thick	380	55	···	···	220	32	14
0.040–0.249 in. thick	395	57	···	···	235	34	15
Plate, T451(b)							
0.500–2.000 in. thick	400	58	···	···	250	36	12–14
0.250–0.499 in. thick	395	57	···	···	250	36	15
0.500–2.000 in. thick	400	58	···	···	250	36	12–14
2.001–3.000 in. thick	395	57	···	···	200	36	8
Sheet and plate, T4							
0.020–0.024 in. thick	370	54	···	···	215	31	14
0.025–0.039 in. thick	380	55	···	···	220	32	14
0.040–0.499 in. thick	395	57	···	···	235	34	15
0.500–1.000 in. thick	400	58	···	···	235	34	14
Sheet, T6							
0.020–0.024 in. thick	425	62	···	···	370	54	7
0.025–0.039 in. thick	435	63	···	···	380	55	7
0.040–0.249 in. thick	440	64	···	···	395	57	8
Plate, T62, T651							
0.250–0.499 in. thick	440	64	···	···	395	57	8
0.500–2.000 in. thick	460	67	···	···	405	59	6
2.001–2.500 in. thick	450	65	···	···	400	58	2
2.501–3.000 in. thick	435	63	···	···	395	57	2
3.001–4.000 in. thick	405	59	···	···	380	55	1
Rolled or cold finished wire (rod and bar)							
T4, T42, T451(b)	380	55	···	···	220	32	16
T6, T62, T651	450	65	···	···	380	55	8
Extruded wire, rod, bar, and shapes							
O	···	···	205	30	125 (max)	12 (max)	12
T4, T4510, T4511	345	50	···	···	240	35	12
T42	345	50	···	···	200	29	12
T6, T6510, T6511							
≤0.499 in. thick	415	60	···	···	365	53	7
0.500–0.749 in. thick	440	64	···	···	400	58	7
0.750 in. thick	470	68	···	···	415	60	7
T62	415	60	···	···	365	53	7(c)
Extruded tube							
O	···	···	205	30	125 (max)	18 (max)	12
T4, T4510, and T4511	345	50	···	···	240	35	12
T42	345	50	···	···	200	29	12
Extruded tube (continued)							
T6, T6510, T6511							
≤0.499 in. thick	415	60	···	···	365	53	7
0.500–0.749 in. thick	440	64	···	···	400	58	7
≥0.750 in. thick	470	68	···	···	415(d)	60(d)	7(d)
T62	415	60	···	···	365	53	7(c)
Drawn tube							
O, 0.18–0.500 in. thick	···	···	220	32	110 (max)	16 (max)	···
T4, T42							
0.018–0.500 in. thick	370	54	···	···	205	30	10–16
Die forgings: axis parallel to direction of grain flow							
T4, ≤4 in. thick	380	55	···	···	205	30	11(e)(f)
T6							
≤2 in. thick	450	65	···	···	385	56	6(e)(g)
>2–3 in. thick	450	65	···	···	380	55	6(e)(g)
≤3–4 in. thick	435	63	···	···	380	55	6(e)(g)
Die forgings: axis not parallel to direction of grain flow							
T6, ≤2 in. thick	440	64	···	···	380	55	3(e)(h)
>2–4 in. thick	435	63	···	···	370	54	2(e)
Hand forgings							
T6							
≤2.000 in. thick longitudinal, long transverse	450	65	···	···	385	56	3–8
2.001–3.000 in. thick							
Longitudinal	440	64	···	···	385	56	8
Long transverse	440	64	···	···	380	55	3
Short transverse	425	62	···	···	380	55	2
3.001–4.000 in. thick							
Longitudinal, long transverse	435	63	···	···	380	55	3–8
Short transverse	420	61	···	···	370	54	2
4.001–5.000 in. thick							
Longitudinal, long transverse	425	62	···	···	370	54	2–7
Short transverse	415	60	···	···	365	53	1
5.001–6.000 in. thick							
Longitudinal, long transverse	420	61	···	···	365	53	2–7
Short transverse	405	59	···	···	365	53	1
6.001–7.000 in. thick							
Longitudinal,							
Long transverse	415	60	···	···	360	52	2–7
Short transverse	400	58	···	···	360	52	1
7.001–8.000 in. thick							
Longitudinal, long transverse	405	59	···	···	350	51	2–7
Short transverse	395	57	···	···	350	51	1
T652							
≤2.000 in. thick							
Longitudinal, long transverse	450	65	···	···	385	56	3–8
2.001–3.000 in. thick							
Longitudinal	440	64	···	···	385	56	8
Long transverse	440	64	···	···	380	55	3
Short transverse	425	62	···	···	360	52	2
3.001–4.000 in. thick							
Longitudinal, long transverse	435	63	···	···	380	55	3–8
Short transverse	420	61	···	···	350	51	2
4.001–5.000 in. thick							
Longitudinal, long transverse	425	62	···	···	370	54	2–7
Short transverse	415	60	···	···	345	50	1
5.001–6.000 in. thick							
Longitudinal, long transverse	420	61	···	···	365	53	2–7
Short transverse	405	59	···	···	345	50	1
6.001–7.000 in. thick							
Longitudinal, long transverse	415	60	···	···	360	52	2–6
Short transverse	400	58	···	···	340	49	1
7.001–8.000 in. thick							
Longitudinal, long transverse	405	59	···	···	350	51	2–6
Short transverse	395	57	···	···	330	48	1
Rolled rings, T6, T652							
≤2.500 in. thick							
Tangential	450	65	···	···	380	55	7
Axial	425	62	···	···	380	55	3
Radial	415	60	···	···	360	52	2
2.501–3.000 in. thick							
Tangential	450	65	···	···	380	55	6
Axial	425	62	···	···	360	52	2

(a) In 50 mm (2 in.) or $4d$, where d is diameter of reduced section of tensile test specimen. Where a range of values appears in this column, specified minimum elongation varies with thickness of the mill product. (b) Upon artificial aging, T451 temper material develops properties applicable to T651 temper. (c) 6% elongation for products over 19 mm (0.750 in.) in diameter or thickness and over 160 through 205 cm^2 (25 through 32 in.^{2}) in cross-sectional area. (d) Value slightly lower for material over 160 through 205 cm^2 (25 through 32 in.^{2}) in cross-sectional area. (e) Test bar machined from sample forging. (f) 16% for test bar taken from separately forged coupon. (g) 8% for test bar taken from separately forged coupon. (h) 2% for forgings over 25 through 50 mm (1 through 2 in.) thick

Table 20 Typical room-temperature mechanical properties of 2017

Property	Temper condition O	T4, T451
Tensile strength, MPa (ksi)	180 (26)	427 (62)
Yield strength (0.2% offset), MPa (ksi)	70 (10)	275 (40)
Elongation in 50 mm (2 in.)(a), %	22	22
Hardness, HB(b)	45	105
Shear strength, MPa (ksi)	125 (18)	262 (38)
Fatigue strength (5×10^8 cycles), MPa (ksi)	90 (13)	125 (18)

(a) Specimens 13 mm (½ in.) diameter. (b) 500 kg load, 10 mm ball

Table 21 Typical tensile properties of 2017 (T4 and T451 tempers) at various temperatures

Test temperature(a) °C	°F	Tensile strength MPa	ksi	Yield strength (0.2% offset) MPa	ksi	Elongation in 50 mm (2 in.), %
−196	−320	550	80	365	53	28
−80	−112	448	65	290	42	24
−28	−18	440	64	283	41	23
24	75	427	62	275	40	22
100	212	393	57	270	39	18
149	300	275	40	207	30	15
204	400	110	16	90	13	35
260	500	62	9	52	7.5	45
316	600	40	6	35	5	65
371	700	30	4.3	24	3.5	70

(a) Tested after holding 10 000 h at temperature

rivets. Used in components for general engineering purposes, structural applications in construction and transportation, screw machine products, and fittings
General characteristics. Age-hardenable wrought aluminum alloy with medium strength and ductility, good machinability, good formability, and fair resistance to atmospheric corrosion. Welding is not recommended unless heat treatment after welding is practicable. Its service temperature is below 100 °C (212 °F).
Forms available. Forgings, extrusions, bars, rods, wire, shapes, and rivets

Mechanical Properties

Tensile properties. See Tables 20 and 21.
Hardness. See Table 20.
Shear strength. See Table 20.
Modulus of elasticity. 72.4 GPa (10.5×10^6 psi) average of tension and compression; modulus is about 2% greater for compression than tension
Modulus of rigidity. 27.5 GPa (4×10^6 psi)
Fatigue strength. See Table 20.

Mass Characteristics

Density. 2.80 g/cm^3 (0.101 lb/in.^{3})

Thermal Properties

Liquidus temperature. 640 °C (1185 °F)
Solidus temperature. 513 °C (955 °F)
Thermal conductivity. At 25 °C (77 °F): 193 W/m · °C (1340 Btu · in./ft^2 · h · °F) with an O temper and 134 W/m · °C (930 Btu · in./ft^2 · h · °F) with a T4 temper
Coefficient of thermal expansion. From 20 to 100 °C (68 to 212 °F): 23.6 μm/m · °C (13.1 μin./in. · °F)

Electrical Properties

Electrical conductivity. At 20 °C (68 °F): 50% IACS on a volume basis (159% IACS on weight basis) with an O temper; 34% IACS on a volume basis (108% IACS on a weight basis) with a T4 temper
Electrical resistivity. At 20 °C (68 °F): 0.035 Ω · mm^2/m (21 Ω · circ mil/ft) with an O temper and 0.05 Ω · mm^2/m (30 Ω · circ mil/ft) with a T4 temper

Fabrication Characteristics

Annealing temperature. 415 °C (775 °F) for a heat-treated anneal and 340 to 350 °C (640 to 660 °F) for cold-work anneal
Solution temperature. 500 to 510 °C (930 to 950 °F)
Aging temperature. Room temperature
Machinability. Fair to good in the annealed condition and excellent in the solution treated and naturally aged condition (T4 temper)
Workability. Has good formability. In the annealed condition (O temper) its formability is equal to or superior to 2024-O. In the T4 temper condition, it forms as readily as 2024-T3 or 2024-T4.
Weldability. Because of the effect of heating on corrosion resistance, welding is rarely recommended except where heat treatment after welding is practicable. The inert gas method and resistance welding have given satisfactory results. Gas welding, brazing, and soldering are not successful. This alloy is so sensitive to cracking during welding that other aluminum alloys, joint design, fixtures, and so on must be arranged so as to put a minimum stress on the joint during cooling. The best filler material is parent metal.

Corrosion Resistance

2017 has a fair resistance to atmospheric corrosion, depending on its thermal treatment. Quenching slowly from the solution-treatment temperature lowers the resistance to corrosion and makes this alloy susceptible to intergranular attack. The same result is obtained by heating the alloy after solution treatment. If, however, the alloy has been slowly quenched, artificial aging tends to restore the normal resistance to attack; in fact, for material that is to be artificially aged, a mild quench may be preferable. For thicker sections, the rate of cooling even by immersion in cold water is not great enough to produce complete freedom from susceptibility to intergranular attack. In thin sections the solution treated material, being aged at room temperature, is more resistant to corrosion than the fully aged material, while in heavy sections the latter is more resistant because of the beneficial effect of artificial aging on more slowly cooled material.

2024, Alclad 2024
4.4Cu-1.5Mg-0.6Mn

Specifications

AMS. See Table 22.
ASME. Rolled or drawn wire, rod, and bar: SB211. Extrusions: SB221
ASTM. See Table 22.
SAE. J454
UNS number. A92024
Government. See Table 22.
Foreign. Austria: Önorm AlCuMg2. Canada: CSA CG42. France: NF A-U4G1. Italy: UNI P-AlCu4.5MgMn; Alclad 2024, P-AlCu4.5MgMn placc. Spain: UNE L-314. Germany: DIN AlCuMg2

Chemical Composition

Composition limits. 0.5 Si max, 0.50 Fe max, 3.8 to 4.9 Cu, 0.30 to 0.9 Mn, 1.2 to 1.8 Mg, 0.10 Cr max, 0.25 Zn max, 0.15 Ti max, 0.05 max other (each), 0.15 max others (total), bal Al
Composition limits of Alclad 2024. 1230 cladding—99.30 Al min, 0.7 Si max + Fe, 0.10 Cu max, 0.05 Mn max, 0.05 Mg max, 0.10 Zn max, 0.05 V max, 0.03 Ti max, 0.03 max other (each)

Applications

Typical uses. Aircraft structures, rivets, hardware, truck wheels, screw machine products, and other miscellaneous structural applications

Mechanical Properties

Tensile properties. See Tables 23, 24, and 25.
Shear strength. See Table 24.
Hardness. See Table 24.
Poisson's ratio. 0.33 at 20 °C (68 °F)
Elastic modulus. Tension, 72.4 GPa (10.5×10^6 psi); shear, 28.0 GPa (4.0×10^6 psi); compression, 73.8 GPa (10.7×10^6 psi)
Fatigue strength. See Table 24.
Elevated-temperature strengths. See Fig. 4.

Mass Characteristics

Density. 2.77 g/cm^3 (0.100 lb/in.^{3}) at 20 °C (68 °F)

Table 22 Standard specifications for alloy 2024

Mill form and condition	Specification number AMS	ASTM	Government
Bare 2024			
Sheet and plate	4033	B 209	QQ-A-250/4
	4035	...	...
	4037	...	...
	4097	...	...
	4098	...	...
	4099	...	...
	4103	...	...
	4104	...	...
	4105	...	...
	4106	...	...
	4192	...	...
	4193	...	...
Wire, rod, and bar (rolled or cold finished)	4112	B 211	QQ-A-225/6
	4119	...	...
	4120	...	...
Wire, rod, bar, shapes, and tube (extruded)	4152	B 221	QQ-A-200/3
	4164	...	...
	4165	...	...
Tube (extruded, seamless)	...	B 241	...
Tube (drawn, seamless)	4087	B 210	WW-T-700/3
	4088	...	MIL-T-50777
Tube (hydraulic)	4086	...	...
Rivet wire and rod	...	B 316	QQ-A-430
Foil	4007	...	MIL-A-81596
Alclad 2024			
Sheet and plate	4034	B 209	QQ-A-250/5
	4040	...	...
	4041	...	...
	4042	...	...
	4060	...	...
	4061	...	...
	4072	...	...
	4073	...	...
	4074	...	...
	4075	...	...
	4194	...	...
	4195	...	...

Thermal Properties

Liquidus temperature. 638 °C (1180 °F)
Solidus temperature. 502 °C (935 °F)
Incipient melting temperature. 502 °C (935 °F)
Coefficient of thermal expansion. Linear:

Temperature range °C	°F	Average coefficient µm/m · K	µin./in. · °F
−50 to 20	−58 to 68	21.1	11.7
20 to 100	68 to 212	22.9	12.7
20 to 200	68 to 392	23.8	13.2
20 to 300	68 to 572	24.7	13.7

Volumetric: 66.0×10^{-6} m³/m³ · K (3.67×10^{-5} in.³/in.³ · °F) at 20 °C (68 °F)
Specific heat. 875 J/kg · K (0.209 Btu/lb · °F) at 20 °C (68 °F)
Thermal conductivity:

Temper	Conductivity W/m · K	Btu/ft · h · °F
O	190	110
T3, T36, T351, T361, T4	120	69
T6, T81, T851, T861	151	88

Table 23 Typical tensile properties of alloy 2024

Temper	Temperature °C	°F	Tensile strength MPa	ksi	Yield strength (0.2% offset) MPa	ksi	Elongation, %
T3 (sheet)	−196	−320	586	85	427	62	18
	−80	−112	503	73	359	52	17
	−28	−18	496	72	352	51	17
	24	75	483	70	345	50	17
	100	212	455	66	331	48	16
	149	300	379	55	310	45	11
	204	400	186	27	138	20	23
	260	500	76	11	62	9	55
	316	600	52	7.5	41	6	75
	371	700	34	5	28	4	100
T4, T351 (plate)	−196	−320	579	84	421	61	19
	−80	−112	490	71	338	49	19
	−28	−18	476	69	324	47	19
	24	75	469	68	324	47	19
	100	212	434	63	310	45	19
	149	300	310	45	248	36	17
	204	400	179	26	131	19	27
	260	500	76	11	62	9	55
	316	600	52	7.5	41	6	75
	371	700	34	5	28	4	100
T6, T651	−196	−320	579	84	469	68	11
	−80	−112	496	72	407	59	10
	−28	−18	483	70	400	58	10
	24	75	476	69	393	57	10
	100	212	448	65	372	54	10
	149	300	310	45	248	36	17
	204	400	179	26	131	19	27
	260	500	76	11	62	9	55
	316	600	52	7.5	41	6	75
	371	700	34	5	28	4	100
T81, T851	−196	−320	586	85	538	78	8
	−80	−112	510	74	476	69	7
	−28	−18	503	73	469	68	7
	24	75	483	70	448	65	7
	100	212	455	66	427	62	8
	149	300	379	55	338	49	11
	204	400	186	27	138	20	23
	260	500	76	11	62	9	55
	316	600	52	7.5	41	6	75
	371	700	34	5	28	4	100
T861	−196	−320	634	92	586	85	5
	−80	−112	558	81	531	77	5
	−28	−18	538	78	510	74	5
	24	75	517	75	490	71	5
	100	212	483	70	462	67	6
	149	300	372	54	331	48	11
	204	400	145	21	117	17	28
	260	500	76	11	62	9	55
	316	600	52	7.5	41	6	75
	371	700	34	5	28	4	100

Electrical Properties

Electrical conductivity. Volumetric, at 20 °C (68 °F):

Temper	Conductivity, %IACS
O	50
T3, T36, T351, T361, T4	30
T6, T81, T851, T861	38

Electrical resistivity:

Temper	Resistivity, nΩ · m
O	34
T3, T36, T351, T361, T4	57
T6, T81, T851, T861	45

Temperature coefficient. 0.1 nΩ · m per K at 20 °C (68 °F)
Electrolytic solution potential. At 25 °C (77 °F) and versus 0.1 *N* calomel electrode in an aqueous solution containing 53 g NaCl plus 3 g H_2O_2 per liter:

Temper	Volts
T3, T4, T361	−0.68
T6, T81, T861	−0.80
Alclad 2024	−0.83

Fabrication Characteristics

Annealing temperature. 413 °C (775 °F)
Solution temperature. 493 °C (920 °F)
Aging temperature. T6 and T8 tempers: 191 °C (375 °F) for 8 to 16 h at temperature

2036
2.6Cu-0.45Mg-0.25Mn

Specifications

UNS number. A92036

Chemical Composition

Composition limits. 0.50 Si max, 0.50 Fe max, 2.2 Cu max, 0.10 to 0.40 Mn, 0.30 to 0.6 Mg, 0.10 Cr max, 0.25 Zn max, 0.15 Ti max, 0.05 max other (each), 0.15 max others (total), bal Al

Applications

Typical uses. Sheet for auto body panels

Mechanical Properties

Tensile properties. Typical, for 0.64 to 3.18 mm (0.025 to 0.125 in.) flat sheet, T4 temper: tensile strength, 340 MPa (49 ksi); yield strength, 195 MPa (28 ksi); elongation, 24% in 50 mm (2 in.). Minimum, for 0.64 to 3.18 mm flat sheet, T4 temper: tensile strength, 290 MPa (42 ksi); yield strength, 160 MPa (23 ksi); elongation, 20% in 50 mm (2 in.)
Hardness. Typical, T4 temper: 80 HR15T
Strain-hardening exponent. 0.23
Elastic modulus. Tension, 70.3 GPa (10.2×10^6 ksi); compression, 71.7 GPa (10.4×10^6 ksi)
Fatigue strength. Typical, T4 temper: 124 MPa (18 ksi) at 10^7 cycles for flat sheet tested in reversed flexure

Mass Characteristics

Density. 2.75 g/cm^3 (0.099 lb/in.^{3}) at 20 °C (68 °F)

Thermal Properties

Liquidus temperature. 650 °C (1200 °F)
Solidus temperature. 554 °C (1030 °F)
Incipient melting temperature. 510 °C (950 °F)
Coefficient of thermal expansion.
Linear:

Temperature range, °C	Temperature range, °F	Average coefficient, μm/m · K	Average coefficient, μin./in. · °F
−50 to 20	−58 to 68	21.6	12.0
20 to 100	68 to 212	23.4	13.0
20 to 200	68 to 392	24.3	13.5
20 to 300	68 to 572	25.2	14.0

Volumetric: 67.5×10^{-6} m^3/m^3 · K (3.75×10^{-5} in.^{3}/in.^{3} · °F) at 20 °C (68 °F)
Specific heat. 882 J/kg · K (0.211 Btu/lb · °F) at 20 °C (68 °F)
Thermal conductivity. At 20 °C (68 °F): O temper, 198 W/m · K (114 Btu/ft · h · °F); T4 temper, 159 W/m · K (91.8 Btu/ft · h · °F)

Electrical Properties

Electrical conductivity. Volumetric, at 20 °C (68 °F): O temper, 52% IACS; T4 temper, 41% IACS
Electrical resistivity. At 20 °C (68 °F): O temper, 33.2 nΩ · m; T4 temper, 42.1 nΩ · m. Temperature coefficient, at 20 °C (68 °F): O and T4 tempers, 0.1 nΩ · m per K
Electrolytic solution potential. At 25 °C (77 °F): −0.75 V versus 0.1 *N* calomel electrode in an aqueous solution containing 53 g NaCl plus 3 g H_2O_2 per liter

Table 24 Typical mechanical properties of alloy 2024

Temper	Tensile strength MPa	Tensile strength ksi	Yield strength MPa	Yield strength ksi	Elongation(a), %	Hardness(b), HB	Shear strength MPa	Shear strength ksi	Fatigue strength(c) MPa	Fatigue strength(c) ksi
Bare 2024										
O	185	27	75	11	20	47	125	18	90	13
T3	485	70	345	50	18	120	285	41	140	20
T4, T351	470	68	325	47	20	120	285	41	140	20
T361	495	72	395	57	13	130	290	42	125	18
Alclad 2024										
O	180	26	75	11	20	...	125	18	...	...
T3	450	65	310	45	18	...	275	40	...	...
T4, T351	440	64	290	42	19	...	275	40	...	...
T361	460	67	365	53	11	...	285	41	...	...
T81, T851	450	65	415	60	6	...	275	40	...	...
T861	485	70	455	66	6	...	290	42	...	...

(a) 1.6 mm (1/16 in.) thick specimen. (b) 500 kg load; 10 mm ball. (c) At 5×10^8 cycles of completely reversed stress; R.R. Moore type test

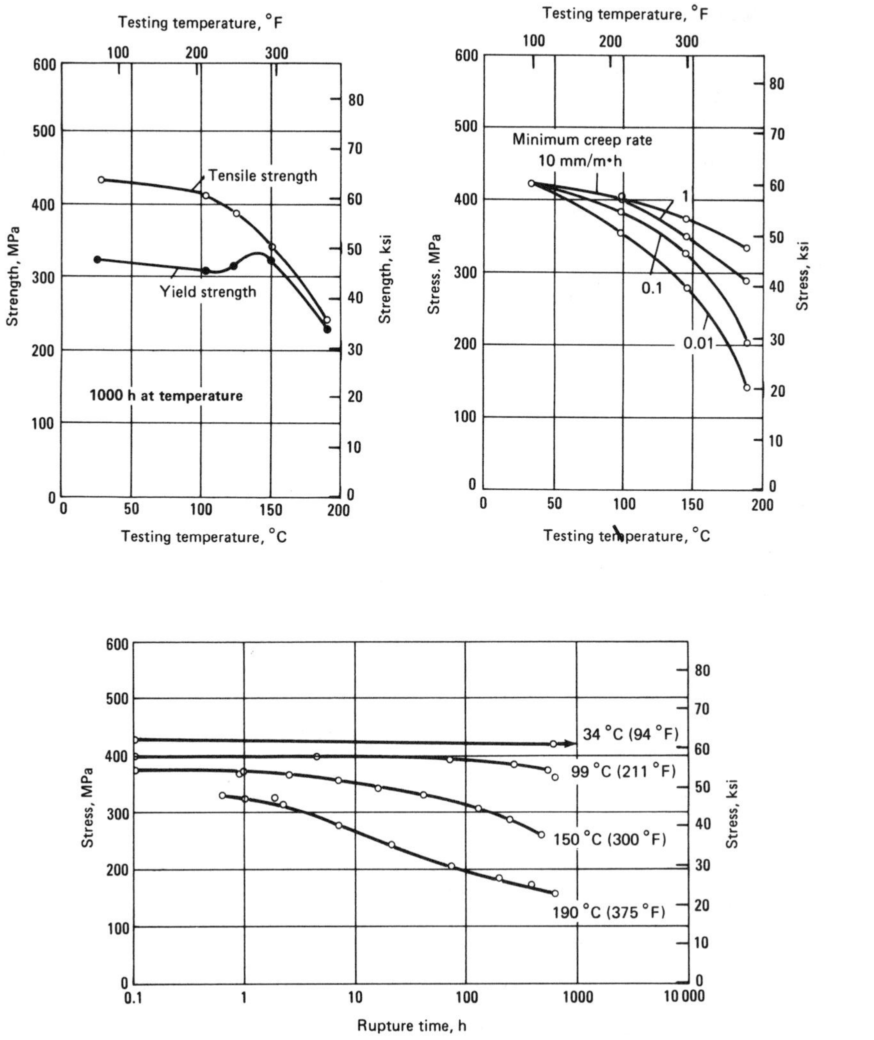

Fig. 4 Effect of temperature on tensile properties of Alclad 2024-T3. Sheet was 1.0 mm (0.04 in.) thick.

Fabrication Characteristics

Weldability. Arc welding with inert gas limited due to crack sensitivity, loss of mechanical properties, and/or loss in resistance to corrosion. When used for automotive parts, can be resistance welded with very good results
Annealing temperature. 385 °C (725 °F); hold 2 to 3 h at temperature for sheet
Solution temperature. 500 °C (930 °F)

Table 25 Tensile property limits for alloy 2024

Temper	Tensile strength (min) MPa	Tensile strength (min) ksi	Yield strength (min) MPa	Yield strength (min) ksi	Elongation (min)(a), %
Sheet and plate					
O	220 (max)	32 (max)	95 (max)	14 (max)	12
T42					
0.010–0.499 in. thick	425	62	260	38	12–15
0.500–1.000 in. thick	420	61	260	38	8
1.001–2.000 in. thick	415	60	260	38	6–7
2.001–3.000 in. thick	400	58	260	38	4
T62					
0.010–0.499 in. thick	440	64	345	50	5
0.500–3.000 in. thick	435	63	345	50	5
T361					
0.020–0.062 in. thick	460	67	345	50	8
0.063–0.249 in. thick	470	68	350	51	9
0.250–0.500 in. thick	455	66	340	49	9–10
T861					
0.020–0.062 in. thick	485	70	425	62	3
0.063–0.249 in. thick	490	71	455	66	4
0.250–0.499 in. thick	485	70	440	64	4
Alclad O					
0.008–0.062 in. thick	205 (max)	30 (max)	95 (max)	14 (max)	10–12
0.063–1.750 in. thick(b)	220 (max)	32 (max)	95 (max)	14 (max)	12
Alclad T42					
0.008–0.009 in. thick	380	55	235	34	10
0.010–0.062 in. thick	395	57	235	34	12–15
0.063–0.499 in. thick	415	60	250	36	12–15
0.500–1.000 in. thick(b)	420	61	260	38	8
1.001–2.000 in. thick(b)	415	60	260	38	6–7
2.001–3.000 in. thick(b)	400	58	260	38	4
Alclad T62					
0.010–0.062 in. thick	415	60	325	47	5
0.063–0.499 in. thick	425	62	340	49	5
Alclad T361					
0.020–0.062 in. thick	420	61	325	47	8
0.063–0.499 in. thick	440	64	330	48	9
0.500 in. thick(b)	445	66	340	49	10
Alclad T861					
0.020–0.062 in. thick	440	64	400	58	3
0.063–0.249 in. thick	475	69	440	64	4
0.250–0.499 in. thick	470	68	425	62	4
0.500 in. thick(b)	485	70	440	64	4
Flat sheet					
T3					
0.008–0.128 in. thick	435	63	290	42	10–15
0.129–0.249 in. thick	440	64	290	42	15
T81	460	67	400	58	5
Alclad T3					
0.008–0.009 in. thick	400	58	270	39	10
0.010–0.062 in. thick	405	59	270	39	12–15
0.063–0.128 in. thick	420	61	275	40	15
0.129–0.249 in. thick	425	62	275	40	15
T81					
0.010–0.062 in. thick	425	62	370	54	5
0.063–0.249 in. thick	450	65	385	56	5
Sheet					
T72	415	60	315	46	5
Alclad T72					
0.010–0.062 in. thick	385	56	295	43	5
0.063–0.249 in. thick	400	58	310	45	5
Coiled sheet					
T4	425	62	275	40	12–15
Alclad T4					
0.010–0.060 in. thick	400	58	250	36	12–15
0.063–0.128 in. thick	420	61	260	38	15
Plate					
T351					
0.250–0.499 in. thick	440	64	290	42	12
0.500–1.000 in. thick	435	63	290	42	8
1.001–2.000 in thick	425	62	290	42	6–7
2.001–3.000 in. thick	415	60	290	42	4
3.001–4.000 in. thick	395	57	285	41	4
T851					
0.250–0.499 in. thick	460	67	400	58	5
Plate (continued)					
0.500–1.000 in. thick	455	66	400	58	5
1.001–1.499 in. thick	455	66	395	57	5
Alclad T351					
0.250–0.499 in. thick	425	62	275	40	12
0.500–1.000 in thick(b)	435	63	290	42	8
1.001–2.000 in. thick(b)	425	62	290	42	6–7
2.001–3.000 in. thick(b)	415	60	290	42	4
3.001–4.000 in. thick(b)	395	57	285	41	4
Alclad T851					
0.250–0.499 in. thick	450	65	385	56	5
0.500–1.000 in. thick(b)	455	66	400	58	5
Wire, rod, and bar (rolled or cold finished)					
O	240 (max)	35 (max)	...	...	16
T36	475	69	360	52	10
T4					
≤0.499 in. thick or in diam	425	62	310(c)	45(c)	10
0.500–4.500 in. thick or in diam	425	62	290(c)	42(c)	10
4.501–6.500 in. thick or in diam	425	62	275(c)	40(c)	10
6.501–8.00 in. in diam	400	58	260	38	10
T42	425	62	275	40	10
T351	425	62	310	45	10
T6	425	62	345	50	5
T62	415	60	315	46	5
T851	455	66	400	58	5
Wire, rod, bar and shapes (extruded)					
O	240 (max)	35 (max)	130 (max)	19 (max)	12
T3, T3510, T3511:					
≤0.249 in. thick or in diam	395	57	290	42	12
0.250–0.749 in. thick or in diam	415	60	305	44	12
0.750–1.499 in. thick or in diam	450	65	315	46	10
≥1.5000 in. thick or in diam:					
≤25 in.^{2} area	485	70	360	52	10
>25–32 in.^{2} area	470	68	330	48	8
T42	395	57	260	38	8–12
T81, T851, T8510, T8511					
0.050–0.249 in. thick or in diam	440	64	385	56	4
0.250–≥1.500 in. thick or in diam: area ≤32 in.^{2}	455	66	400	58	5
Extruded tube					
O	240 (max)	35 (max)	130 (max)	19 (max)	12
T3, T3510, T3511					
≤0.249 in. thick	395	57	290	42	10
0.250–0.749 in. thick	415	60	305	44	10
0.750–1.499 in. thick	450	65	315	46	10
≥1.500 and over in. thick					
Area ≤25 in.^{2}	485	70	330	48	10
Area >25–32 in.^{2}	470	68	315	46	8
T42	395	57	260	38	12–8
T81, T8510, T8511					
0.050–0.249 in. thick	440	64	385	56	4
0.250–≥1.500; area ≤32 in.^{2}	455	66	400	58	5
Drawn tube					
O	220 (max)	32 (max)	105 (max)	15 (max)	...
T3	440	64	290	42	10–16(e)
T42	440	64	275	40	10–16(e)
Rivet and cold-heading wire and rod					
O	240 (max)	35 (max)	...	...	...
H13	220	32	...	...	...
	290 (max)	42 (max)	...	...	...
T4	425	62	275	40	10

(a) In 50 mm (2 in.) or 4*d*, where *d* is diameter of reduced section of tension-test specimen. Where a range of values appears in this column, the specified minimum elongation varies with thickness of the mill product. (b) For plate 12.7 mm (0.500 in.) or over in thickness, listed properties apply to core material only. Tensile and yield strengths of composite plate are slightly lower than the listed value, depending on thickness of the cladding. (c) Minimum yield strength of coiled wire and rod, 276 MPa (40 ksi). (d) Applicable to rod only. (e) Full section specimen; minimum elongation is 10 to 12% for cut-out specimen

Table 26 Typical mechanical properties of alloy 2048 plate, 75 mm (3 in.) thick

	At room temperature	At 120 °C (250 °F)	At 175 °C (350 °F)	At 260 °C (500 °F)
Tensile strength, MPa (ksi)				
Longitudinal	457 (66)	414 (60)	350 (51)	234 (34)
Transverse	465 (67)	414 (60)	345 (50)	230 (33)
Short transverse	463 (67)	...	...	...
Yield strength, MPa (ksi)				
Longitudinal	416 (60)	392 (57)	338 (49)	220 (32)
Transverse	420 (61)	388 (56)	338 (49)	220 (32)
Short transverse	406 (59)	...	...	...
Elongation, %				
Longitudinal	8	13	14	10
Transverse	7	13	...	8
Short transverse	6	...	...	...
Reduction in area, %				
Longitudinal	16	32	37	23
Transverse	12	28	34	15
Short transverse	9	...	...	...
Compressive yield strength, MPa (ksi)				
Longitudinal	420 (61)	391 (57)	350 (51)	243 (35)
Transverse	420 (61)	386 (56)	350 (51)	227 (33)
Elastic moduli, GPa (10^6 psi)				
In tension				
Longitudinal	70 (10)	68 (9.9)	64 (9.3)	57 (8.3)
Transverse	72 (10.4)	68 (9.9)	64 (9.3)	53 (8.7)
Short transverse	77 (11.1)	...	...	...
In compression				
Longitudinal	78 (11.3)	70 (10)	66 (9.6)	65 (9.4)
Transverse	77 (11.1)	71 (10.3)	67 (9.7)	66 (9.6)
Axial fatigue (longitudinal), MPa (ksi)				
Unnotched, R = 0.1				
10^3 cycles	469 (68)	469 (68)	469 (68)	...
10^5 cycles	262 (38)	255 (37)	241 (35)	...
10^7 cycles	221 (32)	193 (28)	172 (25)	...
Notched, K_t = 3.0, R = 0.1				
10^3 cycles	372 (54)	372 (54)	344 (50)	...
10^5 cycles	152 (22)	145 (21)	131 (19)	...
10^7 cycles	110 (16)	97 (14)	82 (12)	...
Creep strength (longitudinal)(a), MPa (ksi)				
100 h	...	303 (44)	241 (35)	60 (9)
1000 h	...	283 (41)	131 (19)	31 (5)
Rupture strength (longitudinal), MPa (ksi)				
100 h	...	345 (50)	269 (39)	90 (13)
1000 h	...	324 (47)	221 (32)	60 (9)

(a) Stress to produce 0.2% plastic extension in the indicated time

2048
3.3Cu-1.5Mg-0.40Mn

Specifications

UNS number. A92048

Chemical Composition

Composition limits. 0.15 Si max, 0.20 Fe max, 2.8 to 3.8 Cu, 0.20 to 0.6 Mn, 1.2 to 1.8 Mg, 0.25 Zn max, 0.10 Ti max, 0.05 max other (each), 0.15 max others (total), bal Al

Applications

Typical uses. Sheet and plate in structural components for aerospace application and military equipment

Mechanical Properties

Tensile properties. See Table 26 and Fig. 5.
Shear strength. Longitudinal, 271 MPa (39.3 ksi); transverse, 270 MPa (39.2 ksi)
Compressive properties. See Table 26 and Fig. 6.
Elastic modulus. See Fig. 5 and 6.
Impact strength. Charpy V-notch: longitudinal, 10.3 J (7.6 ft · lbf); transverse, 6.1 J (4.5 ft · lbf)
Fatigue strength. See Table 26 and Fig. 7 to 10.
Plane-strain fracture toughness. L-T crack orientation, 35.2 MPa$\sqrt{m}$ (32.0 ksi$\sqrt{in.}$); T-L crack orientation, 31.9 MPa$\sqrt{m}$ (29.1 ksi$\sqrt{in.}$)
Creep-rupture characteristics. See Table 26 and Fig. 11.

Mass Characteristics

Density. 2.75 g/cm^3 (0.099 lb/in.^{3}) at 20 °C (68 °F)

Thermal Properties

Coefficient of thermal expansion. Linear, 23.5 μm/m · K (13.0 μin./in. · °F) at 21 to 104 °C (70 to 220 °F)
Specific heat. 926 J/kg · K (0.221 Btu/lb · °F) at 100 °C (212 °F)
Thermal conductivity. T851 temper, 159 W/m · K (92 Btu/ft · h · °F)

Electrical Properties

Electrical conductivity. Volumetric, T851 temper: 42% IACS at 20 °C (68 °F)
Electrical resistivity. T851 temper, 40 nΩ · m at 20 °C (68 °F)

2124
4.4Cu-1.5Mg-0.6Mn

Specifications

AMS. 4101
ASTM. B 209
UNS number. A92124
Government. QQ-A-250/29

Chemical Composition

Composition limits. 0.20 Si max, 0.30 Fe max, 3.8 to 4.9 Cu, 0.30 to 0.9 Mn, 1.2 to 1.8 Mg, 0.10 Cr max, 0.25 Zn max, 0.15 Ti max, 0.05 max other (each), 0.15 max others (total), bal Al

Consequence of exceeding impurity limits. Degrades fracture toughness

Applications

Typical uses. Plate in thicknesses of 40 to 150 mm (1.5 to 6.0 in.) for aircraft structures

Mechanical Properties

Tensile properties. See Tables 27 and 28.
Poisson's ratio. 0.33 at 20 °C (68 °F)
Elastic modulus. See Table 28.
Plane-strain fracture toughness. T851 temper, plate: L-T, 31.9 MPa$\sqrt{m}$ (29.0 ksi$\sqrt{in.}$); T-L, 27.5 MPa$\sqrt{m}$ (25.0 ksi$\sqrt{in.}$); S-L, 24.2 MPa$\sqrt{m}$ (22.0 ksi$\sqrt{in.}$)
Creep-rupture characteristics. See Table 29.

Mass Characteristics

Density. 2.77 g/cm^3 (0.100 lb/in.^{3}) at 20 °C (68 °F)

Thermal Properties

Liquidus temperature. 638 °C (1180 °F)
Solidus temperature. 502 °C (935 °F)
Incipient melting temperature. 502 °C (935 °F)
Coefficient of thermal expansion. Linear:

Temperature range °C	Temperature range °F	Average coefficient μm/m · K	Average coefficient μin./in. · °F
−50 to 20	−58 to 68	21.1	11.7
20 to 100	68 to 212	22.9	12.7
20 to 200	68 to 392	23.8	13.2
20 to 300	68 to 572	24.7	13.7

Volumetric: 66.0 × 10^{-6} m^3/m^3 · K (3.6 × 10^{-5} in.^{3}/in.^{3} · °F) at 20 °C (68 °F)

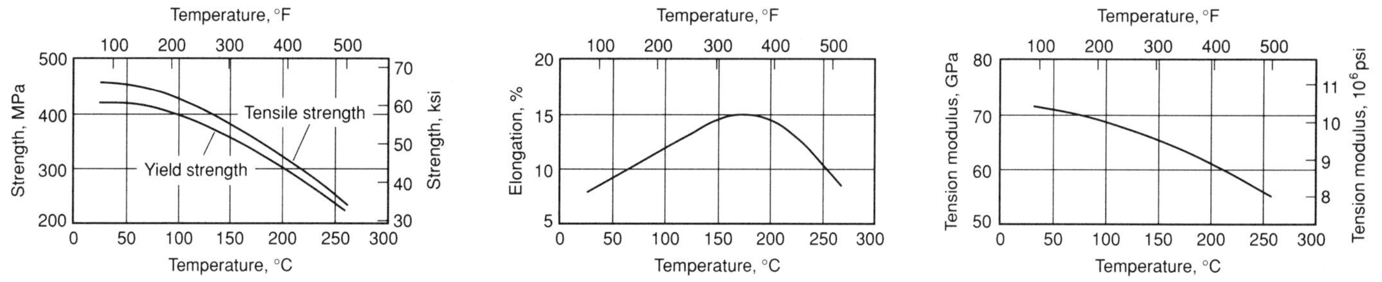

Fig. 5 Typical tensile properties of alloy 2048-T851 plate

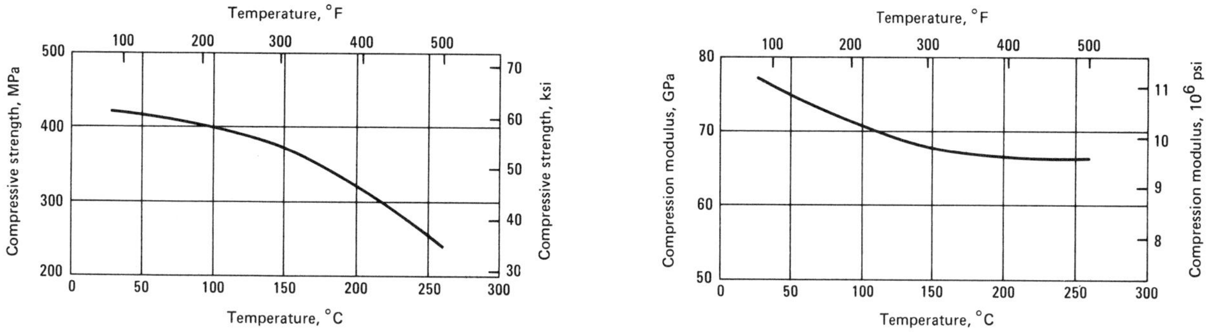

Fig. 6 Typical compressive properties of alloy 2048-T851 plate

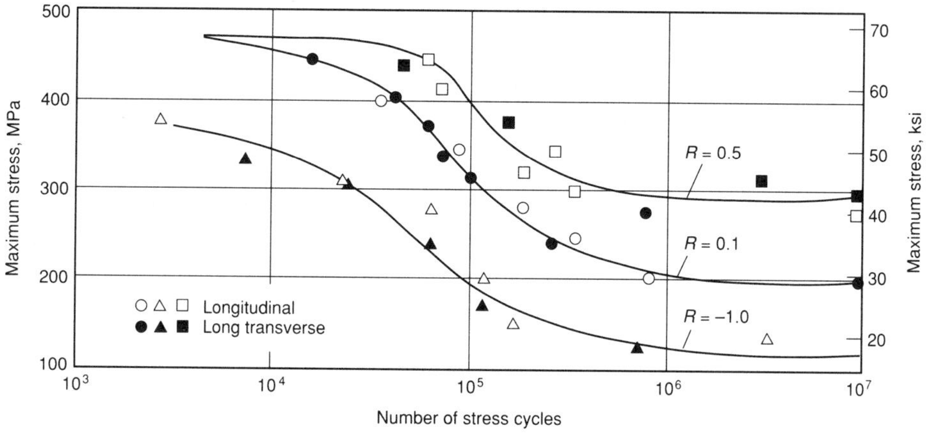

Fig. 7 Axial fatigue curves for unnotched specimens of alloy 2048-T851 plate

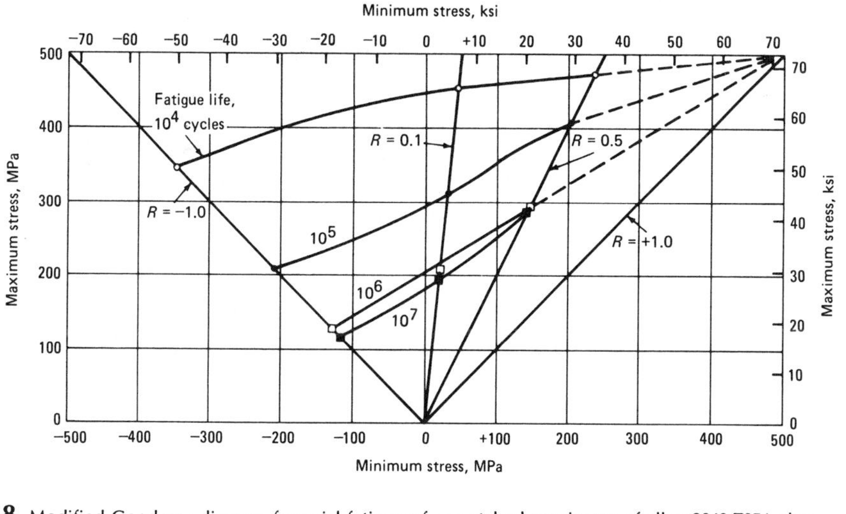

Fig. 8 Modified Goodman diagram for axial fatigue of unnotched specimens of alloy 2048-T851 plate

Specific heat. 882 J/kg · K (0.210 Btu/lb · °F) at 20 °C (68 °F)
Thermal conductivity. At 20 °C (68 °F): O temper, 191 W/m · K (110 Btu/ft · h · °F); T851, 152 W/m · K (87.8 Btu/ft · h · °F)

Electrical Properties

Electrical conductivity. Volumetric, at 20 °C (68 °F): O temper, 50% IACS; T851, 39% IACS
Electrical resistivity. At 20 °C (68 °F): O temper, 34.5 nΩ · m. Temperature coefficient, O and T851 tempers: 0.1 nΩ · m per K at 20 °C (68 °F)
Electrolytic solution potential. T851 temper, −0.80 V versus 0.1 *N* calomel electrode in an aqueous solution containing 53 g NaCl plus 3 g H_2O_2 per liter at 25 °C (77 °F)

Fabrication Characteristics

Annealing temperature. 413 °C (775 °F)
Solution temperature. 493 °C (920 °F)
Aging temperature. 191 °C (375 °F)

2218
4.0Cu-2.0Ni-1.5Mg

Specifications

AMS. Forgings and forging stock: 4142
SAE. J454
UNS. A92218
Government. Forgings and forging stock: QQ-A-367
Foreign. France: NF A-U4N. Spain: UNE L-315. Switzerland: VSM Al-Cu-Ni

Chemical Composition

Composition limits. 0.9 Si max, 1.0 Fe max, 3.5 to 4.5 Cu, 0.20 Mn max, 1.2 to 1.8 Mg,

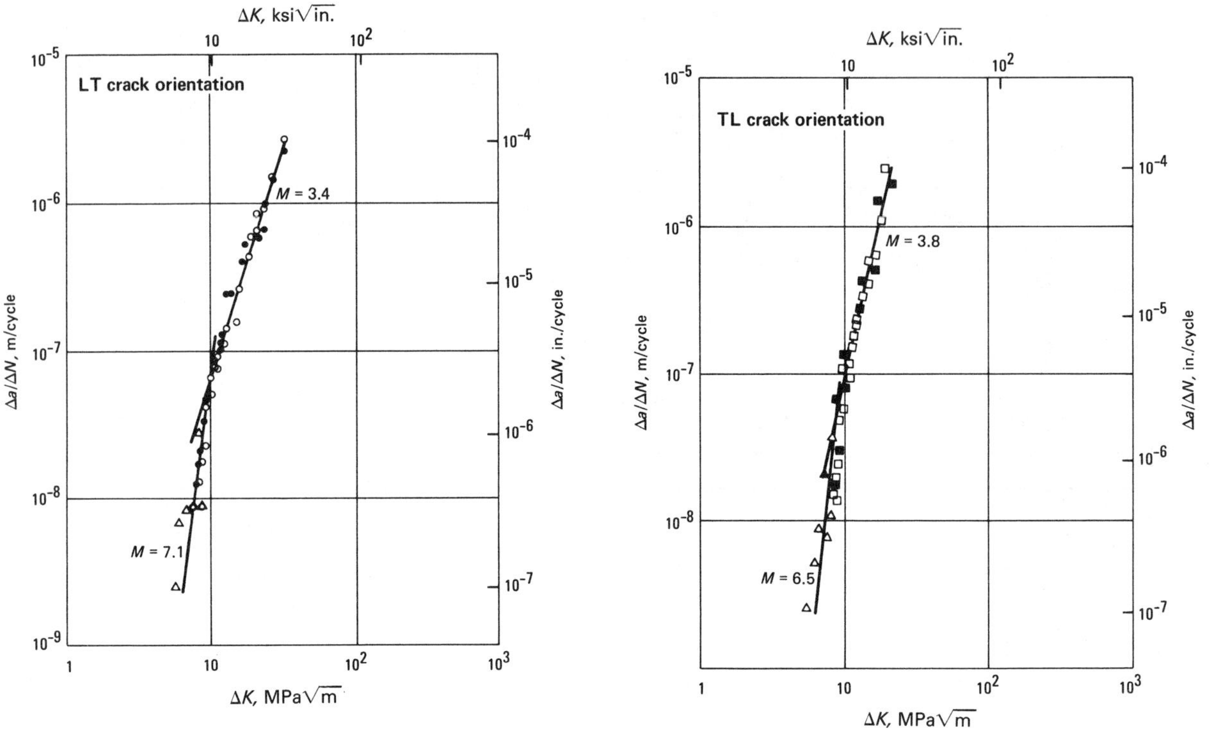

Fig. 9 Fatigue-crack propagation in alloy 2048-T851 plate

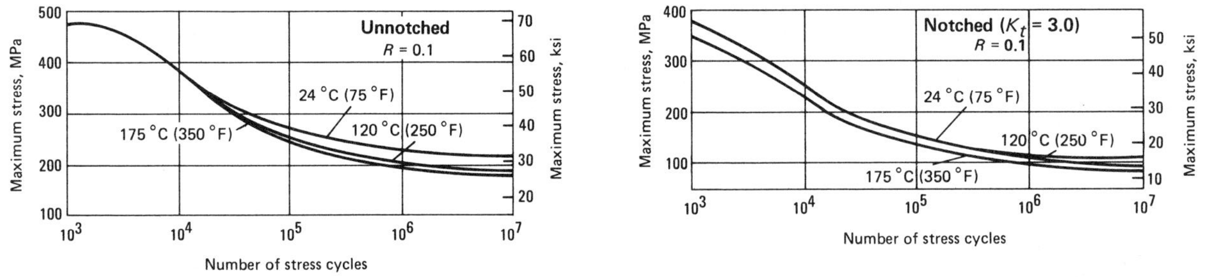

Fig. 10 Axial fatigue of alloy 2048-T851 plate

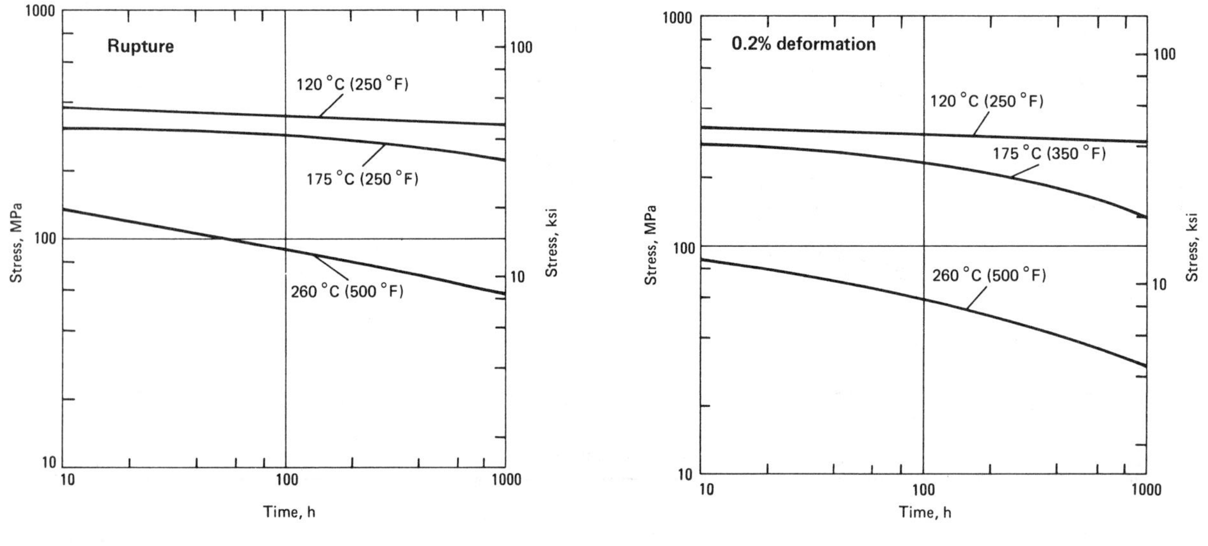

Fig. 11 Creep-rupture curves for alloy 2048-T851 plate, longitudinal orientation

Table 27 Typical tensile properties of alloy 2124-T851

Specimen orientation	Tensile strength MPa	Tensile strength ksi	Yield strength MPa	Yield strength ksi	Elongation, %
1.500–2.000 in. thick					
Longitudinal	490	71	440	64	9
Long transverse	490	71	435	63	9
Short transverse	470	68	420	61	5
2.000–3.000 in. thick					
Longitudinal	480	70	440	64	9
Long transverse	470	68	435	63	8
Short transverse	465	67	420	61	4

0.10 Cr max, 1.7 to 2.3 Ni, 0.25 Zn max, 0.05 max other (each), 0.15 max others (total), bal Al

Applications

Typical uses. Forgings; aircraft and diesel engine pistons; aircraft engine cylinder heads; jet engine impellers and compressor rings

Mechanical Properties

Tensile properties. See Tables 30 and 31.
Shear strength. T72 temper, 205 MPa (30 ksi)
Compressive yield strength. Approximately the same as tensile yield strength
Hardness. See Table 30.

Table 28 Mechanical properties of alloy 2124-T851 plate, 70 mm (2.75 in.) thick

Temperature		Time at tempera-ture, h	At indicated temperature							At room temperature after heating				
			Tensile strength		Yield strength		Elongation, %	Modulus of elasticity		Tensile strength		Yield strength		Elongation, %
°C	°F		MPa	ksi	MPa	ksi		GPa	10^6 psi	MPa	ksi	MPa	ksi	
−269	−452	...	705	102	620	90	10	...	...	...	...	...	...	...
−195	−320	...	595	86	545	79	9	81	11.8	...	...	...	...	...
−80	−112	...	525	76	490	71	8	76	11.0	...	...	...	...	...
−28	−18	...	505	73	470	68	8	74	10.7	...	...	...	...	...
24	75	...	485	70	450	65	8	72	10.5	485	70	450	65	8
100	212	0.1–10 000	455	66	420	61	9	71	10.3	485	70	450	65	8
		100 000	450	65	415	60	9	71	10.3	...	...	...	...	...
150	300	0.1–10	415	60	395	57	10	68	9.9	485	70	450	65	8
		100	405	59	395	57	10	68	9.9	485	70	440	64	8
		1 000	400	58	380	55	11	68	9.9	475	69	435	63	8
		10 000	370	54	330	48	13	68	9.9	460	67	405	59	8
		100 000	345	50	295	43	15	68	9.9	...	...	...	...	...
175	350	0.1	397	57	370	54	12	66	9.6	485	70	450	65	8
		0.5	385	56	365	53	12	66	9.6	485	70	450	65	8
		10	380	55	360	52	12	66	9.6	485	70	435	63	8
		100	360	52	340	49	12	66	9.6	470	68	420	61	8
		1 000	330	48	305	44	14	66	9.6	455	66	400	58	8
		10 000	295	43	250	36	16	66	9.6	405	59	305	44	10
		100 000	220	32	180	26	23	66	9.6	...	...	...	...	...
205	400	0.1	365	53	340	49	13	63	9.2	...	...	...	...	...
		0.5	360	52	330	48	13	63	9.2	475	69	435	63	8
		10	330	48	310	45	14	63	9.2	460	67	405	59	8
		100	305	44	270	39	15	63	9.2	435	63	370	54	8
		1 000	260	38	220	32	19	63	9.2	395	57	305	44	9
		10 000	185	27	140	20	28	63	9.2	290	42	165	24	12
		100 000	125	18	90	13	40	63	9.2	...	...	...	...	...
230	450	0.1	325	47	295	43	15	61	8.9	...	...	...	...	...
		0.5	310	45	285	41	15	61	8.9	470	68	425	62	8
		10	275	40	250	36	17	61	8.9	425	62	360	52	8
		100	235	34	200	29	20	61	8.9	370	54	275	40	10
		1 000	170	25	125	18	30	61	8.9	290	42	170	25	12
		10 000	110	16	76	11	45	61	8.9	215	31	90	13	18
		100 000	83	12	59	8.5	55	61	8.9	...	...	...	...	...
260	500	0.1	270	39	240	35	17	59	8.5	...	...	...	...	...
		0.5	255	37	230	33	17	59	8.5	455	66	400	58	9
		10	205	30	185	27	20	59	8.5	385	56	295	43	10
		100	150	22	125	18	29	59	8.5	290	42	170	25	12
		1 000	105	15	76	11	45	59	8.5	235	34	110	16	17
		10 000	76	11	55	8	60	59	8.5	195	28	83	12	22
		100 000	62	9	45	6.5	65	59	8.5	...	...	...	...	...
315	600	0.1	160	23	145	21	23	53	7.7	...	...	...	...	...
		0.5	140	20	115	17	26	53	7.7	340	49	230	33	10
		10	83	12	69	10	40	53	7.7	270	39	130	19	13
		100	69	10	55	8	50	53	7.7	240	35	105	15	17
		1 000	62	9	45	6.5	65	53	7.7	215	31	83	12	22
		10 000	52	7.5	41	6	75	53	7.7	185	27	76	11	22
		100 000	45	6.5	38	5.5	80	53	7.7	...	...	...	...	...
370	700	0.1	76	11	69	10	35	45	6.5	...	...	...	...	...
		0.5	59	8.5	45	6.5	50	45	6.5	275	40	130	19	13
		10	48	7	34	5	75	45	6.5	255	37	105	15	18
		100	41	6	31	4.5	85	45	6.5	235	34	90	13	22
		1 000	38	5.5	28	4.1	90	45	6.5	205	30	83	12	22
		10 000	34	5	28	4.1	95	45	6.5	185	27	76	11	22
		100 000	34	5	28	4.1	100	45	6.5	...	...	...	...	...
425	800	0.1	34	5	28	4.1	65	...	...	...	...	...	...	...
		0.5	30	4.4	24	3.5	85	...	...	...	...	...	...	...
480	900	...	16	2.3	12	1.8	65	...	...	...	...	...	...	...
535	1000	...	2	0.3	2	0.3	2	...	...	...	...	...	...	...

Table 29 Creep-rupture properties of alloy 2124-T851 plate, 70 mm thick

Temperature °C	Temperature °F	Time under stress, h	Rupture stress MPa	Rupture stress ksi	Stress for creep of 1.0% MPa	1.0% ksi	0.5% MPa	0.5% ksi	0.2% MPa	0.2% ksi	0.1% MPa	0.1% ksi
24	75	0.1	485	70	470	68	455	66	...	...	...	...
		1	475	69	460	67	450	65	...	...	...	...
		10	475	69	455	66	...	...	...	...	...	...
		100	470	68	...	...	...	...	...	...	...	...
		1 000	470	68	...	...	...	...	...	...	...	...
100	212	0.1	455	66	435	63	425	62	420	61	415	60
		1	435	63	420	61	415	60	405	59	395	57
		10	420	61	405	59	395	57	380	55	370	54
		100	400	58	385	56	380	55	360	52	345	50
		1 000	380	55	370	54	360	52	340	49	325	47
150	300	0.1	400	58	380	55	370	54	360	52	345	50
		1	370	54	360	52	345	50	330	48	310	45
		10	345	50	340	49	325	47	310	45	285	41
		100	315	46	310	45	305	44	290	42	250	36
		1 000	290	42	285	41	270	39	235	34	205	30
		10 000	235	34	...	...	...	...	...	...	...	...
		100 000	170	25	...	...	...	...	...	...	...	...
175	350	0.1	365	53	345	50	340	49	325	47	305	44
		1	340	49	325	47	310	45	290	42	260	38
		10	305	44	290	42	275	40	255	37	230	33
		100	270	39	255	37	240	35	205	30	170	25
		1 000	205	30	195	28	170	25	140	20	105	15
		10 000	145	21	...	...	...	...	...	...	...	...
		100 000	90	13	...	...	...	...	...	...	...	...
205	400	0.1	325	47	310	45	295	43	285	41	260	38
		1	290	42	275	40	270	39	250	36	220	32
		10	255	37	240	35	235	34	205	30	170	25
		100	200	29	185	27	180	26	150	22	115	17
		1 000	130	19	125	18	115	17	90	13	52	7.5
		10 000	83	12	69	10	59	8.5	...	...	...	...
		100 000	52	7.5	...	...	...	...	...	...	...	...
230	450	0.1	275	40	260	38	250	36	235	34	215	31
		1	240	35	235	34	220	32	205	30	170	25
		10	195	28	185	27	180	26	150	22	115	17
		100	130	19	125	18	115	17	...	...	...	...
		1 000	76	11	...	...	...	...	...	...	...	...
		10 000	48	7	...	...	...	...	...	...	...	...
		100 000	34	4.9	...	...	...	...	...	...	...	...
260	500	0.1	215	31	205	30	200	29	180	26	170	25
		1	185	27	180	26	170	25	150	22	130	19
		10	140	20	130	19	125	18	97	14	76	11
		100	83	12	76	11	69	10	...	...	...	...
		1 000	48	7	...	...	...	...	...	...	...	...
		10 000	32	4.7	...	...	...	...	...	...	...	...
		100 000	23	3.4	...	...	...	...	...	...	...	...
315	600	0.1	110	16	110	16	105	15	97	14	90	13
		1	97	14	90	13	83	12	69	10	59	8.5
		10	59	8.5	55	8	52	7.5	45	6.5	38	5.5
		100	34	5	34	5	30	4.4	25	3.6	21	3
		1 000	21	3	20	2.9	18	2.6	...	...	...	...

Poisson's ratio. 0.33 at 20 °C (68 °F)
Elastic modulus. Tension, 74.4 GPa (10.8×10^6 psi); shear, 27.5 GPa (4.0×10^6 psi)
Fatigue strength. See Table 32.
Creep-rupture characteristics. See Table 33.

Mass Characteristics

Density. 2.80 g/cm³ (0.101 lb/in.³) at 20 °C (68 °F)

Thermal Properties

Liquidus temperature. 635 °C (1175 °F)
Solidus temperature. 532 °C (990 °F)
Incipient melting temperature. 504 °C (940 °F)
Coefficient of thermal expansion.
Linear:

Temperature range °C	Temperature range °F	Average coefficient μm/m · K	Average coefficient μin./in. · °F
−50 to 20	−58 to 68	20.7	11.5
20 to 100	68 to 212	22.4	12.4
20 to 200	68 to 392	23.3	12.9
20 to 300	68 to 572	24.2	13.4

Volumetric: 6.5×10^{-5} m³/m³ · K (3.6×10^{-5} in.³/in.³ · °F) at 20 °C (68 °F)
Specific heat. 871 J/kg · K (0.208 Btu/lb · °F) at 20 °C (68 °F)
Thermal conductivity. At 20 °C (68 °F): T61 temper, 148 W/m · K (85.5 Btu/ft · h · °F); T72 temper, 155 W/m · K (89.6 Btu/ft · h · °F)

Table 30 Typical mechanical properties of alloy 2218

Temper	Tensile strength MPa	Tensile strength ksi	Yield strength MPa	Yield strength ksi	Elongation, %	Hardness(a), HB
T61	407	59	303	44	13	115
T71	345	50	276	40	11	105
T72	331	48	255	37	11	95

(a) 500 kg load; 10 mm diam ball

Electrical Properties

Electrical conductivity. Volumetric: T61 temper, 38% IACS; T72 temper, 40% IACS
Electrical resistivity. T61 temper, 45.0 nΩ · m; T72 temper, 43.0 nΩ · m. Temperature coefficient, T61 and T72 tempers: 0.1 nΩ · m per K at 20 °C (68 °F)

Fabrication Characteristics

Solution temperature. 510 °C (950 °F)
Aging temperature. T61 temper, 170 °C (340 °F) for 10 h at temperature; T72 temper, 240

Table 31 Tensile properties of alloy 2218-T61

Temperature °C	Temperature °F	Tensile strength(a) MPa	Tensile strength(a) ksi	Yield strength(a) MPa	Yield strength(a) ksi	Elongation, %
−195	−320	495	72.0	360	52.0	15
−80	−112	420	61.0	310	45.0	14
−30	−18	405	59.0	305	44.0	13
25	75	405	59.0	305	44.0	13
100	212	385	56.0	290	42.0	15
150	300	285	41.0	240	35.0	17
205	400	150	22.0	110	16.0	30
260	500	70	10.0	40	6.0	70
315	600	40	5.5	20	3.0	85
370	700	30	4.0	17	2.5	100

(a) Lowest strength determined for representative lot during 10 000 h exposure at temperature under no load

Table 32 Fatigue strength of alloy 2218-T61

Temperature °C	Temperature °F	No. of cycles	Fatigue strength(a) MPa	Fatigue strength(a) ksi
23	75	10^5	270	39.0
		10^6	215	31.0
		10^7	170	25.0
		10^8	135	20.0
		5×10^8	125	18.0
150	300	10^5	...	...
		10^6	170	25.0
		10^7	130	19.0
		10^8	105	15.0
		5×10^8	100	14.0
205	400	10^5	...	...
		10^6	150	22.0
		10^7	105	15.0
		10^8	69	10.0
		5×10^8	59	8.5
260	500	10^5	145	21.0
		10^6	105	15.0
		10^7	72	10.0
		10^8	48	7.0
		5×10^8	41	6.0
315	600	10^5	90	13.0
		10^6	69	10.0
		10^7	48	7.0
		10^8	34	5.0
		5×10^8	31	4.5

(a) R.R. Moore type test

°C (460 °F) for 6 h at temperature

2219, Alclad 2219 6.3Cu-0.3Mn-0.18Zr-0.10V-0.06Ti

Specifications

AMS. Sheet and plate: 4031. Extruded wire, rod, bar, shapes, and tube: 4162, 4163. Forgings: 4143, 4144. Alclad 2219, sheet and plate: 4094, 4095, 4096

ASTM. Sheet and plate: B 209. Rolled or cold finished wire, rod, and bar: B 211. Extruded wire, rod, bar, shapes, and tube: B 221. Extruded, seamless tube: B 241. Forgings: B 247. Alclad 2219, sheet and plate: B 209

SAE. J454

UNS. A92219

Government. Sheet and plate: QQ-A-250/30. Forgings: QQ-A-367, MIL-A-22771. Armor plate: MIL-A-46118. Rivet wire and rod: QQ-A-430

Foreign. France: NF A-U6MT. United Kingdom: DTD 5004

Chemical Composition

Composition limits for 2219. 0.20 Si max, 0.30 Fe max, 5.8 to 6.8 Cu, 0.20 to 0.40 Mn, 0.02 Mg max, 0.10 Zn max, 0.05 to 0.15 V, 0.02 to 0.10 Ti, 0.10 to 0.25 Zr, 0.05 max other (each), 0.15 max others (total), bal Al

Composition limits for Alclad 2219. 7072 cladding—0.10 Cu max, 0.10 Mn max, 0.70 Si max + Fe, 0.80 to 1.3 Zn, 0.10 Mg max, 0.05 max other (each), 0.15 max others (total)

Applications

Typical uses. Welded space booster oxidizer and fuel tanks, supersonic aircraft skin and structure components. Readily weldable and useful for applications over temperature range of −270 to 300 °C (−450 to 600 °F). Has high fracture toughness, and the T8 temper is highly resistant to stress-corrosion cracking

Mechanical Properties

Tensile properties. See Tables 34 to 36.

Poisson's ratio. 0.33 at 20 °C (68 °F)

Elastic modulus. Tension, 73.8 GPa (10.7 × 10^6 psi); compression, 75.2 GPa (10.9 × 10^6 psi)

Fatigue strength. 103 MPa (15 ksi) at 5 × 10^8 cycles, R.R. Moore type test

Creep-rupture characteristics. See Tables 37 and 38.

Mass Characteristics

Density. 2.84 g/cm³ (0.103 lb/in.³) at 20 °C (68 °F)

Thermal Properties

Liquidus temperature. 643 °C (1190 °F)

Incipient melting temperature. 543 °C (1010 °F)

Coefficient of thermal expansion. Linear:

Temperature range °C	Temperature range °F	Average coefficient µm/m · K	Average coefficient µin./in. · °F
−50 to 20	−58 to 68	20.8	11.5
20 to 100	68 to 212	22.5	12.5
20 to 200	68 to 392	23.4	13.0
20 to 300	68 to 572	24.4	13.6

Volumetric: 6.5 × 10^{-5} m³/m³ · K (3.62 × 10^{-5} in.³/in.³ · °F)

Specific heat. 864 J/kg · K (0.206 Btu/lb · °F) at 20 °C (68 °F)

Thermal conductivity. O temper, 170 W/m · K (98.2 Btu/ft · h · °F); T31, T37 tempers, 116 W/m · K (67.0 Btu/ft · h · °F); T62, T81, T87 tempers, 130 W/m · K (75.1 Btu/ft · h · °F)

Electrical Properties

Electrical conductivity. Volumetric, at 20 °C (68 °F): O temper, 44% IACS; T31, T37,

Table 33 Creep-rupture properties of alloy 2218-T61

Temperature °C	Temperature °F	Time under stress, h	Rupture stress MPa	Rupture stress ksi	Stress for creep of: 1% MPa	1% ksi	0.5% MPa	0.5% ksi	0.2% MPa	0.2% ksi	0.1% MPa	0.1% ksi
100	212	Up to 1000	385	56.0	...	...	...	...	...	...	...	...
150	300	0.1	360	52.0	350	51.0	330	48.0	315	46.0	290	42.0
		1	350	51.0	345	50.0	325	47.0	310	45.0	285	41.0
		10	350	51.0	340	49.0	315	46.0	305	44.0	275	40.0
		100	330	48.0	325	47.0	310	45.0	295	43.0	230	33.0
		1000	290	42.0	290	42.0	290	42.0	270	39.0	140	20.0
205	400	0.1	325	47.0	315	46.0	290	42.0	275	40.0	255	37.0
		1	310	45.0	305	44.0	275	40.0	260	38.0	235	34.0
		10	255	37.0	250	36.0	240	35.0	220	32.0	160	23.0
		100	185	27.0	180	26.0	170	25.0	140	20.0	105	15.0
		1000	115	17.0	110	16.0	105	15.0	105	15.0	...	...
315	600	0.1	55	8.0	52	7.5	48	7.0	45	6.5	41	6
		1	48	7.0	48	7.0	45	6.5	41	6.0	38	5.5
		10	45	6.5	41	6.0	47	6.9	34	5.0	21	3.0
		100	27	3.9	23	3.4	21	3.0	14	2.1	...	...

Table 34 Typical tensile properties of alloy 2219

Temper	Tensile strength MPa	Tensile strength ksi	Yield strength MPa	Yield strength ksi	Elongation, %
O	172	25	76	11	18
T42	359	52	186	27	20
T31, T351	359	52	248	36	17
T37	393	57	317	46	11
T62	414	60	290	42	10
T81, T851	455	66	352	51	10
T87	476	69	393	57	10

T351 tempers, 28% IACS; T62, T81, T87, T851 tempers, 30% IACS
Electrical resistivity. At 20 °C (68 °F); O temper, 39 nΩ · m; T31, T37, T351 tempers, 62 nΩ · m; T62, T81, T87, T851 tempers, 57 nΩ · m. Temperature coefficient, all tempers: 0.1 nΩ · m per K at 20 °C (68 °F)
Electrolytic solution potential. T31, T37, T351 tempers, −0.64 V and T62, T81, T87, T851 tempers, −0.80 V versus 0.1 *N* calomel electrode in an aqueous solution containing 53 g NaCl plus 3 g H_2O_2 per liter at 25 °C (77 °F)

Fabrication Characteristics

Annealing temperature. 415 °C (775 °F)
Solution temperature. 535 °C (995 °F)
Aging temperature. 165 to 190 °C (325 to 375 °F) from 18 to 36 h at temperature. Appropriate combination of aging time and temperature is different for different tempers.

2319
5.3Cu-0.3Mn-0.18Zr-0.15Ti-0.10V

Specifications

UNS. A92319
Government. QQ-R-566, MIL-E-16053

Chemical Composition

Composition limits. 5.8 to 6.8 Cu, 0.20 to 0.40 Mn, 0.10 to 0.25 Zr, 0.10 to 0.20 Ti, 0.05 to 0.15 V, 0.20 Si max, 0.30 Fe max, 0.02 Mg max, 0.10 Zn max, 0.0008 Be max, 0.05 max other (each), 0.15 max others (total)

Applications

Typical uses. Electrodes and filler wire for welding 2219

Table 35 Tensile-property limits for alloy 2219

Temper	Tensile strength (min) MPa	Tensile strength (min) ksi	Yield strength (min) MPa	Yield strength (min) ksi	Elongation(a), %
Sheet and plate					
O	220 (max)	32 (max)	110 (max)	16 (max)	12
Alclad O	220 (max)	32 (max)	110 (max)	16 (max)	12
T31(b)					
0.020–0.039 in. thick	315	46	200	29	8
0.040–0.249 in. thick	315	46	195	28	10
Alclad T31(b)					
0.040–0.099 in. thick	290	42	170	25	10
0.100–0.249 in. thick	305	44	180	26	10
T351(c)					
0.250–2.000 in. thick	315	46	195	28	10
2.100–3.000 in. thick	305	44	195	28	10
3.100–4.000 in. thick	290	42	185	27	9
4.100–5.000 in. thick	275	40	180	26	9
5.001–6.000 in. thick	270	39	170	25	8
Alclad T351(c)	305	44	180	26	10
T37					
0.020–0.039 in. thick	340	49	260	38	6
0.040–2.500 in. thick	340	49	255	37	6
2.501–3.000 in. thick	325	47	250	36	6
3.001–4.000 in. thick	310	45	240	35	5
4.001–5.000 in. thick	295	43	235	34	4
Alclad T37					
0.040–0.099 in. thick	310	45	235	34	6
0.100–0.499 in. thick	325	47	240	35	6
T62	370	54	250	36	6–8
Alclad T62					
0.020–0.039 in. thick	305	44	200	29	6
0.040–0.099 in. thick	340	49	220	32	7
0.100–0.499 in. thick	350	51	235	34	7–8
0.500–2.000 in. thick(c)	370	54	250	36	7–8
T81(b)	425	62	315	46	6–7
Alclad T81(b)					
0.020–0.039 in. thick	340	49	255	37	6
0.040–0.099 in. thick	380	55	285	41	7
0.100–0.249 in. thick	400	58	295	43	7
T851(d)					
0.250–2.000 in. thick	425	62	315	46	7–8
2.001–3.000 in. thick	425	62	310	45	6
3.001–4.000 in. thick	415	60	305	44	5
4.001–5.000 in. thick	405	59	295	43	5
5.001–6.000 in. thick	395	57	290	42	4
Alclad T851(d)	400	58	290	42	8
T87					
0.020–0.249 in. thick	440	64	360	52	5–6
0.250–3.000 in. thick	440	64	350	51	6–7
3.001–4.000 in. thick	425	62	345	50	4
4.001–5.000 in. thick	420	61	340	49	3
Alclad T87					
0.040–0.099 in. thick	395	57	315	46	6
0.100–0.499 in. thick	415	60	330	48	6–7
Wire, rod, and bar (rolled or cold finished)					
T851					
0.500–2.000 in. thick or in diam	400	58	275	40	4
2.001–4.000 in. thick or in diam	395	57	270	39	4
Wire, rod, bar, and shapes (extruded)					
O	221 (max)	32 (max)	125 (max)	18 (max)	12
T31, T3510, T3511					
Up thru 0.499 in. thick or in diam.	290	42	180	26	14
0.500–2.999 in. thick or in diam	310	45	185	27	14
T62	370	54	250	36	6
T81, T8510, T8511	400	58	290	42	6
Extruded tube					
O	220 (max)	32 (max)	125 (max)	18 (max)	12
T31, T3510, T3511					
Up thru 0.499 in. thick or in diam.	290	42	180	26	14
0.500–2.999 in. thick or in diam	310	45	185	27	14
T62	370	54	250	36	6
T81, T8510, T8511	400	58	290	42	6
Die forgings					
T6					
Specimen axis parallel to grain flow	400	58	260	38	8(e)(f)
Specimen axis not parallel to grain flow	385	56	250	36	4(e)
Hand forgings(g)					
T6					
Longitudinal axis	400	58	275	40	6
Long transverse axis	380	55	255	37	4
Short transverse axis	365	53	240	35	2
Mechanical property limits					
T852					
Longitudinal axis	425	62	345	50	6
Long transverse axis	425	62	340	49	4
Short transverse axis	415	60	315	46	3
Rolled rings(h)					
T6					
Tangential axis	385	56	275	40	6
Axial axis	380	55	255	37	4
Radial axis	365	53	240	35	2

(a) In 50 mm (2 in.) or 4*d*, where *d* is diameter of reduced section of tensile test specimen. Where a range of values appears in this column, specified minimum elongation varies with thickness of the mill product. (b) Sheet only. (c) For plate 12.7 mm (0.500 in.) or greater in thickness, property limits apply to core material only. Tensile and yield strengths of composite plate slightly lower depending on thickness of cladding. (d) Plate only. (e) Specimen taken from forging. (f) 10% for specimen taken from separately forged coupon. (g) Maximum cross-sectional area 1650 cm^2 (256 $in.^2$). These properties not applicable to upset biscuit forgings or rolled rings. (h) Only applicable to rings having ratio of outside diameter to wall thickness equal to or greater than 10

Table 36 Typical tensile properties of alloy 2219 at various temperatures

Temper	Temperature °C	Temperature °F	Tensile strength(a) MPa	Tensile strength(a) ksi	Yield strength (0.2% offset)(a) MPa	Yield strength (0.2% offset)(a) ksi	Elongation, %
T62	−196	−320	503	73	338	49	16
	−80	−112	434	63	303	44	13
	−28	−18	414	60	290	42	12
	24	75	400	58	276	40	12
	100	212	372	54	255	37	14
	149	300	310	45	227	33	17
	204	400	234	34	172	25	20
	260	500	186	27	133	20	21
	316	600	69	10	55	8	40
	371	700	30	4.4	26	3.7	75
T81, T851	−196	−320	572	83	421	61	15
	−80	−112	490	71	372	54	13
	−28	−18	476	69	359	52	12
	24	75	455	66	345	50	12
	100	212	414	60	324	47	15
	149	300	338	49	276	40	17
	204	400	248	36	200	29	20
	260	500	200	29	159	23	21
	316	600	48	7	41	6	55
	371	700	30	4.4	26	3.7	75

(a) Lowest strength for exposures up to 10 000 h at temperature under no load; test load applied at 35 MPa/min (5 ksi/min) to yield strength and then at strain rate of 5%/min to fracture

Mass Characteristics

Density. 2.83 g/cm³ (0.103 lb/in.³) at 20 °C (68 °F)

Thermal Properties

Liquidus temperature. 643 °C (1190 °F)
Incipient melting temperature. 543 °C (1010 °F)
Coefficient of thermal expansion. Linear:

Temperature range °C	Temperature range °F	Average coefficient µm/m · K	Average coefficient µin./in. · °F
−50 to 20	−58 to 68	20.8	11.5
20 to 100	68 to 212	22.5	12.5
20 to 200	68 to 392	23.4	13.0
20 to 300	68 to 572	24.4	13.6

Volumetric: 6.5×10^{-5} m³/m³ · K (3.62×10^{-5} in.³/in.³ · °F) at 20 °C (68 °F)
Specific heat. 864 J/kg · K (0.206 Btu/lb · °F)
Thermal conductivity. O temper: 170 W/m · K (98.2 Btu/ft · h · °F)

Electrical Properties

Electrical conductivity. Volumetric: O temper, 44% IACS at 20 °C (68 °F)
Electrical resistivity. O temper, 39 nΩ · m at 20 °C (68 °F)
Temperature coefficient. 2.94×10^{-3}/K

Fabrication Characteristics

Annealing temperature. 413 °C (775 °F)

2618
2.3Cu-1.6Mg-1.1Fe-1.0Ni-0.18Si-0.07Ti

Specifications

AMS. Forgings and forging stock: 4132
ASTM. Forgings: B 247
SAE. J454
Government. Forgings: QQ-A-367; MIL-A-22771
Foreign. France: NF A-U2GN. United Kingdom: BS H12

Chemical Composition

Composition limits. 0.10 to 0.25 Si, 0.9 to 1.3 Fe, 1.9 to 2.7 Cu, 1.3 to 1.8 Mg, 0.9 to 1.2 Ni, 0.10 Zn max, 0.04 to 0.10 Ti, 0.05 max other (each), 0.15 others (total), bal Al

Applications

Typical uses. Die and hand forgings. Pistons and rotating aircraft engine parts for operation at elevated temperatures. Tire molds

Mechanical Properties

Tensile properties. See Tables 39 and 40 and Fig. 12.
Shear strength. T61 temper, 260 MPa (38 ksi)
Compressive yield strength. Approximately the same as the tensile yield strength. See also Fig. 13.
Hardness. Die forgings, T61 temper: 115 HB min
Poisson's ratio. 0.33 at 20 °C (68 °F)
Elastic modulus. Tension, 74.4 GPa (10.8×10^6 psi); shear, 28.0 GPa (4.0×10^6 psi)
Fatigue strength. T61 temper, 125 MPa (18 ksi) at 5×10^8 cycles; R.R. Moore type test
Creep-rupture characteristics. See Table 41.

Mass Characteristics

Density. 2.76 g/cm³ (0.100 lb/in.³) at 20 °C (68 °F)

Thermal Properties

Liquidus temperature. 638 °C (1180 °F)
Solidus temperature. 549 °C (1020 °F)
Incipient melting temperature. 502 °C (935 °F)

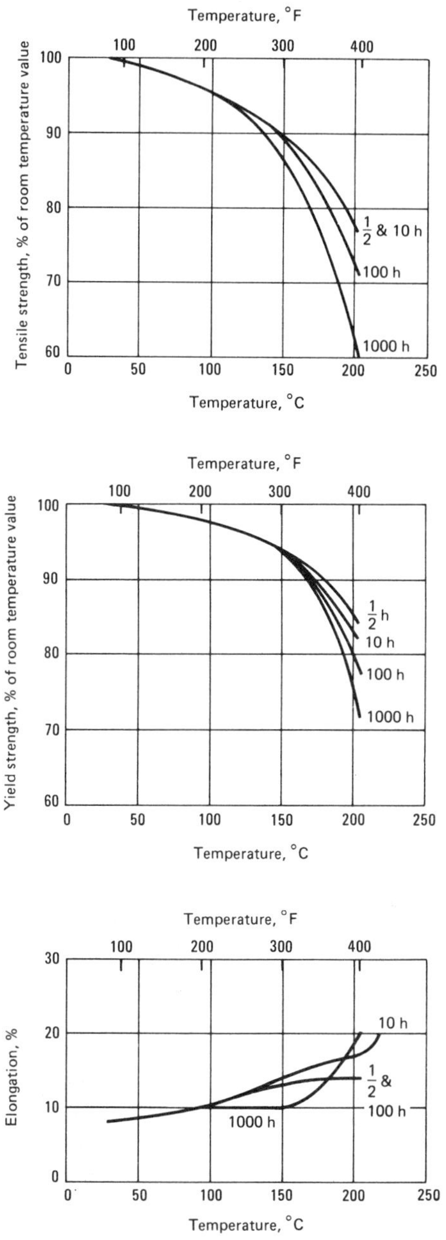

Fig. 12 Influence of prolonged holding at elevated temperature on tensile properties of alloy 2618-T61 hand-forged billets. Properties determined at temperature after holding for the indicated time under no load. Tensile and yield strengths plotted as percentage of corresponding room-temperature value. Elongation plotted as value determined at temperature

Coefficient of thermal expansion. Linear:

Temperature range °C	Temperature range °F	Average coefficient µm/m · K	Average coefficient µin./in. · °F
−50 to 20	−58 to 68	20.6	11.4
20 to 100	68 to 212	22.3	12.4
20 to 200	68 to 392	23.2	12.9
20 to 300	68 to 572	24.1	13.4

Volumetric: 6.45×10^{-5} m³/m³ · K (3.6×10^{-5} in.³/in.³ · °F) at 20 °C (68 °F)
Specific heat. 875 J/kg · K at 20 °C (68 °F)

Table 37 Creep-rupture properties of alloy 2219-T851 plate

Temperature		Time under	Rupture stress		Stress for creep of							
					1.0%		0.5%		0.2%		0.1%	
°C	°F	stress, h	MPa	ksi	MPa	ksi	MPa	ksi	MPa	ksi	MPa	ksi
24	75	0.1	455	66	435	63	415	60	365	53	350	51
		1	450	65	420	61	385	56	360	52	345	50
		10	435	63	400	58	365	53	345	50	330	48
		100	425	62	380	55	360	52	340	49	325	47
		1 000	420	61	365	53	350	51	330	48	315	46
100	212	0.1	395	57	360	52	340	49	315	46	305	44
		1	370	54	340	49	325	47	305	44	285	41
		10	350	51	325	47	310	45	290	42	275	40
		100	330	48	310	45	295	43	275	40	270	39
		1 000	315	46	295	43	285	41	270	39	260	38
150	300	0.1	340	49	305	44	295	43	275	40	260	38
		1	315	46	290	42	275	40	255	37	235	34
		10	290	42	270	39	255	37	235	34	205	30
		100	260	38	250	36	235	34	200	29	170	25
		1 000	235	34	220	32	200	29	165	24	150	22
		10 000	205	30	...	...	...	...	...	...	...	...
		100 000	170	25	...	...	...	...	...	...	...	...
175	350	0.1	305	44	275	40	260	38	250	36	230	33
		1	275	40	255	37	240	35	220	32	200	29
		10	250	36	230	33	215	31	195	28	165	24
		100	220	32	200	29	185	27	160	23	130	19
		1 000	185	27	170	25	160	23	140	20	105	15
		10 000	160	23	...	...	...	...	...	...	...	...
		100 000	130	19	...	...	...	...	...	...	...	...
205	400	0.1	270	39	240	35	235	34	215	31	195	28
		1	235	34	220	32	205	30	180	26	160	23
		10	205	30	195	28	180	26	150	22	130	19
		100	180	26	165	24	150	22	130	19	110	16
		1 000	150	22	140	20	125	18	115	17	90	13
		10 000	125	18	125	18	...	...	...	...	...	...
		100 000	97	14	...	...	...	...	...	...	...	...
230	450	0.1	230	33	205	30	200	29	180	26	165	24
		1	200	29	185	27	170	25	150	22	140	20
		10	170	25	160	23	150	22	130	19	110	16
		100	150	22	140	20	130	19	110	16	90	13
		1 000	125	18	115	17	110	16	90	13	69	10
		10 000	97	14	97	14	97	14	...	...	...	...
		100 000	66	9.5	...	...	...	...	...	...	...	...
260	500	0.1	180	26	170	25	165	24	160	23	145	21
		1	165	24	160	23	150	22	140	20	115	17
		10	150	22	140	20	130	19	110	16	90	13
		100	130	19	125	18	110	16	90	13	69	10
		1 000	105	15	97	14	83	12	69	10	59	8.5
		10 000	69	10	69	10	...	...	...	...	...	...
		100 000	45	6.5	...	...	...	...	...	...	...	...
315	600	0.1	130	19	125	18	125	18	115	17	110	16
		1	115	17	115	17	110	16	105	15	90	13
		10	105	15	97	14	90	13	76	11	62	9
		100	69	10	69	10	62	9	52	7.5	38	5.5
		1 000	41	6	41	6	38	5.5	28	4.1	23	3.4
		10 000	22	3.2	...	...	...	...	...	...	...	...
370	700	0.1	69	10	69	10	69	10	66	9.5	66	9.5
		1	62	9	62	9	59	8.5	45	6.5	32	4.7
		10	32	4.7	30	4.3	27	3.9	23	3.3	18	2.6
		100	22	3.2	20	2.9	18	2.6	13	1.9	...	...
		1 000	14	2.1	13	1.9	11	1.6	...	...	...	...

Thermal conductivity. T61 temper, 146 W/m · K (84 Btu/ft · h · °F) at 20 °C (68 °F)

Electrical Properties

Electrical conductivity. Volumetric, T61 temper, 37% IACS at 20 °C (68 °F)
Electrical resistivity. T61 temper, 47 nΩ · m at 20 °C (68 °F); temperature coefficient, T61 temper: 0.1 nΩ · m per K at 20 °C (68 °F)
Electrolytic solution potential. At 25 °C (77 °F): T61 temper, −0.80 V versus 0.1 *N* calomel electrode in an aqueous solution containing 53 g NaCl plus 3 g H_2O_2 per liter

Fabrication Characteristics

Solution temperature. 530 °C (985 °F)
Aging temperature. T61, 200 °C (390 °F) for 20 h at temperature

3003, Alclad 3003
1.2Mn-0.12Cu

Specifications

AMS. See Table 42.
ASME. See Table 42.
ASTM. See Table 42.
SAE. J454
UNS number. 3003: A93003
Government. See Table 42.
Foreign. Canada: CSA MC10. France: NF A-M1. United Kingdom: BS N3. West Germany: DIN AlMn. ISO: AlMn1Cu

Chemical Composition

Composition limits of 3003: 0.6 Si max, 0.7 Fe max, 0.05 to 0.20 Cu, 1.0 to 1.5 Mn, 0.10 Zn max, 0.05 max other (each), 0.15 max others (total), bal Al
Composition limits of Alclad 3003. 7072 cladding—0.10 Cu max, 0.10 Mg max, 0.10 Mn max, 0.7 Fe max + Si, 0.8 to 1.3 Zn, 0.05 max other (each), 0.15 max others (total), bal Al

Applications

Typical uses of 3003. Applications where good formability, very good resistance to corrosion or good weldability, or all three, are required, and where more strength is

Table 38 Creep-rupture properties of alloy 2219-T87 plate

Temperature		Time under	Rupture stress		Stress for creep of							
					1.0%		0.5%		0.2%		0.1%	
°C	°F	stress, h	MPa	ksi	MPa	ksi	MPa	ksi	MPa	ksi	MPa	ksi
24	75	0.1	460	67	450	65	420	61	385	56	370	54
		1	455	66	425	62	400	58	380	55	365	53
		10	450	65	405	59	385	56	370	54	360	52
		100	435	63	395	57	380	55	365	53	350	51
		1 000	420	61	380	55	370	54	360	52	345	50
100	212	0.1	400	58	365	53	350	51	340	49	325	47
		1	380	55	345	50	340	49	325	47	310	45
		10	350	51	330	48	325	47	305	44	290	42
		100	330	48	315	46	310	45	290	42	260	38
		1 000	315	46	305	44	295	43	260	38	240	35
150	300	0.1	345	50	315	46	310	45	290	42	270	39
		1	315	46	295	43	290	42	260	38	240	35
		10	290	42	275	40	260	38	235	34	205	30
		100	260	38	250	36	235	34	200	29	165	24
		1 000	235	34	230	33	205	30	165	24	140	20
205	400	0.1	255	37	240	35	230	33	200	29	165	24
		1	230	33	215	31	200	29	165	24	140	20
		10	205	30	185	27	170	25	140	20	110	16
		100	180	26	165	24	145	21	115	17	97	14
		1 000	150	22	145	21	130	19	97	14	83	12
230	450	0.1	206	30	195	28	180	26	165	24	140	20
		1	185	27	170	25	165	24	140	20	115	17
		10	170	25	160	23	145	21	115	17	97	14
		100	150	22	140	20	125	18	105	15	83	12
		1 000	130	19	125	18	115	17	83	12	69	10
260	500	0.1	170	25	160	23	150	22	140	20	125	18
		1	160	23	145	21	140	20	125	18	105	15
		10	145	21	130	19	125	18	105	15	83	12
		100	125	18	115	17	110	16	90	13	69	10
		1 000	105	15	105	15	97	14	69	10	59	8.5
315	600	0.1	115	17	110	16	105	15	97	14	83	12
		1	105	15	105	15	97	14	83	12	66	9.5
		10	90	13	83	12	76	11	62	9	52	7.5
		100	62	9	55	8	52	7.5	45	6.5	34	5
		1 000	34	5	31	4.5	28	4	26	3.8	23	3.3
370	700	0.1	59	8.5	55	8	52	7.5	48	7	34	5
		1	48	7	45	6.5	41	6	32	4.7	18	2.6
		10	34	5	30	4.4	26	3.8	17	2.4	12	1.7
		100	23	3.4	20	2.9	17	2.5	11	1.6	8	1.2
		1 000	17	2.4	13	1.9	11	1.6	8	1.2	6	0.9

desired than is provided by unalloyed aluminum. Cooking utensils, food and chemical handling and storage equipment, tanks, trim in transportation equipment, lithographic sheet pressure vessels and piping

Typical uses of Alclad 3003. Farm roofing and siding

Mechanical Properties

Tensile properties. See Tables 43 and 44. Directional characteristics: tensile strength and elongation of sheet in any of the H tempers are slightly lower in transverse direction

Compressive yield strength. Approximately the same as tensile yield strength

Shear yield strength. Approximately 55% of the tensile strength

Hardness. See Table 43.

Poisson's ratio. 0.33 at 20 °C (68 °F)

Elastic modulus. Tension, 70 GPa (10.2 × 10^6 psi); shear, 25 GPa (3.6 × 10^6 psi)

Fatigue strength. See Table 43.

Mass Characteristics

Density. 2.73 g/cm^3 (0.099 lb/in.^{3}) at 20 °C (68 °F)

Thermal Properties

Liquidus temperature. 654 °C (1210 °F)

Solidus temperature. 643 °C (1190 °F)

Coefficient of thermal expansion. Linear:

Temperature range		Average coefficient	
°C	°F	μm/m · K	μin./in. · °F
−50 to 20	−58 to 68	21.5	11.9
20 to 100	68 to 212	23.2	12.9
20 to 200	68 to 392	24.1	13.4
20 to 300	68 to 572	25.1	13.9

Volumetric: 67 × 10^{-6} m^3/m^3 · K (3.72 × 10^{-5} in.^{3}/in.^{3} · °F) at 20 °C (68 °F)

Specific heat. 893 J/kg · K (0.213 Btu/lb · °F) at 20 °C (68 °F)

Thermal conductivity. At 20 °C (68 °F):

Temper	Conductivity	
	W/m · K	Btu/ft · h · °F
O	193	112
H12	163	94.1
H14	159	91.9
H18	155	89.6

Electrical Properties

Electrical conductivity. Volumetric, at 20 °C (68 °F):

Temper	Conductivity, %IACS
O	50
H12	42
H14	41
H18	40

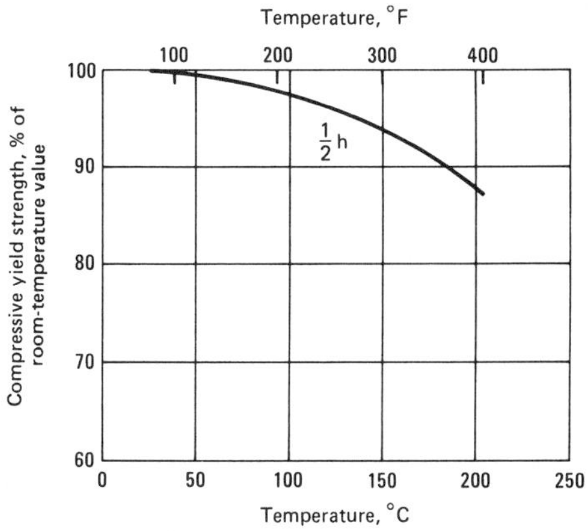

Fig. 13 Influence of temperature on compressive yield strength of alloy 2618-T61 hand-forged billets. Compressive yield strength determined at temperature after holding ½ h under no load. Value plotted as percentage of corresponding room-temperature value

Table 39 Tensile properties of alloy 2618-T61

Product and orientation	Tensile strength MPa	Tensile strength ksi	Yield strength MPa	Yield strength ksi	Elongation(a), %
Typical					
All products	440	64	372	54	10(b)
Property limits					
Die forgings, thickness ≤4 in.(c)					
Axis parallel to grain flow	400	58	310	45	4(d)(e)
Axis not parallel to grain flow	380	55	290	42	4(d)
Hand forgings					
Thickness ≤2.000 in.(c)(f)					
Longitudinal	400	58	325	47	7
Long transverse	380	55	290	42	5
Short transverse	360	52	290	42	4
2.001–3.000 in.					
Longitudinal	395	57	315	46	7
Long transverse	380	55	290	42	5
Short transverse	360	52	290	42	4
3.001–4.000 in.					
Longitudinal	385	56	310	45	7
Long transverse	365	53	275	40	5
Short transverse	350	51	270	39	4
Rolled rings, thickness ≤2.500 in.(g)					
Tangential	380	55	285	41	6
Axial	380	55	285	41	5

(a) In 50 mm (2 in.) or 4 *d*, where *d* is diameter of reduced section of tensile test specimen. (b) 12.5 mm (½ in.) diameter specimen. (c) Properties also apply to forgings machined prior to heat treatment, provided machined thickness is not less than ½ original (as-forged) thickness. (d) Specimen taken from forgings. (e) Elongation 6% min for specimen taken from separately forged coupon. (f) Maximum cross-sectional area 930 cm^2 (1 ft^2). Not applicable to upset biscuit forgings or to rolled rings. (g) Applicable only to rings having ratio of outside diameter to wall thickness equal to or greater than 10

Table 40 Typical tensile properties of alloy 2618-T61 at various temperatures

Temperature °C	Temperature °F	Tensile strength MPa	Tensile strength ksi	Yield strength (0.2% offset) MPa	Yield strength (0.2% offset) ksi	Elongation, %
−196	−320	538	78.0	421	61.0	12
−80	−112	462	67.0	379	55.0	11
−28	−18	441	64.0	372	54.0	10
24	75	441	64.0	372	54.0	10
100	212	427	62.0	372	54.0	10
149	300	345	50.0	303	44.0	14
204	400	221	32.0	179	26.0	24
260	500	90	13.0	62	9.0	50
316	600	52	7.5	31	4.5	80
371	700	34	5.0	24	3.5	120

Electrical resistivity. At 20 °C (68 °F):

Temper	Resistivity, nΩ · m
O	34
H12	41
H14	42
H18	43

Temperature coefficient, all tempers: 0.1 nΩ · m per K at 20 °C (68 °F)
Electrolytic solution potential. 3003 and core of Alclad 3003, −0.83 V; 7072 cladding, −0.96 V versus 0.1 *N* calomel electrode in an aqueous solution containing 53 g NaCl plus 3 g H_2O_2 per liter

Magnetic Properties

Magnetic susceptibility. Mass: 0.8 × 10^{-6} (cgs/g) at 25 °C (77 °F)

Fabrication Characteristics

Annealing temperature. 415 °C (775 °F). Commercial practice: 400 to 600 °C (750 to 1100 °F); higher temperatures used only for flash annealing

3004, Alclad 3004
1.2Mn-1.0Mg

Specifications

ASTM. 3004: sheet and plate, B 209; extruded tube, B 221; welded tube, B 313, B 547. Alclad 3004: sheet and plate, B 209; welded tube, B 313; culvert pipe, B 547
SAE. J454
UNS number. A93004
Government. Culvert pipe: WW-P-402
Foreign. Australia: A3004. France: NF A-M1G. West Germany: DIN AlMn1Mg1

Table 41 Creep-rupture properties of alloy 2618

Temperature °C	Temperature °F	Time under stress, h	Rupture stress MPa	Rupture stress ksi	Stress for creep of 1.0% MPa	1.0% ksi	0.5% MPa	0.5% ksi	0.2% MPa	0.2% ksi	0.1% MPa	0.1% ksi
150	300	0.1	380	55	345	50	345	50	330	48	315	46
		1	360	52	340	49	330	48	315	46	290	42
		10	340	49	325	47	315	46	295	43	270	39
		100	305	44	305	44	290	42	270	39	240	35
		1000	255	37	255	37	250	36	240	35	205	30
177	350	0.1	340	49	325	47	315	46	295	43	285	41
		1	310	45	305	44	295	43	275	40	255	37
		10	285	41	275	40	260	38	250	36	220	32
		100	250	36	240	35	235	34	220	32	185	27
		1000	205	30	200	29	195	28	185	27	150	22
205	400	0.1	290	42	285	41	270	39	255	37	240	35
		1	260	38	255	37	250	36	235	34	205	30
		10	230	33	220	32	215	31	200	29	170	25
		100	195	28	185	27	180	26	165	24	140	20
		1000	160	23	150	22	145	21	130	19	90	13
260	500	0.1	185	27	170	25	165	24	160	23	145	21
		1	165	24	150	22	145	21	140	20	115	17
		10	140	20	130	19	125	18	110	16	83	12
		100	105	15	97	14	90	13	69	10	52	7.5
		1000	62	9	62	9	55	8	48	7	...	...
315	600	0.1	97	14	83	12	69	10	55	8	48	7
		1	69	10	62	9	55	8	45	6.5	41	6
		10	52	7.5	45	6.5	41	6	38	5.5	26	3.8
		100	32	4.6	28	4.1	26	3.7	19	2.8	15	2.2
		1000	20	2.9	17	2.5	14	2.1	...	...	...	...

Table 42 Standard specifications for alloy 3003

Mill form and condition	Specification number AMS	ASME	ASTM	Government
Bare 3003				
Sheet and plate	4006	SB209	B 209	QQ-A-250/2
	4008	. . .	. . .	. . .
Wire, rod, and bar (rolled or cold finished)	. . .	. . .	B 211	QQ-A-225/2
Wire, rod, bar, shapes, and tube (extruded)	. . .	SB221	B 221	QQ-A-200/1
Tube				
Extruded, seamless	. . .	SB241	B 241	. . .
Extruded, coiled	. . .	. . .	B 491	. . .
Drawn	. . .	. . .	B 483	. . .
Drawn, seamless	4065	SB210	B 210	WW-T-700/2
	4067	. . .	. . .	. . .
Condenser	. . .	SB234	B 234	. . .
Condenser with integral fins	. . .	. . .	B 404	. . .
Welded	. . .	. . .	B 313	. . .
	. . .	. . .	B 547	. . .
Pipe: seamless	. . .	. . .	B 241	MIL-P-25995
Gas and oil transmission	. . .	. . .	B 345	. . .
Rivet wire and rod	. . .	. . .	B 316	QQ-A-430
Forgings	. . .	SB247	B 247	. . .
Foil	4010	. . .	. . .	MIL-A-81596
Alclad 3003				
Sheet and plate			B 209	
Tube				
Drawn, seamless			B 210	
Extruded			B 221	
Extruded, seamless			B 241	
Condenser			B 234	
Condenser with integral fins			B 404	
Welded			B 547	
Pipe (gas and oil transmission)			B 345	

Chemical Composition

Composition limits of 3004. 0.25 Cu max, 0.30 Si max, 0.70 Fe max, 1.0 to 1.5 Mn, 0.8 to 1.3 Mg, 0.25 Zn max, 0.05 max other (each), 0.15 max others (total), bal Al

Composition limits of Alclad 3004. 7072 cladding—0.10 Cu max, 0.10 Mg max, 0.10 Mn max, 0.7 Fe max + Si, 0.8 to 1.3 Zn, 0.05 max other (each), 0.15 max others (total), bal Al

Applications

Typical uses of 3004. Drawn and ironed rigid containers (cans), chemical handling and storage equipment, sheet metal work, builders' hardware, incandescent and fluorescent lamp bases and similar applications requiring good formability and higher strength than provided by 3003

Typical uses of Alclad 3004. Siding, culvert pipe, industrial roofing

Mechanical Properties

Tensile properties. See Tables 45 and 46.

Compressive yield strength. Approximately the same as tensile yield strength

Shear yield strength. Approximately 55% of tensile strength

Hardness. See Table 45.

Poisson's ratio. 0.35 at 20 °C (68 °F)

Table 43 Mechanical properties of alloy 3003

Temper	Tensile strength MPa	ksi	MPa	ksi	Yield strength MPa	ksi	Elongation, %	Hardness HB(a)	HR	Shear strength MPa	ksi	Fatigue strength(b) MPa	ksi
Typical properties													
O	110	16	. . .	. . .	42	6	30–40	28	45–65	76	11	48	7
H12	130	19	. . .	. . .	125	18	10–20	35	55–75	83	12	55	8
H14	150	22	. . .	. . .	145	21	8–16	40	70–90	97	14	62	9
H16	175	25	. . .	. . .	175	25	5–14	47	75–92	105	15	69	10
H18	200	29	. . .	. . .	185	27	4–10	55	84–95	110	16	69	10
Property limits	Minimum		Maximum		Minimum								
O (0.006–3.000 in. thick)	97	14	130	19	34	5	14–25	. . .	. . .	. . .	. . .	. . .	. . .
H12 (0.017–2.000 in. thick)	115	17	160	23	83	12	3–10	. . .	. . .	. . .	. . .	. . .	. . .
H14 (0.009–1.000 in. thick)	140	20	180	26	115	17	1–10	. . .	. . .	. . .	. . .	. . .	. . .
H16 (0.006–0.162 in. thick)	165	24	205	30	145	21	1–4	. . .	. . .	. . .	. . .	. . .	. . .
H18 (0.006–0.128 in. thick)	185	27	. . .	. . .	165	24	1–4	. . .	. . .	. . .	. . .	. . .	. . .
H112													
(0.250–0.499 in. thick)	115	17	. . .	. . .	69	10	8	. . .	. . .	. . .	. . .	. . .	. . .
(0.500–2.000 in. thick)	105	15	. . .	. . .	41	6	12	. . .	. . .	. . .	. . .	. . .	. . .
(2.000–3.000 in. thick)	100	14.5	. . .	. . .	41	6	18	. . .	. . .	. . .	. . .	. . .	. . .
Property limits, Alclad 3003(c)													
O (0.006–0.499 in. thick)	90	13	125	18	31	4.5	14–25	. . .	. . .	. . .	. . .	. . .	. . .
(0.500–3.000 in. thick)	97	14	130	19	34	5.0	23	. . .	. . .	. . .	. . .	. . .	. . .
H12													
(0.017–0.499 in. thick)	110	16	150	22	77	11	4–9	. . .	. . .	. . .	. . .	. . .	. . .
(0.500–2.000 in. thick)	115	17	160	23	83	12	10	. . .	. . .	. . .	. . .	. . .	. . .
H14													
(0.009–0.499 in. thick)	130	19	170	25	110	16	1–8	. . .	. . .	. . .	. . .	. . .	. . .
(0.500–2.000 in. thick)	140	20	180	26	115	17	10	. . .	. . .	. . .	. . .	. . .	. . .
H16 (0.006–0.162 in. thick)	160	23	200	29	140	20	1–4	. . .	. . .	. . .	. . .	. . .	. . .
H18 (0.006–0.128 in. thick)	180	26	. . .	. . .	. . .	. . .	1–4	. . .	. . .	. . .	. . .	. . .	. . .
H112													
(0.250–0.499 in. thick)	110	16	. . .	. . .	62	9	8	. . .	. . .	. . .	. . .	. . .	. . .
(0.500–2.000 in. thick)	105	15	. . .	. . .	41	6	12	. . .	. . .	. . .	. . .	. . .	. . .
(2.000–3.000 in. thick)	100	14.5	. . .	. . .	41	6	18	. . .	. . .	. . .	. . .	. . .	. . .

(a) 500 kg load, 10 mm ball, 30 s duration of loading. (b) At 5×10^8 cycles. R.R. Moore type test. (c) Mechanical properties of 3003 clad with 7072 are practically the same as for bare material, except that hardness and fatigue resistance tend to be slightly lower for the clad product.

Table 44 Typical mechanical properties of alloy 3003 at various temperatures

Temperature °C	Temperature °F	Tensile strength(a) MPa	Tensile strength(a) ksi	Yield strength(a) MPa	Yield strength(a) ksi	Elongation, %
O temper						
−200	−328	230	33	60	8.6	46
−100	−148	150	22	52	7.5	43
−30	−22	115	17	45	6.5	41
25	77	110	16	41	6	40
100	212	90	13	38	5.5	43
200	392	60	8.6	30	4.3	60
300	572	29	4.2	17	2.5	70
400	752	18	2.6	12	1.7	75
H14 temper						
−200	−328	250	36	170	25	30
−100	−148	175	25	155	22.5	19
−30	−22	150	22	145	21	16
25	77	150	22	145	21	16
100	212	145	21	130	19	16
200	392	96	14	62	9	20
300	572	29	4.2	17	2.5	70
400	752	18	2.6	12	1.7	75
H18 temper						
−200	−328	290	42	230	33	23
−100	−148	230	33	210	30	12
−30	−22	210	30	190	38	10
25	77	200	29	185	27	10
100	212	180	26	145	21	10
200	392	96	14	62	9	18
300	572	29	4.2	17	2.5	70
400	752	18	2.6	12	1.7	75

(a) Lowest strengths for exposures up to 10 000 h at temperature, no load; test load applied at 35 MPa/min (5 ksi/min) to yield strength and then at strain rate of 5%/min to fracture

Elastic modulus. Tension, 70 GPa (10.2×10^6 psi); shear, 25 GPa (3.6×10^6 psi)
Fatigue strength. See Table 45.

Mass Characteristics

Density. 2.72 g/cm^3 (0.098 lb/in.^{3}) at 20 °C (68 °F)

Thermal Properties

Liquidus temperature. 654 °C (1210 °F)
Solidus temperature. 629 °C (1165 °F)
Coefficient of thermal expansion. Linear:

Temperature range °C	Temperature range °F	Average coefficient µm/m · K	Average coefficient µin./in. · °F
−50 to 20	−58 to 68	21.5	11.9
20 to 100	68 to 212	23.2	12.9
20 to 200	68 to 392	24.1	13.4
20 to 300	68 to 572	25.1	13.9

Volumetric: 67×10^{-6} m^3/m^3 · K (3.72×10^{-5} in.^{3}/in.^{3} · °F) at 20 °C (68 °F)
Specific heat. 893 J/kg · K (0.213 Btu/lb · °F) at 20 °C (68 °F)
Thermal conductivity. O temper: 162 W/m · K (93.6 Btu/ft · h · °F) at 20 °C (68 °F)

Electrical Properties

Electrical conductivity. Volumetric, O temper: 42% IACS at 20 °C (68 °F)
Electrical resistivity. O temper: 41.0 nΩ · m at 20 °C (68 °F); temperature coefficient, 0.1 nΩ · m per K at 20 °C (68 °F)
Electrolytic solution potential. −0.84 V; 3004 and core of Alclad 3004, 7072 cladding, −0.96 V (cladding) versus 0.1 *N* calomel electrode in an aqueous solution containing 53 g NaCl plus 3 g H_2O_2 per liter

Magnetic Properties

Magnetic susceptibility. Mass: 0.8×10^{-6} (cgs/g) at 25 °C (68 °F)

Table 45 Mechanical properties of alloy 3004

Temper	Tensile strength MPa	Tensile strength ksi	Tensile strength MPa	Tensile strength ksi	Yield strength MPa	Yield strength ksi	Elongation, %	Hardness, HB(a)	Shear strength MPa	Shear strength ksi	Fatigue strength(b) MPa	Fatigue strength(b) ksi
Typical properties												
O	180	26	...	...	69	10	20–25	45	110	16	97	14
H32	215	31	...	...	170	25	10–17	52	115	17	105	15
H34	240	35	...	...	200	29	9–12	63	125	18	105	15
H36	260	38	...	...	230	33	5–9	70	140	20	110	16
H38	285	41	...	...	250	36	4–6	77	145	21	110	16
Property limits	Minimum		Maximum		Minimum							
O (0.006–3.000 in. thick)	150	22	200	29	59	8.5	10–18	...	...	...	...	...
H32 (0.017–2.000 in. thick)	195	28	240	35	145	21	1–6	...	...	...	...	...
H34 (0.009–1.000 in. thick)	220	32	260	38	170	25	1–5	...	...	...	...	...
H36 (0.006–0.162 in. thick)	240	35	285	41	195	28	1–4	...	...	...	...	...
H38 (0.006–0.128 in. thick)	260	38	...	...	215	31	1–4	...	...	...	...	...
H112 (0.250–3.000 in. thick)	160	23	...	...	62	9	7	...	...	...	...	...
Property limits, Alclad 3004(c)												
O												
(0.006–0.499 in. thick)	145	21	195	28	55	8	10–18	...	...	...	...	...
(0.500–3.000 in. thick)	150	22	200	29	59	8.5	16	...	...	...	...	...
H32												
(0.017–0.499 in. thick)	185	27	235	34	140	20	1–6	...	...	...	...	...
(0.500–2.000 in. thick)	195	28	240	35	145	21	6	...	...	...	...	...
H34												
(0.009–0.499 in. thick)	215	31	255	37	165	24	1–5	...	...	...	...	...
(0.500–1.000 in. thick)	220	32	260	38	170	25	5	...	...	...	...	...
H36 (0.006–0.162 in. thick)	235	34	275	40	185	27	1–4	...	...	...	...	...
H38 (0.006–0.128 in. thick)	255	37	...	...	...	...	1–4	...	...	...	...	...
H112												
(0.250–0.499 in. thick)	150	22	...	...	59	8.5	7	...	...	...	...	...
(0.500–3.000 in. thick)	160	23	...	...	62	9	7	...	...	...	...	...

(a) 500 kg load, 10 mm ball, 30 s duration of loading. (b) At 5×10^8 cycles, R.R. Moore type test. (c) Mechanical properties of 3004 clad with 7072 are practically the same as for bare material, except that hardness and fatigue resistance tend to be slightly lower for the clad product.

Table 46 Typical mechanical properties of alloy 3004 at various temperatures

Temperature °C	Temperature °F	Tensile strength(a) MPa	Tensile strength(a) ksi	Yield strength(a) MPa	Yield strength(a) ksi	Elongation, %
O temper						
−200	−328	290	42.5	90	13.2	38
−100	−148	200	29	80	11.5	31
−30	−22	180	26	69	10	26
25	77	180	26	69	10	25
100	212	180	26	69	10	25
200	392	96	14	65	9.5	55
300	572	50	7.2	34	4.9	80
400	752	30	4.4	9	2.8	90
H34 temper						
−200	−328	360	52	235	34	26
−100	−148	270	39	212	31	17
−30	−22	245	36	200	29	13
25	77	240	35	200	29	12
100	212	240	35	200	29	12
200	392	145	21	105	15	35
300	572	50	7.2	34	4.9	80
400	752	30	4.4	19	2.8	90
H38 temper						
−200	−328	400	58	295	43	20
−100	−148	310	45	267	39	10
−30	−22	290	42	245	36	7
25	77	280	41	245	36	6
100	212	275	40	245	36	7
200	392	150	22	105	15	30
300	572	50	7.2	34	4.9	80
400	752	30	4.4	19	2.8	90

(a) Lowest strength for exposures up to 10 000 h at temperature, no load; test loading applied at 35 MPa/min (5 ksi/min) to yield strength and then at strain rate of 5%/min to fracture

Fabrication Characteristics

Annealing temperature. 415 °C (775 °F)

3105
0.55Mn-0.50Mg

Specifications

ASTM. B 209
SAE. J454

Chemical Composition

Composition limits. 0.6 Si max, 0.7 Fe max, 0.30 Cu max, 0.20 to 0.80 Mn, 0.20 to 0.80 Mg, 0.20 Cr max, 0.40 Zn max, 0.10 Ti max, 0.05 max other (each), 0.15 max others (total), bal Al

Applications

Typical uses. Residential siding, mobile home sheet, gutters and downspouts, sheet metal work, bottle caps and closures

Mechanical Properties

Tensile properties. See Table 47.
Poisson's ratio. 0.33
Elastic modulus. Tension, 69 GPa (10×10^6 psi); shear, 25 GPa (3.6×10^6 psi)

Mass Characteristics

Density. 2.71 g/cm^3 (0.098 lb/in.^{3}) at 20 °C (68 °F)

Thermal Properties

Liquidus temperature. 657 °C (1215 °F)
Solidus temperature. 638 °C (1180 °F)
Coefficient of thermal expansion. Linear:

Temperature range °C	Temperature range °F	Average coefficient µm/m · K	Average coefficient µin./in. · °F
−50 to 20	−58 to 68	21.8	12.1
20 to 100	68 to 212	23.6	13.1
20 to 200	68 to 392	24.5	13.6
20 to 300	68 to 572	25.5	14.2

Volumetric: 68×10^{-6} m^3/m^3 · K (3.77×10^{-5} in.^{3}/in.^{3} · °F) at 20 °C (68 °F)
Specific heat. 897 J/kg · K (0.214 Btu/lb · °F) at 20 °C (68 °F)
Thermal conductivity. 173 W/m · K (99.9 Btu/ft · h · °F) at 20 °C (68 °F)

Electrical Properties

Electrical conductivity. Volumetric, O temper: 45% IACS at 20 °C (68 °F)
Electrical resistivity. O temper: 38.3 nΩ · m at 20 °C (68 °F); temperature coefficient, 0.1 nΩ · m per K at 20 °C (68 °F)
Electrolytic solution potential. −0.84 V versus 0.1 *N* calomel electrode in an aqueous solution containing 53 g NaCl plus 3 g H_2O_2 per liter

Magnetic Properties

Magnetic susceptibility. Mass: 0.7×10^{-6} (cgs/g) at 20 °C (68 °F)

Fabrication Characteristics

Annealing temperature. 345 °C (650 °F)

4032
12.2Si-1.0Mg-0.9Cu-0.9Ni

Specifications

AMS. Forgings and forging stock: 4145
ASTM. Forgings: B 247
SAE. J454
UNS number. A94032
Government. Forgings: QQ-A-367
Foreign. Canada: CSA SG121. France: NF A-S12UN. Italy: UNI P-AlSi12MgCuNi

Table 47 Mechanical properties of alloy 3105 sheet

Temper	Tensile strength MPa	Tensile strength ksi	Tensile strength MPa	Tensile strength ksi	Yield strength MPa	Yield strength ksi	Elongation, %	Shear strength MPa	Shear strength ksi
Typical properties									
O	115	17	...	...	55	8	24	83	12
H12	150	22	...	...	130	19	7	97	14
H14	170	25	...	...	150	22	5	105	15
H16	195	28	...	...	170	25	4	110	16
H18	215	31	...	...	195	28	3	115	17
H25	180	26	...	...	160	23	8	105	15
Property limits	Minimum		Maximum		Minimum				
O (0.013–0.080 in. thick)	97	14	145	21	34	5	16–20	...	...
H12 (0.017–0.080 in. thick)	130	19	180	26	105	15	1–3	...	...
H14 (0.013–0.080 in. thick)	150	22	200	29	125	18	1–2	...	...
H16 (0.013–0.080 in. thick)	170	25	220	32	145	21	1–2	...	...
H18 (0.013–0.080 in. thick)	195	28	...	...	165	24	1–2	...	...
H25 (0.013–0.080 in. thick)	160	23	...	...	130	19	2–6	...	...

Table 48 Typical mechanical properties of alloy 4032-T6 at various temperatures

Temperature °C	Temperature °F	Tensile strength MPa	Tensile strength ksi	Yield strength MPa	Yield strength ksi	Elongation, %
−200	−328	460	67	337	49	11
−100	−148	415	60	325	47	10
−30	−22	385	56	315	46	9
25	77	380	55	315	46	9
100	212	345	50	300	44	9
200	392	90	13	62	9	30
300	572	38	5.5	24	3.5	70
400	752	21	3.1	12	1.8	90

Chemical Composition

Composition limits. 11.0 to 13.5 Si, 1.0 Fe max, 0.50 to 1.30 Cu, 0.8 to 1.3 Mg, 0.10 Cr max, 0.50 to 1.3 Ni, 0.25 Zn max, 0.05 max other (each), 0.15 max others (total), bal Al

Applications

Typical uses. Pistons and other high-temperature service parts

Mechanical Properties

Tensile properties. T6 temper: tensile strength, 380 MPa (55 ksi); yield strength, 315 MPa (46 ksi); elongation, 9% in 50 mm (2 in.). For typical properties at various temperatures, see Table 48.
Hardness. T6 temper: 120 HB at 500 kg load, 10 mm ball
Poisson's ratio. 0.33
Elastic modulus. Tension, 79 GPa (11.4 × 10^6 psi). Shear, 26 GPa (3.8 × 10^6 psi)
Fatigue strength. T6 temper: 110 MPa (16 ksi) at 5 × 10^8 cycles, R.R. Moore type test. At various temperatures, see Table 49.
Creep-rupture characteristics. See Table 50.

Mass Characteristics

Density. 2.68 g/cm³ (0.097 lb/in.³) at 20 °C (68 °F)

Thermal Properties

Liquidus temperature. 571 °C (1060 °F)
Eutectic temperature. 532 °C (990 °F)
Incipient melting temperature. 532 °C (990 °F)
Coefficient of thermal expansion.

Linear:

Temperature range °C	Temperature range °F	Average coefficient μm/m · K	Average coefficient μin./in. · °F
−50 to 20	−58 to 68	18.0	10.0
20 to 100	68 to 212	19.5	10.8
20 to 200	68 to 392	20.2	11.2
20 to 300	68 to 572	21.0	11.7

Volumetric: 56 × 10^{-6} m³/m³ · K (3.11 × 10^{-5} in.³/in.³ · °F) at 20 °C (68 °F)
Specific heat. 864 J/kg · K (0.206 Btu/lb · °F) at 20 °C (68 °F)
Thermal conductivity. At 20 °C (68 °F): O temper, 155 W/m · K (89.6 Btu/ft · h · °F); T6 temper, 141 W/m · K (81.5 Btu/ft · h · °F)

Electrical Properties

Electrical conductivity. Volumetric, at 20 °C (68 °F): O temper, 40% IACS; T6 temper, 36% IACS
Electrical resistivity. At 20 °C (68 °F): O temper, 43.1 nΩ · m; T6 temper, 47.9 nΩ · m. Temperature coefficient, 0.1 nΩ · m per K at 20 °C (68 °F)

Fabrication Characteristics

Annealing temperature. 415 °C (775 °F); 2 to 3 h at temperature then furnace cooled to 260 °C (500 °F) at 25 °C (50 °F) per h max
Solution temperature. 505 to 515 °C (940 to 960 °F). Hold 4 min at temperature then quench in cold water; for heavy or complicated forgings, quench in water at 65 to 100 °C (150 to 212 °F)
Aging temperature. 170 to 175 °C (335 to 345 °F); 8 to 12 h at temperature

Table 49 Fatigue strength of alloy 4032-T6 at various temperatures

Temperature °C	Temperature °F	No. of cycles	Stress(a) MPa	Stress(a) ksi
24	75	10^4	359	52
		10^5	262	38
		10^6	207	30
		10^7	165	24
		10^8	124	18
		5 × 10^8	114	16.5
149	300	10^5	207	30
		10^6	165	24
		10^7	124	18
		10^8	90	13
		5 × 10^8	79	11.5
204	400	10^5	186	27
		10^6	138	20
		10^7	90	13
		10^8	55	8
		5 × 10^8	48	7
260	500	10^5	131	19
		10^6	83	12
		10^7	55	8
		10^8	34	5
		5 × 10^8	34	5

(a) Based on rotating beam tests at room temperature and cantilever beam tests at elevated temperatures

Hot-working temperature. 315 to 480 °C (600 to 900 °F)

4043
5.2Si

Specifications

AMS. Bare welding rod and electrodes: 4190
SAE. J454
Government. Bare welding rod and electrodes: QQ-R-566, MIL-E-16053; spray gun wire: MIL-W-6712
Foreign. Australia: B4043. Canada: CSA S5. France: NF A-S5. United Kingdom: BS N21. Germany: DIN AlSi5, Werstoff-Nr. 3.2245

Chemical Composition

Composition limits. 4.5 to 6.0 Si, 0.8 Fe max, 0.30 Cu max, 0.05 Mn max, 0.05 Mg max, 0.10 Zn max, 0.20 Ti max, 0.05 max other (each), 0.15 max others (total), 0.0008 Be max for welding electrode only, bal Al

Applications

Typical uses. General purpose weld filler alloy (rod or wire) for welding all wrought and foundry alloys except those rich in magnesium

Mechanical Properties

Tensile properties. See Table 51.

Mass Characteristics

Density. 2.68 g/cm³ (0.097 lb/in.³)

Thermal Properties

Liquidus temperature. 630 °C (1170 °F)
Solidus temperature. 575 °C (1065 °F)
Coefficient of thermal expansion. Linear, 22.0 μm/m · K (12.2 μin./in. · °F) at 20 to 100 °C (68 to 212 °F)

Table 50 Creep-rupture properties of alloy 4032

Temperature °C	Temperature °F	Time under stress, h	Rupture stress MPa	Rupture stress ksi	Stress for creep of 1.0% MPa	1.0% ksi	0.5% MPa	0.5% ksi	0.2% MPa	0.2% ksi
100	212	0.1	331	48	283	41	269	39	...	...
		1	317	46	283	41	262	38	...	...
		10	303	44	283	41	262	38	...	...
		100	296	43	276	40	262	38	...	...
		1000	296	43	276	40	255	37	...	...
149	300	0.1	290	42	276	40	248	36	...	...
		1	276	40	269	39	241	35	...	...
		10	269	39	255	37	234	34	...	...
		100	248	36	241	35	221	32	...	...
		1000	207	30	200	29	186	27	...	...
204	400	0.1	234	34	228	33	221	32	138	20
		1	214	31	207	30	200	29	131	19
		10	186	27	179	26	165	24	103	15
		100	138	20	131	19	124	18	59	8.5
		1000	83	12	76	11	69	10	...	...

Table 51 Typical tensile properties of alloy 4043 welding wire

Wire diameter mm	in.	Temper	Tensile strength MPa	ksi	Yield strength (0.2% offset) MPa	ksi	Elongation, %
5.0	0.20	H16	205	30	180	26	1.7
3.2	0.12	H14	170	25	165	24	1.3
1.6	0.06	H18	285	41	270	39	0.5
1.2	0.05	H16	200	29	185	27	0.4
5.0	0.20	O	130	19	50	7	25
3.2	0.12	O	115	17	55	8	31
1.6	0.06	O	145	21	65	10	22
1.2	0.05	O	110	16	55	8	29

Electrical Properties

Electrical conductivity. Volumetric, O temper: 42% IACS at 20 °C (68 °F)
Electrical resistivity. O temper: 41 nΩ · m at 20 °C (68 °F)

Fabrication Characteristics

Annealing temperature. 350 °C (660 °F)

5005
0.8Mg

Specifications

ASTM. Sheet and plate: B 209. Wire, H19 temper: B 396. Stranded conductor: B 397. Rivet wire and rod: B 316. Rolled rod: B 531. Drawn tube: B 210, B 483
SAE. J454
UNS number. A95005
Government. Rivet wire and rod: QQ-A-430
Foreign. France: NF A-G0.6. United Kingdom: BS N41. Germany: DIN AlMg1. ISO: AlMg1

Chemical Composition

Composition limits. 0.30 Si max, 0.7 Fe max, 0.20 Cu max, 0.20 Mn max, 0.50 to 1.1 Mg, 0.10 Cr max, 0.25 Zn max, 0.05 max other (each), 0.15 max others (total), bal Al

Applications

Typical uses. Electrical conductor wire, cooking utensils, and architectural applications. Medium strength and good resistance to corrosion are two characteristics of 5005 similar to those of 3003. When anodized, film on 5005 is clearer and lighter than on 3003 and gives better color match with 6063 architectural extrusions.

Table 52 Typical mechanical properties of alloy 5005

Temper	Tensile strength(a) MPa	ksi	Yield strength(a) MPa	ksi	Elongation(a)(b), %	Hardness(c), HB	Shear strength MPa	ksi
O	124	18	41	6	25	28	76	11
H12	138	20	131	19	10	...	97	14
H14	159	23	152	22	6	...	97	14
H16	179	26	172	25	5	...	103	15
H18	200	29	193	28	4	...	110	16
H32	138	20	117	17	11	36	97	14
H34	159	23	138	20	8	41	97	14
H36	179	26	165	24	6	46	103	15
H38	200	29	186	27	5	51	110	16

(a) Strengths and elongations unchanged or improved at low temperatures. (b) 1.6 mm (1/16 in.) thick specimen. (c) 500 kg load; 10 mm diam ball

Table 53 Mechanical property limits for alloy 5005 sheet and plate

Temper	Tensile strength, Minimum MPa	ksi	Tensile strength, Maximum MPa	ksi	Yield strength (min) MPa	ksi	Elongation (min), %(a)
O	105	15	145	21	35	5	12–22
H12	125	18	165	24	95	14	2–9
H14	145	21	185	27	115	17	1–8
H16	165	24	205	30	135	18	1–3
H18	185	27	...	...	...	...	1–3
H32	120	17	160	23	85	12	3–10
H34	140	20	180	26	105	15	2–8
H36	160	23	200	29	125	18	1–4
H38	180	26	...	...	...	...	1–4
H112							
0.250–0.492 in. thick	115	17	...	...	...	...	8
0.492–1.60 in. thick	105	15	...	...	...	...	10
1.60–3.20 in. thick	100	15	...	...	...	...	16

(a) In 50 mm (2 in.) or 5*d*, where *d* is diameter or reduced section of tensile test specimen. Where a range of values appears in this column, the specified minimum elongation varies with thickness of the mill product.

Mechanical Properties

Tensile properties. See Tables 52 and 53. Tensile strength and elongation are slightly lower in transverse direction than in longitudinal direction.
Shear yield strength. Approximately 55% of tensile yield strength
Compressive yield strength. Approximately the same as tensile yield strength
Hardness. See Table 52.
Poisson's ratio. 0.33
Elastic modulus. Tension. 68.2 GPa (9.90×10^6 psi); shear, 25.9 GPa (3.75×10^6 psi); compression, 69.5 GPa (10.1×10^6 psi)

Mass Characteristics

Density. 2.70 g/cm^3 (0.097 lb/in.^{3}) at 20 °C (68 °F)

Thermal Properties

Liquidus temperature. 652 °C (1205 °F)
Solidus temperature. 632 °C (1170 °F)
Coefficient of thermal expansion. Linear:

Temperature range °C	°F	Average coefficient μm/m · K	μin./in. · °F
−50 to 20	−58 to 68	21.9	12.2
20 to 100	68 to 212	23.7	13.2
20 to 200	68 to 392	24.6	13.7
20 to 300	68 to 572	25.6	14.2

Volumetric: 68×10^{-6} m^3/m^3 · K (3.77×10^{-5} in.^{3}/in.^{3} · °F) at 20 °C (68 °F)
Specific heat. 900 J/kg · K (0.215 Btu/lb · °F) at 20 °C (68 °F)
Thermal conductivity. 205 W/m · K (118 Btu/ft · h · °F) at 20 °C (68 °F)

Electrical Properties

Electrical conductivity. Volumetric, O and H38 tempers: 52% IACS at 20 °C (68 °F)
Electrical resistivity. O and H38 tempers: 33.2 nΩ · m at 20 °C (68 °F); temperature coefficient, 0.1 nΩ · m per K at 20 °C (68 °F)
Electrolytic solution potential. −0.83 V versus 0.1 *N* calomel electrode in an aqueous solution containing 53 g NaCl plus 3 g H_2O_2 per liter

Fabrication Characteristics

Annealing temperature. 345 °C (650 °F); holding at temperature not required
Hot-working temperature. 260 to 510 °C (500 to 950 °F)

5050
1.4Mg

Specifications

ASTM. Sheet and plate: B 209. Drawn, seamless tube: B 210. Drawn tube: B 483. Welded tube: B 313, B 547
SAE. J454
UNS number. A95050
Foreign. France: NF A-Gl. Italy: P-AlMg 1.5. Switzerland: Al1.5Mg. United Kingdom: BS 3L44. ISO: AlMg1.5

Table 54 Typical mechanical properties of alloy 5050

Temper	Tensile strength(a) MPa	ksi	Yield strength(a) MPa	ksi	Elongation(a)(b), %	Hardness(c), HB	Shear strength MPa	ksi	Fatigue strength(d) MPa	ksi
O	145	21	55	8	24	36	105	15	83	12
H32	170	25	145	21	9	46	115	17	90	13
H34	190	28	165	24	8	53	123	18	90	13
H36	205	30	180	26	7	58	130	19	97	14
H38	220	32	200	29	6	63	138	20	97	14

(a) Strengths and elongation generally unchanged or improved at low temperatures. (b) 1.6 mm (1/16 in.) thick sheet specimen. (c) 500 kg load; 10 mm diam ball. (d) At 5 × 10^8 cycles; R.R. Moore type test

Chemical Composition

Composition limits. 0.40 Si max, 0.7 Fe max, 0.20 Cu max, 0.10 Mn max, 1.1 to 1.8 Mg, 0.10 Cr max, 0.25 Zn max, 0.05 max other (each), 0.15 max others (total), bal Al

Applications

Typical uses. Sheet used as trim in refrigerator applications; tube for automotive gas and oil lines; welded irrigation pipe; also available as plate, tube, rod, bar, and wire

Mechanical Properties

Tensile properties. See Tables 54 to 56. Tensile strength and yield strength are approximately the same in both the transverse and longitudinal directions; however, elongation is slightly lower in the transverse direction than in the longitudinal direction.
Shear yield strength. Approximately 55% of the tensile yield strength
Compressive yield strength. Approximately the same as tensile yield strength
Hardness. See Table 54.
Poisson's ratio. 0.33

Table 55 Typical tensile properties of alloy 5050

Temperature °C	°F	Tensile strength(a) MPa	ksi	Yield strength (0.2% offset)(a) MPa	ksi
−196	−320	255	37	70	10
−80	−112	150	22	60	8.5
−28	−18	145	21	55	8
24	75	145	21	55	8
100	212	145	21	55	8
149	300	130	19	55	8
204	400	95	14	50	7.5
260	500	60	9	41	6
316	600	41	6	29	4.2
371	700	27	3.9	18	2.6
−196	−320	305	44	205	30
−80	−112	205	30	170	25
−28	−18	195	28	165	24
24	75	195	28	165	24
100	212	195	28	165	24
149	300	170	25	150	22
204	400	95	14	50	7.5
260	500	60	9	41	6
316	600	41	6	29	4.2
371	700	27	3.9	18	2.6
−196	−320	315	46	250	36
−80	−112	235	34	205	30
−28	−18	220	32	200	29
24	75	220	32	200	29
100	212	215	31	200	29
219	300	185	27	170	25
204	400	95	14	50	7.5
260	500	60	9	41	6
316	600	41	6	29	4.2
371	700	27	3.9	18	2.6

(a) Lowest strengths for exposures up to 10 000 h at temperature; no load; test loading applied at 35 MPa/min (5 ksi/min) to yield strength and then at strain rate of 5%/min to fracture

Table 56 Tensile-property limits for alloy 5050

Temper	Tensile strength (min) MPa	ksi	Yield strength (min) MPa	ksi	Elongation (min), %(a)
O	125	18	41	6	16–20
H32	150	22	110	16	4–6
H34	170	25	138	20	3–5
H36	185	27	151	22	2–4
H38	200	29	. . .	. . .	2–4

(a) Where a range of values appears in this column, specified minimum elongation varies with thickness of the mill product.

Elastic modulus. Tension, 68.9 GPa (10.0 × 10^6 psi); shear, 25.9 GPa (3.75 × 10^6 psi)

Mass Characteristics

Density. 2.69 g/cm³ (0.097 lb/in.³) at 20 °C (68 °F)

Thermal Properties

Liquidus temperature. 652 °C (1205 °F)
Solidus temperature. 627 °C (1160 °F)
Coefficient of thermal expansion. Linear:

Temperature range °C	°F	Average coefficient µm/m · K	µin./in. · °F
−50 to 20	−58 to 68	21.8	12.1
20 to 100	68 to 212	23.8	13.2
20 to 200	68 to 392	24.7	13.7
20 to 300	68 to 572	25.6	14.2

Specific heat. 900 J/kg · K (0.215 Btu/lb · °F) at 20 °C (68 °F)
Thermal conductivity. 191 W/m · K (110 Btu/ft · h · °F) at 20 °C (68 °F)

Electrical Properties

Electrical conductivity. Volumetric, O and H38 tempers: 50% IACS at 20 °C (68 °F)
Electrical resistivity. O and H38 tempers: 34 nΩ · m at 20 °C (68 °F); temperature coefficient, 0.1 nΩ · m per K at 20 °C (68 °F)
Electrolytic solution potential. −0.83 V versus 0.1 *N* calomel electrode in an aqueous solution containing 53 g NaCl plus 3 g H_2O_2 per liter

Fabrication Characteristics

Annealing temperature. 345 °C (650 °F); holding at temperature not required
Hot-working temperature. 260 to 510 °C (500 to 950 °F)

5052
2.5Mg-0.25Cr

Specifications

AMS. See Table 57.
ASTM. See Table 57.
SAE. J454
UNS number. A95052
Government. Sheet and plate: QQ-A-250/8 Foil: MIL-A-81596. Rolled or cold finished wire, rod, and bar: QQ-A-225/7. Drawn, seamless tube: WW-T-700/4. Rivet wire and rod: QQ-A430. Rivets: MIL-R-24243
Foreign. Canada: CSA GR20. France: NF A-G2.5C. Italy: UNI P-AlMg2.5. Germany: DIN AlMg2.5. ISO: AlMg2.5

Chemical Composition

Composition limits. 0.25 Si max, 0.40 Fe max, 0.10 Cu max, 0.10 Mn max, 2.2 to 2.8 Mg, 0.15 to 0.35 Cr, 0.10 Zn max, 0.05 max other (each), 0.15 max others (total), bal Al

Applications

Typical uses. Aircraft fuel and oil lines, fuel tanks, miscellaneous marine and transport

Table 57 Standard specifications for alloy 5052

Mill form	Specification No. AMS	ASTM
Sheet and plate	4015	B 209
Sheet, plate, bar, and shapes (extruded)	4016, 4017	B 221
Wire, rod, and bar (rolled or cold finished)	4114	B 221
Tube		
Drawn	4069	B 483
Drawn, seamless	4070	B 210
Hydraulic	4071	...
Extruded	...	B 221
Extruded, seamless	...	B 241
Condenser	...	B 234
Condenser with integral fins	...	B 404
Welded	...	B 313, B 547
Rivet wire and rod	...	B 316
Foil	4004	...

applications, sheet metal work, appliances, street light standards, rivets, and wire. Applications where good workability, very good resistance to corrosion, high fatigue strength, weldability, and moderate static strength are desired

Mechanical Properties

Tensile properties. See Tables 58 and 59.
Shear yield strength. Approximately 55% of tensile yield strength
Compressive yield strength. Approximately the same as tensile yield strength
Hardness. See Table 58.
Poisson's ratio. 0.33
Elastic modulus. Tension, 69.3 GPa (10.1×10^6 psi); shear, 25.9 GPa (3.75×10^6 psi); compression, 70.7 GPa (10.3×10^6 psi)

Table 58 Typical mechanical properties of alloy 5052

Temper	Tensile strength(a) MPa	ksi	Yield strength(a) MPa	ksi	Elongation, %(a) 1.6 mm (1/16 in.) thick	Elongation, %(a) 12.5 mm (1/2 in.) diam	Hardness, HB(b)	Shear strength MPa	ksi	Fatigue strength(c) MPa	ksi
O	195	28	90	13	25	27	47	125	18	110	16
H32	230	33	195	28	12	16	60	140	20	115	17
H34	260	38	215	31	10	12	68	145	21	125	18
H36	275	40	240	35	8	9	73	160	23	130	19
H38	290	42	255	37	7	7	77	165	24	140	20

(a) Strengths and elongations unchanged or improved at low temperatures. (b) 500 kg load; 10 mm diam ball. (c) At 5×10^8 cycles; R.R. Moore type test

Mass Characteristics

Density. 2.68 g/cm^3 (0.097 lb/in.^{3}) at 20 °C (68 °F)

Thermal Properties

Liquidus temperature. 649 °C (1200 °F)
Solidus temperature. 607 °C (1125 °F)
Coefficient of thermal expansion. Linear:

Temperature range °C	°F	Average coefficient µm/m · K	µin./in. · °F
−50 to 20	−58 to 68	22.1	12.3
20 to 100	68 to 212	23.8	13.2
20 to 200	68 to 392	24.8	13.8
20 to 300	68 to 572	25.7	14.3

Volumetric: 69×10^{-6} m^3/m^3 · K (3.83×10^{-5} in.^{3}/in.^{3} · °F) at 20 °C (68 °F)

Electrical Properties

Electrical conductivity. Volumetric, O and H38 tempers: 35% IACS at 20 °C (68 °F)
Electrical resistivity. O and H38 tempers: 49.3 nΩ · m at 20 °C (68 °F); temperature coefficient, 0.1 nΩ · m per K at 20 °C (68 °F)
Electrolytic solution potential. −0.85 V versus 0.1 *N* calomel electrode in an aqueous solution containing 53 g NaCl plus 3 g H_2O_2 per liter

Fabrication Characteristics

Annealing temperature. 345 °C (650 °F); holding at temperature not required
Hot-working temperature. 260 to 510 °C (500 to 950 °F)

Table 59 Typical tensile properties of alloy 5052 at various temperatures

Temper	Temperature °C	°F	Tensile strength MPa	ksi	Yield strength (0.2% offset) MPa	ksi	Elongation, %
O	−196	−320	303	44	110	16	46
	−80	−112	200	29	90	13	35
	−28	−18	193	28	90	13	32
	24	75	193	28	90	13	30
	100	212	193	28	90	13	36
	149	300	159	23	90	13	50
	204	400	117	17	76	11	60
	260	500	83	12	52	7.5	80
	316	600	52	7.5	38	5.5	110
	371	700	34	5	21	3	130
H34	−196	−320	379	55	248	36	28
	−80	−112	276	40	221	32	21
	−28	−18	262	38	214	31	18
	24	75	262	38	214	31	16
	100	212	262	38	214	31	18
	149	300	207	30	186	27	27
	204	400	165	24	103	15	45
	260	500	83	12	52	7.5	80
	316	600	52	7.5	38	5.5	110
	371	700	34	5	21	3	130
H38	−196	−320	414	60	303	44	25
	−80	−112	303	44	262	38	18
	−28	−18	290	42	255	37	15
	24	75	290	42	255	37	14
	100	212	276	40	248	36	16
	149	300	234	34	193	28	24

5056, Alclad 5056
5.0Mg-0.1Mn-0.1Cr

Specifications

AMS. Rolled or cold finished wire, rod, and bar: 4182. Foil: 4005
ASTM. Rivet wire and rod: B 316. Rolled or cold finished wire, rod, and bar: B 211. Alclad, rolled or cold finished wire, rod, and bar: B 211
SAE. J454
UNS number. A95056
Government. Rivet wire and rod: QQ-A430. Foil: MIL-A-81596
Foreign. Austria: AlMg5. Canada: CSA-GM50R. United Kingdom: BS N6 2L.58. Germany: DIN AlMg5. ISO: AlMg5

Chemical Composition

Composition limits of 5056. 0.30 Si max, 0.40 Fe max, 0.10 Cu max, 0.05 to 0.20 Mn, 4.5 to 5.6 Mg, 0.20 Cr max, 0.10 Zn max, 0.05 max other (each), 0.15 max others (total), bal Al
Composition limits of Alclad 5056. 6253 cladding—Si, 45 to 65% of Mg content, 0.50 Fe max, 0.10 Cu max, 1.0 to 1.5 Mg, 0.15 to 0.35 Cr, 1.6 to 2.4 Zn, 0.05 max other (each), 0.15 max others (total), bal Al

Applications

Typical uses. Rivets for use with magnesium alloy and cable sheathing; zipper stock, nails; also Alclad wire is extensively used in fabrication of insect screens and other applications where wire products with good resistance to corrosion are required

Mechanical Properties

Tensile properties. See Tables 60, 61, and 62. Elongation, O temper: 20% in 50 mm (2 in.) or 4*d*, where *d* is diameter of reduced section of tension test specimen
Shear yield strength. Approximately 55% of tensile yield strength

Table 60 Typical mechanical properties of alloy 5056

Temper	Tensile strength(a) MPa	Tensile strength(a) ksi	Yield strength(a) MPa	Yield strength(a) ksi	Elongation(a)(b), %	Hardness(c), HB	Shear strength MPa	Shear strength ksi	Fatigue strength(d) MPa	Fatigue strength(d) ksi
O	290	42	152	22	35	65	179	26	138	20
H18	434	63	407	59	10	105	234	34	152	22
H38	414	60	345	50	15	100	221	32	152	22

(a) Strengths and elongations are unchanged or improved at low temperatures. (b) 12.5 mm (½ in.) diam; round specimen. (c) 500 kg load; 10 mm diam ball. (d) At 5 × 10^8 cycles, R.R. Moore type test

Compressive yield strength. Approximately the same as the tensile yield strength
Hardness. See Table 60.
Poisson's ratio. 0.33
Elastic modulus. Tension, 71.7 GPa (10.4 × 10^6 psi); shear, 25.9 GPa (3.75 × 10^6 psi); compression, 73.1 GPa (10.6 × 10^6 psi)

Mass Characteristics

Density. 2.64 g/cm³ (0.095 lb/in.³) at 20 °C (68 °F)

Thermal Properties

Liquidus temperature. 638 °C (1180 °F)
Solidus temperature. 568 °C (1055 °F)
Coefficient of thermal expansion.
Linear, O temper:

Temperature range °C	Temperature range °F	Average coefficient µm/m · K	Average coefficient µin./in. · °F
−50 to 20	−58 to 68	22.5	12.5
20 to 100	68 to 212	24.1	13.7
20 to 200	68 to 392	25.2	14.0
20 to 300	68 to 572	26.1	14.5

Volumetric: 70 × 10^{-6} m³/m³ · K (3.89 × 10^{-5} in.³/in.³ · °F) at 20 °C (68 °F)
Specific heat. 904 J/kg · K (0.216 Btu/lb · °F) at 20 °C (68 °F)
Thermal conductivity. At 20 °C (68 °F): O temper, 120 W/m · K (69.3 Btu/ft · h · °F); H38 temper, 112 W/m · K (64.7 Btu/ft · h · °F)

Electrical Properties

Electrical conductivity. Volumetric, at 20 °C (68 °F): O temper, 29% IACS; H38 temper, 27% IACS
Electrical resistivity. At 20 °C (68 °F): O temper, 59 nΩ · m, H38 temper, 64 nΩ · m. Temperature coefficient, O and H38 tempers: 0.1 nΩ · m per K at 20 °C (68 °F)
Electrolytic solution potential. −0.87 V versus 0.1 *N* calomel electrode in an aqueous solution containing 53 g NaCl plus 3 g H_2O_2 per liter

Fabrication Characteristics

Annealing temperature. 415 °C (775 °F); holding at temperature not required
Hot-working temperature. 315 to 480 °C (600 to 900 °F)

5083
4.4Mg-0.7Mn-0.15Cr

Specifications

AMS. Sheet and plate: 4056, 4057, 4058, 4059
ASTM. Sheet and plate: B 209. Extruded wire, rod, bar, shapes, and tube: B 221. Extruded seamless tube: B 241. Drawn seamless tube: B 210. Welded tube: B 547. Forgings: B 247. Gas and oil transmission pipe: B 345
SAE. J454
UNS number. A95083
Government. Sheet and plate: QQ-A-250/6. Extruded wire, rod, bar, shapes, and tube: QQ-A-200/4. Forgings: QQ-A-367. Armor plate: MIL-A-46027. Extruded armor: MIL-A-46083. Forged armor: MIL-A-45225
Foreign. Canada: CSA GM41. United Kingdom: BS N8. Germany: DIN AlMg4.5Mn; Werstoff-Nr. 3.3547. ISO: AlMg4.5Mn

Chemical Composition

Composition limits. 0.40 Si max, 0.40 Fe max, 0.10 Cu max, 0.40 to 1.0 Mn, 4.0 to 4.9 Mg, 0.05 to 0.25 Cr, 0.25 Zn max, 0.15 Ti max, 0.05 max other (each), 0.15 max others (total), bal Al

Applications

Typical uses. Marine, auto, and aircraft applications, unfired welded pressure vessels, cryogenics, TV towers, drilling rigs, transportation equipment, missile components, armor plate. Applications requiring a weldable moderate-strength alloy having good corrosion resistance

Mechanical Properties

Tensile properties. See Tables 63, 64, and 65.
Shear properties. O temper: shear strength, 172 MPa (25 ksi); shear yield strength, approximately 55% of tensile yield strength
Compressive yield strength. Approximately the same as tensile yield strength
Elastic modulus. Tension, 70.3 GPa (10.2 × 10^6 psi); shear, 26.4 GPa (3.83 × 10^6 psi); compression, 71.7 GPa (10.4 × 10^6 psi)
Fatigue strength. H321 and H116 tempers: 160 MPa (23 ksi) at 5 × 10^8 cycles; R.R. Moore type test

Mass Characteristics

Density. 2.66 g/cm³ (0.096 lb/in.³) at 20 °C (68 °F)

Thermal Properties

Liquidus temperature. 638 °C (1180 °F)
Solidus temperature. 574 °C (1065 °F)
Coefficient of thermal expansion.
Linear:

Temperature range °C	Temperature range °F	Average coefficient µm/m · K	Average coefficient µin./in. · °F
−50 to 20	−58 to 68	22.3	12.4
20 to 100	68 to 212	24.2	13.4
20 to 200	68 to 392	25.0	13.9
20 to 300	68 to 572	26.0	14.4

Table 61 Typical tensile properties of alloy 5056

Temper	Temperature °C	Temperature °F	Tensile strength(a) MPa	Tensile strength(a) ksi	Yield strength(a) MPa	Yield strength(a) ksi	Elongation, %
O	24	75	290	42	150	22	35
	149	300	214	31	117	17	55
	204	400	152	22	90	13	65
	260	500	110	16	69	10	80
	316	600	76	11	48	7	100
	371	700	41	6	28	4	130
H38	24	75	414	60	345	50	15
	149	300	262	38	214	31	30
	204	400	179	26	124	18	50
	260	500	110	16	69	10	80
	316	600	76	11	48	7	100
	371	700	41	6	28	4	130

(a) Lowest strengths for exposures up to 10 000 h at temperature, no load; test loading applied at 35 MPa/min (5 ksi/min) to yield strength and then at strain rate of 5%/min to fracture

Table 62 Mechanical-property limits for alloy 5056—rolled or cold finished wire, rod, and bar

Temper	Tensile strength (min) MPa	Tensile strength (min) ksi
Bare 5056		
O	315 (max)	46 (max)
H111	305	44
H12	315	46
H32	305	44
H14	360	52
H34	345	50
H18	400	58
H38	380	55
H192	415	60
H392	400	58
Alclad 5056		
H192	360	52
H392	345	50
H393	370(a)	54

(a) Yield strength (min), 325 MPa (47 ksi)

Table 63 Typical tensile properties of alloy 5083

Temper	Tensile strength(a) MPa	ksi	Yield strength MPa	ksi	Elongation(a)(b), %
O	290	42	145	21	22
H112	303	44	193	28	16
H116	317	46	228	33	16
H321	317	46	228	33	16
H323, H32	324	47	248	36	10
H343, H34	345	50	283	41	9

(a) Strengths and elongations are unchanged or improved at low temperatures. (b) 1.6 mm (1/16 in.) thick specimen

Volumetric: 70 × 10^{-6} m^3/m^3 · K (3.89 × 10^{-5} $in.^3/in.^3$ · °F) at 20 °C (68 °F)
Specific heat. 900 J/kg · K (0.215 Btu/lb · °F) at 20 °C (68 °F)
Thermal conductivity. 120 W/m · K (69.3 Btu/ft · h · °F) at 20 °C (68 °F)

Electrical Properties

Electrical conductivity. Volumetric, average of all tempers: 29% IACS at 20 °C (68 °F)
Electrical resistivity. 59.5 nΩ · m at 20 °C (68 °F); temperature coefficient, 0.1 nΩ · m per K at 20 °C (68 °F)
Electrolytic solution potential. −0.91 V versus 0.1 *N* calomel electrode in an aqueous solution containing 53 g NaCl plus 3 g H_2O_2 per liter

Fabrication Characteristics

Annealing temperature. 415 °C (775 °F); holding at temperature not required
Hot-working temperature. 315 to 480 °C (600 to 900 °F)

5086, Alclad 5086 4.0Mg-0.4Mn-0.15Cr

Specifications

ASTM. Sheet and plate; B 209. Extruded wire, rod, bar, shapes, and tube: B 221. Extruded seamless tube: B 241. Drawn, seamless tube: B 210. Welded tube: B 313, B 547. Gas and oil transmission pipe: B 345. Alclad 5086, sheet and plate: B 209
SAE. J454
UNS number. A95086

Table 64 Mechanical-property limits for alloy 5083

Temper	Tensile strength Minimum MPa	ksi	Tensile strength Maximum MPa	ksi	Yield strength Minimum MPa	ksi	Yield strength Maximum MPa	ksi	Elongation (min), %(a)
O									
0.051–1.5000 in. thick	275	40	350	51	125	18	200	29	16
1.501–3.000 in. thick	270	39	345	50	115	17	200	29	16
3.001–5.000 in. thick	260	38	...	...	110	16	...	...	14–16
5.001–7.000 in. thick	255	37	...	...	105	15	...	...	14
7.001–8.000 in. thick	250	36	...	...	95	14	...	...	12
H112									
0.250–1.500 in. thick	275	40	...	...	125	18	...	...	12
1.501–3.000 in. thick	270	39	...	...	115	17	...	...	12
H116									
0.063–1.500 in. thick	305	44	...	...	215	31	...	...	12
1.501–3.000 in. thick	285	41	...	...	200	29	...	...	12
H321									
0.188–1.500 in. thick	305	44	385	56	215	31	295	43	12
1.501–3.000 in. thick	285	41	385	56	200	29	295	43	12
H323	310	45	370	54	235	34	305	44	8–10
H343	345	50	405	59	270	39	340	49	6–8

(a) In 50 mm (2 in.) or 4*d*, where *d* is diameter of reduced section of tensile test specimen. Where a range of values appears in this column, the specified minimum elongation varies with thickness of the mill product.

Table 65 Typical tensile properties of alloy 5083-O at various temperatures

Temperature °C	°F	Tensile strength(a) MPa	ksi	Yield strength (0.2% offset)(a) MPa	ksi	Elongation, %
−195	−315	405	59	165	24	36
−80	−112	295	43	145	21	30
−30	−22	290	42	145	21	27
25	80	290	42	145	21	25
100	212	275	40	145	21	36
150	302	215	31	130	19	50
205	400	150	22	115	17	60
260	500	115	17	75	11	80
315	600	75	11	50	7.5	110
370	698	41	6	29	4.2	130

(a) Lowest strength for exposures up to 10 000 h at temperature, no load; test loading applied at 35 MPa/min (5 ksi/min) to yield strength and then at strain rate of 10%/min to fracture

Government. Sheet and plate: QQ-A-250/7, QQ-A-250/19. Extruded wire, rod, bar, shapes, and tube: QQ-A-200/5. Drawn, seamless tube: WW-T-700/5
Foreign. France: NF A-G4MC. Germany: DIN AlMg4. ISO: AlMg4

Chemical Composition

Composition limits. 0.40 Si max, 0.50 Fe max, 0.20 to 0.7 Mn, 3.5 to 4.5 Mg, 0.25 Zn max, 0.15 Ti max, 0.05 max other (each), 0.15 max others (total), bal Al

Applications

Typical uses. Marine, automotive, and aircraft parts, cryogenics, TV towers, drilling rigs, transportation equipment, missile components, armor plate. Applications requiring weldable moderate-strength alloy having comparatively good corrosion resistance

Mechanical Properties

Tensile properties. See Tables 66 and 67. Tensile strength and elongation are approximately equal in the longitudinal and transverse directions.
Shear properties. Shear strength: O temper, 160 MPa (23 ksi); H34 temper, 185 MPa (27 ksi). Shear yield strength: approximately 55% of tensile yield strength
Compressive yield strength. Approximately the same as tensile yield strength
Poisson's ratio. 0.33
Elastic modulus. Tension, 71.0 GPa (10.3 × 10^6 psi); shear, 26.4 GPa (3.83 × 10^6 psi); compression, 72.4 GPa (10.5 × 10^6 psi)

Mass Characteristics

Density. 2.66 g/cm^3 (0.096 $lb/in.^3$) at 20 °C (68 °F)

Thermal Properties

Liquidus temperature. 640 °C (1184 °F)
Solidus temperature. 585 °C (1085 °F)
Coefficient of thermal expansion. Linear:

Temperature range °C	°F	Average coefficient μm/m · K	μin./in. · °F
−50 to 20	−58 to 68	22.0	12.2
20 to 100	68 to 212	23.8	13.2
20 to 200	68 to 392	24.7	13.7
20 to 300	68 to 572	25.8	14.3

Volumetric: 69 × 10^{-6} m^3/m^3 · K (3.83 × 10^{-5} $in.^3/in.^3$ · °F)
Specific heat. 900 J/kg · K (0.215 Btu/lb · °F) at 20 °C (68 °F)
Thermal conductivity. 127 W/m · K (73.4 Btu/ft · h · °F) at 20 °C (68 °F)

Electrical Properties

Electrical conductivity. Volumetric, average of all tempers: 31% IACS at 20 °C (68 °F)
Electrical resistivity. Average of all tempers: 56 nΩ · m at 20 °C (68 °F); temperature coefficient, 0.1 nΩ · m per K at 20 °C (68 °F)
Electrolytic solution potential. −0.88 V ver-

Table 66 Tensile properties of alloy 5086

Temper	Tensile strength MPa	ksi	MPa	ksi	Yield strength MPa	ksi	Elongation(a), %
Typical properties							
O	260	38	...	...	115	17	22
H32, H116	290	42	...	...	205	30	12
H34	325	47	...	...	255	37	10
H112	270	39	...	...	130	19	14
Property limits	Minimum		Maximum		Minimum		Minimum
O (0.020–2.000 in. thick)	240	35	305	44	95	14	15–18
H32 (0.020–2.000 in. thick)	275	40	325	47	195	28	6–12
H34 (0.009–1.000 in. thick)	305	44	350	51	235	34	4–10
H36 (0.006–0.162 in. thick)	325	47	370	54	260	38	3–6
H38 (0.006–0.020 in. thick)	345	50	...	...	285	41	3
H112							
(0.188–0.499 in. thick)	250	36	...	...	125	18	8
(0.500–1.000 in. thick)	240	35	...	...	110	16	10
(1.001–3.000 in. thick)	240	35	...	...	95	14	14
(2.001–3.000 in. thick)	235	34	...	...	95	14	14
H116 (0.063–2.000 in. thick)	275	40	...	...	195	28	8–10

(a) In 50 mm (2 in.) or 4*d*, where *d* is diameter of reduced section of tensile test specimen. Where a range of values appears in this column, specified minimum elongation varies with thickness of the mill product.

Table 67 Typical tensile properties of alloy 5086-O at various temperatures

Temperature °C	°F	Tensile strength(a) MPa	ksi	Yield strength (0.2% offset)(a) MPa	ksi	Elongation, %
−196	−320	379	55	131	19	46
−80	−112	269	39	117	17	35
−28	−18	262	38	117	17	32
24	75	262	38	117	17	30
100	212	262	38	117	17	36
149	300	200	29	110	16	50
204	400	152	22	103	15	60
260	500	117	17	76	11	80
316	600	76	11	52	7.5	110
371	700	41	6	29	4.2	130

(a) Lowest strengths for exposures up to 10 000 h at temperature, no load; test loading applied at 35 MPa/min (5 ksi/min) to yield strength and then at strain rate of 5%/min to fracture

sus 0.1 *N* calomel electrode in an aqueous solution containing 53 g NaCl plus 3 g H_2O_2 per liter

Fabrication Characteristics

Annealing temperature. 345 °C (650 °F); holding at temperature not required
Hot-working temperature. 315 to 480 °C (600 to 900 °F)

5154
3.5Mg-0.25Cr

Specifications

AMS. Sheet and plate: 4018, 4019
ASTM. Sheet and plate: B 209. Rolled or cold finished wire, rod, and bar: B 211. Extruded wire, rod, bar, shapes, and tube: B 221. Drawn, seamless tube: B 210. Welded tube: B 313, B 547
SAE. J454
UNS number. A95154
Foreign. Canada: CSA GR40. France: NF A-G3C. United Kingdom: BS N5. ISO: AlMg3.5

Chemical Composition

Composition limits. 0.25 Si max, 0.40 Fe max, 0.10 Cu max, 0.10 Mn max, 3.1 to 3.9 Mg, 0.15 to 0.35 Cr, 0.20 Zn max, 0.20 Ti max, 0.05 max other (each), 0.15 max others (total), bal Al

Applications

Typical uses. Welded structures, storage tanks, pressure vessels, marine structures, transportation trailer tanks

Mechanical Properties

Tensile properties. See Tables 68 and 69. Tensile strength and elongation are approximately equal in the longitudinal and transverse directions.
Shear properties. Shear strength: see Table 68. Shear yield strength: approximately 55% of tensile yield strength
Compressive yield strength. Approximately the same as tensile yield strength
Hardness. See Table 68.
Poisson's ratio. 0.33
Elastic modulus. Tension, 69.3 GPa (10.1×10^6 psi); shear, 25.9 GPa (3.75×10^6 psi); compression, 70.7 GPa (10.3×10^6 psi)
Fatigue strength. See Table 68.

Mass Characteristics

Density. 2.66 g/cm³ (0.096 lb/in.³) at 20 °C (68 °F)

Thermal Properties

Liquidus temperature. 643 °C (1190 °F)
Solidus temperature. 593 °C (1100 °F)

Table 68 Mechanical properties of alloy 5154

Temper	Tensile strength MPa	ksi	MPa	ksi	Yield strength MPa	ksi	Elongation(a), %	Hardness(b), HB	Shear strength MPa	ksi	Fatigue strength(c) MPa	ksi
Typical properties												
O	240	35	...	...	117	17	27	58	152	22	117	17
H32	270	39	...	...	207	30	15	67	152	22	124	18
H34	290	42	...	...	228	33	13	73	165	24	131	19
H36	310	45	...	...	248	36	12	78	179	26	138	20
H38	330	48	...	...	269	39	10	80	193	28	145	21
H112	240	35	...	...	117	17	25	63	...	...	117	17
Property limits	Minimum		Maximum		Minimum							
O (0.020–3.000 in. thick)	205	30	285	41	75	11	12 to 18	...	...	...	...	...
H32 (0.020–2.000 in. thick)	250	36	295	43	180	26	5 to 12	...	...	...	...	...
H34 (0.009–1.000 in. thick)	270	39	315	46	200	29	4 to 10	...	...	...	...	...
H36 (0.006–0.162 in. thick)	290	42	340	49	220	32	3 to 5	...	...	...	...	...
H38 (0.006–0.128 in. thick)	310	45	...	...	240	35	3 to 5	...	...	...	...	...
H112												
(0.250–0.499 in. thick)	220	32	...	...	125	18	8	...	...	...	...	...
(0.0500–3.000 in. thick)	205	30	...	...	75	11	11 to 15	...	...	...	...	...

(a) In 50 mm (2 in.) or 4*d*, where *d* is diameter of tensile test specimen. Where a range of values appears in this column, specified minimum elongation varies with thickness of the mill product. (b) 500 kg load; 10 mm ball. (c) At 5×10^8 cycles of completely reversed stress; R.R. Moore type test

Table 69 Typical tensile properties of alloy 5154-O at various temperatures

Temperature °C	Temperature °F	Tensile strength MPa	Tensile strength ksi	Yield strength (0.2% offset) MPa	Yield strength (0.2% offset) ksi	Elongation(a), %
−196	−320	360	52	130	19	46
−80	−112	250	36	115	17	35
−28	−18	240	35	115	17	32
24	75	240	35	115	17	30
100	212	240	35	115	17	30
149	300	200	29	110	16	50
204	400	150	22	105	15	60
260	500	115	17	75	11	80
316	600	75	11	50	7.5	110
371	700	41	6	29	4.2	130

(a) In 50 mm (2 in.) or 4*d*, where *d* is diameter of reduced section of tensile test specimen

Coefficient of thermal expansion. Linear:

Temperature range °C	Temperature range °F	Average coefficient µm/m · K	Average coefficient µin./in. · °F
−50 to 20	−58 to 68	22.1	12.3
20 to 100	68 to 212	23.9	13.3
20 to 200	68 to 392	24.9	13.8
20 to 300	68 to 572	25.9	14.4

Volumetric: 69 × 10^{-6} m^3/m^3 · K (3.83 × 10^{-5} $in.^3/in.^3$ · °F) at 20 °C (68 °F)
Specific heat. 900 J/kg · K (0.215 Btu/lb · °F) at 20 °C (68 °F)
Thermal conductivity. 127 W/m · K (73.3 Btu/ft · h · °F) at 20 °C (68 °F)

Electrical Properties

Electrical conductivity. Volumetric, average of all tempers: 32% IACS at 20 °C (68 °F)
Electrical resistivity. Average of all tempers: 53.9 nΩ · m at 20 °C (68 °F); temperature coefficient, 0.1 nΩ · m per K at 20 °C (68 °F)
Electrolytic solution potential. −0.86 V versus 0.1 *N* calomel electrode in an aqueous solution containing 53 g NaCl plus 3 g H_2O_2 per liter

Fabrication Characteristics

Annealing temperature. 345 °C (650 °F); holding at temperature not required
Hot-working temperature. 260 to 510 °C (500 to 950 °F)

5182
4.5Mg-0.35Mn

Specifications

UNS number. A95182

Chemical Composition

Composition limits. 0.20 Si max, 0.35 Fe max, 0.15 Cu max, 0.20 to 0.50 Mn, 4.0 to 5.0 Mg, 0.10 Cr max, 0.25 Zn max, 0.10 Ti max, 0.05 max other (each), 0.15 max others (total), bal Al

Applications

Typical uses. Sheet used for container ends, automotive body panels and reinforcement members, brackets, and parts

Mechanical Properties

Tensile properties. See Table 70.
Shear properties. Shear strength: O temper, 152 MPa (22 ksi). Shear yield strength: approximately 55% of tensile yield strength
Compressive yield strength. Approximately the same as tensile yield strength
Hardness. O temper, 58 HB with 200 kg load, 10 mm diam ball
Poisson's ratio. 0.33
Elastic modulus. Tension, 69.6 GPa (10.1 × 10^6 psi); compression, 70.9 GPa (10.3 × 10^6 psi)
Fatigue strength. O temper, 138 MPa (20 ksi) at 5 × 10^8 cycles in an R.R. Moore type rotating-beam test

Table 71 Tensile properties of alloy 5252

Temper	Tensile strength MPa	Tensile strength ksi	Tensile strength MPa	Tensile strength ksi	Yield strength MPa	Yield strength ksi	Elongation, %
Typical properties							
H25	235	34	...	...	170	25	11(a)
H28, H38	283	41	...	...	240	35	5(a)
Property limits for 0.75–2.3 mm (0.030–0.090 in.) thick sheet							
	Minimum		Maximum				Minimum
H24	205	30	260	38	...	...	10
H25	215	31	270	39	...	...	9
H28	260	38	...	...	...	...	3

(a) 1.6 mm (1/16 in.) thick specimen

Table 70 Typical tensile properties of alloy 5182

Temper	Tensile strength(a) MPa	Tensile strength(a) ksi	Yield strength(a) MPa	Yield strength(a) ksi	Elongation(a)(b), %
O	276	40	138	19	25
H32	317	46	234	34	12
H34	338	49	283	41	10
H19(c)	421	61	393	57	4

(a) Strengths and elongations are unchanged or increased at low temperatures. (b) 1.6 mm (1/16 in.) thick specimen. (c) Properties of this temper are for container end stock 0.25 to 0.38 mm (0.010 to 0.015 in.) thick.

Mass Characteristics

Density. 2.65 g/cm^3 (0.096 $lb/in.^3$) at 20 °C (68 °F)

Thermal Properties

Liquidus temperature. 638 °C (1180 °F)
Solidus temperature. 577 °C (1070 °F)
Coefficient of thermal expansion. Linear:

Temperature range °C	Temperature range °F	Average coefficient µm/m · K	Average coefficient µin./in. · °F
−50 to 20	−58 to 68	22.2	12.3
20 to 100	68 to 212	24.1	13.4
20 to 200	68 to 392	25.0	13.9
20 to 300	68 to 572	26.0	14.4

Volumetric: 70 × 10^{-6} m^3/m^3 · K (3.89 × 10^{-5} $in.^3/in.^3$ · °F) at 20 °C (68 °F)
Specific heat. 904 J/kg · K (0.216 Btu/lb · °F) at 20 °C (68 °F)
Thermal conductivity. 123 W/m · K (71.1 Btu/ft · h · °F) at 20 °C (68 °F)

Electrical Properties

Electrical conductivity. Volumetric, 31% IACS at 20 °C (68 °F)
Electrical resistivity. 55.6 nΩ · m at 20 °C (68 °F); temperature coefficient, 0.1 nΩ · m per K at 20 °C (68 °F)

Fabrication Characteristics

Annealing temperature. 345 °C (650 °F)
Hot-working temperature. 260 to 510 °C (500 to 950 °F)

5252
2.5Mg

Specifications

ASTM. Sheet: B 209
SAE. J454
UNS number. A95252

Chemical Composition

Composition limits. 0.08 Si max, 0.10 Fe max, 0.10 Cu max, 0.10 Mn max, 2.2 to 2.8 Mg, 0.05 Zn max, 0.05 V max, 0.03 max other (each), 0.10 max others (total), bal Al

Applications

Typical uses. Automotive and appliance trim where greater strength is required than in other trim alloys. Can be bright dipped or anodized to give a bright, clear finish

Table 72 Mechanical properties of alloy 5254

Temper	Tensile strength MPa	ksi	MPa	ksi	Yield strength MPa	ksi	Elongation, %	Hardness(a), HB	Shear strength MPa	ksi	Fatigue strength(b) MPa	ksi
Typical properties(c)												
O	240	35	...	...	115	17	27	58	150	22	115	17
H32	270	39	...	...	205	30	15	67	150	22	125	18
H34	290	42	...	...	230	33	13	73	165	24	130	19
H36	310	45	...	...	250	36	12	78	180	26	140	20
H38	330	48	...	...	270	39	10	80	195	28	145	21
H112	240	35	...	...	115	17	25	63	...	...	115	17
Property limits	**Minimum**		**Maximum**		**Minimum**		**Minimum(d)**					
O	205	30	285	41	75	11	12–18	...	...	...	...	...
H32	250	36	295	43	180	26	5–12	...	...	...	...	...
H34	270	39	315	46	200	29	4–10	...	...	...	...	...
H36	290	42	340	49	220	32	3–5	...	...	...	...	...
H38	310	45	...	...	240	35	3–5	...	...	...	...	...
H112												
6–12.5 mm (0.250–0.499 in.) thick	220	32	...	...	125	18	8	...	...	...	...	...
13–75 mm (0.500–3.000 in.) thick	205	30	...	...	75	11	11–15	...	...	...	...	...

(a) 500 kg load; 10 mm ball. (b) At 5×10^8 cycles; R.R. Moore type test. (c) Strengths and elongations are unchanged or increased at low temperatures. (d) In 50 mm (2 in.) or 4*d*, where *d* is diameter of reduced section of test specimen. Where a range of values appears in this column, specified minimum elongation varies with thickness of the mill product.

Mechanical Properties

Tensile properties. See Table 71.
Shear strength. H25 temper: 145 MPa (21 ksi); H28, H38 tempers: 160 MPa (23 ksi)
Compressive yield strength. Approximately the same as tensile yield strength
Hardness. H25 temper: 68 HB. H28, H38 tempers: 75 HB. Brinell hardness determined using 500 kg load, 10 mm ball, 30 s duration of loading
Elastic modulus. Tension, 68.3 GPa (9.90×10^6 psi); compression, 69.7 GPa (10.1×10^6 psi)

Mass Characteristics

Density. 2.67 g/cm³ (0.097 lb/in.³) at 20 °C (68 °F)

Thermal Properties

Liquidus temperature. 649 °C (1200 °F)
Solidus temperature. 607 °C (1125 °F)
Coefficient of thermal expansion.
Linear:

Temperature range °C	°F	Average coefficient µm/m · K	µin./in. · °F
−50 to 20	−58 to 68	23.0	12.2
20 to 100	68 to 212	23.8	13.2
20 to 200	68 to 392	24.7	13.7
20 to 300	68 to 572	25.8	14.3

Volumetric: 69×10^{-6} m³/m³ · K (3.83×10^{-5} in.³/in.³ · °F)
Specific heat. 900 J/kg · K (0.215 Btu/lb · °F at 20 °C (68 °F)
Thermal conductivity. 138 W/m · K (80 Btu/ft · h · °F) at 20 °C (68 °F)

Electrical Properties

Electrical conductivity. Volumetric, average of all tempers: 35% IACS at 20 °C (68 °F)
Electrical resistivity. Average of all tempers: 49 nΩ · m at 20 °C (68 °F); temperature coefficient, 0.1 nΩ · m per K at 20 °C (68 °F)

Fabrication Characteristics

Annealing temperature. 345 °C (650 °F); holding at temperature not required
Hot-working temperature. 260 to 510 °C (500 to 950 °F)

5254
3.5Mg-0.25Cr

Specifications

ASTM. Sheet and plate: B 209. Extruded, seamless tube: B 241
SAE. J454
UNS number. A95254
Foreign. Canada: CSA GR40

Table 73 Typical tensile properties of alloy 5254-O at various temperatures

Temperature °C	°F	Tensile strength(a) MPa	ksi	Yield strength(a) MPa	ksi	Elongation, %
−196	−320	360	52	130	19	46
−80	−112	250	36	115	17	35
−28	−18	240	35	115	17	32
24	75	240	35	115	17	30
100	212	240	35	115	17	36
149	300	200	29	110	16	50
204	400	150	22	105	15	60
260	500	115	17	75	11	80
316	600	75	11	50	7.5	110
371	700	41	6	29	4.2	130

(a) Lowest strengths for exposure up to 10 000 h at temperature, no load; test loading applied at 35 MPa/min (5 ksi/min) to yield strength and then at strain rate of 5%/min to fracture

Chemical Composition

Composition limits. 0.45 Si max + Fe, 0.05 Cu max, 0.01 Mn max, 3.1 to 3.9 Mg, 0.15 to 0.35 Cr, 0.20 Zn max, 0.05 Ti max, 0.05 max other (each), 0.15 max others (total), bal Al

Applications

Typical uses. Storage vessels for hydrogen peroxide and other chemicals

Mechanical Properties

Tensile properties. See Tables 72 and 73.
Shear yield strength. Approximately 55% of tensile yield strength
Compressive yield strength. Approximately the same as tensile yield strength
Hardness. See Table 72.
Elastic modulus. 70.3 GPa (10.2×10^6 psi); compression, 70.9 GPa (10.3×10^6 psi)
Fatigue strength. See Table 72.

Mass Characteristics

Density. 2.66 g/cm³ (0.096 lb/in.³) at 20 °C (68 °F)

Thermal Properties

Liquidus temperature. 643 °C (1190 °F)
Solidus temperature. 593 °C (1100 °F)
Coefficient of thermal expansion.
Linear:

Temperature range °C	°F	Average coefficient µm/m · K	µin./in. · °F
−50 to 20	−58 to 68	22.1	12.3
20 to 100	68 to 212	24.0	13.3
20 to 200	68 to 392	24.9	13.8
20 to 300	68 to 572	25.9	14.4

Volumetric: 69×10^{-6} m³/m³ · K (3.83×10^{-5} in.³/in.³ · °F) at 20 °C (68 °F)
Specific heat. 900 J/kg · K (0.215 Btu/lb · °F) at 20 °C (68 °F)
Thermal conductivity. 127 W/m · K (73.4 Btu/ft · h · °F) at 20 °C (68 °F)

Table 74 Mechanical properties of alloy 5454

Temper	Tensile strength				Yield strength		Elongation, %	Hardness(a), HB	Shear strength	
	MPa	ksi	MPa	ksi	MPa	ksi			MPa	ksi
Typical properties										
O	250	36	...	...	117	17	22	62	159	23
H32	275	40	...	...	207	30	10	73	165	24
H34	305	44	...	...	241	35	10	81	179	26
H36	340	49	...	...	276	40	8	...	...	...
H38	370	54	...	...	310	45	8	...	...	...
H111	260	38	...	...	179	26	14	70	159	23
H112	250	36	...	...	124	18	18	62	159	23
H311	260	38	...	...	179	26	18	70	159	23
Property limits	**Minimum**		**Maximum**		**Minimum**					
O	215	31	285	41	85	12	12–18(b)	...	...	...
H32	250	36	305	44	180	26	5–12(b)	...	...	...
H34	270	39	325	47	200	29	4–10(b)	...	...	...
H112										
6–12.5 mm (0.250–0.499 in.) thick	220	32	...	...	125	18	8	...	...	...
13–75 mm (0.500–3.00 in.) thick	215	31	...	...	85	12	11–15(b)	...	...	...

(a) 500 kg load; 10 mm ball. (b) Range of values indicates that specified minimum elongation varies with thickness of mill product.

Electrical Properties

Electrical conductivity. Volumetric, 32% IACS at 20 °C (68 °F)
Electrical resistivity. 54 nΩ · m at 20 °C (68 °F); temperature coefficient, 0.1 nΩ · m per K at 20 °C (68 °F)
Electrolytic solution potential. −0.86 V versus 0.1 *N* calomel electrode in an aqueous solution containing 53 g NaCl plus 3 g H_2O_2 per liter

Fabrication Characteristics

Annealing temperature. 345 °C (650 °F); holding at temperature not required
Hot-working temperature. 260 to 510 °C (500 to 950 °F)

5356
5.0Mg-0.12Mn-0.12Cr

Specifications

UNS number. A95356
Government. QQ-R-566, MIL-E-16053
Foreign. Canada: CSA GM50P. France: NF A-G5

Chemical Composition

Composition limits. 0.25 Si max, 0.40 Fe max, 0.10 Cu max, 0.05 to 0.20 Mn, 4.5 to 5.5 Mg, 0.05 to 0.20 Cr, 0.10 Zn max, 0.06 to 0.20 Ti, 0.05 max other (each), 0.15 max others (total), 0.0008 Be max, bal Al

Applications

Typical uses. Welding electrodes and filler wire for base metals with high magnesium content (>3% Mg)

Mass Characteristics

Density. 2.64 g/cm³ (0.0954 lb/in.³) at 20 °C (68 °F)

Thermal Properties

Liquidus temperature. 638 °C (1180 °F)
Solidus temperature. 574 °C (1065 °F)
Coefficient of thermal expansion.
Linear:

Temperature range °C	Temperature range °F	Average coefficient µm/m · K	Average coefficient µin./in. · °F
−50 to 20	−58 to 68	22.3	12.3
20 to 100	68 to 212	24.2	13.4
20 to 200	68 to 392	25.1	13.9
20 to 300	68 to 572	26.1	14.5

Volumetric: 70 × 10^{-6} m³/m³ · K (3.89 × 10^{-5} in.³/in.³ · °F) at 20 °C (68 °F)
Specific heat. 904 J/kg · K (0.216 Btu/lb · °F) at 20 °C (68 °F)
Thermal conductivity. 116 W/m · K (67 Btu/ft · h · °F) at 20 °C (68 °F)

Electrical Properties

Electrical conductivity. Volumetric, O temper: 29% IACS at 20 °C (68 °F)
Electrical resistivity. O temper: 59.4 nΩ · m at 20 °C (68 °F). Temperature coefficient, 0.1 nΩ · m per K at 20 °C (68 °F)
Electrolytic solution potential. −0.87 V versus 0.1 *N* calomel electrode in an aqueous solution containing 53 g NaCl plus 3 g H_2O_2 per liter

Fabrication Characteristics

Annealing temperature. 345 °C (650 °F); holding at temperature not required
Hot-working temperature. 260 to 510 °C (500 to 950 °F)

5454
2.7Mg-0.8Mn-0.12Cr

Specifications

ASTM. Sheet and plate: B 209. Extruded wire, rod, bar, shapes, and tube: B 221. Extruded seamless tube: B 241. Condenser tube: B 234. Condenser tube with integral fins: B 404. Welded tube: B 547
SAE. J454
UNS number. A95454
Government. Sheet and plate: QQ-A-250/10. Extruded wire, rod, bar, shapes, and tube: QQ-A-200/6
Foreign. Canada: CSA GM31N. France: NF A-G2.5MC. United Kingdom: BS N51. Germany: DIN AlMg2.7Mn. ISO: AlMg3Mn

Chemical Composition

Composition limits. 0.25 Si max, 0.40 Fe max, 0.10 Cu max, 0.50 to 1.0 Mn, 2.4 to 3.0 Mg, 0.05 to 0.20 Cr, 0.25 Zn max, 0.20 Ti max, 0.05 max other (each), 0.15 max others (total), bal Al

Applications

Typical uses. Welded structures, pressure vessels, tube for marine service

Mechanical Properties

Tensile properties. See Tables 74 and 75.
Shear yield strength. Approximately 55% of tensile yield strength
Compressive yield strength. Approximately the same as tensile yield strength
Hardness. See Table 74.
Elastic modulus. Tension, 69.6 GPa (10.1 × 10^6 psi); compression, 71.0 GPa (10.3 × 10^6 psi)

Mass Characteristics

Density. 2.68 g/cm³ (0.097 lb/in.³) at 20 °C (68 °F)

Thermal Properties

Liquidus temperature. 646 °C (1195 °F)
Solidus temperature. 602 °C (1115 °F)
Coefficient of thermal expansion.
Linear:

Temperature range °C	Temperature range °F	Average coefficient µm/m · K	Average coefficient µin./in. · °F
−50 to 20	−58 to 68	21.9	12.2
20 to 100	68 to 212	23.7	13.2
20 to 200	68 to 392	24.6	13.7
20 to 300	68 to 572	25.6	14.2

Volumetric: 68 × 10^{-6} m³/m³ · K (3.77 × 10^{-5} in.³/in.³ · °F) at 20 °C (68 °F)
Specific heat. 900 J/kg · K (0.215 Btu/lb · °F) at 20 °C (68 °F)
Thermal conductivity. 134 W/m · K (77.4 Btu/ft · h · °F) at 20 °C (68 °F)

Electrical Properties

Electrical conductivity. Volumetric, average of all tempers: 34% IACS at 20 °C (68 °F)
Electrical resistivity. Average of all tempers: 51 nΩ · m at 20 °C (68 °F). Temperature coefficient, 0.1 nΩ · m per K at 20 °C (68 °F)
Electrolytic solution potential. −0.86 V versus 0.1 *N* calomel electrode in an aqueous solution containing 53 g NaCl plus 3 g H_2O_2 per liter

Fabrication Characteristics

Annealing temperature. 345 °C (650 °F); holding at temperature not required

Table 75 Typical tensile properties of alloy 5454 at various temperatures

Temperature °C	Temperature °F	Tensile strength(a) MPa	Tensile strength(a) ksi	Yield strength(a) MPa	Yield strength(a) ksi	Elongation, %
O temper						
−196	−320	370	54	130	19	39
−80	−112	255	37	115	17	30
−28	−18	250	36	115	17	27
24	75	250	36	115	17	25
100	212	250	36	115	17	31
149	300	200	29	110	16	50
204	400	150	22	105	15	60
260	500	115	17	75	11	80
316	600	75	11	50	7.5	110
371	700	41	6	29	4.2	130
H32 temper						
−196	−320	405	59	250	36	32
−80	−112	290	42	215	31	23
−28	−18	285	41	205	30	20
24	75	275	40	205	30	18
100	212	270	39	200	29	20
149	300	220	32	180	26	37
204	400	170	25	130	19	45
260	500	115	17	75	11	80
316	600	75	11	50	7.5	110
371	700	41	6	29	4.2	130
H34 temper						
−196	−320	435	63	285	41	30
−80	−112	315	46	250	36	21
−28	−18	305	44	240	35	18
24	75	305	44	240	35	16
100	212	295	43	235	34	18
149	300	235	34	195	28	32
204	400	180	26	130	19	45
260	500	115	17	75	11	80
316	600	75	11	50	7.5	110
371	700	41	6	29	4.2	130

(a) Lowest strengths for exposures up to 10 000 h at temperature, no load, test loading applied at 35 MPa/min (5 ksi/min) to yield strength and then at strain rate of 5%/min to fracture

Hot-working temperature. 260 to 510 °C (500 to 950 °F)

5456
5.1Mg-0.8Mn-0.12Cr

Specifications

ASTM. Sheet and plate: B 209. Extruded wire, rod, bar, shapes, and tube: B 221. Extruded, seamless tube: B 241. Drawn, seamless tube: B 210

SAE. J454

UNS number. A95456

Government. Sheet and plate: QQ-A-250/9, QQ-A-250/20. Extruded wire, rod, bar, shapes, and tube: QQ-A-200/7. Armor plate: MIL-A-46027. Extruded armor: MIL-A-46083. Forged armor: MIL-A-45225

Chemical Composition

Composition limits. 0.25 Si max, 0.40 Fe max, 0.10 Cu max, 0.50 to 1.0 Mn, 4.7 to 5.5 Mg, 0.05 to 0.20 Cr, 0.25 Zn max, 0.20 Ti max, 0.05 max other (each), 0.15 max others (total), bal Al

Applications

Typical uses. Armor plate, high strength welded structures, storage tanks, pressure vessels, marine service

Mechanical Properties

Tensile properties. See Table 76.

Shear strength. H321, H116 tempers: 207 MPa (30 ksi)

Table 76 Tensile properties of alloy 5456

Temper	Tensile strength MPa	ksi	MPa	ksi	Yield strength MPa	ksi	MPa	ksi	Elongation, %	
Typical properties										
O	310	45	...	...	159	23	...	...	24(a)	
H111	324	47	...	...	228	33	...	...	18(a)	
H112	310	45	...	...	165	24	...	...	22(a)	
H321(b), H116(c)	352	51	...	...	255	37	...	...	16(a)	
									Minimum(d)	
Property limits	**Minimum**		**Maximum**		**Minimum**		**Maximum**		**In 50 mm**	**In 5d(5.65 $\sqrt{A}$)**
O										
1.20–6.30 mm thick	290	42	365	53	130	19	205	30	16	...
6.30–80.00 mm thick	285	41	360	52	125	18	205	30	16	14
80.00–120.00 mm thick	275	40	...	...	120	17	...	...	...	12
120.00–160.00 mm thick	270	39	...	...	115	17	...	...	...	12
160.00–200.00 mm thick	265	38	...	...	105	15	...	...	...	10
H112										
6.30–40.00 mm thick	290	42	...	...	130	19	...	...	12	10
40.00–80.00 mm thick	285	41	...	...	125	18	...	...	...	10
H116(c)(e)										
1.60–30.00 mm thick	315	46	...	...	230	33	...	...	10	10
30.00–40.00 mm thick	305	44	...	...	215	31	...	...	...	10
40.00–80.00 mm thick	285	41	...	...	200	29	...	...	...	10
80.00–110.00 mm thick	275	40	...	...	170	25	...	...	...	10
H321										
4.00–12.50 mm thick	315	46	405	59	230	33	315	46	12	...
12.50–40.00 mm thick	305	44	385	56	215	31	305	44	...	10
40.00–80.00 mm thick	285	41	385	56	200	29	295	43	...	10
H323										
1.20–6.30 mm thick	330	48	400	58	250	36	315	46	6 to 8	...
H343										
1.20–6.30 mm thick	365	53	435	63	285	41	350	51	6 to 8	...

(a) 12.5 mm (½ in.) diam specimen. (b) Material in this temper not recommended for applications requiring exposure to seawater. (c) H116 designation also applies to the condition previously designated H117. (d) Elongations in 50 mm (2 in.) apply to thicknesses through 12.5 mm (½ in.); elongation in $5d$ (5.65 $\sqrt{A}$), where d is diameter and A is cross-sectional area of tensile test specimen, apply to material over 12.5 mm (½ in.) thick. (e) Material in this temper required to pass an exfoliation corrosion test administered by the purchaser

Table 77 Typical mechanical properties of alloy 5457

Temper	Tensile strength(a) MPa	ksi	Yield strength(a) MPa	ksi	Elongation(a)(b), %	Hardness(c), HB	Shear strength MPa	ksi
O	130	19	50	7	22	32	85	12
H25	180	26	160	23	12	48	110	16
H38, H28	205	30	185	27	6	55	125	18

(a) Strengths and elongations are unchanged or improved at lower temperatures. (b) 1.6 mm (1/16 in.) thick specimen. (c) 500 kg load; 10 mm ball

Hardness. H321, H116 tempers: 90 HB
Elastic modulus. Tension, 70.3 GPa (10.2 × 10^6 psi); compression, 71.7 GPa (10.4 × 10^6 psi)

Mass Characteristics

Density. 2.66 g/cm³ (0.096 lb/in.³) at 20 °C (68 °F)

Thermal Properties

Liquidus temperature. 638 °C (1180 °F)
Solidus temperature. 570 °C (1055 °F)
Coefficient of thermal expansion. Linear:

Temperature range °C	°F	Average coefficient µm/m · K	µin./in. · °F
−50 to 20	−58 to 68	22.1	12.3
20 to 100	68 to 212	23.9	13.3
20 to 200	68 to 392	24.8	13.8
20 to 300	68 to 572	25.9	14.4

Volumetric: 69 × 10^{-6} m³/m³ · K (3.83 × 10^{-5} in.³/in.³ · °F) at 20 °C (68 °F)
Specific heat. 900 J/kg · K (0.215 Btu/lb · °F) at 20 °C (68 °F)
Thermal conductivity. 116 W/m · K (67 Btu/ft · h · °F) at 20 °C (68 °F)

Electrical Properties

Electrical conductivity. Volumetric, average of all tempers: 29% IACS at 20 °C (68 °F)
Electrical resistivity. Average of all tempers: 59.5 nΩ · m at 20 °C (68 °F); temperature coefficient, 0.1 nΩ · m per K at 20 °C (68 °F)
Electrolytic solution potential. −0.87 V versus 0.1 *N* calomel electrode in an aqueous solution containing 53 g NaCl plus 3 g H_2O_2 per liter

Fabrication Characteristics

Annealing temperature. 343 °C (650 °F); holding at temperature not required
Hot-working temperature. 260 to 510 °C (500 to 950 °F)

5457
1.0Mg-0.30Mn

Specifications

ASTM. Sheet: B 209
UNS number. A95457

Chemical Composition

Composition limits. 0.08 Si max, 0.10 Fe max, 0.20 Cu max, 0.15 to 0.45 Mn, 0.08 to 1.2 Mg, 0.05 Zn max, 0.05 V max, 0.03 max other (each), 0.10 max others (total), bal Al

Applications

Typical uses. Brightened and anodized automotive and appliance trim
Precautions in use. Fine grain size required for most applications of this alloy

Mechanical Properties

Tensile properties. Tensile strength: min, 110 MPa (16 ksi); max, 150 MPa (22 ksi). Elongation, 20% in 50 mm (2 in.). See also Table 77.
Shear strength. See Table 77.
Compressive yield strength. Approximately the same as tensile yield strength
Hardness. See Table 77.
Poisson's ratio. 0.33 at 20 °C (68 °F)
Elastic modulus. Tension, 68.2 GPa (10.0 × 10^6 psi); shear, 25.9 GPa (3.75 × 10^6 psi); compression, 69.6 GPa (10.1 × 10^6 psi)

Mass Characteristics

Density. 2.69 g/cm³ (0.0972 lb/in.³) at 20 °C (68 °F)

Thermal Properties

Liquidus temperature. 654 °C (1210 °F)
Solidus temperature. 629 °C (1165 °F)
Coefficient of thermal expansion. Linear:

Temperature range °C	°F	Average coefficient µm/m · K	µin./in. · °F
−50 to 20	−58 to 68	21.9	12.2
20 to 100	68 to 212	23.7	13.2
20 to 200	68 to 392	24.6	13.7
20 to 300	68 to 572	25.6	14.2

Volumetric: 68 × 10^{-6} m³/m³ · K (3.77 × 10^{-5} in.³/in.³ · °F) at 20 °C (68 °F)
Specific heat. 900 J/kg · K (0.215 Btu/lb · °F) at 20 °C (68 °F)
Thermal conductivity. 177 W/m · K (102 Btu/ft · h · °F) at 20 °C (68 °F)

Electrical Properties

Electrical conductivity. Volumetric, average of all tempers: 46% IACS at 20 °C (68 °F)
Electrical resistivity. 37.5 nΩ · m at 20 °C (68 °F); temperature coefficient, 0.1 nΩ · m per K at 20 °C (68 °F)
Electrolytic solution potential. −0.84 V versus 0.1 *N* calomel electrode in an aqueous solution containing 53 g NaCl plus 3 g H_2O_2 per liter

Fabrication Characteristics

Formability. Readily formed in both annealed and H25 tempers

Table 78 Mechanical properties of alloy 5652

Temper	Tensile strength MPa	ksi	MPa	ksi	Yield strength MPa	ksi	Elongation(a), %	Hardness(b), HB	Shear strength MPa	ksi	Fatigue strength(c) MPa	ksi
Typical properties												
O	195	28	...	...	90	13	25	47	124	18	110	16
H32	230	33	...	...	195	28	12	60	138	20	117	17
H34	260	38	...	...	215	31	10	68	145	21	124	18
H36	275	40	...	...	240	35	8	73	158	23	131	19
H38	290	42	...	...	255	37	7	77	165	24	138	20
Property limits	Minimum		Maximum		Minimum		Minimum					
O	170	25	215	31	65	9.5	14–18	...	...	...	...	...
H32	215	31	260	38	160	23	4–12	...	...	...	...	...
H34	235	34	285	41	180	26	3–10	...	...	...	...	...
H36	255	37	305	44	200	29	2–4	...	...	...	...	...
H38	270	39	...	...	220	32	2–4	...	...	...	...	...
H112												
(0.250–0.499 in. thick)	195	28	...	...	110	16	7	...	...	...	...	...
(0.500–3.000 in. thick)	170	25	...	...	65	9.5	12–16	...	...	...	...	...

(a) In 50 mm (2 in.) or 4*d*, where *d* is diameter of reduced section of tension-test specimen. Where a range of values appears in this column, the specified minimum elongation varies with thickness of the mill product. (b) 500 kg load; 10 mm ball. (c) At 5 × 10^8 cycles; R.R. Moore type test

Table 79 Tensile properties of alloy 5657

Temper	Tensile strength MPa	ksi	MPa	ksi	Yield strength MPa	ksi	Elongation(a), %
Typical properties(b)							
H25	160	23	...	...	140	20	12
H28, H38	195	28	...	...	165	24	7
Property limits	**Minimum**		**Maximum**				**Minimum**
H241(c)	125	18	180	26	...	...	13
H25	140	20	195	28	...	...	8
H26	150	22	205	30	...	...	7
H28	170	25	...	...	...	...	5

(a) In 50 mm (2 in.) or 4*d*, where *d* is diameter of reduced section of tension-test specimen. (b) Strengths and elongations are unchanged or increased at low temperatures. (c) Material in this temper subject to some recrystallization and attendant loss of brightness

Annealing temperature. 343 °C (650 °F); holding temperature not required
Hot-working temperature. 260 to 510 °C (500 to 950 °F)

5652
2.5Mg-0.25Cr

Specifications

ASTM. Sheet and plate: B 209. Extruded, seamless tube: B 241
SAE. J454
UNS number. A95652

Chemical Composition

Composition limits. 0.40 Si max + Fe, 0.04 Cu max, 0.01 Mn max, 2.2 to 2.8 Mg, 0.15 to 0.35 Cr, 0.10 Zn max, 0.05 max other (each), 0.15 max others (total), bal Al

Applications

Typical uses. Storage vessels for hydrogen peroxide and other chemicals

Mechanical Properties

Tensile properties. See Table 78.
Shear strength. See Table 78.
Compressive yield strength. Approximately the same as tensile yield strength
Hardness. See Table 78.
Poisson's ratio. 0.33
Elastic modulus. Tension, 68.2 GPa (9.89 × 10^6 psi); shear, 25.9 GPa (3.75 × 10^6 psi); compression, 69.6 GPa (10.1 × 10^6 psi)
Fatigue strength. See Table 78.

Mass Characteristics

Density. 2.68 g/cm^3 (0.097 lb/in.^{3}) at 20 °C (68 °F)

Thermal Properties

Liquidus temperature. 649 °C (1200 °F)
Solidus temperature. 607 °C (1125 °F)
Coefficient of thermal expansion.
Linear:

Temperature range °C	°F	Average coefficient μm/m · K	μin./in. · °F
−50 to 20	−58 to 68	22.0	12.2
20 to 100	68 to 212	23.8	13.2
20 to 200	68 to 392	24.7	13.7
20 to 300	68 to 572	25.8	14.3

Volumetric: 69 × 10^{-6} m^3/m^3 · K (3.83 × 10^{-5} in.^{3}/in.^{3} · °F) at 20 °C (68 °F)
Specific heat. 900 J/kg · K (0.215 Btu/lb · °F) at 20 °C (68 °F)
Thermal conductivity. 137 W/m · K (79.1 Btu/ft · h · °F) at 20 °C (68 °F)

Electrical Properties

Electrical conductivity. Volumetric, average of all tempers: 35% IACS at 20 °C (68 °F)
Electrical resistivity. 49 nΩ · m at 20 °C (68 °F); temperature coefficient, 0.1 nΩ · m per K at 20 °C (68 °F)
Electrolytic solution potential. −0.85 V versus 0.1 *N* calomel electrode in an aqueous solution containing 53 g NaCl plus 3 g H_2O_2 per liter

Fabrication Characteristics

Annealing temperature. 345 °C (650 °F); holding at temperature not required
Hot-working temperature. 260 to 510 °C (500 to 950 °F)

5657
0.8Mg

Specifications

ASTM. B 209
UNS number. A95657
Foreign. Italy: P-AlMg0.9

Chemical Composition

Composition limits. 0.08 Si max, 0.10 Fe max, 0.10 Cu max, 0.03 Mn max, 0.6 to 1.0 Mg, 0.05 Zn max, 0.03 Ga max, 0.05 V max, 0.02 max other (each), 0.05 max others (total), bal Al

Applications

Typical uses. Brightened and anodized automotive and appliance trim
Precautions in use. Fine grain size essential for almost all applications of this alloy

Mechanical Properties

Tensile properties. See Table 79.
Shear strength. H25 temper: 95 MPa (14 ksi); H28, H38 tempers: 105 MPa (15 ksi)
Compressive yield strength. Approximately the same as tensile yield strength
Hardness. H25 temper: 40 HB. H28 and H38 tempers: 50 HB. All hardness values obtained with 500 kg load, 10 mm diam ball, and 30 s duration of loading
Poisson's ratio. 0.33
Elastic modulus. Tension, 68.2 GPa (9.89 × 10^6 psi); shear, 25.9 GPa (3.75 × 10^6 psi); compression, 69.6 GPa (10.1 × 10^6 psi)

Mass Characteristics

Density. 2.69 g/cm^3 (0.097 lb/in.^{3}) at 20 °C (68 °F)

Thermal Properties

Liquidus temperature. 657 °C (1215 °F)
Solidus temperature. 638 °C (1180 °F)
Coefficient of thermal expansion.
Linear:

Temperature range °C	°F	Average coefficient μm/m · K	μin./in. · °F
−50 to 20	−58 to 68	21.9	12.2
20 to 100	68 to 212	23.7	13.2
20 to 200	68 to 392	24.6	13.7
20 to 300	68 to 572	25.6	14.2

Volumetric: 68 × 10^{-6} m^3/m^3 · K (3.77 × 10^{-5} in.^{3}/in.^{3} · °F) at 20 °C (68 °F)
Specific heat. 900 J/kg · K (0.215 Btu/lb · °F)

Electrical Properties

Electrical conductivity. Volumetric, 54% IACS at 20 °C (68 °F)
Electrical resistivity. 32 nΩ · m at 20 °C (68 °F); temperature coefficient, 0.1 nΩ · m per K at 20 °C (68 °F)

Fabrication Characteristics

Annealing temperature. 345 °C (650 °F); holding at temperature not required
Hot-working temperature. 260 to 510 °C (500 to 950 °F)

6005
0.8Si-0.5Mg

Specifications

ASTM. Extruded wire, rod, bar, shapes, and tube: B 221
SAE. J454
UNS. A96005

Chemical Composition

Composition limits. 0.6 to 0.9 Si, 0.35 Fe max, 0.10 Cu max, 0.10 Mn max, 0.40 to 0.6 Mg, 0.10 Cr max, 0.10 Zn max, 0.10 Ti max, 0.05 max other (each), 0.15 max others (total), bal Al

Applications

Typical uses. Extruded shapes and tubing for commercial applications requiring strength greater than that of 6063; ladders and TV antennas are among the more common products
Precautions in use. Not recommended for applications requiring resistance to impact loading

Table 80 Typical tensile properties of alloy 6009 automobile body sheet

Orientation	Tensile strength, MPa	Tensile strength, ksi	Yield strength, MPa	Yield strength, ksi	Elongation, %
T4 temper					
Longitudinal	234	34	131	19	24
Transverse and 45°	228	33	124	18	25
T6 temper					
Longitudinal	345	50	324	47	12
Transverse and 45°	338	49	296	43	13

Table 81 Typical tensile properties of alloy 6010 automobile body sheet

Orientation	Tensile strength, MPa	Tensile strength, ksi	Yield strength, MPa	Yield strength, ksi	Elongation, %
T4 temper					
Longitudinal	296	43	186	27	23
Transverse and 45°	290	42	172	25	24
T6 temper					
Longitudinal	386	56	372	54	11
Transverse and 45°	379	55	352	51	12

Mechanical Properties

Tensile properties. Tensile strength (minimum): T1 temper, 172 MPa (25 ksi); T5 temper, 262 MPa (38 ksi). Yield strength (minimum): T1 temper, 103 MPa (15 ksi); T5 temper: 241 MPa (35 ksi). Elongation (minimum): T1 temper, 16%; T5 temper, 8 to 10%, specific value varies with thickness of mill product
Shear strength. T5 temper: 205 MPa (30 ksi)
Hardness. T5 temper: 95 HB
Elastic modulus. Tension, 69 GPa (10×10^6 psi)
Fatigue strength. (minimum). 97 MPa (14 ksi) at 5×10^8 cycles; R.R. Moore type test

Mass Characteristics

Density. 2.7 g/cm^3 (0.098 lb/in.^{3}) at 20 °C (68 °F)

Thermal Properties

Liquidus temperature. 654 °C (1210 °F)
Solidus temperature. 607 °C (1125 °F)
Coefficient of thermal expansion. Linear, 23.4 μm/m · K (13.0 μin./in. · °F) at 20 to 100 °C (68 to 212 °F)
Thermal conductivity. T5 temper: 167 W/m · K (97 Btu/ft · h · °F) at 25 °C (77 °F)

Electrical Properties

Electrical conductivity. Volumetric, T5 temper: 49% IACS at 20 °C (68 °F)
Electrical resistivity. T5 temper: 35 nΩ · m at 20 °C (68 °F)

Fabrication Characteristics

Annealing temperature. 415 °C (778 °F); hold at temperature for 2 to 3 h
Solution temperature. 547 °C (1015 °F)
Aging temperature. 175 °C (346 °F), hold at temperature for 8 h

6009
0.80Si-0.60Mg-0.50Mn-0.35Cu

Specifications

UNS. A96009

Chemical Composition

Composition limits. 0.6 to 1.0 Si, 0.50 Fe max, 0.15 to 0.6 Cu, 0.20 to 0.8 Mn, 0.40 to 0.8 Mg, 0.10 Cr max, 0.25 Zn max, 0.10 Ti max, 0.05 max other (each), 0.15 max others (total), bal Al

Applications

Typical uses. Automobile body sheet

Mechanical Properties

Tensile properties. See Table 80.
Yield stretch. Following simulated forming and a paint bake cycle consisting of 1 h at 175 °C (350 °F). T4 temper: no stretch, 228 MPa (33 ksi); 5% stretch, 262 MPa (38 ksi); 10% stretch, 290 MPa (42 ksi)
Shear strength. Auto body sheet, T4 temper: 152 MPa (22 ksi)
Hardness. T4 temper, auto body sheet: 70 HR15T
Poisson's ratio. 0.33
Elastic modulus. Tension, 69 GPa (10×10^6 psi); shear, 25.4 GPa (3.75×10^6 psi)
Fatigue strength. T4 temper: 117 MPa (17 ksi) at 10×10^6 cycles; sheet flexural specimens

Mass Characteristics

Density. 2.71 g/cm^3 (0.098 lb/in.^{3}) at 20 °C (68 °F)

Thermal Properties

Liquidus temperature. 650 °C (1202 °F)
Solidus temperature. 560 °C (1040 °F)
Coefficient of thermal expansion. Linear:

Temperature range, °C	Temperature range, °F	Average coefficient, μm/m · K	Average coefficient, μin./in. · °F
−50 to 20	−58 to 68	21.6	12.0
20 to 100	68 to 212	23.4	13.0
20 to 200	68 to 392	24.3	13.5
20 to 300	68 to 572	25.2	14.0

Volumetric: 67×10^{-6} m^3/m^3 · K (3.72×10^{-5} in.^{3}/in.^{3} · °F) at 20 °C (68 °F)
Specific heat. 897 J/kg · K (0.214 Btu/lb · °F) at 20 °C (68 °F)
Thermal conductivity. At 20 °C (68 °F): O temper, 205 W/m · K (118 Btu/ft · h · °F); T4 temper, 172 W/m · K (99 Btu/ft · h · °F); T6 temper, 180 W/m · K (104 Btu/ft · h · °F)

Electrical Properties

Electrical conductivity. Volumetric, at 20 °C (68 °F): O temper, 54% IACS; T4 temper, 44% IACS; T6 temper: 47% IACS
Electrical resistivity. At 20 °C (68 °F): O temper, 31.9 nΩ · m; T4 temper, 39.2 nΩ · m; T6 temper, 36.7 nΩ · m. Temperature coefficient, 0.1 nΩ · m per K at 20 °C (68 °F)

Fabrication Characteristics

Formability. Auto body sheet, T4 temper. ½t radius required for 90° bending or for flanging material 0.80 to 1.30 mm (0.032 to 0.050 in.) thick. Standard hems, which are made by bending 180° over 1t interface thickness, also can be made in auto body sheet 0.80 to 1.30 mm thick. Olsen cup height, typically 0.38 in. when tested using a 25 mm (1 in.) diam top die, 15 MPa (2.2 ksi) hold-down pressure and polyethylene film as a lubricant. Strain-hardening exponent (n) typically 0.23; plastic strain ratio (r) typically 0.70
Annealing temperature. 415 °C (775 °F)
Solution temperature. 555 °C (1030 °F)
Aging temperature. 175 °C (350 °F)

6010
1.0Si-0.8Mg-0.5Mn-0.35Cu

Specifications

UNS. A96010

Chemical Composition

Composition limits. 0.8 to 1.2 Si, 0.50 Fe max, 0.15 to 0.6 Cu, 0.20 to 0.8 Mn, 0.60 to 1.0 Mg, 0.10 Cr max, 0.25 Zn max, 0.10 Ti max, 0.05 max other (each), 0.15 max others (total), bal Al

Applications

Typical uses. Automobile body sheet

Mechanical Properties

Tensile properties. Typical. T4 temper: tensile strength, 290 MPa (42 ksi); yield strength, 172 MPa (25 ksi); elongation, 24% in 50 mm (2 in.). See also Table 81.
Yield stretch. Following simulated forming and a paint bake cycle consisting of 1 h at

Table 82 Standard specifications for alloy 6061

Mill form and condition	Specification No. AMS	ASTM	Government
Bare 6061			
Sheet and plate	4025	B 209	QQ-A-250/11
	4026	...	...
	4027	...	...
	4043	...	...
	4053	...	...
Tread plate	...	B 632	MIL-F-17132
Wire, rod, and bar (rolled or cold finished)	4115	B 211	QQ-A-225/8
	4116	...	...
	4117	...	...
	4128	...	...
	4129	...	...
Rod, bar, shapes, and tube (extruded)	4150	B 221	QQ-A-200/8
	4160	...	...
	4161	...	...
	4172	...	...
	4173	...	...
Structural shapes	4113	B 808	QQ-A-200/8
Tube (extruded, seamless)	...	B 241	...
Tube (drawn)	...	B 483	...
Tube (seamless)	4079	B 210	WW-T-700/6
	4080	...	...
	4082	...	...
Tube (hydraulic)	4081	...	MIL-T-7081
	4083	...	...
Tube (condenser)	...	B 234	...
Tube (condenser with integral fins)	...	B 404	...
Tube (welded)	...	B 313	...
	...	B 549	...
Tube (wave guide)	...	...	MIL-W-85
	...	...	MIL-W-23068
	...	...	MIL-W-23351
Pipe	...	B 241	MIL-P-25995
Pipe (gas and oil transmission)	...	B 345	...
Forgings	4127	B 247	QQ-A-367, MIL-A-22771
	4146		
Forging stock	4127	...	QQ-A-367
	4146	...	...
Rivet wire	...	B 316	QQ-A-430
Impacts	...	...	MIL-A-12545
Structural pipe and tube (extruded)	...	B 429	MIL-P-25995
Alclad 6061			
Sheet and plate	4020	B 209	...
	4021	...	...
	4022	...	...
	4023	...	...

Table 83 Typical mechanical properties of alloy 6061

Temper	Tensile strength MPa	Tensile strength ksi	Yield strength MPa	Yield strength ksi	Elongation, % 1.6 mm (1/16 in.) thick specimen	Elongation, % 13 mm (½ in.) diam specimen	Shear strength MPa	Shear strength ksi
Bare 6061								
O	124	18	55	8	25	30	83	12
T4, T451	241	35	145	21	22	25	165	24
T6, T651	310	45	276	40	12	17	207	30
Alclad 6061								
O	117	17	48	7	25	...	76	11
T4, T451	228	33	131	19	22	...	152	22
T6, T651	290	42	255	37	12	...	186	27

175 °C (350 °F). T4 temper: no stretch, 255 MPa (37 ksi); 5% stretch, 295 MPa (43 ksi); 10% stretch, 324 MPa (47 ksi)
Hardness. T4 temper: 76 HR15T
Poisson's ratio. 0.33
Elastic modulus. Tension, 69 GPa (10×10^6 psi); shear, 25.4 GPa (3.75×10^6 psi)
Fatigue strength. T4 temper: 117 MPa (17 ksi) at 10×10^6 cycles; sheet flexural specimens

Mass Characteristics

Density. 2.70 g/cm^3 (0.098 lb/in.^{3}) at 20 °C (68 °F)

Thermal Properties

Liquidus temperature. 650 °C (1200 °F)
Solidus temperature. 585 °C (1085 °F)
Incipient melting temperature. 577 °C (1070 °F)
Coefficient of thermal expansion.

Linear:

Temperature range °C	°F	Average coefficient µm/m · K	µin./in. · °F
−50 to 20	−58 to 68	21.5	11.9
20 to 100	68 to 212	23.2	12.9
20 to 200	68 to 392	24.1	13.4
20 to 300	68 to 572	25.1	13.9

Volumetric: 67×10^{-6} m^3/m^3 · K (3.72×10^{-5} in.^{3}/in.^{3} · °F) at 20 °C (68 °F)
Specific heat. 897 J/kg · K (0.214 Btu/lb · °F) at 20 °C (68 °F)
Thermal conductivity. At 20 °C (68 °F): O temper, 202 W/m · K (117 Btu/ft · h · °F); T4 temper, 151 W/m · K (87.3 Btu/ft · h · °F); T6 temper, 180 W/m · K (104 Btu/ft · h · °F)

Electrical Properties

Electrical conductivity. Volumetric, at 20 °C (68 °F): O temper, 53% IACS; T4 temper, 39% IACS; T6 temper, 44% IACS
Electrical resistivity. At 20 °C (68 °F): O temper, 32.5 nΩ · m; T4 temper, 44.2 nΩ · m; T6 temper, 39.2 nΩ · m. Temperature coefficient, 0.1 nΩ · m per K at 20 °C (68 °F)

Fabrication Characteristics

Formability. Auto body sheet, T4 temper. 1*t* radius required for 90° bending, 1*t* for flanging material 0.80 to 1.30 mm (0.032 to 0.050 in.) thick. Only roped hems, which are made by bending 180° over 2*t* interface thickness, can be made in auto body sheet 0.80 to 1.30 mm (0.0315 to 0.05 in.) thick. Olsen cup height, typically 9.1 mm (0.36 in.) when tested using a 25 mm (1 in.) diam top die, 15 MPa (2200 psi) hold-down pressure and polyethylene film as a lubricant. Strain-hardening exponent (*n*) typically 0.22; plastic strain ratio (*r*) typically 0.70
Annealing temperature. 415 °C (775 °F)
Solution temperature. 565 °C (1050 °F)
Aging temperature. 175 °C (350 °F)

6061 Alclad 6061
1.0Mg-0.6Si-0.30Cu-0.20Cr

Specifications

AMS. See Table 82.
ASTM. See Table 82.
UNS. A96061
Government. See Table 82.
Foreign. Canada: CSA GS11N. France: NF A-G5UC. United Kingdom: BS H20. ISO: AlMg1SiCu

Chemical Composition

Composition limits of 6061. 0.40 to 0.8 Si, 0.7 Fe max, 0.15 to 0.40 Cu, 0.15 Mn max, 0.8 to 1.2 Mg, 0.04 to 0.35 Cr, 0.25 Zn max, 0.15 Ti max, 0.05 max other (each), 0.15 max others (total), bal Al
Composition limits of Alclad 6061. 7072 cladding—0.7 Si max + Fe, 0.10 Cu max, 0.10 Mn max, 0.10 Mg max, 0.8 to 1.3 Zn, 0.05 max other (each), 0.15 max others (total), bal Al

Table 84 Typical tensile properties of alloy 6061-T6 or T651 at various temperatures

Temperature °C	Temperature °F	Tensile strength(a) MPa	Tensile strength(a) ksi	Yield strength (0.2% offset)(a) MPa	Yield strength (0.2% offset)(a) ksi	Elongation, %
−196	−320	414	60	324	47	22
−80	−112	338	49	290	42	18
−28	−18	324	47	283	41	17
24	75	310	45	276	40	17
100	212	290	42	262	38	18
149	300	234	34	214	31	20
204	400	131	19	103	15	28
260	500	51	7.5	34	5	60
316	600	32	4.6	19	2.7	85
371	700	24	3	12	1.8	95

(a) Lowest strength for exposures up to 10 000 h at temperature, no load; test loading applied at 35 MPa/min (5 ksi/min) to yield strength and then at strain rate of 5%/min to fracture

Applications

Typical uses. Trucks, towers, canoes, railroad cars, furniture, pipelines, and other structural applications where strength, weldability, and corrosion resistance are needed

Mechanical Properties

Tensile properties. See Tables 83 and 84.
Shear strength. See Table 83.
Hardness. O temper: 30 HB; T4, T451 tempers: 65 HB; T6, T651 tempers: 95 HB. Data obtained using 500 kg load, 10 mm diam ball, and 30 s duration of loading
Elastic modulus. Tension, 68.9 GPa (10.0 × 10^6 psi); compression, 69.7 GPa (10.1 × 10^6 psi)
Fatigue strength. O temper: 62 MPa (9 ksi). T4, T451, T6, and T651 tempers: 97 MPa (14 ksi). Data correspond to 5 × 10^8 cycles of completely reversed stress in R.R. Moore type tests.

Mass Characteristics

Density. 2.70 g/cm³ (0.098 lb/in.³) at 20 °C (68 °F)

Thermal Properties

Liquidus temperature. 652 °C (1206 °F)
Solidus temperature. 582 °C (1080 °F)
Coefficient of thermal expansion. Linear, 23.6 μm/m · K (13.1 μin./in. · °F) at 20 to 100 °C (68 to 212 °F)
Specific heat. 896 J/kg · K (0.214 Btu/lb · °F) at 20 °C (68 °F)
Thermal conductivity. At 25 °C (77 °F): O temper, 180 W/m · K (104 Btu/ft · h · °F); T4 temper, 154 W/m · K (89.0 Btu/ft · h · °F); T6 temper, 167 W/m · K (96.5 Btu/ft · h · °F)

Electrical Properties

Electrical conductivity. Volumetric at 20 °C (68 °F): O temper, 47% IACS; T4 temper, 40% IACS; T6 temper: 43% IACS
Electrical resistivity. At 20 °C (68 °F): O temper, 37 nΩ · m; T4 temper, 43 nΩ · m; T6 temper, 40 nΩ · m

Fabrication Characteristics

Solution temperature. 530 °C (985 °F)
Aging temperature. Rolled or drawn products: 160 °C (320 °F); hold at temperature for 18 h. Extrusions or forgings: 175 °C (350 °F); hold at temperature for 8 h

6063
0.7Mg-0.4Si

Specifications

AMS. Extruded wire, rod, bar, shapes, and tube: 4156
ASME. Extruded wire, rod, bar, shapes, and tube: SB221. Pipe: SB241
ASTM. See Table 85.
SAE. J454
UNS. A96063
Government. QQ-A-200/9, MIL-P-25995
Foreign. Austria: Önorm AlMgSi0,5. Canada: CSA GS10. France: NF A-GS. Italy: UNI P-AlSi0.4Mg. United Kingdom: BS H19; DTD 372B. Germany: DIN AlMgSi0.5; Werkstoff-Nr. 3.3206. ISO: AlMgSi

Table 85 ASTM specifications for alloy 6063

Mill form and condition	ASTM No.
Wire, rod, bar, shapes, and tube (extruded)	B 221
Tube (extruded, seamless); pipe	B 241
Tube (extruded, coiled)	B 491
Tube (drawn)	B 483
Tube (drawn, seamless)	B 210
Pipe (gas and oil transmission)	B 345
Structural pipe and tube (extruded)	B 429

Chemical Composition

Composition limits. 0.20 to 0.6 Si, 0.35 Fe max, 0.10 Cu max, 0.10 Mn max, 0.45 to 0.9 Mg, 0.10 Cr max, 0.10 Zn max, 0.10 Ti max, 0.05 max other (each), 0.15 max others (total), bal Al

Applications

Typical uses. Pipe, railings, furniture, architectural extrusions, truck and trailer flooring, doors, windows, irrigation pipes

Mechanical Properties

Tensile properties. See Tables 86 and 87.
Hardness. See Table 86.
Poisson's ratio. 0.33
Elastic modulus. Tension, 68.3 GPa (9.91 × 10^6 psi); shear, 25.8 GPa (3.75 × 10^6 psi); compression, 69.7 GPa (10.1 × 10^6 psi)

Table 86 Typical mechanical properties of alloy 6063

Temper	Tensile strength MPa	Tensile strength ksi	Yield strength MPa	Yield strength ksi	Elongation, %	Hardness(a), HB	Shear strength MPa	Shear strength ksi	Fatigue strength MPa	Fatigue strength ksi
O	90	13	48	7	...	25	69	10	55	8
T1(c)	152	22	90	13	20	42	97	14	62	9
T4	172	25	90	13	22	...	...	...	...	...
T5	186	27	145	21	12	60	117	17	69	10
T6	241	35	214	31	12	73	152	22	69	10
T83	255	37	241	35	9	82	152	22	...	...
T831	207	30	186	27	10	70	124	18	...	...
T832	290	42	269	39	12	95	186	27	...	...

(a) 500 kg load; 10 mm diam ball. (b) At 5 × 10^8 cycles; R.R. Moore type test. (c) Formerly T42 temper

Mass Characteristics

Density. 2.69 g/cm³ (0.097 lb/in.³)

Thermal Properties

Liquidus temperature. 655 °C (1211 °F)
Solidus temperature. 615 °C (1139 °F)
Coefficient of thermal expansion. Linear:

Temperature range °C	Temperature range °F	Average coefficient μm/m · K	Average coefficient μin./in. · °F
−50 to 20	−58 to 68	21.8	12.1
20 to 100	68 to 212	23.4	13.0
20 to 200	68 to 392	24.5	13.6
20 to 300	68 to 572	25.6	14.2

Specific heat. 900 J/kg · K (0.215 Btu/lb · °F) at 20 °C (68 °F)
Thermal conductivity. At 25 °C (77 °F):

Temper	Conductivity W/m · K	Conductivity Btu/ft · h · °F
O	218	126
T1 (formerly T42)	193	112
T5	209	121
T6	201	116

Electrical Properties

Electrical conductivity. At 20 °C (68 °F):

Temper	Conductivity, %IACS Equal volume	Conductivity, %IACS Equal weight
O	58	191
T1 (formerly T42)	50	165
T5	55	181
T6, T83	53	175

Table 87 Typical tensile properties of alloy 6063 at various temperatures

Temperature °C	Temperature °F	Tensile strength(a) MPa	Tensile strength(a) ksi	Yield strength (0.2% offset) MPa	Yield strength (0.2% offset) ksi	Elongation, %
T1 temper(b)						
−196	−320	234	34	110	16	44
−80	−112	179	26	103	15	36
−28	−18	165	24	97	14	34
24	75	152	22	90	13	33
100	212	152	22	97	14	18
149	300	145	21	103	15	20
204	400	62	9	45	6.5	40
260	500	31	4.5	24	3.5	75
316	600	23	3.2	17	2.5	80
371	700	16	2.3	14	2	105
T5 temper						
−196	−320	255	37	165	24	28
−80	−112	200	29	152	22	24
−28	−18	193	28	152	22	23
24	75	186	27	145	21	22
100	212	165	24	138	20	18
149	300	138	20	124	18	20
204	400	62	9	45	6.5	40
260	500	31	4.5	24	3.5	75
316	600	23	3.2	17	2.5	80
371	700	16	2.3	14	2	105
T6 temper						
−196	−320	324	47	248	36	24
−80	−121	262	38	228	33	20
−28	−18	248	36	221	32	19
24	75	241	35	214	31	18
100	212	214	31	193	28	15
149	300	145	21	133	20	20
204	400	62	9	45	6.5	40
260	500	31	4.5	24	3.5	75
316	600	23	3.3	17	2.5	80
371	700	16	2.3	14	2	105

(a) Lowest strength for exposures up to 10 000 h at temperature, no load; test loading applied at 35 MPa/min (5 ksi/min) to yield strength and then at strain rate of 5%/min to fracture. (b) T1 temper formerly T42

Table 88 Tensile properties of alloy 6066

Temper	Tensile strength MPa	Tensile strength ksi	Yield strength (0.2% offset) MPa	Yield strength (0.2% offset) ksi	Elongation(a), %
Typical properties					
O	150	22	83	12	18
T4, T451	360	52	207	30	18
T6, T651	395	57	359	52	12
Property limits (extrusions)					
O	200 max	29 max	125 max	18 max	16 min
T4, T4510, T4511	275 min	40 min	170 min	25 min	14 min
T42	275 min	40 min	165 min	24 min	14 min
T6, T6510, T6511	345 min	50 min	310 min	45 min	8 min
T62	345 min	50 min	290 min	42 min	8 min
Property limits (die forgings)					
T6	345 min	50 min	310 min	45 min	...

(a) In 50 mm (2 in.) or 4*d*, where *d* is diameter of reduced section of tensile test specimen

Electrical resistivity. At 20 °C (68 °F):

Temper	Resistivity, nΩ · m
O	30
T1 (formerly T42)	35
T5	32
T6, T83	33

Chemical Properties

General corrosion resistance. Highly resistant to all types of corrosion

Fabrication Characteristics

Machinability. Fair, depending on temper
Weldability. For all commercial processes, excellent weldability and brazability
Annealing temperature. 415 °C (775 °F); hold at temperature 2 to 3 h; cool at 28 °C (50 °F) per h from 415 °C (775 °F) to 260 °C (500 °F)
Solution temperature. 520 °C (970 °F)
Aging temperature. T5 temper: 205 °C (400 °F), hold at temperature for 1 h; or 182 °C (360 °F), hold at temperature for 1 h. All other artificially aged tempers: 175 °C (350 °F), hold at temperature for 8 h

6066
1.4Si-1.1Mg-1.0Cu-0.8Mn

Specifications

ASTM. Extruded wire, rod, bar, shapes, and tube: B 221
SAE. J454
UNS number. A96066
Government. Extruded wire, rod, bar, shapes, and tube: QQ-A-200/10. Forgings: QQ-A-367
Foreign. United Kingdom: BS H11

Chemical Composition

Composition limits. 0.9 to 1.8 Si, 0.50 Fe max, 0.7 to 1.2 Cu, 0.6 to 1.1 Mn, 0.8 to 1.4 Mg, 0.40 Cr max, 0.25 Zn max, 0.20 Ti max, 0.50 max other (each), 0.15 max others (total), bal Al

Applications

Typical uses. Forgings and extrusions for welded structures

Mechanical Properties

Tensile properties. See Table 88.
Shear strength. Typical. O temper: 97 MPa (14 ksi); T4 and T451 tempers: 200 MPa (29 ksi); T6 and T651 tempers: 234 MPa (34 ksi)
Hardness. O temper: 43 HB; T4 and T451 tempers: 90 HB; T6 and T651 tempers: 120 HB
Elastic modulus. Tension, 69 GPa (10×10^6 psi)
Fatigue strength. T6 and T651 tempers, 110 MPa (16 ksi). Data correspond to 5×10^8 cycles in R.R. Moore type test.

Mass Characteristics

Density. 2.71 g/cm^3 (0.098 lb/in.^{3}) at 20 °C (68 °F)

Thermal Properties

Liquidus temperature. 645 °C (1195 °F)
Solidus temperature. 563 °C (1045 °F)
Coefficient of thermal expansion. Linear, 23.2 μm/m · K (12.9 μin./in. · °F) at 20 to 100 °C (68 to 212 °F)
Specific heat. 887 J/kg · K (0.212 Btu/lb · °F) at 20 °C (68 °F)
Thermal conductivity. T6 temper, 147 W/m · K (85 Btu/ft · h · °F) at 20 °C (68 °F)

Electrical Properties

Electrical conductivity. Volumetric, at 20 °C (68 °F): O temper, 40% IACS; T6 temper, 37% IACS
Electrical resistivity. At 20 °C (68 °F): O temper, 43 nΩ · m; T6 temper, 47 nΩ · m

Fabrication Characteristics

Annealing temperature. 415 °C (778 °F); hold at temperature 2 to 3 h
Solution temperature. 530 °C (990 °F); followed by quenching

Table 89 Typical tensile properties of alloy 6101-T6 at various temperatures

Temperature, °C	Temperature, °F	Tensile strength(a), MPa	Tensile strength(a), ksi	Yield strength (0.2% offset)(a), MPa	Yield strength (0.2% offset)(a), ksi	Elongation(b), %
−196	−320	296	43	228	33	24
−80	−112	248	36	207	30	20
−28	−18	234	34	200	29	19
24	75	221	32	193	28	19
100	212	193	28	172	25	20
149	300	145	21	131	19	20
204	400	69	10	48	7	40
260	500	33	4.8	23	3.3	80
316	600	24	3	16	2.3	100
371	700	17	2.5	12	1.8	105

(a) Lowest strength for exposures up to 10 000 h at temperature, no load; test loading applied at 35 MPa/min (5 ksi/min) to yield strength and then at strain rate of 5%/min to fracture. (b) In 50 mm (2 in.)

Table 90 Property limits for alloy 6101 extrusions

Temper	Tensile strength(a), MPa	Tensile strength(a), ksi	Yield strength(a), MPa	Yield strength(a), ksi	Electrical conductivity(a), %IACS
H111	83	12	55	8	59
T6	200	29	172	25	55
T61					
0.125–0.749 in. thick	138	20	103	15	57
0.750–1.499 in. thick	124	18	76	11	57
1.500–2.000 in. thick	103	15	55	8	57
T63	186	27	152	22	56
T64	103	15	55	8	59.5
T65	172–221	25–32	138–186	20–27	56.5

(a) Single entries are minimum values.

Aging temperature. 175 °C (350 °F); hold at temperature 8 h

6070
1.4Si-0.8Mg-0.7Mn-0.3Cu

Specifications

ASTM. Gas and oil transmission pipe: B 345
SAE. J454
Government. Extruded rod, bar, shapes, and tube: MIL-A-46104. Impacts: MIL-A-12545

Chemical Composition

Composition limits. 1.0 to 1.7 Si, 0.50 Fe max, 0.15 to 0.40 Cu, 0.40 to 1.0 Mn, 0.50 to 1.2 Mg, 0.10 Cr max, 0.25 Zn max, 0.15 Ti max, 0.05 max other (each), 0.15 max others (total), bal Al

Applications

Typical uses. Heavy duty welded structures, pipelines, extruded structural components for automobiles

Mechanical Properties

Tensile properties. Typical. Tensile strength: O temper, 145 MPa (21 ksi); T4 temper, 317 MPa (46 ksi); T6 temper, 379 MPa (55 ksi). Yield strength: O temper, 69 MPa (10 ksi); T4 temper, 172 MPa (25 ksi); T6 temper, 352 MPa (51 ksi). Elongation: O and T4 tempers, 20%; T6 temper, 10%
Shear strength. Typical. O temper: 97 MPa (14 ksi); T4 temper: 206 MPa (30 ksi); T6 temper: 234 MPa (34 ksi)
Hardness. O temper: 35 HB; T4 temper: 90 HB; T6 temper: 120 HB. Data obtained using 500 kg load, 10 mm diam ball, and 30 s duration of loading.
Elastic modulus. Tension, 68 GPa (9.9 × 10^6 psi)
Fatigue strength. O temper: 62 MPa (9 ksi); T4 temper: 90 MPa (13 ksi); T6 temper: 97 MPa (14 ksi). Data correspond to 5 × 10^8 cycles of completely reversed stress in an R.R. Moore type test

Mass Characteristics

Density. 2.71 g/cm^3 (0.098 lb/in.^{3})

Thermal Properties

Liquidus temperature. 649 °C (1200 °F)
Solidus temperature. 566 °C (1050 °F)
Specific heat. 891 J/kg · K (0.213 Btu/lb · °F) at 20 °C (68 °F)
Thermal conductivity. T6 temper: 172 W/m · K (99.1 Btu/ft · h · °F) at 20 °C (68 °F)

Electrical Properties

Electrical conductivity. Volumetric, T6 temper: 44% IACS at 20 °C (68 °F)
Electrical resistivity. 39 nΩ · m at 20 °C (68 °F)

Fabrication Characteristics

Solution temperature. 545 °C (1015 °F); followed by quenching
Annealing temperature. T4 temper: 545 °C (1015 °F)
Aging temperature. 160 °C (320 °F); hold at temperature for 18 h

6101
0.6Mg-0.5Si

Specifications

ASTM. Bus conductor: B 317
SAE. J454
UNS number. A96101
Foreign. Austria: Önorm E-AlMgSi. France: NF A-GS/L. Italy: UNI P-AlSi0.5Mg. Switzerland: VSM Al-Mg-Si. United Kingdom: BS 91E. Germany: E-AlMgSi0.5; Werkstoff-Nr. 3.3207

Chemical Composition

Composition limits. 0.30 to 0.7 Si, 0.50 Fe max, 0.10 Cu max, 0.03 Mn max, 0.35 to 0.8 Mg, 0.03 Cr max, 0.10 Zn max, 0.06 B max, 0.03 max other (each), 0.10 max others (total), bal Al

Applications

Typical uses. High strength bus bars, electrical conductors, heat sinks

Mechanical Properties

Tensile properties. Typical. Tensile strength, 221 MPa (32 ksi); yield strength, 193 MPa (28 ksi); elongation, 15%. See also Tables 89 and 90.
Shear strength. 138 MPa (20 ksi)
Hardness. 71 HB with 500 kg load, 10 mm diam ball
Elastic modulus. Tension, 68.9 GPa (10.0 × 10^6 psi); compression, 70.3 GPa (10.2 × 10^6 psi)

Mass Characteristics

Density. 2.69 g/cm^3 (0.097 lb/in.^{3}) at 20 °C (68 °F)

Thermal Properties

Liquidus temperature. 654 °C (1210 °F)
Solidus temperature. 621 °C (1150 °F)
Coefficient of thermal expansion. Linear:

Temperature range, °C	Temperature range, °F	Average coefficient, μm/m · K	Average coefficient, μin./in. · °F
−50 to 20	−58 to 68	21.7	12.0
20 to 100	68 to 212	23.5	13.0
20 to 200	68 to 392	24.4	13.5
20 to 300	68 to 572	25.4	14.1

Specific heat. 895 J/kg · K (0.214 Btu/lb · °F) at 20 °C (68 °F)
Thermal conductivity. 218 W/m · K (126 Btu/ft · h · °F) at 25 °C (77 °F)

Electrical Properties

Electrical conductivity and resistivity at 20 °C (68 °F):

Temper	Electrical conductivity, %IACS	Electrical resistivity, nΩ · m
T6	57	30.2
T61	59	29.2
T63	58	29.7
T64	60	28.7
T65	58	29.7

Table 91 Tensile-property limits for alloy 6151

Temper	Tensile strength MPa	ksi	Yield strength MPa	ksi	Elongation(a), %
Die forgings, T6					
Axis parallel to grain flow	303	44	255	37	14 (coupon) 10 (forging)
Axis not parallel to grain flow	303	44	255	37	6 (forging)
Rolled rings, T6 and T652					
Tangential	303	44	255	37	5
Axial	303	44	241	35	4
Radial	290	42	241	35	2

(a) In 50 mm (2 in.) or 4*d*, where *d* is diameter of reduced section of tensile test specimen

Table 92 Typical tensile properties of alloy 6151

Temperature °C	°F	Tensile strength(a) MPa	ksi	Yield strength (0.2% offset)(a) MPa	ksi	Elongation, %
−196	−321	395	57	345	50	20
−80	−112	345	50	315	46	17
−28	−18	340	49	310	45	17
24	76	330	48	298	43	17
100	212	295	43	275	40	17
149	300	195	28	185	27	20
204	400	95	14	85	12	30
260	500	45	6.5	34	5	50
316	600	34	5	27	3.9	43
371	700	28	4	22	3.2	35

(a) Lowest strength for exposures up to 10 000 h at temperature, no load; test loading applied at 35 MPa/min (5 ksi/min) to yield strength and then at strain rate of 5%/min to fracture

Fabrication Characteristics

Solution temperature. 510 °C (950 °F); hold for 1 h at temperature
Aging temperature. 175 °C (350 °F); hold for 6 to 8 h at temperature
Hot-working temperature. 260 to 510 °C (500 to 950 °F)

6151
0.9Si-0.6Mg-0.25Cr

Specifications

AMS. Forgings: 4125
SAE. J454
UNS number. A96151
Government. Forgings and forging stock: QQ-A-367; MIL-A-22771
Foreign. Canada: CSA SG11P

Chemical Composition

Composition limits. 0.6 to 1.2 Si, 1.0 Fe max, 0.35 Cu max, 0.20 Mn max, 0.45 to 0.8 Mg, 0.15 to 0.35 Cr, 0.25 Zn max, 0.15 Ti max, 0.05 max other (each), 0.15 max others (total), bal Al

Applications

Typical uses. Die forgings and rolled rings for crank cases, fuses, and machine parts. Applications requiring good forgeability, good strength, and resistance to corrosion

Mechanical Properties

Tensile properties. See Tables 91 and 92.
Hardness. T6 temper: 90 HB with 500 kg load, 10 mm diam ball

Mass Characteristics

Density. 2.70 g/cm^3 (0.098 lb/in.^{3}) at 20 °C (68 °F)

Thermal Properties

Liquidus temperature. 650 °C (1200 °F)
Solidus temperature. 588 °C (1090 °F)
Coefficient of thermal expansion. Linear:

Temperature range °C	°F	Average coefficient µm/m · K	µin./in. · °F
−50 to 20	−58 to 68	21.8	12.1
20 to 100	68 to 212	23.0	12.8
20 to 200	68 to 392	24.1	13.4
20 to 300	68 to 572	25.0	13.9

Specific heat. 895 J/kg · K (0.214 Btu/lb · °F)
Thermal conductivity. At 20 °C (68 °F): O temper, 205 W/m · K (118 Btu/ft · h · °F); T4 temper, 163 W/m · K (94 Btu/ft · h · °F); T6 temper, 175 W/m · K (101 Btu/ft · h · °F)

Electrical Properties

Electrical conductivity. Volumetric, at 20 °C (68 °F): O temper, 54% IACS; T4 temper, 42% IACS; T6 temper, 45% IACS
Electrical resistivity. At 20 °C (68 °F): O temper, 32 nΩ · m; T4 temper, 41 nΩ · m; T6 temper, 38 nΩ · m
Electrolytic solution potential. −0.83 V versus 0.1 *N* calomel electrode in an aqueous solution containing 53 g NaCl plus 3 g H_2O_2 per liter

Fabrication Characteristics

Annealing temperature. 413 °C (775 °F); hold at temperature 2 to 3 h; furnace cool to 260 °C (500 °F) at 27 °C (50 °F) per h max
Solution temperature. 510 to 525 °C (950 to 975 °F); hold at temperature 4 min, quench in cold water; heavy or complicated forgings, quench in water at 65 to 100 °C (150 to 212 °F)
Aging temperature. 165 to 175 °C (300 to 345 °F); hold at temperature 8 to 12 h
Hot-working temperature. 260 to 480 °C (500 to 900 °F)

6201
0.7Si-0.8Mg

Specifications

ASTM. Wire, B 398. Stranded conductor, T81 temper: B 399
SAE. J454
UNS. A96201

Chemical Composition

Composition limits. 0.50 to 0.95 Si, 0.50 Fe max, 0.10 Cu max, 0.03 Mn max, 0.6 to 0.9 Mg, 0.03 Cr max, 0.10 Zn max, 0.06 B max, 0.03 max other (each), 0.10 max others (total), bal Al

Applications

Typical uses. Rod and wire for high strength electrical conductors

Mechanical Properties

Tensile properties. Typical. T81 temper: tensile strength, 331 MPa (48 ksi); yield strength, 310 MPa (45 ksi); elongation, 6% in 250 mm (10 in.)
Property limits for T81 temper wire with 1.6 to 3.2 mm (1/16 to 1/8 in.) diameter. Min tensile strength (individual), 315 MPa (46 ksi); min tensile strength (average), 330 MPa (48 ksi)
Property limits for T81 temper wire with 3.2 to 4.8 mm (1/8 to 3/16 in.) diameter. Min tensile strength (individual), 305 MPa (44 ksi); min tensile strength (average), 315 MPa (46 ksi). Min elongation, 3% in 250 mm (10 in.) for all diameters

Mass Characteristics

Density. 2.69 g/cm^3 (0.097 lb/in.^{3}) at 20 °C (68 °F)

Thermal Properties

Liquidus temperature. 654 °C (1210 °F)
Solidus temperature. 607 °C (1125 °F)
Coefficient of thermal expansion:

Temperature range °C	°F	Average coefficient µm/m · K	µin./in. · °F
−50 to 20	−58 to 68	21.6	12.0
20 to 100	68 to 212	23.4	13.0
20 to 200	68 to 392	24.3	13.5
20 to 300	68 to 572	25.2	14.0

Specific heat. 895 J/kg · K (0.214 Btu/lb · °F) at 20 °C (68 °F)
Thermal conductivity. T8 temper: 205 W/m · K (118 Btu/ft · h · °F) at 25 °C (77 °F)

Table 93 Typical tensile properties of alloy 6262 at various temperatures

Temperature, °C	Temperature, °F	Tensile strength(a), MPa	Tensile strength(a), ksi	Yield strength (0.2% offset)(a), MPa	Yield strength (0.2% offset)(a), ksi	Elongation, %
T651 temper						
−196	−320	414	60	324	47	22
−80	−112	338	49	290	42	18
−28	−18	324	47	283	41	17
24	75	310	45	276	40	17
100	212	290	42	262	38	18
149	300	234	34	214	31	20
T9 temper						
−196	−320	510	74	462	67	14
−80	−112	427	62	400	58	10
−28	−18	414	60	386	56	10
24	75	400	58	379	55	10
100	212	365	53	359	52	10
149	300	262	38	255	37	14
204	400	103	15	90	13	34
260	500	59	8.5	41	6	48
316	600	32	4.6	19	2.7	85
371	700	24	3	12	1.8	95

(a) Lowest strength for exposures up to 10 000 h at temperature, no load; test loading applied at 35 MPa/min (5 ksi/min) to yield strength and then at strain rate of 5%/min to fracture

Electrical Properties

Electrical conductivity. Volumetric, T81 temper: 54% IACS at 20 °C (68 °F)
Electrical resistivity. T81 temper: 32 nΩ · m at 20 °C (68 °F)

Fabrication Characteristics

Solution temperature. 510 °C (950 °F)
Aging temperature. 150 °C (300 °F); hold at temperature approximately 4 h

6205
0.8Si-0.5Mg-0.10Mn-0.10Cr-0.10Zr

Specifications

UNS. A96205

Chemical Composition

Composition limits. 0.6 to 0.9 Si, 0.7 Fe max, 0.20 Cu max, 0.05 to 0.15 Mn, 0.40 to 0.6 Mg, 0.05 to 0.15 Cr, 0.25 Zn max, 0.05 to 0.15 Zr, 0.15 Ti max, 0.05 max other (each), 0.15 max others (total), bal Al

Applications

Typical uses. Plate, tread plate, and extrusions for applications requiring high impact strength

Mechanical Properties

Tensile properties. Typical. T1 temper: tensile strength, 262 MPa (38 ksi); yield strength, 138 MPa (20 ksi); elongation, 19%. T5 temper: tensile strength, 310 MPa (45 ksi); yield strength, 290 MPa (42 ksi); elongation, 11%
Shear strength. T5 temper: 207 MPa (30 ksi)
Hardness. T1 temper: 65 HB; T5 temper: 95 HB
Fatigue strength. T5 temper: 103 MPa (15 ksi) at 5 × 10[8] cycles in R.R. Moore type test

Mass Characteristics

Density. 2.70 g/cm³ (0.098 lb/in.³)

Thermal Properties

Liquidus temperature. 645 °C (1210 °F)
Solidus temperature. 613 °C (1135 °F)
Coefficient of thermal expansion. Linear, 23.0 μm/m · K (12.8 μin./in. · °F)
Thermal conductivity. At 25 °C (77 °F): T1 temper, 172 W/m · K (99.1 Btu/ft · h · °F); T5 temper, 188 W/m · K (109 Btu/ft · h · °F)

Electrical Properties

Electrical conductivity. Volumetric, at 20 °C (68 °F): T1 temper, 45% IACS; T5 temper, 49% IACS
Electrical resistivity. At 20 °C (68 °F): T1 temper, 37 nΩ · m per K; T5 temper, 35 nΩ · m

Fabrication Characteristics

Solution temperature. 525 °C (980 °F)
Aging temperature. 175 °C (350 °F); hold at temperature approximately 6 h

6262
1.0Mg-0.6Si-0.3Cu-0.09Cr-0.6Pb-0.6Bi

Specifications

ASTM. Rolled or cold finished wire, rod, and bar: B 211. Extruded wire, rod, bar, shapes, and tube: B 221. Drawn, seamless tube: B 210. Drawn tube: B 483
SAE J454
UNS. A96262
Government. Rolled or cold finished wire, rod, and bar: QQ-A-225/10

Chemical Composition

Composition limits. 0.40 to 0.8 Si, 0.7 Fe max, 0.15 to 0.40 Cu, 0.15 Mn max, 0.8 to 1.2 Mg, 0.04 to 0.14 Cr, 0.25 Zn max, 0.15 Ti max, 0.40 to 0.7 Bi, 0.40 to 0.7 Pb, 0.05 max other (each), 0.15 max others (total), bal Al

Applications

Typical uses. High-stress screw machine products requiring corrosion resistance superior to 2011 and 2017

Mechanical Properties

Tensile properties. Typical, T9 temper: tensile strength, 400 MPa (58 ksi); 0.2% yield strength, 379 MPa (55 ksi); see also Table 93.
Shear strength. Typical, T9 temper: 241 MPa (35 ksi)
Hardness. Typical, T9 temper: 120 HB with 500 kg load, 10 mm diam ball
Fatigue strength. Typical, T9 temper: 90 MPa (13 ksi) at 5 × 10[8] cycles; R.R. Moore type test

Mass Characteristics

Density. 2.71 g/cm³ (0.098 lb/in.³) at 20 °C (68 °F)

Thermal Properties

Liquidus temperature. 650 °C (1205 °F)
Solidus temperature. 585 °C (1080 °F)
Coefficient of thermal expansion. Linear, 23.4 μm/m · K (13.0 μin./in. · °F) at 20 to 100 °C (68 to 212 °F)
Thermal conductivity. T9 temper: 172 W/m · K (99.1 Btu/ft · h · °F) at 20 °C (68 °F)

Electrical Properties

Electrical conductivity. Volumetric, T9 temper: 44% IACS at 20 °C (68 °F)
Electrical resistivity. T9 temper: 39 nΩ · m at 20 °C (68 °F)

Fabrication Characteristics

Annealing temperature. 415 °C (780 °F); hold at temperature 2 to 3 h
Solution temperature. 540 °C (1000 °F); hold at temperature 8 to 12 h
Aging temperature. 170 °C (340 °F); hold at temperature 8 to 12 h

6351
1.0Si-0.6Mg-0.6Mn

Specifications

ASTM. Gas and oil transmission pipe: B 345. Extruded wire, rod, bar, shapes, and tube: B 221
UNS. A96351

Chemical Composition

Composition limits. 0.7 to 1.3 Si, 0.50 Fe max, 0.10 Cu max, 0.40 to 0.8 Mn, 0.40 to 0.8 Mg, 0.20 Zn max, 0.20 Ti max, 0.05 max other (each), 0.15 others (total), bal Al

Applications

Typical uses. Extruded structures used in road vehicles and railroad stock; tubing and pipe for carrying water, oil, or gasoline

Mechanical Properties

Tensile properties. Typical. T4 temper: tensile strength, 248 MPa (36 ksi); 0.2% yield

Table 94 Minimum mechanical properties of alloy 7005

Temper	Tensile strength MPa	Tensile strength ksi	Yield strength MPa	Yield strength ksi	Elongation(a), %	Compressive yield strength MPa	Compressive yield strength ksi	Shear strength MPa	Shear strength ksi	Shear yield strength MPa	Shear yield strength ksi	Bearing strength MPa	Bearing strength ksi	Bearing yield strength MPa	Bearing yield strength ksi
Extrusions															
T53															
L direction	345	50	303	44	10	296	43	193	28	172	25	655(b) 496(c)	95(b) 72(c)	503(b) 407(c)	73(b) 59(c)
L-T direction	331	48	290	42	...	303	44	...	...	...	...	...	...	...	...
Sheet and plate															
T6(d), T63(e), T6351(e)	324	47	262	38	...	269	39	186	27	152	22	634(b) 483(c)	92(b) 70(c)	448(b) 365(c)	65(b) 53(c)

(a) In 50 mm (2 in.) or 4*d*, where *d* is diameter of reduced section of tensile test specimen. (b) *e*/*d* = 2.0, where *e* is edge distance and *d* is pin diameter. (c) *e*/*d* = 1.5. (d) Up to 6.35 mm (0.250 in.) thick. (e) 6.35 to 75 mm (0.250 to 3.00 in.) thick

strength, 152 MPa (22 ksi); elongation, 20%. T6 temper: tensile strength, 310 MPa (45 ksi); 0.2% yield strength, 283 MPa (41 ksi); elongation, 14%. Property limits for extrusions, T54 temper: tensile strength (min), 207 MPa (30 ksi); 0.2% yield strength (min), 138 MPa (20 ksi); elongation (min), 10%
Shear strength. T6 temper, 200 MPa (29 ksi)
Hardness. T6 temper, 95 HB with 500 kg load, 10 mm diam ball
Fatigue strength. Typical, T6 temper: 90 MPa (13 ksi) at 5 × 10^8 cycles in R.R. Moore type test

Mass Characteristics

Density. 2.71 g/cm^3 (0.098 lb/in.^{3})

Thermal Properties

Liquidus temperature. 650 °C (1202 °F)
Solidus temperature. 555 °C (1030 °F)
Coefficient of thermal expansion. Linear, 23.4 μm/m · K (13.0 μin./in. · °F) at 20 to 80 °C (68 to 176 °F)
Thermal conductivity. 176 W/m · K (102 Btu/ft · h · °F) at 25 °C (77 °F)

Electrical Properties

Electrical conductivity. Volumetric, T6 temper: 46% IACS at 20 °C (68 °F)
Electrical resistivity. 38 nΩ · m at 20 °C (68 °F)

Fabrication Characteristics

Annealing temperature. 350 °C (660 °F); hold at temperature for about 4 h
Solution temperature. 505 °C (940 °F)
Aging temperature. 170 °C (338 °F); hold at temperature 6 h

6463
0.40Si-0.7Mg

Specifications

ASTM. Extruded wire, rod, bar, shapes, and tube: B 221
SAE. J454
UNS number. A96463
Foreign. United Kingdom: BS E6

Chemical Composition

Composition limits. 0.20 to 0.6 Si, 0.15 Fe max, 0.20 Cu max, 0.05 Mn max, 0.45 to 0.9 Mg, 0.05 Zn max, 0.05 max other (each), 0.15 max others (total), bal Al

Applications

Typical uses. Architectural, appliance, and bright anodized automotive extrusions

Mechanical Properties

Tensile properties. Typical. Tensile strength: T1 temper, 152 MPa (22 ksi); T5 temper, 186 MPa (27 ksi); T6 temper, 241 MPa (35 ksi). 0.2% yield strength: T1 temper, 90 MPa (13 ksi); T5 temper, 145 MPa (21 ksi); T6 temper: 214 MPa (31 ksi). *Elongation*: T1 temper, 20%; T5 and T6 tempers: 12%
Shear strength. T1 temper, 97 MPa (14 ksi); T5 temper, 117 MPa (17 ksi); T6 temper, 152 MPa (22 ksi)
Hardness. T1 temper, 42 HB; T5 temper, 60 HB; T6 temper, 74 HB. Values obtained with 500 kg load and 10 mm diam ball
Fatigue strength. All tempers: 69 MPa (10 ksi) at 5 × 10^8 cycles; R.R. Moore type test

Mass Characteristics

Density. 2.69 g/cm^3 (0.097 lb/in.^{3})

Thermal Properties

Liquidus temperature. 654 °C (1210 °F)
Solidus temperature. 621 °C (1150 °F)
Coefficient of thermal expansion. Linear, 23.4 μm/m · K (13.0 μin./in. · °F) at 20 to 100 °C (68 to 212 °F)
Thermal conductivity. At 25 °C (77 °F): T1 temper, 192 W/m · K (111 Btu/ft · h · °F); T5 temper, 209 W/m · K (121 Btu/ft · h · °F); T6 temper, 201 W/m · K (116 Btu/ft · h · °F)

Electrical Properties

Electrical conductivity. Volumetric, at 20 °C (68 °F): T1 temper, 50% IACS; T5 temper, 55% IACS; T6 temper, 53% IACS
Electrical resistivity. At 20 °C (68 °F): T1 temper, 34 nΩ · m; T5 temper, 31 nΩ · m; T6 temper, 33 nΩ · m

Fabrication Characteristics

Annealing temperature. 415 °C (780 °F)
Solution temperature. 520 °C (968 °F)
Aging temperature. To produce T6 temper: 175 °C (350 °F), hold at temperature 8 h; can also use 180 °C (360 °F), hold at temperature 6 h. To produce T5 temper: 205 °C (400 °F), hold at temperature 1 h; can also use 180 °C (360 °F), hold at temperature 3 h

7005
4.6Zn-1.4Mg-0.5Mn-0.1Cr-0.1Zr-0.03Ti

Specifications

ASTM. Extruded wire, rod, bar, shapes, and tube: B 221
UNS number. A97005

Chemical Composition

Composition limits. 0.10 Cu max, 1.0 to 1.8 Mg, 0.20 to 0.70 Mn, 0.35 Si max, 0.40 Fe max, 0.06 to 0.20 Cr, 0.01 to 0.06 Ti, 4.0 to 5.0 Zn, 0.08 to 0.20 Zr, 0.05 max other (each), 0.15 max others (total), bal Al

Applications

Typical uses. Extruded structural members such as frame rails, cross members, corner posts, side posts, and stiffeners for trucks, trailers, cargo containers, and rapid transit cars. Welded or brazed assemblies requiring moderately high strength and high fracture toughness, such as large heat exchangers, especially where solution heat treatment after joining is impractical. Sports equipment such as tennis racquets and softball bats
Precautions in use. To avoid stress-corrosion cracking, stresses in the transverse direction should be avoided at exposed machined or sawed surfaces. Parts should be cold formed in O temper, then heat treated; alternatively, parts may be cold formed in W temper, followed by artificial aging. In parts intended for service in aggressive electrolytes such as seawater, selective attack along the heat-affected zone in a weldment or torch-brazed assembly can be avoided by postweld aging. When the service environment is conducive to galvanic corrosion, 7005 should be coupled or joined only to aluminum alloy components having similar electrolytic solution potentials;

Table 95 Typical tensile properties at various temperatures for alloy 7005-T53 extrusions

Temperature, °C	Temperature, °F	Tensile strength(a), MPa	Tensile strength(a), ksi	Yield strength(a), MPa	Yield strength(a), ksi	Elongation, %
−269	−452	641	93	483	70	16
−196	−320	538	78	421	61	16
−80	−112	441	64	379	55	13
−28	−18	421	61	359	52	14
24	75	392	57	345	50	15
100	212	303	44	283	41	20
149	300	165	24	145	21	35
204	400	97	14	83	12	60
260	500	76	11	66	9.5	80

(a) Lowest strength for exposures up to 10 000 h at temperature, no load; test loading applied at 35 MPa/min (5 ksi/min) to yield strength and then at strain rate of 5%/min to fracture

alternatively, joint surfaces should be protected or insulated.

Mechanical Properties

Tensile properties. Typical. Tensile strength: O temper, 193 MPa (28 ksi); T53 temper, 393 MPa (57 ksi); T6, T63, T6351 tempers, 372 MPa (54 ksi). Yield strength: O temper, 83 MPa (12 ksi); T53 temper, 345 MPa (50 ksi); T6, T63, T6351 tempers, 317 MPa (46 ksi). Elongation in 50 mm (2 in.) or 4*d* where *d* is diameter of tensile test specimen: O temper, 20%; T53 temper, 15%; T6, T63, T6351 tempers, 12%. See also Tables 94 and 95.
Shear strength. Typical. O temper: 117 MPa (17 ksi); T53 temper: 221 MPa (32 ksi); T6, T63, T6351 tempers: 214 MPa (31 ksi); see also Table 94.
Compressive strength. See Table 94.
Elastic modulus. Tension, 71 GPa (10.3 × 10^6 psi); shear, 26.9 GPa (3.9 × 10^6 psi); compression, 72.4 GPa (10.5 × 10^6 psi)
Fatigue strength. Rotating beam at 10^8 cycles. T6351 plate: smooth specimens, 115 to 130 MPa (17 to 19 ksi); 60° notched specimens, 20 to 50 MPa (3 to 7 ksi). T53 extrusions: smooth specimens, 130 to 150 MPa (19 to 22 ksi); 60° notched specimens, 24 to 40 MPa (3.5 to 6 ksi). Axial (R = 0) at 10^8 cycles, smooth specimens. T6351 plate: 195 MPa (28 ksi). T53 extrusions: 231 MPa (33.5 ksi)
Plane-strain fracture toughness. Typical, T6351 temper. L-T orientation: 51.3 MPa$\sqrt{m}$ (46.7 ksi$\sqrt{in.}$); data from 75 mm (3 in.) thick notch bend specimens. T-L orientation: 44 MPa$\sqrt{m}$ (40 ksi$\sqrt{in.}$); data from 75 mm (3 in.) thick notch bend specimens. S-L orientation: 30.3 MPa$\sqrt{m}$ (27.6 ksi$\sqrt{in.}$); data from 25 to 32 mm (1 to 1¼ in.) thick compact tensile specimens

Mass Characteristics

Density. 2.78 g/cm^3 (0.100 lb/in.^{3}) at 20 °C (68 °F)

Thermal Properties

Liquidus temperature. 643 °C (1190 °F)
Solidus temperature. 604 °C (1120 °F)
Coefficient of thermal expansion. Linear:

Temperature range, °C	Temperature range, °F	Average coefficient, μm/m · K	Average coefficient, μin./in. · °F
−50 to 20	−58 to 68	21.4	11.9
20 to 100	68 to 212	23.1	12.8
20 to 200	68 to 392	24.0	13.3
20 to 300	68 to 572	25.0	13.9

Volumetric: 67.0 × 10^{-6} m^3/m^3 · K (3.72 × 10^{-5} in.^{3}/in.^{3} · °F) at 20 °C (68 °F)
Specific heat. 875 J/kg · K (0.209 Btu/lb · °F) at 20 °C (68 °F)
Thermal conductivity. At 20 °C (68 °F); O temper, 166 W/m · K (96 Btu/ft · h · °F); T53, T5351, T63, T6351, T63, T6351 tempers, 148 W/m · K (86 Btu/ft · h · °F); T6 temper, 137 W/m · K (79 Btu/ft · h · °F)

Electrical Properties

Electrical conductivity. Volumetric, at 20 °C (68 °F): O temper, 43% IACS; T53, T5351, T63, T6351 tempers, 38% IACS; T6 temper, 35% IACS
Electrical resistivity. At 20 °C (68 °F): O temper, 40.1 nΩ · m; T53, T5351, T63, T6351 tempers, 45.4 nΩ · m; T6 temper, 49.3 nΩ · m. Temperature coefficient, all tempers: 0.1 nΩ · m per K at 20 °C (68 °F)

Fabrication Characteristics

Annealing temperature. 345 °C (650 °F)
Solution temperature. 400 °C (750 °F)
Heat treatment. T53: Press quench from hot working temperature, naturally age 72 h at room temperature, then two-stage artificially age 8 h at 100 to 110 °C (212 to 230 °F) plus 16 h at 145 to 155 °C (290 to 310 °F)

7039
4Zn-2.8Mg-0.25Mn-0.20Cr

Specifications

Military. MIL-A-22771, MIL-A-45225, MIL-A-46063
UNS number. A97039

Chemical Composition

Composition limits. 2.3 to 3.3 Mg, 3.5 to 4.5 Zn, 0.10 to 0.40 Mn, 0.15 to 0.25 Cr, 0.30 max Si, 0.10 max Cu, 0.40 max Fe, 0.10 max Ti, 0.50 max other (each), 0.15 max others (total), bal Al

Applications

Typical uses. Cryogenic storage tanks, unfired pressure vessels, ordnance tanks, armor

Table 96 Typical mechanical properties of 7039

Property	Property value(a) at temper: T64	T61	O
Tensile strength, MPa (ksi)			
Longitudinal	450 (65)	400 (58)	227 (33)
Transverse	450 (65)	400 (58)	227 (33)
0.2% tensile yield strength, MPa (ksi)			
Longitudinal	380 (55)	330 (48)	103 (15)
Transverse	380 (55)	330 (48)	103 (15)
Elongation in 50 mm (2 in.), %			
Longitudinal	13	14	22
Transverse	13	14	22
0.2% compressive yield strength, MPa (ksi)			
Longitudinal	400 (58)	380 (55)	...
Transverse	415 (60)	407 (59)	...
Shear strength, MPa (ksi)			
Longitudinal	270 (39)	...	...
Transverse	255 (37)	235 (34)	...
Bearing strength(b), MPa (ksi)			
Longitudinal	910 (132)	...	...
Transverse	910 (132)	827 (120)	...
Brinell hardness (1500 kg), HB	133	123	61

(a) Property values for 6 to 75 mm (0.25 to 3.0 in.) thick plate. (b) *e*/*d* = 2, where *e* is the edge distance and *d* is the pin diameter

Table 97 Transverse impact toughness of 7039-T64 plate

Plate thickness, mm	Plate thickness, in.	Test temperature, °C	Test temperature, °F	Elongation in 50 mm (2 in.), %	Unnotched impact toughness, J	Unnotched impact toughness, ft · lbf	Notched impact toughness, J	Notched impact toughness, ft · lbf
45	1.75	24	75	12	66.2	48.8	7.6	5.6
		−195	−320	12	87.5	64.5	6.5	4.8
38	1.50	24	75	11	75.3	55.5	7.5	5.5
		−195	−320	11	96.7	71.3	8.3	6.1

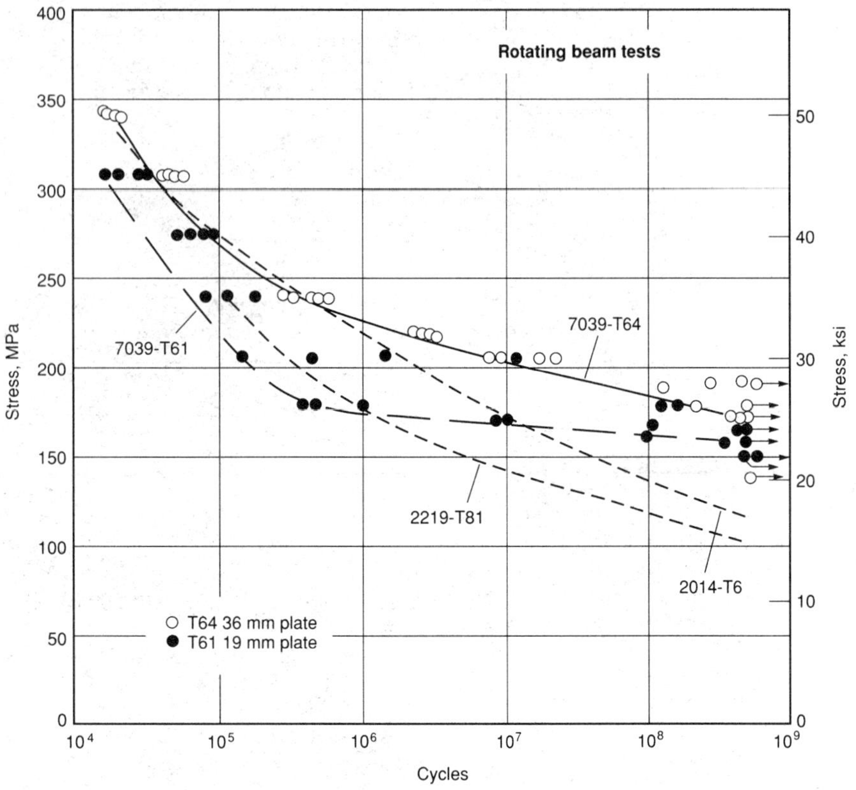

Fig. 14 Rotating beam fatigue data of 7039 plate compared with fatigue characteristics of 2014 and 2219. Data for 7039 are based on least-of-four results in the longitudinal direction with a 7.5 mm (0.3 in.) diam smooth specimen. Curves for 2014 and 2219 are mean values from published literature.

plate, missile structures, low-temperature processing equipment, and storage tanks.
Forms available. Plates, forgings, extrusions, and sometimes sheet

Mechanical Properties

Tensile properties. See Table 96.
Hardness. See Table 96.
Compressive yield strength. See Table 96.
Shear strength. See Table 96.
Elastic modulus. 69.6 GPa (10.1×10^6 psi)
Fatigue strength. See Fig. 14.
Impact toughness. See Table 97.

Mass Characteristics

Density. 2.73 g/cm³ (0.0988 lb/in.³)

Thermal Properties

Liquidus temperature. 638 °C (1180 °F)
Solidus temperature. 482 °C (900 °F)
Coefficient of thermal expansion. From 20 to 100 °C (68 to 212 °F): 23.4 μm/m · °C (13 μin./in. · °F)
Thermal conductivity. 125 to 155 W/m · °C (0.30 to 0.37 cal/cm · s · °C)

Electrical Properties

Electrical conductivity. 32 to 40% IACS (volumetric)

Fabrication Characteristics

Solution treatment. Heat to 460 to 500 °C (860 to 930 °F), soak 2 h, quench in cold water. Sheet stock should be quenched from 490 to 500 °C (910 to 930 °F), while extruded stock should be quenched from 460 to 470 °C (860 to 880 °F).
Aging treatment. T6 temper: reheat to 120 °C (250 °F), hold at temperature for 20 to 24 h, air cool
Annealing treatment. O temper: heat to 415 to 455 °C (775 to 850 °F), soak for 2-3 h, air cool, reheat at 230 °C (450 °F), hold at temperature for 4 h, air cool. Or heat to 355 to 370 °C (670 to 700 °F), air cool
Stress-relief anneal. Heat to 355 to 370 °C (670 to 700 °F), soak for 2 h, air cool to room

Table 98 Mechanical properties of alloy 7049

Size and direction	Tensile strength(a) MPa	ksi	Yield strength (0.2% offset)(a) MPa	ksi	Elongation(a)(b), %	Compressive yield strength MPa	ksi	Shear strength MPa	ksi	Bearing strength(c) MPa	ksi	Bearing yield strength(a) MPa	ksi
Die forgings (AMS 4111), T73 temper													
Parallel to grain flow													
Up to 2 in., incl	496	72	427	62	7	441	64	283	41	917	133	662	96
Over 2–4 in., incl	490	71	421	61	7	434	63	276	40	903	131	655	95
Over 4–5 in., incl	483	70	414	60	7	427	62	269	39	890	129	641	93
Across grain flow													
Up to 1 in., incl	490	71	421	61	3	434	63	283	41	917	133	662	96
Over 1–4 in., incl	483	70	414	60	3–2	427	62	276	40	903	131	655	95
Over 4–5 in., incl	469	68	400	58	2	414	60	269	39	890	129	641	93
Extrusions (AMS 4157), T73511 temper													
Up to 2.999 in., incl													
Longitudinal	510	74	441	64	7	448	65	276	40	758	110	...	...
Long transverse	483	70	414	60	5	420	61	276	40	993	144	...	...
Over 2.999–5.000 in., incl													
Longitudinal	496	72	427	62	7	435	63	269	39	738	107	...	...
Long transverse	469	68	400	58	5	407	59	269	39	965	140	...	...
Extrusions (AMS 4159), T76511 temper													
Up to 2.999 in., incl													
Longitudinal	538	78	483	70	7	490	71	290	42	...	...	586	85
Long transverse	524	76	469	68	5	475	69	290	42	...	...	724	105
Over 2.999–5.000 in., incl													
Longitudinal	524	76	469	68	7	475	69	283	41	...	...	572	83
Long transverse	510	74	455	66	5	462	67	283	41	...	...	696	101

(a) Single values are minimum values. (b) In 50 mm (2 in.) or 4*d*, where *d* is diameter of reduced section of tensile test specimen. Where a range appears in this column, the specified minimum elongation varies with thickness of mill product. (c) *e*/*d* = 2.0, where *e* is edge distance and *d* is pin diameter

Table 99 Minimum mechanical properties of alloy 7050-T736 (or −T74) die forgings

Property	Thickness, in. Up to 2.000	2.001–4.000	4.001–5.000	5.001–6.000
Tensile strength, MPa (ksi)				
Longitudinal direction	496 (72)	490 (71)	483 (70)	483 (70)
Transverse direction	469 (68)	462 (67)	455 (66)	455 (66)
Yield strength, MPa (ksi)				
Longitudinal direction	427 (62)	421 (61)	414 (60)	405 (59)
Transverse direction	386 (56)	379 (55)	372 (54)	372 (54)
Compressive yield strength, MPa (ksi)				
Longitudinal direction	434 (63)	434 (63)	434 (63)	427 (62)
Transverse direction	400 (58)	393 (57)(a)	379 (55)	372 (54)
Shear strength	290 (42)	283 (41)	283 (41)	283 (41)
Bearing strength, MPa (ksi)				
e/*d* = 1.5	683 (99)	676 (98)	669 (97)	669 (97)
e/*d* = 2.0	903 (131)	889 (129)	876 (127)	876 (127)
Bearing yield strength, MPa (ksi)				
e/*d* = 1.5	565 (82)	558 (81)	545 (79)	538 (78)
e/*d* = 2.0	662 (96)	655 (95)	641 (93)	634 (92)
Elongation(b), %				
Longitudinal direction	7	7	7	7
Transverse direction	5	4	3	3

(a) For material 3.001 to 4.000 in. thick, 386 MPa (56 ksi). (b) In 50 mm (2 in.)

temperature

Weldability. Readily weldable by the direct-current inert-gas tungsten-arc (TIG) and by the metal-arc-inert-gas (MIG) process, using a weld-filler alloy of aluminum X5039 or 5183 rod. Has considerably better weld strength and ductility than 5083. Readily welded over a wide range of thicknesses with no decrease in weld ductility. Shows very good crack resistance in restrained plate weldments when joined with X5039 filler wire. Room temperature weld strength averages 360 MPa (52 ksi) and increases to 448 MPa (65 ksi)) at −195 °C (−320 °F). No special pre-weld or post-weld heat treatment is required.

Machinability. Good machinability in the annealed state. Soluble oil emulsions, kerosene, and kerosene-lard oil mixtures are recommended for most machining operations, but high viscosity lubricants are recommended for tapping operations.

Workability. Best formed in its freshly quenched condition. In the soft temper, the alloy can be successfully formed on all types of equipment. Because of its higher strength, a greater allowance for springback will have to be made than when working with other aluminum alloys. Use of heat up to 120 °C (250 °F) during forming in the annealed condition is beneficial in certain swaging, spinning, and drop hammer operations. In the solution treated condition the properties are intermediate between those of O and T6 temper, but definitely higher than O temper condition during the first few hours after quenching. Then formability gradually lessens as age-hardening increases. In the solution treated and aged T6 temper condition, the material exhibits very poor forming qualities. Due to the elaborate annealing and stabilizing treatment required, severe forming in its annealed O temper condition would be impractical. Rubber forming or streaking is usually conducted at 120 to 230 °C (250 to 450 °F).

Corrosion Resistance

The general corrosion resistance characteristics of 7039-T64 are comparable to such highly resistant aluminum-magnesium alloys as 5052, 5086, and 5083. Resistance to general corrosion is very much superior to that of most heat-treatable alloys. Under standard 6% NaCl immersion test for 6 mo or 5% NaCl salt fog, the alloy evidenced a slight superficial staining and a mild and shallow pitting attack with no measurable loss in strength. In a sodium chloride-hydrogen peroxide test, no evidence of intergranular corrosion was observed.

7049
7.6Zn-2.5Mg-1.5Cu-0.15Cr

Specifications

AMS. Extrusions: 4157, 4159. Forgings: 4111

UNS number. A97049

Government. Forgings: QQ-A-367, MIL-H-6088

Chemical Composition

Composition limits. 1.2 to 1.9 Cu, 2.0 to 2.9 Mg, 0.20 Mn max, 0.25 Si max, 0.35 Fe max, 0.10 to 0.22 Cr, 7.2 to 8.2 Zn, 0.10 Ti max, 0.05 max other (each), 0.15 max others (total), bal Al

Applications

Typical uses. Forged aircraft and missile fittings, landing gear cylinders, and extruded sections. Used where static strengths approximately the same as forged 7079-T6

Table 100 Minimum mechanical properties of alloy 7050-T73652 hand forgings

Property	Thickness, in. Up to 2.000	2.001–3.000	3.001–4.000	4.001–5.000	5.001–6.000	6.001–7.000	7.001–8.000
Tensile strength, MPa (ksi)							
Longitudinal direction	496 (72)	496 (72)	490 (71)	483 (70)	476 (69)	469 (68)	462 (67)
L-T direction	490 (71)	483 (70)	483 (70)	476 (69)	469 (68)	462 (67)	455 (66)
S-T direction	...	462 (67)	462 (67)	455 (66)	455 (66)	448 (65)	441 (64)
Yield strength, MPa (ksi)							
Longitudinal direction	434 (63)	427 (62)	421 (61)	414 (60)	407 (59)	400 (58)	393 (57)
L-T direction	421 (61)	414 (60)	407 (69)	400 (58)	386 (56)	372 (54)	359 (52)
S-T direction	...	379 (55)	379 (55)	372 (54)	365 (53)	352 (51)	345 (50)
Compressive yield strength, MPa (ksi)							
Longitudinal direction	441 (64)	434 (63)	427 (62)	421 (61)	414 (60)	407 (59)	400 (58)
L-T direction	448 (65)	441 (64)	434 (63)	427 (62)	414 (60)	400 (58)	386 (56)
S-T direction	...	421 (61)	421 (61)	414 (60)	407 (59)	393 (57)	379 (55)
Shear strength, MPa (ksi)	290 (42)	283 (41)	283 (41)	283 (41)	276 (40)	269 (39)	269 (39)
Bearing strength, MPa (ksi)							
e/*d* = 1.5	689 (100)	683 (99)	683 (99)	669 (97)	662 (96)	655 (95)	641 (93)
e/*d* = 2.0	903 (131)	896 (130)	896 (130)	883 (128)	869 (126)	855 (124)	841 (122)
Bearing yield strength, MPa (ksi)							
e/*d* = 1.5	593 (86)	586 (85)	572 (83)	565 (82)	545 (79)	524 (76)	503 (73)
e/*d* = 2.0	696 (10)	689 (100)	676 (98)	662 (96)	641 (93)	621 (90)	593 (86)
Elongation, %							
Longitudinal direction	9	9	9	9	9	9	9
L-T direction	5	5	5	4	4	4	4
S-T direction	...	4	4	3	3	3	3

Table 101 Typical mechanical properties of alloy 7050

Temperature, °C	Temperature, °F	Time at temp, h	At indicated temperature: Tensile strength, MPa	ksi	Yield strength, MPa	ksi	Elongation(a), %	At room temperature after heating: Tensile strength, MPa	ksi	Yield strength, MPa	ksi	Elongation(a), %
T73651 plate												
24	75	...	510	74	455	66	11	510	74	455	66	11
100	212	0.1–10	441	64	427	62	13	510	74	455	66	11
		100	448	65	434	63	13	510	74	462	67	12
		1 000	441	64	427	62	14	510	74	455	66	12
		10 000	441	64	421	61	15	510	74	441	64	12
149	300	0.1	393	57	386	56	16	510	74	455	66	11
		0.5	393	57	386	56	17	510	74	448	65	12
		10	393	57	386	56	18	503	74	441	64	12
		100	359	52	332	51	19	483	70	407	59	13
		1 000	290	42	276	40	21	407	59	317	46	13
		10 000	221	32	193	28	29	331	48	228	33	14
177	350	0.1	359	52	345	50	19	510	74	448	65	12
		0.5	352	51	345	50	20	496	72	441	64	12
		10	324	47	310	45	22	469	68	400	58	13
		100	248	36	234	34	25	386	56	296	43	13
		1 000	193	28	172	25	31	317	46	214	31	14
		10 000	159	23	124	18	40	248	36	152	22	15
204	400	0.1	303	44	290	42	22	490	71	434	63	12
		0.5	290	42	276	40	23	469	68	421	61	12
		10	221	32	207	30	27	386	56	283	41	13
		100	165	24	152	22	32	317	46	200	29	14
		1 000	131	19	110	16	45	262	38	138	20	16
		10 000	117	17	90	13	54	234	34	117	17	19
T73652 forgings												
−196	−320	...	662	96	572	83	13	...	...	...	...	
−80	−112	...	586	85	503	73	14	...	...	...	...	
−28	−18	...	552	80	476	69	15	...	...	...	...	
24	75	...	524	76	455	66	15	524	76	455	66	15
100	212	0.1–10	462	67	427	62	16	524	76	455	66	15
		100	469	68	434	63	16	524	76	462	67	15
		1 000	462	67	427	62	17	524	76	524	76	16
		10 000	462	67	421	61	17	517	75	517	75	16
149	300	0.1	414	60	386	56	17	517	75	455	66	15
		0.5	414	60	386	56	17	510	74	448	65	15
		10	407	59	386	56	18	503	73	441	64	16
		100	365	53	352	51	20	483	70	407	59	16
		1 000	290	42	276	40	23	407	59	317	46	17
		10 000	221	32	193	28	29	331	48	228	33	17
177	350	0.1	379	55	345	50	19	510	74	448	65	15
		0.5	365	53	345	50	20	496	72	441	64	15
		10	324	47	310	45	22	469	68	400	58	16
		100	248	36	234	34	25	386	56	296	43	17
		1 000	193	28	172	25	31	317	46	214	31	17
		10 000	159	23	124	18	40	248	36	152	22	18
204	400	0.1	324	47	290	42	22	503	73	434	63	15
		0.5	296	43	276	40	23	483	70	421	61	15
		10	221	32	207	30	27	386	56	283	41	16
		100	165	24	152	22	32	317	46	200	29	17
		1 000	131	19	110	16	45	262	38	138	20	19
		10 000	117	17	90	13	54	234	34	117	17	22

(a) In 50 mm (2 in.)

and high resistance to stress-corrosion cracking are required. Fatigue characteristics about equal to those of 7075-T6 products, toughness somewhat higher

Precautions in use. Poor general corrosion resistance

Mechanical Properties

Tensile property limits. See Table 98.
Shear strength. See Table 98.
Compressive strength. See Table 98.
Bearing strength. See Table 98.
Hardness. 135 HB min with 500 kg load, 10 mm diam ball
Poisson's ratio. 0.33
Elastic modulus. Forgings, typical: tension, 70 GPa (10.2×10^6 psi). Extrusions, typical: tension, 72.5 GPa (10.5 ksi); shear, 27.6 GPa (4.0 ksi); compression, 76 GPa (11 ksi)
Fatigue strength. Axial fatigue at stress ratio R of 1.0 for material in the T73 temper. Smooth specimens from 125 mm (5 in.) thick forgings: 275 to 315 MPa (40 to 46 ksi) at 10^7 cycles for temperatures from room temperature to 175 °C (350 °F). Notched specimens from 75 mm (3 in.) thick forgings: 390 MPa (56 ksi) for K_t of 1.0; 115 MPa (17 ksi) for K_t of 3.0; both at 10^7 cycles
Plane-strain fracture toughness. K_Q values from compact tension tests of 7049-T73 die forgings: L-S orientation, 32 to 36 MPa$\sqrt{m}$ (29 to 33 ksi$\sqrt{in.}$); L-T orientation, 31 to 40 MPa$\sqrt{m}$ (28 to 37 ksi$\sqrt{in.}$); S-L orientation, 21 to 27 MPa$\sqrt{m}$ (19 to 25 ksi$\sqrt{in.}$)

Mass Characteristics

Density. 2.82 g/cm^3 (0.102 lb/in.^{3}) at 20 °C (68 °F)

Thermal Properties

Liquidus temperature. 627 °C (1160 °F)
Solidus temperature. 477 °C (890 °F)
Coefficient of thermal expansion. Linear, 23.4 μm/m · K (13.0 μin./in. · °F) at 20 to 100 °C (68 to 212 °F)
Specific heat. 960 J/kg · K (0.23 Btu/lb · °F) at 100 °C (212 °F)
Thermal conductivity. 154 W/m · K (89 Btu/ft · h · °F) at 25 °C (77 °F)

Table 102 Typical axial fatigue strength at 10^7 cycles for alloy 7050

Product and temper	Stress ratio, R	Fatigue strength (max stress): Smooth specimens, MPa	Smooth specimens, ksi	Notched specimens(a), MPa	Notched specimens(a), ksi
Plate, 25–150 mm (1 to 6 in.) thick					
T6 type tempers	0.0	190–290	28–42	...	...
T73*xxx* tempers	0.0	170–300	24–44	50–90	7.5–13
Extrusions, 29.5 mm (1.16 in.) thick					
T76511 temper	0.5	320–340	46–50	110–125	16–18
	0.0	180–210	26–30	70–80	10–12
	−1.0	130–150	19–22	35–50	5–7
Die forgings, 25–150 mm (1 to 6 in.) thick					
T736 temper	0.0	210–275	30–40	75–115	11–17
Hand forgings, 144 × 559 × 2130 mm (4½ × 22 × 84 in.)					
T73652 temper					
Longitudinal	0.5	325	47	145	21
	0.0	225	33	90	13
	−1.0	145	21	50	7
Long transverse	0.5	275	40	115	17
	0.0	170	25	90	13
	−1.0	125	18	50	7
Short transverse	0.5	260	38	115	17
	0.0	170	25	60	9
	−1.0	115	17	50	7

(a) Notch fatigue factor, K_t, of 3.0

Table 103 Plane-strain fracture toughness of alloy 7050

Temper and orientation	Minimum, MPa$\sqrt{m}$	Minimum, ksi$\sqrt{in.}$	Average, MPa$\sqrt{m}$	Average, ksi$\sqrt{in.}$
Plate				
T73651				
L-T	26.4	24	35.2	32
T-L	24.2	22	29.7	27
S-L	22.0	20	28.6	26
Extrusions				
T7651X				
L-T	...	...	30.8	28
T-L	...	...	26.4	24
S-L	...	...	20.9	19
T7351X				
L-T	...	...	45.1	41
T-L	...	...	31.9	29
S-L	...	...	26.4	24
Die forgings				
T736				
L-T	27.5	25	36.3	33
T-L, S-L	20.9	19	25.3	23
Hand forgings				
T73652				
L-T	29.7	27	36.3	33
T-L	18.7	17	23.1	21
S-L	17.6	16	22.0	20

Electrical Properties

Electrical conductivity. Volumetric, 40% IACS min at 20 °C (68 °F)
Electrical resistivity. 43 nΩ · m

7050
6.2Zn-2.3Mg-2.3Cu-0.12Zr

Specifications

AMS. 4050, 4107, 4108
UNS number. A97050

Chemical Composition

Composition limits. 2.0 to 2.6 Cu, 1.9 to 2.6 Mg, 0.10 Mn max, 0.12 Si max, 0.15 Fe max, 0.04 Cr max, 0.08 to 0.15 Zr, 5.7 to 6.7 Zn, 0.06 Ti max, 0.05 max other (each), 0.15 max others (total)
Consequence of exceeding impurity limits. Excess Fe and Si degrade fracture toughness. Increased sensitivity to quenching rate due to excess Mn and Cr results in low strength in thick sections.

Applications

Typical uses. Plate, extrusions, hand and die forgings in aircraft structural parts. Other applications requiring very high strength coupled with high resistance to exfoliation corrosion and stress-corrosion cracking, high fracture toughness and fatigue resistance

Mechanical Properties

Tensile properties. See Tables 99 to 101.
Shear properties. See Tables 99 and 100.
Compressive properties. See Tables 99 and 100.
Bearing properties. See Tables 99 and 100.
Poisson's ratio. 0.33
Elastic modulus. Tension, 70.3 GPa (10.2 × 10^6 psi); shear, 26.9 GPa (3.9 × 10^6 psi); compression, 73.8 GPa (10.7 × 10^6 psi)
Fatigue strength. See Table 102.
Plane-strain fracture toughness. See Table 103.
Creep-rupture characteristics. See Table 104.

Table 104 Creep and rupture properties of alloy 7050-T3651 plate

Temperature, °C	Temperature, °F	Time under stress, h	Rupture stress, MPa	Rupture stress, ksi	Creep 1.0%, MPa	1.0%, ksi	0.5%, MPa	0.5%, ksi	0.2%, MPa	0.2%, ksi	0.1%, MPa	0.1%, ksi
24	75	0.1	510	74	496	72	476	69	455	66	448	65
		1	503	73	483	70	462	67	448	65	441	64
		10	490	71	469	68	455	66	441	64	441	64
		100	476	69	455	66	448	65	441	64	434	63
		1000	469	68	448	65	441	64	...	...	...	...
100	212	0.1	441	64	434	63	427	62	421	61	414	60
		1	427	62	414	60	407	59	400	58	386	56
		10	407	59	393	57	386	56	372	54	359	52
		100	379	55	372	54	365	53	345	50	331	48
		1000	359	52	352	51	345	50	317	46	...	...
149	300	0.1	372	54	365	53	359	52	345	50	324	47
		1	345	50	338	49	324	47	303	44	290	42
		10	310	45	303	44	290	42	269	39	228	33
		100	262	38	255	37	241	35	193	28	152	22
		1000	179	26	179	26	165	24	145	21	124	18

Mass Characteristics

Density. 2.83 g/cm³ (0.102 lb/in.³) at 20 °C (68 °F)

Table 105 Mechanical-property limits for alloy 7072 fin stock

Temper	Tensile strength, Minimum, MPa	Minimum, ksi	Maximum, MPa	Maximum, ksi	Yield strength (min), MPa	Yield strength (min), ksi	Elongation (min), %(a)
O	55	8.0	90	13.0	21	3	15–20
H14	97	14.0	131	19.0	83	12	1–3
H18	131	19.0	...	...	...	...	1–2
H19	145	21.0	...	...	...	...	1
H25	107	15.5	148	21.5	83	12	2–3
H111, H211	62	9.0	97	14.0	41	6.0	12

(a) In 50 mm (2 in.). Where a range of values appears in this column, specified minimum elongation varies with thickness of the mill product.

Thermal Properties

Liquidus temperature. 635 °C (1175 °F)
Solidus temperature. 465 °C (870 °F)
Incipient melting temperature. 488 °C (910 °F) for homogenized (solution treated) wrought material
Eutectic temperature. 465 °C (870 °F) for

Table 106 Standard specifications for alloy 7075

Mill form and condition	AMS	ASTM	Government
Bare products			
Sheet and plate	4038	B 209	QQ-A-250/2
	4044	...	...
	4045	...	...
	4078	...	...
Wire, rod, and bar (rolled or cold finished)	4122	B 211	QQ-A-225/9
	4123	...	...
	4124	...	...
Rod, bar, shapes, and tube (extruded)	4154	B 221	QQ-A-200/11
	4167	...	...
	4168	...	...
	4169	...	...
Tube (extruded, seamless)	...	B 241	...
Tube (drawn, seamless)	...	B 210	...
Forgings and forging stock	4139	B 247	QQ-A-367
	...	...	MIL-A-22771
Impacts	4170	...	MIL-A-12545
Rivets	...	B 316	QQ-A-430
Alclad products			
Sheet and plate	4039	B 209	QQ-A-250/13
	4048	...	...
	4049	...	...
Tapered sheet and plate	4047	...	...
Alclad one side products			
Sheet and plate	4046	B 209	QQ-A-250/18

Table 107 Typical tensile properties for alloy 7075 at various temperatures

Temperature °C	Temperature °F	Tensile strength(a) MPa	Tensile strength(a) ksi	Yield strength (0.2% offset)(a) MPa	Yield strength (0.2% offset)(a) ksi	Elongation(b), %
T6, T651 tempers						
−196	−320	703	102	634	92	9
−80	−112	621	90	545	79	11
−28	−18	593	86	517	75	11
24	75	572	83	503	73	11
100	212	483	70	448	65	14
149	300	214	31	186	27	30
204	400	110	16	87	13	55
260	500	76	11	62	9	65
316	600	55	8	45	6.5	70
271	700	41	6	32	4.6	70
T73, T7351 tempers						
−196	−320	634	92	496	72	14
−80	−112	545	79	462	67	14
−28	−18	524	76	448	65	13
24	75	503	73	434	63	13
100	212	434	63	400	58	15
149	300	214	31	186	27	30
204	400	110	16	90	13	55
260	500	76	11	62	9	65
316	600	55	8	45	6.5	70
371	700	41	6	32	4.6	70

(a) Lowest strength for exposures up to 10 000 h at temperature, no load; test loading applied at 35 MPa/min (5 ksi/min) to yield strength and then at strain rate of 5%/min to fracture. (b) In 50 mm (2 in.)

unhomogenized wrought or as-cast material
Coefficient of thermal expansion. Linear:

Temperature range °C	Temperature range °F	Average coefficient μm/m · K	Average coefficient μin./in. · °F
−50 to 20	−58 to 68	21.7	12.1
20 to 100	68 to 212	23.5	13.1
20 to 200	68 to 392	24.4	13.6
20 to 300	68 to 572	25.4	14.1

Volumetric: 68.0 × 10^{-6} m³/m³ · K (3.78 × 10^{-5} in.³/in.³ · °F) at 20 °C (68 °F)
Specific heat. 860 J/kg · K (0.206 Btu/lb · °F) at 20 °C (68 °F)
Thermal conductivity. At 20 °C (68 °F): O temper, 180 W/m · K (104 Btu/ft · h · °F); T76, T7651 tempers, 154 W/m · K (89 Btu/ft · h · °F); T736, T73651 tempers, 157 W/m · K (91 Btu/ft · h · °F)

Electrical Properties

Electrical conductivity. Volumetric, at 20 °C (68 °F): O temper, 47% IACS; T76, T7651 tempers, 39.5% IACS; T736, T73651 tempers, 40.5% IACS
Electrical resistivity. At 20 °C (68 °F): O temper, 36.7 nΩ · m; T76, T7651 tempers, 43.6 nΩ · m; T736, T73651 tempers, 42.6 nΩ · m. Temperature coefficient, all tempers: 0.1 nΩ · m per K at 20 °C (68 °F)

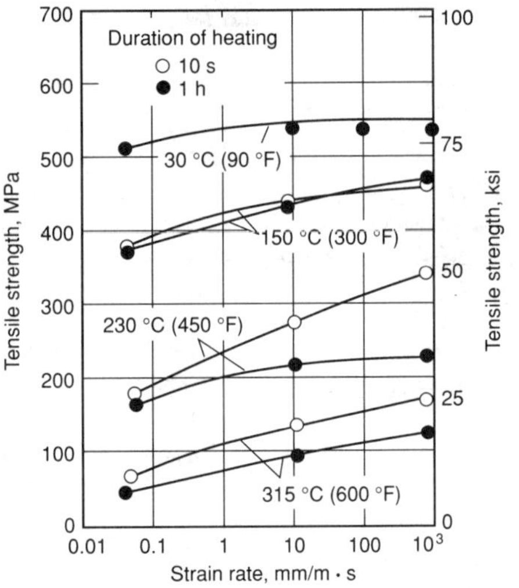

Fig. 15 Effect of strain rate and temperature on tensile strength of alloy 7075-T6

Fabrication Characteristics

Annealing temperature. 415 °C (775 °F)
Solution temperature. 475 °C (890 °F)
Aging temperature. 120 to 175 °C (250 to 350 °F)

7072
1.0Zn

Specifications

ASTM. B 209
SAE. J454
UNS number. A97072

Chemical Composition

Composition limits. 0.10 Cu max, 0.10 Mg max, 0.10 Mn max, 0.7 Si max + Fe, 0.8 to 1.3 Zn, 0.05 max other (each), 0.15 max others (total), bal Al

Applications

Typical uses. Fin stock. Cladding alloy for Alclad sheet, plate, and tube products with the following core alloys: 2219, 3003, 3004, 5050, 5052, 5154, 6061, 7075, 7475, 7178

Mechanical Properties

Tensile properties. See Table 105.
Shear strength. O temper, 55 MPa (8 ksi); H12 temper, 62 MPa (9 ksi); H14 temper, 69 MPa (10 ksi)
Hardness. O temper, 20 HB; H12 temper, 28 HB; H14 temper, 32 HB; all values obtained with 500 kg load, 10 mm diam ball, and 30 s duration of loading
Poisson's ratio. 0.33
Elastic modulus. Tension, 68 GPa (9.9 × 10^6 psi); compression, 70 GPa (10.1 × 10^6 psi)

Mass Characteristics

Density. 2.72 g/cm³ (0.098 lb/in.³) at 20 °C (68 °F)

Table 108 Tensile properties of alloy 7075

Temper	Tensile strength, MPa	Tensile strength, ksi	Yield strength, MPa	Yield strength, ksi	Elongation(a), %
Typical properties					
O	228	33	103	15	17
T6, T651	572	83	503	73	11
T73	503	73	434	63	...
Alclad O	221	32	97	14	17
T6, T651	524	76	462	67	11
Property Limits	Minimum		Minimum		Minimum
Sheet and plate					
O	276 (max)	40 (max)	145 (max)	21 (max)	10
Sheet					
T6, T62					
0.008–0.011 in. thick	510	74	434	63	5
0.012–0.039 in. thick	524	76	462	67	7
0.040–0.125 in. thick	538	78	469	68	8
0.126–0.249 in. thick	538	78	476	69	8
T73	462	67	386	56	8
T76	503	73	427	62	8
Plate					
T62, T651					
0.250–0.499 in. thick	538	78	462	67	9
0.500–1.000 in. thick	538	78	469	68	7
1.001–2.000 in. thick	531	77	462	67	6
2.001–2.500 in. thick	524	76	441	64	5
2.501–3.000 in. thick	496	72	421	61	5
3.001–3.500 in. thick	490	71	400	58	5
3.501–4.000 in. thick	462	67	372	54	3
T7351					
0.250–2.000 in. thick	476	69	393	57	6–7
2.001–2.500 in. thick	455	66	359	52	6
2.501–3.000 in. thick	441	64	338	49	6
T7651					
0.250–0.499 in. thick	496	72	421	61	8
0.500–1.000 in. thick	490	71	414	60	6
Alclad sheet and plate					
O					
0.008–0.062 in. thick	248 (max)	36 (max)	138 (max)	20 (max)	9–10
0.063–0.187 in. thick	262 (max)	38 (max)	138 (max)	20 (max)	10
0.188–0.499 in. thick	269 (max)	39 (max)	145 (max)	21 (max)	10
0.500–1.000 in. thick	276 (max)	40 (max)	...	...	10
Alclad sheet					
T6, T62					
0.008–0.011 in. thick	469	68	400	58	5
0.012–0.039 in. thick	483	70	414	60	7
0.040–0.062 in. thick	496	72	427	62	8
0.063–0.187 in. thick	503	73	434	63	8
0.188–0.249 in. thick	517	75	441	64	8
T73					
0.040–0.062 in. thick	434	63	352	51	8
0.063–0.187 in. thick	441	64	359	52	8
0.188–0.249 in. thick	455	66	372	54	8
T76					
0.125–0.187 in. thick	469	68	393	57	8
0.188–0.249 in. thick	483	70	407	59	8
Alclad plate					
T62, T651					
0.250–0.499 in. thick	517	75	448	65	9
0.500–1.000 in. thick	538(b)	78(b)	469(b)	68(b)	7
1.001–2.000 in. thick	531(b)	77(b)	462(b)	67(b)	6
2.001–2.500 in. thick	524(b)	76(b)	441(b)	64(b)	5
2.501–3.000 in. thick	496(b)	72(b)	421(b)	61(b)	5
3.001–3.500 in. thick	490(b)	71(b)	400(b)	58(b)	5
3.501–4.000 in. thick	462(b)	67(b)	372(b)	54(b)	3
T7351					
0.250–0.499 in. thick	455	66	372	54	8
0.500–1.000 in. thick	476	69	393	57	7
T7651					
0.250–0.499 in. thick	476	69	400	58	8
0.500–1.000 in. thick	490(b)	71(b)	414(b)	60(b)	6

(a) In 50 mm (2 in.) or 4*d*, where *d* is diameter of reduced section of tensile test specimen. Where a range appears in this column, the specified minimum elongation varies with thickness of the mill product. (b) For plate 13 mm (0.500 in.) or over in thickness, listed properties apply to core material only. Tensile and yield strengths of composite plate are slightly lower than listed value, depending on thickness of cladding.

Thermal Properties

Liquidus temperature. 657 °C (1215 °F)
Solidus temperature. 641 °C (1185 °F)
Coefficient of thermal expansion. Linear:

Temperature range, °C	Temperature range, °F	Average coefficient, μm/m · K	Average coefficient, μin./in. · °F
−50 to 20	−58 to 68	21.8	12.1
20 to 100	68 to 212	23.6	13.1
20 to 200	68 to 392	24.5	13.6
20 to 300	68 to 572	25.5	14.2

Volumetric: 68×10^{-3} m³/m³ · K (3.78×10^{-5} in.³/in.³ · °F) at 20 °C (68 °F)
Specific heat. 893 J/kg · K (0.213 Btu/lb · °F) at 20 °C (68 °F)
Thermal conductivity. O temper: 227 W/m · K (131 Btu/ft · h · °F) at 20 °C (68 °F)

Electrical Properties

Electrical conductivity. Volumetric, O temper: 60% IACS at 20 °C (68 °F)
Electrical resistivity. 28.7 nΩ · m at 20 °C (68 °F); temperature coefficient, 0.1 nΩ · m per K at 20 °C (68 °F)
Electrolytic solution potential. −0.96 V versus 0.1 *N* calomel electrode in an aqueous solution containing 53 g NaCl plus 3 g H_2O_2 per liter at 25 °C (77 °F)

Chemical Properties

General corrosion behavior. High resistance to general corrosion. Provides galvanic protection when used as cladding on several different alloys

Fabrication Characteristics

Annealing temperature. 345 °C (650 °F)

7075, Alclad 7075
5.6Zn-2.5Mg-1.6Cu-0.23Cr

Specifications

AMS. See Table 106.
ASTM. See Table 106.
SAE. J454
UNS number. A97075
Government. See Table 106.
Foreign. Austria: Önorm AlZnMg-Cu1.5. Canada: CSA ZG62, ZG62Alclad. France: NF A-Z5GU. Spain: UNE L-371. Switzerland: VSM Al-Zn-Mg-Cu; Alclad, Al-Zn-Mg-Cu-pl. United Kingdom: BS L.95, L.96. Germany: DIN AlZnMgCu1.5; Werkstoff-Nr. 3.4365. ISO: AlZn6MgCu

Chemical Composition

Composition limits of 7075. 1.20 to 2.0 Cu, 2.1 to 2.9 Mg, 0.30 Mn max, 0.40 Si max, 0.50 Fe max, 0.18 to 0.28 Cr, 5.1 to 6.1 Zn, 0.20 Ti max, 0.05 max other (each), 0.15 max others (total), bal Al
Composition limits of Alclad 7075. 7072 cladding—0.10 Cu max, 0.10 Mg max, 0.10 Mn max, 0.7 Si max + Fe, 0.8 to 1.3 Zn, 0.05 max other (each), 0.15 max others (total), bal Al

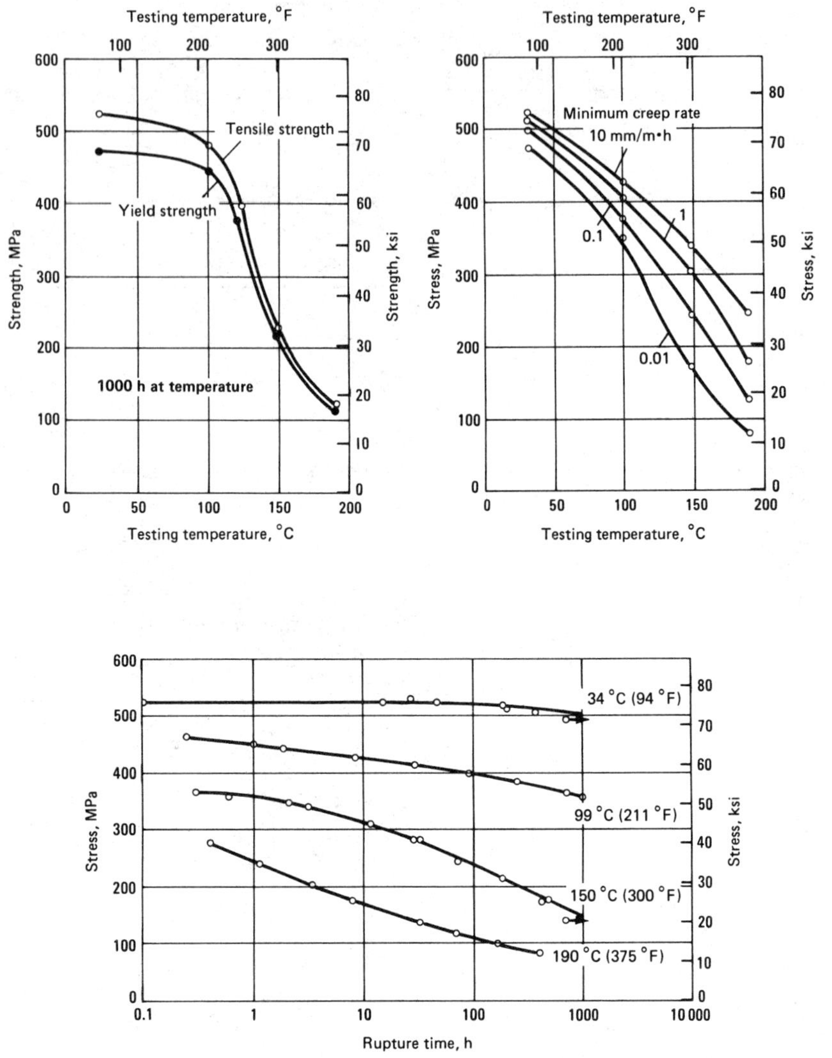

Fig. 16 Effect of temperature on tensile properties of Alclad 7075-T6

Applications

Typical uses. Aircraft structural parts and other highly stressed structural applications where very high strength and good resistance to corrosion are required
Precautions in use. Caution should be exercised in T6 temper applications where sustained tensile stresses are encountered, either residual or applied, particularly in the transverse grain direction. In such instances, the T73 temper should be considered, at some sacrifice in tensile strength.

Mechanical Properties

Tensile properties. See Tables 107 and 108.
Shear strength. Bare and Alclad products, O temper: 152 MPa (22 ksi). Bare products—T6, T651 tempers: 331 MPa (48 ksi); Alclad T6, T651: 317 MPa (46 ksi)
Hardness. O temper, 60 HB; T6, T651 temper, 150 HB; data obtained using 500 kg load, 10 mm diam ball, and 30 s duration of loading
Poisson's ratio. 0.33
Elevated-temperature effects. See Fig. 15 and 16.
Elastic modulus. Tension, 71.0 GPa (10.3 × 10^6 psi); shear, 26.9 GPa (3.9 × 10^6 psi); compression, 72.4 GPa (10.5 × 10^6 psi)
Fatigue strength. T6, T651, T73 tempers: 159 MPa (23 ksi) at 5 × 10^8 cycles in R.R. Moore type test of smooth (unnotched) specimens
Plane-strain fracture toughness. See Table 109.
Directional properties. Transverse mechanical properties of many products, particularly tensile strength and ductility in the short transverse direction, are less than those in the longitudinal direction.

Mass Characteristics

Density. 2.80 g/cm^3 (0.101 lb/in.^{3}) at 20 °C (68 °F)

Thermal Properties

Liquidus temperature. 635 °C (1175 °F)
Solidus temperature. 477 °C (890 °F); eutectic temperature for nonhomogeneous as-cast or wrought material that has not been solution heat treated
Incipient melting temperature. 532 °C (990 °F) for homogenized (solution heat treated) wrought material
Coefficient of thermal expansion. Linear:

Temperature range °C	Temperature range °F	Average coefficient µm/m · K	Average coefficient µin./in. · °F
−50 to 20	−58 to 68	21.6	12.0
20 to 100	68 to 212	23.4	13.0
20 to 200	68 to 392	24.3	13.5
20 to 300	68 to 572	25.2	14.0

Volumetric: 68 × 10^{-6} m^3/m^3 · K (3.78 × 10^{-5} in.^{3}/in.^{3} · °F) at 20 °C (68 °F)
Specific heat. 960 J/kg · K (0.23 Btu/lb · °F) at 100 °C (212 °F)
Thermal conductivity. At 20 °C (68 °F). T6, T62, T651, T652 tempers: 130 W/m · K (75 Btu/ft · h · °F). T76, T7651 tempers: 150 W/m · K (87 Btu/ft · h · °F). T73, T7351, T7352 tempers: 155 W/m · K (90 Btu/ft · h · °F)

Electrical Properties

Electrical conductivity. Volumetric, at 20 °C (68 °F). T6, T62, T651, T652 tempers: 33% IACS. T76, T7651 tempers: 38.5% IACS. T73, T7351, T7352 tempers: 40% IACS
Electrical resistivity. At 20 °C (68 °F). T6, T62, T651, T652 tempers: 52.2 nΩ · m. T76, T7651 tempers: 44.8 nΩ · m. T73, T7351, T7352 tempers: 43.1 nΩ · m. Temperature coefficient, all tempers: 0.1 nΩ · m per K at 20 °C (68 °F)

Fabrication Characteristics

Annealing temperature. 415 °C (775 °F)
Solution temperature. 465 to 480 °C (870 to 900 °F), depending on product
Aging temperature. T6 temper: 120 °C (250 °F); T7 temper: two-stage treatment—107 °C (225 °F) followed by 163 to 177 °C (325 to 350 °F), depending on product

7076
7.5Zn-1.6Mg-0.55Mn-0.65Cu

Specifications

AMS. 4137
ASTM. B 247
Government. QQ-A-367, MIL-A-8097

Chemical Composition

Composition limits. 7.0 to 8.0 Zn, 1.2 to 2.0 Mg, 0.30 to 0.80 Mn, 0.3 to 1.0 Cu, 0.40 Si max, 0.60 Fe max, 0.20 Ti max, 0.05 max other (each), 0.15 max others (total), bal Al

Applications

Typical uses. Aircraft propellers
Available forms. Forgings

Table 109 Typical plane-strain fracture toughness of alloy 7075

Product and temper	Minimum MPa$\sqrt{m}$	Minimum ksi$\sqrt{in.}$	Average MPa$\sqrt{m}$	Average ksi$\sqrt{in.}$	Maximum MPa$\sqrt{m}$	Maximum ksi$\sqrt{in.}$
L-T orientation						
Plate						
T651	27.5	25	28.6	26	29.7	27
T7351	...	...	33.0	30	...	...
Extruded shapes						
T6510,1	28.6	26	30.8	28	35.2	32
T7310,1	34.1	31	36.3	33	37.4	34
Forgings						
T652	26.4	24	28.6	26	30.8	28
T7352	29.7	27	34.1	31	38.5	35
T-L orientation						
Plate						
T651	22.0	20	24.2	22	25.3	23
T7351	27.5	25	31.9	29	36.3	33
Extruded shapes						
T6510,1	20.9	19	24.2	22	28.6	26
T7310,1	24.2	22	26.4	24	30.8	28
Forgings						
T652	...	...	25.3	23	...	...
T7352	25.3	23	27.5	25	28.6	26
S-L orientation						
Plate						
T651	16.5	15	17.6	16	19.8	18
T7351	20.9	19	22.0	20	23.1	21
Extruded shapes						
T6510,1	19.8	18	20.9	19	24.2	22
T7310,1	...	...	22.0	20	...	...
Forgings						
T651	...	...	18.7	17	...	...
T7351	20.9	19	23.1	21	27.5	25

Mechanical Properties

Tensile properties. T61 temper: tensile strength of 485 MPa (70 ksi), yield strength (0.2% offset) of 415 MPa (60 ksi), and elongation of 14% in 50 mm (2 in.)
Hardness. T61 temper: 140 HB (500 kg load).
Modulus of elasticity. 67 GPa (9.7×10^6 psi)

Physical Properties

Density. 2.82 g/cm^3 (0.102 lb/in.^{3})
Coefficient of thermal expansion. 21.6 μm/m · °C (12 μin./in. · °F) from 21 to 100 °C (70 to 212 °F)
Electrical conductivity. 35% IACS (volumetric)

Fabrication Characteristics

Solution anneal. T4 temper: heat to 493 °C (920 °F), quench in water.
Precipitation treatment. T6 temper: after solution anneal, heat at 120 °C (250 °F) for 24 h, air cool
Annealing treatment. O temper; heat to 415 to 455 °C (775 to 850 °F), soak for 2 h, air cool, reheat at 232 °C (450 °F), hold at temperature for 4 h, air cool. Or heat to 355 to 370 °C (670 to 700 °F), soak for 2 h, air cool to 232 °C (450 °F), soak 4 h at 232 °C (450 °F), air cool

Table 110 Typical mechanical properties of alloy 7175-T736 die forgings up to 75 mm (3 in.) thick

Temperature °C	Temperature °F	Time at temperature, h	At indicated temperature: Tensile strength MPa	Tensile strength ksi	Yield strength MPa	Yield strength ksi	Elongation(a), %	At room temperature after heating: Tensile strength MPa	Tensile strength ksi	Yield strength MPa	Yield strength ksi	Elongation(a), %
−253	−423	...	876	127	745	108	12	...	...	...	...	...
−196	−320	...	731	106	676	98	13	...	...	...	...	...
−80	−112	...	621	90	572	83	14	...	...	...	...	...
−28	−18	...	600	87	552	80	16	...	...	...	...	...
24	75	...	552	80	503	73	14	552	80	503	73	14
100	212	0.1	490	71	476	69	14	552	80	503	73	14
		0.5	490	71	462	67	15	552	80	503	73	14
		10	496	72	476	69	16	552	80	510	74	14
		100	503	73	483	70	16	558	81	510	74	14
		1 000	503	73	483	70	17	565	82	517	75	14
		10 000	496	72	476	69	17	558	81	503	73	14
149	300	0.1	427	62	414	60	20	552	80	503	73	15
		0.5	427	62	414	60	18	552	80	503	73	15
		10	427	62	414	60	20	552	80	496	72	15
		100	393	57	372	54	25	524	76	462	67	16
		1 000	310	45	296	43	30	441	64	359	52	17
		10 000	241	35	214	31	30	352	51	248	36	18
176	350	0.1	365	53	345	50	20	538	78	490	71	14
		0.5	379	55	345	50	25	538	78	483	70	14
		10	338	49	324	47	25	496	72	427	62	16
		100	262	38	241	35	25	421	61	331	48	16
		1 000	200	29	179	26	35	331	48	228	33	18
		10 000	165	24	131	19	55	262	38	152	22	20
204	400	0.1	324	47	303	44	20	524	76	469	68	16
		0.5	310	45	283	41	30	503	73	427	62	14
		10	228	33	214	31	35	393	57	296	43	16
		100	165	24	221	32	35	317	46	207	30	18
		1 000	124	18	103	15	45	255	37	138	20	20
		10 000	124	18	90	13	65	234	34	110	16	25
232	450	0.1	262	38	241	35	20	510	74	441	64	16
		0.5	228	33	214	31	25	448	65	359	52	16
		10	159	23	145	21	35	338	49	228	33	17
		100	117	17	103	15	40	269	39	145	21	19
		1 000	97	14	83	12	45	234	34	103	15	25
		10 000	90	13	76	11	50	221	32	97	14	25

(a) In 50 mm (2 in.)

Table 111 Plane-strain fracture toughness of alloy 7175-T736 forgings

Temper and orientation	Plane-strain fracture toughness: Minimum, MPa√m	Minimum, ksi√in.	Average, MPa√m	Average, ksi√in.
Die forgings				
T736				
L-T	29.7	27	33.0	30
T-L, S-L	23.1	21	28.6	26
Hand forgings				
T736				
L-T	33.0	30	37.4	34
T-L	27.5	25	29.7	27
S-L	23.1	21	26.4	24

Stress-relief anneal. Heat to 355 to 370 °C (670 to 700 °F), soak for 2 h, air cool to room temperature

7175
5.6Zn-2.5Mg-1.6Cu-0.23Cr

Commercial Names

Trade name. AA7175

Specifications

AMS. 4109, 4148, 4149, 4179
UNS number. A97175

Chemical Composition

Composition limits. 1.2 to 2.0 Cu, 2.1 to 2.9 Mg, 0.10 Mn max, 0.15 Si max, 0.20 Fe max, 0.18 to 0.28 Cr, 5.1 to 6.1 Zn, 0.10 Ti max, 0.05 max other (each), 0.15 max others (total)
Consequence of exceeding impurity limits. Degraded fracture toughness

Applications

Typical uses. Die and hand forgings for structural parts requiring very high strength, such as aircraft components. T736 tempers supply high strength, resistance to exfoliation corrosion and stress-corrosion cracking, high fracture toughness, and good fatigue resistance.

Mechanical Properties

Tensile properties. Typical. Tensile strength: T66 temper, 593 MPa (86 ksi); T736 temper, 524 MPa (76 ksi). Yield strength: T66 temper, 524 MPa (76 ksi); T736 temper, 455 MPa (66 ksi). Elongation: 11% in 50 mm (2 in.). See also Table 110.
Shear strength. Typical. T66 temper: 324 MPa (47 ksi); T736 temper: 290 MPa (42 ksi)
Hardness. Typical. T66 temper, 150 HB; T736 temper, 145 HB; data obtained with 500 kg load, 10 mm diam ball, and 30 s duration of loading
Poisson's ratio. 0.33
Elastic modulus. Tension, 72 GPa (10.4 × 10^6 psi)
Fatigue strength. Typical. T66 and T736 tempers: 159 MPa (23 ksi)

Table 113 Typical tensile properties of alloy 7178

Temperature, °C	Temperature, °F	Tensile strength(a), MPa	Tensile strength(a), ksi	Yield strength (0.2% offset)(a), MPa	Yield strength (0.2% offset)(a), ksi	Elongation(b), %
T6, T651 tempers						
−196	−320	730	106	650	94	5
−80	−112	650	94	580	84	8
−28	−18	625	91	560	81	9
24	75	605	88	540	78	11
100	212	505	73	470	68	14
149	300	215	31	185	27	40
204	400	105	15	83	12	70
260	500	76	11	62	9	76
316	600	59	8.5	48	7	80
371	700	45	6.5	38	5.5	80
T76, T7651 tempers						
−196	−320	730	106	615	89	10
−80	−112	625	91	540	78	10
−28	−18	605	88	525	76	10
24	75	570	83	505	73	11
100	212	475	69	440	64	17
149	300	215	31	185	27	40
204	400	105	15	83	12	70
260	500	76	11	62	9	76
316	600	59	8.5	48	7	80
371	700	45	6.5	38	5.5	80

(a) Lowest strength for exposures up to 10 000 h at temperature, no load; test loading applied at 35 MPa/min (5 ksi/min) to yield strength and then at strain rate of 5%/min to fracture. (b) In 50 mm (2 in.)

Table 112 Standard specifications for alloy 7178

Mill form and condition	Specification number: ASTM	Government
Sheet and plate	B 209	QQ-A-250/14
	...	QQ-A-250/21
Wire, rod, bar, shapes, and tube (extruded)	B 221	QQ-A-200/13
	...	QQ-A-200/14
Rivet wire	B 316	...
Tube (extruded, seamless)	B 241	...
Alclad sheet and plate	B 209	QQ-A-250/15
	...	QQ-A-250/22
	...	QQ-A-250/28

Plane-strain fracture toughness. See Table 111.

Mass Characteristics

Density. 2.80 g/cm³ (0.101 lb/in.³)

Thermal Properties

Liquidus temperature. 635 °C (1175 °F)
Incipient melting temperature. 532 °C (990 °F) for homogenized (solution heat treated) wrought material
Eutectic temperature. 477 °C (890 °F) for nonhomogeneous as cast or wrought material that has not been solution heat treated
Coefficient of thermal expansion. Linear:

Temperature range, °C	Temperature range, °F	Average coefficient, μm/m · K	Average coefficient, μin./in. · °F
−50 to 20	−58 to 68	21.6	12.0
20 to 100	68 to 212	23.4	13.0
20 to 200	68 to 392	24.3	13.5
20 to 300	68 to 572	25.2	14.0

Volumetric: 68 × 10^{-6} m³/m³ · K (3.78 × 10^{-5} in.³/in.³ · °F) at 20 °C (68 °F)
Specific heat. 864 J/kg · K (0.206 Btu/lb · °F) at 20 °C (68 °F)
Thermal conductivity. At 20 °C (68 °F): O temper, 177 W/m · K (102 Btu/ft · h · °F); T66 temper, 142 W/m · K (82 Btu/ft · h · °F) T736, T73652 tempers, 155 W/m · K (90 Btu/ft · h · °F)

Electrical Properties

Electrical conductivity. Volumetric, at 20 °C (68 °F): O temper, 46% IACS; T66 temper, 36% IACS; T736, T73652 tempers, 40% IACS
Electrical resistivity. At 20 °C (68 °F): O temper, 37.5 nΩ · m; T66 temper, 47.9 nΩ · m; T736, T73652 tempers, 43.1 nΩ · m. Temperature coefficient, all tempers: 0.1 nΩ · m per K at 20 °C (68 °F)

Fabrication Characteristics

Annealing temperature. 415 °C (775 °F)
Solution temperature. 515 °C (960 °F); must be preceded by soak at 477 to 485 °C (890 to 905 °F). Quench from lower temperature.
Aging temperature. 120 to 175 °C (250 to 350 °F)

Table 114 Creep-rupture properties of alloy 7178-T6

Temperature		Time under	Rupture stress		Stress for creep of:							
					1.0%		0.5%		0.2%		0.1%	
°C	°F	stress, h	MPa	ksi	MPa	ksi	MPa	ksi	MPa	ksi	MPa	ksi
150	300	0.1	440	64	420	61	415	60	395	57	365	53
		1	415	60	395	57	380	55	360	52	315	46
		10	370	54	345	50	340	49	310	45	250	36
		100	285	41	270	39	255	37	235	34	185	27
		1000	180	26	180	26	170	25	150	22	130	19
205	400	0.1	275	40	260	38	255	37	235	34	205	30
		1	215	31	205	30	200	29	180	26	145	21
		10	150	22	145	21	145	21	130	19	97	14
		100	105	15	97	14	97	14	83	12	76	11
		1000	69	10	69	10	69	10	59	8.5	55	8
260	500	0.1	110	16	110	16	110	16	105	15	97	14
		1	97	14	97	14	90	13	83	12	66	9.5
		10	69	10	69	10	66	9.5	55	8	41	6
		100	55	8	52	7.5	45	6.5	34	5	...	...
		1000	41	6	34	5	29	4.2	...	...	...	...
315	600	0.1	62	9	52	7.5	48	7	45	6.5	38	5.5
		1	52	7.5	45	6.5	41	6	34	5	26	3.7
		10	41	6	38	5.5	34	4.9	26	3.8	...	...
		100	34	5	30	4.3	26	3.8	...	...	...	...
		1000	28	4	23	3.4	...	...	...	...	...	...

7178, Alclad 7178
6.8Zn-2.7Mg-2.0Cu-0.3Cr

Specifications

AMS. Extruded wire, rod, bar, shapes, and tube; 4158. Alclad 7178, sheet and plate: 4051, 4052
ASTM. See Table 112.
SAE. J454
UNS number. A97178
Government. See Table 112.

Chemical Composition

Composition limits of 7178. 1.6 to 2.4 Cu, 2.4 to 3.1 Mg, 0.30 Mn max, 0.40 Si max, 0.50 Fe max, 0.18 to 0.35 Cr, 6.3 to 7.3 Zn, 0.20 Ti max, 0.05 max other (each), 0.15 max others (total), bal Al
Composition limits of Alclad 7178. 7011 cladding—0.05 Cu max, 1.0 to 1.6 Mg, 0.10 to 0.30 Mn, 0.15 Si max, 0.20 Fe max, 0.08 to 0.20 Cr, 4.0 to 5.5 Zn, 0.05 Ti max, 0.05 max other (each), 0.15 max others (total), bal Al. 7072 cladding—0.10 max Cu, 0.10 Mg max, 0.10 Mn max, 0.70 Si max + Fe, 0.8 to 1.3 Zn, 0.05 max other (each), 0.15 max others (total), bal Al

Applications

Typical uses. Aircraft and aerospace applications where high compressive yield is design criteria
Precautions in use. T6 temper is highly susceptible to exfoliation corrosion. T76 temper has mechanical properties comparable to 7075-T6 and provides improved resistance to exfoliation corrosion.

Mechanical Properties

Tensile properties. See Table 113.
Shear strength. T6, T6510, T6511 tempers: 305 MPa (44 ksi). T76, T76510, T76511 tempers: 295 MPa (43 ksi)
Compressive strength. T6, T6510, T6511 tempers: 530 MPa (77 ksi) at 0.1% permanent set. T76, T76510, T76511 tempers: 460 MPa (67 ksi) at 0.1% permanent set
Bearing properties. T6, T6510, T6511 tempers: bearing strength, 1035 to 1100 MPa (150 to 160 ksi); bearing yield strength, 680 to 730 MPa (99 to 106 ksi). T76, T76510, T76511 tempers: bearing strength, 965 MPa (140 ksi); bearing yield strength, 740 MPa (107 ksi). All data for *e*/*d* ratio of 2.0, where *e* is edge distance and *d* is pin diameter
Poisson's ratio. 0.33
Elastic modulus. Tension, 71.7 GPa (10.4 × 10^6 psi); shear, 27.5 GPa (4.0 × 10^6 psi); compression, 73.7 GPa (10.7 × 10^6 psi)
Fatigue strength. T76 type tempers: 200 to 290 MPa (29 to 42 ksi) at 10^7 cycles in axial fatigue tests (R = 0.0) of smooth specimens; 130 to 195 MPa (19 to 28 ksi) at 10^8 cycles in rotating beam tests (R = −1.0) of polished specimens; 28 to 55 MPa (4 to 8 ksi) at 10^8 cycles in rotating beam tests (R = −1.0) of 60° V-notched specimens (K_t = 3.0)
Creep-rupture characteristics. See Table 114.

Mass Characteristics

Density. 2.83 g/cm³ (0.102 lb/in.³) at 20 °C (68 °F)

Thermal Properties

Liquidus temperature. 629 °C (1165 °F)
Eutectic temperature. 477 °C (890 °F)
Coefficient of thermal expansion. Linear:

Temperature range		Average coefficient	
°C	°F	μm/m · K	μin./in. · °F
−50 to 20	−58 to 68	21.7	12.1
20 to 100	68 to 212	23.5	13.1
20 to 200	68 to 392	24.4	13.6
20 to 300	68 to 572	25.4	14.1

Volumetric: 68 × 10^{-6} m³/m³ · K (3.78 × 10^{-5} in.³/in.³ · °F)
Specific heat. 856 J/kg · K (0.205 Btu/lb · °F) at 20 °C (68 °F)
Thermal conductivity. At 20 °C (68 °F): O temper, 180 W/m · K (104 Btu/ft · h · °F); T6, T651 tempers, 127 W/m · K (73 Btu/ft · h · °F); T76, T7651 tempers, 152 W/m · K (88 Btu/ft · h · °F)

Electrical Properties

Electrical conductivity. Volumetric, at 20 °C (68 °F): O temper, 46% IACS; T6, T651 tempers, 32% IACS; T76, T7651 tempers, 39% IACS
Electrical resistivity. At 20 °C (68 °F): O temper, 37.5 nΩ · m; T6, T651 tempers, 53.9 nΩ · m; T76, T7651 tempers, 44.2 nΩ · m. Temperature coefficient, all tempers: 0.1 nΩ · m per K at 20 °C (68 °F)
Electrolytic solution potential. T6 temper: −0.81 V versus 0.1 *N* calomel electrode in an aqueous solution containing 53 g NaCl plus 3 g H_2O_2 per liter

Fabrication Characteristics

Annealing temperature. 415 °C (775 °F)
Solution temperature. 468 °C (875 °F)
Aging temperature. T6 and T7 tempers, 121 °C (250 °F) for 24 h

7475
5.7Zn-2.3Mg-1.5Cu-0.22Cr

Specifications

AMS. 4084, 4085, 4089, 4090
UNS number. A94475

Chemical Composition

Composition limits. 1.2 to 1.9 Cu, 1.9 to 2.6 Mg, 0.06 Mn max, 0.18 to 0.25 Cr, 0.12 Fe max, 0.10 Si max, 5.2 to 6.2 Zn, 0.06 Ti max, 0.05 max other (each), 0.15 max others (total), bal Al
Consequence of exceeding impurity limits. Degrades fracture toughness

Applications

Typical uses. Bare and Alclad sheet and plate for aircraft fuselage and wing skins,

Table 115 Typical tensile properties of alloy 7475 at various temperatures

Temperature		Time at	At indicated temperature					At room temperature after heating				
			Tensile strength		Yield strength		Elongation(a),	Tensile strength		Yield strength		Elongation(a),
°C	°F	temperature, h	MPa	ksi	MPa	ksi	%	MPa	ksi	MPa	ksi	%
T61 sheet, to 6.35 mm (0.040–0.249 in.) thick												
−196	−320	...	683	99	600	87	10	...	...	...	...	...
−80	−112	...	607	88	545	79	12	...	...	...	...	...
−28	−18	...	579	84	517	75	12	...	...	...	...	...
24	75	...	552	80	496	72	12	552	80	496	72	12
100	212	0.1–0.5	496	72	462	67	14	552	80	496	72	12
		10	496	72	462	67	14	558	81	496	72	12
		100	503	73	469	68	13	558	81	503	73	12
		1 000	503	73	476	69	13	565	82	510	74	12
		10 000	483	70	448	65	14	552	80	490	71	13
149	300	0.1–0.5	434	63	414	60	18	552	80	496	72	12
		10	434	63	414	60	17	545	79	490	71	12
		100	379	55	372	54	19	510	74	434	63	12
		1 000	262	38	255	37	23	400	58	310	45	13
		10 000	207	30	179	26	28	310	45	207	30	14
177	350	0.1	386	56	365	53	19	545	79	490	71	12
		0.5	379	55	365	53	19	538	78	483	70	12
		10	324	47	310	45	21	490	71	414	60	12
		100	228	33	221	32	23	386	56	290	42	12
		1 000	172	25	159	23	30	303	44	193	28	14
		10 000	131	19	110	16	40	234	34	124	18	15
204	400	0.1	331	48	317	46	17	531	77	469	68	12
		0.5	296	43	283	41	19	496	72	427	62	12
		10	200	29	193	28	26	372	54	276	40	12
		100	145	21	138	20	35	296	43	186	27	13
		1 000	110	16	97	14	45	234	34	117	17	15
		10 000	97	14	76	11	55	207	30	97	14	18
232	450	0.1	234	34	221	32	19	490	71	414	60	12
		0.5	200	29	186	27	21	421	61	331	48	12
		10	138	20	131	19	30	303	44	193	28	13
		100	97	14	90	13	45	241	35	124	18	14
		1 000	83	12	76	11	60	214	31	97	14	18
		10 000	83	12	62	9	65	193	28	76	11	22
260	500	0.1	159	23	152	22	20	407	59	310	45	12
		0.5	131	19	124	18	25	338	49	221	32	12
		10	90	13	83	12	45	255	37	131	19	15
		100	76	11	69	10	60	228	33	97	14	19
		1 000	69	10	59	8.5	70	207	30	83	12	21
		10 000	66	9.5	48	7	70	186	27	69	10	22
316	600	0.1	76	11	69	10	35	317	46	193	28	13
		0.5	69	10	62	9	45	269	39	131	19	15
		10	48	7	41	6	65	241	35	90	13	19
		100	45	6.5	38	5.5	75	221	32	83	12	20
		1 000	45	6.5	38	5.5	80	207	30	76	11	21
		10 000	45	6.5	38	5.5	80	186	27	69	10	...
371	700	0.1	41	6	34	5	70	276	40	117	17	17
		0.5	38	5.5	32	4.7	70	...	...	...	...	...
		10–10 000	34	5	27	3.8	85	...	...	...	...	...
427	800	0.1	24	3.5	20	2.8	85	...	...	...	...	...
		0.5	23	3.3	19	2.7	85	...	...	...	...	...
482	900	...	18	2.6	15	2.2	50	...	...	...	...	...
538	1000	...	11	1.6	9	1.3	3	...	...	...	...	...
T761 sheet, 1 to 6.35 mm (0.040 to 0.249 in.) thick												
−196	−320	...	655	95	565	82	11	...	...	...	...	...
−80	−112	...	579	84	503	73	12	...	...	...	...	...
−28	−18	...	552	80	483	70	12	...	...	...	...	...
24	75	...	524	76	462	67	12	524	76	462	67	12
100	212	0.1–10	455	66	434	63	14	524	76	462	67	12
		100–1 000	455	66	434	63	13	531	77	469	68	12
		10 000	441	64	421	61	14	524	76	462	67	13
149	300	0.1–0.5	400	58	386	56	18	524	76	462	67	12
		10	393	57	379	55	17	524	76	455	66	12
		100	359	52	345	50	19	490	71	421	61	12
		1 000	362	38	255	37	23	400	58	303	44	13
		10 000	207	30	179	26	28	310	45	207	30	14
177	350	0.1	352	51	338	49	19	517	75	455	66	12
		0.5	352	51	331	48	19	517	75	455	66	12
		10	303	44	290	42	21	469	68	393	57	12
		100	228	33	221	32	23	379	55	283	41	12
		1 000	172	25	159	23	30	303	44	193	28	14
		10 000	131	19	110	16	40	234	34	124	18	15
204	400	0.1	290	42	269	39	17	503	73	434	63	12
		0.5	276	40	262	38	19	483	70	414	60	12
		10	200	29	193	28	26	372	54	276	40	12

(continued)

(a) In 50 mm (2 in.)

Table 115 (continued)

Temperature °C	Temperature °F	Time at temperature, h	At indicated temperature: Tensile strength MPa	ksi	Yield strength MPa	ksi	Elongation(a), %	At room temperature after heating: Tensile strength MPa	ksi	Yield strength MPa	ksi	Elongation(a), %
T761 sheet, 1 to 6.35 mm (0.040 to 0.249 in.) thick (continued)												
		100	145	21	138	20	35	296	43	186	27	13
		1 000	110	16	97	14	45	234	34	117	17	15
		10 000	97	14	76	11	55	207	30	97	14	18
232	450	0.1	221	32	207	30	19	462	67	386	56	12
		0.5	193	28	179	26	21	414	60	324	47	12
		10	138	20	131	19	30	303	44	193	28	13
		100	97	14	90	13	45	241	35	124	18	14
		1 000	83	12	76	11	60	214	31	97	14	18
		10 000	83	12	62	9	65	193	28	76	11	22
260	500	0.1	159	23	152	22	20	386	56	283	41	12
		0.5	131	19	124	18	25	338	49	221	32	12
		10	90	13	83	12	45	255	37	131	19	15
		100	76	11	69	10	60	228	33	97	14	19
		1 000	69	10	59	8.5	70	207	30	83	12	21
		10 000	66	9.5	48	7	70	186	27	69	10	22
316	600	0.1	76	11	69	10	35	310	45	186	27	13
		0.5	69	10	62	9	45	269	39	131	19	15
		10	48	7	41	6	65	241	35	90	13	19
		100	45	6.5	38	5.5	75	221	32	83	12	20
		1 000	45	6.5	38	5.5	80	207	30	76	11	21
		10 000	45	6.5	38	5.5	80	186	27	69	10	...
371	700	0.1	41	6	34	5	70	276	40	117	17	17
		0.5	38	5.5	32	4.7	70	...	...	...	...	...
		10	34	5	27	3.9	80	...	...	...	...	...
		100–10 000	34	5	27	3.8	85	...	...	...	...	...

(a) In 50 mm (2 in.)

Table 116 Typical fracture-toughness values for alloy 7475

Temper	L-T MPa$\sqrt{m}$	L-T ksi$\sqrt{in.}$	T-L MPa$\sqrt{m}$	T-L ksi$\sqrt{in.}$	S-L MPa$\sqrt{m}$	S-L ksi$\sqrt{in.}$
High-strength plate (K_{Ic})(a)						
T651	42.9	39	37.4	34	29.7	27
T7651	47.3	43	38.5	35	30.8	28
T7351	52.7	48	41.8	38	35.2	32
High-strength sheet (K_c)(b)						
T761						
1.2 mm (0.047 in.) thick, room temperature	143	130	...	...		
−54 °C (−65 °F)	90	82	...	...		
1.4 mm (0.055 in.) thick, room temperature	136	123	...	...		
−54 °C (−65 °F)	87	79	...	...		
1.6 mm (0.063 in.) thick, room temperature	122	112	...	...		
−54 °C (−65 °F)	102	93	...	...		
1.6 mm (0.063 in.) thick, room temperature	150	137	...	...		
−54 °C (−65 °F)	111	101	...	...		
1.6 mm (0.063 in.) thick, room temperature	147	134	...	...		
−54 °C (−65 °F)	109	99	...	...		
1.8 mm (0.071 in.) thick, room temperature	149	136	...	...		
−54 °C (−65 °F)	125	114	...	...		

(a) Determined using standard compact tension specimen. (b) Determined using 400 × 1120 mm (16 × 44 in.) center cracked panel with antibuckling guides

spars, and bulkheads. Other structural applications requiring a combination of high strength and high fracture toughness

Mechanical Properties

Tensile properties. See Table 115.
Shear strength. Plate: T651 temper, 296 MPa (43 ksi); T7351, T7651 tempers, 269 MPa (39 ksi)
Compressive strength. At 0.1% permanent set. Plate: T651 temper, 476 MPa (69 ksi); T7351 temper, 379 MPa (55 ksi); T7651 temper, 414 MPa (60 ksi)
Bearing properties. Plate, all data for *e*/*d* ratio of 2.0, where *e* is edge distance and *d* is pin diameter. T761 temper: bearing strength, 990 MPa (144 ksi); bearing yield strength, 730 MPa (106 ksi). T7351 temper: bearing strength, 875 MPa (127 ksi); bearing yield strength, 640 MPa (93 ksi). T7651 temper: bearing strength, 925 MPa (134 ksi); bearing yield strength, 655 MPa (95 ksi)
Poisson's ratio. 0.33
Elastic modulus. Tension, 70 GPa (10.2 × 10^6 psi); shear, 27 GPa (3.9 × 10^6 psi); compression, 73 GPa (10.6 × 10^6 psi)
Fatigue strength. At 10^7 cycles in axial fatigue tests of smooth specimens from T7351 plate. Longitudinal or transverse orientation: 205 to 235 MPa (30 to 34 ksi) for R = 0.0. Transverse orientation: 315 MPa (46 ksi) for $R = +0.5$; 165 MPa for $R = -1.0$
Plane-strain fracture toughness. See Table 116.
Creep-rupture characteristics. See Table 117.

Mass Characteristics

Density. 2.80 g/cm³ (0.101 lb/in.³) at 20 °C (68 °F)

Thermal Properties

Liquidus temperature. 635 °C (1175 °F)
Incipient melting temperature. 538 °C (1000 °F) for homogenized (solution heat treated) wrought material
Eutectic temperature. 477 °C (890 °F) for as-cast or inhomogeneous wrought material that has not been solution heat treated
Coefficient of thermal expansion.
Linear:

Temperature range °C	Temperature range °F	Average coefficient µm/m · K	Average coefficient µin./in. · °F
−50 to 20	−58 to 68	21.6	12.0
20 to 100	68 to 212	23.4	13.0
20 to 200	68 to 392	24.3	13.5
20 to 300	68 to 572	25.2	14.2

Volumetric: 68 × 10^{-6} m³/m³ · K (3.78 × 10^{-5} in.³/in.³ · °F) at 20 °C (68 °F)
Specific heat. 865 J/kg · K (0.207 Btu/lb · °F) at 20 °C (68 °F)
Thermal conductivity. At 20 °C (68 °F):

Temper	Conductivity W/m · K	Conductivity Btu/ft · h · °F
O	177	102
T61, T651	142	82
T761, T7651	155	90
T7351	163	94

Table 117 Creep-rupture properties of alloy 7475 sheet 1 to 6.35 mm (0.040 to 0.25 in.) thick

Temperature		Time under stress, h	Rupture stress		Stress for creep of: 1.0%		0.5%		0.2%		0.1%	
°C	°F		MPa	ksi	MPa	ksi	MPa	ksi	MPa	ksi	MPa	ksi
T61 sheet												
24	75	0.1	552	80	538	78	524	76	517	75	510	74
		1	545	79	531	77	517	75	510	74	503	73
		10	545	79	517	75	510	74	503	73	496	72
		100	538	78	510	74	503	73	496	72	...	...
		1000	524	76	503	73	496	72	...	...	...	...
100	212	0.1	490	71	476	69	469	68	455	66	448	65
		1	476	69	455	66	448	65	434	63	421	61
		10	455	66	434	63	427	62	414	60	393	57
		100	427	62	414	60	400	58	386	56	365	53
		1000	386	56	379	55	365	53	352	51	...	...
149	300	0.1	414	60	400	58	393	57	379	55	365	53
		1	386	56	372	54	365	53	345	50	310	45
		10	352	51	338	49	317	46	283	41	241	35
		100	262	38	248	39	241	35	214	31	193	28
		1000	186	27	179	26	179	26	165	24	159	23
T761 sheet												
24	75	0.1	524	76	503	73	483	70	476	69	469	68
		1	517	75	490	71	476	69	469	68	462	67
		10	510	74	483	70	469	68	462	67	462	67
		100	496	72	476	69	469	68	462	67	455	66
		1000	490	71	462	67	462	67	455	66	448	65
100	212	0.1	441	64	421	61	414	60	414	60	400	58
		1	421	61	407	59	400	58	393	57	379	55
		10	400	58	386	56	386	56	372	54	359	52
		100	379	55	372	54	365	53	352	51	324	47
		1000	359	52	352	51	345	50	324	47	...	...
149	300	0.1	372	54	365	53	365	53	352	51	324	47
		1	345	50	338	49	331	48	310	45	276	40
		10	310	45	303	44	290	42	255	37	234	34
		100	248	36	234	34	228	33	207	30	193	28
		1000	186	27	179	26	179	26	165	24	159	23

Electrical Properties

Electrical conductivity. Volumetric, at 20 °C (68 °F):

Temper	Conductivity, %IACS
O	46
T61, T651	36
T761, T7651	40
T7351	42

Electrical resistivity. At 20 °C (68 °F):

Temper	Resistivity, nΩ · m
O	37.5
T61, T651	47.9
T761, T7651	43.1
T7351	41.1

Temperature coefficient. All tempers: 0.1 nΩ · m per K at 20 °C (68 °F)

Fabrication Characteristics

Annealing temperature. 415 °C (775 °F)
Solution temperature. 515 °C (960 °F); must be preceded by soak at 465 to 477 °C (870 to 890 °F)
Aging temperature. 120 to 175 °C (250 to 350 °F)

Aluminum Foundry Products

Revised by A. Kearney, Avery Kearney & Company
Elwin L. Rooy, Aluminum Company of America

ALUMINUM CASTING ALLOYS are the most versatile of all common foundry alloys and generally have the highest castability ratings. As casting materials, aluminum alloys have the following favorable characteristics:

- Good fluidity for filling thin sections
- Low melting point relative to those required for many other metals
- Rapid heat transfer from the molten aluminum to the mold, providing shorter casting cycles
- Hydrogen is the only gas with appreciable solubility in aluminum and its alloys, and hydrogen solubility in aluminum can be readily controlled by processing methods
- Many aluminum alloys are relatively free from hot-short cracking and tearing tendencies
- Chemical stability
- Good as-cast surface finish with lustrous surfaces and little or no blemishes

Aluminum alloy castings are routinely produced by pressure-die, permanent-mold, green- and dry-sand, investment, and plaster casting. Aluminum alloys are also readily cast with vacuum, low-pressure, centrifugal, and pattern-related processes such as lost foam. Total shipments of aluminum foundry products (all types of castings exclusive of ingot) in the United States for 1988 were about 10^6 Mg (10^6 tons), of which about 68% was accounted for by die castings (see the shipment statistics in the article "Introduction to Aluminum and Aluminum Alloys" in this Volume).

Alloy Systems

Aluminum casting alloys are based on the same alloy systems as those of wrought aluminum alloys, are strengthened by the same mechanisms (with the exception of strain hardening), and are similarly classified into non-heat-treatable and heat-treatable types. The major difference is that the casting alloys used in the greatest volumes contain alloying additions of silicon far in excess of that found (or used) in most wrought alloys. Aluminum casting alloys must contain, in addition to strengthening elements, sufficient amounts of eutectic-forming elements (usually silicon) in order to have adequate fluidity to feed the shrinkage that occurs in all but the simplest castings.

The phase behavior of aluminum-silicon compositions (Fig. 1) provides a simple eutectic-forming system, which makes possible the commercial viability of most high-volume aluminum casting. Silicon contents, ranging from about 4% to the eutectic level of about 12% (Fig. 1), reduce scrap losses, permit production of much more intricate designs with greater variations in section thickness, and yield castings with higher surface and internal quality. These benefits derive from the effects of silicon in increasing fluidity, reducing cracking, and improving feeding to minimize shrinkage porosity.

The required amounts of eutectic formers depend in part on the casting process. Alloys for sand casting generally are lower in eutectics than those for casting in metal molds because sand molds can tolerate a degree of hot shortness that would lead to extensive cracking in nonyielding metal molds. Resistance to cracking during casting is favored by a small range of solidification temperature, which drops from about 78 °C (140 °F) at 1% Si to zero at about 12% Si. Good feeding characteristics to minimize shrinkage porosity are benefited by a profile of volume fraction solidified versus temperature that is weighted toward the lower portion of the temperature range—that is, toward increased eutectic concentration.

In the binary aluminum-silicon system, under the nonequilibrium conditions of casting, the volume fraction of eutectic increases linearly from about 0 to 1 as silicon content increases from 1 to 12%.

Alloy Groupings and Designations

Although the systems used to identify and group aluminum casting alloys are not internationally standardized, each nation (and in many cases individual firms) has developed its own alloy nomenclature. In the United States and North America, for example, cast aluminum alloys are grouped according to composition limits registered with the Aluminum Association (see Table 3 in the article "Alloy and Temper Designation Systems for Aluminum and Aluminum Alloys"). Comprehensive listings are also maintained by general procurement specifications issued through government agencies (federal, military, and so on) and by technical societies such as the American Society for Testing and Materials and the Society of Automotive Engineers (see Table 1 for examples). In addition, aluminum casting alloys are sometimes grouped according to their quality level or intended end-use application (see the section "Selection of Casting Alloys"). Of these various methods, the grouping of aluminum casting alloys in terms of chemical compositions is discussed below.

The Aluminum Association designation system attempts alloy family recognition by the following scheme:

- 1*xx.x*: Controlled unalloyed compositions
- 2*xx.x*: Aluminum alloys containing copper as the major alloying element
- 3*xx.x*: Aluminum-silicon alloys also containing magnesium and/or copper
- 4*xx.x*: Binary aluminum-silicon alloys
- 5*xx.x*: Aluminum alloys containing magnesium as the major alloying element
- 6*xx.x*: Currently unused
- 7*xx.x*: Aluminum alloys containing zinc as the major alloying element, usually also containing additions of either copper, magnesium, chromium, manganese, or combinations of these elements
- 8*xx.x*: Aluminum alloys containing tin as the major alloying element
- 9*xx.x*: Currently unused

Designations in the form *xxx*.1 and *xxx*.2 include the composition of specific alloys in remelt ingot form suitable for foundry use. Designations in the form *xxx*.0 in all cases define composition limits applicable to castings. Further variations in specified compositions are denoted by prefix letters used primarily to define differences in impurity limits. Accordingly, one of the most common gravity cast alloys, 356, has variations

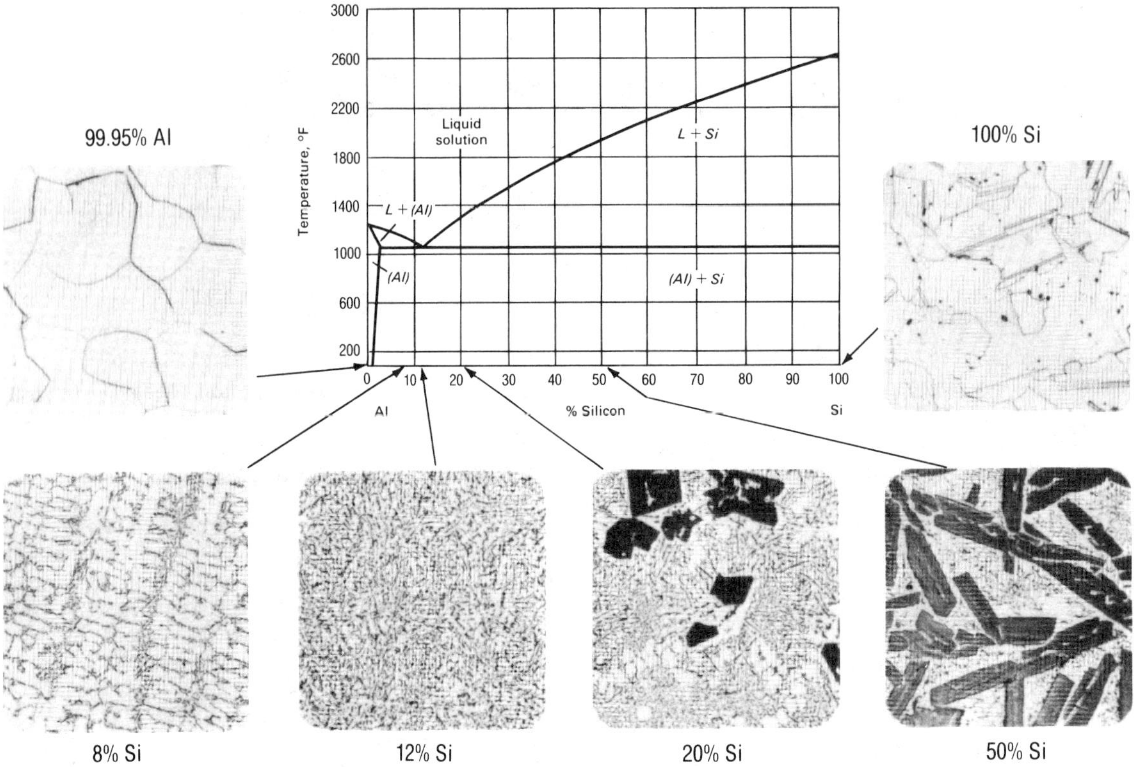

Fig. 1 Aluminum-silicon phase diagram and cast microstructures of pure components and of alloys of various compositions. Alloys with less than 12% Si are referred to as hypoeutectic, those with close to 12% Si as eutectic, and those with over 12% Si as hypereutectic.

A356, B356, and C356; each of these alloys has identical major alloy contents but has decreasing specification limits applicable to impurities, especially iron content.

Composition limits of casting alloys registered with the Aluminum Association are given in Table 3 in the article "Alloy and Temper Designation Systems for Aluminum and Aluminum Alloys."

In designations of the 1*xx.x* type, the second and third digits indicate minimum aluminum content (99.00% or greater); these digits are the same as the two to the right of the decimal point in the minimum aluminum percentage expressed to the nearest 0.01%. The fourth digit in 1*xx.x* designations, which is to the right of the decimal point, indicates product form: 0 denotes castings (such as electric motor rotors), and 1 denotes ingot.

In 2*xx.x* through 8*xx.x* designations for aluminum alloys, the second and third digits have no numerical significance but only identify the various alloys in the group. The digit to the right of the decimal point indicates product form: 0 denotes castings, 1 denotes standard ingot, and 2 denotes ingot having composition ranges narrower than but within those of standard ingot. Alloy modifications, as previously mentioned, are identified by a capital letter preceding the numerical designation.

Alloying-element and impurity limits for ingot are the same as those for castings of the same alloy except that, when the ingot is remelted for making castings, iron and zinc contents tend to increase and magnesium content tends to decrease.

General Composition Groupings. Although the nomenclature and designations for various casting alloys are standardized in North America, many important alloys have been developed for engineered casting production worldwide. For the most part, each nation (and in many cases the individual firm) has developed its own alloy nomenclature. Excellent references are available that correlate, cross reference, or otherwise define significant compositions in international use (Ref 1, 2).

Although a large number of aluminum alloys has been developed for casting, there are six basic types:

- Aluminum-copper
- Aluminum-copper-silicon
- Aluminum-silicon
- Aluminum-magnesium
- Aluminum-zinc-magnesium
- Aluminum-tin

Aluminum-copper alloys that contain 4 to 5% Cu, with the usual impurities iron and silicon and sometimes with small amounts of magnesium, are heat treatable and can reach quite high strengths and ductilities, especially if prepared from ingot containing less than 0.15% Fe. The aluminum-copper alloys are single-phase alloys. Unlike the silicon alloys, there is no highly fluid second phase available at the late stages of solidification. When available, a second phase will aid the required feeding of shrinkage areas and will help compensate for solidification stresses.

When these alloys, and other single-phase alloys, are cast using permanent mold or other rigid mold casting methods, special techniques are required to relieve solidification stresses. Careful techniques are also usually needed to promote the progress of the metal solidification from the remote areas of the casting to the hotter and more liquid casting areas, to the risers, and then to the riser feeders. When these necessary and more exacting casting techniques are used, the aluminum-copper alloys can and have been successfully used to produce high-strength and high-ductility castings. More exacting casting techniques are also helpful and required when casting other single-phase aluminum casting alloys or alloy systems.

In terms of alloy additions, manganese in small amounts may be added, mainly to combine with the iron and silicon and reduce the embrittling effect of essentially insoluble phases. However, these alloys demonstrate poor castability and require more carefully designed gating and more extensive risering if sound castings are to be obtained. Such alloys are used mainly in sand casting; when they are cast in metal molds, silicon must be added to increase fluidity and curtail hot shortness, and this addition of silicon substantially reduces ductility.

Al-Cu alloys with somewhat higher copper contents (7 to 8%), formerly the most

Table 1 Cross-reference chart of frequently used specifications for aluminum alloy sand and permanent mold (PM) castings

Alloy		Federal		ASTM(a)			
AA No.	Former designation	QQ-A-601E (sand)	QQ-A-596d (PM)	B26 (sand)	B108 (PM)	SAE(b)	AMS or MIL-21180c
208.0	108	108	...	CS43A	CS43A	...	...
213.0	C113	...	113	CS74A	CS74A	33	...
222.0	122	122	122	CG100A	CG100A	34	...
242.0	142	142	142	CN42A	CN42A	39	4222
295.0	195	195	...	C4A	...	38	4231
296.0	B295.0	...	B195	...	...	380	...
308.0	A108	...	A108	...	...	...	...
319.0	319, Allcast	319	319	SC64D	SC64D	326	...
328.0	Red X-8	Red X-8	...	SC82A	...	327	...
332.0	F332.0	...	F132	...	SC103A	332	...
333.0	333	...	333	...	...	...	...
336.0	A332.0	...	A132	...	SN122A	321	...
354.0	354	...	...	...	...	...	C354(c)
355.0	355	355	355	SC51A	SC51A	322	4210
C355.0	C355	...	C355	...	SC51B	355	C355(c)
356.0	356	356	356	SG70A	SG70A	323	(d)
A356.0	A356	...	A356	...	SG70B	336	A356(c)
357.0	357	...	357	...	...	...	4241
A357.0	A357	...	...	...	...	...	A357(c)
359.0	359	...	...	...	...	...	359(c)
B443.0	43	43	43	S5A	S5A	...	...
512.0(e)	B514.0	B214	...	GS42A	GS42A	...	...
513.0	A514.0	...	A214	...	GZ42A	...	...
514.0	214	214	...	G4A	...	320	...
520.0	220	220	...	G10A	...	324	4240
535.0	Almag 35	Almag 35	...	GM70B	GM70B	...	4238
705.0	603, Ternalloy 5	Ternalloy 5	Ternalloy 5	ZG32A	ZG32A	311	...
707.0	607, Ternalloy 7	Ternalloy 7	Ternalloy 7	ZG42A	ZG42A	312	...
710.0	A712.0	A612	...	ZG61B	...	313	...
712.0	D712.0	40E	...	ZG61A	...	310	...
713.0	613, Tenzaloy	Tenzaloy	...	ZC81A	...	315	...
771.0	Precedent 71A	Precedent 71A	...	...	...	...	...
850.0	750	750	750	...	...	...	...
851.0	A850.0	A750	A750	...	...	...	...
852.0	B850.0	B750	B750	...	...	...	...

(a) Former designations. ASTM adopted the Aluminum Association designation system in 1974. (b) Former designations used in SAE specifications J452 and/or J453. In 1990, SAE J452 adopted the ANSI/Aluminum Association numbering system for alloys. SAE J453-1986 has also superceded SAE J452. (c) Designation in MIL-21180c. (d) Alloy 356.0 is specified in AMS 4217, 4260, 4261, 4284, 4285, and 4286. (e) Alloy 512.0 is no longer active; it is included for reference purposes only.

commonly used aluminum casting alloys, have steadily been replaced by Al-Cu-Si alloys and today are used to a very limited extent. The best attribute of higher-copper Al-Cu alloys is their insensitivity to impurities. However, these alloys display very low strength and only fair castability.

Also in limited use are Al-Cu alloys that contain 9 to 11% Cu, whose high-temperature strength and wear resistance are attractive for use in aircraft cylinder heads and in automotive (diesel) pistons and cylinder blocks.

Very good high-temperature strength is an attribute of alloys containing copper, nickel, and magnesium, sometimes with iron in place of part of the nickel.

Aluminum-Copper-Silicon Alloys. The most widely used aluminum casting alloys are those that contain silicon and copper. The amounts of both additions vary widely, so that the copper predominates in some alloys and the silicon in others. In these alloys, the copper contributes to strength, and the silicon improves castability and reduces hot shortness; thus, the higher-silicon alloys normally are used for more complex castings and for permanent mold and die casting processes, which often require the use of more exacting casting techniques to avoid problems with hot-short alloys.

Al-Cu-Si alloys with more than 3 to 4% Cu are heat treatable, but usually heat treatment is used only with those alloys that also contain magnesium, which enhances their response to heat treatment. High-silicon alloys (>10% Si) have low thermal expansion, an advantage in some high-temperature operations. When silicon content exceeds 12 to 13% (silicon contents as high as 22% are typical), primary silicon crystals are present, and if properly distributed, impart excellent wear resistance. Automotive engine blocks and pistons are major uses of these hypereutectic alloys.

Aluminum-silicon alloys that do not contain copper additions are used when good castability and good corrosion resistance are needed. Metallographic structures of the pure components and of several intermediate compositions are shown in Fig. 1. The intermediate compositions are mixtures of aluminum containing about 1% Si in solid solution as the continuous phase, with particles of essentially pure silicon. Alloys with less than 12% Si are referred to as hypoeutectic, those with close to 12% Si as eutectic, and those with over 12% Si as hypereutectic.

If high strength and hardness are needed, magnesium additions make these alloys heat treatable. Alloys with silicon contents as low as 2% have been used for casting, but silicon content usually is between 5 and 13%. Strength and ductility of these alloys, especially those with higher silicon, can be substantially improved by modification of the Al-Si eutectic. Modification of hypoeutectic alloys (<12% Si) is particularly advantageous in sand castings and can be effectively achieved through the addition of a controlled amount of sodium or strontium, which refines the eutectic phase. Calcium and antimony additions are also used. Pseudomodification, in which the fineness of the eutectic but not the structure is affected, may be achieved by control of solidification rates.

In hypereutectic Al-Si alloys, refinement of the proeutectic silicon phase by phosphorus additions is essential for casting and product performance.

Aluminum-magnesium casting alloys are essentially single-phase binary alloys with moderate-to-high strength and toughness properties. High corrosion resistance, especially to seawater and marine atmospheres, is the primary advantage of castings made of Al-Mg alloys. Best corrosion resistance requires low impurity content (both solid and gaseous), and thus alloys must be prepared from high-quality metals and handled with great care in the foundry. These alloys are suitable for welded assemblies and are often used in architectural and other decorative or building needs. Aluminum-magnesium alloys also have good machinability and an attractive appearance when anodized. In comparison to the aluminum-silicon alloys, all the aluminum-magnesium alloys require more care in gating and greater temperature gradients to produce sound castings. This often means more chilling and larger risers. Also, careful melting and pouring practices are needed to compensate for the greater oxidizing tendency of these alloys when molten. This care is also needed because many of the applications of these alloys require polishing and/or fine surface finishing, where defects caused by oxide inclusions are particularly undesirable. The relatively poor castability of Al-Mg alloys and the tendency of the magnesium to oxidize increase handling difficulties, and therefore, cost.

Aluminum-zinc-magnesium alloys naturally age, achieving full strength by 20 to 30 days at room temperature after casting. This strengthening process can be accelerated by artificial aging. The high-temperature solution heat treatment and drastic quenching required by other alloys (Al-Cu and Al-Si-Mg alloys, for example) are not necessary for optimum properties in most Al-Zn-Mg alloy castings. However, microsegregation of Mg-Zn phases can occur in these alloys, which reverses the accepted rule that faster solidification results in higher as-cast properties. When it is found in an Al-Zn-Mg alloy casting that the strength of the thin or highly chilled sections are lower than the thick or slowly cooled sections, the weaker sections can be strengthened to the required level by solution heat treatment and quenching, followed by natural or artificial (furnace) aging.

These alloys have moderate to good tensile properties in the as-cast condition. With annealing, good dimensional stability in use is developed. The eutectic melting points of alloys of this group are high, an advantage in castings that are to be assembled by brazing. The alloys have good machinability (and resistance to general corrosion, despite some susceptibility to stress corrosion). They are not generally recommended for service at elevated temperatures. The tensile properties of these alloys develop at room temperatures during the first few weeks after casting due to precipitation hardening. This process continues thereafter at a progressively slower rate. Heat treatments of the T6 and T7 type may be applied to the 707.0, 771.0, and 772.0 alloys.

Castability of Al-Zn-Mg alloys is poor, and careful control of solidification conditions is required to produce sound, defect-free castings. Moderate to steep temperature gradients are required to assure adequate feeding to prevent shrinkage defects. Hot tear cracking where resistance to contraction during solidification and cooling is resisted. However, good foundry techniques and control have enabled well-qualified sand foundries to produce relatively intricate castings. Permanent mold castings, except for relatively simple designs, can be difficult.

Aluminum-tin alloys that contain about 6% Sn (and small amounts of copper and nickel for strengthening) are used for cast bearings because of the excellent lubricity imparted by tin. These tin-containing alloys were developed for bearing applications (in which load-carrying capacity, fatigue strength, and resistance to corrosion by internal-combustion lubricating oil are important criteria). Bearings of aluminum-tin alloys are superior overall to bearings made using most other materials.

Aluminum-tin casting Alloys 850.0, 851.0, and 852.0 can be cast in sand or permanent molds. However, 850.0 (6.3Sn-1Cu-1Ni) and 852.0 (6.3Sn-2Cu-1.2Ni-0.8Mg) usually are cast in permanent molds. Major applications are for connecting rods and crankcase bearings for diesel engines. Sand cast bearings, such as large rolling mill bearings, usually are made of Alloy 851.0 (6.3Sn-2.5Si-1Cu-0.5Ni).

Bearing performance of Al-Sn alloys is strongly affected by casting method. Fine interdendritic distribution of tin, which is necessary for optimum bearing properties, requires small interdendritic spacing, and small spacing is obtained only with casting methods in which cooling is rapid.

From a foundry standpoint, the aluminum-tin alloy system is unique. In the mold, the solidification starts at about 650 °C (1200 °F), and the tin constituents of the alloy are liquid until 229 °C (444 °F). This extremely large solidification range presents unique problems. Rapid solidification rates are recommended to avoid excessive macrosegregation.

Some 850.0 T5 castings are cold worked by reducing the axial dimension 4% (T101 temper). This provides a substantial increase in the compressive yield strength.

Aluminum-lithium casting alloys may offer the same benefits as their wrought counterparts but have not been developed or commercialized as have the wrought aluminum-lithium alloys. (See the article "Aluminum-Lithium Alloys" in this Volume.) The wrought and cast aluminum-lithium alloys consist of aluminum-copper alloys with lithium additions to reduce weight and improve strength.

The behavior of both cast and wrought Al-Li alloys differs from conventional aluminum alloys in terms of fracture mechanisms and temperature effects on mechanical properties. Unlike conventional aluminum alloys, the toughness of Al-Li alloys does not increase with increasing aging temperature beyond the point of overaging (that is, the point required for peak strength). The benefits of reduced density and improved modulus of elasticity are the main incentives in the development of Al-Li alloys.

Aluminum-Base Metal-Matrix Composite (MMC) for Casting. Although aluminum-ceramic composites offer exceptional specific stiffness (elastic modulus-to-weight ratio), the initial development (prior to 1986) of aluminum-base MMC materials required energy- or labor-intensive methods, such as powder metallurgy, thermal spray, diffusion bonding, and high-pressure squeeze casting. None of the composites produced from these methods could be remelted and shape cast, and each proved to be prohibitively expensive for most applications, even in the aerospace/defense sector.

An ingot-metallurgical method for producing a castable aluminum-base MMC material with the trade name of Duralcan was introduced in 1986. The product consists, in foundry ingot form, of foundry alloys to which 10, 15, or 20 vol% of particulate silicon carbide (SiC) had been added. The most attractive feature of this ingot-metallurgical product is its low cost, especially at industrial production levels, and its ability to be remelted without an impairment of properties. The Al-SiC composite foundry ingot can be remelted and shape cast easily using standard aluminum foundry practices and equipment. The casting methods successfully demonstrated to date have been sand, permanent mold, low-pressure permanent mold, high-pressure die casting, and investment casting, both shell and plaster. Additional information on castable aluminum-base MMC materials is given in the section "Discontinuous Aluminum MMC" in the article "Metal-Matrix Composites" in this Volume.

Selection of Casting Alloys

The major factors that influence alloy selection for casting applications include casting process to be used, casting design, required properties, and economic (and availability) considerations. These five factors consist of the following characteristics of an alloy:

- *Casting process considerations*: fluidity, resistance to hot tearing, solidification range
- *Casting design considerations*: solidification range, resistance to hot tearing, fluidity, die soldering (die casting)
- *Mechanical-property requirements*: strength and ductility, heat treatability, hardness
- *Service requirements*: pressure tightness characteristics, corrosion resistance, surface treatments, dimensional stability, thermal stability
- *Economics*: machinability, weldability, ingot and melting costs, heat treatment

Several of these characteristics for various alloys are listed in Tables 2 and 3. Table 2 gives the solidification range and thermal properties, while Table 3 ranks various alloys in terms of castability, corrosion resistance, machinability, and weldability. Note that the rankings in Table 3 can vary with the casting process. Each casting process may also require specific metal characteristics. For example, die and permanent mold casting generally require alloys with good fluidity and resistance to hot tearing, whereas these properties are less critical in sand, plaster, and investment casting, where molds and cores offer less resistance to shrinkage.

Also note that Table 2 groups aluminum casting alloys into the following nine categories:

- Rotor alloys
- Commercial Duralumin alloys
- Premium casting alloys
- Piston and elevated-temperature alloys
- Standard, general-purpose alloys
- Die castings
- Magnesium alloys (see the earlier section "General Composition Groupings" in this article)
- Aluminum-zinc-magnesium alloys (see the section "General Composition Groupings")
- Bearing alloys

Table 2 Typical physical properties of aluminum casting alloys

Alloy	Temper and product form(a)	Specific gravity(b)	Density(b)		Approximate melting range		Electrical conductivity, %IACS	Thermal conductivity at 25 °C (77 °F), cal/cm·s·°C	Coefficient of thermal expansion, per °C × 10^{-6} (per °F × 10^{-6})	
			kg/m³	lb/in.³	°C	°F			20–100 °C (68–212 °F)	20–300 °C (68–570 °F)
Aluminum rotor alloys (pure aluminum)										
Pure aluminum 99.996% Al	0 °F	2.71	2713	0.098	660.2–660.2	1220.4–1220.4	64.94	0.57	23.86 (13.25)	25.45 (14.14)
EC Alloy 99.45% Al, similar to 150.0 alloy	0 °F	2.70	2713	0.098	657–643	1215–1190	57	0.53	23.5 (13)	25.6 (14.2)
Commercial Duralumin alloys (Al-Cu)										
222.0	F(P)	2.95	2962	0.107	520–625	970–1160	34	0.32	22.1 (12.3)	23.6 (13.1)
	O(S)	2.95	2962	0.107	520–625	970–1160	41	0.38	. . .	. . .
	T61(S)	2.95	2962	0.107	520–625	970–1160	33	0.31	22.1 (12.3)	23.6 (13.1)
224.0	T62(S)	2.81	2824	0.102	550–645	1020–1190	30	0.28	. . .	. . .
238.0	F(P)	2.95	1938	0.107	510–600	950–1110	25	0.25	21.4 (11.9)	22.9 (12.7)
240.0	F(S)	2.78	2768	0.100	515–605	960–1120	23	0.23	22.1 (12.3)	24.3 (13.5)
242.0	O(S)	2.81	2823	0.102	530–635	990–1180	44	0.40	. . .	. . .
	T77(S)	2.81	2823	0.102	525–635	980–1180	38	0.36	22.1 (12.3)	23.6 (13.1)
	T571(P)	2.81	2823	0.102	525–635	980–1180	34	0.32	22.5 (12.5)	24.5 (13.6)
	T61(P)	2.81	2823	0.102	525–635	980–1180	33	0.32	22.5 (12.5)	24.5 (13.6)
Premium casting alloys (high strength and toughness alloys)										
201.0	T6(S)	2.80	2796	0.101	570–650	1000–1200	27–32	0.29	19.3 (10.7)	24.7 (13.7)
	T7(P)	2.80	2796	0.101	570–650	1000–1200	32–34	0.29	19.3 (10.7)	24.7 (13.7)
206.0	T4(S)	2.8	2796	0.101	570–650	1000–1200	30	0.29	19.3 (10.7)	24.7 (13.7)
204.0	T4(S)	2.8	2800	0.101	570–650	1060–1200	29	0.29	19.3 (10.7)	. . .
204.0	T6(S)(P)	2.8	2800	0.101	570–650	1060–1200	34	0.29	19.3 (10.7)	. . .
224.0	T62(S)	2.81	2824	0.102	550–645	1020–1190	30	0.28	. . .	. . .
295.0	T4(S)	2.81	2823	0.102	520–645	970–1190	35	0.33	22.9 (12.7)	24.8 (13.8)
	T62(S)	2.81	2823	0.102	520–645	970–1190	35	0.34	22.9 (12.7)	24.8 (13.8)
296.0	T4(P)	2.80	2796	0.101	520–630	970–1170	33	0.32	22.0 (12.2)	23.9 (13.3)
	T6(P)	2.80	2796	0.101	520–630	970–1170	33	0.32	22.0 (12.2)	23.9 (13.3)
	T62(S)	2.80	2796	0.101	520–630	970–1170	33	0.32	. . .	. . .
C355.0	T61(S)	2.71	2713	0.098	550–620	1020–1150	39	0.35	22.3 (12.4)	24.7 (13.7)
A356.0	T6(S)	2.69	2713	0.098	560–610	1040–1130	40	0.36	21.4 (11.9)	23.4 (13.0)
A357.0	T6(S)	2.69	2713	0.098	555–610	1030–1130	40	0.38	21.4 (11.9)	23.6 (13.1)
Piston and elevated-temperature alloys										
332.0	T5(P)	2.76	2768	0.100	520–580	970–1080	26	0.25	20.7 (11.5)	22.3 (12.4)
360.0	F(D)	2.68	2685	0.097	570–590	1060–1090	37	0.35	20.9 (11.6)	22.9 (12.7)
A360.0	F(D)	2.68	2685	0.097	570–590	1060–1090	37	0.35	21.1 (11.7)	22.9 (12.7)
364.0	F(D)	2.63	2630	0.095	560–600	1040–1110	30	0.29	20.9 (11.6)	22.9 (12.7)
380.0	F(D)	2.76	2740	0.099	520–590	970–1090	27	0.26	21.2 (11.8)	22.5 (12.5)
A380.0	F(D)	2.76	2740	0.099	520–590	970–1090	27	0.26	21.1 (11.7)	22.7 (12.6)
384.0	F(D)	2.70	2713	0.098	480–580	900–1080	23	0.23	20.3 (11.3)	22.1 (12.3)
390.0	F(D)	2.73	2740	0.099	510–650	950–1200	25	0.32	18.5 (10.3)	. . .
	T5(D)	2.73	2740	0.099	510–650	950–1200	24	0.32	18.0 (10.0)	. . .
Standard general-purpose alloys										
208.0	F(S)	2.79	2796	0.101	520–630	970–1170	31	0.29	22.0 (12.2)	23.9 (13.3)
308.0	F(P)	2.79	2796	0.101	520–615	970–1140	37	0.34	21.4 (11.9)	22.9 (12.7)
319.0	F(S)	2.79	2796	0.101	520–605	970–1120	27	0.27	21.6 (12.0)	24.1 (13.4)
	F(P)	2.79	2796	0.101	520–605	970–1120	28	0.28	21.6 (12.0)	24.1 (13.4)
324.0	F(P)	2.67	2658	0.096	545–605	1010–1120	34	0.37	21.4 (11.9)	23.2 (12.9)
238.0	F(P)	2.95	2962	0.107	510–600	950–1110	25	0.25	21.4 (11.9)	22.9 (12.7)
240.0	F(S)	2.78	2768	0.100	515–605	960–1120	23	0.23	22.1 (12.3)	24.3 (13.5)
242.0	O(S)	2.81	2823	0.102	530–635	990–1180	44	0.40	. . .	. . .
	T77(S)	2.81	2823	0.102	525–635	980–1180	38	0.36	22.1 (12.3)	23.6 (13.1)
	T571(P)	2.81	2823	0.102	525–635	980–1180	34	0.32	22.5 (12.5)	24.5 (13.6)
	T61(P)	2.81	2823	0.102	525–635	980–1180	33	0.32	22.5 (12.5)	24.5 (13.6)
295.0	T4(S)	2.81	2823	0.102	520–645	970–1190	35	0.33	22.9 (12.7)	24.8 (13.8)
	T62(S)	2.81	2823	0.102	520–645	970–1190	35	0.34	22.9 (12.7)	24.8 (13.8)
296.0	T4(P)	2.80	2796	0.101	520–630	970–1170	33	0.32	22.0 (12.2)	23.9 (13.3)
	T6(P)	2.80	2796	0.101	520–630	970–1170	33	0.32	22.0 (12.2)	23.9 (13.3)
	T62(S)	2.80	2796	0.101	520–630	970–1170	33	0.32	. . .	. . .
308.0	F(P)	2.79	2796	0.101	520–615	970–1140	37	0.34	21.4 (11.9)	22.9 (12.7)
319.0	F(S)	2.79	2796	0.101	520–605	970–1120	27	0.27	21.6 (12.0)	24.1 (13.4)
	F(P)	2.79	2796	0.101	520–605	970–1120	28	0.28	21.6 (12.0)	24.1 (13.4)
324.0	F(P)	2.67	2658	0.096	545–605	1010–1120	34	0.37	21.4 (11.9)	23.2 (12.9)
333.0	F(P)	2.77	2768	0.100	520–585	970–1090	26	0.25	20.7 (11.5)	22.7 (12.6)
	T5(P)	2.77	2768	0.100	520–585	970–1090	29	0.29	20.7 (11.5)	22.7 (12.6)
	T6(P)	2.77	2768	0.100	520–585	970–1090	29	0.28	20.7 (11.5)	22.7 (12.6)
	T7(P)	2.77	2768	0.100	520–585	970–1090	35	0.34	20.7 (11.5)	22.7 (12.6)
336.0	T551(P)	2.72	2713	0.098	540–570	1000–1060	29	0.28	18.9 (10.5)	20.9 (11.6)
354.0	F(P)	2.71	2713	0.098	540–600	1000–1110	32	0.30	20.9 (11.6)	22.9 (12.7)

(continued)

(a) S, sand cast; P, permanent mold; D, die cast. (b) The specific gravity and weight data in this table assume solid (void-free) metal. Because some porosity cannot be avoided in commercial castings, their specific gravity or weight is slightly less than the theoretical value.

Table 2 (continued)

Alloy	Temper and product form(a)	Specific gravity(b)	Density(b) kg/m^3	Density(b) lb/in.^{3}	Approximate melting range °C	Approximate melting range °F	Electrical conductivity, %IACS	Thermal conductivity at 25 °C (77 °F), cal/cm·s·°C	Coefficient of thermal expansion, per °C × 10^{-6} (per °F × 10^{-6}) 20–100 °C (68–212 °F)	Coefficient of thermal expansion, per °C × 10^{-6} (per °F × 10^{-6}) 20–300 °C (68–570 °F)
Standard general-purpose alloys (continued)										
355.0	T51(S)	2.71	2713	0.098	550–620	1020–1150	43	0.40	22.3 (12.4)	24.7 (13.7)
	T6(S)	2.71	2713	0.098	550–620	1020–1150	36	0.34	22.3 (12.4)	24.7 (13.7)
	T61(S)	2.71	2713	0.098	550–620	1020–1150	37	0.35	22.3 (12.4)	24.7 (13.7)
	T7(S)	2.71	2713	0.098	550–620	1020–1150	42	0.39	22.3 (12.4)	24.7 (13.7)
	T6(P)	2.71	2713	0.098	550–620	1020–1150	39	0.36	22.3 (12.4)	24.7 (13.7)
356.0	T51(S)	2.68	2685	0.097	560–615	1040–1140	43	0.40	21.4 (11.9)	23.4 (13.0)
	T6(S)	2.68	2685	0.097	560–615	1040–1140	39	0.36	21.4 (11.9)	23.4 (13.0)
	T7(S)	2.68	2685	0.097	560–615	1040–1140	40	0.37	21.4 (11.9)	23.4 (13.0)
	T6(P)	2.68	2685	0.097	560–615	1040–1140	41	0.37	21.4 (11.9)	23.4 (13.0)
A356.0	T6(S)	2.69	2713	0.098	560–610	1040–1130	40	0.36	21.4 (11.9)	23.4 (13.0)
357.0	T6(S)	2.68	2713	0.098	560–615	1040–1140	39	0.36	21.4 (11.9)	23.4 (13.0)
A357.0	T6(S)	2.69	2713	0.098	555–610	1030–1130	40	0.38	21.4 (11.9)	23.6 (13.1)
358.0	T6(S)	2.68	2658	0.096	560–600	1040–1110	39	0.36	21.4 (11.9)	23.4 (13.0)
359.0	T6(S)	2.67	2685	0.097	565–600	1050–1110	35	0.33	20.9 (11.6)	22.9 (12.7)
392.0	F(P)	2.64	2630	0.095	550–670	1020–1240	22	0.22	18.5 (10.3)	20.2 (11.2)
443.0	F(S)	2.69	2685	0.097	575–630	1070–1170	37	0.35	22.1 (12.3)	24.1 (13.4)
	O(S)	2.69	2685	0.097	575–630	1070–1170	42	0.39	. . .	. . .
	F(D)	2.69	2685	0.097	575–630	1070–1170	37	0.34	. . .	. . .
	F(P)	2.68	2685	0.097	575–630	1070–1170	41	0.38	21.8 (12.1)	23.8 (13.2)
Die casting alloys										
360.0	F(D)	2.68	2685	0.097	570–590	1060–1090	37	0.35	20.9 (11.6)	22.9 (12.7)
A360.0	F(D)	2.68	2685	0.097	570–590	1060–1090	37	0.35	21.1 (11.7)	22.9 (12.7)
364.0	F(D)	2.63	2630	0.095	560–600	1040–1110	30	0.29	20.9 (11.6)	22.9 (12.7)
380.0	F(D)	2.76	2740	0.099	520–590	970–1090	27	0.26	21.2 (11.8)	22.5 (12.5)
A380.0	F(D)	2.76	2740	0.099	520–590	970–1090	27	0.26	21.1 (11.7)	22.7 (12.6)
384.0	F(D)	2.70	2713	0.098	480–580	900–1080	23	0.23	20.3 (11.3)	22.1 (12.3)
390.0	F(D)	2.73	2740	0.099	510–650	950–1200	25	0.32	18.5 (10.3)	. . .
	T5(D)	2.73	2740	0.099	510–650	950–1200	24	0.32	18.0 (10.0)	. . .
413.0	F(D)	2.66	2657	0.096	575–585	1070–1090	39	0.37	20.5 (11.4)	22.5 (12.5)
A413.0	F(D)	2.66	2657	0.096	575–585	1070–1090	39	0.37	. . .	. . .
443.0	F(S)	2.69	2685	0.097	575–630	1070–1170	37	0.35	22.1 (12.3)	24.1 (13.4)
	O(S)	2.69	2685	0.097	575–630	1070–1170	42	0.39	. . .	. . .
	F(D)	2.69	2685	0.097	575–630	1070–1170	37	0.34	. . .	. . .
518.0	F(D)	2.53	2519	0.091	540–620	1000–1150	24	0.24	24.1 (13.4)	26.1 (14.5)
A535.0	F(D)	2.54	2547	0.092	550–620	1020–1150	23	0.24	24.1 (13.4)	26.1 (14.5)
Aluminum-magnesium alloys										
511.0	F(S)	2.66	2657	0.096	590–640	1090–1180	36	0.34	23.6 (13.1)	25.7 (14.3)
512.0	F(S)	2.65	2657	0.096	590–630	1090–1170	38	0.35	22.9 (12.7)	24.8 (13.8)
513.0	F(P)	2.68	2685	0.097	580–640	1080–1180	34	0.32	23.9 (13.3)	25.9 (14.4)
514.0	F(S)	2.65	2657	0.096	600–640	1110–1180	35	0.33	23.9 (13.3)	25.9 (14.4)
518.0	F(D)	2.53	2519	0.091	540–620	1000–1150	24	0.24	24.1 (13.4)	26.1 (14.5)
520.0	T4(S)	2.57	2574	0.093	450–600	840–1110	21	0.21	25.2 (14.0)	27.0 (15.0)
535.0	F(S)	2.62	2519	0.091	550–630	1020–1170	23	0.24	23.6 (13.1)	26.5 (14.7)
A535.0	F(D)	2.54	2547	0.092	550–620	1020–1150	23	0.24	24.1 (13.4)	26.1 (14.5)
B535.0	F(S)	2.62	2630	0.095	550–630	1020–1170	24	0.23	24.5 (13.6)	26.5 (14.7)
Aluminum-zinc alloys (Al-Zn-Mg and Al-Zn)										
705.0	F(S)	2.76	2768	0.100	600–640	1110–1180	25	0.25	23.6 (13.1)	25.7 (14.3)
707.0	F(S)	2.77	2768	0.100	585–630	1090–1170	25	0.25	23.8 (13.2)	25.9 (14.4)
710.0	F(S)	2.81	2823	0.102	600–650	1110–1200	35	0.33	24.1 (13.4)	26.3 (14.6)
711.0	F(P)	2.84	2851	0.103	600–645	1110–1190	40	0.38	23.6 (13.1)	25.6 (14.2)
712.0	F(S)	2.82	2823	0.102	600–640	1110–1180	40	0.38	23.6 (13.1)	25.6 (14.2)
Bearing alloys (aluminum-tin)										
713.0	F(S)	2.84	2879	0.104	595–630	1100–1170	37	0.37	23.9 (13.3)	25.9 (14.4)
850.0	T5(S)	2.87	2851	0.103	225–650	440–1200	47	0.44	. . .	. . .
851.0	T5(S)	2.83	2823	0.102	230–630	450–1170	43	0.40	22.7 (12.6)	. . .
852.0	T5(S)	2.88	2879	0.104	210–635	410–1180	45	0.42	23.2 (12.9)	. . .

(a) S, sand cast; P, permanent mold; D, die cast. (b) The specific gravity and weight data in this table assume solid (void-free) metal. Because some porosity cannot be avoided in commercial castings, their specific gravity or weight is slightly less than the theoretical value.

This grouping of alloys is useful in the selection of alloys because many foundries are dedicated to a particular type of casting alloy. Each group, with the exception of the magnesium and the Al-Zn-Mg alloy groups, is discussed below.

Rotor Castings. Most cast aluminum motor rotors are produced in the carefully controlled pure-alloy conditions 100.0, 150.0, and 170.0 (99.0, 99.5, and 99.7% Al, respectively). Impurities in these alloys are controlled to minimize variations in electrical performance based on conductivity and to minimize the occurrence of microshrinkage and cracks during casting. Minimum and typical conductivities for each alloy grade are:

Alloy	Minimum conductivity, %IACS	Typical conductivity, %IACS
100.1	54	56
150.1	57	59
170.1	59	60

IACS, International Copper Annealed Standard

Table 3 Ratings of castability, corrosion resistance, machinability, and weldability for aluminum casting alloys

1, best; 5, worst. Individual alloys may have different ratings for other casting processes.

Alloy	Resistance to hot cracking(a)	Pressure tightness	Fluidity(b)	Shrinkage tendency(c)	Corrosion resistance(d)	Machinability(e)	Weldability(f)
Sand casting alloys							
201.0	4	3	3	4	4	1	2
208.0	2	2	2	2	4	3	3
213.0	3	3	2	3	4	2	2
222.0	4	4	3	4	4	1	3
240.0	4	4	3	4	4	3	4
242.0	4	3	4	4	4	2	3
A242.0	4	4	3	4	4	2	3
295.0	4	4	4	3	3	2	2
319.0	2	2	2	2	3	3	2
354.0	1	1	1	1	3	3	2
355.0	1	1	1	1	3	3	2
A356.0	1	1	1	1	2	3	2
357.0	1	1	1	1	2	3	2
359.0	1	1	1	1	2	3	1
A390.0	3	3	3	3	2	4	2
A443.0	1	1	1	1	2	4	4
444.0	1	1	1	1	2	4	1
511.0	4	5	4	5	1	1	4
512.0	3	4	4	4	1	2	4
514.0	4	5	4	5	1	1	4
520.0	2	5	4	5	1	1	5
535.0	4	5	4	5	1	1	3
A535.0	4	5	4	4	1	1	4
B535.0	4	5	4	4	1	1	4
705.0	5	4	4	4	2	1	4
707.0	5	4	4	4	2	1	4
710.0	5	3	4	4	2	1	4
711.0	5	4	5	4	3	1	3
712.0	4	4	3	3	3	1	4
713.0	4	4	3	4	2	1	3
771.0	4	4	3	3	2	1	...
772.0	4	4	3	3	2	1	...
850.0	4	4	4	4	3	1	4
851.0	4	4	4	4	3	1	4
852.0	4	4	4	4	3	1	4
Permanent mold casting alloys							
201.0	4	3	3	4	4	1	2
213.0	3	3	2	3	4	2	2
222.0	4	4	3	4	4	1	3
238.0	2	3	2	2	4	2	3
240.0	4	4	3	4	4	3	4
296.0	4	3	4	3	4	3	4
308.0	2	2	2	2	4	3	3
319.0	2	2	2	2	3	3	2
332.0	1	2	1	2	3	4	2
333.0	1	1	2	2	3	3	3
336.0	1	2	2	3	3	4	2
354.0	1	1	1	1	3	3	2
355.0	1	1	1	2	3	3	2
C355.0	1	1	1	2	3	3	2
356.0	1	1	1	1	2	3	2
A356.0	1	1	1	1	2	3	2
357.0	1	1	1	1	2	3	2
A357.0	1	1	1	1	2	3	2
359.0	1	1	1	1	2	3	1
A390.0	2	2	2	3	2	4	2
443.0	1	1	2	1	2	5	1
A444.0	1	1	1	1	2	3	1
512.0	3	4	4	4	1	2	4
513.0	4	5	4	4	1	1	5
711.0	5	4	5	4	3	1	3
771.0	4	4	3	3	2	1	...
772.0	4	4	3	3	2	1	...
850.0	4	4	4	4	3	1	4
851.0	4	4	4	4	3	1	4
852.0	4	4	4	4	3	1	4
Die casting alloys							
360.0	1	1	2	2	3	4	
A360.0	1	1	2	2	3	4	
364.0	2	2	1	3	4	3	
380.0	2	1	2	5	3	4	
A380.0	2	2	2	4	3	4	
384.0	2	2	1	3	3	4	
390.0	2	2	2	2	4	2	
413.0	1	2	1	2	4	4	
C443.0	2	3	3	2	5	4	
515.0	4	5	5	1	2	4	
518.0	5	5	5	1	1	4	

(a) Ability of alloy to withstand stresses from contraction while cooling through hot short or brittle temperature range. (b) Ability of liquid alloy to flow readily in mold and to fill thin sections. (c) Decrease in volume accompanying freezing of alloy and a measure of amount of compensating feed metal required in form of risers. (d) Based on resistance of alloy in standard salt spray test. (e) Composite rating based on ease of cutting, chip characteristics, quality of finish, and tool life. (f) Based on ability of material to be fusion welded with filler rod of same alloy

Rotor alloy 100.0 contains a significantly larger amount of iron and other impurities, and this generally improves castability. With higher iron content crack resistance is improved, and a lower tendency toward shrinkage formation will be observed. This alloy is recommended when the maximum dimension of the part is greater than 125 mm (5 in.). For the same reasons, Alloy 150.0 is preferred over 170.0 in casting performance.

For motor rotors requiring high resistivity (for example, motors with high starting torque) the more highly alloyed die casting compositions are commonly used. The most popular are Alloys 443.2 and A380.2. By choosing alloys such as these, conductivities from 25 to 35% IACS can be obtained; in fact, highly experimental alloys with even higher resistivities have been developed for motor rotor applications.

Although gross casting defects may adversely affect electrical performance, the conductivity of alloys employed in rotor manufacture is more exclusively controlled by composition. Table 4 lists the effects of the various elements in and out of solution on the resistivity of aluminum. Simple calculation using these values accurately predicts total resistivity and its reciprocal conductivity for any composition. A more general and easy-to-use formula for conductivity that offers sufficient accuracy for most purposes is:

$$\text{Conductivity, \%IACS} = 63.50 - 6.9x - 83y$$

where 63.5% is the conductivity of very pure aluminum in %IACS, x = iron + silicon (in wt%), and y = titanium + vanadium + manganese + chromium (in wt%).

References to specific composition limits and manufacturing techniques for rotor alloys show the use of composition controls that reflect electrical considerations. The peritectic elements are limited because their presence is harmful to electrical conductivity. The prealloyed ingots produced to these specifications control conductivity by making boron additions, which form complex precipitates with these elements before casting. In addition the iron and silicon contents are subject to control with the objective of promoting the alpha Al-Fe-Si phase intermetallics least harmful to castability. Ignoring these important relationships results in variable electrical performance, and of at least equal importance, variable casting results.

Commercial Duralumin Alloys. These alloys were first produced and were named by Durener Metallwerke Aktien Gesellschaft in the early 1900s. They were the first heat-treatable aluminum alloys.

The Duralumin alloys have been used extensively as cast and wrought products where high strength and toughness are required. Being essentially a single-phase alloy, improved ductility at higher strengths is inherent as compared to the two-phase silicon alloys. However, this difference also makes these alloys more difficult to cast.

After World War I, the European aluminum casting community developed AU5GT

Table 4 Effect of elements in and out of solid solution on the resistivity of aluminum

Element	Maximum solubility in Al, %	Average increase(a) in resistivity per wt%, microhm-cm	
		In solution	Out of solution(b)
Chromium	0.77	4.00	0.18
Copper	5.65	0.344	0.030
Iron	0.052	2.56	0.058
Lithium	4.0	3.31	0.68
Magnesium	14.9	0.54(c)	0.22(c)
Manganese	1.82	2.94	0.34
Nickel	0.05	0.81	0.061
Silicon	1.65	1.02	0.088
Titanium	1.0	2.88	0.12
Vanadium	0.5	3.58	0.28
Zinc	82.8	0.094(d)	0.023(d)
Zirconium	0.28	1.74	0.044

(a) Add above increase to the base resistivity for high-purity aluminum, 2.65 microhm-cm at 20 °C (68 °F) or 2.71 microhm-cm at 25 °C (77 °F). (b) Limited to about twice the concentration given for the maximum solid solubility, except as noted. (c) Limited to approximately 10%. (d) Limited to approximately 20%

(204 type) and similar Al-Cu-Mg alloys. In the United States, alloys 195 and B195 of the Al-Cu-Si composition were popularized. Between World Wars I and II, and in both communities, these alloys served well in the special situations in which strength and toughness were required. This came at the expense of the extra production costs required because of the poorer castability.

Since World War II, the higher-purity aluminum available from the smelters has enabled the foundryman to make substantial improvements in the mechanical properties of highly castable Al-Si, Al-Si-Cu, and Al-Si-Mg alloys. As a result, the use of the Duralumin alloys has dramatically decreased.

The more recently developed Al-Cu-Mg alloys and applications include many that emphasize the unusual strength and toughness achievable with impurity controls. New developments in foundry equipment and control techniques also have helped some foundries to solve the castability problems.

Premium-quality castings provide higher levels of quality and reliability than are found in conventionally produced parts. These castings may display optimum properties in one or more of the following characteristics: mechanical properties (determined by test coupons machined from representative parts), soundness (determined radiographically), dimensional accuracy, and finish. However, castings of this classification are notable primarily for the mechanical property attainment that reflects extreme soundness, fine dendrite-arm spacing, and well-refined grain structure. These technical objectives require the use of chemical compositions competent to display the premium engineering properties. Alloys considered to be premium engineered compositions appear in separately negotiated specifications or in those such as military specification MIL-A-21180, which is extensively used in the United States for premium casting procurement. Mechanical properties of premium aluminum castings are given in the section "Properties of Aluminum Casting Alloys" in this article.

Alloys considered premium by definition and specification are A201.0, A206.0, 224.0, 249.0, 354.0, A356.0 (D356.0), A357.0 (D357.0), and 358.0. All alloys employed in premium casting engineering work are characterized by optimum concentrations of hardening elements and restrictively controlled impurities. Although any alloy can be produced in cast form with properties and soundness conforming to a general description of premium values relative to corresponding commercial limits, only those alloys demonstrating yield strength, tensile strength, and especially elongation in a premium range belong in this grouping. They fall into two categories: high-strength aluminum-silicon compositions, and those alloys of the 2*xx* series, which by restricting impurity element concentrations provide outstanding ductility, toughness, and tensile properties with notably poorer castability.

In all premium casting alloys, impurities are strictly limited for the purposes of improving ductility. In aluminum-silicon alloys, this translates to control iron at or below 0.01% Fe with measurable advantages to the range of 0.03 to 0.05%, the practical limit of commercial smelting capability.

Beryllium is present in A357 and 158 alloys, not to inhibit oxidation (although that is a corollary benefit), but to alter the form of the insoluble phase to a more nodular form less detrimental to ductility.

The development of hot isostatic pressing is pertinent to the broad range of premium castings but is especially relevant for the more difficult-to-cast aluminum-copper series.

Piston and Other Elevated-Temperature Alloys. The universal acceptance of aluminum pistons by all gasoline engine manufacturers in the United States can be attributed to their light weight and high thermal conductivity. The effect of the lower inertia of the aluminum pistons on the bearing loading permits higher engine speeds and reduced crankshaft counterweighting.

Aluminum automotive pistons generally are permanent mold castings. This design usually is superior in economy and design flexibility. The alloy most commonly used for passenger car pistons, 332.0-T5, has a good combination of foundry, mechanical, and physical characteristics, including low thermal expansion. Heat treatment improves hardness for improved machinability and eliminates any permanent changes in dimensions from residual growth due to aging at operating temperatures.

Piston alloys for heavy-duty engines include the low-expansion alloys 336.0-T551 (A132-T551) and 332.0-T5 (F132-T5). Alloy 242-T571 (142-T571) is also used in some heavy-duty pistons because of its higher thermal conductivity and superior properties at elevated temperatures.

Other applications of aluminum alloys for elevated-temperature use include air-cooled cylinder heads for airplanes and motorcycles. The 10% Cu Alloy 222.0-T61 was used extensively for this purpose prior to the 1940s but has been replaced by the 242.0 and 243.0 compositions because of their better properties at elevated temperatures.

For use at moderate elevated temperatures (up to 175 °C, or 350 °F), Alloys 355 and C355 have been extensively used. These applications include aircraft motor and gear housings. Alloy A201.0 and the A206.0 type alloys have also been used in this temperature range when the combination of high strength at room temperatures and elevated temperatures is required.

Standard General-Purpose Aluminum Casting Alloys. Alloys with silicon as the major alloying constituent are by far the most important commercial casting alloys, primarily because of their superior casting characteristics. Binary aluminum-silicon alloys (443.0, 444.0, 413.0, and A413.0) offer further advantages of high resistance to corrosion, good weldability, and low specific gravity. Although castings of these alloys are somewhat difficult to machine, larger quantities are machined successfully with sintered carbide tools and flood application of lubricant. Application areas are:

- *Alloy 443* (Si at 7%) is used with all casting processes for parts where high strength is less important than good ductility, resistance to corrosion, and pressure tightness
- *Permanent mold Alloys 444 and A444* (Si at 7%) have especially high ductility and are used where impact resistance is a primary consideration (for example, highway bridge-rail support castings)
- *Alloys 413.0 and A413.0* (Si at 12%) are close to the eutectic composition, and as a result, have very high fluidity. They are useful in die casting and where cast-in lettering or other high-definition casting surfaces are required

In the silicon-copper alloys (213.0, 308.0, 319.0, and 333.0), the silicon provides good casting characteristics, and the copper imparts moderately high strength and improved machinability with reduced ductility and lower resistance to corrosion. The silicon range is 3 to 10.5%, and the copper content is 2 to 4.5%. These and similar general-purpose alloys are used mainly in the F temper. The T5 temper can be added to some of these alloys to improve hardness and machinability.

Alloy 356.0 (7 Si, 0.3 Mg) has excellent casting characteristics and resistance to corrosion. This justifies its use in large quantities for sand and permanent mold castings. Several heat treatments are used and provide the various combinations of tensile and physical properties that make it attractive for many applications. This includes many parts in both the auto and aerospace industries. The companion alloy of 356.0 with lower iron content affords higher tensile properties in the premium-quality sand and permanent mold castings. Even higher tensile properties are obtained using this premium casting process using 357.0, A357.0, 358.0, and 359.0 alloys. The high properties of these alloys, attained by T6-type heat treatments, are of special interest to aerospace and military applications.

The 355.0 type alloys, or Al-Si-Mg-Cu alloys, offer greater response to the heat treatment because of the copper addition. This gives the higher strengths with some sacrifice in ductility and resistance to corrosion. Representative sand and permanent mold alloys include 355.0 (5 Si, 1.3 Cu, 0.4 Mg, 0.4 Mn) and 328.0 (8 Si, 1.5 Cu, 0.4 Mg, 0.4 Mn). Some applications include cylinder blocks for internal combustion engines, jet engine compressor cases, and accessory housings.

Alloy C355.0 with low iron is a higher-tensile version of 355, for heat-treated, premium-quality, sand, and permanent mold castings. Some of the applications include tank engine cooling fans, high-speed rotating parts such as impellers. When the premium-strength casting processes are used, even higher tensile properties can be obtained with heat-treated Alloy 354.0 (9 Si, 1.8 Cu, 0.5 Mg). This is also of interest in aerospace applications.

The 390.0 (17 Si, 4.5 Cu, 0.5 Mg) type alloys have enjoyed much growth in recent years. These alloys have high wear resistance and a low thermal expansion coefficient but somewhat poorer casting and machining characteristics than the other alloys in this group. B390.0 is a low-iron version of 390.0 that can be used to advantage for sand and permanent mold casting. Some uses and applications include auto engine cylinder blocks, pistons, and so forth.

Die Casting Alloys. In terms of product tonnage, the use of aluminum alloys for die casting is almost twice as large as the usage of aluminum alloys in all other casting methods combined. In addition, alloys of aluminum are used in die castings more extensively than for any other base metal. Aluminum die castings usually are not heat treated, but occasionally are given dimensional and metallurgical stabilization treatments (variations of aging and annealing processes).

Compositions. The highly castable Al-Si family of alloys is the most important group of alloys for die casting. Of these, alloy 380.0 and its modifications constitute about 85% of aluminum die cast production. The 380.0 family of alloys provides a good combination of cost, strength, and corrosion resistance, together with the high fluidity and freedom from hot shortness that are required for ease of casting. Where better corrosion resistance is required, alloys lower in copper, such as 360.0 and 413.0, must be used. Rankings of these alloys in terms of die soldering and die filling capacity are given in Table 5. The hypereutectic aluminum-silicon alloy 390.0 type has found many useful applications in recent years. In heavy-wear uses, the increased hardness has given it a substantial advantage over normal 380.0 alloys (without any significant problems related to castability). Hypereutectic aluminum-silicon alloys are growing in importance as their valuable characteristics and excellent die casting properties are exploited in automotive and other applications.

Table 5 Characteristics of aluminum die casting alloys

See Table 3 for other characteristics.

Alloy	Resistance to die soldering(a)	Die filling capacity
360.0	2	3
A360.0	2	3
380.0	1	2
A380.0	1	2
383.0	2	1
384.0	2	1
413.0	1	1
A413.0	1	1
C443.0	4	4
518.0	5	5

(a) Rankings from ASTM B 85. Relative rating of die casting alloys from 1 to 5; 1 is the highest or best possible rating. A rating of 5 in one or more categories does not rule an alloy out of commercial use if other attributes are favorable; however, ratings of 5 may present manufacturing difficulties.

Magnesium content is usually controlled at low levels to minimize oxidation and the generation of oxides in the casting process. Nevertheless, alloys containing appreciable magnesium concentrations are routinely produced. Alloy 518.0, for example, is occasionally specified when the highest corrosion resistance is required. This alloy, however, has low fluidity and some tendency to hot shortness. It is difficult to cast, which is reflected in higher costs per casting.

Iron content of 0.7% or greater is preferred in most die casting operations to maximize elevated-temperature strength, to facilitate ejection, and to minimize soldering to the die face. Iron content is usually 1 ± 0.3%. Improved ductility through reduced iron content has been an incentive resulting in widespread efforts to develop a tolerance for iron as low as approximately 0.25%. These efforts focus on process refinements and improved die lubrication.

Additions of zinc are sometimes used to enhance the fluidity of 380.0, and at times, other die casting alloys.

Aluminum-base bearing alloys are primarily alloyed with tin. These alloys are discussed in the section "Aluminum-Tin Alloys" in this article. Aluminum-tin bearing alloys are also discussed in the article "Tin and Tin Alloys" in this Volume.

Effects of Alloying

Antimony. At concentration levels equal to or greater than 0.05%, antimony refines eutectic aluminum-silicon phase to lamellar form in hypoeutectic compositions. The effectiveness of antimony in altering the eutectic structure depends on an absence of phosphorus and on an adequately rapid rate of solidification. Antimony also reacts with either sodium or strontium to form coarse intermetallics with adverse effects on castability and eutectic structure.

Antimony is classified as a heavy metal with potential toxicity and hygiene implications, especially as associated with the possibility of stibine gas formation and the effects of human exposure to other antimony compounds. In cases of direct exposure, OSHA Safety and Health Standards 2206 specifies the following 8-h weighted average exposure limits for antimony and other selected metals:

- Antimony, 0.5 mg/m^3
- Chromium, 0.5 mg/m^3
- Copper, 0.1 mg/m^3
- Lead, 0.2 mg/m^3
- Manganese, 1.0 mg/m^3
- Nickel, 1.0 mg/m^3
- Silver, 0.01 mg/m^3
- Zinc, 5.0 mg/m^3
- Beryllium, 2.0 μg/m^3
- Cadmium, 0.2 mg/m^3

As an additive for aluminum alloys, there is no indication of danger of antimony in aluminum alloys, particularly at the 0.08 to 0.15% levels of antimony in alloys that have been produced for years.

Beryllium additions of as low as a few parts per million may be effective in reducing oxidation losses and associated inclusions in magnesium-containing compositions. Studies have shown that proportionally increased beryllium concentrations are required for oxidation suppression as magnesium content increases.

At higher concentrations (>0.04%), beryllium affects the form and composition of iron-containing intermetallics, markedly improving strength and ductility. In addition to changing beneficially the morphology of the insoluble phase, beryllium changes its composition, rejecting magnesium from the Al-Fe-Si complex and thus permitting its full use for hardening purposes.

Beryllium-containing compounds are, however, numbered among the known carcinogens that require specific precautions in the melting, molten metal handling, dross handling and disposition, and welding of alloys. Standards define the maximum beryllium in welding rod and weld base metal as 0.008 and 0.010%, respectively.

Bismuth improves the machinability of cast aluminum alloys at concentrations greater than 0.1%.

Boron combines with other metals to form borides, such as Al_2 and TiB_2. Titanium boride forms stable nucleation sites for interaction with active grain-refining phases such as $TiAl_3$ in molten aluminum.

Metallic borides reduce tool life in machining operations, and in coarse particle form they consist of objectionable inclusions with detrimental effects on mechanical properties and ductility. At high boron concentrations, borides contribute to furnace sludging, particle agglomeration, and increased risk of casting inclusions. However, boron treatment of aluminum-containing peritectic elements is practiced to improve purity and electrical conductivity in rotor casting. High rotor alloy grades may specify boron to exceed titanium and vanadium contents to ensure either the complexing or precipitation of these elements for improved electrical performance (see the section "Rotor Castings" in this article).

Cadmium in concentrations exceeding 0.1% improves machinability. Precautions that acknowledge volatilization at 767 °C (1413 °F) are essential.

Calcium is a weak aluminum-silicon eutectic modifier. It increases hydrogen solubility and is often responsible for casting porosity at trace concentration levels. Calcium concentrations greater than approximately 0.005% also adversely affect ductility in aluminum-magnesium alloys.

Chromium additions are commonly made in low concentrations to room-temperature aging and thermally unstable compositions in which germination and grain growth are known to occur. Chromium typically forms the compound $CrAl_7$, which displays extremely limited solid-state solubility and is therefore useful in suppressing grain growth tendencies. Sludge that contains iron, manganese, and chromium is sometimes encountered in die casting compositions, but it is rarely encountered in gravity casting alloys. Chromium improves corrosion resistance in certain alloys and increases quench sensitivity at higher concentrations.

Copper. The first and most widely used aluminum alloys were those containing 4 to 10% Cu. Copper substantially improves strength and hardness in the as-cast and heat-treated conditions. Alloys containing 4 to 6% Cu respond most strongly to thermal treatment. Copper generally reduces resistance to general corrosion, and in specific compositions and material conditions, stress-corrosion susceptibility. Additions of copper also reduce hot tear resistance and decrease castability.

Iron improves hot tear resistance and decreases the tendency for die sticking or soldering in die casting. Increases in iron content are, however, accompanied by substantially decreased ductility. Iron reacts to form a myriad of insoluble phases in aluminum alloy melts, the most common of which are $FeAl_3$, $FeMnAl_6$, and $\alpha AlFeSi$. These essentially insoluble phases are responsible for improvements in strength, especially at elevated temperature. As the fraction of insoluble phase increases with increased iron content, casting considerations such as flowability and feeding characteristics are adversely affected. Iron participates in the formation of sludging phases with manganese, chromium, and other elements.

Lead is commonly used in aluminum casting alloys at greater than 0.1% for improved machinability.

Magnesium is the basis for strength and hardness development in heat-treated Al-Si alloys and is commonly used in more complex Al-Si alloys containing copper, nickel, and other elements for the same purpose. The hardening-phase Mg_2Si displays a useful solubility limit corresponding to approximately 0.70% Mg, beyond which either no further strengthening occurs or matrix softening takes place. Common premium-strength compositions in the Al-Si family employ magnesium in the range of 0.40 to 0.070% (see the section "Premium-Quality Castings" in this article).

Binary Al-Mg alloys are widely used in applications requiring a bright surface finish and corrosion resistance, as well as attractive combinations of strength and ductility. Common compositions range from 4 to 10% Mg, and compositions containing more than 7% Mg are heat treatable. Instability and room-temperature aging characteristics at higher magnesium concentrations encourage heat treatment.

Manganese is normally considered an impurity in casting compositions and is controlled to low levels in most gravity cast compositions. Manganese is an important alloying element in wrought compositions through which secondary foundry compositions may contain higher manganese levels. In the absence of work hardening, manganese offers no significant benefits in cast aluminum alloys. Some evidence exists, however, that a high-volume fraction of $MnAl_6$ in alloys containing more than 0.5% Mn may beneficially influence internal casting soundness. Manganese can also be employed to alter response in chemical finishing and anodizing.

Mercury. Compositions containing mercury were developed as sacrificial anode materials for cathodic protection systems, especially in marine environments. The use of these optimally electronegative alloys, which did not passivate in seawater, was severely restricted for environmental reasons.

Nickel is usually employed with copper to enhance elevated-temperature properties. It also reduces the coefficient of thermal expansion.

Phosphorus. In AlP_3 form, phosphorus nucleates and refines primary silicon-phase formation in hypereutectic Al-Si alloys. At parts per million concentrations, phosphorus coarsens the eutectic structure in hypoeutectic Al-Si alloys. Phosphorus diminishes the effectiveness of the common eutectic modifiers sodium and strontium.

Silicon. The outstanding effect of silicon in aluminum alloys is the improvement of casting characteristics. Additions of silicon to pure aluminum dramatically improve fluidity, hot tear resistance, and feeding characteristics. The most prominently used compositions in all casting processes are those of the aluminum-silicon family. Commercial alloys span the hypoeutectic and hypereutectic ranges up to about 25% Si.

In general, an optimum range of silicon content can be assigned to casting processes. For slow cooling-rate processes (such as plaster, investment, and sand), the range is 5 to 7%, for permanent mold 7 to 9%, and for die casting 8 to 12%. The bases for these recommendations are the relationship between cooling rate and fluidity and the effect of percentage of eutectic on feeding. Silicon additions are also accompanied by a reduction in specific gravity and coefficient of thermal expansion.

Silver is used in only a limited range of aluminum-copper premium-strength alloys at concentrations of 0.5 to 1.0%. Silver contributes to precipitation hardening and stress-corrosion resistance.

Sodium modifies the aluminum-silicon eutectic. Its presence is embrittling in aluminum-magnesium alloys. Sodium interacts with phosphorus to reduce its effectiveness in modifying the eutectic and that of phosphorus in the refinement of the primary silicon phase.

Strontium is used to modify the aluminum-silicon eutectic. Effective modification can be achieved at very low addition levels, but a range of recovered strontium of 0.008 to 0.04% is commonly used. Higher addition levels are associated with casting porosity, especially in processes or in thick-section parts in which solidification occurs more slowly. Degassing efficiency may also be adversely affected at higher strontium levels.

Tin is effective in improving antifriction characteristics and is therefore useful in bearing applications. Casting alloys may contain up to 25% Sn. Additions can also be made to improve machinability. Tin may influence precipitation-hardening response in some alloy systems.

Titanium is extensively used to refine the grain structure of aluminum casting alloys, often in combination with smaller amounts of boron. Titanium in excess of the stoichiometry of TiB_2 is necessary for effective grain refinement. Titanium is often employed at concentrations greater than those required for grain refinement to reduce cracking tendencies in hot-short compositions.

Zinc. No significant technical benefits are obtained by the addition of zinc to aluminum.

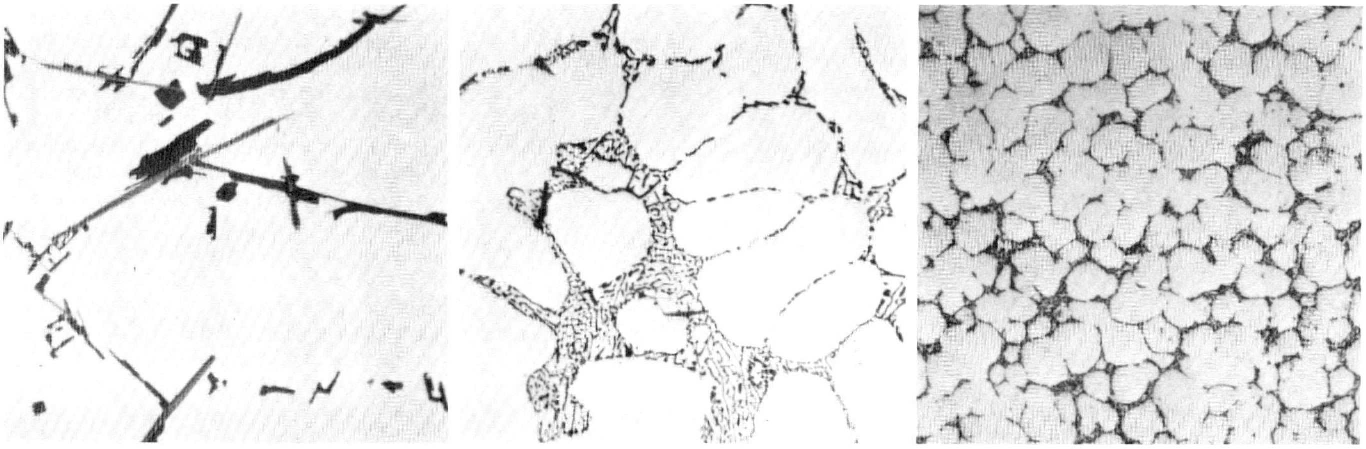

(a) Sand cast 443.0-F

Typical tensile properties				
Tensile strength		Yield strength		Elongation,
MPa	ksi	MPa	ksi	%
130	19	55	8	8

(b) Permanent mold cast B443.0-F

Typical tensile properties				
Tensile strength		Yield strength		Elongation,
MPa	ksi	MPa	ksi	%
160	23	60	9	10

(c) Die cast C443.0-F

Typical tensile properties				
Tensile strength		Yield strength		Elongation,
MPa	ksi	MPa	ksi	%
230	33	110	16	9

Fig. 2 Aluminum, 5% Si alloy microstructures resulting from different solidification rates characteristic of different casting processes. Dendrite cell size and constituent particle size decrease with increasing cooling rate, from sand cast to permanent mold cast to die cast. Etchant, 0.5% hydrofluoric acid. 500×

Accompanied by the addition of copper and/or magnesium, however, zinc results in attractive heat-treatable or naturally aging compositions. A number of such compositions are in common use. Zinc is also commonly found in secondary gravity and die casting compositions. In these secondary alloys, tolerance for up to 3% zinc allows the use of lower grade scrap aluminum to make these alloys and thus lowers cost.

Structure Control

The microstructural features that most strongly affect mechanical properties are:

- Grain size and shape
- Dendrite-arm spacing
- Size and distribution of second-phase particles and inclusions

Some of these microstructural features, such as grain size and dendrite-arm spacing, are primarily controlled by cooling and solidification rates. Figure 2, for example, shows the variation in microstructures and mechanical properties resulting from the different solidification rates associated with different casting processes.

Like grain size and interdendritic spacing, the finer the dispersion of inclusions and second-phase particles, the better the properties of the casting. Fine dispersion requires small particles; large masses of oxides or intermetallic compounds produce excessive brittleness. Controlling size and shape of microconstituents can be done to some extent by controlling composition, but is accomplished more efficiently by minimizing the period of time during which microconstituents can grow. Like minimizing grain size and interdendritic spacing, minimizing time for growth of microconstituents calls for rapid cooling. Thus, it is evident that high cooling rate is of paramount importance in obtaining good casting quality.

Microstructural features such as the size and distribution of primary and intermetallic phases are considerably more complex to control by chemistry. However, chemistry control (particularly control of impurity element concentrations), control of element ratios based on the stoichiometry of intermetallic phases, and control of solidification conditions to ensure uniform size and distribution of intermetallics are all useful. The use of modifiers and refiners to influence eutectic and hypereutectic structures in aluminum-silicon alloys is also an example of the manner in which microstructures and macrostructures can be optimized in foundry operations.

Dendrite-Arm Spacing. In all commercial processes, solidification takes place through the formation of dendrites in the liquid solution. The cells contained within the dendrite structure correspond to the dimensions separating the arms of primary dendrites and are controlled for a given composition primarily by solidification rate. Another factor that may affect interdendritic spacing is the presence of second-phase particles and oxide or gas inclusions. During freezing, inclusions and second-phase particles can segregate to the spaces between dendrite arms and thus increase the spacing.

The farther apart the dendrite arms are, the coarser the distribution of microconstituents and the more pronounced their adverse effects on properties. Thus, small interdendritic spacing is necessary for high casting quality. Figure 3, for example, illustrates the improvement in mechanical properties achievable by the change in dendrite formation controlled by solidification rate. Although several factors affect spacing to some extent, the only efficient way of ensuring fine spacing is use of rapid cooling.

In premium engineered castings and in many other casting applications, careful attention is given to obtaining solidification rates corresponding to optimum mechanical property development. Solidification rate affects more than dendrite cell size, but dendrite cell size measurements are becoming increasingly important.

Grain Refinement. A fine, equiaxed grain structure is normally desired in aluminum castings, because castings with fine, equiaxed grains offer the best combination of strength and ductility. The type and size of grains formed are determined by alloy composition, solidification rate, and the addition of master alloys (grain refiners) containing intermetallic phase particles, which provide sites for heterogeneous grain nucleation.

Grain size is refined by increasing the solidification rate but is also dependent on the presence of grain-refining elements (principally titanium and boron) in the alloy. To some extent, size and shape of grains can be controlled by addition of grain refiners, but use of low pouring temperatures and high cooling rates are the preferred methods.

All aluminum alloys can be made to solidify with a fully equiaxed, fine grain structure through the use of suitable grain-refining additions. The most widely used grain

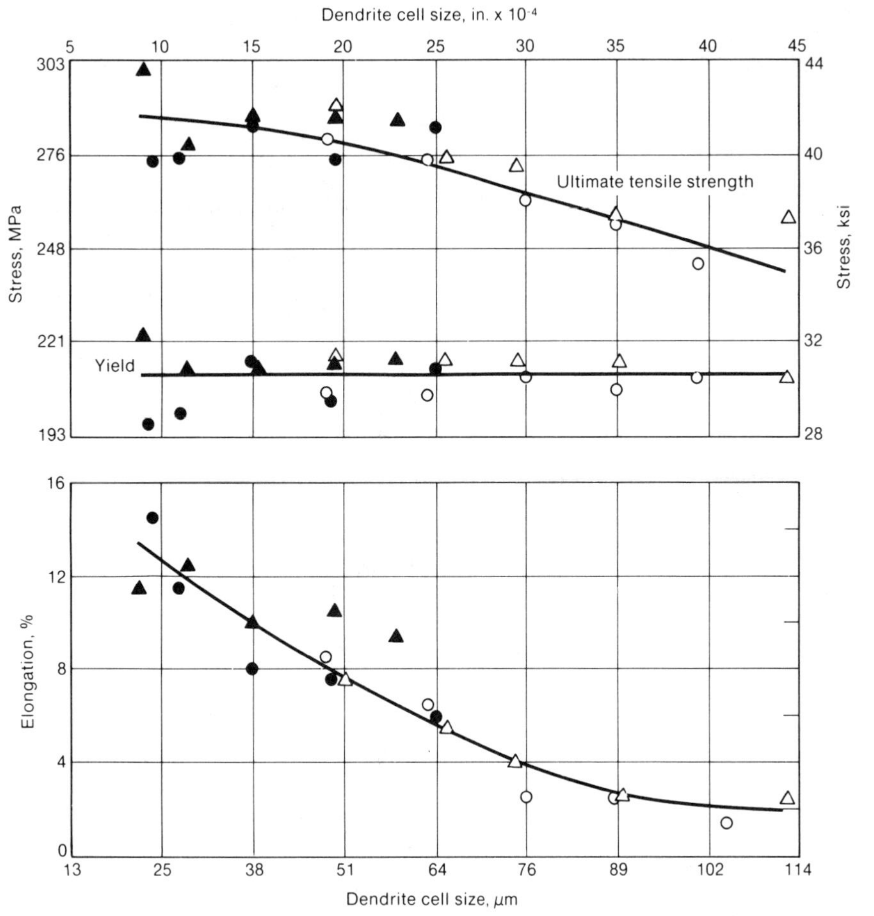

Fig. 3 Tensile properties versus dendrite cell size for four heats of aluminum alloy A356-T62 plaster cast plates

refiners are master alloys of titanium, or of titanium and boron, in aluminum. Aluminum-titanium refiners generally contain from 3 to 10% Ti. The same range of titanium concentrations is used in Al-Ti-B refiners with boron contents from 0.2 to 1% and titanium-to-boron ratios ranging from about 5 to 50. Although grain refiners of these types can be considered conventional hardeners or master alloys, they differ from master alloys added to the melt for alloying purposes alone. To be effective, grain refiners must introduce controlled, predictable, and operative quantities of aluminides (and borides) in the correct form, size, and distribution for grain nucleation. Wrought refiner in rod form, developed for the continuous treatment of aluminum in primary operations, is available in sheared lengths for foundry use. The same grain-refining compositions are furnished in waffle form. In addition to grain-refining master alloys, salts (usually in compacted form) that react with molten aluminum to form combinations of $TiAl_3$ and TiB_2 are also available.

Modification of hypoeutectic aluminum-silicon alloys involves the improvement of properties by inducing structural modification of the normally occurring eutectic. Modification is achieved by the addition of certain elements such as calcium, sodium, strontium, and antimony. It is also understood that increased solidification is useful in achieving modified structures.

In general, the greatest benefits are achieved in alloys containing from 5% Si to the eutectic concentration. The addition of modifying elements (such as calcium, sodium, strontium, and antimony) to these hypoeutectic aluminum-silicon alloys results in a finer lamellar or fibrous eutectic network (Fig. 4). Although there is no agreement on the mechanisms involved, the most popular explanations suggest that modifying additions suppress the growth of silicon crystals within the eutectic, providing a finer distribution of lamellae relative to the growth of the eutectic. It has also been well established that phosphorus interferes with the modification mechanism. Phosphorus reacts with sodium and probably with strontium and calcium to form phosphides that nullify the intended modification additions. It is therefore desirable to use low-phosphorus metal when modification is a process objective and to make larger modifier additions to compensate for phosphorus-related losses.

Effects of Modification. Typically, modified structures display somewhat higher tensile properties and appreciably improved ductility when compared to similar but unmodified structures. Figure 5 illustrates the desirable effects on mechanical properties that can be achieved by modification. Improved performance in casting is characterized by improved flow and feeding as well as by superior resistance to elevated-temperature cracking.

Refinement of Hypereutectic Aluminum-Silicon Alloys. The elimination of large, coarse primary silicon crystals that are harmful in the casting and machining of hypereutectic silicon alloy compositions is a function of primary silicon refinement. Phosphorus added to molten alloys containing more than the eutectic concentration of silicon, made in the form of metallic phosphorus or phosphorus-containing compounds such as phosphor-copper and phosphorus pentachloride, has a marked effect on the distribution and form of the primary silicon phase. Investigations have shown that retained trace concentrations as low as 0.0015 through 0.03% P are effective in achieving the refined structure. Disagreements on recommended phosphorus ranges and addition rates have been caused by the extreme difficulty of accurately sampling and analyzing for phosphorus. More recent developments employing vacuum stage spectrographic or quantometric analysis now provide rapid and accurate phosphorus measurements.

Following melt treatment by phosphorus-containing compounds, refinement can be expected to be less transient than the effects of conventional modifiers on hypoeutectic modification. Furthermore, the solidification of phosphorus-treated melts, cooling to room temperature, reheating, remelting, and resampling in repetitive tests have shown that refinement is not lost; however, primary silicon particle size increases gradually, responding to a loss in phosphorus concentration. Common degassing methods accelerate phosphorus loss, especially when chlorine or freon is used. In fact, brief inert gas fluxing is frequently employed to reactive aluminum phosphide nuclei, presumably by resuspension.

Practices that are recommended for melt refinement are as follows:

- Melting and holding temperature should be held to a minimum
- The alloy should be thoroughly chlorine or freon fluxed before refining to remove phosphorus-scavenging impurities such as calcium and sodium
- Brief fluxing after the addition of phosphorus is recommended to remove the hydrogen introduced during the addition and to distribute the aluminum phosphide nuclei uniformly in the melt

Hydrogen Porosity. In general, two types of porosity may occur in cast aluminum: gas porosity and shrinkage porosity. Gas poros-

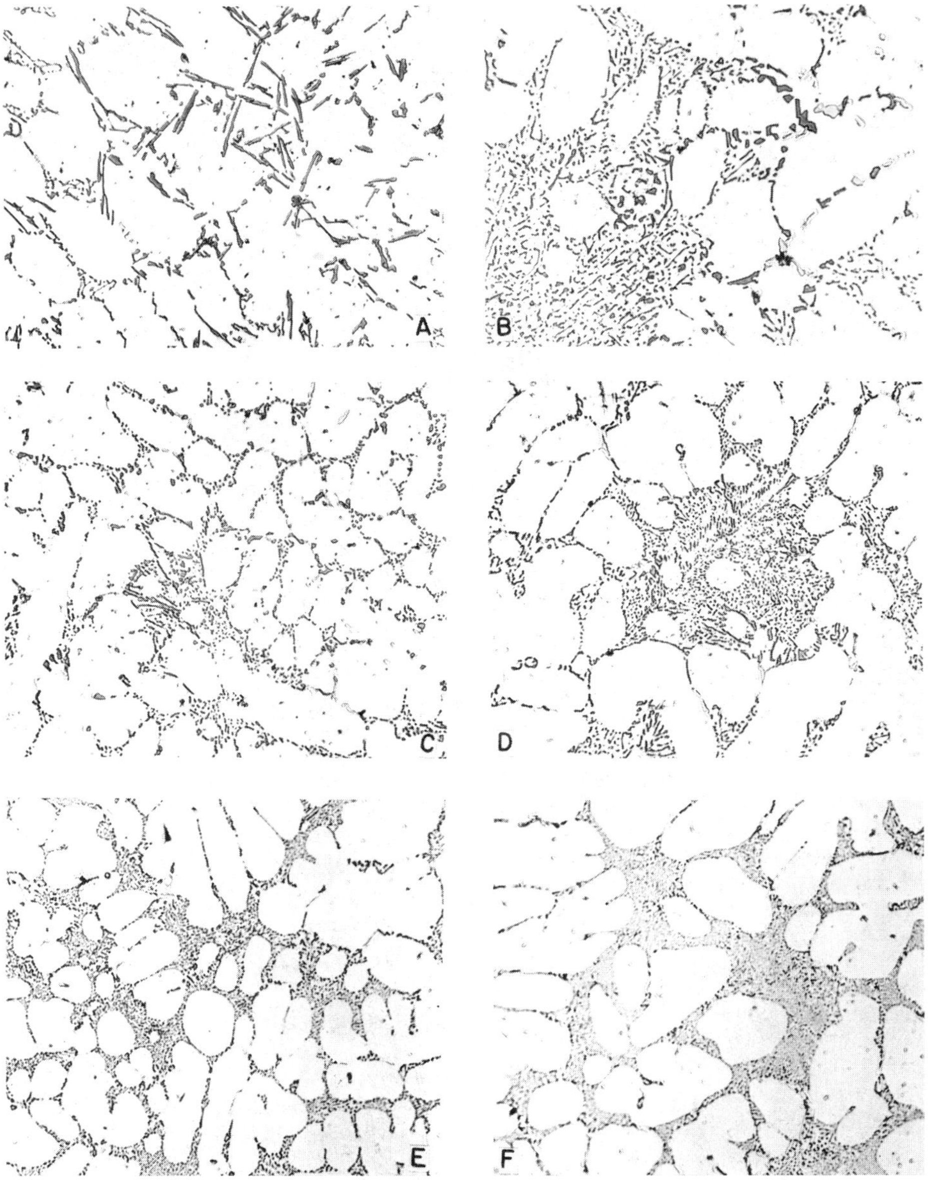

Fig. 4 Varying degrees of aluminum-silicon eutectic modification ranging from unmodified (A) to well modified (F). These are as-cast structures before any solution heat treatment.

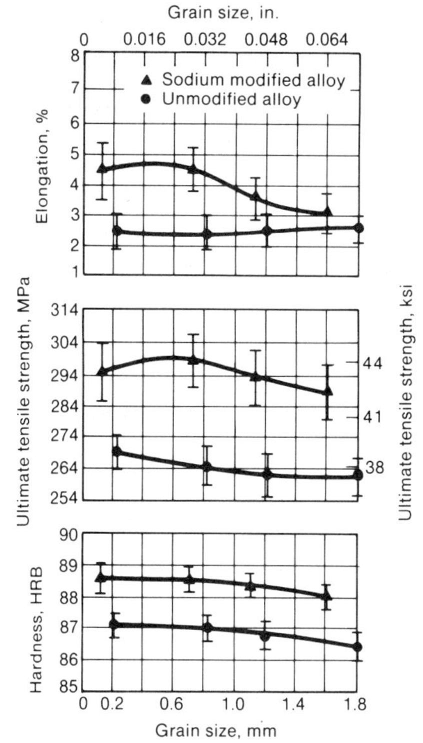

Fig. 5 Mechanical properties of as-cast A356 alloy tensile specimens as a function of modification and grain size

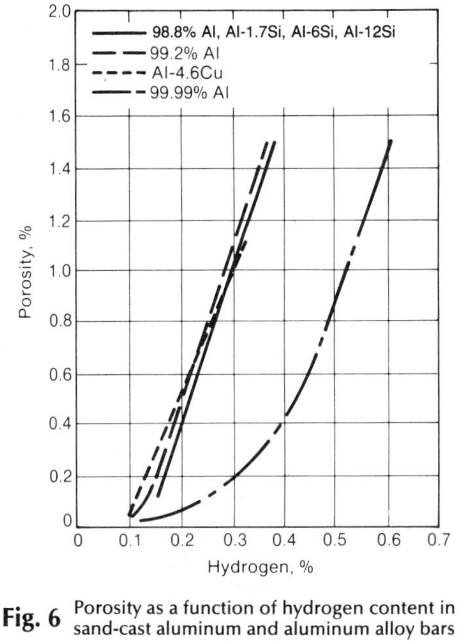

Fig. 6 Porosity as a function of hydrogen content in sand-cast aluminum and aluminum alloy bars

ity, which generally is fairly spherical in shape, results either from precipitation of hydrogen during solidification (because the solubility of this gas is much higher in the molten metal than in the solid metal) or from occlusion of gas bubbles during the high-velocity injection of molten metal in die casting.

Two types or forms of hydrogen porosity may occur in cast aluminum when the precipitation of molecular hydrogen during the cooling and solidification of molten aluminum results in the formation of primary and/or secondary voids. Of greater importance is interdendritic porosity, which is encountered when hydrogen contents are sufficiently high that hydrogen rejected at the solidification front results in solution pressures above atmospheric. Secondary (micron-size) porosity occurs when dissolved hydrogen contents are low, and void formation is characteristically subcritical.

Finely distributed hydrogen porosity may not always be undesirable. Hydrogen precipitation may alter the form and distribution of shrinkage porosity in poorly fed parts or part sections. Shrinkage is generally more harmful to casting properties. In isolated cases, hydrogen may actually be intentionally introduced and controlled in specific concentrations compatible with the application requirements of the casting in order to promote superficial soundness.

Nevertheless, hydrogen porosity adversely affects mechanical properties in a manner that varies with the alloy. Figure 6 shows the relationship between actual hydrogen content and observed porosity. Figure 7 defines the effect of porosity on the ultimate tensile strength of selected compositions.

It is often assumed that hydrogen may be desirable or tolerable in pressure-tight applications. The assumption is that hydrogen porosity is always present in the cast structure as integrally enclosed rounded voids. In fact, hydrogen porosity may occur as rounded or elongated voids and in the presence of shrinkage may decrease rather than increase resistance to pressure leakage.

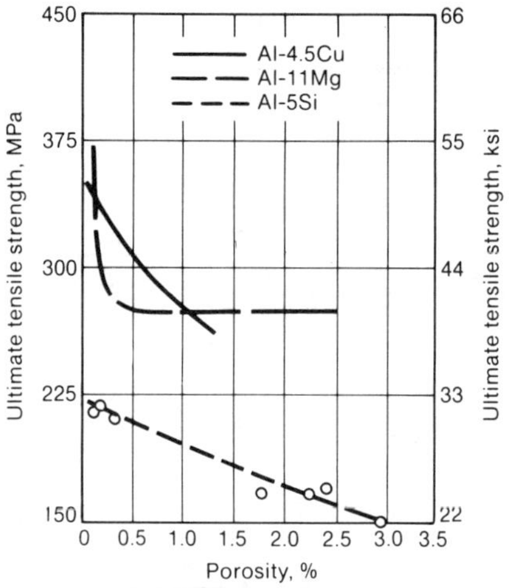

Fig. 7 Ultimate tensile strength versus hydrogen porosity for sand-cast bars of three aluminum alloys. The difference in tensile strength among the three alloys may be a function of heat treatment. The Al-11Mg alloy is typically used in the T4 temper (high toughness and ductility), while the other alloys are typically in the T6 condition (highest strength with acceptable ductility).

Shrinkage Porosity. The other source of porosity is the liquid-to-solid shrinkage that frequently takes the form of interdendritically distributed voids. These voids may be enlarged by hydrogen, and because larger dendrites result from slower solidification, the size of such porosity also increases as solidification rate decreases. It is not possible to establish inherent ratings with respect to anticipated porosity because castings made by any process can vary substantially in soundness—from nearly completely sound to very unsound—depending on casting size and design as well as on foundry techniques.

Heat Treatment. The metallurgy of aluminum and its alloys fortunately offers a wide range of opportunities for employing thermal treatment practices to obtain desirable combinations of mechanical and physical properties. Through alloying and temper selection, it is possible to achieve an impressive array of features that are largely responsible for the current use of aluminum alloy castings in virtually every field of application. Although the term heat treatment is often used to describe the procedures required to achieve maximum strength in any suitable composition through the sequence of solution heat treatment, quenching, and precipitation hardening, in its broadest meaning heat treatment comprises all thermal practices intended to modify the metallurgical structure of products in such a way that physical and mechanical characteristics are controllably altered to meet specific engineering criteria. In all cases, one or more of the following objectives form the basis for temper selection:

- Increase hardness for improved machinability
- Increase strength and/or produce the mechanical properties associated with a particular material condition
- Stabilize mechanical and physical properties
- Ensure dimensional stability as a function of time under service conditions
- Relieve residual stresses induced by casting, quenching, machining, welding, or other operations

To achieve these objectives, parts can be annealed, solution heat treated, quenched, precipitation hardened, overaged, or treated with combinations of these practices. In some simple shapes (for example, bearings), thermal treatment can also include plastic deformation in the form of cold work. Typical heat treatments for various aluminum casting alloys are given in Table 6.

Casting Processes

Aluminum is one of the few metals that can be cast by all of the processes used in casting metals. These processes, in decreasing order of amount of aluminum cast, are: die casting, permanent mold casting, sand casting (green sand and dry sand), plaster casting, and investment casting. Aluminum also is continuous cast. Each of these processes, and the castings produced by them, are discussed below. Other processes such as lost foam, squeeze casting, and hot isostatic pressing are also mentioned.

There are many factors that affect selection of a casting process for producing a specific aluminum alloy part. Some of the important factors in sand, permanent mold, and die casting are discussed in Table 7. The most important factors for all casting processes are:

- Feasibility and cost factors
- Quality factors

In terms of feasibility, many aluminum alloy castings can be produced by any of the available methods. For a considerable number of castings, however, dimensions or design features automatically determine the best casting method. Because metal molds weigh from 10 to 100 times as much as the castings they are used in producing, most very large cast products are made as sand castings rather than as die or permanent mold castings. Small castings usually are made with metal molds to ensure dimensional accuracy. Some parts can be produced much more easily if cast in two or more separate sections and bolted or welded together. Complex parts with many undercuts can be made easily by sand, plaster, or investment casting, but may be practically impossible to cast in metal molds even if sand cores are used.

When two or more casting methods are feasible for a given part, the method used very often is dictated by costs. As a general rule, the cheaper the tooling (patterns, molds, and auxiliary equipment), the greater the cost of producing each piece. Therefore, number of pieces is a major factor in the choice of a casting method. If only a few pieces are to be made, the method involving the least expensive tooling should be used, even if the cost of casting each piece is very high. For very large production runs, on the other hand, where cost of tooling is shared by a large number of castings, use of elaborate tooling usually decreases cost per piece and thus is justified. In mass production of small parts, for example, costs often are minimized by use of elaborate tooling that allows several castings to be poured simultaneously. Die castings are typical of this category.

Quality factors are also important in the selection of a casting process. When applied to castings, the term quality refers to both degree of soundness (freedom from porosity, cracking, and surface imperfections) and levels of mechanical properties (strength and ductility). From the discussions in the section "Control of Structure," it is evident that high cooling rate is of paramount importance in obtaining good casting quality. The tabulation below presents characteristic ranges of cooling rate for the various casting processes.

Casting processes	Cooling rate, °C/s	Dendrite-arm spacing, mm
Plaster, dry sand	0.05–0.2	0.1–1
Green sand, shell	0.1–0.5	0.05–0.5
Permanent mold	0.3–1	0.03–0.07
Die	50–500	0.005–0.015
Continuous	0.5–2	0.03–0.07

However, it should be kept in mind that in die casting, although cooling rates are very high, air tends to be trapped in the casting, which gives rise to appreciable amounts of porosity at the center. Extensive research has been conducted to find ways of reducing such porosity; however, it is difficult if not impossible to eliminate completely, and die castings often are lower in strength than low-pressure or gravity-fed permanent mold castings, which are more sound in spite of slower cooling.

Die Casting. Alloys of aluminum are used in die casting more extensively than alloys of any other base metal. In the United States alone, about 2.5 billion dollars worth of aluminum alloy die castings is produced each year. The die casting process consumes almost twice as much tonnage of aluminum alloys as all other casting processes combined.

Die casting is especially suited to production of large quantities of relatively small parts. Aluminum die castings weighing up to about 5 kg (10 lb) are common, but castings weighing as much as 50 kg (100 lb) are produced when the high tooling and casting-machine costs are justified.

Table 6 Typical heat treatments for aluminum alloy sand and permanent mold castings

Alloy	Temper	Type of casting(a)	Solution heat treatment(b) Temperature(c) °C	°F	Time, h	Aging treatment Temperature(c) °C	°F	Time, h
201.0(d)	T4	S or P	490–500(e)	910–930(e)	2	...	...	...
			+525–530	+980–990	14–20	Minimum of 5 days at room temperature		
	T6	S	510–515(e)	950–960(e)	2	...	...	...
			+525–530	+980–990	14–20	155	310	20
	T7	S	510–515(e)	950–960(e)	2	...	...	...
			+525–530	+980–990	14–20	190	370	5
	T43(f)	...	525	980	20	24 h at room temperature + ½ to 1 h at 160 °C		
	T71	...	490–500(e)	910–930(e)	2	...	...	...
			+525–530	+980–990	14–20	200	390	4
204.0(d)	T4	S or P	530	985	12	Minimum of 5 days at room temperature		
	T4	S or P	520	970	10	...	...	...
	T6(g)	S or P	530	985	12	(g)	(g)	...
206.0(d)	T4	S or P	490–500(e)	910–930(e)	2	...	...	...
			+525–530	+980–990	14–20	Minimum of 5 days at room temperature		
	T6	S or P	490–500(e)	910–930(e)	2	...	...	...
			+525–530	+980–990	14–20	155	310	12–24
	T7	S or P	490–500(e)	910–930(e)	2	...	...	...
			+525–530	+980–990	14–20	200	390	4
	T72	S or P	490–500(e)	910–930(e)	2	...	...	...
			+525–530	+980–990	14–20	243–248	470–480	
208.0	T55	S	...	...	...	155	310	16
222.0	O(h)	S	...	...	...	315	600	3
	T61	S	510	950	12	155	310	11
	T551	P	...	...	...	170	340	16–22
	T65	...	510	950	4–12	170	340	7–9
242.0	O(i)	S	...	...	...	345	650	3
	T571	S	...	...	...	205	400	8
		P	...	...	...	165–170	330–340	22–26
	T77	S	515	960	5(j)	330–355	625–675	2 (minimum)
	T61	S or P	515	960	4–12(j)	205–230	400–450	3–5
295.0	T4	S	515	960	12	...	...	...
	T6	S	515	960	12	155	310	3–6
	T62	S	515	960	12	155	310	12–24
	T7	S	515	960	12	260	500	4–6
296.0	T4	P	510	950	8	...	...	...
	T6	P	510	950	8	155	310	1–8
	T7	P	510	950	8	260	500	4–6
319.0	T5	S	...	...	...	205	400	8
	T6	S	505	940	12	155	310	2–5
		P	505	940	4–12	155	310	2–5
328.0	T6	S	515	960	12	155	310	2–5
332.0	T5	P	...	...	...	205	400	7–9
333.0	T5	P	...	...	...	205	400	7–9
	T6	P	505	950	6–12	155	310	2–5
	T7	P	505	940	6–12	260	500	4–6
336.0	T551	P	...	...	...	205	400	7–9
	T65	P	515	960	8	205	400	7–9
354.0	...	(k)	525–535	980–995	10–12	(h)	(h)	(l)
355.0	T51	S or P	...	...	...	225	440	7–9
	T6	S	525	980	12	155	310	3–5
		P	525	980	4–12	155	310	2–5
	T62	P	525	980	4–12	170	340	14–18
	T7	S	525	980	12	225	440	3–5
		P	525	980	4–12	225	440	3–9
	T71	S	525	980	12	245	475	4–6
		P	525	980	4–12	245	475	3–6
C355.0	T6	S	525	980	12	155	310	3–5
	T61	P	525	980	6–12	Room temperature		8 (minimum)
						155	310	10–12
356.0	T51	S or P	...	...	...	225	440	7–9
	T6	S	540	1000	12	155	310	3–5
		P	540	1000	4–12	155	310	2–5
	T7	S	540	1000	12	205	400	3–5
		P	540	1000	4–12	225	440	7–9
	T71	S	540	1000	10–12	245	475	3
		P	540	1000	4–12	245	475	3–6
A356.0	T6	S	540	1000	12	155	310	3–5
	T61	P	540	1000	6–12	Room temperature		8 (minimum)
						155	310	6–12

(continued)

(a) S, sand; P, permanent mold. (b) Unless otherwise indicated, solution treating is followed by quenching in water at 65–100 °C (150–212 °F). (c) Except where ranges are given, listed temperatures are ±6 °C or ±10 °F. (d) Casting wall thickness, solidification rate, and grain refinement affect the solution heat-treatment cycle in alloys 201.0, 204.0, and 206.0, and care must be taken in approaching the final solution temperature. Too rapid an approach can result in the occurrence of incipient melting. (e) For castings with thick or other slowly solidified sections, a pre-solution heat treatment ranging from about 490 to 515 °C (910 to 960 °F) may be needed to avoid too rapid a temperature rise to the solution temperature and the melting of $CuAl_2$. (f) Temper T43 for 201.0 was developed for improved impact resistance with some decrease in other mechanical properties. Typical Charpy value is 20J (15 ft · lb). (g) The French precipitation treatment technology for the heat treatment of 204.0 alloy requires 12 h at temperature. The aging temperatures of 140, 160, or 180 °C (285, 320, or 355 °F) are selected to meet the required combination of properties. (h) Stress relieve for dimensional stability as follows: hold 5 h at 413 ± 14 °C (775 ± 25 °F); furnace cool to 345 °C (650 °F) over a period of 2 h or more; furnace cool to 230 °C (450 °F) over a period of not more than ½ h; furnace cool to 120 °C (250 °F) over a period of approximately 2 h; cool to room temperature in still air outside the furnace. (i) No quench required; cool in still air outside furnace. (j) Air-blast quench from solution-treating temperature. (k) Casting process varies (sand, permanent mold, or composite) depending on desired mechanical properties. (l) Solution heat treat as indicated, then artificially age by heating uniformly at the temperature and for the time necessary to develop the desired mechanical properties. (m) Quench in water at 65–100 °C (150–212 °F) for 10–20 s only. (n) Cool to room temperature in still air outside the furnace.

Table 6 (continued)

Alloy	Temper	Type of casting(a)	Solution heat treatment(b) Temperature(c) °C	°F	Time, h	Aging treatment Temperature(c) °C	°F	Time, h
357.0	T6	P	540	1000	8	175	350	6
	T61	S	540	1000	10–12	155	310	10–12
A357.0	. . .	(k)	540	1000	8–12	(h)	(h)	(h)
359.0	. . .	(k)	540	1000	10–14	(h)	(h)	(h)
A444.0	T4	P	540	1000	8–12	. . .	. . .	. . .
520.0	T4	S	430	810	18(m)	. . .	. . .	. . .
535.0	T5(h)	S	400	750	5	. . .	. . .	. . .
705.0	T5	S	. . .	. . .	. . .	Room temperature		21 days
						100	210	8
		P	. . .	. . .	. . .	Room temperature		21 days
						100	210	10
707.0	T5	S	. . .	. . .	. . .	155	310	3–5
		P	. . .	. . .	. . .	Room temperature, or		21 days
						100	210	8
	T7	S	530	990	8–16	175	350	4–10
		P	530	990	4–8	175	350	4–10
710.0	T5	S	. . .	. . .	. . .	Room temperature		21 days
711.0	T1	P	. . .	. . .	. . .	Room temperature		21 days
712.0	T5	S	. . .	. . .	. . .	Room temperature, or		21 days
						155	315	6–8
713.0	T5	S or P	. . .	. . .	. . .	Room temperature, or		21 days
						120	250	16
771.0	T53(h)	S	415(n)	775(n)	5(n)	180(n)	360(n)	4(n)
	T5	S	. . .	. . .	. . .	180(n)	355(n)	3–5(n)
	T51	S	. . .	. . .	. . .	205	405	6
	T52	S	. . .	. . .	. . .	(h)	(h)	(h)
	T6	S	590(n)	1090(n)	6(n)	130	265	3
	T71	S	590(i)	1090(i)	6(i)	140	285	15
850.0	T5	S or P	. . .	. . .	. . .	220	430	7–9
851.0	T5	S or P	. . .	. . .	. . .	220	430	7–9
	T6	P	480	900	6	220	430	4
852.0	T5	S or P	. . .	. . .	. . .	220	430	7–9

(a) S, sand; P, permanent mold. (b) Unless otherwise indicated, solution treating is followed by quenching in water at 65–100 °C (150–212 °F). (c) Except where ranges are given, listed temperatures are ±6 °C or ± 10 °F. (d) Casting wall thickness, solidification rate, and grain refinement affect the solution heat-treatment cycle in alloys 201.0, 204.0, and 206.0, and care must be taken in approaching the final solution temperature. Too rapid an approach can result in the occurrence of incipient melting. (e) For castings with thick or other slowly solidified sections, a pre-solution heat treatment ranging from about 490 to 515 °C (910 to 960 °F) may be needed to avoid too rapid a temperature rise to the solution temperature and the melting of $CuAl_2$. (f) Temper T43 for 201.0 was developed for improved impact resistance with some decrease in other mechanical properties. Typical Charpy value is 20J (15 ft · lb). (g) The French precipitation treatment technology for the heat treatment of 204.0 alloy requires 12 h at temperature. The aging temperatures of 140, 160, or 180 °C (285, 320, or 355 °F) are selected to meet the required combination of properties. (h) Stress relieve for dimensional stability as follows: hold 5 h at 413 ± 14 °C (775 ± 25 °F); furnace cool to 345 °C (650 °F) over a period of 2 h or more; furnace cool to 230 °C (450 °F) over a period of not more than ½ h; furnace cool to 120 °C (250 °F) over a period of approximately 2 h; cool to room temperature in still air outside the furnace. (i) No quench required; cool in still air outside furnace. (j) Air-blast quench from solution-treating temperature. (k) Casting process varies (sand, permanent mold, or composite) depending on desired mechanical properties. (l) Solution heat treat as indicated, then artificially age by heating uniformly at the temperature and for the time necessary to develop the desired mechanical properties. (m) Quench in water at 65–100 °C (150–212 °F) for 10–20 s only. (n) Cool to room temperature in still air outside the furnace.

Typical applications of die cast aluminum alloys include:

Alloy 380.0	Lawnmower housings, gear cases, cylinder heads for air-cooled engines
Alloy A380.0	Streetlamp housings, typewriter frames, dental equipment
Alloy 360.0	Frying skillets, cover plates, instrument cases, parts requiring corrosion resistance
Alloy 413.0	Outboard motor parts such as pistons, connecting rods, and housings
Alloy 518.0	Escalator parts, conveyor components, aircraft and marine hardware and fittings

With die casting, it is possible to maintain close tolerances and produce good surface finishes; aluminum alloys can be die cast to basic linear tolerances of ±4 mm/m (±4 mils/in.) and commonly have finishes as fine as 1.3 μm (50 μin.). Die castings are best designed with uniform wall thickness; minimum practical wall thickness for aluminum alloy die castings is dependent on casting size. Small parts are cast as thin as 1.0 mm (0.040 in.). Cores, which are made of metal, are restricted to simple shapes that permit straight-line removal.

Die castings are made by injection of molten metal into metal molds under substantial pressure. Rapid injection (due to the high pressure) and rapid solidification under high pressure (due to the use of bare metal molds) combine to produce a dense, fine-grain surface structure, which results in excellent wear and fatigue properties. Air entrapment and shrinkage, however, may result in porosity, and machine cuts should be limited to 1.0 mm (0.040 in.) to avoid exposing it. Mold coatings are not practical in die casting, which is done at pressures of 2 MPa (300 psi) or higher, because the violence of the rapid injection of molten metal would remove the coating (production of thin-section die castings may involve cavity fill times as brief as 20 ms).

Aluminum alloy die castings usually are not heat treated but occasionally are given dimensional and metallurgical stabilization treatments.

Die castings are not easily welded or heat treated because of entrapped gases. Special techniques and care in production are required for pressure-tight parts. The selection of an alloy with a narrow freezing range also is helpful. The use of vacuum for cavity venting is practiced in some die casting foundries for production of parts for some special applications. In the "pure free" process, the die cavity is purged with oxygen before injection. The entrapped oxygen reacts with the molten aluminum to form oxide particles rather than gas pores.

Approximately 85% of aluminum alloy die castings are produced in aluminum-silicon-copper alloys (alloy 380.0 and its several modifications). This family of alloys provides a good combination of cost, strength, and corrosion resistance, together with the high fluidity and freedom from hot shortness that are required for ease of casting. Where better corrosion resistance is required, alloys lower in copper, such as 360.0 and 413.0, must be used.

Alloy 518.0 is occasionally specified when highest corrosion resistance is required. This alloy, however, has low fluidity and some tendency to hot shortness. It is difficult to cast, which is reflected in higher cost per casting.

The physical and mechanical properties of the most commonly used aluminum die casting alloys are given in the section

Table 7 Factors affecting selection of casting process for aluminum alloys

Factor	Casting process: Sand casting	Permanent mold casting	Die casting
Cost of equipment	Lowest cost if only a few items required	Less than die casting	Highest
Casting rate	Lowest rate	11 kg/h (25 lb/h) common; higher rates possible	4.5 kg/h (10 lb/h) common; 45 kg/h (100 lb/h) possible
Size of casting	Largest of any casting method	Limited by size of machine	Limited by size of machine
External and internal shape	Best suited for complex shapes where coring required	Simple sand cores can be used, but more difficult to insert than in sand castings	Cores must be able to be pulled because they are metal; undercuts can be formed only by collapsing cores or loose pieces
Minimum wall thickness	3.0–5.0 mm (0.125–0.200 in.) required; 4.0 mm (0.150 in.) normal	3.0–5.0 mm (0.125–0.200 in.) required; 3.5 mm (0.140 in.) normal	1.0–2.5 mm (0.100–0.040 in.); depends on casting size
Type of cores	Complex baked sand cores can be used	Reuseable cores can be made of steel, or nonreuseable baked cores can be used	Steel cores; must be simple and straight so they can be pulled
Tolerance obtainable	Poorest; best linear tolerance is 300 mm/m (300 mils/in.)	Best linear tolerance is 10 mm/m (10 mils/in.)	Best linear tolerance is 4 mm/m (4 mils/in.)
Surface finish	6.5–12.5 μm (250–500 μin.)	4.0–10 μm (150–400 μin.)	1.5 μm (50 μin.); best finish of the three casting processes
Gas porosity	Lowest porosity possible with good technique	Best pressure tightness; low porosity possible with good technique	Porosity may be present
Cooling rate	0.1–0.5 °C/s (0.2–0.9 °F/s)	0.3–1.0 °C/s (0.5–1.8 °F/s)	50–500 °C/s (90–900 °F/s)
Grain size	Coarse	Fine	Very fine on surface
Strength	Lowest	Excellent	Highest, usually used in the as-cast condition
Fatigue properties	Good	Good	Excellent
Wear resistance	Good	Good	Excellent
Overall quality	Depends on foundry technique	Highest quality	Tolerance and repeatability very good
Remarks	Very versatile as to size, shape, internal configurations	...	Excellent for fast production rates

"Properties of Aluminum Casting Alloys" in this article. Other characteristics of aluminum die casting alloys are presented in Tables 3 and 5. Final selection of an aluminum alloy for a specific application can best be established by consultation with die casting suppliers.

Permanent mold (gravity die) casting, like die casting, is suited to high-volume production. Permanent mold castings typically are larger than die castings. Maximum weight of permanent mold castings usually is about 10 kg (25 lb), but much larger castings sometimes are made when costs of tooling and casting equipment are justified by the quality required for the casting.

Surface finish of permanent mold castings depends on whether or not a mold wash is used; generally, finishes range from 3.8 to 10 μm (150 to 400 μin.). Basic linear tolerances of about ±10 mm/m (±0.10 in./in.), and minimum wall thicknesses of about 3.6 mm (0.140 in.), are typical. Tooling costs are high, but lower than those for die casting. Because sand cores can be used, internal cavities can be fairly complex. (When sand cores are used, the process usually is referred to as semipermanent mold casting.)

Permanent mold castings are gravity-fed and pouring rate is relatively low, but the metal mold produces rapid solidification. Permanent mold castings exhibit excellent mechanical properties. Castings are generally sound, provided that the alloys used exhibit good fluidity and resistance to hot tearing.

Mechanical properties of permanent mold castings can be further improved by heat treatment. If maximum properties are required, the heat treatment consists of a solution treatment at high temperature followed by a quench (usually in hot water) and then natural or artificial aging. For small castings in which the cooling rate in the mold is very rapid or for less critical parts, thc solution treatment and quench may be eliminated and the fast cooling in the mold relied on to retain in solution the compounds that will produce age hardening.

In low-pressure casting (also called low-pressure die casting or pressure permanent mold casting), molten metal is injected into the metal molds at pressures of 170 kPa (25 psi) or less. Gating systems are used to introduce this metal into the mold inlet at the bottom of the mold so as to aid smooth and nonturbulent flow of the molten metal into the casting cavity. Filling of the mold and control of solidification are aided by application of refractory mold coating to selected areas of the die cavity, which slows down cooling in those areas. Thinner walls can be cast by low-pressure casting than by regular permanent mold casting. Low-pressure casting also has the economic advantage in that it can be highly automated.

Some common aluminum permanent mold casting alloys, and typical products cast from them, are presented below.

Alloy 366.0	Automotive pistons
Alloys 355.0, C355.0, A357.0	Timing gears, impellers, compressors, and aircraft and missile components requiring high strength
Alloys 356.0, A356.0	Machine tool parts, aircraft wheels, pump parts, marine hardware, valve bodies
Alloy B443.0	Carburetor bodies, waffle irons
Alloy 513.0	Ornamental hardware and architectural fittings

Other aluminum alloys commonly used for permanent mold castings include 296.0, 319.0, and 333.0.

Sand casting, which in a general sense involves the forming of a casting mold with sand, includes conventional sand casting and evaporative pattern (lost-foam) casting. This section focuses on conventional sand casting, which uses bonded sand molds. Evaporative pattern casting, which uses unbonded sand molds, is discussed in the next section.

In conventional sand casting, the mold is formed around a pattern by ramming sand, mixed with the proper bonding agent, onto the pattern. Then the pattern is removed, leaving a cavity in the shape of the casting to be made. If the casting is to have internal cavities or undercuts, sand cores are used to make them. Molten metal is poured into the mold, and after it has solidified the mold is broken to remove the casting. In making molds and cores, various agents can be used for bonding the sand. The agent most often used is a mixture of clay and water. (Sand bonded with clay and water is called green sand.) Sand bonded with oils or resins, which is very strong after baking, is used mostly for cores. Water glass (sodium silicate) hardened with CO_2 is used extensively as a bonding agent for both molds and cores.

The main advantages of sand casting are versatility (a wide variety of alloys, shapes, and sizes can be sand cast) and low cost of minimum equipment when a small number of castings is to be made. Among its disadvantages are low dimensional accuracy and poor surface finish; basic linear tolerances of ±30 mm/m (±0.030 in./in.) and surface finishes of 7 to 13 μm, or 250 to 500 μin., as well as low strength as a result of slow cooling, are typical for aluminum sand castings. Use of dry sands bonded with resins or water glass results in better surface finishes and dimensional accuracy, but with a corresponding decrease in cooling rate.

Casting quality is determined to a large extent by foundry technique. Proper metal-handling and gating practice is necessary for obtaining sound castings. Complex castings with varying wall thickness will be sound only if proper techniques are used. A minimum wall thickness of 4 mm (0.15 in.) normally is required for aluminum sand castings.

Typical products made from some common aluminum sand casting alloys include:

Alloy C355.0	Air-compressor fittings, crankcases, gear housings
Alloy A356.0	Automobile transmission cases, oil pans, and rear-axle housings
Alloy 357.0	Pump bodies, cylinder blocks for water-cooled engines
Alloy 443.0	Pipe fittings, cooking utensils, ornamental fittings, marine fittings
Alloy 520.0	Aircraft fittings, truck and bus frame components, levers, brackets
Alloy 713.0	General-purpose casting alloy for applications that require strength without heat treatment or that involve brazing

Other aluminum alloys commonly used for sand castings include 319.0, 355.0, 356.0, 514.0, and 535.0.

Evaporative (lost-foam) pattern casting (EPC) is a sand casting process that uses an unbonded sand mold with an expendable polystyrene pattern placed inside of the mold. This process is somewhat similar to investment casting in that an expendable material can be used to form relatively intricate patterns in a surrounding mold material. Unlike investment casting, however, evaporative pattern casting (EPC) involves a polystyrene foam pattern that vaporizes during the pouring of molten metal into a surrounding mold of unbonded sand. With investment casting, a wax or plastic pattern is encased in a ceramic mold and removed by heat prior to the filling of the mold with molten metal.

The EPC process (also known as lost foam or evaporative foam casting) originated in 1958 when H.F. Shroyer was granted a patent (2,830,343) for a cavityless casting method using a polystyrene foam pattern embedded in traditional green sand. A polystyrene foam pattern left in the sand mold is decomposed by the molten metal, thus replacing the foam pattern and duplicating all of the features of the pattern. Early use of the process was limited to one-of-a-kind rough castings because the foam material was coarse and hand fabricated and because the packed green sand mold would not allow the gases from the decomposing foam pattern to escape rapidly from the mold (the trapped gases usually resulted in porous castings). Later, in 1964, T.R. Smith was granted a patent (3,157,924) for the utilization of loose, unbonded sand as a casting medium. With this important breakthrough, the EPC became an emerging subject of investigation in automotive company research facilities. Use of the process has been increasing rapidly and many casting facilities are now dedicated to the EPC process.

The major difference between sand castings and castings made by the EPC process is in subsequent machining and cleaning operations. The castings in the EPC process are consistently poured at closer tolerances with less stock for grinding and finishing. Dimensional variability associated with core setting, mating of cope, and drag are eliminated.

The use of untreated, unbonded sand makes the sand system economical and easy to manage. Casting cleaning is also greatly reduced and (except for removal of the wash coating) is sometimes eliminated because of the absence of flash, sand, and resin.

Casting yield can be considerably increased by pouring into a three-dimensional flask with the castings gated to a center sprue. An EPC casting facility also has the ability to produce a variety of castings in a continuous and timely manner. Foundries with EPC can pour diverse metals with very few changeover problems, and this adds to the versatility of the foundry.

Further benefits of the EPC process result from the freedom in part design offered by the process. Assembled patterns can be used to make castings that cannot be produced by any other high-production process. Part-development costs can be reduced because of the ability to prototype with the foam. Product and process development can be kept in-house.

The major concern in the EPC process is shrinkage of the foam pattern. The major difference between traditional methods of foundry tooling and evaporative pattern tooling is the continual heating and cooling of the tool and the subsequent stresses and geometrical considerations that this condition implies.

Shell Mold Casting. In shell mold casting, the molten metal is poured into a shell of resin-bonded sand only 10 to 20 mm (0.4 to 0.8 in.) thick—much thinner than the massive molds commonly used in sand foundries. Shell mold castings surpass ordinary sand castings in surface finish and dimensional accuracy and cool at slightly higher rates; however, equipment and production costs are higher, and size and complexity of castings that can be produced are limited.

Plaster Casting. In this method, either a permeable (aerated) or impermeable plaster is used for the mold. The plaster in slurry form is poured around a pattern, the pattern is removed and the plaster mold is baked before the casting is poured. The high insulating value of the plaster allows castings with thin walls to be poured. Minimum wall thickness of aluminum plaster castings typically is 1.5 mm (0.060 in.). Plaster molds have high reproducibility, permitting castings to be made with fine details and close tolerances; basic linear tolerances of ±5 mm/m (±0.005 in./in.) are typical for aluminum castings. Surface finish of plaster castings also is very good; aluminum castings attain finishes of 1.3 to 3.2 μm (50 to 125 μin.). For castings of certain complex shapes, such as some precision impellers and electronic parts, mold patterns made of rubber are used because their flexibility makes them easier to withdraw from the molds than rigid patterns.

Mechanical properties and casting quality depend on alloy composition and foundry technique. Slow cooling due to the highly insulating nature of plaster molds tends to magnify solidification-related problems, and thus solidification must be controlled carefully to obtain good mechanical properties.

Plaster casting is sometimes used to make prototype parts before proceeding to make tooling for production die casting of the part.

Cost of basic equipment for plaster casting is low; however, because plaster molding is slower than sand molding, cost of operation is high. Aluminum alloys commonly used for plaster casting are 295.0, 355.0, C355.0, 356.0, and A356.0.

Investment casting of aluminum most commonly employs plaster molds and expendable patterns of wax or other fusible materials. A plaster slurry is "invested" around patterns for several castings, and the patterns are melted out as the plaster is baked.

Investment casting produces precision parts; aluminum castings can have walls as thin as 0.40 to 0.75 mm (0.015 to 0.030 in.), basic linear tolerances as narrow as ±5 mm/m (±5 mils/in.) and surface finishes of 1.5 to 2.3 μm (60 to 90 μin.). Some internal porosity usually is present, and it is recommended that machining be limited to avoid exposing it. However, investment molding is often used to produce large quantities of intricately shaped parts requiring no further machining so internal porosity seldom is a problem. Because of porosity and slow solidification, mechanical properties are low.

Investment castings usually are small, and thus gating techniques are limited. Christmas-tree gating systems often are em-

Table 8 Effect of squeeze casting on tensile properties

Alloy	Process	Tensile strength MPa	Tensile strength ksi	Yield strength MPa	Yield strength ksi	Elongation, %
356-T6 aluminum	Squeeze casting	309	44.8	265	38.5	3
	Permanent mold	262	38.0	186	27.0	5
	Sand casting	172	25.0	138	20.0	2
535 aluminum (quenched)	Squeeze casting	312	45.2	152	22.1	34.2
	Permanent mold	194	28.2	128	18.6	7
6061-T6 aluminum	Squeeze casting	292	42.3	268	38.8	10
	Forging	262	38.0	241	35.0	10
A356 T4 aluminum	Squeeze casting	265	38.4	179	25.9	20
A206 T4 aluminum	Squeeze casting	390	56.5	236	34.2	24
CDA 377 forging brass	Squeeze casting	379	55.0	193	28.0	32.0
	Extrusion	379	55.0	145	21.0	48.0
CDA 624 aluminum bronze	Squeeze casting	783	113.5	365	53.0	13.5
	Forging	703	102.0	345	50.0	15.0
CDA 925 leaded tin bronze	Squeeze casting	382	55.4	245	35.6	19.2
	Sand casting	306	44.4	182	26.4	16.5
Type 357 (annealed)	Squeeze casting	614	89.0	303	44.0	46
	Sand casting	400	58.0	241	35.0	20
	Extrusion	621	90.0	241	35.0	50
Type 321 (heat treated)	Squeeze casting	1063	154.2	889	129.0	15
	Forging	1077	156.2	783	113.6	7

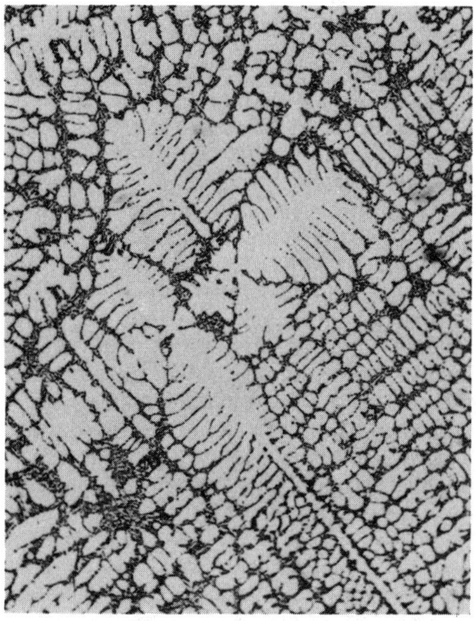
(a)

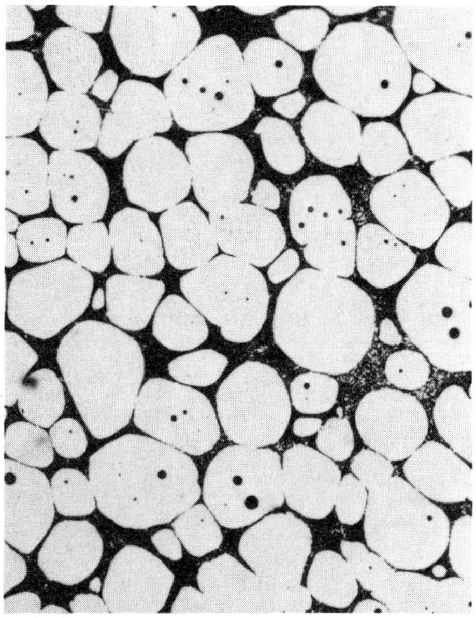
(b)

Fig. 8 Comparison of aluminum alloy 357 (Al-7Si-0.5Mg). (a) A dendritic microstructure from conventional casting. (b) A nondendritic microstructure formed during rheocasting or thixocasting. Both 200×

ployed to produce many parts per mold. Investment casting is especially suited to production of jewelry and parts for precision instruments. Recent strong interest by the aerospace industry in the investment casting process has resulted in limited use of improved technology to produce premium quality castings. The "near-net-shape" requirements of aerospace parts are often attainable using the investment casting techniques. Combining this accurate dimensional control with the high and carefully controlled mechanical properties can, at times, justify casting costs and prices normally not considered practical.

Aluminum alloys commonly used for investment castings are 208.0, 295.0, 308.0, 355.0, 356.0, 443.0, 514.0, 535.0, and 712.0.

Centrifugal Casting. Centrifuging is another method of forcing metal into a mold. Steel, baked sand, plaster, cast iron, or graphite molds and cores are used for centrifugal casting of aluminum. Metal dies or molds provide rapid chilling, resulting in a level of soundness and mechanical properties comparable or superior to that of gravity-poured permanent mold castings. Baked sand and plaster molds are commonly used for centrifuge casting because multiple mold cavities can be arranged readily around a central pouring sprue. Graphite has two major advantages as a mold material: its high heat conductivity provides rapid chilling of the cast metal, and its low specific gravity, compared to ferrous mold materials, reduces the power required to attain the desired speeds.

Centrifugal casting has the advantage over other casting processes in that, if molds are properly designed, inclusions such as gases or oxides tend to be forced into the gates, and thus castings have properties that closely match those of wrought products. Limitations on shape and size are severe, and cost of castings is very high.

Wheels, wheel hubs, and papermaking or printing rolls are examples of aluminum parts produced by centrifugal casting. Aluminum alloys suitable for permanent mold, sand, or plaster casting can be cast centrifugally.

Continuous Casting. Long shapes of simple cross section (such as round, square, and hexagonal rods) can be produced by continuous casting, which is done in a short, bottomless, water-cooled metal mold. The casting is continuously withdrawn from the bottom of the mold; because the mold is water cooled, cooling rate is very high. As a result of continuous feeding, castings generally are free of porosity. In most instances, however, the same product can be made by extrusion at approximately the same cost and with better properties, and thus use of continuous casting is limited. The largest application of continuous casting is production of ingot for rolling, extrusion, or forging.

Composite-Mold Casting. Many of the molding methods described above can be combined to obtain greater flexibility in casting. Thus, dry sand cores often are used in green sand molds, and metal chills can be used in sand molds to accelerate local cooling. Semipermanent molds, which comprise metal molds and sand cores, take advantage of the better properties obtainable with metal molds and the greater flexibility in shape of internal cavities that results from use of cores that can be extracted piecemeal.

Hot isostatic pressing of aluminum castings reduces porosity and can thus decrease the scatter in mechanical properties. The method also makes possible the salvaging of castings that have been scrapped for reasons of internal porosity, thereby achieving improved foundry recovery. This advantage is of more significant importance in the manufacture of castings subject to radiographic inspection when required levels of soundness are not achieved in the casting process. The development of hot isostatic pressing is pertinent to the broad range of premium castings, but is especially relevant for the more difficult-to-cast aluminum-copper series.

Hybrid Permanent Mold Processes. Although die casting, centrifugal casting, and gravity die casting constitute, on a volume basis, the major permanent mold processes, there are also some hybrid processes that use permanent molds. This includes squeeze casting and semisolid metal processing.

Squeeze casting, also known as liquid-metal forging, is a process by which molten metal solidifies under pressure within closed dies positioned between the plates of a hydraulic press. The applied pressure and the instant contact of the molten metal with

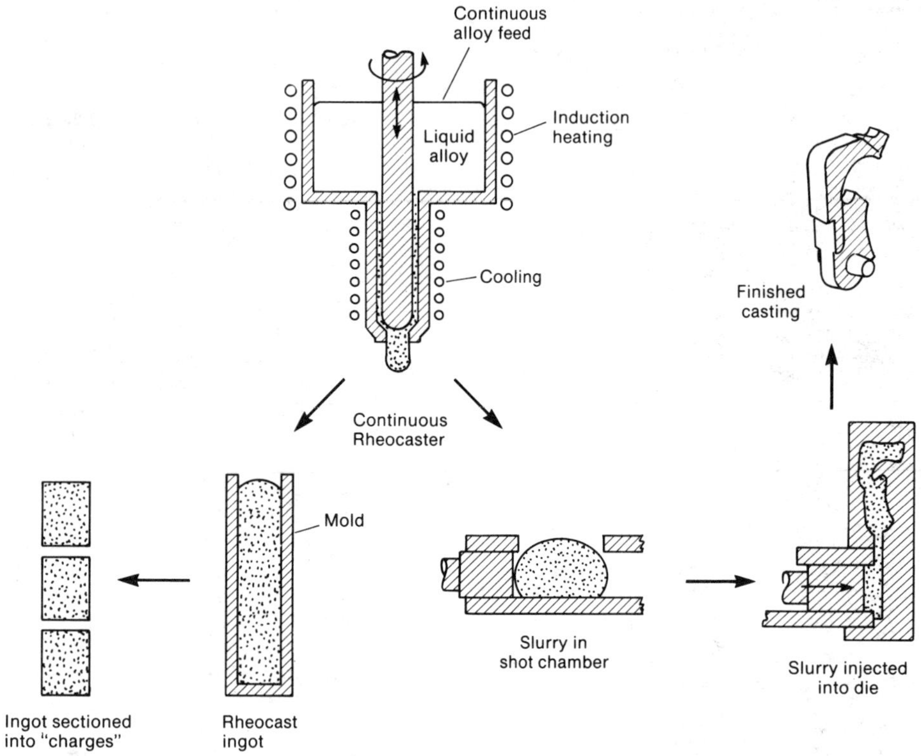

Fig. 9 Semisolid-metal processing with a rheocaster. Commercial semisolid-metal processing is based on thixocasting.

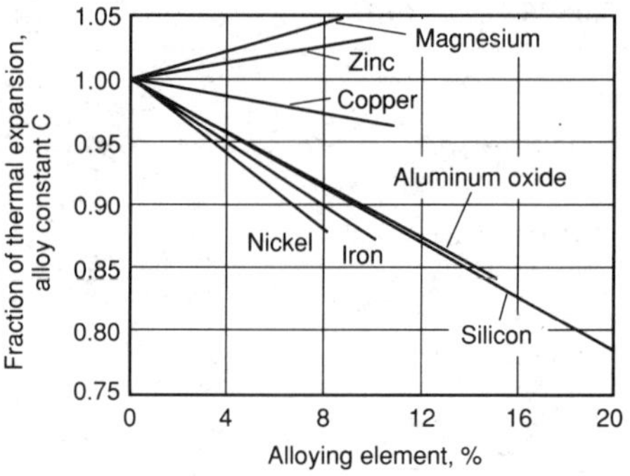

Fig. 10 Effects of alloying elements on the thermal expansion of aluminum. Fraction is based on a value of 1.00 for 99.996 Al. Source: L.A. Willey, Alcoa

Table 9 Tensile properties and hardness of typical semisolid forged aluminum parts

Aluminum alloy	Temper	Ultimate tensile strength, MPa	Ultimate tensile strength, ksi	Tensile yield strength, MPa	Tensile yield strength, ksi	Elongation, %	Hardness, HB
206	T7	386	56.0	317	46.0	6.0	103
2017	T4	386	56.0	276	40.0	8.8	89
2219	T8	352	51.0	310	45.0	5.0	89
6061	T6	330	47.8	290	42.1	8.2	104
6262	T6	365	52.9	330	47.9	10.0	82
7075	T6	496	72.0	421	61.0	7.0	135
356	T5	234	34.0	172	25.0	11.0	89
356	T6	296	43.0	193	28.0	12.0	90
357	T5	296	43.0	207	30.0	11.0	90
357	T6	358	52.0	290	42.0	10.0	100

Table 10 Comparison of semisolid forging and permanent mold casting for the production of aluminum automobile wheels

See Example 1.

Process	Weight direct from die or mold, kg	Weight direct from die or mold, lb	Finished part weight, kg	Finished part weight, lb	Production rate per die or mold, pieces per h	Aluminum alloy	Heat treatment	Ultimate tensile strength, MPa	Ultimate tensile strength, ksi	Yield strength, MPa	Yield strength, ksi	Elongation, %
Semisolid forging	7.5	16.5	6.1	13.5	90	357	T5	290	42	214	31	10
Permanent mold casting	11.1	24.5	8.6	19.0	12	356	T6	221	32	152	22	8

the die surface produces a rapid heat transfer condition that yields a pore-free fine-grain casting with excellent mechanical properties (Table 8). The squeeze casting process is easily automated to produce near-net to net-shape high-quality components.

Squeeze casting has been successfully applied to a variety of ferrous and nonferrous alloys in traditionally cast and wrought compositions. Applications of squeeze-cast aluminum alloys include pistons for engines, disk brakes, automotive wheels, truck hubs, barrel heads, and hubbed flanges. Squeeze casting is simple and economical, efficient in its use of raw material, and has excellent potential for automated operation at high rates of production. The process generates the highest mechanical properties attainable in a cast product. The microstructural refinement and integrity of squeeze-cast products are desirable for many critical applications.

Semisolid-Metal Processing. Semisolid metalworking, also known as semisolid forming, is a hybrid manufacturing method that incorporates elements of both casting and forging. It involves a two-step process for the near-net shape forming of metal parts using a semisolid raw material that incorporates a unique nondendritic microstructure (Fig. 8).

The basic process of semisolid-metal processing is shown schematically in Fig. 9. The key (and first step) to the process involves vigorous agitation of the melt during earlier stages of solidification so as to break up the solid dendrites into small spherulites. There are two general approaches to this process: rheocasting and thixocasting. Rheocasting is a term coined by the researchers at the Massachusetts Institute of Technology (MIT) who initially discovered the techniques of semisolid-metal processing during research on hot tearing undertaken at MIT in the early 1970s. Seeking to understand the magnitude of the forces involved in deforming and fragmenting dendritic growth structures, MIT researchers constructed a high-temperature viscometer. They poured molten lead-tin alloys into the annular space created by two concentric cylinders and measured the forces transmitted through the freezing alloy when the outer cylinder was rotated. During the course of these experiments, it was discovered that when the outer cylinder was continuously rotated, the semisolid alloy exhibited remarkably low shear strength even at relatively high fractions solidified. This unique property was attributed to a novel nondendritic (that is, spheroidal) microstructure.

As these ideas unfolded, research into the nature of semisolid alloys progressed, and it became apparent that bars could be cast from semisolid fluids possessing the rheocast nondendritic microstructure. The final freezing of these bars captures this micro-

Table 11 Typical (and minimum) tensile properties of aluminum casting alloys

Alloy	Temper	Ultimate tensile strength(a) MPa	Ultimate tensile strength(a) ksi	0.2% offset yield strength(a) MPa	0.2% offset yield strength(a) ksi	Elongation(a) in 50 mm (2 in.), %
Rotor alloys (pure aluminum)						
100.1 ingot	· · ·	70	10	40	6	20
150.1 ingot	· · ·	70	10	40	6	20
170.1 ingot	· · ·	70	10	40	6	20
Sand casting alloys						
201.0	T43	414	60	255	37	17.0
	T6	448	65	379	55	8.0
	T7	467	68	414	60	5.5
204.0	T4	372	54	255	37	14
		(295)	(43)	(185)	(27)	(5)
206.0	T4	345	50	193	28	10
		(275)	(40)	(165)	(24)	(6)
	T6	380	55	240	35	10
		(345)	(50)	(205)	(30)	(6)
A206.0	T4	380	55	250	36	5–7
		(345)	(50)	(205)	(30)	(· · ·)
	T71	400	58	330	48	5
		(372)	(54)	(310)	(45)	(3)
208.0	F	145	21	97	14	2.5
		(130)	(19)	(· · ·)	(· · ·)	(1.5)
	T55	(145, min)	(21, min)	· · ·	· · ·	· · ·
A206.0	T4	354	51	250	36	7.0
208.0	F	145	21	97	14	2.5
213.0	F	165	24	103	15	1.5
222.0	O	186	27	138	20	1.0
	T61	283	41	276	40	<0.5
	T62	421	61	331	48	4.0
224.0	T72	380	55	276	40	10.0
240.0	F	235	34	200	28	1.0
242.0	F	214	31	217	30	0.5
	O	186	27	124	18	1.0
	T571	221	32	207	30	0.5
	T77	207	30	159	23	2.0
A242.0	T75	214	31	· · ·	· · ·	2.0
295.0	T4	221	32	110	16	8.5
		(200)	(29)	(· · ·)	(· · ·)	(6)
	T6	250	36	165	24	5.0
		(220)	(32)	(138)	(20)	(3)
	T62	283	41	220	32	2.0
		(248)	(36)	(· · ·)	(· · ·)	(· · ·)
	T7	(200, min)	(29, min)	· · ·	· · ·	(3, min)
319.0	F	186	27	124	18	2.0
	T5	207	30	179	26	1.5
	T6	250	36	164	24	2.0
		(215)	(31)	(· · ·)	(· · ·)	(1.5)
355.0	F	159	23	83	12	3.0
	T51	193	28	159	23	1.5
	T6	241	35	172	25	3.0
		(220)	(32)	(138)	(20)	(2)
	T61	269	39	241	35	1.0
	T7	264	38	250	26	0.5
	T71	240	35	200	29	1.5
	T77	240	35	193	28	3.5
C355.0	T6	270	39	200	29	5.0
		(248)	(36)	(172)	(25)	(2)
356.0	F	164	24	124	18	6.0
	T51	172	25	138	20	2.0
	T6	228	33	164	24	3.5
		(207)	(30)	(138)	(20)	(3)
	T7	235	34	207	30	2.0
		(214)	(31)	(200)	(29)	(· · ·)
	T71	193	28	145	21	3.5
A356.0	F	159	23	83	12	6.0
	T51	179	26	124	18	3.0
	T6	278	40	207	30	6.0
	T71	207	30	138	20	3.0
357.0	F	172	25	90	13	5.0
	T51	179	26	117	17	3.0
	T6	345	50	296	43	2.0
	T7	278	40	234	34	3.0
A357.0	T6	317	46	248	36	3.0
A390.0	F	179	26	179	26	<1.0
	T5	179	26	179	26	<1.0

Alloy	Temper	Ultimate tensile strength(a) MPa	Ultimate tensile strength(a) ksi	0.2% offset yield strength(a) MPa	0.2% offset yield strength(a) ksi	Elongation(a) in 50 mm (2 in.), %
Sand casting alloys (continued)						
	T6	278	40	278	40	<1.0
	T7	250	36	250	36	<1.0
443.0	F	131	19	55	8	8.0
		(117)	(17)	(· · ·)	(· · ·)	(3)
A444.0	F	145	21	62	9	9.0
	T4	159	23	62	9	12.0
511.0	F	145	21	83	12	3.0
512.0	F	138	20	90	13	2.0
		(117)	(17)	(70)	(10)	(· · ·)
514.0	F	172	25	83	12	9.0
		(150)	(22)	(· · ·)	(· · ·)	(6)
520.0	T4	331	48	179	26	16.0
		(290)	(42)	(150)	(22)	(12)
535	F	275	40	145	21	13
		(240)	(35)	(125)	(18)	(9)
A535.0	F	250	36	124	18	9.0
B535.0	F	262	38	130	19	10
705.0	F/T5	(205)	(30, min)	(117)	(17, min)	(5, min)
707.0	F/T5	(227)	(33, min)	(152)	(22, min)	(2, min)
	F/T7	(255)	(37, min)	(207)	(30, min)	(1, min)
710.0	F	241	35	172	25	5.0
		(220)	(32)	(138)	(20)	(2)
712.0	F	240	35	172	25	5.0
		(235)	(34)	(172)	(25)	(4)
713.0	F	240	35	172	25	5.0
		(220)	(32)	(152)	(22)	(3)
771.0	F	303	44	248	36	3
		(270)	(39)	(228)	(33)	(2)
	T2	(248)	(36, min)	(185)	(27, min)	(2, min)
	T5	(290)	(42, min)	(262)	(38, min)	(2, min)
	T6	330	48	262	38	9
		(275)	(40)	(240)	(35)	(5)
772.0	F	275	40	220	32	7
		(255)	(37)	(193)	(28)	(5)
	T6	310	45	240	35	10
		(303)	(44)	(220)	(32)	(6)
850.0	T5	138	20	76	11	8.0
		(110)	(16)	(· · ·)	(· · ·)	(5)
851.0	T5	138	20	76	11	5.0
		(117)	(17)	(· · ·)	(· · ·)	(3)
852.0	T5	186	27	152	22	2.0
		(165)	(24)	(124)	(18)	(· · ·)
Permanent mold casting alloys						
201.0	T43	414	60	255	37	17.0
	T6	448	65	379	55	8.0
	T7	469	68	414	60	5.0
204.0	T4	325	47	200	29	7
		(248)	(36)	(193)	(28)	(5)
206.0	T4	345	50	207	30	10
		(275)	(40)	(165)	(24)	(6)
	T6	385	56	262	38	12
		(345)	(50)	(207)	(30)	(6)
A206.0	T4	430	62	265	38	17
	T71	415	60	345	50	5
		(372)	(54)	(310)	(45)	(3)
	T7	436	63	347	50	11.7
213.0	F	207	30	165	24	1.5
222.0	T52	241	35	214	31	1.0
	T551	255	37	241	35	<0.5
	T65	331	48	248	36	<0.5
238.0	F	207	30	165	24	1.5
242.0	T571	276	40	234	34	1.0
	T61	324	47	290	42	0.5
249.0	T63	476	69	414	60	6.0
	T7	427	62	359	52	9.0
296.0	T4	255	37	131	19	9.0
	T6	276	40	179	26	5.0
		(240)	(35)	(152)	(22)	(2)
	T7	270	39	138	20	4.5
308.0	F	193	28	110	16	2.0
319.0	F	185	27	125	18	2
	T5	207	30	180	26	2
	T6	248	36	165	24	2

(continued)

(a) Minimum values are shown in parenthesis and are listed below their typical values.

Table 11 (continued)

Alloy	Temper	Ultimate tensile strength(a) MPa	Ultimate tensile strength(a) ksi	0.2% offset yield strength(a) MPa	0.2% offset yield strength(a) ksi	Elongation(a) in 50 mm (2 in.), %
Permanent mold casting alloys (continued)						
		(214)	(31)	(· · ·)	(· · ·)	(1.5)
324.0	F	207	30	110	16	4.0
	T5	248	36	179	26	3.0
	T62	310	45	269	39	3.0
332.0	T5	248	36	193	28	1.0
333.0	F	234	34	131	19	2.0
	T5	234	34	172	25	1.0
	T6	290	42	207	30	1.5
	T7	255	37	193	28	2.0
		(215)	(31)	(· · ·)	(· · ·)	(· · ·)
336.0	T551	248	36	193	28	0.5
	T65	324	47	296	43	0.5
354.0	T6	380	55	283	41	6
	T62	393	57	317	46	3
355.0	T51	(185, min)	(27, min)	· · ·	· · ·	· · ·
	T6	290	42	185	27	4
		(255)	(37)	(· · ·)	(· · ·)	(1.5)
	T62	310	45	275	40	1.5
		(290)	(42)	(· · ·)	(· · ·)	(· · ·)
	T71	(235, min)	(34, min)	· · ·	· · ·	· · ·
356.0	F	179	26	124	18	5.0
	T51	186	27	138	20	2.0
	T6	262	38	186	27	5.0
		(207)	(30)	(138)	(20)	(3)
	T7	221	32	165	24	6.0
A356.0	T61	283	41	207	30	10.0
		(255)	(37)	(· · ·)	(· · ·)	(5)
357.0	F	193	28	103	15	6.0
	T51	200	29	145	21	4.0
	T6	360	52	295	43	5.0
		(310)	(45)	(· · ·)	(· · ·)	(3)
A357.0	T61	359	52	290	42	5.0
358.0	T6	345	50	290	42	6
	T62	365	53	317	46	3.5
359.0	T61	325	47	255	37	7
	T62	345	50	290	42	5
A390.0	F	200	29	200	29	<1.0
	T5	200	29	200	29	<1.0
	T6	310	45	310	45	<1.0
	T7	262	38	262	38	<1.0
Permanent mold casting alloys (continued)						
443.0	F	160	23	62	9	10.0
B443.0	F	160	23	62	9	10
444.0	T4	193	28	83	12	25
A444.0	F	165	24	76	11	13.0
	T4	160	23	70	10	21
513.0	F	186	27	110	16	7.0
		(150)	(22)	(· · ·)	(· · ·)	(2.5)
705.0	T5	240	35	103	15	22
707.0	T5	(290, min)	(42, min)			(4, min)
711.0	F	248	36	130	19	8
713.0	T5	275	40	185	27	6
850.0	T5	160	23	76	11	12.0
		(124)	(18)	(· · ·)	(· · ·)	(8)
	T101	160	23	76	11	12
851.0	T5	138	20	76	11	5.0
852.0	T5	221	32	159	23	5.0
		(185)	(27)	(· · ·)	(· · ·)	(3)
Die casting alloys						
360.0	F	324	47	172	25	3.0
A360.0	F	317	46	165	24	5.0
364.0	F	296	43	159	23	7.5
380.0	F	330	48	165	24	3.0
A380.0	F	324	47	160	23	4.0
383.0	F	310	45	150	22	3.5
384.0	F	325	47	172	25	1.0
A384.0	F	330	48	165	24	2.5
390.0	F	279	40.5	241	35	1.0
	T5	296	43	265	38.5	1.0
A390.0	F	283	41	240	35	1.0
B390.0	F	317	46	248	36	· · ·
392.0	F	290	42	262	38	<0.5
413.0	F	296	43	145	21	2.5
A413.0	F	241	35	110	16	3.5
443.0	F	228	33	110	16	9.0
C443.0	F	228	33	95	14	9
513.0	F	276	40	152	22	10.0
515.0	F	283	41	· · ·	· · ·	10.0
518.0	F	310	45	186	27	8.0

(a) Minimum values are shown in parenthesis and are listed below their typical values.

structure. The bars then represented a raw material that could be heated at a later time or a remote location to the semisolid temperature range to reclaim the special rheological characteristics. This process, using semisolid alloys heated from specially cast bars, was termed thixocasting (Ref 3). This distinguished it from rheocasting, which has come to be known as the process used for producing semisolid structures and/or forming parts from slurry without an intermediate freezing step.

A number of alternative approaches to the production of the semisolid raw material have been developed. Although several of these techniques build upon the mechanical agitation approach (Ref 4, 5), others utilize a passive stirring technique for stimulating turbulent flow through cooling channels (Ref 6, 7). At least one approach uses isothermal holding to induce particle coarsening. Most of these alternatives appear to be confined to the laboratory, although one or two have been demonstrated at a pilot production level. To date, none has shown economic viability.

There have been several attempts in the United States and abroad to commercialize rheocasting, but none of these ventures is known to have been commercially successful (Ref 4, 8). On the other hand, semisolid forging, which exploits the manufacturing advantages of thixotropic semisolid alloy bars, began commercial production in 1981 and is now a rapidly expanding commercial process. The production of raw material has been brought to full commercial realization, and the use of semisolid forged parts is broadening in the aerospace, automotive, military, and industrial sectors.

The advantages of semisolid forging have enabled it to compete effectively with a variety of conventional processes in a number of different applications. Semisolid forged parts have replaced conventional forgings, permanent mold and investment castings, impact extrusions, machined extrusion profiles, parts produced on screw machines, and in unusual circumstances, die castings and stampings. Applications include automobile wheels, master brake cylinders, antilock brake valves, disk brake calipers, power steering pump housings, power steering pinion valve housings, engine pistons, compressor housings, steering column mechanical components, airbag containment housings, power brake proportioning valves, electrical connectors, and various covers and housings that require leak-tight integrity. Table 9 lists mechanical properties of selected aluminum alloys used in these components.

There are several potential advantages of semisolid alloys. First, and particularly significant for higher-melting alloys, semisolid metalworking afforded lower operating temperatures and reduced metal heat content (reduced enthalpy of fusion). Second, the viscous flow behavior could provide for a more laminar cavity fill than could generally be achieved with liquid alloys. This could lead to reduced gas entrainment. Third, solidification shrinkage would be reduced in direct proportion to the fraction solidified

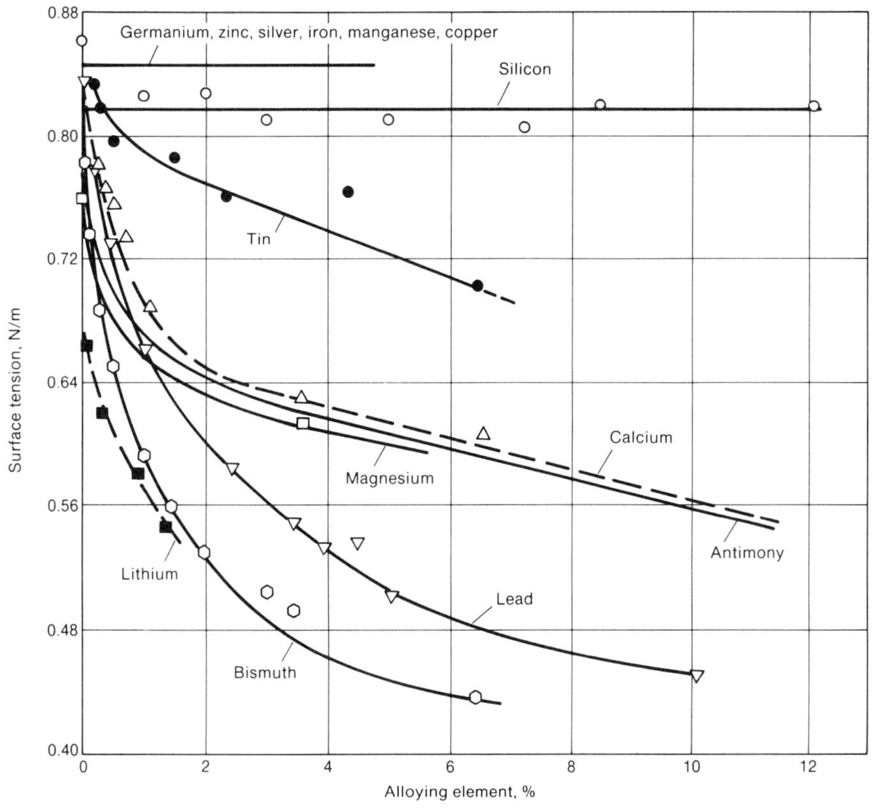

Fig. 11 Effect of various elements on surface tension of 99.99% Al in argon at 700 to 740 °C (1290 to 1365 °F)

within the semisolid metalworking alloy, which should reduce both shrinkage porosity and the tendency toward hot tearing. In addition, the viscous nature of semisolid alloys provides a natural environment for the incorporation of third-phase particles in the preparation of particulate-reinforced metal-matrix composites. The semisolid state also allows greater use of automation in material handling.

Example 1: Comparison of Semisolid Forging and Permanent Mold Casting in the Production of Aluminum Automobile Wheels. Aluminum automobile wheels have been produced by permanent mold casting (gravity and low pressure), squeeze casting, and fabrications of castings or stampings welded to rolled rims. Semisolid forging is a more recent process. Table 10 compares the characteristics of aluminum automobile wheels produced by semisolid forging and permanent mold casting. In addition to an economic advantage, semisolid forging offers other advantages that are discussed below.

Lighter Weight. The ability to form thinner sections without heavy ribs to aid in filling the cavity allows a wheel to be semisolid formed nearer to net size with light ribs on the brake side. This results in a finished wheel that is up to 30% lighter than a cast wheel of the same style.

Consistent Quality. The forging process employs a high-quality, specially prepared (magnetohydrodynamic casting) billet with an engineered metallurgical structure, closely controlled chemistry, and consistent casting variables, supplying an extremely consistent raw material with complete traceability. The wheel-forming process is computer controlled and automated with precise control of the heating and forging process variables, making the entire process adaptable to statistical process control.

Structure and Properties. The semisolid forged wheel is fine grained, dense structured, and formed to close tolerances in precision tooling in which the temperature is controlled to provide consistent forging conditions. This provides consistency in part dimensions and metallurgical properties. Forging in the semisolid state avoids the entrapment of air or mold gas, and the high fraction of solid material, together with the high pressure after forming, reduces the microporosity due to liquid/solid shrinkage. Unlike conventional forgings, the wheel properties are isotropic, reflecting the nondendritic structure of the high-performance aluminum alloy 357 used in the billet.

Design Versatility. The ability to form thin sections (roughly one-quarter to one-half the thickness of casting) permits not only a reduction in the weight of the wheel, but also allows the designer to style the wheel with thinner ribs/spokes and finer detail. Forming in the semisolid state under very high final pressure provides part surfaces and details that reflect the die surfaces. Therefore, the designer has a selection of surface conditions to enhance the style and can obtain exact replication of the fine detail designed in the die.

Properties of Aluminum Casting Alloys

Although the physical and mechanical properties of aluminum casting alloys are well documented, the data given in this section should only be used for alloy comparison and not for design purposes. Properties for design must be obtained from pertinent specifications or design standards or by negotiation with the producer. Additional information on properties is also available in the "Selected References" listed at the end of this article and in the next article "Properties of Cast Aluminum Alloys" in this Volume.

Physical Properties

Table 2 gives typical values for some of the important physical properties of various aluminum casting alloys, which are grouped into the nine alloy categories mentioned earlier in the section "Selection of Casting Alloys." The effects of alloying elements on electrical conductivity and thermal expansion is shown in Table 4 and Fig. 10, respectively. Other important physical properties related to castability are fluidity and shrinkage.

Factors Affecting Fluidity. Fluidity depends on two major factors: the intrinsic fluid properties of the molten metal, and casting conditions. The properties usually thought to influence fluidity are viscosity, surface tension, the character of the surface oxide film, inclusion content, and manner in which the particular alloy solidifies.

Casting conditions that influence fluidity include part configuration; physical measures of the fluid dynamics of the system such as liquidstatic pressure drops, casting head, and velocities; mold material; mold surface characteristics; heat flux; rate of pouring; and degree of superheat.

Viscosity. The measured viscosities of molten aluminum alloys are quite low and fall within a relatively narrow range. Kinematic viscosity (viscosity/specific gravity) is less than that of water. It is evident on this basis that viscosity is not strongly influential in determining casting behavior and therefore is an unlikely source of variability in casting results.

Surface Tension and Oxide Film. A high surface tension has the effect of increasing the pressure required for liquid metal flow. A number of elements influence surface tension, primarily through their effects on the surface tension of the oxide. Figure 11 illustrates the effect of selected elements on surface tension. In aluminum alloys, the true effect of surface tension is overpow-

Table 12 Typical values of hardness, shear strength, fatigue strength, and compressive yield strength of various aluminum casting alloys

Alloy	Temper	Shear MPa	Shear ksi	Fatigue(a) MPa	Fatigue(a) ksi	Hardness, HB(b)	Compressive yield strength MPa	Compressive yield strength ksi
Sand castings								
204.0	T4	110	16	77	11	90	...	...
206.0	T4	...	...	...	...	95	...	...
A206.0	T4	255	37	...	...	100	...	...
A206.0	T71	...	...	160	23	110	...	...
206.0	T6	...	...	...	...	100	...	...
208.0	F	117	17	76	11	55	103	15
295.0	T4	179	26	48	7	60	...	...
	T6	207	30	50	7.5	75	172	25
	T62	227	33	55	8	90	...	...
	T7	...	...	...	...	...	...	...
208.0	F	117	17	76	11	55	103	15
	T55	...	...	...	...	...	...	...
319.0	F	152	22	70	10	70	131	19
	T5	165	24	76	11	80	...	...
	T6	200	29	76	11	80	172	25
355.0	F	...	...	...	...	70	...	...
	T51	152	22	55	8	65	165	24
	T6	193	28	62	9	80	179	26
	T61	248	36	70	10	100	255	37
	T7	193	28	70	10	85	248	36
	T71	241	35	70	10	75	248	36
	T77	179	26	70	10	80	200	29
C355.0	T6	193	28	70	10	90	...	...
356.0	F	...	...	...	...	...	...	...
	T51	138	20	55	8	60	145	21
	T6	179	26	59	8.5	70	172	25
	T7	165	24	62	9	75	214	31
	T71	138	20	59	8.5	60	...	...
357.0	F	...	...	...	...	...	...	...
	T51	...	...	...	...	...	...	...
	T6	164	24	62	9	90	214	31
	T7	...	...	...	...	60	...	...
A357.0	T6	...	...	...	...	85	...	...
A390.0	F, Fs	...	...	70	10	100	...	...
	T6	...	...	90	13	140	...	...
	T7	...	...	...	...	115	...	...
443.0	F	96	14	55	8	40	62	9
511.0	F	117	17	55	8	50	90	13
512.0	F	...	...	...	...	50	96	14
514.0	F	138	20	48	7	50	83	12
535.0	F	193	28	70	10	70	165	24
B535.0	F	207	30	62	9	65	...	...
520.0	T4	234	34	55	8	75	186	27
705.0	F, T5	...	...	...	...	...	...	...
707.0	F, T5	...	...	...	...	...	...	...
	F, T7	...	...	...	...	...	...	...
710.0	F, T5	179	26	55	8	75	172	25
712.0	F, T5	179	26	179	26	9	518	75
713.0	F, T5	179	26	63	9	74	...	...
771.0	F	...	...	...	...	...	...	...
	T2	...	...	...	...	...	...	...
	T5	...	...	...	...	...	...	...
	T6	...	...	...	...	...	...	...
772.0	F	...	...	...	...	...	...	...
	T6	...	...	...	...	...	...	...
850.0	T5	96	14	55	8	45	76	11
851.0	T5	96	14	...	...	45	...	...
852.0	T5	124	18	70	10	65	...	...
Permanent mold								
204.0	T4	...	...	...	...	90	...	...
206.0	T4	...	...	...	...	...	...	...
206.0	T6	255	37	...	...	110	...	...
A206.0	T71	255	37	207	30	110	...	...
Permanent mold (continued)								
296.0	T4	...	...	...	...	...	...	...
296.0	T6	220	32	70	10	90	179	26
296.0	T62	...	...	...	...	...	...	...
296.0	T7	...	...	...	...	...	...	...
213.0	F	...	...	...	...	85	...	...
308.0	F	152	22	89	13	70	117	17
319.0	F	186	27	83	12	85	138	20
	T6	220	32	83	12	95	193	28
333.0	F	186	27	96	14	90	131	19
	T5	186	27	83	12	100	172	25
	T6	228	33	103	15	105	207	30
	T7	193	28	83	12	90	193	28
354.0	T6	262	38	117	17	100	289	42
	T62	276	40	117	17	110	324	47
355.0	T51	...	...	...	...	90	...	...
	T6	234	34	70	10	90	186	27
	T62	248	36	70	10	105	276	40
	T71	...	...	...	...	...	...	...
356.0	F	...	...	...	...	...	...	...
	T51	...	...	...	...	...	...	...
	T6	207	30	90	13	80	186	27
	T7	172	25	76	11	70	165	24
A356.0	T61	193	28	90	13	90	220	32
357.0	T6	241	35	90	13	100	303	44
A357.0	T61	241	35	103	15	100	296	43
358.0	T6	296	43	...	...	105	289	42
	T62	317	46	...	...	...	317	46
359.0	T61	220	32	103	15	90	262	38
	T62	234	34	103	15	100	303	44
A390.0	F, T5	...	...	...	...	110	...	...
	T6	...	...	117	17	145	413	60
	T7	...	...	103	15	120	352	51
393.0	F	...	...	...	...	...	...	...
B443.0	F	110	16	55	8	45	62	9
444.0	T4	...	...	...	...	50	77	11
513	F	152	22	70	10	50	96	14
705.0	T5	152	22	...	...	55	124	18
707.0	T5	...	...	...	...	...	...	...
707.0	T	...	...	...	...	...	...	...
711.0	F	193	28	76	11	70	138	20
713.0	T5	179	26	62	9	75	172	25
850.0	T5	103	15	62	9	45	76	11
851.0	T5	96	14	62	9	45	76	11
852.0	T5	145	21	76	11	70	158	23
850.0	T101	103	15	62	9	45	145	21
Die casting alloys								
360.0	F	207	30	131	19	...	...	...
A360.0	F	200	29	124	18	...	...	...
364.0	F	200	29	124	18	...	...	...
380.0	F	214	31	145	21	...	...	...
A380.0	F	207	30	138	20	...	...	...
383.0	F	...	...	...	...	...	...	...
A384.0	F	200	29	138	20	...	...	...
390.0	F	...	...	76	11	...	...	...
392.0	F	...	...	...	...	...	...	...
A390.0	F	...	...	...	...	...	...	...
B390.0	F	...	...	...	...	...	...	...
392.0	F	...	...	...	...	...	...	...
A413.0	F	172	25	130	19	...	...	...
A413.0	F	159	23	130	19	...	...	...
C443.0	F	130	19	110	16	...	...	...
513.0	F	179	26	124	18	...	...	...
515.0	F	186	27	130	19	...	...	...
518.0	F	200	29	138	20	...	...	...

(a) Strength for 5×10^8 cycles with R.R. Moore rotating beam test. (b) 10 mm (6.4 in.) ball with 500 kgf (1100 lbf) load

ered by the influence of surface oxide film characteristics. The oxide film on pure aluminum, for example, triples apparent surface tension.

Inclusions in the form of suspended insoluble nonmetallic particles dramatically reduce the fluidity of molten aluminum.

Solidification. It has been shown that fluidity is inversely proportional to freezing range (that is, fluidity is highest for pure metals and eutectics, and lowest for solid-solution alloys). The manner in which solidification occurs may also influence fluidity.

Shrinkage. For most metals, the transformation from the liquid to the solid state is accompanied by a decrease in volume. In aluminum alloys, volumetric solidification shrinkage can range from 3.5 to 8.5%. The tendency for formation of shrinkage porosity is related to both the liquid/solid volume

Table 13 Typical mechanical properties of premium-quality aluminum alloy castings and elevated-temperature aluminum casting alloys

Alloy and temper	Hardness, HB(a)	Ultimate tensile strength MPa	Ultimate tensile strength ksi	Tensile yield strength MPa	Tensile yield strength ksi	Elongation in 50 mm (2 in.), %	Compressive yield strength MPa	Compressive yield strength ksi	Shear strength MPa	Shear strength ksi	Fatigue strength(b) MPa	Fatigue strength(b) ksi
Premium-quality castings(c)												
A201.0-T7	...	495	72	448	65	6	...	...	...	...	97	14
A206.0-T7	...	445	65	405	59	6	...	...	...	...	90	13
224.0-T7	...	420	61	330	48	4	...	...	...	...	86	12.5
249.0-T7	...	470	68	407	59	6	...	...	...	...	75	11
354.0-T6	...	380	55	283	41	6	...	...	...	...	135(d)	19.5(d)
C355.0-T6	...	317	46	235	34	6	...	...	...	...	97	14
A356.0-T6	...	283	41	207	30	10	...	...	...	...	90	13
A357.0-T6	...	360	52	290	42	8	...	...	...	...	90	13
Piston and elevated-temperature sand cast alloys												
222.0-T2	80	185	27	138	20	1	...	...	...	...	...	...
222.0-T6	115	283	41	275	40	<0.5	...	...	...	...	...	...
242.0-T21	70	185	27	125	18	1	...	...	145	21	55	8
242.0-T571	85	220	32	207	30	0.5	...	...	180	26	75	11
242.0-T77	75	207	30	160	23	2	165	24	165	24	72	10.5
A242.0-T75	...	215	31	...	...	...	...	...	...	...	...	...
243.0	95	207	30	160	23	2	200	29	70	10	70	10
328.0-F	...	220	32	130	19	2.5	...	...	...	...	...	...
328.0-T6	85	290	42	185	27	4.0	180	26	193	28	...	...
Piston and elevated-temperature alloys (permanent mold castings)												
222.0-T55	115	255	37	240	35	...	295	43	207	30	59	8.5
222.0-T65	...	...	...	...	...	...	...	...	...	...	...	...
242.0-T571	105	275	40	235	34	1	...	...	207	30	72	10.5
242.0-T61	110	325	47	290	42	0.5	...	...	240	35	65	9.5
332.0-T551	105	248	36	193	28	0.5	193	28	193	28	90	13
332.0-T5	105	248	36	193	28	1	200	29	193	28	90	13
336.0-T65	125	325	47	295	43	0.5	193	28	248	36	...	...
336.0-T551	105	248	36	193	28	0.5	193	28	193	28	...	...

(a) 10 mm (0.4 in.) ball with 500 kgf (1100 lbf) load. (b) Rotating beam test at 5×10^8 cycles. (c) Typical values of premium-quality casting are the same regardless of class or the area from which the specimen is cut; see Table 14 for minimum values. (d) Fatigue strength for 10^6 cycles

fraction and the solidification temperature range of the alloy. Riser requirements relative to the casting weight can be expected to increase with increasing solidification temperature range. Requirements for the establishment of more severe thermal gradients, such as by the use of chills or antichills, also increase.

Mechanical Properties

Typical mechanical properties of various aluminum casting alloys are given in Tables 11, 12, and 13. These typical values should be used only for assessing the suitability of an alloy for a particular application, and not for design purposes. Design-stress values are significantly below typical properties as discussed in the section on "Mechanical Test Methods" later in this article. Actual design strength depends on several factors, including:

- Section size
- Expected degree of porosity
- Presence of sharp corners
- Probability of cyclic loading in service

Minimum mechanical property limits are usually defined by the terms of general procurement specifications, such as those developed by government agencies and technical societies. These documents often specify testing frequency, tensile bar type and design, lot definitions, testing procedures, and the limits applicable to test results. By references to general process specifications, these documents also invoke standards and limits for many additional supplier obligations, such as specific practices and controls in melt preparation, heat treatment, radiographic and liquid penetrant inspection, and test procedures and interpretation. Tables 11 and 14 include minimum tensile properties of various casting alloys.

Mechanical Test Methods. Typical and minimum mechanical-property values commonly reported for castings of particular aluminum alloys are determined using separately cast test bars that are ½ in. diameter (for sand and permanent mold castings) or ¼ in. diameter (for die castings). As such, these values represent properties of sound castings, 13 or 6 mm (½ or ¼ in.) in section thickness, made using normal casting practice; they do not represent properties in all sections and locations of full-size production castings. Typical and minimum properties of test bars, however, are useful in determining relative strengths of the various alloy/temper combinations. Minimum properties—those values listed in applicable specifications—apply, except where otherwise noted, only to separate cast test bars. These values, unlike minimum values based on bars cut from production castings, are not usable as design limits for production castings. However, they can be useful in quality assurance. Actual mechanical properties, whether of separately cast test bars or of full-size castings, are dependent on two main factors:

- Alloy composition and heat treatment
- Solidification pattern and casting soundness

Some specifications for sand, permanent mold, plaster, and investment castings have defined the correlation between test results from specimens cut from the casting and separately cast specimens. A frequent error is the assumption that test values determined from these sources should agree. Rather, the properties of separately cast specimens should be expected to be superior to those of specimens machined from the casting. In the absence of more specific guidelines, one rule of thumb defines the average tensile and yield strengths of machined specimens as not less than 75% of the minimum requirements for separately cast specimens, and elongation as not less than 25% of the minimum requirement. These relations may be useful in establishing the commercial acceptability of parts in dispute.

Test Specimens. Accurate determination of mechanical properties of aluminum alloy castings (or of castings of any other metal) requires proper selection of test specimens. For most wrought products, a small piece of the material often is considered typical of the rest, and mechanical properties deter-

Table 14 Minimum tensile properties of premium-quality aluminum alloy castings

These mechanical property values are attainable in favorable casting configurations and must be negotiated with the foundry for the particular configuration desired.

Alloy	Class	Ultimate tensile strength (min), MPa	ksi	0.2% offset yield strength (min) MPa	ksi	Elongation in 50 mm (2 in.), %
Specimens cut from designated casting areas						
A201.0-T7(a)	Class 1	414	60	345	50	5
	Class 2	414	60	345	50	3
224.0-T7	Class 1	345	50	255	37	3
	Class 2	379	55	255	37	5
249.0-T7	Class 1	345	50	276	40	2
	Class 2	379	55	310	45	3
	Class 3	414	60	345	50	5
354.0-T6(a)	Class 1	324	47	248	36	3
	Class 2	345	50	290	42	2
C355.0-T6(a)	Class 1	283	41	214	31	3
	Class 2	303	44	228	33	3
	Class 3	345	50	276	40	2
A356.0-T6(a)	Class 1	262	38	193	28	5
	Class 2	276	40	207	30	3
	Class 3	310	45	234	34	3
A357.0-T6(a)	Class 1	310	45	241	35	3
	Class 2	345	50	276	40	5
224.0	Class 1	345	50	255	37	3
	Class 2	379	55	255	37	5
Specimens cut from any area						
A201.0-T7(a)	Class 10	386	56	331	48	3
	Class 11	379	55	331	48	1.5
224.0-T7	Class 10	310	45	241	35	2
	Class 11	345	50	255	37	3
249.0-T7	Class 10	379	55	310	45	3
	Class 11	345	50	276	40	2
354.0-T6(a)	Class 10	324	47	248	36	3
	Class 11	296	43	228	33	2
355.0-T6(a)	Class 10	283	41	214	31	3
	Class 11	255	37	207	30	1
	Class 12	241	35	193	28	1
A356.0-T6(a)	Class 10	262	38	193	28	5
	Class 11	228	33	186	27	3
	Class 12	221	32	152	22	2
A357.0-T6(a)	Class 10	262	38	193	28	5
	Class 11	283	41	214	31	3
224.0	Class 10	310	45	241	35	2
	Class 11	345	50	255	37	3

(a) Values from specification MIL-A-21180

mined from that small piece also are considered typical. Properties of castings, however, vary substantially from one area of a given casting to another, and may vary from casting to casting in a given heat.

If castings are small, one from each batch can be sacrificed and cut into test bars. If castings are too large to be economically sacrificed, test bars can be molded as an integral part of each casting, or can be cast in a separate mold.

Usually, test bars are cast in a separate mold. When this is done, care must be taken to ensure that the metal poured into the test-bar mold is representative of the metal in the castings that the test bars are supposed to represent. In addition, differences in pouring temperature and cooling rate, which can make the properties of separately cast test bars different from those of production castings, must be avoided.

For highly stressed castings, integrally cast test bars are preferable to separately cast bars. When integrally cast bars are selected, however, gating and risering must be designed carefully to ensure that test bars and castings have equivalent microstructure and integrity. Also, if there are substantial differences between test-bar diameter and wall thickness in critical areas of the casting, use of integrally cast test panels equal in thickness to those critical areas, instead of standard test bars, should be considered.

ASTM E8 defines the test bars suitable for evaluation of aluminum castings. The use of test bars cut from die castings is not recommended; simulated service (proof) testing is considered more appropriate.

Chemical Composition and Heat Treatment. Mechanical properties of castings depend not only on choice of alloy but also depend somewhat on other considerations linked with the alloy. Variations in chemical composition, even within specified limits, can have measurable effects. Metallurgical considerations such as coring, phase segregation, and modification also can alter properties. Modification is commonly used for those aluminum alloys with 5% or more silicon.

In hypoeutectic Al-Si alloys, the coarse silicon eutectic has been refined and dispersed by modification. The modified structure increases both ductility and mechanical strength. Modification is accomplished by addition of small amounts (0.02%) of sodium or strontium. Making those additions often introduces gas into the melt. Their use, therefore, must be weighed against applicable radiographic specifications. In the hypereutectic aluminum-silicon alloys (silicon greater than 11.7%), refinement of primary silicon in sand and permanent mold castings is accomplished by adding 0.05% P. (Phosphorus modification is required in only those die castings that have thick walls.) In these alloys, the phosphorus addition provides moderate improvements in strength and machinability.

Where heat treatment is required, choice of temper affects properties. Heat treating variables such as solution time and temperature, temperature of quenching medium, and quench delay also can alter properties.

Casting variables also contribute to mechanical-property variations. The differences between typical test-bar properties and mechanical properties of full-size castings result from differences in soundness and solidification characteristics. Casting soundness depends on the amount of porosity or other imperfections present in the casting. The presence of dross, shrinkage porosity, and gas porosity will all decrease properties. Dross inclusions and gas porosity are minimized by proper melting and pouring techniques.

Properties of production castings vary depending on two aspects of the solidification characteristics in each section of the casting—solidification rate and location and form of shrinkage. Shrinkage results when supply of molten metal is not adequate throughout solidification. Shrinkage may be apparent as sponge shrinkage, centerline shrinkage, shrinkage porosity, or a large shrinkage cavity. Shrinkage, however, can be controlled by proper use of directional solidification.

Directional solidification in a casting section is accomplished by starting solidification at a selected point and allowing it to progress toward a riser. If solidification also starts at a second point (such as a thinner region), then shrinkage between these two points results, as illustrated in Fig. 12. In designing a gating system for a casting, each section is examined in an attempt to establish directional solidification by proper use of chills, risers, and insulating materials. Casting design also is very important in ensuring that the necessary thermal gradients are established.

When the solidification characteristics of the entire casting are examined, solidification may be faster in some areas of the casting than in others; for example, the rate

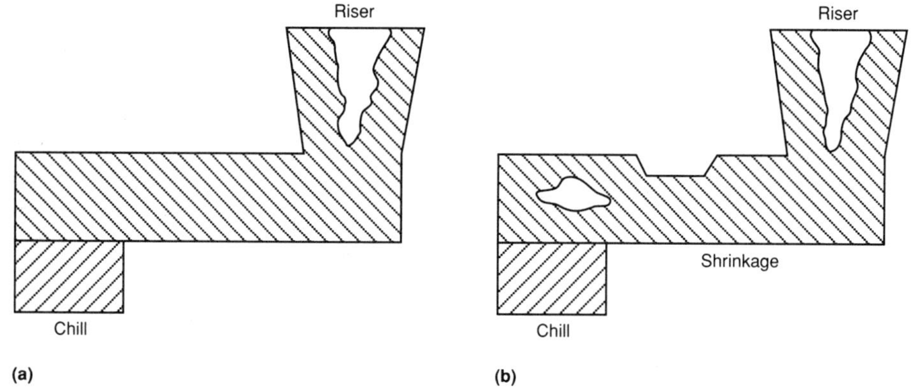

Fig. 12 Effect of gating system on formation of shrinkage cavities. Solidification starts at the chill and progresses toward the riser. In (a), molten metal can easily feed from the riser into the entire length of the casting. In (b), the narrow portion of the casting can freeze shut before solidification of the left portion of the casting is completed, and thus the riser can no longer feed that portion and a shrinkage cavity develops.

will be faster at a chilled area than at a riser area. The solidification rate of a casting section can be determined using a metallographic technique that measures dendrite-arm spacing. High solidification rates produce relatively small dendrite-arm spacing.

As explained in the section "Structure Control" in this article, casting processes all are characterized by solidification mode. In die casting, metal is rapidly injected and thus is subject to high chill rates. This results in rapid solidification and a fine metallographic structure at the surface. The fast chill rate and rapid rejection, however, usually result in centerline shrinkage. It has generally been impractical to achieve directional solidification in die castings to overcome this centerline shrinkage. In permanent mold casting chill rate also is high, but slower pouring allows longer feed times and more sound castings; design is very important in ensuring the proper solidification pattern. Because of high solidification rates, the properties of die and permanent mold castings are relatively insensitive to casting variables.

Due to relatively long solidification times, properties of sand and plaster castings are very sensitive to variations in casting technique; chills, risers, and gates significantly affect properties. Plaster molds, with their insulating properties, keep solidification rate low and consequently produce castings with relatively low properties.

The fatigue strength of castings is normally lower than that of wrought materials when the specimens are smooth. However, because castings are less notch sensitive than wrought products, castings may be beneficial in applications with multidirectional loading and/or notch susceptibility. Table 12 gives typical fatigue-strength limits of various casting alloys.

Premium (high-strength, high-toughness) castings combine a number of conventional casting procedures in a selective manner to provide premium strength, soundness, dimensional tolerances, surface finish, or a combination of these characteristics. As a result, premium engineered castings typically have somewhat higher fatigue resistance than conventional castings. Table 13 gives fatigue limits on various premium castings. Figure 13 compares premium castings and wrought plate in terms of notch-yield ratio and propagation energy values.

The production of premium-quality castings is an example of understanding and using the solidification process to good advantage. In production of premium-quality castings, composite molds combining several mold materials are used to take advantage of the special properties of each casting process. High mechanical properties in designated areas are obtained by use of special chills. Plaster sections and risers may be used to extend the feeding range of the casting.

Premium-quality castings are made to the tight radiographic and mechanical specifications required for aerospace and other critical applications. Basic linear tolerances of ±15 mm/m (±15 mils/in.) are possible with aluminum alloy castings, depending on mold material, equipment, and available fixtures, and minimum wall thickness of 3.8 mm (0.150 in.) is typical. Aluminum alloys commonly poured as premium-quality castings include C355.0, A356.0, and A357.

Table 14 gives mechanical property limits normally applied to premium engineered castings. It must be emphasized that the negotiation of limits for specific parts is usual practice, with higher as well as lower specific limits based on part design needs and foundry capabilities. Properties of the premium casting alloys at elevated temperatures are shown in Fig. 14, 15, and 16.

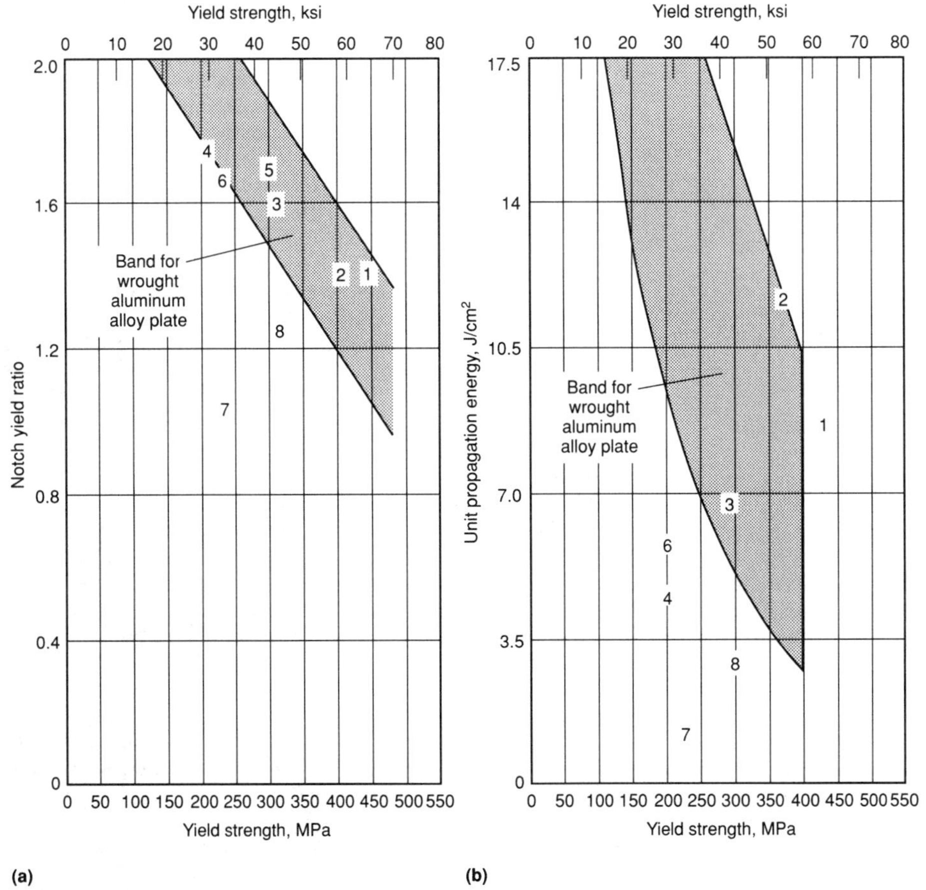

Fig. 13 Comparison of (a) notch yield ratio and (b) unit propagation energy versus yield strength for various aluminum alloy premium-quality castings and wrought aluminum alloy plate. 1, XA201.0-T7; 2, 249.0-T7; 3, 224.0-T7; 4, C355.0-T6; 5, 354.0-T6; 6, A356.0-T6; 7, 356.0-T6; 8, A357.0-T6

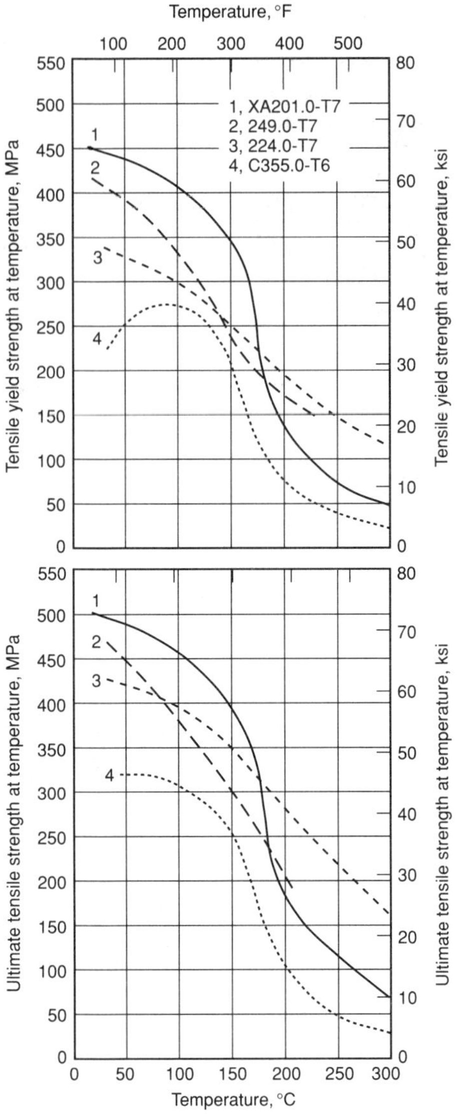

Fig. 14 Elevated-temperature tensile strengths of various premium casting alloys suggested for elevated-temperature service. Duration of temperature exposure was 10 000 h.

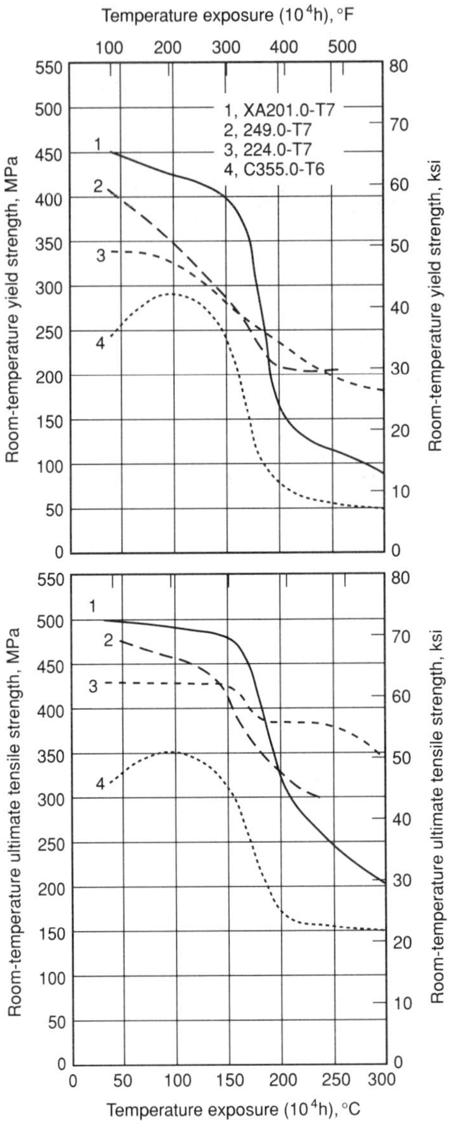

Fig. 15 Room-temperature tensile strengths of various aluminum alloy premium-quality castings after 10 000-h temperature exposure

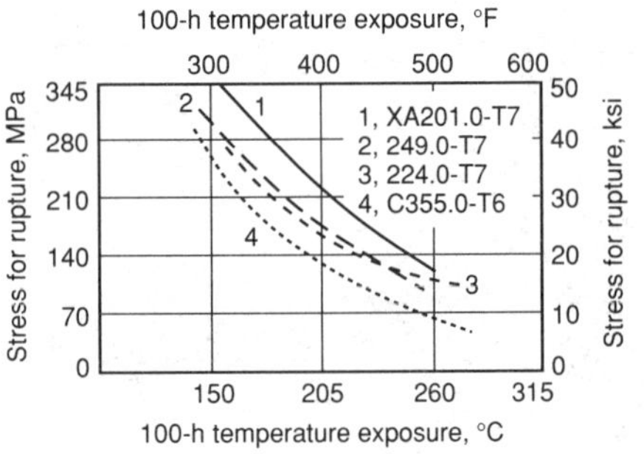

Fig. 16 100-h stress rupture of premium-quality aluminum alloy castings for elevated-temperature service

Other Engineering Characteristics

The modulus of elasticity of cast aluminum alloys varies somewhat with alloy contents, but 71 GPa (10.3×10^6 psi) is an appropriate average value for most design considerations.

Poisson's Ratio. The ratio of transverse contraction to the longitudinal elongation of a strain specimen is about 0.33 for aluminum casting alloys.

Corrosion Resistance. Aluminum and aluminum alloys have a highly protective, tightly adherent, invisible oxide film on its surface. Because of this film there is an inherent ability to resist corrosion. When this film is broken in most environments, it begins to reform immediately, the protective cover is restored, and deterioration by corrosion is resisted. Resistance to corrosion is greatly dependent on alloy content and its effect on the oxide film. For example, copper additions lower corrosion resistance, whereas silicon and magnesium enhance corrosion resistance. The most widely used test for corrosion resistance is the standard salt spray test. Relative rankings are given in Table 3.

Machinability. Aluminum exhibits better machinability than most metals and alloys. Its high thermal conductivity fosters long tool life by quickly diffusing heat away from the cutting tool. The typically low cutting forces permit the use of almost limitless machining speeds. In most cases the surface quality is exceptional. The casting alloys containing silicon (3*xx*.0 and 4*xx*.0 series) are less of a problem with regard to large chips because the cast structure is less homogeneous than wrought products. The grain-boundary segregation and second-phase silicon constituent aid chip breakage. To overcome problems with long chips of the single-phase alloys (2*xx*.0, 5*xx*.0, 7*xx*.0, and 8*xx*.0 series), the cutting tools are often ground or fitted with chip breakers. The main problem in machining aluminum is the tendency to form long continuous chips.

Coarse grain and large silicon constituent size in a casting can detract from the machinability because they accelerate tool wear. In the same manner alloys containing high contents of Fe, Mn, Cr, B, or Ti tend to form very hard second phases and may also cause excessive tool wear.

Alloys that contain small amounts of the low-melting-point metals like Sn, Pb, and Bi may exhibit free-machining characteristics.

Weldability. The weldability ratings given in Table 2 are based on the ability to be fusion welded with filler metal of the same alloy. This relates to the salvage problem of surface and dimensional defects and surface irregularities. The inert gas-shielded arc welding processes are normally used in preference to salt fluxes or flux-coated filler rods. The welding techniques used on castings are similar to those used on wrought products. However, special consideration must be made for the thicker as-cast surface aluminum oxide film and the gas porosity content of the castings. During welding this gas expands and "bubbles" up through the weld pool, causing defects in the weld. From this one must recognize that the quality of the weld is closely related to the casting quality or soundness in the area of the weld.

The large amount of oxides and porosity usually found in commercial die castings makes welding of these castings impractical in most cases. Recent quality-enhancing die casting techniques have, in a few cases, overcome this limitation.

Welding of castings to extrusions or other wrought aluminum products is often done. This part-assembly technique has been useful in many applications. Here the techniques normally used for welding wrought products apply, keeping in mind the effects of casting discontinuities on the weld quality.

REFERENCES

1. *Woldman's Engineering Alloys*, 6th ed., R.C. Gibbons, Ed., American Society for Metals, 1979

2. *Handbook of International Alloy Compositions and Designations*, Metals and Ceramics Information Center, Batelle Memorial Institute, 1976
3. R.G. Riek, A. Vrachnos, K.P. Young, and R. Mehrabian, *Trans. AFS*, Vol 83
4. J. Collot, Gircast—A New Stir-Casting Process Applied to Cu-Sn and Zn-Al Alloys, Castability and Mechanical Properties, in *Proceedings of International Symposium on Zinc-Aluminum Alloy*, Canadian Institute of Mining and Metallurgy, 1986, p 249
5. A.C. Arruda and M. Prates, *Solidification Technology in the Foundry and Cast House*, The Metals Society, 1983
6. R.L. Antona and R. Moschini, *Metall. Sci. Technol.*, Vol 4 (No. 2), Aug 1986, p 49-59
7. G.B. Brook, *Mater. Des.*, Vol 3, Oct 1982, p 558-565
8. U. Feurer and H. Zoller, Effect of Licensed Consection on the Structure of D.C. Cast Aluminum Ingots, Paper presented at the 105th AIME Conference (Las Vegas), The Metallurgical Society, Inc., 1976

SELECTED REFERENCES

- G. Bouse and M. Behrendt, Metallurgical and Mechanical Property Characterization of Premium Quality Vacuum Investment Cast 200 and 300 Series Aluminum Alloys, *Adv. Cast. Technol.*, Nov 1986
- Directionally Solidified Aluminum Foundry Alloys, Paper 69, *Trans. AFS*, 1987
- M. Holt and K. Bogardus, The "Hot" Aluminum Alloys, *Prod. Eng.*, Aug 1965
- F. Mollard, Understanding Fluidity, Paper 33, *Trans. AFS*, 1987
- E. Rooy, Improved Casting Properties and Integrity With Hot Isostatic Processing, *Mod. Cast.*, Dec 1983
- G. Scott, D. Granger, and B. Cheney, Fracture Toughness and Tensile Properties of Directionally Solidified Aluminum Foundry Alloys, *Trans. AFS*, 1987, p 69
- J. Tirpak, "Elevated Temperature Properties of Cast Aluminum Alloys A201-T7 and A357-T6," AFWAL-TR-85-4114, Air Force Wright Aeronautical Laboratories, 1985

Properties of Cast Aluminum Alloys

Revised by A.L. Kearney, Avery Kearney and Company

201.0
4.6Cu-0.7Ag-0.35Mn-0.35Mg-0.25Ti

Commercial Names

Trade name. KO-1

Specifications

Former ASTM. CQ 51A
SAE. 382
UNS number. A02010
AMS. 4228 and 4229

Chemical Composition

Composition limits. 4.0 to 5.2 Cu, 0.15 to 0.55 Mg, 0.20 to 0.50 Mn, 0.10 Si max, 0.15 Fe max, 0.15 to 0.35 Ti, 0.40 to 1.0 Ag, 0.05 other (each) max, 0.10 others (total) max, bal Al

Consequence of exceeding impurity limits. High iron or silicon decreases tensile properties.

Applications

Typical uses. Sand castings, permanent mold and investment castings. Structural casting members, aerospace housings, electrical transmission line fittings, insulator caps, truck and trailer castings, other applications requiring highest tensile and yield strengths with moderate elongation. Gasoline engine cylinder heads and pistons, turbine and supercharger impellers, rocker arms, connecting rods, missile fins, other applications where strength at elevated temperatures is important. Structural gear housings, aircraft landing gear castings, ordnance castings, pump housings, other applications where high strength and high energy-absorption capacity are required

Precautions in use. 201.0 castings should be specified in the T7 temper wherever resistance to stress-corrosion cracking is an important consideration (exceeds requirement of 60 days alternate immersion exposure—10 min/h to 3.5% NaCl and stressed to 75% of yield strength). T6 temper unsuitable for such applications

Mechanical Properties

Tensile properties. See Table 1.
Hardness. T4 temper, 95 HB; T6 temper, 135 HB; T7 temper, 130 HB (500 kg load, 10 mm ball)
Poisson's ratio. 0.33
Elastic modulus. Tension, 71 GPa (10.3 × 10^6 psi); shear, 23 GPa (3.4 × 10^6 psi)
Impact strength. Charpy V-notch: T4 temper, 21.7 J (16 ft · lbf); T6 temper, 10 ± 5 J

Table 1 Tensile properties of alloy 201.0

Condition	Exposure °C	Exposure °F	Exposure time, h	Tensile strength MPa	Tensile strength ksi	Yield strength(a) MPa	Yield strength(a) ksi	Elongation(b), %
Typical properties(c)								
T4	...	...	...	365	53	215	31	20
T6	...	...	...	485	70	435	63	7
T7	...	...	...	460	67	415	60	4.5
Short-time elevated-temperature properties(c)								
T7	150	300	0.5–100	380	55	360	52	6–8.5
			1 000	360	52	345	50	8
			10 000	315	46	275	40	7
	205	400	0.5	325	47	310	45	9
			100	285	41	270	39	10
			1 000	250	36	230	33	9
			10 000	185	27	150	22	14
	260	500	0.5	195	28	185	27	14
			100	150	22	140	20	17
			1 000	125	18	110	16	18
	315	600	0.5	140	20	130	19	12
			100	85	12	75	11	30
			1 000	70	10	60	9	39
			10 000	60	9	55	8	43
Average (avg) and minimum (min) values for T43 temper(d)								
Integral test bars(e)	...	...	...	416(avg)	60.3(avg)	257(avg)	37.3(avg)	17.4(avg)
	...	...	...	372(f)	54(min)(f)	220(f)	32(min)(f)	...
	...	...	...	393(g)	57(min)(g)	235(g)	34(min)(g)	...
Test bars cut from castings(h)	...	...	...	391(avg)	56.7(avg)	241(avg)	35.0(avg)	15.2(avg)
	...	...	...	338(f)	49(min)(f)	193(f)	28(min)(f)	...
	...	...	...	360(g)	52(min)(g)	215(g)	31(min)(g)	...

(a) 0.2% offset. (b) In 50 mm or 2 in. Where a range appears in this column, specified elongation varies with exposure time. (c) Properties of separately sand cast test bars. (d) The T43 temper was developed to provide increased impact resistance with some decrease in other mechanical properties. (e) Property values from 210 tests. (f) Minimum value above which 99% of the population values is expected to fall with a 95% confidence level. (g) Minimum value above which 90% of the population values is expected to fall with a 95% confidence level. (h) Property values from 117 tests

Table 2 Creep-rupture properties of separately sand cast test bars of alloy 201.0

Temperature, °C	Temperature, °F	Time under stress, h	Rupture stress, MPa	Rupture stress, ksi	Minimum creep rate at rupture stress, % per h	Stress for creep of: 1.0%, MPa	1.0%, ksi	0.5%, MPa	0.5%, ksi	0.25%, MPa	0.25%, ksi
150	300	10	Above yield		...	...	...	...	...	...	...
		100	Above yield		...	...	...	...	...	...	...
		1000	270	39	0.00013	260	38	250	36	250	36
		10 000	195	28	0.000023	195	28	185	27	180	76
175(a)	350(b)	10	Above yield		...	...	...	...	...	...	...
		100	250	36	0.00095	250	36	235	34	230	33
		1000	180	26	0.000175	180	26	170	25	170	24
		10 000	130	19	0.000035	130	19	125	18	125	18
205	400	10	250	36	0.0145	240	35	230	33	220	32
		100	180	26	0.0024	180	26	170	24	170	24
		1000	130	19	0.00046	130	19	125	18	110	16
		10 000	95	14	0.000088	95	14	90	13	85	12
230	450	10	185	27	0.028	185	27	170	25	170	24
		100	140	20	0.0048	140	20	130	19	115	17
		1000	95	14	0.00083	95	14	90	13	85	12
		10 000	70	10	0.000175	70	10	65	9.5	60	9
260	500	10	140	20	0.047	140	20	125	18	110	16
		100	95	14	0.0080	95	14	95	14	85	12
		1000	70	10	0.00130	70	10	70	9.8	60	8.6
		10 000	50	7.5	0.00028	50	7.5	50	7.2	45	6.3
290	550	10	90	13	0.062	83	12	83	12	69	10
		100	66	9.5	0.0118	63	9.2	58	8.4	54	7.8
		1000	48	7.0	0.0022	46	6.7	42	6.1	39	5.7
		10 000	34	5.0	0.00037	33	4.8	30	4.4	29	4.2
315	600	10	59	8.6	0.073	55	8.0	51	7.4	47	6.8
		100	43	6.2	0.0126	39	5.7	37	5.4	34	4.9
		1000	31	4.5	0.0023	29	4.2	27	3.9	24	3.5
		10 000	23	3.3	0.00043	21	3.0	20	2.9	18	2.6

(a) For this temperature, properties are interpolations.

(8 ± 4 ft · lbf); T43 temper, typically 20 J (15 ft · lbf)
Creep characteristics. Creep rupture in Table 2 and creep rates in Table 3
Fatigue characteristics. Fatigue limits for temper T7 in Fig. 1
Shear strength. Typical values of 290 ± 30 MPa (42 ± 4 ksi)

Mass Characteristics

Density. 2.80 g/cm³ (0.101 lb/in.³) at 20 °C (68 °F)

Thermal Properties

Liquidus temperature. 650 °C (1200 °F)
Solidus temperature. 535 °C (995 °F)
Coefficient of linear thermal expansion. T6 and T7 conditions:

Temperature range, °C	Temperature range, °F	Average coefficient, μm/m · K	Average coefficient, μin./in. · °F
20–100	68–212	19.3	10.7
20–200	68–392	22.7	12.6
20–300	68–572	24.7	13.7

Specific heat. 920 J/kg · K (0.22 Btu/lb · °F) at 100 °C (212 °F)
Latent heat of fusion. 389 kJ/kg (167 Btu/lb)
Thermal conductivity. 121 W/m · K (70 Btu/ft · h · °F) at 25 °C (77 °F)

Electrical Properties

Electrical conductivity. Volumetric, T6 condition: 27 to 32% IACS at 20 °C (68 °F)
Electrical resistivity. T6 condition: 54 to 64 nΩ · m at 20 °C (68 °F)
Electrolytic solution potential. Versus 0.1 *N* calomel electrode in an aqueous solution containing 53 g NaCl plus 3 g H_2O_2 per liter: as-cast, −0.70 V; T4 condition, −0.59 V; T6 condition, −0.68 V; T7 condition, −0.73 V

Fabrication Characteristics

Solution temperature. See Table 4.
Aging temperature. See Table 4.
Castability. Resistance to both hot cracking and solidification shrinkage is only rated fair for this alloy. Pressure tightness is rated good. Successful production of high-quality castings requires close control of alloy composition, grain refining, melt temperature, fluxing, and heat-treating practices. Melt temperature should not exceed 790 °C (1450 °F).
Machinability. Rating is excellent. Moderate to fast speeds and feeds are recommended.
Weldability. Very good results can be obtained when joining castings of alloy 201.0 to components of similar composition, using arc or resistance welding methods. Weld repair methods are not simple and require close control of the temperature of the casting during welding to prevent hot tearing and solidification cracking.
Finishing. Electroplating imparts an excellent finish to this alloy. Mechanical polishing also gives excellent results. Anodizing produces very good appearance and good protection.

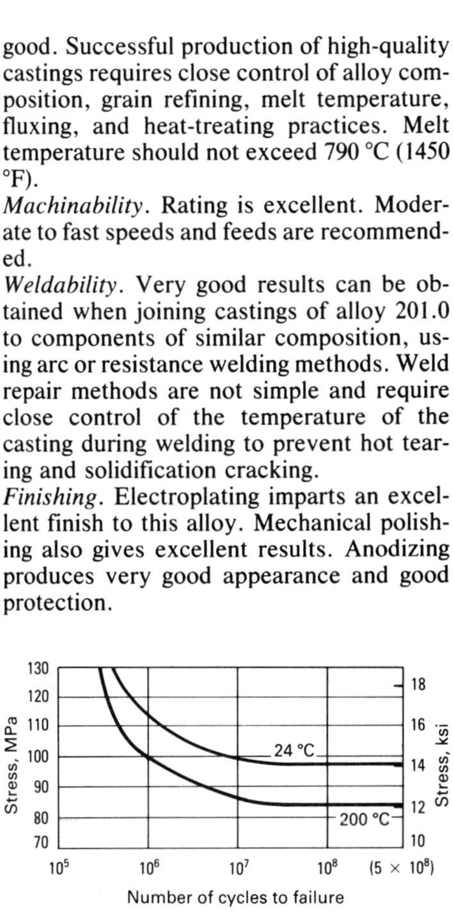

Fig. 1 *S-N* fatigue curves for alloy 201.0-T7 at 24 °C (75 °F) and 200 °C (400 °F). Curves are from R.R. Moore smooth specimens with completely reversed bending.

Table 3 Creep rates of 201.0 at various temperatures

Minimum creep rate, %/h	Stress to produce minimum creep rate at: 150 °C (300 °F), MPa	150 °C, ksi	205 °C (400 °F), MPa	205 °C, ksi	260 °C (500 °F), MPa	260 °C, ksi	315 °C (600 °F), MPa	315 °C, ksi
1×10^{-5}	165	24.0	64	9.3	28	4.0	11	1.6
1×10^{-4}	255	37.5	98	14.2	43	6.2	17	2.5
1×10^{-3}	(a)	(a)	155	22.5	66	9.5	26	3.8
0.01	(a)	(a)	234	34.0	103	15	41	6.0
0.1	(a)	(a)	(a)	(a)	155	22.5	62	9.0
1.0	(a)	(a)	(a)	(a)	241	35.0	97	14.0

(a) Above the minimum yield strength specification value of 345 MPa (50 ksi)

Table 4 Heat treatment practice for alloys 201.0 and 206.0

Temper designation	Solution treatment	Aging treatment
T4	510–515 °C (950–960 °F) for 2 h followed by 525–530 °C (980–990 °F)(a) for 14–20 h(b) and a water quench(c)	None
T6	Same as T4	Room temperature for 12–24 h, or 150–155 °C (305–315 °F) for 20 h
T7	Same as T4	Room temperature for 12–24 h, or 185–190 °C (365–375 °F) for 5 h
T43(d)	525 °C (980 °F) for 20 h and water quenched(c)	24 h at room temperature plus ½ to 1 h at 160 °C (320 °F)

(a) Careful composition and temperature control must be maintained during solution heat treatment in order to attain both adequate solution and to prevent incipient melting. (b) Soaking time periods required for average sand castings after load has reached specified temperature. Time changes may be required. Permanent mold and thin wall castings, in general, take less time. (c) At 65 to 100 °C (150 to 212 °F). (d) Temper T43 developed for alloy 201.0 for improved impact resistance with some decrease in other properties

204.0
4.6Cu-0.25Mg-0.17Fe-0.17Ti

Commercial Name

A-U5GT (Pechiney)
ELT-204 (French AFNOR)

Applications

Typical uses. Extensively used in France for high-performance sand and permanent mold castings
Permanent mold. Break calipers in Bendix brakes for European cars 1958 through early 1980s (many million castings produced and used without problems)
Sand and permanent mold. Light- and heavy-duty impellers, structural parts for aerospace and auto industry. Ordnance parts for tanks, off-the-highway trucks and other equipment, light-weight, high-strength hand tools, light-weight power-train castings in auto and truck industry

Mechanical Properties

Tensile properties. See Table 5 and Fig. 2.
The effects of solution treatment time depend to a considerable extent on the rate of solidification of the casting. A casting that has solidified rapidly permits a short solution treatment, whereas one that has solidified slowly requires a longer solution treatment time. The graphs in Fig. 2 compare mechanical property test results of permanent mold test specimens with sand cast test specimens as related to the solidification rate. Note the much longer times at the solution time required for the slowly solidified sand cast bars as compared with the rapidly solidified permanent mold test pieces. This difference is most pronounced in 204.0-type alloys (Fig. 2). However, some degree of these solution heat-treatment effects on mechanical properties occurs in all aluminum casting alloys.

206.0, A206.0
4.5Cu-0.30Mn-0.25Mg-0.22Ti

Specifications

AMS. 4235, 4236, 4237

Chemical Composition

Composition limits. 4.2 to 5.0 Cu, 0.15 to 0.35 Mg, 0.20 to 0.50 Mn, 0.10 Si max, 0.15 Fe max, 0.10 Zn max, 0.15 to 0.30 Ti, 0.05 Ni max, 0.05 Sn max, 0.05 other (each) max, 0.15 others (total) max, bal Al. A206.0: 4.2 to 5.0 Cu, 0.15 to 0.35 Mg, 0.20 to 0.50 Mn, 0.05 Si max, 0.10 Fe max, 0.10 Zn max, 0.15 to 0.30 Ti, 0.05 Ni max, 0.05 Sn max, 0.05 other (each) max, 0.15 others (total) max, bal Al

Applications

Typical uses. Structural castings in heat-treated temper for automotive, aerospace, and other applications where high tensile and yield strength and moderate elongation are needed. Gear housings, truck spring hanger castings, and other applications where high fracture toughness is required. Cylinder heads for gasoline and diesel motors, turbine and supercharger impellers, other applications where high strength at elevated temperatures and special aging treatment are required
Precautions in use. Subject to corrosion problems due to copper content of alloy. T4 and T7 heat treatments qualify and meet federal test requirements for stress-corrosion cracking. T6 temper should not be used where stress-corrosion cracking could be a problem.

Mechanical Properties

Tensile properties. Separately cast test bars. Tensile strength and yield strength, see Table 6 and Fig. 3. Elongation in 50 mm or 2 in. (typical for A206.0-T7): 11.7% at room temperature, 14.0% at 120 °C (250 °F), 17.7% at 175 °C (350 °F). Reduction in area (typical for A206.0-T7): 26.0% at room temperature, 40.4% at 120 °C (250 °F), 53.7% at 175 °C (350 °F)
Shear strength. See Table 6 and Fig. 3.
Compressive strength. See Table 6 and Fig. 3.
Bearing properties. See Table 6 and Fig. 3.
Hardness. T4 temper, 118 HV; T7 temper, 137 HV
Poisson's ratio. 0.33 at 20 °C (68 °F)
Elastic modulus. Tension: 70 GPa (10.2 × 10^6 psi) at room temperature, 69 GPa (10.0 × 10^6 psi) at 120 °C (250 °F), 65 GPa (9.4 × 10^6 psi) at 175 °C (350 °F). See also Fig. 3.
Impact strength. Charpy V-notch, 9.5 J (7.0 ft · lbf) at 20 °C (68 °F)
Fatigue strength. See Table 6 and Fig. 3.
Plane-strain fracture toughness. 43 MPa$\sqrt{m}$ (39 ksi$\sqrt{in.}$) for A206.0-T7 (not a true K_{Ic} value per ASTM E399). See also Fig. 4.

Table 5 Tensile properties and heat treatment of alloy 204.0

			Mean property values and estimated standard deviation				
			Tensile strength		0.2% yield strength		
Condition	Precipitation treatment(a)	Hardness, HB	MPa	ksi	MPa	ksi	Elongation in 50 mm (2 in.), %
Permanent molds							
Y34 (T4)	At least 5 days at 20 °C (68 °F)	110	400 ± 10	58 ± 1.5	250 ± 10	36 ± 1.5	21 ± 2
Y33 (T6)	12 h at 140 °C (285 °F)	105	395 ± 10	57 ± 1.5	230 ± 10	33 ± 1.5	20 ± 2
	12 h at 160 °C (320 °F)	115	405 ± 20	59 ± 3	290 ± 20	42 ± 3	16 ± 1
	12 h at 180 °C (355 °F)	125	420 ± 20	61 ± 3	380 ± 20	55 ± 3	8 ± 1
Sand castings							
Y24 (T4)	At least 5 days at 20 °C (68 °F)	110	400 ± 10	58 ± 1.5	265 ± 10	38.5 ± 1.5	14 ± 1
Y23 (T6)	12 h at 140 °C (285 °F)	105	395 ± 10	57 ± 1.5	250 ± 10	36 ± 1.5	13.5 ± 1
	12 h at 160 °C (320 °F)	115	400 ± 10	58 ± 1.5	300 ± 20	43.5 ± 3	9 ± 1
	12 h at 180 °C (355 °F)	125	420 ± 10	61 ± 1.5	395 ± 20	57 ± 3	3 ± 0.5

(a) Precipitation treatment preceded by a solution treatment of 12 h at 530 °C (985 °F) and a water quench

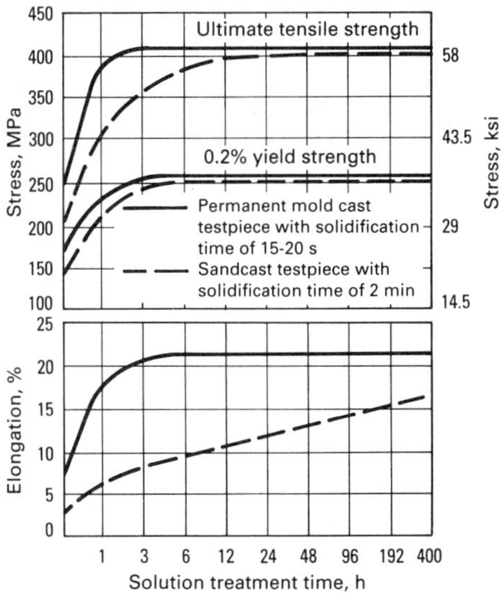

Fig. 2 Effect of solution treatment time on tensile properties of alloy 204.0. An AFNOR testpiece was cast in a permanent mold (solidification time 15 to 20 s). An Aluminium Pechiney testpiece was cast in sand (solidification time 2 min). Solution treatment at 530 °C (985 °F) followed by cold water quench and natural aging

Mass Characteristics

Density. 2.80 g/cm³ (0.101 lb/in.³) at 20 °C (68 °F)

Thermal Properties

Liquidus temperature. 650 °C (1202 °F)
Solidus temperature. 570 °C (1058 °F)
Incipient melting temperature. 542 °C (1008 °F)
Coefficient of linear thermal expansion. 19.3 μm/m · K (10.7 μin./in. · °F) at 20 to 100 °C (68 to 212 °F)
Specific heat. 920 J/kg · K (0.22 Btu/lb · °F) at 100 °C (212 °F)
Thermal conductivity. 121 W/m · K (70.1 Btu/ft · h · °F)

Electrical Properties

Electrical resistivity. At 20 °C (68 °F): T6 temper, 54 to 64 nΩ · m; T7 temper, 50 to 54 nΩ · m

Chemical Properties

General corrosion behavior. Comparable to other wrought or cast aluminum alloys containing equivalent amounts of copper

Fabrication Characteristics

Weldability. Fair repair welding characteristics
Solution temperature. See Table 4.
Aging temperature. See Table 4 or the following:

- T7 temper, 200 °C (390 °F), hold at temperature for 8 h
- T6 temper, 155 °C (310 °F), hold at temperature for 8 h
- T4 temper, room temperature

Table 6 Typical mechanical properties for separately cast test bars of alloy A206.0-T7

Property	Room temperature MPa	Room temperature ksi	120 °C (250 °F) MPa	120 °C (250 °F) ksi	175 °C (350 °F) MPa	175 °C (350 °F) ksi
Tensile strength	436	63	384	56	333	48
Tensile yield strength	347	50	316	46	302	44
Shear strength	257	37	232	34	208	30
Compressive yield strength	372	54	347	50	318	46
Bearing ultimate strength (BUS)						
e/*d* = 1.5(a)	692	100	632	92	545	79
e/*d* = 2.0(a)	960	131	784	114	635	92
Bearing yield strength (BYS)						
e/*d* = 1.5(a)	544	79	507	73	477	69
e/*d* = 2.0(a)	658	95	628	91	566	82
Axial fatigue strength						
Unnotched, *R* = 0.1						
10^3 cycles	435	63	...	...	350	51
10^5 cycles	290	42	...	...	250	36
10^7 cycles	205	30	...	...	160	23
Notched, K_t = 3.0, *R* = 0.1						
10^3 cycles	370	54	...	...	370	54
10^5 cycles	115	17	...	...	115	17
10^7 cycles	90	10	...	...	70	10

(a) Where *e* is edge distance and *d* is pin diameter

208.0
4Cu-3Si

Commercial Names

Former designation. 108

Specifications

Former SAE. 380
Former ASTM. CS43A
UNS number. A 02080
Government. QQ-A-601, class 8

Chemical Composition

Composition limits. 3.5 to 4.5 Cu, 0.10 Mg max, 0.5 Mn max, 2.5 to 3.5 Si, 1.2 Fe max, 1.0 Zn max, 0.35 Ni max, 0.25 Ti max, 0.50 others (total) max, bal Al
Consequence of exceeding impurity limits. High iron decreases mechanical properties, especially ductility. Zinc or tin decreases mechanical properties. Magnesium reduces ductility.

Applications

Typical uses. Manifolds, valve bodies, and similar castings requiring pressure tightness. Other applications where good casting characteristics, good weldability, pressure tightness, and moderate strength are required

Mechanical Properties

Tensile properties. Typical for separately cast test bars. F temper: tensile strength, 145 MPa (21 ksi); yield strength, 95 MPa (14 ksi); elongation, in 50 mm or 2 in., 2.5%
Shear strength. 115 MPa (17 ksi)
Compressive yield strength. 105 MPa (15 ksi)
Hardness. 55 HB (500 kg load, 10 mm ball)
Poisson's ratio. 0.33
Elastic modulus. Tension, 71 GPa (10.3 × 10^6 psi); shear, 26.5 GPa (3.85 × 10^6 psi)
Fatigue strength. 76 MPa (11 ksi) at 5 × 10^8 cycles (R.R. Moore type test)

Mass Characteristics

Density. 2.79 g/cm³ (0.101 lb/in.³) at 20 °C (68 °F)

Thermal Properties

Liquidus temperature. 625 °C (1160 °F)
Solidus temperature. 520 °C (970 °F)
Coefficient of linear thermal expansion.

Temperature range °C	Temperature range °F	Average coefficient μm/m · K	Average coefficient μin./in. · °F
20–100	68–212	22.0	12.2
20–200	68–392	23.0	12.7
20–300	68–572	24.0	13.3

Specific heat. 963 J/kg · K (0.230 Btu/lb · °F) at 100 °C (212 °F)
Latent heat of fusion. 389 kJ/kg (167 Btu/lb)
Thermal conductivity. At 25 °C (77 °F): as-cast, 121 W/m · K (70 Btu/ft · h · °F); annealed, 146 W/m · K (84.4 Btu/ft · h · °F)

Electrical Properties

Electrical conductivity. Volumetric, at 20 °C (68 °F): as-cast, 31% IACS; annealed, 38% IACS
Electrical resistivity. At 20 °C (68 °F): as-cast, 55.6 nΩ · m; annealed, 45.4 nΩ · m
Electrolytic solution potential. −0.77 V versus 0.1 *N* calomel electrode in an aqueous solution containing 53 g NaCl plus 3 g H_2O_2 per liter

Fabrication Characteristics

Melting temperature. 675 to 815 °C (1250 to 1500 °F)
Casting temperature. 675 to 790 °C (1250 to 1450 °F)
Joining. Rivet compositions: 2117-T4, 2017-T4. Soft solder with Alcoa No. 802; no flux. Rub-tin with Alcoa No. 33 flux: flame either

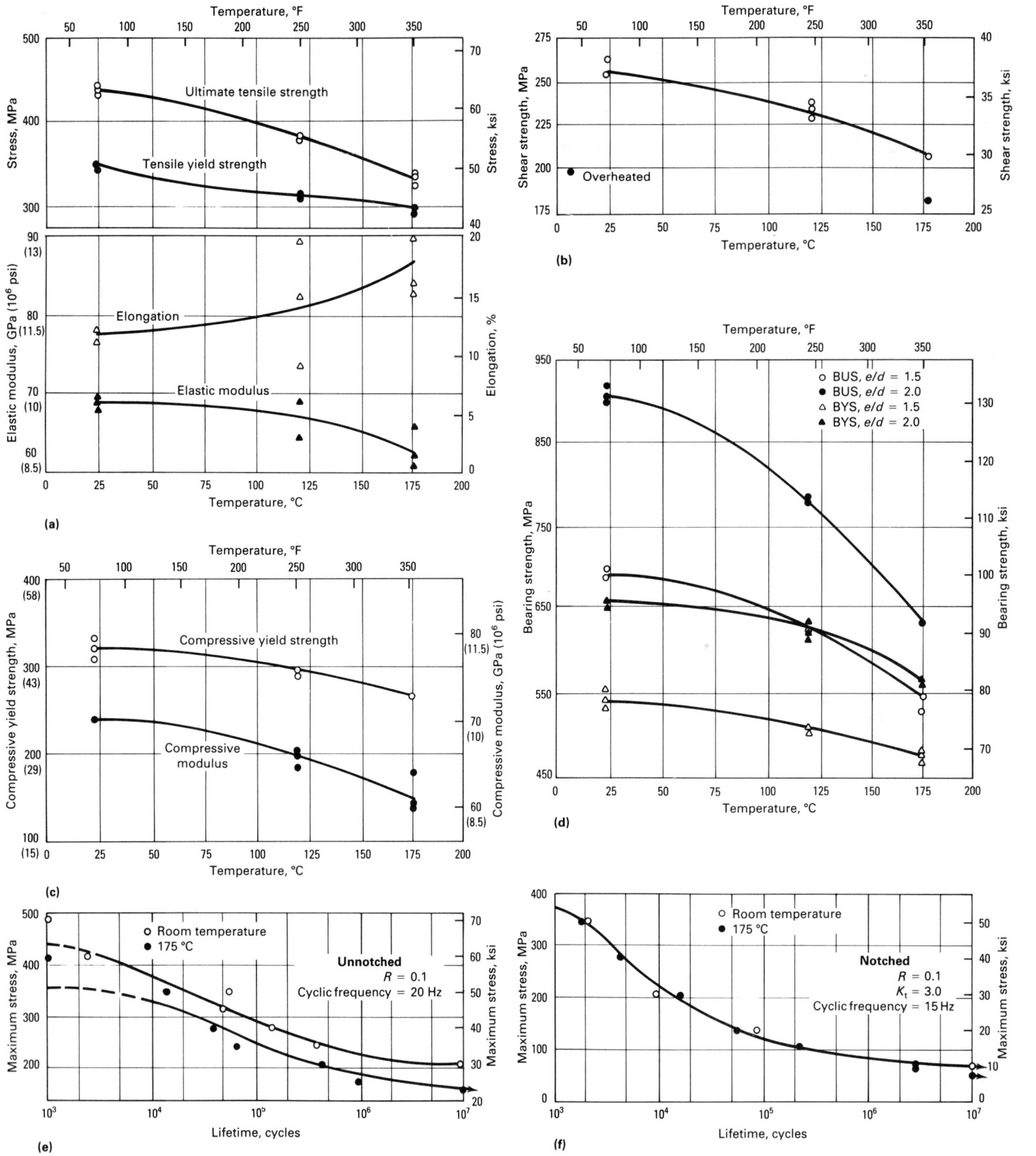

Fig. 3 Effect of temperature on the strength of A206.0-T7. (a) Tensile properties. (b) Shear strength. (c) Compressive properties. (d) Bearing strengths. (e) Unnotched fatigue limits. (f) Notched fatigue limits

reducing oxyacetylene or reducing oxyhydrogen. Metal-arc weld with 4043 alloy: Alcoa No. 27 flux. Carbon-arc weld with 4043 alloy: Alcoa No. 24 flux (automatic), Alcoa No. 27 flux (manual). Tungsten-arc argon-atmosphere weld with 4043 alloy: no flux. Resistance welding: spot, seam, and flash methods

Aging temperature. T55 temper: 150 to 160 °C (300 to 320 °F) for 16 h

238.0
10.0%Cu-4.0%Si-0.3%Mg

Commercial Names

138

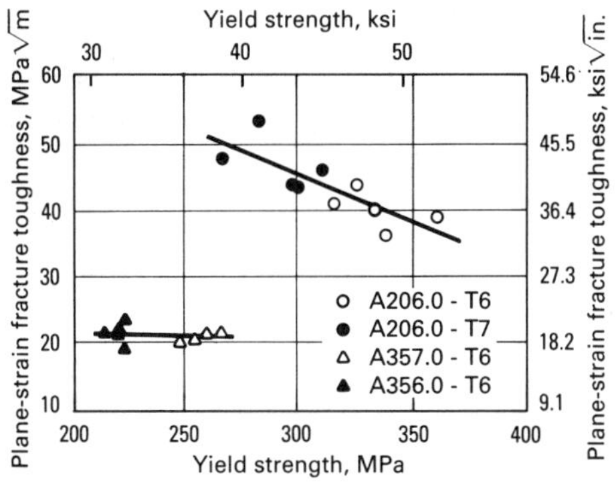

Fig. 4 Variation in plain strain-fracture toughness with yield strength of alloys A206.0, A357.0, and A356.0

Specifications

Former. CS104A, 138

Applications

Applications where a combination of high hardness in the as-cast condition, good casting characteristics, and good machinability are required. Soleplated for electric hand irons is typical.

242.0
4Cu-2Ni-2.5Mg

Commercial Names

Former designation. 142

Specifications

AMS. 4220, 4222
Former ASTM. CN42A
Former SAE. 39
UNS number. A02420
Government. QQ-A-601, class 6 (sand); QQ-A-596, class 3 (permanent mold)
Foreign. Canada: CSA CN42. France: NF A-U4NT. ISO: AlCu4Ni2Mg2

Chemical Composition

Composition limits. 3.5 to 4.5 Cu, 1.2 to 1.8 Mg, 0.35 Mn max, 0.7 Si max, 1.0 Fe max, 0.25 Cr max, 0.35 Zn max, 0.25 Ti max, 1.7 to 2.3 Ni, 0.05 other (each) max, 0.15 others (total) max, bal Al

Consequence of exceeding impurity limits. High iron may cause shrinkage difficulties. High silicon decreases mechanical properties. Chromium decreases thermal conductivity.

Applications

Typical uses. Motorcycle, diesel, and aircraft pistons, air-cooled cylinder heads, aircraft generator housings, other applications where excellent high-temperature strength is required

Mechanical Properties

Tensile properties. See Tables 7 to 9.
Shear strength. See Table 7.
Compressive yield strength. See Table 7.
Hardness. See Table 7.
Poisson's ratio. 0.33
Elastic modulus. Tension, 71 GPa (10.3 × 10^6 psi); shear, 26.5 GPa (3.85 × 10^6 psi)
Fatigue strength. See Table 7.

Mass Characteristics

Density. 2.823 g/cm^3 (0.102 lb/in.^{3}) at 20 °C (68 °F)

Thermal Properties

Liquidus temperature. 635 °C (1175 °F)
Solidus temperature. 530 °C (990 °F)
Coefficient of linear thermal expansion.

Temperature range °C	Temperature range °F	Average coefficient µm/m · K	Average coefficient µin./in. · °F
20–100	68–212	22.5	12.5
20–200	68–392	23.5	13.1
20–300	68–572	24.5	13.6

Specific heat. 963 J/kg · K (0.230 Btu/lb · °F) at 100 °C (212 °F)
Latent heat of fusion. 389 kJ/kg (167 Btu/lb)
Thermal conductivity. At 25 °C (77 °F):

Temper and product	Conductivity W/m · K	Conductivity Btu/ft · h · °F
T21, sand	167	96.7
T571, sand	134	77.4
T77, sand	146	84.6
T61, permanent mold	130	74.9

Electrical Properties

Electrical conductivity. Volumetric:

Temper and product	Conductivity, %IACS
T21, sand	44
T571, sand	34
T77, sand	38
T61, permanent mold	33

Electrical resistivity. At 20 °C (68 °F):

Temper and product	Resistivity, nΩ · m
T21, sand	39.2
T571, sand	50.7
T77, sand	45.4
T61, permanent mold	52.2

Fabrication Characteristics

Melting temperature. 675 to 815 °C (1250 to 1500 °F)
Casting temperature. Sand and permanent mold castings: 675 to 790 °C (1250 to 1450 °F)
Annealing temperature. See Table 10.
Solution temperature. See Table 10.
Aging temperature. See Table 10.
Joining. Rivet compositions: 2117-T4, 2017-T4. Soft solder with Alcoa No. 802; no flux. Rub-tin with Alcoa No. 802. Metal-arc weld with 4043 alloy: Alcoa No. 27 flux. Carbon-arc weld with 4043 alloy: Alcoa No. 24 flux (automatic), Alcoa No. 27 flux (manual). Tungsten-arc argon-atmosphere weld with 4043 alloy: no flux. Resistance welding: spot, seam, and flash welds

295.0
4.5Cu-1.1Si

Commercial Names

Former designation. 195

Specifications

AMS. 4230, 4231
Former ASTM. C4A
SAE. 38
UNS number. A02950
Government. QQ-A-601, class 4
Foreign. Canada: CSA-C4. ISO: AlCu4Si

Table 7 Typical mechanical properties of separately cast test bars of alloy 242.0

Temper	Tensile strength(a) MPa	Tensile strength(a) ksi	Tensile yield strength(a) MPa	Tensile yield strength(a) ksi	Elongation(a), %	Hardness(b), HB	Shear strength MPa	Shear strength ksi	Fatigue strength(c) MPa	Fatigue strength(c) ksi	Compressive yield strength(a) MPa	Compressive yield strength(a) ksi
Sand cast												
T21	185	27	125	18	1.0	70	145	21	55	8.0	125	18
T571	220	32	205	30	0.5	85	180	26	75	11.0	235	34
T77	205	30	160	23	2.0	75	165	24	70	10.5	165	24
Permanent mold cast												
T571	275	40	235	34	1.0	105	205	30	70	10.5	235	34
T61	325	47	290	42	0.5	110	240	35	65	9.5	305	44

(a) Strengths and elongations remain unchanged or improve at low temperatures. (b) 500 kg load; 10 mm ball. (c) At 5 × 10^8 cycles; R.R. Moore type test

Table 8 Typical tensile properties of separately cast test bars of alloy 242.0-T571 at elevated temperature

Temperature °C	Temperature °F	Tensile strength MPa	Tensile strength ksi	Yield strength MPa	Yield strength ksi	Elongation(a), %
Sand cast						
24	75	220	32	205	30	0.5
150	300	205	30	195	28	0.5
205	400	180	26	145	21	1.0
260	500	90	13	55	8	8.0
315	600	55	8	30	4	20.0
Permanent mold cast						
24	75	275	40	235	34	1.0
150	300	255	37	230	33	1.0
205	400	195	28	150	22	2.0
260	500	90	13	55	8	15.0
315	600	55	8	30	4	35.0

(a) In 50 mm or 2 in.

Table 9 Tensile properties of 242.0–T77 alloy at various temperatures

Temperature °C	Temperature °F	Tensile strength MPa	Tensile strength ksi	Yield strength(a) MPa	Yield strength(a) ksi	Elongation, %
Sand castings						
−195	−320	255	37	193	28	2
−80	−112	220	32	172	25	2
−28	−18	220	32	158	23	2
24	75	207	30	158	23	2
100	212	207	30	158	23	2
150	300	186	27	145	21	2
205	400	138	20	103	15	3
260	500	90	13	55	8	6
315	600	55	8	28	4	20
371	700	35	5	21	3	40
Permanent mold casting						
−195	−320	290	42	270	39	1
−80	−112	275	40	248	36	1
−28	−18	275	40	235	34	1
24	75	275	40	235	34	1
100	212	275	40	235	34	1
150	300	255	37	227	33	1
205	400	193	28	152	22	2
260	500	90	13	55	8	15
315	600	55	8	28	4	35
371	700	35	5	21	3	60

(a) 0.2% offset

Table 10 Heat treatment for separately cast test bars of alloy 242.0

Purpose (and resulting temper)	Temperature °C	Temperature °F	Time, h
Sand castings			
Annealing, T21	340–345	645–655	2–4
Aging, T571(a)	170–175	335–345	40–48
Solution heat treatment	520–525	965–975	6(b)(c)
Aging. T77(d)(e)	340–345	645–655	1–3
Permanent mold castings			
Aging, T571(b)	170–175	335–345	40–48
Solution heat treatment	515–520	955–965	4(b)(f)
Aging, T61(e)	200–205	395–405	3–5

(a) No solution heat treatment. (b) Soaking-time periods required for average castings after load has reached specified temperature. Time can be decreased or may have to be increased, depending on experience with particular castings. (c) Still-air cooling. (d) Start with solution heat-treated material. (e) U.S. Patent 1,822,877. (f) Cool in water at 65 to 100 °C (150 to 212 °F).

Chemical Composition

Composition limits. 4.0 to 5.0 Cu, 0.03 Mg max, 0.35 Mn max, 0.7 to 1.5 Si, 1.0 Fe max, 0.35 Zn max, 0.25 Ti max, 0.05 other (each) max, 0.15 others (total) max, bal Al

Consequence of exceeding impurity limits. High iron or silicon decreases tensile properties, especially ductility. Manganese or magnesium decreases ductility. Tin reduces strength, hardness, and resistance to corrosion.

Applications

Typical uses. Flywheel housings, rear-axle housings, bus wheels, aircraft wheels, fittings, crankcases and other applications where a combination of high tensile properties and good machinability is required, but pressure tightness is not needed

Mechanical Properties

Tensile properties. Typical for separately cast test bars. Tensile strength: T4 temper, 220 MPa (32 ksi); T6 temper, 250 MPa (36 ksi); T62 temper, 285 MPa (41 ksi). Yield strength: T4 temper, 110 MPa (16 ksi); T6 temper, 165 MPa (24 ksi); T62 temper, 220 MPa (32 ksi). Elongation in 50 mm or 2 in.: T4 temper, 8.5%; T6 temper, 5.0%; T62 temper, 2.0%. Strengths and elongations remain unchanged or improve at low temperatures. See also Table 11.

Shear strength. T4 temper: 180 MPa (26 ksi). T6 temper: 205 MPa (30 ksi). T62 temper: 230 MPa (33 ksi)

Compressive yield strength. T4 temper: 115 MPa (17 ksi); T6 temper: 170 MPa (25 ksi); T62 temper: 235 MPa (34 ksi)

Hardness. T4 temper, 60 HB; T6 temper, 75 HB; T62 temper, 90 HB (500 kg load, 10 mm ball)

Poisson's ratio. 0.33

Elastic modulus. Tension, 70 GPa (10.0 × 10^6 psi); shear, 25.9 GPa (3.75 × 10^6 psi)

Fatigue strength. At 5 × 10^8 cycles: T4 temper, 48 MPa (7 ksi); T6 temper, 52 MPa (7.5 ksi); T62 temper, 55 MPa (8.0 ksi) (R.R. Moore type test)

Mass Characteristics

Density. 2.823 g/cm^3 (0.102 lb/in.^{3}) at 20 °C (68 °F)

Thermal Properties

Liquidus temperature. 645 °C (1190 °F)

Solidus temperature. 520 °C (970 °F)

Coefficient of linear thermal expansion.

Temperature range °C	Temperature range °F	Average coefficient µm/m · K	Average coefficient µin./in. · °F
20–100	68–212	23.0	12.8
20–200	68–392	24.0	13.3
20–300	68–572	25.0	13.9

Specific heat. 963 J/kg · K (0.230 Btu/lb · °F) at 100 °C (212 °F)

Latent heat of fusion. 389 kJ/kg (167 Btu/lb)

Thermal conductivity. T4 and T62 tempers: 138 W/m · K (79.8 Btu/ft · h · °F) at 25 °C (77 °F)

Electrical Properties

Electrical conductivity. Volumetric: T4 temper, 35% IACS; T62 temper, 37% IACS

Electrical resistivity. T4 and T6 tempers: 49.3 nΩ · m at 20 °C (68 °F)

Electrolytic solution potential. Versus 0.1 *N* calomel electrode in an aqueous solution containing 53 g NaCl plus 3 g H_2O_2 per liter:

Table 11 Typical tensile properties for separately cast test bars of alloy 295.0 at elevated temperatures

Temperature °C	Temperature °F	Tensile strength MPa	Tensile strength ksi	Yield strength MPa	Yield strength ksi	Elongation(a), %
T4 temper						
24	75	220	32	110	16	8.5
150	300	195	28	105	15	5.0
205	400	105	15	60	9	15.0
260	500	60	9	40	6	25.0
315	600	30	4	20	3	75.0
T6 temper						
24	75	250	36	165	24	5.0
150	300	195	28	140	20	5.0
205	400	105	15	60	9	15.0
260	500	60	9	40	6	25.0
315	600	30	4	20	3	75.0

(a) In 50 mm or 2 in.

Table 12 Typical tensile properties of separately permanent mold cast test bars of alloy 296.0 at elevated temperature

Temperature °C	Temperature °F	Tensile strength MPa	Tensile strength ksi	Yield strength MPa	Yield strength ksi	Elongation(a), %
T4 temper						
24	75	255	37	130	19	9
150	300	200	29	160	23	5
205	400	115	17	75	11	15
260	500	50	7	30	4	25
315	600	25	3.5	15	2.5	75
T6 temper						
24	75	275	40	180	26	5
150	300	200	29	160	23	5
205	400	115	17	75	11	15
260	500	50	7	30	4	25
315	600	25	3.5	15	2.5	75

(a) In 50 mm or 2 in.

T4 temper, −0.70 V; T6 temper, −0.71 V; T62 temper, −0.73 V

Fabrication Characteristics

Melting temperature. 675 to 815 °C (1250 to 1500 °F)
Casting temperature. 675 to 790 °C (1250 to 1450 °F)
Solution temperature. 515 to 520 °C (955 to 965 °F); hold at temperature for 12 h; cool in water at 65 to 100 °C (150 to 212 °F)
Aging temperature. 150 to 155 °C (305 to 315 °F). To obtain T6 temper from solution heat-treated material, hold at temperature for 3 to 5 h; for T62 temper, hold at temperature for 12 to 16 h.
Joining. Rivet compositions: 2117-T4, 2017-T4. Soft solder with Alcoa No. 802; no flux. Rub-tin with Alcoa No. 802. Atomic-hydrogen weld with 4043 alloy; Alcoa No. 22 flux. Oxyacetylene weld with 4043 alloy; Alcoa No. 22 flux; flame neutral. Metal-arc weld with 4043 alloy; Alcoa No. 27 flux. Carbon-arc weld with 4043 alloy; Alcoa No. 24 flux (automatic), Alcoa No. 27 flux (manual). Tungsten-arc argon-atmosphere weld with 4043 alloy; no flux. Resistance welding: spot, seam, and flash methods

296.0
4.5Cu-2.5Si

Commercial Names

Former designations. B 295.0, B 195

Specifications

AMS. 4282, 4283
SAE. 380
UNS number. A-22950
Government. QQ-A-596, class 4

Chemical Composition

Composition limits. 4.0 to 5.0 Cu, 0.05 Mg max, 0.35 Mn max, 2.0 to 3.0 Si, 1.2 Fe max, 0.50 Zn max, 0.25 Ti max, 0.35 Ni max, 0.35 others (total) max, bal Al
Consequence of exceeding impurity limits. High iron decreases tensile properties, especially ductility. Zinc or tin decreases tensile properties. Manganese or magnesium reduces ductility.

Applications

Typical uses. Aircraft fittings, aircraft gun control parts, aircraft wheels, railroad car seat frames, compressor connecting rods, full pump bodies, other applications requiring a combination of high-tensile properties and good machinability

Mechanical Properties

Tensile properties. Typical for separately cast test bars. Tensile strength: T4 temper, 255 MPa (37 ksi); T6 temper, 275 MPa (40 ksi); T7 temper, 270 MPa (39 ksi). Yield strength: T4 temper, 130 MPa (19 ksi); T6 temper, 180 MPa (26 ksi); T7 temper, 140 MPa (20 ksi). Elongation, in 50 mm or 2 in.: T4 temper, 9%; T6 temper, 5%; T7 temper, 4.5%. Strengths and elongations remain unchanged or improve at low temperatures. See also Table 12.
Shear strength. T4 and T7 tempers, 205 MPa (30 ksi). T6 temper, 220 MPa (32 ksi)
Compressive yield strength. T4 and T7 tempers: 140 MPa (20 ksi); T6 temper: 180 MPa (26 ksi)
Hardness. T4 temper: 75 HB; T6 temper: 90 HB; T7 temper: 80 HB (500 kg load, 10 mm ball)
Poisson's ratio. 0.33
Elastic modulus. Tension, 69 GPa (10.0×10^6 psi); shear, 26.2 GPa (3.80×10^6 psi)
Fatigue strength. At 5×10^8 cycles: T4 temper, 65 MPa (9.5 ksi); T6 temper, 70 MPa (10 ksi); T7 temper, 60 MPa (9 ksi) (R.R. Moore type test)

Mass Characteristics

Density. 2.796 g/cm^3 (0.101 lb/in.^{3}) at 20 °C (68 °F)

Thermal Properties

Liquidus temperature. 635 °C (1170 °F)
Solidus temperature. 530 °C (990 °F)
Coefficient of linear thermal expansion.

Temperature range °C	Temperature range °F	Average coefficient μm/m · K	Average coefficient μin./in. · °F
20–100	68–212	22	12.2
20–200	68–392	23	12.7
20–300	68–572	24	13.3

Specific heat. 963 J/kg · K (0.230 Btu/lb · °F) at 100 °C (212 °F)
Latent heat of fusion. 389 kJ/kg (167 Btu/lb)
Thermal conductivity. T4 and T6 tempers: 130 W/m · K (75 Btu/ft · h · °F)

Electrical Properties

Electrical conductivity. Volumetric, T4 and T6 tempers: 33% IACS
Electrical resistivity. T4 and T6 tempers: 52.2 nΩ · m at 20 °C (68 °F)
Electrolytic solution potential. T4 temper: −0.71 V versus 0.1 *N* calomel electrode in an aqueous solution containing 53 g NaCl plus 3 g H_2O_2 per liter

Fabrication Characteristics

Melting temperature. 675 to 815 °C (1250 to 1500 °F)
Casting temperature. 675 to 815 °C (1250 to 1500 °F)
Solution temperature. 505 to 515 °C (945 to 955 °F); hold at temperature for 8 h; cool in water at 65 to 100 °C (150 to 212 °F)
Aging temperature. To obtain T6 temper

from solution heat-treated material, 150 to 155 °C (305 to 315 °F) and hold at temperature 5 to 7 h; for T7 temper (U.S. Patent 1,822,877) 255 to 265 °C (495 to 505 °F) and hold at temperature 4 to 6 h
Joining. Same as alloy 295.0

308.0
5.5Si-4.5Cu

Commercial Names

Former designation. A108

Specifications

Former ASTM. SC64A
SAE. 330
UNS number. A03080
Government. QQ-A-596, class 6

Chemical Composition

Composition limits. 4.0 to 5.0 Cu, 0.10 Mg max, 0.50 Mn max, 5.0 to 6.0 Si, 1.0 Fe max, 1.0 Zn max, 0.25 Ti max, 0.50 others (total) max, bal Al
Consequence of exceeding impurity limits. High iron, zinc, or tin decreases mechanical properties. Magnesium decreases ductility.

Applications

Typical uses. Ornamental grills, reflectors, general-purpose castings, and other applications where good casting characteristics, good weldability, pressure tightness, and moderate strength are required

Mechanical Properties

Tensile properties. Typical for separately cast test bars. F temper: tensile strength, 195 MPa (28 ksi); yield strength, 110 MPa (16 ksi); elongation in 50 mm or 2 in., 2.0%
Shear strength. 150 MPa (22 ksi)
Hardness. 70 HB (500 kg load, 10 mm ball)
Poisson's ratio. 0.33
Elastic modulus. Tension, 71 GPa (10.3 × 10^6 psi); shear, 26.5 GPa (3.85 × 10^6 psi)

Mass Characteristics

Density. 2.79 g/cm^3 (0.101 lb/in.^{3}) at 20 °C (68 °F)

Thermal Properties

Liquidus temperature. 615 °C (1135 °F)
Solidus temperature. 520 °C (970 °F)
Coefficient of linear thermal expansion.

Temperature range °C	°F	Average coefficient µm/m · K	µin./in. · °F
20–100	68–212	21.5	11.9
20–200	68–392	22.5	12.5
20–300	68–572	23.0	12.8

Specific heat. 963 J/kg · K (0.230 Btu/lb · °F) at 100 °C (212 °F)
Latent heat of fusion. 389 kJ/kg (167 Btu/lb)
Thermal conductivity. 142 W/m · K (82.2 Btu/ft · h · °F) at 25 °C (77 °F)

Electrical Properties

Electrical conductivity. Volumetric, 37% IACS at 20 °C (68 °F)
Electrical resistivity. 46.6 nΩ · m at 20 °C (68 °F)
Electrolytic solution potential. −0.75 versus 0.1 *N* calomel electrode in an aqueous solution containing 53 g NaCl plus 3 g H_2O_2 per liter

Fabrication Characteristics

Melting temperature. 675 to 815 °C (1250 to 1500 °F)
Casting temperature. 675 to 790 °C (1250 to 1450 °F)
Joining. Rivet compositions: 2117-T4, 2017-T4. Soft solder with Alcoa No. 802. Braze with Alcoa No. 717; Alcoa No. 33 flux; flame either reducing oxyacetylene or reducing oxyhydrogen. Metal-arc weld with 4043 alloy; Alcoa No. 27 flux. Carbon-arc weld with 4043 alloy; Alcoa No. 24 flux (automatic), Alcoa No. 27 flux (manual). Tungsten-arc argon-atmosphere weld with 4043 alloy, no flux. Resistance welding: spot, seam, and flash

319.0
6Si-3.5Cu

Commercial Names

Former designations. 319, Allcast

Specifications

Former ASTM. SC64D
SAE. 326
UNS number. A03190
Foreign. ISO: AlSi6Cu4

Chemical Composition

Composition limits. 3.0 to 4.0 Cu, 0.10 Mg max, 0.50 Mn max, 5.5 to 6.5 Si, 1.0 Fe max, 1.0 Zn max, 0.25 Ti max, 0.35 Ni max, 0.50 others (total) max, bal Al
Consequence of exceeding impurity limits. Mechanical properties are relatively insensitive to impurities.

Applications

Typical uses. Automotive cylinder heads, internal combustion engine crankcases, typewriter frames, piano plates, and other applications where good casting characteristics and weldability, pressure tightness, and moderate strength are required

Mechanical Properties

Tensile properties. See Table 13.
Shear strength. See Table 13.
Compressive yield strength. See Table 13.
Hardness. See Table 13.
Poisson's ratio. 0.33
Elastic modulus. Tension, 74 GPa (10.7 × 10^6 psi); shear, 28 GPa (4.0 × 10^6 psi)
Fatigue strength. See Table 13.

Mass Characteristics

Density. 2.79 g/cm^3 (0.101 lb/in.^{3}) at 20 °C (68 °F)

Thermal Properties

Liquidus temperature. 605 °C (1120 °F)
Solidus temperature. 515 °C (960 °F)
Coefficient of linear thermal expansion.

Temperature range °C	°F	Average coefficient µm/m · K	µin./in. · °F
20–100	68–212	21.5	11.9
20–200	68–392	23.0	12.8
20–300	68–572	23.5	13.1

Specific heat. 963 J/kg · K (0.230 Btu/lb · °F) at 100 °C (212 °F)
Latent heat of fusion. 389 kJ/kg (167 Btu/lb)
Thermal conductivity. 109 W/m · K (62.9 Btu/ft · h · °F) at 25 °C (77 °F)

Electrical Properties

Electrical conductivity. Volumetric, 27% IACS at 20 °C (68 °F)
Electrical resistivity. Sand: 63.9 nΩ · m at 20 °C (68 °F)
Electrolytic solution potential. −0.81 V (sand) and −0.76 V (permanent mold) versus 0.1 *N* calomel electrode in an aqueous solution containing 53 g NaCl plus 3 g H_2O_2 per liter

Fabrication Characteristics

Melting temperature. 675 to 815 °C (1250 to 1500 °F)
Casting temperature. Sand: 675 to 790 °C (1250 to 1450 °F)
Solution temperature. 500 to 505 °C (935 to

Table 13 Typical mechanical properties for separately cast test bars of alloy 319.0

Temper	Tensile strength(a) MPa	ksi	Tensile yield strength(a) MPa	ksi	Elongation(a)(b), %	Hardness(c), HB	Shear strength MPa	ksi	Fatigue strength(d) MPa	ksi	Compressive yield strength(a) MPa	ksi
Sand cast												
As-cast	185	27	125	18	2.0	70	150	22	70	10	130	19
T6	250	36	165	24	2.0	80	200	29	75	11	170	25
Permanent mold												
As-cast	235	34	130	19	2.5	85	165	24	70	10	130	19
T6	280	40	185	27	3.0	95	185	27	...	...	...	...

(a) Strengths and elongations are unchanged or improved at low temperatures. (b) In 50 mm or 2 in. (c) 500 kg load; 10 mm ball. (d) At 5 × 10^8 cycles; R.R. Moore type test

Table 14 Tensile properties of permanent mold 332.0-T5 alloy at various temperatures

Temperature °C	Temperature °F	Tensile strength MPa	Tensile strength ksi	Yield strength(a) MPa	Yield strength(a) ksi	Elongation, %
24	75	248	36	193	28	1
100	212	227	33	185	27	1
150	300	215	31	165	24	2
205	400	172	25	110	16	3
260	500	130	19	83	12	6
315	600	83	12	55	8	15
371	700	55	8	41	6	25

(a) 0.2% offset

Table 15 Typical tensile properties for separately cast test bars of alloy 336.0 at elevated temperature

Temperature °C	Temperature °F	Tensile strength MPa	Tensile strength ksi	Yield strength MPa	Yield strength ksi	Elongation(a), %
25	75	250	36	195	28	0.5
150	300	215	31	150	22	1.0
205	400	180	26	105	15	2.0
260	500	125	18	70	10	5.0
315	600	70	10	30	4	10.0

(a) In 50 mm or 2 in.

Table 16 Tensile properties of sand cast 336.0-T551 alloy at various temperatures

Temperature °C	Temperature °F	Tensile strength MPa	Tensile strength ksi	Yield strength(a) MPa	Yield strength(a) ksi	Elongation, %
−195	−320	310	45	270	39	1
−80	−112	275	40	235	34	1
−28	−18	262	38	215	31	1
24	75	248	36	193	28	0.5
100	212	240	35	172	25	1
150	300	215	31	152	22	1
205	400	180	26	103	15	2
260	500	125	18	70	10	5
315	600	70	10	28	4	10
371	700	35	5	21	3	45

(a) 0.2% offset

945 °F); hold at temperature 12 h (sand), 8 h (permanent mold); cool in water at 65 to 100 °C (150 to 212 °F)
Aging temperature. To obtain T6 temper from solution-treated material, 150 to 155 °C (305 to 315 °F) and hold at temperature 2 to 5 h
Joining. Same as for alloy 208.0

332.0
9.5%Si-3.0%Cu-1.0%Mg

Commercial Names

F332, F132

Specifications

Former ASTM. SC103A
Former SAE. 332
UNS. A03320

Applications

Typical uses are applications where good high-temperature strength, low coefficient of thermal expansion, and good resistance to wear are required (for example, automotive and diesel pistons, pulleys, sheaves, and so forth).

Mechanical Properties

Tensile properties. See Table 14.

336.0
12Si-2.5Ni-1Mg-1Cu

Commercial Names

Former designations. A332.0, A132

Specifications

Former ASTM. SN122A
SAE. 321
UNS number. A13320
Government. QQ-A-596, class 9
Foreign. Canada: CSA SN122. France: NF A-S12N2G

Chemical Composition

Composition limits. 0.5 to 1.5 Cu, 1.3 Mg max, 0.35 Mn max, 11.0 to 13.0 Si, 1.2 Fe max, 0.35 Zn max, 0.25 Ti max, 0.05 other (each) max, 0.15 others (total) max, bal Al
Consequence of exceeding impurity limits. High iron or chromium promotes shrinkage difficulties.

Applications

Typical uses. Automotive and diesel pistons, pulleys, sheaves, and other applications where good high-temperature strength, low coefficient of thermal expansion, and good resistance to wear are required

Mechanical Properties

Tensile properties. Typical for separately cast test bars. Tensile strength: T551 temper, 248 MPa (36 ksi); T65 temper, 324 MPa (47 ksi). Yield strength: T551 temper, 193 MPa (28 ksi); T65 temper, 296 MPa (43 ksi). Elongation, in 50 mm or 2 in.: T551 and T65 tempers, 0.5%. Strengths and elongations remain unchanged or improve at low temperatures. See also Tables 15 and 16.
Shear strength. T551 temper, 193 MPa (28 ksi); T65 temper, 248 MPa (36 ksi)
Compressive yield strength. T551 temper, 193 MPa (28 ksi); T65 temper, 296 MPa (43 ksi)
Hardness. T551 temper, 105 HB; T65 temper, 125 HB (500 kg load, 10 mm ball)
Poisson's ratio. 0.33
Elastic modulus. Tension, 73 GPa (10.6×10^6 psi); shear, 30 GPa (4.35×10^6 psi)

Mass Characteristics

Density. 2.71 g/cm^3 (0.098 lb/in.^{3}) at 20 °C (68 °F)

Thermal Properties

Liquidus temperature. 565 °C (1050 °F)
Solidus temperature. 540 °C (1000 °F)
Melting temperature. 677 to 815 °C (1250 to 1500 °F)
Coefficient of linear thermal expansion.

Temperature range °C	Temperature range °F	Average coefficient µm/m · K	Average coefficient µin./in. · °F
20–100	68–212	19	10.6
20–200	68–392	20	11.1
20–300	68–572	21	11.7

Specific heat. 963 J/kg · K (0.230 Btu/lb · °F) at 100 °C (212 °F)
Latent heat of fusion. 389 kJ/kg (167 Btu/lb)
Thermal conductivity. T551 temper: 117 W/m · K (67.7 Btu/ft · h · °F) at 25 °C (77 °F)

Electrical Properties

Electrical conductivity. Volumetric, T551 temper: 29% IACS at 20°C (68 °F)
Electrical resistivity. T551 temper: 59.5 nΩ · m at 20 °C (68 °F)

Fabrication Characteristics

Melting temperature. 675 to 815 °C (1250 to 1500 °F)
Casting temperature. 675 to 788 °C (1250 to 1450 °F)
Solution temperature. 515 to 520 °C (955 to 965 °F); hold 8 h at temperature; cool in water at 65 to 100 °C (150 to 212 °F)
Aging temperature. 170 to 175 °C (335 to 345 °F); hold at temperature 14 to 18 h to obtain T5 temper from as-cast material; 12 to 26 h to obtain T6 temper from solution heat-treated material

Table 17 Minimum mechanical properties for castings of alloy 354.0-T61

Class(a)	Tensile strength(b) MPa	Tensile strength(b) ksi	Tensile yield strength(b)(c) MPa	Tensile yield strength(b)(c) ksi	Elongation(b)(d), %	Compressive yield strength(e) MPa	Compressive yield strength(e) ksi
1	324	47	248	36	3	248	36
2	345	50	290	42	2	290	42
10	324	47	248	36	3	248	36
11	296	43	227	33	2	227	33

(a) Classes 1 and 2 (levels of properties) obtainable only at designated areas of casting; classes 10 and 11 may be obtained at any location in casting. (b) Specified in MIL-A-21180. (c) 0.2% offset. (d) In 50 mm, 2 in. or 4*d*, where *d* is diameter of reduced section of tensile-test specimen. (e) Design values; not specified

Joining. Rivet compositions: 6053-T4, 6053-T6, 6053-T61. Soft solder with Alcoa No. 802; no flux. Rub-tin with Alcoa No. 802. Metal-arc weld with 4043 alloy; Alcoa No. 27 flux. Carbon-arc weld with 4043 alloy; Alcoa No. 24 flux (automatic); Alcoa No. 27 flux (manual). Tungsten-arc argon-atmosphere weld with 4043 alloys; no flux. Resistance welding: spot, seam, and flash methods

339.0
12.0%Si-1.0%Ni-1.0%Mg-2.25%Cu

Commercial Names

Z332.0, Z132

Applications

A lower cost alloy quite similar to 336.0 alloy. Applications similar to those for 336.0 alloy and not needing the higher elevated-temperature property available in 336.0 alloy

354.0
9Si-1.8Cu-0.5Mg

Commercial Name

354

Specifications

Former ASTM. SC92A
UNS number. AC3540
Government. MIL-A-21180

Chemical Composition

Composition limits. 1.6 to 2.0 Cu, 0.4 to 0.6 Mg, 0.10 Mn max, 8.6 to 9.5 Si, 0.2 Fe max, 0.1 Zn max, 0.2 Ti max, 0.05 other (each) max, 0.15 others (total) max, bal Al

Applications

Typical uses. Permanent mold castings used in applications requiring high strengths and heat treatability

Mechanical Properties

Tensile properties. See Tables 17 and 18.
Compressive yield strength. See Table 17.
Elastic modulus. Tension, 73.1 GPa (10.6 × 10^6 psi); shear, 27.6 GPa (4.0 × 10^6 psi); compression, 74.5 GPa (10.8 × 10^6 psi)
Fatigue strength. See Table 19.
Creep-rupture characteristics. See Table 20.

Mass Characteristics

Density. 2.71 g/cm^3 (0.098 lb/in.^{3})

Thermal Properties

Coefficient of linear thermal expansion. 20.9 μm/m · K (11.6 μin./in. · °F) at 20 to 100 °C (68 to 212 °F)
Specific heat. 963 J/kg · K (0.230 Btu/lb · °F) at 100 °C (212 °F)

Table 18 Typical mechanical properties for separately cast test bars of alloy 354.0-T61 at various temperatures

Temperature °C	Temperature °F	Time at temperature, h	At indicated temperature: Tensile strength MPa	ksi	Yield strength MPa	ksi	Elongation(a), %	At room temperature after heating: Tensile strength MPa	ksi	Yield strength MPa	ksi	Elongation(a), %
−196	−320	...	470	68	340	49	5	...	...	...	...	...
−80, −28	−112, −18	...	400	58	290	42	5	...	...	...	...	...
24	75	...	380	55	285	41	6	380	55	285	41	6
100	212	0.5	345	50	285	41	6	380	55	285	41	6
		10	350	51	285	41	6	385	56	290	42	6
		100	360	52	290	42	6	400	58	295	43	6
		1000	370	54	310	45	6	420	61	310	45	6
		10 000	415	60	340	49	6	435	63	350	51	5
150	300	0.5	325	47	275	40	6	380	55	290	42	6
		10	345	50	295	43	6	395	57	305	44	5
		100	350	51	315	46	6	425	62	345	50	5
		1000	340	49	305	44	6	405	59	360	52	4
		10 000	290	42	240	35	6	340	49	275	40	6
175	350	0.5	310	45	270	39	6	380	55	295	43	6
		10	325	47	290	42	6	405	59	340	49	4
		100	295	43	260	38	8	405	59	350	51	5
		1000	230	33	195	28	13	325	47	255	37	8
		10 000	130	19	95	14	24	205	30	115	17	16
205	400	0.5	290	42	270	39	6	405	59	340	49	5
		10	270	39	250	36	9	400	58	340	49	5
		100	205	30	180	26	17	330	48	255	37	7
		1000	130	19	105	15	30	220	32	125	18	14
		10 000	105	15	75	11	45	185	27	90	13	20
230	450	0.5	255	37	240	35	9	400	58	345	50	5
		10	195	28	170	25	15	315	46	250	36	8
		100	125	18	95	14	25	240	35	140	20	11
		1000	95	14	75	11	40	195	28	95	14	17
		10 000	80	12	60	8.5	55	170	25	75	11	22
260	500	0.5	195	28	170	25	16	360	52	290	42	6
		10	115	17	105	15	22	250	36	150	22	11
		100	80	12	65	9.5	35	205	30	105	15	15
		1000	65	9.5	50	7.5	50	185	27	80	12	19
		10 000	60	8.5	40	6	65	165	24	70	10	11
315	600	0.5	90	13	80	12	29	260	38	145	21	13
		10	60	8.5	50	7	60	205	30	90	13	17
		100	40	6	35	5	85	185	27	75	11	19
		1000	...	...	...	...	...	170	25	65	9.5	21
		10 000	...	...	...	...	...	160	23	60	8.5	23

(a) In 50 mm, 2 in. or 4*d*, where *d* is diameter of reduced section of tensile-test specimen

Table 19 Fatigue strengths for separately cast test bars of alloy 354.0-T61

Temperature		Cycles											
		10^4		10^5		10^6		10^7		10^8		5×10^8	
°C	°F	MPa	ksi	MPa	ksi	MPa	ksi	MPa	ksi	MPa	ksi	MPa	ksi
24	75	345	50	275	40	215	31	175	25.5	145	21	135	19.5
150	300	...	...	255	37	200	29	150	21.5	115	17	110	16
205	400	...	...	215	31	150	22	105	15	70	10	60	9
260	500	195	28	140	20.5	96	14	60	9	40	6	40	6
315	600	...	...	75	11	55	8	40	6	30	4	30	4

Note: R.R. Moore type test

Table 20 Creep-rupture properties for separately cast test bars of alloy 354.0-T61

Temperature		Time under	Rupture stress		Stress for creep of:							
					1%		0.5%		0.2%		0.1%	
°C	°F	stress, h	MPa	ksi	MPa	ksi	MPa	ksi	MPa	ksi	MPa	ksi
177	350	0.1	305	44	295	43	285	41	290	39	255	37
		1.0	295	43	290	42	285	41	290	39	255	37
		10	285	41	285	41	275	40	255	37	240	35
		100	240	35	235	34	230	33	205	30	115	17
		1000	170	25	165	24	165	24	138	20	76	11
205	400	0.1	285	41	275	40	270	39	255	37	240	35
		1.0	255	37	250	36	250	36	235	34	215	31
		10	220	32	215	31	205	30	180	26	125	18
		100	160	23	160	23	150	22	125	18	69	10
		1000	90	13	90	13	83	12	83	12	48	7

Table 21 Minimum mechanical properties for alloy C355.0-T61 castings

	Tensile strength		Tensile yield strength(b)(c)		Elonga-	Compressive yield strength(e)	
Class(a)	MPa	ksi	MPa	ksi	tion(d), %	MPa	ksi
1	285	41	215	31	3	215	31
2	305	44	230	33	3	230	33
3	345	50	275	40	2	275	40
10	285	41	215	31	3	215	31
11	255	37	205	30	1	205	30
12	240	35	195	28	1	195	28

(a) Classes 1, 2, and 3 (levels of properties) obtainable only at designated areas of casting; classes 10, 11, and 12 may be obtained from any location in casting. (b) Specified in MIL-A-21180. High properties are obtained by advanced foundry techniques and by careful control of trace elements at lower levels than specified for alloy 355.0 castings. (c) 0.2% offset. (d) In 4*d*, where *d* is diameter of reduced section of tensile-test specimen. (e) Design values; not specified

Thermal conductivity. 128 W/m · K (74 Btu/ft · h · °F)

Fabrication Characteristics

Solution temperature. 525 °C (980 °F); hold at temperature 10 to 12 h; quench in hot water 60 to 80 °C (140 to 176 °F)
Aging temperature. To obtain T61 temper from solution heat-treated material, room temperature for 8 to 16 h; 155 °C (310 °F); hold at temperature for 10 to 12 h

355.0, C355.0
5Si-1.3Cu-0.5Mg

Specifications

AMS. 4210, 4212, 4214, 4280, 4281
Former ASTM. 355.0: SC51A. C355.0: SC51B
SAE. 322
UNS number. A03550
Government. 355.0: sand castings, QQ-A-601, class 10; permanent mold castings, QQ-A-596, class 6. C355.0: MIL-A-21180
Foreign. Canada: CSA SC51

Chemical Composition

Composition limits. 355.0: 1.0 to 1.5 Cu, 0.40 to 0.60 Mg, 0.50 Mn max, 4.5 to 5.5 Si, 0.6 Fe max, 0.25 Cr max, 0.35 Zn max, 0.25 Ti max, 0.05 other (each) max, 0.15 others (total) max, bal Al. (If Fe exceeds 0.45, Mn content may not be less than ½ Fe content). C355.0: 1.0 to 1.5 Cu, 0.40 to 0.60 Mg, 0.10 Mn max, 4.5 to 5.5 Si, 0.20 Fe max, 0.10 Zn max, 0.20 Ti max, 0.05 other (each) max, 0.15 others (total) max, bal Al
Consequence of exceeding impurity limits. High iron decreases ductility. Nickel decreases resistance to corrosion. Tin reduces mechanical properties.

Applications

Typical uses. Aircraft supercharger covers, fuel-pump bodies, air-compressor pistons, liquid-cooled cylinder heads, liquid-cooled aircraft engine crankcases, water jackets, and blower housings. Other applications where good castability, weldability, and pressure tightness are required. The presence of copper in 355.0 increases strength but reduces corrosion resistance and ductility.

Mechanical Properties

Tensile properties. See Tables 21 through 25.
Compressive yield strength. See Table 21.
Poisson's ratio. 0.33
Elastic modulus. 355.0: tension, 70.3 GPa (10.2×10^6 psi) at 25 °C (75 °F), 67.6 GPa (9.8×10^6 psi) at 150 °C (300 °F), 64.1 GPa (9.3×10^6 psi) at 204 °C (400 °F), 56.5 GPa (8.2×10^6 psi) at 260 °C (500 °F); shear, 26.2 GPa (3.8×10^6 psi). C355.0: tension, 69.6 GPa (10.1×10^6 psi); shear, 26.5 GPa (3.85×10^6 psi); compression, 71 GPa (10.3×10^6 psi)
Fatigue strength. See Table 26.
Creep-rupture characteristics. See Table 27.

Mass Characteristics

Density. 2.71 g/cm³ (0.098 lb/in.³) at 20 °C (68 °F)

Thermal Properties

Liquidus temperature. 620 °C (1150 °F)
Solidus temperature. 545 °C (1015 °F)
Coefficient of linear thermal expansion.

Temperature range		Average coefficient	
°C	°F	μm/m · K	μin./ in. · °F
20–100	68–212	22.4	12.4
20–200	68–392	23	12.8
20–300	68–572	24	13.3

Specific heat. 963 J/kg · K (0.230 Btu/lb · °F) at 100 °C (212 °F)
Thermal conductivity. At 25 °C (77 °F):

	Conductivity	
Temper and form	W/m · K	Btu/ft · h · °F
T51, sand	167	96
T6, T61, sand	152	88
T7, sand	163	94
T6 permanent mold	151	87

Table 22 Typical mechanical properties for separately cast test bars of alloy 355.0

Temper	Tensile strength MPa	Tensile strength ksi	Tensile yield strength MPa	Tensile yield strength ksi	Elongation, %	Hardness(a), HB	Shear strength MPa	Shear strength ksi	Fatigue strength(b) MPa	Fatigue strength(b) ksi	Compressive yield strength MPa	Compressive yield strength ksi
Sand cast												
T51	195	28	160	23	1.5	65	150	22	55	8.0	165	24
T6	240	35	170	25	3.0	80	195	28	62	9.0	180	26
T61	270	39	240	35	1.0	90	215	31	66	9.5	255	37
T7	260	38	250	36	0.5	85	195	28	69	10.0	260	38
T71	240	35	200	29	1.5	75	180	26	69	10.0	205	30
Permanent mold cast												
T51	205	30	165	24	2.0	75	165	24	...	...	165	24
T6	290	42	185	27	4.0	90	235	34	69	10	185	27
T62	310	45	275	40	1.5	105	250	36	69	10	275	40
T7	275	40	205	30	2.0	85	205	30	69	10	205	30
T71	250	36	215	31	3.0	85	185	27	69	10	215	31

(a) 500 kg load; 10 mm ball. (b) At 5×10^8 cycles; R.R. Moore type test

Table 23 Typical tensile properties of separately cast test bars of alloy 355.0 at elevated temperatures

Temperature °C	Temperature °F	Tensile strength(a) MPa	Tensile strength(a) ksi	Yield strength(a) MPa	Yield strength(a) ksi	Elongation(a)(b), %
T6 temper, sand cast						
25	75	240	35	170	25	3
150	300	230	33	170	25	1.5
205	400	115	17	90	13	8
260	500	65	9.5	35	5	16
315	600	40	6	20	3	36
T6 temper, permanent mold cast						
25	75	290	42	185	27	4
150	300	220	32	170	25	10
205	400	130	19	90	13	20
260	500	65	9.5	35	5	40
315	600	40	6	20	3	50
T51 temper, sand cast						
25	75	195	28	160	23	1.5
150	300	165	24	130	19	3
205	400	95	14	70	10	8
260	500	65	9.5	35	5	16
315	600	40	6	20	3	36

(a) Strengths and elongations remain unchanged or improve at low temperatures. (b) In 50 mm or 2 in.

Table 24 Tensile properties of alloy 355.0-T71 at various temperatures

Temperature °C	Temperature °F	Tensile strength MPa	Tensile strength ksi	Yield strength(a) MPa	Yield strength(a) ksi	Elongation, %
Sand castings						
−195	−320	282	41	235	34	1.5
−80	−112	255	37	220	32	1.5
−28	−18	248	36	215	31	1.5
24	75	240	35	200	29	1.5
100	212	235	34	193	28	2
150	300	207	30	180	26	3
205	400	117	17	90	13	8
260	500	67	9.5	35	5	16
315	600	41	6	21	3	36
371	700	25	3.5	14	2	50
Permanent mold castings						
−195	−320	317	46	262	38	1.5
−80	−112	345	50	235	34	2
−28	−18	262	38	227	33	2.5
24	75	248	36	215	31	3
100	212	227	33	200	29	4
150	300	200	29	180	26	8
205	400	130	19	90	13	20
260	500	67	9.5	35	5	40
315	600	41	6	21	3	50
371	700	25	3.5	14	2	60

(a) 0.2% offset

Electrical Properties

Electrical conductivity. Volumetric:

Temper and form	Conductivity, %IACS
T51, sand	43
T6, sand	36
T61, sand	39
T7, sand	42
T6, permanent mold	39

Electrical resistivity. At 20 °C (68 °F):

Temper and form	Resistivity, nΩ · m
T51, sand	40.1
T6, sand	47.9
T61, sand	44.2
T7, sand	41.0
T6, permanent mold	44.2

Electrolytic solution potential. T4 temper, −0.78 V and T6 temper, −0.79 V versus 0.1 *N* calomel electrode in an aqueous solution containing 53 g NaCl plus 3 g H_2O_2 per liter

Fabrication Characteristics

Melting temperature. 675 to 815 °C (1250 to 1500 °F)
Casting temperature. 675 to 790 °C (1250 to 1450 °F)
Solution temperature. See Table 28.
Aging temperature. See Table 28.
Joining. Same as alloy 514.0

356.0, A356.0
7Si-0.3Mg

Specifications

AMS. 356.0: 4217, 4260, 4261, 4284, 4285, 4286. A356.0: 4218
Former ASTM. 356.0, SG70A; A356.0, SG70B
SAE. 356.0: J452, 323
UNS number. 356.0: A03560. A356.0: A13560
Government. 356.0: QQ-A-601, QQ-A-596. A356.0: MIL-C-21180 (class 12)
Foreign. ISO: AlSi7Mg

Chemical Composition

Composition limits. 356.0: 0.25 Cu max, 0.20 to 0.45 Mg, 0.35 Mn max, 6.5 to 7.5 Si, 0.6 Fe max, 0.35 Zn max, 0.25 Ti max, 0.05 other (each) max, 0.15 others (total) max, bal Al. A356.0: 0.20 Cu max, 0.25 to 0.45 Mg, 0.10 Mn max, 6.5 to 7.5 Si, 0.20 Fe max, 0.10 Zn max, 0.20 Ti max, 0.05 other (each) max, 0.15 others (total) max, bal Al
Consequence of exceeding impurity limits. High copper or nickel decreases ductility and resistance to corrosion. High iron decreases strength and ductility.

Applications

Typical uses. 356.0: aircraft pump parts, automotive transmission cases, aircraft fittings and control parts, water-cooled cylin-

Table 25 Tensile properties of alloy 355.0-T51 at various temperatures

Temperature °C	Temperature °F	Tensile strength MPa	Tensile strength ksi	Yield strength(a) MPa	Yield strength(a) ksi	Elongation, %
Sand castings						
−195	−320	227	33	185	27	1.5
−80	−112	200	29	165	24	1.5
−28	−18	193	28	160	23	1.5
24	75	193	28	160	23	1.5
100	212	193	28	152	22	2
150	300	165	24	130	19	3
205	400	95	14	70	10	8
260	500	67	9.5	35	5	16
315	600	40	6	21	3	36
371	700	25	3.5	14	2	50
Permanent mold						
−195	−320	255	37	185	27	1
−80	−112	240	35	172	25	1.5
−28	−18	215	31	165	24	1.5
24	75	207	30	165	24	2
100	212	193	28	165	24	3
150	300	160	23	138	20	4
205	400	103	15	70	10	19
260	500	67	9.5	35	5	33
315	600	41	6	21	3	38
371	700	25	3.5	14	2	60

(a) 0.2% offset

der blocks. Other applications where excellent castability and good weldability, pressure tightness, and good resistance to corrosion are required. A356.0: aircraft structures and engine controls, nuclear energy installations, and other applications where high-strength permanent mold or investment castings are required

Mechanical Properties

Tensile properties. See Tables 29 through 33.
Compressive yield strength. See Table 29.
Poisson's ratio. 0.33
Elastic modulus. Tension, 72.4 GPa (10.5 × 10^6 psi); shear, 27.2 GPa (3.95 × 10^6 psi)
Creep-rupture characteristics. See Table 34.

Mass Characteristics

Density. 2.685 g/cm^3 (0.097 lb/in.^{3}) at 20 °C (68 °F)

Thermal Properties

Liquidus temperature. 615 °C (1135 °F)
Solidus temperature. 555 °C (1035 °F)
Coefficient of linear thermal expansion.

Temperature range °C	Temperature range °F	Average coefficient µm/m · K	Average coefficient µin./in. · °F
20–100	68–212	21.5	11.9
20–200	68–392	22.5	12.5
20–300	68–572	23.5	13.1

Specific heat. 963 J/kg · K (0.230 Btu/lb · °F) at 100 °C (212 °F)
Latent heat of fusion. 389 kJ/kg
Thermal conductivity. At 25 °C (77 °F):

Temper and form	Conductivity W/m · K	Conductivity Btu/ft · h · °F
T51, sand	167	96
T6, sand	151	87
T7, sand	155	90
T6, permanent mold	159	92

Table 26 Fatigue properties for separately cast test bars of alloy C355.0-T61

Temperature °C	Temperature °F	Number of cycles	Fatigue strength(a) MPa	Fatigue strength(a) ksi
24	75	10^5	195	28.0
		10^6	130	19.0
		10^7	110	16.0
		10^8	100	14.5
		5×10^8	95	14.0
260	500	10^5	125	18.0
		10^6	80	11.5
		10^7	50	7.5
		10^8	40	5.5
		5×10^8	35	5.0

(a) Based on rotating-beam tests at room temperature and cantilever beam (rotating load) tests at elevated temperature

Electrical Properties

Electrical conductivity. Volumetric:

Temper and form	IACS, %
T51, sand	43
T6, sand	39
T7, sand	40
T6, permanent mold	41

Electrical resistivity. At 20 °C (68 °F):

Temper and form	Resistivity, nΩ · m
T51, sand	40.1
T6, sand	44.2
T7, sand	43.1
T6, permanent mold	42.1

Electrolytic solution potential. T6 temper (sand): −0.82 V versus 0.1 *N* calomel electrode in an aqueous solution containing 53 g NaCl plus 3 g H_2O_2 per liter

Radiation Effect on Properties

Effect of neutron irradiation. See Table 35.

Fabrication Characteristics

Melting temperature. 675 to 815 °C (1250 to 1500 °F)
Casting temperature. 675 to 790 °C (1250 to 1450 °F)
Solution temperature. See Table 36.
Aging temperature. See Table 36.
Joining. Same as alloy 514.0

Table 27 Creep-rupture properties for separately cast test bars of alloy C355.0-T61

Temperature °C	Temperature °F	Time under stress, h	Rupture stress MPa	Rupture stress ksi	Stress for creep of: 1% MPa	1% ksi	0.5% MPa	0.5% ksi	0.2% MPa	0.2% ksi	0.1% MPa	0.1% ksi
150	300	0.1	285	41	275	40	270	39	240	35	230	33
		1	285	41	270	39	260	38	235	34	220	32
		10	275	40	260	38	250	36	230	33	205	30
		100	260	38	250	36	235	34	215	31	170	25
		1000	220	32	215	31	206	30	185	27	140	20
205	400	0.1	250	36	250	36	240	35	230	33	170	25
		1	230	33	220	32	205	30	170	25	140	20
		10	180	26	120	25	160	23	130	19	110	16
		100	130	19	130	19	125	18	97	14	...	...
		1000	97	14	90	13	83	12	...	...	...	...
260	500	0.1	165	24	145	21	130	19	105	15	83	12
		1	125	18	110	16	97	14	83	12	59	8.5
		10	90	13	83	12	76	11	59	8.5	41	6
		100	62	9	62	9	55	8	41	6	...	...
		1000	45	6.5	45	6.5	41	6	...	...	...	...

Table 28 Heat treatments for separately cast test bars of alloy 355.0

Purpose (and resulting temper)	Temperature °C	°F	Time at temperature, h
Sand castings			
Solution	520–530	970–990	12(a)(b)
Aging			
T51(c)	225–230	435–445	7–9
T6(d)	150–155	300–315	3–5
T61(d)	150–160	300–320	8–10
T7(d)(e)	225–230	435–445	7–9
T71(d)(e)	245–250	470–480	4–6
Permanent mold castings			
Solution	520–530	970–980	8(a)(b)
Aging(f)			
T62(d)	170–175	335–345	14–18

(a) Soaking-time periods required for average castings after load has reached specified temperature. Time can be decreased or may have to be increased, depending on experience with particular castings. (b) Cool in water at 65 to 100 °C (150 to 212 °F). (c) No solution heat treatment. (d) Start with solution heat-treated material. (e) U.S. Patent 1,822,877. (f) Except for temper listed under this head, temperature values for all tempers are the same as for sand castings.

Table 29 Minimum mechanical properties for alloy A356.0-T61 castings

Class(a)	Tensile strength(b) MPa	ksi	Tensile yield strength(b)(c) MPa	ksi	Elongation(d), %	Compressive yield strength(e) MPa	ksi
1	260	38	195	28	5	195	28
2	275	40	205	30	3	205	30
3	310	45	235	34	3	235	34
10	260	38	195	28	5	195	28
11	230	33	185	27	3	185	27
12	220	32	150	22	2	150	22

(a) Classes 1, 2, and 3 (levels of properties) obtainable only at designated areas of casting; classes 10, 11, and 12 may be specified at any location in casting. (b) Specified in MIL-A-21180. (c) 0.2% offset. (d) In 4*d*, where *d* is diameter of reduced section of tensile-test specimen. (e) Design values; not specified

357.0, A357.0
7Si-0.5Mg

Specifications

UNS number. 357.0: A03570. A357.0: A13570
Government. A357.0: MIL-A-21180

Chemical Composition

Composition limits. 357.0: 0.05 Cu max; 0.45 to 0.6 Mg, 0.03 Mn max, 6.5 to 7.5 Si, 0.15 Fe max, 0.05 Zn max, 0.20 Ti max, 0.05 other (each) max, 0.15 others (total) max, bal Al. A357.0: 0.20 Cu max, 0.40 to 0.7 Mg, 0.10 Mn max, 6.5 to 7.5 Si, 0.20 Fe max, 0.10 Zn max, 0.10 to 0.20 Ti, 0.04 to 0.07 Be, 0.05 other (each) max, 0.15 others (total) max, bal Al

Applications

Typical uses. Critical aerospace applications and other uses requiring heat-treatable permanent mold casting that combines ready weldability with high strength and good toughness

Mechanical Properties

Tensile properties. See Tables 37 and 38.
Compressive yield strength. See Table 37.
Hardness. A357.0, T61 temper: 100 HB
Elastic modulus. A357.0: tension, 71.7 GPa (10.4×10^6 psi); shear, 26.8 GPa (3.9×10^6 psi); compression, 72.4 GPa (10.5×10^6 psi)
Fatigue strength. A357.0-T62 (rotating-beam tests):

Number of cycles	Stress MPa	ksi
10^5	255	37
10^6	195	28
10^7	145	21
10^8	115	17
5×10^8	110	16

Mass Characteristics

Density. 2.68 g/cm³ (0.097 lb/in.³)

Thermal Properties

Liquidus temperature. A357.0: 615 °C (1135 °F)
Solidus temperature. 555 °C (1035 °F)
Coefficient of linear thermal expansion. 21.6 μm/m · K (12.0 μin./in. · °F) at 17 to 100 °C (63 to 212 °F)
Specific heat. 963 J/kg · K (0.230 Btu/lb · °F) at 100 °C (212 °F)
Thermal conductivity. 152 W/m · K (88 Btu/ft · h · °F) at 25 °C (77 °F)

Fabrication Characteristics

Solution temperature. 540 °C (1005 °F); hold at temperature for 8 h; hot water quench
Aging temperature. T6 temper: 170 °C (340 °F); hold at temperature 3 to 5 h
Joining. Because of the beryllium content, care should be taken not to inhale fumes during welding.

359.0
9Si-0.6Mg

Specifications

Former ASTM. SG91A
UNS number. A03590
Government. MIL-A-21180

Chemical Composition

Composition limits. 0.20 Cu max, 0.50 to 0.7 Mg, 0.10 Mn max, 8.5 to 9.5 Si, 0.20 Fe max, 0.10 Zn max, 0.20 Ti max, 0.05 other (each) max, 0.15 others (total) max, bal Al

Applications

Typical uses. A moderately high-strength permanent mold casting alloy having superior casting characteristics

Table 30 Typical mechanical properties for separately cast test bars of alloy 356.0

Temper	Tensile strength MPa	ksi	Yield strength MPa	ksi	Elongation(a), %	Hardness(b), HB	Shear strength MPa	ksi	Fatigue strength(c) MPa	ksi	Compressive yield strength MPa	ksi
Sand cast												
T51	172	25	140	20	2.0	60	140	20	55	8.0	145	21
T6	228	33	165	24	3.5	70	180	26	60	8.5	170	25
T7	234	34	205	30	2.0	75	165	24	62	9.0	215	31
T71	193	28	145	21	3.5	60	140	20	60	8.5	150	22
Permanent mold												
T6	262	38	185	27	5.0	80	205	30	90	13	185	27
T7	221	32	165	24	6.0	70	170	25	75	11	165	24

(a) In 50 mm or 2 in. (b) 500 kg load; 10 mm ball. (c) At 5×10^8 cycles; R.R. Moore type test

Table 31 Typical tensile properties of separately cast test bars of alloy 356.0-T6

Temperature °C	Temperature °F	Tensile strength(a) MPa	Tensile strength(a) ksi	Yield strength(a) MPa	Yield strength(a) ksi	Elongation(a)(b), %
24	75	230	33	165	24	3.5
150	300	160	23	140	20	6.0
205	400	85	12	60	8.5	18
260	500	50	7.5	35	5.0	35
315	600	30	4.0	20	3.0	60

(a) Strengths and elongations remain unchanged or improve at low temperatures. (b) In 50 mm or 2 in.

Table 32 Tensile properties of alloy 356.0-T6 at various temperatures

Temperature °C	Temperature °F	Tensile strength MPa	Tensile strength ksi	Yield strength(a) MPa	Yield strength(a) ksi	Elongation, %
Sand castings						
−195	−320	275	40	193	28	3.5
−80	−112	240	35	172	25	3.5
−28	−18	227	33	165	24	3.5
24	75	227	33	165	24	3.5
100	212	220	32	165	24	4
150	300	160	23	138	20	6
205	400	83	12	58	8.5	18
260	500	53	7.5	35	5	35
315	600	28	4	21	3	60
371	700	17	2.5	14	2	80
Permanent mold castings						
−195	−320	330	48	220	32	5
−80	−112	275	40	193	28	5
−28	−18	270	39	185	27	5
24	75	262	38	185	27	5
100	212	207	30	172	25	6
150	300	145	21	117	17	10
205	400	83	12	58	8.5	30
260	500	53	7.5	34	5	55
315	600	28	4	21	3	70
371	700	17	2.5	14	2	80

(a) 0.2% offset

Table 33 Tensile properties of alloy 356.0-T7 at various temperatures

Temperature °C	Temperature °F	Tensile strength MPa	Tensile strength ksi	Yield strength(a) MPa	Yield strength(a) ksi	Elongation, %
Sand castings						
−195	−320	283	41	240	35	2
−80	−112	248	36	220	32	2
−28	−18	235	34	215	31	2
24	75	235	34	207	30	2
100	212	207	30	193	28	2
150	300	160	23	138	20	6
205	400	83	12	58	8.5	18
260	500	53	7.5	34	5	35
315	600	28	4	21	3	60
371	700	17	2.5	14	2	80
Permanent mold castings						
−195	−320	275	40	207	30	6
−80	−112	248	36	180	26	6
−28	−18	165	34	172	25	6
24	75	220	32	165	24	6
100	212	185	27	160	23	10
150	300	160	23	138	20	20
205	400	83	12	58	8.5	40
260	500	50	7	34	5	55
315	600	28	4	21	3	70
371	700	17	2.5	14	2	80

(a) 0.2% offset

Mechanical Properties

Tensile properties. See Tables 39 and 40.
Compressive yield strength. See Table 39.
Elastic modulus. Tension, 72.4 GPa (10.5 $\times$ 10^6 psi); shear, 27.6 GPa (4.0 $\times$ 10^6 psi); compression, 73.8 GPa (10.7 $\times$ 10^6 psi)
Fatigue strength. Rotating-beam tests, T61 temper:

Number of cycles	Stress MPa	Stress ksi
10^5	255	37
10^6	195	28
10^7	145	21
10^8	115	17
5×10^8	110	16

Mass Characteristics

Density. 2.685 g/cm^3 (0.097 lb/in.^{3})

Thermal Properties

Liquidus temperature. 615 °C (1135 °F)
Solidus temperature. 555 °C (1035 °F)
Coefficient of linear thermal expansion. 20.9 μm/m · K (11.6 μin./in. · °F) at 20 to 100 °C (68 to 212 °F)
Specific heat. 963 J/kg · K (0.230 Btu/lb · °F)
Thermal conductivity. 138 W/m · K (80 Btu/ft · h · °F)

Fabrication Characteristics

Solution temperature. 540 °C (1000 °F); hold at temperature 10 to 14 h; hot water quench 60 to 80 °C (140 to 175 °F)
Aging temperature. Room temperature for 8 to 16 h after solution treatment, then 155 °C (310 °F) for 10 to 12 h (T61 temper), or 170 °C (340 °F) for 6 to 10 h (T62 temper)

360.0, A360.0
9.5Si-0.5Mg

Specifications

AMS. 360.0: 4290F
Former ASTM. 360.0: SG100B. A360.0: SG100A
SAE. A360.0: J452, 309
UNS number. 360.0: A03600. A360.0: A13600
Government. 360.0: QQ-A-591

Chemical Composition

Composition limits. 360.0: 0.6 Cu max, 0.40 to 0.6 Mg, 0.35 Mn max, 9.0 to 10.0 Si, 2.0 Fe max, 0.50 Ni max, 0.50 Zn max, 0.15 Sn max, 0.25 other (total) max, bal Al. A360.0: 0.6 Cu max, 0.40 to 0.6 Mg, 0.35 Mn max, 9.0 to 10.0 Si, 1.3 Fe max, 0.50 Ni max, 0.50 Zn max, 0.15 Sn max, 0.25 other (total) max, bal Al
Consequence of exceeding impurity limits. Increasing copper limits lowers resistance to corrosion; increasing iron lowers ductility. Decreasing silicon reduces castability.

Applications

Typical uses. Die castings requiring improved corrosion resistance compared to 3800. Other applications where excellent castability, pressure tightness, resistance to hot cracking, strength at elevated temperatures, and ability to be electroplated are required. Poor weldability and brazeability. General-purpose casting alloy for such items as cover plates and instrument cases

Table 34 Creep-rupture properties for separately cast test bars of alloy A356.0-T61

Temperature °C	Temperature °F	Time under stress, h	Rupture stress MPa	Rupture stress ksi	Stress for creep of: 1% MPa	1% ksi	0.5% MPa	0.5% ksi	0.2% MPa	0.2% ksi	0.1% MPa	0.1% ksi
150	300	0.1	235	34	215	31	205	30	195	28	185	27
		1	235	34	215	31	200	29	185	27	180	26
		10	230	33	205	30	195	28	180	26	170	25
		100	200	29	195	28	185	27	170	25	165	24
		1000	165	24	165	24	160	23	...	...	...	...

Table 35 Effect of neutron radiation on tensile properties of alloy A356.0-T61

Fast neutron flux, n/cm²	Tensile strength MPa	Tensile strength ksi	Yield strength MPa	Yield strength ksi	Elongation, %
Control sample	230	33	180	26	4
2.0×10^{19}	255	37	200	29	6
1.2×10^{20}	290	42	230	33	6
5.6×10^{20}	315	46	290	42	6
9.8×10^{20}	375	54	360	52	3

Note: Separately cast test bars; irradiation temperature, 50 °C (120 °F)

Table 36 Heat treatments for separately cast test bars of alloys 356.0 and A356.0

Purpose (and resulting temper)	Temperature °C	Temperature °F	Time at temperature, h
Sand castings			
Solution	535–540	995–1005	12(a)(b)
Aging			
T51(c)	225–230	435–445	7–9
T6(d)	150–155	305–315	2–5
T7(d)(e)	225–230	435–445	7–9
T71(d)	245–250	470–480	2–4
Permanent mold castings			
Solution	535–540	995–1005	8(a)(b)
Aging(f)			
T6(d)	150–155	305–315	3–5

(a) Soaking-time periods required for average casting after load has reached specified temperature. Time can be decreased or may have to be increased, depending on experience with particular castings. (b) Cool in water at 65 to 100 °C (150 to 212 °F). (c) No solution heat treatment. (d) Start with solution heat-treated material. (e) U.S. Patent 1,822,877. (f) Except for temper listed under this head, temperature values for all tempers are the same as for sand castings.

Table 37 Minimum mechanical properties for alloy A357.0 castings

Class(a)	Tensile strength(b) MPa	Tensile strength(b) ksi	Tensile yield strength(b)(c) MPa	Tensile yield strength(b)(c) ksi	Elongation(b)(d), %	Compressive yield strength(e) MPa	Compressive yield strength(e) ksi
T61, permanent mold castings							
1	317	46	248	36	3	...	...
10	283	41	214	31	3	...	...
T62 castings							
1	310	45	241	35	3	241	35
2	345	50	276	40	5	276	40
10	262	38	193	28	5	193	28
11	283	41	214	31	3	214	31

(a) Classes 1 and 2 (levels of properties obtainable only at designated areas of casting); classes 10 and 11 may be obtained from any location in castings. (b) Specified in MIL-A-21180. (c) 0.2% offset. (d) In 4*d*, where *d* is diameter of reduced section of tensile-test specimen. (e) Design values; not specified

Mechanical Properties

Tensile properties. Typical for separately cast test bars, as-cast. 360.0: tensile strength, 305 MPa (44 ksi); yield strength, 170 MPa (25 ksi); elongation, 2.5% in 50 mm or 2 in. A360.0: tensile strength, 320 MPa (46 ksi); yield strength, 170 MPa (25 ksi); elongation, 3.5% in 50 mm or 2 in. See also Table 41.

Shear strength. 360.0: 190 MPa (28 ksi). A360.0: 180 MPa (26 ksi)

Poisson's ratio. 0.33

Elastic modulus. Tension, 71.0 GPa (10.3×10^6 psi); shear, 26.5 GPa (3.85×10^6 psi)

Fatigue strength. At 5×10^8 cycles, 360.0: 140 MPa (20 ksi). A360.0: 120 MPa (18 ksi) (R.R. Moore type test)

Mass Characteristics

Density. 2.630 g/cm³ (0.095 lb/in.³) at 20 °C (68 °F)

Thermal Properties

Liquidus temperature. 595 °C (1105 °F)

Solidus temperature. 555 °C (1035 °F)

Coefficient of linear thermal expansion.

Temperature range °C	Temperature range °F	Average coefficient μm/m · K	Average coefficient μin./in. · °F
20–100	68–212	21	11.6
20–200	68–392	22	12.2
20–300	68–572	23	12.8

Specific heat. 963 J/kg · K (0.230 Btu/lb · °F) at 100 °C (212 °F)

Latent heat of fusion. 389 kJ/kg (167 Btu/lb)

Thermal conductivity. 113 W/m · K (65.3 Btu/ft · h · °F) at 25 °C (77 °F)

Electrical Properties

Electrical conductivity. Volumetric: 360.0, 28% IACS; A360.0, 30% IACS

Electrical resistivity. 61.6 nΩ · m at 20 °C (68 °F) for alloy 360.0

Fabrication Characteristics

Melting temperature. 650 to 760 °C (1200 to 1400 °F)

Die casting temperature. 635 to 705 °C (1175 to 1300 °F)

Joining. Same as alloys 413.0 and A413.0

380.0, A380.0
8.5Si-3.5Cu

Specifications

AMS. A380.0: 4291

Former ASTM. 380.0: SC84B. A380.0: SC84A

SAE. 380.0: 308. A380.0: 306

UNS number. 380.0: A03800. A380.0: A13800

Government. A380.0: QQ-A-591

Foreign. 380.0: Canada, CSA SC84

Chemical Composition

Composition limits. 380.0: 3.0 to 4.0 Cu, 0.10 Mg max, 0.50 Mn max, 7.5 to 9.5 Si, 2.0 Fe max, 0.50 Ni max, 3.0 Zn max, 0.35 Sn max, 0.50 others (total) max, bal Al. A380.0: 3.0 to 4.0 Cu, 0.10 Mg max, 0.50 Mn max, 7.5 to 9.5 Si, 1.3 Fe max, 0.50 Ni max, 3.0 Zn max, 0.35 Sn max, 0.50 others (total) max, bal Al

Consequence of exceeding impurity limits. Increasing iron will lower ductility. Relatively large quantities of impurities may be present before serious effects are detected.

Applications

Typical uses. Vacuum cleaners, floor polishers, parts for automotive and electrical industries such as motor frames and housings. Most widely used aluminum die casting alloy. Poor weldability and brazeability; fair strength at elevated temperatures

Mechanical Properties

Tensile properties. Typical for separately

Table 38 Typical mechanical properties of separately cast test bars of alloy A357.0-T62 at various temperatures

Temperature °C	Temperature °F	Time at temperature, h	Tensile strength MPa	Tensile strength ksi	Yield strength MPa	Yield strength ksi	Elongation(a), %
−196	−320	...	425	62	330	48	6
−80	−112	...	380	55	310	45	6
−28	−18	...	370	54	305	44	6
24	75	...	360	52	290	42	8
100	212	0.5–100	315	46	270	39	10
		1000	315	46	275	49	8
		10 000	330	48	310	45	6
150	300	0.5	270	39	240	35	10
		10	285	41	255	37	9
		100	290	42	275	40	7
		1000	260	38	250	36	7
		10 000	160	23	145	21	20
175	350	0.5	255	37	235	34	7
		10	275	40	260	38	6
		100	240	35	230	33	7
		1000	150	22	140	20	19
		10 000	90	13	75	11	35
205	400	0.5	250	36	240	35	6
		10	205	30	195	28	7
		100	160	23	145	21	23
		1000	85	12	70	10	40
		10 000	70	10	50	7.5	50
230	450	0.5	215	31	205	30	9
		10	130	19	125	18	13
		100	95	14	90	13	45
260	500	0.5	160	23	150	22	16
		10	85	12	75	11	23
		100	55	8	50	7	55
315	600	0.5	70	10	65	9.5	35

(a) In 4*d*, where *d* is diameter of reduced section of tensile-test specimen

Table 39 Minimum mechanical properties for alloy 359.0-T61

Class(a)	Tensile strength(b) MPa	Tensile strength(b) ksi	Tensile yield strength(b)(c) MPa	Tensile yield strength(b)(c) ksi	Elongation(b)(d), %	Compressive yield strength(e) MPa	Compressive yield strength(e) ksi
1	310	45	241	35	4	241	35
2	324	47	262	38	3	262	38
10	310	45	234	34	4	234	34
11	276	40	207	30	3	207	30

(a) Classes 1 and 2 (levels of properties) obtainable only from designated areas of casting; classes 10 and 11 may be obtained from any location in casting. (b) Specified in MIL-A-21180. (c) 0.2% offset. (d) In 4*d*, where *d* is diameter of reduced section of tensile-test specimen. (e) Design values; not specified

Table 40 Typical tensile properties of separately cast test bars of alloy 359.0-T6 at various temperatures

Temperature °C	Temperature °F	Time at temperature, h	Tensile strength MPa	Tensile strength ksi	Yield strength MPa	Yield strength ksi	Elongation(a), %
−196	−320	...	435	63	325	47	4
−80	−112	...	380	55	325	47	5
−28	−18	...	360	52	310	45	6
150	300	100	290	42	260	38	10
		1000	250	36	235	34	11
		10 000	125	18	95	14	30
260	500	0.5	125	18	115	17	25
		10	65	9.5	60	8.5	40
		100	60	8.5	50	7	50
		1000	50	7.5	40	6	55
		10 000	50	7	35	5	60
315	600	0.5	50	7.5	45	6.5	50
		10	40	6	40	5.5	60
		100	40	5.5	30	4.4	65
370	700	0.5	30	4.4	28	4	55

(a) In 4*d*, where *d* is diameter of reduced section of tensile test specimen

cast test bars, as-cast. 380.0: tensile strength, 330 MPa (48 ksi); yield strength, 165 MPa (24 ksi); elongation, 3% in 50 mm or 2 in. A380.0: tensile strength, 325 MPa (47 ksi); yield strength, 160 MPa (23 ksi); elongation, 4% in 50 mm or 2 in. See also Tables 42 and 43.
Shear strength. 380.0: 195 MPa (28 ksi). A380.0: 185 MPa (27 ksi)
Poisson's ratio. 0.33
Elastic modulus. 71.0 GPa (10.3 × 10^6 psi); shear, 26.5 GPa (3.85 × 10^6 psi)
Fatigue strength. At 5 × 10^8 cycles, 380.0 and A380.0: 138 MPa (20 ksi) (R.R. Moore type test)

Mass Characteristics

Density. 2.71 g/cm³ (0.098 lb/in.³) at 20 °C (68 °F)

Thermal Properties

Liquidus temperature. 595 °C (1100 °F)
Solidus temperature. 540 °C (1000 °F)
Coefficient of linear thermal expansion. At 20 to 200 °C (68 to 392 °F). 380.0: 22.0 μm/m · K (12.2 μin./in. · °F). A380.0: 21.8 μm/m · K (12.1 μin./in. · °F)
Specific heat. 963 J/kg · K (0.230 Btu/lb · °F) at 100 °C (212 °F)
Latent heat of fusion. 389 kJ/kg (167 Btu/lb)
Thermal conductivity. 96.2 W/m · K (55.6 Btu/ft · h · °F) at 25 °C (77 °F)

Electrical Properties

Electrical conductivity. Volumetric, 27% IACS at 20 °C (68 °F)
Electrical resistivity. 65 nΩ · m at 20 °C (68 °F)

Fabrication Characteristics

Melting temperature. 650 to 760 °C (1200 to 1400 °F)
Die casting temperature. 635 to 705 °C (1175 to 1300 °F)
Annealing temperature. For increased ductility, 260 to 370 °C (500 to 700 °F); hold at temperature 4 to 6 h; furnace cool or cool in still air
Stress relief temperature. 175 to 260 °C (350 to 500 °F); hold at temperature 4 to 6 h; cool in still air
Joining. Same as alloy 413.0 and A413.0

383.0
10.5Si-2.5Cu

Specifications

Former ASTM. SC102A
SAE. 383
UNS number. A03830

Chemical Composition

Composition limits. 2.0 to 3.0 Cu, 0.10 Mg max, 0.50 Mn max, 9.5 to 11.5 Si, 1.3 Fe max, 0.30 Ni max, 3.0 Zn max, 0.15 Sn max, 0.50 others (total) max, bal Al

Applications

Typical uses. Applications requiring good die filling capacity, fair pressure tightness, electroplating and machining characteristics, and strength at elevated temperature, poor weldability, and brazeability; anodizing quality is poor

Mechanical Properties

Tensile properties. Typical for separately cast test bars, as-cast: tensile strength, 310

Table 41 Typical tensile properties for separately cast test bars of alloys 360.0-F and A360.0-F at elevated temperature

Temperature		Tensile strength		Yield strength(a)		Elongation(b),
°C	°F	MPa	ksi	MPa	ksi	%
360.0 aluminum						
24	75	325	47	170	25	3
100	212	305	44	170	25	2
150	300	240	35	165	24	4
205	400	150	22	95	14	8
250	500	85	12	50	7.5	20
315	600	50	7	30	4.5	35
370	700	30	4.5	20	3	40
A360.0 aluminum						
24	75	315	46	165	24	5
100	212	295	43	165	24	3
150	300	235	34	160	23	5
205	400	145	21	90	13	14
250	500	75	11	45	6.5	30
315	600	45	6.5	28	4	45
370	700	30	4	15	2.5	45

(a) 0.2% offset. (b) In 50 mm or 2 in.

Table 42 Typical tensile properties for separately cast test bars of alloy 380.0-F at elevated temperature

Temperature		Tensile strength		Yield strength		Elongation,
°C	°F	MPa	ksi	MPa	ksi	%
24	75	330	48	165	24	3
100	212	310	45	165	24	4
150	300	235	34	150	22	5
205	400	165	24	110	16	8
260	500	90	13	55	8	20
315	600	50	7	30	4	30
370	700	30	4	15	2.5	35

Table 43 Tensile properties of die cast alloy 380.0-F at various temperatures

Temperature		Tensile strength		Yield strength(a)		Elongation,
°C	°F	MPa	ksi	MPa	ksi	%
−195	−320	407	59	207	30	2.5
−80	−112	338	49	165	24	2.5
−26	−18	338	49	165	24	3
24	75	330	48	165	24	3
100	212	310	45	165	24	4
150	300	235	34	152	22	5
205	400	165	24	110	16	8
260	500	90	13	55	8	20
315	600	49	7	28	4	30
371	700	28	4	17	2.5	35

(a) 0.2% offset

MPa (45 ksi); yield strength, 150 MPa (22 ksi); elongation, 3.5% in 50 mm or 2 in.
Hardness. 75 HB (500 kg load, 10 mm ball)
Poisson's ratio. 0.33
Fatigue strength. 145 MPa (21 ksi) at 5 × 10^8 cycles
Impact strength. Charpy V-notch: 4 J (3 ft · lbf)

Mass Characteristics

Density. 2.74 g/cm^3 (0.099 lb/in.^{3})

Thermal Properties

Liquidus temperature. 580 °C (1080 °F)
Solidus temperature. 515 °C (960 °F)
Coefficient of linear thermal expansion. 21.1 μm/m · K (11.7 μin./in. · °F) at 20 to 100 °C (68 to 212 °F)
Thermal conductivity. 96.2 W/m · K (55.6 Btu/ft · h · °F)

Electrical Properties

Electrical conductivity. Volumetric, 23% IACS at 20 °C (68 °F)

Fabrication Characteristics

Die casting temperature. 615 to 700 °C (1140 to 1290 °F)
Stress relief temperature. 175 to 260 °C (350 to 500 °F); hold at temperature 4 to 6 h; cool in still air
Annealing temperature. For increased ductility, 260 to 370 °C (500 to 700 °F); hold at temperature 4 to 6 h; furnace cool or cool in still air

384.0, A384.0
11.2Si-3.8Cu

Specifications

Former ASTM. SC114A
SAE. 303
UNS number. 384.0: A03840. A384.0: A13840
Government. 384.0: QQ-A-591

Chemical Composition

Composition limits. 384.0: 3.0 to 4.5 Cu, 0.10 Mg max, 0.5 Mn max, 10.5 to 12.0 Si, 1.3 Fe max, 0.50 Ni max, 3.0 Zn max, 0.35 Sn max, 0.50 others (total) max, bal Al. A384.0: 3.0 to 4.5 Cu, 0.10 Mg max, 0.50 Mn max, 10.5 to 12.0 Si, 1.3 Fe max, 0.50 Ni max, 1.0 Zn max, 0.35 Sn max, 0.50 others (total) max, bal Al
Consequence of exceeding impurity limits. Generally insensitive to minor variations in composition, but resistance to corrosion is reduced and lowers as copper increases

Applications

Typical uses. Die casting applications where fair pressure tightness and fair strength at elevated temperatures are required. Better die filling than 380.0. Poor weldability and brazeability

Mechanical Properties

Tensile properties. Typical for separately cast test bars, as-cast, 384.0 and A384.0: tensile strength, 330 MPa (48 ksi); yield strength, 165 MPa (24 ksi); elongation, 2.5% in 50 mm or 2 in.
Shear strength. 384.0: 200 MPa (29 ksi)
Hardness. 384.0 and A384.0: 85 HB (500 kg load, 10 mm ball)
Fatigue strength. 384.0: 140 MPa (20 ksi)

Mass Characteristics

Density. 384.0: 2.823 g/cm^3 (0.102 lb/in.^{3}). A384.0: 2.768 g/cm^3 (0.100 lb/in.^{3})

Thermal Properties

Liquidus temperature. 580 °C (1080 °F)
Solidus temperature. 515 °C (960 °F)
Coefficient of linear thermal expansion. 384.0: 20.8 μm/m · K (11.6 μin./in. · °F). A384.0: 20.7 μm/m · K (11.5 μin./in. · °F)
Thermal conductivity. 384.0: 92 W/m · K (53 Btu/ft · h · °F). A384.0: 96 W/m · K (56 Btu/ft · h · °F)

Electrical Properties

Electrical conductivity. Volumetric, at 20 °C (68 °F). 384.0: 22% IACS. A384.0: 23% IACS

Fabrication Characteristics

Die casting temperature. 615 to 700 °C (1140 to 1290 °F)

Table 44 Typical room-temperature mechanical properties for separately cast test bars of alloys 390.0 and A390.0

Temper	Tensile strength(a) MPa	Tensile strength(a) ksi	Yield strength(b) MPa	Yield strength(b) ksi	Hardness(a)(c), HB	Fatigue strength(d) MPa	Fatigue strength(d) ksi
A390.0, sand castings							
F, T5	180	26	180	26	100	...	...
T6	275	40	275	40	140	105	15
T7	250	36	250	36	115	...	...
A390.0, permanent mold castings							
F, T5	200	29	200	29	110	...	...
T6	310	45	310	45	145	115	17
T7	260	38	260	38	120	100	14.5
390.0, conventional die castings							
F	280	40.5	240	35	120	140	20
T5	295	43	260	38	125	...	...
390.0, Acurad castings							
F	205	30	195	28	110	90	13
T5	205	30	200	29	110	95	14
T6	365	53	365	53	150	115	17
T7	275	40	275	40	125	110	16

(a) Tensile properties and hardness are determined from standard cast-to-size tensile specimens 12.7 mm (½ in.) diameter for sand, permanent mold, and Acurad castings and 6.4 mm (¼ in.) diameter for die castings and tested without machining the surface. (b) 0.2% offset. For 390.0 and A390.0 castings, yield strength normally equals tensile strength because 0.2% offset is not reached prior to fracture. (c) 500 kg load; 10 mm ball. (d) At 5×10^8 cycles; R.R. Moore type test

Table 45 Typical elevated-temperature tensile yield strength for separately cast test bars of alloy 390.0

Temper	Temperature °C	Temperature °F	Yield strength(a) MPa	Yield strength(a) ksi
Acurad castings				
F	38	100	195	28
	95	200	195	28
	150	300	180	26
	205	400	155	22
	260	500	100	14
T5	38	100	210	30
	95	200	225	32
	150	300	195	28
	205	400	160	23
	260	500	85	12
T6	38	100	365	52
	95	200	335	48
	150	300	305	44
	205	400	235	34
	260	500	70	10
T7	38	100	280	40
	95	200	270	39
	150	300	245	35
	205	400	195	28
	260	500	70	10
Die castings				
F	38	100	260	37
	95	200	285	41
	150	300	265	38
	205	400	210	30
	260	500	125	18

(a) Based on cast-to-size test specimens tested after 1000 h holding at test temperature

Stress relief temperature. 175 to 260 °C (350 to 500 °F); hold at temperature 4 to 6 h; cool in still air
Annealing temperature. For increased ductility, 260 to 370 °C (500 to 700 °F); hold at temperature 4 to 6 h; furnace cool or cool in still air

390.0, A390.0
17.0Si-4.5Cu-0.6Mg

Specifications

UNS number. 390.0 die castings, A03900. A390.0: sand and permanent mold castings, A13900

Chemical Composition

Composition limits. 390.0: 4.0 to 5.0 Cu, 0.45 to 0.65 Mg, 0.10 Mn max, 16.0 to 18.0 Si, 1.3 Fe max, 0.10 Zn max, 0.20 Ti max, 0.10 other (each) max, 0.20 others (total) max, bal Al. A390.0: 4.0 to 5.0 Cu, 0.45 to 0.65 Mg, 0.10 Mn max, 16.0 to 18.0 Si, 0.5 Fe max, 0.10 Zn max, 0.20 Ti max, 0.10 other (each) max, 0.20 others (total) max, bal Al

Applications

Typical uses. Automotive cylinder block, four cycle air-cooled engines, air compressors, Freon compressors, pumps requiring abrasive resistance, pulleys, and brake shoes. Other applications where high wear resistance, low coefficient of thermal expansion, good elevated-temperature strength, and good fluidity are required

Mechanical Properties

Tensile properties. See Tables 44 and 45. Typical elongation. 390.0: die and Acurad castings (F and T5 tempers), 1.0% in 50 mm or 2 in.; Acurad castings (T6 and T7 tempers), <1.0% in 50 mm or 2 in. A390.0: sand castings (all tempers) and permanent mold castings (T6 and T7 tempers), <1.0% in 50 mm or 2 in.; permanent mold castings (F and T5 tempers), 1.0% in 50 mm or 2 in.
Hardness. See Table 44.
Elastic modulus. Tension, 81.2 GPa (11.8×10^6 psi); compression, 82.8 GPa (12.0×10^6 psi)
Fatigue strength. See Table 44.

Mass Characteristics

Density. 2.73 g/cm³ (0.099 lb/in.³) at 20 °C (68 °F)

Thermal Properties

Liquidus temperature. 650 °C (1200 °F)
Solidus temperature. 505 °C (945 °F)
Coefficient of linear thermal expansion. 18.0 μm/m · K (10.0 μin./in. · °F) at 20 to 100 °C (68 to 212 °F)
Thermal conductivity. 134 W/m · K (77.4 Btu/ft · h · °F) at 25 °C (77 °F)

Electrical Properties

Electrical conductivity. Volumetric, at 20 °C (68 °F). F temper: 27% IACS. T5 temper: 25% IACS

Fabrication Characteristics

Solution temperature. 495 °C (925 °F)
Aging temperature. T5 and T7 tempers: 230 °C (450 °F). T6 temper: 175 °C (350 °F); hold at temperature for 8 h

413.0, A413.0
12Si

Commercial Names

Former designation. 413.0: 13. A413.0: A13

Specifications

Former ASTM. 413.0: S12B. B85 S12A
SAE. A413.0: J453, 305
UNS number. 413.0: A04130. A413.0: A14130
Government. A413.0: QQ-A-591 (class 2)
Foreign. Canada: A413.0, CSA S12P. France: NF A-S13. ISO: AlSi12

Chemical Composition

Composition limits. 413.0: 1.0 Cu max, 0.10 Mg max, 0.35 Mn max, 11.0 to 13.0 Si, 2.0 Fe max, 0.50 Ni max, 0.50 Zn max, 0.15 Sn max, 0.25 others (total) max, bal Al. A413.0: 1.0 Cu max, 0.10 Mg max, 0.35 Mn max, 11.0 to 13.0 Si, 1.3 Fe max, 0.50 Ni max, 0.50 Zn max, 0.15 Sn max, 0.25 others (total) max, bal Al
Consequence of exceeding impurity limits. Content of impurities may be quite high before serious effects are detected. Increasing copper lowers corrosion resistance; increasing iron and magnesium lowers ductility; increasing silicon content may lead to machining problems.

Applications

Typical uses. Miscellaneous thin-walled and intricately designed castings. Other applications where excellent castability, resistance to corrosion, and pressure tightness are required

Mechanical Properties

Tensile properties. Typical for separately

Table 46 Typical tensile properties for separately cast test bars of alloy 413.0-F at elevated temperature

Temperature °C	°F	Tensile strength MPa	ksi	Yield strength(a) MPa	ksi	Elongation(b), %
−195	−320	360	52	160	23	1.5
−80	−112	310	45	145	21	2
−28	−18	303	44	145	21	2
24	75	295	43	145	21	2.5
100	212	255	37	140	20	5
150	300	220	32	130	19	8
205	400	165	24	105	15	15
260	500	90	13	60	9	30
315	600	50	7	30	4.5	35
370	700	30	4.5	15	2.5	40

(a) 0.2% offset. (b) In 50 mm or 2 in.

Table 47 Typical tensile properties for separately cast test bars of alloys 443.0, 443.0-F, B443.0-F, and C443.0-F

Temperature °C	°F	Tensile strength MPa	ksi	Yield strength(a) MPa	ksi	Elongation(b), %
443.0-F sand castings						
24	75	130	19	55	8	8
B443.0-F permanent mold castings						
24	75	160	23	60	9	10
C443.0-F die castings						
24	75	230	33	110	16	9
100	212	195	28	110	16	9
150	300	150	22	105	15	10
205	400	110	16	85	12	25
260	500	60	9	40	6	30
315	600	35	5	25	3.5	35
370	700	25	3.5	15	2.5	35

(a) 0.2% offset. (b) In 50 mm or 2 in.

cast test bars, as-cast. 413.0: tensile strength, 300 MPa (43 ksi); yield strength, 145 MPa (21 ksi); elongation, 2.5% in 50 mm or 2 in. A413.0: tensile strength, 290 MPa (42 ksi); yield strength, 130 MPa (19 ksi); elongation, 3.5% in 50 mm or 2 in. See also Table 46.
Shear strength. 170 MPa (25 ksi)
Fatigue strength. At 5 × 10^8 cycles, 130 MPa (19 ksi) (R.R. Moore type test)

Mass Characteristics

Density. 2.657 g/cm³ (0.096 lb/in.³) at 20 °C (68 °F)

Thermal Properties

Coefficient of linear thermal expansion.

Temperature range °C	°F	Average coefficient µm/m · K	µin./in. · °F
20–100	68–212	20.4	11.3
20–200	68–392	21.4	11.8
20–300	68–572	22.4	12.4

Specific heat. 963 J/kg · K (0.230 Btu/lb · °F)
Latent heat of fusion. 389 kJ/kg (167 Btu/lb)
Thermal conductivity. 121 W/m · K (70 Btu/ft · h · °F) at 25 °C (77 °F)

Electrical Properties

Electrical conductivity. Volumetric, 31% IACS at 20 °C (68 °F)
Electrical resistivity. 55.6 nΩ · m at 20 °C (68 °F)

Fabrication Characteristics

Melting temperature. 650 to 760 °C (1200 to 1400 °F)
Die casting temperature. 635 to 705 °C (1175 to 1300 °F)
Joining. Rivet compositions: 6053-T4, 6053-T6, 6053-T61. Soft solder: After copper plating, then use methods applicable to copper-base alloys. Resistance welding: flash method

443.0, A443.0, B443.0, C443.0 5.2Si

Commercial Names

Former designation. 43

Specifications

Former ASTM. 443.0: S5B. B443.0: S5A. C443.0: S5C
SAE. C443.0: 304
UNS number. 443.0: A04430. A443.0: A14430. B443.0: A24430. C443.0: A34430
Government. B443.0: QQ-A-601 (class 2). C443.0: QQ-A-591
Foreign. Canada: CSA S5

Chemical Composition

Composition limits. 443.0: 0.6 Cu max, 0.05 Mg max, 0.50 Mn max, 4.5 to 6.0 Si, 0.8 Fe max, 0.25 Cr max, 0.50 Zn max, 0.25 Ti max, 0.35 others (total) max, bal Al. A443.0: 0.30 Cu max, 0.05 Mg max, 0.50 Mn max, 4.5 to 6.0 Si, 0.8 Fe max, 0.25 Cr max, 0.50 Zn max, 0.25 Ti max, 0.35 others (total) max, bal Al. B443.0: 0.15 Cu max, 0.05 Mg max, 0.35 Mn max, 4.5 to 6.0 Si, 0.8 Fe max, 0.35 Zn max, 0.25 Ti max, 0.25 others (total) max, bal Al. C443.0: 0.6 Cu max, 0.10 Mg max, 0.35 Mn max, 4.5 to 6.0 Si, 2.0 Fe max, 0.50 Ni max, 0.50 Zn max, 0.15 Sn max, 0.25 others (total) max, bal Al
Consequence of exceeding impurity limits. For die cast alloy, relatively large quantities of impurities may be present before serious effects are detected. Increasing copper tends to lower resistance to corrosion; increasing iron and magnesium tends to lower ductility. For sand and permanent mold cast alloys, high copper, iron, or nickel decreases ductility and resistance to corrosion. Increasing magnesium reduces ductility.

Applications

Typical uses. Cooking utensils, food-handling equipment, marine fittings, miscellaneous thin-section castings. Die castings: applications where good pressure tightness, above-average ductility, and excellent resistance to corrosion are required. Sand and permanent mold castings: applications where very good castability and resistance to corrosion with moderate strength are required

Mechanical Properties

Tensile properties. See Table 47.
Shear strength. F temper: 443.0 (sand castings): 95 MPa (14 ksi). B443.0 (permanent mold castings): 110 MPa (16 ksi). C443.0 (die castings): 145 MPa (21 ksi)
Hardness. F temper: 443.0 (sand castings): 40 HB. B443.0 (permanent mold castings): 45 HB. C443.0 (die castings): 65 HB (500 kg load, 10 mm ball)
Poisson's ratio. 0.33
Elastic modulus. Tension, 71.0 GPa (10.3 × 10^6 psi); shear, 26.5 GPa (3.85 × 10^6 psi)
Fatigue strength. F temper, at 5 × 10^8 cycles. 443.0 (sand castings) and B443.0 (permanent mold castings): 55 MPa (8 ksi). C443.0 (die castings): 115 MPa (17 ksi) (R.R. Moore type test)

Mass Characteristics

Density. 2.69 g/cm³ (0.097 lb/in.³) at 20 °C (68 °F)

Thermal Properties

Liquidus temperature. 630 °C (1170 °F)
Solidus temperature. 575 °C (1065 °F)
Coefficient of linear thermal expansion.

Temperature range °C	°F	Average coefficient µm/m · K	µin./in. · °F
20–100	68–212	22	12.2
20–200	68–392	23	12.8
20–300	68–572	24	13.3

Specific heat. 963 J/kg · K (0.230 Btu/lb · °F) at 100 °C (212 °F)
Latent heat of fusion. 389 kJ/kg (167 Btu/lb)
Thermal conductivity. As-cast: 142 W/m · K (82.2 Btu/ft · h · °F). Annealed: 163 W/m · K (94.3 Btu/ft · h · °F)

Electrical Properties

Electrical conductivity. Volumetric at 20 °C (68 °F). As-cast (sand, permanent mold, and die castings): 37% IACS
Electrical resistivity. At 20 °C (68 °F). As-cast (sand, permanent mold, and die castings): 46.6 nΩ · m. Annealed (sand and permanent mold): 41.0 nΩ · m
Electrolytic solution potential. −0.83 V (sand cast) and −0.82 V (permanent mold cast) versus 0.1 *N* calomel electrode in an aqueous solution containing 53 g NaCl plus 3 g H_2O_2 per liter

Fabrication Characteristics

Melting temperature. Die castings: 650 to 760 °C (1200 to 1400 °F). Sand and permanent mold castings: 675 to 815 °C (1250 to 1500 °F)
Casting temperature. Die castings: 635 to 705 °C (1175 to 1300 °F). Sand and permanent mold castings: 675 to 790 °C (1250 to 1450 °F)
Joining. Rivet compositions: 6053-T4, 6053-T6, 6053-T61. Soft solder with copper plate and use methods applicable to copper-base alloys for die castings. Use Alcoa No. 802, no flux or rub-tin with Alcoa No. 802 for sand and permanent mold castings. Sand and permanent mold casting alloys (unless otherwise noted): braze with Alcoa No. 717; Alcoa No. 33 flux; flame either reducing oxyacetylene or reducing oxyhydrogen. Atomic-hydrogen weld with 4043 alloy; Alcoa No. 22 flux. Oxyacetylene weld with 4043 alloy; Alcoa No. 22 flux; neutral flame. Metal-arc weld with 4043 alloy; Alcoa No. 27 flux. Carbon-arc weld with 4043 alloy; Alcoa No. 24 flux (automatic), Alcoa No. 27 flux (manual). Tungsten-arc argon-atmosphere weld with 4043 alloy; no flux. Resistance weld: flash method for die cast alloys; spot, seam, and flash methods for sand and permanent mold cast alloys

514.0
4Mg

Commercial Names

Former designation. 214

Specifications

Former ASTM. G4A
SAE. 320
UNS number. A05140
Government. QQ-A-601 (class 5)
Foreign. Canada: CSA G4. United Kingdom: DTD 165. ISO: AlMg3

Table 48 Typical tensile properties for separately cast test bars of alloy 514.0-F

Temperature		Tensile strength		Yield strength(a)		Elongation,
°C	°F	MPa	ksi	MPa	ksi	%
24	75	170	25	85	12	9
150	300	150	22	85	12	7
205	400	125	18	85	12	9
260	500	90	13	55	8	12
315	600	60	9	30	4	17

(a) 0.2% offset

Chemical Composition

Composition limits. 0.15 Cu max, 3.5 to 4.5 Mg, 0.35 Mn max, 0.35 Si max, 0.50 Fe max, 0.15 Zn max, 0.25 Ti max, 0.05 other (each) max, 0.15 others (total) max, bal Al
Consequence of exceeding impurity limits. High copper or nickel greatly decreases resistance to corrosion and decreases ductility. High iron, silicon, or manganese decreases strength and ductility. Tin reduces resistance to corrosion.

Applications

Typical uses. Dairy and food-handling applications, cooking utensils, fittings for chemical and sewage use. Other applications where excellent resistance to corrosion and tarnish are required

Mechanical Properties

Tensile properties. Typical, F temper. Tensile strength, 145 MPa (21 ksi); yield strength, 95 MPa (14 ksi); elongation, 3.0%. See also Table 48.
Shear strength. 140 MPa (20 ksi)
Compressive yield strength. 85 MPa (12 ksi)
Hardness. 50 HB (500 kg load, 10 mm ball)
Poisson's ratio. 0.33
Elastic modulus. Tension, 71.0 GPa (10.3 × 10^6 psi); shear, 26.5 GPa (3.85 × 10^6 psi)
Fatigue strength. 50 MPa (7 ksi) at 5 × 10^8 cycles (R.R. Moore type test)

Mass Characteristics

Density. 2.650 g/cm³ (0.096 lb/in.³) at 20 °C (68 °F)

Thermal Properties

Liquidus temperature. 630 °C (1170 °F)
Solidus temperature. 585 °C (1090 °F)
Coefficient of linear thermal expansion.

Temperature range		Average coefficient	
°C	°F	μm/m · K	μin./in. · °F
20–100	68–212	24	13.3
20–200	68–392	25	13.9
20–300	68–572	26	14.4

Specific heat. 963 J/kg · K (0.230 Btu/lb · °F) at 100 °C (212 °F)
Latent heat of fusion. 389 kJ/kg (167 Btu/lb)
Thermal conductivity. 146 W/m · K (84.6 Btu/ft · h · °F) at 25 °C (77 °F)

Electrical Properties

Electrical conductivity. Volumetric, 35% IACS at 20 °C (68 °F)
Electrical resistivity. 49.3 nΩ · m at 20 °C (68 °F)
Electrolytic solution potential. −0.87 V versus 0.1 *N* calomel electrode in an aqueous solution containing 53 g NaCl plus 3 g H_2O_2 per liter

Fabrication Characteristics

Melting temperature. 675 to 815 °C (1250 to 1500 °F)
Casting temperature. 675 to 790 °C (1250 to 1450 °F)
Joining. Rivet compositions: 6053-T4, 6053-T6, 6053-T61. Soft solder with Alcoa No. 802; no flux. Rub-tin with Alcoa No. 802. Braze with Alcoa No. 717; Alcoa No. 33 flux; flame either reducing oxyacetylene or reducing oxyhydrogen. Atomic-hydrogen weld with 4043 alloy; Alcoa No. 22 flux. Oxyacetylene weld with 4043 alloy; Alcoa No. 22 flux; flame neutral. Metal-arc weld with 4043 alloy; Alcoa No. 27 flux. Carbon-arc weld with 4043 alloy; Alcoa No. 24 flux (automatic), Alcoa 27 flux (manual). Tungsten-arc argon-atmosphere weld with 4043; no flux. Resistance welding: spot, seam, and flash welds

518.0
8Mg

Commercial Names

Former designation. 218

Specifications

Former ASTM. G8A
UNS number. A05180
Government. QQ-A-591

Chemical Composition

Composition limits. 0.25 Cu max, 7.5 to 8.5 Mg, 0.35 Mn max, 0.35 Si max, 1.8 Fe max, 0.15 Ni max, 0.15 Zn max, 0.15 Sn max, 0.25 others (total) max, bal Al

Applications

Typical uses. Alloy has excellent corrosion resistance and machinability; high ductility; poor castability (is hot short). Takes a high polish; difficult to attain a uniform appearance after anodizing. Non-heat treatable. Poor weldability and brazeability. Used for die cast marine fittings, ornamental hardware, ornamental automotive parts, and other applications requiring the highest corrosion resistance

Table 49 Typical tensile properties for separately cast test bars of alloy 520.0-F at elevated temperature

Temperature °C	Temperature °F	Tensile strength MPa	Tensile strength ksi	0.2% yield strength MPa	0.2% yield strength ksi	Elongation(a), %
24	75	315	46	170	25	14
150	300	240	35	130	19	16
205	400	150	22	80	11.5	40
260	500	105	15	50	7.5	55
315	600	70	10.5	25	3.5	70

(a) In 50 mm or 2 in.

Mechanical Properties

Tensile properties. Typical, F temper. Tensile strength, 310 MPa (45 ksi); yield strength, 190 MPa (28 ksi); elongation, 5 to 8% in 50 mm or 2 in.
Shear strength. 205 MPa (30 ksi)
Hardness. 80 HB (500 kg, 10 mm load)
Impact strength. Charpy V-notch: 9 J (7 ft · lbf)
Fatigue strength. 160 MPa (23 ksi) at 5×10^8 cycles (R.R. Moore type test)

Mass Characteristics

Density. 2.57 g/cm^3 (0.093 lb/in.^{3})

Thermal Properties

Liquidus temperature. 620 °C (1150 °F)
Solidus temperature. 535 °C (995 °F)
Coefficient of linear thermal expansion. 24.1 μm/m · K (13.4 μin./in. · °F) at 20 to 100 °C (68 to 212 °F)
Thermal conductivity. 96.2 W/m · K (55.6 Btu/ft · h · °F)

Electrical Properties

Electrical conductivity. Volumetric, 25% IACS at 20 °C (68 °F)

520.0
10Mg

Commercial Names

Former designation. 220

Specifications

AMS. 4240
Former ASTM. G10A
SAE. 324
UNS number. A05200
Government. QQ-A-601 (class 16)
Foreign. Canada: CSA G10. France: NF A-G10. ISO: AlMg10

Chemical Composition

Composition limits. 0.25 Cu max, 9.5 to 10.6 Mg, 0.15 Mn max, 0.25 Si max, 0.30 Fe max, 0.15 Zn max, 0.25 Ti max, 0.05 other (each) max, 0.15 others (total) max, bal Al
Consequence of exceeding impurity limits. High copper or nickel greatly decreases resistance to corrosion. High iron, silicon, or manganese contents adversely affect mechanical properties.

Applications

Typical uses. Aircraft fittings, railroad passenger-car frames, miscellaneous castings requiring strength and shock resistance. Other applications where excellent machinability and resistance to corrosion with highest strength and elongation of any aluminum sand casting alloy are desired

Mechanical Properties

Tensile properties. Typical. T4 temper: tensile strength, 330 MPa (48 ksi); yield strength, 180 MPa (26 ksi); elongation in 50 mm or 2 in., 16%. See also Table 49.
Shear strength. 235 MPa (34 ksi)
Compressive yield strength. 2 MPa (27 ksi)
Hardness. 75 HB (500 kg load, 10 mm ball)
Poisson's ratio. 0.33
Elastic modulus. Tension, 66 GPa (9.5×10^6 psi); shear, 24.5 GPa (3.55×10^6 psi)
Fatigue strength. 55 MPa (8 ksi) at 5×10^8 cycles (R.R. Moore type test)

Mass Characteristics

Density. 2.57 g/cm^3 (0.093 lb/in.^{3}) at 20 °C (68 °F)

Thermal Properties

Liquidus temperature. 605 °C (1120 °F)
Solidus temperature. 450 °C (840 °F)
Coefficient of linear thermal expansion.

Temperature range °C	Temperature range °F	Average coefficient μm/m · K	Average coefficient μin./in. · °F
20–100	68–212	25	13.9
20–200	68–392	26	14.4
20–300	68–572	27	15.0

Specific heat. 963 J/kg · K (0.230 Btu/lb · °F) at 100 °C (212 °F)
Latent heat of fusion. 389 kJ/kg (167 Btu/lb)
Thermal conductivity. T4 temper: 87.9 W/m · K (50.8 Btu/ft · h · °F) at 25 °C (77 °F)

Electrical Properties

Electrical conductivity. Volumetric, T4 temper: 21% IACS at 20 °C (68 °F)
Electrical resistivity. T4 temper: 82.1 nΩ · m at 20 °C (68 °F)
Electrolytic solution potential. T4 temper: −0.89 V versus 0.1 *N* calomel electrode in an aqueous solution containing 53 g NaCl plus 3 g H_2O_2 per liter

Fabrication Characteristics

Melting temperature. 675 to 815 °C (1250 to 1500 °F)
Casting temperature. 675 to 788 °C (1250 to 1450 °F)
Joining. Rivet compositions: 6053-T4, 6053-T6, 6053-T61. Soft solder with Alcoa No. 802; no flux. Rub-tin with Alcoa No. 802. Resistance welding: spot, seam, and flash methods

535.0, A535.0, B535.0
7Mg

Commercial Names

Former designations. 535.0: Almag35. A535.0: A218. B535.0: B218

Specifications

Former AMS. 4238A, 4239
Former ASTM. 535.0: GM70B
UNS number. 535.0: A05350. A535.0: A15350. B535.0: A25350
Government. 535.0: QQ-A-601, QQ-A-371

Chemical Composition

Composition limits. 535.0: 0.05 Cu max, 6.2 to 7.5 Mg, 0.10 to 0.25 Mn, 0.15 Si max, 0.15 Fe max, 0.10 to 0.25 Ti, 0.003 to 0.007 Be, 0.002 B max, bal Al. A535.0: 0.10 Cu max, 6.5 to 7.5 Mg, 0.10 to 0.25 Mn, 0.20 Si max, 0.20 Fe max, 0.25 Ti max, 0.05 other (each) max, 0.15 others (total) max, bal Al. B535.0: 0.10 Cu max, 6.5 to 7.5 Mg, 0.05 Mn max, 0.15 Si max, 0.15 Fe max, 0.10 to 0.25 Ti, 0.05 other (each) max, 0.15 others (total) max, bal Al

Applications

Typical uses. Maximum properties are available immediately after casting without the aid of heat treatment or natural aging. Used in parts in computing devices, aircraft and missile guidance systems, and electric equipment where dimensional stability is essential. Highly useful in marine and other corrosive-prone applications

Mechanical Properties

Tensile properties. F and T5 tempers: 535.0: Tensile strength: typical, 275 MPa (40 ksi); minimum, 240 MPa (35 ksi). Yield strength: typical, 140 MPa (20 ksi); minimum, 125 MPa (18 ksi). Elongation in 50 mm or 2 in.: typical, 13%; minimum, 8.0%. See also Table 50.
Shear strength. 190 MPa (27.5 ksi)
Compressive yield strength. Typical: 162 MPa (23.5 ksi)
Hardness. Typical: 60 HB. Minimum: 70 HB
Elastic modulus. Tension, 71.0 GPa (10.3×10^6 psi)
Impact strength. Charpy: 90° notch specimen, 14.2 J (10.5 ft · lbf); keyhole specimen, 6.7 J (4.95 ft · lbf); unnotched specimen, 77.0 J (56.8 ft · lbf)

Mass Characteristics

Density. 2.62 g/cm^3 (0.095 lb/in.^{3})

Table 50 Typical tensile properties for separately cast test bars of alloy 535.0-F at elevated temperature

Temperature °C	Temperature °F	Tensile strength MPa	Tensile strength ksi	Elongation(a), %
150	300	260	37.5	11
175	350	235	34	14
205	400	220	32	14
260	500	180	26.5	13
315	600	140	20.5	13
370	700	105	15.5	12

(a) In 50 mm or 2 in.

Table 51 Typical tensile properties for separately cast test bars of alloy 712.0-F at elevated temperature

Temperature °C	Temperature °F	Tensile strength MPa	Tensile strength ksi	0.2% yield strength MPa	0.2% yield strength ksi	Elongation(a), %
79	175	235	34	210	30.5	3
120	250	205	29.5	175	25	2
175	350	135	19.5	115	17	6

(a) In 50 mm or 2 in.

Thermal Properties

Liquidus temperature. 630 °C (1165 °F)
Solidus temperature. 550 °C (1020 °F)
Coefficient of linear thermal expansion.

Temperature range °C	Temperature range °F	Average coefficient μm/m · K	Average coefficient μin./in. · °F
−60 to 20	−76 to 68	21.6	12.0
20–100	68–212	23.6	13.1
20–200	68–392	25.6	14.2
20–300	68–572	26.6	14.8

Electrical Properties

Electrical conductivity. Volumetric, 23% IACS at 20 °C (68 °F)
Electrical resistivity. 75 nΩ · m at 20 °C (68 °F)

Chemical Properties

General corrosion behavior. 535.0 has the highest resistance to corrosion of any of the common aluminum casting alloys.

Fabrication Characteristics

Machinability. Superior, can be milled at speeds four times faster than other aluminum casting alloys. High microfinishes can be achieved at high speeds. 535.0 takes a very high mirror polish. Normally this alloy is used for sand and permanent mold castings, but it can also be used for die casting. Where high dimensional tolerance is required, the following procedure should be used: rough machine parts; heat at 200 °C (400 °F) for 14 h; cycle between −73 to 100 °C (100 to 212 °F) five times (30 h/cycle); finish machine; heat 10 h at 200 °C (400 °F); cycle between −73 to 100 °C (−100 to 212 °F) 25 times (30 h/cycle). 535.0 may be stress relieved at approximately 370 °C (700 °F) for 5 h; air cool. Creep resistance at 370 °C (700 °F) is very low, permitting plastic flow under the load of locked-up stresses and resulting in stress-free castings. On air cooling from 370 °C (700 °F), 535.0 will have full hard and physical properties and will be stable. After being stress relieved, most castings from 535.0, A535.0, B535.0 can be rough and finish machined without breaking into the machining sequence.
Weldability. Can be welded by any inert gas shielded-arc systems using filler material of 5356 or 535.0 aluminum. Welding fluxes should be avoided if possible. Because of the beryllium content in alloy 535.0, care should be taken not to inhale fumes during welding.
Anodizing. Use sulfuric acid process to produce a pure satin white finish capable of being dyed to brilliant pastel colors.

712.0
5.8Zn-0.6Mg-0.5Cr-0.2Ti

Commercial Names

Former designations. D712.0, D612, 40E

Specifications

Former ASTM. ZG61A
SAE. 310
UNS number. A47120
Government. QQ-A-601 (class 17)

Chemical Composition

Composition limits. 0.25 Cu max, 0.50 to 0.65 Mg, 0.10 Mn max, 0.30 Si max, 0.50 Fe max, 0.40 to 0.6 Cr, 5.0 to 6.5 Zn, 0.15 to 0.25 Ti, 0.05 other (each) max, 0.20 others (total) max, bal Al

Applications

Typical uses. Applications where a good combination of mechanical properties is required without heat treatment: shock and corrosion resistance, machinability, dimensional stability, no distortion in heat treating

Mechanical Properties

Tensile properties. F or T5 temper. Typical tensile strength, 240 MPa (35 ksi); yield strength, 170 MPa (25 ksi); elongation, 5%. Low-temperature strength after 24 h at −70 °C (−94 °F): tensile strength, 265 MPa (38.4 ksi); elongation in 50 mm or 2 in., 5%. See also Table 51.
Shear strength. 180 MPa (26 ksi)
Compressive proportional limit. 95 MPa (14 ksi)
Hardness. 70 HB (500 kg load, 10 mm ball)
Poisson's ratio. 0.33
Elastic modulus. Tension, 71.0 GPa (10.3×10^6 psi); shear, 26.5 GPa (3.85×10^6 psi)
Impact strength. Charpy V-notch: 2.7 to 4.0 J (2 to 3 ft · lbf)
Fatigue strength. 62 MPa (9 ksi) at 5×10^8 cycles (R.R. Moore type test)

Mass Characteristics

Density. 2.81 g/cm^3 (0.101 lb/in.^{3}) at 20 °C (68 °F)

Thermal Properties

Liquidus temperature. 615 °C (1140 °F)
Solidus temperature. 570 °C (1060 °F)
Coefficient of linear thermal expansion. 24.7 μm/m · K (13.7 μin./in. · °F) at 20 to 93 °C (68 to 199 °F)
Specific heat. 963 J/kg · K (0.230 Btu/lb · °F) at 100 °C (212 °F)
Latent heat of fusion. 389 kJ/kg (167 Btu/lb)
Thermal conductivity. 138 W/m · K (79.8 Btu/ft · h · °F) at 25 °C (77 °F)

Electrical Properties

Electrical conductivity. Volumetric, 35% IACS at 20 °C (68 °F)
Electrical resistivity. 49.3 nΩ · m at 20 °C (68 °F)

Fabrication Characteristics

Melting temperature. F temper: 610 to 650 °C (1130 to 1200 °F)
Aging temperature. T5 temper: room temperature for 21 days or at 157 °C (315 °F) for 6 to 8 h

713.0
7.5Zn-0.7Cu-0.35Mg

Commercial Names

Former designation. 613, Tenzaloy

Specifications

Former ASTM. Sand castings, B26 ZC81A. Permanent mold castings, B108 ZC81B
Former SAE. 315
UNS number. A07130
Government. Sand castings, QQ-A-601 (class 22). Permanent mold castings: QQ-A-596 (class 12)

Chemical Composition

Composition limits. 0.40 to 1.0 Cu, 0.20 to 0.50 Mg, 0.6 Mn max, 0.25 Si max, 1.1 Fe max, 0.35 Cr max, 0.15 Ni max, 7.0 to 8.0 Zn, 0.25 Ti max, 0.10 other (each) max, 0.25 others (total) max, bal Al

Applications

Typical uses. Cast aluminum furniture and other very large casting applications that require high strength without heat treatment. 713.0 ages at room temperature to produce mechanical properties equivalent to those of common heat-treated aluminum cast alloys. These properties develop in 10 to 14 days at room temperature or in 12 h at 120 °C (250 °F).

Mechanical Properties

Tensile properties. Typical for T5 temper, aged at room temperature for 21 days or artificially aged at 120 ± 5.5 °C (250 ± 10 °F) for 16 h. Sand casting: tensile strength, 205 MPa (30 ksi); yield strength: 150 MPa (22 ksi); elongation, 4.0% in 50 mm or 2 in. Permanent mold casting: tensile strength, 220 MPa (32 ksi); yield strength: 150 MPa (22 ksi); elongation, 3.0% in 50 mm or 2 in.
Shear strength. 180 MPa (26 ksi)
Compressive yield strength. 170 MPa (25 ksi)
Impact strength. Charpy V-notch: sand castings, 3.4 J (2.5 ft · lbf); permanent mold castings, 4 J (3 ft · lbf). Unnotched: sand castings, 16.3 J (12 ft · lbf); permanent mold castings, 27 J (20 ft · lbf)
Fatigue strength. 60 MPa (9 ksi) at 5×10^8 cycles (R.R. Moore type test)

Mass Characteristics

Density. 2.81 g/cm^3 (0.102 lb/in.^{3})

Thermal Properties

Coefficient of linear thermal expansion. 24.1 μm/m · K (13.4 μin./in. · °F) at 20 to 200 °C (68 to 392 °F)
Thermal conductivity. 140 W/m · K (80 Btu/ft · h · °F) at 25 °C (77 °F)

Electrical Properties

Electrical conductivity. Volumetric, 35% IACS at 20 °C (68 °F)

Chemical Properties

General corrosion behavior. Good resistance to corrosion, equivalent to aluminum-silicon alloys. A typical corrosion test showed no loss in mechanical properties after immersion for 90 days in aerated 3% salt-water solution. Not subject to acceleration of corrosion by stress or to stress-corrosion cracking as determined by the standard test of exposure for 14 days to the corrosive medium while under a continuous load of 75% of yield strength

Fabrication Characteristics

Melting temperature. Approximate, 595 to 640 °C (1100 to 1185 °F)
Machinability. Good machinability and polishing characteristics. Very good dimensional stability. Fully aged material shows a decrease in length of less than 0.1 min/in. of length. If 713.0 is given a stress-relief treatment of 6 h at 450 °C (850 °F) and air cooled, it ages naturally. The resulting product is a stress-free, full-strength casting. This is not possible with any heat-treatable aluminum alloy.
Weldability. For high-strength welds, shielded-arc methods can be used with filler alloys 5154 and 5356.
Brazeability. Readily brazed at 540 to 595 °C (1000 to 1100 °F) using any of the common brazing methods

Table 52 Minimum mechanical properties for separately cast test bars of alloy 771.0

Temper	Tensile strength (min) MPa	Tensile strength (min) ksi	Yield strength(min)(a) MPa	Yield strength(min)(a) ksi	Elongation(b), %	Hardness(c), HB
T5	290	42	260	38	1.5	100
T51	220	32	185	27	3.0	85
T52	250	36	205	30	1.5	85
T6	290	42	240	35	5.0	90
T71	330	48	310	45	2.0	120

(a) 0.2% offset. (b) In 50 mm or 2 in. (c) 500 kg load, 10 mm ball

771.0
7Zn-0.9Mg-0.13Cr

Commercial Names

Former designation. Precedent 71A

Specifications

Former ASTM. 771.0: ZG71B
UNS number. A07710
Government. 771.0: QQ-A-601E

Chemical Composition

Composition limits. 0.10 Cu max, 0.8 to 1.0 Mg, 0.10 Mn max, 0.15 Si max, 0.15 Fe max, 0.06 to 0.20 Cr, 6.5 to 7.5 Zn, 0.10 to 0.20 Ti, 0.05 other (each) max, 0.15 others (total), bal Al

Applications

Typical uses. Applications where free machine and dimension stability are important. Polishes to a high luster; anodizes with good clean appearance. Good corrosion resistance

Mechanical Properties

Tensile properties. See Table 52.
Compressive properties. Compressive strength, T71 temper, 925 MPa (134 ksi); compressive yield strength, 370 MPa (54 ksi)
Elastic modulus. Tension, 71.0 GPa (10.3×10^6 psi)

Mass Characteristics

Density. 2.823 g/cm^3 (0.102 lb/in.^{3})

Thermal Properties

Liquidus temperature. 645 °C (1190 °F)
Solidus temperature. 605 °C (1120 °F)
Coefficient of linear thermal expansion. 24.7 μm/m · K (13.7 μin./in. · °F) at 20 to 100 °C (68 to 212 °F)
Thermal conductivity. 138 W/m · K (79.8 Btu/ft · h · °F)

Electrical Properties

Electrical conductivity. Volumetric, 27% IACS at 20 °C (68 °F)

Fabrication Characteristics

Machinability. 771.0-T5 has good stability and machinability. It can be milled five times faster and hole worked at twice the speed of alloys such as 356.0 and 319.0. It can be finished machined in one clamping operation to flatness tolerance of 0.001 in. This reduces total cost of machining over most casting alloys, which require two clamping operations to obtain this type of flatness tolerance.
Welding. Can be welded by either gas tungsten-arc or gas metal-arc welding using 5356 rod or wire. Special procedure should be followed in welding to ensure good results.

If parts are to be welded, the operation should be made part of the heat-treating cycle. If welding is to be done on T6 or T71 parts, the castings are heated to 580 °C (1080 °F), removed from the heat-treating furnace, and welded while hot. The parts are then returned to the furnace and the T6 and T71 heat treatments continued. If the parts are to be used in the T52 or T2 temper, they are heated to 415 °C (775 °F), taken from the furnace, welded hot, then returned to the furnace and the heat treatment continued. Items to be used in the T51 temper are heated to 205 °C (405 °F), taken from the furnace, welded hot, then returned to the furnace and T51 treatment continued. Repair weld parts should be heated and welded as described above.

The T5 temper should not be welded but can be welded if the procedure for T51 is used.
Heat treatments. See Table 53.

850.0
6.2Sn-1Cu-1Ni

Commercial Names

Former designation. 750

Specifications

AMS. Permanent mold casting: 4275
UNS number. A08500
Government. QQ-A-596 (class 15)

Chemical Composition

Composition limits. 0.7 to 1.3 Cu, 0.10 Mg max, 0.10 Mn max, 0.7 Si max, 0.7 Fe max,

Table 53 Heat treatments for alloy 771.0

Temper	Treatment
T2	Hold at 415 ± 14 °C (775 ± 25 °F) for 5 h; cool outside furnace in still air to room temperature; harden by reheating to 180 ± 3 °C (360 ± 5 °F) for 4 h; cool in air
T5	Hold at 180 ± 3 °C (355 ± 5 °F) for 3 to 5 h; cool outside furnace in still air to room temperature
T6	Hold at 580 to 595 °C (1080 to 1100 °F) for 6 h; cool outside furnace to room temperature in still air; age by holding for 3 h at 130 °C (265 °F) followed by cooling in still air
T51	Age by holding at 205 °C (405 °F) for 6 h; cool in still air
T52	Hold at 415 °C (775 ± 25 °F) for 5 h; cool from 415 to 345 °C (775 to 650 °F) in 2 h or more; cool from 345 to 230 °C (650 to 450 °F) in not more than ½ h (20 min desirable); cool from 230 to 120 °C (450 to 250 °F) in approximately 2 h; cool from 120 °C (250 °F) to room temperature in still air outside of furnace; harden by reheating to 165 °C (330 °F) for 6 to 16 h and cooling outside of furnace in still air
T71	Hold at 580 to 595 °C (1080 to 1100 °F) for 6 h; cool outside furnace to room temperature in still air; age by holding at 140 °C (285 °F) for 15 h followed by cooling in still air. Similar properties can be obtained by aging at 155 °C (310 °F) for 3 h

5.5 to 7.0 Sn, 0.7 to 1.3 Ni, 0.20 Ti max, 0.30 others (total) max, bal Al
Consequence of exceeding impurity limits. High iron, manganese, or magnesium decreases ductility and increases hardness. High silicon modifies bearing characteristics.

Applications

Typical uses. Applications where excellent bearing qualities are required

Mechanical Properties

Tensile properties. Typical for T5 temper: tensile strength, 160 MPa (23 ksi); yield strength, 75 MPa (11 ksi); elongation in 50 mm or 2 in., 10%
Shear strength. 103 MPa (15 ksi)
Compressive yield strength. 75 MPa (11 ksi)
Hardness. T5 temper: 45 HB (500 kg load, 10 mm ball)
Poisson's ratio. 0.33
Elastic modulus. Tension, 71.0 GPa (10.3 × 10^6 psi); shear, 26.5 GPa (3.85 × 10^6 psi)
Fatigue strength. 60 MPa (9 ksi) at 5 × 10^8 cycles (R.R. Moore type test)

Mass Characteristics

Density. 2.880 g/cm^3 (0.104 lb/in.^{3}) at 20 °C (68 °F)

Thermal Properties

Liquidus temperature. 650 °C (1200 °F)
Solidus temperature. 225 °C (435 °F)
Coefficient of linear thermal expansion.

Temperature range °C	°F	Average coefficient µm/m · K	µin./in. · °F
20–100	68–212	23.1	12.8
20–200	68–392	24.3	13.5

Specific heat. 963 J/kg · K (0.230 Btu/lb · °F) at 100 °C (212 °F)
Latent heat of fusion. 389 kJ/kg (167 Btu/lb)
Thermal conductivity. 180 W/m · K (104 Btu/ft · h · °F)

Electrical Properties

Electrical conductivity. Volumetric, 47% IACS at 20 °C (68 °F)
Electrical resistivity. 36.7 nΩ · m at 20 °C (68 °F)

Fabrication Characteristics

Melting temperature. 650 to 730 °C (1200 to 1350 °F)
Casting temperature. 650 to 705 °C (1200 to 1300 °F)
Aging temperature. 230 °C (450 °F); hold at temperature for 8 h

Aluminum-Lithium Alloys

Richard S. James, Aluminum Company of America

ALUMINUM-LITHIUM ALLOYS have been developed primarily to reduce the weight of aircraft and aerospace structures; more recently, they have been investigated for use in cryogenic applications (for example, liquid oxygen and hydrogen fuel tanks for aerospace vehicles).

The major development work began in the 1970s, when aluminum producers accelerated the development of aluminum-lithium alloys as replacements for conventional airframe alloys. The lower-density aluminum-lithium alloys were expected to reduce the weight and improve the performance of aircraft. The goal was to introduce ingot aluminum-lithium alloys that could be fabricated on the existing equipment of aluminum producers and then used by airframe manufacturers as direct replacements for the conventional aluminum alloys (which typically have constituted 70 to 80% of the weight of current aircraft). This development work led to the introduction of commercial alloys 8090, 2090, and 2091 in the mid-1980s; Weldalite 049 and CP276 were introduced shortly thereafter. These alloys are characterized by the following approximate nominal (wt%) compositions (balance aluminum):

- *Weldalite 049*: 5.4 Cu, 1.3 Li, 0.4 Ag, 0.4 Mg, 0.14 Zr
- *Alloy 2090*: 2.7 Cu, 2.2 Li, 0.12 Zr
- *Alloy 2091*: 2.1 Cu, 2.0 Li, 0.10 Zr
- *Alloy 8090*: 2.45 Li, 0.12 Zr, 1.3 Cu, 0.95 Mg
- *Alloy CP276*: 2.7 Cu, 2.2 Li, 0.5 Mg, 0.12 Zr

Commercial aluminum-lithium alloys are targeted as advanced materials for aerospace technology primarily because of their low density, high specific modulus, and excellent fatigue and cryogenic toughness properties. The superior fatigue crack propagation resistance of aluminum-lithium alloys, in comparison with that of traditional 2*xxx* and 7*xxx* alloys, is primarily due to high levels of crack tip shielding, meandering crack paths, and the resultant roughness-induced crack closure. However, the fact that these alloys derive their superior properties extrinsically from the above mechanisms has certain implications with respect to small-crack and variable-amplitude behavior. For example, aluminum-lithium alloys lose their fatigue advantage over conventional aluminum alloys in compression-dominated variable-amplitude fatigue spectra tests. However, in tension-dominated spectra, aluminum-lithium alloys show greater retardations on the application of single-peak tensile overloads.

The principal disadvantages of peak-strength aluminum-lithium alloys are reduced ductility and fracture toughness in the short-transverse direction, anisotropy of in-plane properties, the need for cold work to attain peak properties, and accelerated fatigue crack extension rates when cracks are microstructurally small (Ref 1). These limitations have precluded the direct substitution of aluminum airframe alloys with aluminum-lithium alloys, although it is possible to group the present aluminum alloys and the current aluminum-lithium alloys in terms of product form and primary design criteria (as is done in Table 1 for present aluminum and aluminum-lithium aerospace alloys). The grouping in Table 1 indicates that certain aluminum-lithium alloys exhibit more damage tolerance, strength, and corrosion resistance than other aluminum-lithium alloys. While this grouping may not be complete, it does provide a start-

Table 1 Groupings of selected aerospace alloys and temper conditions according to primary design requirements

The groupings do not imply that aluminum-lithium alloys can be directly substituted for aluminum alloys. See text for discussion.

Primary design criteria	Aluminum aerospace alloys				Aluminum-lithium aerospace alloys			
	Sheet	Plate	Forgings	Extrusions	Sheet	Plate	Forgings	Extrusions
High strength	7075-T6	7075-T651 7150-T651 7475-T651 7150-T7751	...	7075-T6511 7175-T6511 7150-T6511 7050-T76511	CP276 8091-T8X 2090-T83	Weldalite 049 8091-T851 2090-T81	Weldalite 049 8091-T652	CP276 Weldalite 049 8090-T8X 2090-T86
Medium strength, corrosion damage tolerance	7075-T76 7075-T73 2214-T6 2014-T6 2219-T87	7010-T7651 2214-T651 2014-T651 7075-T651 7075-T7351 7050-T7651 2124-T851 2219-T852	7050-T74 7075-T6 7075-T73 2014-T652 2024-T852	7050-T3511 2224-T3511 2219-T851 2024-T8511 2014-T6511 7075-T73511	2091-T8X 8090-T8X 2090-T8	8090-T7E20 8090-T8771	8090-T652 2091-T852	8090-T8X 8091-T8551
High damage tolerance	2219-T39 2024-T3	2324-T39 2124-T351 2024-T351	2219-T3511 2024-T352	...	2091-CPHK 2091-T3 2091-T84 8090-T81	8090-T8E57 8090-T8151 2091-T351 2091-T851	...	2091-T8 8090-T81
Weldability	...	2219	2219	2219	Weldalite 049	Weldalite 049	Weldalite 049	Weldalite 049
Cryogenic properties	...	2519	2519	2519	2090/8090	2090/8090	8090	2090
Superplastic forming	7475	...	...	...	2090/8090 Weldalite 049	...	...	...

Source: Ref 2

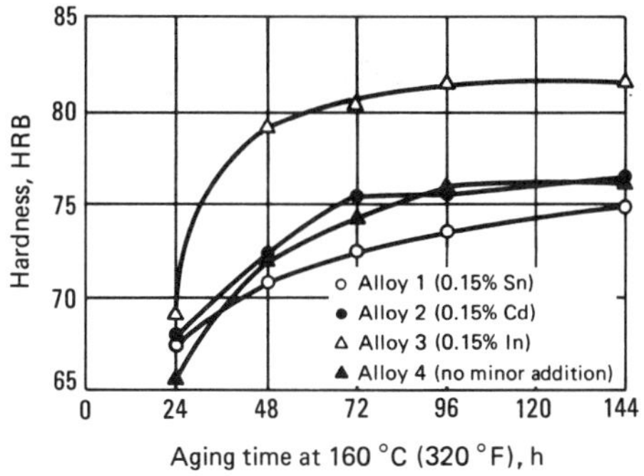

Fig. 1 Effect of minor additions (0.15 wt%) of cadmium, iridium, and tin on the age-hardening response of aluminum-lithium alloy 2090 (2.3 Cu, 2.3 Li, 0.15 Zr)

ing guideline when considering potential applications for aluminum-lithium alloys, even though they are generally unsuitable for direct comparison with or replacement of the more commonly used aluminum aerospace alloys included in Table 1.

Physical Metallurgy

In the development of low-density alloys, the simplest approach to reducing the weight of an alloy is to add elements with low atomic weights as alloying elements. In the case of aluminum alloys, lithium and beryllium are the most effective metallic additions for lowering density. Lithium is the lightest metallic element, and each 1% of lithium (up to the 4.2% Li solubility limit) reduces alloy density by about 3% and increases modulus by about 5% (Ref 3, 4). In addition, lithium in small amounts allows the precipitation strengthening of aluminum when a homogeneous distribution of coherent, spherical δ′ (Al_3Li) precipitates is formed during heat treatment. Beryllium additions, on the other hand, do not give rise to significant precipitation strengthening in aluminum. The combined density-reducing and precipitation-strengthening characteristics of lithium were the main reasons for its choice as the alloying element for the development of low-density aluminum-base alloys.

Like other age-hardened aluminum alloys, aluminum-lithium alloys achieve precipitation strengthening by thermal aging after a solution heat treatment. The precipitate structure is sensitive to a number of processing variables, including, but not limited to, the quenching rate following the solution heat treatment, the degree of cold deformation prior to aging, and the aging temperature and time. Minor alloying elements can also have a significant effect on the aging process (Fig. 1) by changing the interface energy of the precipitate, by increasing the vacancy concentration, and/or by raising the critical temperature for homogeneous precipitation. In addition, heterogeneous precipitation at interfaces and grain boundaries (which occurs in addition to the homogeneous precipitation of the strengthening phase) can have an adverse effect on fracture behavior. Depending on the composition and temperature, the relative size and volume fraction of the different precipitates can be systematically varied.

Al-Li Alloys. The age hardening of aluminum-lithium alloys involves the continuous precipitation of δ′ (Al_3Li) from a supersaturated solid solution. The aluminum and lithium in the δ′ precipitates are positioned at specific locations. The eight shared corner sites are occupied by lithium, and the six shared faces are occupied by aluminum. This gives rise to the aluminum-lithium composition of δ′ precipitates. The geometrical similarity between the lattice of the precipitates and the face-centered cubic lattice of the solid solution facilitates the observed cube/cube orientation dependence (Ref 5-7). The lattice parameters of the precipitate are closely matched to those of the matrix. Consequently, the microstructure of an aluminum-lithium alloy solution heat treated and aged for short times below the δ′ solvus is characterized by a homogeneous distribution of coherent, spherical δ′ precipitates.

Aluminum-lithium-base alloys are microstructurally unique. They differ from most of the aluminum alloys in that once the major strengthening precipitate (δ′) is homogeneously precipitated, it remains coherent even after extensive aging. In addition, extensive aging at high temperatures (>190 °C, or 375 °F) can result in the precipitation of icosahedral grain-boundary precipitates with five-fold symmetry. Although the quasi-crystalline structure and the composition of these grain boundary precipitates are not yet exactly known, it has been suggested that both the precipitates and the precipitate-free zones (PFZs) near the grain boundaries might play a major role in the fracture process.

The low ductility and toughness of binary aluminum-lithium alloys can be traced, at least in part, to the inhomogeneous nature of their slip, resulting from coherent-particle hardening of spherical δ′ precipitates. The presence of equilibrium δ (AlLi) precipitates at grain boundaries can also cause PFZs, which can induce further strain localization and promote intergranular failure. Consequently, for the development of commercial alloys, slip has been homogenized by introducing dispersoids (manganese, zirconium) and semicoherent/incoherent precipitates, such as T_1 (Al_2CuLi), θ′ (Al_2Cu), or S (Al_2LiMg), through copper or magnesium additions. Concurrent developments in thermomechanical processing have optimized aluminum-lithium microstructures for the best combinations of strength and toughness. The resulting material tends to be highly textured where zirconium additions are used to inhibit recrystallization. Texture increases the variability of properties with orientation.

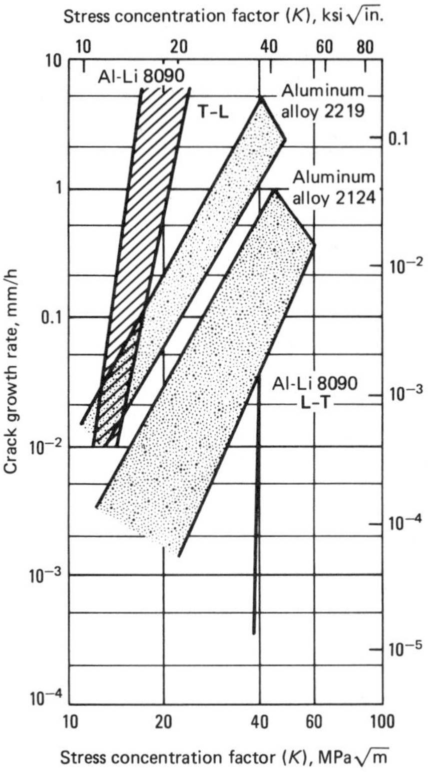

Fig. 2 Comparison of creep crack growth rates for aluminum-lithium alloy extrusions with those for other aluminum alloys. Alloy 8090 contains 2.5% Li, 1.5% Cu, 1.0% Mg, 0.12% Zr, and a balance of aluminum. T-L, crack plane and growth directions parallel to extrusion direction; L-T, crack plane and growth directions perpendicular to extrusion direction. Source: Ref 8

Al-Li-*X* Alloys. As mentioned above, various modifications in alloy chemistry and fabrication techniques have been used in an attempt to improve the ductility and toughness of aluminum-lithium alloys while maintaining a high strength. Copper, magnesium, and zirconium solute additions have been shown to have beneficial effects. Magnesium and copper improve the strength of aluminum-lithium alloys through solid-solution and precipitate strengthening, and they can minimize the formation of PFZs near grain boundaries. Zirconium, which forms the cubic Al_3Zr coherent dispersoid, stabilizes the subgrain structure and suppresses recrystallization.

Compared with traditional high-strength aluminum alloys, Al-Li-*X* alloys show 7 to 12% higher stiffness, generally superior fatigue crack propagation resistance, and improved toughness at cryogenic temperatures. On the negative side, however, they can suffer from poor short-transverse properties, and they have been shown to display significantly accelerated fatigue crack extension rates when cracks are microstructurally small (Ref 1).

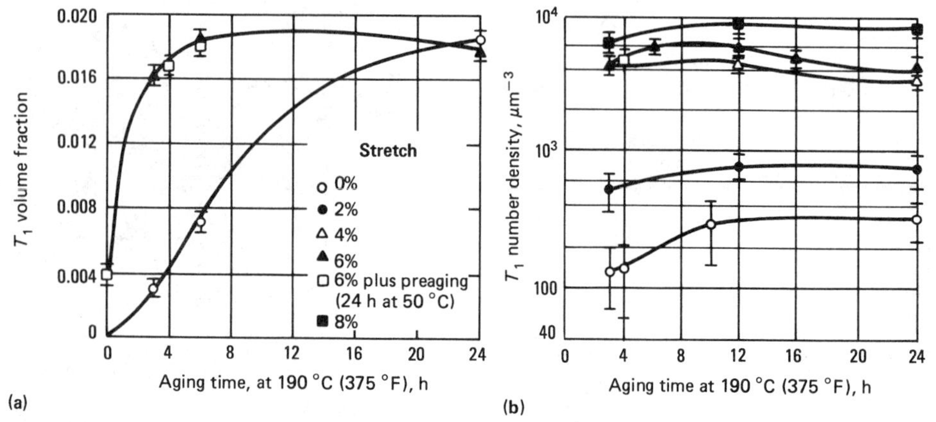

Fig. 3 Effect of prior deformation and aging time on the amount of T_1 precipitates in aluminum-lithium alloy 2090 (2.4% Li, 2.4% Cu, 0.18% Zr, bal aluminum). (a) Volume fraction of T_1. (b) Number density of T_1

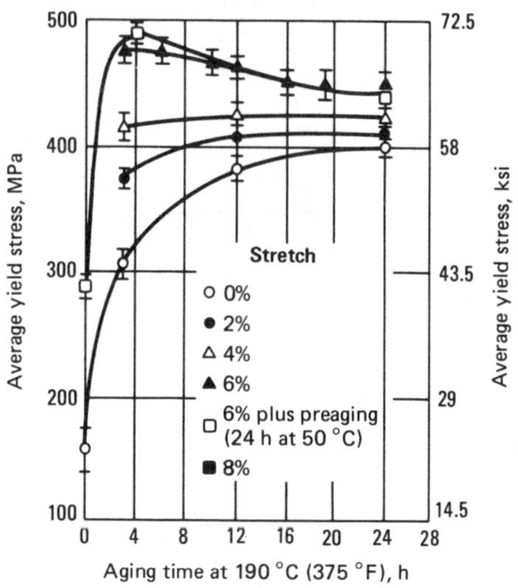

Fig. 4 Average yield stress versus aging time for aluminum-lithium alloy 2090 (2.4% Li, 2.4% Cu, 0.18% Zr, balance aluminum) with various amounts of prior deformation

The creep crack growth rates of Al-Li-*X* alloys may be slower or greater than those of conventional aluminum aerospace alloys, depending on orientation. Figure 2, for example, compares crack growth rates in an extruded Al-Li-Cu-Mg-Zr alloy (Al-Li 8090) with those of other aluminum alloys. Crack growth rates for Al-Li 8090 in the T-L orientation (crack plane and growth direction parallel to the extrusion direction) are on the average much higher than those for aluminum alloys 2219 and 2124. In contrast, the growth rates for Al-Li 8090 in the L-T orientation (crack plane and growth direction parallel to the extrusion direction) are much lower than those for other aluminum alloys.

In terms of δ′ precipitation, the only effect of magnesium appears to be a reduction in the solubility of lithium. The microstructure of an aluminum-magnesium-lithium alloy in the early stages of aging is similar to that of an aluminum-lithium alloy (Ref 9-11). Precipitation in the aluminum-copper-lithium system is more complicated than that in either the aluminum-lithium or aluminum-magnesium-lithium systems.

Thermomechanical Effects. In addition to precipitation hardening, aluminum-lithium alloys derive part of their strength from a controlled grain microstructure generated through hot and cold deformation. Alloying additions of ancillary metallic elements such as manganese, chromium, and zirconium are made to control the grain microstructure during thermomechanical operations. Of these elements, zirconium has improved the combination of strength and fracture toughness in aluminum-lithium-base alloys.

Unlike aluminum-zinc-magnesium-base alloys, aluminum-lithium-base alloys gain increased strength and toughness from deformation prior to aging. This unusual phenomenon has given rise to a number of thermomechanical processing steps for aluminum-lithium alloys aimed at optimizing mechanical properties after artificial aging.

Deformation prior to aging also affects the extent of precipitation strengthening. For example, in aluminum-lithium-copper alloys such as alloy 2090, the T_1 (Al_2CuLi) strengthening precipitates have large coherency strains that are minimized when the precipitates are nucleated on dislocations. In fact, it is believed that dislocations are necessary for the nucleation of the T_1 phase. Similar effects also apply to the S' (Al_2LiMg) strengthening precipitate in aluminum-lithium-copper-magnesium alloys such as 8090.

Because of the effect of dislocations on the precipitation strengthening of these alloys, deformation prior to aging is often used to increase the dislocation density and thus the nucleation sites for the strengthening precipitates. This deformation is normally applied by a tensile stress for sheet, plate, and extruded material, and by compression for forgings. The tensile deformation is usually fairly uniform; however, the compression strain may vary considerably from the surface to the centerline of the product. Therefore, it is important to understand the relationship between the magnitude of strain and the aging response of the alloy.

The effect of deformation on the number, density, and volume fraction of T_1 in an aluminum-lithium-copper-zirconium alloy is shown in Fig. 3. Figure 4, which shows the effect of deformation (stretch) on the yield strength of samples aged for various times at 190 °C (375 °F), correlates very well with Fig. 3 on the effect of stretch on the volume fraction of strengthening precipitates. It is well established that the strength of age-hardened aluminum alloys depends on both the size and volume fraction of the strengthening precipitates. This example shows that the aging response is very sensitive to the degree of deformation prior to artificial aging. An important point to note is that an inhomogeneous distribution of strain can have a pronounced effect on the aging response and thus on the local strength in Al-Li-*X* alloys. This effect of strain on aging response diminishes as the amount of work goes beyond 4%. Consequently, when sections are stretched or compressed to develop a T8 temper, large variations in properties may result if the strain is not homogeneous or at least in excess of some minimum value around 4%.

Table 2 Major producers of aluminum-lithium alloys

Producer	Alloys	Products
Producers with casting facilities		
British Alcan	8090	Sheet, plate, extrusions, forgings
Alcoa	2090, 2091, 8090	Sheet, plate, extrusions, forgings
Pechiney	2091, 8090, CP276	Sheet, plate, extrusions, forgings
Reynolds	Weldalite	Sheet, plate, extrusions, forgings
Producers without casting facilities		
ILM	8090, 2090	Extrusions
Otto Fuchs	8090	Extrusions, forgings
Menziken	8090	Extrusions
VAW	8090	Extrusions
HyDuty	8090	Forgings
Hoogevens	8090	Sheet, plate

Alloy Development

The development of aluminum-lithium alloys has spanned over 65 years. Some of the major milestones in the development of

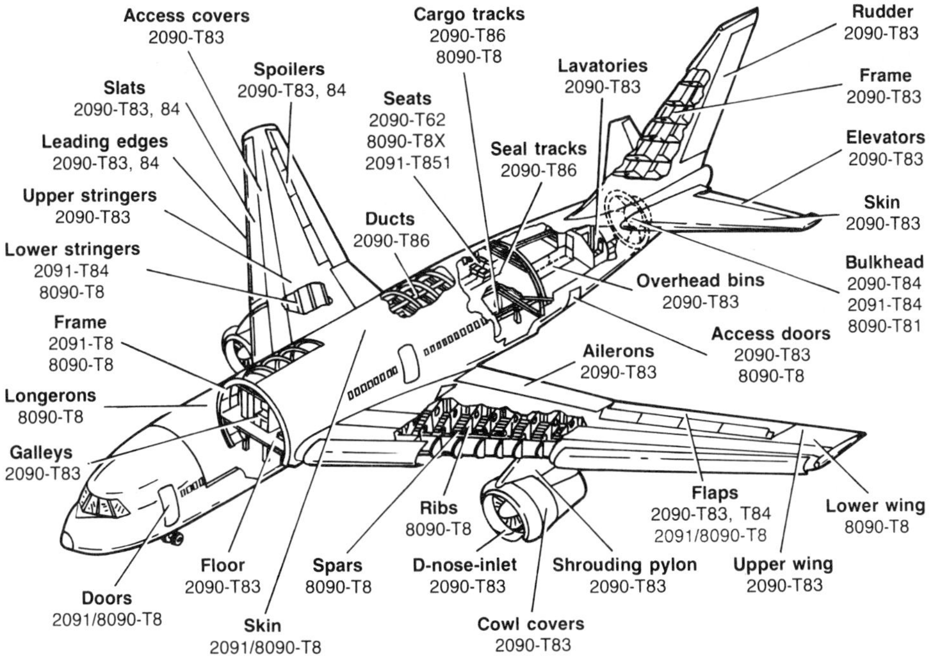

Fig. 5 Use of aluminum-lithium alloys in a commercial aircraft

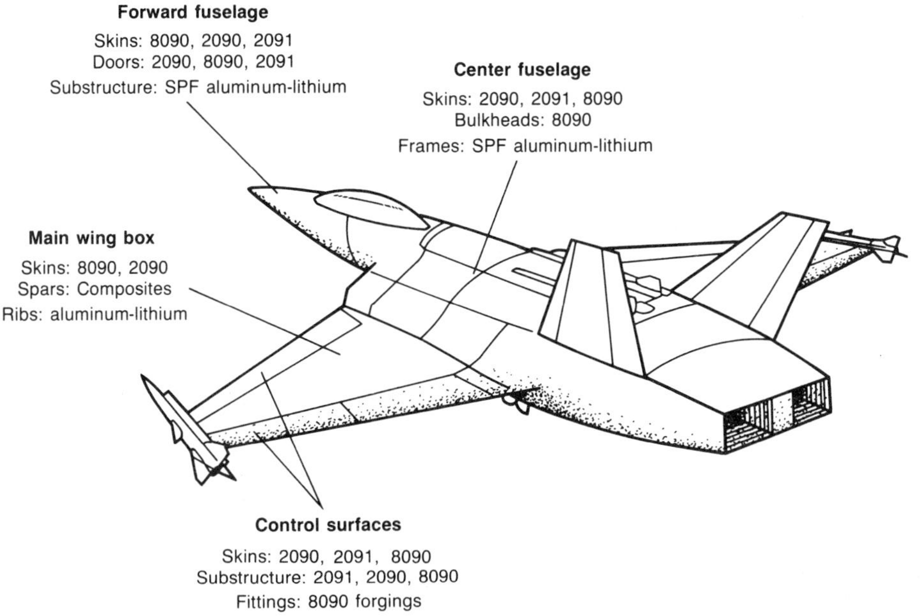

Fig. 6 Use of aluminum-lithium alloys and superplastic-forming (SPF) aluminum-lithium alloys in a fighter aircraft

these alloys include the introduction of the first aluminum-lithium alloy (Scleron, Al-Zn-Cu-Li) by the Germans in the 1920s (Ref 12), the introduction of alloy 2020 (Al-Cu-Li-Cd) in the late 1950s (Ref 13), and the introduction of alloy 1420 (Al-Mg-Li) in the Soviet Union in the mid-1960s (Ref 14). There were no major commercial applications for Scleron; the only applications for 2020 were the wings and horizontal stabilizers for the RA5C Vigilante aircraft (Ref 15). Alloy 1420, while still being used, has not gained widespread acceptance outside of the Soviet Union. There were a number of reasons for the limited applications for each of these early aluminum-lithium alloys, including shortcomings in properties (primarily fracture toughness) and continuing improvements to existing aluminum alloys.

The development of aluminum-lithium alloys continued into the 1970s; these alloys were intended for use as low-density materials for reducing the weight of aircraft. Studies have shown that reducing the density of structural materials is the most influential factor in reducing aircraft weight (Ref 16) and is more important than increasing the strength, modulus, or damage-tolerant properties of these materials. The goal of aluminum-lithium development work was to introduce ingot metallurgy alloys that could be fabricated on the existing equipment of aluminum producers and then used by airframe manufacturers in the same way the conventional aluminum alloys were used. In addition to the lower material costs of aluminum-lithium alloys relative to composites, it was postulated that these alloys would require no basic changes in airframe production methods. A direct-substitution scenario implied that property equivalence would translate directly into weight savings by the replacement of existing aluminum parts with identical aluminum-lithium parts. This direct substitution has not come to pass; however, the properties of aluminum-lithium alloys have been found to be very attractive in a number of applications, and the alloys that have been developed have proved to be very usable in these cases.

Current Al-Li-*X* Alloys and Their Applications. In the early 1980s, aluminum manufacturers were working on aluminum-lithium alloys that they planned to introduce as direct substitutes for aluminum alloys. This work led to the introduction of commercial alloys 8090, 2090, and 2091 in the mid-1980s, and the introduction of Weldalite 049 and CP276 toward the end of the decade. Properties of these alloys are discussed in the section "Commercial Aluminum-Lithium Alloys" in this article.

At the present time, a number of producers (Table 2) are making aluminum-lithium alloys that are registered and well characterized. However, quite a bit of confusion still exists concerning temper designations, alloys, and supply, and any list of tempers, alloys, and producers is subject to change. Therefore, it is suggested that the reader contact the Aluminum Association and the Aerospace Materials specifications for information about current alloy registration and specification properties.

Although the direct replacement of conventional aluminum aerospace alloys with aluminum-lithium alloys has not been possible, aluminum-lithium alloys have gained acceptance in a number of aerospace applications for both primary and secondary structures. Table 1 provides a guideline to start with when considering aluminum-lithium alloys for an existing or new design. However, because the properties of aluminum-lithium alloys are unique and because these alloys cannot be directly substituted for existing aluminum aerospace alloys, most applications for aluminum-lithium alloys are found in new programs. These limitations have slowed the market development of these products. In addition, the cost of Al-Li alloys is typically three to five times that of conventional aluminum alloys because of the specialized equipment re-

Table 3 Aluminum-lithium alloy compositions

Element	Composition, wt% 2090(a)	2091(a)	8090(a)	Weldalite 049(b)
Silicon	0.10	0.20	0.20	...
Iron	0.12	0.30	0.30	...
Copper	2.4–3.0	1.8–2.5	1.0–1.6	5.4
Manganese	0.05	0.10	0.10	...
Magnesium	0.25	1.1–1.9	0.6–1.3	0.4
Chromium	0.05	0.10	0.10	...
Zinc	0.10	0.25	0.25	...
Lithium	1.9–2.6	1.7–2.3	2.2–2.7	1.3
Zirconium	0.08–0.15	0.04–0.16	0.04–0.16	0.14
Titanium	0.15	0.10	0.10	...
Other, each	0.05	0.05	0.05	(Ag 0.4)
Other, total	0.15	0.15	0.15	...
Aluminum	bal	bal	bal	bal

(a) Registered limits. (b) Nominal composition

quired for processing and the high cost of lithium; therefore, application of these alloys has been limited to programs where weight reduction is an overriding concern.

Commercial applications in production include aircraft parts such as leading and trailing edges, access covers, seat tracks, and wing skins. Figure 5 shows present and potential uses of aluminum-lithium materials on a generic commercial transport, and thereby serves as a summary of alloys and tempers that are in production or under evaluation. Figure 5 is not intended as a recommendation for the use of aluminum-lithium for specific applications, nor is it a complete listing of possible applications for these alloys; however, it does provide some indication of the literally thousands of pounds of aluminum-lithium parts that can be designed into a new or existing aircraft. In addition to the commercial transport uses for these alloys, helicopter, commuter, and business aviations applications are in production or under study (Ref 17-24). The viability of using aluminum-lithium for parts can be evaluated on the basis of primary design criteria, such as requirements for strength, stiffness, minimum gage, damage tolerance, and corrosion resistance; also, product form classifications (Table 1) can be consulted to identify suitable aluminum-lithium materials to investigate for use in a particular application.

Military Applications. In many cases, military applications are treated as classified information; for the most part, however, they can be generically viewed in much the same way as commercial applications. Certain types of military aircraft parts (Fig. 6) are likely candidates for the use of aluminum-lithium alloys; in particular, minimum-gage and stiffness-dominated designs are good candidates for the replacement of conventional aluminum alloys with existing aluminum-lithium alloys. Steady progress has been made in using superplastic-formed aluminum-lithium sheet parts (Ref 25), conventionally formed 2090, 2091, and 8090 sheet (Ref 22, 26-32) for structures and wing skins, and plate parts. Representative parts include access covers and wing skins. The lack of a material with good short-transverse stress-corrosion cracking resistance and good short-transverse ductility and toughness has kept aluminum-lithium from capturing a major metal bulkhead application, but such applications should become possible with continuing progress in research and development.

Space Applications. Of all the benefits offered by the use of aluminum-lithium alloys, weight savings is the most critical in space applications. Applications under production and evaluation include chemically milled aluminum-lithium alloys for integrally stiffened primary structures and tankage, as well as products for sheet and stringer construction of shrouds, formings, and adapters. Aluminum-lithium alloy 2090-T81 is a candidate alloy for the cryogenic tankage of booster systems. The cryogenic properties of 2090-T81 have been extensively studied and are summarized in Ref 33. Early test data revealed that there may be an oxygen compatibility problem with aluminum-lithium alloys. However, the latest testing by the National Institute of Science and Technology (Ref 34) has shown that aluminum-lithium alloys are as compatible as 2219 (the presently used cryogenic tankage alloy) with liquid-oxygen environments.

Manufacturing. The rolling, extruding, and forging operations for aluminum-lithium alloys are comparable to those for conventional aluminum alloys. However, there are a number of differences in other areas, particularly ingot production and recycling, that require attention.

Explosion Potential With Water. The explosion potential of molten aluminum-lithium alloys in contact with water is significantly greater than that of conventional molten aluminum alloys. A number of variables affect the explosion potential, including the lithium content of the alloy, the depth and containment of the water, the metal temperature, and the size (diameter) and velocity of the molten metal stream being introduced to the water. Any operation that generates molten aluminum-lithium poses a potential explosion hazard in the presence of water; the melting range for aluminum-lithium alloys is between 500 and 600 °C (930 and 1100 °F). Safe casting practices have been developed by primary producers, but caution is still necessary (Ref 35-37).

Fabricating Ingot Quality. The control of ingot quality—in particular, the control of inclusions, impurities, hydrogen content, grain size, and composition—is critical to attaining good properties following rolling, extrusion, or forging and heat treatment. The reactivity of lithium with air and refractories has necessitated the development of specialized equipment and processes to melt, remelt, or cast aluminum-lithium ingots of sufficient size and quality. The need for specialized equipment to meet quality requirements has contributed to the higher cost of aluminum-lithium alloys (Ref 2).

Explosion Potential With Salt Bath. Like other aluminum alloys, aluminum-lithium alloys may cause severe exothermic reactions or explosions if heat treated in salt baths under conditions in which melting or incipient melting occurs. Limited testing and user experience indicate that aluminum-lithium alloys can be safely handled in salt baths so long as melting is precluded and normal precautions are observed.

Table 4 Plane-strain fracture toughness (K_{IC}) of Weldalite 049 extruded bar

Temperature °C	°F	Temper designation	Orientation(a)	K_{Ic} MPa$\sqrt{m}$	ksi$\sqrt{in.}$	Tensile strength MPa	ksi	Yield strength MPa	ksi
21	70	T3	L-T	36.9	33.6	530	77	405	59
21	70	T3	T-L	30.9	28.1	485	70	350	51
21	70	T3	T-L	29.8	27.1	485	70	350	51
21	70	T6E4	L-T	30	27.3	650	94	605	88
21	70	T6E4	L-T	29	26.4	650	94	605	88
−195	−320	T3	T-L	31.8	28.9	615	89	455	66
−195	−320	T3	T-L	30.9	28.1	615	89	455	66

(a) L-T, crack plane perpendicular to extrusion direction; T-L, crack plane parallel to extrusion direction

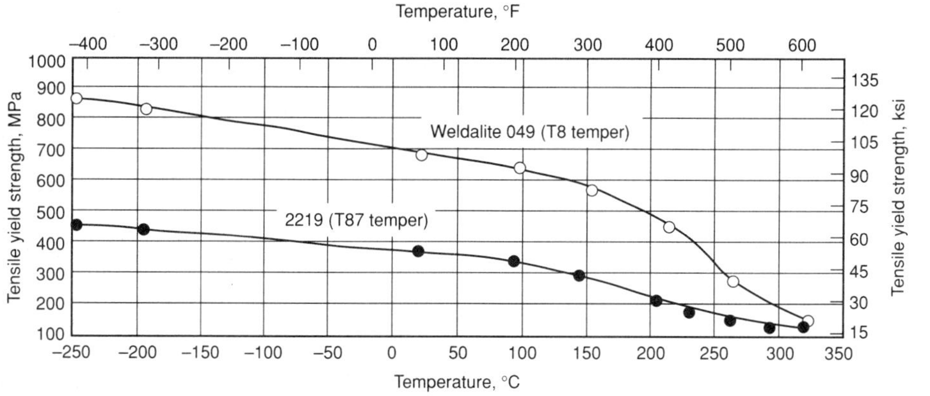

Fig. 7 Yield strengths of two aluminum-lithium candidate alloys for cryogenic tankage applications. Strain rate, 4×10^{-4}/s with a 0.5-h hold at temperature

Table 5 Mean longitudinal tensile properties of Weldalite 049 in various tempers and product forms

Data for 2219-T8 provided for comparison

Description and/or temper	Yield strength MPa	Yield strength ksi	Ultimate tensile strength MPa	Ultimate tensile strength ksi	Elongation in 50 mm (2 in.), %
Extruded products(a)					
T3	407	59.0	529	76.7	16.6
T4	438	63.5	591	85.7	15.7
Reversion	331	48.0	484	70.2	24.2
T6	680	98.7	720	104.4	3.7
T8	692	100.4	713	103.5	5.3
2219-T81					
Minimum	303	44.0	420	61.0	6.0
Typical	352	51.0	455	66.0	10.0
Rolled products(b)					
T8, 5 mm (0.2 in.) thick(c)	643	93.3	664	96.3	5.7
T6, 5 mm (0.2 in.) thick(c)	625	90.7	660	95.8	5.2
T6A, 5 (0.2 in.) mm thick	642	93.1	665	96.5	5.2
T6B, 5 (0.2 in.) mm thick	662	96.0	686	99.5	3.7
T6A, 6.35 mm (0.25 in.) thick	671	97.3	700	101.6	5.3
T6B, 6.35 mm (0.25 in.) thick	668	96.7	692	100.3	5.6
T6A, 9.5 mm (0.38 in.) thick	650	94.3	672	97.4	5.1
Forging(d)					
T4, naturally aged for 1000 h	392	56.9	559	81.1	18.5
Slightly underaged (170 °C, or 340 °F, for 20 h)	658	95.5	702	101.8	5.0

(a) Most are 100 × 9.5 mm (4 × 0.375 in.) extruded plate. (b) Rolled from 180 kg (400 lb) pilot commercial ingots. (c) For tankage. (d) Commercial hook forging

Explosion Potential (Fires). Like the dust of any aluminum alloy, finely divided aluminum-lithium dust in sufficient concentration is explosive in the presence of an ignition source. The finer the dust, the greater the chance of ignition and the more severe the explosion. Aluminum-lithium fires, like those involving conventional aluminum alloys, can generate hydrogen gas in the presence of water, and this gas may be explosive in the presence of an ignition source. Hydrogen generation in confined spaces is of particular concern.

Heat Treating. During the heating of aluminum-lithium alloys to elevated temperatures (>260 °C, or 500 °F), surface oxidation occurs, and lithium oxide and hydroxide are formed. The formation of these compounds is of concern because of the possible skin, eye, and upper respiratory tract irritation or damage they can cause. Lithium oxide/hydroxide is a strong alkali and is potentially corrosive to any tissue with which it comes in contact. Inhalation of the dust may result in upper respiratory tract irritation. Contact with the skin or eyes may cause irritation and, with greater exposure, possible burns.

The American Industrial Hygiene Association has recommended a Workplace Environmental Exposure Level (WEEL) for lithium oxide and hydroxide particulate of 1 milligram per cubic meter (mg/m^3) of air for a maximum exposure time of 1 min. The potential for exceeding the recommended ceiling exposure limit is most likely to exist when these alloys are heated to temperatures greater than ~260 °C (~500 °F) to form significant amounts of lithium oxide/hydroxide on metallic surfaces and when this surface material becomes airborne during operations such as dry machining (Ref 36).

Recycling. The recovery of finished aluminum-lithium products from semifabrication facilities varies with the product form, but it is generally somewhat lower than that of conventional aluminum alloy products. The buy-to-fly ratios of finished aircraft parts in the fabrication processes of aircraft manufacturers also generate large quantities of scrap. Some of the scrap produced by these processes is heavy, relatively easily segregated, and can be taken back for direct recycling into aluminum-lithium ingots. However, the bulk is in the form of machining swarf, which is unsuitable for direct recycling because it is contaminated with other aluminum alloys and other aircraft materials such as stainless steel and titanium.

For conventional aircraft aluminum alloys of the 2*xxx* and 7*xxx* series, the recycling of process scrap from both metal suppliers and aircraft manufacturers is a well-established worldwide industry. For the most part, primary aluminum suppliers internally recycle their own scrap back into aircraft products. Their customers sell off their scrap to secondary aluminum smelters, who convert it into aluminum-silicon foundry alloys.

The introduction of aluminum-lithium alloys will not eliminate conventional recycling activities. However, additional processing steps will have to be introduced, and existing commercial arrangements will need to be modified. The need for these changes stems from three attributes of lithium metal: First, its high price; second, its high reactivity; and, third, its absence from the specifications of all registered alloys other than the currently rather limited number of alloys based on aluminum-copper-lithium (such as 2090) and aluminum-lithium-copper-magnesium (8090 and 2091).

Secondary recyclers must be aware of the problems that can occur if they remelt scraps that contain aluminum-lithium alloys. The major problem areas are:

- Metallurgical property changes that can affect recovered alloys
- Adverse effects on refractories
- Adverse effects on metal cleanliness
- Increased melt losses
- Safety and industrial hygiene
- Environmental issues
- Higher costs

Each of these problem areas has specific technological definitions. Also, many of them overlap and have interactive reactions. Potential solutions to these problems have been reviewed many times in the literature and have been implemented for the numerous existing applications. However, the scrap loop should be taken into consideration as part of the review for deciding

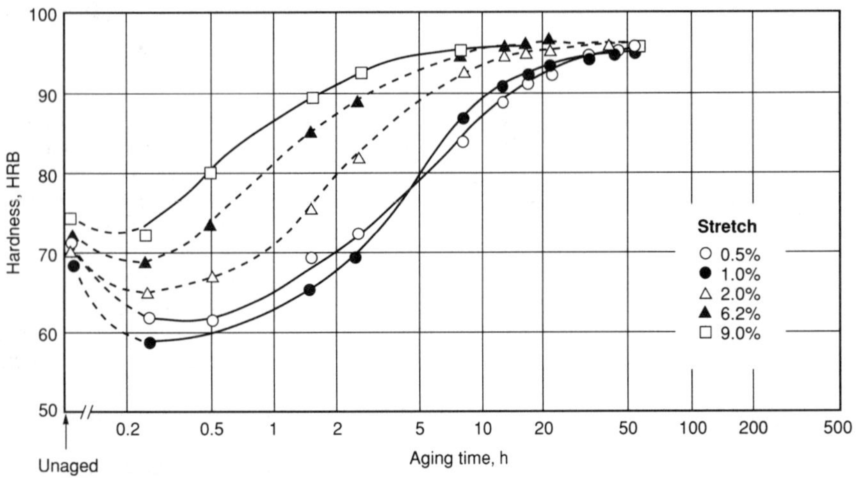

Fig. 8 Aging response of Weldalite 049 with various amounts of deformation prior to aging. Approximate aging temperature, 170 °C (340 °F)

whether or not to use aluminum-lithium for a specific program (Ref 36, 38-41).

Commercial Aluminum-Lithium Alloys

Development of commercially available aluminum-lithium-base alloys was started by adding lithium to aluminum-copper, aluminum-magnesium, and aluminum-copper-magnesium alloys. These alloys were chosen to superimpose the precipitation-hardening characteristics of aluminum-copper-, aluminum-copper-magnesium-, and aluminum-magnesium-base precipitates to the hardening of lithium-containing precipitates. Proceeding in this manner, alloys 2020 (Al-Cu-Li-Cd), 01429 (Al-Mg-Li), 2090 (Al-Cu-Li), and 2091 and 8090 (Al-Cu-Mg-Li) evolved. Besides these registered alloys, other commercial aluminum-lithium alloys include Weldalite 049 and CP276.

This section focuses on alloys 2090, 2091, 8090, and Weldalite 049. Compositions of these alloys are listed in Table 3. General application characteristics and producers are listed in Tables 1 and 2, respectively.

Weldalite 049

The trade name Weldalite refers to alloys (Ref 42-46) designed and developed at Martin Marietta Laboratories for welded aerospace applications. The leading alloy, Weldalite 049, was designed to replace mainstay alloys 2219 and 2014 for launch systems applications. In these applications, Weldalite is a candidate for propellant tankage, which constitutes the bulk of the dry weight of space launch systems. These tanks are most often fabricated by welding because the propellants are usually contained under pressure. Liquid hydrogen and liquid oxygen are the fuel/oxidizer combination of choice; therefore, cryogenic properties are also an important factor in the overall compatibility of these alloys for launch systems applications.

The effect of cryogenic temperatures on the strengths of Weldalite 049 and one other aluminum-lithium alloy is shown in Fig. 7. The toughness of Weldalite 049 remains relatively constant at cryogenic temperatures (Table 4); other aluminum-lithium alloys, on the other hand, show a marked increase in in-plane toughness at cryogenic temperatures. Weldalite has a density of 2.7 g/cm^3 (0.098 lb/in.^{3}) and an elastic modulus of 76 GPa (11 × 10^6 psi). Other physical properties have not yet been determined.

Weldalite 049 shows high strength in a variety of products and tempers (Table 5). Its natural aging response is extremely strong with cold work (temper T3), and even stronger without cold work (T4); in fact, it has a stronger natural aging response than that of any other known aluminum alloy. Weldalite 049 undergoes reversion during the early stages of artificial aging, and its ductility increases significantly up to 24%. Tensile strengths of 700 MPa (100 ksi) have been attained in both T6 and T8 tempers produced in the laboratory. As shown in Fig. 8, specimens of Weldalite 049 in the peak-aged condition all have essentially the same level of hardness despite varying degrees (from 0.5 to 9%) of cold work prior to aging; the yield strength of Weldalite 049 is also relatively unaffected by prior cold work.

The ability of Weldalite 049 to attain high strength without cold work is particularly beneficial for forgings, where the uniform introduction of cold work is often impractical (Table 5). Weldalite 049 small-scale forgings and commercial Boeing hook forgings have displayed tensile strengths of greater than 700 MPa (100 ksi).

Weldalite 049 has very good weldability; for example, it displays no discernable hot cracking in highly restrained weldments made by gas tungsten arc, gas metal arc, and variable polarity plasma arc (VPPA) welding. Extremely high weldment strengths have been reported using conventional 2319 filler, and even higher weldment strengths have been obtained with the use

Table 6 Mean tensile properties of Weldalite 049, 2090, and 2219 weldments with conventional and Weldalite filler

Base metal/filler	Temperature(a)	Thickness, mm	Thickness, in.	Postweld temper	Weld position	Ultimate tensile strength, MPa	Ultimate tensile strength, ksi	Yield strength, MPa	Yield strength, ksi	Elongation, %, in 25 mm (1 in.)	Elongation, %, in 50 mm (2 in.)
VPPA square butt weldments(b)											
2219/2319	RT	9.5	0.375	As-welded	60° horizontal	273	39.6	140	20.4	7.9	4.6
2219/049	RT	9.5	0.375	As-welded	60° horizontal	283	41.1	154	22.3	7.1	4.7
2219/049	RT	5.8	0.230	As-welded	60° horizontal	325	47.1	161	23.4	9.0	5.0
2090/2319	RT	13	0.500	As-welded	Vertical	252	36.5	156	22.7	8.6	4.7
2090/049	RT	6.5	0.255	As-welded	60° horizontal	285	41.3	147	21.3	7.1	3.8
049/2319	RT	9.5	0.375	As-welded	Vertical	274	39.8	248	36.0	1.5	1.0
049/049	RT	9.5	0.375	As-welded	60° horizontal	315	45.7	249	36.1	1.5	1.5
049/049	RT	9.5	0.375	Naturally aged for 800 h	60° horizontal	372	54.0	290	42.1	3.0	...
VPPA weldments of extruded plate(c)											
049/049	175 °C (350 °F)	9.5	0.375	As-welded	...	287	41.6	188	27.3	5.4	...
049/049	RT	9.5	0.375	As-welded	...	372	54.0	290	42.0	3.0	...
049/049	−195 °C (−320 °F)	9.5	0.375	As-welded	...	413	59.9	360	52.2	1.9	...
049/049	−253 °C (−423 °F)	9.5	0.375	As-welded	...	505	73.2	427	61.9	1.7	...

(a) RT, room temperature. (b) All fractures occurred in the heat-affected zone. (c) 100 × 9.5 mm (4 × 0.375 in.) plate. Source: Martin Marietta Manned Space Systems

Table 7 Tempers and corresponding products forms for alloy 2090

Temper	Characteristics	Product forms
O	Annealed, lowest strength, maximum formability	Sheet, plate
T31(a)	Good formability, will approach T83 or T84 properties after aging by user	Sheet, extrusions
T3(a)	Moderate formability, can be aged to T83 and T84 properties by customer	Sheet, extrusions
T86	Strengths similar to those of 7075-T6511	Extrusions
T83(a)	Strengths similar to those of 7075-T6	Sheet
T81(a)	Strengths similar to those of 7075-T651	Plate
T84(a)	Strength and toughness similar to those of 7075-T76	Sheet, plate
T6(a)	Solution heat treated and aged by user	Sheet, plate

(a) Registered with Aluminum Association

Table 8 Typical physical properties of selected aluminum-lithium alloys

Property	2090	2091	8090
Density, g/cm³ (lb/in.³)	2.59 (0.094)	2.58 (0.093)	2.55 (0.092)
Melting range, °C (°F)	560–650 (1040–1200)	560–670 (1040–1240)	600–655 (1110–1210)
Electrical conductivity, %IACS	17–19	17–19	17–19
Thermal conductivity at 25 °C (77 °F), W/m · K (Btu · in./ft² · °F · h)	84–92.3 (580–640)	84 (580)	93.5 (648)
Specific heat at 100 °C (212 °F), J/kg · K (cal/g · °C)	1203 (0.2875)	860 (0.205)	930 (0.22)
Average coefficient of thermal expansion from 20 to 100 °C (68 to 212 °F), μm/m · °C (μin./in. · °F)	23.6 (13.1)	23.9 (13.3)	21.4 (11.9)
Solution potential, mV(a)	−740	−745	−742
Elastic modulus, GPa (10⁶ psi)	76 (11.0)	75 (10.9)	77 (11.2)
Poisson's ratio	0.34	...	...

(a) Measured per ASTM G 60 using a saturated calomel electrode

of a proprietary Weldalite filler (Table 6). As shown by the data of Table 6, a mean VPPA weldment strength of 370 MPa (54 ksi) has been obtained by welding Weldalite 049 with 049 filler. High strengths (310 MPa, or 45 ksi, ultimate tensile strength) have also been attained with tungsten inert-gas welds.

Weldalite 049 has been used in the fabrication of a subscale prototype cryogenic tank. Because of this successful fabrication and the combination of properties offered by Weldalite 049 (such as ultrahigh strength at room and cryogenic temperatures, and good weldability), the alloy has been proposed as a baseline structural material for the now-defunct Advanced Launch System. In addition, one of the Weldalite alloys is under consideration for use in the National Aerospace Plane and in the Titan family of missiles.

Alloy 2090

Alloy 2090 was developed to be a high-strength alloy with 8% lower density and 10% higher elastic modulus than 7075-T6, a major high-strength alloy used in current aircraft structures (Ref 19, 47). Alloy 2090 is intended for use in applications where high- and medium-strength sheet, plate, and extrusions are used. Its excellent weldability and cryogenic properties make it suitable for cryogenic tankage structures. In addition, 2090 is suited for superplastic-forming (SPF) applications.

The chemical composition of alloy 2090 (Table 3) was registered with the Aluminum Association in 1984. A variety of tempers (Table 7) are being developed to offer useful combinations of strength, toughness, corrosion resistance, damage tolerance, and fabricability. Physical properties of 2090 are given in Table 8. The microstructure of 2090 products (Fig. 9) is controlled and is primarily unrecrystallized.

Strength and Toughness. Because alloy 2090 and its tempers are relatively new and in different phases of registration and char-

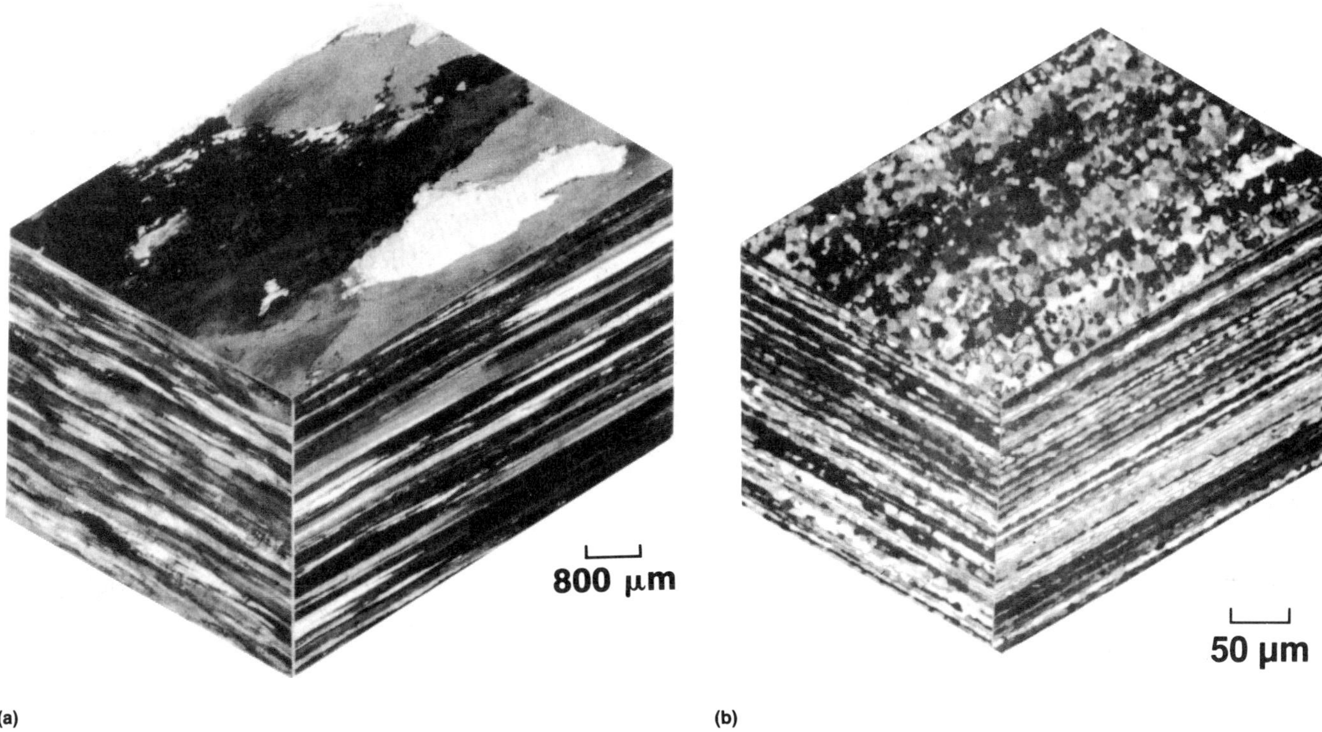

Fig. 9 Unrecrystallized microstructures of alloy 2090. (a) 45 mm (1.75 in.) thick 2090 plate. (b) 1.6 mm (0.063 in.) thick 2090 sheet

Table 9 Tentative mechanical property limits of 2090 products

Typical values are given in parentheses. Data for alloy 7075-T6 are included for comparison.

2090 temper	Thickness, mm	Thickness, in.	Specification	Direction(a)	Tensile properties: Ultimate tensile strength, MPa	Ultimate tensile strength, ksi	Yield strength, MPa	Yield strength, ksi	Elongation in 50 min (2 in.), %	Toughness: Direction(b) and (K_c) or (K_{Ic})(c)	K_{Ic} or K_c, MPa $\sqrt{m}$	K_{Ic} or K_c, ksi $\sqrt{in.}$
Sheet												
T83	0.8–3.175	0.032–0.125	AMS 4351	L	530 (550)	77 (80)	517 (517)	75 (75)	3 (6)	L-T (K_c)	(44)(d)	(40)(d)
				LT	505	73	503	73	5	...	...	...
				45°	440	64	440	64	...	...	...	...
T83	3.2–6.32	0.126–0.249	AMS 4351	L	483	70	483	70	4	...	...	...
				LT	455	66	455	66	5	...	...	...
				45°	385	56	385	56	...	...	...	...
T84	0.8–6.32	0.032–0.249	AMS Draft D89	L	495 (525)	72 (76)	455 (470)	66 (68)	3 (5)	L-T (K_c)	49 (71)(d)	45 (65)(d)
				LT	475	69	415	60	5	T-L (K_c)	49(d)	45(d)
				45°	427	62	345	50	7			
T3(e)	...	...	(f)	LT	317 min	46 min	214 min	31 min	6 min	...	...	...
O	...	...	(f)	LT	213 max	31 max	193 max	28 max	11 min	...	...	...
7075-T6	...	...	...	L	(570)	(83)	(517)	(75)	(11)	L-T (K_c)	(71)(d)	(65)(d)
Extrusions												
T86(g)	0.0–3.15(h)	0.000–0.124(h)	AMS Draft D88BE	L	517	75	470	68	4	...	...	...
	3.175–6.32(h)	0.125–0.249(h)		L	545	79	510	74	4	...	...	...
	6.35–12.65(h)	0.250–0.499(h)		L	550	80	517	75	5	...	...	...
				LT	525	76	483	70	...	...	...	...
Plate												
7075-T6	...	...	...	L	(565)	(82)	(510)	(74)	(11)	L-T (K_{Ic})	(27)	(25)
T81	13–38	0.50–1.50	AMS 4346	L	517 (550)	75 (80)	483 (517)	70 (75)	4 (8)	L-T (K_{Ic})	≥27 (71)	≥25 (65)
				LT	517	75	470	68	3	L-T (K_{Ic})	≥22	≥20

(a) L, longitudinal; LT, long transverse. (b) L-T, crack plane and direction perpendicular to the principal direction of metalworking (rolling or extrusion); T-L, crack plane and direction parallel to the direction of metalworking. (c) K_c, plane-stress fracture toughness; K_{Ic}, plane-strain fracture toughness. (d) Toughness limits based on limited data and typical values (in parentheses) for 405 × 1120 mm (16 × 44 in.) sheet panel. (e) The T3 temper can be aged to the T83 or T84 temper. (f) No end user specification. (g) Temper registration request made to the Aluminum Association. (h) Nominal diameter or least thickness (bars, rod, wire, shapes) or nominal wall thickness (tube)

acterization, data may be incomplete for some forms; however, current capabilities are given with the specifications as they now exist in Table 9. Properties of selected tempers of 2090 at various temperatures are shown in Fig. 10 to 12. Their behavior is similar to that of other 2*xxx* and 7*xxx* alloys, although the 2090 tempers show a higher strength at elevated temperatures than other alloys. Changes in strength and toughness at cryogenic temperatures are more pronounced in 2090 than in conventional aluminum alloys; alloy 2090 has a substantially higher toughness at cryogenic temperatures (Fig. 13).

Design Considerations. In general, the engineering characteristics of aluminum-lithium alloys are similar to those of the current 2*xxx* and 7*xxx* high-strength alloys used by the aerospace industry. However, some material features of the 2090 products vary somewhat from those of the conventional aluminum alloys and should be considered during the design and material selection phase. These distinct characteristics of 2090 include:

- An in-plane anisotropy of tensile properties that is higher than in conventional alloys. In 2090, more importance is placed on 45° and shear properties. These properties of 2090 are well characterized and can easily be checked for individual design application
- An elevated-temperature exposure for the peak-aged tempers (T86, T81, and T83) that shows good stability within 10%

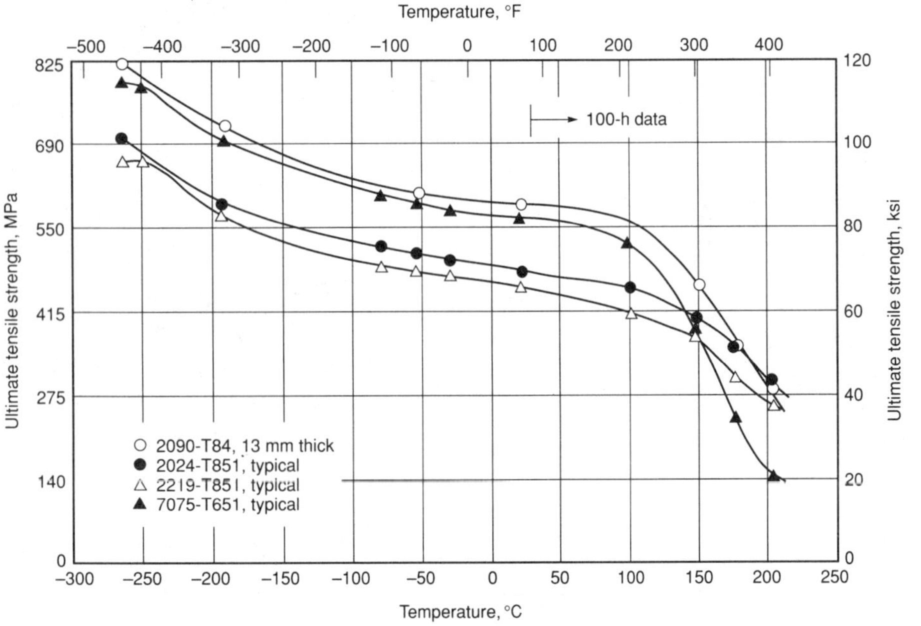

Fig. 10 Longitudinal tensile strength versus temperature for aluminum-lithium alloy 2090-T84 and various other aluminum plate alloys

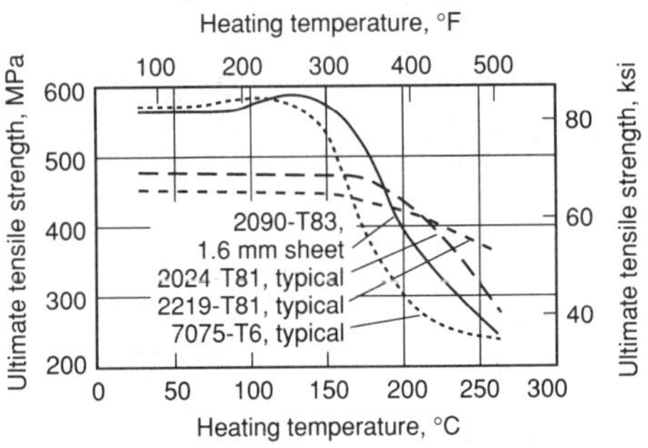

Fig. 11 Room-temperature tensile strength of 2090-T83 sheet and various aluminum sheet alloys after heating (for 100 h)

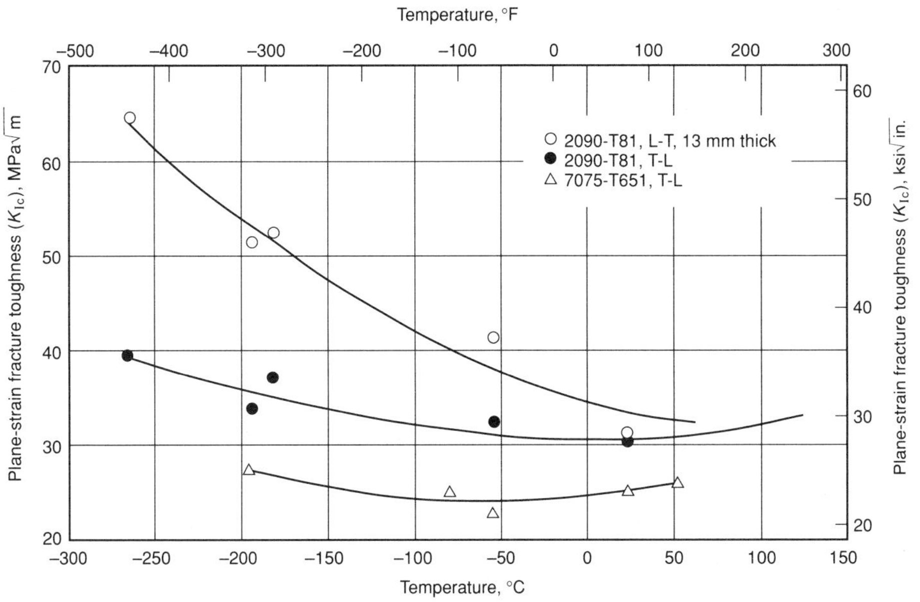

Fig. 12 Effect of temperature on the plane-strain fracture toughness (K_{Ic}) of 2090-T81 and 7075-T651 plate alloys. Exposure of ½ h at temperature. L-T, crack plane and direction perpendicular to the principal direction of rolling; T-L, crack plane and direction parallel to the principal direction of rolling

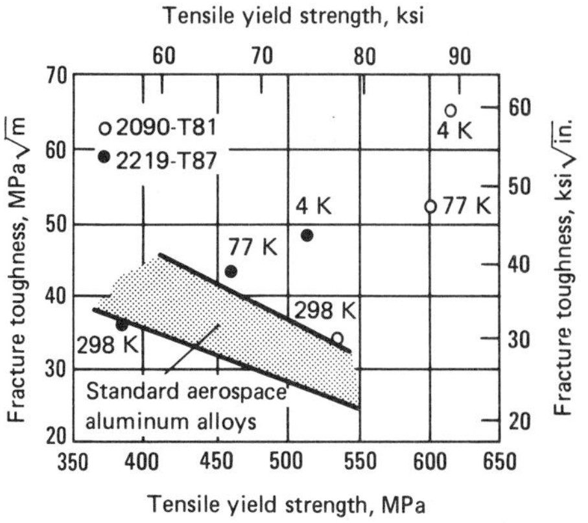

Fig. 13 Yield strength/fracture toughness combination as a function of temperature for 2090-T81 alloy. The shaded region represents the range of strength/toughness combinations for typical aerospace aluminum alloys. Data for alloy 2219-T87 as a function of temperature are shown for comparison. Source: Ref 33

of original properties. However, for the underaged temper (T84), there will be significant additional aging

- Excellent fatigue crack growth behavior. However, the large advantage 2090 has over conventional alloys is reduced with high overloads or compression-dominated regimes
- The need for cold work to achieve optimum properties. In this characteristic, 2090 is similar to 2219 and 2024
- Shape-dependent behavior for extrusions with very high strengths. Extrusions with a low-aspect-ratio section may exhibit shape-dependent behavior when only a small difference exists between the yield and ultimate tensile strengths. This may be a result of a change in the strain-hardening coefficient in these sections. The ductility may remain at acceptable levels

Corrosion. Alloy 2090 sheet and plate, and 2090-T86 extrusions have demonstrated excellent resistance to exfoliation corrosion (Ref 48) in extensive seacoast exposure tests (Table 10). The resistance of these alloys and tempers is superior to that of 7075-T6, which, in some product forms, can suffer very severe exfoliation during a two-year seacoast exposure.

The stress-corrosion cracking (SCC) resistance of 2090 is strongly influenced by artificial aging. Tempers that are underaged, such as T84, may be more susceptible to SCC than the near-peak-aged T83, T81, and T86 tempers. This is particularly true for products subjected to sustained tensile loading in the short-transverse direction. Peak-aged tempers are resistant to SCC in the short-transverse direction with up to 170 MPa (25 ksi) applied stress in alternate immersion and atmospheric exposure (Table 11).

Sheet, thin plate (<25 mm, or 1 in., thick) and thinner extrusions have been extensively tested to evaluate stress-corrosion resistance in the long-transverse direction. The grain structure of wrought unrecrystallized aluminum products makes the long-transverse direction much less sensitive to factors that cause SCC than the short-transverse direction. Both alternate immersion and seacoast exposure tests indicate that 2090-T83, -T84, and -T86 have excellent long-transverse SCC resistance. All three tempers survive alternate immersion testing at stress levels equivalent to 75% of tensile yield strength specification minimums.

Table 12 lists the corrosion potentials of several tempers of 2090 along with those of several other high-strength alloys. The addition of up to between 2 and 2.5% Li to an aluminum alloy does not significantly affect the electrochemical properties of the alloy. For 2090, the amount of copper in solid solution controls the corrosion potential values. The potentials of 2090-T83, -T84, -T81, and -T86 coincide closely enough with those of artificially aged tempers of 2*xxx* and 7*xxx* alloys that galvanic corrosion is not expected to be a major problem when the different alloys are connected to each other. Pitting corrosion results show that 2090 is subject to slightly greater pitting than 7075 (Table 13).

Fatigue. The results of smooth and notched axial stress fatigue tests for alloy 2090 are shown in Fig. 14(a), 14(b), and 14(c) for extrusions, plate, and sheet, respectively. The performance of 2090 is comparable to that of 7075-T6 up to 10^5 cycles for smooth tests and through failure for notched tests. The results of constant-amplitude fatigue crack growth tests (Fig. 15) show that for most stress ratios (R) the fatigue crack growth rate of 2090 is better than that of 7075. However, there may be a

Table 10 Exfoliation results for 2090-T83 sheet, 2090-T81 plate, and 2090-T86 extrusions

					Rating(b)				
					Salt spray(c)			Seacoast exposure(d)	
		Thickness		Plane	1	2	4	Exposure	
Product	Source	mm	in.	location(a)	week	weeks	weeks	time, months	Rating
Sheet	Plant	1.6	0.063	...	...	...	P	(e)	(e)
Plate	Plant	1.3	0.05	T/2	P	P	P	24	N
				T/10	P	P	P	24	N
Extrusion	Plant	20	0.8	T/2	N	P	P	24	N
				T/10	P	P	P	24	N
Plate	Laboratory	6.4	0.25	T/2	P	P	P	48	N
		6.4	0.25	T/2	P	P	P	48	N
		6.4	0.25	T/2	P	P	P	48	N
		6.4	0.25	T/2	P	P	P	48	N
		25	1.0	T/2	P	P	P	48	EA
		25	1.0	T/2	P	P	P	48	EA
		25	1.0	T/2	P	P	P	48	EA
		25	1.0	T/2	P	P	P	48	EA

(a) T/2 and T/10 represent location as a function of thickness (T). (b) Exfoliation rating per ASTM G 34 exfoliation corrosion (EXCO) test: N, no appreciable attack; P, pitting; EA, superficial—tiny blisters, thin slivers, flakes, or powder, with only slight separation of metal. (c) Modified ASTM acetic acid salt intermittent spray. (d) Seacoast exposure at Point Judith, RI. (e) Samples exposed 26 January 1988 and will be evaluated after 1, 2, and 4 years.

Table 11 Stress-corrosion cracking results for 2090-T81 plate and 2090-T86 extrusions tested in the short-transverse direction

Product	Source	Thickness		Specimen(a)	Alternate immersion (30 days)				Seacoast exposure(d)			
					172 MPa (25 ksi)		241 MPa (35 ksi)		172 MPa (25 ksi)		241 MPa (35 ksi)	
		mm	in.		F/N(b)	Days(c)	F/N(b)	Days(c)	F/N(b)	Days(c)	F/N(b)	Days(c)
Plate	Laboratory	25	1.0	C-ring	0/3	. . .	3/3	3, 3, 5	0/3	. . .	2/3	367, 367
		25	1.0	C-ring	0/3	. . .	3/3	9	0/3	. . .	1/3	966
		25	1.0	C-ring	1/3	5	3/3	3, 3, 3	0/3	. . .	3/3	367, 839, 966
		25	1.0	C-ring	0/3	. . .	3/3	3, 9, 9	0/3	. . .	0/3	. . .
Plate	Plant	38	1.5	Tensile bar	0/5	. . .	3/5	9, 12, 14	. . .	. . .	. . .	. . .
Extrusion	Plant	20	0.8	C-ring	0/5	. . .	0/5	. . .	0/5	(e)	0/5	(e)

(a) C-rings were 19 mm (0.75 in.) in diameter; tensile bars were 3.2 mm (⅛ in.) in diameter. (b) Number of failed specimens/total specimens. (c) Number of days to failure. (d) Seacoast exposure at Point Judith, RI. (e) Exposed at Point Judith for 1000 days

Table 12 Typical corrosion potentials for alloy 2090 and other selected aluminum alloys

Alloy and temper	Corrosion potential, mV(a)
2024-T3	−600
2090-T3	−640
2090-T84	−710
2024-T81	−710
2090-T83	−740
2090-T86	−740
7075-T6	−740
1100	−745

(a) Corrosion potentials measured per ASTM G 69 using a saturated calomel reference electrode

Table 13 Pitting corrosion results for 2090 and 7075 sheet

Alloy	Thickness		Exposure period	Maximum pit depth, μm	Average pit depth, μm	Pit number density, pits/mm
	mm	in.				
Alternate immersion in 3.5% NaCl solution (per ASTM G 44)						
7075-T6	1.3	0.050	30 days	149.6	79.2	14.8
2090-T83	1.3	0.050	30 days	198.1	123.4	12.9
	1.6	0.063	30 days	193.5	107.4	5.8
Salt spray, neutral 5% salt fog cabinet (per ASTM B 117)						
7075-T6	1.3	0.050	1000 h	20.5	16.8	6.2
	1.3	0.050	1000 h	24.6	20.3	3.8
2090-T83	1.0	0.040	1000 h	182.9	116.8	4.9
	1.3	0.050	1000 h	209.0	65.3	4.3
	1.6	0.063	1000 h	198.1	101.6	4.8

considerable variation among the fatigue crack growth rates for 2090, depending on crack size (Fig. 16).

Forming. Tests have been conducted to evaluate the formability of 2090-O, -T31, -T3, and -T83 sheet in the bending, stretch-bending, and biaxial-stretching modes (Ref 49). Results for 2090 were compared with those for 2024-T3 and 7075-T6.

Table 14 gives the 90° down flange minimum bend test results expressed as a bend radius/thickness (*R*/*t*) ratio; smaller *R*/*t* ratios suggest better formability. According to these preliminary test results, the O, T31, and T3 tempers of 2090 and 7075-T6 performed better than alloy 2090-T83. Generally, smaller bend radii were achieved when the bend line was oriented parallel to the rolling direction of the material.

The stretch-bend test was used to measure formability in the bending under tension mode. Comparative results of tests performed on various tempers of 2090, 2024, and 7075 are given in Table 15. When the principal axis of stress is parallel to the rolling direction, the tests showed that 2024-O had the best stretch-bend capability followed by, in decreasing order, 2024-T3, 2090-O, 2090-T31, 2090-T3, 7075-T6, and 2090-T83.

In the limiting dome height (LDH) test, rectangular blanks are rigidly clamped in the longer direction and stretched by a hemispherical punch. The height of the dome at fracture, which indicates the combined effect of strain hardening characteristics and limiting strain capability of the material, may be considered as a measure of stretch formability. The results of single-lot LDH tests for 2090, 2024, and 7075 sheet are shown in Table 16. The T31 temper provided the best results of all those tested for 2090. Further, the plane-strain limiting dome height was about 50% lower for 2090-T83 than for 7075-T6. Results for 2090-O are not shown in Table 16, but they can be expected to be slightly better than those for 2090-T31.

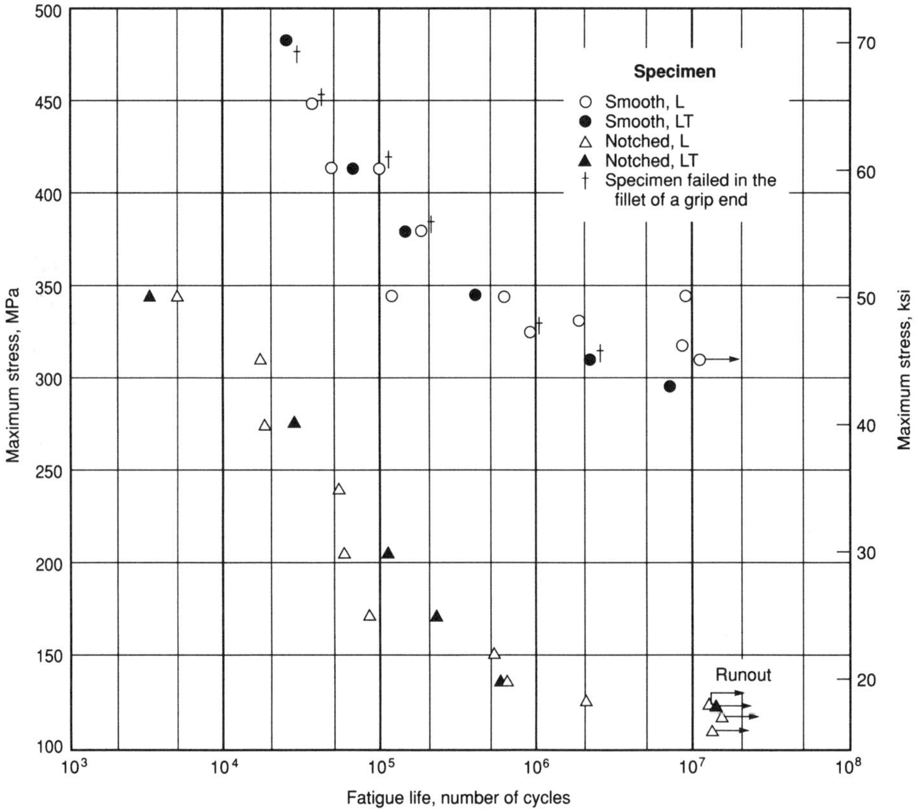

Fig. 14(a) Smooth and notched axial stress fatigue data for commercially produced 2090-T86 extruded tee shapes. Extrusion thickness, 19 mm (0.75 in.); stress ratio (*R*), 0.1; specimen location, T-2. For notched specimen, theoretical stress concentration factor (K_t) = 3. L, longitudinal orientation; LT, long-transverse orientation. Testing conducted in an ambient air atmosphere

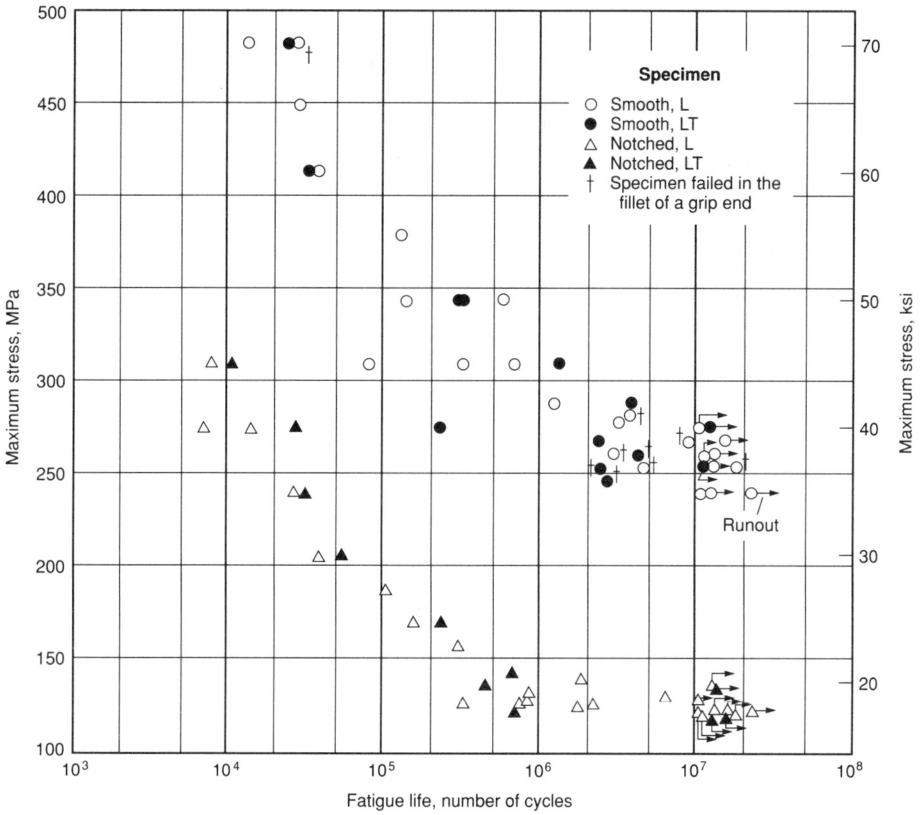

Fig. 14(b) Smooth and notched axial stress fatigue data for commercially produced 2090-T81 plate. Plate thickness, 13 mm (0.5 in.); stress ratio (*R*), 0.1; specimen location, T-2. For notched specimen, theoretical stress concentration factor (K_t) = 3. L, longitudinal orientation; LT, long-transverse orientation. Testing conducted in an ambient air atmosphere

Finishing Characteristics (Ref 50-52). The deoxidizing response of alloy 2090 in the T31 and T83 tempers is similar to that of 2024-T3 and 7075-T6 with regard to weight loss and roughness. Metallographic examinations show smaller but more densely populated pits on the 2090. Detailed weight loss and roughness data have been obtained using conventional deoxidizers such as chromic sulfuric acid, triacid, sulfuric acid, proprietary nonchromated iron sulfate solution, nitric acid, and nitric-hydrofluoric acid. After prolonged treatments (20 min) in hot (71.7 °C, or 161 °F) chromic sulfuric acid, 2090 showed end grain and intergranular attack less than that for 2024-T3 and to about the same depth as on 7075-T6.

Coating weights produced on 2090 by anodizing treatments with various agents, including sulfuric acid, 40 and 20 V chromic acid, and phosphoric acid, are about the same if not somewhat heavier than those produced on 2024-T3 and 7075-T6 (Table 17). However, when these alloys are anodized simultaneously in sulfuric acid, 2024 produces the lightest coating weight, 7075 produces the heaviest, and 2090 falls in between the two. This exception is due to the different voltage requirements of these alloys. The conversion coating response of 2090 is also quite good, with coating weights comparable to those on 7075-T6 for identical processing. Salt spray testing of the anodic coatings (applied and tested per MIL 8625) and conversion coatings (MIL C5541) show that the corrosion resistance of 2090 is as good as or better than that of 2024-T3 and 7075-T6 (Table 17).

Chemical milling studies conducted by airframe manufacturers indicate that an acceptable finish can be obtained by using a standard chemical milling solution. The 2090 alloy develops a slightly rougher finish than 7075 and 2024 from the standard chemical milling process (Table 18).

Table 14 90° bend test results for various tempers of 2024, 2090, 2091, and 7075 sheet

Results expressed as the ratio of bend radius to thickness (*R*/*t*)

Alloy and temper	Specimen gage		Parallel to grain				Across grain			
			R min				R min			
	mm	in.	mm	in.	R/t(a)	Springback	mm	in.	R/t(a)	Springback
2024-O	1.6	0.063	0	0	0	2°	0	0	0	2°
	3.2	0.125	0–3.2	0–0.125	0–1.00	. . .	0–3.2	0–0.125	0–1.00	. . .
7075-O	1.6	0.063	0–1.6	0–0.063	0–1.00	2°	0–1.6	0–0.063	0–1.00	2°
	3.2	0.125	0–3.2	0–0.125	0–1.00	. . .	0–3.2	0–0.125	0–1.00	. . .
2090-O	1.6	0.063	2.3	0.09	1.5	5°	2.3	0.09	1.5	5°
	3.2	0.125	4.8	0.19	1.5	3°	4.8	0.19	1.5	3°
2090-T3E27	0.8	0.032	0.5	0.02	0.7	. . .	1.5	0.06	1.8	. . .
	2.2	0.086	2.3	0.09	1.0	. . .	4.3	0.17	2.0	. . .
	3.2	0.125	6.35	0.25	2.0	7°	9.65	0.38	3.0	6°
2091-T4	1.6	0.063	2.3	0.09	1.5	1°	2.3	0.09	1.5	1°
2091-T3	1.6	0.063	3.3	0.13	2.0	2°	3.3	0.13	2.0	2°
	3.2	0.125	7.9	0.31	2.5	1°	6.35	0.25	2.0	2°
2090-T3E28	1.25	0.049	1.85	0.0735	1.5	. . .	0.373	0.147	3.0	. . .
	1.6	0.063	3.2	0.125	2.0	7°	3.2	0.125	2.0	7°
	2.05	0.081	6.17	0.243	3.0	. . .	6.17	0.243	3.0	. . .
2024-T3	1.6	0.063	3–6	0.12–0.24	2–4	. . .	3–6	0.12–0.24	2–4	. . .
	3.2	0.125	9.65–15	0.38–0.59	3–5	. . .	9.65–15	0.38–0.59	3–5	. . .
2091-T8	3.2	0.125	12.7	0.50	4.0	12°	16	0.63	5.0	11°
7075-T6	1.6	0.063	4.8–8.1	0.19–0.32	3–5	. . .	4.8–8.1	0.19–0.32	3–5	. . .
	3.2	0.125	6.35–9.65	0.25–0.38	4–6	. . .	6.35–9.65	0.25–0.38	4–6	. . .
2090-T8E41	1.2	0.047	4.77	0.188	4.0	. . .	14.32	0.564	12.0	. . .
	1.6	0.063	9.65	0.38	6.0	21°	11.2	0.44	6.9	20°
	2.15	0.085	13	0.51	6.0	. . .	17.3	0.68	8.0	. . .
	3.2	0.125	>25	>1	>8	. . .	25	1	8.0	25–35°

(a) Lower *R*/*t* ratios suggest better formability.

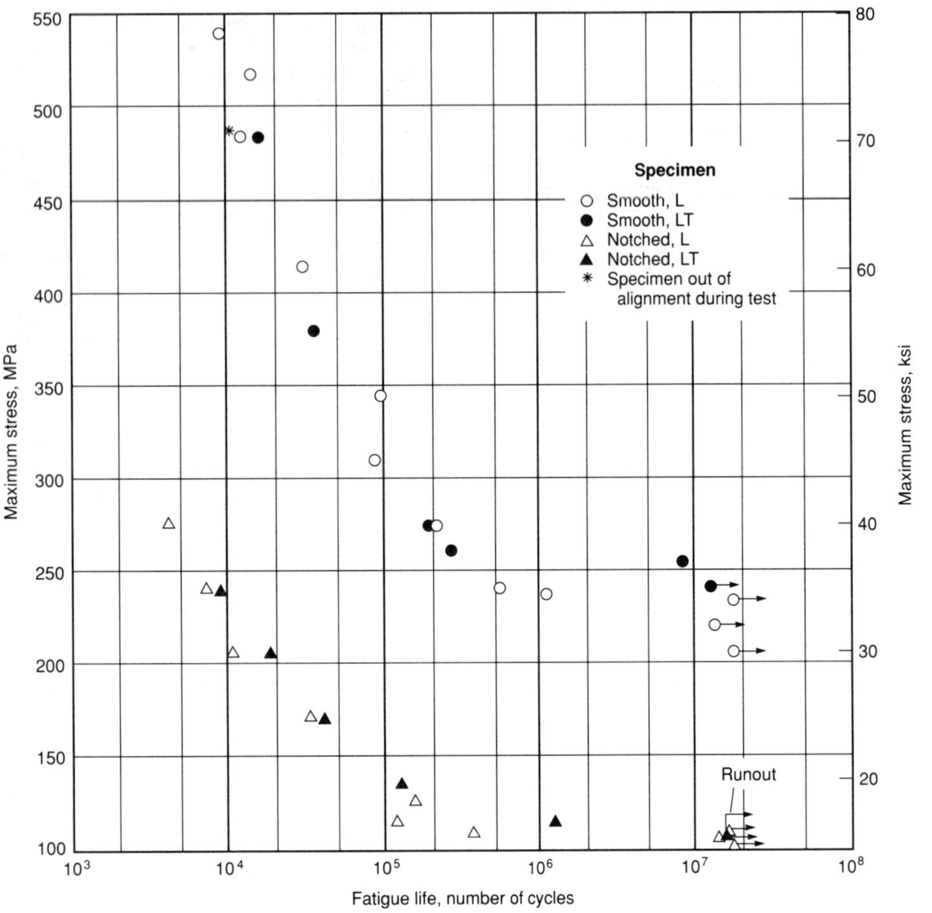

Fig. 14(c) Smooth and notched axial stress fatigue data for commercially produced 2090-T83 sheet. Sheet thickness, 1.6 mm (0.063 in.); stress ratio (R), 0.1. For notched specimen, theoretical stress concentration factor (K_t) = 3. L, longitudinal orientation; LT, long-transverse orientation. Testing conducted in an ambient air atmosphere

Table 15 Stretch-bend test results for 2090, 2091, 2024, and 7075 sheet

Alloy and temper	Gage		Bendline orientation: Parallel to grain, Punch travel			Across grain, Punch travel		
	mm	in.	mm	in.	Punch travel/gage	mm	in.	Punch travel/gage
7075-O	1.6	0.063	41.6	1.64	26.03	43.9	1.73	27.46
	3.2	0.125	41.6	1.64	14.32	40.9	1.61	12.88
2024-O	1.6	0.063	28.7	1.13	17.90	30.5	1.20	19.10
2090-O	1.6	0.063	24.6	0.97	15.37	33.2	1.31	20.79
	3.2	0.125	25.9	1.02	8.17	37.3	1.47	11.76
2024-T3	1.6	0.063	23.4	0.92	14.60	20.8	0.82	13.00
	1.75	0.069	18.8	0.74	10.76	21.1	0.83	12.00
2090-T3E27	9.4	0.037	12.4	0.49	13.24	21.8	0.86	23.19
	1.24	0.049	17.5	0.69	14.06	26.2	1.03	21.12
	2.2	0.086	15.7	0.62	7.22	22.1	0.87	10.11
	3.2	0.125	14.0	0.55	4.40	29.7	1.17	9.34
2090-T3E28	1.3	0.05	15.5	0.61	12.10	27.4	1.08	21.56
	2.2	0.086	16.5	0.65	7.54	19.05	0.75	8.74
7075-T6	8.1	0.032	9.1	0.36	11.25	8.6	0.34	10.78
	1.2	0.049	14.0	0.55	11.18	13.0	0.51	10.45
	1.8	0.07	45.5	1.79	7.87	40.9	1.61	7.74
2090-T8E41	1.2	0.048	7.1	0.28	5.92	8.6	0.34	7.10
	1.6	0.063	8.4	0.33	5.24	9.4	0.37	5.81
	2.15	0.085	6.1	0.24	2.77	6.6	0.26	3.00
	3.2	0.125	8.9	0.35	2.80	7.9	0.31	2.48
2091-T3	1.6	0.063	35.5	1.40	22.22	22.4	0.88	13.95
	3.15	0.124	36.0	1.42	11.45	...	...	...
2091-T4	1.6	0.063	31.75	1.25	19.84	24.4	0.96	15.27
2091-T8	3.2	0.126	15.0	0.59	4.68	...	...	...

Machining. Machinability tests conducted on 2090 plate indicate that the material has machinability comparable to that of the B-rated aerospace alloys, 2024-T351 and 7075-T651. The machinability of B-rated alloys is characterized by curled or easily broken chips and good-to-excellent finish. Recent field machining trials have shown that certain practices should be followed when machining 2090:

- The part should be well supported and braced by adequate fixturing, and vertical flanges should be supported during machining whenever possible
- Sharp positive-rake tooling specifically designed for machining aluminum should be used
- Adequate amounts of coolant should be continuously directed to the cutting area

Welding. Gas metal arc, gas tungsten arc, and electron beam welding processes have been used to evaluate the fusion weldability characteristics of alloy 2090. All three methods show good results (Table 19).

Alloy 2091

Alloy 2091 was developed to be a damage-tolerant alloy with 8% lower density and 7% higher modulus than 2024-T3, a major high-toughness damage-tolerant alloy currently used for most aircraft structures (Ref 53). Alloy 2091 is also suitable for use in secondary structures where high strength is not critical.

The chemical composition of alloy 2091 has been registered with the Aluminum Association (Table 3). A variety of tempers (Table 20) are being developed to offer useful combinations of strength, corrosion resistance, damage tolerance, and fabricability. Typical physical properties of 2091 are given in Table 8. The microstructure of 2091 varies according to product thickness and producer; in general, gages above 3.5 mm (0.140 in.) have an unrecrystallized microstructure, and lighter gages feature an elongated recrystallized grain structure.

Because alloy 2091 and its tempers are relatively new and in different phases of registration and characterization, data may be incomplete for some forms. In the underaged (T8) temper, alloy 2091 has a plane-stress fracture toughness (K_c) of about 130 to 140 MPa$\sqrt{m}$ (120 to 130 ksi$\sqrt{in.}$) and a compressive modulus of 76 to 80 GPa (11 to 11.6 × 10^6 psi). Properties for 2091 in Preliminary European Normale (prEN) specifications are given in Table 21.

In general, the behavior of 2091 is similar to that of other 2*xxx* and 7*xxx* alloys. Material characteristics that have been cause for concern in other aluminum-lithium alloys are of less concern in 2091. Alloy 2091 depends less on cold work to attain its properties than does 2024 (Fig. 17). The properties of 2091 after elevated-temperature (≦125 °C, or 260 °F) exposure are relatively stable (Fig. 18) in that changes in properties during the lifetime of a component are acceptable for most commercial applications.

Corrosion (Ref 48, 53, 54). The exfoliation resistance of 2091-T84, like that of 2024,

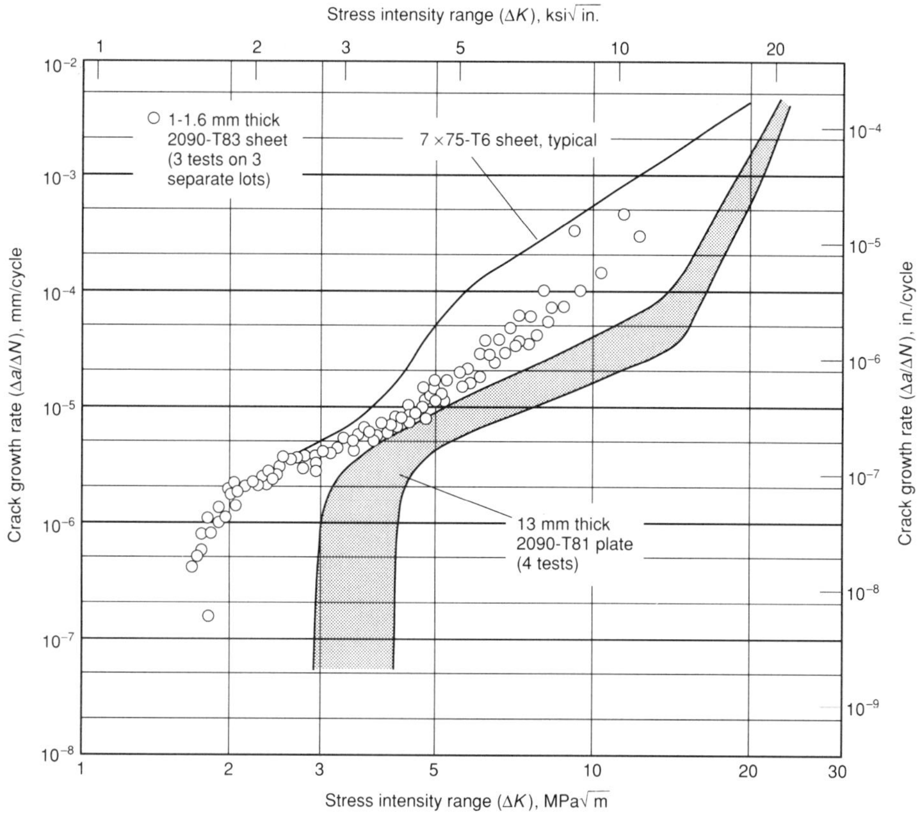

Fig. 15 Fatigue crack propagation data for 2090-T83 sheet, 2090-T81 plate, and 7*x*75-T6 sheet. Crack orientation, L-T (crack plane and propagation direction perpendicular to the principal direction of metalworking); stress ratio (*R*), 0.33. Testing conducted in high-humidity air (>90% relative humidity)

varies depending on the microstructure of the product and its quench rate. The more unrecrystallized the structure, the more even the exfoliation attack. However, the exfoliation resistance of 2091 (Table 22) is generally comparable to that of similar gages of 2024-T3.

The microstructural relationship for stress-corrosion cracking in sheet products is the converse of that for exfoliation. As the microstructure becomes more fibrous, the SCC threshold increases. For thicker unrecrystallized structures and thinner elongated recrystallized structures, it is possible to attain an SCC threshold of 240 MPa (35 ksi), which is quite good compared to that of 2024-T3 (Ref 54). For thinner products, the threshold varies by gage and producer; it may be as low as 50 to 60% of the yield strength or as high as 75% of the yield strength.

Fatigue. Although fatigue testing on 2091 has been done by a number of labs, producers, and users (Ref 53, 55, 56), the results have been difficult to interpret. The results for 2091 have been superior to those for 2024 (Fig. 19), roughly equivalent to those for 2024 (Ref 55, 56), or inferior to those for 2024. In general, the consensus is that under controlled and similar circumstances, the fatigue properties of 2091-T84 are sufficient to allow it to be used as a substitute for 2024.

Finishing. The responses of 2091 to chemical milling, anodizing, conversion coating, and deoxidizing are similar to those of other aluminum or aluminum-lithium alloys (Tables 17 and 18). The same bath that is used for 2024 or 7075 can be used for 2091, or the bath can be adjusted to give improved surface roughness to 2091. The adjustment may be necessary because the etch rates for 2091 may be different than those for other aluminum alloys, or because batch processing of 2091 parts may be necessary. Results from salt spray tests indicate that 2091 gains better protection from anodic and conversion coatings than do the conventional aluminum alloys.

Forming. As a general rule, the as-quenched condition provides the best forming ability for aluminum-lithium alloys. In this temper, the forming ability of 2091 recrystallized sheets exceeds that of 2024. The *n* and *r* (Lankford) coefficients are respectively 0.33 and 1.0 for 2091 and 0.22 and 0.70 for 2024. The outstanding behavior indicated by the values for 2091 facilitates forming operations: It allows a decrease in the number of forming steps for 2091 as compared with 2024. This benefit has been confirmed in bending, drawing, and rubber-forming tests by airframe manufacturers.

Because of grain coarsening, the critical lower work-hardening strain (e_c) limits the maximum intermediate strain in the forming sequence for 2091 before the last solution heat treatment. The same limitation exists on conventional alloys: the actual e_c values are 5 to 13% for 2091, 6% for 2091-CPHK, and 6% for 2024 sheets. The stretch-bend and 90° bend test results given in Tables 14 and 15 indicate that, in similar applications, the formability of 2091 compares favorably with that of 2024. The relatively slow natural aging response of 2091 (Fig. 20) allows considerably more latitude in forming parts on the shop floor than is possible with conventional alloys.

Table 16 Limiting dome height test results for 2090, 2024, and 7075 sheet

Alloy and temper	Gage mm	Gage in.	Longitudinal tensile strength MPa	Longitudinal tensile strength ksi	Longitudinal yield strength MPa	Longitudinal yield strength ksi	Elongation, %	Punch travel mm	Punch travel in.
2090-T3E27	1.24	0.049	348	50.4	248	36.0	9.20	13.39	0.527
	2.18	0.086	298	43.2	215	31.2	9.50	19.81	0.780
	2.24	0.088	316	45.9	256	37.2	8.20	17.04	0.671
2090-T3E28	1.24	0.049	345	50.1	259	37.6	8.50	13.16	0.518
2090-T8E41	2.16	0.085	552	80.1	508	73.7	6.50	10.82	0.426
7075-T6	1.24	0.049	...	...	...	...	...	18.47	0.727
	1.78	0.070	...	...	...	...	...	21.13	0.832

Table 17 Anodizing, conversion coating, and deoxidizing responses of aluminum-lithium alloys 2090 and 2091 compared with those of aluminum alloys 2024 and 7075

Treatment characteristic	2090	2024/7075	2091
Sulfuric anodizing coating weight, g/m²	7.23–9.91	4.57–21.0	Similar to 2090
Chromic anodizing coating weight, g/m²	3.30–4.00	3.12–3.76	...
Phosphoric anodizing coating weight, g/m²	0.34–0.75	0.42–0.74	...
Conversion coating, g/m²	0.65–0.90	0.58–0.86	...
Triacid deoxidizing weight loss, mg/cm²	...	11	17
Chromic/sulfuric acid deoxidizing weight loss, mg/cm²	4.8–5.0	2.8–5.1	5.2
Salt spray results with anodic and conversion coatings	82–89% passed	36–75% passed	...

Alloy 8090

Alloy 8090 was developed to be a damage-tolerant medium-strength alloy with about 10% lower density and 11% higher modulus than 2024 and 2014, two commonly used aluminum alloys. Its use is aimed at

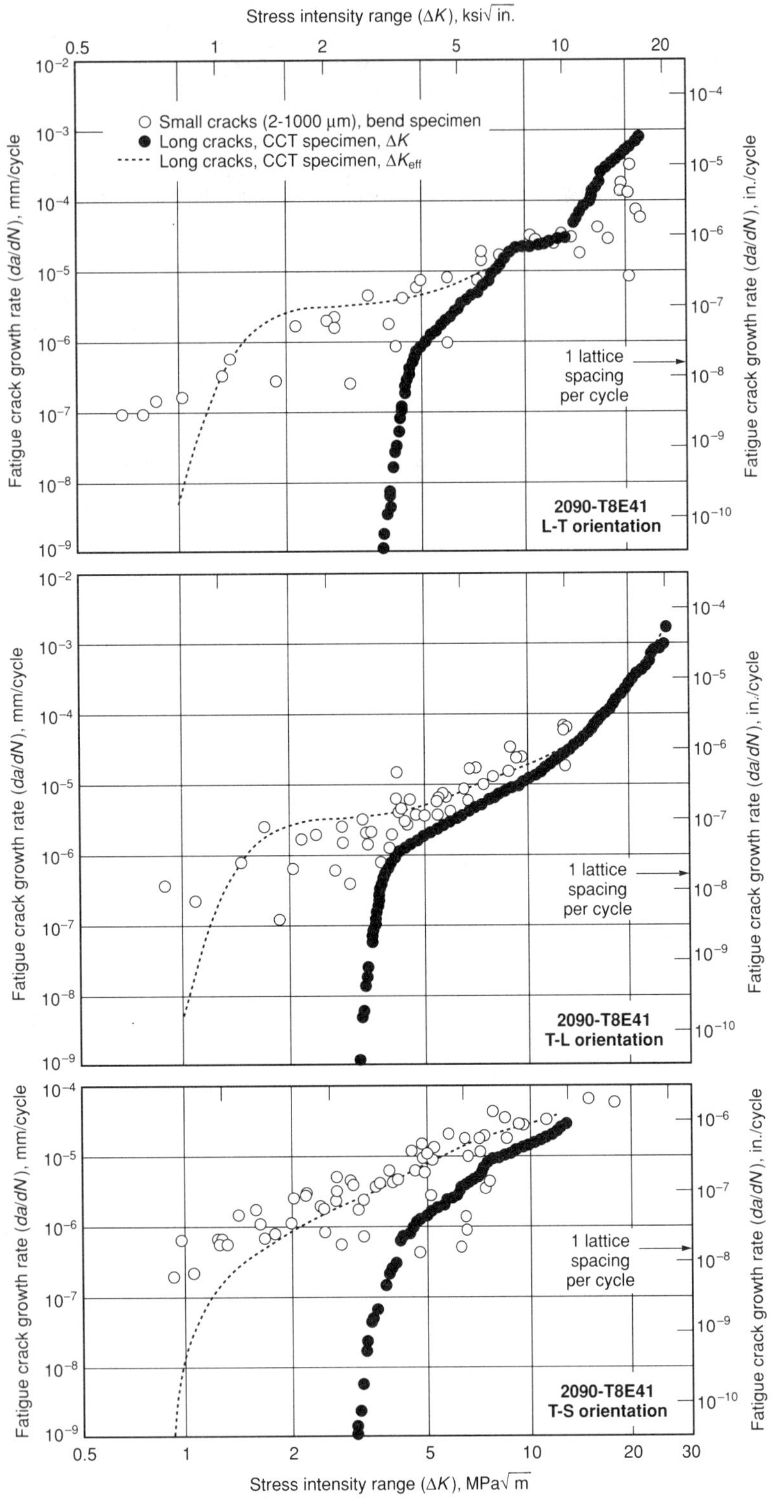

Fig. 16 Comparison of the growth rates of long (greater than ~5 mm) and naturally occurring small (2 to 1000 μm) fatigue cracks in 2090-T8E41 alloy as a function of the nominal and effective stress intensity ranges, ΔK and ΔK_{eff}, respectively. Data are presented for L-T, T-L, and T-S orientations at a stress ratio (R) of 0.1. Growth rates of small cracks exceed those of long cracks by several orders of magnitude when compared on the basis of ΔK; however, they show close correspondence when characterized in terms of ΔK_{eff}. CCT, center-cracked tension. Source: Ref 1

applications where damage tolerance and the lowest possible density are critical. The alloy is available as sheet, plate, extrusions, and forgings, and it can also be used for welded applications.

The chemical composition of 8090 has been registered with the Aluminum Association (Table 3). A variety of tempers have been developed that offer useful combinations of strength, corrosion resistance, dam-

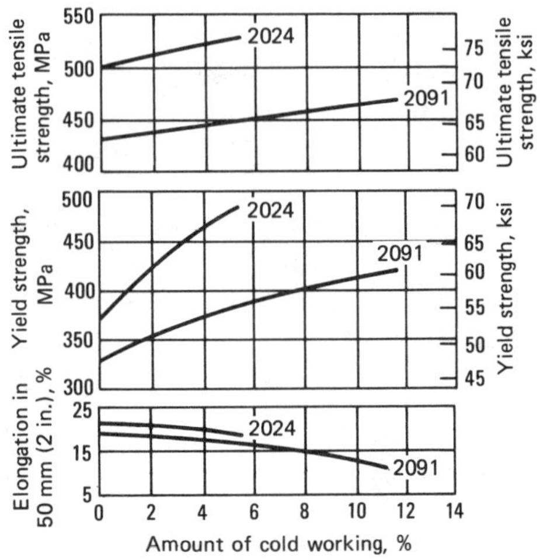

Fig. 17 Longitudinal tensile properties of alloys 2091 and 2024 as a function of cold working prior to aging. Aluminum alloy 2024 is naturally aged; aluminum-lithium alloy 2091 is aged to temper T8X.

age tolerance, and fabricability. Unfortunately, there has been no official temper registration in the United States for any of these tempers for any of the variety of product forms. Descriptions of commonly used unofficial temper designations are given in Table 23. Typical physical properties of 8090 are given in Table 8. Plate, extrusions, and forgings have an unrecrystallized microstructure; damage-tolerant sheet has a recrystallized microstructure. Higher-strength sheet is available with a recrystallized or unrecrystallized microstructure (Fig. 21).

Strength and Toughness. Because alloy 8090 and its tempers and product forms are relatively new and unregistered, property data are incomplete. However, available data for the current capabilities of 8090 products and tempers are given in Table 24. The medium-strength products of alloy 8090 are aged to near-peak strength and show small changes in properties after elevated-temperature exposure (Fig. 22). The very underaged (damage-tolerant) products will undergo additional aging upon exposure to elevated temperatures. Changes in strength and toughness at cryogenic temperatures are more pronounced in 8090 than in conventional aluminum alloys; 8090 has a substantially higher strength and toughness at cryogenic temperatures (Table 25).

The improving quality of commercially available aluminum-lithium alloys such as 8090 has resulted in significant improvements in short-transverse ductility and, consequently, short-transverse tensile strength. Research on the short-transverse fracture toughness of 8090 has shown that the property reaches a minimum plateau (Fig. 23) at an aging temperature of 190 °C (375 °F). The level of the plateau toughness is affected by impurity content. Sodium content, for example, has been claimed to control toughness; however, once the sodi-

Table 18 Response of aluminum-lithium alloys 2090 and 2091 to chemical milling

Response characteristic	Alloy response: 2090	2024/7075	2091
Standard chemical milling solution (NaOH + Na_2S)			
Roughness, μm (μin.)	3.4 –3.8 (135–150)	0.9–1.0 (36–41)	1.8 (70)
Etch rate per side, μm/min (mils/min)	38 –46 (1.5–1.8)	40 (1.6)	63.5 (2.5)
TEA-modified chemical milling solution(a)			
Roughness, μm (μin.)	2.25 –2.35 (89–93)	0.80–0.85 (31–34)	1.65 (65)
Etch rate per side, μm/min (mils/min)	43 –50 (1.7–2.0)	40 (1.6)	66 (2.6)

(a) Sodium hydroxide, triethanolamine (TEA), and sodium sulfide

Table 19 Typical mechanical properties of gas metal arc welds of 13 mm (0.50 in.) thick 2090 plate with 2319 filler alloy

Alloy, temper, and condition	Ultimate tensile strength, MPa	ksi	Yield strength, MPa	ksi	Elongation in 50 mm (2 in.), %
2090-T81, as-welded	232	33.6	204	29.6	5.2
2090-T4, postweld aged	258	37.5	(a)	(a)	0
2090-T81, postweld solution heat treated and aged	386	56.1	(a)	(a)	0

Note: Weld bead reinforcements were removed prior to testing. Failure occurred through the weld fusion zone or at the interface. (a) Failure occurred before reaching 0.2% offset, indicating nil elongation.

Table 20 Registered temper designations of alloy 2091 bare and aluminum-clad sheet

Temper	Characteristics	Forms
O	Annealed, lowest strength, maximum formability	Sheet, plate
T3	Solution heat treated and stretched; can be aged to T84 temper	Sheet, plate
T8, T84(a)	Underaged temper; has the best combination of strength, toughness, and corrosion resistance for damage-tolerant applications	Sheet
T851	Medium-strength product	Plate
T8X51	Underaged damage-tolerant product	Thick sheet and plate

Only T84 is registered with the Aluminum Association at this time.

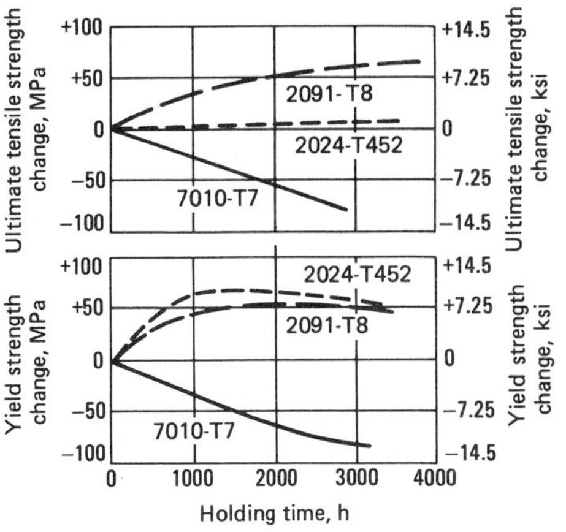

Fig. 18 Variation in room-temperature ultimate tensile strength and yield strength for aluminum-lithium alloy 2091 after holding at 130 °C (265 °F) at indicated times. Data for aluminum alloys 2024 and 7010 are included for comparison.

um level in lithium additions is reduced to a practical level of typical battery grade lithium, the fracture toughness of the 8090 alloy achieves a constant value.

Design Considerations. In general, the engineering considerations involved in the application of aluminum-lithium alloys are similar to those for the 2*xxx* and 7*xxx* alloys used by the aerospace industry. However, just as with 2090, there are some unique factors that need to be considered when using 8090:

- The in-plane anisotropy of tensile properties for unrecrystallized products (plate,

Table 21 Preliminary European Normale (prEN) specifications for minimum tensile properties of bare and aluminum-clad 2091 sheet and light-gage plate

Product thickness, mm	in.	Longitudinal: 0.2% yield strength, MPa	ksi	Ultimate tensile strength, MPa	ksi	Elongation, %(a)	Long-transverse: 0.2% yield strength, MPa	ksi	Ultimate tensile strength, MPa	ksi	Elongation, %(a)	45°: 0.2% yield strength, MPa	ksi	Ultimate tensile strength, MPa	ksi	Elongation, %(a)
Aluminum-clad products (prEN 6003)(b)																
0.79–3.45	0.031–0.136	265	38.5	364	52.8	10	265	38.5	384	55.7	10	236	34.2	350	50.7	15
3.45–6.0	0.136–0.236	334	48.5	418	60.7	8	290	42.1	418	60.7	10	256	37.1	364	52.8	15
Bare products (prEN 6005)(b)																
0.81–3.3	0.032–0.130	290	42.1	394	57.1	10	295	42.8	408	59.2	10	265	38.5	379	55	15
3.3–6.0	0.130–0.236	359	52.1	448	65	8	325	47.1	359	52.1	10	285	41.4	398	57.8	15
6.0–12	0.236–0.472	359	52.1	448	65	8	325	47.1	359	52.1	10	285	41.4	398	57.8	15
12–40	0.472–1.575	354	51.4	438	63.5	7	320	46.4	423	61.4	8	275	40	394	57.2	13

(a) Elongation in 50 mm (2 in.). (b) prEN specifications issued by the AECMA standards organization in Europe

Table 22 Alloy 2091 exfoliation from seacoast exposure and EXCO testing

Product	Gage, mm	in.	Temper	Plane	EXCO testing(a) 4-day exfoliation rating(b)	Seacoast exposure(c): Months	Exfoliation rating(b)
Sheet(d)	1.0–3.5	0.04–0.14	...	T/2	EA (superficial)	...	...
Sheet(e)	6.3	0.25	T8	T/2	EB (moderate)	6	EB (moderate)(f)
Sheet(e)	3.2	0.125	T3	T/2	EB (moderate)	12	P (pitting)
Sheet(e)	3.2	0.125	T8	T/2	EB (moderate)	12	EB (moderate)
Sheet(e)	4.8	0.19	T8	T/2	EA (superficial)	...	...
			T8	T/10	EA (superficial)	...	...
Sheet(g)	1.2	0.047	T3	T/2	P (pitting)	...	...
Plate(e)	12.7	0.5	T8	T/2	EA (superficial)	...	...

(a) Exfoliation corrosion (EXCO) testing per ASTM G 34. (b) Exfoliation rating per ASTM G 34: P, pitting; EA (superficial), tiny blisters, thin slivers, flakes, or powders with only slight separation of metal; EB (moderate), notable layering and penetration into metal. (c) Seacoast exposure at Point Judith, RI. (d) Pechiney data. (e) Sheet and plate fabricated at Davenport, IA facility. (f) Exfoliation more advanced on Point Judith panel than on EXCO panel. (g) Fokker data. Source: Ref 53, 54

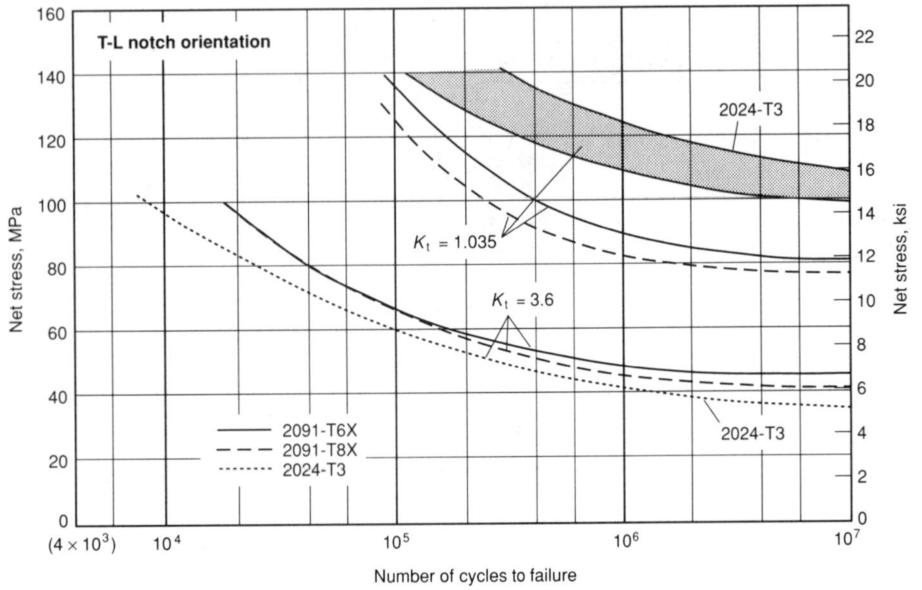

Fig. 19 Longitudinal fatigue resistance of notched 2091-T6X, 2091-T8X, and 2024-T3 sheet. Alloys 2091-T8X and 2024-T3 were stretched 2% before aging. Stress ratio (R), 0.1. K_t, theoretical stress concentration factor. T-L notch orientation as defined in Fig. 23

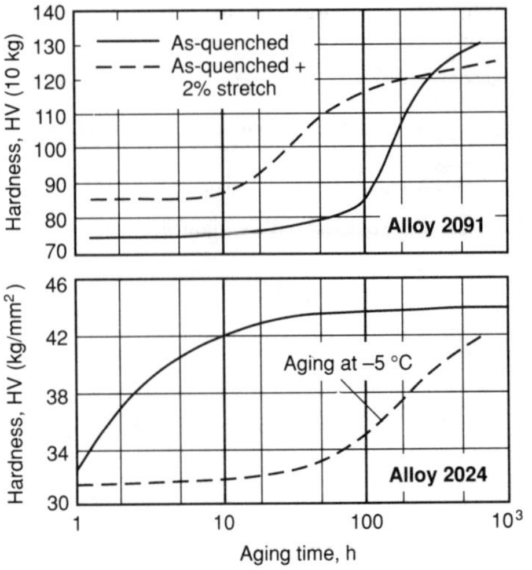

Fig. 20 Natural aging of 2091 and 2024 aluminum alloys. Aging done at room temperature (22 °C, or 71 °F) except where indicated

extrusions, forgings, and some medium-strength sheet) is higher in 8090 than in conventional alloys, placing more importance on 45° and shear properties

- Recrystallized damage-tolerant sheet and recrystallized medium-strength sheet show much less anisotropy of tensile properties than do the unrecrystallized products
- Cold work is required to achieve good properties in 8090 products, although to a lesser extent than with 2090
- Fatigue crack growth behavior is excellent, but the advantage 8090 has over conventional alloys is reduced in overload situations

Corrosion performance for 8090 is a strong function of the degree of artificial aging and the microstructure. Table 26 summarizes corrosion performance by product and temper for various types of corrosion tests. Care needs to be taken when selecting an accelerated test for judging the corrosion resistance of 8090. The modified ASTM acetic acid salt intermittent spray (MASTMAASIS) test seems

Table 23 Temper designations for aluminum-lithium alloy 8090

Temper(a)	Characterization	Forms
T8, T8X	Near-peak-aged medium-strength sheet product	Sheet
T81	Underaged damage-tolerant sheet	Sheet
T8771, T651, T7E20	Near-peak-aged plate	Plate
T8151, T8E57	Underaged damage-tolerant plate	Plate
T6511, T8511/10	Medium-to-high-strength peak-aged extrusions	Extrusions
T8771, T852	Medium-strength peak-aged forgings	Forgings

(a) Temper designations are not registered; this listing is a recap of designations that producers and users have used.

Table 24 Tensile properties and fracture toughness of aluminum-lithium alloy 8090

Temper	Product form	Grain structure(a)	Direction	Minimum and typical(b) tensile properties: Ultimate tensile strength, MPa	Ultimate tensile strength, ksi	0.2% yield strength, MPa	0.2% yield strength, ksi	Elongation in 50 mm (2 in.), %	Minimum and typical(b) fracture toughness values: Fracture orientation(c) and toughness type (K_c or K_{Ic})(d)	Toughness value(b), MPa $\sqrt{m}$	Toughness value(b), ksi $\sqrt{in.}$
8090-T81 (underaged)	Damage-tolerant bare sheet <3.55 mm (0.140 in.) thick	R	Longitudinal	345–440	50–64	295–350	43–51	8–10 typ	L-T (K_c)	94–165	86–150
			Long transverse	385–450	56–65	290–325	42–47	10–12	T-L (K_c)	85 min	77 min
			45°	380–435	55–63	265–340	38.5–49	14 typ	S-L (K_c)	. . .	. . .
8090-T8X (peak aged)	Medium-strength sheet	UR	Longitudinal	470–490	68–71	380–425	55–62	4–5	L-T (K_c)	75 typ	68 typ
			Long transverse	450–485	65–70	350–440	51–64	4–7	T-L (K_c)	. . .	. . .
			45°	380–415	55–60	305–345	44–50	4–11	S-L (K_c)	. . .	. . .
8090-78X	Medium-strength sheet	R	Longitudinal	420–455	61–66	325–385	47–56	4–8	L-T (K_c)	. . .	. . .
			Long transverse	420–440	61–64	325–360	47–52	4–8	T-L (K_c)	. . .	. . .
			45°	420–425	61–62	325–340	47–49	4–10	S-L (K_c)	. . .	. . .
8090-T8771, 8090-T651 (peak aged)	Medium-strength plate	UR	Longitudinal	460–515	67–75	380–450	55–65	4–6 min	L-T (K_{Ic})	20–35	18–32
			Long transverse	435 min	63 min	365 min	53 min	4 min	T-L (K_{Ic})	13–30	12–27
			Short transverse	465 typ	67 typ	360 typ	52 typ	. . .	S-L (K_{Ic})	16 typ	14.5 typ
			45°	420 min	61 min	340 min	49 min	1–1.5 min			
8090-T8151 (underaged)	Damage-tolerant plate	UR	Longitudinal	435–450	63–65	345–370	50–54	5 min	L-T (K_{Ic})	35–49	32–45
			Long transverse	435 min	63 min	325 min	47 min	5 min	T-L (K_{Ic})	30–44	27–40
			45°	425 min	61.5 min	275 min	40 min	8 min	S-L (K_{Ic})	25 typ	23 typ
8090-T852	Die forgings with cold work, or hand forgings	UR	Longitudinal	425–495	62–72	340–415	49–60	6–8	L-T (K_{Ic})	30 typ	27 typ
			Long transverse	405–475	59–69	325–395	47–57	3–6	T-L (K_{Ic})	20 typ	18 typ
			45°	405–450	59–65	305–395	44–57	2–6	S-L (K_{Ic})	15 typ	14 typ
8090-T8511, 8090-T6511	Extrusions	UR	Longitudinal	460–510	67–74	395–450	57–65	3–6	. . .	. . .	. . .

(a) R, recrystallized; UR, unrecrystallized. (b) Unless otherwise specified as only a minimum (min) or a typical (typ) value, the two values given for a property represent its minimum and typical value. The minimum values are proposed by various customer and national specifications and do not reflect a uniform registration. (c) See Fig. 23 for a diagram of fracture orientations. (d) K_c, plane-stress fracture toughness; K_{Ic}, plane-strain fracture toughness

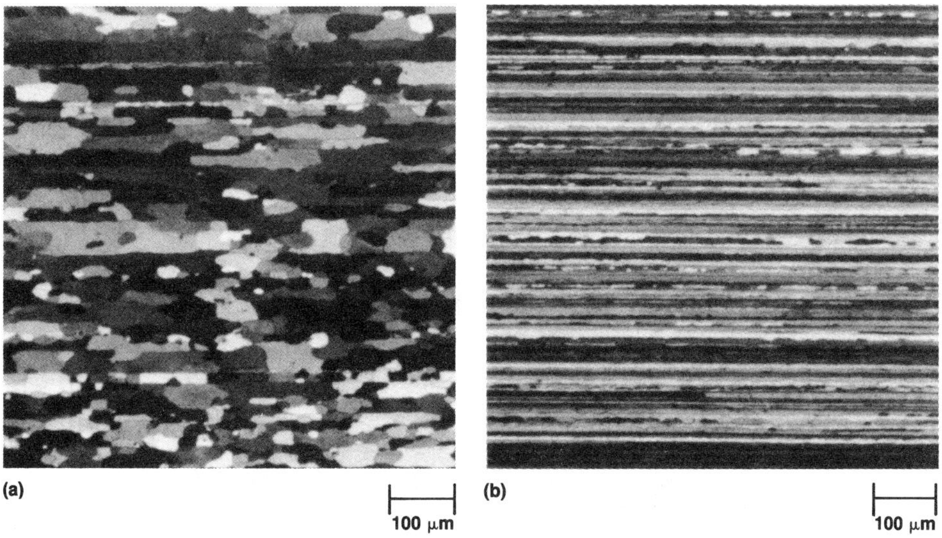

Fig. 21 Microstructures of 8090 sheet. (a) Recrystallized grain structure. (b) Unrecrystallized grain structure

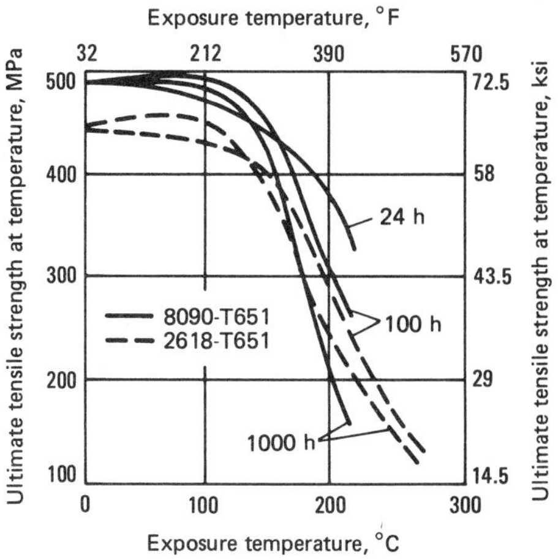

Fig. 22 Ultimate tensile strength for 8090-T651 plate tested at temperature after different exposure times. Data for 2618-T651 provided for comparison

Table 25 Cyrogenic tensile and toughness properties of 8090-T3

Test temperature, K	Direction	Tensile properties: Yield strength MPa	ksi	Tensile strength MPa	ksi	Elongation in 38 mm (1.5 in.), %	Reduction in area, %	Toughness MPa $\sqrt{m}$	ksi $\sqrt{in.}$
295	Longitudinal	217	31.5	326	47	12	18	...	...
	Transverse	208	30	348	50.5	14	26	...	...
76	Longitudinal	248	36	458	66.5	22	27	97(a)	88(a)
	Transverse	241	35	450	65	20	37	60(b)	55(b)
20	Longitudinal	272	39.5	609	88.3	28	28	...	...
	Transverse	268	39	592	86	25	27	...	...
4	Longitudinal	280	41	605	88	26	28	74(a)	67(a)
	Transverse	270	39	597	86.5	24	29	50(b)	45(b)

(a) Toughness with an L-T crack orientation (crack plane and growth direction perpendicular to the rolling direction). (b) Toughness with a T-L crack orientation (crack plane and growth direction parallel to the rolling direction)

to be a better indicator of exfoliation resistance than the ASTM G 34 exfoliation corrosion (EXCO) test. Alloy 8090 has displayed generally good exfoliation resistance in atmospheric exposure. For thick products, short-transverse SCC resistance is best in the peaked-aged temper. Thick products with unrecrystallized microstructures have good SCC resistance in the long-transverse direction, whereas those with recrystallized structures have a lower SCC threshold.

Fatigue. The improved fatigue life of aluminum-lithium alloys as compared with conventional aluminum alloys is viewed by most researchers to be due to crack branching and closure effects. While this improved fatigue performance is valuable, the application designer needs to consider the impact that overloads and negative stress ratios have on the fatigue life of aluminum-lithium alloys to determine the overall advantage that can be obtained with these alloys.

Fatigue of 8090 Sheet. The results of fatigue testing for damage-tolerant 8090 sheet (8090-T81) are shown in Fig. 24. The fatigue crack growth rate of 8090-T81 is up to an order of magnitude slower than that of 2024-T3 at a stress ratio of 0.1 (Fig. 25). The advantage that 8090 has in this area diminishes at higher stress ratios. Both the smooth and notched fatigue strengths of 8090-T81 sheet are higher than those of 2024. Microstructure appears to have some effect on fatigue life: An elongated recrystallized structure provides the best fatigue life results (Fig. 26). There is little fatigue data available for 8090 medium-strength sheet product, but what little there is suggests that an elongated recrystallized microstructure provides the best combination of strength and damage tolerance.

Fatigue of 8090 Plate. Alloy 8090 in plate form also comes in both the damage-tolerant and medium-strength tempers. Fatigue data for the medium-strength temper (T651) are shown in Fig. 27. Data for the damage-tolerant temper are not available. A comparison of the results for the medium-strength tempers to those for 2024 indicates that 8090 has the superior fatigue crack growth rate (Fig. 28),

Table 26 Exfoliation ratings and SCC thresholds for aluminum-lithium alloy 8090

Temper	Product	Microstructure	Exfoliation rating(a): EXCO test(b)	MASTMAASIS test(c)	Atmospheric exposure	SCC threshold
8090-T81 (underaged)	Sheet	Recrystallized	EA	EA	P, EA	60% of yield strength in the L-T direction
8090-T8 (peak aged)	Sheet	Recrystallized	ED	EA	P	
8090-T8510/11 (peak aged)	Extrusions	Unrecrystallized	...	...	...	75% of yield strength in the L-T direction
8090-T8771, 8090-T651 (peak aged)	Plate	Unrecrystallized	Surface P	...	Surface P	105–140 MPa (15–20 ksi) short-transverse threshold
8090-T851	Plate	Unrecrystallized	EC(d)	EB(d)	P, EA	
8090-T8 (peak aged)	Sheet	Unrecrystallized	EC	EB	...	75% of yield strength in the L-T direction
8090 (peak aged)	Forgings	Unrecrystallized	...	...	...	140 MPa (20 ksi) short-transverse threshold

(a) Exfoliation rating per ASTM G 34: P, pitting; EA, superficial—tiny blisters, thin slivers, flakes or powders with only slight separation of metal; EB, moderate, notable layering and penetration in metal; ED, very severe—penetration to a considerable depth and loss of metal. (b) Exfoliation corrosion test per ASTM G 34. (c) MASTMAASIS, modified ASTM acetic acid salt intermittent spray. (d) Rating at a plane location of T/2, where T is plate thickness

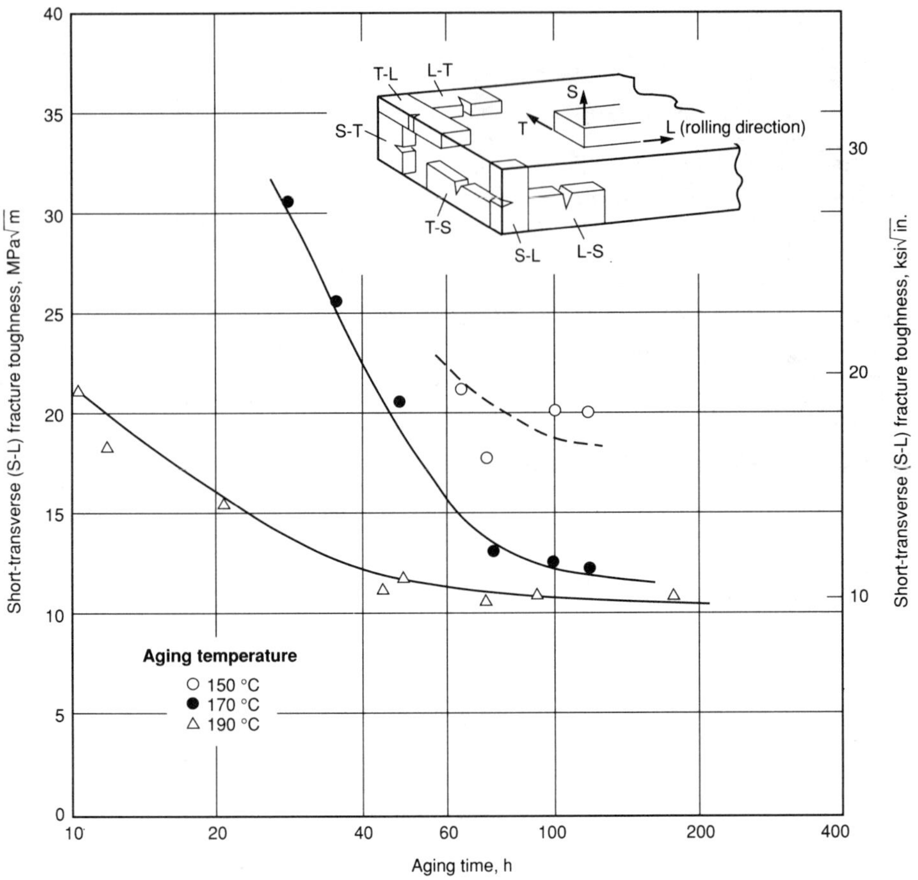

Fig. 23 Effect of aging temperature and time on the short-transverse (S-L) toughness of 8090-T651 plate. The S-L designation indicates a crack plane perpendicular to the S axis and a crack growth in the longitudinal (L) direction.

with a higher fatigue stress for both smooth and notched specimens.

The fatigue performance of 8090 forgings is similar to that of 8090 sheet.

Forming. Bend tests indicate that 8090 has a lower material springback than that associated with conventional aluminum alloys (Fig. 29). Materials in the as-received condition were subjected to a bend of 90°, approaching at 0°, 45°, and 90° to the rolling direction. The average results for 8090 were lower than those for the other two alloys. Tests on 8090 going down to $2t$ bend radius were successfully completed without damage to the testpiece. Tests with an unaged (solutionized) temper of 8090 produced springback results averaging 2° lower than those for traditional alloy 2014. However, 2014 shows a relatively constant springback with regard to orientation to the rolling direct, whereas the 8090 alloy shows a marked increase after 45°.

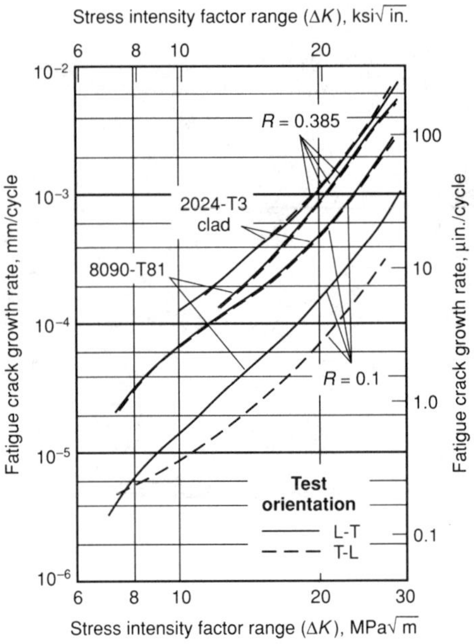

Fig. 25 Effect of test orientation and stress ratio (R), on the fatigue crack growth rates of 8090-T81 and clad 2024-T3 sheet. L-T, crack plane and direction perpendicular to the principal direction of rolling; T-L, crack plane and direction parallel to the principal direction of rolling

Careful preparation of flanged hole surfaces is required to minimize edge cracking when using 8090 in the T8 temper condition. With satisfactory edge preparation, holes can be formed with good results.

Stretch forming is a process in which sheet is wrapped around a forming block and then control stretched to form items such as airfoil sections. The process has been carried out with sufficient success to indicate that it can be used to produce typical aircraft forms. The following factors need to be considered when stretch forming 8090:

- 8090 forms better in the solutionized condition
- $2t$ bends appear to be practical
- Bending springback values may be less than those for conventional alloys

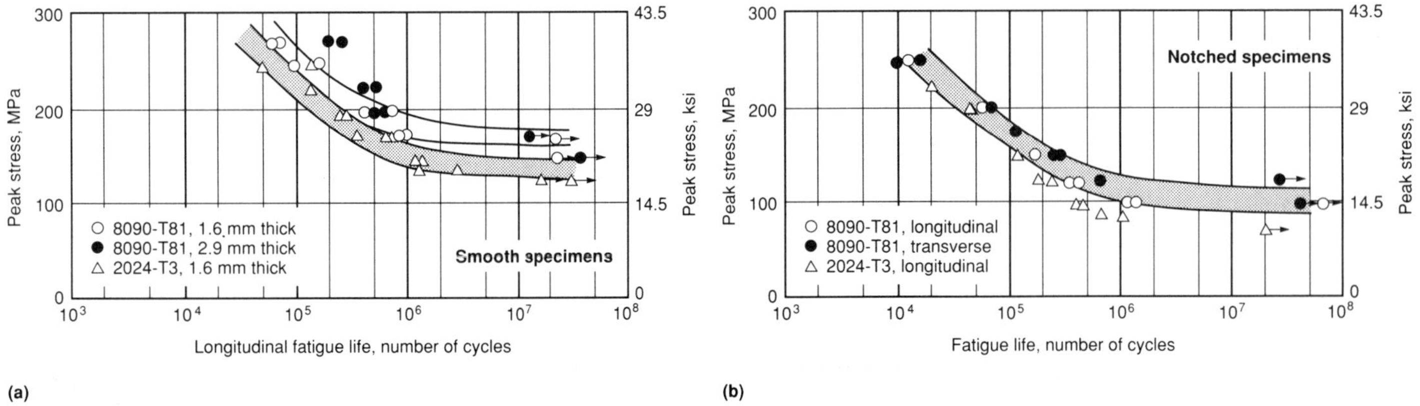

Fig. 24 *S-N* fatigue life curves for damage-tolerant 8090 sheet (8090-T81) and clad 2024-T3 sheet. Stress ratio (R), 0.1; cyclic frequency, 80 Hz. (a) Smooth specimens. Theoretical stress concentration factor (K_t) = 1. (b) Notched specimens. K_t = 2.6

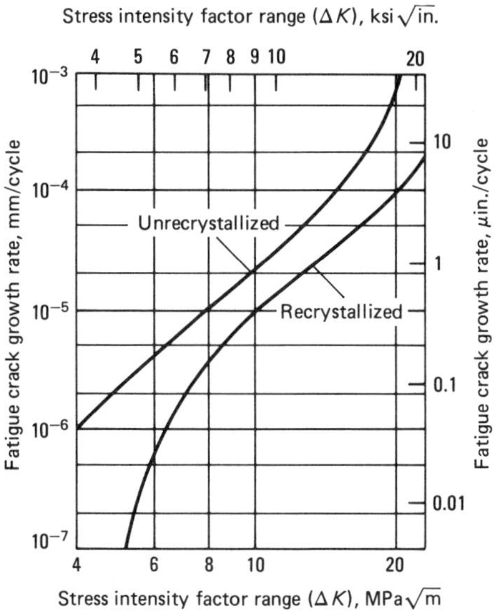

Fig. 26 Effect of microstructure on the fatigue crack growth rate of 8090 sheet under constant-amplitude loading

- Flanged holes require a fine edge finish

Finishing Characteristics. The finishing work done to date indicates that 8090 can be chemically milled with existing solutions to achieve a satisfactory surface roughness (Table 27). The higher etch rate of 8090 necessitates batch processing of parts; however, better protection against corrosive environments is obtained with anodized 8090 than with conventional alloys (Fig. 30).

Welding. Gas metal arc, gas tungsten arc, and electron beam welding processes have been used to evaluate the fusion weldability characteristics of 8090. Results with all three methods indicate that 8090 is commercially weldable (Table 28).

Table 27 Response of aluminum-lithium alloy 8090 to chemical milling

Data for 2024 provided for comparison

Alloy	Etch rate, mm/min	Etch rate, mils/min	Undercut ratio	Setback	Surface finish, RHR(a): Before chemical milling	Surface finish, RHR(a): After chemical milling
8090	0.084	3.3	1.3	0.3	140	55
2024	0.066	2.6	1.0	0.4	20	35

(a) RHR, roughness height rating

REFERENCES

1. K.T. Venkateswara Rao, W. Yu, and R.O. Ritchie, Fatigue Crack Propagation in Aluminum-Lithium Alloy 2090: Part II. Small Crack Behavior, *Metall. Trans. A*, Vol 19A, March 1988, p 563-569
2. R.S. James, Commercial Development of Al-Li Products, *AGARD (NATO)*, No. 444, 1988, p 3
3. E.A. Starke, Jr., T.H. Sanders, Jr., and I.G. Palmer, New Approaches to Alloy Development in the Al-Li System, *J. Met.*, 1981, p 24-33
4. K.K. Sankaran and N.J. Grant, Structure and Properties of Splat Quenched 2024-Aluminum Alloy Containing Lithium Additions, in *Aluminum-Lithium Alloys*, The Metallurgical Society of AIME, 1981, p 205-227
5. B. Noble and G.E. Thompson, *Met. Sci. J.*, Vol 5, 1971, p 114
6. D.B. Williams and J.W. Edington, *Met. Sci. J.*, Vol 9, 1974, p 529
7. T.H. Sanders, Jr., *Mater. Sci. Eng.*, Vol 43, 1980, p 247
8. K. Sadananda and K.V. Jata, Creep Crack Growth Rate of Two Al-Li Alloys, *Metall. Trans. A*, Vol 19A, 1988, p 847-854
9. T.H. Sanders, Jr., Final Report, Contract N62269-74-C-0438, Naval Air Development Center, June 1976

Table 28 Tensile behavior of 8090-T6 weldments

Data for two other weldable aluminum alloys provided for comparison

Alloy	Filler	Heat treatment	Tensile strength, MPa	Tensile strength, ksi	Yield strength, MPa	Yield strength, ksi	Elongation in 50 mm (2 in.), %	Specific tensile strength(a), MPa/g/cm³	Specific tensile strength(a), ft × 10³	Specific yield strength(b), MPa/g/cm³	Specific yield strength(b), ft × 10³
Unwelded 8090-T6	...	...	504	73	429	62.2	6	198	795	168	674
8090-T6	Al	As-welded	165	24	137	19.9	5	65	260	54	215
8090-T6	Al-5Si	As-welded	205	30	165	24	3	81	325	65	260
8090-T6	Al-5Mg	As-welded	228	33	176	25.5	4	90	360	70	280
8090-T6	Al-5Mg	As-welded + T6 temper	302	43.8	245	35.5	4	119	478	97	390
7017-T6	Al-5Mg	30 days natural aging	340	49	220	32	8	122	490	79	317
8090-T6	Al-5Mg (+Zr)	As-welded	235	34	183	26.5	4	93	373	72	290
8090-T6	8090	As-welded	310	45	285	41	2	123	494	113	454
8090-T6	8090	As-welded + T6 temper	367	53.2	315	45.5	4	145	582	124	498
2219-T851	2319	As-welded	300	43.5	185	27	5	105	420	65	260

(a) Specific tensile strength is the ratio of tensile strength to density. (b) Specific yield strength is the ratio of yield strength to density.

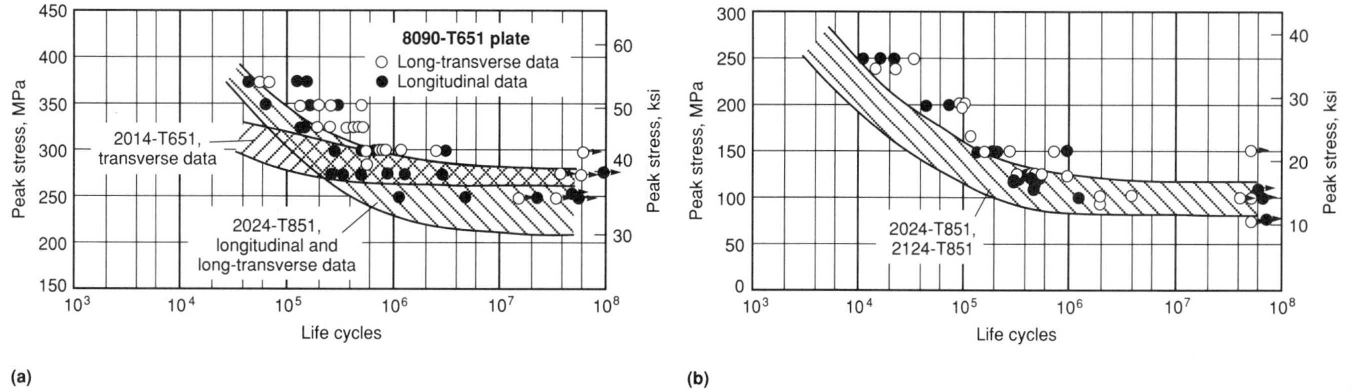

Fig. 27 *S-N* fatigue data for 8090-T651 and selected 2xxx alloys. Stress ratio (R), 0.1. Specimens tested in laboratory air. (a) Smooth specimens. (b) Notched specimens (K_t = 3)

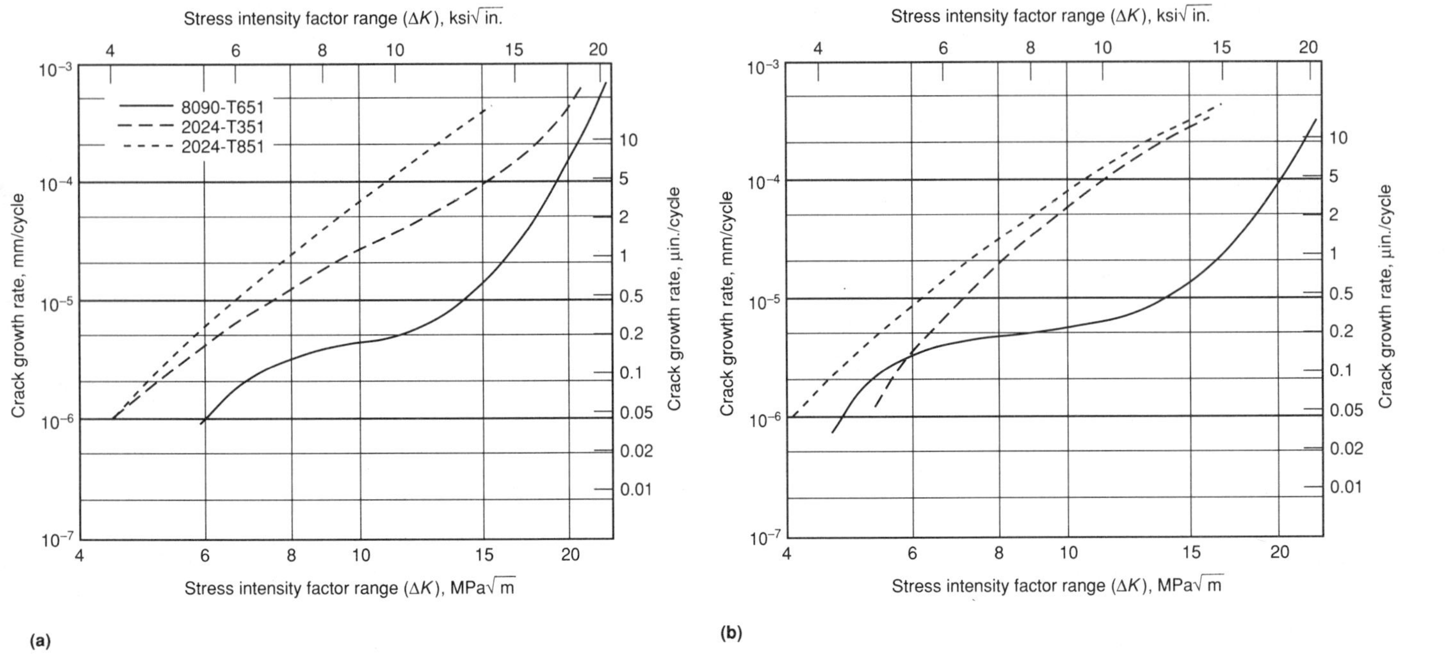

Fig. 28 Fatigue crack growth curves for 8090-T651 and two tempers of 2024 plate. Specimens tested in laboratory air. Stress ratio (*R*), 0.1. (a) L-T specimens (crack plane and direction perpendicular to the principal direction of rolling). (b) T-L specimens (crack plane and direction parallel to the principal direction of rolling)

10. T.H. Sanders, Jr., Final Report, Contract N62269-74-0271, Naval Air Development Center, June 1979
11. G.E. Thompson and B. Noble, *J. Inst. Met.*, Vol 101, 1973, p 111
12. O. Realeaux, Scleron Alloys, *J. Inst. Met.*, Vol 33, 1925, p 346
13. E.H. Spuhler, Alcoa Alloy X-2020, *Alcoa Green Lett.*, 1958, p 156-9-58
14. I.N. Fridlyander, V.F. Shamray, and N.V. Shiryayera, Phase Composition and Mechanical Properties of Aluminum Alloys Containing Magnesium and Lithium, *Russ. Metall.*, No. 2, 1965, p 83-90; translated from *Izv. Akad. Nauk SSSR, Met.*
15. E.S. Balmuth and R. Smith, A Perspective on the Development of Aluminum-Lithium Alloys, in *Aluminum-Lithium Alloys*, The Metallurgical Society of AIME, 1981, p 69-88
16. J.C. Ekvall, J.E. Rhodes, and G.G. Wald, Methodology for Evaluating Weight Savings From Basic Material Properties, in *Design of Fatigue and Fracture Resistant Structures*, STP 761, American Society for Testing and Materials, 1982, p 341
17. J. Koshorst, Point of View of a Civil Aircraft Manufacturer on Al-Li Alloys, *AGARD (NATO)*, No. 444, 1988, p 18-1 to 18-5
18. A.F. Smith, A Comparison of Large AA8090, AA8091 and AA7010 Die Forgings for Helicopter Structural Applications, in *Aluminum-Lithium Alloys*, Vol III, Materials and Component Engineering Publications, 1989, p 1587-1596
19. "Cooperative Test Program for the Evaluation of Engineering Properties of Al-Li Alloy 2090-T8X Sheet, Plate and Extrusion Products," Final Report, Contract N60921-84-C-0078, U.S. Department of the Navy, Naval Surface Warfare Center, 15 Sept 1989
20. Y. Barbaux, Properties of Commercially Available Al-Li Alloys for Possible Use on Civil Aircraft, in *Aluminum-Lithium Alloys*, Vol III, Materials and Component Engineering Publications, 1989, p 1667-1676
21. G.J.H. Vaessen, C. Van Tilborgh, and H.W. van Rooijen, Fabrication of Test Articles From Al-Li 2091 for Fokker 100, *AGARD (NATO)*, No. 444, 1988, p 13-1 to 13-12
22. P.O. Wakeling, P.E. Bretz, G. LeRoy, C.J. Peel, and W.E. Quist, "Aerospace Applications of Al-Li Alloys," Paper presented at the Symposium of the World Materials Congress (Chicago, IL), Sept 1988

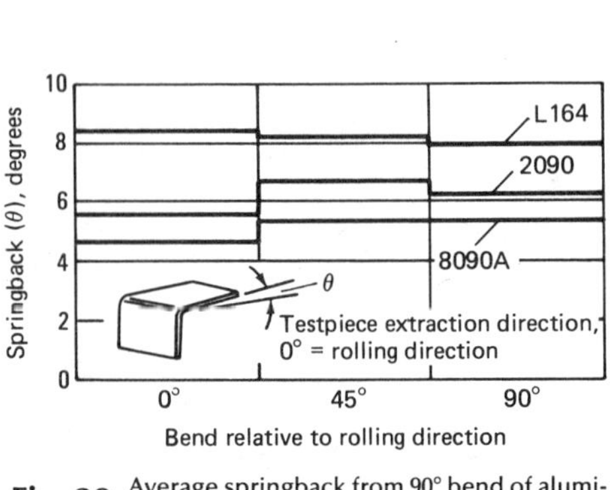

Fig. 29 Average springback from 90° bend of aluminum-lithium alloy 8090 and two conventional alloys. All three alloys were tested in the as-received condition.

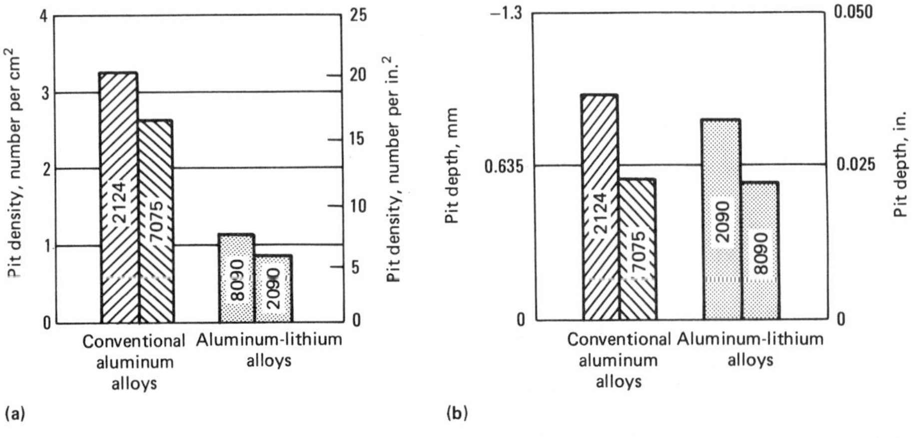

Fig. 30 Pitting characteristics of selected aluminum and aluminum-lithium alloys exposed to an SO_2 salt fog for 32 days. (a) Pit density. (b) Pit depth

23. A.F. Smith, in *Aluminum-Lithium—Development, Applications and Superplastic Forming*, American Society for Metals, 1986, p 19-20
24. C.J. Peel, Current Status of the Application of Conventional Aluminum-Lithium Alloys and the Potential for Future Developments, *AGARD (NATO)*, No. 444, 1988, p 21-1 to 21-9
25. *Aluminum-Lithium—Development, Application and Superplastic Forming*, American Society for Metals, 1986, p 188-226
26. E.J. Tuegel, V.M. Vasey-Glandon, M.O. Pruitt, and K.K. Sankaran, Forming of Aluminum-Lithium Sheet for Fighter Aircraft Applications, in *Aluminum-Lithium Alloys*, Materials and Component Engineering Publications, Vol III, 1989, p 1597-1606
27. J.C. Johnson, "Aluminum-Lithium Applications for Hornet 2000," Paper presented at the Southeastern Regional Conference of the Society of Allied Weight Engineers (Dayton, OH), 1988
28. B.A. Davis, Aluminum-Lithium: Application of Plate and Sheet to Fighter Aircraft, *AGARD (NATO)*, No. 444, 1988, p 20-1 to 20-5
29. P.O. Wakeling, S.D. Forness, and E.A.W. Heckman, The Use of 8090 in the McDonnell-Douglas F15 SMTD Aircraft, in *Aluminum-Lithium Alloys—Design, Development and Application Update*, ASM INTERNATIONAL, 1988, p 339-340
30. J. Waldman, R. Mahapatra, C. Neu, and A.P. Divecha, Aluminum-Lithium Alloys for Naval Aircraft, in *Aluminum-Lithium Alloys—Design, Development and Application Update*, ASM INTERNATIONAL, 1988, p 341-356
31. K.K. Sankaran, V.M. Vasey-Glandon, and E.J. Tuegel, Application of Aluminum-Lithium Alloys to Fighter Aircraft, in *Aluminum-Lithium Alloys*, Vol III, Materials and Component Engineering Publications, 1989, p 1625-1631
32. S.L. Langenbeck, I.F. Sakata, and R.R. Sawtell, Aerospace Structural Application of Aluminum-Lithium, in *Aluminum-Lithium—Development, Application and Superplastic Forming*, American Society for Metals, 1986, p 188-226
33. J. Glazer *et al.*, *Metall. Trans. A*, Vol 18A, 1987, p 1695-1701
34. R.P. Reed, N.J. Simon, J.D. McColskey, J.R. Berger, C.N. McCowan, E.S. Drexler, and R.P. Walsh, "Aluminum Alloys for ALS Cryogenic Tanks: Oxygen Compatibility Interim Report," National Institute of Standards and Technology, Jan 1990
35. F.M. Page, A.T. Chamberlain, and R. Grimes, The Safety of Molten Aluminum-Lithium Alloys in the Presence of Coolants, *J. Phys. (Orsay)*, Sept 1987
36. C.L. Laszcz-Davis, S.G. Epstein, R.P. Hancock, D.R. Hudgins, and R.M. James, "The Safety, Health and Recycling Aspects of Aluminum-Lithium Alloys," Aluminum Association, p 1-3
37. "Aluminum Alloys Containing Lithium," Material Safety Data Sheet 337N, Hazardous Materials Control Committee
38. W.R. Wilson, J. Worth, E.P. Short, and C.F. Pygall, Recycling of Aluminum-Lithium Process Scrap, *J. Phys. (Orsay)*, Sept 1987
39. W.R. Wilson, D.J. Allan, O. Stenzel, M. Lorke, K.W. Krone, and C. Seebauer, Aluminum-Lithium Scrap Recycling by Vacuum Distillation, *Aluminum-Lithium Alloys*, Vol III, Materials and Component Engineering Publications, 1989, p 473-496
40. Alcoa Studies Plan to Recycle Customer's Aluminum-Lithium Scrap, *To Aerosp. Ind.*, Vol 5, Dec 1986
41. T.J. Robare, J.J. Witters, G.M. Kallmeyer, and R.H. Keenan, "Recycling of Aluminum-Lithium Alloy Scrap," Alcoa, 1989
42. T.J. Langan and J.R. Pickens, Identification of Strengthening Phases of Al-Cu-Li Alloy Weldalite^{049}, in *Aluminum-Lithium Alloys*, Vol II, Materials and Component Engineering Publications, 1989, p 691-700
43. F.W. Gayle, F.H. Heubaum, and J.R. Pickens, Natural Ageing and Reversion Behaviour of Al-Cu-Li-Ag-Mg Alloy Weldalite^{049}, in *Aluminum-Lithium Alloys*, Vol II, Materials and Component Engineering Publications, 1989, p 701-710
44. K. Moore, T.J. Langa, F.H. Heuaum, and J.R. Pickens, Effect of Cu Content on the Corrosion and Stress-Corrosion Behavior of Al-Cu-Li Weldalite™ Alloys, in *Aluminum-Lithium Alloys*, Vol III, Materials and Component Engineering Publications, 1989, p 1281-1287
45. A. Cho, R.F. Ashton, G.W. Steele, and J.L. Kirby, Status of Weldable Al-Li Alloys (Weldalite^{049}) Development at Reynolds Metals Company, in *Aluminum-Lithium Alloys*, Vol III, Materials and Component Engineering Publications, 1989, p 1377-1386
46. J.R. Pickens, F.H. Heubaum, T.J. Langan, and L.S. Kramer, Al-(4.5-6.3)Cu-1.3Li-0.4Ag-0.4Mg-0.14Zr Alloy Weldalite^{049}, in *Aluminum-Lithium Alloys*, Volume III, Materials and Component Engineering Publications, 1989, p 1397-1414
47. M.D. Goodyear, Alcoa Alloy 2090, *Alcoa Green Lett.*, April 1989
48. E.L. Colvin and S.J. Murtha, Exfoliation Corrosion Testing of Al-Li Alloys 2090 and 2091, in *Aluminum-Lithium Alloys*, Vol III, Materials and Component Engineering Publications, 1989, p 1251-1260
49. C.J. Warren and R.J. Rioja, Forming Characteristics and Post-Formed Properties of Al-Li Alloys, in *Aluminum-Lithium Alloys*, Vol I, Materials and Component Engineering Publications, 1989, p 417-430
50. J.H. Powers, "Chem Milling Response of Aluminum-Lithium Alloys," Alcoa Laboratories Technical Report 52-88-10, Oct 1988
51. J.H. Powers, "Finishing Response of Aluminum-Lithium Alloys," Alcoa Laboratories Technical Report 52-89-04, Jan 1989
52. W.F. Johnson, "Finishing Response of Aluminum-Lithium 2090 and 2091 Alloys," Alcoa Laboratories Technical Report 52-88-3, July 1988
53. M. Doudeau, P. Meyer, and D. Constant, Al-Li Alloys Developed by Pechiney, *AGARD (NATO)*, No. 444, Oct 1988, p 2
54. C.J.E. Smith, Corrosion and Stress Corrosion of Aluminum-Lithium Alloys, *AGARD (NATO)*, No. 444, Oct 1988, p 7
55. W. Zink, J. Weilke, L. Schwarmann, and K.H. Rendigs, Investigation on Sheet Material of 8090 and 2091 Aluminum-Lithium Alloys, *AGARD (NATO)*, No. 444, Oct 1988, p 9
56. Y. Barbaux, Properties Des Alliages Al-Li, *AGARD (NATO)*, No. 444, Oct 1988, p 8

High-Strength Aluminum P/M Alloys

J.R. Pickens, Martin Marietta Laboratories

POWDER METALLURGY (P/M) technology provides a useful means of fabricating net-shape components that enables machining to be minimized, thereby reducing costs. Aluminum P/M alloys can therefore compete with conventional aluminum casting alloys, as well as with other materials, for cost-critical applications. In addition, P/M technology can be used to refine microstructures compared with those made by conventional ingot metallurgy (I/M), which often results in improved mechanical and corrosion properties. Consequently, the usefulness of aluminum alloys for high-technology applications, such as those in aircraft and aerospace structures, is extended. This article describes and reviews the latter of these two areas of aluminum P/M technology where high strength and improved combinations of properties are obtained by exploiting the inherent advantages of P/M for alloy design.

The metallurgical reasons for the microstructural refinement made possible by P/M are discussed. The two broad high-strength P/M technologies—rapid solidification (RS) and mechanical attrition (mechanical alloying/dispersion strengthening)—are described. The various steps in aluminum P/M technology are explained to produce an appreciation for the interrelationship between powder processing and resultant properties. Finally, the major thrust areas of P/M alloy design and development are reviewed and some properties of the leading aluminum P/M alloys discussed. No attempt is made to provide design-allowable mechanical properties, and the data presented for the various P/M alloys may not be directly comparable because of differences in product forms. Nevertheless, the properties presented will enable the advantages of aluminum P/M alloys to be appreciated. Greater details of aluminum P/M alloys are provided in other reviews (Ref 1 to 7). Conventional pressed and sintered aluminum P/M alloys for less demanding applications are described in the Appendix to this article.

Advantages of Aluminum P/M Technology

Aluminum alloys have numerous technical advantages that have enabled them to be one of the dominant structural material families of the 20th century. Aluminum has low density (2.71 g/cm^3) compared with competitive metallic alloy systems, good inherent corrosion resistance because of the continuous, protective oxide film that forms very quickly in air, and good workability that enables aluminum and its alloys to be economically rolled, extruded, or forged into useful shapes. Major alloying additions to aluminum such as copper, magnesium, zinc, and lithium—alone, or in various combinations—enable aluminum alloys to attain high strength. Designers of aircraft and aerospace systems generally like using aluminum alloys because they are reliable, reasonably isotropic, and low in cost compared to more exotic materials such as organic composites.

Aluminum alloys do have limitations compared with competitive materials. For example, Young's modulus of aluminum (about 70 GPa, or 10 × 10^6 psi) is significantly lower than that of ferrous alloys (about 210 GPa, or 30 × 10^6 psi) and titanium alloys (about 112 GPa, or 16 × 10^6 psi). This lower modulus is almost exactly offset by the density advantage of aluminum compared to iron- and titanium-base alloys. Nevertheless, designers could exploit higher-modulus aluminum alloys in many stiffness-critical applications.

Although aluminum alloys can attain high strength, the strongest such alloys have often been limited by stress-corrosion cracking (SCC) susceptibility in the highest-strength tempers. For example, the high-strength 7*xxx* alloys (Al-Zn-Mg and Al-Zn-Mg-Cu) can have severe SCC susceptibility in the highest-strength (T6) tempers. To remedy this problem, overaged (T7) tempers have been developed that eliminate SCC susceptibility, but with a 10 to 15% strength penalty.

The melting point of aluminum, 660 °C (1220 °F), is lower than that of the major competitive alloy systems: iron-, nickel-, and titanium-base alloys. As might be expected, the mechanical properties of aluminum alloys at elevated temperatures are often not competitive with these other systems. This limitation of aluminum alloys is of particular concern to designers of aircraft and aerospace structures, where high service temperatures preclude the use of aluminum alloys for certain structural components.

The number of alloying elements that have extensive solid solubility in aluminum is relatively low. Consequently, there are not many precipitation-hardenable aluminum alloy systems that are practical by conventional I/M. This can be viewed as a limitation when alloy developers endeavor to design improved alloys. Aluminum P/M technology enables the aforementioned limitations of aluminum alloys to be overcome to various extents, while still maintaining most of the inherent advantages of aluminum.

Structure/Property Benefits. The advantages of P/M stem from the ability of small particles to be processed. This enables:

- The realization of RS rates
- The uniform introduction of strengthening features, that is, barriers to dislocation motion, from the powder surfaces

The powder processes of rapid solidification and mechanical attrition lead to microstructural grain refinement and, in general, better mechanical properties of the alloy. Specifically, the smaller the mean free path between obstacles to dislocation motion, the greater the strengthening. In addition, finer microstuctural features are also less apt to serve as fracture-initiating flaws, thereby increasing toughness.

The RS rates made possible by P/M enable microstructural refinement by several methods. For example, grain size can be reduced because of the short time available for nuclei to grow during solidification. Finer grain size results in a smaller mean free path between grain boundaries, which are effective barriers to dislocation motion, leading to increased "Hall-Petch" strengthening. In addition, RS can extend the alloying limits in aluminum by enhancing supersaturation and thereby enabling greater precipitation hardening without the harmful segregation effects from overalloyed I/M alloys. Moreover, elements that are essentially insoluble in the solid state, but have

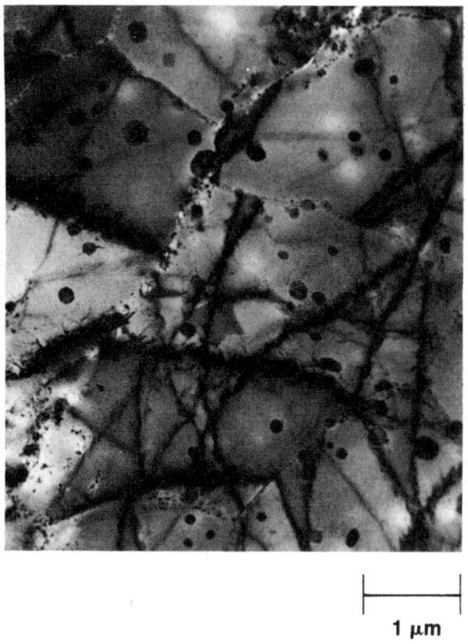

Fig. 1 Experimental Al-Zn-Mg-Cu-Co RS alloy that contains a fine distribution of spherical Co_2Al_9 particles and fine grain size (2 to 5 μm). Courtesy of L. Christodoulou, Martin Marietta Laboratories

significant solubility in liquid aluminum, can be uniformly dispersed in the powder particles during RS. This can lead to the formation of novel strengthening phases that are not possible by conventional I/M, while also suppressing the formation of equilibrium phases that are deleterious to toughness and corrosion resistance. The photomicrograph shown in Fig. 1 exemplifies the microstructural refinement possible by RS-P/M processing. This experimental Al-6.7Zn-2.4Mg-1.4Cu-0.8Co alloy has a grain size of 2 to 5 μm and a fine distribution of Co_2Al_9 particles of about 0.1 to 0.4 μm. Comparable I/M alloys have a grain size that is almost an order of magnitude larger. In addition, making this alloy by I/M would lead to very coarse cobalt-containing particles because of the low solid-solubility of cobalt in aluminum. These coarse particles would significantly degrade toughness. Thus, RS processes constitute one of the two major classes of high-strength P/M technology.

The other class of high-strength P/M technology relies on the introduction of strengthening features from powder surfaces, which can be accomplished on a fine scale because of the high surface-area-to-volume ratio of the powder particles. Most, but not all, such processes involve ball milling and are called mechanical attrition processes.

Oxides can be easily introduced from powder surfaces by consolidating the powder and hot working the product (see the section "Mechanical Attrition Process" in this article). The fine oxides can improve strength by oxide dispersion strengthening (ODS), and in addition, by substructural strengthening resulting from the dislocations created during working, which are generated by dislocation-oxide interactions. However, ODS can be increased by mechanically attriting the powder particles to more finely disperse the oxides. Ball milling in the presence of organic surfactants allows carbides to be dispersed in a similar fashion. Finally, ball milling aluminum and other powders can enable fine intermetallic dispersoids to form during milling or subsequent thermomechanical processing.

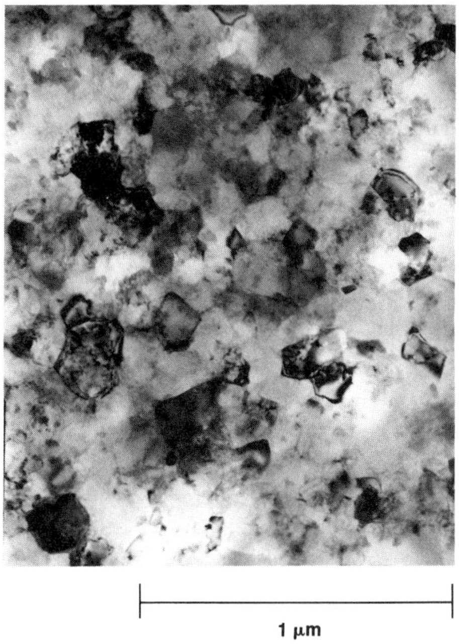

Fig. 2 Experimental mechanically alloyed Al-1.5Li-0.9O-0.6C alloy featuring an ultrafine grain/subgrain size. Courtesy of J.R. Pickens, unpublished research

The interplay between dispersed oxides, carbides (or other ceramics), intermetallics, and dislocations during hot, warm, or cold working can enable great refinement in grain and subgrain size. This can result in significant strengthening because grain and subgrain boundaries can be effective barriers to dislocation motion. For example, an experimental Al-1.5Li-0.9O-0.6C mechanically alloyed material had an extremely fine grain/subgrain size of 0.1 μm (Fig. 2). The ultrafine oxides and carbides, coupled with this fine grain size, enabled the alloy to have a tensile strength of nearly 800 MPa (115 ksi) with 3% elongation. An I/M alloy containing this much oxygen and carbon that is effectively contained in finely dispersed particles is not viable.

P/M processing, including both rapid solidification and mechanical attrition processes, provides alloy designers with additional flexibility resulting from the inherent advantage of working with small bits of matter. In addition to refinement of strengthening features, different metallic powders can be mixed to form duplex microstructures. Furthermore, ceramic powders may be mixed with aluminum alloy powders to form metal-matrix composites (MMCs). However, the fine particles of matter must be consolidated and formed into useful shapes. The potential benefits of P/M technology can be realized, or lost, in the critical consolidation-related and forming processes.

Aluminum P/M Processing

There are several steps in aluminum P/M technology that can be combined in various ways, but they will be conveniently described in three general steps:

- Powder production
- Powder processing (optional)
- Degassing and consolidation

Powder can be made by various RS processes including atomization, splat quenching to form particulates, and melt spinning to form ribbon. Alternatively, powder can be made by non-RS processes such as by chemical reactions including precipitation, or by machining bulk material. Powder-processing operations are optional and include mechanical attrition (for example, ball milling) to modify powder shape and size or to introduce strengthening features, or comminution such as that used to cut melt-spun ribbon into powder flakes for subsequent handling.

Aluminum has a high affinity for moisture, and aluminum powders readily adsorb water. The elevated temperatures generally required to consolidate aluminum powder causes the water of hydration to react and form hydrogen, which can result in porosity in the final product, or under confined conditions, can cause an explosion. Consequently, aluminum powder must be degassed prior to consolidation. This is often performed immediately prior to consolidation at essentially the same temperature as that for consolidation to reduce fabrication costs. Consolidation may involve forming a billet that can be subsequently rolled, extruded, or forged conventionally, or the powder may be consolidated during hot working directly to finished-product form. The various steps in aluminum P/M technology will now be discussed in greater detail.

Powder Production

Atomization is the most widely used process to produce aluminum powder. Aluminum is melted, alloyed, and sprayed through a nozzle to form a stream of very fine particles that are rapidly cooled—most often by an expanding gas. Atomization techniques have been reviewed extensively in Ref 8 and 9 and in *Powder Metallurgy*, Volume 7 of the 9th Edition of *Metals Handbook*. Cooling rates of 10^3 to 10^6 K/s are typically obtained.

Splat cooling is a process that enables cooling rates even greater than those obtained in atomization. Aluminum is melted and alloyed, and liquid droplets are sprayed or dropped against a chilled surface of high thermal conductivity—for example, a copper wheel that is water cooled internally. The resultant splat particulate is removed from the rotating wheel to allow subsequent droplets to contact the bare, chilled surface. Cooling rates of 10^5 K/s are typical, with rates up to 10^9 K/s reported.

Melt-spinning techniques are somewhat similar to splat cooling. The molten aluminum alloy rapidly impinges a cooled, rotating wheel, producing rapidly solidified product that is often in ribbon form. The leading commercial melt-spinning process is the planar flow casting (PFC) process developed by M.C. Narisimhan and co-workers at Allied-Signal Inc. (Ref 10). The liquid stream contacts a rotating wheel at a carefully controlled distance to form a thin, rapidly solidified ribbon and also to reduce oxidation. The ribbon could be used for specialty applications in its PFC form but is most often comminuted into flake powder for subsequent degassing and consolidation.

In the aforementioned RS powder-manufacturing processes, partitionless (that is, no segregation) solidification can occur, and with supersaturation RS can ultimately lead to greater precipitation strengthening. Furthermore, with the highest solidification rates, crystallization can be suppressed. The novelties of RS aluminum microstructures and nonequilibrium phase considerations have been reviewed (Ref 11, 12).

Other Methods. Powder can also be made from machining chips or via chemical reactions (Ref 1, 13). Such powders should be carefully cleaned before degassing and consolidation.

Mechanical Attrition Process

Mechanical attrition processes often involve ball milling in various machines and environments. Such processes can be used to control powder size and distribution to facilitate flow or subsequent consolidation, introduce strengthening features from powder surfaces, and enable intermetallics or ceramic particles to be finely dispersed. The two leading mechanical attrition processes today—mechanical alloying in the United States, and reaction milling in Europe—are improvements on sinter-aluminum-pulver (SAP) technology developed by Irmann in Austria (Ref 14, 15).

SAP Technology. In 1946, Irmann and co-workers were preparing rod specimens for spectrographic analysis by hot pressing mixtures of pure aluminum and other metal powders. They noticed the unexpectedly high hardness of the resulting rods. Mechanical property evaluations revealed that the hot-pressed material had strength approaching that of aluminum structural alloys. Based on microstructural evaluations, Irmann attributed the high strength of the hot-pressed compacts to the breakdown of the surface oxide film on the powder particles during hot pressing. Irmann performed mechanical tests at elevated temperatures and showed that these alloys not only had surprisingly high elevated-temperature strength, but retained much of their room-temperature strength after elevated-temperature exposure.

Irmann also ball milled aluminum powder and noticed that it was similar to the flaky powder used in paint pigment. Hüttig studied the formation of the flaky powder during ball milling and noticed the competition between comminution and welding of the powder particles (Ref 16). This fracture and welding of the powder particles caused the surface oxide film to be ruptured and become somewhat dispersed in the powder particles. Irmann called these hot-pressed materials sinter-aluminum-pulver (SAP), later referred to as sintered aluminum powder or sintered aluminum product in the United States. Various SAP alloys were developed in the 1950s that displayed excellent elevated-temperature properties (Ref 1).

The mechanical alloying process is a high-energy ball-milling process that employs a stirred ball mill called an attritor, a shaken ball mill, or a conventional rotating ball mill (Ref 17, 18). The process is performed in the presence of organic surfactants, for example, methanol, stearic acid, and graphite, to control the cold welding of powder particles and provide oxygen and carbon for dispersion strengthening (Ref 19, 20). Mechanical alloying claims the advantage of milling under "dry" conditions—that is, not in mineral oil, as in some SAP processing, which must be removed after milling—although dry SAP ball milling in the presence of stearic acid was reported in the 1950s (Ref 1).

Elemental powders may be milled with aluminum powders to effect solid-solution strengthening or to disperse intermetallics. The dispersed oxides, carbides, and/or intermetallics create effective dislocation sources during the milling process and suppress dislocation annihilation during subsequent working operations, which result in greatly increased dislocation density. Thus, mechanical alloying enables the effective superimposition of numerous strengthening mechanisms (Ref 2, 19-21), including:

- Oxide dispersion
- Carbide dispersion
- Fine grain size
- High dislocation density and substructure
- Solid-solution strengthening

Consequently, high strength can be obtained without reliance on precipitation strengthening, which may introduce problems such as corrosion and SCC susceptibility, and propensity for planar slip. Nevertheless, the aforementioned five strengthening contributions can be augmented by precipitation strengthening as well as intermetallic dispersion strengthening.

Reaction milling is another mechanical attrition process derived from SAP technology (Ref 22-24). It is extremely similar to mechanical alloying, and subtle differences between the two processes appear to have little appreciable effect on the compositions that can be processed and the resulting microstructures produced. Investigations in Europe have also successfully superimposed numerous strengthening features in aluminum P/M alloys, as described above.

Powder Degassing and Consolidation

The water of hydration that forms on aluminum powder surfaces must be removed to prevent porosity in the consolidated product. Although solid-state degassing has been used to reduce the hydrogen content of aluminum P/M wrought products, it is far easier and more effective to remove the moisture from the powder. Degassing is often performed in conjunction with consolidation, and the most commonly used techniques are described below. The various aluminum fabrication schemes are summarized in Fig. 3.

Can Vacuum Degassing. This is perhaps the most widely used technique for aluminum degassing because it is relatively non-capital intensive. Powder is encapsulated in a can, usually aluminum alloys 3003 or 6061, as shown schematically in Fig. 4. A spacer is often useful to increase packing and to avoid safety problems when the can is welded shut. The author has found that packing densities are typically 60% of theoretical density when utilizing this method on mechanically alloyed powders. Care must be used to allow a clear path for evolved gases through the spacer to prevent pressure buildup and explosion.

To increase packing density, the powder is often cold isostatically pressed (CIP) in a reusable polymeric container before insertion into the can. Powder densities in the CIPed compact of 75 to 80% theoretical density are preferred because they have increased packing density with respect to loose-packed powder, yet allow sufficient interconnected porosity for gas removal. At packing densities of about 84% and higher, effective degassing is not possible for several atomized aluminum-alloy powders (Ref 25, 26). Furthermore, one must control CIP parameters to avoid inhomogeneous load transfer through the powder, which can lead to excessive density in the outer regions of the cylindrical compact and much lower densities in the center. Such CIP parameters often must be developed for a specific powder and compact diameter (see the article "Cold Isostatic Pressing of Metal Powders" in Volume 7 of the 9th Edition of *Metals Handbook*).

The canned powder is sealed by welding a cap that contains an evacuation tube as shown in Fig. 4. After ensuring that the can contains no leaks, the powder is vacuum

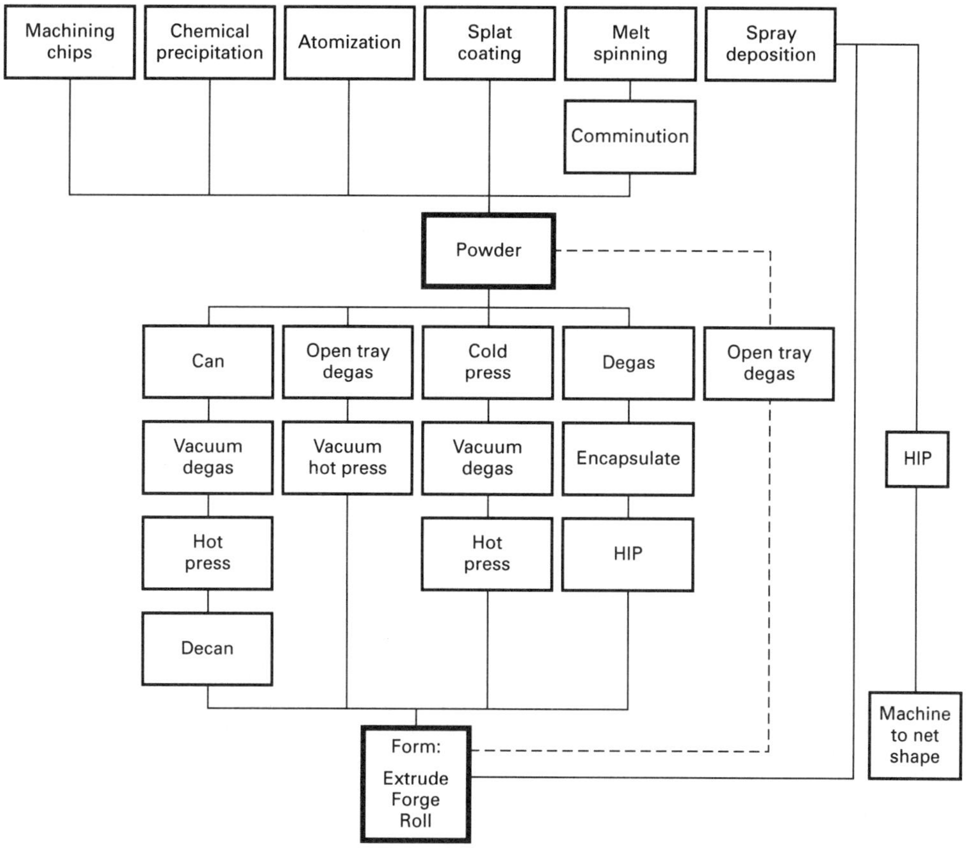

Fig. 3 Aluminum P/M fabrication schemes

degassed while heating to elevated temperatures. The rate in gas evolution as a function of degassing temperature depends on powder size, distribution, and composition. The ultimate degassing temperature should be selected based on powder composition, considering tradeoffs between resulting hydrogen content and microstructural coarsening. For example, an RS-P/M precipitation-hardenable alloy that is to be welded would likely be degassed at a relatively high temperature to minimize hydrogen content (hotter is not always better). Coarsening would not significantly decrease the strength of the resulting product because solution treatment and aging would be subsequently performed and provide most of the strengthening. On the other hand, a mechanically attrited P/M alloy that relies on substructural strengthening and will serve in a mechanically fastened application might be degassed at a lower temperature to reduce the annealing out of dislocations and coarsening of substructure.

When a suitable vacuum is achieved (for example, <5 millitorr), the evacuation tube is sealed by crimping. The degassed powder compact can then be immediately consolidated to avoid the costs of additional heating. An extrusion press using a blind die (that is, no orifice) is often a cost-effective means of consolidation.

Dipurative Degassing. Roberts and coworkers at Kaiser Aluminum & Chemical Corporation have developed an improved degassing method called "dipurative" degassing (Ref 27). In this technique, the vacuum-degassed powder, which is often canned, is backfilled with a dipurative gas (that is, one that effectively removes water of hydration) such as extra-dry nitrogen, and then re-evacuated. Several backfills and evacuations can be performed resulting in lower hydrogen content. In addition, the degassing can often be performed at lower temperatures to reduce microstructural coarsening.

Vacuum Degassing in a Reusable Chamber. The cost of canning and decanning adversely affects the competitiveness of aluminum P/M alloys. This cost can be alleviated somewhat by using a reusable chamber for vacuum hot pressing. The powder or CIPed compact can be placed in the chamber and vacuum degassed immediately prior to compaction in the same chamber. Alternatively, the powder can be "open tray" degassed, that is, degassed in an unconfined fashion, prior to loading into the chamber. The processing time to achieve degassing in an open tray can be much less than that required in a chamber or can, thereby increasing productivity. Care must be exercised to load the powder into the

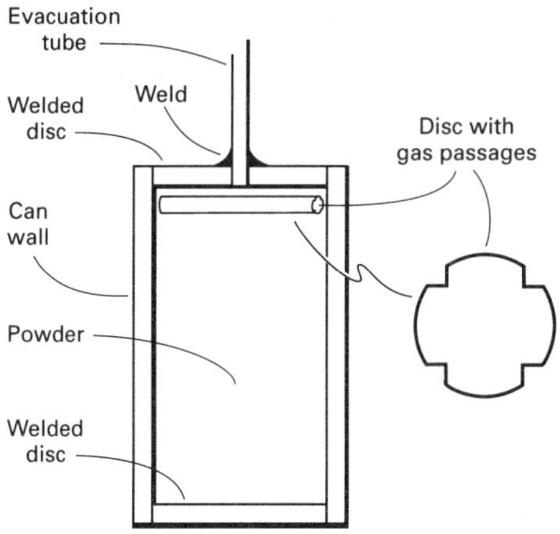

Fig. 4 Degassing can used for aluminum P/M processing. Source: Ref 2

chamber using suitable protection from ambient air and moisture. The compacted billet can then be formed by conventional hot-working operations.

Direct Powder Forming. One of the most cost-effective means of powder consolidation is direct powder forming. Degassed powder, or powder that has been manufactured with great care to avoid contact with ambient air, can be consolidated directly during the hot-forming operation. Direct powder extrusion and rolling have been successfully demonstrated numerous times over the past two decades (Ref 28, 29). It still remains an attractive means for decreasing the cost of aluminum P/M.

Hot Isostatic Pressing (HIP). In HIP, degassed and encapsulated powder is subjected to hydrostatic pressure in a HIP apparatus (see the article "Hot Isostatic Pressing of Metal Powders" in Volume 7 of the 9th Edition of *Metals Handbook*). Can vacuum degassing is often used as the precursor step to HIP. Furthermore, net-shape encapsulation of degassed powder can be used to produce certain near net-shape parts. Relatively high HIP pressures (~200 MPa, or 30 ksi) are often preferred. Unfortunately, the oxide layer on the powder particle surfaces is not sufficiently broken up for optimum mechanical properties. A subsequent hot-working operation that introduces shear-stress components is often necessary to improve ductility and toughness.

Rapid Omnidirectional Consolidation. Engineers at Kelsey-Hayes Company have developed a technique to use existing commercial forging equipment to consolidate powders in several alloy systems (Ref 30). Called rapid omnidirectional consolidation (ROC), it is a lower-cost alternative to HIP. In ROC processing, degassed powder is loaded into a thick-walled "fluid die" that is made of a material that plastically flows at the consolidation temperature and pressure, and which enables the transfer of hydrostatic stress to the powder. Early fluid dies were made of mild steels or a Cu-10Ni alloy,

Table 1 Nominal compositions of aluminum P/M alloys for ambient-temperature service

Alloy	Composition, wt% Zn	Mg	Cu	Co	Zr	Ni	Cr	Li	O	C
7090	8.0	2.5	1.0	1.5	...	...	...	...	...	...
7091	6.5	2.5	1.5	0.4	...	...	...	...	...	...
CW67	9.0	2.5	1.5	...	0.14	0.1	...	...	...	...
7064	7.4	2.4	2.1	0.75	0.3	...	0.15	...	0.2	...
Al-9052	...	4.0	...	...	...	...	...	...	0.5	1.1
Al-905XL	...	4.0	...	...	...	...	...	1.3	0.4	1.1

Table 2 Longitudinal tensile properties from experimental extrusions of aluminum P/M alloys designed for ambient-temperature service

Alloy	Temper	Yield strength, MPa (ksi)	Ultimate tensile strength, MPa (ksi)	Elongation, %	Reference
7091	T6E192	600 (87)	640 (93)	13	41(b)
	T7	545 (79)	595 (86)	11	41(b)
7090	T6511	640 (93)	675 (98)	10	41(b)
	T7	580 (84)	620 (90)	9	41(b)
CW67	T7X1	580 (84)	614 (89)	12	40(c)
7064(a)	T6	635 (92.1)	683 (99.0)	12	42
	T7	621 (90.0)	650 (95)	9	
Al-9052	F	380 (55.0)	450 (65.0)	13	21(c)
	F	630 (91.0)	635 (92.0)	4	41(d)

(a) Formerly called PM-64. (b) Pilot production extruded bar purchased from Alcoa. (c) Typical values. (d) High-strength experimental 23 kg (50 lb) heat made by carefully controlled processing

with subsequent dies made from ceramics, glass, or composites.

The preheated die that contains powder can be consolidated in less than 1 s in a forging press, thereby reducing thermal exposure that can coarsen RS microstructures. In addition, productivity is greatly increased by minimizing press time. Depending upon the type of fluid die material being used, the die can be machined off, chemically leached off, melted off, or designed to "pop off" the net-shape component while cooling from the consolidation temperature. For aluminum-alloy powders, unconfined degassing is critical to optimize the cost effectiveness of the ROC process. An electrodynamic degasser was developed for this purpose.

Just as in the case of HIP, the stress state during ROC is largely hydrostatic. Consequently, there may not be sufficient shear stresses to break the oxide layer and disperse the oxides, which can lead to prior powder-particle boundary (PPB) failure. Consequently, a subsequent hot-working step of the ROC billet is often necessary for demanding applications. More detailed information on this technique can be found in the article "Rapid Omnidirectional Compaction" in Volume 7 of the 9th Edition of *Metals Handbook*.

Dynamic Compaction. Various ultrahigh strain-rate consolidation techniques, that is, dynamic compaction, have been developed and utilized for aluminum alloys (Ref 31, 32). In dynamic compaction, a high-velocity projectile impacts the degassed powders that are consolidated by propagation of the resultant shock wave through the powder. The bonding between the powder particles is believed to occur by melting of a very thin layer on the powder surfaces, which is caused by the heat resulting from friction between the powder particles that occurs during impact. The melted region is highly localized and self-quenched by the powder interiors shortly after impact. Thus, dynamic compaction has the advantage of minimizing thermal exposure and microstructural coarsening, while breaking up the PPBs.

New Directions in Powder Consolidation. Perhaps the most promising area in powder consolidation to have matured over the past decade is actually a bridge between RS-P/M technology and I/M technology. Spray-forming techniques such as the Osprey process (Ref 33), liquid dynamic compaction (Ref 34, 35), and vacuum plasma structural deposition (Ref 36) build consolidated product directly from the atomized stream. Solidification rates greater than those in conventional I/M are attained, but they are not as high as those in RS-P/M processes such as atomization or splat cooling. Spray-forming technology, which will be discussed briefly later, is also described in the article "Spray Deposition of Metal Powders" in Volume 7 of the 9th Edition of *Metals Handbook*.

Alloy Design Research

Aluminum P/M technology is being used to improve the limitations of aluminum alloys and also to push the inherent advantages of aluminum alloys to new limits. The alloy design efforts can be described in three areas (Ref 4):

- High ambient-temperature strength with improved corrosion and SCC resistance
- Improved elevated-temperature properties so aluminum alloys can more effectively compete with titanium alloys
- Increased stiffness and/or reduced density for aluminum alloys to compete with organic composites

This third area includes aluminum-lithium alloys, aluminum-beryllium-lithium alloys, and metal-matrix composites.

High Ambient-Temperature Strength and Improved Corrosion/SCC Resistance

RS Alloys. Rapid-solidification processing has been used to develop improved aluminum alloys for ambient-temperature service for over 25 years. Much of the early work has been conducted by investigators at the Aluminum Company of America (Alcoa) (Ref 37-39). The most successful work in this area was in the Al-Zn-Mg alloy subsystem (7*xxx* alloys), where more highly alloyed 7*xxx* alloy variants with dispersed transition-metal intermetallic phases were investigated. The leading alloys developed have been cobalt-containing alloys 7091 and 7090 (Ref 39), and a subsequent nickel- and zirconium-containing alloy CW67 (Ref 40). Compositions of aluminum alloys for ambient-temperature service are provided in Table 1. Mechanical properties, of course, depend upon mill product form and thermomechanical history. Tensile properties of extrusions are provided in Table 2 to allow an appreciation for the strengths that are possible.

The SCC resistance of 7090 and 7091 is improved with respect to I/M 7*xxx* alloys. In fact, SCC resistance generally increases with cobalt content from 0 to 1.6 wt%, but general corrosion resistance, as assessed by weight-loss tests, decreases (Ref 41). For example, in the peak-strength condition, alloy 7091 with 0.4 wt% Co has a similar SCC plateau crack velocity to I/M 7075, but with 1.5 wt% Co, 7090 has a lower plateau velocity (Fig. 5). Alloy CW67 also has improved combinations of strength and SCC resistance with respect to conventional 7*xxx* I/M alloys and it has replaced 7090 and 7091 as Alcoa's leading RS-P/M aluminum alloy for ambient-temperature service (Ref 40).

The effect of P/M processing on the fatigue behavior of aluminum alloys is complex. In general, resistance to fatigue-crack initiation is improved by P/M processing, in part because of the refinement of constituent particles at which fatigue cracks can nucleate. On the other hand, resistance to fatigue-crack growth can be decreased by P/M processing because of the refinement in grain size. For fine-grain P/M aluminum alloys, the plastic zone size may span several grains, thereby enabling the transfer of deformation across grain boundaries. In coarser-grain I/M alloys, the plastic zone may be contained within one grain.

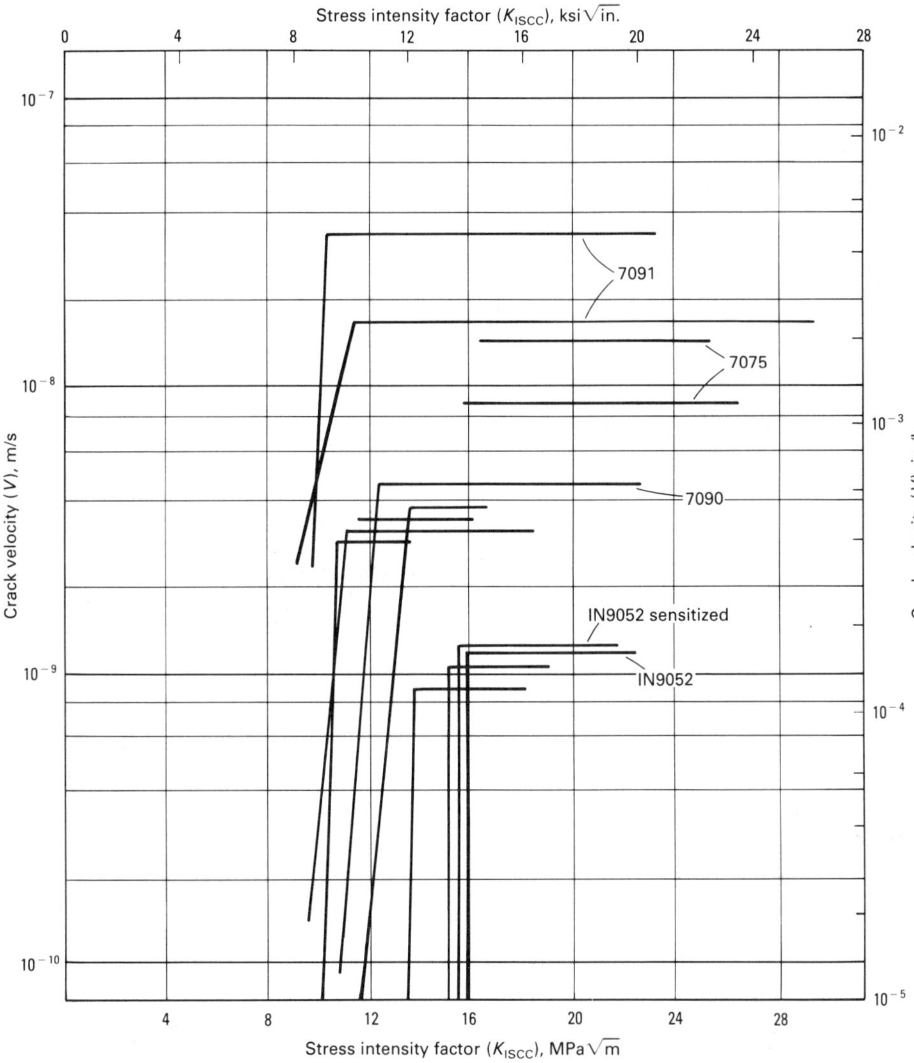

Fig. 5 Stress-corrosion cracking characteristics for the three P/M alloys and I/M 7075, all in their highest-strength (peak aged) conditions. Source: Ref 41

Voss compared the fatigue-crack growth rate under constant load amplitude of 7075-T6510 made by I/M and P/M and enhanced-purity I/M alloy of similar composition, 7475-T651 (Ref 42). Typical results are shown in Fig. 6. The I/M alloys generally displayed better resistance to fatigue-crack propagation in laboratory air. This behavior can change under spectrum fatigue testing, that is, testing designed to simulate service conditions by altering amplitude, where P/M aluminum alloys can perform better than I/M counterparts. The prospective user should exercise care in selecting the proper fatigue testing for aluminum alloys by performing tests that are relevant to service.

Scientists at Kaiser Aluminum & Chemical Corporation have also developed RS-P/M 7*xxx* alloys (Ref 43). The leading such alloy, 7064 (formerly called PM-64), is in part strengthened by zirconium-, chromium-, and cobalt-containing dispersoids (see Table 1). It also displays attractive combinations of strength and SCC resistance, and in addition, has been shown to be superplastically formable.

Wear-Resistant RS Alloys. Japanese researchers have also been developing alloys for ambient-temperature service by RS-P/M. However, they have been more interested in wear resistance than corrosion resistance. For example, Honda Motor Company, Ltd. (Ref 44) has been developing duplex powder alloys that contain atomized aluminum-silicon or aluminum-iron "hard alloy" powders with more ductile aluminum powders (Ref 44). The wear resistance is provided by the hard alloy powders, and good forgeability is provided by the more ductile powders.

Japanese researchers at Sumitomo Electric Industries and Kobe Steel are also looking at RS-P/M Al-Si alloys for wear-resistant applications (Ref 45, 46). For example, Kobe Steel has patented an Al-Si-Cu-Mg family of alloys that contain dispersoid-forming elements for compressor pistons and connecting rods. The leading composition is an Al-20Si-5Fe-2Cu-1Mg-1Mn (wt%) alloy.

Researchers at Tokoku University in Japan have developed RS amorphous alloys that can be produced in films, fibers, plates, or pipes (Ref 47). These alloys contain various combinations of yttrium, nickel, and lanthanum, producing unprecedented strengths as high as 1140 MPa (165 ksi) on, presumably, small samples. Although it is doubtful that such alloys could be scaled up to large wrought products that have such strengths, the alloys are aimed at wear-resistant applications in pistons, valves, gears, brakes, bearings, and rotors (Ref 47).

Mechanically Alloyed Materials. Mechanical attrition processes have also been used to develop improved alloys for ambient-temperature service. Mechanically alloyed Al-4Mg-1.1C-0.5O alloy Al-9052 (formerly IN-9052) (Ref 19-21) and Al-4Mg-1.3Li-1.1C-0.4O alloy Al-905XL (formerly IN-905XL) (Ref 48) were designed to obtain high strength from dispersion and substructural strengthening, while minimizing corrosion and SCC susceptibility that might be introduced by certain precipitates. In addition, Al-905XL was designed to have increased specific stiffness, introduced by the lithium addition. Each of these alloys has been fabricated over a wide range of strength levels by controlling dispersoid content and thermomechanical processing to vary substructural strengthening. In general, these alloys are hot worked at relatively high values of temperature-compensated strain rate, referred to as the Zener Holloman parameter, which is defined by:

$$Z = \dot{\epsilon} \exp (Q/RT) \qquad \text{(Eq 1)}$$

where $\dot{\epsilon}$ is the mean strain rate, Q is the activation energy of the rate-controlling process in the deformation mechanism, R is the universal gas constant, and T is absolute temperature. Such strain rates lead to increased substructural strengthening. For example, extruding at a relatively low temperature and high speed produces a high Z, which can lead to a fine array of subgrains. Hot working at too high a value of Z can result in sufficient stored energy to cause recrystallization. However, the relatively high-volume fraction of finely dispersed oxides and carbides tends to stabilize the fine grain or subgrain structure and inhibit the formation of coarse recrystallized grains.

Alloy Al-9052 is dispersion strengthened by magnesium oxides, aluminum oxides, and aluminum carbide; solid-solution strengthened by magnesium; and fine grain/subgrain strengthened (Ref 19, 20). Alloy Al-905XL is strengthened by similar features, although it does display a slight artificial aging response from lithium-containing precipitates (Ref 49-51). The SCC behavior of a high-strength variant of Al-9052 was compared with RS-P/M 7091 and 7090 in their

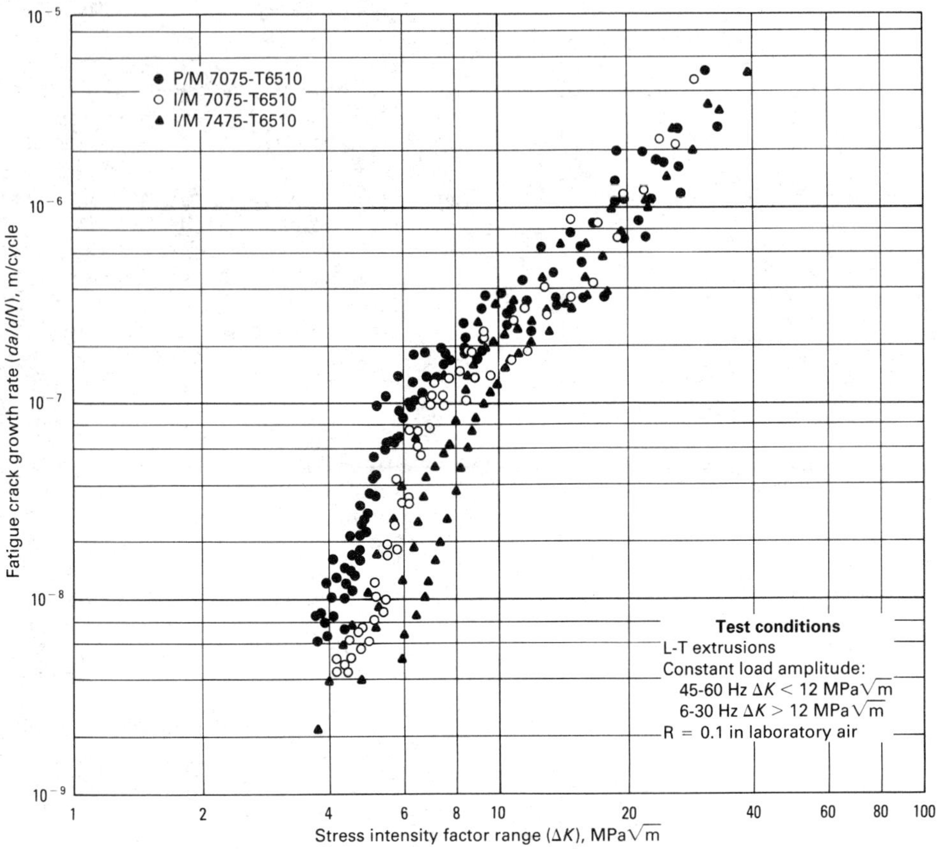

Fig. 6 Fatigue crack growth behavior of equivalent yield strength P/M 7075-T6510, I/M 7075-T6510, and I/M 7475-T6510 in a laboratory air environment. Source: Ref 43

peak strength (T6) conditions (Ref 41). The nonheat-treatable mechanically alloyed material displayed the lowest susceptibility in these peak strength conditions (see Fig. 5). However, the RS-P/M alloys were immune to SCC in the overaged T7 conditions, which resulted in a 10% decrease in strength. The higher cobalt-containing RS-P/M alloy 7090 displayed lower susceptibility and higher strength than 7091, although the higher-volume fraction of cobalt-containing dispersoids did increase pitting susceptibility with respect to 7091. The three P/M alloys showed clear SCC resistance advantages over competitive I/M alloys (Fig. 5). This work concluded with the following generalization concerning the SCC susceptibility of aluminum P/M alloys.

SCC Susceptibility. P/M alloys made by RS or by mechanical alloying achieve excellent combinations of high strength and superb SCC resistance because they enable the uniform introduction of microstructural features that either improve SCC resistance (for example, Co_2Al_9 intermetallic particles), or increase strength without degrading SCC resistance (for example, oxide and carbide dispersion strengthening) (Ref 41).

Al-905XL is currently the leading mechanically alloyed material under commercialization. The alloy is primarily aimed at forging applications where its attractive strength (Table 3), low density, and good corrosion and SCC resistance offer advantages over conventional aluminum alloys. A mechanically alloyed 2*xxx* alloy, Al-9021, is also under development, but its corrosion resistance is not as good as that of A1-9052 and Al-905XL.

Concluding Remarks. Attractive alloys have been developed by both RS-P/M and mechanical attrition processes for ambient-temperature service. Their primary advantage is improved combinations of strength and SCC/corrosion resistance. However, their advantage over high-strength aluminum I/M alloys is often viewed as not significant enough to effect alloy changes in existing applications. Furthermore, the extensive, successful efforts in developing aluminum-lithium I/M alloys has provided additional competition for these P/M alloys.

P/M Alloys With Improved Elevated-Temperature Properties

Powder metallurgy technology is inherently suited to alloy design for elevated-temperature service. Rapid solidification technology allows formation of finely dispersed intermetallic strengthening phases that resist coarsening and are not practical, or in some cases are not even possible, by

Table 3 Typical mechanical properties of Al-905XL P/M forgings

Material property	Direction: Longitudinal	Direction: Transverse
Ultimate tensile strength		
MPa	517	483
ksi	75	70
Yield strength (0.2% offset)		
MPa	448	414
ksi	65	60
Elongation, %	9	6
Fracture toughness, K_{Ic}		
MPa$\sqrt{m}$	30	30
ksi$\sqrt{in.}$	27	27
Modulus of elasticity		
GPa	80	...
10^6 psi	11.6	...

Source: Ref 50

conventional I/M. In addition, mechanical attrition processes can similarly disperse intermetallics and also disperse oxides and carbides on an extremely fine scale. These oxides and carbides are extremely resistant to coarsening at the service temperatures envisioned for aluminum alloys.

Early efforts focused on extending the service temperature of aluminum alloys to 315 to 345 °C (600 to 650 °F). In the mid-1980s, a U.S. Air Force initiative sought to extend the possible service temperature to an extremely challenging 480 °C (900 °F).

RS-P/M Alloys. Much of the development in this area sought to disperse slow-diffusivity transition metals as intermetallic phases in aluminum. Most of the successful alloys developed contain iron.

Scientists at Alcoa (Ref 52-54) investigated numerous such RS-P/M alloys, and Al-Fe-Ce alloys CU78 and CZ42 were the leading alloys developed (compositions of elevated-temperature service alloys are given in Table 4; a typical microstructure is shown in Fig. 7). The alloys displayed good tensile strength up to 315 °C (600 °F) as indicated in Tables 5 and 6, good room-temperature properties after elevated-temperature exposure (Fig. 8), and surprisingly good resistance to environmentally assisted cracking (Ref 54). No failures were observed in 180 days in a 3.5% NaCl solution under conditions of alternate immersion at a stress of 275 MPa (40 ksi). Pratt & Whitney has found that molybdenum additions to RS-P/M Al-Fe produce attractive properties also (Ref 56).

Perhaps the leading RS-P/M aluminum alloys developed are the Al-Fe-V-Si alloys by Allied-Signal Inc. that are made using planar flow casting (Ref 57, 58). For example, alloys FVS-0812 and FVS-1212 display exceptionally high strengths at 315 °C (600 °F) and usable strengths at 425 °C (800 °F) (see Tables 7 and 8). Furthermore, the alloys contain high-volume fractions of silicides that increase modulus. Both alloys have good corrosion resistance in salt-fog environments and good resistance to SCC

Table 4 Nominal compositions of aluminum P/M alloys for elevated-temperature service

Alloy	Supplier	Composition, wt%(a) Fe	Ce	Cr	V	Si	Zr	Mn	Mo	Ti
CU78	Alcoa	8.3	4.0	...	...	...	...	...	...	...
CZ42	Alcoa	7.0	6.0	...	...	...	...	...	...	...
...	Pratt & Whitney	8.0	...	...	...	...	...	...	2	...
FVS-0812	Allied-Signal	8.5	...	...	1.3	1.7	...	...	...	...
FVS-1212	Allied-Signal	11.7	...	...	1.2	2.4	...	...	...	...
FVS-0611	Allied-Signal	6.5	...	...	0.6	1.3	...	...	...	...
...	Alcan	...	...	5	...	...	2	...	...	...
...	Alcan	...	...	5	...	...	2	1	...	...
...	Inco	...	...	...	...	...	...	...	...	4(b)
...	Inco	...	...	...	...	...	...	...	...	8(b)
...	Inco	...	...	...	...	...	...	...	...	12(b)

(a) Oxygen also present. (b) Oxide- and carbide-dispersion strengthening also present

Table 5 Modified property goals (minimum values) for shaped extrusions of P/M alloy CZ42

Material property	Value
Tensile strength, MPa (ksi)	
At room temperature	448 (65)
At 166 °C (330 °F)	365 (53)
At 232 °C (450 °F)	310 (45)
At 260 °C (500 °F)	283 (41)
At 316 °C (600 °F)	221 (32)
Yield strength, MPa (ksi)	
At room temperature	379 (55)
At 166 °C (330 °F)	345 (50)
At 232 °C (450 °F)	296 (43)
At 260 °C (500 °F)	262 (38)
At 316 °C (600 °F)	200 (29)
Modulus of elasticity, GPa (10^6 psi)	
At room temperature	78.6 (11.4)
At 166 °C (330 °F)	68.9 (10.0)
At 232 °C (450 °F)	64.1 (9.3)
At 260 °C (500 °F)	62.1 (9.0)
At 316 °C (600 °F)	56.5 (8.2)
Elongation, at room temperature to 316 °C (600 °F), %	5.0
Fracture toughness (K_{Ic}), MPa$\sqrt{m}$ (ksi$\sqrt{in.}$)	
L-T at room temperature	23 (21)
T-L at room temperature	18 (16)

Source: Ref 54

in saline environments. The dispersed silicides are so fine that they contribute greatly to strength over a wide temperature range (Fig. 9), but do not degrade corrosion and SCC resistance. The silicides are apparently so stable that alloy FVS-0812 can be exposed to 425 °C (800 °F) for up to 1000 h without degradation of room-temperature properties.

Alloy FVS-0812 has been forged into aircraft wheels and is now under consideration as a replacement for conventional alloy 2014, the mainstay aircraft wheel alloy. In addition, the stability of the alloy after elevated-temperature exposure, coupled with its high strength at 300 °C (575 °F), is enabling the alloy to challenge titanium alloys in certain engine components.

Allied-Signal, Inc. is also developing RS-P/M alloy FVS-0611 for applications where good formability is necessary. It is lower in alloying content (Table 4) than the other PFC alloys for elevated-temperature service and is aimed for applications such as rivets and thin-walled tubing.

Scientists at Alcan International Limited took a different approach to designing RS-P/M alloys for elevated-temperature service (Ref 59). They pursued the Al-Cr-Zr system to develop an alloy that could attain attractive elevated-temperature properties from powder that is less sensitive to cooling rate and also has better hot workability. Both chromium and zirconium can produce thermally stable solid solutions in aluminum at modest solidification rates (10^3 K/s). The consolidated billet could then be extruded or rolled with a relatively low flow stress at the lower end of the hot-working range and then aged at a higher temperature to form stable intermetallics that provide elevated-temperature strength.

This alloy design approach provides a potential advantage over RS-P/M Al-Fe-*X* alloys, which often attain their highest hardness just after RS (see Fig. 10). Hardness, and consequently yield strength, can be lost in each thermomechanical processing step from degassing/consolidation through hot working. In addition, elevated-temperature flow stress of the as-compacted billet can make extrusion, forging, and rolling difficult and expensive. Another advantage over RS-P/M Al-Fe-*X* alloys is that the leading Al-Cr-Zr compositions have lower density.

Palmer *et al.* (Ref 59) present analysis by Ekvall *et al.* (Ref 60) and other data comparing the modulus and density of various RS-P/M alloys for elevated-temperature service (Table 9). They compare two parameters that are related to weight-savings ability—specific modulus $E \div \rho$ and the parameter $E^{1/3} \div \rho$. Specific modulus is the weight-controlling parameter where the predominant failure mode involves aeroelastic stiffness, and the parameter $E^{1/3} \div \rho$ applies in certain buckling-limited applications. The leading Al-Cr-Zr alloys, Al-5Cr-2Zr and Al-5Cr-2Zr-1Mn, compare favorably to other RS-P/M alloys and have a substantial advantage over leading I/M alloys such as 2219.

Mechanically Attrited Alloys. Reaction milling is an SAP-derived process developed by Jangg (Ref 22, 23) and co-workers in Europe. It is claimed to be an attritor milling process that mills without a process control agent (PCA) (Ref 24), such as the stearic acid often used in mechanical alloying. Reaction milling is often performed in the presence of lampblack, or graphite, the latter of which the author has found to act as a PCA (Ref 19, 20) that introduces carbon. Oxygen is introduced by careful con-

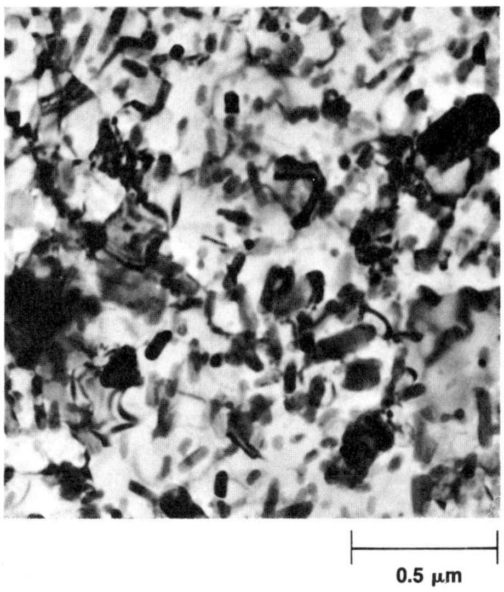

Fig. 7 TEM micrograph of an RS-P/M Al-8Fe-4Ce alloy showing fine Al-Fe-Ce phases that strengthen the 1 mm (0.040 in.) thick sheet. Courtesy of G.J. Hildeman and L. Angers, Alcoa Laboratories

Table 6 Tensile and fracture properties of Al-Fe-*X* alloy thin-sheet specimens tested in the L-T orientation

Material	Temperature °C	°F	Yield strength MPa	ksi	Tensile strength MPa	ksi	Elongation, %	Fracture toughness (K_{Ic}) MPa$\sqrt{m}$	ksi$\sqrt{in.}$
Al-8Fe-7Ce	25	77	418.9	60.8	484.9	70.3	7.0	8.5	7.7
	316	600	178.1	25.8	193.8	28.1	7.6	7.9	7.2
Al-8Fe-2Mo-1V	25	77	323.5	46.9	406.6	59.0	6.7	9.0	8.2
	316	600	170.0	24.6	187.5	27.2	7.2	8.1	7.4
Al-10.5Fe-2.5V	25	77	464.1	67.3	524.5	76.1	4.0	5.7	5.2
	316	600	206.3	29.9	240.0	34.8	6.9	8.1	7.4
Al-8Fe-1.4V-1.7Si	25	77	362.5	52.6	418.8	60.7	6.0	36.4	33.1
	316	600	184.4	26.7	193.8	28.1	8.0	14.9	13.6

Note: Tensile and compact-tension (CT) specimens were prepared according to ASTM specifications E399 and E813. The tensile specimens were 203 mm (8.0 in.) in total length, 50.8 mm (2 in.) in gage length, 12.7 mm (0.5 in.) in width, and 1.27 mm (0.05 in.) in thicknes[illegible] specimens were 38.1 mm (1½ in.) in width and 7.6 to 10.2 mm (0.3 to 0.4 in.) in thickness. Source: Ref 55

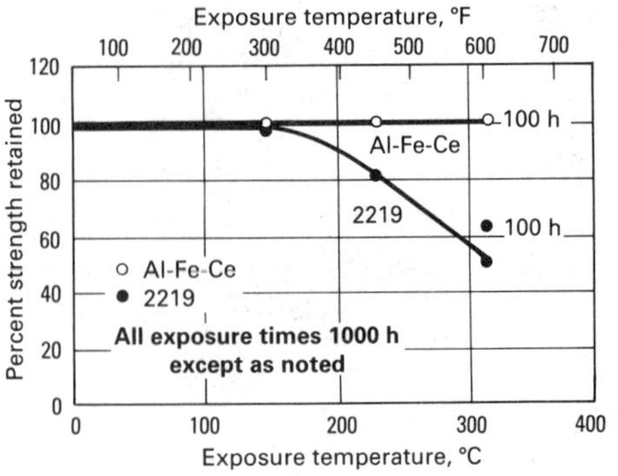

Fig. 8 Percent of room-temperature strength retained after elevated-temperature exposure of Al-Fe-Ce alloys

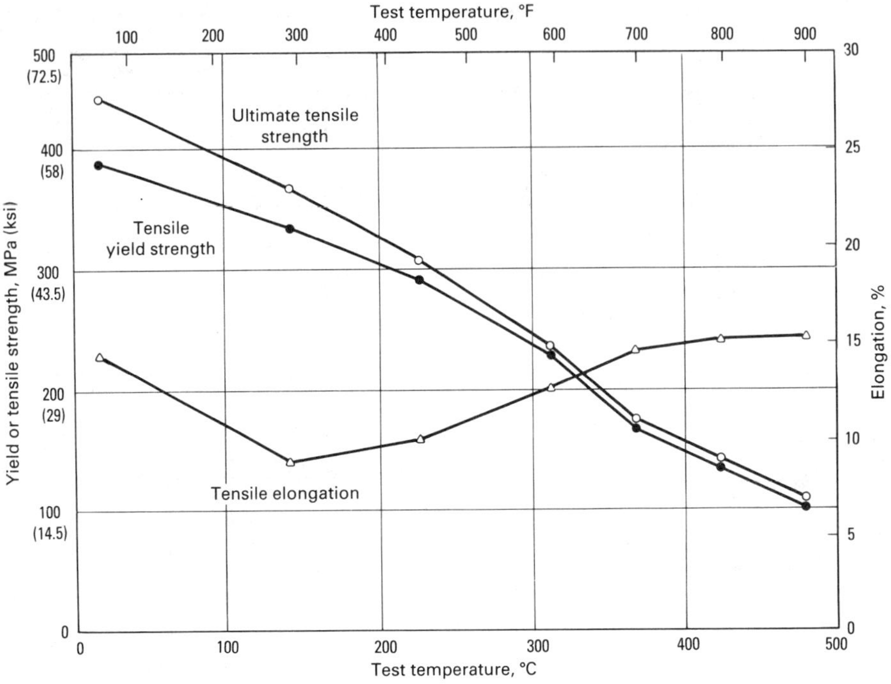

Fig. 9 Tensile properties of FVS-0812 alloy extrusions

trol of the milling atmosphere and forms oxides that are finely dispersed by the reaction milling process. Oxygen in the milling atmosphere may be considered a PCA by reducing the degree of cold welding when powder particles are impacted during ball milling.

The leading reaction-milled alloys (DISPAL alloys) are not as strong as most other elevated-temperature P/M alloys at ambient or intermediate temperatures (Ref 24). However, their decrease in strength with increasing temperature is rather flat, so they have reasonable strength and also good stability at higher temperatures, for example, 150 MPa (22 ksi) tensile strength at 400 °C (750 °F).

Several mechanically alloyed aluminum alloys for elevated-temperature service were designed in 1978 by the author while working for the International Nickel Company (now Inco Alloys International, Inc.) (Ref 61). Various low-solid solubility transition-metal elements were mechanically alloyed to potentially form stable dispersoids, as were electropositive solid-solution elements (for example, lithium and magnesium) to potentially form stable oxides. The leading alloy developed in the program was an Al-4Ti alloy that contained carbon and oxygen from the mechanical alloying process. A yield strength of 159 MPa (23 ksi) and a tensile strength of 186 MPa (27 ksi) were attained at 345 °C (650 °F) on the first heat investigated. In addition, room-temperature strengths (325 MPa, or 47 ksi yield strength and 383 MPa, or 56 ksi ultimate tensile strengths) were essentially unchanged after 100 h at 345 °C (650 °F). In recent work in the mechanically alloyed Al-Ti system by Wilsdorf *et al.* (Ref 62), higher titanium levels and yttria additions were investigated that improved strength at 345 °C (650 °F). In addition, yield and tensile strengths at 425 °C (800 °F), as high as 113 MPa (16 ksi) and 133 MPa (19 ksi), respectively, were attained.

Concluding Remarks. This is the area where aluminum P/M has perhaps the best chance for replacing conventional aluminum or titanium alloys. In fact, the attractive room-temperature properties of these alloys and their surprisingly good corrosion resistance will enable them to compete for warm- and ambient-temperature applications. RS-P/M alloys generally attain higher strength than those made by mechanical attrition processes up to about 350 °C (660 °F), with the Allied-Signal Inc. FVS series alloys showing particular promise. The mechanically attrited materials display superb stability—that is, room-temperature strength after elevated-temperature exposure—and high strengths for aluminum alloys at 425 °C (800 °F).

Table 7 Room- and elevated-temperature tensile properties of planar flow cast alloy FVS-0812

Test temperature		Yield strength		Ultimate tensile strength		Elongation, %	Young's modulus	
°C	°F	MPa	ksi	MPa	ksi		GPa	10^6 psi
24	75	413	60	462	67	12.9	88.4	12.8
149	300	345	50	379	55	7.2	83.2	12.0
232	450	310	45	338	49	8.2	73.1	10.6
316	600	255	37	276	40	11.9	65.5	9.5
427	800	138	20	155	22	15.1	61.4	8.9

Table 8 Mechanical properties of planar flow cast alloy FVS-1212

Test temperature		Yield strength		Ultimate tensile strength		Elongation, %	Young's modulus	
°C	°F	MPa	ksi	MPa	ksi		GPa	10^6 psi
24	75	531	77	559	81	7.2	95.5	13.9
150	300	455	66	469	68	4.2	...	...
230	450	393	57	407	59	6.0	...	...
315	600	297	43	303	44	6.8	...	...

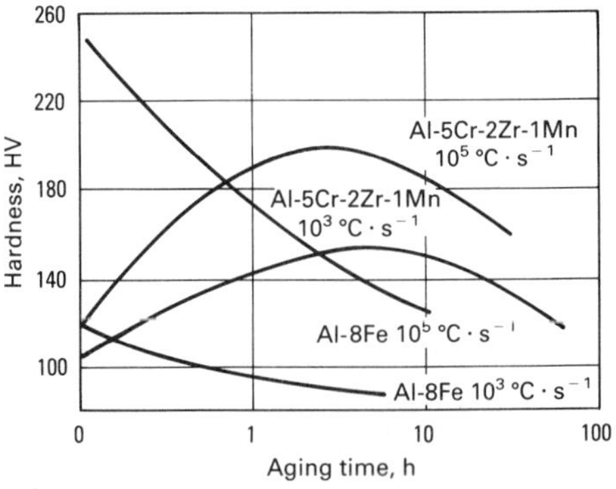

Fig. 10 Variation of microhardness with aging time at 400 °C for Al-5.2Cr-1.9Zr-1Mn alloy compared to Al-8Fe alloy. Source: Ref 59

Table 9 Elastic modulus, density, and weight-saving parameters for thermally stable RS-P/M aluminum alloys

Alloy designation	Composition	Elastic modulus (E) GPa	Elastic modulus (E) 10^6 psi	Density (ρ), g/cm^3	E/ρ $MN \cdot m \cdot kg^{-1}$	$E^{1/3}/\rho$ $N^{1/3}m^{7/3}kg^{-1}$
...	Al-5Cr-2Zr	80.8	11.72	2.82	28.7	1.53
...	Al-5Cr-2Zr-1Mn	86.5	12.54	2.86	30.2	1.55
CU78	Al-8.3Fe-4.0Ce	79.6	11.54	2.95	27.0	1.46
CZ42	Al-7.0Fe-6.0Ce	80.0	11.60	3.01	26.6	1.43
...	Al-8Fe-2Mo	86.2	12.50	2.91	29.6	1.52
FVS-0812	Al-8.5Fe-1.3V-1.7Si	88.4	12.82	3.02	29.3	1.48
FVS-1212	Al-12.4Fe-1.2V-2.3Si	95.5	13.85	3.07	31.1	1.49
RAE 72	Al-7.5Cr-1.2Fe	89.0	12.91	2.89	30.8	1.54
I/M 2219	Al-6.3Cu-0.3Mn-0.06Ti-0.1V-0.18Zr	73.0	10.59	2.86	25.5	1.46

Note: E/ρ, specific modulus; $E^{1/3}/\rho$, buckling parameter. Source: Ref 60

Table 10 Properties of RS-P/M aluminum-beryllium-lithium alloys

Composition, wt%	Temper	Yield strength MPa	Yield strength ksi	Ultimate tensile strength MPa	Ultimate tensile strength ksi	Elongation, %	Young's modulus GPa	Young's modulus 10^6 psi	Density g/cm^3
Al-20.5Be-2.4Li	As-quenched	321	46.5	451	65	6.4	123	18	2.298
	Peak aged	483	70	531	77	3.3	123	18	2.298
Al-29.6Be-1.3Li	Underaged	434	63	494	72	5.0	142	21	...
	Peak aged	497	72	536	78	2.6	142	21	...

Source: Ref 66

High-Modulus and/or Low-Density Alloys

Aluminum P/M is useful to attain stiffer and/or lighter alloys in several ways. First, the alloying limits of elements that increase modulus (E) or decrease density (ρ) can be extended by either RS or mechanical attrition processes. Lithium and beryllium are the only elements that generally do both in aluminum, and RS aluminum-lithium and aluminum-lithium-beryllium alloys have been made that contain alloying levels not practical by I/M. In addition, attempts to mechanically alloy aluminum-lithium alloys have been fruitful at modest lithium levels, but not so at very high levels (3.5 to 4 wt%) (Ref 48). Powder metallurgy technology is also useful in producing metal-matrix composites (MMCs).

The development of P/M alloys that have increased stiffness and/or decreased density is not receiving as much emphasis as it did less than a decade ago (early 1980s). This is largely due to the success of aluminum-lithium alloys made by I/M, which have the advantage of being producible using modified conventional equipment (see the article "Aluminum-Lithium Alloys" in this Volume). In addition, several technologies have arisen that enable MMCs to be made by I/M—for example, Alcan Aluminum Limited's Dural technology for SiC in aluminum, and Martin Marietta's XD technology for TiB_2 or TiC in aluminum. Several high $E \div \rho$ aluminum P/M alloys will be briefly described to show how P/M technology can be utilized in this area.

Al-Li and Al-Li-Be Alloys. Scientists at Allied-Signal Inc. investigated 3 to 4 wt% Li aluminum-lithium alloys made by the PFC technique. The rapid cooling rate allows lithium levels to be attained that are higher than those practical by conventional I/M using commercial direct-chill casting technology. The practical limit for lithium in aluminum by commercial-scale direct-chill casting is about 2.7 wt%. The decreased thermal conductivity caused by lithium additions results in slower cooling rates, which lead to unacceptable segregation at high lithium levels. In addition, the potential explosive hazard increases with lithium content.

Several of the high-lithium-containing PFC alloy compositions were assessed in an Air Force program in cooperation with The Boeing Company (Ref 63). The leading such alloy, 644B, is an Al-3.2Li-1Cu-0.5Mg-0.5Zr alloy with a density of 2.537 g/cm^3. The high zirconium content, which is over 3 times the effective maximum (about 0.14 wt%) in commercial-scale direct-chill casting, was added to form composite Al_3Zr-Al_3Li precipitates that would hopefully be less shearable than Al_3Li (δ').

The alloy attained a mean yield strength of 424 MPa (62 ksi), tensile strength of 521 MPa (76 ksi), with 7% elongation in the longitudinal direction in extruded form. Unfortunately, toughness was low, so subsequent alloy variations contained lower lithium content (Ref 64). The alloy also displayed SCC susceptibility.

As mentioned earlier, mechanically alloyed Al-905XL attains moderate to high strength and good corrosion resistance. The 1.3 wt% Li in the alloy coupled with lithium-, magnesium-, and aluminum-containing oxides also increases elastic modulus, which is 80 GPa (11.6×10^6 psi). Lithium levels of up to 4% were investigated in the early development of mechanically alloyed aluminum-lithium alloys (Ref 48), but oxygen contamination caused ductility to be low. More recently, various other lithium-containing compositions were integrated into the aforementioned Air Force program (Ref 63).

As pointed out by Perepezko, the aluminum-beryllium alloy system is one that can benefit greatly from the RS-P/M approach (Ref 65). The solid-solubility limit of beryllium in aluminum is low (0.3 at%), as is the eutectic composition (2.5 wt%), so the extension of solubility by RS-P/M techniques is particularly useful.

Lewis and co-workers at Lockheed Corporation have examined Al-Be-Li alloys made by RS-P/M and have attained values of specific modulus that are significantly higher than those attained in conventional aluminum alloys (Ref 66). They used melt spinning to examine Al-Be-Li and Al-Be alloys designed to be age hardenable and to have ultralow density. The alloys were designed to be strengthened by δ' and possibly α-Be dispersoids. The ability of RS-P/M technology to extend solubility limits can be appreciated by the Al-20.5Be-2.4Li and Al-29.6Be-1.3Li alloys that displayed perhaps the best properties (Table 10).

Metal-Matrix Composites. A significant amount of work has been undertaken to produce aluminum MMCs using P/M technology (Ref 2-6). Much of this work has emphasized SiC for reinforcement, although work with Al_2O_3, TiB_2, and other ceramics has received attention. P/M technology enables unique matrix alloys to be designed to accept specific ceramic particles. The alloy designer has the flexibility to select compositions that might not be practical by I/M to improve wetability and/or the nature of the ceramic matrix interface. In addition, P/M allows fine-scale mixing of ceramic and alloy powder (or powders for duplex matrices) and provides the option of "slushy state" consolidation, that is, at a consolidation temperature between the solidus and liquidus of the matrix alloy. DWA Corporation and Arco Metals (formerly Silag Corporation) have extended significant effort in this area throughout the past decade and each produced alloys with high strength and stiffness. Some of the more interesting aluminum MMC work has involved reinforcing high-strength P/M alloy matrices. For example, Agarwala *et al.* examined 7091-SiC composites and found attractive mechanical properties, but decreased corrosion resistance (Ref 67). Unfortunately, no one alloy has gained widespread commercial acceptance.

Mechanical alloying can also be used to make MMCs. A low-volume fraction of Al_2O_3 can be easily dispersed by the attrition process (Ref 19, 20). More recently, Nieh *et al.* demonstrated that 15 vol% SiC could be introduced into Al-4Cu-1.5Mg-1.1C-0.8O alloy Al-9021 and produce an MMC with a modulus of 81.8 GPa (11.8 × 10^6 psi) (Ref 68).

It appears that MMC research by RS-P/M is losing favor to spray-deposition forming technologies, which, as mentioned earlier, are a bridge between P/M and I/M.

Spray Deposition. In spray-deposition processes, the atomized stream of alloyed particles is deposited onto a chilled substrate and builds up a compact directly. With proper control of atmosphere, this obviates degassing and consolidation, thereby reducing costs. For more demanding applications, a subsequent hot-working step is often necessary. The as-sprayed compact can be hot worked conventionally into useful shapes that exhibit improved ductility. The details of the leading such processes, Osprey (Ref 33) and liquid dynamic compaction (LDC) (Ref 34, 35, 69), have been described in detail.

Very high strengths have been obtained on LDC-processed 7075 containing 1% Ni and 0.8% Zr. For extrusions, an ultimate tensile strength of 816 MPa (118 ksi) and yield strength of 740 MPa (107 ksi) with 9% elongation have been achieved. More recently, preliminary LDC work on the Al-4-6.3Cu-1.3Li-0.4Ag-0.4Mg-0.14Zr alloy (Weldalite 049)—which was designed to be a greater than 700 MPa (102 ksi) tensile strength I/M alloy (Ref 70)—showed extremely high ductility in very underaged tempers by refinement of constituent particles (42% elongation and 484 MPa, or 70 ksi, tensile strength).

The vacuum plasma structural deposition process is currently being explored by Pechiney Aluminum Company to produce properties equivalent to those in isothermally forged products. The process is being used for titanium alloys and titanium aluminide matrix systems aimed for elevated-temperature applications.

Scientists at Alcan Aluminum Limited have combined Osprey Corporation technology with proprietary ceramic powder-spray technology to make MMCs (Ref 71). They call their technology COSPRAY, and have introduced SiC or B_4C into aluminum-lithium matrix alloy 8090 and I/M alloy 2618. The strength of the 8090 + 10 vol% SiC was slightly higher than that of unreinforced 8090, but ductility decreased. However, modulus increased from 79.5 to 95.9 GPa (11.5 to 14 × 10^6 psi). Similar behavior was observed in 2618 + 14 vol% SiC, but with a greater strength increment. Unfortunately, the B_4C reacted with aluminum, and possibly with lithium, and degraded mechanical properties.

Concluding Remarks. Aluminum-lithium I/M alloys and new techniques of making MMCs by I/M are now competing with high $E \div \rho$ alloys made by P/M. It is likely that research and development resources will be decreased in this area. However, spray-forming technologies like Osprey and liquid dynamic compaction will probably receive increased research and development resources and may be cost competitive for near net-shape components.

New Directions in Aluminum P/M Research

Intermetallics. The need for improved elevated-temperature properties over the range 300 to 1500 °C (570 to 2730 °F) has generated considerable interest in ordered intermetallics. With the inherent low density of aluminum-rich aluminide intermetallics, and some promising compositions that have cubic structures, it is likely that RS and mechanical attrition processes will be extended to aluminides to an increasing extent (see the article "Ordered Intermetallics" in this Volume). In addition, the RS-P/M aluminum alloys have a good chance to displace conventional aluminum I/M alloys, and perhaps titanium alloys in certain applications because the elevated-temperature properties required by designers likely cannot be attained by I/M. Combinations of P/M processes will likely improve properties such as mechanically alloying RS powder or attrited PFC ribbon, both with and without ceramic reinforcement (Ref 72, 73).

Superplastic forming (SPF) is a process attractive for reducing costs in aerospace structures (Ref 74, 75). Superplasticity is generally attainable with a fine array of high-angle grain boundaries that are stabilized by fine dispersoids. Such microstructures have been attained readily by P/M, both with and without ceramic reinforcement. It is likely that research and development in this promising area will continue.

Appendix: Conventionally Pressed and Sintered Aluminum P/M Alloys

CONVENTIONALLY PRESSED AND SINTERED aluminum powder metal parts have been commercially available for many years. Sintered aluminum P/M parts are competitive with many aluminum castings, extrusions, and screw machine products that require expensive and time-consuming finishing operations. In addition, sintered aluminum P/M parts compete with other metal powder parts in applications where some of the attractive physical and mechanical properties of aluminum can be used.

Commercially available aluminum powder alloy compositions (Table 11) consist of blends of atomized aluminum powders mixed with powders of various alloying elements such as zinc, copper, magnesium, and silicon. The most common heat-treatable grades are comparable to the 2*xxx* and 6*xxx* series wrought aluminum alloys. Alloys 201AB and MD-24 are most similar to wrought alloy 2014. They develop high strength and offer moderate corrosion resistance. Alloys 601AB and MD-69 are similar to wrought alloy 6061. These alloys offer high strength, good ductility, corrosion resistance, and can be specified for anodized parts. Alloy 601AC is the same as 601AB, but does not contain an admixed lubricant. It is used for isostatic and die-wall-lubricated compaction. When high conductivity is required, alloy 602AB often is used. Conductivity of 602AB ranges from 24 × 10^6 to 28 × 10^6 S/m (42.0 to 49% IACS), depending on the type of heat treatment selected.

Aluminum P/M Part Processing

Basic design details for aluminum P/M parts involve the same manufacturing operations, equipment, and tooling that are used for iron, copper, and other metal-powder compositions. Detailed information on P/M design and processing can be found in *Powder Metallurgy*, Volume 7 of the 9th Edition of *Metals Handbook*.

Compacting. Aluminum P/M parts are compacted at low pressures and are adaptable to all types of compacting equipment. The pressure density curve, which compares the compacting characteristics of aluminum with other metal powders, indicates that aluminum is simpler to compact. Figure 11 shows the relative difference in compacting characteristics for aluminum and sponge iron or copper.

The lower compacting pressures required for aluminum permit wider use of existing presses. Depending on the press, a larger part often can be made by taking advantage

Table 11 Compositions of typical aluminum P/M alloy powders

Grade	Composition, % Cu	Mg	Si	Al	Lubricant
601AB	0.25	1.0	0.6	bal	1.5
201AB	4.4	0.5	0.8	bal	1.5
602AB	· · ·	0.6	0.4	bal	1.5
202AB	4.0	· · ·	· · ·	bal	1.5
MD-22	2.0	1.0	0.3	bal	1.5
MD-24	4.4	0.5	0.9	bal	1.5
MD-69	0.25	1.0	0.6	bal	1.5
MD-76	1.6	2.5	· · ·	bal	1.5

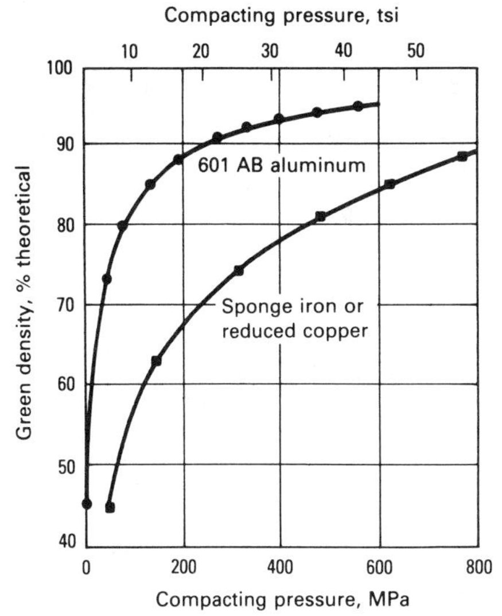

Fig. 11 Relationship of green density and compacting pressure

of maximum press force. For example, a part with a 130 cm^2 (20 $in.^2$) surface area and 50 mm (2 in.) depth is formed readily on a 4450 kN (500 ton) press. The same part in iron would require a 5340 kN (600 ton) press. In addition, because aluminum responds better to compacting and moves more readily in the die, more complex shapes having more precise and finer detail can be produced.

Sintering. Aluminum P/M parts can be sintered in a controlled, inert atmosphere or in vacuum. Sintering temperatures are based on alloy composition and generally range from 595 to 625 °C (1100 to 1160 °F). Sintering time varies from 10 to 30 min. Nitrogen, dissociated ammonia, hydrogen, argon, and vacuum have been used for sintering aluminum; however, nitrogen is preferred because it results in high as-sintered mechanical properties (Table 12). It is also economical in bulk quantities. If a protective atmosphere is used, a dew point of −40 °C (−40 °F) or below is recommended. This is equivalent to a moisture content of 120 mL/m^3 (120 ppm) maximum.

Aluminum preforms can be sintered in batch furnaces or continuous radiant tube mesh or cast belt furnaces. Optimum dimensional control is best attained by maintaining furnace temperature at ±2.8 °C (±5 °F). Typical heating cycles for aluminum parts sintered in various furnaces are illustrated in Fig. 12.

Mechanical properties are directly affected by thermal treatment. All compositions respond to solution heat treating, quenching, and aging in the same manner as conventional heat-treatable alloys. More detailed information on sintering of aluminum can be found in the article "Production Sintering Practices for P/M Materials" in Volume 7 of the 9th Edition of *Metals Handbook*.

Re-Pressing. The density of sintered compacts may be increased by re-pressing. When re-pressing is performed primarily to improve the dimensional accuracy of a compact, it usually is termed "sizing"; when performed to improve configuration, it is termed "coining." Re-pressing may be followed by resintering, which relieves stress due to cold work in re-pressing and may further consolidate the compact. By pressing and sintering only, parts of over 80% theoretical density can be produced. By re-pressing, with or without resintering, parts of 90% theoretical density or more can be produced. The density attainable is limited by the size and shape of the compact.

Forging of aluminum is a well-established technology. Wrought aluminum alloys have been forged into a variety of forms, from small gears to large aircraft structures, for many years (see the article "Forging of Aluminum Alloys" in *Forming and Forging*, Volume 14 of the 9th Edition of *Metals Handbook*). Aluminum lends itself to the forging of P/M preforms to produce structural parts.

In forging of aluminum preforms, the sintered aluminum part is coated with a graphite lubricant to permit proper metal flow during forging. The part is either hot or cold forged; hot forging at 300 to 450 °C (575 to 850 °F) is recommended for parts requiring critical die fill. Forging pressure usually does not exceed 345 MPa (50 ksi). Forging

Table 12 Typical properties of nitrogen-sintered aluminum P/M alloys

Alloy	Compacting pressure MPa	Compacting pressure tsi	Green density %	Green density g/cm^3	Green strength MPa	Green strength psi	Sintered density %	Sintered density g/cm^3	Temper	Tensile strength(a) MPa	Tensile strength(a) ksi	Yield strength(a) MPa	Yield strength(a) ksi	Elongation, %	Hardness
601AB	96	7	85	2.29	3.1	450	91.1	2.45	T1	110	16	48	7	6	55–60 HRH
									T4	141	20.5	96	14	5	80–85 HRH
									T6	183	26.5	176	25.5	1	70–75 HRE
	165	12	90	2.42	6.55	950	93.7	2.52	T1	139	20.1	88	12.7	5	60–65 HRH
									T4	172	24.9	114	16.6	5	80–85 HRH
									T6	232	33.6	224	32.5	2	75–80 HRE
	345	25	95	2.55	10.4	1500	96.0	2.58	T1	145	21	94	13.7	6	65–70 HRH
									T4	176	25.6	117	17	6	85–90 HRH
									T6	238	34.5	230	33.4	2	80–85 HRE
602AB	165	12	90	2.42	6.55	950	93.0	2.55	T1	121	17.5	59	8.5	9	55–60 HRH
									T4	121	17.5	62	9	7	65–70 HRH
									T6	179	26	169	24.5	2	55–60 HRE
	345	25	95	2.55	10.4	1500	96.0	2.58	T1	131	19	62	9	9	55–60 HRH
									T4	134	19.5	65	9.5	10	70–75 HRH
									T6	186	27	172	25	3	65–70 HRE
201AB	110	8	85	2.36	4.2	600	91.0	2.53	T1	169	24.5	145	24	2	60–65 HRE
									T4	210	30.5	179	26	3	70–75 HRE
									T6	248	36	248	36	0	80–85 HRE
	180	13	90	2.50	8.3	1200	92.9	2.58	T1	201	29.2	170	24.6	3	70–75 HRE
									T4	245	35.6	205	29.8	3.5	75–80 HRE
									T6	323	46.8	322	46.7	0.5	85–90 HRE
	413	30	95	2.64	13.8	2000	97.0	2.70	T1	209	30.3	181	26.2	3	70–75 HRE
									T4	262	38	214	31	5	80–85 HRE
									T6	332	48.1	327	47.5	2	90–95 HRE
202AB															
Compacts	180	13	90	2.49	5.4	780	92.4	2.56	T1	160	23.2	75	10.9	10	55–60 HRH
									T4	194	28.2	119	17.2	8	70–75 HRH
									T6	227	33	147	21.3	7.3	45–50 HRE
Cold-formed parts (19% strain)	180	13	90	2.49	5.4	780	92.4	2.56	T2	238	33.9	216	31.4	2.3	80 HRE
									T4	236	34.3	148	21.5	8	70 HRE
									T6	274	39.8	173	25.1	8.7	85 HRE
									T8	280	40.6	250	36.2	3	87 HRE

(a) Tensile properties determined using powder metal flat tension bar (MPIF standard 10-63), sintered 15 min at 620 °C (1150 °F) in nitrogen

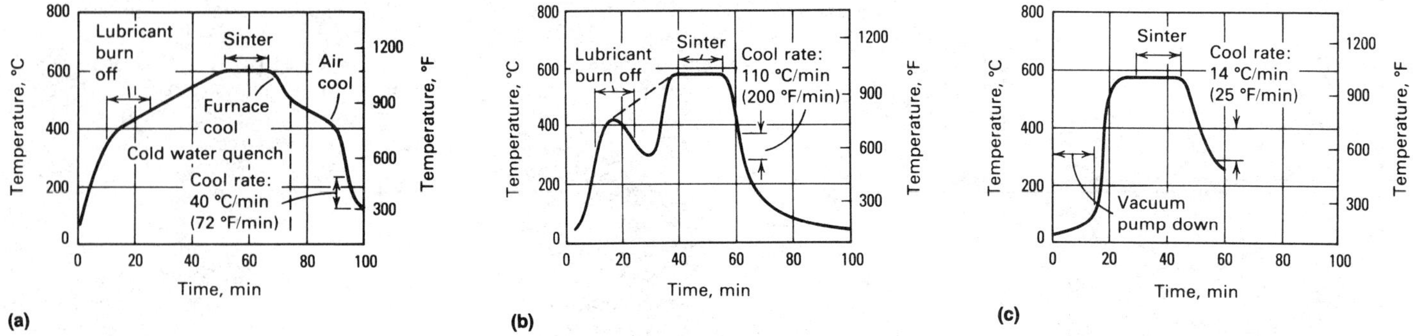

Fig. 12 Typical heating cycles for aluminum P/M parts sintered in (a) A batch furnace. (b) A continuous furnace. (c) A vacuum furnace

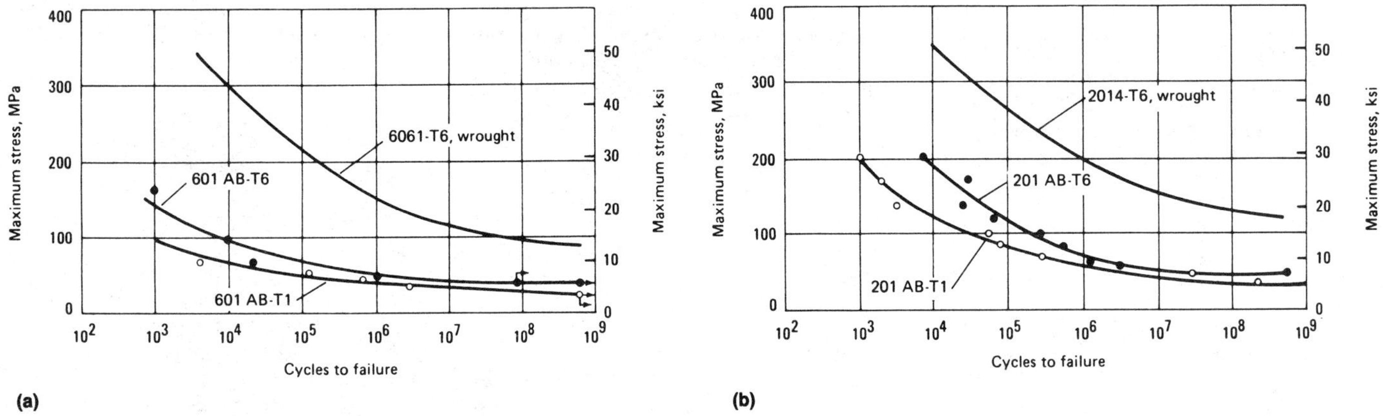

Fig. 13 Fatigue curves for (a) P/M 601AB. (b) P/M 201AB

normally is performed in a confined die so that no flash is produced and only densification and lateral flow result from the forging step. Scrap loss is less than 10% compared to conventional forging, which approaches 50%. Forged aluminum P/M parts have densities of over 99.5% of theoretical density. Strengths are higher than nonforged P/M parts, and in many ways, are similar to conventional forging. Fatigue endurance limit is doubled over that of nonforged P/M parts.

Alloys 601AB, 602AB, 201AB, and 202AB are designed for forgings. Alloy 202AB is especially well suited for cold forging. All of the aluminum powder alloys respond to strain hardening and precipitation hardening, providing a wide range of properties. For example, hot forging of alloy 601AB-T4 at 425 °C (800 °F) followed by heat treatment gives ultimate tensile strengths of 221 to 262 MPa (32 to 38 ksi), and a yield strength of 138 MPa (20 ksi), with 6 to 16% elongation in 25 mm (1 in.).

Heat treated to the T6 condition, 601AB has ultimate tensile strengths of 303 to 345 MPa (44 to 50 ksi). Yield strength is 303 to 317 MPa (44 to 46 ksi), with up to 8% elongation. Forming pressure and percentage of reduction during forging influence final properties.

Ultimate tensile strengths of 358 to 400 MPa (52 to 58 ksi), and yield strengths of 255 to 262 MPa (37 to 38 ksi), with 8 to 18% elongation, are possible with 201AB heat treated to the T4 condition. When heat treated to the T6 condition, the tensile strength of 201AB increases from 393 to 434 MPa (57 to 63 ksi). Yield strength for this condition is 386 to 414 MPa (56 to 60 ksi), and elongation ranges from 0.5 to 8%.

Properties of cold-formed aluminum P/M alloys are increased by a combination of strain-hardened densification and improved interparticle bonding. Alloy 601AB achieves 257 MPa (37.3 ksi) tensile strength and 241 MPa (34.9 ksi) yield strength after forming to 28% upset. Properties for the T4 and T6 conditions do not change notably between 3 and 28% upset. Alloy 602AB has moderate properties with good elongation. Strain hardening (28% upset) results in 221 MPa (32 ksi) tensile and 203 MPa (29.4 ksi) yield strength. The T6 temper parts achieve 255 MPa (37 ksi) tensile strength and 227 MPa (33 ksi) yield strength. Highest cold-formed properties are achieved by 201AB. In the as-formed condition, yield strength increases from 209 MPa (30.3 ksi) for 92.5% density, to 281 MPa (40.7 ksi) for 96.8% density.

Table 13 Typical heat-treated properties of nitrogen-sintered aluminum P/M alloys

Heat-treated variables and properties	Grades			
	MD-22	MD-24	MD-69	MD-76
Solution treatment				
Temperature, °C (°F)	520 (970)	500 (930)	520 (970)	475 (890)
Time, min	30	60	30	60
Atmosphere	Air	Air	Air	Air
Quench medium	H_2O	H_2O	H_2O	H_2O
Aging				
Temperature, °C (°F)	150 (300)	150 (300)	150 (300)	125 (257)
Time, h	18	18	18	18
Atmosphere	Air	Air	Air	Air
Heat-treated (T_6) properties(a)				
Transverse-rupture strength, MPa (ksi)	550 (80)	495 (72)	435 (63)	435 (63)
Yield strength, MPa (ksi)	200 (29)	195 (28)	195 (28)	275 (40)
Tensile strength, MPa (ksi)	260 (38)	240 (35)	205 (30)	310 (45)
Elongation, %	3	3	2	2
Rockwell hardness, HRE	74	72	71	80
Electrical conductivity, %IACS	36	32	39	25

(a)T_6, solution heat treated, quenched, and artificially age hardened

Table 14 Electrical and thermal conductivity of sintered aluminum alloys, wrought aluminum, brass, bronze, and iron

Material	Temper	Electrical conductivity(a) at 20 °C (68 °F), %IACS	Thermal conductivity(b) at 20 °C (68 °F), cgs units
601AB	T4	38	0.36
	T6	41	0.38
	T61	44	0.41
201AB	T4	32	0.30
	T6	35	0.32
	T61	38	0.36
602AB	T4	44	0.41
	T6	47	0.44
	T61	49	0.45
6061 wrought aluminum	T4	40	0.37
	T6	43	0.40
Brass (35% Zn)	Hard	27	0.28
	Annealed	27	0.28
Bronze (5% Sn)	Hard	15	0.17
	Annealed	15	0.17
Iron (wrought plate)	Hot rolled	16	0.18

(a) Determined with FM-103 Magnatester. (b) Converted from electrical conductivity values

Alloy 202AB is best suited for cold forming. Treating to the T2 condition, or as-cold formed, increases the yield strength significantly. In the T8 condition, 202AB develops 280 MPa (40.6 ksi) tensile strength and 250 MPa (36.2 ksi) yield strength, with 3% elongation at the 19% upset level.

Properties of Sintered Parts

Mechanical Properties. Sintered aluminum P/M parts can be produced with strength that equals or exceeds that of iron or copper P/M parts. Tensile strengths range from 110 to 345 MPa (16 to 50 ksi), depending on composition, density, sintering practice, heat treatment, and repressing procedures. Table 12 lists typical properties of four nitrogen-sintered P/M alloys. Properties of heat-treated, pressed, and sintered grades are provided in Table 13.

Impact tests are used to provide a measure of toughness of powder metal materials, which are somewhat less ductile than similar wrought compositions. Annealed specimens develop the highest impact strength, whereas fully heat-treated parts have the lowest impact values. Alloy 201AB generally exhibits higher impact resistance than alloy 601AB at the same percent density, and impact strength of 201AB increases with increasing density. A desirable combination of strength and impact resistance is attained in the T4 temper for both alloys. In the T4 temper, 95% density 201AB develops strength and impact properties exceeding those for as-sintered 99Fe-1C alloy, a P/M material frequently employed in applications requiring tensile strengths under 345 MPa (50 ksi).

Fatigue is an important design consideration for P/M parts subject to dynamic stresses. Fatigue strengths of pressed and sintered P/M parts may be expected to be about half those of the wrought alloys of corresponding compositions (see comparisons of two P/M alloys with two wrought alloys in Fig. 13). These fatigue-strength levels are suitable for many applications.

Fig. 14 Machining chips from a wrought aluminum alloy (right) and from a P/M aluminum alloy (left)

Electrical and Thermal Conductivity. Aluminum has higher electrical and thermal conductivities than most other metals. Table 14 compares the conductivities of sintered aluminum alloys with wrought aluminum, brass, bronze, and iron.

Machinability. Secondary finishing operations such as drilling, milling, turning, or grinding can be performed easily on aluminum P/M parts. Aluminum P/M alloys provide excellent chip characteristics; compared to wrought aluminum alloys, P/M chips are much smaller and are broken more easily with little or no stringer buildup, as can be seen in Fig. 14. This results in improved tool service life and higher machinability ratings.

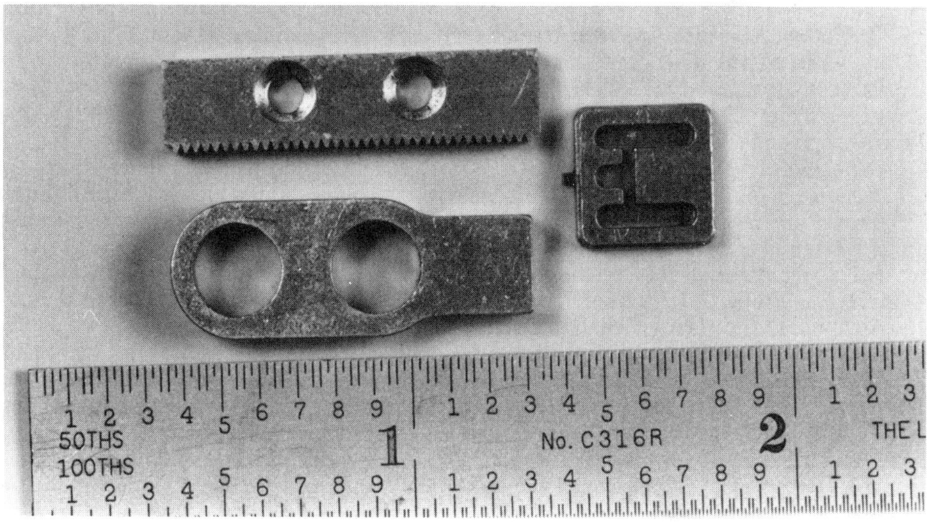

Fig. 15 Typical pressed and sintered aluminum P/M parts made from alloy 601AB. Top: gear rack used on a disc drive. Bottom: link flexure used on a print tip for a typewriter. Right: header/cavity block used on a high-voltage vacuum capacitor. Courtesy of D. Burton, Perry Tool & Research Company

Applications for Sintered Parts

Aluminum P/M parts are used in an increasing number of applications. The business machine market currently uses the greatest variety of aluminum P/M parts. Other markets that indicate growth potential include automotive components, aerospace components, power tools, appliances, and structural parts. Due to their mechanical and physical properties, aluminum P/M alloys provide engineers with flexibility in material selection and design. These factors, coupled with the economic advantages of this technology, should continue to expand the market for aluminum P/M parts. A variety of pressed and sintered aluminum P/M parts are shown in Fig. 15.

REFERENCES

1. E.A. Bloch, *Metall. Rev.*, Vol 6 (No. 22), 1961
2. J.R. Pickens, A Review of Aluminum Powder Metallurgy for High-Strength Applications, *J. Mater. Sci.*, Vol 16, 1981, p 1437-1457
3. T.E. Tietz and I.G. Palmer, in *Advanc-*

es in Powder Technology, G.Y. Chin, Ed., American Society for Metals, 1981, p 189-224
4. J.R. Pickens and E.A. Starke, Jr., The Effect of Rapid Solidification on the Microstructures and Properties of Aluminum Powder Metallurgy Alloys, in *Rapid Solidification Processing: Principle and Technologies*, Vol III, R. Mehrabian, Ed., National Bureau of Standards, p 150-170
5. F.H. Froes and J.R. Pickens, Powder Metallurgy of Light Metal Alloys for Demanding Applications, *J. Met.*, Vol 36 (No. 1), 1984, p 14-28
6. J.R. Pickens, High-Strength Aluminum Alloys Made by Powder Metallurgy: A Brief Overview, in *Proceedings of the Fifth International Conference on Rapidly Quenched Metals (RQ5)* (Wurzburg, West Germany), 3-7 Sept 1984, p 1711
7. J.R. Pickens, K.S. Kumar, and T.J. Langan, High-Strength Aluminum Alloy Development, in the *Proceedings of the 33rd Sagamore Army Materials Research Conference, Corrosion Prevention and Control* (Burlington, VT), Vol 33, 28-31 July 1986
8. N.J. Grant, in *Proceedings of the International Conference on the Rapid Solidification Process*, Claitor's Law Books & Publishing Division, Inc., p 230
9. A. Lawley, Preparation of Metal Powders, *Annu. Rev. Mater. Sci.*, Vol 8, 1978, p 49-71
10. M.C. Narisimhan, U.S. Patent 4,142,571, 1979
11. H. Jones, Phenomenology and Fundamentals of Rapid Solidification of Aluminum Alloys, in *Dispersion Strengthened Aluminum Alloys*, Y.-W. Kim and W.M. Griffith, Ed., TMS, 1988
12. H. Jones, *Mater. Sci. Eng.*, Vol 5 (No. 1), 1969-1970
13. S. Storchheim, *Light Met. Age*, Vol 36 (No. 5), 1978
14. R. Irmann, *Metallurgia*, Vol 46 (No. 125), 1952
15. A. von Zeerleder, *Mod. Met.*, Vol 8, 12, and 40, 1953
16. G.F. Hüttig, *Z. Metallkunde*, Vol 48, 1957, p 352
17. J.S. Benjamin, *Sci. Am.*, Vol 234 (No. 40), 1976
18. J.S. Benjamin and M.J. Bomford, *Metall. Trans. A*, Vol 8A (No. 1301), 1977
19. J.R. Pickens, R.D. Schelleng, S.J. Donachie, and T.J. Nichol, High-Strength Aluminum Alloy and Process, U.S. Patent 4,292,079, 29 Sept 1981
20. J.R. Pickens, R.D. Schelleng, S.J. Donachie, and T.J. Nichol, High-Strength Aluminum Alloy and Process, U.S. Patent 4,297,136, 27 Oct 1981
21. IncoMAP Alloy AL-9052, *Alloy Dig.*, IncoMAP Light Alloys, Inco Alloys International, Inc., May 1989
22. G. Jangg and F. Kutner, *Powder Metall. Int.*, Vol 9 (No. 24), 1977
23. G. Jangg, Powder Forging of Dispersion Strengthened Aluminum Alloys, *Met. Powder Rep.*, Vol 35, May 1980, p 206-208
24. V. Arnhold and K. Hummert, DISPAL-Aluminum Alloys With Non-Metallic Dispersoids, in *Dispersion Strengthened Aluminum Alloys*, Y.-W. Kim and W.M. Griffith, Ed., TMS, p 483-500
25. J.R. Pickens, J.S. Ahearn, R.O. England, and D.C. Cooke, High-Strength Weldable Aluminum Alloys Made From Rapidly Solidified Powder, in *Proceedings of the AIME-TMS Symposium on Powder Metallurgy* (Toronto), G.J. Hildeman and M.J. Koczak, Ed., 15-16 Oct 1985, p 105-133
26. J.R. Pickens, J.S. Ahearn, R.O. England, and D.C. Cooke, "Welding of Rapidly Solidified Powder Metallurgy Aluminum Alloys," End-of-year report on contract N00167-84-C-0099, Martin Marietta Laboratories and DARPA, Dec 1985
27. S.G. Roberts, U.S. Patent 4,104,061
28. G. Naeser, *Modern Developments in PM*, Vol 3, Hausner, Ed., 1965
29. D.H. Ro, "Direct Rolling of Aluminum Powder Metal Strip," Report for contract F33615-80-C-5161, 1 Sept 1980 to March 1981
30. J.R. Lizenby, "Rapid Omnidirectional Compaction," Kelsey-Hayes Company, Powder Technology Center, 1982
31. D. Raybould, Dynamic Powder Compaction, *Proceedings of the Eighth International HERF Conference* (New York), American Society of Mechanical Engineers, I. Berman and J.W. Schroeder, Ed., 1984; also, in *Carbide Tool J.*, March/April 1984
32. D. Raybould and T.Z. Blazynski, Dynamic Compaction of Powders, in *Materials at High Strain Rates*, Elsevier, 1987, p 71-130
33. P. Mathur, D. Appelian, and A. Lawley, Analysis of the Spray Deposition Process, *Acta. Metall.*, Vol 37 (No. 2), Feb 1989, p 429-443
34. E.J. Lavernia and N.J. Grant, Spray Deposition of Metals: A Review, *Proceedings of the Sixth International Conference on Rapidly Quenched Metals (RQ6)*
35. E.J. Lavernia and N.J. Grant, Structure and Property of a Modified 7075 Aluminum Alloy Produced by Liquid Dynamic Compaction, *Int. J. Rapid Solidification*, Vol 2, 1986, p 93-106
36. H.J. Heine, *33 Met. Prod.*, Vol 26 (No. 12), Dec 1988, p 40
37. A.P. Haarr, Frankford Arsenal Report 13-64-AP59S, U.S. Army, Oct 1964
38. Idem, Frankford Arsenal Report 13-65-AP59S, U.S. Army, Dec 1965
39. W.S. Cebulak, E.W. Johnson, and H. Markus, *Int. J. Powder Metall. Powder Technol.*, Vol 12 (No. 299), 1976
40. G.J. Hildeman, L.C. Laberre, A. Hefeez, and L.M. Angers, Microstructural Mechanical Property and Corrosion Evaluations of 7*xxx* P/M Alloy CW67, in *High-Strength Powder Metallurgy Aluminum Alloys II*, G.J. Hildeman and M.J. Koczak, Ed., TMS, 1986, p 25-43
41. J.R. Pickens and L. Christodoulou, The Stress-Corrosion Cracking Behavior of High-Strength Aluminum Powder Metallurgy Alloys, *Metall. Trans. A*, Vol 18A, Jan 1987, p 135-149
42. D.P. Voss, "Structure and Mechanical Properties of Powder Metallurgy 2024 and 7075 Aluminum Alloys," Report EOARD-TR-80-1 on AFOSR Grant 77-3440, 31 Oct 1979
43. S.W. Ping, Developments in Premium High-Strength Powder Metallurgy Alloys by Kaiser Aluminum, in STP 890, American Society for Testing and Materials, 1984, p 369–380
44. Japanese Patent JP8975605A-Kokai, 18 Sept 1987
45. Y. Hirai, K. Kanayama, M. Nakamura, H. Sano, and K. Kubo, *Rep. Gov. Ind. Res. Inst. Nagoya*, Vol 37 (No. 5), May 1988, p 139-146
46. Japanese Patent JP88312942A-Kokai, 20 Dec 1988
47. Development of High-Strength Aluminum Alloys, *Alutopia*, Vol 18 (No. 7), July 1988, p 47-49
48. J.R. Pickens, Mechanically Alloyed Dispersion Strengthened Aluminum-Lithium Alloy, U.S. Patent 4,532,106, 30 July 1985; European Examiner Application 45622A, 10 Feb 1982; Japanese Examiner Application 57857/82, 7 April 1982
49. S.J. Donachie and P.S. Gilman, The Microstructure and Properties of Al-Mg-Li Alloys Prepared by Mechanical Alloying, in *Aluminum Lithium Alloys II*, T.H. Sanders, Jr. and E.A. Starke, Jr., Ed., TMS, 1984, p 507-515
50. IncoMAP Alloy Al-905XL, *Alloy Dig.*, IncoMAP Light Alloys, Inco Alloys International, Inc.
51. R.D. Schelleng, P.S. Gilman, and S.J. Donachie, Aluminum Magnesium-Lithium Forging Alloy Made by Mechanical Alloying, in *Overcoming Material Boundaries*, proceedings of National SAMPE Technical Conference Series, Vol 17, p 106
52. R.E. Sanders, Jr., G.J. Hildeman, and D.L. Lege, Contract F33615-77-C-5086 Report, U.S. Air Force Materials Laboratories, March 1979
53. G.J. Hildeman and R.E. Sanders, Jr., U.S. Patent 4,379,719
54. Alloys CU78 and CZ42 for Elevated Temperature Applications, in *Wrought P/M Alloys*, Aluminum Company of America

55. K.S. Chan, Evidence of a Thin Sheet Toughening Mechanism in Al-Fe-X Alloys, *Met. Trans. A* , Vol 20A, Jan 1989, p 155-164
56. A.R. Cox, Report FR 100754, U.S. Air Force Materials Laboratories, Aug 1978
57. P.S. Gilman, M.S. Zedalis, J.M. Peltier, and S.K. Das, Rapidly Solidified Aluminum, in *Transition Metal Alloys for Aerospace Applications*
58. Rapidly Solidified Aluminum Al-Fe-V-Si Alloy, *Alloy Dig.*, FVS 0812, Allied-Signal Inc., Aug 1989
59. I.G. Palmer, M.P. Thomas, and G.J. Marshall, Development of Thermally Stable Al-Cr-Zr Alloys Using Rapid Solidification Technology, in *Dispersion Strengthened Aluminum Alloys*, Y.-W. Kim and W.M. Griffith, Ed., TMS, p 217-242
60. J.C. Ekvall, J.E. Rhodes, and G.G. Wald, Methodology for Evaluating Weight Savings From Basic Materials Properties, in *Design of Fatigue and Fracture Resistant Structures*, STP 761, American Society for Testing and Materials, 1982, p 328-341
61. D.L. Erich, "Development of a Mechanically Alloyed Aluminum Alloy for 450-650 °F Service," Report AFML-TR-79-4210, U.S. Air Force, 1980
62. H.G.F. Wilsdorf, "Very High Temperature Aluminum Materials' Concepts, U.S. Air Force Contract F33615-86-C-5074," Report UVA/525661/MS88/104, Aug 1987
63. G.H. Narayanan, W.E. Quist, *et al.*, "Low Density Aluminum Alloy Development," Final Report on Contract F33615-81-C-5053, Report WRDC-TR-89-4037, May 1989
64. N.J. Kim, D.J. Skinner, K. Okazaki, and C.M. Adam, Development of Low Density Aluminum-Lithium Alloys Using Rapid Solidification Technology, in *Aluminum-Lithium Alloys III*, C. Baker, P.J. Gregson, S.J. Harris, C.J. Peel, Ed., Institute of Metals, 1986, p 78-84
65. J.H. Perepezko, D.U. Furrer, B.A. Mueller, Undercooling of Aluminum Alloys, in *Dispersion Strengthened Aluminum Alloys*, Y.-W. Kim and W.M. Griffith, Ed., TMS, p 77-102
66. R.E. Lewis, D.L. Yaney, and L.E. Tanner, The Effect of Lithium in Al-Be Alloys, in *Aluminum-Lithium Alloys V*, T.H. Sanders and E.A. Starke, Jr., Ed., Materials and Components, p 731-740
67. R.C. Paciej and V.S. Agarwala, Influence of Processing Variables on the Corrosion Susceptibility of Metal Matrix Composites, *Corrosion*, Vol 44 (No. 10), p 680-684
68. T.G. Nieh, C.M. McNally, J. Wadsworth, D.L. Yaney, and P.S. Gilman, Mechanical Properties of a SiC Reinforced Aluminum Composite Prepared by Mechanical Alloying, in *Dispersion Strengthened Aluminum Alloys*, Y.-W. Kim and W.M. Griffith, Ed., TMS, p 681-692
69. J. Marinkovich, F.A. Mohammed, E.J. Lavernia, and J.R. Pickens, Spray Atomization and Deposition Processing of Al-Cu-Li-Ag-Mg-Zr Alloy Weldalite 049, *J. Met.*, 1989
70. J.R. Pickens, F.H. Heubaum, T.J. Langan, and L.S. Kramer, Al-4.5-6.3Cu-1.3Li-0.4Ag-0.4Mg-0.14Zr Alloy Weldalite™ 049," in *Aluminum-Lithium Alloys*, T.H. Sanders and E.A. Starke, Jr., Ed., Materials and Components, p 1397-1414
71. S.A. Court, I.R. Hughes, I.G. Palmer, and J. White, Microstructure and Properties of Aluminum-Particulate Reinforced Composites Produced by Spray Deposition, presented at 1989 AIME-TMS fall meeting (in press)
72. S.S. Ezz, A. Lawley, and M.J. Koczak, Dispersion Strengthened Al-Fe-Ni: A Dual Rapid Solidification—Mechanical Alloying Approach, in *Aluminum Alloys, Their Physical and Mechanical Properties*, Vol II, L.A. Starke, Jr. and T.H. Sanders, Jr., Ed., Engineering Materials Advisory Services, Ltd., p 1013-1028
73. D.L. Yaney, M.L. Ovecogla, and W.D. Nix, The Effect of Mechanical Alloying on the Deformation Behavior of Rapidly Solidified Al-Fe-Ce Alloy, in *Dispersion Strengthened Aluminum Alloys*, Y.-W. Kim and W.M. Griffith, Ed., TMS, p 619-630
74. R. Crooks, Microstructural Evolution in a Superplastic P/M 7*xxx* Aluminum Alloy, in *Superplasticity in Aerospace*, TMS, Jan 1988, p 25-28
75. G. Gonzalez-Dancel, S.D. Karmarkar, A.P. Divecha, and O.D. Sherby, Influence of Anisotropic Distribution of Whiskers on the Superplastic Behavior of Aluminum in a Back-Extruded 6061 Al-70Si C_w Composite, *Compos. Sci. Technol.*, Vol 35 (No. 2), 1989, p 105-120

Introduction to Copper and Copper Alloys

Derek E. Tyler, Olin Corporation, and William T. Black, Copper Development Association Inc.

COPPER and copper alloys constitute one of the major groups of commercial metals. They are widely used because of their excellent electrical and thermal conductivities, outstanding resistance to corrosion, ease of fabrication, and good strength and fatigue resistance. They are generally nonmagnetic. They can be readily soldered and brazed, and many coppers and copper alloys can be welded by various gas, arc, and resistance methods. For decorative parts, standard alloys having specific colors are readily available. Copper alloys can be polished and buffed to almost any desired texture and luster. They can be plated, coated with organic substances, or chemically colored to further extend the variety of available finishes.

Pure copper is used extensively for cables and wires, electrical contacts, and a wide variety of other parts that are required to pass electrical current. Coppers and certain brasses, bronzes, and cupronickels are used extensively for automobile radiators, heat exchangers, home heating systems, panels for absorbing solar energy, and various other applications requiring rapid conduction of heat across or along a metal section. Because of their outstanding ability to resist corrosion, coppers, brasses, some bronzes, and cupronickels are used for pipes, valves, and fittings in systems carrying potable water, process water, or other aqueous fluids.

In all classes of copper alloys, certain alloy compositions for wrought products have counterparts among the cast alloys; this enables the designer to make an initial alloy selection before deciding on the manufacturing process. Most wrought alloys are available in various cold-worked conditions, and the room-temperature strengths and fatigue resistances of these alloys depend on the amount of cold work as well as the alloy content. Typical applications of cold-worked wrought alloys (cold-worked tempers) include springs, fasteners, hardware, small gears, cams, electrical contacts, and components.

Certain types of parts, most notably plumbing fittings and valves, are produced by hot forging simply because no other fabrication process can produce the required shapes and properties as economically. Copper alloys containing 1 to 6% Pb are free-machining grades. These alloys are widely used for machined parts, especially those produced in screw machines.

Although fewer alloys are produced now than in the 1930s, new alloys continue to be developed and introduced, in particular to meet the challenging requirements of the electronics industry. Information on the use of copper alloys for lead frames, conductors, and other electronic components can be found in *Packaging*, Volume 1 of the *Electronic Materials Handbook* published by ASM INTERNATIONAL.

Properties and applications of wrought copper alloys are presented in Tables 1 and 2. Similar data for cast copper alloys are presented in Table 3. More detailed information on the properties and applications of both wrought and cast copper alloys is presented in the articles that follow in this Section.

Properties of Importance

Along with strength, fatigue resistance, and ability to take a good finish, the primary selection criteria for copper and copper alloys are:

- Corrosion resistance
- Electrical conductivity
- Thermal conductivity
- Color
- Ease of fabrication

Corrosion Resistance. Copper is a noble metal but, unlike gold and other precious metals, can be attacked by common reagents and environments. Pure copper resists attack quite well under most corrosive conditions. Some copper alloys, however, sometimes have limited usefulness in certain environments because of hydrogen embrittlement or stress-corrosion cracking (SCC).

Hydrogen embrittlement is observed when tough pitch coppers, which are alloys containing cuprous oxide, are exposed to a reducing atmosphere. Most copper alloys are deoxidized and thus are not subject to hydrogen embrittlement.

Stress-corrosion cracking most commonly occurs in brass that is exposed to ammonia or amines. Brasses containing more than 15% Zn are the most susceptible. Copper and most copper alloys that either do not contain zinc or are low in zinc content generally are not susceptible to SCC. Because SCC requires both tensile stress and a specific chemical species to be present at the same time, removal of either the stress or the chemical species can prevent cracking. Annealing or stress relieving after forming alleviates SCC by relieving residual stresses. Stress relieving is effective only if the parts are not subsequently bent or strained in service; such operations reintroduce stresses and resensitize the parts to SCC.

Dealloying is another form of corrosion that affects zinc-containing copper alloys. In dealloying, the more active metal is selectively removed from an alloy, leaving behind a weak deposit of the more noble metal.

Copper-zinc alloys containing more than 15% Zn are susceptible to a dealloying process called dezincification. In the dezincification of brass, selective removal of zinc leaves a relatively porous and weak layer of copper and copper oxide. Corrosion of a similar nature continues beneath the primary corrosion layer, resulting in gradual replacement of sound brass by weak, porous copper. Unless arrested, dealloying eventually penetrates the metal, weakening it structurally and allowing liquids or gases to leak through the porous mass in the remaining structure.

Corrosion ratings for wrought and cast copper alloys in a variety of media are given in Tables 4 and 5, respectively. An extensive review on the corrosion properties of copper can be found in *Corrosion*, Volume 13 of the 9th Edition of *Metals Handbook*.

Table 1 Properties of wrought copper and copper alloys

Alloy number (and name)	Nominal composition, %	Commercial forms(a)	Mechanical properties(b): Tensile strength, MPa	Tensile strength, ksi	Yield strength, MPa	Yield strength, ksi	Elongation in 50 mm (2 in.), %(b)	Machinability rating, %(c)
C10100 (oxygen-free electronic copper)	99.99 Cu	F, R, W, T, P, S	221–455	32–66	69–365	10–53	55–4	20
C10200 (oxygen-free copper)	99.95 Cu	F, R, W, T, P, S	221–455	32–66	69–365	10–53	55–4	20
C10300 (oxygen-free extra-low-phosphorus copper)	99.95 Cu, 0.003 P	F, R, T, P, S	221–379	32–55	69–345	10–50	50–6	20
C10400, C10500, C10700 (oxygen-free silver-bearing copper)	99.95 Cu(d)	F, R, W, S	221–455	32–66	69–365	10–53	55–4	20
C10800 (oxygen-free low-phosphorus copper)	99.95 Cu, 0.009 P	F, R, T, P	221–379	32–55	69–345	10–50	50–4	20
C11000 (electrolytic tough pitch copper)	99.90 Cu, 0.04 O	F, R, W, T, P, S	221–455	32–66	69–365	10–53	55–4	20
C11100 (electrolytic tough pitch anneal-resistant copper)	99.90 Cu, 0.04 O, 0.01 Cd	W	455	66	. . .	. . .	1.5 in 1500 mm (60 in.)	20
C11300, C11400, C11500, C11600 (silver-bearing tough pitch copper)	99.90 Cu, 0.04 O, Ag(e)	F, R, W, T, S	221–455	32–66	69–365	10–53	55–4	20
C12000, C12100	99.9 Cu(f)	F, T, P	221–393	32–57	69–365	10–53	55–4	20
C12200 (phosphorus-deoxidized copper, high residual phosphorus)	99.90 Cu, 0.02 P	F, R, T, P	221–379	32–55	69–345	10–50	45–8	20
C12500, C12700, C12800, C12900, C13000 (fire-refined tough pitch with silver)	99.88 Cu(g)	F, R, W, S	221–462	32–67	69–365	10–53	55–4	20
C14200 (phosphorus-deoxidized arsenical copper)	99.68 Cu, 0.3 As, 0.02 P	F, R, T	221–379	32–55	69–345	10–50	45–8	20
C14300	99.9 Cu, 0.1 Cd	F	221–400	32–58	76–386	11–56	42–1	20
C14310	99.8 Cu, 0.2 Cd	F	221–400	32–58	76–386	11–56	42–1	20
C14500 (phosphorus-deoxidized tellurium-bearing copper)	99.5 Cu, 0.50 Te, 0.008 P	F, R, W, T	221–386	32–56	69–352	10–51	50–3	85
C14700 (sulfur-bearing copper)	99.6 Cu, 0.40 S	R, W	221–393	32–57	69–379	10–55	52–8	85
C15000 (zirconium-copper)	99.8 Cu, 0.15 Zr	R, W	200–524	29–76	41–496	6–72	54–1.5	20
C15100	99.82 Cu, 0.1 Zr	F	262–469	38–68	69–455	10–66	36–2	20
C15500	99.75 Cu, 0.06 P, 0.11 Mg, Ag(h)	F	276–552	40–80	124–496	18–72	40–3	20
C15710	99.8 Cu, 0.2 Al_2O_3	R, W	324–724	47–105	268–689	39–100	20–10	. . .
C15720	99.6 Cu, 0.4 Al_2O_3	F, R	462–614	67–89	365–586	53–85	20–3.5	. . .
C15735	99.3 Cu, 0.7 Al_2O_3	R	483–586	70–85	414–565	60–82	16–10	. . .
C15760	98.9 Cu, 1.1 Al_2O_3	F, R	483–648	70–94	386–552	56–80	20–8	. . .
C16200 (cadmium-copper)	99.0 Cu, 1.0 Cd	F, R, W	241–689	35–100	48–476	7–69	57–1	20
C16500	98.6 Cu, 0.8 Cd, 0.6 Sn	F, R, W	276–655	40–95	97–490	14–71	53–1.5	20
C17000 (beryllium-copper)	99.5 Cu, 1.7 Be, 0.20 Co	F, R	483–1310	70–190	221–1172	32–170	45–3	20
C17200 (beryllium-copper)	99.5 Cu, 1.9 Be, 0.20 Co	F, R, W, T, P, S	469–1462	68–212	172–1344	25–195	48–1	20
C17300 (beryllium-copper)	99.5 Cu, 1.9 Be, 0.40 Pb	R	469–1479	68–200	172–1255	25–182	48–3	50
C17400	99.5 Cu, 0.3 Be, 0.25 Co	F	620–793	90–115	172–758	25–110	12–4	20
C17500 (copper-cobalt-beryllium alloy)	99.5 Cu, 2.5 Co, 0.6 Be	F, R	310–793	45–115	172–758	25–110	28–5	. . .
C18200, C18400, C18500 (chromium-copper)	99.5 Cu(i)	F, W, R, S, T	234–593	34–86	97–531	14–77	40–5	20
C18700 (leaded copper)	99.0 Cu, 1.0 Pb	R	221–379	32–55	69–345	10–50	45–8	85
C18900	98.75 Cu, 0.75 Sn, 0.3 Si, 0.20 Mn	R, W	262–655	38–95	62–359	9–52	48–14	20
C19000 (copper-nickel-phosphorus alloy)	98.7 Cu, 1.1 Ni, 0.25 P	F, R, W	262–793	38–115	138–552	20–80	50–2	30
C19100 (copper-nickel-phosphorus-tellurium alloy)	98.15 Cu, 1.1 Ni, 0.50 Te, 0.25 P	R, F	248–717	36–104	69–634	10–92	27–6	75
C19200	98.97 Cu, 1.0 Fe, 0.03 P	F, T	255–531	37–77	76–510	11–74	40–2	20
C19400	97.5 Cu, 2.4 Fe, 0.13 Zn, 0.03 P	F	310–524	45–76	165–503	24–73	32–2	20
C19500	97.0 Cu, 1.5 Fe, 0.6 Sn, 0.10 P, 0.80 Co	F	552–669	80–97	448–655	65–95	15–2	20
C19700	99 Cu, 0.6 Fe, 0.2 P, 0.05 Mg	F	344–517	50–75	165–503	24–73	32–2	20
C21000 (gilding, 95%)	95.0 Cu, 5.0 Zn	F, W	234–441	34–64	69–400	10–58	45–4	20
C22000 (commercial bronze, 90%)	90.0 Cu, 10.0 Zn	F, R, W, T	255–496	37–72	69–427	10–62	50–3	20
C22600 (jewelry bronze, 87.5%)	87.5 Cu, 12.5 Zn	F, W	269–669	39–97	76–427	11–62	46–3	30
C23000 (red brass, 85%)	85.0 Cu, 15.0 Zn	F, W, T, P	269–724	39–105	69–434	10–63	55–3	30
C24000 (low brass, 80%)	80.0 Cu, 20.0 Zn	F, W	290–862	42–125	83–448	12–65	55–3	30
C26000 (cartridge brass, 70%)	70.0 Cu, 30.0 Zn	F, R, W, T	303–896	44–130	76–448	11–65	66–3	30
C26800, C27000 (yellow brass)	65.0 Cu, 35.0 Zn	F, R, W	317–883	46–128	97–427	14–62	65–3	30
C28000 (Muntz metal)	60.0 Cu, 40.0 Zn	F, R, T	372–510	54–74	145–379	21–55	52–10	40
C31400 (leaded commercial bronze)	89.0 Cu, 1.75 Pb, 9.25 Zn	F, R	255–414	37–60	83–379	12–55	45–10	80
C31600 (leaded commercial bronze, nickel-bearing)	89.0 Cu, 1.9 Pb, 1.0 Ni, 8.1 Zn	F, R	255–462	37–67	83–407	12–59	45–12	80
C33000 (low-leaded brass tube)	66.0 Cu, 0.5 Pb, 33.5 Zn	T	324–517	47–75	103–414	15–60	60–7	60
C33200 (high-leaded brass tube)	66.0 Cu, 1.6 Pb, 32.4 Zn	T	359–517	52–75	138–414	20–60	50–7	80

(continued)

(a) F, flat products; R, rod; W, wire; T, tube; P, pipe; S, shapes. (b) Ranges are from softest to hardest commercial forms. The strength of the standard copper alloys depends on the temper (annealed grain size or degree of cold work) and the section thickness of the mill product. Ranges cover standard tempers for each alloy. (c) Based on 100% for C36000. (d) C10400, 250 g/Mg (8 oz/ton) Ag; C10500, 310 g/Mg (10 oz/ton); C10700, 780 g/Mg (25 oz/ton). (e) C11300, 250 g/Mg (8 oz/ton) Ag; C11400, 310 g/Mg (10 oz/ton); C11500, 500 g/Mg (16 oz/ton); C11600, 780 g/Mg (25 oz/ton). (f) C12000, 0.008 P; C12100, 0.008 P and 125 g/Mg (4 oz/ton) Ag. (g) C12700, 250 g/Mg (8 oz/ton) Ag; C12800, 500 g/Mg (10 oz/ton); C12900, 500 g/Mg (16 oz/ton); C13000, 780 g/Mg (25 oz/ton). (h) 260 g/Mg (8.30 oz/ton) Ag. (i) C18200, 0.9 Cr; C18400, 0.8 Cr; C18500, 0.7 Cr. (j) Values are for as-hot-rolled material. (k) Values are for as-extruded material. (l) Rod, 61.0 Cu min. Source: Copper Development Association Inc.

Table 1 (continued)

Alloy number (and name)	Nominal composition, %	Commercial forms(a)	Mechanical properties(b): Tensile strength, MPa	Tensile strength, ksi	Yield strength, MPa	Yield strength, ksi	Elongation in 50 mm (2 in.), %(b)	Machinability rating, %(c)
C33500 (low-leaded brass)	65.0 Cu, 0.5 Pb, 34.5 Zn	F	317–510	46–74	97–414	14–60	65–8	60
C34000 (medium-leaded brass)	65.0 Cu, 1.0 Pb, 34.0 Zn	F, R, W, S	324–607	47–88	103–414	15–60	60–7	70
C34200 (high-leaded brass)	64.5 Cu, 2.0 Pb, 33.5 Zn	F, R	338–586	49–85	117–427	17–62	52–5	90
C34900	62.2 Cu, 0.35 Pb, 37.45 Zn	R, W	365–469	53–68	110–379	16–55	72–18	50
C35000 (medium-leaded brass)	62.5 Cu, 1.1 Pb, 36.4 Zn	F, R	310–655	45–95	90–483	13–70	66–1	70
C35300 (high-leaded brass)	62.0 Cu, 1.8 Pb, 36.2 Zn	F, R	338–586	49–85	117–427	17–62	52–5	90
C35600 (extra-high-leaded brass)	63.0 Cu, 2.5 Pb, 34.5 Zn	F	338–510	49–74	117–414	17–60	50–7	100
C36000 (free-cutting brass)	61.5 Cu, 3.0 Pb, 35.5 Zn	F, R, S	338–469	49–68	124–310	18–45	53–18	100
C36500 to C36800 (leaded Muntz metal)(j)	60.0 Cu(l), 0.6 Pb, 39.4 Zn	F	372	54	138	20	45	60
C37000 (free-cutting Muntz metal)	60.0 Cu, 1.0 Pb, 39.0 Zn	T	372–552	54–80	138–414	20–60	40–6	70
C37700 (forging brass)(k)	59.0 Cu, 2.0 Pb, 39.0 Zn	R, S	359	52	138	20	45	80
C38500 (architectural bronze)(k)	57.0 Cu, 3.0 Pb, 40.0 Zn	R, S	414	60	138	20	30	90
C40500	95 Cu, 1 Sn, 4 Zn	F	269–538	39–78	83–483	12–70	49–3	20
C40800	95 Cu, 2 Sn, 3 Zn	F	290–545	42–79	90–517	13–75	43–3	20
C41100	91 Cu, 0.5 Sn, 8.5 Zn	F, W	269–731	39–106	76–496	11–72	13–2	20
C41300	90.0 Cu, 1.0 Sn, 9.0 Zn	F, R, W	283–724	41–105	83–565	12–82	45–2	20
C41500	91 Cu, 1.8 Sn, 7.2 Zn	F	317–558	46–81	117–517	17–75	44–2	30
C42200	87.5 Cu, 1.1 Sn, 11.4 Zn	F	296–607	43–88	103–517	15–75	46–2	30
C42500	88.5 Cu, 2.0 Sn, 9.5 Zn	F	310–634	45–92	124–524	18–76	49–2	30
C43000	87.0 Cu, 2.2 Sn, 10.8 Zn	F	317–648	46–94	124–503	18–73	55–3	30
C43400	85.0 Cu, 0.7 Sn, 14.3 Zn	F	310–607	45–88	103–517	15–75	49–3	30
C43500	81.0 Cu, 0.9 Sn, 18.1 Zn	F, T	317–552	46–80	110–469	16–68	46–7	30
C44300, C44400, C44500 (inhibited admiralty)	71.0 Cu, 28.0 Zn, 1.0 Sn	F, W, T	331–379	48–55	124–152	18–22	65–60	30
C46400 to C46700 (naval brass)	60.0 Cu, 39.25 Zn, 0.75 Sn	F, R, T, S	379–607	55–88	172–455	25–66	50–17	30
C48200 (naval brass, medium-leaded)	60.5 Cu, 0.7 Pb, 0.8 Sn, 38.0 Zn	F, R, S	386–517	56–75	172–365	25–53	43–15	50
C48500 (leaded naval brass)	60.0 Cu, 1.75 Pb, 37.5 Zn, 0.75 Sn	F, R, S	379–531	55–77	172–365	25–53	40–15	70
C50500 (phosphor bronze, 1.25% E)	98.75 Cu, 1.25 Sn, trace P	F, W	276–545	40–79	97–345	14–50	48–4	20
C51000 (phosphor bronze, 5% A)	95.0 Cu, 5.0 Sn, trace P	F, R, W, T	324–965	47–140	131–552	19–80	64–2	20
C51100	95.6 Cu, 4.2 Sn, 0.2 P	F	317–710	46–103	345–552	50–80	48–2	20
C52100 (phosphor bronze, 8% C)	92.0 Cu, 8.0 Sn, trace P	F, R, W	379–965	55–140	165–552	24–80	70–2	20
C52400 (phosphor bronze, 10% D)	90.0 Cu, 10.0 Sn, trace P	F, R, W	455–1014	66–147	193 (Annealed)	28	70–3	20
C54400 (free-cutting phosphor bronze)	88.0 Cu, 4.0 Pb, 4.0 Zn, 4.0 Sn	F, R	303–517	44–75	131–434	19–63	50–16	80
C60800 (aluminum bronze, 5%)	95.0 Cu, 5.0 Al	T	414	60	186	27	55	20
C61000	92.0 Cu, 8.0 Al	R, W	483–552	70–80	207–379	30–55	65–25	20
C61300	92.65 Cu, 0.35 Sn, 7.0 Al	F, R, T, P, S	483–586	70–85	207–400	30–58	42–35	30
C61400 (aluminum bronze, D)	91.0 Cu, 7.0 Al, 2.0 Fe	F, R, W, T, P, S	524–614	76–89	228–414	33–60	45–32	20
C61500	90.0 Cu, 8.0 Al, 2.0 Ni	F	483–1000	70–145	152–965	22–140	55–1	30
C61800	89.0 Cu, 1.0 Fe, 10.0 Al	R	552–586	80–85	269–293	39–42.5	28–23	40
C61900	86.5 Cu, 4.0 Fe, 9.5 Al	F	634–1048	92–152	338–1000	49–145	30–1	. . .
C62300	87.0 Cu, 3.0 Fe, 10.0 Al	F, R	517–676	75–98	241–359	35–52	35–22	50
C62400	86.0 Cu, 3.0 Fe, 11.0 Al	F, R	621–724	90–105	276–359	40–52	18–14	50
C62500 (k)	82.7 Cu, 4.3 Fe, 13.0 Al	F, R	689	100	379	55	1	20
C63000	82.0 Cu, 3.0 Fe, 10.0 Al, 5.0 Ni	F, R	621–814	90–118	345–517	50–75	20–15	30
C63200	82.0 Cu, 4.0 Fe, 9.0 Al, 5.0 Ni	F, R	621–724	90–105	310–365	45–53	25–20	30
C63600	95.5 Cu, 3.5 Al, 1.0 Si	R, W	414–579	60–84	. . .	. . .	64–29	40
C63800	95.0 Cu, 2.8 Al, 1.8 Si, 0.40 Co	F	565–896	82–130	372–786	54–114	36–4	. . .
C64200	91.2 Cu, 7.0 Al	F, R	517–703	75–102	241–469	35–68	32–22	60
C65100 (low-silicon bronze, B)	98.5 Cu, 1.5 Si	R, W, T	276–655	40–95	103–476	15–69	55–11	30
C65400	95.44 Cu, 3 Si, 1.5 Sn, 0.06 Cr	F	276–793	40–115	130–744	20–108	40–3	20
C65500 (high-silicon bronze, A)	97.0 Cu, 3.0 Si	F, R, W, T	386–1000	56–145	145–483	21–70	63–3	30
C66700 (manganese brass)	70.0 Cu, 28.8 Zn, 1.2 Mn	F, W	315–689	45.8–100	83–638	12–92.5	60–2	30
C67400	58.5 Cu, 36.5 Zn, 1.2 Al, 2.8 Mn, 1.0 Sn	F, R	483–634	70–92	234–379	34–55	28–20	25
C67500 (manganese bronze, A)	58.5 Cu, 1.4 Fe, 39.0 Zn, 1.0 Sn, 0.1 Mn	R, S	448–579	65–84	207–414	30–60	33–19	30
C68700 (aluminum brass, arsenical)	77.5 Cu, 20.5 Zn, 2.0 Al, 0.1 As	T	414	60	186	27	55	30
C68800	73.5 Cu, 22.7 Zn, 3.4 Al, 0.40 Co	F	565–889	82–129	379–786	55–114	36–2	. . .

(continued)

(a) F, flat products; R, rod; W, wire; T, tube; P, pipe; S, shapes. (b) Ranges are from softest to hardest commercial forms. The strength of the standard copper alloys depends on the temper (annealed grain size or degree of cold work) and the section thickness of the mill product. Ranges cover standard tempers for each alloy. (c) Based on 100% for C36000. (d) C10400, 250 g/Mg (8 oz/ton) Ag; C10500, 310 g/Mg (10 oz/ton); C10700, 780 g/Mg (25 oz/ton). (e) C11300, 250 g/Mg (8 oz/ton) Ag; C11400, 310 g/Mg (10 oz/ton); C11500, 500 g/Mg (16 oz/ton); C11600, 780 g/Mg (25 oz/ton). (f) C12000, 0.008 P; C12100, 0.008 P and 125 g/Mg (4 oz/ton) Ag. (g) C12700, 250 g/Mg (8 oz/ton) Ag; C12800, 500 g/Mg (10 oz/ton); C12900, 500 g/Mg (16 oz/ton); C13000, 780 g/Mg (25 oz/ton). (h) 260 g/Mg (8.30 oz/ton) Ag. (i) C18200, 0.9 Cr; C18400, 0.8 Cr; C18500, 0.7 Cr. (j) Values are for as-hot-rolled material. (k) Values are for as-extruded material. (l) Rod, 61.0 Cu min. Source: Copper Development Association Inc.

Table 1 (continued)

Alloy number (and name)	Nominal composition, %	Commercial forms(a)	Mechanical properties(b): Tensile strength, MPa	Tensile strength, ksi	Yield strength, MPa	Yield strength, ksi	Elongation in 50 mm (2 in.), %(b)	Machinability rating, %(c)
C69000	73.3 Cu, 3.4 Al, 0.6 Ni, 22.7 Zn	F	496–896	72–130	345–807	50–117	40–2	. . .
C69400 (silicon red brass)	81.5 Cu, 14.5 Zn, 4.0 Si	R	552–689	80–100	276–393	40–57	25–20	30
C70250	96.2 Cu, 3 Ni, 0.65 Si, 0.15 Mg	F	586–758	85–110	552–784	80–105	40–3	20
C70400	92.4 Cu, 1.5 Fe, 5.5 Ni, 0.6 Mn	F, T	262–531	38–77	276–524	40–76	46–2	20
C70600 (copper-nickel, 10%)	88.7 Cu, 1.3 Fe, 10.0 Ni	F, T	303–414	44–60	110–393	16–57	42–10	20
C71000 (copper-nickel, 20%)	79.0 Cu, 21.0 Ni	F, W, T	338–655	49–95	90–586	13–85	40–3	20
C71300	75 Cu, 25 Ni	F	338–655	49–95	90–586	13–85	40–3	20
C71500 (copper-nickel, 30%)	70.0 Cu, 30.0 Ni	F, R, T	372–517	54–75	138–483	20–70	45–15	20
C71700	67.8 Cu, 0.7 Fe, 31.0 Ni, 0.5 Be	F, R, W	483–1379	70–200	207–1241	30–180	40–4	20
C72500	88.2 Cu, 9.5 Ni, 2.3 Sn	F, R, W, T	379–827	55–120	152–745	22–108	35–1	20
C73500	72.0 Cu, 10.0 Zn, 18.0 Ni	F, R, W, T	345–758	50–110	103–579	15–84	37–1	20
C74500 (nickel silver, 65–10)	65.0 Cu, 25.0 Zn, 10.0 Ni	F, W	338–896	49–130	124–524	18–76	50–1	20
C75200 (nickel silver, 65–18)	65.0 Cu, 17.0 Zn, 18.0 Ni	F, R, W	386–710	56–103	172–621	25–90	45–3	20
C75400 (nickel silver, 65–15)	65.0 Cu, 20.0 Zn, 15.0 Ni	F	365–634	53–92	124–545	18–79	43–2	20
C75700 (nickel silver, 65-12)	65.0 Cu, 23.0 Zn, 12.0 Ni	F, W	359–641	52–93	124–545	18–79	48–2	20
C76200	59.0 Cu, 29.0 Zn, 12.0 Ni	F, T	393–841	57–122	145–758	21–110	50–1	. . .
C77000 (nickel silver, 55-18)	55.0 Cu, 27.0 Zn, 18.0 Ni	F, R, W	414–1000	60–145	186–621	27–90	40–2	30
C72200	82.0 Cu, 16.0 Ni, 0.5 Cr, 0.8 Fe, 0.5 Mn	F, T	317–483	46–70	124–455	18–66	46–6	. . .
C78200 (leaded nickel silver, 65-8-2)	65.0 Cu, 2.0 Pb, 25.0 Zn, 8.0 Ni	F	365–627	53–91	159–524	23–76	40–3	60

(a) F, flat products; R, rod; W, wire; T, tube; P, pipe; S, shapes. (b) Ranges are from softest to hardest commercial forms. The strength of the standard copper alloys depends on the temper (annealed grain size or degree of cold work) and the section thickness of the mill product. Ranges cover standard tempers for each alloy. (c) Based on 100% for C36000. (d) C10400, 250 g/Mg (8 oz/ton) Ag; C10500, 310 g/Mg (10 oz/ton); C10700, 780 g/Mg (25 oz/ton). (e) C11300, 250 g/Mg (8 oz/ton) Ag; C11400, 310 g/Mg (10 oz/ton); C11500, 500 g/Mg (16 oz/ton); C11600, 780 g/Mg (25 oz/ton). (f) C12000, 0.008 P; C12100, 0.008 P and 125 g/Mg (4 oz/ton) Ag. (g) C12700, 250 g/Mg (8 oz/ton) Ag; C12800, 500 g/Mg (10 oz/ton); C12900, 500 g/Mg (16 oz/ton); C13000, 780 g/Mg (25 oz/ton). (h) 260 g/Mg (8.30 oz/ton) Ag. (i) C18200, 0.9 Cr; C18400, 0.8 Cr; C18500, 0.7 Cr. (j) Values are for as-hot-rolled material. (k) Values are for as-extruded material. (l) Rod, 61.0 Cu min. Source: Copper Development Association Inc.

Electrical and Thermal Conductivity. Copper and its alloys are relatively good conductors of electricity and heat. In fact, copper is used for these purposes more often than any other metal. Alloying invariably decreases electrical conductivity and, to a lesser extent, thermal conductivity. The amount of reduction due to alloying does not depend on the conductivity or any other bulk property of the alloying element, but only on the effect that the particular solute atoms have on the copper lattice. For this reason, coppers and high-copper alloys are preferred over copper alloys containing more than a few percent total alloy content when high electrical or thermal conductivity is required for the application.

Color. Copper and certain copper alloys are used for decorative purposes alone, or when a particular color and finish is combined with a desirable mechanical or physical property of the alloy. Table 6 lists the range of colors that can be obtained with standard copper alloys.

Ease of Fabrication. Copper and its alloys are generally capable of being shaped to the required form and dimensions by any of the common fabricating processes. They are routinely rolled, stamped, drawn, and headed cold; they are rolled, extruded, forged, and formed at elevated temperature. Copper alloys are readily stamped and formed into components. Most countries in the world employ copper alloys for coinage. There are casting alloys for all of the generic families of coppers and copper alloys. Copper metals can be polished, textured, plated, or coated to provide a wide variety of functional or decorative surfaces.

Copper and copper alloys are readily assembled by any of the various mechanical or bonding processes commonly used to join metal components. Crimping, staking, riveting, and bolting are mechanical means of maintaining joint integrity. Soldering, brazing, and welding are the most widely used processes for bonding copper metals. Selection of the best joining process is governed by service requirements, joint configuration, thickness of the components, and alloy composition(s). These factors are discussed in detail in *Welding, Brazing, and Soldering*, Volume 6 of the 9th Edition of *Metals Handbook*.

Mechanical Working

High-purity copper is a very soft metal. It is softest in its undeformed single-crystal form and requires a shear stress of only 3.9 MPa (570 psi) on {111} crystal planes for slip. Annealed tough pitch copper is almost as soft as high-purity copper, but many of the copper alloys are much harder and stiffer, even in annealed tempers.

Copper is easily deformed cold. Once flow has been started, it takes little energy to continue, and thus extremely large changes in shape or reductions in section are possible in a single pass. The only limitation appears to be the ability to design and build the necessary tools. Very heavy reductions are possible, especially with continuous flow. Rolling reductions of more than 90% in one pass are used for rolling strip.

Copper and many of its alloys also respond well to sequential cold working. Tandem rolling and gang-die drawing are common. Some copper alloys work harden rapidly; therefore, the number of operations that can be performed before annealing to resoften the metal is limited.

Copper can be cold reduced almost limitlessly without annealing, but heavy deformation (more than about 80 to 90%) may induce preferred crystal orientation, or texturing. Textured metal has different properties in different directions, which is undesirable for some applications.

Both the cold-working and the hot-working characteristics of copper are described below. For more detailed information, see *Forming and Forging*, Volume 14 of the 9th Edition of *Metals Handbook*.

Cold working increases both tensile strength and yield strength, but it has a more pronounced effect on the latter. For most coppers and copper alloys, the tensile strength of the hardest cold-worked temper is approximately twice the tensile strength of the annealed temper. For the same alloys, the yield strength of the hardest cold-worked temper can be as much as five to six times that of the annealed temper.

Hardness as a measure of temper is inaccurate: The relation between hardness and strength is different for different alloys. Usually, hardness and strength for a given alloy can be correlated only over a rather narrow range of conditions. Also, the range

Table 2 Fabrication characteristics and typical applications of wrought copper and copper alloys

Alloy number (and name)	Fabrication characteristics and typical applications
C10100 (oxygen-free electronic copper)	Excellent hot and cold workability; good forgeability. Fabricated by coining, coppersmithing, drawing and upsetting, hot forging and pressing, spinning, swaging, stamping. Uses: busbars, bus conductors, waveguides, hollow conductors, lead-in wires and anodes for vacuum tubes, vacuum seals, transistor components, glass-to-metal seals, coaxial cables and tubes, klystrons, microwave tubes, rectifiers
C10200 (oxygen-free copper)	Fabrication characteristics same as C10100. Uses: busbars, waveguides
C10300 (oxygen-free, extra-low-phosphorus copper)	Fabrication characteristics same as C10100. Uses: busbars, electrical conductors, tubular bus, and applications requiring good conductivity and good welding or brazing properties
C10400, C10500, C10700 (oxygen-free, silver-bearing copper)	Fabrication characteristics same as C10100. Uses: auto gaskets, radiators, busbars, conductivity wire, contacts, radio parts, winding, switches, terminals, commutator segments, chemical process equipment, printing rolls, clad metals, printed circuit foil
C10800 (oxygen-free, low-phosphorus copper)	Fabrication characteristics same as C10100. Uses: refrigerators; air conditioners; gas and heater lines; oil burner tubes; plumbing pipe and tube; brewery tubes; condenser and heat exchanger tubes; dairy and distiller tubes; pulp and paper lines; tanks; air, gasoline, hydraulic, and oil lines
C11000 (electrolytic tough pitch copper)	Fabrication characteristics same as C10100. Uses: downspouts, gutters, roofing, gaskets, auto radiators, busbars, nails, printing rolls, rivets, radio parts, flexible circuits
C11100 (electrolytic tough pitch anneal-resistant copper)	Fabrication characteristics same as C10100. Uses: electrical power transmission where resistance to softening under overloads is desired
C11300, C11400, C11500, C11600 (silver-bearing tough pitch copper)	Fabrication characteristics same as C10100. Uses: gaskets, radiators, busbars, windings, switches, chemical process equipment, clad metals, printed circuit foil
C12000, C12100	Fabrication characteristics same as C10100. Uses: busbars, electrical conductors, tubular bus, and applications requiring welding or brazing
C12200 (phosphorus-deoxidized copper, high residual phosphorus)	Fabrication characteristics same as C10100. Uses: gas and heater lines; oil burner tubing; plumbing pipe and tubing; condenser, evaporator, heat exchanger, dairy, and distiller tubing; steam and water lines; air, gasoline, and hydraulic lines
C12500, C12700, C12800, C12900, C13000 (fire-refined tough pitch with silver)	Fabrication characteristics same as C10100. Uses: same as C11000
C14200 (phosphorus-deoxidized arsenical copper)	Fabrication characteristics same as C10100. Uses: plates for locomotive fireboxes, staybolts, heat exchanger and condenser tubes
C14300	Fabrication characteristics same as C10100. Uses: anneal-resistant electrical applications requiring thermal softening and embrittlement resistance, lead frames, contacts, terminals, solder-coated and solder-fabricated parts, furnace-brazed assemblies and welded components, cable wrap
C14310	Same as C14300
C14500 (phosphorus-deoxidized, tellurium-bearing copper)	Fabrication characteristics same as C10100. Uses: forgings and screw machine products and parts requiring high conductivity, extensive machining, corrosion resistance, copper color, or a combination of these properties; electrical connectors, motor and switch parts, plumbing fittings, soldering coppers, welding torch tips, transistor bases, and furnace-brazed articles
C14700 (sulfur-bearing copper)	Fabrication characteristics same as C10100. Uses: screw machine products and parts requiring high conductivity, extensive machining, corrosion resistance, copper color, or a combination of these properties; electrical connectors; motor and switch components; plumbing fittings; cold-headed and machined parts; cold forgings; furnace-brazed articles; screws; soldering coppers; rivets; and welding torch tips
C15000, C15100 (zirconium-copper)	Fabrication characteristics same as C10100. Uses: switches, high-temperature circuit breakers; commutators, stud bases for power transmitters, rectifiers, soldering welding tips, lead frames
C15500	Fabrication characteristics same as C10100. Uses: high-conductivity light-duty springs, electrical contacts, fittings, clamps, connectors, diaphragms, electronic components, resistance welding electrodes
C15710	Excellent cold workability. Fabricated by extrusion, drawing, rolling, impacting, heading, swaging, bending, machining, blanking, roll threading. Uses: electrical connectors, light-duty current-carrying springs, inorganic insulated wire, thermocouple wire, lead wire, resistance welding electrodes for aluminum, heat sinks
C15720	Excellent cold workability. Fabricated by extrusion, drawing, rolling, impacting, heading, swaging, machining, blanking. Uses: relay and switch springs, lead frames, contact supports, heat sinks, circuit breaker parts, rotor bars, resistance welding electrodes and wheels, connectors, high-strength high-temperature parts
C15735	Excellent cold workability. Fabricated by extrusion, drawing, heading, impacting, machining. Uses: resistance welding electrodes, circuit breakers, feed-through conductors, heat sinks, motor parts, high-strength high-temperature parts
C15760	Excellent cold workability. Fabricated by extrusion and drawing. Uses: resistance welding electrodes, circuit breakers, electrical connectors, wire feed contact tips, plasma spray nozzles, high-strength high-temperature parts
C16200 (cadmium-copper)	Excellent cold workability; good hot formability. Uses: trolley wire, heating pads, electric-blanket elements, spring contacts, rail bands, high-strength transmission lines, connectors, cable wrap, switch gear components, and waveguide cavities
C16500	Fabrication characteristics same as C16200. Uses: electrical springs and contacts, trolley wire, clips, flat cable, resistance welding electrodes
C17000 (beryllium-copper)	Fabrication characteristics same as C16200. Commonly fabricated by blanking, forming and bending, turning, drilling, tapping. Uses: bellows, Bourdon tubing, diaphragms, fuse clips, fasteners, lock washers, springs, switch parts, roll pins, valves, welding equipment
C17200 (beryllium-copper)	Similar to C17000, particularly for its nonsparking characteristics
C17300 (beryllium-copper)	Combines superior machinability with the good fabrication characteristics of C17200
C17500 (copper-cobalt-beryllium alloy)	Fabrication characteristics same as C16200. Uses: fuse clips, fasteners, springs, switch and relay parts, electrical conductors, welding equipment
C18200, C18400, C18500 (chromium-copper)	Excellent cold workability, good hot workability. Uses: resistance welding electrodes, seam welding wheels, switch gear, electrode holder jaws, cable connectors, current-carrying arms and shafts, circuit breaker parts, molds, spot-welding tips, flash welding electrodes, electrical and thermal conductors requiring strength, switch contacts
C18700 (leaded copper)	Good cold workability; poor hot formability. Uses: connectors, motor and switch parts, screw machine parts requiring high conductivity
C18900	Fabrication characteristics same as C10100. Uses: welding rod and wire for inert-gas tungsten arc and metal arc welding and oxyacetylene welding of copper

(continued)

Table 2 (continued)

Alloy number (and name)	Fabrication characteristics and typical applications
C19000 (copper-nickel-phosphorus alloy)	Fabrication characteristics same as C10100. Uses: springs, clips, electrical connectors, power tube and electron tube components, high-strength electrical conductors, bolts, nails, screws, cotter pins, and parts requiring some combination of high strength, high electrical or thermal conductivity, high resistance to fatigue and creep, and good workability
C19100 (copper-nickel-phosphorus-tellurium alloy)	Good hot and cold workability. Uses: forgings and screw machine parts requiring high strength, hardenability, extensive machining, corrosion resistance, copper color, good conductivity, or a combination of these properties; bolts, bushings, electrical connectors, gears, marine hardware, nuts, pinions, tie rods, turnbuckle barrels, welding torch tips
C19200	Excellent hot and cold workability. Uses: automotive hydraulic brake lines, flexible hose, electrical terminals, fuse clips, gaskets, gift hollowware, applications requiring resistance to softening and stress corrosion, air conditioning and heat exchanger tubing
C19400, C19700	Fabrication characteristics same as C19200. Uses: electrical terminals, cable wrap, electronic connectors, lead frames, applications requiring resistance to softening and stress relaxation at greater-than-ambient temperatures
C19400	Excellent hot and cold workability. Uses: circuit breaker components, contact springs, electrical clamps, electrical springs, electrical terminals, flexible hose, fuse clips, gaskets, gift hollowware, plug contacts, rivets, welded condenser tubes
C19500	Excellent hot and cold workability. Uses: electrical springs, sockets, terminals, connectors, clips, and other current-carrying parts having strength
C21000 (gilding, 95%)	Excellent cold workability, good hot workability for blanking, coining, drawing, piercing and punching, shearing, spinning, squeezing and swaging, stamping. Uses: coins, medals, bullet jackets, fuse caps, primers, plaques, jewelry base for gold plate
C22000 (commercial bronze, 90%)	Fabrication characteristics same as C21000, plus heading and upsetting, roll threading and knurling, hot forging and pressing. Uses: etching bronze, grillwork, screen cloth, weather stripping, lipstick cases, compacts, marine hardware, screws, rivets
C22600 (jewelry bronze, 87.5%)	Fabrication characteristics same as C21000, plus heading and upsetting, roll threading and knurling. Uses: angles, channels, chain, fasteners, costume jewelry, lipstick cases, compacts, base for gold plate
C23000 (red brass, 85%)	Excellent cold workability, good hot formability. Uses: weather stripping, conduit, sockets, fasteners, fire extinguishers, condenser and heat exchanger tubing, plumbing pipe, radiator cores
C24000 (low brass, 80%)	Excellent cold workability. Fabrication characteristics same as C23000. Uses: battery caps, bellows, musical instruments, clock dials, pump lines, flexible hose
C26000 (cartridge brass, 70%)	Excellent cold workability. Fabrication characteristics same as C23000, except for coining, roll threading, and knurling. Uses: radiator cores and tanks, flashlight shells, lamp fixtures, fasteners, locks, hinges, ammunition components, plumbing accessories, pins, rivets
C26800, C27000 (yellow brass)	Excellent cold workability. Fabrication characteristics same as C23000. Uses: same as C26000 except not used for ammunition
C28000 (Muntz metal)	Excellent hot formability and forgeability for blanking, forming and bending, hot forging and pressing, hot heading and upsetting, shearing. Uses: architectural panel sheets, large nuts and bolts, brazing rod, condenser plates, heat exchanger and condenser tubing, hot forgings
C31400 (leaded commercial bronze)	Excellent machinability. Uses: screws, machine parts, pickling crates
C31600 (leaded commercial bronze, nickel-bearing)	Good cold workability; poor hot formability. Uses: electrical connectors, fasteners, hardware, nuts, screws, screw machine parts
C33000 (low-leaded brass tube)	Combines good machinability and excellent cold workability. Fabricated by forming and bending, machining, piercing, punching. Uses: pump and power cylinders and liners, ammunition primers, plumbing accessories
C33200 (high-leaded brass tube)	Excellent machinability. Fabricated by piercing, punching, machining. Uses: general-purpose screw machine parts
C33500 (low-leaded brass)	Similar to C33200. Commonly fabricated by blanking, drawing, machining, piercing and punching, stamping. Uses: butts, hinges, watch backs
C34000 (medium-leaded brass)	Similar to C33200. Fabricated by blanking, heading and upsetting, machining, piercing and punching, roll threading and knurling, stamping. Uses: butts, gears, nuts, rivets, screws, dials, engravings, instrument plates
C34200 (high-leaded brass)	Combines excellent machinability with moderate cold workability. Uses: clock plates and nuts, clock and watch backs, gears, wheels, channel plate
C34900	Good cold workability, fair hot workability for bending and forming, heading and upsetting, machining, roll threading and knurling. Uses: building hardware, rivets and nuts, plumbing goods, and parts requiring moderate cold working combined with some machining
C35000 (medium-leaded brass)	Fair cold workability; poor hot formability. Uses: bearing cages, book dies, clock plates, engraving plates, gears, hinges, hose couplings, keys, lock parts, lock tumblers, meter parts, nuts, sink strainers, strike plates, templates, type characters, washers, wear plates
C35300 (high-leaded brass)	Similar to C34200
C35600 (extra-high-leaded brass)	Excellent machinability. Fabricated by blanking, machining, piercing and punching, stamping. Uses: same as C34200 and C35300
C36000 (free-cutting brass)	Excellent machinability. Fabricated by machining, roll threading and knurling. Uses: gears, pinions, automatic high-speed screw machine parts
C36500 to C36800 (leaded Muntz metal)	Combines good machinability with excellent hot formability. Uses: condenser tube plates
C37000 (free-cutting Muntz metal)	Fabrication characteristics similar to C36500 to C36800. Uses: automatic screw machine parts
C37700 (forging brass)	Excellent hot workability. Fabricated by heading and upsetting, hot forging and pressing, hot heading and upsetting, machining. Uses: forgings and pressings of all kinds
C38500 (architectural bronze)	Excellent machinability and hot workability. Fabricated by hot forging and pressing, forming, bending and machining. Uses: architectural extrusions, store fronts, thresholds, trim, butts, hinges, lock bodies, forgings
C40500	Excellent cold workability. Fabricated by blanking, forming, drawing. Uses: meter clips, terminals, fuse clips, contact and relay springs, washers
C40800	Excellent cold workability. Fabricated by blanking, stamping, shearing. Uses: electrical connectors
C41100	Excellent cold workability, good hot formability. Fabricated by blanking, forming, drawing. Uses: bushings, bearing sleeves, thrust washers, terminals, connectors, flexible metal hose, electrical conductors
C41300	Excellent cold workability; good hot formability. Uses: plater bar for jewelry products, flat springs for electrical switchgear
C41500	Excellent cold workability. Fabricated by blanking, drawing, bending, forming, shearing, stamping. Uses: spring applications for electrical switches
C42200	Excellent cold workability; good hot formability. Fabricated by blanking, piercing, forming, drawing. Uses: sash chains, fuse clips, terminals, spring washers, contact springs, electrical connectors
C42500	Excellent cold workability. Fabricated by blanking, piercing, forming, drawing. Uses: electrical switches, springs, terminals, connectors, fuse clips, pen clips, weather stripping

(continued)

Table 2 (continued)

Alloy number (and name)	Fabrication characteristics and typical applications
C43000	Excellent cold workability; good hot formability. Fabricated by blanking, coining, drawing, forming, bending, heading, upsetting. Uses: same as C42500
C43400	Excellent cold workability. Fabricated by blanking, drawing, bonding, forming, stamping, shearing. Uses: electrical switch parts, blades, relay springs, contacts
C43500	Excellent cold workability for fabrication by forming and bending. Uses: Bourdon tubing and musical instruments
C44300, C44400, C44500 (inhibited admiralty)	Excellent cold workability for forming and bending. Uses: condenser, evaporator, and heat exchanger tubing; condenser tubing plates; distiller tubing; ferrules
C46400 to C46700 (naval brass)	Excellent hot workability and hot forgeability. Fabricated by blanking, drawing, bending, heading and upsetting, hot forging, pressing. Uses: aircraft turnbuckle barrels, balls, bolts, marine hardware, nuts, propeller shafts, rivets, valve stems, condenser plates, welding rod
C48200 (naval brass, medium-leaded)	Good hot workability for hot forging, pressing, and machining operations. Uses: marine hardware, screw machine products, valve stems
C48500 (leaded naval brass)	Combines excellent hot forgeability and machinability. Fabricated by hot forging and pressing, machining. Uses: marine hardware, screw machine parts, valve stems
C50500 (phosphor bronze, 1.25% E)	Excellent cold workability; good hot formability. Fabricated by blanking, bending, heading and upsetting, shearing and swaging. Uses: electrical contacts, flexible hose, pole-line hardware
C51000 (phosphor bronze, 5% A)	Excellent cold workability. Fabricated by blanking, drawing, bending, heading and upsetting, roll threading and knurling, shearing, stamping. Uses: bellows, Bourdon tubing, clutch discs, cotter pins, diaphragms, fasteners, lock washers, wire brushes, chemical hardware, textile machinery, welding rod
C51100	Excellent cold workability. Uses: bridge bearing plates, locator bars, fuse clips, sleeve bushings, springs, switch parts, truss wire, wire brushes, chemical hardware, perforated sheets, textile machinery, welding rod
C52100 (phosphor bronze, 8% C)	Good cold workability for blanking, drawing, forming and bending, shearing, stamping. Uses: generally for more severe service conditions than C51000
C52400 (phosphor bronze, 10% D)	Good cold workability for blanking, forming and bending, shearing. Uses: heavy bars and plates for severe compression; bridge and expansion plates and fittings; articles requiring good spring qualities, resiliency, fatigue resistance, and good wear and corrosion resistance
C54400 (free-cutting phosphor bronze)	Excellent machinability; good cold workability. Fabricated by blanking, drawing, bending, machining, shearing, stamping. Uses: bearings, bushings, gears, pinions, shafts, thrust washers, valve parts
C60800 (aluminum bronze, 5%)	Good cold workability; fair hot formability. Uses: condenser, evaporator, and heat exchanger tubes; distiller tubes; ferrules
C61000	Good hot and cold workability. Uses: bolts, pump parts, shafts, tie rods, overlay on steel for wearing surfaces
C61300	Good hot and cold formability. Uses: nuts, bolts, stringers and threaded members, corrosion-resistant vessels and tanks, structural components, machine parts, condenser tube and piping systems, marine protective sheathing and fastening, munitions mixing troughs and blending chambers
C61400 (aluminum bronze, D)	Similar to C61300
C61500	Good hot and cold workability. Fabrication characteristics similar to C52100. Uses: hardware, decorative metal trim, interior furnishings, and other articles requiring high tarnish resistance
C61800	Fabricated by hot forging and hot pressing. Uses: bushings, bearings, corrosion-resistant applications, welding rods
C61900	Excellent hot formability for fabricating by blanking, forming, bending, shearing, and stamping. Uses: springs, contacts, switch components
C62300	Good hot and cold formability. Fabricated by bending, hot forging, hot pressing, forming, welding. Uses: bearings, bushings, valve guides, gears, valve seats, nuts, bolts, pump rods, worm gears, and cams
C62400	Excellent hot formability for fabrication by hot forging and hot bending. Uses: bushings, gears, cams, wear strips, nuts, drift pins, tie rods
C62500	Excellent hot formability for fabrication by hot forging and machining. Uses: guide bushings, wear strips, cams, dies, forming rolls
C63000	Good hot formability. Fabricated by hot forming and forging. Uses: nuts, bolts, valve seats, plunger tips, marine shafts, valve guides, aircraft parts, pump shafts, structural members
C63200	Good hot formability. Fabricated by hot forming and welding. Uses: nuts, bolts, structural pump parts, shafting requiring corrosion resistance
C63600	Excellent cold workability; fair hot formability. Fabricated by cold heading. Uses: components for pole-line hardware, cold-headed nuts for wire and cable connectors, bolts and screw products
C63800	Excellent cold workability and hot formability. Uses: springs, switch parts, contacts, relay springs, glass sealing, porcelain enameling
C64200	Excellent hot formability. Fabricated by hot forming, forging, machining. Uses: valve stems, gears, marine hardware, pole-line hardware, bolts, nuts, valve bodies and components
C65100 (low-silicon bronze, B)	Excellent hot and cold workability. Fabricated by forming and bending, heading and upsetting, hot forging and pressing, roll threading and knurling, squeezing and swaging. Uses: hydraulic pressure lines, anchor screws, bolts, cable clamps, cap screws, machine screws, marine hardware, nuts, pole-line hardware, rivets, U-bolts, electrical conduits, heat exchanger tubing, welding rod
C65400	Excellent hot and cold workability. Fabricated by forming, bending, blanking. Uses: springs, switch parts, contacts and relay springs in above-ambient-temperature conditions demanding superior stress relaxation
C65500 (high-silicon bronze, A)	Excellent hot and cold workability. Fabricated by blanking, drawing, forming and bending, heading and upsetting, hot forging and pressing, roll threading and knurling, shearing, squeezing, swaging. Uses: similar to C65100 including propeller shafts
C66700 (manganese brass)	Excellent cold formability. Fabricated by blanking, bending, forming, stamping, welding. Uses: brass products resistance welded by spot, seam, and butt welding
C67400	Excellent hot formability. Fabricated by hot forging and pressing, machining. Uses: bushings, gears, connecting rods, shafts, wear plates
C67500 (manganese bronze, A)	Excellent hot workability. Fabricated by hot forging and pressing, hot heading and upsetting. Uses: clutch discs, pump rods, shafting, balls, valve stems and bodies
C68700 (aluminum brass, arsenical)	Excellent cold workability for forming and bending. Uses: condenser, evaporator, and heat exchanger tubing; condenser tubing plates; distiller tubing; ferrules
C68800	Excellent hot and cold formability. Fabricated by blanking, drawing, forming and bending, shearing and stamping. Uses: springs, switches, contacts, relays, drawn parts
C69000	Fabricating characteristics same as C68800. Uses: wiring devices, relays, switches, springs, high-strength shells
C69400 (silicon red brass)	Excellent hot formability for fabrication by forging, screw machine operations. Uses: valve stems where corrosion resistance and high strength are critical

(continued)

Table 2 (continued)

Alloy number (and name)	Fabrication characteristics and typical applications
C70250	Excellent hot and cold workability. Fabricated by blanking, forming, bending. Uses: relays, switches, springs, lead frames for use at service temperatures above ambient where superior stress relaxation is required
C70400	Excellent cold workability; good hot formability. Fabricated by forming, bending, welding. Uses: condensers, evaporators, heat exchangers, ferrules, saltwater piping, lithium bromide absorption tubing, shipboard condenser intake systems
C70600 (copper-nickel, 10%)	Good hot and cold workability. Fabricated by forming and bending, welding. Uses: condensers, condenser plates, distiller tubing, evaporator and heat exchanger tubing, ferrules, saltwater piping
C71000 (copper-nickel, 20%)	Good hot and cold formability. Fabricated by blanking, forming and bending, welding. Uses: communication relays, condensers, condenser plates, electrical springs, evaporator and heat exchanger tubes, ferrules, resistors
C71300	Good hot and cold formability. Fabricated by blanking. Uses: U.S. 5-cent coin and, when clad to C11000, U.S. 10-cent, 25-cent, 50-cent, and $1 coins
C71500 (copper-nickel, 30%)	Similar to C70600
C71700	Good hot and cold formability. Uses: high-strength constructional parts for seawater corrosion resistance, hydrophone cases, mooring cable wire, springs, retainer rings, bolts, screws, pins for ocean telephone cable applications
C72500	Excellent cold and hot formability. Fabricated by blanking, brazing, coining, drawing, etching, forming and bending, heading and upsetting, roll threading and knurling, shearing, spinning, squeezing, stamping and swaging. Uses: relay and switch springs, connectors, brazing alloy, lead frames, control and sensing bellows
C73500	Fabrication characteristics same as C74500. Uses: hollowware, medallions, jewelry, base for silver plate, cosmetic cases, musical instruments, nameplates, contacts
C74500 (nickel silver, 65-10)	Excellent cold workability. Fabricated by blanking, drawing, etching, forming and bending, heading and upsetting, roll threading and knurling, shearing, spinning, squeezing and swaging. Uses: rivets, screws, slide fasteners, optical parts, etching stock, hollowware, nameplates, platers' bars
C75200 (nickel silver, 65-18)	Fabrication characteristics similar to C74500. Uses: rivets, screws, table flatware, truss wire, zippers, bows, camera parts, core bars, temples, base for silver plate, costume jewelry, etching stock, hollowware, nameplates, radio dials
C75400 (nickel silver, 65-15)	Fabrication characteristics similar to C74500. Uses: camera parts, optical equipment, etching stock, jewelry
C75700 (nickel silver, 65-12)	Fabrication characteristics similar to C74500. Uses: slide fasteners, camera parts, optical parts, etching stock, nameplates
C76200	Fabrication characteristics same as C77000. Uses: electrical terminals, contact springs, release brackets, ornamental bits and spurs, optical parts, surgical instruments, electrical contacts
C77000 (nickel silver, 55-18)	Good cold workability. Fabricated by blanking, forming and bending, shearing. Uses: optical goods, springs, resistance wire
C72200	Good hot and cold workability. Fabricated by forming, bending and welding. Uses: condenser and heat exchanger tubing, saltwater piping
C78200 (leaded nickel silver, 65-8-2)	Good cold formability. Fabricated by blanking, milling and drilling. Uses: key blanks, watch plates, watch parts

of correlation is often different for different methods of hardness determination.

Hot Working. Not all shaping is confined to cold deformation. Hot working is commonly used for alloys that remain ductile above the recrystallization temperature. Hot working permits more extensive changes in shape than cold working, and thus a single operation often can replace a sequence of forming and annealing operations. To avoid preferred orientation and textures, and to achieve processing economy, copper and many copper alloys are hot worked to nearly finished size. Hot working reduces the as-cast grain size from about 1 to 10 mm (0.04 to 0.4 in.) to about 0.1 mm (0.004 in.) or less and yields a soft texture-free structure suitable for cold finishing.

Some hot-working operations may produce strengths that exceed that of the annealed temper. However, property control by hot working is very difficult and is rarely attempted.

Heat Treating

Work-hardened metal can be returned to a soft state by heating, or annealing. During the annealing of simple single-phase alloys, deformed and highly stressed crystals are transformed into unstressed crystals by recovery, recrystallization, and grain growth. In severely deformed metal, recrystallization occurs at lower temperatures than in lightly deformed metal. Also, the grains are smaller and more uniform in size when severely deformed metal is recrystallized.

Grain size can be controlled by proper selection of cold-working and annealing practices. Large amounts of prior cold work, fast heating to annealing temperature, and short annealing times favor fine grain sizes. Larger grain sizes are normally produced by a combination of limited deformation and long annealing times. In normal commercial practice, annealed grain sizes are controlled to about a median value in the range of 0.01 to 0.10 mm (0.0004 to 0.004 in.).

Variations in annealed grain size produce variations in hardness and other mechanical properties that are smaller than those that occur in cold-worked material, but these variations are by no means negligible. Fine grain sizes often are required to enhance end-product characteristics such as load-carrying capacity, fatigue resistance, resistance to SCC, and surface quality for polishing and buffing of either annealed or cold-formed parts.

Heat-treating processes can also be applied to copper and copper alloys to achieve homogenization, stress relieving, solutionizing, precipitation hardening, and quench hardening and tempering. These aspects are referred to throughout the articles that follow in this Section, and they are reviewed in more detail in *Heat Treating*, Volume 4 of the 9th Edition of *Metals Handbook*.

Temper Designations

The temper designations for wrought copper and copper alloys were traditionally specified on the basis of cold reduction imparted by rolling or drawing. This scheme related the nominal temper designations to the amount of reduction stated in Brown & Sharpe (B & S) gage numbers for rolled sheet and drawn wire (Table 7). Heat-treatable alloys and product forms such as rod, tube, extrusions, and castings were not readily described by this system. To remedy this situation, ASTM B 601, "Standard Practice for Temper Designations for Copper and Copper Alloys—Wrought and Cast," was developed. This standard established an alphanumeric code that can be assigned to each of the standard descriptive temper designations (Table 8).

Electrical Coppers

Commercially pure copper is represented by UNS numbers C10100 to C13000. The various coppers within this group have different degrees of purity and therefore different characteristics. Fire-refined tough pitch copper C12500 is made by deoxidizing anode copper until the oxygen content has been lowered to a value of 0.02 to 0.04%. Both the traditional method of poling (or pitching) a bath of molten anode copper and the more modern method of deoxidizing

Table 3 Properties and applications of cast copper and copper alloys

UNS designa-tion(a)	Nominal composition, %(a)	Typical mechanical properties, as-cast (heat treated)(b)								Machinability rating, %(c)	Casting types(d)	Typical applications
		Tensile strength		Yield strength		Elongation in 50 mm (2 in.), %	Hardness					
		MPa	ksi	MPa	ksi		Rockwell	Brinell 500 kg	Brinell 3000 kg			
C80100	99.95 Cu + Ag min, 0.05 other max	172	25	62	9	40	...	44	...	10	C, T, I, M, P, S	Electrical and thermal conductors; corrosion- and oxidation-resistant applications
C80300	99.95 Cu + Ag min, 0.034 Ag min, 0.05 other max	172	25	62	9	40	...	44	...	10	C, T, I, M, P, S	Electrical and thermal conductors; corrosion- and oxidation-resistant applications
C80500	99.75 Cu + Ag min, 0.034 Ag min, 0.02 B max, 0.23 other max	172	25	62	9	40	...	44	...	10	C, T, I, M, P, S	Electrical and thermal conductors; corrosion- and oxidation-resistant applications
C80700	99.75 Cu + Ag min, 0.02 B max, 0.23 other max	172	25	62	9	40	...	44	...	10	C, T, I, M, P, S	Electrical and thermal conductors; corrosion- and oxidation-resistant applications
C80900	99.70 Cu + Ag min, 0.034 Ag min, 0.30 other max	172	25	62	9	40	...	44	...	10	C, T, I, M, P, S	Electrical and thermal conductors; corrosion- and oxidation-resistant applications
C81100	99.70 Cu + Ag min, 0.30 other max	172	25	62	9	40	...	44	...	10	C, T, I, M, P, S	Electrical and thermal conductors; corrosion- and oxidation-resistant applications
High-copper alloys												
C81300	98.5 Cu min, 0.06 Be, 0.80 Co, 0.40 other max	(365)	(53)	(248)	(36)	(11)	...	(39)	...	20	C, T, I, M, P, S	Higher-hardness electrical and thermal conductors
C81400	98.5 Cu min, 0.06 Be, 0.80 Cr, 0.40 other max	(365)	(53)	(248)	(36)	(11)	(B 69)	...	...	20	C, T, I, M, P, S	Higher-hardness electrical and thermal conductors
C81500	98.0 Cu min, 1.0 Cr, 0.50 other max	(352)	(51)	(276)	(40)	(17)	...	(105)	...	20	C, T, I, M, P, S	Electrical and/or thermal conductors used as structural members where strength and hardness greater than that of C80100-C81100 are required
C81700	94.25 Cu min, 1.0 Ag, 0.4 Be, 0.9 Co, 0.9 Ni	(634)	(92)	(469)	(68)	(8)	...	...	(217)	30	C, T, I, M, P, S	Electrical and/or thermal conductors used as structural members where strength and hardness greater than that of C80100-C81100 are required. Also used in place of C81500 where electrical and/or thermal conductivities can be sacrificed for hardness and strength
C81800	95.6 Cu min, 1.0 Ag, 0.4 Be, 1.6 Co	345 (703)	50 (102)	172 (517)	25 (75)	20 (8)	B 55 (B 96)	...	...	20	C, T, I, M, P, S	Resistance welding electrodes, dies
C82000	96.8 Cu, 0.6 Be, 2.6 Co	345 (689)	50 (100)	138 (517)	20 (75)	20 (8)	B 55 (B 95)	...	(195)	20	C, T, I, M, P, S(e)	Current-carrying parts, contact and switch blades, bushings and bearings, soldering iron, resistance welding tips
C82100	97.7 Cu, 0.5 Be, 0.9 Co, 0.9 Ni	(634)	(92)	(469)	(68)	(8)	...	...	(217)	30	C, T, I, M, P, S	Electrical and/or thermal conductors used as structural members where strength and hardness greater than that of C80100-C81100 are required. Also used in place of C81500 where electrical and/or thermal conductivities can be sacrificed for hardness and strength
C82200	96.5 Cu min, 0.6 Be, 1.5 Ni	393 (655)	57 (95)	207 (517)	30 (75)	20 (8)	B 60 (B 96)	...	...	20	C, T, I, M, P, S	Clutch rings, brake drums, seam welder electrodes, projection welding dies, spot welding tips, beam welder shapes, bushings, water-cooled holders
C82400	96.4 Cu min, 1.70 Be, 0.25 Co	496 (1034)	72 (150)	255 (965)	37 (140)	20 (1)	B 78 (C 38)	...	...	20	C, I, M, P, S(e)	Safety tools, molds for plastic parts, cams, bushings, bearings, valves, pump parts, gears
C82500	97.2 Cu, 2.0 Be, 0.5 Co, 0.25 Si	552 (1103)	80 (160)	310	45	20 (1)	B 82 (C 40)	...	...	20	C, I, M, P, S(e)	Safety tools, molds for plastic parts, cams, bushings, bearings, valves, pump parts
C82600	95.2 Cu min, 2.3 Be, 0.5 Co, 0.25 Si	565 (1138)	82 (165)	324 (1069)	47 (155)	20 (1)	B 83 (C 43)	...	...	20	C, I, M, P, S(e)	Bearings and molds for plastic parts
C82700	96.3 Cu, 2.45 Be, 1.25 Ni	(1069)	(155)	(896)	(130)	(0)	(C 39)	...	...	20	C, I, M, P, S	Bearings and molds for plastic parts

(continued)

(a) Nominal composition, unless otherwise noted. For seldom-used alloys, only compositions are available. (b) Values for C82700, C84200, C96200, and C96300 are minimum, not typical. As-cast values are for sand casting except C93900, continuous cast; and C85800, C87800, C87900, die cast. Heat-treated values, in parentheses, indicate that the alloy responds to heat treatment. If heat-treated values are not shown, the copper or copper alloy does not respond. (c) Based on a value of 100% for free-cutting brass. (d) C, centrifugal; T, continuous; D, die; I, investment; M, permanent mold; P, plaster; S, sand. (e) Also pressure cast. (f) Property and application data not available from the Copper Development Association Inc. (g) As-heat-treated value for C94700, 20; for C94800, 40. Source: Copper Development Association Inc.

Table 3 (continued)

UNS designation(a)	Nominal composition, %(a)	Typical mechanical properties, as-cast (heat treated)(b): Tensile strength, MPa	Tensile strength, ksi	Yield strength, MPa	Yield strength, ksi	Elongation in 50 mm (2 in.), %	Hardness: Rockwell	Hardness: Brinell, 500 kg	Hardness: Brinell, 3000 kg	Machinability rating, %(c)	Casting types(d)	Typical applications
High-copper alloys (continued)												
C82800	96.6 Cu, 2.6 Be, 0.5 Co, 0.25 Si	669 (1138)	97 (165)	379 (1000)	55 (145)	20 (1)	B 85 (C 45)	...	...	10	C, I, M, P, S(e)	Molds for plastic parts, cams, bushings, bearings, valves, pump parts, sleeves
Red brasses and leaded red brasses												
C83300	93 Cu, 1.5 Sn, 1.5 Pb, 4 Zn	221	32	69	10	35	...	35	...	35	S	Terminal ends for electrical cables
C83400	90 Cu, 10 Zn	241	35	69	10	30	F 50	...	...	60	C, S	Moderate-strength, moderate-conductivity castings; rotating bands
C83600	85 Cu, 5 Sn, 5 Pb, 5 Zn	255	37	117	17	30	...	60	...	84	C, T, I, S	Valves, flanges, pipe fittings, plumbing goods, pump castings, water pump impellers and housings, ornamental fixtures, small gears
C83800	83 Cu, 4 Sn, 6 Pb, 7 Zn	241	35	110	16	25	...	60	...	90	C, T, S	Low-pressure valves and fittings, plumbing supplies and fittings, general hardware, air-gas-water fittings, pump components, railroad catenary fittings
Semired brasses and leaded semired brasses												
C84200	80 Cu, 5 Sn, 2.5 Pb, 12.5 Zn	193	28	103	15	27	...	60	...	80	C, T, S	Pipe fittings, elbows, Ts, couplings, bushings, lock nuts, plugs, unions
C84400	81 Cu, 3 Sn, 7 Pb, 9 Zn	234	34	103	15	26	...	55	...	90	C, T, S	General hardware, ornamental castings, plumbing supplies and fixtures, low-pressure valves and fittings
C84500	78 Cu, 3 Sn, 7 Pb, 12 Zn	241	35	97	14	28	...	55	...	90	C, T, S	Plumbing fixtures, cocks, faucets, and stops; waste, air, and gas fittings; low-pressure valve fittings
C84800	76 Cu, 3 Sn, 6 Pb, 15 Zn	248	36	97	14	30	...	55	...	90	C, S	Plumbing fixtures, cocks, faucets, stops, waste, air and gas fittings, general hardware, and low-pressure valve fittings
Yellow brasses and leaded yellow brasses												
C85200	72 Cu, 1 Sn, 3 Pb, 24 Zn	262	38	90	13	35	...	45	...	80	C, T	Plumbing fittings and fixtures, ferrules, valves, hardware, ornamental brass, chandeliers, and irons
C85400	67 Cu, 1 Sn, 3 Pb, 29 Zn	234	34	83	12	35	...	50	...	80	C, T, M, P, S	General-purpose yellow casting alloy not subject to high internal pressure. Furniture hardware, ornamental castings, radiator fittings, ship trimmings, cocks, battery clamps, valves, and fittings
C85500	61 Cu, 0.8 Al, bal Zn	414	60	159	23	40	B 55	85	...	80	C, S	Ornamental castings
C85700	63 Cu, 1 Sn, 1 Pb, 34.7 Zn, 0.3 Al	345	50	124	18	40	...	75	...	80	C, M, P, S	Bushings, hardware fittings, ornamental castings
C85800	58 Cu, 1 Sn, 1 Pb, 40 Zn	379	55	207	30	15	B 55	...	...	80	D	General-purpose die casting alloy having moderate strength
Manganese and leaded manganese bronze alloys												
C86100	67 Cu, 21 Zn, 3 Fe, 5 Al, 4 Mn	655	95	345	50	20	...	...	180	30	C, I, P, S	Marine castings, gears, gun mounts, bushings and bearings, marine racing propellers
C86200	64 Cu, 26 Zn, 3 Fe, 4 Al, 3 Mn	655	95	331	48	20	...	...	180	30	C, T, D, I, P, S	Marine castings, gears, gun mounts, bushings and bearings
C86300	63 Cu, 25 Zn, 3 Fe, 6 Al, 3 Mn	793	115	572	83	15	...	...	225	8	C, I, P, S	Extra-heavy-duty high-strength alloy. Large valve stems, gears, cams, slow-speed heavy-load bearings, screwdown nuts, hydraulic cylinder parts
C86400	59 Cu, 1 Pb, 40 Zn	448	65	172	25	20	...	90	105	65	C, D, M, P, S	Free-machining manganese bronze. Valve stems, marine fittings, lever arms, brackets, light-duty gears

(continued)

(a) Nominal composition, unless otherwise noted. For seldom-used alloys, only compositions are available. (b) Values for C82700, C84200, C96200, and C96300 are minimum, not typical. As-cast values are for sand casting except C93900, continuous cast; and C85800, C87800, C87900, die cast. Heat-treated values, in parentheses, indicate that the alloy responds to heat treatment. If heat-treated values are not shown, the copper or copper alloy does not respond. (c) Based on a value of 100% for free-cutting brass. (d) C, centrifugal; T, continuous; D, die; I, investment; M, permanent mold; P, plaster; S, sand. (e) Also pressure cast. (f) Property and application data not available from the Copper Development Association Inc. (g) As-heat-treated value for C94700, 20; for C94800, 40. Source: Copper Development Association Inc.

Table 3 (continued)

UNS designation(a)	Nominal composition, %(a)	Typical mechanical properties, as-cast (heat treated)(b): Tensile strength, MPa	Tensile strength, ksi	Yield strength, MPa	Yield strength, ksi	Elongation in 50 mm (2 in.), %	Hardness: Rockwell	Hardness: Brinell, 500 kg	Hardness: Brinell, 3000 kg	Machinability rating, %(c)	Casting types(d)	Typical applications
Manganese and leaded manganese bronze alloys (continued)												
C86500	58 Cu, 0.5 Sn, 39.5 Zn, 1 Fe, 1 Al	490	71	193	28	30	. . .	100	130	26	C, I, P, S	Machinery parts requiring strength and toughness, lever arms, valve stems, gears
C86700	58 Cu, 1 Pb, 41 Zn	586	85	290	42	20	B 80	. . .	155	55	C, S	High-strength free-machining manganese bronze. Valve stems
C86800	55 Cu, 37 Zn, 3 Ni, 2 Fe, 3 Mn	565	82	262	38	22	. . .	. . .	80	30	S	Marine fittings, marine propellers
Silicon bronzes and silicon brasses												
C87200	89 Cu min, 4 Si	379	55	172	25	30	. . .	85	. . .	40	C, I, M, P, S	Bearings, bells, impellers, pump and valve components, marine fittings, corrosion-resistant castings
C87400	83 Cu, 14 Zn, 3 Si	379	55	165	24	30	. . .	70	100	50	C, D, I, M, P, S	Bearings, gears, impellers, rocker arms, valve stems, clamps
C87500	82 Cu, 14 Zn, 4 Si	462	67	207	30	21	. . .	115	134	50	C, D, I, M, P, S	Bearings, gears, impellers, rocker arms, valve stems, small boat propellers
C87600	90 Cu, 5.5 Zn, 4.5 Si	455	66	221	32	20	B 76	110	135	40	S	Valve stems
C87800	82 Cu, 14 Zn, 4 Si	586	85	345	50	25	B 85	. . .	. . .	40	D	High-strength thin-wall die castings, brush holders, lever arms, brackets, clamps, hexagonal nuts
C87900	65 Cu, 34 Zn, 1 Si	483	70	241	35	25	B 70	. . .	. . .	80	D	General-purpose die casting alloy having moderate strength
Tin bronzes												
C90200	93 Cu, 7 Sn	262	38	110	16	30	. . .	70	. . .	20	C, S	Bearings and bushings
C90300	88 Cu, 8 Sn, 4 Zn	310	45	145	21	30	. . .	70	. . .	30	C, T, I, P, S	Bearings, bushings, pump impellers, piston rings, valve components, seal rings, steam fittings, gears
C90500	88 Cu, 10 Sn, 2 Zn	310	45	152	22	25	. . .	75	. . .	30	C, T, I, S	Bearings, bushings, pump impellers, piston rings, valve components, steam fittings, gears
C90700	89 Cu, 11 Sn	303 (379)	44 (55)	152 (207)	22 (30)	20 (16)	. . .	80 (102)	. . .	20	C, T, I, M, S	Gears, bearings, bushings
C90800	87 Cu, 12 Sn											
C90900	87 Cu, 13 Sn	276	40	138	20	15	. . .	90	. . .	20	C, S	Bearings and bushings
C91000	85 Cu, 14 Sn, 1 Zn	221	32	172	25	2	. . .	105	. . .	20	C, T, I, S	Piston rings and bearings
C91100	84 Cu, 16 Sn	241	35	172	25	2	. . .	. . .	135	10	S	Piston rings, bearings, bushings, bridge plates
C91300	81 Cu, 19 Sn	241	35	207	30	0.5	. . .	. . .	170	10	C, T, M, S	Piston rings, bearings, bushings, bridge plates, bells
C91600	88 Cu, 10.5 Sn, 1.5 Ni	303 (414)	44 (60)	152 (221)	22 (32)	16 (16)	. . .	85 (106)	. . .	20	C, T, M, S	Gears
C91700	86.5 Cu, 12 Sn, 1.5 Ni	303 (414)	44 (60)	152 (221)	22 (32)	16 (16)	. . .	85 (106)	. . .	20	C, T, I, M, S	Gears
Leaded tin bronzes												
C92200	88 Cu, 6 Sn, 1.5 Pb, 4.5 Zn	276	40	138	20	30	. . .	65	. . .	42	C, T, I, M, P, S	Valves, fittings, pressure-containing parts for use up to 290 °C (550 °F)
C92300	87 Cu, 8 Sn, 4 Zn	276	40	138	20	25	. . .	70	. . .	42	C, T, S	Valves, pipe fittings, and high-pressure steam castings. Superior machinability to C90300
C92400 (f)	88 Cu, 10 Sn, 2 Pb, 2 Zn	. . .	. . .	. . .	. . .	. . .	. . .	. . .	. . .	. . .	. . .	. . .
C92500	87 Cu, 11 Sn, 1 Pb, 1 Ni	303	44	138	20	20	. . .	80	. . .	30	C, T, M, S	Gears, automotive synchronizer rings
C92600	87 Cu, 10 Sn, 1 Pb, 2 Zn	303	44	138	20	30	F 78	70	. . .	40	C, T, S	Bearings, bushings, pump impellers, piston rings, valve components, steam fittings, gears. Superior machinability to C90500
C92700	88 Cu, 10 Sn, 2 Pb	290	42	145	21	20	. . .	77	. . .	45	C, T, S	Bearings, bushings, pump impellers, piston rings, valve components, steam fittings, gears. Superior machinability to C90500

(continued)

(a) Nominal composition, unless otherwise noted. For seldom-used alloys, only compositions are available. (b) Values for C82700, C84200, C96200, and C96300 are minimum, not typical. As-cast values are for sand casting except C93900, continuous cast; and C85800, C87800, C87900, die cast. Heat-treated values, in parentheses, indicate that the alloy responds to heat treatment. If heat-treated values are not shown, the copper or copper alloy does not respond. (c) Based on a value of 100% for free-cutting brass. (d) C, centrifugal; T, continuous; D, die; I, investment; M, permanent mold; P, plaster; S, sand. (e) Also pressure cast. (f) Property and application data not available from the Copper Development Association Inc. (g) As-heat-treated value for C94700, 20; for C94800, 40. Source: Copper Development Association Inc.

Table 3 (continued)

UNS designation(a)	Nominal composition, %(a)	Typical mechanical properties, as-cast (heat treated)(b): Tensile strength, MPa	Tensile strength, ksi	Yield strength, MPa	Yield strength, ksi	Elongation in 50 mm (2 in.), %	Hardness: Rockwell	Hardness: Brinell, 500 kg	Hardness: Brinell, 3000 kg	Machinability rating, %(c)	Casting types(d)	Typical applications
Leaded tin bronzes (continued)												
C92800	79 Cu, 16 Sn, 5 Pb	276	40	207	30	1	B80	...	...	70	C, S	Piston rings
C92900	84 Cu, 10 Sn, 2.5 Pb, 3.5 Ni	324 (324)	47 (47)	179 (179)	26 (26)	20 (20)	...	80 (80)	...	40	C, T, M, S	Gears, wear plates, guides, cams, parts requiring machinability superior to that of C91600 or C91700
High-leaded tin bronzes												
C93200	83 Cu, 7 Sn, 7 Pb, 3 Zn	241	35	124	18	20	...	65	...	70	C, T, M, S	General-utility bearings and bushings
C93400	84 Cu, 8 Sn, 8 Pb	221	32	110	16	20	...	60	...	70	C, T, S	Bearings and bushings
C93500	85 Cu, 5 Sn, 9 Pb	221	32	110	16	20	...	60	...	70	C, T, S	Small bearings and bushings, bronze backing for babbit-lined automotive bearings
C93700	80 Cu, 10 Sn, 10 Pb	241	35	124	18	20	...	60	...	80	C, T, M, S	Bearings for use at high speed and heavy pressures, pumps, impellers, corrosion-resistant applications, pressure-tight castings
C93800	78 Cu, 7 Sn, 15 Pb	207	30	110	16	18	...	55	...	80	C, T, M, S	Bearings for general service at moderate pressures, pump impellers and bodies for use in acid mine water
C93900	79 Cu, 6 Sn, 15 Pb	221	32	152	22	7	...	63	...	80	T	Continuous castings only. Bearings for general service, pump bodies and impellers for mine waters
C94000 (f)	70.5 Cu, 13.0 Sn, 15.0 Pb, 0.50 Zn, 0.75 Ni, 0.25 Fe, 0.05 P, 0.35 Sb	...	...	...	...	...	...	...	...	...	...	...
C94100 (f)	70.0 Cu, 5.5 Sn, 18.5 Pb, 3.0 Zn, 1.0 other max	...	...	...	...	...	...	...	...	...	...	...
C94300	70 Cu, 5 Sn, 25 Pb	186	27	90	13	15	...	48	...	80	C, S	High-speed bearings for light loads
C94400	81 Cu, 8 Sn, 11 Pb	221	32	110	16	18	...	55	...	80	C, T, S	General-utility alloy for bushings and bearings
C94500	73 Cu, 7 Sn, 20 Pb	172	25	83	12	12	...	50	...	80	C, S	Locomotive wearing parts, high-speed low-load bearings
Nickel-tin bronzes												
C94700	88 Cu, 5 Sn, 2 Zn, 5 Ni	345 (586)	50 (85)	159 (414)	23 (60)	35 (10)	...	85	(180)	30(g)	C, T, I, M, S	Valve stems and bodies, bearings, wear guides, shift forks, feeding mechanisms, circuit breaker parts, gears, piston cylinders, nozzles
C94800	87 Cu, 5 Sn, 5 Ni	310 (414)	45 (60)	159 (207)	23 (30)	35 (8)	...	80 (120)	...	50(g)	M, S	Structural castings, gear components, motion translation devices, machinery parts, bearings
C94900 (f)	80 Cu, 5 Sn, 5 Pb, 5 Zn, 5 Ni	...	...	...	...	...	...	...	...	...	...	...
Aluminum bronzes												
C95200	88 Cu, 3 Fe, 9 Al	552	80	186	27	35	...	...	125	50	C, T, M, P, S	Acid-resisting pumps, bearings, gears, valve seats, guides, plungers, pump rods, bushings
C95300	89 Cu, 1 Fe, 10 Al	517 (586)	75 (85)	186 (290)	27 (42)	25 (15)	...	...	140 (174)	55	C, T, M, P, S	Pickling baskets, nuts, gears, steel mill slippers, marine equipment, welding jaws
C95400	85 Cu, 4 Fe, 11 Al	586 (724)	85 (105)	241 (372)	35 (54)	18 (8)	...	...	170 (195)	60	C, T, M, P, S	Bearings, gears, worms, bushings, valve seats and guides, pickling hooks
C95410	85 Cu, 4 Fe, 11 Al, 2 Ni											
C95500	81 Cu, 4 Ni, 4 Fe, 11 Al	689 (827)	100 (120)	303 (469)	44 (68)	12 (10)	...	...	192 (230)	50	C, T, M, P, S	Valve guides and seats in aircraft engines, corrosion-resistant parts, bushings, gears, worms, pickling hooks and baskets, agitators

(continued)

(a) Nominal composition, unless otherwise noted. For seldom-used alloys, only compositions are available. (b) Values for C82700, C84200, C96200, and C96300 are minimum, not typical. As-cast values are for sand casting except C93900, continuous cast; and C85800, C87800, C87900, die cast. Heat-treated values, in parentheses, indicate that the alloy responds to heat treatment. If heat-treated values are not shown, the copper or copper alloy does not respond. (c) Based on a value of 100% for free-cutting brass. (d) C, centrifugal; T, continuous; D, die; I, investment; M, permanent mold; P, plaster; S, sand. (e) Also pressure cast. (f) Property and application data not available from the Copper Development Association Inc. (g) As-heat-treated value for C94700, 20; for C94800, 40. Source: Copper Development Association Inc.

Table 3 (continued)

UNS designa-tion(a)	Nominal composition, %(a)	Typical mechanical properties, as-cast (heat treated)(b): Tensile strength MPa	ksi	Yield strength MPa	ksi	Elongation in 50 mm (2 in.), %	Hardness: Rockwell	Brinell 500 kg	Brinell 3000 kg	Machinability rating, %(c)	Casting types(d)	Typical applications
Aluminum bronzes (continued)												
C95600	91 Cu, 7 Al, 2 Si	517	75	234	34	18	...	...	140	60	C, T, M, P, S	Cable connectors, terminals, valve stems, marine hardware, gears, worms, pole-line hardware
C95700	75 Cu, 2 Ni, 3 Fe, 8 Al, 12 Mn	655	95	310	45	26	...	...	180	50	C, T, M, P, S	Propellers, impellers, stator clamp segments, safety tools, welding rods, valves, pump casings
C95800	81 Cu, 5 Ni, 4 Fe, 9 Al, 1 Mn	655	95	262	38	25	...	...	159	50	C, T, M, P, S	Propeller hubs, blades, and other parts in contact with saltwater
Copper-nickels												
C96200	88.6 Cu, 10 Ni, 1.4 Fe	310	45	172	25	20	...	...	...	10	C, S	Components of items being used for seawater corrosion resistance
C96300	79.3 Cu, 20 Ni, 0.7 Fe	517	75	379	55	10	...	150	...	15	C, S	Centrifugally cast tailshaft sleeves
C96400	69.1 Cu, 30 Ni, 0.9 Fe	469	68	255	37	28	...	...	140	20	C, T, S	Valves, pump bodies, flanges, elbows used for seawater corrosion resistance
C96600	68.5 Cu, 30 Ni, 1 Fe, 0.5 Be	(758)	(110)	(482)	(70)	(7)	...	...	(230)	20	C, T, I, M, S	High-strength constructional parts for seawater corrosion resistance
C96700	67.6 Cu, 30 Ni, 0.9 Fe, 1.15 Be, 0.15 Zr, 0.15 Ti	(1207)	(175)	(552)	(80)	(10)	C26	...	...	40	I, M, S	Corrosion-resistant molds for plastics, high-strength constructional parts for seawater use
Nickel silvers												
C97300	56 Cu, 2 Sn, 10 Pb, 12 Ni, 20 Zn	241	35	117	17	20	...	55	...	70	I, M, S	Hardware fittings, valves and valve trim, statuary, ornamental castings
C97400	59 Cu, 3 Sn, 5 Pb, 17 Ni, 16 Zn	262	38	117	17	20	...	70	...	60	C, I, S	Valves, hardware, fittings, ornamental castings
C97600	64 Cu, 4 Sn, 4 Pb, 20 Ni, 8 Zn	310	45	165	24	20	...	80	...	70	C, I, S	Marine castings, sanitary fittings, ornamental hardware, valves, pumps
C97800	66 Cu, 5 Sn, 2 Pb, 25 Ni, 2 Zn	379	55	207	30	15	...	...	130	60	I, M, S	Ornamental and sanitary castings, valves and valve seats, musical instrument components
Leaded coppers												
C98200 (f)	76.0 Cu, 24.0 Pb	...	...	...	...	...	...	...	...	...	...	...
C98400 (f)	70.5 Cu, 28.5 Pb, 1.5 Ag	...	...	...	...	...	...	...	...	...	...	...
C98600 (f)	65.0 Cu, 35.0 Pb, 1.5 Ag	...	...	...	...	...	...	...	...	...	...	...
C98800 (f)	59.5 Cu, 40.0 Pb, 5.5 Ag	...	...	...	...	...	...	...	...	...	...	...
Special alloys												
C99300	71.8 Cu, 15 Ni, 0.7 Fe, 11 Al, 1.5 Co	655	95	379	55	2	...	200	20	20	T, S	Glass-making molds, plate glass rolls, marine hardware
C99400	90.4 Cu, 2.2 Ni, 2.0 Fe, 1.2 Al, 1.2 Si, 3.0 Zn	455 (545)	66 (79)	234 (372)	34 (54)	25	...	...	125 (170)	50	C, T, I, S	Valve stems, marine and other uses requiring resistance to dezincification and dealuminification, propeller wheels, electrical parts, mining equipment gears
C99500	87.9 Cu, 4.5 Ni, 4.0 Fe, 1.2 Al, 1.2 Si, 1.2 Zn	483	70	276	40	12	...	145	50	50	C, T, S	Same as C99400, but where higher yield strength is required
C99600	58 Cu, 2 Al, 40 Mn	558 (558)	81 (81)	248 (303)	36 (44)	34 (27)	B 72	...	130	...	C, T, M, S	Damping alloys to reduce noise and vibration
C99700	56.5 Cu, 1 Al, 1.5 Pb, 12 Mn, 5 Ni, 24 Zn	379	55	172	25	25	...	...	110	80	C, D, I, M, P, S	
C99750	58 Cu, 1 Al, 1 Pb, 20 Mn, 20 Zn	448 (517)	65 (75)	221 (276)	32 (40)	30 (20)	B77 (B82)	110 (119)	...	...	D, I, M, P, S	

(a) Nominal composition, unless otherwise noted. For seldom-used alloys, only compositions are available. (b) Values for C82700, C84200, C96200, and C96300 are minimum, not typical. As-cast values are for sand casting except C93900, continuous cast; and C85800, C87800, C87900, die cast. Heat-treated values, in parentheses, indicate that the alloy responds to heat treatment. If heat-treated values are not shown, the copper or copper alloy does not respond. (c) Based on a value of 100% for free-cutting brass. (d) C, centrifugal; T, continuous; D, die; I, investment; M, permanent mold; P, plaster; S, sand. (e) Also pressure cast. (f) Property and application data not available from the Copper Development Association Inc. (g) As-heat-treated value for C94700, 20; for C94800, 40. Source: Copper Development Association Inc.

Table 4 Corrosion ratings of wrought copper alloys in various corrosive media

The letters E, G, F, and P have the following significance: E (excellent), resists corrosion under almost all conditions of service; G (good), some corrosion will take place, but satisfactory service can be expected under all but the most severe conditions; F (fair), corrosion rates are higher than for the G classification, but the metal can be used if needed for a property other than corrosion resistance and if either the amount of corrosion does not cause excessive maintenance expense or the effects of corrosion can be lessened, such as by use of coatings or inhibitors; P (poor), corrosion rates are high, and service is generally unsatisfactory.

Corrosive medium	Coppers	Low-zinc brasses	High-zinc brasses	Special brasses	Phosphor bronzes	Aluminum bronzes	Silicon bronzes	Copper-nickels	Nickel silvers
Acetate solvents	E	E	G	E	E	E	E	E	E
Acetic acid(a)	E	E	P	P	E	E	E	E	G
Acetone	E	E	E	E	E	E	E	E	E
Acetylene(b)	P	P	(b)	P	P	P	P	P	P
Alcohols(a)	E	E	E	E	E	E	E	E	E
Aldehydes	E	E	F	F	E	E	E	E	E
Alkylamines	G	G	G	G	G	G	G	G	G
Alumina	E	E	E	E	E	E	E	E	E
Aluminum chloride	G	G	P	P	G	G	G	G	G
Aluminum hydroxide	E	E	E	E	E	E	E	E	E
Aluminum sulfate and alum	G	G	P	G	G	G	G	E	G
Ammonia, dry	E	E	E	E	E	E	E	E	E
Ammonia, moist(c)	P	P	P	P	P	P	P	F	P
Ammonium chloride(c)	P	P	P	P	P	P	P	F	P
Ammonium hydroxide(c)	P	P	P	P	P	P	P	F	P
Ammonium nitrate(c)	P	P	P	P	P	P	P	F	P
Ammonium sulfate(c)	F	F	P	P	F	F	F	G	F
Aniline and aniline dyes	F	F	F	F	F	F	F	F	F
Asphalt	E	E	E	E	E	E	E	E	E
Atmosphere									
Industrial(c)	E	E	E	E	E	E	E	E	E
Marine	E	E	E	E	E	E	E	E	E
Rural	E	E	E	E	E	E	E	E	E
Barium carbonate	E	E	E	E	E	E	E	E	E
Barium chloride	G	G	F	F	G	G	G	G	G
Barium hydroxide	E	E	G	E	E	E	E	E	E
Barium sulfate	E	E	E	E	E	E	E	E	E
Beer(a)	E	E	G	E	E	E	E	E	E
Beet sugar syrup(a)	E	E	G	E	E	E	E	E	E
Benzene, benzine, benzol	E	E	E	E	E	E	E	E	E
Benzoic acid	E	E	E	E	E	E	E	E	E
Black liquor, sulfate process	P	P	P	P	P	P	P	G	P
Bleaching powder, wet	G	G	P	G	G	G	G	G	G
Borax	E	E	E	E	E	E	E	E	E
Bordeaux mixture	E	E	G	E	E	E	E	E	E
Boric acid	E	E	G	E	E	E	E	E	E
Brines	G	G	P	G	G	G	G	E	E
Bromine, dry	E	E	E	E	E	E	E	E	E
Bromine, moist	G	G	P	F	G	G	G	G	G
Butane(d)	E	E	E	E	E	E	E	E	E
Calcium bisulfate	G	G	P	G	G	G	G	G	G
Calcium chloride	G	G	F	G	G	G	G	G	G
Calcium hydroxide	E	E	G	E	E	E	E	E	E
Calcium hypochlorite	G	G	P	G	G	G	G	G	G
Cane sugar syrup(a)	E	E	E	E	E	E	E	E	E
Carbolic acid (phenol)	F	G	P	G	G	G	G	G	G
Carbonated beverages(a)(e)	E	E	E	E	E	E	E	E	E
Carbon dioxide, dry	E	E	E	E	E	E	E	E	E
Carbon dioxide, moist(a)(e)	E	E	E	E	E	E	E	E	E
Carbon tetrachloride, dry	E	E	E	E	E	E	E	E	E
Carbon tetrachloride, moist	G	G	F	G	E	E	E	E	E
Castor oil	E	E	E	E	E	E	E	E	E
Chlorine, dry(f)	E	E	E	E	E	E	E	E	E
Chlorine, moist	F	F	P	F	F	F	F	G	F
Chloracetic acid	G	F	P	F	G	G	G	G	G
Chloroform, dry	E	E	E	E	E	E	E	E	E
Chromic acid	P	P	P	P	P	P	P	P	P
Citric acid(a)	E	E	F	E	E	E	E	E	E
Copper chloride	F	F	P	F	F	F	F	F	F
Copper nitrate	F	F	P	F	F	F	F	F	F
Copper sulfate	G	G	P	G	G	G	G	E	G
Corn oil(a)	E	E	G	E	E	E	E	E	E
Cottonseed oil(a)	E	E	G	E	E	E	E	E	E
Creosote	E	E	G	E	E	E	E	E	E
Dowtherm A	E	E	E	E	E	E	E	E	E
Ethanol amine	G	G	G	G	G	G	G	G	G
Ethers	E	E	E	E	E	E	E	E	E
Ethyl acetate (esters)	E	E	G	E	E	E	E	E	E
Ethylene glycol	E	E	G	E	E	E	E	E	E
Ferric chloride	P	P	P	P	P	P	P	P	P
Ferric sulfate	P	P	P	P	P	P	P	P	P
Ferrous chloride	G	G	P	G	G	G	G	G	G
Ferrous sulfate	G	G	P	G	G	G	G	G	G
Formaldehyde (aldehydes)	E	E	G	E	E	E	E	E	E
Formic acid	G	G	P	F	G	G	G	G	G
Freon, dry	E	E	E	E	E	E	E	E	E
Freon, moist	E	E	E	E	E	E	E	E	E
Fuel oil, light	E	E	E	E	E	E	E	E	E
Fuel oil, heavy	E	E	G	E	E	E	E	E	E
Furfural	E	E	F	E	E	E	E	E	E
Gasoline	E	E	E	E	E	E	E	E	E
Gelatin(a)	E	E	E	E	E	E	E	E	E
Glucose(a)	E	E	E	E	E	E	E	E	E
Glue	E	E	G	E	E	E	E	E	E
Glycerin	E	E	G	E	E	E	E	E	E
Hydrobromic acid	F	F	P	F	F	F	F	F	F
Hydrocarbons	E	E	E	E	E	E	E	E	E
Hydrochloric acid (muriatic)	F	F	P	F	F	F	F	F	F
Hydrocyanic acid, dry	E	E	E	E	E	E	E	E	E
Hydrocyanic acid, moist	P	P	P	P	P	P	P	P	P
Hydrofluoric acid, anhydrous	G	G	P	G	G	G	G	G	G
Hydrofluoric acid, hydrated	F	F	P	F	F	F	F	F	F
Hydrofluosilicic acid	G	G	P	G	G	G	G	G	G
Hydrogen(d)	E	E	E	E	E	E	E	E	E
Hydrogen peroxide up to 10%	G	G	F	G	G	G	G	G	G
Hydrogen peroxide over 10%	P	P	P	P	P	P	P	P	P
Hydrogen sulfide, dry	E	E	E	E	E	E	E	E	E
Hydrogen sulfide, moist	P	P	F	F	P	P	P	F	F
Kerosine	E	E	E	E	E	E	E	E	E
Ketones	E	E	E	E	E	E	E	E	E
Lacquers	E	E	E	E	E	E	E	E	E
Lacquer thinners (solvents)	E	E	E	E	E	E	E	E	E
Lactic acid(a)	E	E	F	E	E	E	E	E	E
Lime	E	E	E	E	E	E	E	E	E
Lime sulfur	P	P	F	F	P	P	P	F	F
Linseed oil	G	G	G	G	G	G	G	G	G
Lithium compounds	G	G	P	F	G	G	G	E	E
Magnesium chloride	G	G	F	F	G	G	G	G	G
Magnesium hydroxide	E	E	G	E	E	E	E	E	E
Magnesium sulfate	E	E	G	E	E	E	E	E	E
Mercury or mercury salts	P	P	P	P	P	P	P	P	P
Milk(a)	E	E	G	E	E	E	E	E	E
Molasses	E	E	G	E	E	E	E	E	E
Natural gas(d)	E	E	E	E	E	E	E	E	E
Nickel chloride	F	F	P	F	F	F	F	F	F
Nickel sulfate	F	F	P	F	F	F	F	F	F
Nitric acid	P	P	P	P	P	P	P	P	P
Oleic acid	G	G	F	G	G	G	G	G	G
Oxalic acid(g)	E	E	P	P	E	E	E	E	E
Oxygen(h)	E	E	E	E	E	E	E	E	E
Palmitic acid	G	G	F	G	G	G	G	G	G
Paraffin	E	E	E	E	E	E	E	E	E
Phosphoric acid	G	G	P	F	G	G	G	G	G
Picric acid	P	P	P	P	P	P	P	P	P
Potassium carbonate	E	G	E	E	E	E	E	E	E
Potassium chloride	G	G	P	F	G	G	G	E	E
Potassium cyanide	P	P	P	P	P	P	P	P	P
Potassium dichromate (acid)	P	P	P	P	P	P	P	P	P
Potassium hydroxide	G	G	F	G	G	G	G	E	E
Potassium sulfate	E	E	G	E	E	E	E	E	E
Propane(d)	E	E	E	E	E	E	E	E	E
Rosin	E	E	E	E	E	E	E	E	E

(continued)

Note: This table is intended to serve only as a general guide to the behavior of copper and copper alloys in corrosive environments. It is impossible to cover in a simple tabulation the performance of a material for all possible variations of temperature, concentration, velocity, impurity content, degree of aeration, and stress. The ratings are based on general performance; they should be used with caution, and then only for the purpose of screening candidate alloys. (a) Copper and copper alloys are resistant to corrosion by most food products. Traces of copper may be dissolved and affect taste or color of the products. In such cases, copper alloys are often tin coated. (b) Acetylene forms an explosive compound with copper when moisture or certain impurities are present and the gas is under pressure. Alloys containing less than 65% Cu are satisfactory; when the gas is not under pressure, other copper alloys are satisfactory. (c) Precautions should be taken to avoid SCC. (d) At elevated temperatures, hydrogen will react with tough pitch copper, causing failure by embrittlement. (e) Where air is present, corrosion rate may be increased. (f) Below 150 °C (300 °F), corrosion rate is very low; above this temperature, corrosion is appreciable and increases rapidly with temperature. (g) Aeration and elevated temperature may increase corrosion rate substantially. (h) Excessive oxidation may begin above 120 °C (250 °F). If moisture is present, oxidation may begin at lower temperatures. (i) Use of high-zinc brasses should be avoided in acids because of the likelihood of rapid corrosion by dezincification. Copper, low-zinc brasses, phosphor bronzes, silicon bronzes, aluminum bronzes, and copper-nickels offer good resistance to corrosion by hot and cold dilute H_2SO_4 and to corrosion by cold concentrated H_2SO_4. Intermediate concentrations of H_2SO_4 are sometimes more corrosive to copper alloys than either concentrated or dilute acid. Concentrated H_2SO_4 may be corrosive at elevated temperatures due to breakdown of acid and formation of metallic sulfides and sulfur dioxide, which cause localized pitting. Tests indicate that copper alloys may undergo pitting in 90–95% H_2SO_4 at about 50 °C (122 °F), in 80% acid at about 70 °C (160 °F), and in 60% acid at about 100 °C (212 °F). (j) Wetting agents may increase corrosion rates of copper and copper alloys slightly to substantially when carbon dioxide or oxygen is present by preventing formation in a film on the metal surface and by combining (in some instances) with the dissolved copper to produce a green, insoluble compound.

Table 4 (continued)

Corrosive medium	Coppers	Low-zinc brasses	High-zinc brasses	Special brasses	Phosphor bronzes	Aluminum bronzes	Silicon bronzes	Copper-nickels	Nickel silvers
Seawater	G	G	F	E	G	E	G	E	E
Sewage	E	E	F	E	E	E	E	E	E
Silver salts	P	P	P	P	P	P	P	P	P
Soap solution	E	E	E	E	E	E	E	E	E
Sodium bicarbonate	E	E	G	E	E	E	E	E	E
Sodium bisulfate	G	G	F	G	G	G	G	E	E
Sodium carbonate	E	E	G	E	E	E	E	E	E
Sodium chloride	G	G	P	F	G	G	G	E	E
Sodium chromate	E	E	E	E	E	E	E	E	E
Sodium cyanide	P	P	P	P	P	P	P	P	P
Sodium dichromate (acid)	P	P	P	P	P	P	P	P	P
Sodium hydroxide	G	G	F	G	G	G	G	E	E
Sodium hypochlorite	G	G	P	G	G	G	G	G	G
Sodium nitrate	G	G	P	F	G	G	G	E	E
Sodium peroxide	F	F	P	F	F	F	F	G	G
Sodium phosphate	E	E	G	E	E	E	E	E	E
Sodium silicate	E	E	G	E	E	E	E	E	E
Sodium sulfate	E	E	G	E	E	E	E	E	E
Sodium sulfide	P	P	F	F	P	P	P	F	F
Sodium thiosulfate	P	P	F	F	P	P	P	F	F
Steam	E	E	F	E	E	E	F	E	E
Stearic acid	E	E	F	E	E	E	E	E	E
Sugar solutions	E	E	G	E	E	E	E	E	E
Sulfur, solid	G	G	E	G	G	G	G	E	G
Sulfur, molten	P	P	P	P	P	P	P	P	P
Sulfur chloride, dry	E	E	E	E	E	E	E	E	E
Sulfur chloride, moist	P	P	P	P	P	P	P	P	P
Sulfur dioxide, dry	E	E	E	E	E	E	E	E	E
Sulfur dioxide, moist	G	G	P	G	G	G	G	F	F
Sulfur trioxide, dry	E	E	E	E	E	E	E	E	E
Sulfuric acid 80–95%(i)	G	G	P	F	G	G	G	G	G
Sulfuric acid 40–80%(i)	F	F	F	P	F	F	F	F	F
Sulfuric acid 40%(i)	G	G	P	F	G	G	G	G	G
Sulfurous acid	G	G	P	G	G	G	G	F	F
Tannic acid	E	E	E	E	E	E	E	E	E
Tartaric acid(a)	E	E	G	E	E	E	E	E	E
Toluene	E	E	E	E	E	E	E	E	E
Trichloracetic acid	G	G	P	F	G	G	G	G	G
Trichlorethylene, dry	E	E	E	E	E	E	E	E	E
Trichlorethylene, moist	G	G	F	G	E	E	E	E	E
Turpentine	E	E	E	E	E	E	E	E	E
Varnish	E	E	E	E	E	E	E	E	E
Vinegar(a)	E	E	P	F	E	E	E	E	G
Water, acidic mine	F	F	P	F	G	F	F	P	F
Water, potable	E	E	G	E	E	E	E	E	E
Water, condensate(c)	E	E	E	E	E	E	E	E	E
Wetting agents(j)	E	E	E	E	E	E	E	E	E
Whiskey(a)	E	E	E	E	E	E	E	E	E
White water	G	G	G	E	E	E	E	E	E
Zinc chloride	G	G	P	G	G	G	G	G	G
Zinc sulfate	E	E	P	E	E	E	E	E	E

Note: This table is intended to serve only as a general guide to the behavior of copper and copper alloys in corrosive environments. It is impossible to cover in a simple tabulation the performance of a material for all possible variations of temperature, concentration, velocity, impurity content, degree of aeration, and stress. The ratings are based on general performance; they should be used with caution, and then only for the purpose of screening candidate alloys. (a) Copper and copper alloys are resistant to corrosion by most food products. Traces of copper may be dissolved and affect taste or color of the products. In such cases, copper alloys are often tin coated. (b) Acetylene forms an explosive compound with copper when moisture or certain impurities are present and the gas is under pressure. Alloys containing less than 65% Cu are satisfactory; when the gas is not under pressure, other copper alloys are satisfactory. (c) Precautions should be taken to avoid SCC. (d) At elevated temperatures, hydrogen will react with tough pitch copper, causing failure by embrittlement. (e) Where air is present, corrosion rate may be increased. (f) Below 150 °C (300 °F), corrosion rate is very low; above this temperature, corrosion is appreciable and increases rapidly with temperature. (g) Aeration and elevated temperature may increase corrosion rate substantially. (h) Excessive oxidation may begin above 120 °C (250 °F). If moisture is present, oxidation may begin at lower temperatures. (i) Use of high-zinc brasses should be avoided in acids because of the likelihood of rapid corrosion by dezincification. Copper, low-zinc brasses, phosphor bronzes, silicon bronzes, aluminum bronzes, and copper-nickels offer good resistance to corrosion by hot and cold dilute H_2SO_4 and to corrosion by cold concentrated H_2SO_4. Intermediate concentrations of H_2SO_4 are sometimes more corrosive to copper alloys than either concentrated or dilute acid. Concentrated H_2SO_4 may be corrosive at elevated temperatures due to breakdown of acid and formation of metallic sulfides and sulfur dioxide, which cause localized pitting. Tests indicate that copper alloys may undergo pitting in 90–95% H_2SO_4 at about 50 °C (122 °F), in 80% acid at about 70 °C (160 °F), and in 60% acid at about 100 °C (212 °F). (j) Wetting agents may increase corrosion rates of copper and copper alloys slightly to substantially when carbon dioxide or oxygen is present by preventing formation in a film on the metal surface and by combining (in some instances) with the dissolved copper to produce a green, insoluble compound.

with hydrocarbons produce metal with essentially the same high ductility and excellent electrical conductivity. Fire-refined tough pitch copper contains a small amount of residual sulfur, normally 10 to 30 ppm, and a somewhat larger amount of cuprous oxide, normally 500 to 3000 ppm.

Electrolytic tough pitch copper C11000 is made from cathode copper, that is, copper that has been refined electrolytically. C11000 is the most common of all the electrical coppers. It has high electrical conductivity, in excess of 100% IACS. It has the same oxygen content as C12500 but contains less than 50 ppm total metallic impurities (including sulfur).

Oxygen-free coppers C10100 and C10200 are made by melting prime-quality cathode copper under nonoxidizing conditions, which are produced by a granulated graphite bath covering and a protective reducing atmosphere that is low in hydrogen. Oxygen-free coppers are particularly suitable for applications requiring high conductivity coupled with exceptional ductility, low gas permeability, freedom from hydrogen embrittlement, or low out-gassing tendency.

If resistance to softening at slightly elevated temperature is required, C11100 is often specified. This copper contains a small amount of cadmium, which raises the temperature at which recovery and recrystallization occur. Oxygen-free copper, electrolytic tough pitch copper, and fire-refined tough pitch copper are available as silver-bearing coppers having specific minimum silver contents. The silver, which may be present as an impurity in anode copper or may be intentionally alloyed to molten cathode copper, also imparts resistance to softening to cold-worked metal. Silver-bearing coppers and cadmium-bearing coppers are used for applications such as automotive radiators and electrical conductors that must operate at temperatures above about 200 °C (400 °F).

If good machinability is required, C14500 (tellurium-bearing copper) or C14700 (sulfur-bearing copper) can be selected. As might be expected, machinability is gained at a modest sacrifice in electrical conductivity.

Adding small amounts of elements such as silver, cadmium, iron, cobalt, and zirconium to deoxidized copper imparts resistance to softening at the times and temperatures encountered in soldering operations, such as those used to join components of automobile and truck radiators, and those used in semiconductor packaging operations. Thermal and electrical conductivities and room-temperature mechanical properties are unaffected by small additions of these elements. However, cadmium-copper and zirconium-copper work harden at higher rates than either silver-bearing copper or electrolytic tough pitch copper (Fig. 1).

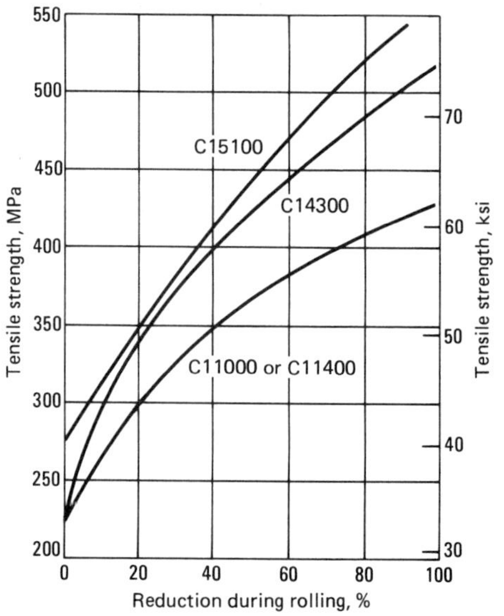

Fig. 1 Tensile strength versus reduction during rolling for cadmium-copper (C14300), zirconium-copper (C15100), and tough pitch copper (C11000)

Table 5 Corrosion ratings of cast copper alloys in various media

The letters A, B, and C have the following significance: A, recommended; B, acceptable; C, not recommended

Corrosive medium	Copper	Tin bronze	Leaded tin bronze	High-leaded tin bronze	Leaded red brass	Leaded semired brass	Leaded yellow brass	Leaded high-strength yellow brass	High-strength yellow brass	Aluminum bronze	Leaded nickel brass	Leaded nickel bronze	Silicon bronze	Silicon brass
Acetate solvents	B	A	A	A	A	A	B	A	A	A	A	A	A	B
Acetic acid														
20%	A	C	B	C	B	C	C	C	C	A	C	A	A	B
50%	A	C	B	C	B	C	C	C	C	A	C	B	A	B
Glacial	A	A	A	C	A	C	C	C	C	A	B	B	A	A
Acetone	A	A	A	A	A	A	A	A	A	A	A	A	A	A
Acetylene(a)	C	C	C	C	C	C	C	C	C	C	C	C	C	C
Alcohols(b)	A	A	A	A	A	A	A	A	A	A	A	A	A	A
Aluminum chloride	C	C	C	C	C	C	C	C	C	B	C	C	C	C
Aluminum sulfate	B	B	B	B	B	C	C	C	C	A	C	C	A	A
Ammonia, moist gas	C	C	C	C	C	C	C	C	C	C	C	C	C	C
Ammonia, moisture-free	A	A	A	A	A	A	A	A	A	A	A	A	A	A
Ammonium chloride	C	C	C	C	C	C	C	C	C	C	C	C	C	C
Ammonium hydroxide	C	C	C	C	C	C	C	C	C	C	C	C	C	C
Ammonium nitrate	C	C	C	C	C	C	C	C	C	C	C	C	C	C
Ammonium sulfate	B	B	B	B	B	C	C	C	C	A	C	C	A	A
Aniline and aniline dyes	C	C	C	C	C	C	C	C	C	B	C	C	C	C
Asphalt	A	A	A	A	A	A	A	A	A	A	A	A	A	A
Barium chloride	A	A	A	A	A	C	C	C	C	A	A	A	A	C
Barium sulfide	C	C	C	C	C	C	C	C	B	C	C	C	C	C
Beer(b)	A	A	B	B	B	C	C	C	A	A	C	A	A	B
Beet sugar syrup	A	A	B	B	B	A	A	A	B	A	A	A	B	B
Benzine	A	A	A	A	A	A	A	A	A	A	A	A	A	A
Benzol	A	A	A	A	A	A	A	A	A	A	A	A	A	A
Boric acid	A	A	A	A	A	A	A	B	A	A	A	A	A	A
Butane	A	A	A	A	A	A	A	A	A	A	A	A	A	A
Calcium bisulfite	A	A	B	B	B	C	C	C	C	A	B	A	A	B
Calcium chloride (acid)	B	B	B	B	B	B	C	C	C	A	C	C	A	C
Calcium chloride (alkaline)	C	C	C	C	C	C	C	C	C	A	C	A	C	B
Calcium hydroxide	C	C	C	C	C	C	C	C	C	B	C	C	C	C
Calcium hypochlorite	C	C	B	B	B	C	C	C	C	B	C	C	C	C
Cane sugar syrups	A	A	B	A	B	A	A	A	A	A	A	A	A	B
Carbonated beverages(b)	A	C	C	C	C	C	C	C	C	A	C	C	A	C
Carbon dioxide, dry	A	A	A	A	A	A	A	A	A	A	A	A	A	A
Carbon dioxide, moist(b)	B	B	B	C	B	C	C	C	C	A	C	A	A	B
Carbon tetrachloride, dry	A	A	A	A	A	A	A	A	A	A	A	A	A	A
Carbon tetrachloride, moist	B	B	B	B	B	B	B	B	B	B	B	A	A	A
Chlorine, dry	A	A	A	A	A	A	A	A	A	A	A	A	A	A
Chlorine, moist	C	C	B	B	B	C	C	C	C	C	C	C	C	C
Chromic acid	C	C	C	C	C	C	C	C	C	C	C	C	C	C
Citric acid	A	A	A	A	A	A	A	A	A	A	A	A	A	A
Copper sulfate	B	A	A	A	A	C	C	C	C	B	B	B	A	A
Cottonseed oil(b)	A	A	A	A	A	A	A	A	A	A	A	A	A	A
Creosote	B	B	B	B	B	C	C	C	C	A	B	B	B	B
Ethers	A	A	A	A	A	A	A	A	A	A	A	A	A	A
Ethylene glycol	A	A	A	A	A	A	A	A	A	A	A	A	A	A
Ferric chloride (sulfate)	C	C	C	C	C	C	C	C	C	C	C	C	C	C
Ferrous chloride (sulfate)	C	C	C	C	C	C	C	C	C	C	C	C	C	C
Formaldehyde	A	A	A	A	A	A	A	A	A	A	A	A	A	A
Formic acid	A	A	A	A	A	B	B	B	B	A	B	B	B	C
Freon	A	A	A	A	A	A	A	A	A	A	A	A	A	B
Fuel oil	A	A	A	A	A	A	A	A	A	A	A	A	A	A
Furfural	A	A	A	A	A	A	A	A	A	A	A	A	A	A
Gasoline	A	A	A	A	A	A	A	A	A	A	A	A	A	A
Gelatin(b)	A	A	A	A	A	A	A	A	A	A	A	A	A	A
Glucose	A	A	A	A	A	A	A	A	A	A	A	A	A	A
Glue	A	A	A	A	A	A	A	A	A	A	A	A	A	A
Glycerin	A	A	A	A	A	A	A	A	A	A	A	A	A	A
Hydrochloric or muriatic acid	C	C	C	C	C	C	C	C	C	B	C	C	C	C
Hydrofluoric acid	B	B	B	B	B	B	B	B	B	A	B	B	B	B

(continued)

Note: This table is intended to serve only as a general guide to the behavior of copper and copper alloys in corrosive environments. It is impossible to cover in a simple tabulation the performance of a material for all possible variations of temperature, concentration, velocity, impurity content, degree of aeration, and stress. The ratings are based on general performance; they should be used with caution, and then only for the purpose of screening candidate alloys. (a) Acetylene forms an explosive compound with copper when moist or when certain impurities are present and the gas is under pressure. Alloys containing less than 65% Cu are satisfactory for this use. When gas is not under pressure, other copper alloys are satisfactory. (b) Copper and copper alloys resist corrosion by most food products. Traces of copper may be dissolved and affect taste or color. In such cases, copper metals are often tin coated.

Table 5 (continued)

Corrosive medium	Copper	Tin bronze	Leaded tin bronze	High-leaded tin bronze	Leaded red brass	Leaded semired brass	Leaded yellow brass	Leaded high-strength yellow brass	High-strength yellow brass	Aluminum bronze	Leaded nickel brass	Leaded nickel bronze	Silicon bronze	Silicon brass
Hydrofluosilicic acid	B	B	B	B	B	C	C	C	C	B	C	C	B	C
Hydrogen	A	A	A	A	A	A	A	A	A	A	A	A	A	A
Hydrogen peroxide	C	C	C	C	C	C	C	C	C	C	C	C	C	C
Hydrogen sulfide, dry	C	C	C	C	C	C	C	C	C	B	C	C	B	C
Hydrogen sulfide, moist	C	C	C	C	C	C	C	C	C	B	C	C	C	C
Lacquers	A	A	A	A	A	A	A	A	A	A	A	A	A	A
Lacquer thinners	A	A	A	A	A	A	A	A	A	A	A	A	A	A
Lactic acid	A	A	A	A	A	C	C	C	C	A	C	C	A	C
Linseed oil	A	A	A	A	A	A	A	A	A	A	A	A	A	A
Liquors														
Black liquor	B	B	B	B	B	C	C	C	C	B	C	C	B	B
Green liquor	C	C	C	C	C	C	C	C	C	B	C	C	C	B
White liquor	C	C	C	C	C	C	C	C	C	A	C	C	C	B
Magnesium chloride	A	A	A	A	A	C	C	C	C	A	C	C	A	B
Magnesium hydroxide	B	B	B	B	B	B	B	B	B	A	B	B	B	B
Magnesium sulfate	A	A	A	A	B	C	C	C	C	A	C	B	A	B
Mercury, mercury salts	C	C	C	C	C	C	C	C	C	C	C	C	C	C
Milk(b)	A	A	A	A	A	A	A	A	A	A	A	A	A	A
Molasses(b)	A	A	A	A	A	A	A	A	A	A	A	A	A	A
Natural gas	A	A	A	A	A	A	A	A	A	A	A	A	A	A
Nickel chloride	A	A	A	A	A	C	C	C	C	B	C	C	A	C
Nickel sulfate	A	A	A	A	A	C	C	C	C	A	C	C	A	C
Nitric acid	C	C	C	C	C	C	C	C	C	C	C	C	C	C
Oleic acid	A	A	B	B	B	C	C	C	C	A	C	A	A	B
Oxalic acid	A	A	B	B	B	C	C	C	C	A	C	A	A	B
Phosphoric acid	A	A	A	A	A	C	C	C	C	A	C	A	A	A
Picric acid	C	C	C	C	C	C	C	C	C	C	C	C	C	C
Potassium chloride	A	A	A	A	A	C	C	C	C	A	C	C	A	C
Potassium cyanide	C	C	C	C	C	C	C	C	C	C	C	C	C	C
Potassium hydroxide	C	C	C	C	C	C	C	C	C	A	C	C	C	C
Potassium sulfate	A	A	A	A	A	C	C	C	C	A	C	C	A	C
Propane gas	A	A	A	A	A	A	A	A	A	A	A	A	A	A
Seawater	A	A	A	A	A	C	C	C	C	A	C	C	B	B
Soap solutions	A	A	A	A	B	C	C	C	C	A	C	C	A	C
Sodium bicarbonate	A	A	A	A	A	A	A	A	A	A	A	A	A	B
Sodium bisulfate	C	C	C	C	C	C	C	C	C	A	C	C	C	C
Sodium carbonate	C	A	A	A	A	C	C	C	C	A	C	C	C	A
Sodium chloride	A	A	A	A	A	B	C	C	C	A	C	C	A	C
Sodium cyanide	C	C	C	C	C	C	C	C	C	B	C	C	C	C
Sodium hydroxide	C	C	C	C	C	C	C	C	C	A	C	C	C	C
Sodium hypochlorite	C	C	C	C	C	C	C	C	C	C	C	C	C	C
Sodium nitrate	B	B	B	B	B	B	B	B	B	A	B	B	A	A
Sodium peroxide	B	B	B	B	B	B	B	B	B	B	B	B	B	B
Sodium phosphate	A	A	A	A	A	A	A	A	A	A	A	A	A	A
Sodium sulfate (silicate)	A	A	B	B	B	B	C	C	C	A	C	C	A	B
Sodium sulfide (thiosulfate)	C	C	C	C	C	C	C	C	C	B	C	C	C	C
Stearic acid	A	A	A	A	A	A	A	A	A	A	A	A	A	A
Sulfur, solid	C	C	C	C	C	C	C	C	C	A	C	C	C	C
Sulfur chloride	C	C	C	C	C	C	C	C	C	C	C	C	C	C
Sulfur dioxide, dry	A	A	A	A	A	A	A	A	A	A	A	A	A	A
Sulfur dioxide, moist	A	A	A	B	B	C	C	C	C	A	C	C	A	B
Sulfur trioxide, dry	A	A	A	A	A	A	A	A	A	A	A	A	A	A
Sulfuric acid														
78% or less	B	B	B	B	B	C	C	C	C	A	C	C	B	B
78% to 90%	C	C	C	C	C	C	C	C	C	B	C	C	C	C
90% to 95%	C	C	C	C	C	C	C	C	C	B	C	C	C	C
Fuming	C	C	C	C	C	C	C	C	C	A	C	C	C	C
Tannic acid	A	A	A	A	A	A	A	A	A	A	A	A	A	A
Tartaric acid	B	A	A	A	A	A	A	A	A	A	A	A	A	A

(continued)

Note: This table is intended to serve only as a general guide to the behavior of copper and copper alloys in corrosive environments. It is impossible to cover in a simple tabulation the performance of a material for all possible variations of temperature, concentration, velocity, impurity content, degree of aeration, and stress. The ratings are based on general performance; they should be used with caution, and then only for the purpose of screening candidate alloys. (a) Acetylene forms an explosive compound with copper when moist or when certain impurities are present and the gas is under pressure. Alloys containing less than 65% Cu are satisfactory for this use. When gas is not under pressure, other copper alloys are satisfactory. (b) Copper and copper alloys resist corrosion by most food products. Traces of copper may be dissolved and affect taste or color. In such cases, copper metals are often tin coated.

Table 5 (continued)

Corrosive medium	Copper	Tin bronze	Leaded tin bronze	High-leaded tin bronze	Leaded red brass	Leaded semired brass	Leaded yellow brass	Leaded high-strength yellow brass	High-strength yellow brass	Aluminum bronze	Leaded nickel brass	Leaded nickel bronze	Silicon bronze	Silicon brass
Toluene	B	B	A	A	A	B	B	B	B	B	B	B	B	A
Trichlorethylene, dry	A	A	A	A	A	A	A	A	A	A	A	A	A	A
Trichlorethylene, moist	A	A	A	A	A	A	A	A	A	A	A	A	A	A
Turpentine	A	A	A	A	A	A	A	A	A	A	A	A	A	A
Varnish	A	A	A	A	A	A	A	A	A	A	A	A	A	A
Vinegar	A	A	B	B	B	C	C	C	C	B	C	C	A	B
Water, acid mine	C	C	C	C	C	C	C	C	C	C	C	C	C	C
Water, condensate	A	A	A	A	A	A	A	A	A	A	A	A	A	A
Water, potable	A	A	A	A	A	A	B	B	B	A	A	A	A	A
Whiskey(b)	A	A	C	C	C	C	C	C	C	A	C	C	A	C
Zinc chloride	C	C	C	C	C	C	C	C	C	B	C	C	B	C
Zinc sulfate	A	A	A	A	A	C	C	C	C	B	C	A	A	C

Note: This table is intended to serve only as a general guide to the behavior of copper and copper alloys in corrosive environments. It is impossible to cover in a simple tabulation the performance of a material for all possible variations of temperature, concentration, velocity, impurity content, degree of aeration, and stress. The ratings are based on general performance; they should be used with caution, and then only for the purpose of screening candidate alloys. (a) Acetylene forms an explosive compound with copper when moist or when certain impurities are present and the gas is under pressure. Alloys containing less than 65% Cu are satisfactory for this use. When gas is not under pressure, other copper alloys are satisfactory. (b) Copper and copper alloys resist corrosion by most food products. Traces of copper may be dissolved and affect taste or color. In such cases, copper metals are often tin coated.

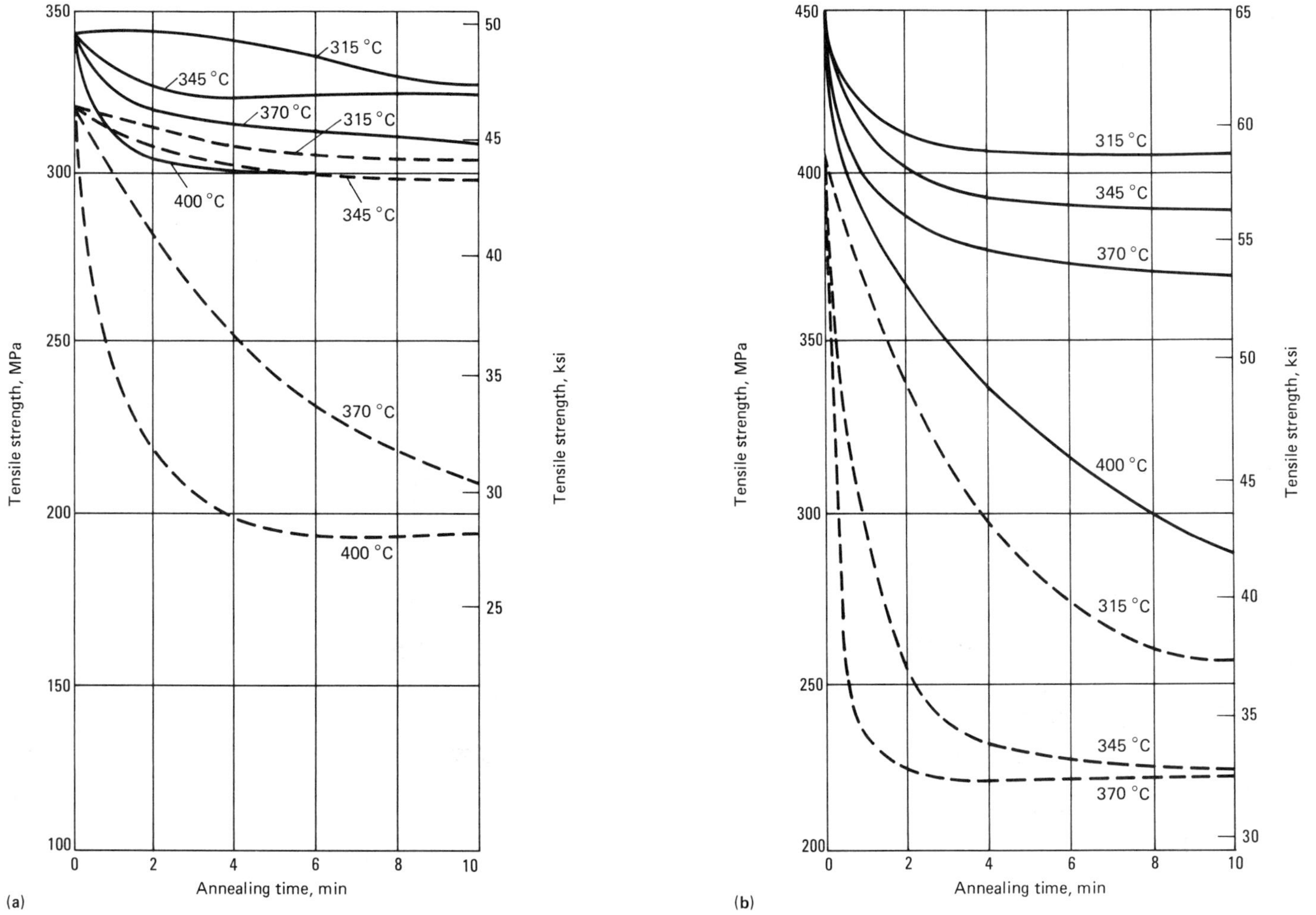

Fig. 2 Softening characteristics of cadmium-bearing copper and silver-bearing tough pitch copper. (a) Softening curves for material cold reduced 21% in area, from 0.1 to 0.075 mm (0.0038 to 0.0030 in.) in thickness. (b) Softening curves for material cold reduced 90% in area, from 0.75 to 0.075 mm (0.0300 to 0.0030 in.) in thickness

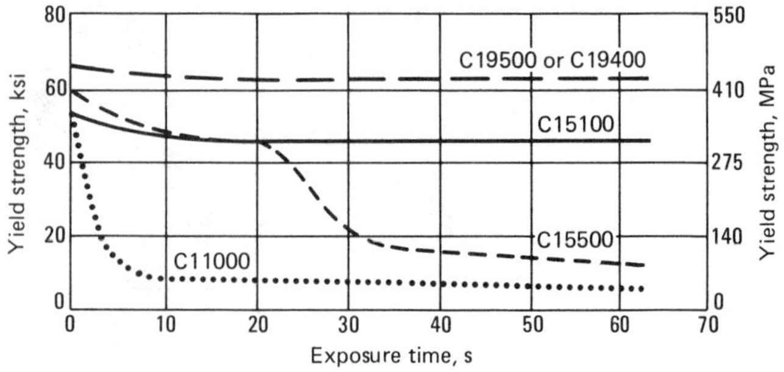

Fig. 3 Softening resistance of lead frame materials at the upper temperature limit (500 °C, or 930 °F)

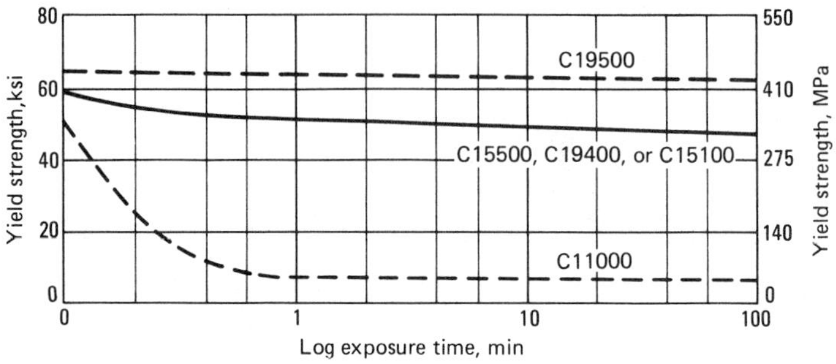

Fig. 4 Softening resistance of lead frame materials at an intermediate temperature level (350 °C, or 660 °F)

Cold-rolled silver-bearing copper is used extensively for automobile radiator fins. Usually such strip is only moderately cold rolled because heavy cold rolling makes silver-bearing copper more likely to soften during soldering or baking operations. Some manufacturers prefer cadmium-copper C14300 because it can be severely cold rolled without making it susceptible to softening during soldering. Figure 2 illustrates the softening characteristics of C14300 and C11400 as measured for several temperatures and two tempers. As shown in Fig. 2(b), C14300 cold rolled to a tensile strength of 440 MPa (64 ksi) retains 91% of its strength after a typical core bake of 3 min at 345 °C (650 °F). Silver-bearing copper C11400 given the same cold reduction retains only 60% of its tensile strength after the same baking schedule.

Another application in which softening resistance is of paramount importance is lead frames for electronic devices, such as plastic dual-in-line packages. During packaging and assembly, lead frames may be subjected to temperatures up to 350 °C (660 °F) for several minutes and up to 500 °C (930 °F) for several seconds. The leads must maintain good strength because they are pressed into socket connectors, often by automated assembly machines; softened leads collapse, causing spoilage. See the article "Lead Frame Materials" in *Packaging*, Volume 1 of the *Electronic Materials Handbook* for additional information.

Alloy C15100 (copper-zirconium), alloy C15500 (copper-silver-magnesium-phosphorus), alloy C19400 (copper-iron-phosphorus-zinc), and alloy C19500 (copper-iron-cobalt-tin-phosphorus) are popular for these applications because they have good conductivity, good strength, and good softening resistance. Figures 3 and 4 compare the softening resistance of these alloys with electrolytic copper C11000.

Copper Alloys

The most common way to catalog copper and copper alloys is to divide them into six families: coppers, dilute-copper (or high-copper) alloys, brasses, bronzes, copper-nickels, and nickel silvers. The first family, the coppers, is essentially commercially pure copper, which ordinarily is soft and ductile and contains less than about 0.7% total impurities. The dilute-copper alloys contain small amounts of various alloying elements, such as beryllium, cadmium, chromium, or iron, each having less than 8 at.% solid solubility; these elements modify one or more of the basic properties of copper. Each of the remaining families contains one of five major alloying elements as its primary alloying ingredient:

Family	Alloying element	Solid solubility, at.%(a)
Brasses	Zinc	37
Phosphor bronzes	Tin	9
Aluminum bronzes	Aluminum	19
Silicon bronzes	Silicon	8
Copper-nickels, nickel silvers	Nickel	100

(a) At 20 °C (70 °F)

A general classification for wrought and cast copper alloys is given in Table 9.

Solid-Solution Alloys. The most compatible alloying elements with copper are those that form solid-solution fields. These include all elements forming useful alloy families (see Table 9) plus manganese. Hardening in these systems is great enough to make useful objects without encountering brittleness associated with second phases or compounds.

Cartridge brass is typical of this group. It consists of 30% Zn in copper and exhibits no β phase except an occasional small amount due to segregation; the β phase normally disappears after the first anneal. Provided that there are no tramp elements such as iron, cold-working and grain growth relationships are easily reproduced in practice.

Modified Solid-Solution Alloys. The solid-solution-strengthened alloys of copper are noted for their strength and formability. Be-

Table 6 Standard color-controlled wrought copper alloys

UNS number	Common name	Color description
C11000	Electrolytic tough pitch copper	Soft pink
C21000	Gilding, 95%	Red-brown
C22000	Commercial bronze, 90%	Bronze-gold
C23000	Red brass, 85%	Tan-gold
C26000	Cartridge brass, 70%	Green-gold
C28000	Muntz metal, 60%	Light brown-gold
C63800	Aluminum bronze	Gold
C65500	High-silicon bronze, A	Lavender-brown
C70600	Copper-nickel, 10%	Soft lavender
C74500	Nickel silver, 65-10	Gray-white
C75200	Nickel silver, 65-18	Silver

Table 7 Temper designations for wrought copper and brass based on cold reduction

Nominal temper designation	Rolled sheet: Increase in B and S gage numbers	Rolled sheet: Reduction in thickness and area, %	Rolled sheet: True strain(a)	Drawn wire: Reduction in diameter, %	Drawn wire: Reduction in area, %	Drawn wire: True strain(a)
¼ hard	1	10.9	0.116	10.9	20.7	0.232
½ hard	2	20.7	0.232	20.7	37.1	0.463
¾ hard	3	29.4	0.347	29.4	50.1	0.694
Hard	4	37.1	0.463	37.1	60.5	0.926
Extra hard	6	50.1	0.696	50.1	75.1	1.39
Spring	8	60.5	0.928	60.5	84.4	1.86
Extra spring	10	68.6	1.16	68.6	90.2	2.32
Special spring	12	75.1	1.39	75.1	93.8	2.78
Super spring	14	80.3	1.62	80.3	96.1	3.25

(a) True strain equals ln A_0/A, where A_0 is the initial cross-sectional area and A is the final area.

Table 8 ASTM B 601 temper designation codes for copper and copper alloys

Temper designation	Temper name or material condition
Cold-worked tempers(a)	
H00	⅛ hard
H01	¼ hard
H02	½ hard
H03	¾ hard
H04	Hard
H06	Extra hard
H08	Spring
H10	Extra spring
H12	Special spring
H13	Ultra spring
H14	Super spring
Cold-worked tempers(b)	
H50	Extruded and drawn
H52	Pierced and drawn
H55	Light drawn; light cold rolled
H58	Drawn general purpose
H60	Cold heading; forming
H63	Rivet
H64	Screw
H66	Bolt
H70	Bending
H80	Hard drawn
H85	Medium-hard-drawn electrical wire
H86	Hard-drawn electrical wire
H90	As-finned
Cold-worked and stress-relieved tempers	
HR01	H01 and stress relieved
HR02	H02 and stress relieved
HR04	H04 and stress relieved
HR08	H08 and stress relieved
HR10	H10 and stress relieved
HR20	As-finned
HR50	Drawn and stress relieved
Cold-rolled and order-strengthened tempers(c)	
HT04	H04 and order heat treated
HT08	H08 and order heat treated
As-manufactured tempers	
M01	As-sand cast
M02	As-centrifugal cast
M03	As-plaster cast
M04	As-pressure die cast
M05	As-permanent mold cast
M06	As-investment cast
M07	As-continuous cast
M10	As-hot forged and air cooled
M11	As-forged and quenched
M20	As-hot rolled
M30	As-hot extruded
M40	As-hot pierced
M45	As-hot pierced and rerolled
Annealed tempers(d)	
O10	Cast and annealed (homogenized)
O11	As-cast and precipitation heat treated
O20	Hot forged and annealed
O25	Hot rolled and annealed
O30	Hot extruded and annealed
O31	Extruded and precipitation heat treated
O40	Hot pierced and annealed
O50	Light annealed
O60	Soft annealed
O61	Annealed
O65	Drawing annealed
O68	Deep-drawing annealed
O70	Dead-soft annealed
O80	Annealed to temper, ⅛ hard
O81	Annealed to temper, ¼ hard
O82	Annealed to temper, ½ hard
Annealed tempers(e)	
OS005	Average grain size, 0.005 mm
OS010	Average grain size, 0.010 mm
OS015	Average grain size, 0.015 mm
OS025	Average grain size, 0.025 mm
OS035	Average grain size, 0.035 mm
OS050	Average grain size, 0.050 mm
OS060	Average grain size, 0.060 mm
OS070	Average grain size, 0.070 mm
OS100	Average grain size, 0.100 mm
OS120	Average grain size, 0.120 mm
OS150	Average grain size, 0.150 mm
OS200	Average grain size, 0.200 mm
Solution-treated temper	
TB00	Solution heat treated
Solution-treated and cold-worked tempers	
TD00	TB00 cold worked to ⅛ hard
TD01	TB00 cold worked to ¼ hard
TD02	TB00 cold worked to ½ hard
TD03	TB00 cold worked to ¾ hard
TD04	TB00 cold worked to full hard
Solution-treated and precipitation-hardened temper	
TF00	TB00 and precipitation hardened
Cold-worked and precipitation-hardened tempers	
TH01	TD01 and precipitation hardened
TH02	TD02 and precipitation hardened
TH03	TD03 and precipitation hardened
TH04	TD04 and precipitation hardened
Precipitation-hardened and cold-worked tempers	
TL00	TF00 cold worked to ⅛ hard
TL01	TF00 cold worked to ¼ hard
TL02	TF00 cold worked to ½ hard
TL04	TF00 cold worked to full hard
TL08	TF00 cold worked to spring
TL10	TF00 cold worked to extra spring
Mill-hardened tempers	
TM00	AM
TM01	¼ HM
TM02	½ HM
TM04	HM
TM06	XHM
TM08	XHMS
Quench-hardened tempers	
TQ00	Quench hardened
TQ50	Quench hardened and temper annealed
TQ55	Quench hardened and temper annealed, cold drawn and stress relieved
TQ75	Interrupted quench hardened
Precipitation-hardened, cold-worked, and thermal-stress-relieved tempers	
TR01	TL01 and stress relieved
TR02	TL02 and stress relieved
TR04	TL04 and stress relieved
Solution-treated and spinodal-heat-treated temper	
TX00	Spinodal hardened
Tempers of welded tubing(f)	
WH00	Welded and drawn to ⅛ hard
WH01	Welded and drawn to ¼ hard
WH02	Welded and drawn to ½ hard
WH03	Welded and drawn to ¾ hard
WH04	Welded and drawn to full hard
WH06	Welded and drawn to extra hard
WM00	As welded from H00 (⅛-hard) strip
WM01	As welded from H01 (¼-hard) strip
WM02	As welded from H02 (½-hard) strip
WM03	As welded from H03 (¾-hard) strip
WM04	As welded from H04 (full-hard) strip
WM06	As welded from H06 (extra-hard) strip
WM08	As welded from H08 (spring) strip
WM10	As welded from H10 (extra-spring) strip
WM15	WM50 and stress relieved
WM20	WM00 and stress relieved
WM21	WM01 and stress relieved
WM22	WM02 and stress relieved
WM50	As welded from annealed strip
WO50	Welded and light annealed
WR00	WM00; drawn and stress relieved
WR01	WM01; drawn and stress relieved
WR02	WM02; drawn and stress relieved
WR03	WM03; drawn and stress relieved
WR04	WM04; drawn and stress relieved
WR06	WM06; drawn and stress relieved

(a) Cold-worked tempers to meet standard requirements based on cold rolling or cold drawing. (b) Cold-worked tempers to meet standard requirements based on temper names applicable to specific products. (c) Tempers produced by controlled amounts of cold work followed by a thermal treatment to produce order strengthening. (d) Annealed to meet specific mechanical property requirements. (e) Annealed to meet prescribed nominal average grain size. (f) Tempers of fully finished tubing that has been drawn or annealed to produce specified mechanical properties or that has been annealed to produce a prescribed nominal average grain size are commonly identified by the appropriate H, O, or OS temper designation.

cause they are single phase and are not transformed by heating or cooling, their maximum strength is developed by cold working methods such as cold rolling or cold drawing. Formability is reduced in proportion to the amount of cold work applied.

Modifications of some solid-solution alloys were developed by adding elements that react to form dispersions of intermetallic particles. These dispersions have a grain-refining and strengthening effect. As a result, higher strengths can be produced with less cold working, resulting in better formability at higher strength levels. Because these modifications do not require large amounts of costly elements, the gains are reasonably economical.

Alloy C63800 (95Cu-2.8Al-1.8Si-0.4Co) is a high-strength alloy with a nominal annealed tensile strength of 570 MPa (82 ksi) and nominal tensile strengths of 660 to 900 MPa (96 to 130 ksi) for the standard-rolled tempers. Cobalt provides the dispersion of strengthening intermetallic particles.

Alloy C68800 (73.5Cu-22.7Zn-3.4Al-0.4Co) is a high-strength modified aluminum brass. Its bend formability parallel to the direction of rolling is outstanding relative to its strength. It owes some of its unique properties to a dispersion of intermetallic particles resulting from the presence of cobalt. Its strength range is essentially the same as that of alloy C63800.

Alloy C65400 (95.44Cu-3Si-1.5Sn-0.06Cr) is a very high-strength alloy that has excellent stress-relaxation resistance at temperatures up to 105 °C (220 °F). Its nominal strength range in rolled tempers is 570 to 945 MPa (82 to 137 ksi). Electrical contact and connector springs are heat treated at 200 to 250 °C (390 to 480 °F) for 1 h to stabilize internal stresses and maximize stress-relaxation resistance.

Alloy C66400 (86.5Cu-11.5Zn-1.5Fe-0.5Co) is a low-zinc brass modified by the addition of iron and cobalt. The dispersion of intermetallic particles resulting from these additions strengthens the alloy. At the same time, conductivity is only moderately reduced, and resistance to SCC is very high. A high-zinc brass of the same strength and conductivity would be subject to SCC unless plated for protection.

Other modified solid-solution-strengthened alloys are probably described in the literature of the brass mill industries. Those described above should serve as examples of this additional class of copper alloys, which is expanding through the development efforts of the producers of brass mill products throughout the world.

Age-Hardenable Alloys. Age hardening produces very high strengths but is limited to those few copper alloys in which the solubility of the alloying element decreases sharply with decreasing temperature. The beryllium-coppers can be considered typical of the age-hardenable copper alloys.

Wrought beryllium-coppers can be precipitation hardened to the highest strength levels attainable in copper-base alloys. There are two commercially significant alloy families employing two ranges of beryllium with additions of cobalt or nickel. The so-called red alloys contain beryllium at levels ranging from approximately 0.2 to 0.7 wt%, with additions of nickel or cobalt totaling 1.4 to 2.7 wt%, depending on the alloy. Alloys C17500 and C17510 are examples of red alloys; these low-beryllium alloys achieve relatively high conductivity (for example, 50% IACS) and retain the pink luster of other low-alloy coppers. The red alloys achieve yield strengths ranging from about 170 to 550 MPa (25 to 80 ksi) with no heat treatment to greater than 895 MPa (130 ksi) after precipitation hardening, depending on degree of cold work.

The more highly beryllium-alloyed systems can contain from 1.6 to 2.0 wt% Be and about 0.25 wt% Co, for example, alloys C17000 and C17200. These alloys frequently are called the gold alloys because of the shiny luster imparted by the substantial amount of beryllium present (~12 at.%). The gold alloys are the high-strength beryllium-coppers because they can attain yield strengths ranging from approximately 205 to 690 MPa (30 to 100 ksi) in the age-hardenable condition to above 1380 MPa (200 ksi) after aging. The conductivity of the gold alloys is lower than that of the red alloy family by virtue of the high beryllium content. However, conductivity ranging from about 20% to higher than 30% IACS is obtained in wrought products depending on the amount of cold work and the heat treatment schedule. For enhanced machinability in rod and wire, lead is added (as in alloy C17300). More detailed information on beryllium-containing copper alloys can be found in the article "Beryllium-Copper and Other Beryllium-Containing Alloys" in this Section.

Other age-hardenable alloys include C15000; C15100 (zirconium-copper); C18200, C18400, and C18500 (chromium-coppers); C19000 and C19100 (copper-nickel-phosphorus alloys); and C64700 and C70250 (copper-nickel-silicon alloys). Some age-hardening alloys have different desirable characteristics, such as high strength combined with better electrical conductivity than the beryllium-coppers.

Alloy C71900 (copper-nickel-tin) and other similar alloys can be hardened by spinodal decomposition. By combining cold working with hot working, these alloys can achieve high strengths that are equivalent to those of the hardenable beryllium-coppers. These alloys are unique in that their forming characteristics are isotropic and thus do not reflect the directionality normally associated with wrought alloys.

Other Alloys. Certain aluminum bronzes, most notably those containing more than about 9% Al, can be hardened by quenching from above a critical temperature. The hardening process is a martensitic-type process, similar to the martensitic hardening that occurs when iron-carbon alloys are quenched. Mechanical properties of aluminum bronzes can be varied somewhat by temper annealing after quenching or by using an interrupted quench instead of a standard quench. Aluminum bronzes alloyed with nickel or zinc use reversible martensitic transformations to provide shape memory effects (see the article "Shape Memory Alloys" in this Volume).

Insoluble Alloying Elements. Lead, tellurium, and selenium are added to copper and copper alloys to improve machinability. These elements, along with bismuth, make hot rolling and hot forming nearly impossible and severely limit the useful range of cold working. The high-zinc brasses avoid these limitations, however, because they become fully β phase at high temperature. The β phase can dissolve lead, thus avoiding a liquid grain-boundary phase at hot forging or extrusion temperatures. Most free-cutting brass rod is made by β extrusion. Alloy C37700, one of leaded high-zinc brasses, is so readily hot forged that it is the standard alloy against which the forgeability of all copper alloys is measured.

Table 9 Generic classification of copper alloys

Generic name	UNS numbers	Composition
Wrought alloys		
Coppers	C10100–C15760	>99% Cu
High-copper alloys	C16200–C19600	>96% Cu
Brasses	C20500–C28580	Cu-Zn
Leaded brasses	C31200–C38590	Cu-Zn-Pb
Tin brasses	C40400–C49080	Cu-Zn-Sn-Pb
Phosphor bronzes	C50100–C52400	Cu-Sn-P
Leaded phosphor bronzes	C53200–C54800	Cu-Sn-Pb-P
Copper-phosphorus and copper-silver-phosphorus alloys	C55180–C55284	Cu-P-Ag
Aluminum bronzes	C60600–C64400	Cu-Al-Ni-Fe-Si-Sn
Silicon bronzes	C64700–C66100	Cu-Si-Sn
Other copper-zinc alloys	C66400–C69900	...
Copper-nickels	C70000–C79900	Cu-Ni-Fe
Nickel silvers	C73200–C79900	Cu-Ni-Zn
Cast alloys		
Coppers	C80100–C81100	>99% Cu
High-copper alloys	C81300–C82800	>94% Cu
Red and leaded red brasses	C83300–C85800	Cu-Zn-Sn-Pb (75–89% Cu)
Yellow and leaded yellow brasses	C85200–C85800	Cu-Zn-Sn-Pb (57–74% Cu)
Manganese bronzes and leaded manganese bronzes	C86100–C86800	Cu-Zn-Mn-Fe-Pb
Silicon bronzes, silicon brasses	C87300–C87900	Cu-Zn-Si
Tin bronzes and leaded tin bronzes	C90200–C94500	Cu-Sn-Zn-Pb
Nickel-tin bronzes	C94700–C94900	Cu-Ni-Sn-Zn-Pb
Aluminum bronzes	C95200–C95810	Cu-Al-Fe-Ni
Copper-nickels	C96200–C96800	Cu-Ni-Fe
Nickel silvers	C97300–C97800	Cu-Ni-Zn-Pb-Sn
Leaded coppers	C98200–C98800	Cu-Pb
Miscellaneous alloys	C99300–C99750	...

Deoxidizers

Lithium, sodium, beryllium, magnesium, boron, aluminum, carbon, silicon, and phosphorus can be used to deoxidize copper. Calcium, manganese, and zinc can sometimes be considered deoxidizers, although they normally fulfill different roles.

The first requirement of a deoxidizer is that it have an affinity for oxygen in molten copper. Probably the second most impor-

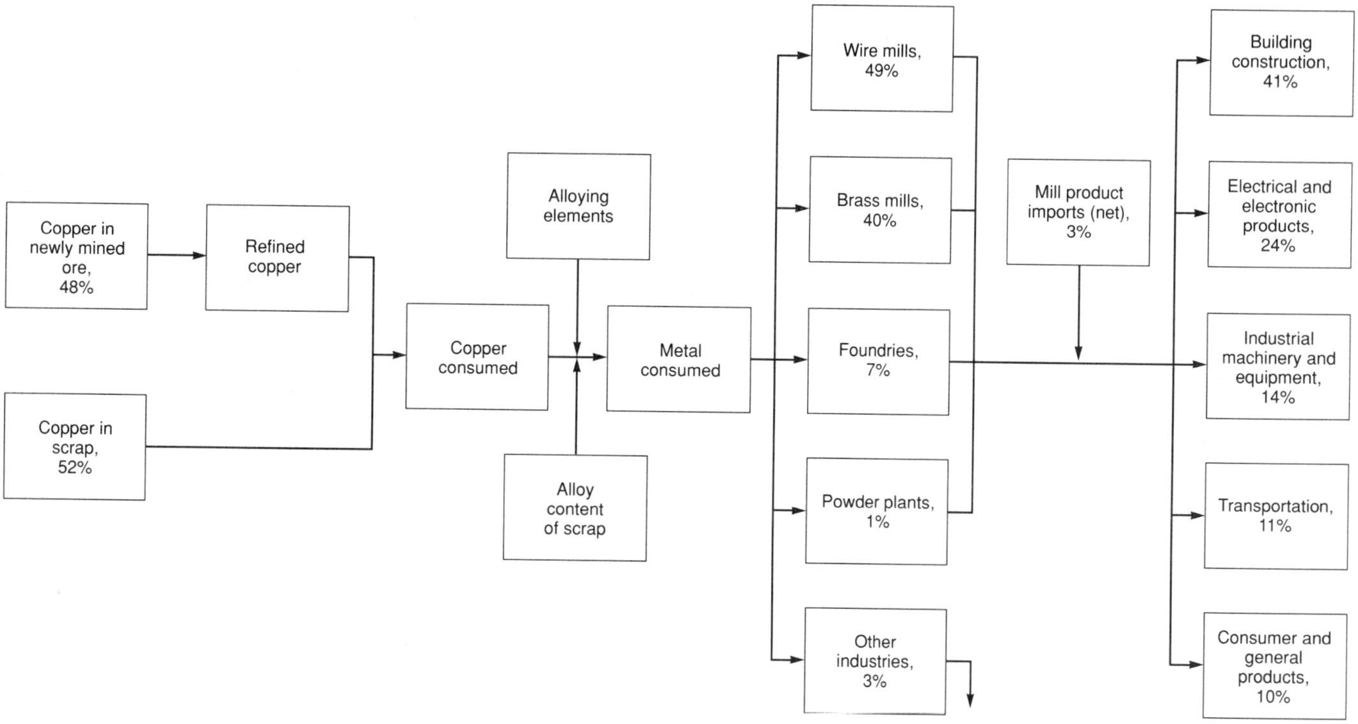

Fig. 5 Flow of copper in the U.S. economy. Percentages are based on 1989 data.

tant requirement is that it be relatively inexpensive compared to copper and any other additions. Thus, although zinc normally functions as a solid-solution strengthener, it is sometimes added in small amounts to function as a deoxidizer because it has a high affinity for oxygen and is relatively low in cost. In tin bronze, phosphorus has traditionally been the deoxidizer, hence the name phosphor bronzes for these alloys. Silicon, rather than phosphorus, is the deoxidizer for chromium-coppers because phosphorus severely reduces electrical conductivity. Most deoxidizers contribute to hardness and other qualities, thus obscuring their role in alloy deoxidization.

Production of Copper Metals

The copper industry in the United States, broadly speaking, is composed of two segments: producers (mining, smelting, and refining companies) and fabricators (wire mills, brass mills, foundries, and powder plants). The end products of copper producers, the most important of which are refined cathode copper and wire rod, are sold almost entirely to copper fabricators. The end products of copper fabricators can be generally described as mill products and foundry products, and they consist of wire and cable, sheet, strip, plate, rod, bar, mechanical wire, tubing, forgings, extrusions, castings, and powder metallurgy shapes. These products are sold to a wide variety of industrial users. Certain mill products—chiefly wire, cable, and most tubular products—are used without further metalworking. On the other hand, most flat-rolled products, rod, bar, mechanical wire, forgings, and castings go through multiple metalworking, machining, finishing, and/or assembly operations before emerging as finished products.

Copper Producers. Figure 5 is a simplified flow chart of the copper industry. The box at upper left represents mining companies, which remove vast quantities of low-grade material, mostly from open-pit mines, to extract copper from the crust of the earth. Approximately 2 tons of overburden must be removed along with each ton of copper ore. (The ratio of overburden to ore is sometimes as high as 5 to 1.) The ore itself averages only about 0.7% Cu.

Copper ore normally is crushed, ground, and concentrated, usually by flotation, to produce a beneficiated ore containing about 25% Cu. The ore concentrates are then reduced to the metallic state, most often by a pyrometallurgical process. Traditionally, the concentrated ore is processed in a primary smelting reactor, such as a reverberatory furnace, to produce a copper sulfide-iron sulfide matte containing up to 60% Cu. Reverberatory technology is rapidly being replaced by oxygen/flash smelting, which greatly reduces the volume of off gases. Sulfuric acid is manufactured from the sulfur dioxide contained in these off gases and is an important co-product of copper smelting. The matte is oxidized in a converter to transform the iron sulfides to iron oxides, which separate out in a slag, and to reduce the copper sulfide to blister copper, which contains at least 98.5% Cu. Current technology combines the converter step with the preceding smelting step. Fire refining of blister copper then removes most of the oxygen and other impurities, leaving a product at least 99.5% pure, which is cast into anodes. Finally, most anode copper is electrolytically refined, usually to a purity of at least 99.95%.

The resulting cathodes are the normal end product of the producer companies and are a common item of commerce. In recent years, many producers have installed continuous casting rod mills to directly convert cathode copper to wire rod (typically, 8 mm, or 5/16 in., in diameter), which is the feed material for the wire mills. Primary producers may also convert the cathode to cakes or billets of copper for sale to brass mills. The consumption of refined copper (mostly cathodes) in the United States was about 2.2×10^6 Mg (2.4×10^6 tons) in 1989, about 26% of the free-world total of 8.5×10^6 Mg (9.3×10^6 tons).

Hydrometallurgical processing is an increasingly important alternative to pyrometallurgy, particularly for nonsulfide ores such as oxides, silicates, and carbonates. In this process, weak acid is percolated through ore or waste dumps of rejected materials. Copper is leached out and recovered from the pregnant leach liquor, most often by solvent extraction, to produce an electrolyte suitable for electrowinning. In electrowinning, copper is extracted electrolytically in much the same way as anode copper is electrorefined. Copper extracted

Table 10 Major wrought copper and copper-base alloy systems used in the United States

Copper or copper alloy group	Designation	Approximate U.S. shipments in 1989 Mg × 10^3	lb × 10^6	Remarks
Wire mill products				
Coppers	C10000–C15900	1522	3356	C11000 is the predominant material.
Brass mill products				
Coppers	C10000–C15900 C16000–C16900 C18000–C18900	570	1257	Includes modified coppers, cadmium copper, and chromium copper
Common brasses	C20000–C29900	219	482	Of this amount, 90% is strip, sheet, and plate.
Leaded brasses	C30000–C39900	347	766	Of this amount, 96% is rod.
Tin bronzes (phosphor bronzes)	C50000–C53900	14	31	Unleaded only
Aluminum bronzes, silicon bronzes, and manganese bronzes	C60000–C68400	13	28	
Copper-nickels	C70000–C72900	32	71	
Nickel silvers	C73000–C79900	4	9	
Others	C17000–C17900 C19000–C19900 C40000–C49900 C54000–C54900 C68500–C69900	49	107	Includes beryllium-coppers, copper-iron alloys, tin brasses, leaded tin bronzes, aluminum brasses, and silicon brasses
Total		**2770**	**6107**	

Source: Copper Development Association Inc.

by electrowinning is equal in quality to that produced by electrolytic refining.

The box at the lower left in Fig. 5 represents the portion of the copper supply provided by scrap. In recent years, well over half the copper consumed in the United States has been derived from recycled scrap, and this percentage has grown somewhat over the last two decades. About 55% of this scrap has been new scrap, such as turnings from screw-machined rod, as opposed to old scrap, such as used electrical cable or auto radiators. Scrap recycled within a particular plant or company (runaround scrap) is not included in these statistics. About one-third of the scrap recycled in the United States is fed into the smelting or refining stream and quickly loses its identity. The remainder is consumed directly by brass mills; by ingot makers, whose main function is to process scrap into alloy ingot for use by foundries; by foundries themselves; by powder plants; and by others, such as the chemical, aluminum, and steel industries.

The box labeled "copper consumed" in Fig. 5 represents the total tonnage of refined copper plus the copper content of scrap consumed directly by fabricator companies. The various alloying elements used in producing copper alloys and the alloy content of the directly consumed scrap are added to obtain the total metal consumed by copper fabricators and other industries.

Copper Fabricators. The four classes of copper fabricators together account for about 97% of the total copper (including alloying metal) consumed each year in the United States (Fig. 5). Other industries, such as steel, aluminum, and chemical producers, consume the remaining 3%.

The share of metal consumed by wire rod mills has grown sharply over the last 20 years to its present (1989) level of about 49%; consumption by brass mills has dropped to 40%. Foundries account for about 7% of fabricated mill products, and powder plants use less than 1% of the U.S. supply of copper.

Wire mill products are destined for use as electrical conductors. Starting with wire rod, these mills cold draw the material (with necessary anneals) to final dimensions through a series of dies. The individual wires may be stranded and normally are insulated before being gathered into cable assemblies.

Brass mills melt and alloy feedstock to make strip, sheet, plate, tube, rod, bar, mechanical wire, forgings, and extrusions. Less than half the copper input to brass mills is refined; the rest is scrap. Fabricating processes such as hot rolling, cold rolling, extrusion, and drawing are employed to convert the melted and cast feedstock into mill products.

About 45% of the output of U.S. brass mills is unalloyed copper and high-copper alloys, chiefly in such forms as plumbing and air-conditioning tube, busbar and other heavy-gage current-carrying flat products, and roofing sheet. Copper alloys make up the remaining 55% (Table 10). Free-cutting brass rod, which exhibits outstanding machinability and good corrosion resistance, and brass strip, which has high strength, good corrosion resistance, excellent formability, and good electrical properties, together constitute about 80% of the total tonnage of copper alloys shipped from U.S. brass mills. Other alloy types of major commercial significance include tin bronzes (phosphor bronzes), which are noted for their excellent cold-forming behavior and strength; tin brasses, known for outstanding corrosion resistance; copper-nickels, which are strong and particularly resistant to seawater; nickel silvers, which combine a silvery appearance with good formability and corrosion resistance; beryllium-coppers, which provide outstanding strength when hardened; and aluminum bronzes, which have high strength along with good resistance to oxidation, chemical attack, and mechanical abrasion.

Foundries use prealloyed ingot, scrap, and virgin metal as raw materials. Their chief products are shaped castings for many different industrial and consumer goods, the most important of which are plumbing products and industrial valves. Centrifugal and continuously cast products find major application as bearings, cylinders, and other symmetrical components (see the article "Selection and Application of Copper Alloy Castings" in this Section). Powder plants produce powder and flake for further fabrication into powder metallurgy parts, chiefly small sintered bronze bushings (see the article "Copper P/M Products" in this Section). Net imports (imports less exports) of mill and foundry products have fluctuated in recent years, from a high of 7% in the period from 1984 to 1986 to less than 3% in 1989.

Industry Structure

The structure of the U.S. copper and copper alloy industry has undergone dramatic changes during the last two decades. In 1966, for example, the United States was easily the largest producer and consumer of newly mined copper. In addition, U.S. companies accounted for most of the output of Chilean mines; thus, the United States had effective control of about 45% of free-world production. However, the U.S.-owned mines in Chile began to be nationalized in the late 1960s, and the government of Chile has greatly expanded the output of these mines. The present U.S. share of world mine production is about 16%. Chile also accounts for 18%, nearly all of which is exported, and has the largest copper reserves in the world. Other important free-world producers are, in order of 1989 mine production, Canada, Zambia, Zaire, and Peru. The centrally controlled economies, including both the Eastern bloc and some free-world countries, today account for perhaps 60% of world mine production of copper, versus about 40% 20 years ago. World copper mine production statistics are provided in Fig. 6.

U.S. Companies. Many of the U.S. copper mining, smelting, and refining companies were purchased by oil companies in the 1970s. A string of unprofitable years in the early 1980s led to divestiture. These events, along with expanding international competition, led managements to institute cost controls and production efficiencies per-

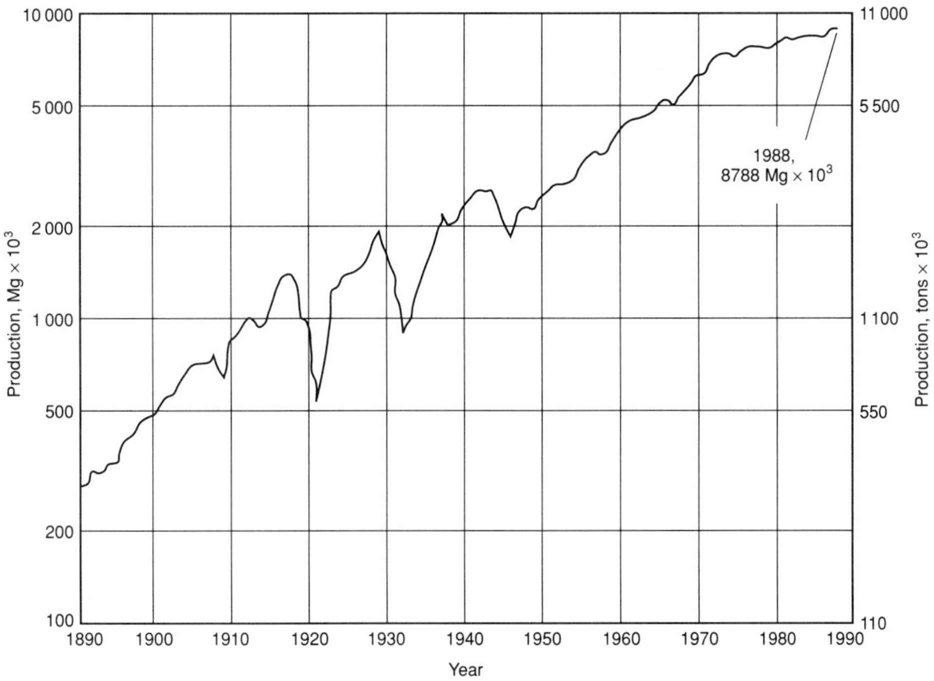

Fig. 6 World copper mine production. Despite the 10 000-year history of copper, about three-quarters of all copper ever consumed has been used in the period since World War II. 1989 production: 9178 Mg × 10^3. Source: Metallgesellschaft and the World Bureau of Metal Statistics

haps unprecedented in the U.S. metals industry. At present, despite the high cost of complying with the strictest environmental standards in the world, U.S. copper companies are well positioned to live with low price levels and to profit from the increasing demand for copper.

The U.S. brass mill industry has also undergone a significant restructuring in the past few years. The same economic factors at work in recent years in the mining industry have squeezed the profit margins of brass mills. In the recent past, this industry had been dominated by large full-line mills producing a wide range of products, including strip, sheet, plate, rod, bar, forgings, mechanical wire, and tubing. At present, only one full-line mill is in operation, as compared with eight in 1970. Instead, single-product operations have come to the forefront, many with the most up-to-date production equipment to enable them to compete on a worldwide scale. Many older mills have been closed, and other companies have been restructured through leveraged buyouts and employee stock ownership plans.

The wire and cable mills have seen some restructuring, particularly in the breakup of some of the large multiproduct operations that had equity ties to mining companies; also, numerous mergers and buyouts have taken place. Overall, however, these changes have not been as profound as those experienced by the copper companies and brass mills.

In contrast to other metal industries, today almost no top-to-bottom integration exists in the U.S. copper industry. The only move toward integration in the last 20 years has been the addition of continuous casting wire rod mills to the end of the production process by several refining companies.

Applications of Copper and Copper Alloys

The five major market categories shown at the far right in Fig. 5 constitute the chief customer industries of the copper fabricators. Of the chief customer industries, the largest is building construction, which purchases large quantities of electrical wire, tubing, and parts for building hardware and for electrical, plumbing, heating, and air-conditioning systems. The second largest category is electrical and electronic products, including those for telecommunications, electronics, wiring devices, electric motors, and power utilities. The industrial machinery and equipment category includes industrial valves and fittings; industrial, chemical, and marine heat exchangers; and various other types of heavy equipment, off-road vehicles, and machine tools. Transportation applications include road vehicles, railroad equipment, and aircraft parts; automobile radiators and wiring harnesses are the most important products in this category. Finally, consumer and general products include electrical appliances, fasteners, ordnance, coinage, and jewelry.

Table 11 is a more detailed listing of the largest markets for copper and copper alloys in the United States and compares 1989 markets with those for 1980. Building wiring and plumbing constitute the two highest

Table 11 Major end-use applications for copper and copper alloys in the United States in 1989

Application	% of total 1989	1980
Building wiring	16.9	10.7
Plumbing and heating	14.9	13.4
Autos, trucks, and buses	9.8	8.7
Telecommunications	8.1	13.1
Power utilities	7.7	7.4
Air conditioning and commercial refrigeration	7.1	6.3
In-plant equipment	7.1	8.4
Electronics	5.7	4.2
Industrial valves and fittings	3.4	3.5
Appliances and extension cords	2.7	2.9
Coinage	0.9	2.7
Other	15.7	18.7
Total	**100.0**	**100.0**

Source: Copper Development Association Inc.

end-use applications. Both have benefited from an extended housing boom and from an increasing intensity of use (that is, more electrical loads and more bathrooms in new homes). Plastics are an ongoing threat to replace copper in plumbing applications, but they are susceptible to permeation by gasoline and other organic materials and to mechanical damage. They also do not exhibit the bactericidal properties of copper.

Automotive applications represent the third largest market for copper. In 1989, about 23 kg (50 lb) of copper and copper alloys were used in an average U.S.-built passenger car; this compares to a use of 16 kg (36 lb) per car in 1980.

Telecommunications applications, on the other hand, have dropped from second to fourth place in less than a decade. The development of technological innovations such as subscriber carrier (that is, the piggybacking of many phone conversations on a single pair of copper wires), fiber optics (to a lesser degree), and the use of wires of smaller cross section have been the cause of this decrease.

In the power utilities market, the greatest competition to copper comes from aluminum, which is used almost exclusively in overhead transmission and distribution cable. An area of potentially significant growth, however, is that of superconductors, which may open up new markets for copper, such as for transmission lines, and energy storage devices. See the article "Superconducting Materials" in this Volume for more detailed information.

The copper end-use market that experienced the largest decrease between 1980 and 1989 was coinage. In 1982, a copper alloy was replaced by zinc in the penny. This caused a significant drop in the use of copper for coinage because roughly three-quarters of all U.S. coins minted are pennies. A move is now afoot to create a new copper-base dollar coin to eventually replace the dollar bill. The dollar of today has a value equal to that of a quarter in the early

Table 12 1988 U.S. mine production compared with estimates of copper reserves and resources

Production or reserves	Amount Mg × 10^6	lb × 10^9
U.S. mine production	1.5	3.3
U.S. reserve base	90	198
World reserve base	566	1248
Total world resources	2300	5070
Land-base resources	1600	3530
Deep-sea nodules	700	1540

Source: U.S. Bureau of Mines

1960s, and Australia, Britain, Canada, and other countries have replaced their smallest-denomination bill with a coin.

Copper Supply and Reserves

In the future, the market for copper has the potential of expanding into such promising areas as superconductivity applications; marine uses, such as ship hulls and sheathing for offshore platforms; electric vehicles and solar energy (both of which will inevitably reemerge at some point when oil supplies tighten); fire sprinkler systems; and agricultural applications. However, the maintenance of present markets for copper and attempts to open up new ones must be balanced against the prospects for future copper availability. Table 12 gives the 1989 U.S. mine production of copper as compared with estimates for the U.S. copper reserve base, the worldwide reserve base, and the total worldwide copper resources. The projected U.S. reserves have grown over time from the 24×10^6 Mg (52×10^9 lb) estimated by the Paley Commission in 1952 to almost 90×10^6 Mg (198×10^9 lb) in 1989, despite the fact that tens of billions of pounds have been mined and put into service in the interim. One reason for this increase is that the continuous development of better extraction techniques has enabled worthless rock to be reclassified as useful ore.

Going into the 1990s, the United States is nearly self-sufficient in copper and has averaged about 89% self-sufficiency over the last ten years; in comparison, the United States is only about 20% self-sufficient in aluminum.

Wrought Copper and Copper Alloy Products

Revised by Derek E. Tyler, Olin Corporation

COPPER AND COPPER ALLOYS are produced in various mill-product forms for a variety of applications (Table 1). High electrical conductivity, corrosion resistance, ease of fabrication (including good machinability) and good heat-transfer properties are typical reasons for selecting copper alloys. About 90% of the total tonnage of wrought copper alloys sold by U.S. manufacturers is represented by the 16 application categories listed in Table 1.

This article describes the manufacturing processes used to produce wrought copper and copper alloys in the form of sheet and strip products, tubular products, and wire or cable. The stress-relaxation characteristics of copper alloys are also discussed in this article. Information on the properties, applications, and comparisons of specific wrought copper alloys is given in the articles "Introduction to Copper and Copper Alloys" and "Properties of Wrought Coppers and Copper Alloys" in this Volume.

Table 1 Major end-use applications for copper and copper alloys in the United States

Application	Mill products	Principal reason(s) for using copper(a)
Telecommunications	Copper wire	Electrical properties
Automotive: automobiles, trucks, and buses	Brass and copper strip, copper wire	Corrosion resistance, heat transfer, electrical properties
Plumbing and heating	Copper tube, brass rod, castings	Corrosion resistance, machinability
Building wiring	Copper wire	Electrical properties
Heavy industrial equipment	All	Corrosion resistance, wear resistance, electrical properties, heat transfer, machinability
Air conditioning and commercial refrigeration	Copper tube	Heat transfer, formability
Industrial valves and fittings	Brass rod, castings	Corrosion resistance, machinability
Power utilities	Copper wire and bar	Electrical properties
Appliances	Copper wire and tube	Electrical properties, heat transfer
Lighting and wiring devices	Alloy strip, copper wire	Electrical properties
Electronics	Alloy strip, copper wire	Electrical properties
Fasteners	Brass wire	Machinability, corrosion resistance
Military and commercial ordnance	Brass strip and tube	Ease of fabrication
Coinage	Alloy and copper strip	Ease of fabrication, corrosion resistance, electrical properties, aesthetics
Builders' hardware	Brass rod and strip	Corrosion resistance, formability, aesthetics
Heat exchangers	Alloy tube and plate	Heat transfer, corrosion resistance

(a) Although not specifically listed as a principal reason in all applications, ease of fabrication is a factor in all application categories.

Sheet and Strip*

OF THE VARIOUS WROUGHT COPPER alloys covered in the article "Introduction to Copper and Copper Alloys" in this Volume, several alloys are covered in ASTM specifications for sheet and strip products (Table 2). Composition limits and mechanical properties of these wrought alloys are given in the applicable ASTM specifications. Additional information on the properties of specific copper alloys used as sheet or strip is given in the article "Properties of Wrought Coppers and Copper Alloys" in this Volume.

This section describes the manufacturing of copper alloy sheet and strip and how properties are controlled in the manufacturing process. The manufacture of sheet and strip in the modern brass mill begins with one of two basic casting operations: either the molten metal is cast in the form of slabs that are subsequently heated and hot rolled to coils of heavy-gage strip, or the metal is directly cast in strip form and coiled. The coils, in either case, then have their surfaces milled to remove any defects from casting or hot rolling. The next set of operations provides the desired final gage and temper by a series of cold-rolling, annealing, and cleaning operations. Finally, the sheet or strip may be slit into narrower widths, leveled, edge rolled or otherwise treated, and packaged for shipment.

*Edited from *Understanding Copper Alloys*, Olin Corporation, 1977

Raw Materials, Melting, and Casting

Raw materials from which the melt is prepared consist primarily of virgin copper, either electrolytic or fire-refined; selected clean scrap of known origin, carefully checked for composition; and special alloy elements such as virgin zinc, lead, tin, or nickel. Scrap is baled to make it dense, so it sinks below the surface of the molten metal in the furnace and melts rapidly. This is the first of many operations in which the quality of the finished product can be affected. In this case the impurities that can alter the processing characteristics of the copper alloys can be avoided or controlled to established tolerance levels. In copper-aluminum alloys, for example, excessive amounts of lead, zinc, silicon, or phosphorus will cause hot shortness and cracking during hot working and welding. In copper-zinc alloys (brasses), the effects of various impurities are as follows:

Table 2 ASTM specifications for copper and copper alloy sheet and strip

Specification	Product description
B 36	Brass plate, sheet, strip, and rolled bar
B 96 and B 96M (metric)	Copper-silicon alloy plate, sheet, strip, and rolled bar for general purposes
B 103	Phosphor bronze plate, sheet, strip, and rolled bar
B 121	Leaded brass plate, sheet, strip, and rolled bar
B 122	Copper-nickel-tin alloy, copper-nickel-zinc alloy (nickel silver), and copper-nickel alloy plate, sheet, strip, and rolled bar
B 152 and B 152M (metric)	Copper sheet, strip, plate, and rolled bar
B 169 and B 169M (metric)	Aluminum bronze plate, sheet, strip, and rolled bar
B 194	Copper-beryllium alloy plate, sheet, strip, and rolled bar
B 291	Copper-zinc-manganese alloy (manganese brass) sheet and strip
B 422	Copper-aluminum-silicon-cobalt alloy, copper-nickel-aluminum-silicon alloy, copper-nickel-aluminum-magnesium alloy sheet and strip
B 465	Copper-iron alloy plate, sheet, strip, and rolled bar
B 534	Copper-cobalt-beryllium alloy, copper-nickel-beryllium alloy plate, sheet, strip, and rolled bar
B 591	Copper-zinc-tin alloy plate, sheet, strip, and rolled bar
B 592	Copper-zinc-aluminum-cobalt or nickel-alloy plate, sheet, strip, and rolled bar
B 694	Copper, copper alloy, and copper-clad stainless steel sheet and strip for electrical cable shielding
B 740	Copper-nickel-tin spinodal alloy strip

- *Lead* should be kept under 0.01% for hot rolling, although additions of lead up to 4% improve machinability in material processed by extrusion and cold working. Lead lowers room-temperature ductility in brass, and also leads to hot shortness above 315 °C (600 °F)
- *Aluminum* as high as 2% has no adverse effect on hot or cold working. However, annealing and grain size are affected
- *Arsenic* does not affect hot or cold working, but tends to refine the grain size, and thus to lower the ductility
- *Cadmium* is controversial; some claim as much as 0.10% has little effect, others would keep it under 0.05%
- *Chromium* affects temperature of anneal and grain size. This condition is aggravated when iron is present
- *Iron* chiefly affects annealing and magnetic properties
- *Nickel* restrains grain growth
- *Phosphorus* has no adverse effect up to 0.04%; it does, however, restrain grain growth, increase tensile strength, and lower ductility to some extent

Additional information on the effect of impurities in some specific copper is given in the article "Properties of Wrought Coppers and Copper Alloys" in this Volume.

Melting. In one melting method, the raw materials are discharged into hoppers that feed electric-induction melting furnaces. With this type of melting, the analysis of the raw materials is carefully controlled and few or no impurities are introduced during melting. The metal is protected from atmospheric oxidation by a cover of carbon or bone ash. When the composition and temperature have been determined to meet the requirements of the alloy being melted, the molten metal is transferred to a holding furnace.

Casting in Book Molds. For many years in brass mills, the molten metal was poured from the melting furnace into a pouring box that distributed it into long, rectangular book molds (split molds, hinged like a book). The slab cast in these molds is about 50 to 100 mm (2 to 4 in.) thick, ½ to ¾ m (2 to 2.5 ft) wide, and about 2.5 m (8 ft) long. The book mold method has some important disadvantages: The maximum weight of a casting, and therefore the length of finished coil, are limited; molten metal is dropped 2.5 m (8 ft) from a pouring box to the base of the mold resulting in oxide generation and entrapment; and metal splash causes the bottom end of the bar to be spongy. The casting also varies from bottom to top in temperature and solidification rate with potential problems from shrinkage cavities, gas entrapment, surface laps, and mold coating defects.

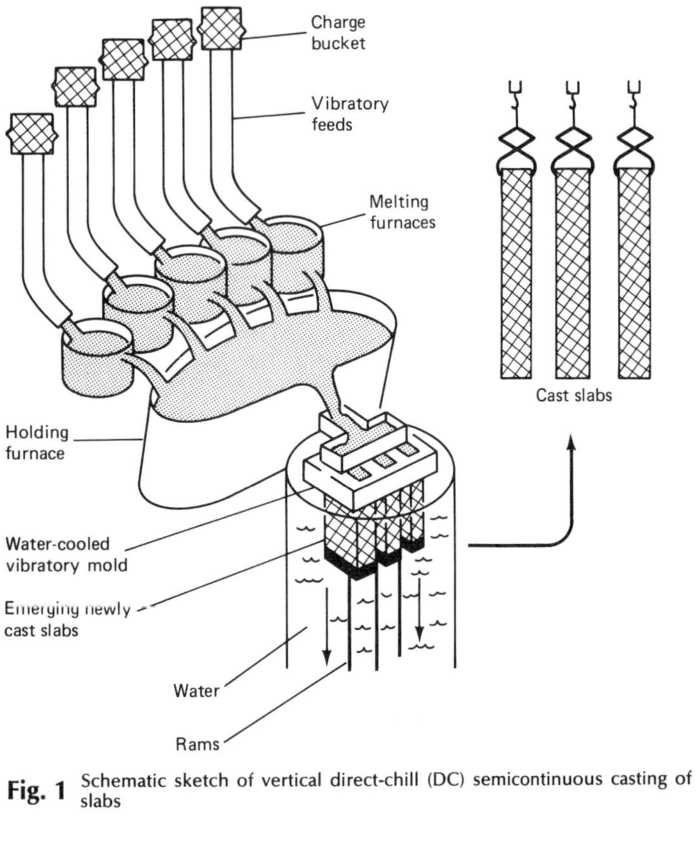

Fig. 1 Schematic sketch of vertical direct-chill (DC) semicontinuous casting of slabs

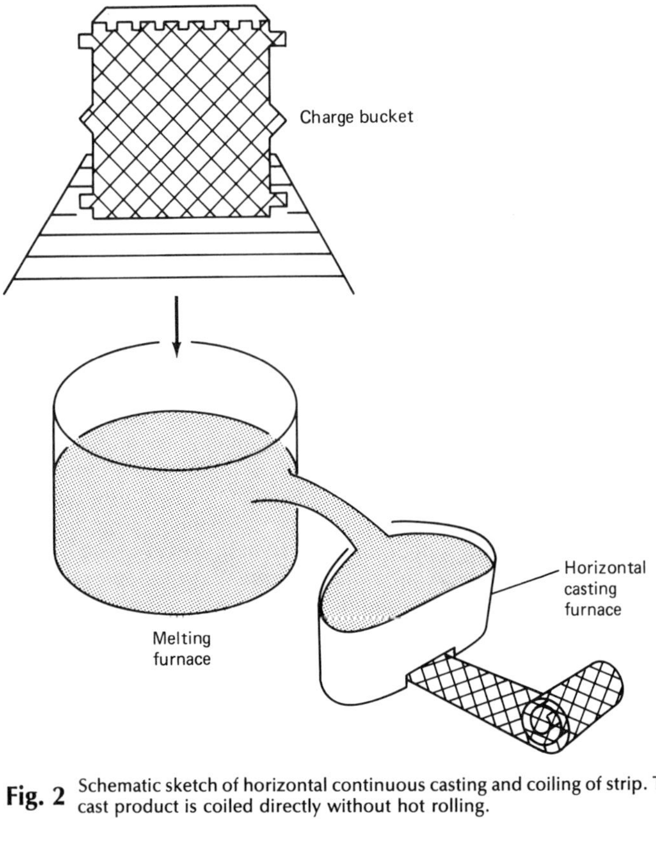

Fig. 2 Schematic sketch of horizontal continuous casting and coiling of strip. The cast product is coiled directly without hot rolling.

Semicontinuous and Continuous Casting. During the period of 1960 to 1975, two other casting processes began to supplant book molds. These two methods are:

- Vertical direct-chill semicontinuous casting
- Horizontal continuous casting

The vertical direct-chill (DC) semicontinuous casting process (Fig. 1) is used to produce slabs of large cross section, which are subsequently reheated, hot rolled into heavy-gage strip, and coiled. The continuous casting process (Fig. 2) uses a horizontal mold and casts a thin, rectangular section in much longer lengths that are coiled directly without hot rolling. Both processes provide a cast product that is fine grained, sound, and generally free of nonmetallic inclusions.

Vertical Semicontinuous Casting. In this method, molten metal flows into a short, rectangular, water-cooled mold, which initially is closed at one end by a plug on a movable ram or a starter bar (Fig. 1). The metal freezes to the plug and forms a shell against the mold surface. The ram is then steadily withdrawn, pulling the shell with it. As the shell exits the bottom of the mold, cold water is sprayed on it, cooling it rapidly and causing the contained molten metal to freeze. In this manner a continuously cast slab of the desired length is produced.

The vertical DC semicontinuous cast method is designed to produce large slabs from which heavy-weight finished coils can be made. Such large coils are the most economical to handle through the subsequent rolling and annealing processes at the brass mill, and later by the user who is fabricating finished parts.

Direct-chill slabs are hot rolled to produce coils, but not all alloys can be rolled by this method. Such alloys must be cold rolled, and the amount of reduction in thickness that can be achieved before annealing becomes necessary is small when compared to hot-rolling reductions. The problem with alloys that are hard to hot work is overcome with the horizontal continuous casting method. It offers a means of producing relatively thin castings in long lengths that can be coiled in the cast state and later reduced by cold rolling. Tedious, costly cold breakdown rolling and the attendant annealing are avoided.

In the DC casting process, more than one slab can be cast from each pour. A typical DC casting station consists of multiple melting furnaces and a large holding furnace. When the metal in each of the melting furnaces has been melted, the composition has been established, and the proper temperature attained, the molten metal is transferred to the holding furnace. Casting then proceeds from the holding furnace. Typically, three 6800 kg (15 000 lb) slabs of uniform composition are cast at one time. Smaller versions of this same process exist in the industry as well.

The horizontal continuous casting process (Fig. 2) also provides a product of excellent quality. Typically, one low-frequency electric-induction furnace is used as a melter. When the proper analysis is established and the pouring temperature attained, part of the metal is poured into a second, smaller electric-induction holding furnace. This furnace is constantly monitored to maintain the metal at the desired casting temperature. The casting mold is attached to the lower front of this furnace. It is a graphite mold contained in a copper, water-cooled jacket. A silicon carbide plate in the front of the furnace contains a slot that opens into the mold. At the beginning of a cast, a starter bar is inserted into the mold and the metal freezes to it. The mold is typically inches long.

The cast bar, frozen to the starter bar, is continuously withdrawn as the metal freezes in the mold. Although it is a simple process, its practice requires that tolerances on mold dimensions be held to a fraction of a millimeter (a few ten-thousandths of an inch) and exceptional melt cleanliness be maintained. Any dross or other foreign material that enters the graphite mold will quickly destroy it. Mold sizes range from 250 to over 600 mm (10 to 24 in.) in width and from about 10 to 15 mm (3/8 to 5/8 in.) in thickness. A saw or shear in the withdrawal line cuts the bars off at the desired length, and they are coiled in preparation for subsequent processing. This process lends itself to in-line coil milling and to maximum coil lengths, dependent only on handling equipment capacity and practical processing of the material itself.

The rapid chilling of the small amount of metal in the horizontal mold produces a fine, equiaxed cast grain structure. The metal drawn from the furnace as it solidifies always has a pool of molten liquid above it where gases and nonmetallic impurities tend to collect. The cast bar is generally free of porosity and defects caused by solid inclusions.

Some smaller mills depend almost entirely on horizontal casting, regardless of alloy, because the process is readily adaptable to the casting of small quantities of several alloys.

Hot Rolling of Direct-Chill Semicontinuous-Cast Slabs

The rolling of slab into sheet or strip products is performed for reduction in thickness and/or grain refinement. The initial rolling of slabs is for grain refinement as well as to begin reduction in thickness. For copper and copper alloys that can be hot worked, the quickest and most economical method of reduction is hot rolling.

To prepare the slab for hot rolling, the top or gate end is trimmed by sawing and then it is conveyed into a furnace for heating. Slabs or bars of the same alloy are grouped together in a lot and processed through the furnace and the hot mill. The furnace temperature and the time for each bar to pass through the furnace are adjusted in order to allow the bar to reach the appropriate temperature throughout its thickness, length, and width by the time it passes through to the exit conveyor.

Temperature control is an important factor in hot rolling. Hot rolling can be accomplished only within a certain temperature range for each alloy. The bars will be damaged and have to be scrapped if hot rolling is attempted at a temperature that is too high or too low. Further, for all alloys, the grain size of the hot-rolled product is determined by the temperature at the last rolling pass. Subsequent processing (that is, cold working and annealing) to meet specified properties, is dependent on this grain size. In some alloys, elements go into solution above certain temperatures and then precipitate out at lower temperatures. By completing hot rolling at a temperature above the precipitation temperature and quenching in a high-pressure water spray, solution heat treatment can be accomplished. This also affects both the physical and the mechanical properties attained in subsequent processing.

The roll stand used for hot rolling is a very sturdy mill having two rolls (two-high) whose direction of rotation can be rapidly reversed so the strip can be passed back and forth between them. The large horizontal rolls that reduce the thickness are supplemented by a pair of vertical edging rolls. The vertical rolls are needed to maintain the proper width by rolling edges because an appreciable spread in width takes place during hot rolling. The rolls are water cooled to avoid overheating, which would cause the surfaces to crack and check. Further, a polishing stone continuously dresses the rolls as they operate. As the thickness is reduced, the bar length increases proportionately. After the final rolling pass, the metal is spray cooled and coiled. Rolling temperatures and the percent of reduction per pass are designed to suit each alloy.

Milling or Scalping

Along with continuous casting, an equipment development that significantly advanced production is the high-speed coil milling machine. All coppers and copper alloys, produced with the good surface expected of brass mill sheet and strip, have their surfaces removed or scalped by a machining operation after breakdown rolling to remove all surface oxides remaining from casting or hot rolling. This operation is accomplished in a specially designed milling machine having rolls with inset blades that cut or mill away the surface layer of metal. The capability of this machine to handle the product in coiled form means that a much longer bar can be conveniently and economically milled.

Following hot rolling the DC cast bars are coil milled, and after careful surface inspection are ready to be applied on orders for

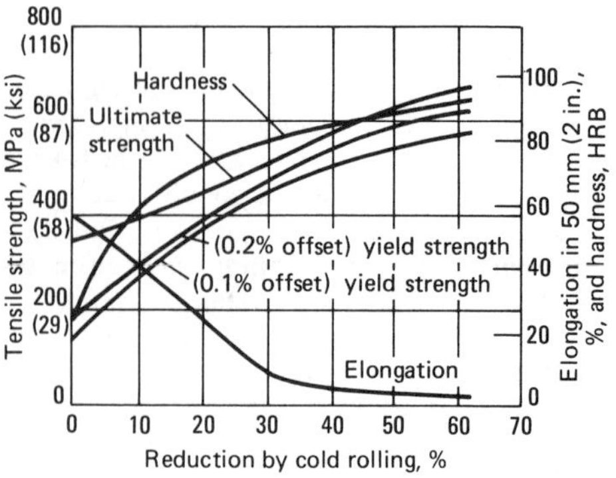

Fig. 3 The effect of cold rolling on the strength, hardness, and ductility of annealed Alloy C26000 when it is cold rolled in varying amounts up to 62% reduction in thickness

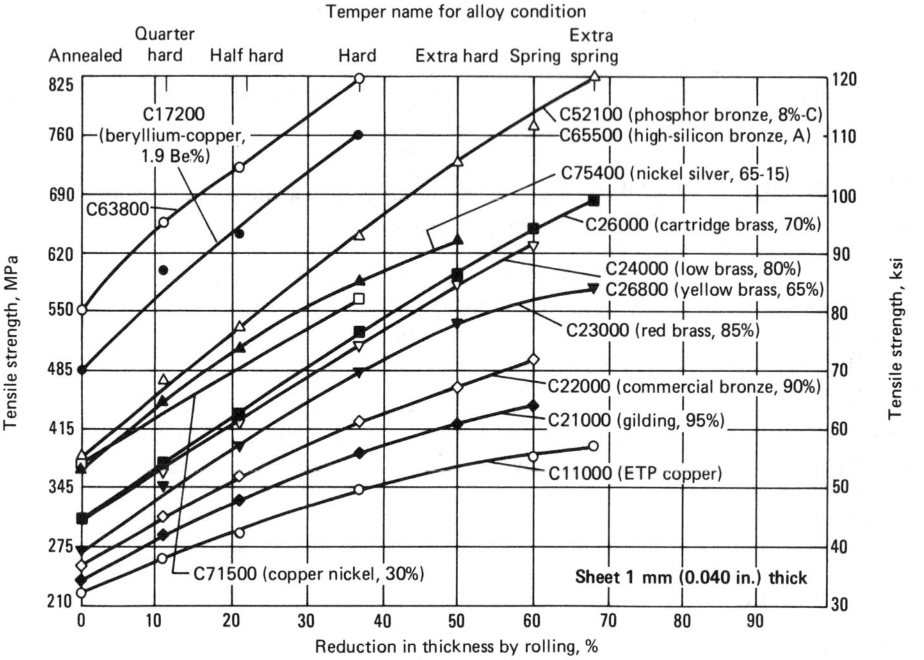

Fig. 4 Tensile strength of single-phase copper alloys as affected by percentage reduction in thickness by rolling (temper). Curves of lesser slope indicate a lower rate of work hardening and a higher capacity for redrawing.

processing to final gage, temper, and width. Horizontally continuous-cast bars arrive at this stage by a somewhat different processing path. The coiled cast bars are annealed to provide a stress-free structure of maximum ductility. They are then cold rolled to work the structure sufficiently, so a fully recrystallized wrought grain structure will develop in the subsequent anneal. The bars are then scalped by milling. Both hot-reduced and cold-reduced milled bars are typically in the thickness range of 7.5 to 10 mm (0.300 to 0.400 in.).

Cold Rolling to Final Thickness

The sequence of operations for processing metal from milled condition to finish thickness or gage is designed to meet specified requirements for each application.

The earliest stages of cold rolling and annealing are designed to achieve the largest practical reduction in thickness (limits to the amount of reduction are discussed in the section "Effects on Properties" in this article). In the final rolling operations, where the strip is brought to finish gage, the cold reductions are designed to meet the specified property (temper) requirement. Meeting the tensile strength requirement, which is the basic mechanical property requirement for rolled tempers, is accomplished by cold rolling to the appropriate ready-to-finish gage, annealing to the desired grain size, and then rolling to finish gage. The percent reduction between ready-to-finish and finish gage is chosen to provide the amount of work hardening needed to produce the tensile strength required. Unavoidable small variations in thickness at both ready-to-finish and finish gages and in grain size from the ready-to-finish anneal require that the tensile strength requirement be given as a range, rather than a single value.

Rolling Mills. All thickness reduction is accomplished by cold rolling, and a variety of rolling mills are used. Cold rolling of coppers and copper alloys into sheet and strip of excellent quality requires a combination of skillful workmanship, knowledge, and good rolling mills. To keep cost as low as possible and competitive, the reduction in thickness to final gage needs to be accomplished in the fewest operations compatible with quality requirements. The basic problem is to reduce the thickness as much as possible in each rolling operation while maintaining uniformity of thickness across the width and length of a coil that is 60 to 180 m (200 to 600 ft) long at the first rolling pass and could be 7500 m (25 000 ft) long if rolled to 0.1 mm (0.004 in.) finish gage. Coupled with the need to maintain uniformity of thickness through all processing stages is the need to maintain flatness across the width and length of the coiled metal. Metal with uniformity of flatness across and along its length is described as having good shape. It is free of humps, waves, and buckles.

A rolling mill is capable of applying a large, but still limited, force upon the surfaces of the metal as it passes between the rolls to reduce its thickness. The applied force is spread across the contact area of the rolls on the metal. The larger the contact area, the smaller the force that is applied per unit of area and the smaller the reduction in thickness that can be achieved per pass through the rolls. Rolls of small diameter will have a small contact area, and greater force per unit of area. Small-diameter work rolls are most desirable for providing maximum utilization of roll force in reducing metal thickness, but they lack the stiffness required. The wider the metal to be rolled, the longer the rolls, and the greater the tendency for the rolls to bend or spring. To overcome the tendency, four-high and cluster rolling mills are used for cold rolling in the brass mill.

Four-high rolling mills contain a pair of work rolls of relatively small diameter (for example, 300 mm, or 12 in.). A second pair of rolls, of large diameter (for example, 900 mm, or 36 in.), is placed above and below the work rolls in the stand to back them up and prevent them from springing. This arrangement allows the advantage of the small contact area of small work rolls and the transmittal of high force through the large backup rolls, while maintaining the rigidity required for gage control. The minimum size of the work rolls is limited by the forces in rolling, which tend to bow them backward or forward during rolling.

Cluster rolling mills (for example, Sendzimir or "Z" mills) were designed to counteract both the vertical and horizontal elements of the rolling forces and thus enable the use of minimum-diameter work rolls. In cluster mills the work rolls are backed up by a cluster of backup rolls placed with respect to the work rolls so they contain the rolling forces and prevent bending or springing of the work rolls. By the use of such rolling mills, the thickness from width edge to edge across the 600 to 915 mm (24 to 36 in.) metal coils can be kept uniform through each gage reduction by rolling. This edge-to-edge gage control contributes to the maintenance of good shape. Good shape contributes to the production of flat, straight metal when slitting to the final specified width needed by the consumer.

The control equipment included in the rolling mill is a feature that bears directly on

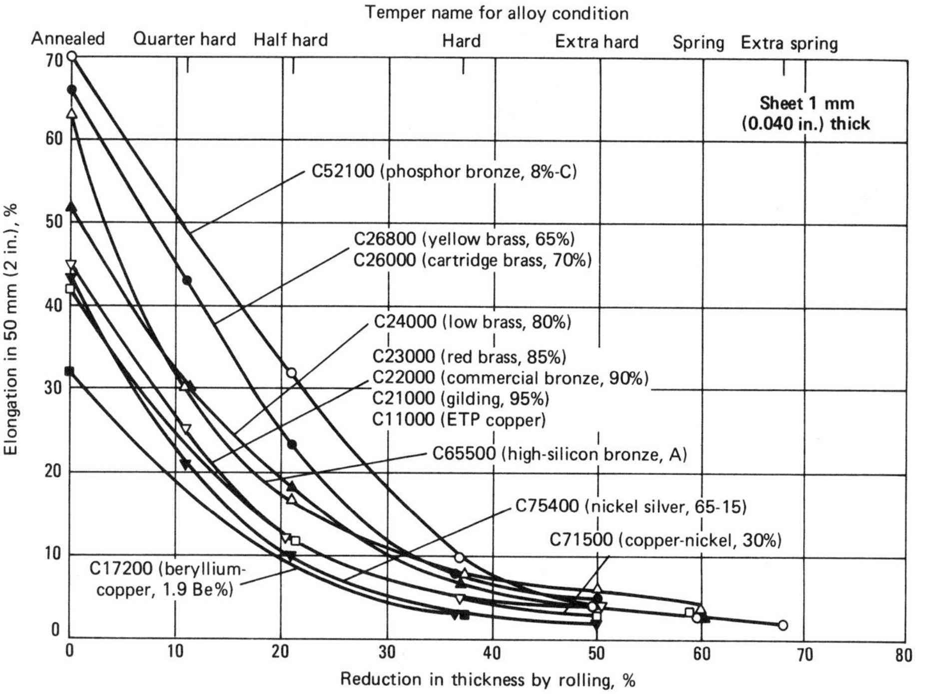

Fig. 5 Elongation of single-phase copper alloys as affected by percentage reduction in thickness by rolling (temper). The elongation values for a given percentage of cold reduction indicate the remaining capacity for deep drawing in a single operation.

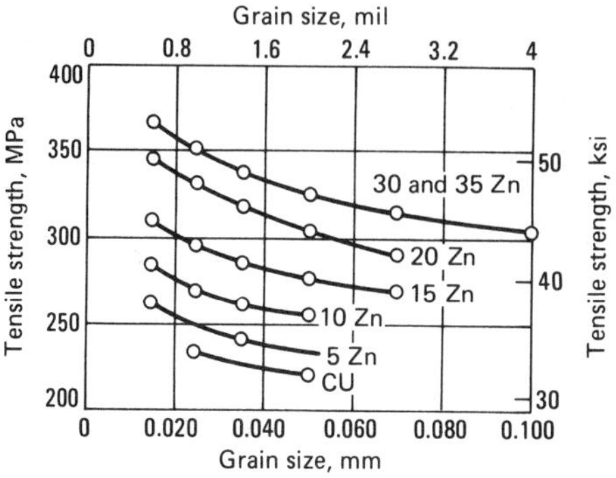

Fig. 6 Effect of grain size on tensile strength of annealed 0.040 in. strip of copper and brasses of designated zinc contents. Source: Ref 1

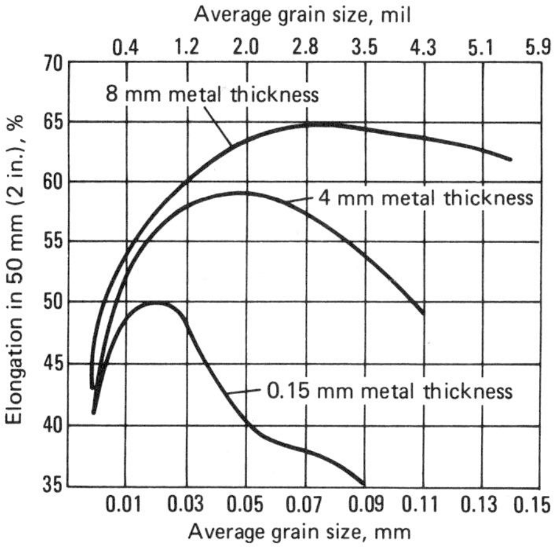

Fig. 7 The relationship between grain size and thickness versus elongation for Alloy C26000

control of the gage from end to end of a coil of metal during rolling. For thickness control during high-speed rolling, continuous measurement of this dimension is a necessity. Rolling mills are equipped with x-ray and beta-ray instruments, which continuously gage the metal and provide a continuous readout of thickness. There are also control devices that actuate the screws in the roll housings and automatically open or close the gap between the work rolls to adjust the thickness being produced as required. These gages may also adjust back tension and forward tension applied by payoff and recoil arbors to effect changes in the thickness of the rolled metal.

Roll Lubricants. Rolling also exerts considerable influence on the surface quality of the metal. Work rolls are made of hardened steel, much harder than the copper alloy being rolled. As the rolls squeeze the metal to reduce its thickness, forward and backward slip between the rolls and the metal surfaces takes place. The frictional forces between the roll and metal surfaces, if direct contact was made, would tear the surface of the metal and load the roll surfaces with bits of the metal. To avoid damaging the surfaces in this manner, the metal and roll surfaces are flooded with cushioning lubricants. The selection of roll lubricants that will provide the protection needed without staining the metal, will be readily removable from the metal surfaces, and will not interfere with the rolling mill performance is an important engineering function that influences the economic production of high-quality copper alloy strip.

Effect on Properties. The more metal is cold worked, the harder and stronger it becomes. The hardening that occurs when copper and copper alloys are cold rolled allows each of them to be produced with a range of strengths or tempers that are suitable for a variety of applications. Starting with annealed temper, the metal will increase in strength approximately proportionally by the amount of reduction by cold rolling. A series of standard cold-rolled tempers for each copper and copper alloy has been established. A typical plot of reduction versus tensile properties and hardness is shown in Fig. 3 for C26000 (cartridge brass). Figures 4 and 5 show, respectively, the variation of tensile strength and elongation for various degrees of reduction (and the associated rolling "temper" name). More information on the properties of rolled tempers for specific copper alloys is given in the article "Properties of Wrought Coppers and Copper Alloys."

For each of the coppers and copper alloys there are limits to the amount of cold reduction that is desirable before annealing the metal to provide a recrystallized soft structure for further cold reduction. Some alloys, such as the phosphor bronzes, the high-zinc-content nickel silvers, and the aluminum-containing high-zinc brasses work harden rapidly. As they are cold rolled, they quickly become too hard for further reduction and must be annealed.

With large amounts of cold reduction prior to annealing, some coppers and copper alloys will develop differences in their strength and ductility when these properties are measured along the direction of rolling, compared to measurements across the direction of rolling. This directionality in mechanical properties arises from the fact that the normal random orientation of the atomic planes from grain to grain is gradually forced into a pattern conforming to the constant working of the metal in one direction. This directionality can affect the fabricability and final performance of the strip or sheet. Its control requires that the amount of reduction between anneals and the temperature of successive anneals be carefully controlled.

Annealing

The basic purpose of annealing is recrystallization and softening to prepare the metal for further cold working in the mill or by the consumer. Anneals are usually designed to produce a chosen grain size for a specified tensile strength, which in annealed copper and copper alloys is largely dependent on grain size, with few exceptions. The effect of grain size on the tensile strength of copper and brass strip is shown in Fig. 6. The effect of grain size on the elongation of C26000 is shown in Fig. 7.

Table 3 Available grain-size ranges and recommended applications

Average grain size, mm	Type of press operation and surface characteristics
0.005–0.015	Shallow forming or stamping. Parts will have good strength and very smooth surface. Also used for very thin metal
0.010–0.025	Stampings and shallow drawn parts. Parts will have high strength and smooth surface. General use for metal under 0.25 mm (0.010 in.) thick
0.015–0.030	Shallow drawn parts, stampings, and deep drawn parts that require buffable surfaces. General use for gages under 0.3 mm (0.012 in.)
0.020–0.035	This grain size range includes the largest average grain that will produce parts essentially free of orange peel. For this reason it is used for all sorts of drawn parts produced from brass up to 0.8 mm (0.032 in.) thick.
0.025–0.040	Brass with 0.040 mm average grain size begins to show some roughening of the surface when severely stretched. Good deep drawing quality for 0.4 to 0.5 mm (0.015 to 0.020 in.) gage range
0.030–0.050	Drawn parts from 0.4 to 0.635 mm (0.015 to 0.025 in.) thick brass requiring relatively good surface, or stamped parts requiring no polishing or buffing
0.040–0.060	Commonly used grain size range for general applications for deep and shallow drawings of parts from brass in 0.5 to 1.0 mm (0.020 to 0.040 in.) gages. Moderate orange peel may develop on drawn surfaces
0.050–0.080, 0.060–0.090, 0.070–0.120	Large average-grain-size ranges are used for deep drawing of difficult shapes or deep drawing parts for gages 1.0 mm (0.040 in.) and greater. Drawn parts will have rough surfaces with orange peel except where smoothed by ironing

Source: Ref 1

Besides strength, grain size also affects workability, the control of directionality, and surface roughness. The consistent performance of the metal in subsequent cold working is dependent on grain-size uniformity. All these factors are considered when selecting the grain size to be established by any of the anneals included in the processing of each coil. Table 3 lists recommended applications of grain size ranges. Uniformity of grain size is influenced by the type of annealing furnaces and the method of operation. Each type of annealing method has certain advantages and disadvantages.

Coil Annealing. When annealing coiled metal, heat from the furnace must be absorbed through the coil surface and then penetrate to the innermost wraps, mostly by conduction. Temperature tends to vary in the coil with distance from the heat-absorbing surfaces. Coil annealing must be carefully controlled, by slowly applying heat at a rate that will avoid overheating the surface, while the temperature of the inner wraps rises and equalizes with that of the outer wraps.

Coil annealing may be done in a roller hearth furnace where the coils are slowly moved through the furnace as they are gradually heated to the annealing temperature. This type of furnace usually does not have a prepared atmosphere, but the products of combustion fill the furnace and reduce the metal oxidation rate. More commonly, coil annealing is done in bell furnaces where a controlled atmosphere can be maintained. The annealing unit consists of a base on which the coils are stacked. Under the base is a fan for circulating the hot gases through the load, to provide more uniform and rapid heating. Surrounding the base is a trough, which may be filled with water, oil, or some other material to seal the inner hood when it is placed over the metal load to enclose it for atmosphere control.

In this type of batch annealing, bell furnaces capable of annealing up to 45 Mg (100 000 lb) of metal at a time are used. After the metal is stacked on the base, temperature-control thermocouples are placed throughout the load to continuously measure the temperature. The inner hood or retort is placed over the load and sealed. Controlled atmosphere begins to flow through the hood, purging the air. The furnace is placed over the hood and heating is begun.

In the well-equipped brass mill, large groups of such annealing units may be connected to a central process-control computer. As the furnace and load thermocouples measure the temperatures and relay them to the control unit, the heat input is constantly adjusted to maintain temperature uniformity in the load. This controlled temperature rise also allows roll lubricants to vaporize and be carried off before the metal gets so hot that surfaces can be harmed. After the metal has reached the annealing temperature, it is held there for a short period or soaked to provide maximum uniformity. Then the furnace is turned off and removed, and the metal cools in the controlled atmosphere under the inner hood. Cooling may be aided by a cooling cover containing a water spray system. The inner hood is not removed until the metal temperature is low enough that no discoloring or oxidation of the metal takes place.

The controlled atmosphere is produced in gas-cracking units. Combustible gases are burned with sufficient air to oxidize all the gaseous elements. The products of this combustion are then refined, and all gases that would be harmful to the metal surfaces are removed by chemical means. Those remaining pass into the annealing hoods, where they expel the air and protect the metal during annealing. For most coppers and copper alloys a slightly oxidizing atmosphere is desirable. For copper Alloy C11000, the atmosphere must be nearly free of hydrogen and the annealing temperature low enough to avoid hydrogen embrittlement. For alloys containing zinc, the small amount of oxygen in the atmosphere combines with the zinc fumes given off and prevents them from attacking the metal parts in the annealing unit. The oxide film that forms on the surface is very thin and readily removed in the subsequent cleaning processes.

Advantages and Disadvantages. One of the advantages of coil annealing in a controlled atmosphere furnace is that the surface of the metal can be readily restored to its natural color by appropriate cleaning following the anneal. The rather rare exception is when an abnormally high annealing temperature is required that causes excessive oxidation or dezincification of a high-zinc brass. Special cleaning methods that remove surface metal are then required to correct this condition. The more common situation is that annealing is done in a well-controlled atmosphere and followed by normal cleaning practices. This produces a metal surface uniform in color and free from detrimental oxides.

A disadvantage of coil annealing is that large coils of some alloys in thinner gages can be easily damaged. When the coiled metal is heated, it expands and the coil wraps get tighter. One wrap can become welded to the next because of the high temperature and pressure encountered, usually making the coil unsuitable for further processing. Coil annealing is also time consuming. A large bell furnace full of metal may require from 24 to 40 h to complete an annealing cycle; additional is the time needed for cleaning, done as a separate operation.

Continuous Strand Annealing. In the late 1940s continuous strand, or strip annealing lines came into use in brass mills. From these early beginnings, the high-speed vertical strip annealers were developed in the 1960s. Annealing lines of this type are now used for annealing copper and copper alloy strip in thicknesses from under 0.25 mm to over 3 mm (0.010 to over 0.125 in.). When several such lines are available, a variety of thickness ranges can be rapidly annealed, providing great flexibility in production scheduling and enabling fast delivery of finished strip.

Because every foot of a coil is exposed to the same temperature as it passes through the strip-annealing furnaces, grain size from end to end should be uniform. Furnace instrumentation continuously records the furnace temperature and controls the heat input. Strip speed through the furnace is similarly monitored. The combination of furnace temperature and speed determines the temperature attained in the metal, and, therefore, the grain size. Samples commonly are cut from each end of each coil after strip or coil annealing, and the grain size or mechanical properties are determined as a further control on the quality uniformity of the product.

The continuous-strip anneal lines include payoff reels, a stitcher for joining the front end of a coil to the trailing end of the preceding one, a degreaser for removing roll lubricants, looping towers for metal storage, a seven-story-high vertical furnace that includes a heating zone, a controlled-atmosphere cooling zone, and a water quench tank. This is followed by acid cleaning tanks, a water rinse, a drying oven, and a reel for recoiling the metal. The fact that the metal is uncoiled before passing through the furnace removes annealing limitations on coil length. Degreasing units remove roll lubricants from the metal surfaces before the metal enters the furnace, so a clean, uniform surface is presented for annealing. The metal passes over a large roller outside the furnace at the top and does not touch anything inside while it is being heated. It then passes under another large roller at the bottom in the cooling water tank. This arrangement avoids any possibility of surface damage to the hot metal, which was common in the earlier horizontal-strip anneal furnaces. Although the furnace temperature is high, the metal is exposed to it for only a few seconds.

The furnace atmosphere may consist of hot burned gases that are blown against the strip surfaces to heat the metal. The metal is rapidly and uniformly raised to the annealing temperature as it passes through the heating zone of the furnace, and is then cooled rapidly by cold burned gases as it passes through the cooling zone, still protected from excessive oxidation.

Following a water quench, which completes the cooling cycle, the metal passes through the cleaning tanks. A normal cleaning solution is dilute sulfuric acid, which dissolves most of the oxide film left on the metal by annealing. As noted earlier, the atmosphere in the furnace must be slightly oxidizing to prevent zinc fumes from attacking the furnace steel framework. For most coppers and copper alloys this small amount of surface oxidation is not detrimental after normal cleaning, and they are regularly strip annealed throughout processing, including finish gage. They have a faintly different color than does bell-annealed and cleaned metal, but the difference is so slight that it is insignificant in most applications. In fact, brasses containing 15% or more of zinc have surfaces that many users feel are better suited for later fabricating if the strip has been continuously annealed. The metal surface holds lubricants well and has a low coefficient of friction against tool steels, making it desirable for press forming and deep drawing. It is likely that some zinc oxide remains on the surface and acts as a natural lubricant. After acid cleaning, rinsing, and drying, the surface is usually coated with a detergent solution or a light sulfur-free oil to protect it during handling in transit.

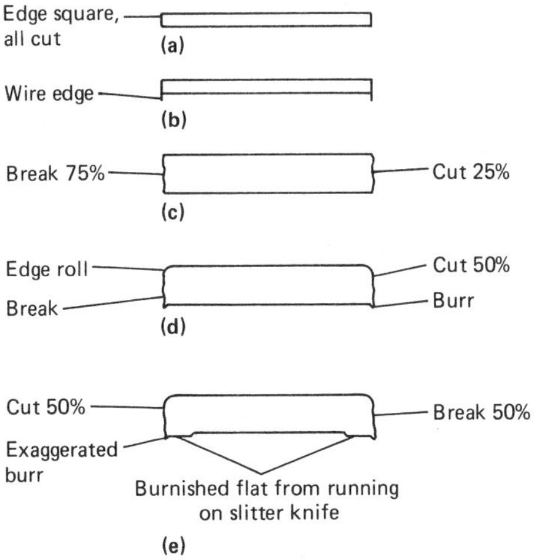

Fig. 8 The different edge contours that can result from slitting, depending on thickness, temper, and alloy. (a) Thin gages; all alloys. Edges square with almost no break. (b) Thin gages. On soft metal, set must be adjusted to avoid wire edges. (c) Heavy gages; hard metal; all alloys. Edges square with 25% cut balanced break. (d) Heavy gages; soft metal; all alloys. Edges square with slight roll. (e) As a rule, the heavy-gage, high-copper alloys have the greatest tendency to roll and burr.

Stress-relief heat treatments are sometimes required after the harder rolling tempers such as Extra Hard, Spring, and Extra Spring (Fig. 5). Although the internal residual stresses left in the strip, from edge to edge and along the length, from this severe working are relatively uniform, small variations sometimes exist that can cause a difference in spring-back during subsequent forming operations. To reduce such residual-stress variations, the metal is heated to a temperature below the recrystallization temperature, usually between 200 and 350 °C (390 and 660 °F), and held there for 0.5 to 1 h. Such treatment results in a product with uniform spring-back.

Heating for stress relief also can change other properties. In phosphor bronzes tensile elongation is increased and strength slightly decreased. These changes are an advantage in the case of difficult-to-form parts requiring maximum strength. In the high-zinc alloys, stress-relief heat treatment increases strength and decreases tensile elongation. In this case, the formability may be decreased.

Cleaning

As noted, following each anneal or heat treatment the metal is cleaned. After cleaning in the appropriate solution the metal is thoroughly washed in water, including brushing with wire or synthetic brushes when needed. The rinse water usually contains a tarnish inhibitor, such as tolutriazole or benzotriazole, to protect the metal. For product at finish gage the rinse tank has a detergent solution added that further protects the metal when dried and also lubricates it slightly to reduce the danger of friction scratches during coiling and uncoiling. Squeegee rolls are used to remove the bulk of the rinse water, and drying ovens in the cleaning lines complete the job.

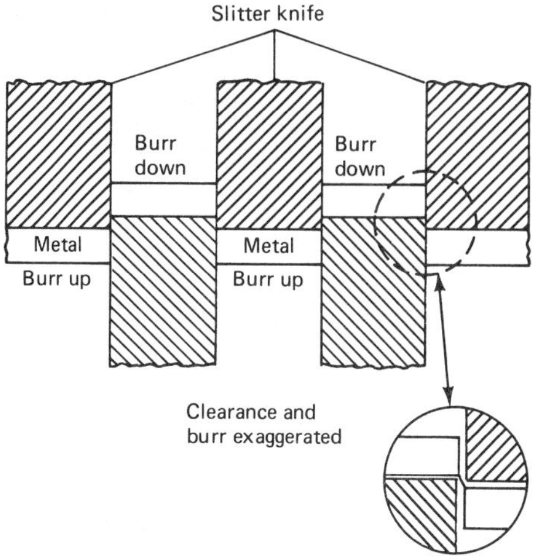

Fig. 9 Burr up/burr down relationship in slitting setup. Such burrs are never excessive on strip released for shipment.

If desired for subsequent working, annealed strip can also be coated with a film of light nontarnishing oil for protection and lubrication. Metal that is finished in a rolled temper will normally contain a light film of rolling lubricant on the surfaces to protect and lubricate it during coiling and uncoiling and in transit.

Slitting, Cutting, and Leveling

Following the final rolling or the final annealing and cleaning, the strip or sheet product is slit to its final width. Slitting is accomplished by opposing rotary discs mounted on rotating arbors. These knife sets mesh as the metal passes between them and shear it into a variety of widths. Slitter knife sets are assembled on arbors. The sets are assemblies of disc knives, cylindrical metal and rubber fillers, and shims. Clearance between the opposing knife edges must be exact for the thickness, alloy, and temper of the metal to be slit. The distance between knife edges on each arbor must be set accurately to cut the specified width within the tolerance allowed. Knife edges must be sharp and continuously lubricated. Dull knives or incorrect clearance between knives for the particular material being slit causes distorted or burred edges.

Camber, that is, departure from edgewise straightness, has often been attributed incorrectly to poor slitting practice. It is true that strips can be pulled crooked when slitting a large number of them from a wide bar, because the slit strips are sometimes fanned out for subsequent coiling using divider plates. This difficulty is diminished on slitters

Table 4 Copper tube alloys and typical applications

UNS number	Alloy type	ASTM specifications	Typical uses
C10200	Oxygen-free copper	B 68, B 75, B 88, B 111, B 188, B 280, B 359, B 372, B 395, B 447	Bus tube, conductors, wave guides
C12200	Phosphorus deoxidized copper	B 68, B 75, B 88, B 111, B 280, B 306, B 359, B 360, B 395, B 447, B 543	Water tubes; condenser, evaporator and heat-exchanger tubes; air conditioning and refrigeration, gas, heater and oil burner lines; plumbing pipe and steam tubes; brewery and distillery tubes; gasoline, hydraulic and oil lines; rotating bands
C19200	Copper	B 111, B 359, B 395, B 469	Automotive hydraulic brake lines; flexible hose
C23000	Red brass, 85%	B 111, B 135, B 359, B 395, B 543	Condenser and heat-exchanger tubes, flexible hose; plumbing pipe; pump lines
C26000	Cartridge brass, 70%	B 135	Plumbing brass goods
C33000	Low-leaded brass (tube)	B 135	Pump and power cylinders and liners; plumbing brass goods
C36000	Free-cutting brass		Screw machine parts; plumbing goods
C43500	Tin brass		Bourdon tubes; musical instruments
C44300, C44400, and C44500	Inhibited admiralty metal	B 111, B 359, B 395	Condenser, evaporator and heat-exchanger tubes; distiller tubes
C46400, C46500, C46600, and C46700	Naval brass		Marine hardware, nuts
C60800	Aluminum bronze, 5%	B 111, B 359, B 395	Condenser, evaporator and heat-exchanger tubes; distiller tubes
C65100	Silicon bronze B	B 315	Heat-exchanger tubes; electrical conduits
C65500	Silicon bronze A	B 315	Chemical equipment, heat-exchanger tubes; piston rings
C68700	Arsenical aluminum brass	B 111, B 359, B 395	Condenser, evaporator and heat-exchanger tubes; distiller tubes
C70600	Copper-nickel, 10%	B 111, B 359, B 395, B 466, B 467, B 543, B 552	Condenser, evaporator and heat-exchanger tubes; salt water piping; distiller tubes
C71500	Copper-nickel, 30%	B 111, B 359, B 395, B 446, B 467, B 543, B 552	Condenser, evaporator and heat-exchanger tubes; distiller tubes; salt water piping

equipped with over-arm separators because strips need not be fanned out as much. This kind of problem can be anticipated, and, if necessary, the bar split at an intermediate stage in processing prior to the final slitting, so fewer cuts are made in this last operation.

Instead of slitting practice, it is the maintenance of good shape during each of the rolling operations that is most important in the control of camber. If good control of thickness across the width is maintained at each rolling operation, the edges and centers of the bar will have elongated uniformly, and when narrow strips are slit they will remain satisfactorily straight.

The shape of the slit edge of strip depends to a great extent on the properties of the metal being slit. The metal may be thick, soft, and ductile, at one extreme of shearing characteristics, to thin, hard, and brittle, at the other extreme. Between these fall all the variations that are characteristic of the gage, alloy, and temper required for the final application. A certain amount of edge distortion cannot be avoided when slitting thick, soft metals (Fig. 8). Even with the best slitter setup, the cross section of a narrow strip will tend to have a "loaf" shape. By contrast, thin, hard phosphor bronze or nickel silver in narrow widths will have a cross section of rectangular shape with square cut edges. Leaded brasses shear cleanly because the lead, present as microscopic globules, lowers the ductility and shear strength. It is for this purpose—ease of cutting and machinability—that lead is added to copper alloys.

As the metal comes from the slitter, both edges of each strip, if distorted, will be distorted in the same direction. The immediately adjacent strips will have edges distorted in the opposite direction. There are some applications for which it is desirable that any edge distortions be in the same direction relative to the part being produced. The user recognizes that the edges of every other coil will be opposite and arranges to uncoil either over or under the coil so the edge condition entering the press is always the same (Fig. 9).

Coil set, the curvature that remains in a strip when it is unwound from a coil, is an inherent characteristic. The degree of this coil set is dependent on a number of factors. The final coiling operation takes place after slitting, and some measure of control over coil set can be exercised at this process stage. However, there are frequently other considerations that also have a bearing. For annealed tempers and the lightly cold-rolled tempers such as ¼ Hard and ½ Hard, coil set may be established during final coiling. The degree of set will be lowest when the largest inside diameter compatible with the specified gage and weight can be used. For the harder rolled tempers and lighter gages, the coil set is actually controlled in the final rolling operation, rather than during coiling, and is usually kept to a minimum.

Processing operations after final slitting are occasionally required. Blanking and edge rolling are two such operations.

Blanking of squares or rectangles is generally done by cutting to length. The metal is first flattened and then cut to length on a flying shear. If the tolerance on length cannot be achieved on the automatic cutting lines, the cut lengths are resheared by hand. When circular blanks are required, they are die cut on a press. The tolerances for the diameter of circular blanks are the same as those for slit metal of corresponding width.

Edge rolling can produce rolled square edges, rounded edges, rounded corners, or rolled full-rounded edges. It can only be done on a limited range of gage, width, and temper combinations. Properties and tolerances are generally the same as those for similar slit-edge products.

Tubular Products

TUBE AND PIPE made of copper or copper alloys are used extensively for carrying potable water in buildings and homes. These products also are used throughout the oil, chemical, and process industries to carry diverse fluids, ranging from various natural and process waters, to seawater, to an extremely broad range of strong and dilute organic and inorganic chemicals. In the automotive and aerospace industries, copper tube is used for hydraulic lines, heat exchangers (such as automotive radiators), air conditioning systems, and various formed or machined fittings. In marine service, copper tube and pipe are used to carry potable water, seawater, and other fluids, but their chief application is in tube bundles for condensers, economizers, and auxiliary heat exchangers. Copper tube and pipe are used in food and beverage industries to carry process fluids for beet and cane sugar refining, for brewing of beer, and for many other food-processing operations. In the building trades, copper tube is used widely for heating and air conditioning systems in homes, commercial buildings, and industrial plants and offices. Table 4 summarizes the copper alloys that are standard tube alloys, and gives ASTM specifications and typical uses for each of the alloys.

Table 5 Typical mechanical properties for copper alloy tube(a)

Temper	Tensile strength MPa	Tensile strength ksi	Yield strength(b) MPa	Yield strength(b) ksi	Elongation(c), %
C10200					
OS050	220	32	69	10	45
OS025	235	34	76	11	45
H55	275	40	220	32	25
H80	380	55	345	50	8
C12200					
OS050	220	32	69	10	45
OS025	235	34	76	11	45
H55	275	40	220	32	25
H80	380	55	345	50	8
C19200					
H55(d)	290	42	205(e)	30(e)	35
C23000					
OS050	275	40	83	12	55
OS015	305	44	125	18	45
H55	345	50	275	40	30
H80	485	70	400	58	8
C26000					
OS050	325	47	105	15	65
OS025	360	52	140	20	55
H80	540	78	440	64	8
C33000					
OS050	325	47	105	15	60
OS025	360	52	140	20	50
H80	515	75	415	60	7
C43500					
OS035	315	46	110	16	46
H80	515	75	415	60	10
C44300, C44400, C44500					
OS025	365	53	150	22	65
C46400, C46500, C46600, C46700(f)					
H80	605	88	455	66	18
C60800					
OS025	415	60	185	27	55
C65100					
OS015	310	45	140	20	55
H80	450	65	275	40	20
C65500					
OS050	395	57	...	...	70
H80	640	93	...	...	22
C68700					
OS025	415	60	185	27	55
C70600					
OS025	305	44	110	16	42
H55	415	60	395	57	10
C71500					
OS025	415	60	170	25	45

(a) Tube size: 25 mm (1 in.) OD by 1.65 mm (0.065 in.) wall. (b) 0.5% extension under load. (c) In 50 mm (2 in.). (d) Tube size: 4.8 mm (0.1875 in.) OD by 0.76 mm (0.030 in.) wall. (e) 0.2% offset. (f) Tube size: 9.5 mm (0.375 in.) OD by 2.5 mm (0.097 in.) wall

Frequently, resistance to corrosion is a critical factor in selecting a tube alloy for a specific application. Information that can help determine the alloy(s) most suitable for a given type of service can be found in the article "Introduction to Copper and Copper Alloys" in this Volume.

Joints in copper tube and pipe are made in various ways. Permanent joints can be made by brazing or welding. Semipermanent joints are made most often by soldering, usually in conjunction with standard socket-type solder fittings, but threaded joints also can be considered semipermanent joints for pipe. Detachable joints are almost always some form of mechanical joint—flared joints, flange-and-gasket joints, and joints made using any of a wide variety of specially designed compression fittings are all common.

Properties of Tube. As with most wrought products, the mechanical properties of copper tube depend on prior processing. With copper, it is not so much the methods used to produce tube, but rather the resulting metallurgical condition that has the greatest bearing on properties. Table 5 summarizes tensile properties for the standard tube alloys in their most widely used conditions. Information on other properties of tube alloys can be found in the data compilations for the individual alloys; see the article "Properties of Wrought Coppers and Copper Alloys" in this Volume.

Production of Tube Shells

Copper tubular products are typically produced from shells made by extruding or piercing copper billets.

Extrusion of copper and copper alloy tube shells is done by heating a billet of material above the recrystallization temperature, and then forcing material through an orifice in a die and over a mandrel held in position with the die orifice. The clearance between mandrel and die determines the wall thickness of the extruded tube shell.

In extrusion, the die is located at one end of the container section of an extrusion press; the metal to be extruded is driven through the die by a ram, which enters the container from the end opposite the die. Tube shells are produced either by starting with a hollow billet or by a two-step operation in which a solid billet is first pierced and then extruded.

Extrusion pressure varies with alloy composition. C36000 (61.5Cu-3Pb-35.5Zn) requires a relatively low pressure, whereas C26000 (70Cu-30Zn) and C44300 (71.5Cu-1Sn-27.5Zn-0.06As) require the highest pressure of all the brasses. Most of the coppers require an extrusion pressure intermediate between those for C26000 and C36000. C71500 (70Cu-30Ni) requires a very high extrusion pressure.

Extrusion pressure also depends on billet temperature, extrusion ratio (the ratio of the cross-sectional area of the billet to that of the extruded section), speed of extrusion, and degree of lubrication. The flow of metal during extrusion depends on many factors, including copper content of the metal, amount of lubricant, and die design.

Rotary piercing on a Mannesmann mill is another method commonly used to produce seamless pipe and tube from copper and certain copper alloys. Piercing is the most severe forming operation customarily applied to metals. The process takes advantage of tensile stresses that develop at the center of a billet when it is subjected to compressive forces around its periphery. In rotary piercing, one end of a heated cylindrical billet is fed between rotating work rolls that lie in a horizontal plane and are inclined at an angle to the axis of the billet (Fig. 10). Guide rolls beneath the billet prevent it from dropping from between the work rolls. Because the work rolls are set at an angle to each other as well as to the billet, the billet is simultaneously rotated and driven forward

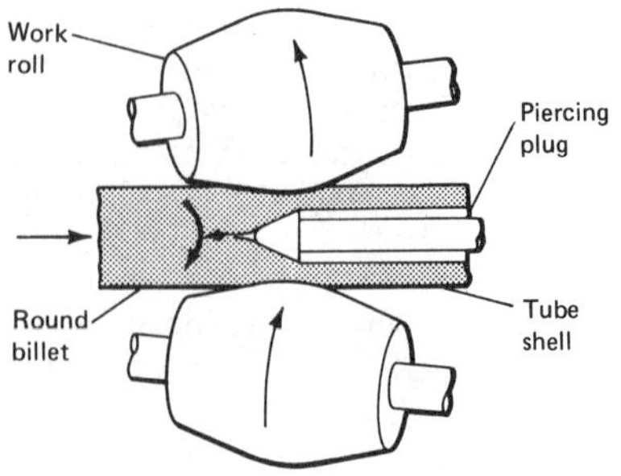

Fig. 10 Schematic diagram of metal piercing. Arrows indicate direction of motion.

toward the piercing plug, which is held in position between the work rolls.

The opening between work rolls is set smaller than the billet, and the resultant pressure acting around the periphery of the billet opens up tensile cracks, and then a rough hole, at the center of the billet just in front of the piercing plug. The piercing plug assists in further opening the axial hole in the center of the billet, smooths the wall of the hole, and controls the wall thickness of the formed tube.

Coppers and plain alpha brasses can be pierced, provided the lead content is held to less than 0.01%. Alpha-beta brasses can tolerate higher levels of lead without adversely affecting their ability to be pierced.

When piercing brass, close temperature control must be maintained because the range in which brass can be pierced is narrow. Each alloy has a characteristic temperature range within which it is sufficiently plastic for piercing to take place. Below this range, the central hole does not open up properly under the applied peripheral forces. Overheating may lead to cracked surfaces. Suggested piercing temperatures for various alloys are given below:

UNS number	Piercing temperature °C	Piercing temperature °F
C11000	815–870	1500–1600
C12200	815–870	1500–1600
C22000	815–870	1500–1600
C23000	815–870	1500–1600
C26000	760–790	1400–1450
C28000	705–760	1300–1400
C46400	730–790	1350–1450

Production of Finished Tubes

Cold drawing of extruded or pierced tube shells to smaller sizes is done on draw blocks for coppers and on draw benches for brasses and other alloys. With either type of machine, the metal is cold worked by pulling the tube through a die that reduces the diameter. Concurrently, wall thickness is reduced by drawing over a plug or mandrel that may be either fixed or floating. Cold drawing increases the strength of the material and simultaneously reduces ductility. Tube size is reduced—outside diameter, inside diameter, wall thickness, and cross-sectional area all are smaller after drawing. Because the metal work hardens, tubes may be annealed at intermediate stages when drawing to small sizes. However, coppers are so ductile that they frequently can be drawn to finished size without intermediate annealing.

Tube reducing is an alternative process for cold sizing of tube. In tube reducing, semicircular grooved dies are rolled or rocked back and forth along the tube while a tapered mandrel inside the tube controls the inside diameter and wall thickness. The process yields tube having very accurate dimensions and better concentricity than can be achieved by tube drawing.

The grooves in the tube-reducing dies are tapered, one end of the grooved section being somewhat larger than the outside diameter of the tube to be sized. As the dies are rocked, the tube is pinched against the tapered mandrel, which reduces wall thickness and increases tube length. The tube is fed longitudinally, and rotated on its axis to distribute the cold work uniformly around the circumference. Feeding and rotating are synchronized with die motion and take place after the dies have completed their forward stroke.

Tube reducing may be used for all alloys that can be drawn on draw benches. Slight changes in die design and operating conditions may be required to accommodate different alloys. Small-diameter tube may be produced by block or bench drawing following tube reducing.

Product Specifications

Copper tube and pipe are available in a wide variety of nominal diameters and wall thicknesses, from small-diameter capillary tube to 300 mm (12 in.) nominal-diameter pipe. To a certain extent, dimensions and tolerances for copper tube and pipe depend on the type of service for which they are intended. The standard dimensions and tolerances for several kinds of copper tube and pipe are given in the ASTM specifications listed in Table 6, along with other requirements for the tubular products. Seamless copper tube for automotive applications (⅛ to ¾ in. nominal diameter) is covered by SAE J528. Requirements for copper tube and pipe to be used in condensers, heat exchangers, economizers, and similar unfired pressure vessels are given in the ASME specifications listed in Table 6. (ASME materials specifications are almost always identical to ASTM specifications having the same numerical designation; for example, ASME SB111 is identical to ASTM B 111.) Certain tube alloys are covered in AMS specifications, which apply to materials for aerospace applications. These are given below:

AMS specification	Product	Copper alloy
4555	Seamless brass tube, light annealed	C26000 C33000
4558	Seamless brass tube drawn	C33200
4625	Phosphor bronze, hard temper	C51000
4640	Aluminum bronze	C63000
4665	Seamless silicon bronze tube, annealed	C65500

Wire and Cable

WIRE made from copper and its alloys has been used since about 3000 to 2000 BC. According to archaeological evidence, the ancient Assyrians, Babylonians, and Egyptians were skilled in producing copper wire for ornamental purposes. Drawing wire through a die is a much more modern development. The earliest evidence of drawn wire comes from sixth century AD Venetian and French artifacts. Theophilus, a German monk, produced the first written records in a treatise on metalworking circa 1110 to 1140 AD. His description of wiredrawing reads, in part, "Two pieces of iron three or four fingers wide, smaller at the top and bottom, rather thin, pierced with three or four holes through which wire may be drawn. . ." By 1270, a set of rules had been passed to govern wiredrawing in Paris, and at least nine wiredrawers were at work in that city.

Development of wiredrawing processes during the Middle Ages concentrated to a large extent on drawing iron and steel wires to make pins and instrument strings. But with the invention of the electric telegraph in 1847 came the requirement for long continuous lengths of electric conductor wire made of copper. In 1850, copper wire was used to make a submarine cable connecting England and France.

At the beginning of the twentieth century, wire was still being drawn through single dies—a process commonly known as "bull-block" drawing. Dies were made by punching a series of holes in a steel plate. These holes were then trimmed to final size with a master punch. Rows of single capstans, power driven from a common drive shaft, were used for drawing single lengths of wire. As each reduction was completed, the steel-plate die was replaced with one containing smaller holes until the final diameter was achieved.

Multiple wire-drawing machines were introduced about 1900. As a result, chilled cast iron plates and dies that could be reamed to size replaced the punch-sized steel-plate dies. Lubricants were introduced because of the considerable heat generated by friction between the wire and draw-capstan and by successive reductions through progressively smaller dies. In turn, use of lubricants permitted wire to be drawn at faster speeds.

Table 6 ASTM and ASME specifications for copper tube and pipe

Tubular product	ASTM	ASME
Seamless pipe and tube		
Seamless copper alloy (C69100) pipe and tube	B 706	. . .
Seamless pipe and tube, copper-nickel alloy(a)	B 466	SB466
	B 466M(a)	
Seamless pipe and tube, copper-silicon alloy	B 315	SB315
Seamless pipe and tube, for electrical conductors	B 188	. . .
Seamless pipe, standard sizes	B 42	. . .
Seamless pipe, threadless	B 302	. . .
Seamless tube		
Seamless copper alloy tubes (C19200 and C70600), for pressure applications	B 469	. . .
Seamless copper-nickel tubes, for desalting plants	B 552	. . .
Seamless tube(a)	B 75	SB75
	B 75M(a)	. . .
Seamless tube, brass(a)	B 135	SB135
	B 135M(a)	
Seamless tube, bright annealed(a)	B 68	. . .
	B 68M(a)	. . .
Seamless tube, capillary, hard drawn	B 360	. . .
Seamless tube, condenser and heat exchanger(a)	B 111, B 395	SB111, SB395
	B 111M(a)	. . .
	B 395M(a)	. . .
Seamless tube, condenser and heat exchanger, with integral fins(a)	B 359	SB359
	B 359M(a)	. . .
Seamless tube, for air conditioning and refrigeration service	B 280	. . .
Seamless tube, drainage	B 306	. . .
Seamless tube, general requirements(a)	B 251	. . .
	B 251M(a)	. . .
Seamless tube, rectangular waveguide	B 372	. . .
Seamless tube, water(a)	B 88	. . .
	B 88M(a)	. . .
Welded pipe and tube		
Hard temper welded copper tube (C21000), for general plumbing and fluid conveyance	B 642	. . .
Welded brass tube, for general application	B 587	. . .
Welded copper tube, for air conditioning-refrigeration	B 640	. . .
Welded pipe and tube, copper-nickel alloy	B 467	SB467
Welded tube, C10800 and 12000(a)	B 543	SB543
	B 543M(a)	. . .
Welded tube, all other coppers	B 447	. . .

(a) Suffix "M" indicates a metric specification.

The prime development during the 1920s was the introduction of drawing dies made of tungsten carbide. High hardness and lack of porosity made tungsten carbide dies ideal for high-speed wiredrawing and provided longer die life than was usually possible with dies made of chilled cast iron. Tungsten carbide dies are standard today. For very fine wire sizes, below about 1.3 mm (0.05 in.), diamond dies are used because they are harder and last longer than tungsten carbide dies. Some wire mills use diamond dies for high-speed drawing of larger wires, up to 8 mm (0.32 in.) in diameter, to reduce the frequency of shutdowns for die replacement.

Classification of Copper for Conductors

Copper metals used for electrical conductors fall into three general categories: high-conductivity coppers, high-copper alloys, and electrical bronzes.

High-conductivity coppers are covered by ASTM specifications B 4, B 5, B 170, B 442, and B 623. ASTM B 4 covers both high-resistance and low-resistance Lake copper. Lake copper is refined from Lake Superior ore deposits. ASTM B 5 covers copper electrolytically refined from blister copper, converter copper, black copper, or Lake copper. ASTM B 170 covers oxygen-free electrical copper.

Oxygen-free copper is produced by special manufacturing techniques and is used to avoid embrittlement where conductors are subjected to hydrogen or other reducing gases at elevated temperatures.

Some specialty coppers are produced by adding minimal amounts of hardening agents (such as chromium, tellurium, beryllium, cadmium, or zirconium). These are used in applications where high anneal resistance is required.

A series of bronzes has been developed for use as conductors; these alloys are covered by ASTM B 105. These bronzes are intended to provide better corrosion resistance and higher tensile strengths than standard conductor coppers. There are nine conductor bronzes, designated 8.5 to 85 in accordance with their electrical conductivities, as given below:

ASTM B 105 alloy designation	Alternative alloy types
8.5	Cu-Si-Fe
	Cu-Si-Mn
	Cu-Si-Zn
	Cu-Si-Sn-Fe
	Cu-Si-Sn-Zn
13	Cu-Al-Sn
	Cu-Al-Si-Sn
	Cu-Si-Sn
15	Cu-Al-Si
	Cu-Al-Sn
	Cu-Al-Si-Sn
	Cu-Si-Sn
20	Cu-Sn
30	Cu-Sn
	Cu-Zn-Sn
40(a)	Cu-Sn
	Cu-Sn-Cd
55(a)	Cu-Sn-Cd
65(a)	Cu-Sn
	Cu-Sn-Cd
80(a)	Cu-Cd
85	Cu-Cd

(a) Normally used for trolley-wire applications in either a round or grooved cross-sectional configuration, as set forth in ASTM B 9

The compositions of these alloys must be within the total limits prescribed in the following table, and no alloy may contain more than the allowed maximum of any constituent other than copper.

Element	Composition limit, % max
Fe	0.75
Mn	0.75
Cd	1.50
Si	3.00
Al	3.50
Sn	5.00
Zn	10.50
Cu	89.00 min
Sum of above elements	**99.50 min**

Classification of Wire and Cable

Round Wire. Standard nominal diameters and cross-sectional areas of solid round copper wires used as electrical conductors are prescribed in ASTM B 258. Wire sizes have almost always been designated in the American Wire Gauge (AWG) system. This system is based on fixed diameters for two wire sizes (4/0 and 36 AWG, respectively) with a geometric progression of wire diameters for the 38 intermediate gages and for gages smaller than 36 AWG (see Table 7). This is an inverse series in which a higher number denotes a smaller wire diameter. Each increase of one AWG number is approximately equivalent to a 20.7% reduction in cross-sectional area.

ASTM B 3 specifies soft (or annealed) copper wire with a maximum volumetric resistivity of 0.017241 $\Omega \cdot mm^2/m$ at 20 °C (68 °F), which corresponds to a maximum weight-basis resistivity of 875.20 $\Omega \cdot lb/mile^2$ when the density is 8.89 g/cm^3. This type of copper is used as the International Annealed Copper Standard (IACS) for electrical conductivity. Table 7 lists some properties of annealed copper wire for various AWG sizes. Tensile strengths are not specified for annealed copper wire.

Hard-drawn copper wire and hard-drawn copper alloy wire for electrical purposes are specified in ASTM B 1 and B 105, respectively. ASTM B 1 specifies hard-drawn round wire that has been reduced at least four AWG numbers (60% reduction in area). Table 8 lists the mechanical properties of hard-drawn copper wire and several hard-drawn copper alloy wires. The electrical resistivity and conductivity of these hard-drawn wires at 20 °C (68 °F) are as follows:

Alloy (hard drawn)	Maximum resistivity, $\Omega \cdot mm^2/m$	Maximum resistivity, $\Omega \cdot lb/mile^2$	Conductivity (volume basis), %IACS
Copper (ASTM B 1) wire with diameter of:			
8.25 to 11.68 mm (0.325 to 0.460 in.)	0.017745	900.77	97.16
1.02 to <8.25 mm (0.0403 to <0.325 in.)	0.017930	910.15	96.16
Copper alloys (ASTM B 105):			
C65100	0.20284	10 169.0	8.5
C51000	0.13263	6649.0	13
C50700	0.057471	2917.3	30
C16500	0.031348	1591.3	55
C19600	0.023299	1182.7	74
C16200	0.021552	1094.0	80

Table 7 Sizes of round wire in the American Wire Gauge (AWG) system and the properties of solid annealed copper wire (ASTM B 3)

					Annealed copper (ASTM B 3)			
	Conductor diameter		Conductor area at 20 °C (68 °F)		Net weight(a)		Elongation(b)	Nominal resistance(c)
Conductor size, AWG	mm	in.	mm^2	circular mils	kg/km	lb/1000 ft	%	Ω/1000 ft (305 m)
4/0	11.684	0.4600	107.0	211 600	953.2	640.5	35	0.0490
3/0	10.404	0.4096	85.0	167 800	755.7	507.8	35	0.06180
2/0	9.266	0.3648	67.4	133 100	599.4	402.8	35	0.07792
1/0	8.252	0.3249	53.5	105 600	475.5	319.5	35	0.09821
1	7.348	0.2893	42.4	83 690	377.0	253.3	30	0.1239
2	6.543	0.2576	33.6	66 360	299.0	200.9	30	0.1563
3	5.827	0.2294	26.7	52 620	237.1	159.3	30	0.1971
4	5.189	0.2043	21.2	41 740	188.0	126.3	30	0.2485
5	4.620	0.1819	16.8	33 090	149.1	100.2	30	0.3134
6	4.115	0.1620	13.3	26 240	118.2	79.44	30	0.3952
7	3.665	0.1443	10.5	20 820	93.8	63.03	30	0.4981
8	3.264	0.1285	8.37	16 510	74.4	49.98	30	0.6281
9	2.906	0.1144	6.63	13 090	59.0	39.62	30	0.7923
10	2.588	0.1019	5.26	10 380	46.8	31.43	25	0.9992
11	2.304	0.0907	4.17	8 230	37.1	24.9	25	1.26
12	2.052	0.0808	3.31	6 530	29.5	19.8	25	1.59
13	1.829	0.0720	2.63	5 180	23.4	15.7	25	2.00
14	1.628	0.0641	2.08	4 110	18.5	12.4	25	2.52
15	1.450	0.0571	1.65	3 260	14.7	9.87	25	3.18
16	1.290	0.0508	1.31	2 580	11.6	7.81	25	4.02
17	1.151	0.0453	1.04	2 050	9.24	6.21	25	5.06
18	1.024	0.0403	0.823	1 620	7.32	4.92	25	6.40
19	0.912	0.0359	0.654	1 290	5.80	3.90	25	8.04
20	0.813	0.0320	0.517	1 020	4.61	3.10	25	10.2
21	0.724	0.0285	0.411	812	3.66	2.46	25	12.8
22	0.643	0.0253	0.324	640	2.89	1.94	25	16.2
23	0.574	0.0226	0.259	511	2.31	1.55	25	20.3
24	0.511	0.0201	0.205	404	1.82	1.22	20	25.7
25	0.455	0.0179	0.162	320	1.44	0.970	20	32.4
26	0.404	0.0159	0.128	253	1.14	0.765	20	41.0
27	0.361	0.0142	0.102	202	0.908	0.610	20	51.4
28	0.320	0.0126	0.081	159	0.716	0.481	20	65.2
29	0.287	0.0113	0.065	128	0.576	0.387	20	81.0
30	0.254	0.0100	0.051	100	0.451	0.303	15	104.0
31	0.226	0.0089	0.040	79.2	0.357	0.240	15	131.0
32	0.203	0.0080	0.032	64.0	0.289	0.194	15	162.0
33	0.180	0.0071	0.026	50.4	0.228	0.153	15	206.0
34	0.160	0.0063	0.020	39.7	0.179	0.120	15	261.0
35	0.142	0.0056	0.016	31.4	0.141	0.0949	15	330.0
36	0.127	0.0050	0.013	25.0	0.113	0.0757	15	415.0
37	0.114	0.0045	0.010	20.2	0.0912	0.0613	15	513.0
38	0.102	0.0040	0.0081	16.0	0.0720	0.0484	15	648.0
39	0.089	0.0035	0.0062	12.2	0.0552	0.0371	15	850.0
40	0.079	0.0031	0.0049	9.61	0.0433	0.0291	15	1079.0
41	0.071	0.0028	0.0040	7.84	0.0353	0.0237	15(d)	1323.0
42	0.0635	0.0025	0.0032	6.25	0.0281	0.0189	15(d)	1659.0
43	0.056	0.0022	0.0023	4.48	0.0219	0.0147	15(d)	2143.0
44	0.050	0.0020	0.0020	4.00	0.0180	0.0121	15(d)	2593.0
45	0.045	0.00176	0.0016	3.10	0.0140	0.00938	(d)	3345.6
46	0.040	0.00157	0.00125	2.46	0.0111	0.00745	(d)	4216.0
47	0.036	0.00140	0.00099	1.96	0.00882	0.00593	(d)	5291.6
48	0.031	0.00124	0.00078	1.54	0.00673	0.00466	(d)	6734.7
49	0.028	0.00111	0.00062	1.23	0.00554	0.00372	(d)	8432.1
50	0.025	0.00099	0.00050	0.980	0.00442	0.00297	(d)	10583
51	0.022	0.00088	0.00039	0.774	0.00348	0.00234	(d)	13400
52	0.020	0.00078	0.00031	0.608	0.00274	0.00184	(d)	17058
53	0.018	0.00070	0.00025	0.490	0.00220	0.00148	(d)	21166
54	0.016	0.00062	0.00019	0.384	0.00173	0.00116	(d)	27009
55	0.014	0.00055	0.00015	0.302	0.00136	0.000914	(d)	34342
56	0.012	0.00049	0.00012	0.240	0.00108	0.000726	(d)	43214

(a) Based on a density of 8.89 g/cm^3 at 20 °C (68 °F). (b) Minimum elongation in 250 mm (10 in.). (c) Based on a resistivity value of 0.017241 Ω · mm^2/m (875.20 Ω · $lb/mile^2$), which is the resistivity for the International Annealed Copper Standard (IACS) of electrical conductivity. (d) Elongation not specified in ASTM B 3

Square and Rectangular Wire. ASTM B 48 specifies soft (annealed) square and rectangular copper wire.

Stranded wire is normally used in electrical applications where some degree of flexing is encountered either in service or during installation. In order of increasing flexibility, the common forms of stranded wire are: concentric lay, unilay, rope lay, and bunched.

Concentric-lay stranded wire and cable are composed of a central wire surrounded by one or more layers of helically laid wires, with the direction of lay reversed in successive layers, and with the length of lay increased for each successive layer. The outer layer usually has a left-hand lay.

ASTM B 8 establishes five classes of concentric-lay stranded wire and cable, from AA (the coarsest) to D (the finest). Details of concentric-lay constructions are given in Table 9.

Unilay stranded wire is composed of a central core surrounded by more than one layer of helically laid wires, all layers having a common lay length and direction. This type of wire sometimes is referred to as "smooth bunch." The layers usually have a left-hand lay.

Rope-lay stranded wire and cable are composed of a stranded member (or members) as a central core, around which are laid one or more helical layers of similar stranded members. The members may be concentric or bunch-stranded. ASTM B 173 and B 172 establish five classes of rope-lay

Table 8 Tensile properties of hard-drawn copper and copper alloy round wire

Conductor size, AWG	Hard-drawn copper wire (ASTM B 1): Nominal tensile strength(a), MPa	ksi	Nominal elongation(b), %	Nominal breaking strength, N	lbf	Minimum tensile strength of hard-drawn copper alloy wire (ASTM B 105): C65100, MPa	ksi	C51000, MPa	ksi	C50700, MPa	ksi	C16500, MPa	ksi	C19600, MPa	ksi	C16200, MPa	ksi	ASTM B 105 Minimum elongation(b), %
4/0	340	49.0	3.8	36220	8143	...	...	...	...	...	...	...	...	...	...	...	...	...
3/0	350	51.0	3.3	29900	6720	...	...	...	...	...	...	...	...	...	...	...	...	...
2/0	365	52.8	2.8	24550	5519	...	...	...	...	...	...	...	...	...	...	...	...	...
1/0	375	54.5	2.4	20095	4518	...	...	...	...	...	...	...	...	...	...	...	...	...
1	385	56.1	2.2	17290	3888	672	97.5	707	102.5	510	74.0	524	76.0	510	74.0	496	72.0	2.2
2	395	57.6	2.0	13350	3002	716	103.8	750	108.8	552	80.0	536	77.8	520	75.5	507	73.5	2.0
3	405	59	1.8	10850	2439	741	107.5	776	112.5	586	85.0	547	79.3	534	77.5	517	75.0	1.8
4	415	60.1	1.7	8762	1970	760	110.2	794	115.2	614	89.0	558	80.9	545	79.0	527	76.4	1.6
5	420	61.2	1.6	7072	1590	774	112.2	808	117.2	638	92.5	568	82.4	552	80.0	534	77.5	1.5
6	430	62.1	1.4	5693	1280	786	114.0	820	119.0	654	94.8	579	84.0	558	81.0	542	78.6	1.4
7	435	63	1.3	4580	1030	795	115.3	829	120.3	665	96.5	590	85.5	568	82.4	550	79.8	1.3
8	440	63.7	1.3	3674	826.1	804	116.6	836	121.6	675	97.9	600	87.0	576	83.5	558	81.0	1.3
9	445	64.3	1.2	2940	660.9	812	117.8	847	122.8	683	99.0	610	88.5	583	84.6	567	82.2	1.2
10	445	64.9	1.2	2354	529.3	820	118.9	854	123.9	690	100.1	620	90.0	590	85.5	575	83.4	1.2
11	450	65.4	1.1	1880	423	826	119.8	860	124.8	698	101.2	630	91.3	597	86.6	583	84.6	1.2
12	455	65.7	1.1	1500	337	832	120.6	866	125.6	705	102.2	638	92.6	605	87.7	591	85.7	1.1
13	455	65.9	1.1	1190	268	836	121.2	870	126.2	710	103.0	647	93.8	612	88.8	598	86.8	1.1
14	455	66.2	1.0	952	214	839	121.7	874	126.7	715	103.7	655	95.0	619	89.8	605	87.8	1.1
15	460	66.4	1.0	756	170	843	122.2	877	127.2	720	104.4	662	96.0	625	90.6	612	88.7	1.0
16	460	66.6	1.0	600	135	845	122.5	879	127.5	725	105.2	669	97.0	634	92.0	617	89.5	1.0
17	460	66.8	1.0	480	108	847	122.8	881	127.8	730	105.9	676	98.0	640	92.8	623	90.3	1.0
18	460	67.0	1.0	380	85.5	848	123.0	883	128.0	735	106.6	680	98.6	645	93.5	627	91.0	0.9
19	463	67.2	...	302	68.0	849	123.2	884	128.2	740	107.3	683	99.0	648	94.0	632	91.6	0.9
20	465	67.4	...	241	54.2	852	123.5	886	128.5	745	108.0	686	99.5	652	94.5	636	92.2	0.9
21	467	67.7	...	192	43.2	...	...	...	...	...	...	...	...	...	...	...	...	...
22	468	67.9	...	152	34.1	...	...	...	...	...	...	...	...	...	...	...	...	...
23	470	68.1	...	121	27.3	...	...	...	...	...	...	...	...	...	...	...	...	...
24	471	68.3	...	96.5	21.7	...	...	...	...	...	...	...	...	...	...	...	...	...
25	473	68.6	...	77.0	17.3	...	...	...	...	...	...	...	...	...	...	...	...	...
26	474	68.8	...	60.9	13.7	...	...	...	...	...	...	...	...	...	...	...	...	...
27	476	69.0	...	48.5	10.9	...	...	...	...	...	...	...	...	...	...	...	...	...
28	478	69.3	...	38.4	8.64	...	...	...	...	...	...	...	...	...	...	...	...	...
29	479	69.4	...	31.0	6.96	...	...	...	...	...	...	...	...	...	...	...	...	...
30	481	69.7	...	24.3	5.47	...	...	...	...	...	...	...	...	...	...	...	...	...
31	482	69.9	...	19.3	4.35	...	...	...	...	...	...	...	...	...	...	...	...	...
32	484	70.2	...	15.7	3.53	...	...	...	...	...	...	...	...	...	...	...	...	...
33	485	70.4	...	12.4	2.79	...	...	...	...	...	...	...	...	...	...	...	...	...
34	487	70.6	...	9.79	2.20	...	...	...	...	...	...	...	...	...	...	...	...	...
35	489	70.9	...	7.78	1.75	...	...	...	...	...	...	...	...	...	...	...	...	...
36	490	71.1	...	6.23	1.40	...	...	...	...	...	...	...	...	...	...	...	...	...
37	492	71.3	...	5.03	1.13	...	...	...	...	...	...	...	...	...	...	...	...	...
38	493	71.5	...	4.39	0.898	...	...	...	...	...	...	...	...	...	...	...	...	...
39	495	71.8	...	3.07	0.691	...	...	...	...	...	...	...	...	...	...	...	...	...
40–44	496	72.0	...	2.42–1.00	0.543–0.226	...	...	...	...	...	...	...	...	...	...	...	...	...

(a) Tensile strengths cannot always be met if wire is drawn into coils of less than 480 mm (19 in.). (b) Elongation in 250 mm (10 in.)

stranded conductors: classes G and H, which have concentric members; and classes I, K, and M, which have bunched members. Construction details are shown in Tables 10 and 11. These cables are normally used to make large, flexible conductors for portable service, such as mining cable or apparatus cable.

Bunch stranded wire is composed of any number of wires twisted together in the same direction without regard to geometric arrangement of the individual strands. ASTM B 174 provides for five classes (I, J, K, L, and M); these conductors are commonly used in flexible cords, hookup wires, and special flexible welding conductors. Typical construction details are given in Table 12.

Tin-Coated Wire. Solid and stranded wires are available with tin coatings. These are manufactured to the latest revisions of ASTM B 33, which covers soft or annealed tinned-copper wires, and B 246, which covers hard-drawn or medium-hard-drawn tinned-copper wires. Characteristics of tinned, round, solid wire are given in Table 13.

Fabrication of Wire Rod

Wire rod is the intermediate product in the manufacture of wire. Although wire rod is the term used in the U.S. for the intermediate product, the term drawing stock is used in international standards and customs documents.

Rolling. The traditional process for converting prime copper into wire rod involves hot rolling of cast wirebar. Almost all drawing stock is rolled to 8 mm (0.32 in.) diameter. Larger sizes, up to 22 mm (0.87 in.) or more in diameter are available on special order.

Some special oxygen-free copper wirebar is produced by vertical casting, but most wirebar is produced by horizontal casting of tough-pitch copper into open molds. The oxygen content is controlled at 0.03 to 0.06% to give a level surface. Cast wirebars weigh about 110 to 135 kg (250 to 300 lb) each. Their ends are tapered to facilitate entry into the first pass of the hot rolling mill.

Prior to rolling, bars are heated to about 925 °C (1700 °F) in a neutral atmosphere and then rolled on a continuous mill through a series of reductions to yield round rod about 6 to 22 mm (¼ to ⅞ in.) in diameter. The hot-rolled rod is coiled, water-quenched, and then pickled to remove the black cupric oxide that forms during rolling. This method can produce rod at rates up to 7.5 kg/s (30 tons/h).

Disadvantages of this process include:

- High capital investment to achieve low operating cost
- Relatively small coils that must be welded together for efficient production,

Table 9 Characteristics of concentric-lay stranded copper conductors specified in ASTM B 8

Conductor size, circular mils or AWG	Nominal weight, lb/1000 ft(a)	Nominal resistance(b), Ω/1000 ft(a)	Class AA		Class A		Class B		Class C		Class D	
			Number of wires	Diameter of individual wires, mils(a)	Number of wires	Diameter of individual wires, mils(a)	Number of wires	Diameter of individual wires, mils(a)	Number of wires	Diameter of individual wires, mils(a)	Number of wires	Diameter of individual wires, mils(a)
5 000 000	15 890	0.002 178	...	...	169	172.0	217	151.8	271	135.8	271	135.8
4 500 000	14 300	0.002 420	...	...	169	163.2	217	144.0	271	128.9	271	128.9
4 000 000	12 590	0.002 696	...	...	169	153.8	217	135.8	271	121.5	271	121.5
3 500 000	11 020	0.003 082	...	...	127	166.0	169	143.9	217	127.0	271	113.6
3 000 000	9 353	0.003 561	...	...	127	153.7	169	133.2	217	117.6	271	105.2
2 500 000	7 794	0.004 278	...	...	91	165.7	127	140.3	169	121.6	217	107.3
2 000 000	6 175	0.005 289	...	...	91	148.2	127	125.5	169	108.8	217	96.0
1 900 000	5 886	0.005 568	...	...	91	144.5	127	122.3	169	106.0	217	93.6
1 800 000	5 558	0.005 877	...	...	91	140.6	127	119.1	169	103.2	217	91.1
1 750 000	6 403	0.006 045	...	...	91	138.7	127	117.4	169	101.8	217	89.8
1 700 000	5 249	0.006 223	...	...	91	136.7	127	115.7	169	100.3	217	88.5
1 600 000	4 940	0.006 612	...	...	91	132.6	127	112.2	169	97.3	217	85.9
1 500 000	4 631	0.007 052	...	...	61	156.6	91	128.4	127	108.7	169	94.2
1 400 000	4 323	0.007 556	...	...	61	151.5	91	124.0	127	105.0	169	91.0
1 300 000	4 014	0.008 137	...	...	61	146.0	91	119.5	127	101.2	169	87.7
1 250 000	3 859	0.008 463	...	...	61	143.1	91	117.2	127	99.2	169	86.0
1 200 000	3 705	0.008 815	...	...	61	140.3	91	114.8	127	97.2	169	84.3
1 100 000	3 396	0.009 617	...	...	61	134.3	91	109.9	127	93.1	169	80.7
1 000 000	3 088	0.010 88	37	164.4	61	128.0	61	128.0	91	104.8	127	88.7
900 000	2 779	0.011 75	37	156.0	61	121.5	61	121.5	91	99.4	127	84.2
800 000	2 470	0.013 22	37	147.0	61	114.5	61	114.5	91	93.8	127	79.4
750 000	2 316	0.014 10	37	142.4	61	110.9	61	110.9	91	90.8	127	76.8
700 000	2 161	0.015 11	37	137.5	61	107.1	61	107.1	91	87.7	127	74.2
650 000	2 007	0.016 27	37	132.5	61	103.2	61	103.2	91	84.5	127	71.5
600 000	1 853	0.017 63	37	127.3	37	127.3	61	99.2	91	81.2	127	68.7
550 000	1 698	0.019 23	37	121.9	37	121.9	61	95.0	91	77.7	127	65.8
500 000	1 544	0.021 16	19	162.2	37	116.2	37	116.2	61	90.5	91	74.1
450 000	1 389	0.023 51	19	153.9	37	110.3	37	110.3	61	85.9	91	70.3
400 000	1 235	0.026 45	19	145.1	19	145.1	37	104.0	61	81.0	91	66.3
350 000	1 081	0.030 22	12	170.8	19	135.7	37	97.3	61	75.7	91	62.0
300 000	926.3	0.035 26	12	158.1	19	125.7	37	90.0	61	70.1	91	57.4
250 000	771.9	0.042 31	12	144.3	19	114.6	37	82.2	61	64.0	91	52.4
4/0	653.3	0.049 99	7	173.9	7	173.9	19	105.5	37	75.6	61	58.9
3/0	518.1	0.063 04	7	154.8	7	154.8	19	94.0	37	67.3	61	52.4
2/0	410.9	0.079 48	7	137.9	7	137.9	19	83.7	37	60.0	61	46.7
1/0	326.0	0.100 2	7	122.8	7	122.8	19	74.5	37	53.4	...	...
1	258.4	0.126 4	3	167.0	7	109.3	19	66.4	37	47.6	...	...
2	204.9	0.159 4	3	148.7	7	97.4	7	57.4	19	59.1	...	...
3	162.5	0.201 0	3	132.5	7	86.7	7	86.7	19	52.6	...	...
4	128.9	0.253 4	3	118.0	7	77.2	7	77.2	19	48.9	...	...
5	102.2	0.319 7	...	...	...	...	7	68.8	19	41.7	...	...
6	81.05	0.403 1	...	...	...	...	7	61.2	19	37.2	...	...
7	64.28	0.508 1	...	...	...	...	7	54.5	19	33.1	...	...
8	50.98	0.640 7	...	...	...	...	7	48.6	19	29.5	...	...
9	40.42	0.808 1	...	...	...	...	7	43.2	19	28.2	...	...

(a) Units used in ASTM B 8 specification; see "Metric Conversion Guide" for conversions. (b) Uncoated wire

- where the welded junctions present potential sources of weakness in subsequent wire-drawing operations
- Unsuitability of rod rolled from cast wirebars for certain specialized wire applications

Because of the disadvantages inherent in producing rolled rod from conventionally cast wirebars, processes have been developed for continuously converting liquid metal directly into wire rod, thus avoiding the intermediate wirebar stage.

Continuous Casting. In 1963, the first plant for continuous casting and rolling of copper wirebar began operation. The plant was operated by the Southwire Company in conjunction with Western Electric Company. The Southwire process was an adaptation of a process developed by Ilario Properzi prior to 1950—a process that had been used for many years by aluminum and zinc producers for conversion of prime metal or scrap into wire rod.

Continuously cast wire rod has come to dominate the copper wire rod market and now accounts for more than 1.8×10^6 Mg (2 million tons) annually, or about 50% of the total amount of wire rod produced.

Other casting systems that have been developed include the Hazelett Strip-Casting Corporation and General Electric Company dip-form systems. Smaller-capacity machines have been developed by Outokumpu Metals Inc., Davy Corporation, Wertli, Lamitref, and others.

Advantages of continuous casting and rolling include:

- Large coil weights, up to 10 Mg (11 tons)
- Ability to reprocess scrap at considerable savings
- Improved rod quality and surface condition
- Homogeneous metallurgical conditions and close process control
- Low capital investment and low operating costs for moderate production rates

The standard feed for continuous casting processes is cathode copper, which is charged directly into a melting furnace. An ASARCO, Inc. shaft furnace is used for the Southwire/Properzi and Hazelett systems, but an electric furnace is preferred for the smaller systems, such as the GE dip-form process.

Southwire/Properzi Continuous Rod System. In this system, the casting machine produces a cast copper bar 2500 to 3000 mm^2 (4 to 5 in.^{2}) in cross-sectional area by pouring molten copper between a grooved wheel made of steel or copper and a steel band. The cast bar is cooled by water sprays as the wheel rotates, and is withdrawn by pinch rolls as it exits from the wheel. Next

Table 10 Characteristics of rope-lay stranded copper conductors having uncoated or tinned concentric members specified in ASTM B 173

Conductor sizes, circular mils or AWG	Class G: Diameter of individual wires, mils(a)	Class G: Number of ropes	Class G: Number of wires each rope	Class G: Net weight, lb/1000 ft(a)	Class H: Diameter of individual wires, mils(a)	Class H: Number of ropes	Class H: Number of wires each rope	Class H: Net weight, lb/1000 ft(a)
5 000 000	65.7	61	19	16 052	53.8	91	19	15 057
4 500 000	62.3	61	19	14 434	51.0	91	19	14 429
4 000 000	58.7	61	19	12 814	48.1	91	19	12 835
3 500 000	55.0	61	19	11 249	45.0	91	19	11 234
3 000 000	50.9	61	19	9 635	41.7	91	19	9 647
2 500 000	59.6	37	19	8 012	46.4	61	19	8 006
2 000 000	53.3	37	19	6 408	41.5	61	19	6 405
1 900 000	52.0	37	19	6 099	40.5	61	19	6 100
1 800 000	50.6	37	19	5 775	39.4	61	19	5 773
1 750 000	49.9	37	19	5 617	38.9	61	19	5 627
1 700 000	49.2	37	19	5 460	38.3	61	19	5 455
1 600 000	47.7	37	19	5 132	37.2	61	19	5 146
1 500 000	59.3	61	7	4 772	46.2	37	19	4 815
1 400 000	57.3	61	7	4 456	44.6	37	19	4 487
1 300 000	55.2	61	7	4 135	43.0	37	19	4 171
1 250 000	54.1	61	7	3 972	42.2	37	19	4 017
1 200 000	53.0	61	7	3 814	41.3	37	19	3 847
1 100 000	50.8	61	7	3 502	39.6	37	19	3 537
1 000 000	48.4	61	7	3 179	37.7	37	19	3 206
900 000	45.9	61	7	2 859	35.8	37	19	2 891
800 000	43.3	61	7	2 544	33.7	37	19	2 562
750 000	41.9	61	7	2 383	32.7	37	19	2 412
700 000	40.5	61	7	2 226	31.6	37	19	2 252
650 000	39.0	61	7	2 064	30.4	37	19	2 085
600 000	37.5	61	7	1 908	29.2	37	19	1 923
550 000	35.9	61	7	1 749	28.0	37	19	1 768
500 000	43.9	37	7	1 579	34.2	61	7	1 587
450 000	41.7	37	7	1 425	32.5	61	7	1 433
400 000	39.3	37	7	1 265	30.6	61	7	1 271
350 000	36.8	37	7	1 109	28.6	61	7	1 110
300 000	34.0	37	7	947.1	26.5	61	7	953.0
250 000	31.1	37	7	792.4	24.2	61	7	794.8
4/0	39.9	19	7	666.6	28.6	37	7	670.1
3/0	35.5	19	7	527.7	25.5	37	7	532.7
2/0	31.6	19	7	418.1	22.7	37	7	422.2
1/0	28.2	19	7	333.0	20.7	37	7	334.3
1	25.1	19	7	263.8	18.0	37	7	265.4
2	36.8	7	7	206.9	22.3	19	7	208.2
3	37.8	7	7	164.4	19.9	19	7	165.8
4	29.2	7	7	130.3	17.7	19	7	131.2
5	26.0	7	7	103.3	15.8	19	7	104.5
6	23.1	7	7	81.52	14.0	19	7	82.06
7	20.6	7	7	64.83	12.5	19	7	65.42
8	18.4	7	7	51.72	11.1	19	7	51.59
9	15.3	7	7	40.59	9.9	19	7	41.04
10	14.6	7	7	32.57	...	...	...	...
12	11.5	7	7	20.20	...	...	...	...
14	9.2	7	7	12.93	...	...	...	...

(a) Units used in ASTM B 173; see "Metric Conversion Guide" for conversion multipliers.

it goes to a bar-conditioning unit, and then to the rolling mill. After passing through a series of reductions in the mill, the rod enters an in-line pickling system, which quenches and cleans the rod prior the final coiling.

Ge Dip-Form Process. The GE dip-form process was introduced in 1964. A seed rod approximately 9 mm (0.35 in.) in diameter is passed at a controlled rate through a bath of molten C10100. Copper freezes onto the seed rod, thickening its diameter to about 16.5 mm (0.65 in.). The rod emerging from the copper bath is hot rolled on a 2-stand mill, cooled, and coiled. The entire operation is performed in a controlled atmosphere, from the time C10100 cathodes enter the melting furnace until the rod emerges. This rod is generally used for production of fine and ultrafine wires. Production rates up to 2.5 kg/s (10 tons/h) can be obtained.

Hazelett Process. This process was originally developed in 1957 for zinc slab, and was further refined in conjunction with Métallurgie Hoboken-Overpelt (Belgium) to make it applicable to copper wire rod. Molten copper is passed between two water-cooled, counter rotating steel belts having specially cooled side dams. The resulting bar is sent through a conventional Krupp rolling mill similar to that used in the Southwire system. Production rates of about 6.3 to 7.5 kg/s (25 to 30 tons/h) have been achieved.

Outokumpu Process. In the Outokumpu process, metal is pulled through a graphite die where it solidifies. One end of the die extends into the melt and the other is surrounded by a water-cooled jacket. The Outokumpu process is unique in that the direction of withdrawal is vertically upwards. A 12-strand plant can produce up to 13,000 Mg (14,000 tons) of oxygen-free rod annually.

Wiredrawing and Wire Stranding

Preparation of Rod. In order to provide a wire of good surface quality, it is necessary to have clean wire rod with a smooth, oxide-free surface. Conventional hot-rolled rod must be cleaned in a separate operation, but with the advent of continuous casting, which provides better surface quality, a separate cleaning operation is not required. Instead, the rod passes through a cleaning station as it exits from the rolling mill.

Table 11 Characteristics of rope-lay stranded copper conductors having uncoated or tinned bunched members specified in ASTM B 172

Conductor size, circular mils or AWG	Class of strand	Construction and wire size, AWG	Total number of wires	Approximate diameter, in.(a)	Net weight, lb/1000 ft(a)	Conductor size, circular mils or AWG	Class of strand	Construction and wire size, AWG	Total number of wires	Approximate diameter, in.(a)	Net weight, lb/1000 ft(a)
1 000 000	I	19×7×19/24	2 527	1.290	3306	250 000	K	7×7×61/30	2 499	0.638	802
	K	37×7×39/30	10 101	1.329	3272		M	19×7×48/34	6 384	0.658	821
	M	61×7×59/34	25 193	1.353	3239	4/0	I	19×28/24	532	0.569	683
900 000	I	19×7×17/24	2 261	1.217	2959		K	7×7×43/30	2 107	0.584	676
	K	37×7×35/30	9 065	1.255	2936		M	19×7×40/34	5 320	0.598	684
	M	61×7×53/34	22 631	1.279	2909	3/0	I	19×22/24	418	0.502	537
800 000	I	19×7×15/24	1 995	1.140	2611		K	7×7×34/30	1 666	0.516	535
	K	19×7×60/30	7 980	1.174	2585		M	19×7×32/34	4 256	0.532	547
	M	61×7×47/34	20 069	1.200	2580	2/0	I	19×18/24	342	0.452	439
750 000	I	19×7×14/24	1 862	1.099	2437		K	7×7×27/30	1 323	0.457	424
	K	19×7×57/30	7 581	1.143	2455		M	19×7×25/34	3 325	0.467	427
	M	61×7×44/34	18 788	1.160	2415	1/0	I	19×14/24	266	0.396	342
700 000	I	19×7×13/24	1 729	1.057	2262		K	19×56/30	1 064	0.408	338
	K	19×7×52/30	6 916	1.089	2240		M	7×7×54/34	2 646	0.414	337
	M	61×7×41/34	17 507	1.117	2251	1	I	7×30/24	210	0.350	267
650 000	I	19×7×12/24	1 596	1.014	2088		K	19×44/30	836	0.359	266
	K	19×7×49/30	6 517	1.056	2111		M	7×7×43/34	2 107	0.368	268
	M	61×7×38/34	16 226	1.074	2086	2	I	7×23/24	161	0.304	205
600 000	I	7×7×30/24	1 470	0.971	1906		K	19×35/30	665	0.319	211
	K	19×7×45/30	5 985	1.010	1938		M	7×7×34/34	1 666	0.325	212
	M	61×7×35/34	14 945	1.028	1921	3	I	7×19/24	133	0.275	169
550 000	I	7×7×28/24	1 372	0.936	1779		K	19×28/30	532	0.283	169
	K	19×7×41/30	5 453	0.961	1766		M	7×7×27/34	1 323	0.288	168
	M	61×7×32/34	13 664	0.981	1757	4	I	7×15/24	105	0.243	134
500 000	I	7×7×25/24	1 225	0.882	1588		K	7×60/30	420	0.250	132
	K	19×7×38/30	5 054	0.924	1637		M	19×56/34	1 064	0.257	134
	M	37×7×49/34	12 691	0.900	1631	5	I	7×12/24	84	0.216	107
450 000	I	7×7×23/24	1 127	0.845	1461		K	7×48/30	336	0.223	106
	K	19×7×34/30	4 522	0.871	1465		M	19×44/34	836	0.226	105
	M	37×7×44/34	11 396	0.892	1465	6	I	7×9/24	63	0.186	80
400 000	I	7×7×20/24	980	0.785	1270		K	7×38/30	266	0.197	84
	K	19×7×30/30	3 990	0.816	1292		M	19×35/34	665	0.201	84
	M	37×7×39/34	10 101	0.837	1298	7	K	7×30/30	210	0.174	66
350 000	I	7×7×18/24	882	0.743	1143		M	19×28/34	532	0.178	67
	K	19×7×26/30	3 458	0.757	1120	8	K	7×/30	168	0.155	53
	M	37×7×34/34	8 806	0.779	1132		M	7×60/34	420	0.158	53
300 000	I	7×7×15/24	735	0.675	953	9	K	7×19/30	133	0.137	42
	K	7×7×61/30	2 989	0.701	959		M	7×48/34	336	0.140	42
	M	19×7×57/34	7 581	0.720	975	10	M	7×37/34	259	0.122	33
250 000	I	7×7×13/24	637	0.626	826	12	M	7×24/34	168	0.097	21

(a) See "Metric Conversion Guide" for conversion into metric units.

The standard method for cleaning copper wire rod is pickling in hot 20% sulfuric acid followed by rinsing in water. When fine wire is being produced, it is necessary to provide rod of even better surface quality. This can be achieved in a number of ways. One is by open-flame annealing of cold-drawn rod—that is, heating to 700 °C (1300 °F) in an oxidizing atmosphere. This eliminates shallow discontinuities. A more common practice, especially for fine magnet-wire applications, is die shaving, where rod is drawn through a circular cutting die made of steel or carbide to remove approximately 0.13 mm (0.005 in.) from the entire surface of the rod. A further refinement of this cleaning operation for rod made from conventionally cast wirebar involves scalping the top surface of cast wirebar and subsequently die shaving the hot-rolled bar.

Wiredrawing. Single-die machines called bull blocks are used for drawing special heavy sections such as trolley wire. Drawing speeds range from about 1 to 2.5 m/s (200 to 500 ft/min). Tallow is generally used as the lubricant, and the wire is drawn through hardened steel or tungsten carbide dies. In some instances, multiple-draft tandem bull blocks (in sets of 3 or 5 passes) are used instead of single-draft machines.

Tandem drawing machines having 10 to 12 dies for each machine are used for breakdown of hot-rolled or continuous-cast copper rod. The rod is reduced in diameter from 8.3 mm (0.325 in.) to about 2 mm (0.08 in.) by drawing it through dies at speeds up to 25 m/s (5000 ft/min). The drawing machine operates continuously; the operator merely welds the end of each rod coil to the start of the next coil.

Several newer continuous extrusion processes for production of wire are currently under development. These include the Conform process. Data indicate so far that copper wire can be produced successfully by these techniques.

Production of Flat or Rectangular Wire. Depending on size and quantity, flat or rectangular wire is drawn on bull block machines or Turk's Head machines, or is rolled on tandem rolling mills with horizontal and vertical rolls. Larger quantities are produced by rolling, smaller quantities by drawing.

Annealing. Wiredrawing, like any other cold-working operation, increases tensile strength and reduces ductility of copper. Although it is possible to cold work copper up to 99% reduction in area, copper wire usually is annealed after 90% reduction.

In some plants, electrical-resistance heating methods are used to fully anneal copper wire as it exits from the drawing machines. Wire coming directly from drawing passes over suitably spaced contact pulleys that carry the electrical current necessary to heat the wire above its recrystallization temperature in less than a second.

In plants where batch annealing is practiced, drawn wire is treated either in a continuous tunnel furnace, where reels travel through a neutral or slightly reducing atmosphere and are annealed during transit, or in batch bell furnaces under a similar protective atmosphere. Annealing temperatures range from 400 to 600 °C (750 to 1100 °F) depending chiefly on wire diameter and reel weight.

Wire Coating. Four basic coatings are used on copper conductors for electrical applications:

Table 12 Characteristics of bunch stranded copper conductors having uncoated or tinned members specified in ASTM B 174

Conductor size, AWG	Class of strand	Number and size of wire, AWG	Approximate diameter, in.(a)	Approximate weight, lb/1000 ft(a)
7	I	52/24	0.168	64.9
8	I	41/24	0.148	51.1
9	I	33/24	0.132	41.2
10	I	26/24	0.117	32.4
	J	65/28	0.118	31.9
	K	104/30	0.120	32.1
12	J	41/28	0.093	20.1
	K	65/30	0.094	20.1
	L	104/32	0.096	20.6
14	J	26/28	0.073	12.7
	K	41/30	0.074	12.7
	L	65/32	0.075	12.8
	M	104/34	0.076	12.7
16	J	16/28	0.057	7.84
	K	26/30	0.058	8.03
	L	41/32	0.059	8.10
	M	65/34	0.059	7.97
18	J	10/28	0.044	4.90
	K	16/30	0.045	4.94
	L	26/32	0.046	5.14
	M	41/34	0.046	5.02
20	J	7/28	0.038	3.43
	K	10/30	0.035	3.09
	L	16/32	0.036	3.16
	M	26/34	0.037	3.19

(a) See "Metric Conversion Guide" for conversion into metric units.

Table 13 Characteristics of tinned solid round copper wire specified in ASTM B 33, B 246, B 258

Conductor size, AWG	Net weight, lb/1000 ft(a)	Soft (annealed) wire: Nominal resistance, Ω/1000 ft(a)	Soft (annealed) wire: Minimum elongation(b), %	Hard-drawn wire: Nominal resistance, Ω/1000 ft(a)	Hard-drawn wire: Minimum breaking strength, lbf(a)
2	200.9	0.1609	25	...	...
3	159.3	0.2028	25	...	...
4	126.3	0.2557	25	0.2680	1773
5	100.2	0.3226	25	0.3380	1432
6	79.44	0.4067	25	0.4263	1152
7	63.03	0.5127	25	0.5372	927.3
8	49.98	0.6465	25	0.6776	743.1
9	39.62	0.8154	25	0.8545	595.1
10	31.43	1.039	20	1.087	476.1
11	24.9	1.31	20	1.37	381.0
12	19.8	1.65	20	1.73	303.0
13	15.7	2.08	20	2.18	241.0
14	12.4	2.62	20	2.74	192.0
15	9.87	3.31	20	3.46	153.0
16	7.81	4.18	20	4.37	121.0
17	6.21	5.26	20	...	...
18	4.92	6.66	20	...	...
19	3.90	8.36	20	...	...
20	3.10	10.6	20	...	...
21	2.46	13.3	20	...	...
22	1.94	16.9	20	...	...
23	1.55	21.1	20	...	...
24	1.22	26.7	15	...	...
25	0.970	34.4	15	...	...
26	0.765	43.5	15	...	...
27	0.610	54.5	15	...	...
28	0.481	69.3	15	...	...
29	0.387	86.1	15	...	...
30	0.303	110.0	10	...	...
31	0.204	141.0	10	...	...
32	0.194	174.0	10	...	...
33	0.153	221.0	10	...	...
34	0.120	281.0	10	...	...

(a) See "Metric Conversion Guide" for conversion into metric units. (b) In 250 mm (10 in.)

- Lead, or lead alloy (80Pb-20Sn), ASTM B 189
- Nickel, ASTM B 355
- Silver, ASTM B 298
- Tin, ASTM B 33

Intermediate and fine wires are drawn on smaller machines that have 12 to 20 or more dies each. The wire is reduced in steps of 20 to 25% in cross-sectional area. Intermediate machines can produce wire as small as 0.5 mm (0.020 in.) in diameter, and fine wire machines can produce wire in diameters from 0.5 mm (0.020 in.) to less than 0.25 mm (0.010 in.). Drawing speeds are typically 25 to 30 m/s (5000 to 6000 ft/min) and may be even higher.

All drawing is performed with a copious supply of lubricant to cool the wire and prevent rapid die wear. Traditional lubricants are soap and fat emulsions, which are fed to all machines from a central reservoir. Breakdown of rod usually requires a lubricant concentration of about 7%; drawing of intermediate and fine wires, concentrations of 2 to 3%. Today, synthetic lubricants are becoming more widely accepted.

Drawn wire is collected on reels or stem packs, depending on the next operation. Fine wire is collected on reels carrying as little as 4.5 kg (10 lb); large-diameter wire, on stem packs carrying up to 450 kg (1000 lb). To ensure continuous operation, many drawing machines are equipped with dual take-up systems. When one reel is filled, the machine automatically flips the wire onto an adjacent empty reel and simultaneously cuts the wire. This permits the operator to unload the full reel and replace it with an empty one without stopping the wiredrawing operation.

Until the early 1970s, hydrostatic extrusion was essentially a batch process and was not considered a competitor to conventional wiredrawing. In 1970, Western Electric Company patented a "viscous drag machine" that uses a pressurized, flowing viscous fluid to feed wire rod into an extrusion chamber and through a die for continuous extrusion of wire. By forcing the viscous fluid to flow along the surface of the wire rod, shear stresses between fluid and rod are used to move the rod through the die. In 1973, a refinement of this process was announced; in the refined system, shear forces transmitted through a viscous medium were used to feed the rod toward the extrusion die. This process has not yet proved economical enough to be a significant commercial process.

Coatings are applied to:

- Retain solderability for hookup-wire applications
- Provide a barrier between the copper and insulation materials such as rubber, that would react with the copper and adhere to it (thus making it difficult to strip insulation from the wire to make an electrical connection)
- Prevent oxidation of the copper during high-temperature service

Tin-lead alloy coatings and pure tin coatings are the most common; nickel and silver are used for specialty and high-temperature applications.

Copper wire can be coated by hot dipping in a molten metal bath, electroplating, or cladding. With the advent of continuous processes, electroplating has become the dominant process, especially because it can be done "on line" following the wiredrawing operation.

Stranded wire is produced by twisting or braiding several wires together to provide a flexible cable. (For a description of various strand constructions, see the section of this article entitled "Classification of Wire and Cable.") Different degrees of flexibility for a given current-carrying capacity can be achieved by varying the number, size, and arrangement of individual wires. Solid wire, concentric strand, rope strand, and bunched strand provide increasing degrees of flexibility; within the last three categories, a larger number of finer wires provides greater flexibility.

Stranded copper wire and cable are made on machines known as bunchers or strand-

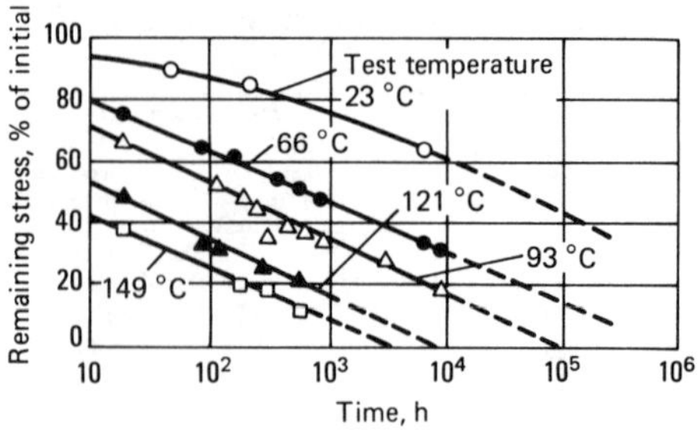

Fig. 11 Tensile-stress-relaxation characteristics of C11000. Data are for tinned 30 AWG (0.25 mm diam) annealed ETP copper wire; initial elastic stress, 89 MPa (13 ksi).

ers. Conventional bunchers are used for stranding small-diameter wires (34 AWG up to 10 AWG). Individual wires are payed off reels located alongside the equipment and are fed over flyer arms that rotate about the take-up reel to twist the wires. The rotational speed of the arm relative to the take-up speed controls the length of lay in the bunch. For small, portable, flexible cables, individual wires are usually 30 to 34 AWG, and there may be as many as 150 wires in each cable.

A tubular buncher has up to 18 wire-payoff reels mounted inside the unit. Wire is taken off each reel while it remains in a horizontal plane, is threaded along a tubular barrel, and is twisted together with other wires by a rotating action of the barrel. At the take-up end, the strand passes through a closing die to form the final bunch configuration. The finished strand is wound onto a reel that also remains within the machine.

Supply reels in conventional stranders for large-diameter wire are fixed onto a rotating frame within the equipment and revolve about the axis of the finished conductor. There are two basic types of machines. In one, known as a rigid-frame strander, individual supply reels are mounted in such a way that each wire receives a full twist for every revolution of the strander. In the other, known as a planetary strander, the wire receives no twist as the frame rotates.

These types of standers are comprised of multiple bays, with the first bay carrying six reels and subsequent bays carrying increasing multiples of six. The core wire in the center of the strand is payed off externally. It passes through the machine center and individual wires are laid over it. In this manner, strands with up to 127 wires are produced in one or two passes through the machine, depending on its capacity for stranding individual wires.

Normally, hard-drawn copper is stranded on a planetary machine so that the strand will not be as springy and will tend to stay bunched rather than spring open when it is cut off. The finished product is wound onto a power-driven external reel that maintains a prescribed amount of tension on the stranded wire.

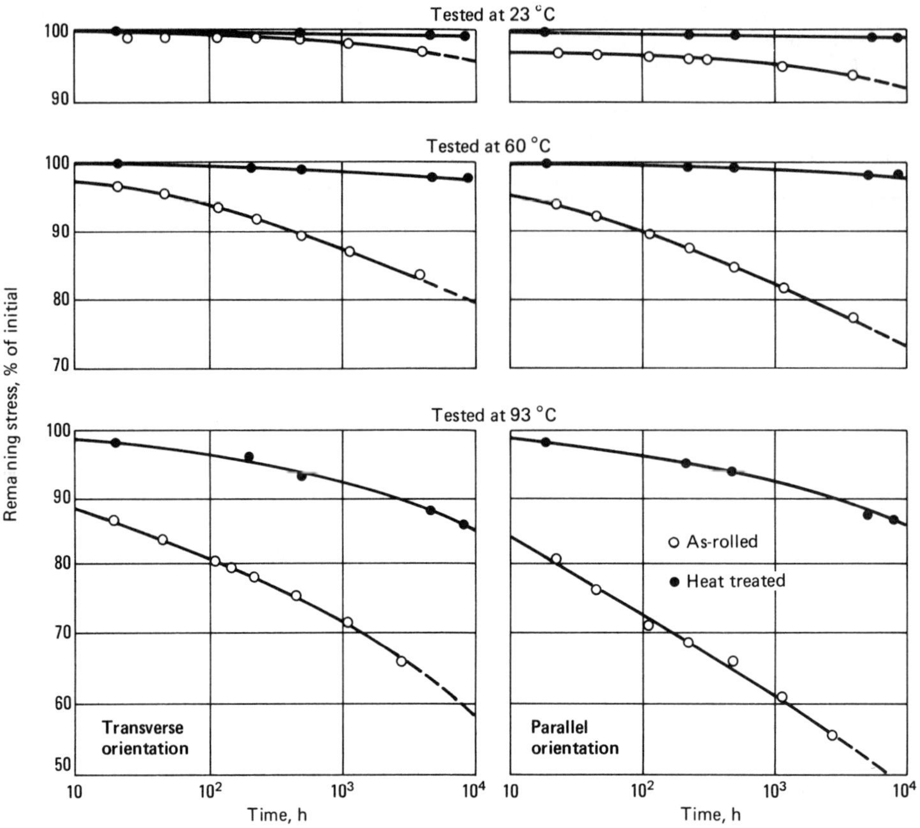

Fig. 12 Anisotropic stress-relaxation behavior in bending for highly cold-worked C51000 strip. Data are for 5% Sn phosphor bronze cold rolled 93% (reduction in area) to 0.25 mm (0.01 in.) and heat treated 2 h at 260 °C (500 °F). Graphs at left are for stress relaxation transverse to the rolling direction; graphs at right, for stress relaxation parallel to the rolling direction. Initial stresses: as rolled, parallel orientation, 607 MPa (88 ksi); as rolled, transverse orientation, 634 MPa (92 ksi); heat treated, parallel orientation, 641 MPa (93 ksi); heat treated, transverse orientation, 738 MPa (107 ksi)

Insulation and Jacketing

Of the three broad categories of insulation—polymeric, enamel, and paper-and-oil—polymeric insulation is the most widely used.

Polymeric Insulation. The most common polymers are polyvinyl chloride (PVC), polyethylene, ethylene propylene rubber (EPR), silicone rubber, polytetrafluoroethylene (PTFE), and fluorinated ethylene propylene (FEP). Polyimide coatings are used where fire resistance is of prime importance, such as in wiring harnesses for manned space vehicles. Until a few years ago, natural rubber was used, but this has now been supplanted by synthetics such as butyl rubber and EPR. Synthetic rubbers are used wherever good flexibility must be maintained, such as in welding or mining cable.

Many varieties of PVC are made, including several that are flame resistant. PVC has good dielectric strength and flexibility, and is one of the least expensive conventional insulating and jacketing materials. It is used mainly for communication wire, control cable, building wire, and low-voltage power cables. PVC insulation is normally selected for applications requiring continuous operation at temperatures up to 75 °C (165 °F).

Polyethylene, because of its low and stable dielectric constant, is specified when better electrical properties are required. It resists abrasion and solvents. It is used chiefly for hookup wire, communication wire, and high-voltage cable. Cross-linked polyethylene (XLPE), which is made by adding organic peroxides to polyethylene and then vulcanizing the mixture, yields better heat resistance, better mechanical properties, better aging characteristics, and freedom from environmental stress cracking. Special compounding can provide flame resistance in cross-linked polyethylene. Typical uses include building wire, control cables, and power cables. The usual maximum sustained operating temperature is 90 °C (200 °F).

PTFE and FEP are used to insulate jet aircraft wire, electronic equipment wire, and specialty control cables, where heat resistance, solvent resistance, and high reliability are important. These electrical cables can operate at temperatures up to 250 °C (480 °F).

All of the polymeric compounds are applied over copper conductors by hot extrusion. The extruders are machines that convert pellets or powders of thermoplastic polymers into continuous covers. The insulating compound is loaded into a hopper that feeds into a long, heated chamber. A contin-

Table 14 Tensile-stress-relaxation data for selected types of copper wire

Material	Temper	Length of test, h	Temperature °C	Temperature °F	Initial stress MPa	Initial stress ksi	Percent of initial stress remaining after: 10 000 h	Percent of initial stress remaining after: 40 years
0.25 mm (0.01 in.) diameter wire								
C10200, tinned	O61	10 000	27	80	41.0	5.95	72	55
		10 000	27	80	61.5	8.92	70	53
		10 000	27	80	82.0	11.9	69	50
		2 850	121	250	82.0	11.9	15	0
		2 850	149	300	82.0	11.9	6	0
C10200, tinned	H04	10 000	27	80	79.9	11.6	82	68
		8 600	66	150	88.9	12.9	78	68
		9 300	93	200	88.9	12.9	67	42
		2 850	121	250	88.9	12.9	55	37
		2 850	149	300	88.9	12.9	42	18
		10 000	27	80	160	23.2	80	68
		8 600	66	150	160	23.2	69	57
		9 300	93	200	160	23.2	59	43
		2 850	121	250	160	23.2	40	14
		2 850	149	300	160	23.2	22	0
C11000, tinned	O61	10 000	23	73	44.8	6.5	60	41
		9 300	66	150	44.8	6.5	47	22
		9 700	93	200	44.8	6.5	32	3
		2 850	121	250	44.8	6.5	12	0
		2 850	149	300	44.8	6.5	12	0
		10 000	23	73	88.9	12.9	60	38
		9 300	66	150	88.9	12.9	30	6
		9 700	93	200	88.9	12.9	20	0
		2 850	121	250	88.9	12.9	8	0
		2 850	149	300	88.9	12.9	8	0
C12000, tinned	O61	10 000	27	80	52.4	7.6	86	80
		10 000	27	80	77.9	11.3	85	79
		10 000	27	80	104	15.1	84	78
C13400, tinned(a)	H00	2 833	93	200	88.9	12.9	50	27
		2 833	93	200	101	14.7	49	28
		2 833	93	200	152	22.1	45	25
		2 833	93	200	203	29.5	42	19
C13700, tinned(b)	H00	9 700	23	73	88.9	12.9	88	83
		9 300	66	150	88.9	12.9	78	67
		9 700	93	200	88.9	12.9	70	52
		2 850	121	250	88.9	12.9	51	27
		2 760	149	300	88.9	12.9	41	8
		9 700	23	73	136	19.7	86	81
		9 300	66	150	136	19.7	77	64
		9 700	93	200	136	19.7	67	48
		2 850	121	250	136	19.7	42	19
		2 760	149	300	136	19.7	28	0
C15000, tinned	H04(c)	9 700	23	73	88.9	12.9	93	92
		9 300	66	150	88.9	12.9	93	89
		9 700	93	200	88.9	12.9	92	82
		2 850	121	250	88.9	12.9	82	78
		2 850	149	300	88.9	12.9	80	76
		9 700	23	73	203	29.5	93	92
		9 800	66	150	203	29.5	93	87
		9 700	93	200	203	29.5	92	82
		2 850	121	250	203	29.5	80	76
		2 850	149	300	203	29.5	78	74
C15000, bare	H04(c)	9 700	23	73	88.9	12.9	96	95
		9 600	66	150	88.9	12.9	96	95
		9 700	93	200	88.9	12.9	96	95
		9 700	23	73	203	29.5	96	95
		9 600	66	150	203	29.5	96	95
		9 700	93	200	203	29.5	86	79
C15000, tinned	H00	2 800	93	200	88.9	12.9	96	91
		2 800	93	200	128	18.6	95	90
		2 800	93	200	192	27.9	94	89
		2 800	93	200	256	37.2	93	89
C15000, silver plated	(d)	9 800	27	80	74.4	10.8	97.9	95
		9 800	27	80	112	16.2	98.8	94
		9 800	27	80	149	21.6	96.7	93
C16200, tinned	H04(e)	9 700	23	73	88.9	12.9	97	94
		9 700	66	150	88.9	12.9	93	92
		9 700	93	200	88.9	12.9	92	87
		2 800	121	250	88.9	12.9	79	71
		2 800	149	300	88.9	12.9	62	40
		9 700	23	73	226	32.8	95	92
		9 700	66	150	226	32.8	91	88
		9 700	93	200	226	32.8	88	84
		2 800	121	250	226	32.8	77	64
		2 800	149	300	226	32.8	60	34

(continued)

(a) Boron-deoxidized copper containing 0.027% Ag. (b) Boron-deoxidized copper containing 0.085% Ag. (c) In-process strand annealed. (d) Proprietary mill processing. (e) Batch annealed

Table 14 (continued)

Material	Temper	Length of test, h	Temperature °C	Temperature °F	Initial stress MPa	Initial stress ksi	Percent of initial stress remaining after: 10 000 h	Percent of initial stress remaining after: 40 years
0.25 mm (0.01 in.) diameter wire (continued)								
C16200, tinned	H00	2 800	93	200	88.9	12.9	91	85
		2 800	93	200	114	16.6	91	84
		2 800	93	200	172	24.9	91	84
		2 800	93	200	229	33.2	91	84
0.5 mm (0.02 in.) diameter wire								
C10200	O61	22 600	27	80	58.6	8.5	81	71
		22 600	27	80	75.8	11.0	81	71
		22 600	27	80	86.2	12.5	81	71
		22 600	27	80	103	15.0	79	70
		22 600	27	80	110	16.0	78	67
C10200	H00	4 060	93	200	68.9	10.0	48	9
		4 060	93	200	142	20.6	42	0
C11000	O61	35 000	27	80	34.5	5.0	60	43
		35 000	27	80	68.9	10.0	55	39
C1100	O61	24 500	27	80	34.5	5.0	60	38
		24 500	27	80	41.2	6.0	60	38
		24 500	27	80	51.7	7.5	59	38
		24 500	27	80	68.9	10.0	57	38
		24 500	27	80	82.7	12.0	56	38
		24 500	27	80	96.5	14.0	55	37
C11000	H00	4 100	93	200	68.9	10.0	35	6
		4 100	93	200	121	17.5	23	0
C11600	H00	4 100	93	200	68.9	10.0	50	20
		4 100	93	200	143	20.7	43	18
C13400	H00	4 100	93	200	68.9	10.0	53	27
		4 100	93	200	148	21.4	38	14
C15500, bare	H00	4 060	93	200	68.9	10.0	78	62
		4 060	93	200	164	23.8	74	60
C16200	H00	4 100	93	200	68.9	10.0	88	82
		4 100	93	200	158	22.9	80	69

(a) Boron-deoxidized copper containing 0.027% Ag. (b) Boron-deoxidized copper containing 0.085% Ag. (c) In-process strand annealed. (d) Proprietary mill processing. (e) Batch annealed

uously revolving screw moves the pellets into the hot zone where the polymer softens and becomes fluid. At the end of the chamber, molten compound is forced out through a small die over the moving conductor, which also passes through the die opening. As the insulated conductor leaves the extruder it is water cooled and taken up on reels. Cables jacketed with EPR and XLPE go through a vulcanizing chamber prior to cooling to complete the cross-linking process.

Enamel Film Insulation. Film-coated wire, usually fine magnet wire, is composed of a metallic conductor coated with a thin, flexible enamel film. These insulated conductors are used for electromagnetic coils in electrical devices and must be capable of withstanding high breakdown voltages. Temperature ratings range from 105 to 220 °C (220 to 425 °F), depending on enamel composition. The most commonly used enamels are based on polyvinyl acetals, polyesters, and epoxy resins.

Equipment for enamel coating of wire often is custom built, but standard lines are available. Basically, systems are designed to insulate large numbers of wires simultaneously. Wires are passed through an enamel applicator that deposits a controlled thickness of liquid enamel onto the wire. Then the wire travels through a series of ovens to cure the coating, and finished wire is collected on spools. In order to build up a heavy coating of enamel, it may be necessary to pass wires through the system several times. In recent years, some manufacturers have experimented with powder-coating methods. These avoid evolution of solvents, which is characteristic of curing conventional enamels, and thus make it easier for the manufacturer to meet Occupational Safety and Health Administration and Environmental Protection Agency standards. Electrostatic sprayers, fluidized beds, and other experimental devices are used to apply the coatings.

Paper-and-Oil Insulation. Cellulose is one of the oldest materials for electrical insulation and is still used for certain applications. Oil-impregnated cellulose paper is used to insulate high-voltage cables for critical power-distribution applications. The paper, which may be applied in tape form, is wound helically around the conductors using special machines in which six to twelve paper-filled pads are held in a cage that rotates around the cable. Paper layers are wrapped alternately in opposite directions, free of twist. Paper-wrapped cables then are placed inside special impregnating tanks to fill the pores in the paper with oil and to ensure that all air has been expelled from the wrapped cable.

The other major use of paper insulation is for flat magnet wire. In this application, magnet-wire strip (with a width-to-thickness ratio greater than 50 to 1) is helically wrapped with one or more layers of overlapping tapes. These may be bonded to the conductor with adhesives or varnishes. The insulation provides highly reliable mechanical separation under conditions of electrical overload.

Stress-Relaxation Characteristics

WHEN AN EXTERNAL STRESS is applied to a piece of metal, it reacts by developing an equal and opposite internal stress. If the metal is held in this strained position, the internal stress will decrease as a function of time. This phenomenon is called stress relaxation and happens because of the transformation of elastic strain in the material to plastic, or permanent strain. The reduction-of-stress rate will be a function of alloy, temper, temperature, and time.

A useful example for visualizing stress relaxation is a bolt clamping two components together. Tightening the bolt results in a compressive force on the two components that is equal and opposite to the tensile force in the bolt. Over a period of time, stress relaxation in the bolt will reduce this tensile stress and cause a loss of clamping force.

Copper alloys are used in many applications where they are subjected to an applied stress at either ambient or moderately elevated temperature. Examples include electrical and electronic connectors, automotive radiators, solar heating panels, and

Table 15 Chemical composition of copper wire tested for stress relaxation

Material	Composition, %				
	Ag	Pb	Fe	Ni	Others
0.25 mm (0.01 in.) diameter					
C10200	0.002	. . .	. . .	. . .	. . .
C11000	0.002	0.001	0.002	0.001	. . .
C12000	0.002	. . .	. . .	. . .	. . .
C13400	0.031	. . .	0.003	. . .	0.02 B, 0.001 Si
C13700	0.090	0.001	0.004	. . .	0.01 B, 0.001 Si
C15000	0.003	0.001	0.002	. . .	0.001 Mg, 0.15 Zr
C16200	0.005	0.01	0.015	0.005	0.75 Cd, 0.01 Sn
0.05 mm (0.02 in.) diameter					
C10200	0.002	. . .	. . .	. . .	. . .
C11000	0.001	. . .	. . .	. . .	0.0355 O
C13000	0.083	. . .	0.0029	0.0079	0.0310 O
C13400	0.031	. . .	0.003	. . .	0.02 B, 0.001 Si
C15500	0.037	. . .	. . .	. . .	0.10 Mg, 0.063 P
C16200	0.003	0.005	. . .	0.005	0.97 Cd, 0.007 Sn, 0.07 Zn

Table 16 Typical mechanical properties of copper wire tested for stress relaxation

Material	Temper	Tensile strength		Yield strength(a)		Elongation(b), %	Conductivity, %IACS
		MPa	ksi	MPa	ksi		
0.25 mm (0.01 in.) diameter							
C10200, tinned	O61	250	36.3	160	23.2	24.6	99
C10200, tinned	H04	271	39.3	228	33.1	2.3	99
C11000, tinned	O61	254	36.9	147	21.3	30.0	99
C12000, tinned	O61	276	40.0	198	28.7	18.1	98
C13400, tinned	H00	365	53.0	359	52.1	0.82	98
C13700, tinned	H00	242	35.1	221	32.1	10.5	99
C15000, tinned	H04	414	60.0	365	53.0	3.7	90
C15000, bare	H04	. . .	. . .	. . .	. . .	. . .	. . .
C15000, tinned	H00	393	57.0	388	56.3	0.82	93
C15000, silver plated	. . .	347	50.3	298	43.3	13.5	. . .
C16200, tinned	H04	473	68.6	404	58.6	6.6	85
C16200, tinned	H00	403	58.5	395	57.3	0.91	85
0.05 mm (0.02 in.) diameter							
C10200, tinned	O61	249	36.1	138	20.0	34.7	101
C10200, bare	H00	278	40.3	261	37.8	11.1	101
C11000, tinned	O61	239	34.7	. . .	. . .	28.8	101
C11000, silver plated	O61	233	33.8	. . .	. . .	24.6	101
C11000, bare	H00	280	40.6	245	35.5	11.7	101
C11600	H00	295	42.8	274	39.8	9.5	98
C13400	H00	258	37.5	240	34.8	9.4	98
C15500, bare	H00	367	53.2	332	48.1	1.9	93.8
C16200	H00	312	45.3	276	40.1	10.9	85

(a) 0.2% offset. (b) In 254 mm (10 in.)

communications cable. Stress relaxation, which can be significant even at ambient temperature, can result in service failures. Data indicating the amount of relaxation that can be expected is, therefore, useful to product designers.

Because of the wide variations in composition and processing among commercial copper alloys, resistance to stress relaxation varies considerably. Of course, selection of an alloy for a given application is based not only on stress-time-temperature response but also on such factors as cost, basic mechanical and physical properties, operating temperature, service environment, and formability. For many applications, electrical conductivity is a primary consideration.

Stress-Relaxation Data

Standard practices for performing stress-relaxation tests are described in ASTM E 328. Testing can be performed in either tension, compression, torsion, or bending. It is advisable to use data generated for the stress state that most closely approximates the application being considered. Since the lifetime of a part is often very long, it is common practice to extrapolate short-time data to longer periods. Methods used include plots like those shown in Fig. 11 (where the dotted portion of the line is an extrapolation), and the Larson-Miller parameter (Ref 2). Use of accelerated tests at temperature in excess of the service temperature can be misleading (Ref 1).

Unalloyed copper C11000 (electrolytic tough pitch) is probably the most inexpensive high-conductivity copper and is used extensively because of its ease of fabrication. The stress-relaxation behavior of this material is rather poor, as demonstrated in Fig. 11, in which relaxed stress is plotted as a function of time and temperature for 0.25 mm (0.010 in.) C11000 wire initially stressed in tension to 89 MPa (13 ksi). Comparison of stress values at a given time for different temperatures illustrated the very sharp dependence of stress relaxation on temperature for this copper. At 93 °C (200 °F), for example, no tension remains after 10^5 h (11.4 years), whereas 40% of the initial stress remains after 40 years at room temperature. For C11000 and for many other copper metals, stress relaxation in a given time period is inversely proportional to absolute temperature (Ref 3).

The stress-relaxation behavior of C10200 (oxygen-free copper) is somewhat better than that of C11000, as shown in Table 14, which also presents stress-relaxation data for many other high-conductivity copper metals. (For compositions of these metals, see Table 15; basic mechanical properties are given in Table 16.) A more extensive comparison of the mechanical behavior of C10200 and C11000 has been presented in Ref 4.

Among the high-conductivity coppers, relaxation is greatest in very-high-purity copper (99.999+%)—a material used mainly in research. Improvement in the stress-relaxation behavior of high-conductivity copper can be achieved by adding alloying elements that cause solid-solution strengthening, age hardening, or dispersion hardening (Ref 5). For example, minute additions of silver significantly reduce stress relaxation in copper (Ref 6, 7).

Besides the strengthened high-conductivity coppers for which stress-relaxation data are given in Table 14, proprietary coppers strengthened with small amounts of cadmium, chromium, zirconium, or a combination of one or more of these elements have been found to have good to superior stress-relaxation resistance (Ref 3). Processing variations that strengthen copper metals, even including internal oxidation, are almost always beneficial.

Lower-conductivity alloys, which are strengthened by alloy additions or precipitation hardening, exhibit improved resistance to stress relaxation, compared to pure copper. The performance of any particular material will be dependent on its chemical composition, condition, and temperature at which it is tested. For materials that are strengthened by cold rolling, several general comments can be made. First, the amount of relaxation that will occur during a given time at a certain temperature will increase with increasing amounts of prior cold work. Second, the performance of these materials may vary depending on the orientation of the test sample to the rolling direction. Finally, the performance of heavily cold-rolled materials can be improved by stress-relief annealing (Ref 8 and 9). Figures 12 and 13 illustrate these effects for alloys C51000 and C72500.

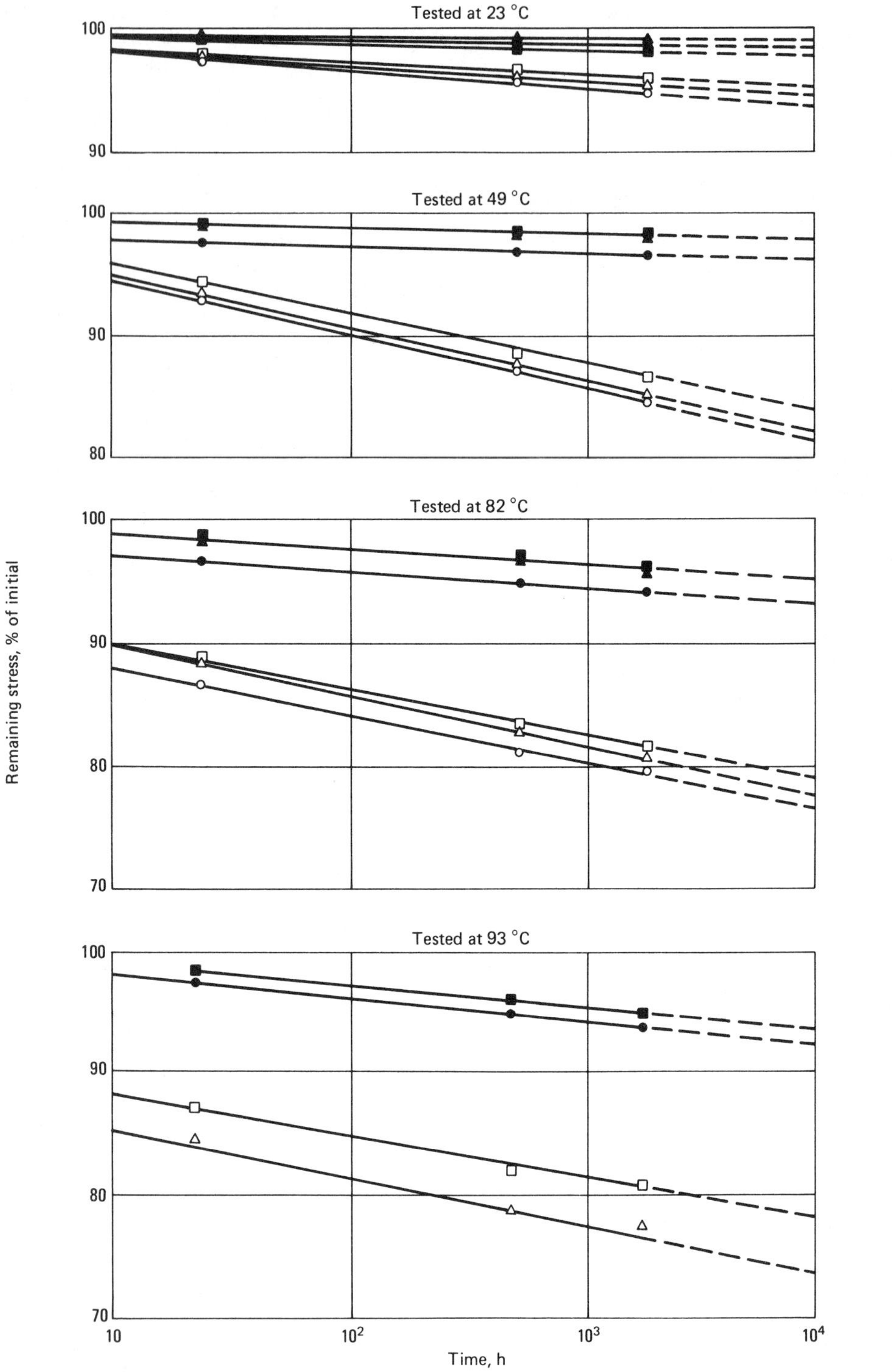

Fig. 13 Anisotropic stress-relaxation behavior in bending for highly cold-worked C72500 strip. Data are for 89Cu-9Ni-2Sn alloy cold rolled 98.7% (reduction in area) to 0.25 mm (0.01 in.) and heat treated 2 h at 357 °C (675 °F). Points represented by circles are for stress relaxation parallel to the rolling direction; triangles, for relaxation at 45° to the rolling direction; squares, for relaxation transverse to the rolling direction. Open points are for as-rolled stock; solid points, for heat-treated stock. Initial stresses: as-rolled, parallel orientation, 524 MPa (76 ksi); as-rolled, 45° orientation, 510 MPa (74 ksi); as-rolled, transverse orientation, 586 MPa (85 ksi); heat treated, parallel orientation, 669 MPa (97 ksi); heat treated, 45° orientation, 552 MPa (80 ksi); heat treated, transverse orientation, 710 MPa (103 ksi)

Since copper alloys are commonly used in electrical and electronic connectors, some manufacturers have published data to assist in the design of these items (see Ref 1 and 10). A compilation of stress-relaxation data from these sources for some of the more common alloys is given in Table 17. It is important to realize that the data from relaxation tests, such as in Table 17, serves as only a comparative ranking of alloy performance because these tests are usually performed on flat samples cut from strip, rather than on finished parts. Other factors that might influence part performance include part geometry, deformation introduced during fabrication, and performance of other materials used in the part (for example, plastic housings used in electrical connectors). Testing of prototype parts is the best way to assess overall performance.

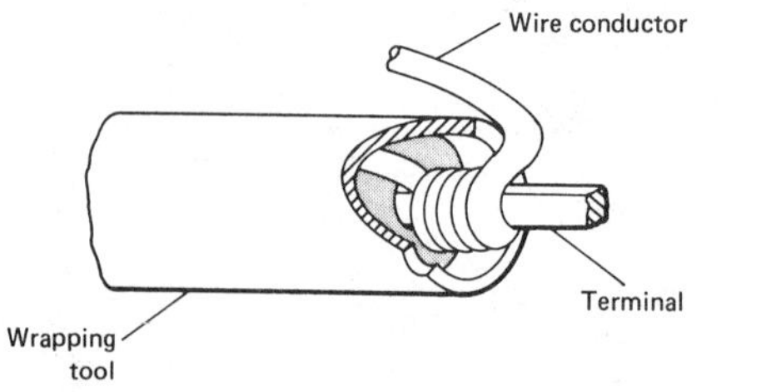

Fig. 14 Typical solderless wrapped connection. Wrapping tool is removed after connection is made.

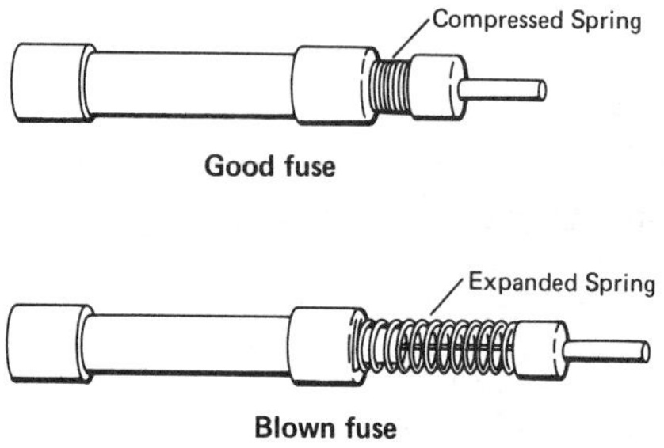

Fig. 15 Typical spring-type alarm fuse

Stress Relaxation in Mechanical Components

A solderless wrapped connection such as the one shown in Fig. 14, in which electrical contact is made by wrapping a wire around a terminal, is a typical application for high-conductivity copper metals where stress relaxation is of concern. Typical operating temperatures can be as high as 85 °C (185 °F); generally, conductivities higher than 98% IACS at 20 °C (68 °F) are desirable. After the connection is made, it is maintained by elastic stresses in the two members. If the wire undergoes stress relaxation, electrical contact between the wire and the terminal may be lost.

The spring in the alarm fuse shown in Fig. 15 is a typical application for copper alloys with room-temperature conductivities of 55 to 85% IACS. This spring conducts relatively high electrical currents and also triggers an alarm circuit if the fuse blows. To perform the latter function reliably, the spring must retain spring force for extended periods of time. But if the spring material undergoes stress relaxation, the device may fail to trigger the alarm when the fuse blows. C16200 (Cu-1Cd) has been used suc-

Table 17 Typical stress-relaxation values for selected copper alloys

Alloy	Temper	Percent of initial stress remaining after specified time at: Room temperature 10^3 h	Room temperature 10^5 h	75 °C (170 °F) 10^3 h	75 °C (170 °F) 10^5 h	105 °C (220 °F) 10^3 h	105 °C (220 °F) 10^5 h	150 °C (300 °F) 10^3 h	150 °C (300 °F) 10^5 h	200 °C (390 °F) 10^3 h	200 °C (390 °F) 10^5 h
C15100	H02	97	95	...	...	86	81	84	80	...	...
	H04	93	89	...	...	80	76	...	...	...	...
	H06	94	92	...	...	78	71	...	...	...	...
C17200	AM	...	...	...	...	97(a)	...	...	...	67	...
	¼HM	...	...	...	...	97(a)	...	...	...	67	...
	½HM	...	...	...	...	97(a)	...	...	...	67	...
	HM	...	...	...	...	98(a)	...	...	...	68	...
	SHM	...	...	...	...	98(a)	...	...	...	68	...
	XHM	...	...	...	...	98(a)	...	...	...	69	...
	XHMS	...	...	...	...	98(a)	...	...	...	69	...
C17410	½HT	...	...	...	...	85(a)	...	...	...	...	...
	HT	...	...	...	...	80(a)	...	76	...	...	...
C19400	H02	95	93	...	...	78	71	...	...	...	...
C26000	H04	95	92	74	63	...	...	...	...	...	...
	H08	87	84	68	53	...	...	36	...	...	...
C51000	H02	94	94	...	...	89	81	...	...	...	...
	H08	95	93	90	81	75	62	44	...	...	...
C52100	H02	97	96	92	89	88	83	...	...	...	...
	H04	99	98	89	84	81	71	...	...	...	...
	H08	98	98	91	87	82	74	45	...	...	...
C63800	H03	89	84	75	66	...	...	...	...	...	...
	H06	87	81	69	56	...	...	...	...	...	...
C65400	H02	...	...	...	...	81	78	...	...	...	...
	H04	...	...	...	...	76	69	...	...	...	...
	H08	...	...	...	...	71	63	...	...	...	...
C68800	H02	97	95	78	67	...	...	...	...	...	...
	H08	95	90	73	61	...	...	...	...	...	...
C70250	TM00	98	98	97	96	91	83	80	65	...	...
C72500	H02	...	...	...	...	86	82	...	...	...	...
	H08	97	96	...	...	83	76	74	...	61	...

(a) Data for C17200 and C17410 in the 105 °C (220 °F) column are actually for tests at 100 °C (212 °F). Source: Ref 2, 10, 11

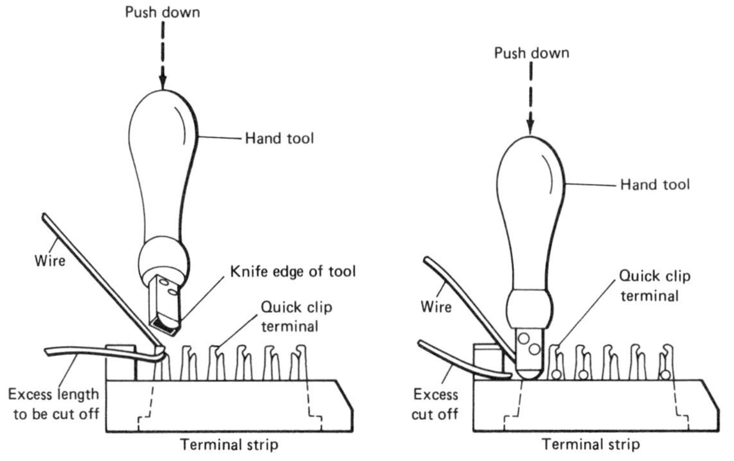

Fig. 16 Typical quick clip connection

cessfully in spring-loaded alarm fuses operating at temperatures below 95 °C (200 °F). (This alloy has an electrical conductivity of 80 to 85% IACS at 20 °C, or 68 °F.) For higher operating temperatures up to 165 °C (330 °F), C19000 (Cu-Ni-P alloy) springs have performed adequately, provided the ratio of nickel to phosphorus is at least 5 to 1 (Ref 12). (The nominal ratio for this alloy is about 3.5 to 1.) For applications where lower electrical conductivity can be tolerated, C17500 (Cu-2.5Co-0.6Be) conductivity, 45% IACS at 20 °C (68 °F), is a satisfactory alternative for temperatures up to 165 °C (330 °F). Both C19000 and C17500 must be age hardened after forming.

A typical application of the lower-conductivity, high-strength copper alloys is the pressure-type, split-beam connector shown in Fig. 16. The knife edges of the connector first cut through the insulation on the conductor and then must maintain electrical contact with it. Materials used for connectors of this type, depending on operating stress and temperature, include some of the phosphor bronzes, nickel silvers, copper-nickels, beryllium-coppers, and some of the newer copper-nickel-tin

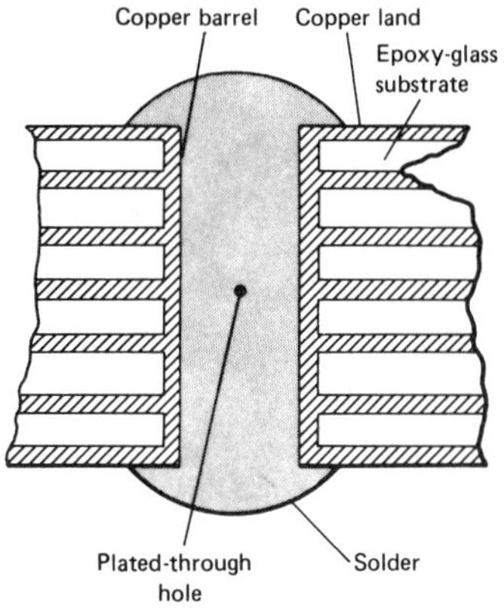

Fig. 17 Cutaway view through a typical multilayer circuit board showing barrel-land construction and a plated-through hole

alloys (Ref 13) strengthened by spinodal decomposition.

Stress relaxation can produce mechanical or thermal ratcheting, which sometimes occurs in multilayer circuit boards such as the one illustrated in Fig. 17. A multilayer board usually consists of an epoxy-glass composite substrate with several lands of electroplated copper and including a plated-through hole. When a leadwire is soldered into the plated-through hole, differential thermal expansion causes the hole to expand more than the substrate, and a tensile stress is applied to the copper barrel. While at temperature, the stressed electroplated copper relaxes according to behavior that varies with the plating system and bath used (Ref 14). If the board is repeatedly heated and cooled, the electroplated copper alternately expands and relaxes during each heating cycle, and permanent strain accumulates in the copper barrel until the assembly fails by buckling or by low-cycle fatigue.

REFERENCES

1. *Understanding Copper Alloys*, J.H. Mendenhall, Ed., Robert E. Krieger, 1977
2. J.B. Conway, Chapter 2, in *Stress-Rupture Parameters: Origin, Calculation and Use*, Gordon and Breach, 1969
3. A. Fox, Stress-Relaxation Characteristics in Tension of High-Strength, High Conductivity Copper and High Copper Alloy Wires, *J. Test. Eval.*, Vol 2 (No. 1), Jan 1974, p 32-39
4. W.R. Opie, P.W. Taubenblat, and Y.T. Hsu, A Fundamental Comparison of the Mechanical Behavior of Oxygen-Free and Tough Pitch Coppers, *J. Inst. Met.*, Vol 98, 1970, p 245
5. A. Fox and J.J. Swisher, Superior Hook-Up Wires for Miniaturized Solderless Wrapped Connections, *J. Inst. Met.*, Vol 100, 1972, p 30

6. A. Fox, "The Creep and Stress Relaxation Behavior of Silver Bearing Copper Wire," Master's thesis, New Jersey Institute of Technology, 1972
7. W.L. Finlay, Silver-Bearing Copper: A Compendium of the Origin, Characteristics, Uses and Future of Copper Containing 12 to 25 Ounces per Ton of Silver, Corinthian Editions, 1968
8. A. Fox, The Effect of Extreme Cold Rolling on the Stress Relaxation Characteristics of CDA Copper Alloy 510 Strip, *J. Mater.*, Vol 6 (No. 2), June 1971, p 422-435
9. A. Fox, Stress Relaxation and Fatigue of Two Electromechanical Spring Materials Strengthened by Thermomechanical Processing, *IEEE Trans. Parts, Mater. Packag.*, Vol PMP-7 (No. 1), March 1971, p 34-47
10. O. Brass, *Spring Designers Data Package*, 2nd ed., T.D. Hann and S.P. Zarlingo, Ed.
11. E.W. Filer and C.R. Scorey, Stress Relaxation in Beryllium Copper Strip, in *Stress Relaxation Testing*, STP 676, American Society for Testing and Materials, 1979, p 89-111
12. A. Fox and R.C. Stoffers, High Conductivity, High Strength Wire Springs for Fuse Applications, Electrical Contacts/1977 Proceedings of the Annual Holm Conference on Electrical Contacts, p 233-240
13. J.T. Plewes, Spinodal Cu-Ni-Sn Alloys Are Strong and Superductile, *Met. Prog.*, July 1974, p 46
14. A. Fox, Mechanical Properties at Elevated Temperature of Cu Bath Electroplated Copper for Multilayer Boards, *J. Test. Eval.*, Vol 4 (No. 1), Jan 1976, p 74-84

Properties of Wrought Coppers and Copper Alloys

Revised by Peter Robinson, Olin Corporation

C10100, C10200

Commercial Names

Previous trade names. C10100: Oxygen-free electronic copper. C10200: Oxygen-free copper
Common name. Oxygen-free copper
Designations. C10100: OFE. C10200: OF

Specifications

ASTM specifications for C10100. Flat products: B 48, B 133, B 152, B 187, B 272, B 432, F 68. Pipe: B 42, B 188, F 68. Rod: B 12, B 49, B 133, B 187, F 68. Shapes: B 133, B 187, F 68. Tubing: B 372, B 68, B 75, B 188, B 280, F 68. Wire: B 1, B 2, B 3, F 68
ASTM specifications for C10200. Flat products: B 48, B 133, B 152, B 187, B 272, B 370, B 432. Pipe: B 42, B 188. Rod: B 12, B 49, B 124, B 133, B 187. Tubing: B 68, B 75, B 88, B 111, B 188, B 280, B 359, B 372, B 395, B 447. Wire: B 1, B 2, B 3, B 33, B 47, B 116, B 189, B 246, B 286, B 298, B 355. Shapes: B 124, B 133, B 187
Government specifications for C10100. Rod: QQ-C-502
Government specifications for C10200. Flat products: QQ-C-576. Rod and shapes: QQ-C-502. Tubing: WW-T-775. Wire: QQ-C-502, QQ-W-343, MIL-W-3318

Chemical Composition

Composition limits. C10100: 99.99 Cu min (there are specific limits in ppm for 17 named elements; refer to ASTM B 170 or the Copper Development Association *Standards Handbook*). C10200: 99.95 Cu + Ag min
Consequence of exceeding impurity limits. C10100 and C10200 are high-conductivity electrolytic coppers produced without use of metal or metalloid deoxidizers. Excessive amounts of impurities reduce conductivity. Excessive oxygen causes the metal to fail the ASTM B 170 bend test after being heated 30 min at 850 °C (1560 °F) in pure hydrogen.

Applications

Typical uses. Busbars, waveguides, lead-in wire, anodes, vacuum seals, transistor components, glass-to-metal seals, coaxial cables, klystrons, microwave tubes
Precautions in use. Avoid heating in oxidizing atmospheres.

Mechanical Properties

Tensile properties. See Table 1, Fig. 1, and Fig. 2.
Shear strength. See Table 1.
Hardness. See Table 1.
Elastic modulus. Tension, 115 GPa (17×10^6 psi); shear, 44 GPa (6.4×10^6 psi)
Impact resistance. See Fig. 2.
Fatigue strength. See Table 1.
Creep-rupture characteristics. See Tables 2 and 3.

Mass Characteristics

Density. 8.94 g/cm^3 (0.323 lb/in.^{3}) at 20 °C (68 °F)

Thermal Properties

Melting point. 1083 °C (1981 °F)
Coefficient of linear thermal expansion. 17.0 μm/m · K (9.4 μin./in. · °F) at 20 to 100 °C (68 to 212 °F); 17.3 μm/m · K (9.6 μin./in. · °F) at 20 to 200 °C (68 to 392 °F); 17.7 μm/m · K (9.8 μin./in. · °F) at 20 to 300 °C (68 to 572 °F)
Specific heat. 385 J/kg · K (0.092 Btu/lb · °F) at 20 °C (68 °F)
Thermal conductivity. 391 W/m · K (226 Btu/ft · h · °F) at 20 °C (68 °F)

Electrical Properties

Electrical conductivity. Annealed: volumetric, 101% IACS min at 20 °C (68 °F)
Electrical resistivity. 17.1 nΩ · m at 20 °C (68 °F)

Chemical Properties

General corrosion behavior. Copper is cathodic to hydrogen in the electromotive series and therefore is the cathode in galvanic couples with other base metals such as iron, aluminum, magnesium, lead, tin, and zinc. C10100 and C10200 have excellent resistance to atmospheric corrosion and to corrosion by most waters, including brackish water and seawater. They have good resistance to nonoxidizing acids but poor resistance to oxidizing acids, moist ammonia, moist halogens, sulfides, and solutions containing ammonium ions.

Fabrication Characteristics

Machinability. 20% of C36000 (free-cutting brass)
Forgeability. 65% of C37700 (forging brass)
Formability. Readily formed by a wide variety of hot and cold methods. Can be easily stamped, bent, coined, sheared, spun, upset, swaged, forged, roll threaded, and knurled
Weldability. Can be readily soldered, brazed, gas tungsten arc welded, gas metal arc welded, or upset welded. Its capacity for being oxyfuel gas welded is fair. Shielded metal arc welding and most resistance welding methods are not recommended.
Annealing temperature. 375 to 650 °C (700 to 1200 °F)
Hot-working temperature. 750 to 875 °C (1400 to 1600 °F)

C10300

Commercial Names

Common name. Oxygen-free extra-low-phosphorus copper
Designation. OFXLP

Specifications

ASTM. Flat products: B 133, B 152, B 187, B 272, B 432. Pipe: B 42, B 188, B 302. Rod: B 12, B 133, B 187. Shapes: B 133, B 187. Tubing: B 68, B 75, B 88, B 111, B 188, B 251, B 280, B 306, B 359, B 372, B 395, B 447

Chemical Composition

Composition limits. 99.95 Cu + Ag + P min, 0.001 to 0.005 P, 0.05 max other (total)

Table 1 Typical mechanical properties of C10100 and C10200

Temper	Tensile strength MPa	Tensile strength ksi	Yield strength(a) MPa	Yield strength(a) ksi	Elongation in 50 mm (2 in.), %	Hardness HRF	Hardness HRB	Hardness HR30T	Shear strength MPa	Shear strength ksi	Fatigue strength(b) MPa	Fatigue strength(b) ksi
Flat products, 1 mm (0.04 in.) thick												
M20	235	34	69	10	45	45	...	...	160	23	...	...
OS025	235	34	76	11	45	45	...	...	160	23	76	11
OS050	220	32	69	10	45	40	...	...	150	22	...	...
H00	250	36	195	28	30	60	10	25	170	25	...	...
H01	260	38	205	30	25	70	25	36	170	25	...	...
H02	240	42	250	36	14	84	40	50	180	26	90	13
H04	345	50	310	45	6	90	50	57	195	28	90	13
H08	380	55	345	50	4	94	60	63	200	29	95	14
H10	395	57	360	53	4	95	62	64	200	29	...	...
Flat products, 6 mm (0.25 in.) thick												
M20	220	32	69	10	50	40	...	...	150	22	...	...
OS050	220	32	69	10	50	40	...	...	150	22	...	...
H00	250	36	195	28	40	60	10	...	170	25	...	...
H01	260	38	205	30	35	70	25	...	170	25	...	...
H04	345	50	310	45	12	90	50	...	195	28	...	...
Flat products, 25 mm (1 in.) thick												
H04	310	45	275	40	20	85	45	...	180	26	...	...
Rod, 6 mm (0.25 in.) in diameter												
H80 (40%)	380	55	345	50	10	94	80	...	200	29	...	...
Rod, 25 mm (1 in.) in diameter												
M20	220	32	69	10	55(c)	40	...	...	150	22	...	...
OS050	220	32	69	10	55(c)	40	...	...	150	22	...	...
H80 (35%)	330	48	305	44	16(d)	87	47	...	185	27	115	17
Rod, 50 mm (2 in.) in diameter												
H80 (16%)	310	45	275	40	20	85	45	...	180	26	...	...
Wire, 2 mm (0.08 in.) in diameter												
OS050	240	35	...	...	35(e)	45	...	...	165	24	...	...
H04	380	55	...	...	1.5(f)	...	...	...	200	29	...	...
H08	455	66	...	...	1.5(f)	...	...	...	230	33	...	...
Tubing, 25 mm (1 in.) outside diameter × 1.65 mm (0.065 in.) wall thickness												
OS025	235	34	76	11	45	45	...	...	160	23	...	...
OS050	220	32	69	10	45	40	...	...	150	22	...	...
H55 (15%)	275	40	220	32	25	77	35	45	180	26	...	...
H80 (40%)	380	55	345	50	8	95	60	63	200	29	...	...
Shapes, 13 mm (0.50 in.) in diameter												
M20	220	32	69	10	50	45	...	...	150	22	...	...
M30	220	32	69	10	50	45	...	...	150	22	...	...
OS050	220	32	69	10	50	45	...	...	150	22	...	...
H80 (15%)	275	40	220	32	30	...	35	...	180	26	...	...

(a) At 0.5% extension under load. (b) At 10^8 cycles. (c) 70% reduction in area. (d) 55% reduction in area. (e) Elongation in 254 mm (10 in.). (f) Elongation in 1500 mm (60 in.)

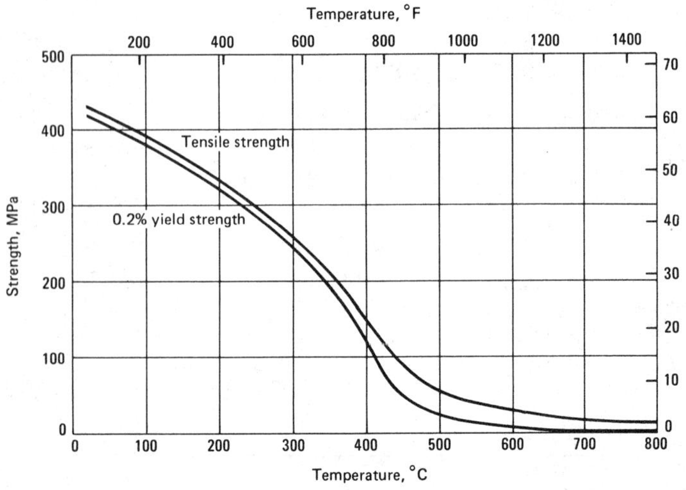

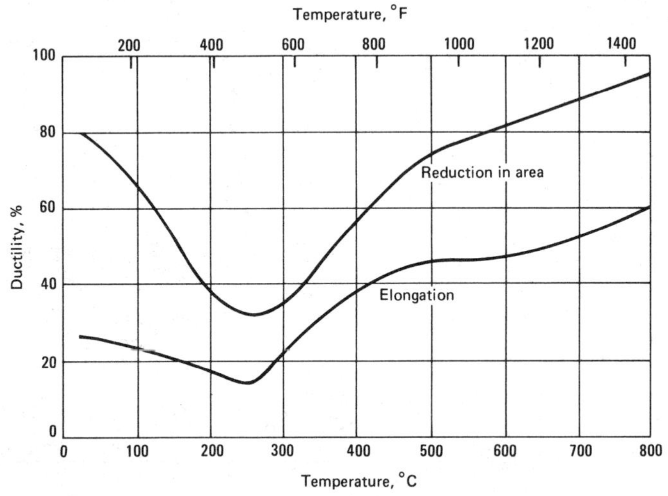

Fig. 1 Elevated-temperature tensile properties of C10100 or C10200 rod, 1180 temper

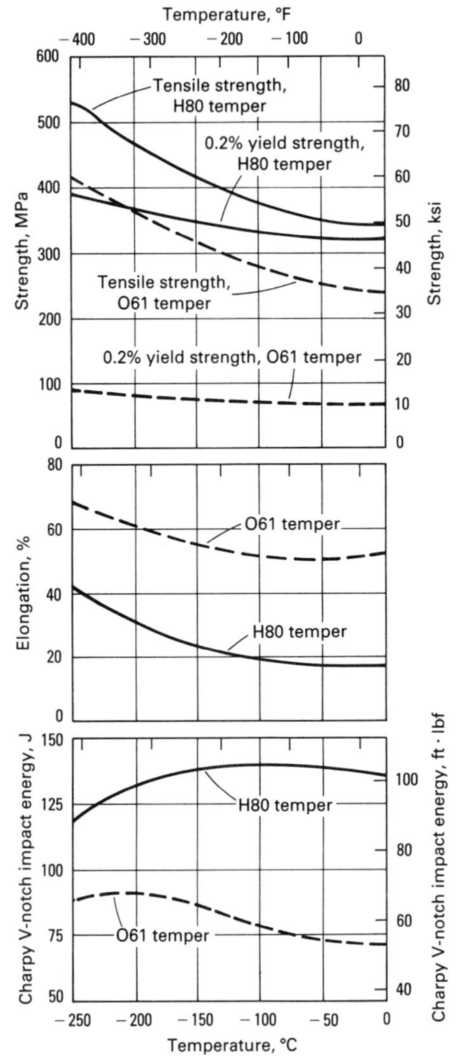

Fig. 2 Low-temperature mechanical properties of C10100 or C10200 bar

Applications

Typical uses. Busbars, electrical conductors and terminals, commutators, tubular busbars, clad products, waveguide tubing, thermostatic control tubing

Mechanical Properties

Tensile properties. See Table 4.
Shear strength. See Table 4.
Hardness. See Table 4.
Elastic modulus. Tension, 115 GPa (17 × 10^6 psi); shear, 44 GPa (6.4 × 10^6 psi)
Fatigue strength. 1 mm (0.04 in.) thick strip: OS025 temper, 76 MPa (11 ksi); H02 or H04 temper, 90 MPa (13 ksi)

Mass Characteristics

Density. 8.94 g/cm^3 (0.323 lb/in.^{3}) at 20 °C (68 °F)

Thermal Properties

Liquidus temperature. 1083 °C (1981 °F)
Solidus temperature. 1083 °C (1981 °F)
Coefficient of linear thermal expansion. 17.0 μm/m · K (9.4 μin./in. · °F) at 20 to 100 °C (68 to 212 °F); 17.3 μm/m · K (9.6 μin./in. · °F) at 20 to 200 °C (68 to 392 °F); 17.7 μm/m · K (9.8 μin./in. · °F) at 20 to 300 °C (68 to 572 °F)
Specific heat. 385 J/kg · K (0.092 Btu/lb · °F) at 20 °C (68 °F)
Thermal conductivity. 386 W/m · K (223 Btu/ft · h · °F) at 20 °C (68 °F)

Electrical Properties

Electrical conductivity. O61 temper: volumetric, 99% IACS at 20 °C (68 °F)
Electrical resistivity. 17.4 nΩ · m at 20 °C (68 °F)

Fabrication Characteristics

Machinability. 20% of C36000 (free-cutting brass)
Annealing temperature. 375 to 650 °C (700 to 1200 °F)
Hot-working temperature. 750 to 875 °C (1400 to 1600 °F)

C10400, C10500, C10700

Commercial Names

Trade name. AMSIL copper
Common name. Oxygen-free silver-copper

Specifications

ASTM. See Table 5.

Chemical Composition

Composition limits. C10400: 99.95 Cu + Ag min, 0.027 Ag min. C10500: 99.95 Cu + Ag min, 0.034 Ag min. C10700: 99.95 Cu + Ag min, 0.085 Ag min

Applications

Typical uses. Busbars, conductivity wire, contacts, radio parts, windings, switches, commutator segments, automotive gaskets and radiators, chemical plant equipment, printing rolls, printed-circuit foil. Many uses are based on the good creep strength at elevated temperatures and the high softening temperature of these alloys.

Mechanical Properties

Tensile properties. See Table 6 and Fig. 3.
Shear strength. See Table 6.
Hardness. See Table 6.
Elastic modulus. Tension, 115 GPa (17 × 10^6 psi); shear, 44 GPa (6.4 × 10^6 psi)

Mass Characteristics

Density. 8.94 g/cm^3 (0.323 lb/in.^{3}) at 20 °C (68 °F)

Thermal Properties

Liquidus temperature. 1083 °C (1981 °F)
Solidus temperature. 1083 °C (1981 °F)
Coefficient of linear thermal expansion. 17.0 μm/m · K (9.4 μin./in. · °F) at 20 to 100 °C (68 to 212 °F); 17.3 μm/m · K (9.6 μin./in. · °F) at 20 to 200 °C (68 to 392 °F); 17.7 μm/m · K (9.8 μin./in. · °F) at 20 to 300 °C (68 to 572 °F)
Specific heat. 385 J/kg · K (0.092 Btu/lb · °F) at 20 °C (68 °F)
Thermal conductivity. 388 W/m · K (224 Btu/ft · h · °F) at 20 °C (68 °F)

Electrical Properties

Electrical conductivity. O61 temper: volumetric, 100% IACS at 20 °C (68 °F)

Table 2 Creep properties of C10100 and C10200

	Test temperature		Stress(a) for creep rate of												
			10^{-6}%/h		10^{-5}%/h		10^{-4}%/h		10^{-3}%/h		10^{-2}%/h		10^{-1}%/h	
Condition and grain size	°C	°F	MPa	ksi	MPa	ksi	MPa	ksi	MPa	ksi	MPa	ksi	MPa	ksi
OS025(b)	43	110	...	...	...	...	...	...	170	25	185	27	200	29
	120	250	...	...	...	...	...	...	125	18	150	22	165	24
	150	300	11	1.6	25	3.6	55	8.0	110	16	130	19	150	22
	205	400	3	0.5	10	1.5	33	4.8	...	...	...	...	...	...
	260	500	0.7	0.1	3	0.4	12	1.7	...	...	...	...	...	...
	370	700	...	...	...	...	...	...	...	...	21	3.1	(40)	(5.8)
	480	900	...	...	...	...	...	...	...	...	9.9	1.45	(23)	(3.3)
Cold drawn 40%(c)	43	110	...	...	...	...	...	...	310	45	330	48	...	...
	120	250	...	...	...	...	...	...	240	35	270	39	(295)	(43)
	150	300	...	...	...	...	...	...	200	29	235	34	250	36
	370	700	...	...	...	...	...	...	11	1.6	26	3.8	(39)	(5.6)
	480	900	...	...	...	...	...	...	...	...	8.3	1.2	(17)	(2.4)
	650	1200	...	...	...	...	...	...	...	...	3	0.5	6	0.9
Cold drawn 84%(d)	150	300	...	...	55	8.0	89.6	13.0	...	...	...	...	...	...
	205	400	(4.5)	(0.65)	12	1.7	35	5.0	...	...	...	...	...	...

(a) Parentheses indicate extrapolated values. (b) Tensile strength, 220 MPa (31.9 ksi) at 21 °C (70 °F). (c) Tensile strength, 352 MPa (51.1 ksi) at 21 °C (70 °F). (d) Tensile strength, 376 MPa (54.5 ksi) at 21 °C (70 °F)

Table 3 Stress-rupture properties of C10100 and C10200

Temper or condition	Test temperature °C	Test temperature °F	Stress(a) for rupture in 10 h, MPa	10 h, ksi	100 h, MPa	100 h, ksi	1000 h, MPa	1000 h, ksi
OS025(b)	150	300	...	...	161	23.4	147	21.3
	200	380	...	...	130	18.9	106	15.3
Cold drawn 40%(c)	120	250	...	...	272	39.4	(245)	(35.6)
	150	300	...	...	241	35.0	(215)	(31.2)
H80(d)	450	840	33	4.8	17	2.4	...	...
	650	1200	9.7	1.4	5.2	0.75	...	...

(a) Parentheses indicate extrapolated values. (b) Tensile strength, 238 MPa (34.5 ksi) at 21 °C (70 °F). (c) Tensile strength, 352 MPa (51.1 ksi) at 21 °C (70 °F). (d) Tensile strength, 426 MPa (61.8 ksi) at 21 °C (70 °F)

Electrical resistivity. 17.2 nΩ · m at 20 °C (68 °F)

Fabrication Characteristics

Machinability. 20% of C36000 (free-cutting brass)
Annealing temperature. 475 to 750 °C (900 to 1400 °F). See also Fig. 3.
Hot-working temperature. 750 to 875 °C (1400 to 1600 °F)

C10800

Commercial Names

Trade name. AMAX-LP copper
Common name. Oxygen-free low-phosphorus copper

Specifications

ASTM. Flat products: B 113, B 152, B 187, B 432. Pipe: B 42, B 302. Rod: B 12, B 133. Shapes: B 133. Tubing: B 68, B 75, B 88, B 111, B 188, B 251, B 280, B 306, B 357, B 360, B 372, B 395, B 447, B 543

Chemical Composition

Composition limits. 99.95 Cu + Ag + P min, 0.005 to 0.012 P

Applications

Typical uses. Refrigerator and air conditioner tubing and terminals, commutators, clad products, gas and burner lines and units, oil burner tubes, condenser and heat exchanger tubes, pulp and paper lines, steam and water lines, tank gage lines, plumbing pipe and tubing, thermostatic control tubing, plate for welded continuous casting molds, tanks, kettles, rotating bands, and similar uses

Mechanical Properties

Tensile properties. See Table 7.
Shear strength. See Table 7.
Hardness. See Table 7.
Elastic modulus. Tension, 115 GPa (17×10^6 psi); shear, 44 GPa (6.4×10^6 psi)
Fatigue strength. See Table 7.

Mass Characteristics

Density. 8.94 g/cm^3 (0.323 lb/in.^{3}) at 20 °C (68 °F)

Thermal Properties

Liquidus temperature. 1083 °C (1981 °F)
Solidus temperature. 1083 °C (1981 °F)
Coefficient of linear thermal expansion. 17.0 μm/m · K (9.4 μin./in. · °F) at 20 to 100 °C (68 to 212 °F); 17.3 μm/m · K (9.6 μin./in. · °F) at 20 to 200 °C (68 to 392 °F); 17.7 μm/m · K (9.8 μin./in. · °F) at 20 to 300 °C (68 to 572 °F)
Specific heat. 385 J/kg · K (0.092 Btu/lb · °F) at 20 °C (68 °F)
Thermal conductivity. 350 W/m · K (202 Btu/ft · h · °F) at 20 °C (68 °F)

Electrical Properties

Electrical conductivity. O61 temper: volumetric, 92% IACS at 20 °C (68 °F)
Electrical resistivity. 18.7 nΩ · m at 20 °C (68 °F)

Table 4 Typical mechanical properties of C10300

Temper	Tensile strength, MPa	Tensile strength, ksi	Yield strength(a), MPa	Yield strength(a), ksi	Elongation in 50 mm (2 in.), %	Hardness HRF	Hardness HRB	Hardness HR30T	Shear strength, MPa	Shear strength, ksi
Flat products, 1 mm (0.04 in.) thick										
OS050	220	32	69	10	45	40	...	...	150	22
OS025	235	34	76	11	45	45	...	...	160	23
H00	250	36	195	28	30	60	10	25	170	25
H01	260	38	205	30	25	70	25	36	170	25
H02	290	42	250	36	14	84	40	50	180	26
H04	345	50	310	45	6	90	50	57	195	28
Flat products, 6 mm (0.25 in.) thick										
OS050	220	32	69	10	50	40	...	...	150	22
H00	250	36	195	28	40	60	10	...	170	25
H04	345	50	310	45	12	90	50	...	195	28
M20	220	32	69	10	50	40	...	...	150	22
Flat products, 25 mm (1 in.) thick										
H04	310	45	275	40	20	85	45	...	180	26
Rod, 6 mm (0.25 in.) in diameter										
H80 (40%)	380	55	345	50	20	94	60	...	200	29
H80 (35%)	330	48	305	44	16	87	47	...	185	27
H80 (16%)	310	45	275	40	20	85	45	...	180	26
Tubing, 25 mm (1 in.) outside diameter × 1.65 mm (0.65 in.) wall thickness										
OS050	220	32	69	10	45	40	...	...	150	22
OS025	235	34	76	11	45	45	...	...	160	23
H80 (15%)	275	40	220	32	25	77	35	45	180	26
H80 (40%)	380	55	345	50	8	95	60	63	200	29
Pipe, ¾ SPS										
H80 (30%)	345	50	310	45	10	90	50	...	195	28
Shapes, 13 mm (0.50 in.) section size										
OS050	220	32	69	10	50	40	...	...	150	22
H80 (15%)	275	40	220	32	30	...	35	...	180	26

(a) At 0.5% extension under load

Table 5 Summary of ASTM and government specifications for C10400, C10500, and C10700

Mill product	C10400	C10500	C10700
ASTM numbers			
Flat products	B 48, B 133, B 152, B 187, B 272	B 152, B 187, B 272	B 152, B 187, B 272
Pipe	B 42, B 188	B 188	B 188
Rod	B 12, B 49, B 133, B 187	B 12, B 49, B 133, B 187	B 12, B 49, B 133, B 187
Shapes	B 133, B 187	B 187	B 187
Tube	B 188	B 188	B 188
Wire	B 1, B 2, B 3	B 1, B 2, B 3	B 1, B 2, B 3
Government numbers			
Flat products	QQ-C-502, QQ-C-576	QQ-C-502, QQ-C-576	QQ-C-576
Rod	QQ-C-502	QQ-C-502	QQ-C-502
Shapes	QQ-C-502, QQ-B-825	QQ-C-502	QQ-B-825, MIL-B-19231
Tube	QQ-B-825	QQ-B-825	QQ-B-825
Wire	QQ-W-343, MIL-W-3318	QQ-W-343, MIL-W-3318	QQ-W-343, MIL-W-3318

Fabrication Characteristics

Machinability. 20% of C36000 (free-cutting brass)
Annealing temperature. 375 to 650 °C (700 to 1200 °F)
Hot-working temperature. 750 to 875 °C (1400 to 1600 °F)

C11000
99.95Cu-0.04O

Commercial Names

Common name. Electrolytic tough pitch copper
Designation. ETP

Specifications

AMS. Sheet and strip: 4500. Wire: 4701
ASME. Plate for locomotive fireboxes: SB11. Rod for locomotive staybolts: SB12
ASTM: See Table 8.
SAE. J463
Government. Federal specifications: See Table 8. Military specifications: Rod, MIL-C-12166; wire, MIL-W-3318, MIL-W-6712

Chemical Composition

Composition limits. 99.90 Cu min (silver counted as copper)
Silver has little effect on mechanical and electrical properties, but does raise the recrystallization temperature and tends to produce a fine-grain copper.
Iron, as present in commercial copper, has no effect on mechanical properties, but even traces of iron can cause C11000 to be slightly ferromagnetic.
Sulfur causes spewing and unsoundness, and is kept below 0.003% in ordinary refinery practice.

Table 6 Typical mechanical properties of C10400, C10500, and C10700

Temper	Tensile strength MPa	Tensile strength ksi	Yield strength(a) MPa	Yield strength(a) ksi	Elongation in 50 mm (2 in.), %	Hardness HRF	Hardness HRB	Hardness HR30T	Shear strength MPa	Shear strength ksi
Flat products, 1 mm (0.04 in.) thick										
OS025	235	34	76	11	45	45	...	...	160	23
H00	250	36	195	28	30	60	10	25	170	25
H01	260	38	205	30	25	70	25	36	170	25
H02	290	42	250	36	14	84	40	50	180	26
H04	345	50	310	45	6	90	50	57	195	28
H08	380	55	345	50	4	94	60	63	200	29
H10	395	57	365	53	4	95	62	64	200	29
M20	235	34	69	10	45	45	...	...	160	23
Flat products, 6 mm (0.25 in.) thick										
OS050	220	32	69	10	50	40	...	...	150	22
H00	250	36	195	28	40	60	10	...	170	25
H01	260	38	205	30	35	70	25	...	170	25
H04	345	50	310	45	12	90	50	...	195	28
M20	220	32	69	10	50	40	...	...	150	22
Flat products, 25 mm (1 in.) thick										
H04	310	45	275	40	20	85	45	...	180	26
Rod, 6 mm (0.25 in.) in diameter										
H80 (40%)	380	55	345	50	10	94	60	...	200	29
Rod, 25 mm (1 in.) in diameter										
OS050	220	32	69	10	55	40	...	...	150	22
H80 (35%)	330	48	305	44	16	87	47	...	185	27
M20	220	32	69	10	55	40	...	...	150	22
Rod, 50 mm (2 in.) in diameter										
H80 (16%)	310	45	275	40	20	85	45	...	180	26
Wire, 2 mm (0.08 in.) in diameter										
OS050	240	35	...	...	35(b)	...	...	...	165	24
H04	380	55	...	...	1.5(c)	...	...	...	200	29
H08	455	66	...	...	1.5(c)	...	...	...	230	33
Shapes, 13 mm (0.50 in.) section size										
OS050	220	32	69	10	50	40	...	...	150	22
H80 (15%)	275	40	220	32	30	...	35	...	180	26
M20	220	32	69	10	50	40	...	...	150	22
M30	220	32	69	10	50	40	...	...	150	22
Tubing, 25 mm (1.0 in.) diameter × 1.65 mm (0.065 in.) wall thickness										
OS050	220	32	69	10	45	40	...	...	150	22
OS025	235	34	76	11	45	45	...	...	160	23
H80 (15%)	275	40	220	32	25	77	35	45	180	26
H80 (50%)	380	55	345	50	8	95	60	62	200	29

(a) At 0.5% extension under load. (b) Elongation in 25 mm (10 in.). (c) Elongation in 1500 mm (60 in.)

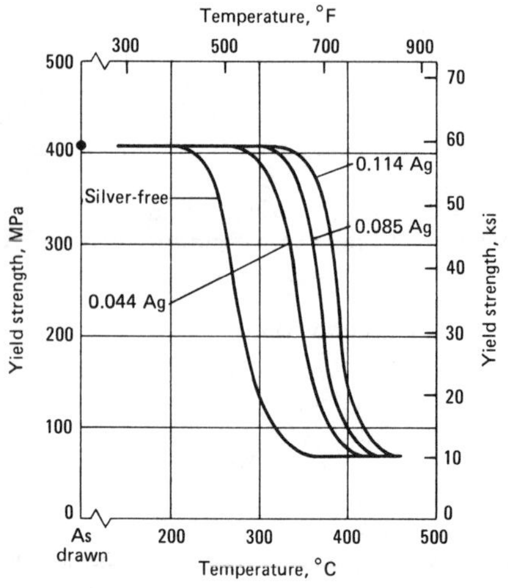

Fig. 3 Softening characteristics of oxygen-free copper containing various amounts of silver. Data are for copper wire cold worked 90% to a diameter of 2 mm (0.08 in.) and then annealed ½ h at various temperatures.

Selenium and tellurium are usually considered undesirable impurities but may be added to improve machinability.
Bismuth creates brittleness in amounts greater than 0.001%.
Lead should not be present in amounts greater than 0.005% if the copper is to be hot rolled.
Cadmium is rarely present; its effect is to toughen copper without much loss in conductivity.
Arsenic decreases the conductivity of copper noticeably, although it is often added intentionally to copper not used in electrical service because it increases the toughness and heat resistance of the metal.
Antimony is sometimes added to the copper when a high recrystallization temperature is desired.

Applications

Typical uses. Produced in all forms except pipe, and used for building fronts, downspouts, flashing, gutters, roofing, screening, spouting, gaskets, radiators, busbars, electrical wire, stranded conductors, contacts, radio parts, switches, terminals, ball floats, butts, cotter pins, nails, rivets, soldering copper, tacks, chemical process equipment, kettles, pans, printing rolls, rotating bands, roadbed expansion plates, vats
Precautions in use. C11000 is subject to embrittlement when heated to 370 °C (700 °F) or above in a reducing atmosphere, as in annealing, brazing, or welding. If hydrogen or carbon monoxide is present in the reducing atmosphere, embrittlement can be rapid.

Mechanical Properties

Tensile properties. See Table 9 and Fig. 4 to 8.
Shear strength. See Table 9.
Hardness. See Table 9 and Fig. 9.
Poisson's ratio. 0.33
Elastic modulus. O60 temper: tension, 115 GPa (17×10^6 psi); shear, 44 GPa (6.4×10^6 psi). Cold-worked (H) tempers: tension, 115 to 130 GPa (17×10^6 to 19×10^6 psi); shear, 44 to 49 GPa (6.4×10^6 to 7.1×10^6 psi)
Impact strength. See Table 10.
Fatigue strength. See Table 9; values shown there are typical of all tough pitch, oxygen-free, phosphorus-deoxidized and arsenical coppers. Copper does not exhibit an endurance limit under fatigue loading and, on the average, will fracture in fatigue at the stated number of cycles when subjected to an alternating stress equal to the corresponding fatigue strength (see Fig. 10).
Creep-rupture characteristics. See Table 11.
Specific damping capacity. The damping capacity of coppers and brasses depends on the amplitude and, in some instances, on the frequency of vibration; it is also affected by the condition of the metal. Up to a point, damping capacity increases with increasing cold work; for example, the damping capacity of 70-30 brass has been reported to increase for reductions up to 60%. When subjected to the same conditions, coppers have about three times the damping capacity of C21000 or C22000. A specific damping capacity of 5×10^{-5} has been recorded for

Table 7 Typical mechanical properties of C10800

Temper	Tensile strength MPa	Tensile strength ksi	Yield strength(a) MPa	Yield strength(a) ksi	Elongation in 50 mm (2 in.), %	Hardness HRF	Hardness HRB	Hardness HR30T	Shear strength MPa	Shear strength ksi	Fatigue strength(b) MPa	Fatigue strength(b) ksi
Flat products, 1 mm (0.04 in.) thick												
OS025	235	34	76	11	45	45	...	...	160	23	76	11
H00	250	36	195	28	30	60	10	25	170	25	...	...
H01	260	38	205	30	25	70	25	36	170	25	...	...
H02	290	42	250	36	14	84	40	50	180	26	90	13
H04	345	50	310	45	6	90	50	57	195	28	90	13
H08	380	55	345	50	4	94	60	63	200	29	97	14
Flat products, 6 mm (0.25 in.) thick												
OS050	220	32	69	10	50	40	...	...	150	22	...	...
H00	250	36	195	28	40	60	10	...	170	25	...	...
H04	345	50	310	45	12	90	50	...	195	28	...	...
M20	220	32	69	10	50	40	...	...	150	22	...	...
Flat products, 25 mm (1 in.) thick												
H04	310	45	275	40	20	85	45	...	180	26	...	...
Rod, 6 mm (0.25 in.) in diameter												
H80 (40%)	380	55	345	50	20	94	60	...	200	29	...	...
Rod, 25 mm (1 in.) in diameter												
H80 (35%)	330	48	305	44	16	87	47	...	185	27	115	17
Rod, 50 mm (2 in.) in diameter												
H80 (16%)	310	45	275	40	20	85	45	...	180	26	...	...
Tubing, 25 mm (1 in.) outside diameter × 1.65 mm (0.065 in.) wall thickness												
OS050	220	32	69	10	45	40	...	...	150	22	...	...
OS025	235	34	76	11	45	45	...	...	160	23	...	...
H55 (15%)	275	40	220	32	25	77	35	45	180	26	...	...
H80 (40%)	380	55	345	50	8	95	60	63	200	29	...	...
Pipe, ¾ SPS												
H80 (30%)	345	50	310	45	10	90	50	...	195	28	...	...

(a) At 0.5% extension under load. (b) At 10^8 cycles

Table 8 ASTM and federal specifications for C11000

Product and condition	Specification number ASTM	Federal
Flat products		
General requirements for copper and copper alloy plate, sheet, strip, and rolled bar	B 248	...
Sheet, strip, plate, and rolled bar	B 152	QQ-C-576
Sheet, lead coated	B 101	...
Sheet and strip for building construction	B 370	...
Strip and flat wire	B 272	QQ-C-502
Foil, strip, and sheet for printed circuits	B 451	...
Rod, bar, and shapes		
General requirements for copper and copper alloy rod, bar, and shapes	B 249	...
Rod, bar, and shapes	B 133	QQ-C-502, QQ-C-576
Rod, hot rolled	B 49	...
Rod, bar, and shapes for forging	B 124	QQ-C-502
Busbars, rods, and shapes	B 187	QQ-B-825
Wire		
General requirements for copper and copper alloy wire	B 250	...
Hard drawn	B 1	QQ-W-343
Tinned	B 246	...
Medium-hard drawn	B 2	QQ-W-343
Tinned	B 246	...
Soft	B 3	QQ-W-343
Lead alloy coated	B 189	...
Nickel coated	B 355	...
Rectangular and square	B 48, B 272	...
Tinned	B 33	...
Silver coated	B 298	...
Trolley	B 47, B 116	...
Conductors		
Bunch stranded	B 174	...
Concentric-lay stranded	B 8, B 226, B 496	...
Conductors for electronic equipment	B 286, B 470	...
Rope-lay stranded	B 172, B 173	...
Composite conductors (copper plus copper-clad steel)	B 229	...
Tubular products		
Bus pipe and tube	B 188	QQ-B-825
Pipe	...	WW-P-377
Welded copper tube	B 447	...
Miscellaneous		
Standard classification of coppers	B 224	...
Electrolytic Cu wirebars, cakes, slabs, billets, ingots, and ingot bars	B 5	...
Anodes	...	QQ-A-673
Die forgings	B 283	...

single-crystal annealed copper. Log decrement: O60 temper, 3.2; cold rolled (H) tempers, 5.0

Coefficient of friction. Values given below apply to any of the unalloyed coppers in contact with the indicated materials without lubrication of any kind between the contacting surfaces:

Opposing material	Coefficient of friction Static	Sliding
Carbon steel	0.53	0.36
Cast iron	1.05	0.29
Glass	0.68	0.53

Mass Characteristics

Density. Solid: 8.89 g/cm^3 (0.321 lb/in.^{3}) at 20 °C (68 °F); 8.32 g/cm^3 (0.301 lb/in.^{3}) at 1083 °C (1981 °F); see also Fig. 11. Liquid: 7.93 g/cm^3 (0.286 lb/in.^{3}) at 1083 °C (1981 °F)

Thermal Properties

Liquidus temperature. 1083 °C (1981 °F)

Solidus temperature. Eutectic point, 1065 °C (1950 °F)

Coefficient of linear thermal expansion. 17 μm/m · K (9.4 μin./in. · °F) at 20 to 100 °C (68 to 212 °F); 17.3 μm/m · K (9.6 μin./in. · °F) at 20 to 200 °C (68 to 392 °F); 17.7 μm/m · K (9.8 μin./in. · °F) at 20 to 300 °C (68 to 572 °F). See also Fig. 12.

Specific heat. 385 J/kg · K (0.092 Btu/lb · °F) at 20 °C (68 °F)

Enthalpy. See Fig. 12.

Latent heat of fusion. 205 kJ/kg

Thermal conductivity. 388 W/m · K (224 Btu/ft · h · °F) at 20 °C (68 °F). For high conductivity coppers, a value of 387 W/m · K (223 Btu/ft · h · °F) is an adjusted value corresponding to an electrical conductivity of 101% IACS:

Temperature K	°C	Thermal conductivity W/m · K	Btu/ft · h · °F
4.2	−268.8	300	170
20	−253	1300	750
77	−196	550	318
194	−79	400	230
273	0	390	225
373	100	380	220
573	300	370	215
973	700	300	170

Electrical Properties

Electrical conductivity. Volumetric: O60 temper: 100 to 101.5% IACS; H14 temper, 97% IACS. See also Fig. 13.

Electrical resistivity. O60 temper: 17.00 to 17.24 nΩ · m; temperature coefficient, 0.00393/K at −100 to 200 °C (−148 to 392 °F) for 100% IACS material, 0.00397/K at −100 to 200 °C for 101% IACS material. H14 temper: 1.78 nΩ · m; temperature coefficient, 0.00381/K at 0 to 100 °C (32 to 212 °F) for 97% IACS material. See also Fig. 13.

Thermoelectrical potential. See Fig. 14.

Electrochemical equivalent. Cu^{2+}, 0.329 mg/C; Cu^{+}, 0.659 mg/C

Electrolytic solution potential. Cu^{2+}, −0.344 V versus standard hydrogen electrode; Cu^{+}, −0.470 V versus standard hydrogen electrode; temperature coefficient, −0.01 mV/K at 20 to 50 °C (68 to 122 °F)

Hydrogen overvoltage. Approximately 0.23 V in dilute sulfuric acid; specific value varies with current density.

Hall effect. Hall coefficient, −52 pV · m/A · T

Optical Properties

Color. Reddish metallic

Spectral reflectivity. 32.7% for λ of 420 nm; 43.7% for λ of 500 nm; 71.8% for λ of 600 nm; 83.4% for λ of 700 nm. See also Fig. 15 and 16.

Chemical Properties

General corrosion behavior. Although many factors influence the corrosion resistance of copper under specific conditions of service, copper is generally less subject to corrosion than other engineering metals. Copper often is used where resistance to corrosion is of prime importance. Sometimes, it is better to use a copper alloy rather than an unalloyed copper.

In general, copper resists nonoxidizing mineral and organic acids, caustic solutions, saline solutions, and various natural waters or process waters. It is suitable for underground service because it resists soil corrosion. Copper is not suitable for service in oxidizing acids such as nitric acid, and it is not recommended for use with ammonia, nitric acid, acid chromate solutions, ferric chloride, mercury salts, perchlorates, or persulfates. Also, copper may corrode in aerated nonoxidizing acids such as sulfuric or acetic acids, even though it is practically immune to these acids in the complete absence of air.

Tough pitch copper is considered to be immune to stress-corrosion cracking in am-

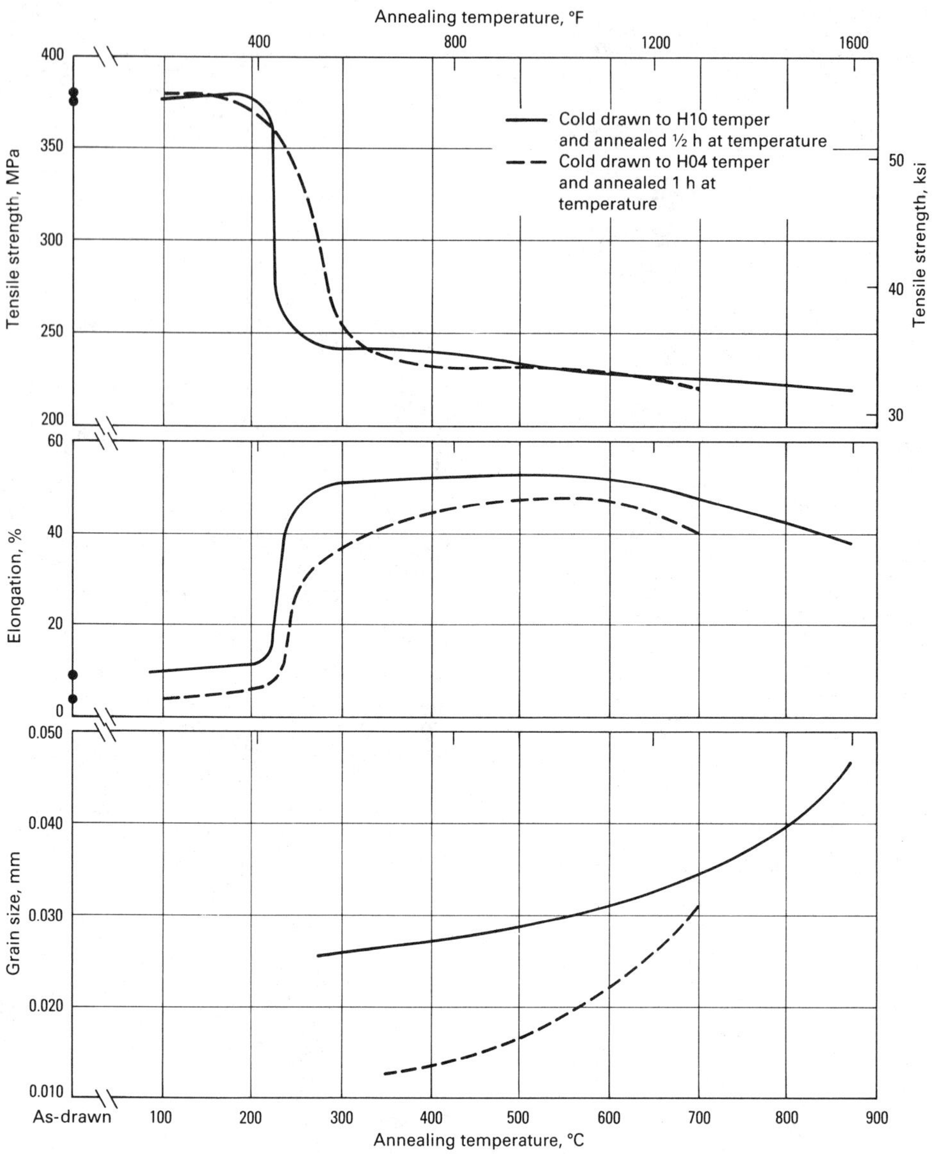

Fig. 4 Variation of tensile properties and grain size of electrolytic tough pitch copper (C11000) and similar coppers

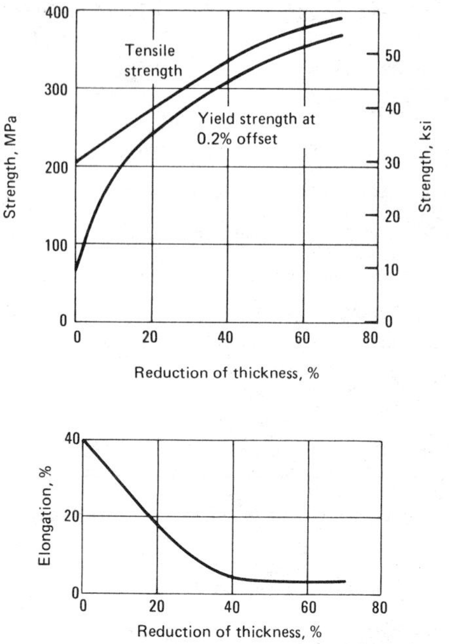

Fig. 5 Variation of tensile properties with amount of cold reduction by rolling for C11000 and similar coppers

monia and the other agents that induce season cracking of brasses. However, tough pitch copper is susceptible to embrittlement in reducing atmospheres, especially those containing hydrogen.

Resistance to specific corroding agents. Depending on concentration and specific conditions of exposure, copper generally resists the following agents.

Acids: mineral acids such as hydrochloric and sulfuric acids; organic acids such as acetic acid (including vinegar and acetates), carbolic acid, citric acid, formic acid, oxalic acid, and tartaric acid; fatty acids; and acidic solutions containing sulfur, such as the sulfurous acid and sulfite solutions used in pulp mills

Alkalies: fused sodium or potassium hydroxide; concentrated or dilute caustic solutions

Salt solutions: aluminum chloride, aluminum sulfate, calcium chloride, copper sulfate, sodium carbonate, sodium nitrate, sodium sulfate, and zinc sulfate

Waters: all types of potable water, many industrial and mine waters, seawater, and brackish water

Fabrication Characteristics

Machinability. 20% of C36000 (free-cutting brass)
Forgeability. 65% of C37700 (forging brass)
Formability. Excellent for cold working and hot forming
Weldability. Soldering: excellent. Brazing and resistance butt welding: good. Gas-shielded arc welding: fair. Oxyfuel gas, shielded metal-arc, resistance spot, and resistance seam welding: not recommended

Annealing temperature. 475 to 750 °C (900 to 1400 °F). See also Fig. 4 and 17.
Hot-working temperature. 750 to 875 °C (1400 to 1600 °F)
Typical softening temperature. 360 °C (675 °F)

C11100
99.95Cu-0.04O-0.01Cd

Commercial Names

Previous trade name. Electrolytic tough pitch copper, anneal resistant
Common name. Anneal-resistant electrolytic copper

Specifications

ASTM. See Table 12.
SAE. Bar, plate, sheet, strip: J461, J463
Government: See Table 12.

Chemical Composition

Composition limits. 99.90 Cu min. Limits on O and Cd or other elements present to make this copper anneal resistant are established by conductivity tests and/or stress relaxation tests rather than by chemical analysis.

Applications

Typical uses. Produced mainly as wire for electrical power transmission where resistance to softening under overloads is desired

Mechanical Properties

Typical tensile properties. Tensile strength, 455 MPa (66 ksi); elongation, 1.5% in 150 cm (60 in.)

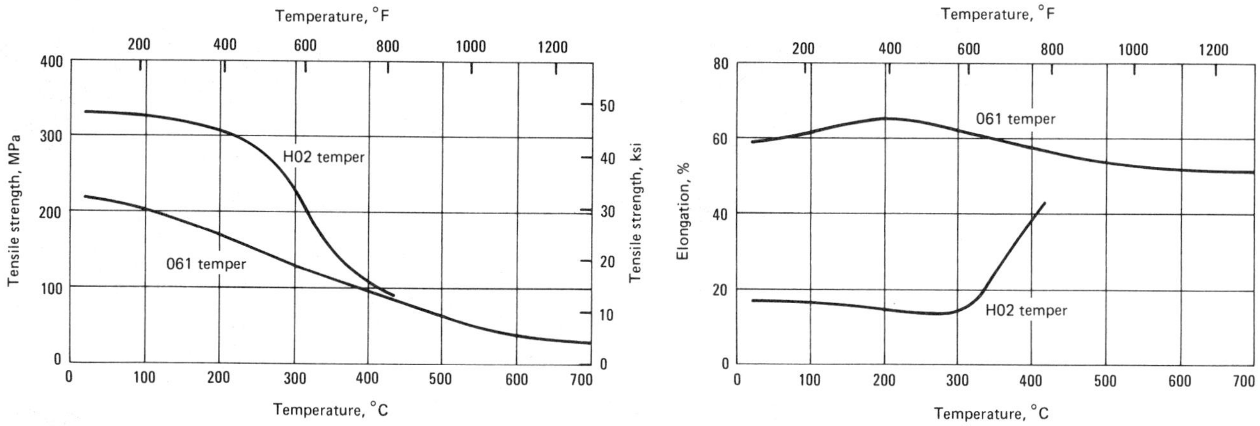

Fig. 6 Short-time elevated-temperature tensile properties of C11000 and similar coppers

Elastic modulus. Tension, 115 GPa (17 × 10^6 psi); shear, 44 GPa (6.4 × 10^6 psi)

Mass Characteristics

Density. 8.89 to 8.94 g/cm³ (0.321 to 0.323 lb/in.³) at 20 °C (68 °F)

Thermal Properties

Liquidus temperature. 1085 °C (1980 °F)
Solidus temperature. 1065 °C (1950 °F)
Coefficient of linear thermal expansion. 17.0 μm/m · K (9.4 μin./in. · °F) at 20 to 100 °C (68 to 212 °F); 17.3 μm/m · K (9.6 μin./in. · °F) at 20 to 200 °C (68 to 392 °F); 17.7 μm/m · K (9.8 μin./in. · °F) at 20 to 300 °C (68 to 572 °F)
Specific heat. 385 J/kg · K (0.092 Btu/lb · °F) at 20 °C (68 °F)
Thermal conductivity. 388 W/m · K (224 Btu/ft · h · °F)

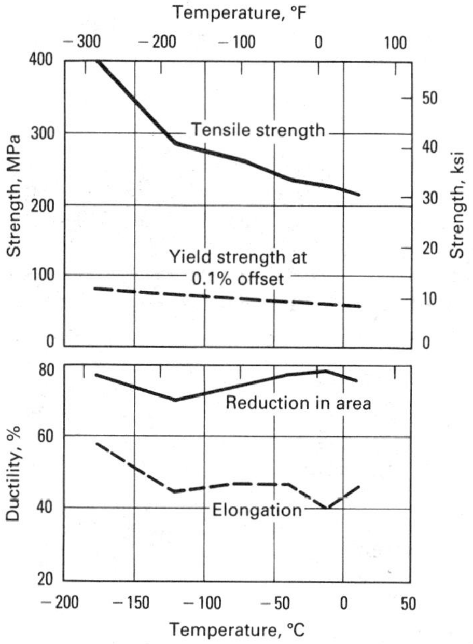

Fig. 7 Low-temperature tensile properties of C11000 and similar coppers

Electrical Properties

Electrical conductivity. Volumetric: 100% IACS at 20 °C (68 °F)
Electrical resistivity. 17.2 nΩ · m at 20 °C (68 °F)

Fabrication Characteristics

Machinability. 20% of C36000 (free-cutting brass)
Forgeability. 65% of C37700 (forging brass)
Formability. Excellent for cold working and hot forming. Common processes include drawing, stranding, and stamping.
Weldability. Soldering: excellent. Brazing and resistance butt welding: good. Gas-shielded arc welding: fair. Oxyacetylene coated metal-arc and resistance spot and seam welding: not recommended
Annealing temperature. 475 to 750 °C (900 to 1400 °F)
Hot-working temperature. 750 to 875 °C (1400 to 1600 °F)
Typical softening temperature. 355 °C (675 °F)

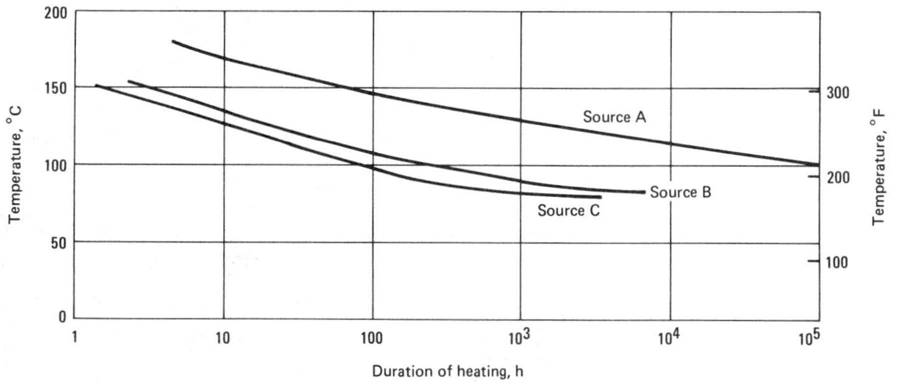

Fig. 8 Stress relaxation curves for C11000 and similar coppers. Data are for H80 temper wire, 2 mm (0.08 in.) in diameter, and represent the time-temperature combination necessary to produce a 5% reduction in tensile strength.

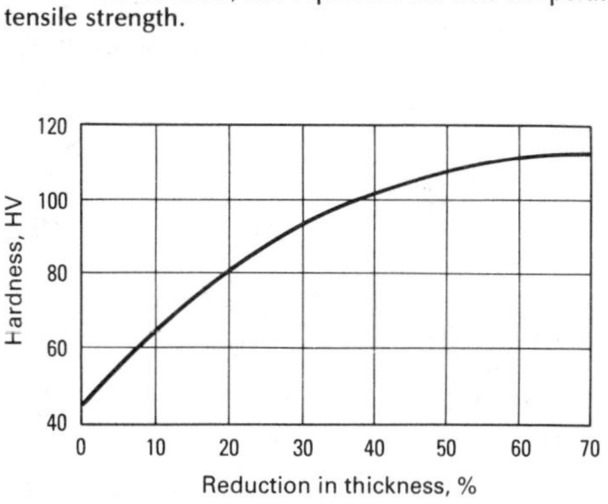

Fig. 9 Variation of hardness with amount of cold reduction by rolling for C11000 and similar coppers

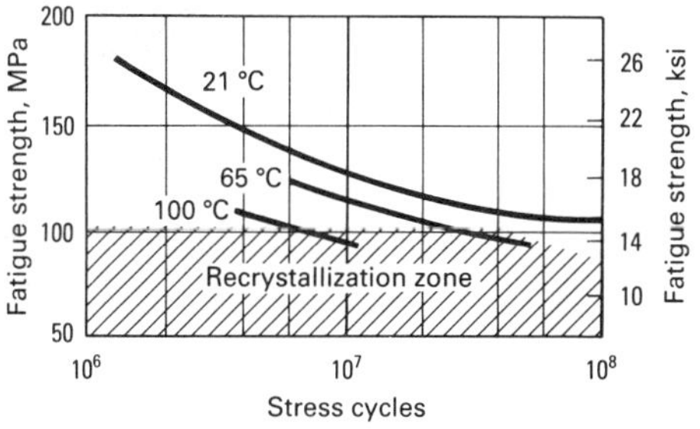

Fig. 10 Rotating-beam fatigue strength of C11000 wire, 2 mm (0.08 in.) in diameter, H80 temper

Table 9 Typical mechanical properties of C11000

Temper	Tensile strength MPa	Tensile strength ksi	Yield strength(a) MPa	Yield strength(a) ksi	Elongation in 50 mm (2 in.), %	Hardness HRF	Hardness HRB	Hardness HR30T	Shear strength MPa	Shear strength ksi	Fatigue strength(b) MPa	Fatigue strength(b) ksi
Flat products, 1 mm (0.04 in.) thick												
OS050	220	32	69	10	45	40	...	...	150	22	...	...
OS025	235	34	76	11	45	45	...	...	160	23	76	11
H00	250	36	195	28	30	60	10	25	170	25	...	...
H01	260	38	205	30	25	70	25	36	170	25	...	...
H02	290	42	250	36	14	84	40	50	180	26	90	13
H04	345	50	310	45	6	90	50	57	195	28	90	13
H08	380	55	345	50	4	94	60	63	200	29	97	14
H10	395	57	365	53	4	95	62	64	200	29	...	...
M20	235	34	69	10	45	45	...	...	160	23	...	...
Flat products, 6 mm (0.25 in.) thick												
OS050	220	32	69	10	50	40	...	...	150	22	...	...
H00	250	36	195	28	40	60	10	...	170	25	...	...
H01	260	38	205	30	35	70	25	...	170	25	...	...
H04	345	50	310	45	12	90	50	...	195	28	...	...
M20	220	32	69	10	50	40	...	...	150	22	...	...
Flat products, 25 mm (1.0 in.) thick												
H04	310	45	275	40	20	85	45	...	180	26	...	...
Rod, 6 mm (0.25 in.) diameter												
H80 (40%)	380	55	345	50	10	94	60	...	200	29	...	...
Rod, 25 mm (1.0 in.) diameter												
OS050	220	32	69	10	55	40	...	...	150	22	...	...
H80 (35%)	330	48	305	44	16	87	47	...	185	27	115(c)	17(c)
M20	220	32	69	10	55	40	...	...	150	22	...	...
Rod, 50 mm (2.0 in.) diameter												
H80 (16%)	310	45	275	40	20	85	45	...	180	26	...	...
Wire, 2 mm (0.08 in.) diameter												
OS050	240	35	...	...	35(d)	...	...	...	165	24	...	...
H04	380	55	...	...	1.5(e)	...	...	...	200	29	...	...
H08	455	66	...	...	1.5(e)	...	...	...	230	33	...	...
Tube, 25 mm (1.0 in.) diameter × 1.65 mm (0.065 in.) wall thickness												
OS050	220	32	69	10	45	40	...	...	150	22	...	...
OS025	235	34	76	11	45	45	...	...	160	23	...	...
H55 (15%)	275	40	220	32	25	77	35	45	180	26		
H80 (40%)	380	55	345	50	8	95	60	63	200	29	...	...
Shapes, 13 mm (0.50 in.) section size												
OS050	220	32	69	10	50	40	...	...	150	22	...	...
H80 (15%)	275	40	220	32	30	...	35	...	180	26	...	...
M20	220	32	69	10	50	40	...	...	150	22	...	...
M30	220	32	69	10	50	40	...	...	150	22	...	...

(a) At 0.5% extension under load. (b) At 10^8 cycles in a reversed bending test. (c) At 3×10^8 cycles in a rotating beam test. (d) Elongation in 250 mm (10 in.). (e) Elongation in 1500 mm (60 in.)

Table 10 Typical impact strength of C11000

Product and condition	Impact strength J	Impact strength ft · lbf
Charpy V-notch		
Hot rolled, annealed	96	71
Charpy keyhole-notch		
As-cast	11	8
As-hot rolled	43	32
Rod		
Annealed	52	38
Commercial temper	35	26
Izod		
Rod		
Annealed and drawn 30%	54	40
Drawn 30%	45	33
Plate		
As-hot rolled	52	38
Annealed	53(a)	39(a)
	39(b)	29(b)
Cold rolled 50%	26(a)	19(a)
	12(b)	9(b)

(a) Parallel to rolling direction. (b) Transverse to rolling direction

C11300, C11400, C11500, C11600 99.96Cu + Ag-0.4O

Commercial Names

Previous trade name. Tough pitch copper with silver
Common name. Silver-bearing tough pitch copper
Designation. STP

Specifications

AMS. Soft wire (all alloys) and trolley wire (C11300 only): 4701
ASME. Strip (C11300 only): SB152
ASTM. See Table 13.
SAE. Bar, sheet, strip (C11300, C11400, and C11600) and plate (C11300 and C11400): J463
Federal. See Table 13.
Military. Soft wire (all alloys) and trolley wire (C11300 only): MIL-W-3318. Commutator bar (11600 only): MIL-B-19231

Chemical Composition

Copper limits. 99.0 to 99.9 Cu
Oxygen limit. 0.04% O max
Silver limits. C11300, 0.027 Ag max; C11400, 0.034 Ag max; C11500, 0.054 Ag max; C11600, 0.085 Ag max. These coppers may be low-resistance lake copper or electrolytic copper to which Ag has been intentionally added.

Applications

Typical uses. All forms except pipe and tubing: gaskets, radiators, busbars, conductivity wire, contacts, radio parts, windings, switches, terminals, commutator segments, chemical process equipment, printing rolls, clad metals, printed circuit foil

Mechanical Properties

Tensile properties. See Table 14.
Shear strength. See Table 14.
Hardness. See Table 14.

Table 11 Creep properties of copper

Temper	Testing temperature °C	Testing temperature °F	Stress MPa	Stress ksi	Duration of test, h	Total extension, %(a)	Intercept, %	Minimum creep rate, % per 1000 h
Strip, 2.5 mm (0.10 in.) thick								
OS030	130	265	55	8	2500	2.6	2.0	0.15
			100	14.5	2600	10.0	7.6	1.2
			140	20	170	29.8(b)	...	39
	175	345	55	8	2000	3.3	2.3	0.65
			100	14.5	350	15(b)	8.0	6.3
H01	130	265	55	8	8250	0.20	0.15	0.01
			100	14.5	8600	0.67	0.26	0.042
			140	20	1750	2.4(b)	0.32	0.45
	175	345	55	8	6850	1.14	0.14	0.088
			100	14.5	1100	2.0	0.22	0.66
H02	130	265	55	8	7200	0.24	0.13	0.01
			100	14.5	8600	1.02	0.25	0.054
			140	20	4680	3.4(b)	0.36	0.27
	175	345	55	8	1050	3.3(b)	...	0.6
H06	130	265	55	8	8250	1.58	0.08	0.035
			100	14.5	8700	7.31	0.16	0.055
			140	20	4030	11(b)	0.24	0.17
Rod, 3.2 mm (0.13 in.) diameter								
OS025	260	500	2.5	0.36	6000	0.08	0.016	0.011
			4.1	0.60	6000	0.19	0.010	0.030
			7.2	1.05	6500	0.64	0.113	0.080
			13.8	2.0	6500	2.88	0.87	0.306
H08	205	400	7.2	1.05	6500	0.06	0.045	0.011
			14.5	2.1	6500	0.20	0.112	0.012
			28	4.05	6500	1.08	0.41	0.097
			50	7.25	6500	5.42	2.47	0.44

Note: Values shown are typical for the tough pitch grades of copper. Oxygen-free, phosphorus-deoxidized, and arsenical coppers have marginally greater resistance to creep deformation. (a) Total extension is initial extension (not given in table) plus intercept (column 8) plus the product of minimum creep rate (column 9) and duration (column 6). (b) Rupture test

Elastic modulus. Tension, 115 GPa (17 × 10⁶ psi); shear, 44 GPa (6.4 × 10⁶ psi)

Mass Characteristics

Density. 8.89 to 8.94 g/cm³ (0.321 to 0.323 lb/in.³) at 20 °C (68 °F)

Thermal Properties

Liquidus temperature. 1080 °C (1980 °F)
Coefficient of linear thermal expansion. 17.7 μm/m · K (9.8 μin./in. · °F) at 20 to 300 °C (68 to 572 °F)
Specific heat. 385 J/kg · K (0.092 Btu/lb · °F) at 20 °C (68 °F)
Thermal conductivity. 388 W/m · K (224 Btu/ft · h · °F) at 20 °C (68 °F)

Electrical Properties

Electrical conductivity. Volumetric: 100% IACS at 20 °C (68 °F)
Electrical resistivity. 17.2 nΩ · m at 20 °C (68 °F)

Fabrication Characteristics

Machinability. 20% of C36000 (free-cutting brass)
Forgeability. 65% of C37700 (forging brass)
Formability. Excellent for cold working and hot forming
Weldability. Soldering: excellent. Brazing and resistance butt welding: good. Gas-shielded arc welding: fair. Oxyacetylene coated metal-arc and resistance spot and seam welding: not recommended
Annealing temperature. 475 to 750 °C (900 to 1400 °F)
Hot-working temperature. 750 to 875 °C (1400 to 1600 °F)

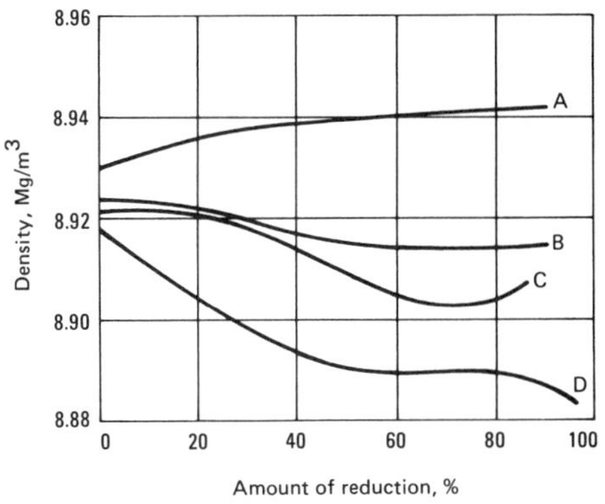

Fig. 11 Variation of density with amount of cold reduction by rolling for C11000 and similar coppers. A, vacuum annealed 12 h at 880 °C (1615 °F) and cold drawn. B, vacuum annealed 12 h at 970 °C (1780 °F) and flat rolled. C, vacuum annealed 12 h at 995 °C (1825 °F) and cold drawn. D, hot rolled, vacuum annealed 4 h at 600 °C (1110 °F), and drawn

C12500, C12700, C12800, C12900, C13000

Commercial Names

Previous trade name. Fire-refined tough pitch copper (C12500); fire-refined tough pitch copper with silver (C12700, C12800, C12900, C13000)
Common name. Fire-refined copper
Designation. C12500; FRTP. Others: FRTSP

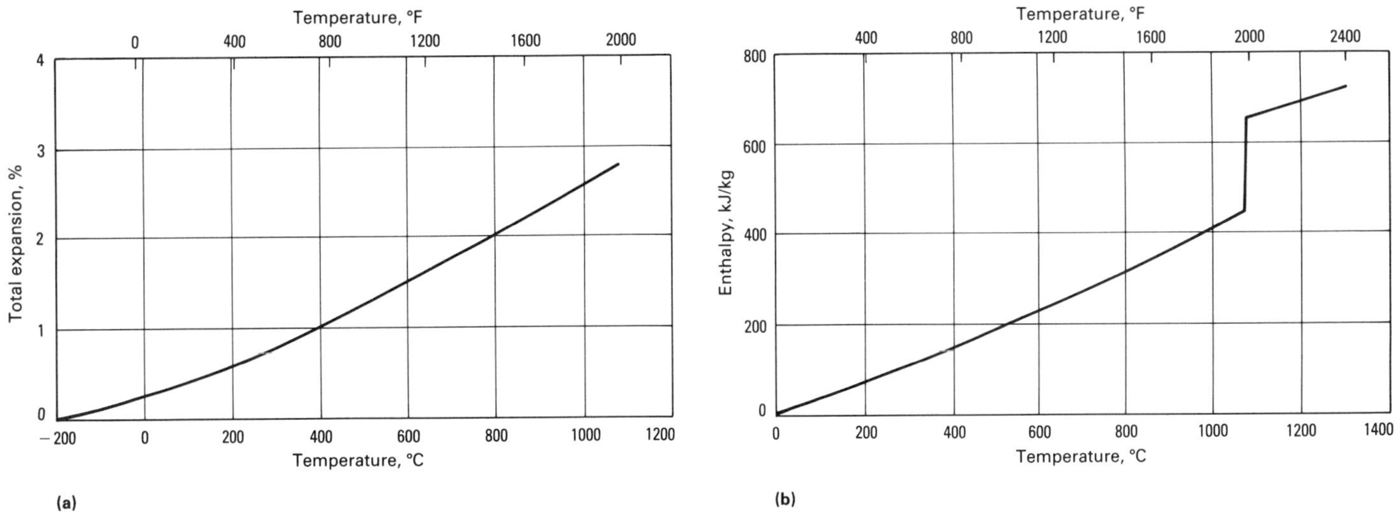

Fig. 12 Thermal expansion and enthalpy of C11000. (a) Total thermal expansion from −190 °C (−310 °F). (b) Enthalpy (heat content) above 0 °C (32 °F)

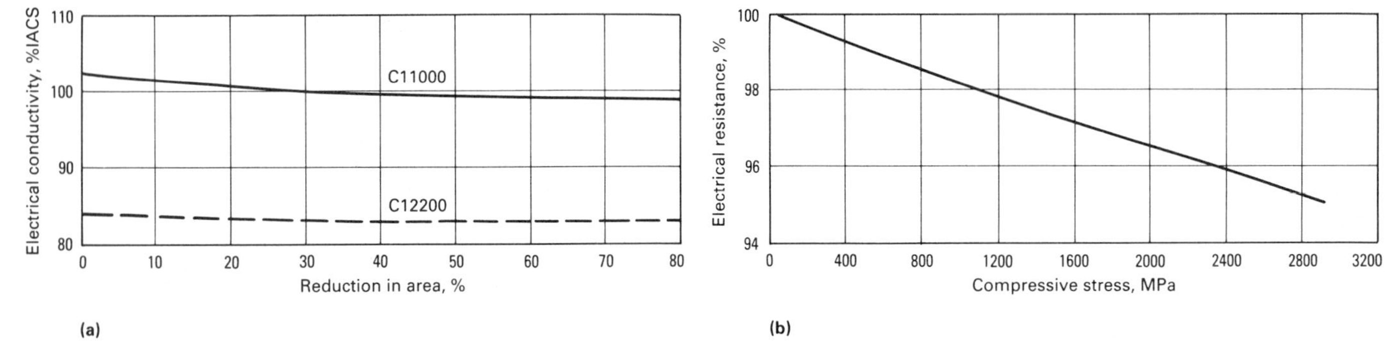

Fig. 13 Electrical properties of copper. (a) Electrical conductivity as a function of amount of cold reduction by drawing. (b) Variation of electrical resistance with applied compressive stress at 30 and 75 °C (86 and 167 °F). Resistance expressed as percent of no load value.

Specifications

ASTM. Flat products: B 11, B 124, B 133, B 152, B 272. Rod: B 12, B 124, B 133. Shapes: B 124, B 133, B 216. Lake copper wirebar, cake, slab, billet, and ingot: B 4
Government. MIL-W-3318

Chemical Composition

Composition limits in ASTM B 216: 99.88 Cu + Ag min (minimum Ag content may be specified by agreement), 0.012 As max, 0.003 Sb max, 0.025 Se + Te max, 0.05 Ni max, 0.003 Bi max, 0.004 Pb max
Consequence of exceeding impurity limits. Bi and Pb can cause hot workability problems if composition limits are exceeded. Se and Te greatly affect recrystallization and grain growth.

Applications

Typical uses. Architectural: Building fronts, downspouts, flashing, gutters, roofing, screening, spouting. Automotive: gaskets, radiators. Electrical: busbars, contacts, radio parts, commutator segments, switches, terminals. Miscellaneous: anodes, chemical process equipment, kettles, pans, printing rolls, rotating bands, roadbed expansion plates, vats. This copper is suitable for use where the high conductivity and low annealing temperature of electrolytic tough pitch copper are not required.
Precautions in use. This copper is subject to embrittlement when heated in a reducing atmosphere, as in annealing, brazing, or welding at temperatures of 370 °C (700 °F) or above. If hydrogen or carbon monoxide is present, embrittlement can be rapid.

Mechanical Properties

Tensile properties. See Table 15.
Shear strength. See Table 15.
Hardness. See Table 15.
Fatigue strength. Strip, OS025 temper, 76 MPa (11 ksi)

Mass Characteristics

Density. 8.89 g/cm^3 (0.321 lb/in.^{3}) at 20 °C (68 °F)

Thermal Properties

Liquidus temperature. 1085 °C (1980 °F)
Coefficient of linear thermal expansion. 16.8 μm/m · K (9.3 μin./in. · °F) at 20 to 100 °C (68 to 212 °F); 17.4 μm/m · K (9.7 μin./in. · °F) at 20 to 200 °C (68 to 392 °F); 17.7 μm/m · K (9.8 μin./in. · °F) at 20 to 300 °C (68 to 572 °F)
Specific heat. 385 J/kg · K (0.092 Btu/lb · °F) at 20 °C (68 °F)
Thermal conductivity. 377 W/m · K (218 Btu/ft · h · °F) at 20 °C (68 °F)

Electrical Properties

Electrical conductivity. Volumetric, 98% IACS at 20 °C (68 °F), annealed
Electrical resistivity. 17.6 nΩ · m at 20 °C (68 °F)

Fabrication Characteristics

Machinability. 20% of C36000 (free-cutting brass)
Formability. Excellent for hot or cold forming but should not be heated for forming or annealed in a reducing atmosphere
Joining. Riveting: use copper rivets. Pressure welding: use Koldweld proprietary method. Soft solder with all grades of solder, commercial solder fluxes, or rosin. Silver braze with all types of flame using copper-phosphorus, silver, or copper-zinc (see ASTM B 260). Satisfactory fluxes are commercially available. Use gas-shielded

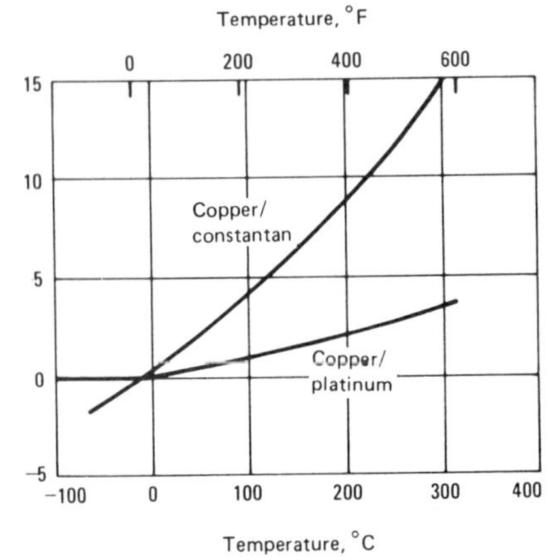

Fig. 14 Thermoelectric properties of copper with cold junctions at 0 °C (32 °F)

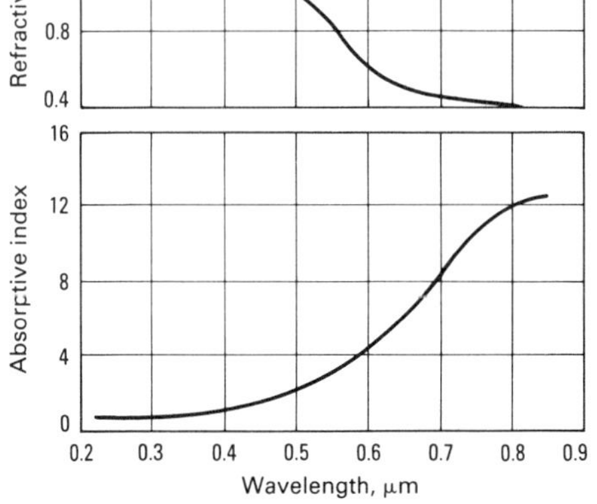

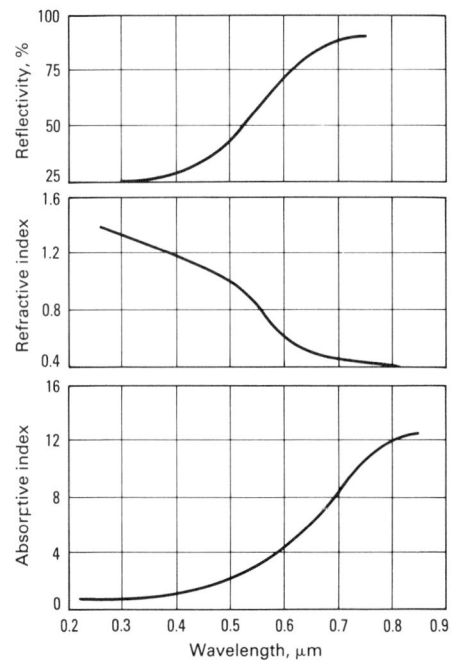

Fig. 15 Optical properties of C11000 and similar coppers at 21 °C (70 °F)

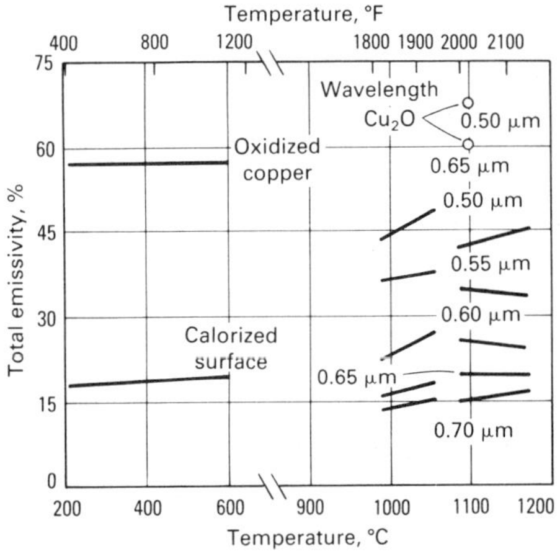

Fig. 16 Emissivity of commercial coppers

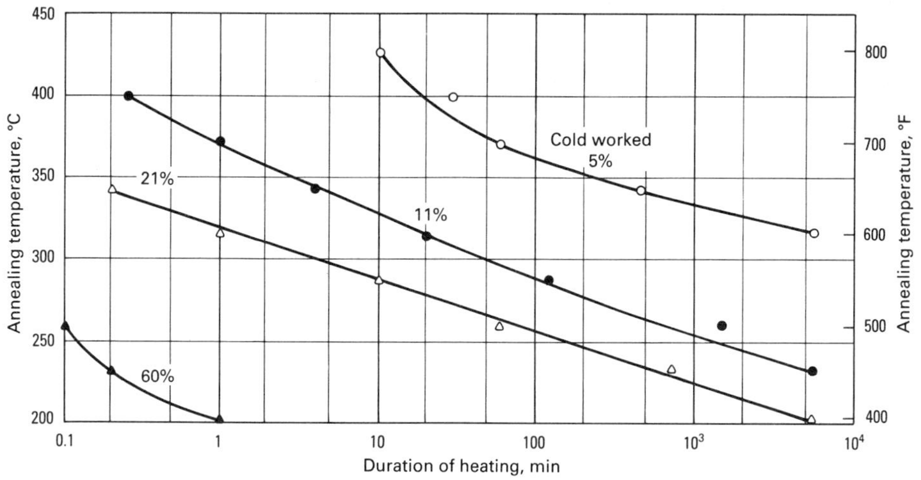

Fig. 17 Time-temperature relationships for annealing C11000 and similar coppers

Table 12 Specifications for C11100

Product	Federal	ASTM
Bar	QQ-C-502, QQ-C-576	. . .
Bar, bus	QQ-B-865	. . .
Pipe, bus	QQ-B-825	. . .
Plate	QQ-C-576	. . .
Rod	QQ-B-502	B 49, B 133
Rod, bus	QQ-B-825	. . .
Shapes	QQ-C-502	. . .
Shapes, bus	QQ-B-825	. . .
Sheet	QQ-C-576	. . .
Strip	QQ-C-502, QQ-C-576	. . .
Tubing, bus	QQ-B-825	. . .
Wire, coated	. . .	B 246
With tin	. . .	B 334
With lead alloy	. . .	B 189
With nickel	. . .	B 355
With silver	. . .	B 298
Wire, flat	QQ-C-502	. . .
Wire, hard drawn	QQ-W-343	B 1
Wire, medium-hard drawn	. . .	B 2
Wire, stranded	. . .	B 8, B 172, B 173, B 174, B 226, B 228, B 229, B 286
Wire, rod	. . .	B 47
Wire, trolley	. . .	B 116

arc welding processes with recommended filler metals, depending on application. Other welding methods generally are not recommended.
Annealing temperature. 400 to 650 °C (750 to 1200 °F)
Hot-working temperature. 750 to 950 °C (1400 to 1750 °F)

C14300, C14310
99.9Cu-0.1Cd; 99.8Cu-0.2Cd

Commercial Names

Previous trade names. C14300: cadmium-copper, deoxidized

Chemical Composition

Copper. 99.8 to 99.9 Cu
Cadmium. C14300, 0.05 to 0.15 Cd; C14310, 0.1 to 0.3 Cd

Applications

Typical uses. Rolled strip for anneal-resistant electrical applications: applications requiring thermal softening and embrittlement resistance, such as lead frames, contacts, terminals, solder-coated and solder-fabricated parts. Furnace brazed assemblies and welded components such as tube or cable wrap

Mechanical Properties

Tensile properties. See Table 16.
Shear strength. See Table 16.
Hardness. See Table 16.
Elastic modulus. Tension, 115 GPa (17 × 10^6 psi); shear, 44 GPa (6.4 × 10^6 psi)

Mass Characteristics

Density. 8.94 g/cm^3 (0.323 lb/in.^{3}) at 20 °C (68 °F)

Thermal Properties

Liquidus temperature. 1080 °C (1976 °F)
Solidus temperature. 1052 °C (1926 °F)
Coefficient of linear thermal expansion. 17.0 μm/m · K (9.4 μin./in. · °F) at 20 to 100 °C (68 to 212 °F); 17.3 μm/m · K (9.6 μin./in. · °F) at 20 to 200 °C (68 to 392 °F); 17.7 μm/m · K (9.8 μin./in. · °F) at 20 to 300 °C (68 to 572 °F)
Specific heat. 385 J/kg · K (0.092 Btu/lb · °F) at 20 °C (68 °F)
Thermal conductivity. C14300, 377 W/m · K (218 Btu/ft · h · °F) at 20 °C (68 °F); C14310, 343 W/m · K (198 Btu/ft · h · °F)

Electrical Properties

Electrical conductivity. Volumetric: C14300, 96% IACS at 20 °C (68 °F); C14310, 85% IACS at 20 °C (68 °F)
Electrical resistivity. C14300, 18 nΩ · m at 20 °C (68 °F); C14310, 20.3 nΩ · m at 20 °C (68 °F)

Fabrication Characteristics

Machinability. 20% of C36000 (free-cutting brass)
Forgeability. 65% of C37700 (forging brass)
Formability. Excellent capacity for cold working and hot forming
Weldability. Soldering, brazing, and gas-shielded arc welding: excellent. Oxyacetylene welding and resistance butt welding: good. Coated metal arc and resistance seam and spot welding: not recommended
Annealing temperature. 535 to 750 °C (1000 to 1400 °F)
Hot-working temperature. 750 to 875 °C (1400 to 1600 °F)

C14500
99.5Cu-0.5Te

Commercial Names

Previous trade name. Phosphorus-deoxidized tellurium-bearing copper
Common name. Free-machining copper
Designation. DPTE

Specifications

ASTM. Flat products and rod: B 124, B 301, Shapes: B 124, B 283

Chemical Composition

Composition limits. 99.90 Cu + Ag + Te min, 0.004 to 0.012 P, 0.40 to 0.60 Te

Applications

Typical uses. Forgings and screw machine products requiring high conductivity, extensive machining, corrosion resistance, copper color, or a combination of these qualities; typical parts include electrical connectors, motor parts, switch parts, plumbing fixtures, soldering tips, welding-torch tips, transistor bases, and parts that are assembled by furnace brazing
Precautions in use. Carbide-tipped tools should be used for machining C14500.

Mechanical Properties

Tensile properties. See Table 17.
Shear strength. See Table 17.

Table 13 Specifications for C11300, C11400, C11500, and C11600

Product	ASTM	Federal
Bar	B 152(a)	QQ-C-576(a), QQ-C-502(b)
Bar, bus	B 187(a)	QQ-B-825(a)
Pipe, bus	B 188(a)	QQ-B-825(c)
Plate	B 152(a)	QQ-B-825(a)
Rod	B 49(e)	QQ-C-502(f)
Rod, bus	B 187(a)	QQ-B-825(f)
Shapes	...	QQ-C-502(f)
Shapes, bus	B 187(a)	QQ-B-825(f)
Sheet	B 152(a)	QQ-C-576(f)
Sheet, clad	B 506(a)	QQ-C-502(c)
Strip	B 152(a), B 272(e)	QQ-C-502(b), QQ-C-576(f)
Strip, clad	B 506(a)	...
Tube, bus	B 188(a)	QQ-B-825(f)
Wire, coated with		
Tin	B 246, B 334(e)	...
Lead alloy	B 189(e)	...
Nickel	B 355(e)	...
Silver	B 298(e)	...
Wire, flat	B 272(e)	QQ-C-502(a)
Wire, hard drawn	B 1(e)	QQ-W-343(e)
Wire, medium-hard drawn	B 2(e)	QQ-W-343(e)
Wire, rod	B 49(e)	...
Wire, soft	B 3(a), B 48(e)	QQ-W-343(e)
Wire, stranded	B 8, B 172, B 173, B 174, B 226, B 228, B 229, B 286(e)	...
Wire, trolley	B 47, B 116(e)	QQ-W-343(g)

(a) C11300, C11400, and C11600. (b) C11600 only. (c) C11400 only. (d) C11300 and C11400. (e) C11300, C11400, C11500, and C11600. (f) C11400 and C11600. (g) C11300 only

Table 14 Typical mechanical properties of C11300, C11400, C11500, and C11600

Temper	Tensile strength MPa	Tensile strength ksi	Yield strength(a) MPa	Yield strength(a) ksi	Elongation in 50 mm (2 in.), %	Hardness HRF	Hardness HRB	Hardness HR30T	Shear strength MPa	Shear strength ksi
Flat products, 1 mm (0.04 in.) thick										
OS025	235	34	75	11	45	45	...	...	160	23
H00	250	36	195	28	30	60	10	25	170	25
H01	260	38	205	30	25	70	25	36	170	25
H02	290	42	250	36	14	84	40	50	180	26
H04	345	50	310	45	6	90	50	57	195	28
H08	380	55	345	50	4	94	60	63	200	29
H10	395	57	365	53	4	95	62	64	200	29
M20	235	34	69	10	45	45	...	...	160	23
Flat products, 6 mm (0.25 in.) thick										
OS050	220	32	69	10	50	40	...	...	150	22
H00	250	36	195	28	40	60	10	...	170	25
H01	260	38	205	30	35	70	25	...	170	25
H04	345	50	310	45	12	90	50	...	195	28
M20	220	32	69	10	50	40	...	...	150	22
Flat products, 25 mm (1.0 in.) thick										
H04	310	45	275	40	20	85	45	...	180	26
Rod, 6 mm (0.25 in.) diameter										
H80 (40%)	380	55	345	50	10	94	60	...	200	29
Rod, 25 mm (1.0 in.) diameter										
OS050	220	32	69	10	55	40	...	...	150	22
H80 (35%)	330	48	305	44	16	87	47	...	185	27
M20	220	32	69	10	55	40	...	...	150	22
Rod, 51 mm (2.0 in.) diameter										
H80 (16%)	310	45	275	40	20	85	45	...	180	26
Wire, 2 mm (0.08 in.) diameter										
OS050	240	35	...	...	35(b)	...	...	...	165	24
H04	380	55	...	...	1.5(c)	...	...	...	200	29
H08	455	66	...	...	1.5(c)	...	...	...	230	33
Shapes, 13 mm (0.50 in.) diameter										
OS050	220	32	69	10	50	40	...	...	150	22
H80 (15%)	275	40	220	32	30	...	35	...	180	26
M20	220	32	69	10	50	40	...	...	150	22
M30	220	32	69	10	50	40	...	...	150	22

(a) At 0.5% extension under load. (b) Elongation in 250 mm (10 in.). (c) Elongation in 1500 mm (60 in.)

Hardness. See Table 17.
Elastic modulus. Tension, 115 GPa (17 × 10^6 psi); shear, 44 GPa (6.4 × 10^6 psi)

Mass Characteristics

Density. 8.94 g/cm^3 (0.323 lb/in.^{3}) at 20 °C (68 °F)

Thermal Properties

Liquidus temperature. 1075 °C (1967 °F)
Solidus temperature. 1051 °C (1924 °F)
Coefficient of linear thermal expansion. 17.1 μm/m · K (9.5 μin./in. · °F) at 20 to 100 °C (68 to 212 °F); 17.4 μm/m · K (9.7 μin./in. · °F) at 20 to 200 °C (68 to 392 °F); 17.8 μm/m · K (9.9 μin./in. · °F) at 20 to 300 °C (68 to 572 °F)
Specific heat. 385 J/kg · K (0.092 Btu/lb · °F) at 20 °C (68 °F)
Thermal conductivity. 355 W/m · K (205 Btu/ft · h · °F) at 20 °C (68 °F)

Electrical Properties

Electrical conductivity. Volumetric: 93% IACS at 20 °C (68 °F)
Electrical resistivity. 18.6 nΩ · m at 20 °C (68 °F)

Fabrication Characteristics

Machinability. 85% of C36000 (free-cutting brass)
Formability. Good capacity for being cold worked, usually by drawing, rolling, or swaging. Excellent capacity for being hot formed, most often by extrusion, forging, or rolling
Weldability. Soldering or brazing: excellent. Arc welding, oxyfuel gas welding, and most resistance welding processes are not recommended.
Annealing temperature. 425 to 650 °C (800 to 1200 °F)
Hot-working temperature. 750 to 875 °C (1400 to 1600 °F)

C14700
99.6Cu-0.4S

Commercial Names

Previous trade name. Sulfur-bearing copper
Common name. Free-machining copper, sulfur copper

Specifications

ASTM. Flat products and rod: B 301

Chemical Composition

Composition limits. 0.20 to 0.50 S, 0.10 max other (total), bal Cu + Ag

Applications

Typical uses. Screw machine products and parts requiring high conductivity, extensive machining, corrosion resistance, copper color, or a combination of these properties; electrical connectors, motors, and switch components; plumbing fittings; furnace-

Table 15 Typical mechanical properties of C12500, C12700, C12800, C12900, and C13000

	Tensile strength		Yield strength: At 0.5% extension under load		Yield strength: At 0.2% offset		Elongation in 50 mm (2 in.), %	Hardness			Shear strength	
Temper	MPa	ksi	MPa	ksi	MPa	ksi		HRF	HRB	H30T	MPa	ksi
Flat products, 1 mm (0.04 in.) thick												
OS025	235	34	76	11	...	...	45	45	...	...	160	23
H00	250	36	195	28	235	34	30	60	10	25	170	25
H02	290	42	250	36	270	39	14	84	40	50	180	26
H04	345	50	310	45	327	47.5	6	90	50	57	195	28
H08	380	55	345	50	360	52	4	94	60	63	200	29
H10	395	57	365	53	370	54	4	95	62	64	200	29
M20	235	34	69	10	...	...	45	45	...	...	160	23
Rod, 25 mm (1 in.) diameter												
OS050	220	32	69	10	...	...	55	40	...	...	150	22
H80 (35%)	330	48	305	44	...	...	16	87	47	...	185	27
M20	220	32	69	10	...	...	55	40	...	...	150	22
Wire, 2 mm (0.08 in.) diameter												
OS050	240	35	...	...	...	...	35(a)	...	...	...	165	24
H04	380	55	...	...	...	...	1.5(b)	...	...	...	200	29
H08	455	66	...	...	...	...	1.5(b)	...	...	...	230	33
Shapes, 13 mm (0.50 in.) diameter												
OS050	220	32	69	10	...	...	50	40	...	...	150	22
H80 (15%)	275	40	220	32	...	...	30	...	35	...	180	26
M20	220	32	69	10	...	...	50	40	...	...	150	22
M30	220	32	69	10	...	...	50	40	...	...	150	22

(a) Elongation in 250 mm (10 in.). (b) Elongation in 1500 mm (60 in.)

Table 16 Typical mechanical properties of C14300 and C14310

	Tensile strength		Yield strength at 0.2% offset		Elongation in 50 mm (2 in.), %
Temper	MPa	ksi	MPa	ksi	
OS025	220	32	75	11	42
H04	310	45	275	40	14
H08	350	51	330	48	7
H10	400	58	385	56	3

Note: Values for strip, 0.3 to 1 mm (0.01 to 0.04 in.) diameter

Table 17 Typical mechanical properties of C14500, C14700, and C18700 rod

	Tensile strength		Yield strength(a)		Elongation in 50 mm (2 in.), %	Hardness, HRB	Shear strength	
Temper	MPa	ksi	MPa	ksi			MPa	ksi
6 mm (0.25 in.) diameter								
H02	295	43	275	40	18	43	180	26
H04	365	53	340	49	10	54	200	29
13 mm (0.50 in.) diameter								
OS015	230	33	76	11	46	43 HRF	150	22
H02	295	43	275	40	20	43	180	26
H04	330	48	305	44	15	48	185	27
25 mm (1 in.) diameter								
OS050	220	32	69	10	50	40 HRF	150	22
H02	290	42	275	40	25	42	170	25
H04	330	48	305	44	20	48	185	27
50 mm (2 in.) diameter								
H02	290	42	270	39	35	42	170	25

(a) At 0.5% extension under load

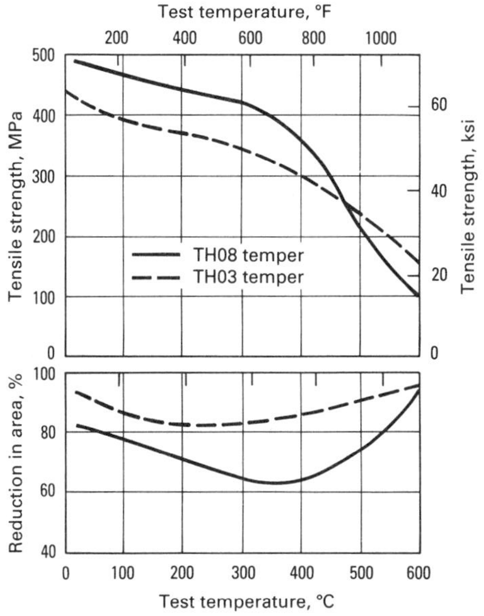

Fig. 18 Short-time elevated-temperature tensile properties of C15000. Material was solution treated 15 min at 900 °C (1650 °F), quenched, cold worked, and aged. The TH03 temper material was cold worked 54%, then aged 1 h at 400 °C (750 °F); the TH08 temper material was cold worked 84%, then aged 1 h at 375 °C (705 °F).

brazed articles; screws; soldering coppers; rivets; and welding-torch tips

Mechanical Properties

Tensile properties. See Table 17, C14500.
Hardness. See Table 17, C14500.
Elastic modulus. Tension, 115 GPa (17 × 10^6 psi); shear, 44 GPa (6.4 × 10^6 psi)

Mass Characteristics

Density. 8.94 g/cm^3 (0.323 lb/in.^{3}) at 20 °C (68 °F)

Thermal Properties

Liquidus temperature. 1076 °C (1969 °F)
Solidus temperature. 1067 °C (1953 °F)
Coefficient of linear thermal expansion. 17.0 μm/m · K (9.4 μin./in. · °F) at 20 to 100 °C (68 to 212 °F); 17.3 μm/m · K (9.6 μin./in. · °F) at 20 to 200 °C (68 to 392 °F); 17.7 μm/m · K (9.8 μin./in. · °F) at 20 to 300 °C (68 to 572 °F)
Specific heat. 385 J/kg · K (0.092 Btu/lb · °F) at 20 °C (68 °F)
Thermal conductivity. 374 W/m · K (216 Btu/ft · h · °F) at 20 °C (68 °F)

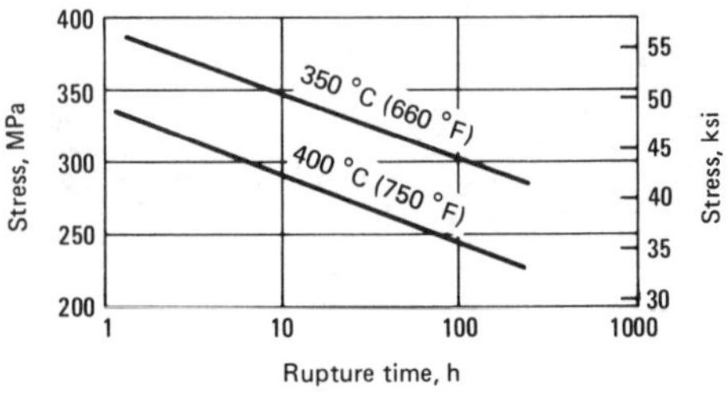

Fig. 19 Stress-rupture properties of C15000, TH08 temper. Material was solution treated 1 h at 950 °C (1740 °F), quenched, cold worked 85%, and aged 1 h at 425 °C (795 °F).

Electrical Properties

Electrical conductivity. Volumetric: O61 temper, 95% IACS at 20 °C (68 °F)
Electrical resistivity. 18.1 nΩ · m at 20 °C (68 °F)

Fabrication Characteristics

Machinability. 85% of C36000 (free-cutting brass)
Annealing temperature. 425 to 650 °C (800 to 1200 °F)
Hot-working temperature. 750 to 875 °C (1400 to 1600 °F)

Table 18 Typical mechanical properties of C15000

Section size mm	Section size in.	Cold work, %, after Solution treating(a)	Cold work, %, after Aging(b)	Tensile strength MPa	Tensile strength ksi	Yield strength(c) MPa	Yield strength(c) ksi	Elongation in 50 mm (2 in.), %
Rod								
5	0.20	...	76	430	62	385	56	8
6	0.25	10(d)	...	285	41	250	36	34
9.5	0.37	80	44	470	68	440	64	11
13	0.50	56	47	460	67	435	63	15
16	0.62	61	31	440	64	430	62	15
19	0.75	50	34	435	63	420	61	15
22	0.87	48	52	430	62	415	60	15
25	1.0	48	47	430	62	415	60	15
32	1.25	32	17	413	60	400	58	18
Wire								
1	0.04	...	98(e)	525	76	495	72	1.5
2.3	0.09	...	62(e)	495	72	470	68	3
		0	...	200	29	40	6	54
		...	0	205	30	90	13	49
6	0.25	0(d)(f)	...	255	37	75	11	50
13	0.50	30(d)	...	365	53	340	49	23

(a) At 900 to 925 °C (1650 to 1695 °F). (b) For 1 h or more at 400 to 425 °C (750 to 795 °F). (c) At 0.5% extension under load. (d) Mill annealed. (e) Solution treated, cold worked the stated amount, then aged. (f) OS025 temper

C15000
99.85Cu-0.15Zr

Commercial Names

Trade name. Amzirc Brand copper; N-4 alloy
Common name. Zirconium-copper

Chemical Composition

Composition limits. 99.95 Cu + Ag + Zr min, 0.13 to 0.20 Zr

Applications

Typical uses. Stud bases for power transmitters and rectifiers, switches and circuit breakers for high-temperature service, commutators, resistance welding tips and wheels, solderless wrapped connectors. Zirconium-copper is heat treatable and retains much of its room-temperature strength up to 450 °C (840 °F).
Precautions in use. During hot working, forging should be discontinued if the temperature falls below 800 °C (1470 °F). The part must be reheated to at least 900 °C (1650 °F) before forging can be resumed.

Mechanical Properties

Tensile properties. See Tables 18 and 19, and Fig. 18.
Hardness. Rod, up to 16 mm (0.62 in.) in diameter, TB04 or TH04 temper: 72 HRB. Wire: 6 mm (0.25 in.) in diameter, OS025 temper, 40 HRB; 13 mm (0.50 in.) in diameter, H01 temper, 90 HRF
Elastic modulus. Tension, 129 GPa (18.7 × 10^6 psi)
Fatigue strength. TH04 temper: 180 MPa (26 ksi) at 10^8 cycles
Impact strength. See Table 19.
Creep-rupture characteristics. See Table 20 and Fig. 19.

Mass Characteristics

Density. 8.89 g/cm^3 (0.321 lb/in.^{3}) at 20 °C (68 °F)

Thermal Properties

Liquidus temperature. 1080 °C (1976 °F)
Solidus temperature. 980 °C (1796 °F)
Coefficient of linear thermal expansion. 16.9 μm/m · K (9.4 μin./in. · °F) at 20 to 100 °C (68 to 212 °F); 17.6 μm/m · K (9.8 μin./in. · °F) at 20 to 300 °C (68 to 572 °F); 20.2 μm/m · K (11.2 μin./in. · °F) at 20 to 650 °C (68 to 1200 °F)
Specific heat. 385 J/kg · K (0.092 Btu/lb · °F) at 20 °C (68 °F)
Thermal conductivity. Solution-treated, cold-worked 84%, and aged material, 367 W/m · K (212 Btu/ft · h · °F) at 20 °C (68 °F)

Electrical Properties

Electrical conductivity. Volumetric: 93% IACS at 20 °C (68 °F)
Electrical resistivity. Solution-treated, cold-worked 84%, and aged material, 18.6 nΩ · m at 20 °C (68 °F)

Fabrication Characteristics

Machinability. 20% of C36000 (free-cutting brass)
Formability. Excellent capacity for being cold worked or hot formed. Most often fabricated by swaging, bending, heading, or forging
Weldability. Soldering: excellent. Brazing or resistance butt welding: good. Other welding processes are not recommended.
Heat treating. Solution treat 5 to 30 min at temperature, then age 1 to 4 h. Aging time and temperature depend on section size and amount of previous cold work.
Annealing temperature. 600 to 700 °C (1110 to 1300 °F)
Solution temperature. 900 to 925 °C (1650 to 1700 °F)

Table 19 Typical low-temperature mechanical properties of C15000

Test temperature °C	Test temperature °F	Tensile strength MPa	Tensile strength ksi	Notched tensile strength(a) MPa	Notched tensile strength(a) ksi	Yield strength(b) MPa	Yield strength(b) ksi	Elongation(c), %	Reduction in area, %	Impact strength(d) J	Impact strength(d) ft · lbf
22	72	445	64.5	673	97.6	411	59.6	16	62	121	89
−78	−108	463	67.2	711	103.1	423	61.3	20	66	142	105
−197	−323	534	77.4	775	112.4	453	65.7	26	71	155	114
−253	−423	587	85.2	820	119.0	458	66.4	37	72	155	114
−269	−452	591	85.7	838	121.6	446	64.7	36	69	...	...

Note: Data are for TH04 temper materials solution treated at 950 °C (1740 °F), cold worked 85 to 90%, and aged 1 h at 450 °C (840 °F). (a) For K_t of 5.0. (b) At 0.2% offset. (c) In 2 diameters. (d) Charpy V-notch, standard 10 mm (0.39 in.) square specimen

Table 20 Typical creep strength of C15000

Test temperature °C	°F	Stress for 1% creep in 1000 h, MPa	1000 h, ksi	10 000 h, MPa	10 000 h, ksi	100 000 h, MPa	100 000 h, ksi
TH01 temper (17% cold work)							
300	570	277	40.2	241	35.0	208	30.2
350	660	217	31.5	166	24.0	185	26.8
400	750	150	21.7	123	17.9	102	14.8
450	840	98	14.2	70	10.2	51	7.4
500	930	88	12.7	39	5.6	16	2.3
600	1110	28	4.1	15	2.2	7.5	1.1
TH02 temper (43% cold work)							
250	480	343	49.7	330	47.8	317	46.0
300	570	325	47.2	297	43.1	272	39.5
350	660	247	35.8	212	30.7	181	26.2
400	750	176	25.6	142	20.6	114	16.5
450	840	100	14.5	74	10.7	51	7.4
500	930	74	10.7	53	7.7	39	5.6
600	1110	18	2.6	12	1.8	8.3	1.2
TH04 temper (82% cold work)							
250	480	321	46.5	312	45.2	303	44.0
300	570	305	44.2	271	39.3	240	34.8
350	660	257	37.3	238	34.5	219	31.8
400	750	201	29.2	161	23.4	139	20.2
450	840	77	11.1	53	7.7	44	6.4
500	930	63	9.2	41	6.0	28	4.0
600	1100	5.2	0.75	2.8	0.41	1.5	0.22
650	1200	3.0	0.44	1.7	0.25	1.0	0.14

Note: Data are for materials solution treated, cold worked the indicated amount, then aged 1 h at 425 °C (795 °F).

Table 21 Nominal mechanical properties of C15100 rolled strip

Temper	Tensile strength, MPa	Tensile strength, ksi	Yield strength, MPa	Yield strength, ksi	Elongation in 50 mm (2 in.), %	Hardness, HRB	Fatigue strength(a), MPa	Fatigue strength(a), ksi
H01	295	43	240	35	22	32	...	...
H02	325	47	295	43	10	38	...	...
H03	360	52	345	50	5	48	...	...
H04	400	58	385	56	3	57	95	14
H06	430	62	415	60	2	60	...	...
H08	470	68	455	66	2	62	...	...

(a) At 10^8 cycles

Aging temperature. Aged only, 500 to 550 °C (930 to 1020 °F); cold worked and aged, 375 to 475 °C (705 to 885 °F)
Hot-working temperature. 900 to 950 °C (1650 to 1740 °F)

C15100
99.9Cu-0.1Zr

Commercial Names

Trade name. ZHC Copper

Chemical Composition

Composition limits. 0.05 to 0.15 Zr, 0.005 Al max, 0.005 Mn max, 0.005 Fe max, 0.01 Al + Mn + Fe max, bal Cu

Applications

Typical uses. Lead frames for high-power electronic circuits, connectors, and switch-blade jaws. Applications requiring high conductivity, moderate strength, good bend formability, and good stress relaxation resistance

Mechanical Properties

Tensile properties. See Table 21.
Hardness. See Table 21.
Elastic modulus. 121 GPa (17.5 × 10^6 psi)
Fatigue strength. See Table 21.

Mass Characteristics

Density. 8.94 g/cm³ (0.323 lb/in.³) at 20 °C (68 °F)

Thermal Properties

Liquidus temperature. 1080 °C (1976 °F)
Solidus temperature. 1065 °C (1949 °F) at 0.05 Zr; 966 °C (1771 °F) at 0.15 Zr
Coefficient of linear thermal expansion. 17.7 μm/m · K (9.8 μin./in. · °F) at 20 to 300 °C (68 to 572 °F)
Thermal conductivity. 360 W/m · K (208 Btu/ft · h · °F) at 20 °C (68 °F)

Electrical Properties

Electrical conductivity. Volumetric at 20 °C (68 °F): annealed, 95% IACS; rolled, 90% IACS
Electrical resistivity. At 20 °C (68 °F): annealed, 18.1 nΩ · m; rolled, 19.2 nΩ · m

Fabrication Characteristics

Machinability. 20% of C36000 (free-cutting brass)
Formability. Excellent capacity for both cold and hot forming
Weldability. Solderability: excellent. Brazing or resistance butt welding: good. All other welding processes are not recommended.
Annealing temperature. 450 to 550 °C (840 to 1025 °F)
Hot-working temperature. 750 to 875 °C (1400 to 1600 °F)

C15500
99.75Cu-0.11Mg-0.06P

Chemical Composition

Composition limits. 99.75 Cu min, 0.027-0.10 Ag, 0.04-0.080 P, 0.08-0.13 Mg

Applications

Typical uses. High-conductivity light-duty springs, electrical contacts, resistance welding electrodes, electrical fittings, clamps, connectors, diaphragms, electronic components

Mechanical Properties

Tensile properties. See Table 22.
Hardness. See Table 22.
Elastic modulus. 115 GPa (17 × 10^6 psi)
Modulus of rigidity. 44 GPa (6.4 × 10^6 psi)
Fatigue strength. See Table 22.

Mass Characteristics

Density. 8.9 g/cm³ (0.322 lb/in.³)

Thermal Properties

Liquidus temperature. 1080 °C (1980 °F)
Solidus temperature. 1078 °C (1972 °F)
Coefficient of linear thermal expansion. 17.7 μm/m · K (9.8 μin./in. · °F) from 20 to 300 °C (68 to 570 °F)
Thermal conductivity. 345 W/m · K (200 Btu/ft · h · °F)
Specific heat. 385 J/kg · K (0.092 Btu/lb · °F) at 20 °C (68 °F)

Electrical Properties

Electrical conductivity. Annealed condition: 90% IACS at 20 °C (68 °F)
Electrical resistivity. Annealed condition after precipitation heat treatment: 1.92 μΩ · cm at 20 °C (68 °F)

Fabrication Characteristics

Machinability. 20% of C36000 (free-cutting brass)
Formability. Excellent capacity for cold working and hot forming
Hot forgeability. 65% of C37700 (forging brass)
Annealing temperature. 485 to 540 °C (900 to 1000 °F)
Hot-working temperature. 750 to 875 °C (1400 to 1600 °F)

Table 22 Mechanical properties of C15500 flat products

Temper	Tensile strength		Yield strength: At 0.5% extension under load		Yield strength: At 0.2% offset		Elongation in 50 mm (2 in.), %	Hardness		Fatigue strength	
	MPa	ksi	MPa	ksi	MPa	ksi		HRF	HRB	MPa	ksi
Section size, 1.0 mm (0.040 in.) thick											
Light anneal (O50)	275	40	125	18	123	17.8	34	70	...	103	15.0
Quarter hard (H01)	310	45	250	36	247	35.8	25	89	...	...	...
Half hard (H02)	365	53	325	47	324	47.0	13	92	...	...	...
Hard (H04)	425	62	393	57	394	57.2	5	97	...	162	23.5
Spring (H08)	460	67	450	65	462	67.0	4	100	...	155	22.5
Extra spring (H10)	495	72	470	68	490	71.0	3	...	80	...	...
Super spring	515	75	480	70	503	73.0	3	...	82	...	...
Special spring	550	80	495	72	517	75.0	3	...	84	...	...
Section size, 5.0 mm (0.200 in.) thick											
Light anneal	275	40	125	18	122	17.7	40	70	...	...	...
Quarter hard	310	45	248	36	246	35.7	28	89	...	...	...

Joining

Weldability. Welding not recommended except for resistance spot welding
Solderability. Excellent
Brazing. Excellent

C15710
99.8Cu-0.2Al_2O_3

Chemical Composition

Composition limits. 99.69 to 99.85 Cu, 0.15 to 0.25 Al_2O_3, 0.01 Fe max, 0.01 Pb max, 0.04 O max

Applications

Typical uses. Rolled strip, rolled flat wire, rod and wire for electrical connectors, light-duty current-carrying springs, inorganic insulated wire, thermocouple wire, lead wire, resistance welding electrodes for aluminum, heat sinks

Mechanical Properties

Tensile properties. See Table 23.
Hardness. See Table 23.
Elastic modulus. Tension, 105 GPa (15 × 10^6 psi)

Mass Characteristics

Density. 8.82 g/cm^3 (0.319 lb/in.^{3}) at 20 °C (68 °F)

Thermal Properties

Liquidus temperature. 1080 °C (1980 °F)
Coefficient of linear thermal expansion. 19.5 μm/m · K (10.8 μin./in. · °F) from 20 to 300 °C (68 to 572 °F)
Specific heat. 380 J/kg · K (0.09 Btu/lb · °F) at 20 °C (68 °F)
Thermal conductivity. 360 W/m · K (208 Btu/ft · h · °F)

Electrical Properties

Electrical conductivity. Volumetric, 90% IACS at 20 °C (68 °F)
Electrical resistivity. 19.2 nΩ · m at 20 °C (68 °F); temperature coefficient, 5.22 nΩ · m per K at 20 °C (68 °F)

Fabrication Characteristics

Formability. Excellent for cold working; poor for hot forming
Weldability. Soldering: excellent. Brazing: good. Resistance butt welding: fair. Resistance spot and seam welding: poor. Oxyacetylene, gas-shielded arc, and coated metal arc welding are not recommended.
Annealing temperature. 650 to 875 °C (1200 to 1600 °F)

Table 23 Typical mechanical properties of C15710

Diameter mm	Diameter in.	Amount of cold working or temper designation	Tensile strength(a) MPa	Tensile strength(a) ksi	Yield strength at 0.2% offset(a) MPa	Yield strength at 0.2% offset(a) ksi	Elongation in 50 mm (2 in.), %	Hardness, HRB
Rod								
24	0.94	0%	325	47	270	39	20	60
22	0.88	13%	345	50	330	48	18	65
19	0.75	39%	415	60	400	58	16	70
16	0.63	56%	450	65	425	62	12	70
10	0.38	82%	510	74	470	68	10	72
6	0.25	93%	530	77	485	70	10	74
		O61	325	47	275	40	20	60
Wire								
2	0.09	98.5%	565	82	540	78	...	...
1	0.05	99.5%	650	94	620	90	...	...
		O61	325	47	275	40	...	...
0.8	0.03	99.8%	685	99	650	94	...	...
		65%	455	66	420	61	...	...
0.5	0.02	99.9%	725	105	690	100	...	...
		85%	475	69	450	65	...	...
		O61	345	50	290	42	...	...

(a) Properties will vary, depending on extrusion ratio and temperature.

C15720
99.6Cu-0.4Al_2O_3

Chemical Composition

Composition limits. 99.49 to 99.6 Cu, 0.35 to 0.45 Al_2O_3, 0.01 Pb max, 0.01 Fe max, 0.04 O max

Applications

Typical uses. Rolled and drawn strip, rolled flat wire, drawn bar, rod, wire, and shapes for relay and switch springs, lead frames, contact supports, heat sinks, circuit breaker parts, rotor bars, resistance welding electrodes and wheels, and connectors. Parts requiring a combination of high strength and conductivity, particularly after exposure to high manufacturing or operating temperatures

Mechanical Properties

Tensile properties. See Table 24.
Hardness. See Table 24.
Elastic modulus. Tension, 113 GPa (16.4 × 10^6 psi)

Mass Characteristics

Density. 8.81 g/cm^3 (0.319 lb/in.^{3}) at 20 °C (68 °F)

Thermal Properties

Liquidus temperature. 1080 °C (1980 °F)
Coefficient of linear thermal expansion. 19.6 μm/m · K (10.9 μin./in. · °F) at 20 to 300 °C (68 to 572 °F)
Specific heat. 380 J/kg · K (0.09 Btu/lb · °F) at 20 °C (68 °F)
Thermal conductivity. 353 W/m · K (204 Btu/ft · h · °F)

Table 24 Typical mechanical properties of C15720

Size	Amount of cold working or temper designation	Tensile strength(a) MPa	Tensile strength(a) ksi	Yield strength at 0.2% offset(a) MPa	Yield strength at 0.2% offset(a) ksi	Elongation in 50 mm (2 in.), %	Hardness, HRB
Flat products							
0.76 mm (0.03 in.) thick	91%	570	83	545	79	7	. . .
0.51 mm (0.02 in.) thick	95%	585	85	565	82	6	. . .
0.25 mm (0.01 in.) thick	97%	605	88	580	84	5	. . .
0.152 mm (0.006 in.) thick	98%	615	89	585	85	3.5	. . .
	O61	485	70	380	55	13	. . .
Rod							
24 mm (0.94 in.) diameter	0%	470	68	365	53	19	74
21 mm (0.81 in.) diameter	26%	495	72	470	68	16	77
18 mm (0.72 in.) diameter	42%	510	74	485	70	14	78
16 mm (0.63 in.) diameter	56%	530	77	495	72	13	79
13 mm (0.50 in.) diameter	72%	540	78	505	73	11	79
10 mm (0.38 in.) diameter	82%	550	80	510	74	10	80
76 mm (3.0 in.) diameter	M30	525	76	510	74	13	78
102 mm (4.0 in.) diameter	M30	460	67	395	57	20	68

(a) Properties will vary, depending on extrusion ratio and temperature.

Table 25 Typical mechanical properties of C15735 rod

Diameter mm	Diameter in.	Amount of cold working or temper designation	Tensile strength(a) MPa	Tensile strength(a) ksi	Yield strength at 0.2% offset(a) MPa	Yield strength at 0.2% offset(a) ksi	Elongation in 50 mm (2 in.), %	Hardness, HRB
24	0.94	0%	485	70	420	61	16	77
19	0.75	39%	550	80	540	78	13	80
16	0.63	56%	585	85	565	82	10	83
64	2.5	M30	490	71	415	60	16	76
76	3.0	M30	565	82	540	78	11	78
102	4.0	M30	515	75	485	70	13	75

(a) Properties will vary, depending on extrusion ratio and temperature.

Electrical Properties

Electrical conductivity. Volumetric, 89% IACS at 20 °C (68 °F)
Electrical resistivity. 19.4 nΩ · m at 20 °C (68 °F)

Fabrication Characteristics

Formability. Excellent for cold working; poor for hot forming
Weldability. Soldering: excellent. Brazing: good. Resistance butt welding: fair. Resistance spot and seam welding: poor. Oxyacetylene, gas-shielded arc, and coated metal arc welding are not recommended.
Annealing temperature. 650 to 925 °C (1200 to 1700 °F)

C15735
99.3Cu-0.7Al_2O_3

Chemical Composition

Composition limits. 99.19 to 99.35 Cu, 0.65 to 0.75 Al_2O_3, 0.01 Fe max, 0.01 Pb max, 0.04 O max

Applications

Typical uses. Rod for resistance welding electrodes, circuit breakers, feed-through conductors, heat sinks, motor parts; parts requiring retention of high strength and conductivity after high-temperature exposure

Mechanical Properties

Tensile properties. See Table 25.
Hardness. See Table 25.
Elastic modulus. Tension, 123 GPa (17.8 × 10^6 psi)

Mass Characteristics

Density. 8.80 g/cm³ (0.318 lb/in.³) at 20 °C (68 °F)

Thermal Properties

Liquidus temperature. 1080 °C (1980 °F)
Coefficient of linear thermal expansion. 20 μm/m · K (11.1 μin./in. · °F) at 20 to 300 °C (68 to 572 °F)
Specific heat. 420 J/kg · K (0.10 Btu/lb · °F) at 20 °C (68 °F)
Thermal conductivity. 339 W/m · K (196 Btu/ft · h · °F)

Electrical Properties

Electrical conductivity. Volumetric, 85% IACS at 20 °C (68 °F)
Electrical resistivity. 20.3 nΩ · m at 20 °C (68 °F)

Fabrication Characteristics

Formability. Excellent for cold working; poor for hot forming
Weldability. Soldering: excellent. Brazing: good. Resistance butt welding: fair. Resistance spot and seam welding: poor. Oxyacetylene, gas-shielded arc, and coated metal arc welding are not recommended.
Annealing temperature. 650 to 925 °C (1200 to 1700 °F)

C16200
99Cu-1Cd

Commercial Names

Previous trade name. Cadmium-copper

Specifications

ASTM. Wire: B 9, B 105
SAE. J463

Chemical Composition

Composition limits. 98.78 to 99.3 Cu, 0.7 to 1.2 Cd, 0.02 Fe max

Applications

Typical uses. Rolled strip, rod, and wire for trolley wire, heating pad and electric blanket elements, spring contacts, rail bands, high-strength transmission lines, connectors, cable wrap, switch gear components, waveguide cavities

Mechanical Properties

Tensile properties. See Table 26.
Shear strength. Rod, 13 mm (0.50 in.) in diameter: OS050 temper, 185 MPa (27 ksi); H04 temper, 385 MPa (56 ksi)
Hardness. See Table 26.
Elastic modulus. Tension, 115 GPa (17 × 10^6 psi); shear, 44 GPa (6.4 × 10^6 psi)
Fatigue strength. At 10^8 cycles for rod 13 mm (0.50 in.) in diameter: OS050 temper, 100 MPa (14.5 ksi); H04 temper, 205 MPa (30 ksi)

Mass Characteristics

Density. 8.89 g/cm³ (0.321 lb/in.³) at 20 °C (68 °F)

Thermal Properties

Liquidus temperature. 1076 °C (1969 °F)
Solidus temperature. 1030 °C (1886 °F)
Coefficient of linear thermal expansion. 17.0 μm/m · K (9.4 μin./in. · °F) at 20 to 100 °C (68 to 212 °F); 17.3 μm/m · K (9.6 μin./in. · °F) at 20 to 200 °C (68 to 392 °F); 17.7 μm/m · K (9.8 μin./in. · °F) at 20 to 300 °C (68 to 572 °F)
Specific heat. 380 J/kg · K (0.09 Btu/lb · °F) at 20 °C (68 °F)
Thermal conductivity. 360 W/m · K (208 Btu/ft · h · °F)

Electrical Properties

Electrical conductivity. Volumetric, 90% IACS at 20 °C (68 °F)
Electrical resistivity. 19.2 nΩ · m at 20 °C (68 °F)

Fabrication Characteristics

Machinability. 20% of C36000 (free-cutting brass)
Formability. Excellent for cold working; good for hot forming

Table 26 Typical mechanical properties of C16200

Temper	Tensile strength MPa	ksi	Yield strength(a) MPa	ksi	Elongation in 50 mm (2 in.), %	Hardness
Flat products, 1 mm (0.04 in.) thick						
OS025 anneal (0.025 mm grain size)	240	35	76	11	52	54 HRF
Hard	415	60	310	45	5	64 HRB
Spring	440	64	...	...	3	73 HRB
Extra spring	495	72	405	59	1	75 HRB
Rod, 13 mm (0.50 in.) diameter						
OS050 anneal (0.050 mm grain size)	240	35	48	7	56	46 HRF
OS025 anneal (0.025 mm grain size)	250	36	83	12	57	46 HRF
Half hard (25%)	400	58	310	45	12	65 HRB
Hard	505	73	474	68.7	9	73 HRB
Wire, 0.25 mm (0.01 in.) diameter						
Drawn (>99%)	690	100	...	...	1.0(b)	...
Wire, 2 mm (0.08 in.) diameter						
OS025 anneal (0.025 mm grain size)	260	38	83	12	50	...
Hard	485	70	380	55	6	...
Spring	550	80	455	66	2	...
Drawn (>96%)	605	88	...	...	1.5(b)	...

(a) At 0.5% extension under load. (b) In 1.5 m (60 in.)

Weldability. Soldering and brazing: excellent. Oxyfuel gas, gas shielded arc, and resistance butt welding: good. Shielded metal arc, resistance spot, and resistance seam welding are not recommended.
Annealing temperature. 425 to 750 °C (800 to 1400 °F)
Hot-working temperature. 750 to 875 °C (1400 to 1600 °F)

C17000
98Cu-1.7Be-0.3Co

Commercial Names

Trade name. Berylco 165
Common name. Beryllium-copper; 165 alloy

Specifications

ASTM. Flat products: B 194. Rod, bars: B 196. Forgings and extrusions: B 570
SAE: J463
Government: QQ-C-533
Resistance Welding Manufacturers' Association. Class IV

Chemical Composition

Composition limits. 1.60 to 1.79 Be, 0.20 Ni + Co min, 0.6 Ni + Fe + Co max, bal Cu

Applications

Typical uses. Bellows, Bourdon tubing; diaphragms, fuse clips, fasteners, lock washers, springs, switch and relay parts, electrical and electronic components, retaining rings, roll pins, valves, pumps, spline shafts, rolling mill parts, welding equipment, nonsparking safety tools
Precautions in use. This material is a potential health hazard. Because it contains beryllium, ventilation must be provided for dry sectioning and grinding, machining, melting, welding, and any other process that produces metal dust or fumes.

Mechanical Properties

Tensile properties. See Tables 27 and 28.
Hardness. See Tables 28 and 29.
Poisson's ratio. 0.30
Elastic modulus. Tension, 115 GPa (17×10^6 psi); shear, 50 GPa (7.3×10^6 psi)
Fatigue strength. See Table 27.

Mass Characteristics

Density. 8.26 g/cm^3 (0.298 lb/in.^{3}) at 20 °C (68 °F)
Volume change on phase transformation. During age hardening: 0.2% maximum decrease in length; 0.6% maximum increase in density

Thermal Properties

Liquidus temperature. 980 °C (1800 °F)
Solidus temperature. 865 °C (1590 °F)
Coefficient of linear thermal expansion. 16.7 μm/m · K (9.3 μin./in. · °F) at 20 to 100 °C (68 to 212 °F); 17.0 μm/m · K (9.4 μin./in. · °F) at 20 to 200 °C (68 to 392 °F); 17.8 μm/m · K (9.9 μin./in. · °F) at 20 to 300 °C (68 to 572 °F)
Specific heat. 420 J/kg · K (0.10 Btu/lb · °F) at 20 °C (68 °F)
Thermal conductivity. 118 W/m · K (69 Btu/ft · h · °F) at 20 °C (68 °F); 145 W/m · K (84 Btu/ft · h · °F) at 200 °C (392 °F)

Electrical Properties

Electrical conductivity. Volumetric, 15 to 33% IACS at 20 °C (68 °F), depending on heat treatment. See also Tables 27 and 28.
Electrical resistivity. Typical, 76.8 nΩ · m at 20 °C (68 °F), but varies with heat treatment

Chemical Properties

General corrosion behavior. Similar to that of other high-copper alloys and basically the same as that of pure copper
Resistance to specific corroding agents. Essentially the same as that of C17200. See also Table 30.

Fabrication Characteristics

Machinability. 20% of C36000 (free-cutting brass)
Formability. This alloy can be formed, drawn, blanked, pierced, and machined in the unhardened condition.
Weldability. Soldering, brazing, gas-shielded arc welding, shielded metal arc welding, and resistance spot welding: good. Resis-

Table 27 Typical mechanical properties and electrical conductivity of C17000 strip

Temper	Tensile strength MPa	ksi	Proportional limit at 0.002% offset MPa	ksi	Yield strength at 0.2% offset MPa	ksi	Elongation in 50 mm (2 in.), %	Electrical conductivity, %IACS	Fatigue strength(a) MPa	ksi
TB00	410–540	60–78	100–140	15–20	190–370	28–53	35–60	17–19	190–230	28–33
TD01	520–610	75–88	280–410	40–60	310–520	45–75	10–35	16–18	200–235	29–34
TD02	590–690	85–100	380–480	55–70	450–620	65–90	5–25	15–17	220–260	32–38
TD04	690–825	100–120	480–590	70–85	550–760	80–110	2–8	15–17	240–270	35–39
TF00(b)	1030–1240	150–180	550–760	80–110	895–1140	130–165	4–10	22–25	240–270	35–39
TH01(c)	1100–1280	160–185	620–795	90–115	930–1170	135–170	3–6	22–25	250–280	36–41
TH02(c)	1170–1340	170–195	660–860	95–125	1000–1210	145–175	2 5	22–25	250–290	36–42
TH04(c)	1240–1380	180–200	690–930	100–135	1070–1240	155–180	1–4	22–25	260–310	38–45
TM00(d)	690–760	100–110	480–590	70–85	520–620	75–90	18–22	20–33	230–255	33–37
TM01(d)	760–825	110–120	520–660	75–95	620–760	90–110	15–19	20–33	230–260	34–38
TM02(d)	825–930	120–135	550–690	80–100	690–860	100–125	12–16	20–33	240–270	35–39
TM04(d)	930–1030	135–150	590–725	85–105	760–930	110–135	9–13	20–33	250–280	36–40
TM06(d)	1030–1100	150–160	590–760	85–110	860–965	125–140	9–12	20–33	255–290	37–42
TM08(d)	1100–1210	160–175	620–795	90–115	965–1140	140–165	3–7	20–33	230–310	33–45

(a) Rotating beam at 10^8 cycles. (b) Aged 3 h at 315 °C (600 °F). (c) Aged 2 h at 315 °C (600 °F). (d) Proprietary mill heat treatment intended to produce the stated tensile properties

Table 28 Typical mechanical properties and electrical conductivities of C17000 rod, bar, plate, tubing, billets, and forgings

Product form	Temper	Tensile strength MPa	Tensile strength ksi	Yield strength at 0.2% offset MPa	Yield strength at 0.2% offset ksi	Elongation in 50 mm (2 in.), %	Hardness	Electrical conductivity, %IACS
Rod, bar, plate, tubing								
All sizes	TB00	415–585	60–85	140–205	20–30	35–60	45–85 HRB	17–19
	TF00(a)	1035–1240	150–180	860–1070	125–155	4–10	32–39 HRC	22–25
<10 mm (<⅜ in.)	TD04	655–895	95–130	515–725	75–105	10–20	92–103 HRB	15–17
	TH04(b)	1205–1380	175–200	930–1140	135–165	2–5	36–41 HRC	22–25
10–25 mm (⅜–1 in.)	TD04	620–825	90–120	515–725	75–105	10–20	91–102 HRB	15–17
	TH04(b)	1170–1345	170–195	930–1140	135–165	2–5	35–40 HRC	22–25
>25 mm (>1 in.)	TD04	585–795	85–115	515–725	75–105	10–20	88–101 HRB	15–17
	TH04(b)	1140–1310	165–190	930–1140	135–165	2–5	34–39 HRC	22–25
Billet	As-cast	515–585	75–85	275–345	40–50	15–30	80–85 HRB	16–22
	Cast and aged(a)	655–690	95–100	485–515	70–75	10–25	18–25 HRC	18–23
	TB00	415–515	60–75	170–205	25–30	25–45	65–75 HRB	13–18
	TF00(a)	965–1170	140–170	725–930	105–135	1–4	30–38 HRC	18–25
Forgings	TB00	415–585	60–85	140–205	20–30	35–60	45–85 HRB	17–19
	TF00(a)	1035–1240	150–180	860–1070	125–155	4–10	32–39 HRC	22–25

(a) Aged 3 h at 350 °C (625 °F). (b) Aged 2 to 3 h at 330 °C (625 °F)

tance seam and resistance butt welding: fair. Oxyfuel gas welding is not recommended.

Recrystallization temperature. 730 °C (1350 °F)

Annealing temperature. Strip, thin rod, wire: 775 to 800 °C (1425 to 1475 °F) for 10 min, water quench. Larger sections: 1 h for each 25 mm (1 in.) of thickness

Solution temperature. 760 to 790 °C (1400 to 1450 °F). All annealing of this material is a solution treatment.

Aging temperature. 260 to 425 °C (500 to 800 °F). Maximum strength is obtained by aging 1 to 3 h at 315 to 345 °C (600 to 650 °F), depending on amount of cold work preceding the aging treatment.

Hot-working temperature. 650 to 825 °C (1200 to 1500 °F)

Hot-shortness temperature. 845 °C (1550 °F)

C17200, C17300

Commercial Names

Previous trade names. C17200: 25 alloy, alloy 25. C17300: alloy M25

Common name. Beryllium-copper

Table 29 Typical hardnesses of C17000 strip

Temper	HV	Standard Rockwell	Superficial Rockwell
TB00	90–160	45–78 HRB	45–67 HR30T
TD01	150–190	68–90 HRB	62–75 HR30T
TD02	185–225	88–96 HRB	74–79 HR30T
TD04	200–260	96–102 HRB	79–83 HR30T
TF00	320 min	33–38 HRC	55–58 HR30N
TH01	343 min	35–39 HRC	55–59 HR30N
TH02	360 min	37–40 HRC	56–80 HR30N
TH04	370 min	39–41 HRC	58–61 HR30N
TM00	200–235	18–23 HRC	37–42 HR30N
TM01	230–265	21–26 HRC	42–46 HR30N
TM02	260–295	25–30 HRC	46–50 HR30N
TM04	290–325	30–35 HRC	50–54 HR30N
TM06	320–350	31–37 HRC	52–56 HR30N
TM08	434–375	32–38 HRC	55–58 HR30N

Specifications

AMS. Flat products: 4530, 4532, Rod, bar, and forgings: 4650. Wire: 4725

ASTM. Flat products: B 194 (C17200 only), B 196. Rod and bar: B 196. Wire: B 197 (C17200 only). Forgings and extrusions: B 570 (C17200 only)

SAE. J463 (C17200)

Government. Strip: QQ-C-533 (C17200 only). Rod and bar: MIL-C-21657, QQ-C-530. Wire: QQ-C-530

Resistance Welding Manufacturers' Association. Class IV

Chemical Composition

Composition limits. 1.80 to 2.00 Be, 0.20 Ni + Co min, 0.6 Ni + Co + Fe max, 0.10 Pb max (C17200) or 0.20 to 0.6 Pb (C17300), 0.5 max other (total), bal Cu

Consequences of exceeding impurity limits. Excessive P and Si decrease electrical conductivity. Excessive Sn and Pb cause hot shortness.

Applications

Typical uses. C17200 and C17300 are used in parts that are subject to severe forming conditions but require high strength, anelasticity, and fatigue and creep resistance (a wide variety of springs, flexible metal hose, Bourdon tubing, bellows, clips, washers, retaining rings); in parts that require high

Table 30 Approximate corrosion resistance of C17000

Good resistance(a)	Fair resistance(b)	Poor resistance(c)
Acetate solvents	Acetic acid, cold, aerated	Acetic acid, hot
Acetic acid, cold, unaerated	Acetic anhydride	Ammonia, moist
Alcohols	Acetylene	Ammonium hydroxide
Ammonia, dry	Ammonium chloride	Ammonium nitrate
Atmosphere, rural, industrial, marine	Ammonium sulfate	Bromine, aerated or hot
Benzene	Aniline	Chlorine, moist or warm
Borax	Bromine, dry	Chromic acid
Boric acid	Carbonic acid	Ferric chloride
Brine	Copper nitrate	Ferric sulfate
Butane	Ferrous chloride	Fluorine, moist or warm
Carbon dioxide	Ferrous sulfate	Hydrochloric acid, over 0.1%
Carbon tetrachloride	Fluorine, dry	Hydrocyanic acid
Chlorine, dry	Hydrochloric acid, up to 0.1%	Hydrofluoric acid, concentrate
Freon	Hydrofluoric acid, dilute	Hydrogen sulfide, moist
Gasoline	Hydrofluosilicic acid	Lactic acid, hot or aerated
Hydrogen	Hydrogen peroxide	Mercuric chloride
Nitrogen	Nitric acid, up to 0.1%	Mercury
Oxalic acid	Phenol	Mercury salts
Potassium chloride	Phosphoric acid, unaerated	Nitric acid, over 0.1%
Potassium sulfate	Potassium hydroxide	Phosphoric acid, aerated
Propane	Sodium hydroxide	Picric acid
Rosin	Sodium hypochlorite	Potassium cyanide
Sodium bicarbonate	Sodium peroxide	Silver chloride
Sodium chloride	Sodium sulfide	Sodium cyanide
Sodium sulfate	Sulfur	Stannic chloride
Sulfur dioxide	Sulfur chloride	Sulfuric acid, aerated
Sulfur trioxide	Sulfuric acid, unaerated	Sulfurous acid
Water, fresh or salt	Zinc chloride	Tartaric acid, hot or aerated

(a) <0.25 mm/year (0.01 in./year) penetration. (b) 0.025–2.5 mm/year (0.001–0.10 in./year) penetration. (c) >0.25 mm/year (0.01 in./year) penetration

Table 31 Tensile property ranges for C17200 and C17300 strip of various tempers

Temper	Tensile strength MPa	Tensile strength ksi	Proportional limit at 0.002% offset MPa	Proportional limit at 0.002% offset ksi	Yield strength at 0.2% offset MPa	Yield strength at 0.2% offset ksi	Elongation in 50 mm (2 in.), %
TB00	415–540	60–78	105–140	15–20	195–380	28–55	35–60
TD01	515–605	75–88	275–415	40–60	415–605	60–88	10–36
TD02	585–690	85–100	380–485	55–70	515–655	75–95	5–25
TD04	690–825	100–120	485–585	70–85	620–770	90–112	2–8
TF00(a)	1140–1310	165–190	690–860	100–125	965–1205	140–175	4–10
TH01(b)	1205–1380	175–200	760–930	110–135	1035–1275	150–185	3–6
TH02(b)	1275–1450	185–210	825–1000	120–145	1105–1345	160–195	2–5
TH04(b)	1310–1480	190–215	860–1070	125–155	1140–1415	165–205	1–4
TM00(c)	690–760	100–110	450–585	65–85	515–620	75–90	18–23
TM01(c)	760–825	110–120	515–655	75–95	620–760	90–110	15–20
TM02(c)	825–930	120–135	585–725	85–105	690–860	100–125	12–18
TM04(c)	930–1035	135–150	655–795	95–115	795–930	115–135	9–15
(c)	1035–1105	150–160	725–825	105–120	860–965	125–140	9–14
TM06(c)	1105–1205	160–175	760–860	110–125	1000–1170	145–170	4–10
TM08(c)	1205–1310	175–190	795–895	115–130	1070–1240	155–180	3–9

(a) Solution treated and aged 3 h at 315 °C (600 °F). (b) Cold rolled and aged 2 h at 315 °C (600 °F). (c) Proprietary mill treatment to produce the indicated tensile properties

strength or wear resistance along with good electrical conductivity and/or magnetic characteristics (navigational instruments, nonsparking safety tools, firing pins, bushings, valves, pumps, shafts, rolling mill parts); and in parts requiring high strength and good corrosion resistance and electrical conductivity (electrochemical springs, diaphragms, contact bridges, bolts, screws).
Precautions in use. Because this alloy contains beryllium, it is a potential health hazard. Adequate ventilation should be provided for dry sectioning, melting, grinding, machining, welding, and any other fabrication or testing process that produces dust or fumes.

Mechanical Properties

Tensile properties. See Tables 31 and 32.
Hardness. See Tables 31 and 32.
Poisson's ratio. 0.30
Elastic modulus. Tension, 125 to 130 GPa (18 to 19 × 10^6 psi); shear, 50 GPa (7.3 × 10^6 psi)
Fatigue strength. Rotating beam: 380 to 480 MPa (55 to 70 ksi) at 10^7 cycles for both TF00 temper rod having a tensile strength of 1140 to 1310 MPa (165 to 190 ksi) and TH04 temper rod having a tensile strength of 1280 to 1480 MPa (185 to 215 ksi). Reversed torsion: 170 to 275 MPa (25 to 40 ksi). See also Table 33.

Structure

Crystal structure. Alpha copper solid solution is face-centered cubic, disordered. At 20 °C (68 °F), the lattice parameter of the parent phase with about 1.8% Be, homogenized at 815 °C (1500 °F), and quenched in water, is 0.3570 nm. The lattice parameter decreases sharply with increasing beryllium content. Age hardening begins with the formation of coherent Guinier-Preston (G-P) zones on {100} planes. The intermediate precipitate γ′ may be nucleated either from the G-P zones or discontinuously at the grain boundaries. In either case, it has a B2 superlattice structure, a lattice parameter of 0.270 nm, and the orientation $(\bar{1}13)_\alpha \parallel (130)_{\gamma'}$, $[110]_\alpha \parallel [001]_{\gamma'}$. The equilibrium precipitate γ′, which requires longer aging times than are normally used commercially, is body-centered cubic of the CsCl type with a B2 superlattice structure, a lattice parameter of 0.270 nm, and the orientation $(\bar{1}11)_\alpha \parallel (110)_\gamma$, $[110]_\alpha \parallel [001]_\gamma$.
Microstructure. Small, mainly spheroidal, uniformly dispersed (Cu,Co)Be beryllides (bluish gray) in a matrix of equiaxed α copper. (Typical grain size is 0.012 to 0.030 nm in wrought product.) There is a strong tendency to form mechanical and annealing twins. In the age-hardened condition, the matrix shows pronounced striations (the so-called tweed structure) caused by G-P zone formation on {110} planes. At long aging times (>8 h), or high aging temperatures (>315 °C, or >600 °F), there is a strong tendency to form continuous bands

Table 32 Property ranges for various mill products of C17200 and C17300

Temper	Thickness or diameter	Tensile strength MPa	Tensile strength ksi	Yield strength at 0.2% offset MPa	Yield strength at 0.2% offset ksi	Elongation in 50 mm (2 in.), %	Hardness	Electrical conductivity, %IACS
Rod, bar, plate, and tubing								
TB00	All sizes	415–585	60–85	140–205	20–30	35–60	45–85 HRB	17–19
TD04	<9.5 mm (3/8 in.)	655–900	95–130	515–725	75–105	10–20	92–103 HRB	15–17
	9.5–25 mm (3/8–1 in.)	620–825	90–120	515–725	75–105	10–20	91–102 HRB	15–17
	>25 mm (1 in.)	585–790	85–115	515–725	75–105	10–20	88–102 HRB	15–17
TF00(a)	All sizes	1140–1310	165–190	1000–1210	145–175	3–10	36–40 HRC	22–25
TH04(b)	<9.5 mm (3/8 in.)	1280–1480	185–215	1140–1380	165–200	2–5	39–45 HRC	22–25
	9.5–25 mm (3/8–1 in.)	1240–1450	180–210	1140–1380	165–200	2–5	38–44 HRC	22–25
	>25 mm (1 in.)	1210–1410	175–205	1030–1340	150–194	2–5	37–43 HRC	22–25
Wire								
TB00	All sizes	400–540	58–78	140–240	20–35	35–55	...	17–19
TD04	<2 mm (0.08 in.)	895–1070	130–155	760–930	110–135	2–8	...	15–17
	2–9.5 mm (0.08–0.38 in.)	655–900	95–130	515–725	75–105	10–35	...	15–17
	>9.5 mm (0.38 in.)	620–825	90–120	515–725	75–105	10–35	...	15–17
TF00(a)	All sizes	1140–1310	165–190	1000–1210	145–175	3–8	...	22–25
TH04(c)	<2 mm (0.08 in.)	1310–1590	190–230	1240–1410	180–205	1–3	...	22–25
TH04(d)	2–9.5 mm (0.08–0.38 in.)	1280–1480	185–215	1210–1380	175–200	2–5	...	22–25
	>9.5 mm (0.38 in.)	1240–1450	180–210	1140–1380	165–200	2–5	...	22–25
Billets								
As cast	...	515–585	75–85	275–345	40–50	15–30	80–85 HRB	16–22
Cast and aged(a)	...	725–760	105–110	515–550	75–80	10–20	20–25 HRC	18–23
TB00	...	415–515	60–75	170–205	25–30	25–45	65–75 HRB	13–18
TF00(a)	...	1070–1210	155–175	860–1030	125–150	1–3	36–42 HRC	18–25
Forgings								
TB00	...	415–585	60–85	140–205	20–30	35–60	45–85 HRB	17–19
TF00(a)	...	1140–1310	165–190	1000–1210	145–175	3–10	36–41 HRC	22–25

(a) Aged 3 h at 330 °C (625 °F). (b) Aged 2 to 3 h at 330 °C (625 °F). (c) Aged 1 h at 330 °C (625 °F). (d) Aged 1½ to 3 h at 330 °C (625 °F)

Table 33 Hardness, conductivity, and fatigue strength for C17200 and C17300 strip of various tempers

Temper	Hardness HV	Hardness HRC	Hardness HR30N	Electrical conductivity, %IACS	Fatigue strength(a) MPa	Fatigue strength(a) ksi
TB00	90–160	45–78 HRB	45–67 HR30T	17–19	205–240	30–35
TD01	150–190	68–90 HRB	62–75 HR30T	16–18	215–250	31–36
TD02	185–225	88–96 HRB	74–79 HR30T	15–17	220–260	32–38
TD04	200–260	96–102 HRB	79–83 HR30T	15–17	240–270	35–39
TF00	343 min	31–41	56–61	22–25	240–260	35–38
TH01	370 min	38–42	58–63	22–25	240–270	35–39
TH02	380 min	39–44	59–65	22–25	270–295	39–43
TH04	385 min	40–45	60–65	22–25	285–315	41–46
TM00	200–235	18–23	37–42	20–28	230–255	33–37
TM01	230–265	21–26	42–47	20–28	235–260	34–38
TM02	260–295	25–30	45–51	20–28	240–295	35–43
TM04	290–325	30–35	50–55	20–28	260–310	38–45
(b)	320–350	31–37	52–56	20–28	260–310	38–45
TM06	343–375	32–38	55–58	20–28	260–310	38–45
TM08	370–400	33–42	56–63	20–28	275–330	40–48

(a) In reversed bending at 10^8 cycles. (b) Proprietary mill heat treatment to produce tensile strength of 1030–1100 MPa (150 to 160 ksi)

Table 34 Approximate corrosion resistance of C17200 and C17300

Good resistance(a)	Fair resistance(b)	Poor resistance(c)
Acetate solvents	Acetic acid, cold, aerated	Acetic acid, hot
Acetic acid, cold, unaerated	Acetic anhydride	Ammonia, moist
Alcohols	Acetylene	Ammonium hydroxide
Ammonia, dry	Ammonium chloride	Ammonium nitrate
Atmosphere, rural, industrial, marine	Ammonium sulfate	Bromine, aerated or hot
Benzene	Aniline	Chlorine, moist or warm
Borax	Bromine, dry	Chromic acid
Boric acid	Carbonic acid	Ferric chloride
Brine	Copper nitrate	Ferric sulfate
Butane	Ferrous chloride	Fluorine, moist or warm
Carbon dioxide	Ferrous sulfate	Hydrochloric acid, over 0.1%
Carbon tetrachloride	Fluorine, dry	Hydrocyanic acid
Chlorine, dry	Hydrochloric acid, up to 0.1%	Hydrofluoric acid, concentrated
Freon	Hydrofluoric acid, dilute	Hydrogen sulfide, moist
Gasoline	Hydrofluosilicic acid	Lactic acid, hot or aerated
Hydrogen	Hydrogen peroxide	Mercuric chloride
Nitrogen	Nitric acid, up to 0.1%	Mercury
Oxalic acid	Phenol	Mercury salts
Potassium chloride	Phosphoric acid, unaerated	Nitric acid, over 0.1%
Potassium sulfate	Potassium hydroxide	Phosphoric acid, aerated
Propane	Sodium hydroxide	Picric acid
Rosin	Sodium hypochlorite	Potassium cyanide
Sodium bicarbonate	Sodium peroxide	Silver chloride
Sodium chloride	Sodium sulfide	Sodium cyanide
Sodium sulfate	Sulfur	Stannic chloride
Sulfur dioxide	Sulfur chloride	Sulfuric acid, aerated
Sulfur trioxide	Sulfuric acid, unaerated	Sulfurous acid
Water, fresh or salt	Zinc chloride	

(a) <0.25 mm/year (0.01 in./year) attack. (b) 0.025–2.54 mm/year (0.001–0.10 in./year) attack. (c) >0.25 mm/year (0.01 in./year) attack

of cellular precipitate at the grain boundaries and along twin boundaries.

Conventional metallographic techniques may be used. Dry sectioning and grinding should be done in a ventilated area.

One of the common etchants for immersion etching is ammonium persulfate: 3 parts concentrated NH_4OH, 1 part 3% H_2O_2, 2 parts 10% $(NH_4)_2S_2O_3$, and 7 to 10 parts H_2O. This etchant reveals general details of the microstructure. The matrix is stained blue to deep lavender, depending on the state of heat treatment, etchant concentration, and etching time. The etchant should be freshly made.

A common etchant for swabbing is potassium dichromate: $K_2Cr_2O_7$, 1.5 g NaCl, 8 ml H_2SO_4, and 100 ml H_2O. This etchant emphasizes grain boundaries, particularly when heavily decorated with discontinuous precipitate. A very effective procedure for studying grain boundaries and discontinuous precipitation is to first etch with ammonium persulfate, then remove the stain with a single wipe of the dichromate etchant.

Mass Characteristics

Density. 8.25 g/cm^3 (0.298 lb/in.^{3}) at 20 °C (68 °F)

Volume change on phase transformation. During age hardening, there is a maximum decrease in length of 0.2% and a maximum increase in density of 0.6%.

Thermal Properties

Liquidus temperature. 980 °C (1800 °F)

Solidus temperature. 865 °C (1590 °F)

Coefficient of linear thermal expansion. 16.7 µm/m · K (9.3 µin./in. · °F) at 20 to 100 °C (68 to 212 °F); 17.0 µm/m · K (9.4 µin./in. · °F) at 20 to 200 °C (68 to 392 °F); 17.8 µm/m · K (9.9 µin./in. · °F) at 20 to 300 °C (68 to 572 °F)

Specific heat. 420 J/kg · K (0.10 Btu/lb · °F) at 20 to 100 °C (68 to 212 °F)

Thermal conductivity. 105 to 130 W/m · K (60 to 75 Btu/ft · h · °F) at 20 °C (68 °F); 130 to 133 W/m · K (75 to 77 Btu/ft · h · °F) at 200 °C (392 °F)

Electrical Properties

Electrical conductivity. Volumetric, 15 to 30% IACS at 20 °C (68 °F), depending on heat treatment. See also Tables 32 and 33.

Electrical resistivity. 57 to 115 nΩ · m at 20 °C (68 °F), depending on heat treatment

Chemical Properties

General corrosion behavior. Similar to that of other high-copper alloys; basically the same as that of pure copper

Resistance to specific corroding agents. See Table 34.

Fabrication Characteristics

Machinability. C17200: 20% of C36000 (free-cutting brass). C17300: 50% of C36000. Both alloys can be readily machined by all conventional methods. Specific machining parameters depend on shapes, machining method, and temper or condition of the metal. The leaded version of this alloy, C17300, is especially intended for machined parts. Other properties are unchanged by the addition of lead to enhance machinability.

Recrystallization temperature. Approximately 730 °C (1350 °F)

Annealing temperature. Strip, thin rod, and wire: 760 to 790 °C (1400 to 1450 °F)/10 min/water quench. Larger sections: 1 h per inch or fraction of an inch of cross section

Solution temperature. 760 to 790 °C (1400 to 1450 °F). All annealing of this material is a solution treatment.

Aging temperature. 260 to 425 °C (500 to 800 °F). Maximum strength is obtained by aging material 1 to 3 h at 315 to 345 °C (600 to 650 °F), depending on the amount of cold work.

Hot-working temperature. 650 to 800 °C (1200 to 1475 °F). C17300 cannot be hot rolled or forged, but it can be hot extruded.

Hot-shortness temperature. 845 °C (1550 °F)

C17410
99.2Cu-0.3Be-0.5Co

Chemical Composition

Composition limits. 0.15 to 0.5 Be, 0.35 to 0.6 Co, 0.2 Al max, 0.2 Si max, 0.2 Fe max, 99.5 min Cu + Ag + named elements

Table 35 Nominal mechanical properties of mill-hardened C17410 strip

Temper	Tensile strength MPa	ksi	Yield strength MPa	ksi	Elongation in 50 mm (2 in.), %	Hardness, HRB	Fatigue strength MPa	ksi
½ HT	725	105	620	90	15	95	...	...
HT	830	120	760	110	12	95 min	300	43

Applications

Typical uses. Strip and wire: fuse clips, fasteners, springs, diaphragms, lead frames, switch parts, and electrical connectors. Rod and plate: resistance spot welding tips, die casting plunger tips, tooling for plastic molding
Precautions in use. Because this alloy contains beryllium, it is a potential health hazard. Adequate safety precautions are mandatory for all melting, welding, grinding, and machining operations.

Mechanical Properties

Tensile properties. See Table 35.
Hardness. See Table 35.
Elastic modulus. 138 GPa (20.0 × 10^6 psi)
Fatigue strength. See Table 35.

Mass Characteristics

Density. 8.80 g/cm^3 (0.318 lb/in.^{3}) at 20 °C (68 °F)

Thermal Properties

Liquidus temperature. 1065 °C (1950 °F)
Solidus temperature. 1025 °C (1875 °F)
Thermal conductivity. 233 W/m · K (135 Btu/ft · h · °F) at 20 °C (68 °F)

Electrical Properties

Electrical conductivity. Volumetric, 45% IACS at 20 °C (68 °F)
Electrical resistivity. 38.2 nΩ · m at 20 °C (68 °F)

Fabrication Characteristics

Machinability. 25% of C36000 (free-cutting brass)
Formability. Excellent capacity for both cold and hot forming
Weldability. Solderability, brazing, or resistance spot welding: good. Oxyacetylene welding is not recommended. Other welding processes: fair
Heat-treating temperature. 450 to 550 °C (840 to 1025 °F)
Hot-working temperature. 650 to 925 °C (1200 to 1700 °F)

C17500
97Cu-0.5Be-2.5Co

Commercial Names

Trade names. 10 alloy, alloy 10, Berylco 10
Common name. Low-beryllium copper

Specifications

ASTM. Flat products: B 534. Rod, bar: B 441
SAE. J463
Government. Rod, bar: MIL-C-46087. Strip: MIL-C-81021
Resistance Welding Manufacturers' Association. Class III

Chemical Composition

Composition limits. 0.40 to 0.7 Be, 2.4 to 2.7 Co, 0.10 Fe max, 0.5 max other (total), bal Cu

Applications

Typical uses. Strip, wire: fuse clips, fasteners, springs, switch parts, electrical connectors, and conductors. Rod, plate: resistance spot welding tips, seam welding discs, die casting plunger tips, tooling for plastic molding
Precautions in use. Because this alloy contains beryllium, it is a potential health hazard. Adequate safety precautions are mandatory for all melting, welding, grinding, and machining operations.

Mechanical Properties

Tensile properties. See Table 36 and Fig. 20.
Hardness. See Table 36.
Elastic modulus. Tension, 125 to 130 GPa (18 to 19 × 10^6 psi)
Fatigue strength. Rod, TF00 temper (rotating-beam tests): 275 to 310 MPa (40 to 45 ksi) at 10^7 cycles. Strip: see Table 36.

Structure

Crystal structure. The α Cu solid solution is face-centered cubic. The beryllide, (Cu, Co)Be, is ordered body-centered cubic of the CsCl (B2) type.
Microstructure. Alpha copper with beryllium in solid solution and with (Cu,Co)Be beryllide inclusions. The appearance of the matrix of the beryllides depends on the extent of deformation and the state of heat treatment.

In the cast condition, the matrix is essentially like pure copper; the beryllides, which are blue-gray, are large and sharply angular in the grain boundaries and small with Widmanstätten orientation within the grains. When such cast shapes are annealed, the cored appearance is reduced slightly, and the matrix becomes slightly cleaner as small amounts of the beryllides are dissolved. As the cast product is reduced by either hot or cold working, the beryllides are broken up and uniformly distributed.

For such products as strip or rod, the microstructure is fine-grain equiaxed α copper with small, mainly spherical, uniformly dispersed beryllides. For all product types, there is little difference in microstructure between the annealed and the aged conditions.

Table 36 Typical mechanical properties and electrical conductivity of C17500

Temper	Tensile strength MPa	ksi	Proportional limit at 0.002% offset MPa	ksi	Yield strength at 0.2% offset MPa	ksi	Elongation in 50 mm (2 in.), %	Hardness, HRB	Electrical conductivity, %IACS	Fatigue strength(a) MPa	ksi
Strip											
TB00	240–380	35–55	69–140	10–20	140–205	20–30	20–35	28–50	20–30	...	...
H04	485–585	70–85	240–450	35–65	380–550	55–80	3–10	70–80	20–30	205	30
TF00	690–825	100–120	380–515	55–75	550–690	80–100	10–20	92–100	45–60	205	30
TH04	760–895	110–130	485–655	70–95	690–825	100–120	8–15	98–102	50–60	240	35
HTR(b)	825–1035	120–150	550–760	80–110	760–965	110–140	1–4	98–103	45–60	240–260	35–38
HTC(c)	515–585	75–85	205–415	30–60	345–515	50–75	8–15	79–88	60 min	205–240	30–35
Rod, bar, plate, tubing											
TB00	240–380	35–55	...	...	140–205	20–30	20–35	20–50	20–30	...	...
H04	450–550	65–80	...	...	380–515	55–75	10–15	60–80	20–30	...	...
TF00	690–825	100–120	...	...	550–690	80–100	10–25	92–100	45–60	...	...
TH04	760–895	110–130	...	...	690–825	100–120	10–20	95–102	50–60	...	...
Forged products											
TB00	240–380	35–55	...	...	140–205	20–30	20–35	20–50	20–30	...	...
TF00	690–825	100–120	...	...	550–690	80–100	10–25	92–100	45–60	...	...

(a) Reversed bending at 10^8 cycles. (b) Proprietary mill hardening for maximum strength. (c) Proprietary mill hardening for maximum electrical conductivity

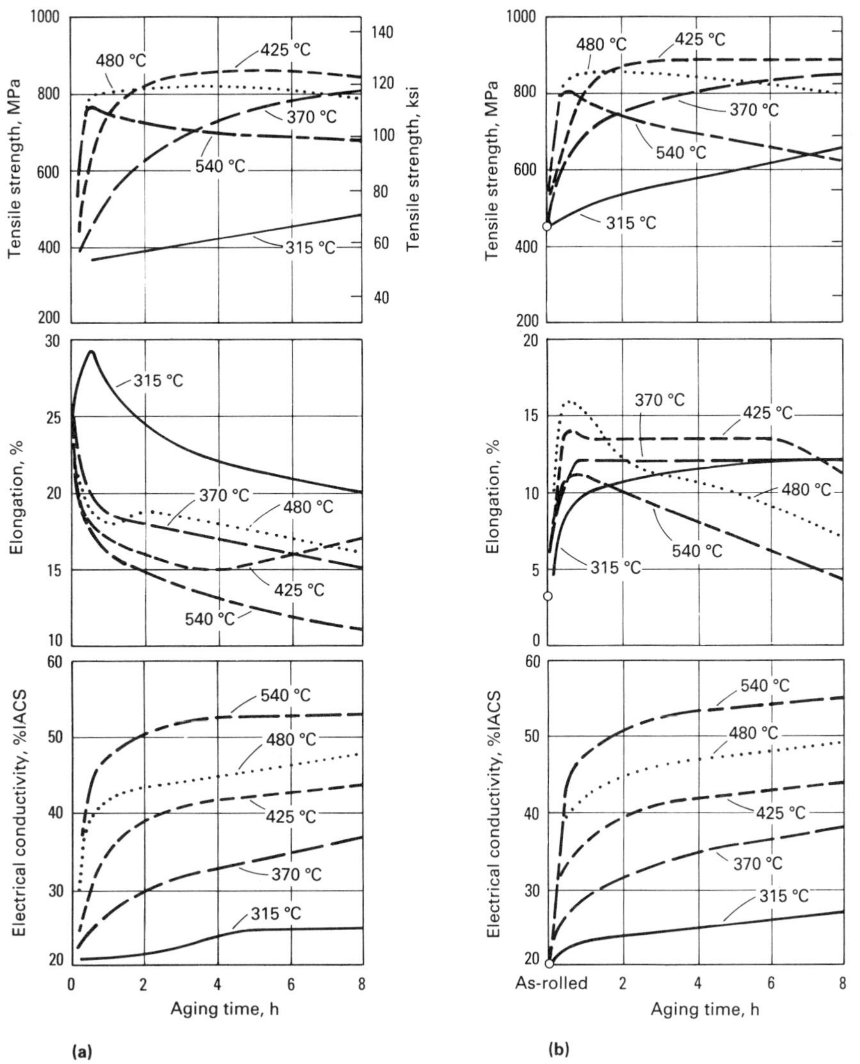

Fig. 20 Aging curves for C17500. (a) TB00 temper. (b) TD02 temper

Metallography is by ordinary metallographic techniques, except that grinding must be performed in a vented area, and all other appropriate OSHA requirements should be strictly observed.

Mass Characteristics

Density. 8.75 g/cm^3 (0.316 lb/in.^{3}) at 20 °C (68 °F)
Volume change on phase transformation. Slight contraction during age hardening; exact amount depends on starting condition of material and on time and temperature of aging.

Thermal Properties

Liquidus temperature. 1070 °C (1955 °F)
Solidus temperature. 1030 °C (1885 °F)
Coefficient of linear thermal expansion. 17.6 μm/m · K (9.8 μin./in. · °F) at 20 to 200 °C (68 to 392 °F)
Specific heat. 420 J/kg · K (0.10 Btu/lb · °F) at 20 °C (68 °F)

Electrical Properties

Electrical conductivity. See Table 36.
Electrical resistivity. 29 to 86 nΩ · m at 20 °C (68 °F), depending on heat treatment

Chemical Properties

General corrosion behavior. Comparable to that of other high-copper alloys. May tarnish in humid or sulfur-bearing atmospheres

Fabrication Characteristics

Machinability. Readily machinable by all common methods. Recommended machining conditions depend greatly on the shape of the part, on the heat treatment of material, and on the type of machining operation.

Because this alloy contains beryllium, OSHA requirements must be strictly observed. Normally, these requirements include flooding and/or special ventilation to prevent personnel from inhaling or ingesting metal dust.

Annealing temperature. All annealing of this alloy is a solution treatment.
Solution temperature. Strip, rod, bar, tubing, wire: 10 min at 900 to 955 °C (1650 to 1750 °F), water quench. Large sections: 1 h per inch or fraction of an inch at 900 to 925 °C (1650 to 1700 °F), water quench
Aging temperature. For maximum strength: 3 to 6 h at 425 °C (800 °F), depending on the degree of cold work. Commercial practice: 2 to 3 h at 470 to 495 °C (875 to 925 °F) to provide a combination of high strength and electrical conductivity. Cooling rate after aging is not critical. See also Fig. 20.
Hot-working temperature. 700 to 925 °C (1300 to 1700 °F)
Hot-shortness temperature. 980 °C (1800 °F)

C17600

Commercial Names

Previous trade name. 50 alloy, alloy 50
Common name. Beryllium-copper

Specifications

SAE. J463 (CA176)
Resistance Welding Manufacturers' Association. Class III

Chemical Composition

Composition limits. 99.5 Cu + Be + additives min, 0.25 to 0.50 Be, 1.40 to 1.70 Co, 0.90 to 1.10 Ag, 1.40 Co + Ni min, 1.90 Co + Ni + Fe max

Applications

Typical uses. A high-conductivity alloy designed especially for resistance welding electrodes for spot, seam, flash, and projection welding methods; electrical connectors, clips
Precautions in use. Ventilation should be used during melting, welding, grinding, and all machining operations.

Mechanical Properties

Tensile properties. See Table 37.
Hardness. See Table 37.
Elastic modulus. Tension, 125 to 130 GPa (18 to 19 × 10^6 psi); shear, 44 GPa (6.4 × 10^6 psi)

Structure

Crystal structure. Alpha copper solid solution is face-centered cubic; the beryllide, (Cu,Co)Be, is ordered body-centered cubic of the CsCl (B2) type.
Microstructure. Matrix of α copper; large and sharply angular blue-gray beryllide inclusions in grain boundaries of cast product, smaller Widmanstätten beryllides within the grain. In wrought products with large amounts of deformation, the beryllides are small, mainly spherical, and uniformly distributed.

Metallography is by conventional techniques. For dry grinding, ventilation

Table 37 Typical mechanical properties and electrical conductivity of C17600 heat treated to various tempers

Temper(a)	Tensile strength, MPa	Tensile strength, ksi	Yield strength at 0.2% offset, MPa	Yield strength at 0.2% offset, ksi	Elongation in 50 mm (2 in.), %	Hardness, HRB	Electrical conductivity, %IACS
Rod, bar, wire, tubing, plate							
TB00	240–380	35–55	140–205	20–30	20–35	20–50	20–30
H04	450–550	65–80	380–515	55–75	10–15	60–80	20–30
TF00	690–825	100–120	550–690	80–100	10–25	92–100	45–60
TH04	760–900	110–130	690–825	100–120	10–20	95–102	50–60
Billet							
As-cast	310–415	45–60	105–240	15–35	15–25	60–65	32–37
Cast and aged	415–515	60–75	205–380	30–55	10–20	65–90	40–50
TB00	275–345	40–50	69–115	10–17	20–40	10–45	22–28
TF00	655–760	95–110	515–550	75–80	3–15	92–100	50–60
Forged products							
TB00	240–380	35–55	140–205	20–30	20–35	25–45	20–30
TF00	690–825	100–120	550–690	80–100	10–25	92–100	50–60

(a) For TB00 temper: solution treat strip, bar, rod, and tubing 10 min at 900 to 955 °C (1650 to 1750 °F) and water quench; solution treat thicker products such as billet 1 h for each 25 mm (1 in.) of thickness or fraction thereof at 900 to 925 °C (1650 to 1700 °F) and water quench. For aging cast billets or producing TF00 temper, age 3 h at 470 to 500 °C (875 to 925 °F). For producing TH04 temper, age 2 h at 470 to 500 °C (875 to 925 °F).

Table 38 Nominal mechanical properties of C18100 strip and wire

Temper	Tensile strength, MPa	Tensile strength, ksi	Yield strength, MPa	Yield strength, ksi	Elongation in 50 mm (2 in.), %	Hardness, HRB
Strip						
Cold worked (40% reduction)	460	67	430	62	6	...
Cold worked (40% reduction), aged	495	72	455	66	10	...
Wire						
Cold worked (60% reduction)	480	70	435	63	6	...
Cold worked (60% reduction), aged	515	75	470	68	11	80
Cold worked (75% reduction)	495	72	455	66	5	...
Cold worked (75% reduction), aged	550	80	475	69	12	...
Cold worked (90% reduction)	500	73	455	66	4	...
Cold worked (90% reduction), aged	585	85	515	75	13	...

should be provided. Some common etchants for immersion etching are 3 parts concentrated NH_4OH, 1 part 3% H_2O_2, 2 parts 10% $(NH_4)_2S_2O_3$, and 7 to 10 parts H_2O. Common etchants for swabbing are 3 g $K_2Cr_2O_7$, 1.5 g NaCl, 8 ml H_2SO_4, and 100 ml H_2O.

Mass Characteristics

Density. 8.75 g/cm³ (0.316 lb/in.³) at 20 °C (68 °F)

Thermal Properties

Liquidus temperature. 1068 °C (1955 °F)
Solidus temperature. 1013 °C (1855 °F)
Coefficient of linear thermal expansion. 16.7 μm/m · K (9.3 μin./in. · °F) at 20 to 200 °C (68 to 392 °F)
Specific heat. 420 J/kg · K (0.10 Btu/lb · °F) at 20 °C (68 °F)
Thermal conductivity. 215 to 245 W/m · K (125 to 140 Btu/ft · h · °F) at 20 °C (68 °F)

Electrical Properties

Electrical conductivity. See Table 37.
Electrical resistivity. 28.7 to 86.2 nΩ · m at 20 °C (68 °F), depending strongly on heat treatment

Fabrication Characteristics

Machinability. Readily machinable by all conventional methods
Annealing temperature. For strip, wire, rod, and bar, 900 to 950 °C (1650 to 1750 °F)/10 min/water quench. For larger sections, anneal 1 h per inch or fraction of an inch at 900 to 925 °C (1650 to 1700 °F) and water quench.
Solution temperature. All annealing for this alloy is solution treatment.
Aging temperature. Maximum strength is obtained by 3 to 6 h at 425 °C (800 °F). Commercial practice is to age material 2 to 3 h at 480 °C (900 °F) to obtain a combination of high strength and electrical conductivity.
Hot-working temperature. 750 to 925 °C (1400 to 1700 °F)
Hot-shortness temperature. 975 °C (1800 °F)

C18100
99Cu-0.8Cr-0.16Zr-0.04Mg

Chemical Composition

Composition limits. 0.4 to 1.2 Cr, 0.05 to 0.3 Zr, 0.03 to 0.06 Mg

Applications

Typical uses. Resistance welding electrodes and wheels, switches, circuit breakers, high-temperature wire, semiconductor bases, heat sinks, and continuous casting molds

Mechanical Properties

Tensile properties. See Table 38.
Hardness. See Table 38.
Elastic modulus. 125 GPa (18.2 × 10⁶ psi)

Mass Characteristics

Density. 8.88 g/cm³ (0.319 lb/in.³) at 20 °C (68 °F)

Thermal Properties

Liquidus temperature. 1075 °C (1967 °F)
Thermal conductivity. 324 W/m · K (187 Btu/ft · h · °F) at 20 °C (68 °F)
Coefficient of linear thermal expansion. 16.7 μm/m · K (9.3 μin./in. · °F) at 20 to 100 °C (68 to 212 °F); 18.4 μm/m · K (10.2 μin./in. · °F) at 20 to 200 °C (68 to 392 °F); 19.3 μm/m · K (10.7 μin./in. · °F) at 20 to 300 °C (68 to 572 °F)

Electrical Properties

Electrical conductivity. Volumetric, 80% IACS at 20 °C (68 °F), annealed
Electrical resistivity. 21.7 nΩ · m at 20 °C (68 °F), annealed

Fabrication Characteristics

Formability. Excellent capacity for both cold and hot forming
Weldability. Solderability, excellent; brazing and gas-shielded arc welding, good; butt resistance welding, fair. Oxyacetylene, spot, and seam resistance welding are not recommended.
Annealing temperature. 600 to 700 °C (1110 to 1300 °F)
Heat-treating temperatures. Solution treatment: 900 to 975 °C (1650 to 1790 °F) for 1 h. Aging treatment: 400 to 500 °C (750 to 930 °F) for 1 h
Hot-working temperature. 790 to 925 °C (1450 to 1700 °F)

C18200, C18400, C18500
99Cu-1Cr

Commercial Names

Previous trade name. CA182, CA184, CA185; Chrome Copper 999 (C18200)
Common name. Chromium-copper

Specifications

ASTM. Wire: F 9
SAE. J463 (C18400 only)
Government. Bar, forgings, rod, strip: MIL-C-19311 (C18400, C18500)

Chemical Composition

Composition limits of C18200. 0.6 to 1.2 Cr, 0.10 Fe max, 0.10 Si max, 0.05 Pb max, 0.5 max other (total), bal Cu + Ag

Table 39 Typical mechanical properties of C18200, C18400, and C18500

Temper	Tensile strength MPa	ksi	Yield strength(a) MPa	ksi	Elongation in 50 mm (2 in.), %	Hardness, HRB
Flat products, 1 mm (0.04 in.) thick						
TB00	235	34	130	19	40	16
TF00(b)	350	51	250	36	22	59
TD04	365	53	350	51	6	66
TH04(c)	460	67	405	59	14	79
Plate, 50 mm (2.0 in.) thick						
TF00	400	58	290	42	25	70
Plate, 75 mm (3.0 in.) thick						
TF00	385	56	275	40	30	68
Rod: 4 mm (0.156 in.) diameter						
TD08	510	74	505	73	5	...
TH08	595	86	530	77	14	...
Rod, 13 mm (0.50 in.) diameter						
TB00	310	45	97	14	40	...
TF00(b)	485	70	380	55	21	70
TD04	395	57	385	56	11	65
TH04(c)	530	77	450	65	16	82
TH03, cold worked 6%	530	77	460	67	19	83
Rod, 25 mm (1.0 in.) diameter						
TF00	495	72	450	65	18	80
Rod, 50 mm (2.0 in.) diameter						
TF00	485	70	450	65	18	75
Rod, 75 mm (3.0 in.) diameter						
TF00	450	65	380	55	18	70
Rod, 100 mm (4.0 in.) diameter						
TF00	380	55	295	43	25	68
Tube, 9.5 mm (3/8 in.) outside diameter × 2.4 mm (0.094 in.) wall thickness						
O60	275	40	105	15	50	59 HRF
Tube, 31.8 mm (1¼ in.) outside diameter × 5.4 mm (0.212 in.) wall thickness						
TD04	405	59	395	57	21	67
TH04, cold-worked 28%	475	69	435	63	26	84

(a) At 0.5% extension under load. (b) Aged 3 h at 500 °C (930 °F). (c) Aged 3 h at 450 °C (840 °F)

Composition limits of C18400. 0.40 to 1.2 Cr, 0.7 Zn max, 0.15 Fe max, 0.10 Si max, 0.05 P max, 0.05 Li max, 0.005 As max, 0.005 Ca max, 0.2 max other (total), bal Cu + Ag
Composition limits of C18500. 0.40 to 1.0 Cr, 0.08 to 0.12 Ag, 0.04 P max, 0.04 Li max, 0.015 Pb max, bal Cu + Ag

Applications

Typical uses. Applications requiring excellent cold workability and good hot workability coupled with medium-to-high conductivity. Uses include resistance welding electrodes, seam welding wheels, switch gears, electrode holder jaws, cable connectors, current-carrying arms and shafts, circuit breaker parts, molds, spot welding tips, flash welding electrodes, electrical and thermal conductors requiring more strength than that provided by unalloyed coppers, and switch contacts

Mechanical Properties

Tensile properties. See Table 39.
Hardness. See Table 39.
Elastic modulus. Tension, 130 GPa (19 × 10^6 psi); shear, 50 GPa (7.2 × 10^6 psi)

Mass Characteristics

Density. 8.89 g/cm^3 (0.321 lb/in.^{3}) at 20 °C (68 °F)

Thermal Properties

Liquidus temperature. 1075 °C (1965 °F)
Solidus temperature. 1070 °C (1960 °F)
Coefficient of linear thermal expansion. 17.6 μm/m · K (9.8 μin./in. · °F) at 20 to 100 °C (68 to 212 °F)
Specific heat. 385 J/kg · K (0.092 Btu/lb · °F) at 20 °C (68 °F)
Thermal conductivity. TB00 temper: 171 W/m · K (99 Btu/ft · h · °F) at 20 °C (68 °F). TH04 temper: 324 W/m · K (187 Btu/ft · h · °F) at 20 °C (68 °F)

Electrical Properties

Electrical conductivity. Volumetric. TB00 temper: 40% IACS at 20 °C (68 °F). TH04 temper: 80% IACS at 20 °C (68 °F)
Electrical resistivity. TH04 temper: 21.6 nΩ · m at 20 °C (68 °F)

Fabrication Characteristics

Machinability. 20% of C36000 (free-cutting brass)
Formability. Suited for hot working by extrusion, rolling, and forging (subsequent solution treatment required) and for cold working (in soft, solution-annealed, or suitable drawn temper) by drawing, rolling, impacting, heading, bending, or swaging
Weldability. Welding and brazing temperatures lower the properties developed by heat treatment; such processes normally are applied to material in the soft condition, followed by necessary heat treatment. Soldering: good. Oxyfuel gas, shielded metal arc, resistance spot, and resistance seam welding are not recommended.
Solution treatment. 980 to 1000 °C (1800 to 1850 °F) for 10 to 30 min, water quench
Aging temperature. 425 to 500 °C (800 to 930 °F) for 2 to 4 h
Hot-working temperature. 800 to 925 °C (1500 to 1700 °F)

C18700
99Cu-1Pb

Commercial Names

Previous trade name. Leaded copper
Common name. Free-machining copper

Specifications

ASTM. Flat products, rod: B 301
SAE. Rod: J463

Chemical Composition

Composition limits. 0.8 to 1.5 Pb, 0.10 max other (total), bal Cu. Oxygen-free grades or grades containing deoxidizers such as P, B, or Li may be specified.

Applications

Typical uses. Electrical connectors, motor parts, switch parts, and screw machine parts requiring high conductivity
Precautions in use. Unless specifically deoxidized, this copper is subject to embrittlement when heated in a reducing atmosphere (as in annealing or brazing) at temperatures of 350 °C (660 °F) or higher. If hydrogen or carbon monoxide is present, embrittlement can be rapid.

Table 40 Typical mechanical properties of C18700 rod, H04 temper

Diameter mm	in.	Tensile strength MPa	ksi	Yield strength MPa	ksi	Elongation in 50 mm (2 in.), %	Hardness, HRB	Shear strength MPa	ksi
6	0.25	415	60	380	55	10	55	200	32
13	0.50	380	55	345	50	11	50	205	30
19	0.75	365	53	330	48	12	50	200	29
25	1.0	350	51	315	46	14	50	195	28

Table 41 Typical mechanical properties of C19200

Temper	Tensile strength MPa	Tensile strength ksi	Yield strength At 0.5% extension under load MPa	Yield strength At 0.5% extension under load ksi	Yield strength At 0.2% offset MPa	Yield strength At 0.2% offset ksi	Elongation in 50 mm (2 in.), %	Hardness, HRB
Strip, 1 mm (0.04 in.) diameter								
O60	310	45	...	...	140 min	20 min	25 min	38
O82	395	57	...	...	305	44	20	55
H02	395	57	...	...	305	44	9	55
H04	450	65	...	...	415	60	7	72
H06	485	70	...	...	460	67	3	75
H08	510	74	...	...	490	71	2 min	76
H10	530	77	...	...	510	74	2 min	77
Tubing, 48 mm (1.88 in.) outside diameter × 3 mm (0.12 in.) wall thickness								
O50	290	42	160	23	150	22	30	...
O60	255	37	83	12	76	11	40	...
H80 (40%)	385	56	360	52	360	52	7	...
Tubing, 5 mm (0.19 in.) outside diameter × 0.8 mm (0.03 in.) wall thickness								
H55	290	42	215	31	205	30	35	...

Table 42 Nominal tensile properties of C19210 sheet

Temper	Tensile strength MPa	Tensile strength ksi	Yield strength MPa	Yield strength ksi	Elongation in 50 mm (2 in.), %
H01	345	50	330	48	13
H02	390	57	385	56	6
H04	440	64	435	63	4
H08	490	71	480	70	2

Mechanical Properties

Tensile properties. See Tables 17 and 40.
Shear strength. See Tables 17 and 40.
Hardness. See Tables 17 and 40.
Elastic modulus. Tension, 115 GPa (17 × 10^6 psi); shear, 44 GPa (6.4 × 10^6 psi)

Mass Characteristics

Density. 8.94 g/cm^3 (0.323 lb/in.^{3}) at 20 °C (68 °F)

Thermal Properties

Liquidus temperature. 1080 °C (1975 °F)
Solidus temperature. 950 °C (1750 °F)
Coefficient of linear thermal expansion. 17.6 μm/m · K (9.8 μin./in. · °F) at 20 to 300 °C (68 to 572 °F)
Specific heat. 385 J/kg · K (0.092 Btu/lb · °F) at 20 °C (68 °F)
Thermal conductivity. 377 W/m · K (218 Btu/ft · h · °F) at 20 °C (68 °F)

Electrical Properties

Electrical conductivity. Volumetric, 96% IACS at 20 °C (68 °F)
Electrical resistivity. 17.9 nΩ · m at 20 °C (68 °F)

Fabrication Characteristics

Machinability. 85% of C36000 (free-cutting brass)
Formability. Good for cold working; poor for hot forming
Weldability. Soldering: excellent. Brazing: good. Most arc, gas, and resistance welding processes are not recommended.
Annealing temperature. 425 to 650 °C (800 to 1200 °F)
Hot-working temperature. 750 to 875 °C (1400 to 1600 °F)

C19200
98.97Cu-1.0Fe-0.03P

Specifications

ASTM. Tubing: B 111, B 359, B 395, B 469

Chemical Composition

Composition limits. 98.7 to 99.19 Cu, 0.8 to 1.2 Fe, 0.01 to 0.04 P

Applications

Typical uses. Rolled strip and tubing for air conditioning and heat exchanger tubing, applications requiring resistance to softening and stress corrosion, automotive hydraulic brake lines, cable wrap, circuit breaker components, contact springs, electrical connectors and terminals, eyelets, flexible hose, fuse clips, gaskets, gift hollowware, lead frames

Mechanical Properties

Tensile properties. See Table 41.
Hardness. See Table 41.
Elastic modulus. Tension, 115 GPa (17 × 10^6 psi); shear, 44 GPa (6.4 × 10^6 psi)

Mass Characteristics

Density. 8.87 g/cm^3 (0.320 lb/in.^{3}) at 20 °C (68 °F)

Thermal Properties

Liquidus temperature. 1084 °C (1983 °F)
Solidus temperature. 1078 °C (1973 °F)
Coefficient of linear thermal expansion. 16.2 μm/m · K (9.0 μin./in. · °F) at 20 to 100 °C (68 to 212 °F)
Specific heat. 380 J/kg · K (0.09 Btu/lb · °F) at 20 °C (68 °F)
Thermal conductivity. Strip, 251 W/m · K (145 Btu/ft · h · °F) at 20 °C (68 °F); tubing, 216 W/m · K (125 Btu/ft · h · °F) at 20 °C (68 °F)

Electrical Properties

Electrical conductivity. Strip, 60% IACS at 20 °C (68 °F); tubing, 50% IACS at 20 °C (68 °F)
Electrical resistivity. Strip, 28.8 nΩ · m at 20 °C (68 °F); tubing, 34.5 nΩ · m at 20 °C (68 °F)

Fabrication Characteristics

Machinability. 20% of C36000 (free-cutting brass)
Forgeability. 65% of C37700 (forging brass)
Weldability. Soldering, brazing, and gas-shielded arc welding: excellent. Oxyacetylene welding: good. Coated metal arc and resistance seam, spot, and butt welding are not recommended.
Annealing temperature. 700 to 815 °C (1300 to 1500 °F)
Hot-working temperature. 825 to 950 °C (1500 to 1750 °F)

C19210
99.87Cu-0.1Fe-0.03P

Chemical Composition

Composition limits. 0.05 to 0.15 Fe, 0.025 to 0.04 P, bal Cu

Applications

Typical uses. Air conditioner and heat exchanger tubing, lead frames, electrical connectors and terminals

Mechanical Properties

Tensile properties. See Table 42.
Elastic modulus. 125 GPa (18.2 × 10^6 psi)

Mass Characteristics

Density. 8.94 g/cm^3 (0.323 lb/in.^{3}) at 20 °C (68 °F)

Thermal Properties

Liquidus temperature. 1082 °C (1980 °F)
Coefficient of linear thermal expansion. 16.9 μm/m · K (9.4 μin./in. · °F) at 20 to 300 °C (68 to 572 °F)

Electrical Properties

Electrical conductivity. Volumetric, 80% IACS at 20 °C (68 °F), annealed
Electrical resistivity. 21.6 nΩ · m at 20 °C (68 °F), annealed

Fabrication Characteristics

Machinability. 20% of C36000 (free-cutting brass)
Formability. Excellent capacity for both cold and hot forming
Weldability. Soldering, brazing, and coated metal arc welding: excellent. Butt, resistance and oxyacetylene welding: good. Gas-shielded arc, spot, and seam resistance welding are not recommended.
Annealing temperature. 450 to 550 °C (840 to 1020 °F)
Hot-working temperature. 700 to 900 °C (1300 to 1650 °F)

Table 43 Typical mechanical properties of C19400

Temper	Tensile strength MPa	Tensile strength ksi	Yield strength at 0.2% offset MPa	Yield strength at 0.2% offset ksi	Elongation in 50 mm (2 in.), %	Hardness HRB	Hardness HR30T	Fatigue strength(a) MPa	Fatigue strength(a) ksi
Flat products, 0.64 mm (0.025 in.) thick									
O60	310	45	150 max	22 max	29 min	38	...	110	16
O50	345	50	160	23	28	45	...	...	...
O82	400	58	255	37	15	...	...	...	...
Flat products, 1 mm (0.04 in.) thick									
H02	400	58	315(b)	46(b)	18	68	66	...	...
H04	450	65	380	55	7	73	69	145	21
H06	485	70	465	67.5	3	74	71	...	...
H08	505	73	486	70.5	3	75	72	148	21.5
H10	530	77	507	73.5	2 max	77	74	141	20.5
H14	550 min	80 min	530 min	77 min	2 max	...	>73	...	...
Tubing, 25 mm (1 in.) outside diameter × 0.9 mm (0.035 in.) wall thickness									
O60	310	45	165	24	28	38	...	...	...
O50	345	50	205	30	16	45	...	...	...
WM02	400	58	365	53	9	61	60	...	...
WM04	450	65	435	63	4	73	66	...	...
WM06	485	70	465	67.5	3	74	68	...	...
WM08	505	73	486	70.5	2	75	69	...	...
WM10	525	76	505	73	1	76	69	...	...
H55 (15%)	400	58	380	55	9	61	60	...	...
H80 (35%)	470	68	455	66	2	73	66	...	...

(a) At 10^8 cycles as determined by the rotating-beam test. (b) At 0.5% extension under load.

Hot forgeability rating. 65% of C37700 (forging brass)

C19400
Cu-2.35Fe-0.03P-0.12Zn

Commercial Names

Previous trade name. High-strength modified copper, HSM copper

Specifications

ASME. Welded tubing: SB543
ASTM. Flat products: B 465. Welded tubing: B 543, B 586

Chemical Composition

Composition limits. 2.1 to 2.6 Fe, 0.05 to 0.20 Zn, 0.015 to 0.15 P, 0.03 Pb max, 0.03 Sn max, 0.15 max other (total), bal Cu

Applications

Typical uses. Applications requiring excellent hot and cold workability as well as high strength and conductivity. Specific uses include circuit breaker components; contact springs; electrical clamps, springs, and terminals; flexible hose; fuse clips; gaskets; gift hollowware; plug contacts; rivets; welded condenser tubes; semiconductor lead frames, and cable shielding

Table 44 Typical room-temperature and low-temperature (cryogenic) properties of C19400

Temper	Tensile strength MPa	Tensile strength ksi	Yield strength at 0.2% offset MPa	Yield strength at 0.2% offset ksi	Elongation in 50 mm (2 in.), %
Room-temperature properties					
O61	325	47	170	25	28
H02	405	59	360	52	15
H04	455	66	405	59	10
Cryogenic properties: −196 °C (−320 °F)					
O61	475	69	195	28	38
H02	570	83	425	62	30
H04	615	89	485	70	23

Mechanical Properties

Tensile properties. See Tables 43, 44, 45, and 46.
Hardness. See Table 43.
Elastic modulus. Tension, 121 GPa (17.5 × 10^6 psi); shear, 45.5 GPa (6.6 × 10^6 psi)
Charpy impact strength. Plate, O61 temper: longitudinal, 144 J (106 ft · lbf) at −196 °C (−320 °F); transverse, 99 J (73 ft · lbf) at −196 °C (−320 °F)
Fatigue strength. See Table 43.
Creep and stress-rupture properties. See Table 45.

Table 45 Typical elevated-temperature properties of annealed C19400 strip

Test temperature °C	Test temperature °F	Tensile strength, min MPa	Tensile strength, min ksi	Yield strength at 0.2% offset, min MPa	Yield strength at 0.2% offset, min ksi	Creep strength, min(a) MPa	Creep strength, min(a) ksi	Stress-rupture stress, min(b) MPa	Stress-rupture stress, min(b) ksi
Ambient		341	49.5	150	22.0	...	...	...	...
65	150	324	47.0	144	20.9	...	...	...	...
95	200	313	45.4	144	20.9	...	...	...	...
120	250	300	43.5	144	20.9	190	27.6	...	...
150	300	289	41.9	139	20.2	171	24.8	171	24.9
175	350	276	40.1	135	19.6	143	20.8	148	21.4
205	400	266	38.6	131	19.0	124	18.0	125	18.1
230	450	253	36.8	131	19.0	110	16.0	105	15.2
260	500	235	34.1	127	18.4	96	13.9	82	11.9
290	550	219	31.8	123	17.8	84	12.2	65	9.4
315	600	203	29.5	116	16.8	74	10.8	47	6.8

(a) Stress causing secondary creep of 0.01% per 1000 h in a 10 000-h test. (b) Stress causing rupture in 100 000 h (extrapolated from 10 000 h)

Table 46 Annealing response of C19400 strip

Annealing temperature °C	°F	Tensile strength MPa	ksi	Yield strength at 0.2% offset MPa	ksi	Elongation in 50 mm (2 in.), %	Electrical conductivity, %IACS
H04 temper							
100	212	460	67	450	65	3	66
205	400	450	65	435	63	5	67
315	600	440	64	415	60	9	68
370	700	415	60	385	56	12	68
425	800	415	60	360	52	14	72
480	900	400	59	345	50	16	71
540	1000	385	56	310	45	17	64(a)
595	1100	350	51	220	32	23	52
650	1200	315	46	140	20	33	51
705	1300	310	45	115	17	34	49
760	1400	305	44	110	16	36	48
815	1500	305	44	110	16	36	48
H10 temper							
100	212	510	74	490	71	3	65
205	400	495	72	460	67	5	66
315	600	485	70	415	60	8	67
370	700	330	48	170	25	25	71
425	800	325	47	145	21	27	74
480	900	315	46	140	20	28	69
540	1000	315	46	140	20	31	64(a)
595	1100	310	45	130	19	33	58
650	1200	305	44	130	19	34	52
705	1300	295	43	115	17	34	49
760	1400	290	42	110	16	35	48
815	1500	285	41	105	15	35	48

(a) Conductivity may be restored to about 70% IACS by holding at 500 °C (925 °F) for 1 h.

Mass Characteristics

Density. 8.78 g/cm^3 (0.317 lb/in.^{3}) at 20 °C (68 °F)

Thermal Properties

Liquidus temperature. 1090 °C (1990 °F)
Solidus temperature. 1080 °C (1980 °F)
Coefficient of linear thermal expansion. 16.3 μm/m · K (9.0 μin./in. · °F) at 20 to 300 °C (68 to 572 °F)
Specific heat. 385 J/kg · K (0.092 Btu/lb · °F) at 20 °C (68 °F)
Thermal conductivity. 260 W/m · K (150 Btu/ft · h · °F) at 20 °C (68 °F)

Electrical Properties

Electrical conductivity. Volumetric, at 20 °C (68 °F). O60 temper: 40% IACS nominal. H14 temper: 50% IACS min. All other tempers: 65% IACS nominal, 60% IACS min. In O50, O80, and H02 tempers, 75% IACS min conductivity may be available depending on mill processing restrictions.
Electrical resistivity. At 20 °C (68 °F). O60 temper: 43.1 nΩ · m nominal. H14 temper: 34.5 nΩ · m max. All other tempers: 26.6 nΩ · m nominal; may be only 23.0 nΩ · m max under certain circumstances

Magnetic Properties

Magnetic permeability. 1.1

Chemical Properties

General corrosion behavior. Very corrosion resistant and essentially immune to stress-corrosion cracking

Fabrication Characteristics

Machinability. 20% of C36000 (free-cutting brass)
Formability. Suited to forming by blanking, coining, coppersmithing, drawing, bending, heading and upsetting, hot forging and pressing, piercing and punching, roll threading and knurling, shearing, spinning, squeezing, and stamping
Weldability. Joining by soldering, brazing, and gas tungsten arc welding: excellent
Annealing temperature. See Table 46.

Table 47 Typical mechanical properties of C19500

Temper	Tensile strength MPa	ksi	Yield strength at 0.2% offset MPa	ksi	Elongation in 50 mm (2 in.), %	Hardness, HRB
O61	360 min	52 min	170 min	25 min	25 min	...
O50	520–590	75–85	395–530	57–77	11–17	81–89
H02	565–620	82–90	505–605	73–88	3–13	85–88
H08	605–670	88–97	585–650	85–94	2–5	87–90
H10	670 min	97 min	650 min	94 min	2 max	90 min

C19500
97Cu-1.5Fe-0.1P-0.8Co-0.6Sn

Commercial Names

Trade name. Strescon

Chemical Composition

Composition limits. 1.3 to 1.7 Fe, 0.6 to 1.0 Co, 0.08 to 0.12 P, 0.40 to 0.7 Sn, 0.20 Zn max, 0.02 Al max, 0.02 Pb max, 0.05 max other (each), 0.10 max other (total), bal Cu

Applications

Typical uses. Electrical springs, sockets, terminals, connectors, clips, and other current-carrying parts requiring strength and exceptional softening resistance. Applications requiring excellent hot and cold workability, high strength, and high conductivity

Mechanical Properties

Tensile properties. See Table 47.
Hardness. See Table 47.
Elastic modulus. Tension, 119 GPa (17.3 × 10^6 psi)

Mass Characteristics

Density. 8.92 g/cm^3 (0.322 lb/in.^{3}) at 20 °C (68 °F)

Thermal Properties

Liquidus temperature. 1090 °C (1995 °F)
Solidus temperature. 1085 °C (1985 °F)
Coefficient of linear thermal expansion. 16.9 μm/m · K (9.4 μin./in. · °F) at 20 to 300 °C (68 to 572 °F)
Thermal conductivity. 199 W/m · K (115 Btu/ft · h · °F) at 20 °C (68 °F)

Electrical Properties

Electrical conductivity. Volumetric, 50% IACS at 20 °C (68 °F), annealed
Electrical resistivity. 34.4 nΩ · m at 20 °C (68 °F)

Fabrication Characteristics

Machinability. 20% of C36000 (free-cutting brass)
Formability. Suited to forming by bending, coining, drawing, and stamping

C19520
97.97Cu-0.75Fe-1.25Sn-0.03P

Chemical Composition

Composition limits. 0.5 to 1.5 Fe, 0.5 to 1.5 Sn, 0.01 to 0.35 P, 96.6 Cu min

Applications

Typical uses. Lead frames

Mechanical Properties

Tensile properties. See Table 48.
Hardness. See Table 48.
Elastic modulus. 117 GPa (17 × 10^6 psi)

Table 48 Nominal mechanical properties of C19520 strip

Temper	Tensile strength MPa	Tensile strength ksi	Elongation in 50 mm (2 in.), %	Hardness, HV
H01	415	60	20	125
H02	440	64	10	140
H04	460	67	4	150
H06	515	75	2	160
H08	585	85	. . .	170
H10	640	93	. . .	180
H12	660	96 min	. . .	190 min

Mass Characteristics

Density. 8.8 g/cm³ (0.318 lb/in.³) at 20 °C (68 °F)

Thermal Properties

Coefficient of linear thermal expansion. 16.7 μm/m · K (9.3 μin./in. · °F) at 20 to 300 °C (68 to 572 °F)
Thermal conductivity. 173 W/m · K (100 Btu/ft · h · °F) at 20 °C (68 °F)

Electrical Properties

Electrical conductivity. Volumetric, 40% IACS at 20 °C (68 °F)
Electrical resistivity. 49.3 nΩ · m at 20 °C (68 °F)

C19700
99.15Cu-0.6Fe-0.2P-0.05Mg

Chemical Composition

Composition limits. 0.3 to 1.2 Fe, 0.1 to 0.4 P, 0.01 to 0.2 Mg, 0.2 max each Sn and Zn, 0.05 max each Co, Mn, Ni, and Pb, 99.8 min Cu + named elements

Applications

Typical uses. Electrical and electronic connectors, circuit breaker components, fuse clips, cable shielding, and lead frames. Generally suited to applications requiring excellent formability combined with high strength and conductivity

Mechanical Properties

Tensile properties. See Table 49.
Hardness. See Table 49.
Elastic modulus. 121 GPa (17.5 × 10⁶ psi)

Mass Characteristics

Density. 8.83 g/cm³ (0.319 lb/in.³) at 20 °C (68 °F)

Table 49 Nominal mechanical properties of C19700 strip

Temper	Tensile strength MPa	Tensile strength ksi	Yield strength MPa	Yield strength ksi	Elongation in 50 mm (2 in.), %	Hardness, HRB
H02	380	55	315	46	10	68
H04	450	65	415	60	6	70
H06	480	70	470	68	3	73
H08	500	73	490	71	2	75

Thermal Properties

Liquidus temperature. 1086 °C (1987 °F)
Solidus temperature. 1069 °C (1956 °F)
Coefficient of linear thermal expansion. 15.8 μm/m · K (8.8 μin./in. · °F) at 20 to 100 °C (68 to 212 °F); 16.8 μm/m · K (9.3 μin./in. · °F) at 20 to 200 °C (68 to 392 °F); 17.3 μm/m · K (9.6 μin./in. · °F) at 20 to 300 °C (68 to 572 °F)
Thermal conductivity. 320 W/m · K (185 Btu/ft · h · °F) at 20 °C (68 °F)

Electrical Properties

Electrical conductivity. Volumetric, 80% IACS at 20 °C (68 °F)
Electrical resistivity. 21.6 nΩ · m at 20 °C (68 °F)

Fabrication Characteristics

Machinability. 20% of C36000 (free-cutting brass)
Formability. Excellent capacity for both cold and hot forming
Weldability. Soldering and brazing: excellent
Annealing temperature. 450 to 600 °C (840 to 1110 °F)
Hot-working temperature. 750 to 950 °C (1400 to 1740 °F)

C21000
95Cu-5Zn

Commercial Names

Previous trade name. Gilding metal, 95%; CA210

Specifications

ASTM. Rolled bar, plate, sheet, and strip: B 36. Wire: B 134
SAE. J463
Government. Wire: QQ-W-321. Sheet and strip: MIL-C-21768

Chemical Composition

Composition limits. 94.0 to 96.0 Cu, 0.05 Pb max, 0.05 Fe max, bal Zn
Effect of zinc content on properties. See Fig. 21.

Applications

Typical uses. Coins, medals, tokens, bullet jackets, firing-pin supports, shells, fuse caps and primers, emblems, jewelry plaques, base for gold plate, base for vitreous enamel

Mechanical Properties

Tensile properties. See Table 50 and Fig. 22.
Shear strength. See Table 50.
Hardness. See Table 50.
Elastic modulus. Tension, 115 GPa (17 × 10⁶ psi); shear, 44 GPa (6.4 × 10⁶ psi)
Velocity of sound. 3.78 km/s at 20 °C (68 °F)

Structure

Crystal structure. Face-centered cubic alpha; lattice parameter, 0.3627 nm
Minimum interatomic distance. 0.2564 nm

Mass Characteristics

Density. 8.86 g/cm³ (0.320 lb/in.³) at 20 °C (68 °F)

Thermal Properties

Liquidus temperature. 1065 °C (1950 °F)
Solidus temperature. 1050 °C (1920 °F)
Coefficient of linear thermal expansion. 18 μm/m · K (10 μin./in. · °F) at 20 to 300 °C (68 to 572 °F)
Specific heat. 380 J/kg · K (0.09 Btu/lb · °F) at 20 °C (68 °F)
Thermal conductivity. 234 W/m · K (135 Btu/ft · h · °F) at 20 °C (68 °F)

Electrical Properties

Electrical conductivity. Volumetric, 56% IACS at 20 °C (68 °F), annealed
Electrical resistivity. 31 nΩ · m at 20 °C (68 °F), annealed; temperature coefficient, 0.0231 nΩ · m per K at 20 °C (68 °F). Liquid phase: 244 nΩ · m at 1100 °C (2010 °F), 266 nΩ · m at 1300 °C (2370 °F)

Magnetic Properties

Magnetic susceptibility. -1.0×10^{-6} to -12.5×10^{-6} (SI units)

Optical Properties

Spectral reflectance. 90% for λ = 578 nm

Table 50 Typical mechanical properties of C21000

Temper	Tensile strength MPa	Tensile strength ksi	Yield strength(a) MPa	Yield strength(a) ksi	Elongation in 50 mm (2 in.), %	Hardness HRB	Hardness HR30T	Shear strength MPa	Shear strength ksi
OS050 anneal (0.050 mm grain size)	235	34	69	10	45	46 HRF	. . .	. . .	. . .
OS035 anneal (0.035 mm grain size)	240	35	76	11	45	52 HRF	4	195	28
OS015 anneal (0.015 mm grain size)	260	38	97	14	42	60 HRF	15	205	30
Quarter hard	290	42	220	32	25	38	44	220	32
Half hard	330	48	275	40	12	52	54	235	34
Hard	385	56	345	50	5	64	60	255	37
Extra hard	420	61	380	55	4	70	64	270	39
Spring	440	64	400	58	4	73	66	275	40

Note: Values for flat products, 1 mm (0.04 in.) thick. (a) At 0.5% extension under load

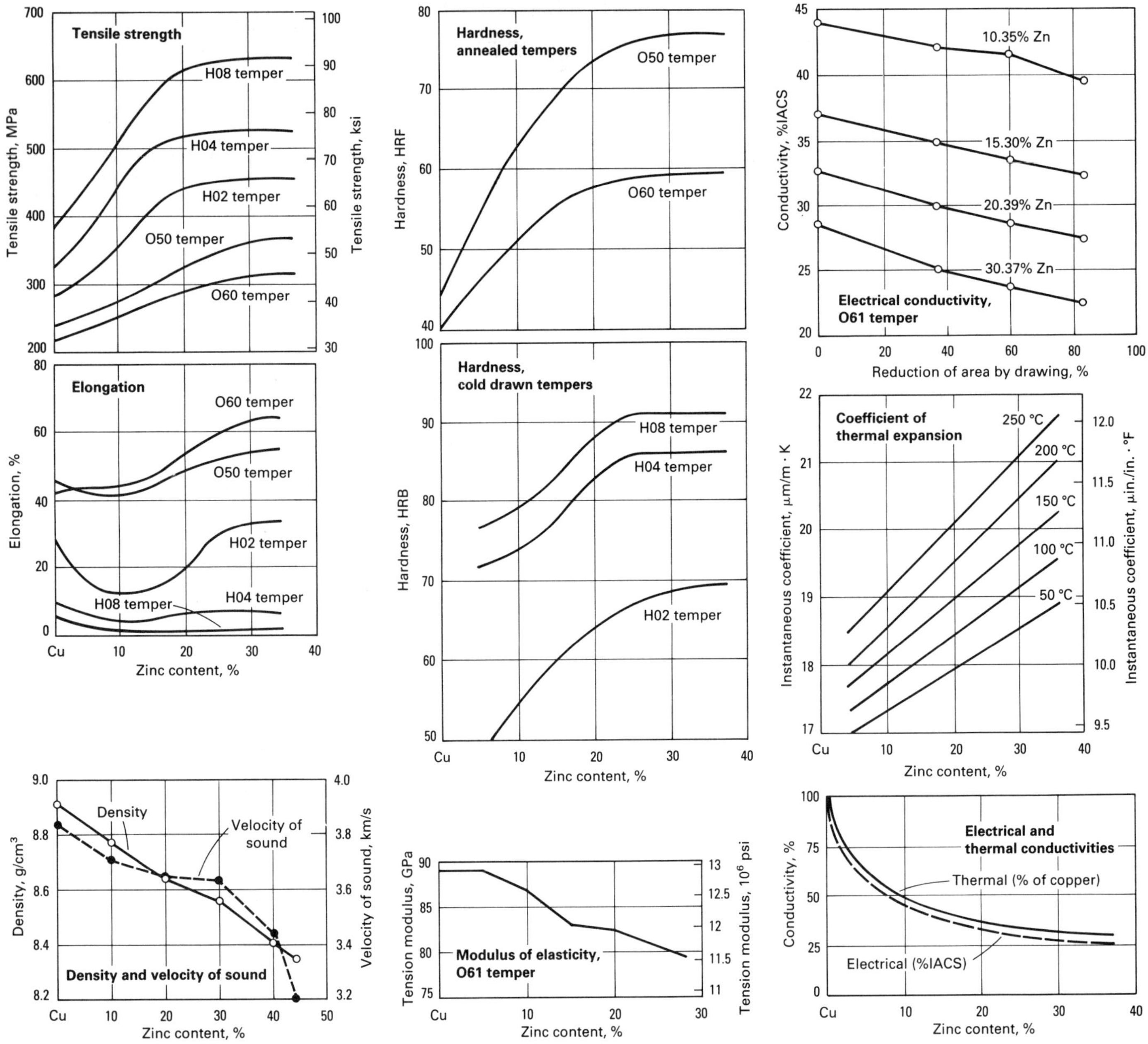

Fig. 21 Variation of properties with zinc content for wrought copper-zinc alloys

Fabrication Characteristics

Machinability. 20% of C36000 (free-cutting brass)
Recrystallization temperature. 370 °C (700 °F) for 50% reduction and 0.015 to 0.070 mm initial grain size. See Fig. 22.
Annealing temperature. 425 to 800 °C (800 to 1450 °F)
Hot-working temperature. 750 to 875 °C (1400 to 1600 °F)

C22000
90Cu-10Zn

Commercial Names

Previous trade name. Commercial bronze, 90%; CA220

Specifications

ASTM. Rolled bar, plate, and sheet: B 36. Strip: B 36 and B 130. Cups, bullet jacket: B 131. Tube, rectangular waveguide: B 372. Seamless tube: B 135. Wire: B 134
SAE. Rolled bar, plate, sheet, strip, and seamless tube: J463 (CA220)
Government. Wire: QQ-W-321; MIL-W-6712. Bands, projectile rotating: MIL-B-18907. Blanks, rotating band for projectiles: MIL-B-20292. Cups, bullet jacket: MIL-C-3383. Sheet and strip: MIL-C-21768. Tube, rectangular waveguide: MIL-W-85. Seamless tube for microwave use: MIL-T-52069

Chemical Composition

Composition limits. 89.0 to 91.0 Cu, 0.05 Pb max, 0.05 Fe max, bal Zn
Consequence of exceeding impurity limits. See general statement for brasses under C26000.
Effect of zinc content on properties. See Fig. 21.

Applications

Typical uses. Architectural: etching bronze, grillwork, screen cloth, weather stripping. Hardware: escutcheons, kickplates, line clamps, marine hardware, rivets, screws, screw shells. Munitions: primer caps, rotating bands. Miscellaneous: compacts, lipstick cases, costume jewelry, ornamental trim, screen wire, base for vitreous enamel, waveguides

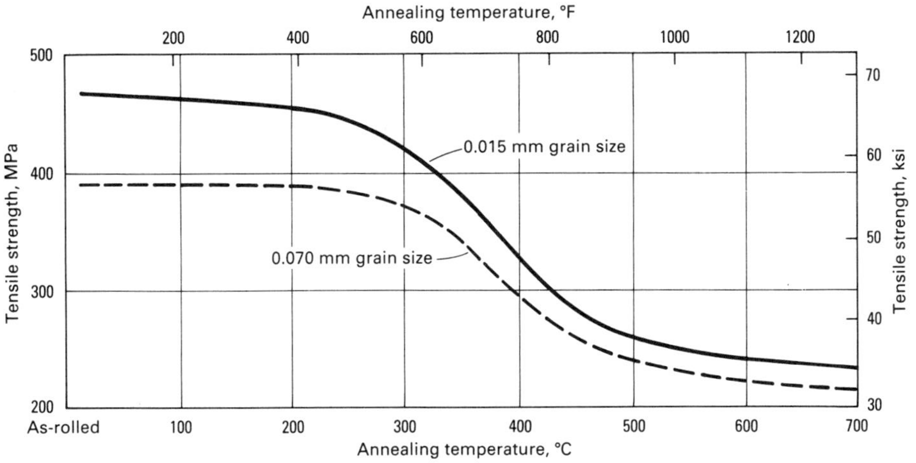

Fig. 22 Variation of tensile strength with annealing temperature for C21000. Data are for 1 mm (0.04 in.) thick ready-to-finish strip that was cold rolled 50% then annealed 1 h at the indicated temperature. Recrystallization temperature, 370 °C (700 °F) for initial grain sizes of 0.015 to 0.070 mm

Mechanical Properties

Tensile properties. See Table 51 and Fig. 23.
Shear strength. See Table 51.
Hardness. See Table 51.
Elastic modulus. Tension, 115 GPa (17 × 10^6 psi); shear, 44 GPa (6.4 × 10^6 psi)
Fatigue strength. Spring temper flat product 1.0 mm (0.40 in.) thick: 145 MPa (21 ksi) at 15 × 10^6 cycles; hard wire 2.0 mm (0.080 in.) in diameter: 160 MPa (23 ksi) at 10^8 cycles
Velocity of sound. 3720 m/s (12 200 ft/s) at 20 °C (68 °F)

Structure

Crystal structure. Face-centered cubic α; lattice parameter, 0.364 nm
Minimum interatomic distance. 0.257 nm

Table 51 Typical mechanical properties of C22000

Temper	Tensile strength MPa	Tensile strength ksi	Yield strength(a) MPa	Yield strength(a) ksi	Elongation in 50 mm (2 in.), %	Hardness HRF	Hardness HR30T	Shear strength MPa	Shear strength ksi
Flat products, 1 mm (0.040 in.) thick									
OS050	255	37	69	10	45	53	6	195	28
OS035	260	38	83	12	45	57	12	205	30
OS025	270	39	97	14	44	60	16	215	31
OS015	280	41	105	15	42	65	26	220	32
H01	310	45	240	35	25	42 HRB	44	230	33
H02	360	52	310	45	11	58 HRB	56	240	35
H04	420	61	370	54	5	70 HRB	63	260	38
H06	460	67	400	58	4	75 HRB	67	275	40
H08	495	72	425	62	3	78 HRB	69	290	42
M20	270	39	97	14	44	60	...	215	31
Flat products, 6 mm (0.250 in.) thick									
OS035	260	38	83	12	50	57	...	205	30
H02	360	52	310	45	15	58 HRB	...	240	35
M20	255	37	69	10	45	53	...	195	28
Wire, 2 mm (0.080 in.) diameter									
OS035	275	40	...	...	50	...	...	205	30
OS015	290	42	...	...	48	...	...	220	32
H00	305	44	...	...	27	...	...	230	33
H01	345	50	...	...	13	...	...	235	34
H02	415	60	...	...	6	...	...	255	37
H04	510	74	...	...	4	...	...	290	42
H06	570	83	...	...	3	...	...	...	...
H08	620	90	...	...	3	...	...	...	...
Tubing, 25 mm (1 in.) outside diameter × 1.65 mm (0.065 in.) wall thickness									
OS025	260	38	83	12	50	57	12	...	...
H80(b)	415	60	365	53	6	69 HRB	62	...	...
Rod, 12.7 mm (0.500 in.) diameter									
OS035	275	40	...	...	50	55	...	220	32
H00	310	45	...	...	25	42 HRB	...	230	33

(a) At 0.5% extension under load. (b) Drawn 35%

Mass Characteristics

Density. 8.80 g/cm³ (0.318 lb/in.³) at 20 °C (68 °F)

Thermal Properties

Liquidus temperature. 1045 °C (1910 °F)
Solidus temperature. 1020 °C (1870 °F)
Boiling point. About 1400 °C (2550 °F) at 101 kPa (1 atm)
Coefficient of linear thermal expansion. 18.4 μm/m · K (10.2 μin./in. · °F) at 20 to 300 °C (68 to 572 °F), cold rolled
Specific heat. 376 J/kg · K (0.09 Btu/lb · °F) at 20 °C (68 °F)
Thermal conductivity. 189 W/m · K (109 Btu/ft · h · °F) at 20 °C (68 °F)

Electrical Properties

Electrical conductivity. Volumetric, 44% IACS at 20 °C (68 °F), annealed
Electrical resistivity. 39.1 nΩ · m at 20 °C (68 °F). Liquid phase, 272 nΩ · m at 1100 °C (2012 °F). Temperature coefficient, 0.00186 nΩ · m per K at 20 °C (68 °F)

Magnetic Properties

Magnetic susceptibility. -0.086×10^{-6} to -1.00×10^{-6} (cgs units)

Fabrication Characteristics

Machinability. 20% of C36000 (free-cutting brass)
Recrystallization temperature. 370 °C (700 °F) for 37% reduction and 0.050 mm (0.002 in.) initial grain size. See Fig. 23.
Annealing temperature. 425 to 800 °C (800 to 1450 °F)
Hot-working temperature. 750 to 875 °C (1400 to 1600 °F)

C22600
87.5Cu-12.5Zn

Commercial Names

Previous trade name. Jewelry bronze, 87½%; CA226
Common name. Jewelry bronze

Chemical Composition

Composition limits. 86.0 to 89.0 Cu, 0.05 Pb max, 0.005 Fe max, bal Zn

Applications

Typical uses. Architectural: angles, channels. Hardware: chain, eyelets, fasteners, slide fasteners. Novelties: compacts, costume jewelry, emblems, etched articles, lipstick containers, plaques, base for gold plate

Mechanical Properties

Tensile properties. See Table 52 and Fig. 24.
Shear strength. See Table 52.
Elastic modulus. Tension, 115 GPa (17 × 10^6 psi); shear, 44 GPa (6.4 × 10^6 psi)

Mass Characteristics

Density. 8.78 g/cm³ (0.317 lb/in.³) at 20 °C (68 °F)

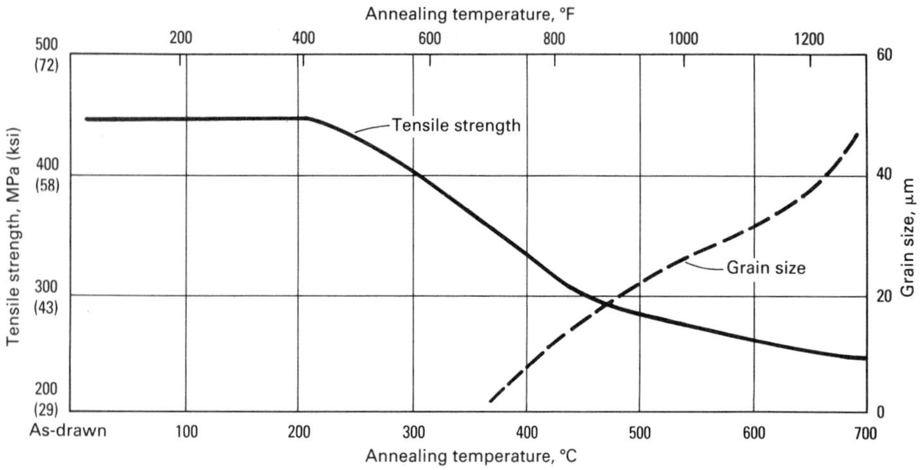

Fig. 23 Variation of tensile strength and grain size with annealing temperature for C22000. Data are for rod less than 25 mm (1 in.) in diameter that was cold drawn to a 37% reduction in area and then annealed 1 h at the indicated temperature. Grain size before annealing was 0.050 mm.

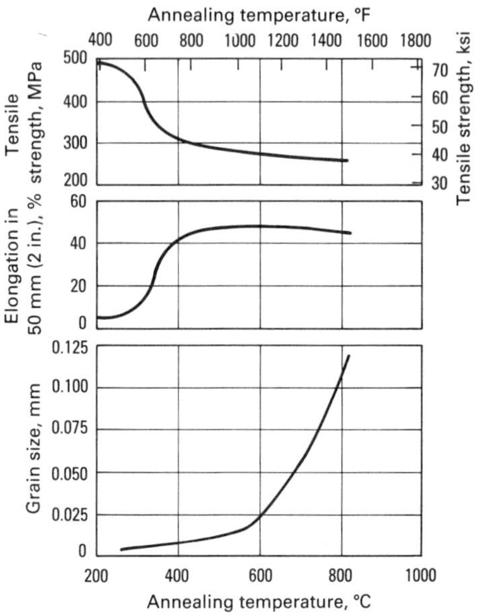

Fig. 24 Annealing characteristics of C22600. Data are for jewelry bronze strip with an initial grain size of 0.035 mm that was cold rolled 50% to a thickness of 1 mm (0.04 in.) and annealed 1 h at various temperatures.

Thermal Properties

Liquidus temperature. 1035 °C (1895 °F)
Solidus temperature. 1005 °C (1840 °F)
Coefficient of linear thermal expansion. 18.6 µm/m · K (10.3 µin./in. · °F) at 20 to 300 °C (68 to 572 °F)
Specific heat. 380 J/kg · K (0.09 Btu/lb · °F) at 20 °C (68 °F)
Thermal conductivity. 173 W/m · K (100 Btu/ft · h · °F) at 20 °C (68 °F)

Electrical Properties

Electrical conductivity. Volumetric, 40% IACS at 20 °C (68 °F), annealed
Electrical resistivity. 43 nΩ · m at 20 °C (68 °F), annealed

Fabrication Characteristics

Machinability. 30% of C36000 (free-cutting brass)
Recrystallization temperature. About 330 °C (625 °F) for 1 mm (0.04 in.) strip rolled six Brown and Sharpe numbers hard from a 0.035 mm (0.001 in.) grain size. See also Fig. 24.
Annealing temperature. 425 to 750 °C (800 to 1400 °F)
Hot-working temperature. 750 to 900 °C (1400 to 1650 °F)

C23000
85Cu-15Zn

Commercial Names

Previous trade name. Red brass, 85%; CA230
Common name. Red brass

Specifications

ASME. Pipe: SB43. Condenser tubing: SB111. Finned tubing: SB359. U-bend tubing: SB395
ASTM. Plate, sheet, strip, hot-rolled bar: B 36. Pipe: B 43. Condenser tubing: B 111. Finned tubing: B 359. Seamless tubing: B 135. U-bend tubing: B 395. Wire: B 134
SAE. Sheet, strip, seamless tube: J463 (CA230)
Government. Bar, forgings, rod, shapes, strip: QQ-B-626. Plate, sheet, strip, hot-rolled bar: QQ-B-613. Pipe: WW-P-351. Seamless tubing: WW-T-791; MIL-T-20168. Wire: QQ-W-321

Chemical Composition

Composition limits. 84.0 to 86.0 Cu, 0.06 Pb max, 0.05 Fe max, bal Zn
Consequence of exceeding impurity limits. See general statement for cartridge brass (C26000).
Effect of zinc on properties. See Fig. 21.

Applications

Typical uses. Architectural: etching parts, trim, weather strip. Electrical: conduit, screw shells, sockets. Hardware: eyelets, fasteners, fire extinguishers. Industrial: condenser and heat exchanger tubes, flexible hose, pickling crates, pump lines, radiator cores. Plumbing: plumbing pipc, J-bcnds, service lines, traps. Miscellaneous: badges, compacts, costume jewelry, dials, etched articles, lipstick containers, nameplates, tags

Mechanical Properties

Tensile properties. See Table 53 and Fig. 25.

Table 52 Typical mechanical properties of C22600

Temper	Tensile strength MPa	Tensile strength ksi	Yield strength(a) MPa	Yield strength(a) ksi	Elongation in 50 mm (2 in.), %	Hardness	Shear strength MPa	Shear strength ksi
Flat products, 1 mm (0.04 in.) thick								
OS050	270	39	76	11	46	55 HRF	200	29
OS035	275	40	90	13	45	59 HRF	205	30
OS025	290	42	105	15	44	64 HRF	215	31
OS015	305	44	110	16	42	68 HRF	220	32
H01	325	47	255	37	25	47 HRB	235	34
H02	370	54	325	47	12	61 HRB	250	36
H04	455	66	385	56	5	73 HRB	275	40
H06	495	72	415	60	4	78 HRB	290	42
H08	545	79	425	62	4	82 HRB	305	44
Wire, 2 mm (0.08 in.) diameter								
OS050	275	40	90	13	44	...	200	29
OS035	285	41	105	15	42	...	205	30
OS025	295	43	115	17	40	...	215	31
OS015	310	45	125	18	38	...	220	32
H00	325	47	240	35	26	...	235	34
H01	385	56	360	52	12	...	250	36
H02	470	68	415	60	7	70 HRB	275	40
H04	570	83	440	64	5	...	...	...
H06	615	89	450	65	4	...	...	...
H08	670	97	455	66	3	...	...	...

(a) At 0.5% extension under load

Table 53 Typical mechanical properties of C23000

Temper	Tensile strength MPa	Tensile strength ksi	Yield strength(a) MPa	Yield strength(a) ksi	Elongation in 50 mm (2 in.), %	Hardness HRF	Hardness HR30T	Shear strength MPa	Shear strength ksi
Flat products, 1 mm (0.04 in.) thick									
OS070	270	39	69	10	48	56	10	215	31
OS050	275	40	83	12	47	59	14	215	31
OS035	285	41	97	14	46	63	22	215	31
OS025	295	43	110	16	44	66	28	220	32
OS015	310	45	125	18	42	71	38	230	33
H01	345	50	270	39	25	55 HRB	54	240	35
H02	395	57	340	49	12	65 HRB	60	255	37
H04	485	70	395	57	5	77 HRB	68	290	42
H06	540	78	420	61	4	83 HRB	72	305	44
H08	580	84	435	63	3	86 HRB	74	315	46
Wire, 2 mm (0.08 in.) diameter									
OS035	285	41	...	...	48	...	...	215	31
OS025	295	43	...	...	...	...	...	220	32
OS015	310	45	...	...	...	...	...	230	33
H00	345	50	...	...	25	...	...	240	35
H01	405	59	...	...	11	...	...	260	38
H02	495	72	...	...	8	...	...	295	43
H04	605	88	...	...	6	...	...	330	48
H08	725	105	...	...	...	...	...	370	54
Tubing, 25 mm (1.0 in.) outside diameter × 1.65 mm (0.065 in.) wall thickness									
OS050	275	40	83	12	55	60	15	...	...
OS015	305	44	125	18	45	71	38	...	...
H55 (15%)	345	50	275	40	30	55 HRB	54	...	...
H80 (35%)	485	70	365	53	8	77 HRB	68	...	...
Pipe, 19 mm (0.75 in.) SPS									
OS015	305	44	125	18	45	71	...	...	...

(a) At 0.5% extension under load

Shear strength. See Table 53.
Hardness. See Table 53.
Impact strength. Izod: cast, 45 J (33 ft · lbf); cast and annealed, 43 J (32 ft · lbf). Charpy keyhole: annealed rod, 69 J (51 ft · lbf). See also Fig. 26.
Elastic modulus. Tension, 115 GPa (17 × 10^6 psi); shear, 44 GPa (6.4 × 10^6 psi)
Fatigue strength. Rod, H00 temper, 140 MPa (20 ksi) at 300 × 10^6 cycles
Creep-rupture characteristics. See Fig. 27.
Velocity of sound. 3660 m/s (12 000 ft/s) at 20 °C (68 °F)

Mass Characteristics

Density. 8.75 g/cm^3 (0.316 lb/in.^{3}) at 20 °C (68 °F)

Thermal Properties

Liquidus temperature. 1025 °C (1880 °F)
Solidus temperature. 990 °C (1810 °F)
Coefficient of linear thermal expansion. 18.7 μm/m · K (10.4 μin./in. · °F) at 20 to 300 °C (68 to 572 °F), cold rolled
Specific heat. 380 J/kg · K (0.09 Btu/lb · °F) at 20 °C (68 °F)
Thermal conductivity. 159 W/m · K (92 Btu/ft · h · °F) at 20 °C (68 °F)

Electrical Properties

Electrical conductivity. Volumetric, 37% IACS at 20 °C (68 °F), annealed
Electrical resistivity. 47 nΩ · m at 20 °C (68 °F), annealed. Liquid: 299 nΩ · m at 1100 °C (2012 °F); 304 nΩ · m at 1200 °C (2192 °F). Temperature coefficient, 0.0016/°C at 20 °C (68 °F)

Magnetic Properties

Magnetic susceptibility. Approximately −1.00 × 10^{-6} (cgs units)

Fabrication Characteristics

Machinability. 30% of C26000 (free-cutting brass)
Recrystallization temperature. About 350 °C (660 °F) for 1 mm (0.04 in.) sheet rolled six Brown and Sharpe numbers hard with a 50% reduction and 0.035 mm (0.001 in.) initial grain size
Annealing temperature. 425 to 725 °C (800 to 1350 °F). See also Fig. 25.
Hot-working temperature. 800 to 900 °C (1450 to 1650 °F)

C24000 80Cu-20Zn

Commercial Names

Trade name. Low brass, 80%; CA240
Common name. Low brass

Specifications

ASTM. Flat products: B 36. Wire: B 134
SAE. Sheet, strip: J463 (CA240)
Government. Finished-edge bar and strip, forgings, rod, shapes: QQ-B-626. Rolled bar, plate, sheet, strip: QQ-B-613. Wire: QQ-W-321. Brazing alloy wire: QQ-B-650

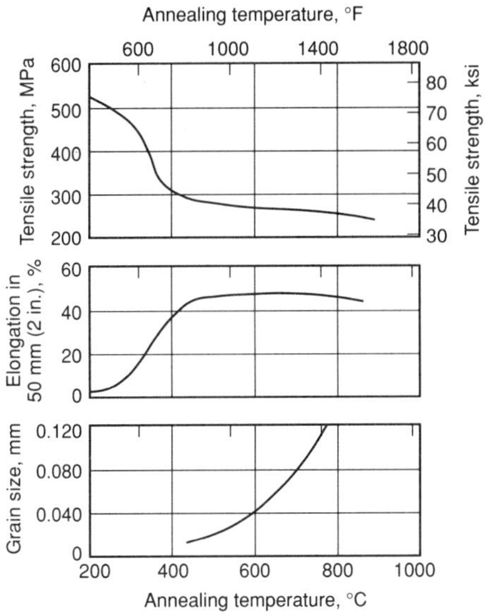

Fig. 25 Annealing characteristics of C23000. Data are for 1 mm (0.04 in.) thick red brass sheet, H06 temper, annealed 1 h at various temperatures.

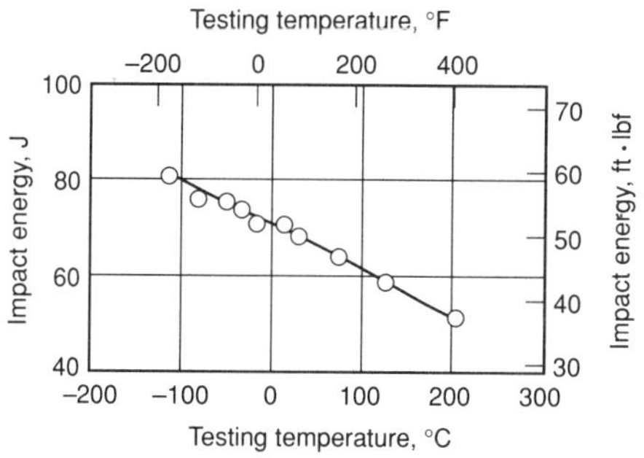

Fig. 26 Impact strength of C23000. Charpy keyhole specimens were machined from O61 temper material, then tested at the indicated temperatures. Impact strengths represent energy absorbed without fracture.

Chemical Composition

Composition limits. 78.5 to 81.5 Cu, 0.05 Pb max, 0.05 Fe max, bal Zn
Effect of zinc content on properties. See Fig. 21.

Applications

Typical uses. Ornamental metal work, medallions, spandrels, electrical battery caps, bellows and musical instruments, clock dials, flexible hose, pump lines, tokens

Mechanical Properties

Tensile properties. See Table 54 and Fig. 28.
Shear strength. See Table 54.
Hardness. See Table 54.
Elastic modulus. Tension, 110 GPa (16 × 10^6 psi); shear, 40 GPa (6 × 10^6 psi)
Fatigue strength. 1 mm (0.04 in.) thick strip, H08 temper: 165 MPa (24 ksi) at 20 × 10^6 cycles

Structure

Crystal structure. Face-centered cubic α;

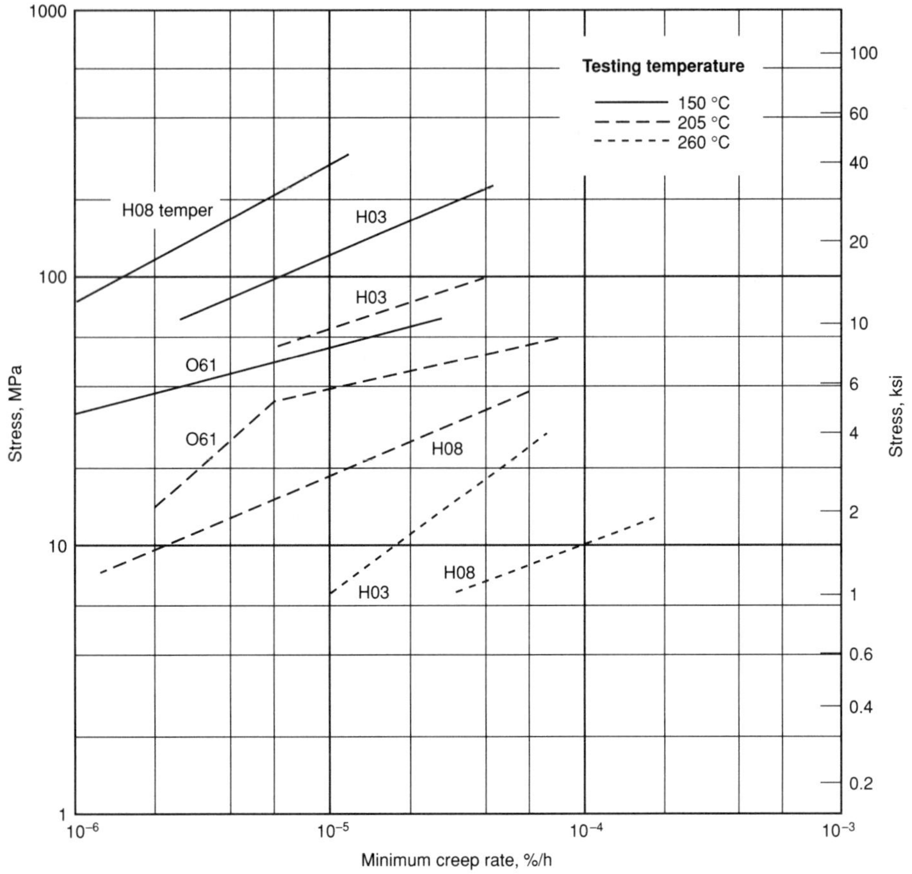

Fig. 27 Minimum creep rates for C23000 wire. Data are for red brass wire, 3.2 mm (0.125 in.) in diameter, that was cold drawn to size, then tested in the as-drawn or annealed condition.

lattice parameter, 0.366 nm
Minimum interatomic distance. 0.259 nm

Mass Characteristics

Density. 8.67 g/cm³ (0.313 lb/in.³) at 20 °C (68 °F)

Solidification shrinkage. 5 to 6%

Thermal Properties

Liquidus temperature. 1000 °C (1830 °F)
Solidus temperature. 965 °C (1770 °F)
Coefficient of linear thermal expansion. 19.1 μm/m · K (10.6 μin./in. · °F) at 20 to 300 °C (68 to 572 °F)
Specific heat. 380 J/kg · K (0.09 Btu/lb · °F) at 20 °C (68 °F)
Thermal conductivity. 140 W/m · K (81 Btu/ft · h · °F) at 20 °C (68 °F)

Table 54 Typical mechanical properties of C24000

Temper	Tensile strength MPa	Tensile strength ksi	Yield strength(a) MPa	Yield strength(a) ksi	Elongation in 50 mm (2 in.), %	Hardness HRF	Hardness HR30T	Shear strength MPa	Shear strength ksi
Flat products, 1 mm (0.04 in.) thick									
OS070	290	42	83	12	52	57	8	. . .	. . .
OS050	305	44	97	14	50	61	16	220	32
OS035	315	46	105	15	48	66	28	. . .	. . .
OS025	330	48	115	17	47	69	32	. . .	. . .
OS015	345	50	140	20	46	75	42	230	33
H01	365	53	275	40	30	55 HRB	54	250	36
H02	420	61	345	50	18	70 HRB	64	270	39
H04	510	74	405	59	7	82 HRB	71	295	43
H08	625	91	450	65	3	91 HRB	77	330	48
Wire, 2 mm (0.08 in.) diameter									
OS050	305	44	. . .	. . .	55	. . .	. . .	220	32
OS035	315	46	. . .	. . .	50	. . .	. . .	. . .	. . .
OS015	345	50	. . .	. . .	47	. . .	. . .	230	33
H00	385	56	. . .	. . .	27	. . .	. . .	255	37
H01	470	68	. . .	. . .	12	. . .	. . .	290	42
H02	565	82	. . .	. . .	8	. . .	. . .	325	47
H04	740	107	. . .	. . .	5	. . .	. . .	365	53
H06	800	116	. . .	. . .	4	. . .	. . .	. . .	. . .
H08	860	125	. . .	. . .	3	. . .	. . .	415	60

(a) At 0.5% extension under load

Electrical Properties

Electrical conductivity. Volumetric, O61 temper: 32% IACS at 20 °C (68 °F)
Electrical resistivity. O61 temper: 54 nΩ · m at 20 °C (68 °F). Liquid: 330 nΩ · m at 1000 °C (1830 °F); 338 nΩ · m at 1200 °C (2190 °F). Temperature coefficient, 0.00154/°C at 20 °C (68 °F)

Magnetic Properties

Magnetic susceptibility. Approximately -1.00×10^{-6} (cgs units)

Fabrication Characteristics

Machinability. 30% of C36000 (free-cutting brass)
Recrystallization temperature. About 400 °C (750 °F) for 37% reduction and 0.060 mm initial grain size
Annealing temperature. 425 to 700 °C (800 to 1300 °F). See also Fig. 28.
Hot-working temperature. 825 to 900 °C (1500 to 1650 °F)

C26000
70Cu-30Zn

Commercial Names

Previous trade name. Cartridge brass, 70%; CA260
Common name. Cartridge brass, 70-30 brass, spinning brass, spring brass, extra-quality brass

Specifications

AMS. Flat products: 4505, 4507. Tube: 4555
ASTM. Flat products: B 19, B 36, B 569. Cups for cartridge cases: B 129. Tube: B 135, B 587. Wire: B 134
SAE. J463
Government. Flat products: QQ-B-613, QQ-B-626, MIL-C-50. Rod, bar, shapes, forgings: QQ-B-626. Tube: MIL-T-6945, MIL-T-20219. Wire: QQ-W-321, QQ-B-650. Shim stock, laminated: MIL-S-22499. Cups for cartridge cases: MIL-C-10375

Chemical Composition

Composition limits. 68.5 to 71.5 Cu, 0.07 Pb max, 0.05 Fe max, 0.15 max other (total), bal Zn
Effect of zinc on properties. See Fig. 21.
Lead should be kept under 0.01% for hot rolling, although additions of lead up to 4% improve machinability in material processed by extrusion and cold working. Lead lowers room-temperature ductility in brass and leads to hot shortness at temperatures above 315 °C (600 °F).
Aluminum at levels as high as 2% has no adverse effect on hot or cold working. How-

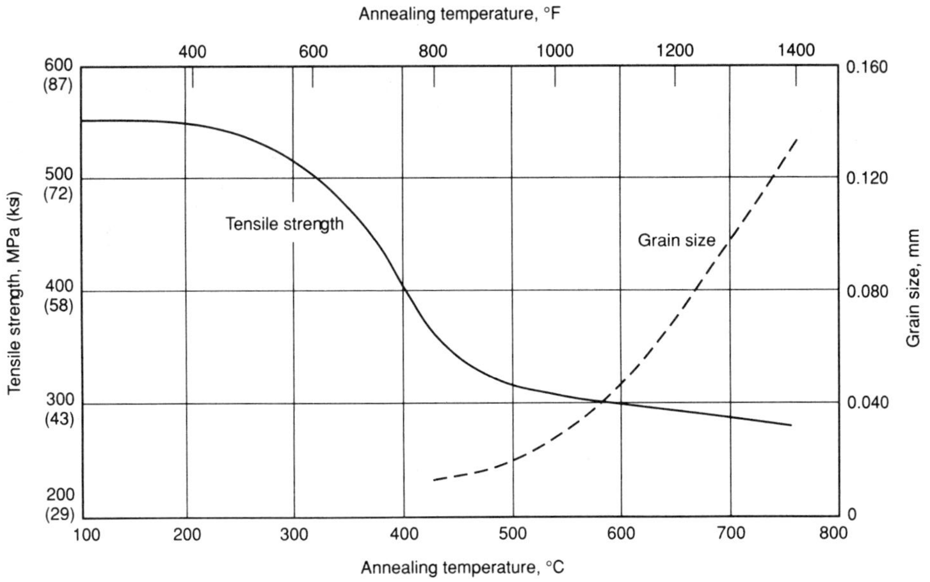

Fig. 28 Tensile strength and grain size versus annealing temperature for C24000, annealed from HC2 temper. Data are for low brass with an initial grain size of 0.060 mm that was cold drawn 37% to a diameter of less than 25 mm (1 in.) and annealed 1 h at the indicated temperature.

ever, annealing and grain size are affected. *Arsenic* does not affect hot or cold working, but it tends to refine the grain size, thereby lowering ductility.
Cadmium. The effects of cadmium are not universally agreed upon; some claim as much as 0.10% has little effect, others maintain that it should be kept below 0.05%.
Chromium affects temperature of anneal and grain size. This condition is aggravated when iron is present.
Iron chiefly affects annealing and magnetic properties.
Nickel restrains grain growth.
Phosphorus has no adverse effect up to 0.04%; it does, however, restrain grain growth, increase tensile strength, and lower ductility to some extent.

Applications

Typical uses. Architectural: grillwork. Automotive: radiator cores and tanks. Electrical: bead chain. flashlight shells, reflectors, lamp fixtures, socket shells, screw shells. Hardware: eyelets, fasteners, pins, hinges, kickplates, locks, rivets, springs, stampings, tubes, etched articles. Munitions: ammunition components, particularly cartridge cases. Plumbing: accessories, fittings. Industrial: pump and power cylinders, cylinder liners
Precautions in use. Highly susceptible to season cracking in ammoniacal environments

Mechanical Properties

Tensile properties. See Tables 55 and 56 and Fig. 29, 30, and 31.
Hardness. See Table 55 and Fig. 31.
Elastic modulus. Tension, 110 GPa (16 × 10^6 psi); shear, 40 GPa (6 × 10^6 psi)
Fatigue strength. See Table 55.
Impact strength. Charpy V-notch: O61 temper, 60 J (44 ft · lbf); M20 temper, 19 J (14 ft · lbf); Izod: O61 temper, 89 J (66 ft · lbf) for notched round specimen
Creep-rupture properties. See Fig. 32.
Velocity of sound. 3660 m/s (12 000 ft/s) at 20 °C (68 °F)

Table 55 Typical mechanical properties of C26000

Temper	Tensile strength MPa	ksi	Yield strength(a) MPa	ksi	Elongation in 50 mm (2 in.), %	Hardness HRF	HR30T	Shear strength MPa	ksi	Fatigue strength(b) MPa	ksi
Flat products, 1 mm (0.04 in.) thick											
OS100	300	44	75	11	68	54	11	215	31	90	13
OS070	315	46	95	14	65	58	15	220	32	90	13
OS050	325	47	105	15	62	64	26	230	33	...	...
OS035	340	49	115	17	57	68	31	235	34	95	14
OS025	350	51	130	19	55	72	36	235	34	...	...
OS015	365	53	150	22	54	78	43	240	35	105	15
H01	370	54	275	40	43	55 HRB	54	250	36	...	...
H02	425	62	360	52	23	70 HRB	65	275	40	125	18
H04	525	76	435	63	8	82 HRB	73	305	44	145	21
H06	595	86	450	65	5	83 HRB	76	315	46	...	...
H08	650	94	...	...	3	91 HRB	77	330	48	160	23
H10	680	99	...	...	3	93 HRB	78	...	...	...	...
Wire, 2 mm (0.08 in.) diameter											
OS050	330	48	110	16	64	...	...	230	33	...	...
OS035	345	50	125	18	60	...	...	235	34	...	...
OS025	360	52	145	21	58	...	...	240	35	...	...
OS015	370	54	160	23	58	...	...	250	36	...	...
H00	400	58	315	46	35	...	...	260	38	...	...
H01	485	70	395	57	20	...	...	290	42	...	...
H06	855	124	...	...	4	...	...	...	...	...	...
H08	895	130	...	...	3	...	...	415	60	150	22
Tube, 25 mm (1 in.) outside diameter × 1.6 mm (0.065 in.) wall thickness											
OS050	325	47	105	15	65	64	26	...	...	...	...
OS025	360	52	140	20	55	75	40	...	...	...	...
H80	540	78	440	64	8	82 HRB	73	...	...	...	...
Rod, 25 mm (1.0 in.) diameter											
OS050	330	48	110	16	65	65	...	235	34	...	...
H00	380	55	275	40	48	60 HRB	...	260	38	...	...
H02	480	70	360	52	30	80 HRB	...	290	42	22(c)	150(c)

(a) At 0.5% extension under load. (b) Reverse bending, at 10^8 cycles. (c) Reverse bending, at 5 × 10^7 cycles

Table 56 Typical tensile properties of cold-rolled and annealed C26000 sheet

Direction in sheet	Tensile strength MPa	ksi	Elongation, %
Parallel to RD	330	48	59
45° to RD	305	44	66
90° to RD	325	47	61

Note: Approximate values for material given a ready-to-finish anneal at 400 °C (750 °F), then cold rolled 70% and annealed 1 h at 575 °C (1070 °F). RD, rolling direction

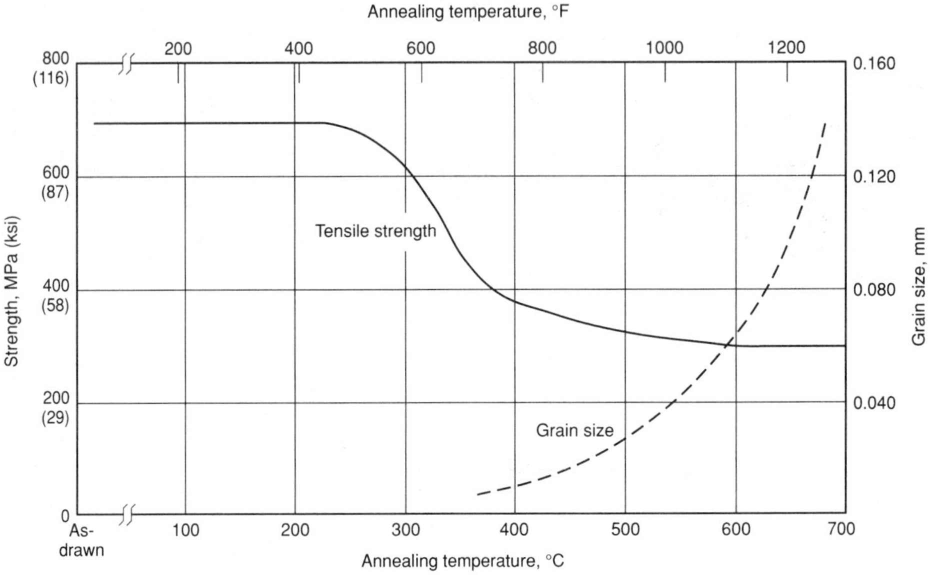

Fig. 29 Tensile strength and grain size as a function of annealing temperature for C26000 rod. Data are for cartridge brass rod less than 25 mm (1 in.) in diameter that was cold drawn 50% (from starting material having a grain size of 0.045 mm), then annealed 1 h at the indicated temperature.

Structure

Crystal structure. Face-centered cubic; lattice parameter, 0.3684 nm
Minimum interatomic distance. 0.2605 nm
Microstructure. Single-phase α usually with extensive pattern of annealing twins
Damping capacity. See Fig. 33.

Mass Characteristics

Density. 8.53 g/cm^3 (0.308 lb/in.^{3}) at 20 °C (68 °F)

Thermal Properties

Liquidus temperature. 955 °C (1750 °F)
Solidus temperature. 915 °C (1680 °F)
Coefficient of linear thermal expansion. Cold-rolled stock: 19.9 μm/m · K (11.1 μin./in. · °F) at 20 to 300 °C (68 to 572 °F). Equation for 20 to 300 °C: $L_t = L_0[1 + (17.75t + 0.00653t^2) \times 10^{-6}]$, where t is temperature difference from 20 °C
Specific heat. 375 J/kg · K (0.09 Btu/lb · °F) at 20 °C (68 °F)
Thermal conductivity. 120 W/m · K (70 Btu/ft · h · °F) at 20 °C (68 °F)

Electrical Properties

Electrical conductivity. Volumetric, O61 temper, 28% IACS at 20 °C (68 °F)
Electrical resistivity. O61 temper, 62 nΩ · m at 20 °C (68 °F), temperature coefficient, 0.092 nΩ · m per K at 20 °C (68 °F)
Hall coefficient. 25 pV · m/A · T

Magnetic Properties

Iron in excess of 0.03% can precipitate from C26000 during suitable low-temperature anneals. Precipitation is slow and occurs chiefly in a nonmagnetic form, which is converted to a ferromagnetic structure on subsequent cold working.
Magnetic susceptibility. -8×10^{-8} to -16×10^{-8} (mks units); susceptibility in α brasses decreases with increasing zinc content.

Chemical Properties

General corrosion behavior. Resists corrosion in a wide variety of waters and chemical solutions; may undergo dezincification in stagnant or slowly moving salt solutions, brackish water, or mildly acidic solutions. Susceptible to stress-corrosion cracking (season cracking), especially in ammoniacal environments

Fabrication Characteristics

Machinability. 30% of C36000 (free-cutting brass)
Formability. Excellent for cold working and forming; fair for hot forming. Directionality in brass is more readily developed with high zinc content, such as in C26000 and higher-zinc brasses. Earing usually occurs 45° to the direction of rolling and is aggravated by heavy final reductions, low ready-to-finish annealing temperatures, and high finish annealing temperatures.
Weldability. Soldering and brazing: excellent. Oxyfuel gas, resistance spot, and resistance butt welding: good. Gas metal arc welding: fair. Other welding processes are not recommended.
Recrystallization temperature. About 300 °C (575 °F) for 0.045 mm initial grain size and a cold reduction of 50%
Annealing temperature. 425 to 750 °C (800 to 1400 °F)
Hot-working temperature. 725 to 850 °C (1350 to 1550 °F)

C26800, C27000
65Cu-35Zn

Commercial Names

Previous trade name. C26800: Yellow brass, 66%. C27000: Yellow brass, 65
Common name. Yellow brass

Specifications

AMS. Wire: 4710, 4712
ASTM. Flat products: B 36 (C26800). Tube: B 135 (C27000), B 587 (C26800 and C27000). Wire: B 134
SAE. J463
Government. Flat products: QQ-B-613, Bar, rod, forgings, shapes: QQ-B-626. Wire: QQ-W-321, MIL-W-6712

Chemical Composition

Composition limits of C26800. 64.0 to 68.5 Cu, 0.15 Pb max, 0.05 Fe max, bal Zn
Composition limits of C27000. 63.0 to 68.5 Cu, 0.10 Pb max, 0.07 Fe max, bal Zn
Effect of zinc on properties. See Fig. 21.

Applications

Typical uses. Architectural grillwork, radi-

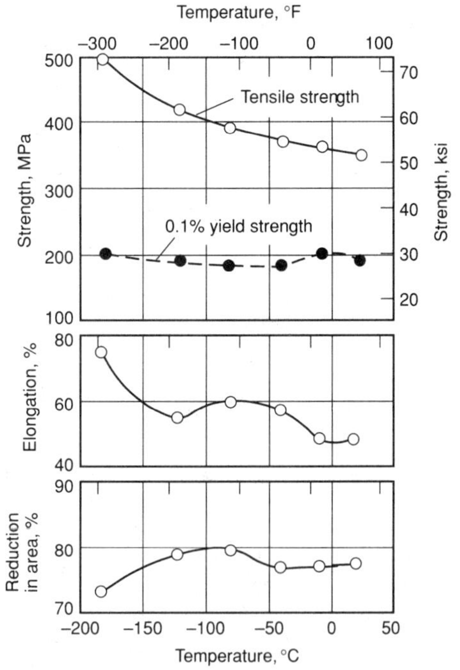

Fig. 30 Low-temperature tensile properties of C26000 rod, O61 temper

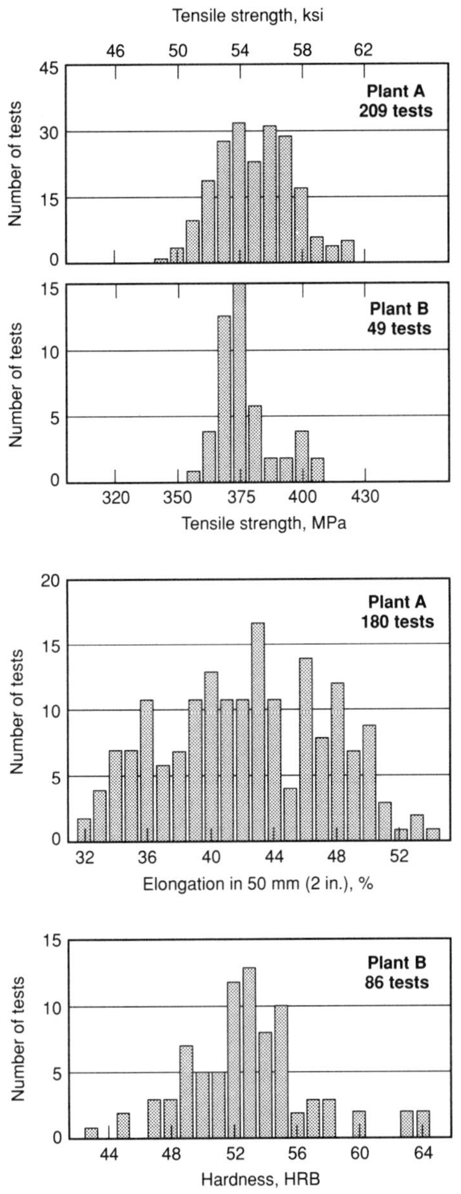

Fig. 31 Typical distribution of tensile properties and hardness for C26000 strip, H01 temper. Data are for cartridge brass strip 0.5 to 1 mm (0.020 to 0.040 in.) thick.

ator cores and tanks, reflectors, flashlight shells, lamp fixtures, screw shells, socket shells, bead chain, chain, eyelets, fasteners, grommets, kickplates, push plates, stencils, plumbing accessories, sink strainers, wire, pins, rivets, screws, springs

Mechanical Properties

Tensile properties. See Table 57.
Shear strength. See Table 57.
Hardness. See Table 57.
Elastic modulus. Tension, 105 GPa (15 × 10^6 psi); shear, 35 GPa (5 × 10^6 psi)
Fatigue strength. Rotating beam tests. At 10^8 cycles, for strip 1 mm (0.04 in.) thick: OS070 temper, 83 MPa (12 ksi); H04 temper, 97 MPa (14 ksi); H08 temper, 140 MPa (20 ksi)

Table 57 Typical mechanical properties of C26800 and C27000

Temper	Tensile strength, MPa	Tensile strength, ksi	Yield strength(a), MPa	Yield strength(a), ksi	Elongation in 50 mm (2 in.), %	Hardness, HRF	Hardness, HR30T	Shear strength, MPa	Shear strength, ksi
Flat products, 1 mm (0.04 in.) thick									
OS070	315	46	97	14	65	58	15	220	32
OS050	325	47	105	15	62	64	26	230	33
OS035	340	49	115	17	57	68	31	235	34
OS025	350	51	130	19	55	72	36	240	35
OS015	365	53	150	22	54	78	43	250	36
H01	370	54	275	40	43	55 HRB	54	250	36
H02	420	61	345	50	23	70 HRB	65	275	40
H04	510	74	415	60	8	80 HRB	70	295	43
H06	585	85	425	62	5	87 HRB	74	310	45
H08	625	91	425	62	3	90 HRB	76	325	47
H10	675	98	435	63	3	91 HRB	77	...	...
Rod, 25 mm (1.0 in.) diameter									
OS050	330	48	110	16	65(b)	65	...	235	34
H00 (6%)	380	55	275	40	48(c)	...	55	...	36
Wire, 2 mm (0.08 in.) diameter									
OS050	330	48	110	16	64	...	...	230	33
OS035	345	50	125	18	60	...	...	235	34
OS025	360	52	145	21	58	...	...	240	35
OS015	370	54	160	23	55	...	...	250	36
H00	400	58	315	46	35	...	...	260	38
H01	485	70	395	57	20	...	...	290	42
H02	605	88	420	61	15	...	...	...	...
H04	760	110	...	...	8	...	...	380	55
H06	825	120	...	...	4	...	...	...	...
H08	885	128	...	...	3	...	...	415	60

(a) At 0.5% extension under load. (b) 75% reduction in area. (c) 70% reduction in area

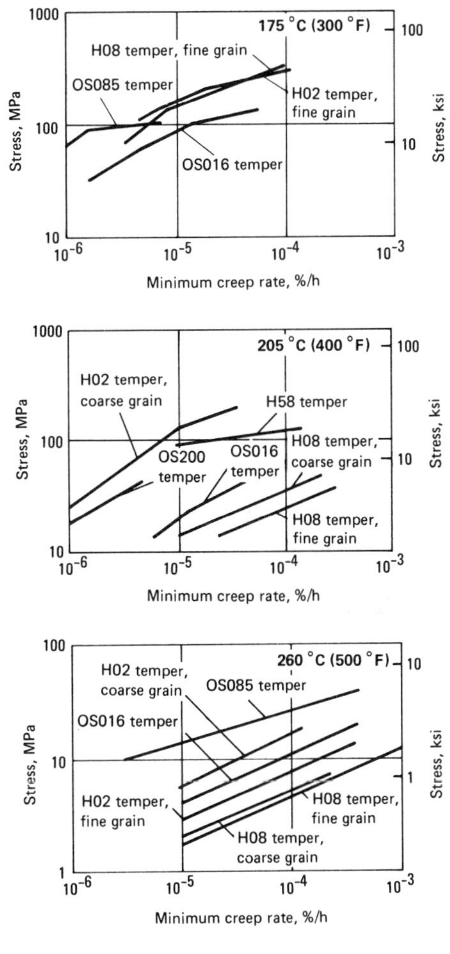

Fig. 32 Minimum creep rates for C26000

Mass Characteristics

Density. 8.47 g/cm³ (0.306 lb/in.³) at 20 °C (68 °F)

Thermal Properties

Liquidus temperature. 930 °C (1710 °F)
Solidus temperature. 905 °C (1660 °F)
Coefficient of linear thermal expansion. 20.3 μm/m · K (11.3 μin./in. · °F) at 20 to 300 °C (68 to 572 °F)
Specific heat. 380 J/kg · K (0.09 Btu/lb · °F) at 20 °C (68 °F)
Thermal conductivity. 116 W/m · K (67 Btu/ft · h · °F) at 20 °C (68 °F)

Electrical Properties

Electrical conductivity. Volumetric, 27% IACS at 20 °C (68 °F), annealed
Electrical resistivity. 64 nΩ · m at 20 °C (68 °F), annealed

Structure

Crystal structure. Face-centered cubic α

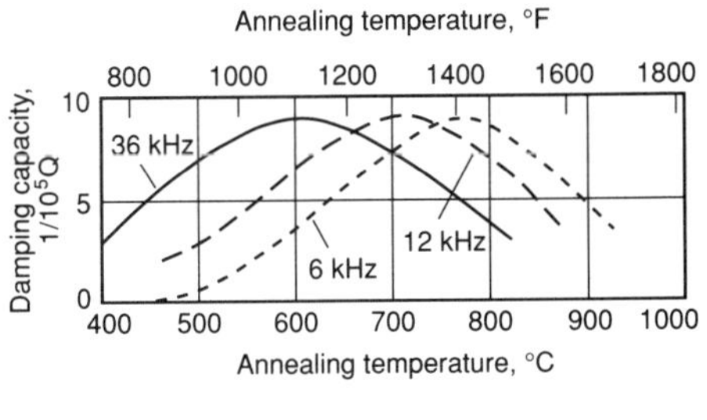

Fig. 33 Damping capacity of annealed C26000

Table 58 Typical mechanical properties of C28000

Temper	Tensile strength MPa	Tensile strength ksi	Yield strength(a) MPa	Yield strength(a) ksi	Elongation in 50 mm (2 in.), %	Hardness, HRF	Shear strength MPa	Shear strength ksi
Flat products, 1 mm (0.04 in.) thick								
M20	370	54	145	21	45	85	275	40
O61	370	54	145	21	45	80	275	40
H00	415	60	240	35	30	55 HRB	290	42
H02	485	70	345	50	10	75 HRB	305	44
Rod, 25 mm (1 in.) diameter								
M30	360	52	140	20	52	78	270	39
O61	370	54	145	21	50	80	275	40
H01	495	72	345	50	25	78	310	45

(a) 0.5% extension under load

Microstructure. Single-phase α

Fabrication Characteristics

Machinability. 30% of C36000 (free-cutting brass)

Recrystallization temperature. About 290 °C (550 °F) for strip cold rolled 50% to 1 mm (0.04 in.) thickness and having an initial grain size of 0.035 mm

Maximum cold reduction between anneals. 90%

Annealing temperature. 425 to 700 °C (800 to 1300 °F)

Hot-working temperature. 700 to 820 °C (1300 to 1500 °F)

C28000 60Cu-40Zn

Commercial Names

Previous trade name. Muntz metal, 60%; CA280

Common name. Muntz metal

Specifications

ASME. Condenser tubing: SB111

ASTM. Tubing: B 111, B 135

Government. Flat products: QQ-B-613. Bar, rod, forgings, shapes: QQ-B-626. Seamless tubing: WW-T-791

Chemical Composition

Composition limits. 59.0 to 63.0 Cu, 0.30 Pb max, 0.07 Fe max, bal Zn

Effect of zinc on properties. See Fig. 21.

Applications

Typical uses. Decoration, as architectural panel sheets; structural, as heavy plates; bolting and valve stems; tubing for heat exchangers; brazing rod (for copper alloys and cast iron); hot forgings

Precautions in use. C28000 has poor cold-drawing and forming properties in comparison with those of higher-copper alloys,

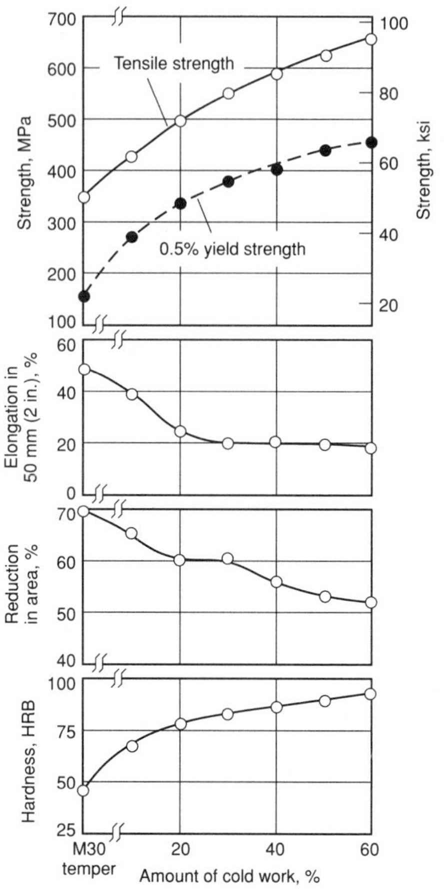

Fig. 34 Typical mechanical properties of extruded and drawn C28000. Data are for Muntz metal rod less than 25 mm (1 in.) in diameter that was extruded and then cold drawn to various percentages of reduction in area.

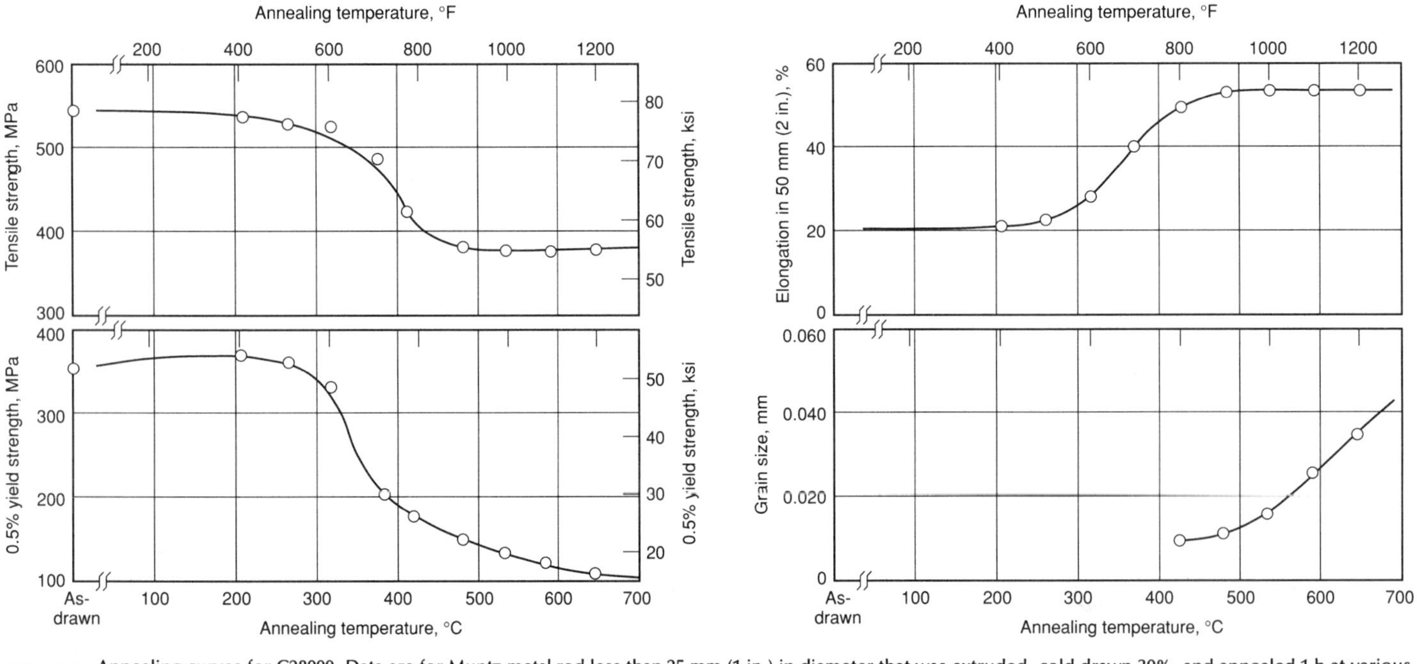

Fig. 35 Annealing curves for C28000. Data are for Muntz metal rod less than 25 mm (1 in.) in diameter that was extruded, cold drawn 30%, and annealed 1 h at various temperatures.

Table 59 Typical mechanical properties of C31600

Temper	Tensile strength MPa	Tensile strength ksi	Yield strength(a) MPa	Yield strength(a) ksi	Elongation in 50 mm (2 in.), %	Hardness, HRB	Shear strength MPa	Shear strength ksi
Drawn bar, 6 mm (0.25 in.) diameter								
H04	435	63	385	56	12	70	...	...
Rod, 13 mm (0.50 in.) diameter								
H04	460	67	405	59	13	72	275	40
Rod, 25 mm (1 in.) diameter								
OS050	255	37	83	12	45	55 HRF	165	24
H04	450	65	395	57	15	70	270	39

(a) 0.5% extension under load

but it has excellent hot-working properties. It is the strongest of the copper-zinc alloys but is less ductile than higher-copper alloys. It is subject to dezincification and stress-corrosion cracking under certain conditions.

Mechanical Properties

Tensile properties. See Table 58 and Fig. 34 and 35.
Shear strength. See Table 58.
Hardness. See Table 58.
Elastic modulus. Tension, 105 GPa (15 × 10^6 psi); shear, 39 GPa (5.6 × 10^6 psi)

Mass Characteristics

Density. 8.39 g/cm^3 (0.303 lb/in.^{3}) at 20 °C (68 °F)

Thermal Properties

Liquidus temperature. 905 °C (1660 °F)
Solidus temperature. 900 °C (1650 °F)
Coefficient of linear thermal expansion. 20.8 μm/m · K (11.6 μin./in. · °F) at 20 to 300 °C (68 to 572 °F)
Specific heat. 375 J/kg · K (0.09 Btu/lb · °F) at 20 °C (68 °F)
Thermal conductivity. 123 W/m · K (71 Btu/ft · h · °F) at 20 °C (68 °F)

Electrical Properties

Electrical conductivity. Volumetric, 28% IACS at 20 °C (68 °F)
Electrical resistivity. 61.6 nΩ · m at 20 °C (68 °F)

Structure

Microstructure. Two phase: face-centered cubic α plus body-centered cubic β. Beta phase appears lemon yellow when etched with ammonia peroxide; it is dark when etched with ferric chloride. In grain size determination, the beta phase should be ignored.

Optical Properties

Color. Reddish compared to C26000 (70-30 cartridge brass). C28000 is used as a good match to the color of C23000 (85-15 red brass).

Chemical Properties

General corrosion behavior. Generally good; similar to copper except as noted below
Resistance to specific corroding agents. Better resistance to sulfur-bearing compounds than that of higher-copper alloys

Fabrication Characteristics

Machinability. 40% of C36000 (free-cutting brass)
Forgeability. 90% of C37700 (forging brass)
Formability. Fair capacity for cold working; excellent capacity for hot forming
Weldability. Soldering or brazing: excellent. Oxyfuel gas welding, resistance spot welding, or resistance butt welding: good. Gas-shielded arc welding: fair
Annealing temperature. 425 to 600 °C (800 to 1100 °F). See also Fig. 35.
Hot-working temperature. 625 to 800 °C (1150 to 1450 °F)

C31400
89Cu-9.1Zn-1.9Pb

Commercial Names

Previous trade name. Leaded commercial bronze; CA314

Specifications

ASTM. B 140

Chemical Composition

Composition limits. 87.5 to 90.5 Cu, 1.3 to 2.5 Pb, 0.10 Fe max, 0.7 Ni max, 0.5 max other (total), bal Zn

Applications

Typical uses. Screws, screw machine parts, pickling racks and fixtures, electrical plug-type connectors, builders' hardware

Mechanical Properties

Tensile properties. Rod, typical. O61 temper: tensile strength, 255 MPa (37 ksi); yield strength, 83 MPa (12 ksi) at 0.5% extension under load; elongation, 45% in 50 mm (2 in.); reduction in area, 70%. H02 temper: tensile strength, 360 MPa (52 ksi); yield strength, 310 MPa (45 ksi); elongation, 18%; reduction in area, 60%
Shear strength. Rod, typical: O61 temper, 165 MPa (24 ksi); H02 temper, 205 MPa (30 ksi)
Hardness. O61 temper, 55 HRF; H02 temper, 58 HRB; H04 temper, 61 to 65 HRB
Elastic modulus. Tension, 115 GPa (17 × 10^6 psi); shear, 45 GPa (6.4 × 10^6 psi)

Mass Characteristics

Density. 8.83 g/cm^3 (0.319 lb/in.^{3}) at 20 °C (68 °F)

Thermal Properties

Liquidus temperature. 1040 °C (1900 °F)
Solidus temperature. 1010 °C (1850 °F)
Coefficient of linear thermal expansion. 18.4 μm/m · K (10.2 μin./in. · °F) at 20 to 300 °C (68 to 572 °F)
Specific heat. 375 J/kg · K (0.09 Btu/lb · °F) at 20 °C (68 °F)
Thermal conductivity. 180 W/m · K (104 Btu/ft · h · °F) at 20 °C (68 °F)

Electrical Properties

Electrical conductivity. Volumetric, 42% IACS at 20 °C (68 °F)
Electrical resistivity. 41 nΩ · m at 20 °C (68 °F)

Optical Properties

Color. Rich bronze

Fabrication Characteristics

Machinability. 80% of C36000 (free-cutting brass)
Formability. Cold working, good; hot forming, poor
Weldability. Soldering: excellent. Brazing: good. Resistance butt welding: fair. All other welding processes are not recommended.
Annealing temperature. 425 to 650 °C (800 to 1200 °F)

C31600
89Cu-8.1Zn-1.9Pb-1Ni

Commercial Names

Previous trade name. Leaded commercial bronze-nickel bearing; CA316

Specifications

ASTM. B 140

Chemical Composition

Composition limits. 87.5 to 90.5 Cu, 1.3 to 2.5 Pb, 0.7 to 1.2 Ni, 0.1 Fe max, 0.04 to 0.10 P, 0.5 max other (total), bal Zn

Applications

Typical uses. Electrical connectors, fasteners, hardware, nuts, screws, screw machine parts. Most commonly used as rod or drawn bar

Mechanical Properties

Tensile properties. See Table 59.
Hardness. See Table 59.
Elastic modulus. Tension, 115 GPa (17 × 10^6 psi)

Table 60 Typical mechanical properties of C33000 tubing

Temper	Tensile strength MPa	Tensile strength ksi	Yield strength(a) MPa	Yield strength(a) ksi	Elongation in 50 mm (2 in.), %	Hardness HRF	Hardness HRB	Hardness HR30T
OS050	325	47	105	15	60	64	...	26
OS025	360	52	135	20	50	75	...	36
H58	450	65	345	50	32	100	70	66
H80	515	75	415	60	7	...	85	76

Note: Values for tubing, 25 mm (1.0 in.) outside diameter × 1.65 mm (0.065 in.) wall thickness. (a) 0.5% extension under load

Table 61 Typical mechanical properties of C33200 tubing

Temper	Tensile strength MPa	Tensile strength ksi	Yield strength(a) MPa	Yield strength(a) ksi	Elongation in 50 mm (2 in.), %	Hardness HRF	Hardness HRB	Hardness HR30T
OS050	325	47	105	15	60	64	...	26
OS025	360	52	135	20	50	75	...	36
H58	450	65	345	50	32	100	70	66
H80	515	75	415	60	7	...	85	76

Note: Values for tubing, 25 mm (1.0 in.) outside diameter × 1.65 mm (0.065 in.) wall thickness. (a) 0.5% extension under load

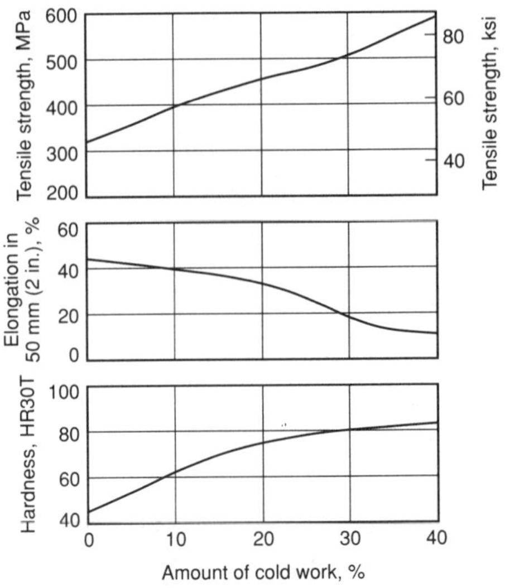

Fig. 36 Typical mechanical properties of cold drawn C33200 copper alloy tubing

Mass Characteristics

Density. 8.86 g/cm³ (0.320 lb/in.³) at 20 °C (68 °F)

Thermal Properties

Liquidus temperature. 1040 °C (1900 °F)
Solidus temperature. 1010 °C (1850 °F)
Coefficient of linear thermal expansion. 18.4 μm/m · K (10.2 μin./in. · °F) at 20 to 300 °C (68 to 572 °F)
Specific heat. 380 J/kg · K (0.09 Btu/lb · °F) at 20 °C (68 °F)
Thermal conductivity. 140 W/m · K (81 Btu/ft · h · °F) at 20 °C (68 °F)

Electrical Properties

Electrical conductivity. Volumetric, 32% IACS at 20 °C (68 °F)
Electrical resistivity. 54 nΩ · m at 20 °C (68 °F)

Optical Properties

Color. Rich bronze

Fabrication Characteristics

Machinability. 80% of C36000 (free-cutting brass)
Formability. Cold working: good. Hot forming, poor
Weldability. Soldering: excellent. Brazing: good. Resistance butt welding: fair. All other welding processes are not recommended.
Annealing temperature. 425 to 650 °C (800 to 1200 °F)

C33000
66Cu-33.5Zn-0.5Pb

Commercial Names

Previous trade name. Low-leaded brass (tube)
Common name. High brass; yellow brass

Specifications

AMS. 4555
ASTM. B 135
SAE. J463
Government. WW-T-791, MIL-T-46072

Chemical Composition

Composition limits. 65 to 68 Cu, 0.2 to 0.8 Pb, 0.07 Fe max, 0.5 max other (total), bal Zn. For tubing with an outside diameter greater than 125 mm (5 in.), Pb content may be less than 0.2%.

Applications

Typical uses. General-purpose use where some degree of machinability is required together with moderate cold-working properties; for example, primers for munitions. Plumbing: J-bends, pump lines, trap lines

Mechanical Properties

Tensile properties. See Table 60.
Hardness. See Table 60.
Elastic modulus. Tension, 105 GPa (15 × 10⁶ psi); shear, 39 GPa (5.6 × 10⁶ psi)

Mass Characteristics

Density. 8.50 g/cm³ (0.31 lb/in.³) at 20 °C (68 °F)

Thermal Properties

Liquidus temperature. 940 °C (1720 °F)
Solidus temperature. 905 °C (1660 °F)
Coefficient of linear thermal expansion. 20.2 μm/m · K (11.2 μin./in. · °F) at 20 to 300 °C (68 to 572 °F)
Specific heat. 380 J/kg · K (0.09 Btu/lb · °F) at 20 °C (68 °F)
Thermal conductivity. 115 W/m · K (67 Btu/ft · h · °F) at 20 °C (68 °F)

Electrical Properties

Electrical conductivity. Volumetric, O61 temper, 26% IACS at 20 °C (68 °F)
Electrical resistivity. 66 nΩ · m at 20 °C (68 °F)

Fabrication Characteristics

Machinability. 60% of C36000 (free-cutting brass)
Formability. Cold working, excellent; hot forming, poor
Weldability. Soldering: excellent. Brazing: good. Oxyfuel gas, gas-shielded arc, resistance spot, and resistance butt welding; fair. All other welding processes are not recommended.
Recrystallization temperature. 290 °C (550 °F)
Annealing temperature. 425 to 650 °C (800 to 1200 °F)

C33200
66Cu-32.4Zn-1.6Pb

Commercial Names

Previous trade name. High-leaded brass (tube)
Common name. Free-cutting tube brass

Specifications

AMS. 4558
ASTM. B 135
Government. MIL-T-46072

Chemical Composition

Composition limits. 65.0 to 68.0 Cu, 1.3 to 2.0 Pb, 0.07 Fe max, 0.5 max other (total), bal Zn

Applications

Typical uses. General-purpose screw machine products

Mechanical Properties

Tensile properties. See Table 61 and Fig. 36.
Hardness. See Table 61 and Fig. 36.
Elastic modulus. Tension, 105 GPa (15 × 10⁶ psi); shear, 39 GPa (5.6 × 10⁶ psi)

Mass Characteristics

Density. 8.53 g/cm³ (0.31 lb/in.³) at 20 °C (68 °F)

Table 62 Typical mechanical properties of C33500

Temper	Tensile strength MPa	Tensile strength ksi	Yield strength(a) MPa	Yield strength(a) ksi	Elongation in 50 mm (2 in.), %	Hardness HRF	Hardness HR30T	Shear strength MPa	Shear strength ksi
OS070	315	46	97	14	65	58	15	220	32
OS050	325	47	105	15	62	64	26	...	...
OS035	340	49	115	17	57	68	31	235	34
OS025	350	51	130	19	55	72	36	...	...
H01	370	54	275	40	43	55 HRB	54	250	36
H02	420	61	345	50	23	70 HRB	65	275	40
H04	510	74	415	60	8	80 HRB	69	295	43
H06	580	84	...	...	...	86 HRB	74	...	...

Note: Values for flat products, 1 mm (0.04 in.) thick. (a) 0.5% extension under load

Table 63 Typical mechanical properties of C34000

Temper	Tensile strength MPa	Tensile strength ksi	Yield strength(a) MPa	Yield strength(a) ksi	Elongation in 50 mm (2 in.), %	Hardness HRB	Hardness HR30T	Shear strength MPa	Shear strength ksi
Flat products, 1 mm (0.04 in.) thick									
OS035	340	49	115	17	54	68 HRF	31	225	33
OS025	350	51	130	19	53	72 HRF	36	235	34
H01	370	54	275	40	41	55	54	250	36
H02	420	61	345	50	21	70	63	275	40
H04	510	74	415	60	7	80	70	295	43
H06	585	85	425	62	5	87	73	310	45
Rod, 25 mm (1.0 in.) diameter									
OS025	345	50	135	20	60	70 HRF	...	235	34
H03	380	55	290	42	40	60	...	250	36
H02	435	63	330	48	30	68	...	275	40
Wire, 2 mm (0.08 in.) diameter									
OS025	345	50	...	...	50	...	...	235	34
H00	400	58	...	...	30	...	...	260	38
H01	485	70	...	...	13	...	...	290	42
H02	605	88	...	...	7	...	...	315	46

(a) 0.5% extension under load

Thermal Properties

Liquidus temperature. 930 °C (1710 °F)
Solidus temperature. 900 °C (1650 °F)
Coefficient of linear thermal expansion. 20.3 μm/m · K (11.3 μin./in. · °F) at 20 to 300 °C (68 to 572 °F)
Specific heat. 380 J/kg · K (0.09 Btu/lb · °F) at 20 °C (68 °F)
Thermal conductivity. 115 W/m · K (67 Btu/ft · h · °F) at 20 °C (68 °F)

Electrical Properties

Electrical conductivity. Volumetric. O61 temper, 26% IACS at 20 °C (68 °F)
Electrical resistivity. 66 nΩ · m at 20 °C (68 °F)

Fabrication Characteristics

Machinability. 80% of C36000 (free-cutting brass)
Formability. Cold working, fair; hot forming, poor
Weldability. Soldering: excellent. Brazing: good. Resistance butt welding: fair. All other welding processes are not recommended.
Recrystallization temperature. 288 °C (550 °F)
Annealing temperature. 425 to 650 °C (800 to 1200 °F)

C33500
65Cu-34.5Zn-0.5Pb

Commercial Names

Previous trade name. Low-leaded brass

Specifications

ASTM. Flat products: B 121. Rod: B 453
Government. Flat products: QQ-B-613. Bar, forgings, rod, shapes, strip: QQ-B-626

Chemical Composition

Composition limits. 62.5 to 66.5 Cu, 0.3 to 0.8 Pb, 0.1 Fe max, 0.5 max other (total), bal Zn

Applications

Typical uses. Hardware such as butts and hinges; watch backs

Mechanical Properties

Tensile properties. See Table 62.
Shear strength. See Table 62.
Hardness. See Table 62.
Elastic modulus. Tension, 105 GPa (15×10^6 psi); shear, 39 GPa (5.6×10^6 psi)

Mass Characteristics

Density. 8.47 g/cm^3 (0.306 lb/in.^{3}) at 20 °C (68 °F)

Thermal Properties

Liquidus temperature. 925 °C (1700 °F)
Solidus temperature. 900 °C (1650 °F)
Coefficient of linear thermal expansion. 20.3 μm/m · K (11.3 μin./in. · °F) at 20 to 300 °C (68 to 572 °F)
Specific heat. 380 J/kg · K (0.09 Btu/lb · °F) at 20 °C (68 °F)
Thermal conductivity. 115 W/m · K (67 Btu/ft · h · °F) at 20 °C (68 °F)

Electrical Properties

Electrical conductivity. Volumetric, 26% IACS at 20 °C (68 °F)
Electrical resistivity. 66 nΩ · m at 20 °C (68 °F)

Fabrication Characteristics

Machinability. 60% of C36000 (free-cutting brass)
Formability. Cold working, good; hot forming, poor. Commonly fabricated by blanking, drawing, machining, piercing, punching, and stamping
Weldability. Soldering: excellent. Brazing: good. Oxyfuel gas, gas-shielded arc, resistance spot, and resistance butt welding: fair. Shielded metal arc and resistance seam welding are not recommended.
Annealing temperature. 425 to 700 °C (800 to 1300 °F)

C34000
65Cu-34Zn-1Pb

Commercial Names

Previous trade name. Medium-leaded brass, 64.5%

Specifications

ASTM. Flat products: B 121. Rod: B 453
Government. Flat products: QQ-B-613. Bar, forgings, rod, shapes, strip: QQ-B-626

Chemical Composition

Composition limits. 62.5 to 66.5 Cu, 0.8 to 1.4 Pb, 0.10 Fe max, 0.5 max other (total), bal Zn

Applications

Typical uses. Flat products: butts, dials, engravings, gears, instrument plates, nuts, or drawn shells, all involving piercing, threading, or machining. Rod, bar, and wire: couplings, free-machining screws and rivets, gears, nuts, tire valve stems, screw machine products involving severe knurling and roll threading or moderate cold heading, flaring, spinning, or swaging

Mechanical Properties

Tensile properties. See Table 63.
Shear strength. See Table 63.
Hardness. See Table 63.
Elastic modulus. Tension, 105 GPa (15×10^6 psi); shear, 39 GPa (5.6×10^6 psi)

Mass Characteristics

Density. 8.47 g/cm^3 (0.306 lb/in.^{3}) at 20 °C (68 °F)

Table 64 Typical mechanical properties of C34200

Temper	Tensile strength MPa	ksi	Yield strength(a) MPa	ksi	Elongation in 50 mm (2 in.), %	Hardness HRB	HR30T	Shear strength MPa	ksi
Flat products, 1 mm (0.04 in.) thick									
OS015	370	54	165	24	45	78 HRF	41	255	37
OS025	360	52	140	20	48	76 HRF	37	250	36
OS035	340	49	115	17	52	68 HRF	32	235	34
OS050	325	47	105	15	55	66 HRF	28	225	33
H01	370	54	275	40	38	55	54	250	36
H02	420	61	345	50	20	70	63	275	40
H04	510	74	415	60	7	80	71	295	43
H06	585	85	425	62	5	87	75	310	45
Rod, 25 mm (1.0 in.) diameter									
O50	325	47	125	18	50(b)	66 HRF	...	...	...
H55	400	58	270	39	28(c)	65	...	...	...
H02	450	65	310	45	23(d)	72	...	...	...

(a) 0.5% extension under load. (b) Reduction in area 65%. (c) Reduction in area 50%. (d) Reduction in area 35%

Table 65 Typical mechanical properties of C34900

Temper	Tensile strength MPa	ksi	Yield strength(a) MPa	ksi	Elongation in 50 mm (2 in.), %	Hardness	Shear strength MPa	ksi
Rod, 6 mm (0.25 in.) diameter								
OS035	365	53	165	24	50	75 HRF	235	34
Rod, 25 mm (1.0 in.) diameter								
H01	385	56	290	42	42	70 HRB	250	36
Wire, 6 mm (0.25 in.) diameter								
OS015	380	55	150	22	48	70 HRF	240	35
H01	470	68	380	55	18	72 HRB	285	410
Wire, 19 mm (0.75 in.) diameter								
OS050	330	48	110	16	72	67 HRF	220	32

(a) At 0.5% extension under load

Thermal Properties

Liquidus temperature. 925 °C (1700 °F)
Solidus temperature. 885 °C (1630 °F)
Coefficient of linear thermal expansion. 20.3 μm/m · K (11.3 μin./in. · °F) at 20 to 300 °C (68 to 572 °F)
Specific heat. 380 J/kg · K (0.09 Btu/lb · °F) at 20 °C (68 °F)
Thermal conductivity. 115 W/m · K (67 Btu/ft · h · °F) at 20 °C (68 °F)

Electrical Properties

Electrical conductivity. Volumetric. O61 temper, 26% IACS at 20 °C (68 °F)
Electrical resistivity. 66 nΩ · m at 20 °C (68 °F)

Fabrication Characteristics

Machinability. 60% of C36000 (free-cutting brass)
Formability. Cold working, good; hot forming, poor
Weldability. Soldering: excellent. Brazing: good. Resistance butt welding: fair. All other welding processes are not recommended.
Recrystallization temperature. 288 °C (550 °F)
Annealing temperature. 425 to 650 °C (800 to 1200 °F)

C34200 64.5Cu-33.5Zn-2Pb C35300 62Cu-36.2Zn-1.8Pb

Commercial Names

Previous trade name. High-leaded brass
Common name. Clock brass, engraver's brass, heavy-leaded brass

Specifications

ASTM. Flat products: B 121. Rod: B 453
SAE. J463
UNS number. C34200, C35300
Government. Flat products: QQ-B-613. Bar, forgings, rod, shapes, strip: QQ-B-626

Chemical Composition

Composition limits of C34200. 62.5 to 66.5 Cu, 1.5 to 2.5 Pb, 0.1 Fe max, 0.5 max other (total), bal Zn
Composition limits of C35300. 59.0 to 64.5 Cu, 1.3 to 2.3 Pb, 0.1 Fe max, 0.5 max other (total), bal Zn

Applications

Typical uses. Flat products: gears, wheels, nuts, plates for clocks, keys, bearing cages, engraver's plates. Rod: gears, pinions, valve stems, automatic screw machine parts that need more severe cold working than can be tolerated with free-cutting brass (for example, processes such as knurling and moderate staking)

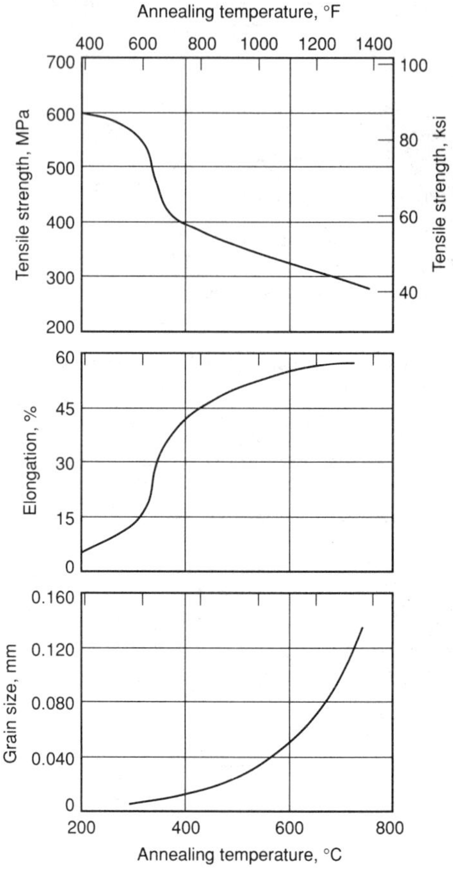

Fig. 37 Annealing behavior of C34200. Curves are for 1 mm (0.04 in.) thick strip cold rolled from OS035 temper starting stock.

Mechanical Properties

Tensile properties. See Table 64.
Hardness. See Table 64.
Elastic modulus. Tension, 105 GPa (15 × 10^6 psi); shear, 39 GPa (5.6 × 10^6 psi)

Structure

Crystal structure. Face-centered cubic α
Microstructure. Two phase, α and lead

Mass Characteristics

Density. 8.5 g/cm^3 (0.307 lb/in.^{3}) at 20 °C (68 °F)

Thermal Properties

Liquidus temperature. 910 °C (1670 °F)
Solidus temperature. 885 °C (1630 °F)
Coefficient of linear thermal expansion. 20.3 μm/m · K (11.3 μin./in. · °F) at 20 to 300 °C (68 to 572 °F)
Specific heat. 380 J/kg · K (0.09 Btu/lb · °F) at 20 °C (68 °F)
Thermal conductivity. 115 W/m · K (67 Btu/ft · h · °F) at 20 °C (68 °F)

Table 66 Typical mechanical properties of C35000

Temper	Tensile strength MPa	Tensile strength ksi	Yield strength, 0.5% extension under load MPa	Yield strength, 0.5% extension under load ksi	Yield strength, 0.2% offset MPa	Yield strength, 0.2% offset ksi	Elongation in 50 mm (2 in.), %	Hardness HRB	Hardness HR30T	Shear strength MPa	Shear strength ksi
Flat products, 1 mm (0.04 in.) thick											
OS050	310	45	90	13	90	13	57	61 HRF	...	...	...
OS035	325	47	110	16	110	16	54	67 HRF	...	...	...
OS025	330	48	135	20	135	20	50	70 HRF	...	...	...
OS015	350	51	170	25	170	25	46	74 HRF	52	...	...
H01	370	54	220	32	235	34	43	66	60	...	...
H02	415	60	310	45	310	45	29	75	68	...	...
H03	460	67	365	53	380	55	17	80	71	...	...
H04	505	73	415	60	415	60	10	86	75	...	...
H06	580	84	450	65	475	69	5				
Rod, 12 mm (0.5 in.) diameter											
OS050	330	48	110	16	...	...	56	65 HRF	25	235	34
OS015	380	55	170	25	...	...	46	85 HRF	50	250	36
H01	400	58	305	44	...	...	42	60	57	260	38
H02	485	70	360	52	...	...	22	80	70	290	42

Table 67 Typical mechanical properties of 1 mm (0.04 in.) thick C35600 sheet and strip

Temper	Tensile strength MPa	Tensile strength ksi	Yield strength(a) MPa	Yield strength(a) ksi	Elongation in 50 mm (2 in.), %	Hardness HRB	Hardness HR30T
OS035	340	49	115	17	50	68 HRF	31
H01	370	54	275	40	35	55	54
H02	420	61	345	50	20	70	65
H04	510	74	415	60	7	80	69

(a) At 0.5% extension under load

Electrical Properties

Electrical conductivity. Volumetric, O61 temper, 26% IACS at 20 °C (68 °F)
Electrical resistivity. 66 nΩ · m at 20 °C (68 °F)

Fabrication Characteristics

Machinability. 90% of C36000 (free-cutting brass)
Formability. Cold working, fair; hot forming, poor
Weldability. Soldering: excellent. Brazing: good. Resistance butt welding: fair. All other welding processes are not recommended.
Recrystallization temperature. 320 °C (600 °F)
Annealing temperature. 425 to 600 °C (800 to 1100 °F). See also Fig. 37.
Hot-working temperature. 785 to 815 °C (1445 to 1500 °F)

C34900
62Cu-37.5Zn-0.3Pb

Chemical Composition

Composition limits. 61.0 to 64.0 Cu, 0.1 to 0.5 Pb, 0.1 Fe max, 0.5 max other (total), bal Zn

Applications

Typical uses. Building hardware, drilled and tapped rivets, plumbing goods, saw nuts, and parts requiring moderate cold working combined with some machining

Mechanical Properties

Tensile properties. See Table 65.
Shear strength. See Table 65.
Hardness. See Table 65.
Elastic modulus. Tension, 105 GPa (15 × 10^6 psi); shear, 39 GPa (5.6 × 10^6 psi)

Mass Characteristics

Density. 8.44 g/cm^3 (0.305 lb/in.^{3}) at 20 °C (68 °F)

Thermal Properties

Liquidus temperature. 910 °C (1670 °F)
Solidus temperature. 895 °C (1640 °F)
Coefficient of linear thermal expansion. 20.3 μm/m · K (11.3 μin./in. · °F) at 20 to 300 °C (68 to 572 °F)
Specific heat. 380 J/kg · K (0.09 Btu/lb · °F) at 20 °C (68 °F)
Thermal conductivity. 115 W/m · K (67 Btu/ft · h · °F) at 20 °C (68 °F)

Electrical Properties

Electrical conductivity. Volumetric, 26% IACS at 20 °C (68 °F)
Electrical resistivity. 66 nΩ · m at 20 °C (68 °F)

Fabrication Characteristics

Machinability. 50% of C36000 (free-cutting brass)
Formability. Cold working, good; hot forming, poor
Weldability. Soldering: excellent. Brazing: good. Oxyfuel gas, gas-shielded arc, and resistance spot and butt welding: fair. Shielded metal arc and resistance seam welding are not recommended.
Annealing temperature. 425 to 650 °C (800 to 1200 °F)
Hot-working temperature. 675 to 800 °C (1250 to 1450 °F)

C35000
62.5Cu-36.4Zn-1.1Pb

Commercial Names

Previous trade name. Medium-leaded brass, 62%

Specifications

ASTM. Flat products: B 121. Rod: B 453
SAE. J463
Government. Flat products: QQ-B-613. Bar, forgings, rod, shapes, strip: QQ-B-626

Chemical Composition

Composition limits. 59.0 to 64.0 Cu, 0.8 to 1.4 Pb, 0.1 Fe max, 0.5 max other (total), bal Zn

Applications

Typical uses. Bearing cages, book dies, clock plates, engraving plates, gears, hinges, hose couplings, keys, lock parts, lock tumblers, meter parts, sink strainers, strike plates, templates, nuts, type characters, washers, wear plates

Mechanical Properties

Tensile properties. See Table 66.
Shear strength. See Table 66.
Hardness. See Table 66.
Elastic modulus. Tension, 105 GPa (15 × 10^6 psi); shear, 39 GPa (5.6 × 10^6 psi)

Mass Characteristics

Density. 8.47 g/cm^3 (0.306 lb/in.^{3}) at 20 °C (68 °F)

Thermal Properties

Liquidus temperature. 915 °C (1680 °F)
Solidus temperature. 895 °C (1640 °F)
Coefficient of linear thermal expansion. 20.3 μm/m · K (11.3 μin./in. · °F) at 20 to 300 °C (68 to 572 °F)
Specific heat. 380 J/kg · K (0.09 Btu/lb · °F) at 20 °C (68 °F)
Thermal conductivity. 115 W/m · K (67 Btu/ft · h · °F) at 20 °C (68 °F)

Electrical Properties

Electrical conductivity. Volumetric, 26% IACS at 20 °C (68 °F)
Electrical resistivity. 66 nΩ · m at 20 °C (68 °F)

Fabrication Characteristics

Machinability. 70% of C36000 (free-cutting brass)
Forgeability. 50% of C37700 (forging brass)
Formability. Fair for cold working and hot forming
Weldability. Soldering: excellent. Brazing: good. Resistance butt welding: fair. All other welding processes are not recommended.
Annealing temperature. 425 to 600 °C (800 to 1100 °F)
Hot-working temperature. 760 to 800 °C (1400 to 1500 °F)

Table 68 Typical mechanical properties of C36000

Temper	Tensile strength MPa	Tensile strength ksi	Yield strength(a) MPa	Yield strength(a) ksi	Elongation in 50 mm (2 in.), %	Reduction in area, %	Hardness, HRB	Shear strength MPa	Shear strength ksi
Rod, 6 mm (0.25 in.) diameter									
H02(b)	470	68	360	52	18	48	80	260	38
Rod, 25 mm (1 in.) diameter									
O61	340	49	125	18	53	58	68 HRF	205	30
H02(c)	400	58	310	45	25	50	78	235	34
Rod, 50 mm (2 in.) diameter									
H02(d)	380	55	305	44	32	52	75	220	32
Shapes									
M30	340	49	125	18	50	...	68 HRF	205	30
H01(e)	385	56	310	45	20	...	62	230	33

(a) 0.5% extension under load. (b) Cold drawn 25%. (c) Cold drawn 20%. (d) Cold drawn 18%. (e) Cold drawn 15%

C35600
62Cu-35.5Zn-2.5Pb

Commercial Names

Previous trade name. Extra-high-leaded brass

Specifications

ASTM. Flat products: B 121. Rod: B 453
Government. Flat products: QQ-B-613. Bar, rod, shapes, strip: QQ-B-626

Chemical Composition

Composition limits. 59.0 to 64.5 Cu, 2.0 to 3.0 Pb, 0.1 Fe max, 0.5 max other (total), bal Zn

Applications

Typical uses. Hardware: clock plates and nuts, clock and watch backs, clock gears and wheels. Industrial: channel plate

Mechanical Properties

Tensile properties. See Table 67.
Hardness. See Table 67.
Elastic modulus. Tension, 97 GPa (14 × 10^6 psi); shear, 37 GPa (5.3 × 10^6 psi)

Mass Characteristics

Density. 8.5 g/cm^3 (0.307 lb/in.^{3}) at 20 °C (68 °F)

Thermal Properties

Liquidus temperature. 905 °C (1660 °F)
Solidus temperature. 885 °C (1630 °F)
Coefficient of linear thermal expansion. 20.5 μm/m · K (11.4 μin./in. · °F) at 20 to 300 °C (68 to 572 °F)
Specific heat. 380 J/kg · K (0.09 Btu/lb · °F) at 20 °C (68 °F)
Thermal conductivity. 115 W/m · K (67 Btu/ft · h · °F) at 20 °C (68 °F)

Electrical Properties

Electrical conductivity. Volumetric, O61 temper, 26% IACS at 20 °C (68 °F)
Electrical resistivity. 66 nΩ · m at 20 °C (68 °F)

Fabrication Characteristics

Machinability. 100% of C36000 (free-cutting brass)
Formability. Cold working, poor; hot forming, fair
Weldability. Soldering: excellent. Brazing: good. Resistance butt welding: fair. All other welding processes are not recommended.
Annealing temperature. 425 to 600 °C (800 to 1100 °F)
Hot-working temperature. 700 to 800 °C (1300 to 1450 °F)

C36000
61.5Cu-35.5Zn-3Pb

Commercial Names

Previous trade name. Free-cutting brass
Common name. Free-turning brass, free-cutting yellow brass, high-leaded brass

Specifications

AMS. 4610
ASTM. B 16
SAE. J463
Government. Flat products: QQ-B-613. Bar, forgings, rod, shapes, strip: QQ-B-626

Chemical Composition

Composition limits. 60.0 to 63.0 Cu, 2.5 to 3.7 Pb, 0.35 Fe max, 0.5 max other (total), bal Zn

Applications

Typical uses. Hardware: gears, pinions. Industrial: automatic high-speed screw machine parts

Mechanical Properties

Tensile properties. See Table 68 and Fig. 38.
Hardness. See Table 68.
Elastic modulus. Tension, 97 GPa (14 × 10^6 psi); shear, 37 GPa (5.3 × 10^6 psi)
Fatigue strength. Rotating-beam tests on 90 mm (0.350 in.) diam specimens taken from 50 mm (2 in.) diam rod. H02 temper (cold drawn 15%): 140 MPa (20 ksi) at 10^8 cycles; 97 MPa (14 ksi) at 3 × 10^8 cycles

Structure

Microstructure. Generally three phase: α, β, and lead

Mass Characteristics

Density. 8.5 g/cm^3 (0.307 lb/in.^{3}) at 20 °C (68 °F)

Thermal Properties

Liquidus temperature. 900 °C (1660 °F)
Solidus temperature. 885 °C (1630 °F)
Coefficient of linear thermal expansion. 20.5 μm/m · K (11.4 μin./in. · °F) at 20 to 300 °C (68 to 572 °F)
Specific heat. 380 J/kg · K (0.09 Btu/lb · °F) at 20 °C (68 °F)

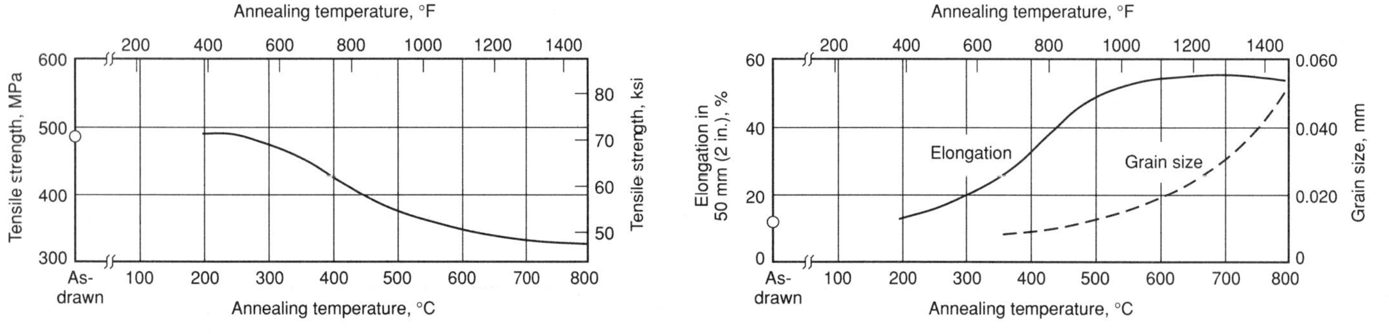

Fig. 38 Annealing curves for C36000. Data are for free-cutting brass rod, cold drawn 30% to 19 mm (0.75 in.) in diameter from M30 temper (as-extruded) starting stock, then annealed 1 h at temperature.

Table 69 Typical mechanical properties of C37000

Temper	Tensile strength MPa	Tensile strength ksi	Yield strength(a) MPa	Yield strength(a) ksi	Elongation in 50 mm (2 in.), %	Hardness HRB	Hardness HR30T
Tube, 38 mm (1.5 in.) outside diameter × 3 mm (0.125 in.) wall thickness							
O50	370	54	140	20	40	80 HRF	43
H80(b)	550	80	415	60	6	85	74
Tube, 50 mm (2 in.) outside diameter × 6 mm (0.25 in.) wall thickness							
H80(c)	485	70	310	45	10	75	67

(a) At 0.5% extension under load. (b) Cold drawn 35%. (c) Cold drawn 25%

Thermal conductivity. 115 W/m · K (67 Btu/ft · h · °F) at 20 °C (68 °F)

Electrical Properties

Electrical conductivity. Volumetric, O61 temper, 26% IACS at 20 °C (68 °F)
Electrical resistivity. 66 nΩ · m at 20 °C (68 °F)

Fabrication Characteristics

Machinability. 100%. This is the standard material against which the machining qualities of all other copper alloys are judged.
Formability. Cold working, poor; hot forming, fair
Weldability. Soldering: excellent. Brazing: good. Resistance butt welding: fair. All other welding processes are not recommended.
Recrystallization temperature. 330 °C (625 °F)
Annealing temperature. 425 to 600 °C (800 to 1100 °F). See also Fig. 38.
Hot-working temperature. 700 to 800 °C (1300 to 1450 °F)

C36500, C36600, C36700, C36800
60Cu-39.4Zn-0.6Pb

Commercial Names

Previous trade name. C36500, uninhibited leaded Muntz metal; C36600, arsenical leaded Muntz metal; C36700, antimonial leaded Muntz metal; C36800, phosphorized leaded Muntz metal
Common name. Leaded Muntz metal; inhibited leaded Muntz metal

Specifications

ASME. Plate, condenser tube: SB171
ASTM. Plate, condenser tube: B 171. Plate, clad: B 432

Chemical Composition

Composition limits. 58.0 to 61.0 Cu, 0.4 to 0.9 Pb, 0.15 Fe max, 0.25 Sn max; As Sb, or P (see below); 0.1 max other (total), bal Zn
Antimony or phosphorus limits. C36500, none specified; C36600, 0.02 to 0.1 As; C36700, 0.02 to 0.1 Sb; C36800, 0.02 to 0.1 P

Applications

Typical uses. Main tube sheets for condensers and heat exchangers; support sheets; baffles

Mechanical Properties

Tensile properties. 25 mm (1 in.) plate, M20 temper: tensile strength, 370 MPa (54 ksi); yield strength (0.5% extension), 140 MPa (20 ksi); elongation, 45% in 50 mm (2 in.)
Shear strength. M20 temper: 275 MPa (40 ksi)
Hardness. M20 temper: 80 HRF
Elastic modulus. Tension, 105 GPa (15 × 10^6 psi); shear, 39 GPa (5.6 × 10^6 psi)

Structure

Crystal structure. Face-centered cubic
Microstructure. Alpha and β with undissolved lead. Beta phase appears lemon yellow with ammonia peroxide etch; it may be darkened with ferric chloride etch. Lead appears as insoluble gray particles randomly distributed throughout the structure.

Mass Characteristics

Density. 8.41 g/cm^3 (0.304 lb/in.^{3}) at 20 °C (68 °F)

Thermal Properties

Liquidus temperature. 900 °C (1650 °F)
Solidus temperature. 885 °C (1630 °F)
Coefficient of linear thermal expansion. 20.8 μm/m · K (11.6 μin./in. · °F) at 20 to 300 °C (68 to 572 °F)
Specific heat. 380 J/kg · K (0.09 Btu/lb · °F) at 20 °C (68 °F)
Thermal conductivity. 123 W/m · K (71 Btu/ft · h · °F) at 20 °C (68 °F)

Electrical Properties

Electrical conductivity. Volumetric, O61 temper: 28% IACS at 20 °C (68 °F)
Electrical resistivity. 62 nΩ · m at 20 °C (68 °F)

Chemical Properties

General corrosion behavior. Good resistance to corrosion in both fresh and salt water. C36500 is the uninhibited alloy and is subject to dezincification; the inhibited alloys each contain 0.02 to 0.10% of an inhibitor element (As, Sb, or P), which imparts high resistance to dezincification.

Fabrication Characteristics

Machinability. 60% of C36000 (free-cutting brass)
Formability. Cold working, fair; hot working, excellent
Weldability. Soldering: excellent. Brazing: good. Oxyfuel gas, gas-shielded arc, and resistance butt welding: fair. All other welding processes are not recommended.
Annealing temperature. 425 to 600 °C (800 to 1100 °F)
Hot-working temperature. 625 to 800 °C (1150 to 1450 °F)

C37000
60Cu-39Zn-1Pb

Commercial Names

Previous trade name. Free-cutting Muntz metal

Specifications

ASTM. Tube: B 135
Government. Flat products: QQ-B-613. Bar, forgings, rod, strip: QQ-B-626. Tube: MIL-T-46072

Chemical Composition

Composition limits. 59.0 to 62.0 Cu, 0.9 to 1.4 Pb, 0.15 Fe max, 0.5 max other (total), bal Zn

Applications

Typical uses. Automatic screw machine parts

Mechanical Properties

Tensile properties. See Table 69.
Hardness. See Table 69.
Elastic modulus. Tension, 105 GPa (15 × 10^6 psi); shear, 39 GPa (5.6 × 10^6 psi)

Mass Characteristics

Density. 8.41 g/cm^3 (0.304 lb/in.^{3}) at 20 °C (68 °F)

Thermal Properties

Liquidus temperature. 900 °C (1650 °F)
Solidus temperature. 885 °C (1630 °F)
Coefficient of linear thermal expansion. 20.8 μm/m · K (11.6 μin./in. · °F) at 20 to 300 °C (68 to 572 °F)
Specific heat. 375 J/kg · K (0.09 Btu/lb · °F) at 20 °C (68 °F)
Thermal conductivity. 120 W/m · K (69 Btu/ft · h · °F) at 20 °C (68 °F)

Electrical Properties

Electrical conductivity. Volumetric, O61 temper: 27% IACS at 20 °C (68 °F)
Electrical resistivity. 63.9 nΩ · m at 20 °C (68 °F)

Fabrication Characteristics

Machinability. 70% of C36000 (free-cutting brass)
Weldability. Soldering: excellent. Brazing: good. Resistance butt welding: fair. All other welding processes are not recommended.
Annealing temperature. 425 to 600 °C (800 to 1100 °F)
Hot-working temperature. 625 to 800 °C (1150 to 1450 °F)

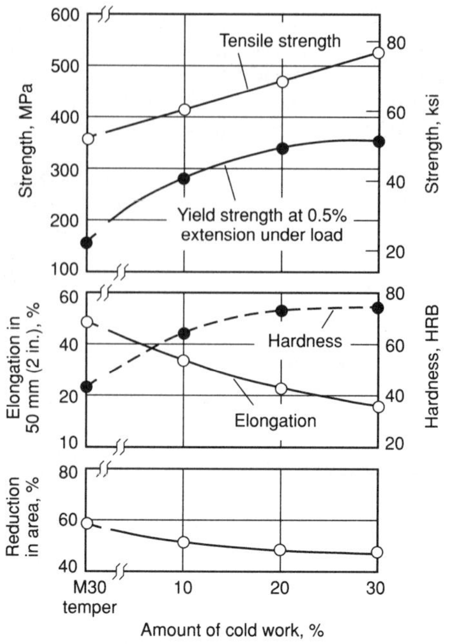

Fig. 39 Typical mechanical properties of extruded and drawn C37700. Data are for forging brass rod less than 25 mm (1 in.) in diameter that was extruded, then cold drawn to various percentages of reduction in area.

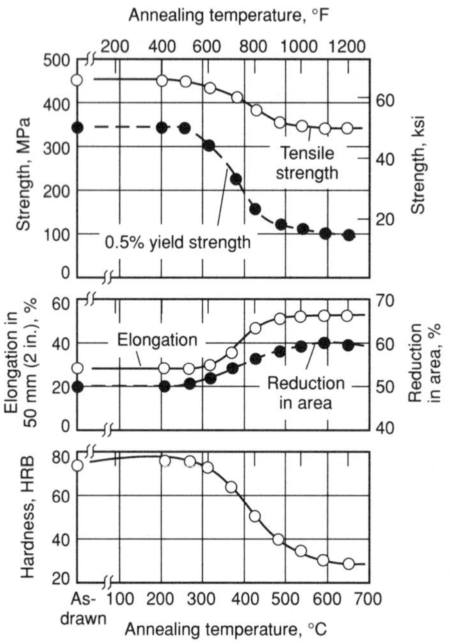

Fig. 40 Annealing curves for C37700. Typical data are for forging brass rod less than 25 mm (1 in.) in diameter that was extruded, cold drawn 18%, and annealed 1 h at various temperatures.

C37700
60Cu-38Zn-2Pb

Commercial Names

Previous trade name. Forging brass

Specifications

AMS. Die forgings, forging rod: 4614
ASME. Die forgings: SB283
ASTM. Bar, forging, rod, shapes: B 124. Die forgings: B 283
SAE. Die forgings: J463
Government. QQ-B-626. Die forgings: MIL-C-13351

Chemical Composition

Composition limits. 58.0 to 62.0 Cu, 1.5 to 2.5 Pb, 0.3 Fe max, 0.5 max other (total), bal Zn

Applications

Typical uses. Forgings and pressings of all kinds

Mechanical Properties

Tensile properties. M30 temper: tensile strength, 360 MPa (52 ksi); yield strength (0.5% extension), 140 MPa (20 ksi); elongation, 45% in 50 mm (2 in.). See Fig. 39 and 40.
Hardness. M30 temper: 78 HRF. See also Fig. 39 and 40.
Elastic modulus. Tension, 105 GPa (15 × 10^6 psi); shear, 39 GPa (5.6 × 10^6 psi)

Structure

Crystal structure. Face-centered cubic
Microstructure. Two phase: α and β, with undissolved lead. Beta phase appears lemon yellow with ammonia peroxide etch. Ferric chloride darkens β phase. Lead appears as gray particles.

Mass Characteristics

Density. 8.44 g/cm^3 (0.305 lb/in.^{3}) at 20 °C (68 °F)

Thermal Properties

Liquidus temperature. 895 °C (1640 °F)
Solidus temperature. 880 °C (1620 °F)
Coefficient of linear thermal expansion. 20.7 μm/m · K (11.5 μin./in. · °F) at 20 to 300 °C (68 to 572 °F)
Specific heat. 380 J/kg · K (0.09 Btu/lb · °F) at 20 °C (68 °F)
Thermal conductivity. 120 W/m · K (69 Btu/ft · h · °F) at 20 °C (68 °F)

Electrical Properties

Electrical conductivity. Volumetric, 27% IACS at 20 °C (68 °F)
Electrical resistivity. 64 nΩ · m at 20 °C (68 °F)

Magnetic Properties

Magnetic susceptibility. Nonmagnetic

Optical Properties

Color. Golden hue compared to yellow of C26000 (cartridge brass)

Fabrication Characteristics

Machinability. 80% of C36000 (free-cutting brass)
Forgeability. 100%. This is the standard material against which the forging qualities of all other copper alloys are judged.
Formability. Cold working, poor; hot forming, excellent
Weldability. Soldering: excellent. Brazing: good. Resistance butt welding: fair. All other welding processes are not recommended.
Annealing temperature. 425 to 600 °C (800 to 1100 °F). See also Fig. 40.
Hot-working temperature. 650 to 825 °C (1200 to 1500 °F)

C38500
57Cu-40Zn-3Pb

Commercial Names

Previous trade name. Architectural bronze

Specifications

ASTM. Shapes: B 455

Chemical Composition

Composition limits. 55.0 to 60.0 Cu, 2.0 to 3.8 Pb, 0.35 Fe max, 0.5 max other (total), bal Zn

Applications

Typical uses. Architectural: extrusions, storefronts, thresholds, and trim. Hardware: butts, hinges, and lock bodies. Industrial: forgings

Mechanical Properties

Tensile properties. M30 temper: tensile strength, 415 MPa (60 ksi); yield strength (0.5% extension), 140 MPa (20 ksi); elongation, 30% in 50 mm (2 in.)
Shear strength. M30 temper: 240 MPa (35 ksi)
Hardness. M30 temper: 65 HRB
Elastic modulus. Tension, 97 GPa (14 × 10^6 psi); shear, 37 GPa (5.3 × 10^6 psi)

Mass Characteristics

Density. 8.47 g/cm^3 (0.306 lb/in.^{3}) at 20 °C (68 °F)

Thermal Properties

Liquidus temperature. 890 °C (1630 °F)
Solidus temperature. 875 °C (1610 °F)
Coefficient of linear thermal expansion. 20.9 μm/m · K (11.6 μin./in. · °F) at 20 to 300 °C (68 to 572 °F)
Specific heat. 380 J/kg · K (0.09 Btu/lb · °F) at 20 °C (68 °F)
Thermal conductivity. 123 W/m · K (71 Btu/ft · h · °F) at 20 °C (68 °F)

Electrical Properties

Electrical conductivity. Volumetric, O61 temper: 28% IACS at 20 °C (68 °F)
Electrical resistivity. 62 nΩ · m at 20 °C (68 °F)

Fabrication Characteristics

Machinability. 90% of C36000 (free-cutting brass)
Formability. Cold working, poor; hot forming, excellent
Weldability. Soldering: excellent. Brazing: good. Resistance butt welding: fair. All other welding processes are not recommended.

Table 70 Typical mechanical properties of C40500

Temper	Tensile strength		Yield strength At 0.5% extension under load		Yield strength At 0.2% offset		Elongation in 50 mm (2 in.), %	Hardness		Shear strength	
	MPa	ksi	MPa	ksi	MPa	ksi		HRB	HR30T	MPa	ksi
Flat products, 1 mm (0.04 in.) thick											
OS035	270	39	83	12	69	10	49	55 HRF	10	215	31
OS025	280	41	83	12	76	11	48	58 HRF	13	230	33
OS015	290	42	90	13	97	14	47	64 HRF	24	230	33
H01	325	47	250	36	250	36	30	46	47	240	35
H02	360	52	295	43	345	50	15	60	56	255	37
H03	400	58	340	49	385	56	12	67	62	260	38
H04	440	64	380	55	425	62	10	72	65	270	39
H06	475	69	415	60	460	67	7	76	69	280	41
H08	510	74	435	63	495	72	4	79	71	295	43
H10	540	78	485	70	525	76	3	82	72	310	45

Annealing temperature. 425 to 600 °C (800 to 1100 °F)
Hot-working temperature. 625 to 725 °C (1150 to 1350 °F)

C40500
95Cu-4Zn-1Sn

Commercial Names

Trade name. High-conductivity bronze
Common name. Penny bronze

Specifications

ASTM. B 591

Chemical Composition

Composition limits. 94 to 96 Cu, 0.7 to 1.3 Sn, 0.05 Pb max, 0.05 Fe max, bal Zn

Applications

Typical uses. Meter clips, terminals, fuse clips, contact springs, relay springs, washers from rolled strip, rolled bar, sheet

Mechanical Properties

Tensile properties. See Table 70.
Shear strength. See Table 70.
Hardness. See Table 70.
Elastic modulus. Tension: hard, 110 GPa (16×10^6 psi); annealed, 125 GPa (18×10^6 psi)

Mass Characteristics

Density. 8.83 g/cm^3 (0.319 lb/in.^{3}) at 20 °C (68 °F)

Thermal Properties

Liquidus temperature. 1060 °C (1940 °F)
Solidus temperature. 1025 °C (1875 °F)
Thermal conductivity. 165 W/m · K (95 Btu/ft · h · °F) at 20 °C (68 °F)

Electrical Properties

Electrical conductivity. Volumetric, 41% IACS at 20 °C (68 °F)
Electrical resistivity. 42 nΩ · m at 20 °C (68 °F)

Fabrication Characteristics

Machinability. 20% of C36000 (free-cutting brass)
Formability. Excellent for cold working; good for hot forming
Weldability. Soldering and brazing: excellent. Gas-shielded arc and resistance spot and butt welding: good. Oxyacetylene and resistance seam welding: fair. Coated metal arc welding is not recommended.
Annealing temperature. 510 to 670 °C (950 to 1240 °F)
Hot-working temperature. 830 to 890 °C (1525 to 1635 °F)

C40800
95Cu-2Sn-3Zn

Specifications

ASTM. Flat products: B 591

Chemical Composition

Composition limits. 94 to 96 Cu, 1.8 to 2.2 Sn, 0.05 Pb max, 0.05 Fe max, bal Zn

Applications

Typical uses. Rolled strip for electrical connectors

Mechanical Properties

Tensile properties. See Table 71.
Shear strength. See Table 71.
Hardness. See Table 71.
Elastic modulus. Tension: hard, 110 GPa (16×10^6 psi); shear, 41 GPa (6×10^6 psi)

Mass Characteristics

Density. 8.86 g/cm^3 (0.320 lb/in.^{3}) at 20 °C (68 °F)

Thermal Properties

Liquidus temperature. 1054 °C (1930 °F)
Solidus temperature. 1038 °C (1900 °F)
Coefficient of linear thermal expansion. 18.2 μm/m · K (10.1 μin./in. · °F) at 20 to 300 °C (68 to 572 °F)
Specific heat. 380 J/kg · K (0.09 Btu/lb · °F) at 20 °C (68 °F)
Thermal conductivity. 160 W/m · K (92 Btu/ft · h · °F) at 20 °C (68 °F)

Electrical Properties

Electrical conductivity. Volumetric, 37% IACS at 20 °C (68 °F)

Table 71 Typical mechanical properties of C40800

Temper	Tensile strength		Yield strength At 0.5% extension under load		Yield strength At 0.2% offset		Elongation in 50 mm (2 in.), %	Hardness		Shear strength	
	MPa	ksi	MPa	ksi	MPa	ksi		HRB	HR30T	MPa	ksi
OS035	290	42	90	13	. . .	. . .	43	60 HRF	22	230	33
OS025	305	44	97	14	. . .	. . .	43	65 HRF	26	235	34
OS015	310	45	105	15	. . .	. . .	42	69 HRF	31	235	34
H01	345	50	270	39	310	45	24	50	54	250	36
H02	370	54	315	46	380	55	12	65	62	260	38
H03	425	62	360	52	415	60	6	72	67	280	41
H04	460	67	395	57	485	70	5	76	70	295	43
H06	505	73	420	61	540	78	4	82	73	310	45
H08	545	79	455	66	565	82	3	85	77	330	48
H10	545 min	79 min	515	75	580	84	3	84 min	75 min	340	49

Note: Values for flat products, 1 mm (0.04 in.) thick

Table 72 Typical mechanical properties of C41100

Temper	Tensile strength MPa	Tensile strength ksi	Yield strength, At 0.5% extension under load MPa	ksi	Yield strength, At 0.2% offset MPa	ksi	Elongation in 50 mm (2 in.), %	Hardness HRB	Hardness HR30T	Shear strength MPa	Shear strength ksi
Flat products, 1 mm (0.04 in.) thick											
OS050	260	38	76	11	62	9	44	58 HRF	. . .	220	32
OS035	270	39	76	11	83	12	43	60 HRF	. . .	230	33
OS025	280	41	83	12	97	14	41	68 HRF	. . .	. . .	. . .
OS015	290	42	83	12	105	15	40	71 HRF	. . .	235	34
H01	330	48	260	38	280	41	23	52	58	. . .	. . .
H02	380	55	325	47	365	53	14	62	60	250	36
H03	415	60	360	52	400	58	6	70	66	. . .	. . .
H04	455	66	380	55	440	64	5	76	69	275	40
H06	495	72	415	60	485	70	4	78	71	. . .	. . .
H08	540	78	485	70	515	75	3	81	72	. . .	. . .
H10	550	80	495	72	525	76	2	83	73	. . .	. . .
Wire, 6 mm (0.25 in.) diameter											
H80 (70%)	560	81	. . .	. . .	. . .	. . .	2(a)	. . .	. . .	. . .	. . .
Wire, 3 mm (0.10 in.) diameter											
H80 (95%)	705	102	. . .	. . .	. . .	. . .	1(b)	. . .	. . .	. . .	. . .
Wire, 1 mm (0.05 in.) diameter											
H80 (98.7%)	730	106	. . .	. . .	. . .	. . .	0.9(b)	. . .	. . .	. . .	. . .

(a) Elongation in 254 mm (10 in.). (b) Elongation in 1500 mm (60 in.)

Electrical resistivity. 46.6 nΩ · m at 20 °C (68 °F)

Fabrication Characteristics

Machinability. 20% of C36000 (free-cutting brass)
Formability. Excellent for cold working; fair for hot forming
Weldability. Soldering and brazing: excellent. Oxyacetylene, gas-shielded arc, and resistance butt welding: good. Resistance spot and seam welding are not recommended.
Annealing temperature. 450 to 675 °C (850 to 1250 °F)
Hot-working temperature. 830 to 890 °C (1525 to 1635 °F)

C41100
91Cu-8.5Zn-0.5Sn

Commercial Names

Previous trade name. Lubaloy

Specifications

ASTM. Flat products: B 508, B 591. Wire: B 105

Chemical Composition

Composition limits. 89 to 93 Cu, 0.3 to 0.7 Sn, 0.1 Pb max, 0.05 Fe max, bal Zn

Applications

Typical uses. Rolled strip; rolled bar, rod, and sheet for bushings, bearing sleeves, thrust washers, terminals, connectors, flexible metal hose, and electrical conductors

Mechanical Properties

Tensile properties. See Table 72.
Hardness. See Table 72.
Elastic modulus. Tension: hard, 115 GPa (16.7 × 10^6 psi); annealed, 125 GPa (18 × 10^6 psi). Shear, 46 GPa (6.7 × 10^6 psi)

Mass Characteristics

Density. 8.80 g/cm^3 (0.318 lb/in.^{3}) at 20 °C (68 °F)

Thermal Properties

Liquidus temperature. 1040 °C (1905 °F)
Solidus temperature. 1020 °C (1870 °F)
Coefficient of linear thermal expansion. 18 μm/m · K (10 μin./in. · °F) at 20 to 100 °C (68 to 212 °F)
Thermal conductivity. 130 W/m · K (75 Btu/ft · h · °F) at 20 °C (68 °F)
Specific heat. 380 J/kg · K (0.09 Btu/lb · °F) at 20 °C (68 °F)

Electrical Properties

Electrical conductivity. Volumetric, 32% IACS at 20 °C (68 °F)
Electrical resistivity. 54 nΩ · m at 20 °C (68 °F)

Fabrication Characteristics

Machinability. 20% of C36000 (free-cutting brass)
Formability. Excellent for cold working; good for hot forming
Weldability. Soldering: excellent. Gas-shielded arc welding and resistance butt

Table 73 Typical mechanical properties of C41500

Temper	Tensile strength MPa	Tensile strength ksi	Yield strength, At 0.5% extension under load MPa	ksi	Yield strength, At 0.2% offset MPa	ksi	Elongation in 50 mm (2 in.), %	Hardness HRB	Hardness HR30T	Fatigue strength MPa	Fatigue strength ksi
OS035	315	46	115	17	125	18	44	64 HRF	24	240	35
OS025	. . .	. . .	. . .	. . .	. . .	. . .	. . .	68 HRF	29	. . .	. . .
OS015	345	50	180	26	185	27	42	74 HRF	36	250	36
H01	345	50	280	41	. . .	. . .	28	62	58	. . .	. . .
H02	385	56	365	53	370	54	16	74	65	280	41
H03	435	63	. . .	. . .	. . .	. . .	. . .	78	68	290	42
H04	485	70	450	65	455	66	5	83	72	305	44
H06	525	76	490	71	515	75	4	86	73	305	44
H08	560	81	505	73	570	83	3	90	75	345	50
H10	560 min	81 min	515	75	605	88	2	89 min	74 min	360	52

Note: Values for flat products, 1 mm (0.04 in.) thick

Table 74 Typical mechanical properties of C41900 strip

Temper	Tensile strength MPa	Tensile strength ksi	Yield strength at 0.2% offset MPa	Yield strength at 0.2% offset ksi	Elongation in 50 mm (2 in.), %	Hardness, HRB
O61	340	49	130	19	42	67 HRF
H01	400	58	315	46	25	64
H02	470	68	395	57	14	73
H03	515	75	450	65	5	78
H04	565	82	510	74	4	87
H06	640	93	530	77	3	92
H08	705	102	550	80	2	95

Table 75 Typical mechanical properties of C42200

Temper	Tensile strength		Yield strength				Elongation in 50 mm (2 in.), %	Hardness	
			At 0.5% extension under load		At 0.2% offset				
	MPa	ksi	MPa	ksi	MPa	ksi		HRB	HR30T
OS035	295	43	105	15	97	14	46	65 HRF	27
OS025	305	44	110	16	105	15	45	70 HRF	31
OS015	315	46	115	17	130	19	44	75 HRF	40
H01	360	52	275	40	270	39	30	56	54
H02	415	60	350	51	395	57	12	70	64
H03	455	66	380	55	440	64	6	77	68
H04	505	73	450	65	485	70	4	81	70
H06	550	80	470	68	525	76	3	84	72
H08	600	87	505	73	560	81	2	87	73
H10	605 min	88 min	515	75	580	84	2	86 min	74 min

Note: Values for flat products, 1 mm (0.04 in.) thick

welding: good. Brazing, oxyacetylene, and resistance spot welding: fair. Coated metal arc and resistance seam welding are not recommended.
Annealing temperature. 500 to 700 °C (930 to 1290 °F)
Hot-working temperature. 830 to 890 °C (1525 to 1635 °F)

C41500
91Cu-7.2Zn-1.8Sn

Specifications

ASTM. B 591

Chemical Composition

Composition limits. 89 to 93 Cu, 1.5 to 2.2 Sn, 0.1 Pb max, 0.05 Fe max, bal Zn

Applications

Typical uses. Rolled strip for spring applications for electrical switches

Mechanical Properties

Tensile properties. See Table 73.
Hardness. See Table 73.
Elastic modulus. Tension: hard, 110 GPa (16 × 10^6 psi); annealed, 125 GPa (18 × 10^6 psi). Shear, 46 GPa (6.7 × 10^6 psi)

Mass Characteristics

Density. 8.80 g/cm^3 (0.318 lb/in.^{3}) at 20 °C (68 °F)

Thermal Properties

Liquidus temperature. 1032 °C (1890 °F)
Solidus temperature. 1010 °C (1850 °F)
Coefficient of linear thermal expansion. 18.6 μm/m · K (10.3 μin./in. · °F) at 20 to 300 °C (68 to 572 °F)
Specific heat. 380 J/kg · K (0.09 Btu/lb · °F) at 20 °C (68 °F)
Thermal conductivity. 123 W/m · K (71 Btu/ft · h · °F) at 20 °C (68 °F)

Electrical Properties

Electrical conductivity. Volumetric, 28% IACS at 20 °C (68 °F)
Electrical resistivity. 62 nΩ · m at 20 °C (68 °F)

Fabrication Characteristics

Machinability. 30% of C36000 (free-cutting brass)
Formability. Excellent for cold working; fair for hot forming
Weldability. Soldering and brazing: excellent. Oxyacetylene, gas-shielded arc, and resistance butt welding: good. Resistance spot welding: fair. Coated metal arc and resistance seam welding are not recommended.
Annealing temperature. 400 to 705 °C (750 to 1300 °F)
Hot-working temperature. 730 to 845 °C (1350 to 1550 °F)

C41900
90.5Cu-4.35Zn-5.15Sn

Commercial Names

Previous trade name. CA419
Common name. Tin brass

Chemical Composition

Composition limits. 89 to 92 Cu, 4.8 to 5.5 Sn, 0.10 Pb max, 0.05 Fe max, bal Zn

Applications

Typical uses. Electrical connectors

Mechanical Properties

Tensile properties. See Table 74.
Hardness. See Table 74.
Elastic modulus. Tension, 125 GPa (18 × 10^6 psi)

Mass Characteristics

Density. 8.80 g/cm^3 (0.318 lb/in.^{3}) at 20 °C (68 °F)

Thermal Properties

Liquidus temperature. 1025 °C (1880 °F)
Solidus temperature. 1000 °C (1830 °F)
Coefficient of linear thermal expansion. 18.7 μm/m · K (10.4 μin./in. · °F) at 20 to 300 °C (68 to 572 °F)
Thermal conductivity. 100 W/m · K (58 Btu/ft · h · °F) at 20 °C (68 °F)

Electrical Properties

Electrical conductivity. Volumetric, 22% IACS at 20 °C (68 °F)
Electrical resistivity. 78 nΩ · m at 20 °C (68 °F)

Fabrication Characteristics

Annealing temperature. 480 to 680 °C (900 to 1250 °F)

C42200
87.5Cu-11.4Zn-1.1Sn

Commercial Names

Previous trade name. Lubronze

Specifications

ASTM. B 591

Chemical Composition

Composition limits. 86.0 to 89.0 Cu, 0.8 to 1.4 Sn, 0.35 P max, 0.05 Pb max, 0.05 Fe max, bal Zn

Applications

Typical uses. Rolled strip, rolled bar and sheet for sash chains, terminals, fuse clips, spring washers, contact springs, and electrical connectors

Mechanical Properties

Tensile properties. See Table 75.
Hardness. See Table 75.
Elastic modulus. Tension: hard, 110 GPa (16 × 10^6 psi); annealed, 125 GPa (18 × 10^6 psi)

Mass Characteristics

Density. 8.80 g/cm^3 (0.318 lb/in.^{3}) at 20 °C (68 °F)

Thermal Properties

Liquidus temperature. 1040 °C (1905 °F)
Solidus temperature. 1020 °C (1870 °F)
Thermal conductivity. 130 W/m · K (75 Btu/ft · h · °F) at 20 °C (68 °F)

Electrical Properties

Electrical conductivity. Volumetric, 31% IACS at 20 °C (68 °F)

Table 76 Typical mechanical properties of C42500

Temper	Tensile strength		Yield strength: At 0.5% extension under load		Yield strength: At 0.2% offset		Elongation in 50 mm (2 in.), %	Hardness	
	MPa	ksi	MPa	ksi	MPa	ksi		HRB	HR30T
OS035	310	45	125	18	105	15	49	70 HRF	32
OS025	315	46	125	18	125	18	48	72 HRF	36
OS015	325	47	135	20	130	19	47	79 HRF	45
H01	370	54	310	45	315	46	35	60	56
H02	435	63	345	50	405	59	20	75	68
H03	470	68	395	57	450	65	15	80	70
H04	525	76	435	63	505	73	9	86	73
H06	565	82	485	70	545	79	7	90	74
H08	615	89	515	75	585	85	4	92	76
H10	635 min	92 min	525	76	615	89	2	92 min	76 min

Note: Values for flat products, 1 mm (0.04 in.) thick

Table 77 Typical mechanical properties of C43000

Temper	Tensile strength		Yield strength(a)		Elongation in 50 mm (2 in.), %	Hardness		
	MPa	ksi	MPa	ksi		HRF	HRB	HR30T
OS035	315	46	125	18	55	69	30	...
OS025	...	...	...	...	...	72	34	...
OS015	...	...	...	...	...	77	39	...
H01	365	53	275	40	44	...	57	57
H02	425	62	380	55	25	...	73	65
H03	495	72	450	65	13	...	79	69
H04	540	78	460	67	10	...	84	73
H06	605	88	485	70	5	...	81	75
H08	650	94	495	72	4	...	91	77
H10	620 min	90 min	505	73	3	...	90 min	75 min

Note: Values for flat products, 1 mm (0.04 in.) thick. (a) At 0.5% extension under load

Table 78 Typical mechanical properties of C43400

Temper	Tensile strength		Yield strength(a)		Elongation in 50 mm (2 in.), %	Hardness		Shear strength	
	MPa	ksi	MPa	ksi		HRB	HR30T	MPa	ksi
OS035	310	45	105	15	49	64 HRF	22	250	36
OS025	315	46	110	16	48	65 HRF	26	255	37
OS015	330	48	115	17	47	70 HRF	30	255	37
H01	360	52	280	41	28	54	55	275	40
H02	405	59	350	51	18	66	63	290	42
H03	470	68	405	59	10	73	68	310	45
H04	510	74	460	67	7	80	71	340	49
H06	580	84	490	71	5	83	74	360	52
H08	620	90	510	74	4	86	76	370	54
H10	605 min	88 min	515	75	3	84 min	74 min	385	56

Note: Values for flat products, 1 mm (0.04 in.) thick. (a) At 0.5% extension under load

Electrical resistivity. 55 nΩ · m at 20 °C (68 °F)

Fabrication Characteristics

Machinability. 30% of C36000 (free-cutting brass)
Formability. Excellent for cold working, good for hot forming
Weldability. Soldering and gas-shielded arc welding: excellent. Resistance spot and butt welding: good. Resistance seam welding and brazing: fair. Oxyacetylene welding is not recommended.
Annealing temperature. 500 to 675 °C (930 to 1250 °F)
Hot-working temperature. 830 to 890 °C (1525 to 1635 °F)

C42500
88.5Cu-9.5Zn-2Sn

Specifications

ASTM. B 591

Chemical Composition

Composition limits. 87 to 90 Cu, 1.5 to 3.0 Sn, 0.35 P max, 0.05 Pb max, 0.05 Fe max, bal Zn

Applications

Typical uses. Rolled strip, rolled bar and sheet for electrical switch springs, terminals, connectors, fuse clips, pen clips, and weather stripping

Mechanical Properties

Tensile properties. See Table 76.
Hardness. See Table 76.
Elastic modulus. Tension: hard, 110 GPa (16×10^6 psi); annealed, 125 GPa (18×10^6 psi)

Mass Characteristics

Density. 8.78 g/cm³ (0.317 lb/in.³) at 20 °C (68 °F)

Thermal Properties

Liquidus temperature. 1030 °C (1890 °F)
Solidus temperature. 1010 °C (1850 °F)
Coefficient of linear thermal expansion. 18.4 μm/m · K (10.2 μin./in. · °F) at 20 to 100 °C (68 to 212 °F)
Specific heat. 380 J/kg · K (0.09 Btu/lb · °F) at 20 °C (68 °F)
Thermal conductivity. 120 W/m · K (69 Btu/ft · h · °F) at 20 °C (68 °F)

Electrical Properties

Electrical conductivity. Volumetric, 28% IACS at 20 °C (68 °F)
Electrical resistivity. 62 nΩ · m at 20 °C (68 °F)

Fabrication Characteristics

Machinability. 30% of C36000 (free-cutting brass)
Formability. Excellent for cold working; fair for hot forming
Weldability. Soldering and brazing: excellent. Oxyacetylene, gas-shielded arc and resistance butt welding: good. Coated metal arc and resistance spot and seam welding are not recommended.
Annealing temperature. 425 to 700 °C (800 to 1300 °F)
Hot-working temperature. 790 to 840 °C (1455 to 1545 °F)

C43000
87Cu-10.8Zn-2.2Sn

Specifications

ASTM. Flat products: B 591

Chemical Composition

Composition limits. 84 to 87 Cu, 1.7 to 2.7 Sn, 0.10 Pb max, 0.05 Fe max, bal Zn

Applications

Typical uses. Rolled strip and sheet for electrical switches, springs, fuse and pen clips, and weather stripping

Mechanical Properties

Tensile properties. See Table 77.
Hardness. See Table 77.
Elastic modulus. Tension, 110 GPa (16×10^6 psi); shear, 119 GPa (17.3×10^6 psi)

Mass Characteristics

Density. 8.75 g/cm³ (0.316 lb/in.³) at 20 °C (68 °F)

Thermal Properties

Liquidus temperature. 1025 °C (1877 °F)
Solidus temperature. 1000 °C (1832 °F)

Table 79 Typical mechanical properties of C43500

Temper	Tensile strength MPa	Tensile strength ksi	Yield strength(a) MPa	Yield strength(a) ksi	Elongation in 50 mm (2 in.), %	Hardness HRF	Hardness HR30T	Shear strength MPa	Shear strength ksi
Flat products, 1 mm (0.04 in.) thick									
OS025	340	49	125	18	46	70	31	250	36
H02	450	65	370	54	16	72 HRB	. . .	286	41.5
H04	550	80	470	68	7	85 HRB	. . .	310	45
Tubing, 25 mm (1.0 in.) outside diameter × 1.65 mm (0.065 in.) wall thickness									
OS035	315	46	110	16	46	69	40	. . .	. . .
H80 (35%)	515	75	415	60	10	. . .	. . .	. . .	. . .

(a) At 0.5% extension under load

Coefficient of linear thermal expansion. 18.4 µm/m · K (10.2 µin./in. · °F) at 20 to 100 °C (68 to 212 °F)
Specific heat. 380 J/kg · K (0.09 Btu/lb · °F) at 20 °C (68 °F)
Thermal conductivity. 119 W/m · K (69 Btu/ft · h · °F)

Electrical Properties

Electrical conductivity. Volumetric, 27% IACS at 20 °C (68 °F)
Electrical resistivity. 64 nΩ · m at 20 °C (68 °F)

Fabrication Characteristics

Machinability. 30% of C36000 (free-cutting brass)
Formability. Excellent for cold working; good for hot forming
Weldability. Soldering and brazing: excellent. Oxyacetylene, gas-shielded arc, and resistance butt welding: good. Resistance spot welding: fair. Coated metal arc and resistance seam welding are not recommended.
Annealing temperature. 425 to 700 °C (800 to 1300 °F)
Hot-working temperature. 790 to 840 °C (1455 to 1545 °F)

C43400
85Cu-14.3Zn-0.7Sn

Specifications

ASTM. Flat products: B 591

Chemical Composition

Composition limits. 84 to 87 Cu, 0.4 to 1.0 Sn, 0.05 Pb max, 0.05 Fe max, bal Zn

Applications

Typical uses. Rolled strip for electrical uses: switch parts, blades, relay springs, contacts

Mechanical Properties

Tensile properties. See Table 78.
Shear strength. See Table 78.
Hardness. See Table 78.
Elastic modulus. Tension: hard, 110 GPa (16×10^6 psi); annealed, 40 GPa (6×10^6 psi)

Mass Characteristics

Density. 8.75 g/cm^3 (0.316 lb/in.^{3}) at 20 °C (68 °F)

Thermal Properties

Liquidus temperature. 1020 °C (1870 °F)
Solidus temperature. 990 °C (1810 °F)
Coefficient of linear thermal expansion. 18.9 µm/m · K (10.5 µin./in. · °F) at 20 to 300 °C (68 to 572 °F)
Specific heat. 380 J/kg · K (0.09 Btu/lb · h · °F) at 20 °C (68 °F)
Thermal conductivity. 137 W/m · K (79 Btu/ft · h · °F) at 20 °C (68 °F)

Electrical Properties

Electrical conductivity. Volumetric, 31% IACS at 20 °C (68 °F)
Electrical resistivity. 56 nΩ · m at 20 °C (68 °F)

Fabrication Characteristics

Machinability. 30% of C36000 (free-cutting brass)
Formability. Excellent for cold working; fair for hot forming
Weldability. Soldering and brazing: excellent. Oxyacetylene, gas-shielded arc, and resistance spot and butt welding: good. Resistance seam welding is not recommended.
Annealing temperature. 425 to 675 °C (800 to 1250 °F)
Hot-working temperature. 815 to 870 °C (1500 to 1600 °F)

C43500
81Cu-18.1Zn-0.9Sn

Chemical Composition

Composition limits. 79 to 83 Cu, 0.6 to 1.2 Sn, 0.1 Pb max, 0.05 Fe max, 0.15 max other (total), bal Zn

Applications

Typical uses. Rolled strip and tubing for Bourdon tubing and musical instruments

Mechanical Properties

Tensile properties. See Table 79.
Shear strength. See Table 79.
Elastic modulus. Tension, 110 GPa (16×10^6 psi); shear, 40 GPa (6×10^6 psi)

Mass Characteristics

Density. 8.66 g/cm^3 (0.313 lb/in.^{3}) at 20 °C (68 °F)

Thermal Properties

Liquidus temperature. 1005 °C (1840 °F)
Solidus temperature. 965 °C (1770 °F)
Coefficient of linear thermal expansion. 19.4 µm/m · K (10.8 µin./in. · °F) at 20 °C (68 °F)
Specific heat. 380 J/kg · K (0.09 Btu/lb · °F) at 20 °C (68 °F)

Electrical Properties

Electrical conductivity. Volumetric, 28% IACS at 20 °C (68 °F)
Electrical resistivity. 62 nΩ · m at 20 °C (68 °F)

Fabrication Characteristics

Machinability. 30% of C36000 (free-cutting brass)
Formability. Excellent for cold working; good for hot forming

Table 80 Typical mechanical properties of C44300, C44400, and C44500

Temper	Tensile strength MPa	Tensile strength ksi	Yield strength(a) MPa	Yield strength(a) ksi	Elongation in 50 mm (2 in.), %	Hardness HRF	Hardness HR15T	Hardness HR30T
Tubing, 25 mm (1 in.) outside diameter × 1.65 mm (0.065 in.) wall thickness								
OS025	365	53	152	22	65	75	. . .	37
H01	434	63	. . .	. . .	45	. . .	86.5	. . .
H02	503	73	. . .	. . .	29	. . .	90	78
H03	565	82	. . .	. . .	15	. . .	90	81
H04	669	97	. . .	. . .	4	. . .	93	84
Plate, 25 mm (1 in.) diameter								
M20	330	48	124	18	65	70	. . .	. . .
Strip, 1 mm (0.04 in.) diameter								
O60 (0.080 mm)	310	45	90	13	69	59	9	20
O60 (0.015 mm)	330	48	97	14	62	60	9	20
H04	607	88	496	72	4	109	90	76

(a) At 0.5% extension under load. Apparent elastic limit (tubing), 125 MPa (18 ksi)

Table 81 Typical Charpy impact strength data for C44300, C44400, or C44500

Test temperature °C	°F	Impact strength J	ft · lbf
20	68	82.4	60.8
3	38	82.2	60.6
−18	0	79.7	58.8
−30	−25	82.4	60.8
−50	−60	79.9	58.9
−80	−110	83.4	61.5
−115	−175	80.3	59.2

Note: Annealed specimens, cut from 19 mm (0.75 in.) diam rod into keyhole-notch bars. Values are averages of data from three tests (specimens did not fracture). Tensile strength at 20 °C (68 °F), 320 MPa (46.5 ksi); yield strength, 92 MPa (13.3 ksi); elongation, 83.5%; hardness, 64 HRF

Weldability. Soldering and brazing: excellent. Oxyacetylene and resistance spot and butt welding: good. Gas-shielded metal arc welding: fair. Coated metal arc and resistance seam welding are not recommended.

C44300, C44400, C44500 71Cu-28Zn-1Sn

Commercial Names

Previous trade names. C44300, arsenical admiralty metal; C44400, antimonial admiralty metal; C44500, phosphorized admiralty metal

Table 82 Typical creep data for C44300, C44400, or C44500

Temperature		Stress required to produce designated creep in 1000 h: Nil(a)		0.01%		0.10%		1.00%	
°C	°F	MPa	ksi	MPa	ksi	MPa	ksi	MPa	ksi
205	400	69	10	90	13	117	17	130	19
315	600	(b)	(b)	6.9	1.0	13.4	1.95	26	3.8
425	800	(b)	(b)	0.37	0.054	1.1	0.16	3.4	0.5

Note: Values for rod, hot rolled to 22.2 mm (0.875 in.), then cold drawn to 19.0 mm (0.750 in.). (a) No measurable flow. (b) Nearly zero

Common names. Inhibited admiralty metal; admiralty brass

Specifications

ASME. Condenser plate: SB171. Tubing: SB111, SB359, SB395, SB543

ASTM. Condenser plate: B 171. Tubing: B 111, B 359, B 395, B 543

Chemical Composition

Composition limits. 70.0 to 73.0 Cu, 0.07 Pb max, 0.06 Fe max, 0.9 to 1.2 Sn (or 0.8 to 1.2 Sn for flat-rolled products); As, Sb, or P (see below); bal Zn

Arsenic, antimony, or phosphorus limits. C44300, 0.02 to 0.10 As; C44400, 0.02 to 0.10 Sb, C44500, 0.02 to 0.10 P

Applications

Typical uses. Condenser, distiller, and heat exchanger tubes, ferrules, strainers, condenser tube plates

Precautions in use. These three alloys are susceptible to stress-corrosion cracking. Whenever possible, they should be used in the annealed condition. Where fabrication results in residual stresses, a suitable stress-relieving heat treatment should be applied.

Mechanical Properties

Tensile properties. See Table 80 and Fig. 41.

Hardness. See Table 80.

Elastic modulus. Tension, 110 GPa (16 × 10^6 psi); shear, 40 GPa (6 × 10^6 psi)

Impact strength. See Table 81.

Fatigue strength. 115 to 125 MPa (17 to 18 ksi) at 10^7 cycles

Creep-rupture characteristics. See Table 82.

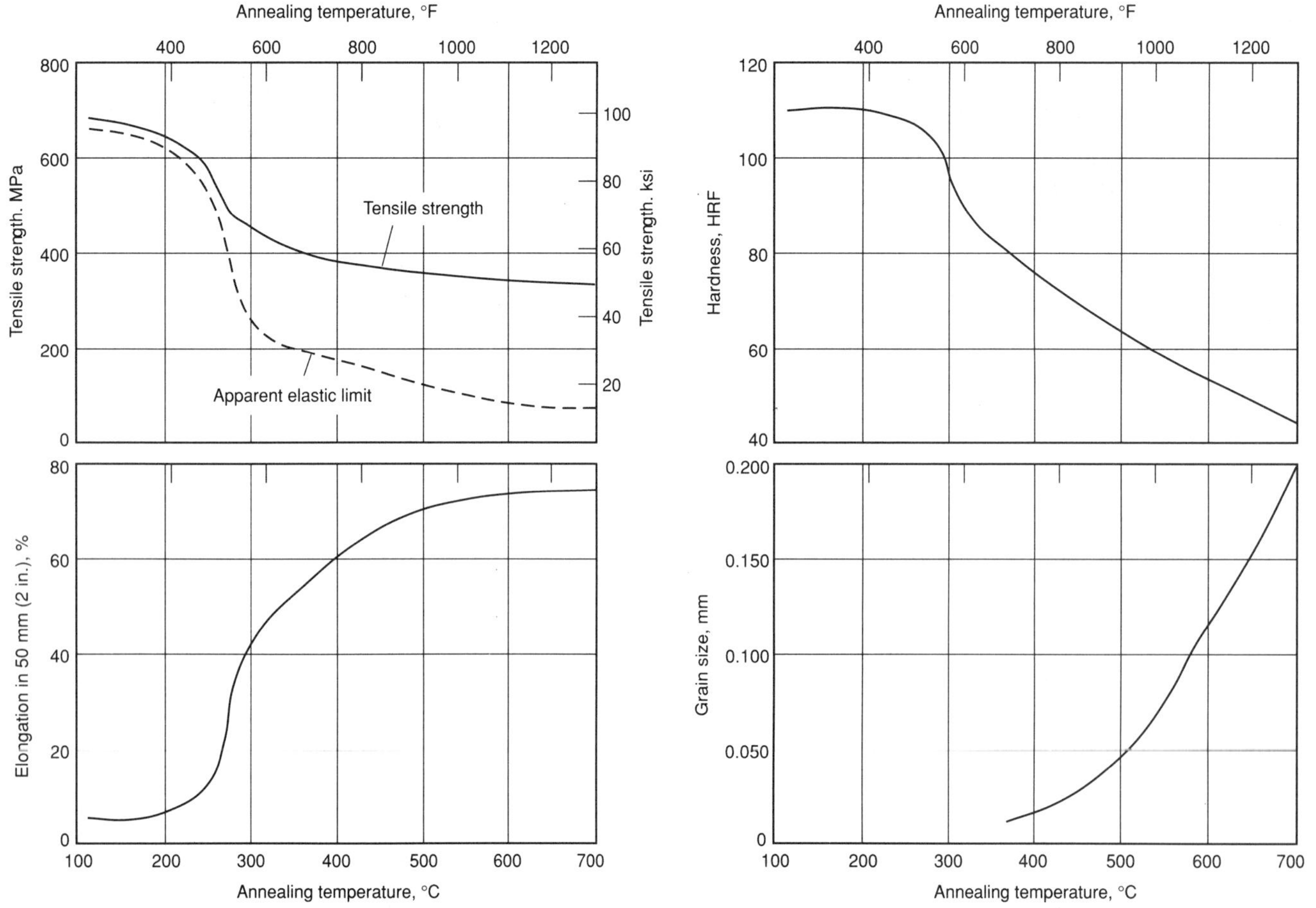

Fig. 41 Variation of properties and grain size with annealing temperature for C44300, C44400, or C44500. Data for inhibited admiralty metal tubing (71Cu-28Zn-1Sn), cold drawn 50% and annealed 1 h at temperature

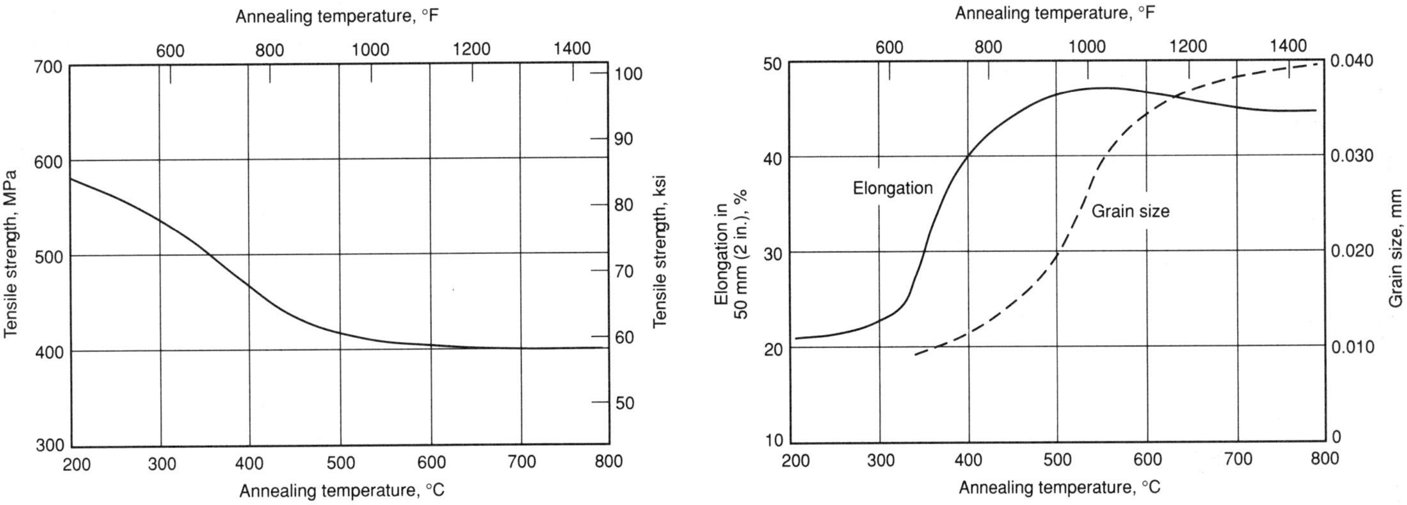

Fig. 42 Variation of strength, ductility, and grain size with annealing temperature for C46400, C46500, C46600, or C46700. Data for 19 mm (0.75 in.) diam naval brass rod (60Cu-39.2Zn-0.8Sn), cold drawn 30% and annealed 1 h at temperature. Grain size before cold drawing, 0.025 mm

Mass Characteristics

Density. 8.53 g/cm^3 (0.308 lb/in.^{3}) at 20 °C (68 °F)

Thermal Properties

Liquidus temperature. 935 °C (1720 °F)
Solidus temperature. 900 °C (1650 °F)
Coefficient of linear thermal expansion. 20.2 μm/m · K (11.2 μin./in. · °F) at 20 to 300 °C (68 to 572 °F)
Specific heat. 380 J/kg · K (0.09 Btu/lb · °F) at 20 °C (68 °F)
Thermal conductivity. 110 W/m · K (64 Btu/ft · h · °F) at 20 °C (68 °F)

Electrical Properties

Electrical conductivity. Volumetric, 25% IACS at 20 °C (68 °F)
Electrical resistivity. 69 nΩ · m at 20 °C (68 °F)

Chemical Properties

General corrosion behavior. Good resistance to salt and fresh waters at low velocities. Water velocities above 1.8 m/s (6 ft/s) give rise to impingement attack. A different inhibitor is added to each alloy to protect against dezincification.

Fabrication Characteristics

Machinability. 30% of C36000 (free-cutting brass)
Formability. Excellent for cold working; fair for hot forming
Weldability. Soft soldering: excellent. Silver alloy brazing, oxyfuel gas welding, resistance spot welding, and flash welding: good. Gas-shielded arc welding: fair. Shielded metal arc welding and resistance seam welding are not recommended.
Recrystallization temperature. 300 °C (575 °F) for 1 mm (0.04 in.) strip cold rolled hard (50% reduction) from a grain size of 0.015 mm. See also Fig. 41.
Annealing temperature. 425 to 600 °C (800 to 1100 °F)
Hot-working temperature. 650 to 800 °C (1200 to 1450 °F)

C46400, C46500, C46600, C46700
60Cu-39.2Zn-0.8Sn

Commercial Names

Previous trade names. C46400, uninhibited naval brass; C46500, arsenical naval brass; C46600, antimonial naval brass; C46700, phosphorized naval brass
Common names. Naval brass; inhibited naval brass

Specifications

AMS. Bar and rod (C46400 only): 4611, 4612
ASME. Condenser plate: SB171

Table 83 Typical mechanical properties of C46400, C46500, C46600, or C46700

Temper	Tensile strength MPa	Tensile strength ksi	Yield strength(a) MPa	Yield strength(a) ksi	Elongation in 50 mm (2 in.), %	Reduction in area, %	Hardness, HRB	Shear strength MPa	Shear strength ksi
Flat products, 1 mm (0.04 in.) thick									
O50	427	62	207	30	40	...	60	283	41
H01	483	70	400	58	17	...	75	296	43
Flat products, 6 mm (0.25 in.) thick									
O60	400	58	172	25	49	...	56	275	40
O50	414	60	193	28	45	...	58	283	41
Flat products, 25 mm (1.0 in.) thick									
M20	379	55	172	25	50	...	55	275	40
Rod and bar, 6 mm (0.25 in.) diameter									
O60	400	58	186	27	45	60	56	275	40
O50	434	63	207	30	40	55	60	290	42
H01 (10%)	482	70	331	48	25	50	80	296	43
H02 (20%)	552	80	393	57	20	45	85	310	45
Rod and bar, 25 mm (1.0 in.) diameter									
O60	393	57	172	25	47	60	55	275	40
O50	434	63	207	30	40	55	60	290	42
H01 (8%)	476	69	317	46	27	50	78	296	43
H02 (20%)	517	75	365	53	20	45	82	303	44
Rod and bar, 51 mm (2.0 in.) diameter									
O60	386	56	172	25	47	60	55	275	40
O50	427	62	193	28	43	55	60	290	42
H01 (8%)	462	67	276	40	35	50	75	296	43
Tubing, 9 mm (0.375 in.) outside diameter × 2.5 mm (0.097 in.) wall thickness									
H80 (35%)	607	88	455	66	18	40	95	...	...
O61	427	62	207	30	45	...	25	...	...
Extruded shapes									
M30	400	58	170	25	40	...	...	275	40

(a) At 0.5% extension under load

Table 84 Typical mechanical properties of C48200

Temper	Tensile strength MPa	Tensile strength ksi	Yield strength(a) MPa	Yield strength(a) ksi	Elongation in 50 mm (2 in.), %	Hardness, HRB	Shear strength MPa	Shear strength ksi
Rod, 25 mm (1.0 in.) diameter								
O60	395	57	170	25	40	55	260	38
O50	435	63	205	30	35	60	270	39
H01 (8%)	475	69	315	46	20	78	275	40
H02 (20%)	515	75	365	53	15	82	285	41
Rod, 51 mm (2.0 in.) diameter								
O60	385	56	170	25	40	55	260	38
O50	425	62	195	28	37	60	270	39
H01 (8%)	460	67	275	40	30	75	275	40
H02 (15%)	485	70	360	52	17	78	285	41
Rod, 76 mm (3.0 in.) diameter								
H01 (4%)	435	63	230	33	43	78	275	40
Bar, 10 mm (0.38 in.) diameter								
M30	435	63	230	33	34	60	270	39
Bar, 38 mm (1.5 in.) diameter								
H01	455	66	275	40	32	75	275	40

(a) At 0.5% extension under load

ASTM. Bar, rod, and shapes (C46400 only): B 21, B 124, Forgings (C46400 only): B 283. Condenser plate: B 171
SAE. Bar, rod, and shapes (C46400 only): J461, J463
Government. QQ-B-626. Bar, rod, shapes, forgings, and wire (C46400 only): QQ-B-637. Bar and flat products (C46400 only): QQ-B-639. Tubing (C46400 only): MIL-T-6945

Chemical Composition

Composition limits. 59.0 to 62.0 Cu; 0.50 to 1.0 Sn; 0.20 Pb max; 0.10 Fe max; As, Sb, or P (see below); bal Zn
Arsenic, antimony, or phosphorus limits. C46400, none specified; C46500, 0.02 to 0.10 As; C46600, 0.02 to 0.10 Sb; C46700, 0.2 to 0.10 P

Applications

Typical uses. Condenser plates, welding rod, marine hardware, propeller shafts, valve stems, airplane turnbuckle barrels, balls, nuts, bolts, rivets, fittings

Mechanical Properties

Tensile properties. See Table 83 and Fig. 42.
Shear strength. See Table 83.
Hardness. See Table 83.
Elastic modulus. Tension, 100 GPa (15 × 10^6 psi); shear, 39 GPa (5.6 × 10^6 psi)
Impact strength. 43 J (32 ft · lbf) at 21 °C (70 °F) for Charpy keyhole specimens 10 mm (0.4 in.) square machined from annealed plate 13 mm (0.5 in.) thick; plate hardness, 96 HRF
Fatigue strength. 100 MPa (15 ksi) at 3 × 10^8 cycles

Structure

Crystal structure. Face-centered-cubic α and body-centered-cubic β
Microstructure. Generally two phases: α and β

Mass Characteristics

Density. 8.41 g/cm^3 (0.304 lb/in.^{3}) at 20 °C (68 °F)

Thermal Properties

Liquidus temperature. 900 °C (1650 °F)
Solidus temperature. 885 °C (1630 °F)
Coefficient of linear thermal expansion. 21.2 μm/m · K (11.8 μin./in. · °F) at 20 to 300 °C (68 to 572 °F)
Specific heat. 380 J/kg · K (0.09 Btu/lb · °F) at 20 °C (68 °F)
Thermal conductivity. 116 W/m · K (67 Btu/ft · h · °F) at 20 °C (68 °F)

Electrical Properties

Electrical conductivity. Volumetric, 26% IACS at 20 °C (68 °F), annealed
Electrical resistivity. 66.3 nΩ · m at 20 °C (68 °F), annealed

Chemical Properties

General corrosion behavior. Good resistance to corrosion in both fresh and salt water; different inhibitor elements are added to C46500, C46600, and C46700 to protect against dezincification.

Fabrication Characteristics

Machinability. 30% of C36000 (free-cutting brass)
Forgeability. 90% of C37700 (forging brass)
Formability. Excellent for hot forming; fair for cold forming
Weldability. Soft soldering and silver alloy brazing: excellent. Oxyfuel gas welding, resistance spot welding, and flash welding: good. Gas-shielded arc welding and resistance seam welding: fair. Shielded arc welding is not recommended.
Recrystallization temperature. About 350 °C (660 °F) for 19 mm (0.75 in.) diam rod cold drawn 30%. See also Fig. 42.
Annealing temperature. 425 to 600 °C (800 to 1100 °F)
Maximum cold reduction between anneals. 30%
Hot-working temperature. 650 to 825 °C (1200 to 1500 °F)

C48200
60.5Cu-38Zn-0.8Sn-0.7Pb

Commercial Names

Previous trade names. Naval brass, medium leaded; CA482
Common name. Leaded naval brass

Specifications

ASTM. Rod, bar, and shapes: B 21 (CA482), B 124 (C48200)
Government. QQ-B-626. Bar, rod, shapes, forgings, and wire: QQ-B-637. Bar and plate: QQ-B-639

Chemical Composition

Composition limits. 59.0 to 62.0 Cu, 0.40 to 1.0 Pb; 0.10 Fe max, 0.50 to 1.0 Sn, bal Zn

Applications

Typical uses. Marine hardware, screw machine products, valve stems

Mechanical Properties

Tensile properties. See Table 84.
Shear strength. See Table 84.
Hardness. See Table 84.
Elastic modulus. Tension, 100 GPa (15 × 10^6 psi); shear, 39 GPa (5.6 × 10^6 psi)

Structure

Microstructure. Generally three phases: α, β, and lead

Mass Characteristics

Density. 8.44 g/cm^3 (0.305 lb/in.^{3}) at 20 °C (68 °F)

Thermal Properties

Liquidus temperature. 900 °C (1650 °F)
Solidus temperature. 885 °C (1625 °F)
Coefficient of linear thermal expansion. 21.2 μm/m · K (11.8 μin./in. · °F) at 20 to 300 °C (68 to 572 °F)
Specific heat. 380 J/kg · K (0.09 Btu/lb · °F) at 20 °C (68 °F)
Thermal conductivity. 116 W/m · K (67 Btu/ft · h · °F)

Electrical Properties

Electrical conductivity. Volumetric, 26% IACS at 20 °C (68 °F)
Electrical resistivity. 66.3 nΩ · m at 20 °C (68 °F), annealed

Chemical Properties

General corrosion behavior. Good resistance to seawater and marine atmospheres

Table 85 Typical mechanical properties of C48500

Temper	Tensile strength MPa	Tensile strength ksi	Yield strength(a) MPa	Yield strength(a) ksi	Elongation in 50 mm (2 in.), %	Hardness, HRB	Shear strength MPa	Shear strength ksi
O60	393	57	172	25	40	55	248	36
H01 (8%)	476	69	317	46	20	78	269	39
H02 (20%)	517	75	365	53	15	82	276	40

Note: Values for rod, 25 mm (1.0 in.) diameter. (a) At 0.5% extension under load.

Fabrication Characteristics

Machinability. 50% of C36000 (free-cutting brass)
Forgeability. 90% of C37700 (forging brass)
Formability. Good for hot working; poor for cold working
Weldability. Soft soldering: excellent. Silver alloy brazing: good. Flash welding: fair. Oxyfuel gas welding, arc welding, and most resistance welding processes are not recommended.
Recrystallization temperature. About 360 °C (680 °F) for 19 mm (0.75 in.) diam rod cold drawn 30%
Annealing temperature. 425 to 600 °C (800 to 1100 °F)
Hot-working temperature. 650 to 760 °C (1200 to 1400 °F)

C48500
60Cu-37.5Zn-1.8Pb-0.7Sn

Commercial Names

Previous trade names. High-leaded naval brass; CA485
Common name. Leaded naval brass

Specifications

ASTM. Rod, bar, and shapes: B 21 (CA482), B 124 (C48500). Forgings: B 283 (CA485)
Government. QQ-B-626. Bar, rod, shapes, forgings, and wire: QQ-B-637. Bar and plate products: QQ-B-639

Chemical Composition

Composition limits. 59.0 to 62.0 Cu, 1.3 to 2.2 Pb; 0.10 Fe max, 0.50 to 1.0 Sn, bal Zn

Applications

Typical uses. Marine hardware, screw-machine products, valve stems

Mechanical Properties

Tensile properties. See Table 85.
Shear strength. See Table 85.
Hardness. See Table 85.
Elastic modulus. Tension, 100 GPa (15 × 10^6 psi); shear, 39 GPa (5.6 × 10^6 psi)

Structure

Microstructure. Generally three phases: α, β, and lead

Mass Characteristics

Density. 8.44 g/cm^3 (0.305 lb/in.^{3}) at 20 °C (68 °F)

Thermal Properties

Liquidus temperature. 900 °C (1650 °F)
Solidus temperature. 885 °C (1625 °F)
Coefficient of linear thermal expansion. 21.2 μm/m · K (11.8 μin./in. · °F) at 20 to 300 °C (68 to 572 °F)
Specific heat. 380 J/kg · K (0.09 Btu/lb · °F) at 20 °C (68 °F)
Thermal conductivity. 116 W/m · K (67 Btu/ft · h · °F)

Electrical Properties

Electrical conductivity. Volumetric, 26% IACS at 20 °C (68 °F)
Electrical resistivity. 66.3 nΩ · m at 20 °C (68 °F)

Chemical Properties

General corrosion behavior. Good resistance to seawater and marine atmospheres

Fabrication Characteristics

Machinability. 70% of C36000 (free-cutting brass)
Forgeability. 90% of C37700 (forging brass)
Formability. Good for hot working; poor for cold working
Weldability. Soft soldering: excellent. Silver alloy brazing: good. Flash welding: fair. Oxyfuel gas welding, arc welding, and most resistance welding processes are not recommended.
Recrystallization temperature. About 360 °C (680 °F) for 19 mm (0.75 in.) diam rod cold drawn 30%. See also Fig. 43.
Annealing temperature. 425 to 600 °C (800 to 1100 °F)
Maximum cold reduction between anneals. 20%
Hot-working temperature. 650 to 760 °C (1200 to 1400 °F)

C50500
98.7Cu-1.3Sn

Commercial Names

Previous names. Phosphor bronze, 1.25% E; CA505
Common name. Phosphor bronze (1.25% Sn)

Specifications

ASTM. Strip: B 105. Wire: B 105

Chemical Composition

Composition limits. 1.0 to 1.7 Sn; 0.05 Pb

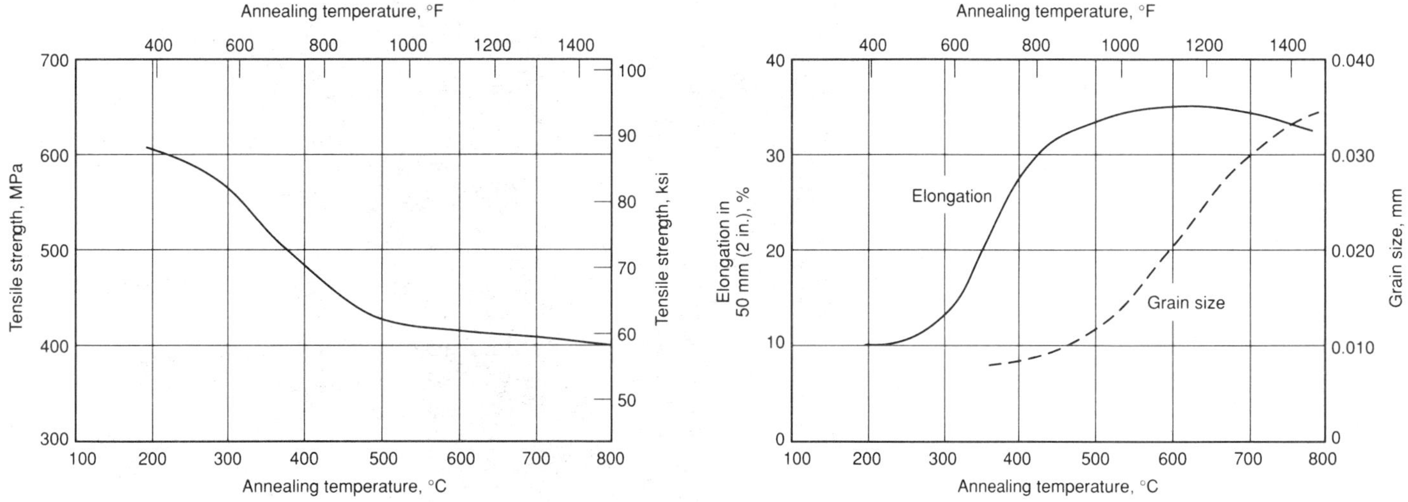

Fig. 43 Variation of strength, ductility, and grain size with annealing temperature for C48500. Data for 19 mm (0.75 in.) diam high-leaded naval brass (60Cu-37.5Zn-1.8Pb-0.7Sn) rod that was cold drawn 30% and annealed 1 h at temperature. Grain size before cold drawing, 0.025 mm

Table 86 Typical mechanical properties of C50500

Temper	Grain size, mm	Tensile strength MPa	Tensile strength ksi	Yield strengh(a) MPa	Yield strengh(a) ksi	Elongation in 50 mm (2 in.), %	Hardness, HRB	Fatigue strengh(b) MPa	Fatigue strengh(b) ksi
OS035	0.035	276	40	76	11	47.0	...	114	16.5
OS075	0.015	290	42	90	13	47.0	...	121	17.5
H02	0.035	365	53	352	51	12.0	59.0	162	23.5
	0.015	372	54	359	52	13.0	60.0	172	25
H04	0.035	421	61	414	60	5.0	67.0	179	26
	0.015	441	64	434	63	5.0	69.0	190	27.5
H06	0.035	462	67	455	66	3.0	73.0	172	25
	0.015	483	70	476	69	3.0	75.0	193	28
H08	0.035	483	70	476	69	3.0	76.0	197	28.5
	0.015	510	74	503	73	3.0	78.0	203	29
H10	0.035	510	74	503	73	3.0	79.0	197	28.5
	0.015	524	76	517	75	3.0	80.0	210	30.5

Note: Values for flat products 1 mm (0.040 in.) thick. Data in this table were interpolated from ASTM STP 1. (a) At 0.2% offset. (b) At 10^8 cycles of fully reversed stress

Table 87 Nominal mechanical properties of C50710 strip

Temper	Tensile strength MPa	Tensile strength ksi	Elongation in 50 mm (2 in.), %	Hardness, HV
H02	455	66	25	150
H04	540	78	11	168
H06	585	85	9	185

max; 0.10 Fe max; 0.30 Zn max; 0.35 P max; bal Cu; 99.5 Cu + Sn + P min

Applications

Typical uses. Electrical contacts, flexible hose, pole line hardware

Mechanical Properties

Tensile properties. See Table 86.
Hardness. See Table 86.
Elastic modulus. Tension, 117 GPa (17 × 10^6 psi); shear, 44 GPa (6.4 × 10^6 psi)
Fatigue strength. See Table 86.

Mass Characteristics

Density. 8.89 g/cm³ (0.321 lb/in.³) at 20 °C (68 °F)

Thermal Properties

Liquidus temperature. 1075 °C (1970 °F)
Solidus temperature. 1035 °C (1900 °F)
Coefficient of linear thermal expansion. 17.8 μm/m · K (9.9 μin./in. · °F) at 20 to 300 °C (68 to 572 °F)
Specific heat. 380 J/kg · K (0.09 Btu/lb · °F) at 20 °C (68 °F)
Thermal conductivity. 208 W/m · K (120 Btu/ft · h · °F) at 20 °C (68 °F)

Electrical Properties

Electrical conductivity. Volumetric, 48% IACS at 20 °C (68 °F)
Electrical resistivity. 36 nΩ · m at 20 °C (68 °F)

Fabrication Characteristics

Machinability. 20% of C36000 (free-cutting brass)
Formability. Cold: excellent. Hot: good. Commonly fabricated by blanking, forming, bending, heading, upsetting, shearing, squeezing, and swaging
Weldability. Flash welding, soldering, and brazing: excellent. Gas metal arc welding: good. Oxyfuel gas welding and shielded metal arc welding: fair. Other processes are not recommended.
Annealing temperature. 475 to 650 °C (900 to 1200 °F)
Hot-working temperature. 800 to 875 °C (1450 to 1600 °F)

C50710
97.7Cu-2.0Sn-0.3Ni

Chemical Composition

Composition limits. 1.7 to 2.3 Sn, 0.1 to 0.4 Ni, 0.35 P max

Applications

Typical uses. Lead frames

Mechanical Properties

Tensile properties. See Table 87.
Hardness. See Table 87.
Elastic modulus. 113 GPa (16.4 × 10^6 psi)

Mass Characteristics

Density. 8.88 g/cm³ (0.321 lb/in.³) at 20 °C (68 °F)

Thermal Properties

Liquidus temperature. 1065 °C (1950 °F)
Solidus temperature. 995 °C (1820 °F)
Coefficient of linear thermal expansion. 17.0 μm/m · K (9.4 μin./in. · °F) from 20 to 550 °C (68 to 1025 °F)
Thermal conductivity. 154 W/m · K (89 Btu/ft · h · °F) at 20 °C (68 °F)

Electrical Properties

Electrical conductivity. Volumetric, 30% IACS at 20 °C (68 °F)
Electrical resistivity. 57.4 nΩ · m at 20 °C (68 °F)

C51000
94.8Cu-5Sn-0.2P

Commercial Names

Previous trade names. Phosphor bronze, 5% A

Specifications

AMS. Flat products: 4510. Bar, rod, tubing: 4625. Wire: 4720
ASTM. Flat products: B 100, B 103. Bar: B 103, B 139. Rod, shapes: B 139. Wire: B 159
SAE. J463
Government. Flat products, bar, shapes: QQ-B-750. Rod: QQ-B-750, MIL-B-13501.

Table 88 Typical mechanical properties of C51000

Temper	Tensile strength MPa	Tensile strength ksi	Yield strength(a) MPa	Yield strength(a) ksi	Elongation in 50 mm (2 in.), %	Hardness, HRB
Flat products, 1 mm (0.04 in.) thick						
OS050	325	47	130	19	64	26
OS035	340	49	140	20	58	28
OS025	345	50	145	21	52	30
OS015	365	53	150	22	50	34
H02	470	68	380	55	28	78
H04	560	81	515	75	10	87
H06	635	92	550	80	6	93
H08	690	100	...	...	4	95
H10	740	107	...	...	3	97
Rod, 13 mm (0.05 in.) diameter						
H02	515	75	450	65	25	80
Rod, 25 mm (1 in.) diameter						
H02	480	70	400	58	25	78
Wire, 2 mm (0.08 in.) diameter						
OS035	345	50	140	20	58	...
H01	470	68	415	60	24	...
H02	585	85	550	80	8	...
H04	760	110	...	...	5	...
H06	895	130	...	...	3	...
H08	965	140	...	...	2	...

(a) At 0.5% extension under load

Table 89 Typical mechanical properties of 1 mm (0.04 in.) thick C51100 strip

Temper	Tensile strength, MPa	Tensile strength, ksi	Yield strength at 0.2% offset, MPa	Yield strength at 0.2% offset, ksi	Elongation in 50 mm (2 in.), %	Hardness, HRB	Hardness, HR30T
OS050	315	46	110	16	48	70 HRF	...
OS035	330	48	130	19	47	73 HRF	...
OS025	345	50	145	21	46	75 HRF	...
OS015	350	51	160	23	46	76 HRF	...
H01	380	55	295	43	36	48	45
H02	425	62	385	56	19	70	65
H03	510	74	495	72	11	84	72
H04	550	80	530	77	7	86	74
H06	635	92	615	89	4	91	78
H08	675	98	655	95	3	93	79
H10	710	103	675	98	2	95	80

Table 90 Typical mechanical properties of C52100

Temper	Tensile strength, MPa	Tensile strength, ksi	Yield strength(a), MPa	Yield strength(a), ksi	Elongation in 50 mm (2 in.), %	Hardness, HRF	Hardness, HRB	Hardness, HR30T
Flat products, 1 mm (0.04 in.) thick								
OS050	380	55	...	...	70	75	...	...
OS035	400	58	...	...	65	80	...	...
OS025	415	60	165	24	63	82	50	...
OS015	425	62	...	...	60	85	...	...
H02	525	76	380	55	32	...	84	73
H04	640	93	495	72	10	...	93	78
H06	730	106	550	80	4	...	96	80
H08	770	112	...	...	3	...	98	81
H10	825	120	...	...	2	...	100	82
Rod, 13 mm (0.5 in.) diameter								
H02	550	80	450	65	33	...	85	...
Wire, 2 mm (0.08 in.) diameter								
OS035	415	60	165	24	65	...	...	...
H01	560	81	...	...	...	...	...	...
H02	725	105	...	...	...	...	...	...
H04	895	130	...	...	...	...	...	...
H06	965	140	...	...	...	...	...	...

(a) At 0.5% extension under load

Bearings: MIL-B-13501. Wire: QQ-B-750, QQ-W-321, MIL-W-6712

Chemical Composition

Composition limits. 93.6 to 95.6 Cu, 4.2 to 5.8 Sn, 0.03 to 0.35 P, 0.05 Pb max, 0.1 Fe max, 0.3 Zn max

Applications

Typical uses. Architectural: bridge bearing plates. Hardware: beater bars, bellows, Bourdon tubing, clutch disks, cotter pins, diaphragms, fuse clips, fasteners, lock washers, sleeve bushings, springs, switch parts, truss wire, wire brushes. Industrial: chemical hardware, perforated sheets, textile machinery, welding rods

Mechanical Properties

Tensile properties. See Table 88.
Hardness. See Table 88.
Elastic modulus. Tension, 110 GPa (16 × 10^6 psi); shear, 41 GPa (6 × 10^6 psi)
Fatigue structure. At 10^8 cycles. Flat products: H04 temper, 170 MPa (25 ksi); H08 temper, 150 MPa (22 ksi). Wire: H04 temper, 185 MPa (27 ksi); H06 temper, 205 MPa (30 ksi)

Mass Characteristics

Density. 8.86 g/cm^3 (0.320 lb/in.^{3}) at 20 °C (68 °F)

Thermal Properties

Liquidus temperature. 1060 °C (1945 °F)
Solidus temperature. 975 °C (1785 °F)
Coefficient of linear thermal expansion. 17.8 μm/m · K (9.9 μin./in. · °F) at 20 to 300 °C (68 to 572 °F)
Specific heat. 380 J/kg · K (0.09 Btu/lb · °F)
Thermal conductivity. 84 W/m · K (48.4 Btu/ft · h · °F) at 20 °C (68 °F)

Electrical Properties

Electrical conductivity. Volumetric, 20% IACS at 20 °C (68 °F)
Electrical resistivity. 87 nΩ · m at 20 °C (68 °F)

Fabrication Characteristics

Machinability. 20% of C36000 (free-cutting brass)
Formability. Excellent capacity for cold working by blanking, drawing, forming, bending, roll threading, knurling, shearing, and stamping. Poor capacity for hot forming
Weldability. Soldering, brazing, and resistance butt welding: excellent. Gas metal arc and resistance spot welding: good. Oxyfuel gas, shielded metal arc, and resistance seam welding: fair
Annealing temperature. 475 to 675 °C (900 to 1250 °F)

C51100
95.6Cu-4.2Sn-0.2P

Specifications

ASTM. Flat products: B 100, B 103

Chemical Composition

Composition limits. 94.5 to 96.3 Cu, 3.5 to 4.9 Sn, 0.003 to 0.35 P, 0.05 Pb max, 0.1 Fe max, 0.3 Zn max

Applications

Typical uses. Architectural: bridge bearing plates. Hardware: beater bars, bellows, clutch disks, connectors, diaphragms, fuse clips, fasteners, lock washers, sleeve bushings, springs, switch parts, terminals. Industrial: chemical hardware, perforated sheets, textile machinery

Mechanical Properties

Tensile properties. See Table 89.
Hardness. See Table 89.
Elastic modulus. Tension, 110 GPa (16 × 10^6 psi); shear, 41 GPa (6 × 10^6 psi)

Mass Characteristics

Density. 8.86 g/cm^3 (0.32 lb/in.^{3}) at 20 °C (68 °F)

Thermal Properties

Liquidus temperature. 1060 °C (1945 °F)
Solidus temperature. 975 °C (1785 °F)
Coefficient of linear thermal expansion. 17.8 μm/m · K (9.9 μin./in. · °F) at 20 to 300 °C (68 to 572 °F)
Specific heat. 380 J/kg · K (0.09 Btu/lb · °F) at 20 °C (68 °F)
Thermal conductivity. 84 W/m · K (48.4 Btu/ft · h · °F) at 20 °C (68 °F)

Electrical Properties

Electrical conductivity. Volumetric, 20% IACS at 20 °C (68 °F)
Electrical resistivity. 87 nΩ · m at 20 °C (68 °F)

Fabrication Characteristics

Machinability. 20% of C36000 (free-cutting brass)
Formability. Excellent capacity for cold working by blanking, drawing, forming, bending, roll threading, knurling, shearing, and stamping. Poor capacity for hot forming
Weldability. Soldering, brazing, and resistance butt welding: excellent. Gas metal arc and resistance spot welding: good. Oxyfuel gas, shielded metal arc, and resistance seam welding: fair
Annealing temperature. 475 to 675 °C (900 to 1250 °F)

Table 91 Typical mechanical properties of C52400

Temper	Tensile strength MPa	ksi	Elongation in 50 mm (2 in.), %	Hardness, HRB
Flat products, 1 mm (0.04 in.) thick				
OS035	455	66	68	55
H02	570	83	32	92
H04	690	100	13	97
H06	795	115	7	100
H08	840	122	4	101
H10	885	128	3	103
Wire, 2 mm (0.08 in.) diameter				
OS035	455	66	70	. . .
H01	640	93	. . .	. . .
H02	815	118	. . .	. . .
H04	1013	147	. . .	. . .

Table 92 Typical mechanical properties of C54400

Temper	Tensile strength MPa	ksi	Yield strength(a) MPa	ksi	Elongation in 50 mm (2 in.), %	Hardness, HRB
Sheet and strip, 1 mm (0.04 in.) thick						
OS050	315	46	. . .	. . .	48	70 HRF
OS035	330	48	. . .	. . .	47	73 HRF
OS025	345	50	. . .	. . .	46	75 HRF
OS015	350	51	. . .	. . .	46	76 HRF
H02	425	62	370	54	19	70
H04	550	80	510	74	7	86
H06	635	92	. . .	. . .	4	91
H08	675	98	550	80	3	93
H10	710	103	. . .	. . .	2	95
Flat products, 8 mm (0.38 in.) thick						
H04	415	60	310	45	20	70
Flat products, 19 mm (0.75 in.) thick						
H04	380	55	240	35	25	. . .
Rod, 13 mm (0.50 in.) diameter						
H04	515	75	435	63	15	83
Rod, 25 mm (1.0 in.) diameter						
H04	470	68	395	57	20	80

(a) At 0.5% extension under load

C52100
92Cu-8Sn

Commercial Names

Previous trade names. Phosphor bronze, 8% C

Specifications

ASTM. Flat products: B 103. Bar: B 103, B 139. Rod, shapes: B 139. Wire: B 159
SAE. J463
Government. MIL-E-23765

Chemical Composition

Composition limits. 90.5 to 92.8 Cu, 7.0 to 9.0 Sn, 0.03 to 0.35 P, 0.05 Pb max, 0.1 Fe max, 0.2 Zn max

Applications

Typical uses. For more severe service conditions than C51000. Architectural: bridge bearing plates. Hardware: beater bars, bellows. Bourdon tubing, clutch disks, cotter pins, diaphragms, fuse clips, fasteners, lock washers, sleeve bushings, springs, switch parts, truss wire, wire brushes. Industrial: chemical hardware, perforated sheets, textile machinery, welding rods

Mechanical Properties

Tensile properties. See Table 90.
Hardness. See Table 90.
Elastic modulus. Tension, 110 GPa (16×10^6 psi); shear, 41 GPa (6×10^6 psi)
Fatigue strength. Strip, 1 mm (0.04 in.) thick, H04 temper: 150 MPa (22 ksi) at 10^8 cycles

Mass Characteristics

Density. 8.8 g/cm^3 (0.318 lb/in.^{3}) at 20 °C (68 °F)

Thermal Properties

Liquidus temperature. 1025 °C (1880 °F)
Solidus temperature. 880 °C (1620 °F)
Coefficient of linear thermal expansion. 18.2 μm/m · K (10.1 μin./in. · °F) at 20 to 300 °C (68 to 572 °F)
Specific heat. 380 J/kg · K (0.09 Btu/lb · °F) at 20 °C (68 °F)
Thermal conductivity. 62 W/m · K (36 Btu/ft · h · °F) at 20 °C (68 °F)

Electrical Properties

Electrical conductivity. Volumetric, 13% IACS at 20 °C (68 °F)
Electrical resistivity. 133 nΩ · m at 20 °C (68 °F)

Fabrication Characteristics

Machinability. 20% of C36000 (free-cutting brass)
Formability. Good capacity for cold working by blanking, drawing, forming, bending, shearing, and stamping. Poor capacity for hot forming
Weldability. Soldering, brazing, and resistance butt welding: excellent. Gas metal arc and resistance spot welding: good. Oxyfuel gas, shielded metal arc, and resistance seam welding: fair
Annealing temperature. 475 to 675 °C (900 to 1250 °F)

C52400
90Cu-10Sn

Commercial Names

Previous trade name. Phosphor bronze, 10% D

Specifications

ASTM. Flat products: B 103. Bar: B 103, B 139. Rod, shapes: B 139. Wire: B 159
Government. Flat products, wire: QQ-B-750

Chemical Composition

Composition limits. 88.3 to 90.07 Cu, 9.0 to 11.0 Sn, 0.03 to 0.35 P, 0.05 Pb max, 0.1 Fe max, 0.2 Zn max

Applications

Typical uses. Heavy bars and plates for severe compression requiring good wear and corrosion resistance; bridge and expansion plates and fittings; and articles requiring extra spring qualities and optimum resiliency, particularly in fatigue

Mechanical Properties

Tensile properties. Tensile strength and elongation, see Table 91. Yield strength, typical, OS035 temper: 195 MPa (28 ksi) at 0.5% extension under load
Hardness. See Table 91.
Elastic modulus. Tension, 110 GPa (16×10^6 psi); shear, 41 GPa (6×10^6 psi)

Mass Characteristics

Density. 8.78 g/cm^3 (0.317 lb/in.^{3}) at 20 °C (68 °F)

Thermal Properties

Liquidus temperature. 1000 °C (1830 °F)
Solidus temperature. 845 °C (1550 °F)
Coefficient of linear thermal expansion. 18.4 μm/m · K (10.2 μin./in. · °F) at 20 to 300 °C (68 to 572 °F)
Specific heat. 380 J/kg · K (0.09 Btu/lb · °F) at 20 °C (68 °F)
Thermal conductivity. 50 W/m · K (29 Btu/ft · h · °F) at 20 °C (68 °F)

Electrical Properties

Electrical conductivity. Volumetric, 11% IACS at 20 °C (68 °F)
Electrical resistivity. 157 nΩ · m at 20 °C (68 °F)

Fabrication Characteristics

Machinability. 20% of C36000 (free-cutting brass)
Formability. Good capacity for cold work-

ing by blanking, forming, bending, and shearing. Poor capacity for hot forming
Weldability. Soldering, brazing, and resistance butt welding: excellent. Gas metal arc and resistance spot welding: good. Oxyfuel gas, shielded metal arc, and resistance seam welding: fair
Annealing temperature. 475 to 675 °C (900 to 1250 °F)

C54400
88Cu-4Pb-4Sn-4Zn

Commercial Names

Previous trade name. Phosphor bronze B-2
Common names. Free-cutting phosphor bronze; 444 bronze; bearing bronze

Specifications

AMS. Strip 4520
ASTM. B 103, B 139
SAE. J463. Bearing alloy: J460 (791)
Government. Bar and rod: QQ-B-750

Chemical Composition

Composition limits. 3.5 to 4.5 Pb, 3.5 to 4.5 Sn, 1.5 to 4.5 Zn, 0.10 Fe max, 0.01 to 0.50 P, bal Cu; 99.5 Cu + Pb + Sn + Zn + P min

Applications

Typical uses. Bearings (sleeve and thrust), bushings, gears, pinions, screw machine products, shafts, thrust washers, valve parts

Mechanical Properties

Tensile properties. See Table 92.
Hardness. See Table 92.
Elastic modulus. Tension, 103 GPa (15×10^6 psi); shear, 39 GPa (5.6×10^6 psi)

Mass Characteristics

Density. 8.89 g/cm^3 (0.321 lb/in.^{3}) at 20 °C (68 °F)

Thermal Properties

Liquidus temperature. 1000 °C (1830 °F)
Solidus temperature. 930 °C (1700 °F)
Coefficient of linear thermal expansion. 17.3 μm/m · K (9.6 μin./in. · °F) at 20 to 300 °C (68 to 572 °F)
Specific heat. 380 J/kg · K (0.09 Btu/lb · °F) at 20 °C (68 °F)
Thermal conductivity. 87 W/m · K (50 Btu/ft · h · °F) at 20 °C (68 °F)

Electrical Properties

Electrical conductivity. Volumetric, 19% IACS at 20 °C (68 °F)
Electrical resistivity. 91 nΩ · m at 20 °C (68 °F)

Fabrication Characteristics

Machinability. 80% of C36000 (free-cutting brass)
Formability. Good cold working characteristics; commonly fabricated by machining, shearing, blanking, drawing, forming, bending. Hot working and hot forming are not recommended.
Weldability. Soldering: excellent. Brazing: good. Flash welding: fair. Other welding processes are not recommended.

C60600
95Cu-5Al

Commercial Names

Previous trade name. Aluminum bronze A; CA606
Common names. Aluminum bronze, 5%

Specifications

ASTM. Flat products: B 169
Government. Bar, rod, forgings, shapes: QQ-C-645. Sheet and plate: QQ-C-450. Strip: QQ-C-450, QQ-C-465

Chemical Composition

Composition limits. 92.0 to 96.0 Cu, 4.0 to 7.0 Al, 0.50 Fe max, 0.50 max other (total)
Consequence of exceeding impurity limits. Excessive amounts of Pb, Zn, and P will cause hot shortness and difficulties in welding.

Applications

Typical uses. Produced as sheet, strip, and rolled bar; used to make fasteners, deep drawn "gold" decoration, and parts requiring corrosion resistance
Precautions in use. Not suitable for use in oxidizing acids

Mechanical Properties

Tensile properties. Typical data for 13 mm (0.5 in.) thick plate. Tensile strength: O61 temper, 310 MPa (45 ksi); H04 temper, 415 MPa (60 ksi). Yield strength: O60 temper, 115 MPa (17 ksi); H04 temper, 165 MPa (24 ksi). Elongation: O60 temper, 40% in 50 mm (2 in.); H04 temper, 25% in 50 mm (2 in.)
Hardness. O60 temper, 42 HRB; H04 temper, 55 HRB
Poisson's ratio. 0.326
Elastic modulus. Tension, 121 GPa (17.5×10^6 psi); shear, 46 GPa (6.6×10^6 psi)
Fatigue strength. Rotating beam, 169 MPa (24.5 ksi) at 10^8 cycles)

Structure

Microstructure. Alpha structure, face-centered cubic

Mass Characteristics

Density. 8.17 g/cm^3 (0.295 lb/in.^{3}) at 20 °C (68 °F)
Volume change on freezing. Approximately 1.6% contraction

Thermal Properties

Liquidus temperature. 1065 °C (1945 °F)
Solidus temperature. 1050 °C (1920 °F)
Coefficient of linear thermal expansion. 18 μm/m · K (10 μin./in. · °F) at 20 to 300 °C (68 to 572 °F)
Specific heat. 375 J/kg · K (0.09 Btu/lb · °F) at 20 °C (68 °F)
Thermal conductivity. 79.5 W/m · K (45.9 Btu/ft · h · °F) at 20 °C (68 °F)

Electrical Properties

Electrical conductivity. Volumetric, 17% IACS at 20 °C (68 °F)
Electrical resistivity. 100 nΩ · m at 20 °C (68 °F)

Magnetic Properties

Magnetic permeability. 1.01

Chemical Properties

General corrosion resistance. See C61400.
Resistance to specific agents. Has been used in sulfuric acid pickling applications where oxygen content is low. Has been used for anhydrous NH_4OH, but the presence of moisture leads to season cracking. Not suitable for use with nitric acid. Oxidizing salts such as chromates and metal salts such as ferric chloride are generally corrosive to C60600.

Fabrication Characteristics

Machinability. 20% of C36000 (free-cutting brass). Tends to form tough, stringy chips. Good lubrication and cooling essential for good finish. Carbide or tool steel cutters may be used.
Recrystallization temperature. 350 °C (660 °F) at 44% reduction and 0.075 mm (0.003 in.) initial grain size
Annealing temperature. 550 to 650 °C (1020 to 1200 °F)
Hot-working temperature. 815 to 870 °C (1500 to 1600 °F)

C60800
95Cu-5Al

Commercial Names

Previous trade name. 5% aluminum bronze
Common name. Aluminum bronze, 5%

Specifications

ASME. Tubing: SB111, SB359, SB395
ASTM. Tubing: B 111, B 359, B 395

Chemical Composition

Composition limits. 92.5 to 94.8 Cu, 5.0 to 6.5 Al, 0.02 to 0.35 As, 0.10 Pb max, 0.10 Fe max
Consequence of exceeding impurity limits. Excessive amounts of Pb, Zn, and P will cause difficulties in welding and hot working.

Applications

Typical uses. Produced as seamless tubing and ferrule stock for heat exchanger tubes, condenser tubes, and other applications requiring corrosion-resistant seamless tubing

Precautions in use. Not suitable for use in oxidizing acids

Mechanical Properties

Tensile properties. Typical for OS025 temper tubing, 25 mm (1.0 in.) outside diameter × 1.65 mm (0.065 in.) wall thickness: tensile strength, 415 MPa (60 ksi); yield strength (0.5% extension under load), 185 MPa (27 ksi); elongation, 55% in 50 mm (2 in.)
Hardness. OS025 temper: 77 HRF
Poisson's ratio. 0.325
Elastic modulus. Tension, 121 GPa (17.5 × 10^6 psi); shear, 46 GPa (6.6 × 10^6 psi)

Structure

Microstructure. Alpha structure, face-centered cubic

Mass Characteristics

Density. 8.17 g/cm^3 (0.295 lb/in.^{3}) at 20 °C (68 °F)
Volume change on freezing. Approximately 1.6% contraction

Thermal Properties

Liquidus temperature. 1065 °C (1945 °F)
Solidus temperature. 1050 °C (1920 °F)
Coefficient of linear thermal expansion. 18 μm/m · K (10 μin./in. · °F) at 20 to 300 °C (68 to 572 °F)
Specific heat. 380 J/kg · K (0.09 Btu/lb · °F) at 20 °C (68 °F)
Thermal conductivity. 79.5 W/m · K (45.9 Btu/ft · h · °F) at 20 °C (68 °F)

Electrical Properties

Electrical conductivity. Volumetric, 17% IACS at 20 °C (68 °F)
Electrical resistivity. 100 nΩ · m at 20 °C (68 °F)

Magnetic Properties

Magnetic permeability. 1.01

Chemical Properties

General corrosion resistance. See C61400.
Resistance to specific agents. Has been used in sulfuric acid pickling applications where oxygen content is low. Has been used for anhydrous NH_4OH, but the presence of moisture leads to season cracking. Not suitable for use with nitric acid. Oxidizing salts such as chromates and metal salts such as ferric chloride are generally corrosive to C60800.

Fabrication Characteristics

Machinability. 20% of C36000 (free-cutting brass). Tends to form tough, stringy chips. Good lubrication and cooling essential for good finishes. Carbide or tool steel cutters may be used.
Formability. Good for cold working; fair for hot forming
Weldability. Arc and resistance welding: good. Brazing: fair. Soldering and oxyfuel gas welding are not recommended.
Recrystallization temperature. 350 °C (660 °F) at 44% reduction and 0.075 mm (0.003 in.) initial grain size
Annealing temperature. 550 to 650 °C (1020 to 1200 °F)
Hot-working temperature. 800 to 875 °C (1470 to 1610 °F)

C61000
92Cu-8Al

Commercial Names

Common name. 8% aluminum bronze

Specifications

ASME. SB169
ASTM. B 169
Government. QQ-C-450; MIL-E-23765

Chemical Composition

Composition limits. 6.0 to 8.5 Al, 0.50 Fe max, 0.02 Pb max, 0.20 Zn max, 0.10 Si max, 0.50 max other (total), bal Cu

Applications

Typical uses. Produced as rod or wire and used to make bolts, shafts, tire rods, and pump parts. Also used as a welded overlay on steel to improve surface wear resistance

Mechanical Properties

Tensile properties. Typical data for rod, 25 mm (1 in.) in diameter. O60 temper: tensile strength 480 MPa (70 ksi); yield strength (0.5% extension under load), 205 MPa (30 ksi); elongation, 65% in 50 mm (2 in.). H04 temper: tensile strength, 550 MPa (80 ksi); yield strength, 380 MPa (55 ksi); elongation, 25%
Hardness. O60 temper: 60 HRB. H04 temper: 85 HRB
Elastic modulus. Tension, 117 GPa (17 × 10^6 psi); shear, 44 GPa (6.4 × 10^6 psi)

Mass Characteristics

Density. 7.78 g/cm^3 (0.281 lb/in.^{3}) at 20 °C (68 °F)

Thermal Properties

Liquidus temperature. 1040 °C (1905 °F)
Coefficient of linear thermal expansion. 17.9 μm/m · K (9.9 μin./in. · °F) at 20 to 300 °C (68 to 572 °F)
Specific heat. 375 J/kg · K (0.09 Btu/lb · °F) at 20 °C (68 °F)
Thermal conductivity. 69 W/m · K (40 Btu/ft · h · °F) at 20 °C (68 °F)

Electrical Properties

Electrical conductivity. Volumetric, 15% IACS at 20 °C (68 °F)
Electrical resistivity. 115 nΩ · m at 20 °C (68 °F)

Fabrication Characteristics

Machinability. 20% of C36000 (free-cutting brass)
Forgeability. 70% of C37700 (forging brass)

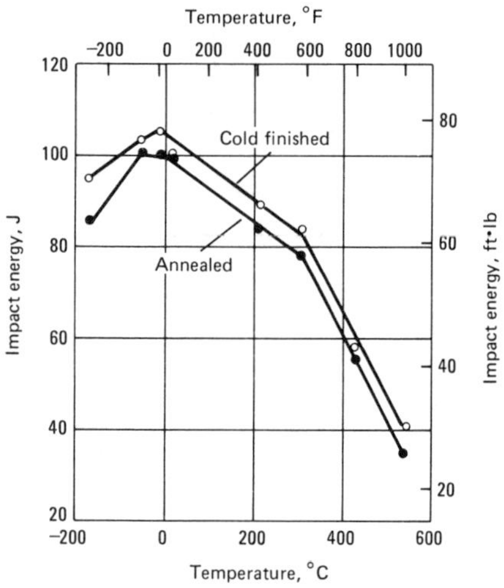

Fig. 44 Variation in Charpy V-notch impact strength with temperature for C61300 and C61400

Formability. Good capacity for being hot formed or cold worked. Common fabrication processes include blanking, drawing, forming, bending, cold heading, and roll threading.
Weldability. Arc welding, resistance spot welding, and resistance butt welding: good. Soldering and resistance seam welding: fair. Brazing and oxyfuel gas welding are not recommended.
Annealing temperature. 600 to 675 °C (1100 to 1250 °F)
Hot-working temperature. 760 to 875 °C (1400 to 1600 °F)

C61300
90Cu-7Al-2.7Fe-0.3Sn

Commercial Names

Common name. Aluminum bronze, 7%

Specifications

Government. Flat products: QQ-C-450

Chemical Composition

Composition limits. 88.5 to 91.5 Cu, 6.0 to 7.5 Al, 0.02 to 0.50 Sn, 2.0 to 3.0 Fe, 0.10 Mn max, 0.15 Ni (+ Co) max, 0.01 Pb max, 0.05 Zn max, 0.05 max other
Consequence of exceeding impurity limits. Excessive amounts of Pb, Zn, P, or Si will cause hot shortness, which can lead to problems during hot working or welding.

Applications

Typical uses. Produced as rod, bar, sheet, plate, seamless tubing and pipe, welded pipe, fasteners, tube sheets, heat exchanger tubes, acid-resistant piping, columns, water boxes, and corrosion-resistant vessels
Precautions in use. Not suitable for use in oxidizing acids

Mechanical Properties

Tensile properties. Typical data for 13 mm

Table 93 Typical mechanical properties of C61300 and C61400 rod at various temperatures

Temperature °C	Temperature °F	Tensile strength MPa	Tensile strength ksi	Yield strength(a) MPa	Yield strength(a) ksi	Elongation in 50 mm (2 in.), %	Reduction in area, %	Modulus of elasticity GPa	Modulus of elasticity 10^6 psi	Hardness, HB(b)
Cold finished										
−182	−295	718	104.1	397	57.6	50	49	156	22.7	186
−60	−75	611	88.6	335	48.7	45	55	149	21.6	170
−29	−20	606	87.9	339	49.2	44	58	172	25.0	162
20	70	590	85.5	318	46.1	42	59	126	18.3	157
204	400	532	77.2	298	43.3	35	32	128	18.5	144
316	600	432	62.6	271	39.3	22	24	88	12.8	137
427	800	170	24.6	105	15.2	52	41	48	6.9	83
538	1000	88	12.8	71	10.3	27	26	45	6.6	49
Annealed										
−182	−295	707	102.6	347	50.3	52	51	139	20.2	185
−60	−75	610	88.4	305	44.3	47	57	176.5	25.6	162
−29	−20	600	87.1	303	43.9	44	56	172	24.9	161
20	70	583	84.6	288	41.8	45	56	136.5	19.8	155
204	400	522	75.7	276	40.1	34	32	130	18.8	142
316	600	427	62.0	256	37.1	30	27	81	11.7	134
427	800	174	25.3	123	17.8	60	55	67	9.7	84
538	1000	92	13.3	68	9.9	36	33	45	6.5	50

(a) At 0.5% extension under load. (b) 3000 kg (6615 lb) load

(0.50 in.) thick plate. Tensile strength: O60 temper, 540 MPa (78 ksi); H04 temper, 585 MPa (85 ksi). Yield strength (0.5% extension): O60 temper, 240 MPa (35 ksi); H04 temper, 400 MPa (58 ksi). Elongation: O60 temper, 42% in 50 mm (2 in.); H04 temper, 35% in 50 mm (2 in.). Reduction in area: O60 temper, 32%; H04 temper, 25%. See also Table 93.
Compressive properties. Typical data for 13 mm (0.50 in.) thick plate. Compressive strength, ultimate: O60 temper, 825 MPa (120 ksi); H04 temper, 860 MPa (125 ksi)
Hardness. O60 temper, 82 HRB; H04 temper, 91 HRB. See also Table 93.
Poisson's ratio. 0.312
Elastic modulus. Tension, 115 GPa (17 × 10^6 psi); shear, 44 GPa (6.4 × 10^6 psi). See also Table 93.
Impact strength. Charpy keyhole, 81 to 88 J (60 to 65 ft · lbf) at −30 to 150 °C (−20 to 300 °F); Izod, 54 to 66 J (40 to 49 ft · lbf) at −30 to 150 °C (−20 to 300 °F). See also Fig. 44.
Fatigue strength. Reverse bending, 180 MPa (26 ksi) at 10^8 cycles

Structure

Microstructure. Alpha structure, single phase, with iron-rich precipitates

Mass Characteristics

Density. 7.89 g/cm³ (0.285 lb/in.³) at 20 °C (68 °F)
Volume change on freezing. Approximately 1.8% contraction

Thermal Properties

Liquidus temperature. 1045 °C (1915 °F)
Solidus temperature. 1040 °C (1905 °F)
Coefficient of linear thermal expansion. 16.2 μm/m · K (9.0 μin./in. · °F) at 20 to 300 °C (68 to 572 °F)
Specific heat. 375 J/kg · K (0.09 Btu/lb · °F) at 20 °C (68 °F)
Thermal conductivity. 56.5 W/m · K (32.7 Btu/ft · h · °F) at 20 °C (68 °F); temperature coefficient, 0.12 W/m · K per K at 20 °C (68 °F)

Electrical Properties

Electrical conductivity. Volumetric, 12% IACS at 20 °C (68 °F)
Electrical resistivity. 144 nΩ · m at 20 °C (68 °F)

Magnetic Properties

Magnetic permeability. 1.16

Chemical Properties

General corrosion resistance. See C61400.
Resistance to specific agents. C61300 is very resistant to neutral and nonoxidizing salts. It has given extended service in potash solutions of potassium chloride, sodium chloride, magnesium chloride, and calcium chloride. The alloy resists nonoxidizing mineral acids and has been used successfully for tanks containing hydrofluoric acid in glass-etching applications. In organic acid service, it has been used to make acetic acid distillation columns.

C61300 is highly resistant to dealloying and to season cracking in steam and in hot oxidizing aqueous solutions and vapors. The presence of tin in 7% aluminum bronze (C61300 contains 0.3% Sn, C61400 does not contain Sn) evidently renders the alloy immune to stress-corrosion cracking in these environments. Like many copper alloys, C61300 is susceptible to season cracking in moist ammonia and mercurous nitrate solutions. However, it is highly resistant to season cracking in anhydrous ammonia, especially when the moisture content is below 500 ppm and the temperature is below 85 °C (180 °F).

Because of its high resistance to corrosion in salt water, C61300 has been specified for a wide variety of components for marine and desalting plant service. Typical uses of C61300 include tube sheets for condensers in both nuclear and fossil fuel power stations, cooling tower transfer piping, seawater piping for secondary cooling systems in nuclear power plants, and piping for geothermal heat transfer systems.

Fabrication Characteristics

Machinability. Fair to poor, with chips tending to be stringy and gummy. Good lubrication and cooling are essential. Tool steel cutters: roughing speed, 90 m/min (300 ft/min) with a feed of 0.3 mm/rev (0.011 in./rev); finishing speed, 350 m/min (1150 ft/min) with a feed of 0.3 mm/rev (0.011 in./rev)
Forgeability. 50% of C37700 (forging brass)
Formability. Good for cold working and hot forming
Weldability. Arc and resistance welding: good. Brazing: fair. Soldering and oxyfuel gas welding are not recommended.
Recrystallization temperature. 785 to 870 °C (1450 to 1600 °F)
Annealing temperature. 600 to 875 °C (1125 to 1600 °F)
Hot-working temperature. 800 to 925 °C (1450 to 1700 °F)
Hot-shortness temperature. 1010 °C (1850 °F)

C61400
91Cu-7Al-2Fe

Commercial Names

Previous trade name. Aluminum bronze D
Common name. Aluminum bronze, 7%

Specifications

ASME. Flat products: SB169, SB171. Bar, rod, shapes: SB150
ASTM. Flat products: B 169, B 171. Bar, rod, shapes: B 150
SAE. J463
Government. Flat products: QQ-C-450,

Table 94 Typical mechanical properties of C61400

Size	Tensile strength MPa	Tensile strength ksi	Yield strength(a) MPa	Yield strength(a) ksi	Elongation in 50 mm (2 in.), %	Shear strength MPa	Shear strength ksi
Flat products, O60 temper							
3 mm (0.12 in.) thick	565	82	310	45	40	310	45
8 mm (0.31 in.) thick	550	80	275	40	40	290	42
13 mm (0.50 in.) thick	535	78	240	35	42	275	40
25 mm (1.00 in.) thick	525	76	230	33	45	275	40
Flat products, H04 temper							
3 mm (0.12 in.) thick	615	89	415	60	32	...	...
8 mm (0.31 in.) thick	585	85	400	58	35	...	...
13 mm (0.50 in.) thick	550	80	370	54	38	...	...
25 mm (1.00 in.) thick	535	78	310	45	40	...	...
Rod, H04 temper							
13 mm (0.50 in.) diameter	585	85	310	45	35	330	48
25 mm (1.00 in.) diameter	565	82	275	40	35	310	45
51 mm (2.00 in.) diameter	550	80	240	35	35	275	40

(a) At 0.5% extension under load

Table 95 Typical mechanical properties of 1 mm (0.04 in.) thick C61500 sheet and strip

Temper	Tensile strength MPa	Tensile strength ksi	Yield strength at 0.2% offset MPa	Yield strength at 0.2% offset ksi	Elongation in 50 mm (2 in.), %	Hardness, HR30T	Fatigue strength at 10^8 cycles MPa	Fatigue strength at 10^8 cycles ksi
O60	485	70	150	22	55	42	...	...
O50	585	85	345	50	36	70	260	38
H02	725	105	515	75	15	81	...	...
H04	860	125	620	90	5	83	...	...
H06	930	135	690	100	4	84	270	39
H08	965	140	725	105	3	84.5	...	...
HR06(a)	1000	145	965	140	1	86.5	275	40

(a) Cold worked 50%, then stress relieved for 1 h at 300 °C (570 °F)

QQ-C-465. Bar, rod, shapes, forgings: QQ-C-465. Flat wire: QQ-C-465

Chemical Composition

Composition limits. 88.0 to 92.5 Cu, 6.0 to 8.0 Al, 1.5 to 3.5 Fe, 1.0 Mn max, 0.20 Zn max, 0.01 Pb max, 0.015 P max, 0.5 max other (total)
Consequence of exceeding impurity limits. Excessive amounts of Pb, Zn, Si, or P will cause hot shortness and cracking during hot working and welding.

Applications

Typical uses. Produced as seamless tubing, welded and seamless pipe, sheet, plate, rod, and bar for condenser and heat exchanger tubes, fasteners, tube sheets, and corrosion-resistant vessels
Precautions in use. Not suitable for use in oxidizing acids. Susceptible to stress-corrosion cracking in moist ammonia or in steam environments, especially when stress levels are high

Mechanical Properties

Tensile properties. See Tables 93 and 94.
Shear strength. See Table 94.
Compressive properties. Compressive strength, ultimate: O60 temper, 825 MPa (120 ksi); H04 temper, 860 MPa (125 ksi)
Hardness. O60 temper, 80 to 84 HRB; H04 temper, 84 to 91 HRB. See also Table 94.
Poisson's ratio. 0.312
Elastic modulus. Tension, 115 GPa (17 × 10^6 psi); shear, 44 GPa (6.4 × 10^6 psi). See also Table 94.
Impact strength. Charpy keyhole, 81 to 88 J (60 to 65 ft · lbf); Izod, 54 to 61 J (40 to 45 ft · lbf). See also Fig. 44.
Fatigue strength. Reverse bending, 180 MPa (26 ksi) at 10^8 cycles

Structure

Microstructure. Alpha solid solution with precipitates of iron-rich phase

Mass Characteristics

Density. 7.89 g/cm^3 (0.285 lb/in.^{3}) at 20 °C (68 °F)
Volume change on freezing. Approximately 1.8% expansion

Thermal Properties

Liquidus temperature. 1045 °C (1915 °F)
Solidus temperature. 1040 °C (1905 °F)
Coefficient of linear thermal expansion. 16.2 μm/m · K (9.0 μin./in. · °F) at 20 to 300 °C (68 to 572 °F)
Specific heat. 375 J/kg · K (0.09 Btu/lb · °F) at 20 °C (68 °F)
Thermal conductivity. 56.5 W/m · K (32.6 Btu/ft · h · °F) at 20 °C (68 °F); temperature coefficient, 0.12 W/m · K per K at 20 °C (68 °F)

Electrical Properties

Electrical conductivity. Volumetric, 14% IACS at 20 °C (68 °F)
Electrical resistivity. 123 nΩ · m at 20 °C (68 °F)

Magnetic Properties

Magnetic permeability. 1.16

Chemical Properties

General corrosion behavior. The aluminum bronzes resist nonoxidizing mineral acids such as sulfuric, hydrochloric, and phosphoric acid. Resistance tends to decrease with increasing concentration of dissolved oxygen or oxidizing agents, particularly as temperatures increase above 55 °C (130 °F). Aluminum bronzes are generally suited for service in alkalies, neutral salts, nonoxidizing acid salts, and many organic acids and compounds. Oxidizing acids, oxidizing salts, and heavy-metal salts are corrosive. Aluminum bronzes resist waters, whether potable water, brackish water, or seawater. Softened water tends to be more corrosive than hard water. Aluminum bronzes resist dealloying, but to different degrees depending on alloy composition. In general, corrosion resistant is influenced most by solution concentration, aeration, temperature, velocity, and the type and amount of any impurities in the solution.

Like many other copper alloys, the aluminum bronzes are susceptible to stress-corrosion cracking in moist ammonia and mercury compounds. When stress levels are high, they may also be susceptible to stress-corrosion cracking in purified steam or in steam containing acidic or salt vapors.
Resistance to specific agents. C61400 has been used successfully to contain mineral acids, alkalies such as sodium or potassium hydroxide, neutral salts such as sodium chloride, and organic acids such as acetic, lactic, or oxalic acid. C61400 resists anhydrous ammonia, but precautions must be taken to exclude moisture and thus avoid season cracking. Similarly, this alloy resists anhydrous chlorinated hydrocarbons such as carbon tetrachloride, but the presence of moisture makes those chemicals corrosive.

Fabrication Characteristics

Machinability. 20% of C36000 (free-cutting brass). Tendency to form continuous, stringy chips. Good lubrication and cooling essential. Tool steel or carbide cutters may be used. Typical conditions, using tool steel cutters: roughing speed, 90 m/min (300 ft/min) with a feed of 0.3 mm/rev (0.011 in./rev); finishing speed, 350 m/min (1150 ft/min) with a feed of 0.3 mm/rev (0.011 in./rev)
Formability. Fair for cold working; good for hot forming
Weldability. Gas-shielded arc, coated metal arc, and resistance welding: good. Brazing: fair. Soldering, oxyacetylene, and carbon arc welding are not recommended.

Table 96 Typical mechanical properties of C62300 rod at various temperatures

Temperature °C	Temperature °F	Tensile strength MPa	Tensile strength ksi	Yield strength(a) MPa	Yield strength(a) ksi	Elongation in 50 mm (2 in.), %	Reduction in area, %	Modulus of elasticity in tension GPa	Modulus of elasticity in tension 10^6 psi	Hardness, HB(b)
Cold finished										
−182	−295	778	112.8	390	56.5	37	41	127	18.4	193
−60	−75	682	98.9	340	49.3	34	41	108	15.7	170
−29	−20	663	96.2	326	47.3	34	44	114	16.5	168
20	70	652	94.5	320	46.4	34	44	111	16.1	165
204	400	550	79.8	296	43.0	22	22	114	16.5	152
316	600	465	67.5	296	43.0	10	13	85	12.4	148
427	800	196	28.5	138	20.0	32	33	54	7.9	98
538	1000	103	15.0	92	13.3	18	29	41	5.9	54
Annealed										
−182	−295	762	110.5	377	54.7	35	38	125	18.1	195
−60	−75	664	96.3	330	47.9	33	39	121	17.5	171
−29	−20	647	93.8	323	46.9	31	38	124	18.0	168
20	70	620	90.0	294	42.6	32	39	120	17.4	161
204	400	534	77.5	302	43.8	20	21	138	20.0	151
316	600	448	65.0	288	41.8	10	13	79	11.5	146
427	800	210	30.5	153	22.2	46	39	73	10.6	97
538	1000	94	13.6	84	12.2	27	32	51	7.4	46

(a) At 0.5% extension under load. (b) 3000 kg (6615 lb) load

Table 97 Typical compressive properties for C62300 rod, H50 temper

	Compressive strength at permanent set of							
	0.1%		1%		10%		20%	
Rod diameter	MPa	ksi	MPa	ksi	MPa	ksi	MPa	ksi
≤25 mm (≤1 in.)	360	52	485	70	825	120	965	140
25–50 mm (1–2 in.)	345	50	450	65	675	98	930	135
50–75 mm (2–3 in.)	315	46	415	60	620	90	895	130

Recrystallization temperature. 785 to 870 °C (1450 to 1600 °F)
Annealing temperature. 600 to 900 °C (1125 to 1650 °F)
Hot-working temperature. 800 to 925 °C (1450 to 1700 °F)
Hot-shortness temperature. 1010 °C (1850 °F)

C61500
90Cu-8Al-2Ni

Commercial Names

Previous trade name. Lusterloy

Chemical Composition

Composition limits. 89.0 to 90.5 Cu, 7.7 to 8.3 Al, 1.8 to 2.2 Ni, 0.015 Pb max

Applications

Typical uses. Hardware, decorative metal trim, interior furnishings, giftware, springs, fasteners, architectural panels and structural sections, deep drawn articles, tarnish-resistant articles

Mechanical Properties

Tensile properties. See Table 95.
Hardness. See Table 95.
Elastic modulus. Tension, 112 GPa (16.6 × 10^6 psi)
Fatigue strength. See Table 95.

Mass Characteristics

Density. 7.65 g/cm^3 (0.278 lb/in.^{3}) at 20 °C (68 °F)

Thermal Properties

Liquidus temperature. 1040 °C (1904 °F)
Solidus temperature. 1030 °C (1890 °F)
Coefficient of linear thermal expansion. 16.8 μm/m · K (9.3 μin./in. · °F) at 20 to 300 °C (68 to 572 °F)
Specific heat. 380 J/kg · K (0.09 Btu/lb · °F) at 20 °C (68 °F)
Thermal conductivity. 58 W/m · K (33.6 Btu/ft · h · °F) at 20 °C (68 °F)

Electrical Properties

Electrical conductivity. Volumetric, 12.6% IACS at 20 °C (68 °F)
Electrical resistivity. 137 nΩ · m at 20 °C (68 °F)

Optical Properties

Color. Gold

Chemical Properties

General corrosion behavior. Excellent; similar to that of other aluminum bronzes

Fabrication Characteristics

Machinability. 30% of C36000 (free-cutting brass)
Forgeability. 50% of C37700 (forging brass)
Formability. Suitable for forming by bending, drawing, deep drawing, forging, extrusion, blanking, and stamping; only slight directionality in bending. Good for cold working and hot forming
Weldability. Gas-shielded arc welding, shielded metal arc welding, and resistance welding: excellent. Soldering and brazing: easily done using mildly aggressive fluxes. Oxyfuel gas welding is not recommended.
Annealing temperature. 620 to 675 °C (1150 to 1250 °F)
Aging temperature. Order strengthening, 300 °C (575 °F) for 1 h
Hot-working temperature. 815 to 870 °C (1500 to 1600 °F)

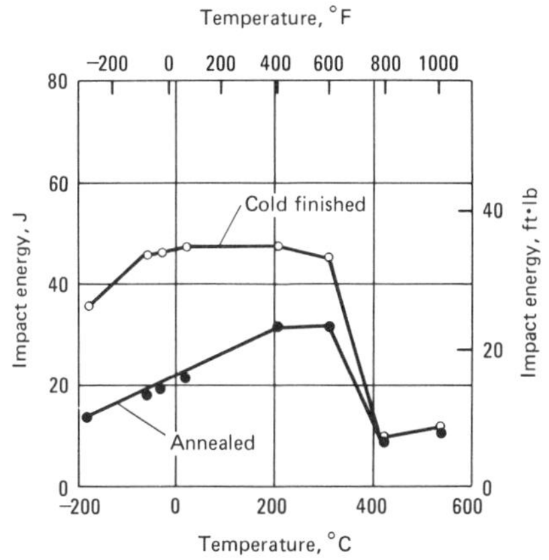

Fig. 45 Variation in Charpy V-notch impact strength with temperature for C62300

C62300
87Cu-10Al-3Fe

Commercial Names

Common name. Aluminum bronze, 9%

Specifications

ASME. Bar, rod, shapes: SB150
ASTM. Bar, rod, shapes: B 150. Forgings: B 283
SAE. J463
Government. Forgings: MIL-B-16166

Table 98 Typical compressive properties for C62400 rod, H50 temper

Rod diameter	Compressive strength at permanent set of: 0.1% MPa	0.1% ksi	1% MPa	1% ksi	10% MPa	10% ksi	Ultimate compressive strength MPa	Ultimate compressive strength ksi
≤25 mm (≤1 in.)	290	42	470	68	885	128	1140	165
25–50 mm (1–2 in.)	220	32	400	58	825	120	1090	158
50–75 mm (2–3 in.)	175	25	330	48	795	115	1090	158

Chemical Composition

Composition limits. 82.2 to 89.5 Cu, 8.5 to 11.0 Al, 2.0 to 4.0 Fe, 1.0 Ni (+ Co) max, 0.6 Sn max, 0.50 Mn max, 0.25 Si max, 0.5 max other (total)
Consequence of exceeding impurity limits. An excessive amount of Pb will cause hot shortness, and excessive Si will cause the alloy to lose ductility. Excessive Al will reduce ductility and corrosion resistance.

Applications

Typical uses. Produced as rod and bar for bearings, bushings, bolts, nuts, gears, valve guides, pump rods, cams, and applications requiring corrosion resistance
Precautions in use. Not suitable for use in oxidizing acids

Mechanical Properties

Tensile properties. Typical. Tensile strength, 605 MPa (88 ksi); yield strength, 305 MPa (44 ksi); elongation, 15% in 50 mm (2 in.); reduction in area, 15%. See Table 96.
Compressive properties. See Table 97.
Hardness. 89 HRB. See Table 96.
Poisson's ratio. 0.328
Elastic modulus. Tension, 115 GPa (17 × 10^6 psi); shear, 44 GPa (6.4 × 10^6 psi). See also Table 96.
Impact strength. Charpy V-notch, 25 to 40 J (18 to 30 ft · lbf); Izod, 43 to 47 J (32 to 35 ft · lbf). See also Fig. 45.
Fatigue strength. Reverse bending, 200 MPa (29 ksi) at 10^8 cycles

Structure

Microstructure. Duplex structure of face-centered-cubic α plus metastable body-centered-cubic β with iron-rich precipitates

Mass Characteristics

Density. 7.65 g/cm³ (0.276 lb/in.³) at 20 °C (68 °F)
Volume change on freezing. Approximately 2% expansion

Thermal Properties

Liquidus temperature. 1045 °C (1915 °F)
Solidus temperature. 1040 °C (1905 °F)
Phase transformation temperature. Eutectoid transformation, 563 to 570 °C (1045 to 1055 °F)
Coefficient of linear thermal expansion. 16.2 μm/m · K (9.0 μin./in. · °F) at 20 to 300 °C (68 to 572 °F)
Specific heat. 375 J/kg · K (0.09 Btu/lb · °F) at 20 °C (68 °F)
Thermal conductivity. 54.4 W/m · K (31.4 Btu/ft · h · °F) at 20 °C (68 °F); temperature coefficient, 0.12 W/m · K per K at 20 °C (68 °F)

Electrical Properties

Electrical conductivity. Volumetric, 12% IACS at 20 °C (68 °F)
Electrical resistivity. 144 nΩ · m at 20 °C (68 °F)

Magnetic Properties

Magnetic permeability. 1.17

Chemical Properties

General corrosion behavior. See C61400.
Resistance to specific agents. C62300 resists nonoxidizing mineral acids, but hydrochloric acid is more corrosive than other nonoxidizing mineral acids. C62300 resists dealloying, but to a lesser extent than C61300 or C61400. Like other aluminum bronzes, C62300 is not suitable for use in an oxidizing acid such as nitric acid.

Fabrication Characteristics

Machinability. Fair, with good surface finish possible. Carbide or tool steel cutters may be used. Typical conditions using tool steel cutters: roughing speed, 107 m/min (350 ft/min) with a feed of 0.3 mm/rev (0.011 in./rev); finishing speed, 350 m/min (1150 ft/min) with a feed of 0.15 mm/rev (0.006 in./rev)
Forgeability. 75% of C37700 (forging brass)
Formability. Good for cold working and hot forming
Weldability. Gas-shielded arc, shielded metal arc, and all types of resistance welding: good. Brazing: fair. Soldering and oxyfuel gas welding are not recommended.
Annealing temperature. 600 to 650 °C (1110 to 1200 °F)
Hot-working temperature. 700 to 875 °C (1290 to 1600 °F)
Hot-shortness temperature. 1010 °C (1850 °F)

C62400
86Cu-11Al-3Fe

Commercial Names

Common name. Aluminum bronze, 11%

Specifications

SAE. J463

Chemical Composition

Composition limits. 82.8 to 88.0 Cu, 10.0 to 11.5 Al, 2.0 to 4.5 Fe, 0.30 Mn max, 0.25 Si max, 0.20 Sn max, 0.5 max other (total)
Consequence of exceeding impurity limits. Excessive amounts of Si and Al decrease ductility.

Applications

Typical uses. Produced as rod and bar for gears, wear plates, cams, bushings, nuts, drift pins, and tie rods
Precautions in use. May lose ductility upon prolonged heating in range from 370 to 565 °C (700 to 1050 °F). Not suitable for use in oxidizing acids

Mechanical Properties

Tensile properties. Typical data for 50 mm (2 in.) diam round rod (half hard). Tensile strength, 655 MPa (95 ksi); yield strength (0.5% extension), 330 MPa (48 ksi); elongation, 14% in 50 mm (2 in.); reduction in area, 11%
Compressive properties. See Table 98.
Hardness. 92 HRB
Poisson's ratio. 0.318
Elastic modulus. Tension, 115 GPa (17 × 10^6 psi); shear, 44 GPa (6.4 × 10^6 psi)
Impact strength. Charpy keyhole, 15 J (11 ft · lbf) at −23 to 27 °C (−10 to 80 °F); Izod, 23 J (17 ft · lbf) at −23 to 27 °C (−10 to 80 °F)
Fatigue strength. Reverse bending, 235 MPa (34 ksi) at 10^8 cycles

Structure

Microstructure. Duplex-structure α plus metastable β phases and iron-rich precipitates

Mass Characteristics

Density. 7.45 g/cm³ (0.269 lb/in.³) at 20 °C (68 °F)
Volume change on freezing. Approximately 2% contraction

Thermal Properties

Liquidus temperature. 1040 °C (1900 °F)
Solidus temperature. 1025 °C (1880 °F)
Phase transformation temperature. Eutectoid, 560 to 570 °C (1045 to 1055 °F)
Coefficient of linear thermal expansion. 16.5 μm/m · K (9.2 μin./in. · °F) at 20 to 300 °C (68 to 572 °F)
Specific heat. 375 J/kg · K (0.09 Btu/lb · °F) at 20 °C (68 °F)
Thermal conductivity. 58.6 W/m · K (33.9 Btu/ft · h · °F) at 20 °C (68 °F); temperature coefficient, 0.12 W/m · K per K at 20 °C (68 °F)

Electrical Properties

Electrical conductivity. Volumetric, 12% IACS at 20 °C (68 °F)
Electrical resistivity. 144 nΩ · m at 20 °C (68 °F)

Magnetic Properties

Magnetic permeability. 1.34

Chemical Properties

General corrosion behavior. See C61400.

Table 99 Typical mechanical properties of C63000 rod at various temperatures

Temperature °C	Temperature °F	Tensile strength MPa	Tensile strength ksi	Yield strength(a) MPa	Yield strength(a) ksi	Elongation in 50 mm (2 in.), %	Reduction in area, %	Modulus of elasticity GPa	Modulus of elasticity 10^6 psi	Hardness, HB(b)
Cold finished										
−182	−295	845	122.5	469	68.1	8	10	128	18.5	238
−60	−75	774	112.3	443	64.3	26	28	131	19.0	216
−29	−20	784	113.7	463	67.1	24	29	132	19.1	209
20	70	776	112.5	407	59.1	20	21	117	16.9	200
204	400	694	100.7	403	58.5	13	15	136	19.7	188
316	600	582	84.4	373	54.1	8	9	85	12.3	181
427	800	245	35.5	166	24.1	51	56	57	8.2	98
538	1000	107	15.5	88	12.8	41	59	44	6.4	47
Annealed										
−182	−295	867	125.8	431	62.5	12	12	130	18.8	235
−60	−75	784	113.7	379	54.9	24	26	123	17.9	212
−29	−20	781	113.3	384	55.7	23	23	139	20.1	209
20	70	766	111.1	370	53.6	21	21	125	18.1	200
204	400	706	102.4	348	50.4	16	15	107	15.5	189
316	600	605	87.7	337	48.9	10	11	110	15.9	176
427	800	232	33.7	158	22.9	41	46	64	9.3	101
538	1000	95	13.7	78	11.3	39	46	48	7.0	50

(a) At 0.5% extension under load. (b) 3000 kg (6615 lb) load

Resistance to specific agents. C62400 resists nonoxidizing mineral acids, but hydrochloric acid is more corrosive than other nonoxidizing mineral acids. C62400 is susceptible to dealloying, but proper heat treatment increases resistance to this type of corrosion. Like other aluminum bronzes, C62400 is not suitable for use in an oxidizing acid such as nitric acid.

Fabrication Characteristics

Machinability. 50% of C36000 (free-cutting brass); chips break readily. Carbide or tool steel cutters may be used. Typical conditions using tool steel cutters: roughing speed, 90 m/min (300 ft/min) with a feed of 0.3 mm/rev (0.011 in./rev); finishing speed, 290 m/min (950 ft/min) with a feed of 0.1 mm/rev (0.004 in./rev). Using carbide cutters with 2.3 to 6.4 mm (0.09 to 0.25 in.) cut: roughing speed, 53 m/min (175 ft/min) with a feed of 0.3 mm/rev (0.011 in./rev); finishing speed, 38 to 45 m/min (125 to 150 ft/min) with a feed of 0.3 mm/rev (0.011 in./rev)
Weldability. Similar to that of C62300
Annealing temperature. 600 to 700 °C (1110 to 1300 °F)
Hot-working temperature. 760 to 925 °C (1400 to 1700 °F)

C62500
82.7Cu-4.3Fe-13Al

Commercial Names

Trade name. Ampco 21, Wearite 4-13

Chemical Composition

Composition limits. 12.5 to 13.5 Al, 3.5 to 5.0 Fe, 2.0 Mn max, 0.5 max other (total), bal Cu
Consequence of exceeding impurity limits. Possibility of hot shortness, reduced wear resistance, increased spalling tendency, and lower strength when elements such as Pb, Zn, P, and Si are present in more than trace quantities

Applications

Typical uses. Guide bushings, wear strips, cams, sheet metal forming dies, forming rolls
Precautions in use. Low ductility and impact resistance make it advisable to provide adequate structural support for components made of C62500 that will be subjected to shock loads or high stress. Corrosion resistance is inferior to that of aluminum bronzes containing less aluminum.

Mechanical Properties

Tensile properties. Typical M30 and O61 tempers: tensile strength, 690 MPa (100 ksi); yield strength (0.5% extension), 380 MPa (55 ksi); elongation, 1% in 50 mm (2 in.); reduction in area, 1%
Compressive properties. Compressive strength, 450 MPa (65 ksi) at a permanent set of 0.1%; 880 MPa (128 ksi) at a permanent set of 1%
Hardness. 27 HRC
Poisson's ratio. 0.312
Elastic modulus. Tension, 110 GPa (16×10^6 psi); shear, 42.3 GPa (6.13×10^6 psi)
Impact strength. Izod or Charpy keyhole, 3 J (2 ft · lbf) at −18 to 100 °C (0 to 212 °F)
Fatigue strength. Rod, M30 temper: 460 MPa (67 ksi) at 10^8 cycles

Structure

Microstructure. Primarily body-centered-cubic metastable β phase with small crystals of ordered close-packed hexagonal γ phase

Magnetic Properties

Magnetic permeability. 1.2

Mass Characteristics

Density. 7.21 g/cm^3 (0.260 lb/in.^{3}) at 20 °C (68 °F)

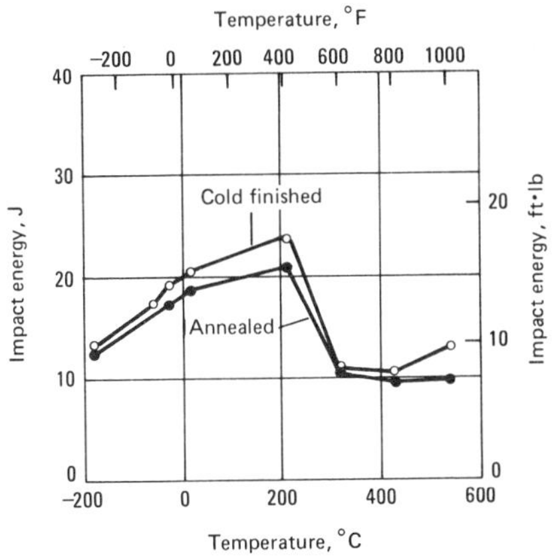

Fig. 46 Variation in Charpy V-notch impact strength with temperature for C63000

Thermal Properties

Liquidus temperature. 1052 °C (1925 °F)
Solidus temperature. 1047 °C (1917 °F)
Coefficient of linear thermal expansion. 16.2 μm/m · K (9.0 μin./in. · °F) at 20 to 300 °C (68 to 572 °F)
Specific heat. 380 J/kg · K (0.09 Btu/lb · °F) at 20 °C (68 °F)
Thermal conductivity. 38.9 W/m · K (22.5 Btu/ft · h · °F) at 20 °C (68 °F); temperature coefficient, 0.093 W/m · K per K at −100 to 150 °C (−150 to 300 °F)

Electrical Properties

Electrical conductivity. Volumetric, 10% IACS at 20 °C (68 °F)
Electrical resistivity. 172 nΩ · m at 20 °C (68 °F)

Chemical Properties

General corrosion behavior. Adequate corrosion resistance to ambient moisture and industrial atmospheres. C62500 is rarely used for its corrosion characteristics in strongly corrosive environments. General corrosion characteristics are inferior to those of C62400 and C62300.

Fabrication Characteristics

Machinability. 20% of C36000 (free-cutting brass)
Formability. Not recommended for cold working; excellent for hot forming
Weldability. Gas-shielded arc and shielded metal arc welding: good. Brazing and resistance welding: fair. Oxyfuel gas welding and soldering are not recommended.
Annealing temperature. 600 to 650 °C (1100 to 1200 °F)
Hot-working temperature. 745 to 850 °C (1375 to 1550 °F)

C63000
82Cu-10Al-5Ni-3Fe

Commercial Names

Previous trade name. Aluminum bronze E
Common name. Nickel-aluminum bronze

Specifications

AMS. Bar, shapes: 4640
ASME. Bar, rod, shapes: SB150. Condenser tube plate: SB171
ASTM. Bar, rod, shapes: B 124, B 150. Condenser tube plate: B 171. Forgings: B 283
SAE. J463
Government. Flat products, rod, shapes: QQ-C-465. Forgings: QQ-C-465, MIL-B-16166

Chemical Composition

Composition limits. 78.0 to 85.0 Cu, 9.0 to 11.0 Al, 2.0 to 4.0 Fe, 4.0 to 5.5 Ni (+ Co), 1.5 Mn max, 0.30 Zn max, 0.25 Si max, 0.20 Sn max, 0.5 max other (total)
Consequence of exceeding impurity limits. Excessive amounts of Zn, Sn, and Pb will cause cracking during hot working and joining. Excessive Si will result in machining difficulties.

Applications

Typical uses. Produced as rod, bar, and forgings for nuts, bolts, shafting, pump parts, valve seats, faucet balls, gears, cams, structural members, and tube sheets for condensers in power stations and desalting units
Precautions in use. Not suitable for use in oxidizing acids

Mechanical Properties

Tensile properties. Typical data for 25 mm (1 in.) diam round rod, HR50 temper: tensile strength, 760 MPa (110 ksi); yield strength, 470 MPa (68 ksi); elongation, 10% in 50 mm (2 in.); reduction in area, 10%. See also Table 99.
Compressive properties. HR50 temper. Compressive strength, ultimate: 1035 MPa (150 ksi)
Hardness. HR50 temper: 94 HRB. See also Table 99.
Poisson's ratio. 0.328
Elastic modulus. Tension, 115 GPa (17×10^6 psi); shear, 44 GPa (6.4×10^6 psi). See also Table 99.
Impact strength. Charpy V-notch, 16 to 21 J (12 to 15 ft · lbf) at 20 °C (68 °F). See also Fig. 46.
Fatigue strength. Reverse bending, 255 MPa (37 ksi) at 10^8 cycles

Structure

Microstructure. Features α, κ, and metastable β phases in various structures, depending on heat treatment and/or thermal history and composition. Normally, α plus α-κ lamellar structure with areas of β

Mass Characteristics

Density. 7.58 g/cm^3 (0.274 lb/in.^{3}) at 20 °C (68 °F)
Volume change on freezing. Approximately 2% contraction

Thermal Properties

Liquidus temperature. 1055 °C (1930 °F)
Solidus temperature. 1035 °C (1895 °F)
Coefficient of linear thermal expansion. 16.2 μm/m · K (9.0 μin./in. · °F) at 20 to 300 °C (68 to 572 °F)
Specific heat. 375 J/kg · K (0.09 Btu/lb · °F) at 20 °C (68 °F)
Thermal conductivity. 37.7 W/m · K (21.8 Btu/ft · h · °F) at 20 °C (68 °F); temperature coefficient, 0.09 W/m · K per K at 20 °C (68 °F)

Electrical Properties

Electrical conductivity. Volumetric, 9% IACS at 20 °C (68 °F)
Electrical resistivity. 192 nΩ · m at 20 °C (68 °F)

Magnetic Properties

Magnetic permeability. 1.05

Chemical Properties

General corrosion behavior. See C61400.

Fabrication Characteristics

Machinability. 30% of C36000 (free-cutting brass), with breaking to slightly stringy chips. Carbide or tool steel cutters may be used, and good lubrication and cooling are essential. Typical conditions using tool steel cutters: roughing speed, 75 m/min (250 ft/min) with a feed of 0.3 mm/rev (0.011 in./rev); finishing speed, 290 m/min (950 ft/min) with a feed of 0.1 mm/rev (0.004 in./rev). Using carbide cutters: roughing speed, 53 m/min (175 ft/min) with a feed of 0.3 mm/rev (0.011 in./rev); finishing speed, 38 m/min (125 ft/min) with a feed of 0.3 mm/rev (0.011 in./rev)
Forgeability. 75% of C37700 (forging brass)
Formability. Poor for cold working; good for hot forming
Weldability. Gas-shielded arc, coated metal arc and spot, seam, and butt resistance welding: good. Brazing: fair. Soldering and oxyacetylene welding are not recommended.
Annealing temperature. 600 to 700 °C (1100 to 1300 °F)
Hot-working temperature. 800 to 925 °C (1450 to 1700 °F)

C63200
82Cu-9Al-5Ni-4Fe

Commercial Names

Common name. Nickel-aluminum bronze

Chemical Composition

Composition limits. 75.9 to 84.5 Cu, 8.5 to 9.5 Al, 3.0 to 5.0 Fe, 4.0 to 5.5 Ni (+ Co), 3.5 Mn max, 0.10 Si max, 0.02 Pb max, 0.5 max other (total)
Iron content shall not exceed Ni content.
Consequence of exceeding impurity limits. Excessive Pb and Si will cause hot shortness in weld joints. Excessive Mn will reduce corrosion resistance.

Table 100 Typical mechanical properties of C63600

Diameter mm	in.	Temper	Tensile strength MPa	ksi	Elongation in 50 mm (2 in.), %	Hardness, HRB
Rod						
16	0.63 ...	O61	415	60	64	...
14	0.56 ...	H01	510	74	31	...
Wire						
10	0.40 ...	O61	415	60	67	...
11	0.42 ...	H00 (7%)	470	68	52	71
12	0.49 ...	H01 (21%)	580	84	29	84

Applications

Typical uses. Produced as rod, bar, and forgings for nuts, bolting, shafts, pump parts, propellers, and miscellaneous uses for corrosion- and spark-resistant parts for industrial, marine, and submarine applications
Precautions in use. Not suitable for use in oxidizing acids

Mechanical Properties

Tensile properties. Depending on amount of cold work or heat treatment. Tensile strength, 640 to 725 MPa (93 to 105 ksi); yield strength, 330 to 380 MPa (48 to 55 ksi); elongation, 18% in 50 mm (2 in.); reduction in area, 18%
Compressive properties. Compressive strength, ultimate: 760 MPa (110 ksi)
Hardness. 92 to 97 HRB
Poisson's ratio. 0.320
Elastic modulus. Tension, 115 GPa (17×10^6 psi); shear, 44 GPa (6.4×10^6 psi)
Impact strength. Charpy V-notch, 23 to 27 J (17 to 20 ft · lbf) at −29 to 20 °C (−20 to 78 °F); Charpy keyhole, 13 to 16 J (10 to 12 ft · lbf) at −29 to 20 °C (−20 to 78 °F)

Structure

Microstructure. Features α, κ, and metastable β phases in various structures, depending on heat treatment and/or thermal history and composition. Normally, α plus α-κ lamellar structure with or without areas of β phase

Mass Characteristics

Density. 7.64 g/cm^3 (0.276 lb/in.^{3}) at 20 °C (68 °F)
Volume change on freezing. Approximately 2% contraction

Thermal Properties

Liquidus temperature. 1060 °C (1940 °F)
Solidus temperature. 1040 °C (1905 °F)
Coefficient of linear thermal expansion. 16.2 μm/m · K (9.0 μin./in. · °F) at 20 to 300 °C (68 to 572 °F)
Specific heat. 439 J/kg · K (0.105 Btu/lb · °F) at 20 °C (68 °F)
Thermal conductivity. 36 W/m · K (21 Btu/ft · h · °F) at 20 °C (68 °F)

Table 101 Typical mechanical properties of C63800 sheet and strip

Temper	Tensile strength MPa	Tensile strength ksi	Yield strength at 0.2% offset MPa	Yield strength at 0.2% offset ksi	Elongation in 50 mm (2 in.), %	Hardness HRB	Hardness HR30T
O61	565	82	385	56	33	...	74
H01	660	96	565	82	17	94	78
H02	730	106	640	93	10	97	80
H03	765	111	680	99	8	98	81
H04	825	120	750	109	5	99	82
H06	855	124	780	113	4	100	82
H08	895	130	800	116	3	100	83
H10	895 min	130 min	820 min	119 min	2 max	100 min	83 min

Electrical Properties

Electrical conductivity. Volumetric, 7% IACS at 20 °C (68 °F)
Electrical resistivity. 246 nΩ · m at 20 °C (68 °F)

Magnetic Properties

Magnetic permeability. 1.04

Fabrication Characteristics

Machinability. Fair. Tendency to form stringy chips and to gall makes good lubrication and cooling essential. Tool steel or carbide cutters may be used. Good finish and fine thread tapping possible. Typical conditions using tool steel cutters: roughing speed, 75 m/min (250 ft/min) with a feed of 0.3 mm/rev (0.011 in./rev); finishing speed, 290 m/min (950 ft/min) with a feed of 0.1 mm/rev (0.004 in./rev). Using carbide cutters: roughing speed, 53 m/min (175 ft/min) with a feed of 0.3 mm/rev (0.011 in./rev); finishing speed, 38 m/min (125 ft/min) with a feed of 0.3 mm/rev (0.011 in./rev)
Annealing temperature. 705 to 880 °C (1300 to 1615 °F)
Hot-working temperature. 705 to 925 °C (1300 to 1700 °F)

C63600
95.5Cu-3.5Al-1.0Si

Chemical Composition

Composition limits. 93.5 to 96.3 Cu, 3.0 to 4.0 Al, 0.7 to 1.3 Si, 0.50 Zn max, 0.20 Sn max, 0.15 Ni max, 0.15 Fe max, 0.05 Pb max

Applications

Typical uses. Rod and wire for components for pole line hardware; cold-headed nuts for wire and cable connectors; bolts; screw machine products

Mechanical Properties

Tensile properties. See Table 100.
Hardness. See Table 100.
Elastic modulus. Tension, 110 GPa (16 × 10^6 psi)

Mass Characteristics

Density. 8.33 g/cm^3 (0.301 lb/in.^{3}) at 20 °C (68 °F)

Thermal Properties

Liquidus temperature. 1035 °C (1890 °F)
Coefficient of linear thermal expansion. 17.2 μm/m · K (9.4 μin./in. · °F) at 20 to 300 °C (68 to 572 °F)
Thermal conductivity. 57 W/m · K (33 Btu/ft · h · °F) at 20 °C (68 °F)

Electrical Properties

Electrical conductivity. Volumetric, 12% IACS at 20 °C (68 °F)
Electrical resistivity. 143 nΩ · m at 20 °C (68 °F)

Fabrication Characteristics

Machinability. 40% of C36000 (free-cutting brass)
Formability. Excellent for cold working; fair for hot forming
Weldability. Gas-shielded arc, shielded metal arc, and resistance welding: fair. Soldering, brazing, and oxyfuel gas welding are not recommended.
Hot-working temperature. 760 to 875 °C (1400 to 1600 °F)

C63800
95Cu-2.8Al-1.8Si-0.40Co

Commercial Names

Trade name. Coronze

Chemical Composition

Composition limits. 2.5 to 3.1 Al, 1.5 to 2.1 Si, 0.25 to 0.55 Co, 0.80 Zn max, 0.10 Ni max, 0.05 Pb max, 0.10 Fe max, 0.10 Mn max, bal Cu

Applications

Typical uses. Springs, switch parts, contacts, relay springs, glass sealing, and porcelain enameling

Mechanical Properties

Tensile properties. See Table 101 and Fig. 47 and 48.
Hardness. See Table 101.
Poisson's ratio. 0.312
Elastic modulus. Tension, 117 GPa (16.7 × 10^6 psi)

Mass Characteristics

Density. 8.28 g/cm^3 (0.299 lb/in.^{3}) at 20 °C (68 °F)

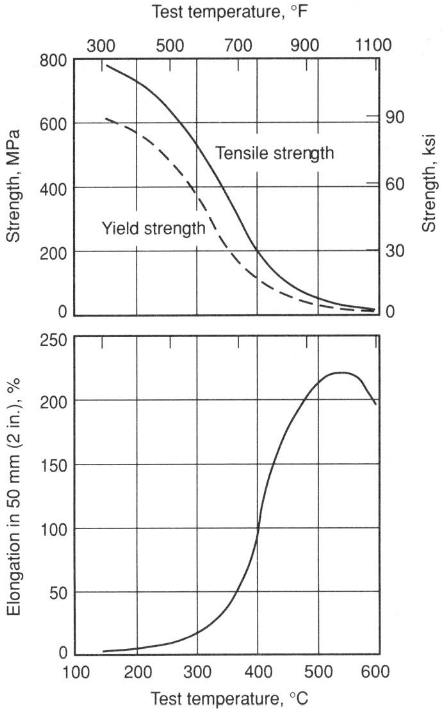

Fig. 47 Typical short-time tensile properties of C63800, H02 temper

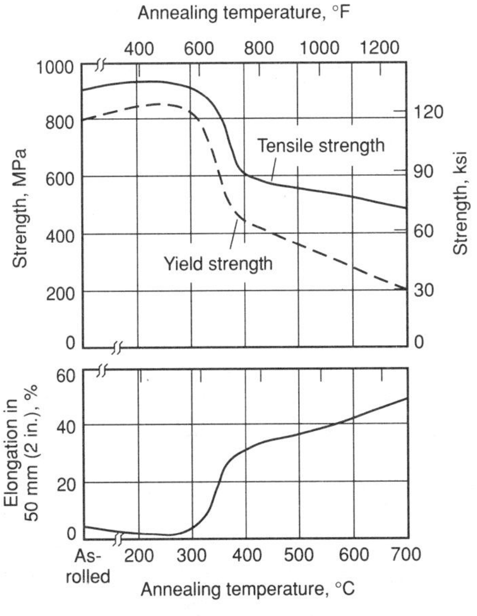

Fig. 48 Anneal resistance of C63800 strip, H08 temper. Typical room-temperature tensile properties for material annealed 1 h at various temperatures

Thermal Properties

Liquidus temperature. 1030 °C (1885 °F)
Solidus temperature. 1000 °C (1830 °F)
Coefficient of linear thermal expansion. 17.1 μm/m · K (9.5 μin./in. · °F) at 20 to 300 °C (68 to 572 °F)
Specific heat. 375 J/kg · K (0.09 Btu/lb · °F) at 20 °C (68 °F)
Thermal conductivity. 42 W/m · K (24 Btu/ft · h · °F) at 20 °C (68 °F)

Table 102 Typical mechanical properties of C65100

Temper	Tensile strength MPa	Tensile strength ksi	Yield strength(a) MPa	Yield strength(a) ksi	Elongation in 50 mm (2 in.), %	Hardness	Shear strength MPa	Shear strength ksi
Rod, 25 mm (1 in.) thick								
OS035	275	40	105	15	50	55 HRF	...	...
H04 (36%)	485	70	380	55	15	80 HRB	310	45
H06 (50%)	620	90	460	67	12	90 HRB	345	50
Wire, 2 mm (0.08 in.) diameter								
H00	380	55	275	40	40	...	250	36
H01	450	65	345	50	25	...	275	40
H02	550	80	435	63	15	...	310	45
H04	690	100	485	70	11	...	345	50
H06	725	105	490	71	10	...	365	53
Wire, 11 mm (0.44 in.) diameter								
H00 (21%)	435	63	...	...	30	...	...	...
H02 (37%)	550	80	...	...	20	...	...	...
H04 (60%)	655	95	...	...	12	...	...	...
Tubing, 25 mm (1.0 in.) outside diameter × 1.65 mm (0.065 in.) wall thickness								
OS015	310	45	140	20	55	68 HRF	...	...
H80 (35%)	450	65	275	40	20	75 HRB	...	...

(a) At 0.5% extension under load

Table 103 Nominal mechanical properties of C65400 strip

Temper	Tensile strength MPa	Tensile strength ksi	Yield strength MPa	Yield strength ksi	Elongation in 50 mm (2 in.), %	Hardness, HRB	Fatigue strength at 10^8 cycles MPa	Fatigue strength at 10^8 cycles ksi
H01	570	83	415	60	30	82	...	...
H02	655	95	585	85	20	92	...	...
H03	725	105	635	92	13	95	...	...
H04	790	115	700	102	6	97	235	34
H06	825	120	760	110	5	99	...	...
H08	890	129	815	118	3	100	255	37
H10	930	135	860	125	2	101	...	...

Electrical Properties

Electrical conductivity. Volumetric, 10% IACS at 20 °C (68 °F), annealed
Electrical resistivity. 174 nΩ · m at 20 °C (68 °F), annealed

Chemical Properties

General corrosion behavior. C63800 is more resistant to stress corrosion than the nickel silvers, approaching the performance of the highly resistant phosphor bronzes. This alloy is far superior to most other copper alloys in resistance to crevice corrosion. At elevated temperature, the oxidation resistance of C63800 is excellent. For example, after heating in air for 2 h, the film thickness on C83800 was 7 nm (0.26 μin.) at 450 °C (840 °F), 12 nm (0.47 μin.) at 600 °C (1100 °F), and 24 nm (0.94 μin.) at 700 °C (1300 °F). On the basis of weight gain after heating 2 to 24 h in air at temperatures of 600 to 800 °C (1100 to 1300 °F), C63800 was consistently superior to Nickel 270, Nichrome (80Ni-20Cr), type 301 stainless steel, Incoloy 800 (ASTM B 408), and C60600. The superiority of C63800 was especially evident at 800 °C (1300 °F).

Fabrication Characteristics

Formability. Suitable for blanking, drawing, bending, shearing, and stamping. Excellent for cold working and hot forming
Weldability. Soft soldering utilizing standard fluxes is normally employed. Brazing, gas-shielded arc welding, and all forms of resistance welding are also commonly used.
Annealing temperature. 400 to 600 °C (750 to 1100 °F). See also Fig. 48.

C65100
98.5Cu-1.5Si

Commercial Names

Previous trade name. Low-silicon bronze B
Common name. Low-silicon bronze

Specifications

ASME. Bar, rod, shapes: SB98. Tubular products: SB315
ASTM. Flat products: B 97. Bar, rod, shapes: B 98. Tubular products: B 315. Wire: B 99
Government. QQ-C-591

Chemical Composition

Composition limits. 0.8 to 2.0 Si, 0.05 Pb max, 0.8 Fe max, 1.5 Zn max, 0.7 Mn max, bal Cu

Applications

Typical uses. Aircraft: hydraulic pressure lines. Hardware: anchor screws, bolts, cable clamps, cap screws, machine screws, marine hardware, nuts, pole line hardware, rivets, U-bolts. Industrial: electrical conduits, heat exchanger tubes, welding rod

Mechanical Properties

Tensile properties. See Table 102.
Shear strength. See Table 102.
Hardness. See Table 102.
Elastic modulus. Tension, 115 GPa (17 × 10^6 psi); shear, 44 GPa (6.4 × 10^6 psi)
Fatigue strength. Reverse bending, H04 temper, 170 MPa (25 ksi) at 10^8 cycles; H06 temper, 195 MPa (28 ksi) at 10^8 cycles

Mass Characteristics

Density. 8.75 g/cm^3 (0.316 lb/in.^{3}) at 20 °C (68 °F)

Thermal Properties

Liquidus temperature. 1060 °C (1940 °F)
Solidus temperature. 1030 °C (1890 °F)
Coefficient of linear thermal expansion. 18 μm/m · K (9.9 μin./in. · °F) at 20 to 300 °C (68 to 572 °F)
Specific heat. 380 J/kg · K (0.09 Btu/lb · °F) at 20 °C (68 °F)
Thermal conductivity. 57 W/m · K (33 Btu/ft · h · °F) at 20 °C (68 °F)

Electrical Properties

Electrical conductivity. Volumetric, 12% IACS at 20 °C (68 °F)
Electrical resistivity. 144 nΩ · m at 20 °C (68 °F)

Fabrication Characteristics

Machinability. 30% of C36000 (free-cutting brass)
Formability. Excellent for cold working and hot forming
Weldability. Soldering, brazing, gas-shielded arc, resistance spot, and resistance butt welding: excellent. Oxyfuel gas and resistance seam welding: good. Shielded metal arc welding: fair
Annealing temperature. 475 to 675 °C (900 to 1250 °F)
Hot-working temperature. 700 to 875 °C (1300 to 1600 °F)

C65400
95.4Cu-3.0Si-1.5Sn-0.1Cr

Chemical Composition

Composition limits. 2.7 to 3.4 Si, 1.2 to 1.9 Sn, 0.01 to 0.12 Cr, 0.5 Zn max, 0.05 Pb max, bal Cu

Applications

Typical uses. Applications where high strength and good formability combined with good stress relaxation resistance is required. Specific uses include contact springs, connectors, and wiring devices.

Mechanical Properties

Tensile properties. See Table 103.

Table 104 Typical mechanical properties of C65500

Temper	Tensile strength MPa	ksi	Yield strength(a) MPa	ksi	Elongation in 50 mm (2 in.), %	Hardness, HRB	Shear strength MPa	ksi
Flat products, 1 mm (0.04 in.) thick								
OS070	385	56	145	21	63	40	290	42
OS035	415	60	170	25	60	62	295	43
OS015	435	63	205	30	55	66	310	45
H01	470	68	240	35	30	75	325	47
H02	540	78	310	45	17	87	345	50
H04	650	94	400	58	8	93	390	57
H06	715	104	415	60	6	96	415	60
H08	760	110	427	62	4	97	435	63
Rod, 25 mm (1.0 in.) diameter								
OS050	400	58	150	22	60	60	295	43
H02 (20%)	540	78	310	45	35	85	360	52
H04 (36%)	635	92	380	55	22	90	400	58
H06 (50%)	745	108	415	60	13	95	425	62
Wire, 2 mm (0.08 in.) diameter								
OS035	415	60	170	25	60	...	295	43
H00	485	70	275	40	35	...	330	48
H01	550	80	330	48	20	...	360	52
H02	675	98	395	57	8	...	400	58
H04	860	125	450	65	5	...	450	65
H08 (80%)	1000	145	485	70	3	...	485	70
Tubing, 25 mm (1.0 in.) outside diameter × 1.65 mm (0.065 in.) wall thickness								
OS050	395	57	...	...	70	45	...	...
H80 (35%)	640	93	...	...	22	92	...	...

(a) At 0.5% extension under load

Hardness. See Table 103.
Elastic modulus. 117 GPa (17.0 × 10^6 psi)
Fatigue strength. See Table 103.

Mass Characteristics

Density. 8.55 g/cm^3 (0.309 lb/in.^{3}) at 20 °C (68 °F)

Thermal Properties

Liquidus temperature. 1020 °C (1865 °F)
Solidus temperature. 955 °C (1755 °F)
Coefficient of linear thermal expansion. 17.5 μm/m · K (9.7 μin./in. · °F) at 20 to 300 °C (68 to 572 °F)
Thermal conductivity. 36 W/m · K (21 Btu/ft · h · °F) at 20 °C (68 °F)

Electrical Properties

Electrical conductivity. Volumetric, 7% IACS at 20 °C (68 °F)
Electrical resistivity. 246 nΩ · m at 20 °C (68 °F)

Fabrication Characteristics

Machinability. 30% of C36000 (free-cutting brass)
Formability. Excellent capacity for both cold working and hot forming
Weldability. Soldering and brazing: good. All forms of resistance welding and gas-shielded arc welding: excellent. Coated metal arc welding: fair
Annealing temperature. 400 to 600 °C (750 to 1100 °F)

C65500
97Cu-3Si

Commercial Names

Previous trade name. High-silicon bronze A
Common name. High-silicon bronze

Specifications

AMS. Bar, rod: 4615. Tubing: 4665
ASME. Flat products: SB96. Bar, rod, shapes: SB98. Tubular products: SB315
ASTM. Flat products: B 96, B 97, B 100. Bar, rod, shapes: B 98, B 124. Forgings: B 283. Tubular products: B 315. Wire: B 99
SAE. J463
Government. QQ-C-591. Tubing: MIL-T-8231

Chemical Composition

Composition limits. 2.8 to 3.8 Si, 0.5 Pb max, 0.8 Fe max, 1.5 Zn max, 1.5 Mn max, 0.6 Ni max, bal Cu

Applications

Typical uses. Aircraft: hydraulic pressure lines. Hardware: bolts, burrs, butts, clamps, cotter pins, hinges, marine hardware, nails, nuts, pole line hardware, screws. Industrial: bearing plates, bushings, cable, channels, chemical equipment, heat exchanger tubes, kettles, piston rings, tanks, rivets, screen cloth and wire, screen plates, shafting. Marine: propeller shafts

Mechanical Properties

Tensile properties. See Table 104.
Shear strength. See Table 104.
Hardness. See Table 104.

Table 105 Typical mechanical properties of C66400

Temper	Tensile strength MPa	ksi	Yield strength at 0.2% offset MPa	ksi	Elongation in 50 mm (2 in.), %
O60	435	63	310	45	25
H01	495	72	455	66	13
H02	545	79	525	76	7
H03	570	83	560	81	6
H04	605	88	585	85	5
H06	650	94	615	89	4
H08	670	97	635	92	3
H10	690	100	640	93	3

Elastic modulus. Tension, 105 GPa (15 × 10^6 psi); shear, 39 GPa (5.6 × 10^6 psi)
Fatigue strength. Reverse bending, H04 temper, 200 MPa (29 ksi) at 10^8 cycles; H08 temper, 205 MPa (30 ksi) at 10^8 cycles

Mass Characteristics

Density. 8.53 g/cm^3 (0.308 lb/in.^{3}) at 20 °C (68 °F)

Thermal Properties

Liquidus temperature. 1025 °C (1880 °F)
Solidus temperature. 970 °C (1780 °F)
Coefficient of linear thermal expansion. 18 μm/m · K (10 μin./in. · °F) at 20 to 300 °C (68 to 572 °F)
Specific heat. 380 J/kg · K (0.09 Btu/lb · °F) at 20 °C (68 °F)
Thermal conductivity. 36 W/m · K (21 Btu/ft · h · °F) at 20 °C (68 °F)

Electrical Properties

Electrical conductivity. Volumetric, 7% IACS at 20 °C (68 °F)
Electrical resistivity. 246 nΩ · m at 20 °C (68 °F)

Fabrication Characteristics

Machinability. 30% of C36000 (free-cutting brass)
Forgeability. 40% of C37700 (forging brass)
Formability. Excellent for cold working and hot forming
Weldability. Brazing, gas-shielded arc, and all forms of resistance welding: excellent. Soldering and oxyfuel gas welding: good. Shielded metal arc welding: fair
Annealing temperature. 475 to 700 °C (900 to 1300 °F)
Hot-working temperature. 700 to 875 °C (1300 to 1600 °F)

C66400
86.5Cu-1.5Fe-0.5Co-11.5Zn

Commercial Names

Previous trade name. Cobron

Chemical Composition

Composition limits. 1.3 to 1.7 Fe, 0.30 to 0.70 Co, 11.0 to 12.0 Zn, 0.05 Sn max, 0.05 Ni max, 0.05 Al max, 0.05 Mn max, 0.05 Si

Table 106 Typical mechanical properties of 1 mm (0.04 in.) thick C68800 strip

Temper	Tensile strength, MPa	Tensile strength, ksi	Yield strength at 0.2% offset, MPa	Yield strength at 0.2% offset, ksi	Elongation in 50 mm (2 in.), %	Hardness, HRB	Hardness, HR30T
O60(a)	565	82	365	53	35	78	69
O50	615	89	475	69	30	. . .	. . .
H01	650	94	525	76	20	90.5	78
H02	725	105	635	92	9	95	81
H04	780	113	705	102	5	97	82.5
H06	825	120	750	109	3	98	83
H08	885	128	785	114	2	99	83.5
H10	895 min	130 min	805 min	117 min	2 max	99 max	84 min

(a) Annealed C68800 usually has a very fine grain size (0.010 mm or less).

Table 107 Typical mechanical properties of C68800 after low-temperature thermal treatment

Temper	As-rolled: Tensile strength, MPa	As-rolled: Tensile strength, ksi	As-rolled: Yield strength at 0.2% offset, MPa	As-rolled: Yield strength at 0.2% offset, ksi	Stabilization treated(a): Tensile strength, MPa	Stabilization treated(a): Tensile strength, ksi	Stabilization treated(a): Yield strength at 0.2% offset, MPa	Stabilization treated(a): Yield strength at 0.2% offset, ksi
H02(b)	695	101	640	93	725	105	690	100
H04	750	109	670	97	780	113	740	107
H06	840	122	760	110	910	132	895	130
H08	890	129	785	114	960	139	925	134
H10	895 min	130 min	805 min	117 min	965	140	945	137

(a) Heated 1 h at 205 to 230 °C (400 to 445 °F). (b) Stabilization treatment is not effective on H00 or H01 temper material.

Table 108 Typical mechanical properties of C69000

Temper	Tensile strength, MPa	Tensile strength, ksi	Yield strength at 0.2% offset, MPa	Yield strength at 0.2% offset, ksi	Elongation in 50 mm (2 in.), %	Hardness, HRB	Hardness, HR30T
OS025	565	82	360	52	35	. . .	69
H01	650	94	525	76	19.5	90.5	. . .
H02	715	105	635	92	9	95	. . .
H04	780	113	700	102	4.5	97	. . .
H06	825	120	750	109	2.5	98	. . .
H08	870	126	785	114	1.5	96	. . .
H10	895 min	130 min	805	117	2 max	99 min	. . .
EHT(a)	930	135	875	127	1	. . .	84.5

(a) Cold rolled 50% and stress relief annealed 1 h at 220 °C (425 °F)

max, 0.05 Ag max, 0.02 P max, 0.015 Pb max, bal Cu. Note: The Fe + Co content shall be 1.8 to 2.0 (total).

Applications

Typical uses. Spring washers, switchblades, fuse clips, contact springs, socket contacts, connectors, terminals, and similar parts for electronic and electromechanical assemblies

Mechanical Properties

Tensile properties. See Table 105.
Elastic modulus. Tension, 112 GPa (16.3 × 10^6 psi)
Fatigue strength. Reverse bending, O60 temper, 165 MPa (24 ksi) at 10^8 cycles; H04 temper, 185 MPa (27 ksi) at 10^8 cycles

Mass Characteristics

Density. 8.74 g/cm^3 (0.317 lb/in.^{3}) at 20 °C (68 °F)

Thermal Properties

Liquidus temperature. 1055 °C (1930 °F)
Solidus temperature. 1035 °C (1895 °F)
Thermal conductivity. 116 W/m · K (67 Btu/ft · h · °F) at 20 °C (68 °F)

Electrical Properties

Electrical conductivity. Volumetric, O61 temper: 30% IACS at 20 °C (68 °F)
Electrical resistivity. O61 temper: 57.5 nΩ · m at 20 °C (68 °F)

C68800
73.5Cu-22.7Zn-3.4Al-0.4Co

Commercial Names

Trade name. Alcoloy

Specifications

ASTM. Flat products: B 592

Chemical Composition

Composition limits. 72.3 to 74.7 Cu, 3.0 to 3.8 Al, 0.25 to 0.55 Co, 0.05 Pb max, 0.05 Fe max, 0.010 max other (total), bal Zn (25.1 to 27.1 Al + Zn)

Applications

Typical uses. Springs, switches, contacts, relays, terminals, plug receptacles, connectors

Mechanical Properties

Tensile properties. See Tables 106 and 107.
Hardness. See Table 106.
Elastic modulus. Tension, 116 GPa (16.8 × 10^6 psi)

Mass Characteristics

Density. 8.20 g/cm^3 (0.296 lb/in.^{3}) at 20 °C (68 °F)

Thermal Properties

Liquidus temperature. 965 °C (1765 °F)
Solidus temperature. 950 °C (1740 °F)
Coefficient of linear thermal expansion. 18 μm/m · K (10 μin./in. · °F) at 20 to 300 °C (68 to 572 °F)
Specific heat. 375 J/kg · K (0.09 Btu/lb · °F) at 20 °C (68 °F)
Thermal conductivity. 69 W/m · K (40 Btu/ft · h · °F) at 20 °C (68 °F)

Electrical Properties

Electrical conductivity. Volumetric: O61 temper: 18% IACS at 20 °C (68 °F); H08 temper, 16.6% IACS at 20 °C (68 °F)
Electrical resistivity. O61 temper, 96 nΩ · m at 20 °C (68 °F); H08 temper, 104 nΩ · m

Magnetic Properties

Magnetic permeability. 1.003

Chemical Properties

General corrosion behavior. C68800 is more resistant than C26000 to both corrosion and stress-corrosion cracking.

Fabrication Characteristics

Formability. Suitable for blanking, drawing, bending, shearing, and stamping. Bending characteristics are nearly nondirectional for all annealed and rolled tempers. Excellent for cold working and hot forming
Weldability. Can be joined by soft soldering when mildly activated commercial fluxes are used and exhibits substantially better tarnish resistance than most other copper alloys. Can also be joined by brazing and resistance welding
Annealing temperature. 400 to 600 °C (750 to 1100 °F)
Order strengthening. When heated to 220 °C (425 °F), temper-rolled C68800 undergoes an ordering reaction that increases strength (see Table 107) and decreases ductility. Because this decrease in ductility would adversely affect formability, parts should be order strengthened after forming. Specific times and temperatures for thermal treatment may vary, depending on cold-worked temper. Susceptibility to stress-corrosion cracking increases dramatically with an increase in the degree of ordering.
Stabilization treatment. Stabilization treatment is performed when enhanced stress relaxation is desired. This treatment causes

Table 109 Typical mechanical properties of copper alloy C69400 rod

Temper	Section size mm	Section size in.	Tensile strength MPa	Tensile strength ksi	Yield strength (0.5% extension under load) MPa	Yield strength (0.5% extension under load) ksi	Elongation in 50 mm (2 in.), %	Hardness, HRB
O60	13	0.5	621	90	310	45	20	85
	25	1.0	586	85	296	43	25	85
	51	2.0	550	80	276	40	25	85
H00	19	0.75	689	100	393	57	21	95

Table 111 Typical mechanical properties of C70400

Temper	Tensile strength MPa	Tensile strength ksi	Yield strength At 0.5% extension under load MPa	ksi	Yield strength At 0.2% offset MPa	ksi	Elongation in 50 mm (2 in.), %	Hardness HRB	Hardness HR30T
Strip									
O61	260	38	83	12	...	...	41	8	...
H01	350	51	...	...	275	40	21	54	57
H02	395	57	...	...	380	55	11	67	65
H04	440	64	...	...	435	63	5	72	68
H06	485	70	...	...	475	69	3	75	69
H08	530	77	...	...	525	76	2 min	76 min	70 min
Tubing, 25 mm (1.0 in.) outside diameter × 1.65 mm (0.065 in.) wall thickness									
OS015	285	41	97	14	...	...	46	58 HRF	...
H55	330	48	250	36	...	...	18	67 HRF	...

Table 110 Nominal tensile properties of C70250 strip

Temper	Tensile strength MPa	Tensile strength ksi	Yield strength MPa	Yield strength ksi	Elongation in 50 mm (2 in.), %
TM00	585	85	655	95	6
TM04	690	100	730	106	2

little change in the 0.2% offset yield strength. Temperature and time of the treatment vary, depending on cold-worked temper; the ranges are 280 to 320 °C (535 to 610 °F) and 10 min to 2 h, respectively. To gain the maximum benefit from a stabilization treatment, parts should be stabilized after forming. There is no increase in stress-corrosion susceptibility as a result of this treatment.

C69000
73.3Cu-22.7Zn-3.4Al-0.6Ni

Chemical Composition

Composition limits. 72 to 74.5 Cu, 3.3 to 3.5 Al, 0.50 to 0.70 Ni, 0.05 Fe max, 0.025 Pb max, bal Zn

Applications

Typical uses. Electrical component parts, contacts, connectors, switches, relays, springs, high-strength shells

Mechanical Properties

Tensile properties. See Table 108.
Hardness. See Table 108.
Elastic modulus. Tension, 115 GPa (16.7 × 10^6 psi)

Mass Characteristics

Density. 8.19 g/cm^3 (0.296 lb/in.^{3}) at 20 °C (68 °F)

Thermal Properties

Liquidus temperature. 960 °C (1760 °F)
Solidus temperature. 950 °C (1745 °F)
Coefficient of linear thermal expansion. 18 μm/m · K (10 μin./in. · °F) at 20 to 300 °C (68 to 572 °F)
Specific heat. 380 J/kg · K (0.09 Btu/lb · °F) at 20 °C (68 °F)
Thermal conductivity. 40 W/m · K (23 Btu/ft · h · °F) at 20 °C (68 °F)

Electrical Properties

Electrical conductivity. Volumetric, O61 temper: 18% IACS at 20 °C (68 °F)
Electrical resistivity. O61 temper: 96 nΩ · m

Chemical Properties

General corrosion behavior. Significantly better corrosion performance than C26000, both in uniform corrosion rate and stress-corrosion resistance

Fabrication Characteristics

Formability. Behavior in blanking, drawing, forming, bending, stamping, and other cold forming operations is similar to that of C26000, but with lower directionality in bending of cold-worked tempers. Excellent for cold working and hot forming
Weldability. Resistance welding: good. Soldering and brazing: fair, provided that active flux is used. Oxyfuel gas and arc welding are not recommended.
Annealing temperature. 400 to 600 °C (750 to 1100 °F); stress relief anneal, 225 °C (435 °F) for 1 h
Hot-working temperature. 790 to 840 °C (1450 to 1550 °F)

C69400
81.5Cu-14.5Zn-4Si

Commercial Names

Previous trade name. Silicon red brass, CA694

Specifications

ASTM. Rod B 371 (CA694)

Chemical Composition

Composition limits. 80.0 to 83.0 Cu, 0.30 Pb max, 0.20 Fe max, 3.5 to 4.5 Si, bal Zn

Applications

Typical uses. Valve stems requiring a combination of corrosion resistance and high strength; forged or screw-machined parts

Mechanical Properties

Tensile properties. See Table 109.
Hardness. See Table 109.
Elastic modulus. Tension, 110 GPa (16 × 10^6 psi)

Mass Characteristics

Density. 8.19 g/cm^3 (0.296 lb/in.^{3}) at 20 °C (68 °F)

Thermal Properties

Liquidus temperature. 920 °C (1685 °F)
Solidus temperature. 820 °C (1510 °F)
Coefficient of linear thermal expansion. 20.2 μm/m · K (11.2 μin./in. · °F) at 20 to 300 °C (68 to 572 °F)
Specific heat. 380 J/kg · K (0.09 Btu/lb · °F) at 20 °C (68 °F)
Thermal conductivity. 26 W/m · K (15 Btu/ft · h · °F) at 20 °C (68 °F)

Electrical Properties

Electrical conductivity. Volumetric, 6.2% IACS at 20 °C (68 °F)
Electrical resistivity. 280 nΩ · m at 20 °C (68 °F)

Fabrication Characteristics

Machinability. 30% of C36000 (free-cutting brass)
Forgeability. 80% of C37700 (forging brass)
Formability. Excellent capacity for being hot formed; poor for being cold formed
Weldability. Soft soldering and silver alloy brazing: excellent. Oxyfuel gas welding and resistance welding: good. Arc welding is not recommended.
Annealing temperature. 425 to 650 °C (800 to 1200 °F)
Hot-working temperature. 650 to 875 °C (1200 to 1600 °F)

C70250
95.4Cu-3.0Ni-0.6Si-0.1Mg

Chemical Composition

Composition limits. 2.2 to 4.2 Ni, 0.25 to 1.2

Table 112 Typical mechanical properties of C70600 and C71000

Temper	Tensile strength MPa	Tensile strength ksi	Yield strength 0.5% extension under load MPa	Yield strength 0.5% extension under load ksi	Yield strength 0.2% offset MPa	Yield strength 0.2% offset ksi	Elongation in 50 mm (2 in.), %	Hardness HRF	Hardness HRB
Flat products, 1 mm (0.04 in.) thick									
OS050	350	51	90	13	90	13	35	72	25
OS035	358	52	98	14	98	14	35	73	27
OS025	365	53	110	16	110	16	35	75	30
H01	415	60	330	48	338	49	20	92	58
H02	468	68	425	62	435	63	8	100	75
H04	518	75	490	71	500	72	5	...	80
H06	540	78	518	75	525	76	4	...	82
H08	565	82	540	78	545	79	3	...	84
H10	585	85	540	78	545	79	3	...	86
Tubing, 25 mm (1 in.) outside diameter × 1.65 mm (0.065 in.) wall thickness									
OS025	338	49	125	18	...	...	40	72	25
H55	468	68	430	62	...	...	14	...	76
Wire, 2 mm (0.080 in.) diameter									
H10	655	95	585	85	...	...	5	...	...

Si, 0.05 to 0.3 Mg, 1.0 Zn max, 0.05 Pb max, 0.1 Mn max, 0.2 Fe max, 99.5 Cu + named elements min

Applications

Typical uses. Applications where high strength and good formability combined with good stress relaxation resistance and moderate conductivity are required. Specific uses include contact springs, connectors, and lead frames.

Mechanical Properties

Tensile properties. See Table 110.
Elastic modulus. 131 GPa (19.0 × 10^6 psi)

Mass Characteristics

Density. 8.80 g/cm^3 (0.318 lb/in.^{3}) at 20 °C (68 °F)

Thermal Properties

Liquidus temperature. 1095 °C (2003 °F)
Solidus temperature. 1075 °C (1967 °F)
Coefficient of linear thermal expansion. 17.6 μm/m · K (9.8 μin./in. · °F) at 20 to 300 °C (68 to 572 °F)
Thermal conductivity. 147 to 190 W/m · K (85 to 110 Btu/ft · h · °F) at 20 °C (68 °F)

Electrical Properties

Electrical conductivity. Volumetric, 35 to 40% IACS at 20 °C (68 °F)
Electrical resistivity. 43.1 to 49.3 nΩ · m at 20 °C (68 °F)

C70400
92.4Cu-5.5Ni-1.5Fe-0.6Mn

Specifications

ASTM. Pipe: B 466. Tubing: B 111, B 359, B 395, B 466, B 543

Chemical Composition

Composition limits. 91.2 Cu min, 4.8 to 6.2 Ni, 1.3 to 1.7 Fe, 0.3 to 0.8 Mn, 1.0 Zn max, 0.05 Pb max

Applications

Typical uses. Rolled strip, sheet, and tubing for industrial uses; condensers, condenser plates, evaporator and heat exchanger tubes, ferrules, saltwater piping, lithium bromide absorption system tubing, shipboard condenser intake systems

Mechanical Properties

Tensile properties. See Table 111.
Hardness. See Table 111.
Elastic modulus. Tension, 115 GPa (17 × 10^6 psi); shear, 44 GPa (6.4 × 10^6 psi)

Mass Characteristics

Density. 8.94 g/cm^3 (0.323 lb/in.^{3}) at 20 °C (68 °F)

Thermal Properties

Liquidus temperature. 1125 °C (2050 °F)
Coefficient of linear thermal expansion. 17.5 μm/m · K (9.7 μin./in. · °F) at 20 to 300 °C (68 to 572 °F)
Specific heat. 380 J/kg · K (0.09 Btu/lb · °F) at 20 °C (68 °F)
Thermal conductivity. 64 W/m · K (37 Btu/ft · h · °F) at 20 °C (68 °F)

Electrical Properties

Electrical conductivity. Volumetric, 14% IACS at 20 °C (68 °F)
Electrical resistivity. 120 nΩ · m at 20 °C (68 °F)

Fabrication Characteristics

Machinability. 20% of C36000 (free-cutting brass)
Formability. Excellent for cold working; good for hot forming
Weldability. Soldering, brazing, and gas-shielded arc welding: excellent. Coated metal arc and resistance spot, seam, and butt welding: good. Oxyacetylene welding: fair
Annealing temperature. 565 to 815 °C (1050 to 1500 °F)

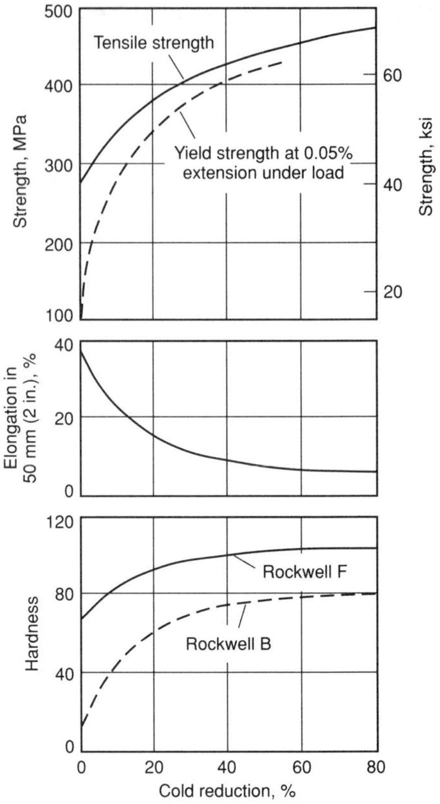

Fig. 49 Mechanical properties of cold drawn C70600 tubing. Data for variation in mechanical properties with amount of cold reduction for tubing with a diameter of 60 mm (2⅜ in.) and a wall thickness of 4.8 mm (3/16) wall

Hot-working temperature. 815 to 950 °C (1500 to 1750 °F)

C70600
90Cu-10Ni

Commercial Names

Previous trade names. Copper-nickel, 10%; CA706
Common name. 90-10 cupronickel

Specifications

ASME. Flat products: SB171, SB402. Pipe: SB466, SB467. Tubing: SB111, SB359, SB395, SB466, SB467, SB543
ASTM. Flat products: B 122, B 171, B 402, B 432. Pipe: B 466, B 467. Rod: B 151. Tubing: B 111, B 359, B 395, B 466, B 467, B 543, B 552
SAE. Plate and tubing: J463
Government. Bar, flat products, forgings, rod: MIL-C-15726 E(2). Tubing: MIL-T-16420 J(3), MIL-T-1368 C(2), MIL-T-23520 A(4). Condenser tubing: MIL-T-15005 F

Chemical Composition

Composition limits. 0.05 Pb max, 1 to 1.8 Fe, 1.0 Zn max, 9 to 11 Ni, 1.0 Mn max, 0.5 max other (total), bal Cu

Table 113 Typical mechanical properties of C71500

Size	Temper	Tensile strength MPa	Tensile strength ksi	Yield strength(a) MPa	Yield strength(a) ksi	Elongation in 50 mm (2 in.), %	Hardness, HRB
Flat products							
25 mm (1 in.) plate	M20 temper	380	55	140	20	45	36
1 mm (0.04 in.) strip	O61 temper(b)	380	55	125	18	36	40
	H80 temper	580	84	545	79	3	86
Rod							
<25 mm (1 in.) diameter	O61 temper(c)	380	55	140	20	45	37
	H80 temper(d)	585	85	540	78	15	81
25 mm (1 in.) diameter	H02 temper(e)	515	75	485	70	15	80
Tubing							
19 mm (0.75 in.) outside diameter × 1.25 mm (0.049 in.) wall thickness	O61 temper(b)	340	49	...	...	50	...
	H80 temper	580	84	...	...	4	...
25 mm (1 in.) outside diameter × 1.65 mm (0.065 in.) wall thickness	OS025 temper	415	60	170	25	45	45
114 mm (4.5 in.) outside diameter × 2.75 mm (0.109 in.) wall thickness	OS035 temper	370	54	...	...	45	36

(a) 0.5% extension under load. (b) Annealed at 705 °C (1300 °F). (c) Annealed at 760 °C (1400 °F). (d) Cold drawn 50%. (e) Cold drawn 20%

Applications

Typical uses. Condensers, condenser plates, distiller tubes, evaporator and heat exchanger tubes, ferrules, saltwater piping, boat hulls

Mechanical Properties

Tensile properties. See Table 112 and Fig. 49.
Elastic modulus. Tension, 140 GPa (20 × 10^6 psi); shear, 52 GPa (7.5 × 10^6 psi)
Fatigue strength. Tubing, H55 temper: 138 MPa (20 ksi) at 10^8 cycles

Mass Characteristics

Density. 8.94 g/cm^3 (0.323 lb/in.^{3}) at 20 °C (68 °F)

Thermal Properties

Liquidus temperature. 1150 °C (2100 °F)
Solidus temperature. 1100 °C (2010 °F)
Coefficient of linear thermal expansion. 17.1 μm/m · K (9.5 μin./in. · °F) at 20 to 300 °C (68 to 572 °F)
Specific heat. 380 J/kg · K (0.09 Btu/lb · °F) at 20 °C (68 °F)
Thermal conductivity. 40 W/m · K (23 Btu/ft · h · °F) at 20 °C (68 °F)

Electrical Properties

Electrical conductivity. Volumetric, 9.1% IACS
Electrical resistivity. 190 nΩ · m at 20 °C (68 °F)

Optical Properties

Color. Pink-silver

Chemical Properties

Resistance to specific agents. Excellent resistance to seawater

Fabrication Characteristics

Machinability. 20% of C36000 (free-cutting brass)
Formability. Good capacity for being both cold worked and hot formed
Weldability. Soldering, brazing, gas-shielded arc, and resistance butt welding: excellent. Shielded metal arc and resistance spot and seam welding: good. Oxyfuel gas welding: fair
Annealing temperature. 600 to 825 °C (1100 to 1500 °F)
Hot-working temperature. 850 to 950 °C (1550 to 1750 °F)

C71000
80Cu-20Ni

Commercial Names

Previous trade names. Copper-nickel, 20%; CA710
Common name. 80-20 cupronickel

Specifications

ASME. Pipe: SB466, SB467. Tubing: SB111, SB359, SB395, SB466, SB467
ASTM. Bar and flat products: B 122. Pipe: B 466, B 467. Tubing: B 111, B 359, B 395, B 466, B 467. Wire: B 206
SAE. Bar, flat products, and tubing: J463

Chemical Composition

Composition limits. 0.05 Pb max, 1.00 Fe max, 1.00 Zn max, 19 to 23 Ni, 1.00 Mn max, 0.5 max other (total), bal Cu

Applications

Typical uses. Communication relays, condensers, condenser plates, electrical springs, evaporator and heat exchanger tubes, ferrules, resistors

Mechanical Properties

Tensile properties. See Table 112.
Fatigue strength. Tubing, H55 temper: 138 MPa (20 ksi) at 10^8 cycles
Elastic modulus. Tension, 140 GPa (20 × 10^6 psi); shear, 52 GPa (7.5 × 10^6 psi)

Mass Characteristics

Density. 8.94 g/cm^3 (0.323 lb/in.^{3}) at 20 °C (68 °F)

Thermal Properties

Liquidus temperature. 1200 °C (2190 °F)
Solidus temperature. 1150 °C (2100 °F)
Coefficient of linear thermal expansion. 16.4 μm/m · K (9.1 μin./in. · °F) at 20 to 300 °C (68 to 572 °F)
Specific heat. 380 J/kg · K (0.09 Btu/lb · °F) at 20 °C (68 °F)
Thermal conductivity. 36 W/m · K (21 Btu/ft · h · °F) at 20 °C (68 °F)

Electrical Properties

Electrical conductivity. Volumetric, O61 temper: 6.5% IACS at 20 °C (68 °F)
Electrical resistivity. O61 temper: 265 nΩ · m at 20 °C (68 °F)

Optical Properties

Color. Pale silver

Fabrication Characteristics

Machinability. 20% of C36000 (free-cutting brass)
Formability. Good capacity for being cold worked by blanking, forming, and bending; good capacity for being hot formed
Weldability. Soldering, brazing, gas-shielded arc welding and resistance welding (all forms): excellent. Shielded metal arc welding: good. Oxyfuel gas welding: fair
Annealing temperature. 650 to 825 °C (1200 to 1500 °F)
Hot-working temperature. 875 to 1050 °C (1600 to 1900 °F)

C71500
70Cu-30Ni

Commercial Names

Previous trade names. Copper-nickel, 30%; CA715
Common name. 70-30 cupronickel

Specifications

ASME. Flat products: SB171, SB402. Pipe: SB466, SB467. Tubing: SB111, SB359, SB395, SB466, SB467, SB543

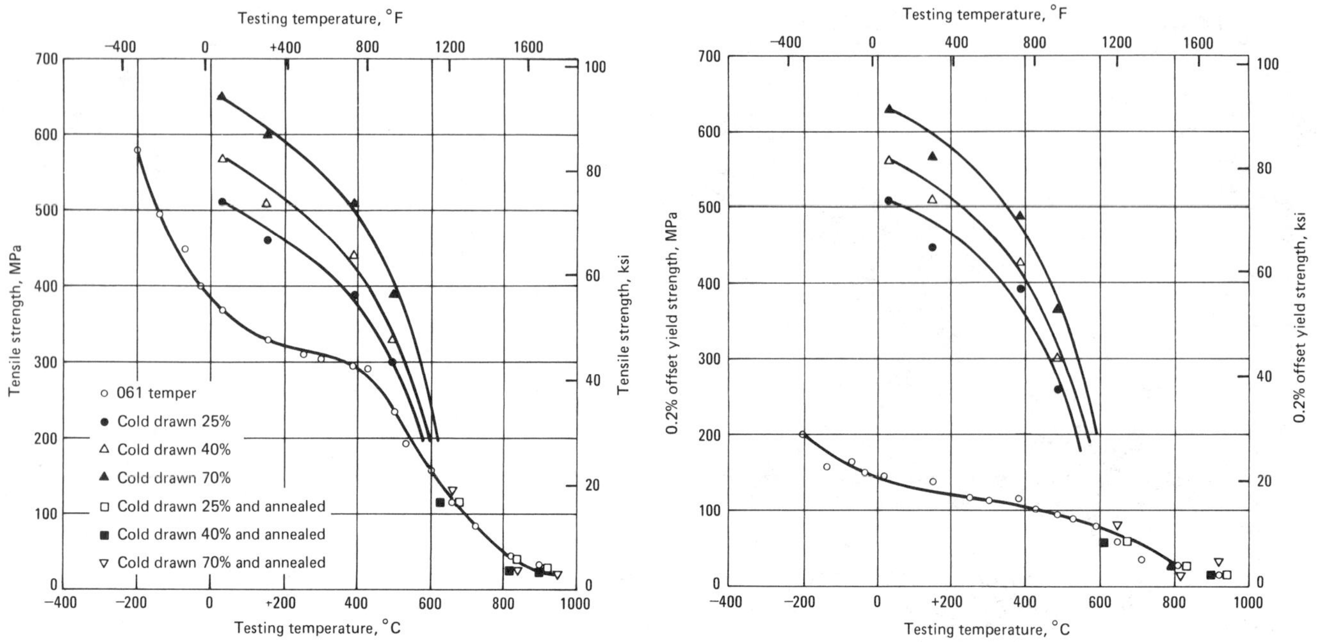

Fig. 50 Typical tensile and yield strengths of C71500 rod

ASTM. Flat products: B 122, B 151, B 171, B 402. Pipe: B 466, B 467. Rod: B 151. Tube: B 111, B 359, B 395, B 466, B 467, B 552
SAE. Bar, flat products, tubing: J463
Government. Bar, flat products, forgings, rod, wire: MIL-C-15726. Tubing: MIL-T-15005, MIL-T-16420, MIL-T-22214

Chemical Composition

Composition limits. 0.05 Pb max, 0.4 to 0.7 Fe, 1.0 Zn max, 29 to 33 Ni, 1.0 Mn max, 0.5 max other (total), bal Cu

Applications

Typical uses. Condensers, condenser plates, distiller tubes, evaporator and heat exchanger tubes, ferrules, saltwater piping
Precautions in use. Stress relieving or full annealing should precede exposure to solders of all kinds.

Mechanical Properties

Tensile properties. See Table 113 and Fig. 50.
Elastic modulus. Tension, 150 GPa (22 × 10^6 psi) at 20 °C (68 °F); shear, 57 GPa (8.3 × 10^6 psi) at 20 °C
Hardness. See Table 113.
Impact strength. Charpy keyhole data: 107 J (79 ft · lbf) at 21 °C (70 °F); 88 to 93 J (65 to 69 ft · lbf) at −73 °C (−100 °F) after holding 30 to 140 days at −87 °C (−125 °F). Data are for 10 mm (0.4 in.) square specimens machined from 25 mm (1 in.) thick plate having a room temperature hardness of 88 HRF. See Table 114 for additional impact data.
Fatigue strength. Rod, H80 temper, drawn 50% to 25 mm (1 in.) in diameter: 220 MPa (32 ksi) at 10^8 cycles. Rod, O61 temper, drawn 50% to 25 mm (1 in.) in diameter, then annealed at 760 °C (1400 °F): 150 MPa (22 ksi) at 10^8 cycles
Creep-rupture characteristics. Creep strength for 3.2 mm (0.125 in.) diam wire, OS020 temper: for a creep rate of 0.001% in 1000 h, 165 MPa (24 ksi) at 150 °C (300 °F) or 110 MPa (16 ksi) at 260 °C (500 °F); for a creep rate of 0.01% in 1000 h, 240 MPa (35 ksi) at 150 °C or 205 MPa (30 ksi) at 260 °C. Creep strength for 19 mm (¾ in.) diam rod, O61 temper, drawn to size and annealed at 550 °C (1020 °F): 63 MPa (9.1 ksi) for a creep rate of 0.01% in 1000 h at 400 °C (750 °F); 130 MPa (18.8 ksi) for a creep rate of 0.1% in 1000 h at 400 °C (750 °F)

Mass Characteristics

Density. 8.94 g/cm^3 (0.323 lb/in.^{3}) at 20 °C (68 °F)

Thermal Properties

Liquidus temperature. 1240 °C (2260 °F)
Solidus temperature. 1170 °C (2140 °F)
Coefficient of linear thermal expansion. 16.2 μm/m · K (9 μin./in. · °F) at 20 to 300 °C (68 to 572 °F)
Specific heat. 380 J/kg · K (0.09 Btu/lb · °F) at 20 °C (68 °F)
Thermal conductivity. 29 W/m · K (17 Btu/ft · h · °F) at 20 °C (68 °F)

Electrical Properties

Electrical conductivity. Volumetric, O61 temper: 4.6% IACS at 20 °C (68 °F)
Electrical resistivity. O61 temper: 375 nΩ · m at 20 °C (68 °F); temperature coefficient, 4.8 × 10^{-5}/K (2.6 × 10^{-5}/ °F) at 20 to 200 °C (68 to 392 °F)

Optical Properties

Color. White

Fabrication Characteristics

Machinability. 20% of C36000 (free-cutting brass)
Formability. Good capacity for being both cold worked and hot formed by bending and forming and welding processes
Weldability. Soldering, brazing, all forms of arc welding, and all forms of resistance welding: excellent. Oxyfuel gas welding: good
Annealing temperature. 650 to 825 °C (1200 to 1500 °F)
Hot-working temperature. 925 to 1050 °C (1700 to 1900 °F)

C71900
67.2Cu-30Ni-2.8Cr

Commercial Names

Previous trade names. Copper-nickel, chromium-bearing; CA719
Common name. Cupronickel with Cr

Table 114 Typical Charpy impact strengths for C71500

Testing temperature °C	°F	Charpy impact strength(a) J	ft · lbf
−115	−175	81	60
−18	0	81	60
3	38	87	64
20	68	89	66
65	150	72	53
120	250	72	53
205	400	68	50

(a) For 10 mm (0.39 in.) square keyhole specimens machined from annealed rod

Table 115 Typical mechanical properties of C71900 strip

Condition	Tensile strength MPa	Tensile strength ksi	Yield strength at 0.2% offset MPa	Yield strength at 0.2% offset ksi	Elongation in 50 mm (2 in.), %	Hardness, HRB
Heat treated(a)	600	87	365	53	32	87
Half-hard temper(b)	730	106	685	99	14	100
Hard temper(c)	780	113	740	107	8	100
Spring temper(d)	835	121	800	116	6	101

(a) Spinodally decomposed by air cooling from 900 °C (1650 °F). (b) Spinodally decomposed, then cold rolled 20%. (c) Spinodally decomposed, then cold rolled 37%. (d) Spinodally decomposed, then cold rolled 60%

Chemical Composition

Composition limits. 28 to 32 Ni, 2.4 to 3.2 Cr, 0.5 Fe max, 0.2 to 1.0 Mn, 0.01 to 0.20 Ti, 0.02 to 0.25 Zr, 0.04 C max, 0.25 Si max, 0.5 max other (total), bal Cu

Applications

Typical uses. Heat exchanger tubes, tube sheets, water boxes, ferrules, saltwater pipe

Mechanical Properties (Spinodally Decomposed Condition)

Tensile properties. Tensile strength, 540 MPa (78 ksi); yield strength, 330 MPa (47 ksi) at 0.2% offset, elongation, 25%. See Table 115.
Elastic modulus. Tension, 150 GPa (22 × 10^6 psi); shear, 59 GPa (8.5 × 10^6 psi)
Fatigue strength. Smooth bar, rotating beam: 275 MPa (40 ksi) at 10^8 cycles for both spinodally decomposed condition and half-hard temper (spinodally decomposed plus 44% cold work)

Mass Characteristics

Density. 8.85 g/cm^3 (0.319 lb/in.^{3}) at 20 °C (68 °F)

Thermal Properties

Liquidus temperature. 1220 °C (2225 °F)
Solidus temperature. 1170 °C (2140 °F)
Coefficient of linear thermal expansion. 16.8 μm/m · K (9.3 μin./in. · °F) at 20 to 200 °C (68 to 392 °F); 17.1 μm/m · K (9.5 μin./in. · °F) at 20 to 300 °C (68 to 572 °F)
Thermal conductivity. 29 W/m · K (16.5 Btu/ft · h · °F) at 20 °C (68 °F)

Electrical Properties (Spinodally Decomposed Condition)

Electrical conductivity. Volumetric, 4.4% IACS at 20 °C (68 °F)
Electrical resistivity. 395 nΩ · m at 20 °C (68 °F)

Magnetic Properties

Magnetic permeability. 1.0003 at magnetic field strength of 16 kA/m

Chemical Properties

Resistance to specific corroding agents. Seawater: C71900 resists both general and localized attack. The corrosion rate is very low (generally less than 0.1 mm/yr, or less than 4 mils/yr) in seawater flowing at velocities above about 1.8 m/s (6 ft/s). This level of corrosion resistance is comparable to or slightly less than the resistance of C71500 exposed to the same conditions. At low velocities or in stagnant seawater, C71900 exhibits slightly higher general weight loss than C71500, and corrosion is a broad, uniform type of attack. C71900 is not quite as resistant to crevice corrosion as C71500 at all velocities; for example, in a 3-month test in seawater flowing at intermediate velocity, C71900 incurred 0.2 to 0.33 mm (8 to 13 mils) penetration compared to nil penetration for C71500. Welding does not have an adverse effect on corrosion resistance; corrosion in seawater is about the same for the weld zone, heat-affected zone, and unaffected base metal. C71900 appears to be immune to stress-corrosion cracking in seawater, even when the seawater is contaminated with 5 ppm H_2S. C71900 is cathodic to carbon steel and Ni-Resist type cast iron, slightly cathodic to C71500, and anodic to austenitic stainless steels.

Fabrication Characteristics

Formability. C71900 can be cold worked in a manner similar to C71500, although C71900 has higher tensile and yield strengths at any given reduction. Hot working is readily accomplished from a starting temperature of 1040 to 1065 °C (1900 to 1950 °F), but working should not be continued below 840 °C (1550 °F) because of reduced ductility. About 25% more extrusion pressure is required for C71900 than for C71500. Because of microsegregation, cast billets should be homogenized 3 to 4 h at 1040 to 1065 °C before being extruded.
Weldability. Soldering, brazing, and all forms of arc welding: excellent. Resistance welding is not normally used for this alloy. Oxyfuel gas welding is not recommended. Material thick enough to require multipass welds develops a minimum yield strength of 345 MPa (50 ksi) as-welded. Single-pass welds develop a minimum yield strength of 275 MPa (40 ksi) as-welded. The yield strength can be raised to 345 MPa (50 ksi) by a postweld heat treatment consisting of 1 h at 480 °C (900 °F).
Heat treatment. Full properties of the spinodally decomposed condition can be achieved by slow cooling (furnace cooling or still air cooling) through the temperature range 760 to 425 °C (1400 to 800 °F) from a soaking temperature of 900 to 1000 °C (1650 to 1850 °F).
Hot-working temperature. 900 to 1065 °C (1650 to 1950 °F)

C72200
83Cu-16.5Ni-0.5Cr

Commercial Names

Previous trade names. Copper-nickel, chromium-bearing; CA722
Common name. Cupronickel with Cr

Chemical Composition

Composition limits. 15 to 18 Ni, 0.3 to 0.7 Cr, 0.5 to 1.0 Fe, 0.4 to 0.9 Mn, 0.03 Si max, 0.03 Ti max, 0.03 C max, 0.5 max other (total), bal Cu

Applications

Typical uses. Condenser and heat exchanger tubing, saltwater pipe

Mechanical Properties

Tensile properties. O61 temper: tensile strength, 315 MPa (46 ksi); yield strength, 125 MPa (18 ksi) at 0.2% offset; elongation, 46%. H04 temper: tensile strength, 480 MPa (70 ksi); yield strength, 455 MPa (66 ksi) at 0.2% offset; elongation, 6%
Elastic modulus. Tension, 135 GPa (20 × 10^6 psi); shear, 55 GPa (8.2 × 10^6 psi)

Mass Characteristics

Density. 8.94 g/cm^3 (0.323 lb/in.^{3}) at 20 °C (68 °F)

Thermal Properties

Liquidus temperature. 1176 °C (2148 °F)
Solidus temperature. 1122 °C (2052 °F)
Coefficient of linear thermal expansion. 15.8 μm/m · K (8.8 μin./in. · °F) at 20 to 300 °C (68 to 572 °F)
Specific heat. 396 J/kg · K (0.094 Btu/lb · °F) at 20 °C (68 °F)
Thermal conductivity. 34.5 W/m · K (19.9 Btu/ft · h · °F) at 20 °C (68 °F)

Electrical Properties

Electrical conductivity. Volumetric, 6.53% IACS
Electrical resistivity. 264 nΩ · m at 20 °C (68 °F)

Fabrication Characteristics

Formability. Good capacity for being cold worked or hot formed
Weldability. Gas-shielded arc welding: excellent. Soldering, brazing, shielded metal arc welding, and resistance welding (all forms): good. Oxyfuel gas welding: fair
Annealing temperature. 730 to 815 °C (1350 to 1500 °F)
Hot-working temperature. 900 to 1040 °C (1650 to 1900 °F)

Table 116 Tensile properties of C72500

Temper	Tensile strength MPa	Tensile strength ksi	Yield strength, 0.5% extension under load MPa	Yield strength, 0.5% extension under load ksi	Yield strength, 0.2% offset MPa	Yield strength, 0.2% offset ksi	Elongation in 50 mm (2 in.), %	Hardness, HRB
Flat products, 1 mm (0.04 in.) thick								
Annealed(a)	380	55	150	22	150	22	35	42
Quarter hard	450	65	365	53	400	58	18	71
Half hard	490	71	450	65	475	69	6	78
Hard	570	83	515	75	555	81	3	85
Extra hard	600	87	555	81	590	86	2	88
Spring	625	91	570	83	620	90	1	90
Super spring	770	112	570	83	740	108	1	99
Wire, 2 mm (0.08 in.) diameter								
Annealed(a)	415	60	170	25	...	...	...	...

(a) Grain size, 0.015 mm

C72500
88.2Cu-9.5Ni-2.3Sn

Commercial Names

Previous trade names. Copper-nickel, tin-bearing; CA725
Common name. Cupronickel with Sn

Chemical Composition

Composition limits. 0.05 Pb max, 0.6 Fe max, 0.5 Zn max, 0.2 Mn max, 8.5 to 10.5 Ni, 1.8 to 2.8 Sn, 0.2 max other (total), bal Cu

Applications

Typical uses. Relay and switch springs, connectors, lead frames, control and sensing bellows, brazing alloy

Mechanical Properties

Tensile properties. See Table 116.
Elastic modulus. Tension, 137 GPa (20×10^6 psi); shear, 52 GPa (7.5×10^6 psi)

Mass Characteristics

Density. 8.89 g/cm^3 (0.321 lb/in.^{3}) at 20 °C (68 °F)

Thermal Properties

Liquidus temperature. 1130 °C (2065 °F)
Solidus temperature. 1060 °C (1940 °F)
Coefficient of linear thermal expansion. 16.5 μm/m · K (9.2 μin./in. · °F) at 20 to 300 °C (68 to 572 °F)
Thermal conductivity. 55 W/m · K (31 Btu/ft · h · °F) at 20 °C (68 °F)

Electrical Properties

Electrical conductivity. Volumetric, 11% IACS at 20 °C (68 °F)
Electrical resistivity. O61 temper, 157 nΩ · m at 20 °C (68 °F)

Optical Properties

Color: Silver

Chemical Properties

Resistance to specific agents. Excellent resistance to seawater

Fabrication Characteristics

Machinability. 20% of C36000 (free-cutting brass)
Formability. Excellent capacity for being both cold worked and hot formed by blanking, coining, drawing, forming, bending, heading, upsetting, roll threading, knurling, shearing, spinning, squeezing, stamping, and swaging
Weldability. Soldering, brazing, and resistance spot and resistance butt welding: excellent. Gas-shielded arc, shielded metal arc, and resistance seam welding: good. Oxyfuel gas welding: fair
Annealing temperature. 650 to 800 °C (1200 to 1475 °F)
Hot-working temperature. 850 to 950 °C (1550 to 1750 °F)

C74500
65Cu-25Zn-10Ni

Commercial Names

Common name. Nickel silver, 65-10

Specifications

ASTM. Flat products: B 122. Bar: B 122, B 151. Rod: B 151. Wire: B 206
Government. Flat products: QQ-C-585. Bar: QQ-C-585, QQ-C-586. Rod, shapes, flat wire: QQ-C-586. Wire: QQ-W-321

Table 117 Typical mechanical properties of C74500

Temper	Tensile strength MPa	Tensile strength ksi	Yield strength(a) MPa	Yield strength(a) ksi	Elongation in 50 mm (2 in.), %	Hardness HRF	Hardness HRB	Hardness HR30T	Shear strength MPa	Shear strength ksi
Flat products, 1 mm (0.04 in.) thick										
OS070	340	49	125	18	49	67	22	30	...	...
OS050	350	51	130	19	46	71	28	34	...	...
OS035	365	53	140	20	43	76	35	38	285	41
OS025	385	56	160	23	40	80	42	44	...	...
OS015	415	60	195	28	36	85	52	51	...	...
H00	415	60	240	35	34	...	60	55	295	43
H01	450	65	310	45	25	...	70	63	310	45
H02	505	73	415	60	12	...	80	70	345	50
H04	590	86	515	75	4	...	89	76	380	55
H06	655	95	525	76	3	...	92	78	405	59
Wire, 2 mm (0.08 in.) diameter										
OS070	345	50	...	...	50	...	...	...	...	...
OS050	360	52	...	...	48	...	...	...	...	...
OS035	385	56	...	...	45	...	...	...	...	...
OS025	400	58	...	...	40	...	...	...	...	...
OS015	435	63	...	...	35	...	...	...	...	...
H00 (10%)	450	65	...	...	25	...	...	...	...	...
H01 (20%)	495	72	...	...	10	...	...	...	...	...
H02 (37%)	585	85	...	...	7	...	...	...	...	...
H04 (60%)	725	105	...	...	5	...	...	...	...	...
H06 (75%)	825	120	...	...	3	...	...	...	...	...
H08 (84%)	895	130	...	...	1	...	...	...	...	...

(a) At 0.5% extension under load

Table 118 Typical mechanical properties of C75200

Temper	Tensile strength MPa	Tensile strength ksi	Yield strength(a) MPa	Yield strength(a) ksi	Elongation in 50 mm (2 in.), %	Hardness HRF	Hardness HRB	Hardness HR30T
Flat products, 1 mm (0.04 in.) thick								
OS035	400	58	170	25	40	85	40	...
OS015	415	60	205	30	32	90	55	...
H01	450	65	345	50	20	...	73	65
H02	510	74	427	62	8	...	83	72
H04	585	85	510	74	3	...	87	75
Rod, 13 mm (0.5 in.) diameter								
OS035	385	56	170	25	42	...	...	...
H02 (20%)	485	70	415	60	20	...	78	...
Wires, 2 mm (0.08 in.) diameter								
OS035	400	58	170	25	45	...	...	...
OS015	415	60	205	30	35	...	...	...
H01	505	73	450	65	16	...	...	...
H02	590	86	550	80	7	...	...	...
H04	710	103	620	90	3	...	...	...

(a) At 0.5% extension under load

Table 119 Typical mechanical properties of C75400 sheet or strip, 1 mm (0.04 in.) thick

Temper	Tensile strength MPa	Tensile strength ksi	Yield strength(a) MPa	Yield strength(a) ksi	Elongation in 50 mm (2 in.), %	Hardness HRF	Hardness HR30T	Shear strength MPa	Shear strength ksi
OS070	365	53	125	18	43	69	27	...	...
OS050	380	55	130	19	42	73	33	...	...
OS035	395	57	145	21	40	79	41	285	41
OS025	405	59	165	24	37	82	46	...	...
OS015	420	61	195	28	34	89	53	...	...
H00	425	62	240	35	30	60 HRB	55	295	43
H01	450	65	340	49	21	70 HRB	63	305	44
H02	510	74	425	62	10	80 HRB	70	325	47
H04	585	85	515	75	3	87 HRB	75	360	52
H06	635	92	545	79	2	90 HRB	77	370	54

(a) At 0.5% extension under load

Chemical Composition

Composition limits. 63.5 to 68.5 Cu, 9.0 to 11.0 Ni, 0.10 Pb max, 0.25 Fe max, 0.5 Mn max, 0.5 max other, bal Zn

Applications

Typical uses. Hardware: rivets, screws, slide fasteners. Optical goods: optical parts. Miscellaneous: etching stock, hollowware, nameplates, platers' bars

Mechanical Properties

Tensile properties. See Table 117.
Hardness. See Table 117.
Elastic modulus. Tension, 120 GPa (17.5 × 10^6 psi); shear, 46 GPa (6.6 × 10^6 psi)

Mass Characteristics

Density. 8.69 g/cm^3 (0.314 lb/in.^{3}) at 20 °C (68 °F)

Thermal Properties

Liquidus temperature. 1020 °C (1870 °F)
Coefficient of linear thermal expansion. 16.4 μm/m · K (9.1 μin./in. · °F) at 20 to 300 °C (68 to 572 °F)
Specific heat. 380 J/kg · K (0.09 Btu/lb · °F) at 20 °C (68 °F)
Thermal conductivity. 45 W/m · K (26 Btu/ft · h · °F) at 20 °C (68 °F)

Electrical Properties

Electrical conductivity. Volumetric, 9.0% IACS at 20 °C (68 °F)
Electrical resistivity. 192 nΩ · m at 20 °C (68 °F)

Fabrication Characteristics

Machinability. 20% of C36000 (free-cutting brass)
Formability. Excellent for cold working; poor for hot forming
Weldability. Soldering and brazing: excellent. Oxyfuel gas, resistance spot, and resistance butt welding: good. Gas metal arc and resistance seam welding: fair. Shielded metal arc welding: not recommended
Annealing temperature. 600 to 750 °C (1100 to 1400 °F)

C75200
65Cu-18Ni-17Zn

Commercial Names

Common name. Nickel silver, 65-18

Specifications

ASTM. Flat products: B 122. Bar: B 122, B 151. Rod: B 151. Wire: B 206
SAE. J463
Government. Flat products: QQ-C-585. Bar: QQ-C-585, QQ-C-586. Rod, shapes, flat wire: QQ-C-586. Wire: QQ-W-321

Chemical Composition

Composition limits. 63.0 to 66.5 Cu, 16.5 to 19.5 Ni, 0.1 Pb max, 0.25 Fe max, 0.5 Mn max, 0.5 max other (total), bal Zn

Applications

Typical uses. Hardware: rivets, screws, table flatware, truss wire, zippers. Optical goods: bows, camera parts, core bars, templates. Miscellaneous: base for silver plate, costume jewelry, etching stock, hollowware, nameplates, radio dials

Mechanical Properties

Tensile properties. See Table 118.
Hardness. See Table 118.
Elastic modulus. Tension, 125 GPa (18 × 10^6 psi); shear, 47 GPa (6.8 × 10^6 psi)

Mass Characteristics

Density. 8.73 g/cm^3 (0.316 lb/in.^{3}) at 20 °C (68 °F)

Thermal Properties

Liquidus temperature. 1110 °C (2030 °F)
Solidus temperature. 1070 °C (1960 °F)
Coefficient of linear thermal expansion. 16.2 μm/m · K (9.0 μin./in. · °F) at 20 to 300 °C (68 to 572 °F)
Specific heat. 380 J/kg · K (0.09 Btu/lb · °F) at 20 °C (68 °F)
Thermal conductivity. 33 W/m · K (19 Btu/ft · h · °F) at 20 °C (68 °F)

Electrical Properties

Electrical conductivity. Volumetric, 6% IACS at 20 °C (68 °F)
Electrical resistivity. 287 nΩ · m at 20 °C (68 °F)

Fabrication Characteristics

Machinability. 20% of C36000 (free-cutting brass)
Formability. Excellent for cold working; poor for hot forming
Weldability. Soldering and brazing: excellent. Oxyfuel gas, resistance spot, and resistance butt welding: good. Gas metal arc and resistance seam welding: fair. Shielded metal arc welding is not recommended.

C75400
65Cu-20Zn-15Ni

Commercial Names

Common name. Nickel silver, 65-15

Chemical Composition

Composition limits. 63.5 to 66.5 Cu, 14.0 to

Table 120 Typical mechanical properties of C75700 sheet or strip, 1 mm (0.04 in.) thick

Temper	Tensile strength, MPa	Tensile strength, ksi	Yield strength(a), MPa	Yield strength(a), ksi	Elongation in 50 mm (2 in.), %	Hardness HRF	Hardness HRB	Hardness HR30T	Shear strength, MPa	Shear strength, ksi
OS070	360	52	125	18	48	69	22	27	...	...
OS050	370	54	130	19	45	73	30	33	...	...
OS035	385	56	145	21	42	78	37	38	285	41
OS025	405	59	165	24	38	82	45	44	...	...
OS015	420	61	195	28	35	88	55	51	...	...
H00	415	60	240	35	32	...	60	55	295	43
H01	450	65	310	45	23	...	70	63	305	44
H02	505	73	415	60	11	...	80	70	325	47
H04	585	85	515	75	4	...	89	75	360	52
H06	640	93	545	79	2	...	92	77	385	56

(a) At 0.5% extension under load

16.0 Ni, 0.1 Pb max, 0.25 Fe max, 0.5 Mn max, 0.5 max other (total), bal Zn

Applications

Typical uses. Camera parts, optical equipment, etching stock, jewelry

Mechanical Properties

Tensile properties. See Table 119.
Hardness. See Table 119.
Elastic modulus. Tension, 125 GPa (18 × 10^6 psi); shear, 47 GPa (6.8 × 10^6 psi)

Mass Characteristics

Density. 8.70 g/cm^3 (0.314 lb/in.^{3}) at 20 °C (68 °F)

Thermal Properties

Liquidus temperature. 1075 °C (1970 °F)
Solidus temperature. 1040 °C (1900 °F)
Coefficient of linear thermal expansion. 16.2 μm/m · K (9.0 μin./in. · °F) at 20 to 300 °C (68 to 572 °F)
Specific heat. 380 J/kg · K (0.09 Btu/lb · °F) at 20 °C (68 °F)
Thermal conductivity. 36 W/m · K (21 Btu/ft · h · °F) at 20 °C (68 °F)

Electrical Properties

Electrical conductivity. Volumetric, 7% IACS at 20 °C (68 °F)
Electrical resistivity. 246 nΩ · m at 20 °C (68 °F)

Fabrication Characteristics

Machinability. 20% of C36000 (free-cutting brass)
Formability. Excellent for cold working by blanking, drawing, forming, bending, heading, upsetting, roll threading, knurling, shearing, spinning, squeezing, or swaging; poor for hot forming
Weldability. Soldering and brazing: excellent. Oxyfuel gas, resistance spot, and resistance butt welding: good. Gas metal arc and resistance seam welding: fair. Shielded metal arc welding is not recommended.
Annealing temperature. 600 to 815 °C (1100 to 1500 °F)

C75700
65Cu-23Zn-12Ni

Commercial Names

Common name. Nickel silver, 65-12

Specifications

ASTM. Bar, rod: B 151. Wire: B 206
Government. Wire: QQ-W-321

Chemical Composition

Composition limits. 63.5 to 66.5 Cu, 11.0 to 13.0 Ni, 0.05 Pb max, 0.25 Fe max, 0.5 Mn max, 0.5 max other (total), bal Zn

Applications

Typical uses. Slide fasteners, camera parts, optical parts, etching stock, nameplates

Mechanical Properties

Tensile properties. See Table 120.
Hardness. See Table 120.
Elastic modulus. Tension, 125 GPa (18 × 10^6 psi); shear, 47 GPa (6.8 × 10^6 psi)

Mass Characteristics

Density. 8.69 g/cm^3 (0.314 lb/in.^{3}) at 20 °C (68 °F)

Thermal Properties

Liquidus temperature. 1040 °C (1900 °F)
Coefficient of linear thermal expansion. 16.2 μm/m · K (9.0 μin./in. · °F) at 20 to 300 °C (68 to 572 °F)
Specific heat. 380 J/kg · K (0.09 Btu/lb · °F) at 20 °C (68 °F)
Thermal conductivity. 40 W/m · K (23 Btu/ft · h · °F) at 20 °C (68 °F)

Electrical Properties

Electrical conductivity. Volumetric, 8% IACS at 20 °C (68 °F)
Electrical resistivity. 216 nΩ · m at 20 °C (68 °F)

Fabrication Characteristics

Machinability. 20% of C36000 (free-cutting brass)
Formability. Excellent for cold working by blanking, drawing, etching, forming, bending, heading, upsetting, roll threading, knurling, shearing, spinning, squeezing, or swaging; poor for hot forming
Weldability. Soldering and brazing: excellent. Oxyfuel gas, resistance spot, and resistance butt welding: good. Gas metal arc and resistance seam welding: fair. Shielded metal arc welding is not recommended.
Annealing temperature. 600 to 825 °C (1100 to 1500 °F)

Table 121 Typical mechanical properties of C77000

Temper	Tensile strength, MPa	Tensile strength, ksi	Yield strength(a), MPa	Yield strength(a), ksi	Elongation in 50 mm (2 in.), %	Hardness HRF	Hardness HRB	Hardness HR30T
Flat products, 1 mm (0.04 in.) thick								
OS035	415	60	185	27	40	90	55	...
H04	690	100	585	85	3	...	91	77
H06	745	108	620	90	2.5	...	96	80
H08	795	115	...	...	2.5	...	99	81
Wire, 2 mm (0.08 in.) diameter								
OS035	415	60	...	...	40	...	...	...
H08 (68%)	1000	145	...	...	2	...	...	...

(a) At 0.5% extension under load

C77000
55Cu-27Zn-18Ni

Commercial Names

Common name. Nickel silver, 55-18

Specifications

ASTM. Flat products: B 122. Bar: B 122, B 151. Rod: B 151. Wire: B 206
SAE. J463
Government. Flat products: QQ-C-585. Bar: QQ-C-585, QQ-C-586. Rod, shapes, flat wire: QQ-C-586. Wire: QQ-W-321

Chemical Composition

Composition limits. 53.5 to 56.5 Cu, 16.5 to

Table 122 Typical mechanical properties of 1 mm (0.04 in.) thick C78200 sheet

Temper	Tensile strength MPa	Tensile strength ksi	Yield strength(a) MPa	Yield strength(a) ksi	Elongation in 50 mm (2 in.), %	Hardness, HRB	Shear strength MPa	Shear strength ksi
OS035	365	53	160	23	40	78 HRF	275	40
OS015	405	59	185	27	32	85 HRF	295	43
H01	425	62	290	42	24	65	305	44
H02	475	69	400	58	12	78	325	47
H03	540	78	435	63	5	84	350	51
H04	585	85	505	73	4	87	370	54
H06	625	91	525	76	3	90	400	58

(a) At 0.5% extension under load

19.5 Ni, 0.1 Pb max, 0.25 Fe max, 0.5 Mn max, 0.5 max other (total), bal Zn

Applications

Typical uses. Optical goods, springs, resistance wire

Mechanical Properties

Tensile properties. See Table 121.
Hardness. See Table 121.
Elastic modulus. Tension, 125 GPa (18 × 10^6 psi); shear, 47 GPa (6.8 × 10^6 psi)

Mass Characteristics

Density. 8.70 g/cm^3 (0.314 lb/in.^{3}) at 20 °C (68 °F)

Thermal Properties

Liquidus temperature. 1055 °C (1930 °F)
Coefficient of linear thermal expansion. 16.7 μm/m · K (9.3 μin./in. · °F) at 20 to 300 °C (68 to 572 °F)
Specific heat. 380 J/kg · K (0.09 Btu/lb · °F) at 20 °C (68 °F)
Thermal conductivity. 29 W/m · K (17 Btu/ft · h · °F) at 20 °C (68 °F)

Electrical Properties

Electrical conductivity. Volumetric, 5.5% IACS at 20 °C (68 °F)
Electrical resistivity. 314 nΩ · m at 20 °C (68 °F)

Fabrication Characteristics

Machinability. 30% of C36000 (free-cutting brass)
Formability. Good for cold working by blanking, forming, bending, and shearing; poor for hot forming
Weldability. Soldering and brazing: excellent. Oxyfuel gas, resistance spot, and resistance butt welding: good. Gas metal arc and resistance seam welding: fair. Shielded metal arc welding is not recommended.
Annealing temperature. 600 to 825 °C (1100 to 1500 °F)

78200 65Cu-25Zn-8Ni-2Pb

Chemical Composition

Composition limits. 63.0 to 67.0 Cu, 1.5 to 2.5 Pb, 7.0 to 9.0 Ni, 0.35 Fe max, 0.50 Mn max, 0.10 max other (total), bal Zn

Applications

Typical uses. Key blanks, watch plates, watch parts

Mechanical Properties

Tensile properties. See Table 122.
Shear strength. See Table 122.
Hardness. See Table 122.
Elastic modulus. Tension, 117 GPa (17 × 10^6 psi); shear, 44 GPa (6.4 × 10^6 psi)

Mass Characteristics

Density. 8.69 g/cm^3 (0.314 lb/in.^{3}) at 20 °C (68 °F)

Thermal Properties

Liquidus temperature. 1000 °C (1830 °F)
Solidus temperature. 970 °C (1780 °F)
Coefficient of linear thermal expansion. 18.5 μm/m · K (10.3 μin./in. · °F) at 20 to 100 °C (68 to 212 °F)
Specific heat. 380 J/kg · K (0.09 Btu/lb · °F) at 20 °C (68 °F)
Thermal conductivity. 48 W/m · K (28 Btu/ft · h · °F) at 20 °C (68 °F)

Electrical Properties

Electrical conductivity. Volumetric, 10.9% IACS at 20 °C (68 °F)
Electrical resistivity. 160 nΩ · m at 20 °C (68 °F)

Fabrication Characteristics

Machinability. 60% of C36000 (free-cutting brass)
Formability. Cold working, good; hot forming, poor. Commonly fabricated by blanking, milling, and drilling
Weldability. Soldering: excellent. Brazing: good. Oxyfuel gas, arc, and resistance welding generally are not recommended.
Annealing temperature. 500 to 620 °C (930 to 1150 °F)

Selection and Application of Copper Alloy Castings

Revised by Robert F. Schmidt, Colonial Metals Co., and Donald G. Schmidt, R. Lavin & Sons, Inc.

COPPER ALLOY CASTINGS are used in applications that require superior corrosion resistance, high thermal or electrical conductivity, good bearing surface qualities, or other special properties. Casting makes it possible to produce parts with shapes that cannot be easily obtained by fabrication methods such as forming or machining. Often, it is more economical to produce a part as a casting than to fabricate it by other means.

Types of Copper Alloys

Because pure copper is extremely difficult to cast and is prone to surface cracking, porosity problems, and the formation of internal cavities, small amounts of alloying elements (such as beryllium, silicon, nickel, tin, zinc, and chromium) are used to improve the casting characteristics of copper. Larger amounts of alloying elements are added for property improvement.

The copper-base castings are designated in the Unified Numbering System (UNS) with numbers ranging from C80000 to C99999. Also, copper alloys in the cast form are sometimes classified according to their freezing range (that is, the temperature range between the liquidus and solidus temperatures). The freezing range of various copper alloys is discussed in the section "Control of Solidification" in this article.

Compositions of copper casting alloys (Table 1) may differ from those of their wrought counterparts for various reasons. Generally, casting permits greater latitude in the use of alloying elements because the effects of composition on hot- or cold-working properties are not important. However, imbalances among certain elements, and trace amounts of certain impurities in some alloys, will diminish castability and can result in castings of lower quality.

Certain cast alloys may also be unsuitable for wrought products. For example, several alloys listed in Table 1 have lead contents of 5% or more. Alloys containing such high percentages of lead are not suited to hot working, but they are ideal for low- or medium-speed bearings, where the lead prevents galling (and excessive wear if a lubricant is not present).

The tolerance for impurities is normally greater in castings than in their wrought counterparts because of the adverse effects certain impurities have on hot or cold workability. On the other hand, impurities that inhibit response to heat treatment must be avoided both in castings and in wrought products. The choice of an alloy for any casting usually depends on five factors: metal cost, castability, machinability, properties, and final cost.

Castability

Castability should not be confused with fluidity, which is only a measure of the distance to which a metal will flow before solidifying. Fluidity is thus one factor determining the ability of a molten alloy to completely fill a mold cavity in every detail. Castability, on the other hand, is a general term relating to the ability to reproduce fine detail on a surface. Colloquially, good castability refers to the ease with which an alloy responds to ordinary foundry practice without requiring special techniques for gating, risering, melting, sand conditioning, or any of the other factors involved in making good castings. High fluidity often ensures good castability, but it is not solely responsible for that quality in a casting alloy.

The castability of alloys is generally influenced by their shrinkage characteristics and their freezing range (which is not necessarily related directly to shrinkage). Classification of copper casting alloys according to a narrow or wide freezing range is discussed in the article "Copper and Copper Alloys" in Volume 15, *Casting*, of the 9th Edition of *Metals Handbook*. The effect of the freezing range on castability is discussed in the section "Control of Solidification" in this article.

Foundry alloys are also classified as high-shrinkage or low-shrinkage alloys. The former class includes the manganese bronzes, aluminum bronzes, silicon bronzes, silicon brasses, and some nickel-silvers. They are more fluid than the low-shrinkage red brasses, more easily poured, and give high-grade castings in the sand, permanent mold, plaster, die, and centrifugal casting processes. With high-shrinkage alloys, careful design is necessary to promote directional solidification, avoid abrupt changes in cross section, avoid notches (by using generous fillets), and properly place gates and risers; all of these design precautions help avoid internal shrinks and cracks. Turbulent pouring must be avoided to prevent the formation of dross becoming entrapped in the casting. Liberal use of risers or exothermic compounds ensures adequate molten metal to feed all sections of the casting. Table 2 presents foundry characteristics of selected standard alloys, including a comparative ranking of both fluidity and overall castability for sand casting; number 1 represents the highest castability or fluidity ranking.

All copper alloys can be successfully cast in sand. Sand casting allows the greatest flexibility in casting size and shape and is the most economical casting method if only a few castings are made (die casting is more economical above ~50 000 units). Permanent mold casting is best suited for tin, silicon, aluminum and manganese bronzes, and yellow brasses. Die casting is well suited for yellow brasses, but increasing amounts of permanent mold alloys are also being die cast. Size is a definite limitation for both methods, although large slabs weighing as much as 4500 kg (10 000 lb) have been cast in permanent molds. Brass die castings generally weigh less than 0.2 kg (0.5 lb) and seldom exceed 0.9 kg (2 lb). The limitation of size is due to the reduced die life with larger castings.

Virtually all copper alloys can be cast successfully by the centrifugal casting process. Castings of almost every size from less than 100 g to more than 22 000 kg (<0.25 to >50 000 lb) have been made.

Table 1 Nominal compositions of principal copper casting alloys

UNS number	Common name	Previous ASTM designation	Composition						
			Cu	Sn	Pb	Zn	Fe	Al	Other
ASTM B 22									
C86300	Manganese bronze	B 22-E	62	...	...	24	3	6	3 Mn
C90500	Tin bronze	B 22-D	88	10	...	2	...	...	...
C91100	Tin bronze	B 22-B	84	16	...	...	...	...	...
C91300	Tin bronze	B 22-A	81	19	...	...	...	...	...
ASTM B 61									
C92200	Valve bronze	...	88	6	1.5	4	...	...	1 Ni max
ASTM B 62									
C83600	Leaded red brass	...	85	5	5	5	...	...	...
ASTM B 66									
C93800	High-lead tin bronze	...	78	7	15	...	...	...	...
C94300	High-lead tin bronze	...	70	5	25	...	...	...	...
C94400	Leaded phosphor bronze	...	81	8	11	...	...	...	0.35 P
C94500	High-lead tin bronze	...	73	7	19	1	...	...	...
ASTM B 67									
C94100	High-lead tin bronze	...	bal	5.5	20	...	...	...	...
ASTM B 148									
C95200	Aluminum bronze	B 148-9A	88	...	...	...	3	9	...
C95300	Aluminum bronze	B 148-9B	89	...	...	...	1	10	...
C95400	Aluminum bronze	B 148-9C	85.5	...	...	...	4	10.5	...
C95410	Aluminum bronze	...	84	...	...	...	4	10	2 Ni
C95500	Nickel-aluminum bronze	B 148-9D	81	...	...	...	4	11	4 Ni
C95600	Silicon-aluminum bronze	B 148-9E	91	...	...	...	...	7	2 Si
C95700	Aluminum bronze	...	75	...	...	...	3	8	2 Ni, 12 Mn
C95800	Nickel-aluminum bronze	...	81.5	...	...	...	4	9	4 Ni, 1.5 Mn
ASTM B 176									
C85700	Yellow brass	...	61	1	1	37	...	...	...
C85800	Yellow brass	Z30A	58	1	1	40	...	...	...
C86500	Manganese bronze	...	58	...	...	39	1	1	0.5 Mn
C87800	Silicon brass	ZS144A	82	...	...	14	...	...	4 Si
C87900	Silicon yellow brass	ZS331A	65	...	...	33	...	...	1 Si
C99700	White manganese bronze	...	58	...	2	22	...	1	5 Ni, 12 Mn
C99750	White manganese bronze	...	58	...	1	20	...	1	20 Mn
ASTM B 584									
C83450	Leaded red brass	...	88	2.5	2	6.5	...	...	1 Ni
C83600	Leaded red brass	B 145-4A	85	5	5	5	...	...	...
C83800	Leaded red brass	B 145-4B	83	4	6	7	...	...	...
C84400	Leaded semired brass	B 145-5A	81	3	7	9	...	...	...
C84800	Leaded semired brass	B 145-5B	76	3	6	15	...	...	...
C85200	Leaded yellow brass	B 146-6A	72	1	3	24	...	...	...
C85400	Leaded yellow brass	B 146-6B	67	1	3	29	...	...	...
C85700	Leaded naval brass	B 146-6C	61	1	1	37	...	...	...
C86200	High-strength manganese bronze	B 147-8B	63	...	...	27	3	4	3 Mn
C86300	High-strength manganese bronze	B 147-8C	62	...	...	26	3	6	3 Mn
C86400	Leaded manganese bronze	B 147-7A	58	1	1	38	1	0.5	0.5 Mn
C86500	Manganese bronze	B 147-8A	58	...	...	39	1	1	1 Mn
C86700	Leaded manganese bronze	B 132-B	58	1	1	34	2	2	2 Mn
C87300	Silicon bronze	B 198-12A	95	...	...	...	...	...	1 Mn, 4 Si
C87400	Leaded silicon brass	B 198-13A	82	...	0.5	14	...	...	3.5 Si
C87500	Silicon brass	B 198-13B	82	...	...	14	...	...	4 Si
C87600	Silicon bronze	B 198-13C	91	...	...	5	...	...	4 Si
C87610	Silicon bronze	B 198-12A	92	...	...	4	...	...	4 Si
C90300	Modified G bronze	B 143-1B	88	8	...	4	...	...	...
C90500	G bronze	B 143-1A	88	10	...	2	...	...	...
C92200	Navy M	B 143-2A	88	6	1.5	3.5	...	...	...
C92300	Leaded tin bronze	B 143-2B	87	8	1	4	...	...	...
C92500	Leaded tin bronze	...	87	10	1	2	...	...	...
C93200	High-lead tin bronze	B 144-3B	83	7	7	3	...	...	...
C93500	High-lead tin bronze	B 144-3C	85	5	9	1	...	...	...
C93700	High-lead tin bronze	B 144-3A	80	10	10	...	...	...	...
C93800	High-lead tin bronze	B 144-3D	78	7	15	...	...	...	...
C94300	High-lead tin bronze	B 144-3E	70	5	25	...	...	...	...
C94700	Nickel-tin bronze	B 292-A	88	5	...	2	...	...	5 Ni
C94800	Leaded nickel-tin bronze	B 292-B	87	5	1	2	...	...	5 Ni
C94900	Leaded nickel-tin bronze	...	80	5	5	5	...	...	5 Ni
C96800	Spinodal alloy	...	82	8	...	...	...	...	10 Ni, 0.2 Nb
C97300	Leaded nickel-silver	B 149-10A	56	2	10	20	...	...	12 Ni
C97600	Leaded nickel-silver	B 149-11A	64	4	4	8	...	...	20 Ni
C97800	Leaded nickel-silver	B 149-11B	66	5	2	2	...	...	25 Ni

Table 2 Foundry properties of the principal copper alloys for sand casting

UNS number	Common name	Shrinkage allowance, %	Approximate liquidus temperature °C	Approximate liquidus temperature °F	Castability rating(a)	Fluidity rating(a)
C83600	Leaded red brass	5.7	1010	1850	2	6
C84400	Leaded semired brass	2.0	980	1795	2	6
C84800	Leaded semired brass	1.4	955	1750	2	6
C85400	Leaded yellow brass	1.5–1.8	940	1725	4	3
C85800	Yellow brass	2.0	925	1700	4	3
C86300	Manganese bronze	2.3	920	1690	5	2
C86500	Manganese bronze	1.9	880	1615	4	2
C87200	Silicon bronze	1.8–2.0	...	...	5	3
C87500	Silicon brass	1.9	915	1680	4	1
C90300	Tin bronze	1.5–1.8	980	1795	3	6
C92200	Leaded tin bronze	1.5	990	1810	3	6
C93700	High-lead tin bronze	2.0	930	1705	2	6
C94300	High-lead tin bronze	1.5	925	1700	6	7
C95300	Aluminum bronze	1.6	1045	1910	8	3
C95800	Aluminum bronze	1.6	1060	1940	8	3
C97600	Nickel-silver	2.0	1145	2090	8	7
C97800	Nickel-silver	1.6	1180	2160	8	7

(a) Relative rating for casting in sand molds. The alloys are ranked from 1 to 8 in both overall castability and fluidity; 1 is the highest or best possible rating.

Because of their low lead contents, aluminum bronzes, yellow brasses, manganese bronzes, low-nickel bronzes, and silicon brasses and bronzes are best adapted to plaster mold casting. For most of these alloys, lead should be held to a minimum because it reacts with the calcium sulfate in the plaster, resulting in discoloration of the surface of the casting and increased cleaning and machining costs. Size is a limitation on plaster mold casting, although aluminum bronze castings that weigh as little as 100 g (0.25 lb) have been made by the lost-wax process, and castings that weigh more than 150 kg (330 lb) have been made by conventional plaster molding.

Control of Solidification. Production of consistently sound castings requires an understanding of the solidification characteristics of the alloys as well as knowledge of relative magnitudes of shrinkage. The actual amount of contraction during solidification does not differ greatly from alloy to alloy. Its distribution, however, is a function of the freezing range and the temperature gradient in critical sections. Manganese and aluminum bronzes are similar to steel in that their freezing ranges are quite narrow—about 40 and 14 °C (70 and 25 °F), respectively. Large castings can be made by the same conventional methods used for steel, as long as proper attention is given to placement of gates and risers—both those for controlling directional solidification and those for feeding the primary central shrinkage cavity.

Tin bronzes have wider freezing ranges (~165 °C, or 300 °F, for C83600). Alloys with such wide freezing ranges form a mushy zone during solidification, resulting in interdendritic shrinkage or microshrinkage. Because feeding cannot take place properly under these conditions, porosity results in the affected sections. In overcoming this effect, design and riser placement, plus the use of chills, are important. Another means of overcoming interdendritic shrinkage is to maintain close temperature control of the metal during pouring and to provide for rapid solidification. These requirements limit section thickness and pouring temperatures, and this practice requires a gating system that will ensure directional solidification. Sections up to 25 mm (1 in.) in thickness are routinely cast. Sections up to 50 mm (2 in.) thick can be cast, but only with difficulty and under carefully controlled conditions. A bronze with a narrow solidification (freezing) range and good directional solidification characteristics is recommended for castings having section thicknesses greater than about 25 mm (1 in.).

It is difficult to achieve directional solidification in complex castings. The most effective and most easily used device is the chill. For irregular sections, chills must be shaped to fit the contour of the section of the mold in which they are placed. Insulating pads and riser sleeves sometimes are effective in slowing down the solidification rate in certain areas to maintain directional solidification. Further information on the casting of copper alloys is given in *Casting*, Volume 15 of the 9th Edition of *Metals Handbook*.

Mechanical Properties

Most copper-base casting alloys containing tin, lead, or zinc have only moderate tensile and yield strengths, low-to-medium hardness, and high elongation. When higher tensile or yield strength is required, the aluminum bronzes, manganese bronzes, silicon brasses, silicon bronzes, and some nickel-silvers are used instead. Most of the higher-strength alloys have better-than-average resistance to corrosion and wear. Mechanical and physical properties of copper-base casting alloys are presented in Table 3. (Throughout this discussion, as well as in Table 3, the mechanical properties quoted are for sand cast test bars. Properties of the castings themselves may be lower, depending on section size and process-design variables.)

Tensile strengths for cast test bars of aluminum bronzes and manganese bronzes range from 450 to 900 MPa (65 to 130 ksi), depending on composition; some aluminum bronzes attain maximum tensile strength only after heat treatment.

Although manganese and aluminum bronzes are often used for the same applications, the manganese bronzes are handled more easily in the foundry. As-cast tensile strengths as high as 800 MPa (115 ksi) and elongations of 15 to 20% can be obtained readily in sand castings; slightly higher values are possible in centrifugal castings. Stresses can be relieved at 175 to 200 °C (350 to 400 °F). Lead can be added to the lower-strength manganese bronzes to increase machinability, but at the expense of tensile strength and elongation. Lead content should not exceed 0.1% in high-strength manganese bronzes. Although manganese bronzes range in hardness from 125 to 250 HB, they are readily machined.

Tin is added to the low-strength manganese bronzes to enhance resistance to dezincification, but it should be limited to 0.1% in high-strength manganese bronzes unless sacrifices in strength and ductility can be accepted.

Manganese bronzes are specified for marine propellers and fittings, pinions, ball bearing races, worm wheels, gear shift forks, and architectural work. Manganese bronzes are also used for rolling mill screw-down nuts and slippers, bridge trunnions, gears, and bearings, all of which require high strength and hardness.

Various cast aluminum bronzes contain 9 to 14% Al and lesser amounts of iron, manganese, or nickel. They have a very narrow solidification range; therefore, they have a greater need for adequate gating and risering than do most other copper casting alloys and thus are more difficult to cast. A wide range of properties can be obtained with these alloys, especially after heat treatment, but close control of composition is necessary. Like the manganese bronzes, aluminum bronzes can develop tensile strengths well over 700 MPa (100 ksi).

Most aluminum bronzes contain from 0.75 to 4% Fe to refine grain structure and increase strength. Alloys containing from 8 to 9.5% Al cannot be heat treated unless other elements (such as nickel or manganese) in amounts over 2% are added as well. They have higher tensile strengths and greater ductility and toughness than any of the ordinary tin bronzes. Applications include valve nuts, cam bearings, impellers,

Table 3 Typical properties of copper casting alloys

UNS number	Tensile strength, MPa	Tensile strength, ksi	Yield strength(a), MPa	Yield strength(a), ksi	Compressive yield strength(b), MPa	Compressive yield strength(b), ksi	Elongation, %	Hardness, HB(c)	Electrical conductivity, % IACS
ASTM B 22									
C86300	820	119	468	68	490	71	18	225(d)	8.0
C90500	317	46	152	22	...	...	30	75	10.9
C91100	241	35	172	25	125 min	18 min	2	135(d)	8.5
C91300	241	35	207	30	165 min	24 min	0.5	170(d)	7.0
ASTM B 61									
C92200	280	41	110	16	105	15	45	64	14.3
ASTM B 62									
C83600	240	35	105	15	100	14	32	62	15.0
ASTM B 66									
C93800	221	32	110	16	83	12	20	58	11.6
C94300	186	27	90	13	76	11	15	48	9.0
C94400	221	32	110	16	...	...	18	55	10.0
C94500	172	25	83	12	...	...	12	50	10.0
ASTM B 67									
C94100	138	20	97	14	...	...	15	44	...
ASTM B 148									
C95200	552	80	200	29	207	30	38	120(d)	12.2
C95300	517	75	186	27	138	20	25	140(d)	15.3
C95400	620	90	255	37	...	...	17	170(d)	13.0
C95400 (HT)(e)	758	110	317	46	...	...	15	195(d)	12.4
C95410	620	90	255	37	...	...	17	170(d)	13.0
C95410 (HT)(e)	793	116	400	58	...	...	12	225(d)	10.2
C95500	703	102	303	44	...	...	12	200(d)	8.8
C95500 (HT)(e)	848	123	545	79	...	...	5	248(d)	8.4
C95600	517	75	234	34	...	...	18	140(d)	8.5
C95700	655	95	310	45	...	...	26	180(d)	3.1
C95800	662	96	255	37	241	35	25	160(d)	7.0
ASTM B 176									
C85700	...	...	...	...	...	...	...	...	...
C85800	380	55	205(f)	30(f)	...	...	15	...	22.0
C86500	...	...	...	...	...	...	...	...	...
C87800	620	90	205(f)	30(f)	...	...	25	...	6.5
C87900	400	58	205(f)	30(f)	...	...	15	...	...
C99700	415	60	180	26	...	...	15	120(d)	3.0
C99750	...	...	...	...	...	...	...	...	...
ASTM B 584									
C83450	255	37	103	15	69	10	34	62	20.0
C83600	241	35	103	15	97	14	32	62	15.1
C83800	241	35	110	16	83	12	28	60	15.3
C84400	234	34	97	14	...	...	28	55	16.8
C84800	262	38	103	15	90	13	37	59	16.4
C85200	262	38	90	13	62	9	40	46	18.6
C85400	234	34	83	12	62	9	37	53	19.6
C85700	352	51	124	18	...	...	43	76	21.8
C86200	662	96	331	48	352	51	20	180(d)	7.4
C86300	820	119	469	68	489	71	18	225(d)	8.0
C86400	448	65	166	24	159	23	20	108(d)	19.3
C86500	489	71	179	26	166	24	40	130(d)	20.5
C86700	586	85	290	42	...	...	20	155(d)	16.7
C87300	400	58	172	25	131	19	35	85	6.1
C87400	379	55	165	24	...	...	30	70	6.7
C87500	469	68	207	30	179	26	17	115	6.1
C87600	456	66	221	32	...	...	20	135(d)	8.0
C87610	400	58	172	25	131	19	35	85	6.1
C90300	310	45	138	20	90	13	30	70	12.4
C90500	317	46	152	22	103	15	30	75	10.9
C92200	283	41	110	16	103	15	45	64	14.3
C92300	290	42	138	20	69	10	32	70	12.3
C92600	303	44	138	20	83	12	30	72	10.0
C93200	262	38	117	17	...	...	30	67	12.4
C93500	221	32	110	16	...	...	20	60	15.0
C93700	269	39	124	18	124	18	30	67	10.1
C93800	221	32	110	16	83	12	20	58	11.6
C94300	186	27	90	13	76	11	15	48	9.0
C94700	345	50	159	23	...	...	35	85	11.5
C94700 (HT)(g)	620	90	483	70	...	...	10	210(d)	14.8
C94800	310	45	159	23	...	...	35	80	12.0
C94900	262 min	38 min	97 min	14 min	...	...	15 min	...	...
C96800	862 min	125 min	689 min(f)	100 min(f)	...	...	3 min	...	...
C97300	248	36	117	17	...	...	25	60	5.9
C97600	324	47	179	26	159	23	22	85	4.8
C97800	379	55	214	31	...	...	16	130(d)	4.5

Note: HT indicates alloy in heat-treated condition. (a) At 0.5% extension under load. (b) At a permanent set of 0.025 mm (0.001 in). (c) 500 kgf (1100 lbf) load. (d) 3000 kgf (6600 lbf) load. (e) Heat treated at 900 °C (1650 °F), water quenched, tempered at 590 °C (1100 °F), water quenched. (f) At 0.2% offset. (g) Solution anneal of 760 °C (1400 °F) for 4 h, water quench, and then age at 315 °C (600 °F) for 5 h and air cool

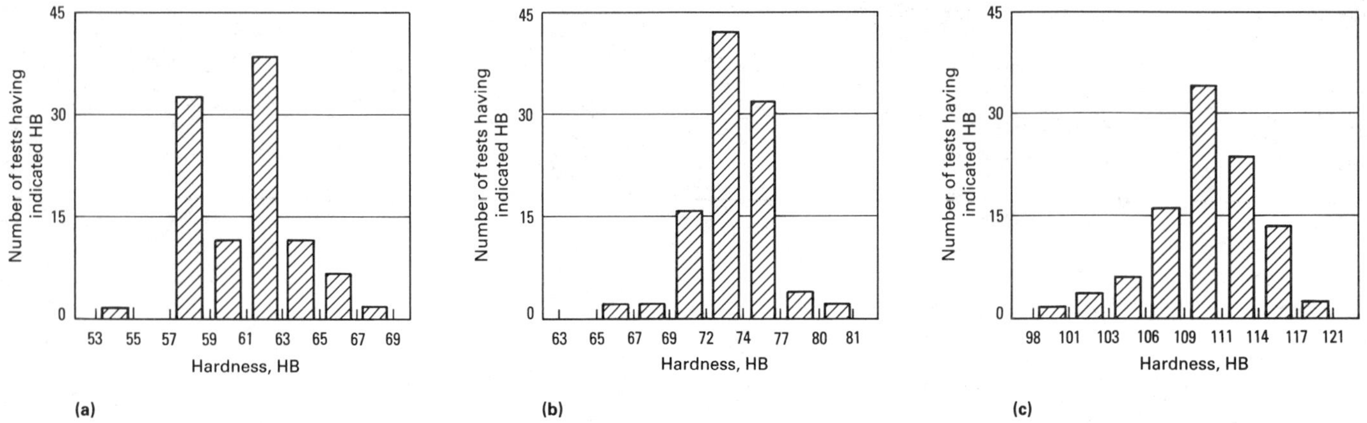

Fig. 1 Distribution of hardness over 100 tests for three copper casting alloys of different tensile strengths. (a) C83600. Tensile strength, 235 to 260 MPa (34 to 38 ksi); 500 kg (1100 lbf) load. (b) C90300. Tensile strength, 275 to 325 MPa (40 to 47 ksi); 500 kg (1100 lbf) load. (c) C87500. Tensile strength, 420 to 500 MPa (61 to 72 ksi); 1500 kg (3300 lbf) load

hangers in pickling baths, agitators, crane gears, and connecting rods.

The heat-treatable aluminum bronzes contain from 9.5 to 11.5% Al; they also contain iron, with or without nickel or manganese. These castings are quenched in water or oil from temperatures between 760 and 925 °C (1400 and 1700 °F) and tempered at 425 to 650 °C (800 to 1200 °F), depending on the exact composition and the required properties.

From the range of properties shown in Table 3, it can be seen that all the maximum properties cannot be obtained in any one aluminum bronze. In general, alloys with higher tensile strengths, yield strengths, and hardnesses have lower values of elongation. Typical applications of the higher-hardness alloys are rolling mill screwdown nuts and slippers, worm gears, bushings, slides, impellers, nonsparking tools, valves, and dies.

Aluminum bronzes resist corrosion in many substances, including pickling solutions. When corrosion occurs, it often proceeds by preferential attack of the aluminum-rich phases. Duplex alpha-plus-beta aluminum bronzes are more susceptible to preferential attack of the aluminum-rich phases than are the all-alpha aluminum bronzes.

Aluminum bronzes have fatigue limits that are considerably greater than those of manganese bronze or any other cast copper alloy. Unlike Cu-Zn and Cu-Sn-Pb-Zn alloys, the mechanical properties of aluminum and manganese bronzes do not decrease with increases in casting cross section. This is because these alloys have narrow freezing ranges, which result in denser structures when castings are properly designed and properly fed.

Whereas manganese bronzes experience hot shortness above 230 °C (450 °F), aluminum bronzes can be used at temperatures as high as 400 °C (750 °F) for short periods of time without an appreciable loss in strength. For example, a room-temperature tensile strength of 540 MPa (78 ksi) declines to 529 MPa (76.7 ksi) at 260 °C (500 °F), 460 MPa (67 ksi) at 400 °C (750 °F), and 400 MPa (58 ksi) at 540 °C (1000 °F). Corresponding elongation values change from 28% to 32, 35, and 25%, respectively.

Unlike manganese bronzes, many aluminum bronzes increase in yield strength and hardness but decrease in tensile strength and elongation upon slow cooling in the mold. Whereas some manganese bronzes precipitate a relatively soft phase during slow cooling, aluminum bronzes precipitate a hard constituent rather rapidly within the narrow temperature range of 565 to 480 °C (1050 to 900 °F). Therefore, large castings, or smaller castings that are cooled slowly, will have properties different from those of small castings cooled relatively rapidly. The same phenomenon occurs upon heat treating the hardenable aluminum bronzes. Cooling slowly through the critical temperature range after quenching, or tempering at temperatures within this range, will decrease elongation. An addition of 2 to 5% Ni greatly diminishes this effect.

Nickel brasses, silicon brasses, and silicon bronzes, although generally higher in strength than red metal alloys, are used more for their corrosion resistance and are discussed in the section "Selection of Alloys for Corrosion Service" in this article. Distributions of hardness and tensile strength data for separately cast test bars of three different alloys are shown in Fig. 1.

Properties of Test Bars. The mechanical properties of separately cast test bars often differ widely from those of production castings poured at the same time, particularly when the thickness of the casting differs markedly from that of the test bar.

The mechanical properties of tin bronzes are particularly affected by variations in casting section size. With increasing section sizes up to about 50 mm (2 in.), the mechanical properties—both strength and elongation—of the castings themselves are progressively lower than the corresponding properties of separately cast test bars. Elongation is particularly affected; for some tin bronzes, elongation of a 50 mm (2 in.) section may be as little as 1/10 that of a 10 mm (0.4 in.) section or of a separately cast test bar.

The metallurgical behavior of many copper alloy systems is complex. The cooling rate (a function of casting section size) directly influences grain size, segregation, and interdendritic shrinkage; these factors, in turn, affect the mechanical properties of the cast metal. Therefore, molding and casting techniques are based on metallurgical characteristics as well as on casting shape.

Dimensional Tolerances

Typical dimensional tolerances are different for castings produced by different molding methods. A molding process involving two or more mold parts requires greater tolerances for dimensions that cross the parting line than for dimensions wholly within one mold part. For castings made in green sand molds, tolerances across the parting line depend on the accuracy of pins and bushings that align the cope with the drag.

Figure 2 shows variations in two important dimensions for 50 production castings of red brass. The larger dimension presented the greatest difficulty: None of the 50 production castings had an actual dimension as large as the nominal design value. Figure 3 shows dimensional variations in two similar cored valve castings. For each design, both the cores and the corresponding cavities in the castings were measured for about 100 castings. For both designs, the castings had actual dimensions less than those of the cores. This indicates that cores may need to have a slightly larger nominal size than is desired in the finished casting in order to ensure proper as-cast hole sizes.

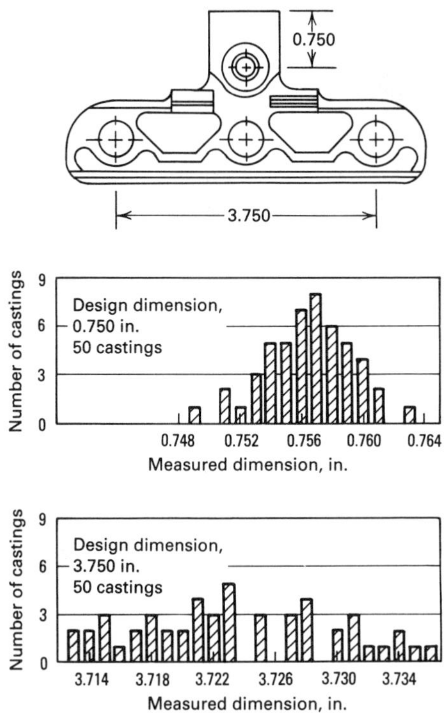

Fig. 2 Variations from design dimensions for a typical red brass casting. Parts were cast in green sand molds made using the same pattern. All dimensions in inches

Machinability

Machinability ratings of copper casting alloys are similar to those of their wrought counterparts. The cast alloys can be separated into three groups. The relative machinability of alloys belonging to the three groups is shown in Table 4.

The first group includes only those containing a single copper-rich phase plus lead. Whether present merely to improve machinability or for some other purpose, lead facilitates chip breakage, thus allowing higher machining speeds with decreased tool wear and improved surface finishes.

Alloys of the second group contain two or more phases. Generally, the secondary phases are harder or more brittle than the matrix. Silicon bronzes, several aluminum bronzes, and the high-tin bronzes belong to this group. Hard and brittle secondary phases act as internal chip breakers, resulting in short chips and easier machining. Manganese bronzes produce a long spiral chip that is smooth on both sides and that does not break. Some aluminum bronzes, on the other hand, produce a long spiral chip that is rough on the underside and that breaks, thus acting like a short chip. Some of the alloys in the second group are classified as moderately machinable because tools wear more rapidly when these alloys are machined, even though chip formation is entirely adequate.

The third group, the most difficult to machine, is composed mainly of the high-strength manganese bronzes and aluminum bronzes that are high in iron or nickel content. Additional information on the machining of copper castings is available in *Machining*, Volume 16 of the 9th Edition of *Metals Handbook.*

Table 4 Machinability ratings of several copper casting alloys

UNS number	Common name	Machinability rating, %(a)
Group 1: free-cutting alloys		
C83600	Leaded red brass	90
C83800	Leaded red brass	90
C84400	Leaded semired brass	90
C84800	Leaded semired brass	90
C94300	High-lead tin bronze	90
C85200	Leaded yellow brass	80
C85400	Leaded yellow brass	80
C93700	High-lead tin bronze	80
C93800	High-lead tin bronze	80
C93200	High-lead tin bronze	70
C93500	High-lead tin bronze	70
C97300	Leaded nickel brass	70
Group 2: moderately machinable alloys		
C86400	Leaded high-strength manganese bronze	60
C92200	Leaded tin bronze	60
C92300	Leaded tin bronze	60
C90300	Tin bronze	50
C90500	Tin bronze	50
C95600	Silicon-aluminum bronze	50
C95300	Aluminum bronze	35
C86500	High-strength manganese bronze	30
Group 3: hard-to-machine alloys		
C86300	High-strength manganese bronze	20
C95200	9% aluminum bronze	20
C95400	11% aluminum bronze	20
C95500	Nickel-aluminum bronze	20

(a) Expressed as a percentage of the machinability of C36000, free-cutting brass. The rating is based on relative speed for equivalent tool life. For example, a material having a rating of 50 should be machined at about half the speed that would be used to make a similar cut in C36000.

General-Purpose Alloys

General-purpose copper casting alloys are often classified as either red or yellow alloys. Nominal compositions and general properties of these alloys are given in Tables 1 and 3.

The leaded red and leaded semired brasses respond readily to ordinary foundry practice and are rated very high in castability. Alloy C83600 is the best known of this group and usually is referred to by one of its common names—85-5-5-5 or ounce metal. Alloy C83600 and its modification, C83800 (83-4-6-7), constitute the largest tonnage of copper-base foundry alloys. They are used where moderate corrosion resistance, good machinability, moderate strength and ductility, and good castability are required. C83800 has lower mechanical properties but better machinability and lower initial metal cost than C83600.

Both C83600 and C83800 are used for plumbing goods, flanges, feed pumps, meter casings and parts, general household and machinery hardware and fixtures, papermaking machinery, hydraulic and steam valves, valve disks and seats, impellers, injectors, memorial markers, plaques, statuary, and similar products.

Alloys C84400 and C84800 are higher in lead and zinc and lower in copper and tin than C83600 and C83800. They are lower in

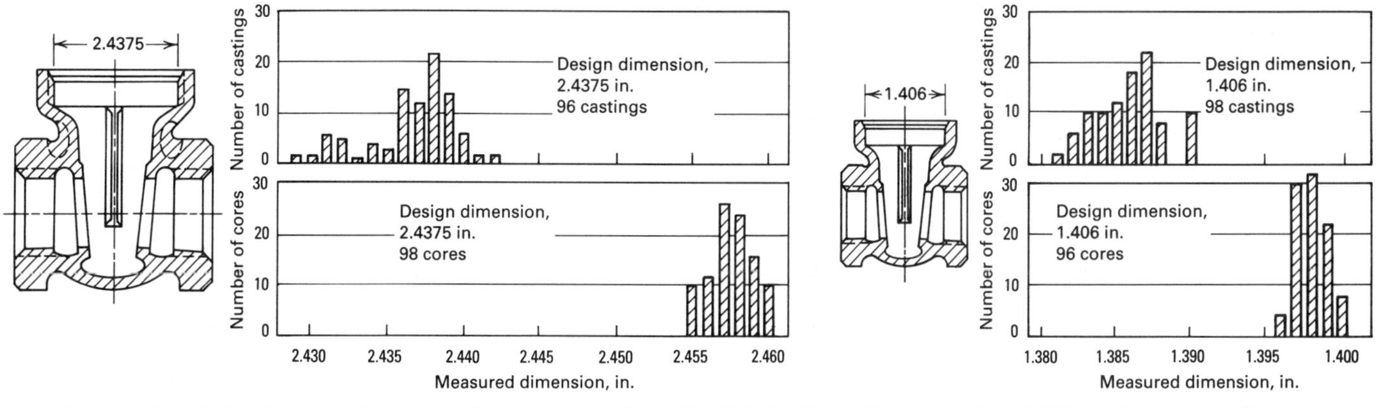

Fig. 3 Variations from design dimensions for two typical cast red brass valve bodies. Valve bodies, similar in design but of different sizes, were made using dry sand cores to shape the internal cavities. The upper histograms indicate dimensional variations for the castings; the lower histograms indicate variations for the corresponding cores. All dimensions in inches

price, and they have lower tensile strengths and hardnesses. Their widest application is in the plumbing industry.

The leaded yellow brasses C85200 and C85700 are even lower in price and mechanical properties. Their main applications are die castings for plumbing goods and accessories, low-pressure valves, air and gas fittings, general hardware, and ornamental castings. In general, they are best suited for small parts; larger parts with thick sections should be avoided. Aluminum (0.15 to 0.25%) is added to yellow brasses to increase fluidity and to give a smoother surface.

All of the red and yellow general-purpose alloys, when properly made and cleaned, can be plated with nickel or chromium.

Alloys that do not contain lead, such as the tin bronzes C90500 (Navy G bronze) and C90300 (modified Navy G bronze), are considerably more difficult to machine than leaded alloys. Alloys containing 10 to 12% Sn, 1 to 2% Ni, and 0.1 to 0.3% P are known as gear bronzes. Up to 1.5% Pb frequently is added to increase machinability. The addition of lead to C90300 increases machinability, but a concurrent decrease in tin is needed to maintain elongation. The leaded tin bronzes include C92200 (known as steam bronze, valve bronze, or Navy M bronze) and C92300 (commercial G bronze).

All of the tin bronzes are suitable wherever corrosion resistance, leak tightness, or greater strength is required at higher operating temperatures than can be tolerated with leaded red or semired brasses. The limiting temperature for long-time operation of C92200 is 290 °C (550 °F); for C90300, C90500, and C92300, it is 260 °C (500 °F) because of the embrittlement caused by the precipitation of a high-tin phase. This reaction does not occur in tin bronzes with tin contents less than about 8%. For elevated-temperature service in handling fluids and gases, Table UNF-23 of the ASME Boiler and Pressure Vessel Code defines allowable working stresses for C92200 (leaded tin bronze, ASTM B 61) and C83600 (leaded red brass, ASTM B 62) at different temperatures (Table 5).

Nickel frequently is added to tin bronzes to increase density and leak tightness. Alloys containing more than 3% Ni are heat treatable, but they must contain less than 0.01% Pb for optimum properties; one example of such an alloy is C94700 (88Cu-5Sn-2Zn-5Ni).

Table 5 Allowable working stresses for C92200 and C83600 castings

Temperature		Working stress			
		ASTM B 61(a)		ASTM B 62(b)	
°C	°F	MPa	ksi	MPa	ksi
38	100	47	6.8	41	6.0
65	150	47	6.8	41	6.0
93	200	47	6.8	40	5.8
120	250	47	6.8	38	5.5
150	300	45	6.5	34	5.0
175	350	41	6.0	31	4.5
205	400	38	5.5	24	3.5
230	450	34	5.0	24	3.5
260	500	28	4.0	24	3.5
290	550	23	3.3	24	3.5

(a) A minimum tensile strength of 235 MPa (34 ksi) is specified for C92200 in ASTM B 61. (b) A minimum tensile strength of 205 MPa (30 ksi) is specified for C83600 in ASTM B 62. Source: ASME Boiler and Pressure Vessel Code, Table UNF-23

Selection of Alloys for Corrosion Service

In Table 6, the relative corrosion resistance in a wide variety of liquids and gases is given for 14 different classes of copper casting alloys. Certain generalizations can be drawn from an examination of these data. In many liquids, the yellow brasses do not have corrosion resistance as high as that of the other copper alloys. However, the high strength of the yellow brasses may make them more desirable even though some corrosion may be encountered. Corrosion of these alloys often takes place by dezincification; if this is a problem, alloys with copper contents of 80% or more must be selected.

Often, experience must be relied on for the proper selection of alloys. Laboratory tests are only guides because they fail to duplicate or approximate the conditions that will be encountered in service. When used in a "recommended" service application (see Table 6), copper metals generally give adequate service life. However, the table can only serve as a guide, and it should be used judiciously. Additional information on the corrosion of copper alloys is given in *Corrosion*, Volume 13 of the 9th Edition of *Metals Handbook*.

Atmospheric Corrosion. Copper alloy castings have been used for centuries for their superior resistance to atmospheric corrosion. Resistance is afforded by the formation of a coating or patina of basic copper sulfate, which ultimately reacts further to form some basic copper carbonate. The sulfate is virtually insoluble in water and thus affords good protection.

Liquid Corrosion. Copper alloy castings are widely used for their superior corrosion resistance in many liquid media. Their resistance to corrosion in liquids, like their resistance to atmospheric corrosion, is increased by the formation of a stable, adherent reaction product. If the coating is removed by chemical or mechanical means, corrosion resistance is reduced, and this reduction often is severe. Thus, rapid corrosion takes place in aerated mineral acids or under conditions of severe agitation, impingement, or high-velocity flow.

Copper metals are attacked by strong organic and inorganic acids and, to some extent, by weak organic acids. Although a copper metal may not visibly corrode, even a minute quantity of copper ions in the solution is not acceptable in certain applications. This is particularly true for food products, in which adverse color or taste can develop.

Ammonium hydroxide attacks all copper alloys severely, and these alloys are not recommended where ammonium ions may be formed. Copper metals are generally satisfactory for applications involving neutral organic compounds, including petroleum products, solvents, and animal and vegetable products. However, in the presence of moisture, certain of these materials may form acids, which in turn may attack a copper metal.

In aqueous solutions, attack is accelerated by dissolved oxygen and carbon dioxide. Thus, although copper alloys are widely used for plumbing goods, they are attacked by many natural waters, especially the very soft waters with high oxygen and carbon dioxide contents. In these waters, carbonic acid is formed, which prevents the development of a resistant layer or dissolves any previously formed layer. Dezincification of high-zinc alloys frequently results if they are used indiscriminately in fresh-water service.

Bearing and Wear Properties

Copper alloys have long been used for bearings because of their combination of moderate-to-high strength, corrosion resistance, and self-lubrication properties. The choice of an alloy depends on the required corrosion resistance and fatigue strength, the rigidity of the backing material, lubrication, the thickness of bearing material, load, the speed of rotation, atmospheric conditions, and other factors. Copper alloys can be cast into plain bearings, cast on steel backs, cast on rolled strip, made into sintered powder metallurgy shapes, or pressed and sintered onto a backing material.

Three groups of alloys are used for bearing and wear-resistant applications: phosphor bronzes (Cu-Sn); copper-tin-lead (low-zinc) alloys; and manganese, aluminum, and silicon bronzes (see Table 1).

Phosphor bronzes (Cu-Sn-P or Cu-Sn-Pb-P alloys) have residual phosphorus ranging from a few hundredths of 1% (for deoxidation and slight hardening) to a maximum of 1%, a level which imparts great hardness. Nickel often is added to refine grain size and disperse the lead. Copper-tin bearings have high resistance to wear, high hardness, and moderately high strength. Alloy C90700 is so widely used for gears that it is commonly called gear bronze.

Phosphor bronzes of higher tin content, such as C91100 and C91300, are used in bridge turntables, where loads are high and rotational movement is slow. The maximum load permitted for C91100 (16% Sn) is 17 MPa (2500 psi); for C91300 (19% Sn) it is 24 MPa (3500 psi). These bronzes are high in

Table 6 Corrosion ratings of cast copper metals in various media

Ratings: A, recommended; B, acceptable; C, not recommended

Corrosive medium	Copper	Tin bronze	Leaded tin bronze	High-lead tin bronze	Leaded red brass	Leaded semired brass	Leaded yellow brass	Leaded high-strength yellow brass	High-strength yellow brass	Aluminum bronze	Leaded nickel brass	Leaded nickel bronze	Silicon bronze	Silicon brass
Acetate solvents	B	A	A	A	A	A	B	A	A	A	A	A	A	B
Acetic acid														
20%	A	C	B	C	B	C	C	C	C	A	C	A	A	B
50%	A	C	B	C	B	C	C	C	C	A	C	B	A	B
Glacial	A	A	A	C	A	C	C	C	C	A	B	B	A	A
Acetone	A	A	A	A	A	A	A	A	A	A	A	A	A	A
Acetylene(a)	C	C	C	C	C	C	C	C	C	C	C	C	C	C
Alcohols(b)	A	A	A	A	A	A	A	A	A	A	A	A	A	A
Aluminum chloride	C	C	C	C	C	C	C	C	C	B	C	C	C	C
Aluminum sulfate	B	B	B	B	B	C	C	C	C	A	C	C	A	A
Ammonia, moist gas	C	C	C	C	C	C	C	C	C	C	C	C	C	C
Ammonia, moisture-free	A	A	A	A	A	A	A	A	A	A	A	A	A	A
Ammonium chloride	C	C	C	C	C	C	C	C	C	C	C	C	C	C
Ammonium hydroxide	C	C	C	C	C	C	C	C	C	C	C	C	C	C
Ammonium nitrate	C	C	C	C	C	C	C	C	C	C	C	C	C	C
Ammonium sulfate	B	B	B	B	B	C	C	C	C	A	C	C	A	A
Aniline and aniline dyes	C	C	C	C	C	C	C	C	C	B	C	C	C	C
Asphalt	A	A	A	A	A	A	A	A	A	A	A	A	A	A
Barium chloride	A	A	A	A	A	C	C	C	C	A	A	A	A	C
Barium sulfide	C	C	C	C	C	C	C	C	B	C	C	C	C	C
Beer(b)	A	A	B	B	B	C	C	C	A	A	C	A	A	B
Beet sugar syrup	A	A	B	B	B	A	A	A	B	A	A	A	B	B
Benzine	A	A	A	A	A	A	A	A	A	A	A	A	A	A
Benzol	A	A	A	A	A	A	A	A	A	A	A	A	A	A
Boric acid	A	A	A	A	A	A	A	B	A	A	A	A	A	A
Butane	A	A	A	A	A	A	A	A	A	A	A	A	A	A
Calcium bisulfite	A	A	B	B	B	C	C	C	C	A	B	A	A	B
Calcium chloride, acid	B	B	B	B	B	B	C	C	C	A	C	C	A	C
Calcium chloride, alkaline	C	C	C	C	C	C	C	C	C	A	C	A	C	B
Calcium hydroxide	C	C	C	C	C	C	C	C	C	B	C	C	C	C
Calcium hypochlorite	C	C	B	B	B	C	C	C	C	B	C	C	C	C
Cane sugar syrups	A	A	B	A	B	A	A	A	A	A	A	A	A	B
Carbonated beverages(b)	A	C	C	C	C	C	C	C	C	A	C	C	A	C
Carbon dioxide, dry	A	A	A	A	A	A	A	A	A	A	A	A	A	A
Carbon dioxide, moist(b)	B	B	B	C	B	C	C	C	C	A	C	A	A	B
Carbon tetrachloride, dry	A	A	A	A	A	A	A	A	A	A	A	A	A	A
Carbon tetrachloride, moist	B	B	B	B	B	B	B	B	B	B	B	A	A	A
Chlorine, dry	A	A	A	A	A	A	A	A	A	A	A	A	A	A
Chlorine, moist	C	C	B	B	B	C	C	C	C	C	C	C	C	C
Chromic acid	C	C	C	C	C	C	C	C	C	C	C	C	C	C
Citric acid	A	A	A	A	A	A	A	A	A	A	A	A	A	A
Copper sulfate	B	A	A	A	A	C	C	C	C	B	B	B	A	A
Cottonseed oil(b)	A	A	A	A	A	A	A	A	A	A	A	A	A	A
Creosote	B	B	B	B	B	C	C	C	C	A	B	B	B	B
Ethers	A	A	A	A	A	A	A	A	A	A	A	A	A	A
Ethylene glycol	A	A	A	A	A	A	A	A	A	A	A	A	A	A
Ferric chloride, sulfate	C	C	C	C	C	C	C	C	C	C	C	C	C	C
Ferrous chloride, sulfate	C	C	C	C	C	C	C	C	C	C	C	C	C	C
Formaldehyde	A	A	A	A	A	A	A	A	A	A	A	A	A	A
Formic acid	A	A	A	A	A	B	B	B	B	A	B	B	B	C
Freon	A	A	A	A	A	A	A	A	A	A	A	A	A	B
Fuel oil	A	A	A	A	A	A	A	A	A	A	A	A	A	A
Furfural	A	A	A	A	A	A	A	A	A	A	A	A	A	A
Gasoline	A	A	A	A	A	A	A	A	A	A	A	A	A	A
Gelatin(b)	A	A	A	A	A	A	A	A	A	A	A	A	A	A
Glucose	A	A	A	A	A	A	A	A	A	A	A	A	A	A
Glue	A	A	A	A	A	A	A	A	A	A	A	A	A	A
Glycerin	A	A	A	A	A	A	A	A	A	A	A	A	A	A
Hydrochloric or muriatic acid	C	C	C	C	C	C	C	C	C	B	C	C	C	C
Hydrofluoric acid	B	B	B	B	B	B	B	B	B	A	B	B	B	B
Hydrofluosilicic acid	B	B	B	B	B	C	C	C	C	B	C	C	B	C
Hydrogen	A	A	A	A	A	A	A	A	A	A	A	A	A	A
Hydrogen peroxide	C	C	C	C	C	C	C	C	C	C	C	C	C	C
Hydrogen sulfide, dry	C	C	C	C	C	C	C	C	C	B	C	C	B	C
Hydrogen sulfide, moist	C	C	C	C	C	C	C	C	C	B	C	C	C	C
Lacquers	A	A	A	A	A	A	A	A	A	A	A	A	A	A
Lacquer thinners	A	A	A	A	A	A	A	A	A	A	A	A	A	A
Lactic acid	A	A	A	A	A	C	C	C	C	A	C	C	A	C
Linseed oil	A	A	A	A	A	A	A	A	A	A	A	A	A	A

(continued)

(a) Acetylene forms an explosive compound with copper when moist or when certain impurities are present and the gas is under pressure. Alloys containing less than 65% Cu are satisfactory under this use. When gas is not under pressure, other copper alloys are satisfactory. (b) Copper and copper alloys resist corrosion by most food products. Traces of copper may be dissolved and affect taste or color. In such cases, copper metals often are tin coated.

Table 6 (continued)

Corrosive medium	Copper	Tin bronze	Leaded tin bronze	High-lead tin bronze	Leaded red brass	Leaded semired brass	Leaded yellow brass	Leaded high-strength yellow brass	High-strength yellow brass	Aluminum bronze	Leaded nickel brass	Leaded nickel bronze	Silicon bronze	Silicon brass
Liquors														
Black	B	B	B	B	B	C	C	C	C	B	C	C	B	B
Green	C	C	C	C	C	C	C	C	C	B	C	C	C	B
White	C	C	C	C	C	C	C	C	C	A	C	C	C	B
Magnesium chloride	A	A	A	A	A	C	C	C	C	A	C	C	A	B
Magnesium hydroxide	B	B	B	B	B	B	B	B	B	A	B	B	B	B
Magnesium sulfate	A	A	A	A	B	C	C	C	C	A	C	B	A	B
Mercury and mercury salts	C	C	C	C	C	C	C	C	C	C	C	C	C	C
Milk(b)	A	A	A	A	A	A	A	A	A	A	A	A	A	A
Molasses(b)	A	A	A	A	A	A	A	A	A	A	A	A	A	A
Natural gas	A	A	A	A	A	A	A	A	A	A	A	A	A	A
Nickel chloride	A	A	A	A	A	C	C	C	C	B	C	C	A	C
Nickel sulfate	A	A	A	A	A	C	C	C	C	A	C	C	A	C
Nitric acid	C	C	C	C	C	C	C	C	C	C	C	C	C	C
Oleic acid	A	A	B	B	B	C	C	C	C	A	C	A	A	B
Oxalic acid	A	A	B	B	B	C	C	C	C	A	C	A	A	B
Phosphoric acid	A	A	A	A	A	C	C	C	C	A	C	A	A	A
Picric acid	C	C	C	C	C	C	C	C	C	C	C	C	C	C
Potassium chloride	A	A	A	A	A	C	C	C	C	A	C	C	A	C
Potassium cyanide	C	C	C	C	C	C	C	C	C	C	C	C	C	C
Potassium hydroxide	C	C	C	C	C	C	C	C	C	A	C	C	C	C
Potassium sulfate	A	A	A	A	A	C	C	C	C	A	C	C	A	C
Propane gas	A	A	A	A	A	A	A	A	A	A	A	A	A	A
Sea water	A	A	A	A	A	C	C	C	C	A	C	C	B	B
Soap solutions	A	A	A	A	B	C	C	C	C	A	C	C	A	C
Sodium bicarbonate	A	A	A	A	A	A	A	A	A	A	A	A	A	B
Sodium bisulfate	C	C	C	C	C	C	C	C	C	A	C	C	C	C
Sodium carbonate	C	A	A	A	A	C	C	C	C	A	C	C	C	A
Sodium chloride	A	A	A	A	A	B	C	C	C	A	C	C	A	C
Sodium cyanide	C	C	C	C	C	C	C	C	C	B	C	C	C	C
Sodium hydroxide	C	C	C	C	C	C	C	C	C	A	C	C	C	C
Sodium hypochlorite	C	C	C	C	C	C	C	C	C	C	C	C	C	C
Sodium nitrate	B	B	B	B	B	B	B	B	B	A	B	B	A	A
Sodium peroxide	B	B	B	B	B	B	B	B	B	B	B	B	B	B
Sodium phosphate	A	A	A	A	A	A	A	A	A	A	A	A	A	A
Sodium sulfate, silicate	A	A	B	B	B	B	C	C	C	A	C	C	A	B
Sodium sulfide, thiosulfate	C	C	C	C	C	C	C	C	C	B	C	C	C	C
Stearic acid	A	A	A	A	A	A	A	A	A	A	A	A	A	A
Sulfur, solid	C	C	C	C	C	C	C	C	C	A	C	C	C	C
Sulfur chloride	C	C	C	C	C	C	C	C	C	C	C	C	C	C
Sulfur dioxide, dry	A	A	A	A	A	A	A	A	A	A	A	A	A	A
Sulfur dioxide, moist	A	A	A	B	B	C	C	C	C	A	C	C	A	B
Sulfur trioxide, dry	A	A	A	A	A	A	A	A	A	A	A	A	A	A
Sulfuric acid														
78% or less	B	B	B	B	B	C	C	C	C	A	C	C	B	B
78% to 90%	C	C	C	C	C	C	C	C	C	B	C	C	C	C
90% to 95%	C	C	C	C	C	C	C	C	C	B	C	C	C	C
Fuming	C	C	C	C	C	C	C	C	C	A	C	C	C	C
Tannic acid	A	A	A	A	A	A	A	A	A	A	A	A	A	A
Tartaric acid	B	A	A	A	A	A	A	A	A	A	A	A	A	A
Toluene	B	B	A	A	A	B	B	B	B	B	B	B	B	A
Trichlorethylene, dry	A	A	A	A	A	A	A	A	A	A	A	A	A	A
Trichlorethylene, moist	A	A	A	A	A	A	A	A	A	A	A	A	A	A
Turpentine	A	A	A	A	A	A	A	A	A	A	A	A	A	A
Varnish	A	A	A	A	A	A	A	A	A	A	A	A	A	A
Vinegar	A	A	B	B	B	C	C	C	C	B	C	C	A	B
Water, acid mine	C	C	C	C	C	C	C	C	C	C	C	C	C	C
Water, condensate	A	A	A	A	A	A	A	A	A	A	A	A	A	A
Water, potable	A	A	A	A	A	A	B	B	B	A	A	A	A	A
Whiskey(b)	A	A	C	C	C	C	C	C	C	A	C	C	A	C
Zinc chloride	C	C	C	C	C	C	C	C	C	B	C	C	B	C
Zinc sulfate	A	A	A	A	A	C	C	C	C	B	C	A	A	C

(a) Acetylene forms an explosive compound with copper when moist or when certain impurities are present and the gas is under pressure. Alloys containing less than 65% Cu are satisfactory under this use. When gas is not under pressure, other copper alloys are satisfactory. (b) Copper and copper alloys resist corrosion by most food products. Traces of copper may be dissolved and affect taste or color. In such cases, copper metals often are tin coated.

phosphorus (1% max) to impart high hardness, and low in zinc (0.25% max) to prevent seizing. They are very brittle, and because of this brittleness are sometimes replaced by manganese bronzes or aluminum bronzes.

High-lead tin bronzes are used where a softer metal is required at slow-to-moderate speeds and at loads not exceeding 5.5 MPa (800 psi). Alloys of this type include C93200 and C93700. The former, also known as 83-7-7-3, is an excellent general bearing alloy; it is especially well suited for applications where lubrication may be deficient. Alloy C93200 is widely used in machine tools, electrical and railroad equipment, steel mill machinery, and automotive applications. Alloy C93200 is produced by the

Table 7 Composition and typical properties of heat-treated copper casting alloys of high strength and conductivity

UNS number	Nominal composition	Tensile strength MPa	Tensile strength ksi	Yield strength MPa	Yield strength ksi	Elongation, %	Hardness	Electrical conductivity, %IACS
C81400	99Cu-0.8Cr-0.06Be	365	53	250	36	11	69 HRB	70
C81500	99Cu-1Cr	350	51	275	40	17	105 HB	85
C81800	97Cu-1.5Co-1Ag-0.4Be	705	102	515	75	8	96 HRB	48
C82000	97Cu-2.5Co-0.5Be	660	96	515	75	6	96 HRB	48
C82200	98Cu-1.5Ni-0.5Be	655	95	515	75	7	96 HRB	48
C82500	97Cu-2Be-0.5Co-0.3Si	1105	160	1035	150	1	43 HRC	20
C82800	96.6Cu-2.6Be-0.5Co-0.3Si	1140	165	1070	155	1	46 HRC	18

continuous casting process and has replaced sand castings for mass-produced bearings of high quality. Alloys C93800 (15% Pb) and C94300 (24% Pb) are used where high loads are encountered under conditions of poor or nonexistent lubrication; under corrosive conditions, such as in mining equipment (pumps and car bearings); or in dusty atmospheres, as in stone-crushing and cement plants. These alloys replace the tin bronzes or low-lead tin bronzes where operating conditions are unsuitable for alloys containing little or no lead. They also are produced by the continuous casting process.

High-strength manganese bronzes have high tensile strength, hardness, and resistance to shock. Large gears, bridge turntables (slow motion and high compression), roller tracks for antiaircraft guns, and recoil parts of cannons are typical applications.

Aluminum bronzes with 8 to 9% Al are widely used for bushings and bearings in light-duty or high-speed machinery. Aluminum bronzes containing 11% Al, either as-cast or heat treated, are suitable for heavy-duty service (such as valve guides, rolling mill bearings, screwdown nuts, and slippers) and precision machinery. As aluminum content increases above 11%, hardness increases and elongation decreases to low values. Such bronzes are well suited for guides and aligning plates, where wear would be excessive. Aluminum bronzes that contain more than 13% Al exceed 300 HB in hardness but are brittle. Such alloys are suitable for dies and other parts not subjected to impact loads.

Aluminum bronze generally has a considerably higher fatigue limit and freedom from galling than manganese bronze. On the other hand, manganese bronze has great toughness for equivalent tensile strength and does not need to be heat treated.

Electrical and Thermal Conductivity

Electrical and thermal conductivity of any casting will invariably be lower than for wrought metal of the same composition. Copper castings are used in the electrical industry for their current-carrying capacity, and they are used for water-cooled parts of melting and refining furnaces because of their high thermal conductivity. However, for a copper casting to be sound and have electrical or thermal conductivity of at least 85%, care must be taken in melting and casting. The ordinary deoxidizers (silicon, tin, zinc, aluminum, and phosphorus) cannot be used because small residual amounts lower electrical and thermal conductivity drastically. Calcium boride or lithium will help to produce sound castings with high conductivity.

Cast copper is soft and low in strength. Increased strength and hardness and good conductivity can be obtained with heat-treated alloys containing silicon, cobalt, chromium, nickel, and beryllium in various combinations. These alloys, however, are expensive and less readily available than the standardized alloys. Table 7 presents some of the properties of these alloys after heat treatment.

Cost Considerations

During the design of a copper alloy casting, foundry personnel or the design engineer must choose a method of producing internal cavities. There is no general rule for choosing between cored and coreless designs. A cost analysis will determine which is the more economical method of producing the casting, although frequently the choice can be decided by experience.

For example, costs were compared for producing a small (13 mm, or ½ in.) valve disk both as a cored casting and as a machined casting (internal cavities made without cores). The machined casting could be produced for about 78% of the cost of making the identical casting using dry sand cores—a savings of 22% in favor of the machined casting. In a similar instance, producing a larger (38 mm, or 1½ in.) valve disk as a cored casting that required only a minimal amount of machining saved more than 8% in overall cost compared to producing the same valve disk without cores. Thus, for two closely related parts, a difference in manufacturing economy may exist when all cost factors are taken into account.

Properties of Cast Copper Alloys

Revised by Arthur Cohen, Copper Development Association, Inc.

C81100

Commercial Names

Previous trade name. CA811

Chemical Composition

Composition limits. 99.70 Cu + Ag min, 0.30 max other (total), 0.01 P + Si max to achieve a conductivity of 92% IACS

Applications

Typical uses. Electrical and thermal conductors, applications requiring resistance to corrosion and oxidation

Mechanical Properties

Tensile properties. Typical data for sand-cast test bars: tensile strength, 170 MPa (25 ksi); yield strength, 62 MPa (9 ksi) at 0.5% extension under load; elongation, 40% in 50 mm (2 in.)
Hardness. 44 HB
Elastic modulus. Tension, 115 GPa (17 × 10^6 psi)
Fatigue strength. 62 MPa (9 ksi) at 10^8 cycles

Mass Characteristics

Density. 8.94 g/cm^3 (0.323 lb/in.^{3}) at 20 °C (68 °F)
Volume change on freezing. 4.92% contraction
Patternmaker's shrinkage. 21 mm/m (¼ in./ft)

Thermal Properties

Liquidus temperature. 1083 °C (1981 °F)
Solidus temperature. 1065 °C (1948 °F)
Coefficient of linear thermal expansion. 16.9 μm/m · K (9.4 μin/in. · °F) at 20 to 300 °C (68 to 572 °F)
Specific heat. 380 J/kg · K (0.09 Btu/lb · °F) at 20 °C (68 °F)
Thermal conductivity. 346 W/m · K (200 Btu/ft · h · °F) at 20 °C (68 °F)

Electrical Properties

Electrical conductivity. Volumetric, 92% IACS at 20 °C (68 °F)

Magnetic Properties

Magnetic permeability. 1.0

Fabrication Characteristics

Machinability. 10% of C36000 (free-cutting brass)

C81300

Commercial Names

Previous trade name. CA813
Common name. Beryllium-copper

Chemical Composition

Composition limits. 98.5 Cu min, 0.20 to 0.10 Be, 0.6 to 1.0 Co. (Cu + sum of named elements shall be 99.5% minimum.)

Applications

Typical uses. Higher-hardness electrical and thermal conductors

Mechanical Properties

Tensile properties. Properties for separately cast heat-treated (TF00 temper) test bars; tensile strength, 365 MPa (53 ksi) min; yield strength, 250 MPa (36 ksi) min at 0.2% offset; elongation, 11% min in 50 min (2 in.)
Hardness. 89 HB (500 kg), typical
Elastic modulus. Tension, 110 GPa (16 × 10^6 psi) at 20 °C (68 °F)

Mass Characteristics

Density. 8.81 g/cm^3 (0.318 lb/in.^{3}) at 20 °C (68 °F)
Volume change on freezing. Patternmaker's shrinkage, 21 mm/m (¼ in./ft)

Thermal Properties

Liquidus temperature. 1093 °C (2000 °F)
Solidus temperature. 1066 °C (1950 °F)
Coefficient of linear thermal expansion. 18 μm/m·K (10.0 μin./in. · °F) at 20 to 300 °C (68 to 572 °F)
Specific heat. 390 J/kg · K (0.093 Btu/lb · °F) at 20 °C (68 °F)
Thermal conductivity. 260 W/m · K (150 Btu/ft · h · °F) at 20 °C (68 °F)

Electrical Properties

Electrical conductivity. Volumetric, 60% IACS at 20 °C (68 °F)

Fabrication Characteristics

Machinability. TF00 temper: 20% of C36000 (free-cutting brass)
Solution heat-treating temperature. 980 to 1010 °C (1800 to 1850 °F)
Aging temperature. 480 °C (900 °F)
Stress-relieving temperature. 260 °C (500 °F)

C81400
99Cu-0.8Cr-0.06Be

Commercial Names

Previous trade name. Beryllium-copper 70C, CA814
Common name. Be-modified chrome copper

Specifications

RWMA. Class II

Chemical Composition

Composition limits. 98.5 Cu min, 0.6 to 1.0 Cr, 0.02 to 0.10 Be

Applications

Typical uses. Electrical parts that meet RWMA Class II standards. The beryllium content of this alloy ensures that the chromium content will be kept under control during melting and casting, thus allowing the production of chrome copper castings of consistently high quality.
Precautions from health hazard. See C82500.

Mechanical Properties

Tensile properties. Typical as-cast: tensile strength, 205 MPa (30 ksi); yield strength, 83 MPa (12 ksi) at 0.2% offset; elongation, 35% in 50 mm (2 in.). TF00 temper: tensile strength, 365 MPa (53 ksi); yield strength, 250 MPa (36 ksi) at 0.2% offset; elongation, 11% in 50 mm (2 in.)
Hardness. As-cast: 62 HRB. TF00 temper: 69 HRB
Elastic modulus. Tension, 110 GPa (16 × 10^6 psi); shear, 41 GPa (5.9 × 10^6 psi)

Mass Characteristics

Density. 8.81 g/cm^3 (0.318 lb/in.^{3}) at 20 °C (68 °F)
Patternmaker's shrinkage. 1.96%

Table 1 Typical mechanical properties of C81800

Temper	Tensile strength MPa	Tensile strength ksi	Yield strength(a) MPa	Yield strength(a) ksi	Elongation(b), %	Hardness, HRB
As-cast	345	50	140	20	20	50
Cast and aged(c)	450	65	275	40	15	70
TB00(c)	310	45	83	12	25	40
TF00(d)(c)	705	102	515	75	8	96

(a) At 0.2% offset. (b) In 50 mm (2 in.). (c) Aged 3 h at 480 °C (900 °F). (d) Solution treated at 900 to 950 °C (1650 to 1750 °F)

Thermal Properties

Liquidus temperature. 1095 °C (2000 °F)
Solidus temperature. 1065 °C (1950 °F)
Coefficient of linear thermal expansion. 18 μm/m · K (10 μin./in. · F) at 20 to 300 °C (68 to 572 °F)
Specific heat. 389 J/kg · K (0.093 Btu/lb · °F) at 20 °C (68 °F)
Thermal conductivity. 259 W/m · K (150 Btu/ft · h · °F) at 20 °C (68 °F)

Fabrication Characteristics

Machinability. As-cast or TB00 temper: 30% of C36000 (free-cutting brass); TF00 temper: 40% of C36000
Melting temperature. 1065 to 1095 °C (1950 to 2000 °F)
Casting temperature. Light castings, 1200 to 1260 °C (2200 to 2300 °F); heavy castings, 1175 to 1230 °C (2150 to 2250 °F)
Solution temperature. 1000 to 1010 °C (1830 to 1850 °F)
Aging temperature. 480 °C (900 °F)

C81500
99Cu-1Cr

Commercial Names

Previous trade names. Chromium-copper; CA815
Common name. Chrome copper

Chemical Composition

Composition limits. 98.0 to 99.6 Cu, 0.40 to 1.50 Cr, 0.015 Pb max, 0.04 P max, 0.15 max other (total)
Consequence of exceeding impurity limits. Elements that contribute to hot shortness must be avoided. Because of the high solution temperatures necessary to develop the desired mechanical properties, elements that enter into solid solution must be held to close limits.

Applications

Typical uses. Electrical and/or thermal conductors used as structural members in applications requiring greater strength and hardness than that of cast coppers C80100 to C81100

Mechanical Properties

Tensile properties. Typical data for sand-cast test bars, heat treated: tensile strength, 350 MPa (51 ksi); yield strength, 275 MPa (40 ksi) at 0.5% extension under load; elongation, 17% in 50 mm (2 in.)
Hardness. Heat treated, 105 HB
Poisson's ratio. 0.32
Elastic modulus. Tension, 115 GPa (17 × 10^6 psi)
Impact strength. Izod, 41 J (30 ft · lbf); Charpy V-notch, 27 J (20 ft · lbf)
Fatigue strength. 105 MPa (15 ksi) at 10^8 cycles

Mass Characteristics

Density. 8.82 g/cm³ (0.319 lb/in.³) at 20 °C (68 °F)
Patternmaker's shrinkage. 21 mm/m (¼ in./ft)

Thermal Properties

Liquidus temperature. 1085 °C (1985 °F)
Solidus temperature. 1075 °C (1967 °F)
Coefficient of linear thermal expansion. 17.1 μm/m · K (9.5 μin./in. · °F) at 20 to 300 °C (68 to 572 °F)
Specific heat. 376 J/kg · K (0.09 Btu/lb · °F) at 20 °C (68 °F)
Thermal conductivity. 315 W/m · K (182 Btu/ft · h · °F) at 20 °C (68 °F)

Electrical Properties

Electrical conductivity. Volumetric: solution heat treated, 40 to 50% IACS at 20 °C (68 °F); precipitation hardened, 80 to 90% IACS at 20 °C (68 °F)
Electrical resistivity. Solution heat treated, 38.3 nΩ · m at 20 °C (68 °F); precipitation hardened, 21 nΩ · m at 20 °C (68 °F). Temperature coefficient: solution heat treated, 0.08 nΩ · m per K at 20 °C (68 °F); precipitation hardened, 0.06 nΩ · m per K at 20 °C (68 °F)

Magnetic Properties

Magnetic permeability. 1.0

Fabrication Characteristics

Machinability. 20% of C36000 (free-cutting brass)
Weldability. Chromium copper can be silver soldered, soft soldered, or brazed; it can be carbon arc welded with copper-chromium filler rod and fused-borax flux.
Solution temperature. 1000 to 1010 °C (1830 to 1850 °F)
Aging temperature. 480 °C (900 °F)

C81800
97Cu-1.5Co-1Ag-0.4Be

Commercial Names

Previous trade name. Beryllium-copper alloy 50C, CA818

Specifications

RWMA. Class III

Chemical Composition

Composition limits. 0.30 to 0.55 Be, 1.4 to 1.7 Co, 0.8 to 0.12 Ag, 0.15 Si max, 0.20 Ni max, 0.10 Fe max, 0.10 Al max, 0.10 Sn max, 0.002 Pb max, 0.10 Zn max, 0.10 Cr max, bal Cu
Consequence of exceeding impurity limits. See C82500.

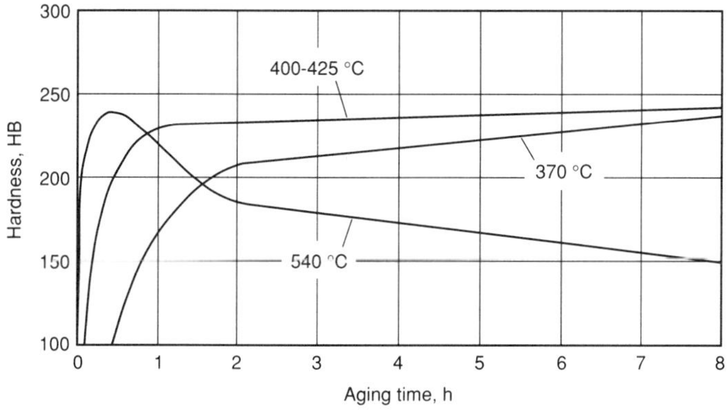

Fig. 1 Aging curves for cast and solution-treated C81800

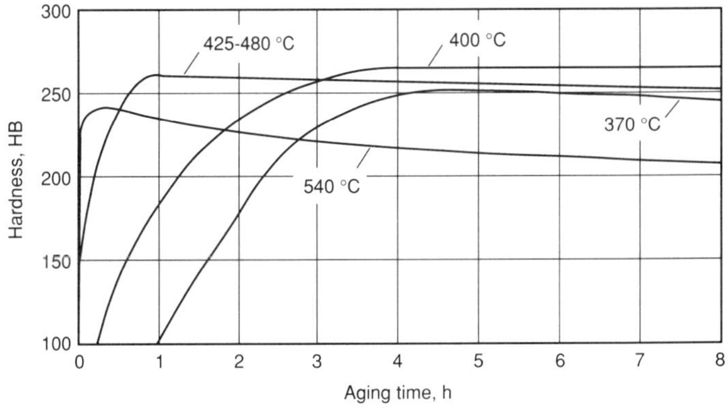

Fig. 2 Aging curves for cast and solution-treated C82000

Table 2 Typical mechanical properties of C82000

Temper	Tensile strength MPa	Tensile strength ksi	Yield strength(a) MPa	Yield strength(a) ksi	Elongation(b), %	Hardness, HRB
As-cast	345	50	140	20	20	52
Cast and aged(c)	450	65	255	37	12	70
TB00(d)	325	47	105	15	25	40
TF00(d)(e)	660	96	515	75	6	96

(a) At 0.2% offset. (b) In 50 mm (2 in.). (c) Aged 2 h at 480 °C (900 °F). (d) Solution treated at 900 to 950 °C (1650 to 1750 °F). (e) Aged 3 h at 480 °C (900 °F)

Applications

Typical uses. The silver content of C81800 provides an improved surface conductivity over other RWMA Class III alloys. Typical uses are resistance welding electrode tips and holders and arms.
Precautions as health hazard. See C82500.

Mechanical Properties

Tensile properties. See Table 1.
Hardness. See Table 1 and Fig. 1.
Poisson's ratio. 0.33
Elastic modulus. Tension, 110 GPa (16×10^6 psi); shear, 41 GPa (6×10^6 psi)

Mass Characteristics

Density. 8.62 g/cm^3 (0.311 lb/in.^{3}) at 20 °C (68 °F)
Patternmaker's shrinkage. 1.56%

Thermal Properties

Liquidus temperature. 1070 °C (1955 °F)
Solidus temperature. 1010 °C (1855 °F)
Coefficient of linear thermal expansion. 18 μm/m · K (10 μin./in.· °F) at 20 to 300 °C (68 to 572 °F)
Specific heat. 420 J/kg · K (0.10 Btu/lb · °F) at 20 °C (68 °F)
Thermal conductivity. 218 W/m · K (126 Btu/ft · h · °F) at 20 °C (68 °F)

Electrical Properties

Electrical conductivity. Volumetric, 48% IACS at 20 °C (68 °F)
Electrical resistivity. 359 μΩ · m at 20 °C (68 °F)

Magnetic Properties

Magnetic susceptibility. See C82000.

Nuclear Properties

Effect of neutron irradiation. See C82500.

Chemical Properties

See C82000.

Fabrication Characteristics

Machinability. As-cast or TB00 temper: 30% of C36000 (free-cutting brass). TF00 temper: 40% of C36000
Melting temperature. 1010 to 1070 °C (1855 to 1955 °F)
Casting temperature. Light castings, 1175 to 1230 °C (2150 to 2250 °F); heavy castings, 1120 to 1175 °C (2050 to 2150 °F)
Solution temperature. 900 to 925 °C (1650 to 1700 °F)
Aging temperature. 480 °C (900 °F) See also Fig. 1.

C82000
97Cu-2.5Co-0.5Be

Commercial Names

Previous trade name. Beryllium-copper alloy 10C, CA820
Common name. Beryllium-copper casting alloy 10C

Specifications

Government. QQ-C-390 (CA820), MIL-C-19464 (Class I)

Chemical Composition

Composition limits. 0.45 to 0.8 Be, 2.4 to 2.7 Co, 0.15 Si max, 0.20 Ni max, 0.10 Fe max, 0.10 Al max, 0.10 Sn max, 0.02 Pb max, 0.10 Zn max, 0.10 Cr max, bal Cu
Consequence of exceeding impurity limits. See C82500.

Applications

Typical uses. C82000 castings are used when a combination of high conductivity and high strength is required. Applications include resistance welding tips, holders and arms, circuit-breaker parts, switch gear parts, plunger tips for die casting, con-casting molds, for continuous casting installations, soldering-iron tips, brake drums, and whenever RWMA Class III properties are required.
Precautions as health hazard. See C82500.

Mechanical Properties

Tensile properties. See Table 2.
Hardness. See Table 2 and Fig. 2 and 3.
Poisson's ratio. 0.33
Elastic modulus. Tension, 115 GPa (17×10^6 psi); shear, 44 GPa (6.4×10^6 psi)
Fatigue strength. Rotating beam, 125 MPa (18 ksi) at 5×10^7 cycles

Mass Characteristics

Density. 8,62 g/cm^3 (0.311 lb/in.^{3}) at 20 °C (68 °F)
Patternmaker's shrinkage. 1.56%

Thermal Properties

Liquidus temperature. 1090 °C (1990 °F)

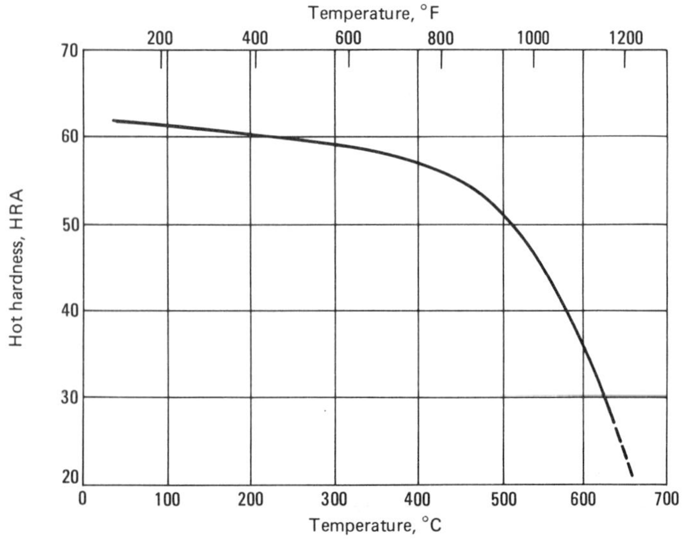

Fig. 3 Hot hardness of C82000, TF00 temper. Cast specimens were solution treated, then aged at 480 °C (900 °F). Useful design range is up to about 400 °C (750 °F).

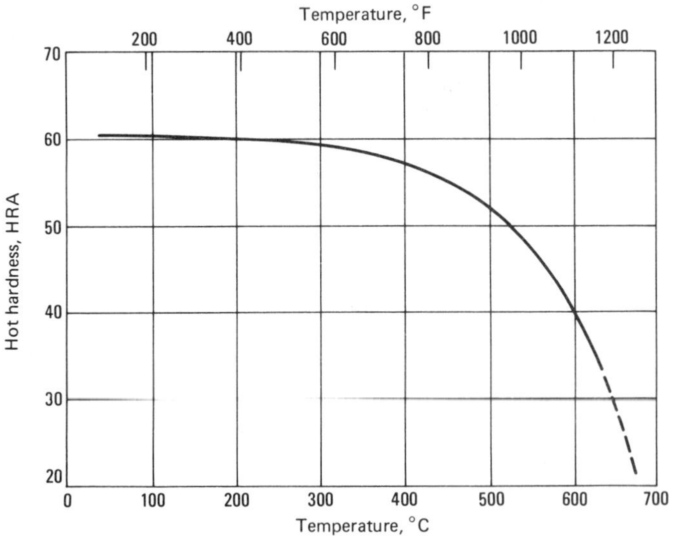

Fig. 4 Hot hardness of C82200, TF00 temper. Aged at 480 °C (900 °F). Useful design range is up to 370 °C (700 °F).

Table 3 Typical mechanical properties of C82200

Temper	Tensile strength MPa	ksi	Yield strength(a) MPa	ksi	Elongation(b), %	Hardness, HRB
As-cast	345	50	170	25	20	55
Cast and aged(c)	450	65	275	40	15	75
TB00(d)	310	45	85	12	30	30
TF00(d)(c)	655	95	515	75	7	96

(a) At 0.2% offset. (b) In 50 mm (2 in.). (c) Aged 3 h at 480 °C (900 °F). (d) Solution treated at 900 to 955 °C (1650 to 1750 °F)

Solidus temperature. 970 °C (1780 °F)
Coefficient of linear thermal expansion. 17.8 μm/m · K (9.9 μin./in. · °F) at 20 to 300 °C (68 to 572 °F)
Specific heat. 420 J/kg · K (0.10 Btu/lb · °F) at 20 °C (68 °F)
Thermal conductivity. 218 W/m · K (126 Btu/ft · h · °F) at 20 °C (68 °F)

Electrical Properties

Electrical conductivity. Volumetric, 48% IACS at 20 °C (68 °F)
Electrical resistivity. 359 μΩ · m at 20 °C (68 °F)

Magnetic Properties

Magnetic susceptibility. Commercial beryllium-copper casting alloys containing 0.02 to 0.8% Be exhibit magnetic susceptibility of +0.001 cgs units or less. Magnetic susceptibility varies principally with iron content—on the high side of the commercial range of iron content (0.10%), magnetic susceptibility is approximately 0.001 cgs units; on the low side of the range (0.05%), the value will be much lower (0.0001 cgs units or less). A high-temperature solution treatment of 950 °C (1750 °F) and a low aging temperature of 425 to 450 °C (800 to 850 °F) will tend to keep iron in solution and keep the alloy nonmagnetic—that is, with a magnetic susceptibility less than 0.0001 cgs units.

Nuclear Properties

Effect of neutron irradiation. See C82500.

Chemical Properties

General corrosion behavior. At elevated temperatures, beryllium is an active oxide former. Beryllium in beryllium-copper alloys will preferentially form BeO when low partial pressures of oxygen are present. This causes intergranular oxidation during solution treating in air, often resulting in surface deterioration up to 0.05 mm (0.002 in.) deep. For resistance to water, chemical solutions, organic chemicals and chemical gases, and for resistance to stress-corrosion cracking, see C82500.

Fabrication Characteristics

Machinability. As-cast or TB00 temper: 30% of C36000 (free-cutting brass). TF00 temper: 40% of C36000
Melting temperature. 970 to 1090 °C (1780 to 1990 °F)
Casting temperature. Light castings, 1175 to 1230 °C (2150 to 2250 °F); heavy castings, 1120 to 1175 °C (2050 to 2150 °F)
Solution temperature. 900 to 925 °C (1650 to 1700 °F)
Aging temperature. 480 °C (900 °F). See also Fig. 2.

C82200
98Cu-1.5Ni-0.5Be

Commercial Names

Previous trade name. Beryllium-copper alloy 30C, CA822
Common name. Beryllium-copper casting alloy 30C, 35C, or 53B

Specifications

RWMA. Class III

Chemical Composition

Composition limits. 0.35 to 0.8 Be, 1.0 to 2.0 Ni, 0.15 Si max, 0.20 Co max, 0.10 Fe max, 0.10 Al max, 0.10 Sn max, 0.02 Pb max, 0.10 Zn max, 0.10 Cr max, bal Cu
Consequence of exceeding impurity limits. See C82500.

Applications

Typical uses. Seam welder electrodes, projection welder dies, spot welding tips, beam welder shapes, water-cooled holders, arms bushings for resistance welding, clutch rings, brake drums
Precautions as health hazard. See C82500.

Mechanical Properties

Tensile properties. See Table 3.
Hardness. See Table 3 and Fig. 4.
Poisson's ratio. 0.33
Elastic modulus. Tension, 114 GPa (16.5 $\times 10^6$ psi); shear, 43 GPa (6.2 $\times 10^6$ psi)

Mass Characteristics

Density. 8.75 g/cm³ (0.316 lb/in.³) at 20 °C (68 °F)
Patternmaker's shrinkage. 1.56%

Thermal Properties

Liquidus temperature. 1115 °C (2040 °F)
Solidus temperature. 1040 °C (1900 °F)
Coefficient of linear thermal expansion. 16.2 μm/m · K (9 μin./in. · °F) at 20 to 200 °C (68 to 392 °F)
Specific heat. 420 J/kg · K (0.10 Btu/lb · °F) at 20 °C (68 °F)
Thermal conductivity. 183 W/m · K (106 Btu/ft · h · °F) at 20 °C (68 °F)

Electrical Properties

Electrical conductivity. Volumetric, 48% IACS at 20 °C (68 °F)
Electrical resistivity. 359 μΩ · m at 20 °C (68 °F)

Magnetic Properties

Magnetic susceptibility. See C82000.

Nuclear Properties

Effect of neutron irradiation. See C82500.

Chemical Properties

See C82000.

Fabrication Characteristics

Machinability. As-cast or TB00 temper: 30% of C36000 (free-cutting brass). TF00 temper: 40% of C36000
Melting temperature. 1035 to 1115 °C (1900 to 2040 °F)
Casting temperature. Light castings, 1200 to 1260 °C (2200 to 2300 °F); heavy castings, 1150 to 1200 °C (2100 to 2200 °F)
Solution temperature. 900 to 955 °C (1650 to 1750 °F)
Aging temperature. 445 to 455 °C (835 to 850 °F)

C82400
98Cu-1.7Be-0.3Co

Commercial Names

Previous trade name. Beryllium-copper alloy 165C; CA824
Common name. Beryllium-copper casting alloy 165C

Specifications

Government. QQ-C-390 (CA824)

Chemical Composition

Composition limits. 1.65 to 1.75 Be, 0.20 to 0.40 Co, 0.10 Ni max, 0.20 Fe max, 0.15 Al max, 0.10 Sn max, 0.02 Pb max, 0.10 Zn max, 0.10 Cr max, bal Cu
Consequence of exceeding impurity limits. See C82500.

Applications

Typical uses. C82400 was developed for use in marine service as a corrosion-resistant, pressure-tight casting material. Its lower beryllium content compared to C82500 makes this alloy the least expensive of the commercial high-strength beryllium-copper alloys. When its hardness is relatively low, C82400 exhibits greater-than-normal toughness. Typical uses include various parts for the submarine telephone cable repeater system and hydrophone, molds for forming plastics, safety tools, plunger tips for die casting, cams, bushings, bearings, valves, pump parts, and gears.
Precautions as health hazard. See C82500.

Mechanical Properties

Tensile properties. See Table 4.

Table 4 Typical mechanical properties of C82400

Temper	Tensile strength MPa	ksi	Yield strength(a) MPa	ksi	Elongation(b), %	Hardness
As-cast	485	70	275	40	15	78 HRB
Cast and aged(c)	690	100	550	80	3	21 HRC
TB00(d)	415	60	140	20	40	59 HRB
TF00(d)(c)	1070	155	1000	145	1	38 HRC

(a) At 0.2% offset. (b) In 50 mm (2 in.). (c) Aged 3 h at 345 °C (650 °F). (d) Solution treated at 800 to 815 °C (1475 to 1500 °F)

Table 5 Typical mechanical properties of C82500

Temper	Tensile strength MPa	ksi	Yield strength(a) MPa	ksi	Elongation(b), %	Hardness
As-cast	515	75	275	40	15	81 HRB
Cast and aged(c)	825	120	725	105	2	30 HRC
TB00(d)	415	60	170	25	35	63 HRB
TF00(d)(c)	1105	160	1035	150	1	43 HRC

(a) At 0.2% offset. (b) In 50 mm (2 in.). (c) Aged 3 h at 345 °C (650 °F). (d) Solution treated at 790 to 800 °C (1450 to 1475 °F)

Hardness. See Table 4.
Poisson's ratio. 0.30
Elastic modulus. Tension, 128 GPa (18.5 $\times 10^6$ psi); shear, 50 GPa (7.3 $\times 10^6$ psi)
Fatigue strength. Rotating beam, 160 MPa (23 ksi) at 5 $\times 10^7$ cycles

Structure

Crystal structure. See C82500.

Mass Characteristics

Density. 8.31 g/cm^3 (0.301 lb/in.^{3}) at 20 °C (68 °F)
Patternmaker's shrinkage. 1.56%
Dilation during aging. Linear, 0.2%
Change in density during aging. 0.6% increase

Thermal Properties

Liquidus temperature. 995 °C (1825 °F)
Solidus temperature. 900 °C (1650 °F)
Incipient melting temperature. 865 °C (1585 °F)
Coefficient of linear thermal expansion. 17.0 μm/m · K (9.4 μin./in. · °F) at 20 to 200 °C (68 to 392 °F)
Specific heat. 420 J/kg · K (0.10 Btu/lb · °F) at 20 °C (68 °F)
Thermal conductivity. 109 W/m · K (63 Btu/ft · h · °F) at 20 °C (68 °F)

Electrical Properties

Electrical conductivity. Volumetric, 25% IACS at 20 °C (68 °F)
Electrical resistivity. 690 μΩ · m at 20 °C (68 °F)

Magnetic Properties

Magnetic susceptibility. See C82500.

Nuclear Properties

Effect of neutron irradiation. See C82500.

Chemical Properties

See C82500.

Fabrication Characteristics

Machinability. As-cast or TB00 temper: 30% of C36000 (free-cutting brass). Cast and aged or TF00 temper: 10 to 20%
Melting temperature. 900 to 1000 °C (1650 to 1825 °F)
Casting temperature. Light castings, 1080 to 1135 °C (1975 to 2075 °F); heavy castings, 1025 to 1080 °C (1875 to 1975 °F)
Solution temperature. 790 to 815 °C (1450 to 1500 °F)
Aging temperature. 345 °C (650 °F)

C82500
97.2Cu-2Be-0.5Co-0.25Si

Commercial Names

Previous trade name. Beryllium-copper 20C, CA825
Common name. Standard beryllium-copper casting alloy

Specifications

AMS. Investment castings: 4890
Government. Sand castings: QQ-C-390, MIL-C-19464 (class 2); centrifugal castings: QQ-C-390; precision castings: MIL-C-11866 (composition 17), MIL-C-17324; investment castings: MIL-C-22087
Other. ICI-Cu-2-10780

Chemical Composition

Composition limits. 95.5 Cu min, 1.90 to 2.15 Be, 0.35 to 0.7 Co, 0.20 to 0.35 Si, 0.20 Ni max, 0.25 Fe max, 0.15 Al max, 0.10 Sn max, 0.02 Pb max, 0.10 Zn max, 0.10 Cr max. Available with or without 0.02 to 0.10% Ti added as a grain refiner.
Consequence of exceeding impurity limits. Generally, electrical conductivity is lowered. High Fe raises magnetic susceptibility. High Sn, Zn, or Pb causes hot shortness. High Cr diminishes response to precipitation hardening.

Applications

Typical uses. Molds for forming plastics, die casting plunger tips, safety tools, cams, bushings, bearings, gears, sleeves, valves, wear parts, structural parts, resistance welding electrodes and inserts, holders, and structural members. Exhibits low casting temperature, good castability, excellent ability to reproduce fine detail in the pattern, high strength, high electrical and thermal conductivity, and excellent resistance to corrosion and wear. Can be sand, shell, ceramic, investment, permanent, pressure, and die cast. Especially suited for investment castings and often replaces ferrous castings having similar mechanical properties. Investment castings are used for communication, textile, aerospace, business machine, firearm, instrument, and ordnance parts.
Precautions in use as health hazard. Melting, casting, abrasive-wheel operations, abrasive blasting, welding, arc cutting, flame cutting, grinding, polishing, and buffing under improper conditions may raise the concentration of beryllium in the air to levels above the limits prescribed by OSHA, thus creating a potential for personnel to contract berylliosis, a chronic lung disease. Exhaust ventilation, the principal means of achieving compliances with these limits, is a specific OSHA requirement for processes involving beryllium alloys. Careful attention to the exhaust-ventilation requirements of these and any other effluent-producing operations is essential. Actual exposure of workers should be continually monitored using prescribed air-sampling and calculation methods to determine compliance or noncompliance with OSHA limits.

Mechanical Properties

Tensile properties. See Table 5 and Fig. 5.
Compressive properties. Compressive yield strength, 1030 to 1200 MPa (150 to 175 ksi) at a permanent set of 0.1%
Hardness. See Table 5 and Fig. 6 and 7.
Poisson's ratio. 0.30
Elastic modulus. Tension, 128 GPa (18.5 $\times 10^6$ psi); shear, 50 GPa (7.3 $\times 10^6$ psi)
Fatigue strength. Rotating beam, 165 MPa (24 ksi) at 5 $\times$ 10^7 cycles
Tensile properties and hardness versus temperature. See Fig. 5 and 7.

Structure

Crystal structure. Alpha phase, face-centered cubic. Lattice parameter; a: solution treated (2.1% Be in solid solution), 0.357 nm; precipitation hardened, 0.361 nm

Mass Characteristics

Density. 8.26 g/cm^3 (0.298 lb/in.^{3}) at 20 °C (68 °F)
Patternmaker's shrinkage. 1.56%
Dilation during aging. Linear, 0.2%
Change in density during aging. 0.6% increase

Thermal Properties

Liquidus temperature. 980 °C (1800 °F)
Solidus temperature. 855 °C (1575 °F)

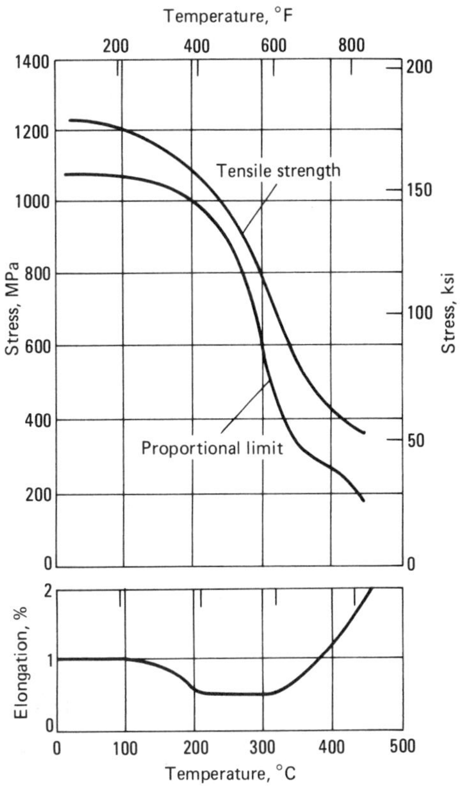

Fig. 5 Elevated-temperature tensile properties of C82500, TF00 temper. Sand cast test bars were solution treated, then aged at 345 °C (650 °F). Useful design range is limited to about 220 °C (425 °F).

Incipient melting temperature. 835 °C (1535 °F)
Coefficient of linear thermal expansion. 17 μm/m · K (9.4 μin./in. · °F) at 20 to 200 °C (68 to 392 °F)
Specific heat. 420 J/kg · K (0.10 Btu/lb · °F) at 20 °C (68 °F)
Thermal conductivity. 105 W/m · K (61 Btu/ft · h · °F) at 20 °C (68 °F)

Electrical Properties

Electrical conductivity. Volumetric, 20% IACS at 20 °C (68 °F)
Electrical resistivity. 862 μΩ · m at 20 °C (68 °F)

Magnetic Properties

Magnetic susceptibility. Commercial beryllium-copper casting alloys with 1.6 to 2.7% Be content exhibit magnetic susceptibility of approximately +0.002 cgs units. Magnetic susceptibility varies principally with iron content. On the high side of the range for iron content (0.25%), magnetic susceptibility is greater than 0.002 cgs units. On the low side of the commercial range (0.05%), the value is much less than 0.001 cgs units. A high-temperature solution treatment of 815 °C (1500 °F) and a low aging temperature of 315 to 345 °C (600 to 650 °F) will tend to keep iron in solution and keep magnetic susceptibility below 0.002 cgs units.

Nuclear Properties

Effect of neutron irradiation. Neutron irradiation causes precipitation hardening because of thermal spikes and induced vacancies. This will affect as-cast or solution-treated tempers but have little effect on material already peak aged or overaged.

Chemical Properties

General corrosion behavior. At elevated temperatures, beryllium is an active oxide former. Beryllium in beryllium-copper alloys will preferentially form BeO when low partial pressures of oxygen are present, especially when the environment is reducing with respect to copper. This leads to preferential formation of BeO films during hot processing of alloys containing 1.6% Be or more. BeO films may be abrasive to fabricating tools, and may be removed mechanically or by pickling. The general corrosion resistance of beryllium-copper alloys is similar to that of deoxidized copper, except as indicated above.
Resistance to specific agents. Beryllium-copper alloys possess excellent resistance to atmospheric corrosion in marine, industrial, and rural environments. They have excellent resistance to organic chemicals such as alcohols, aldehydes, esters, and ketones. They are slightly more resistant to seawater than tough pitch or deoxidized copper. Resistance is good with respect to: fresh water; most organic acids, hot or cold dilute sulfuric acid, cold concentrated sulfuric acid, and cold dilute hydrochloric acid; hot or cold dilute alkalis and cold concentrated alkalis; salts, including most sulfates and chlorides. Resistance is only fair towards sulfides, especially at elevated temperatures. Resistance is poor towards: mercury and mercury compounds; nitric acid; ferric chloride, ferric sulfate, and other heavy-metal salts with oxidizing cations and strong acid anions; acid chromates; and halogens (fluorine, chlorine, bromine, and iodine), particularly at elevated temperatures.
Stress-corrosion cracking. Beryllium-copper alloys resist stress-corrosion cracking in marine and most chemical environments, even when stressed up to 90% of their 0.2% offset yield strengths. They are susceptible to stress-corrosion cracking in ammonia

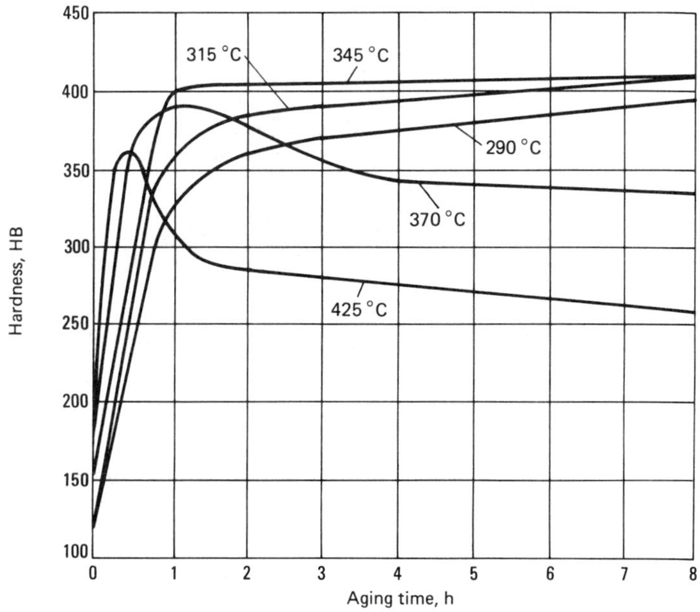

Fig. 6 Aging curves for solution-treated C82500 or beryllium-copper alloy 21C

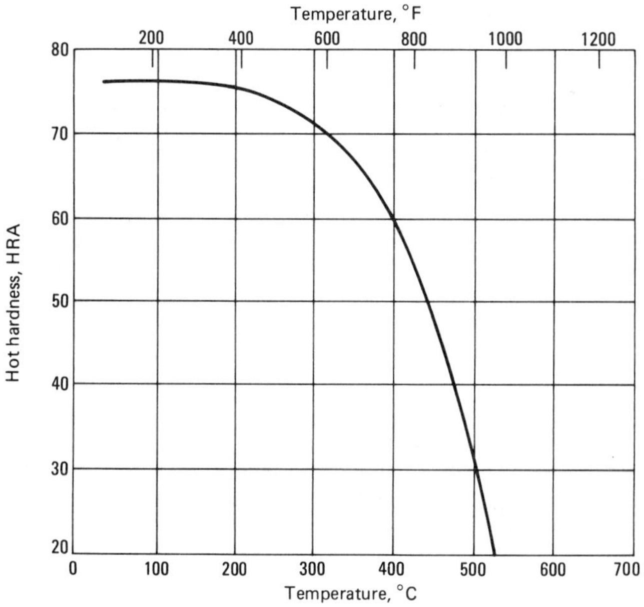

Fig. 7 Hot hardness of C82500, TF00 temper. Specimens were solution treated, then aged at 345 °C (650 °F).

Table 6 Typical mechanical properties of C82600

Temper	Tensile strength MPa	Tensile strength ksi	Yield strength(a) MPa	Yield strength(a) ksi	Elongation(b), %	Hardness
As-cast	550	80	345	50	10	86 HRB
Cast and aged(b)	825	120	725	105	2	31 HRC
TB00(c)	485	70	205	30	12	75 HRB
TF00(d)(c)	1140	165	1070	155	1	45 HRC

(a) At 0.2% offset. (b) In 50 mm (2 in.). (c) Aged 3 h at 345 °C (650 °F). (d) Solution treated at 790 to 800 °C (1450 to 1475 °F)

Table 7 Typical mechanical properties of C82800 sand cast test bars

Temper	Tensile strength MPa	Tensile strength ksi	Yield strength(a) MPa	Yield strength(a) ksi	Elongation(b), %	Hardness
As-cast	550	80	345	50	10	88 HRB
Cast and aged(c)	860	125	760	110	2	31 HRC
TB00(d)	550	80	240	35	10	85 HRB
TF00(d)(c)	1140	165	1070	155	1	46 HRC

(a) At 0.2% offset. (b) In 50 mm (2 in.). (c) Aged 3 h at 345 °C (650 °F). (d) Solution treated at 790 to 800 °C (1450 to 1475 °F)

and halogen gas environments, especially at elevated temperature.

Fabrication Characteristics

Machinability. As-cast or TB00 temper: 30% of C36000 (free-cutting brass). Cast and aged or TF00 temper: 10 to 20%
Melting temperature. 850 to 980 °C (1575 to 1800 °F)
Casting temperature. Light castings, 1065 to 1175 °C (1950 to 2150 °F); heavy castings, 1010 to 1065 °C (1850 to 1950 °F)
Solution temperature. 790 to 800 °C (1450 to 1475 °F)
Aging temperature. 345 °C (650 °F). See also Fig. 6.

C82600
97Cu-2.4Be-0.5Co

Commercial Names

Previous trade name. Beryllium-copper 245C
Common name. Beryllium-copper casting alloy 245C

Specifications

Government. QQ-C-390

Chemical Composition

Composition limits. 2.25 to 2.45 Be, 0.35 to 0.7 Co, 0.20 to 0.35 Si, 0.20 Ni max, 0.25 Fe max, 0.15 Al max, 0.10 Sn max, 0.02 Pb max, 0.10 Zn max, 0.10 Cr max, bal Cu
Consequence of exceeding impurity limits. See C82500.

Applications

Typical uses. C82600 is a beryllium-copper casting alloy intermediate in beryllium content between C82500 and C82800. It exhibits better fluidity, castability, and hardness than C82500 and better toughness and lower cost than C82800. C82600 is used primarily to produce molds for plastic parts. In pressure castings, the lower pouring temperature results in longer tool life than for similar castings of C82500.
Precautions as health hazard. See C82500.

Mechanical Properties

Tensile properties. See Table 6.
Hardness. See Table 6.
Poisson's ratio. 0.30
Elastic modulus. Tension, 130 GPa (19 $\times 10^6$ psi); shear, 50 GPa (7.3 $\times 10^6$ psi)

Structure

Crystal structure. Alpha phase, face-centered cubic

Mass Characteristics

Density. 8.16 g/cm^3 (0.295 lb/in.^{3}) at 20 °C (68 °F)
Patternmaker's shrinkage. 1.56%
Dilation during aging. Linear, 0.2%
Change in density during aging. 0.6% increase

Thermal Properties

Liquidus temperature. 955 °C (1750 °F)
Solidus temperature. 855 °C (1575 °F)
Incipient melting temperature. 835 °C (1535 °F)
Coefficient of linear thermal expansion. 17 μm/m · K (9.4 μin./in. · °F) at 20 to 200 °C (68 to 392 °F)
Specific heat. 420 J/kg · K (0.10 Btu/lb · °F) at 20 °C (68 °F)
Thermal conductivity. 100 W/m · K (58 Btu/ft · h · °F) at 20 °C (68 °F)

Electrical Properties

Electrical conductivity. Volumetric, 19% IACS at 20 °C (68 °F)
Electrical resistivity. 907 μΩ · m at 20 °C (68 °F)

Magnetic Properties

Magnetic susceptibility. See C82500.

Nuclear Properties

Effect of neutron irradiation. See C82500.

Chemical Properties

See C82500.

Fabrication Characteristics

Machinability. As-cast or TB00 temper: 30% of C36000 (free-cutting brass). Cast and aged or TF00 temper: 10 to 20% of C36000
Melting temperature. 855 to 955 °C (1575 to 1750 °F)
Casting temperature. Light castings, 1040 to 1150 °C (1900 to 2100 °F); heavy castings, 980 to 1040 °C (1800 to 1900 °F)
Solution temperature. 790 to 800 °C (1450 to 1475 °F)
Aging temperature. 345 °C (650 °F)

C82800
96.6Cu-2.6Be-0.5Co-0.3Si

Commercial Names

Previous trade name. Beryllium-copper alloy 275C, CA828
Common name. Beryllium-copper casting alloy 275C

Specifications

Government. QQ-C-390, MIL-T-16243, MIL-C-19464 (Class IV)
Other. ICI-Cu-2-10785

Chemical Composition

Composition limits. 94.8 Cu min, 2.50 to 2.75 Be, 0.37 to 0.7 Co, 0.20 to 0.35 Si, 0.20 Ni max, 0.25 Fe max, 0.15 Al max, 0.10 Sn max, 0.02 Pb max, 0.10 Zn max, 0.10 Cr max
Consequence of exceeding impurity limits. See C82500.

Applications

Typical uses. C82800 is a special-purpose, high-fluidity casting alloy developed for molds for forming plastics and other applications where the casting process should replicate finest detail with maximum fidelity and the resultant part must exhibit maximum hardness and wear resistance for a cast beryllium-copper alloy. The relative slow pouring temperature results in increased tool life during pressure casting and permanent molding. Typical uses are molds for forming plastics, cams, bushings, bearings, valves, pump parts, sleeves, and precision cast parts for the communications, textile, aerospace, business machine, firearm, instrument, ordnance, and other industries.
Precautions in use. See C82500.

Mechanical Properties

Tensile properties. See Table 7 and Fig. 8.
Hardness. See Table 7 and Fig. 9 and 10.
Poisson's ratio. 0.30
Elastic modulus. Tension, 133 GPa (19.3 $\times 10^6$ psi); shear, 51 GPa (7.4 $\times 10^6$ psi)

Structure

Crystal structure. See C82500.

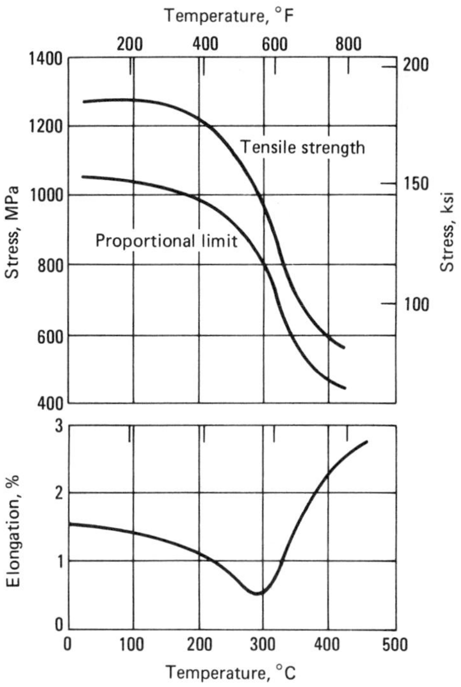

Fig. 8 Typical tensile properties of C82800, TF00 temper. Sand cast test bars were solution treated, then aged at 345 °C (650 °F).

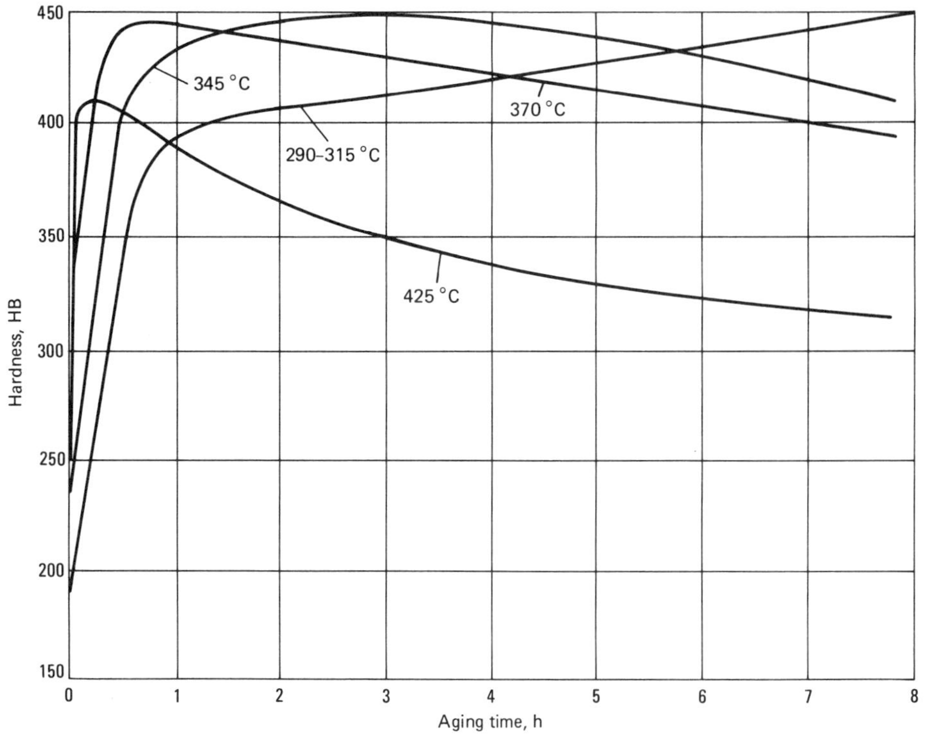

Fig. 9 Aging curves for solution-treated C82800

Mass Characteristics

Density. 8.09 g/cm^3 (0.292 lb/in.^{3}) at 20 °C (68 °F)
Patternmaker's shrinkage. 1.56%
Linear dilation during aging. 0.2%
Change in density during aging. 0.6% increase

Thermal Properties

Liquidus temperature. 930 °C (1710 °F)
Solidus temperature. 835 °C (1535 °F)
Incipient melting temperature. 855 °C (1575 °F)

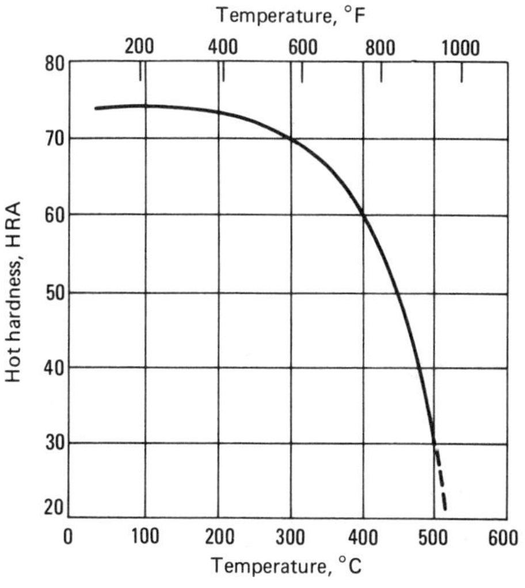

Fig. 10 Hot hardness of C82800, TF00 temper. Specimens were solution treated, then aged at 345 °C (650 °F).

Coefficient of linear thermal expansion. 17 µm/m · K (9.4 µin./in. · °F) at 20 to 200 °C (68 to 392 °F)
Specific heat. 420 J/kg · K (0.10 Btu/lb · °F) at 20 °C (68 °F)
Thermal conductivity. 95 W/m · K (55 Btu/ft · h · °F) at 20 °C (68 °F)

Electrical Properties

Electrical conductivity. Volumetric, 18% IACS at 20 °C (68 °F)
Electrical resistivity. 958 µΩ · m at 20 °C (68 °F)

Magnetic Properties

Magnetic susceptibility. See C82500.

Nuclear Properties

Effect of neutron irradiation. See C82500.

Chemical Properties

See C82500.

Fabrication Characteristics

Machinability. As-cast or TB00 temper: 30% of C36000 (free-cutting brass). Cast and aged or TF00 temper: 10 to 20% of C36000
Melting temperature. 860 to 930 °C (1575 to 1710 °F)
Casting temperature. Light castings, 1040 to 1150 °C (1900 to 2100 °F); heavy castings, 965 to 1040 °C (1770 to 1900 °F)
Solution temperature. 790 to 800 °C (1450 to 1475 °F)
Aging temperature. 345 °C (650 °F). See also Fig. 9.

C83300

Commercial Names

Previous trade name. CA833
Common name. Contact metal

Specification

Ingot code number. 131

Chemical Composition

Composition limits. 92.0 to 94.0 Cu, 1.0 to 2.0 Pb, 1.0 to 2.0 Sn, 2.0 to 6.0 Zn
Copper specification. In reporting chemical analyses by the use of instruments such as spectrograph, x-ray, and atomic absorption, copper may be indicated as balance. In reporting chemical analyses obtained by wet methods, zinc may be indicated as balance on those alloys with over 2% Zn.

Applications

Typical use. Terminal ends for electrical cables

Mechanical Properties

Tensile properties. Typical data for as-sand-cast separately cast test bars (M01 temper): tensile strength, 220 MPa (32 ksi); yield strength, 70 MPa (10 ksi) at 0.5% extension under load; elongation, 35% in 50 mm (2 in.)
Hardness. 35 HB (500 kg), typical
Elastic modulus. Tension: 105 GPa (15 × 10^6 psi) at 20 °C (68 °F)

Mass Characteristics

Density. 8.8 g/cm^3 (0.318 lb/in.^{3}) at 20 °C (68 °F)

Patternmaker's shrinkage. 16 to 21 mm/m (3⁄16 to 1⁄4 in./ft)

Thermal Properties

Liquidus temperature. 1060 °C (1940 °F)
Solidus temperature. 1030 °C (1886 °F)
Specific heat. 380 J/kg · K (0.09 Btu/lb · °F) at 20 °C (68 °F)

Electrical Properties

Electrical conductivity. Volumetric, 32% IACS at 20 °C (68 °F)

Fabrication Characteristics

Machinability. M01 temper, 35% of C36000 (free-cutting brass)
Stress-relieving temperature. 260 °C (500 °F)

C83600
85Cu-5Sn-5Pb-5Zn

Commercial Names

Previous trade names. Leaded red brass; CA836
Common names. Ounce metal; 85-5-5-5; composition metal

Specifications

AMS. 4855
ASTM. B 30, B 62, B 271, B 505, B 584
SAE. J462 (CA836)
Ingot identification number. 115
Government. QQ-C-390 (CA836), MIL-C-15345 (Alloy 1)

Chemical Composition

Composition limits. 84.0 to 86.0 Cu, 4.0 to 6.0 Sn, 4.0 to 6.0 Pb, 4.0 to 6.0 Zn, 0.30 Fe max, 0.25 Sb max, 1.0 Ni max, 0.05 P max (1.5 max for continuous castings), 0.08 S max, 0.005 Al max, 0.005 Si max. In determining Cu min, Cu may be calculated as Cu + Ni.
Consequence of exceeding impurity limits. Aluminum and/or silicon in excess of 0.005% will adversely affect mechanical properties and pressure tightness.

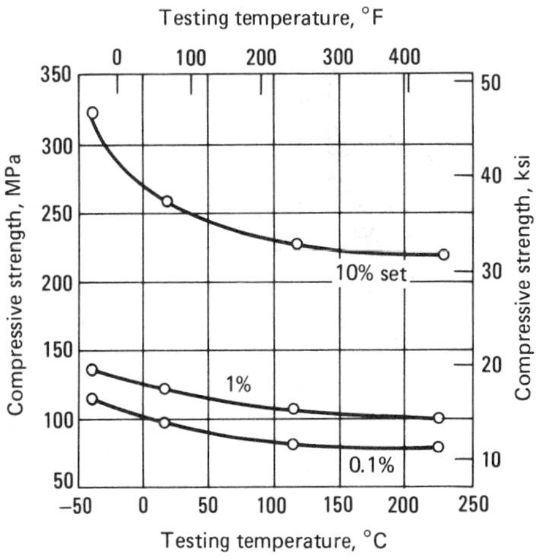

Fig. 11 Typical compressive strength for C83600

Applications

Typical uses. Good general-purpose casting alloy. For castings requiring moderate strength, soundness, and good machinability, such as low-pressure valves, pipe fittings, gasoline- and oil-line fittings, fire-equipment fittings, small gears, small pump parts, general plumbing hardware

Mechanical Properties

Tensile properties. Typical data for separately cast test bars: tensile strength, 255 MPa (37 ksi); yield strength, 117 MPa (17 ksi) at 0.5% extension under load; elongation, 30% in 50 mm (2 in.)
Compressive properties. Compressive strength at room temperature: 97 MPa (14 ksi) at permanent set of 0.1%; 120 MPa (17.4 ksi) at permanent set of 1%; 258 MPa (37.5 ksi) at permanent set of 10%. See also Fig. 11.
Hardness. 60 HB, typical
Elastic modulus. Tension, 83 GPa (12×10^6 psi) at 20 °C (68 °F). See also Fig. 12.
Impact strength. Izod, 14 J (10 ft · lbf); Charpy V-notch, 15 J (11 ft · lbf)
Fatigue strength. 76 MPa (11 ksi) at 10^8 cycles. See also Fig. 13.
Creep strength. For 0.1% creep in 10 000 h: 86 MPa (12.5 ksi) at 180 °C (350 °F); 77 MPa (11.1 ksi) at 230 °C (450 °F); 48 MPa (7 ksi) at 290 °C (550 °F)

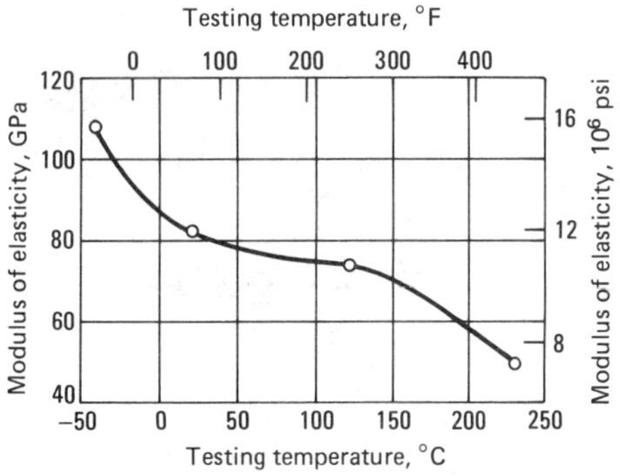

Fig. 12 Typical modulus of elasticity in tension for C83600

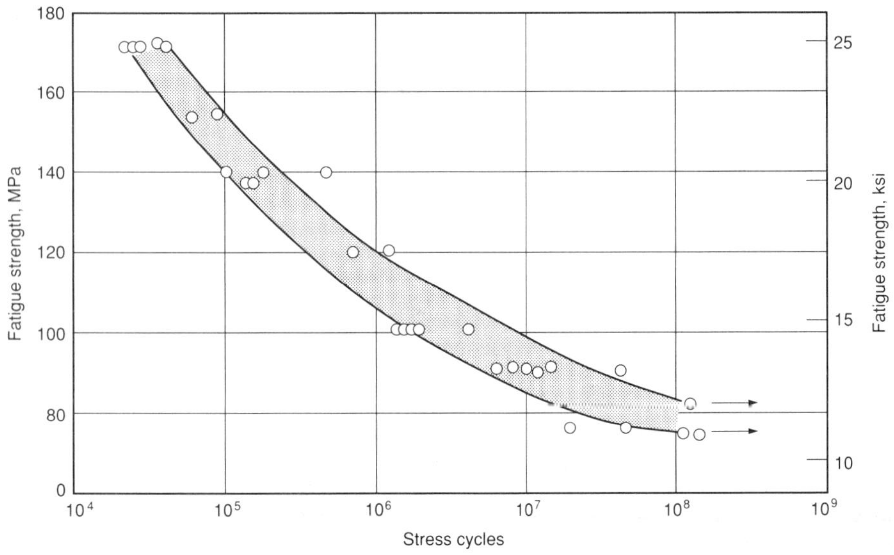

Fig. 13 Fatigue strength of C83600

Mass Characteristics

Density. 8.83 g/cm³ (0.318 lb/in.³) at 20 °C (68 °F)
Volume change on freezing. 10.6% contraction
Patternmaker's shrinkage. 13 to 16 mm/m (5⁄32 to 3⁄16 in./ft)

Thermal Properties

Liquidus temperature. 1010 °C (1850 °F)
Solidus temperature. 855 °C (1570 °F)
Coefficient of linear thermal expansion. 18.0 μm/m · K (10.0 μin./in. · °F) at 20 to 205 °C (68 to 400 °F). See also Fig. 14.
Specific heat. 380 J/kg · K (0.09 Btu/lb · °F) at 20 °C (68 °F)
Thermal conductivity. 72.0 W/m · K (41.6 Btu/ft · h · °F) at 20 °C (68 °F)

Electrical Properties

Electrical conductivity. Volumetric, 15% IACS

Magnetic Properties

Magnetic permeability. 1.0

Fabrication Characteristics

Machinability. 84% of C36000 (free-cutting brass)

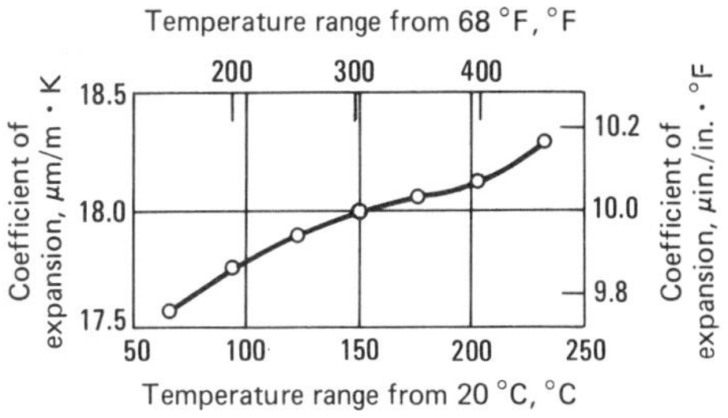

Fig. 14 Mean thermal expansion of C83600

C83800
83Cu-4Sn-6Pb-7Zn

Commercial Names

Previous trade name. CA838
Common names. Hydraulic bronze; 83-4-6-7

Specifications

ASTM. B 30 (CA838), B 271 (CA838), B 505 (CA838), B 584 (CA838)
SAE. J462
Ingot identification number. 120
Government. QQ-C-390

Chemical Composition

Composition limits. 82.0 to 83.8 Cu, 3.3 to 4.2 Sn, 5.0 to 7.0 Pb, 5.0 to 8.0 Zn, 0.30 Fe max, 0.25 Sb max, 1.0 Ni max, 0.03 P max (1.5 max for continuous castings), 0.08 S max, 0.005 Al max, 0.005 Si max. In determining Cu min, Cu may be calculated as Cu + Ni.
Consequence of exceeding impurity limits. Aluminum and/or silicon in excess of 0.005% will adversely affect mechanical properties and pressure tightness.

Applications

Typical uses. General-purpose free-machining alloy. For air, gas, and water fittings; plumbing supplies and fittings; pumps and pump fittings; hardware; carburetors; injectors; railroad catenary; and overhead fittings

Mechanical Properties

Tensile properties. Typical data for separately cast test bars: tensile strength, 240 MPa (35 ksi); yield strength, 110 MPa (16 ksi) at 0.5% extension under load; elongation, 25% in 50 mm (2 in.)
Compressive properties. Compressive strength: 79 MPa (11.5 ksi) at permanent set of 0.1%; 200 MPa (29 ksi) at permanent set of 10%
Hardness. 60 HB
Elastic modulus. Tension, 92 GPa (13.3 $\times 10^6$ psi)
Impact strength. Izod, 11 J (8 ft · lbf)

Mass Characteristics

Density. 8.6 g/cm^3 (0.312 lb/in.^{3}) at 20 °C (68 °F)
Patternmaker's shrinkage. 15.6 mm/m (3/16 in./ft)

Thermal Properties

Liquidus temperature. 1005 °C (1840 °F)
Solidus temperature. 845 °C (1550 °F)
Coefficient of linear thermal expansion. 18 µm/m · K (10 µin./in. · °F) at 20 to 232 °C (68 to 450 °F)
Specific heat. 380 J/kg · K (0.09 Btu/lb · °F) at 20 °C (68 °F)
Thermal conductivity. 72.5 W/m · K (41.9 Btu/ft · h · °F) at 20 °C (68 °F)

Electrical Properties

Electrical conductivity. Volumetric, 15% IACS

Magnetic Properties

Magnetic permeability. 1.0

Fabrication Characteristics

Machinability. 90% of C36000 (free-cutting brass)

C84400
81Cu-3Sn-7Pb-9Zn

Commercial Names

Previous trade names. Leaded semi-red brass; CA844
Common names. Valve metal; 81-3-7-9

Specifications

ASTM. B 30 (CA844), B 271 (CA844), B 505 (CA844), B 584 (CA844)
Ingot identification number. 123
Government. QQ-C-390

Chemical Composition

Composition limits. 78.0 to 82.0 Cu, 2.3 to 3.5 Sn, 6.0 to 8.0 Pb, 7.0 to 10.0 Zn, 0.40 Fe max, 0.25 Sb max, 1.0 Ni max, 0.02 P max (1.5 max for continuous castings), 0.08 S max, 0.005 Al max, 0.005 Si max. In determining Cu min, Cu may be calculated as Cu + Ni.
Consequence of exceeding impurity limits. Aluminum and/or silicon in excess of 0.005% will adversely affect mechanical properties and pressure tightness.

Applications

Typical uses. Low-pressure valves and fittings, general hardware fittings, plumbing supplies and fixtures, ornamental fixtures

Mechanical Properties

Tensile properties. Typical data for separately cast test bars: tensile strength, 235 MPa (34 ksi); yield strength, 105 MPa (15 ksi) at 0.5% extension under load; elongation, 26% in 50 mm (2 in.)
Hardness. 55 HB
Elastic modulus. Tension, 90 GPa (13.0 $\times 10^6$ psi)
Impact strength. Izod, 11 J (8 ft · lbf)

Mass Characteristics

Density. 8.70 g/cm^3 (0.314 lb/in.^{3}) at 20 °C (68 °F)
Patternmaker's shrinkage. 15.6 mm/m (3/16 in./ft)

Thermal Properties

Liquidus temperature. 1005 °C (1840 °F)
Solidus temperature. 840 °C (1540 °F)
Coefficient of linear thermal expansion. 18 µm/m · K (10 µin./in. · °F) at 20 to 260 °C (68 to 500 °F)
Specific heat. 380 J/kg · K (0.09 Btu/lb · °F) at 20 °C (68 °F)
Thermal conductivity. 72.5 W/m · K (41.9 Btu/ft · h · °F) at 20 °C (68 °F)

Electrical Properties

Electrical conductivity. Volumetric, 16.4% IACS

Magnetic Properties

Magnetic permeability. 1.0

Fabrication Characteristics

Machinability. 90% of C36000 (free-cutting brass)

C84800
76Cu-2½Sn-6½Pb-15Zn

Commercial Names

Common name. Leaded semi-red brass, plumbing goods brass, 76-2½-6½-15

Specifications

ASTM. B 30, B 271, B 505, B 584
Government. QQ-C-390, CA848
Other. Ingot code number 130

Chemical Composition

Composition limits. 75.0 to 77.0 Cu, 2.0 to 3.0 Sn, 5.5 to 7.0 Pb, 13.0 to 17.0 Zn, 0.40 Fe max, 0.25 Sb max, 1.0 Ni max, 0.02 P max (1.5 P max for continuous castings), 0.08 S max, 0.005 Al max, 0.005 Si max
Copper specification. In determining Cu, minimum may be calculated as Cu + Ni.
Consequence of exceeding impurity limits. Aluminum and/or silicon in excess of 0.005% will adversely affect mechanical properties and pressure tightness.

Applications

Typical uses. Plumbing fixtures, cocks, faucets, stops, wastes, air- and gas-line fittings, general hardware fittings, low-pressure valves and fittings

Mechanical Properties

Tensile properties. Typical data for separately cast test bars: tensile strength, 255 MPa (37 ksi); yield strength, 97 MPa (14 ksi) at 0.5% extension under load; elongation, 35% in 50 mm (2 in.)
Compressive properties. Typical compressive strength: 88.3 MPa (12.8 ksi) at a permanent set of 0.1%; 109 MPa (15.8 ksi) at a permanent set of 1%; 236 MPa (34.3 ksi) at a permanent set of 10%
Hardness. 55 HB

Elastic modulus. Tension, 105 GPa (15 × 10^6 psi)
Impact strength. Charpy V-notch, 16 J (12 ft · lbf)
Fatigue strength. 76 MPa (11 ksi) at 10^8 cycles
Creep-rupture characteristics. Limiting creep stress for 10^{-5}%/h: 82.0 MPa (11.9 ksi) at 177 °C (350 °F); 55 MPa (8 ksi) at 204 °C (400 °F); 20 MPa (3 ksi) at 288 °C (550 °F)

Mass Characteristics

Density. 8.58 g/cm³ (0.310 lb/in.³) at 20 °C (68 °F)
Patternmaker's shrinkage. 16 mm/m (3/16 in./ft)

Thermal Properties

Liquidus temperature. 954 °C (1750 °F)
Solidus temperature. 832 °C (1530 °F)
Coefficient of linear thermal expansion. 18.7 μm/m · K (10.4 μin./in. · °F) at 20 to 260 °C (68 to 500 °F)
Specific heat. 376 J/kg · K (0.09 Btu/lb · °F) at 20 °C (68 °F)
Thermal conductivity. 72.0 W/m · K (41.6 Btu/ft · h · °F) at 20 °C (68 °F)

Electrical Properties

Electrical conductivity. Volumetric, 16.4% IACS at 20 °C (68 °F)

Magnetic Properties

Magnetic permeability. 1.0

Fabrication Characteristics

Machinability. 90% of C36000 (free-cutting brass)

C85200
72Cu-1Sn-3Pb-24Zn

Commercial Names

Previous trade names. Leaded yellow brass; CA852
Common names. High-copper yellow brass; 72-1-3-24

Specifications

ASTM. B 30 (CA852), B 271 (CA852), B 584 (CA852)
SAE. J462
Ingot identification number. 400
Government. QQ-C-390 (CA852), MIL-C-15345 (Alloy 28)

Chemical Composition

Composition limits. 70.0 to 74.0 Cu, 0.7 to 2.0 Sn, 1.5 to 3.8 Pb, 20.0 to 27.0 Zn, 0.6 Fe max, 0.20 Sb max, 1.0 Ni max, 0.02 P max, 0.05 S max, 0.005 Al max, 0.05 Si max

Applications

Typical uses. Plumbing fittings and fixtures, ferrules, low-pressure valves, hardware fittings, ornamental brass, chandeliers, andirons

Mechanical Properties

Tensile properties. Typical data for separately cast test bars: tensile strength, 260 MPa (38 ksi); yield strength, 90 MPa (13 ksi) at 0.5% extension under load; elongation, 35% in 50 mm (2 in.)
Hardness. 45 HB
Elastic modulus. Tension, 76 GPa (11 ×10^6 psi)

Mass Characteristics

Density. 8.50 g/cm³ (0.307 lb/in.³) at 20 °C (68 °F)
Volume change on freezing. 12.4% contraction
Patternmaker's shrinkage. 16 mm/m (3/16 in./ft)

Thermal Properties

Liquidus temperature. 940 °C (1725 °F)
Solidus temperature. 925 °C (1700 °F)
Coefficient of linear thermal expansion. 21 μm/m · K (11.5 μin./in. · °F) at 20 to 100 °C (68 to 212 °F)
Specific heat. 380 J/kg · K (0.09 Btu/lb · °F) at 20 °C (68 °F)
Thermal conductivity. 83.9 W/m · K (48.5 Btu/ft · h · °F) at 20 °C (68 °F)

Electrical Properties

Electrical conductivity. Volumetric, 18.6% IACS

Magnetic Properties

Magnetic permeability. 1.0

Fabrication Characteristics

Machinability. 80% of C36000 (free-cutting brass)

C85400
67Cu-1Sn-3Pb-29Zn

Commercial Names

Previous trade names. Leaded yellow brass; CA854
Common names. No. 1 yellow brass; 67-1-3-29

Specifications

ASTM. B 30 (CA854), B 271 (CA854), B 584 (CA854)
SAE. J462 (CA854)
Ingot identification number. 403
Government. QQ-C-390 (CA854), MIL-C-15345 (Alloy 23)

Chemical Composition

Composition limits. 65.0 to 70.0 Cu, 0.50 to 1.5 Sn, 1.5 to 3.5 Pb, 24.0 to 32.0 Zn, 0.7 Fe max, 1.0 Ni max, 0.35 Al max, 0.05 Si max
Aluminum. Addition of 0.20 to 0.30% Al improves castability.

Applications

Typical uses. General-purpose casting alloy. For lightweight castings not subject to high internal pressure, such as furniture hardware, ornamental castings, radiator fittings, ship trimmings, gas cocks, light fixtures, battery clamps

Mechanical Properties

Tensile properties. Typical data for separately cast test bars: tensile strength, 235 MPa (34 ksi); yield strength, 83 MPa (12 ksi) at 0.5% extension under load; elongation, 35% in 50 mm (2 in.)
Hardness. 50 HB

Mass Characteristics

Density. 8.45 g/cm³ (0.305 lb/in.³) at 20 °C (68 °F)
Patternmaker's shrinkage. 16 mm/m (3/16 in./ft)

Thermal Properties

Liquidus temperature. 940 °C (1725 °F)
Solidus temperature. 925 °C (1700 °F)
Coefficient of linear thermal expansion. 20.2 μm/m · K (11.2 μin./in. · °F) at 20 to 100 °C (68 to 212 °F)
Specific heat. 380 J/kg · K (0.09 Btu/lb · °F) at 20 °C (68 °F)
Thermal conductivity. 88 W/m · K (51 Btu/ft · h · °F) at 20 °C (68 °F)

Electrical Properties

Electrical conductivity. Volumetric, 19.6% IACS

Magnetic Properties

Magnetic permeability. 1.0

Fabrication Characteristics

Machinability. 80% of C36000 (free-cutting brass)

C85700, C85800
63Cu-1Sn-1Pb-35Zn

Commercial Names

Previous trade names. CA857, CA858
Common names. Leaded yellow brass; 63-1-1-35

Specifications

ASTM. B 30 (CA857, CA858), B 176 (CA858), B 271 (CA857), B 584 (CA857)
SAE. J462
Ingot identification number. 406
Government. QQ-C-390 (CA857), MIL-C-15345 (Alloy 3)

Chemical Composition

Composition limits. See Table 8. Addition of 0.20 to 0.30% Al improves castability.

Applications

Typical uses. Bushings, hardware fittings, ornamental castings, lock hardware

Mechanical Properties

Tensile properties. Typical data for separately cast test bars. Sand castings or centrifugal castings (C85700): tensile strength, 345 MPa (50 ksi); yield strength, 125 MPa

Table 8 Composition limits of C85700 and C85800

Sand castings or centrifugal castings (C85700)			
Cu	58.0–64.0	Fe	0.7 max
Sn	0.50–1.50	Ni	1.0 max
Pb	0.80–1.50	Al	0.55 max
Zn	32.0–40.0	Si	0.05 max
Die castings (C85800)			
Cu	58.0 min	Fe	0.50 max
Sn	1.5 max	Al	(b)
Pb	1.5 max	Mn	0.25 max
Zn	(a)	Other	0.50 max (c)

(a) ASTM B 176, 31 to 41; SAE J462, 31.0 to 34.0. (b) ASTM B 176, 0.25 max; SAE J462, 0.50 max. (c) SAE J462 allows 0.05 Sb max, 0.50 Ni max, 0.05 As max, 0.05 S max, 0.01 P max, and 0.25 Si max before determination of total unnamed elements.

(18 ksi) at 0.5% extension under load; elongation, 40% in 50 mm (2 in.). Die castings (C85800): tensile strength, 380 MPa (55 ksi); yield strength, 205 MPa (30 ksi) at 0.5% extension under load; elongation, 15% in 50 mm (2 in.)
Hardness. Sand castings or centrifugal castings (C85700), 75 HB; die castings (C85800), 102 HB
Elastic modulus. Tension: sand castings or centrifugal castings (C85700), 97 GPa (14 × 10^6 psi); die castings (C85800), 105 GPa (15 ×10^6 psi)

Mass Characteristics

Density. 8.41 g/cm^3 (0.304 lb/in.^{3}) at 20 °C (68 °F)
Patternmaker's shrinkage. 16 mm/m (3/16 in./ft)

Thermal Properties

Liquidus temperature. 920 °C (1688 °F)
Solidus temperature. 903 °C (1657 °F)
Coefficient of linear thermal expansion. 22 μm/m · K (12 μin./in. · °F) at 20 to 260 °C (68 to 500 °F)
Specific heat. 376 J/kg · K (0.09 Btu/lb · °F) at 20 °C (68 °F)
Thermal conductivity. 83.9 W/m · K (48.5 Btu/ft · h · °F) at 20 °C (68 °F)

Electrical Properties

Electrical conductivity. Volumetric, 22% IACS

Magnetic Properties

Magnetic permeability. 1.0

Fabrication Characteristics

Machinability. 80% of C36000 (free-cutting brass)

C86100, C86200
64Cu-24Zn-3Fe-5Al-4Mn

Commercial Names

Common names. Manganese bronze (90 000 psi); High-strength yellow brass; CA861; CA862

Specifications

ASTM. C86100: none. C86200: Ingot, B 30; centrifugal castings, B 271; sand castings, B 584; continuous castings, B 505
SAE. J462. (Former alloy number: 430A)
Government. QQ-C-390, QQ-C-523. C86100: centrifugal castings, MIL-C-15345 (Alloy 5); investment castings, MIL-C-22087 (composition 7); sand castings, MIL-C-22229 (composition 10). C86200: investment castings, MIL-C-22087 (composition 9); precision castings, MIL-C-11866 (composition 20); sand castings, MIL-C-22229 (composition 9)
Ingot identification number. 423

Chemical Composition

Composition limits. C86100: 66.0 to 68 Cu, 4.5 to 5.5 Al, 2.0 to 4.0 Fe, 2.5 to 5.0 Mn, 1.0 Ni max, 0.2 Sn max, 0.2 Pb max, bal Zn. C86200: 60.0 to 68.0 Cu, 3.0 to 7.5 Al, 2.0 to 4.0 Fe, 2.5 to 5.0 Mn, 1.0 Ni max, 0.2 Sn max, 0.2 Pb max, bal Zn

Applications

Typical uses. Marine castings, gears, gun mounts, bushings, and bearings

Mechanical Properties

Tensile properties. Nominal. Tensile strength, 655 MPa (95 ksi); yield strength, 330 MPa (48 ksi); elongation, 20% in 50 mm (2 in.)
Compressive properties. Compressive strength, 345 MPa (50 ksi) at a permanent set of 0.1%
Hardness. 180 HB
Elastic modulus. Tension, 105 GPa (15 ×10^6 psi)
Impact strength. Izod, 16 J (12 ft · lbf)

Mass Characteristics

Density. 7.9 g/cm^3 (0.285 lb/in.^{3}) at 20 °C (68 °F)
Volume change on freezing. 2%

Thermal Properties

Liquidus temperature. 940 °C (1725 °F)
Solidus temperature. 900 °C (1650 °F)
Coefficient of linear thermal expansion. 22 μm/m · K (12 μin./in. · °F) at 20 to 260 °C (68 to 500 °F)
Specific heat. 376 J/kg · K (0.09 Btu/lb · °F) at 20 °C (68 °F)
Thermal conductivity. 35 W/m · K (20 Btu/ft · h · °F) at 20 °C (68 °F)

Electrical Properties

Electrical conductivity. Volumetric, 7.5% IACS at 20 °C (68 °F)

Magnetic Properties

Magnetic permeability. 1.24 at field strength of 16 kA/m

Fabrication Characteristics

Machinability. 30% of C36000 (free-cutting brass)
Annealing temperature. 260 °C (500 °F)

C86300
64Cu-26Zn-3Fe-3Al-4Mn

Commercial Names

Common names. Manganese bronze (110 000 psi); High-strength yellow brass; CA863

Specifications

AMS. 4862
ASTM. Sand castings: B 22, B 584; centrifugal castings: B 271; continuous castings: B 505, ingot: B 30
SAE. J462
Government. QQ-C-390, QQ-C-523. Centrifugal castings, MIL-C-15345 (Alloy 6); investment castings, MIL-C-22087 (composition 9); precision castings, MIL-C-11866 (composition 21); sand castings, MIL-C-22229 (composition 8)
Ingot identification number. 424

Chemical Composition

Composition limits. 60.0 to 68.0 Cu, 2.5 to 5.0 Mn, 3.0 to 7.5 Al, 2.0 to 4.0 Fe, 0.2 Pb max, 0.2 Sn max, bal Zn
Consequence of exceeding impurity limits. Excessive Sn causes brittleness; excessive Pb or Ni decreases elongation.

Applications

Typical uses. Extra-heavy duty, high-strength alloy for gears, cams, bearings, screw-down nuts, bridge parts, hydraulic cylinder parts
Precautions in use. Not to be used in marine atmospheres, ammonia, or high-corrosive atmospheres

Mechanical Properties

Tensile properties. Nominal. Tensile strength, 820 MPa (119 ksi); yield strength, 460 MPa (67 ksi); elongation, 18% in 50 mm (2 in.)
Compressive properties. Compressive strength: 415 MPa (60 ksi) at permanent set of 0.1%; 670 MPa (97 ksi) at permanent set of 1%
Hardness. 225 HB
Elastic modulus. Tension, 105 GPa (15.5 ×10^6 psi)
Fatigue strength. Rotating beam, 170 MPa (25 ksi) at 100 million cycles
Impact strength. Izod, 20 J (15 ft · lbf). Charpy V-notch, 16 J (12 ft · lbf)

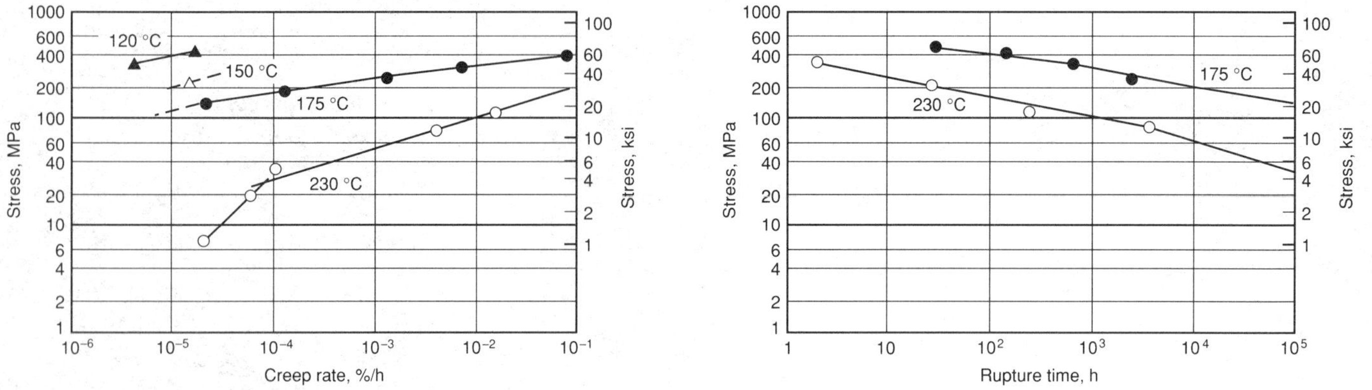

Fig. 15 Creep-rupture properties of C86300

Creep-rupture characteristics. Stress for 0.17% creep in 10 000 h: 390 MPa (56.5 ksi) at 120 °C (250 °F); 225 MPa (32.5 ksi) at 150 °C (300 °F); 130 MPa (19 ksi) at 175 °C (350 °F); 3 MPa (0.5 ksi) at 230 °C (450 °F). See also Fig. 15 and 16.

Mass Characteristics

Density. 7.7 g/cm^3 (0.278 lb/in.^{3}) at 20 °C (68 °F)
Volume change on freezing. 2%

Thermal Properties

Liquidus temperature. 923 °C (1693 °F)
Solidus temperature. 885 °C (1625 °F)
Coefficient of linear thermal expansion. 22 μm/m · K (12 μin./in. · °F) at 20 to 260 °C (68 to 500 °F)
Specific heat. 376 J/kg · K (0.09 Btu/lb · °F) at 20 °C (68 °F)
Thermal conductivity. 36 W/m · K (21 Btu/ft · h · °F) at 20 °C (68 °F)

Electrical Properties

Electrical conductivity. Volumetric, 9% IACS at 20 °C (68 °F)

Magnetic Properties

Magnetic permeability. 1.09 at field strength of 16 kA/m

Fabrication Characteristics

Machinability. 8% of C36000 (free-cutting brass)
Annealing temperature. 260 °C (500 °F)

C86400
59Cu-0.75Sn-0.75Pb-37Zn-1.25Fe-0.75Al-0.5Mn

Commercial Names

Previous trade name. Leaded high-strength yellow brass; stem manganese bronze
Common name. Manganese bronze (60 000 psi)

Specifications

ASTM. Sand castings: B 584; centrifugal castings: B 271; ingot: B 30
Government. QQ-C-390, QQ-C-523
Ingot identification number. 420

Chemical Composition

Composition limits. 56.0 to 62.0 Cu, 1.5 Sn max, 0.5 to 1.5 Pb, 2.0 Fe max, 1.5 Al max, 1.5 Mn max, 1.0 Ni max, bal Zn

Applications

Typical uses. Free-machining manganese bronze for valve stems, marine castings and fittings, pump bodies

Mechanical Properties

Tensile properties. Typical tensile strength, 450 MPa (65 ksi); yield strength, 170 MPa (25 ksi); elongation, 20% in 50 mm (2 in.)
Compressive properties. Compressive strength: 150 MPa (22 ksi) at 0.1% perma-

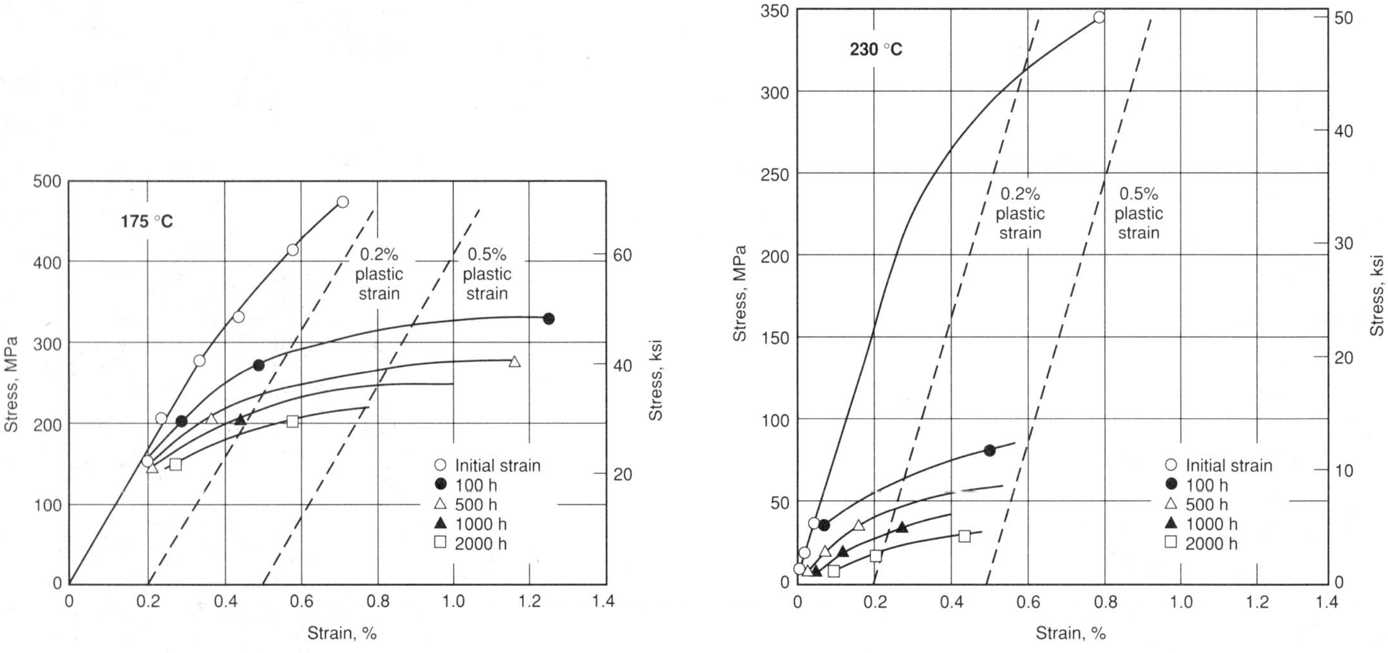

Fig. 16 Isochronous stress-strain curves for C86300

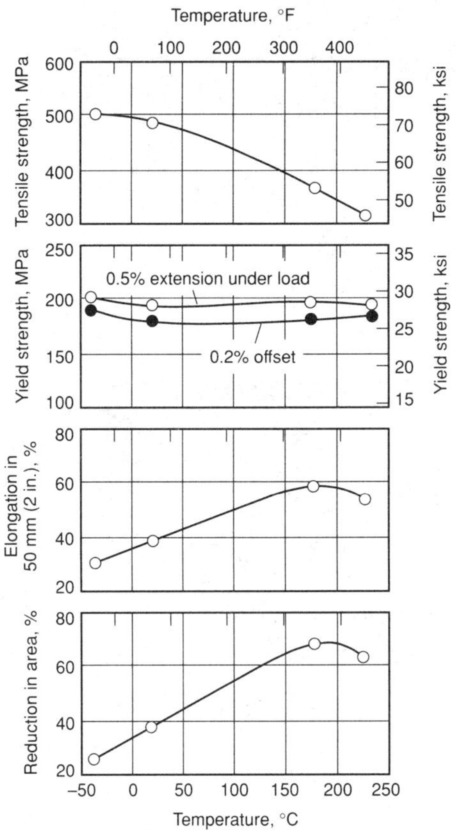

Fig. 17 Typical tensile properties of C86500

nent set; 600 MPa (87 ksi) at 10% permanent set
Hardness. 105 HB
Elastic modulus. Tension, 96 GPa (14 $\times 10^6$ psi)
Impact strength. Izod, 40 J (30 ft · lbf). Charpy V-notch, 34 J (25 ft · lbf)

Mass Characteristics

Density. 8.32 g/cm³ (0.301 lb/in.³) at 20 °C (68 °F)
Volume change on freezing. 2%

Thermal Properties

Liquidus temperature. 880 °C (1615 °F)
Solidus temperature. 860 °C (1585 °F)
Coefficient of linear thermal expansion. 20 μm/m · K (11.4 μin./in. · °F) at 21 to 204 °C (70 to 400 °F)
Specific heat. 376 J/kg · K (0.09 Btu/lb · °F) at 20 °C (68 °F)

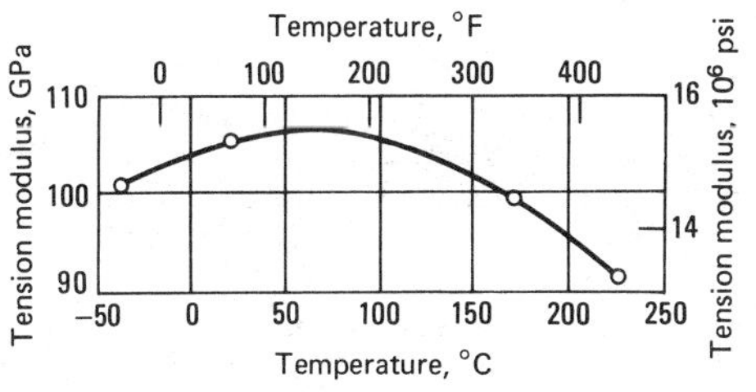

Fig. 20 Elastic modulus in tension for C86500

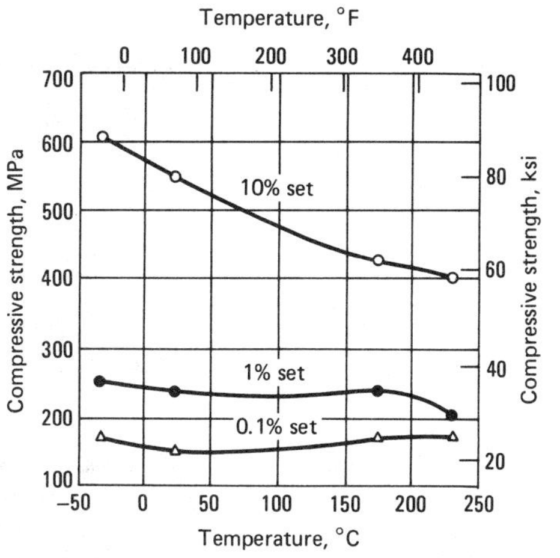

Fig. 18 Typical compressive strength of C86500

Thermal conductivity. 88 W/m · K (51 Btu/ft · h · °F) at 20 °C (68 °F)

Electrical Properties

Electrical conductivity. Volumetric, 22% IACS at 20 °C (68 °F)

Fabrication Characteristics

Machinability. 60% of C36000 (free-cutting brass)
Casting temperature range. Light castings, 1040 to 1120 °C (1900 to 2050 °F); heavy castings, 955 to 1040 °C (1750 to 1900 °F)
Annealing temperature. 260 °C (500 °F)

C86500
58Cu-39Zn-1.3Fe-1Al-0.5Mn

Commercial Names

Previous trade name. High-strength yellow brass
Common name. Manganese bronze (65 000 psi)

Specifications

AMS. 4860A
ASTM. Sand castings: B 584; centrifugal castings: B 271, ingot: B 30
SAE. J462
Government. QQ-C-390. Sand castings, MIL-C-22229 (composition 7); centrifugal castings, MIL-C-15345 (Alloy 4); investment castings, MIL-C-22087 (composition 5)
Ingot identification number. 421

Chemical Composition

Composition limits. 55.0 to 60.0 Cu, 0.4 to 2.0 Fe, 0.5 to 1.5 Al, 1.5 Mn max, 0.4 Pb max, 1.0 Sn max, 1.0 Ni max, bal Zn

Applications

Typical uses. Propeller hubs, blades, and other parts in contact with salt and fresh water, gears, liners

Mechanical Properties

Tensile properties. Typical. Tensile

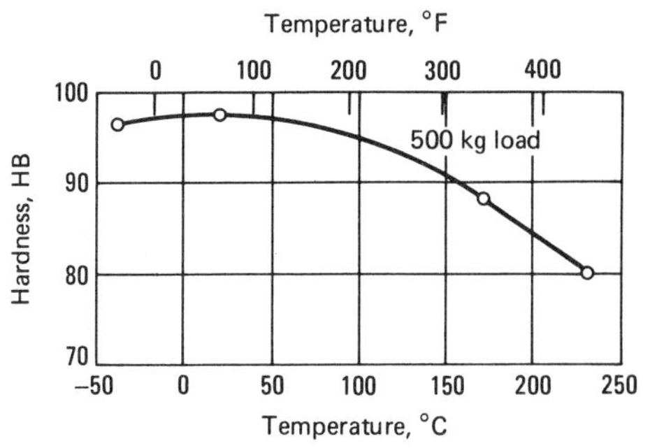

Fig. 19 Typical Brinell hardness of C86500

strength, 490 MPa (71 ksi); yield strength, 195 MPa (28 ksi); elongation, 30% in 50 mm (2 in.). See also Fig. 17.
Compressive properties. Compressive strength: 165 MPa (24 ksi) at permanent set of 0.1%; 240 MPa (35 ksi) at permanent set of 1%; 545 MPa (79 ksi) at permanent set of 10%. See also Fig. 18.
Hardness. 130 HB. See also Fig. 19.
Elastic modulus. Tension, 105 GPa (15 $\times 10^6$ psi). See also Fig. 20.
Fatigue strength. Reverse bending, 145 MPa (21 ksi) at 10^8 cycles. See also Fig. 21.
Impact strength. Charpy, 42 J (31 ft · lbf). See also Fig. 22.
Creep-rupture characteristics. Stress for 0.1% creep in 10 000 h: 190 MPa (28 ksi) at 120 °C (250 °F); 43 MPa (6.2 ksi) at 175 °C (350 °F); 12 MPa (1.7 ksi) at 230 °C (450 °F). See also Fig. 23 and 24.

Mass Characteristics

Density. 8.3 g/cm³ (0.299 lb/in.³) at 20 °C (68 °F)
Patternmaker's shrinkage. 1.65 to 2.15% for pouring temperature of 905 °C (1665 °F)

Thermal Properties

Liquidus temperature. 880 °C (1616 °F)
Solidus temperature. 862 °C (1583 °F)
Coefficient of linear thermal expansion. 20.3 μm/m · K (11.3 μin./in. · °F) at 21 to 93 °C (70 to 200 °F). See also Fig. 25.
Specific heat. 373 J/kg · K (0.089 Btu/lb · °F) at 20 °C (68 °F)
Thermal conductivity. 87 W/m · K (50.2 Btu/ft · h · °F) at 20 °C (68 °F). See also Fig. 25.

Electrical Properties

Electrical conductivity. Volumetric, 20.5% IACS at 20 °C (68 °F). See also Fig. 26.
Electrical resistivity. See Fig. 26.

Magnetic Properties

Magnetic permeability. 1.09 at field strength of 16 kA/m

Fabrication Characteristics

Machinability. 26% of C36000 (free-cutting brass)
Annealing temperature. 260 °C (500 °F)

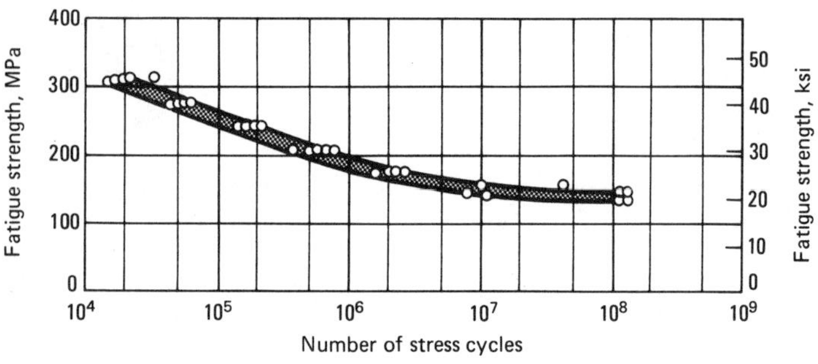

Fig. 21 Typical reverse bending fatigue curve at room temperature for C86500

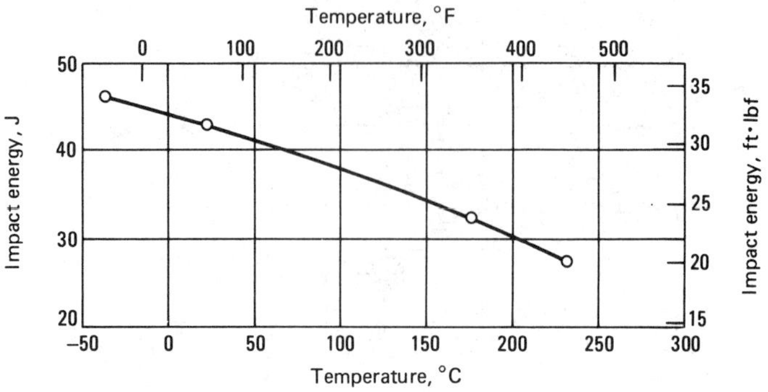

Fig. 22 Typical Charpy V-notch impact strength for C86500

C86700

Commercial Names

Previous trade name. CA867
Common names. Leaded high-strength yellow brass; 80 000 psi tensile manganese bronze

Specifications

ASTM. Centrifugal, B 271; ingot, B 30; sand, B 584, B 763

Chemical Composition

Composition limits. 55.0 to 60.0 Cu, 1.0 to 3.0 Al, 1.0 to 3.0 Fe, 0.5 to 1.5 Pb, 1.0 to 3.5 Mn, 1.0 Ni max, 1.5 Sn max, 30.0 to 38.0 Zn. Ingot for remelting specifications may vary from the ranges shown.

Copper and zinc specifications. In reporting chemical analyses by the use of instruments such as spectrograph, x-ray, and atomic absorption, copper may be indicated as balance. In reporting chemical analyses obtained by wet methods, zinc may be indicated as balance on those alloys with over 2% Zn.

Applications

Typical uses. High-strength free-machining manganese bronze valve stems

Mechanical Properties

Tensile properties. Typical data for as-sand-cast separately cast test bar (M01 temper): tensile strength, 585 MPa (85 ksi); yield strength, 290 MPa (42 ksi) at 0.5% extension under load; elongation, 20% in 50 mm (2 in.)
Hardness. Typically 80 HRB or 155 HB (3000 kg)
Elastic modulus. Tension: 105 GPa (15 × 10^6 psi) at 20 °C (68 °F)

Mass Characteristics

Density. 8.32 g/cm^3 (0.301 lb/in.^{3}) at 20 °C (68 °F)
Patternmaker's shrinkage. 21 mm/m (¼ in./ft)

Thermal Properties

Liquidus temperature. 880 °C (1616 °F)
Solidus temperature. 862 °C (1583 °F)
Coefficient of linear thermal expansion. 19 μm/m · K (11 μin./in. · °F) at 20 to 200 °C (68 to 392 °F)
Specific heat. 376 J/kg · K (0.09 Btu/lb · °F) at 20 °C (68 °F)

Electrical Properties

Electrical conductivity. Volumetric, 32% IACS at 20 °C (68 °F)

Fabrication Characteristics

Machinability. M01 temper; 55% of C36000 (free-cutting brass)
Stress-relieving temperature. 260 °C (500 °F)

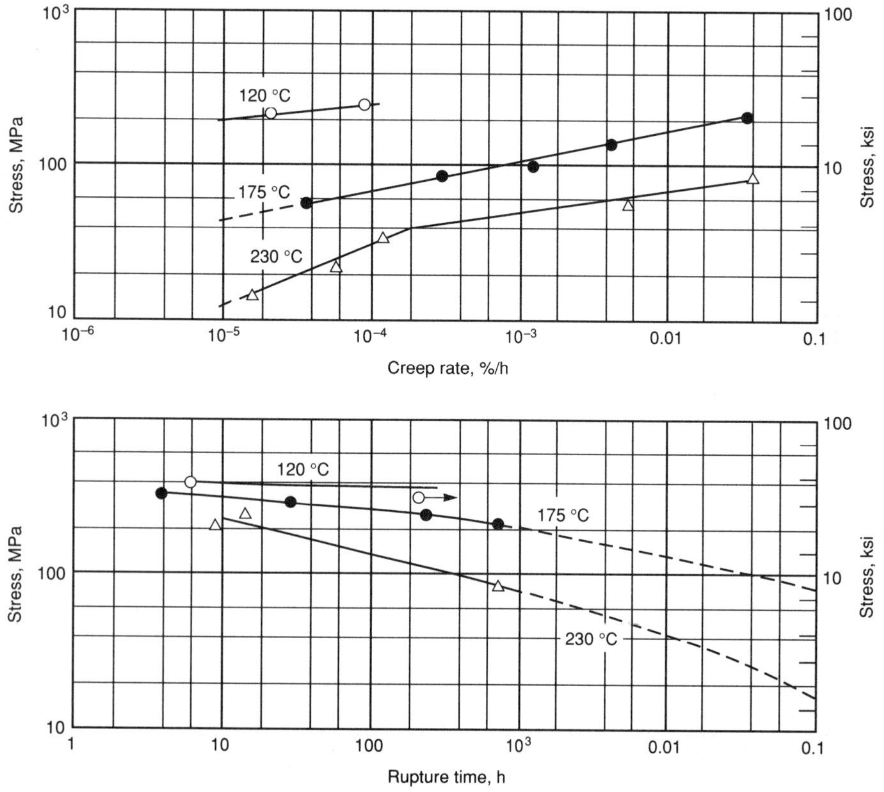

Fig. 23 Typical creep-rupture properties of C86500

C86800

Commercial Names

Previous trade name. CA868
Common name. Nickel-manganese bronze

Specifications

ASTM. Die, B 176
Government. Sand, QQ-C-390; valves, WW-V-1967

Chemical Composition

Composition limits. 53.5 to 57.0 Cu, 2.0 Al max, 1.0 to 2.5 Fe, 0.20 Pb max, 2.5 to 4.0 Mn, 2.5 to 4.0 Ni, 1.0 Sn max, bal Zn. Ingot for remelting specifications may vary from the ranges shown.
Copper and zinc specifications. In reporting chemical analyses by the use of instruments such as spectrograph, x-ray, and atomic absorption, copper may be indicated as balance.

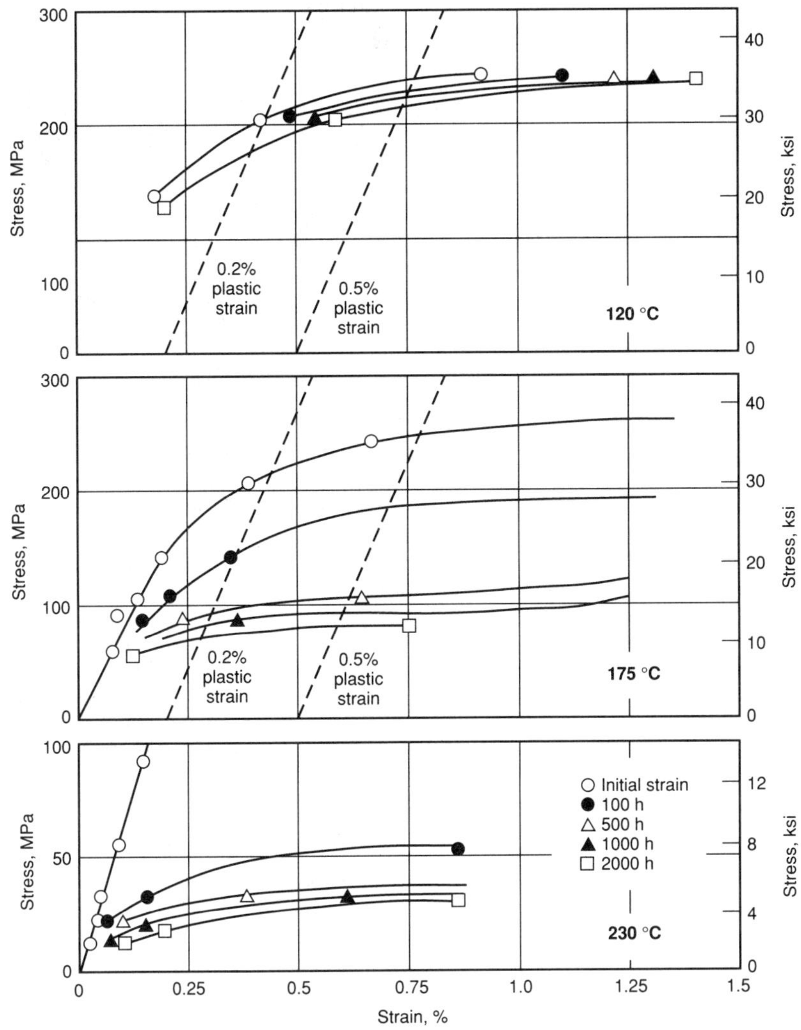

Fig. 24 Isochronous stress-strain curves for C86500

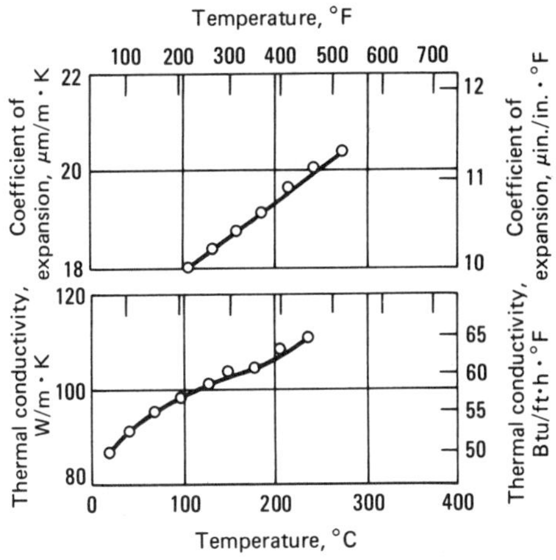

Fig. 25 Typical thermal properties of C86500

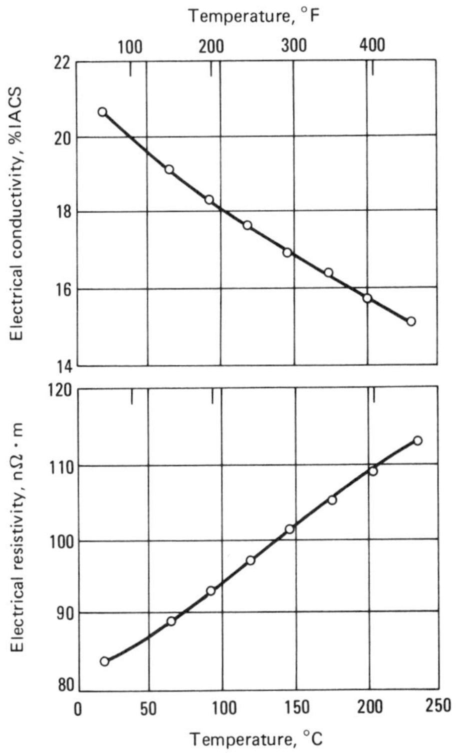

Fig. 26 Variation of electrical properties with temperature for C86500

In reporting chemical analyses obtained by wet methods, zinc may be indicated as balance on those alloys with over 2% Zn.

Applications

Marine fittings and propellers

Mechanical Properties

Tensile properties. Typical data for as-sand-cast separately cast test bars (M01 temper); tensile strength, 565 MPa (82 ksi); yield strength, 260 MPa (38 ksi) at 0.5% extension under load; elongation, 22% in 50 mm (2 in.)
Hardness. Typically 80 HB (3000 kg)
Elastic modulus. Tension, 105 GPa (15 × 10^6 psi) at 20 °C (68 °F)

Mass Characteristics

Density. 8.0 g/cm^3 (0.29 lb/in.^{3}) at 20 °C (68 °F)
Volume change on freezing. Patternmaker's shrinkage, 21 mm/m (¼ in./ft)

Thermal Properties

Liquidus temperature. 900 °C (1652 °F)
Solidus temperature. 880 °C (1616 °F)
Specific heat. 376 J/kg · K (0.09 Btu/lb · °F) at 20 °C (68 °F)

Electrical Properties

Electrical conductivity. Volumetric, 9.0% IACS at 20 °C (68 °F)

Fabrication Characteristics

Machinability. M01 temper, 30% of C36000 (free-cutting brass)
Stress-relieving temperature. 260 °C (500 °F)

C87300 (formerly C87200)

Commercial Names

Trade name. Everdur, Herculor, Navy Tombasil
Common name. Silicon bronze, 95-1-4, 92-4-4, 89-6-5

Specifications

ASTM. Centrifugal, B 271; ingot, B 30; sand, B 585, B 763
SAE. J461, J462
Government. QQ-C-390, WW-V-1967
Military. MIL-C-11866 (composition 19); MIL-C-22229
Other. Ingot code number 500A

Chemical Composition

Composition limits. 94.0 Cu min, 0.20 Pb max, 0.25 Zn max, 0.20 Fe max, 3.5 to 4.5 Si, 0.8 to 1.5 Mn
Cu + sum of named elements. 99.5 min

Applications

Typical uses. As a substitute for tin bronze where good physical and corrosion resistance are required. Bearings, bells, impellers, pump and valve components, marine fittings, statuary and art castings

Mechanical Properties

Tensile properties. Typical data for separately cast test bars: tensile strength, 380 MPa (55 ksi); yield strength, 170 MPa (25 ksi) at 0.5% extension under load; elongation, 30% in 50 mm (2 in.)
Compressive properties. Typical compressive strength, 125 MPa (18 ksi) at permanent set of 0.1%, 415 MPa (60 ksi) at permanent set of 10%
Hardness. 85 HB
Elastic modulus. Tension, 105 GPa (15 × 10^6 psi)
Impact strength. Izod, 45 J (33 ft · lbf)

Mass Characteristics

Density. 8.36 g/cm^3 (0.302 lb/in.^{3}) at 20 °C (68 °F)
Patternmaker's shrinkage. 21 mm/m (¼ in./ft)

Thermal Properties

Liquidus temperature. 916 °C (1680 °F)
Solidus temperature. 821 °C (1510 °F)
Coefficient of linear thermal expansion. Linear, 19.6 μm/m · K (10.9 μin./in. · °F) at 20 to 260 °C (68 to 500 °F)
Thermal conductivity. 28 W/m · K (16 Btu/ft · h · °F) at 20 °C (68 °F)

Electrical Properties

Electrical conductivity. Volumetric, 6.7% IACS at 20 °C (68 °F)

Magnetic Properties

Magnetic permeability. 1.0

Fabrication Characteristics

Machinability. 50% of C36000 (free-cutting brass)
Stress-relieving temperature. 260 °C (500 °F)

C87600

Commercial Names

Common names. Low-zinc silicon brass, CA876

Specifications

ASTM. Ingot, B 30; sand, B 584, B 763
Ingot code number. 500D

Chemical Composition

Composition limits. 88.0 Cu min, 0.50 Pb max, 4.0 to 7.0 Zn, 0.20 Fe max, 3.5 to 5.5 Si, 0.25 Mn max
Cu + sum of named elements. 99.5 min

Applications

Typical uses. Valve stems

Mechanical Properties

Tensile properties. Typical data for as-sand-cast separately cast test bars (M01 temper): tensile strength, 455 MPa (66 ksi); yield strength, 220 MPa (32 ksi) at 0.5% extension under load; elongation, 20% in 50 mm (2 in.)
Compressive strength. Typically, 415 MPa (60 ksi) at 0.1 mm/mm (0.1 in./in.) set
Elastic modulus. Tension, 115 GPa (17 × 10^6 psi) at 20 °C (68 °F)

Mass Characteristics

Density. 8.3 g/cm^3 (0.300 lb/in.^{3}) at 20 °C (68 °F)
Patternmaker's shrinkage. 16 mm/m (3/16 in./ft)

Thermal Properties

Liquidus temperature. 971 °C (1780 °F)
Solidus temperature. 860 °C (1580 °F)
Specific heat. 376 J/kg · K (0.09 Btu/lb · °F) at 20 °C (68 °F)

Electrical Properties

Electrical conductivity. Volumetric, 6.0% IACS at 20 °C (68 °F)

Fabrication Characteristics

Machinability. M01 temper; 40% of C36000 (free-cutting brass)
Stress-relieving temperature. 260 °C (500 °F)

C87610, Silicon Bronze

Specifications

ASTM. Ingot: B 30
Ingot code number. 500E

Chemical Composition

Composition limits. 90.0 Cu min, 0.20 Pb max, 3.0 to 5.0 Zn, 0.20 Fe max, 3.0 to 5.0 Si, 0.25 Mn max
Cu + sum of named elements. 99.5 min

Applications

Typical uses. Bearings, bells, impellers, pump and valve components, marine fittings, corrosion-resistant castings

Mechanical Properties

Tensile properties. Typical data for as-sand-cast separately cast test bars (M01 temper): tensile strength, 380 MPa (55 ksi); yield strength, 170 MPa (25 ksi) at 0.5% extension under load; elongation, 30% in 50 mm (2 in.)
Compressive strength. Typically 125 MPa (18 ksi) at 0.001 mm/mm (0.001 in./in.) set and 415 MPa (60 ksi) at 0.01 mm/mm (0.01 in./in.) set
Shear strength. Typically 193 MPa (28 ksi)
Hardness. Typically 85 HB (500 kg)
Impact strength. Izod: 45 J (33 ft · lbf)
Elastic modulus. Tension, 105 GPa (15 × 10^6 psi) at 20 °C (68 °F)

Mass Characteristics

Density. 8.4 g/cm^3 (0.302 lb/in.^{3}) at 20 °C (68 °F)
Patternmaker's shrinkage. 21 mm/m (¼ in./ft)

Thermal Properties

Specific heat. 376 J/kg · K (0.09 Btu/lb · °F) at 20 °C (68 °F)

Electrical Properties

Electrical conductivity. Volumetric, 6.0% IACS at 20 °C (68 °F)

Fabrication Characteristics

Machinability. M01 temper; 40% of C36000 (free-cutting brass)
Stress-relieving temperature. 260 °C (500 °F)

C87500, C87800
82Cu-4Si-14Zn

Commercial Names

Trade name. Tombasil
Common name. Silicon brass, 82-4-14

Specifications

ASTM. C87500: ingots, B 30; centrifugal castings, B 271; sand castings, B 584. C87800: die castings, B 176
SAE. J462
Government. C 87500: sand castings, QQ-C-390; investment castings, MIL-C-22087 (composition 4). C87800: die castings, MIL-B-15894 (class 3)
Other. Ingot code 500T

Chemical Composition

Composition limits. C87500: 79.0 min Cu, 0.50 Pb max, 12.0 to 16.0 Zn, 0.50 Al max, 3.0 to 5.0 Si. C87800: 80.0 to 83.0 Cu, 0.25 Sn max, 0.15 Pb max, 0.15 Fe max, 0.15 Mn max, 0.15 Al max, 3.75 to 4.25 Si, 0.01 Mg max, 0.25 max others (total), bal Zn, but As, Sb, and S not to exceed 0.05 each, and P not to exceed 0.01

Applications

Typical uses. Bearings, gears, impellers, rocker arms, valve stems, brush holders, bearing races, small boat propellers

Mechanical Properties

Tensile properties. Typical data for separately cast test bars. Sand castings: tensile strength, 460 MPa (67 ksi); yield strength, 205 MPa (30 ksi) at 0.5% extension under load; elongation, 21% in 50 mm (2 in.). Die castings: tensile strength, 585 MPa (85 ksi); yield strength, 310 MPa (45 ksi) at 0.5% extension under load; elongation, 25% in 50 mm (2 in.)
Compressive properties. Compressive strength, 183 MPa (26.5 ksi) at a permanent set of 0.1%; 515 MPa (75 ksi) at a permanent set of 10%
Hardness. Sand cast, 134 HB; die cast, 163 HB
Elastic modulus. Tension: sand cast, 106 GPa (15.4 × 10^6 psi); die cast, 138 GPa (20.0 × 10^6 psi)
Impact strength. Charpy V-notch, 43 J (32 ft · lbf)

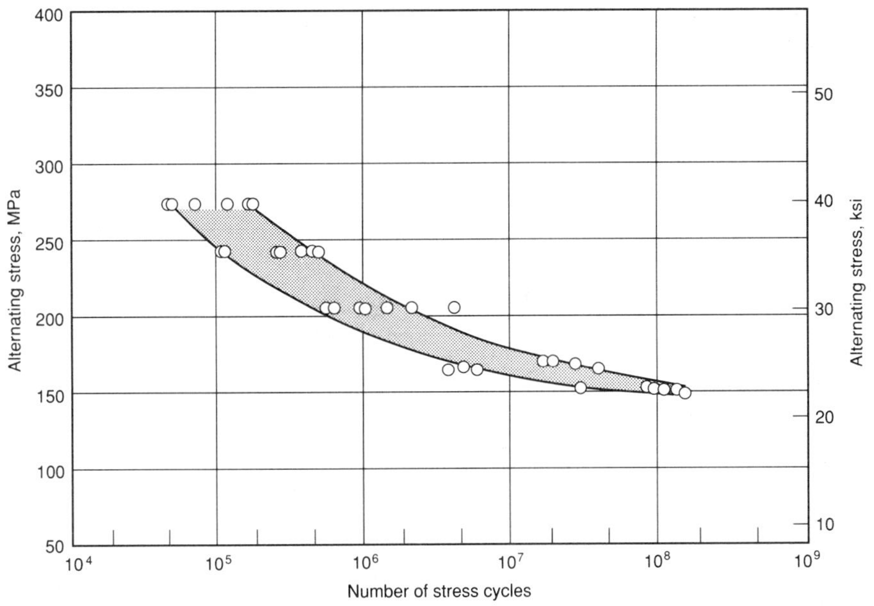

Fig. 27 Fatigue curve for C87500 and C87800

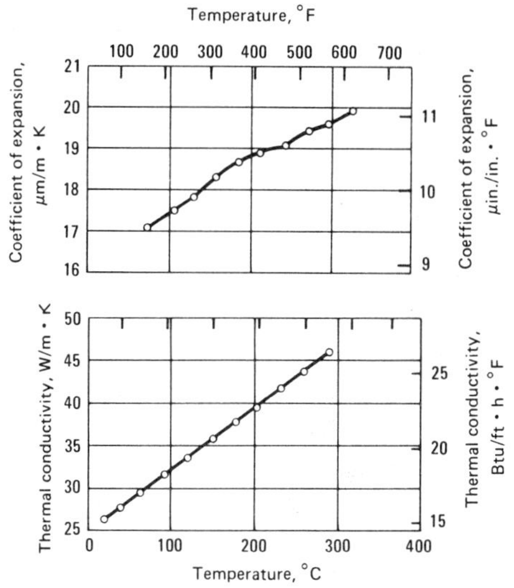

Fig. 28 Selected thermal properties of C87500 and C87800

Fatigue strength. Rotating beam, 150 MPa (22 ksi) at 10^8 cycles. See also Fig. 27.
Creep-rupture characteristics. Limiting creep stress for 10^{-5}%/h: 195 MPa (28 ksi) at 175 °C (350 °F); 75 MPa (11 ksi) at 230 °C (450 °F); 9.5 MPa (1.4 ksi) at 290 °C (550 °F). Stress for rupture in 100 000 h: 125 MPa (18 ksi) at 230 °C (450 °F); 20 MPa (3 ksi) at 290 °C (550 °F)

Mass Characteristics

Density. 8.28 g/cm³ (0.299 lb/in.³) at 20 °C (68 °F)
Patternmaker's shrinkage. 1.5 to 1.9%

Thermal Properties

Liquidus temperature. 917 °C (1683 °F)
Solidus temperature. 821 °C (1510 °F)
Coefficient of linear thermal expansion. 19.6 μm/m · K (10.9 μin./in. · °F) at 20 to 260 °C (68 to 500 °F). See also Fig. 28.
Specific heat. 375 J/kg · K (0.09 Btu/lb · °F) at 20 °C (68 °F)
Thermal conductivity. 28 W/m · K (16 Btu/ft · h · °F) at 20 °C (68 °F). See also Fig. 28.

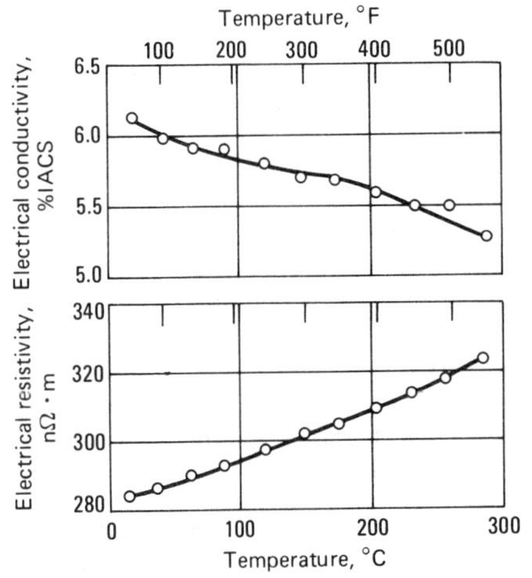

Fig. 29 Electrical conductivity and resistivity of C87500 and C87800

Electrical Properties

Electrical conductivity. Volumetric, 6.7% IACS at 20 °C (68 °F). See also Fig. 29.
Electrical resistivity. 284 nΩ · m at 20 °C (68 °F). See also Fig. 29.

Magnetic Properties

Magnetic permeability. 1.0

Fabrication Characteristics

Machinability. C87500: 50% of C36000 (free-cutting brass). C87800: 40% of C36000
Casting temperature. 980 to 955 °C (1800 to 1750 °F)
Stress-relieving temperature. 260 °C (500 °F)

C87900

Commercial Names

Common names. Silicon yellow brass, CA879

Specifications

ASTM. Ingot: B 30; die: B 176
Government. MIL-B-15894
SAE. J461, J462
Ingot identification number. 500G

Chemical Composition

Composition limits. 63.0 Cu min, 0.25 Sn max, 0.25 Pb max, 30.0 to 36.0 Zn, 0.40 Fe max, 0.15 Al max, 0.8 to 1.2 Si, 0.15 Mn max, 0.50 Ni (including Co) max, 0.05 S max, 0.01 P max, 0.05 As max, 0.05 Sb max. Total named elements shall be 99.5% minimum.
Copper and zinc specifications. In reporting chemical analyses by the use of instruments such as spectrograph, x-ray, and atomic absorption, copper may be indicated as balance. In reporting chemical analyses obtained by wet methods, zinc may be indicated as balance on those alloys with over 2% zinc. In determining Cu min, copper may be calculated as Cu + Ni.

Applications

Typical uses. General-purpose die-casting alloy having moderate strength

Mechanical Properties

Tensile properties. Typical data for as-die-cast test bars (M04 temper): tensile strength, 485 MPa (70 ksi); yield strength, 240 MPa (35 ksi) at 0.2% offset; elongation, 25% in 50 mm (2 in.)
Hardness. Typically 70 HRB
Impact strength. Charpy unnotched: 68 J (50 ft · lbf)
Elastic modulus. Tension, 105 GPa (15 × 10^6 psi) at 20 °C (68 °F)

Mass Characteristics

Density. 8.5 g/cm³ (0.308 lb/in.³) at 20 °C (68 °F)
Patternmaker's shrinkage. 15.6 mm/m (3/16 in./ft)

Thermal Properties

Liquidus temperature. 926 °C (1700 °F)
Solidus temperature. 900 °C (1650 °F)
Specific heat. 376 J/kg · K (0.09 Btu/lb · °F) at 20 °C (68 °F)

Electrical Properties

Electrical conductivity. Volumetric, 15% IACS at 20 °C (68 °F)

Fabrication Characteristics

Machinability. M04 temper; 80% of C36000 (free-cutting brass)
Stress-relieving temperature. 260 °C (500 °F)

C90300
88Cu-8Sn-4Zn

Commercial Names

Common name. Tin bronze; 88-8-0-4; "G"-bronze

Specifications

ASTM. Sand castings: B 584; centrifugal castings: B 271; continuous castings: B 505; ingot: B 30
SAE. J462
Government. QQ-C-390, QQ-C-525. Sand castings: MIL-C-22229, composition 1; centrifugal castings: MIL-C-15345, alloy 8; investment castings: MIL-C-22087, composition 3; precision castings: MIL-C-11866, composition 26
Ingot identification number. 225

Chemical Composition

Composition limits. 86.0 to 89.0 Cu, 7.5 to 9.0 Sn, 3.0 to 5.0 Zn, 1.0 Ni max, 0.30 Pb max, 0.15 Fe max, 0.05 P max. (For continuous castings, 1.5 P max), 0.2 Sb max, 0.05 S max, 0.005 Si max, 0.005 Al max

Applications

Typical uses. Bearings, bushings, pump impellers, piston rings, valve components, seal rings, steam fittings, gears

Mechanical Properties

Tensile properties. Typical tensile strength, 310 MPa (45 ksi); yield strength, 145 MPa (21 ksi); elongation, 30% in 50 mm (2 in.)
Compressive properties. Compressive strength, 90 MPa (13 ksi)
Hardness. 70 HB
Elastic modulus. Tension, 97 GPa (14 × 10^6 psi)
Impact strength. Charpy V-notch, 19 J (14 ft · lbf)

Mass Characteristics

Density. 8.80 g/cm^3 (0.318 lb/in.^{3}) at 20 °C (68 °F)
Volume change on freezing. 1.6%

Thermal Properties

Liquidus temperature. 1000 °C (1830 °F)
Solidus temperature. 854 °C (1570 °F)
Coefficient of linear thermal expansion. 18 μm/m · K (10 μin./in. · °F) at 20 to 177 °C (68 to 340 °F)
Specific heat. 376 J/kg · K (0.09 Btu/lb · °F) at 20 °C (68 °F)
Thermal conductivity. 74 W/m · K (43 Btu/ft · h · °F)

Electrical Properties

Electrical conductivity. Volumetric, 12% IACS at 20 °C (68 °F)

Magnetic Properties

Magnetic permeability. 1.0

Fabrication Characteristics

Machinability. 30% of C36000 (free-cutting brass)

C90500
88Cu-10Sn-2Zn

Commercial Names

Common name. Tin bronze; Gun metal; 88-10-0-2

Specifications

AMS. 4845
ASTM. Sand castings: B 22, B 584; centrifugal castings: B 271; continuous castings: B 505; ingot: B 30
SAE. J462
Government. QQ-C-390
Ingot identification number. 210

Chemical Composition

Composition limits. 86.0 to 89.0 Cu, 9.0 to 11.0 Sn, 1.0 to 3.0 Zn, 1.0 Ni max, 0.3 Pb max, 0.15 Fe max, 0.05 P max. (For continuous castings, 1.5 max P), 0.2 Sb max, 0.05 S max, 0.005 Si max, 0.005 Al max

Applications

Typical uses. Bearings, bushings, pump impellers, piston rings, pump bodies, valve components, steam fittings, gears

Mechanical Properties

Tensile properties. Typical tensile strength, 310 MPa (45 ksi); yield strength, 150 MPa (22 ksi); elongation, 25% in 50 mm (2 in.); reduction in area, 40%
Compressive properties. Compressive strength, 275 MPa (40 ksi)
Elastic modulus. Tension, 105 GPa (15 × 10^6 psi)
Fatigue strength. Rotating beam, 90 MPa (13 ksi) at 10^8 cycles
Impact strength. Izod, 14 J (10 ft · lbf)

Mass Characteristics

Density. 8.72 g/cm^3 (0.315 lb/in.^{3}) at 20 °C (68 °F)
Volume change on freezing. 1.6%

Thermal Properties

Liquidus temperature. 1000 °C (1830 °F)
Solidus temperature. 854 °C (1570 °F)
Coefficient of linear thermal expansion. 20 μm/m · K (11μin./in. · °F) at 20 to 300 °C (68 to 572 °F)
Specific heat. 376 J/kg · K (0.09 Btu/lb · °F) at 20 °C (68 °F)
Thermal conductivity. 74 W/m · K (43 Btu/ft · h · °F)

Electrical Properties

Electrical conductivity. Volumetric, 11% IACS at 20 °C (68 °F)

Magnetic Properties

Magnetic permeability. 1.0

Fabrication Characteristics

Machinability. 30% of C36000 (free-cutting brass)

C90700
89Cu-11Sn

Commercial Names

Common name. Tin bronze, 65; Phosphor gear bronze

Specifications

ASTM. Continuous castings: B 505; ingot: B 30
Ingot identification number. 205

Chemical Composition

Composition limits. 88.0 to 90.0 Cu, 10.0 to 12.0 Sn, 0.15 Fe max, 0.1 to 0.3 P, 0.005 Al max, 0.30 Pb max, 0.50 Zn max, Pb + Zn + Ni, 1.0 max
Consequence of exceeding impurity limits. Ductility decreases rapidly with tin contents over 12%, with 13% a practical limit for gear applications.

Applications

Typical uses. Worm wheels and gears; bearings expected to carry heavy loads at relatively low speeds

Mechanical Properties

Tensile properties. Typical. Sand castings: tensile strength, 305 MPa (44 ksi); yield strength, 150 MPa (22 ksi); elongation, 20% in 50 mm (2 in.). Permanent mold castings: tensile strength, 380 MPa (55 ksi); yield strength, 205 MPa (30 ksi); elongation, 16% in 50 mm or 2 in.
Hardness. Sand castings, 80 HB; permanent mold castings, 102 HB
Elastic modulus. Tension, 105 GPa (15 × 10^6 psi)
Fatigue strength. Rotating beam, 170 MPa (25 ksi) at 10^8 cycles

Mass Characteristics

Density. 8.77 g/cm^3 (0.317 lb/in.^{3}) at 20 °C (68 °F)
Volume change on freezing. 1.6%

Thermal Properties

Liquidus temperature. 1000 °C (1830 °F)
Solidus temperature. 832 °C (1530 °F)
Coefficient of linear thermal expansion. 18 μm/m · K (10 μin./in. · °F) at 20 to 200 °C (68 to 392 °F)
Specific heat. 376 J/kg · K (0.09 Btu/lb · °F) at 20 °C (68 °F)
Thermal conductivity. 71 W/m · K (41 Btu/ft · h · °F)

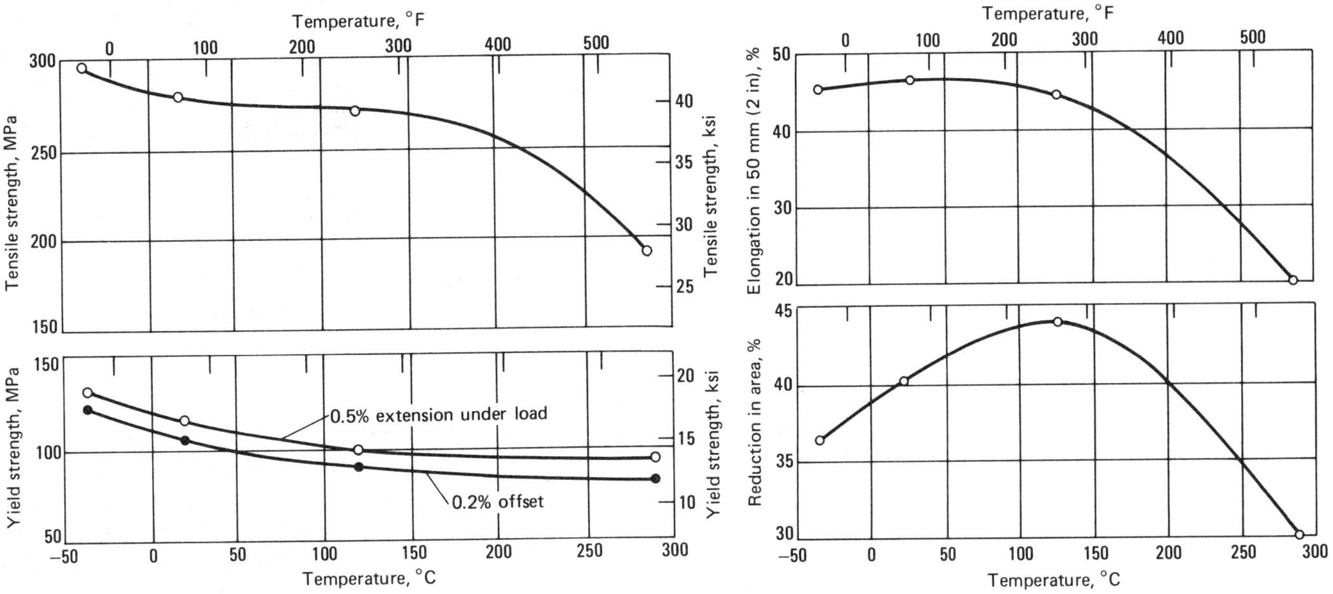

Fig. 30 Tensile properties of C92200

Electrical Properties

Electrical conductivity. Volumetric, 9.6% IACS at 20 °C (68 °F)
Electrical resistivity. 15 nΩ · m at 20 °C (68 °F)

Magnetic Properties

Magnetic permeability. 1.0

Fabrication Characteristics

Machinability. 20% of C36000 (free-cutting brass)

C91700
86½Cu-12Sn-1½Ni

Commercial Names

Common name. Nickel gear bronze, 86½-12-0-0-1½

Specifications

ASTM. Ingot: B 30; sand castings: B 427
Other. Ingot code number 205

Chemical Composition

Composition limits. 85.0 to 87.5 Cu, 11.3 to 12.5 Sn, 0.25 Pb max, 1.3 to 2.0 Ni, 0.30 P max

Applications

Typical uses. Worm wheels and gears, bearings with heavy loads and relatively low speeds

Mechanical Properties

Tensile properties. Typical data for sand-cast test bars: tensile strength, 305 MPa (44 ksi); yield strength, 150 MPa (22 ksi) at 0.5% extension under load; elongation, 16% in 50 mm (2 in.). Typical data for centrifugal or permanent mold test bars: tensile strength, 415 MPa (60 ksi); yield strength, 220 MPa (32 ksi) at 0.5% extension under load; elongation, 16% in 50 mm (2 in.)
Hardness. Sand cast, 85 HB; centrifugal or permanent mold cast, 106 HB
Elastic modulus. Tension, 105 GPa (15 × 10^6 psi)

Mass Characteristics

Density. 8.75 g/cm^3 (0.316 lb/in.^{3}) at 20 °C (68 °F)
Patternmaker's shrinkage. 16 mm/m (3/16 in./ft)

Thermal Properties

Liquidus temperature. 1015 °C (1860 °F)
Solidus temperature. 850 °C (1565 °F)
Coefficient of linear thermal expansion. 16.2 μm/m · K (9.0 μin./in. · °F) at 20 to 200 °C (68 to 392 °F)
Specific heat. 376 J/kg · K (0.09 Btu/lb · °F) at 20 °C (68 °F)
Thermal conductivity. 71 W/m · K (41 Btu/ft · h · °F) at 20 °C (68 °F)

Electrical Properties

Electrical conductivity. Volumetric, 10% IACS at 20 °C (68 °F)

Magnetic Properties

Magnetic permeability. 1.0

Fabrication Characteristics

Machinability. 20% of C36000 (free-cutting brass)

C92200
88Cu-6Sn-1½Pb-4½Zn

Commercial Names

Common name. Navy "M" bronze, steam bronze, 88-6-1½-4½

Specifications

ASTM. B 584, B 61, B 271, B 505, B 30
SAE. J462 (C92200)
Government. CA922, QQ-B-225 (Alloy number 1), MIL-B-16541, MIL-B-15345
Other. Ingot code number 245

Chemical Composition

Composition limits. 86.0 to 90.0 Cu, 5.5 to 6.5 Sn, 1.0 to 2.0 Pb, 3.0 to 5.0 Zn, 1.0 Ni max, 0.25 Fe max, 0.05 P max. (1.5 P max for continuous castings), 0.05 S max, 0.005 Si max, 0.25 Sb max

Applications

Typical uses. Component castings of valves, flanges and fittings, oil pumps, gears, bushings, bearings, backing for babbitt-lined bearings, pressure-containing parts at temperatures up to 290 °C (550 °F), and stresses up to 20 MPa (3 ksi)

Mechanical Properties

Tensile properties. Typical data for sand-cast test bars: tensile strength, 275 MPa (40 ksi); yield strength, 140 MPa (20 ksi) at 0.5% extension under load; elongation, 30% in 50 mm (2 in.). See also Fig. 30.
Compressive properties. Compressive strength, 105 MPa (15 ksi) at permanent set of 10%; 260 MPa (38 ksi) at permanent set of 0.1%. See also Fig. 31.
Hardness. 65 HB (500 kg load). See also Fig. 32.
Elastic modulus. Tension, 97 GPa (14 × 10^6 psi). See also Fig. 33.
Fatigue strength. Rotating beam, 76 MPa (11 ksi) at 10^8 cycles. See also Fig. 34.
Creep-rupture characteristics. Limiting creep stress for 10^{-5}%/h: 110 MPa (16.0 ksi) at 177 °C (350 °F); 77.2 MPa (11.2 ksi) at 232

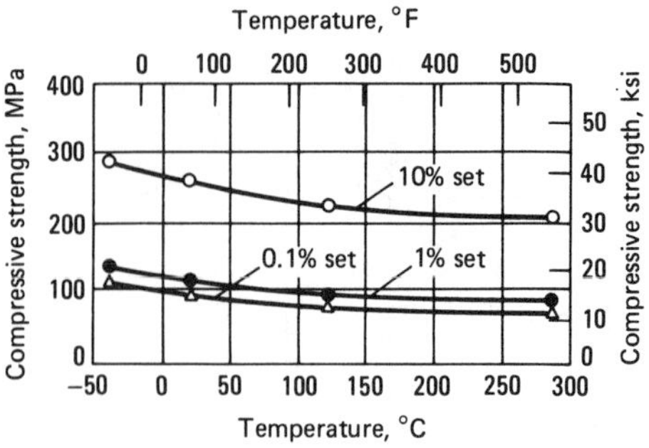

Fig. 31 Compressive strength of C92200

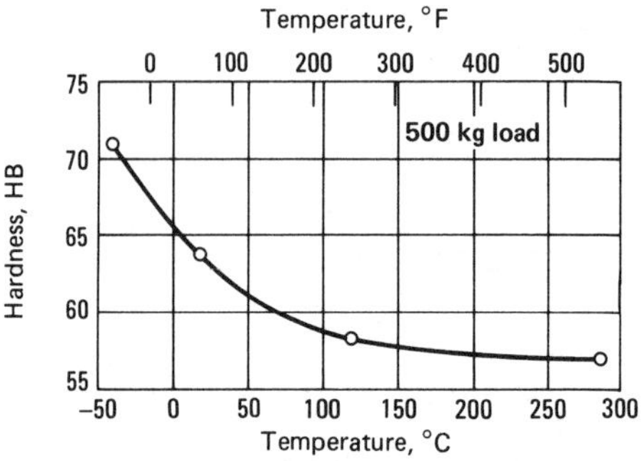

Fig. 32 Brinell hardness of C92200

°C (450 °F); 43 MPa (6.2 ksi) at 288 °C (550 °F). See also Fig. 35.

Mass Characteristics

Density. 8.64 g/cm^3 (0.312 lb/in.^{3}) at 20 °C (68 °F)
Patternmaker's shrinkage. 16 mm/m (3/16 in./ft)

Thermal Properties

Liquidus temperature. 990 °C (1810 °F)
Solidus temperature. 825 °C (1520 °F)
Incipient melting temperature. Pb, 315 °C (600 °F)
Coefficient of linear thermal expansion. See Fig. 36.
Specific heat. 376 J/kg · K (0.09 Btu/lb · °F) at 20 °C (68 °F)
Thermal conductivity. 70 W/m · K (40 Btu/ft · h · °F) at 20 °C (68 °F). See also Fig. 36.

Electrical Properties

Electrical conductivity. Volumetric, 14.3% IACS at 20 °C (68 °F)
Electrical resistivity. 120 nΩ · m at 20 °C (68 °F)

Magnetic Properties

Magnetic permeability. 1.0

Fabrication Characteristics

Machinability. 42% of C36000 (free-cutting brass)
Weldability. Soldering: excellent. Brazing: excellent, but strain must be avoided during brazing and subsequent cooling because brazing is performed at temperatures within the hot short range. Oxyfuel gas welding and all forms of arc welding are not recommended.
Stress-relieving temperature. 260 °C (500 °F)

C92300
87Cu-8Sn-1Pb-4Zn

Commercial Names

Common names. Leaded tin bronze, leaded Navy "G"-bronze, 87-8-1-4

Specifications

ASTM. Sand castings: B 584; centrifugal castings: B 271; continuous castings: B 505; ingot: B 30
SAE. J462
Government. QQ-C-390. Centrifugal castings: MIL-C-15345 (Alloy 10)
Other. Ingot code number 230

Chemical Composition

Composition limits. 85.0 to 89.0 Cu, 7.0 to 9.0 Sn, 1.0 Pb max, 2.5 to 5.0 Zn, 1.0 Ni max, 0.25 Fe max, 0.05 P max (1.5 P max for continuous castings), 0.25 Sb max, 0.05 S max, 0.005 Si max, 0.005 Al max

Applications

Typical uses. Strong general-utility structural bronze for use under severe conditions; valves, expansion joints, special high-pressure pipe fittings, steam pressure castings

Mechanical Properties

Tensile properties. Typical data for sand-cast test bars: tensile strength, 275 MPa (40 ksi); yield strength, 140 MPa (20 ksi) at 0.5% extension under load; elongation, 25% in 50 mm (2 in.)
Compressive properties. Compressive strength, 69 MPa (10 ksi) at permanent set of 0.1%; 240 MPa (35 ksi) at permanent set of 10%
Hardness. 70 HB
Elastic modulus. Tension, 97 GPa (14 × 10^6 psi)
Impact strength. Izod, 18.3 J (13.5 ft · lbf)

Mass Characteristics

Density. 8.8 g/cm^3 (0.317 lb/in.^{3}) at 20 °C (68 °F)
Patternmaker's shrinkage. 16 mm/m (3/16 in./ft)

Thermal Properties

Liquidus temperature. 1000 °C (1830 °F)
Solidus temperature. 855 °C (1570 °F)
Incipient melting temperature. Pb, 315 °C (600 °F)
Coefficient of linear thermal expansion. 18 μm/m · K (10 μin./in. · °F) at 20 to 177 °C (68 to 350 °F)
Specific heat. 376 J/kg · K (0.09 Btu/lb · °F) at 20 °C (68 °F)
Thermal conductivity. 75 W/m · K (43 Btu/ft · h · °F) at 20 °C (68 °F)

Electrical Properties

Electrical conductivity. Volumetric, 12% IACS at 20 °C (68 °F)

Fabrication Characteristics

Machinability. 42% of C36000 (free-cutting brass)
Weldability. Soldering: excellent. Brazing: good, but strain must be avoided during brazing and subsequent cooling because brazing is performed at temperatures within the hot short range. Oxyfuel gas welding and all forms of arc welding are not recommended.
Stress-relieving temperature. 260 °C (500 °F)

C92500
87Cu-11Sn-1Pb-1Ni

Commercial Names

Common name. Leaded tin bronze, 640; 87-11-1-0-1

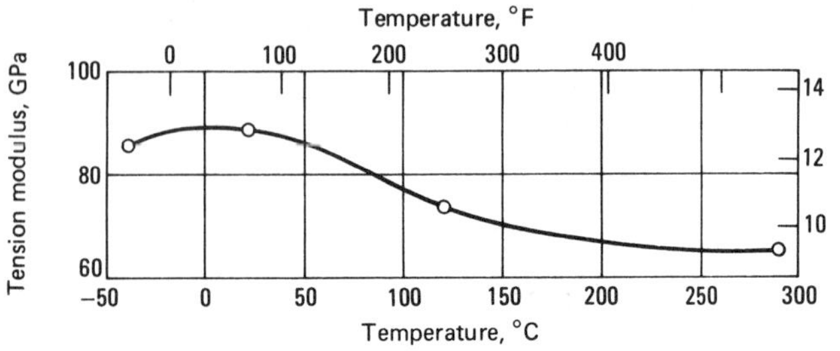

Fig. 33 Elastic modulus in tension for C92200

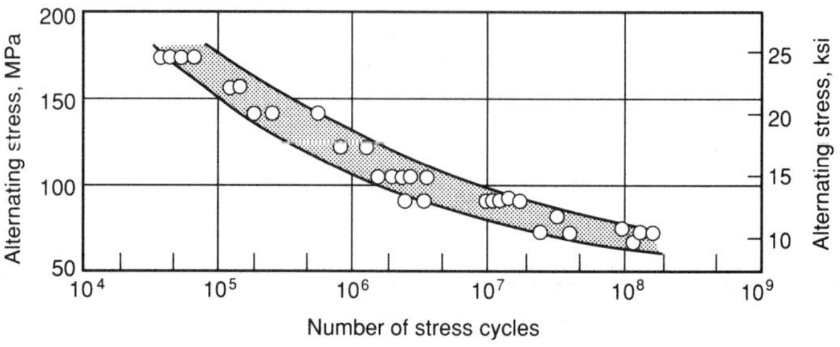

Fig. 34 Fatigue strength of C92200

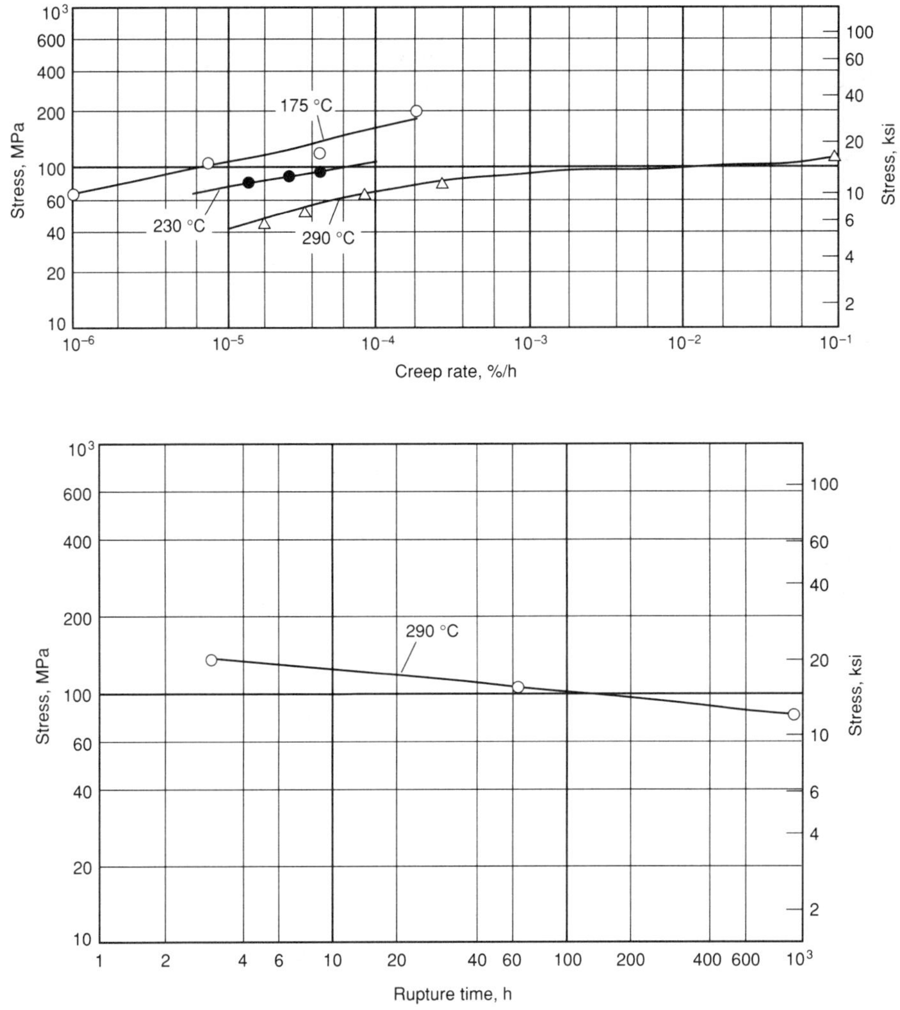

Fig. 35 Creep-rupture properties of C92200

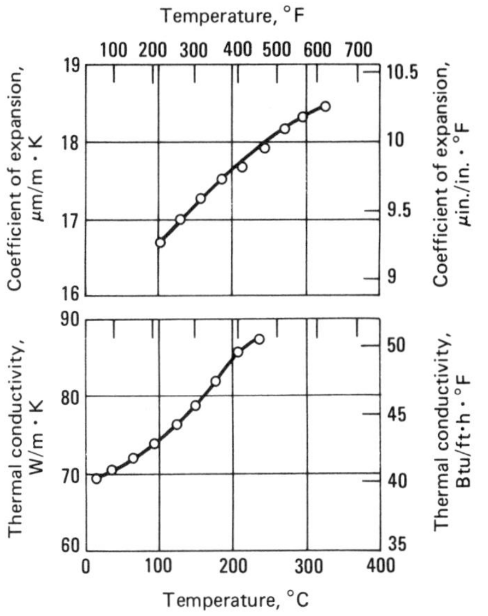

Fig. 36 Thermal properties of C92200

Specifications

ASTM. Continuous castings: B 505; ingot: B 30
SAE. J462
Other. Ingot code number 250

Chemical Composition

Composition limits. 85.0 to 88.0 Cu, 10.0 to 12.0 Sn, 1.0 to 1.5 Pb, 0.5 Zn max, 0.8 to 1.5 Ni, 0.15 Fe max, 0.20 to 0.30 P, 0.005 Al max

Applications

Typical uses. Gears, automotive synchronizer rings

Mechanical Properties

Tensile properties. Typical data for sand-cast test bars: tensile strength, 305 MPa (44 ksi); yield strength, 140 MPa (20 ksi) at 0.5% extension under load; elongation, 20% in 50 mm (2 in.)
Hardness. 80 HB
Elastic modulus. Tension, 110 GPa (16 × 10^6 psi)

Mass Characteristics

Patternmaker's shrinkage. 16 mm/m (3/16 in./ft)

Thermal Properties

Incipient melting temperature. Pb, 315 °C (600 °F)
Specific heat. 376 J/kg · K (0.09 Btu/lb · °F) at 20 °C (68 °F)

Fabrication Characteristics

Machinability. 30% of C36000 (free-cutting brass)
Weldability. Soldering: excellent. Brazing: good, but strain must be avoided during brazing and subsequent cooling because brazing is performed at temperatures within the hot short range. Oxyfuel gas welding and all forms of arc welding are not recommended.
Stress-relieving temperature. 260 °C (500 °F)

C92600
87Cu-10Sn-1Pb-2Zn

Commercial Names

Common name. Leaded tin bronze

Specifications

Ingot code number. 215

Chemical Composition

Composition limits. 86.0 to 88.5 Cu, 9.3 to 10.5 Sn, 0.8 to 1.2 Pb, 1.3 to 2.5 Zn, 0.75 Ni max, 0.15 Fe max, 0.25 Sb max, 0.05 S max, 0.005 Si max, 0.03 P max, 0.005 Al max

Applications

Typical uses. Commercial bronze for high-duty bearings where wear resistance is essential; strong general-utility structural bronze for use under severe conditions; bolts, nuts, gears; heavy-pressure bearings and bushings to use against hardened steel; valves, expansion joints, special high-pressure pipe fittings; pump pistons; elevator components; steam pressure castings

Mechanical Properties

Tensile properties. Typical data for sand-cast test bars: tensile strength, 304 MPa (44 ksi); yield strength, 140 MPa (20 ksi) at 0.5% extension under load; elongation, 30% in 50 mm (2 in.)
Compressive properties. Compressive strength, 85 MPa (12 ksi) at permanent set of 0.1%; 275 MPa (40 ksi) at permanent set of 10%
Hardness. 78 HRF, 72 HB
Elastic modulus. Tension, 105 GPa (15 × 10^6 psi)
Impact strength. Izod, 9 J (7 ft · lbf)

Mass Characteristics

Density. 8.70 g/cm^3 (0.315 lb/in.^{3}) at 20 °C (68 °F)
Patternmaker's shrinkage. 16 mm/m (3/16 in./ft)

Thermal Properties

Liquidus temperature. 980 °C (1800 °F)

Solidus temperature. 845 °C (1550 °F)
Incipient melting temperature. Pb, 315 °C (600 °F)
Specific heat. 376 J/kg · K (0.09 Btu/lb · °F) at 20 °C (68 °F)

Electrical Properties

Electrical conductivity. Volumetric, 9% IACS at 20 °C (68 °F)

Fabrication Characteristics

Machinability. 40% of C36000 (free-cutting brass)
Weldability. Soldering: excellent. Brazing: good, but strain must be avoided during brazing and subsequent cooling because brazing is done at temperatures within the hot short range. Oxyfuel gas welding and all forms of arc welding are not recommended.
Stress-relieving temperature. 260 °C (500 °F)

C92700
88Cu-10Sn-2Pb

Commercial Names

Common name. Leaded tin bronze, 88-10-2-0

Specifications

ASTM. Continuous castings: B 505; ingot: B 30
SAE. J462
Other. Ingot code number 206

Chemical Composition

Composition limits. 86.0 to 89.0 Cu, 9.0 to 11.0 Sn, 1.0 to 2.5 Pb, 0.7 Zn max, 1.0 Ni max, 0.15 Fe max, 0.25 P max, 0.005 Al max

Applications

Typical uses. Bearings, bushings, pump impellers, piston rings, valve components, steam fittings, gears

Mechanical Properties

Tensile properties. Typical data for sand-cast test bars: tensile strength, 290 MPa (42 ksi); yield strength, 145 MPa (21 ksi) at 0.5% extension under load; elongation, 20% in 50 mm (2 in.)
Hardness. 77 HB
Elastic modulus. Tension, 110 GPa (16×10^6 psi)

Mass Characteristics

Density. 8.8 g/cm^3 (0.317 lb/in.^{3}) at 20 °C (68 °F)
Patternmaker's shrinkage. 16 mm/m (3/16 in./ft)

Thermal Properties

Liquidus temperature. 980 °C (1800 °F)
Solidus temperature. 845 °C (1550 °F)
Incipient melting temperature. Pb, 315 °C (600 °F)
Coefficient of linear thermal expansion. 18 μm/m · K (10 μin./in. · °F) at 20 to 177 °C (68 to 350 °F)
Specific heat. 376 J/kg · K (0.09 Btu/lb · °F) at 20 °C (68 °F)

Electrical Properties

Electrical conductivity. Volumetric, 11% IACS at 20 °C (68 °F)

Fabrication Characteristics

Machinability. 45% of C36000 (free-cutting brass)
Weldability. Soldering: excellent. Brazing: good, but parts must not be strained during brazing or subsequent cooling because brazing is done at temperatures within the hot short range. Oxyfuel gas welding and all forms of arc welding are not recommended.
Stress-relieving temperature. 260 °C (500 °F)

C92900
84Cu-10Sn-2½Pb-3½Ni

Commercial Names

Common name. Leaded nickel-tin bronze, 84-10-2½-0-3½

Specifications

ASTM. Sand and centrifugal castings: B 427; continuous castings: B 505; ingot: B 30
SAE. J462

Chemical Composition

Composition limits. 81.0 to 85.5 Cu, 9.0 to 11.0 Sn, 2.0 to 3.2 Pb, 2.8 to 4.0 Ni, 0.50 P max, 0.50 max other (total)

Applications

Typical uses. Gears, wear plates and guides, cams

Mechanical Properties

Tensile properties. Typical data for sand-cast test bars: tensile strength, 325 MPa (47 ksi); yield strength, 180 MPa (26 ksi) at 0.5% extension under load; elongation, 20% in 50 mm (2 in.)
Hardness. 80 HB
Elastic modulus. Tension, 97 GPa (14×10^6 psi)
Impact strength. Izod, 16 J (12 ft · lbf)

Mass Characteristics

Density. 8.79 g/cm^3 (0.318 lb/in.^{3}) at 20 °C (68 °F)
Patternmaker's shrinkage. 16 mm/m (3/16 in./ft)

Thermal Properties

Liquidus temperature. 1030 °C (1887 °F)
Solidus temperature. 860 °C (1575 °F)
Incipient melting temperature. Pb, 315 °C (600 °F)
Coefficient of linear thermal expansion. 17 μm/m · K (9.5 μin./in. · °F) at 20 to 200 °C (68 to 392 °F)
Specific heat. 376 J/kg · K (0.09 Btu/lb · °F) at 20 °C (68 °F)
Thermal conductivity. 58.2 W/m · K (33.6 Btu/ft · h · °F) at 20 °C (68 °F)

Electrical Properties

Electrical conductivity. Volumetric, 9.2% IACS at 20 °C (68 °F)

Fabrication Characteristics

Machinability. 40% of C36000 (free-cutting brass)
Weldability. Soldering: excellent. Brazing: good, but parts must not be strained during brazing or subsequent cooling because brazing is done at temperatures within the hot short range. Oxyfuel gas welding and all forms of arc welding are not recommended.
Stress-relieving temperature. 260 °C (500 °F)

C93200
83Cu-7Sn-7Pb-3Zn

Commercial Names

Common name. High-leaded tin bronze; bearing bronze 660; 83-7-7-3

Specifications

ASTM. Sand castings: B 584; centrifugal castings, B 271; continuous castings: B 505; ingot: B 30
SAE. J462
Government. QQ-C-390; QQ-C-525; QQ-L-225 (Alloy 12); MIL-C-15345 (Alloy 17); MIL-C-11553 (Alloy 12); MIL-B-16261 (Alloy VI)
Other. Ingot code number 315

Chemical Composition

Composition limits. 81.0 to 85.0 Cu, 6.3 to 7.5 Sn, 6.0 to 8.0 Pb, 2.0 to 4.0 Zn, 0.50 N max, 0.20 Fe max, 0.15 P max, 0.35 Sb max, 0.08 S max, 0.003 Si max. In determining Cu, minimum may be calculated as Cu + Ni.
Other phosphorus specifications. 1.5 P max for continuous castings; 0.50 P max for permanent mold castings

Applications

Typical uses. General-utility bearings and bushings, automobile fittings

Mechanical Properties

Tensile properties. Typical data for sand-cast test bars: tensile strength, 240 MPa (35 ksi); yield strength, 125 MPa (18 ksi) at 0.5% extension under load; elongation, 20% in 50 mm (2 in.)
Compressive properties. Compressive strength, 315 MPa (46 ksi) at permanent set of 10%
Hardness. 65 HB
Elastic modulus. Tension, 100 GPa (14.5×10^6 psi)
Impact strength. Izod, 8 J (6 ft · lbf)
Fatigue strength. Reverse bending, 110 MPa (16 ksi) at 10^8 cycles

Mass Characteristics

Density. 8.93 g/cm^3 (0.322 lb/in.^{3}) at 20 °C (68 °F)
Patternmaker's shrinkage. 18 mm/m (7/32 in./ft)

Thermal Properties

Liquidus temperature. 975 °C (1790 °F)
Solidus temperature. 855 °C (1570 °F)
Incipient melting temperature. Pb, 315 °C (600 °F)
Coefficient of linear thermal expansion. 18 μm/m · K (10 μin./in. · °F) at 0 to 100 °C (32 to 212 °F)
Specific heat. 376 J/kg · K (0.09 Btu/lb · °F) at 20 °C (68 °F)
Thermal conductivity. 59 W/m · K (34 Btu/ft · h · °F) at 20 °C (68 °F)

Electrical Properties

Electrical conductivity. Volumetric, 12% IACS at 20 °C (68 °F)

Fabrication Characteristics

Machinability. 70% of C36000 (free-cutting brass)
Weldability. Soldering: excellent. Brazing: good, but parts must not be strained during brazing or subsequent cooling because brazing is done at temperatures within the hot short range. Oxyfuel gas welding and all forms of arc welding are not recommended.
Stress-relieving temperature. 260 °C (500 °F)

C93400

Commercial Names

Common name. High-leaded tin bronze, CA934, 84-8-8-0

Specifications

ASTM. Continuous, B 505; ingot, B 30
Government. QQ-C-390; MIL-C-22087; MIL-C-22229
Ingot identification number. 310

Chemical Composition

Composition limits. 82.0 to 85.0 Cu, 7.0 to 9.0 Sn, 7.0 to 9.0 Pb, 0.8 Zn max, 0.20 Fe max, 0.50 Sb max, Ni (including Co) 1.0 max, 0.08 S max, 0.50 P max (for continuous castings, phosphorus shall be 1.5% maximum), 0.005 Al max, 0.005 Si max. Ingot for remelting specifications vary from the ranges given.
Copper and zinc specifications. In reporting chemical analyses by the use of instruments such as spectrograph, x-ray, and atomic absorption, copper may be indicated as balance. In reporting chemical analyses obtained by wet methods, zinc may be indicated as balance on those alloys with over 2% zinc.

Applications

Typical uses. Bearings and bushings

Mechanical Properties

Tensile properties. Typical data for as-sand-cast separately cast test bars (M01 temper): tensile strength, 220 MPa (32 ksi); yield strength, 110 MPa (16 ksi) at 0.5% extension under load; elongation, 20% in 50 mm (2 in.)
Hardness. Typically 60 HB (500 kg)
Compressive strength. 330 MPa (48 ksi) at 0.1 mm/mm (0.1 in./in.) set
Impact strength. Izod, 6.8 J (5 ft · lbf)
Proportional limit. 55 MPa (8 ksi)
Fatigue strength. 100 MPa (15 ksi) at 10^8 cycles
Elastic modulus. Tension, 76 GPa (11 × 10^6 psi) at 20 °C (68 °F)

Mass Characteristics

Density. 8.87 g/cm^3 (0.320 lb/in.^{3}) at 20 °C (68 °F)
Volume change on freezing. Patternmaker's shrinkage, 16 mm/m (3/16 in./ft)

Thermal Properties

Specific heat. 376 J/kg · K (0.09 Btu/lb · °F) at 20 °C (68 °F)

Electrical Properties

Electrical conductivity. Volumetric, 12% IACS at 20 °C (68 °F)

Fabrication Characteristics

Machinability. M01 temper; 70% of C36000 (free-cutting brass)
Stress-relieving temperature. 260 °C (500 °F)

C93500
85Cu-5Sn-9Pb-1Zn

Commercial Names

Common name. High-leaded tin bronze, 85-5-9-1

Specifications

ASTM. Sand castings: B 584; centrifugal castings, B 271; continuous castings, B 505; ingot, B 30
SAE. J462
Government. QQ-C-390; QQ-L-225 (Alloy 14); MIL-B-11553B (Alloy 14)
Other. Ingot code number 326

Chemical Composition

Composition limits. 83.0 to 86.0 Cu, 4.5 to 6.0 Sn, 8.0 to 10.0 Pb, 2.0 Zn max, 0.50 Ni max, 0.20 Fe max, 0.02 P max (1.5 P max for continuous castings), 0.30 Sb max, 0.08 S max, 0.003 Si max. In determining Cu, minimum may be calculated as Cu + Ni.

Applications

Typical uses. Small bearings and bushings, bronze backings for babbitt-lined automotive bearings

Mechanical Properties

Tensile properties. Typical data for sand-cast test bars: tensile strength, 220 MPa (32 ksi); yield strength, 110 MPa (16 ksi) at 0.5% extension under load; elongation, 20% in 50 mm (2 in.)
Compressive properties. Compressive strength, 90 MPa (13 ksi) at permanent set of 0.1%
Hardness. 60 HB
Elastic modulus. Tension, 100 GPa (14.5 × 10^6 psi)
Impact strength. Charpy V-notch or Izod, 11 J (8 ft · lbf)

Mass Characteristics

Density. 8.87 g/cm^3 (0.320 lb/in.^{3}) at 20 °C (68 °F)
Patternmaker's shrinkage. 16 mm/m (3/16 in./ft)

Thermal Properties

Liquidus temperature. 1000 °C (1830 °F)
Solidus temperature. 855 °C (1570 °F)
Incipient melting temperature. Pb, 315 °C (600 °F)
Coefficient of linear thermal expansion. 18 μm/m · K (10 μin./in. · °F) at 20 to 200 °C (68 to 392 °F)
Specific heat. 376 J/kg · K (0.09 Btu/lb · °F) at 20 °C (68 °F)
Thermal conductivity. 71 W/m · K (41 Btu/ft · h · °F) at 20 °C (68 °F)

Electrical Properties

Electrical conductivity. Volumetric, 15% IACS at 20 °C (68 °F)

Magnetic Properties

Magnetic permeability. 1.0

Fabrication Characteristics

Machinability. 70% of C36000 (free-cutting brass)
Weldability. Soldering: good. Brazing: good, but parts must not be strained during brazing or subsequent cooling because brazing is done at temperatures within the hot short range. Oxyfuel gas welding and all forms of arc welding are not recommended.
Stress-relieving temperature. 260 °C (500 °F)

C93700
80Cu-10Sn-10Pb

Commercial Names

CDA and UNS number. C93700
Common names. High-leaded tin bronze; bushing and bearing bronze; 80-10-10

Specifications

AMS. Sand and centrifugal castings: 4842
ASTM. Sand castings: B 22, B 584; centrifugal castings: B 271; continuous castings: B 505; ingot: B 30
SAE. J462
Government. QQ-C-390; MIL-B-13506 (Alloy A2)
Other. Ingot code number 305

Chemical Composition

Composition limits. 78.0 to 82.0 Cu, 9.0 to 11.0 Sn, 8.0 to 11.0 Pb, 0.70 Zn max, 0.70 Ni

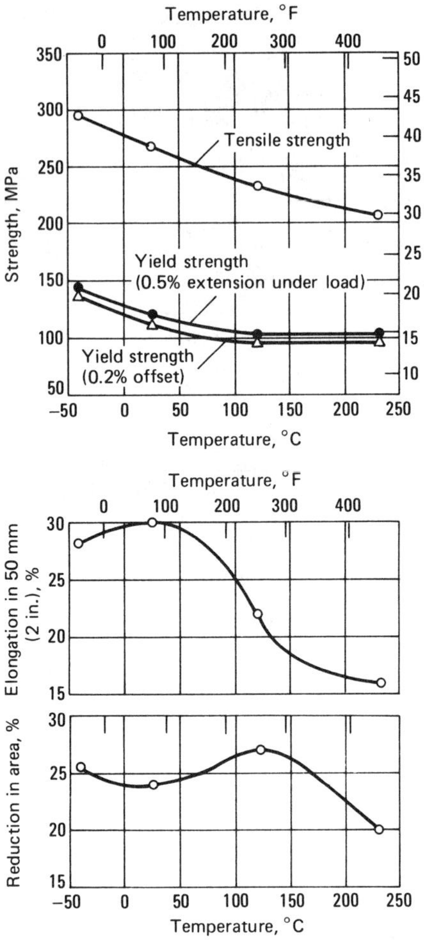

Fig. 37 Typical tensile properties of C93700 at various temperatures

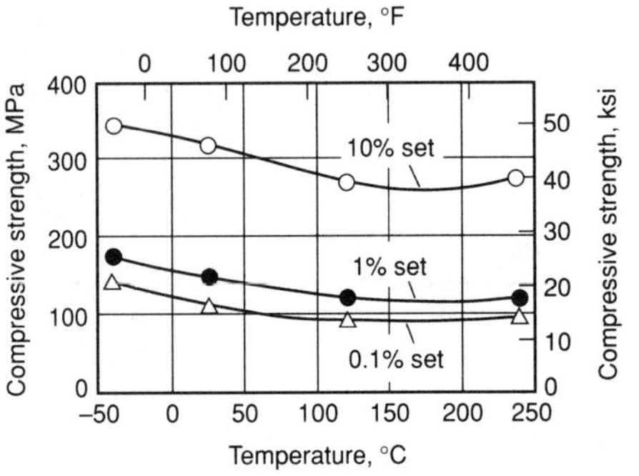

Fig. 38 Variation of compressive strength with temperature for C93700

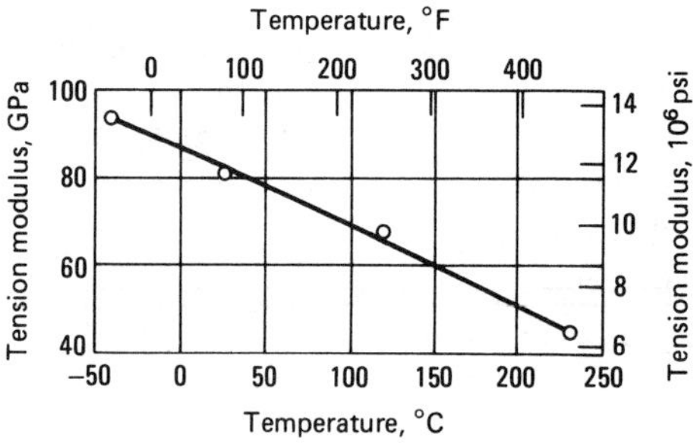

Fig. 39 Variation of elastic modulus with temperature for C93700

max, 0.15 Fe max, 0.05 P max, 0.50 Sb max, 0.08 S max, 0.003 Si max

Applications

Typical uses. Bearings for high speed and heavy pressure, pumps, impellers, applications requiring corrosion resistance, pressure-tight castings

Mechanical Properties

Tensile properties. Typical data for sand-cast test bars: tensile strength, 240 MPa (35 ksi); yield strength, 125 MPa (18 ksi) at 0.5% extension under load; elongation, 20% in 50 mm (2 in.). See also Fig. 37.
Compressive properties. Compressive strength, 90 MPa (13 ksi) at permanent set of 0.1%; 325 MPa (47 ksi) at permanent set of 10%. See also Fig. 38.
Hardness. 60 HB
Elastic modulus. See Fig. 39.
Impact strength. Izod, 7 J (5 ft · lbf); Charpy V-notch, 15 J (11 ft · lbf)
Fatigue strength. Reverse bending, 90 MPa (13 ksi) at 10^8 cycles. See also Fig. 40.
Creep-rupture characteristics. Limiting creep stress for 10^{-5}%/h: 71.7 MPa (10.4 ksi) at 177 °C (350 °F); 51 MPa (7.4 ksi) at 232 °C (450 °F); 12 MPa (1.8 ksi) at 288 °C (550 °F). See also Fig. 41.

Mass Characteristics

Density. 8.95 g/cm³ (0.323 lb/in.³) at 20 °C (68 °F)
Volume change on freezing. 7.3%
Patternmaker's shrinkage. 11 mm/m (⅛ in./ft)

Thermal Properties

Liquidus temperature. 930 °C (1705 °F)
Solidus temperature. 762 °C (1403 °F)
Incipient melting temperature. Pb, 315 °C (600 °F)
Coefficient of linear thermal expansion. 18.5 μm/m · K (10.3 μin./in. · °F) at 20 to 200 °C (68 to 392 °F). See also Fig. 42.
Specific heat. 376 J/kg · K (0.09 Btu/lb · °F) at 20 °C (68 °F)
Thermal conductivity. 46.9 W/m · K (27.1 Btu/ft · h · °F) at 20 °C (68 °F). See also Fig. 42.

Electrical Properties

Electrical conductivity. See Fig. 43.
Electrical resistivity. 170 nΩ · m at 20 °C (68 °F)

Magnetic Properties

Magnetic permeability. 1.0

Fabrication Characteristics

Machinability. 80% of C36000 (free-cutting brass)
Weldability. Soldering: good. Brazing: good, but parts must not be strained during brazing or subsequent cooling because brazing is done at temperatures in the hot short range. Oxyfuel gas welding and all forms of arc welding are not recommended.
Stress-relieving temperature. 260 °C (500 °F)

C93800
78Cu-7Sn-15Pb

Commercial Names

Common names. High-leaded tin bronze, anti-acid metal, 78-7-15

Specifications

ASTM. Sand castings: B 66, B 584; centrifugal castings: B 271; continuous castings: B 505; ingot: B 30
SAE. J462
Government. QQ-C-390; QQ-C-525 (Alloy 7); QQ-L-225 (Alloys 19 and 7); MIL-B-16261 (Alloy IV)
Other. Ingot code number 319

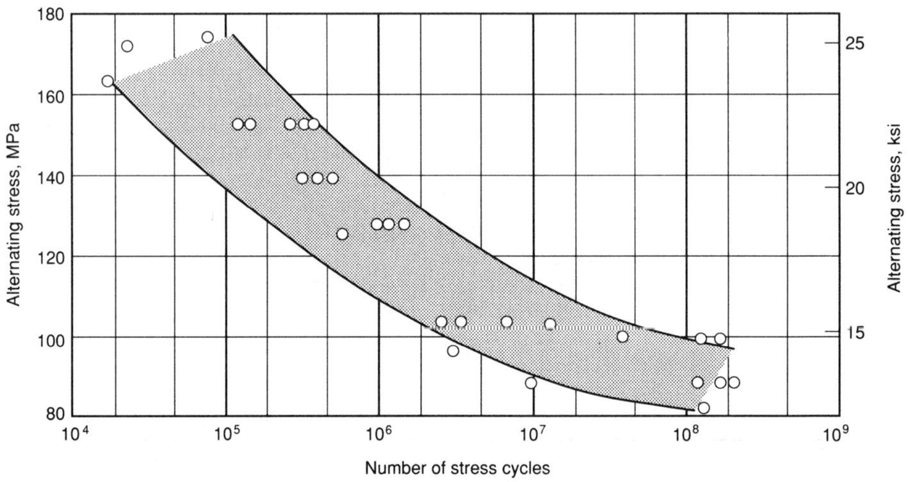

Fig. 40 Typical reverse bending fatigue curve for C93700

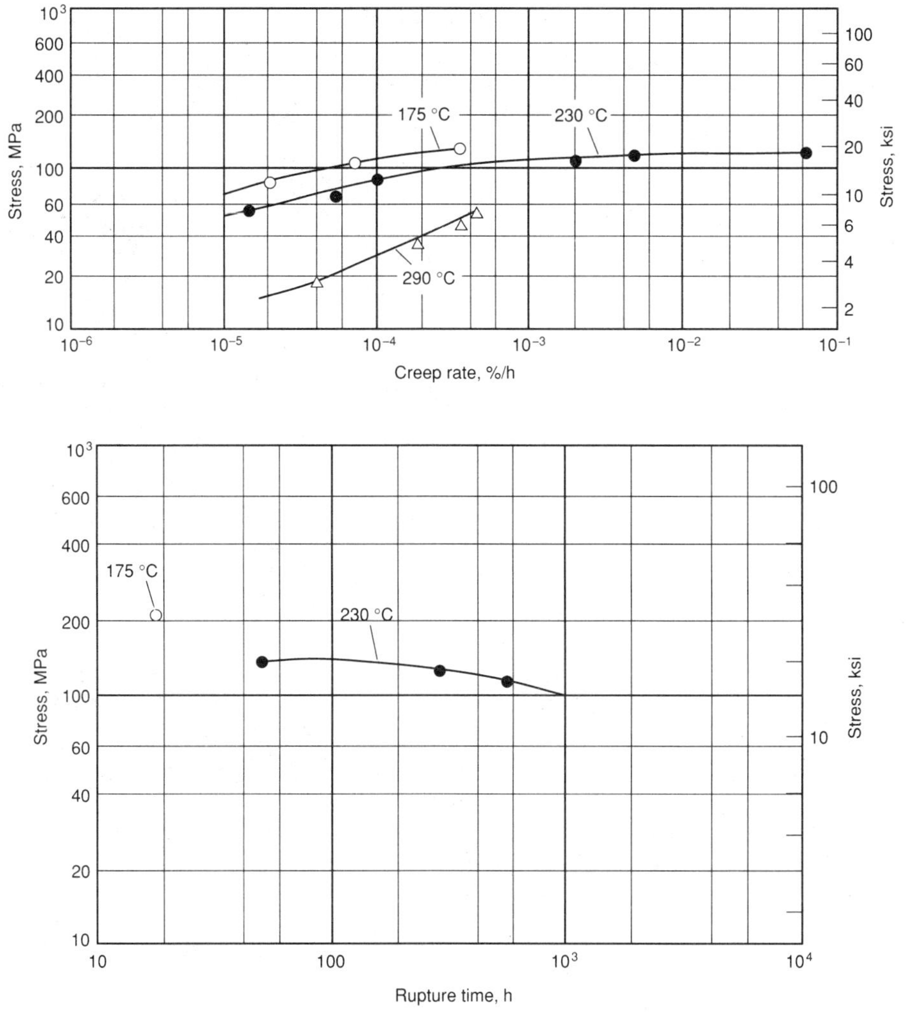

Fig. 41 Typical creep-rupture properties of C93700

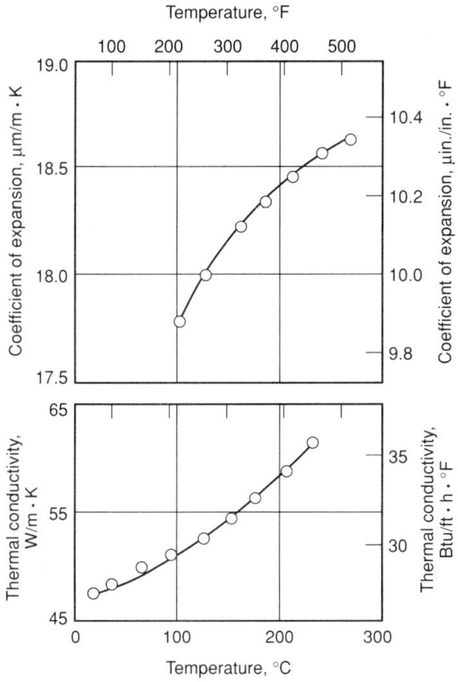

Fig. 42 Selected thermal properties of C93700

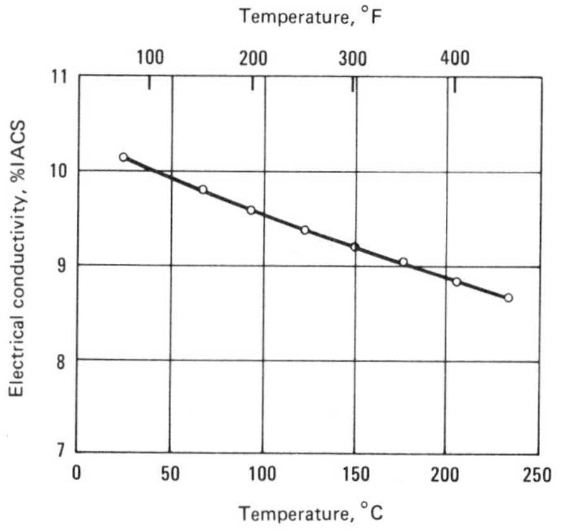

Fig. 43 Variation of electrical conductivity with temperature for C93700

Chemical Composition

Composition limits. 75.0 to 79.0 Cu, 6.3 to 7.5 Sn, 13.0 to 16.0 Pb, 0.70 Zn max, 0.70 Ni max, 0.15 Fe max, 0.05 P max, 0.70 Sb max, 0.08 S max, 0.003 Si max, 0.005 Al max. In determining Cu, minimum may be calculated as Cu + Ni.

Consequence of exceeding impurity limits. Aluminum or silicon causes lead sweating during solidification and may cause a substantial portion of castings to be unsound.

Applications

Typical uses. Locomotive engine castings and general-service bearings for moderate pressure; general-purpose wearing metal for rod bushings, shoes, and wedges; freight car bearings; backs for lined journal bearings for locomotive tenders and passenger cars; pump impellers, and bodies for use in acid mine water

Mechanical Properties

Tensile properties. Typical data for sand-cast test bars: tensile strength, 205 MPa (30 ksi); yield strength, 110 MPa (16 ksi) at 0.5% extension under load; elongation, 18% in 50 mm (2 in.). Typical data for chilled centrifugally cast test bars: tensile strength, 230 MPa (33 ksi); yield strength, 140 MPa (20 ksi) at 0.5% extension under load; elongation, 12% in 50 mm (2 in.)

Shear strength. 105 MPa (15 ksi)

Compressive properties. Compressive strength: sand cast: 83 MPa (12 ksi) at permanent set of 0.1%; 260 MPa (38 ksi) at permanent set of 10%. Centrifugally cast: 130 MPa (19 ksi) at permanent set of 0.1%

Hardness. Sand-cast: 55 HB

Elastic modulus. Sand-cast test bars: tension, 72.4 GPa (10.5 × 10^6 psi)

Impact strength. Sand cast: Charpy V-notch or Izod, 7 J (5 ft · lbf)

Fatigue strength. Reverse bending, sand-cast test bars: 69 MPa (10 ksi) at 10^8 cycles

Mass Characteristics

Density. 9.25 g/cm^3 (0.334 lb/in.^{3}) at 20 ° C (68 °F)

Patternmaker's shrinkage. 11 mm/m (1/8 in./ft)

Thermal Properties

Liquidus temperature. 945 °C (1730 °F)

Solidus temperature. 855 °C (1570 °F)

Incipient melting temperature. Pb, 315 °C (600 °F)

Coefficient of linear thermal expansion. 18.5 µm/m · K (10.3 µin./in. · °F) at 20 to 205 °C (68 to 400 °F)

Specific heat. 376 J/kg · K (0.09 Btu/lb · °F) at 20 °C (68 °F)

Thermal conductivity. 52 W/m · K (30 Btu/ft · h · °F) at 20 °C (68 °F)

Electrical Properties

Electrical conductivity. Volumetric, 11.5% IACS at 20 °C (68 °F)

Magnetic Properties

Magnetic permeability. 1.0

Fabrication Characteristics

Machinability. 80% of C36000 (free-cutting brass)
Weldability. Soldering: good. Brazing: poor. Oxyfuel gas welding and all forms of arc welding are not recommended.
Stress-relieving temperature. 260 °C (500 °F)

C93900
79Cu-6Sn-15Pb

Commercial Names

Common name. High-leaded tin bronze, 79-6-15

Specifications

ASTM. B 505, B 30

Chemical Composition

Composition limits. 76.5 to 79.5 Cu, 5.0 to 7.0 Sn, 14.0 to 18.0 Pb, 1.5 Zn max, 0.80 Ni max, 0.40 Fe max, 0.05 P max. 1.5 P max for continuous castings

Applications

Typical uses. Continuous castings only; common products include bearings for general service, pump bodies and impellers for mine use

Mechanical Properties

Tensile properties. Typical tensile strength, 220 MPa (32 ksi); yield strength, 150 MPa (22 ksi) at 0.5% extension under load; elongation, 7% in 50 mm (2 in.)
Hardness. 63 HB, typical
Elastic modulus. Tension, 76 GPa (11 × 10^6 psi)

Mass Characteristics

Density. 9.25 g/cm^3 (0.334 lb/in.^{3}) at 20 °C (68 °F)
Patternmaker's shrinkage. 11 mm/m (⅛ in./ft)

Thermal Properties

Liquidus temperature. 943 °C (1730 °F)
Solidus temperature. 854 °C (1570 °F)
Incipient melting temperature. Pb, 315 °C (600 °F)
Coefficient of linear thermal expansion. 18.5 μm/m · K (10.3 μin./in. · °F) at 20 to 204 °C (68 to 400 °F)
Specific heat. 376 J/kg · K (0.09 Btu/lb · °F)
Thermal conductivity. 52 W/m · K (30 Btu/ft · h · °F) at 20 °C (68 °F)

Electrical Properties

Electrical conductivity. Volumetric, 11.5% IACS at 20 °C (68 °F)

Magnetic Properties

Magnetic permeability. 1.0

Fabrication Characteristics

Machinability. 80% of C36000 (free-cutting brass)

C94300
70Cu-5Sn-25Pb

Commercial Names

Common name. High-leaded tin bronze, soft bronze, 70-5-25

Specifications

ASTM. B 584, B 66, B 271, B 505, B 30
SAE. J462 (CA943)
Government. QQ-L-225, Alloy 18; MIL-B-16261, Alloy V
Other. Ingot code number 322

Chemical Composition

Composition limits. 68.5 to 73.5 Cu, 4.5 to 6.0 Sn, 22.0 to 25.0 Pb, 0.50 Zn max, 0.70 Ni max, 0.15 Fe max, 0.70 Sb max, 0.05 P max, 0.08 S max
Supplementary composition limits. In determining Cu, minimum may be calculated as Cu + Ni. 0.35 Fe max when used for steel-backed bearings. 1.5 P max for continuous castings

Applications

Typical uses. Bearings under light loads and high speed, driving boxes, railroad bearings

Mechanical Properties

Tensile properties. Typical data for sand-cast test bars: tensile strength, 185 MPa (27 ksi); yield strength, 90 MPa (13 ksi) at 0.5% extension under load; elongation, 10% in 50 mm (2 in.); reduction in area, 8%
Compressive properties. Typical compressive strength: 76 MPa (11 ksi) at permanent set of 0.1%; 160 MPa (23 ksi) at permanent set of 10%
Hardness. 48 HB
Elastic modulus. Tension, 72.4 GPa (10.5 × 10^6 psi)
Impact strength. Izod, 7 J (5 ft · lbf)

Mass Characteristics

Density. 9.29 g/cm^3 (0.336 lb/in.^{3}) at 20 °C (68 °F)
Patternmaker's shrinkage. 11 mm/m (⅛ in./ft)

Thermal Properties

Solidus temperature. 900 °C (1650 °F)
Incipient melting temperature. Pb, 315 °C (600 °F)
Specific heat. 376 J/kg · K (0.09 Btu/lb · °F) at 20 °C (68 °F)
Thermal conductivity. 62.7 W/m · K (36.2 Btu/ft · h · °F) at 20 °C (68 °F)

Electrical Properties

Electrical conductivity. Volumetric, 9% IACS at 20 °C (68 °F)

Magnetic Properties

Magnetic permeability. 1.0

Fabrication Characteristics

Machinability. 80% of C36000 (free-cutting brass)

C94500
73Cu-7Sn-20Pb

Commercial Names

Common name. Medium bronze

Specifications

ASTM. Sand castings: B 66; ingot: B 30
Government. QQ-L-225, Alloy 15; MIL-B-16261, Alloy I

Chemical Composition

Composition limits. 6.0 to 8.0 Sn, 16 to 22 Pb, 1.2 Zn max, 1.0 Ni max, 0.8 Sb max, 0.005 Al max, 0.15 Fe max, 0.5 P max (1.5 P max for continuous castings), 0.08 S max, 0.005 Si max, bal Cu

Applications

Typical uses. Locomotive wearing parts, high-load low-speed bearings

Mechanical Properties

Tensile properties. Typical. Tensile strength, 170 MPa (25 ksi); yield strength, 83 MPa (12 ksi); elongation, 12% in 50 mm (2 in.)
Compressive properties. Compressive strength, 250 MPa (36 ksi)
Hardness. 50 HB
Elastic modulus. Tension, 72 GPa (10.5 × 10^6 psi); shear, 90 GPa (13 × 10^6 psi)
Fatigue strength. Rotating beam, 69 MPa (10 ksi) at 10^8 cycles
Impact strength. Izod, 5.4 J (4.0 ft · lbf)

Mass Characteristics

Density. 9.4 g/cm^3 (0.34 lb/in.^{3}) at 20 °C (68 °F)
Volume change on freezing. 1.1%

Thermal Properties

Liquidus temperature. 940 °C (1725 °F)
Solidus temperature. 800 °C (1475 °F)
Incipient melting temperature. 315 °C (600 °F)
Coefficient of linear thermal expansion. 18.5 μm/m · K (10.3 μin./in. · °F) at 20 to 200 °C (68 to 392 °F)
Specific heat. 376 J/kg · K (0.09 Btu/lb · °F) at 20 °C (68 °F)
Thermal conductivity. 52 W/m · K (30 Btu/ft · h · °F) at 20 °C (68 °F)

Electrical Properties

Electrical conductivity. Volumetric, 10% IACS at 20 °C (68 °F)

Magnetic Properties

Magnetic permeability. 1.0

Fabrication Characteristics

Machinability. 80% of C36000 (free-cutting brass)

C95200
88Cu-3Fe-9Al

Commercial Names

Previous trade name. Ampco Al

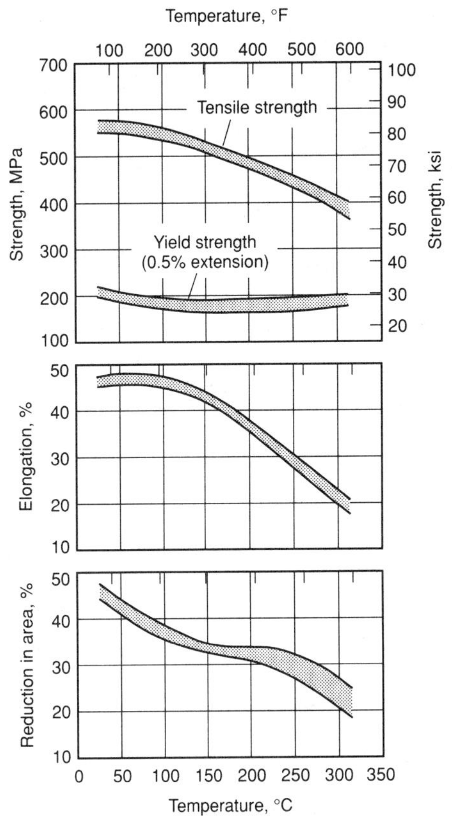

Fig. 44 Typical short-time tensile properties of C95200, as-cast

Common name. Aluminum bronze 9A; 88-3-9

Specifications

ASME. Sand castings: SB148; centrifugal castings: SB271
ASTM. Sand castings: B 148; centrifugal castings: B 271; continuous castings: B 505; ingot: B 30
SAE. J462
Government. Centrifugal, sand, and continuous castings: QQ-C-390; sand castings: MIL-C-22229
Other. Ingot code number 415

Chemical Composition

Composition limits. 86 Cu min, 8.5 to 9.5 Al, 2.5 to 4.0 Fe, 1.0 max other (total)
Consequence of exceeding impurity limits. Possible hot shortness and/or hot cracking, embrittlement, and reduced soundness of castings

Applications

Typical uses. Acid-resisting pumps, bearings, bushings, gears, valve seats, guides, plungers, pump rods, pickling hooks, nonsparking hardware
Precautions in use. Not suitable for use in oxidizing acids

Mechanical Properties

Tensile properties. Typical data for sand-cast test bars: tensile strength, 550 MPa (80 ksi); yield strength, 185 MPa (27 ksi); elongation, 35% in 50 mm (2 in.). See also Fig. 44.
Hardness. 64 HRB; 125 HB (3000 kg load)
Poisson's ratio. 0.31
Elastic modulus. Tension, 105 GPa (15×10^6 psi); shear, 39 GPa (5.7×10^6 psi)
Impact strength. Charpy keyhole, 27 J (20 ft · lbf) at −18 to 38 °C (0 to 100 °F); Izod, 40 J (30 ft · lbf) at −18 to 38 °C (0 to 100 °F)
Fatigue strength. Rotating beam, 150 MPa (22 ksi) at 10^8 cycles
Creep-rupture characteristics. Limiting creep stress for 10^{-5}%/h: 145 MPa (21 ksi) at 230 °C (450 °F); 54 MPa (7.9 ksi) at 315 °C (600 °F). See also Fig. 45.

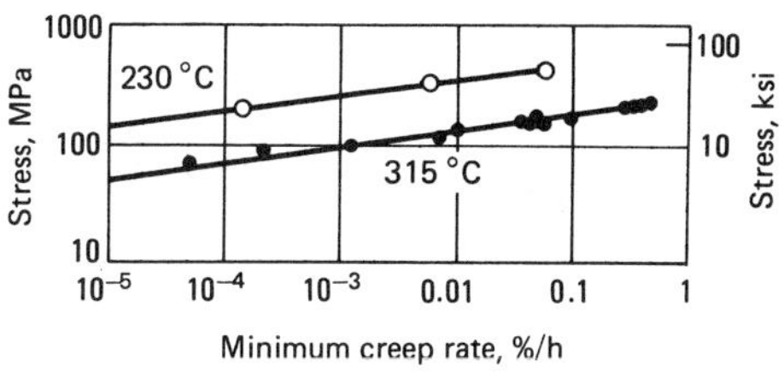

Fig. 45 Typical creep properties of C95200, as-cast

Structure

Microstructure. As cast, the microstructure is primarily fcc alpha, with precipitates of iron-rich alpha in the form of rosettes and spheres. Depending on the cooling rate, small amounts of metastable cph beta or alpha-gamma eutectoid decomposition products may be present. Annealing followed by rapid cooling reduces the amount of residual beta to about 5% of the apparent volume.
Metallographic etchant. Acid ferric chloride (10% HCl, 5% $FeCl_3$)

Mass Characteristics

Density. 7.64 g/cm^3 (0.276 lb/in.^{3}) at 20 °C (68 °F)
Volume change on freezing. Approximately 1.7% contraction
Patternmaker's shrinkage. 2%

Thermal Properties

Liquidus temperature. 1045 °C (1915 °F)
Solidus temperature. 1040 °C (1905 °F)
Coefficient of linear thermal expansion. 16.2 μm/m · K (9.0 μin./in. · °F) at 20 to 300 °C (68 to 572 °F)
Specific heat. 380 J/kg · K (0.091 Btu/lb · °F) at 20 °C (68 °F)
Thermal conductivity. 50 W/m · K (29.1 Btu/ft · h · °F) at 20 °C (68 °F)

Electrical Properties

Electrical conductivity. Volumetric, 12% IACS at 20 °C (68 °F)
Electrical resistivity. 144 nΩ · m at 20 °C (68 °F)

Magnetic Properties

Magnetic permeability. 1.20 at 16 000 A/m (200 oersteds)

Chemical Properties

General corrosion behavior. C95200 has generally fair resistance to attack in nonoxidizing mineral acids such as sulfuric, hydrochloric, and phosphoric, and in alkalies such as sodium and potassium hydroxide. Cast components are used successfully in systems for seawater, brackish water, and potable water. The alloy resists many organic acids, including acetic and lactic, plus all esters and ethers. Moist ammonia atmospheres can cause stress-corrosion cracking.

Fabrication Characteristics

Machinability. 20% of C36000 (free-cutting brass). Carbide or tool steel cutters may be used. Good surface finish and precision attainable with all conventional methods. Typical conditions using tool steel cutters: roughing speed, 105 m/min (350 ft/min) with a feed of 0.3 mm/rev (0.011 in./rev); finishing speed, 350 m/min (1150 ft/min) with a feed of 0.15 mm/rev (0.006 in./rev)
Annealing temperature. 650 to 745 °C (1200 to 1375 °F)

C95300
89Cu-1Fe-10A1

Commercial Names

Trade name. Ampco B2
Common names. Aluminum bronze 9B; 89-1-10

Specifications

ASTM. Sand castings: B 148; centrifugal castings: B 271; continuous castings: B 505; ingots: B 30
SAE. J462
Government. Centrifugal and sand castings: QQ-C-390; precision castings: MIL-C-11866, composition 22
Ingot identification number. 415

Chemical Composition

Composition limits. 86 Cu min, 9.0 to 11.0 Al, 0.8 to 1.5 Fe, 1.0 max other (total)
Consequence of exceeding impurity limits. Possible hot shortness, loss of casting soundness, embrittlement, reduced response to heat treatment

Applications

Typical uses. Pickling baskets, nuts, gears, steel mill slippers, marine equipment, welding jaws, nonsparking hardware
Precautions in use. Not suitable for exposure to oxidizing acids. Prolonged heating in the 320 to 565 °C (610 to 1050 °F) range can result in a loss of ductility and notch toughness.

Mechanical Properties

Tensile properties. Minimum values. As-cast: tensile strength, 450 MPa (65 ksi);

yield strength, 170 MPa (25 ksi); elongation, 20% in 50 mm (2 in.); reduction in area, 25%. TQ50 temper: tensile strength, 550 MPa (80 ksi); yield strength, 275 MPa (40 ksi); elongation, 12% in 50 mm (2 in.); reduction in area, 14%
Compressive properties. Compressive ultimate strength: as-cast, 760 MPa (110 ksi); TQ50 temper, 825 MPa (120 ksi). Elastic limit: as-cast, 125 MPa (18 ksi); TQ50 temper, 205 MPa (30 ksi)
Hardness. As-cast, 67 HRB; TQ50 temper, 81 HRB
Poisson's ratio. 0.314
Elastic modulus. Tension, 110 GPa (16 × 10^6 psi); shear, 42 GPa (6.1 × 10^6 psi)
Impact strength. Cast and annealed: Charpy keyhole, 31 J (23 ft · lbf); Izod, 38 J (28 ft · lbf) at −20 to 100 °C (−5 to 212 °F). TQ50 temper: Charpy keyhole, 37 J (27 ft · lbf) at −20 to 100 °C (−5 to 212 °F)

Structure

Crystal structure. Alpha phase, face-centered cubic; beta phase, close-packed hexagonal
Microstructure. As-cast and properly cooled or annealed, the structure is approximately 70% alpha and 30% metastable beta. Quenched and tempered (TQ50 temper), the structure is largely tempered metastable beta martensite, but also contains both primary alpha and reprecipitated acicular alpha.

Mass Characteristics

Density. 7.53 g/cm³ (0.272 lb/in.³) at 20 °C (68 °F)
Patternmaker's shrinkage. 1.6%

Thermal Properties

Liquidus temperature. 1045 °C (1915 °F)
Solidus temperature. 1040 °C (1905 °F)
Coefficient of linear thermal expansion. 16.2 μm/m · K (9.0 μin./in. · °F) at 20 to 300 °C (68 to 572 °F)
Specific heat. 375 J/kg · K (0.09 Btu/lb · °F) at 20 °C (68 °F)
Thermal conductivity. 63 W/m · K (36 Btu/ft · h · °F) at 20 °C (68 °F); temperature coefficient, 0.12 W/m · K per K at 20 °C (68 °F)

Electrical Properties

Electrical conductivity. Volumetric, 13% IACS at 20 °C (68 °F)
Electrical resistivity. 133 nΩ · m at 20 °C (68 °F)

Magnetic Properties

Magnetic permeability. 1.07 at field strength of 8 kA/m

Chemical Properties

General corrosion behavior. Corrosion characteristics of C95300 are slightly inferior to those of C95200, primarily because C95300 has more and larger beta areas.

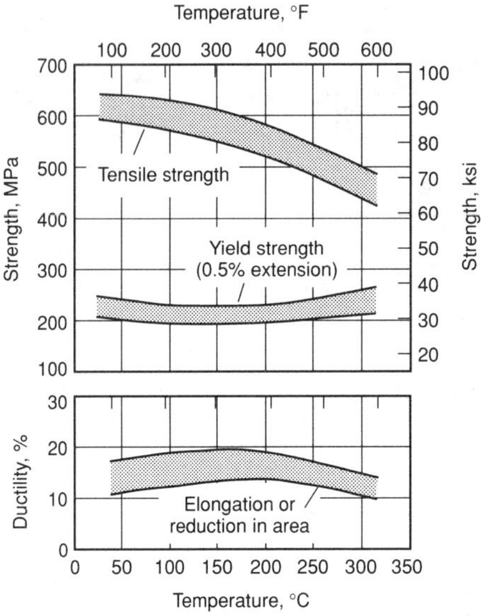

Fig. 46 Typical short-time tensile properties of C95400, as-cast

Heat treatment enhances corrosion resistance, particularly in mediums that promote dealloying. The alloy shows characteristic resistance to nonoxidizing mineral acids, neutral salt solutions, seawater, brackish water, and some organic acids.

Fabrication Characteristics

Machinability. 55% of C36000 (free-cutting brass). Tool steel or carbide cutters may be used. Good surface and precision finish may be obtained in the as-cast, cast and annealed, and TQ50 tempers. Typical conditions using tool steel cutters: roughing speed, 90 m/min (300 ft/min) at a feed of 0.2 mm/rev (0.009 in./rev); finishing speed, 290 m/min (950 ft/min) at a feed of 0.1 mm/rev (0.004 in./rev)
Annealing temperature. 595 to 650 °C (1100 to 1200 °F)

C95400 (85Cu-4Fe-11Al) and C95410

Commercial Names

Trade name. Ampco C3
Common names. Aluminum bronze 9C; G5; 85-4-11

Specifications

ASME. Sand castings: SB148
ASTM. Sand castings: B 148; centrifugal castings: B 271; continuous castings: B 505; ingots: B 30
Government. QQ-C-390. Sand castings, MIL-C-22229 (composition 6); investment castings, MIL-C-15345 (Alloy 13); centrifugal castings, MIL-C-22087 (composition 8)
Ingot identification number 415

Chemical Composition

Composition limits of C95400. 83 min Cu, 10.0 to 11.5 Al, 3.0 to 5.0 Fe, 0.50 Mn max, 2.5 Ni max (+ Co), 0.5 max other (total)
Composition limits of C95410. 83.0 Cu min, 3.0 to 5.0 Fe, 1.5 to 2.5 Ni (including Co), 10.0 to 11.5 Al, 0.50 Mn max
Cu + sum of named elements. 99.5 min
Consequence of exceeding impurity limits. Possible hot shortness, reduced casting soundness, embrittlement and loss of heat treating response

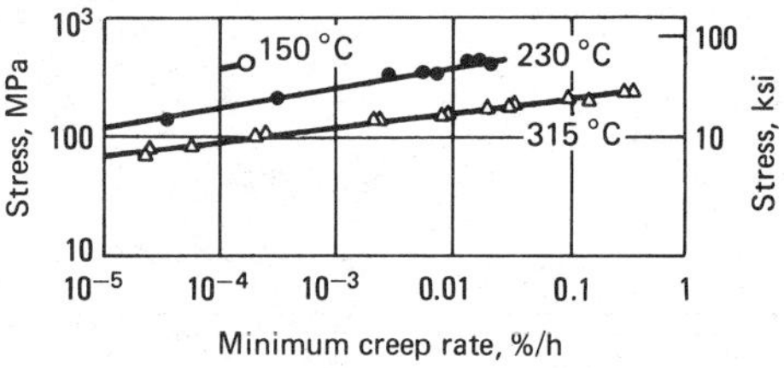

Fig. 47 Typical creep properties of C95400, as-cast

Applications

Typical uses. Pump impellers, bearings, gears, worms, bushings, valve seats and guides, rolling mill slippers, slides, nonsparking hardware
Precautions in use. Not suitable for use in oxidizing acids. Prolonged heating in the 320 to 565 °C (610 to 1050 °F) range can result in loss of ductility and notch toughness.

Mechanical Properties

Tensile properties. Minimum values. As cast: tensile strength, 515 MPa (75 ksi); yield strength, 205 MPa (30 ksi); elongation, 12% in 50 mm (2 in.); reduction in area, 12%. TQ50 temper: tensile strength, 620 MPa (90 ksi); yield strength, 310 MPa (45 ksi); elongation, 6% in 50 mm (2 in.), reduction in area, 6%. See also Fig. 46.
Compressive properties. Compressive strength, ultimate: as-cast, 940 MPa (136 ksi); TQ50 temper, 1070 MPa (155 ksi)
Hardness. As-cast, 83 HRB; TQ50 temper: 94 HRB
Poisson's ratio. 0.316
Elastic modulus. Tension, 110 GPa (16 × 10^6 psi); shear, 41 GPa (6.1 × 10^6 psi)
Impact strength. As-cast: Charpy keyhole, 15 J (11 ft · lbf); Izod, 22 J (16 ft · lbf) at 20 °C (68 °F). TQ50 temper: Charpy keyhole, 9 J (7 ft · lbf); Izod, 15 J (11 ft · lbf) at 20 °C (68 °F)
Fatigue strength. Reverse bending, 240 MPa (35 ksi) at 10^8 cycles (TQ50 temper)
Creep-rupture characteristics. Limiting creep stress at a strain rate of 10^{-5}%/h: 115 MPa (17 ksi) at 230 °C (450 °F); 51 MPa (7.4 ksi) at 315 °C (600 °F); 30 MPa (4.4 ksi) at 370 °C (700 °F); 20 MPa (2.9 ksi) at 425 °C (800 °F). See also Fig. 47.

Structure

Crystal structure. Alpha, face-centered cubic; beta, close-packed hexagonal

Microstructure. As-cast and annealed material normally consists of approximately 50% alpha and 50% metastable beta. Under some conditions, eutectoid decomposition may produce an alpha-gamma-2 structure instead of the beta phase. Quenched-and-tempered structures consist of fine acicular alpha crystals in a tempered beta matrix.

Mass Characteristics

Density. 7.45 g/cm^3 (0.269 lb/in.^{3}) at 20 °C (68 °F)
Patternmaker's shrinkage. 1.6%

Thermal Properties

Liquidus temperature. 1040 °C (1900 °F)
Solidus temperature. 1025 °C (1880 °F)
Coefficient of linear thermal expansion. 16.2 μm/m · K (9.0 μin./in. · °F) at 20 to 300 °C (68 to 572 °F)
Specific heat. 420 J/kg · K (0.10 Btu/lb · °F) at 20 °C (68 °F)
Thermal conductivity. 59 W/m · K (34 Btu/ft · h · °F) at 20 °C (68 °F); temperature coefficient, 0.117 W/m · K per K at 20 °C (68 °F)

Electrical Properties

Electrical conductivity. Volumetric, 13% IACS at 20 °C (68 °F)
Electrical resistivity. 133 nΩ · m at 20 °C (68 °F)

Magnetic Properties

Magnetic permeability. As-cast, 1.27 at field strength of 16 kA/m; TQ50 temper, 1.20 at field strength of 16 kA/m

Chemical Properties

General corrosion behavior. C95400 has fair resistance to attack by nonoxidizing solutions of mineral acids such as sulfuric and phosphoric, as well as to neutral salts such as sodium chloride. The alloy also resists acetic, lactic, and oxalic acids; organic solvents such as esters and ethers; and seawater, brackish water, and potable waters. In some environments, C95400 can undergo dealloying caused by corrosive attack on the beta phase. Heat treatment improves resistance to dealloying. Moist ammonia environments may cause stress-corrosion cracking under high levels of applied stress.

Fabrication Characteristics

Machinability. 60% of C36000 (free-cutting brass). C95400, in either as-cast or TQ50 temper, is easily machined by all standard operations using high-strength tool steel or carbide cutters. Typical conditions using tool steel cutters: roughing speed, 90 m/min (300 ft/min) at a feed of 0.3 mm/rev (0.011 in./rev); finishing speed, 290 m/min (950 ft/min) at a feed of 0.1 mm/rev (0.004 in./rev)
Annealing temperature. 620 °C (1150 °F)

C95500
81Cu-4Fe-4Ni-11Al

Commercial Names

Previous trade name. Ampco D4
Common names. Aluminum bronze 9D; 415; 81-4-4-11

Specifications

AMS. 4880
ASTM. Sand castings: B 148; centrifugal castings: B 271; continuous castings: B 505; ingots: B 30
SAE. J462
Government. QQ-C-390; centrifugal castings, MIL-C-15345 (Alloy 14); sand castings, MIL-C-22229 (composition 6); investment castings, MIL-C-22087 (composition 8)
Ingot identification number. 415

Chemical Composition

Composition limits. 78 Cu min, 10.0 to 11.5 Al, 3.0 to 5.0 Fe, 3.5 Mn max, 3.0 to 5.5 Ni (+ Co), 0.5 max other (total)
Consequence of exceeding impurity limits. Possible hot shortness in welding, embrittlement, increased quench-cracking susceptibility, possible loss of heat-treating response. Excessive Si can cause machining difficulties.

Applications

Typical uses. Valve guides and seats in aircraft engines, corrosion-resistant parts, bushings, gears, worms, pickling hooks and baskets, agitators
Precautions in use. Not suitable for use in strong oxidizing acids

Mechanical Properties

Tensile properties. Typical. As-cast: tensile strength, 620 MPa (90 ksi); yield strength, 275 MPa (40 ksi); elongation, 6% in 50 mm (2 in.); reduction in area, 7%. TQ50 temper: tensile strength, 760 MPa (110 ksi); yield strength, 415 MPa (60 ksi); elongation, 5% in 50 mm (2 in.); reduction in area, 5%
Compressive properties. As-cast: compressive strength, 895 MPa (130 ksi); compressive yield strength, 825 MPa (120 ksi) at a permanent set of 10%; elastic limit, 310 MPa (45 ksi). TQ50 temper: compressive strength, 1140 MPa (165 ksi); compressive yield strength, 1030 MPa (150 ksi) at a permanent set of 10%; elastic limit, 415 MPa (60 ksi)
Hardness. As-cast, 87 HRB; TQ50 temper, 96 HRB
Poisson's ratio. 0.32
Elastic modulus. Tension: as-cast, 110 GPa (16 × 10^6 psi); TQ50 temper, 115 GPa (17 × 10^6 psi). Shear: as-cast, 42 GPa (6.1 × 10^6 psi); TQ50 temper, 44 GPa (6.4 × 10^6 psi)
Impact strength. Charpy keyhole, 14 J (10 ft · lbf); Izod, 18 J (13 ft · lbf) at 20 °C (68 °F)
Fatigue strength. Rotating beam, as-cast, 215 MPa (31 ksi) at 10^8 cycles; TQ50 temper, 260 MPa (38 ksi) at 10^8 cycles
Creep-rupture characteristics. Limiting creep stress at a strain rate of 10^{-5}%/h: 72 MPa (10.5 ksi) at 315 °C (600 °F); 38 MPa (5.5 ksi) at 370 °C (700 °F); 17 MPa (2.5 ksi) at 425 °C (800 °F)

Structure

Crystal structure. Alpha, face-centered cubic; beta, close-packed hexagonal; kappa, ordered face-centered cubic
Microstructure. As-cast or annealed structures consist of alpha crystals plus kappa precipitates, forming a pearlitic appearance. Small areas of metastable beta may exist. Heat-treated structures consist of tempered beta martensite with very fine reprecipitated alpha needles. Some undissolved equiaxed alpha crystals may be evident, depending on the actual composition and quenching temperature.

Mass Characteristics

Density. 7.53 g/cm^3 (0.272 lb/in.^{3}) at 20 °C (68 °F)
Patternmaker's shrinkage. 1.6%

Thermal Properties

Liquidus temperature. 1055 °C (1930 °F)
Solidus temperature. 1040 °C (1900 °F)
Coefficient of linear thermal expansion. 16.2 μm/m · K (9.0 μin./in. · °F) at 20 to 300 °C (68 to 572 °F)
Specific heat. 418 J/kg · K (0.10 Btu/lb · °F) at 20 °C (68 °F)
Thermal conductivity. 42 W/m · K (24 Btu/ft · h · °F) at 20 °C (68 °F)

Electrical Properties

Electrical conductivity. Volumetric, 8.5% IACS at 20 °C (68 °F)
Electrical resistivity. 203 nΩ · m at 20 °C (68 °F)

Magnetic Properties

Magnetic permeability. As-cast, 1.30 at field strength of 16 kA/m; TQ50 temper, 1.20 at field strength of 16 kA/m

Chemical Properties

General corrosion behavior. Good cavitation resistance in salt water and fresh tap water. Avoid nitric acid and strong aeration when using other acids.

Fabrication Characteristics

Machinability. 50% of C36000 (free-cutting brass). Heat treating reduces machinability in drilling and tapping operations. Tool steel or carbide cutters may be used. Typical conditions using tool steel cutters follow. Roughing speed: as-cast, 76 m/min (250 ft/min) at a feed of 0.3 mm/rev (0.011 in./rev); TQ50 temper, 90 m/min (300 ft/min) at a feed of 0.2 mm/rev (0.009 in./rev). Finishing speed: as-cast and TQ50 temper, 290 m/min (950 ft/min) at a feed of 0.1 mm/rev (0.004 in./rev)
Annealing temperature. 620 to 705 °C (1150 to 1300 °F)

C95600
91Cu-2Si-7Al

Commercial Names

Common name. Aluminum-silicon bronze

Specifications

ASTM. Ingot: B 30; sand castings: B 148, B 763
Government. QQ-B-675, MIL-V-11 87
Ingot identification number. 415E

Chemical Composition

Composition limits. 88.0 Cu min, 0.25 Ni (including Co) max, 6.0 to 8.0 Al, 1.8 to 3.3 Si.
Cu + sum of named elements. 99.0% min

Applications

Typical uses. Cable connectors, terminals, valve stems, marine hardware, gears, worms, pole-line hardware

Mechanical Properties

Tensile properties. Typical data for as-sand-cast separately cast test bars (M01 temper): tensile strength, 515 MPa (75 ksi); yield strength, 235 MPa (34 ksi) at 0.5% extension under load; elongation, 18% in 50 mm (2 in.)
Hardness. Typically, 140 HB (3000 kg)
Elastic modulus. Tension, 105 GPa (15 × 10^6 psi) at 20 °C (68 °F)

Mass Characteristics

Density. 7.69 g/cm^3 (0.278 lb/in.^{3}) at 20 °C (68 °F)
Volume change on freezing. Patternmaker's shrinkage, 16 mm/m (3⁄16 in./ft)

Thermal Properties

Liquidus temperature. 1005 °C (1840 °F)
Solidus temperature. 982 °C (1800 °F)
Specific heat. 376 J/kg · K (0.09 Btu/lb · °F) at 20 °C (68 °F)

Electrical Properties

Electrical conductivity. Volumetric, 8.5% IACS at 20 °C (68 °F)

Fabrication Characteristics

Machinability. M01 temper; 60% of C36000 (free-machining brass)
Stress-relieving temperature. 260 °C (500 °F)

C95700
75Cu-3Fe-8Al-2Ni-12Mn

Commercial Names

Previous trade name. Superstone 40, Novoston, Ampcoloy 495
Common name. Manganese-aluminum bronze; 75-3-8-2-12

Specifications

ASTM. Sand castings: B 148; ingot: B 30
Government. Sand castings: MIL-B-24480

Chemical Composition

Composition limits. 71.0 Cu min, 11.0 to 14.0 Mn, 7.0 to 8.5 Al, 2.0 to 4.0 Fe, 1.5 to 3.0 Ni, 0.10 Si max, 0.03 Pb max, 0.5 max others (total)
Consequence of exceeding impurity limits. Possible hot shortness and reduced cast strength

Applications

Typical uses. Propellers, impellers, stator clamp segments, safety tools, welding rods, valves, pump casings, marine fittings
Precautions in use. Slow cooling or prolonged heating in the 350 to 565 °C (660 to 1050 °F) range may cause embrittlement. Not suitable for use in oxidizing acids

Mechanical Properties

Tensile properties. Typical data for sand-cast test bars: tensile strength, 620 MPa (90 ksi); yield strength, 275 MPa (40 ksi); elongation, 20% in 50 mm (2 in.); reduction in area, 24%
Compressive properties. Compressive strength, as-cast: 1035 MPa (150 ksi) at a permanent set of 0.1%
Hardness. As-cast, or cast and annealed: 85 to 90 HRB
Poisson's ratio. 0.326
Elastic modulus. Tension, 125 GPa (18 × 10^6 psi); shear, 44 GPa (6.4 × 10^6 psi)
Impact strength. Izod, 27 J (20 ft · lbf) at 20 °C (68 °F)
Fatigue strength. Reverse bending, 231 MPa (33.5 ksi) at 10^8 cycles
Creep-rupture characteristics. Limiting creep stress for 10^{-5}%/h: 66 MPa (9.6 ksi) at 205 °C (400 °F); 31 MPa (4.5 ksi) at 290 °C (550 °F). Rupture stress for 10^5 h life: 470 MPa (68 ksi) at 205 °C (400 °F); 232 MPa (33.6 ksi) at 260 °C (500 °F); 39 MPa (5.7 ksi) at 370 °C (700 °F)

Structure

Microstructure. As-cast and annealed tempers: fcc alpha crystals with cph beta phase in various amounts, typically, 25% by volume

Mass Characteristics

Density. 7.53 g/cm^3 (0.272 lb/in.^{3}) at 20 °C (68 °F)
Patternmaker's shrinkage. 1.6%

Thermal Properties

Liquidus temperature. 990 °C (1815 °F)
Solidus temperature. 950 °C (1740 °F)
Coefficient of linear thermal expansion. 17.6 μm/m · K (9.8 μin./in. · °F) at 20 to 300 °C (68 to 572 °F)
Specific heat. 440 J/kg · K (0.105 Btu/lb · °F) at 20 °C (68 °F)
Thermal conductivity. 12.1 W/m · K (7.0 Btu/ft · h · °F) at 20 °C (68 °F)

Electrical Properties

Electrical conductivity. Volumetric, 3.1% IACS at 20 °C (68 °F)
Electrical resistivity. 556 nΩ · m at 20 °C (68 °F)

Magnetic Properties

Magnetic permeability. Magnetic condition (as-cast, slow cooled): 2.2 to 15.0. Demagnetized (annealed, fast cooled): 1.03

Chemical Properties

General corrosion behavior. Generally comparable to that of the aluminum bronzes and nickel-aluminum bronzes. See C95200.

Fabrication Characteristics

Machinability. 50% of C36000 (free-cutting brass). Tool steel or carbide cutters may be used. Good surface finishes and tolerance are possible in all conventional machining operations. Typical conditions using tool steel cutters: roughing speed, 75 m/min (250 ft/min) with a feed of 0.3 mm/rev (0.011 in./rev); finishing speed, 290 m/min (950 ft/min) with a feed of 0.1 mm/rev (0.004 in./rev)
Annealing temperature. 620 °C (1150 °F)

C95800
82Cu-4Fe-9Al-4Ni-1Mn

Commercial Names

Common names. Alpha nickel-aluminum bronze; propeller bronze

Specifications

ASTM. Sand castings: B 148; centrifugal castings: B 271; continuous castings: B 505; ingots, B 30
SAE. J462
Government. Sand and centrifugal castings: QQ-C-390; MIL-B-24480; centrifugal castings only: MIL-C-15345, Alloy 28
Ingot identification number. 415

Chemical Composition

Composition limits. 79.0 Cu min, 0.03 Pb max, 3.5 to 4.5 Fe, 4.0 to 5.0 Ni (+ Co), 0.8 to 1.5 Mn, 8.5 to 9.5 Al, 0.10 Si max
Consequence of exceeding impurity limits. Hard spots, embrittlement, possible hot shortness, possible weld cracking

Applications

Typical uses. Propeller blades and hubs for fresh- and salt-water service, fittings, gears, worm wheels, valve guides and seals, structural applications
Precautions in use. Not suitable for use in oxidizing acids or strong alkalies

Mechanical Properties

Tensile properties. Typical. Cast and annealed: tensile strength, 585 MPa (85 ksi); yield strength, 240 MPa (35 ksi); elongation, 15% in 50 mm (2 in.); reduction in area, 16%
Compressive properties. Compressive strength, cast and annealed: 240 MPa (35 ksi) at a permanent set of 0.1%; 330 MPa (48 ksi) at a permanent set of 1%; 690 MPa (100 ksi) at a permanent set of 10%
Hardness. Cast and annealed, 84 to 89 HRB
Poisson's ratio. 0.32

Elastic modulus. Tension, 110 GPa (16 × 10^6 psi); shear, 42 GPa (6.1 × 10^6 psi)
Impact strength. Charpy keyhole, 13 J (10 ft · lbf) at −23 to 66 °C (−10 to 150 °F); Charpy V-notch, 22 J (16 ft · lbf) at −23 to 66 °C (−10 to 150 °F)
Fatigue strength. Rotating beam, 230 MPa (33 ksi) at 10^8 cycles

Structure

Crystal structure. Alpha, face-centered cubic; beta, close-packed hexagonal; kappa, ordered face-centered cubic
Microstructure. As-cast or annealed structures are generally continuous equiaxed alpha crystals with small areas of metastable beta phase. Kappa phase precipitates are found in the alpha phase, in grain boundaries, and in beta areas. Quench-and-temper treatments result in refinement and redistribution of the kappa phase throughout a matrix of tempered beta martensite and alpha-kappa eutectoid decomposition product. Some undissolved primary alpha crystals may also be present.

Mass Characteristics

Density. 7.64 g/cm³ (0.276 lb/in.³) at 20 °C (68 °F)
Patternmaker's shrinkage. 1.6%

Thermal Properties

Liquidus temperature. 1060 °C (1940 °F)
Solidus temperature. 1045 °C (1910 °F)
Coefficient of linear thermal expansion. 16.2 μm/m · K (9.0 μin./in. · °F) at 20 to 300 °C (68 to 572 °F)
Specific heat. 440 J/kg · K (0.105 Btu/lb · °F) at 20 °C (68 °F)
Thermal conductivity. 36 W/m · K (21 Btu/ft · h · °F) at 20 °C (68 °F)

Electrical Properties

Electrical conductivity. Volumetric, 7.1% IACS at 20 °C (68 °F)
Electrical resistivity. 243 nΩ · m at 20 °C (68 °F)

Magnetic Properties

Magnetic permeability. 1.05 at field strength of 16 kA/m

Chemical Properties

General corrosion behavior. Corrosion properties of C95800 are similar to those of other nickel-aluminum bronzes, except that C95800 has better resistance to cavitation and seawater fouling attack. Resists dealloying in most mediums

Fabrication Characteristics

Machinability. 50% of C36000 (free-cutting brass). Excellent surface finish and tolerances possible in all standard machining operations. Carbide or tool steel cutters may be used. Typical conditions using tool steel cutters: roughing speed, 76 m/min (250 ft/min) at a feed of 0.3 mm/rev (0.011 in./rev); finishing speed, 290 m/min (950 ft/min) at a feed of 0.1 mm/rev (0.004 in./rev)
Annealing temperature. 650 to 705 °C (1200 to 1300 °F)

C96200
90Cu-10Ni

Commercial Names

Common name. 90 Cu-10 Ni

Specifications

ASTM. Centrifugal, B 369; sand, B 369; ingot, B 30
Government. Centrifugal: QQ-C-390; MIL-C-15345, Alloy 25; MIL-C-20159, type II. Sand: QQ-C-390; MIL-C-20159, type II; MIL-V-18436
SAE. Centrifugal and sand: J461, J462

Chemical Composition

Composition limits. 84.5 to 87.0 Cu, 1.0 to 1.8 Fe, 9 to 11.0 Ni, 0.15 C max, 0.03 Pb max (0.01 Pb max for welding grades), 1.5 Mn max, 1.0 Nb max, 0.30 Si max

Applications

Typical uses. Component parts of items being used for seawater corrosion resistance

Mechanical Properties

Tensile properties. Properties for as-sand-cast separately cast (M01 temper) test bars: Tensile strength, 310 MPa min (45 ksi min); yield strength, 172 MPa min (25 ksi min); elongation, 20% min in 50 mm (2 in.)
Compressive strength. Typically, 255 MPa (37 ksi) at 0.1 mm/mm (0.1 in./in.) set
Impact strength. Charpy V-notch, 135 J (100 ft · lbf)
Elastic modulus. Tension, 124 GPa (18 × 10^6 ksi) at 20 °C (68 °F)

Mass Characteristics

Density. 8.94 g/cm³ (0.323 lb/in.³)
Patternmaker's shrinkage. 16 mm/m (3/16 in./ft)

Thermal Properties

Liquidus temperature. 1150 °C (2100 °F)
Solidus temperature. 1100 °C (2010 °F)
Specific heat. 376 J/kg · K (0.09 Btu/lb · °F) at 20 °C (68 °F)
Thermal conductivity. 45 W/m · K (26 Btu/ft · h · °F) at 20 °C (68 °F)

Electrical Properties

Electrical conductivity. Volumetric, 11% IACS at 20 °C (68 °F)

Fabrication Characteristics

Machinability. M01 temper; 10% of C36000 (free-cutting brass)
Weldability. Soldering and brazing: excellent. Gas shielded arc welding: poor. Metal arc welding: good, using R, Cu, Ni, or E, Cu, Ni filler metal. Oxyacetylene and carbon arc welding are not recommended.

C96400
70Cu-30Ni

Commercial Names

Previous trade name. 70-30 copper-nickel

Specifications

ASTM. Centrifugal castings: B 369; continuous castings: B 505; sand castings, ingot: B 30
Government. Centrifugal castings: MIL-C-15345 (Alloy 24); sand castings: QQ-C-390, MIL-C-20159 (type 1)

Chemical Composition

Composition limits. 65.0 to 69.0 Cu, 28.0 to 32.0 Ni, 0.50 to 1.5 Nb, 0.25 to 1.5 Fe, 1.5 Mn max, 0.50 Si max, 0.15 C max, 0.03 Pb max (0.01 Pb max for welding applications)

Applications

Typical uses. Centrifugal, continuous, and sand castings for valves, pump bodies, flanges, and elbows for applications requiring resistance to seawater corrosion

Mechanical Properties

Tensile properties. Typical data for sand-cast test bars: tensile strength, 470 MPa (68 ksi); yield strength, 255 MPa (37 ksi) at 0.5% extension under load; elongation, 28% in 50 mm (2 in.)
Hardness. Typical, 140 HB using 3000 kg load
Elastic modulus. Tension, 145 GPa (21 × 10^6 psi)
Impact strength. Charpy V-notch, 106 J (78 ft · lbf)
Fatigue strength. Reverse bending, 125 MPa (18 ksi) at 10^8 cycles

Mass Characteristics

Density. 8.94 g/cm³ (0.323 lb/in.³) at 20 °C (68 °F)
Patternmaker's shrinkage. 19 mm/m (7/32 in./ft)

Thermal Properties

Liquidus temperature. 1240 °C (2260 °F)
Solidus temperature. 1170 °C (2140 °F)
Coefficient of linear thermal expansion. 16 μm/m · K (9.0 μin./in. · °F) at 20 to 300 °C (68 to 572 °F)
Specific heat. 375 J/kg · K (0.09 Btu/lb · °F) at 20 °C (68 °F)
Thermal conductivity. 29 W/m · K (17 Btu/ft · h · °F) at 20 °C (68 °F)

Electrical Properties

Electrical conductivity. Volumetric, as-cast tempers: 5% IACS at 20 °C (68 °F)

Fabrication Characteristics

Machinability. 20% of C36000 (free-cutting brass)
Weldability. Soldering, brazing: excellent. Gas-shielded arc and shielded metal-arc welding: good, using RCuNi or ECuNi filler

metal. Oxyfuel gas and carbon arc welding are not recommended.

C96600
69.5Cu-30Ni-0.5Be

Commercial Names

Previous trade name. Beryllium cupro-nickel alloy 71C; CA966
Common name. Beryllium cupro-nickel

Specifications

Government. Sand castings: MIL-C-81519

Chemical Composition

Composition limits. 0.40 to 0.7 Be, 29.0 to 33.0 Ni, 0.8 to 1.1 Fe, 1.0 Mn max, 0.15 Si max, 0.01 Pb max, bal Cu
Consequence of exceeding impurity limits. An excessive amount of Si will increase as-cast hardness and lower ductility. High Pb will cause hot shortness.

Applications

Typical uses. C96600 is a high-strength version of the well-known cupro-nickel alloy C96400, possessing twice the strength. Like C96400, C96600 exhibits excellent corrosion resistance to seawater. Typical uses are high-strength constructional parts for marine service; pressure housings for long, unattended submergence; pump bodies; valve bodies; seawater line fittings; marine low-tide hardware; gimbal assemblies; and release mechanisms.
Precautions in use. See C82500.

Mechanical Properties

Tensile properties. Typical data for separately cast test bars. TB00 temper: tensile strength, 515 MPa (75 ksi); yield strength, 260 MPa (38 ksi); elongation in 50 mm (2 in.), 12%. TF00 temper: tensile strength, 825 MPa (120 ksi); yield strength, 515 MPa (75 ksi); elongation, 12%
Hardness. TB00 temper: 74 HRB. TF00 temper: 24 HRC
Poisson's ratio. 0.33
Elastic modulus. Tension, 150 GPa (22×10^6 psi); shear, 57 GPa (8.3×10^6 psi)

Mass Characteristics

Density. 8.80 g/cm^3 (0.320 lb/in.^{3}) at 20 °C (68 °F)
Patternmaker's shrinkage. 1.8%

Thermal Properties

Liquidus temperature. 1180 °C (2160 °F)
Solidus temperature. 1100 °C (2010 °F)
Coefficient of linear thermal expansion. 16 μm/m · K (9 μin./in. · °F) at 20 to 300 °C (68 to 572 °F)
Specific heat. 377 J/kg · K (0.091 Btu/lb · °F) at 20 °C (68 °F)
Thermal conductivity. 30 W/m · K (17.3 Btu/ft · h · °F) at 20 °C (68 °F)

Electrical Properties

Electrical conductivity. Volumetric, 4.3% IACS at 20 °C (68 °F)
Electrical resistivity. 4 nΩ · m at 20 °C (68 °F)

Chemical Properties

General corrosion behavior. Essentially identical to that of C96400

Fabrication Characteristics

Machinability. TF00 temper, 40% of C36000 (free-cutting brass)
Melting temperature. 1100 to 1180 °C (2010 to 2160 °F)
Casting temperature. 1260 to 1370 °C (2300 to 2500 °F)
Solution temperature. 995 °C (1825 °F)
Aging temperature. 510 °C (950 °F). Typical aging time, 3 h

C97300
56Cu-2Sn-10Pb-20Zn-12Ni

Commercial Names

Previous trade name. 12% nickel silver
Common name. Leaded nickel brass; 56-2-10-20-12

Specifications

ASTM. Centrifugal castings: B 271; sand castings: B 584; ingot: B 30

Chemical Composition

Composition limits. 53.0 to 58.0 Cu, 1.5 to 3.0 Sn, 8.0 to 11.0 Pb, 17.0 to 25.0 Zn, 11.0 to 14.0 Ni, 1.5 Fe max, 0.50 Mn max, 0.35 Sb max, 0.15 Si max, 0.08 S max, 0.05 P max, 0.005 Al max

Applications

Typical uses. Investment, centrifugal, permanent mold, and sand castings for hardware fittings; valves and valve trim; statuary, and ornamental castings

Mechanical Properties

Tensile properties. Typical data for sand-cast test bars: tensile strength, 240 MPa (35 ksi); yield strength, 115 MPa (17 ksi) at 0.5% extension under load; elongation, 20% in 50 mm (2 in.)
Hardness. Typical, 55 HB using 500 kg load
Elastic modulus. Tension, 110 GPa (16×10^6 psi)

Mass Characteristics

Density. 8.95 g/cm^3 (0.321 lb/in.^{3}) at 20 °C (68 °F)
Patternmaker's shrinkage. 16 mm/m (3/16 in./ft)

Thermal Properties

Liquidus temperature. 1040 °C (1904 °F)
Solidus temperature. 1010 °C (1850 °F)
Coefficient of linear thermal expansion. 16.2 μm/m · K (9.0 μin./in. · °F) at 20 to 260 °C (68 to 500 °F)
Specific heat. 375 J/kg · K (0.09 Btu/lb · °F) at 20 °C (68 °F)
Thermal conductivity. 28.5 W/m · K (16.5 Btu/ft · h · °F) at 20 °C (68 °F)

Electrical Properties

Electrical conductivity. Volumetric, as-cast tempers: 5.7% IACS at 20 °C (68 °F)

Fabrication Characteristics

Machinability. 70% of C36000 (free-cutting brass)
Weldability. Soldering, brazing: excellent. Welding: not recommended
Stress-relieving temperature. 260 °C (500 °F), 1 h for each 25 mm (1 in.) of section thickness
Casting temperature. Light castings, 1200 to 1315 °C (2200 to 2400 °F); heavy castings, 1090 to 1200 °C (2000 to 2200 °F). Melt rapidly at no more than 55 to 85 °C (100 to 150 °F) above maximum casting temperature.

C97600
64Cu-4Sn-4Pb-8Zn-20Ni

Commercial Names

Previous trade name. 20% nickel silver
Common name. Dairy metal, leaded nickel bronze, 64-4-4-8-20

Specifications

ASME. Sand castings: SB584
ASTM. Centrifugal castings: B 271; sand castings: B 584; ingot: B 30
Government. Sand castings: MIL-C-17112
Other. Ingot code number 412

Chemical Composition

Composition limits. 63.0 to 67.0 Cu, 3.5 to 4.5 Sn, 3.0 to 5.0 Pb, 3.0 to 9.0 Zn, 19.0 to 21.5 Ni, 1.5 Fe max, 1.0 Mn max, 0.25 Sb max, 0.15 Si max, 0.08 S max, 0.05 P max, 0.005 Al max

Applications

Typical uses. Centrifugal, investment, and sand castings for marine castings; sanitary fittings; ornamental hardware; valves, and pumps

Mechanical Properties

Tensile properties. Typical data for sand-cast test bars: tensile strength, 310 MPa (45 ksi); yield strength, 165 MPa (24 ksi) at 0.5% extension under load; elongation, 20% in 50 mm (2 in.)
Compressive properties. Compressive strength, 205 MPa (30 ksi) at a permanent set of 1%; 395 MPa (57 ksi) at a permanent set of 10%
Hardness. Typical, 80 HB using 500 kg load
Elastic modulus. Tension, 130 GPa (19×10^6 psi)
Impact strength. Charpy V-notch, 15 J (11 ft · lbf)
Fatigue strength. Reverse bending, 107

MPa (15.5 ksi) at 10^8 cycles
Creep-rupture characteristics. Limiting stress for creep of 10^{-5}%/h: 224 MPa (32.5 ksi) at 230 °C (450 °F); 153 MPa (22.2 ksi) at 290 °C (550 °F)

Mass Characteristics

Density. 8.90 g/cm^3 (0.321 lb/in.^{3}) at 20 °C (68 °F)
Patternmaker's shrinkage. 11 mm/m (1/8 in./ft)

Thermal Properties

Liquidus temperature. 1143 °C (2089 °F)
Solidus temperature. 1108 °C (2027 °F)
Coefficient of linear thermal expansion. 17 μm/m · K (9.3 μin./in. · °F) at 20 to 300 °C (68 to 572 °F)
Specific heat. 375 J/kg · K (0.90 Btu/lb · °F) at 20 °C (68 °F)
Thermal conductivity. 22 W/m · K (13 Btu/ft · h · °F) at 20 °C (68 °F)

Electrical Properties

Electrical conductivity. Volumetric, as-cast tempers: 5% IACS at 20 °C (68 °F)

Fabrication Characteristics

Machinability. 70% of C36000 (free-cutting brass)
Weldability. Soldering, brazing: excellent. Welding: not recommended
Stress-relieving temperature. 260 °C (500 °F), 1 h for each 25 mm (1 in.) of section thickness
Casting temperature. Light castings, 1260 to 1430 °C (2300 to 2600 °F); heavy castings, 1230 to 1320 °C (2250 to 2400 °F). Melt rapidly at no more than 55 to 85 °C (100 to 150 °F) above casting temperature range.

C97800
66.5Cu-5Sn-1.5Pb-2Zn-25Ni

Commercial Names

Previous trade name. 25% nickel silver
Common name. Leaded nickel bronze; 66-5-2-2-25

Specifications

ASTM. Centrifugal castings: B 271; sand castings: B 584; ingot, B 30

Chemical Composition

Composition limits. 64.0 to 67.0 Cu, 4.0 to 5.5 Sn, 1.0 to 2.5 Pb, 1.0 to 4.0 Zn, 24.0 to 27.0 Ni, 1.5 Fe max, 1.0 Mn max, 0.20 Sb max, 0.15 Si max, 0.08 S max, 0.05 P max, 0.005 Al max

Applications

Typical uses. Investment, permanent mold, and sand castings for ornamental castings; sanitary fittings; valve bodies; valve seats; and musical instrument components

Mechanical Properties

Tensile properties. Typical data for sand-cast test bars: tensile strength, 380 MPa (55 ksi); yield strength, 205 MPa (30 ksi) at 0.5% extension under load; elongation, 15% in 50 mm (2 in.)
Hardness. Typical, 130 HB using 3000 kg load
Elastic modulus. Tension, 130 GPa (19 × 10^6 psi)

Mass Characteristics

Density. 8.86 g/cm^3 (0.320 lb/in.^{3}) at 20 °C (68 °F)
Patternmaker's shrinkage. 16 mm/m (3/16 in./ft)

Thermal Properties

Liquidus temperature. 1180 °C (2156 °F)
Solidus temperature. 1140 °C (2084 °F)
Coefficient of linear thermal expansion. 17.5 μm/m · K (9.7 μin./in. · °F) at 20 to 260 °C (68 to 500 °F)
Specific heat. 375 J/kg · K (0.09 Btu/lb · °F) at 20 °C (68 °F)
Thermal conductivity. 25.4 W/m · K (14.7 Btu/ft · h · °F) at 20 °C (68 °F)

Electrical Properties

Electrical conductivity. Volumetric, as-cast tempers: 4.5% IACS at 20 °C (68 °F)

Fabrication Characteristics

Machinability. 60% of C36000 (free-cutting brass)
Weldability. Soldering, brazing: excellent. Welding: not recommended
Stress-relieving temperature. 260 °C (500 °F)

C99400
90.4Cu-2.2Ni-2.0Fe-1.2Al-1.2Si-3.0Zn

Commercial Names

Common name. Nondezincification alloy, NDZ

Chemical Composition

Composition limits. 0.25 Pb max, 1.0 to 3.5 Ni, 1.0 to 3.0 Fe, 0.50 to 2.0 Al, 0.50 to 2.0 Si, 0.50 to 5.0 Zn, 0.50 Mn max, bal Cu

Applications

Typical uses. Centrifugal, continuous, investment, and sand castings for valve stems; propeller wheels; electrical parts; gears for mining equipment; outboard motor parts; marine hardware; and other environmental uses where resistance to dezincification and dealuminification is required

Mechanical Properties

Tensile properties. Typical. M01 temper: tensile strength, 455 MPa (66 ksi); yield strength, 235 MPa (34 ksi) at 0.5% extension under load; elongation, 25% in 50 mm (2 in.). TF00 temper: tensile strength, 545 MPa (79 ksi); yield strength, 370 MPa (54 ksi) at 0.5% extension under load
Shear strength. M01 temper, 330 MPa (48 ksi)
Hardness. M01 temper, 125 HB; TF00 temper, 170 HB. Determined using 3000 kg load
Elastic modulus. Tension, 133 GPa (19.3 × 10^6 psi)

Mass Characteristics

Density. 8.30 g/cm^3 (0.30 lb/in.^{3}) at 20 °C (68 °F)
Patternmaker's shrinkage. 16 mm/m (3/16 in./ft)

Electrical Properties

Electrical conductivity. Volumetric, TF00 temper: 16.8% IACS at 20 °C (68 °F)

Fabrication Characteristics

Machinability. 50% of C36000 (free-cutting brass)
Weldability. Shielded metal-arc welding: poor
Solution temperature. 885 °C (1625 °F), 1 h for each 25 mm (1 in.) of section thickness
Aging temperature. 480 °C (900 °F), 1 h at temperature
Stress-relieving temperature. 315 °C (600 °F), 1 h for each 25 mm (1 in.) of section thickness

C99500

Commercial Name

Trade name. NDZ-S

Specifications

ASTM. Sand castings: B 763

Chemical Composition

Composition limits. 0.25 Pb max, 3.5 to 5.5 Ni, 3.0 to 5.0 Fe, 0.50 to 2.0 Al, 0.50 to 2.0 Si, 0.50 Mn max, 0.50 to 2.0 Zn, bal Cu

Applications

Typical uses. Valve stems, marine, and other environmental uses where resistance to dezincification and dealuminification is required, propeller wheels, electrical parts, gears for mining equipment and outboard marine industry; same as C99400 but used where higher yield strength is required

Mechanical Properties

Tensile properties. Properties for as-sand-cast separately cast (M01 temper) test bars: tensile strength, 483 MPa min (70 ksi min); yield strength, 275 MPa min (40 ksi min); elongation, 12% min in 50 mm (2 in.)
Hardness. Typically, 145 HB (500 kg); 50 HB (3000 kg)
Proportional limit. Typically, 145 MPa (21 ksi)

Mass Characteristics

Density. 8.3 g/cm^3 (0.30 lb/in.^{3})
Volume change on freezing. Patternmaker's shrinkage, 16 mm/m (3/16 in./ft)

Table 9 Typical mechanical properties of beryllium-copper alloy 21C

Temper	Tensile strength		Yield strength		Elongation in 50 mm (2 in.), %	Hardness
	MPa	ksi	MPa	ksi		
As-cast	515	75	275	40	25	75 HRB
Cast and aged(a)	825	120	725	105	5	30 HRC
Solution treated(b)	415	60	170	25	40	63 HRB
Solution treated(b) and aged(a)	1105	160	1035	150	1	42 HRC

(a) Aged 3 h at 345 °C (650 °F). (b) At 790 to 800 °C (1450 to 1475 °F)

Electrical Properties

Electrical conductivity. Volumetric, 13.7% IACS at 20 °C (68 °F)

Fabrication Characteristics

Machinability. M01 temper; 50% of C36000 (free-cutting brass)
Stress-relieving temperature. 315 °C (600 °F)

C99700 56.5Cu-5Ni-1Al-1.5Pb-12Mn-24Zn

Commercial Names

Common name. White manganese brass
Trade name. White Tombasil

Chemical Composition

Composition limits. 54.0 Cu min, 19.0 to 25.0 Zn, 11.0 to 15.0 Mn, 4.0 to 6.0 Ni, 2.0 Pb max, 1.0 Sn max, 1.0 Fe max, 0.50 to 3.0 Al

Applications

Typical uses. Building hardware (interior and exterior), architectural and ornamental fittings, marine hardware, floor drain covers, food handling equipment, swimming pool hardware, valves

Mechanical Properties

Tensile properties. Typical data for separately cast test bars. Sand cast: tensile strength, 380 MPa (55 ksi); yield strength, 170 MPa (25 ksi) at 0.5% extension under load; elongation, 25% in 50 mm (2 in.). Die cast: tensile strength, 450 MPa (65 ksi); yield strength, 185 MPa (27 ksi) at 0.5% extension under load; elongation, 15% in 50 mm (2 in.)
Hardness. Sand cast: 110 HB (3000 kg load); die cast: 125 HB
Elastic modulus. Tension, 114 GPa (16.5 × 10^6 psi)

Mass Characteristics

Density. 8.19 g/cm^3 (0.296 lb/in.^{3}) at 20 °C (68 °F)
Patternmaker's shrinkage. 21 mm/m (0.25 in./ft)

Thermal Properties

Liquidus temperature. 900 °C (1655 °F)
Solidus temperature. 880 °C (1615 °F)

Electrical Properties

Electrical conductivity. Volumetric, 3% IACS at 20 °C (68 °F)

Fabrication Characteristics

Machinability. 80% of C36000 (free-cutting brass)

C99750

Specifications

ASTM. Die castings: B 176

Chemical Composition

Composition limits. 0.25 to 3.0 Al, 55.0 to 61.0 Cu, 0.50 to 2.5 Pb, 17.0 to 23.0 Mn, 5.0 Ni max, 17.0 to 23.0 Zn, 1.0 Fe max (iron content shall not exceed nickel content)

Mechanical Properties

Tensile properties. Typical properties for as-sand-cast separately cast (M01 temper) test bars: tensile strength, 448 MPa (65 ksi); yield strength, 220 MPa (32 ksi) at 0.2% offset; elongation, 30% in 50 mm (2 in.)
Hardness. Typically, 77 HRB, 110 HB (500 kg)
Compressive strength. Typically, 193 MPa (28 ksi) at 0.001 mm/mm (0.001 in./in.) set, 262 MPa (38 ksi) at 0.01 mm/mm (0.01 in./in.) set, and 495 MPa (72 ksi) at 0.1 mm/mm (0.1 in./in.) set
Impact strength. Charpy V-notch, 100 J (75 ft · lbf)
Fatigue strength. 128 MPa (18.5 ksi) at 10^8 cycles
Elastic modulus. Tension, 117 GPa (17 × 10^6 psi) at 20 °C (68 °F)

Mass Characteristics

Density. 8.0 g/cm^3 (0.29 lb/in.^{3})

Thermal Properties

Liquidus temperature. 843 °C (1550 °F)
Solidus temperature. 819 °C (1505 °F)
Specific heat. 376 J/kg · K (0.09 Btu/lb · °F) at 20 °C (68 °F)

Electrical Properties

Electrical conductivity. Volumetric, 2% IACS at 20 °C (68 °F)

Fabrication Properties

Stress-relieving temperature. 260 °C (500 °F)

Beryllium copper 21C 97Cu-2Be-1Co

Commercial Names

Common name. Grain-refined beryllium-copper casting alloy 21C

Chemical Composition

Composition limits. 2.00 to 2.25 Be, 1.0 to 1.2 Co, 0.20 to 0.40 Si, 0.20 Ni max, 0.25 Fe max, 0.15 Al max, 0.10 Sn max, 0.02 Pb max, 0.10 Zn max, 0.10 Cr max
Consequence of exceeding impurity limits. See C82500.

Applications

Typical uses. The 1% Co content is a strong grain refiner, and as a result, this alloy is used instead of beryllium-copper alloys C82500 and C82400 when thin sections must be cast at high temperatures or when thick and thin sections are present within the same casting in order to achieve a uniform fine-grained structure. The higher cobalt content imparts better wear resistance but less desirable polishability and machinability. Typical uses are comparable to those of beryllium-copper alloys C82400 and C82500.
Precautions in use. See C82500.

Mechanical Properties

Tensile properties. See Table 9.
Hardness. See Table 9.
Poisson's ratio. 0.30
Elastic modulus. Tension, 128 GPa (18.5 × 10^6 psi); shear, 50 GPa (7.3 × 10^6 psi)

Mass Characteristics

Density. 8.26 g/cm^3 (0.298 lb/in.^{3}) at 20 °C (68 °F)
Dilation during aging. Linear, 0.2%
Change in density during aging. 0.6% increase
Patternmaker's shrinkage. 1.56%

Thermal Properties

Liquidus temperature. 980 °C (1800 °F)
Solidus temperature. 860 °C (1575 °F)
Incipient melting temperature. 835 °C (1535 °F)
Coefficient of linear thermal expansion. 10 μm/m · K (5.5 μin./in. · °F) at 20 to 200 °C (68 to 392 °F)
Specific heat. 419 J/kg · K (0.10 Btu/lb · °F) at 20 °C (68 °F)
Thermal conductivity. 105 W/m · K (61 Btu/ft · h · °F) at 20 °C (68 °F)

Electrical Properties

Electrical conductivity. Volumetric, 20% IACS at 20 °C (68 °F)
Electrical resistivity. 862 μΩ · m at 20 °C (68 °F)

Magnetic Properties

Magnetic susceptibility. See C82500.

Nuclear Properties

Effect of irradiation. See C82500.

Chemical Properties

Same as C82500

Table 10 Typical mechanical properties of cast beryllium cupro-nickel alloy 72C

Temper	Tensile strength MPa	Tensile strength ksi	Yield strength MPa	Yield strength ksi	Elongation in 50 mm (2 in.), %	Hardness
As-cast and aged(a)	555	81	310	45	15	90 HRB
Solution treated(b) and aged(a)	860	125	550	80	70	26 HRC

(a) Aged 3 h at 510 °C (950 °F). (b) Water quenched from 995 °C (1825 °F)

Fabrication Characteristics

Machinability. As-cast or solution treated, 30% of C36000 (free-cutting brass). Cast and aged, or solution treated and aged, 10 to 20% of C36000
Solution temperature. 790 to 800 °C (1450 to 1475 °F)
Aging temperature. 340 °C (650 °F)
Melting temperature. 860 to 980 °C (1575 to 1800 °F)
Casting temperature. Light castings, 1065 to 1175 °C (1950 to 2150 °F); heavy castings, 1000 to 1065 °C (1850 to 1950 °F)

Beryllium copper nickel 72C 68.8Cu-30Ni-1.2Be

Commercial Names

Common name. Modified beryllium cupronickel alloy 72C

Chemical Composition

Composition limits. 1.1 to 1.2 Be, 29.0 to 33.0 Ni, 0.7 to 1.0 Fe, 0.10 to 0.20 Zr, 0.10 to 0.20 Ti, 0.7 Mn max, 0.15 Si max, 0.1 Pb max, bal Cu
Consequence of exceeding impurity limits. High silicon will raise as-cast hardness and lower ductility. High lead will cause hot shortness. High carbon will result in undesirable carbides.

Applications

Typical uses. Alloy 72C is a modified version of beryllium cupro-nickel alloy 71C, its increased beryllium content providing improved castability. Its field of application is the plastic tooling industry. Alloy 72C generally is ceramic mold cast into tooling used for molding flame-retardant plastics containing bromine, bromine-boron, chlorinated paraffins and phosphates, and other halogens. Additionally, alloy 72C tooling is resistant to corrosion by the foaming agents used in structural plastics that generate ammonia at elevated temperatures, as well as to decompositional products of PVC that contain HCl. The good castability of 72C allows it to be cast into tooling of fine detail.
Precautions in use. See C82500.

Mechanical Properties

Tensile properties. See Table 10.
Hardness. See Table 10.
Poisson's ratio. 0.33
Elastic modulus. Tension, 150 GPa (22 × 10^6 psi); shear, 57 GPa (8.3 × 10^6 psi)

Mass Characteristics

Density. 8.60 g/cm^3 (0.311 lb/in.^{3}) at 20 °C (68 °F)
Patternmaker's shrinkage. 1.8%

Thermal Properties

Liquidus temperature. 1155 °C (2110 °F)
Solidus temperature. 1065 °C (1950 °F)
Coefficient of linear thermal expansion. 16 μm/m · K (9 μin./in. · °F) at 20 to 300 °C (68 to 572 °F)
Specific heat. 337 J/kg · K (0.08 Btu/lb · °F) at 20 °C (68 °F)
Thermal conductivity. 30 W/m · K (17 Btu/ft · h · °F) at 20 °C (68 °F)

Electrical Properties

Electrical conductivity. Volumetric, 43% IACS at 20 °C (68 °F)
Electrical resistivity. 4 mΩ · m at 20 °C (68 °F)

Chemical Properties

General corrosion behavior. Essentially the same as C96400

Fabrication Characteristics

Machinability. Solution treated and aged, 40% of C36000 (free-cutting brass)
Solution temperature. 995 °C (1825 °F)
Aging temperature. 510 °C (950 °F)
Melting temperature. 1065 to 1155 °C (1950 to 2110 °F)
Casting temperature. 1200 to 1300 °C (2200 to 2400 °F)

Copper P/M Products

Erhard Klar and David F. Berry, SCM Metal Products, Inc.

COPPER-BASE POWDER-METALLURGY (P/M) products rank second after iron- and steel-base P/M products in terms of volume, with an estimated weight consumption in 1988 of 15 500 Mg (17 000 short tons) in North America. This consumption is associated with the major copper-base P/M applications, which include bronze bearings, copper and copper alloy structural parts, friction materials, copper carbon brushes, and high-electrical-conductivity copper. Another 4500 Mg (5000 tons) of copper and copper alloy powders were consumed as additions to iron and steel powders and for infiltration of iron and steel P/M parts. Figure 1 shows the total consumption of copper and copper alloy powders since 1950, including powders for other than P/M uses, which during the past ten years accounted for about 10% of the total. A gradual decline in total powder consumption since the early 1970s can be seen. This is attributed in part to the decline of bronze self-lubricating bearings through substitution of less-expensive dilute bronze (that is, iron containing), and iron-base bearings, or, particularly in the low-performance end of the market, by plastic bearings. Use of sintered metallic friction materials also decreased because of substitution by other materials.

This article briefly reviews the subject of copper-base P/M products in terms of powder production methods and the product properties/consolidation practices of the major applications mentioned above. Additional information on copper-base P/M is contained in *Powder Metallurgy*, Volume 7 of the 9th Edition of *Metals Handbook*.

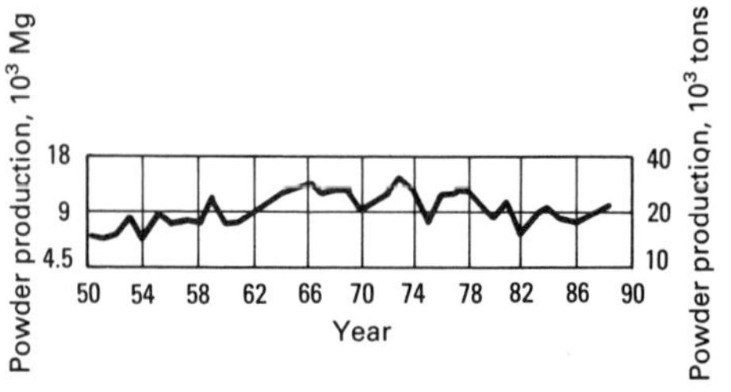

Fig. 1 Copper and copper alloy powder production in North America

Powder Production

Of the four major methods for making copper and copper alloy powders, atomization and oxide reduction are presently practiced on a large scale in North America; electrolytic and hydrometallurgical copper powders have not been manufactured in the United States since the early 1980s.

Table 1 shows a comparison of some of the typical fundamental powder characteristics of commercial copper powders made by various production processes. These powders are used in all major P/M copper-base products mentioned above except for brass and nickel silver structural parts, which are made exclusively from atomized prealloyed powders.

Atomization. The disintegration of a liquid metal stream by means of an impinging jet of liquid or gas is known as atomization, or more specifically, twin-fluid atomization. This process was originally used for the production of alloy powders but is now used increasingly for the production of plain iron and copper powders. Figure 2 is a schematic representation of the processing stages involved in atomization.

Figure 3 shows copper powders produced from a gas-atomizing medium and a water-atomizing medium. For plain copper powders, water is the preferred atomizing medium. The atomized powder is often subjected to an elevated temperature reduction and agglomeration treatment that improves its compacting properties. Table 2 shows powder properties of commercial grades of water-atomized copper powders.

The particle size of atomized powders can be varied greatly by adjusting processing conditions. Increases in gas or water pressure result in smaller particle sizes. For plain copper powders, average particle sizes less than 325 mesh (45 μm) are feasible.

The particle shape of water-atomized copper powder also can be controlled from almost spherical (Fig. 3c) to irregular (Fig. 3b) by modifying the water jet/metal stream interaction. Greater particle irregularity is possible through the use of small alloy additions to the copper melt that lower its surface tension.

Prealloyed Powders of Brass and Nickel Silver. Air atomization is used mainly for making prealloyed powders of brass and nickel silver, and to a lesser degree bronze powders, for use in high-density (>7.0 g/cm^3) components. The low surface tension of the molten alloys of these compositions renders the particle shape sufficiently irregular to make the powders compactible (Fig. 4). Reduction of oxides is not necessary for the standard P/M grades. Table 3 shows typical properties of commercial grades of brass, bronze, and nickel silver powders.

Commercial prealloyed brass powders are available in leaded and nonleaded compositions. Commercial brass alloys range from 90Cu-10Zn to 65Cu-35Zn; however, leaded versions of 80Cu-20Zn and 70Cu-30Zn are most commonly used for the manufacture of sintered structural parts that may require secondary machining operations. The only commercially available nickel silver powder has a nominal composition of 65Cu-18Ni-17Zn, which is modified by addition of lead when improved machinability is required.

Prealloyed atomized bronze powders are not used widely for structural parts fabrication because their nodular particle form and high apparent density result in low green strength. However, blends of such powders

Table 1 Characteristics of commercial copper powders

Type of powder	Composition, %: Copper	Oxygen	Acid insolubles	Particle shape	Surface area
Electrolytic	99.1–99.8	0.1–0.8	0.03 max	Dendritic	Medium to high
Oxide reduced	99.3–99.6	0.2–0.6	0.03–0.1	Irregular; porous	Medium
Water atomized	99.3–99.7	0.1–0.3	0.01–0.03	Irregular to spherical; solid	Low
Hydrometallurgical	97–99.5	0.2–0.8	0.03–0.8	Irregular agglomerates	Very high

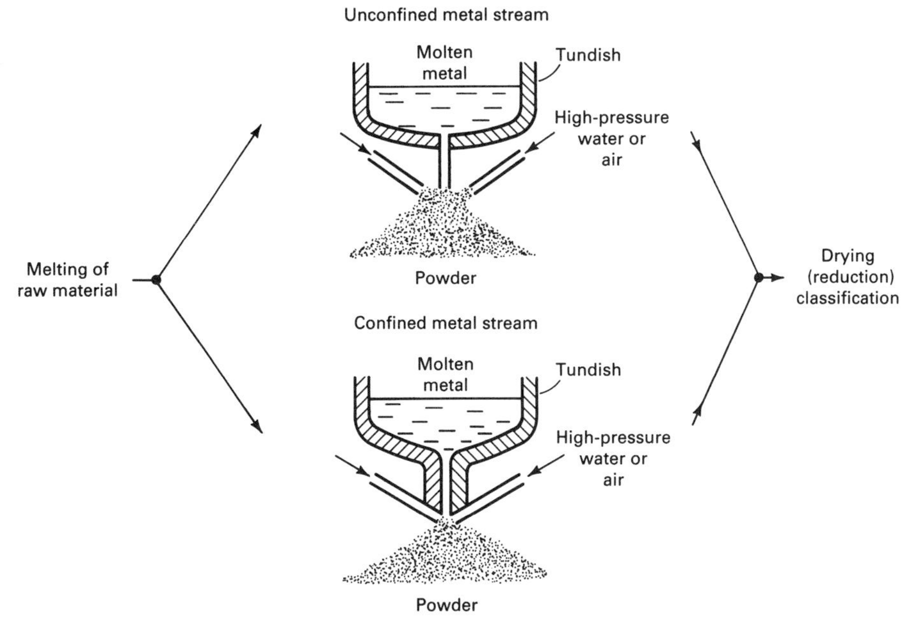

Fig. 2 Processing stages in production of metal powders by atomization

with irregular copper powders and phosphorus-copper yield sintered parts with good mechanical properties.

Reduction of Oxide. In this process particulate copper oxide is reduced with solid or gaseous reducing agents at elevated temperatures. The resulting sintered porous copper cake is milled to a powder. The raw material for this process was originally copper mill scale, but as demand exceeded supply, copper oxide had to be produced specifically from copper for the reduction process. Sources of raw material include particulate copper scrap, electrolytic copper, and atomized copper. Selection of raw material is based on purity requirements and end use. Table 4 shows properties of commercial grades of copper powder produced by copper oxide reduction.

Particle size is controlled through milling of the starting oxide and the reduced sinter cake. The milled copper powder particles are irregular and porous (Fig. 5). Through control of reduction conditions it is possible to obtain a broad range of pore characteristics. Compacting and other properties may be varied through control of the pore characteristics of the spongy particles.

Electrolysis. In this process the variables are adjusted so that a spongy or brittle polycrystalline deposit, rather than a smooth deposit, is formed at the cathode with electroplating. Aqueous electrolytes and soluble anodes are used. The low-metal overvoltage of copper permits its economic deposition as sponge. Electrolytic copper powder normally is of high purity; however, impurities more noble than copper are codeposited.

Typical processing conditions include an electrolyte concentration of 5 to 8 g/L of copper and 100 to 160 g/L of sulfuric acid, a bath temperature of about 50 °C (120 °F), a current density of 0.05 to 0.1 A/cm^2 (50 to 100 A/ft^2), and a cell voltage of about 1 V. A broad range of particle sizes is possible through control of processing conditions. The particle shape of as-deposited powder is dendritic or fernlike (Fig. 6). For applications requiring a powder with a low apparent density and a high surface area, additions to the electrolytic bath can be made that decrease dendrite arm thickness. A final furnace treatment can alter this shape to such an extent that the processing history of the powder is difficult to determine.

Hydrometallurgy. The basic processing steps include preparing a pregnant liquor by leaching ore or another suitable raw material, followed by precipitation of the metal from its solution. For copper the most important precipitation methods are cementation, reduction with hydrogen, and electrolysis.

In cementation, the copper-bearing solution is passed over scrap iron, which results in precipitation of copper according to:

$$Fe + CuSO_4 \rightarrow Cu + FeSO_4$$

Fig. 3 Scanning electron micrographs of gas- and water-atomized copper powders. (a) Nitrogen atomized. (b) Water atomized, apparent density of 3.04 g/cm^3. (c) Water atomized, apparent density of 4.60 g/cm^3

Table 2 Properties of commercial grades of water-atomized copper powders

	Chemical properties, %		Physical properties						
					Tyler sieve analysis, %				
Copper, %	Hydrogen loss	Acid insolubles	Hall flow rate, s/50 g	Apparent density, g/cm^3	+100	−100+150	−150+200	−200+325	−325
99.65(a)	0.28	...	...	2.65	Trace	0.31	8.1	28.2	63.4
99.61(a)	0.24	...	...	2.45	0.2	27.3	48.5	21.6	2.4
99.43(a)	0.31	...	...	2.70	tr	0.9	3.2	14.2	81.7
>99.1(b)	<0.35	<0.2	~50	2.4	<8	17–22	18–30	22–26	18–38

(a) Water atomized plus reduced. (b) Contains magnesium

Table 3 Physical properties of typical brass, bronze, and nickel silver alloy compositions

Property	Brass(a)	Bronze(a)	Nickel silver(a)(b)
Sieve analysis, %			
+100 mesh	2.0 max	2.0 max	2.0 max
−100+200	15–35	15–35	15–35
−200+325	15–35	15–35	15–35
−325	60 max	60 max	60 max
Physical properties			
Apparent density	3.0–3.2	3.3–3.5	3.0–3.2
Flow rate, s/50 g	24–26	...	...
Mechanical properties			
Compressibility (c) at 415 MPa (30 tsi), g/cm³	7.6	7.4	7.6
Green strength(c) at 415 MPa (30 tsi), MPa (psi)	10–12 (1500–1700)	10–12 (1500–1700)	9.6–11 (1400–1600)

(a) Nominal mesh sizes: brass, −60 mesh; bronze, −60 mesh; nickel silver, −100 mesh. (b) Contains no lead. (c) Compressibility and green strength data of powders lubricated with 0.5% lithium stearate

Fig. 4 Prealloyed air-atomized nickel silver powder (63Cu-18Ni-17Zn-2Pb). 165×

Subsequent separation, washing, thermal reduction, and pulverizing usually produce a copper powder that contains significant amounts of iron and acid insolubles such as alumina and silica. Contamination with gangue varies and depends on the nature of the pregnant liquor. Low purity in general, and high-iron content in particular, restrict the use of cement copper in P/M applications. Its irregular particle shape and high specific surface area (Fig. 7), however, impart good green strength and make it useful in friction applications.

Self-Lubricating Sintered Bronze Bearings

Mechanism of Lubrication. The function of a bearing is to guide a moving part with as little friction as possible. For sintered self-lubricating bearings this is accomplished by using the interconnected porosity of the bearing as an oil reservoir. Figure 8 shows schematically the mechanism of this type of lubrication for a rotating shaft. As the shaft begins to rotate, metal-to-metal friction between the shaft and the bearing causes the temperature of the bearing assembly to rise. As a result, the oil contained in the pores of the bearing expands, and the oil wedge (that is, the space between the shaft and the bearing) is partially filled with oil.

Rotation of the shaft develops a so-called hydrodynamic pressure, p, within the oil film that with correct clearance, shaft velocity, and pore structure of the bearing is able to lift the shaft so that it rides on a liquid film of oil. This is known as hydrodynamic lubrication and is a condition of lowest friction. During operation, the oil that passes into the pores of the bearing is being recirculated to the unloaded region. With low shaft velocities and during startup, the hydrodynamic pressure is insufficient to separate shaft and bearing. This leads to so-called "mixed" or even to "boundary" lubrication with attendant friction increase, temperature rise, oil loss, wear, and reduced bearing life. When the shaft ceases to rotate, the temperature of the assembly decreases and the oil within the oil wedge is drawn back into the porous bearing by capillary forces. Thus, the oil can be reused many times.

Uses. For light-duty applications, these bearings are designed to last for the life of the equipment or machine in which they are used. For medium- and heavy-duty applications, relubrication is usually necessary. Table 5 shows examples of applications. These bearings run quietly and may be used in vertical positions, whereas solid bearings would normally be impractical because of lubricant run-out. They are particularly useful if it is difficult to lubricate the part, such as in a refrigerator motor, or where oil splashing may interfere with the operation of the machine.

Composition. The most widely used bearing material is 90Cu-10Sn bronze, often with the addition of up to 1.5% graphite. So-called dilute bronze bearings contain various amounts of iron. Dilution with iron reduces the cost of a bearing at the expense of some loss in performance.

Although bronze bearings can be produced from partially or fully prealloyed powders, they are predominantly made from elemental powder blends. The tin powders used in these blends are typically made by air atomization.

Compaction pressures range from about 10 to 30 tsi. Figure 9 shows an assortment of P/M bronze bearings. The most common shapes are simple or flanged bushings, but self-aligning bearings with spherical external surfaces are also used. Sizes range from about 0.8 to 75 mm (1/32 to 3 in.) in diameter.

Sintering is typically done in a continuous-mesh belt furnace at temperatures between 815 to 870 °C (1500 to 1600 °F) for about 3 to 8 min at temperature. Typical furnace atmospheres are dissociated ammonia or endothermic gas. To obtain reproducible sintering results it is important to carefully control time and temperature because of their influence upon the kinetics of the liquid-phase alloying process, which in turn determines the dimensional changes taking place during sintering. The desired microstructure is an alpha-bronze such as shown in Fig. 10.

Sizing. Most bearings are sized for improved dimensional accuracy. Sizing pres-

Table 4 Properties of commercial grades of copper powder produced by the copper oxide process

	Chemical properties, %					Physical properties							Compacted properties		
								Tyler sieve analysis, %						Green strength, MPa (psi), at:	
Copper	Tin	Graphite	Lubricant	Hydrogen loss	Acid insolubles	Apparent density, g/cm³	Hall flow rate, s/50 g	+100	+150	+200	+325	−325	Green density, g/cm³	165 MPa (12 tsi)	6.30 g/cm³
99.53	...	...	...	0.23	0.04	2.99	23	0.3	11.1	26.7	24.1	37.8	6.04	6.15 (890)	...
99.64	...	...	...	0.24	0.03	2.78	24	...	0.6	8.7	34.1	56.6	5.95	7.85 (1140)(a)	...
99.62	...	...	...	0.26	0.03	2.71	27	...	0.3	5.7	32.2	61.8	5.95	9.3 (1350)(a)	...
99.36	...	...	...	0.39	0.12	1.56	...	0.1	1.0	4.9	12.8	81.2	5.79	21.4 (3100)(a)	...
99.25	...	...	...	0.30	0.02	2.63	30	0.08	7.0	13.3	16.0	63.7	...	...	8.3 (1200)(a)
90	10	...	0.75	...	...	3.23	30.6	0.0	1.4	9.0	32.6	57.0	6.32	...	3.80 (550)
88.5	10	0.5	0.80	...	...	3.25	12(b)	...	...	...	...	...	...	...	3.6 (525)

(a) Measured with die wall lubricant only. (b) Carney flow

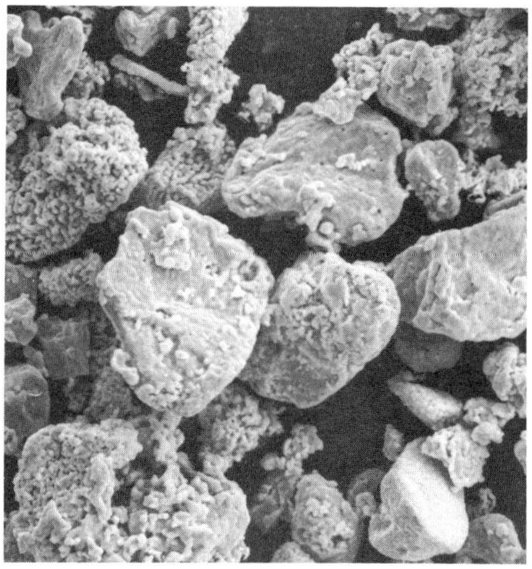

Fig. 5 Oxide-reduced copper powder. 500×

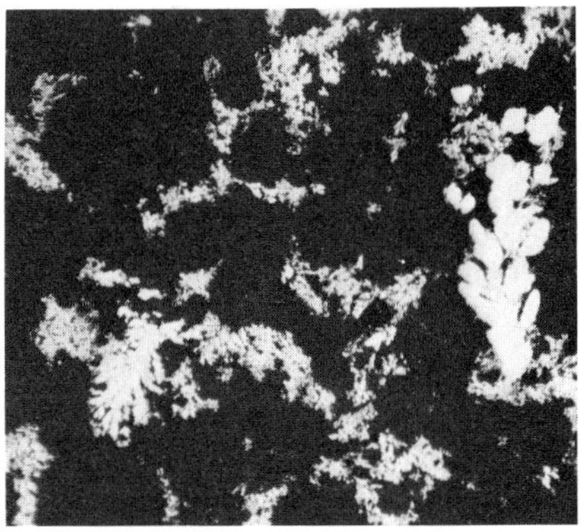

Fig. 6 Electrolytic copper powder showing dendritic structure. 85×

Table 5 Applications of self-lubricating sintered bronze bearings (fractional horsepower electric motors)

Automotive components	**Consumer electronics**
Starters	Phonographs
Light generators	Record changers
Oil and water pumps	Tape recorders
Windshield wipers	**Business machines**
Hood and window raisers	Typewriters
Heaters	Computers
Air conditioners	Copiers
Power antennae	**Industrial equipment**
Power seat adjusters	Textile machines
Home appliances	Packaging machines
Dishwashers	Electric fans
Clothes dryers	**Portable power tools**
Washing machines	Drills
Sewing machines	Saws
Vacuum cleaners	
Refrigerators	
Food mixers	
Farm and lawn equipment	
Tractors	
Combines	
Cotton pickers	
Lawn mowers	
String cutters	
Chain saws	

sures range from about 200 to 550 MPa (15 to 40 tsi). During sizing the density increases slightly.

Impregnation. Bearings are marketed either dry or saturated with oil, usually by a vacuum impregnation process that enables oil efficiencies of 90% or more to be achieved (that is, 90% or more of the available porosity is filled with oil). For use, these bearings are force-fitted into a housing.

Load-Carrying Capacity and Bearing Life. Tables 6 and 7 show minimum strength, oil content, densities, and typical loads for two low-graphite bronze compositions. The product of shaft surface velocity, *V*, times specific bearing load, *P*, the so-called *PV* factor, is a useful parameter for describing bearing performance. Strictly speaking, the concept of a permissible or maximum *PV* value means that a bearing should operate under the hydrodynamic (lowest friction) mode of lubrication for any combination of load and velocity not exceeding that maximum *PV* value. For this to be valid it is assumed that other bearing running conditions (shaft clearance, shaft alignment, housing design, oil viscosity, and so forth) have been optimized. For low velocities or for frequent start/stop operation, however, a coherent load-bearing oil film may not be formed. Also, at high velocities, oil losses may be excessive. Under these conditions, higher bearing temperatures and reduced bearing life may apply even at low *PV* values.

For a given configuration the bearing temperature increases with increasing *PV*. With increasing bearing temperature the life of the bearing decreases steeply. Use of a bearing below its maximum *PV* value results in a large increase in bearing life as a result of the lower temperature, which in turn reduces oil losses from evaporation and decomposition. A reduction in temperature by 10 °C (20 °F) will approximately double the life of the oil (Fig. 11). Under conditions where the lubrication mode

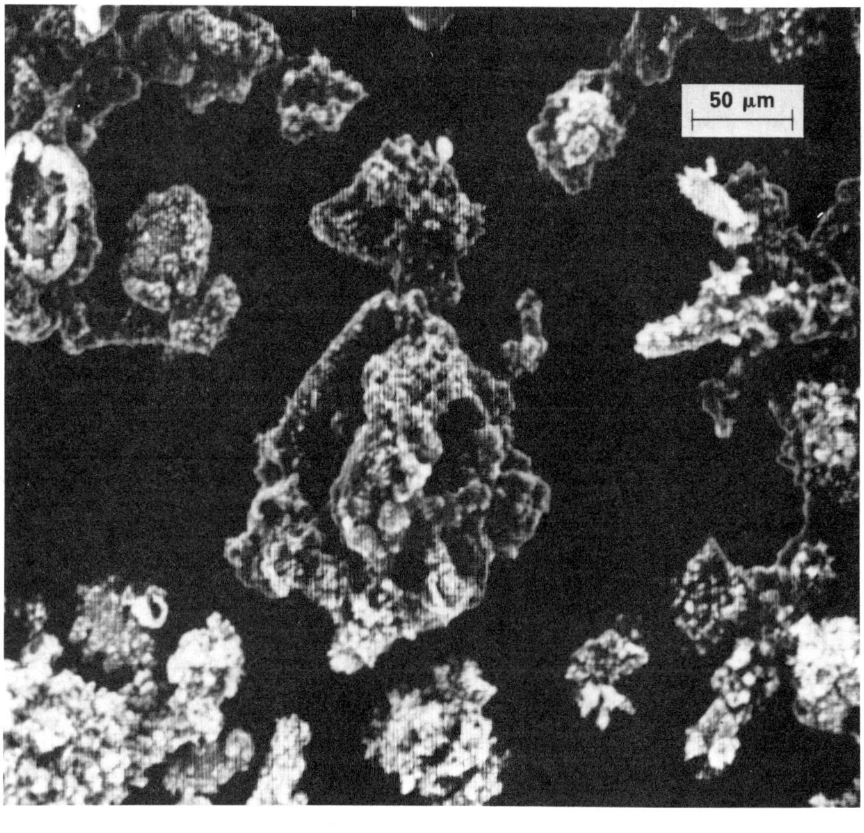

Fig. 7 Scanning electron micrograph of hydrometallurgically produced copper powder (cement copper)

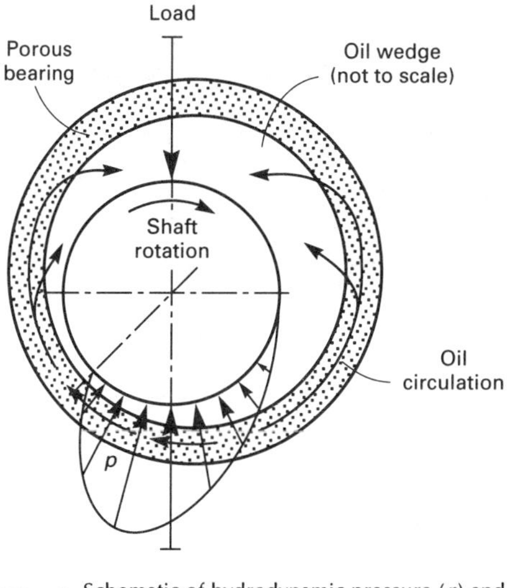

Fig. 8 Schematic of hydrodynamic pressure (*p*) and oil circulation in an oil-impregnated porous bearing

Fig. 9 Assorted P/M bronze bearings

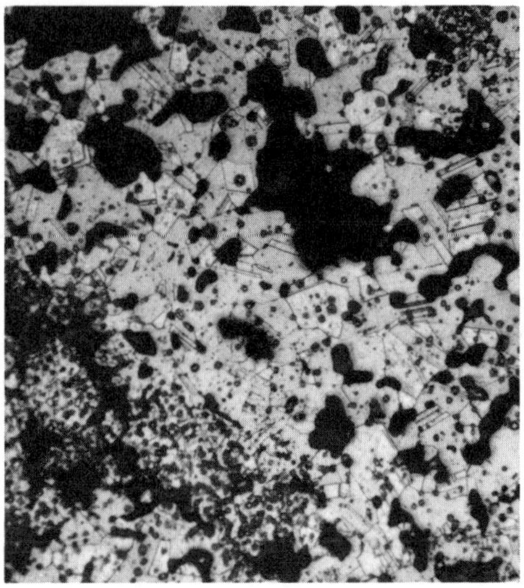

Fig. 10 Microstructure of P/M 90Cu-10Sn bronze

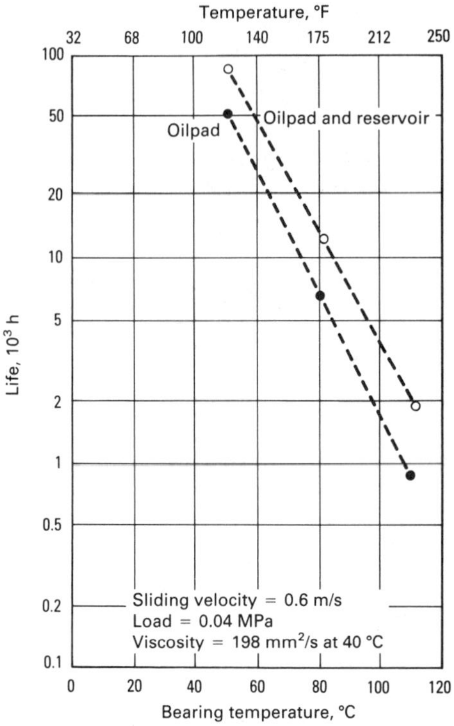

Fig. 11 Life of sintered bronze bearings MKZ (Sint-B50) in fan motors with different lubrication as a function of temperature using increased volume of supplementary lubrication. Source: Ref 2

changes from hydrodynamic to mixed, the friction coefficient rises with an attendant rise of the oil temperature and decreases in bearing life. For mixed lubrication conditions, addition of up to 1.5% graphite is often made; this has been shown to decrease bearing temperature and increase bearing life in applications including intermittent stop/start operation.

New Developments. In recent years there have been several attempts to improve and expand the performance of sintered bronze bearings. In 1965, Youssef and Eudier (Ref 3) described porous bearings that contained a layer of very fine powder at the inside bore of the bearing, permitting load capacity to be increased by a factor of 10 to 20. This radical improvement was attributed to the fine porous layer that prevented air introduction and subsequent oil loss. In 1980, Kohno *et al.* (Ref 4) described phosphorus, molybdenum disulfide, and graphite-containing bronze bearings that showed excellent performance under both boundary and hydrodynamic lubrication conditions at velocities of up to 800 m/min (2600 ft/min) and *PV* factors of 250 MPa · m/min (120 ksi · ft/min). In 1983, Eudier and Youssef (Ref 5) obtained a patent for P/M bronze bearings containing dispersed hard faces with varying amounts of antimony, bismuth, and nickel. Improved *PV* factors (3 to >6 MPa · m/s, or 85 to 170 ksi · ft/min) and suppression of the temperature peak during the running-in period were attributed to the formation of low-melting glasslike intermetallic phases. In 1985, Shikata *et al.* (Ref 6) described molybdenum disulfide and graphite-containing bronze bearings that exhibited lower and more stable friction coefficients at low speeds (about 0.02 m/s, or 4 ft/min). In the above cases much of the improvement was attributed to the strengthening of the matrix with phosphorus and nickel, respectively, and the presence of MoS_2, which improved lubrication and decreased wear under boundary conditions.

Bearing life is affected by a large number of factors, from powder properties, compaction, sintering, sizing, and choice of oil, to design and thermal properties of the bearing housing. The section "P/M Self-Lubricating Bearings" of *MPIF Standard 35* (Ref 1) provides information on press fits, interference fits, running clearances, and dimensional tolerances, in addition to chemical and mechanical properties. Some bearing manufacturers offer proprietary bearing compositions. It is therefore recommended that the designer consult with the bearing manufacturer for more specific advice. References 7 and 8 by V.T. Morgan give the most comprehensive descriptions of this subject available in the literature.

Table 6 Properties of sintered bronze (low-graphite) bearings

Material description code	Chemical composition, wt% Element	Min	Max	Minimum strength constant (*K*) MPa	ksi	Minimum oil content, vol%	Density (D_{wet}), g/cm³ Min	Max
CT-1000-K19	Copper	87.2	90.5					
	Tin	9.5	10.5	130	19	24	6.0	6.4
	Graphite	0	0.3					
	Other	...	2.0					
CT-1000-K26	Copper	87.2	90.5					
	Tin	9.5	10.5	180	26	19	6.4	6.8
	Graphite	0	0.3					
	Other	...	2.0					

Source: Ref 1

Copper-Base Structural Parts

Applications of copper-base P/M materials that rely mainly on the load-bearing capacities of the sintered parts are commonly classified as structural applications. The most important copper-base structural-part compositions include brass, nickel silver, and bronze. Structural P/M parts also include pure-copper P/M products and oxide-dispersion-strengthened (ODS) copper for applications where good electrical and thermal conductivity is important. Oxide-dispersion-strengthened copper is described in a separate section so-named in this article.

P/M Structural Parts From Brass, Nickel Silver, or Bronze. The use of P/M techniques for producing these parts is due to economic advantages such as cost savings in labor and materials. These classes of P/M materials are characterized by their combination of mechanical strength, ductility, and corrosion resistance.

Compositions and Properties. Compositions (Table 8) and properties (Table 9) of structural parts of brass, bronze, and nickel silver are covered in *MPIF Standard 35* (Ref 1). The leaded compositions are used whenever secondary machining operations are required.

Table 7 Recommended loads and shaft velocities for sintered bronze bearings

Bearing material	Loading, MPa (ksi), for a shaft velocity of: Static	Slow and intermittent	7.5 m/min (25 ft/min)	15–30 m/min (50–100 ft/min)	30–45 m/min (100–150 ft/min)	45–60 m/min (150–200 ft/min)	60–150 m/min (200–500 ft/min)	150–300 m/min (500–1000 ft/min)
CT-1000-K19	38	22	14	3.8	2.5	1.9	1245/v(a)	1355/v(a)
	(5.5)	(3.2)	(2.0)	(0.55)	(0.365)	(0.280)	55/V(b)	60/V(b)
CT-1000-K26	59	27.5	14	3.4	2.25	1.7	1130/v(a)	...
	(8.5)	(4.0)	(2.0)	(0.50)	(0.325)	(0.250)	50/V(b)	...
CT-1000-K37	77.5	31	12.5	3.1	2.05	1.55	1020/v(a)	...
	(11.25)	(4.5)	(1.8)	(0.45)	(0.300)	(0.225)	45/V(b)	...
F-0000-K15	52	25	12.5	2.75	1.60	1.2	790/v(a)	...
	(7.5)	(3.6)	(1.8)	(0.40)	(0.235)	(0.175)	35/V(b)	...
FC-1000-K23	103	55	20	4.8	2.75	2	905/v(a)	...
	(15)	(8.0)	(3.0)	(0.70)	(0.400)	(0.300)	40/V(b)	...

(a) Load in MPa with v expressed in m/min. (b) Load in ksi with V expressed in ft/min. Source: Ref 1

Uses. Typical applications for brass and nickel silver parts include latch bolts and cylinders for locks; shutter mechanism components for cameras; gears, cams, and actuator bars in timing assemblies and in small-generator drive assemblies; and decorative trim and medallions. In many of these applications, corrosion resistance, wear resistance, and aesthetic appearance play important roles. The surface finish may be improved by burnishing. An assortment of brass P/M parts is shown in Fig. 12.

Bronze P/M structural parts, which generally are produced by methods similar to those used for self-lubricating bearings, frequently are selected because of the corrosion and wear resistance of bronze. Bronze P/M components are used in applications such as automobile clutches, copiers, outboard motors, and paint-spraying equipment. Figure 13 illustrates an assortment of bronze P/M parts.

Compaction. Brass and nickel silver alloys are usually blended with lubricants in amounts from 0.5 to 1.0 wt%. Lithium stearate is the preferred lubricant because of its cleansing and scavenging action during sintering. However, bilubricant systems are common, such as lithium stearate and zinc stearate, on a 50/50 basis to minimize the surface staining attributed to excessive lithium stearate. Lubricated powders compact to 75% of theoretical density at 207 MPa (30 ksi) and to 85% of theoretical density at 415 MPa (60 ksi).

Sintering of brass and nickel silver compacts is normally performed in protective atmospheres (that is, dissociated ammonia, endothermic gas, and nitrogen-base atmospheres) at temperatures ranging from 815 to 925 °C (1500 to 1700 °F) depending on alloy composition. To avoid distortion and/or blistering of the compacts, sintering temperatures should not exceed the solidus temperature of the alloy. Through multiple pressing and sintering, yield strength and hardness may approach those of the wrought alloy counterparts. To minimize zinc losses during sintering, yet allow for adequate lubricant removal, protective-sintering-tray arrangements are used. Figure 14 shows the marked effect different lubricants have upon various properties after sintering. Figure 15 shows the effect of sintering time upon sintered strength and dimensional change of a 70Cu-30Zn leaded brass.

Pure copper P/M parts are used mainly in electrical and electronic applications, although ODS copper has also been developed for electrical/electronic applications (see the section "Oxide-Dispersion-Strengthened Copper" in this article). Oxide-dispersion-strengthened copper offers better strength at room and elevated temperatures.

In the production of pure-copper P/M parts, it is essential to use very pure copper

Table 8 Compositions of copper-base P/M structural materials (brass, bronze, and nickel silver)

Material designation	Chemical composition, %: Cu	Zn	Pb	Sn	Ni	Element
CZ-1000	88.0	rem	...	...	...	Min
	91.0	rem	...	...	...	Max
CZP-1002	88.0	rem	1.0	...	...	Min
	91.0	rem	2.0	...	...	Max
CZP-2002	77.0	rem	1.0	...	...	Min
	80.0	rem	2.0	...	...	Max
CZ-3000	68.5	rem	...	...	...	Min
	71.5	rem	...	...	...	Max
CZP-3002	68.5	rem	1.0	...	...	Min
	71.5	rem	2.0	...	...	Max
CNZ-1818	62.5	rem	...	...	16.5	Min
	65.5	rem	...	...	19.5	Max
CNZP-1816	62.5	rem	1.0	...	16.5	Min
	65.5	rem	2.0	...	19.5	Max
CT-1000	87.5	rem	...	9.5	...	Min
	90.5	rem	...	10.5	...	Max

Note: Total by difference equals 2.0% max, which may include other minor elements added for specific purposes. Source: Ref 1

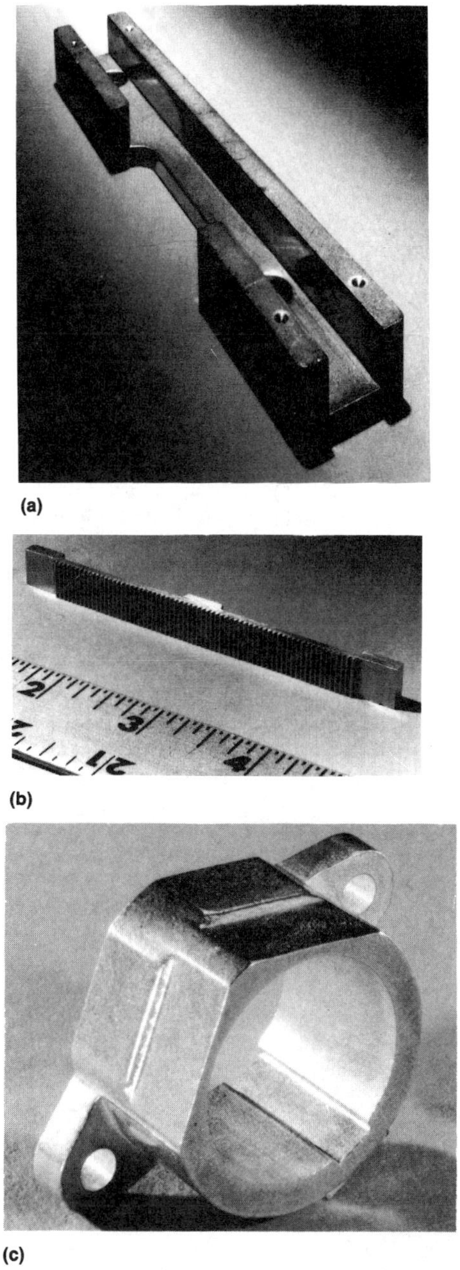

Fig. 12 P/M brass components. (a) Brass rack guide for rack-and-pinion steering column of an electric outdoor motor. (b) Leaded brass rack for a stereo three-dimensional microscope. (c) Leaded brass objective mounts for a microscope. Courtesy of Metal Powder Industries Federation

Fig. 13 Assorted P/M bronze parts. Courtesy of Norddeutsche Affinerie

Table 9 Properties of copper-base P/M structural materials (brass, bronze, and nickel silver)

Mechanical property data derived from laboratory-prepared test specimens sintered under commercial manufacturing conditions

Material designation code(a)	Minimum yield strength MPa	Minimum yield strength ksi	Typical values: Ultimate tensile strength MPa	Ultimate tensile strength ksi	Yield strength (0.2%) MPa	Yield strength (0.2%) ksi	Elongation in 25 mm (1 in.) %	Young's modulus GPa	Young's modulus 10^6 psi	Transverse rupture strength MPa	Transverse rupture strength ksi	Unnotched Charpy impact strength J	Unnotched Charpy impact strength ft · lbf	Density g/cm³	Compressive yield strength (0.1%) MPa	Compressive yield strength (0.1%) ksi	Apparent hardness, HRH
CZ-1000-9	62	9	124	18.0	65	9.5	9.0	52	7.5	270	39	(b)	(b)	7.60	(b)	(b)	65
CZ-1000-10	70	10	138	20.0	76	11.0	10.5	69	10.0	315	46	(b)	(b)	7.90	(b)	(b)	72
CZ-1000-11	75	11	159	23.0	83	12.0	12.0	(b)	(b)	360	52	(b)	(b)	8.10	(b)	(b)	80
CZP-1002-	(b)	(b)	(b)	(b)	(b)	(b)	(b)	(b)	(b)	(b)	(b)	(b)	(b)	(b)	(b)	(b)	(b)
CZP-2002-11	75	11	159	23.0	93	13.5	12.0	69	10.0	345	50	38	28.0	7.60	103	15.0	75
CZP-2002-12	83	12	207	30.0	110	16.0	14.5	83	12.0	480	70	76	56.0	8.00	110	16.0	84
CZ-3000-14	97	14	193	28.0	110	16.0	14.0	62	9.0	425	62	31	23.0	7.60	83	12.0	84
CZ-3000-16	110	16	234	34.0	131	19.0	17.0	69	10.0	590	86	51.5	38.0	8.00	90	13.0	92
CZP-3002-13	90	13	186	27.0	103	15.0	14.0	62	9.0	395	57	(b)	(b)	7.60	(b)	(b)	80
CZP-3002-14	97	14	217	31.5	115	16.5	16.0	69	10.0	490	71	(b)	(b)	8.00	(b)	(b)	88
CNZ-1818-17	117	17	234	34.0	140	20.0	11.0	75	11.0	500	73	32.5	24.0	7.90	172	25.0	90
CNZP-1816-	(b)	(b)	(b)	(b)	(b)	(b)	(b)	(b)	(b)	(b)	(b)	(b)	(b)	(b)	(b)	(b)	(b)
CT-1000-13 (repressed)	90	13	152	22.0	110	16.0	4.0	38	5.5	310	45	5.4	4.0	7.20	186	27.0	82

(a) Suffix numbers represent minimum yield-strength values in ksi. (b) Additional data in preparation will appear in subsequent editions of *MPIF Standard 35*. Source: Ref 1

powders (≧99.95% purity) or to bring about the precipitation of soluble impurities during sintering. As little as 0.023% Fe in solid solution in copper lowers its conductivity to 86% of that of pure copper. Small amounts of iron mechanically mixed with the copper powder lower the conductivity much less, unless the iron dissolves in the copper during sintering. If high-purity copper is used, or if soluble impurities are precipitated during sintering, it is possible to obtain the values of strength and conductivity shown in Fig. 16.

Conductivity is directly related to porosity; the greater the void content, the lower the conductivity. Electrical conductivity of as-pressed and sintered pure copper parts varies from 80 to 90% IACS. Full-density properties, as shown above, are reached or approached by compacting at moderate pressure of 205 to 250 MPa (15 to 18 tsi), sintering at temperatures 50 to 150 °C (90 to 270 °F) below the melting point of copper (1083 °C, or 1981 °F), followed by re-pressing, coining, or forging.

Typical applications of pure copper parts in which high electrical conductivity is required include commutator rings, contacts, shading coils, nose cones, and electrical twist-type plugs. Copper powders also are used in copper-graphite compositions that have low contact resistance, high current-carrying capacity, and high thermal conductivity. Typical applications include brushes for motors and generators and moving parts for rheostats, switches, and current-carrying washers.

Friction Materials

Sintered-metal friction materials were developed in the 1920s and commercialized in the early 1930s by Wellman. They are used in applications involving the transmission of motion through friction (clutches) and for deceleration and braking. In these processes mechanical energy is converted into frictional heat, which is absorbed and dissipated by the friction material (that is, the brake linings or clutch facings).

Metal-base friction materials are strong and heat resistant and were developed in

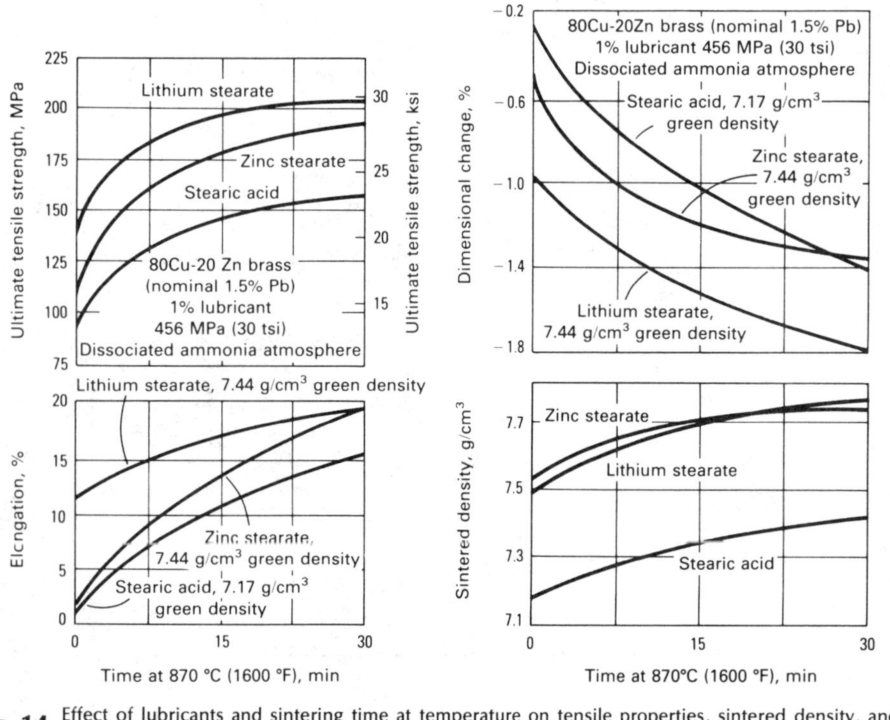

Fig. 14 Effect of lubricants and sintering time at temperature on tensile properties, sintered density, and dimensional change of brass compacts

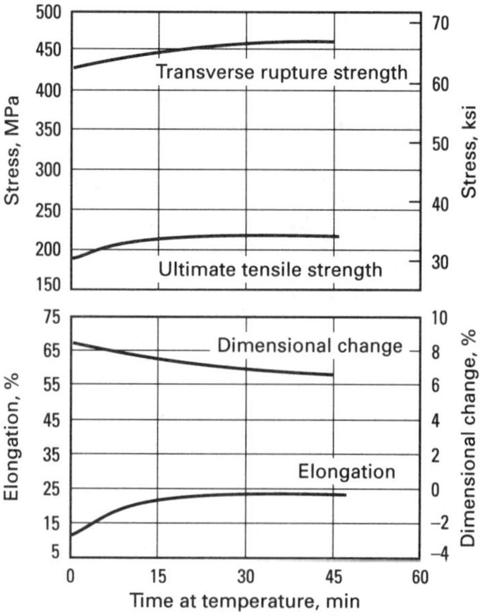

Fig. 15 Effect of varying sintering time on properties of prealloyed 70Cu-30Zn leaded brass (nominal 1.5% Pb). Lubricant: 0.375% lithium stearate and 0.375% zinc stearate; compaction pressure: 415 MPa (30 tsi); green density: 7.3 g/cm³; sintering temperature and atmosphere: 870 °C (1600 °F) in dissociated ammonia

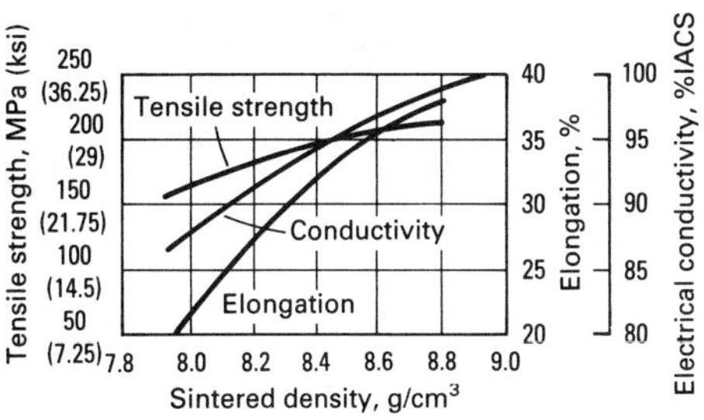

Fig. 16 Effect of density on electrical conductivity and tensile properties of P/M copper

response to energy inputs and temperatures that exceeded the capabilities of the organic-base friction materials used in the 20s and 30s. World War II, with its demands for large quantities of heavy-duty friction materials in military vehicles and aircraft, contributed much to that industry. More recently, improved organic-base friction materials for light- and medium-duty applications have grown at the expense of metal-base friction materials.

Uses. Sintered-metal friction-materials applications or operating conditions may be classified in terms of dry/wet and mild/moderate/severe, as shown in Fig. 17. The majority of the clutch applications are for wet (oil) operation. In oil applications the coefficient of friction is lower, but part life is longer. Also, there are big differences in finishing operations depending on whether parts must perform dry or in oil. Parts operating in oil have surface grooves (Fig. 18a) that help remove oil from the interface and raise the coefficient of friction. Figure 18 shows friction elements used as brake linings and clutch facings.

Composition. Early P/M friction materials were solely copper-base materials. Today, copper-base materials are still being used in all applications, but lower-cost iron-base compounds have been developed for moderate- to severe-duty dry applications. Some typical friction-material compositions are shown in Table 10 for both dry and wet applications.

Copper-base materials are used mainly where semifluid friction occurs. For dry friction they are suitable only where operating conditions are relatively mild (less than 350 °C, or 660 °F).

Processing. The mixtures of metal and ceramic powders (Table 11) are carefully blended. Fine metal powders with high surface area are necessary to provide a strong and thermally conductive matrix for the nonmetallic components.

Compacting pressures range from 165 to 275 MPa (12 to 20 tsi). Properties are very sensitive to production conditions. Seemingly minor changes in raw materials or processing may lead to drastic changes in performance characteristics of the final product.

Bell-type sintering furnaces usually are used where the friction facing is bonded to a supporting steel backing plate such as in clutch disks. The green disks are placed on the copper-plated steel plates and stacked. Pressure is applied on the vertical stack of disks. Sintering temperatures range from 550 to 950 °C (1020 to 1740 °F) in a protective atmosphere. Typical sintering times are from 30 to 60 min. The sintered parts are typically machined for dimensional accuracy and surface parallelism.

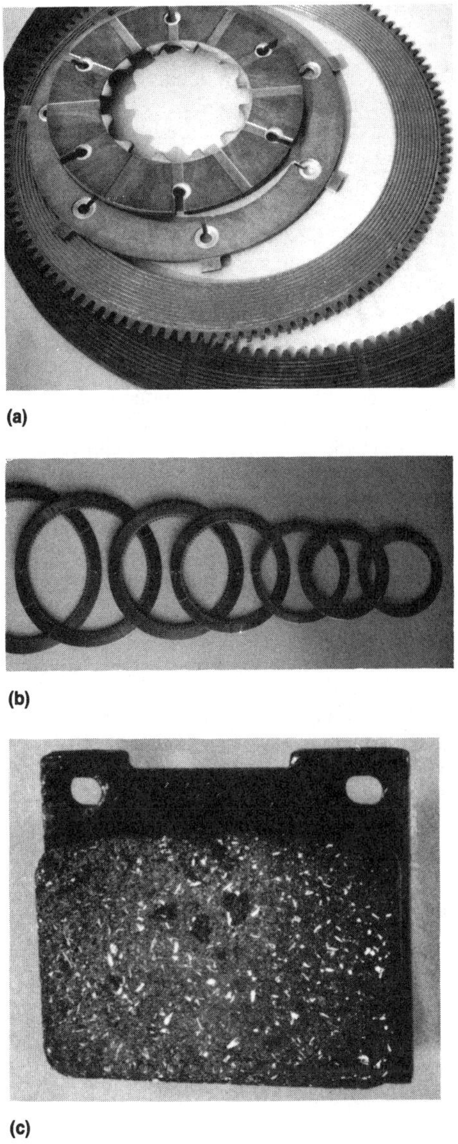

Fig. 18 Copper-base P/M friction elements. (a) Grooved P/M friction elements for wet applications. (b) Copper-base P/M clutch plates (280 to 500 mm OD) used in power-shift transmissions for tractors. (c) Copper-base P/M friction pad

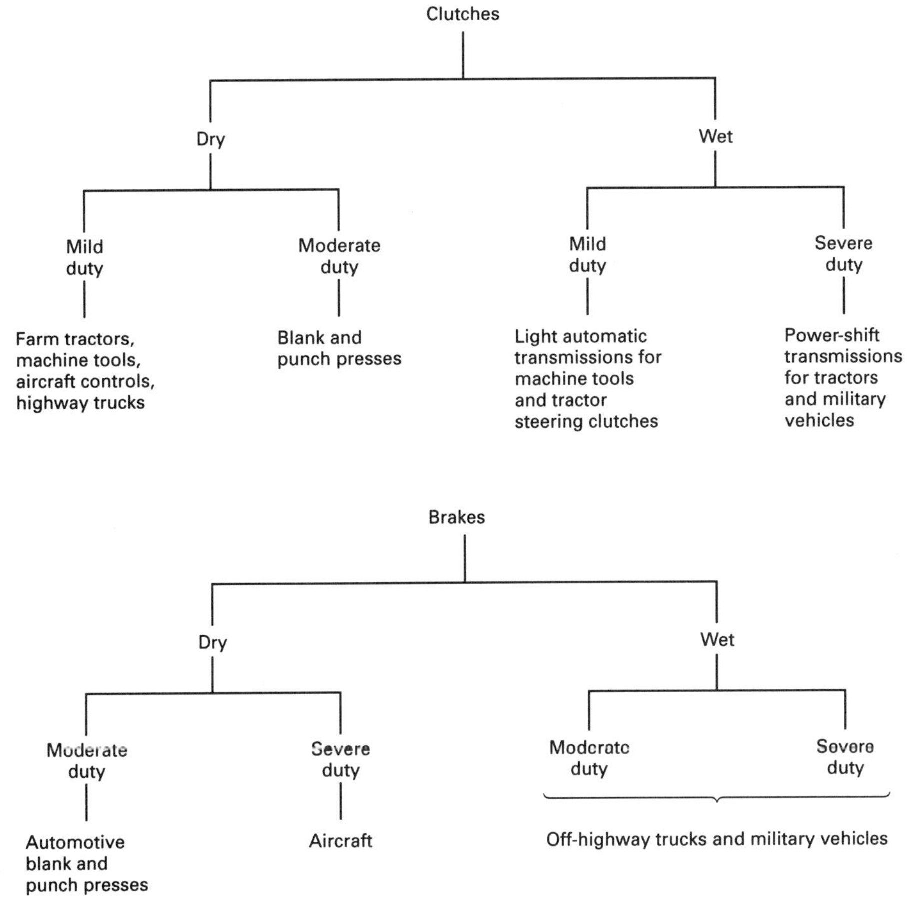

Fig. 17 Applications of sintered-metal friction materials. Source: Ref 9

Table 10 Compositions of sintered copper-base materials for wet and dry applications

Country	Composition, wt% Cu	Sn	Fe	Pb	Graphite	MoS_2	Other	Use(a)
USSR	65–80	7–9	4–7	5–10	3–8	. . .	2–4 SiO_2	W, D
	70	9	4	6	4	. . .	3 SiO_2, 3 asbestos	W
	60	10	4	5	4	. . .	9 asbestos, 8 bakelite powder	W
East Germany	81.5	4.5	. . .	5	4	. . .	5 mullite	W
	rem	. . .	. . .	5	12	. . .	8 MgO; 5 Ti	W, D
USA	60–75	4–10	5–10	. . .	3–10	3–12	2–7 SiO_2	D
	52.5	. . .	. . .	7.5	. . .	. . .	5 SiO_2; 15 Bi	W
	72	4.7	3.3	3.5	8.7	1.4	1.9 SiO_2; 0.2 Al_2O_3	W, D
	72	7	3	6	6	. . .	3 SiO_2; 4 MoO_3	D
	62	7	8	12	7	. . .	4 sand	D
	74	3.5	. . .	. . .	16	. . .	2 Sb; 4.5 SiO_2	D
United Kingdom	rem	3–10	5–10	1–10	0.8	≤4	1.5–4 SiO_2	W
West Germany	67.7	5.1	8	1.5	6.2	5	2.5 SiO_2; 3 Al_2O_3	D
	rem	4–15	5–30		20–30	. . .	3–10 Al_2O_3	W
Sweden	68.5	5.2	4.5	1.8	6.5	≤4	3.3 SiO_2; 3 Al_2O_3	W, D
	68.5	8	4.5	3	6	6	4 SiO_2	W, D
Italy	68	5.5	7	9	6	. . .	4.5 SiO_2	W, D
Austria	68	5	8	1.5	6.2	≤3	2.5 SiO_2; 3 Al_2O_3	W
	54.4	0.8	3.7	21.4	19	. . .	0.5 S; 0.04 Mn	D

(a) W, wet; D, dry. Source: Ref 10

Table 11 P/M friction material components used for various functions

Function	Components
Friction, strength, heat conductivity	Matrix/binder: Cu- or Fe-base (Sn, Zn, Pb additions)
Lubrication (seizure prevention; stability)	Dispersed lubricants: graphite, MoS_2, Pb
Abrasion/friction	Abrasive (frictional) components: SiO_2, mullite, Al_2O_3, Si_3N_4
Wear resistance	Cementite, cast iron grit, spinels
Filler	Carbon, minerals

The friction segments usually are brazed, welded, riveted, or mechanically fastened to supporting steel members or are pressure bonded directly to the assembly.

Function of Components. Only multiphase composites are capable of fulfilling the diverse requirements of high-performance friction materials (Table 12). Also, in developing friction materials compatibility of the opposing member is important. Typical opposing-member materials are cast iron, and hardened and unhardened low-alloy steels.

The matrix or binder, which is usually an iron-base or copper-base material, accounts for about 50 to 80% of total weight (greater than 40 vol%). About 5 to 15% consists of a low-melting-point metal such as tin or zinc that alloys with the major constituent through liquid-phase sintering. For maximum friction, soft metals with high coefficients of friction are preferred. To avoid gross seizure between friction liner and pad, lubricants such as graphite, lead, and molybdenum disulfide are added.

While lubricants (5 to 25 wt%) prevent gross seizure, they do not prevent local welding and metal transfer. To minimize these, up to 20% of an abrasive (often called the frictional component) is added. Because these abrasive components also produce wear, the amount added depends on how much wear can be tolerated in a specific application.

Table 12 Critical performance characteristics of friction materials

Characteristics
Dynamic coefficient of friction
Static coefficient of friction
Static to dynamic coefficients ratio
Durability
Energy capacity
Engagement characteristics
Cost
Wear of opposing member
Fabricability
Temperature coefficient of friction
Time coefficient of friction

An important requirement is thermal stability, which means that the coefficient of friction and the wear rate do not appreciably change up to a specific temperature. Maximum operating temperatures are around 350 °C (660 °F) for copper-base friction materials and range from 600 to 1100 °C (1100 to 2000 °F) for iron-base materials. The wear-resistant components account for up to 10 wt%, essentially for dry applications. Some of these components, such as spinels and mixed metal oxide solutions, may be formed during sintering. Finally, fillers are used, in amounts up to 15 wt%, to decrease costs.

The coefficient of friction is dependent not only on speed, pressure, and temperature of operation, but also on composition and powder characteristics of the components. Because of this complexity, optimum compositions are still derived empirically.

Oxide-Dispersion-Strengthened Copper

The use of pure-copper P/M parts in electrical applications is limited because of the low strength of copper at room and elevated temperatures. Oxide-dispersion-strengthened copper overcomes these limitations and is finding many uses. Basically, in ODS copper a fine and uniform dispersion of aluminum oxide particles (3 to 12 nm) in the copper matrix hardens and strengthens the material and retards recrystallization. Thus, mechanical properties are retained up to very high temperatures. Precipitation-hardened copper alloys lose much of their strength above 400 to 550 °C (750 to 1000 °F).

Manufacture. Oxide-dispersion-strengthened copper can be made by simple mechanical mixing of the metallic and oxidic constituents, by coprecipitation from salt solutions, by mechanical alloying, and by selective or internal oxidation. Dispersion quality and cost vary substantially among these methods; internal oxidation produces the finest and most uniform dispersion.

In internal oxidation, an atomized copper-aluminum alloy is internally oxidized at elevated temperature. This process converts the aluminum into aluminum oxide. Size and uniformity of dispersion of the aluminum oxide depend on several process parameters. Consolidation of the powder to full density and/or various mill forms is accomplished through any of the conventional consolidation processes. Properties of the fully dense material depend upon the amount of deformation introduced during consolidation. Finished parts can be made from consolidated shapes by cold forming, machining, brazing, and soldering. Flash welding and electron-beam welding have also been used successfully.

Properties. Figure 19 shows the ranges in tensile strength, elongation, hardness, and electrical conductivity as a function of aluminum-oxide content. These properties are typical for wrought stock in the hot extruded condition. Cold work broadens these ranges with only minimal effect on conductivity. The three commercial grades of ODS copper are designated as C15760, C15725, and C15715. Other grades can be produced to specified requirements. Oxygen-free compositions immune to hydrogen embrittlement are also available. Rod, bar, tube, wire, strip, plate, and assorted large shapes are available in a wide range of sizes with varying amounts of cold work.

Table 13 gives physical properties of the three grades of ODS copper. Melting point, density, modulus of elasticity, and coefficient of thermal expansion are similar to those of pure copper. Figures 20 and 21 show the fatigue strength and the 100-h stress-rupture strength, respectively, of the C15760 and C15715 ODS coppers produced by SCM Metal Products, Inc. Figure 21 shows the superior strength of ODS copper above 400 °C (750 °F) in comparison to other high-conductivity copper alloys.

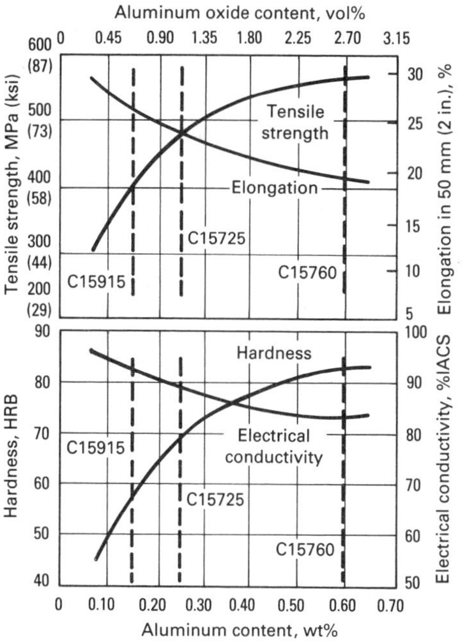

Fig. 19 Properties of three ODS coppers. Source: SCM Metal Products, Inc.

Uses. The combination of high electrical and thermal conductivity, outstanding corrosion resistance, ease of fabrication, and retainment of high strength at elevated temperatures make dispersion-strengthened copper useful in many applications. Dispersion-strengthened copper enhances the current-carrying or heat-dissipating capabilities for a given section size and structural strength. Alternatively, it enables reduction of section sizes for component miniaturization.

Welding Electrodes. Important applications of dispersion-strengthened copper include resistance welding electrodes in automotive, appliance, and other sheet metal industries. They outperform Cu-Cr, Cu-Cr-Zr, and Cu-Zr electrodes. For use on galvanized steel, where the latter materials encounter severe sticking problems, and in automatic press and robot welding applications, they minimize downtime from dressing and changing operations. Seam-welding wheels of ODS copper have also proved beneficial in high-speed welding of coated steels.

Lead Wires. As lead wire for incandescent lamps, ODS copper supports the tungsten filament and facilitates pressing of glass stems without undue softening of the leads. This eliminates the need for expensive molybdenum support wires. Higher light output at reduced wattage and reduced heat losses results from the use of thinner lead wires.

Relay Blades and Contact Supports. Strip products of ODS copper are used in relay blades and contact supports where strength retention after exposure to elevated temperature from brazing is important. In these applications it has replaced phosphor bronze and beryllium-copper.

Lead Frames. The use of ODS copper strip is also being evaluated in several high-performance, integrated-circuit lead-frame applications. Its high thermal conductivity effectively dissipates heat from the integrated circuit chips. The high strength improves the integrity of the leads during handling, that is, it minimizes bending during insertion in the circuit board.

Table 13 Physical properties of three ODS coppers and oxygen-free (OF) copper

Property	Material: C15715(a)	C15725(a)	C15760(a)	OF Copper(a)
Melting point, °C (°F)	1083 (1981)	1083 (1981)	1083 (1981)	1083 (1981)
Density, g/cm^3 (lb/in.)	8.90 (0.321)	8.86 (0.320)	8.81 (0.318)	8.94 (0.323)
Electrical resistivity at 20 °C (68 °F), Ω · mm^2/m (Ω · circular mil/ft)	0.0186 (11.19)	0.0198 (11.91)	0.0221 (13.29)	0.017 (10.20)
Electrical conductivity at 20 °C (68 °F), M mho/m (%IACS)	54 (92)	50 (87)	45 (78)	58 (101)
Thermal conductivity at 20 °C (68 °F), W/m · K (Btu/ft · h · °F)	365 (211)	344 (199)	322 (186)	391 (226)
Linear coefficient of thermal expansion for 20 to 1000 °C (68 to 1830 °F), ppm/°C (ppm/°F)	16.6 (9.2)	16.6 (9.2)	16.6 (9.2)	17.7 (9.8)
Modulus of elasticity, GPa (10^6 psi)	130 (19)	130 (19)	130 (19)	115 (17)

(a) Glidcop grades. Source: SCM Products, Inc.

Porous Bronze Filters

Porous P/M parts are made from various types of metal powders depending on the particular application. The most commonly used powders include bronze, stainless steel, nickel and nickel-base alloys, titanium, and aluminum. Materials used less frequently include the refractory metals (tungsten, molybdenum, and tantalum) and the noble metals (silver, gold, and platinum).

Filters constitute one of the major applications of porous metals. The ability to achieve close control of porosity and pore size is the main reason metal powders are used in filter applications. Most producers of nonferrous filters prefer atomized spherical powder of closely controlled particle size to allow production of filters within the desired pore range. The effective pore size of filters generally ranges from 5 to 125 μm.

Tin bronze is the most widely used P/M filter material, but nickel silver, stainless steel, copper-tin-nickel alloys, and nickel-base alloys also are used. The major advantage of P/M bronze materials over other porous metals is cost. Porous P/M bronze

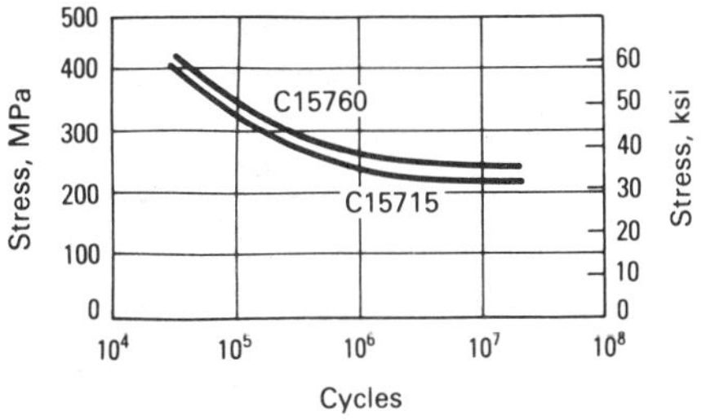

Fig. 20 Fatigue resistance of dispersion-strengthened copper. Tests conducted at room temperature in a Krause cantilever bending-rotating beam made at a frequency of 10 000 cpm. C15760 underwent 14% cold work, and C15715 underwent 94% cold work prior to test. Source: SCM Metal Products, Inc.

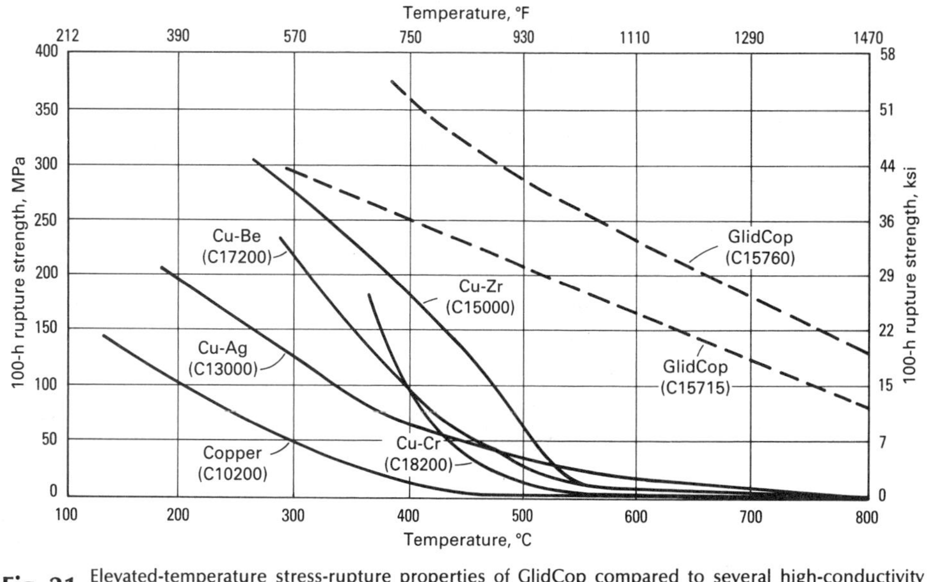

Fig. 21 Elevated-temperature stress-rupture properties of GlidCop compared to several high-conductivity copper alloys. Source: SCM Metal Products, Inc.

Fig. 22 Assorted filters made from P/M bronze. Courtesy of Arrow Pneumatics, Inc.

Table 14 Properties of four grades of filter materials produced by loose powder sintering spherical powders

Particle size of spherical powder particles: Mesh range	Range in μm	Tensile strength: MPa	ksi	Recommended minimum filter thickness: mm	in.	Largest dimensions of particles retained, μm	Viscous permeability coefficient, m^2
20–30	850–600	20–22	2.9–3.2	3.2	0.125	50–250	2.5×10^{-4}
30–40	600–425	25–28	3.6–4.1	2.4	0.095	25–50	1×10^{-4}
40–60	425–250	33–35	4.8–5.1	1.6	0.063	12–25	2.7×10^{-5}
80–120	180–125	33–35	4.8–5.1	1.6	0.063	2.5–12	9×10^{-6}

Source: Ref 11

filters can be obtained with tensile strengths ranging from 20 to 140 MPa (3 to 20 ksi) and appreciable ductility, up to 20% elongation. Also, P/M bronze has the same corrosion resistance as cast bronze of the same composition and thus can be used in a wide range of environments. Figure 22 shows assorted product forms of bronze P/M filters.

Fabrication. Bronze filters usually are made by gravity sintering of spherical bronze powders, which are generally made from the atomization of molten prealloyed bronze. These powders typically contain 90 to 92% Cu and 8 to 10% Sn. Filters made from atomized bronze have sintered densities ranging from 5.0 to 5.2 g/cm³. To produce filters with the highest permeability for a given maximum pore size, powder particles of a uniform particle size must be used.

Although not widely used, coarser powders for bronze filters can be obtained by chopping copper wire and tumbling the choppings. Filters made from tin-coated cut copper wire with tin contents ranging from 2.5 to 8% are also used to a lesser extent. Filters made from these materials have sintered densities ranging from 4.6 to 5.0 g/cm³.

During sintering the filters shrink slightly—as much as 8%. To avoid excessive shrinkage, filters from powders with fine particle size require lower sintering temperatures in the neighborhood of 815 °C (1500 °F). Because of the shrinkage during sintering, filters must be designed with a slight draft, so they can be removed from the mold.

Properties of four grades of bronze filter materials are presented in Table 14. By far the most common of these grades is the third. The two coarsest grades are no longer widely used.

Applications. Powder metallurgy bronze filters are used to filter gases, oils, refrigerants, and chemical solutions. They have been used in fluid systems of space vehicles to remove particles as small as 1 μm. Bronze diaphragms can be used to separate air from liquids or mixtures of liquids that are not emulsified. Only liquids capable of wetting the pore surface can pass through the porous metal part.

Bronze filter materials can be used as flame arrestors on electrical equipment operating in flammable atmospheres, where the high thermal conductivity of the bronze prevents ignition. They can also be used as vent pipes on tanks containing flammable liquids. In these applications, heat is conducted away rapidly so that the ignition temperature is not reached. Additional information on the manufacture, properties, performance characteristics, and applications of P/M bronze filters can be found in the article "P/M Porous Parts" in *Powder Metallurgy*, Volume 7 of the 9th Edition of *Metals Handbook*.

Other Applications

Flaked (ball milled) and other forms of copper are used in combination with graphited materials to form carbon brushes, which are used extensively as sliding electrical contacts in electrical motor units (see the article "Electrical Contact Materials" in this Volume).

Cupronickel powders are widely used in the production of coins, tokens, and medallions. They also find use in components for marine applications because of their good corrosion resistance in sea water.

Age- (precipitation-) hardening alloys based on the systems Cu-Cr, Cu-Co-Be, Cu-Be, Cu-Ti, and Cu-Ni-Sn have attractive combinations of strength, wear, and corrosion resistance. In recent years there have been studies to extend the solubility of copper alloys by rapid solidification processing.

REFERENCES

1. *MPIF Standard 35*, Metal Powder Industries Federation, 1986-1987
2. A.E. Kindler and H. Stein, Determination of the Life of Sintered Bearings, *Met. Powder Rep.*, 1985, p 342-346
3. H. Youssef and M. Eudier, Production and Properties of a New Porous Bearing, in *Modern Developments in Powder Metallurgy*, Vol 3, 1966, p 129-137
4. T. Kohno and Y. Nishino, Development of Sintered Bearings for High Speed Revolution Applications, in *Modern Developments in Powder Metallurgy*, Vol 12, 1981, p 855-870
5. Sintered self-lubricating bearing and process to produce it, French Patent 2,555,682, 1983
6. H. Shikata, H. Funabashi, Y. Ebine, and T. Hayasaka, Performance of Sintered Cu-Sn-Ni Bearings Containing MoS_2, *Met. Powder Rep.*, June 1985, p 351-357
7. V.T. Morgan, Porous Metal Bearings, in *Perspectives in Powder Metallurgy*, Vol 4, *Friction and Antifriction Materials*, Plenum Press, 1970, p 187-210
8. V.T. Morgan, Copper Powder Metallurgy for Bearings, in *New Perspectives in P/M*, Vol 7, *Copper Base Powder Metallurgy*, Metal Powder Industries Federation, 1980, p 39-63
9. B.T. Collins, The U.S. Friction Materials Industry, in *Perspectives in Powder Metallurgy*, Vol 4, 1970, p 3-7
10. W. Schatt, *Pulvermetallurgie Sinter und Verbundwerkstoffe*, VEB Deutscher Verlag fur Grundstoffindustrie, 1979, p 315
11. F.R. Lenel, *Powder Metallurgy Principles and Applications*, Metal Powder Industries Federation, 1980

Beryllium-Copper and Other Beryllium-Containing Alloys

John C. Harkness, William D. Spiegelberg, and W. Raymond Cribb, Brush Wellman Inc.

BERYLLIUM ADDITIONS, up to about 2 wt%, produce dramatic effects in several base metals. In copper and nickel, this alloying addition promotes strengthening through precipitation hardening. In aluminum alloys, a small addition improves oxidation resistance, castability, and workability. Other advantages are produced in magnesium, gold, zinc, and other base metals.

The most widely used beryllium-containing alloys by far are the wrought beryllium-coppers. They rank high among copper alloys in attainable strength while retaining useful levels of electrical and thermal conductivity. Applications for these alloys include:

- Electronic components, where the strength, formability, and favorable elastic modulus of these alloys make them well suited for use as electronic connector contacts
- Electrical equipment, where their fatigue strength, conductivity, and stress relaxation resistance lead to their use as switch and relay blades
- Control bearings, where antigalling features are important
- Housings for magnetic sensing devices, where low magnetic susceptibility is critical
- Resistance welding systems, where hot hardness and conductivity are important in structural and consumable welding components

Precipitation hardening is a critical attribute for the cast beryllium-copper alloys. Hardness, thermal conductivity, and castability are important in most of their applications. For example, they are used in molds for plastic component production where fine cast-in detail such as wood or leather grain is desired. Cast alloys are also used for thermal management in welding equipment, for waveguides, and for mold components such as core pins. High-strength alloys are used in sporting equipment such as investment cast golf club heads.

Master alloys of beryllium in copper, nickel, and aluminum are available for foundry use in preparing casting alloys or otherwise treating alloy melts. Beryllium-copper atomized powder is used in several applications, notably as a conductive matrix for carbide or diamond cutters and as permeable electric contacts.

Because beryllium-copper and other beryllium-containing alloys are precipitation hardenable, they can be tailored across a wide range of property combinations. Recent advances in composition control, processing techniques, and recycling technology have broadened their capabilities and expanded their range of application. This article describes the important features of this alloy group, including information on safe handling.

Beryllium-Copper Alloys

Beryllium-copper alloys are available in all common commercial mill forms, including strip, wire, rod, bar, tube, plate, casting ingot, and cast billet. Free-machining beryllium-copper is offered as rod. Beryllium-nickel alloys are supplied primarily as strip, rod, and casting ingot, although other wrought forms are obtainable.

Beryllium-copper alloys respond readily to conventional forming, plating, and joining processes. Depending on mill form and condition (temper), the wrought materials can be stamped, cold formed by a variety of conventional processes, or machined. Cast billet can be hot forged, extruded, or machined, and castings can be produced by a variety of foundry techniques. Finished components can be conventionally plated with tin, nickel, semiprecious metals, or precious metals. Alternatively, strip can be clad or inlayed with other metals. Surfaces can also be modified by various techniques to enhance performance or appearance. Beryllium-copper alloys are solderable with standard fluxes and, if care is taken to preserve the properties achieved by heat treament, can be joined by normal brazing and many fusion welding processes.

Composition

Commercial beryllium-copper alloys are classified as high-copper alloys. Wrought products fall in the nominal range 0.2 to 2.00 wt% Be, 0.2 to 2.7 wt% Co (or up to 2.2 wt% Ni), with the balance consisting essentially of copper. Casting alloys are somewhat richer, with up to 2.85 wt% Be. Within this compositional band, two distinct classes of commercial materials have been developed, the high-strength alloys and the high-conductivity alloys. Compositions of the commercial alloys are listed in Table 1.

The wrought high-strength alloys (C17000 and C17200) contain 1.60 to 2.00 wt% Be and nominal 0.25 wt% Co. A free-machining version of C17200, which is modified with a small lead addition and available only as rod and wire, is designated C17300. The traditional wrought high-conductivity alloys (C17500 and C17510) contain 0.2 to 0.7 wt% Be and nominal 2.5 wt% Co (or 2 wt% Ni). The leanest and most recently developed high-conductivity alloy is C17410, which contains somewhat less than 0.4 wt% Be and 0.6 wt% Co.

The high-strength casting alloys (C82400, C82500, C82600, and C82800) contain 1.60 to 2.85 wt% Be, nominal 0.5 wt% Co, and a small silicon addition. Grain refinement in these foundry products is achieved by a minor titanium addition to the casting ingot or by increased cobalt content (up to a nominal content of 1 wt% Co) as in C82510. The high-conductivity casting alloys (C82000, C82100, and C82200) contain up to 0.8 wt% Be.

The beryllium in the high-strength alloys, at a level of close to 12 at.%, imparts a gold luster to these copper-base materials. The lower atomic fraction in the high-conductivity alloys produces a reddish or coral-gold color.

Physical Metallurgy

The binary beryllium-copper phase diagram in Fig. 1 is a useful, although somewhat simplified, tool for understanding the metallurgy of these alloys. The diagram

Table 1 Composition of commercial beryllium-copper alloys

UNS number	Composition, wt%							
	Be	Co	Ni	Co + Ni	Co + Ni + Fe	Si	Pb	Cu
Wrought alloys								
C17200	1.80–2.00	...	...	0.20 min	0.6 max	...	...	bal
C17300	1.80–2.00	...	...	0.20 min	0.6 max	...	0.20–0.6	bal
C17000	1.60–1.79	...	...	0.20 min	0.6 max	...	...	bal
C17510	0.2–0.6	...	1.4–2.2	...	...	...	...	bal
C17500	0.4–0.7	2.4–2.7	...	...	...	...	...	bal
C17410	0.15–0.50	0.35–0.60	...	...	...	...	...	bal
Cast alloys								
C82000	0.45–0.80	...	...	2.40–2.70	...	...	...	bal
C82200	0.35–0.80	...	1.0–2.0	...	...	...	...	bal
C82400	1.60–1.85	...	...	0.20–0.65	...	...	...	bal
C82500	1.90–2.25	...	...	0.35–0.70	...	0.20–0.35	...	bal
C82510	1.90–2.15	...	...	1.00–1.20	...	0.20–0.35	...	bal
C82600	2.25–2.55	...	...	0.35–0.65	...	0.20–0.35	...	bal
C82800	2.50–2.85	...	...	0.35–0.70	...	0.20–0.35	...	bal

Note: Copper plus additions, 99.5% min

shows that the solid solubility of beryllium in the α copper matrix decreases as the temperature is lowered, and thus the beryllium-copper alloys are precipitation hardenable. Heat treatment typically consists of solution annealing followed by precipitation treatment (also known as age hardening). Cold work can be performed on wrought products between annealing and age hardening to enhance the magnitude of the age-hardening response.

The precipitation sequence in C17200 commences with homogeneous nucleation of Guinier-Preston (G-P) zones. As age hardening progresses, coherent metastable γ″ and subsequently γ′ precipitates form from the G-P zones. Strength increases with aging time as a result of the coherency strains that develop as the copper matrix attempts to accommodate the growing submicroscopic precipitates. At certain age-hardening time-temperature combinations, the optically resolvable equilibrium γ phase develops, either homogeneously in the matrix or heterogeneously at grain boundaries. This phase is partially coherent with the copper matrix. The associated loss of coherency strains results in a decrease in strength compared to that developed by the formation of the metastable precipitates.

Commercial beryllium-copper alloys contain a third element addition, either of cobalt or of nickel. This addition to the binary alloy system restricts grain growth during annealing by establishing a dispersion of beryllide particles in the matrix. The addition also enhances the magnitude of the age-hardening response and retards the tendency to overage or soften at extended aging times and higher aging temperatures. In C17500 and C17200, the beryllides are (Cu,Co)Be with an ordered body-centered cubic CsCl (B2) superlattice. The beryllides in C17510 are (Cu,Ni)Be; they also display the B2 superlattice.

Microstructure

Distinctive features in the microstructure of beryllium-copper alloys are easily revealed by conventional metallographic and scanning electron microscope techniques. Beryllides, other phases, and surface effects can be examined on as-polished specimens; however, etchants (Table 2) must be used to reveal other features of interest.

Cast beryllium-copper alloy microstructures exhibit α copper dendrites and blue-gray intermetallic beryllide particles of the order of 10 μm in the longest dimension. Primary beryllides formed during solidification display a Chinese script morphology. Secondary beryllides formed after solidifi-

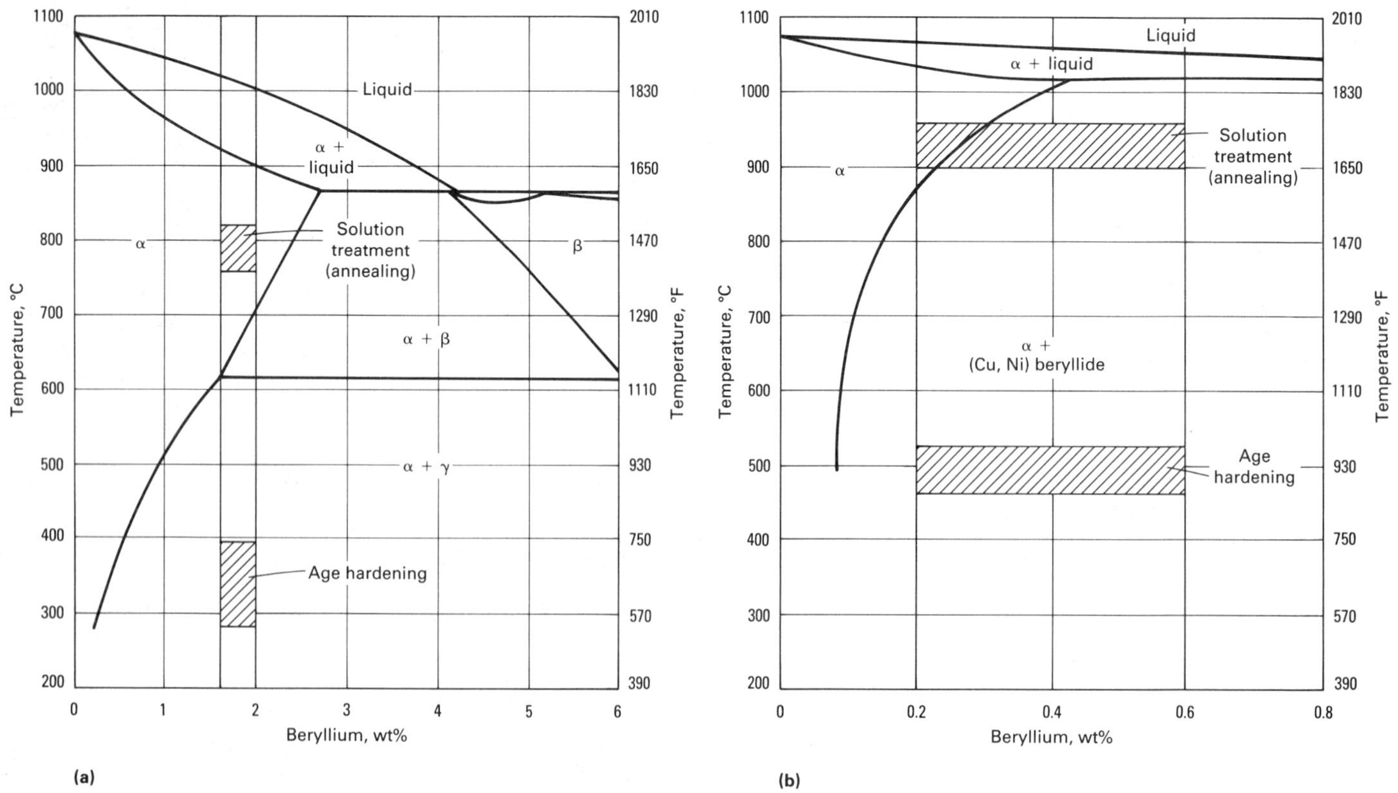

Fig. 1 Phase diagrams for beryllium-copper alloys. (a) Binary composition for high-strength alloys such as C17200. (b) Pseudobinary composition for C17510, a high-conductivity alloy

Table 2 Recommended etching reagents for beryllium-copper alloys

Etchant	Composition(a)	Comments
1. Ammonium persulfate hydroxide	1 part NH_4OH (concentrated) and 2 parts $(NH_4)_2S_2O_8$ (ammonium persulfate) 2.5% in H_2O	Used for observation of the general structure of all beryllium-copper alloys. Preheat sample in hot water (optional); swab etch 2–20 s; use fresh.
2. Ammonium persulfate hydroxide (variation)	2 parts 10% $(NH_4)_2S_2O_8$, 3 parts NH_4OH (concentrated), 1 part 3% H_2O_2, and 5–7 parts H_2O	Used for all beryllium-copper alloys. Offers improved grain boundary delineation in unaged material. A, ¼ H, ½ H, H tempers (unaged, use less H_2O. AT through HT and aged, use more H_2O). Use fresh; swab or immerse 5–60 s. Preheat specimen in hot H_2O if etching rate is slow.
3. Dichromate	2 g $K_2Cr_2O_2$ (potassium dichromate), 8 mL H_2SO_4 (concentrated), 1 drop HCl per 25 mL of solution, and 100 mL H_2O	Used for observation of the grain structure of wrought C17000, C17200, C17300. Use for AT through HT and mill hardened (aged) tempers. Etch first with ammonium persulfate hydroxide (No. 1 or 2); wipe dichromate 1–2 times over specimen to remove dark etch color. Do not overetch; sample may pit. Can be used with laboratory aging of annealed or as-rolled material at 370 °C (700 °F) for 15–20 min to enhance grain boundary delineation for grain size determination
4. Hydroxide/peroxide	5 parts NH_4OH (concentrated), 2–5 parts 3% H_2O_2, and 5 parts H_2O	Common etchant for copper and brass, also applicable to beryllium-copper alloys. Use fresh.
5. Ferric chloride	5 g $FeCl_3$ (ferric chloride), 50 mL HCl, and 100 mL H_2O	Common etchant for copper alloys, also applicable to cold-rolled tempers of beryllium-copper alloys C17500 and C17510 to show grain structure. Immerse 3–12 s.
6. Cyanide	1 g KCN (potassium cyanide) and 100 mL H_2O	General structure of beryllium-copper alloys C17500, C17510 (No. 6). Immerse 1–5 min; stir slowly while etching; use etchant 7 if others are too weak to bring out structure. A two-step technique for improved results on C17510 includes immersion in etchant 6 followed by swabbing with etchant 8. *Caution: Poison fumes.* Use fume hood. Do not dispose of used solutions directly into drains. Pour used solution into beaker containing chlorine bleach. Let stand 1 h, then flush down drain with plenty of running water.
7. Persulfate hydroxide/cyanide	4 parts ammonium persulfate hydroxide etchant (etchant 1 or 2) and 1 part cyanide etchant (etchant 6)	
8. Cyanide peroxide hydroxide	20 mL KCN, 5 mL H_2O_2, and 1–2 mL NH_4OH	
9. Phosphoric acid electrolyte	20 mL H_2O (tap, not distilled), 58 mL 3% H_2O_2, 48 mL H_3PO_4, and 48 mL ethyl alcohol	For deep etching of beryllium-copper. Polish specimen through 1 μm or finer Al_2O_3. Use 0.5-1 cm^2 (0.08–0.16 $in.^2$) mask. 0.1 A to etch (higher amperes to polish). Low-to-moderate flow rate. 3 to 6 s to etch, up to 60 s to polish

(a) Where H_2O is indicated, use distilled water unless otherwise noted.

cation of the primary phase exhibit a rodlike morphology and preferred orientation. The β phase forms peritectically from the liquid and can be observed in high-strength alloy castings as an interdendritic network surrounding the primary copper-rich α phase. The β phase decomposes to α and γ phases by eutectoid transformation on cooling to room temperature. The transformed β phase exists as angular milky-white patches surrounded by a dark outline in the as-polished microstructure. Subsequent thermomechanical processing of wrought products refines the primary beryllides to a population of smaller, roughly spherical, blue-gray particles and normally dissolves the transformed β phase.

The coherent precipitates responsible for age hardening in both the high-conductivity and the high-strength beryllium-copper alloys are too small to be resolved optically and can be detected only by transmission electron microscopy. Age-hardened microstructures of the high-strength alloys are distinguishable from unaged material by a dark etching response associated with striations resulting from surface relief accompanying metastable precipitation. Overaged and slack-quenched unaged high-strength alloys exhibit colonies of equilibrium γ phase at grain boundaries. The fine lamellar morphology of these cellular precipitates can be observed by transmission electron microscopy or by scanning electron microscopy of an etched metallographic specimen. In the age-hardened state, these γ phase colonies are softer than the coherent precipitate-strengthened matrix.

Age-hardened microstructures of the high-conductivity alloys are indistinguishable from unaged microstructures in the optical microscope. Coherency strains associated with metastable precipitates are insufficient to cause a dark etching response. In these alloys, the equilibrium γ phase forms not by a discontinuous reaction at the grain boundaries, but instead by a continuous transformation in the matrix.

Heat Treatment

Solution annealing is performed by heating the alloy to a temperature slightly below the solidus to dissolve a maximum amount of beryllium, then rapidly quenching the material to room temperature to retain the beryllium in a supersaturated solid solution. Users of beryllium-copper alloys are seldom required to perform solution annealing; this operation is almost always done by the supplier.

Typical annealing temperature ranges are 760 to 800 °C (1400 to 1475 °F) for the high-strength alloys and 900 to 955 °C (1650 to 1750 °F) for the high-conductivity alloys. Temperatures below the minimum can result in incomplete recrystallization. Too low a temperature can also result in the dissolution of an insufficient amount of beryllium for satisfactory age hardening. Annealing at temperatures above the maximum can cause excessive grain growth or induce incipient melting.

Once the set temperature is reached, it is not necessary to hold the metal at the annealing temperature for more than a few minutes to accomplish solution treatment. In general, thin strip or wire can be annealed in less than 2 min; heavy-section products usually are held at the annealing temperature for 30 min or less. It is important to be sure to reach the set temperature; as a guide, heat-up time is usually estimated as ½ to 1 h per inch of thickness. The use of thermal measurement equipment is helpful in establishing these parameters because they depend on the size and quantity of parts being treated. Prolonged annealing time does not increase the solution of beryllium at a given annealing temperature. At the high end of the annealing temperature range, extended dwell time can promote undesirable secondary grain growth.

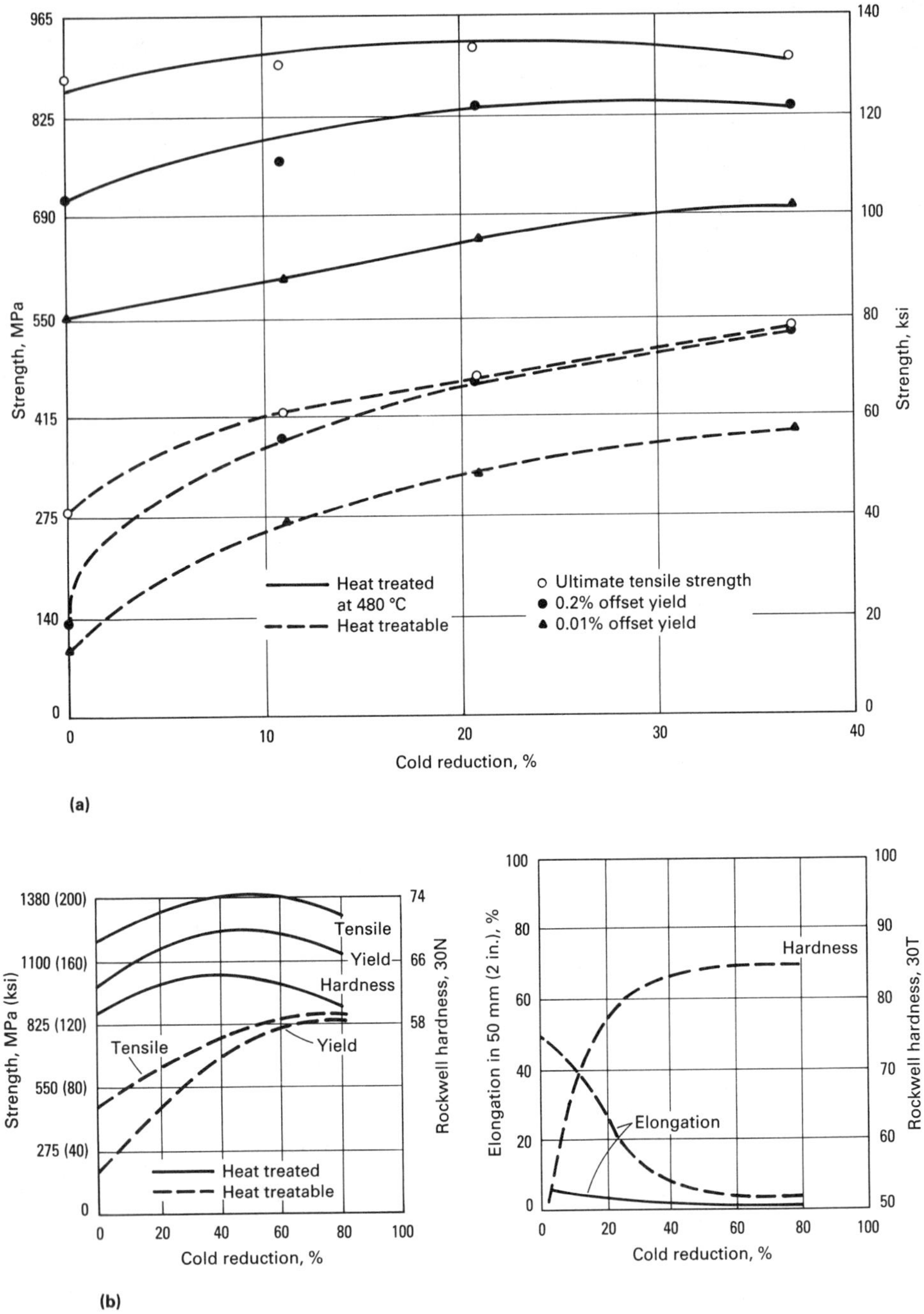

Fig. 2 Influence of cold reduction and age hardening on the mechanical properties of beryllium-copper alloys. (a) C17510 aged at 480 °C (895 °F) for 2 or 3 h. (b) C17200 aged at 315 °C (600 °F) for 2 or 3 h

Interrupted or slow quenching rates should be avoided because they permit precipitation of beryllium during cooling, resulting in an unacceptably high level of as-quenched hardness and an inadequate final age hardening response. This is caused by the annealing out of quenched-in vacancies at these lower quenching rates. The annealing practice for beryllium-copper is in distinct contrast to that for many copper alloys that do not strengthen by heat treatment. These alloys are typically subjected to lower-temperature and longer-time annealing for recovery of cold-working strains and control of recrystallized grain size.

Age hardening involves reheating the solution-annealed material to a temperature below the equilibrium solvus for a time sufficient to nucleate and grow the beryllium-rich precipitates responsible for hardening. For the high-strength alloys, age hardening is typically performed at temperatures of 260 to 400 °C (500 to 750 °F) for 0.1 to 4 h. The high-conductivity alloys are age hardened at 425 to 565 °C (800 to 1050 °F) for 0.5 to 8 h.

Within limits, cold working the alloy between solution annealing and age hardening increases both the rate and the magnitude of the age-hardening response in wrought products. As cold work increases to about a 40% reduction in area, the maximum peak-age hardness increases. Further cold work beyond this point is nonproductive and results in decreased hardness after age hardening and diminished ductility in the unaged condition (Fig. 2). Commercial alloys intended for user age hardening are therefore limited to a maximum of about 37% cold work in strip (H temper). For wire, the maximum amount of cold work is commonly somewhat greater.

Electrical conductivity is lowest when the alloy is in the solution-annealed condition because of the large amount of beryllium dissolved in the copper matrix. During age hardening, electrical conductivity increases as dissolved beryllium precipitates from solid solution. Conductivity increases monotonically with both aging time and temperature; aging temperature has the more pronounced effect (Fig. 3). In the high-conductivity alloys, electrical conductivity is 20 to 30% IACS in the unaged condition and 45 to 60% IACS in the peak-aged condition. The conductivity of unaged high-strength alloys is 15 to 19% IACS, increasing to 22 to 28% IACS after peak aging.

Mill-hardened high-strength alloys can be produced either slightly underaged or moderately overaged. Electrical conductivity of these products thus can range from about 17 to 28% IACS.

High-Strength Wrought Alloys. Typical aging response curves for solution-annealed and annealed and cold worked C17200 are shown in Fig. 4. When age hardened at 315 to 335 °C (600 to 635 °F), strength increases to a plateau in about 3 h for annealed material or about 2 h for cold-worked material and remains essentially constant thereafter. At lower age-hardening temperatures, longer aging times are required to reach an aging response plateau.

At higher temperatures, such as 340 °C (640 °F) or above, a relative maximum appears in the age-hardening response curve. At constant cold work, the strength associated with this relative maximum and its time of occurrence diminish with increasing age-hardening temperature. If the alloy is kept at a constant aging temperature but is subjected to an increasing amount of cold work (to about 40% reduction in cross section), the magnitude of the relative maximum strength increases slightly and its time of occurrence diminishes. Below about 330 °C (625 °F), age hardening results almost exclusively from formation of the metastable coherent precipitates. Above this temperature, both metastable and equilibrium precipitates form; the latter concentrate at grain boundaries.

High-Conductivity Wrought Alloys. Typical aging response curves for solution-an-

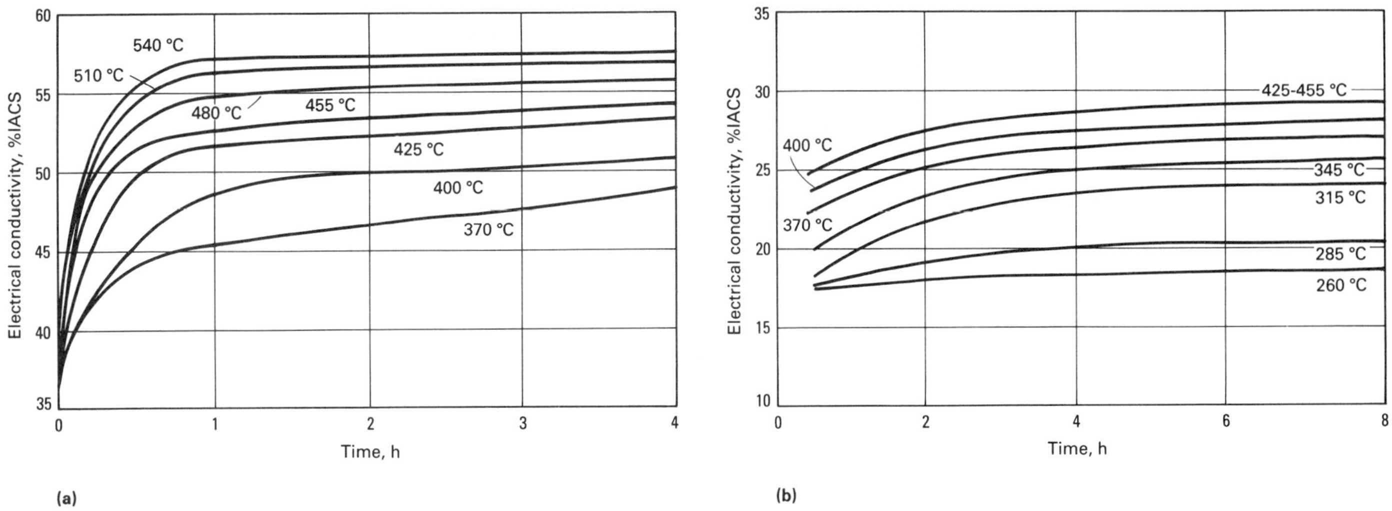

Fig. 3 Effect of aging temperature and time on the electrical conductivity of beryllium-copper. (a) Roll-hardened (TD04 temper) C17510. (b) Composite data for C17200 in the annealed, ¼ hard, ½ hard, and hard conditions (TB00 and TD04 tempers)

nealed C17500 and C17510 are shown in Fig. 5. Aging at 450 to 480 °C (840 to 900 °F) for 2 to 3 h is commonly recommended. Overaging is less pronounced than in the high-strength alloys and can be employed to advantage because the appreciable cobalt or nickel content of these alloys increases the thermal stability of the age-hardening precipitates.

Underage, Peak-Age, and Overage Treatments. Material that has been aged for an insufficient amount of time to attain the maximum possible hardness at a particular temperature is said to be underaged. Material aged at time-temperature combinations resulting in maximum attainable hardness is said to be peak aged. Material aged beyond the relative maximum in the aging response curve is said to be overaged. Underaged material retains the capacity to increase in hardness through additional age hardening; overaged material does not.

Considerable latitude exists for achieving target strength levels with combinations of cold work and age-hardening temperature and time. When strength less than maximum is desired, for example, for increased ductility, cold work can be reduced and underaging (lower-temperature/longer-time, higher-temperature/shorter-time) or overaging (higher-temperature/longer-time) heat treatments can be employed to attain the desired properties.

If parts are inadvertently overaged to lower-than-desired hardness, they require re-solution annealing to restore the age-hardening response. In this case, the strengthening contribution of any cold work imparted before the original age-hardening treatment is erased, and the maximum strength attainable in the salvaged components is that available from solution-annealed and peak-aged material.

From a process control standpoint, peak aging at intermediate temperatures is relatively insensitive to minor fluctuations in temperature. Appreciable extension in time beyond that to attain the aging response plateau is tolerable. The low sensitivity of final strength to aging conditions once this plateau has been attained accounts for the recommendation of age-hardening temperatures of 315 °C (600 °F) for the high-strength alloys and 480 °C (900 °F) for the high-conductivity alloys. Peak-aging treatments are ideally suited for hardening large lots of components on reels or in baskets or trays. Greater precision is needed to select the correct temperature and time to achieve the desired properties when using (in order of

Table 3 Physical properties of beryllium-copper alloys

Tabulated properties apply to age-hardened products.

Alloy	Density		Elastic modulus		Thermal expansion coefficient from 20–200 °C (70–390 °F)		Thermal conductivity		Melting range	
	g/cm³	lb/in.³	GPa	10⁶ psi	10⁻⁶/°C	10⁻⁶/°F	W/m · °C	Btu/ft · h · °F	°C	°F
Wrought alloys										
C17200 (a)	8.36	0.302	131	19	17	9.4	105	60	870–980	1600–1800
C17300 (a)	8.36	0.302	131	19	17	9.4	105	60	870–980	1600–1800
C17000 (a)	8.41	0.304	131	19	17	9.4	105	60	890–1000	1635–1830
C17510 (b)	8.83	0.319	138	20	18	10	240	140	1000–1070	1830–1960
C17500 (b)	8.83	0.319	138	20	18	10	200	115	1000–1070	1830–1960
C17410	8.80	0.318	138	20	18	10	230	133	1020–1070	1870–1960
Casting alloys										
C82000	8.83	0.319	140	20.3	18	10	195	113	. . .	. . .
C82200	8.83	0.319	140	20.3	18	10	250	145	. . .	. . .
C82400	8.41	0.304	130	18.9	18	10	100	58	. . .	. . .
C82500	8.30	0.300	130	18.9	18	10	97	56	. . .	. . .
C82510	8.30	0.300	130	18.9	18	10	97	56	. . .	. . .
C82600	8.22	0.297	130	18.9	18	10	93	54	. . .	. . .
C82800	8.14	0.294	130	18.9	18	10	90	52	. . .	. . .

(a) Density before age hardening, 8.25 g/cm³ (0.298 lb/in.³). (b) Density before age hardening, 8.75 g/cm³ (0.316 lb/in.³)

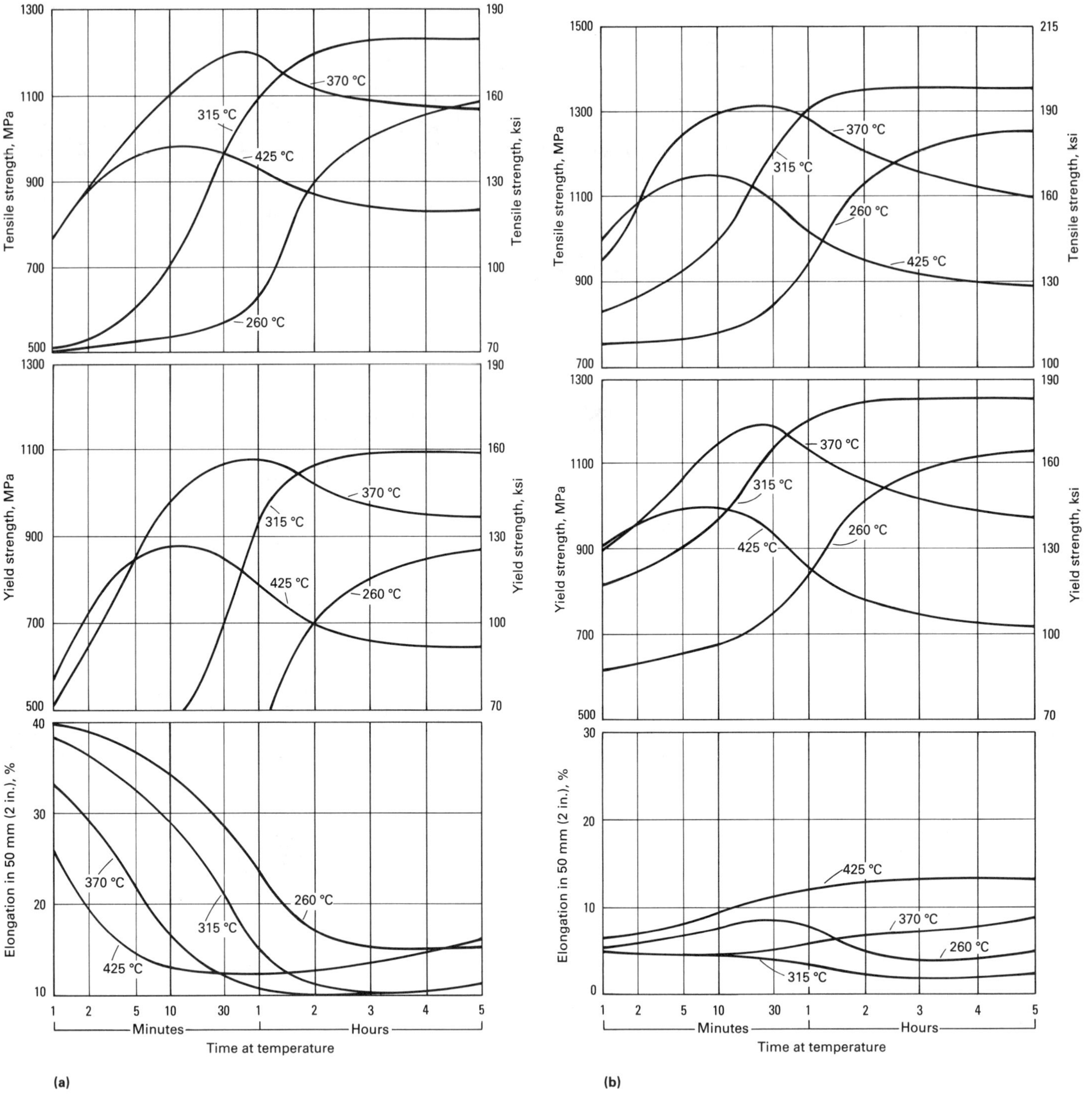

Fig. 4 Age-hardening response curves for the tensile strength, yield strength, and elongation of C17200. (a) Annealed (TB00) temper. (b) Roll-hardened (TD04) temper

increasing need for precision) overaging, low-temperature underaging, high-temperature underaging, and aging to the relative maximum hardness at higher temperatures.

Age-hardening treatments involving precise temperature and time combinations pose problems for batch-type heat-treating processes. Furnace loads must be evenly distributed to ensure uniform heating rates and soaking times in all components for a consistent part-to-part aging response. The use of vacuum furnaces for age-hardening necessitates the shielding of parts from direct radiation. The furnace should be backfilled with an inert gas to provide a more uniform convective heat transfer to the load than that which can be achieved by radiation alone.

Physical Properties

Beryllium and ternary elements in beryllium-copper modify physical properties, but in most cases the effects are not as dramatic as those they produce in mechanical properties. Data for selected physical properties of beryllium-copper alloys are given in Table 3. These data, when examined in tandem with the composition data from Fig. 1, show that beryllium reduces density and lowers liquidus and solidus temperatures. Thermal expansion is relatively unaffected by beryllium content; thermal and electrical conductivities are reduced in proportion to the amount of alloying additions.

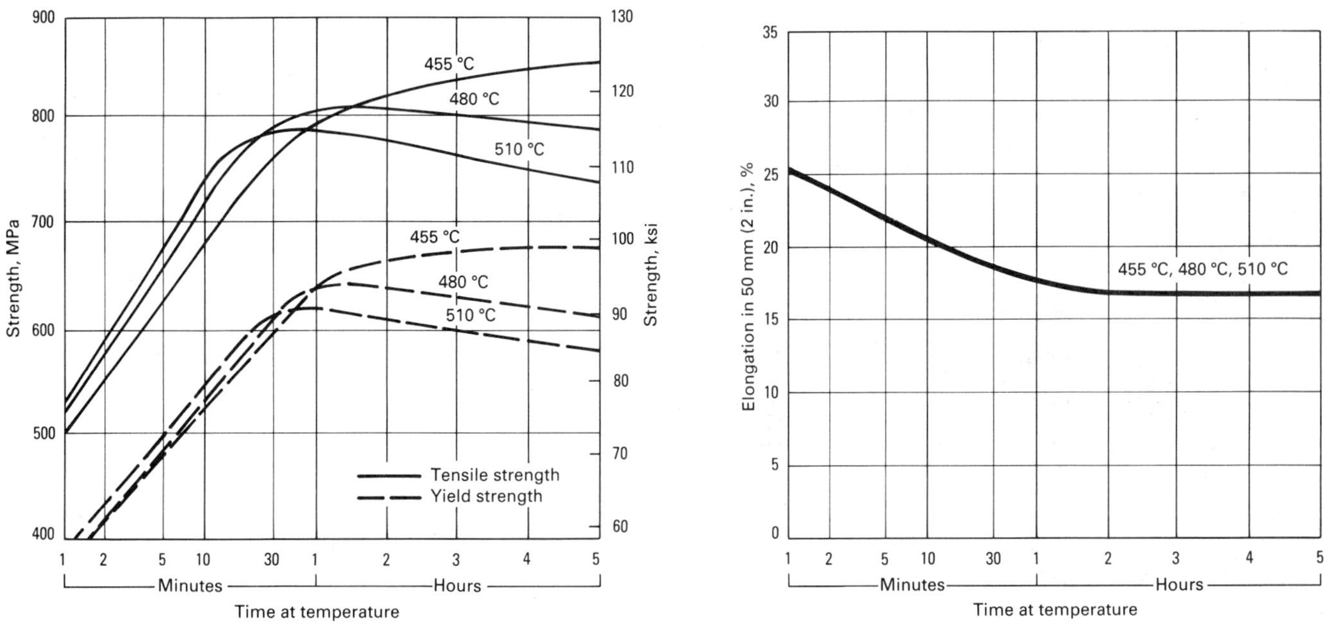

Fig. 5 Age-hardening response curves for annealed (TB00 temper) C17510

Comparing the high-strength and high-conductivity alloys reveals differences in density, thermal conductivity, and melting behavior, but little difference in modulus or thermal expansion coefficient. The thermal expansion coefficients of both alloy families are similar to those of steel. This means that beryllium-coppers and steels are compatible in the same assemblies over wide temperature ranges.

The specific heat of beryllium-copper increases with temperature. For both the high-strength and high-conductivity alloys it ranges from 375 J/kg · °C (0.09 Btu/lb · °F) at room temperature to 420 J/kg · °C (0.10 Btu/lb · °F) at 90 °C (200 °F). The magnetic permeability of beryllium-copper is very close to unity, meaning that these alloys are nearly perfectly transparent to slowly varying magnetic fields. All beryllium-copper alloys and product forms have a Poisson's ratio of 0.3.

Mechanical Properties

Strength, hardness, and ductility data for the various tempers of beryllium-copper alloy strip and selected tempers of other wrought products are shown in Tables 4, 5, and 6.

Wrought products are supplied in a range of both heat-treatable and mill-hardened conditions (tempers). The heat-treatable conditions include the solution-annealed temper (commercial designation A, or ASTM designation TB00) and a range of annealed and cold-worked tempers (¼ H through H, or TD01 through TD04) that must be age hardened by the user after forming. Increasing cold work, within limits, increases the strength obtained during age hardening. Heat-treatable tempers are the softest and generally most ductile materials in the as-shipped condition, and they can be formed into components of varying complexity depending upon the level of cold work. Age hardening these heat-treatable tempers develops strength levels that range higher than those in any other copper-base alloys. After age hardening by the user, the solution-annealed material is redesignated AT, or TF00, and the annealed and cold-worked tempers are redesignated ¼ HT through HT, or TH01 through TH04.

Mill-hardened tempers, designated AM through XHMS, or TM00 through TM08, receive proprietary cold-working and age-hardening treatments from the supplier prior to shipment, and they do not require heat treatment by the user after forming. Mill-hardened tempers exhibit intermediate-to-high strength and good-to-moderate ductility; these property levels satisfy many component fabrication requirements.

Strip. Wrought high-strength beryllium-copper alloy C17200 strip attains ultimate tensile strengths as high as 1520 MPa (220 ksi) in the peak-age-hardened HT (TH04) condition; the corresponding electrical conductivity is on the order of 20% IACS (Table 4). Because of its slightly lower beryllium content, alloy C17000 achieves maximum age-hardened strengths slightly lower than those of C17200. Mill-hardened C17200 strip is supplied in a range of tempers that have ultimate tensile strengths from 680 to 1320 MPa (99 to 190 ksi).

Ductility varies inversely with strength. It decreases with increasing cold work in the heat-treatable tempers and with increasing strength in the mill-hardened tempers. Beryllium-copper C17500 and C17510 strip can be age hardened to tensile strengths up to 940 MPa (136 ksi) and electrical conductivities in excess of 45% IACS. Mill-hardened strip tempers of these high-conductivity alloys span the tensile strength range of 510 to 1040 MPa (74 to 150 ksi) and include one specially processed temper with a minimum electrical conductivity of 60% IACS.

Other Wrought Products. Plate, bar, wire, rod, and tube also are available in the solution-annealed temper, the annealed and cold-worked heat-treatable temper, and the mill-hardened temper (Tables 5 and 6). Strength and ductility combinations in wire are similar to those of corresponding alloys in strip form. Age-hardened strengths of plate, bar, and tube products range somewhat lower than those of strip or wire and, to a minor degree, vary inversely with section thickness. In addition to these traditional heavy-section product properties, unique property combinations often can be developed by proprietary mill-hardening treatments in response to the changing requirements of emerging applications.

Forgings and hot-finished extruded products are available in the solution-annealed temper and the annealed and age-hardened temper. Cold work is not imparted prior to age hardening. Mechanical properties of beryllium-copper forgings and extrusions are shown in Table 7.

Cast Products. Typical mechanical property ranges for the beryllium-copper casting alloys are shown in Table 8. Four conditions exist for castings:

- As-cast (C temper, or ASTM M01 through M07; the ASTM temper designation depends upon the casting practice, such as sand, permanent mold, investment, continuous casting, and so on)

Table 4 Temper designations and properties for beryllium-copper strip in various conditions

Temper designations		Initial	Aging	Tensile strength		Yield strength at 0.2% offset		Elongation,		Electrical conductivity,
ASTM B 601	Commercial	condition(a)	treatment(b)	MPa	ksi	MPa	ksi	%	Rockwell hardness	%IACS
C17000 (97.9Cu-1.7Be)										
TB00	A	Annealed	. . .	410–530	59–77	190–250	28–36	35–65	45–78 HRB	15–19
TB00	A (planish)	Annealed	. . .	410–540	59–78	200–380	29–55	35–60	45–78 HRB	15–19
TD01	¼ H	¼ hard	. . .	510–610	74–88	410–560	59–81	20–45	68–90 HRB	15–19
TD02	½ H	½ hard	. . .	580–690	84–100	510–660	74–96	12–30	88–96 HRB	15–19
TD04	H	Hard	. . .	680–830	99–120	620–800	90–116	2–10	96–102 HRB	15–19
TF00 (c)	AT	Annealed	3 h at 315 °C	1030–1250	149–181	890–1140	129–165	3–20	33–38 HRB	22–28
			3 h at 345 °C	1105–1275	160–185	860–1140	125–165	4–10	34–40 HRC	22–28
TH01 (c)	¼ HT	¼ hard	2 h at 315 °C	1100–1320	160–191	930–1210	135–175	3–15	35–40 HRC	22–28
			3 h at 330 °C	1170–1345	170–195	895–1170	130–170	3–6	36–41 HRC	22–28
TH02 (c)	½ HT	½ hard	2 h at 315 °C	1170–1380	170–200	1030–1250	149–181	1–10	37–42 HRC	22–28
			2 h at 330 °C	1240–1380	180–200	965–1240	140–180	2–5	38–42 HRC	22–28
TH04 (c)	HT	Hard	2 h at 315 °C	1240–1380	180–200	1060–1250	154–181	1–6	38–44 HRC	22–28
			2 h at 330 °C	1275–1415	185–205	1070–1345	155–195	2–5	39–43 HRC	22–28
TM00	AM	Annealed	M	680–760	99–110	480–660	70–96	18–30	98 HRB–23 HRC	18–33
TM01	¼ HM	¼ hard	M	750–830	109–120	550–760	80–110	15–25	20–26 HRC	18–33
TM02	½ HM	½ hard	M	820–940	119–136	650–870	94–126	12–22	24–30 HRC	18–33
TM04	HM	Hard	M	930–1040	135–151	750–940	109–136	9–20	28–35 HRC	18–33
TM05	SHM	Hard	M	1030–1110	149–161	860–970	125–141	9–18	31–37 HRC	18–33
TM06	XHM	Hard	M	1060–1210	154–175	930–1140	135–165	3–10	32–38 HRC	18–33
C17200 (98.1Cu-1.9Be)										
TB00	A	Annealed	. . .	410–530	59–77	190–250	28–36	35–65	45–78 HRB	15–19
TB00	A (planish)	Annealed	. . .	410–540	59–78	200–380	29–55	35–60	45–78 HRB	15–19
TD01	¼ H	¼ hard	. . .	510–610	74–88	410–560	59–81	20–45	68–90 HRB	15–19
TD02	½ H	½ hard	. . .	580–690	84–100	510–660	74–95	12–30	88–96 HRB	15–19
TD04	H	Hard	. . .	680–830	99–120	620–800	90–116	2–18	96–102 HRB	15–19
TF00 (c)	AT	Annealed	3 h at 315 °C	1130–1350	164–195	960–1205	139–175	3–15	36–42 HRC	22–28
			½ h at 370 °C	1105–1310	160–190	895–1205	130–175	3–10	34–40 HRC	22–28
TH01 (c)	¼ HT	¼ hard	2 h at 315 °C	1200–1420	174–206	1030–1275	149–185	3–10	36–43 HRC	22–28
			¼ h at 370 °C	1170–1380	170–200	965–1275	140–185	2–6	36–42 HRC	22–28
TH02 (c)	½ HT	½ hard	2 h at 315 °C	1270–1490	184–216	1100–1350	159–196	1–8	38–44 HRC	22–28
			¼ h at 370 °C	1240–1450	180–210	1035–1345	150–195	2–5	38–44 HRC	22–28
TH04 (c)	HT	Hard	2 h at 315 °C	1310–1520	190–220	1130–1420	164–206	1–6	38–45 HRC	22–28
			¼ h at 370 °C	1275–1480	185–215	1105–1415	160–205	1–4	39–45 HRC	22–28
TM00	AM	Annealed	M	680–760	99–110	480–660	70–96	16–30	95 HRB–23 HRC	17–28
TM01	¼ HM	¼ hard	M	750–830	109–120	550–760	80–110	15–25	20–26 HRC	17–28
TM02	½ HM	½ hard	M	820–940	119–136	650–870	94–126	12–22	23–30 HRC	17–28
TM04	HM	Hard	M	930–1040	135–150	750–940	109–136	9–20	28–35 HRC	17–28
TM05	SHM	Hard	M	1030–1110	149–160	860–970	125–140	9–18	31–37 HRC	17–28
TM06	XHM	Hard	M	1060–1210	154–175	930–1180	135–171	4–15	32–38 HRC	17–28
TM08	XHMS	Hard	M	1200–1320	174–191	1030–1250	149–181	3–12	33–42 HRC	17–28
C17400 (99.5Cu(min)-0.30Be-0.25Co) and C17410 (99.5 Cu(min)-0.3Be-0.5Co)										
. . .	HT	Hard	M	750–900	109–130	650–870	94–126	7–17	95 HRB–27 HRC	45–55
C17500 (96.9Cu-0.55Be-2.55Co) and C17510 (97.8Cu-0.4Be-1.8Ni)										
TB00	A	Annealed	. . .	240–380	35–55	130–210	19–30	20–40	20–45 HRB	20–30
TB00	A (planish)	Annealed	. . .	240–380	35–55	170–320	25–46	20–40	20–45 HRB	20–30
TD04	H	Hard	. . .	480–590	70–85	370–560	54–81	2–10	78–88 HRB	20–30
TF00 (c)	AT	Annealed	3 h at 455 °C	725–825	105–120	550–725	80–105	8–12	93–100 HRB	45–60
			3 h at 480 °C	680–900	99–130	550–690	80–100	10–25	92–100 HRB	45–60
TM00	AM	Annealed	M	680–900	99–130	550–690	80–100	10–25	92–100 HRB	45–60
TH04 (c)	HT	Hard	2 h at 455 °C	792–950	115–138	725–860	105–125	5–8	97–104 HRB	45–52
			2 h at 480 °C	750–940	109–136	650–830	94–120	8–20	95–102 HRB	48–60
TM04	HM	Hard	M	750–940	109–136	650–830	94–120	8–20	95–102 HRB	48–60
. . .	HTR	Hard	M	820–1040	119–150	750–970	109–140	1–5	98–103 HRB	48–60
. . .	HTC	Hard	M	510–590	74–85	340–520	49–75	8–20	79–88 HRB	60 min

(a) All annealing is solution treating, and all alloys are annealed prior to roll hardening and/or heat treatment where applicable. (b) M, mill hardened with special mill processing and precipitation treatment. (c) Two heat treatments given for comparison

- As-cast plus age hardened (CT temper, no ASTM designation)
- As-cast plus solution annealed (A temper, or ASTM TB00)
- As-cast plus solution annealed and age hardened (AT temper, or ASTM TF00)

The solution-annealing temperature range for the high-strength casting alloys, C82400 through C82800, is 760 to 790 °C (1400 to 1450 °F); these alloys are age hardened at 340 °C (640 °F). The high-conductivity casting alloys, C82000 and C82200, are annealed at 870 to 900 °C (1600 to 1650 °F) and age hardened at 480 °C (895 °F). Annealing times of 1 h per inch of casting section thickness are recommended, with a minimum soak of 3 h for the high-strength alloys to ensure maximum property uniformity. An age-hardening time of 3 h is recommended for the temperatures indicated.

Maximum strength is obtained from the casting alloys in the AT (TF00) temper. These alloys reach strength levels slightly lower than those of the corresponding wrought AT temper beryllium-coppers. The CT temper produces strengths slightly lower than those of the AT temper; however, the lower strength is offset by reduced processing costs. In addition, CT temper components experience less shrinkage and age-hardening distortion than do the AT temper castings.

The CT temper strengths shown in Table 8 apply to castings poured in metal molds. The slower solidification and cooling rates associated with sand or ceramic molds or heavy sections can result in lower CT temper strength. Castings in the solution-an-

Table 5 Mechanical and electrical properties of beryllium-copper wire

Temper designations			Wire diameter		Tensile strength		Yield strength		Elongation,	Electrical conductivity,
ASTM	Commercial	Aging treatment	mm	in.	MPa	ksi	MPa	ksi	%	%IACS
C17200 and C17300										
TB00	A	...	1.3–12.7	0.05–0.5	410–540	59–78	130–210	19–30	30–60	15–19
TD01	¼ H	...	1.3–12.7	0.05–0.5	620–800	90–116	510–730	74–106	3–25	15–19
TD02	½ H	...	1.3–12.7	0.05–0.5	750–940	110–136	620–870	90–126	2–15	15–19
TD03	¾ H	...	1.3–2.0	0.05–0.08	890–1070	130–155	790–1040	115–151	2–8	15–19
TD04	H	...	1.3–2.0	0.05–0.08	960–1140	140–165	890–1110	129–161	1–6	15–19
TF00	AT	3 h at 315–330 °C	1.3–12.7	0.05–0.5	1100–1380	160–200	990–1250	144–181	3 min	22–28
TH01	¼ HT	2 h at 315–330 °C	1.3–12.7	0.05–0.5	1200–1450	175–210	1130–1380	164–200	2 min	22–28
TH02	½ HT	1.5 h at 315–330 °C	1.3–12.7	0.05–0.5	1270–1490	184–216	1170–1450	170–210	2 min	22–28
TH03	¾ HT	1 h at 315–330 °C	1.3–2.0	0.05–0.08	1310–1590	190–230	1200–1520	174–220	2 min	22–28
TH04	HT	1 h at 315–330 °C	1.3–2.0	0.05–0.08	1340–1590	194–230	1240–1520	180–220	1 min	22–28
C17510 and C17500										
TB00	A	...	1.3–12.7	0.05–0.5	240–380	35–55	60–210	8.7–30	20–60	20–30
TD04	H	...	1.3–12.7	0.05–0.5	440–560	64–81	370–520	54–75	2–20	20–30
TF00	AT	3 h at 480–495 °C	1.3–12.7	0.05–0.5	680–900	99–130	550–760	80–110	10 min	45–60
TH04	HT	2 h at 480–495 °C	1.3–12.7	0.05–0.5	750–970	109–140	650–870	94–126	10 min	48–60

nealed and age-hardened (AT) temper are less susceptible to the effects of a slow cooling rate or variable section size. Water quenching of annealed temper castings with a large cast grain size may cause cracking. Slowing the cooling rate during quenching is recommended in such cases; however, this will reduce the AT temper aging response of the materials.

Fabrication Characteristics

Formability. Strip products can be fabricated, depending on their temper, into components by stamping, coining, deep drawing, or hydroforming. The severe strains associated with the latter two cold-forming processes generally confine their application to the solution-annealed (TB00) or ¼ hard (TD01) tempers.

Bend formability is commonly measured by 90° or 180° plane-strain bend tests and is reported as the ratio of minimum bend radius for no cracking to the strip thickness. As indicated by the data given in Table 9, formability is highest and most isotropic in the annealed (TB00) and ¼ hard (TD01) tempers. Slightly anisotropic but good formability is retained as cold work increases to the hard (TD04) temper; these formability characteristics are also exhibited by the low-to-intermediate strength mill-hardened tempers (TM00 through TM04). Moderate-to-limited, more anisotropic formability is displayed by the high-strength mill-hardened tempers through TM08 and in mill-hardened C17410 TH04 strip. Formability data provided by strip producers are typically based on test results for strip as thick as 1.2 mm (0.05 in.); the reported figures generally are conservative. Users will experience better forming characteristics in thinner strip and in components with smaller width-to-thickness ratios, such as those

Table 6 Mechanical and electrical properties of beryllium-copper rod, bar, tube, and plate

Temper designations			Outside diameter or across flats		Tensile strength		Yield strength		Elongation,		Electrical conductivity,
ASTM	Commercial	Aging treatment	mm	in.	MPa	ksi	MPa	ksi	%	Hardness	%IACS
C17200											
TB00	A	...	All sizes		410–590	59–86	130–250	19–36	20–60	45–85 HRB	15–19
TD04	H	...	≤9.5	≤3/8	620–900	90–130	510–730	74–106	8–30	92–103 HRB	15–19
			9.5–25	3/8–1	620–870	90–126	510–730	74–106	8–30	88–102 HRB	15–19
			25–50	1–2	580–830	84–120	510–730	74–106	8–20	88–101 HRB	15–19
			50–75	2–3	580–830	84–120	510–730	74–106	8–20	88–101 HRB	15–19
TF00	AT	3 h at 315–330 °C	All sizes		1130–1380	164–200	890–1210	129–175	3–10	36–41 HRC	22–28
TH04	HT	2–3 h at 315–330 °C	≤9.5	≤3/8	1270–1560	184–226	1100–1380	160–200	2–9	39–45 HRC	22–28
			9.5–25	3/8–1	1240–1520	180–220	1060–1350	154–196	2–9	38–44 HRC	22–28
			25–50	1–2	1200–1490	174–216	1030–1320	149–191	4–9	37–44 HRC	22–28
			50–75	2–3	1200–1490	174–216	990–1280	144–186	4–9	37–44 HRC	22–28
C17000											
TB00	A	...	All sizes		410–590	60–86	130–250	19–36	20–60	45–85 HRB	15–19
TD04	H	...	≤9.5	≤3/8	620–900	90–130	510–730	74–106	8–30	92–103 HRB	15–19
			9.5–25	3/8–1	620–870	90–126	510–730	74–106	8–30	92–102 HRB	15–19
			25–50	1–2	580–830	84–120	510–730	74–106	8–20	88–101 HRB	15–19
			50–75	2–3	580–830	84–120	510–730	74–106	8–20	88–101 HRB	15–19
TF00	AT	3 h at 315–330 °C	All sizes		1030–1320	150–191	860–1070	125–155	3–10	32–39 HRC	22–28
TH04	HT	2–3 h at 315–330 °C	≤9.5	≤3/8	1170–1450	170–210	990–1280	144–186	2–5	35–41 HRC	22–28
			9.5–25	3/8–1	1170–1450	170–210	990–1280	144–186	2–5	35–41 HRC	22–28
			25–50	1–2	1130–1380	164–200	960–1250	139–181	2–5	34–39 HRC	22–28
			50–75	2–3	1130–1380	164–200	930–1210	135–175	2–6	34–39 HRC	22–28
C17500 and C17510											
TB00	A	...	All sizes		240–380	35–55	60–210	8.7–30	20–35	20–50 HRB	20–30
TD04	H	...	≤5	≤3	440–560	64–81	340–520	49–75	10–15	60–80 HRB	20–30
TF00	AT	3 h at 480 °C	All sizes		680–900	99–130	550–690	80–100	10–25	92–100 HRB	45–60
TH04	HT	2 h at 480 °C	≤75	≤3	750–970	109–140	650–870	94–126	5–25	95–102 HRB	48–60

Table 7 Mechanical and electrical properties of beryllium-copper forgings and extrusions

Alloy(a)	Heat treatment	Tensile strength, MPa	Tensile strength, ksi	Yield strength, MPa	Yield strength, ksi	Elongation, %	Hardness	Electrical conductivity, %IACS
C17200 (TB00)	...	410–590	59–85	130–280	19–41	35–60	45–85 HRB	15–19
	3 h at 330 °C	1130–1320	164–191	890–1210	129–175	3–10	36–42 HRC	22–28
C17000 (TB00)	...	410–590	59–85	130–280	19–41	35–60	45–85 HRB	15–19
	3 h at 330 °C	1030–1250	149–181	860–1070	125–155	4–10	32–39 HRC	22–28
C17500 (TB00) and C17510 (TB00)	...	240–380	35–55	130–280	19–41	20–35	20–50 HRB	20–35
	3 h at 480 °C	680–830	99–120	550–690	80–100	10–25	92–100 HRB	45–60

(a) ASTM temper designations in parentheses; all alloys in the annealed condition prior to heat treatment

Table 8 Mechanical properties of beryllium-copper casting alloys

UNS designation	Temper	Yield strength at 0.2% offset, MPa	Yield strength at 0.2% offset, ksi	Tensile strength, MPa	Tensile strength, ksi	Elongation in 50 mm (2 in.), %	Hardness
C82000	As-cast	105–170	15–25	310–380	45–55	15–25	50–60 HRB
	As-cast and aged	170–310	25–45	380–480	55–70	10–15	65–75 HRB
	Solution annealed and aged	480–550	70–80	620–760	90–110	3–15	92–100 HRB
C82200	As-cast	170–240	25–35	380–410	55–60	15–25	55–65 HRB
	As-cast and aged	280–380	40–55	410–520	60–75	10–20	75–90 HRB
	Solution annealed and aged	480–550	70–80	620–690	90–100	5–10	92–100 HRB
C82400	As-cast	240–280	35–40	450–520	65–75	20–25	74–82 HRB
	As-cast and aged	450–520	65–75	655–720	95–105	10–20	20–24 HRC
	Solution annealed and aged	930–1000	135–145	1000–1070	145–155	2–4	34–39 HRC
C82500 and C82510	As-cast	280–345	40–50	520–590	75–85	15–30	80–85 HRB
	As-cast and aged	480–520	70–75	690–720	100–105	10–20	20–24 HRC
	Solution annealed and aged	830–1030	120–150	1030–1210	150–175	1–3	38–43 HRC
C82600	As-cast	310–345	45–50	550–590	80–85	15–25	81–86 HRB
	As-cast and aged	410–450	60–65	650–720	95–105	10–15	20–25 HRC
	Solution annealed and aged	1070–1170	155–170	1140–1240	165–180	1–2	40–45 HRC
C82800	As-cast	345–410	50–60	590–620	85–90	5–25	80–90 HRB
	As-cast and aged	410–480	60–70	655–720	95–105	10–15	20–25 HRC
	Solution annealed and aged	1140–1240	165–180	1240–1340	180–195	0.5–3	43–47 HRC

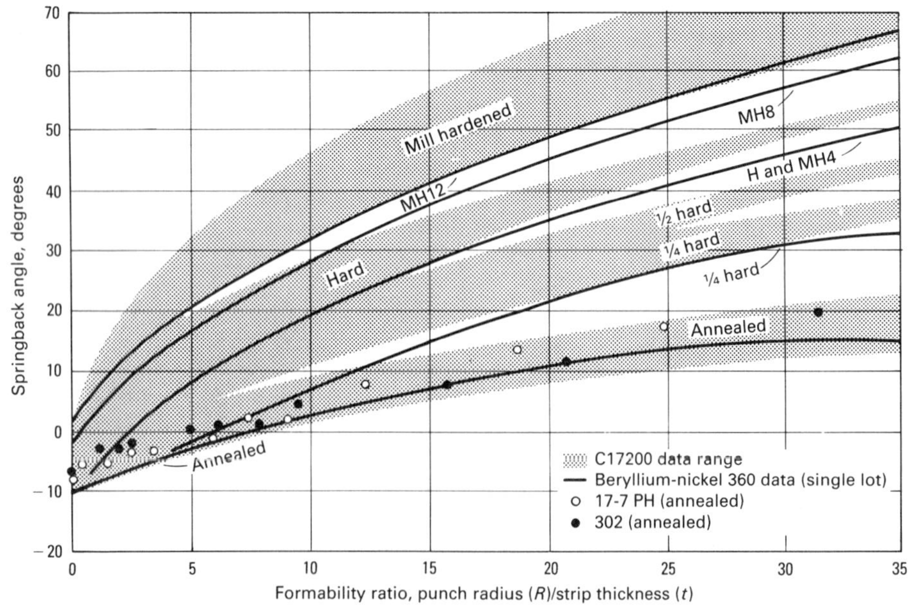

Fig. 6 Angular springback of heat-treatable and mill-hardened tempers of beryllium-copper C17200 and beryllium-nickel N03360 strip (90° V-block plane-strain bends)

typical of electronic connector contact springs.

Elastic Springback. As with any metallic material, cold-formed beryllium-copper alloys exhibit elastic springback upon release of the forming tool pressure. Compensation is often incorporated into forming tools to enable the material to attain target dimensions. Angular springback curves for heat-treatable and mill-hardened strip formed longitudinally in a 90° V-block (plane-strain bends) are shown in Fig. 6.

Springback of C17200 increases with increasing yield strength. It is lowest in the solution-annealed (TB00) temper and highest in the high-strength mill-hardened tempers. Springback also increases with the forming ratio (punch radius divided by strip thickness) as elastic strain becomes a larger fraction of the total forming strain. Negative angular springback (collapse of the formed part against the punch) is evident at very small forming ratios in these V-block bends and is associated with a substantial shift of the neutral bend axis toward the inner bend radius.

Figure 6 shows that consistency of elastic springback in a component formed with a tool of fixed punch radius depends more on uniformity of thickness than on yield strength within a given temper. Springback also depends on the strains imparted during forming and varies with factors such as punch clearance and method of forming. Wiping bends and coining deformations produce less springback than plane-strain V-block bends.

Age-Hardening Dimensional Change. Selecting mill-hardened tempers allows the user to avoid the effects of dimensional change upon age hardening because these tempers do not need to be heat treated after stamping and forming. However, for those users who form parts before age hardening, this section contains information about factors that can contribute to dimensional change during age hardening and suggestions on how to control these factors.

Causes. The coherent age-hardening precipitates have a slightly higher density than the copper-rich matrix, and their formation is accompanied by shrinkage of the matrix. The magnitude of this shrinkage depends on the beryllium content, temper, and age-

Table 9 Relative formability of beryllium-copper strip

		Suitable alloy condition for specified formability rating and approximate formability ratio (*R*/*t*) for a 90° bend(a)								
		Alloy C17000			Alloy C17200			Alloys C17500 and C17510		
Formability rating	Specific formability	Alloy condition(b)	Transverse(c) *R*/*t* ratio	Longitudinal(d) *R*/*t* ratio	Alloy condition(b)	Transverse(c) *R*/*t* ratio	Longitudinal(d) *R*/*t* ratio	Alloy condition(b)	Transverse(c) *R*/*t* ratio	Longitudinal(d) *R*/*t* ratio
Excellent	Used for deep-drawn and severely cupped or formed parts	TB00	0.0	0.0	TB00	0.0	0.0	TB00	0.0	0.0
	As formable as the annealed (TB00) temper but easier to blank	TD01	0.0	0.0	TD01	0.0	0.0	...	...	...
					TM00(e)	0.0	0.0			
					TM02(e)	0.0	0.0			
Very good	Used for moderately drawn or cupped parts	TD02	1.0	0.5	TD02	1.0	0.5	TD04	0.6	0.5
		TM00	1.0	1.0	TM00	0.8	0.8	TF00	1.0	1.0
					TM01	1.0	1.0	HTC	1.0	1.0
					TM04(e)	1.0	1.0			
Good	Formable to a 90° bend around a radius <3× stock thickness	TD04	2.9	1.0	TD04	2.9	1.0	TH04	2.0	2.0
		TM01	1.7	1.5	TM02	1.3	1.3			
		TM02	2.2	1.9	TM04	2.5	2.5			
					TM06(e)	2.0	2.5			
Moderate(f)	Suitable for light drawing; used for springs	...	...	...	TM05	3.2	2.8	HTR	3.5	2.8
					TM06	3.8	3.0			
					TM08(e)	3.0	3.5			
Limited	For essentially flat parts; forming requires very generous punch radii	TM04	5.1	3.8	TM08	6.0	4.1	...	...	...
		TM05	7.7	5.0						
		TM06	10.4	6.1						

(a) Formability ratios of punch radius (*R*) to stock thickness (*t*) are valid for strip up to 1.3 mm (0.050 in.) thick. Strip less than 0.25 mm (0.010 in.) thick will form somewhat better than shown. Values reflect the smallest punch radius that forms a strip sample into a 90° vee-shaped die without failure. (b) See Table 4 for descriptions of the alloy condition designations. (c) Transverse bend direction has a bend axis parallel to the rolling direction (see Fig. 8). (d) Longitudinal bend direction has a bend axis perpendicular to the rolling direction. (e) Special mill processing for high formability. (f) Moderately formable strip includes alloy C17410 in the TH04 condition with a longitudinal *R*/*t* ratio of 1.0 and a transverse *R*/*t* ratio of 6.0.

hardening conditions, but it is nominally only 0.6% in volume or 0.2% in linear dimension in the high-strength alloys (C17000, C17200, and C17300). The high-conductivity alloys (C17500 and C17510) have a negligible volume change on age hardening.

The rate of precipitation and accompanying shrinkage depend upon residual stresses in the matrix. Compressive stresses locally intensify the aging process, whereas tensile residual stresses slow the process. When the residual stress distribution is nonuniform, as in many formed components, nonuniform shrinkage can promote distortion from as-formed dimensions. In a 90° bend, for example, the inner bend radius is under compression and exhibits a greater aging response and more shrinkage than the outer bend radius. The included angle decreases during age hardening in this case. Distortion is greater in flatter components than in other shapes unless stiffening features are included in the design. Distortion tends to be magnified over long dimensions. Bending, twist, waviness, and angular changes can occur in unsupported parts during age hardening, but each of these problems can be controlled with stiffening in the design or with one or more of the approaches suggested below.

Controls. Opportunities for distortion control are present in the forming operation itself. For example, a two-step bending process promotes a more uniform residual stress distribution in the formed part. The part can be initially bent past the intended angle and then bent back to the desired position. Rather than having compressive stresses on the inner bend radius and tensile stresses on the outer bend radius (as in a single-stage bend), the two-stage bend component will exhibit compressive stresses on both the inner and outer bend radii, thus promoting more uniform through-thickness shrinkage.

Larger bend radii promote more uniform residual stresses, but bends with an excessively large radius exhibit elastic springback sufficiently great to reverse the sign of residual stresses on the inner and outer bend surfaces. When this occurs, aging distortion appears in the direction opposite to the tighter-radius bends. Dull tooling or excessive die clearances will increase material distortion at the stamped edge, imparting localized residual stresses that can increase age-hardening distortion.

Temper selection is another means of distortion control. The product temper that results in the lowest nonuniformity of final residual stress distribution in the component to be aged will show the least distortion. For example, distortion in parts machined from large-diameter cold-worked rod can be controlled by substituting an annealed rod temper that is free from the residual stress of cold drawing. On the other hand, many components formed from thin strip benefit from the selection of more heavily cold-worked tempers in which the nonuniformity of residual stress from forming is less significant than in lightly cold-worked strip. Selecting the hardest heat-treatable strip temper consistent with the formability requirements of the part will result in the least aging distortion.

Fixturing can be employed to constrain parts during age hardening; the exact technique varies substantially with component shape. Long, flat rod or tubular parts can be secured to rigid beams or between flat plates; other parts can be packed tightly in a supporting and heat-distributing medium such as sand. Thin, flat parts can be self-fixtured by incorporating in-plane stiffening features into the design.

Special heat-treating conditions can be employed to lessen age-hardening distortion in the high-strength alloys as compared with that obtained by peak aging at 315 °C (600 °F) for 2 or 3 h. Aging at a higher temperature for a shorter time corresponding to the relative maximum in the aging response curve will reduce distortion by increasing the proportion of incoherent precipitates relative to coherent precipitates in the age-hardening process, thereby reducing the net shrinkage tendency of the material. For example, the magnitude of distortion reduction achieved by aging at 370 °C (700 °F) for about 30 min is significant, but it depends on the initial residual stress distribution in the part. Some sacrifice in peak strength also accompanies this high-temperature short-time age-hardening approach. In C17200, the decrease in strength is about 100 MPa (14 ksi).

The desired strength level and the process control techniques used during aging (which are influenced by lot size, furnace type, and part configuration) will dictate the optimal age-hardening temperature for this approach. Thermal stress-relieving treatments at temperatures below the age-hardening range may reduce some residual stress. A treatment at 175 °C (350 °F) for 3 to 4 h is a simple and effective stress-relief process.

Cleaning. Beryllium-copper products are thoroughly cleaned and provided with a tarnish inhibitor before delivery. They therefore need no cleaning or other preparation prior to use. After stamping and forming, age hardening, or any type of thermal treatment, these products should be cleaned as a preparatory step to plating, coating, or joining processes.

The first step in the surface preparation of beryllium-copper is the removal of soils, oils, and grease. These are normally present as residual traces of lubricants used in forming or from exposure of the material to shop atmospheres laden with oil mist. Surface soils such as fingerprints also should be removed. Conventional cleaners, such as organic solvents and alkaline solutions, are normally adequate for removing organic residues. Vapor degreasing is effective, as is ultrasonic agitation for augmenting the cleaning medium action.

Like all copper alloys, beryllium-copper forms a thin surface oxide, or tarnish, when exposed to air. Tarnish formation is accelerated by moisture and by elevated temperature. Even when protective atmospheres are used, heat treatment usually oxidizes the surface to such a degree that cleaning is required for materials intended for precision applications.

The surfaces of user-heat-treated beryllium-copper components can be prepared with a procedure such as:

- Immerse parts in an aqueous solution of 20 to 25 vol% sulfuric acid plus 2 to 3 vol% hydrogen peroxide held at 45 to 60 °C (115 to 140 °F). Immerse for sufficient time to remove any dark coloration
- Rinse thoroughly and dry. Joining or plating should immediately follow the cleaning process

Pretesting should be done to avoid removal of measurable amounts of metal caused by a high acid concentration or an excessive immersion time. Special cases such as the removal of oxides left by annealing or welding processes should be reviewed with the material supplier.

Soldering is specified when the service temperature is below about 150 °C (300 °F), where higher joining temperatures might damage components, and where electrical and thermal continuity require more strength than a mechanical bond can provide. Soldering is the most common joining technique for beryllium-copper in electrical and electronic applications, where component thickness is typically less than 0.3 mm (0.012 in.). Soldering can be performed with localized heating by resistance, induction, infrared, and flame techniques; by selective application techniques such as wave soldering; or by general techniques such as bulk immersion and vapor phase deposition.

Parts to be soldered must first be thoroughly cleaned to remove oil, grease, tarnish, and oxides. Soldering should immediately follow cleaning. The mildest flux suitable for the job should be used. Noncorrosive (rosin) fluxes that are active only when heated and need only a warm-water cleanup are the most common choice for beryllium-copper. Gas fluxes (containing hydrazine in nitrogen or argon) can also be used and leave no deleterious reaction products.

Tin-lead compositions are the most commonly used solders for joining beryllium-copper. A 63Sn-37Pb solder generally is selected for high-volume electronic work. The 50Sn-50Pb type is typically chosen for hand soldering. Special silver- or indium-containing solders are used where higher bond strength and ductility are required or where silver coatings on electronic components might otherwise be dissolved by tin-lead solders. Melting temperatures for the indicated tin-lead solders range from 180 °C (350 °F) to about 240 °C (460 °F). These temperatures are low enough in the beryllium-copper age-hardening range that mechanical properties will not be affected during soldering.

The solderability of bare beryllium-copper is adequate for manual- or moderate-speed automated joining. High-speed soldering is aided by precoating with a minimum of 0.007 mm (0.275 mil) of 60-40 solder or pure tin, applied by hot dipping or electroplating. Overheating must be avoided to minimize oxidation, to prevent flux degradation, to retard formation of undesirable intermetallic compounds at the solder/substrate interface, and to prevent metallurgical changes in the substrate. Soldering beryllium-copper to itself poses no special problems, but the high conductivity of these alloys may necessitate the use of heat sinks to concentrate heat at joints made with lower-conductivity materials.

Brazing provides stronger, more heat-resistant joints than those formed by soldering. For beryllium-copper, brazing temperatures are higher than age-hardening levels. It is therefore preferable that brazing be performed before age hardening; however, with a rapid cycle, hardened beryllium-copper can be brazed effectively. The brazing temperature should not exceed the solution-annealing temperature. Braze integrity depends on joint design and assembly size, as well as on heat input and dissipation rates. Heat sinks are recommended to confine increases in temperature to the joint area.

Two general brazing techniques are used for joining beryllium-copper. A low-temperature approach employs brazing filler metals that melt at temperatures below 620 °C (1150 °F), such as American Welding Society (AWS) BAg-1. It is performed after age hardening and is accompanied by a very minor loss of hardness. This technique is recommended for joining small parts of similar size; it requires a localized and rapid application of heat to the joint area for a dwell time of 1 min maximum, followed by rapid forced-air cooling or water quenching. Small brazed assemblies may allow the incorporation of short-time high-temperature age hardening into the brazing cycle.

A high-temperature technique is better suited to larger parts or joints between components of dissimilar size. The process is applied after solution annealing but before age hardening. It uses brazing filler metals that melt near the annealing temperature of the beryllium-copper alloy being joined. Where practical, large furnace-brazed parts can be quenched after brazing in preparation for subsequent age-hardening treatments.

The high-strength beryllium-coppers require a brazing alloy that melts at a temperature near 760 °C (1400 °F) (for example, AWS BAg-8). The high-conductivity alloys require a filler metal that will flow at temperatures between 900 and 950 °C (1650 and 1740 °F), such as AWS RBCuZn-D. Heat can be applied by torch, induction, resistance, or furnace methods. Heating must be followed by rapid cooling, preferably water quenching, to preserve the age-hardening response. As with soldering, the surfaces to be joined must be thoroughly cleaned immediately before brazing to remove oils, grease, dirt, or oxides. Flux residues should be removed after joining by hot water and brushing or with warm dilute sulfuric acid.

Welding is useful for joining beryllium-copper, and a variety of processes are available. Common techniques include electron beam and laser welding, resistance welding for face-to-face sheet joining, gas tungsten arc welding for overlay or thick section joining, and friction welding for tubular sections.

Careful metallurgical process planning is essential for any welding process. Consideration must be given to joint design, preheat (which must be held below the age-hardening temperature), weld technique, and postweld practice. It is best to weld beryllium-copper in the annealed condition with subsequent precipitation hardening. Previously age-hardened material can be welded, but the higher thermal conductivity of the base metal and the complex metallurgical conditions in the heat-affected zone make postweld heat treatments desirable for these materials.

Adhesive bonding is being used with increasing frequency for beryllium-copper assemblies because of its low cost and good performance at temperatures up to 150 °C

Table 10 Mechanical properties of beryllium-copper alloys at hot-working temperatures

Alloy	Temperature °C	Temperature °F	Yield strength in tension MPa	Yield strength in tension ksi	Elongation, %	Deformation resistance MPa	Deformation resistance ksi
High-strength beryllium-copper alloys (C17000 and C17200)(a)	705	1300	48	7	60	70–193	10–28
	760	1400	14	2	105	55–138	8–20
	815	1500	7	1	130	48–117	7–17
High-conductivity beryllium-copper alloys (C17500 and C17510)	705	1300	76	11	35	110–228	16–33
	760	1400	55	8	45	83–186	12–27
	815	1500	34	5	55	69–159	10–23
	870	1600	20	3	70	41–138	6–20
	925	1700	14	2	85	34–110	5–16

(a) Alloy C17300 cannot be forged. Extrusion of C17300 requires special conditions and processes.

(300 °F). Good bond strength can be obtained with a variety of adhesive formulations without complex surface preparation. Joint design is important, and loads should be transmitted along rather than across the bond joint where possible.

Thermosetting epoxy formulations achieve lap shear strengths of 20 MPa (3 ksi) or more and peel strengths of 0.8 to 1.4 MPa per millimeter of width (3 to 5 ksi per inch of width) with a 35 °C (95 °F) cure. A two-component modified epoxy applied to degreased mill-hardened strip can achieve this level of strength. Higher cure temperatures, special resin additives, and surface-etching techniques are sometimes used to increase strength even further when needed.

Hot-Working Processes. Beryllium-copper alloys are readily hot worked to produce dimensionally accurate and sound parts that respond well to subsequent age hardening. The forgeability of these alloys is considered good. The principal variables affecting product integrity are preheating time and temperature, temperature control during processing, and deformation ratio.

Forging. For the high-conductivity alloys C17500 and C17510, the forging range is 760 to 925 °C (1400 to 1700 °F). The forging range for the high-strength alloys C17000 and C17200 is 705 to 775 °C (1300 to 1425 °F). Free-machining alloy C17300 is not hot forgeable, but it can be hot extruded.

The amount and rate of deformation dictate forging temperature selection, with larger reductions and faster rates (which increase adiabatic heating) requiring a starting temperature low in the indicated range. Forging must be conducted at a temperature sufficiently high to recrystallize the material without promoting excessive grain growth. Reheating is called for when the surface temperature cools because of radiation loss and die contact and when adiabatic heating is insufficient to maintain surface temperature above the working range minimum.

Open die forging processes (including ring rolling and roll forging) are applicable to short-run items or relatively large parts of simple geometry. Closed die forging is better suited to longer runs or more complex and precision component designs. During either type of forging process, the workpiece must be maintained within a controlled temperature range to avoid forging defects associated with incipient melting on the high-temperature side and surface cracking on the low-temperature side of the range.

Room-temperature billets should be brought to temperature at a rate of 1 h per inch of section and then soaked at the forging temperature for a short time to ensure temperature uniformity. On reheating, 0.5 h per inch of section is sufficient heating time. Furnace temperature uniformity and control should be within ±5 °C (±10 °F). Furnace atmospheres should be neutral or only slightly oxidizing to prevent excessive scale formation. The use of high-sulfur fuel in gas-fired furnaces should be avoided.

Reduction ratios should be sufficiently large to promote deformation penetration through the entire workpiece section. Partial penetration of deformation from light hammer blows, particularly on the final passes, can result in undesirable nonuniform dynamic recrystallization and associated nonuniformity in microstructure and mechanical properties after age hardening. The desirable range of reduction ratios is 3:1 to 5:1, but reduction ratios as large as 8:1 to 10:1 can be obtained in some cases. Deformation should not exceed the elevated-temperature ductility limit of the material (Table 10).

Extrusion. The temperature ranges for extruding beryllium-copper alloys are:

- C17000 and C17200, 705 to 775 °C (1300 to 1425 °F)
- C17500 and C17510, 815 to 900 °C (1500 to 1650 °F)

Free-machining C17300 is more difficult to extrude than the other beryllium-coppers and requires special control of process parameters.

Key variables in the extrusion process are temperature, speed, extrusion ratio, die design, die material, dummy block configuration, and lubrication. A temperature above the recommended range can promote oxidation of the extruded product. Cracking, coarse grain size, or incipient melting are possible under extreme conditions. The somewhat wider extrusion temperature range of alloys C17500 and C17510 renders these alloys less susceptible to heat checking.

Too low a billet temperature causes high press loads and may promote nonuniform deformation. Near-neutral preheating furnace atmospheres retard oxidation that might increase die wear. Billets can be economically heated by induction with minimum oxidation. The press speed and extrusion ratio depend on die design and the ability of the tooling and metal system to distribute adiabatic heat generated during deformation. Reduction ratios as high as 10:1 are feasible. Dummy blocks are typically flat, and a die half angle of 35° is satisfactory for the high-strength beryllium-coppers. Die materials include tool steels, high-temperature cobalt- or nickel-base alloys, and zirconia or alumina ceramics. Ceramic dies have the advantages of a lower coefficient of friction with beryllium-coppers and lower wear rates. Mandrels for tube extrusion are typically made of high-strength materials such as Inconel 718. Lubrication reduces friction and die wear and is typically provided by graphite-base, sulfur-free formulations.

Heading and Forming. Heat-treatable tempers of rod and wire can be formed by cold heading. Mill-hardened C17500 and C17510 rod and bar products, by virtue of their moderate strength and good ductility, can also be cold formed by bending.

Heat Treatment and Properties (Hot-Worked Products). Forgings and extrusions require heat treatment to attain maximum strength (Table 7); solution annealing, rapid quenching, and age hardening at temperatures and times recommended for other wrought beryllium-copper products are the methods generally employed. As an alternative to solution annealing, hot-finished beryllium-copper components (extrusions in particular) can be quenched directly from the press by immersion in a well-agitated medium. Spray quenching is usually inadequate, and the resultant age-hardening response depends on both the hot finishing temperature and the rapidity of the quench.

Machining. Beryllium-copper can be machined at metal removal rates comparable to those used for other high-performance copper alloys and stainless steels. Machining can be done on annealed or cold-worked products, but tough, continuous chips may lead to difficulty in some instances. Machining in the age-hardened condition (AT or HT) eliminates this difficulty and alleviates the need for postaging cleaning. Alloy C17300 is generally used for automated machining operations because the minor lead additive is an effective lubricant and chip breaker.

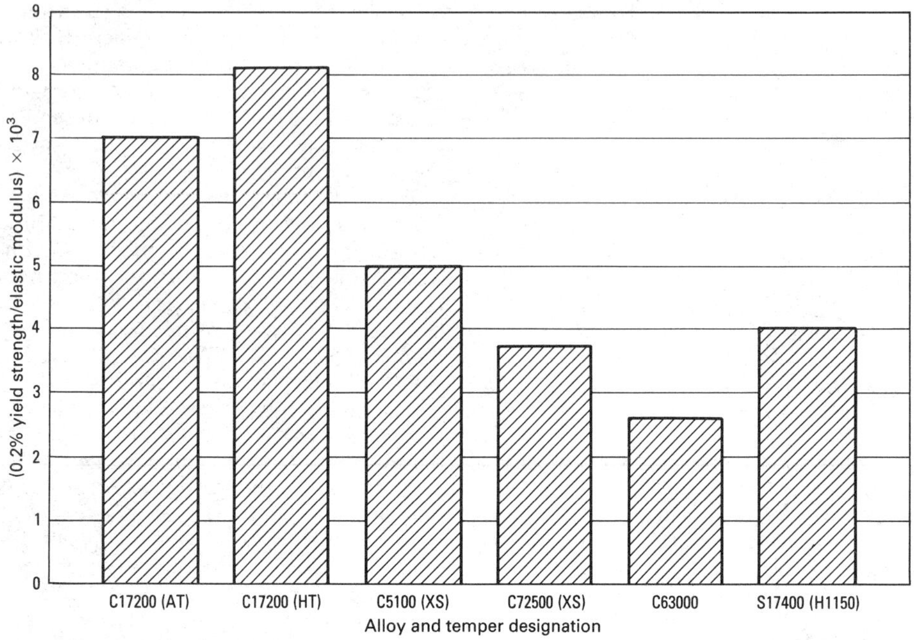

Fig. 7 Ratio of yield strength to elastic modulus for several connector alloys

Cutting tools should be kept sharp and should have a positive rake angle between 5° and 20° for best performance. Chip breakers are recommended for turning. Cutting fluid is recommended as a coolant; water-soluble oils and synthetic emulsions are commonly used for this purpose. Although the best finishes are obtained from sulfurized oils, these oils will discolor the surface. The stain is not harmful, but should be removed after machining, particularly if the parts are to undergo subsequent age hardening.

Surface grinding is also commonly used for beryllium-copper alloys. Guidelines on wheels, speeds, and metal removal rates are available from abrasives manufacturers. Grinding should always be done with a coolant. Other machining methods, such as photochemical etching, electrical discharge machining, and electrochemical machining, are practiced by numerous vendors.

Design and Alloy Selection Considerations

To perform effectively in an engineering system, an alloy often must satisfy a set of complex requirements. These requirements consist of a list of mechanical and physical properties in which two or three predominate in importance. The combination of properties must provide measurable benefits over the lifetime of the system to justify the choice of one alloy over another. The beryllium-copper family of alloys, by virtue of its ability to be tailored to fit a wide variety of strength, conductivity, and fabricability parameters, provides benefits in a number of diverse industries. This section presents examples of the beneficial properties of beryllium-copper alloys and quantifies some of the major reasons for their selection for particular applications.

Resilience and Formability (Electronic Contact Spring Applications). The ratio of yield strength to elastic modulus measures the ability of a spring to apply a high force from a relatively large deflection without taking a permanent set. The property, which can be termed resilience, is shown in Fig. 7 for two tempers of beryllium-copper, several other electronic component alloys, and one steel. Mechanical properties shown in this and other graphs in this section are the minimum values from property ranges provided in current alloy specifications. The striking feature of this chart is the high ratio displayed by the beryllium-copper tempers, meaning that they can grip the edge of a printed circuit board or a mating pin member with positive action, and this action will be retained through repeated insertion cycles.

Electronic connector contact springs require resilience, and, because of high demands on space in miniaturized electronics, they also require the ability to be formed into complex spring configurations. In electronic connectors, these springs are most often formed from strip, less frequently from wire. In either case, formability is important. Bend formability in strip, discussed earlier in this article (Table 9), is the minimum ratio of bend radius to strip thickness that avoids visible cracking at the outside surface of the bend. Smaller values of this ratio represent an improved ability to form a small, tightly formed part and thus to conserve both space and material. The bend formabilities of selected copper-base alloys are compared in Fig. 8, which shows the formability ratio (R/t) as a function of the 0.2% yield strength. Alloy selection from this chart can be performed by plotting a point representing the R/t of the tightest bend in a contact spring design against the yield strength (or resilience) determined from normal force/stress analysis of that design. The alloy selected should be the one corresponding to the curve immediately below the plotted point.

Other attributes influence designer selection of a particular connector alloy for electronic contacts. Electrical conductivity plays a role; however, because these connectors operate at extremely low (signal level) currents, they are almost always plated, clad, or otherwise provided with a coating that has a high electrical conductivity. This coating can be a precious or semiprecious metal or a conductive polymer, but in all cases it must be oxidation resistant and soft enough to present a nascent metal surface under the wiping action of connector insertion. Because of this coating, electrical conductivity is a secondary consideration in alloy selection. Electronic contact alloys are chosen on the basis of resilience and formability properties from among those alloys having an electrical conductivity of at least 15% IACS. These selection criteria have resulted in the evolution of a relatively small number of alloy families (Table 11a) commonly used for electrical connectors.

Typical connector design requirements call for longitudinal and transverse formabilities of the order of 1 to 2 R/t and 0.2% yield strengths of 940 MPa (136 ksi) and more. Many exceptions to these guidelines exist among the various connector classes. Insulation-displacement connectors need isotropic formability and somewhat higher hardness to maintain a sharp knife edge to cut through insulation. Surface-mount connectors need stress relaxation resistance to withstand soldering cycles. In general, automotive electronic connectors require less formability, but because they must operate in hot environments, alloys such as C17510 and C17410 frequently are specified for these applications.

Thermal Stability of Spring Properties. Many electrical and electronic connectors must operate reliably for extended periods at elevated temperatures. Others must carry appreciable currents that may cause a temperature rise in the device. In either circumstance, stability of spring properties over the operating life of the connector is important for satisfactory performance. Beryllium-containing connector alloys are suited to a wide range of elevated-temperature or moderately high-current applications.

Exposure of mechanically stressed springs to elevated temperatures can cause relaxation of spring force and permanent set, even at stresses below the yield

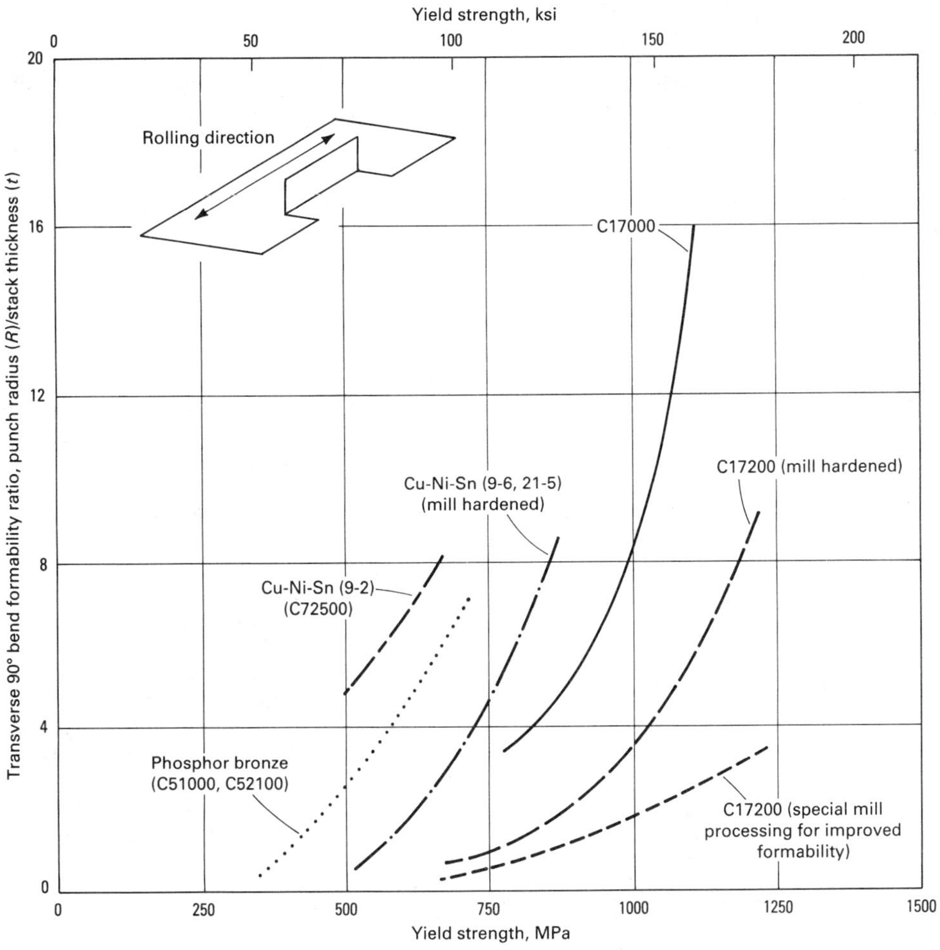

Fig. 8 Strength and transverse bend formability relationships in selected connector alloys (90° plane-strain bends)

Table 11(a) Copper alloys used for electrical connectors

See Tables 11(b) and 11(c) for mechanical properties.

Alloy name	UNS designation	ASTM specifications	Nominal composition, wt%(a)	Density g/cm³	Density lb/in.³	Elastic modulus GPa	Elastic modulus 10⁶ psi	Electrical conductivity, %IACS
Alloys hardened by rolling								
Brass	C26000	B 36	30 Zn	8.53	0.308	110	16.0	28(b)
Tin brass	C42500	B 591	9.5 Zn, 2.0 Sn, 0.2 P	8.77	0.317	110	16.0	28
Phosphor bronze	C51000	B 103	5.0 Sn, 0.2 P	8.86	0.320	110	16.0	15
Copper-silicon	C65400	B 96	3.1 Si, 1.5 Sn, 0.1 Cr	8.55	0.309	117	17.0	7
Ni-Sn copper	C72500	B 122	9.5 Ni, 2.3 Sn	8.89	0.321	130	19.0	11
Alloys hardened by thermal treatment								
Spinodal Ni-Sn copper	C72900	B 740	8 Sn, 15 Ni	8.94	0.323	128	18.5	7.8
Beryllium-copper	C17200	B 194	1.9 Be, 0.2 Co	8.25	0.298	128	18.5	22

(a) Alloying elements plus copper and trace elements total 100%. (b) Annealed

strength of the material. The ability of a material to resist such loss of spring force over time at elevated temperature is called stress relaxation resistance. The typical stress relaxation behavior of several copper-base connector alloys and beryllium-nickel is shown in Fig. 9. Stress relaxation increases slightly with increases in the initial stress level at constant temperature; it becomes more pronounced as the exposure temperature increases. The copper alloys strengthened by work hardening (for example, C51000, C68800, and C72500) exhibit appreciably lower resistance to stress relaxation than do the copper alloys strengthened by heat treatment (for example, C72400, the beryllium-coppers, and spinodal alloys such as C72700 and C72900).

Because of their higher aging temperatures, the high-conductivity beryllium-copper alloys are proportionately more relaxation resistant than the high-strength beryllium-coppers (C17000 and C17200). However, the yield strengths of the latter alloys are generally higher than those of the high-conductivity alloys, permitting greater initial stress levels in the connector design. Peak-age-hardened C17000 and C17200 in the AT to HT tempers (TF00 to TD04) are slightly more relaxation resistant than the mill-hardened tempers of these alloys (TM01 to TM08). The stress relaxation resistance of beryllium-nickel N03360 exceeds that of the beryllium-copper and spinodal alloys.

The increase in temperature in a connector beam of fixed length and cross-sectional area is directly proportional to the square of the applied current and inversely proportional to the electrical and thermal conductivities of the material. Because of their lower solute contents, beryllium-copper alloys, the high conductivity compositions in particular, exhibit a markedly lower temperature rise than do other copper alloys of similar or lower strength. The consequences of increased temperature include increased loss of spring force through accelerated stress relaxation, increased contact resistance from heat-induced oxidation at the connector interface, and, in extreme cases, distortion and melting of the plastic connector housing.

Strength and Electrical Conductivity in Interference Grounding. Electronic device containers, from connector shells to cabinets, must be shielded from the emission of radio frequency electromagnetic energy that can cause interference. These devices also must seal tightly to minimize the possibility of picking up minute quantities of atmospheric dust or particulates. The solution is to line the device at sealing surfaces with a conductive strip that has been cut into formed fingerlike spring members. Upon closure, the strip fingers fold elastically into a very thin configuration. High fracture strength in bending is required to eliminate cracking at the compressed bend, and electrical conductivity is needed to effectively and reliably ground out the small interference signal. Beryllium-copper thin strip is a common choice for this application. Typical strength and electrical conductivity combinations of commercial conductive spring alloys are compared in Fig. 10.

Fatigue Strength and Resilience (Mechanical Spring and Electrical Switch Applications). Beryllium-copper has long been noted for having the highest fatigue strength among copper alloys. This property and the high resilience of beryllium-copper in strip form have resulted in the use of the alloys as blades in many different types of switches, thermostatic controls, and electromechanical relays. A switch, for example a snap-acting type, undergoes a high flexural load with each cycle. The lifetime of the device

Table 11(b) Yield strengths and formability ratios (*R*/*t*) for the roll-hardened electrical connector copper alloys in Table 11(a)

Alloy condition	ASTM temper designation	0.2% yield strength MPa	ksi	Longitudinal *R*/*t*(a) for a 90° bend	Transverse *R*/*t*(b) for a 90° bend	Stress relaxation(c) after 1000 h at 105 °C (220 °F)
C26000						
Annealed	TB00	69–227	10–33	0	0	. . .
Half hard	H02	290–415	42–60	0	0.6	. . .
Hard	H04	460–538	67–78	0.6	1.3	76(d)
Spring	H08	565–627	82–91	1.5	3.0	. . .
Extra spring	H10	593–640	86–93	1.7	3.5	. . .
C42500						
Annealed	TB00	90–150	13–22	0	0.5	. . .
Half hard	H02	352–455	51–66	0	1.3	. . .
Hard	H04	455–545	66–79	0.5	2.0	80
Spring	H08	558–613	81–89	1.5	8.0	. . .
Extra spring	H10	600 min	87 min	. . .	. . .	. . .
C51000						
Annealed	TB00	130–200	19–29	0	0	. . .
Half hard	H02	325–470	47–68	0.4	2.0	. . .
Hard	H04	510–605	74–88	0.6	3.0	74
Spring	H08	635–745	92–108	1.5	7.5	. . .
Extra spring	H10	675–758	98–110	2.3	13	. . .
C65400						
Annealed	TB00	310	45	0	0.5	. . .
Half hard	H02	455–600	66–87	0.5	1.5	. . .
Hard	H04	648–750	94–109	1.4	3.5	83
Spring	H08	780–848	113–123	3.0	5.0	. . .
Extra spring	H10	835–895	121–130	4.0	6.0	. . .
C72500						
Annealed	TB00	125–172	18–25	1.0	1.0	. . .
Half hard	H02	407–538	59–78	1.0	1.0	. . .
Hard	H04	503–607	73–88	1.1	1.2	85
Spring	H08	572–670	83–97	1.5	1.7	. . .
Extra spring	H10	607–703	88–102	. . .	. . .	. . .

(a) Longitudinal bend has a bend axis perpendicular to the rolling direction; *R*, punch radius; *t*, stock thickness. (b) Transverse bend has a bend axis parallel to the rolling direction; *R*, punch radius; *t*, stock thickness. (c) Percentage of 0.2% yield strength remaining after exposure. (d) 75 °C (170 °F) exposure

Table 11(c) Yield strengths and formability ratios (*R*/*t*) for a spinodal Ni-Sn copper (C72900) and a beryllium-copper (C17200)

Alloy condition	ASTM temper designation	0.2% yield strength MPa	ksi	Longitudinal *R*/*t*(a) for a 90° bend	Transverse *R*/*t*(b) for a 90° bend
C72900					
Annealed and hardened	TF00	415–703(c)	60–102(c)	. . .	. . .
Annealed, rolled, and hardened	TH02	725–883(c)	105–128(c)	. . .	. . .
	TH04	895–1048(c)	130–152(c)	1.0	4.0
Mill hardened	TM00	517–655(d)	75–95(d)	0	0
	TM04	725–860(d)	105–125(d)	0.5	1.0
C17200					
Annealed and hardened	TF00	965–1205	140–175	0	0
Annealed, rolled, and hardened	TH02	1100–1345	160–195	0.5	1.0
	TH04	1138–1413	165–205	1.0	2.9
Mill hardened	TM00	483–655	70–95	0	0
	TM04	760–930(e)	110–135(e)	1.0	1.0

(a) Longitudinal bend has a bend axis perpendicular to the rolling direction; *R*, punch radius; *t*, stock thickness. (b) Transverse bend has a bend axis parallel to the rolling direction; *R*, punch radius; *t*, stock thickness. (c) 0.05 offset yield strength. (d) 93% of yield strength remaining after 1000-h exposure at 105 °C (220 °F). (e) 98% of yield strength remaining after 1000-h exposure at 105 °C (220 °F)

is ultimately limited by the fatigue life of the switch or relay contact blade. The *S-N* curves shown in Fig. 11 illustrate the fatigue behavior of beryllium-copper for two different values of the stress ratio *R*, the ratio of minimum stress to maximum stress. Switch action almost always involves unidirectional bending ($R = 0$), and therefore a very large number of cycles can be sustained by even a miniaturized switch or relay configuration. Switch designers often specify beryllium-copper in the HT condition to achieve the highest strength possible. If mill-hardened alloys are used, it is common to choose the highest-strength SHM or XHMS (TM05 through TM08) tempers when forming requirements are not severe. Relay and thermostatic control users typically select the high-conductivity alloys such as C17510 or C17410 because their current levels are higher, and thermal management is critical in these applications.

Other types of devices also use the spring resilience and fatigue strength of beryllium-copper to advantage. Detectors for seismic, ultrasonic, or other types of vibratory energy, for example, must have very high sensitivity to small signals. Springs for these devices are produced in foil thicknesses and are usually designed to have very high stiffness to vibration modes other than those in the direction of greatest interest.

Resilience and High Galling Stress (Dynamically Loaded Bearing Applications). A high ratio of yield strength to elastic modulus (Fig. 7) is a desirable attribute for an alloy used in journal bearings. The ratio measures the ability of a sleeve bearing to take a radial load from a journal (or a rod-end bearing to take a thrust load applied to a race) and to distribute this load elastically without deforming permanently. As the chart shows, beryllium-copper is higher in resilience than other coppers and steels used for this purpose in, for example, aircraft landing gear. Resilience in itself is not enough, however; the alloy also must resist galling and have high wear resistance. The wear characteristics of a variety of common alloys are compared in Fig. 12. Although beryllium-copper is not the highest on this chart, when galling resistance and resilience are also taken into account, its combination of properties usually makes it a leading candidate for bearing applications.

Low Magnetic Susceptibility and Galling Resistance (Magnetic Sensor Housing Applications). The sensing of low-level magnetic fields, such as by oil field, biomedical, and navigational instruments, requires tubular housings that are transparent to magnetic fields. Thin-wall instrument tubes house sensitive magnetometers that are responsive to fields at up to 10^{-4} tesla (1 G) or less. For example, in oil and gas exploration and logging, devices are lowered to depths of 300 to 6000 m (1000 to 20 000 ft) to determine the azimuth and inclination of a wellbore in drilling or production. Instrument tubes for this application can be thin-wall devices lowered on wire lines that transmit data electrically, or they can be heavy-wall tubes that make up part of the bottom-hole assembly of the drilling string near the drill-ing bit. The latter application is important in the recently developed technology known as measurement while drilling.

Alloys selected for service in these instruments require corrosion resistance, especially resistance to chloride stress cracking, and galling resistance to minimize the maintenance expense associated with galled

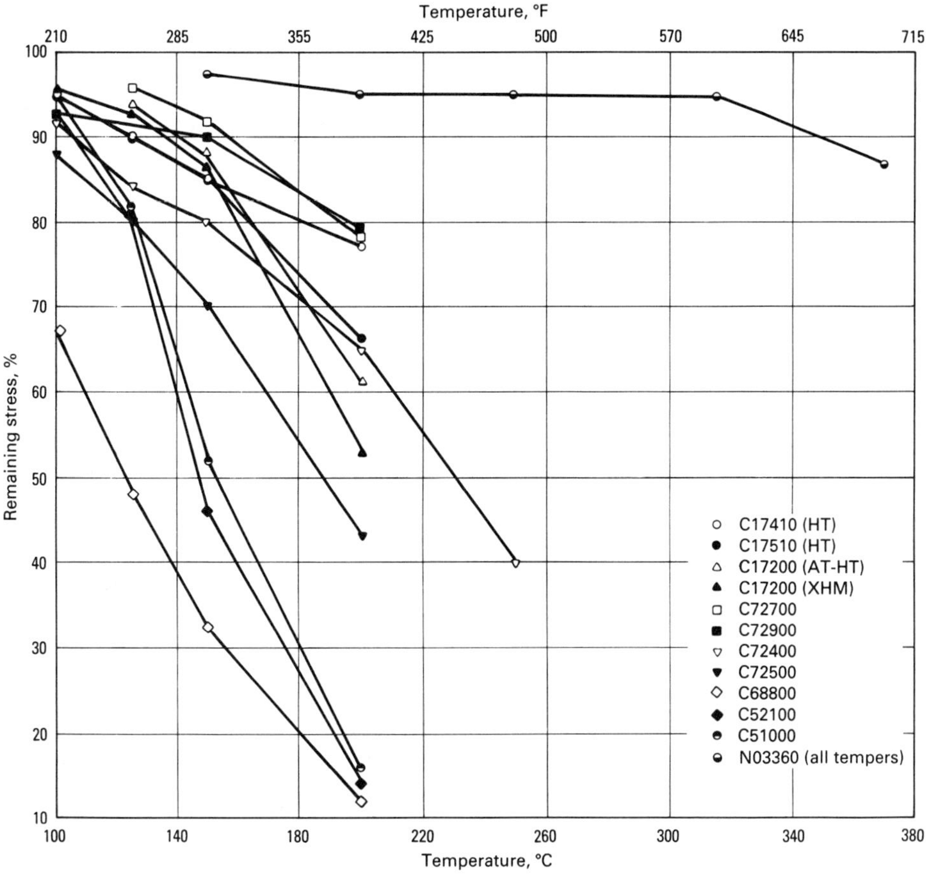

Fig. 9 Isochronal (1000-h) stress relaxation behavior of selected connector alloys at an initial stress level of 50 to 75% of the room-temperature 0.2% offset yield strength

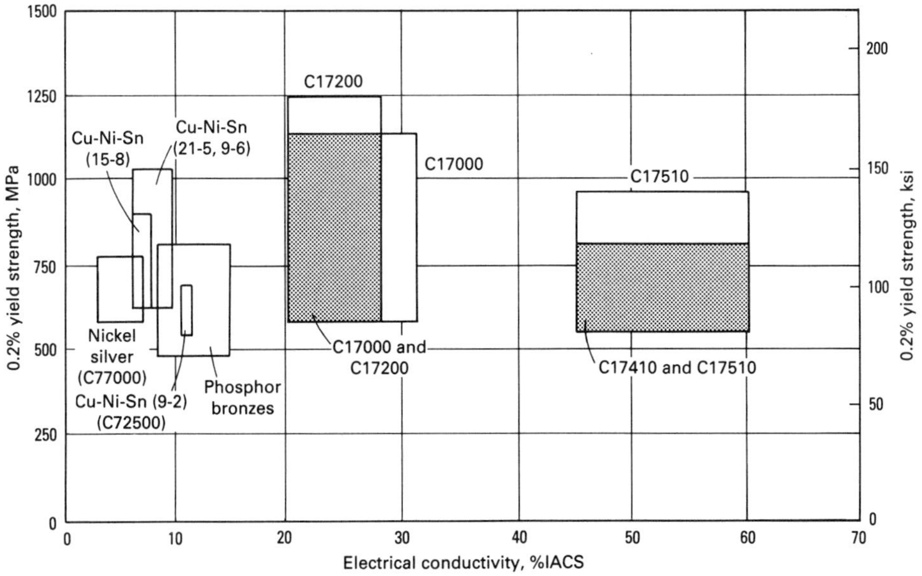

Fig. 10 Strength and electrical conductivity relationships in selected copper alloys. Each box represents the range of properties spanned by available tempers of the indicated alloy.

threads in frequently disassembled instrument packages. Freedom from magnetic hot spots is essential; as Fig. 13 shows, beryllium-copper has low initial susceptibility that remains stable and is not influenced by cold work resulting from rigorous service and rugged handling.

Thermal and Electrical Conductivity Product and Hardness (Resistance Welding Applications). The product of electrical and thermal conductivity is a measure of the ability of an alloy to manage Joule heating in an electrical system. An alloy with a high value for this product, which is plotted in Fig. 14 for a variety of copper-base alloys, minimizes the amount of $I^2 \cdot R$ heat generated with its low electrical resistivity and distributes the heat for effective convection and radiation cooling with its high thermal conductivity.

An alloy with a high product of thermal and electrical conductivity provides good thermal management, but this property by itself is not enough to make the material well suited for high-productivity resistance welding applications such as in autobody steel assembly systems. These welding systems require hardness at moderately elevated temperatures to resist tooling deformation from the sheet steel clamping pressure while pulsed current is applied. A minimum room-temperature hardness of 95 HRB is required for the demands of this application, and a substantial fraction of this hardness must be retained at a temperature of 200 °C (400 °F) and above. Typical elevated-temperature strength and hardness values for beryllium-copper alloys are shown in Fig. 15.

Thermal Conductivity and Fatigue Strength (Injection Molding Tooling Applications). Modern plastic injection molding systems require high thermal conductivity for productivity. Tooling designers continually strive for lower cycle time by developing increasingly complex die cavities to nest as many parts as possible. The life of the tooling component is a major economic factor also, and this life is often limited by erosion or heat checking caused by thermal fatigue. Modern design techniques allow the injection molder to control the flow of the plastic resin through the management of die and insert temperatures.

Alloys C17200 and C17510 are frequently selected for mold components because they provide the thermal stress resistance required for long life as well as the thermal conductivity needed for effective heat control. Typically, wrought alloys in the hardness range from 36 to 42 HRC with a thermal conductivity equivalent to that of aluminum are specified for these applications. This grade has mechanical properties typical of alloy C17200 and is usually machined into mold components from plate or heavy-section round products. Higher thermal conductivity is available in a wrought version with a specified hardness of 90 to 102 HRB; this material has mechanical properties equivalent to those of alloy C17510.

Cast alloys have long been used for castability advantages in molding because they are able to accurately replicate intricate pattern details. The alloy most commonly used for this work is C82200, but each of the casting alloys has found application in the field of plastic molding.

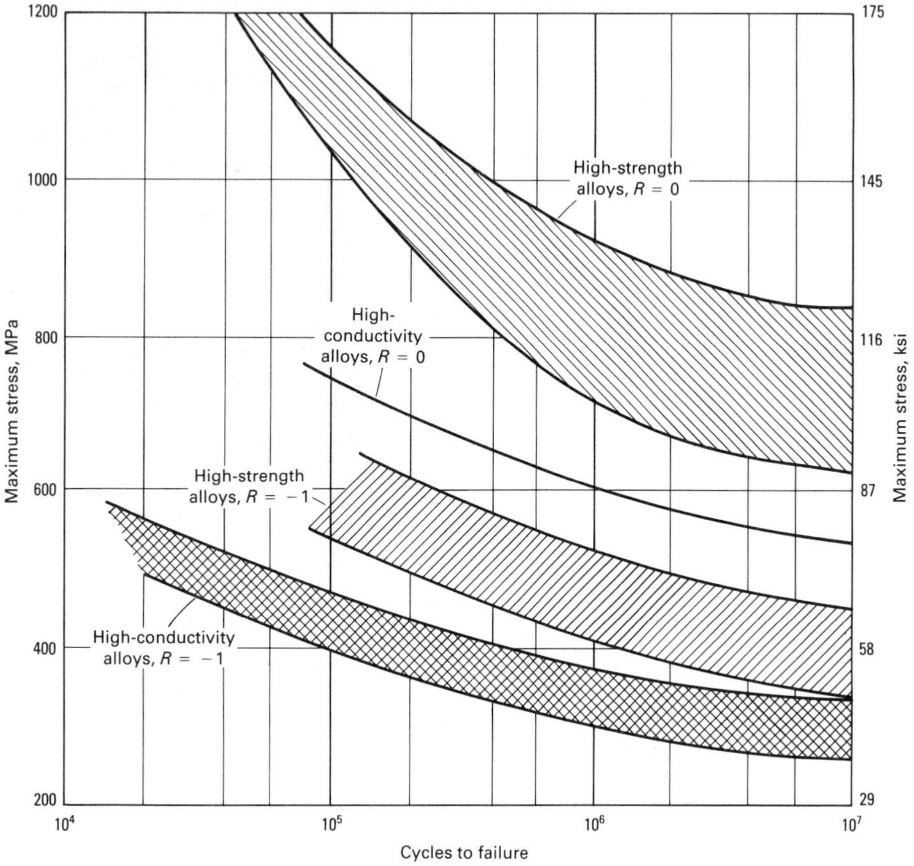

Fig. 11 Fatigue behavior of beryllium-copper strip according to the stress ratio R in unidirectional ($R = 0$) and fully reversed ($R = -1$) bending

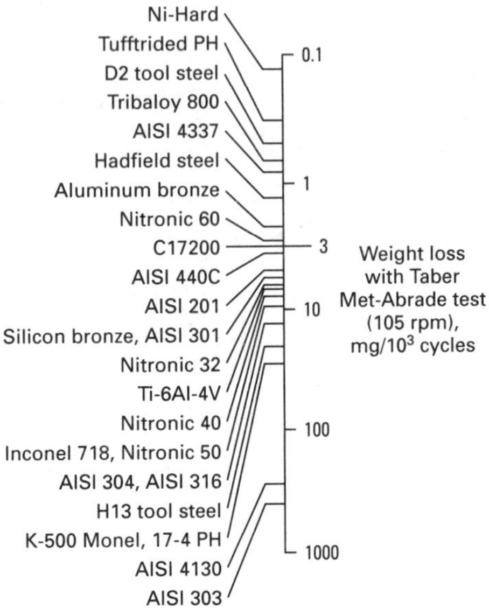

Fig. 12 Wear characteristics of C17200 beryllium-copper compared to those of other materials

Thermal and Electrical Conductivity Product at Cryogenic Temperatures and Strength (Magnetic Coil Applications). A copper solenoid or toroidal coil attached to a current source generates a magnetic field at the bore proportional to the electrical current flowing through the coil. In an air core coil the magnitude of the attainable magnetic field is not limited by a saturization magnetization as an iron core electromagnet would be. Thus it is possible to generate a very large magnetic field in the open bore of a resistive magnet. The field is limited only by the magnitude of the electric current, the heat generated in the coil turns of the magnet, and the magnetic stresses applied to the coils by the magnetic field itself. Thermal management in this type of magnet is provided by cryogenic cooling. Therefore, the thermal conductivity (Fig. 16) and the electrical conductivity at cryogenic temperatures must both be reasonably high. However, it is not sufficient to consider just the product of thermal and electrical conductivity when selecting an alloy for magnet design. A pure copper would have a very high conductivity product, but it would not perform well as a magnet because it undergoes creep deformation under the stress applied in a high-field magnet device. Many systems require a relatively high yield strength for successful operation.

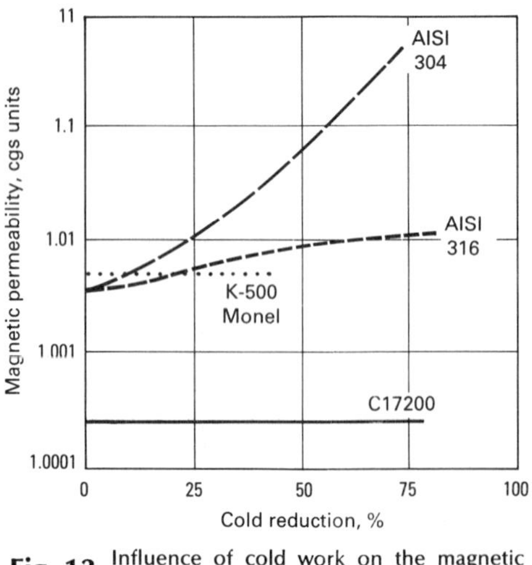

Fig. 13 Influence of cold work on the magnetic permeability of C17200 and selected other materials

For example, alloy C17510 can be tailored for an electrical conductivity of 60% IACS minimum and a minimum yield strength of 725 MPa (105 ksi). This property combination is satisfactory for many designs, but others might benefit from higher conductivity. For these needs, it is feasible to produce a version of the same alloy with an electrical conductivity of 70% IACS and a yield strength of 510 MPa (74 ksi) or more. Magnets of this type are important in research on field effects and for development of advanced electrical power generation systems. Analytical chemistry and medical systems also can benefit from this rapidly developing technology.

Corrosion Resistance for Demanding Service. Many electronic, marine, structural, industrial, and hydrocarbon applications benefit from the combination of corrosion resistance and desirable mechanical and physical properties of beryllium-copper alloys. In electronics, for example, the shelf life of unprotected beryllium-copper strip exposed for 2 years is comparable to that of C51000 or C72500; all of these alloys retain limited solderability with an activated rosin flux. Surface inhibition with benzotriazole extends the shelf life of beryllium-copper beyond this 2-year period.

Good machinability, an ability to withstand rigorous handling, low corrosion rates, and resistance to biofouling qualify beryllium-copper for various marine applications. One such use is submarine cable repeater housings. The corrosion (penetration) rate of C17200 in seawater is nominally on the order of 0.025 to 0.050 mm per year (1 to 2 mil/year) for short-term exposure at a low-to-moderate flow velocity; the rate diminishes with longer exposure times because of the formation of corrosion products and microorganism films.

Beryllium-copper is immune to cracking in chloride and sulfide environments. In hydrogen sulfide, the alloy shows general rather than localized corrosion. For example, at 150 °C (300 °F) and a concentration of 1%, the penetration rate is less than 0.050 mm per year (2 mil/yr), and the alloy maintains structural integrity. Moisture is re-

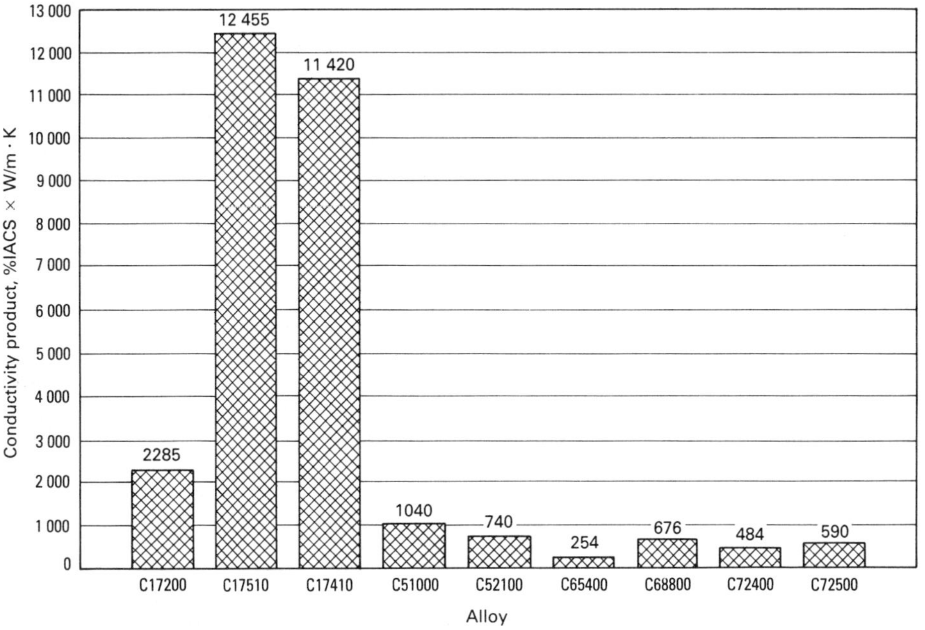

Fig. 14 Product of electrical and thermal conductivity for selected connector alloys. The product is inversely proportional to the temperature increase in a current-carrying contact beam.

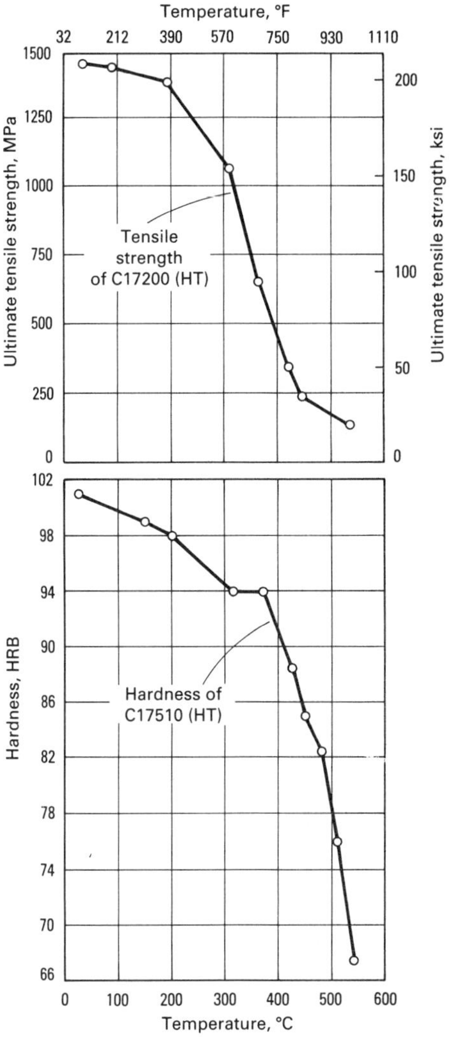

Fig. 15 Hardness of C17510 (HT) beryllium-copper and strength of C17200 (HT) beryllium-copper at elevated temperatures

quired for the attack of beryllium-copper and other copper alloys by halogens such as fluorine gas. Copper alloys will exhibit stress-corrosion cracking only in the combined presence of ammonia, high relative humidity, and oxygen.

Structural and industrial applications expose materials to a wide variety of atmospheric, chemical, and gaseous environments. Table 12 summarizes the corrosion behavior of beryllium-copper in various environments. Factors such as temperature, concentration, velocity, and the presence of impurities affect actual performance.

Plastics vary in their effect on beryllium-copper, depending on the nature of volatiles emitted. In polymer molding processes, resins cause no attack. Upon combustion, however, polyvinyl chloride and room-temperature vulcanized silicone, for example, are reported to produce fumes that corrode copper alloys. Other plastics, such as acetal, nylon 6/6, and polytetrafluoroethylene (PTFE) do not. In addition, commercial polymers can contain flame retardants such as brominated organic compounds that attack copper alloys under some circumstances. Ionic species, particularly alkali metals, can be leached from some polymers and cause corrosion.

Production Metallurgy

Melting, Casting, and Hot Working. The first step performed by commercial manufacturers of beryllium-copper alloys is the production of a nominal copper and 4 wt% Be master alloy by the carbothermic reduction of BeO in a bath of molten copper in an arc furnace. The master alloy is remelted in coreless induction furnaces and diluted with additional copper, cobalt (or nickel), and recycled mill scrap to adjust the final composition. Melts are semicontinuously cast into rectangular or round billets for hot working into wrought product forms, or they are poured as small casting ingots for foundry use.

Semicontinously cast rectangular billets are hot rolled into plate or into coils of hot band for conversion to strip. Round billets are hot extruded into bar, seamless tube, or rod coil. As-cast billets exhibit an inverse segregation layer, rich in beryllium, that cannot be eliminated by normal thermal homogenizing treatments and must be removed mechanically. This operation is typically performed by turning round billets prior to extrusion and by slab milling hot-rolled flat products prior to intermediate cold working. The hot-working temperatures typically coincide with the solution-annealing temperatures for the respective alloys, that is, about 705 to 815 °C (1300 to 1500 °F) for the high-strength alloys and about 815 to 925 °C (1500 to 1700 °F) for the high-conductivity alloys.

Hot-worked products are softened as needed by solution annealing before further processing. Heavy-section mill products and coiled hot band or rod can be annealed by heating in car-bottom furnaces equipped with transfer racks and water quenching pits. Hot band coils also can be strand annealed. Accumulated mill scale from hot working and annealing is then removed by chemical cleaning or mechanical surface conditioning.

Cold Working and Age Hardening. Subsequent processing of wrought beryllium-copper alloys typically includes one or more cycles of cold working and intermediate solution annealing until a ready-to-finish size is reached. At the ready-to-finish size, a final solution anneal is applied that establishes the final grain size and age-hardening response of the alloy. Chemical cleaning is necessary after each anneal because, even in the best of commercial vacuums or protective atmospheres, a tenacious oxide film will form. If not removed, this oxide can cause roll wear in cold working, die wear in stamping, and poor adhesion of solder or plating layers in finished components. The high-conductivity alloys are prone to subsurface oxidation unless they are annealed in protective atmospheres. This internally oxidized surface layer lacks age-hardening response and should be suppressed with the use of a protective atmosphere or subsequently removed by chemical or mechanical cleaning.

Processing after the final anneal can include cold rolling, heat treatment to speci-

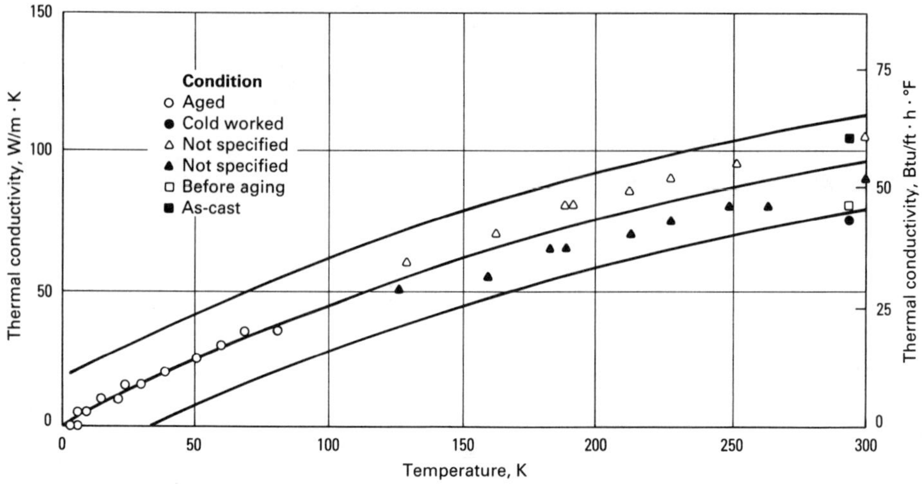

Fig. 16 Thermal conductivity of C17200 at cryogenic temperatures. Source: Ref 1

fied strength levels (mill hardening), and, for strip, slitting to specified width. Also, wrought products are treated with corrosion-inhibiting films to extend shelf life. During manufacture, mill products are monitored for stringent control of as-cast composition, nonmetallic inclusion content, intermediate and finish annealed grain size, dimensional consistency, as-shipped mechanical properties, age-hardening response, and surface condition.

Cast Products. In addition to the strengths afforded by age-hardening response, the beryllium-coppers as casting alloys display appreciable fluidity and ability to replicate very fine pattern detail. Melting and casting procedures recommended for these alloys are designed to minimize beryllium loss through oxidation and maintain the excellent castability.

The beryllium-copper casting alloys can be melted in most commercial resistance, gas, coreless induction, and arc furnaces. Coreless induction furnaces afford greater control; consequently, they reduce hydrogen pickup, beryllium loss, and dross contamination as compared with channel-type furnaces. The alloys can be air or vacuum melted. Furnace refractories suitable for melting beryllium-copper casting alloys include clay graphite, silicon carbide, alumina, magnesia, and zirconia. High-silica refractories may react with beryllium-copper melts.

Charge materials should be clean, dry, and free of contamination. Impurities derived from scrap input can affect the finished casting. Zinc, tin, phosphorus, lead, and chrome impurities can induce brittleness and loss of strength. Aluminum and iron can reduce age-hardening response and degrade electrical or thermal conductivity and corrosion resistance. Where low magnetic permeability is desired in the finished casting, iron must be held to as low a level as possible. Silicon is normally added at the 0.20 to 0.35 wt% level to many of the high-strength beryllium-copper casting alloys, and it is added at the nominal 0.15 wt% level or below to the high-conductivity alloys to promote fluidity and control drossing. Excess silicon, however, increases brittleness.

Castings for critical applications usually employ virgin ingot. However, where acceptable, up to 50% clean scrap returns, such as gates and risers, can be added to the melt. Grain refinement is afforded by the optional addition of a small amount of titanium to C82400 through C82800 casting ingot. The increased cobalt content of C82510 also provides grain refinement.

Care must be exercised in any melting system to avoid extreme superheating, which can increase beryllium loss, drossing, and gas absorption. The high affinity of beryllium for oxygen also mandates that melt agitation and melt hold times be kept to a minimum. The recommended pouring temperature range decreases with increasing beryllium content:

- C82000 and C82200, 1090 to 1180 °C (2000 to 2150 °F)
- C82400 through C82510, 1010 to 1120 °C (1850 to 2050 °F)
- C82600, 970 to 1070 °C (1775 to 1960 °F)
- C82800, 960 to 1040 °C (1760 to 1900 °F)

Drossing can be minimized by melting under an inert gas or graphite cover or by pouring the instant the pouring temperature is reached. Drosses are easily removed, but the use of a commercial fluoride-containing flux can assist the separation of dross from entrapped metal. Melts can be degassed by bubbling with dry nitrogen or argon (nominal −40 °C, or −40 °F, dew point) or with commercial solid degassers such as the PTFE plastic types.

No special pouring rate requirements exist for beryllium-copper casting alloys. Strainer cores can be a source of dross-promoting turbulence and are not recommended. Nonmetallic inclusions can be controlled, however, by the use of ceramic foam filters.

Most common casting methods suitable for copper-base alloys are applicable to beryllium-copper. These include pressure casting, investment casting, centrifugal casting, the Shaw process, die casting, and casting in permanent, ceramic, and various types of sand molds. The solidification shrinkage of beryllium-copper is similar to

Table 12 Room-temperature corrosion properties of beryllium-copper

Type of environment	Acceptable application	Application not recommended
Atmosphere	Industrial	
	Marine	
	Rural	
Water	Fresh	
	Softened	
	Brine	
	Sewage	
Gas (dry)	Chlorine	Acetylene
	Oxygen/ozone	
	Carbon dioxide	
	Sulfur dioxide	
	Ammonia	
	Fuel gases	
Organic compounds	Most organic acids	Pyridine
	Alcohols	
	Ketones	
	Chlorinated solvents	
	Fuels	
	Lubricating/hydraulic oils	
Inorganic chemicals	Nonoxidizing acids	Ammonium hydroxide
	Acetic acid	Sodium hydroxide
	Hydrochloric acid	Hydrogen peroxide
	Dilute sulfuric acid	Oxidizing acids/salts
	Phosphoric acid	Chromic acid
	Aqueous solutions	Nitric acid
	Chlorides	Ferric chloride
	Carbonates	Mercury
	Sulfates	

Table 13 Nominal compositions of commercial beryllium-nickel alloys

Product form	Alloy	Composition, wt% Be	Cr	Other	Ni
Wrought	N03360	1.85–2.05	...	0.4–0.6 Ti	bal(a)
Cast	M220C	2.0	...	0.5 C	bal
Cast	41C	2.75	0.5	...	bal(b)
Cast	42C	2.75	12.0	...	bal(b)
Cast	43C	2.75	6.0	...	bal(b)
Cast	44C	2.0	0.5	...	bal(b)
Cast	46C	2.0	12.0	...	bal(b)
Cast	Master	6	...	...	bal(c)

(a) 99.4 Ni + Be + Ti + Cu min, 0.25 Cu max. (b) 0.1 C max. (c) Master alloys with 10, 25, and 50 wt% Be are also available.

Table 14 Recommended etchants for beryllium-nickel alloys

Etchant	Composition	Comments
1. Nitric acid and water	30 ml HNO_3 (concentrated) and 70 ml H_2O(a)	Used for observation of the general structure of all tempers of beryllium-nickel. Swab etch
2. Modified Marble's etchant	4 g $CuSO_4$ (copper sulfate), 20 ml HCl (concentrated), and 20 ml H_2O(a)	Used for observation of the general structure of all tempers of beryllium-nickel. Swab etch. Can also be used with sensitive tint illumination to reveal the grain structure of hot-worked or annealed material
3. Nitric and acetic acids	50 ml HNO_3 (concentrated) and 50 ml glacial CH_3COOH (acetic acid), optionally diluted with 25–50 ml CH_3COCH_3 (acetone)	Used for observation of the general structure of beryllium-nickel casting alloys. Swab etch

(a) Use distilled water.

that of tin bronze and less than that of aluminum bronze, silicon bronze, or manganese bronze. Metal or graphite chills can be placed in sand molds to promote directional solidification and reduce shrinkage porosity.

Beryllium-Nickel Alloys

Beryllium-nickel alloys, like their beryllium-copper counterparts, are age hardenable. The alloys are distinguished by very high strength, excellent formability, and excellent resistance to fatigue, elevated temperature softening, stress relaxation, and corrosion. Wrought beryllium-nickel is available in strip, rod, and wire forms. The wrought product is used primarily as mechanical and electrical/electronic components that must exhibit good spring properties at elevated temperatures (for example, thermostats, bellows, diaphragms, burn-in connectors, and sockets).

A variety of beryllium-nickel casting alloys exhibit strengths nearly as high as those of the wrought products, and they have the advantage of excellent castability. Many of the casting alloys are used in molds and cores for glass and polymer molding, other glass-forming tools, diamond drill bit matrices, and cast turbine parts. Some casting alloys are also used in jewelry and dental applications by virtue of their high replication of detail in the investment casting process.

Compositions of the wrought and cast beryllium-nickel alloys are shown in Table 13. Only one composition is supplied in wrought form, UNS N03360, which contains 1.85 to 2.05 wt% Be, 0.4 to 0.6 wt% Ti, and a balance of nickel. Commercially available beryllium-nickel casting alloys include a 6 wt% Be master alloy, a series with 2.2 to 2.6 wt% Be that includes one alloy with a minor carbon addition for enhanced machinability, and a series of ternary nickel-base alloys with up to 2.75 wt% Be and 12 wt% Cr.

Physical Metallurgy. The metallurgy of beryllium-nickel alloys is analogous to that of the high-strength beryllium-copper alloys. The alloys are solution annealed at a temperature high in the α nickel region to dissolve a maximum amount of beryllium, then rapidly quenched to room temperature to create a supersaturated solid solution. Precipitation hardening involves heating the alloy to a temperature below the equilibrium solvus to nucleate and grow metastable beryllium-rich precipitates, which harden the matrix. In the high-strength beryllium-coppers, the equilibrium γ precipitate forms at grain boundaries only at higher age-hardening temperatures; commercial beryllium-nickel alloys, on the other hand, exhibit a degree of equilibrium grain-boundary precipitate formation at all temperatures in the age-hardening range.

Microstructure. Metallographic sample preparation of beryllium-nickel is identical to the technique used for beryllium-copper. Descriptions of the swab etchants suitable for revealing the microstructure of all tempers of wrought and cast beryllium-nickel alloys are shown in Table 14.

Wrought unaged beryllium-nickel microstructures exhibit nickel-beryllide intermetallic compound particles containing titanium in a nickel-rich matrix of equiaxed or deformed grains, depending on whether the alloy is in the solution-annealed or a cold-worked temper. After age hardening, a small volume fraction of equilibrium nickel-beryllium phase is generally observed at the grain boundaries. In other respects, unaged and aged beryllium-nickel microstructures are essentially indistinguishable when viewed in an optical microscope. Cast beryllium-nickel alloys containing carbon exhibit graphite nodules in a matrix of nickel-rich dendrites with an interdendritic nickel-beryllium phase. Cast chromium-containing alloys exhibit primary dendrites of nickel-chromium-beryllium solid solution and an interdendritic nickel-beryllium phase. Solution annealing cast beryllium-nickel partially spheroidizes, but does not appreciably dissolve, the interdendritic nickel-beryllium phase.

Heat Treatment. Wrought UNS N03360 is typically solution annealed at about 1000 °C (1830 °F). Cold work up to about 40% can be imparted between solution annealing and aging to increase the rate and magnitude of the age-hardening response. Aging to peak strength is performed at 510 °C (950 °F) for up to 2.5 h for annealed material and for up to 1.5 h for cold-worked material. Aging response curves for hard-temper beryllium-nickel strip are presented in Fig. 17. The underaging, peak-aging, and overaging behavior of N03360 is similar to that of C17200. The cast binary alloys are solution annealed at about 1065 °C (1950 °F) and aged at 510 °C (950 °F) for 3 h. Cast ternary alloys are annealed at a temperature of approximately 1090 °C (1990 °F) and given the same aging treatment. Castings are typically used in the solution-annealed and aged (AT) temper for maximum strength. The cast plus aged (CT) temper is not employed.

Mechanical and Physical Properties of Wrought Beryllium-Nickel. Annealed beryllium-nickel is designated the A temper, and cold worked material is designated ¼ H through H temper. As with the wrought beryllium-copper alloys, increasing cold work through about a 40% reduction in area increases the rate and magnitude of the age-hardening response. User age-hardened materials are designated the AT through HT tempers. As with the high-strength beryllium-coppers, beryllium-nickel strip is processed by proprietary cold-working and age-hardening techniques to provide a series of ascending-strength mill-hardened tempers designated MH2 through MH12; these tempers do not require heat treatment by the user after stamping and forming.

Mechanical properties of beryllium-nickel strip and casting alloys are given in Tables 15 and 16, respectively. The ultimate tensile strengths of wrought materials range from a minimum of 1480 MPa (215 ksi) in the annealed and aged AT temper to a

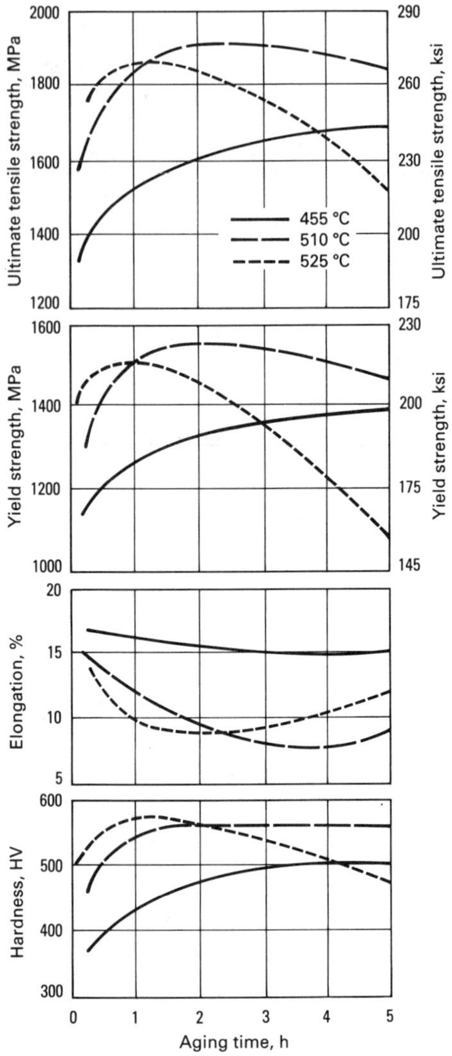

Fig. 17 Aging response curves for beryllium-nickel alloy N03360 strip

minimum of 1860 MPa (270 ksi) in the cold-rolled and aged HT temper. Tensile strengths of mill-hardened strip range from 1065 MPa (155 ksi) to over 1790 MPa (260 ksi). Ductility decreases with increasing strength in both the heat-treatable and age-hardened conditions. In addition to high strength in tension, beryllium-nickel strip exhibits high fatigue strength in fully reversed bending (Fig. 18). A significant fraction of room-temperature strength is maintained through short exposure to temperatures as high as 540 °C (1000 °F) (Fig. 19).

Physical and electrical properties of selected beryllium-nickel alloys are given in Table 17. Electrical conductivity is about 6% IACS in the age-hardened condition. Beryllium-nickel displays only a fraction of the conductivity of the beryllium-coppers, but its conductivity exceeds that of stainless steel.

Fabrication Characteristics of Wrought Products. Beryllium-nickel strip exhibits excellent to good formability in all heat-treatable and mill-hardened tempers (Table 18). The anisotropy of formability in the unaged cold-rolled and mill-hardened tempers is less pronounced than in beryllium-copper alloys. The strength and formability combinations available in beryllium-nickel strip surpass those of the beryllium-coppers and stainless steels.

Despite the significantly greater yield strength available in mill-hardened beryllium-nickel, these beryllium-nickel tempers do not display any greater amounts of elastic springback than beryllium-copper (Fig. 6). This parity is due to the higher elastic modulus of beryllium-nickel. The total strain associated with a given bend is the sum of the plastic deformation (permanent set) and elastic springback. The elastic component of this total strain, from the stress-strain curve, is equal to the yield strength divided by elastic modulus. Consequently, for a fixed bend, different materials with comparable ratios of yield strength to elastic modulus will exhibit comparable springback. This ratio is about 0.009 for the strongest mill-hardened tempers of both beryllium-copper C17200 (TM08) and beryllium-nickel strip (MH12).

As with the high-strength beryllium-copper alloys, beryllium-nickel alloys can have their densities increased by the age-hardening process. Precipitation hardening is accompanied by lineal shrinkage of about 0.2%. Fixture heat treatment, increased preage-hardening cold work, or two-stage bending to provide more uniform through-thickness residual stress distribution can be used to reduce aging distortion in wrought products. The aging response mechanism of beryllium-nickel does not afford users the option of age hardening at temperatures above the recommended 510 °C (950 °F) peak-aging temperature to reduce distortion while preserving near-peak strength.

Beryllium-nickel strip, either heat-treatable or mill-hardened, is harder than similarly designated tempers of beryllium-copper or other copper-base alloys. Tool wear in stamping will accordingly be greater than that for the copper alloys as a class and will approximate die life associated with the stamping of stainless steel or other nickel-base alloys.

Cleaning. No special atmosphere is required for the age hardening of beryllium-nickel, but oxidation can be minimized through the use of a protective atmosphere such as nitrogen. Thin oxide films acquired during aging can be removed by one of two methods:

- Immersion for 1 h in a 70 °C (160 °F) aqueous solution of 50 vol% sulfuric acid, followed by a thorough water rinse
- Immersion for 1 to 2 min in a 54 to 77 °C (130 to 170 °F) aqueous solution of 15 vol% hydrochloric acid, 15 vol% sulfuric acid, and 75 to 110 g/L (10 to 15 oz/gal) of ferric nitrate

Heavier oxide films formed during aging can be removed by brief application of a solution of 2 parts acetic acid, 1 part nitric acid, a few drops of hydrochloric acid (for activation), and a suitable wetting agent. Precautions are necessary because this solution becomes extremely hot during use.

Joining. Wrought or cast beryllium-nickel can be satisfactorily joined to itself or to other metals by welding, brazing, or solder-

Table 15 Mechanical properties of beryllium-nickel alloy N03360 strip

Temper designations: ASTM	Temper designations: Commercial	Heat treatment(a)	Tensile strength: MPa	Tensile strength: ksi	Yield strength at 0.2% offset: MPa	Yield strength at 0.2% offset: ksi	Minimum elongation in 50 mm (2 in.), %	Rockwell hardness
TB00	A	...	655–895	95–130	275–485	40–70	30	39–57 HRA
TD01	¼ H	...	760–1035	110–150	445–860	65–125	15	50–65 HRA
TD02	½ H	...	895–1205	130–175	790–1170	115–170	4	51–70 HRA
TD04	H	...	1065–1310	154–190	1035–1310	150–190	1	55–75 HRA
TF00	AT	2.5 h at 510 °C	1480 min	215 min	1035 min	150 min	12	78–86 HRN
TH01	¼ HT	2.5 h at 510 °C	1585 min	230 min	1205 min	175 min	10	80–88 HRN
TH02	½ HT	1.5 h at 510 °C	1690 min	245 min	1380 min	200 min	9	81–90 HRN
TH04	HT	1.5 h at 510 °C	1860 min	270 min	1585 min	230 min	8	83–90 HRN
...	MH2	M	1065–1240	154–180	690–860	100–125	14	...
...	MH4	M	1240–1415	180–205	825–1065	120–154	12	...
...	MH6	M	1380–1550	200–225	1035–1205	150–175	10	...
...	MH8	M	1515–1690	220–245	1170–1415	170–205	9	...
...	MH10	M	1655–1860	240–270	1380–1550	200–225	8	...
...	MH12	M	1790–2000	260–290	1515–1690	220–245	8	...

(a) M, heat treatment performed at mill

Table 16 Typical mechanical properties of selected beryllium-nickel casting alloys

Alloy	Condition	Ultimate tensile strength MPa	ksi	0.2% yield strength MPa	ksi	Elongation in 50 mm (2 in.), %	Rockwell hardness
M220C	Annealed(a)	760	110	345	50	35	95 HRC
	Annealed and aged(b)	1620	235	1380	200	4	54 HRC
41C	Annealed and aged(b)	1585	230	...	...	...	55 HRC
42C	Annealed and aged(c)	1035	150	...	...	6	38 HRC
43C	Annealed and aged(b)	1310	190	...	...	...	45 HRC
44C	Annealed and aged(b)	1310	190	...	...	...	48 HRC
46C	Annealed and aged(b)	1035	150	...	...	...	35 HRC

(a) Solution annealed at 1065 °C (1950 °F) for 1 h and water quenched. (b) Solution annealed and aged at 510 °C (950 °F) for 3 h. (c) Solution annealed at 1093 °C (2000 °F) for 10 h, water quenched, and then aged at 510 °C (950 °F) for 3 h

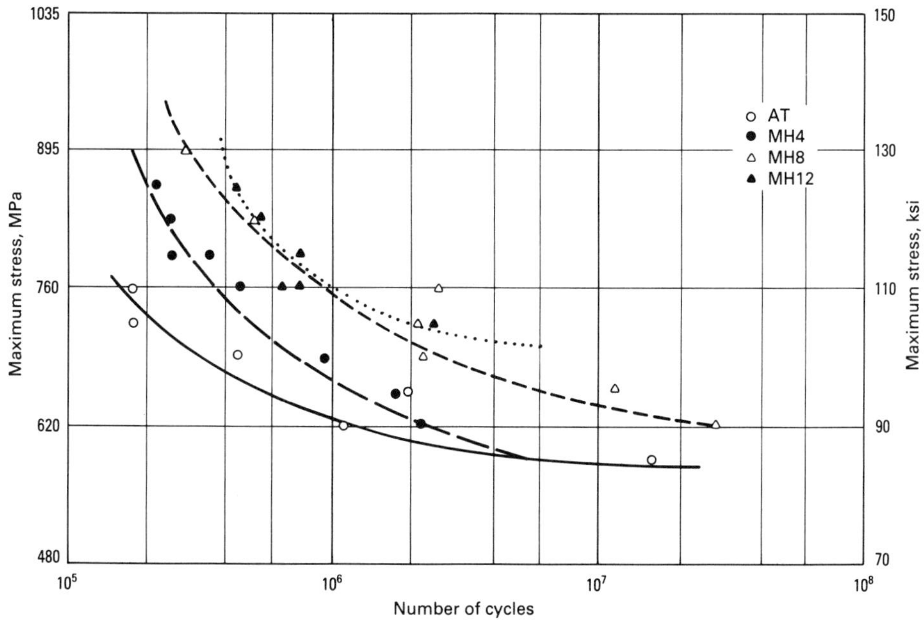

Fig. 18 Fatigue behavior of beryllium-nickel alloy N03360 strip in fully reversed bending (stress ratio, $R = -1$)

ing. As with any other alloy, surfaces to be joined must be thoroughly cleaned to remove dirt, oil, and oxide films. Soldering is typically performed with 50Sn-50Pb alloy and a flux of 50% zinc chloride and 50% ammonium chloride. Brazing can be performed with AWS BAg-1a filler metal, using a short heating cycle after solution annealing but before age hardening. Applicable welding processes include metal arc, tungsten inert-gas, and gas welding; procedures and commercial nickel alloy filler rods recommended for the joining of other nickel-base alloys should be used. These filler metals will mix with molten beryllium-nickel in the weld pool and will exhibit a limited postjoining age-hardening response. An improved aging response in the welded joints can be obtained by using beryllium-nickel filler metal.

Melting and Casting Characteristics (Foundry Products). Beryllium-nickel casting alloys are readily air melted, preferably in electric or induction furnaces. Gas- or oil-fired furnaces should use fuel with a very low sulfur content. The melt surface forms a protective beryllium oxide film, which retards beryllium loss. Additional melt surface protection can be supplied by a blanket of argon gas or an alumina-base slag cover. Furnace linings or crucibles of magnesia are preferred; Alundum, zirconium silicate, or mullite are adequate.

Charges should be clean, dry, and free of oil or grease, particularly leaded or sulfurized lubricants. Scrap additions should not exceed about 50% of the charge weight and should be free of low-melting-point film-forming elements such as antimony, arsenic, bismuth, boron, lead, lithium, selenium, sulfur, tellurium, tin, and phosphorus. Small amounts of these tramp elements render beryllium-nickel (and other nickel-base) casting alloys red short and susceptible to hot tearing during solidification. Beryllium in the composition is an effective deoxidizer and scavenger of sulfur and nitrogen. Additional deoxidation by silicon or magnesium is not usually necessary, but a small amount of titanium can be added to stabilize residual nitrogen as TiN. Degassing of melts that have absorbed hydrogen from moisture can be accomplished by bubbling dry argon through the melt or by the addition of a PTFE-type solid degassing agent. A carbon boil, commonly used to degas molten nickel-base alloys, should not be used. With limited available oxygen, addition of nickel oxide to initiate a carbon boil will only deplete the beryllium content.

The pouring temperature for most of the beryllium-nickel casting alloys is about 1370 °C (2500 °F). Castings should be poured soon after this temperature is attained, with a minimum of turbulence and exposure of the pouring stream to air. Mold design should incorporate appropriate numbers, sizes, and positions of chills, risers, and exothermic or insulating sleeves. The use of these design elements will ensure progressive directional solidification and thus minimize shrinkage porosity. Sand, shell, investment, ceramic, graphite, and permanent mold materials are appropriate for these alloys. The excellent castability of beryllium-nickel alloys makes them particularly well suited for precision casting processes.

Production Metallurgy. The production process for beryllium-nickel alloys parallels that for beryllium-copper in many respects. The primary differences lie in higher processing temperatures and higher mill loads imposed by the wrought nickel-base alloy. Processing commences with induction melting in air of charges consisting of a 6Be-Ni master alloy, additive elements, and mill scrap. The 6Be-Ni master is produced by induction melting commercial-purity beryllium and nickel rather than by carbothermic reduction of BeO as for the 4Be-Cu master alloy. Casting alloys are poured as small ingots for foundry use. The wrought alloy is semicontinuously cast into rectangular or round billets for hot working into strip or round products.

Hot rolling or extrusion is performed in the vicinity of the solution-annealing temperature, about 980 °C (1800 °F). The elevated-temperature deformation resistance of beryllium-nickel is appreciably greater than that of beryllium-copper; as a result, the hot mill loads or extrusion pressures for beryllium-nickel are substantially higher than those for the copper-base alloys. The hot-worked products are then brought to a ready-to-finish size by one or more iterations of solution annealing and cold working. Chemical cleaning is required after each anneal because of the propensity of beryllium-nickel to form oxide films even in protective commercial furnace atmospheres.

A final solution anneal establishes the finished grain size and age-hardening response of the alloy. Cold working and mill hardening can follow the final anneal.

Other Beryllium-Nickel Casting Alloys. Nonprecious-metal dental alloys for precision investment cast crown and bridgework include several proprietary beryllium-containing nickel-base compositions. These alloys contain from 0.5 to 2 wt% Be and include varying amounts of chromium, mo-

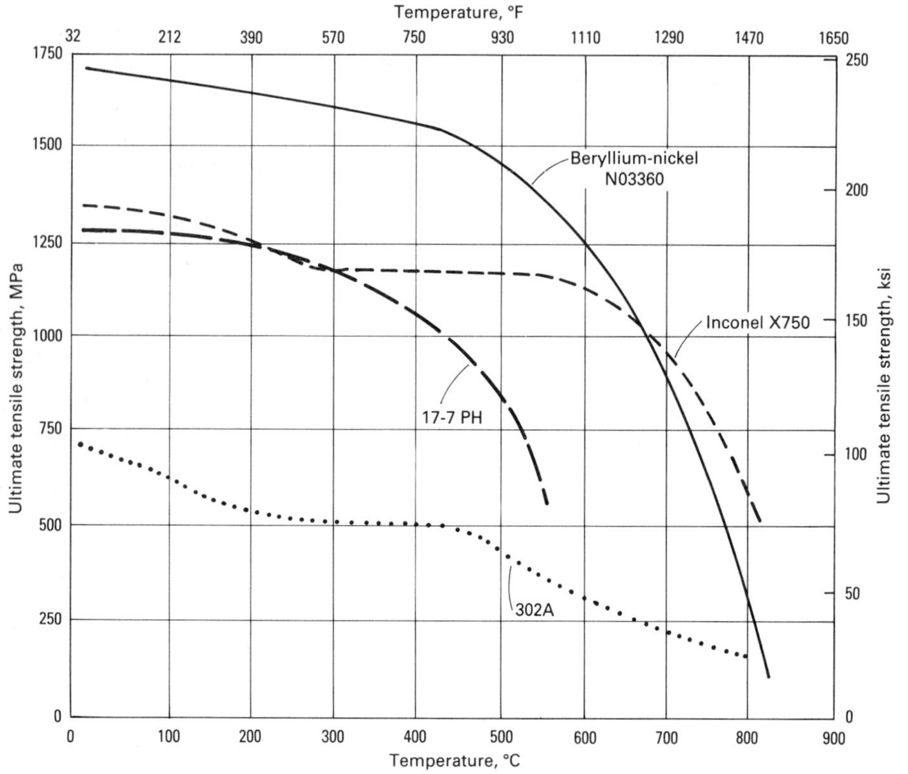

Fig. 19 Elevated-temperature strength of beryllium-nickel alloy N03360 strip compared with that of selected stainless steels

lybdenum, aluminum, cobalt, and titanium, depending upon the supplier. Beryllium enhances the castability of these alloys for replication of fine detail and provides a favorable oxide film morphology on the casting surface that enhances the adhesion of porcelain enamel coatings. The overall compositions are formulated to match the thermal expansion of the porcelain enamel.

Other Beryllium-Containing Alloys

Beryllium in Aluminum Alloys. Small additions of beryllium to aluminum and magnesium systems improve casting consistency. When as little as 0.005 to 0.05 wt% Be is added to an aluminum-base alloy in the liquid state, a protective oxide film forms on the surface. This film reduces drossing, increases metal yield and cleanliness, and improves fluidity; the result is cleaner, higher-quality castings with improved surface finish, consistent strength, and improved ductility. The protective film reduces the absorption of hydrogen into the melt and thus decreases the propensity of the material for internal gas-related defects. Also, the film tends to reduce metal reaction with sand molds during casting. Preferentially oxidizable alloy additions to aluminum, such as magnesium and sodium, also withstand composition fade because of oxidation during melting and casting. Vaporization of the base metal is also positively affected by beryllium additions.

End-use benefits imparted to aluminum alloys by beryllium include a reduction in tarnishing, an improved buffing and polishing response, and a consistent aging response, particularly in alloys containing magnesium or silicon. Applications for beryllium-containing aluminum alloys include aircraft premium castings (ASTM A 357, for example) and certain wrought age-hardenable alloys.

Beryllium in Magnesium. Magnesium systems are particularly sensitive to oxidation and are even susceptible to ignition when oxygen is readily available. Small additions of beryllium, as little as 10 ppm residual, increase the ignition temperature of magnesium by as much as 200 °C (360 °F) and can provide a safety factor during melting and casting. Fluxes accomplish similar results but can add to the system cost through increased equipment corrosion and the need for increased melt cleanliness. Beryllium is one of the few elements that is more reactive to oxygen than magnesium is, and it can be used in place of fluxes for further benefits. Beryllium is known to getter iron and similar impurities from the melt, thus providing some degree of refinement in purity where needed. As in aluminum alloys, the film-forming characteristics of beryllium reduce moisture interaction and mold reactions.

End-use benefits of adding beryllium to magnesium alloys include resistance to high-temperature oxidation and corrosion. Applications include aircraft-quality castings and automotive engine blocks.

Safe Handling of Beryllium-Containing Alloys

Beryllium-containing materials, like many other materials used in industry, can present a health risk unless precautions are employed. These alloys present a hazard only if a person who has become sensitized to beryllium inhales a sufficiently large concentration of beryllium-containing dust, mist, or fumes. Excessive inhalation of beryllium particulates can cause serious chronic pulmonary illness. Although only a small percentage of people (of the order of 1 in 100) ever become sensitized to beryllium, there is no way to identify these sensitive people in advance. As a result, everyone must be protected.

Controls. While the hazard associated with dilute beryllium alloys is not to be minimized, years of experience have shown that risks in the occupational environment can be controlled with proper ventilation and other safeguards. Care must be exercised in the processing and

Table 17 Typical physical and electrical properties of selected beryllium-nickel alloys

Product form	Alloy	Condition	Density, g/cm^3	Thermal expansion from 20–550 °C (70–1020 °F), 10^{-6}/°C	Thermal conductivity, W/m · K	Electrical resistivity, μΩ · cm	Electrical conductivity, %IACS	Elastic modulus GPa	Elastic modulus 10^6 psi
Wrought	N03360	Aged(a)	8.27	4.5	28 (at 20 °C)	28.7 max	6 min	193–206	28–30
		Mill hardened(b)	. . .	. . .	. . .	34.5 max	5 min	. . .	. . .
		Unaged(c)	. . .	. . .	. . .	43.1 max	4 min	. . .	. . .
Cast	M220C	Aged(a)	8.08–8.19	4.8	36.9 (at 38 °C) 51.1 (at 538 °C)	21.0	. . .	179–193	26–28
Cast	42C	Aged(a)	7.8	. . .	34.6 (at 93 °C)	34.5	5 min	193	28

(a) Solution annealed and aged at 510 °C (950 °F) for 3 h. (b) Heat treated by producing mill. (c) Solution annealed with 0 to 37% cold work

Table 18 Relative formability of beryllium-nickel alloy N03360 strip

Formability rating	Specific formability	Alloy condition: Rolled tempers	Alloy condition: Mill-hardened tempers	Formability ratio (R/t) for 90° bend(a): Longitudinal	Formability ratio (R/t) for 90° bend(a): Transverse
Excellent	Excellent formability, used for deep-drawn and severely cupped or formed parts; can be bent flat through 180° angle in any direction.	A	...	0	0
		¼ H	...	0	0
		...	MH2	0	0
		...	MH4	0.5	0.5
Very good	Very good formability, used for moderately drawn and cupped parts; formable to 90° bend around a radius.	½ H	...	0.7	1.2
		...	MH6	1.0	1.2
		...	MH8	1.2	1.6
Good	Slightly reduced formability, formable to 90° bend around a radius.	H	...	1.2	2.0
		...	MH10	1.5	2.2
		...	MH12	2.0	3.0

(a) For strip ≤1.3 mm (≤0.050 in.) thick. Strip <0.25 mm (<0.010 in.) thick will exhibit formability somewhat better than shown. This chart should not be used for part design because punch radius, not part dimension, is used to calculate formability, and springback is not considered.

fabrication of these alloys to avoid inhalation of airborne particles, such as dust, mist, or fumes, in excess of the permissible exposure level.

Permissible Levels. The U.S. Occupational Safety and Health Administration (OSHA) has adopted the following in-plant permissible exposure level standards designed to keep airborne concentrations well below the levels known to cause health problems:

- Daily time-weighted average exposure over an 8-h day is not to exceed 2 μg of beryllium per cubic meter of air
- Short-term exposure above 5 but not more than 25 μg of beryllium per cubic meter of air is to be limited to 30 min or less during an 8-h work period

As demonstrated by over 40 years of documented experience, even the most sensitive person is safe if exposure is below these threshold values.

To protect the general public from environmental exposure to airborne beryllium, the U.S. Environmental Protection Agency has established an emission standard of 10 g of elemental beryllium per day as a permissible emission into the air surrounding a plant.

REFERENCE

1. N. Simon and R. Reed, *Cryogenic Properties of Copper and Copper Alloys*, Vol II, National Bureau of Standards, 1987

Nickel and Nickel Alloys

W.L. Mankins and S. Lamb, Inco Alloys International, Inc.

NICKEL in elemental form or alloyed with other metals and materials has made significant contributions to our present-day society and promises to continue to supply materials for an even more demanding future. This article provides a historical overview of nickel, highlights alloy developments achieved to the present, and provides a discussion of the physical metallurgy of nickel alloys. Emphasis has been placed on alloys developed for corrosion-resistant applications. Alloys developed for heat-resistant applications, low-expansion alloys, electrical resistance alloys, and mechanically alloyed/dispersion-strengthened alloys are only briefly reviewed as these materials are described elsewhere in this Volume or in Volume 1, 10th Edition, of *Metals Handbook*. Nickel-base alloy castings, which like their wrought counterparts are also used extensively in corrosive-media and high-temperature applications, will not be described in the present article. Detailed information on cast alloys can be found, however, in *Casting*, Volume 15 of the 9th Edition of *Metals Handbook* and in the aforementioned Volume 1 of the 10th Edition.

Changes in supply and demand in a technology-driven economy have spawned a diverse range of nickel-base alloys with various properties. Commercial nickel-base alloys, which account for approximately 13% of all nickel consumed, can be divided into groups or families by their major elemental constituents. Compositions, typical mechanical and physical properties, and applications are shown for representative alloys in each group. Other uses for nickel are:

Use	Amount consumed, %
Stainless steel	57
Alloy steel	9.5
Nickel-base alloys	13
Copper-base alloys	2.3
Plating	10.4
Foundry	4.4
Other	3.3

Source: Nickel Development Institute

Historical Development

Nickel has been used in alloys that date back to the dawn of civilization. Chemical analysis of artifacts has shown that weapons, tools, and coins contain nickel in varying amounts. Perhaps the earliest nickel-containing white alloy was Pai-Thong or white copper. This Chinese-made material added zinc to nickel-copper ores. Many decorative pieces such as candlesticks were produced and were ultimately brought to Europe by the East India Company in the seventeenth century (Ref 1).

Early in the eighteenth century, miners in the Saxony region of Germany tried to smelt some newly discovered copper-appearing ores, only to discover that the white metal they produced was too hard to be hammered into useful items. The miners thought the material was cursed, calling it "Old Nick's Copper" or "Kupfer-Nickel." Similar ores were discovered in the following years in other locations, and these were also called nickel because the hard, white metal resisted efforts to deform it.

A.F. Cronstedt, working for the Swedish Department of Mines, worked five years with these curious ores and was finally able to separate and identify a new element that he named nickel. Five years after the identification of elemental nickel in 1751, another Swedish scientist, Von Engestrom, found that nickel was a major component of Pai-Thong and this led to the invention of German silver or nickel-silver (Cu-Sn-Pb-Zn alloys containing 12 to 25 wt% Ni).

Nickel plating, coinage, and nickel-silver were the main applications for the small quantity of nickel that was produced through the second half of the nineteenth century. Additional nickel was provided from Norwegian mines beginning in the mid-1800s, and significant ones were discovered in the South Pacific island of New Caledonia.

The chance finding of the huge nickel-containing ore deposits in the Sudbury district of Ontario, Canada, during the construction of the trans-Canada railroad aroused the interest of prospectors, miners, industrialists, and investors. Initially, the ores were sought for the contained copper that was in demand. The sulfide ores were found to contain nickel, copper, cobalt, iron, and precious metals (gold, silver, and the platinum-group metals).

Attempts to separate the metals were plagued with many problems. Bessemer-type converters were explored but the heavy, acrid, sulfur-dioxide-laden fumes rendered the process impractical. Researchers continued to find ways to separate the nickel from the more needed copper. The selling price for nickel was ten times that of copper, and once a separation process was perfected, there would be a glut on the market. This prompted the need for markets other than coinage, plating, and nickel-silver to be found.

While searching for an alternative material to substitute for cast iron in an ammonia refrigeration unit, inventor John Gamgee discovered nickel steel (Ref 2). This led to other nickel-iron combinations resulting in alloys for cryogenic applications and tough nickel steel armor plate. Further experimental work in America and Europe convinced governments that the nickel steel plate was superior to other known materials in resisting penetration from armor-piercing projectiles. Consequently, the demand for nickel to produce armor plate for the world's navies would exceed the supply unless improved methods for producing nickel could be found.

Early Refining Advances. The Orford Tops and Bottoms process in which sodium sulfate was added to molten nickel-copper-sulfide ores in the smelter was the solution to the problem. This process caused the nickel sulfides and the copper sulfides to separate into distinct portions when slow cooled in a mold. Nickel sulfides comprised the shiny "bottoms" while the copper sulfide "tops" remained in the upper end of the mold. A hammer blow separated the two fractions. Refinement of the two sulfide mattes using different techniques provided both high-quality nickel and copper. In the case of the nickel-containing bottoms, these were crushed, leached with sulfuric acid, dried, sintered, crushed again, fused in reverberatory furnaces with low ash coal, and cast as anodes (Ref 3). Final purification of these anodes was achieved through electrolytic refining, resulting in a pure nickel cathode sheet.

The Orford process was replaced by more efficient refining techniques in 1948. With much improved ore beneficiation processes, resulting from more efficient and effective methods of separation of the metallic ores and rock (gangue), it became more economical to process lower-grade ores. In the case of the sulfide deposits, the ore can be separated into distinct particles so that after crushing, mechanical separation is practical. Froth flotation, followed by thickening and dewatering, allowed the separation of both nickel and copper concentrates, as well as iron concentrates. Pyrometallurgical, hydrometallurgical, and vapometallurgical processes were and are still used to refine the concentrates to the elemental metal. Each of these processes is briefly described below. Refining practices vary among nickel producers. The methods selected are dictated by the composition of the ores and sources of economic energy available, as well as by technical and historical factors.

Pyrometallurgy. The concentrates are calcined in a roasting furnace, smelted in a reverberatory or other similar furnace using coal or natural gas for fuel, and blown with air in a converter. Nickel and copper concentrates are further refined to increase the purity of the metals. Nickel is refined to lower to an acceptable level those elements that would have a deleterious effect on subsequent metal performance. The elements removed include antimony, arsenic, bismuth, copper, iron, lead, phosphorus, sulfur, tin, and zinc. The major and minor impurities removed are also recovered if economically viable. Precious metals present at minute levels in sulfide ores are ultimately recovered to complete the total refining/recovery process.

Hydrometallurgy. Electrorefining and electrowinning both are electrolytic processes used to produce pure nickel cathode sheet. In the former process, a nickel electrolytic refining cell consisting of a slab of crude nickel to be refined (anode) and a thin nickel starting sheet (cathode) are immersed in an aqueous electrolyte, and direct current is passed through the cell. Nickel and some impurities are dissolved in the electrolyte (anolyte). The solution is pumped from the cell and the impurities are chemically removed. Purified electrolyte is returned to the cell as catholyte and deposited on the cathode as pure nickel. Electrowinning, also an electrolytic process, can electrolyze soluble anodes of nickel sulfide or utilize insoluble anodes to extract the nickel from a leach liquor.

Vapometallurgy. The carbonyl process uses gas-to-metal transformation to extract pure nickel from an impure nickel oxide. In this process, the oxide is reduced with hydrogen and the nickel reacts selectively with carbon monoxide to form gaseous nickel carbonyl (Ref 4). The gas is decomposed by heat to yield pure nickel. This is acknowledged as the best method of refining nickel that is available. The powder or pellet produced is of a very high purity, the process is energy efficient, and there are no polluting waste by-products.

Alloy and Market Developments. There have been a number of significant developments in nickel technology that have served to shape the present industry. A number of these are listed below. Additional information can be found in the cited references.

- The discovery in 1905 of Monel, a high tensile strength nickel-copper alloy that was found to be highly resistant to atmospheric corrosion, salt water, and various acid and alkaline solutions
- Developmental work by Marsh (Ref 5) on nickel-chromium alloys that led to the discovery of the Nimonic (Ni-Cr+Ti) series of alloys, which are primarily used for creep resistance, high strength, and stability at high temperature
- The work of Elwood Haynes on binary nickel-chromium and cobalt-chromium alloys used for oxidation-resistant and wear-resistant applications (Ref 6)
- The work of Paul D. Merica on the use of nickel in cast irons, bronzes, and steel as well as his significant discovery that aluminum and titanium led to precipitation hardening of nickel-base alloys (Ref 7). This mechanism continues to provide the basis for material strengthening in today's superalloys
- The work of William A. Mudge on a precipitation hardening nickel-copper alloy (K-Monel)
- The establishment of the Kure Beach, and Harbor Island, NC, corrosion testing facilities by F.L. LaQue. These two facilities, established in 1935, comprise the LaQue Center for Corrosion Technology
- The addition of ferrochrome (70Cr-30Fe) to nickel to create Inconel alloys known for their high strength at high temperatures, oxidation resistance, and carburization resistance
- Developmental work during the 1920s of nickel-molybdenum alloys that led to the discovery of the Hastelloy series of alloys, known for their high corrosion resistance
- Further advances in high-temperature alloys used for aircraft applications led to the development of Nimonic alloy 80 and Nimonic alloy 80A during the 1940s (Ref 8)
- The development of the turbo-supercharger for aircraft engines, which operated at temperatures ranging from 650 to 815 °C (1200 to 1500 °F) as the rotor turned at speeds of 20 000 to 30 000 rpm, led to improved precipitation-hardened alloys Hastelloy alloy B and Hastelloy alloy X
- The production of the first gas turbine engine led to developments in new alloys for blades, vanes, and disks with improved creep and fatigue resistance
- The introduction of a new family of Fe-Ni-Cr alloys (Incoloy series) based on a lower nickel content (20 to 40 wt%) designed to meet the need for high-temperature oxidation resistance and aqueous corrosion protection
- Advances in powder metallurgy (P/M) processing that led to the development of mechanically alloyed dispersion-strengthened superalloys
- New melting technologies (vacuum induction melting, electron beam melting, plasma melting/refining and vacuum arc skull melting) in concert with net shape investment casting led to the development of fine grained equiaxed castings as well as directionally solidified and single-crystal superalloys (Ref 9, 10)

Physical Metallurgy of Nickel and Nickel Alloys

Nickel is a versatile element and will alloy with most metals. Complete solid solubility exists between nickel and copper. Wide solubility ranges between iron, chromium, and nickel make possible many alloy combinations. The face-centered cubic structure of the nickel matrix (γ) can be strengthened by solid-solution hardening, carbide precipitation, or precipitation hardening. An overview of each of these hardening mechanisms as they apply to wrought alloys, follows. Cast alloy physical metallurgy is described in Volume 1 of the 10th Edition of *Metals Handbook* (see Ref 9 and 10). Detailed reviews on the metallurgy and microstructures of wrought nickel alloys can be found elsewhere in the *Metals Handbook* series (see Ref 11 and 12).

Solid-Solution Hardening. Cobalt, iron, chromium, molybdenum, tungsten, vanadium, titanium, and aluminum are all solid-solution hardeners in nickel. The elements differ with nickel in atomic diameter from 1 to 13%. Lattice expansion related to atomic diameter oversize can be related to the hardening observed (Ref 13 and 14). Above 0.6 T_m (melting temperature), which is the range of high-temperature creep, strengthening is diffusion dependent and large slow diffusing elements such as molybdenum and tungsten are the most effective hardeners (Ref 15).

Carbide Strengthening. Nickel is not a carbide former. Carbon reacts with other elements alloyed with nickel to form carbides that can be either a bane or a blessing to the designer of alloys. An understanding of the carbide class and its morphology is beneficial to the alloy designer.

The carbides most frequently found in nickel-base alloys are MC, M_6C, M_7C_3, and $M_{23}C_6$ (where M is the metallic carbide-forming element or elements). MC is usually a large blocky carbide, random in distribution, and generally not desired. M_6C carbides are also blocky; formed in grain boundaries they can be used to control grain size, or precipitated in a Widmanstätten pattern throughout the grain these carbides can impair ductility and rupture life. M_7C_3 carbides (predominately Cr_7C_3) form intergranularly and are beneficial if precipitated as discrete particles. They can cause embrittlement if they agglomerate, forming continuous grain-boundary films. This condition will occur over an extended period of time at high temperatures. $M_{23}C_6$ carbides show a propensity for grain-boundary precipitation. The $M_{23}C_6$ carbides are influential in determining the mechanical properties of nickel-base alloys. Discrete grain-boundary particles enhance rupture properties. Long time exposure at 760 to 980 °C (1400 to 1800 °F) will cause precipitation of angular intragranular carbides as well as particles along twin bands and twin ends.

Heat treatment provides the alloy designer with a means of creating desired carbide structures and morphologies before placing the material in service. The alloy chemistry, its prior processing history, and the heat treatment given to the material influence carbide precipitation and ultimately performance of the alloy. Each new alloy must be thoroughly examined to determine its response to heat treatment or high temperature.

Precipitation Hardening. The precipitation of γ', $Ni_3(Al,Ti)$ in a high-nickel matrix provides significant strengthening to the material. This unique intermetallic phase has a face-centered cubic structure similar to that of the matrix and a lattice constant having 1% or less mismatch in the lattice constant with the γ matrix (Ref 15). This close matching allows low surface energy and long time stability.

Precipitation of the γ' from the supersaturated matrix yields an increase in strength with increasing precipitation temperature, up to the overaging or coarsening temperature. Strengthening of alloys by γ' precipitation is a function of γ' particle size. The hardness of the alloy increases with particle size growth, which is a function of temperature and time. Several factors contribute to the magnitude of the hardening, but a discussion of the individual factors is beyond the scope of this paper.

The volume percent of γ' precipitated is also important because high-temperature strength increases with amount of the phase present. The amount of gamma prime formed is a function of the hardener content of the alloy. Aluminum, titanium, niobium, and tantalum are strong γ' formers. Effective strengthening by γ' decreases above about 0.6 T_m as the particles coarsen. To retard coarsening, the alloy designer can add elements to increase the volume percent of γ' or add high-partitioning, slow-diffusing elements such as niobium or tantalum to form the desired precipitate.

Table 1 Role of elements in iron-base and nickel-base superalloys

Effect	Iron base	Nickel base
Solid-solution strengtheners	Cr, Mo	Co, Cr, Fe, Mo, W, Ta
Fcc matrix stabilizers	C, W, Ni	...
Carbide form		
MC type	Ti	W, Ta, Ti, Mo, Nb
M_7C_3 type	...	Cr
$M_{23}C_6$ type	Cr	Cr, Mo, W
M_6C type	Mo	Mo, W
Carbonitrides		
M(CN) type	C, N	C, N
Forms γ' Ni_3 (Al,Ti)	Al, Ni, Ti	Al, Ti
Retards formation of hexagonal η (Ni_3Ti)	Al, Zr	...
Raises solvus temperature of γ'	...	Co
Hardening precipitates and/or intermetallics	Al, Ti, Nb	Al, Ti, Nb
Forms γ'' (Ni_3Nb)	...	Nb
Oxidation resistance	Cr	Al, Cr
Improves hot corrosion resistance	La, Y	La, Th
Sulfidation resistance	Cr	Cr
Increases rupture ductility	B	B(a), Zr
Causes grain-boundary segregation	...	B, C, Zr
Facilitates working	...	...

(a) If present in large amounts, borides are formed. Source: Ref 12

The γ' phase can transform to other (Ni_3X) precipitates if the alloy is supersaturated in titanium, niobium, or tantalum. Titanium-rich metastable γ' can transform to (Ni_3Ti), or eta phase (η), a hexagonal close-packed phase. Formation of η-phase can alter mechanical properties, and effects of the phase must be determined on an individual alloy basis. Excess niobium results in metastable η transforming to γ'' (body-centered tetragonal phase) and ultimately to the equilibrium phase Ni_3Nb (orthorhombic phase). Both γ' and γ'' can be present at peak hardness, whereas transformation to the coarse, elongated Ni_3Nb (orthorhombic) results in a decrease in hardness. The phases precipitated are functions of alloy chemistry and the heat treatment given the material prior to service or the temperature/time exposure of in-service application.

Table 1 summarizes the effects of adding various elements to nickel-base and iron-base superalloys. Table 2 provides additional detail on the various phases precipitated in wrought heat-resistant alloys.

Applications and Characteristics of Nickel Alloys

Nickel and nickel alloys are used for a wide variety of applications, the majority of which involve corrosion resistance and/or heat resistance. Some of these include:

- *Aircraft gas turbines*: disks, combustion chambers, bolts, casings, shafts, exhaust systems, cases, blades, vanes, burner cans, afterburners, thrust reversers (Ref 9, 10, and 12)
- *Steam turbine power plants*: bolts, blades, stack gas reheaters (Ref 16)
- *Reciprocating engines*: turbochargers, exhaust valves, hot plugs, valve seat inserts
- *Metal processing*: hot-work tools and dies (Ref 17)
- *Medical applications*: dentistry uses (Ref 18), prosthetic devices (Ref 19)
- *Space vehicles*: aerodynamically heated skins, rocket engine parts (Ref 20)
- *Heat-treating equipment*: trays, fixtures, conveyor belts, baskets, fans, furnace mufflers (Ref 21)
- *Nuclear power systems*: control rod drive mechanisms, valve stems, springs, ducting (Ref 22)
- *Chemical and petrochemical industries*: bolts, fans, valves, reaction vessels, piping, pumps (Ref 23, 24, and 25)
- *Pollution control equipment*: scrubbers, flue gas desulfurization equipment (liners, fans, stack gas reheaters, ducting) (Ref 16, 26)
- *Metals processing mills*: ovens, afterburners, exhaust fans
- *Coal gasification and liquefaction systems*: heat exchangers, reheaters, piping
- *Pulp and paper mills*: tubing, doctor blades, bleaching circuit equipment, scrubbers (Ref 27)

A number of other applications for nickel alloys involve the unique physical properties of special-purpose nickel-base or high-nickel alloys. These include:

- Low-expansion alloys
- Electrical resistance alloys
- Soft magnetic alloys
- Shape memory alloys

Each of these special-purpose alloys, which are briefly discussed below, are also described in separate articles elsewhere in this Handbook.

Heat-Resistant Applications

Nickel-base alloys are used in many applications where they are subjected to harsh environments at high temperatures. Nickel-chromium alloys or alloys that contain more than about 15% Cr are used to provide both oxidation and carburization resistance at temperatures exceeding 760 °C (1400 °F). The chromium in the alloys promotes the formation of a protective surface oxide, and the nickel provides good retention of the protective coating, especially during cyclic exposure to high temperatures.

In atmospheres that are oxidizing to chromium but reducing to nickel, nickel-chromi-

Table 2 Constituents observed in wrought heat-resistant alloys

Phase	Crystal structure	Lattice parameter, nm	Formula	Comments
γ'	fcc (ordered $L1_2$)	0.3561 for pure Ni_3Al to 0.3568 for $Ni_3(Al_{0.5}Ti_{0.5})$	Ni_3Al $Ni_3(Al,Ti)$	Principal strengthening phase in many nickel- and nickel-iron-base superalloys; crystal lattice varies slightly in size (0–0.5%) from that of austenite matrix; shape varies from spherical to cubic; size varies with exposure time and temperature
η	hcp ($D0_{24}$)	$a_0 = 0.5093$ $c_0 = 0.8276$	Ni_3Ti (no solubility for other elements)	Found in iron-, cobalt-, and nickel-base superalloys with high titanium/aluminum ratios after extended exposure; may form intergranularly in a cellular form or intragranularly as acicular platelets in a Widmanstätten pattern
γ''	bct (ordered $D0_{22}$)	$a_0 = 0.3624$ $c_0 = 0.7406$	Ni_3Nb	Principal strengthening phase in Inconel 718; γ'' precipitates are coherent disk-shaped particles that form on the {100} planes (average diameter approximately 60 nm, thickness approximately 5–9 nm); metastable phase
Ni_3Nb (δ)	Orthorhombic (ordered Cu_3Ti)	$a_0 = 0.5106–0.511$ $b_0 = 0.421–0.4251$ $c_0 = 0.452–0.4556$	Ni_3Nb	Observed in overaged Inconel 718; has an acicular shape when formed between 815 and 980 °C (1500 and 1800 °F); forms by cellular reaction at low aging temperatures and by intragranular precipitation at high aging temperatures
MC	Cubic	$a_0 = 0.430–0.470$	TiC NbC HfC	Titanium carbide has some solubility for nitrogen, zirconium, and molybdenum; composition is variable; appears as globular, irregularly shaped particles that are gray to lavender; M elements can be titanium, tantalum, niobium, hafnium, thorium, or zirconium
$M_{23}C_6$	fcc	$a_0 = 1.050–1.070$ (varies with composition)	$Cr_{23}C_6$ $(Cr,Fe,W,Mo)_{23}C_6$	Form of precipitation is important; it can precipitate as films, globules, platelets, lamellae, and cells; usually forms at grain boundaries; M element is usually chromium, but nickel-cobalt, iron, molybdenum, and tungsten can substitute
M_6C	fcc	$a_0 = 1.085–1.175$	Fe_3Mo_3C $Fe_3W_3C–Fe_4W_2C$ Fe_3Nb_3C Nb_3Co_3C Ta_3Co_3C Cr_7C_3	Randomly distributed carbide; may appear pinkish; M elements are generally molybdenum or tungsten; there is some solubility for chromium, nickel-niobium, tantalum, and cobalt
M_7C_3	Hexagonal	$a_0 = 1.398$ $c_0 = 0.4523$		Generally observed as a blocky intergranular shape; observed only in alloys such as Nimonic 80A after exposure above 1000 °C (1830 °F), and in some cobalt-base alloys
M_3B_2	Tetragonal	$a_0 = 0.560–0.620$ $c_0 = 0.300–0.330$	Ta_3B_2 V_3B_2 Nb_3B_2 $(Mo,Ti,Cr,Ni,Fe)_3B_2$ Mo_2FeB_2	Observed in iron-nickel- and nickel-base alloys with about 0.03% B or greater; borides appear similar to carbides, but are not attacked by preferential carbide etchants; M elements can be molybdenum, tantalum, niobium, nickel, iron, or vanadium
MN	Cubic	$a_0 = 0.4240$	TiN (Ti,Nb,Zr)N (Ti,Nb,Zr) (C,N) ZrN NbN	Nitrides are observed in alloys containing titanium, niobium, or zirconium; they are insoluble at temperatures below the melting point; easily recognized as-polished, having square to rectangular shapes and ranging from yellow to orange
μ	Rhombohedral	$a_0 = 0.475$ $c_0 = 2.577$	Co_7W_6 $(Fe,Co)_7(Mo,W)_6$	Generally observed in alloys with high levels of molybdenum or tungsten; appears as coarse, irregular Widmanstätten platelets; forms at high temperatures
Laves	Hexagonal	$a_0 = 0.475–0.495$ $c_0 = 0.770–0.815$	Fe_2Nb Fe_2Ti Fe_2Mo Co_2Ta Co_2Ti	Most common in iron-base and cobalt-base superalloys; usually appears as irregularly shaped globules, often elongated, or as platelets after extended high-temperature exposure
σ	Tetragonal	$a_0 = 0.880–0.910$ $c_0 = 0.450–0.480$	FeCr FeCrMo CrFeMoNi CrCo CrNiMo	Most often observed in iron- and cobalt-base superalloys, less commonly in nickel-base alloys; appears as irregularly shaped globules, often elongated; forms after extended exposure between 540 and 980 °C (1005 and 1795 °F)

Source: Ref 11

um alloys may be subject to internal oxidation. The addition of iron to these alloys greatly reduces the susceptibility to internal oxidation.

Increasing the nickel content provides good resistance to carburizing environments. High nickel contents in nickel-chromium-iron alloys also improve resistance to nitriding. In addition, the high percentage of nickel improves resistance to thermal fatigue and maintains an austenitic structure so the alloy remains ductile.

Sulfidation resistance at high temperatures is enhanced in nickel-base alloys with the addition of chromium. While the addition of chromium is beneficial to properties at high temperatures, it also makes the alloys more difficult to produce. More detailed information on the high-temperature properties of nickel-base alloys can be found in Volume 1 of the 10th Edition of *Metals Handbook* (see Ref 9, 10, and 12) and in *Corrosion*, Volume 13 of the 9th Edition of *Metals Handbook*.

Corrosion Resistance

This section will briefly review the types of corrosion resulting from exposure of nickel alloys to aqueous environments. More detailed information can be found in Ref 28 and in *Corrosion*, Volume 13 of the 9th Edition of *Metals Handbook*.

General Corrosion. Nickel-base alloys offer excellent corrosion resistance to a wide range of corrosive media. However, as with all types of corrosion, many factors influence the rate of attack. The corrosive media itself is the most important factor governing corrosion of a particular metal. Acidity, temperature, concentration, motion relative to metal surface, degree of oxidizing power and aeration, and presence or absence of inhibitors or accelerators should always be considered. Most of these factors interact, and often this interaction is very complex. For instance, sulfuric acid is generally considered a reducing acid, but in high concentrations the acid becomes oxidizing and this shift usually overshadows other factors in the corrosion behavior of the acid.

As a general rule, increases in temperature increase reaction rates, but increasing temperature also tends to drive dissolved gases out of solution so that a reaction that requires dissolved oxygen can often be slowed down by heating.

Chemical Pitting. Although pitting can arise from various causes, certain chemicals—mainly halide salts and particularly chlorides—are recognized as well-known pit producers. The passive metals are particularly susceptible to pitting in chloride environments, especially oxidizing chlorides such as ferric, cupric, and mercuric chloride. It seems that the chloride ions accumulate at anodic areas and either penetrate or dissolve the passive film at these points. Since the chloride corrosion product is hydrolyzed to hydrochloric acid, the acidity at the anode increases as more chloride migrates to the anode, and the corrosion rate increases with time. Self-accelerating reactions of this kind are described as autocatalytic reactions.

The effect of molybdenum added to nickel-containing alloys offers higher degrees of nobility and resistance to these chloride types of attack. Figure 1 shows these effects on the corrosion resistance of various alloys.

Intergranular Corrosion. The austenitic stainless steels containing ~8 to 40 wt% Ni comprise a class of materials in which this form of attack is most common. It is usually caused by an improper heat treatment or heat from welding that causes the precipitation of certain alloy components at the grain boundary. Chromium combines with carbon in solution to form a chromium carbide. This precipitation causes a depletion of corrosion-resisting elements in the area surrounding the grain boundary, and this area becomes anodic to the remainder of the grain.

This condition occurs when these alloys are held at temperatures between 425 and 760 °C (800 and 1400 °F). The depleted chromium region becomes sensitized to the corrosive environment resulting in attack at this point. The condition is known as sensitization.

There are three methods of combating intergranular corrosion in cases where susceptible materials must be heated in the

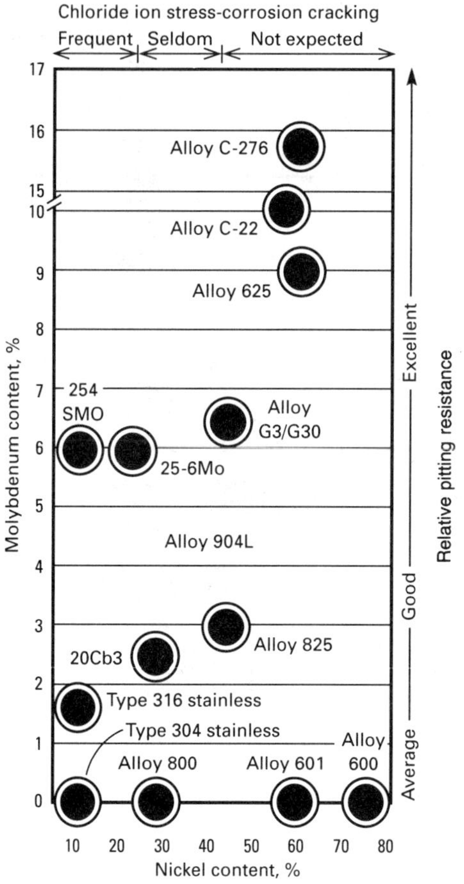

Fig. 1 Effect of molybdenum and nickel contents on the corrosion resistance of selected commercial alloys

sensitizing range. The first method is to reheat the metal to a temperature high enough to redissolve the precipitated phase and then to cool it quickly enough to maintain this phase in solution. The second method, called stabilization, is to add certain elements such as niobium, tantalum, and titanium in order to make use of their ability to combine more readily than chromium with carbon. In this way, chromium is not depleted and the metal retains its corrosion resistance. This technique is used in redesigning austenitic stainless and nickel-base alloys to suitably resist various acidic aqueous environments. The third is to restrict the amount of one of the constituents of the precipitate—usually carbon, for example, 316L stainless steel—and to thereby reduce the extent of the precipitation and resulting alloy depletion.

Corrosion Fatigue. Metals that fail as a result of being alternately or cyclicly stressed are said to fatigue. Failure is by transgranular cracking and is usually only a single crack. (High-temperature fatigue is intergranular, since above the equicohesive temperature, grain boundaries are weaker than the grains.) Endurance limit and fatigue strength are measures of a metal's ability to withstand cyclic stressing in air.

Fatigue data determined in air are useless as design criteria for a part to be placed in service in a corrosive environment. When the metal is cyclicly stressed in corrosive environments, the joint action of corrosion and fatigue greatly intensifies the damage. Cracking is again transgranular, but there are usually multiple cracks and they quite often begin at the base of a corrosion pit. Unfortunately, corrosion fatigue data for environments other than water or sea water are almost totally lacking.

The most important consideration in selecting a metal for resistance to corrosion fatigue is the resistance of the metal to the corrosive environment. One of the major characteristics of the nickel-base alloys is that the higher the nickel content, the better the corrosion fatigue resistance.

Stress-Corrosion Cracking (SCC). In contrast to corrosion fatigue, which will occur in any corrosive media, stress-corrosion cracking requires a specific combination of alloy and environment. Austenitic stainless steels that contain surface tensile stresses, either locked in during fabrication or externally applied, may fail by transgranular cracking in *chloride* solutions. Similarly, Alloy 400 that is in a stressed condition may fail by intergranular cracking when exposed in *mercury* or *mercury salts*. The following table lists some of the common alloy systems and the specific media in which they are subject to this attack:

Alloy	Environment	Cracking
Aluminum-magnesium	Cl^-	Intergranular
Brass	NH_4^+	Intergranular
Steel	NO_3^-, OH^-	Intergranular
300 series stainless steel	Cl^-, OH	Transgranular (can be intergranular if sensitized)
Alloy 400	Hg, Hg^+, chromic acid, aerated hydrofluoric acid vapor	Intergranular and transgranular
Alloy 600	Fused caustic	Intergranular
Nickel 200, 201	Hg, Hg_2^+, molten metals	Intergranular

Since stress-corrosion cracking requires three simultaneous factors—surface tensile stress, alloy, and environment—the alteration or elimination of any one of them can prevent this attack. Where it is possible, the alteration of the environment or the choice of a different alloy is the best solution. Elimination of stress is usually attempted through heat treatment, but it is often difficult or impossible to completely eliminate stresses on complex fabricated equipment, and the procedure is always costly.

The most common form of stress-corrosion cracking is that involving chloride and halide ions, which may be present in a wide variety of water and process streams. Stan-

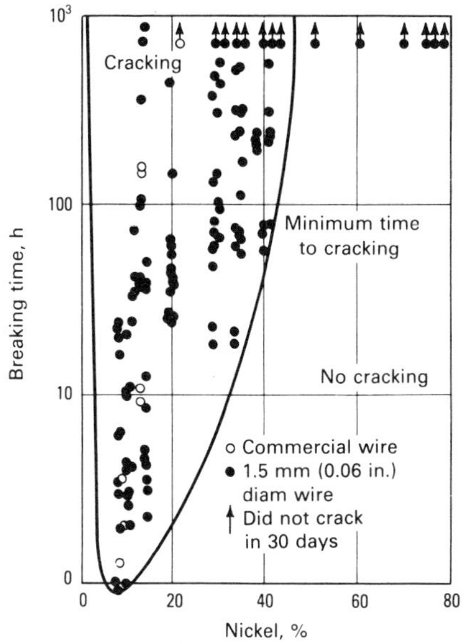

Fig. 2 Effect of nickel additions to a 17 to 24% Cr steel on resistance to SCC in boiling 42% magnesium chloride. 1.5 mm (0.06 in.) diam wire specimens deadweight loaded to 228 or 310 MPa (33 or 45 ksi) Source: Ref 30

dard ASTM tests have been established to evaluate the sensitivity of nickel alloys to cracking under these types of conditions, using boiling magnesium chloride solutions, U-bend tests, and slow strain tensile tests (Ref 29). Iron-chromium-nickel alloys with nickel contents greater than 50% are immune to cracking in boiling 42% magnesium chloride (Fig. 2). However, SCC of nickel and high-nickel alloys has been experienced in high-temperature caustic soda and caustic potash solutions and in molten caustic.

Cracking of some nickel-base alloys has also occurred under special conditions in fluosilicic acid, hydrofluoric acid, mercuric salt solutions, and high-temperature water and steam that are contaminated with trace amounts of oxygen, lead, fluorides, or chlorides. Sensitized alloys are susceptible to SCC in sulfur compounds such as sodium sulfite, sodium thiosulfate, and polythionic acids (Ref 29).

Low-Expansion Alloys

French physicist C.E. Guillaume discovered in 1896 that alloys of nickel-iron had low thermal expansion characteristics. Nickel was found to have a profound effect on the thermal expansion of iron. Alloys can be designed to have a very low thermal expansion or display uniform and predictable expansion over certain temperature ranges.

Iron-36% Ni alloy (Invar) has the lowest expansion of the Fe-Ni alloys and maintains nearly constant dimensions during normal variations in atmospheric temperature, as shown in Fig. 3. Higher nickel results in greater thermal expansion, which allows for

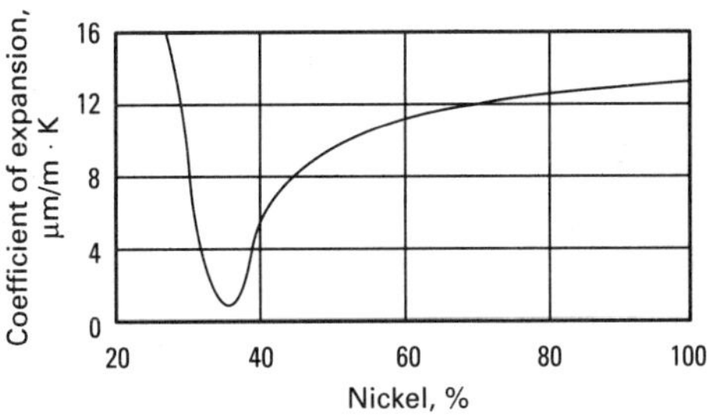

Fig. 3 Coefficient of linear expansion at 20 °C (68 °F) versus nickel content for iron-nickel alloys containing 0.4% Mn and 0.1% C

specific expansion rates to be selected by adjusting the nickel content. It can be seen in Fig. 4 that uniform expansion is exhibited up to the inflection point, which also increases with nickel content. Nickel-iron alloys are ferromagnetic at room temperature, and their inflection point is closely associated with the Curie temperature.

The addition of cobalt to the nickel-iron matrix produces alloys with a low coefficient of expansion, a constant modulus of elasticity, and high strength. The alloys can also be strengthened by precipitation-hardening heat treatments made possible by the addition of niobium and titanium. Additional information on low-expansion alloys can be found later in this article (see the following section on Commercial Nickel and Nickel Alloys) and in the article "Low-Expansion Alloys" in this Volume.

Electrical Resistance Alloys

Several alloy systems based on nickel or containing high nickel contents are used in instruments and control equipment to measure and regulate electrical characteristics (resistance alloys) or are used in furnaces and appliances to generate heat (heating alloys).

The primary requirements for resistance alloys are uniform resistivity, stable resistance (no time-dependent aging effects), reproducible temperature coefficient of resistance, and low thermoelectrical potential versus copper. Properties of secondary importance are coefficient of expansion, mechanical strength, ductility, corrosion resistance, and the ability to be joined to other metals by welding, brazing, or soldering.

Types of resistance alloys containing nickel include:

- Cu-Ni alloys containing 2 to 45% Ni
- Ni-Cr-Al alloys containing 35 to 95% Ni
- Ni-Cr-Fe alloys containing 35 to 60% Ni
- Ni-Cr-Si alloys containing 70 to 80% Ni

The primary requirements of materials used for heating elements are high melting point, high electrical resistivity, reproducible temperature coefficient of resistance, good oxidation resistance in furnace environments, absence of volatile components, and resistance to contamination. Other desirable properties are good elevated-temperature creep strength, high emissivity, low thermal expansion and low modulus (both of which help minimize thermal fatigue), good resistance to thermal shock, and good strength and ductility at fabrication temperatures.

Types of resistance heating alloys containing nickel include:

- Ni-Cr alloys containing 65 to 80% Ni with 1.5% Si
- Ni-Cr-Fe alloys containing 35 to 70% Ni with 1.5% Si + 1% Nb

Additional information on electrical resistance alloys can be found later in this article (see the following section on Commercial Nickel and Nickel Alloys) and in the article "Electrical Resistance Alloys" in this Volume.

Soft Magnetic Alloys

Two broad classes of magnetically soft materials have been developed in the Fe-Ni system. The high-nickel alloys (about 79% Ni with 4 to 5% Mo; bal Fe) have high initial permeability and low saturation induction. The low-nickel alloys (about 50% Ni) are lower in initial permeability but higher in saturation induction (see the article "Magnetically Soft Materials" in this Volume for specific property data). Additional information on these alloys can also be found later in this article (see the following section on Commercial Nickel and Nickel Alloys).

Shape Memory Alloys

Metallic materials that demonstrate the ability to return to their previously defined shape when subjected to the appropriate heating schedule are referred to as shape memory alloys. Nickel-titanium alloys (50Ni-50Ti) are one of the few commercially important shape memory alloys. The physical metallurgy, properties, and applications of equiatomic NiTi alloys are described in the article "Shape Memory Alloys" in this Volume.

Commercial Nickel and Nickel Alloys

The commercial forms of nickel and nickel-base alloys are fully austenitic and are used/selected mainly for their resistance to high temperature and aqueous corrosion. From a tariff definition, these are alloys where nickel is in the predominance. From a nomenclature perspective, they are alloys which contain greater than 30% Ni in their content.

Figure 5 categorizes these alloys by nickel content, while relating the alloys to the more common austenitic stainless steels. All of these materials are characterized by having 15 to 23% Cr and, therefore, can be ranked by nickel content. The austenitic steels, thus, are iron-chromium-nickel al-

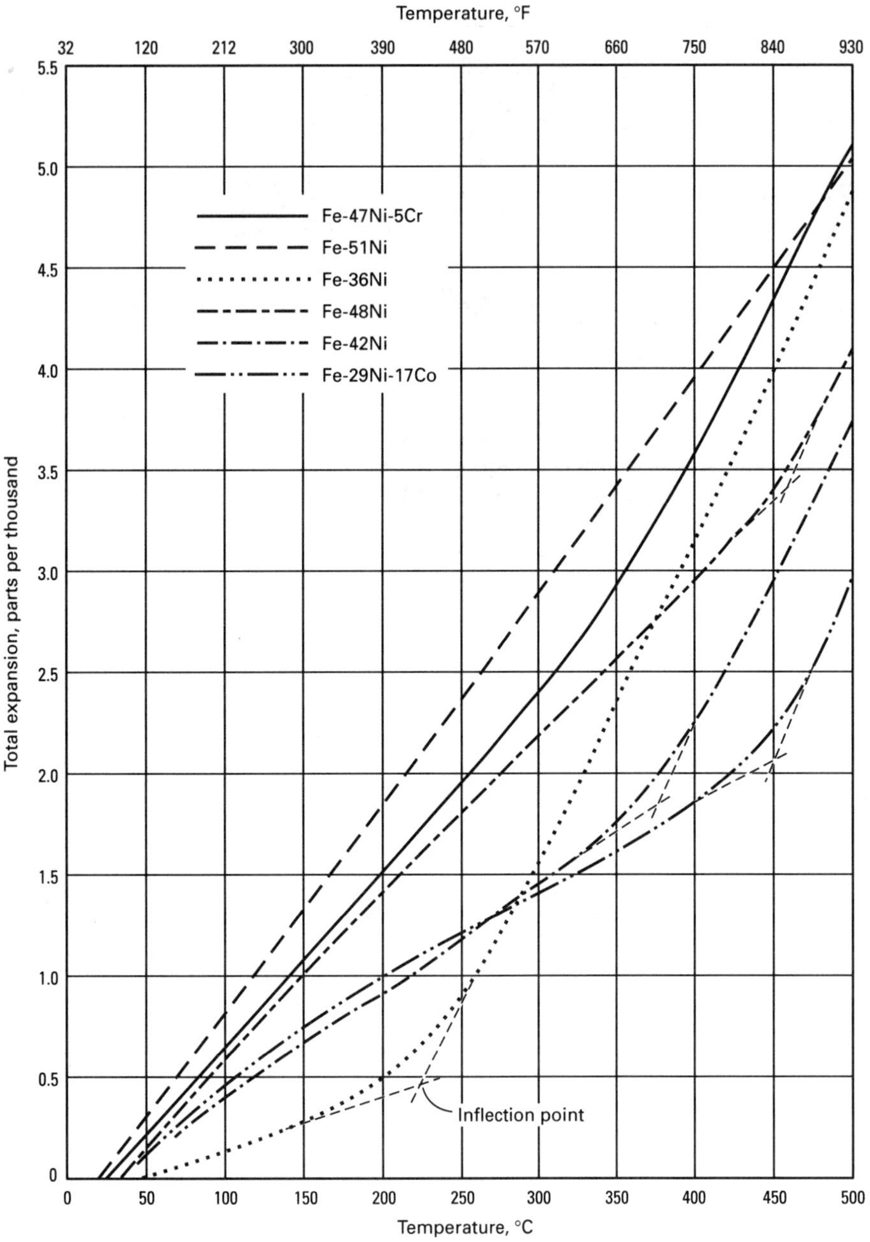

Fig. 4 Total thermal expansion of iron-nickel alloys showing the effect of third elements

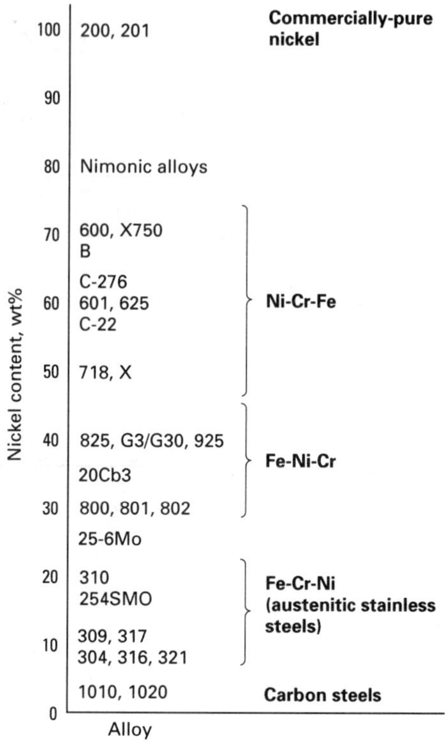

Fig. 5 Nickel-base alloy chart showing alloys containing varying amounts of nickel and iron. Chromium contents are constant at approximately 18 to 20%.

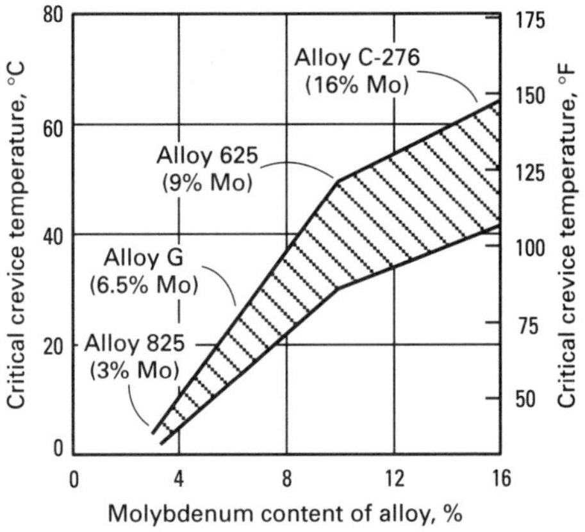

Fig. 6 Molybdenum content versus critical crevice temperature range for several nickel-base alloys. The range of critical crevice temperatures was obtained by varying test procedure.

loys (ranking elements by declining predominance). Any nickel-chromium-iron containing alloys can therefore be identified and related by family to one another and their general corrosion characteristics identified. Omitted from Fig. 5 is the Ni-Cu (Monel) series of alloys.

Maximum suggested operating temperatures for each of the characterized groupings are as follows:

- Fe-Cr-Ni 1050 °C (1920 °F)
- Fe-Ni-Cr 1150 °C (2100 °F)
- Ni-Cr-Fe 1200 °C (2200 °F)

Variations exist for each of these series to enhance their corrosion characteristics or their mechanical properties. Hence, molybdenum and nitrogen may be added to this series of alloys to improve their resistance to pitting and crevice corrosion; chromium additions improve sulfidation resistance; aluminum, titanium, and niobium allow strengthening through precipitation hardening (age hardening). But in all cases, the nickel content allows improvements in fatigue strength and high-temperature performance, especially in reducing environments such as those involving carburization and nitriding.

Figure 6 compares the corrosion characteristics of several nickel-base alloys using the Materials Technology Institute (MTI) test procedure, which involves the use of 6% ferric chloride solution for determining the relative resistance of alloy to crevice corrosion in oxidizing chloride environments (Ref 31). This method proposes alloy rankings on the merits of increased critical crevice temperature. As shown in Fig. 6, increased molybdenum content is the major contributor to crevice corrosion resistance in aggressive chloride-containing environments.

Hastelloy Alloy B-2, which contains 27% Mo, offers the best resistance to hot hydrochloric acid environments of any of the nickel-base alloys. Selection of the other molybdenum-containing alloys will depend upon the operating conditions and severity of the environment in question.

Table 3 General corrosion resistance of nickel-base alloys

Alloy series	Applications	Aqueous environments(a)
Nickel 200; Alloys 400, 600	Good general fabricable alloys for vessels and pipelines handling complex chemical derivatives from petrochemical (organic) feedstock	Reducing
Alloys C-276, 625, G3/G30, C-22/622, 825	Molybdenum-containing alloys for pitting and crevice corrosion resistance	Neutral, reducing, oxidizing
Alloys 800, 904L; type 304, 316, 317 stainless steels	Alloys commonly used in food processing, pulp and paper industries, and chemical transportation	Oxidizing

(a) Reducing environments: caustic soda (NaOH), hydrochloric acid, sulfuric acid (dilute solutions), hydrofluoric acid; hydrochloric acid requires high-molybdenum alloys. Neutral environments: organic acid salts (NaCl, bisulfates). Oxidizing environments: sulfuric acid (concentrated), phosphoric acid, nitric acid; nitric acid requires high-chromium alloys.

Table 3 provides a general perspective of corrosion resistance for the groups of alloys. Generally, the higher the nickel in an alloy, the greater is its inherent resistance to reducing environment, both acid and alkali. By contrast, the austenitic stainless steels (Fe-Cr-Ni alloys) rely on oxygen or oxidizing conditions to assist in maintaining a protective oxide film for their corrosion resistance. Breakdown of these films makes them susceptible to pitting and crevice corrosion attack. The transition group of alloys offers degrees of resistance to both oxidizing and reducing conditions. These alloys generally contain varying amounts of molybdenum in their content, and they offer a broad range of resistance to these mixed environments.

Figures 7 and 8 show the high-temperature performance and capabilities for the nickel-base alloys. The higher the nickel content, the better the resistance. In Fig. 7, the nickel-chromium-iron alloys exhibit excellent oxidation resistance. This characteristic is most desired in alloys specified for aerospace jet engine applications and for thermal processing requirements. Alloy 230 with tungsten additions (Ni-22Cr-14W-2Mo-3Fe-5Co) combines excellent high-temperature strength with outstanding resistance to oxidizing environments up to 1150 °C (2100 °F). Nickel-chromium-iron alloys also exhibit excellent carburization resistance (Fig. 8).

Summaries of wrought nickel-base alloy compositions, typical properties, and applications are provided in Tables 4, 5, and 6.

Commercially Pure and Low-Alloy Nickels. Nickel is supplied to the producers of nickel alloys in powder, pellets, or anode forms. Purity is important, and these products constitute Grade 1 nickel in the primary metals marketplace. The feedstock is either melted and cast into ingots, or direct rolled or powder compacted into semifinished forms. This has led to a whole series of alloy modifications, with controlled compositions having nickel contents ranging from about 94% to virtually 100%.

These materials are characterized by high density, offering magnetic and electronic property capabilities. They also offer excellent corrosion resistance to reducing environments, along with reasonable thermal transfer characteristics. Hence, they find utilization in heat exchangers, evaporators, and various food processing applications, where purity of finished products is important. Some nickels of commercial importance include:

- Nickel 200. 99.5% Ni min, 0.10% C max, other minor additions include manganese and silicon
- Nickel 201. Low-carbon version (0.02% C max) for service temperatures above 290 °C (550 °F)
- Nickel 205. Controlled magnesium levels (0.01 to 0.08%) for electronic characteristics
- Nickel 270 and 290. High-purity powder alloys (99.95% Ni) made from water-atomized and carbonyl powders
- Permanickel Alloy 300. Age-hardenable nickel containing 0.2 to 0.6 Ti: spring properties with higher thermal and electrical properties
- Duranickel Alloy 301. Age-hardenable nickel containing 4.00 to 4.75 Al and 0.25 to 1.00 Ti: spring applications for electrical hardware

Applications for Nickel 200 and 201 are found in the chemical processing industry. These alloys are important in handling hot-concentrated caustic soda and dry chlorine. Caustic is a key ingredient in dissolving wood products for production of paper and in the extraction of aluminum from its ores.

Originally, Nickel 200 was used for food processing, kitchen hardware, and roofing. It continues to be selected for use as a coinage material. Today, it finds major applications in the electronics industry (plated pins for printed circuit board interconnects), battery applications (molten lithium, long-life storage batteries), and synthetic diamond production. In all instances, corrosion resistance and the need for end-product purity have played prime roles in its selection and utilization.

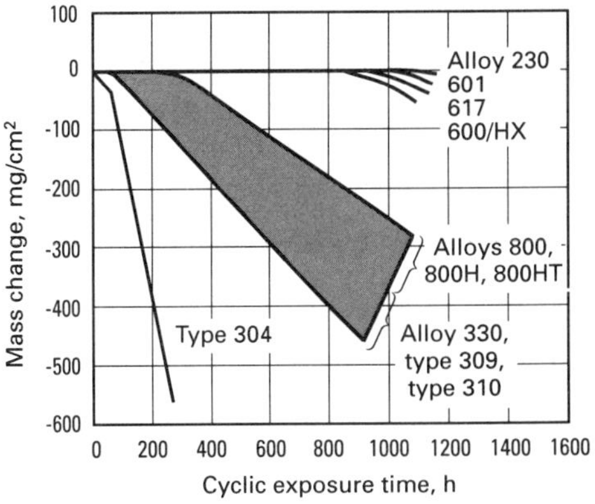

Fig. 7 Cyclic oxidation resistance at 1095 °C (2000 °F). Each cycle consisted of 15 min heating followed by 5 min of cooling in air.

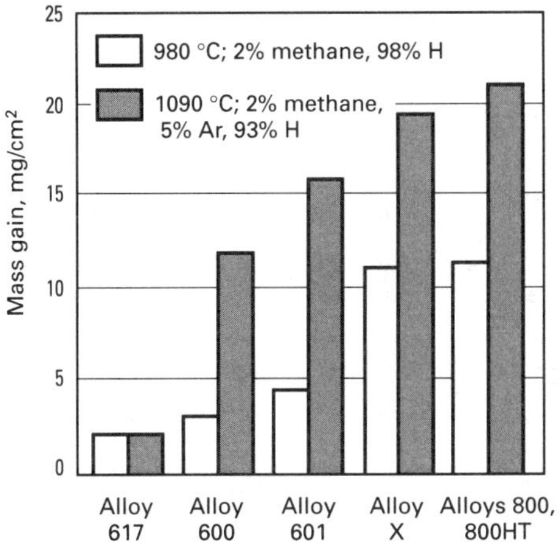

Fig. 8 Resistance to gas carburization at 980 and 1090 °C (1800 and 2000 °F). Test duration, 100 h

Nickel-copper alloys have been found to possess excellent corrosion resistance in reducing chemical environments and in sea water, where they deliver excellent service in nuclear submarines and various surface vessels.

These alloys have excellent ductility and can be readily fabricated and formed into a variety of shapes. By changing the various proportions of nickel and copper in the alloy, a whole series of alloys with different electrical resistivities and Curie points (magnetic/nonmagnetic transition temperatures) can be created. Some nickel-copper alloys of commercial importance include:

- Alloy 400 (66% Ni, 33% Cu). The base alloy in the series; can be magnetic depending upon composition and previous work history (Curie point is 20 to 50 °C, or 70 to 120 °F)
- Alloy R-405. Controlled sulfur addition (0.025 to 0.06% S) for improved machinability
- Alloy K-500. Added aluminum and titanium for age hardening; totally nonmagnetic and spark resistant

This series of alloys finds major applications in the fastener industry serving the

Table 4 Compositions of selected nickel and nickel-base alloys

Alloy	Composition, wt%(a) Ni	Cu	Fe	Mn	C	Si	S	Other
Commercially pure and low-alloy nickels								
Nickel 200	99.0 min	0.25	0.40	0.35	0.15	0.35	0.01	...
Nickel 201	99.0 min	0.25	0.40	0.35	0.02	0.35	0.01	...
Nickel 205	99.0 min(b)	0.15	0.20	0.35	0.15	0.15	0.008	0.01–0.08 Mg, 0.01–0.05Ti
Nickel 211	93.7 min(b)	0.25	0.75	4.25–5.25	0.20	0.15	0.015	...
Nickel 212	97.0 min	0.20	0.25	1.5–2.5	0.10	0.20	...	0.20 Mg
Nickel 222	99.0 min(b)	0.10	0.10	0.30	...	0.10	0.008	0.01–0.10 Mg, 0.005 Ti
Nickel 270	99.9 min	0.01	0.05	0.003	0.02	0.005	0.003	0.005 Mg, 0.005 Ti
Duranickel 301	93.00 min	0.25	0.60	0.50	0.30	1.00	0.01	4.00–4.75 Al, 0.25–1.00 Ti
Nickel-copper alloys								
Alloy 400	63.0 min(b)	28.0–34.0	2.5	0.20	0.3	0.5	0.024	...
Alloy 401	40.0–45.0(b)	bal	0.75	2.25	0.10	0.25	0.015	...
Alloy R-405	63.0 min(b)	28.0–34.0	2.5	2.0	0.3	0.5	0.025–0.060	...
Alloy 450	29.0–33.0	bal	0.4–1.0	1.0	...	...	0.02	1.0 Zn, 0.05 Pb, 0.02 P
Alloy K-500	63.0 min(b)	27.0–33.0	2.0	1.5	0.25	0.5	0.01	2.30–3.15 Al, 0.35–0.85 Ti

Alloy	Composition, wt%(a) Ni	Cr	Fe	Co	Mo	W	Nb	Ti	Al	C	Mn	Si	B	Other
Nickel-chromium and nickel-chromium-iron alloys														
Alloy 230	bal	22.0	3.0	5.0	2.0	14.0	...	...	0.3	0.10	0.5	0.4	0.005	0.02 La
Alloy 600	72.0 min(b)	14.0–17.0	6.0–10.0	...	...	...	...	...	...	0.15	1.0	0.5	...	0.5 Cu
Alloy 601	58.0–63.0	21.0–25.0	bal	...	...	...	...	...	1.0–1.7	0.10	1.0	0.50	...	1.0 Cu
Alloy 617	44.5 min	20.0–24.0	3.0	10.0–15.0	8.0–10.0	...	...	0.6	0.8–1.5	0.05–0.15	1.0	1.0	0.006	0.5 Cu
Alloy 625	58.0 min	20.0–23.0	5.0	1.0	8.0–10.0	...	3.15–4.15(c)	0.40	0.40	0.10	0.50	0.50	...	...
Alloy 690	58.0 min	27.0–31.0	7.0–11.0	...	...	...	...	...	...	0.05	0.50	0.50	...	0.50 Cu
Alloy 718	50.0–55.0(b)	17.0–21.0	bal	1.0	2.80–3.30	...	4.75–5.50(c)	0.65–1.15	0.20–0.80	0.08	0.35	0.35	0.006	0.30 Cu
Alloy X750	70.0 min(b)	14.0–17.0	5.0–9.0	1.0	...	...	0.70–1.20(c)	2.25–2.75	0.40–1.00	0.08	1.00	0.50	...	0.50 Cu
Alloy 751	70.0 min(b)	14.0–17.0	5.0–9.0	...	...	...	0.7–1.2(c)	2.0–2.6	...	0.10	1.0	0.5	...	0.5 Cu
Alloy MA 754(d)	78.0	20	1.0	...	...	...	...	0.5	0.3	0.05	...	...	...	0.6 Y_2O_3
Alloy C-22	51.6	21.5	5.5	2.5	13.5	4.0	...	...	...	0.01	1.0	0.1	...	0.3 V
Alloy C-276	bal	14.5–16.5	4.0–7.0	2.5	15.0–17.0	3.0–4.5	...	...	...	0.01	1.0	0.08	...	0.35 V
Alloy G3	bal	21.0–23.5	18.0–21.0	5.0	6.0–8.0	1.5	0.50(c)	...	...	0.015	1.0	1.0	...	1.5–2.5 Cu
Alloy HX	bal	20.5–23.0	17.0–20.0	0.5–2.5	8.0–10.0	0.2–1.0	...	...	...	0.05–0.15	1.0	1.0	...	...
Alloy S	bal	14.5–17.0	3.0	2.0	14.0–16.5	1.0	...	...	0.10–0.50	0.02	0.30–1.0	0.20–0.75	0.015	0.01–0.10 La, 0.35 Cu
Alloy W	63.0	5.0	6.0	2.5	24.0	...	...	...	...	0.12	1.0	1.0	...	...
Alloy X	bal	20.50–23.00	17.0–20.0	0.5–2.5	8.0–10.0	0.2–1.0	...	0.15	0.50	0.05–0.15	1.0	1.0	0.008	0.5 Cu
Iron-nickel-chromium alloys														
Alloy 556	20.0	22.0	bal	18.0	3.0	2.5	...	...	0.2	0.10	1.0	0.4	...	0.6 Ta, 0.02 La, 0.02 Zr
Alloy 800	30.0–35.0	19.0–23.0	39.5 min	...	...	...	...	0.15–0.60	0.15–0.60	0.10	1.5	1.0	...	...
Alloy 800HT	30.0–35.0	19.0–23.0	39.5 min	...	...	...	...	0.15–0.60	0.15–0.60	0.06–0.10	1.5	1.0	...	0.85–1.20 Al + Ti
Alloy 825	38.0–46.0	19.5–23.5	22.0 min	...	2.5–3.5	...	...	0.6–1.2	0.2	0.05	1.0	0.5	...	...
Alloy 925	44.0	21.0	28.0	...	3.0	...	...	2.1	0.3	0.01	...	...	...	...
20Cb3	32.0–38.0	19.0–21.0	bal	...	2.0–3.0	...	1.0	...	...	0.07	2.0	1.0	...	3.0–4.0 Cu
20Mo-4	35.0–40.0	22.5–25.0	bal	...	3.5–5.0	...	0.15–0.35	...	...	0.03	1.0	0.5	...	0.5–1.5 Cu
20Mo-6	33.0–37.20	22.0–26.0	bal	...	5.0–6.7	...	...	...	...	0.03	1.0	0.5	...	2.0–4.0 Cu
Controlled-expansion alloys (Fe-Ni-Cr, Fe-Ni-Co)														
Alloy 902	41.0–43.5(b)	4.9–5.75	bal	...	...	...	...	2.2–2.75	0.3–0.8	0.06	0.8	1.0	...	...
Alloy 903	38.0	...	42.0	15.0	...	...	3.0	1.4	0.9	...	...	...	...	...
Alloy 907	38.0	...	42.0	13.0	...	...	4.7	1.5	0.03	...	...	0.15	...	...
Alloy 909	38.0	...	42.0	13.0	...	...	4.7	1.5	0.03	0.01	...	0.4	...	...
Nickel-iron alloys														
Alloy 36	35.0–38.0	0.50	bal	1.0	0.5	...	...	...	...	0.10	0.60	0.35	...	...
Alloy 42	42.0(e)	0.50	bal	1.0	0.5	...	...	...	0.15	0.05	0.80	0.30	...	...
Alloy 48	48.0(e)	0.25	bal	1.0	...	...	...	...	0.10	0.05	0.80	0.30	...	...

(a) Single values are maximum values unless otherwise indicated. (b) Nickel plus cobalt content. (c) Niobium plus tantalum content. (d) Mechanically alloyed, dispersion-strengthened, powder metallurgy alloy. (e) Nominal value; adjusted to meet expansion requirements

marine, aerospace, and chemical processing industries. The nonmagnetic characteristics of Alloy K-500 are used for gyroscope application and anchor cable aboard minesweepers. Alloy K-500 is also used for propeller shafts on a wide variety of vessels and exhibits high fatigue strength in seawater.

This series of alloys also finds application in chemical process applications for handling of organic acids, caustic, and dry chlorine. The age-hardenable Alloy K-500 is used in valves and pumps—both as castings and wrought forms. The oil and gas industry has used this alloy extensively for sucker rods and associated Christmas tree wellhead applications, especially in sour gas environments.

The nickel-chromium and nickel-chromium-iron series of alloys led the way to higher strength and resistance to elevated temperatures. While these series of alloys were finding their way into the early European jet engine problem, these alloys found their initial applications in North America in thermal process equipment and the chemical process industry, where carburizing environments and elevated temperatures were service limiting to stainless steels. Today they also form the basis for both commercial and military power systems. Two of the earliest developed Ni-Cr and Ni-Cr-Fe alloys were:

- Alloy 600 (76Ni-15Cr-8Fe). The basic alloy in the Ni-Cr-Fe system; high nickel content makes it resistant to reducing environments

Table 5 Room-temperature mechanical properties and characteristics of selected nickel-base alloys

Properties are for annealed sheet unless otherwise indicated.

Alloy	Ultimate tensile strength		Yield strength (0.2% offset)		Elongation in 50 mm (2 in.), %	Elastic modulus (tension)		Hardness	Description/major applications
	MPa	ksi	MPa	ksi		GPa	10^6 psi		
Commercially pure and low-alloy nickels									
Nickel 200	462	67	148	21.5	47	204	29.6	109 HB	Commercially pure wrought nickel with good mechanical properties and excellent resistance to many corrosives. Nickel 201 has low carbon (0.02% max) for applications over 315 °C (600 °F). Used for food processing equipment, chemical shipping drums, caustic handling equipment and piping, electronic parts, aerospace and missile components, rocket motor cases, and magnetostrictive devices
Nickel 201	403	58.5	103	15	50	207	30	129 HB	
Nickel 205	345	50	90	13	45	...	...	...	Wrought nickel similar to Nickel 200 but with compositional adjustments to enhance performance in electrical and electronic applications. Used for the anodes and grids of electronic valves, magnetostrictive transducers, lead wires, transistor housings, and battery cases
Nickel 211	530	77	240	35	40	...	...	...	Nickel-manganese alloy that is slightly harder than Nickel 200. The manganese addition provides resistance to sulfur compounds at elevated temperatures. Used as fuses in light bulbs, as grids in vacuum tubes, and in assemblies where sulfur is present in heating flames
Nickel 212	483	70	...	...	...	...	...	...	Wrought nickel strengthened with an addition of manganese. Used for electrical and electronic applications such as lead wires, supporting components in lamps and cathode ray tubes, and electrodes in glow-discharge lamps
Nickel 222	380	55	...	...	...	...	...	...	Wrought nickel with an addition of magnesium for electronic applications. The magnesium provides activation for cathodes of thermionic devices. Used for sleeves of indirectly heated oxide-coated cathodes
Nickel 270	345	50	110	16	50	...	...	30 HRB	A high-purity grade of nickel made by powder metallurgy. It has a low base hardness and high ductility. Its extreme purity is useful for components of hydrogen thyratrons. Also used for electrical resistance thermometers
Duranickel 301 (precipitation hardened)	1170	170	862	125	25	207	30	30–40 HRC	Nickel-aluminum-titanium alloy used for applications that require the corrosion resistance of commercially pure nickel but with greater strength or spring properties. These applications include diaphragms, springs, clips, press components for extrusion of plastics, and molds for production of glass articles
Nickel-copper alloys									
Alloy 400	550	80	240	35	40	180	26	110–150 HB	A nickel-copper alloy with high strength and excellent corrosion resistance in a range of media, including seawater, hydrofluoric acid, sulfuric acid, and alkalies. Used for marine engineering, chemical and hydrocarbon processing equipment, valves, pumps, shafts, fittings, fasteners, and heat exchangers
Alloy 401	440	64	134	19.5	51	...	...	...	A copper-nickel alloy designed for specialized electrical and electronic applications. It has a very low temperature coefficient of resistance and medium-range electrical resistivity. Used for wire-wound precision resistors and bimetal contacts
Alloy R-405	550	80	240	35	40	180	26	110–140 HB	The free-machining version of Alloy 400. A controlled amount of sulfur is added to the alloy to provide sulfide inclusions that act as chip breakers during machining. Used for meter and valve parts, fasteners, and screw machine products
Alloy 450	385	56	165	24	46	...	...	...	A copper-nickel alloy of the 70-30 type having superior weldability. It is resistant to corrosion and biofouling in seawater, has good fatigue strength, and has relatively high thermal conductivity. Used for seawater condensers, condenser plates, distiller tubes, evaporator and heat exchanger tubes, and saltwater piping
Alloy K-500 (precipitation hardened)	1100	160	790	115	20	180	26	300 HB	A precipitation-hardenable nickel-copper alloy that combines the corrosion resistance of Alloy 400 with greater strength and hardness. It also has low permeability and is nonmagnetic to under −100 °C (−150 °F). Used for pump shafts, oil well tools and instruments, doctor blades and scrapers, springs, valve trim, fasteners, and marine propeller shafts

(continued)

(a) Cold rolled and solution annealed at 1230 °C (2250 °F). Sheet thickness, 1.2 to 1.6 mm (0.048 to 0.063 in.). (b) Annealed at 980 °C (1800 °F) for 30 min, air cooled, and aged at 760 °C (1400 °F) for 8 h, furnace cooled at a rate of 55 °C (100 °F)/h, heated to 620 °C (1150 °F) for 8 h, air cooled

Table 5 (continued)

Properties are for annealed sheet unless otherwise indicated.

Alloy	Ultimate tensile strength MPa	Ultimate tensile strength ksi	Yield strength (0.2% offset) MPa	Yield strength (0.2% offset) ksi	Elongation in 50 mm (2 in.), %	Elastic modulus (tension) GPa	Elastic modulus (tension) 10^6 psi	Hardness	Description/major applications
Nickel-chromium and nickel-chromium-iron alloys									
Alloy 230 (a)	860	125	390	57	47.7	211	30.6	92.5 HRB	Nickel-chromium-tungsten alloy that combines excellent high-temperature strength with resistance to oxidizing environments up to 1150 °C (2100 °F) and resistance to nitriding environments. Used for aerospace gas turbine components, chemical processing equipment, and heat-treating equipment
Alloy 600	655	95	310	45	40	207	30	75 HRB	A nickel-chromium alloy with good oxidation resistance at high temperatures and resistance to chloride ion stress-corrosion cracking, corrosion by high-purity water, and caustic corrosion. Used for furnace components, in chemical and food processing, in nuclear engineering, and for sparking electrodes
Alloy 601	620	90	275	40	45	207	30	65–80 HRB	A nickel-chromium alloy with an addition of aluminum for outstanding resistance to oxidation and other forms of high-temperature corrosion. It also has high mechanical properties at elevated temperatures. Used for industrial furnaces; heat-treating equipment such as baskets, muffles, and retorts; petrochemical and other process equipment; and gas turbine components
Alloy 617 (solution annealed)	755	110	350	51	58	211	30.6	173 HB	A nickel-chromium-cobalt-molybdenum alloy with an exceptional combination of metallurgical stability, strength, and oxidation resistance at high temperatures. Resistance to oxidation is enhanced by an aluminum addition. The alloy also resists a wide range of corrosive aqueous environments. Used in gas turbines for combustion cans, ducting, and transition liners; for petrochemical processing; for heat-treating equipment; and in nitric acid production
Alloy 625	930	135	517	75	42.5	207	30	190 HB	A nickel-chromium-molybdenum alloy with an addition of niobium that acts with the molybdenum to stiffen the alloy's matrix and thereby provide high strength without a strengthening heat treatment. The alloy resists a wide range of severely corrosive environments and is especially resistant to pitting and crevice corrosion. Used in chemical processing, aerospace and marine engineering, pollution-control equipment, and nuclear reactors
Alloy 690	725	105	348	50.5	41	211	30.6	88 HRB	A high-chromium-nickel alloy with excellent resistance to many aqueous media and high-temperature atmospheres. Used for various applications involving nitric or nitric/hydrofluoric acid solutions. Also useful for high-temperature service in gases containing sulfur
Alloy 718 (precipitation hardened)	1240	180	1036	150	12	211	30.6	36 HRC	A precipitation-hardenable nickel-chromium alloy containing significant amounts of iron, niobium, and molybdenum along with lesser amounts of aluminum and titanium. It combines corrosion resistance and high strength with outstanding weldability, including resistance to postweld cracking. The alloy has excellent creep-rupture strength at temperatures up to 700 °C (1300 °F). Used in gas turbines, rocket motors, spacecraft, nuclear reactors, pumps, and tooling
Alloy X750 (precipitation hardened)	1137	165	690	100	20	207	30	330 HB	A nickel-chromium alloy similar to Alloy 600 but made precipitation hardenable by additions of aluminum and titanium. The alloy has good resistance to corrosion and oxidation along with high tensile and creep-rupture properties at temperatures up to about 700 °C (1300 °F). Its excellent relaxation resistance is useful for high-temperature springs and bolts. Used in gas turbines, rocket engines, nuclear reactors, pressure vessels, tooling, and aircraft structures
Alloy 751 (precipitation hardened)	1310	190	976	141.5	22.5	214	31	352 HB	A nickel-chromium alloy similar to Alloy X750 but with increased aluminum content for greater precipitation hardening. Designed for use as exhaust valves in internal-combustion engines. In that application, the alloy offers high strength at operating temperatures, high hot hardness for wear resistance, and corrosion resistance in hot exhaust gases containing lead oxide, sulfur, bromine, and chlorine
Alloy MA 754	965	140	585	85	22	...	...	...	A mechanically alloyed nickel-chromium alloy with oxide dispersion strengthening. The strength, corrosion reistance, and microstructural stability of the alloy make it useful for gas turbine vanes and other extreme-service applications.

(continued)

(a) Cold rolled and solution annealed at 1230 °C (2250 °F). Sheet thickness, 1.2 to 1.6 mm (0.048 to 0.063 in.). (b) Annealed at 980 °C (1800 °F) for 30 min, air cooled, and aged at 760 °C (1400 °F) for 8 h, furnace cooled at a rate of 55 °C (100 °F)/h, heated to 620 °C (1150 °F) for 8 h, air cooled

Table 5 (continued)

Properties are for annealed sheet unless otherwise indicated.

Alloy	Ultimate tensile strength MPa	Ultimate tensile strength ksi	Yield strength (0.2% offset) MPa	Yield strength (0.2% offset) ksi	Elongation in 50 mm (2 in.), %	Elastic modulus (tension) GPa	Elastic modulus (tension) 10^6 psi	Hardness	Description/major applications
Nickel-chromium and nickel-chromium-iron alloys (continued)									
Alloy C-22	785	114	372	54	62	...	...	209 HB	A nickel-chromium-molybdenum alloy with outstanding resistance to pitting, crevice corrosion, and stress-corrosion cracking. Also exhibits high resistance to oxidizing media, including wet chlorine and mixtures containing nitric and oxidizing acids. Used for pollution control and pulp and paper equipment
Alloy C-276	790	115	355	52	61	205	29.8	90 HRB	A nickel-molybdenum-chromium alloy with an addition of tungsten having excellent corrosion resistance in a wide range of severe environments. The high molybdenum content makes the alloy especially resistant to pitting and crevice corrosion. The low carbon content minimizes carbide precipitation during welding to maintain corrosion resistance in as-welded structures. Used in pollution control, chemical processing, pulp and paper production, and waste treatment
Alloy G3	690	100	320	47	50	199	28.9	79 HRB	A nickel-chromium-iron alloy with additions of molybdenum and copper. It has good weldability and resistance to intergranular corrosion in the welded condition. The low carbon content helps prevent sensitization and consequent intergranular corrosion of weld heat-affected zones. Used for flue gas scrubbers and for handling phosphoric and sulfuric acids
Alloy HX (solution annealed)	793	115	358	52	45.5	205	29.7	90 HRB	A nickel-chromium-iron-molybdenum alloy with outstanding strength and oxidation resistance at temperatures up to 1200 °C (2200 °F). Matrix stiffening provided by the molybdenum content results in high strength in a solid-solution alloy having good fabrication characteristics. Used in gas turbines, industrial furnaces, heat-treating equipment, and nuclear engineering
Alloy S (solution annealed)	835	121	445	64.5	49	212	30.8	52 HRA	High-temperature alloy with excellent thermal stability, low thermal expansion, and oxidation resistance to 1095 °C (2000 °F). Retains strength and ductility after aging at temperatures of 425 to 870 °C (800 to 1600 °F). Developed for applications involving severely cyclical heating conditions. Used extensively as seal rings in gas turbine engines
Alloy W (solution annealed)	850	123	370	53.5	55	...	...	...	A solid-solution-strengthened alloy that was developed primarily for the welding of dissimilar alloys. It is available as straight cut-length wire for gas tungsten arc welding, layer-wound wire for gas metal arc welding, and coated electrodes for shielded metal arc welding. It has also been produced in the form of sheet and plate for structural applications up to 760 °C (1400 °F).
Alloy X (solution annealed)	785	114	360	52.5	43	196	28.5	89 HRB	A nickel-chromium-iron-molybdenum alloy that possesses an exceptional combination of oxidation resistance, fabricability, and high-temperature strength. It has also been found to be exceptionally resistant to stress-corrosion cracking in petrochemical applications. Exhibits good ductility after prolonged exposure at temperatures of 650, 760, and 870 °C (1200, 1400, and 1600 °F) for 16 000 h
Iron-nickel-chromium alloys									
Alloy 556	815	118.1	410	59.5	47.7	205	29.7	91 HB	An iron-nickel-chromium-cobalt alloy that combines effective resistance to sulfidizing, carburizing, and chlorine-bearing environments at high temperatures with good oxidation resistance, good fabricability, and excellent high-temperature strength. It has also been found to resist corrosion by molten salts and is resistant to corrosion from molten zinc. Used for waste incinerator, chemical process, and pulp and paper mill equipment
Alloy 800	600	87	295	43	44	193	28	138 HB	An iron-nickel-chromium alloy with good strength and excellent resistance to oxidation and carburization in high-temperature atmospheres. It also resists corrosion by many aqueous environments. The alloy maintains a stable, austenitic structure during prolonged exposure to high temperatures. Used for process piping, heat exchangers, carburizing equipment, heating-element sheathing, and nuclear steam-generator tubing

(continued)

(a) Cold rolled and solution annealed at 1230 °C (2250 °F). Sheet thickness, 1.2 to 1.6 mm (0.048 to 0.063 in.). (b) Annealed at 980 °C (1800 °F) for 30 min, air cooled, and aged at 760 °C (1400 °F) for 8 h, furnace cooled at a rate of 55 °C (100 °F)/h, heated to 620 °C (1150 °F) for 8 h, air cooled

Table 5 (continued)

Properties are for annealed sheet unless otherwise indicated.

Alloy	Ultimate tensile strength MPa	Ultimate tensile strength ksi	Yield strength (0.2% offset) MPa	Yield strength (0.2% offset) ksi	Elongation in 50 mm (2 in.), %	Elastic modulus (tension) GPa	Elastic modulus (tension) 10^6 psi	Hardness	Description/major applications
Iron-nickel-chromium alloys (continued)									
Alloy 800HT	See Alloy 800.								An iron-nickel-chromium alloy having the same basic composition as Alloy 800 but with significantly higher creep-rupture strength. The higher strength results from close control of the carbon, aluminum, and titanium contents in conjunction with a high-temperature anneal. Used in chemical and petroleum processing, in power plants for superheater and reheater tubing, in industrial furnaces, and for heat treating equipment.
Alloy 825	690	100	310	45	45	206	29.8	...	An iron-nickel-chromium alloy with additions of molybdenum and copper. It has excellent resistance to both reducing and oxidizing acids, to stress-corrosion cracking, and to localized attack such as pitting and crevice corrosion. The alloy is especially resistant to sulfuric and phosphoric acids. Used for chemical processing, pollution-control equipment, oil and gas well piping, nuclear fuel reprocessing, acid production, and pickling equipment
Alloy 925 (b)	1210	176	815	118	24	...	...	36.5 HRC	A precipitation-hardenable iron-nickel-chromium alloy with additions of molybdenum and copper. The alloy has outstanding resistance to general corrosion, pitting, crevice corrosion, and stress-corrosion cracking in many aqueous environments, including those containing sulfides and chlorides. Used for surface and down-hole hardware in sour gas wells and for oil production equipment
20Cb3	550	80	240	35	30	...	...	90 HRB	A high-nickel austenitic stainless steel with excellent resistance to chemicals containing chlorides and sulfuric, phosphoric, and nitric acids. Resists pitting, crevice corrosion, and intergranular attack used for tanks, piping, heat exchangers, pumps, valves, and other chemical process equipment
20Mo-4	615	89	262	38	41	186	27	80 HRB	Alloy is designed for applications requiring improved resistance to pitting and crevice corrosion. It should be considered for acid environments where pitting and crevice corrosion problems are encountered. Applications include heat exchangers, chemical process piping and equipment, mixing tanks, and metal-cleaning and pickling tanks
20Mo-6	607	88	275	40	50	186	27	...	An austenitic stainless steel that is resistant to corrosion in hot chloride environments with low pH. It has good resistance to pitting, crevice corrosion, and stress-corrosion cracking in chloride environments. It is also resistant to oxidizing media. Applications include commercial fume scrubbers, offshore platforms, and equipment for pulp and paper mills.
Controlled-expansion alloys									
Alloy 902 (precipitation hardened)	1210	175	760	110	25	...	...	...	A nickel-iron-chromium alloy made precipitation hardenable by additions of aluminum and titanium. The titanium content also helps provide a controllable thermoelastic coefficient, which is the outstanding characteristic of the alloy. The alloy can be processed to have a constant modulus of elasticity at temperatures from −45 to 65 °C (−50 to 150 °F). Used for precision springs, mechanical resonators, and other precision elastic components
Alloy 903 (precipitation hardened)	1310	190	1100	160	14	...	...	...	A nickel-iron-cobalt alloy with additions of niobium, titanium, and aluminum for precipitation hardening. The alloy combines high strength with a low and constant coefficient of thermal expansion at temperatures up to about 430 °C (800 °F). It also has a constant modulus of elasticity and is highly resistant to thermal fatigue and thermal shock. Used in gas turbines for rings and casings
Alloy 907	See Alloy 903.								A nickel-iron-cobalt alloy with additions of niobium and titanium for precipitation hardening. It has the low coefficient of expansion and high strength of Alloy 903 but with improved notch-rupture properties at elevated temperatures. Used for components of gas turbines, including seals, shafts, and casings
Alloy 909 (precipitation hardened)	1275	185	1035	150	15	159	23	...	A nickel-iron-cobalt alloy with a silicon addition and containing niobium and titanium for precipitation hardening. It is similar to Alloys 903 and 907 in that it has low thermal expansion and high strength. However, the silicon addition results in improved notch-rupture and tensile properties that are achieved with less-restrictive processing and significantly shorter heat treatments. Used for gas turbine casings, shrouds, vanes, and shafts

(a) Cold rolled and solution annealed at 1230 °C (2250 °F). Sheet thickness, 1.2 to 1.6 mm (0.048 to 0.063 in.). (b) Annealed at 980 °C (1800 °F) for 30 min, air cooled, and aged at 760 °C (1400 °F) for 8 h, furnace cooled at a rate of 55 °C (100 °F)/h, heated to 620 °C (1150 °F) for 8 h, air cooled

Table 6 Room-temperature physical properties of selected nickel-base alloys

Properties are for annealed sheet unless otherwise indicated.

Alloy	Density, g/cm^3	Melting point/range		Specific heat		Average coefficient of thermal expansion		Thermal conductivity		Electrical resistivity, nΩ · m	Curie temperature	
		°C	°F	J/kg · K	Btu/lb · °F	µm/m · K	µin./in. · °F	W/m · K	Btu · in./ft² · h · °F		°C	°F
Commercially pure and low-alloy nickels												
Nickel 200	8.89	1435–1445	2615–2635	456	0.109	13.3	7.4	70	485	95	360	680
Nickel 201	8.89	1435–1445	2615–2635	456	0.109	13.1	7.3	79.3	550	85	360	680
Nickel 205	8.89	1435–1445	2615–2635	456	0.109	13.3	7.4	75.0	520	95	360	680
Nickel 211	8.72	1427	2600	532	0.127	13.3	7.4	44.7	310	169	310	590
Nickel 212	8.86	1435–1445	2615–2635	430	0.103	12.9	7.2	44.0	305	109	...	...
Nickel 222	8.89	1435–1445	2615–2635	460	0.110	13.3	7.4	75	520	88	Ferromagnetic	
Nickel 270	8.89	1455	2650	460	0.110	13.3	7.4	86	595	75	Ferromagnetic	
Duranickel 301 (precipitation hardened)	8.25	1438	2620	435	0.104	13.0	7.2	23.8	165	424	16–50	60–120
Nickel-copper alloys												
Alloy 400	8.80	1300–1350	2370–2460	427	0.102	13.9	7.7	21.8	151	547	20–50	70–120
Alloy 401	8.89	...	...	...	...	13.7	7.6	19.2	133	489	<−196	<−320
Alloy R-405	8.80	1300–1350	2370–2460	427	0.102	13.7	7.6	21.8	151	510	20–50	70–120
Alloy 450	8.91	1170–1240	2140–2260	...	...	15.5	8.6	29.4	204	412	...	...
Alloy K-500 (precipitation hardened)	8.44	1315–1350	2400–2460	419	0.100	13.7	7.6	17.5	121	615	−134	−210
Nickel-chromium and nickel-chromium-iron alloys												
Alloy 230 (a)	8.83	1300–1370	2375–2500	397	0.095	12.6	7.0	8.9	62	1250	...	...
Alloy 600	8.47	1355–1413	2470–2575	444	0.106	13.3	7.4	14.9	103	1030	−124	−192
Alloy 601	8.11	1360–1411	2480–2571	448	0.107	13.75	7.6	11.2	78	1190	−196	−320
Alloy 617 (solution annealed)	8.36	1330–1380	2430–2510	419	0.100	11.6(b)	6.4(b)	13.6	94	1220	...	...
Alloy 625	8.44	1290–1350	2350–2460	410	0.098	12.8	7.1	9.8	68	1290	...	...
Alloy 690	8.19	1343–1377	2450–2510	450	0.107	14.06(b)	7.80(b)	13.5	93	1148	...	...
Alloy 718 (precipitation hardened)	8.19	1260–1336	2300–2437	435	0.104	13.0	7.2	11.4	79	1250	−112	−170
Alloy X750	8.28	1390–1430	2540–2600	431	0.103	12.6	7.0	12.0	83	1220	−125	−193
Alloy 751	8.22	1390–1430	2540–2600	431	0.103	12.6	7.0	12.0	83	1220	−125	−193
Alloy C-276	8.89	1325–1370	2415–2500	427	0.102	11.2(b)	6.2(b)	9.8	67.9	1300	...	...
Alloy G3	8.14	1260–1345	2300–2450	452	0.108	14.6	8.1	10.0	69	...	...	...
Alloy HX (solution annealed)	8.23	1260–1355	2300–2470	461	0.110	13.3	7.4	11.6	80.4	1160	...	...
Alloy S (solution annealed)	8.75	1335–1380	2435–2516	398(c)	0.095(c)	11.5	6.4	14.0(d)	97(d)	1280	...	...
Alloy X (solution annealed)	8.22	1260–1355	2300–2470	486	0.116	13.9	7.7	9.1	63	1180	...	...
Iron-nickel-chromium alloys												
Alloy 556	8.23	1330–1415	2425–2580	464	0.111	14.6	8.1	11.1	77	952	...	...
Alloy 800	7.94	1357–1385	2475–2525	460	0.110	14.4	7.9	11.5	80	989	−115	−175
Alloy 800HT	See Alloy 800.											
Alloy 825	8.14	1370–1400	2500–2550	440	0.105	14.0	7.8	11.1	77	1130	−196	−320
Alloy 925 (e)	8.14	1311–1366	2392–2490	435	0.104	13.2	7.32	...	...	1166	...	...
20Cb3	8.08	...	...	500	0.12	14.69	8.16	12.2(f)	84.6(f)	1082	...	...
20Mo-4	8.106	...	...	458	0.109	14.92(g)	8.29(g)	12.1(f)	83.9(f)	1056	...	...
20Mo-6	8.133	...	...	460	0.11	14.8(b)	8.22(b)	12.1(f)	83.9(f)	1082	...	...
Controlled-expansion alloys												
Alloy 902 (precipitation hardened)	8.05	1455–1480	2650–2700	500	0.12	7.6	4.2	12.1	83.9	1020	190	380
Alloy 903 (precipitation hardened)	8.25	1318–1393	2405–2539	435	0.104	7.65(h)	4.25(h)	16.7	116	610	415–470	780–880
Alloy 907 (precipitation hardened)	8.33	1335–1400	2440–2550	431	0.103	7.7(h)	4.3(h)	14.8	103	697	400–455	750–850
Alloy 909 (precipitation hardened)	8.30	1395–1430	2540–2610	427	0.102	7.7(h)	4.3(h)	14.8	103	728	400–455	750–850

(a) Cold rolled and solution annealed at 1230 °C (2250 °F). Sheet thickness, 1.2–1.6 mm (0.048–0.063 in.). (b) Average value at 25–100 °C (75–200 °F). (c) Average value at 0 °C (32 °F). (d) Average value at 200 °C (390 °F). (e) Annealed at 980 °C (1800 °F) for 30 min, air cooled, and aged at 760 °C (1400 °F) for 8 h, furnace cooled at a rate of 55 °C (100 °F)/h, heated to 620 °C (1150 °F) for 8 h, air cooled. (f) Average value at 50 °C (120 °F). (g) Average value at 25 to 200 °C (75 to 390 °F). (h) Average value at 25 to 425 °C (75 to 800 °F)

- Nimonic alloys (80Ni-20Cr + Ti/Al). The basic alloy for jet engine development (see Ref 12 for compositions and properties)

Some high-temperature variants include:

- Alloy 601. Lower nickel (61%) content with aluminum and silicon additions for improved oxidation and nitriding resistance
- Alloy X750. Aluminum and titanium additions for age hardening: originally used for skin of Bell X-1 experimental aircraft
- Alloy 718. Titanium and niobium additions to overcome strain-age cracking problems during welding and weld repair
- Alloy X (48Ni-22Cr-18Fe-9Mo + W). High-temperature flat-rolled product for aerospace applications
- Waspaloy (60Ni-19Cr-4Mo-3Ti-1.3Al). Proprietary alloy for jet engine applications

To achieve higher design strengths, dispersion-strengthened powder metallurgy alloys have been developed. Mechanical alloying represents the technology by which these materials can be made. In this process, a controlled mixture of alloy powder and about 1 vol% oxides (typically Y_2O_3) are charged into a high-energy ball mill. The metal particle size initially is in the 2 to 200 µm range, whereas the oxide particles are less than 10 µm in size. The milling operation, carried out dry, causes the superalloy particles to weld repeatedly to the oxide particles and then break apart. The resultant acicular or platelike powders are composites with extremely fine, homogeneous microstructures. Consolidation is carried out by placing the powders in steel cylinders, aligning the ends closed, and either extruding to bar or rolling to plate or sheet.

Alloy MA 754 (Ni-20Cr-1.0Fe-0.6Y_2O_3) was the initial alloy produced by this new technology and specifically developed for gas turbine engine vanes. It is superior to cast cobalt-base alloys in dimensional stability, thermal fatigue, and strength at 1100 °C (2010 °F). More detailed information on mechanically alloyed materials can be found in the article "Dispersion-Strengthened Nickel-Base and Iron-Base Alloys" in this Volume and Volume 1 of the 10th Edition of *Metals Handbook* (see Ref 12).

Some corrosion-resistant variants in the Ni-Cr-Fe system include:

- Alloy 625. The addition of 9% Mo plus 3% Nb offers both high-temperature and wet corrosion resistance; resists pitting and crevice corrosion
- Alloy G3/G30 (Ni-22Cr-19Fe-7Mo-2Cu). The increased molybdenum content in these alloys offers improved pitting and crevice corrosion resistance
- Alloy C-22 (Ni-22Cr-6Fe-14Mo-4W). Superior corrosion resistance in oxidizing acid chlorides, wet chlorine, and other severe corrosive environments
- Alloy C-276 (17% Mo plus 3.7W). Good seawater corrosion resistance and excellent pitting and crevice corrosion resistance
- Alloy 690 (27% Cr addition). Excellent oxidation and nitric acid resistance; specified for nuclear waste disposal by the vitreous encapsulation method

Throughout most of the nuclear power era, Alloy 600 has been the preferred alloy for all steam generator tubing, both in commercial and military reactor design. With the recognition of the potential for Alloy 600 to suffer stress-corrosion cracking in superheated pure waters, Alloy 690 has become the alternate material for future replacement steam generator tubing and new designs.

All of the high-temperature variants and the Nimonic series of alloys find application in many of the hot sections of aircraft gas turbine engines—blades, turbine rings, fasteners, and so forth. The alloys are also used for thermal processing equipment, much of which is used in annealing and heat treating of the age-hardenable aerospace alloys.

The molybdenum-containing systems are notable for their multipurpose capabilities in chemical processing, pulp and paper production (bleachers and washers), and pollution-control equipment (flue gas desulfurization, scrubbers, precipitators). These alloys also find use in oil country tubular goods for handling highly corrosive sour gases. In critical applications, this series of alloys has found use in various instrumentation applications (sensors and safety diaphragms). Bellows-quality material is used in honeycomb applications and for improved low cycle fatigue capabilities.

As described earlier in this article, a series of alloys has also been designed for electrical resistance heating applications. These nickel-chromium alloys offer oxidation and thermal shock resistance in on-off service, with the electrical resistance characteristics being realized through variations of chemistry, wire diameters, and processing history (grain size effects). Some alloys of commercial importance include:

- 80Ni-20Cr (plus 1.5 Si). Operating conditions up to 1150 to 1175 °C (2100 to 2150 °F). High electrical resistivity characteristics
- 60Ni-24Fe-16Cr. Suitable for less exacting applications and operating conditions up to 950 °C (1750 °F); for example, clothes dryer elements
- 35Ni-45Fe-20Cr. Cost effective consideration for operating conditions up to 1065 °C (1950 °F). However, large temperature coefficient of resistance must be taken into effect during element design

A common application today is the spiral-wound electrical element contained within a metal sheath, for use as an appliance heating element. However, these alloys have also been used extensively for heating in other applications such as domestic fan heaters and thermal storage units. They exhibit good resistance to oxidation up to their recommended maximum operating temperatures.

Iron-Nickel-Chromium Alloys. The development of this series of alloys took place in the early 1950s with their introduction during the Korean War period. Incoloy Alloy 800 was designed as a leaner nickel version of the nickel-chromium series of materials. It offered good oxidation resistance and was introduced as a sheathing material for electric stove elements. Since that time, even leaner nickel alloys have been designed for selected sheathing applications.

This series of alloys has also found extensive use in the high-temperature petrochemical environments, where sulfur-containing feedstocks (naphtha and heavy oils) are cracked into component distillate parts. Not only were they resistant to chloride-ion stress-corrosion cracking, but they also offered resistance to polythionic acid cracking. Some alloys of commercial importance include:

- Alloy 800 (Fe-32Ni-21Cr). The basic alloy in the Fe-Ni-Cr system; resistant to oxidation and carburization at elevated temperatures
- Alloy 800H. Modification with controlled carbon (0.05 to 0.10%) and grain size (>ASTM 5), to optimize stress-rupture properties
- Alloy 800HT. Similar to 800H with further modification to combined titanium and aluminum levels (0.85 to 1.2%) to ensure optimum high-temperature properties
- Alloy 801. Increased titanium content (0.75 to 1.5%); exceptional resistance to polythionic acid cracking
- Alloy 802. High-carbon version (0.2 to 0.5%) for improved strength at high temperatures
- Alloy 825 (Fe-42Ni-21.5Cr-2Cu). Stabilized with titanium addition (0.6 to 1.2%). Also contains molybdenum (3%) for pitting resistance in aqueous corrosion applications. Copper content bestows resistance to sulfuric acid
- Alloy 925. Addition of titanium and aluminum to 825 composition for strengthening through age hardening

The 800 alloy series offers excellent strength at elevated temperature (creep and stress rupture). These alloys are consequently useful for catalytic cracking tubes, pigtails, and reformer tubes. For optimum performance in the new millisecond design of crackers, Alloy 800HT internally finned tubing is being used. This offers greater heat exchanger wall area. For environments where polythionic acid can form during downtime periods, Alloy 801 offers optimum resistance. Alloy 802 offers competitive high-temperature strength capabilities because of its high carbon level and has been used extensively in sinter deck plate applications for handling abrasive, high-temperature environments.

Some corrosion variants in the Fe-Ni-Cr system include:

- 20Cb3 (Fe-35Ni-20Cr-3.5Cu-2.5Mo + Nb). This alloy was developed for the handling of sulfuric acid environments
- 20Mo-4 and 20Mo-6 (Fe-36Ni-23Cr-5Mo + Cu). Increased corrosion resistance in pulp and paper industry environments

The corrosion-resistant series of alloys (825, 925, and 20Cb3) all contain molybdenum for enhanced corrosion resistance. This series of alloys has excellent resistance to sulfuric acid. The alloys have found extensive use as tubing and plate for production of fertilizer and associated products.

Alloy 925 has found use for downhole components in sour gas wells around the world. It has been forged into block master valves for well-head applications, along with associated Christmas tree well-head components. Alloy 825 is used for downhole tubular components where hydrogen sulfide, carbon dioxide, and sodium chlo-

Table 7 Compositions and uses of coated nickel-base welding electrodes

AWS class or proprietary name	Composition, wt%(a) Ni + Co	C	Mn	Fe	S	Si	Cu	Cr	Al	Ti	P	Other	Major uses
ENi-1	92.0 min	0.10	0.75	0.75	0.02	1.25	0.25	· · ·	1.0	1.0–4.0	0.03	(b)	Joining Nickel 200 and Nickel 201; the clad side of nickel-clad steel; joining steels to nickel alloys
ENiCu-7	62.0–69.0	0.15	4.00	2.5	0.015	1.5	bal	· · ·	0.75	1.0	0.02	(b)	Joining Alloy 400 to itself, to low-alloy and carbon steels, to copper and copper-nickel alloys; surfacing of steels
ENiCrFe-1	62.0 min	0.08	3.5	11.0	0.015	0.75	0.50	13.0–17.0	· · ·	· · ·	0.03	1.5–4.0 Nb + Ta, (b)	Joining Alloy 600; Alloy 330
ENiCrFe-3	59.0 min	0.10	5.0–9.5	10.0	0.015	1.0	0.50	13.0–17.0	· · ·	1.0	0.03	1.0–2.5 Nb + Ta, (b)	Joining Alloys 600 and 601; surfacing of steel; dissimilar combinations of steels and nickel alloys
ENiCrMo-3	55.0 min	0.10	1.0	7.0	0.02	0.75	0.50	20.0–23.0	· · ·	· · ·	0.03	3.15–4.15 Nb + Ta, 8.0–10.0 Mo, (b)	Joining Alloys 625 and 601; surfacing of steel; dissimilar combinations of steels and nickel alloys
ENiCrCoMo-1	bal	0.05	1.0–2.0	18.0–21.0	0.03	1.0	1.5–2.5	21.0–23.5	· · ·	· · ·	0.04	12.0 Co, 9.0 Co	Joining Alloy 617; Alloy 800HT (for temperatures above 760 °C, or 1400 °F); dissimilar combinations of steels and nickel alloys
ENiCrMo-4	bal	0.02	1.0	4.0–7.0	0.03	0.2	0.50	14.5–16.5	· · ·	· · ·	0.04	2.50 Co, 15.0–17.0 Mo, 0.35 V, 3.0–4.5 W, (b)	Joining Alloy C-276; other pit-resistant alloys; surfacing of steels
ENiCrMo-2	bal	0.05–0.15	1.0	17.0–20.0	0.03	1.0	0.50	20.5–23.0	· · ·	· · ·	0.04	8.0–10.0 Mo, 0.20–1.0 W, (b)	Joining Alloys 800 and 800HT (for temperatures below 675 °C, or 1250 °F); dissimilar combinations of steels and nickel alloys; 9% Ni steel; surfacing of steels
Incoloy 135	35.40 min	0.08	1.25–1.50	bal	0.03	0.75	1.0–2.5	26.5–30.5	· · ·	· · ·	· · ·	6.0–8.0 Mo, 5.0 Co, 0.50 Nb + Ti, 1.5 W, (b)	Joining Alloy 825
ENiCrMo-9	bal	0.02	1.0	18.0–21.0	0.03	1.0	1.5–2.5	21.0–23.5	· · ·	· · ·	0.04	(b)	Joining Alloys G3 and G30; other pit-resistant alloys; dissimilar combinations of steels and nickel alloys; surfacing of steels
ENiCI	95.0 min	1.0	0.20	3.0	0.005	0.70	0.10	· · ·	· · ·	· · ·	· · ·	· · ·	Joining cast irons, especially for thin sections
ENiFeCI	53.0 min	1.20	0.30	45.0	0.005	0.70	0.10	· · ·	· · ·	· · ·	· · ·	· · ·	Joining cast irons, especially thick sections and high-phosphorus irons

(a) Single values are minimum values unless otherwise indicated. (b) Also contains 0.50% total other elements (unspecified)

ride (salt) are at elevated temperatures. For very aggressive corrosion sour gas environments, the nickel-chromium-iron-molybdenum series of alloys has to be considered.

Controlled-expansion alloys include alloys in both the Fe-Ni-Cr and Fe-Ni-Co series. Some alloys of commercial importance include:

- Alloy 902 (Fe-42Ni-5Cr with 2.2 to 2.75% Ti and 0.3 to 0.8% Al). This is an alloy with a controllable thermoelastic coefficient
- Alloys 903, 907, 909 (42Fe-38Ni-13Co with varying aging elements such as niobium, titanium, and aluminum). These alloys offer high strength and low coefficient of thermal expansion

The 900 alloy series offers very unusual characteristics and properties. Alloys 903, 907, and 909 were all designed to provide high strength and low coefficient of thermal expansion for applications up to 650 °C (1200 °F). These advantages have been used by the aerospace industry to design near net-shape components and to provide closer clearance between the tips of rotating turbine blades and retainer rings. This allows for greater power output and fuel efficiencies. These high-strength alloys also allow increased strength-to-weight ratios in engine design, resulting in weight savings. Alloy 909 offers attractive properties for rocket engine thrust chambers, ordnance hardware, springs, gage blocks, and instrumentation.

Alloy 902 is used extensively in spring, pressure sensor, and instrumentation applications for its thermoelastic coefficient. It is particularly useful in the design of pressure-sensing devices, and can be used in both low-frequency components (Bourdon tubing, aneroid capsules, and springs) and high-frequency components (tuning forks, mechanical filters, and vibrating reeds).

Nickel-Iron Low-Expansion Alloys. This series of alloys plays a very important role in both the lamp industry and electronics, where glass-to-metal seals in encapsulated components are important. The nickel alloys are chosen for a variety of reasons:

- They have readily reducible oxides and offer a capacity for easy outgassing
- They are readily fabricated and retain structural integrity at temperature
- They offer good thermal conductivity and low electrical resistance

Some alloys of commercial importance include:

- Invar (Fe-36Ni). This alloy has the lowest thermal expansion of any metal from ambient to 230 °C (450 °F)
- Alloy 42 (Fe-42Ni). This alloy has the closest thermal expansion match to alumina, beryllia, and vitreous glass
- Alloy 426. Additions of 6% Cr are added to this alloy for vacuum-tight sealing applications
- Alloy 52 (Fe-51.5Ni). This alloy has a thermal expansion that closely matches vitreous potash-soda-lead glass

Alloy 42, along with Kovar, an Fe-29Ni-17Co alloy, formed the core of the lead frame market in servicing the computer industry in its early days. They offered, and still offer, high-integrity electronic components that are required for military packages. Alloy 42, aluminum striped, remains in prime demand for ceramic dual-in-line packages (CERDIP) components, while

Table 8 Compositions and uses of nickel-base filler metals

AWS class or proprietary name	Composition, wt%(a)											
	Ni + Co	C	Mn	Fe	S	Si	Cu	Cr	Al	Ti	P	Other
ERNi-1	93.0 min	0.15	1.0	1.0	0.015	0.75	0.25	...	1.5	2.0–3.5	0.03	(b)
ERNiCu-7	62.0–69.0	0.15	4.0	2.5	0.015	1.25	bal	...	1.25	1.5–3.0	0.02	(b)
ERNiCr-3	67.0 min	0.10	2.5–3.5	3.0	0.015	0.50	0.50	18.0–22.0	...	0.75	0.03	2.0–3.0 Nb + Ta, (b)
Inconel 601	58.0–63.0	0.10	1.0	bal	0.015	0.50	1.0	21.0–25.0	1.0–1.7	...	...	...
ERNiCrCoMo-1	bal	0.05–0.15	1.0	3.0	0.015	1.0	0.50	20.0–24.0	0.80–1.50	0.60	0.03	10.0–15.0 Co, 8.0–10.0 Mo, (b)
ERNiCrMo-3	58.0 min	0.10	0.50	5.0	0.015	0.50	0.50	20.0–23.0	0.40	0.40	0.02	3.15–4.15 Nb + Ta, 8.0–10.0 Mo, (b)
ERNiFeCr-2	50.0–55.0	0.08	0.35	bal	0.015	0.35	0.30	17.0–21.0	0.20–0.80	0.65–1.15	0.015	4.75–5.50 Nb + Ta, 2.80–3.30 Mo, (b)
ERNiCrMo-4	bal	0.02	1.0	4.0–7.0	0.03	0.08	0.50	14.5–16.5	...	...	0.04	2.50 Co, 15.0–17.0 Mo, 0.35 V, 3.0–4.5 W, (b)
ERNiCrMo-2	bal	0.05–0.15	1.0	17.0–20.0	0.03	1.0	0.50	20.5–23.0	...	...	0.04	0.50–2.50 Co, 8.0–10.0 Mo, 0.20–1.0 W, (b)
NC 80/20 Filler Metal	bal	0.26	1.2	0.5	...	0.5	0.2	18.0–21.0	...	...	...	1.0 Co
ERNiFeCr-1	38.0–46.0	0.05	1.0	22.0	0.03	0.50	1.5–3.0	19.5–23.5	0.20	0.60–1.20	0.03	2.5–3.5 Mo
NI-ROD Filler Metal 44	44 min	0.3	11	45	...	...	...	...	...	...	...	...
NI-ROD FC 55 Cored Wire	50.0 min	1.0	4.2	44.0	...	0.6	...	...	...	...	...	...

AWS class or proprietary name	Major uses
ERNi-1	Joining Nickel 200 and 201; dissimilar combinations of nickel alloys and steels; surfacing of steels
ERNiCu-7	Joining Alloys 400, R-405, and K-500; surfacing of steel
ERNiCr-3	Joining Alloys 600 and 601; Alloys 800 and 800HT (for temperatures below 760 °C, or 1400 °F); Alloy 330; dissimilar combinations of steels and nickel alloys; surfacing of steels
Inconel 601	Joining Alloy 601
ERNiCrCoMo-1	Joining Alloy 617; Alloy 800HT (for temperatures above 760 °C, or 1400 °F); dissimilar combinations of high-temperature alloys
ERNiCrMo-3	Joining Alloys 625 and 601; pit-resistant alloys; dissimilar combinations of steels and nickel alloys; surfacing of steels
ERNiFeCr-2	Joining Alloys 718 and X750
ERNiCrMo-4	Joining Alloy C-276; other pit-resistant alloys; surfacing of steels
ERNiCrMo-2	Joining Alloy HX
NC 80/20 Filler Metal	Joining electrical-resistance alloys
ERNiFeCr-1	Joining Alloy 825
NI-ROD Filler Metal 44	Joining cast irons, especially robotic and automatic welding
NI-ROD FC 55 Cored Wire	Automatic welding of cast irons by flux-cored arc welding

(a) Single values are maximum values unless otherwise indicated. (b) Also contains 0.50% total other elements (unspecified)

Kovar is used in lids and closures for hybrid electronic packages today.

Invar (Alloy 36) also finds use in the electronic industry for printed circuit boards. Used in conjunction with copper, it offers a clad composite product that can be designed with controlled-expansion characteristics depending upon the proportions of each alloy present in the composite. The copper allows excellent thermal and electrical transference, while the Invar offers very low expansion to provide constraint to the copper, which otherwise wants to expand rapidly. The low expansion of Invar with other alloys of differing expansion can provide a series of thermomechanical control and switchgear devices. Other applications for this alloy have been found in cryogenic tanks for handling liquid gases and a variety of instrumentation and fixturing applications.

The chromium-bearing grades find a major use in television shadow mask screens and for headlight ferrules, where vacuum-tight sealing is required for performance. In less demanding glass sealing applications, copper-coated Alloy 42 (Dumet) is used for traditional incandescent lamp production. More detailed information can be found in the article "Low-Expansion Alloys" in this Volume.

Soft Magnetic Alloys. The nickel-iron alloys also offer an interesting set of magnetic permeability properties, which have played an important part in switchgear and for direct current (dc) motor and generator designs.

As described earlier in this article, such nickel-iron alloys are known as "soft" magnetic alloys as compared with the "hard" iron and silicon-iron materials (differences in magnetic materials are described in the articles "Magnetically Soft Materials" and "Permanent Magnet Materials" in this Volume). The magnetically soft materials found initial major applications in telegraph applications and later in telephone equipment.

With high initial and maximum permeability, these products offer excellent magnetic shielding. The lower-nickel alloys (~50% Ni) offer fairly constant permeabilities over a narrower range of flux densities and have found their primary use in rotors, armatures, and low level transformers. High-nickel alloys (~77% Ni) are useful for applications in which power requirements must be minimized such as transformers, inductors, magnetic amplifiers, magnetic shields, tape recorder heads, and memory storage devices.

Welding Alloys. Welding products for nickel alloys have similar compositions to the base metals, although additions of aluminum, titanium, magnesium, and other elements are made to the filler metals and welding electrodes to ensure proper deoxidation of the molten weld pool and to overcome any hot-short cracking and malleability problems.

Nickel-base welding products are also used for welding of dissimilar materials, for example, stainless steels to cast iron, as in a drive shaft and yoke for automotive assemblies. In other cases, overmatching filler materials (high-molybdenum products) can be used to ensure adequate corrosion resistance in welds where matching-composition electrodes may not provide enough molybdenum in the final weld deposit to resist pitting and crevice corrosion. Alloy 625 welding materials, with 9% Mo, are widely used for joining many corrosion-resistant materials with less molybdenum.

Pure nickel and nickel-iron welding electrodes are established products for welding of cast iron, especially ductile iron. The nickel-iron composition is more suitable for welding the higher-sulfur grades of cast iron and for repair welding. Nickel-iron filler metals and flux-core wire are also used, especially for automatic, single-pass welding.

Tables 7 and 8 provide compositions of nickel-base welding electrodes and consumables, and their major uses.

Welding procedures for nickel alloys are similar to those used for stainless steel. However, the nickel-base molten weld deposit is less fluid than that for stainless steel, so joint configurations and joint angles must be opened up to accommodate manipulation of the weld rod or torch during placement of the weld metal in the joint area. Cleaning of the base metal is also important before welding the nickel-base alloys.

These materials can be joined by any one of the basic welding processes—gas-tungsten arc (GTAW), metal gas arc (GMAW), shielded metal-arc (SMAW), brazing, and soldering. In more sophisticated applications, laser welding, electron beam welding, and ultrasonic welding have all been used. The choice of welding processes should be based upon the thickness of the metal to be joined, design of the unit, design of the joint, position in which the weld is to be made, need for jigs and fixtures, service conditions and corrosive environments, and any special shop or field construction conditions.

REFERENCES

1. *The Romance of Nickel*, International Nickel Company, 1957, p 61
2. J.F. Thompson and N. Beasley, *For the Years to Come*, G.P. Putnam's Sons, 1960, p 25-27
3. C.R. Hayward, *Outline of Metallurgical Practice*, 3rd ed., Van Nostrand, 1952, p 299
4. J.R. Boldt, Jr. and P. Queneau, *The Winning of Nickel*, Van Nostrand, 1967, p 359
5. A.L. Marsh, British Patent 2129, 1906
6. R.D. Gray, *Stellite(R), A History of the Haynes Stellite Company 1912-1972*, Cabot Corporation, 1981, p 20
7. F.B. Howard-White, *Nickel, An Historical Review*, Van Nostrand, 1963, p 169
8. W. Betteridge, *The NIMONIC Alloys*, Edward Arnold, 1959, p 6
9. G.L. Erickson, Polycrystalline Cast Superalloys, in *Properties and Selection: Irons, Steels, and High-Performance Alloys*, Vol 1, 10th ed., *Metals Handbook*, ASM INTERNATIONAL, 1990, p 981-994
10. K. Harris, G.L. Erickson, and R.E. Schwer, Directionally Solidified and Single-Crystal Superalloys, in *Properties and Selection: Irons, Steels, and High-Performance Alloys*, Vol 1, 10th ed., *Metals Handbook*, ASM INTERNATIONAL, 1990, p 995-1006
11. G.F. Vander Voort and H.M. James, Wrought Heat-Resistant Alloys, in *Metallography and Microstructures*, Vol 9, 9th ed., *Metals Handbook*, American Society for Metals, 1985, p 305-329
12. N.S. Stoloff, Wrought and P/M Superalloys, in *Properties and Selection: Irons, Steels, and High-Performance Alloys*, Vol 1, 10th ed., *Metals Handbook*, ASM INTERNATIONAL, 1990, p 950-977
13. R.M.N. Pelloux and N.J. Grant, *Trans. AIME*, Vol 218, 1960, p 232
14. E.R. Parker and T.H. Hazlett, Principles of Solution Hardening, in *Relation of Properties to Microstructures*, American Society for Metals, 1954, p 30
15. R.F. Decker, "Strengthening Mechanisms in Nickel-Base Superalloys," Paper presented at the Steel Strengthening Mechanism Symposium (Zurich), May 1969
16. B.C. Syrett *et al.*, Corrosion in Fossil Fuel Power Plants, in *Corrosion*, Vol 13, 9th ed., *Metals Handbook*, ASM INTERNATIONAL, 1987, p 985-1010
17. S. Shaw, Isothermal and Hot-Die Forging, in *Forming and Forging*, Vol 14, 9th ed., *Metals Handbook*, ASM INTERNATIONAL, 1988, p 150-157
18. H.J. Mueller, Tarnish and Corrosion of Dental Alloys, in *Corrosion*, Vol 13, 9th ed., *Metals Handbook*, ASM INTERNATIONAL, 1987, p 1336-1366
19. A.C. Fraker, Corrosion of Metallic Implants and Prosthetic Devices, in *Corrosion*, Vol 13, 9th ed., *Metals Handbook*, ASM INTERNATIONAL, 1987, p 1324-1335
20. L.J. Korb, Corrosion of Manned Spacecraft, in *Corrosion*, Vol 13, 9th ed., *Metals Handbook*, ASM INTERNATIONAL, 1987, p 1058-1100
21. G.Y. Lai and C.R. Patriarca, Corrosion of Heat-Treating Furnace Accessories, in *Corrosion*, Vol 13, 9th ed., *Metals Handbook*, ASM INTERNATIONAL, 1987, p 1310-1315
22. J.C. Danko *et al.*, Corrosion in the Nuclear Power Industry, in *Corrosion*, Vol 13, 9th ed., *Metals Handbook*, ASM INTERNATIONAL, 1987, p 927-984
23. T.F. Degnan *et al.*, Corrosion in the Chemical Processing Industry, in *Corrosion*, Vol 13, 9th ed., *Metals Handbook*, ASM INTERNATIONAL, 1987, p 1134-1185
24. J.E. Donham *et al.*, Corrosion in Petroleum Production Operations, in *Corrosion*, Vol 13, 9th ed., *Metals Handbook*, ASM INTERNATIONAL, 1987, p 1232-1261
25. J. Gutzeit, R.D. Merrick, and L.R. Scharfstein, Corrosion in Petroleum Refining and Petrochemical Operations, in *Corrosion*, Vol 13, 9th ed., *Metals Handbook*, ASM INTERNATIONAL, 1987, p 1262-1287
26. W.J. Gilbert and R.J. Chironna, Corrosion of Emission-Control Equipment, in *Corrosion*, Vol 13, 9th ed., *Metals Handbook*, ASM INTERNATIONAL, 1987, p 1367-1370
27. A. Garner *et al.*, Corrosion in the Pulp and Paper Industry, in *Corrosion*, Vol 13, 9th ed., *Metals Handbook*, ASM INTERNATIONAL, 1987, p 1186-1220
28. A.I. Asphahani *et al.*, Corrosion of Nickel-Base Alloys, in *Corrosion*, Vol 13, 9th ed., *Metals Handbook*, ASM INTERNATIONAL, 1987, p 641-657
29. D.O. Sprowls, Evaluation of Stress-Corrosion Cracking, in *Corrosion*, Vol 13, 9th ed., *Metals Handbook*, ASM INTERNATIONAL, 1987, p 245-290
30. H.R. Copson, Effect of Composition on Stress Corrosion Cracking of Some Alloys Containing Nickel, in *Physical Metallurgy of Stress Corrosion Fracture*, T.N. Rhodin, Ed., Interscience, 1959, p 247-272
31. R.M. Kain, Evaluation of Crevice Corrosion, in *Corrosion*, Vol 13, 9th ed., *Metals Handbook*, ASM INTERNATIONAL, 1987, p 303-313

Cobalt and Cobalt Alloys

Paul Crook, Haynes International, Inc.

COBALT is a tough silver-gray magnetic metal that resembles iron and nickel in appearance and in some properties. Cobalt is useful in applications that utilize its magnetic properties, corrosion resistance, wear resistance, and/or its strength at elevated temperatures. Some cobalt-base alloys are also biocompatible, which has prompted their use as orthopedic implants (see the article "Corrosion of Cobalt-Base Alloys" in Volume 13 of the 9th Edition of *Metals Handbook*).

This article provides a general overview of cobalt-base alloys as wear-resistant, corrosion-resistant, and/or heat-resistant materials. Particular emphasis is placed on cobalt-base alloys for wear resistance, because this is the single largest application area of cobalt-base alloys. In heat-resistant applications, cobalt is more widely used as an alloying element in nickel-base alloys with cobalt tonnages in excess of those used in cobalt-base heat-resistant alloys.

Elemental Cobalt

Mining and Processing. Much of cobalt today derives from copper and copper-nickel rich sulfide deposits in Zaire and Zambia. Other countries where the mining of cobalt is significant include Canada and Finland.

The largest deposits, in the Shaba Province of Zaire, are mined using both open pit and underground methods. Here the ore is subjected to crushing, grinding, and flotation, prior to a magnetic concentrating process. This concentrate is then leached in sulfuric acid and the cobalt and copper extracted by electrolysis (Ref 1).

Alternative future sources of cobalt include the manganese-rich nodules discovered on the floor of the Pacific Ocean.

Physical Properties. With an atomic number of 27, cobalt falls between iron and nickel on the periodic table. The density of cobalt is 8.8 g/cm^3 similar to that of nickel. Its thermal expansion coefficient (Fig. 1) lies between those of iron and nickel. At temperatures below 417 °C (783 °F), cobalt exhibits a hexagonal close-packed structure. Between 417 °C (783 °F) and its melting point of 1493 °C (2719 °F), cobalt has a face-centered cubic structure.

Typical tensile properties of pure cobalt are given in Table 1. The elastic modulus of cobalt is about 210 GPa (30 × 10^6 psi) in tension and about 183 GPa (26.5 × 10^6 psi) in compression. Electrical and magnetic properties of pure cobalt are summarized in Table 2. More information on the properties of cobalt is contained in the article "Properties of Pure Metals" in this Volume.

Uses of Cobalt. As well as forming the basis of the cobalt-base alloys discussed in this article, cobalt is also an important ingredient in other materials:

- Paint pigments
- Nickel-base superalloys
- Cemented carbides and tool steels
- Magnetic materials
- Artificial γ-ray sources

Of these applications, paint pigment represents the single largest use of cobalt.

In the nickel-base superalloys, cobalt (which is present typically in the range 10 to 15 wt%) provides solid-solution strengthening and decreases the solubility of aluminum and titanium, thereby increasing the volume fraction of gamma prime (γ′) precipitate. Cobalt in nickel-base superalloys also reduces the tendency for grain boundary carbide precipitation, thus reducing chromium depletion at the grain boundaries (Ref 2).

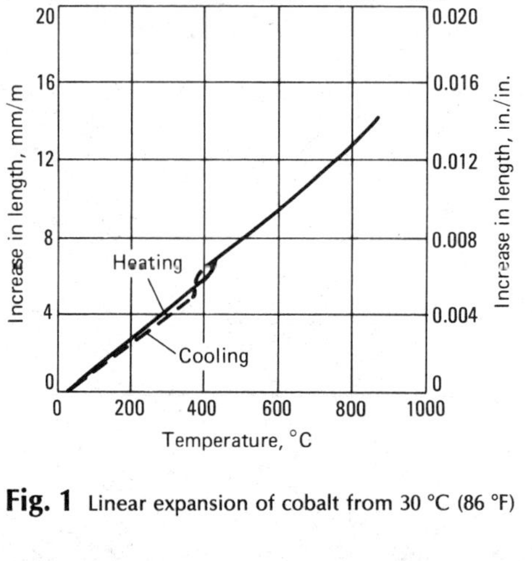

Fig. 1 Linear expansion of cobalt from 30 °C (86 °F)

In Cemented Carbides. The role of cobalt in cemented carbides is to provide a ductile bonding matrix for tungsten-carbide particles. Cobalt is used as a bonding matrix with tungsten carbide because its wetting or capillary action during liquid phase sintering allows the achievement of high densities (Ref 3). The commercially significant cemented carbides contain cobalt in the range of 3 to 25 wt%. As cutting tool materials, cemented carbides with 3 to 12 wt% Co are commonly used (see the article "Cemented Carbides" in this Volume for more details). Cobalt is also used in tool steels, which are covered in Volume 1 of the 10th Edition of *Metals Handbook*.

In Magnetic Materials. Cobalt, which is naturally ferromagnetic, provides resistance to demagnetization in several groups of permanent magnet materials. These include the aluminum-nickel-cobalt alloys (in which cobalt ranges from about 5 to 35 wt%), the iron-cobalt alloys (approximately 5 to 12 wt%), and the cobalt rare-earth intermetallics (which have some of the highest magnetic properties of all known materials). Further information is contained in the article "Permanent Magnet Materials" in this Volume.

As Radioactive Source. The artificial isotope cobalt-60 is an important γ-ray source in medical and industrial applications. Table 3 compares the characteristics of cobalt-60 γ-ray sources with other γ-ray sources used in industrial radiography.

Cobalt-Base Alloys

As a group, the cobalt-base alloys may be generally described as wear resistant, corrosion resistant, and heat resistant (strong even at high temperatures). Table 4 lists typical compositions of present-day cobalt-base alloys in these three application areas. Many of the properties of the alloys arise from the crystallographic nature of cobalt (in particular its response to stress), the solid-solution-strengthening effects of chromium, tungsten, and molybdenum, the for-

Table 1 Typical mechanical properties of pure cobalt

Form and purity	Tensile strength MPa	Tensile strength ksi	0.2% yield strength MPa	0.2% yield strength ksi	Compressive yield strength MPa	Compressive yield strength ksi
As-cast (99.9%)	235	34	...	...	290	42
Annealed (99.9%)	255	37	...	...	385	56
Swaged (99.9%)	690	100	...	...	...	...
Zone refined (99.8%)	945	137	760	111	...	...

Table 2 Electrical and magnetic properties of pure cobalt

Property	Value
Electrical properties	
Electrical conductivity, IACS at 20 °C (68 °F), %	27.6
Electrical resistivity, nΩ · m	52.5
Electrical resistivity temperature coefficient at 20 °C (68 °F), nΩ · m per K (nΩ · in. per °F)	5.31 (116)
Magnetic properties	
Magnetic permeability	
Initial	68
Maximum	245
Coercive force for H_{max} = 0.1 T (1000 G), A/m (Oe)	708 (8.9)
Saturation magnetization ($4\pi I_s$), T (kG)	1.87 (18.7)
Residual induction for H_{max} = 0.1 T (1000 G), T (G)	0.49 (4900)
Hysteresis loss for B_{max} = 0.5 T (5000 G), J/m^3 · cycle	690
Curie temperature, °C (°F)	1121 (2050)

mation of metal carbides, and the corrosion resistance imparted by chromium. Generally, the softer and tougher compositions are used for high-temperature applications such as gas-turbine vanes and buckets. The harder grades are used for resistance to wear.

Historically, many of the commercial cobalt-base alloys are derived from the cobalt-chromium-tungsten and cobalt-chromium-molybdenum ternaries first investigated by Elwood Haynes at the turn of the century. He discovered the high strength and stainless nature of the binary cobalt-chromium alloy, and he later identified tungsten and molybdenum as powerful strengthening agents within the cobalt-chromium system. When he discovered these alloys, Haynes named them the Stellite alloys after the Latin, *stella*, for star because of their star-like luster. Having discovered their high strength at elevated temperatures, Haynes also promoted the use of Stellite alloys as cutting tool materials.

Following the success of cobalt-base tool materials during World War I, they were then used from about 1922 in weld overlay form to protect surfaces from wear. These early cobalt-base "hardfacing" alloys were used on plowshares, oil well drilling bits, dredging cutters, hot trimming dies, and internal combustion engine valves and valve seats. Since the 1920s, some of these applications have ceased, some have continued, and many more have been added. In 1982, 1360 Mg (1500 tons) of cobalt-base alloys were sold for the purpose of hardfacing, one-third of this quantity being used to protect valve seating surfaces (both fluid control and engine valves).

Later in the 1930s and early 1940s, cobalt-base alloys for corrosion and high-temperature applications were developed in a series of related events involving the Austenal Laboratories and the Haynes Stellite Division of Union Carbide (Ref 4). Of the corrosion-resistant alloys, a cobalt-chromium-molybdenum alloy with a moderately low carbon content was developed to satisfy the need for a suitable investment cast dental material. This bicompatible material, which has the tradename Vitallium, is in use today for surgical implants. In the 1940s this same alloy also underwent investment casting trials for World War II aircraft turbocharger blades, and, with modifications to enhance structural stability, was used successfully for many years in this and other elevated-temperature applications. This early high-temperature material, Stellite alloy 21, is still in use today, but predominantly as an alloy for wear resistance.

Cobalt-Base Wear-Resistant Alloys

The cobalt-base wear alloys of today are little changed from the early alloys of Elwood Haynes. The most important differences relate to the control of carbon and silicon (which were impurities in the early alloys). Indeed, the main differences in the current Stellite alloy grades are carbon and tungsten contents (hence the amount and type of carbide formation in the microstructure during solidification). As will be discussed, carbon content influences hardness, ductility, and resistance to abrasive wear. Tungsten also plays an important role in these properties.

Types of Wear. There are several distinct types of wear which generally fall into three main categories:

- Abrasive wear
- Sliding wear
- Erosive wear

The type of wear encountered in a particular application is an important factor that influences the selection of a wear-resistant material.

Abrasive wear is encountered when hard particles, or hard projections (on a counterface) are forced against, and moved relative to, a surface. The terms high and low stress abrasion relate to the condition of the abrasive medium (be it hard particles or projections) after interaction with the surface. If the abrasive medium is crushed, then the high stress condition is said to prevail. If the abrasive medium remains intact, the process is described as low stress abrasion. Typically, high stress abrasion results from the entrapment of hard particles between metallic surfaces (in relative motion), while low stress abrasion is encountered when moving surfaces come into contact with packed abrasives, such as soil and sand.

In alloys such as the cobalt-base wear alloys, which contain a hard phase, the abrasion resistance generally increases as the volume fraction of the hard phase increases. Abrasion resistance is, however, strongly influenced by the size and shape of the hard phase precipitates within the microstructure, and the size and shape of the abrading species.

Sliding Wear. Of the three major types of wear, sliding is perhaps the most complex, not in concept, but in the way different materials respond to sliding conditions. Sliding wear is a possibility whenever two surfaces are forced together and moved relative to one another. The chances of damage are increased markedly if the two surfaces are metallic in nature, and if there is little or no lubrication present.

Sliding wear generally occurs by one or more of three mechanisms. In the first mechanism, oxide control of the sliding wear process, and low wear rates, are experienced when surface temperatures are high, by virtue of either a high ambient temperature or frictional heating. This is because oxide growth rates increase dramatically with temperature. In some cases, so-called "oxide glazes" are formed on the surfaces. These "oxide glazes" are very smooth, highly reflective regions, caused by the shearing of oxide asperities (peaks) and redistribution of the oxide debris in the

Table 3 Characteristics of γ-ray sources used in industrial radiography

γ-ray source	Half-life	Photon energy, MeV	Radiation output, RHM/Ci(a)	Penetrating power of steel, mm (in.)
Thulium-170	128 days	0.054 and 0.084(b)	0.003	13 (½)
Iridium-192	74 days	12 rays from 0.21–0.61	0.48	75 (3)
Cesium-137	33 years	0.66	0.32	75 (3)
Cobalt-60	5.3 years	1.17 and 1.33	1.3	230 (9)

(a) Output for typical unshielded, encapsulated sources; RHM/Ci, roentgens per hour at 1m per curie. (b) Against strong background of higher-MeV radiation

Table 4 Nominal compositions of various cobalt-base alloys

Alloy tradename	Nominal composition, %									
	Co	Cr	W	Mo	C	Fe	Ni	Si	Mn	Others
Cobalt-base wear-resistant alloys										
Stellite 1	bal	31	12.5	1 (max)	2.4	3 (max)	3 (max)	2 (max)	1 (max)	...
Stellite 6	bal	28	4.5	1 (max)	1.2	3 (max)	3 (max)	2 (max)	1 (max)	...
Stellite 12	bal	30	8.3	1 (max)	1.4	3 (max)	3 (max)	2 (max)	1 (max)	...
Stellite 21	bal	28	...	5.5	0.25	2 (max)	2.5	2 (max)	1 (max)	...
Haynes alloy 6B	bal	30	4	1	1.1	3 (max)	2.5	0.7	1.5	...
Tribaloy T-800	bal	17.5	...	29	0.08 (max)	...	...	3.5	...	...
Stellite F	bal	25	12.3	1 (max)	1.75	3 (max)	22	2 (max)	1 (max)	...
Stellite 4	bal	30	14.0	1 (max)	0.57	3 (max)	3 (max)	2 (max)	1 (max)	...
Stellite 190	bal	26	14.5	1 (max)	3.3	3 (max)	3 (max)	2 (max)	1 (max)	...
Stellite 306	bal	25	2.0	...	0.4	...	5	...	...	6 Nb
Stellite 6K	bal	31	4.5	1.5 (max)	1.6	3 (max)	3 (max)	2 (max)	2 (max)	...
Cobalt-base high-temperature alloys										
Haynes alloy 25 (L605)	bal	20	15	...	0.10	3 (max)	10	1 (max)	1.5	...
Haynes alloy 188	bal	22	14	...	0.10	3 (max)	22	0.35	1.25	0.05 La
MAR-M alloy 509	bal	22.5	7	...	0.60	1.5 (max)	10	0.4 (max)	0.1 (max)	3.5 Ta, 0.2 Ti, 0.5 Zr
Cobalt-base corrosion-resistant alloys										
MP35N, Multiphase alloy	bal	20	...	10	...	...	35	...	...	...
Haynes alloy 1233	bal	25.5	2	5	0.08 (max)	3	9	...	...	0.1N (max)

bal, balance

surface valleys. There are times when the oxide debris, if trapped as discrete particles or flakes between the sliding surfaces, can become abrasive. This is an important consideration when the surfaces oscillate with respect to one another over small amplitudes. The combined sliding/abrasive wear mechanism set up under these conditions is known as fretting.

The second mechanism of sliding wear is normally associated with high contact stresses and assumes breakdown of the oxide films to the point where true metal-to-metal contact is established. Under these conditions, there is an opportunity for cold welding of the surfaces to occur, and for subsequent movement to result in fracture of small pieces away from the original interface (normally in the weaker of the two mating materials). Damage caused by this mechanism is termed galling. Substantial metal transfer from one surface to the other and gross deformation of surface materials are typical of this condition.

The third mechanism of sliding wear, which can also produce substantial metallic damage, is one of subsurface fatigue. This mechanism is associated with cyclic stress conditions caused by materials periodically pressing on one another. Material is lost through fatigue crack nucleation and growth at a specific depth.

The metallic materials which perform well under sliding conditions do so either by virtue of their oxidation behavior or their ability to resist deformation and fracture. Little is known of the influence of metal-to-metal bond strength during cold welding. For materials such as the cobalt-base wear alloys with a hard phase dispersed throughout a softer matrix, the sliding-wear properties are controlled predominantly by the matrix. Indeed, within the cobalt alloy family, resistance to galling is generally independent of hard particle volume fraction and overall hardness.

Erosive Wear. Four distinct forms of erosive wear have been identified:

- Solid-particle erosion
- Liquid-droplet erosion
- Cavitation erosion
- Slurry erosion

Solid-particle erosion is caused by the impingement of small, solid particles against a surface. The solid particles themselves are typically airborne or entrained in some other gaseous environment. Particle sizes typically range from 5 to 500 μm. Typical velocities associated with solid-particle erosion range from 2 m/s (6 ft/s) in fluid bed combustors to 500 m/s (1650 ft/s). The rate of solid-particle erosion is dependent upon the velocity of the particles, their impingement angle (the angle dependency generally being different for ductile materials than it is for brittle materials), and the nature of the erodent (shape, size, strength).

Slurry erosion, or liquid-solid particle erosion, is similar to solid-particle erosion, except that there are differences in the viscosity of the carrier fluid (gas in solid-particle erosion, liquid carrier in slurry erosion). Slurry erosion occurs at the surfaces impinged by the solid particles in the liquid stream. The similarity to abrasion arises from the fact that the particles are hydrodynamically forced against the surface.

Although quite different mechanistically, liquid-droplet erosion and cavitation erosion have a similar effect upon a surface. They both result in a succession of shock (or stress) waves into the surface. For this reason, those materials which resist liquid-droplet erosion also perform well under cavitation conditions and vice versa.

Liquid-droplet erosion is easily envisaged, whereas cavitation erosion is a more complex phenomenon. For the latter to occur, the surface must be in contact with a liquid undergoing pressure changes. Surface damage results from the collapse of near-surface bubbles in the liquid, or, more precisely, from the action of liquid jets which arise during bubble implosion. The bubbles themselves are created when the pressure in the liquid falls below its vapor pressure. Collapse is induced by subsequent pressure increases.

The abrasion resistance of cobalt-base alloys, like other wear-resistant alloys, generally depends on the hardness of the carbide phases and/or the metal matrix. With the complex mechanisms of solid-particle and slurry erosion, however, such generalizations may not be warranted. In solid-particle erosion, for example, ductility may also be a factor (see the discussion on solid-particle erosion in the section "Wear Data" of this article).

As for liquid-droplet or cavitation erosion, the performance of a material is largely dependent upon its ability to absorb the shock (stress) waves without, essentially, microscopic fracture. In cobalt-base wear alloys, it has been found that carbide volume fraction (hence, bulk hardness) has very little effect upon resistance to liquid-droplet and cavitation erosion (Ref 5). Much more important are the properties of the matrix.

Alloy Compositions and Product Forms. The nominal compositions of various cobalt-base wear-resistant alloys are listed in Table 4, with six popular cobalt-base wear alloys listed first. Stellite alloys 1, 6, and 12 are derivatives of the original cobalt-chro-

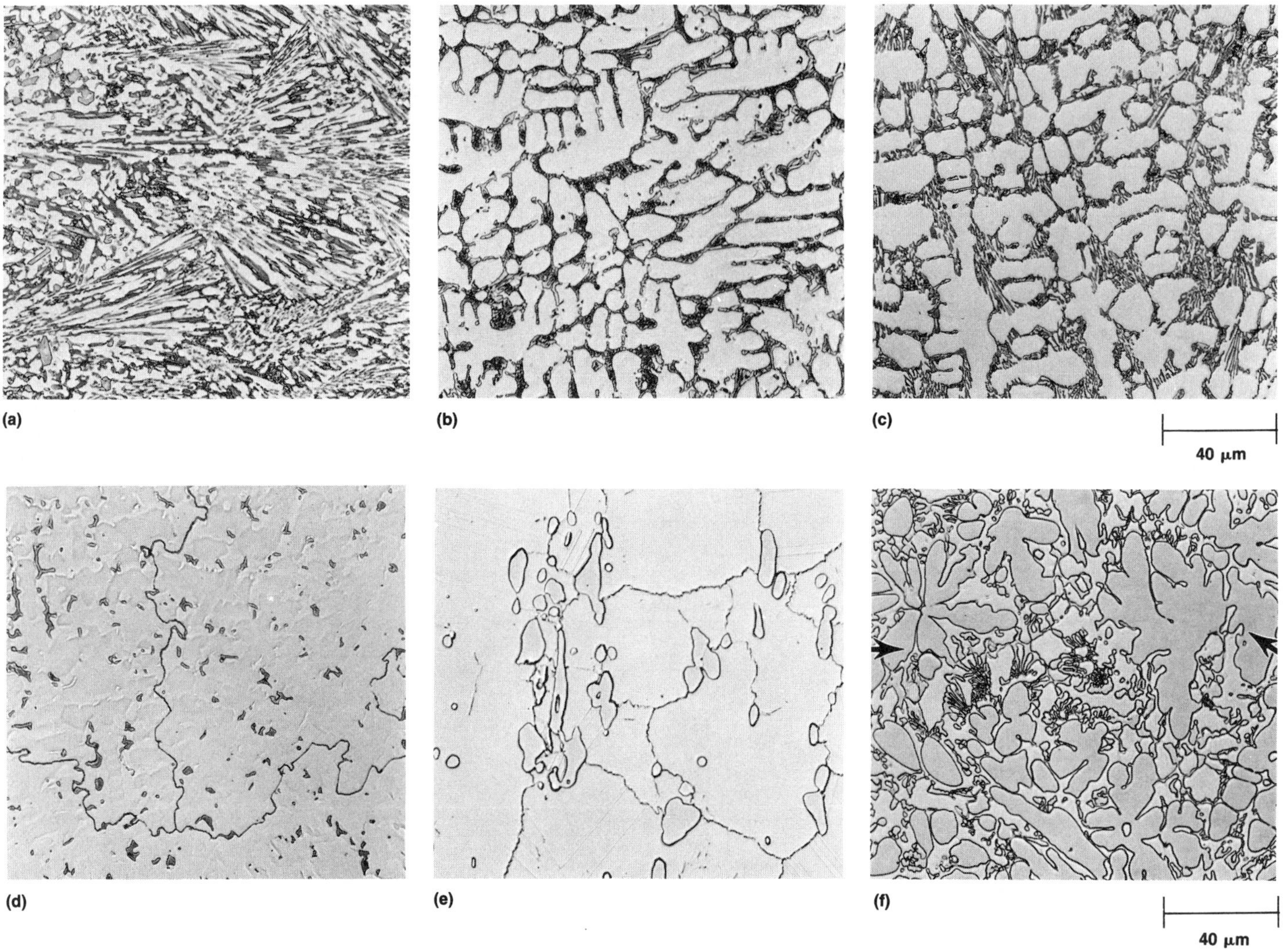

Fig. 2 Microstructures of various cobalt-base wear-resistant alloys. (a) Stellite 1, two-layer GTA deposit. (b) Stellite 6, two-layer GTA deposit. (c) Stellite 12, two-layer GTA deposit. (d) Stellite 21, two-layer GTA deposit. (e) Haynes alloy 6B, 13 mm (0.5 in.) plate. (f) Tribaloy alloy (T-800) showing the Laves precipitates (the largest continuous precipitates some of which are indicated with arrows). All 500×

mium-tungsten alloys developed by Haynes. These alloys are characterized by their carbon and tungsten contents, with Stellite alloy 1 being the hardest, most abrasion resistant, and least ductile. Their microstructures (in weld overlay form) are presented in Fig. 2(a) to 2(c) and illustrate the extent of carbide precipitation. These carbides are generally of the chromium-rich M_7C_3 type, although in high-tungsten alloys (such as Stellite alloy 1) tungsten-rich M_6C carbides usually are present also.

Stellite alloy 21 (Fig. 2d) differs from the first three alloys in that it employs molybdenum, rather than tungsten, to strengthen the solid solution. Stellite alloy 21 also contains considerably less carbon. By virtue of the high molybdenum content, and the fact that most of the chromium is in solution (rather than in Cr_7C_3 carbides), the alloy is more resistant to corrosion than Stellite alloys 1, 6, and 12.

Unlike these four alloys described above, which are generally used in the form of castings and weld overlays, Haynes alloy 6B is a wrought product available in plate, sheet, and bar form. Subtle compositional differences between alloy 6B and Stellite alloy 6 (such as silicon control) facilitate processing. The advantages of wrought processing include greatly enhanced ductility, chemical homogeneity, and resistance to abrasion by virtue of the coarse, blocky carbides within the microstructure (Fig. 2e).

The sixth wear-resistant alloy in Table 2 is the Tribaloy alloy (T-800), which is from an alloy family developed by DuPont in the early 1970s. In the search for resistance to abrasion and corrosion, workers at DuPont took the unprecedented step of alloying with excessive amounts of molybdenum and silicon to induce the formation during solidification of a hard and corrosion-resistant intermetallic compound, known as Laves phase. The carbon in T-800 is held as low as possible to discourage the precipitation of carbides (which, if encouraged to form, would tie up the vital elements chromium and molybdenum). The extent and nature of the Laves precipitates in T-800 are illustrated in Fig. 2(f). The Laves precipitates confer outstanding resistance to abrasion, but limit ductility. As a result of this limited ductility, the alloy is now generally used in the form of plasma-sprayed coatings.

Powder metallurgy (P/M) versions of several Stellite alloys (typically containing low levels of boron to enhance sintering) are available for applications where the P/M process is cost effective (that is, high volumes of relatively small components).

Wear Data. Abrasion data are presented for the six popular wear alloy compositions in Fig. 3, along with data for 316L stainless steel and D2 tool steel (60 HRC) for comparison. These data were generated using the ASTM G 65B (dry sand/rubber wheel) test and, except in the case of Haynes alloy 6B (samples of which were prepared from solution annealed plates with a thickness of 13 mm, or ½ in.), samples were prepared

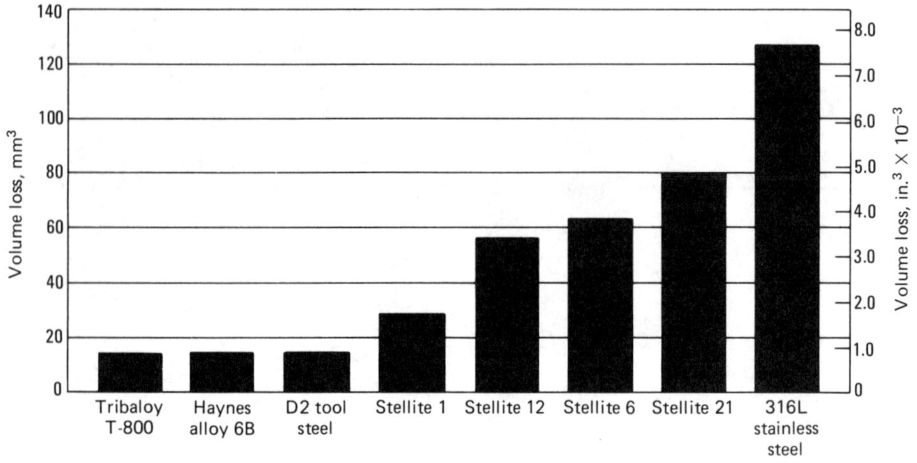

Fig. 3 Abrasion data of various cobalt-base alloys tested per ASTM G 65B

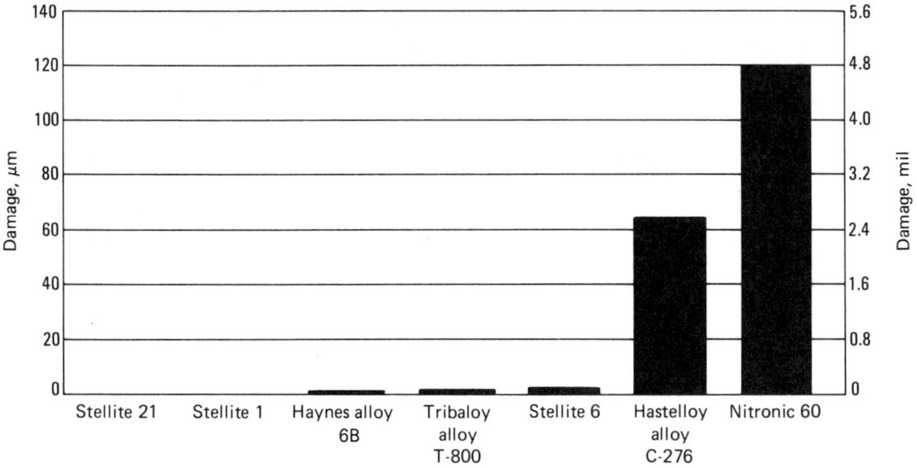

Fig. 4 Galling data of various cobalt-base alloys, Hastelloy C-276, and Nitronic-60 stainless steel. Data are from a 120°-10 stroke test with a 26.7 kN (6000 lbf) load.

from two-layer gas-tungsten-arc (GTA) deposits. Within the Stellite alloy family, it is evident from Fig. 3 that abrasion resistance is a function of carbon and tungsten content (Table 4). As the carbon content increases in the chromium-tungsten Stellite alloys, so does the tungsten content. This results in an increase in carbide content and thus hardness. In Fig. 3, the benefits of wrought processing in alloy 6B and the effectiveness of the Laves phase in T-800 are also evident.

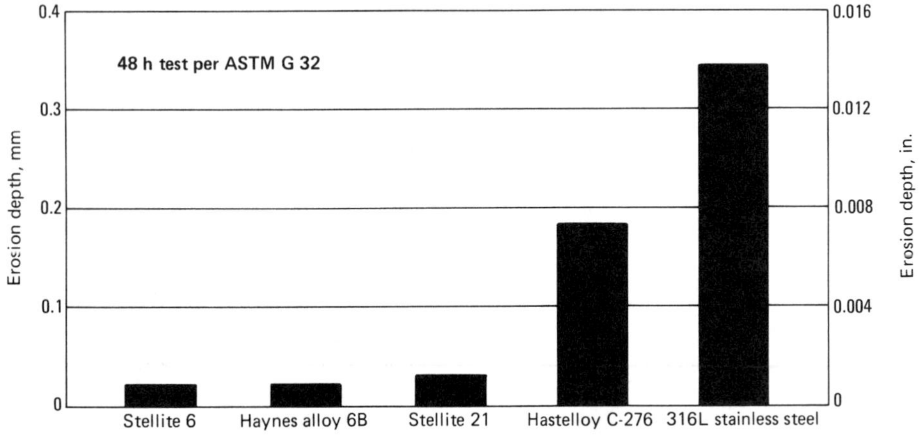

Fig. 5 Cavitation erosion data on various cobalt-base alloys, Hastelloy alloy C-276, and 316L stainless steel

Galling data (under self-mated conditions) are presented for five of the popular alloys in Fig. 4, along with data for corrosion-resistant Ni-Cr-Mo alloy (Hastelloy alloy C-276) and a stainless steel noted for its resistance to galling (Nitronic 60). The test, which was too severe for 316L stainless steel in that it failed by total seizure, involved twisting cylindrical pins with diameters of 16 mm (or 0.63 in.) back and forth 10 times through an arc of 120° at a load of 26.7 kN (6000 lbf). The amount of damage to the test surfaces was determined by use of a surface profilometer. The Haynes 6B, Nitronic 60, and Hastelloy C-276 samples were prepared from solution-annealed plates; the Stellite alloy samples were prepared from two-layer GTA deposits; the Tribaloy alloy (T-800) samples were prepared from two-layer plasma-transferred-arc deposits. From Fig. 4 it is evident that the cobalt wear alloys are exceptionally resistant to galling when coupled against like materials, and that this characteristic is largely independent of the compositional variations between the commonly used alloys. It should be noted that the self-mated galling resistance of Nitronic 60 equals that of the cobalt alloys at lower loads.

Solid-Particle Erosion Data. As previously mentioned in the section "Types of Wear" in this article, resistance against solid-particle erosion may not be as closely correlated with hardness or carbon content as with other types of wear mechanisms. This is evident from the solid-particle erosion data of Ref 6, which compares the solid particle of Haynes alloy 6B with 304 stainless steel and two heat-resistant cobalt-base alloys (Haynes alloys 25 and 188) listed in Table 4. The tests involved solid-particle erosion with 127 μm alumina particles impacting the test specimens at a 90° impact angle and with a speed of 170 m/s (560 ft/s). The weight loss of each alloy, as a fraction of the weight loss for Haynes alloy 6B, were as follows at room temperature:

Alloy	Erosion weight loss compared with alloy 6B
304 stainless steel	Same
Haynes alloy 25	96% of alloy 6B
Haynes alloy 188	97% of alloy 6B

From this data, there is little difference in the solid-particle erosion resistance of the wear-resistant 6B alloy and the two heat-resistant cobalt alloys. The two heat-resistant alloys (which have lower carbon contents and are more ductile than 6B) even exhibit slightly more erosion resistance than alloy 6B in this particular test. At temperatures of 700 °C (1290 °F), the advantage of the more ductile cobalt alloys became more pronounced, with weight loss fractions of 0.85 and 0.83 for Haynes alloys 25 and 188, respectively.

Cavitation Erosion Data. The outstanding cavitation erosion properties of the

Table 5 Mechanical and physical properties of cobalt-base wear-resistant alloys

Property	Alloy 1	Alloy 6	Alloy 12	Alloy 21	Alloy 6B	Alloy T-800
Hardness, HRC	55	40	48	32	37(a)	58
Yield strength, MPa (ksi)	...	541 (78.5)	649 (94.1)	494 (71.6)	619 (89.8)(a)	...
Ultimate tensile strength, MPa (ksi)	618 (89.6)	896 (130)	834 (135.5)	694 (100)	998 (145)(a)	...
Elongation, %	<1	1	<1	9	11	...
Thermal expansion coefficient, μm/m · °C						
From 20 to 100 °C (68–212 °F)	10.5	11.4	11.5	11.0	13.9(b)	...
From 20 to 500 °C (68–930 °F)	12.5	14.2	13.3	13.1	15.0(b)	12.6
From 20 to 1000 °C (68–1830 °F)	14.8	...	15.6	...	17.4(b)	15.1
Thermal conductivity, W/m · K	...	...	...	...	14.8	14.3
Specific gravity	8.69	8.46	8.56	8.34	8.39	8.64
Electrical resistivity, μΩ · m	0.94	0.84	0.88	...	0.91	...
Melting range, °C (°F)						
Solidus	1255 (2291)	1285 (2345)	1280 (2336)	1186 (2167)	1265 (2309)	1288 (2350)
Liquidus	1290 (2354)	1395 (2543)	1315 (2400)	1383 (2521)	1354 (2470)	1352 (2465)

(a) 3.2 mm (1/8 in.) thick sheet. (b) Starting temperature of 0 °C (32 °F)

cobalt-base wear alloys as compared with Hastelloy alloy C-276 and 316L stainless steel are illustrated in Fig. 5. This information was generated using ASTM G-32 procedures. The samples were prepared from solution-annealed plates (in the case of Haynes alloy 6B, Hastelloy alloy C-276, and 316L stainless steel) or from two-layer GTA deposits (in the case of Stellite alloys 6 and 21). The remaining three wear alloy compositions (from Table 4) lacked sufficient ductility for this test because the samples were of an intricate nature and subjected to high mechanical stresses during attachment to, and detachment from, the ultrasonic horn.

The physical and mechanical properties of six commonly used cobalt wear alloys are presented in Table 5. In the case of the Stellite and Tribaloy alloys, this information pertains to sand castings. Notable are the moderately high yield strengths and hardnesses of the alloys, the inverse relationship between carbon content and ductility (in the case of the Stellite alloys), and the enhanced ductility imparted to alloy 6B by wrought processing. A list of typical applications of the cobalt wear-resistant alloys of Table 4 is given in Table 6. Generally, the alloys are used in moderately corrosive and/or elevated-temperature environments.

Cobalt-Base High-Temperature Alloys

For many years, the predominant user of high-temperature alloys was the gas turbine industry. In the case of aircraft gas turbine power plants, the chief material requirements were elevated-temperature strength, resistance to thermal fatigue, and oxidation resistance. For land-base gas turbines, which typically burn lower grade fuels and operate at lower temperatures, sulfidation resistance was the major concern. Today, the use of high-temperature alloys is more diversified, as more efficiency is sought from the burning of fossil fuels and waste, and as new chemical processing techniques are developed.

Although cobalt-base alloys are not as widely used as nickel and nickel-iron alloys in high-temperature applications, cobalt-base high-temperature alloys nevertheless play an important role, by virtue of their excellent resistance to sulfidation and their strength at temperatures exceeding those at which the gamma-prime- and gamma-double-prime-precipitates in the nickel and nickel-iron alloys dissolve. Cobalt is also used as an alloying element in many nickel-base high-temperature alloys. The various types of iron-base, nickel-base, and cobalt-base alloys for high-temperature application are discussed in the article "Wrought and P/M Superalloys" in Volume 1 of the 10th Edition of *Metals Handbook*. Nickel-base and cobalt-base castings for high-temperature service are also covered in the article "Polycrystalline Cast Superalloys" in Volume 1 of the 10th Edition of *Metals Handbook*.

Alloy Compositions and Product Forms. As previously noted, Stellite 21 was an early type of cobalt-base high-temperature alloy that is used now primarily for wear resistance. Since the early use of Stellite 21, cobalt-base high-temperature materials have gone through various stages of development to increase their high-temperature capability. The use of tungsten rather than molybdenum, moderate nickel contents, lower carbon contents, and rare-earth additions typify cobalt-base high-temperature alloys of today.

Typical wrought and cast cobalt alloy compositions developed for high-temperature use are presented in Table 4. Haynes

Table 6 Typical applications of various cobalt-base wear-resistant alloys

Applications	Stellite alloys from Table 4	Forms	Mode of degradation
Automotive industry			
Engine valve seating surfaces	6, F	Weld overlay	Solid particle erosion, hot corrosion
Power industry			
Control valve seating surfaces	6, 21	Weld overlay	Sliding wear, cavitation erosion
Steam turbine erosion shields	Haynes alloy 6B	Wrought sheet	Liquid droplet erosion, particulate erosion
Marine industry			
Rudder bearings	306	Weld overlay	Sliding wear
Steel industry			
Hot shear edges	6	Weld overlay	Sliding wear, impact, abrasion
Bar mill guide rolls	12	Weld overlay	Sliding wear, impact, abrasion
Chemical processing industry			
Control valve seating surfaces	6	Weld overlay	Sliding wear, cavitation erosion
Plastic extrusion screw flights	1, 6, 12	Weld overlay	Sliding wear, abrasion
Pump seal rings	6, 12	Weld overlay	Sliding wear
Dry battery molds	4	Casting	Abrasion
Pulp and paper industry			
Chain saw guide bars	6, Haynes alloy 6B	Wrought sheet, weld overlay	Sliding wear, abrasion
Textile industry			
Carpet knives	6K, 12	Wrought sheet, weld overlay	Abrasion
Oil and gas industry			
Rotary drill bearings	190	Weld overlay	Abrasion, sliding wear

Table 7 Mechanical and physical properties of selected cobalt-base high-temperature alloys

Property	Alloy 25	Alloy 188	Alloy MAR-M 509
Yield strength, MPa (ksi)			
At 21 °C (70 °F)	445 (64.5)(a)	464 (67.3)(b)	585 (85)(c)
At 540 °C (1000 °F)	. . .	305 (44)(d)	400 (58)(c)
Tensile strength, MPa (ksi)			
At 21 °C (70 °F)	970 (141)(a)	945 (137)(b)	780 (113)(c)
At 540 °C (1000 °F)	800 (116)(e)	740 (107)(d)	570 (83)(c)
1000-h rupture strength, MPa (ksi)			
At 870 °C (1600 °F)	75 (11)	70 (10)	140 (20)
At 980 °C (1800 °F)	30 (4)	30 (4)	90 (13)
Elongation, %	62(a)	53(b)	3.5(c)
Thermal expansion coefficient, μm/m · K			
From 21 to 93 °C (70–200 °F)	12.3	11.9	. . .
From 21 to 540 °C (70–1000 °F)	14.4	14.8	. . .
From 21 to 1090 °C (70–2000 °F)	17.7	18.5	. . .
Thermal conductivity, W/m · K			
At 20 °C (68 °F)	9.8(f)	10.8	. . .
At 500 °C (930 °F)	18.5(g)	19.9	. . .
At 900 °C (1650 °F)	26.5(h)	25.1	. . .
Specific gravity	9.13	8.98	8.86
Electrical resistivity, μΩ · m	0.89	1.01	. . .
Melting range, °C (°F)			
Solidus	1329 (2424)	1302 (2375)	1290 (2350)
Liquidus	1410 (2570)	1330 (2426)	1400 (2550)

(a) Sheet 3.2 mm (1/8 in.) thick. (b) Sheet 0.75–1.3 mm (0.03–0.05 in.) thick. (c) As-cast. (d) Sheet, heat treated at 1175 °C (2150 °F) for 1 h with rapid air cool. (e) Sheet, heat treated at 1230 °C (2250 °F) for 1 h with rapid air cool. (f) At 38 °C (100 °F). (g) At 540 °C (1000 °F). (h) At 815 °C (1500 °F)

alloys 25 (also known as L605) and 188 are wrought alloys available in the form of sheets, plates, bars, pipes, and tubes (together with a range of matching welding products for joining purposes). MAR-M alloy 509 is an alloy designed for vacuum investment casting. Selected mechanical properties of the three alloys are given in Table 7. Stress rupture data are presented in Fig. 6. Other cobalt-base alloys for high-temperature service are covered in Volume 1 of the 10th Edition of *Metals Handbook*.

In the three high-temperature alloys of Table 4, chromium and tungsten are the chief solid-solution strengtheners, and nickel is present to stabilize the face-centered cubic structure (that is, to reduce the allotropic phase transformation temperature). During the design of Haynes alloy 188, consideration was also given to control of the electron hole number (N_v) to restrict the formation of topologically close-packed phases (for example, Laves) in the material (Ref 7). Oxidation resistance is enhanced in the case of Haynes alloy 188 by the rare-earth addition, lanthanum.

Carbides serve to increase high-temperature strength and, in the wrought alloys, control grain size. In the solution-annealed condition, the carbides in Haynes alloy 188 are of the M_6C type. Aging in the temperature range 650 to 1175 °C (1200 to 2150 °F) promotes secondary carbide precipitation (M_6C at the higher aging temperatures and $M_{23}C_6$ at the lower temperatures). In the MAR-M alloy 509 (which has a higher carbon content) the active carbide-forming elements, tantalum, titanium, and zirconium, exhibit a predominant Chinese-script MC carbide in the as-cast condition (Ref 8). Further information on the carbides in cobalt-base superalloys is provided in Volume 1 of the 10th Edition of *Metals Handbook*.

Resistance to Oxidation and Sulfidation. Dynamic oxidation data for Haynes alloys 25 and 188 relative to the nickel-base alloys X and 601 are presented in Fig. 7. The values represent the total depth of metal affected (that is, the depth of metal turned to oxide plus the depth subjected to internal oxidation). The 100-hour test involved cooling the samples every half hour to 540 °C (1000 °F). The test atmosphere contained the combustion products of A-640 aviation kerosene, using an air-to-fuel ratio of 40 to 1. The test gas velocity was approximately 390 km/h (355 ft/s). The improved performance of Haynes alloy 188 in this test is attributable to effects of lanthanum upon oxide scale adherence.

Sulfidation test results at 980 °C (1800 °F) for alloys 25 and 188, relative to the nickel-base alloys, X, 601, and Waspaloy alloy, are presented in Fig. 8. These results, which were generated in an environment with a sulfur partial pressure of 0.4 Pa (4×10^{-6} atmosphere) and an oxygen partial pressure of 3×10^{-12} Pa (3×10^{-17} atm), show the outstanding sulfidation resistance of the cobalt-base high-temperature alloys relative to both gamma-prime-strengthened and solid-solution-strengthened high-temperature nickel alloys. Of the alloys tested under these conditions, only Haynes alloy 6B (Table 4) and Haynes alloy HR-160 (a high-silicon, cobalt-containing alloy) exhibit superior properties.

Applications. Haynes alloys 25 and 188, and MAR-M alloy 509 are well established in the gas turbine industry. As a casting alloy, 509 is generally used for complex shapes such as nozzle guide vanes. As wrought alloys, 25 and 188 are used for fabricated assemblies and ductwork. In particular, Haynes alloy 188 is the alloy of choice for combustor cans and afterburner liners in high-performance aircraft gas turbines. Haynes alloy 25 has also been used successfully in a variety of indus-

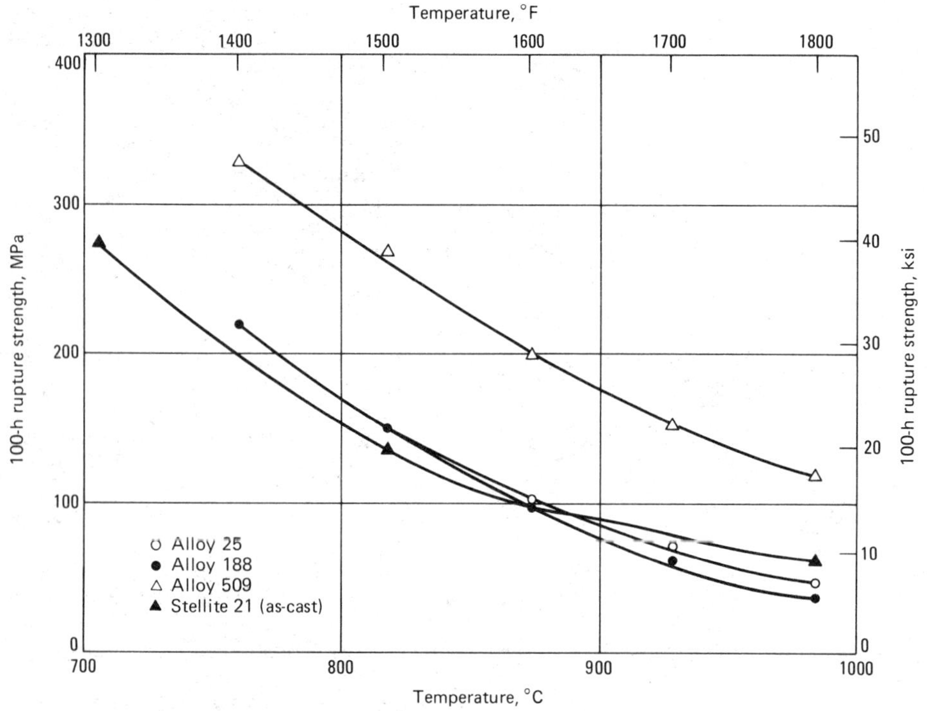

Fig. 6 100-h stress-rupture strengths of selected cobalt-base superalloys. See Table 7 for 1000-h rupture strengths. Stellite 21 included for comparison with MAR-M alloy 509

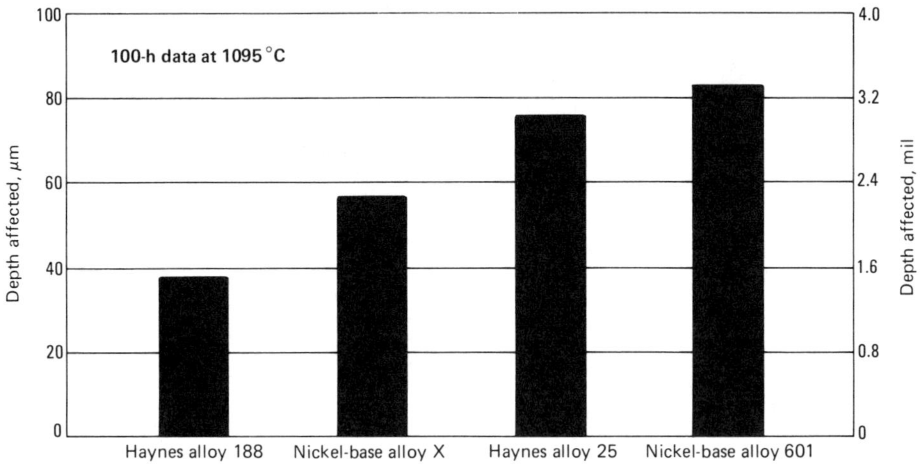

Fig. 7 Dynamic oxidation data for selected high-temperature cobalt alloys and nickel-base alloys at 1095 °C (2000 °F)

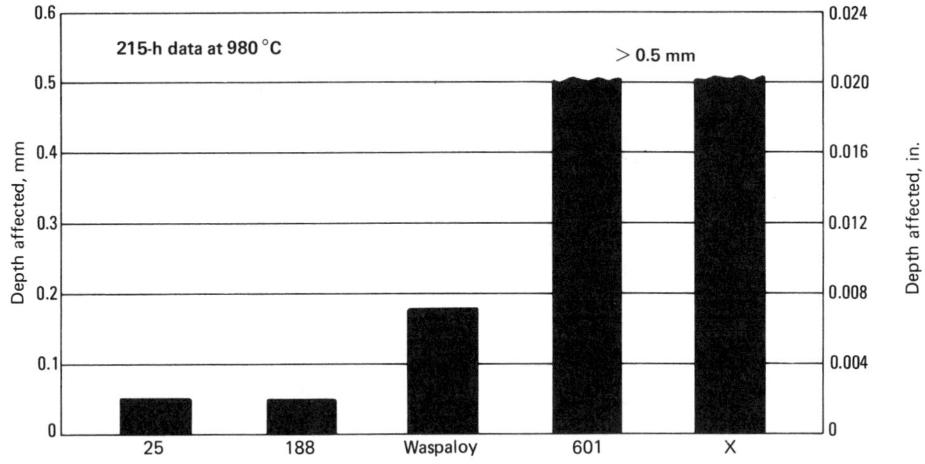

Fig. 8 Sulfidation data of Haynes alloys 25 and 188 relative to selected nickel-base alloys at 980 °C (1800 °F)

trial furnace applications (for example, muffles and liners).

Cobalt-Base Corrosion-Resistant Alloys

Although the cobalt-base wear-resistant alloys possess some resistance to aqueous corrosion, they are limited by grain boundary carbide precipitation, the lack of vital alloying elements in the matrix (after formation of the carbides or Laves precipitates) and, in the case of the cast and weld overlay materials, by chemical segregation in the microstructure.

By virtue of their homogeneous microstructures and lower carbon contents, the wrought cobalt-base high-temperature alloys (which typically contain tungsten rather than molybdenum) are even more resistant to aqueous corrosion, but still fall well short of the nickel-chromium-molybdenum alloys in corrosion performance.

To satisfy the industrial need for alloys which exhibit outstanding resistance to aqueous corrosion, yet share the attributes of cobalt as an alloy base (resistance to various forms of wear, and high strength over a wide range of temperatures), several low-carbon, wrought cobalt-nickel-chromium-molybdenum alloys are produced. The compositions of two of these are presented in Table 4. In addition, the cobalt-chromium-molybdenum Vitallium alloy is still widely used today for prosthetic devices and implants on account of its excellent compatibility with body fluids and tissues (see the article "Corrosion of Cobalt-Base Alloys" in Volume 13 of the 9th Edition of *Metals Handbook*).

Types of Aqueous Corrosion. Just as there are several types of wear, there are several types of aqueous corrosion. Of concern in this article are:

- Uniform corrosion
- Localized corrosion (pitting)
- Stress-corrosion cracking

Other forms of attack, such as galvanic corrosion and intergranular corrosion, can be overcome by judicious design and by appropriate alloy heat treatments.

Uniform attack is the most common form of corrosion and is so named because loss of surface material is of a uniform nature. From an electrochemical standpoint, uniform corrosion occurs because of the continual shifting of anodic and cathodic surface sites.

Localized attack, or pitting, occurs when anodic surface sites remain stationary. Deep pits may form on such surfaces, giving rise to rapid and unpredictable failures of components. Chloride ions are particularly strong promoters of this type of attack.

Stress-corrosion cracking is caused by the combined effects of corrosion and tensile stresses. As with localized corrosion, failures of components by this mechanism can be rapid and unpredictable. The generally held view of stress-corrosion cracking is that it is initiated by localized corrosion, then progresses by virtue of a combination of mechanical crack propagation and corrosion at the crack tip (this becoming anodic with respect to surrounding surfaces).

Alloy Compositions and Product Forms. The two corrosion-resistant alloys presented in Table 4 rely for their corrosion resistance on chromium and molybdenum. The corrosion properties of alloy 1233 are also enhanced by tungsten. Both alloys are available in a variety of wrought product forms (plates, sheets, bars, tubes, and so forth). They are also available in the form of welding consumables for joining purposes.

Corrosion Properties. The uniform corrosion properties of alloy 1233, relative to a number of well known nickel- and iron-base corrosion alloys, are summarized in Table 8. The outstanding pitting resistance of alloy 1233 is illustrated in Fig. 9. Although not as resistant to pitting as alloy 1233, MP35N alloy is nevertheless excellent in a wide range of mineral acid environments and chloride solutions.

Table 8 Comparison of corrosion rates for selected cobalt-base, iron-base, and nickel-base alloys in various solutions

	Corrosion rate, mm/year							
Alloy	Boiling 99% acetic acid	Boiling 65% nitric acid	Boiling 1% hydrochloric acid	Boiling 2% hydrochloric acid	54% P_2O_5 at 116 °C (240 °F)	Boiling 10% sulfuric acid	Boiling ASTM-G28A solution	Boiling ASTM-G28B solution
1233	<0.01	0.15	0.01	13.49	0.19	2.52	0.20	0.02
C-276	<0.01	21.51	0.52	1.90	0.58	0.51	8.05	0.86
625	0.01	0.51	0.03	14.15	0.30	0.64	0.43	71.08
20CB-3	0.11	0.19	1.80	5.77	0.92	0.40	0.25	69.08
316L	0.19	0.24	13.31	25.15	5.11	47.46	0.94	80.51

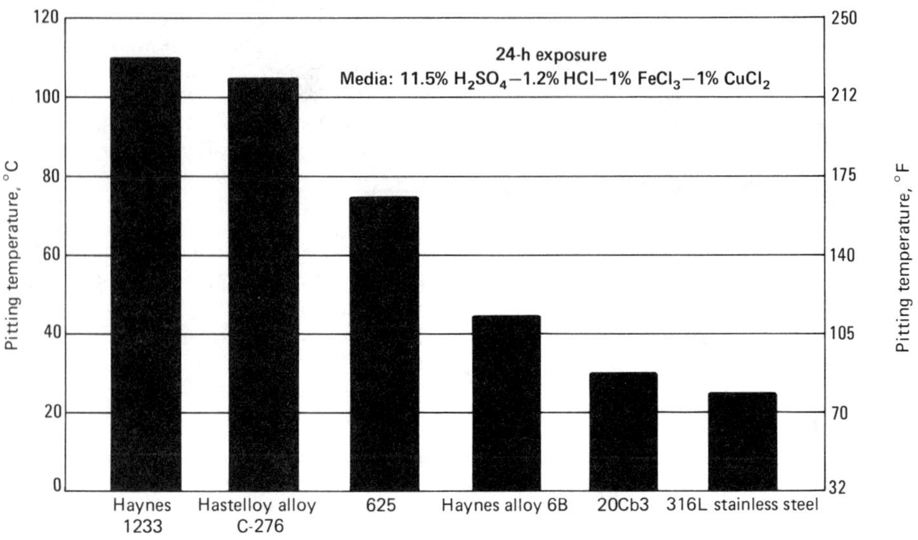

Fig. 9 Critical pitting temperature of the cobalt-base corrosion-resistant alloy 1233 compared with various other alloys

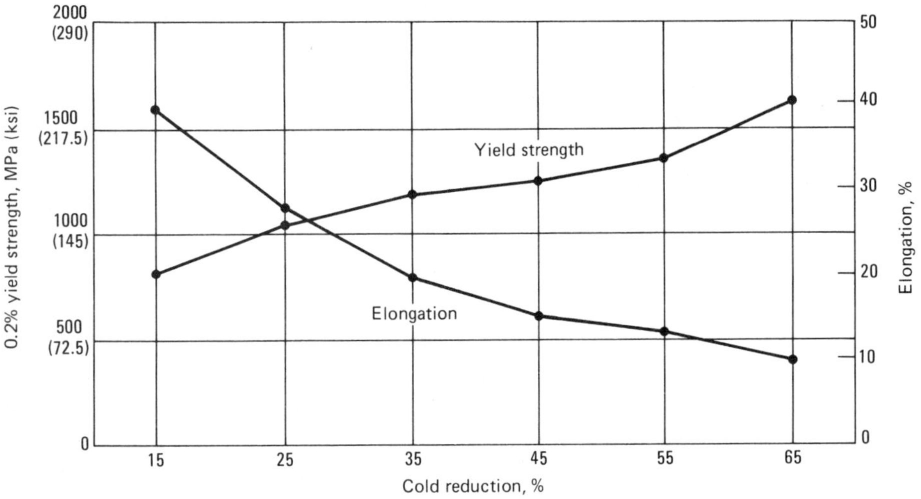

Fig. 10 Yield strength and elongation versus cold reduction for the corrosion-resistant MP35N multiphase alloy

Alloy 1233 is less resistant to stress-corrosion cracking than the nickel-chromium-molybdenum Hastelloy alloys; however, it offers better resistance than 316L stainless steel to this form of failure. By virtue of its higher nickel content, MP35N alloy is very resistant to stress-corrosion cracking and is suitable for hydrogen sulfide service.

Mechanical Properties. An advantage of the two corrosion-resistant alloys in Table 4 is that they may be strengthened considerably by cold working. In the case of MP35N alloy, the alloy is intended for use in the work-hardened (or work-hardened and aged) condition, and the manufacturers have supplied considerable data concerning the mechanical properties of the alloy at different levels of cold work. Some of these data (0.2% offset yield strengths and elongation values) are plotted in Fig. 10. Other mechanical and physical properties of the two alloys are given in Table 9.

Table 9 Mechanical and physical properties of selected cobalt-base corrosion-resistant alloys

Property	Alloy 1233(a)	MP35N alloy
Hardness	28 HRC	90 HRB(b)
Yield strength, MPa (ksi)	558 (81)	380 (55)(b)
Ultimate tensile strength, MPa (ksi)	1020 (148)	895 (130)(b)
Elongation, %	33	65(b)
Thermal expansion coefficient, μm/m · K		
21–93 °C (70–200 °F)	...	12.8(c)
21–315 °C (70–600 °F)	...	14.8(c)
21–540 °C (70–1000 °F)	...	15.7(c)
Thermal conductivity, W/m · K	...	11.2(c)
Electrical resistivity, μΩ · m	...	1.03(c)
Melting range, °C (°F)		
Solidus	1333 (2431)	1315 (2400)
Liquidus	1355 (2471)	1440 (2625)

(a) 13 mm (½ in.) plate, solution annealed. (b) Cold-drawn bar, solution annealed. (c) Work-strengthened and aged

Applications of both these alloys include pump and valve components and spray nozzles. MP35N alloy is also popular for fasteners, cables, and marine hardware.

REFERENCES

1. S.F. Sibley, *Cobalt*, MCP-5, U.S. Department of the Interior, 1977
2. L. Habraken and D. Coutsouradis, *Cobalt*, Vol 26, 1965, p 10
3. J. Hinnuber and O. Rudiger, *Cobalt*, Vol 19, 1963, p 57
4. R.D. Gray, *A History of the Haynes Stellite Company*, Cabot Corporation, 1974
5. K.C. Antony and W.L. Silence, *ELSI-5 Proceedings*, University of Cambridge, 1979, p 67
6. J.S. Hansen, J.E. Kelly, and F.W. Wood, U.S. Bureau of Mines Report RI8335, 1979
7. R.B.H. Herchenroeder, S.J. Matthews, J.W. Tackett, and S.T. Wlodek, *Cobalt*, Vol 54, 1972, p 3
8. H.L. Wheaton, *Cobalt*, Vol 29, 1965, p 163

Selection and Application of Magnesium and Magnesium Alloys

Revised by Susan Housh and Barry Mikucki, Dow Chemical U.S.A., and Archie Stevenson, Magnesium Elektron, Inc.

MAGNESIUM and magnesium alloys are used in a wide variety of structural and nonstructural applications. Structural applications include automotive, industrial, materials-handling, commercial, and aerospace equipment. The automotive applications include clutch and brake pedal support brackets, steering column lock housings, and manual transmission housings. In industrial machinery, such as textile and printing machines, magnesium alloys are used for parts that operate at high speeds and thus must be lightweight to minimize inertial forces. Materials-handling equipment includes dockboards, grain shovels, and gravity conveyors. Commercial applications include hand-held tools, luggage, computer housings, and ladders. Magnesium alloys are valuable for aerospace applications because they are lightweight and exhibit good strength and stiffness at both room and elevated temperatures.

Magnesium is also employed in various nonstructural applications. It is used as an alloying element in alloys of aluminum, zinc, lead, and other nonferrous metals. It is used as an oxygen scavenger and desulfurizer in the manufacture of nickel and copper alloys; as a desulfurizer in the iron and steel industry; and as a reducing agent in the production of beryllium, titanium, zirconium, hafnium, and uranium. Another important nonstructural use of magnesium is in the Grignard reaction in organic chemistry. In finely divided form, magnesium finds some use in pyrotechnics, both as pure magnesium and alloyed with 30% or more aluminum. The relative position of magnesium in the electromotive series allows it to be used for cathodic protection of other metals from corrosion and in construction of dry-cell, seawater, and reserve-cell batteries. Gray iron foundries use magnesium and magnesium-containing alloys as ladle addition agents introduced just before the casting is poured. The magnesium makes the graphite particles nodular and greatly improves the toughness and ductility of the cast iron. Because of its rapid but controllable response to etching and its light weight, magnesium is also used in photoengraving.

Table 1 presents figures for the global shipments of primary magnesium from 1983 to 1988, broken down into its major uses. Primary magnesium is furnished to ASTM B 92, grade 9980A, with a specified minimum magnesium content of 99.8%. Also available are special grades of primary magnesium in which manganese, aluminum, and iron impurities are held to especially low levels. These special grades are employed in chemical and metallurgical applications, such as the preparation of uranium metal and other reactive metals.

Aluminum and zinc are relatively soluble in solid magnesium, but their solubilities decrease at low temperatures. The solubility of aluminum is 12.7% by weight at 437 °C (819 °F) and 3.0% at 93 °C (200 °F); solubility of zinc is 6.2% at 340 °C (644 °F) and 2.8% at 204 °C (400 °F). Solubilities of manganese, zirconium, and cerium are less than 1.0% by weight at 482 °C (900 °F). At the eutectic temperature, 4.5% thorium is soluble in magnesium. Manganese is effective in improving the corrosion stability of magnesium alloys that contain aluminum and zinc. Several of these alloys, which employ manganese to control both the iron content and activity in the alloy, are available and have excellent corrosion resistance. Other alloys not containing aluminum and zinc, but containing yttrium, are also available and exhibit good corrosion resistance.

Designations. A standard system of alloy and temper designations, adopted in 1948, is explained in Table 2. As an example of how the system works, consider magnesium alloy AZ91E-T6, the nominal composition and typical properties of which are given in Table 3. The first part of the designation, AZ, signifies that aluminum and zinc are the two principal alloying elements. The second

Table 1 Primary magnesium shipments

Application	Shipments, metric tons: 1983	1984	1985	1986	1987	1988
Structural						
Die castings	27 900	30 400	29 700	28 800	26 600	28 500
Gravity castings(a)	2 000	1 300	1 200	1 600	1 800	2 100
Wrought products(b)	7 100	6 600	4 800	5 400	8 400	7 400
Total structural	**37 000**	**38 300**	**35 700**	**35 800**	**36 800**	**38 000**
Nonstructural						
Aluminum alloying	110 800	113 500	121 000	122 100	122 100	134 300
Desulfurization	13 400	17 400	19 100	20 300	21 900	28 600
Nodular iron	8 900	9 800	11 300	12 300	14 200	15 800
Metal reduction	9 200	12 200	10 300	9 600	8 800	10 200
Chemical(c)	8 200	7 800	8 000	8 000	7 200	8 100
Electrochemical(d)	7 600	7 700	9 100	8 300	8 000	8 000
Other(e)	9 300	9 300	10 300	10 000	17 000	8 200
Total nonstructural	**167 400**	**177 700**	**189 100**	**190 600**	**199 200**	**213 200**
Total all uses	**204 400**	**216 000**	**224 800**	**226 400**	**236 000**	**251 200**

(a) Includes sand, permanent mold, investment, and plaster castings. (b) Includes extrusions, forgings, sheet, plate, and photoengraving stock. (c) Grignard reaction or pyrotechnic applications. (d) Cast or extruded anodes and batteries. (e) Shipments into the U.S.S.R., the People's Republic of China, and Comecon countries. Source: International Magnesium Association

Table 2 Standard four-part ASTM system of alloy and temper designations for magnesium alloys

See text for discussion.

First part	Second part	Third part	Fourth part
Indicates the two principal alloying elements	Indicates the amounts of the two principal alloying elements	Distinguishes between different alloys with the same percentages of the two principal alloying elements	Indicates condition (temper)
Consists of two code letters representing the two main alloying elements arranged in order of decreasing percentage (or alphabetically if percentages are equal)	Consists of two numbers corresponding to rounded-off percentages of the two main alloying elements and arranged in same order as alloy designations in first part	Consists of a letter of the alphabet assigned in order as compositions become standard	Consists of a letter followed by a number (separated from the third part of the designation by a hyphen)
A—aluminum B—bismuth C—copper D—cadmium E—rare earth F—iron G—magnesium H—thorium K—zirconium L—lithium M—manganese N—nickel P—lead Q—silver R—chromium S—silicon T—tin W—yttrium Y—antimony Z—zinc	Whole numbers	Letters of alphabet except I and O	F—as fabricated O—annealed H10 and H11—slightly strain hardened H23, H24, and H26—strain hardened and partially annealed T4—solution heat treated T5—artificially aged only T6—solution heat treated and artificially aged T8—solution heat treated, cold worked, and artificially aged

part of the designation, 91, gives the rounded-off percentages of aluminum and zinc (9 and 1, respectively). The third part, E, indicates that this is the fifth alloy standardized with 9% Al and 1% Zn as the principal alloying additions. The fourth part, T6, denotes that the alloy is solution treated and artificially aged.

Casting Alloys

High-Pressure Die Casting Alloys. There are three systems of magnesium alloys used commercially for high-pressure die casting: magnesium-aluminum-zinc-manganese (AZ), magnesium-aluminum-manganese (AM), and magnesium-aluminum-silicon-manganese (AS). Nominal compositions and typical properties of these alloys are given in Table 3. General characteristics of high-pressure die casting alloys are given in Table 4.

The most commonly used magnesium die casting alloy is AZ91D. The AZ91D alloy exhibits good mechanical and physical properties in combination with excellent castability and saltwater corrosion resistance. The excellent corrosion resistance of high-purity AZ91D is essentially the result of controlling the impurity level of three critical contaminants: iron, nickel, and copper. High-purity AZ91D alloy has replaced less-pure AZ91B as the workhorse for die casting. The AZ91B alloy is still used because it can be easily produced from scrap or secondary metal. This cost-saving alternative can be used for applications in which corrosion resistance is not an important consideration, for example, painted parts in a noncorrosive environment. Die castings are used in the as-cast condition.

For applications requiring greater ductility than available with AZ91D, high-purity die cast alloy AM60B is used. The AM60B alloy has better elongation and toughness than AZ91D. Despite the decrease in aluminum content, the tensile and yield strengths of AM60B are only slightly lower than those of AZ91D. The AM60B alloy is used in the production of die cast automobile wheels and in some archery and other sports equipment. As with AZ91D, AM60B exhibits excellent saltwater corrosion resistance.

Die cast alloy AS41A has creep strength much superior to that of the AZ91D or AM60B alloys at temperatures up to 175 °C (350 °F); it also has good elongation, yield strength, and tensile strength. The AS41A alloy was used in crankcases of air-cooled automotive engines. A high-purity version of AS41A, which exhibits excellent saltwater corrosion resistance, is being introduced as AS41XB.

The AS21 alloy exhibits even better creep strength than AS41A. However, AS21 has lower room-temperature tensile and yield strengths, and it is somewhat more difficult to cast.

Sand and Permanent Mold Casting Alloys. Several systems of magnesium alloys are available for sand and permanent mold castings:

- Magnesium-aluminum-manganese with and without zinc (AM and AZ)
- Magnesium-zirconium (K)
- Magnesium-zinc-zirconium with and without rare earths (ZK, ZE, and EZ)
- Magnesium-thorium-zirconium with and without zinc (HK, HZ, and ZH)
- Magnesium-silver-zirconium with rare earths or thorium (QE and QH)
- Magnesium-yttrium-rare-earth-zirconium (WE)
- Magnesium-zinc-copper-manganese (ZC)

Nominal compositions and typical properties of these alloys are given in Table 3. A summary of general characteristics of the sand and permanent mold casting alloys is given in Table 5.

Magnesium-Aluminum Casting Alloys. The magnesium sand and permanent mold casting alloys that contain aluminum as the primary alloying ingredient (AM100A, AZ63A, AZ81A, AZ91C, AZ91E, and AZ92A) exhibit good castability, good ductility, and moderately high yield strength at temperatures up to approximately 120 °C (~250 °F). Of these alloys, AZ91E has become prominent; it has almost completely replaced AZ91C because it has superior corrosion performance. In AZ91E, the iron, nickel, and copper contaminants are controlled to very low levels. As a result, it exhibits excellent saltwater corrosion resistance.

In any of the magnesium-aluminum-zinc alloys, an increase in aluminum content raises yield strength but reduces ductility for comparable heat treatment. Final selection of the specific composition may be based on tests of the finished castings.

The K1A alloy is used primarily where high damping capacity is required. It has low tensile and yield strength.

Magnesium alloys that contain high levels of zinc (ZK51A, ZK61A, ZK63A, and ZH62A) develop the highest yield strengths of the casting alloys and can be cast into complicated shapes. However, these grades are more costly than the alloys of the AZ

Table 3 Nominal compositions and typical room-temperature mechanical properties of magnesium alloys

Alloy	Composition, %						Tensile strength		Yield strength						Elongation in 50 mm (2 in.), %	Shear strength		Hardness, HRB(c)
	Al	Mn(a)	Th	Zn	Zr	Other(b)			Tensile		Compressive		Bearing					
							MPa	ksi	MPa	ksi	MPa	ksi	MPa	ksi		MPa	ksi	
Sand and permanent mold castings																		
AM100A-T61	10.0	0.1	···	···	···	···	275	40	150	22	150	22	···	···	1	···	···	69
AZ63A-T6	6.0	0.15	···	3.0	···	···	275	40	130	19	130	19	360	52	5	145	21	73
AZ81A-T4	7.6	0.13	···	0.7	···	···	275	40	83	12	83	12	305	44	15	125	18	55
AZ91C- and E-T6(d)	8.7	0.13	···	0.7	···	···	275	40	145	21	145	21	360	52	6	145	21	66
AZ92A-T6	9.0	0.10	···	2.0	···	···	275	40	150	22	150	22	450	65	3	150	22	84
EQ21A-T6	···	···	···	···	0.7	1.5 Ag, 2.1 Di	235	34	195	28	195	28	···	···	2	···	···	65–85
EZ33A-T5	···	···	···	2.7	0.6	3.3 RE	160	23	110	16	110	16	275	40	2	145	21	50
HK31A-T6	···	···	3.3	···	0.7	···	220	32	105	15	105	15	275	40	8	145	21	55
HZ32A-T5	···	···	3.3	2.1	0.7	···	185	27	90	13	90	13	255	37	4	140	20	57
K1A-F	···	···	···	···	0.7	···	180	26	55	8	···	···	125	18	1	55	8	···
QE22A-T6	···	···	···	···	0.7	2.5 Ag, 2.1 Di	260	38	195	28	195	28	···	···	3	···	···	80
QH21A-T6	···	···	1.0	···	0.7	2.5 Ag, 1.0 Di	275	40	205	30	···	···	···	···	4	···	···	···
WE43A-T6	···	···	···	···	0.7	4.0 Y, 3.4 RE	250	36	165	24	···	···	···	···	2	···	···	75–95
WE54A-T6	···	···	···	···	0.7	5.2 Y, 3.0 RE	250	36	172	25	172	25	···	···	2	···	···	75–95
ZC63A-T6	···	0.25–0.75	···	6.0	···	2.7 Cu	210	30	125	18	···	···	···	···	4	···	···	55–65
ZE41A-T5	···	···	···	4.2	0.7	1.2 RE	205	30	140	20	140	20	350	51	3.5	160	23	62
ZE63A-T6	···	···	···	5.8	0.7	2.6 RE	300	44	190	28	195	28	···	···	10	···	···	60–85
ZH62A-T5	···	···	1.8	5.7	0.7	···	240	35	170	25	170	25	340	49	4	165	24	70
ZK51A-T5	···	···	···	4.6	0.7	···	205	30	165	24	165	24	325	47	3.5	160	23	65
ZK61A-T5	···	···	···	6.0	0.7	···	310	45	185	27	185	27	···	···	···	170	25	68
ZK61A-T6	···	···	···	6.0	0.7	···	310	45	195	28	195	28	···	···	10	180	26	70
Die castings																		
AM60A- and B-F(e)	6.0	0.13	···	···	···	···	205	30	115	17	115	17	···	···	6	···	···	···
AS21X1	1.7	0.4	···	···	···	1.1 Si	240	35	130	19	130	19	···	···	9	···	···	···
AS41A-F(f)	4.3	0.35	···	···	···	1.0 Si	220	32	150	22	150	22	···	···	4	···	···	···
AZ91A,B, and D-F(g)	9.0	0.13	···	0.7	···	···	230	33	150	22	165	24	···	···	3	140	20	63
Extruded bars and shapes																		
AZ10A-F	1.2	0.2	···	0.4	···	···	240	35	145	21	69	10	···	···	10	···	···	···
AZ21X1-F(h)	1.8	0.02	···	1.2	···	···	···	···	···	···	···	···	···	···	···	···	···	···
AZ31 B and C-F(i)	3.0	···	···	1.0	···	···	260	38	200	29	97	14	230	33	15	130	19	49
AZ61A-F	6.5	···	···	1.0	···	···	310	45	230	33	130	19	285	41	16	140	20	60
AZ80A-T5	8.5	···	···	0.5	···	···	380	55	275	40	240	35	···	···	7	165	24	82
HM31A-F	···	1.2	3.0	···	···	···	290	42	230	33	185	27	345	50	10	150	22	···
M1A-F	···	1.2	···	···	···	···	255	37	180	26	83	12	195	28	12	125	18	44
ZC71-F	···	0.5-1.0	···	6.5	···	1.2 Cu	360	52	340	49	···	···	···	···	5	···	···	70–80
ZK21A-F	···	···	···	2.3	0.45(a)	···	260	38	195	28	135	20	···	···	4	···	···	···
ZK40A-T5	···	···	···	4.0	0.45(a)	···	276	40	255	37	140	20	···	···	4	···	···	···
ZK60A-T5	···	···	···	5.5	0.45(a)	···	365	53	305	44	250	36	405	59	11	180	26	88
Sheet and plate																		
AZ31B-H24	3.0	···	···	1.0	···	···	290	42	220	32	180	26	325	47	15	160	23	73
HK31A-H24	···	···	3.0	···	0.6	···	255	37	200	29	160	23	285	41	9	140	20	68
HM21A-T8	···	0.6	2.0	···	···	···	235	34	170	25	130	19	270	39	11	125	18	···
PE(j)	3.3	···	···	0.7	···	···	···	···	···	···	···	···	···	···	···	···	···	···

(a) Minimum. (b) RE, rare earth; Di, didymium (a mixture of rare-earth elements made up chiefly of neodymium and praseodymium). (c) 500 kg load, 10 mm ball. (d) Properties of C and E are identical, but AZ91E castings have maximum contaminant levels of 0.005% Fe, 0.0010% Ni, and 0.015% Cu. (e) Properties of A and B are identical, but AM60B castings have maximum contaminant levels of 0.005% Fe, 0.002% Ni, and 0.010% Cu. (f) Properties of A and XB are identical, but AS41XB castings have maximum contaminant levels of 0.0035% Fe, 0.002% Ni, and 0.020% Cu. (g) Properties of A, B, and D are identical, except that 0.30% max residual Cu is allowable in AZ91B, and AZ91D castings have maximum contaminant levels of 0.005% Fe, 0.002% Ni, and 0.030% Cu. (h) For battery applications. (i) Properties of B and C are identical, but AZ31C has 0.15% min Mn, 0.1% max Cu, and 0.03% max Ni. (j) Photoengraving grade

series. Therefore, these alloys are used where exceptionally good yield strengths are required. They are intended primarily for use at room temperature.

Because ZK61A has a higher zinc content, it has significantly greater strength than ZK51A (Table 3). Both alloys maintain high ductility after an artificial aging treatment (T5). The strength of ZK61A can be further increased (3 to 4%) by solution treatment plus artificial aging (T6), without impairing ductility. Both of these alloys have fatigue strengths equal to those of the magnesium-aluminum-zinc alloys, but they are more susceptible to microporosity and hot cracking, and are less weldable. Addition of either thorium or rare-earth metals overcomes these deficiencies. The strength properties of ZE63A are equivalent to those of ZK61A, and those of ZH62A are equivalent to or better than those of ZK51A (Table 3).

Alloy ZE63A is a high-strength grade with excellent tensile strength and yield strength; these superior properties are obtained by heat treating in a hydrogen atmosphere. Because hydriding proceeds from the surface, heat-treating time, wall thickness, and penetrability are limiting factors. This alloy has excellent casting characteristics.

The ZE41A alloy was developed to meet the growing need for an alloy with medium strength, good weldability, and improved castability in comparison with AZ91C and AZ92A. It has good fatigue and creep properties and maximum freedom from microshrinkage. Unlike the AZ alloys, there is a very close relationship between separately cast test bar properties and those obtained from the casting itself, even where relatively thick cast sections are involved. Alloy ZE41A is used at temperatures up to 160 °C (320 °F) in such applications as aircraft engines, helicopter and airframe components, and wheels and gear boxes.

Alloy ZC63 is a member of a new family of magnesium alloys containing neither aluminum nor zirconium. The alloy exhibits good castability, and it is pressure tight and weldable. No grain refining or hardeners are required to obtain its properties, but a heat treatment must be used to achieve the full properties. The alloy has attractive room-temperature and moderately elevated temperature properties. The corrosion resistance of the alloy is similar to that of AZ91C, but it is less than that of AZ91E.

Table 4 Characteristics of high-pressure die cast alloys

Alloy(a)	General characteristics
AZ91D	Most commonly used die casting alloy. Good strength at room temperature, good castability, good atmospheric stability, excellent saltwater corrosion resistance
AM60B	Good elongation and toughness, excellent saltwater corrosion resistance, good yield and tensile properties
AS21X1	Best creep resistance of die casting alloys, good room-temperature properties, useful in high-temperature applications
AS41XB	Good creep resistance up to 175 °C (350 °F), good room-temperature properties, excellent saltwater corrosion resistance, useful in high-temperature applications

(a) All alloys are in the as-cast condition.

Table 5 Characteristics of sand and permanent mold cast alloys

Alloy	Temper	General characteristics
AM100A	T4, T6	Permanent mold alloy. Pressure tight, weldable, good atmospheric stability
AZ63A	T4, T6	Good saltwater corrosion resistance even with a high iron level, good toughness, difficult to cast. Very seldom used today
AZ91E	T6	General-purpose alloy. Good strength at room temperature, useful properties up to 175 °C (350 °F), good atmospheric stability, and excellent saltwater corrosion resistance. The most commonly used alloy in the Mg-Al-Zn family
AZ92C	T6	General-purpose alloy. Excellent strength at room temperature, useful properties up to 175 °C (350 °F), good atmospheric stability
EQ21A	T6	Heat-treated alloy. High yield strength up to 250 °C (480 °F), pressure tight, weldable
EZ33A	T5	Creep resistant up to 250 °C (480 °F), excellent castability, pressure tight, weldable
HK31A	T6	Creep resistant up to 345 °C (650 °F) for short-time applications, pressure tight, weldable
HZ32A	T5	Creep resistant up to 345 °C (650 °F), pressure tight, weldable
QE22A	T6	Heat-treated alloy. High yield strength up to 250 °C (480 °F), pressure tight, weldable
QH21A	T6	Good creep resistance, high yield strength up to 300 °C (570 °F), pressure tight, weldable
WE43	T6	Heat-treated alloy. Good properties up to 250 °C (480 °F) for extended periods of time, pressure tight, weldable, good corrosion resistance
WE54A	T6	The first of a new family of alloys containing yttrium. Exceptional strength at both room and elevated temperatures
ZC63A	T6	Good room-temperature properties, useful strength at moderately elevated temperatures, excellent castability, pressure tight, weldable
ZE41A	T5	Easily cast, weldable, pressure tight, useful strength at elevated temperatures
ZE63A	T6	Excellent castability, pressure tight, weldable, highly developed properties in thin-wall castings
ZH62A	T5	Stronger than, but as castable as, ZE41A. Weldable, pressure tight
ZK51A	T5	Good strength at room temperature
ZK61A	T6	Excellent strength at room temperature. Only fair castability but capable of developing excellent properties in castings

Source: Ref 1

The magnesium-rare-earth-zirconium alloys are used at temperatures between 175 and 260 °C (350 and 500 °F). Because their high-temperature strengths exceed those of the magnesium-aluminum-zinc alloys, thinner walls can be used, and a savings in weight is possible.

The magnesium-rare-earth-zinc-zirconium alloy EZ33A has good strength stability when exposed to elevated temperatures. (Strength stability is the ability to resist deterioration of strength from extended exposure to elevated temperatures.) This alloy is more difficult to cast in some designs than magnesium-aluminum-zinc alloys. Castings of EZ33A have excellent pressure tightness. Alloy ZE41A, discussed earlier, is similar to EZ33A, but it has higher tensile and yield strengths because of its higher zinc content. Some sacrifice is made in castability and weldability in ZE41A to obtain the higher mechanical properties.

When the operating temperature of an engine housing was increased from 120 to 205 °C (250 to 400 °F), alloy EZ33A-T5 was successfully substituted for AZ92A-T6. The change was based on creep tests of separately cast bars of the two alloys; stress values for 0.1% creep in 1000 h were:

Alloy	Temperature °C	Temperature °F	Stress MPa	Stress ksi
AZ92A-T6	205	400	6.9	1.0
	260	500	2.1	0.3
EZ33A-T5	205	400	58	8.4
	260	500	26	3.7
	315	600	8.3	1.2

The magnesium-thorium-zirconium alloys HK31A and HZ32A are intended primarily for use at temperatures of 200 °C (400 °F) and higher; at these temperatures, properties superior to those of EZ33A are required. For full development of properties, HK31A requires the T6 treatment (solution heat treatment plus artificial aging), whereas HZ32A, which contains zinc, requires only the T5 treatment (artificial aging). Castings of HK31A and HZ32A have been used at temperatures as high as 345 to 370 °C (650 to 700 °F) in a few applications. The magnesium-zinc-thorium-zirconium alloy ZH62A differs from other magnesium-thorium-zirconium alloys in that it is intended primarily for use at room temperature.

Magnesium-thorium-zirconium alloys are more difficult to cast than EZ33A because they are more susceptible to the formation of inclusions and defects as a result of gating turbulence. The tendency for inclusions to form in the magnesium-thorium-zirconium alloys is particularly marked in thin-wall parts that require rapid pouring rates. These alloys have adequate castability for production of complex parts of moderate-to-heavy wall thickness.

At 260 °C (500 °F) and slightly higher, HZ32A is equal to or better than HK31A in short-time and long-time creep strength at all extensions. The HK31A alloy has higher tensile, yield, and short-time creep strengths up to 370 °C (700 °F). However, HZ32A has greater strength stability at elevated temperatures, and it has much better foundry characteristics than does HK31A.

Magnesium-Silver Casting Alloys. The presence of silver improves the room-temperature strength of magnesium alloys. When rare-earth elements or thorium is present, along with the silver, elevated-temperature strength is also increased. The QE22A and EQ21A grades are high tensile strength and yield strength alloys with fairly good properties at temperatures up to 205 °C (400 °F). Alloy QH21A has similar properties to QE22A and EQ21 at room temperature, but it exhibits superior properties at temperatures from 205 °C (400 °F) up to 260 °C (500 °F). The alloys QE22A, EQ21A, and QH21A have good castability and weldability. They do require solution and aging heat treatments to achieve the higher mechanical properties. The QE22A and QH21A alloys are relatively expensive because of their silver contents; EQ21A, which has a lower silver content, is less expensive.

Alloys WE54 and WE43 have high tensile strengths and yield strengths, and they exhibit good properties at temperatures up to 300 °C (570 °F) and 250 °C (480 °F), respectively. The WE54 alloy retains its properties at high temperature for up to 1000 h, whereas WE43 retains its properties at high temperature in excess of 5000 h. Both WE54 and WE43 have good castability and weldability, but they require solution and aging

Table 6 Forgeability of four magnesium alloys

Alloy	Transverse ductility, %(a)	Forging characteristics
ZK60A	7.0	Excellent on hydraulic or mechanical presses for small forgings; large forgings confined to hydraulic presses. Properties nearly equivalent to those of AZ80A
AZ80A	5.0	Forgings have maximum strength. Forging limited to hydraulic presses
HK31A	5.0	Readily forged if proper temperature is maintained. Recommended for elevated-temperature applications
HM21A	5.0	Rolled ring and die forgings. Recommended for elevated-temperature applications

(a) For a minimum web thickness of approximately 3 mm (1/8 in.) and a minimum draft of 1°

heat treatments to optimize their mechanical properties. They are relatively expensive because of their yttrium content. Both alloys are corrosion resistant, with corrosion rates similar to those of the common aluminum-base casting alloys.

Wrought Alloys

Wrought magnesium alloys are produced as bars, billets, shapes, wire, sheet, plate, and forgings.

Extruded bars and shapes are made of several types of magnesium alloys (Table 3). For normal strength requirements, one of the magnesium-aluminum-zinc (AZ) alloys is usually selected. The strength of these alloys increases as aluminum content increases. Alloy AZ31B is a widely used moderate-strength grade with good formability; it is used extensively for cathodic protection. Alloy AZ31C is a lower-purity commercial variation of AZ31B for lightweight structural applications that do not require maximum corrosion resistance. The M1A and ZM21A alloys can be extruded at higher speeds than AZ31B, but they have limited use because of their lower strength. Alloy AZ10A has a low aluminum content and thus is of lower strength than AZ31B, but it can be welded without subsequent stress relief. The AZ61A and AZ80A alloys can be artificially aged for additional strength (with a sacrifice in ductility); AZ80A is not available in hollow shapes. Alloy AZ21X1 is designed specially for use in battery applications.

Alloy ZK60A is used where high strength and good toughness are required. This alloy is heat treatable and is normally used in the artificially aged (T5) condition. ZK21A and ZK40A alloys are of lower strength and are more readily extrudable than ZK60A; they have had limited use in hollow tubular strength requirements.

Alloy ZC71 is a member of a new family of magnesium alloys containing neither aluminum nor zirconium. The alloy can be extruded at high rates and exhibits good strength properties. The corrosion resistance of ZC71 is similar to that of AZ91C, but it falls far short of that of AZ91E.

Alloy HM31A is of moderate strength. It is suitable for use in applications requiring good strength and creep resistance at temperatures in the range of 150 to 425 °C (300 to 800 °F).

Forgings are made of AZ31B, AZ61A, AZ80A, M1A, and ZK60A; the compositions and properties of these alloys are listed under extruded bars and shapes in Table 3. Alloy HM21A, which is listed under sheet and plate alloys in Table 3, is also a good forging alloy. Alloys M1A and AZ31B may be used for hammer forgings (whereas the other alloys are almost always press forged); however, there has been a gradual decline in the use of the magnesium-manganese alloy M1A. The AZ80A alloy has greater strength than AZ61A and requires the slowest rate of deformation of the magnesium-aluminum-zinc alloys. Alloy ZK60A has essentially the same strength as AZ80A but with greater ductility. To develop maximum properties, both AZ80A and ZK60A are heat treated to the artificially aged (T5) condition; AZ80A may be given the T6 solution heat treatment, followed by artificial aging to provide maximum creep stability. Alloy HM21A is given the T5 temper. It is useful at elevated temperatures up to 370 to 425 °C (700 to 800 °F) for applications in which good creep resistance is needed.

Table 7 Recommended minimum radii for 90° bends in magnesium sheet

The numerical values for bend radii are given as multiples of sheet thickness.

Alloy and temper	Forming temperature(a): 20 °C (70 °F)	95 °C (200 °F)	150 °C (300 °F)	205 °C (400 °F)	260 °C (500 °F)	315 °C (600 °F)	370 °C (700 °F)	425 °C (800 °F)
AZ31B-O	5.5*t*	5.5*t*	4*t*	3*t*	2*t*	...	...	...
AZ31B-H24	8*t*	8*t*	6*t*	3*t*	2*t*	...	...	...
HK31A-O	6*t*	6*t*	6*t*	5*t*	4*t*	3*t*	2*t*	1*t*
HK31A-H24	13*t*	13*t*	13*t*	9*t*	8*t*	5*t*	3*t*	...
HM21A-T8	9*t*	9*t*	9*t*	9*t*	9*t*	8*t*	6*t*	4*t*

(a) See Table 18 for maximum time at temperature.

Hydraulic and mechanical processes are both used for the forging of magnesium. A slow and controlled rate of deformation is desirable because it facilitates control of the plastic flow of metal; therefore, hydraulic press forging is the most commonly used process. Magnesium, which has a hexagonal crystal structure, is more easily worked at elevated temperatures. Consequently, forging stock (ingot or billet) is heated to a temperature between 350 and 500 °C (650 and 950 °F) prior to forging. Forgeability ratings of four magnesium alloys are given in Table 6. Forgeability was measured by transverse ductility, assuming a web thickness of approximately 3 mm (1/8 in.) and a minimum draft angle of 1°. The ZK60A alloy is indicated as having slightly higher forgeability than the other three high-strength alloys.

Sheet and plate are rolled from magnesium-aluminum-zinc (AZ and photoengraving grade, or PE) and magnesium-thorium (HK and HM) alloys (Table 3).

Alloy AZ31B is the most widely used alloy for sheet and plate and is available in several grades and tempers. It can be used at temperatures up to 100 °C (200 °F). The HK31A and HM21A alloys are suitable for use at temperatures up to 315 and 345 °C (600 and 650 °F), respectively. However, HM21A has superior strength and creep resistance. For example, an air impeller manufactured from thick plate of alloy HK31A failed as a result of excessive creep. The same item manufactured from HM21A provided satisfactory performance and service life. Test coupons machined from the two materials gave the following stress values:

Alloy	Stress for 0.1% creep in 100 h: MPa	ksi
At 205 °C (400 °F)		
HM21A	86.2	12.5
HK31A	41	6.0
At 260 °C (500 °F)		
HM21A	72.4	10.5
HK31A	28	4.0
At 315 °C (600 °F)		
HM21A	52	7.5
HK31A	14	2.0

Alloy PE is a special-quality sheet with excellent flatness, corrosion resistance, and etchability. It is used in photoengraving.

Good formability is an important requirement for most sheet materials. The approximate formability of magnesium alloy sheet is indicated by its ability to withstand 90° bending over a mandrel without cracking. The minimum size of the mandrel (minimum radius) over which the sheet can be bent without cracking depends on alloy composition and temper, material thickness, and temperature (Table 7). When correct temperatures and forming conditions are employed, all magnesium alloys can be deep drawn to about equal reduction.

Metal-Matrix Composites. Magnesium serves as an excellent matrix for metal-matrix composites. It has an excellent affinity for bonding to reinforcing ceramic materials, which include continuous and discontinuous ceramic fibers along with ceramic particles. Magnesium composites can be manufactured by liquid metal infiltration processes, continuous casting, squeeze casting, diffusion bonding, powder metallurgy, and proprietary casting techniques.

Continuous fiber reinforced magnesium composites have excellent room- and elevated-temperature stiffness and strength. However, they are relatively expensive because of the high cost of the continuous reinforcements, which include alpha Al_2O_3, boron, graphite, SiC, and steel fibers. Magnesium alloys reinforced with continuous graphite fibers have the best combination of specific stiffness and thermal-deformation resistance of any common engineering material. Magnesium alloys reinforced with discontinuous alumina, alumina-silica, or mineral fibers are being investigated for use in commercial automobile pistons because their wear properties are superior to those of typical magnesium alloys. Magnesium-ceramic particle composites have shown excellent specific stiffness, strength, and increased abrasive wear resistance. These properties, along with the low cost of ceramic particles such as SiC, B_4C, and Al_2O_3, make magnesium-ceramic particle composites a competitive candidate for many commercial automotive applications. A more detailed explanation of reinforced magnesium alloys can be found in the article "Metal-Matrix Composite Materials" in this Volume.

Mechanical Properties

Typical values for the mechanical and physical properties of magnesium alloys are given in the article "Properties of Magnesium Alloys" in this Volume. For castings, these values are obtained by testing separately cast specimens. Tensile strengths of investment mold and shell mold castings compare favorably with those of sand and permanent mold castings. Yield strength, tensile strength, and percentage elongation may vary with cooling rate and generally are lower than those of separately cast sand mold test bars.

Figure 1 shows the effect of specimen location on the tensile properties of specimens machined from two representative sand castings that have sections of varying thickness. Some specifications permit a 25% reduction in tensile strength and a 75% reduction in elongation for specimens machined from castings, as compared with requirements for separately cast bars.

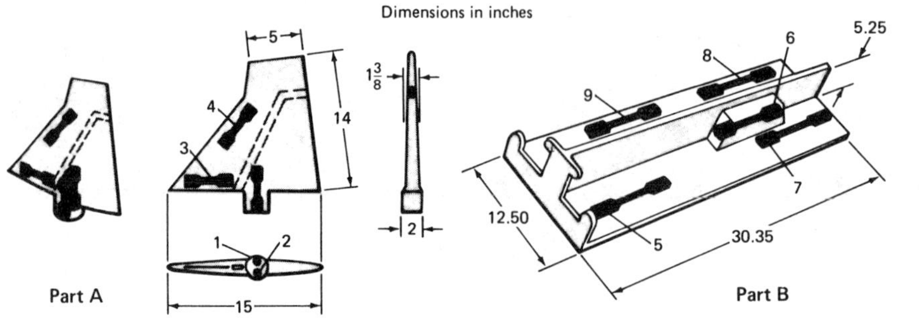

Specimen location	Tensile strength MPa	Tensile strength ksi	Yield strength MPa	Yield strength ksi	Elongation in 50 mm (2 in.), %
Part A					
1	202.4	29.35	116.5	16.90	2.8
2	199.6	28.95	123.1	17.85	2.8
3	227.5	33.00	133.1	19.30	2.5
4	265.4	38.50	141.0	20.45	4.0
Part B					
5	242.7	35.20	140.8	20.42	2.8
6	214.8	31.15	135.5	19.65	3.1
7	248.6	36.05	142.1	20.61	2.5
8	250.6	36.35	141.8	20.57	2.2
9	244.4	35.45	140.0	20.30	22.8

Fig. 1 Effect of specimen location on the tensile properties of two AZ91A-T6 production sand castings. Part A, average mechanical properties for parts made from 11 different heats. Specimens 1 and 2 were machined 13 mm (0.50 in.) in diameter; specimens 3 and 4, 6.4 × 13 mm (0.25 × 0.50 in.) flat. Part B, average mechanical properties for parts made from 16 heats. Castings weighed approximately 7 kg (15 lb), and all sections were 6.4 to 9.5 mm (¼ to ⅜ in.) thick, except specimen 6, which was taken from a section about 25 mm (1.0 in.) thick. Test specimens were 6.4 × 13 mm (¼ × ½ in.) flat, except number 6, which was 13 mm (0.50 in.) in diameter. Customer required a minimum of 176 MPa (25.5 ksi) tensile strength, 110 MPa (16.0 ksi) yield strength, and 1.0% elongation in 50 mm (2 in.).

Most magnesium alloys have ratios of tensile strength to density and tensile yield strength to density that are comparable to those of other common structural metals.

The direction, temperature, and speed at which an alloy is fabricated have a significant effect on the mechanical properties of wrought parts. For example, extrusions produced at higher temperatures and speeds have lower strength than those produced under normal operating conditions. Mechanical properties of forgings depend on the orientation of the tested specimen in relation to the flow patterns developed during forging.

Compressive Strength. Compressive yield strength is defined as the stress required to produce a deviation or offset of 0.2% from the modulus line. For castings, compressive yield strength is approximately equal to tensile yield strength. For wrought alloys, however, yield strength in compression may be considerably less than yield strength in tension. The ratio of yield strength in compression to yield strength in tension varies from about 0.4 for alloy M1A to an average value of about 0.7 for the other wrought magnesium alloys. Typical compressive yield strength values for various magnesium alloys are given in Table 3.

Maximum design stresses for magnesium alloy columns that are loaded axially and that have sufficient stability to prevent local failure may be determined, for columns in the long-column range, by using the Euler column formula. (A long column is one in which the length and cross section are such that the stress at which it will buckle does not exceed the elastic limit of the column material.) Maximum design stresses for magnesium alloy columns in the short-column range depend on the strengths and the forms of the alloys being tested. (A short column is any column of such length and cross section that it fails under compressive loading by plastic yielding and/or crushing, rather than by buckling.) In practical application, the maximum design stress of a column is considered to be the minimum compressive yield stress of the material. Various formulas have been developed for determining the maximum design stresses for columns of intermediate length (those that fail by elastic buckling in combination with plastic yielding and/or crushing). Column strength curves for several magnesium extrusion alloys are shown in Fig. 2.

Bearing strength is the resistance to a stress applied by a pin in a hole; it is particularly important in the design of bolted and riveted joints. Bearing yield strength is defined as the stress required to produce an offset from the initial straight portion of the curve equal to 2% of the hole diameter. Bearing strength values listed in Table 3 were determined using specimens with an

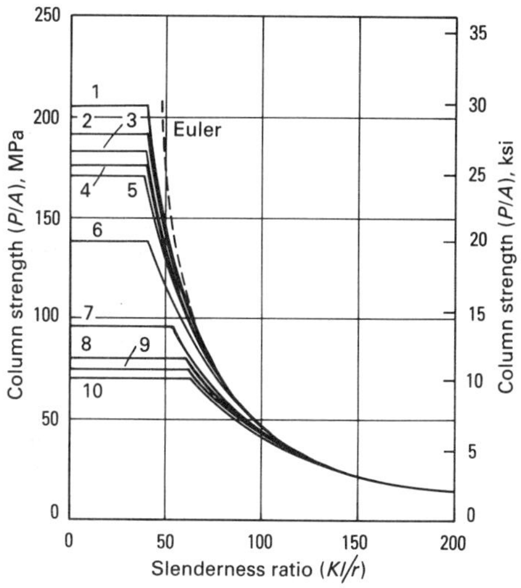

Alloy	Curve number	Least dimension of area range	
ZK60A-T5......	1	<1290 mm^2	<2.000 $in.^2$
	2	1290–1935 mm^2	2.000–2.999 $in.^2$
	5	1935–3225 mm^2	3.000–4.999 $in.^2$
AZ80A-T5......	2	6.35–38.09 mm	0.250–1.499 in.
	3	38.10–63.49 mm	1.500–2.499 in.
	4	63.50–127.00 mm	2.500–5.000 in.
ZK21A-F.......	6	<3226 mm^2	<5.000 $in.^2$
AZ61A-F.......	7	6.35–127.00 mm	0.250–5.000 in.
AZ31B-F.......	8	≤6.35 mm	≤0.250 in.
	9	6.35–63.49 mm	0.250–2.499 in.
	10	63.50–127.00 mm	2.500–5.000 in.

Fig. 2 Minimum column strength curves for several magnesium extrusion alloys. *P*, ultimate column load; *A*, cross-sectional area; *K*, constant that depends on end conditions, *l*, column length; and *r*, minimum radius of gyration of column cross section

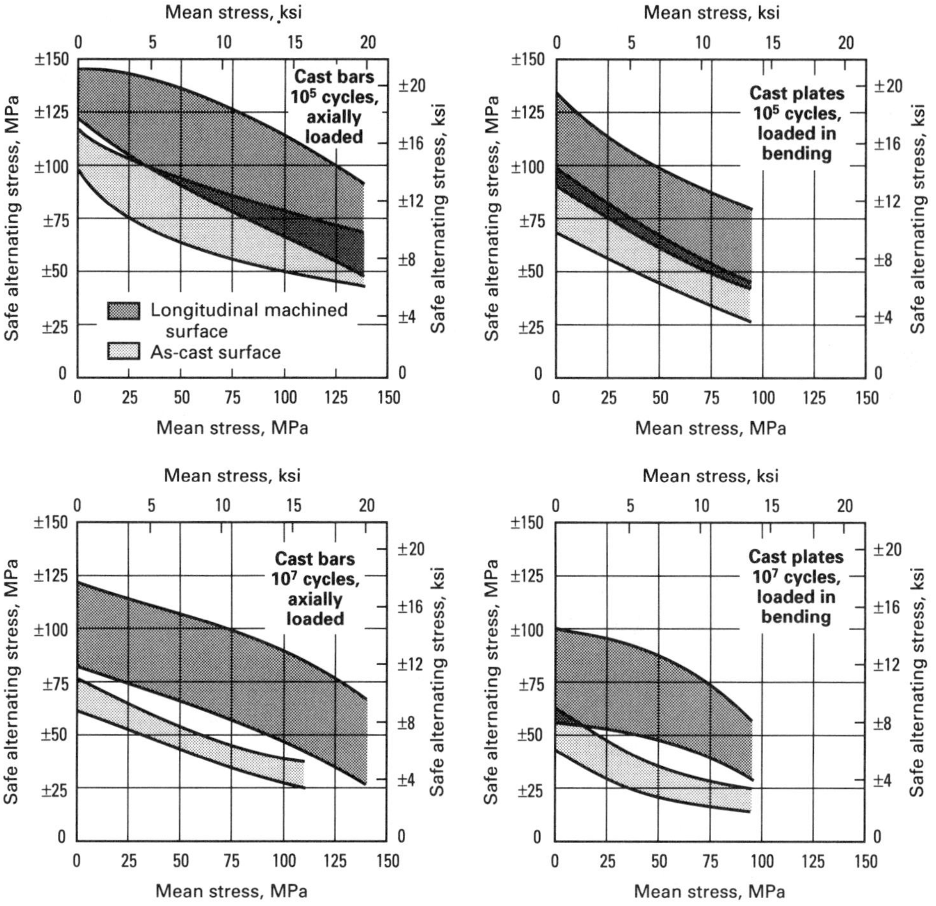

Fig. 3 Effect of surface type on the fatigue properties of cast magnesium-aluminum-zinc alloys

edge distance (from the center of the hole) of 2½ times the pin diameter and a width of 8 times the pin diameter. Increasing edge distance to more than about twice the pin diameter had little effect on bearing strength values. Sheet thicknesses in a wide range were tested, and the ratio of pin diameter to sheet thickness had no observable effect, except when buckling occurred. A pin diameter not greater than four times the sheet thickness prevented buckling.

Shear strength is an important consideration in the design of joints in magnesium parts, such as threaded joints and spot welds. Values for castings and extrusions given in Table 3 were obtained by the conventional double-shear method, using solid rods. Values for sheet and AZ80A-T5 structural shapes were obtained by the punch method, using flat specimens.

Hardness and Wear Resistance. Magnesium alloys have sufficient hardness for all structural applications except those involving severe abrasion. Hardness values are given in Table 3 and in the article "Properties of Magnesium Alloys" in this Volume. Although rather wide variations in hardness are observed in magnesium alloys, resistance of the alloys to abrasion varies by only about 15 to 20%. When subjected to wear by rubbing, by frequent removal of studs, or by heavy bearing loads, magnesium can be protected by inserts of steel, bronze, or nonmetallic materials; these materials can be attached as sleeves, liners, plates, or bushings. Such inserts may be attached mechanically by pressing, shrinking, riveting, bolting, or bonding; in castings, inserts may be cast in place.

Magnesium alloys perform satisfactorily as bearing materials for applications in which:

- Loads do not exceed 14 MPa (2 ksi)
- Shafts are hardened (350 to 600 HB)
- Lubrication is ample
- Speeds are low (5 m/s, or 1000 ft/min, max)
- Operating temperatures do not exceed 105 °C (220 °F)

Fatigue strength of magnesium alloys, as determined using laboratory test samples, covers a relatively wide scatter band, which is characteristic of other metals as well. The *S-N* curves have a gradual change in slope and become essentially parallel to the horizontal axis at 10 to 100 million cycles.

Fatigue strengths are higher for wrought products than for cast test bars. The fatigue strengths of several alloys are listed in the article "Properties of Magnesium Alloys" in this Volume. Increasing surface smoothness improves resistance to fatigue failure. For example, removing the relatively rough as-cast surfaces of castings by machining improves the fatigue properties of the castings (Fig. 3). Sharp notches, small radii, fretting, and corrosion are more likely to reduce fatigue life than are variations in chemical composition or heat treatment.

A specimen of alloy ZK60A-T5 (static yield strength, 290 MPa, or 42 ksi) with a machined 60° notch of 0.025 mm (0.001 in.) radius has a fatigue limit of 28 MPa (4 ksi) at 500 million cycles, compared with 110 MPa (16 ksi) for an unnotched specimen. This is a notch factor of about 0.25. For a shorter life of 100 000 cycles, the notch factor is about 0.48. As the severity of the notch decreases, its effect on fatigue limit decreases rapidly. For instance, a semicircular notch with radius of 1.2 mm (0.047 in.) reduces fatigue strength by only 20%, compared with 75% for the sharp V-notch cited above.

When fatigue is the controlling factor in design, every effort should be made to decrease the severity of stress raisers. Use

Table 8 Damping capacity of selected magnesium alloys and other metals

Alloy	Temper	Specific damping capacity, %, at 7.0 MPa (1.0 ksi)	14 MPa (2.0 ksi)	20 MPa (3.0 ksi)	25 MPa (3.5 ksi)	35 MPa (5.0 ksi)
Magnesium alloys						
AM60A,B	F	5.33	13.33	24.0	32.0	52.0
AS21A	F	16.0	33.33	48.0	53.33	60.0
AS41A,XB	F	5.33	13.33	21.33	28.0	44.0
AZ31B	F	1.04	1.57	2.04	2.38	2.72
AZ91A,B,D	F	2.67	5.33	12.0	16.0	29.33
AZ92	F	0.17	0.45	2.09	5.54	...
	T4	0.50	1.04	1.29	2.62	3.78
	T6	0.35	0.70	1.64	3.08	4.78
EZ33A	T5	...	4.88	12.55	18.15	22.42
HK31	T6	0.37	0.66	1.12	...	...
HZ32A	T5	1.93	7.81	11.64	...	...
K1A	F	40.0	48.8	56.0	61.7	66.1
M1A	F	0.35	1.28	2.22	3.14	3.92
ZE41A	T5	1.86	1.94	2.02	2.06	2.19
Aluminum alloys						
355	T6	...	0.51	0.67	1.0	...
356	T6	0.3	0.48	0.62	0.82	1.2
Cast iron	...	...	5.0	12.2	14.2	16.5

of generous fillets in reentrant corners and gradual changes of section greatly increase fatigue life. Situations in which the effects of one stress raiser overlap those of another should be eliminated. Further improvement in fatigue strength can be obtained by inducing stress patterns conducive to long life. Cold working the surfaces of critical regions by rolling or peening to achieve appreciable plastic deformation produces residual compressive surface stress and increases fatigue life.

Surface rolling of radii is especially beneficial to fatigue resistance because radii generally are the locations of higher-than-normal stresses. In surface rolling, the size and shape of the roller, as well as the feed and pressure, are controlled to obtain definite plastic deformation of the surface layers for an appreciable depth (0.25 to 0.38 mm, or 0.010 to 0.015 in.). In all surface working processes, caution must be exercised to avoid surface cracking, which decreases fatigue life. For example, if shot peening is used, the shot must be smooth and round. The use of broken shot or grit can result in surface cracks.

Damping capacity is the ability of a metal to elastically absorb vibrational energy and keep the vibrations from transmitting through the metal. Magnesium and magnesium alloys have excellent damping capacity, giving them the ability to decrease vibration and noise in many applications. These applications include vibration testing fixtures and mounting brackets for electronic equipment that is sensitive to vibrations. The percent damping capacity of magnesium alloys is compared with that of other metals in Table 8.

Low-Temperature Properties. With decreasing temperature, magnesium alloys increase in tensile strength, yield strength, and hardness but generally decrease in ductility. Tables 9(a) and 9(b) give the results of tensile tests at both room and low temperatures.

Elevated temperatures have adverse effects on ultimate and yield strengths, as demonstrated by the bearing strength data given for two magnesium alloys in Table 10. The effect of elevated temperatures on the mechanical properties of magnesium alloys is evaluated by considering:

- The strength as determined by bringing the test specimen up to temperature and testing immediately (short-time test)
- The strength at temperature after prolonged heating at elevated temperature
- The effect on room-temperature properties of heating at elevated temperature for short and long times
- The deformation produced by prolonged heating under load (creep test)

Data showing the effects of elevated temperature on the mechanical properties of several magnesium alloys are presented in Tables 11 and 12. Elevated-temperature properties, including isochronous stress-strain curves, are given in greater detail in the data compilations in the article "Properties of Magnesium Alloys" in this Volume.

Designs of many parts for use at elevated temperatures under continuous loads are based on maximum allowable deformation. The limiting creep-stress values given in Table 12 are based on 0.2% total extension. The isochronous curves in the data compilations cover a wide range of deformations. The alloys that contain thorium have the greatest resistance to creep at 205 and 315 °C (400 and 600 °F), and the magnesium-aluminum-zinc alloys have the lowest resistance. The decrease in modulus of elasticity with increase in temperature, a characteristic of the magnesium-aluminum-zinc alloys, is considerably less for thorium-containing alloys.

Selection of Product Form

Selection of a particular product form for a structural application is based on mechanical property requirements and on cost, availability, and fabricability. Requirements for production and design may change under operating conditions or as need arises. A part originally machined from bar stock may subsequently be made by extrusion or forging. Assemblies built up by joining sheets and extrusions may be redesigned as castings with equivalent performance at lower cost.

Castings. Parts too intricate to fabricate economically by other methods can be produced as castings. Sand, permanent mold, and die castings are more widely used than investment and shell mold castings. The choice of casting method is determined primarily by the size, shape, quantity, cost, and desired mechanical properties of the casting. The cost of magnesium alloy castings is governed largely by ingot price, alloy castability, and required heat treatment. Ingot price increases with additions of rare-earth metals, zirconium, silver, yttrium, and thorium. Small changes in composition can affect the cost of heat treatment. Comparative costs for casting an aircraft engine part from three different magnesium alloys are given in Table 13.

Magnesium alloys are cast by the permanent mold process when the number of parts required justifies the very high cost of equipment. The mechanical properties of sand and permanent mold castings are comparable, but the permanent mold process normally provides closer control of dimensions and produces better cast surfaces. Because of the slow solidification rate inherent in the processes, sand and permanent mold castings usually require a heat treatment to obtain the required mechanical properties. Temper designations are shown in Table 2.

The cost of castings is also influenced by such factors as required tolerances, mold and die costs, and machining costs. The quantity of a part to be produced is an important factor affecting cost and must be considered in seeking the most economical method of production. For example, in making the part illustrated in Fig. 4, tooling costs would have made investment casting more expensive if only 30 pieces were produced; however, because more than 30 pieces were produced, investment casting was cheaper than machining from bar stock. Although rejection rates vary from one casting method to another (Fig. 5), they are high for all methods when tolerances are close.

Figure 6 illustrates a casting that was produced at lower cost by shell mold casting than by sand casting and machining when more than 700 castings were made. The sand mold casting required one extra machining operation to obtain a dimension

Table 9(a) Low-temperature tensile properties of selected sheet and plate magnesium alloys and magnesium weldments

Alloy	Thickness mm	Thickness in.	Tensile strength MPa	Tensile strength ksi	Yield strength MPa	Yield strength ksi	Elongation, %
Transverse tests of plate alloys at 24 °C (75 °F)							
HK31A-H24	6.35	0.250	243	35.2	180	25.9	21.0
HK31A-O	6.35	0.250	200	29.0	125	18.0	30.5
HM21A-T8	6.35	0.250	241	35.0	170	24.8	13.7
Longitudinal tests of sheet and plate alloys at 24 °C (75 °F)							
HK31A-H24	1.63	0.064	250	36.3	200	29.0	7.5
HK31A-H24	6.35	0.250	238	34.5	190	27.3	14.2
Welded(a)	6.35	0.250	200	28.8	150	21.7	2.4
HK31A-O	1.63	0.064	205	29.7	123	17.9	27.5
HK31A-O	6.35	0.250	200	28.9	122	17.7	29.7
Welded(a)	6.35	0.250	160	23.4	119	17.3	3.2
HM21A-T5	(b)	(b)	210	30.4	107	15.5	8.0
HM21A-T8	1.63	0.064	222	32.2	160	23.1	7.2
HM21A-T8	6.35	0.250	223	32.4	173	25.1	5.6
Welded(a)	6.35	0.250	197	28.6	128	18.6	2.7
Longitudinal tests of sheet and plate alloys at −54 °C (−65 °F)							
HK31A-H24	1.63	0.064	300	43.3	220	32.0	5.0
	6.35	0.250	280	40.8	230	33.4	9.0
HK31A-O	1.63	0.064	275	39.9	147	21.4	20.7
	6.35	0.250	265	38.3	148	21.5	18.0
HM21A-T5	(b)	(b)	272	39.5	110	15.8	9.3
HM21A-T8	1.63	0.064	273	39.6	177	25.6	6.2
	6.35	0.250	265	38.4	205	29.7	4.7
Longitudinal tests of sheet and plate alloys at −72 °C (−98 °F)							
HK31A-H24	1.63	0.064	295	42.7	210	30.6	4.2
HK31A-H24	6.35	0.250	292	42.4	235	33.8	11.5
Welded(a)	6.35	0.250	195	28.1	163	23.6	0.5
HK31A-O	1.63	0.064	285	41.3	145	21.1	17.5
HK31A-O	6.35	0.250	277	40.2	150	21.9	20.2
Welded(a)	6.35	0.250	203	29.4	145	21.0	2.2
HM21A-T5	(b)	(b)	275	40.0	112	16.3	8.3
HM21A-T8	1.63	0.064	280	40.8	152	22.1	17.5
	6.35	0.250	275	40.1	215	31.3	5.0
	6.35	0.250	200	29.2	120	17.5	1.5
Longitudinal tests of sheet and plate alloys at −196 °C (−320 °F)							
HK31A-H24	1.63	0.064	372	54.0	227	33.0	6.2
HK31A-H24	6.35	0.250	365	52.9	240	34.7	8.0
Welded(a)	6.35	0.250	232	33.7	180	25.9	1.5
HK31A-O	1.63	0.064	330	47.9	168	24.3	12.7
HK31A-O	6.35	0.250	325	47.2	170	24.7	12.5
Welded(a)	6.35	0.250	205	29.7	150	21.6	2.2
HM21A-T5	(b)	(b)	320	46.6	125	18.1	8.0
HM21A-T8	1.63	0.064	328	47.6	172	24.9	4.0
HM21A-T8	6.35	0.250	325	47.3	210	30.6	4.2
Welded(a)	6.35	0.250	228	33.1	145	20.9	1.5

Note: Values for wrought alloys are averages of two to four tests at room temperature (50 mm, or 2 in. gage length). Values of duplicate tests at low temperatures are also averages (25 mm, or 1 in. gage length). Values for cast alloys (Table 9b) are averages of two to four tests on separately cast bars. (a) Welding rod was EZ33A; weld bead intact. (b) Specimens machined from a forging

that could be held within the tolerance of the shell mold.

Die castings made of magnesium alloys may be selected in preference to aluminum die castings of the same design because of the savings in weight. Magnesium die castings may also replace assembled steel parts, zinc die castings, and plastics. Service requirements and size may govern whether a magnesium alloy is selected for use in a die casting, but quantity is the most important factor because die castings are high-production items. The effects of both quantity and weight are shown in Fig. 7. Magnesium alloy die castings, like castings in general, are always priced and purchased on a per piece basis. The cost per pound varies, depending primarily on complexity of design, wall thickness, number of cavities in the mold, and quality level.

Magnesium has several processing advantages over aluminum in die casting. The low heat content of magnesium produces rapid solidification of molten metal and short die holding times; therefore, production rates can be much higher for magnesium. This low heat content also reduces thermal fatigue and leads to longer die life when casting magnesium as opposed to aluminum. The low reactivity of magnesium alloys with iron allows the use of steel processing equipment and reduces die erosion, which again leads to longer die life. Depending on the application, magnesium can be cast in both hot and cold chamber die casting machines, whereas aluminum is most commonly processed only in cold chamber machines. Hot chamber machines can offer significantly increased production rates over cold chamber machines.

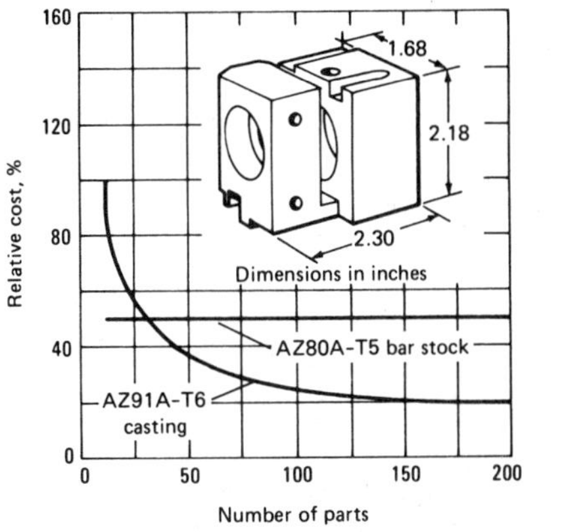

Fig. 4 Comparison of cost-quantity relationships for producing a part by investment casting and by machining from bar stock

Extrusions. Magnesium alloys are extruded as round rods and as a variety of bars, tubes, and shapes. A wide variety of special shapes also can be extruded. Extrusion is selected as a means of producing certain shapes when:

- Several small extrusions or a combination of extrusions and sheet can be joined to form an assembly
- Shapes are desired that are uneconomical to machine from castings
- Pieces cut from extrusions can replace individually cast or forged parts

The extrusion process offers many design possibilities not economically attainable by other production methods. These include reentrant angles and undercuts, thin-wall tubing of large diameter, and almost unrestricted variations in section thickness. Probably the most important factor in determining whether a magnesium alloy shape will extrude well is good symmetry, preferably around both axes.

Very thin and wide sections with large circumscribing circles should be avoided. The optimum width-to-thickness (*w*/*t*) ratio for magnesium extrusions normally is less than 20. Parts with higher ratios can be extruded, but require more generous tolerances. A thick section tapering to a thin wedge must always be modified by rounding the edge, or the die may not fill properly. A thin leg attached to a thick body of an extrusion should be limited to a length not exceeding ten times the leg thickness. Semiclosed shapes requiring long, thin die tongues should be avoided. For best extrudability, the length of the tongue should not exceed three times its width, although it is possible to extrude lengths five times the width. Similarly, shapes requiring unbalanced die tongues do not constitute good extrusion design. Hollow shapes that contain unsymmetrical voids, or voids separat-

Table 9(b) Low-temperature tensile properties of selected cast magnesium alloys

Alloy	Tensile strength MPa	Tensile strength ksi	Yield strength MPa	Yield strength ksi	Elongation, %	Charpy impact energy, Notched specimens J	Charpy impact energy, Notched specimens ft · lbf	Charpy impact energy, Unnotched specimens J	Charpy impact energy, Unnotched specimens ft · lbf
At 24 °C (75 °F)									
AZ91C-T6	290	41.8	132	19.2	6.3	7.96	5.87	1.36	1.00
AZ92A-T6	290	41.8	160	23.4	4.0	7.62	5.62	0.68	0.50
EZ33A-T5	190	27.5	115	16.9	7.6	7.46	5.50	0.84	0.62
HK31A-T6	225	32.7	112	16.3	9.5	16.61	12.25	3.80	2.81
QE22A-T6	280	40.6	213	30.9	4.4	23.5	17.3	1.36	1.0
ZH62A-T5	275	39.9	192	27.9	5.7	15.02	11.08	1.02	0.75
At −78 °C (−109 °F)									
AZ91C-T6	305	44.3	150	21.6	5.1	6.26	4.62	1.36	1.00
AZ92A-T6	295	42.7	170	24.6	2.3	6.44	4.75	0.76	0.56
EZ33A-T5	190	27.6	125	18.0	3.1	4.83	3.56	0.68	0.50
HK31A-T6	300	43.3	120	17.5	8.6	16.43	12.12	3.21	2.37
QE22A-T6	334	48.4	219	31.7	4.4	18.7	13.8	1.5	1.1
ZH62A-T5	330	47.6	200	29.2	2.7	18.99	14.00	1.02	0.75
At −196 °C (−321 °F)									
AZ91C-T6	310	44.9	180	26.0	1.7	4.06	3.00	1.02	0.75
AZ92A-T6	320	46.5	195	28.5	0.8	4.57	3.37	0.68	0.50
EZ33A-T5	200	29.0	140	20.3	2.2	5.00	3.69	0.68	0.50
HK31A-T6	330	48.1	135	19.6	6.1	13.72	10.12	3.05	2.25
QE22A-T6	359	52.1	233	33.8	2.4	14.4	10.6	1.1	0.8
ZH62A-T5	320	46.6	235	34.1	1.0	8.56	6.31	1.02	0.75

Note: Values for wrought alloys (Table 9a) are averages of two to four tests at room temperature (50 mm, or 2 in. gage length). Values of duplicate tests at low temperatures are also averages (25 mm, or 1 in. gage length). Values for cast alloys are averages of two to four tests on separately cast bars.

ed by sections of inadequate thickness, are undesirable. Sharp outside corners result in excessive stress concentration and die breakage and thus should be avoided. Inside corners should be filleted to reduce stress concentration in the part and to ensure complete filling of the die during extrusion. Regardless of the shape being extruded, it is difficult to hold distances between thin sections to close tolerances.

Many shapes can be extruded economically. Extrusion dies are relatively inexpensive, and dimensions can be held closely enough so that machining is often unnecessary. Part A in Fig. 8 is an example of a shape that lends itself to extrusion. However, because more metal had to be machined from this part as an extrusion than as a die casting, die casting proved to be cheaper. A slight design change in the bracket shown as part B in Fig. 8 permitted the part to be made more economically by extrusion than by casting.

Impact extrusions are tubular parts of symmetrical shape. The impact type of hot extrusion is particularly applicable when:

- It is not practical to make the part by any other method, such as with parts requiring very thin walls, parts in which thin walls having high strength are essential or parts that must incorporate irregular profiles
- High production rates are required, where scrap loss from machining would be excessive if the part were made by other means, where strength requirements cannot be met by die castings, where the number of manufacturing operations or the number of parts in an assembly can be reduced by the use of impact extrusions, where portions of the part require zero draft, and/or where closer tolerances are required

In designing impact extrusions, the following factors should be taken into consideration:

- A wide variety of symmetrical shapes is possible
- Variations in wall thickness are possible (thin sidewall, thick bottom; thick sidewall, thin bottom)
- Ribs, flanges, bosses, and indentations can be incorporated
- Length-to-diameter ratios can range from 1.5:1 to 15:1. Ratios from 6:1 to 8:1 are considered good working ratios

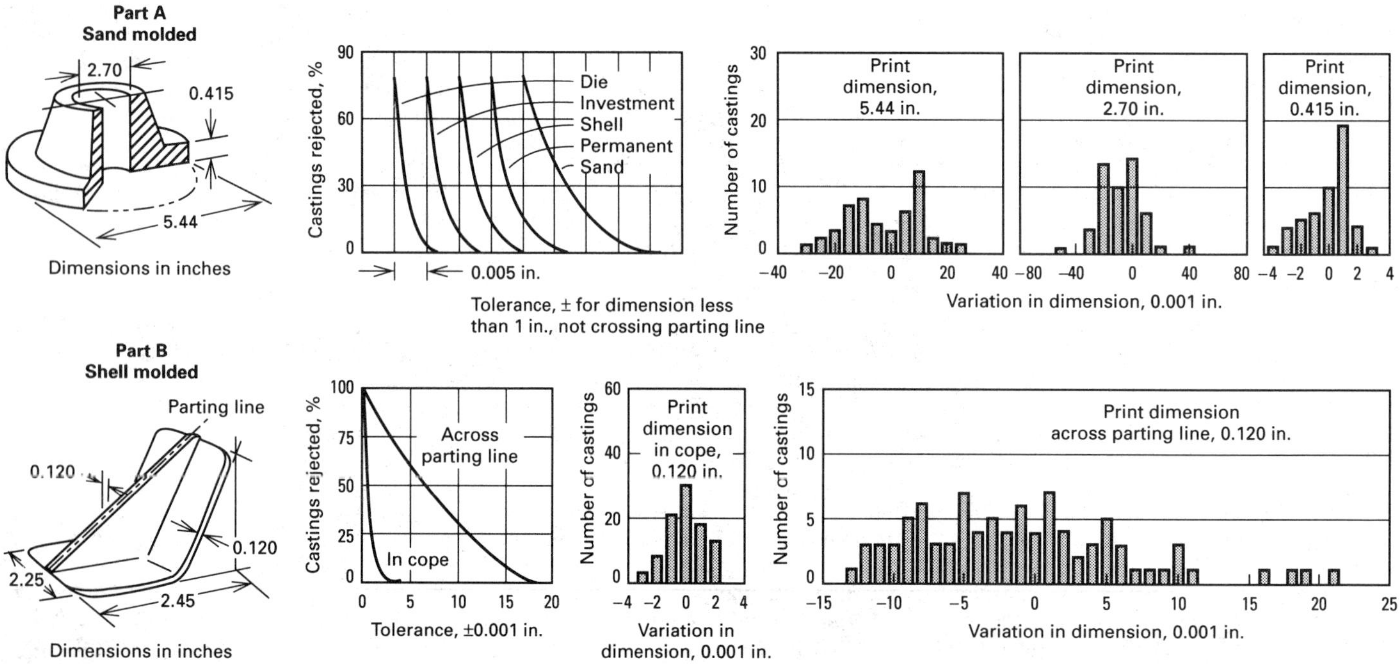

Fig. 5 Dimensional variations, tolerances, and rejection rates for two magnesium-aluminum-zinc alloy castings

Table 10 Typical bearing strengths of two magnesium alloys at elevated temperatures

Temperature		Bearing strength			
		Ultimate		Yield	
°C	°F	MPa	ksi	MPa	ksi
HK31A-H24 sheet					
20	70	420	61	285	41
205	400	285	41.4	210	30.3
260	500	225	32.4	190	27.8
315	600	170	25	145	21
HM21A-T8 sheet					
20	70	415	60	275	40
205	400	270	39	185	27
260	500	250	36	180	26
315	600	200	29	150	22
425	800	160	23	115	17

- Reduction in area varies with the alloy being extruded and is limited by the size of available equipment. Parts with reduction in area up to 95% have been made. In general, extrusions of alloys M1A and AZ31B can have thinner walls than AZ61A, AZ80A, and ZK60A extrusions
- Sharp corner radii are possible in some areas of impact extrusions. This is not true of other product forms
- The average properties of impact extrusions are slightly higher than the typical properties of the hot-extruded stock from which the parts are made

Forgings. Magnesium forgings can be produced in the same variety of shapes and sizes as forgings of other metals. Maximum size is limited primarily by the size of available equipment. Tolerances can be held to the same values as in normal forging of other metals; they vary somewhat with forging size and design.

Forgings have the best combination of strength characteristics of all forms of magnesium. They are used where light weight coupled with rigidity and high strength are required. Magnesium forgings are sometimes used because of their pressure tightness, machinability, and lack of warpage rather than because of their high strength-to-weight ratio.

Forging is used for parts produced in quantities sufficient to amortize die costs and for parts requiring high strength and ductility; greater uniformity and soundness than can be obtained with castings. For small quantities, hand forgings may be used, but die forgings have better mechanical properties and are less expensive in larger quantities.

The ease with which magnesium can be worked greatly reduces the number of forging operations needed to produce finished parts. Many of the steps commonly required in forging brass, bronze, and steel (such as punching, planishing, drawing and ironing, sizing and coining, and edging and rolling) are unnecessary in forging magnesium. Bending, blocking, and finishing are the principal steps used in the forging process. Recommended corner and fillet radii for magnesium forgings (Fig. 9) are given in Tables 14(a) and 14(b).

Die design and resulting metal flow cause variations in tensile properties at different sections of large forgings. An example involving aircraft wheels forged from magnesium alloys is illustrated in Fig. 10. Test specimens were taken at several locations in forgings made from alloy AZ80A-T5 and from alloy ZK60A in the T5 and T6 conditions.

Mechanical properties of magnesium alloy forgings or castings, as determined by testing of separately cast or forged bars, are useful for evaluating certain characteristics on a comparative basis and serve as a means of control. However, test results for these bars may vary significantly from properties of specimens taken from various locations in production castings or forgings. The amount of this variation is affected by section thickness and direction of metal flow. The wide variations in properties between forged brackets of the same design made from two different heat-treated alloys are shown in Fig. 11, and a comparison of the properties of separately cast and forged test specimens with those of specimens cut from castings and forgings are presented in Fig. 12. Forgings made from magnesium-

Table 11 Effect of elevated temperature on the tensile strengths of magnesium alloys

	Tested at exposure temperature										Tested at room temperature			
			Exposed 10 min at				Exposed 1000 h at				Exposed 1000 h at			
	20 °C (70 °F)		150 °C (300 °F)		315 °C (600 °F)		205 °C (400 °F)		315 °C (600 °F)		205 °C (400 °F)		315 °C (600 °F)	
Alloy	MPa	ksi	MPa	ksi	MPa	ksi	MPa	ksi	MPa	ksi	MPa	ksi	MPa	ksi
Castings														
AZ63A-T6	275	40	165	24	55	8	110	16	...	...	255	37	...	...
AZ92A-T6	275	40	195	28	55	8	115	17	...	...	270	39	...	...
EQ21A-T6	261	37.9	211	30.6	132	19.1	...	...	...	...	246	35.7	...	...
EZ33A-T5	160	23	145	21	83	12	130	19	76	11	170	25	180	26
HK31A-T6	215	31	195	28	125	18	180	26	62	9	240	35	180	26
HZ32A-T5	200	29	145	21	83	12	115	17	76	11	220	32	235	34
QE22A-T6	266	38.6	208	30.2	80	11.6	...	...	...	...	...	...	...	...
QH21A-T6	275	40	235	34	97	14	...	...	...	...	...	...	...	...
ZC63A-T6	242	35.1	179	26	...	...	...	...	...	...	...	...	...	...
ZE41A-T5	218	31.6	167	24.2	77	11.2	...	...	...	...	...	...	...	...
ZH62A-T5	290	42	195	28	69	10	...	...	...	...	...	...	...	...
WE43-T6	265	38.4	243	35.2	163	23.6	...	...	...	...	250	36.3	...	...
WE54-T6	280	40.6	255	37.0	184	26.7	235	34.1	...	...	272	39.5	217	31.5
Extrusions														
AZ80A-T5	380	55	235	34	69	10	...	...	...	...	...	...	...	...
ZK60A-T5	365	53	180	26	41	6	...	...	...	...	315	46	315	46
HM31A-F	275	40	195(a)	28(a)	115	17	...	...	...	...	...	...	...	...
Sheet														
AZ31B-H24	285	41	145	21	48	7	90	13	62(a)	9(a)	255	37	260	38
HK31A-T6	255	37	180	26	115	17	...	...	55	8	255	37	215	31
HM21A-T8	235	34	140	20	97	14	...	...	...	...	...	...	...	...

(a) Tested at 260 °C (500 °F)

Table 12 Effect of elevated temperatures on values of creep stress and elastic modulus for magnesium alloys

	Creep stress(a) at				Elastic modulus at			
	205 °C (400 °F)		315 °C (600 °F)		205 °C (400 °F)		315 °C (600 °F)	
Alloy	MPa	ksi	MPa	ksi	GPa	10^6 psi	GPa	10^6 psi
Castings								
AZ92A-T6	3.4	0.5	...	...	31	4.5	21	3.0
EQ21A-T6	62	9.0	...	...	41	6.0	33	4.8
EZ33A-T5	38	5.5	6.9	1.0	40	5.8	38	5.5
HK31A-T6	64	9.3	14	2.0	40	5.8	39	5.6
HZ32A-T5	52	7.5	22	3.2	40	5.8	39	5.6
QE22A-T6	55	8.0	...	...	37	5.4	31	4.5
ZC63A-T6	49	7.1	...	...	...	...	...	...
ZE41A-T5	31	4.5	...	...	41	6.0	24	3.5
ZH62A-T5	17	2.5	...	...	40	5.8	38	5.5
WE43A-T6	96	13.9	...	...	39	5.6	37	5.4
WE54A-T6	132	19.1	...	...	41	6.0	36	5.3
Extrusions								
ZK60A-T5	7	1.0(b)	...	...	...	...	...	...
HM31A-F	83	12.0	41	6.0	40	5.8	38	5.5
Sheet								
AZ31B-H24	7	1.0(b)	...	...	30	4.3	17	2.5
HK31A-T6	69	10.0	17	2.5	40	5.8	25	3.6
HM21A-T8	76	11.0	34	5.0	40	5.8	34	5.0

(a) Stress to produce 0.2% creep strain in 1000 h for cast alloys and 100 h for wrought alloys. (b) Tested at 150 °C (300 °F)

thorium alloys are considerably more expensive than those of magnesium-zinc-zirconium alloys. Comparing the costs of forging and casting requires careful analysis for each design and specification.

Inserts

Magnesium surfaces that are subjected to heavy bearing loads or severe wear require protection. Inserts that provide such protection to the surfaces of holes can be made of various materials and can be attached in numerous ways.

Cast-In Inserts. Inserts in magnesium may be fixed in place by casting the magnesium around them. Cast-in inserts for use in magnesium can be made of steel, brass, bronze, or other metals. Nonferrous inserts can be plated with chromium to prevent alloying with magnesium, although such plating is seldom used. Tinning of ferrous inserts prevents galvanic corrosion of the magnesium.

Cast-in inserts become securely fixed when the cast metal shrinks around them. The insert is even more securely fixed if the outside of the insert is knurled or grooved. Care should be taken to ensure that stress concentrations are not set up by sharp corners or insufficient metal around the insert.

Shrinkage of the magnesium alloy around cast-in inserts can cause high residual stress in the metal surrounding the insert. This possibility can be significantly reduced by preheating the insert, although inserts with a wall thickness of 1.3 mm (0.050 in.) or less will preheat sufficiently from contact with the hot die and the molten magnesium. If these design and manufacturing conditions are not followed, the high residual stresses may lead to failure of the part in service brought about by stress-corrosion cracking. Safe design dimensions for cast-in steel inserts with a wall thickness greater than 1.3 mm (0.050 in.) are shown in Fig. 13.

In highly stressed castings, it may be desirable to cast in a pilot to which the insert can be attached, thus avoiding high shrink stresses around the insert. Cast-in inserts can also be used for providing design details not otherwise feasible, such as lubrication lines of steel or copper, or appendages that otherwise could not be attached conveniently. Cast-in inserts complicate manufacture of castings and should be used only if other methods are not feasible.

Press-Fit and Shrink-Fit Inserts. Because of the low modulus of elasticity of magnesium, greater interference must be used for press-fit or shrink-fit inserts in magnesium than for inserts in other metals in order to obtain sufficient gripping force. An interference of 0.5 to 1.0 mm/m (0.0005 to 0.001

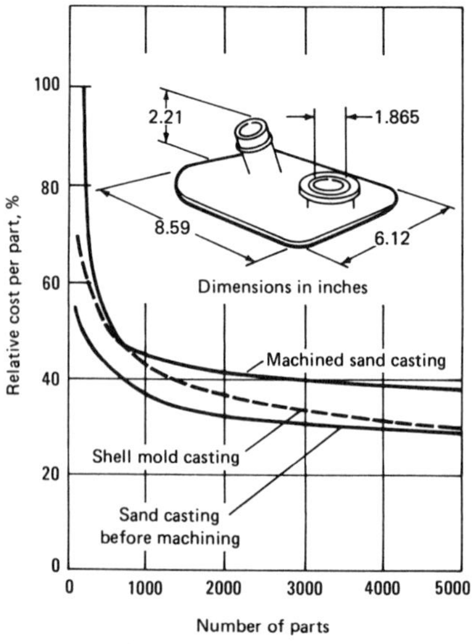

Fig. 6 Comparison of cost-quantity relationships for two methods of casting a magnesium-aluminum-zinc alloy part. The sand casting required an extra machining operation to meet a dimensional limit that could be held in the shell mold casting without machining. Thus, the curve for the machined sand casting should be used in comparing the total costs of the two casting methods.

Table 13 Comparison of costs for making a typical aircraft engine casting from three different magnesium alloys

Item	AZ91C	ZE41A	QE22A
Cores	$ 41.18	$ 41.18	$ 41.18
Molding	79.44	79.44	79.44
Metal	76.33	125.83	310.77
Cleaning	56.66	56.66	56.66
Heat treatment	18.28	1.39	8.03
Visual inspection	11.00	11.00	11.00
Nondestructive testing	94.68	94.68	94.68
Fixturing	8.80	8.80	8.80
Total cost per casting	**$386.37**	**$418.98**	**$610.56**

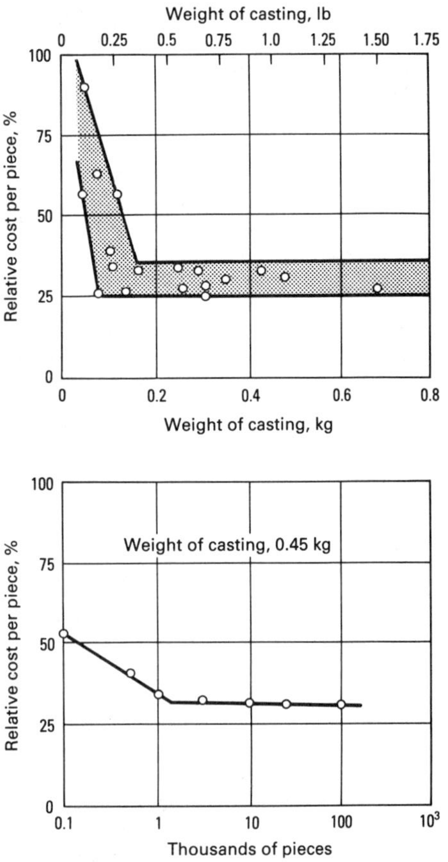

Fig. 7 Typical effects of weight and quantity on the cost of magnesium die castings. On a weight basis, the cost of die castings heavier than about 0.15 kg (⅓ lb) is fairly constant. For lighter castings, those weighing less than about 0.10 or 0.15 kg (4 or 5 oz), the cost per pound may vary by a factor of three or four, depending on design, wall thickness, and quality level.

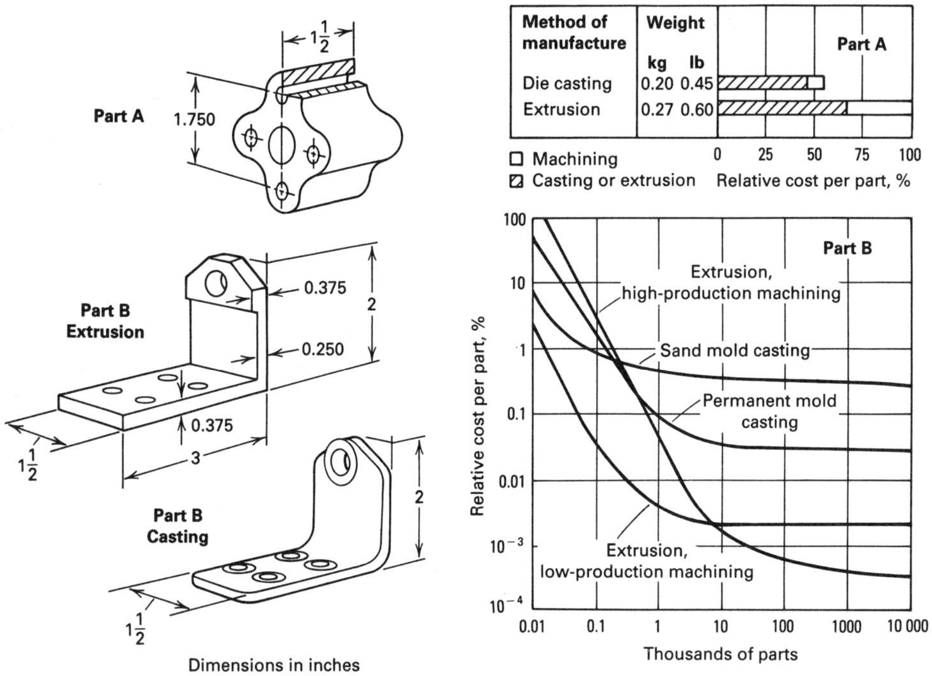

Fig. 8 Effect of manufacturing method on the cost of two magnesium alloy parts. Costs are based on AZ91B die castings and AZ31B extrusions.

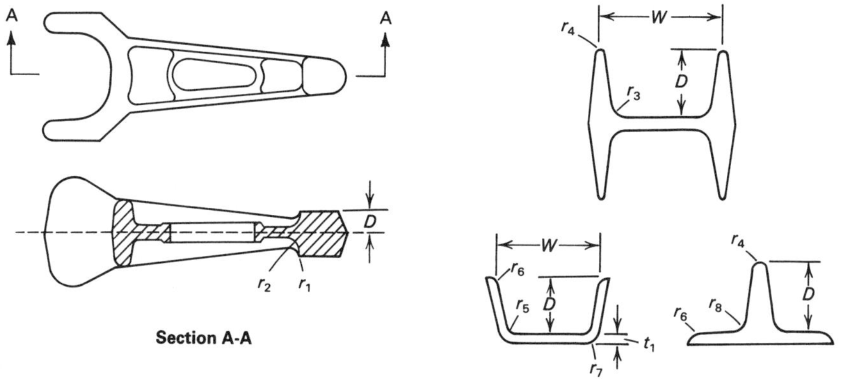

Fig. 9 Large magnesium forgings. Corner and fillet radii are given in Tables 14(a) and 14(b).

in./in.) is usually satisfactory, but may be increased appreciably where high-torque loads are likely to be encountered. Table 15 presents recommended interferences for steel and bronze inserts in normal service at various temperatures. The values given serve only as a guide because service temperature, type of insert material, severity of service, thickness of insert, and sensitivity of the alloy to stress-corrosion cracking all influence the correct amount of interference. Differences in thermal expansion, yield strength, and modulus of elasticity must also be considered for service at elevated temperature.

Inserts are more easily assembled by shrinking than by pressing. It is relatively easy to heat a magnesium part and thus expand a hole in the part so that it is large enough to receive an insert. Where necessary, the insert may be cooled to facilitate insertion. On the other hand, assembly by pressing requires careful machining of both insert and hole, and proper lubrication. When large interferences are required, it is best to specify shrink-fit inserts because press-fits may score the magnesium.

Screwed-In Inserts. Various types of screwed-in (and other mechanically attached) inserts may also be used successfully in magnesium parts. When screwed-in inserts are used, locations of threaded holes must be chosen carefully in order to avoid stress concentrations and to provide sufficient engagement length for the thread.

Formability

Magnesium alloys, like other alloys with hexagonal crystal structures, are much more workable at elevated temperatures than at room temperature. Consequently, magnesium alloys are usually formed at elevated temperatures, and cold forming is used only for mild deformations around generous radii. The methods and equipment used in forming magnesium alloys are the same as those commonly employed in forming alloys of other metals, except for differences in tooling and technique that are required when forming is done at elevated temperatures.

Working of metals at elevated temperature has several advantages over cold working. Magnesium parts usually are drawn at elevated temperature in one operation without repeated annealing and redrawing, thus reducing the time involved for making the part and also eliminating the necessity of additional die equipment for extra stages. Hardened dies are unnecessary for most types of forming. Hot-formed parts can be made to closer dimensional tolerances than cold-formed parts because they experience less springback. Suggested maximum forming temperatures and times for various wrought magnesium alloys are given in Table 16.

Sheet and Plate. Rolled magnesium alloy products include flat sheet and plate, coiled sheet, circles, tooling plate, and tread plate. These products are supplied in a variety of standard and nonstandard sizes (Table 17).

The ability to use increased section thickness without weight penalty is of particular importance in designs that employ magnesium sheet. Thick-sheet construction provides the rigidity necessary in a structure, without the need for costly assembly of ribs and similar reinforcing members.

Rolled magnesium alloy products can be worked by most conventional methods. For severe forming, sheet in the annealed (O temper) condition is preferred. However, sheet in the partially annealed (H24 temper) condition can be formed to a considerable extent. Because heat has significant effects on properties of hard-rolled magnesium, properties of the metal after exposure to elevated temperature must be considered in forming. The design curves shown in Fig. 14 give minimum values suitable for design use. Although the curves are based primarily on tests of sheet 1.63 mm (0.064 in.) thick or less, check tests indicate reasonable applicability for gages up to 6.35 mm (0.250 in.).

Figure 14 shows how the properties of AZ31B-H24 vary with exposure time at several temperatures. The curves have been extrapolated above the typical property levels of AZ31B-H24 sheet. Thus, if the value selected from a curve exceeds the actual property level of the material before exposure, the actual figure must be used.

Tests indicate that the effects of multiple exposures at elevated temperature are cumulative. Using Fig. 14 as an example, suppose that a part is held at 195 °C (380 °F) for 10 min; the resultant design compressive

Table 14(a) Corner and fillet radii (in millimeters) for large magnesium forgings

See Fig. 9 for identification of symbols.

D, mm	Minimum r_1, mm	Minimum r_2, mm
<4.76	2.38	6.35
4.76–10.32	4.76	10.32
10.32–25.4	7.94	15.87
25.4–44.45	12.70	25.4
44.45–63.50	19.04	31.75
63.50–101.6	26.99	41.28
>101.6	31.75	50.80

D, mm	r_3, mm	r_4, mm
9.52	6.35	1.59
12.70	7.14	1.59
15.87	8.73	1.59
19.04	9.52	1.98
22.22	11.11	2.38
25.40	11.91	2.78
28.58	12.70	3.17
31.75	14.28	3.97
38.10	15.87	4.76
44.45	18.25	5.56
50.80	20.64	6.35
57.15	22.22	7.14
63.50	23.81	7.93
69.85	26.99	8.73
76.20	28.57	9.52

D, mm	Minimum r_5, mm
W, 15.87 mm max	
10.32 max	2.78
W, 15.87–20.64	
10.32 max	2.78
10.32–12.70	3.97
W, 20.64–25.4 mm	
10.32 max	3.97
10.32–12.70	3.97
12.70–15.87	3.97
W, 25.4–32.54 mm	
10.32 max	3.97
10.32–12.70	3.97
12.70–15.87	3.97
15.87–20.64	6.35
W, 32.54–41.28 mm	
10.32 max	3.97
10.32–12.70	6.35
12.70–15.87	6.35
15.87–20.64	6.35
20.64–25.4	6.35
W, 41.28–50.80 mm	
10.32 max	6.35
10.32–12.70	6.35
12.70–15.87	6.35
15.87–20.64	10.32
20.64–25.4	10.32
25.4–33.34	10.32
W, 50.80–63.50 mm	
10.32 max	6.35
W, 50.80–63.50 mm (continued)	
10.32–12.70	6.35
12.70–15.87	6.35
15.87–20.64	10.32
20.64–25.4	10.32
25.4–33.34	10.32
W, 63.50–76.2 mm	
10.32 max	10.32
10.32–12.70	10.32
12.70–15.87	10.32
15.87–20.64	10.32
20.64–25.4	10.32
25.4–33.34	10.32
33.34–50.8	15.87
W, 76.2–101.6 mm	
10.32 max	10.32
10.32–12.70	10.32
12.70–15.87	10.32
15.87–20.64	10.32
20.64–25.4	10.32
25.4–33.34	15.87
33.34–50.8	15.87
W, 101.6–127 mm	
10.32 max	10.32
10.32–12.70	10.32
12.70–15.87	10.32
15.87–20.64	15.87
20.64–25.4	15.87
25.4–33.34	15.87
33.34–50.8	20.64

D, mm	r_6, mm
10.32 max	1.59
10.32–12.70	1.59
12.70–15.87	2.78
15.87–20.64	2.78
20.64–25.4	2.78
25.4–33.34	3.97
33.34–50.8	6.35

D, mm	r_8, mm
9.52 max	1.59
12.70	6.35
15.87	7.94
19.04	8.73
22.22	9.52
25.40	10.32
28.58	11.11
31.75	12.70
38.10	15.08
44.45	16.67
50.80	19.04
57.15	21.43
63.50	23.02
69.85	25.40
76.20	26.99

Note: The values given for r_3 and r_4, the corner and fillet radii for I sections in magnesium forgings, apply when the W/D ratio is approximately 2 to 1. For W/D ratios greater than 2 to 1, the radii should be increased; for lower W/D ratios the radii may be decreased. In corner and fillet radii for channel sections, $r_7 = r_5 + t_1$.

yield stress is 159 MPa (23.1 ksi). Then suppose the part is subsequently exposed for 300 min at 170 °C (340 °F). Because 200 min at 170 °C (340 °F) and 10 min at 195 °C (380 °F) both result in a minimum compressive yield stress of 159 MPa (23.1 ksi), they are equivalent exposures. The total equivalent exposure time at 170 °C (340 °F) then is 300 plus 200, or 500 min, which according to the data given in Fig. 14 results in a compressive yield stress of 155 MPa (22.5 ksi).

AZ31B-H24 sheet is commonly hot formed at temperatures below 160 °C (325 °F) to avoid annealing it to room-temperature property levels lower than the specified minimums. Annealing is a function of both time and temperature of exposure; thus, temperatures higher than 160 °C (325 °F) can be tolerated if exposure is carefully controlled. Table 18 shows the maximum permissible combination of time and temperature that will ensure that the specified minimum room-temperature properties of AZ31B-H24, HK31A-H24, and HM21A-T8 can be retained. This table is used for establishing limits of time and temperature for single exposures in normal forming operations. Whenever the sheet must endure multiple exposures, or whenever time of exposure at a given temperature must exceed the value indicated in Table 18, the data compilations in the article "Properties of Magnesium Alloys" in this Volume should be consulted.

Deep Drawing. Magnesium alloys can be cold drawn to a maximum reduction of 15 to 25% in the annealed condition. The cold drawability limit of alloy AZ31B-O is about 20%. The drawability, or percentage reduction in blank diameter, is calculated by the formula:

$$\text{Percentage reduction} = \frac{D - d}{D} \times 100$$

where D is the blank diameter before drawing and d is the diameter of the punch.

The time and temperature for annealing AZ31B-O are 15 min at 260 °C (500 °F). Cold-drawn parts are stress relieved at 150 °C (300 °F) for 1 h after the final draw, to eliminate the danger of cracking from residual stresses.

Both hydraulic and mechanical presses can be used in drawing operations. Hydraulic presses are utilized most often because they can operate at slower, more uniform speeds, thus providing accurate control when intricate draws are made. Deep draws also require the use of a hydraulic press because mechanical presses are not capable of extremely long strokes. Mechanical presses can be used for moderate draws at higher production rates.

The technique for hot drawing of magnesium alloy sheet has been developed largely so that drawing can be completed in a single operation. Heating magnesium alloys increases their drawability to such an extent that most parts can be made in a single draw. The amount of possible reduction increases as temperature increases up to about 230 °C (450 °F) for alloy AZ31B. It is usual practice to draw annealed AZ31B sheet to reductions as high as 68% in a single draw. When heated, magnesium can be drawn to higher reduction in a single draw than other metals. Maximum single-draw reduction as a function of temperature is shown in Fig. 15. These values were determined for 1.63 mm (0.064 in.) sheet using a cupping die 38 mm (1½ in.) in diameter. The radii on the draw ring and punch were both $5t$, and drawing speeds were 38 mm (1½ in.) and 500 mm (20 in.) per minute. The data in Fig. 15 show that the amount of reduction possible at a given temperature is also influenced by drawing speed.

The possibilities inherent in two-step draws are illustrated by the following parts: In the first operation, 610 mm (24 in.) blanks of 0.64 mm (0.025 in.) annealed sheet were drawn to a cup 200 mm (8 in.) in diameter by 400 mm (16 in.) deep; they were redrawn to

Table 14(b) Corner and fillet radii (in inches) for large magnesium forgings

See Fig. 9 for identification of symbols.

D, in.	Minimum r_1, in.	Minimum r_2, in.
<3/16	3/32	1/4
3/16–13/32	3/16	13/32
13/32–1	5/16	5/8
1–1 3/4	1/2	1
1 3/4–2 1/2	3/4	1 1/4
2 1/2–4	1 1/16	1 5/8
>4	1 1/4	2

D, in.	r_3, in.	r_4, in.	*D*, in.	r_3, in.	r_4, in.
3/8	1/4	1/16	1 1/2	5/8	3/16
1/2	9/32	1/16	1 3/4	23/32	7/32
5/8	11/32	1/16	2	13/16	1/4
3/4	3/8	5/64	2 1/4	7/8	9/32
7/8	7/16	3/32	2 1/2	15/16	5/16
1	15/32	7/64	2 3/4	1 1/16	11/32
1 1/8	1/2	1/8	3	1 1/8	3/8
1 1/4	9/16	5/32			

D, in.	Minimum r_5, in.
***W*, 5/8 in. max**	
13/32 max	7/64
***W*, 5/8–13/16 in.**	
13/32 max	7/64
13/32–1/2	5/32
***W*, 13/16–1 in.**	
13/32 max	5/32
13/32–1/2	5/32
1/2–5/8	5/32
***W*, 1–1 9/32 in.**	
13/32 max	5/32
13/32–1/2	5/32
1/2–5/8	5/32
5/8–13/16	1/4
***W*, 1 9/32–1 5/8 in.**	
13/32 max	5/32
13/32–1/2	1/4
1/2–5/8	1/4
5/8–13/16	1/4
13/16–1	1/4
***W*, 1 5/8–2 in.**	
13/32 max	1/4
13/32–1/2	1/4
1/2–5/8	1/4
5/8–13/16	13/32
13/16–1	13/32
1–1 5/16	13/32
***W*, 2–2 1/2 in.**	
13/32 max	1/4
13/32–1/2	1/4
1/2–5/8	1/4
5/8–13/16	13/32
***W*, 2–2 1/2 in. (continued)**	
13/16–1	13/32
1–1 5/16	13/32
***W*, 2 1/2–3 in.**	
13/32 max	13/32
13/32–1/2	13/32
1/2–5/8	13/32
5/8–13/16	13/32
13/16–1	13/32
1–1 5/16	13/32
1 5/8–2	5/8
***W*, 3–4 in.**	
13/32 max	13/32
13/32–1/2	13/32
1/2–5/8	13/32
5/8–13/16	13/32
13/16–1	13/32
1–1 5/16	5/8
1 5/8–2	5/8
***W*, 4–5 in.**	
13/32 max	13/32
13/32–1/2	13/32
1/2–5/8	13/32
5/8–13/16	5/8
13/16–1	5/8
1–1 5/16	5/8
1 5/8–2	13/16

D, in.	r_6, in.
13/32 max	1/16
13/32–1/2	1/16
1/2–5/8	7/64
5/8–13/16	7/64
13/16–1	7/64
1–1 5/16	5/32
1 5/8–2	1/4

D, in.	r_8, in.
3/8 max	1/16
1/2	1/4
5/8	5/16
3/4	11/32
7/8	3/8
1	13/32
1 1/8	7/16
1 1/4	1/2
1 1/2	19/32
1 3/4	21/32
2	3/4
2 1/4	27/32
2 1/2	29/32
2 3/4	1
3	1 1/16

Note: The values given for r_3 and r_4, the corner and fillet radii for I sections in magnesium forgings, apply when the *W*/*D* ratio is approximately 2 to 1. For *W*/*D* ratios greater than 2 to 1, the radii should be increased; for lower *W*/*D* ratios the radii may be decreased. In corner and fillet radii for channel sections, $r_7 = r_5 + t_1$.

a cup 140 mm (5½ in.) in diameter by 585 mm (23 in.) deep. Starting with a rectangular blank of 1.3 mm (0.051 in.) AZ31B-O, 455 × 485 mm (18 × 19 in.), a rectangular box 110 × 275 × 165 mm (4⅜ × 10¾ × 6½ in.) deep was drawn in the first operation. This box was then redrawn into a rectangular box 90 × 255 × 170 mm (3½ × 10 × 6¾ in.) deep, having 5.6 mm (7/32 in.) corner radii.

For most parts, however, depth of draw is not a primary consideration, and usually no trouble is experienced in drawing to the depth required. More trouble is encountered in keeping the metal free from puckers in parts with rounded corners or contours. Temperatures above those required for maximum drawability often are necessary to eliminate these puckers. On unusual or difficult jobs, it may be necessary to vary the procedure to obtain minimum scrap.

Choice of die materials is influenced chiefly by the severity of the operation and the number of parts to be produced. For most applications, unhardened low-carbon steel boiler plate or cast iron is satisfactory. For runs of 10 000 parts or more, for maximum surface smoothness, or for close tolerances where no significant die wear can be tolerated, hardened tool steels are recommended. W1 or O1 tool steels are satisfactory for extremely long runs (one million parts). For the most severe draws, however, the more abrasion-resistant tool steels, such as A2 or D2, will probably be more satisfactory and more economical. For room-temperature drawing, it is usually desirable that die steels be heat treated to obtain near-maximum hardness in service. However, for elevated-temperature drawing, the maximum temperature to which the dies will be exposed in drawing must also be considered. In this situation, the dies must be tempered slightly above the maximum service temperature, even though some hardness may be sacrificed.

Dimensional allowances must be considered for pieces of the punch and die that will affect part size. For example, magnesium, aluminum, and zinc alloys can be used in the design of punches to minimize the difference in the coefficients of thermal expansion between magnesium and the punch. However, if mild steel or cast iron are used in the design of a punch, dimensional allowances must be given to account for the great difference in the coefficients of thermal expansion.

Parts with smooth bottoms are usually drawn with open dies or dies in which the punch does not bottom on the female, or bottom, plate. These parts are formed by the pull exerted on the blank by the pressure ring and the draw ring. Mating dies are used with magnesium only in forming parts having reentrant portions that cannot be made by other means.

Stretch Forming. Both magnesium sheet and magnesium extrusions can be stretch formed. The temper of the alloy has no effect on the techniques employed. Sheet is usually heated to 165 to 290 °C (325 to 550 °F) and slowly stretched to the desired contour. Annealed sheet can be stretched at room temperature to a limited extent. However, the formabilities of alloys of any temper are so much greater at elevated temperatures than at room temperature that elevated-temperature forming is preferred for most operations. The percentage of differential stretching (*P*) during stretch forming is expressed:

$$P = \frac{L - S}{S} \times 100$$

where *L* is the longest stretched length in the part and *S* is the shortest comparable length parallel to the longest stretched length. *L* and *S* are assumed to be of equal length before stretching.

Hot stretch forming results in minimum springback; the little springback that may

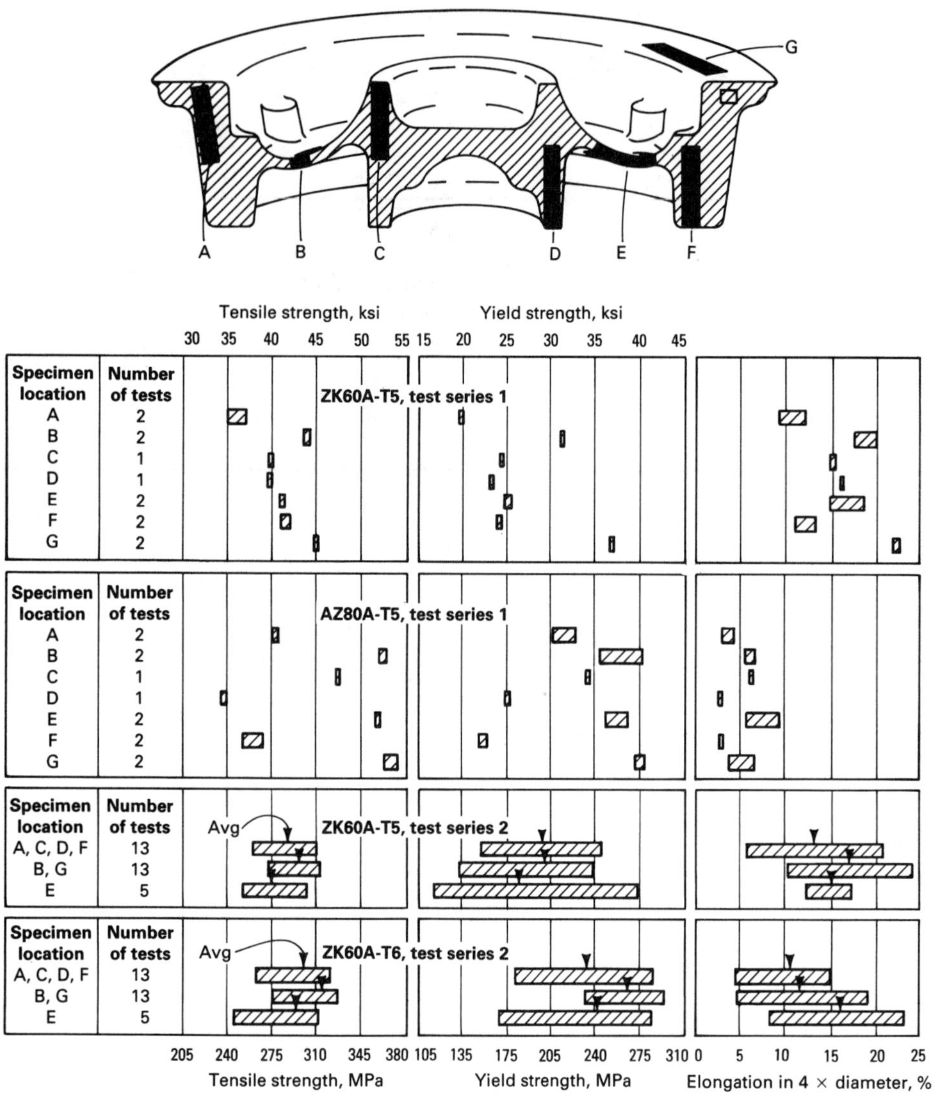

Fig. 10 Effect of alloy, heat treatment, and specimen location on mechanical properties of forged magnesium aircraft wheels

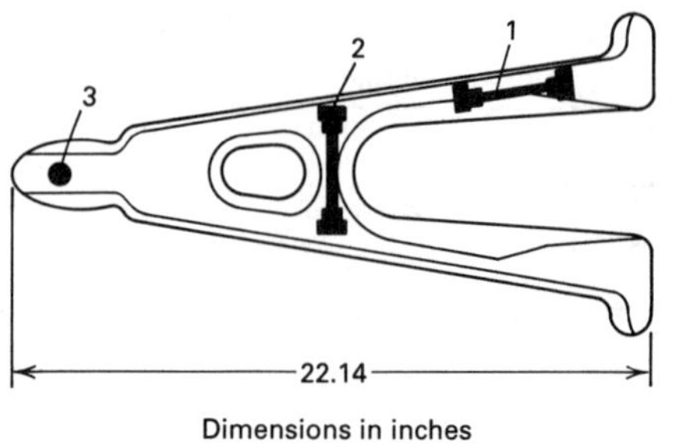

Specimen location	Tensile strength MPa	Tensile strength ksi	Yield strength MPa	Yield strength ksi	Elongation in 50 mm (2 in.), %
ZK60A-T					
1	305	44.1	238	34.5	8.5
2	302	43.8	203	29.5	23.5
3	260	37.7	131	19.0	12.5
AZ80A-T5					
1	368	53.4	279	40.5	4.0
2	333	48.3	243	35.2	2.7
3	283	41.1	163	23.7	3.2

Fig. 11 Comparison of the mechanical properties of forged brackets made of two different magnesium alloys

occur is controlled by adding about 1% to the total stretch. Contour and springback control are good for positive contour radii curvature up to 6 m (20 ft).

For differential stretching of sheet over dies of low curvature, the maximum practical limit is about 15%. A 12% maximum is considered more desirable, however, because it permits enough overstretch for springback control.

Wrinkling is often a problem in the stretch forming of magnesium sheet, particularly in making asymmetrical parts of low curvature. The best way to control wrinkling is to build the proper restraints in the dies.

Dies can be made of a variety of materials, including magnesium alloys, aluminum alloys, zinc alloys, iron, or steel. One aircraft company has reported the use of dies made of concrete cast over a wire mesh and heated by electrical resistance. Zinc alloy blocks should not be used above 230 °C (450 °F). Grippers should not have sharp serrated edges, which tear magnesium alloys; the use of emery paper between the grips and the magnesium sheet helps to reduce the possibility of tearing.

Heat control is important, and proper arrangement of heating units provides correct heat distribution. In addition to electrical-resistance heating, infrared radiant heating units can be employed. Thermostatic control is essential and should be used on all heaters. The resistance heaters can be placed at various points where critical forming occurs. It is important that heating be sufficient to induce plastic flow, yet not great enough to cause excessive elongation and rupture of the part.

Die temperature varies with the material being formed. AZ31B-O sheet is usually formed at 290 °C (550 °F) without loss of mechanical properties. Hard-rolled AZ31B-H24 can withstand a temperature of 165 °C (325 °F) for 1 h and higher temperatures for shorter periods. The maximum time at tem-

Table 15 Approximate interferences suggested for press-fit and shrink-fit inserts in magnesium

Outside diameter mm	Outside diameter in.	Wall thickness mm	Wall thickness in.	Steel inserts, operating temperatures 21 °C (70 °F)	38 °C (100 °F)	93 °C (200 °F)	149 °C (300 °F)	Bronze inserts, operating temperatures 21 °C (70 °F)	38 °C (100 °F)	93 °C (200 °F)	149 °C (300 °F)
12.70	0.50	1.6	0.06	0.0004	0.0005	0.0009	0.0013	0.0004	0.0005	0.0007	0.0009
25.40	1.0	2.2	0.09	0.0006	0.0010	0.0017	0.0025	0.0006	0.0008	0.0012	0.0016
50.80	2.0	3.3	0.13	0.0010	0.0015	0.0031	0.0048	0.0010	0.0013	0.0019	0.0028
76.20	3.0	4.0	0.16	0.0015	0.0024	0.0047	0.0072	0.0015	0.0019	0.0028	0.0044
101.6	4.0	4.8	0.19	0.0021	0.0034	0.0062	0.0096	0.0021	0.0026	0.0039	0.0060
127.0	5.0	5.6	0.22	0.0028	0.0044	0.0080	0.0121	0.0028	0.0034	0.0053	0.0079
152.4	6.0	6.4	0.25	0.0036	0.0053	0.0098	0.0147	0.0036	0.0043	0.0070	0.0098

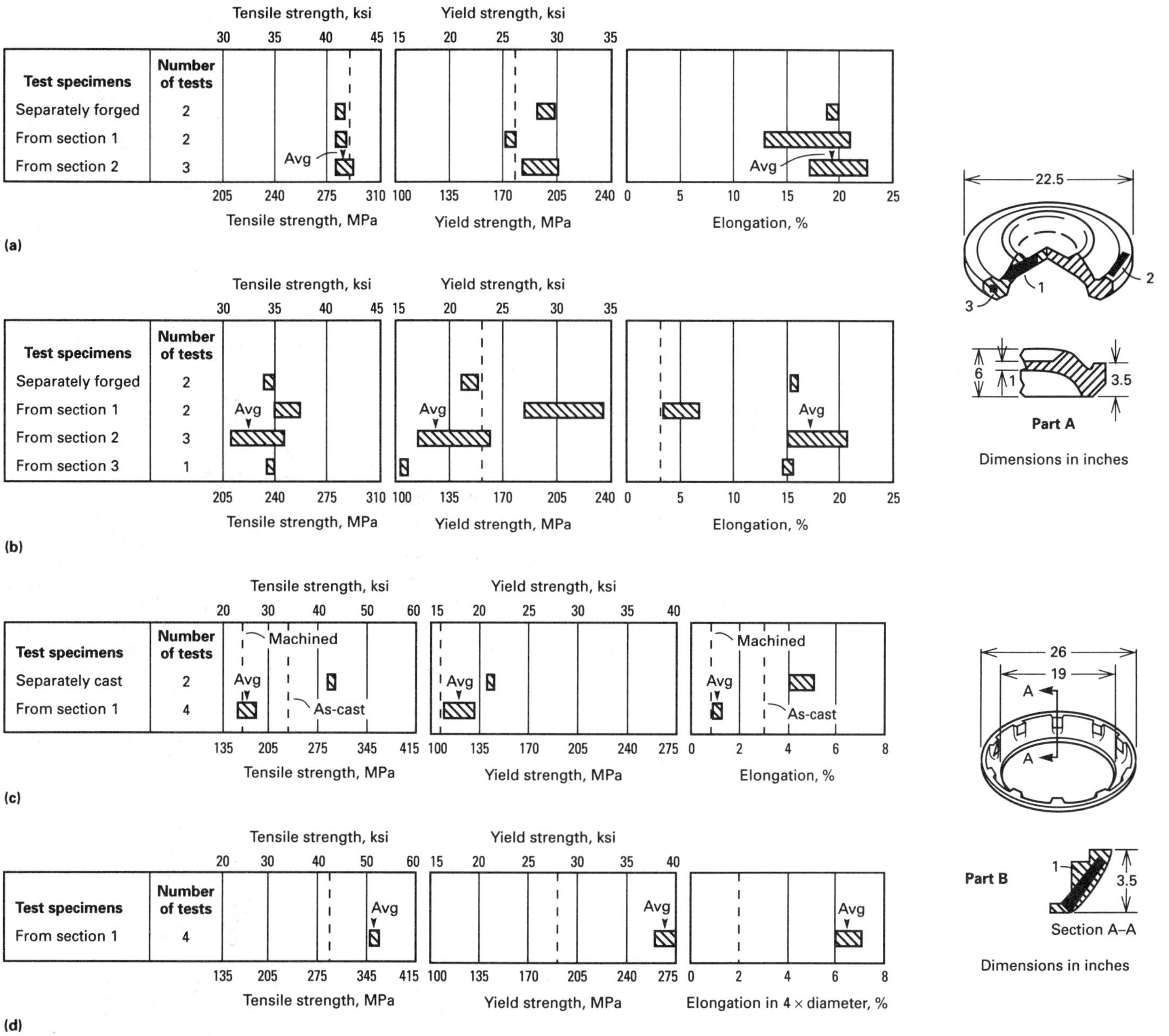

Fig. 12 Comparison of mechanical properties of separately forged or cast specimens with those of specimens cut from forgings or castings. The dashed lines indicate the specified minimums for each property. (a) Part A, ZK60A-T5 forgings. (b) Part A, HM21A-T5 forgings. Approximate weight of part A, 34 kg (75 lb). (c) Part B, AZ91C-T6 castings. Approximate weight of part B, 10 kg (22 lb). (d) Part B, AZ80A-T5 forgings. Approximate weight of part B, 9 kg (20 lb)

perature to maintain original properties is given in Table 18. Extruded magnesium alloys can be heated to 315 °C (600 °F) in most instances. For most contours, only heating of the die is necessary because the sheet will pick up heat quickly from the hot die. More complicated parts may require additional heating of the sheet by radiant heat or thermal blankets. Thick material may require the same treatment.

Lubricants are normally recommended; colloidal graphite is the best for high-temperature forming. Lower-temperature operation (≦260 °C, or 500 °F) may permit the use of heat-resisting waxes and greases, or a dry-film stearate-type lubricant instead of colloidal graphite. Some shops have used heat-resisting synthetic rubber or fiberglass cloth between the magnesium sheet and the forming block.

Annealed magnesium sheet can be shrunk satisfactorily at room temperature. The amount of shrinkage possible can be greatly increased by heating. Standard shrinking machines are employed.

Extrusion Formability. Production bending of extrusions can be done on standard angle rolls, in mating dies, in stretch-forming machines, or in other specialized bending equipment. If the forming is severe, extrusions are heated to approximately 260 to 345 °C (500 to 650 °F) and formed hot.

Some bend radii data for flat extruded magnesium strip, as well as maximum forming temperatures and times, are given in Table 19. Bending of more complex extrusions is more difficult, and the bend radii required must be established for a given shape. The minimum bend radii for round magnesium tubing varies with the ratio of outer diameter to wall thickness (Table 20).

Joining of Magnesium Alloys

Welding. Magnesium alloys can be readily welded by gas metal arc welding and by resistance spot welding. Rods of approximately the same composition as the base

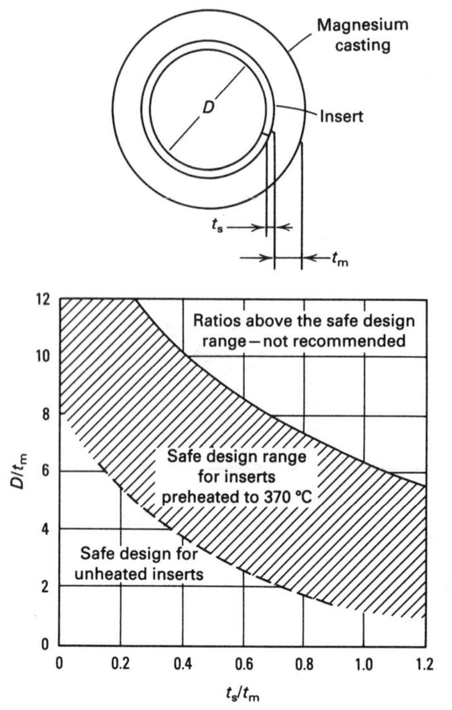

Fig. 13 Safe design dimensions for cast-in steel inserts in magnesium alloy castings. Values apply to inserts having a wall thickness (t_s) greater than 1.3 mm (0.050 in.).

metal are generally satisfactory. With alloys HM21A and HM31A, EZ33A rods give higher joint efficiencies (Table 21).

Table 16 Maximum forming temperatures and times for wrought magnesium alloys

Alloy	Temperature °C	Temperature °F	Time(a)
Sheet			
AZ31B-O	288	550	1 h
AZ31B-H24	163	325	1 h
HK31A-H24	343	650	15 min
	371	700	5 min
	399	750	3 min
Extrusions			
AZ61A-F	288	550	1 h
AZ31B-F	288	550	1 h
M1A-F	371	700	1 h
AZ80A-F	288	550	½ h
AZ80A-T5	193	380	1 h
ZK60A-F	288	550	½ h
ZK60A-T5	204	400	½ h

(a) Maximum time the alloy can be held at temperature without adverse effects on properties

Butt and fillet joints are preferred in magnesium because they are the easiest to make by arc welding, and they provide more consistent results than other types of joints. Lap joints are used sometimes, but they are generally less satisfactory than butt joints for load-carrying applications.

Arc welded joints in annealed magnesium alloy sheet and plate have room-temperature tensile strengths less than 10% lower than those of the base metal (joint efficiencies of greater than 90%). Tensile strengths of arc welds in hard-rolled material, however, are significantly lower than those of the base metal (joint efficiencies of only 60 to 85%) as a result of the annealing effect of welding. Consequently, room-temperature strengths of arc welded joints in magnesium alloy sheet and plate are about the same regardless of the temper of the base metal.

Joint efficiencies also are affected by service temperatures. For example, arc welds in HK31A-H24 sheet exhibit joint efficiencies of 75 to 80% at room temperature, but these increase to nearly 100% at 260 °C (500 °F). Joint efficiencies of arc welds in HM21A-T8 sheet range from about 80% at room temperature to 100% at 200 °C (400 °F). HM31-T5 extrusions exhibit joint efficiencies of 75 to 85% from room temperature to about 370 °C (700 °F), and 100% at 425 °C (800 °F) and above. There are no appreciable differences in properties between welds made with alternating current and those made with direct current.

Stress Relieving. Arc welds in some magnesium alloys—specifically the magnesium-aluminum-zinc series and alloys containing more than 1% Al—are subject to stress-corrosion cracking, and thermal treatment must be used to remove the residual stresses that cause this condition. This treatment consists of placing the parts in a jig or clamping plate and heating them at the temperatures indicated in Table 22 for the specified times. After heating, the parts are cooled in still air. The use of jigs

Table 17 Sizes of flat-rolled products available in magnesium alloys

Thickness range, mm	Thickness range, in.	Width, Standard, mm	Width, Standard, in.	Width, Maximum, mm	Width, Maximum, in.	Length, Standard, m	Length, Standard, ft	Length, Maximum(a), m	Length, Maximum(a), ft
Flat sheet									
0.25–0.41	0.010–0.016	...	...	610	24	...	...	5.5	18
0.41–0.51	0.016–0.020	915	36	915	36	3.66	12	5.5	18
0.51–0.81	0.020–0.032	1220	48	1220	48	3.66	12	5.5	18
0.81–6.35	0.032–0.250	1220	48	1525	60	3.66	12	5.5	18
Coiled sheet(b)									
0.81–1.02	0.032–0.040	...	...	1525	60	...	...	...	...
1.02–6.35	0.040–0.250	...	...	1830	72	...	...	...	...
Plate									
6.35–50.8	0.250–2.000	1220	48	1830	72	3.66	12	5.5	18
50.8–76.2	2.000–3.000	1220	48	1830	72	3.66	12	3.9	13
Tooling plate									
6.35, 9.52	0.250, 0.375	1220	48	1830	72	2.44, 3.66	8, 12	5.5	18
12.7–50.8(c)	0.500–2.000(c)	1220, 1525	48, 60	1830	72	2.44, 3.66	8, 12	5.5	18
63.5	2.500	1220	48	1830	72	2.44	8	4.9	16
76.2	3.000	1220	48	1830	72	2.44	8	3.9	13
88.9	3.500	1220	48	1830	72	2.44	8	3.4	11
101.6	4.000	1245	49	1830	72	2.1	7	3.0	10
127.0	5.000	1245	49	1830	72	1.8	6	2.4	8
152.4	6.000	1245	49	1830	72	1.5	5	2.0	6.5
Tread plate (raised-pattern floor plate)									
3.18	0.125	1220, 1525	48, 60	1830	72	3.66	12	5.5	18
4.76–9.52(d)	0.188–0.375(d)	1220, 1525	48, 60	1830	72	3.66	12	5.5	18
11.11–15.88(d)	0.438–0.625(d)	1220, 1525	48, 60	1830	72	3.66	12	5.5	18
19.05	0.750	1220, 1525	48, 60	1830	72	3.66	12	5.5	18

(a) Maximum length for maximum width. Size of a single piece is limited to 998 kg (2200 lb). (b) Coiled sheet up to 610 mm (24 in.) wide can be produced in thicknesses down to 0.25 mm (0.10 in.). (c) In steps of 6.35 mm (0.250 in.). (d) In steps of 1.588 mm (0.0625 in.)

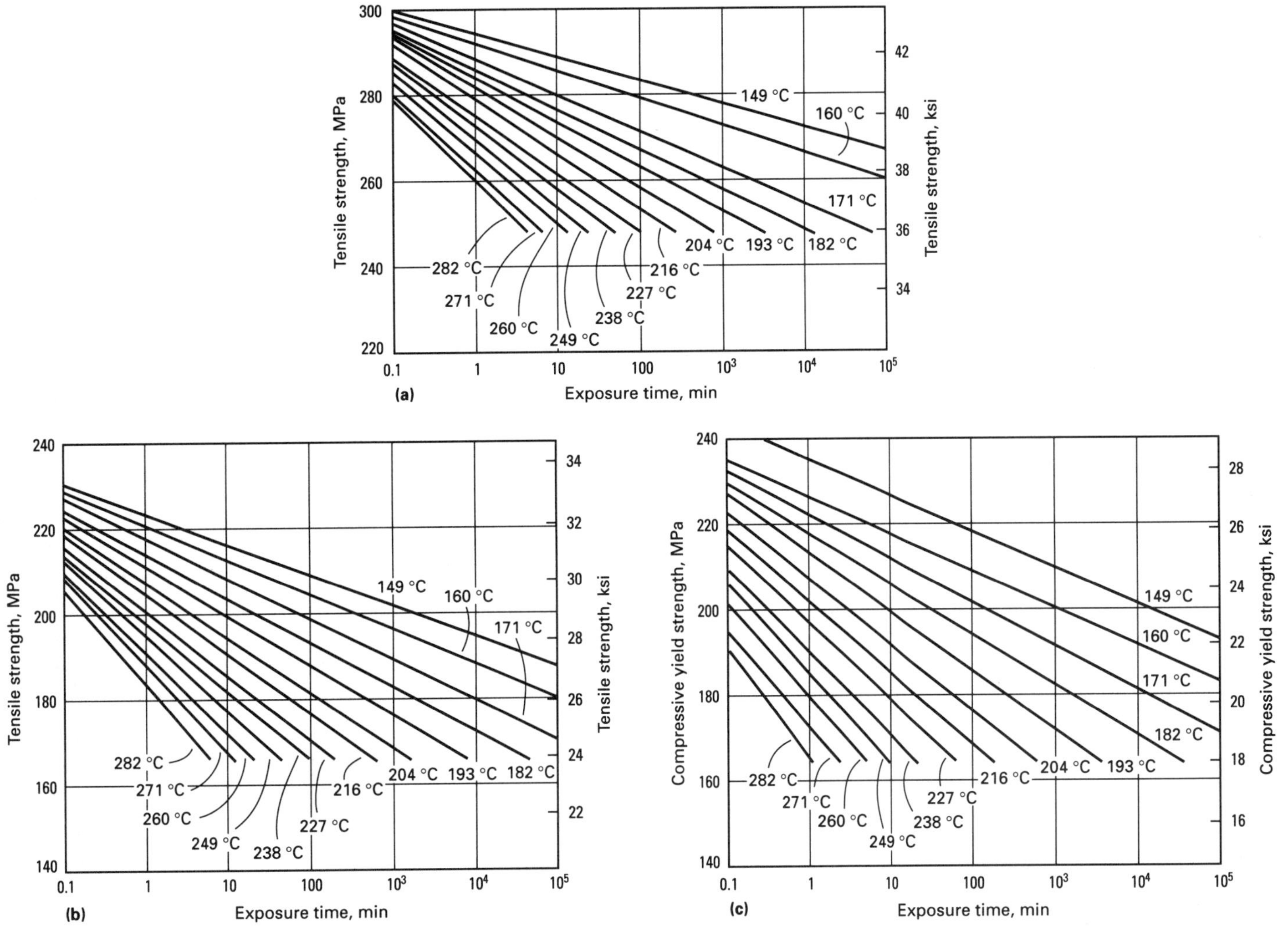

Fig. 14 Effect of exposure time at elevated temperature on the mechanical properties of AZ31B-H24 at room temperature. Data are based on sheet 1.63 mm (0.064 in.) thick. Check tests indicate reasonable applicability for thicknesses up to 6.35 mm (0.250 in.).

Table 18 Maximum time at temperature to maintain properties of magnesium alloy sheet

Maximum time, min	Temperature °C	°F
AZ31B-H24		
0.3	260	500
1	224	435
2	210	410
3	202	395
4	196	385
5	188	370
10	182	360
30	174	345
60	163	325
HK31A-H24		
15	343	650
5	371	700
3	385	725
3	399	750
HM21A-T8(a)		
60	399	750
10	427	800

Note: Annealed sheet will endure much higher temperatures than heat-treated sheet for short periods without significant reduction of properties at room temperature. (a) Based on limited data obtained in laboratory tests

is sometimes necessary so that relief of stresses does not result in warpage of the assembly.

The other types of magnesium alloys, including those containing manganese, rare earths, thorium, zinc, or zirconium, are not sensitive to stress corrosion and normally do not require stress relief after welding.

Repaired castings are generally heat treated again after welding. All alloys which require full (T6) heat treatment are best welded in the solution-treated (T4) condition. After welding, a short solution treatment followed by the normal aging treatment is necessary. The postweld heat treatment of magnesium casting alloys depends on the desired final temper of the castings (Table 23). However, if complete solution heat treatment is not desired, welded castings should always be stress relieved as described in Table 22.

Spot welds in magnesium have good static strength, but their fatigue strength is lower than for either riveted or adhesive-bonded joints. Spot-welded assemblies are used mainly for low-stress applications and are not recommended where joints are subject to vibration. Typical shear strengths of spot welds in three alloys are given in Table 24. Shear strengths of welds AZ61A and HK31A are about the same as those in AZ31B.

Recommended spot spacings and edge distances for spot welds are given in Table 25. Where magnesium sheets of unequal thickness are to be spot welded, the thickness ratio should not exceed 2½ to 1.

Seam welds of the continuous or intermittent types have strength properties comparable to those of spot welds. Shear strengths of about 19.2 to 40.2 kg/linear mm (1075 to 2250 lb/linear in.) of welded seam can be obtained in AZ31B sheet from 1 to 3 mm (0.040 to 0.12 in.) thick.

The cost of weldments is less likely to vary significantly with quantity than the cost of other methods of fabrication. Therefore, weldments are used most often where quantities are small or where fabrication of specific designs is impractical or impossible by other methods. For a dozen parts of the design shown in Fig. 16, sand castings cost

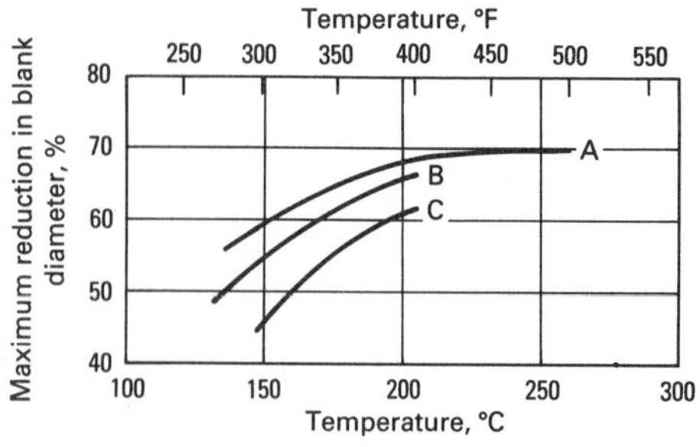

Fig. 15 Effect of temperature on the drawability of 1.63 mm (0.064 in.) AZ31B sheet. Cupping die diameter, 38 mm (1½ in.); radii on draw ring and punch, 5*t*. A, temper O; drawing speed, 38 mm (1½ in.)/min. B, temper H24; drawing speed 38 mm (1½ in.)/min. C, temper H24; drawing speed, 500 mm (20 in.)/min

twice as much as weldments; at about 35 pieces, the tooling cost for casting was absorbed, and casting was more economical for larger lots.

For the electronic mounting base shown in Fig. 17, the die casting was superior to the weldment in mechanical properties, although properties of both were above minimum requirements. Die castings were less expensive than weldments in quantities of 5000 (including cost of tooling); in quantities of 100, weldments were less expensive.

Adhesive bonding of magnesium has become an important fabrication technique. The fatigue characteristics of adhesive-bonded lap joints are better than those of other types of joints. The probability of stress concentration failure in adhesive-bonded joints is minimal. Adhesive bonding permits the use of thinner materials than can be effectively riveted. The adhesive fills the spaces between the contacting surfaces and thus acts as an insulator between any dissimilar metals in the joint. It also permits manufacture of assemblies having surfaces smoother than those associated with riveting.

Adhesive bonding has been limited almost exclusively to lap joints. A few general factors should be considered when designing adhesive-bonded joints:

- Joint strengths vary with the lap width, metal thickness, direction in which loads are applied, and type of adhesive used
- The joint should be designed so that it provides a sufficiently large bonded area
- The adhesive layer should be uniform in thickness
- The adhesive layer should be as thin as possible, yet applied in sufficient quantity so that no joints are starved
- Joints should be designed so that pressure and heat can be readily applied
- The curing temperatures of the common structural adhesives are below the temperatures at which the properties of hard-rolled magnesium sheet are affected, and thus they do not significantly reduce the properties of magnesium alloys in the annealed (O) condition

The effect of lap width on the shear strength of joints bonded with phenolic rubber-base resin adhesive is shown in Fig. 18(a). The effect of temperature on the shear strength of adhesive-bonded joints in magnesium and aluminum is shown in Fig. 18(b).

The characteristics and properties of some adhesives used with magnesium are given in Table 26. These adhesives cannot be utilized in assemblies operating above 80 °C (180 °F) because of low shear strength.

Riveting. Essentially the same procedures employed in riveting other materials are used in riveting magnesium alloys. Standard procedures are used for drilling and countersinking holes. Both dimpling and machine countersinking are used in flush riv-

Table 19 Suggested limits for bending flat magnesium alloy extrusions

As established for 2.29 × 22.2 mm (0.090 × 0.875 in.) extruded flat strip. Numerical values for bend radii are given as multiples of extrusion thickness.

Alloy	Typical bend radius at 21 °C (70 °F)	Limits for hot bending: At temperature, °C	°F	Time, h	Typical bend radius
AZ61A-F	1.9*t*	288	550	1	1.0*t*
AZ80A-F	2.4*t*	288	550	½	0.7*t*
AZ31B-F	2.4*t*	288	550	1	1.5*t*
M1A-F	4.8*t*	371	700	1	2.0*t*
AZ80A-T5	8.3*t*	193	380	1	1.7*t*
ZK60A-F	12*t*	288	550	½	2.0*t*
ZK60A-T5	12*t*	204	400	½	6.6*t*

Table 20 Form bending parameters for magnesium tubing

Alloy	Forming temperature, °C	°F	Bend radius(a)
AZ31B-F	21	70	4*D*
	93	200	3*D*
AZ61A-F	21	70	4*D*
	−7	20	3*D*
M1A-F	21	70	6*D*
	204	400	4*D*
ZK60A-F	21	70	5*D*

Alloy	Minimum bend radius at 21 °C (70 °F)(b): *D*/*t* = 17	*D*/*t* = 6	*D*/*t* = 3
AZ61A-F(c)	5*D*	2½*D*	2*D*
AZ61A-F(d)	2½*D*	2½*D*	2½*D*
AZ31B-F(c)	6*D*	4*D*	3*D*
AZ31B-F(d)	3*D*	2*D*	2*D*
M1A-F(c)	6*D*	3*D*	2½*D*
M1A-F(d)	6*D*	6*D*	2½*D*

(a) *D*, tube outside diameter. Bend radius taken to axis of tube. (b) Minimum bend radius for various *D*/*t* ratios at 21 °C (70 °F). *D*, tube outside diameter; *t*, wall thickness. (c) Tubing unfilled before bending. (d) Tubing filled with low-melting alloy (50% Bi, 26.7% Pb, 13.3% Sn, 10% Cd) before bending

Table 21 Weldability of magnesium alloys

Alloy	Thickness, mm	in.	Welding rod	Joint efficiency, %	Joint ductility(a)
AZ31B-O	1.63	0.064	AZ61A, AZ92A	97	12.0
AZ31B-H24	1.63	0.064	AZ61A, AZ92A	88	10.0
ZE10A-O	1.63	0.064	AZ61A, AZ92A	94	7.0
ZE10A-H24	1.63	0.064	AZ61A, AZ92A	87	3.0
M1A-F	3.17	0.125	M1A	55	2.0
AZ31B-F	3.17	0.125	AZ61A, AZ92A	92	12.0
AZ61A-F	3.17	0.125	AZ61A, AZ92A	89	8.0
AZ80A-F	3.17	0.125	AZ61A, AZ92A	86	4.0
AZ63A-F	12.70	0.5	AZ63A	83	2.5
AZ63A-T4	12.70	0.5	AZ63A	70	5.0
AZ63A-T6	12.70	0.5	AZ63A	75	2.0
AZ92A-F	12.70	0.5	AZ92A	100	2.5
AZ92A-T4	12.70	0.5	AZ92A	70	4.0
AZ92A-T6	12.70	0.5	AZ92A	75	2.0
AZ91C-F	12.70	0.5	AZ92A	100	2.5
AZ91C-T4	12.70	0.5	AZ92A	78	4.0
AZ91C-T6	12.70	0.5	AZ92A	75	2.0
AZ81A-F	12.70	0.5	AZ92A	100	2.5
AZ81A-T4	12.70	0.5	AZ92A	85	8.0
EK41A-T5	12.70	0.5	EK41A	100	1.0
EK41A-T6	12.70	0.5	EK41A	93	6.2
EZ33A-T5	12.70	0.5	EZ33A	100	1.1
HK31A-T6	12.70	0.5	HK31A	100	9.5
HK31A-H24	...	...	EZ33A	83	1.0
HZ32A-T5	12.70	0.5	HZ32A	93	3.8
HM21A-T8	1.63	0.064	EZ33A	88	1.5
	...	...	HM31A	74	1.5
HM31A-F	15.88	0.625	EZ33A	71	1.8
			HM31A	58	2.5

(a) Percentage elongation across the weld over a 50 mm (2 in.) gage length from tension tests

Table 22 Times and temperatures for stress relieving arc welds in magnesium alloys

Alloy	Temperature °C	°F	Time, min
Sheet			
AZ31B-H24(a)	150	300	60
AZ31B-O(a)	260	500	15
Extrusions(b)			
AZ31B-F(a)	260	500	15
AZ61A-F(a)	260	500	15
AZ80A-F(a)	260	500	15
AZ80A-T5(a)	204	400	60
HM31A-T5	425	800	60
ZK60A-F(c)	260	500	15
ZK60A-T5(c)	150	300	60
Castings(d)			
AM100A	260	500	60
AZ63A	260	500	60
AZ81A	260	500	60
AZ91C	260	500	60
AZ92A	260	500	60
EZ33A	250	480	600
HZ32A	350	660	120
K1A(e)	...	...	...
ZE41A	330	625	120
ZH62A	330	625	120

(a) Postweld stress relief is required to prevent possible stress-corrosion cracking in this alloy. Postweld heat treatment of other alloys is used primarily for straightening or for stress relieving prior to machining. (b) When extrusions are welded to sheet, distortion may be minimized by using a lower stress-relieving temperature and a longer time. For example, 60 min at 150 °C (300 °F) instead of 15 min at 260 °C (500 °F). (c) ZK60 has limited weldability. (d) These stress-relief schedules for casting alloys will not develop maximum joint strength. For maximum strength, use the postweld heat treatments shown in Table 23. (e) No stress relief is necessary after welding this alloy.

eting. With machine countersinking, it is desirable to have a cylindrical land with a minimum depth of 0.38 mm (0.015 in.) at the bottom of the hole. Thus, machine countersinking is limited to sheet thick enough to permit lands of this depth with a given size of rivet. Dimpling of magnesium alloy sheet is a hot-forming operation; to prevent reduction of properties during dimpling, the sheet must not be heated to excessively high temperatures or for long periods.

Only aluminum rivets should be used if galvanic incompatibility is to be minimized, and those up to 8 mm (5⁄16 in.) in diameter can be driven cold. The ease of driving rivets of alloy 5056 will vary with the temper. Quarter-hard temper (5056-H32) is satisfactory for all normal riveting.

Machinability

Magnesium and its alloys can be machined at extremely high speeds using greater depths of cut and higher rates of feed than can be used in machining other structural metals. There are no significant differences in machinability among magnesium alloys. Therefore, a specific magnesium alloy rarely, if ever, is selected in place of another magnesium alloy solely on the basis of machinability.

Because of the free-cutting characteristic of magnesium, chips produced in machining are well broken. Dimensional tolerances of about ±0.1 mm (a few thousandths of an inch) can be obtained using standard operations.

The power required to remove a given amount of metal is lower for magnesium than for any other commonly machined metal. Based on the volume of metal removed per minute, the comparative power requirements of various metals are:

Metal	Relative power
Magnesium alloys	1.0
Aluminum alloys	1.8
Brass	2.3
Cast iron	3.5
Low-carbon steel	6.3
Nickel alloys	10.0

Tool wear is also reduced when machining magnesium because of the high thermal conductivity of the metal, which allows rapid dissipation of heat, and the low cutting pressures required. Ordinary carbon steel tools can be used in machining magnesium, but high-speed tools and carbide-tipped tools can be used for high production rate jobs.

An outstanding machining characteristic of magnesium alloys is their ability to acquire an extremely fine finish. Often, it is unnecessary to grind and polish magnesium to obtain a smooth finished surface. Surface smoothness readings of about 0.1 μm (3 to 5 μin.) have been reported for machined magnesium and are attainable at both high and low speeds, with or without cutting fluids.

Cutting Fluids (Coolants). In the machining of magnesium alloys, cutting fluids provide far smaller reductions in friction than they provide in the machining of other metals; thus, they are of little use in improving surface finish and tool life. Most machining of magnesium alloys is done dry, but cutting fluids sometimes are used for cooling the work.

Although less heat is generated during machining of magnesium alloys than during machining of other metals, higher cutting speeds and the low heat capacity and relatively high thermal expansion characteristics of magnesium may make it necessary to dissipate the small amount of heat that is generated. Heat generation can be minimized by the use of correct tooling and machining techniques, but cutting fluids are sometimes needed to reduce the possibilities of distortion of the work and ignition of fine chips. Because they are used primarily to dissipate heat, cutting fluids are referred to as coolants when used in the machining of magnesium alloys.

Numerous mineral oil cutting fluids of relatively low viscosity are satisfactory for use as coolants in the machining of magnesium. Suitable coolants represent a compromise between cooling power and flash point. Additives designed to increase wetting power are usually beneficial. Only mineral oils should be used as coolants; animal and vegetable oils are not recommended.

Water-soluble oils, oil-water emulsions, or water solutions of any kind should not be used on magnesium. Water reduces the scrap value of magnesium turnings and introduces potential fire hazards during shipment and storage of machine shop scrap.

Safe Practice. The possibility of chips or turnings catching fire must be considered when magnesium is to be machined. Chips must be heated close to their melting point before ignition can occur. Roughing cuts and medium finishing cuts produce chips too large to be readily ignited during machining. Fine finishing cuts, however, produce fine chips that can be ignited by a spark. Stopping the feed and letting the tool dwell before disengagement, and letting the tool or tool holder rub on the work, produce extremely fine chips and should be avoided.

Factors that increase the probability of chip ignition are:

- Extremely fine feeds
- Dull or chipped tools
- Improperly designed tools
- Improper machining techniques
- Sparks caused by tools hitting iron or steel inserts

Feeds less than 0.02 mm (0.001 in.) per revolution and cutting speeds higher than 5 m/s (1000 ft/min) increase the risk of fire. Even under the most adverse conditions—with dull tools and fine feeds—chip fires are very unlikely at cutting speeds below 3.5 m/s (700 ft/min).

Any fire hazard connected with machining of magnesium is easy to control, and large quantities of magnesium are machined without difficulty. Following these rules will reduce the fire hazard:

- Keep all cutting tools sharp and ground with adequate relief and clearance angles
- Use heavy feeds to produce thick chips
- Use mineral oil coolants (15 to 19 L/min, or 4 to 5 gal/min) whenever possible; when not possible, avoid fine cuts
- Do not allow chips to accumulate on machines or on the clothing of operators. Remove dust and chips at frequent intervals and store in clean, plainly labeled, covered metal cans
- Keep an adequate supply of a recommended magnesium fire extinguisher within reach of operators

If dry chips are ignited, they will burn with a brilliant white light, but the fire will not flare up unless disturbed. Burning chips should be extinguished as follows:

- Scatter a generous layer of clean, dry cast iron chips or metal extinguishing powder over the burning magnesium
- Cover actively burning fires on combustible surfaces like wood floors with a layer of the extinguishant, then shovel the

Table 23 Weld preheat and postweld heat treatment of magnesium castings

Alloy	Metal temper before welding(a)	Desired temper after welding(a)	Weld preheat(b)	Postweld heat treatment (time after reaching temperature)(c)
AZ63A	T4	T4	None to 380 °C (720 °F) max(d)	½ h at 390 °C (730 °F)(d)
AZ63A	T4 or T6	T6	None to 380 °C (720 °F) max(d)	½ h at 390 °C (730 °F)(d) + 5 h at 220 °C (425 °F)
AZ63A	T5	T5	None to 260 °C (500 °F); 1½ h max at 260 °C (500 °F)	5 h at 220 °C (425 °F)
AZ81A	T4	T4	None to 400 °C (750 °F) max(d)	½ h at 415 °C (780 °F)(d)
AZ91C	T4	T4	None to 400 °C (750 °F) max(d)	½ h at 415 °C (780 °F)(d)
AZ91C	T4 or T6	T6	None to 400 °C (750 °F) max(d)	½ h at 415 °C (780 °F)(d) + either 4 h at 215 °C (420 °F) or 16 h at 170 °C (335 °F)
AZ92A	T4	T4	None to 400 °C (750 °F) max(d)	½ h at 415 °C (780 °F)(d)
AZ92A	T4 or T6	T6	None to 400 °C (750 °F) max(d)	½ h at 415 °C (780 °F)(d) + either 4 h at 260 °C (500 °F) or 5 h at 220 °C (425 °F)
AM100A	T6	T6	None to 400 °C (750 °F) max(d)	½ h at 415 °C (780 °F)(d) + 5 h at 220 °C (425 °F)
EK41A	T4 or T6	T6	None to 260 °C (500 °F); 1½ h max at 260 °C (500 °F)	16 h at 205 °C (400 °F)
EK41A	T5	T5	None to 260 °C (500 °F); 1½ h max at 260 °C (500 °F)	16 h at 205 °C (400 °F)
EZ33A	F or T5	T5	None to 260 °C (500 °F); 1½ h max at 260 °C (500 °F)	5 h at 215 °C (420 °F) (optional)(e) 2 h at 345 °C (650 °F) + 5 h at 215 °C (420 °F)
HK31A	T4 or T6	T6	None to 260 °C (500 °F)	16 h at 205 °C (400 °F) (optional)(e) 1 h at 315 °C (600 °F) + 16 h at 205 °C (400 °F)
HZ32A	F or T5	T5	None to 260 °C (500 °F)	16 h at 315 °C (600 °F)
K1A	F	F	None	None
ZE41A	F or T5	T5	None to 315 °C (600 °F)	2 h at 330 °C (625 °F) (optional)(e) 2 h at 330 °C (625 °F) + 16 h at 175 °C (350 °F)
ZH62A	F or T5	T5	None to 315 °C (600 °F)	16 h at 250 °C (480 °F) (optional)(e) 2 h at 330 °C (625 °F) + 16 h at 175 °C (350 °F)
ZK51A	F or T5	T5	None to 315 °C (600 °F)	16 h at 175 °C (350 °F) (optional)(e) 2 h at 330 °C (625 °F) + 16 h at 175 °C (350 °F)
ZK61A	F or T5	T5	None to 315 °C (600 °F)	48 h at 150 °C (300 °F)
ZK61A	T4 or T6	T6	None to 315 °C (600 °F)	2-5 h at 500 °C (930 °F)(d) + 48 h at 130 °C (265 °F)

(a) Temper T4, solution heat treated; T6, solution heat treated and aged; T5, artificially aged; F, as-cast. (b) Heavy and unrestrained sections usually need no preheat; thin and restrained sections may need to be preheated to indicated temperatures to avoid weld cracking. (c) Temperatures listed are maximum allowable; furnace controls should be set so that temperature does not cycle above indicated maximum. (d) SO_2 or CO_2 atmosphere recommended when heating temperature exceeds 370 °C (700 °F). (e) Optional postweld heat treatment serves to induce greater stress relief.

entire mass into an iron container or onto a piece of iron plate
- Do not use water or any of the common liquid or foam-type extinguishers, which intensify magnesium chip fires

Distortion of magnesium parts during machining occurs rarely and usually can be attributed to excessive heating or improper chucking or clamping.

Heating of the work is increased by use of dull or improperly designed tools, extremely high machining feeds and speeds, or very fine cuts. Because magnesium has a relatively high coefficient of thermal expansion, such excessive heating results in substantial increases in dimensions—particularly in thin sections, where heating causes relatively large increases in temperature. Use of sharp, properly designed tools; mineral oil coolants; and relatively coarse feeds and depths of cut reduces excessive heating. Wide variations in room temperature during machining can also cause sufficient dimensional change to affect machining tolerances.

Clamping should always be done on heavier sections of magnesium castings, and clamping pressures should not be high enough to cause distortion. Special care should be taken with light parts that could be distorted easily by the chuck or by use of heavy cuts.

Distortion of magnesium parts is seldom caused by stresses during casting, forging, or extruding, but it may result from stresses caused by straightening or welding. Such stresses can be relieved prior to machining by heating at 260 °C (500 °F) for 2 h and slowly cooling. However, such treatment causes some loss of strength in AZ31B-H24 sheet products. If distortion of parts is observed after rough machining, the cutting tool should be inspected to ensure that it is sharp and properly ground. If so, the size of cut should be decreased. With complex parts or parts machined to extremely close tolerances, it may be advisable to stress relieve or, if time permits, to store parts for 2 or 3 days between rough machining and finishing.

Design and Weight Reduction

By substituting magnesium alloys for heavier metals such as steel and aluminum alloys, many structural parts can be substantially reduced in weight with little or no redesign. This is possible because manufacturing limitations make many parts heavier than necessary. For example, for successful filling of the mold, a casting may require a minimum wall thickness greater than that dictated by service requirements and the strength of the metal used. Similarly, forgings and extrusions sometimes must be made thicker than necessary, and the light weight of magnesium can be used to advantage. In many instances, a casting, forging, or extrusion for which magnesium is substituted for a heavier metal can have adequate strength with no increase in wall thickness.

In other parts, substitution of magnesium may require greater wall thickness, and substantial redesign may be necessary in order to realize maximum weight savings. Because strength and stiffness in bending of many structural sections increase approximately as the square and cube of the section depth, respectively, it is possible to obtain large increases in strength and stiffness with moderate increases in depth and cross-sectional area. When such increases in depth are permissible, it usually is economical to redesign the part for magnesium. The greater bulk of the redesigned part reduces local instability, and although the saving in weight is less than maximum, the reduction in instability allows design simplification and thus reduces manufacturing costs.

The room-temperature thickness, strength, stiffness, and weight of magnesium alloys are compared with those of aluminum alloys and steel in Table 27. Bending strength is defined as the product of yield strength and section modulus.

Bending. Rectangular steel, aluminum, and magnesium sections of equal thickness have rigidities in the ratio of their moduli of elasticity. The magnesium section weighs about 63% as much as the aluminum section and about 22% as much as the steel section.

The rigidity in bending of a rectangular section is proportional both to the cube of its depth and to its modulus of elasticity. If the section thicknesses of a magnesium

Table 24 Typical shear strengths of spot welds in magnesium alloys

Material thickness, sheet		Average spot diameter		Single-spot shear strength for			
				AZ31B-O		HK31A-H24	
mm	in.	mm	in.	kg	lb	kg	lb
0.508	0.020	3.56	0.14	100	220	. . .	. . .
0.635	0.025	4.06	0.16	120	270	. . .	. . .
0.813	0.032	4.57	0.18	150	330	135	300
1.016	0.040(a)	5.08	0.20	185	410	170	375
1.270	0.050	5.84	0.23	240	530	250	550
1.600	0.063(b)	6.86	0.27	340	750	325	720
2.032	0.080	7.87	0.31	405	890	. . .	. . .
2.540	0.100	8.64	0.34	535	1180	. . .	. . .
3.175	0.125(c)	9.65	0.38	695	1530	675	1490

Material thickness, extrusions		Average spot diameter		Single-spot shear strength for M1A-F	
mm	in.	mm	in.	kg	lb
0.508	0.020	3.05	0.12	50	105
0.635	0.025	3.56	0.14	70	150
0.813	0.032	4.06	0.16	95	210
1.016	0.040	4.57	0.18	130	285
1.295	0.051	5.33	0.21	175	385
1.626	0.064	6.10	0.24	225	500
2.057	0.081	7.11	0.28	305	670
2.591	0.102	7.87	0.31	400	885
3.175	0.125	8.89	0.35	515	1135

(a) Single-spot shear strength for HM21A-T8 alloy is 165 kg (360 lb). (b) Single-spot shear strength for HM21A-T8 alloy is 300 kg (660 lb). (c) Single-spot shear strength for HM21A-T8 alloy is 555 kg (1220 lb).

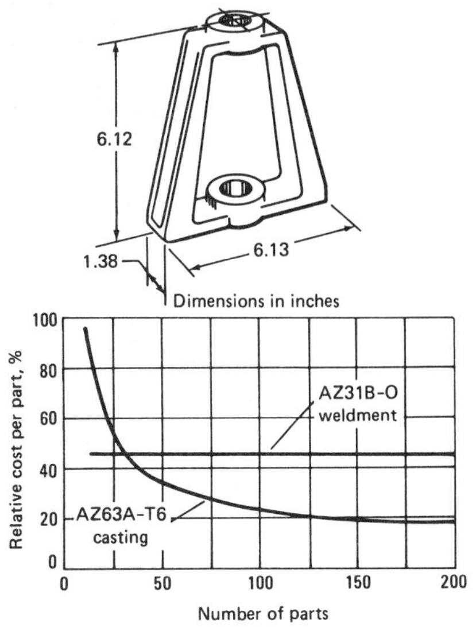

Fig. 16 Effect of quantity on the cost of magnesium alloy sand castings compared with the same parts made as weldments

Table 25 Recommended spot spacing and edge distance for spot welds in magnesium alloy sheet

Sheet thickness		Spot spacing				Edge distance			
		Minimum		Nominal		Minimum		Nominal	
mm	in.	mm	in.	mm	in.	mm	in.	mm	in.
0.508	0.020	6.35	0.25	12.70	0.50	3.81	0.15	6.35	0.25
0.635	0.025	6.35	0.25	12.70	0.50	4.06	0.16	6.35	0.25
0.813	0.032	7.87	0.31	15.75	0.62	4.57	0.18	6.35	0.25
1.015	0.040	9.65	0.38	19.05	0.75	5.08	0.20	6.35	0.25
1.296	0.051	10.41	0.41	19.05	0.75	5.84	0.23	7.87	0.31
1.626	0.064	12.70	0.50	25.40	1.00	6.85	0.27	9.65	0.38
2.057	0.081	15.75	0.62	31.75	1.25	7.87	0.31	10.41	0.41
2.591	0.102	15.75	0.62	31.75	1.25	9.40	0.37	12.70	0.50
3.175	0.125	19.05	0.75	38.10	1.50	11.18	0.44	15.75	0.62

section, an aluminum section, and a steel section are adjusted until their rigidities are equal, the magnesium section will weigh about 71% of the aluminum and about 40% of the steel. If the section thickness of the magnesium is increased to about twice that of the steel, the magnesium will be more than 70% more rigid than the steel and less than 50% as heavy. Magnesium supporting its own weight shows no more deflection than other metals under the same conditions.

At high temperatures, the difference between short-time ultimate and yield strengths of certain magnesium alloys decreases significantly. Creep properties that depend on time must also be considered in evaluating materials for long-time operation at elevated temperature. Creep-strength values of several magnesium alloys are given in the data compilations in the article "Properties of Magnesium Alloys" in this Volume.

Plate Buckling. Structures subjected to compressive loads may be limited in efficiency (load carried versus weight of structure) by buckling at relatively low stresses.

A structural index is a valuable aid to designers in the selection of optimum mate-

Table 26 Characteristics of adhesives used for bonding magnesium

These adhesives are for service at temperatures up to 82 °C (180 °F).

General type of composition	Curing conditions					Adhesive thickness		Shear strength	
	Temperature			Pressure					
	°C	°F	Time, min	MPa	ksi	mm	in.	MPa	ksi
Phenol formaldehyde plus polyvinyl formal powder(a)	132	270	32	0.34–3.44	0.05–0.5	0.0–0.152	0.0–0.006	11–18	1.6–2.6
Phenolic rubber-base resin(a)	163	325	20	1.38	0.2	0.076–0.152	0.003–0.006	15–18	2.2–2.6
Phenolic synthetic rubber base plus thermosetting resin	177	350	10	0.048–0.310	0.007–0.045	0.127–0.508	0.005–0.020	7–17	1.0–2.5
			60	0.689	0.1	0.127–0.508	0.005–0.020	14–20	2.1–2.9
Ethoxyline resin liquid, powder, or stick used like solder	199	390	60	Contact	Contact	0.025–0.152	0.001–0.006	10–15	1.5–2.2
Ethoxyline resin (two liquids)	Room	Room	24 h	Contact	Contact	0.025–0.152	0.001–0.006	. . .	. . .
Epoxy-type resin paste plus liquid activator	93	200	60	Contact	Contact	0.076–0.127	0.003–0.005	21 max	3.0 max
Rubber base	204	400	8	1.38	0.2	0.254–0.381	0.010–0.015	12–16	1.7–2.3
Vinyl phenolic	149	300	8 preheat	1.38	0.2	0.102–0.305	0.004–0.012	7–12	1.0–1.7
	135–204	275–400	70–4	. . .	. . .	. . .	. . .	. . .	. . .
Epoxy-type resin	93	200	45	Contact	Contact	0.254–0.762	0.010–0.030	8–12	1.2–1.8
	93	200	45	0.096	0.014	(tape)	(tape)	10–12	1.4–1.7
Phenolic	149	300	15	0.193	0.028	0.051–0.102	0.002–0.004	17 max	2.4 max

(a) Known to meet USAF specifications

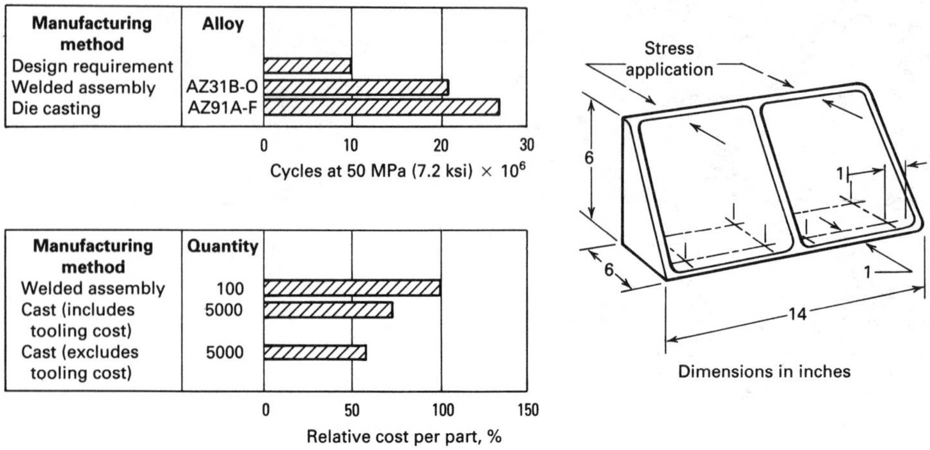

Fig. 17 Comparison of a magnesium alloy electronic mounting base as manufactured by welding and by casting. Weight of part, 1.25 kg (2.75 lb)

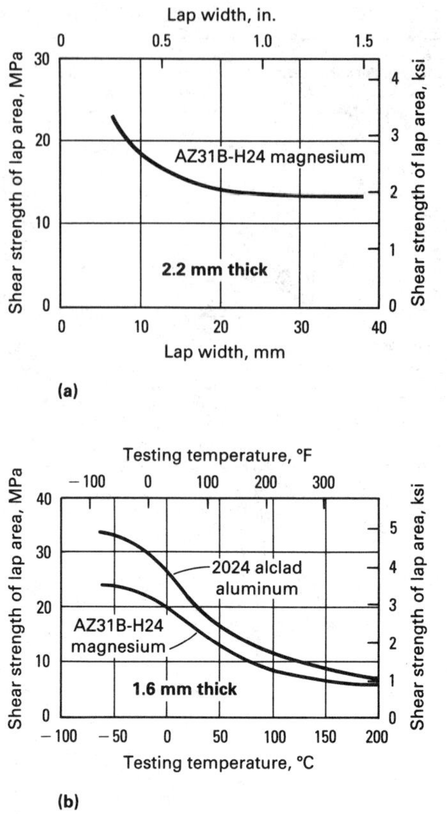

Fig. 18 Effect of (a) lap width and (b) temperature on the shear strength of joints bonded with a phenolic rubber-base resin adhesive

Table 27 Relative bending strength, stiffness, and weight of selected structural metals

Material	Thickness	Bending strength	Stiffness	Weight
For equal thickness				
1025 steel	100	100.0	100.0	100.0
6061-T6 aluminum sheet and extrusions	100	97.2	34.5	34.5
AZ31B magnesium extrusions	100	47.2	22.4	22.5
ZK60A-T5 magnesium extrusions	100	88.9	22.4	22.5
AZ31B-H24 magnesium sheet	100	73.4	22.4	22.5
For equal bending strength				
1025 steel	100	100	100.0	100.0
6061-T6 aluminum sheet and extrusions	101	100	35.8	34.8
AZ31B magnesium extrusions	146	100	69.2	32.9
ZK60A-T5 magnesium extrusions	106	100	26.7	23.9
AZ31B-H24 magnesium sheet	117	100	35.6	26.3
For equal stiffness				
1025 steel	100	100	100	100.0
6061-T6 aluminum sheet and extrusions	143	199	100	49.2
AZ31B magnesium extrusions	165	129	100	37.2
ZK60A-T5 magnesium extrusions	165	242	100	37.2
AZ31B-H24 magnesium sheet	165	200	100	37.2
For equal weight				
1025 steel	100	100	100	100
6061-T6 aluminum sheet and extrusions	290	817	841	100
AZ31B magnesium extrusions	444	930	1962	100
ZK60A-T5 magnesium extrusions	444	1753	1962	100
AZ31B-H24 magnesium sheet	444	1451	1962	100

Note: Comparison made at room temperature for rectangular beams of constant width with the following minimum yield strengths: 1025 steel, 250 MPa (36 ksi); 6061-T6 aluminum, 240 MPa (35 ksi); magnesium alloys, average of minimum tensile yield and compressive yield strengths. All comparisons expressed in percent

rials for plate structures that are critical in compression loading. A structural index is nondimensional; that is, equivalent designs give the same value of structural index regardless of the size of the actual part. The plate-buckling index is computed from the maximum edge load (P_{cr}) that will not cause crippling, the width (b) of the plate and a factor, K, determined by the amount of restraint or clamping along the unloaded edges for a simply supported edge, K = 4.0). The formula is:

$$\text{Index} = \frac{P_{cr}K^{0.5}}{b^2}$$

Using this index, the efficiency of various structural materials can be directly compared for given conditions of loading and structural configuration. For example, the efficiencies of three materials at room temperature and at 260 °C (500 °F) for plate-buckling indexes up to 4000 are shown in Fig. 19. Comparisons are based on typical properties after short-time exposure at temperature.

A low value of plate-buckling index means either that the critical edge load is low or that the plate is wide, corresponding in either instance to a more lightly stressed structure. As the index value increases, it represents a transition to narrower plates and/or heavier edge loads and, at high values, corresponds to a condition of pure prismatic compression.

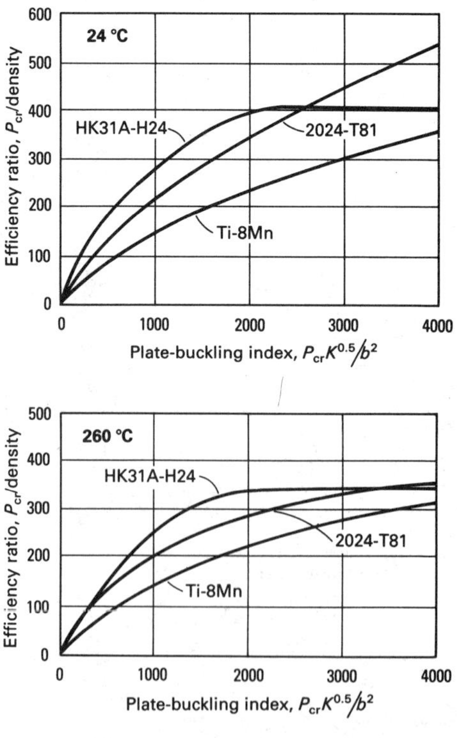

Fig. 19 Effect of plate-buckling index and temperature on the structural efficiency of magnesium, aluminum, and titanium alloys. See text for discussion.

The ratio of working stress to density is an inverse measure of structural weight—the higher the ratio, the lighter the structure. Figure 19 shows the expected advantage in efficiency of the lower-density magnesium alloy HK31A-H24 over the higher-density aluminum and titanium alloys. This advantage fades as the index increases, and the stress condition moves from elastic buckling toward prismatic compression. Comparison of the two charts shows that the range over which the magnesium alloy is the most efficient of the three alloys (magnesium, aluminum, and titanium) is higher at 260 °C (500 °F) than at room temperature.

REFERENCE

1. R.S. Busk, *Magnesium Products Design*, Vol 1, Marcel Dekker, 1987, p 180

Properties of Magnesium Alloys

Revised by Susan Housh and Barry Mikucki, Dow Chemical U.S.A., and Archie Stevenson, Magnesium Elektron, Inc.

Wrought Magnesium Alloys

AZ10A

Specifications

UNS. M11100
Government. Extruded rods, bars, and shapes: QQ-M-31

Chemical Composition

Composition limits. 1.0 to 1.5 Al, 0.2 to 0.6 Zn, 0.2 Mn min, 0.1 Si max, 0.1 Cu max, 0.005 Ni max, 0.005 Fe max, 0.04 Ca max, bal Mg

Applications

Typical uses. Low-cost extrusion alloy with moderate mechanical properties and high elongation. Used in as-extruded (F) temper

Mechanical Properties

Tensile properties. See Table 1.
Compressive properties. See Table 1.
Poisson's ratio. 0.35
Elastic modulus. Tension, 45 GPa (6.5 × 10^6 psi)

Mass Characteristics

Density. 1.76 g/cm^3 (0.064 lb/in.^{3}) at 20 °C (68 °F)

Thermal Properties

Liquidus temperature. 645 °C (1190 °F)
Solidus temperature. 630 °C (1170 °F)
Coefficient of linear thermal expansion. 26.6 μm/m · K (14.8 μin./in. · °F) at 21 to 204 °C (70 to 400 °F)
Thermal conductivity. 110 W/m · K (64 Btu/ft · h · °F) at 20 °C (68 °F)

Electrical Properties

Electrical resistivity. 64 nΩ · m at 20 °C (68 °F)

Fabrication Characteristics

Weldability. Good; does not require stress relief after welding

Table 1 Typical mechanical properties of AZ10A at room temperature

Size and shape	Tensile strength, MPa	Tensile strength, ksi	Yield strength, MPa	Yield strength, ksi	Elongation, %	Compressive yield strength, MPa	Compressive yield strength, ksi
Solid shapes with least dimension up to 6.4 mm (0.025 in.)	240	35	145	21	10	69	10
Solid shapes with least dimension to 6.4 to 38 mm (0.025 to 1.5 in.)	240	35	150	22	10	76	11
Hollow and semihollow shapes	230	33	145	21	8	69	10
Tube (152 mm, or 6 in. OD max) with 0.7 to 6.4 mm (0.028 to 0.25 in.) wall	230	33	145	21	8	69	10

AZ21X1

Specifications

UNS. M11210

Chemical Composition

Composition limits. 1.6 to 2.5 Al, 0.8 to 1.6 Zn, 0.1 to 0.25 Ca, 0.15 Mn max, 0.05 Si max, 0.05 Cu max, 0.005 Fe max, 0.002 Ni max, 0.3 max other, bal Mg

Applications

Typical uses. Impact-extruded battery anodes. Used in as-extruded (F) temper

AZ31B, AZ31C

Specifications

AMS. AZ31B sheet: O temper, 4357; H24 temper, 4376
ASTM: Sheet: B 90. Extruded rod, bar, shapes, tubing, and wire: B 107. AZ31B forgings: B 91
SAE. AZ31B: J466. Former SAE alloy number: 510
UNS numbers. AZ31B: M11311. AZ31C: M11312
Government. AZ31B: forgings, sheet, and plate, QQ-M-40; extruded bar, rod, and shapes, QQ-M-31B; extruded tubing, WW-T-825B
Foreign. Elektron AZ31 (extruded bar and tubing). British: sheet, BS 3370 MAG111; extruded bar and tubing, BS 3373 MAG111. German: DIN 9715 3.5312. French: AFNOR G-A371

Chemical Composition

Composition limits of AZ31B. 2.5 to 3.5 Al, 0.20 Mn min, 0.60 to 1.4 Zn, 0.04 Ca max, 0.10 Si max, 0.05 Cu max, 0.005 Ni max, 0.005 Fe max, 0.30 max other (total); bal Mg
Composition limits of AZ31C. 2.4 to 3.6 Al, 0.15 Mn min, 0.50 to 1.5 Zn, 0.10 Cu max, 0.03 Ni max, 0.10 Si max, bal Mg
Consequence of exceeding impurity limits. Excessive Cu, Ni, or Fe degrades corrosion resistance.

Applications

Typical uses. AZ31B and AZ31C: forgings and extruded bar, rod, shapes, structural sections, and tubing with moderate mechanical properties and high elongation; AZ31C is the commercial grade, with the same properties as AZ31B but higher impurity limits. AZ31B only: sheet and plate with good formability and strength, high resistance to corrosion, and good weldability. AZ31B and AZ31C are used in the as-fabricated (F), annealed (O), and hard-rolled (H24) tempers.

Mechanical Properties

Tensile properties. See Tables 2 and 3.
Shear strength. See Table 2.
Compressive yield strength. See Table 2.
Bearing properties. See Table 2.
Hardness: See Table 2.
Poisson's ratio. 0.35
Elastic modulus. Tension, 45 GPa (6.5 × 10^6 psi); shear, 17 GPa (2.4 × 10^6 psi)

Table 2 Typical mechanical room-temperature mechanical properties of AZ31B

Product form	Tensile strength MPa	Tensile strength ksi	Tensile yield strength(a) MPa	Tensile yield strength(a) ksi	Elongation, %(b)	Hardness HB(c)	Hardness HRE	Shear strength MPa	Shear strength ksi	Compressive yield strength(a) MPa	Compressive yield strength(a) ksi	Ultimate bearing strength(d) MPa	Ultimate bearing strength(d) ksi	Bearing yield strength(d) MPa	Bearing yield strength(d) ksi
Sheet, annealed	255	37	150	22	21	56	67	145	21	110	16	485	70	290	42
Sheet, hard rolled	290	42	220	32	15	73	83	160	23	180	26	495	72	325	47
Extruded bar, rod, and solid shapes	255	37	200	29	12	49	57	130	19	97	14	385	56	230	33
Extruded hollow shapes and tubing	241	35	165	24	16	46	51	...	...	83	12	...	...	...	...
Forgings	260	38	170	25	15	50	59	130	19	...	...	...	...	...	...

(a) At 0.2% offset. (b) In 50 mm (2 in.). (c) 500 kg load, 10 mm ball. (d) 4.75 mm (3/16 in.) pin diameter

Impact strength. Forgings and extruded bar, rod, and solid shapes: Charpy V-notch, 4.3 J (3.2 ft · lbf)
Directional properties. See Table 4.

Mass Characteristics

Density: 1.77 g/cm^3 (0.064 lb/in.^{3}) at 20 °C (68 °F)

Thermal Properties

Liquidus temperature. 630 °C (1170 °F)
Solidus temperature. 605 °C (1120 °F)
Coefficient of linear thermal expansion. 26 μm/m · K (14 μin./in. · °F)
Specific heat versus temperature. $C_p = 0.2441 + 0.000105T - 2783T^{-2}$
Latent heat of fusion. 330 to 347 kJ/kg (142 to 149 Btu/lb)
Thermal conductivity. 96 W/m · K (56 Btu/ft · h · °F) at 100 to 300 °C (212 to 572 °F)

Electrical Properties

Electrical conductivity. 18.5% IACS
Electrical resistivity. 92 nΩ · m at 20 °C (68 °F)
Electrolytic solution potential. 1.59 V versus saturated calomel electrode

Fabrication Characteristics

Weldability. Gas-shielded arc welding with AZ61A or AZ92A rod (AZ61A preferred), excellent; stress relief required. Resistance welding, excellent
Recrystallization temperature. Recrystallizes after 1 h at 205 °C (400 °F) following 15% cold work
Annealing temperature. 345 °C (650 °F)
Hot-working temperature. 230 to 425 °C (450 to 800 °F)

AZ61A

Specifications

AMS. Extrusions: 4350. Forgings: 4358
ASTM. Extrusions: B 107. Forgings: B 91
SAE. J466. Former SAE alloy numbers: 520 (extrusions) and 531 (forgings)
UNS number. M11610
Government. Extruded bar, rod, and shapes: QQ-M-31B. Extruded tubing: WW-T-825A. Forgings: QQ-M-40B
Foreign. Elektron AZ61 (extruded bar, sections, and tubing). British: extruded bar, sections, and tubing, BS 3373 MAG121; forgings, BS 3372 MAG121. German: DIN 9715 3.5612; castings, DIN 1729 3.5612. French: AFNOR G-A6Z1

Chemical Composition

Composition limits. 5.8 to 7.2 Al, 0.15 Mn min, 0.40 to 1.5 Zn. 0.10 Si max, 0.05 Cu max, 0.005 Ni max, 0.005 Fe max, 0.30 max other (total), bal Mg
Consequence of exceeding impurity limits. Excessive Cu, Ni, or Fe degrades corrosion resistance.

Applications

Typical uses. General-purpose extrusions with good properties and moderate cost, and forgings with good mechanical properties; used in the as-fabricated (F) temper. This alloy is used in sheet form for battery applications only.

Mechanical Properties

Tensile properties. See Tables 5 and 6.
Shear strength. See Table 5.
Compressive yield strength. See Table 5.
Bearing properties. See Table 5.
Hardness. See Table 5.
Poisson's ratio. 0.35
Elastic modulus. Tension, 45 GPa (6.5 × 10^6 psi); shear, 17 GPa (2.4 × 10^6 psi)
Impact strength. Charpy V-notch: forgings, 3 J (2.2 ft · lbf); extruded rod, bar, and shapes, 4.1 J (3.0 ft · lbf)

Mass Characteristics

Density: 1.8 g/cm^3 (0.065 lb/in.^{3}) at 20 °C (68 °F)

Thermal Properties

Liquidus temperature. 620 °C (1145 °F)
Solidus temperature. 525 °C (975 °F)
Incipient melting temperature. 418 °C (785 °F)
Coefficient of linear thermal expansion. 26 μm/m · K (14 μin./in. · °F) at 20 °C (68 °F)
Specific heat. 1.05 kJ/kg · K (0.25 Btu/lb · °F) at 25 °C (78 °F)

Table 3 Typical tensile properties of AZ31B at various temperatures

Testing temperature °C	Testing temperature °F	Tensile strength MPa	Tensile strength ksi	Yield strength MPa	Yield strength ksi	Elongation in 50 mm (2 in.), %
Sheet, hard rolled						
−80	−112	331	48.0	234	34.0	...
−27	−18	310	45.0	234	34.0	...
21	70	290	42.0	221	32.0	15
100	212	207	30.0	145	21.0	30
150	300	152	22.0	90	13.0	45
200	400	103	15.0	59	8.5	55
260	500	76	11.0	31	4.5	75
315	600	41	6.0	21	3.0	125
370	700	28	4.0	14	2.0	140
Extrusions, as fabricated						
−185	−300	434	63.0	338	49.0	6.0
−130	−200	359	52.0	303	44.0	7.5
−73	−100	314	45.5	262	38.0	9.5
−18	0	283	41.0	228	33.0	12.5
21	70	262	38.0	200	29.0	15.0
93	200	238	34.5	148	21.5	23.5
120	250	217	31.5	117	17.0	29.5
150	300	179	26.0	100	14.5	37.5

Table 4 Typical directional properties of AZ31B

Condition	Tensile strength MPa	Tensile strength ksi	Yield strength MPa	Yield strength ksi	Elongation, %(a)
Parallel to rolling direction					
Annealed	255	37	150	22	21
Hard rolled	290	42	220	32	15
Perpendicular to rolling direction					
Annealed	270	39	170	25	19
Hard rolled	295	43	235	34	19

(a) In 50 mm (2 in.)

Table 5 Typical room-temperature mechanical properties of AZ61A-F

Form and condition	Tensile strength MPa	Tensile strength ksi	Tensile yield strength(a) MPa	Tensile yield strength(a) ksi	Elongation, %(b)	Hardness HB(c)	Hardness HRE	Shear strength MPa	Shear strength ksi	Compressive yield strength(a) MPa	Compressive yield strength(a) ksi	Ultimate bearing strength(d) MPa	Ultimate bearing strength(d) ksi	Bearing yield strength(d) MPa	Bearing yield strength(d) ksi
Forgings	295	43	180	26	12	55	66	145	21	125	18	...	...	...	...
Extruded bar, rod, and shapes	305	44	205	30	16	60	72	140	20	130	19	470	68	285	41
Extruded tubing and hollow shapes	285	41	165	24	14	50	60	...	...	110	16	...	...	...	...
Sheet	305	44	220	32	8	...	...	...	...	150	22	...	...	...	...

(a) At 0.2% offset. (b) In 50 mm (2 in.). (c) 500 kg load, 10 mm ball. (d) 4.75 mm (3/16 in.) pin diameter

Latent heat of fusion. 373 kJ/kg (160 Btu/lb)
Thermal conductivity. 80 W/m · K (46 Btu/ft · h · °F)

Electrical Properties

Electrical conductivity. 11.6% IACS at 20 °C (68 °F)
Electrical resistivity. 125 nΩ·m at 20°C (68 °F)
Electrolytic solution potential. 1.58 V versus saturated calomel electrode

Fabrication Characteristics

Weldability. Gas-shielded arc welding with AZ61A or AZ92A rod (AZ61A preferred), good; stress relief required. Resistance welding, excellent
Recrystallization temperature. Recrystallizes after 1 h at 288 °C (550 °F) following 20% cold work
Annealing temperature. 345 °C (650 °F)
Hot-working temperature. 230 to 400 °C (450 to 750 °F)
Hot-shortness temperature. 415 °C (780 °F)

AZ80A

Specifications

AMS. Forgings: 4360
ASTM. Extruded rod, bar, and shapes: B 107. Forgings: B 91
SAE. J466. Former SAE alloy numbers: 523 (extrusions) and 532 (forgings)
UNS number. M11800
Government. Extruded bar, rod, and shapes: QQ-M-31B. Extruded tubing: WW-T-825. Forgings: QQ-M-40B

Chemical Composition

Composition limits. 7.8 to 9.2 Al, 0.20 to 0.80 Zn, 0.12 Mn min, 0.10 Si max, 0.05 Cu max, 0.005 Ni max, 0.005 Fe max, 0.30 max other (total), bal Mg
Consequence of exceeding impurity limits. Excessive Si, Cu, Ni, or Fe degrades corrosion resistance.

Applications

Typical uses. Extruded products and press forgings. This alloy can be heat treated.

Mechanical Properties

Tensile properties. See Tables 7 and 8.
Shear strength. See Table 7.
Compressive yield strength. See Table 7.
Bearing properties. See Table 7.
Hardness. See Table 7.
Poisson's ratio. 0.35
Elastic modulus. Tension, 45 GPa (6.5 × 10[6] psi); shear, 17 GPa (2.4 × 10[6] psi)

Mass Characteristics

Density. 1.8 g/cm³ (0.065 lb/in.³) at 20 °C (68 °F)

Thermal Properties

Liquidus temperature. 610 °C (1130 °F)
Solidus temperature. 490 °C (915 °F)
Incipient melting temperature. 427 °C (800 °F)
Coefficient of linear thermal expansion. 26 μm/m · K (14 μin./in. · °F) at 20 °C (68 °F)
Specific heat. 1.05 kJ/kg · K (0.25 Btu/lb · °F) at 25 °C (78 °F)
Thermal conductivity. 76 W/m · K (44 Btu/ft · h · °F) at 100 to 300 °C (212 to 572 °F)

Electrical Properties

Electrical conductivity. Extruded condition, 10.6% IACS at 20 °C (68 °F)
Electrical resistivity. 145 nΩ · m at 20 °C (68 °F)
Electrolytic solution potential. 1.57 V versus saturated calomel electrode

Fabrication Characteristics

Weldability. Gas-shielded arc welding with AZ61A or AZ92A rod (AZ61A preferred), good; stress relief required. Resistance welding, excellent
Recrystallization temperature. Recrystallizes after 1 h at 345 °C (650 °F) following 10% cold work
Annealing temperature. 385 °C (725 °F)
Hot-working temperature. 320 to 400 °C (600 to 750 °F)
Hot-shortness temperature. 415 °C (775 °F)

Table 6 Typical properties of AZ61A-F extrusions at various temperatures

Temperature °C	Temperature °F	Tensile strength MPa	Tensile strength ksi	Yield strength MPa	Yield strength ksi	Elongation in 50 mm (2 in.), %
−185	−300	379	55.0	317	46.0	4
−130	−200	355	51.5	296	43.0	6.5
−73	−100	331	48.0	265	38.5	9.5
−18	0	317	46.0	238	34.5	13
21	70	310	45.0	228	33.0	16
93	200	286	41.5	179	26.0	23
150	300	217	31.5	134	19.5	32
200	400	145	21.0	97	14.0	48.5
315	600	52	7.5	34	5.0	70

Table 7 Typical room-temperature mechanical properties of AZ80A

Form and condition	Tensile strength MPa	Tensile strength ksi	Tensile yield strength(a) MPa	Tensile yield strength(a) ksi	Elongation, %(b)	Hardness HB(c)	Hardness HRE	Shear strength MPa	Shear strength ksi	Compressive yield strength MPa	Compressive yield strength ksi	Ultimate bearing strength MPa	Ultimate bearing strength ksi	Bearing yield strength MPa	Bearing yield strength ksi
Forgings															
As-forged	330	48	230	33	11	69	80	150	22	170	25	...	...	...	...
Aged (T5 temper)	345	50	250	36	6	72	82	160	23	195	28	...	...	...	...
Bar, rod, and shapes															
As-extruded	340	49	250	36	11	67	77	150	22	...	...	550	80	350	51
Aged (T5 temper)	380	55	275	40	7	80	88	165	24	240	35	...	...	...	...

(a) At 0.2% offset. (b) In 50 mm (2 in.). (c) 500 kg load, 10 mm ball

Table 8 Typical mechanical properties of AZ80A-F at various temperatures

Testing temperature °C	°F	Tensile strength MPa	ksi	Yield strength MPa	ksi	Elongation in 50 mm (2 in.), %
−73	−100	386	56.0	269	39.0	8.5
−18	0	355	51.5	252	36.5	10.5
21	70	338	49.0	248	36.0	11.0
93	200	307	44.5	221	32.0	18.0
150	300	241	35.0	176	25.5	25.5
200	400	197	28.5	121	17.5	35.0
260	500	110	16.0	76	11.0	57.0

Table 9 Typical tensile properties of HK31A-H24 sheet at elevated temperatures

Testing temperature °C	°F	Tensile strength MPa	ksi	Yield strength MPa	ksi	Elongation in 50 mm (2 in.), %
21	70	260	38	205	30	8
150	300	180	26	165	24	20
200	400	165	24	145	21	21
260	500	140	20	115	17	19
315	600	89	13	48	7	70
345	650	55	8	28	4	>100

HK31A

See also cast alloy HK31A.

Specifications

AMS. Annealed sheet and plate: 4384E
ASTM. Sheet and plate: B 90
SAE. J465. Former SAE alloy number: 507
UNS number. M13310
Government. Sheet and plate: MIL-M-26075

Chemical Composition

Composition limits. 2.5 to 4.0 Th, 0.4 to 1.0 Zr, 0.3 Zn max, 0.1 Cu max, 0.01 Ni max, 0.3 max other (total), bal Mg

Applications

Typical uses. Sheet and plate with excellent weldability and formability, and with high strength up to 315 °C (600 °F)

Mechanical Properties

Tensile properties. Tensile strength: H24 temper, 260 MPa (38 ksi); O temper, 230 MPa (33 ksi). Yield strength: H24 temper, 205 MPa (30 ksi); O temper, 140 MPa (20 ksi). Elongation in 50 mm (2 in.): O temper, 23%; H24 temper, 9%
Tensile properties versus temperature. See Table 9 and Fig. 1 and 2.
Compressive yield strength. O temper: 97 MPa (14 ksi) at 21 °C (70 °F). H24 temper: 160 MPa (23 ksi) at 21 °C (70 °F); 150 MPa (22 ksi) at 204 °C (400 °F). See also Fig. 1 and 2.
Bearing properties. H24 temper: ultimate bearing strength, 420 MPa (61 ksi); bearing yield strength, 285 MPa (41 ksi)
Hardness. H24 temper, 68 HRE: O temper, 55 HRE
Poisson's ratio. 0.35
Elastic modulus. Tension, 45 GPa (6.5 × 10^6 psi); shear, 17 GPa (2.4 × 10^6 psi)
Impact strength. Charpy V-notch, at 20 °C (68 °F): H24 temper, 4.1 J (3.0 ft · lbf); O temper, 5.4 J (4.0 ft · lbf)
Creep characteristics. See Fig. 3 and 4.

Mass Characteristics

Density. 1.8 g/cm^3 (0.065 lb/in.^{3}) at 20 °C (68 °F)

Table 10 Thermal conductivity of HK31A sheet and plate at various temperatures

Testing temperature °C	°F	Thermal conductivity W/m · K	Btu/ft · h · °F
H24 temper			
18	65	114	66
38	100	114	66
93	200	119	69
150	300	123	71
200	400	128	74
260	500	132	76
O temper			
18	65	107	62
38	100	107	62
93	200	110	64
150	300	114	66
200	400	119	69
260	500	123	71

Thermal Properties

Liquidus temperature. 650 °C (1200 °F)
Solidus temperature. 590 °C (1090 °F)
Incipient melting temperature. 627 to 632 °C (1160 to 1170 °F) in circulating air
Specific heat versus temperature. $C_p = 1374 + 0.0002306T + 3370T^{-2}$
Latent heat of fusion. 318 to 335 kJ/kg (137 to 144 Btu/lb)
Thermal conductivity. See Table 10.

Electrical Properties

Electrical resistivity. At 20 °C (68 °F): H24 temper, 61 nΩ · m; O temper, 60 nΩ · m

Fabrication Characteristics

Weldability. Gas-shielded arc welding with HK 31A or EZ33A rod (EZ33A preferred), excellent; stress relief can be used for sheet and plate, but is not required. Resistance welding, excellent

Table 11 Typical tensile and compressive properties of HM21A at elevated temperatures

Testing temperature °C	°F	Tensile strength MPa	ksi	Tensile yield strength MPa	ksi	Compressive yield strength MPa	ksi	Elongation, %(a)
HM21A-T8 sheet(b)								
21	70	235	34	170	25	130	19	8
200	400	125	18	115	17	105	15	30
260	500	110	16	105	15	105	15	25
315	600	97	14	83	12	83	12	15
370	700	76	11	55	8	55	8	50
HM21A-T5 forgings(c)								
21	70	230	33	140	20	115	17	15
200	400	110	16	90	13	...	...	49(d)
315	600	90	13	76	11	...	...	37(d)
370	700	76	11	55	8	...	...	43(d)

(a) In 50 mm (2 in.). (b) Under 100-h exposure. (c) Rapid heating, except for forging tested at 21 °C (70 °F). (d) In 25 mm (1 in.)

HM21A

Specifications

AMS. Sheet and plate: 4390. Forgings: 4363
ASTM. Sheet and plate: B 90. Forgings: B 91
UNS number. M13210
Government. Sheet and plate: MIL-M-8917. Forgings: QQ-M-40

Chemical Composition

Composition limits. 1.5 to 2.5 Th, 0.45 to 1.1 Mn, 0.30 max other (total), bal Mg

Applications

Typical uses. Sheet, plate, and forgings in the solution-heat-treated, cold-worked, and annealed condition (T8 temper), usable to 343 °C (650 °F) and above

Mechanical Properties

Tensile properties. T8 temper: tensile strength, 235 MPa (34 ksi); yield strength at 0.2% offset, 170 MPa (25 ksi)
Shear strength. 125 MPa (18 ksi)

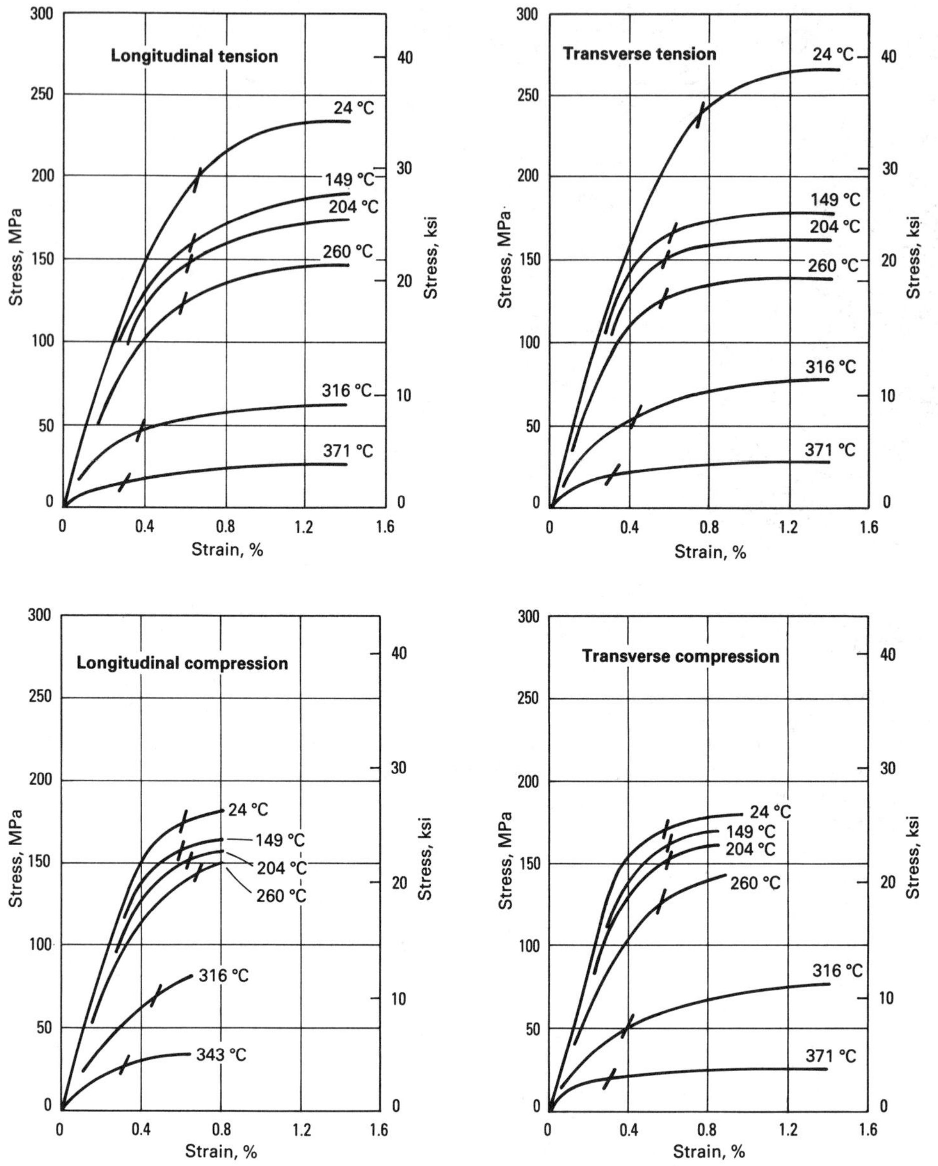

Fig. 1 Typical stress-strain curves for 1.63 mm (0.064 in.) thick HK31A-H24 sheet

Compressive yield strength. 130 MPa (19 ksi)
Tensile and compressive properties versus temperature. See Table 11 and Fig. 5.
Bearing properties. Ultimate bearing strength, 415 MPa (60 ksi); bearing yield strength, 270 MPa (39 ksi)
Poisson's ratio: 0.35
Elastic modulus. Tension, 45 GPa (6.5 × 10^6 psi); shear, 17 GPa (2.4 × 10^6 psi)
Creep characteristics. See Table 12 and Fig. 6(a) and 6(b).

Mass Characteristics

Density. 1.78 g/cm^3 (0.064 lb/in.^{3}) at 20 °C (68 °F)

Thermal Properties

Liquidus temperature. 650 °C (1200 °F)
Solidus temperature. 605 °C (1120 °F)
Specific heat versus temperature. $C_p = 0.1412 + 0.0002294T + 3068T^{-2}$
Latent heat of fusion. 343 kJ/kg (148 Btu/lb)
Thermal conductivity. H24 temper, 134 W/m · K (77 Btu/ft · h · °F); O temper, 138 W/m · K (80 Btu/ft · h · °F)

Electrical Properties

Electrical resistivity. At 20 °C (68 °F): H24 temper, 52 nΩ · m; O temper, 50 nΩ · m

Fabrication Characteristics

Weldability. Gas-shielded arc welding with EZ33A rod, excellent; resistance welding, very good
Annealing temperature. 455 °C (850 °F)
Hot-working temperature. 455 to 595 °C (850 to 1100 °F)

HM31A

Specifications

AMS. As-extruded: 4388. Extruded and aged: 4389
SAE. J466
UNS number. M13312
Government. MIL-M-8916

Chemical Composition

Composition limits. 2.5 to 3.5 Th, 1.2 Mn min, 0.30 max other (total), bal Mg

Applications

Typical uses. Weldable alloy developed primarily for elevated-temperature structural service in the form of extruded bar, rod, shapes, and tubing. Exposure to temperatures up to 315 °C (600 °F) for 1000 h causes virtually no change in short-time room- and elevated-temperature properties. Superior elastic modulus, particularly at elevated temperatures. Although certain extruded sections develop optimum properties in the as-extruded (F) temper, other sections require aging to the T5 temper.

Mechanical Properties

Tensile properties. See Table 13 and Fig. 7.
Shear strength. Punch, 150 MPa (22 ksi) at 21 °C (70 °F)
Compressive yield strength. See Table 13.
Bearing properties. At 21 °C (70 °F): ultimate bearing strength, 480 MPa (70 ksi); bearing yield strength, 345 MPa (50 ksi)
Poisson's ratio. 0.35
Elastic modulus. Tension: 45 GPa (6.5 × 10^6 psi) at 21 °C (70 °F); 42 GPa (6.1 × 10^6 psi) at 150 °C (300 °F); 40 GPa (5.9 × 10^6 psi) at 200 °C (400 °F); 39 GPa (5.6 × 10^6 psi) at 315 °C (600 °F). Shear: 17 GPa (2.4 × 10^6 psi) at 21 °C (70 °F)
Creep characteristics. See Table 14.

Mass Characteristics

Density. 1.8 g/cm^3 (0.065 lb/in.^{3})

Thermal Properties

Liquidus temperature. 650 °C (1200 °F)
Solidus temperature. 605 °C (1120 °F)
Incipient melting temperature. 482 °C (900 °F)
Coefficient of linear thermal expansion. 26 μm/m · K (14.5 μin./in. · °F) at 20 to 93 °C (68 to 200 °F); 28 μm/m · K (15.6 μin./in. · °F) at 20 to 316 °C (68 to 600 °F); 30 μm/m · K (16.8 μin./in. · °F) at 20 to 540 °C (68 to 1000 °F)
Specific heat versus temperature. $C_p = 0.0982 + 0.0002894T + 5300T^{-2}$
Latent heat of fusion. 331 kJ/kg (143 Btu/lb)
Thermal conductivity. 104 W/m · K (60 Btu/ft · h · °F)

Electrical Properties

Electrical conductivity. F temper at 20 °C (68 °F): volumetric, 26% IACS; mass, 135% IACS

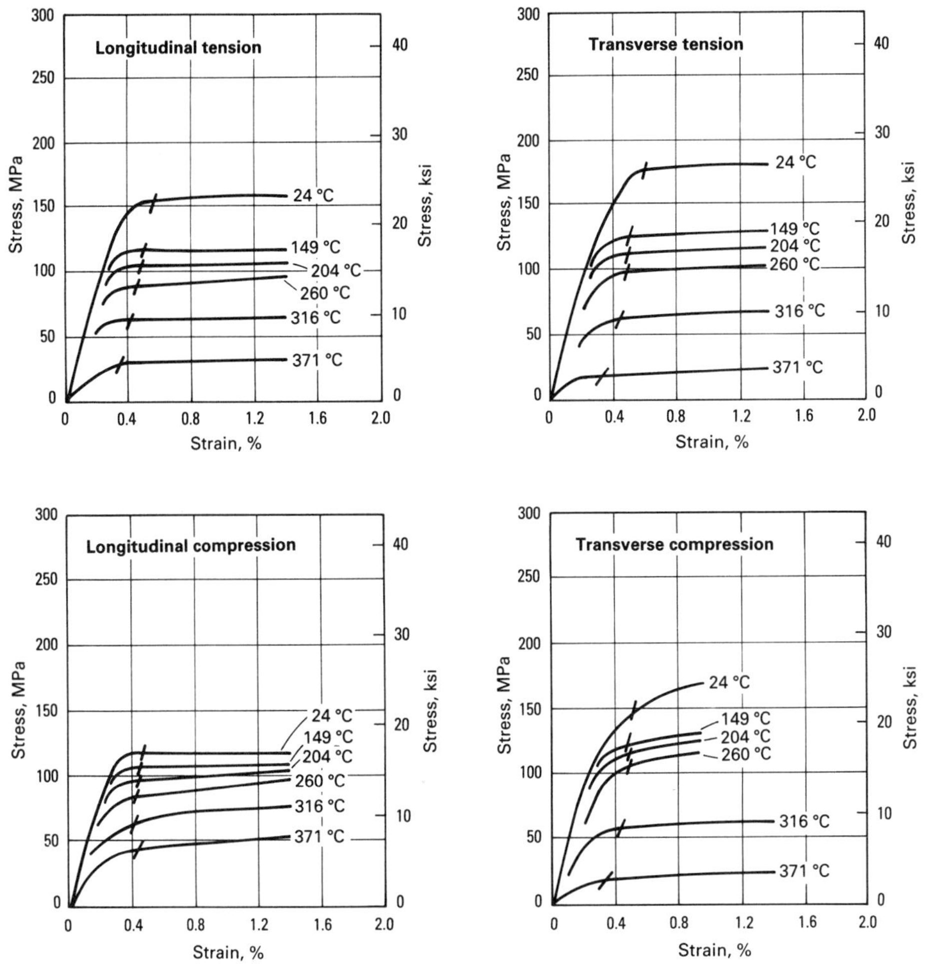

Fig. 2 Typical stress-strain curves for 1.63 mm (0.064 in.) thick HK31A-0 sheet

Table 12 Typical creep properties of HM21A-T8 sheet

Testing temperature		Stress to produce, in 100 h, extension of: 0.1% (creep)		0.2% (total)		0.5% (total)	
°C	°F	MPa	ksi	MPa	ksi	MPa	ksi
150	300	103	14.9	80	11.5	108	15.6
200	400	92	13.3	72	10.5	93	13.5
260	500	55	8.0	48	7.0	62	9.0
315	600	34	5.0	34	5.0	41	6.0
370	700	16	2.3	18	2.6	24	3.5

Table 13 Typical tensile and compressive properties of HM31A extrusions up to 2600 mm^2 (4 in.^{2}) in area

Testing temperature		Tensile strength		Tensile yield strength		Compressive yield strength		Elongation in 50 mm (2 in.), %
°C	°F	MPa	ksi	MPa	ksi	MPa	ksi	
21	70	283	41	230	33	165	24	10
150	300	195	28	180	26	170	25	30
200	400	165	24	160	23	160	23	32
260	500	145	21	140	20	140	20	25
315	600	115	17	110	16	110	16	22
370	700	90	13	83	12	...	...	35
425	800	55	8	48	7	...	...	60
480	900	14	2	7	1	...	...	100

Table 14 Typical creep properties of HM31A extrusions

Testing temperature		Stress to produce, in 100 h, extension of: 0.1% (creep)		0.2% (total)		0.5% (total)	
°C	°F	MPa	ksi	MPa	ksi	MPa	ksi
200	400	110	16	83	12	115	17
260	500	76	11	69	10	83	12
315	600	41	6	41	6	48	7

Electrical resistivity. F temper: 66 nΩ · m at 20 °C (68 °F); 79 nΩ · m at 93 °C (200 °F); 97 nΩ · m at 200 °C (400 °F); 115 nΩ · m at 315 °C (600 °F)
Temperature coefficient of electrical resistivity. 0.18 nΩ · m per K

Fabrication Characteristics

Weldability. Gas-shielded arc welding with EZ33A rod, excellent; no stress relief is necessary. Resistance welding, very good
Recrystallization temperature. Recrystallizes after 1 h at 400 °C (750 °F) following 50% cold work
Hot-working temperature. 370 to 540 °C (700 to 1000 °F)

M1A

Specifications

ASTM. Extruded rod, bar, shapes, and tubing: B 107
SAE. J466. Former SAE alloy numbers: 522 (extrusions) and 533 (forgings)
UNS number. M15100
Government. Extruded bar, rod, and shapes: QQ-M-31. Extruded tubing: WW-T-825. Forgings: QQ-M-40. Sheet and plate: QQ-M-54
Foreign. Elektron AM503. British: BS 3370 MAG101. German: DIN 9715 3.5200

Chemical Composition

Composition limits. 1.2 Mn min, 0.30 Ca max, 0.05 Cu max, 0.01 Ni max, 0.10 Si max, 0.30 max others (total), bal Mg
Consequence of exceeding impurity limits. Excessive Si tends to precipitate Mn. Excessive Cu or Ni degrades corrosion resistance in salt water.

Applications

Typical uses. Wrought products with moderate mechanical properties as well as excellent weldability, corrosion resistance, and hot formability; not heat treatable

Mechanical Properties

Tensile properties. See Tables 15 and 16.
Shear strength. See Table 15.
Compressive properties. See Table 15.
Bearing properties. See Table 15.
Hardness. See Table 15.
Directional properties. See Table 17.
Poisson's ratio. 0.35
Elastic modulus. Tension, 45 GPa (6.5 × 10^6 psi); shear, 17 GPa (2.4 × 10^6 psi)

Mass Characteristics

Density. 1.77 g/cm^3 (0.064 lb/in.^{3}) at 20 °C (68 °F)

Thermal Properties

Liquidus temperature. 649 °C (1200 °F)
Solidus temperature. 648 °C (1198 °F)
Coefficient of linear thermal expansion. 26 μm/m · K (14 μin./in. · °F) at 20 to 100 °C (68 to 212 °F)

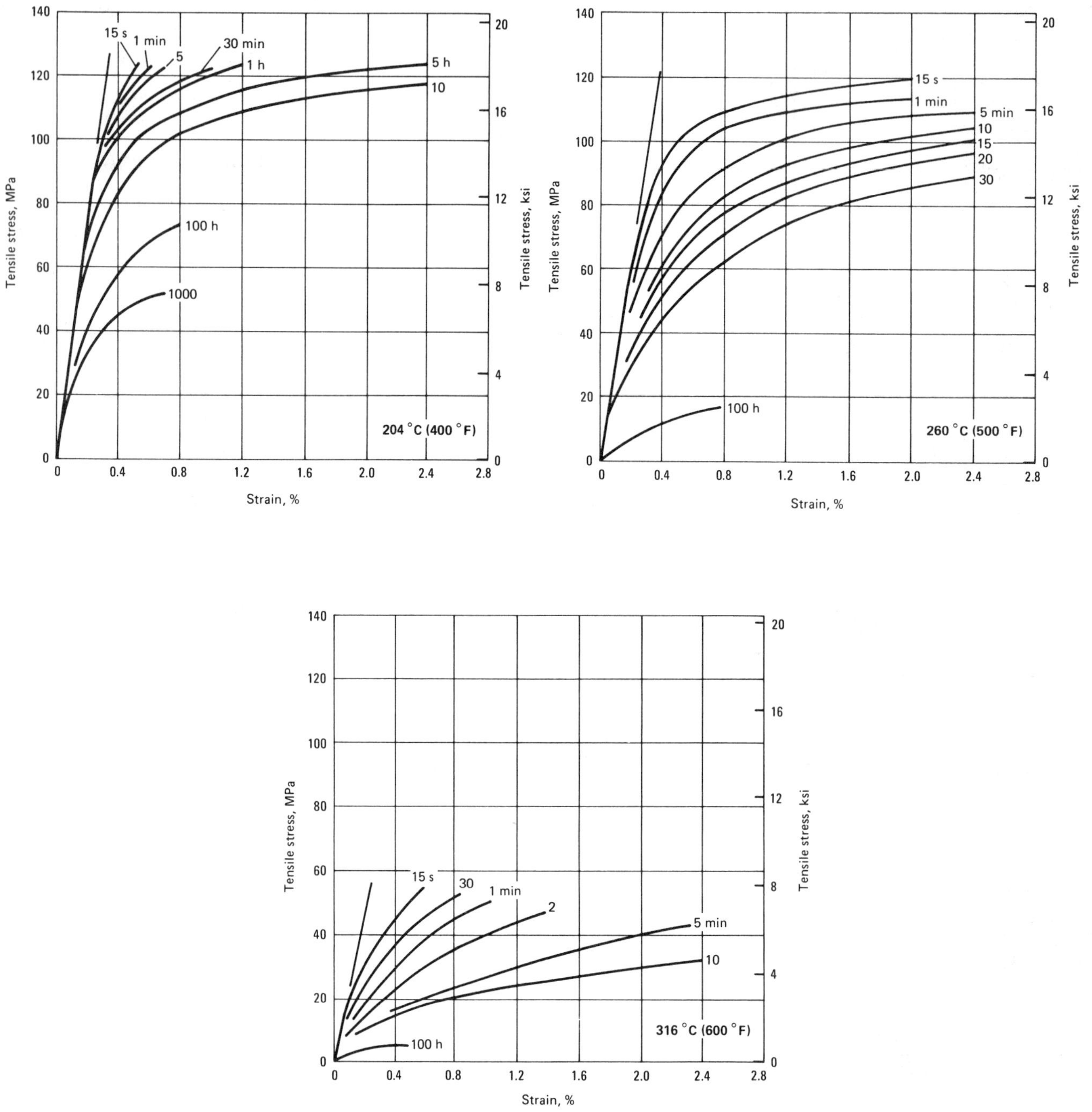

Fig. 3 Isochronous stress-strain curves for 1.63 mm (0.064 in.) thick HK31A-H24 sheet. Specimens exposed at testing temperatures for 3 h before loading.

Table 15 Typical room-temperature mechanical properties of M1A

Product form	Tensile strength		Tensile yield strength(a)		Elongation, %(b)	Hardness		Shear strength		Compressive yield strength(a)		Ultimate bearing strength(d)		Bearing yield strength(d)	
	MPa	ksi	MPa	ksi		HB(c)	HRE	MPa	ksi	MPa	ksi	MPa	ksi	MPa	ksi
Sheet, annealed	230	33	125	18	17	48	55	115	17	76	11	350	51	200	29
Sheet, hard rolled	240	35	180	26	7	54	65	115	17	125	18	395	57	270	39
Extruded bar and shapes	255	37	180	26	12	44	45	125	18	83	12	350	51	195	28
Extruded tubing and hollow shapes	240	35	145	21	9	42	41	...	...	62	9	...	...	...	...
Forgings	250	36	160	23	7	47	54	110	16	...	...	...	...	...	...

(a) At 0.2% offset. (b) In 50 mm (2 in.). (c) 500 kg load, 10 mm ball. (d) 4.75 mm (3/16 in.) pin diameter

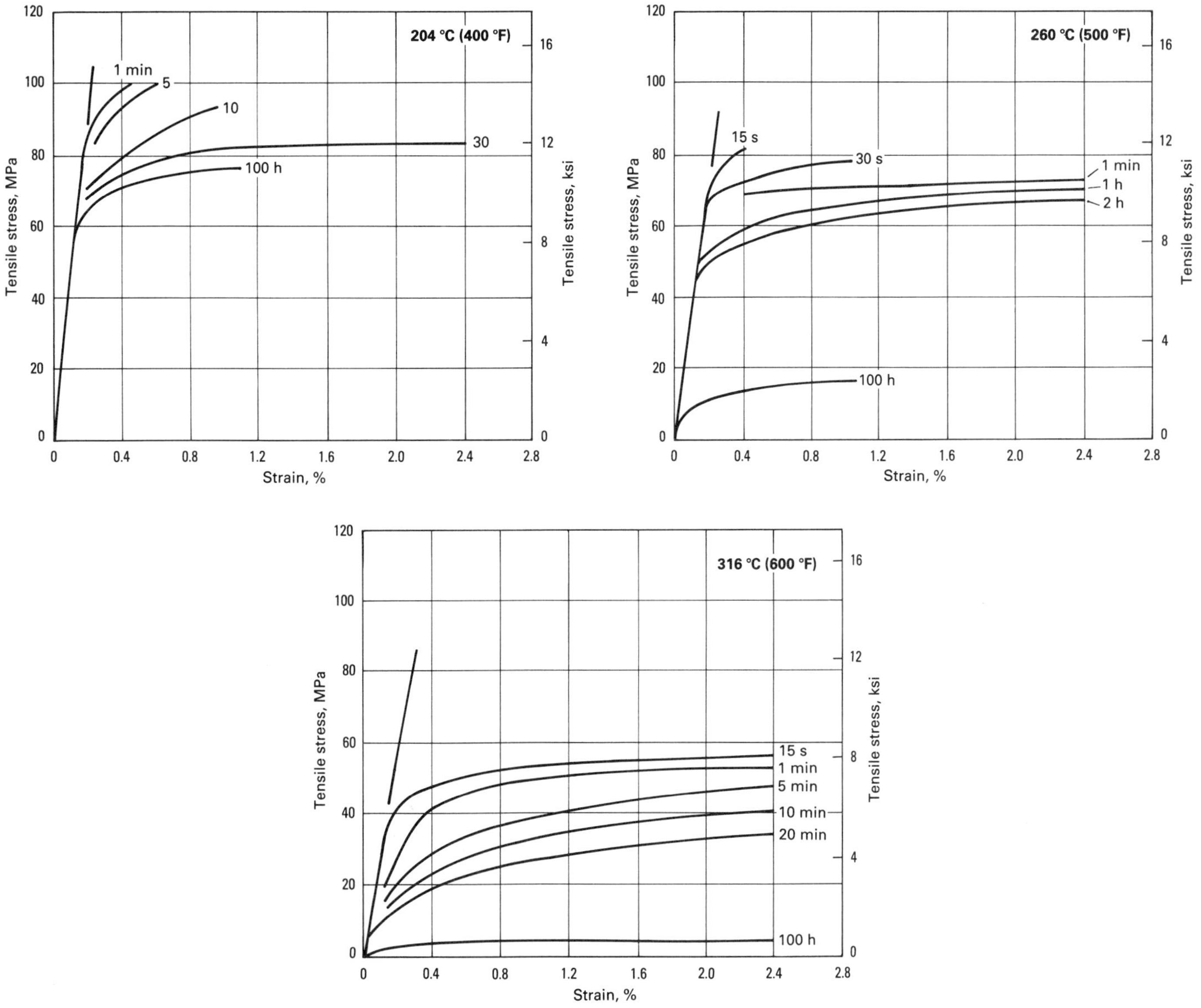

Fig. 4 Isochronous stress-strain curves for 1.63 mm (0.064 in.) thick HK31A-0 sheet. Specimens exposed at testing temperatures for 3 h before loading.

Specific heat. 1.05 kJ/kg · K (0.25 Btu/lb · °F)
Latent heat of fusion. 373 kJ/kg (160 Btu/lb)
Thermal conductivity. 138 W/m · K (79.8 Btu/ft · h · °F)

Electrical Properties

Electrical conductivity. 34.5% IACS at 20 °C (68 °F)
Electrical resistivity. 50 nΩ · m at 20 °C (68 °F)
Electrolytic solution potential. 1.64 V versus saturated calomel electrode

Fabrication Characteristics

Weldability. Gas-shielded arc welding with AZ61A, AZ92A, or M1A rod (AZ61A preferred), excellent; stress relief not required but may be used. Resistance welding, good. Oxyacetylene welding, if necessary, can be done with M1A rod, magnesium flux, and neutral flame.
Recrystallization temperature. Recrystallizes after 1 h at 260 °C (500 °F) following 20% cold work
Annealing temperature. 370 °C (700 °F)
Hot-working temperature. 295 to 540 °C (560 to 1000 °F)

PE

Chemical Composition

Composition limits. 2.5 to 4.0 Al, 0.08 Mn max, 0.7 to 1.6 Zn, 0.05 Si max, 0.05 Cu max, 0.005 Ni max, 0.005 Fe max, 0.04 Ca max, 0.03 max other impurities (total), bal Mg
Consequence of exceeding impurity limits. Poor etch quality

Applications

Typical uses. Photoengraving

Mass Characteristics

Density. 1.76 g/cm^3 (0.064 lb/in.^{3}) at 20 °C (68 °F)

Thermal Properties

Liquidus temperature. 632 °C (1170 °F)
Solidus temperature. 605 °C (1120 °F)
Incipient melting temperature. 532 °C (990 °F)
Coefficient of linear thermal expansion. 26 μm/m · K (14 μin./in. · °F)
Specific heat. 1047 J/kg · K (0.25 Btu/lb · °F) at 20 °C (68 °F)
Latent heat of fusion. 330 to 347 kJ/kg (142 to 149 Btu/lb)

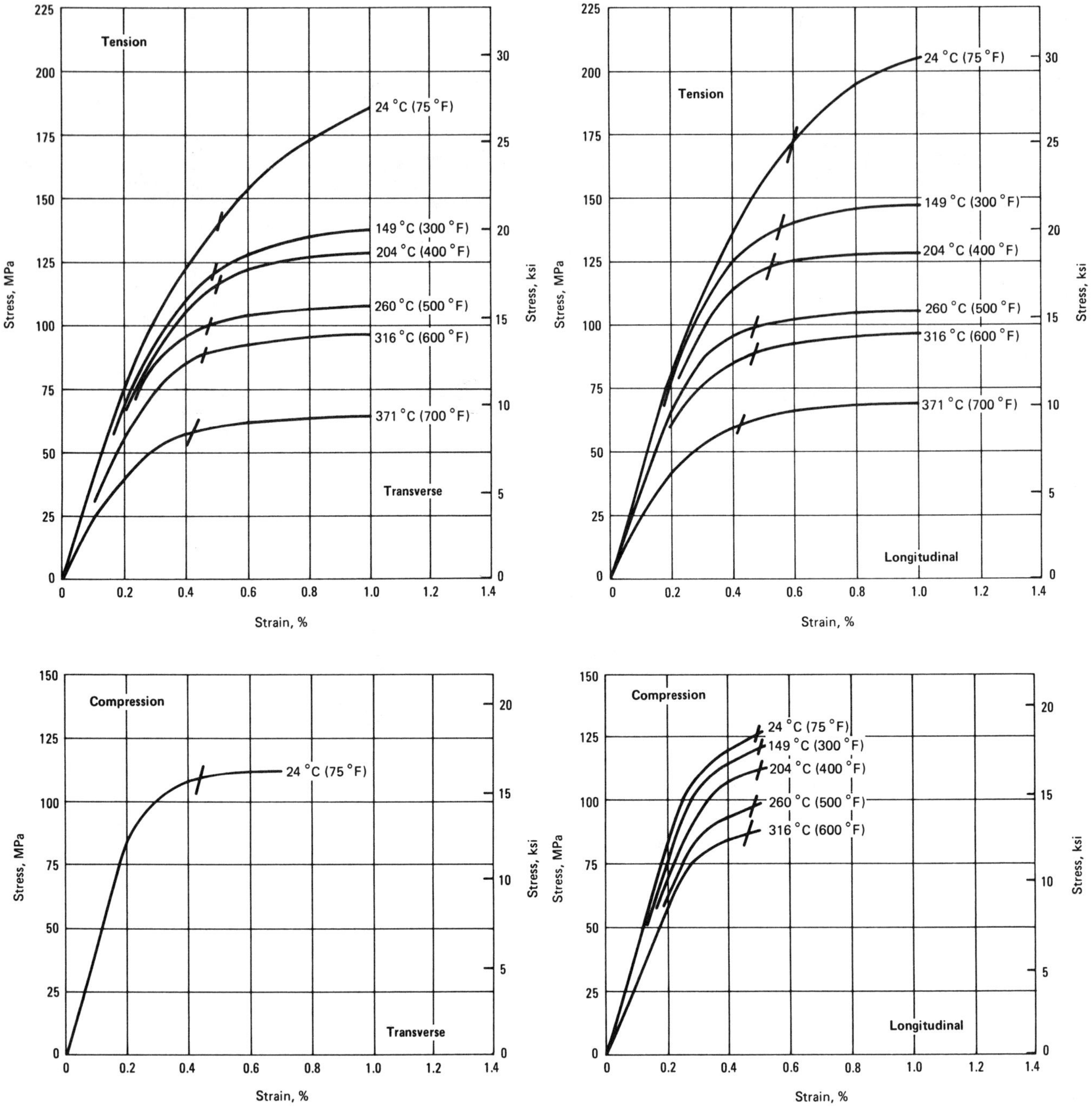

Fig. 5 Typical stress-strain curves for HM21A-T8 sheet. Specimens held at test temperature 3 h before testing.

Fabrication Characteristics

Annealing temperature. 345 °C (650 °F)
Hot-working temperature. 230 to 425 °C (450 to 800 °F)
Hot-shortness temperature. 345 °C (650 °F)

ZC71

Specifications

ASTM. Extrusions: B 107
UNS number. M16710

Chemical Composition

Composition limits. 6.0 to 7.0 Zn, 1.0 to 1.5 Cu, 0.5 to 1.0 Mn, 0.20 Si max, 0.010 Ni max, 0.30 max other (total), bal Mg

Applications

Typical uses. Medium-cost extrusion alloy with good mechanical properties and high elongation. Used in the solution-heat-treated and artificially aged (T6) condition

Mechanical Properties

Tensile properties. See Table 18.
Elastic modulus. Tension, 44.2 GPa (6.4 × 10^6 psi) at 20 °C (68 °F)
Hardness. 70 to 80 HB

Mass Characteristics

Density. 1.83 g/cm^3 (0.066 lb/in.^{3}) at 20 °C (68 °F)

Thermal Properties

Liquidus temperature. 635 °C (1175 °F)
Solidus temperature. 455 °C (850 °F)
Thermal conductivity. 122 W/m · K (70.5 Btu · ft · °F) at 20 °C (68 °F)

Electrical Properties

Electrical resistivity. 54 nΩ · m at 20 °C (68 °F)

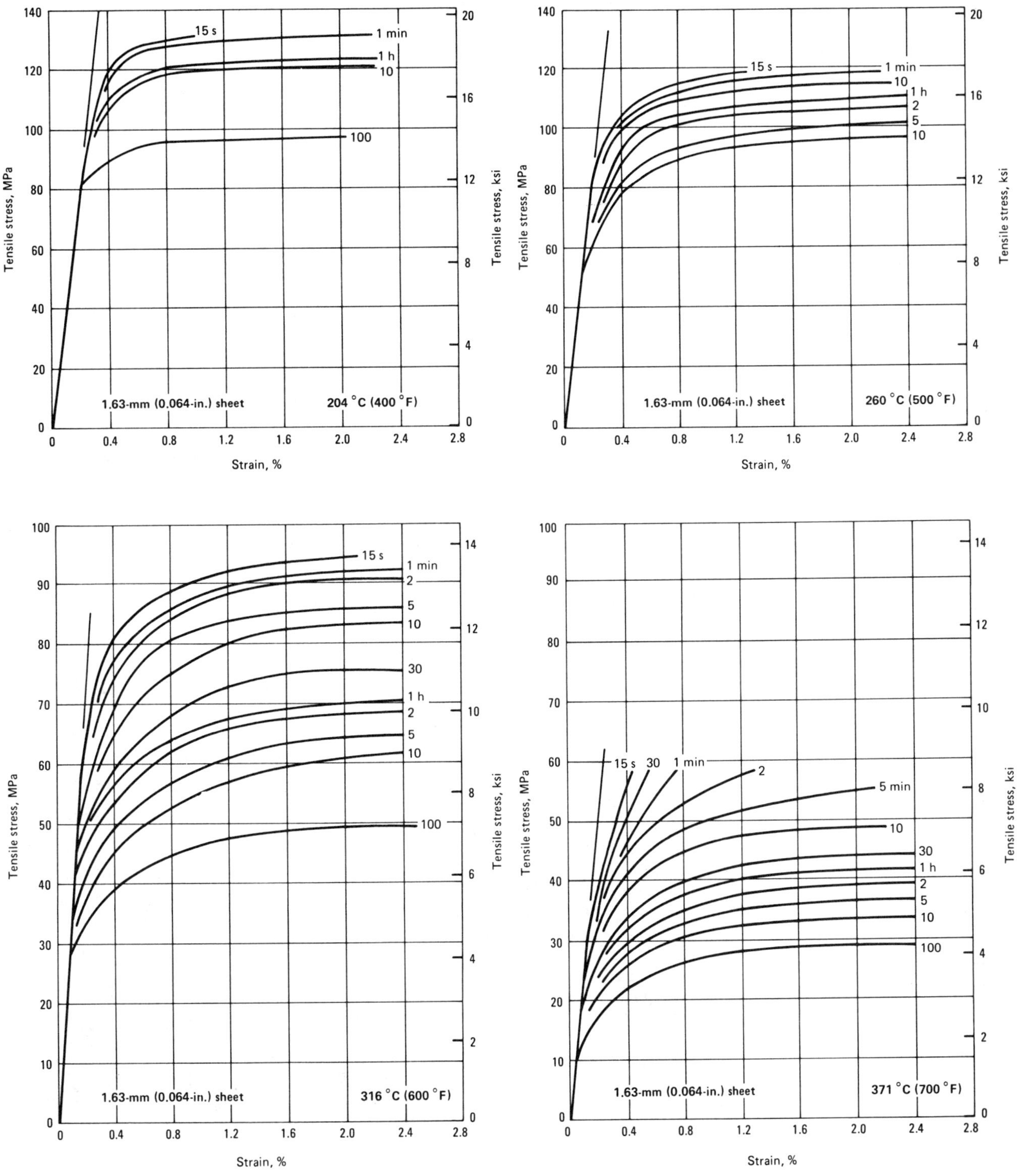

Fig. 6(a) Isochronous stress-strain curves for HM21A-T8 sheet tested at 204, 260, 316, and 371 °C (400, 500, 600, and 700 °F)

Fabrication Characteristics

Weldability. Gas-shielded arc welding with weld rod of same base metal composition

ZK21A

Specifications

AMS. Extruded tubes, bars, rods, and shapes: 4387

UNS. M16210

Government. Extrusions: MIL-M-46039

Chemical Composition

Composition limits. 2.0 to 2.6 Zn, 0.45 to 0.8 Zr, 0.3 max impurities (total), bal Mg

Applications

Typical uses. Moderate-strength extrusion alloy with good weldability. Stress relief is not required. Used in as-extruded (F) temper

Mechanical Properties

Tensile properties. See Table 19.

Compressive properties. See Table 19.

Fabrication Characteristics

Weldability. Gas-shielded arc welding with

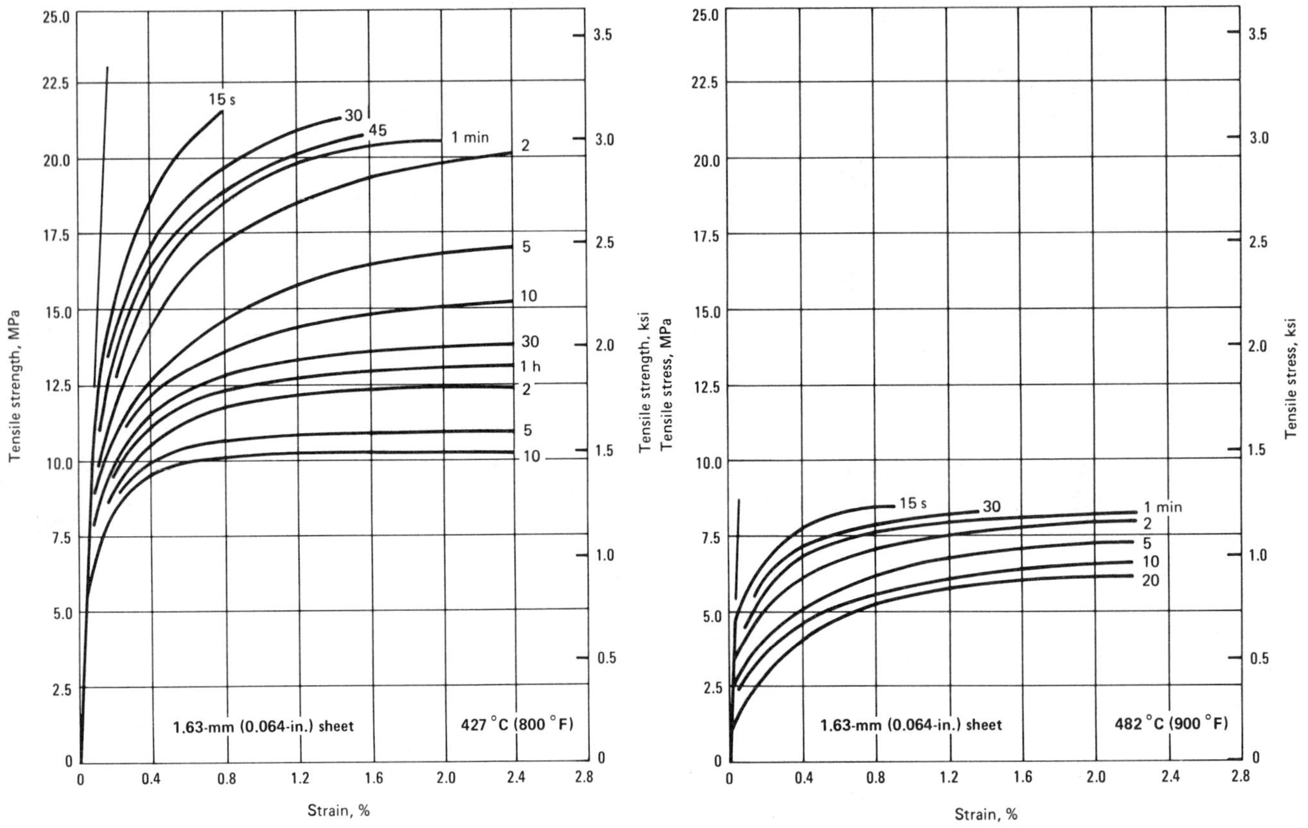

Fig. 6(b) Isochronous stress-strain curves for HM21A-T8 sheet tested at 427 and 482 °C (800 and 900 °F). Specimens held at test temperature 3 h before testing.

AZ61A or AZ92A rod, satisfactory. Resistance welding, satisfactory

ZK40A

Specifications

ASTM. Extrusions: B 107
UNS. M16400
Foreign. Canadian, CSA HG.5 ZK40A

Chemical Composition

Composition limits. 3.5 to 4.5 Zn, 0.45 Zr min, 0.30 max other (total), bal Mg

Applications

Typical uses. High yield strength extrusion alloy, available in as-extruded (F) and artificially aged (T5) tempers. Not as sensitive to stress concentration at thread roots as other high-strength alloys. Can be heat treated. Can replace ZK60A, especially for diamond drill rod, and is more readily extruded

Mechanical Properties

Tensile properties. See Table 20.
Poisson's ratio. 0.35
Elastic modulus. Tension, 45 GPa (6.5 × 10^6 psi); shear, 17 GPa (2.4 × 10^6 psi)

Mass Characteristics

Density. 1.83 g/cm^3 (0.066 lb/in.^{3}) at 20 °C (68 °F)

ZK60A

Specifications

AMS. Extrusions: 4352. Forgings: 4362
ASTM. Extrusions: B 107. Forgings: B 91
SAE. J466. Former SAE alloy number: 524
UNS number. M16600
Government. Extruded rod, bar, and shapes: QQ-M-31. Extruded tubing: WW-T-825. Forgings: QQ-M-40
Foreign. Elektron ZW6. British: BS 3373 MAG161. German: DIN 9715 3.5161. French: AFNOR G-Z5Zr

Chemical Composition

Composition limits. 4.8 to 6.2 Zn, 0.45 Zr min, 0.30 max other (total), bal Mg

Applications

Typical uses. Extruded products and press forgings with high strength and good ductility; can be artificially aged to T5 temper

Mechanical Properties

Tensile properties. See Table 21.
Shear strength. See Table 21.
Compressive yield strength. See Table 21.
Bearing properties. See Table 21.
Hardness. See Table 21.
Poisson's ratio. 0.35
Elastic modulus. Tension, 45 GPa (6.5 × 10^6 psi); shear, 17 GPa (2.4 × 10^6 psi)

Mass Characteristics

Density. 1.83 g/cm^3 (0.066 lb/in.^{3}) at 20 °C (68 °F)

Thermal Properties

Liquidus temperature. 635 °C (1175 °F)
Solidus temperature. 520 °C (970 °F)
Incipient melting temperature. 518 °C (965 °F)
Coefficient of linear thermal expansion. 26 μm/m · K (14 μin./in. · °F) at 20 °C (68 °F)
Specific heat versus temperature. $C_p = 0.1233 + 0.0002566T + 3939T^{-2}$
Latent heat of fusion. 300 to 335 kJ/kg (129 to 144 Btu/lb)
Thermal conductivity. F temper, 117 W/m · K (68 Btu/ft · h · °F) at 20 °C (68 °F); T5 temper, 121 W/m · K (70 Btu/ft · h · °F) at 20 °C (68 °F)

Electrical Properties

Electrical conductivity. At 20 °C (68 °F): F temper, 29% IACS; T5 temper, 30% IACS

Fabrication Characteristics

Weldability. Gas-shielded arc welding with AZ92A welding rod is possible but not recommended because these alloys are prone to hot-shortness cracking; when welds free of cracks are obtained, they

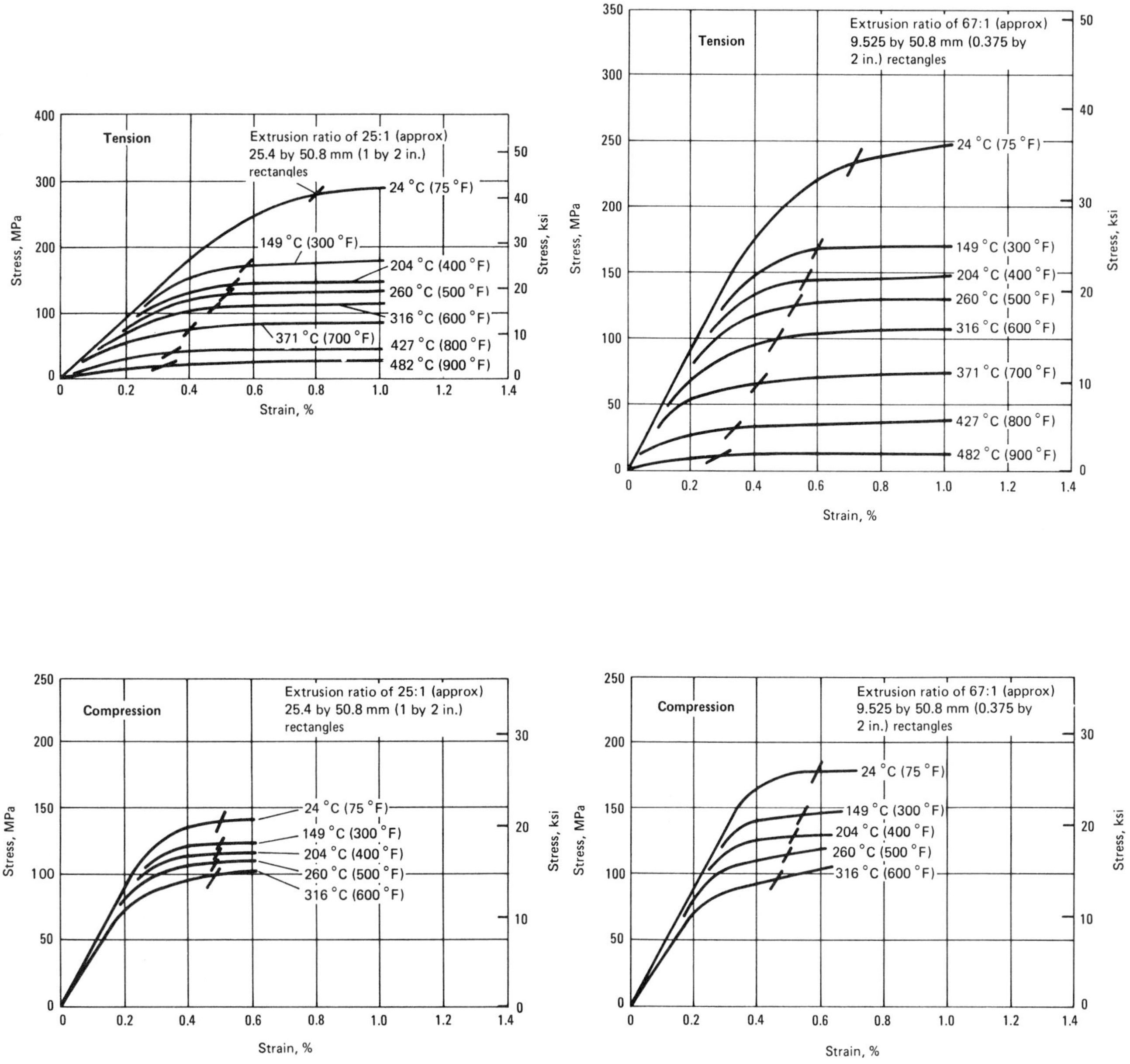

Fig. 7 Typical stress-strain curves for HM31A extrusions. Tested in longitudinal direction

exhibit high weld efficiencies. Resistance welding, excellent
Aging temperature. 150 °C (300 °F) for 24 h in the air, followed by air cooling
Hot-working temperature. 315 to 400 °C (600 to 750 °F)
Hot-shortness temperature. Cast, 315 °C (600 °F); wrought, 510 °C (950 °F)

Cast Magnesium Alloys

AM60A, AM60B

Specifications

ASTM. Die castings: B 94
UNS numbers. AM60A: M10600. AM60B: M10603
Foreign. German: DIN 1729 3.5662

Chemical Composition

Composition limits of AM60A. 5.5 to 6.5 Al, 0.13 Mn min, 0.50 Si max, 0.35 Cu max, 0.22 Zn max, 0.03 Ni max, bal Mg
Composition limits of AM60B. 5.5 to 6.5 Al, 0.25 Mn min, 0.10 Si max, 0.22 Zn max, 0.005 Fe max, 0.010 Cu max, 0.002 Ni max, 0.003 max other (total), bal Mg. If the Mn content is less than 0.25% or the Fe content in AM60B exceeds 0.005%, then the Fe-Mn ratio will not exceed 0.010, and corrosion resistance will rapidly decrease.
Consequence of exceeding impurity limits. Corrosion resistance decreases with increasing Fe, Cu, or Ni content.

Applications

Typical uses. Die casting alloy used in as-cast (F) temper for production of automotive wheels and other parts requiring good elongation and toughness combined with reasonable yield and tensile properties

Mechanical Properties

Tensile properties. F temper: tensile

Table 16 Typical tensile properties of M1A at elevated temperatures

Testing temperature °C	°F	Tensile strength MPa	ksi	Yield strength MPa	ksi	Elongation, %
Bar and shapes, extruded						
93	200	186	27.0	145	21.0	16
120	250	165	24.0	131	19.0	18
150	300	145	21.0	110	16.0	21
200	400	117	17.0	83	12.0	27
315	600	62	9.0	34	5.0	53
Sheet, annealed						
93	200	169	24.5	110	16.0	31
120	250	148	21.5	100	14.5	41
150	300	133	19.3	86	12.5	44
Sheet, hard rolled						
93	200	203	29.5	183	26.5	11
120	250	190	27.5	169	24.5	13
150	300	172	25.0	145	21.0	15
Forgings						
93	200	165	24.0	121	17.5	25
120	250	145	21.0	107	15.5	26
150	300	131	19.0	93	13.5	31
200	400	114	16.5	69	10.0	34
260	500	83	12.0	45	6.5	67
315	600	41	6.0	28	4.0	140

Table 17 Typical directional properties of M1A sheet

Condition	Tensile strength MPa	ksi	Yield strength MPa	ksi	Elongation, %
Parallel to rolling direction					
Annealed	230	33	125	18	17
Hard rolled	250	36	180	26	7
Perpendicular to rolling direction					
Annealed	220	32	115	17	17
Hard rolled	255	37	185	27	13

strength, 220 MPa (32 ksi); yield strength, 130 MPa (19 ksi); elongation, 6% in 50 mm (2 in.)
Compressive yield strength. F temper: 130 MPa (19 ksi)
Poisson's ratio. 0.35
Elastic modulus. Tension, 45 GPa (6.5 × 10^6 psi)

Mass Characteristics

Density. 1.8 g/cm^3 (0.065 lb/in.^{3}) at 20 °C (68 °F)

Thermal Properties

Liquidus temperature. 615 °C (1140 °F)
Solidus temperature. 540 °C (1005 °F)
Coefficient of linear thermal expansion. 25.6 μm/m · K (14.2 μin./in. · °F) at 20 to 100 °C (68 to 212 °F)
Thermal conductivity. 62 W/m · K (36 Btu/ft · h · °F) at 20 °C (68 °F)

Fabrication Characteristics

Casting temperature. 650 to 695 °C (1200 to 1280 °F)
Weldability. Not weldable

Corrosion Resistance

ASTM B 177 salt spray test. AM60B: <0.13 mg/cm^2/day (<20 mils/yr)

AM100A

Specifications

AMS. Permanent mold castings: 4483. Investment castings: 4455
ASTM. Sand castings: B 80. Ingot for sand, permanent mold, and die castings: B 93. Permanent mold castings: B 199. Investment castings: B 403
SAE. J465. Former SAE alloy number: 502
UNS number. M10100
Government. Permanent mold castings: QQ-M-55

Chemical Composition

Composition limits. 9.3 to 10.7 Al, 0.10 Mn min, 0.30 Zn max, 0.30 Si max, 0.10 Cu max, 0.01 Ni max, 0.30 max other (total), bal Mg
Consequence of exceeding impurity limits. Corrosion resistance decreases with increasing amounts of Cu, Ni, and Fe. Increased amounts of Zn decrease pressure tightness. More than 0.5% Si decreases elongation.

Applications

Typical uses. Pressure-tight sand and permanent mold castings with good combinations of tensile strength, yield strength, and elongation

Mechanical Properties

Tensile properties. See Tables 22 and 23, and Fig. 8.
Shear strength. See Table 22.
Compressive yield strength. See Table 22.
Hardness. At room temperature: See Table 22. At −78 °C (−108 °F): F temper, 63 HB or 75 HRE; T4 temper, 60 HB or 73 HRE; T6 temper, 85 HB or 90 HRE
Bearing properties. Ultimate bearing strength: T4 temper, 475 MPa (69 ksi); T61 temper, 560 MPa (81 ksi). Bearing yield strength: T4 temper, 310 MPa (45 ksi); T61 temper, 470 MPa (68 ksi)
Poisson's ratio. 0.35
Impact strength. Charpy V-notch. At 20 °C (68 °F): F temper, 0.8 J (0.6 ft · lbf); T4 temper, 2.7 J (2.0 ft · lbf); T61 temper, 0.9 J (0.7 ft · lbf). At −78 °C (−108 °F): F temper, 1.1 J (0.8 ft · lbf); T4 temper, 3.4 J (2.5 ft · lbf); T6 temper, 1.1 J (0.8 ft · lbf)
Fatigue strength. R.R. Moore type test. At 5 × 10^8 cycles: F and T61 tempers, 70 MPa (10 ksi); T4 temper, 75 MPa (11 ksi)
Elastic modulus. Tension, 45 GPa (6.5 × 10^6 psi); shear, 17 GPa (2.4 × 10^6 psi)

Mass Characteristics

Density. 1.83 g/cm^3 (0.066 lb/in.^{3}) at 20 °C (68 °F)

Thermal Properties

Liquidus temperature. 595 °C (1100 °F)
Solidus temperature. 463 °C (865 °F)
Incipient melting temperature. 430 °C (810 °F)
Coefficient of linear thermal expansion. 25 μm/m · K (14 μin./in. · °F) at 18 to 100 °C (65 to 212 °F)
Specific heat. 1.05 kJ/kg · K (0.25 Btu/lb · °F) at 25 °C (77 °F)
Thermal conductivity. 73 W/m · K (42 Btu/ft · h · °F) at 100 to 300 °C (212 to 572 °F)
Latent heat of fusion. 372 kJ/kg (160 Btu/lb)

Electrical Properties

Electrical conductivity. F temper, 11.5% IACS; T4 temper, 9.9% IACS; T6 temper, 12.3% IACS
Electrical resistivity. F temper, 150 nΩ · m; T4 temper, 175 nΩ · m; T6 temper, 140 nΩ · m; at 20 °C (68 °F)
Electrolytic solution potential. 1.57 V versus saturated calomel electrode
Hydrogen overvoltage. 0.27 V for extrusions; 0.06 V for castings

Fabrication Characteristics

Weldability. Gas-shielded arc welding with AM100A rod, very good
Casting temperatures. Sand castings, 735 to 845 °C (1350 to 1550 °F); permanent mold castings, 650 to 815 °C (1200 to 1500 °F); ingot, 650 to 705 °C (1200 to 1300 °F)

AS41A, AS41XB

Specifications

ASTM. Die castings: AS41A, B 94

Table 18 Typical and minimum tensile properties of magnesium alloy ZC71

Specimen	Condition	0.2% yield strength MPa	ksi	Tensile strength MPa	ksi	Elongation, %
Round bar with 13–125 mm (½–5 in.) diameter	ZCM 711-F (as-extruded)	158	23	240	35	7
	ZCM 711-T5 (precipitation treated)	200	29	248	36	5
	ZCM 711-T6 (fully heat treated)	295	43	324	47	3
16 mm (⅝ in.) diam bar	As-extruded	180–190	26.1–27.6	280–290	40.6–42.1	10–13
	T5 condition	240–265	34.8–38.4	305–320	44.2–46.4	6–10
	T6 condition	340–350	49.3–50.8	360–375	52.2–54.4	4–6
125 mm (5 in.) diam bar	As-extruded	170–190	24.7–27.6	255–275	37.0–40.0	12–15
	T5 condition	215–235	31.2–34.1	275–295	39.9–42.8	8–10
	T6 condition	315–335	45.7–48.6	340–360	49.3–52.2	5–7

Table 19 Minimum mechanical properties at room temperature of ZK21A-F extrusions

Form	Tensile strength MPa	ksi	Yield strength MPa	ksi	Compressive yield strength MPa	ksi	Elongation, %
Rods, bars, and shapes	260	38	195	28	135	20	4
Tubing	235	34	180	26	97	14	4

Table 20 Minimum mechanical properties of ZK40A-T5 at room temperature

Form	Tensile strength MPa	ksi	Yield strength MPa	ksi	Elongation, %	Compressive yield strength MPa	ksi
Extruded bars and shapes	275	40	255	37	4	140	20
Extruded tubes	275	40	250	36	4	140	20

UNS numbers. AS41A, M10410
Foreign. German: DIN 1729 3.5470

Chemical Composition

Composition limits of AS41A. 3.5 to 5.0 Al, 0.50 to 1.5 Si, 0.20 to 0.50 Mn, 0.12 Zn max, 0.06 Cu max, 0.03 Ni max, 0.30 max other (total), bal Mg
Composition limits of AS41XB. 3.5 to 5.0 Al, 0.50 to 1.50 Si, 0.35 Mn min, 0.12 Zn max, 0.0035 Fe max, 0.020 Cu max, 0.002 Ni max, bal Mg
Consequence of exceeding impurity limits. Corrosion resistance decreases with increasing Fe, Cu, or Ni content. If the Mn content is less than 0.35% or the Fe content in AS41XB exceeds 0.0035%, then the Fe-Mn ratio will not exceed 0.010, and corrosion resistance will rapidly decrease.

Applications

Typical uses. Die castings used in the as-cast condition (F temper), with creep resistance superior to that of AZ91A, AZ91B, AZ91D, or AM60A up to 175 °C (350 °F), and with good tensile strength, tensile yield strength, and elongation

Mechanical Properties

Tensile properties. F temper: tensile strength, 210 MPa (31 ksi); yield strength, 140 MPa (20 ksi); elongation, 6% in 50 mm (2 in.)
Compressive yield strength. F temper, 140 MPa (20 ksi)
Poisson's ratio. 0.35
Elastic modulus. Tension, 45 GPa (6.5 × 10^6 psi)

Mass Characteristics

Density. 1.77 g/cm^3 (0.064 lb/in.^{3}) at 20 °C (68 °F)

Thermal Properties

Liquidus temperature. 620 °C (1150 °F)
Solidus temperature. 565 °C (1050 °F)
Coefficient of linear thermal expansion. 26.1 μm/m · K (14.5 μin./in. · °F) at 20 to 100 °C (68 to 212 °F)
Specific heat. 1.0 kJ/kg · K (0.24 Btu/lb · °F) at 20 °C (68 °F)
Thermal conductivity. 68 W/m · K (40 Btu/ft · h · °F) at 20 °C (68 °F)

Fabrication Characteristics

Casting temperature. 660 to 695 °C (1220 to 1280 °F)
Weldability. Not weldable

Corrosion Resistance

ASTM B 117 salt spray test. AS41XB: <0.25 mg/cm^2/day (<20 mils/yr)

AZ63A

Specifications

AMS. Sand castings: F temper, 4420; T4 temper, 4422; T5 temper, 4424
ASTM. Ingot: B 93. Sand castings: B 80
SAE. J465. Former SAE alloy number: 50
UNS number. M11630
Government: Sand castings: QQ-M-56. Permanent mold castings: QQ-M-55
Foreign. Elektron AZG

Chemical Composition

Composition limits. 5.3 to 6.7 Al, 2.5 to 3.5 Zn, 0.15 Mn min, 0.30 Si max, 0.25 Cu max, 0.01 Ni max, 0.30 other (total), bal Mg
Consequence of exceeding impurity limits. Excessive Si causes brittleness. Excessive Cu degrades mechanical properties and corrosion resistance. Excessive Ni degrades corrosion resistance.

Applications

Typical uses. Sand castings with good strength, ductility, and toughness

Table 21 Typical mechanical properties of ZK60A at room temperature

Form and condition	Tensile strength MPa	ksi	Tensile yield strength(a) MPa	ksi	Elongation, %	Hardness HB(b)	Hardness HRE	Shear strength MPa	ksi	Compressive yield strength MPa	ksi	Ultimate bearing strength MPa	ksi	Bearing yield strength MPa	ksi
Extruded bars, rod, and shapes															
ZK60A-F	340	49	260	38	11	75	84	185	27	230	33	550	80	380	55
ZK60A-T5	350	51	285	41	11	82	88	180	26	250	36	585	85	405	59
Extruded hollow shapes and tubing															
ZK60A-F	315	46	235	34	12	75	84	...	...	170	25	...	...	...	...
ZK60A-T5	345	50	275	40	11	82	88	...	...	200	29	...	...	...	...
Forgings															
ZK60A-T5	305	44	215	31	16	65	77	165	24	160	23	420	61	285	41

(a) 0.2% offset. (b) 500 kg load, 10 mm ball

Table 22 Typical mechanical properties of AM100A sand castings at room temperature

Temper	Tensile strength MPa	Tensile strength ksi	Tensile or compressive yield strength(a) MPa	Tensile or compressive yield strength(a) ksi	Elongation in 50 mm (2 in.), %	Hardness HB	Hardness HRE	Shear strength MPa	Shear strength ksi
F	150	22	83	12	2	53	61	125	18
T4	275	40	90	13	10	52	62	140	20
T61	275	40	150	22	1	69	80	145	21
T5	150	22	110	16	2	58	70	...	...
T7	260	38	125	18	1	67	78	...	...

(a) Values are the same for tensile and compressive yield strengths.

Table 23 Typical tensile properties of AM100A sand castings at elevated and subzero temperatures

Testing temperature °C	Testing temperature °F	Tensile strength MPa	Tensile strength ksi	Tensile yield strength MPa	Tensile yield strength ksi	Elongation in 50 mm (2 in.), %
F temper						
−78	−108	150	22	125	18.0	1
T4 temper						
−78	−108	260	38	125	18.0	7
93	200	235	34	...	...	1.5
150	300	160	23	...	...	9
260	500	83	12	...	...	22
T6 temper(a)						
−78	−108	270	39	180	26.0	2
150	300	165	24	62	9.0	4
200	400	115	17	45	6.5	25
260	500	83	12	28	4.0	45
315	600	59	8.5	17	2.5	60
370	700	38	5.5	10	1.5	100

(a) Elevated-temperature properties were determined after prolonged heating.

Mechanical Properties

Tensile properties. Tensile strength: F and T5 tempers, 200 MPa (29 ksi); T4, T6, and T7 tempers, 275 MPa (40 ksi). Yield strength: F and T4 tempers, 97 MPa (14 ksi); T5 temper, 105 MPa (15 ksi); T6 temper, 130 MPa (19 ksi); T7 temper, 115 MPa (17 ksi). Elongation in 50 mm (2 in.): F and T7 tempers, 6%, T4 temper, 12%; T5 temper, 4%; T6 temper, 5%. See also Fig. 9.
Tensile properties versus temperature. See Table 24.
Shear strength. F and T4 tempers, 125 MPa (18 ksi); T5 temper, 130 MPa (19 ksi); T6 and T7 tempers, 140 MPa (20 ksi)
Compressive yield strength. F, T4, and T5 tempers, 97 MPa (14 ksi); T6 temper, 130 MPa (19 ksi); T7 temper, 115 MPa (17 ksi)
Bearing properties. Ultimate bearing strength: F, T4, and T6 tempers, 415 MPa (60 ksi); T5 temper, 455 MPa (66 ksi); T7 temper, 515 MPa (75 ksi). Bearing yield strength: F and T5 tempers, 275 MPa (40 ksi); T4 temper, 305 MPa (44 ksi); T6 temper, 360 MPa (52 ksi); T7 temper, 325 MPa (47 ksi)
Hardness. F temper, 50 HB or 59 HRE; T4 and T5 tempers, 55 HB or 66 HRE; T6 temper, 73 HB or 83 HRE; T7 temper, 64 HB or 76 HRE
Poisson's ratio. 0.35
Elastic modulus. Tension, 45 GPa (6.5 × 10^6 psi); shear, 17 GPa (2.4 × 10^6 psi)

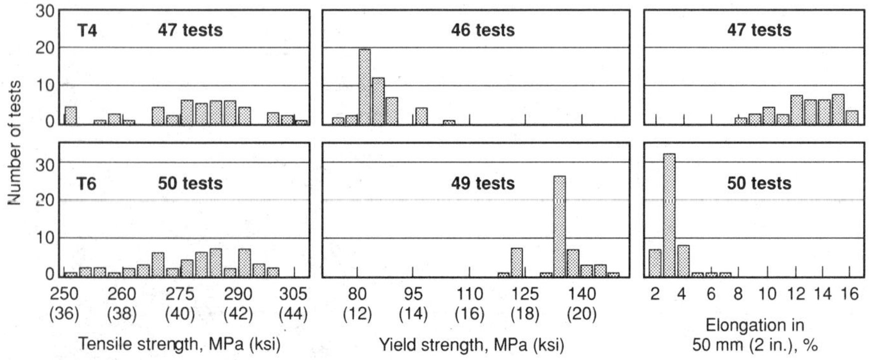

Fig. 8 Distribution of tensile properties for separately sand cast test bars of AM100A

Impact strength. Charpy V-notch: F temper, 1.4 J (1.0 ft · lbf); T4 temper, 3.4 J (2.5 ft · lbf); T5 temper, 3.5 J (2.6 ft · lbf); T6 temper, 1.5 J (1.1 ft · lbf)
Fatigue strength. R.R. Moore type test. At 5 × 10^8 cycles: F, T5, and T6 tempers, 76 MPa (11 ksi); T4 temper, 83 MPa (12 ksi); T7 temper, 115 MPa (17 ksi)

Mass Characteristics

Density. 1.83 g/cm³ (0.066 lb/in.³) at 20 °C (68 °F)

Thermal Properties

Liquidus temperature. 610 °C (1130 °F)
Solidus temperature. 455 °C (850 °F)
Coefficient of linear thermal expansion. 26.1 μm/m · K (14.5 μin./in. · °F) at 20 to 100 °C (68 to 212 °F)
Specific heat. 1.05 kJ/kg · K (0.25 Btu/lb · °F) at 25 °C (77 °F)
Latent heat of fusion. 373 kJ/kg (160 Btu/lb)
Thermal conductivity. 77 W/m · K (44.3 Btu/ft · h · °F) at 100 to 300 °C (212 to 572 °F)

Electrical Properties

Electrical conductivity. At 20 °C (68 °F): F temper, 15% IACS; T4 temper, 12.3% IACS; T5 temper, 13.8% IACS
Electrical resistivity. At 20 °C (68 °F): F temper, 115 nΩ · m; T4 temper, 140 nΩ · m; T5 temper, 125 nΩ · m
Electrolytic solution potential. 1.57 V versus saturated calomel electrode
Hydrogen overvoltage. As cast, 0.34 V

Fabrication Characteristics

Casting temperature. Sand castings, 705 to 845 °C (1300 to 1550 °F)
Weldability. Gas-shielded arc welding with AZ63A or AZ92A rod (AZ63A preferred), fair

AZ81A

Specifications

ASTM. Sand castings: B 80. Ingot: B 93. Permanent mold castings: B 199. Investment castings: B 403
SAE. J465. Former SAE alloy number: 505
UNS number. M11810
Government. Sand castings: QQ-M-56. Permanent mold castings: QQ-M-55
Foreign. Elektron A8. British: BS 2970 MAG1. German: DIN 1729 3.5812. French: AIR 3380 G-A9

Chemical Composition

Composition limits. 7.0 to 8.1 Al, 0.4 to 1.0 Zn, 0.13 Mn min, 0.30 Si max, 0.10 Cu max, 0.01 Ni max, 0.30 max other (total), bal Mg
Consequence of exceeding impurity limits. Excessive Si causes brittleness. Excessive Cu degrades mechanical properties and corrosion resistance. Excessive Ni degrades corrosion resistance.

Table 24 Typical tensile properties of AZ63A sand castings at elevated temperatures

Tested as soon as specimens reached testing temperature

Testing temperature °C	Testing temperature °F	Tensile strength MPa	Tensile strength ksi	Yield strength MPa	Yield strength ksi	Elongation in 50 mm (2 in.), %
F temper						
24	75	197	28.6	94	13.7	4.5
65	150	210	30.5	...	...	3.0
93	200	208	30.1	...	...	4.5
120	250	191	27.7	...	...	7.5
150	300	166	24.1	...	...	20.5
200	400	105	15.3	...	...	50.5
260	500	71	10.3	...	...	38.0
T4 temper						
24	75	254	36.8	94	13.6	10.0
65	150	253	36.7	...	...	9.0
93	200	236	34.3	...	...	7.0
120	250	207	30.0	...	...	9.0
150	300	154	22.4	...	...	33.2
200	400	101	14.6	...	...	38.0
260	500	75	10.9	...	...	26.0
T6 temper						
35	95	232	33.7	122	17.7	5.5
93	200	248	36.0	119	17.3	11.0
120	250	223	32.4	114	16.5	11.0
150	300	169	24.5	103	15.0	15.0
200	400	121	17.5	83	12.0	17.0
260	500	83	12.0	61	8.8	15.0
315	600	57	8.2	39	5.6	20.0

Applications

Typical uses. Sand and permanent mold castings used in the solution-treated condition (T4 temper), with good strength and excellent ductility and toughness. This alloy is readily castable, with a low microshrinkage tendency.

Mechanical Properties

Tensile properties. T4 temper: tensile strength, 275 MPa (40 ksi); yield strength, 83 MPa (12 ksi); elongation, 15% in 50 mm (2 in.). See also Fig. 10.

Tensile properties versus temperature. See Table 25.

Shear strength. T4 temper, 145 MPa (21 ksi)

Bearing properties. Ultimate bearing strength, 400 MPa (58 ksi); bearing yield strength, 240 MPa (35 ksi)

Compressive yield strength. 83 MPa (12 ksi)

Hardness. 55 HB or 66 HRE

Poisson's ratio. 0.35

Elastic modulus. Tension, 45 GPa (6.5 × 10^6 psi); shear, 17 GPa (2.4 × 10^6 psi)

Impact strength. Charpy V-notch, 6.1 J (4.5 ft · lbf)

Creep characteristics. See Table 26.

Mass Characteristics

Density. 1.80 g/cm³ (0.065 lb/in.³) at 20 °C (68 °F)

Thermal Properties

Liquidus temperature. 610 °C (1130 °F)

Solidus temperature. 490 °C (915 °F)

Coefficient of linear thermal expansion. 25 μm/m · °C (14 μin./in. · °F)

Thermal conductivity. 51.1 W/m · K (29.5 Btu/ft · h · °F) at 20 °C (68 °F)

Electrical Properties

Electrical conductivity. 12% IACS at 20 °C (68 °F)

Electrical resistivity. 13 nΩ · m

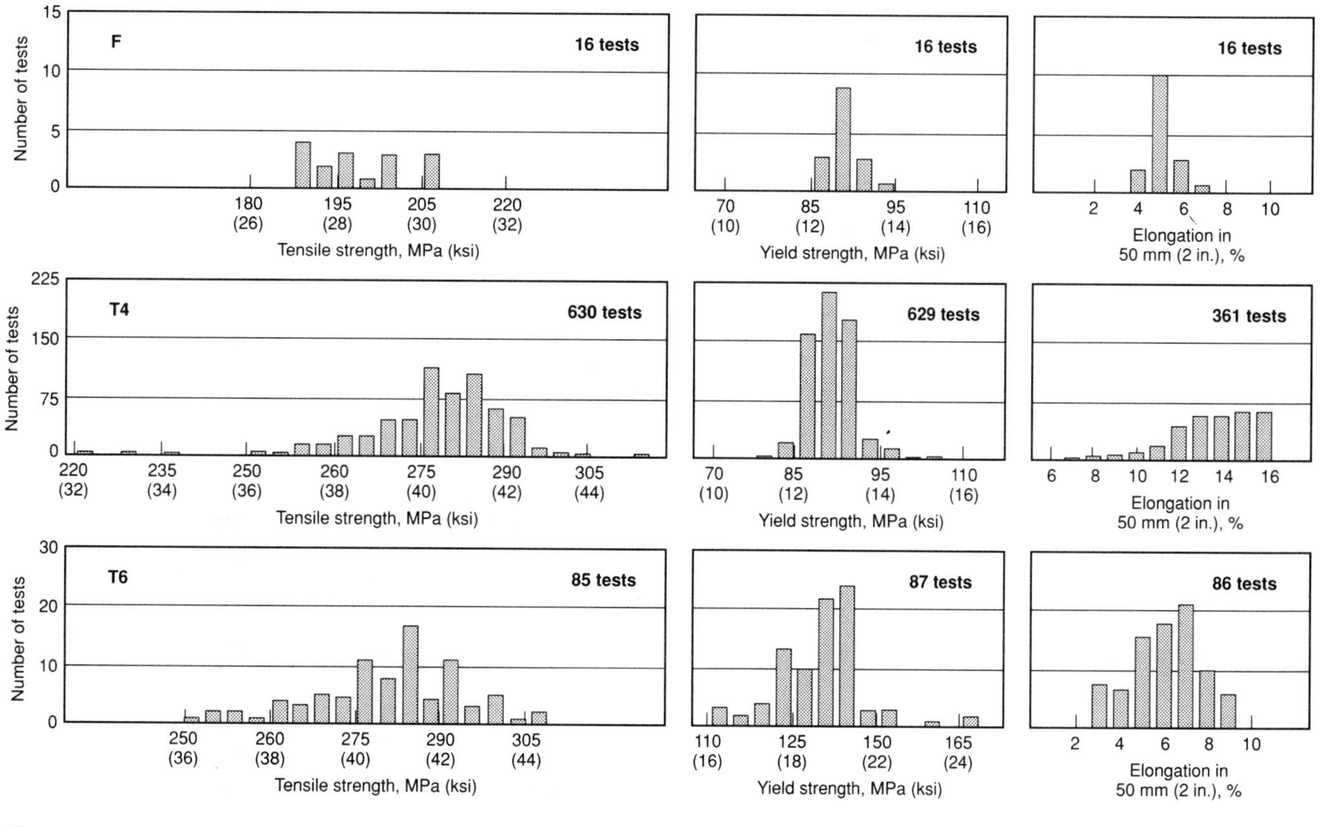

Fig. 9 Distribution of tensile properties for separately cast test bars of AZ63A

Table 25 Typical tensile properties of AZ81A-T4 sand castings at elevated temperatures

Properties determined using separately cast test bars

Testing temperature °C	°F	Tensile strength MPa	ksi	Yield strength MPa	ksi	Elongation in 50 mm (2 in.), %
21	70	275	40.0	83	12.0	15.0
93	200	260	37.5	83	12.0	20.0
150	300	190	27.5	80	11.5	24.5
200	400	140	20.0	76	11.0	29.0
260	500	97	14.0	72	10.5	35.0

Table 26 Typical creep properties of AZ81A-T4 sand castings

Properties determined using separately cast test bars

Time under load, h	Tensile stress resulting in total extension(a) of 0.1% MPa	ksi	0.2% MPa	ksi	0.5% MPa	ksi
At 93 °C (200 °F)						
1	39	5.6	58	8.4	86	12.5
10	37	5.4	55	8.0	83	12.0
100	36	5.2	51	7.4	81	11.8
At 150 °C (300 °F)						
1	37	5.4	53	7.7	...	...
10	28	4.0	45	6.5	62	9.0
100	15	2.2	24	3.5	46	6.6
At 200 °C (400 °F)						
1	23	3.4	41	6.0	...	...
10	12	1.7	21	3.1	...	...
100	7	1.0	12	1.7	21	3.0

(a) Total extension equals initial extension plus creep extension.

Fabrication Characteristics

Casting temperature. 705 to 845 °C (1300 to 1550 °F)
Weldability. Gas-shielded arc welding with AZ92A rod, very good

AZ91A, AZ91B, AZ91C, AZ91D, AZ91E

Specifications

AMS. Die castings: AZ91A, 4490. Sand castings: AZ91C, 4437; AZ91E, 4446
ASTM. Die castings: AZ91A, AZ91B, and AZ91D, B 94. Sand castings: AZ91C and AZ91E, B 80. Permanent mold castings: AZ91C and AZ91E, B 199. Investment castings: AZ91C and AZ91E, B 403. Ingot: B 93
SAE. J465. Former SAE alloy numbers: AZ91A, 501; AZ91B, 501A; AZ91C, 504
UNS numbers. AZ91A: M11910. AZ91B: M11912. AZ91C: M11914. AZ91D: M11916. AZ91E: M11921
Government. Die castings: AZ91A, QQ-M-38. Permanent mold castings: AZ91C, QQ-M-55 and MIL-M-46062. Sand castings: AZ91C, QQ-M-56, and MIL-M-46062
Foreign. Elektron AZ91. British: BS 2970 MAG3. French: AIR 3380 G-AZ91. German: DIN 1729 3.5912

Chemical Composition

Composition limits of AZ91A. 8.3 to 9.7 Al, 0.13 Mn min, 0.35 to 1.0 Zn, 0.50 Si max, 0.10 Cu max, 0.03 Ni max, 0.30 max other, bal Mg
Composition limits of AZ91B. 8.3 to 9.7 Al, 0.13 Mn min, 0.35 to 1.0 Zn, 0.50 Si max, 0.35 Cu max, 0.03 Ni max, 0.30 max other, bal Mg
Composition limits of AZ91C. 8.1 to 9.3 Al, 0.13 Mn min, 0.40 to 1.0 Zn, 0.30 Si max, 0.10 Cu max, 0.01 Ni max, 0.3 max other (total), bal Mg
Composition limits of AZ91D. 8.3 to 9.7 Al, 0.15 Mn min, 0.35 to 1.0 Zn, 0.10 Si max, 0.005 Fe max, 0.030 Cu max, 0.002 Ni max, 0.02 max other (each), bal Mg
Composition limits of AZ91E. 8.1 to 9.3 Al, 0.17 to 0.35 Mn, 0.4 to 1.0 Zn, 0.20 Si max, 0.005 Fe max, 0.015 Cu max, 0.0010 Ni max, 0.01 max other (each), 0.30 max other (total)
Consequence of exceeding impurity limits. Corrosion resistance decreases with increasing Fe, Cu, or Ni content. More than 0.5% Si decreases elongation. If Fe content exceeds 0.005% in AZ91D or AZ91E, the permissible Fe-Mn ratio will not exceed 0.032, and corrosion resistance will rapidly decrease.

Applications

Typical uses. AZ91A, AZ91B, and AZ91D, which have the same nominal composition except for iron, copper, and nickel contents, are die casting alloys used in the as-cast condition (F temper). AZ91D is a high-purity alloy which has excellent corrosion resistance; it is the most commonly used magnesium die casting alloy. AZ91A and AZ91B can be made from secondary metal, reducing the cost of the alloy; they must be used when maximum corrosion resistance is not required. AZ91E is a high-purity alloy with excellent corrosion resistance used in pressure-tight sand and permanent mold castings with high tensile strength and moderate yield strength. AZ91C is used in sand and permanent mold castings when maximum corrosion resistance is not required.

Corrosion Resistance

ASTM B 117 salt spray test. AZ91D: <0.13 mg/cm²/day (<10 mils/yr). AZ91E-T6: <0.63 mg/cm²/day (<50 mils/yr)

Mechanical Properties

Tensile properties. See Tables 27 and 28, and Fig. 11, 12, and 13.
Shear strength. AZ91A, AZ91B, and AZ91D: F temper, 140 MPa (20 ksi)
Compressive yield strength. See Table 27.
Bearing properties. See Table 27.
Hardness. See Table 27.
Poisson's ratio. 0.35
Elastic modulus. Tension, 45 GPa (6.5×10^6 psi); shear, 17 GPa (2.4×10^6 psi)
Impact strength. See Table 27.
Fatigue strength. R.R. Moore type tests. At 5×10^8 cycles: AZ91A, AZ91B, and AZ91D (F temper): 97 MPa (14 ksi) at 5×10^8 cycles. At 1×10^8 cycles: AZ91C and AZ91E, 80 to 95 MPa (12 to 14 ksi)

Mass Characteristics

Density. 1.81 g/cm³ (0.066 lb/in.³) at 20 °C (68 °F)

Thermal Properties

Liquidus temperature. 595 °C (1105 °F)
Solidus temperature. 470 °C (875 °F)
Coefficient of linear thermal expansion. 26 μm/m · K (14 μin./in. · °F) at 20 to 100 °C (68 to 212 °F)
Specific heat. 1.05 kJ/kg · K (0.25 Btu/lb · °F) at 20 °C (68 °F)
Latent heat of fusion. 373 kJ/kg (160 Btu/lb)
Thermal conductivity. 72 W/m · K (41.8 Btu/ft · h · °F) at 100 to 300 °C (212 to 572 °F)
Incipient melting temperature. 421 °C (790 °F)

Electrical Properties

Electrical conductivity. AZ91A: F temper, 10.1% IACS. AZ91C and AZ91E: F temper, 11.5% IACS; T4 temper, 9.9% IACS; T6 temper, 11.2% IACS
Electrical resistivity. AZ91A, AZ91B, and AZ91D: F temper, 170 nΩ · m; AZ91C and AZ91E: F temper, 150 nΩ · m; T4 temper, 175 nΩ · m; T6 temper, 151.5 nΩ · m
Electrolytic solution potential. 1.58 V versus saturated calomel electrode
Hydrogen overvoltage. As-cast, 0.40 V

Fabrication Characteristics

Casting temperature. AZ91C and AZ91E: sand castings, 705 to 845 °C (1300 to 1550 °F); permanent mold castings, 650 to 815 °C (1200

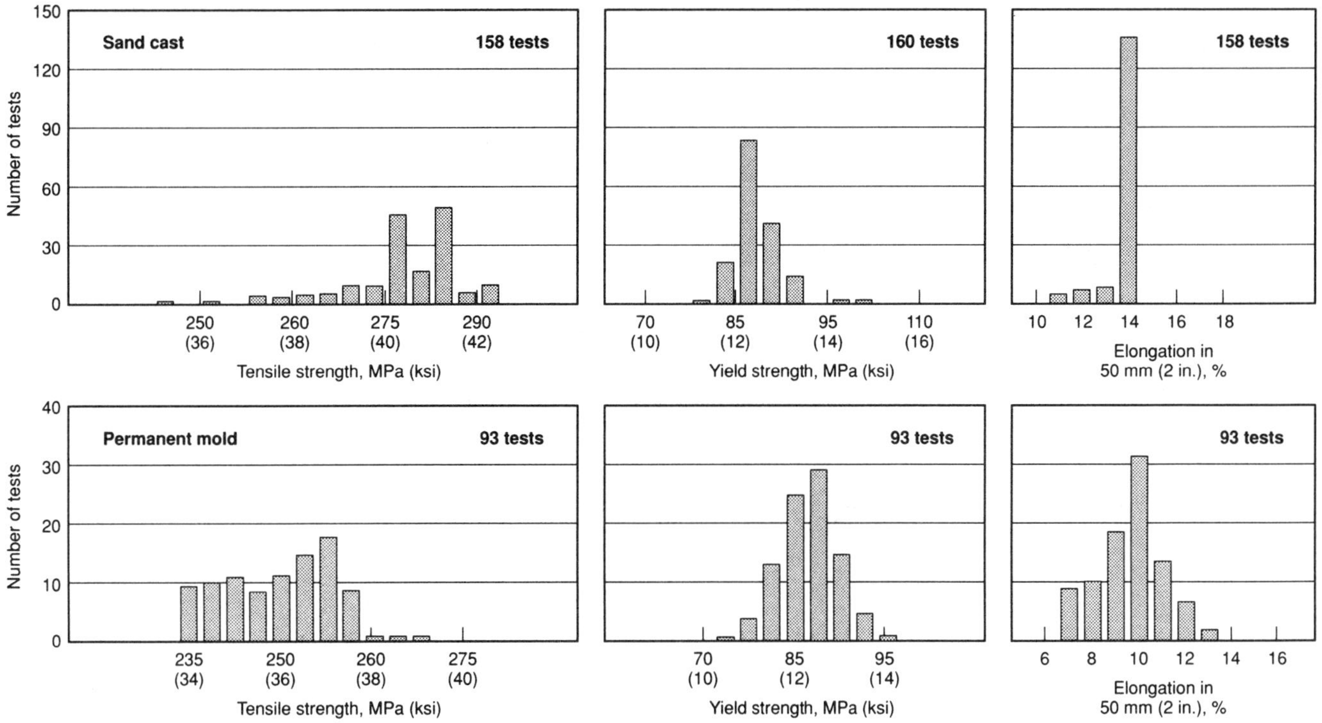

Fig. 10 Distribution of tensile properties for separately cast test bars of AZ81A-T4

to 1500 °F). AZ91A, AZ91B, and AZ91D: die castings, 625 to 700 °C (1160 to 1290 °F)
Weldability. AZ91C and AZ91E can be readily welded by the gas-shielded arc process using AZ91C or AZ92A rod; stress relief required. AZ91A, AZ91B, and AZ91D not weldable
Hot-shortness temperature. 400 °C (750 °F)

AZ92A

Specifications

AMS. Sand castings: 4434. Investment castings: 4453. Permanent mold castings: 4484
ASTM. Ingot: B 93. Sand castings: B 80. Permanent mold castings: B 199. Investment castings: B 403
SAE. J465. Former SAE alloy number: 500
UNS number. M11920
Government. Sand castings: QQ-M-56 and MIL-M-46062. Permanent mold castings: QQ-M-55 and MIL-M-46062

Chemical Composition

Composition limits. 8.3 to 9.7 Al, 0.10 Mn min, 1.6 to 2.4 Zn, 0.30 Si max, 0.25 Cu max, 0.01 Ni max, 0.30 max other (total), bal Mg

Table 27 Typical room-temperature mechanical properties of AZ91A, AZ91B, AZ91C, AZ91D, and AZ91E castings

Property	AZ91A, AZ91B, AZ91D, F temper	AZ91C and AZ91E F temper	AZ91C and AZ91E T4 temper	AZ91C and AZ91E T6 temper
Tensile strength, MPa (ksi)	230 (33)	165 (24)	275 (40)	275 (40)
Tensile yield strength, MPa (ksi)	150 (22)	97 (14)	90 (13)	145 (21)
Elongation in 50 mm (2 in.), %	3	2.5	15	6
Compressive yield strength at 0.2% offset, MPa (ksi)	165 (24)	97 (14)	90 (13)	130 (19)
Ultimate bearing strength, MPa (ksi)	...	415 (60)	415 (60)	515 (75)
Bearing yield strength, MPa (ksi)	...	275 (40)	305 (44)	360 (52)
Hardness				
HB	63	60	55	70
HRE	75	66	62	77
Charpy V-notch impact strength, J (ft · lbf)	2.7 (2.0)	0.79 (0.58)	4.1 (3.0)	1.4 (1.0)

Table 28 Typical tensile properties of AZ91C-T6 sand castings at elevated temperatures

Testing temperature °C	Testing temperature °F	Tensile strength MPa	Tensile strength ksi	Tensile yield strength MPa	Tensile yield strength ksi	Elongation in 50 mm (2 in.), %
150	300	185	27	97	14	40
200	400	115	17	83	12	40

Consequence of exceeding impurity limits. Excessive Cu or Ni degrades corrosion resistance. More than 0.5% Si decreases elongation.

Applications

Typical uses. Pressure-tight sand and permanent mold castings with high tensile strength and good yield strength

Mechanical Properties

Tensile properties. See Table 29 and Fig. 14.
Tensile properties versus temperature. See Table 30.
Shear strength. F temper, 125 MPa (18 ksi); T4 and T5 tempers, 140 MPa (20 ksi); T6 temper, 145 MPa (21 ksi); T7 temper, 150 MPa (22 ksi)
Compressive yield strength. F and T4 tempers, 97 MPa (14 ksi); T5 temper, 115 MPa (17 ksi); T6 temper, 150 MPa (22 ksi); T7 temper, 145 MPa (21 ksi)
Bearing properties. Ultimate bearing strength: F and T5 tempers, 345 MPa (50 ksi); T4 temper, 470 MPa (68 ksi); T6 temper, 550 MPa (80 ksi). Bearing yield strength: F, T4, and T5 tempers, 315 MPa (46 ksi); T6 temper, 450 MPa (65 ksi)
Hardness. F temper: 65 HB or 76 HRE. T4 temper: 63 HB or 75 HRE. T5 temper: 69 HB or 80 HRE. T6 temper: 81 HB or 88 HRE. T7 temper: 78 HB or 86 HRE
Poisson's ratio. 0.35
Elastic modulus. Tension, 45 GPa (6.5 × 10^6 psi); shear, 17 GPa (2.4 × 10^6 psi)

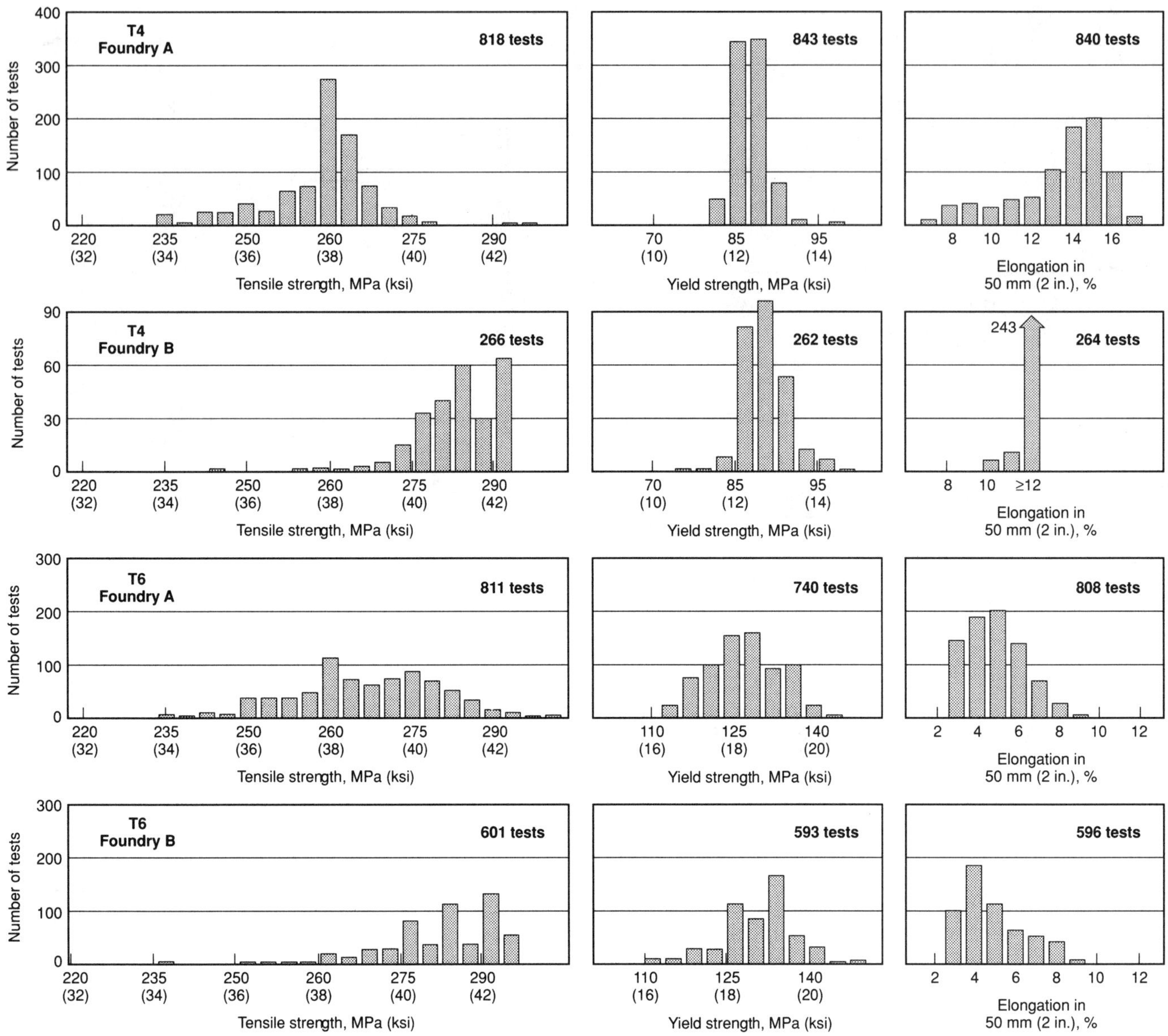

Fig. 11 Distribution of tensile properties for separately sand cast test bars of AZ91C

Impact strength. Charpy V-notch: F temper, 0.7 J (0.5 ft · lbf); T4 temper, 2.7 J (2.0 ft · lbf); T6 temper, 1.1 J (0.8 ft · lbf)
Fatigue strength. R.R. Moore type test. At 5×10^8 cycles: F and T6 tempers, 83 MPa (12 ksi); T4 and T7 tempers, 90 MPa (13 ksi); T5 temper, 76 MPa (11 ksi)

Table 29 Typical tensile properties of AZ92A sand castings

Properties determined using separately cast test bars

Temper	Tensile strength, MPa	Tensile strength, ksi	Yield strength, MPa	Yield strength, ksi	Elongation, %(a)
F	170	25	97	14	2
T4	275	40	97	14	10
T5	170	25	115	17	1
T6	275	40	150	22	3
T7	275	40	145	21	3

(a) In 50 mm (2 in.)

Mass Characteristics

Density. 1.83 g/cm³ (0.066 lb/in.³) at 20 °C (68 °F)

Thermal Properties

Liquidus temperature. 595 °C (1100 °F)
Solidus temperature. 445 °C (830 °F)
Coefficient of linear thermal expansion. 26 μm/m · K (14 μin./in. · °F) at 18 to 100 °C (65 to 212 °F)
Incipient melting temperature. 410 °C (770 °F)
Specific heat. 1.05 kJ/kg · K (0.25 Btu/lb · °F) at 25 °C (78 °F)
Latent heat of fusion. 373 kJ/kg (160 Btu/lb)
Thermal conductivity. 72 W/m · K (41.8 Btu/ft · h · °F) at 100 to 300 °C (212 to 572 °F)

Electrical Properties

Electrical conductivity. At 20 °C (68 °F): F temper, 12.3% IACS; T4 temper, 10.5% IACS; T6 temper, 12.3% IACS
Electrical resistivity. At 20 °C (68 °F): F temper, 140 nΩ · m; T4 temper, 165 nΩ · m; T6 temper, 140 nΩ · m
Electrolytic solution potential. 1.56 V versus saturated calomel electrode
Hydrogen overvoltage. As-cast, 0.3 V

Fabrication Characteristics

Casting temperature. Sand castings, 705 to 845 °C (1300 to 1550 °F); permanent mold castings, 650 to 815 °C (1200 to 1500 °F)
Weldability. Gas-shielded arc welding with AZ92A rod, good; stress relief required

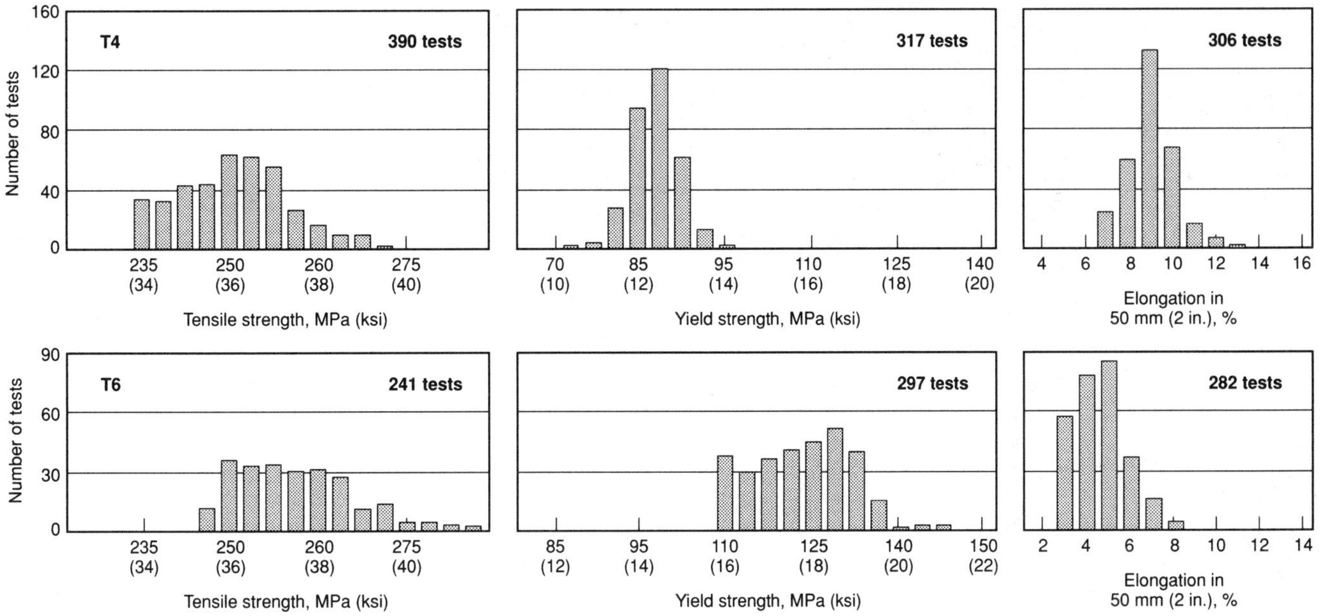

Fig. 12 Distribution of tensile properties for separately cast permanent mold test bars of AZ91C

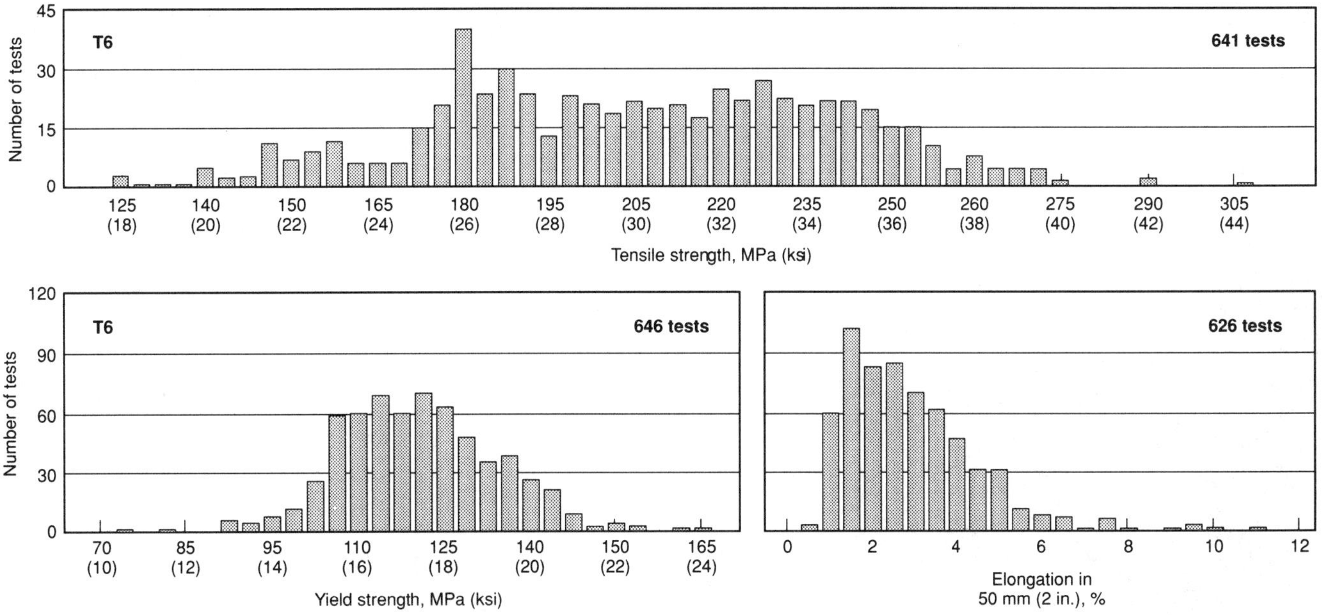

Fig. 13 Distribution of tensile properties for specimens cut from AZ91C sand castings

EQ21

Specifications

AMS. 4417
ASTM. Sand castings: B 80. Permanent mold castings: B 199. Investment castings: B 403
UNS number. M16330
Government. Sand and permanent mold castings: MIL-M-46062
Foreign. British: BS 2970 MAG13

Chemical Composition

Composition limits. 1.3 to 1.7 Ag, 1.75 to 2.5 Nd-rich rare earths, 0.4 to 1.0 Zr, 0.05 to 0.10 Cu, 0.01 Ni max, 0.3 max other (total), bal Mg
Consequence of exceeding impurity limits. Zr content below 0.5% may result in somewhat coarser as-cast grains and lower mechanical properties.

Applications

Typical uses. Sand and permanent mold castings used in the solution-treated and artificially aged condition (T6 temper), with high yield strengths up to 200 °C (390 °F). Castings have excellent short-time elevated-temperature mechanical properties and are pressure tight and weldable.

Mechanical Properties

Tensile properties. T6 temper: tensile strength, 235 MPa (34 ksi); yield strength, 170 MPa (25 ksi); elongation, 2% in 50 mm (2 in.)
Tensile properties versus temperature. See Table 31 and Fig. 15 and 16.
Compressive properties. T6 temper: compressive strength, 345 MPa (50 ksi); compressive yield strength, 195 MPa (28 ksi)
Elastic modulus. Tension, 45 GPa (6.5 ×

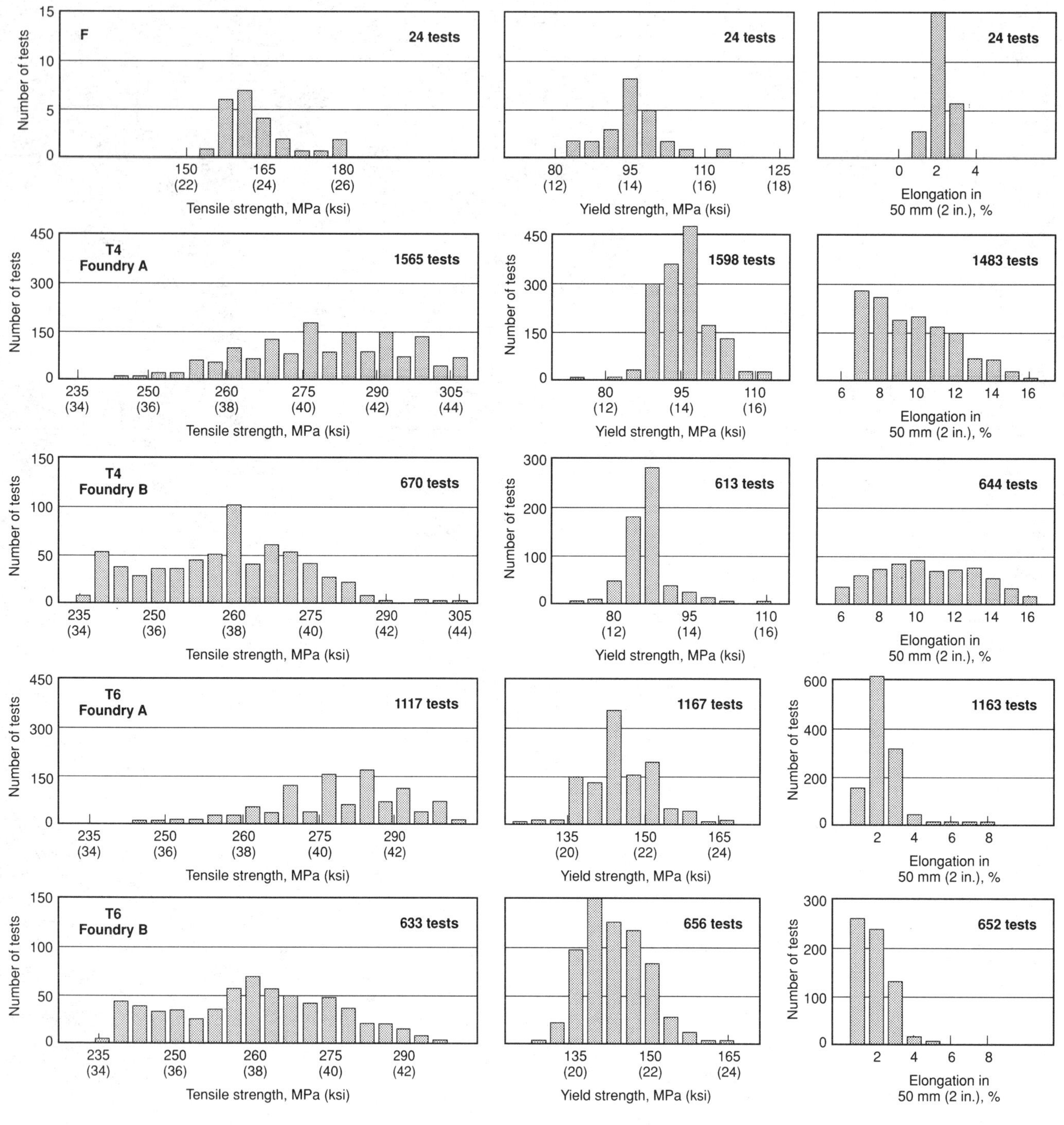

Fig. 14 Distribution of tensile properties for separately sand cast test bars of AZ92A

10^6 psi) (See also Fig. 17); shear, 17 GPa (6.4×10^6 psi)
Creep characteristics. See Table 32.
Fatigue strength. See Fig. 18.
Poisson's ratio. 0.35
Hardness. 65 to 85 HB
Specific damping capacity. 0.4 at stress equal to 10% of tensile yield strength

Mass Characteristics

Density. 1.81 g/cm^3 (0.065 lb/in.^{3}) at 20 °C (68 °F)

Thermal Properties

Liquidus temperature. 640 °C (1184 °F)
Solidus temperature. 540 °C (1004 °F)
Thermal conductivity. 113 W/m · K (65.3 Btu/ft · h · °F)
Coefficient of linear thermal expansion. 26.7 μm/m · K (14.8 μin./in. · °F) from 20 to 200 °C (68 to 212 °F)
Specific heat. 1.00 kJ/kg · K (0.24 Btu/lb · °F) at 20 to 100 °C (68 to 212 °F)
Latent heat of fusion. 373 kJ/kg (160 Btu/lb)

Electrical Properties

Electrical conductivity. 25.2% IACS at 20 °C (68 °F)
Electrical resistivity. 68.5 nΩ · m at 20 °C (68 °F)

Fabrication Characteristics

Casting temperature. Sand castings, 750 to 820 °C (1380 to 1510 °F)
Weldability. Gas-shielded arc welding with welding rod of base metal composition

Table 30 Typical tensile properties of AZ92A sand castings at elevated temperatures

Testing temperature(a) °C	°F	Tensile strength MPa	ksi	Elongation in 50 mm (2 in.), %	Time at temperature, days(b)
F temper					
93	200	170	25	2	80
150	300	150	22	3	160
200	400	110	16	36	160
260	500	83	12	34	40
T4 temper					
93	200	275	40	8	160
150	300	180	26	40	160
200	400	115	17	41	160
260	500	76	11	52	40
T6 temper					
93	200	260	38	7	160
150	300	170	25	40	160
200	400	115	17	43	160
260	500	76	12	47	40

(a) Tested after prolonged heating at testing temperature. (b) Prior to testing

Solution temperature. 515 to 525 °C (960 to 980 °F)
Aging temperature. 200 °C (390 °F)

EZ33A

Specifications

AMS. Sand castings: 4442
ASTM. Sand castings: B 80. Permanent mold castings: B 199. Investment castings: B 403
SAE. J465. Former SAE alloy number: 506
UNS number. M12330
Government. Sand castings: QQ-M-56. Permanent mold castings: QQ-M-55. Welding rod: MIL-R-6944
Foreign. Elektron ZRE1. British: BS 2970 MAG6. German: DIN 1729 3.5103. French: AIR 3380 ZRE1

Chemical Composition

Composition limits. 2.5 to 4.0 rare earths, 2.0 to 3.1 Zn, 0.50 to 1.0 Zr, 0.10 Cu max, 0.01 Ni max, 0.30 max other (total), bal Mg

Applications

Typical uses. Pressure-tight sand and permanent mold castings relatively free from microporosity, used in T5 condition for applications requiring good strength properties up to 260 °C (500 °F)

Mechanical Properties

Tensile properties. T5 temper: tensile strength, 160 MPa (23 ksi); yield strength, 110 MPa (16 ksi); elongation, 3% in 50 mm (2 in.). See also Fig. 19.
Tensile properties versus temperature. See Table 33 and Fig. 20.
Shear strength. T5 temper, 135 MPa (19.8 ksi)
Compressive yield strength. 110 MPa (16 ksi)
Bearing properties. T5 temper: ultimate bearing strength, 395 MPa (57 ksi); bearing yield strength, 275 MPa (39.9 ksi)
Hardness: 50 HB or 59 HRE
Creep characteristics. See Table 34 and Fig. 21(a) and 21(b).

Table 32 Long-term creep properties of EQ21A sand castings

Time under load, h	Tensile stress resulting in creep extension(a) of 0.1% MPa	0.1% ksi	0.2% MPa	0.2% ksi	0.5% MPa	0.5% ksi
At 150 °C (300 °F)						
10	149	21.6	...	...	...	...
100	138	20.0	155	22.5	...	...
1000	123	17.8	134	19.4	152	22.0
At 200 °C (390 °F)						
10	109	15.8	...	...	...	...
100	78	11.3	95	13.8	116	16.8
1000	...	...	62	9.0	76	11.0
At 250 °C (480 °F)						
10	46	6.7	...	...	...	...
100	29	4.2	36	5.2	42	6.1
1000	...	...	19	2.8	24	3.5

(a) Does not include initial extension

Table 31 Typical tensile properties of EQ21A sand castings at various temperatures

Testing temperature °C	°F	Tensile strength MPa	ksi	Yield strength MPa	ksi
20	68	261	37.8	195	28.3
100	212	230	33.4	189	27.4
200	390	191	27.7	170	24.6
300	570	132	19.1	117	17.0

Mass Characteristics

Density. 1.83 g/cm^3 (0.066 lb/in.^{3}) at 20 °C (68 °F)

Thermal Properties

Liquidus temperature. 645 °C (1190 °F)
Solidus temperature. 545 °C (1010 °F)
Coefficient of linear thermal expansion. 26.1 μm/m · K (14.5 μin./in. · °F) from 20 to 100 °C (68 to 212 °F)
Specific heat. 1.05 kJ/kg · K (0.25 Btu/lb · °F) at 20 °C (68 °F)
Latent heat of fusion. 373 kJ/kg (160 Btu/lb)
Thermal conductivity. 100 W/m · K (58 Btu/ft · h · °F)

Electrical Properties

Electrical conductivity. 25% IACS at 20 °C (68 °F)
Electrical resistivity. 70 nΩ · m at 20 °C (68 °F)

Fabrication Characteristics

Casting temperature. Sand and permanent mold castings, 750 to 820 °C (1380 to 1510 °F)
Weldability. Gas-shielded arc welding with EZ33A rod, excellent; preheating not necessary but may be used; postweld heat treatment required

HK31A

See also wrought alloy HK31A.

Specifications

AMS. Sand castings: 4445
ASTM. Sand castings: B 80. Permanent mold castings: B 199. Investment castings: B 403
SAE. J465. Former SAE alloy number: 507
UNS number. M13310
Government. Sand castings: QQ-M-56 and MIL-M-46062. Permanent mold castings: QQ-M-55 and MIL-M-46062

Chemical Composition

Composition limits. 2.5 to 4.0 Th, 0.40 to 1.0 Zr, 0.30 Zn max, 0.10 Cu max, 0.01 Ni max, 0.30 max other (total), bal Mg

Applications

Typical uses. Sand castings for use at temperatures up to 345 °C (650 °F)

Mechanical Properties

Tensile properties. T6 temper: tensile strength, 220 MPa (32 ksi); yield strength,

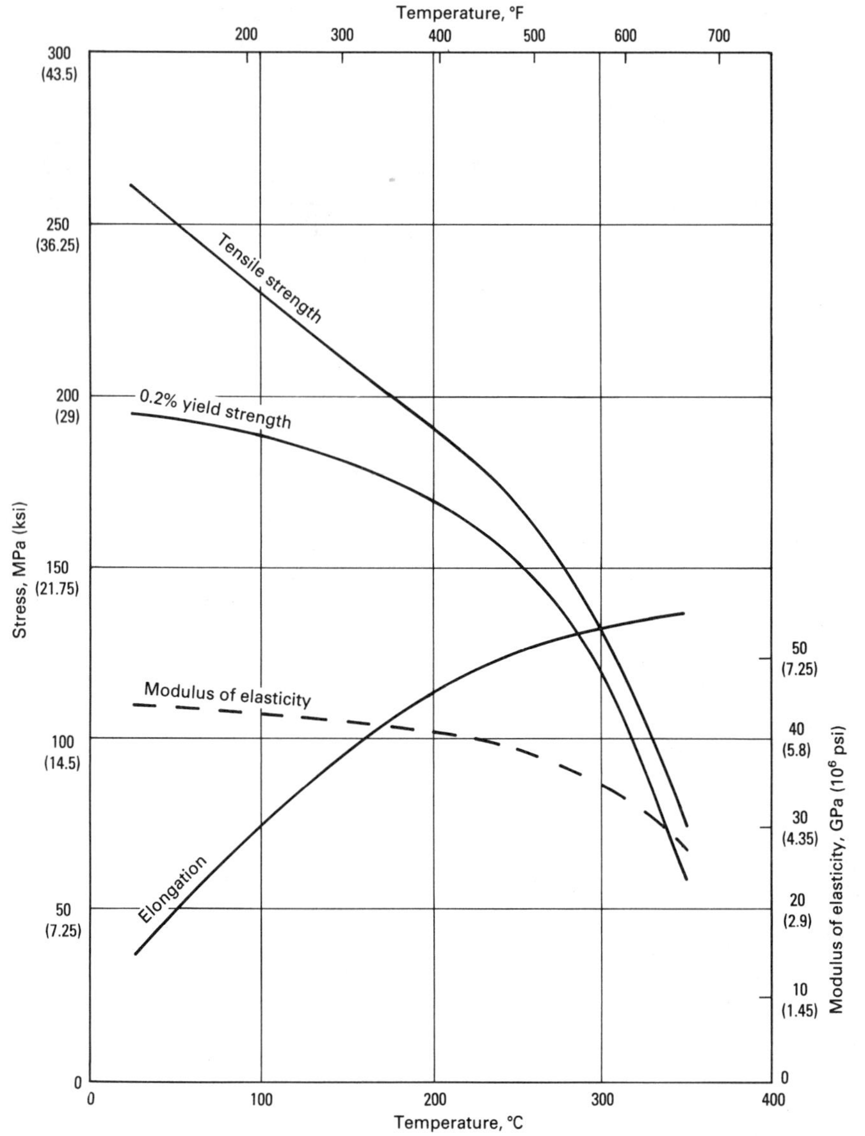

Fig. 15 Effect of temperature on the strength of EQ21A-T6 sand castings

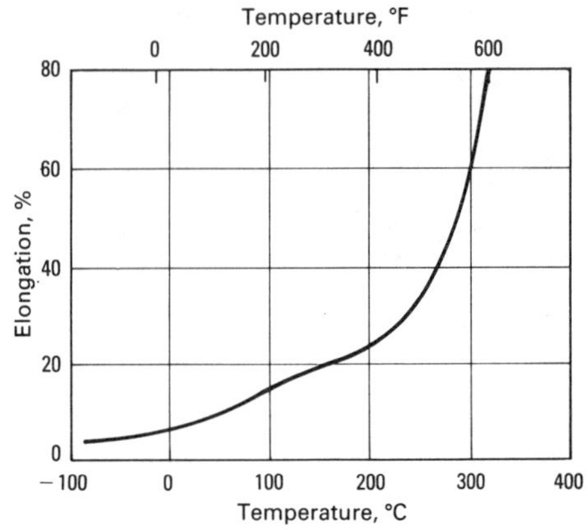

Fig. 16 Effect of temperature on the elongation of EQ21A-T6 sand castings

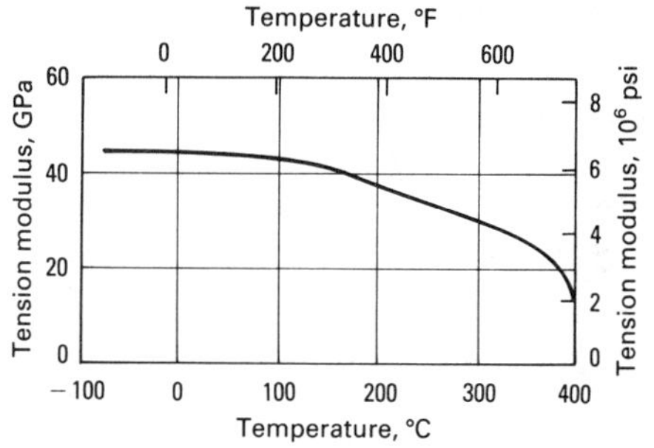

Fig. 17 Effect of temperature on the elastic modulus of EQ21-T6 sand castings

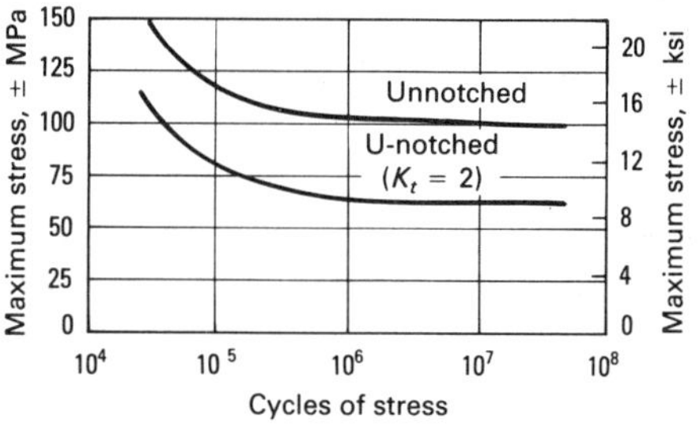

Fig. 18 Fatigue characteristics of EQ21A-T6 sand castings. Rotating beam (Wohler) tests; machine speed, 2960 Hz

105 MPa (15 ksi); elongation, 8% in 50 mm (2 in.). See also Fig. 22.
Tensile properties versus temperature. See Table 35 and Fig. 23.
Compressive yield strength. T6 temper, 105 MPa (15 ksi)
Bearing properties. T6 temper: ultimate bearing strength, 420 MPa (61 ksi); bearing yield strength, 275 MPa (40 ksi)
Hardness. T6 temper, 66 HRE
Poisson's ratio. 0.35

Elastic modulus. Tension, 45 GPa (6.5×10^6 psi); shear, 17 GPa (2.4×10^6 psi)
Creep characteristics. See Table 36 and Fig. 24.

Mass Characteristics

Density. 1.8 g/cm^3 (0.065 lb/in.^{3}) at 20 °C (68 °F)

Thermal Properties

Liquidus temperature. 650 °C (1205 °F)

Table 33 Typical tensile properties of EZ33A-T5 sand castings at elevated temperatures

Properties determined using separately cast test bars

Testing temperature		Tensile strength		Yield strength		Elongation in 50 mm (2 in.), %
°C	°F	MPa	ksi	MPa	ksi	
24	75	160	23	110	16	3
150	300	150	22	97	14	10
200	400	145	21	76	11	20
260	500	125	18	69	10	31
315	600	83	12	55	8	50

Solidus temperature. 590 °C (1090 °F)
Incipient melting temperature. 627 to 632 °C (1160 to 1170 °F) in circulating air
Specific heat versus temperature. $C_p = 1374 + 0.0002306T + 3370T^{-2}$
Latent heat of fusion. 318 to 335 kJ/kg (137 to 144 Btu/lb)
Thermal conductivity. At 20 °C (68 °F): T6 temper, 92 W/m · K (53 Btu/ft · h · °F); H24 temper, 113 W/m · K (65 Btu/ft · h · °F); O temper, 105 W/m · K (61 Btu/ft · h · °F)
Thermal conductivity versus temperature. See Table 37.

Electrical Properties

Electrical conductivity. T6 temper, 22% IACS at 20 °C (68 °F)
Electrical resistivity. At 20 °C (68 °F): T6 temper, 77 nΩ · m; H24 temper, 61 nΩ · m; O temper, 60 nΩ · m

Table 34 Typical creep properties of EZ33A-T5 sand castings

Properties determined using separately cast test bars

Time under load, h	Tensile stress resulting in total extension(a) of: 0.1%, MPa	0.1%, ksi	0.2%, MPa	0.2%, ksi	0.5%, MPa	0.5%, ksi	1.0%, MPa	1.0%, ksi
At 200 °C (400 °F)								
1	41	6	69	10	89	13	105	15
10	41	6	62	9	83	12	89	13
100	34	5	55	8	69	10	76	11
1000	28	4	41	6	48	7	55	8
At 260 °C (500 °F)								
1	34	5	55	8	69	10	83	12
10	28	4	34	5	48	7	55	8
100	14	2	21	3	28	4	34	5
1000	14	2	14	2	14	2	21	3
At 315 °C (600 °F)								
1	14	2	21	3	28	4	34	5
10	14	2	14	2	21	3	21	3
100	14	2	7	1	14	2	14	2
1000	7	1	7	1	7	1	7	1

(a) Total extension equals initial extension plus creep extension.

Table 35 Typical tensile properties of HK31A-T6 sand castings at elevated temperatures

Properties determined using separately cast test bars

Testing temperature, °C	Testing temperature, °F	Tensile strength, MPa	Tensile strength, ksi	Yield strength, MPa	Yield strength, ksi	Elongation in 50 mm (2 in.), %
24	75	215	31	110	16	6
200	400	165	24	97	14	17
260	500	160	23	89	13	19
315	600	140	20	83	12	22
370	700	89	13	55	8	26

Fabrication Characteristics

Casting temperature. Sand and permanent mold castings, 750 to 820 °C (1380 to 1510 °F)
Weldability. Gas-shielded arc welding with EZ33A or HK31A rod (EZ33A preferred), very good; stress relief required for sand castings

HZ32A

Specifications

AMS. Sand castings: 4447
ASTM. Sand castings: B 80
UNS number. M13320
Government. Sand castings: QQ-M-56, MIL-M-46062
Foreign. Elektron ZT1. British: BS 2970 MAG8. German: DIN 1729 3.5105

Chemical Composition

Composition limits. 1.7 to 2.5 Zn, 2.5 to 4.0 Th, 0.10 rare earths max, 0.50 to 1.0 Zr, 0.10 Cu max, 0.01 Ni max, 0.30 max other (total), bal Mg
Consequence of exceeding impurity limits. More than 0.1% rare earths causes a loss in creep resistance.

Applications

Typical uses. Sand castings used in the artificially aged condition (T5 temper), with moderate strength and an optimum combination of properties for medium- and long-time exposure at temperatures above 260 °C (500 °F). Castings are pressure tight, and under long-time exposure can withstand higher stresses and higher temperatures than any other commercially available magnesium alloy.

Mechanical Properties

Tensile properties. T5 temper: tensile strength, 185 MPa (27 ksi); yield strength, 90 MPa (13 ksi); elongation in 50 mm (2 in.), 4%. See also Fig. 25.
Tensile properties versus temperature. See Table 38 and Fig. 26.
Compressive properties. T5 temper: compressive yield strength, 110 MPa (16 ksi)
Hardness: 55 HB
Poisson's ratio. 0.3
Elastic modulus. Tension, 45 GPa (6.5×10^6 psi); shear, 17 GPa (2.4×10^6 psi)
Creep characteristics. See Fig. 27.

Mass Characteristics

Density. 1.83 g/cm³ (0.066 lb/in.³) at 20 °C (68 °F)

Thermal Properties

Liquidus temperature. 650 °C (1200 °F)
Solidus temperature. 550 °C (1025 °F)
Coefficient of linear thermal expansion. 26.7 μm/m · K (14.8 μin./in. · °F) at 20 to 200 °C (68 to 390 °F)
Specific heat. 0.96 kJ/kg · K (0.23 Btu/lb · °F)
Latent heat of fusion. 373 kJ/kg (160 Btu/lb)
Thermal conductivity. 110 W/m · K (64 Btu/ft · h · °F) at 20 °C (68 °F)

Electrical Properties

Electrical conductivity. 26.5% IACS at 20 °C (68 °F)

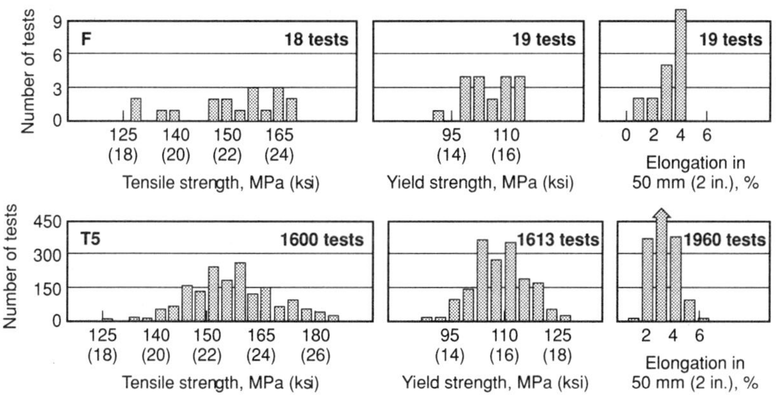

Fig. 19 Distribution of tensile properties for separately sand cast test bars of EZ33A

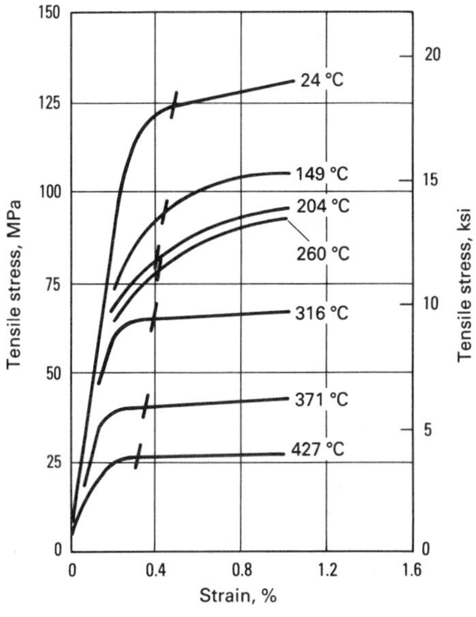

Fig. 20 Typical stress-strain curves for separately sand cast test bars of EZ33A-T5

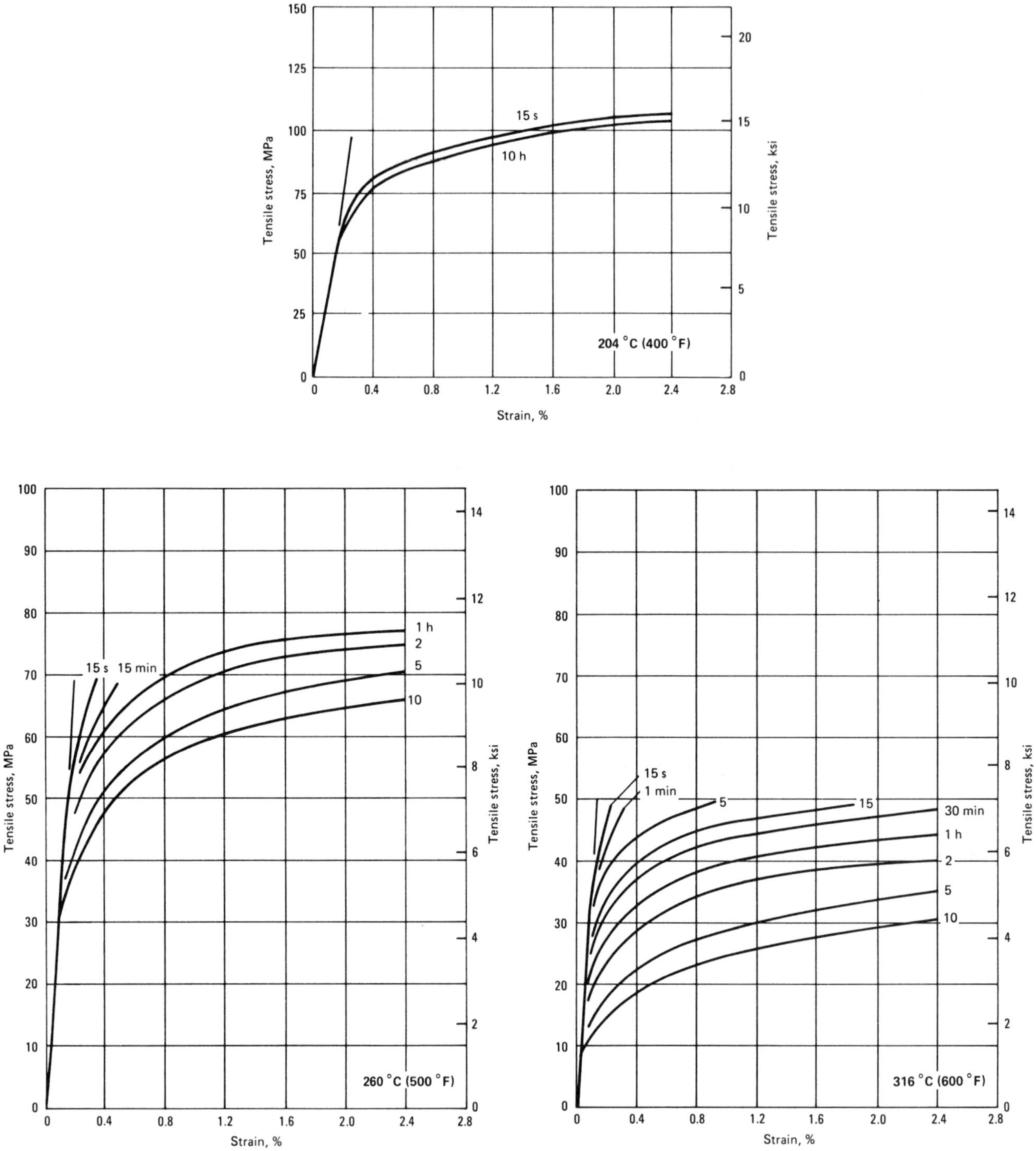

Fig. 21(a) Isochronous stress-strain curves for separately sand cast test bars of EZ33A-T5 tested at 204, 260, and 316 °C (400, 500, and 600 °F). Specimens exposed at testing temperatures for 3 h before loading.

Electrical resistivity. T5 temper: 65 nΩ · m at 20 °C (68 °F)

Fabrication Characteristics

Casting temperature. Sand castings, 750 to 820 °C (1380 to 1510 °F)
Weldability. Gas-shielded arc welding with HZ32A or EZ33A welding rod, fair; heavy-section castings require stress relief after welding.

K1A

Specifications

ASTM. Sand castings: B 80
UNS number. M18010

Chemical Composition

Composition limits. 0.40 to 1.0 Zr, 0.30 max other (total), bal Mg

Applications

Typical uses. K1A is used in the as-cast condition (F temper) for its high damping capacity. It has slightly better mechanical properties as die cast than as sand cast.

Mechanical Properties

Tensile properties. Sand castings, F temper: tensile strength, 180 MPa (26 ksi); yield strength, 55 MPa (8 ksi); elongation, 19%.

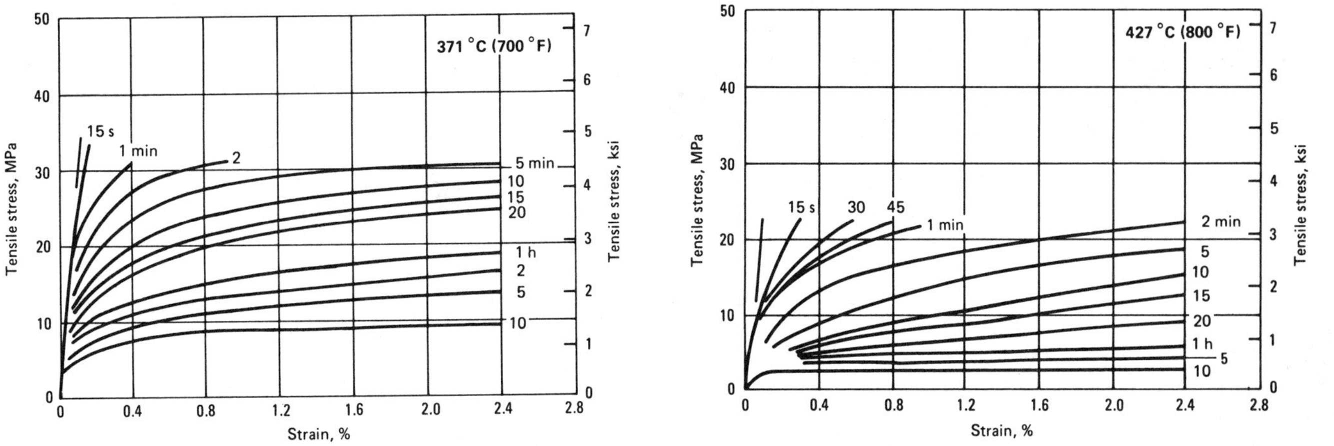

Fig. 21(b) Isochronous stress-strain curves for separately sand cast test bars of EZ33A-T5 tested at 371 and 427 °C (700 and 800 °F). Specimens exposed at testing temperatures for 3 h before loading.

Die castings, F temper: tensile strength, 165 MPa (24 ksi); yield strength, 83 MPa (12 ksi); elongation, 8%
Tensile properties versus temperature. See Table 39.
Shear strength. Sand castings, F temper: 55 MPa (8 ksi)
Bearing properties. Sand castings, F temper: ultimate bearing strength, 317 MPa (46 ksi); bearing yield strength, 125 MPa (18 ksi)

Mass Characteristics

Density. 1.74 g/cm^3 (0.063 lb/in.^{3}) at 20 °C (68 °F)

Thermal Properties

Liquidus temperature. 650 °C (1200 °F)
Solidus temperature. 650 °C (1200 °F)
Coefficient of linear thermal expansion. 27 μm/m · K (15 μin./in. · °F) at 21 to 200 °C (70 to 400 °F)
Thermal conductivity. 122 W/m · K (71 Btu/ft · h · °F) at 20 °C (68 °F)
Latent heat of fusion. 343 to 360 kJ/kg (148 to 155 Btu/lb)

Electrical Properties

Electrical resistivity. 57 nΩ · m

Fabrication Characteristics

Casting temperature. Sand castings, 750 to 820 °C (1380 to 1510 °F)
Weldability. Can be readily welded and soldered

QE22A

Specifications

AMS. Sand castings: 4418C
ASTM. Sand castings: B 80. Permanent mold castings: B 199. Investment castings: B 403
UNS number. M18220
Government. Sand castings: QQ-M-56B. Sand and permanent mold castings: MIL-M-46062B. Permanent mold castings: QQ-M-55
Foreign. Elektron MSR-B. British: DTD 5055. French: MSR-B AECMA MG-C-51. German: DIN 1729 3.5164

Chemical Composition

Composition limits. 2.0 to 3.0 Ag, 1.75 to 2.5 Nd-rich rare earths, 0.4 to 1.0 Zr, 0.1 Cu max, 0.01 Ni max, 0.3 max other (total), bal Mg
Consequence of exceeding impurity limits. Zr content below 0.5% may result in somewhat coarser as-cast grains and lower mechanical properties.

Applications

Typical uses. Sand and permanent mold castings used in the solution-treated and artificially aged condition (T6 temper), with high yield strengths at temperatures up to 200 °C (390 °F). Castings have excellent short-time elevated-temperature mechanical properties and are pressure tight and weldable.

Mechanical Properties

Tensile properties. T6 temper: tensile strength, 260 MPa (38 ksi); yield strength, 195 MPa (28 ksi); elongation, 3% in 50 mm (2 in.)
Tensile properties versus temperature. See Table 40 and Fig. 28 and 29.
Compressive properties. T6 temper: compressive strength, 345 MPa (50 ksi); compressive yield strength, 195 MPa (28 ksi)
Hardness. 65 to 85 HB
Poisson's ratio. 0.35
Elastic modulus. Tension, 45 GPa (6.5 × 10^6 psi) (see also Fig. 30); shear, 17 GPa (2.5 × 10^6 psi); compression, 44 GPa (6.4 × 10^6 psi)
Impact strength. Charpy, at 20 °C (68 °F): unnotched, 6.8 to 13.6 J (5 to 10 ft · lbf); V-notched, 1.4 to 2.7 J (1 to 2 ft · lbf)
Fatigue strength. See Fig. 31.
Creep-rupture characteristics. See Table 41 and Fig. 32.
Specific damping capacity. 0.4 at stress equal to 10% of tensile yield strength

Mass Characteristics

Density. 1.8 g/cm^3 (0.065 lb/in.^{3}) at 20 °C (68 °F)

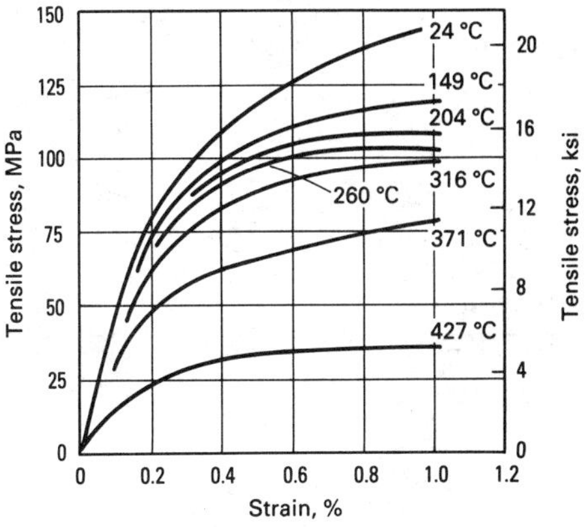

Fig. 23 Typical stress-strain curves for separately sand cast test bars of HK31A

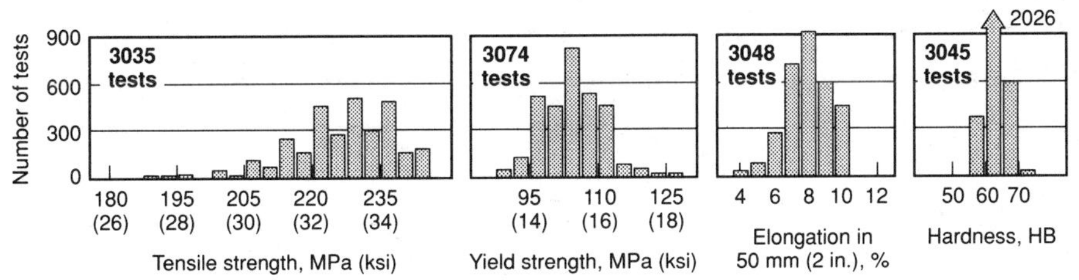

Fig. 22 Distribution of mechanical properties for separately sand cast test bars of HK31A-T6

Table 36 Creep properties of HK31A-T6 sand castings

Properties determined using separately cast test bars

Time under load, h	Tensile stress resulting in total extension(a) of: 0.1%, MPa	0.1%, ksi	0.2%, MPa	0.2%, ksi	0.5%, MPa	0.5%, ksi	1.0%, MPa	1.0%, ksi
At 200 °C (400 °F)								
1	41	6.0	71	10.3	103	15.0	110	16.0
10	40	5.8	68	9.8	103	15.0	110	16.0
100	39	5.6	66	9.5	103	15.0	110	16.0
1000	37	5.4	63	9.1	97	14.0	109	15.8
At 260 °C (500 °F)								
1	36	5.25	69	10.0	97	14.0	107	15.5
10	30	4.4	59	8.6	88	12.7	100	14.4
100	24	3.5	43	6.3	67	9.7	84	12.2
1000	21	3.1	29	4.2	47	6.8	52	7.6
At 290 °C (550 °F)								
1	...	...	54	7.8	85	12.3	...	...
10	...	...	44	6.4	66	9.5	...	...
100	...	...	31	4.5	43	6.3	...	...
1000	...	...	17	2.5	22	3.2	...	...
At 315 °C (600 °F)								
1	29	4.15	43	6.2	72	10.4	85	12.3
10	22	3.25	33	4.75	50	7.2	60	8.7
100	15	2.15	20	2.9	24	3.5	28	4.1
1000	6	0.94	8	1.1	10	1.4	11	1.55
At 350 °C (660 °F)								
1	...	...	30	4.4	41	6.0	...	...
10	...	...	16	2.3	22	3.2	...	...
100	...	...	7	1.0	9	1.3	...	...
1000	...	...	4	0.63	5	0.72	...	...

(a) Total extension equals initial extension plus creep extension.

Thermal Properties

Liquidus temperature. 645 °C (1190 °F)
Solidus temperature. 550 °C (1020 °F)
Coefficient of linear thermal expansion. 26.7 μm/m · K (14.8 μin./in. · °F) at 20 to 200 °C (68 to 390 °F)
Specific heat. 1.00 kJ/kg · K (0.24 Btu/lb · °F) at 20 to 100 °C (68 to 212 °F)
Latent heat of fusion. 373 kJ/kg (160 Btu/lb)
Thermal conductivity. 113 W/m · K (65.3 Btu/ft · h · °F)

Electrical Properties

Electrical conductivity. 25.2% IACS at 20 °C (68 °F)
Electrical resistivity. 68.5 nΩ · m at 20 °C (68 °F)

Fabrication Characteristics

Casting temperature. Sand and permanent mold castings, 750 to 820 °C (1380 to 1510 °F)
Weldability. Gas-shielded arc welding with welding rod of base metal composition, good
Solution temperature. 520 to 530 °C (970 to 990 °F)
Aging temperature. 200 °C (390 °F)

QH21A

Specifications

Foreign. Elektron QH21A

Chemical Composition

Composition limits. 2.0 to 3.0 Ag, 0.6 to 1.6 Th, 0.6 to 1.5 rare earths (composed of at least 70% Nd), 0.40 to 1.0 Zr, 0.20 Zn max, 0.10 Cu max, 0.01 Ni max, 0.30 max other (total), bal Mg. Optimum total for Th plus rare earths, 1.6 to 2.2

Applications

Typical uses. Castings used in solution-heat-treated and artificially aged condition (T6 temper). Ideally suited for aircraft and aerospace components, especially where pressure tightness is required. This alloy is of particular interest to designers and stress engineers for highly stressed components operating at temperatures up to 250 °C (480 °F).

Mechanical Properties

Tensile properties. See Table 42 and Fig. 33.
Fatigue strength. See Fig. 34.
Creep characteristics. See Fig. 35.

Mass Characteristics

Density. 1.83 g/cm^3 (0.066 lb/in.^{3}) at 21 °C (70 °F)

Thermal Properties

Liquidus temperature. 640 °C (1185 °F)
Solidus temperature. 535 °C (995 °F)
Coefficient of linear thermal expansion. 26.7 μm/m · K (14.8 μin./in. · °F) at 20 to 100 °C (68 to 212 °F)
Specific heat. 1.00 kJ/kg · K (0.23 Btu/lb · °F)

Table 37 Thermal conductivity of HK31A-T6 sand castings at various temperatures

Temperature °C	°F	Thermal conductivity W/m · K	Btu/ft · h · °F
20	68	92	53
38	100	92	53
93	200	100	58
150	300	105	61
200	400	109	63
260	500	113	65

Table 38 Typical tensile properties of HZ32A-T5 sand castings at elevated temperatures

Testing temperature °C	°F	Tensile strength MPa	ksi	Yield strength MPa	ksi	Elongation in 50 mm (2 in.), %
24	75	200	29	105	15	6
93	200	180	26	97	14	15
150	300	150	22	83	12	23
200	400	115	17	69	10	33
260	500	97	14	63	9	33
315	600	83	12	55	8	28
370	700	69	10	48	7	29

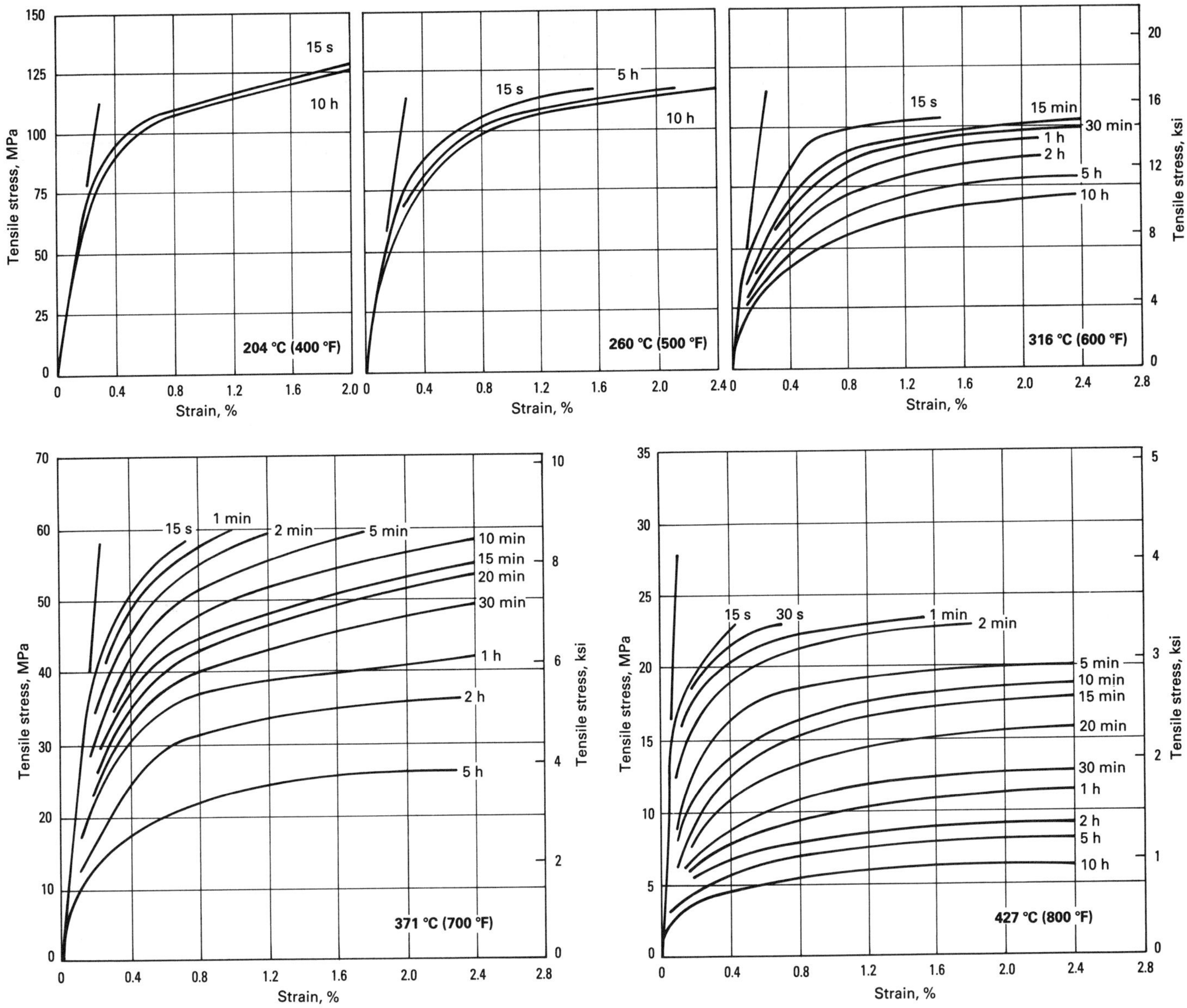

Fig. 24 Isochronous stress-strain curves for separately cast test bars of HK31A-T6. Specimens exposed at testing temperature for 3 h before loading.

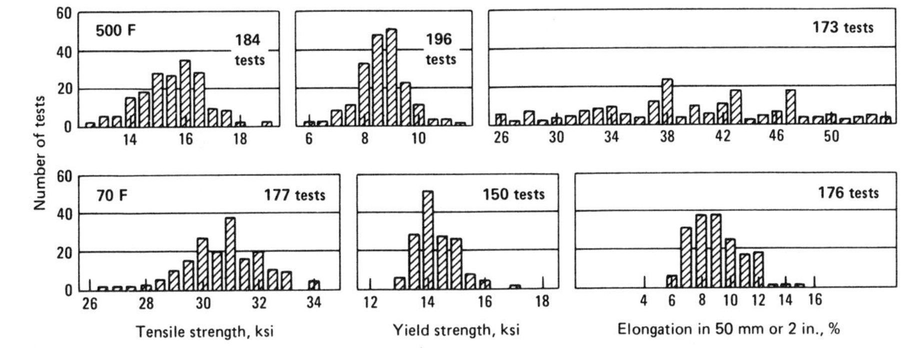

Fig. 25 Distribution of tensile properties for separately sand cast test bars of HZ32A-T5

Thermal conductivity. 113 W/m · K (65 Btu/ft · h · °F) at 21 °C (70 °F)

Electrical Properties

Electrical resistivity. 68.5 nΩ · m

Fabrication Characteristics

Castability. Fine-grain alloy with good casting characteristics

Casting temperature. Sand castings, 750 to 820 °C (1380 to 1510 °F)

Weldability. Fully weldable by the gas tungsten arc process, using rod of the base metal composition

WE43

Specifications

ASTM. Sand castings: B 80. Permanent mold castings: B 93. Investment castings: B 199

UNS number. M18430

Chemical Composition

Composition limits. 3.7 to 4.3 Y, 2.4 to 4.4 rare earths, 0.4 to 1.0 Zr, 0.15 Mn max, 0.2 Zn max, 0.03 Cu max, 0.01 Si max, 0.005 Ni max, 0.2 Li max, bal Mg. Rare earths consist of 2.0 to 2.5% Nd with the remainder comprising heavy rare earths (HRE), principally Tb, Er, Dy, and Gd. The HRE fraction is directly related to the Y content of the alloy

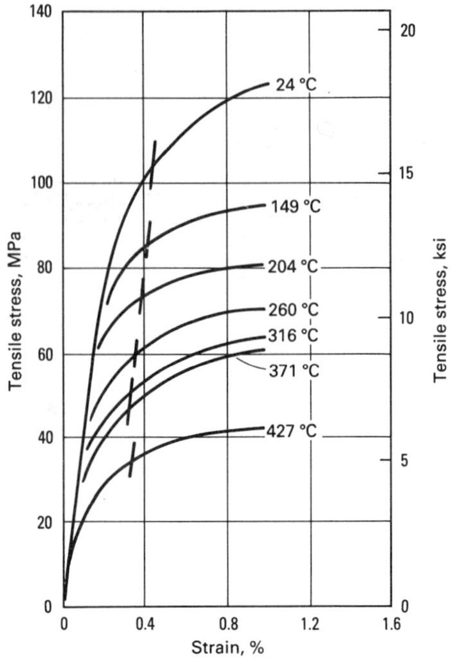

Fig. 26 Typical stress-strain curves for separately sand cast test bars of HZ32A-T5

(that is, Y is present in a nominal 80Y-20HRE mixture).
Consequence of exceeding impurity limits. Zr content below 0.5% may result in somewhat coarser as-cast grains and lower mechanical properties.

Applications

Typical uses. Sand castings used in the solution-heat-treated and artificially aged condition (T6). Castings retain properties at elevated temperatures (≤250 °C, or 480 °F) for extended periods of time (>5000 h), and they are pressure tight and weldable.

Mechanical Properties

Tensile properties. T6 temper: tensile strength, 250 MPa (36.3 ksi); yield strength, 162 MPa (23.5 ksi), elongation, 2%
Tensile properties versus temperature. See Table 43 and Fig. 36.
Elastic modulus. Tension, 44.2 GPa (6.4 × 10^6 psi) at 20 °C (68 °F)
Creep characteristics. See Table 44.
Poisson's ratio. 0.27
Hardness. 75 to 95 HB

Mass Characteristics

Density. 1.84 g/cm^3 (0.0665 lb/in.^{3}) at 20 °C (68 °F)

Thermal Properties

Liquidus temperature. 640 °C (1185 °F)
Solidus temperature. 540 to 550 °C (1005 to 1020 °F)
Thermal conductivity. 51.3 W/m · K (29.6 Btu/h · ft · °F) at 20 °C (68 °F)

Table 39 Typical tensile properties of K1A-F sand castings at elevated temperatures

Testing temperature °C	°F	Tensile strength MPa	ksi	Tensile yield strength MPa	ksi	Elongation, %
93	200	115	17	48	7	30
200	400	55	8	34	5	71
315	600	28	4	14	2	78

Electrical Properties

Electrical resistivity. 148 nΩ · m at 20 °C (68 °F)
Electrical conductivity. 6.8 MS/m (6.8 × 10^4 mho/cm^3)

Fabrication Characteristics

Casting temperature. Sand castings, 750 to 820 °C (1380 to 1510 °F)
Weldability. Gas-shielded arc welding with weld rod of base metal composition

Corrosion Resistance

ASTM B 117 salt fog test. 0.1 to 0.2 mg/cm^2/day

WE54

Specifications

AMS. 4426
ASTM. Sand castings: B 80. Permanent mold castings: B 199. Investment castings: B 403
UNS number. M18410

Chemical Composition

Composition limits. 4.75 to 5.5 Y, 2.0 to 4.0 rare earths, 0.4 to 1.0 Zr, 0.15 Mn max, 0.2 Zn max, 0.03 Cu max, 0.01 Si max, 0.005 Ni max, 0.2 Li max, bal Mg. Rare earths consist of 1.5 to 2.0% Nd with the remainder comprising heavy rare earths (HRE), principally Tb, Er, Dy, and Gd. The HRE fraction is directly related to the Y content of the alloy (that is, Y is present in a nominal 80Y-20HRE mixture).
Consequence of exceeding impurity limits. Zr content below 0.5% may result in somewhat coarser as-cast grains and lower mechanical properties.

Applications

Typical uses. Sand castings used in the solution-heat-treated and artificially aged (T6) condition. Castings retain properties at high temperatures (300 °C, or 570 °F) for short-term applications (up to 1000 h) and are pressure tight and weldable.

Mechanical Properties

Tensile properties. T6 temper: tensile strength, 250 MPa (36.5 ksi); yield strength, 172 MPa (24.9 ksi); elongation, 2%
Tensile properties versus temperature. See Fig. 37.

Table 40 Typical tensile properties of QE22A sand castings at various temperatures

Testing temperature °C	°F	Tensile strength MPa	ksi	Yield strength MPa	ksi
20	68	263	38.1	208	30.2
100	212	235	34.1	193	28.0
200	392	193	28.0	166	24.0
300	572	83	12.0	69	10

Table 41 Long-time creep properties of QE22A sand castings

Time under load, h	Tensile stress resulting in creep extension(a) of: 0.05% MPa	ksi	0.1% MPa	ksi	0.2% MPa	ksi	0.5% MPa	ksi	1.0% MPa	ksi
At 150 °C (300 °F)										
10	150	21.6	...	...	...	...	...	...	...	...
100	120	17.4	140	20.5	165	23.8	...	...	...	...
1000	90	13.0	105	15.5	125	18.0	150	21.7		
At 200 °C (390 °F)										
10	83	12.0	105	15.0	...	...	...	...	...	...
100	55	8.0	73	10.6	87	12.6	105	15.0	110	16.0
1000	...	...	...	...	55	8.0	72	10.5	78	11.3
At 250 °C (480 °F)										
10	32	4.7	41	6.0	...	...	...	...	...	...
100	17	2.5	26	3.7	32	4.7	40	5.8	...	...
1000	...	...	10	1.4	16	2.3	22	3.2	26	3.8

(a) Does not include initial extension

Table 42 Typical tensile properties of QH21A-T6 castings before and after exposure to elevated temperature

Exposure condition	Tensile strength MPa	ksi	Tensile yield strength MPa	ksi	Elongation in 50 mm (2 in.), %
Unexposed(a)	276	40.0	207	30.0	4
Exposed 500 h at 200 °C (390 °F)	284	41.2	205	29.7	8
Exposed 1000 h at 200 °C (390 °F)	282	40.9	200	29.0	8

(a) Room-temperature values determined using separately cast test bars

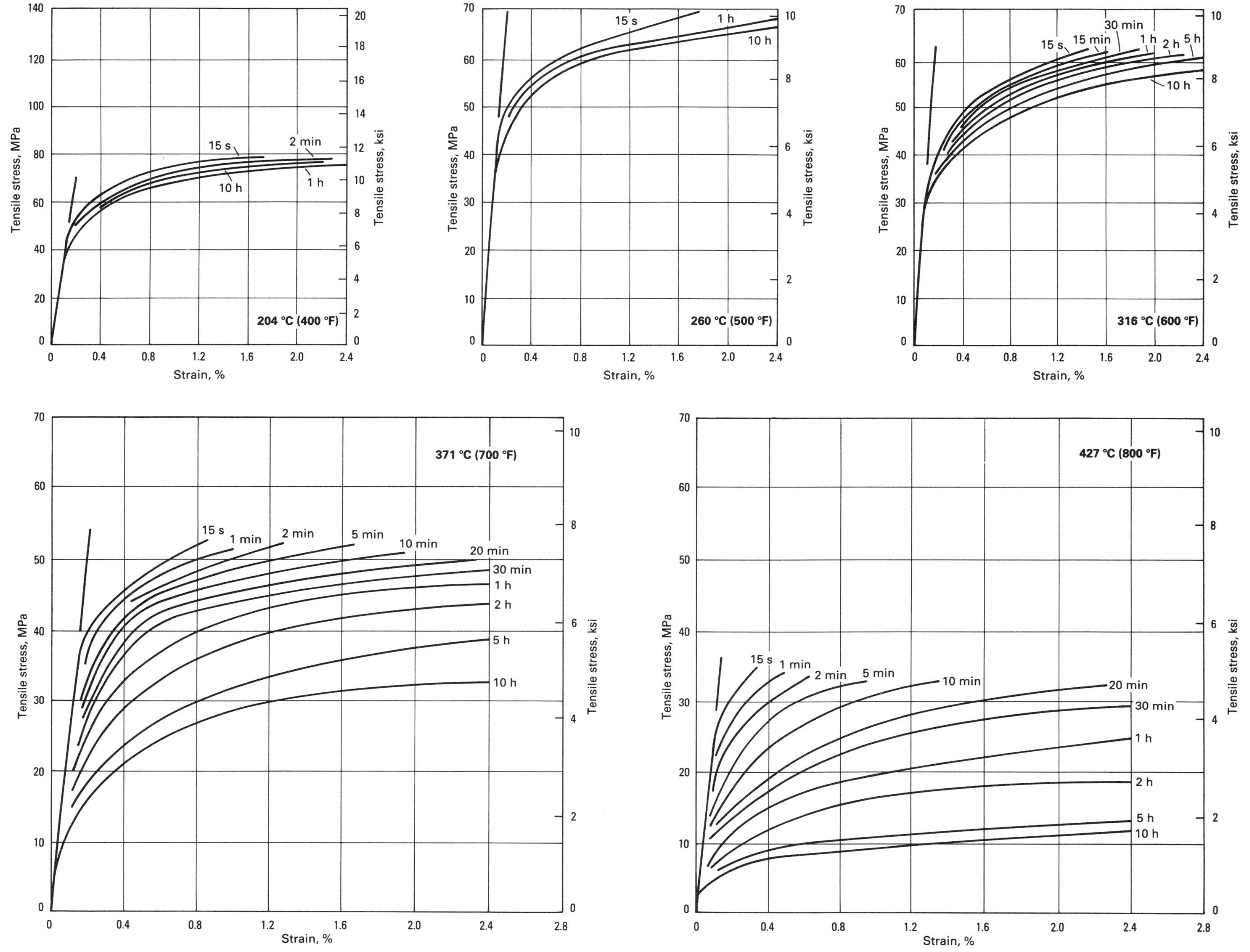

Fig. 27 Isochronous stress-strain curves for separately sand cast test bars of HZ32A-T5. Specimens exposed at testing temperatures for 3 h before loading.

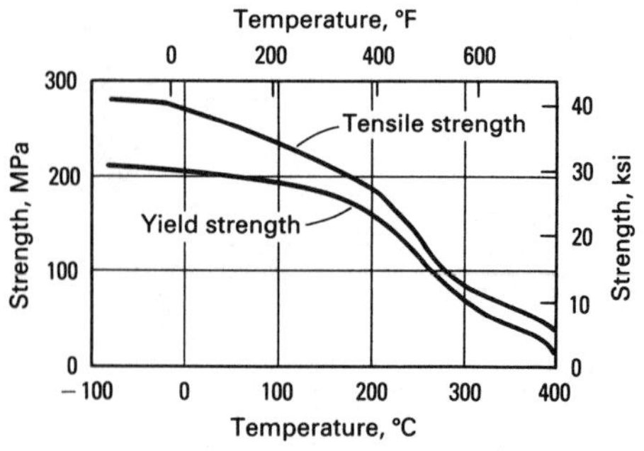

Fig. 28 Effect of temperature on the strength of QE22A-T6 sand castings

Compressive properties. T6 temper: compressive strength, 410 MPa (59.5 ksi); compressive yield strength, 172 MPa (24.9 ksi)
Elastic modulus. Tension, 44.4 GPa (6.4 × 10^6 psi)
Creep characteristics. See Table 45.
Poisson's ratio. 0.27
Hardness. 75 to 95 HB

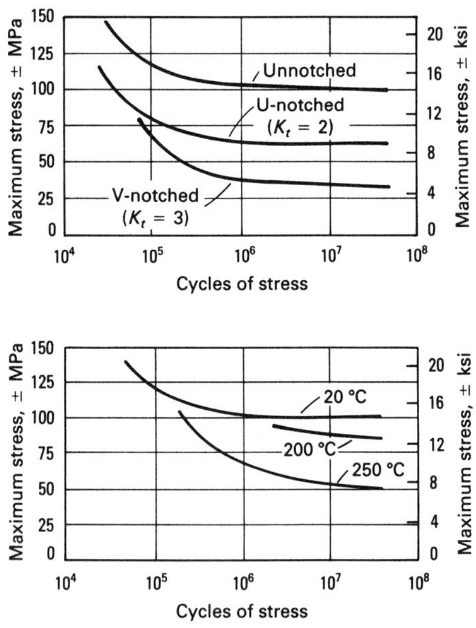

Fig. 31 Fatigue characteristics of QE22A-T6 sand castings. Rotating beam (Wohler) tests; machine speed, 2960 Hz

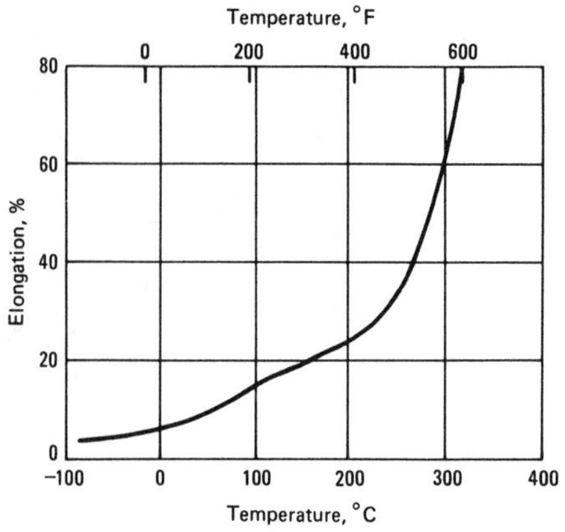

Fig. 29 Effect of temperature on the elongation of QE22A-T6 sand castings

Mass Characteristics

Density. 1.85 g/cm^3 (0.067 lb/in.^{3}) at 20 °C (68 °F)

Thermal Properties

Liquidus temperature. 640 °C (1185 °F)
Solidus temperature. 545 to 555 °C (1015 to 1030 °F)
Thermal conductivity. 52 W/m · K (30 Btu/ft · h · °F) at 20 °C (68 °F)

Electrical Properties

Electrical resistivity. 173 nΩ · m at 20 °C (68 °F)
Electrical conductivity. 5.8 MS/m (5.8 × 10^4 mho/cm^3)

Fabrication Characteristics

Casting temperature. Sand castings, 750 to 820 °C (1380 to 1510 °F)
Weldability. Gas-shielded arc welding with weld rod of base metal composition

Corrosion Resistance

ASTM B 117 salt fog test. 0.1 to 0.2 mg/cm^2/day

ZC63

Specifications

ASTM. Sand castings: B 80. Permanent mold castings: B 199. Investment castings: B 403
UNS number. M16631

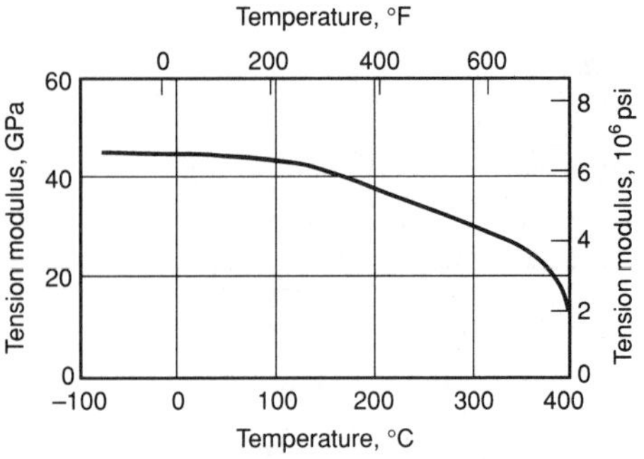

Fig. 30 Effect of temperature on the elastic modulus of QE22A-T6 sand castings

Chemical Composition

Composition limits. 5.5 to 6.5 Zn, 2.4 to 3.0 Cu, 0.25 to 0.75 Mn, 0.20 Si max, 0.010 Ni max, 0.30 max other (total), bal Mg

Applications

Typical uses. Sand castings used in the solution-heat-treated and artificially aged (T6) condition. Superior properties to AZ91C-applications with better castability. Useful in pressure-tight applications. Can be welded

Mechanical Properties

Tensile properties. T6 temper: tensile strength, 210 MPa (30.5 ksi); yield strength, 125 MPa (18.1 ksi); 3 to 5% elongation
Tensile properties versus temperature. See Fig. 38.
Elastic modulus. Tension, 45 GPa (6.5 × 10^6 psi) at 20 °C (68 °F)
Creep characteristics. See Fig. 39 and 40.
Poisson's ratio. 0.27
Hardness. 55 to 65 HB

Mass Characteristics

Density. 1.87 g/cm^3 (0.068 lb/in.^{3}) at 20 °C (68 °F)

Thermal Properties

Liquidus temperature. 635 °C (1175 °F)

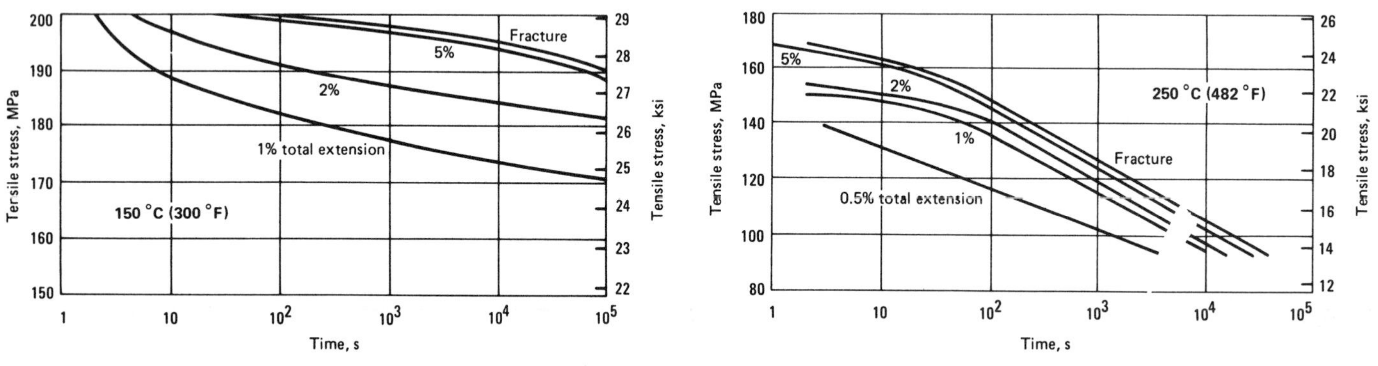

Fig. 32 Short-time creep-rupture properties of QE22A-T6 sand castings

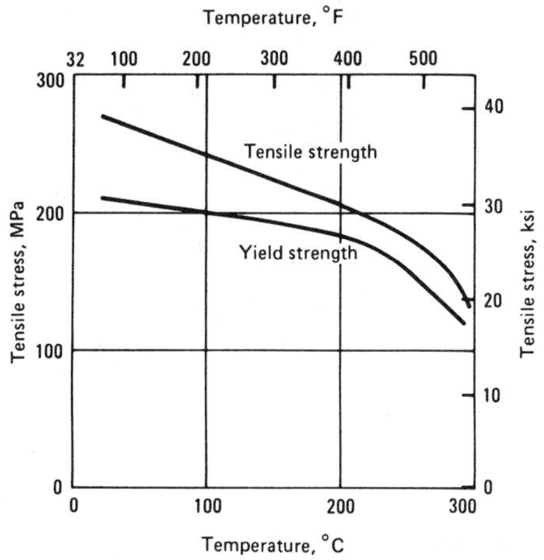

Fig. 33 Effect of temperature on the strength of QH21A-T6 sand castings

Solidus temperature. 465 °C (870 °F)
Thermal conductivity. 122 W/m · K (70.5 Btu/ft · h · °F) at 20 °C (68 °F)

Electrical Properties

Electrical resistivity. 54 nΩ · m at 20 °C (68 °F)

Fabrication Characteristics

Weldability. Gas-shielded arc welding with weld rod of same base metal composition

ZE41A

Specifications

ASTM. Sand castings: B 80
UNS number. M16410
Foreign. Elektron RZ5. British: BS 2970 MAG5. German: DIN 1729 3.5101. French: AIR 3380 RZ5

Chemical Composition

Composition limits. 3.5 to 5.0 Zn, 0.75 to 1.75 rare earths (as mischmetal), 0.40 to 1.0 Zr, 0.15 Mn max, 0.10 Cu max, 0.01 Ni max, 0.30 max other (total), bal Mg

Table 43 Elevated-temperature properties taken from specimens of sand cast 25 mm (1 in.) thick plate of WE43 magnesium alloy

Test temperature		Young's modulus		0.2% yield strength		Tensile strength		Elongation,
°C	°F	GPa	10^6 psi	MPa	ksi	MPa	ksi	%
150	300	47	6.8	170–180	25–26	240–250	35–36	4–8
200	390	39	5.7	160–180	23–26	240–260	35–38	8–14
250	480	36	5.2	150–170	22–25	210–230	30–33	15–20
300	570	36	5.2	110–130	16–19	150–170	22–25	30–50

Consequence of exceeding impurity limits. Content of less than 0.6% soluble Zr may increase grain size and thus reduce mechanical properties; weldability also may decrease.

Applications

Typical uses. Sand castings used in the artificially aged condition (T5 temper), with better castability than ZK51A and good strength up to 93 °C (200 °F). Useful in pressure-tight applications. Can be welded. Stress relieved at 345 °C (650 °F)

Mechanical Properties

Tensile properties. T5 temper: tensile strength, 205 MPa (30 ksi); yield strength, 140 MPa (20 ksi); elongation, 3.5%
Tensile properties versus temperature. See Table 46 and Fig. 41.
Compressive properties. T5 temper: compressive strength, 345 MPa (50 ksi); compressive yield strength, 140 MPa (20 ksi)
Impact strength. Charpy V-notch, 1.4 J (1 ft · lbf)
Elastic modulus. Tension, 45 GPa (6.5 × 10^6 psi); shear, 17 GPa (2.4 × 10^6 psi)
Creep characteristics. See Table 47.
Poisson's ratio. 0.35
Bearing properties. Ultimate bearing strength, 485 MPa (70 ksi); bearing yield strength, 350 MPa (51 ksi)
Hardness. 62 HB or 72 HRE

Mass Characteristics

Density. 1.82 g/cm³ (0.066 lb/in.³) at 20 °C (68 °F)

Thermal Properties

Liquidus temperature. 645 °C (1190 °F)
Solidus temperature. 525 °C (975 °F)
Thermal conductivity. 113 W/m · K (65.3 Btu/ft · h · °F)

Electrical Properties

Electrical resistivity. T5 temper: 60 nΩ · m at 20 °C (68 °F)

Fabrication Characteristics

Casting temperature. Sand castings, 750 to 820 °C (1380 to 1510 °F)
Weldability. Gas-shielded arc welding with weld rod of base metal composition, good; complete all welding before hydrogen treatment; stress relief required

ZE63A

Specifications

AMS. Sand castings: 4425
UNS number. M16630
Government. Sand castings: MIL-M-46062B
Foreign. Elektron ZE63A. British: DTD 5045

Chemical Composition

Composition limits. 5.5 to 6.0 Zn, 2.1 to 3.0 rare earths, 0.40 to 1.0 Zr, 0.10 Cu max, 0.01 Ni max, 0.30 max other (total), bal Mg

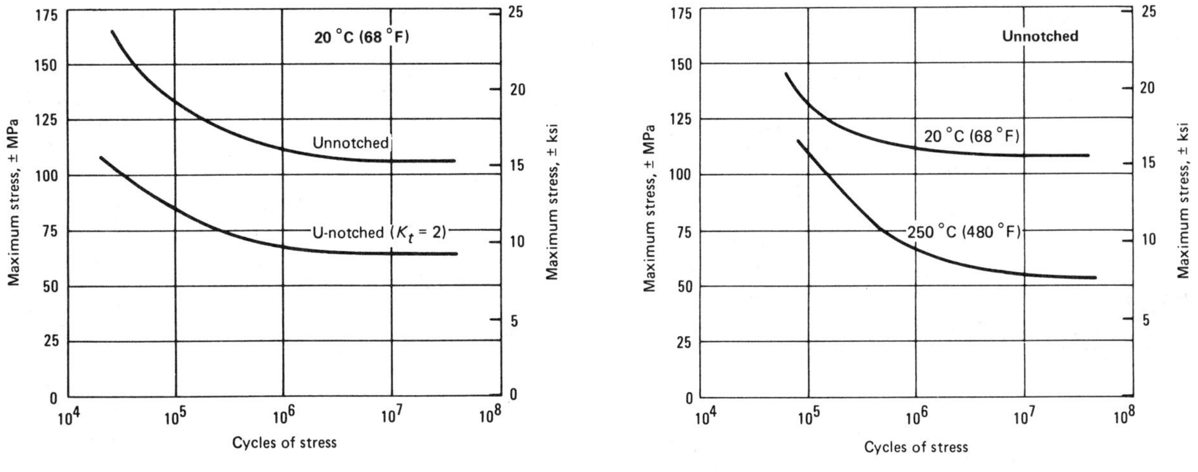

Fig. 34 Fatigue characteristics of QH21A-T6 sand castings. Rotating beam (Wohler) tests; machine speed, 2960 Hz

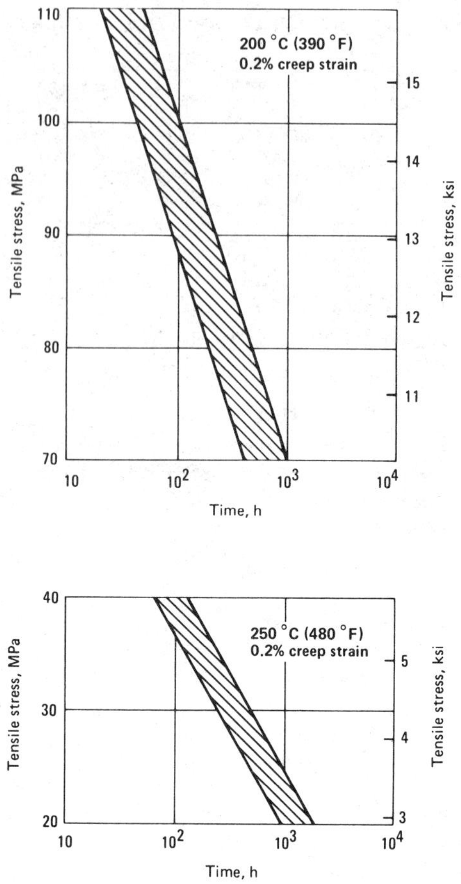

Fig. 35 Creep properties of QH21A-T6 sand castings

Applications

Typical uses. Sand and investment castings used in solution-heat-treated and artificially aged condition (T6 temper). Especially useful in thin-section castings for applications requiring high mechanical strength and freedom from porosity. Special heat treatment in hydrogen is required to develop properties.

Mechanical Properties

Tensile properties. T6 temper: tensile strength, 300 MPa (44 ksi); yield strength, 190 MPa (28 ksi); elongation, 10% in 5.65 $\sqrt{A}$

Tensile properties versus temperature. See Table 48.

Table 44 Creep properties of WE43

Time, h	Stress to produce creep strain of 0.1% MPa	0.1% ksi	0.2% MPa	0.2% ksi	0.5% MPa	0.5% ksi
At 200 °C (390 °F)						
10	170	24.6	176	25.5	185	26.8
100	148	21.5	161	23.4	173	25
1000	...	...	96	13.9	139	20
At 250 °C (480 °F)						
10	69	10	75	10.8	...	...
100	44	6.4	61	8.8	...	...
1000	...	...	39	5.7	...	...

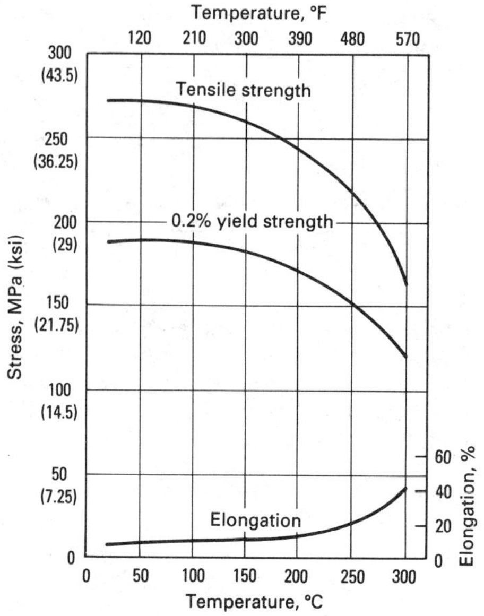

Fig. 36 Effect of test temperature on the tensile properties of WE43

Compressive properties. T6 temper: compressive strength, 450 MPa (65 ksi); compressive yield strength, 195 MPa (28 ksi)

Hardness. 60 to 85 HB

Poisson's ratio. 0.35

Elastic modulus. Tension, 45 GPa (6.5×10^6 psi); shear, 17 GPa (2.5×10^6 psi); compression, 44 GPa (6.4×10^6 psi)

Impact strength. Unnotched, 0.33 to 0.55 J (0.24 to 0.41 ft · lbf); notched, 0.063 to 0.084 J (0.046 to 0.062 ft · lbf)

Plane-strain fracture toughness. 27 MPa$\sqrt{m}$ (24.5 ksi$\sqrt{in.}$)

Creep characteristics. See Table 49.

Mass Characteristics

Density. 1.87 g/cm^3 (0.067 lb/in.^{3}) at 20 °C (68 °F)

Table 45 Creep properties of WE54 magnesium alloy

Type and amount of strain	Stress to produce indicated creep or total strain in 10 h MPa	10 h ksi	100 h MPa	100 h ksi	1000 h MPa	1000 h ksi
Creep strain at 200 °C (390 °F), %						
0.05	...	...	131	19.0	80	11.6
0.1	...	...	160	23.2	102	14.8
0.2	...	...	170	24.7	132	19.2
Total strain at 200 °C (390 °F), %						
0.5	...	...	...	...	126	18.3
Creep strain at 250 °C (480 °F), %						
0.1	90	13.1	47	6.8	16	2.3
0.2	110	16	61	8.8	32	4.6
0.5	135	19.6	81	11.7	48	7.0
Total strain at 250 °C (480 °F), %						
0.2	56	8.1	41	5.9	26	3.8
0.5	108	15.7	75	10.9	45	6.5
1	140	20.3	90	13.1	60	8.7

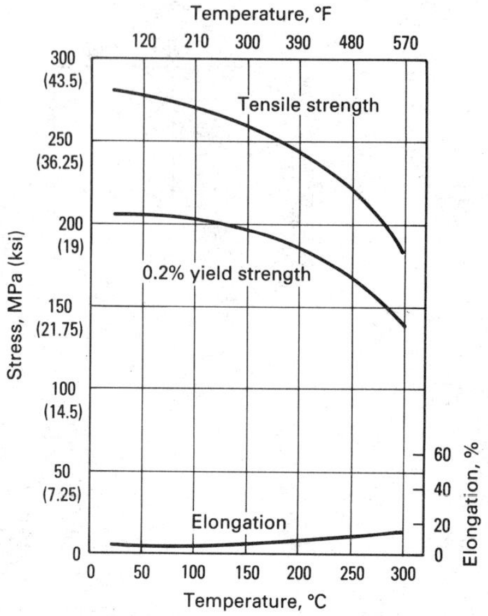

Fig. 37 Effect of test temperature on the tensile properties of WE54

Thermal Properties

Liquidus temperature. 635 °C (1175 °F)

Solidus temperature. 510 °C (950 °F)

Coefficient of linear thermal expansion. 26.5 μm/m · K (14.7 μin./in. · °F)

Specific heat. 0.96 kJ/kg · K (0.23 Btu/lb · °F)

Thermal conductivity. 109 W/m · K (63 Btu/ft · h · °F)

Electrical Properties

Electrical conductivity. 30.9% IACS at 20 °C (68 °F)

Electrical resistivity. 56 nΩ · m at 20 °C (68 °F)

Fabrication Characteristics

Casting temperature. Sand castings, 750 to 820 °C (1380 to 1510 °F)

Weldability. Gas-shielded arc welding with ZE63A welding rod, very good. Must be welded prior to heat treatment

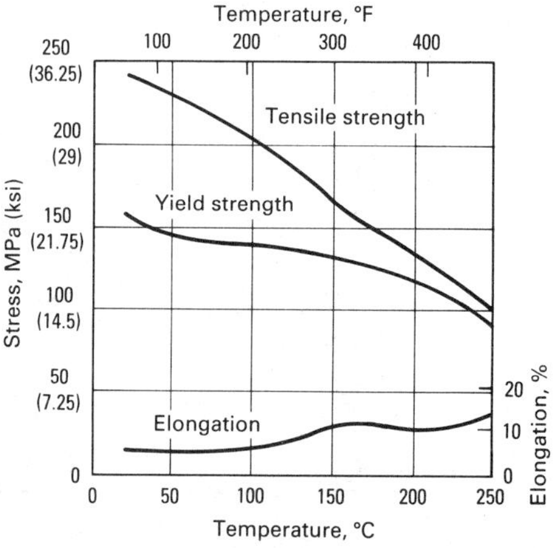

Fig. 38 Elevated-temperature tensile properties of magnesium alloy ZC63

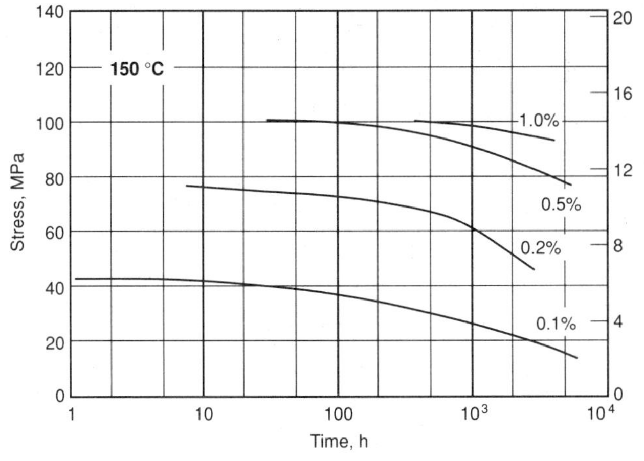

(a)

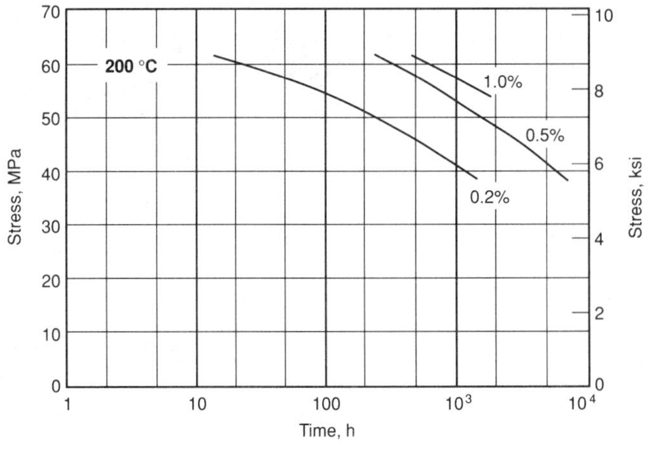

(b)

Fig. 39 Stress-time relationships for specified total strains of magnesium alloy ZC63. (a) At 150 °C (300 °F). (b) At 200 °C (390 °F)

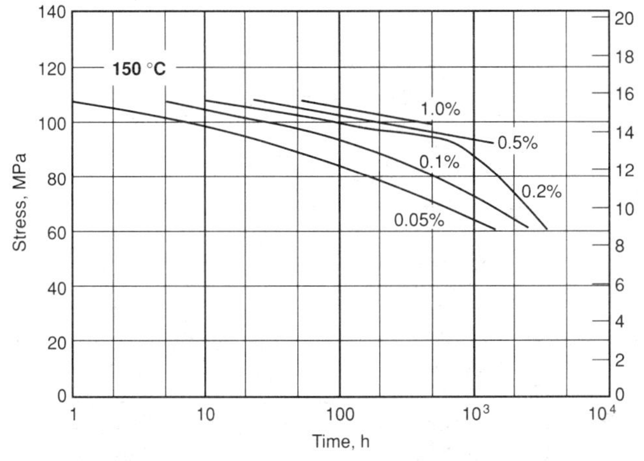

(a)

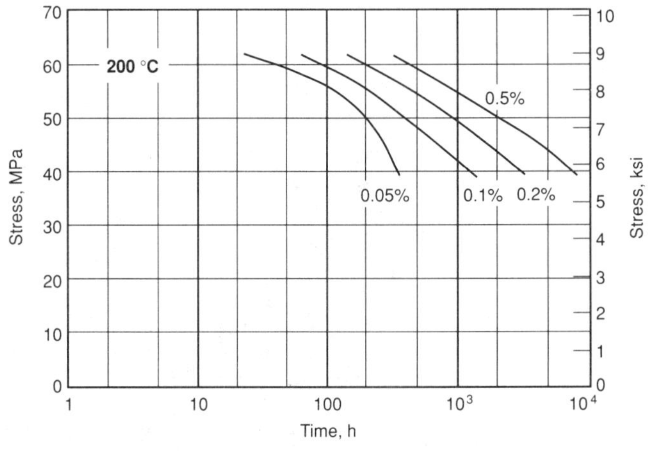

(b)

Fig. 40 Stress-time relationships for specified creep strains of magnesium alloy ZC63. (a) At 150 °C (300 °F). (b) At 200 °C (390 °F)

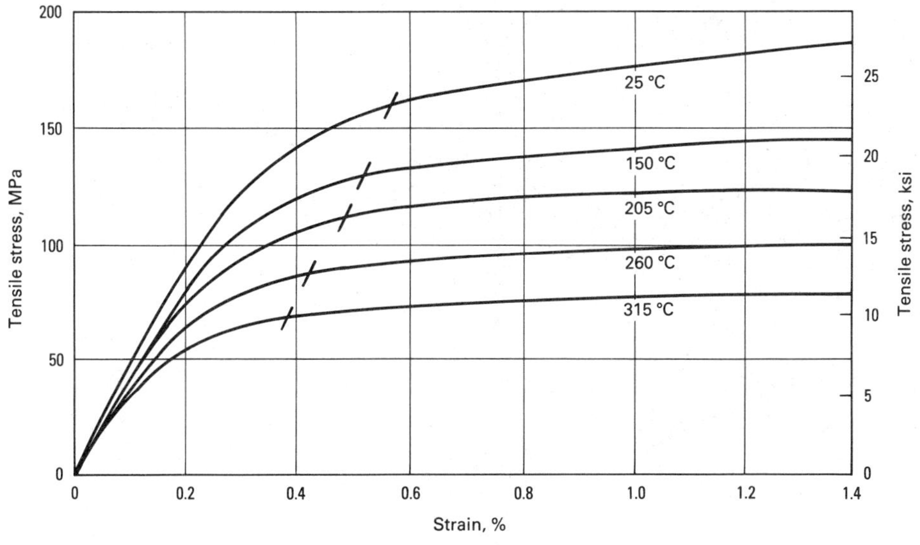

Fig. 41 Typical stress-strain curves for separately cast test bars of ZE41A-T5

ZH62A

Specifications

AMS. Sand castings: 4448
ASTM. Sand castings: B 80
SAE. J465. Former SAE alloy number: 508
UNS number. M16620
Government. Sand castings: QQ-M-56, MIL-M-46062
Foreign. Elektron TZ6. British: BS 2970 MAG9. German: DIN 1729 3.5102. French: AIR 3380 TZ6

Chemical Composition

Composition limits. 5.2 to 6.2 Zn, 1.4 to 2.2 Th, 0.50 to 1.0 Zr, 0.10 Cu max, 0.01 Ni max, 0.30 max other (total), bal Mg

Applications

Typical uses. Sand and permanent mold castings used in artificially aged condition (T5 temper) for room-temperature service. Highest in yield strength of all magnesium

Table 46 Typical tensile properties of ZE41A-T5 sand castings at elevated temperatures

Properties determined on separately cast test bars

Testing temperature °C	°F	Tensile strength MPa	ksi	Tensile yield strength MPa	ksi	Elongation in 50 mm (2 in.), %
93	200	193	28.0	138	20.0	8
150	300	172	25.0	130	18.8	12
200	400	141	20.5	114	16.5	31
260	500	106	15.4	88	12.7	40
315	600	82	11.9	69	10.0	45

Table 47 Creep properties of ZE41A-T5 sand castings

Properties determined using separately cast test bars

Time under load, h	Tensile stress resulting in total extension(a) of: 0.1% MPa	0.1% ksi	0.2% MPa	0.2% ksi	0.5% MPa	0.5% ksi	1.0% MPa	1.0% ksi
At 93 °C (200 °F)								
1	47	6.8	85	12.3	135	20.0	...	...
10	46	6.6	83	12.0	130	19.0	...	...
100	42	6.1	76	11.0	125	18.1	...	...
1000	37	5.4	68	9.8	115	16.5	...	...
At 150 °C (300 °F)								
1	43	6.3	74	10.7	112	16.3	...	...
10	43	6.2	71	10.3	105	15.2	...	...
100	41	6.0	68	9.9	99	14.3	...	...
1000	34	5.0	63	9.1	86	12.5	...	...
At 200 °C (400 °F)								
1	38	5.5	67	9.7	104	15.1	...	...
10	33	4.8	56	8.1	91	13.2	...	...
100	23	3.4	41	6.0	74	10.7	...	...
1000	14	2.1	23	3.3	37	5.4	...	...
At 260 °C (500 °F)								
1	28	4.1	39	5.6	55	8.0	66	9.5
10	16	2.3	23	3.4	35	5.1	43	6.2
100	7	1.0	12	1.8	21	3.0	25	3.6
1000	6	0.84	7	1.0	10	1.4	12	1.7

(a) Total extension equals initial extension plus creep extension.

casting alloys except ZK61A-T6 and QE22A-T6

Mechanical Properties

Tensile properties. T5 temper: tensile strength, 240 MPa (35 ksi); yield strength, 150 MPa (22 ksi); elongation, 4% in 50 mm (2 in.). See also Fig. 42.
Compressive yield strength. T5 temper, 150 MPa (22 ksi)
Poisson's ratio. 0.3
Elastic modulus. Tension, 45 GPa (6.5 × 10^6 psi); shear, 17 GPa (2.5 × 10^6 psi)
Hardness. 70 HB
Impact strength. Notched Izod, 3.4 J (2.5 ft · lbf) at 20 °C (68 °F)

Table 48 Typical tensile properties of ZE63A sand castings at various temperatures

Testing temperature °C	°F	Tensile strength MPa	ksi	Tensile yield strength MPa	ksi
20	68	289	41.9	173	25.1
100	212	235	34.1	131	19.0
150	302	187	27.1	111	16.1
200	392	131	19.0	97	14.1

Mass Characteristics

Density. 1.86 g/cm³ (0.067 lb/in.³) at 20 °C (68 °F)

Thermal Properties

Liquidus temperature. 630 °C (1170 °F)
Solidus temperature. 520 °C (970 °F)
Coefficient of linear thermal expansion. 27.1 μm/m · K (15 μin./in. · °F) at 20 to 200 °C (68 to 390 °F)
Specific heat. 0.96 kJ/kg · K (0.23 Btu/lb · °F)

Table 49 Creep properties of ZE63A sand castings

Stress MPa	ksi	Time, h, to reach total extension(a) of: 0.15%	0.2%	0.25%	0.3%	0.5%	0.75%	1.0%	2.0%	3.0%	4.0%
At 100 °C (212 °F)											
46	6.7	25	650	1440	...	...	...	...	...	...	...
62	9.0	...	120	480	960	...	...	...	...	...	...
77	11.1	...	...	30	135	1250	...	...	...	...	...
92	13.3	...	...	...	15	280	1400	...	...	...	...
At 150 °C (300 °F)											
39	5.7	70	530	...	...	...	...	...	...	...	...
46	6.7	20	156	...	912	...	...	...	...	...	...
54	7.8	...	8	...	50	720	...	...	...	...	...
62	9.0	...	5	...	35	335	840	1200	...	...	...
70	10.1	...	...	...	12	135	350	550	920	...	...
77	11.1	...	...	...	5	35	90	145	290	350	390

(a) Total extension equals initial extension plus creep extension.

Latent heat of fusion. 373 kJ/kg (160 Btu/lb)
Thermal conductivity. 110 W/m · K (63 Btu/ft · h · °F) at 20 °C (68 °F)

Electrical Properties

Electrical conductivity. 26.5% IACS at 20 °C (68 °F)
Electrical resistivity. 65 nΩ · m at 20 °C (68 °F)

Fabrication Characteristics

Casting temperature. Sand castings, 750 to 820 °C (1380 to 1510 °F)
Weldability. Gas-shielded arc welding with EZ33A or ZH62A welding rod, poor; castings should be heat treated after welding.

ZK51A

Specifications

AMS. Sand castings: 4443
ASTM. Sand castings: B 80
SAE. J465. Former SAE alloy number: 509
UNS number. M16510
Government. Sand castings: QQ-M-56A, MIL-M-46062
Foreign. Elektron Z5Z. British: BS 2970 MAG4. French: AIR 3380 Z5Z

Chemical Composition

Composition limits. 3.6 to 5.5 Zn, 0.50 to 1.0 Zr, 0.10 Cu max, 0.01 Ni max, 0.30 max other (total), bal Mg

Applications

Typical uses. Sand castings used in artificially aged condition (T5 temper), with high yield strength and good ductility. This alloy is suggested for highly stressed parts that are small or relatively simple in design. Solution treatment is not required.

Mechanical Properties

Tensile properties. T5 temper: tensile strength, 205 MPa (30 ksi); yield strength, 140 MPa (20 ksi); elongation, 3.5% in 50 mm (2 in.)
Tensile properties versus temperature. See Table 50.
Shear strength. T5 temper, 150 MPa (22 ksi)

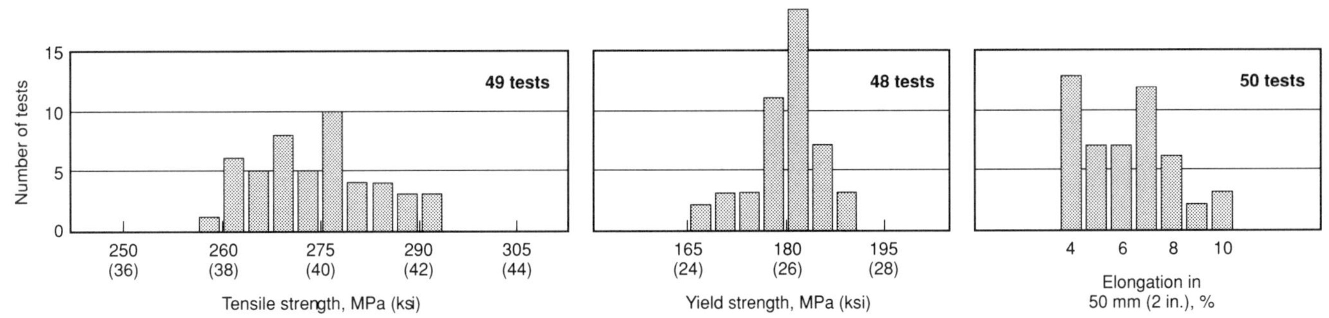

Fig. 42 Distribution of tensile properties for separately cast test bars of ZH62A-T5

Compressive properties. T5 temper: compressive strength, 345 MPa (50 ksi); compressive yield strength, 140 MPa (20 ksi)
Bearing properties. T5 temper: ultimate bearing strength, 485 MPa (70 ksi); bearing yield strength, 350 MPa (51 ksi)
Hardness. 62 HB or 72 HRE
Creep characteristics. See Table 51.

Mass Characteristics

Density. 1.83 g/cm^3 (0.066 lb/in.^{3}) at 20 °C (68 °F)
Solidification shrinkage. 4.2%
Volume change during cooling. 5% contraction from 600 to 20 °C (1110 to 68 °F)

Thermal Properties

Liquidus temperature. 640 °C (1185 °F)
Solidus temperature. 560 °C (1040 °F)
Nonequilibrium solidus temperature. 550 °C (1020 °F)
Coefficient of linear thermal expansion. 26 μm/m · K (14.5 μin./in. · °F) at 20 °C (68 °F)
Specific heat. 1.02 kJ/kg · K (0.244 Btu/lb · °F) at 20 °C (68 °F)
Latent heat of fusion. 318 kJ/kg (137 Btu/lb)
Thermal conductivity. 110 W/m · K (63 Btu/ft · h · °F) at 20 °C (68 °F)

Electrical Properties

Electrical conductivity. 28% IACS at 20 °C (68 °F)
Electrical resistivity. 62 nΩ · m at 20 °C (68 °F). Temperature coefficient, 0.16 nΩ · m per K at 20 °C (68 °F)

Fabrication Characteristics

Casting temperature. Sand castings, 750 to 820 °C (1380 to 1510 °F)
Weldability. Gas-shielded arc welding with EZ33A or ZK51A rod (EZ33A preferred), limited; preheating not necessary, but may be used; postweld heat treatment required

Table 50 Typical tensile properties of ZK51A-T5 sand castings at elevated temperatures
Properties determined using separately cast test bars

Testing temperature		Tensile strength		Yield strength		Elongation in 50 mm (2 in.), %
°C	°F	MPa	ksi	MPa	ksi	
25	75	275	40	180	26	8
95	200	205	30	145	21	12
150	300	160	23	115	17	14
205	400	115	17	90	13	17
260	500	83	12	62	9	16
315	600	55	8	41	6	16

Table 51 Creep properties of ZK51A-T5 sand castings
Properties determined using separately cast test bars

Time under load, h	Tensile stress resulting in total extension(a) of 0.1% MPa	0.1% ksi	0.2% MPa	0.2% ksi	0.5% MPa	0.5% ksi	1.0% MPa	1.0% ksi
At 95 °C (200 °F)								
1	47	6.8	85	12.3	138	20.0	...	...
10	46	6.6	83	12.0	131	19.0	...	...
100	42	6.1	76	11.0	125	18.1	...	...
1000	37	5.4	68	9.8	114	16.5	...	...
At 150 °C (300 °F)								
1	43	6.3	74	10.7	112	16.3	...	...
10	43	6.2	71	10.3	105	15.2	...	...
100	41	6.0	68	9.9	99	14.3	...	...
1000	34	5.0	63	9.1	86	12.5	...	...
At 205 °C (400 °F)								
1	38	5.5	67	9.7	104	15.1	...	...
10	33	4.8	56	8.1	91	13.2	...	...
100	23	3.4	41	6.0	74	10.7	...	...
1000	14	2.1	23	3.3	37	5.4	...	...
At 260 °C (500 °F)								
1	28	4.1	39	5.6	55	8.0	66	9.5
10	16	2.3	23	3.4	35	5.1	43	6.2
100	7	1.0	12	1.8	21	3.0	25	3.6
1000	6	0.84	7	1.0	10	1.4	12	1.7

(a) Total extension equals initial extension plus creep extension.

ZK61A

Specifications

ASTM. Sand castings: B 80
SAE. J465. Former SAE alloy number: 513
UNS number. M16610
Government. Sand castings: QQ-M-56B

Chemical Composition

Composition limits. 5.5 to 6.5 Zn, 0.6 to 1.0 Zr, 0.10 Cu max, 0.01 Ni max, 0.30 max other (total), bal Mg

Applications

Typical uses. Simple, highly stressed castings of uniform cross section. High in cost. Intricate castings subject to microporosity and cracking due to shrinkage. Not readily welded. Sometimes used in the artificially aged condition (T5 temper) but usually in the solution-heat-treated and artificially aged condition (T6 temper) to develop properties fully

Mechanical Properties

Tensile properties. T6 temper: tensile strength, 310 MPa (45 ksi); yield strength, 195 MPa (28 ksi); elongation in 50 mm (2 in.), 10%

Fatigue properties. At least equal to those of the Mg-Al-Zn alloys

Mass Characteristics

Density. 1.83 g/cm^3 (0.066 lb/in.^{3}) at 20 °C (68 °F)

Thermal Properties

Liquidus temperature. 635 °C (1175 °F)
Solidus temperature. 530 °C (985 °F)
Coefficient of linear thermal expansion. 27.0 μm/m · K (15.0 μin./in. · °F) at 20 to 200 °C (68 to 390 °F)

Fabrication Characteristics

Weldability. Not readily weldable. Addition of Th or rare earths decreases porosity and improves weldability.
Casting temperature. Sand castings, 705 to 815 °C (1300 to 1500 °F)

Tin and Tin Alloys

Revised by William B. Hampshire, Tin Research Institute, Inc.

TIN was one of the first metals known to man. Throughout ancient history, various cultures recognized the virtues of tin in coatings, alloys, and compounds, and the use of the metal increased with advancing technology. Today, tin is an important metal in industry even though the annual tonnage used is much smaller than those of many other metals. One reason for the small tonnage is that, in most applications, only very small amounts of tin are used at a time.

Tin Production and Consumption

Tin is produced from both primary and secondary sources. Secondary tin is produced from recycled materials (see the article "Recycling of Tin" in this Volume). Figure 1 shows the consumption of primary and secondary tin in the United States during recent years. Figure 2 shows 1988 data for the relative consumption of tin in the United States by application.

Primary Production. Tin ore generally is centered in areas far distant from centers of consumption. The leading tin-producing countries (excluding the USSR and China) are, in descending order, Brazil, Indonesia, Malaysia, Thailand, Bolivia, and Australia (1988 totals). These countries supply over 85% of total world production.

Cassiterite, a naturally occurring oxide of tin, is by far the most economically important tin mineral. The bulk of the world's tin ore is obtained from low-grade placer deposits of cassiterite derived from primary ore bodies or from veins associated with granites or rocks of granitic composition.

Primary ore deposits can contain very low percentages of tin (0.01%, for example), and thus large amounts of soil or rock must be worked to provide recoverable amounts of tin minerals. Unlike ores of other metals, cassiterite is very resistant to chemical and mechanical weathering, but extended erosion of primary lodes by air and water has resulted in deposition of the ore as eluvial and alluvial deposits.

Underground lode deposits of tin ores are worked by sinking shafts and driving adits, and the rock is broken from the working face by drilling and blasting. Cassiterite is recovered from eluvial and alluvial deposits by dredging, gravel pumping, and hydraulicking. In open-pit mining, a much less widely employed mining method, mechanical and manual methods are used to move tin-bearing materials. After ball mill concentration of the ore, a final culling is provided at dressing stations.

The final concentrates, which contain 70 to 77% tin, are then sent to the smelter, where they are mixed with anthracite and limestone. This charge is heated in a reverberatory furnace to about 1400 °C (2550 °F) to reduce the tin oxide to impure tin metal, which is again heated in huge cast iron melting pots to refine the metal. Steam or compressed air is introduced into the molten metal, and this treatment, plus addition of controlled amounts of other elements that combine with the impurities, results in tin of high purity (99.75 to 99.85%). This high-purity tin often is treated again by liquating or electrolytic refining, which provides tin with a purity level approaching 99.99%.

After the tin is refined, it is cast into ingots weighing 12 to 25 kg (26 to 56 lb) or bars in weights of 1 kg (2 lb) and upwards. Tin normally is sold by brand name, and the choice of brand is determined largely by the amounts of impurities that can be tolerated in each end product. High-purity brands of tin may contain small amounts of lead, antimony, copper, arsenic, iron, bismuth, nickel, cobalt, and silver. Total impurities in commercially pure tin rarely exceed 0.25%.

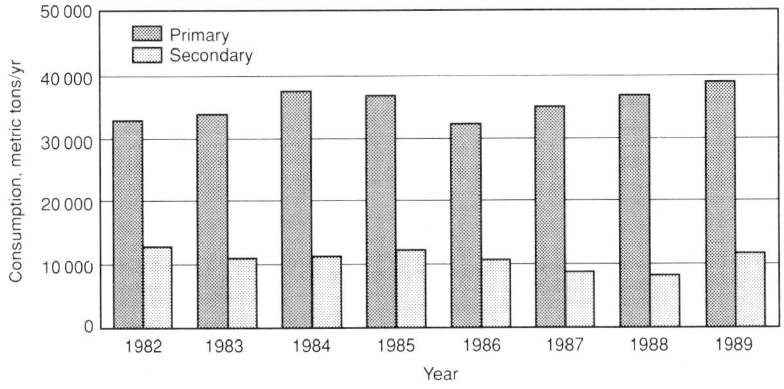

Fig. 1 U.S. consumption of primary and secondary tin in recent years

Tin in Coatings

Tinplate. The largest single application of tin worldwide is in the manufacture of tinplate (steel sheet coated with tin), which accounts for about 40% of total world tin consumption. Since 1940, the traditional hot dip method of making tinplate has been largely replaced by electrodeposition of tin on continuous strips of rolled steel. Electrolytic tinplate can be produced with either equal or unequal amounts of tin on the two surfaces of the steel base metal. Nominal coating thicknesses for equally coated tinplate range from 0.38 to 1.5 μm (15 to 60 μin.) on each surface. The thicker coating on tinplate with unequal coatings (differential tinplate) rarely exceeds 2.0 μm (80 μin.). Tinplate is produced in thicknesses from 0.15 to 0.60 mm (0.006 to 0.024 in.).

Over 90% of world production of tinplate is used for containers (tin cans). Traditional tinplate cans are made of three pieces of tin-coated steel: two ends and a body with a soldered side seam. Innovations in can manufacture have produced two-piece cans made by drawing and ironing. Tinplate cans find their most important use in the packaging of food products, beer, and soft drinks, but they are also used for holding paint,

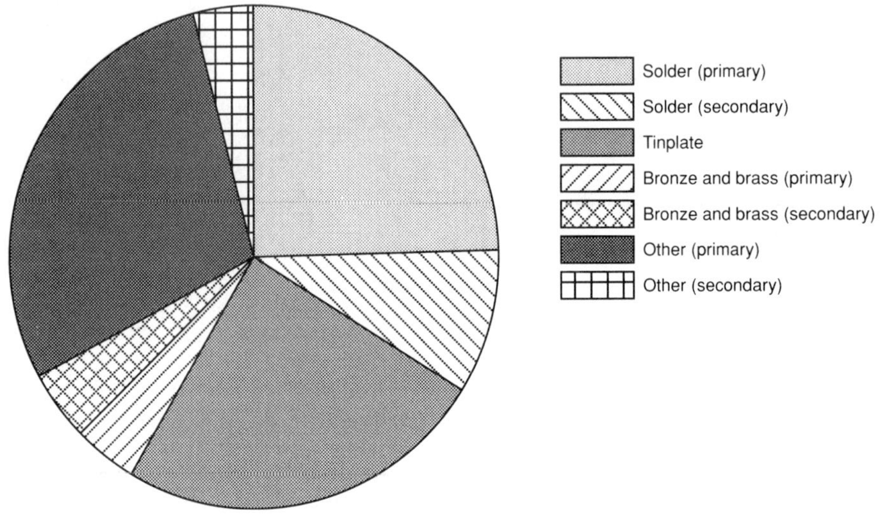

Fig. 2 Relative consumption of tin in the United States by application. 1988 data. Source: U.S. Bureau of Mines

motor oil, disinfectants, detergents, and polishes. Other applications of tinplate include signs, filters, batteries, toys, and gaskets, and containers for pharmaceuticals, cosmetics, fuels, tobacco, and numerous other commodities.

Electroplating accounts for one of the major uses of tin and tin chemicals. Tin is used in anodes, and tin chemicals are used in formulating various electrolytes and for coating a variety of substrates. Tin electroplating can be performed in either acid or alkaline solutions. Sodium or potassium stannates form the bases of alkaline tinplating electrolytes that are very efficient and capable of producing high-quality deposits. Advantages of these alkaline stannate baths are that they are not corrosive to steel and that they do not require additional agents. Acid electrotinning solutions operate at higher current densities and higher plating rates and require additions of organic compounds.

A number of alloy coatings can be electroplated from mixed stannate-cyanide baths, including coatings of tin-zinc and tin-cadmium alloys and a wide range of tin-copper alloys (bronzes). The bronzes range in tin content from 7 to 98%. Red bronze deposits contain up to 20% tin; high-tin bronzes, called speculum, usually contain about 40% tin.

Tin-nickel and tin-lead electrodeposits are plated from acid electrolytes and are important coatings for printed circuits and electronic components. Tin-cobalt plate is used in applications requiring an attractive finish and good corrosion resistance.

Two ternary alloy electrodeposits are used by industry. These are the copper-tin-lead for bearing surfaces and the copper-tin-zinc alloy for coatings in certain electronic applications.

Hot Dip Coatings. Coating steel with lead-tin alloys produces a material called terneplate (see the article "Lead and Lead Alloys" in this Volume). Terneplate is easily formed and easily soldered. It is used as a roofing and weather-sealing material and in the construction of automotive gasoline tanks, signs, radiator header tanks, brackets, chassis and covers for electronic equipment, and sheathing for cable and pipe.

Hot dip tin coatings are used both on wire for component leads and on food-handling and food-processing equipment. In addition, hot dip tin coatings are used to provide the bonding layer for the babbitting of bearing shells.

Pure Tin

Commercial tin is considered to be pure when it contains a minimum of 99.8% Sn. Of the various types of commercially pure tin, about 80 to 90% is a high-purity commercial tin known as Grade A tin as specified in ASTM B 339. According to this specification, Grade A tin must have a minimum tin purity of 99.85% Sn and maximum residual impurities of 0.04% Sb, 0.05% As, 0.030% Bi, 0.001% Cd, 0.04% Cu, 0.015% Fe, 0.05% Pb, 0.01% S, 0.005% Zn, and 0.01% (Ni + Co). Other specifications for commercially pure tin include:

- U.S. government specification QQT-371, Grade A (99.75% Sn)
- British specification BS 3252, Grade T (99.8% Sn)
- German specification DIN 1704, Grade A2 (99.75% Sn)

Table 1 summarizes selected physical, thermal, electrical, and optical properties of pure tin. Further information is contained in the article "Properties of Pure Metals" in this Volume. General applications of Grade A tin include tinplate foil, collapsible tubes, block tin products, and pewter.

Mechanical Properties. Typical tensile properties of commercially pure tin are given in Table 2. Hardness and elasticity values are given in Table 1.

Creep Characteristics. Like lead, tin is subject to creep deformation and rupture even at room temperature. Consequently, tensile strength may not be an important design criterion because creep rupture can occur at stresses even below the yield strengths in Table 2. For example, one series of tests on a commercially pure tin resulted in the following creep characteristics at room temperature:

Initial stress, MPa	Initial stress, psi	Time, days	Extension, %
1.083	157.0	551	3.5
1.351	196.0	551	7
2.256	327.1	173(a)	101
2.772	402.1	79(a)	132
3.227	468.1	21(a)	119
4.214	611.2	4.6	105
7.069	1025.2	0.5(a)	78

(a) Specimen failed

Fatigue Strength. Rotating-cantilever fatigue tests on a commercially pure tin resulted in fatigue strength levels of 2.9 MPa (430 psi) for 10^7 cycles at 15 °C (59 °F) and 2.6 MPa (380 psi) for 10^8 cycles at 100 °C (212 °F). Because creep deformation of tin occurs at room temperature, fatigue strengths may be influenced by creep-fatigue interaction and thus may depend on the frequency and/or waveform of stress cycling.

Impact Strength. Charpy V-notch tests on commercially pure tin at various temperatures resulted in the following impact strengths:

Temperature, °C	Temperature, °F	Charpy V-notch impact energy, J	Charpy V-notch impact energy, ft · lbf
−80	−112	3.7	2.75
−60	−76	11.5	8.5
−15	5	28.5	21.0
0	32	44.1	32.5
150	302	22.7	16.75
190	374	20.3	15.0
215	419	2.7	2.0

Specific Damping Capacity. Tests on bars vibrating at audio frequencies in the free-free mode produced these results:

Temperature, °C	Temperature, °F	Logarithmic decrement, Polycrystalline	Logarithmic decrement, Single crystals
25	77	0.022	0.0010
50	122	0.045	0.0013
75	167	0.060	0.0015
100	212	0.054	0.0018
125	257	0.045	0.0024
150	302	0.060	0.0032

Chemical Properties and Corrosion Behavior. Tin reacts with both strong acids and strong alkalies, but it is relatively resistant to near-neutral solutions. Oxygen

Table 1 Physical, thermal, electrical, and optical properties of commercially pure tin

Property	Value
Physical properties	
Atomic number	50
Atomic weight	118.69
Crystal structure	α phase or β phase
Density, g/cm³ (lb/in.³)	
α phase at 1 °C (33.8 °F)	5.765 (0.2083)
β phase at 20 °C (68 °F)	7.168 (0.2590)
Liquid surface tension at 400–800 °C (750–1470 °F), mN/m	700–0.17 × T + (25 + 0.015 × T)(a)
Hardness, HB	
At 20 °C (68 °F)	3.9
At 60 °C (140 °F)	3.0
At 100 °C (212 °F)	2.3
Modulus of elasticity, GPa (10^6 psi)	
Cast (coarse grain)	41.6 (6.03)
Self-annealed (fine grain)	44.3 (6.43)
Poisson's ratio	0.33
Volume change on freezing, %	2.8%
Volume change on phase transformation, %	~27%
Thermal properties	
Melting point, °C (°F)	231.9 (449.4)
Boiling point, °C (°F)	2270 (4118)
Phase transformation temperature on cooling (β phase to α phase), °C (°F)	13.2 (55.8)
Latent heat of fusion, J/g (Btu/lb)	59.5 (25.6)
Latent heat of phase transformation, J/g (Btu/lb)	17.6 (7.57)
Latent heat of vaporization, kJ/g (Btu/lb)	2.4 (1.03 × 10^3)
Specific heat, J/kg · K (Btu/lb · °F)	
α phase at 10 °C (50 °F)	205 (49 × 10^{-3})
β phase at 25 °C (77 °F)	222 (53 × 10^{-3})
Linear coefficient of thermal expansion, 10^{-6}/K	
α phase at −100 °C (−150 °F)	18.1
α phase at −50 °C (−60 °F)	19.2
β phase at 100 °C (212 °F)	23.8
β phase at 150 °C (300 °F)	26.7
Thermal conductivity, W/m · K	
β phase at 100 °C (212 °F)	60.7
β phase at 200 °C (390 °F)	56.5
Electrical properties	
Electrical conductivity (volumetric) at 20 °C (68 °F)	15.6% IACS
Electrical resistivity, μΩ · m	
At 0 °C (32 °F)	0.110
At 100 °C (212 °F)	0.155
At 200 °C (390 °F)	0.200
Optical properties (546.1 nm wavelength)	
Reflectance index	
Film, 42–200 nm thick	0.70
Bulk solid	0.80
Refractive index	
Film, 42–200 nm thick	2.4
Bulk solid	1.0
Absorptive index	
Film, 42–200 nm thick	1.9
Bulk solid	4.2

(a) T, temperature in degrees Kelvin

greatly accelerates corrosion in aqueous solutions. In general, with mineral acids the rate of attack increases with the temperature and concentration. Dilute solutions of weak alkalies have little effect on tin, but strong alkalies are corrosive even in cold dilute solutions. Salts with an acid reaction attack tin in the presence of oxidizers or air. Tin resists demineralized waters but is slightly attacked near the water line by hard tap waters. The corrosion resistance of tin in specific environments is summarized in Table 3. Additional information on the corrosion of tin is given in *Corrosion*, Volume 13 of the 9th Edition of *Metals Handbook*.

Applications of Unalloyed Tin. There are only a few applications where tin is used unalloyed with other metals. Unalloyed tin is the most practical lining material for handling high-purity water in distillation plants because it is chemically inert to pure water and will not contaminate the water in any way.

In the manufacture of plate glass, the molten glass is fed from the furnace onto the surface of a molten tin bath, which is protected from oxidation by an atmosphere that contains nitrogen and some hydrogen. The natural forces of surface tension and gravity within the bath ordinarily produce plate glass about 6 mm (¼ in.) thick, but the thickness of the glass can be varied by adjusting the speed at which the molten glass is drawn from the float bath and the temperature of the tin. With this process, glass ribbons are formed with flat and parallel surfaces. The surfaces of the glass are so smooth that surface polishing is not required.

Table 2 Tensile properties of commercially pure tin

Temperature °C	Temperature °F	Yield strength MPa	Yield strength ksi	Elongation in 25 mm (1 in.), %	Reduction in area, %
Strained at 0.2 mm/m · min (0.0002 in./in. · min)					
−200	−328	36.2	5.25	6	6
−160	−256	90.3	13.10	15	10
−120	−184	87.6	12.71	60	97
−80	−112	38.9	5.64	89	100
−40	−40	20.1	2.92	86	100
0	32	12.5	1.81	64	100
23	73	11.0	1.60	57	100
Strained at 0.4 mm/m · min (0.0004 in./in. · min)					
15	59	14.5	2.10	75	...
50	122	12.4	1.80	85	...
100	212	11.0	1.60	55	...
150	302	7.6	1.10	55	...
200	392	4.5	0.65	45	...

Note: It is uncertain if the inconsistencies among these data are due to differences in purity or the difference in straining rate.

Powder Applications. Much of the supply of tin powders is used in making sintered bronze or sintered iron parts. However, tin powders are also increasingly employed in making paste solders and creams used in the plumbing and electronic manufacturing industries. Tin and tin alloy powders find minor uses in sprayed coatings for food-handling equipment, metallizing of nonconductors, and bearing repairs. Tin particles can also be used in food can lacquers to decrease the dissolution of iron and any exposed lead-base solder by the food product.

Additions of 2% tin powder and 3% copper powder aid the sintering of iron compacts. The tin provides a low-melting-point phase, which in turn provides diffusion paths for the iron. Iron-tin-copper compacts sintered at 950 °C (1740 °F) have mechanical properties comparable to those of iron-copper powder metallurgy parts containing 7 to 10% Cu sintered at 1150 °C (2100 °F). In addition, closer control of finished dimensions is afforded by the iron-tin-copper mixture, and this control results in improved quality and cost effectiveness.

Sintered compacts made from mixtures of iron and tin-lead solder powders are suitable for certain low-stress engineering applications. Warm compressing of these compacts (at ~450 °C, or 840 °F) provides cohesion of the iron solder mixtures but

Table 3 Resistance of tin to specific corroding agents

Corrosive agent	Resistance	Remarks
Acid, acetic	Slight attack	Increased by air
Acid, butyric	Resistant	...
Acid, citric	Moderate attack	At water line
Acids, fatty	Moderate attack	...
Acid, hydrochloric	Severe attack	In presence of air
Acid, hydrofluoric	Severe attack	In presence of air
Acid, lactic	Moderate attack	Increased by air
Acid, nitric	Severe attack	...
Acid, oxalic	Moderate attack	(a)
Acid, phosphoric	Resistant	...
Acid, salts	Severe attack	Air present
Acid, sulfuric	Severe attack	(b)
Acid, tartaric	Slight attack	...
Air	Resistant	...
Ammonia	Resistant	...
Bromine	Severe attack	...
Carbon tetrachloride	Resistant	...
Chlorine	Severe attack	...
Iodine	Severe attack	...
Milk	Resistant	...
Motor fuel	Resistant	...
Petroleum products	Resistant	...
Potassium hydroxide	Severe attack	Increased by air
Sodium carbonate	Slight attack	...
Sodium hydroxide	Severe attack	Increased by air
Water, distilled	Resistant	...
Water, sea	Slight attack	...

(a) Most corrosive of common organic acids. (b) Increased with concentration and in the presence of air

does not recrystallize the iron powder; therefore, any work hardening obtained during compaction is retained. Different properties can be obtained in the pressed-and-sintered compacts by varying the pressing conditions and the relative amounts of the iron and solder powders.

Tin in Chemicals

The manufacture of inorganic and organic chemicals containing tin constitutes one of the major uses of metallic tin. The use of tin compounds has grown so rapidly over the past quarter century that the tin chemicals industry has been transformed from one based mainly on recovered secondary tin to one that consumes significant amounts of primary ingot tin.

Tin chemicals are used for such widely diversified applications as electrolyte solutions for depositing tin and its alloys; pigments and opacifiers for ceramics and glazes; catalysts and stabilizers for plastics; pesticides, fungicides, and antifouling agents in agricultural products, paints, and adhesives; and corrosion-inhibiting additives for lubricating oils.

Solders

Solders account for the largest use of tin in the United States (Fig. 2). Tin is an important constituent in solders because it wets and adheres to many common base metals at temperatures considerably below their melting points. Tin is alloyed with lead to produce solders with melting points lower than those of either tin or lead (see the article "Lead and Lead Alloys" in this Volume). Small amounts of various metals, notably antimony and silver, are added to tin-lead solders to increase their strength. These solders can be used for joints subjected to high or even subzero service temperatures.

Solder compositions and the applications of joining by soldering are many and varied (Table 4). Commercially pure tin is used for soldering side seams of cans for special food products and aerosol sprays. The electronics and electrical industries employ solders containing 40 to 70% Sn that provide strong and reliable joints under a variety of environmental conditions. High-tin solders are used for joining parts of electrical apparatuses because their electrical conductivity is higher than that of high-lead solders. High-tin solders are also used where lead may be a hazard, for example, in contact with foodstuffs or in potable-water plumbing applications.

General-purpose solders (50Sn-50Pb and 40Sn-60Pb) are used for light engineering applications, plumbing, and sheet metal work. Lower-tin solders (20 to 35% Sn, balance Pb) are used in joining cable and in the production of automobile radiators and heat exchangers. Some solders are used to fill crevices at seams and welds in automotive bodies, thereby providing smooth joints and contours.

Tin-zinc solders are used to join aluminum. Tin-antimony and tin-silver solders are employed in applications requiring joints with high creep resistance, and in applications requiring a lead-free solder composition, such as potable-water plumbing. Also, tin solders that contain 5% Sb (or 5% Ag) are suitable for use at higher temperatures than are the tin-lead solders. Further information on solders is provided in Ref 1 and in Volume 6 of the 9th Edition of *Metals Handbook*.

Impurities in solders can affect wetting properties, flow within the joint, melting temperature of the solder, strength capabilities of joints, and oxidation characteristics of the solder alloys. The most common impurity elements and their principal levels and effects are discussed below.

Aluminum. Traces of aluminum in a tin-lead solder bath can seriously affect soldering qualities. More than 0.005% Al can cause grittiness, lack of adhesion, and surface oxidation of the solder alloy. A deterioration in the surface brightness of a molten bath sometimes is an indication of the presence of aluminum.

Antimony is slightly detrimental to wetting properties, but it can be used as an intentional addition for strengthening. As an impurity, antimony tends to reduce the effective spread of a solder alloy. High-lead solder specifications usually require a maximum limit of 0.5% Sb. The general rule is that antimony should not exceed 6% of the tin content, although in some applications this rule can be invalid. In various high-lead solders (such as Sn40B, Sn30B, Sn35B, Sn25B, and Sn20B in ASTM B 32), the presence of antimony is used to ensure that a transformation from β tin to α tin does not take place. Such a transformation would result in a volume change and a drastic loss in solder strength.

Arsenic. A progressive deterioration in the quality of the solder is observed with increases in arsenic content. As little as 0.005% As induces some dewetting, and dewetting becomes more severe as the percentage of arsenic is increased to 0.02%. Arsenic levels should be kept within this range. At the maximum allowable level of 0.03%, arsenic can cause dewetting problems when soldering brass.

Bismuth. Low levels of bismuth in the solder alloy generally do not cause any difficulties, although some discoloration of soldered surfaces occurs at levels above 0.5%.

Cadmium. A progressive decrease in wetting capability occurs with additions of cadmium to tin-lead solders. While there is no significant change in the molten appearance, small amounts of cadmium can increase the risk of bridging and icicle formation in printed circuits. For this reason, and for health reasons, cadmium levels should be kept to a minimum.

Copper. Although copper levels above about 0.25% can cause grittiness of solder, for the most part, the role of copper as a solder contaminant appears to be variable and related to the particular product. A molten tin-lead solder bath is capable of dissolving copper at a high rate, and the level of copper in the bath can easily reach 0.3%. Copper in liquid solder does not appear to have any deleterious effect upon the wetting rate or joint formation. Excess copper settles to the bottom of a solder bath as an intermetallic compound sludge. New solder alloy allows a maximum copper content of 0.08%.

Iron and nickel are not naturally present in solder alloy. The presence of iron-tin compounds in tin-lead solders can be identified as a grittiness. Generally, iron is limited to a maximum of 0.02% in new solder. There are no specification limits for nickel, but levels as low as 0.02% can produce some reduction in wetting characteristics. Iron levels above about 0.1% cause grittiness of solder.

Phosphorus and Sulfur. Phosphorus at a level of 0.01% is capable of producing dewetting and some grittiness. At higher levels, surface oxidation occurs, and some

Table 4 Applications, specifications, and nominal compositions of selected tin-base solder materials

Common name	Specifications				Nominal composition, %	Liquidus temperature		Solidus temperature		Typical applications
	ASTM	Government	British	German		°C	°F	°C	°F	
Commercially pure tin	B 339, Grade A	QQ-T-371, Grade A	BS 3252, Grade T	DIN 1704, Grade A2	(a)	...	...	...	...	Soldering sideseams of cans for foods or aerosols
Antimonial-tin solder	B 32, Grade S65	...	...	...	95 Sn, 5 Sb	240	464	234	452	Soldering of electrical equipment, joints in copper tubing, and cooling coils for refrigerators. Resistant to SO_2
Tin-silver solder	B 32, Grade Sn95	...	...	...	95 Sn, 5 Ag	245	473	221	430	Soldering of components for electrical and high-temperature service
Tin-silver eutectic alloy	B 32, Grade Sn96	QQ-S-571, Grade Sn96	...	...	96 Sn, 3.5 Ag	221	430	221	430	Popular choice with properties similar to those of ASTM B 32, Grade Sn95
Soft solder (70–30 solder)	B 32, Grade Sn70	QQ-S-571, Grade Sn70	...	...	70 Sn, 30 Pb	192	378	183	361	Joining and coating of metals
Eutectic solder (63–37 soft solder)	B 32, Grade Sn63	QQ-S-571, Grade Sn63	...	DIN 1707, LSn 63Pb	63 Sn, 37 Pb	183	361	183	361	Lowest-melting (eutectic) solder for electronics
Soft solder (60–40 solder)	B 32, Grade Sn60	QQ-S-571, Grade Sn60	BS 219, Grade K	DIN 1707, LSn 60Pb(Sb)	60 Sn, 40 Pb	190	374	183	361	Solder for electronic and electrical work, especially mass soldering of printed circuits

(a) See the section "Pure Tin" in this article for minimum tin contents.

identifiable problems such as grittiness and dewetting become readily discernible. Sulfur causes grittiness in solders at a very low level and should be held to 0.001%. Discrete particles of tin-sulfide can be formed. Both of these elements are detrimental to good soldering.

Zinc. The ASTM new solder alloy specification states that zinc content must be kept to a maximum of 0.005% in tin-lead solders. At this maximum limit, even with new solders in a molten bath, some surface oxidation can be observed, and oxide skins may form, encouraging icicles and bridging. Up to 0.01% Zn has been identified as the cause of dewetting on copper surfaces. Excessive zinc causes oxidation of solder to be more noticeable.

The combined effects of the above impurity elements can be significant. Excessive contamination in solder baths or dip pots generally can be identified through surface oxidation, changes in the product quality, and the appearance of grittiness or frostiness in joints made in this bath. A general sluggishness of the solder also may indicate excessive impurities. In addition to analysis, experience with solder bath operation is helpful in determining the point at which the material should be renewed for good solder joint production. The ASTM solder specifications, which specify maximum allowable impurity concentrations, are useful when purchasing solder for general use (Table 5). In particular applications, specific contaminants or a combination of elements may be detrimental to a particular soldered product. On occasion, determining a revised or limited specification for solder materials is required.

Impurities of a metallic and nonmetallic nature can be found in raw materials and in the scrap solder that is sometimes used by reclaimers. Reclaimed solder is used in many industrial applications where impurities may not be detrimental. However, correct selection of solder grade is important for economical production. Manufacturing problems can result from inappropriate solder selection, from the use of solder baths for longer periods than contamination buildup will tolerate, or from processing methods that rapidly contaminate a solder bath. Determination of suitable specifications, of allowable impurities in new materials, and of allowable impurities in the solder bath through its deterioration to the point at which it is discarded should be included in any soldering quality control program.

Electrical and mechanical property data for selected tin-base solders are given in Table 6. The effects of elevated temperatures on the tensile strength and elongation of 60-40 solder are listed in Table 7.

When measuring the tensile properties of bulk solder, the results depend greatly on

Table 5 Impurity limits in ASTM specifications for the tin-base solders listed in Table 4

Common name	Nominal composition, %	Impurity limits, %(a)											
		Sb	Ag	Al	As	Bi	Cd	Cu	Fe	Pb	S	Zn	Other
Commercially pure tin (ASTM B 339, Grade A)	99.85 Sn min	0.04	...	...	0.05	0.015	0.001	0.04	0.015	0.05	0.01	0.005	(b)
Antimonial-tin solder	95 Sn, 5 Sb	4.5–5.5	0.015	0.005	0.05	0.15	0.03	0.08	0.04	0.2	...	0.005	...
Tin-silver solder	95 Sn, 5 Ag	0.12	4.4–4.8	0.005	0.01	0.15	0.005	0.08	0.02	0.10	...	0.005	...
Tin-silver eutectic alloy	96 Sn, 3.5 Ag	0.12	3.4–3.8	0.005	0.01	0.15	0.005	0.08	0.02	0.10	...	0.005	...
70–30 solder	70 Sn, 30 Pb	0.50	0.015	0.005	0.03	0.25	0.001	0.08	0.02	30 nom	...	0.005	...
Eutectic solder (63–37 solder)	63 Sn, 37 Pb	0.50	0.015	0.005	0.03	0.25	0.001	0.08	0.02	37 nom	...	0.005	...
60–40 solder	60 Sn, 40 Pb	0.50	0.015	0.005	0.03	0.25	0.001	0.08	0.02	40 nom	...	0.005	...

(a) Maximum unless a range or nominal (nom) is specified. (b) Ni + Co, 0.01% max

Table 6 Electrical and mechanical properties of selected tin-base solders

Antimonial-tin solder (95Sn–5Sb)

Tensile properties. Cast: typical tensile strength, 40.7 MPa (5.9 ksi); elongation in 100 mm (4 in.), 38%. Soldered copper joint: typical tensile strength, 97.9 MPa (14.2 ksi)
Shear strength. Cast, 41.4 MPa (6.0 ksi). Soldered copper joint, 76.5 MPa (11.1 ksi)
Impact strength. Cast (Izod test), 27 J (20 ft · lbf)
Electrical conductivity. Volumetric, 11.9% IACS at 20 °C (68 °F)
Electrical resistivity. 145 nΩ · m at 25 °C (77 °F)

Tin-silver solder (95Sn–5Ag)

Tensile properties. Sheet, 1.02 mm (0.040 in.) thick, aged 14 days at room temperature: typical tensile strength, 31.7 MPa (4.6 ksi); yield strength, 24.8 MPa (3.6 ksi); elongation in 50 mm (2 in.), 49%. Soldered copper joint: typical tensile strength, 96.5 MPa (14 ksi)
Shear strength. Soldered copper joint, 73.1 MPa (10.6 ksi)
Electrical conductivity. Volumetric, 16.6% IACS at 20 °C (68 °F)
Electrical resistivity. 104 nΩ · m at 0 °C (32 °F)
Temperature coefficient of electrical resistivity. 0–100 °C (32–212 °F), 42.3 pΩ · m/K

70–30 soft solder (70Sn–30Pb)

Tensile properties. Cast: typical tensile strength, 46.9 MPa (6.8 ksi)
Hardness. 12 HB
Electrical conductivity. Volumetric, 11.8% IACS
Electrical resistivity. 146 nΩ · m

Eutectic solder (63Sn–37Pb)

Tensile properties. Cast: typical tensile strength, 51.7 MPa (7.5 ksi); elongation in 100 mm (4 in.), 32%. Soldered copper joint: typical tensile strength, 200 MPa (2g ksi)
Shear strength. Cast, 42.7 MPa (6.2 ksi); soldered copper joint, 55.2 MPa (8 ksi)
Hardness. Cast, 14 HB
Impact strength. Cast (Izod test), 20 J (15 ft · lbf)
Creep characteristics. Minimum creep rate: at room temperature and 2.3 MPa (335 psi), 0.1 mm/m (100 μin./in.) per day; at 80 °C (176 °F) and 467 MPa (68 psi), 0.1 mm/m (100 μin./in.) per day
Dynamic viscosity. 1.33 mPa · s (0.0133 poise) at 280 °C (536 °F)
Liquid surface tension. 0.490 N/m at 280 °C (536 °F)
Electrical conductivity. Volumetric, 11.9% IACS
Electrical resistivity. 145 nΩ · m

60–40 soft solder (60Sn–40Pb)

Tensile properties. Bulk solder at room temperature (measurements depend greatly on conditions of casting and testing): mean tensile strength, 52.5 MPa (7.61 ksi); elongation, 30–60%.
Shear strength. Mean, 37.1 MPa (5.38 ksi) (depends greatly on conditions of casting and testing)
Hardness. 16 HV (depends on casting conditions)
Elastic modulus. Tension (bulk solder), 30.0 GPa (4.35 × 10^6 psi)
Creep-rupture characteristics. Limiting creep stress, 2.2–3.0 MPa (320–430 psi) for a strain rate of 10^{-4} m/m per day at room temperature. Rupture life: 1000 h under stress of 4.5 MPa (650 psi) at 26 °C (79 °F); 1000 h under stress of 1.4 MPa (200 psi) at 80 °C (176 °F)
Dynamic liquid viscosity. Estimated, 2.0 mPa · s (0.020 poise) at the liquidus temperature
Liquid surface tension. Estimated: 468 mN/m at 330 °C (626 °F), 461 mN/m at 430 °C (806 °F)
Electrical conductivity. Volumetric, 11.5% IACS
Electrical resistivity. 149.9 nΩ · m
Thermoelectric potential. Same as pure tin when measured against copper
Temperature of superconductivity. 7.05 K. Critical field, 83.2 mT at 1.3 K

the casting and testing conditions. For example, eutectic and near-eutectic 60-40 solder compositions were examined for superplasticity. It was found that the strain rate sensitivity *m* has a value of about 0.4 at a strain rate of 10^{-4} m/m · s, increasing to a relative maximum of about 0.5 at a strain rate of 10^{-3} m/m · s, then decreasing to a value near 0.2 at a strain rate of 10^{-1} m/m · s.

Thermal Properties. Solidus and liquidus points of various solder compositions are given in Table 4 and in the article "Lead and Lead Alloys" in this Volume. Other thermal properties include:

Solder	Linear thermal expansion at 15–110 °C (60–230 °F), 10^{-6}/K	Thermal conductivity at 0–180 °C (32–355 °F), W/m · K
70Sn–30Pb	21.6	...
63Sn–37Pb	24.7	50
60Sn–40Pb	24	...

The 60-40 solder has a specific heat of 176 J/kg · K (0.042 Btu/lb · °F) and an estimated heat of fusion of 37 J/g (16 Btu/lb).

Pewter

Pewter is a tin-base white metal containing antimony and copper. Originally, pewter was defined as an alloy of tin and lead, but to avoid toxicity and dullness of finish, lead is excluded from modern pewter. These modern compositions contain 1 to 8% Sb and 0.25 to 3.0% Cu. Pewter casting alloys usually are lower in copper than pewters used for spinning hollowware and thus have greater fluidity at casting temperatures.

Modern pewter consists of a cored solid solution of antimony in tin within which are distributed fine crystals of η (Cu_6Sn_5) phase. Pewter is malleable and ductile, and it is easily spun or formed into intricate designs and shapes. Pewter parts do not require annealing during fabrication. Much of the costume jewelry produced today is made of pewter alloys centrifugally cast in rubber or silicone molds. Typical pewter products include coffee and tea services, trays, steins, mugs, candy dishes, jewelry, bowls, plates, vases, candlesticks, compotes, decanters, and cordial cups.

Chemical Composition. Although a wide range of compositions has been called pewter, the usual modern alloys contain 90 to 95% Sn and 1 to 3% Cu, with the balance consisting of antimony. Some pewterlike materials are sand cast or spun aluminum alloys, which are traditionally not considered to be pewter. Although some pewter contains lead as an alloying constituent, a considerable portion of lead is undesirable for applications in which the material may be in contact with food or beverages. In addition, lead may impart a dullness to the ware.

Composition limits of modern pewter are shown in Table 8.

Physical Properties. Typical tensile properties and hardnesses of pewter are given in Table 9. The effect of processing variables on the mechanical properties of pewter is covered in Table 10. In addition to those properties given in Table 9, pewter has:

- *An elastic modulus* of 53 GPa (7.7 × 10^6 psi)
- *A density* of 7.28 g/cm³ (0.263 lb/in.³)
- *A liquidus temperature* of 295 °C (563 °F)
- *A solidus temperature* of 244 °C (471 °F)

Chemical Properties and Corrosion Resistance. Pewter tarnishes in soft water, with the production of a visible film of interference-tint thickness. It does not tarnish in hard water, but localized attack can occur at the water line and sometimes elsewhere if a chalky deposit is formed from the water. Pewter is attacked by dilute hydrochloric and citric acids in the presence of air.

Fabrication Characteristics. Pewter has good solderability. Casting temperatures of pewter range from 315 to 330 °C (600 to 625 °F).

Pewter can be formed by rolling, hammering, spinning, or drawing. The earing of pewter sheet can be reduced by an intermediate cross-rolling operation or heat treatment; rolling can then be continued down to final thickness.

Bearing Alloys

The primary consideration in the selection of a bearing alloy is that the material

Table 7 Effect of temperature on properties of 60–40 solder cast at 300 °C (570 °F) in steel molds (specimens not machined)

Temperature °C	Temperature °F	Tensile strength MPa	Tensile strength ksi	Elongation, %
Cast in 150 °C (300 °F) molds				
19	66	56.4	8.18	60(a)
50	122	45.4	6.58	80(a)
75	167	41.7	6.05	90(a)
100	212	30.9	4.48	110(a)
125	257	19.3	2.80	180(a)
150	302	12.4	1.80	180(a)
Cast in 200 °C (390 °F) molds				
0	32	59	8.6	50(b)
−40	−40	76	11.0	50(b)
−80	−112	97	14.1	55(b)
−120	−184	119	17.3	30(b)
−160	−256	112	16.2	10(b)
−200	−328	109	15.8	5(b)

(a) In 22.5 mm (0.89 in.). (b) In 25.4 mm (1.00 in.)

Table 8 Chemical composition limits for modern pewter

Specification	Composition, % Sn	Sb	Cu	Pb max	As max	Fe max	Zn max	Cd max
ASTM B 560								
Type 1(a)	90–93	6–8	0.25–2.0	0.05	0.05	0.015	0.005	. . .
Type 2(b)	90–93	5–7.5	1.5–3.0	0.05	0.05	0.015	0.005	. . .
Type 3(c)	95–98	1.0–3.0	1.0–2.0	0.05	0.05	0.015	0.005	. . .
BS 5140	bal	5–7	1.0–2.5	0.5	. . .	. . .	. . .	0.05
		3–5	1.0–2.5	0.5	. . .	. . .	. . .	0.05
DIN 17810	bal	1–3	1–2	0.5	. . .	. . .	. . .	. . .
		3.1–7.0	1–2	0.5	. . .	. . .	. . .	. . .

(a) Casting alloy, nominal composition 92Sn-7.5Sb-0.5Cu. (b) Sheet alloy, nominal composition 91Sn-7Sb-2Cu. (c) Special-purpose alloy

Table 9 Typical mechanical properties of pewter

Form and condition	Section thickness, mm	Section thickness, in.	Tensile strength, MPa	Tensile strength, ksi	Elongation in 50 mm (2 in.), %	Hardness, HB
Chill cast(a)	19.05	0.750	. . .	. . .	. . .	23.8
Sheet, annealed 1 h at 205 °C (400 °F), air cooled	6.12	0.241	59	8.6	40	9.5
Sheet, cold rolled, 32% reduction	6.12	0.241	52	7.6	50	8.0

(a) Modulus of elasticity, 53 GPa (7.7 × 10^6 psi)

must have a low coefficient of friction. Bearing alloys also must maintain a balance between softness and strength. Aluminum-tin bearing alloys, for example, provide an excellent compromise between the requirement for high fatigue strength and the need for good surface properties such as softness, seizure resistance, and embeddability. Tin-base bearing alloys are specified in ASTM B 23, AMS 4800, and U.S. Government specification QQ-M-161.

Compositions. Table 11 lists the chemical compositions of various tin-base bearing alloys specified in ASTM and SAE standards. Tin has a low coefficient of friction and thus meets the primary requirement of a bearing material. Tin is structurally a weak metal; therefore, when it is used in bearing applications it is alloyed with copper and antimony for increased hardness, tensile strength, and fatigue resistance. Normally, the quantity of lead in these alloys, called tin-base babbitts, is limited to 0.35 to 0.5% to avoid formation of the tin-lead eutectic, which would significantly reduce strength properties at operating temperatures.

The presence of zinc in tin-base bearing metals generally is not favored. Arsenic increases resistance to deformation at all temperatures; zinc has a similar effect at 38 °C (100 °F), but causes little or no change at room temperature. Zinc has a marked effect on the microstructures of some of these alloys. Small quantities of aluminum (even less than 1%) will modify their microstructures. Bismuth is objectionable because, in combination with tin, it forms a eutectic that melts at 137 °C (279 °F). At temperatures above this eutectic, alloy strength is appreciably decreased.

In high-tin alloys, such as ASTM grades 1, 2, and 3, and SAE 11 and 12, lead content is limited to 0.50% or less because of the deleterious effect of higher percentages on the strength of these alloys at temperatures of 150 °C (300 °F) and above. Lead and tin form a eutectic that melts at 183 °C (361 °F). At higher temperatures, bearings become fragile as a result of the formation of a liquid phase within them.

Lead-base bearing alloys, called lead-base babbitts, contain up to 10% Sn and 12 to 18% Sb. In general, these alloys are inferior in strength to tin-base babbitts, and this must be equated with their lower cost. Segregation of the constituents of these alloys may provide some difficulties during

Table 10 Effect of processing variables on the mechanical properties of pewter sheet and on the amount of earing during drawing

Properties are mean values of three determinations each on 1 mm (0.04 in.) thick sheets of Sn-6Sb-2Cu alloy that were cold rolled from 25 mm (1.00 in.) thick cast slabs.

Processing	Delay between processing and testing	Tensile strength at angle to rolling direction of 0°, MPa	0°, ksi	55°, MPa	55°, ksi	90°, MPa	90°, ksi	Elongation, % at angle to rolling direction of 0°	55°	90°	Hardness, HV	Earing, %
Cross rolling from intermediate thickness	12 months	64	9.3	62	9.0	64	9.3	56	49	53	15	10
	24 h	48	7.0	48	7.0	50	7.3	92	136	122	13	4½
Unidirectional rolling, with heat treatment(a) at intermediate thickness	24 h	68	9.9	69	10.0	73	10.6	47	36	17	20	2½

(a) About 150–200 °C (302–392 °F). Source: Ref 2

Table 11 Compositions of tin-base bearing alloys

Designation	Nominal composition, % Sn(a)	Sb	Pb max(b)	Cu	Fe max	As max	Bi max	Zn max	Al max	Total other max
ASTM B 23 alloys										
Alloy 1	91.0	4.5	0.35	4.5	0.08	0.10	0.08	0.005	0.005	0.05 Cd(c)
Alloy 2	89.0	7.5	0.35	3.5	0.08	0.10	0.08	0.005	0.005	0.05 Cd(c)
Alloy 3	84.0	8.0	0.35	8.0	0.08	0.10	0.08	0.005	0.005	0.05 Cd(c)
Alloy 11	87.5	6.8	0.50	5.8	0.08	0.10	0.08	0.005	0.005	0.05 Cd(c)
SAE alloys										
SAE 11	86.0	6.0–7.5	0.50	5.0–6.5	0.08	0.10	0.08	0.005	0.005	0.20
SAE 12	88.0	7.0–8.0	0.50	3.0–4.0	0.08	0.10	0.08	0.005	0.005	0.20
Intermediate lead-tin alloys										
Lead-tin babbitt	75	12	9.3–10.7	3	0.08	0.15	. . .	. . .	. . .	. . .
ASTM B 102, Alloy PY1815A	65	15	17–19	2	0.08	0.15	. . .	0.01	0.01	. . .

(a) Desired minimum in ASTM alloys; specified minimum in SAE alloys. (b) Maximum unless a range is specified. (c) Total named elements, 99.80%

centrifugal casting of linings. During casting, careful selection of rotational speed in relation to bearing size is necessary. Additions of cerium, arsenic, or nickel also assist in controlling segregation of these alloys. Lead-base babbitt alloys are discussed in more detail in the article "Lead and Lead Alloys" in this Volume.

Intermediate Lead-Tin Babbitt Alloys. In addition to the tin-base and lead-base babbitts, there is a series of intermediate lead-tin bearing alloys. These alloys have tin and lead contents between 20 and 65%; in addition, they contain various amounts of antimony and copper. Increasing the tin content of these alloys provides higher hardness and greater ease of casting. These alloys are less prone to segregation during melting than lead-base babbitts. Cast intermediate bearing alloys, however, exhibit lower strength values than either tin-base or lead-base babbitts.

Aluminum-tin bearing alloys represent an excellent compromise between the requirement for high fatigue strength and the need for good surface properties such as softness, seizure resistance, and embeddability. Aluminum-tin bearing alloys are usually employed in conjunction with hardened-steel or ductile-iron crankshafts, and they allow significantly higher loading than tin- or lead-base bearing alloys.

Low-tin aluminum-base alloys (5 to 7% Sn) containing small amounts of strengthening elements, such as copper and nickel, are often used for connecting-rod and thrust bearings in high-duty engines. Strict dimensional tolerances must be adhered to, and oil contamination should be avoided. Alloys containing 20 to 40% Sn and a balance of aluminum show excellent resistance to corrosion by products of oil breakdown; they also exhibit good embeddability, particularly in dusty environments. The higher-tin alloys have adequate strength and better surface properties, which make them useful for crosshead bearings in high-power marine diesel engines.

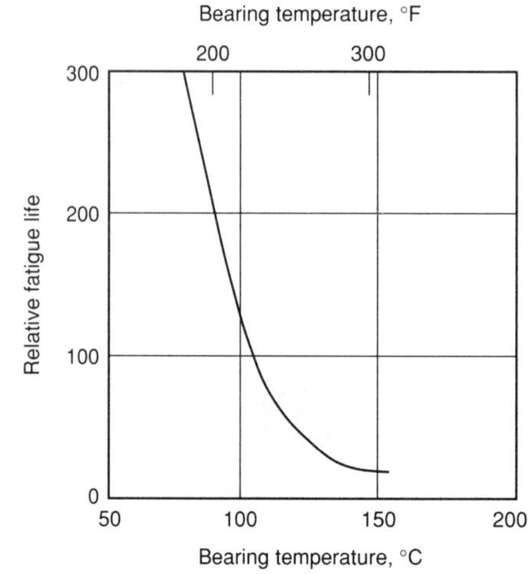

Fig. 3 Variation of bearing life with temperature for SAE 12 bimetal bearings. Thickness of alloy lining, 0.05 to 0.13 mm (0.002 to 0.005 in.); bearing load, 14 MPa (2000 psi)

Properties of Tin-Base Bearing Alloys. The mechanical properties of selected tin-base bearing alloys are shown in Tables 12 and 13. The mechanical-property values obtained from massive cast specimens are dependent on temperature. Also, hardness and compression tests are sensitive to the duration of the load because of the plastic nature of these materials. Bulk properties may be of some value in initial screening of materials, but they do not accurately predict the behavior that the material will exhibit when it is in the form of a thin layer bonded to a strong backing, which is the manner in which the babbitts are normally used. The relationship that exists between bearing life and the thickness of the babbitt is shown in Fig. 3, which also shows the marked influence of operating temperature.

Compared with other bearing materials, tin alloys have low resistance to fatigue, but their strength is sufficient to warrant their use under low-load conditions. These alloys are easy to bond and handle, and they have excellent antiseizure qualities. In addition, they are much more resistant to corrosion than lead-base bearing alloys.

Microstructures. Tin-base bearing alloys vary in microstructure in accordance with their composition. Alloys that contain about 0.5 to 8% Cu and less than about 8% Sb are characterized by a solid-solution matrix in which needles of a copper-rich constituent and fine, rounded particles of precipitated SbSn are distributed. The proportion of the copper-rich constituent increases with copper content. SAE 12 (ASTM Grade 2) has a structure of this type in which the needles often assume a characteristic hexagonal starlike pattern. Alloys that contain about 0.5 to 8% Cu and more than about 8% Sb exhibit primary cuboids of SbSn and needles of the copper-rich constituent in the solid-solution matrix. In alloys with about 8% Sb and about

Table 12 Physical properties and compressive strengths of selected tin-base bearing alloys

Designation	Specific gravity	Compressive yield strength(a)(b) At 20 °C (68 °F) MPa	ksi	At 100 °C (212 °F) MPa	ksi	Compressive ultimate strength(a)(c) At 20 °C (68 °F) MPa	ksi	At 100 °C (212 °F) MPa	ksi	Hardness, HB(d) At 20 °C (68 °F)	At 100 °C (212 °F)	Solidus temperature °C	°F	Liquidus temperature °C	°F	Pouring temperature °C	°F
ASTM B 23, Alloy 1	7.34	30.3	4.40	18.3	2.65	88.6	12.85	47.9	6.95	17.0	8.0	223	433	371	700	440	825
ASTM B 23, Alloy 2	7.39	42.1	6.10	20.7	3.00	102.7	14.90	60.0	8.70	24.5	12.0	241	466	354	669	425	795
ASTM B 23, Alloy 3	7.46	45.5	6.60	21.7	3.15	121.3	17.60	68.3	9.90	27.0	14.5	240	464	422	792	490	915
Lead-tin babbitt from Table 11	7.53	38.3	5.55	14.8	2.15	111.4	16.15	47.6	6.9	24(e)	12	184	363	306	583	...	...
ASTM B 102, Alloy PY1815A (die cast)	7.75	34	5	14	2.1	103	15	46	6.7	23	10	181	358	296	565	...	...

(a) The compression test specimens were cylinders 38 mm (1½ in.) long and 13 mm (½ in.) in diameter, machined from chill castings 50 mm (2 in.) long and 20 mm (¾ in.) in diameter. (b) Values for yield point were taken from stress-strain curves at a deformation of 0.125% reduction of gage length. (c) Values for ultimate strength were taken as the unit load necessary to produce a deformation of 25% of the length of the specimen. (d) Tests were made on the bottom face of parallel machined specimens cast at room temperature in a steel mold 50 mm (2 in.) in diameter by 16 mm (⅝ in.) deep. The Brinell hardness values listed are the averages of three impressions on each alloy, using a 10 mm ball and applying a 500 kg load for 30 s. (e) Chill cast hardness of 27 HB

Table 13 Mechanical properties of selected tin-base babbitt alloys

See Table 12 for compressive strengths.

ASTM B 23 alloy	Condition	Typical tensile strength MPa	ksi	Elongation, %	Elastic modulus GPa	10^6 psi	Izod impact strength J	ft · lbf	Fatigue strength MPa	ksi
Alloy 1	Chill cast	64	9.3	2(a)	50	7.3	3.4(b)	2.5(b)	26(c)	3.8(c)
	Die cast	62	9	2(a)	...	...	...	...	...	...
Alloy 2	Chill cast	77(d)	11.2(d)	18(e)	...	...	...	...	33(c)	4.8(c)
	Die cast	87(f)	12.6(f)	...	52(g)	7.6(g)	...	...	...	...
Alloy 3	Die cast	69	10	1(a)	...	...	...	...	...	...

(a) Elongation in 50 mm (2 in.). (b) Izod impact energy of 0.9 J (0.7 ft · lbf) at 200 °C (390 °F). (c) Fatigue strength for 2×10^7 cycles, R.R. Moore-type test. (d) Tensile strength of 45 MPa (6.5 ksi) at 100 °C (212 °F) and 20 MPa (2.9 ksi) at 175 °C (345 °F). (e) Gage length equals $4\sqrt{\text{area}}$. (f) Cast from 315 °C (600 °F) into mold at 150 °C (300 °F). (g) Cast from 400 °C (750 °F) into a mold at 100 °C (212 °F)

Table 14 Mechanical properties of white metal

Form and condition	Section size mm	in.	Tensile strength MPa	ksi	Elongation, %
Chill cast, tested 2 months after casting(a)	50×13	2 × ½	50	7.2	...
Chill cast, annealed at 225 °C (437 °F)(b)	50×13	2 × ½	45	6.5	50
Cast(c)	...	...	...	...	...
Annealed sheet	2.5	0.1	46	6.7	70(d)
Sheet, quenched from 220 °C (428 °F)	2.5	0.1	51	7.4	28(d)
Sheet, aged 150 °C (302 °F)	2.5	0.1	61	8.8	28(d)
Wire, extruded	3.5	0.14	59	8.5	63
Wire, extruded and annealed 24 h at 225 °C (437 °F)	3.5	0.14	54	7.8	10

(a) Brinell hardness, 20. (b) Brinell hardness, 17. (c) Izod impact value, 30 J (22 ft · lbf); shear strength, 46 MPa (6.7 ksi). (d) In 50 mm (2 in.)

Table 15 Creep-rupture characteristics of white metal

Tests conducted at room temperature (9 to 27 °C, or 48 to 81 °F) on rolled material 2.5 mm (0.1 in.) thick

Stress MPa	ksi	Time to fracture, days	Final extension, %
9.7	1.4	19	66
8.3	1.2	54	54
7.6	1.1	71	37
6.9	1.0	155	42
6.2	0.9	198	49
5.5	0.8	360	98
4.1	0.6	339(a)	4.12(a)
2.8	0.4	337(a)	1.06(a)

(a) Specimen did not fracture.

0.5 to 8% Cu, rapid cooling suppresses formation of the SbSn cuboids; this is particularly true of alloys containing lower percentages of copper.

Other Tin-Base Alloys

Alloys for Organ Pipes. Tin-lead alloys are used in the manufacture of organ pipes. These materials are commonly called spotted metal because they develop large nucleated crystals, or spots, when solidified as strip on casting tables. The pipes that produce the diapason tones of organs generally are made of alloys with tin contents varying from 20 to 90% according to the tone required. Broad tones generally are produced by alloys rich in lead; as tin content increases, the tone becomes brighter. Cold-rolled tin-copper-antimony alloys (95% Sn) also have been used successfully in the manufacture of pipes, and the adoption of these alloys has improved the efficiency and speed of fabrication of finished pipes. This composition provides for a bright appearance that is more tarnish resistant than the tin-lead alloys.

Type metals are cast alloys containing various proportions of lead, antimony, and tin. They do not readily segregate on solidification from the melt, but they are subject to porosity in the central regions of type characters and slugs because air in molds escapes with difficulty. When these alloys are used, good fill of the mold should be ensured by rapid injection, and the temperature of the metal should be high enough to avoid premature solidification and entrapment of gases. Further information on type metals is given in the article "Lead and Lead Alloys" in this Volume.

Tin-base casting alloys are included in ASTM specification B 102. Alloy CY44A in this specification is similar to Alloy 1 in ASTM B 23 for sleeve bearings (Tables 11 and 12). Composition limits of the die casting version (Alloy CY44A in ASTM B 102) are 90 to 92% Sn, 4 to 5% Sb, 4 to 5% Cu, 0.35% Pb max, 0.08% Fe max, 0.08% As max, 0.01% Zn max, and 0.01% Al max.

Alloy PY1815A in ASTM B 102 is another alloy used for die castings and sleeve bearings. This alloy which has nominal contents of 82% Sn, 13% Sb, and 5% Cu, is included in Tables 11 and 12 with the other tin-base bearing alloys.

Alloy YC135A in ASTM B 102 has nominal composition contents of 65% Sn, 18% Pb, 15% Sb, and 2% Cu and is typically used for die castings. This alloy has a typical tensile strength of 69 MPa (10 ksi), an elongation value of 1% in 50 mm (2 in.), and a hardness of 29 HB. Alloy YC135A has creep-rupture strengths of about 17 MPa (2.5 ksi) for 1 year and 13 MPa (1.875 ksi) for a 10-year life.

White metal (92Sn-8Sb) is a tin-base alloy used for jewelry. Typical mechanical properties of white metal are listed in Tables 14 and 15. During cold rolling, the alloy hardens at first, and maximum hardness is reached at a reduction of about 40 to 45%. Further working causes progressive softening until, at about 80% reduction, the hardness approaches that of the cast alloy; annealing at 200 to 225 °C (392 to 437 °F) causes the severely worked alloy to harden slightly.

The solidus temperature of white metal is 246 °C (475 °F). White metal has a volumetric electrical conductivity of about 11.1% IACS and an electrical resistivity of about 155 nΩ · m at 25 °C (77 °F).

Fusible alloys are any of the more than 100 white metal alloys that melt at relatively low temperatures. Most commercial fusible alloys contain bismuth, lead, tin, cadmium, indium, and antimony, and special alloys of this class may also contain significant amounts of zinc, silver, thallium, or gallium. Further information on fusible alloys is contained in the article "Indium and Bismuth" in this Volume.

Many of the fusible alloys used in industrial applications are based on eutectic compositions. These alloys find important uses in automatic safety devices such as fire sprinklers, boiler plugs, and furnace controls. Under ambient temperature, these alloys have sufficient strength to hold parts together, but at a specific elevated temperature the fusible-alloy link will melt, thus disconnecting the parts. Examples of tin-base eutectic fusible alloys are:

Alloy composition, %	Melting temperature °C	°F
51.2 Sn, 30.6 Pb, 18.2 Cd	142	288
67.75 Sn, 32.25 Cd	177	351
61.86 Sn, 38.14 Pb	183	362
91 Sn, 9.0 Zn	199	390
96 Sn, 3.5 Ag	221	430

Collapsible tubes and tin foil are forms of tin metal that are still in use, although they are not as common as they once were. Collapsible tubes of tin are used for certain pharmaceutical products and for some premium artist paints. Tin foil is used for packaging some premium products and for wine bottle capsules. Two common tin alloys for these types of applications are described below.

Hard tin (99.6Sn-0.4Cu) is used for collapsible tubes and foils. Hard tin is resistant to attack by foodstuffs, medicinal products, cosmetics, and artist's colors. The liquidus temperature of hard tin is 230 °C (446 °F); the solidus temperature is 227 °C (441 °F).

Typical tensile strengths of 2.5 mm (0.1 in.) thick strip hard tin in various conditions are:

- 23 MPa (3.3 ksi) for strip annealed for 3 h at 100 °C (212 °F)
- 21 MPa (3.1 ksi) for strip annealed for 3 h at 200 °C (390 °F)
- 28 MPa (4.0 ksi) for cold-rolled strip (80% reduction)

Bursting of a tube 25 mm (1 in.) in diameter and 0.1 mm (0.004 in.) in wall thickness occurred with an internal pressure of 320 kPa (46 psi). In a bend test, a flattened impact-extruded collapsible tube 0.1 mm (0.004 in.) in wall thickness survived 21 bends over 90° jaws (1 kg load).

Tin foil (92Sn-8Zn) is used for food packaging. Its suitability for this application is indicated by, for example, immersion and bottle-capping tests with milk that showed that this alloy is only slightly soluble and has no effect on the milk (Ref 3).

Typical tensile properties of tin foil include a tensile strength of 60 MPa (8.7 ksi), a yield strength of 41 MPa (6.0 ksi), and an elongation of 40%. The solidus temperature is about 200 °C (390 °F).

Other Alloys Containing Tin

Battery Grid Alloys. Lead-calcium-tin alloys have been developed for storage-battery grids, largely as replacements for antimonial-lead alloys. The use of ternary lead-base alloys containing up to 1.3% Sn has substantially reduced gassing, and thus batteries with grids made of these alloys do not require periodic water additions during their working life. Two chief methods of grid manufacture are casting and fabrication of wrought alloys; fabrication of wrought alloys includes punching, roll forging, and expanded-metal processes.

Copper Alloys. Copper-tin bronzes were some of the first alloys used by man, and these alloys continue to be used for structural and decorative purposes. True bronzes contain tin in amounts up to 10% as well as very small amounts of phosphorus. Quaternary bronzes containing 5% Sn, 5% Zn, 5% Pb, and a balance of copper are used for general-purpose castings for applications requiring reasonable strength and soundness, such as gears, pumps, and automotive fittings. Special copper-base alloys with 20 to 24% Sn have historically been used for cast bells of excellent tonal quality. Spinodal copper-nickel-tin alloys containing 2 to 8.5% Sn have excellent elastic properties and have replaced tin-free copper-nickel alloys in some spring and electrical-contact applications. In addition to these uses in copper-base alloys, small quantities of tin (0.75 to 1.0%) are added to copper-zinc alloys (brasses) for increased corrosion resistance. Cast leaded brasses may contain up to 4% Sn.

Dental alloys for making amalgams contain silver, tin, mercury, and some copper and zinc. The copper increases hardness and strength, and the zinc acts as a scavenger during alloy manufacture, protecting major constituents from oxidation. Most of the dental alloys presently available contain 25 to 27% Sn and consist mainly of the intermetallic compound Ag_3Sn. When porcelain veneers are added to gold alloys for high-grade dental restoration, 1% Sn is added to the gold alloy to ensure bonding with the porcelain.

Cast Irons. The presence of about 0.1% Sn in flake or ductile iron castings ensures a completely pearlitic structure, and this pearlite is retained even at elevated temperatures. Commercially pure tin is added to the cast iron in the form of shot, bars, or cast pieces; in cupola melting, the tin is commonly added to the ladle or to the cupola spout during tapping. Tin is also added to special mixing chambers along with suitable inoculant materials in the production of ductile iron castings. Because the mixing chambers are an integral part of the mold, this technique allows one-step treatment of the molten metal as it enters the mold, and it prevents fading (that is, the loss of effectiveness of inoculating additions before the metal is cast). In addition, the mixing chamber provides immediate dissolution of the tin in the iron and ensures uniform distribution in the casting.

Titanium Alloys. Tin strengthens titanium alloys by forming solid solutions. Titanium can exist in the low-temperature α phase or the higher temperature β phase, which remains stable up to the melting point. In titanium alloys, relative amounts of α and β phases present at the service temperature have profound effects on properties. Aluminum additions raise the transformation temperature and stabilize the α phase, but they can cause embrittlement in amounts greater than 7%. However, with tin additions, increased strength without embrittlement can be obtained in aluminum-stabilized α titanium alloys. Optimum strength and workability can be obtained with 5% Al and 2.5% Sn; in addition, this alloy has the advantage of being weldable. Alpha-beta titanium alloys contain aluminum as an α stabilizer and combinations of β stabilizers (such as chromium, iron, molybdenum, manganese, or vanadium), as well as tin and zirconium as substitutional solid-solution strengthening elements. Such alloys have good strength and creep resistance at elevated temperatures. Strength and forming properties of many of these alloys can be optimized by various heat treatments.

Zirconium alloys are similar to titanium alloys in that the elements they contain can be divided into two classes: α stabilizers, which raise the transformation temperature, and β stabilizers, which lower it. Tin and aluminum are α stabilizers in zirconium alloys and enhance high-temperature strength. A commercial series of corrosion-resistant zirconium alloys containing 0.15 to 2.5% Sn has been developed for nuclear service.

REFERENCES

1. R.J. Klein Wassink, *Soldering in Electronics*, 2nd ed., Electrochemical Publications, 1989
2. R. Duckett and P.A. Ainsworth, *Sheet Met. Ind.*, Vol 50 (No. 7), 1973, p 412
3. R. Kerr, The Behavior of Some Metal Foils in Contact with Milk, *J. Soc. Chem. Ind.*, Vol 61, 1942, p 128

SELECTED REFERENCE

- *The Properties of Tin*, Publication No. 218, International Tin Research Institute, 1954

Zinc and Zinc Alloys

Robert J. Barnhurst, Noranda Technology Centre

ZINC AND ZINC ALLOYS for decorative and functional applications are described in this article. Zinc and zinc alloys are used in the form of coatings, castings, rolled sheets, drawn wire, forgings, and extrusions. Other uses of zinc are as a major constituent in brasses (see the articles on copper-base alloys in this Volume) and as a sacrificial anode for marine environments.

In its purer form, zinc is available as slabs, ingots, shot, powder, and dust; combined with oxygen, it is available as zinc oxide powder. Slab zinc is produced in three grades (Table 1). Impurity limits are very important when zinc is used for alloying purposes. Exceeding impurity limits can result in poor mechanical and corrosion properties. Pure zinc shot is used primarily for additions to electrogalvanizing baths, and zinc powder and dust are used in batteries and in enhanced corrosion-resistant paints. Zinc oxide is used as a pigment in primers and finish paint, as a reducing agent in chemical processes, and as a common additive in the production of rubber products.

Zinc Products

Coating of steel constitutes the largest single use of zinc, but it is used in large tonnages in zinc alloy castings, as zinc dust and oxide (for zinc-rich organic and inorganic coatings), and in wrought zinc products. This section will review the various zinc product forms as well as provide references to articles in other *Metals Handbook* volumes that contain more detailed information. The mechanical and physical properties of both wrought and cast alloys are described in the section "Properties of Zinc Alloys" in this article.

Zinc Coatings

The use of zinc as a coating to protect steel and iron from corrosion is the largest single application for the metal worldwide. Metallic zinc coatings are applied to steels:

- From a molten metal bath (hot dip galvanizing)
- By electrochemical means (electrogalvanizing)
- From a spray of molten metal (metallizing)
- In the form of zinc powder by chemical/mechanical means (mechanical galvanizing)

Zinc coatings are applied to many different types of products, ranging in size from small fasteners to continuous strip to large structural shapes and assemblies.

Hot Dip Galvanizing. The hot dip galvanizing industry is currently the largest consumer of zinc in the coatings field. It is divided into two segments:

- Production of continuously galvanized steel strip
- Galvanizing of structural shapes and products after fabrication

Galvanized products can be joined by conventional techniques such as welding and bolting.

Conventional strip galvanizing makes use of an alloy with a nominal content of 0.20% Al and a balance of zinc. The coating thickness is generally less than 25 μm (0.001 in.), or approximately 175 g/m^2 (0.573 oz/ft^2) of steel surface (one-side total). The coating is characterized by excellent adhesion and formability. These attributes, along with good weldability by conventional welding techniques, make strip galvanizing particularly attractive for automobile manufacturing. Galvanized strip is also used in the building industry, where significant tonnages are used in prepainted condition. The appliance industry is also a large consumer of both painted and unpainted galvanized strip. Some galvanized strip is subjected to a heat treatment known as galvannealing that converts the coating to an iron-zinc alloy. Galvannealing has been used for building products for a number of years and, more recently, for automotive parts. In recent years, new strip coatings with improved corrosion resistance, namely Galfan (5% Al) and Galvalume (55% Al), have been introduced. Galfan has been incorporated into ASTM B 750 (Table 2) and Galvalume into ASTM A 792. Additional information on strip galvanizing is available in the article "Precoated Steel Sheet" in *Properties and Selection: Irons, Steels, and High-Performance Alloys*, Volume 1 of the 10th Edition of *Metals Handbook* and in the articles "Hot Dip Coatings," "Organic Coatings and Linings," and "Corrosion of Zinc" in *Corrosion*, Volume 13 of the 9th Edition of *Metals Handbook*.

After-Fabrication Galvanizing. An aluminum-free grade of zinc that contains up to 1 wt% Pb and a balance of zinc is used for after-fabrication galvanizing. Most specifications call for a minimum coating thickness in the range of 85 to 100 μm (0.0034 to 0.004 in.), or 500 to 600 g/m^2 (1.6 to 2.0 oz/ft^2). Coating thickness is controlled by immersion time, which in turn is governed by the substrate thickness; it can be much higher with some reactive grades of steel containing even small amounts of silicon (silicon-killed steels). Proprietary galvanizing processes that use small additions of aluminum (Polygalva) or nickel (Technigalva) in concentrations of 0.04 and 0.08 wt%, respectively, have been developed in attempts to control coating thickness.

Traditional markets for after-fabrication coatings include electric utility and microwave transmission towers; highway-related products such as guard rails, signs, and lighting standards; structural applications in the industrial sector (for example, chemical, petrochemical, agricultural, and pulp and paper

Table 1 Grades and compositions of slab zinc (ASTM B 6)

Grade	UNS number	Composition, % Pb	Fe max	Cd max	Al max	Cu max	Sn max	Total nonzinc max	Zn min by difference
Special high grade	Z13001	0.003 max	0.003	0.003	0.002	0.002	0.001	0.010	99.990
High grade	Z15001	0.03 max	0.02	0.02	0.01	...	...	0.10	99.90
Prime western	Z19001	0.5–1.4	0.05	0.20	0.01	0.20	...	2.0	98.0

Table 2 Zn-5Al-MM alloy ingot chemical requirements for hot dip coatings (Galfan, or UNS Z38510) per ASTM B 750

Element	Composition, %
Aluminum(a)	4.2–6.2
Cerium plus lanthanum	0.03–0.10
Iron max	0.075
Silicon max	0.015
Lead max(b)	0.005
Cadmium max(b)	0.005
Tin max	0.002
Other max each(c)	0.02
Other max total(c)	0.04
Zinc	bal

Note: For purposes of acceptance and rejection, the observed value or calculated value obtained from analysis should be rounded to the nearest unit in the last right-hand place of figures used in expressing the specified limit, in accordance with the rounding procedure prescribed in Section 3 of ASTM E 29. By agreement between purchaser and supplier, analysis may be required and limits established for elements or compounds not specified in the table of chemical composition. Zn-5Al-MM alloy ingot for hot dip coatings may contain antimony, copper, and magnesium in amounts of up to 0.002, 0.1, and 0.05%, respectively. No harmful effects have ever been noted from the presence of these elements up to these concentrations; therefore, analyses are not required for these elements. Magnesium may be specified by the buyer up to 0.1% max. Zirconium and titanium may each be specified by the buyer up to 0.02% max. (a) Aluminum may be specified by the buyer up to 7.2% max. (b) Lead and cadmium and, to a lesser extent, tin and antimony are known to cause intergranular corrosion in zinc-aluminum alloys. Therefore, it is important to maintain the levels of these elements below the limits specified. (c) Except antimony, copper, magnesium, zirconium, and titanium

industries); drainage products; pipe for potable drinking water; heat exchangers; and reinforcing bar for concrete structures.

Electrogalvanizing. Strip-applied electrogalvanized coatings are becoming increasingly important for automotive applications. These coatings are applied at high-speed, high-current-density electroplating lines. Pure zinc as well as zinc-nickel and zinc-iron coatings are produced. The coatings are generally more uniform, smoother, and thinner than hot dip coatings. Corrosion resistance, coating-to-steel adhesion, formability, weldability, and paintability are critical properties for automotive applications of electrogalvanized steel.

Metallizing, also known as thermal spraying, is used in applications where heavy coatings are specified for corrosion protection. The process is amenable to field applications and is used in refurbishing existing structures. Very long service lives are possible with composite systems, often with the use of thinner coatings (plus suitable organic paint coats) than those that are required with conventional metallizing alloys.

In a typical metallizing procedure, either pure zinc (special high grade) or Zn-15Al is sprayed onto the steel surface to be protected. The zinc alloy is provided either in dust form or as rods that are atomized by a flame or electric arc and then propelled onto the substrate by a high-speed gas jet. Additional information is available in the article "Thermal Spray Coating" in *Surface Cleaning, Finishing, and Coating*, Volume 5 of the 9th Edition of *Metals Handbook*.

Mechanical galvanizing is a batch process that is carried out in rotating drums. During processing, the workpiece is tumbled in a mixture of zinc dust, chemicals, and glass beads, and the coating is impacted onto the surface of the workpiece by the tumbling action. It is used for coating fasteners fabricated from specialty spring or case-hardened steels, or both materials, because the properties of such fasteners might be adversely affected by the high temperature of a hot dip bath. Mechanical galvanizing is also used for applications where relatively heavy coating weights are specified.

Zinc Alloy Castings

Zinc alloys are used extensively in both gravity and pressure die castings. When used as general casting alloys, zinc alloys can be cast using such processes as high-pressure die casting, low-pressure die casting, sand casting, permanent mold casting (iron, graphite, or plaster molds), spin casting (silicone rubber molds), investment (lost-wax) casting, continuous or semicontinuous casting, and centrifugal casting. A newer process involves semisolid casting, of which several techniques can be employed. A detailed treatment of casting processes for zinc alloys is included in the article "Zinc and Zinc Alloys" in *Casting*, Volume 15 of the 9th Edition of *Metals Handbook*. Corrosion is of no concern for most applications. However, for castings under moderate-to-severe corrosive attack, some loss of properties is to be expected. Long-term aging also may cause some small loss of properties; the effects will vary from alloy to alloy and depend upon the casting method used.

Pressure Die Castings. Zinc alloys have been used for die casting for over 60 years. Until recently, all zinc alloys were based on a hypoeutectic composition, that is, they contained less aluminum (close to 4.0% Al) than the eutectic chemistry of 5.0% Al. Recently, a family of hypereutectic zinc-aluminum alloys with higher aluminum contents (>5.0% Al), have become widely used as die casting alloys. These alloys were originally designed as gravity casting alloys (see the section "Gravity Castings" in this article). They possess higher strength than the hypoeutectic zinc alloys. The compositions of current die casting alloys are included in Tables 3 and 4.

Zinc casting alloys have dendritic/eutectic microstructures. The hypoeutectic alloys solidify with zinc-rich (η) dendrites, whereas the hypereutectic alloys solidify with aluminum-rich dendrites. The ZA-8 and ZA-12 alloys solidify with cored β dendrites, whereas ZA-27 solidifies with α dendrites. The microstructures of zinc alloys are discussed in detail in the article "Zinc and Zinc Alloys" in *Metallography and Microstructures*, Volume 9 of the 9th Edition of *Metals Handbook*.

It is critically important that all zinc-aluminum casting alloys be carefully han-

Table 3 Nominal compositions of common zinc alloy die castings and zinc alloy ingot for die casting

Alloy(a) UNS number	ASTM designation	Common designation	Composition, % Cu	Al	Mg	Fe max	Pb max	Cd max	Sn max	Ni	Zn
Castings (ASTM B 86)											
Z33520 (b)	AG40A	No. 3	0.25 max(d)	3.5–4.3	0.020–0.05(e)	0.100	0.005	0.004	0.003	...	bal
Z33523 (b)	AG40B	No. 7	0.25 max	3.5–4.3	0.005–0.020	0.075	0.0030	0.0020	0.0010	0.005–0.020	bal
Z35531 (b)	AC41A	No. 5	0.75–1.25	3.5–4.3	0.03–0.08(e)	0.100	0.005	0.004	0.003	...	bal
Z35541	AC43A	No. 2	2.5–3.0	3.5–4.3	0.020–0.050	0.100	0.005	0.004	0.003	...	bal
Ingot form (ASTM B 240)											
Z33521 (c)	AG40A	No. 3	0.10 max	3.9–4.3	0.025–0.05	0.075	0.004	0.003	0.002	...	bal
Z33522 (c)	AG40B	No. 7	0.10 max	3.9–4.3	0.010–0.02	0.075	0.002	0.002	0.001	0.005–0.020	bal
Z35530 (c)	AC41A	No. 5	0.75–1.25	3.9–4.3	0.03–0.06	0.075	0.004	0.003	0.002	...	bal
Z35540	AC43A	No. 2	2.6–2.9	3.9–4.3	0.025–0.05	0.075	0.004	0.003	0.002	...	bal

Note: For purposes of acceptance and rejection, the observed value or calculated value obtained from analysis should be rounded to the nearest unit in the last right-hand place of figures used in expressing the specified limit, in accordance with the rounding procedure prescribed in ASTM E 29. (a) ASTM alloy designations were established in accordance with ASTM B 275. UNS designations were established in accordance with ASTM E 527. The last digit of a UNS number differentiates between alloys of similar composition. UNS designations for ingot and casting versions of an alloy were not assigned in the same sequence for all alloys. (b) Zinc alloy die castings may contain nickel, chromium, silicon, and manganese in amounts of 0.02, 0.02, 0.035, and 0.06%, respectively. No harmful effects have ever been noted from the presence of these elements in these concentrations; therefore, analyses are not required for these elements. (c) Zinc alloy ingot for die casting may contain nickel, chromium, silicon, and manganese in amounts of up to 0.02, 0.02, 0.035 and 0.05%, respectively. No harmful effects have ever been noted from the presence of these elements up to these concentrations; therefore, analyses are not required for these elements, except that nickel analysis is required for Z33522. (d) For the majority of commercial applications, a copper content in the range of 0.25–0.75% will not adversely affect the serviceability of die castings and should not serve as a basis for rejection. (e) Magnesium may be as low as 0.015% provided that the lead, cadmium, and tin do not exceed 0.003, 0.003, and 0.001%, respectively.

Table 4 Nominal compositions of zinc-aluminum foundry and die casting alloys directly poured to produce castings and in ingot form for remelting to produce castings

Alloy		Composition, %							
		Additions				Impurities(c)			
Common designation	UNS number(a)	Al	Cu	Mg	Zn(b)	Fe max	Pb max	Cd max	Sn max
Castings (ASTM B 791)									
ZA-8	Z35636	8.0–8.8	0.8–1.3	0.015–0.030	bal	0.075	0.006	0.006	0.003
ZA-12	Z35631	10.5–11.5	0.5–1.2	0.015–0.030	bal	0.075	0.006	0.006	0.003
ZA-27	Z35841	25.0–28.0	2.0–2.5	0.010–0.020	bal	0.075	0.006	0.006	0.003
Ingot form (ASTM B 669)									
ZA-8	Z35635	8.2–8.8	0.8–1.3	0.020–0.030	bal	0.065	0.005	0.005	0.002
ZA-12	Z35630	10.8–11.5	0.5–1.2	0.020–0.030	bal	0.065	0.005	0.005	0.002
ZA-27	Z35840	25.5–28.0	2.0–2.5	0.012–0.020	bal	0.072	0.005	0.005	0.002

(a) UNS alloy designations have been established in accordance with ASTM E 527. (b) Determined arithmetically by difference. (c) Zinc-aluminum ingot for foundry and pressure die casting may contain chromium, manganese, or nickel in amounts of up to 0.01% each or 0.03% total. No harmful effects have ever been noted from the presence of these elements in these concentrations; therefore, analyses are not required for these elements.

dled to prevent excessive pickup of harmful impurity elements such as lead, cadmium, tin, and iron, among others. Cross contamination caused by melting the alloys in furnaces used for casting copper and aluminum alloys or iron is particularly troublesome because these alloys contain elements harmful to zinc alloys. Purity concerns have led producers in many countries (those belonging to the European Economic Community, for example) to require that only 100% virgin material be used in the production of zinc foundry alloys. This requirement does not apply in North America, but alloyed ingots obtained from external suppliers are expected to meet strict impurity limits. A maximum 50% remelt of foundry returns to the melting furnace is acceptable during the making of castings.

Zinc alloys have low melting points, require relatively low heat input, do not require fluxing or protective atmospheres, and are nonpolluting; the last is a particularly important advantage. The rapid chilling rate inherent in zinc die castings results in minor property and dimensional changes with time, particularly if the casting is quenched from the die rather than air cooled. Although this is rarely a problem, a stabilizing heat treatment can be applied prior to service if rigid dimensional tolerances are to be met. The higher the heat treatment temperature, the shorter the stabilizing time required; 100 °C (212 °F) is a practical limit to prevent blistering of the casting or other problems. A common treatment consists of 3 to 6 h at 100 °C (212 °F), followed by air cooling. The time extends to 10 to 20 h for a treatment temperature of 70 °C (158 °F).

Because of their high fluidity, zinc alloys can be cast in much thinner walls than other die casting alloys, and they can be die cast to tighter dimensional tolerances. Zinc alloys allow the use of very low draft angles; in some cases, a zero draft angle is possible.

Alloy No. 2 has the highest tensile strength, creep strength, and hardness of all alloys in the hypoeutectic Zamak series of die casting alloys. The high copper content (3.0% Cu) causes some dimensional instability and leads to a net expansion of approximately 0.0014% after 20 years. It also causes some loss of impact strength and ductility. Alloy No. 2 has good bearing properties.

Alloy No. 3 is the most widely used zinc die casting alloy in the United States. It provides the best overall combination of strength, castability, dimensional stability, ease of finishing, and cost.

Alloy No. 5 produces castings that are both harder and stronger than those made from alloy No. 3. However, these properties improvements come at the expense of ductility, and postforming operations such as riveting, swaging, or crimping must be done with additional care. The creep resistance of alloy No. 5 is second only to that of alloy No. 2 among the hypoeutectic zinc-aluminum alloys.

Alloy No. 7 is essentially a high-purity version of alloy No. 3. Because of its lower magnesium content, alloy No. 7 has even better castability than alloy No. 3, enabling excellent reproduction of surface detail in castings. Alloy No. 7 has the highest ductility among the hypoeutectic alloys.

Alloy ZA-8 is the only member of the hypereutectic alloys that can be hot chamber die cast along with the hypoeutectic alloys. It is equivalent to alloy No. 2 in many respects, but ZA-8 has higher tensile, fatigue, and creep strengths, is more dimensionally stable, and has lower density. Alloy ZA-8 castings can be readily finished, thereby combining their high structural strength with excellent appearance.

Alloy ZA-12 has very good castability in cold chamber die casting machines. It is lower in density than all other zinc alloys except ZA-27, and it is frequently specified for castings that must combine casting quality with optimum performance. The plating quality of ZA-12 is lower than that of ZA-8, but it has excellent bearing and wear properties.

Alloy ZA-27 is the lightest, hardest, and strongest of all the zinc alloys, but it has relatively low ductility and impact strength when pressure die cast. Because of the wide freezing range of ZA-27, casting quality can suffer unless care is taken. The secondary creep strength of ZA-27 is better than that of all other zinc alloys except for the now rarely used ILZRO (International Lead-Zinc Research Organization) 16; however, ZA-8 has better primary creep strength. Alloy ZA-27 demonstrates the highest sound and vibration damping properties of all the zinc casting alloys; as a group, zinc alloys have a damping resistance equal to that of cast irons at elevated temperatures.

Alloy ILZRO 16 was developed specifically for optimum creep resistance, particularly at elevated temperatures. It does have the highest creep resistance of all zinc alloys, but it is difficult to manufacture and suffers from melt instability; for these reasons, ZA-8 often is used in its place.

It should be noted that the strength performance of zinc alloys drops significantly with increases in temperature. At 100 °C (212 °F), tensile and yield strengths are typically 65 to 75% of those at room temperature, and creep strength is similarly reduced.

Gravity Castings. With the exceptions of forming die alloys, slush casting alloys, and specialty alloys developed and used for bearings, no general-purpose gravity casting zinc alloys existed until the 1960s. In the 1960s and 1970s, a new family of hypereutectic zinc-aluminum alloys was developed. Alloy ILZRO-12 (now ZA-12) was the first to appear, beginning in 1962; ZA-8 and ZA-27 were quickly added. Alloy ZA-12 was developed first as a prototyping alloy for alloy No. 3 pressure die castings. Alloy ZA-27 was developed specifically as a sand casting alloy, and ZA-8 was turned into a permanent mold casting alloy. All three alloys are now used more extensively in pressure die castings.

The performance of the ZA alloys when they are gravity cast varies markedly from that of the same alloys when they are pressure die cast. The compositions of the gravity casting alloys are given in Table 4. The same requirements concerning impurities,

Table 5 Nominal compositions of zinc casting alloys used for sheet metal forming dies and for slush casting alloys in ingot form

Alloy		Composition, %							
Common designation	UNS number	Al	Cd max	Cu	Fe max	Pb max	Mg	Sn max	Zn
Forming die alloys (ASTM B 793)									
Alloy A	Z35543	3.5–4.5	0.005	2.5–3.5	0.100	0.007	0.02–0.10	0.005	bal
Alloy B	Z35542	3.9–4.3	0.003	2.5–2.9	0.075	0.003	0.02–0.05	0.001	bal
Slush casting alloys (ASTM B 792)									
Alloy A	Z34510	4.50–5.00	0.005	0.2–0.3	0.100	0.007	...	0.005	bal
Alloy B	Z30500	5.25–5.75	0.005	0.1 max	0.100	0.007	...	0.005	bal

melt cross contamination, and general handling described for the die casting alloys apply equally to the gravity casting alloys. As with the die casting alloys, microstructural changes with time can alter the properties and dimensions of cast parts. However, property changes are normally very small over the normal life span of a component, and dimensional changes, except in ZA-27, are negligible. A stabilizing heat treatment of 12 h at 250 °C (482 °F), followed by furnace cooling, effectively eliminates three-fourths of the dimensional changes that occur upon long-term aging.

Alloy ZA-8 is used mostly with ferrous permanent mold casting, but it is also used with graphite molds. Alloy ZA-8 can also be sand cast if needed, although sand casting is not used extensively for this alloy. With the exception of creep resistance, the strength of a permanent mold casting is lower than that of a pressure die casting due to the coarser microstructure of the former. The plating quality of ZA-8 is excellent, and its surface detail reproducibility is better than that of all other ZA alloys.

Alloy ZA-12 is more versatile than ZA-8 because it can be either sand cast or permanent mold cast. Its strength properties are high, and its ductility and impact strength properties are acceptable. It is clearly the alloy of choice for graphite mold casting. The bearing and damping properties of ZA-12 are both very high. Alloy ZA-12 can be readily semicontinuous cast in solid and hollow rounds for machining bushings and industrial bearings.

Alloy ZA-27 develops its optimum properties when it is sand cast. However, care should be taken when producing heavy-section castings to ensure maximum soundness and minimal underside shrinkage. Underside shrinkage, caused by gravity segregation of the aluminum-rich phase during solidification, causes a roughening on the drag surface of the casting as zinc liquid is drawn up into the casting. Both a reduction in underside shrinkage and a sound casting can be ensured when chills are used to promote directional solidification and to increase the solidification rate. The addition of rare earth elements has also been reported to reduce underside shrinkage.

In a sound gravity casting, ZA-27 produces ductility and impact strength properties much higher than those found in many die castings. Alloy ZA-27 has excellent bearing and wear properties, and it demonstrates the best damping resistance of any zinc alloy. Although it is very rarely required, a simple heat treatment of 3 h at 320 °C (608 °F), followed by furnace cooling, can increase the ductility and impact strength of ZA-27 castings. Dimensional stability is enhanced by a stabilizing heat treatment of 12 h at 250 °C (482 °F), followed by furnace cooling.

Kirksite alloy is used as a forming die alloy, and is capable of being sand cast to shape rapidly. It has an almost identical composition to the No. 2 die casting alloy. It is mainly used in the construction of cast two-piece dies for forming sheet metal parts such as components for use in the transportation and aerospace industries. Kayem 1 and Kayem 2 are similar alloys used extensively in Europe. Cast-to-size molds made of Kirksite are being used for plastic injection molding for both short-run prototyping and production operations. Two die forming die alloys are included in ASTM B 793 (Table 5).

Slush casting alloys are used extensively for the production of hollow castings such as table lamp bases. The molten alloy is poured into the mold until it is full or nearly full, and then the mold is inverted, allowing the unsolidified metal (slush) to run out. The solidified shell that is left is then removed. The thickness of the shell depends on the time interval between pouring and inverting the mold, the melt and mold temperatures, and the mold material. Two slush casting alloys are currently available (Table 5).

Specialty Alloys. Main Metal alloy and Alzen alloy are still used in Europe for the production of continuously cast bearing stock. They are also used for sliding elements, hydraulic components, worm wheels, roller bearing cages, and several other products. A series of Cosmal alloys, specifically formulated for applications requiring high damping, have been developed in Japan.

Cast Product Applications. Zinc is used extensively in the transportation industry for parts such as carburetors, fuel pump bodies, wiper parts, speedometer frames, grilles, horns, shift levers, load-bearing transmission cases, heater components, brake parts, radio bodies, electronic heat sinks, lamp and instrument bezels, steering wheel hubs, alternator brackets, exterior and interior hardware, instrument panels, and body moldings. Zinc castings are also extensively used in general hardware and electronic and electrical fittings of all kinds, including parts for domestic appliances (for example, washing machines, vacuum cleaners, mixers, and so on), oil burners, motor housings, locks, and clocks. Zinc castings are frequently and increasingly being specified for hardware used in the computer industry, in business machines (photocopiers, facsimile machines, cash registers, and typewriters), and in such items as recording machines, projectors, vending machines, cameras, gasoline pumps, many hand tools, and machinery such as larger drill presses and lathes. The ZA alloys are increasingly being specified for bearings and bushings in low-speed high-load applications.

Finishing and Secondary Operations for Zinc Alloy Castings. Many of the finishes applied to other types of metal products can be applied to zinc die castings and gravity castings. Suitable finishing treatments include:

- Mechanical buffing, polishing, brushing, and tumbling
- Plating with materials such as copper, nickel, silver and black nickel, chromium, electroless nickel, and brass
- Chemical finishing such as chromating, enameling, lacquering, painting, varnishing, anodizing, and vacuum aluminizing
- Plastic (powder coat) finishing

Phosphating of the cast surface is generally recommended to provide good adhesion for subsequent paint or powder coatings. Additional information on surface preparation techniques and coatings for zinc is available in the article "Zinc Alloys" in *Surface Cleaning, Finishing, and Coating*, Volume 5 of the 9th Edition of *Metals Handbook*.

Although zinc alloys have good natural corrosion resistance (provided that impurity limits are not exceeded), chromating and anodizing provide added corrosion protection in moderate-to-severe corrosive environments. The white rust that can form on zinc castings stored in damp environments is effectively prevented or delayed by chromating the surface. Detailed information is available in the article "Corrosion of Zinc" in *Corrosion*, Volume 13 of the 9th Edition of *Metals Handbook*.

All zinc casting alloys have excellent machining properties, with long tool life, low cutting forces, good surface finish, low tool wear, and small chip formation. Common machining operations performed on these alloys include drilling, tapping, reaming,

Table 6 Nominal compositions of rolled zinc alloys per ASTM B 69

Alloy: Common designation	Alloy: UNS number	Composition, %: Cu	Pb	Cd	Fe max	Al max	Other max	Zn
Zn–0.08Pb	Z21210	0.001 max	0.10 max	0.005 max	0.012	0.001	0.001 Sn	bal
Zn–0.06Pb–0.06Cd	Z21220	0.005 max	0.05–0.10	0.05–0.08	0.012	0.001	0.001 Sn	bal
Zn–0.3Pb–0.3Cd	Z21540	0.005 max	0.25–0.50	0.25–0.45	0.002	0.001	0.001 Sn	bal
Zn–1Cu	Z44330	0.85–1.25	0.10 max	0.005 max	0.012	0.001	0.001 Sn	bal
Zn–1Cu–0.010Mg	Z45330	0.85–1.25	0.15 max	0.04 max	0.015	0.001	0.006–0.016 Mg 0.001 Sn	bal
Zn–0.8Cu–0.15Ti	Z41320	0.50–1.50	0.10 max	0.05 max	0.012	0.001	0.12–0.50 Ti 0.001 Sn	bal
Zn–0.8Cu	Z40330	0.70–0.90	0.02 max	0.02 max	0.01	0.005	0.02 Ti	bal

Table 7 Typical applications of wrought zinc and zinc alloys

Alloy	Applications
Pure zinc	Deep-drawn hardware, expanded metal
Zn-Cu	Building construction materials, deep-drawn hardware, coinage
Zn-Cu-Ti	Roofing, gutters, and downspouts; building construction materials; deep-drawn hardware; address plates; solar collectors
Zn-Pb-Cd-Fe	Building construction materials, dry-cell battery cans, deep-drawn hardware, address plates, electrical components
Zn-Al (superplastic zinc)	Shaped components such as typewriter casings, computer panels, and covers

Source: *Engineering Properties of Zinc Alloys*, International Lead-Zinc Research Organization, 1989

broaching, routing, turning, milling, die threading, and sawing. Detailed information is available in the article "Machining of Zinc Alloy Die Castings" in *Machining*, Volume 16 of the 9th Edition of *Metals Handbook*.

Zinc alloy castings can be conveniently joined by soldering or brazing, or by certain welding techniques using zinc-base fillers. Cadmium-, tin-, or lead-base solders are not recommended because they can promote intergranular corrosion problems unless the castings are plated with heavy coatings of nickel or copper prior to soldering. Newer zinc-base solders are becoming available. Detailed information about these joining techniques is available in *Welding, Brazing, and Soldering*, Volume 6 of the 9th Edition of *Metals Handbook*.

Adhesive bonding or mechanical fasteners are also excellent methods for joining castings. Zinc castings can be riveted, staked, and crimped. Threaded fasteners, including self-tapping screws, should not be overtightened but rather tightened to recommended torques. Up to 40% loss of torque should be incorporated into the design for parts operating at elevated temperatures of 50 °C (122 °F) or higher. Significant torque loss can be avoided by using special fasteners, including cone (spring or Belleville) or star washers of the correct size. Joining two or more parts can be accomplished by die casting a joint to properly align and join the parts.

Additional Properties of Zinc Castings. In addition to their excellent physical and mechanical properties, zinc alloys offer:

- Good corrosion resistance
- Excellent vibration- and sound-damping properties that increase exponentially with temperature (because of these damping characteristics, zinc alloys can be designated HIDAMETS, or high-damping metals)
- Excellent bearing and wear properties
- Spark (incendivity) resistance (with the exception of the high-aluminum ZA-27 alloy)

Wrought Zinc and Zinc Alloys

Zinc in pure form or with small alloying additions is used in three main types of wrought products: flat-rolled products, wire-drawn products, and extruded and forged products. Wrought zinc is readily machined, joined, and finished.

Flat-Rolled Products. Zinc is usually cast into 25 to 100 mm (1.0 to 4.0 in.) thick flat slabs that are suitable for rolling; these slabs are preheated and then rough and finish rolled. Schedules for finish rolling of zinc strip vary and depend on the product required. Strip is produced in various widths up to 2 m (79 in.) and in thicknesses down to 0.1 mm (0.004 in.). Foil in thicknesses of 0.025 mm (0.001 in.) or less is produced in special mills. For a bright surface combined with high ductility, finish rolling is performed at 120 to 150 °C (250 to 300 °F).

Rolled zinc can be readily formed into many different shapes by bending, spinning, deep drawing, roll forming, coining, and impact extrusion. Joining is easily achieved by soldering and resistance welding. When alloyed with copper and titanium, zinc sheet is very creep resistant and can be used in functional applications; examples include architectural applications such as in roofing and siding. Rolled zinc is produced in seven basic alloys and also as pure zinc (Table 6). Variations in chemistry and rolling conditions produce a variety of properties.

Superplastic zinc, which contains 21 to 23% Al and a small amount of copper (0.4 to 0.6%), can be easily formed into complex shapes and displays the characteristics of plastics or molten glass at temperatures of 250 to 270 °C (480 to 520 °F). The very fine grain size produced by processing at 275 to 375 °C (525 to 705 °F) followed by quenching and aging gives superplastic zinc its unique properties. Different grades of superplastic zinc, such as air-cooled or furnace-cooled varieties, have different levels of strength. When superplastic zinc is reheated to above 275 °C (527 °F) and slowly cooled to room temperature, the superplastic properties disappear. Detailed information about superplasticity is available in the article "Sheet Formability Testing" in *Mechanical Testing*, Volume 8 of the 9th Edition of *Metals Handbook*.

Wire-Drawn Products. Zinc is easily rolled or extruded into rod and then drawn into wire. The manufacture of zinc alloy wire is normally continuous and follows casting, rolling, and drawing operations. Special finishes or lubricants are sometimes applied following drawing. Wire sizes vary from 1.0 to 6.35 mm (0.004 to 0.25 in.).

Zinc alloy wire is widely used in thermal spraying, or metallizing, where the wire is melted and sprayed onto a substrate using a special gun. This process is used primarily for the corrosion protection of steel (see the section "Zinc Coatings" in this article). In addition to pure zinc, zinc alloys containing 15% Al are used in thermal spraying because the zinc-aluminum alloy provides increased corrosion protection. In addition to its thermal spraying applications, zinc alloy wire is used in nail and screw production and in zinc-base solders.

Extruded and Forged Products. Two zinc forging alloys are currently in commercial use; one has a zinc-aluminum base, and the other contains copper and titanium. Korloy 2573 (Zn-14.5Al-0.02Mg-0.75Cu) has high impact strength at low temperatures. The titanium-containing alloy, Korloy 3130 (Zn-1.0Cu-0.1Ti), is a more general-purpose alloy and has better creep strength. Both alloys have excellent machining, joining, and finishing characteristics, although machining of the titanium-containing alloy is best performed with carbide-tipped tooling.

Zinc alloys are generally capable of being extruded but require higher pressures and lower speeds than other nonferrous metals. Alloy ZA-27 has been fabricated from continuously cast billets into extruded stock for bushings. The tensile properties of the extruded ZA-27 are improved over those of the cast stock, especially elongation. Extruding offers near-net-shape capability with minimal or no machining. Temperatures of the order of 250 to 300 °C (480 to 570 °F) are required to extrude zinc alloys.

Wrought Product Applications. In addition to the use of wrought zinc in roofing and flashing, rolled zinc is drawn into many different products, including dry-cell battery cans, batter cups, handrails, eyelets, meter cases, buckles, ferrules, gaskets, and electrical components such as lamp parts. Examples of typical end uses for the various grades of wrought zinc are given in Table 7.

Properties of Zinc Alloys

AG40A
Zn-4Al-0.04Mg

Commercial Names

Trade name. No. 3 die casting alloy
Previous trade name. Zamak 3
Foreign. Mazak 3

Specifications

ASTM. B 86: alloy AG40A (die castings). B 240: alloy AG40A (ingot)
SAE. J468, alloy 903
UNS. Z33521 (ingot), Z33520 (castings)
U.S. Government. QQ-Z-363
Foreign. AFNOR-ZA4G; BS 1004A; CSA HZ-3 (ingot), HZ-11 (castings); DIN-1743; JIS H 2201 class 1 (ingot), H 5301 ZDC 2 (castings); SAA-AS1881

Chemical Composition

Composition limits. ASTM B 86: 3.5 to 4.3 Al, 0.020 to 0.05 Mg, 0.25 Cu max, 0.100 Fe max, 0.005 Pb max, 0.004 Cd max, 0.003 Sn max, bal Zn
Consequence of exceeding impurity limits. Alloy becomes subject to intergranular corrosive attack and fails prematurely by warping and cracking.

Applications

Typical uses. Die castings such as automotive parts, household appliances and fixtures, office and computer equipment, building hardware

Mechanical Properties

Tensile properties. Die cast specimen, 6.35 mm (0.25 in.) in diameter: tensile strength, 283 MPa (41 ksi); elongation, 10% in 50 mm (2 in.) gage length. Tensile and other mechanical properties are affected by operating temperature and time (see Table 8 and Fig. 1).
Shear strength. 214 MPa (31.0 ksi)
Compressive strength. 414 MPa (60.0 ksi)
Hardness. 82 HB (500 kg load, 10 mm hardened steel ball, 30-s duration)
Impact strength. Charpy, unnotched 6.35 mm (0.25 in.) square bar: 58 J (43 ft · lbf). See Table 8.
Fracture toughness. Pressure die cast, 12.3 MPa$\sqrt{m}$ (11.2 ksi$\sqrt{in.}$) at 24 °C (75 °F). See Table 8.
Fatigue strength. Reverse bending, 48 MPa (7.0 ksi) at 5×10^8 cycles
Creep strength. Stress to give 1% secondary creep extension in 10^5 h, per ASME boiler code: 21 MPa (3.0 ksi) at 20 °C (68 °F). See Fig. 2.

Mass Characteristics

Density. 6.6 g/cm^3 (0.24 lb/in.^{3}) at 21 °C (70 °F)
Volume change on freezing. 1.17% shrinkage

Thermal Properties

Liquidus temperature. 387 °C (728 °F)
Solidus temperature. 381 °C (718 °F)
Coefficient of linear thermal expansion. 27.4 μm/m · K (15.2 μin./in. · °F) at 20 to 100 °C (68 to 212 °F)
Specific heat. 0.419 kJ/kg · K (0.10 Btu/lb · °F) at 20 to 100 °C (68 to 212 °F)
Thermal conductivity. 113.0 W/m · K (784 Btu · in./ft^2 · h · °F)

Electrical Properties

Electrical conductivity. 27% IACS at 20 °C (68 °F)
Electrical resistivity. 6.4 μΩ · cm (2.52 μΩ · in.) at 20 °C (68 °F)

Chemical Properties

General corrosion behavior. This alloy has better resistance than pure zinc and can

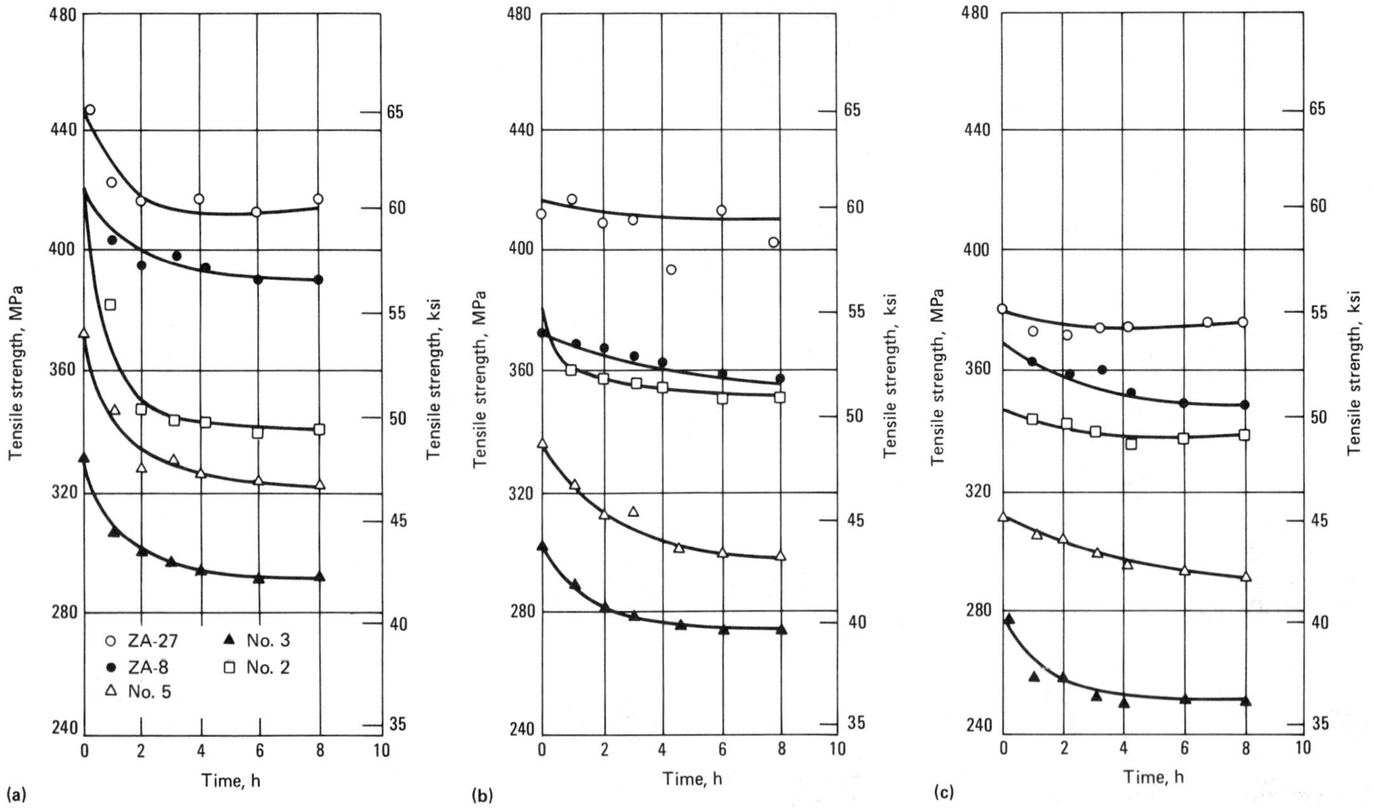

Fig. 1 Effect of aging time on the tensile strengths of five zinc alloys. Aging temperature, 100 °C (212 °F). (a) 0.76 mm (0.030 in.) casting wall thickness. (b) 1.52 mm (0.060 in.) casting wall thickness. (c) 2.54 mm (0.100 in.) casting wall thickness. Source: Noranda Technology Centre

Table 8 Effect of temperature on the mechanical properties of zinc-alloy and zinc-aluminum alloy castings

Alloy designation	Temperature		Tensile strength(a)		Impact energy(b)		Fracture toughness (average K_{Ic})			
							Pressure die cast		Sand cast	
	°C	°F	MPa	ksi	J	ft · lbf	MPa $\sqrt{m}$	ksi $\sqrt{in.}$	MPa $\sqrt{m}$	ksi $\sqrt{in.}$
Conventional die casting alloys										
No. 2	21	70	359	52.1	47.5	35	...	...	...	...
No. 3	−40	−40	308.9	44.8	2.7	2	10.1	9.2	...	...
	−20	−4	301.3	43.8	5.4	4	...	...	...	...
	0	32	284.8	41.3	31.2	23	...	...	...	...
	21	70	282.7	41.0	58.3	43	...	...	...	...
	24	75	...	...	...	...	12.3	11.2	...	...
	40	104	244.8	35.5	57.0	42	...	...	...	...
	95	203	195.1	28.3	54	40	...	...	...	...
No. 5	−40	−40	337.2	48.9	2.7	2	...	...	...	...
	−20	−4	340.6	49.4	5.4	4	...	...	...	...
	0	32	333.0	48.3	55.6	41	...	...	...	...
	21	70	328.2	47.6	65.1	48	...	...	...	...
	40	104	295.8	42.9	62.4	46	...	...	...	...
	95	203	242.0	35.1	58.3	43	...	...	...	...
No. 7	−40	−40	308.9	44.8	1.4	1.0	...	...	...	...
	−20	−4	299.2	43.4	1.9	1.4	...	...	...	...
	−10	14	...	...	2.4	1.8	...	...	...	...
	0	32	282.7	41.0	3.8	2.8	...	...	...	...
	20	68	...	...	54.2	40	...	...	...	...
	50	122	232.4	33.7	58.3	43	...	...	...	...
	95	203	193.1	28.0	54.2	40	...	...	...	...
	150	302	120.0	17.4	43.4	32	...	...	...	...
Zinc-aluminum casting alloys										
ZA-8	−40	−40	409.6	59.4	1	1	10.2	9.3	...	...
	−20	−4	402.7	58.4	1	1	...	...	...	...
	−10	14	...	...	2	1.5	...	...	...	...
	0	32	382.7	55.5	2	1.5	...	...	...	...
	20	68	373.7	54.2	42	31	...	...	...	...
	24	75	...	...	...	...	12.6	11.5	...	...
	40	104	...	...	54	40	...	...	...	...
	50	122	328.2	47.6	...	...	...	...	...	...
	60	144	...	...	56	41	...	...	...	...
	80	176	...	...	65	48	...	...	...	...
	100	212	224.1	32.5	63	46	27.7	25.2	...	...
ZA-12	−40	−40	450.2	65.3	1.5	1	11.2	10.2	9.8	8.9
	−20	−4	...	...	1.5	1	...	...	...	...
	0	32	434.4	63.0	3	1.7	...	...	...	...
	20	68	403.4	58.5	29	21	...	...	...	...
	24	75	...	...	...	...	14.4	13.1	14.5	13.2
	40	104	...	...	35	26	...	...	...	...
	50	122	349.6	50.7	...	...	...	...	...	...
	60	140	...	...	40	29	...	...	...	...
	80	176	...	...	46	33	...	...	...	...
	100	212	228.9	33.2	46	34	29.0	26.4	29.1	26.5
	150	302	119.3	17.3	...	...	...	...	...	...
ZA-27	−40	−40	520.6	75.5	2	1.5	11.9	10.8	16.4	14.9
	−20	−4	500.6	72.6	3	2.5	...	...	...	...
	−10	14	...	...	7	5	...	...	...	...
	0	32	497.1	72.1	...	...	...	...	...	...
	24	75	...	...	...	...	20.2	18.4	23.7	21.6
	20	68	425.4	61.7	13	9.5	...	...	...	...
	40	104	...	...	15	11	...	...	...	...
	50	122	397.8	57.7	...	...	...	...	...	...
	60	140	...	...	16	12	...	...	...	...
	80	176	...	...	16	12	...	...	...	...
	100	212	259.3	37.6	16	12	35.2	32.0	42.1	38.3
	150	302	129.0	18.7	...	...	...	...	...	...

(a) As-cast. (b) As-cast, unnotched 6.35 mm (0.25 in.) square specimen. Source: *Engineering Properties of Zinc Alloys*. International Lead-Zinc Research Organization, 1989 and the Noranda Technology Centre

safely be used wherever zinc, zinc-coated iron, or zinc-coated steel has been used successfully in the past.

Other Properties

Damping. Q^{-1} (× 10^3) at 100 Hz, 3.6 at 20 °C (68 °F), 6.4 at 100 °C (212 °F). See Table 9.

Fabrication Characteristics

Joining. Mechanical fasteners, adhesives, tungsten inert-gas (TIG) or metal inert-gas (MIG) welding. Solder with lead-tin solder over nickel-plated surface; acidulated zinc chloride flux. Oxyacetylene weld with alloy No. 3, no flux, soft flame. Resistance welding pulsation technique can be used.

Recommended casting temperature range. 395 to 425 °C (740 to 800 °F)

Precautions in melting. Stir well after meltdown and allow a 15-min holding period. Skim surface dross. Avoid high melt temperatures to prevent magnesium loss.

Precautions in casting. Strict adherence to chemical composition limits and proper die design are necessary to ensure sound castings and high strength and ductility.

AC41A
Zn-4Al-1Cu-0.05Mg

Commercial Names

Trade name. No. 5 die casting alloy

Previous trade name. Zamak 5

Foreign. Mazak 5

Specifications

ASTM. B 86: alloy AC41A (die castings). B 240: alloy AC41A (ingot)
SAE. J468, alloy 925
UNS. Z35530 (ingot), Z35531 (castings)
Government. QQ-Z-363
Foreign. AFNOR-ZA4UIG; BS 1004A; CSA HZ-3 (ingot), HZ-11 (castings); DIN-1743; JIS H 2201 class 2 (ingot), H 5301 ZDC 1 (castings); SAA-AS1881

Chemical Composition

Composition limits. ASTM B 86: 3.5 to 4.3 Al, 0.75 to 1.25 Cu, 0.030 to 0.08 Mg, 0.100 Fe max, 0.005 Pb max, 0.004 Cd max, 0.003 Sn max, bal Zn
Consequence of exceeding impurity limits. Alloy becomes subject to intergranular corrosive attack and fails prematurely by warping and cracking.

Applications

Typical uses. Die castings such as automotive parts, household appliances and fixtures, office and computer equipment, building hardware

Mechanical Properties

Tensile properties. Die cast specimen, 6.35 mm (0.25 in.) in diameter: tensile strength, 328 MPa (47.6 ksi), elongation, 7% in 50 mm (2 in.) gage length. Tensile and other mechanical properties are affected by operating temperature and time (see Table 8 and Fig. 1).
Shear strength. 262 MPa (38.0 ksi)
Compressive strength. 600 MPa (87.0 ksi)
Hardness. 91 HB (500 kg load, 10 mm hardened steel ball, 30-s duration)
Impact strength. Charpy, unnotched 6.35 mm (0.25 in.) square bar: 65 J (48 ft · lbf). See Table 8.
Fatigue strength. Reverse bending, 56.5 MPa (8.20 ksi) at 5×10^8 cycles

Mass Characteristics

Density. 6.7 g/cm^3 (0.24 lb/in.^{3}) at 21 °C (70 °F)
Volume change on freezing. 1.17% shrinkage

Thermal Properties

Liquidus temperature. 386 °C (727 °F)
Solidus temperature. 380 °C (717 °F)

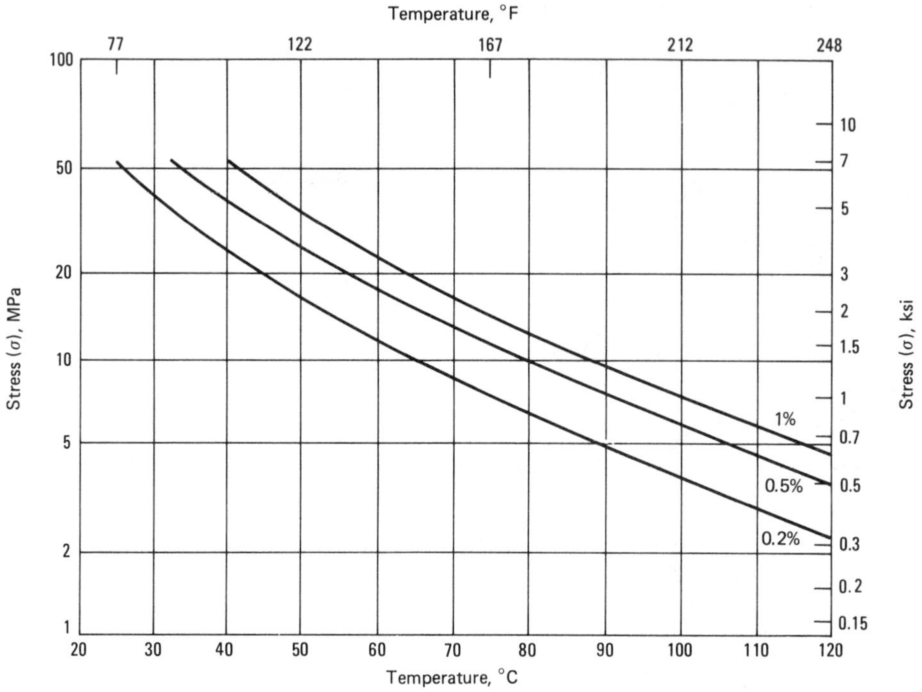

Fig. 2 Elongation of zinc alloy No. 3 at various combinations of stress and temperature for a service life of 3×10^3 h. Source: AM&S Europe Ltd.

Coefficient of linear thermal expansion. 27.4 μm/m · K (15.2 μin./in. · °F) at 20 to 100 °C (68 to 212 °F)
Specific heat. 0.419 kJ/kg · K (0.10 Btu/lb · °F) at 20 to 100 °C (68 to 212 °F)
Thermal conductivity. 109.0 W/m · K (755 Btu · in./ft^2 · h · °F)

Electrical Properties

Electrical conductivity. 26% IACS at 20 °C (68 °F)
Electrical resistivity. 6.5 μΩ · cm (2.56 μΩ · in.) at 20 °C (68 °F)

Chemical Properties

General corrosion behavior. This alloy has better resistance than pure zinc and can safely be used wherever zinc, zinc-coated iron, or zinc-coated steel has been used successfully in the past.

Fabrication Characteristics

Joining. Mechanical fasteners, adhesives, TIG or MIG welding. Solder with lead-tin solder over nickel-plated surface; acidulated zinc chloride flux. Oxyacetylene weld with alloy No. 3, no flux, soft flame. Resistance welding pulsation technique can be used.
Recommended casting temperature range. 395 to 425 °C (740 to 800 °F)
Precautions in melting. Stir well after meltdown and allow a 15-min holding period. Skim surface dross. Avoid high melt temperatures to prevent magnesium loss.
Precautions in casting. Strict adherence to chemical composition limits and proper die design are necessary to ensure sound castings and high strength and ductility.

AG40B Zn-4Al-0.015Mg

Commercial Names

Trade name. No. 7 die casting alloy
Previous trade name. Zamak 7
Foreign. Mazak 7

Specifications

ASTM. B 86: alloy AG40B (die castings). B 240: alloy AG40B (ingot)

Table 9 Damping properties of selected zinc alloys at 20 °C (68 °F) and 100 °C (212 °F)

	$Q^{-1} \times 10^3$(a)											
	10 Hz		50 Hz		100 Hz		500 Hz		10^3 Hz		5×10^3 Hz	
Alloy	20 °C (68 °F)	100 °C (212 °F)	20 °C (68 °F)	100 °C (212 °F)	20 °C (68 °F)	100 °C (212 °F)	20 °C (68 °F)	100 °C (212 °F)	20 °C (68 °F)	100 °C (212 °F)	20 °C (68 °F)	100 °C (212 °F)
No. 3 (b)	3.3	6.9	4.8	8.2	3.6	6.4	1.2	3.1	0.8	2.4	0.4	1.6
ZA-8 (b)	3.2	7.0	5.4	9.3	5.2	8.8	1.9	4.1	1.2	3.0	0.5	1.8
ZA-12 (b)	3.5	9.3	5.6	10.8	5.1	9.8	1.9	4.8	1.2	3.7	0.5	2.3
ZA-27 (b)	5.0	12.8	6.6	13.5	5.0	10.9	1.7	5.8	1.2	4.8	0.6	3.3
SPZ (c)	9.2	35.5	9.3	31.3	7.8	27.8	4.2	19.9	3.4	17.7	2.2	13.8

(a) Specific damping capacity = $200\pi\ Q^{-1}$. (b) As die cast. (c) Cold-rolled superplastic zinc sheet. Source: Atomic Energy of Canada and the International Lead-Zinc Research Organization

UNS. Z33522 (ingot), Z33523 (castings)

Chemical Composition

Composition limits. ASTM B 86: 3.5 to 4.3 Al, 0.25 Cu max, 0.005-0.020 Mg, 0.075 Fe max, 0.003 Pb max, 0.002 Cd max, 0.001 Sn max, 0.005-0.020 Ni, bal Zn
Consequence of exceeding impurity limits. Alloy becomes subject to intergranular corrosive attack and fails prematurely by warping and cracking.

Applications

Typical uses. Die castings such as automotive parts, household appliances and fixtures, office and computer equipment, building hardware

Mechanical Properties

Tensile properties. Die cast specimen, 6.35 mm (0.25 in.) in diameter: tensile strength, 283 MPa (41.0 ksi); elongation, 13% in 50 mm (2 in.) gage length. Tensile and other mechanical properties are affected by operating temperature and time (see Table 8).
Shear strength. 214 MPa (31.0 ksi)
Compressive strength. 414 MPa (60.0 ksi)
Hardness. 80 HB (500 kg load, 10 mm hardened steel ball, 30-s duration)
Impact strength. Charpy, unnotched 6.35 mm (0.25 in.) square bar: 58 J (43 ft · lbf). See Table 8.
Fatigue strength. Reverse bending, 47 MPa (6.80 ksi) at 5×10^8 cycles

Mass Characteristics

Density. 6.6 g/cm^3 (0.24 lb/in.^{3}) at 21 °C (70 °F)
Volume change on freezing. 1.17% shrinkage

Thermal Properties

Liquidus temperature. 387 °C (728 °F)
Solidus temperature. 381 °C (718 °F)
Coefficient of linear thermal expansion. 27.4 μm/m · K (15.2 μin./in. · °F) at 20 to 100 °C (68 to 212 °F)
Specific heat. 0.419 kJ/kg · K (0.10 Btu/lb · °F) at 20 to 100 °C (68 to 212 °F)
Thermal conductivity. 113.0 W/m · K (784 Btu · in./ft^2 · h · °F)

Electrical Properties

Electrical conductivity. 27% IACS at 20 °C (68 °F)

Chemical Properties

General corrosion behavior. This alloy has better resistance than pure zinc and can safely be used wherever zinc, zinc-coated iron, or zinc-coated steel has been used successfully in the past.

Fabrication Characteristics

Joining. Mechanical fasteners, adhesives, TIG or MIG welding. Solder with lead-tin solder over nickel-plated surface; acidulated zinc chloride flux. Oxyacetylene weld with alloy No. 3, no flux, soft flame. Resistance welding pulsation technique can be used.
Recommended casting temperature range. 395 to 425 °C (740 to 800 °F)
Precautions in melting. Stir well after meltdown and allow a 15-min holding period. Skim surface dross.
Precautions in casting. Strict adherence to chemical composition limits and proper die design are necessary to ensure sound castings and high strength and ductility.

AC43A
Zn-4Al-2.5Cu-0.04Mg

Commercial Names

Trade name. No. 2 die casting alloy
Previous trade name. Zamak 2
Foreign. Mazak 2

Specifications

ASTM. B 86: alloy AC43A (die castings). B 240: alloy AC43A (ingot)
SAE. Alloy 921
UNS. Z35540 (ingot), Z35541 (castings)
Foreign. AFNOR-ZA4U3G; BS 1004; DIN-1743

Chemical Composition

Composition limits. ASTM B 86: 3.5 to 4.3 Al, 2.5 to 3.0 Cu, 0.020 to 0.05 Mg, 0.100 Fe max, 0.005 Pb max, 0.004 Cd max, 0.003 Sn max, bal Zn
Consequence of exceeding impurity limits. Alloy becomes subject to intergranular corrosive attack and fails prematurely by warping and cracking.

Applications

Typical uses. Die castings such as automotive parts, household appliances and fixtures, office and computer equipment, building hardware

Mechanical Properties

Tensile properties. Die cast specimen, 6.35 mm (0.25 in.) in diameter: tensile strength, 358 MPa (52.0 ksi); elongation, 7% in 50 mm (2 in.) gage length. Tensile and other mechanical properties are affected by operating temperature and time (see Table 8 and Fig. 1).
Shear strength. 317 MPa (46.0 ksi)
Compressive strength. 641 MPa (93.0 ksi)
Hardness. 100 HB (500 kg load, 10 mm hardened steel ball, 30-s duration)
Impact strength. Charpy, unnotched 6.35 mm (0.25 in.) square bar: 47 J (35 ft · lbf). See Table 8.
Fatigue strength. Reverse bending, 59 MPa (8.50 ksi) at 5×10^8 cycles

Mass Characteristics

Density. 6.6 g/cm^3 (0.24 lb/in.^{3}) at 21 °C (70 °F)
Volume change on freezing. 1.25% shrinkage

Thermal Properties

Liquidus temperature. 390 °C (734 °F)
Solidus temperature. 379 °C (715 °F)
Coefficient of linear thermal expansion. 27.8 μm/m · K (15.4 μin./in. · °F) at 20 to 100 °C (68 to 212 °F)
Specific heat. 0.419 kJ/kg · K (0.10 Btu/lb · °F) at 20 to 100 °C (68 to 212 °F)
Thermal conductivity. 105.0 W/m · K (726 Btu · in./ft^2 · h · °F)

Electrical Properties

Electrical conductivity. 25% IACS at 20 °C (68 °F)

Chemical Properties

General corrosion behavior. This alloy has better resistance than pure zinc and can safely be used wherever zinc, zinc-coated iron, or zinc-coated steel has been used successfully in the past.

Fabrication Characteristics

Joining. Mechanical fasteners, adhesives, TIG or MIG welding. Solder with lead-tin solder over nickel-plated surface; acidulated zinc chloride flux. Oxyacetylene weld with alloy No. 3, no flux, soft flame. Resistance welding pulsation technique can be used.
Recommended casting temperature range. 395 to 425 °C (740 to 800 °F)
Precautions in melting. Stir well after meltdown and allow a 15-min holding period. Skim surface dross. Avoid high melt temperatures to prevent magnesium loss.
Precautions in casting. Strict adherence to chemical composition limits and proper die design are necessary to ensure sound castings and high strength and ductility.

ZA-8
Zn-8Al-1Cu-0.02Mg

Commercial Names

Trade name. ZA-8, zinc foundry alloy

Specifications

ASTM. B 791 (castings), B 669 (ingot)
UNS. Z35635 (ingot), Z35636 (castings)
Foreign. BS DD 139 (castings)

Chemical Composition

Composition limits. ASTM B 669 (ingot): 8.2 to 8.8 Al, 0.8 to 1.3 Cu, 0.020 to 0.030 Mg, 0.065 Fe max, 0.005 Pb max, 0.005 Cd max, 0.002 Sn max, bal Zn. ASTM B 791 (castings): 8.0 to 8.8 Al, 0.8 to 1.3 Cu, 0.015 to 0.030 Mg, 0.075 Fe max, 0.006 Pb max, 0.006 Cd max, 0.003 Sn max, bal Zn
Consequence of exceeding impurity limits. Alloy becomes subject to intergranular corrosive attack and fails prematurely by warping and cracking.

Applications

Typical uses. For pressure die castings and gravity castings, wherever high strength is

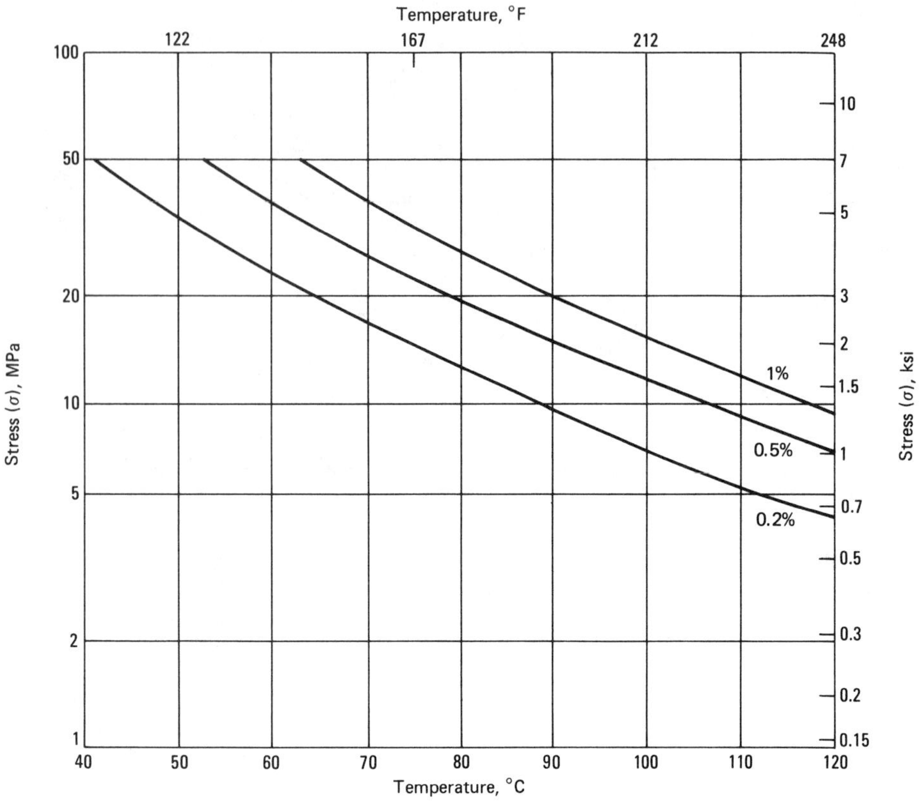

Fig. 3 Elongation of pressure die cast zinc alloy ZA-8 at various combinations of stress and temperature for a service life of 3×10^3 h. Source: AM&S Europe Ltd.

required. Automobiles, general hardware, agricultural equipment, electronic and electrical fittings, domestic and garden appliances, computer hardware, business machines, recording machines, radios, and hand tools

Mechanical Properties

Tensile properties. Die cast specimen, 6.35 mm (0.25 in.) in diameter: tensile strength, 374 MPa (54.2 ksi); yield strength (0.2% offset), 290 MPa (42.1 ksi); elongation, 8% in 50 mm (2 in.) gage length. Permanent mold cast specimen, 12.7 mm (0.5 in.) in diameter: tensile strength, 240 MPa (34.8 ksi); yield strength (0.02% offset), 208 MPa (30.2 ksi); elongation, 1.3% in 50 mm (2 in.). Elastic modulus, all processes: 85.5 GPa (12.4×10^6 psi). Tensile and other mechanical properties are affected by operating temperature and time (see Table 8 and Fig. 1).
Shear strength. Die cast, 275 MPa (39.9 ksi); permanent mold cast, 242 MPa (35.1 ksi)
Compressive strength. 0.1% offset: die cast, 252 MPa (36.5 ksi); permanent mold cast, 210 MPa (30.5 ksi)
Hardness. 500 kg load, 10 mm ball, 30-s duration: die cast, 103 HB; permanent mold cast, 87 HB
Impact strength. Charpy, unnotched: die cast 6.35 mm (0.25 in.) square bar, 42 J (31 ft · lbf); permanent mold cast 10.0 mm (0.394 in.) square bar, 20 J (15 ft · lbf). See Table 8.
Fracture toughness. Pressure die cast, 12.6 MPa$\sqrt{m}$ (11.5 ksi$\sqrt{in.}$) at 24 °C (75 °F). See Table 9.
Fatigue strength. Reverse bending at 5×10^8 cycles: die cast, 103 MPa (15.0 ksi); permanent mold cast, 52 MPa (7.5 ksi)
Creep strength. Stress to give 1% secondary creep extension in 10^5 h, per ASME boiler code: 70 MPa (10.1 ksi) at 20 °C (68 °F). See Fig. 3.

Mass Characteristics

Density. 6.30 g/cm^3 (0.227 lb/in.^{3}) at 21 °C (70 °F)
Volume change on freezing. 1% shrinkage

Thermal Properties

Liquidus temperature. 404 °C (759 °F)
Solidus temperature. 375 °C (707 °F)
Coefficient of linear thermal expansion. 23.2 μm/m · K (12.9 μin./in. · °F) at 20 to 100 °C (68 to 212 °F)
Latent heat of fusion. 112 kJ/kg (48 Btu/lb)
Specific heat. 0.435 kJ/kg · K (0.104 Btu/lb · °F) at 20 to 100 °C (68 to 212 °F)
Thermal conductivity. 115 W/m · K (795 Btu · in./ft^2 · h · °F)

Electrical Properties

Electrical conductivity. 27.7% IACS at 20 °C (68 °F)
Electrical resistivity. 6.2 μΩ · cm (2.44 μΩ · in.) at 20 °C (68 °F)

Chemical Properties

General corrosion behavior. This alloy has better resistance than pure zinc and hypoeutectic zinc-aluminum alloys and can be safely used wherever zinc, zinc-coated iron, or zinc-coated steel has been used successfully in the past.

Other Properties

Damping. Q^{-1} ($\times 10^3$) at 100 Hz: pressure die cast, 5.2 at 20 °C (68 °F), 8.8 at 100 °C (212 °F). See Table 9.

Fabrication Characteristics

Joining. Mechanical fasteners, adhesives, TIG or MIG welding. Lead-tin or zinc-cadmium soft solder. Resistance welding pulsation technique can be used.
Recommended casting temperature range. 435 to 460 °C (815 to 860 °F)
Precautions in melting. Stir well after meltdown and allow a 15-min holding period. Skim surface dross. Avoid high melt temperatures to prevent magnesium loss.
Precautions in casting. Strict adherence to chemical composition limits and proper pattern or die design are necessary to ensure sound castings and high strength and ductility.

ZA-12
Zn-11Al-1Cu-0.025Mg

Commercial Names

Trade name. ZA-12, zinc foundry alloy
Previous trade name. ILZRO 12

Specifications

ASTM. B 791 (castings), B 669 (ingot)
UNS. Z35630 (ingot), Z35631 (castings)
Foreign. BS DD 139 (castings); SAA-AS1881

Chemical Composition

Composition limits. ASTM B 669 (ingot): 10.8 to 11.5 Al, 0.5 to 1.2 Cu, 0.020 to 0.030 Mg, 0.065 Fe max, 0.005 Pb max, 0.005 Cd max, 0.002 Sn max, bal Zn. ASTM B 791 (castings): 10.5 to 11.5 Al, 0.5 to 1.2 Cu, 0.015 to 0.030 Mg, 0.075 Fe max, 0.006 Pb max, 0.006 Cd max, 0.003 Sn max, bal Zn
Consequence of exceeding impurity limits. Alloy becomes subject to intergranular corrosive attack and fails prematurely by warping and cracking. High iron causes excessive tool wear.

Applications

Typical uses. For pressure die castings and gravity castings, wherever high strength is required. Automobiles, general hardware, agricultural equipment, electronic and electrical fittings, domestic and garden appliances, computer hardware, business machines, recording machines, radios, and hand tools. This alloy is used in bearings and bushings for high-load low-speed applications.

Mechanical Properties

Tensile properties. Die cast specimen, 6.35 mm (0.25 in.) in diameter: tensile strength, 404 MPa (58.5 ksi); yield strength (0.2% offset), 320 MPa (46.4 ksi); elongation, 5% in 50 mm (2 in.) gage length. Permanent mold cast specimen, 12.7 mm (0.5 in.) in diameter: tensile strength, 328 MPa (47.5 ksi); yield strength (0.02% offset), 268 MPa (38.9 ksi); elongation, 2.2% in 50 mm (2 in.). Sand cast specimen, 12.7 mm (0.5 in.) in diameter: tensile strength, 299 MPa (43.4 ksi); elongation, 1.5% in 50 mm (2 in.). Elastic modulus, all processes: 82.7 GPa (12.0×10^6 psi). Tensile and other mechanical properties are affected by operating temperature and time (see Table 8 and Fig. 1).

Shear strength. Die cast, 296 MPa (42.9 ksi); sand cast, 253 MPa (36.7 ksi)

Compressive strength. 0.1% offset: die cast, 269 MPa (39.0 ksi); permanent mold cast, 234 MPa (34.0 ksi); sand cast, 230 MPa (33.3 ksi)

Hardness. 500 kg load, 10 mm hardened steel ball, 30-s duration: die cast, 100 HB; permanent mold, 89 HB; sand cast, 94 HB

Impact strength. Charpy, unnotched. Die cast, 6.35 mm (0.25 in.) square bar: 29 J (21 ft · lbf); permanent mold, machined 10.0 mm (0.394 in.) square bar: 20 J (15 ft · lbf); sand cast, machined 10.0 mm (0.394 in.) square bar: 26 J (19 ft · lbf). See Table 8.

Fracture toughness. At 24 °C (75 °F): sand cast, 14.5 MPa$\sqrt{m}$ (13.2 ksi$\sqrt{in.}$); pressure die cast, 14.4 MPa$\sqrt{m}$ (13.1 ksi$\sqrt{in.}$). See Table 8.

Fatigue strength. Reverse bending at 5×10^8 cycles: die cast, 117 MPa (17.0 ksi); sand cast, 103 MPa (15.0 ksi)

Creep strength. Stress to give 1% secondary creep extension in 10^5 h, per ASME boiler code: 69 MPa (10.0 ksi) at 20 °C (68 °F)

Mass Characteristics

Density. 6.03 g/cm^3 (0.218 lb/in.^{3}) at 21 °C (70 °F)

Volume change on freezing. 1.3% shrinkage

Thermal Properties

Liquidus temperature. 432 °C (810 °F)

Solidus temperature. 377 °C (710 °F)

Coefficient of linear thermal expansion. 24.1 μm/m · K (13.4 μin./in. · °F) at 20 to 100 °C (68 to 212 °F)

Latent heat of fusion. 118 kJ/kg (52 Btu/lb)

Specific heat. 0.450 kJ/kg · K (0.107 Btu/lb · °F) at 20 to 100 °C (68 to 212 °F)

Thermal conductivity. 116 W/m · K (805 Btu · in./ft^2 · h · °F)

Electrical Properties

Electrical conductivity. 28.3% IACS at 20 °C (68 °F)

Electrical resistivity. 6.1 μΩ · cm (2.4 μΩ · in.) at 20 °C (68 °F)

Chemical Properties

General corrosion behavior. This alloy has better resistance than pure zinc and hypoeutectic zinc-aluminum alloys, and it can safely be used wherever zinc, zinc-coated iron, or zinc-coated steel has been used successfully in the past.

Other Properties

Damping. Q^{-1} ($\times 10^3$) at 100 Hz, pressure die cast: 5.1 at 20 °C (68 °F); 9.8 at 100 °C (212 °F). See Table 9.

Fabrication Characteristics

Joining. Mechanical fasteners, adhesives, TIG or MIG welding. Lead-tin or zinc-cadmium soft solder. Resistance welding pulsation technique can be used.

Recommended casting temperature range. 460 to 490 °C (860 to 915 °F)

Precautions in melting. Stir well after meltdown and allow a 15-min holding period. Skim surface dross. Avoid high melt temperatures to prevent magnesium loss.

Precautions in casting. Strict adherence to chemical composition limits and proper pattern or die design are necessary to ensure sound castings and high strength and ductility. Not recommended for hot chamber die casting unless shot end components and pot are made with special materials

ZA-27
Zn-27Al-2Cu-0.015Mg

Commercial Names

Trade name. ZA-27, zinc foundry alloy

Specifications

ASTM. B 791 (castings), B 669 (ingot)

UNS. Z35840 (ingot), Z35841 (castings)

Foreign. BS DD 139 (castings); SAA-AS1881

Chemical Composition

Composition limits. ASTM B 669 (ingot): 25.5 to 28.0 Al, 2.0 to 2.5 Cu, 0.012 to 0.020 Mg, 0.072 Fe max, 0.005 Pb max, 0.005 Cd max, 0.002 Sn max, bal Zn. ASTM B 791 (castings): 25.0 to 28.0 Al, 2.0 to 2.5 Cu, 0.01 to 0.02 Mg, 0.075 Fe max, 0.006 Pb max, 0.006 Cd max, 0.003 Sn max, bal Zn

Consequence of exceeding impurity limits. Alloy becomes subject to intergranular corrosive attack and fails prematurely by warping and cracking. High iron causes excessive tool wear.

Applications

Typical uses. For pressure die castings and gravity castings, wherever very high strength is required. In automobile engine mounts and drive trains, general hardware, agricultural equipment, domestic and garden appliances, and heavy-duty hand and work tools. This alloy is extensively used in bearings and bushings for high-load low-speed applications.

Mechanical Properties

Tensile properties. Die cast specimen, 6.35 mm (0.25 in.) in diameter: tensile strength, 426 MPa (61.8 ksi); yield strength (0.2% offset), 371 MPa (53.8 ksi); elongation, 2.5% in 50 mm (2 in.) gage length. Permanent mold cast specimen, 12.7 mm (0.5 in.) in diameter: tensile strength, 424 MPa (61.5 ksi), elongation, 2% in 50 mm (2 in.). Sand cast specimen, 12.7 mm (0.5 in.) in diameter: tensile strength, 421 MPa (61.0 ksi); elongation, 4.5% in 50 mm (2 in.). Elastic modulus, all processes: 77.9 GPa (11.3×10^6 psi). Tensile and other mechanical properties are affected by operating temperature and time (see Table 8 and Fig. 1).

Shear strength. Die cast, 325 MPa (47.1 ksi); sand cast, 292 MPa (42.3 ksi)

Compressive strength. 0.1% offset: die cast, 359 MPa (52.1 ksi); sand cast, 330 MPa (47.8 ksi)

Hardness. 500 kg load, 10 mm hardened steel ball, 30-s duration: die cast, 119 HB; permanent mold, 113 HB; sand cast, 90 HB

Impact strength. Charpy, unnotched. Die cast, 6.35 mm (0.25 in.) square bar: 12 J (9 ft · lbf); sand cast, machined 10.0 mm (0.394 in.) square bar: 48 J (35 ft · lbf). See Table 8.

Fracture toughness. At 24 °C (75 °F): sand cast, 23.7 MPa$\sqrt{m}$ (21.6 ksi$\sqrt{in.}$); pressure die cast, 20.2 MPa$\sqrt{m}$ (18.4 ksi$\sqrt{in.}$). See Table 8.

Fatigue strength. Reverse bending at 5×10^8 cycles: die cast, 117 MPa (17.0 ksi); sand cast, 172 MPa (25.0 ksi)

Creep strength. Stress to give 1% secondary creep extension in 10^5 h, per ASME boiler code: pressure die cast, 69 MPa (10.0 ksi) at 20 °C (68 °F); sand cast, 76 MPa (11.0 ksi); sand cast and homogenized (320 °C, or 608 °F, furnace cooled), 95 MPa (13.8 ksi). See Fig. 4.

Mass Characteristics

Density. 5.00 g/cm^3 (0.181 lb/in.^{3}) at 21 °C (70 °F)

Volume change on freezing. 1.3% shrinkage

Thermal Properties

Liquidus temperature. 484 °C (903 °F)

Solidus temperature. 375 °C (708 °F)

Coefficient of linear thermal expansion. 26.0 μm/m · K (14.4 μin./in. · °F) at 20 to 100 °C (68 to 212 °F)

Latent heat of fusion. 128 kJ/kg (55 Btu/lb)

Specific heat. 0.525 kJ/kg · K (0.125 Btu/lb · °F) at 20 to 100 °C (68 to 212 °F)

Thermal conductivity. 125.5 W/m · K (870 Btu · in./ft^2 · h · °F)

Electrical Properties

Electrical conductivity. 29.7% IACS at 20 °C (68 °F)

Electrical resistivity. 5.8 μΩ · cm (2.3 μΩ · in.) at 20 °C (68 °F)

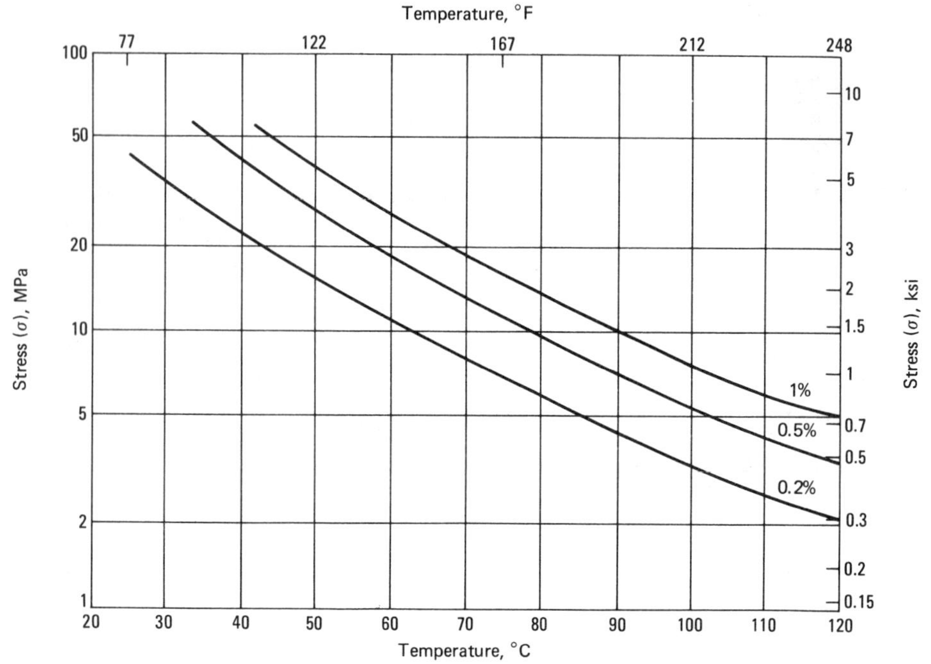

Fig. 4 Elongation of pressure die cast zinc alloy ZA-27 at various combinations of stress and temperature for a service life of 3 × 10³ h. Source: AM&S Europe Ltd.

Chemical Properties

General corrosion behavior. This alloy has better resistance than pure zinc and hypoeutectic zinc-aluminum alloys, and it can safely be used wherever zinc, zinc-coated iron, or zinc-coated steel has been used successfully in the past.

Other Properties

Damping. Q^{-1} (× 10^3) at 100 Hz: pressure die cast, 5.0 at 20 °C (68 °F); 10.9 at 100 °C (212 °F). See Table 9.

Fabrication Characteristics

Joining. Mechanical fasteners, adhesives, TIG or MIG welding. Lead-tin or zinc-cadmium soft solder. Resistance welding pulsation technique can be used.
Recommended casting temperature range: 515 to 545 °C (960 to 1015 °F)
Precautions in melting. Stir well after meltdown and allow a 15-min holding period. Skim surface dross. Avoid high melt temperatures to prevent magnesium loss.
Precautions in casting. Strict adherence to chemical composition limits and proper pattern or die design are necessary to ensure sound castings and high strength and ductility. Alloy cannot be hot chamber cast without special shot end and pot materials.

ILZRO 16
Zn-1.25Cu-0.2Ti-0.15Cr

Commercial Names

Trade name. ILZRO 16

Chemical Composition

Composition limits. 1.0 to 1.5 Cu, 0.15 to 0.25 Ti, 0.10 to 0.20 Cr, 0.01 to 0.04 Al, 0.02 Mg max, 0.04 Fe max, 0.005 Pb max, 0.004 Cd max, 0.003 Sn max, bal Zn

Applications

Typical uses. Sustained high-load bearing components for elevated-temperature service

Mechanical Properties

Tensile properties. Die cast specimen, 6.35 mm (0.25 in.) in diameter: tensile strength, 230 MPa (33.5 ksi); yield strength (0.2% offset), 140 MPa (20.5 ksi); elongation, 6% in 50 mm (2 in.) gage length. Elastic modulus, 97.0 GPa (14.0 × 10^6 psi)
Hardness. 500 kg load, 10 mm hardened steel ball, 30-s duration: die cast, 76 HB; permanent mold, 113 HB; sand cast, 90 HB
Impact strength. Charpy, unnotched. Die cast, 6.35 mm (0.25 in.) square bar: 25 J (18 ft · lbf)
Creep strength. Stress to give 1% secondary creep extension in 10^5 h, per ASME boiler code: 95 MPa (13.8 ksi) at 20 °C (68 °F), 28 MPa (4.0 ksi) at 100 °C (212 °F). See Fig. 5.

Mass Characteristics

Density. 7.1 g/cm³ (0.256 lb/in.³) at 21 °C (70 °F)

Thermal Properties

Liquidus temperature. 416 °C (781 °F)
Solidus temperature. 418 °C (785 °F)
Coefficient of linear thermal expansion. 27.0 μm/m · K (15.0 μin./in. · °F) at 20 to 100 °C (68 to 212 °F)
Specific heat. 0.402 kJ/kg · K (0.096 Btu/lb · °F) at 20 to 100 °C (68 to 212 °F)
Thermal conductivity. 104.7 W/m · K (726 Btu · in./ft² · h · °F)

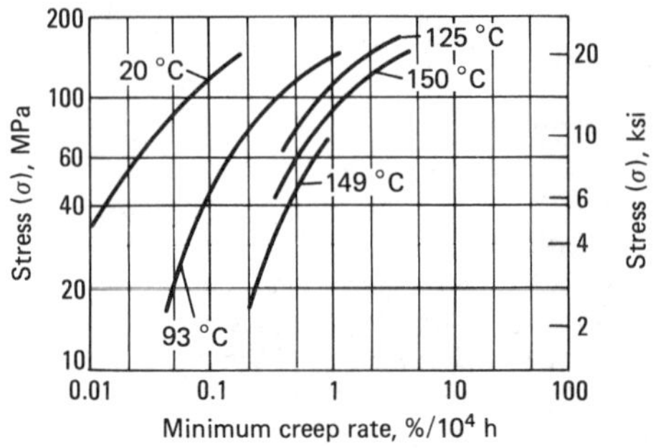

Fig. 5 Tensile creep properties of zinc alloy ILZRO 16 at various temperatures. (0.1%/10⁴ h) = (1%/10⁵ h). Source: *Engineering Properties of Zinc Alloys,* International Lead-Zinc Research Organization, 1989

Electrical Properties

Electrical resistivity. 8.4 μΩ · cm (3.3 μΩ · in.) at 20 °C (68 °F)

Fabrication Characteristics

Recommended casting temperature range. 460 to 470 °C (860 to 880 °F)
Precautions in melting. Use refractory crucibles. Stir well after meltdown.
Precautions in casting. Strict adherence to chemical composition limits and proper pattern or die design are necessary to ensure sound castings and high strength. Alloy cannot be hot chamber die cast without special shot end and pot materials.

Slush Casting Alloy
Zn-4.75Al-0.25Cu

Commercial Names

Trade name. Slush casting Alloy A

Specifications

ASTM. B 792 (ingot)
UNS. Z34510 (ingot)

Chemical Composition

Composition limits. ASTM B 792 (ingot): 4.5 to 5.0 Al, 0.2 to 0.3 Cu, 0.10 Fe max, 0.007 Pb max, 0.005 Cd max, 0.005 Sn max, bal Zn
Consequence of exceeding impurity limits. Alloy becomes hot short, subject to intergranular corrosive attack, and fails prematurely by warping and cracking.

Applications

Typical uses. For all slush and permanent mold castings, chiefly in the manufacture of lighting fixtures and statues

Mechanical Properties

Tensile properties. Chill cast specimen, 12.7 mm (0.50 in.) in diameter: tensile strength,

193 MPa (28.0 ksi); elongation, 1.0% in 50 mm (2 in.) gage length
Impact strength. Charpy, unnotched: chill cast 6.35 mm (0.25 in.) square bar, 4 J (3.0 ft · lbf)

Thermal Properties

Liquidus temperature. Approximately 390 °C (734 °F)
Solidus temperature. 380 °C (716 °F)

Chemical Properties

General corrosion behavior. This alloy has better resistance than pure zinc and can be safely used wherever zinc, zinc-coated iron, or zinc-coated steel has been used successfully in the past.

Fabrication Characteristics

Joining. Mechanical fasteners, adhesives, TIG or MIG welding. Lead-tin or zinc-cadmium soft solder. Resistance welding pulsation technique can be used.
Precautions in casting. Avoid contamination with lead, tin, or cadmium.

Slush Casting Alloy Zn-5.5Al

Commercial Names

Trade name. Slush casting Alloy B, unbreakable metal

Specifications

ASTM. B 792 (ingot)
UNS. Z30500 (ingot)

Chemical Composition

Composition limits. ASTM B 792 (ingot): 5.25 to 5.75 Al, 0.1 Cu max, 0.10 Fe max, 0.007 Pb max, 0.005 Cd max, 0.005 Sn max, bal Zn
Consequence of exceeding impurity limits. Alloy becomes hot short, subject to intergranular corrosive attack, and fails prematurely by warping and cracking.

Applications

Typical uses. For all slush and permanent mold castings, chiefly in the manufacture of lighting fixtures and statues

Mechanical Properties

Tensile properties. Chill cast specimen, 12.7 mm (0.50 in.) in diameter: tensile strength, 172 MPa (25.0 ksi); elongation, 1.0% in 50 mm (2 in.) gage length
Impact strength. Charpy, unnotched: chill cast 6.35 mm (0.25 in.) square bar, 1.4 J (1.0 ft · lbf)

Thermal Properties

Liquidus temperature. Approximately 395 °C (745 °F)
Solidus temperature. 380 °C (716 °F)

Chemical Properties

General corrosion behavior. This alloy has better resistance than pure zinc and can be safely used wherever zinc, zinc-coated iron, or zinc-coated steel has been used successfully in the past.

Fabrication Characteristics

Joining. Mechanical fasteners, adhesives, TIG or MIG welding. Lead-tin or zinc-cadmium soft solder. Resistance welding pulsation technique can be used.
Precautions in casting. Avoid contamination with lead, tin, or cadmium.

Commercial Rolled Zinc Zn-0.08Pb

Commercial Names

Previous trade name. Deep-drawing zinc

Specifications

ASTM. B 69
UNS. Z21210

Chemical Composition

Composition limits. ASTM B 69: 0.10 Pb max, 0.012 Fe max, 0.005 Cd max, 0.001 Cu max, 0.001 Al max, 0.001 Sn max, bal Zn
Consequence of exceeding impurity limits. Iron or cadmium increases hardness and reduces ductility. Alloy becomes hot short, subject to intergranular corrosive attack, and fails prematurely by warping and cracking.

Applications

Typical uses. Generally drawn, formed, or spun articles requiring some rigidity, such as drawn battery cans and formed eyelets and grommets
Precautions in use. Deforms under light continuous load, particularly at elevated temperatures

Mechanical Properties

In wrought materials, properties often are anisotropic (dependent on direction). Those properties that are strongly anisotropic are given in longitudinal and transverse direction (relative to rolling direction), respectively; for example, tensile strength, 134 MPa (19.4 ksi) (longitudinal) to 159 MPa (23.0 ksi) (transverse).
Tensile properties. Hot-rolled specimen: tensile strength, 134 to 159 MPa (19.5 to 23.0 ksi); elongation, 65 to 50%. Cold-rolled specimen: tensile strength, 145 to 186 MPa (21.0 to 27.0 ksi); elongation, 50 to 40%
Hardness. Hot rolled, 42 HB
Fatigue strength. Rotating beam, hot-rolled strip: 17 MPa (2.5 ksi) at 10^8 cycles

Mass Characteristics

Density. 7.14 g/cm^3 (0.258 lb/in.^{3}) at 21 °C (70 °F)

Thermal Properties

Some properties are anisotropic (dependent on direction). Those properties that are strongly anisotropic are given in longitudinal and transverse direction (relative to rolling direction), respectively; for example, coefficient of thermal expansion, 32.5 μm/m · K (18.05 μin./in. · °F) (longitudinal) to 23.0 μm/m · K (12.7 μin./in. · °F) (transverse).
Melting point. 419 °C (786 °F)
Coefficient of linear thermal expansion. 32.5 to 23 μm/m · K (18.05 to 12.7 μin./in. · °F) at 20 to 40 °C (68 to 105 °F)
Specific heat. 0.395 kJ/kg · K (0.094 Btu/lb · °F) at 20 to 100 °C (68 to 212 °F)
Thermal conductivity. 108 W/m · K (749 Btu · in./ft^2 · h · °F)

Electrical Properties

Electrical conductivity. 28.4% IACS at 20 °C (68 °F)
Electrical resistivity. 6.2 μΩ · cm (2.4 μΩ · in.) at 20 °C (68 °F)

Chemical Properties

General corrosion behavior. This alloy has excellent resistance to atmospheric corrosion.

Fabrication Characteristics

Forming methods. Suited to drawing, bending, roll forming, spinning, swaging, and impact extrusion
Precautions in forming. Keep temperature above 21 °C (70 °F). Use pure soapy water or noncorrosive mineral oil as lubricant.
Precautions in finishing. Air-dried coatings are preferred over baked finishes because many enamels must be baked at temperatures high enough to change the mechanical properties of the rolled zinc.
Joining. Mechanical fasteners, adhesives. Lead-tin or zinc-cadmium soft solder. Resistance welding pulsation technique can be used.
Hot-working temperature. 120 to 275 °C (250 to 525 °F). Hot shortness occurs at 300 to 420 °C (570 to 785 °F).

Commercial Rolled Zinc Zn-0.06Pb-0.06Cd

Specifications

ASTM. B 69
UNS. Z21220

Chemical Composition

Composition limits. ASTM B 69: 0.05 to 0.10 Pb, 0.012 Fe max, 0.05 to 0.08 Cd, 0.005 Cu max, 0.001 Al max, 0.001 Sn max, bal Zn
Consequence of exceeding impurity limits. Iron or cadmium increases hardness and reduces ductility. Alloy becomes hot short, subject to intergranular corrosive attack, and fails prematurely by warping and cracking.

Applications

Typical uses. Generally drawn, formed, or spun articles requiring some rigidity, such

as drawn battery cans and formed eyelets and grommets
Precautions in use. Deforms under light continuous load, particularly at elevated temperatures

Mechanical Properties

Some properties are anisotropic (dependent on direction). Those properties that are strongly anisotropic are given in longitudinal and transverse direction (relative to rolling direction), respectively; for example, tensile strength, 150 MPa (21.8 ksi) (longitudinal) to 170 MPa (24.7 ksi) (transverse).
Tensile properties. Hot-rolled specimen: tensile strength, 145 to 173 MPa (21.0 to 25.0 ksi); elongation, 52 to 30%. Cold-rolled specimen: tensile strength, 152 to 201 MPa (22.0 to 29.0 ksi); elongation, 40 to 30%
Hardness. Hot rolled: 43 HB
Fatigue strength. Rotating beam, hot-rolled strip: 26 MPa (3.8 ksi) at 10^8 cycles
Shear strength. Approximately 124 to 138 MPa (18.0 to 20.0 ksi)

Mass Characteristics

Density. 7.14 g/cm^3 (0.258 lb/in.^{3}) at 21 °C (70 °F)

Thermal Properties

Some properties are anisotropic (dependent on direction). Those properties that are strongly anisotropic are given in longitudinal and transverse direction (relative to rolling direction), respectively; for example, coefficient of thermal expansion, 32.5 μm/m · K (18.1 μin./in. · °F) (longitudinal) to 23.0 μm/m · K (12.7 μin./in. · °F) (transverse).
Melting point. 419 °C (786 °F)
Coefficient of linear thermal expansion. 32.5 to 23 μm/m · K (18.05 to 12.7 μin./in. · °F) at 20 to 40 °C (68 to 105 °F)
Specific heat. 0.395 kJ/kg · K (0.094 Btu/lb · °F) at 20 to 100 °C (68 to 212 °F)
Thermal conductivity. 108 W/m · K (749 Btu · in./ft^2 · h · °F)

Electrical Properties

Electrical conductivity. 32% IACS at 20 °C (68 °F)
Electrical resistivity. 6.06 μΩ · cm (2.4 μΩ · in.) at 20 °C (68 °F)

Chemical Properties

General corrosion behavior. This alloy has excellent resistance to atmospheric corrosion.

Fabrication Characteristics

Forming methods. Suited to drawing, bending, roll forming, spinning, swaging, and impact extrusion
Precautions in forming. Keep temperature above 21 °C (70 °F). Use pure soapy water or noncorrosive mineral oil as lubricant.
Precautions in finishing. Air-dried coatings are preferred over baked finishes because many enamels must be baked at temperatures high enough to change the mechanical properties of the rolled zinc.
Joining. Mechanical fasteners, adhesives. Lead-tin or zinc-cadmium soft solder. Resistance welding pulsation technique can be used.
Hot-working temperature. 120 to 275 °C (250 to 525 °F). Hot shortness occurs at 300 to 420 °C (570 to 785 °F).

Commercial Rolled Zinc Zn-0.3Pb-0.03Cd

Specifications

ASTM. B 69
UNS. Z21540

Chemical Composition

Composition limits. ASTM B 69: 0.25 to 0.50 Pb, 0.25 to 0.45 Cd, 0.002 Fe max, 0.005 Cu max, 0.001 Al max, 0.001 Sn max, bal Zn
Consequence of exceeding impurity limits. Iron or cadmium increases hardness and reduces ductility. Alloy becomes hot short, subject to intergranular corrosive attack, and fails prematurely by warping and cracking.

Applications

Typical uses. Plates and strip for soldered battery cans; photoengraver's and lithographer's sheet
Precautions in use. Deforms under continuous load, particularly at elevated temperatures

Mechanical Properties

Some properties are anisotropic (dependent on direction). Those properties that are strongly anisotropic are given in longitudinal and transverse direction (relative to rolling direction), respectively; for example, tensile strength, 160 MPa (23.2 ksi) (longitudinal) to 200 MPa (29.0 ksi) (transverse).
Tensile properties. Hot-rolled specimen: tensile strength, 160 to 200 MPa (23.2 to 29.0 ksi); elongation, 50 to 32%. Cold-rolled specimen: tensile strength, 170 to 210 MPa (24.7 to 30.5 ksi); elongation, 45 to 28%
Hardness. Hot rolled: 47 HB
Fatigue strength. Rotating beam, hot-rolled strip: 28 MPa (4.1 ksi) at 10^8 cycles

Mass Characteristics

Density. 7.14 g/cm^3 (0.258 lb/in.^{3}) at 21 °C (70 °F)

Thermal Properties

Some properties are anisotropic (dependent on direction). Those properties that are strongly anisotropic are given in longitudinal and transverse direction (relative to rolling direction), respectively; for example, coefficient of thermal expansion, 39.9 μm/m · K (22.2 μin./in. · °F) (longitudinal) to 23.4 μm/m · K (13.0 μin./in. · °F) (transverse).
Melting point. 419 °C (786 °F)
Coefficient of linear thermal expansion. 39.9 to 23.4 μm/m · K (22.2 to 13.0 μin./in. · °F) at 20 to 40 °C (68 to 105 °F)
Specific heat. 0.395 kJ/kg · K (0.094 Btu/lb · °F) at 20 to 100 °C (68 to 212 °F)
Thermal conductivity. 108 W/m · K (749 Btu · in./ft^2 · h · °F)

Electrical Properties

Electrical conductivity. Approximately 32% IACS at 20 °C (68 °F)
Electrical resistivity. Approximately 6.06 μΩ · cm (2.4 μΩ · in.) at 20 °C (68 °F)

Chemical Properties

General corrosion behavior. This alloy has excellent resistance to atmospheric corrosion.

Fabrication Characteristics

Forming methods. Suited to drawing, bending, roll forming, spinning, swaging, and impact extrusion
Precautions in forming. Keep temperature above 21 °C (70 °F). Use pure soapy water or noncorrosive mineral oil as lubricant.
Annealing temperature. 105 °C (220 °F)
Precautions in finishing. Air-dried coatings are preferred over baked finishes because many enamels must be baked at temperatures high enough to change the mechanical properties of the rolled zinc.
Joining. Mechanical fasteners, adhesives. Lead-tin or zinc-cadmium soft solder. Resistance welding pulsation technique can be used.
Hot-working temperature. 120 to 275 °C (250 to 525 °F). Hot shortness occurs at 300 to 420 °C (570 to 785 °F).

Copper-Hardened Rolled Zinc Zn-1.0Cu

Specifications

ASTM. B 69
UNS. Z44330

Chemical Composition

Composition limits. ASTM B 69: 0.85 to 1.25 Cu, 0.10 Pb max, 0.012 Fe max, 0.005 Cd max, 0.001 Al max, 0.001 Sn max, bal Zn
Consequence of exceeding impurity limits. Iron or cadmium increases hardness and reduces ductility. Alloy becomes hot short, subject to intergranular corrosive attack, and fails prematurely by warping and cracking.

Applications

Typical uses. Weather stripping, nameplates, ferrules, and drawn, formed, or spun articles requiring stiffness
Precautions in use. Deforms under heavy

continuous load, particularly at elevated temperatures

Mechanical Properties

Some properties are anisotropic (dependent on direction). Those properties that are strongly anisotropic are given in longitudinal and transverse direction (relative to rolling direction), respectively; for example, tensile strength, 170 MPa (24.7 ksi) (longitudinal) to 210 MPa (30.5 ksi) (transverse).
Tensile properties. Hot-rolled specimen: tensile strength, 170 to 210 MPa (24.7 to 30.5 ksi); elongation, 50 to 35%. Cold-rolled specimen: tensile strength, 210 to 280 MPa (30.5 to 40.6 ksi); elongation, 40 to 25%
Hardness. Hot rolled, 52 HB; cold rolled, 60 HB
Fatigue strength. Rotating beam, hot-rolled strip: 28 MPa (4.1 ksi) at 10^8 cycles
Shear strength. 138 to 152 MPa (20.0 to 22.0 ksi)

Mass Characteristics

Density. 7.17 g/cm^3 (0.259 lb/in.^{3}) at 21 °C (70 °F)

Thermal Properties

Some properties are anisotropic (dependent on direction). Those properties that are strongly anisotropic are given in longitudinal and transverse direction (relative to rolling direction), respectively; for example, coefficient of thermal expansion, 32.5 μm/m · K (18.1 μin./in. · °F) (longitudinal) to 23.0 μm/m · K (12.9 μin./in. · °F) (transverse).
Liquidus temperature. 422 °C (792 °F)
Solidus temperature. 419 °C (786 °F)
Coefficient of linear thermal expansion. 34.7 to 21.1 μm/m · K (19.3 to 11.7 μin./in. · °F) at 20 to 40 °C (68 to 105 °F)
Specific heat. 0.402 kJ/kg · K (0.096 Btu/lb · °F) at 20 to 100 °C (68 to 212 °F)
Thermal conductivity. 104.7 W/m · K (726 Btu · in./ft^2 · h · °F)

Electrical Properties

Electrical conductivity. Approximately 28% IACS at 20 °C (68 °F)
Electrical resistivity. Approximately 6.2 μΩ · cm (2.45 μΩ · in.) at 20 °C (68 °F)

Chemical Properties

General corrosion behavior. This alloy has excellent resistance to atmospheric corrosion.

Fabrication Characteristics

Forming methods. Suited to drawing, bending, roll forming, spinning, and swaging
Precautions in forming. Keep temperature above 21 °C (70 °F). Use pure soapy water or noncorrosive mineral oil as lubricant.
Precautions in finishing. Air-dried coatings are preferred over baked finishes because many enamels must be baked at temperatures high enough to change the mechanical properties of the rolled zinc.
Joining. Mechanical fasteners, adhesives. Lead-tin or zinc-cadmium soft solder. Resistance welding pulsation technique can be used.
Hot-working temperature. 175 to 300 °C (345 to 570 °F). Hot shortness occurs at 300 to 420 °C (570 to 785 °F).

Rolled Zinc Alloy Zn-1.0Cu-0.010Mg

Specifications

ASTM. B 69
UNS. Z45330

Chemical Composition

Composition limits. ASTM B 69: 0.85 to 1.25 Cu, 0.006 to 0.016 Mg, 0.15 Pb max, 0.012 Fe max, 0.04 Cd max, 0.001 Al max, 0.001 Sn max, bal Zn
Consequence of exceeding impurity limits. Iron or cadmium increases hardness and reduces ductility. Alloy becomes hot short, subject to intergranular corrosive attack, and fails prematurely by warping and cracking.

Applications

Typical uses. Corrugated roofing and flat, drawn, or mildly formed articles requiring maximum stiffness
Precautions in use. Deforms under heavy continuous load, particularly at elevated temperatures

Mechanical Properties

Some properties are anisotropic (dependent on direction). Those properties that are strongly anisotropic are given in longitudinal and transverse direction (relative to rolling direction), respectively; for example, tensile strength, 200 MPa (29.0 ksi) (longitudinal) to 276 MPa (40.0 ksi) (transverse).
Tensile properties. Hot-rolled specimen: tensile strength, 200 to 276 MPa (29.0 to 40.0 ksi); elongation, 20 to 10%. Cold-rolled specimen: tensile strength, 248 to 317 MPa (36.0 to 46.0 ksi); elongation, 25 to 10%
Hardness. Hot rolled, 61 HB; cold rolled, 80 HB
Fatigue strength. Rotating beam, hot-rolled strip: 47 MPa (6.8 ksi) at 10^8 cycles

Mass Characteristics

Density. 7.17 g/cm^3 (0.259 lb/in.^{3}) at 21 °C (70 °F)

Thermal Properties

Some properties are anisotropic (dependent on direction). Those properties that are strongly anisotropic are given in longitudinal and transverse direction (relative to rolling direction), respectively; for example, coefficient of thermal expansion, 34.8 μm/m · K (19.3 μin./in. · °F) (longitudinal) to 21.1 μm/m · K (11.7 μin./in. · °F) (transverse).
Liquidus temperature. 422 °C (792 °F)
Solidus temperature. 419 °C (786 °F)
Coefficient of linear thermal expansion. 34.8 to 21.1 μm/m · K (19.3 to 11.7 μin./in. · °F) at 20 to 40 °C (68 to 105 °F)
Specific heat. 0.401 kJ/kg · K (0.096 Btu/lb · °F) at 20 to 100 °C (68 to 212 °F)
Thermal conductivity. 105 W/m · K (728 Btu · in./ft^2 · h · °F)

Electrical Properties

Electrical conductivity. Approximately 27% IACS at 20 °C (68 °F)
Electrical resistivity. Approximately 63 μΩ · cm (2.5 μΩ · in.) at 20 °C (68 °F)

Chemical Properties

General corrosion behavior. This alloy has excellent resistance to atmospheric corrosion.

Fabrication Characteristics

Forming methods. Suited to drawing, bending, and roll forming
Precautions in forming. Keep temperature above 21 °C (70 °F). Use pure soapy water or noncorrosive mineral oil as lubricant.
Precautions in finishing. Air-dried coatings are preferred over baked finishes because many enamels must be baked at temperatures high enough to change the mechanical properties of the rolled zinc.
Joining. Mechanical fasteners, adhesives. Lead-tin or zinc-cadmium soft solder. Resistance welding pulsation technique can be used.
Annealing temperature. 175 °C (345 °F)
Hot-working temperature. 175 to 300 °C (345 to 570 °F). Hot shortness occurs at 300 to 420 °C (570 to 785 °F).

Zn-Cu-Ti Alloy Zn-0.8Cu-0.15Ti

Specifications

ASTM. B 69
UNS. Z41320

Chemical Composition

Composition limits. ASTM B 69: 0.50 to 1.50 Cu, 0.12 to 0.50 Ti, 0.10 Pb max, 0.012 Fe max, 0.05 Cd max, 0.001 Al max, 0.001 Sn max, bal Zn
Consequence of exceeding impurity limits. Iron or cadmium increases hardness and reduces ductility. Alloy becomes hot short, subject to intergranular corrosive attack, and fails prematurely by warping and cracking.

Applications

Typical uses. Corrugated roofing, leaders and gutters, and formed articles requiring maximum creep resistance
Precautions in use. Creep resistance decreases with increasing temperature of use. Should be heat treated after cold working for maximum creep resistance

Mechanical Properties

Some properties are anisotropic (dependent on direction). Those properties that are strongly anisotropic are given in longitudinal and transverse direction (relative to rolling direction), respectively; for example, tensile strength, 221 MPa (32.0 ksi) (longitudinal) to 290 MPa (42.0 ksi) (transverse).

Tensile properties. Hot-rolled specimen: tensile strength, 221 to 290 MPa (32.0 to 42.0 ksi); yield strength (0.2% offset), 125 to 177 MPa (18.1 to 25.7 ksi); elastic modulus, 63.5 to 88.0 GPa (9.2 to 12.8 × 10^6 psi); elongation, 38 to 21%. Cold-rolled specimen: tensile strength, 200 to 260 MPa (29.0 to 37.7 ksi); elongation, 60 to 44%

Shear strength. 138 to 152 MPa (20.0 to 22.0 ksi)

Hardness. Hot rolled, 61 HB; cold rolled, 80 HB

Fatigue strength. Rotating beam, hot-rolled strip: 47 MPa (6.8 ksi) at 10^8 cycles

Mass Characteristics

Density. 7.17 g/cm^3 (0.259 lb/in.^{3}) at 21 °C (70 °F)

Thermal Properties

Some properties are anisotropic (dependent on direction). Those properties that are strongly anisotropic are given in longitudinal and transverse direction (relative to rolling direction), respectively; for example, coefficient of thermal expansion, 24.0 μm/m · K (13.3 μin./in. · °F) (longitudinal) to 19.4 μm/m · K (10.8 μin./in. · °F) (transverse).

Liquidus temperature. 422 °C (792 °F)

Solidus temperature. 419 °C (786 °F)

Coefficient of linear thermal expansion. 24.0 to 19.4 μm/m · K (13.3 to 10.8 μin./in. · °F) at 20 to 40 °C (68 to 105 °F)

Specific heat. 0.402 kJ/kg · K (0.096 Btu/lb · °F) at 20 to 100 °C (68 to 212 °F)

Thermal conductivity. 105 W/m · K (727 Btu · in./ft^2 · h · °F)

Electrical Properties

Electrical conductivity. 27% IACS at 20 °C (68 °F)

Electrical resistivity. 6.24 μΩ · cm (2.5 μΩ · in.) at 20 °C (68 °F)

Chemical Properties

General corrosion behavior. This alloy has excellent resistance to atmospheric corrosion.

Fabrication Characteristics

Forming methods. Suited to drawing, bending, and roll forming

Precautions in forming. Keep temperature above 21 °C (70 °F). Use pure soapy water or noncorrosive mineral oil as lubricant.

Joining. Mechanical fasteners, adhesives. Lead-tin or zinc-cadmium soft solder. Resistance welding pulsation technique can be used.

Annealing temperature. High creep resistance can be restored after cold working by annealing for 45 min at 250 °C (480 °F).

Hot-working temperature. 150 to 300 °C (300 to 570 °F). Hot shortness occurs at 300 to 420 °C (570 to 785 °F).

Superplastic Zinc
Zn-22Al

Commercial Names

Trade names. Super Z300, Formetal 22 Alloy, Korloy 2684

Chemical Composition

Composition limits. 21 to 23 Al, 0.40 to 0.60 Cu, 0.008 to 0.012 Mg, 0.01 Pb max, 0.002 Fe max, 0.01 Cd max, 0.001 Sn max, bal Zn

Consequence of exceeding impurity limits. Iron or cadmium increases hardness and reduces ductility. Alloy becomes hot short, subject to intergranular corrosive attack, and fails prematurely by warping and cracking.

Applications

Typical uses. Supplied as sheet for thermal forming. Especially useful for low-volume applications where tooling costs must be kept low. Used for electronic enclosures, cabinets and panels, business machine parts, and medical and other laboratory instruments and tools

Precautions in use. Subject to creep if highly stressed, particularly at elevated temperature

Mechanical Properties

Some properties are anisotropic (dependent on direction). Those properties that are strongly anisotropic are given in longitudinal and transverse direction (relative to rolling direction), respectively; for example, tensile strength, 310 MPa (45.0 ksi) (longitudinal) to 380 MPa (55.1 ksi) (transverse).

Tensile properties. As-rolled: tensile strength, 310 to 380 MPa (45.0 to 55.0 ksi); yield strength (0.2% offset), 255 to 297 MPa (37.0 to 43.0 ksi); elongation, 27 to 25%. Annealed at 315 °C (600 °F), air cooled: tensile strength, 400 to 441 MPa (58 to 64.0 ksi); yield strength (0.2% offset), 352 to 386 MPa (51.0 to 56.0 ksi); elongation, 11 to 9 %. Elastic modulus, 68 to 93 GPa (9.9 to 13.5 × 10^6 psi)

Hardness. As-rolled, 70 to 79 HRB; annealed, 84 to 85 HRB

Creep strength. Stress for 1% extension in 10^5 h at 20 °C (68 °F): as-rolled, 20 to 25 MPa (3.0 to 3.6 ksi); annealed, 40 to 69 MPa (5.8 to 10.0 ksi)

Impact strength. Rolled and annealed: 9.5 to 27 J (7 to 20 ft · lbf) at 20 °C (68 °F), depending on direction of testing

Damping. Q^{-1} (× 10^3) at 100 Hz: cold rolled, 7.8 at 20 °C (68 °F); 27.8 at 100 °C (212 °F). See Table 9.

Mass Characteristics

Density. 5.20 g/cm^3 (0.188 lb/in.^{3}) at 21 °C (70 °F)

Thermal Properties

Some properties are anisotropic (dependent on direction). Those properties that are strongly anisotropic are given in longitudinal and transverse direction (relative to rolling direction), respectively; for example, coefficient of thermal expansion, 22.0 μm/m · K (12.2 μin./in. · °F) (longitudinal) to 21.5 μm/m · K (11.9 μin./in. · °F) (transverse).

Coefficient of linear thermal expansion. At 20 to 40 °C (68 to 105 °F). As-rolled, 22.0 to 21.5 μm/m · K (12.2 to 11.9 μin./in. · °F); annealed, 26.6 to 26.8 μm/m · K (14.8 to 14.9 μin./in. · °F)

Electrical Properties

Electrical conductivity. At 20 °C (68 °F). As rolled, 32% IACS; annealed, 28% IACS

Electrical resistivity. 6.0 μΩ · cm (2.4 μΩ · in.) at 20 °C (68 °F)

Chemical Properties

General corrosion behavior. This alloy has excellent resistance to atmospheric corrosion.

Fabrication Characteristics

Forming methods. Suited to deep drawing, compressing molding, bending, stretch forming, and roll forming

Hot-working temperature. 250 to 270 °C (480 to 520 °F)

Annealing temperature. When superplastic zinc is reheated to above 275 °C (527 °F) and slowly cooled to room temperature, the superplastic properties disappear.

Joining. Mechanical fasteners, adhesives. Lead-tin or zinc-cadmium soft solder. Resistance welding pulsation technique can be used.

Lead and Lead Alloys

Revised by Anthony W. Worcester and John T. O'Reilly, The Doe Run Company

LEAD was one of the first metals known to man. Probably the oldest lead artifact is a figure made about 3000 BC, which was found in the temple of Osiris near Abydos. All civilizations, beginning with the ancient Egyptians, Assyrians, and Babylonians, have used lead for many ornamental and structural purposes. Many magnificent buildings erected in the 15th and 16th centuries still stand under their original lead roofs.

Pipe was one of the earliest applications of lead. The Romans produced 15 standard sizes of water pipe in regular 3 m (10 ft) lengths. In the old Roman baths at Bath, England, the lead pipe installed by the Romans 1900 years ago is still in use. The Romans made their pipe by folding heavy sheets of cast lead and fusing the seams together. Presently, battery applications constitute more than 80% of lead alloy use.

Lead Processing

Sources of Lead. Although there are at least 60 known lead-containing minerals, by far the most important as a source of primary lead is galena (PbS). Recycling of scrap lead (from batteries, lead sheet, and cable sheathing) is also a major source, providing more than half of the lead used in the United States. Antimonial lead, soft lead, and lead-calcium alloys are produced from recycled lead. Considerable tonnages of scrap solder and bearing metals are recovered and used again. Lead recycling is discussed in more detail in the article "Recycling of Lead" in this Volume.

Galena is generally associated with substantial amounts of zinc and copper minerals and is not normally mined independently. Whenever possible, galena is separated from other minerals through comminution followed by differential flotation. Sometimes, however, the mineral particles are so small and so tightly locked together that separation by flotation is not efficient. In such instances, bulk concentrates containing lead, zinc, and copper are produced. These concentrates are then separated by the Imperial Smelting Process or at the facilities of an integrated polymetallic smelter.

The major lead mining centers of the world are, in alphabetical order, Australia, Canada, Mexico, Peru, the Soviet Union, and the United States. Total annual world production of lead from ores is about 3.5×10^6 Mg (3.8×10^6 tons).

Smelting and Refining. Concentrates leaving the flotation mills contain at least 40% Pb; they generally contain about 70%. The concentrates are sintered to remove sulfur, convert the lead to oxides, and agglomerate the fine flotation product, which is not physically desirable in a blast furnace. The roasted sinter is charged into the top of a blast furnace along with suitable fluxes and coke; the resulting impure lead-base bullion, containing copper, silver, gold, and various impurities, is shipped to the refinery for further treatment before the metal is suitable for industrial use.

In the initial refining, copper is removed by cooling the bullion and the addition of sulfur (or sulfur in combination with pyrite) to the molten furnace bullion (usually in the smelter). The resulting copper-rich drosses are re-treated in a reverberatory furnace to produce a high-copper matte, which is further treated elsewhere to recover the copper.

In the next step, antimony, tin, and arsenic are removed, usually by the Harris process. In this process, additions of sodium hydroxide and sodium nitrate react with the antimony, tin, and arsenic to form sodium salts, which are then removed and treated to recover the metals. Alternatively, the antimony, tin, and arsenic can be oxidized by air and/or oxygen blowing (usually in a continuous softener reverberatory furnace) and later recovered from the resulting oxide dross or slag.

Precious metals are removed next. In the Parkes process, zinc is added to form intermetallic compounds with the precious metals. These compounds are skimmed from the molten bullion and further processed to recover the gold and silver. Residual zinc generally is removed by the vacuum dezincing process. In electrolytic practice, the drossed lead bullion is cast into anodes, which are placed in electrolytic cells. Lead dissolves from the anode and is deposited on the cathode. Precious metals, along with antimony, arsenic, tin, bismuth, and other nonferrous elements, remain on the anode as a slimy residue. This residue is further treated to recover its metallic values.

Bismuth, if present, can be removed either by electrolytic refining or by the Betterton-Kroll process, in which a calcium-magnesium alloy is added to the molten lead. The calcium and magnesium selectively react with the bismuth to form a floating intermetallic dross.

The purity of commercial lead refined by these processes varies with ore composition and processing conditions. Final purification, using caustic or nitre, attains a purity of greater than or equal to 99.99%. Pig lead generally is cast into bars weighing roughly 30 to 45 kg (60 to 100 lb), although blocks weighing about 900 kg (1 ton) have become increasingly popular in recent years, particularly among large users.

Compositions and Grades

Table 1 lists the Unified Numbering System (UNS) designations for various pure lead grades and lead-base alloys. The designations are grouped according to general nominal chemical contents.

Grades of Lead. Composition limits of the four grades of pig lead covered in ASTM B 29 (1979) are given in Table 2. These grades are pure lead (also called corroding lead) and common lead (both containing 99.94% min lead), and chemical lead and acid-copper lead (both containing 99.90% min lead). Lead of higher specified purity (99.99%) is also available in commercial quantities.

Specifications other than ASTM B 29 for grades of pig lead include federal specification QQ-L-171, German standard DIN 1719, British specification BS 334, Canadian Standard CSA-HP2, and Australian Standard 1812.

Corroding Lead. Most lead produced in the United States is pure (or corroding) lead (99.94% min Pb). Corroding lead, which exhibits the outstanding corrosion resistance typical of lead and its alloys, is named

Table 1 UNS categories and nominal compositions of various lead grades and lead-base alloys

Lead alloy type(a)	UNS designations
Pure leads (UNS L50000–L50099)	
Zone-refined lead (99.9999% Pb min)	L50001
Refined soft lead (99.999% Pb min)	L50005
Refined soft lead (99.99% Pb min)	L50011, L50012, L50013, L50014
Corroding lead (99.94% Pb min)	L50042
Common lead (99.94% Pb min)	L50045
Lead-silver alloys (UNS L50100–L50199)	
Cable-sheathing alloy (0.2% Ag, 99.8% Pb)	L50101
Electrowinning alloys (0.5–1.0% Ag, 99.5–99% Pb)	L50110, L50115, L50120
Electrowinning alloy (1.0% Ag, 1.0% As, 98% Pb)	L50122
Cathodic protection anode alloy (2.0% Ag, 98% Pb)	L50140
Solder alloys (1.0–1.5% Ag, 1.0 Sn, bal Pb)	L50121, L50131
Solder alloys (1.5–2.5% Ag, with no tin)	L50132, L50150, L50151
Solder alloy (1.5% Ag, 5.0% Sn, 93.5% Pb)	L50134
Solder alloy (2.5% Ag, 2.0% Sn, 95.5% Pb)	L50152
Solder alloy (5.0% Ag, 95% Pb)	L50170
Solder alloys (5.0% Ag, with 5% Sn or 5% In)	L50171, L50172
Solder alloy (5.5% Ag)	L50180
Lead-arsenic alloys (UNS L50300–L50399)	
Arsenical lead cable-sheathing alloy (0.15% As, 0.10% Bi, 0.10% Sn, 99.6% Pb)	L50310
Lead-barium alloys (UNS L50500–L50599)	
Lead-barium alloy (0.05% Ba, 99.9% Pb)	L50510
Lead-tin-barium alloys (0.05–0.10% Ba, 1.0–2.0% Sn, 97.9–99% Pb)	L50520–L50522, L50530, L50535
Frary metal (0.4–1.2% Ba, 0.5–0.8% Ca, 97.2–98.8% Pb)	L50540–L50543
Lead-calcium alloys (UNS L50700–L50899)	
Lead-calcium alloys (99.9% Pb, 0.008–0.03% Ca)	L50710, L50720
Cable-sheathing alloys (0.025% Ca, 99.7–99.9% Pb, 0.0–0.025% Sn)	L50712, L50713
Lead-copper-calcium alloy (99.9% Pb, 0.06% Cu, 0.03% Ca)	L50722
Electrowinning anode alloy (0.5% Ag, 99.4% Pb, 0.05% Ca)	L50730
Battery grid alloy (99.9% Pb, 0.06% Ca)	L50735
Battery grid alloys (0.065% Ca, 0.2–1.5% Sn, 99.7–98.4% Pb)	L50736, L50737, L50740, L50745, L50750, L50755
Battery grid alloys (0.07% Ca, 0.0–0.7% Sn, 99.2–99.9% Pb)	L50760, L50765
Battery grid alloys (0.10% Ca, 0.0–1.0% Sn, 98.9–99.9% Pb)	L50770, L50775, L50780, L50790
Battery grid alloys (0.12% Ca, 0.3% Sn, 99.6% Pb)	L50795, L50800
Bearing metal (0.02% Al, 0.04% Li, 0.7% Ca, 0.6% Na, 98.7% Pb)	L50810
Bearing metal (0.02% Al, 0.04% Li, 0.7% Ca, 0.2% Na, 0.4% Ba, 98.7% Pb)	L50820
Lead-calcium alloys (1.0–6.0% Ca, 94.0–99.0% Pb)	L50840, L50850, L50880
Lead-cadmium alloys (UNS L50900–L50999)	
Lead-cadmium eutectic alloy (17.0% Cd, 83.0% Pb)	L50940
Lead-copper alloys (UNS L51100–L51199)	
Copperized lead (0.05% Cu, 99.9% Pb)	L51110
Chemical lead (see Table 2)	L51120
Copper-bearing lead (0.06% Cu, 99.90% Pb min)	L51121
Lead-tellurium-copper alloys (0.06% Cu, 0.045–0.055% Te, 99.82–99.85% Pb min)	L51123, L51124
Copperized soft lead (0.06% Cu, 99.9% Pb min)	L51125
Copper-bearing alloy (51% Pb, 3.0% Sn, other 0.8% max, bal Cu) (alloy 485 in SAE J460)	L51180
Lead-indium alloys (UNS L51500–L51599)	
Lead-indium-silver solder alloys (2.38–2.5% Ag, 4.76–5.0% In, 92.5–92.8% Pb)	L51510, L51512
Lead-indium solder alloys (5.0% In, 95.0% Pb)	L51511
Lead-indium alloys (19.0–70% In, 30–81% Pb)	L51530, L51532, L51535, L51540, L51550, L51560, L51570
Indium-tin-lead alloy (40% In, 40% Sn, 20% Pb)	L51545
Indium-silver-lead alloy (80% In, 5% Ag, 15% Pb)	L51585
Lead-lithium alloys (UNS L51700–L51799)	
Lead-lithium alloys (0.01–0.07% Li, 99.9% Pb)	L51705, L51708, L51710, L51720, L51730
Lead-tin-lithium alloys (0.02–0.04% Li, 0.35–0.7% Sn, 99.2–99.9% Pb)	L51740, L51748
Lead-tin-lithium-calcium alloys (0.08–0.065% Li, 1–2% Sn, 0.02–0.15% Ca, 97.8–99.6% Pb)	L51770, L51775, L51778, L51780, L51790
Lead-antimony alloys (UNS L52500–L53799)(b)	
Lead-antimony alloys (<1.0% Sb)	L52500–L52599
Lead-antimony alloys (1.0–1.99% Sb)	L52600–L52699
Lead-antimony alloys (2.0–2.99% Sb)	L52700–L52799
Lead-antimony alloys (3.0–3.99% Sb)	L52800–L52899
Lead-antimony alloys (4.0–4.99% Sb)	L52900–L52999
Lead-antimony alloys (5.0–5.99% Sb)	L53000–L53099
Lead-antimony alloys (6.0–6.99% Sb)	L53100–L53199
Lead-antimony alloys (7.0–8.99% Sb)	L53200–L53299
Lead-antimony alloys (9.0–10.99% Sb)	L53300–L53399
Lead-antimony alloys (11.0–12.99% Sb)	L53400–L53499
Lead-antimony alloys (13.0–15.99% Sb)	L53500–L53599
Lead-antimony alloys (16.0–19.99% Sb)	L53600–L53699
Lead-antimony alloys (>20% Sb)	L53700–L53799
Lead-tin alloys (UNS L54000–L55099)(c)	
Lead-tin alloys (<1.0% Sn)	L54000–L54099
Lead-tin alloys (1.0–1.99% Sn)	L54100–L54199
Lead-tin alloys (2.0–3.99% Sn)	L54200–L54299
Lead-tin alloys (4.0–7.99% Sn)	L54300–L54399
Lead-tin alloys (8.0–11.99% Sn)	L54400–L54499
Lead-tin alloys (12.0–15.99% Sn)	L54500–L54599
Lead-tin alloys (16.0–19.99% Sn)	L54600–L54699
Lead-tin alloys (20.0–27.99% Sn)	L54700–L54799
Lead-tin alloys (28.0–37.99% Sn)	L54800–L54899
Lead-tin alloys (38.0–47.99% Sn)	L54900–L54999
Lead-tin alloys (48.0–57.99% Sn)	L55000–L55099
Lead-strontium alloys (UNS L55200–L55299)	
Battery alloys (0.06–0.2% Sr, 0.0–0.03% Al, 0.0–0.08% Sn, 0.0–0.6% Ca, 99–99.8% Pb)	L55210, L55230, L55260
Lead-strontium alloy (2% Sr, 98% Pb)	L55290

(a) Unless otherwise specified as a minimum (min) or balance (bal), the listed compositions represent nominal values (or the range of nominal values when several alloy designations are grouped together). (b) See Table 3 for compositions of specific lead-antimony alloys. (c) See Table 4 for compositions of specific lead-tin alloys.

not for a characteristic of the metal but rather for a process in which it was formerly used. Corroding lead is used in making pigments, lead oxides, and a wide variety of other lead chemicals. The required high purity is necessary to avoid problems, such as unwanted colors in white lead pigment, caused by impurities during and after processing.

Chemical Lead. Refined lead with a residual copper content of 0.04 to 0.08% and a residual silver content of 0.002 to 0.02% is particularly desirable in the chemical industries and thus is called chemical lead. Most chemical lead produced in the United States is refined from the lead ores of southeastern Missouri, which contain small amounts of copper and silver. The copper content significantly improves corrosion resistance and mechanical strength, making chemical lead the next most frequently used grade

Table 2 ASTM B 29 present and proposed pig lead composition specifications

Lead type	Composition, %(a) Ag	Bi	Cd	Cu	Fe	Ni	Pb	Te	Zn	Other
Pure lead (UNS L50042)(b)										
Present specification	0.0015(c)	0.050	NR	0.0015(c)	0.002	NR	99.94 min	NR	0.001	(d)
Proposed specification	0.0025	0.030 or 0.025	0.0010	0.0010	0.001 or NR	0.0002	99.96 min	0.00005	0.0005	(e)
Common lead (UNS L50045)(f)										
Present specification	0.005	0.05(g)	NR	0.0015	0.002	NR	99.94 min	NR	0.001	(d)
Proposed specification	0.005	0.05	0.001	0.0015	0.002 or NR	0.001	99.94 min	...	0.001	(h)
Chemical lead(i) (UNS L51120)	0.002–0.020	0.005	NR	0.040–0.080	0.002	NR	99.90 min	NR	0.001	(d)
Acid-copper lead(j) (UNS L51121)	0.002	0.025	NR	0.040–0.080	0.002	NR	99.90 min	NR	0.001	(d)

(a) Compositions are maximums unless a range or minimum (min) is specified; by agreement between the purchaser and the supplier, analyses may be required and limits established for elements (or compounds) not specified here. NR, not required. (b) Corroding lead is a designation used in the trade to describe lead that has been refined to a high degree of purity. (c) Silver plus copper, 0.0025% max total. (d) Arsenic plus antimony plus tin, 0.002% max total. (e) Arsenic, antimony, and tin each at 0.0005% max. (f) Common lead is fully refined desilverized lead. (g) By agreement between the purchaser and the supplier, bismuth levels up to 0.150% may be allowed. (h) Arsenic, antimony, and tin each at 0.001% max. (i) Chemical lead designates the undesilverized lead produced from southeastern Missouri ores. (j) Copper-bearing lead is made by adding copper to fully refined lead.

after corroding lead. The silver content also improves corrosion resistance in some applications.

Copper-bearing lead provides corrosion protection comparable to that of chemical lead in most applications that require high corrosion resistance. This grade is made by adding copper to fully refined lead. It differs from chemical lead primarily in its higher allowed bismuth content.

Common lead, which contains higher amounts of silver and bismuth than does corroding lead, is used for battery oxide and general alloying.

Lead-Base Alloys. Because lead is very soft and ductile, it is normally used commercially as lead alloys. Antimony, tin, arsenic, and calcium are the most common alloying elements.

Antimony generally is used to give greater hardness and strength, as in storage battery grids, sheet, pipe, and castings. Antimony contents of lead-antimony alloys can range from 0.5 to 25%, but they are usually 2 to 5%. Table 3 lists the compositions of various lead-antimony alloys.

Within the past decade, lead-calcium alloys have replaced lead-antimony alloys in a number of applications, in particular, storage battery grids and casting applications. These alloys contain 0.03 to 0.15% Ca (see, for example, some of the lead-calcium alloys in Table 1). More recently, aluminum has been added to calcium-lead and calcium-tin-lead alloys as a stabilizer for calcium.

Adding tin to lead or lead alloys increases hardness and strength, but lead-tin alloys are more commonly used for their good melting, casting, and wetting properties, as in type metals and solders. Tin gives the alloy the ability to wet and bond with metals such as steel and copper; unalloyed lead has poor wetting characteristics. Tin combined with lead and bismuth or cadmium forms the principal ingredient of many low-melting alloys. Table 4 lists the compositions of various lead-tin alloys.

Arsenical lead (UNS L50310) is used for cable sheathing. Arsenic is often used to harden lead-antimony alloys and is essential to the production of round dropped shot.

Properties of Lead

The properties of lead that make it useful in a wide variety of applications are density, malleability, lubricity, flexibility, electrical conductivity, and coefficient of thermal expansion, all of which are quite high; and elastic modulus, elastic limit, strength, hardness, and melting point, all of which are quite low. Lead also has good resistance to corrosion under a wide variety of conditions. Lead is easily alloyed with many other metals and casts with little difficulty.

Pouring temperature and rate of cooling markedly influence the microstructures and properties of lead alloys. High pouring temperatures and low cooling rates, such as those that result from the use of overly hot molds, promote segregation and the formation of a coarse structure. A coarse structure can cause brittleness, low compressive strength, and low hardness.

Density. The high density of lead (11.35 g/cm^3, at room temperature) makes it very effective in shielding against x-rays and gamma radiation (Fig. 1). In very large installations, it is often used for lining concrete structures to greatly reduce the thickness of concrete that otherwise would be required.

The combination of high density, high limpness (low stiffness), and high damping capacity (Fig. 2) makes lead an excellent material for deadening sound and for isolating equipment and structures from mechanical vibrations.

Because of its high density, lead generally is excluded from use in applications where light weight is important. However, even when light weight is desirable, the high density of lead sometimes can be used to advantage. For example, the use of lead in aircraft counterweights often reduces the total weight, because the high density of lead allows more mass to be concentrated at the point of greatest effect. In addition, lead has a low melting point and is relatively easy to cast; thus it is possible to fit lead weights into irregular and out-of-the-way spaces.

Malleability, softness, and lubricity are three related properties that account for the extensive use of lead in many applications. For example, high malleability is largely responsible for the use of lead as a caulking material, enabling it to fill caulked joints completely. The softness and self-lubricating properties of lead account in substantial part for its use in bearing alloys, gaskets, and washers. As a coating on wire or sheet metal, lead acts as a drawing lubricant, and in the form of powder, it imparts lubricity to antiseize compounds and engine bearings. The malleability of lead is used to greatest advantage in the manufacture of foil; lead foil often is rolled as thin as 0.01 mm (0.0005 in.).

On the other hand, the softness of lead requires that care be taken in designing for many applications. For example, excessive stream velocity in lead pipes can result in severe erosion if proper precautions are not taken in system design.

Strength. The low tensile strength (Table 5) and low creep strength (Table 6) of lead must always be considered when designing lead components. The principal limitation on the use of lead as a structural material is not its low tensile strength but its susceptibility to creep. Lead continuously deforms at low stresses (Fig. 3), and this deformation ultimately results in failure at stresses far below the ultimate tensile strength.

The low strength of lead does not necessarily preclude its use. Lead products can be designed to be self-supporting, or inserts or supports of other materials can be provided. Alloying with other metals, notably calcium or antimony, is a common method of strengthening lead for many applications. In general, consideration should always be given to supporting lead structures by lead-covered steel straps. When lead is used as a lining in a structure made of a stronger material, the lining can be supported by bonding it to the structure. With the development of improved bonding and adhesive techniques, composites of lead with other materials can be made. Composites have improved strength yet also retain the desirable properties of lead.

Thermal Expansion. The relatively high coefficient of thermal expansion of lead (29.3 μm/m/K, or 16.3 μin./in./°F, for pure lead) is another important design parame-

Table 3 Nominal compositions of selected lead-antimony alloys

UNS designation	Description and/or specification	Composition, %(a)				
		Pb	Sb	Sn	As	Other
<1% Sb						
L52505	Lead-antimony alloy	99.9	0.1	. . .	. . .	. . .
L52515	Cable-sheathing alloy	99.8	0.2	. . .	0.015	. . .
L52520	Cable-sheathing alloy	99.4	0.2	0.4	. . .	. . .
L52535	Cable-sheathing alloy	99.6	0.4	. . .	0.03	. . .
L52560	Bullet alloy	99.2	0.75	. . .	. . .	. . .
L52565	Overhead cable alloy	99.2	0.75	. . .	. . .	. . .
1–1.9% Sb						
L52605	1% antimonial lead	99.0	1.0	. . .	. . .	. . .
L52615	Lead-base die casting alloy	98.6	1.0	0.3	0.1	0.003 S
L52620	Battery alloy	97.0	1.5	. . .	. . .	1.45 Cd
L52625	Shot alloy	98.0	1.55	0.0005 max	0.45	. . .
L52630	Battery alloy	98	1.6	0.1	0.3	0.02 Se, 0.005 S
2–2.9% Sb						
L52705	2% antimonial lead	98.0	2.0	. . .	. . .	. . .
L52710	Battery alloy	97.5	2.0	0.3	0.15	0.003 S
L52720	Battery alloy	97.2	2.25	0.2	0.3	0.008 S, 0.02 Se
L52725	Bullet alloy	97.5	2.5	. . .	. . .	. . .
L52730	Electrotype (general)	95	2.5	2.5	. . .	. . .
L52750	Battery alloy	96.5	2.75	0.3	0.4	0.005 S, 0.075 Ca
L52760	Battery alloy	96.8	2.75	0.2	0.18	0.008 S, 0.075 Cu
L52770	Battery alloy	96.6	2.9	0.3	0.15	0.004 S, 0.15 Cu
3–3.9% Sb						
L52805	3% antimonial lead	97.0	3.0	. . .	. . .	. . .
L52810	Battery alloy	96.5	3.0	0.3	0.15	0.003 S
L52815	Shot alloy	96.4	3.0	0.0005 max	0.6	. . .
L52830	Electrotype (general)	94	3	3	. . .	. . .
L52840	Battery alloy	95.7	3.25	0.4	0.5	0.06 Ca, 0.12 Cu
L52860	Bearing alloy	bal	3.0–4.0	3.5–4.7	0.05 max	0.005 Al max, 0.10 Bi max, 0.005 Cd max, 0.10 Cu max, 0.005 Zn max, 0.40 max total of others
4–5% Sb						
L52901	4% antimonial lead	96	4	. . .	. . .	. . .
L52905	Battery alloy	95.5	4.0	0.3	0.15	0.003 S
L52915	Type metal alloy	93	4	3	. . .	. . .
L52922	Anode alloy	95	4.5	. . .	. . .	0.5 S
L52930	Battery alloy	94.6	4.75	0.3	0.3	0.007 S, 0.05 Cu
L53020	Bullet alloy	90.0	5.0	5.0	. . .	. . .
6–7% Sb						
L53105	6% antimonial lead	94	6	. . .	. . .	. . .
L53115	Rolled sheet alloy	93.7	6.0	0.3	. . .	. . .
L53120	Electrowinning anode alloy	93.6	6.0	. . .	0.4	. . .
L53125	Creep-resistant pipe and sheet	93.3	6.0	. . .	0.65	. . .
L53122	High-strength sheet lead	93.6	6.0	. . .	0.4	. . .
L53130	Lead alloy	92.8	6.0	0.6	0.6	. . .
L53135	Battery alloy	93.4	6.0	0.3	0.3	0.006 S, 0.07 Cu
L53140	Hard shot alloy	92.6	6.2	0.0005 max	1.2	. . .
L53220	Lead alloy	91.8	7.0	0.6	0.6	. . .
8–9% Sb						
L53230	8% antimonial lead	92	8	. . .	. . .	. . .
L53235	Hard shot alloy	90.7	8.0	0.0005 max	1.25	. . .
L53260	Spin casting alloy	88.9	8.0	3.1	. . .	. . .
L53265	Type metal alloy	88	8.0	4.0	. . .	. . .
L53305	9% antimonial lead	91	9.0	. . .	. . .	. . .
L53310	Lead alloy	90	9	1	. . .	. . .
L53320	White metal bearing alloy	86	9	5	. . .	. . .
10–11% Sb						
L53340	Lead-base die casting alloy (ASTM B 102)	bal	10	. . .	0.15 max	0.50 Cu max, 0.01 Zn max
L53345	Lead-base bearing alloy (SAE J460)	bal	10	6.0	0.25 max	0.005 Al max, 0.01 Bi max, 0.05 Cd max, 0.50 Cu max, 0.005 Zn max, 0.20 max total of others
L53346	Lead-base bearing alloy (ASTM B23)	bal	10	6.0	0.25 max	0.005 Al max, 0.10 Bi max, 0.05 Cd max, 0.50 Cu max, 0.10 Fe max, 0.005 Zn max
L53405	11% antimonial lead	89	11	. . .	. . .	. . .
L53420	Linotype alloy	86	11	3	. . .	. . .
L53425	Special linotype alloy	84	11	5	. . .	. . .

(continued)

(a) Nominal compositions unless otherwise specified. bal, balance. Source: Ref 1

Table 3 (continued)

UNS designation	Description and/or specification	Composition, %(a) Pb	Sb	Sn	As	Other
12–14% Sb						
L53454	Type metal alloy	85	12	3	...	...
L53455	Linotype B (eutectic) alloy	84	12	4	...	...
L53460	Type metal alloy	82	12	6	...	...
L53465	Lead alloy	77.5	12.5	10.0	...	0.05 Cu
L53480	Arsenical babbitt	83.5	12.75	0.75	3.0	...
L53510	General stereotype alloy	80.5	13	6.5	...	...
L53530	Flat stereotype alloy	80	14	6	...	...
15% Sb						
L53550	15% antimonial lead	85	15	...	...	...
L53555	Lead alloy	83	15	2	...	...
L53558	Type metal alloy	81	15	4	...	...
L53560	Lead-base die casting alloy (ASTM B 102)	80	15	5	0.15 max	0.01 Al max, 0.50 Cu max, 0.01 Zn max
L53565	Lead-base white metal bearing alloy (ASTM B 23)	bal	15	5.0	0.45	0.005 Al max, 0.1 Bi max, 0.05 Cd max, 0.50 Cu max, 0.1 Fe max, 0.005 Zn max

(a) Nominal compositions unless otherwise specified. bal, balance. Source: Ref 1

ter. In lead roofing and flashing, thermal expansion must always be considered. It is provided for by using small sheets and loose-locking each sheet to the next, thus minimizing both individual and cumulative expansion. In pipelines subject to wide variations in temperature, allowance must be made for free expansion. The excellent flexibility of lead can be used to advantage in designing such systems.

Fatigue Properties. Because lead and lead-base alloys are susceptible to creep even at room temperature, the fatigue properties of lead-base alloys can be affected by creep-fatigue interaction. The effect of creep on fatigue life is manifested by frequency effects (Fig. 4). Lower cyclic frequencies are associated with a reduction in fatigue life.

Stress limits for a fatigue life of 10^7 cycles generally range from about 3 MPa (0.435 ksi) for corroding lead (UNS L50042) to about 30 MPa (4.35 ksi) for lead babbit (ASTM B 23, 1983) in the chill cast condition. Typical fatigue strengths in bending for two cable-sheathing lead alloys are shown in Fig. 5.

Corrosion Resistance. Lead is highly resistant to corrosion by the atmosphere, by waters, and by a wide range of chemicals in common use. Where resistance to corrosion must be combined with long service life, the limitations imposed by the mechanical properties of lead must be carefully considered in the final design. A general description of the resistance of chemical lead to specific corroding agents is given in Table 7. For a detailed description of the corrosion resistance of lead and lead alloys, see the article "Corrosion Resistance of Lead and Lead Alloys" in Volume 13 of the 9th Edition of *Metals Handbook*.

Extrusion Characteristics. Lead and lead-base alloys exhibit high ductility and are easy to extrude. The addition of alloying elements increases the force required for extruding, but the process is carried out with little difficulty using billets heated to a maximum temperature of about 230 °C (450 °F). Principal applications include pipes, wire, tubes, munitions, and sheathing for cable. Molten lead is used in continuous extruders for cable sheathing and hose curing. Vertical extrusion presses are sometimes used to produce protective sheathings of lead on electrical conductors. Arsenical lead used for sheathing (UNS L50310) has a typical extrusion temperature of about 200 to 230 °C (400 to 450 °F).

Casting temperatures for various lead alloys are:

Table 4 Compositions of selected lead-tin alloys

UNS designation	Description and/or specification	Composition, %(a) Pb	Sn	Sb	As	Other
L54030	Cable-sheathing alloy	99.7	0.2	...	...	0.075 Cd nom
L54050	Cable-sheathing alloy	99.4	0.4	...	...	0.15 Cd nom
L54210	2% tin solder (ASTM B 32 and DIN 1707)	98	2.0	0.12	0.02	(b)
L54250	SAE solder alloy 9B (SAE J473)	bal	2.50–2.75	4.90–5.40	0.40–0.60	(b)(c)
L54320	5/95 solder (ASTM B 32)	95	4.5–5	0.12	0.02	(b)
L54370	Plated overlay for bearings (SAE J460)	bal	5.0–9.0	...	...	Total others, 0.08 max
L54410	8% Sn solder (DIN 1707)	92	8	0.3 nom	...	...
L54510	Plated overlay for bearings (SAE J460)	bal	8.0–12.0	...	...	Total others, 3.5 max
L54520	10/90 solder (ASTM B 32)	90	10	0.20–0.50	0.02	(b)
L54525	88/10/2 solder (Federal specification QQ-S-571)	bal	9.0–11.0	0.20	0.02	(d), 1.7–2.4 Ag
L54560	15/85 solder (ASTM B 32)	85	15	0.20–0.50	0.02	(b)
L54710	20% Sn solder (British standard 219)	80	20	0.2 nom	...	...
L54720	25/75 solder (ASTM B 32)	75	25	0.25	0.02	(b)
L54727	Lead-base bearing alloy (DIN 1741)	59	25	13 nom	...	3 Cu nom
L54820	30/70 solder (ASTM B 32 and British standard 219)	70	30	0.25	0.02	(b)
L54850	35/65 solder (ASTM B 32)	65	35	0.25	0.02	(b)
L54855	Silver-loaded solder	61.5	35.5	...	...	3.0 Ag nom
L54915	40/60 solder (ASTM B 32 and DIN 1707)	60	40	0.12	0.02	(b)
L54925	Lead-base bearing alloy (DIN 1741)	46	40	12 nom	...	2 Cu nom
L54950	45/55 solder (ASTM B 32 and AMS 4750)	55	45	0.12	0.03	(b)
L55030	50/50 solder (ASTM B 32 and DIN 1707)	50	50	0.20–0.50	0.03	(b)

(a) Unless otherwise noted, lead and tin are nominal (nom) values, and antimony and arsenic are maximum (max) values. (b) Maximums on residual elements: 0.005 Al, 0.25 Bi, 0.08 Cu, 0.02 Fe, and 0.005 Zn. (c) Maximum on other elements, 0.08 total. (d) Maximums on residual elements: 0.005 Al, 0.03 Bi, 0.001 Cd, 0.08 Cu, 0.005 Zn, 0.10 max total on all others. Source: Ref 1

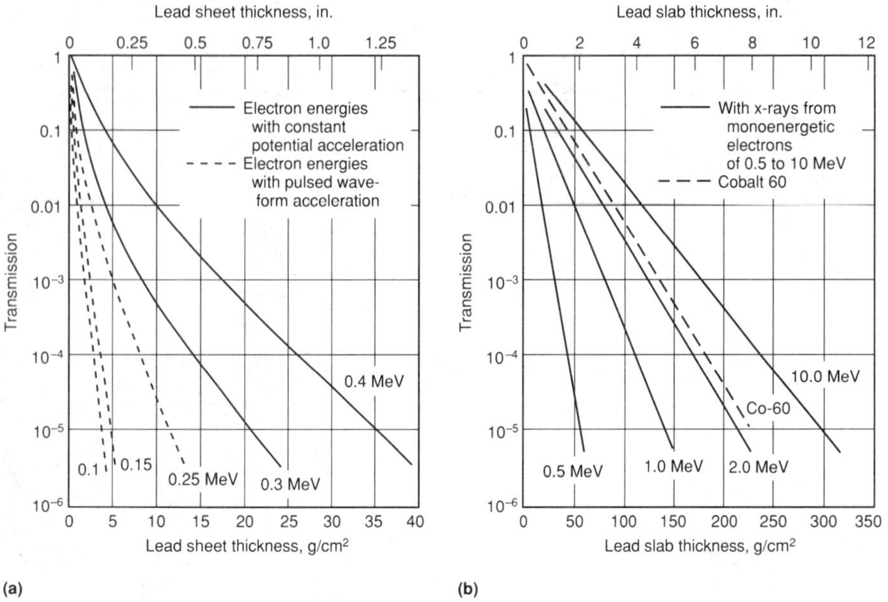

Fig. 1 Broad-beam transmission of x-rays through (a) lead sheet and (b) lead slab. The energy designations (in MeV) on each curve refer to the energy of electrons impinging upon a thick x-ray producing target. The curves represent transmission as a dose-equivalent index ratio, and the bottom scale indicates the required mass thickness (g/cm²), which is useful in selecting a material when weight is a factor. Transmission of cobalt-60 gamma rays is included for comparison. Source: Adapted from Ref 2

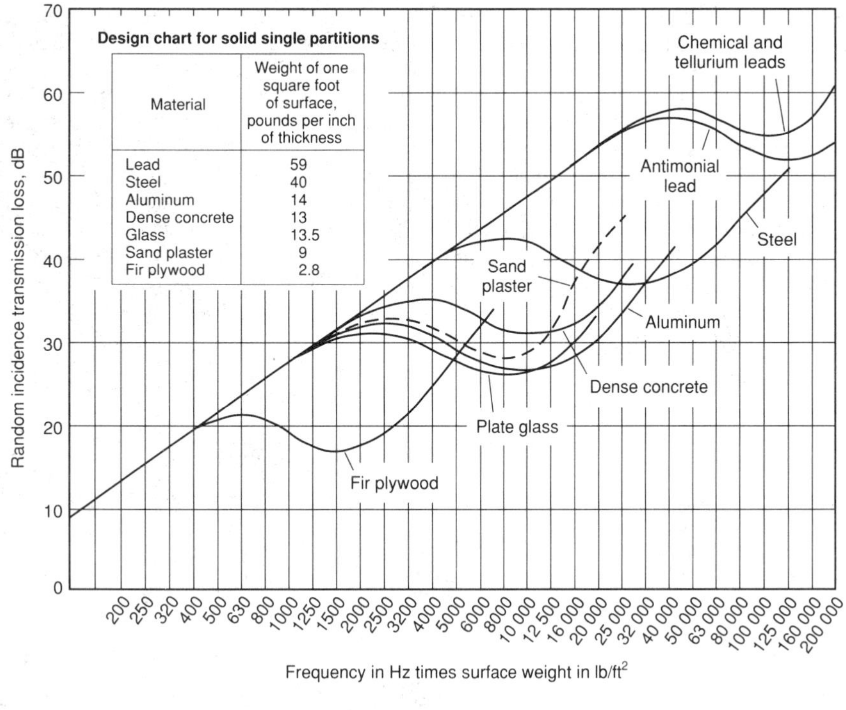

Fig. 2 Damping capacity of lead compared with that of other materials. Source: Ref 3

- 420 to 445 °C (790 to 830 °F) for chemical lead (UNS L51120)
- 400 °C (750 °F) for arsenical lead (UNS L50310)
- 425 to 500 °C (800 to 930 °F) for 1% Sb lead (UNS L52605), 4% Sb lead (UNS L52901), 6% Sb lead (UNS L53105), and 8% Sb lead (UNS L53230)
- 325 to 400 °C (617 to 750 °F) for lead babbitt alloys 7 and 13 in ASTM B 23 (UNS alloys L53581 and L53346, respectively)
- 480 to 540 °C (900 to 1000 °F) for lead babbitt alloy 15 in ASTM B 23 (UNS L53620)
- 340 to 425 °C (645 to 800 °F) for lead babbitt alloy 8 in ASTM B 23 (UNS L53565)

The recrystallization temperature for lead is below 0 °C (32 °F).

Products and Applications

The most significant applications of lead and lead alloys are lead-acid storage batteries (in the grid plates, posts, and connector straps), ammunition, cable sheathing, and building construction materials (such as sheet, pipe, solder, and wool for caulking). Other important applications include counterweights, battery clamps and other cast products such as: bearings, ballast, gaskets, type metal, terneplate, and foil. Lead in various forms and combinations is finding increased application as a material for controlling sound and mechanical vibrations. Also, in many forms it is important as shielding against x-rays and, in the nuclear industry, gamma rays. In addition, lead is used as an alloying element in steel and in copper alloys to improve machinability and other characteristics, and it is used in fusible (low-melting) alloys for fire sprinkler systems.

Although most lead is used in metallic form, substantial amounts are used in the form of lead compounds. These include tetraethyl and tetramethyl lead (used as antiknock compounds in gasoline), litharge (PbO), and various corrosion-inhibiting lead pigments such as red lead (Pb_3O_4), lead chromates, lead silicochromates, and lead silicates. Litharge is used in paste mixtures for grid plates of lead-acid storage batteries, in cements, glasses, and ceramics, and as a starting material for preparation of many other lead compounds.

Red lead has long been one of the most important rust-inhibiting pigments used in primers and undercoats for the protection of steel structures. Commercially important white corrosion-inhibiting pigments are basic lead carbonate, dibasic lead phosphite, dibasic lead phosphosilicate, and basic lead silicate. The most important colored pigments are tribasic lead chromosilicate, basic lead silicochromate, and normal lead silicochromate. The color of tribasic lead chromosilicate varies from red to orange, whereas the color of lead silicochromate is generally yellow. Lead silicochromate is used in yellow paint for marking pavement.

Battery Grids. The largest use of lead is in the manufacture of lead-acid storage batteries. These batteries consist of a series of grid plates made from either cast or wrought calcium lead or antimonial lead that is pasted with a mixture of lead oxides and immersed in sulfuric acid.

The length of the active life of a battery depends on the resistance of the lead alloy grids to corrosion under repeated cycling (charge and discharge) in the sulfuric acid. Automotive positive battery plates are usually made from antimony-lead alloys containing 1.5 to 3% antimony and other ele-

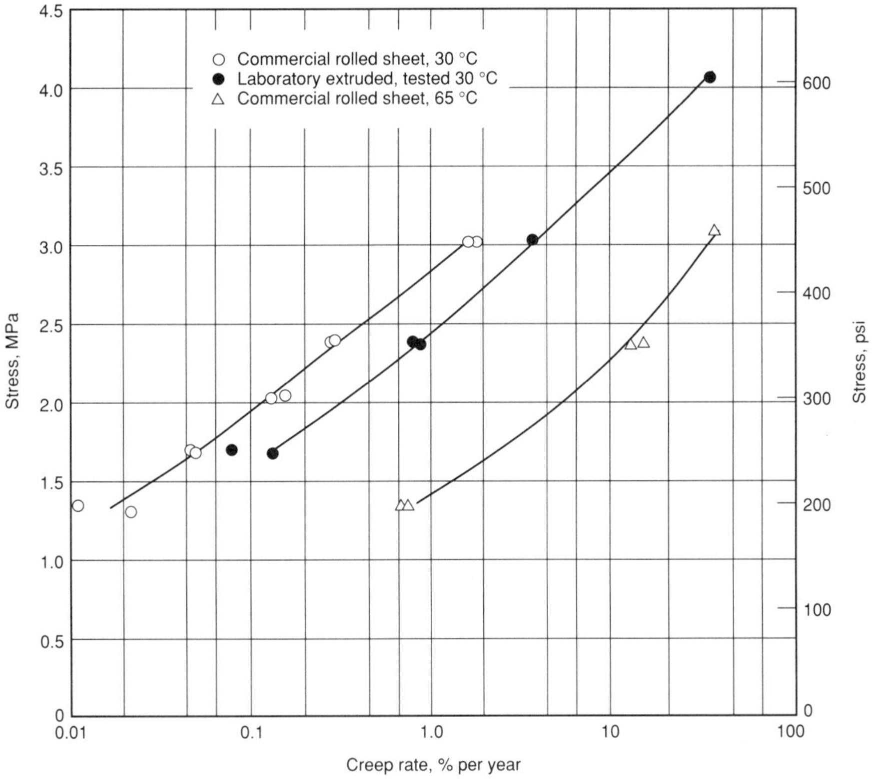

Fig. 3 Creep rate of chemical lead (+99.90% Pb). Test specimens were 19 by 32 mm (¾ by ⅛ in.) with a 250 mm (10 in.) gage length. Length of specimen parallel to the direction of rolling or extrusion

ments such as tin, arsenic, copper, sulfur, and selenium. Other automotive battery grids are made from lead-calcium-tin-aluminum alloys. The exact composition used varies with the manufacturer. Table 8 lists some typical compositions of lead alloys used for battery applications. Hybrid automotive batteries are made from 1 to 2% Sb-Pb alloys for the positive grid and lead-calcium alloys (0.04 to 0.15% Ca) for the negative grid. Battery grid alloys may also include 0.1 to 0.8% Sn (Table 8). In addition, some lead-calcium alloys contain 0.01 to 0.03% Al. Industrial batteries usually are made from alloys containing 5 to 8% Sb and various other elements. For all of these alloys, long battery life requires close control of impurities. Large standby stationary batteries can be made with grids of relatively pure lead of special design. This type of battery normally contains lead calcium.

Type metals, a class of metals used in the printing industry, generally consist of lead-antimony and tin alloys. Small amounts of copper are added to increase hardness for some applications. Compositions of type metals in present commercial use are given in Table 9. The lead base provides low cost, a low melting point, and ease of casting—properties that are desirable for all type metals. Additions of antimony harden the alloy, make it more resistant to compressive impact and wear, lower the casting temperature, and minimize contraction during freezing. Tin adds fluidity, improves castability, reduces brittleness, and imparts a finer structure—a characteristic that helps type reproduce fine detail.

Electrotype metal contains the lowest percentages of tin and antimony (see Table 9) because it is used as a backing metal only and is not required to resist wear. Unlike electrotype metal, stereotype metal ordinarily is used directly for printing; therefore, it must be harder and more wear resistant than electrotype metal, necessitating higher contents of tin and antimony. For greater resistance to wear, stereotypes can be lightly electroplated with chromium or nickel.

Linotype, or slug-casting metal, is used for high-speed composition of newspaper type. For this purpose, a low melting point and short temperature range during solidification are of greatest importance. The ternary eutectic alloy containing 84% Pb, 4% Sn, and 12% Sb, or an alloy of similar composition, is favored for linotype.

Like linotype metal, monotype metal is machine die cast. In monotype casting, only one type character is cast at a time. A rapid cooling rate is therefore possible, permitting the use of harder alloys of higher melting range than can be used for linotype metal, which is die cast an entire line at a time.

Foundry type metal is used exclusively to cast type for hand composition. The cast type is used over and over again instead of being melted before reuse, as is the case for other type metals. If not used to print directly, foundry type metal is subjected to heavy pressure in forming molds for electrotypes, stereotypes, and other duplicate plates. Such service requires the hardest,

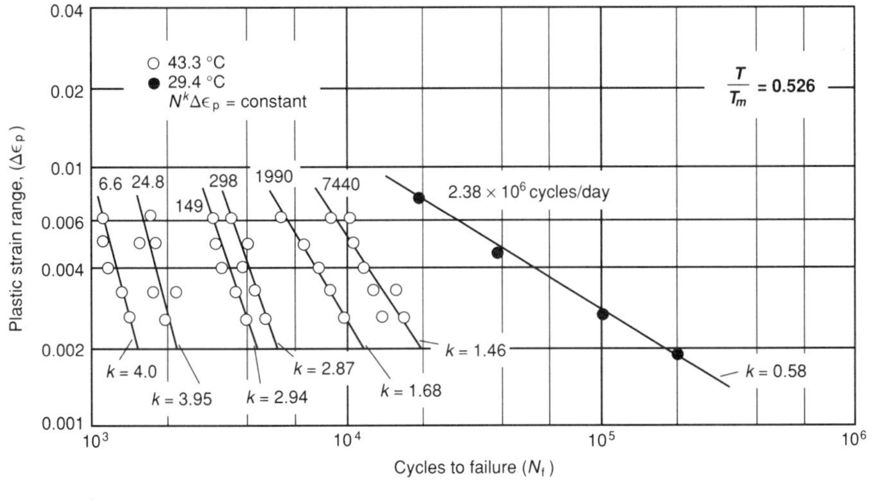

Fig. 4 Effect of frequency of cycling on the low-cycle fatigue behavior of lead

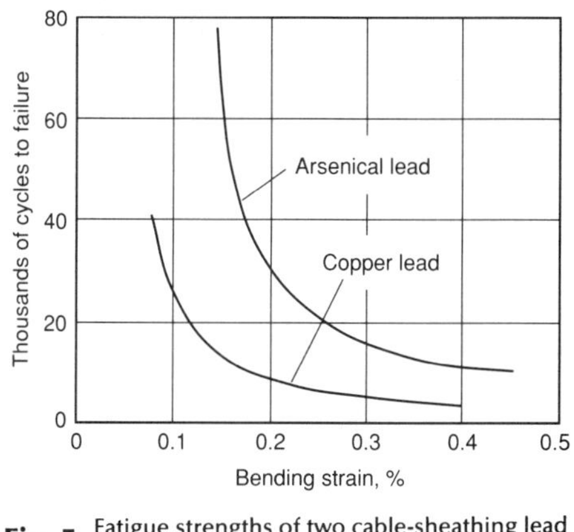

Fig. 5 Fatigue strengths of two cable-sheathing lead alloys in bending. Bending was at 25 °C (77 °F), one cycle per minute.

Table 5 Typical room-temperature tensile properties of selected lead alloys

UNS designation	Typical product forms or name	Condition	Tensile strength MPa	Tensile strength ksi	Yield strength at 0.2% offset MPa	Yield strength at 0.2% offset ksi	Elongation, %(a)	Hardness
50042	Corroding lead	Sand cast	12–13	1.7–1.9	5.5	0.8	30	3.2–4.5 HB(b)
	Corroding lead	Chill cast	14	2.0	. . .	. . .	47	4.2 HB(b)
50131	Solder	. . .	. . .	. . .	. . .	. . .	. . .	13.0 HB
50132	Solder	. . .	35	5	. . .	. . .	28	13 HB
50134	Solder	. . .	30	4⅓	. . .	. . .	20	. . .
50172	Solder	. . .	40	5.7	. . .	. . .	. . .	. . .
50180	Solder	. . .	32	4.6	. . .	. . .	28	. . .
50310	As-Pb cable sheath	Extruded at any temperature(c)	17.2	2.5	. . .	. . .	40	. . .
	As-Pb pipe	Extruded at any temperature(c)	16.2	2.35	. . .	. . .	40	4.9 HB
	As-Pb cable sheath	Extruded at 425 °C (795 °F)(d)	19.9	2.9	. . .	. . .	35	. . .
	As-Pb pipe	Extruded at 425 °C (795 °F)(d)	20.6	3.0	. . .	. . .	35	6.6 HB
	As-Pb cable sheath	Extruded at 450 °C (840 °F)(d)	20.6	3.0	. . .	. . .	35	. . .
	As-Pb pipe	Extruded at 450 °C (840 °F)(d)	22.0	3.2	. . .	. . .	35	7.0 HB
	As-Pb pipe	Extruded at 550 °C (1022 °F) and quenched	31.7	4.6	. . .	. . .	25	10.0 HB
50737	Sheet, strip, wire	. . .	45–48	6.5–7.0	. . .	. . .	15	. . .
50740	Sheet, strip, wire	Rolled strip(e)	62	9	55	8	10	. . .
50750	Sheet, strip, wire	Rolled strip(e)	70	10	66	9.5	10	. . .
50760	Cast battery grids	Fully aged, air-cooled castings	36–39	5.2–5.6	. . .	. . .	30–45	70–80 HR(f)
50775	Cast battery grids	Fully aged, air-cooled casting	41–45	6.0–6.5	. . .	. . .	20–35	90–95 HR(f)
50780	Cast parts requiring high strength	Fully aged, air-cooled castings	45–52	6.5–7.5	. . .	. . .	25–30	85–90 HR(f)
50790	Cast parts requiring high strength	Fully aged, air-cooled castings	52–55	7.5–8.0	. . .	. . .	20–35	90–95 HR(f)
51120	Chemical lead	Unspecified	16–19	2.3–2.7	6–8	0.9–1.2	30–60	4–6 HB(b), or 80–85 HRR(g)
	Laboratory-rolled chemical lead	<30 days after rolling	20	2.96	. . .	. . .	42	84 HR(f)
	Laboratory-rolled chemical lead	30 days after rolling	19	2.8	. . .	. . .	44	80 HR(f)
	Laboratory-rolled chemical lead	3 years after rolling	19	2.75	. . .	. . .	35	75 HR(f)
	Commercially rolled sheet	Age unknown	18	2.6	. . .	. . .	52	75 HR(f)
	Commercially rolled sheet	Age unknown	17	2.5	9	1.3	27(h)	75 HR(f)
	Extruded sheet	Age unknown	18	2.6	. . .	. . .	57	78 HR(f)
51121	Copper lead	. . .	16–19	2.3–2.7	. . .	. . .	30–60	. . .
51123	Lead-tellurium-copper alloy	. . .	21	3	. . .	. . .	. . .	5.8 HB
51125	Copperized soft lead	. . .	16–19	2.3–2.7	. . .	. . .	30–60	. . .
51510	Solder alloy	. . .	32	4.6	. . .	. . .	. . .	. . .
51535	Lead-indium alloy	. . .	38	5.5	. . .	. . .	. . .	. . .
51570	Lead-indium alloy	. . .	24	3.5	. . .	. . .	. . .	. . .
52605	1% Sb sheathing	Extruded, aged 1 month	20	3.0	. . .	. . .	50	7 HB
52901	Hard lead (96-4)	Cold-rolled sheet (95% reduced)	27.5	4.0	. . .	. . .	48	8 HB(i)
	Hard lead (96-4)	Heat treated(j)	80	11.7	. . .	. . .	6.3	24 HB(i)
53105	Hard lead (94-6)	Chill cast	47.2	6.84	. . .	. . .	24	13 HB(i)
	Hard lead (94-6)	Cold rolled (95%)	28.3	4.10	. . .	. . .	47	. . .
	Hard lead (94-6)	Extruded	22.8	3.30	. . .	. . .	65	10.7 HB(i)
53230	Heavy-duty battery grids	Cold rolled (95%)	32	4.65	. . .	. . .	31	9.5 HB(i)
	Heavy-duty battery grids	Heat treated(k)	85	12.35	. . .	. . .	4.7	26.3 HB(i)
53305	Heavy-duty battery grids	Chill cast	52	7.5	. . .	. . .	17	15.4 HB
53346	Lead-base babbitt	Chill cast	70	10	. . .	. . .	5	19 HB
53565	Lead-base babbitt	Chill cast	70	10	. . .	. . .	5	20 HB
53581/53585(l)	Lead-base babbitt	Chill cast	72	10.5	. . .	. . .	4	22 HB
53620	Lead-base babbitt	Chill cast	71	10.4	. . .	. . .	2	20 HB
54320	5/95 solder	. . .	23	3.4	10	1.5	50	8 HB
54321	5% Sn antimonial solder	. . .	28	4.06	10	1.45	55	8 HB
54520	10/90 solder	. . .	30	4.35	. . .	. . .	10	. . .
54711	20/80 solder	. . .	40	5.8	25	3.6	16	11.3 HB
54280	30/70 solder	. . .	34	4.9	. . .	. . .	18	12 HB
54915	40/60 solder	. . .	37	5.4	. . .	. . .	25	. . .
55030	50/50 solder	. . .	42	6.1	33	4.8	60	14.5 HB

(a) Elongation in 50 mm (2 in.) unless otherwise specified. (b) Brinell hardness indication using a 10 mm ball and a 100 kg load for a 30-s duration. (c) Air cooled after extruding. (d) Water quenched after extruding. (e) Longitudinal properties of rolled strip. (f) Unspecified Rockwell scale for soft metals that is similar to the Rockwell R scale; see alloy L51120 for a comparison of the Brinell scale and the Rockwell R scale. (g) Rockwell R scale (30-s duration). (h) Elongation in 200 mm (8 in.). (i) Brinell test with 1/16 in. ball and a 9.85 kg load for a 30-s duration. (j) Heat treated at 235 °C (455 °F), quenched, and then aged 150 days. (k) Heat treated at 235 °C (455 °F), quenched, and then aged 1 day at room temperature. (l) Alloy 7 in ASTM B 23; designated as L53581 in the ASTM specification but designated as L53585 in other listings

most wear-resistant alloy that is practical to use. Small additions of copper as a hardener are feasible for foundry type.

Because type metals other than foundry type are remelted and recast repeatedly, there is always a possibility of contamination by unwanted metals, as well as by oxide and dross formed during the melting and handling of the molten metal. Copper, zinc, nickel, aluminum, and arsenic are the principal metallic impurities that can impair the castability of type metals. Iron is also present in very small amounts, principally because of the action of molten tin on the steel equipment used in melting and casting. Iron usually is not considered a harmful impurity, except that it increases the amount of dross.

Cable Sheathing. Lead sheathing extruded around electrical power and communication cables gives the most durable protection against moisture and corrosion damage, and provides mechanical protection of the insulation. Chemical lead, 1% antimonial lead, and arsenical lead are most commonly employed for this purpose. The additional stiffness imparted to lead by antimony is advantageous for overhead cables. The additional resistance to bending and creep imparted by arsenic is desirable in applications involving severe vibration. Lead alloyed with 0.03% Ca or with tellurium has also been used with satisfactory results.

Lead-sheathed cables used underground or under water are usually protected against mechanical damage to the sheathing. Sheathing on underground cable generally is protected from contact with the ground

Table 6 Creep characteristics of selected lead alloys

Alloy	Creep rate characteristics: Applied stress, MPa	Applied stress, ksi	Minimum creep rate, 10^{-4}%/h	Room-temperature creep-rupture characteristics: Applied stress, MPa	Applied stress, ksi	Time to rupture, h
Chemical lead (L51120)	2	0.3	3.4 at 20 °C (68 °F)	...	...	...
Arsenical lead (L50310)	2	0.3	0.15 at 24 °C (76 °F)	...	...	...
			0.36 at 43 °C (110 °F)	...	...	...
			2.0 at 65 °C (150 °F)	...	...	...
Rolled calcium-tin-lead (L50740)	...	...	...	28	4	50
	...	...	...	21	3	500
Rolled calcium-tin-lead (L50750)	6.9	1.0	0.06 at 20 °C (68 °F)	28	4	1000
1% Sb (L52605) extruded cable sheath	4.5	0.65	0.1 at 0 °C (32 °F)	...	...	...
	2.4	0.35	0.1 at 30 °C (86 °F)	...	...	...
	1.0	0.15	0.1 at 66 °C (150 °F)	...	...	...
Cold-rolled 6% Sb (L53105) sheet	2.8	0.4	0.1 at 30 °C (86 °F)	...	...	...
	0.34	0.05	0.1 at 100 °C (212 °F)	...	...	...
Cold-rolled 8% Sb (L53230) sheet	2.9	0.425	0.1 at 30 °C (86 °F)	...	...	...
Cable-sheathing alloy (L54030)	...	...	...	5.0	0.725	1200
10/90 solder (L54520)	...	...	...	3.5	0.5	1000
				1.1(a)	0.16(a)	1000(a)
30/70 solder (L54820)	0.79	0.115	0.01 per day at 20 °C (68 °F)	...	...	...
40/60 solder (L54915)	...	...	...	2.1	0.3	1000
Solder alloy (40C of ASTM B 32, or L54918)	...	...	...	4.9	0.7	1000
				0.6(a)	0.09(a)	1000(a)

(a) At 100 °C (212 °F)

Table 7 Resistance of chemical lead to specific corroding agents

Corrosive agent	Resistance	Corrosive agent	Resistance
Acetone	Resistant	Ammonium hydroxide	Resistant
Acetylene	Resistant	Ammonium phosphate	Resistant
Acid, acetic(a)	Moderate general attack	Benzol	Resistant
Acid, chromic	Resistant	Bromine(i)	Resistant
Acid, citric	Moderate general attack	Carbon dioxide	Resistant
Acid, hydrochloric(b)	Moderate general attack	Carbon tetrachloride(j)	Resistant
Acid, hydrofluoric	Resistant	Chlorine(k)	Resistant
Acids, mixed(c)	Resistant	Dyestuffs	Generally resistant
Acid, nitric(d)	Severe general attack	Formaldehyde	Moderate general attack
Acid, phosphoric(e)	Resistant	Magnesium chloride	Severe general attack
Acid, sulfuric(f)	Resistant	Magnesium sulfate	Resistant
Acid, sulfurous	Resistant	Motor fuel	Resistant
Acid, tartaric	Moderate general attack	Nickel sulfate	Resistant
Air	Resistant	Oxygen	Resistant
Alcohol, ethyl	Resistant	Phenols	Resistant
Alcohol, methyl	Resistant	Photographic solutions	Generally resistant
Aluminum sulfate	Resistant	Sodium carbonate	Resistant
Ammonia(g)	Resistant	Sodium chloride(l)	Resistant
Ammonium azide	Resistant	Sodium hydroxide(m)	Resistant
Ammonium chloride(h)	Resistant	Sodium sulfate(n)	Resistant
		Sulfur dioxide(o)	Resistant
		Water, chlorinated	Resistant
		Water, sea	Resistant

(a) Used to handle acetic anhydride and glacial acetic acid. (b) Use generally not recommended. (c) At ordinary temperatures with 30% H_2O. (d) Used at normal temperatures above 80% concentration. (e) Up to 80% concentration at 200 °C (390 °F). (f) Up to 96% concentration at room temperature, or 85% concentration at 220 °C (430 °F). (g) Unless sodium or potassium is dissolved in it. (h) Up to 10% concentration at ordinary temperatures. (i) When cold and free from acid. (j) At ordinary temperatures. (k) Moist to 110 °C (230 °F), or dry. (l) Dilute solutions at ordinary temperatures. (m) Up to 26% concentration and 80 °C (175 °F). (n) Up to 10% concentration boiling. (o) Moist up to 200 °C (390 °F), or dry

Table 8 Compositions of selected lead alloys for battery grids

UNS designation	Composition, %: As max	Ag max	Ca	Pb	Sb	Sn	Other
Calcium-lead alloys							
L50760	0.0005	0.001	0.06–0.08	bal	0.0005 max	0.0005 max	(a)
L50770	0.0005	0.001	0.10 nom	bal	0.0005 max	0.0005 max	(a)
L50775	0.0005	0.001	0.08–0.11	bal	0.0005 max	0.2–0.4	(a)
L50780	0.0005	0.001	0.08–0.11	bal	0.0005 max	0.4–0.6	(a)
L50790	0.0005	0.001	0.08–0.10	bal	0.0005 max	0.9–1.1	(a)
Antimony-lead alloys							
L52760	0.18 nom	...	...	bal	2.75 nom	0.2 nom	...
L52765	0.3 nom	...	...	bal	2.75 nom	0.3 nom	...
L52770	0.15 nom	...	...	bal	2.9 nom	0.3 nom	...
L52840	0.15 nom	...	...	bal	2.9 nom	0.3 nom	...

(a) 0.005% max Bi and 0.0005% max each for Cu, Zn, Cd, Ni, and Fe

by wood, cement, clay, or fiber. Where scoring of the sheathing or a severely corrosive environment is likely to be encountered, a polyethylene or neoprene jacket is applied over the lead. Underwater lead-sheathed cables are protected with asphalt-impregnated jute and galvanized steel wire.

Sheet. Lead sheet is a construction material of major importance in chemical and related industries because lead resists attack by a wide range of chemicals (for example, see Table 7, which describes the corrosion properties of chemical lead). Lead sheet is also used in building construction for roofing and flashing, shower pans, flooring, x-ray and gamma-ray protection, and vibration damping and soundproofing. Sheet for use in chemical industries and building construction is made from either pure lead or 6% antimonial lead. Calcium-lead and calcium-lead-tin alloys are also suitable for many of these applications.

Lead sheet is rolled in widths up to 3.6 m (11¾ ft) and in any thickness desired. Thickness often is designated by weight per unit area; lead weighs approximately 5 kg/m^2 (1 lb/ft^2) for each 0.4 mm (1/64 in.) of thickness. This approximation, however, should be used with care for thicknesses exceeding 6 mm (¼ in.); lead sheet 13 mm (½ in.) thick weighs 145 kg/m^2 (30 lb/ft^2).

Roofing and flashing for general purposes are made of 3 lb lead sheet 1.2 mm (3/64 in.) thick. Flashing installed in contact with fresh cement, mortar, or concrete should be coated with black asphalt.

Pans placed beneath the concrete flooring of shower and bath stalls are made of at least 4 lb lead sheet. They should be coated on both sides with asphalt or covered with tar paper. As flooring, lead sheet offers a corrosion-resistant surface; it also is non-sparking, which is required for some specialized applications. Because of its excel-

Table 9 Typical compositions and properties of type metals

Item	Composition, % Pb	Sn	Sb	Hardness, HB(a)	Liquidus temperature °C	°F	Solidus temperature °C	°F
Electrotype								
General	95	2.5	2.5	...	303	578	246	475
General	94	3	3	12.4	298	568	246	475
Curved plates	93	4	3	12.5	294	561	245	473
Stereotype								
Flat plate	80	6	14	23	256	493	239	462
General	80.5	6.5	13	22	252	485	239	462
Curved plates	77	8	15	25	263	505	239	462
Linotype								
Standard	86	3	11	19	247	477	239	462
Special	84	5	11	22	246	475	239	462
Ternary eutectic alloy	84	4	12	22	239	463	239	462
Monotype								
Ordinary	78	7	15	24	262	503	239	462
Display	75	8	17	27	271	520	239	462
Case type(b)	72	9	19	28.5	286	546	239	462
Case type	64	12	24	33	330	626	239	462
Rules	75	10	15	26	270	518	239	462
Foundry type								
Hard (1.5% Cu)	60.5	13	25	...	...	...	...	...
Hard (1.5% Cu)	58.5	20	20	...	...	...	...	...
Hard (2.0% Cu)	61	12	25	...	...	...	...	...

(a) 10 mm ball, 250 kg load. (b) Lanston standard

lent absorption characteristics, lead sheet is widely used as radiation shielding for medical and industrial installations.

Lead sheet is used in many applications where its vibration-damping characteristics are advantageous. For example, vibration-damping pads of lead sheet and steel are placed under the column footings of buildings to prevent the transmission of underground vibrations, such as those that originate from subway and railroad trains. The lead serves as a moistureproof envelope in addition to absorbing vibration. Hangers for rigid pipes often are lined with lead, which acts as a vibration and movement absorber. The soundproof abilities of lead sheet are discussed in the section "Sound Control Materials" in this article.

Pipe. Seamless pipe made from lead and lead alloys is readily fabricated by extrusion. Because of its corrosion resistance and flexibility, lead pipe finds many uses in the chemical industry and in plumbing and water distribution systems. Pipe for these applications is made from either chemical lead or 6% antimonial lead. Sizes range from fine tubing to pipes 300 mm (12 in.) or more in diameter, with almost any wall thickness.

In the chemical industry, horizontal runs of exposed lead pipe are usually supported continuously in troughs or sheet metal shells. Unbonded lead-lined steel pipe can sometimes be used as a simple solution to the problem of pipe support. Lengths of pipe can also be fabricated with welded-on lead hanger bars; hanger hooks can be attached through the hanger bars. Vertical runs are supported at intervals of approximately 460 mm (18 in.). Lengths of pipe are joined by welding or by bolting through welded-on flanges. Expansion bends are provided for pipe that will operate at elevated or fluctuating temperatures.

Heating and cooling coils are important uses of lead pipe in the chemical industry. They usually are in the form of helixes or return-bend banks of coils. Lead spacer supports are welded between turns at about 460 mm (18 in.) intervals.

For lead pipe used in the chemical industry, the appropriate wall thickness depends on operating pressures and allowances for corrosion and abrasion. Pipe 40 mm (1½ in.) in diameter with 13 mm (½ in.) walls is commonly used with steam pressures up to 310 kPa (45 psi). For nonpressure service, pipes up to 50 mm (2 in.) in diameter are usually no lighter than the class known as B or M weight, and the minimum wall thickness of larger pipes generally is 6 mm (¼ in.). For pressurized lead pipe, safe working pressure is calculated using the formula:

$$P = \frac{2St}{D}$$

where P is the working pressure, S is the maximum allowable fiber stress, t is the wall thickness, and D is the inside diameter of the pipe. For chemical lead, the maximum allowable fiber stress ranges from 1400 kPa (200 psi) at room temperature to 550 kPa (80 psi) at 150 °C (300 °F). Proper design of steam lines allows for condensate drainage, thus eliminating water hammer damage.

When optimum strength is essential, lead-lined steel pipe can be used, or coils can be made of copper tubing completely covered with an adherent layer of lead.

Lead pipes and traps have had a long history of use in water and waste service because of the excellent corrosion resistance of lead and its ability to adjust to ground settlement without damage. Joints in service pipes have been successfully made by wiping, by welding, by cupping and soldering, and through the use of compression-type couplings. Service pipe should be laid with goosenecks to allow for settlement, and a cast iron sleeve should be provided where the pipe passes through

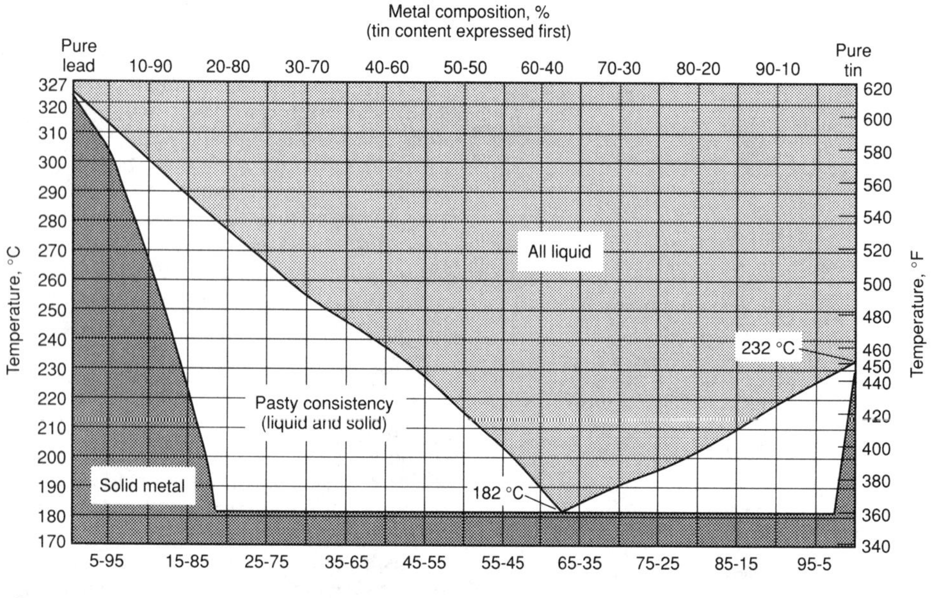

Fig. 6 Tin-lead phase diagram. Source: Lead Industries Association, Inc.

Table 10 Tin-lead solders

Composition, %		Solidus temperature		Liquidus temperature		Pasty range		
Tin	Lead	°C	°F	°C	°F	Δ °C	Δ °F	Uses
2	98	316	601	322	611	6	10	Side seams for can manufacturing
5	95	305	581	312	594	7	13	Coating and joining metals
10	90	268	514	302	576	34	62	Sealing cellular automobile radiators, filling seams or dents
15	85	227	440	288	550	61	110	Sealing cellular automobile radiators, filling seams or dents
20	80	183	361	277	531	94	170	Coating and joining metals, or filling dents or seams in automobile bodies
25	75	183	361	266	511	83	150	Machine and torch soldering
30	70	183	361	255	491	72	130	
35	65	183	361	247	477	64	116	General-purpose and wiping solder
40	60	183	361	238	460	55	99	Wiping solder for joining lead pipes and cable sheaths; also for automobile radiator cores and heating units
45	55	183	361	227	441	44	80	Automobile radiator cores and roofing seams
50	50	183	361	216	421	33	60	Most popular general-purpose solder
60	40	183	361	190	374	7	13	Primarily for electronic soldering applications where low soldering temperatures are required
63	37	183	361	183	361	0	0	Lowest-melting (eutectic) solder for electronic applications

foundation walls. Where electrolysis, free lime, or cinder fill is encountered, lead pipe should be suitably protected.

Solders in the tin-lead system are the most widely used of all joining materials. The low melting range of tin-lead solders (Fig. 6) makes them ideal for joining most metals by convenient heating methods with little or no damage to heat-sensitive parts. Tin-lead solder alloys can be obtained with melting temperatures as low as 182 °C (360 °F) and as high as 315 °C (600 °F). Except for the pure metals and the eutectic solder with 63% Sn and 37% Pb, all tin-lead solder alloys melt within a temperature range that varies according to the alloy composition.

Industrial solder alloys include a wide variety of material combinations, from 100% Pb to 100% Sn, as demanded by the particular application. Table 10 gives the melting characteristics of some common tin-lead solders and their typical applications. The solders containing less than 5% Sn are used for sealing precoated containers, for coating and joining metals, and for applications where the service temperatures exceed 120 °C (250 °F). At those temperatures, the solder functions primarily as a seal. The 10/90, 15/85, and 20/80 solders are used for sealing cellular automobile radiators and for filling seams and dents in automobile bodies. The general-purpose solders are 40/60 and 50/50. They typically are used for soldering automobile radiator cores; electrical, and electronic connections; and roofing seams and heating units. Plumber's wiping solder used to be used on water pipe but is no longer permitted, for health reasons.

Other solders contain additional alloy additions, such as antimony or silver. For the electronics industry, silver is added to tin-lead solders to reduce the dissolution of silver from silver alloy coatings. Silver can also be added to improve creep resistance. Tin-silver-lead alloys exhibit good tensile, creep, and shear strengths. Some are used for higher-temperature bonds in sequential soldering operations. Fatigue properties are increased by the addition of silver to the solder. Lead solder with 1% Sn and 1.5% Ag is used in cryogenic equipment because it does not embrittle at low temperatures.

Lead-base bearing alloys, which are called lead-base babbitt metals, vary widely in composition but can be categorized into two groups:

- Alloys of lead, tin, antimony, and, in many instances, arsenic
- Alloys of lead, calcium, tin, and one or more of the alkaline earth metals

Many alloys of the first group have been used for centuries as type metals. They most likely were chosen for use as bearing materials because of the properties they were known to possess. The advantages of arsenic additions in this type of bearing alloy have been generally recognized since 1938. Alloys of the second type were developed early in the 20th century.

Pouring temperature and rate of cooling markedly influence the microstructures and properties of lead alloys, particularly when they are used in the form of heavy liners for railway journals. High pouring temperatures and low cooling rates, such as those that result from the use of overly hot mandrels, promote segregation and the formation of a coarse structure. A coarse structure can cause brittleness, low compressive strength, and low hardness. Therefore, low pouring temperatures (325 to 345 °C, or 620 to 650 °F) are usually recommended. Because these alloys remain relatively fluid almost to the point of complete solidification (about 240 °C, or 465 °F, for most compositions), they are easy to manipulate and can be handled with no great loss of metal from the formation of dross.

Typical compositions of lead-base bearing alloys covered by ASTM specifications, and the corresponding SAE designations, are listed in Table 11 along with compositions of selected proprietary alloys. Additional information on the mechanical properties of some of these alloys is given in Table 12.

In the absence of arsenic, the microstructures of these alloys comprise cuboid primary crystals of SbSn or of antimony embedded in a ternary mixture of Pb-Sb-SbSn in which lead forms the matrix. The number of these cuboids per unit volume of alloy increases as antimony content increases. If antimony content is more than about 15%, the total amount of the hard constituents increases to such an extent that the alloys become too brittle to be useful as bearing materials.

Arsenic is added to lead babbitts to improve their mechanical properties, particularly at elevated temperatures. All lead babbitts are subject to softening or loss of strength during prolonged exposure to the temperatures (95 to 150 °C, 200 to 300 °F) at which they serve as bearings in internal-combustion engines. The addition of arsenic minimizes such softening. Under suitable casting conditions, the arsenical lead babbitts—for example, SAE 15 (ASTM grade 15)—develop remarkably fine and uniform structures. They also have better fatigue strength than arsenic-free alloys.

Arsenical babbitts give satisfactory service in many applications. The use of these alloys increased greatly during the Second World War, particularly in the automobile industry and in the manufacture of diesel engines. The most widely used alloy is SAE 15 (ASTM grade 15), which contains 1% arsenic. Automobile bearings of this alloy are usually made from continuously cast bimetal (steel and babbitt) strip. When properly handled, this alloy can withstand the considerable strain that results from forming the bimetal strip into bearings.

Diesel engine bearings often are cast as individual bearing shells by either centrifugal or gravity methods. An alloy that contains 3% arsenic (alloy B in Table 11) has been used successfully for applications where higher hardness is required and where formability requirements are less severe (rolling mill bearings, for example).

For many years, lead-base bearing alloys were considered to be only inferior low-cost substitutes for tin alloys. However, the two groups of alloys do not differ greatly in antiseizure characteristics, and when lead-base alloys are used with steel backs and in

Table 11 Nominal compositions of lead-base babbitt alloys

Designation	Nominal composition, % Pb	Sb	Sn	Cu max	Fe max	As	Bi max	Zn max	Al max	Cd max	Other
ASTM B 23 alloys											
Alloy 7 (a)	bal	15.0	10.0	0.50	0.1	0.45	0.10	0.005	0.005	0.05	...
Alloy 8	bal	15.0	5.0	0.50	0.1	0.45	0.10	0.005	0.005	0.05	...
Alloy 13 (b)	bal	10.0	6.0	0.50	0.1	0.25	0.10	0.005	0.005	0.05	...
Alloy 15 (c)	bal	16.0	1.0	0.50	0.1	1.10	0.10	0.005	0.005	0.05	...
Other alloys											
SAE 16	bal	3.5	4.5	0.10	...	0.05	0.10	0.005	0.005	0.05	...
AAR M501 (d)	bal	8.75	3.5	0.50	...	0.20	...	...	...	...	...
SAE 19	bal	...	10.0	...	...	...	...	...	...	...	...
SAE 190	bal	...	7.0	3.0	...	...	...	...	...	...	...
Proprietary alloys											
A	95.65	...	3.35	0.08	...	...	...	...	...	...	0.67 Ca
B	83.30	12.54	0.84	0.10	...	3.05	...	...	...	...	...
C	bal	10.0	3.0	0.20	...	...	...	...	...	...	2.0 Ag

(a) Also SAE 14. (b) Also SAE 13. (c) Also SAE 5. (d) Association of American Railroads specification M501; also ASTM B 67

Table 12 Properties of selected ASTM B 23 lead-base babbitt alloys

Designation	Specific gravity	Compressive yield strength(a)(b) At 20 °C (68 °F) MPa	ksi	At 100 °C (212 °F) MPa	ksi	Compressive ultimate strength(a)(c) At 20 °C (68 °F) MPa	ksi	At 100 °C (212 °F) MPa	ksi	Hardness, HB(d) At 20 °C	At 100 °C	Solidus temperature °C	°F	Liquidus temperature °C	°F	Pouring temperature °C	°F
Alloy 7	9.73	24.5	3.55	11.0	1.60	107.9	15.65	42.4	6.15	22.5	10.5	240	464	268	514	338	640
Alloy 8	10.04	23.4	3.40	12.1	1.75	107.6	15.60	42.4	6.15	20.0	9.5	237	459	272	522	340	645
Alloy 15	10.05	...	...	...	...	...	...	...	...	21.0	13.0	248	479	281	538	350	662

(a) The compression test specimens were cylinders 38 mm (1.5 in.) long × 13 mm (0.5 in.) in diameter that were machined from chill castings 50 mm (2 in.) long × 19 mm (0.75 in.) in diameter. (b) Values were taken from stress-strain curves at a deformation of 0.125% reduction of gage length. (c) Values were taken as the unit load necessary to produce a deformation of 25% of the length of the specimen. (d) Tests were made on the bottom face of parallel-machined specimens that had been cast at room temperature in a steel mold, 50 mm (2 in.) in diameter × 16 mm (0.625 in.) deep. Values listed are the averages of three impressions on each alloy, using a 10 mm ball and applying a 500 kg load for 30 s.

thicknesses below 0.75 mm (0.03 in.), they have a fatigue resistance that is equal to, if not better than, that of tin alloys. Bearings of any of these alloys remain serviceable longest when they are no more than 0.13 mm (0.005 in.) thick (Fig. 7). The superiority of lead alloys over tin alloys becomes more marked as operating temperatures increase. For this reason, automotive engineers generally favor lead-base alloys of compositions that approximate ASTM alloys 7 and 15, and SAE alloy 16. The SAE alloy is cast into and on a porous sintered matrix, usually of copper-nickel, that is bonded to steel. The surface layer of the babbitt is 0.025 to 0.13 mm (0.001 to 0.005 in.) thick.

The use of lead babbitts containing calcium and alkaline earth metals is confined almost entirely to railway applications, although these babbitts also are employed to some extent in certain diesel engine bearings. One of the more widely used alloys contains 1.0 to 1.5% Sn, 0.50 to 0.75% Ca, and small amounts of various other elements. The strength of this alloy approximates that of a tin alloy containing 90% Sn, 8% Sb, and 2% Cu. The hardness of this lead alloy is about 20 HB, the solidus temperature is 321 °C (610 °F), and the liquidus temperature is probably near 338 °C (640 °F). The pouring temperature, which varies from 500 to 520 °C (930 to 970 °F), is relatively high. The high temperature and reactive nature of calcium accounts for the formation of a much larger volume of dross than that encountered in the melting of lead-antimony-tin alloys. Care must be taken to avoid contamination of the alloy with antimonial lead babbitts, and vice versa. Deformability and resistance to wear are of the same order as those of the other lead babbitts. Most alloys of this type are subject to corrosion by acidic oils.

The fatigue resistance of bearing materials depends to a great extent on the design of the bearing. The strength and rigidity of the supporting structure, the thickness of the backing metal (steel or bronze), the thickness of the bearing material, and the character of the bond between the bearing material and the backing are all factors of consequence in bearings for use in high-speed reciprocating engines, such as the main and connecting-rod bearings of automobile and aircraft engines.

Resistance to fatigue is somewhat less important in bearings that operate under static load, for example, journal bearings in traction motor supports for diesel locomotives and in railway freight cars. In such bearings, antiseizure characteristics, conformability, compressive strength, and resistance to abrasion and corrosion are of greater significance. The lining metal generally employed in such journal bearings is the low-arsenic Association of American Railroads alloy M501 (ASTM B 67) (Table 11) cast onto a leaded-bronze back.

Ammunition. Large quantities of lead are used in ammunition for both military and sporting purposes. Alloys used for shot contain up to 8% Sb and 2% As; those used for bullet cores contain up to 2% Sb.

Terne Coatings. Long terne steel sheet is carbon steel sheet that has been continuously coated by various hot dip processes with terne metal (lead with 3 to 15% Sn). This coated sheet is duller in appearance than conventional tin-coated sheet; this accounts for the name terne, which means dull or tarnished in French. The smooth, dull coating gives the sheet corrosion resistance, formability, excellent solderability, and

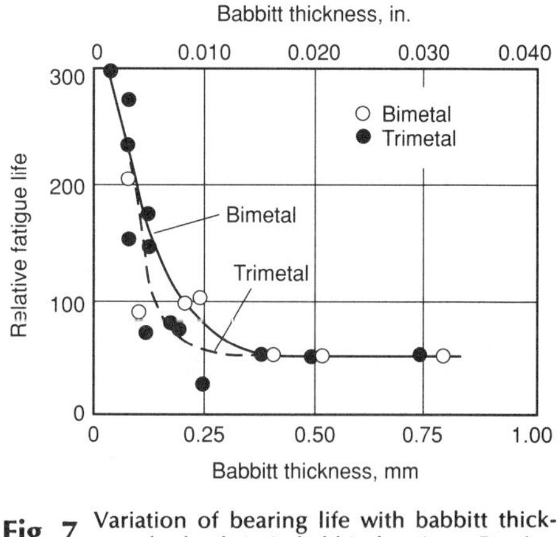

Fig. 7 Variation of bearing life with babbitt thickness for lead or tin babbitt bearings. Bearing load, 14 MPa (2000 psi) for all tests

paintability. The term long terne is used to describe terne-coated sheet, whereas short terne is used for terne-coated plate.

Because of its unusual properties, long terne sheet has been adapted to a wide variety of applications. Its principal use is in automotive gasoline tanks. Its excellent solderability and special corrosion resistance make the product well-suited for this application. Other typical applications include:

- Automotive parts such as air conditioners, air filters, cylinder head covers, distributor tubes, oil filters, oil pans, radiator parts, and valve rocker arm covers
- Caskets
- Electronic chassis and parts for radios, tape recorders, and television sets
- File drawer tracks
- Fire doors and frames
- Furnace and heating equipment parts
- Railroad switch lamps
- Small fuel tanks for lawn mowers, power saws, tractors, and outboard motors

Long terne sheet is often produced in accordance with ASTM A 308. For applications requiring good formability, the coating is applied over commercial quality, drawing quality, or drawing quality special killed low-carbon steel sheet. The terne coating acts as a lubricant and facilitates forming, and the strong bond of the terne metal allows it to be formed along with the basis metal. When higher strength is required, the coating can be applied over low-carbon steel sheet of structural (physical) quality, although this will result in some loss in ductility. The mechanical properties of long terne sheet are essentially the same as those of hot dip galvanized or aluminized steel sheet.

Lead has excellent corrosion resistance, and terne metal is principally lead, with 3 to 15% tin added to react with the steel to form a tight intermetallic bond. However, because lead does not offer galvanic protection to the steel basis metal, care must be exercised to avoid scratches and pores in the coating. Small openings can be sealed by corrosion products of iron, lead, and oxygen, but larger ones can corrode in an environment unfavorable to the steel basis metal.

Long terne sheet can be readily soldered with noncorrosive fluxes using normal procedures because the sheet is already presoldered. This makes it a good choice for applications in which ease of solderability is important, such as television and radio chassis and gasoline tanks. It also can be readily welded by either resistance seam welding or spot welding; however, when the coating is subjected to high temperatures, significant concentrations of lead fumes can be released. Therefore, the U.S. Occupational Safety and Health Administration and similar state agencies have promulgated standards that must be followed when welding, cutting, or brazing metals containing lead or metals coated with lead or lead alloys.

Long terne sheet has excellent paint adherence, which allows it to be painted using conventional systems; however, it is not usually painted. When painting is done, no prior special surface treatment or primer is necessary, except for the removal of ordinary dirt, oil, and grease. Oiled sheet, however, should be thoroughly cleaned to remove the oil. Alternatively, a wash primer treatment or a paint that will tolerate a slight residue of manufacturing oil can be used.

Long terne sheet normally is furnished dry and requires no special handling. It should be stored indoors in a warm, dry place. Unprotected outdoor storage of coils or bundles can result in white or gray staining of the terne coating. Also, if pores are present in the terne coating, rust staining can occur.

Lead foil, generally known as composition metal foil, is usually made by rolling a sandwich of lead between two sheets of tin, producing a tight union of the metals. Thicknesses of 0.01 mm (0.0005 in.) or less are common. Lead foil is used for moisture protection in the construction industry and for oxygen barriers on wine and champagne bottles. Lead-tin composite foils also are used in the electronics industry.

Fusible Alloys. Lead alloyed with tin, bismuth, cadmium, indium, or other elements, either alone or in combination, forms alloys with particularly low melting points. Some of these alloys, which melt at temperatures even lower than the boiling point of water, are referred to as fusible alloys. They are used for automatic sprinkler systems, electric fuses, and boiler plugs. Additional information on fusible alloys is contained in the article "Indium and Bismuth" in this Volume.

Anodes made of lead alloys are used in the electrowinning and plating of metals such as manganese, copper, nickel, and zinc. Rolled lead-calcium-tin and lead-silver alloys are the preferred anode materials in these applications, because of their high resistance to corrosion in the sulfuric acid used in electrolytic solutions. Lead anodes also have high resistance to corrosion by seawater, making them economical to use in systems for the cathodic protection of ships and offshore rigs. Anodes for these purposes are sometimes made of unalloyed lead, but they are usually made of lead alloyed with silver, tin, or antimony. These anodes are produced not only in cast form but also as extruded bars or supported sheet. Lead-7% tin anodes are used in chrome plating.

Structures

In many applications, lead is combined with stiffer and stronger materials to make structures that have the best qualities of both materials. An example of this type of structure is the series of lead and lead-coated structures described below, which are used for corrosion-resistant equipment. However, lead also is combined with plastics having relatively low stiffness and strength to make structures with superior sound control characteristics.

The plumbum series is a group of material combinations, each of which features lead as a major constituent. These materials are used in applications requiring good strength and high resistance to corrosion. Lead or lead alloy coatings provide the corrosion resistance; the strength is provided by steel, concrete, wood, brick, or another suitable material.

The six members of the plumbum series are described below, along with examples of typical uses for each type. They are presented in roughly increasing order of cost and strength:

- *Basic plumbum*: Lead or lead alloys in cast or extruded form with limited support. Used for cast antimonial lead valves, pipe fittings, pumps, anodes, and vessels
- *Supported plumbum*: Lead or lead alloys in sheet, pipe, or other extruded forms that are mechanically fastened to supporting structures of steel, wood, concrete, copper, or other metals. Used for concrete cells lined with lead sheet (loosely lined or cage supported) that are needed for the electrolytic refining of metals; also used for flues, ducts, towers, floors, expanded lead-lined pipe, cable sheathing, roofing, and anodes
- *Adhesive plumbum*: Lead or lead alloys in sheet, pipe, or other forms that are joined with an adhesive to steel, concrete, wood, or any other material that can provide extensive support. Used for acid storage tanks made of lead sheet joined with an adhesive to a steel outer shell
- *Bonded plumbum*: A heavy lead or lead alloy layer metallurgically bonded to steel, copper, or another metal. Used for homogeneously bonded lead-lined steel reaction vessels and lead-clad copper heating and cooling coils
- *Brick plumbum*: Lead or lead alloy sheet sandwiched between an outer shell of concrete or steel and an inner layer of chemical-resistant ceramic brick or masonry (usually acid brick). The sheet is metallurgically or chemically bonded to the outer shell and the inner layer; for some applications a layer of cushioning material is placed between the sheet and the inner layer. Used for sulfuric acid mist scrubbers, precipitators, concentrators, and storage tanks
- *Plumbum coatings*: Thin lead or lead alloy coatings metallurgically or mechanically bonded to equipment to protect it from corrosion. Used for lead-tin alloy

Table 13 Sound control materials containing lead

Material	Description	Uses
Sheet lead	Usual weight, 0.25 to 2 kg (½ to 4 lb)	Used alone or laminated to substrates of various types
Lead-foam composites	Lead sheet with a usual weight of 0.25 or 0.5 kg (½ or 1 lb) sandwiched between layers of polyurethane foam	Laminated to enclosures
Leaded plastic sheets	Lead-loaded vinyl or neoprene sheet with or without fabric reinforcement	As a curtain or to line enclosures
Damping tile	Lead-loaded epoxy or urethane tiles	Damping heavy machinery
Casting compounds	Lead-loaded epoxy	Potting; filling complex voids
Troweling compounds	Lead-loaded epoxy or urethane	Damping enclosures, surfaces, resonating members, and rattling panels
Lead-fiberglass composites	Sandwich composite of lead and fiberglass with a usual weight of 0.25 to 0.5 kg (½ to 1 lb)	Damping enclosures; sound isolation between walls and rooms

coatings on steel for roofing, gutters, and downspouts

The five major characteristics of the plumbum series are relatively low material costs, low-to-moderate installation and maintenance costs, inherently high corrosion resistance, long service life, and adaptability to a wide range of operating conditions.

The relatively low material costs of plumbum series equipment are due to the fact that the three basic component materials—steel, lead, and concrete—are comparatively inexpensive. A fourth component, wood, is used only in applications such as the manufacture of explosives, where its nonsparking property makes it essential. The fifth major component, chemical-resistant masonry, is not low in cost. However, it is used only where high-temperature strength or abrasion resistance is required. Under those conditions, the cost of using the only other suitable materials or material combinations is usually comparable or higher.

All plumbum series equipment can be used in cold climates without failure due to embrittlement. Temperatures as high as 1000 °C (1830 °F) have been handled successfully by brick plumbum. Both brick plumbum and bonded plumbum have high resistance to damage from thermal shock caused by large and rapid fluctuations in temperature. High heat conductivity is a normal feature of bonded plumbum, basic plumbum, and most types of plumbum coatings. A strong barrier to heat transfer is provided by brick plumbum, adhesive plumbum, and a few types of supported plumbum. Thus, bonded plumbum is used to make heating and cooling coils, and brick plumbum is widely used as insulation in ducts handling very hot gases.

Basic plumbum performs reliably under pressures up to 0.3 MPa (3 atm). All other plumbums can withstand substantially higher pressures; bonded plumbum is especially resistant to damage. Both bonded plumbum and brick plumbum can be used to handle vacuums as well as pressures that fluctuate both above and below atmospheric level. Adequate abrasion resistance often is provided by using the harder alloys of lead. However, for extremely abrasive conditions, the hard masonry of brick plumbum is required. The pliability and malleability of soft lead make possible the manufacture of a wide variety of intricately shaped items such as corrugated helical heating and cooling coils. These qualities also allow the manufacture of glassy smooth lead pipes that minimize friction energy losses.

Plumbum coatings and the basic, brick, and supported plumbums are electrically conductive. However, they can be made into insulating barriers by incorporating a layer of nonconducting material. By choosing the appropriate adhesive, adhesive plumbum can be made either insulating or conductive. Bonded plumbum is always electrically conductive. Equipment with nonsparking surfaces can be constructed using supported or adhesive plumbum combinations that contain only wood and lead as the major components.

Sound Control Materials. Lead is an excellent barrier to sound transmission. Essentially, a good sound barrier should have high density and low stiffness, and it should be impermeable. Lead and lead composites more than satisfy these requirements. In addition, the high internal damping capacity of lead and lead composites make them even more effective in controlling sound. Several examples of sound control materials that contain lead are given in Table 13. An advantage of these products is that they are unaffected by coolants, cutting oils, drawing compounds, and similar industrial fluids.

REFERENCES

1. *Properties of Lead and Lead Alloys*, Lead Industries Association, 1984
2. *Radiography & Radiation Testing*, Vol 3, *Nondestructive Testing Handbook*, 2nd ed., American Society for Nondestructive Testing, 1985, p 750, 753
3. Bolt, Beranek, and Newman, "Improved Sound Barriers Employing Lead," Lead Industries Association, 1960

Refractory Metals and Alloys

Chairman: John B. Lambert, Fansteel Inc.

Introduction

John B. Lambert and John J. Rausch, Fansteel Inc.

THE REFRACTORY METALS include niobium (also known as columbium), tantalum, molybdenum, tungsten, and rhenium. With the exception of two of the platinum-group metals, osmium and iridium, they have the highest melting temperatures and lowest vapor pressures of all metals. The refractory metals are readily degraded by oxidizing environments at moderately low temperatures, a property that has restricted the applicability of the metals in low-temperature or nonoxidizing high-temperature environments. Protective coating systems have been developed, mostly for niobium alloys, to permit their use in high-temperature oxidizing aerospace applications.

Refractory metals at one time were limited to use in lamp filaments, electron tube grids, heating elements, and electrical contacts; however, they have since found widespread application in the aerospace, electronics, nuclear and high-energy physics, and chemical process industries. Each of the refractory metals, with the exception of rhenium, is consumed in quantities exceeding 900 Mg (1000 tons) annually on a worldwide basis. In 1988, consumption of refractory metals in the United States was:

Metal	Amount, Mg(a)
Niobium	2 665(b)(c)
Tantalum	422(d)
Tungsten	8 298
Molybdenum	17 422
Rhenium	8(b)

(a) 1 Mg = 1 metric ton. (b) Estimated. (c) Used primarily in the form of ferrocolumbium; includes nickel-niobium and a small quantity of other niobium materials. (d) Apparent

Applications

Most niobium is consumed as a ferroalloy used in the production of high-strength low-alloy and stainless steels; the consumption of niobium-base metals and alloys accounts for about 6% of the total. The single largest use for tantalum is as powder and anodes for electronic capacitors, representing about 50% of total consumption. Mill products—sheet and plate, rod and bar, and tubing—constitute nearly 25% of tantalum consumption. The major end use for tungsten is in cemented carbides, which are used for cutting tools and wear-resistant materials. Tungsten carbides make up nearly 60% of tungsten consumption; mill products account for approximately 25%. Most molybdenum is used as an alloying addition in steels, irons, and superalloys. Molybdenum-base mill products represent less than 5% of usage. Platinum-rhenium reforming catalysts account for 85% of rhenium consumption.

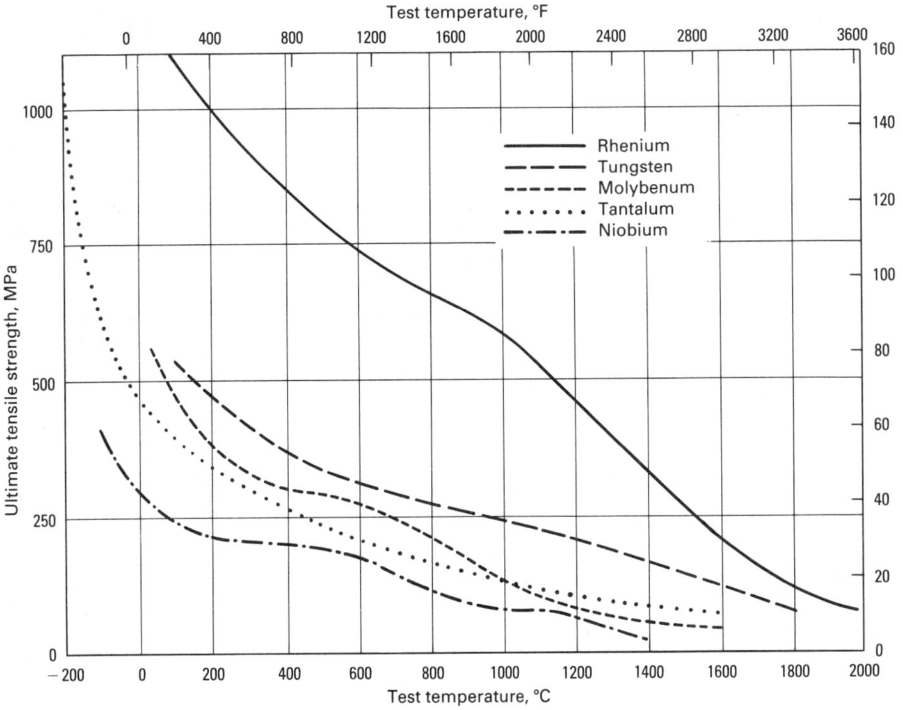

Fig. 1 Test temperature versus ultimate tensile strength for pure refractory metals

As a result primarily of experience gained in aerospace programs, applications for refractory metals now encompass almost every type of industry. Table 1 summarizes the commercially significant uses of these metals. Table 2 compares the physical, thermal, electrical, magnetic, and optical properties of pure refractory metals. Figures 1 and 2 compare the temperature-dependent ultimate tensile strengths and elastic moduli of the refractory metals. The values for hexagonal close-packed (hcp) rhenium are quite different from those of the other metals, which are body-centered cubic (bcc). Table 3 lists nominal compositions of commercially prominent refractory metal alloys.

Selection of a specific alloy from the refractory metal group often is based on fabricability rather than on strength or corrosion resistance. Niobium, tantalum, and their alloys are the most easily fabricated

Table 1 Commercial applications of refractory metals and alloys by industry

Application	Material
Aerospace and nuclear industries	
Counterweights (aircraft, inertial guidance systems)	Tungsten alloys
Solid-propellant rockets	
2650–2750 °C (4800–4980 °F) flame temperature	Molybdenum, tungsten
3425–3550 °C (6195–6420 °F) flame temperature	Silver and copper-infiltrated tungsten
Lifting and guidance structures for glide reentry vehicles	Cb-752(a), FS-85(a), C-129Y(a)
Leading edges and nose caps for hypersonic flight vehicles	Cb-752(a), FS-85(a), Ta-10W(a)
Thrust chambers	C-103(a)
Radiation nozzle extensions	C-103(a), FS-85
Jet engine components	
Augmenter liners	C-103(a)
Center body	Cb-752(a)
Flame holders	Cb-752(a)
Rocket nozzles	FS-85(a), Ta-10W (Ta-Hf clad)
Thermal shields	C-129Y(a)
Porous ionizer plates	Tungsten
Heat shields and cesium vapor inlet tubes (ion engine)	Tantalum
Fasteners	Nb-Ti
Honeycomb structures	Molybdenum, Cb-752, Ta-10W
Hot gas tubing	Ta-10W(a)
Hot gas bellows	C-103(a)
Solid propellant expansion nozzle	C-103(a)
Nuclear and high-energy physics	
Linear accelerators, microwave cavities	Niobium
Superconductors	Nb-Ti, Nb_3Sn
Liquid metal containers and piping	Nb-1Zr
Electronics industry	
Capacitors	Tantalum powder, foil, wire
Capacitor cases	Tantalum strip
Rectifiers, railway signals	Tantalum
Battery chargers	Tantalum
Transducers	Molybdenum, tungsten
Electron tube parts	
Heaters	Tungsten, W-Re
Supports	Molybdenum
Cathodes	Tantalum
Anodes	Molybdenum, tungsten
Superconducting wire	Nb-Ti, Nb_3Sn
X-ray targets	Tungsten, molybdenum, rhenium, composite W-Mo
Electrodes (mercury switches)	Molybdenum, tungsten
Thin-film substrates	Molybdenum, tungsten
Electrical contacts	Tungsten, rhenium, W-Ag, W-Cu
Heat sinks	Molybdenum, tungsten
Backing wafers, semiconductors	Molybdenum, tungsten
Filaments, ion gages, photoflash	Rhenium, W-Re
Process industries	
Heating and cooling coils	Tantalum, Ta-Nb
Shell and tube heat exchangers	Tantalum
Condensers	Tantalum
Tantalum-clad steel vessels	Tantalum
Distillation towers	Tantalum
Valves for hot sulfuric acid service	Molybdenum, tantalum, Ta-Nb
Expansion joints (bellows)	Tantalum
Glass-processing equipment	Tantalum
Crucibles, all sizes up to 1 m (3 ft) diameter × 1.3 m (4 ft) high	Tungsten, tantalum, Ta-40Nb
Spinnerettes, textile industry	Tantalum, niobium
Thermocouple protection tubes	Tantalum-coated copper or steel
Rupture discs	Tantalum
Thermowells	Tantalum-clad copper, tantalum
Spargers, funnels, jet ejectors	Tantalum
Bayonet heaters	Tantalum
Pumps for hydrogen chloride service at 200 kPa (30 psi) and 150 °C (300 °F)	Tantalum (exposed parts)
Cathodic protection electrodes	Niobium
Special equipment	
Furnace parts	
Heating elements, shields, boats, trays, platens, fixtures	Tungsten, molybdenum, tantalum
Susceptors (induction furnace)	Tungsten
Extrusion dies	Tungsten, molybdenum
Piercing points, hot punches	Tungsten, molybdenum
Cups	Tungsten, molybdenum, tantalum
Fasteners (nuts, screws, studs, rivets)	Tantalum, molybdenum, C-3009, C-129Y
Die casting molds, cores	Molybdenum, tungsten
Vacuum-metallizing coils, boats	Tungsten, molybdenum
Springs	Tungsten, molybdenum, tantalum
Boring bars	Tungsten, molybdenum
Surgical implants	Tantalum
Instruments	Tantalum
Electroplating equipment	Tantalum
Thermocouples, spot weld electrodes	W, W-Re alloys
Cathodes, plasma generator	W-1Ni
Rapid-fire gun barrels	C-3009(a)
Sodium vapor lamp electrodes	Nb-1Zr

(a) Parts are silicide-coated in use.

refractory metals. They can be formed, machined, and joined by conventional methods. They are ductile in the pure state and have high interstitial solubilities for carbon, nitrogen, oxygen, and hydrogen. Because of the high solubilities in niobium and tantalum, these embrittling contaminants normally do not present problems in fabrication. However, tantalum and niobium dissolve sufficient amounts of oxygen at elevated temperatures to destroy ductility at normal operating temperatures. Therefore, elevated-temperature fabrication of these metals is used only when necessary. Protective coatings or atmospheres are mandatory unless some contamination can be tolerated. The allowable level of contamination, in turn, determines the maximum permissible exposure time in air at elevated temperature.

Molybdenum, molybdenum alloys, tungsten, and tungsten alloys require special fabrication techniques. Fabrication involving mechanical working should be performed below the recrystallization temperature. These materials have limited solubilities for carbon, nitrogen, oxygen, and hydrogen. Because the residual levels of these elements required to prevent embrittlement are impractically low, the microstructure must be controlled to ensure a sufficiently low ductile-to-brittle transition temperature (DBTT).

The resistance of refractory metals to corrosion by liquid metals and aggressive acid solutions can cut maintenance and downtime if high initial costs can be accepted. Systems for containing liquid metals such as lithium and cesium at high temperatures have been fabricated of Nb-1Zr alloy tubing; tantalum and tantalum-clad steel processing equipment have performed well in high-temperature sulfuric acid service.

Most refractory metals and alloys are available as wire. Tungsten wire, for example, which comes in diameters as small as 0.0102 mm (0.0004 in.), is used as fiber reinforcement in composite materials in which the matrix is any one of various ductile alloys. Tantalum wire is used extensively in capacitor manufacture and in surgical applications.

In the nuclear field, tungsten crucibles that are pressed and sintered, shear spun, chemical vapor deposited, or plasma sprayed and sintered are used in recovering uranium and plutonium from spent reactor fuel.

Tantalum and, to a lesser extent, molybdenum have been used for many years in the chemical process industries. The severe corrosion problems accompanying many chemical processes have given impetus to greater use of refractory alloys. Recently, chemical equipment has been fabricated from steel plate explosively clad with tantalum. Forming and welding methods have

Table 2 Mechanical and physical properties of pure refractory metals

Property	Niobium	Tantalum	Molybdenum	Tungsten	Rhenium
Structure and atomic properties					
Atomic number	41	73	42	74	75
Atomic weight	92.9064	180.95	95.94	183.85	186.31
Density at 20 °C (70 °F), g/cm³ (lb/in.³)	8.57 (0.310)	16.6 (0.600)	10.22 (0.369)	19.25 (0.695)	21.04 (0.760)
Crystal structure	bcc	bcc	bcc	bcc	hcp
Lattice constants, nm					
a	0.3294	0.3303	0.3147	0.3165	0.27609
c	...	...	...	...	0.45829
Slip plane at room temperature	110	110	112	...	0001–1010
Thermal properties					
Melting temperature, °C (°F)	2468 (4474)	2996 (5425)	2610 (4730)	3410 (6170)	3180 (5755)
Boiling temperature, °C (°F)	4927 (8901)	5427 (9801)	5560 (10040)	5700 (10290)	5760 (10400)
Vapor pressure at 2500 K, mPa (torr)	5.3 (4 × 10^{-5})	0.11 (8 × 10^{-7})	80 (6 × 10^{-4})	0.0093 (7 × 10^{-8})	0.17 (1.3 × 10^{-6})
Coefficient of expansion, near RT(a), μm/m · K (μin./in. · °F)	7.3 (4.1)	6.5 (3.6)	4.9 (2.7)	4.6 (2.6)	6.7 (3.7)
Specific heat at 20 °C (70 °F), kJ/kg · K (Btu/lb · °F)	0.268 (0.0643)	0.139 (0.0333)	0.276 (0.0662)	0.138 (0.0331)	0.138 (0.0331)
Latent heat of fusion, kJ/kg (Btu/lb)	290 (125)	145–174 (62–75)	270 (115)	220 (95)	177 (76)
Latent heat of vaporization, kJ/kg (Btu/lb)	7490 (3202)	4160–4270 (1790–1840)	5123 (2160)	4680 (2010)	3415 (1470)
Thermal conductivity, W/m · K (Btu/ft · h · °F)					
At 20 °C (70 °F)	52.7 (30.4)	54.4 (31.4)	142 (81.9)	155 (89.4)	71 (41)
At 500 °C (930 °F)	63.2 (36.5)	66.6 (38.4)	123 (71.0)	130 (75)	...
Electrical properties					
Electrical conductivity at 18 °C (64 °F), %IACS(b)	13.2	13.0	33.0	30.0	8.1
Electrical resistivity, at 20 °C (70 °F), nΩ · m	160	135	52	53	193
Electrochemical equivalent, mg/C	0.1926	0.375	0.166	0.318	0.276
Hall coefficient, nV · m/A · T	0.09	0.095	...	...	...
Magnetic properties					
Magnetic susceptibility (volume) at 25 °C (75 °F), mks system	28 × 10^{-6}	10.4 × 10^{-6}	1.17 × 10^{-8}	4.1 × 10^{-8}	0.37 × 10^{-6}
Optical properties					
Total emissivity at 1500 °C (2730 °F), %	0.19	0.21	0.19	0.23	...
Spectral emittance at λ = 650 nm, %	0.37	0.49	0.37	0.43	...
Additional properties					
Poisson's ratio at 25 °C (75 °F)	0.38	0.35	0.32	0.28	0.49
Elastic modulus, GPa	103	185	324	400	469
Ductile-to-brittle transition temperature (DBTT), K	<147(c)	<25(c)	...	250(d)	...

(a) RT, room temperature. (b) IACS, International Annealed Copper Standard. (c) Viscous with iron purity. (d) Value for as-drawn material; DBTT for annealed tungsten is 325 K.

been developed for fabrication of the clad plate into reactor vessels, tanks, and other types of chemical equipment. Explosive bonding produces a metallurgical bond at the tantalum/steel interface. Bond efficiency is over 98%, and bond shear strength exceeds the American Society of Mechanical Engineers (ASME) minimum acceptable value for clad material.

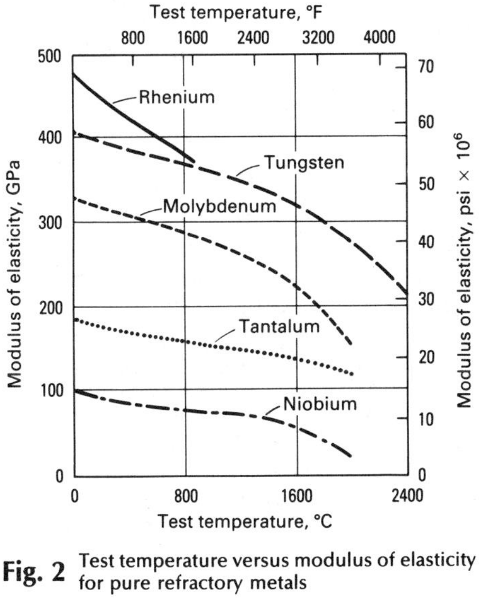

Fig. 2 Test temperature versus modulus of elasticity for pure refractory metals

Electronic applications constitute one of the major uses for refractory metals. The largest use for tantalum is in electrolytic capacitors. Porous sintered powder metallurgy (P/M) anodes are used in both solid and wet electrolytic capacitors, and, to a lesser extent, precision tantalum foil is used in foil capacitors. The dielectric film of tantalum oxide is electrolytically formed on the tantalum surface in the manufacturing process. It has been reported that commercial quantities of P/M niobium capacitors are being produced in the Soviet Union. Although the dielectric constant of niobium oxide is greater than that of tantalum oxide (41.4 versus 25.3), niobium powder is not extensively used for capacitors. The problem with niobium is that the amorphous anodic oxide film, formed as the dielectric, crystallizes at a relatively low temperature, thereby causing performance decay in the capacitor.

Tantalum cases (or cans) are used for hermetically sealing wet electrolytic tantalum capacitors. The package consists of a cold drawn can having a porous sintered tantalum powder lining that serves as a cathode, a cap or a header, and a glass-to-metal seal in the header; the seal insulates a tantalum lead wire that passes through the header and connects to the porous tantalum powder capacitor anode. The lead is either embedded in the anode during powder pressing or spot welded after the anode is sintered. After the slug has been inserted and the can filled with electrolyte, the header is resistance or laser welded to the tantalum can, forming a hermetic seal. Headers are formed in progressive dies, although recent designs also supply a stamped tantalum washer for the end seal. Cans approximately 9.52 mm (0.375 in.) in diameter and 19.0 mm (0.750 in.) in length are usually made either by drawing on a transfer press or by spinning.

Other electronic components in which refractory metals are used include a composite x-ray target, which consists of a forged or spun molybdenum substrate with a plasma-sprayed optical track of tungsten-rhenium alloy; molybdenum cathode supports for radar devices; and magnetron end hats.

In vacuum metallizing equipment, evaporation boats are commonly fabricated by coating formed molybdenum substrates with plasma-sprayed refractory oxides.

Table 3 Nominal compositions of commercially important refractory metal alloys

Alloy designation	Composition, %														
	Nb	Ta	Mo	W	Re	Zr	Hf	Ti	Y	C	ThO_2	Si	K	Al	O
Niobium alloys															
Nb-1Zr	bal	...	...	...	...	1	...	...	...	...	...	...	...	...	...
FS-85	bal	27.5	...	11	...	1	...	...	...	...	...	...	...	...	...
Cb-752	bal	...	...	10	...	2.5	...	...	...	...	...	...	...	...	...
C-103	bal	...	...	...	...	...	10	1	...	...	...	...	...	...	...
C-129Y	bal	...	...	10	...	...	10	...	0.15	...	...	...	...	...	...
C-3009 (a)	bal	...	...	10	...	...	30	...	...	...	...	...	...	...	...
Cb-Ti superconductor		...	...	...	...	...	...	46.5	...	...	...	...	...	...	...
Tantalum alloys															
63 metal	0.15	bal	...	2.5	...	...	...	...	...	...	...	...	...	...	...
Ta-10W	...	bal	...	10	...	...	...	...	...	...	...	...	...	...	...
T-111	...	bal	...	8	...	...	2	...	...	...	...	...	...	...	...
T-222	...	bal	...	10	...	...	10	...	...	0.01	...	...	...	...	...
Ta-40Nb	40	bal	...	...	...	...	...	...	...	...	...	...	...	...	...
61 metal (P/M)		...	bal	...	7.5	...	...	...	...	...	...	...	...	...	...
Molybdenum alloys															
Mo-0.5Ti	...	...	bal	0.02	...	...	...	0.5	...	...	...	...	...	...	...
TZM	...	...	bal	0.02	...	0.1	...	0.5	...	...	...	...	...	...	...
Tungsten alloys															
W-ThO_2 alloys															
W-1 ThO_2	...	...	...	bal	...	...	...	...	...	...	1	...	...	...	...
W-2 ThO_2	...	...	...	bal	...	...	...	...	...	...	2	...	...	...	...
W-Mo alloys(b)															
W-2 Mo	...	...	2	bal	...	...	...	...	...	...	...	...	...	...	...
W-15 Mo	...	...	15	bal	...	...	...	...	...	...	...	...	...	...	...
W-Re alloys(c)															
W-1.5 Re	...	...	...	bal	1.5	...	...	...	...	...	...	...	...	...	...
W-3 Re	...	...	...	bal	3	...	...	...	...	...	...	...	...	...	...
W-25 Re	...	...	...	bal	25	...	...	...	...	...	...	...	...	...	...
Doped W(d)	...	...	...	bal	...	...	...	...	...	...	...	50 ppm	90 ppm	15 ppm	35 ppm

(a)C-3009 is in the literature as a family of alloys ranging from 9–15% W and ≤5% Ti. The hafnium content is constant at 30%. (b) Various molybdenum contents; two most common alloys listed. (c) Various rhenium contents to ≤26%; three most common alloys listed. (d) See also Table 22.

Chemical vapor deposition has proved useful for fabricating free-standing refractory metal parts for advanced electronics applications. Tungsten emitters of preferred crystal orientation, tungsten collectors, and sandwich insulators (Nb-Al_2O_3-Nb, for example) are used in nuclear thermionic conversion devices.

Production of Refractory Metals

The refractory metals, except for niobium, are produced exclusively as metal powders, which are consolidated by sintering and/or melting. The process for niobium differs only in that the metal is most commonly reduced by aluminothermic reduction of oxide. In this process, oxide impurities slag from the molten niobium.

For tantalum and niobium, electron beam (EB) melting is widely used for further purification. Powders can also be produced for these metals from ingot by the hydride-crush-dehydride process. Alloys are made by adding alloying agents during melting. Low-volatility metals such as tungsten and tantalum can be added during EB melting. More volatile agents such as titanium, hafnium, or zirconium are frequently added in vacuum arc remelting.

Hot forging or extrusion is used for breaking down ingots into rounds or rectangular sheet bar. These bars, as well as sintered products, are processed into sheet, plate, foil, tubing, and bar. Table 4 gives typical mill-processing temperatures for the refractory metals.

Fabrication

A general flow sheet for the fabrication of refractory metals is shown in Fig. 3.

Machining

Equipment for machining refractory metals must be rigid and powerful to ensure optimum results. Carbide and, on occasion, cast cobalt tools give acceptable tool life and cutting properties. Detailed information is available in the articles "Cemented Carbides" and "Cast Cobalt Alloys" in *Machining*, Volume 16 of the 9th Edition of *Metals Handbook*.

Niobium alloys and tantalum alloys are readily machined using high-speed steel (see the articles "High-Speed Tool Steels" and "P/M High-Speed Tool Steels" in *Machining*, Volume 16 of the 9th Edition of *Metals Handbook*) or carbide tools. The machining and grinding characteristics of these alloys vary from being similar to those of soft copper to those of annealed stainless steel. Tooling recommendations are summarized in Table 5.

Molybdenum is machined using carbide tools of the same configurations as those used for machining 1040 and 4340 steel because the machining characteristics of these two metals are similar to those of molybdenum. Machining speeds for molybdenum alloys (TZM, for example) are about 40% higher than those for type 302 stainless steel. Finish grinding of molybdenum requires a heavy coolant flow and the use of aluminum oxide wheels to prevent heat checking. Tool configurations and grinding techniques are similar to those for grinding cast iron; conventional machines with standard feeds and speeds are satisfactory.

Turning is a problem only with tungsten. For tungsten, the use of carbide tools ground with a negative back rake, 15° lead, and 0° side rake are mandatory. All turning is done at room temperature. However, machinable tungsten heavy-metal alloys bonded with copper, nickel, and iron are produced (see the section "Tungsten" in this article for additional information).

For grinding tungsten, wheels of 60-grit silicon carbide or 46-grit alumina are recommended. Normal precautions, extra-light pressures, and heavy coolant flow are required.

Tungsten and molybdenum must be punched and sheared at temperatures above their ductile-to-brittle transition temperatures. Sheets over 1.3 mm (0.050 in.) thick must have an excess thickness of 1.6 to 3.2 mm (1/16 to 1/8 in.) to allow for belt sanding to final dimensions. Cutting can be done using abrasive (60-grit silicon carbide) cutoff wheels.

Table 4 Mill-processing temperatures for refractory metals

Metal or alloy	Forging: Temperature(a) °C	Forging: Temperature(a) °F	Forging: Typical total reduction, %	Extrusion: Temperature(a) °C	Extrusion: Temperature(a) °F	Extrusion: Typical reduction ratio	Rolling: Temperature(a) °C	Rolling: Temperature(a) °F	Rolling: Typical total reduction between anneals, %
Niobium and niobium alloys									
Niobium	980–650	1800–1200	50–80	1095–650	2000–1200	10:1	315–205	600–400	50 breakdown
							20	70	90 finish
Nb-1Zr	1205–980	2200–1800	50–80	1205–980	2200–1800	10:1	315–205	600–400	50 breakdown
							20	70	80 finish
FS-85	1315–980	2400–1800	50	1315–980	2400–1800	4:1	370–205	700–400	40 breakdown
							20	70	50–65 finish
Cb-752	1205–980	2200–1800	30	1315–980	2400–1800	4:1	370–260	700–500	50 breakdown
							20	70	60–75 finish
C-103	1315–980	2400–1800	50	1315–980	2400–1800	8:1	205	400	50 breakdown
							20	70	60–70 finish
C-129Y	1315–980	2400–1800	50	1315–980	2400–1800	4:1	425	800	50 breakdown
							20	70	60–70 finish
Tantalum and tantalum alloys									
Tantalum	<500	<930	50–80	1095	2000	10:1	370–260	700–500	80 breakdown
	20	70	Finish				20	70	90 finish
Ta-10W	1260–980	2300–1800	50	1650–1425	3000–2600	10:1	370–260	700–500	80 breakdown
	1095–815	2000–1500	Finish				20	70	90 finish
T-222	1260–1205	2300–2200	50	2040–1650	3700–3000	10:1	370–260	700–500	75 breakdown
							20	70	50–75 finish
Molybdenum and molybdenum alloys									
Molybdenum	1315–1150	2400–2100	50	1760–1370	3200–2500	8:1	1205	2200	50 breakdown
	925–815	1700–1500	Finish				870	1600	90–75 finish
Mo-0.5Ti	1425–1260	2600–2300	50	1815–1480	3300–2700	8:1	1205	2200	50 breakdown
	1315–1150	2400–2100	Finish				870	1600	75 finish
TZM	1480–1315	2700–2400	50	1815–1540	3300–2800	8:1	1350–1205	2460–2200	50 breakdown
	1370–1205	2500–2200	Finish				1000–980	1830–1800	60
							315	600	10 finish
Tungsten									
Tungsten	1815–1595	3300–2900	20	1925–1650	3500–3000	9:1	1450–1400	2640–2550	50 breakdown
	1315–1010	2400–1850	Finish				1370–980	2500–1800	90 finish

(a) Where a range is given, the higher temperature is the typical starting temperature and the lower temperature is the minimum working temperature for that process.

Electrical discharge machining can be used for shaping niobium and tantalum. Some limited work has been done to apply electrochemical machining to these metals, but the formation of the tenacious anodic oxide layer on the metals, which inhibits the process, and hydrogen embrittlement are problems. Both electrochemical machining and electrical discharge machining are suitable for molybdenum and tungsten.

Photoetching and chemical blanking have been used on molybdenum. In these processes, photographic masking is followed by etching in a solution of HNO_3 and HF. Such techniques are used for applications in which complex integral shapes or weight reductions are required. Chemical blanking is a particularly attractive process for simultaneously cutting parts with many different shapes from the same plate or sheet.

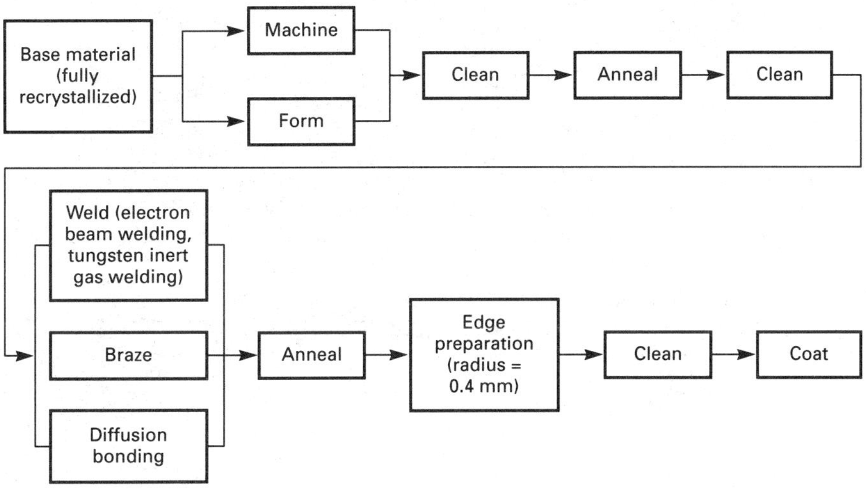

Fig. 3 Typical sequence of operations for the fabrication of refractory metals

Band or circular saws can be used to cut molybdenum sheet. Sheet thicknesses in the range of 1.0 to 1.5 mm (0.039 to 0.059 in.) require band speeds of 37 m/min (120 sfm); thinner sections with thicknesses from 0.4 to 0.75 mm (0.016 to 0.030 in.) can be cut at band speeds of 76 to 91 m/min (250 to 300 sfm). Sheets in the thickness range of 0.50 to 1.52 mm (0.020 to 0.060 in.) can be cut with abrasive wheels rotating at 1000 to 1400 rev/min.

Molybdenum sheet can be effectively drilled using high-speed steel drills and conventional oil lubricants. Table 6 shows some typical drill parameters for the use of automatic drill machines.

Acceptable machining techniques for rhenium include electrical discharge machin-

Table 5 Tooling recommendations for machining niobium and tantalum

Approach angle	15–20°
Side rake	30–35°
Side and end clearance	5°
Plan relief angle	15–20°
Nose radius	0.50–0.75 mm (0.020–0.030 in.)
Cutting speed	
High-speed steel tools	0.3–0.4 m/s (60–80 sfm)
Carbide tools	1.3–1.5 m/s (250–300 sfm)
Feed	
Roughing	0.23–0.30 mm/rev (0.009–0.012 in./rev)
Finishing	0.13 mm/rev (0.005 in./rev) max
Depth of cut	0.75–3.2 mm (0.030–0.125 in.)

Table 6 Parameters for drilling holes in molybdenum sheet

Drill size			Speed,	Feed	
Number	mm	in.	rev/min	mm/rev	in./rev
40	2.49	0.0980	1200	0.127	0.0050
30	3.25	0.1285	900	0.178	0.0070
10	4.90	0.1935	600	0.254	0.0100

ing, electrochemical milling, abrasive cutting, and grinding. Rhenium is very difficult to machine with carbide tools and other conventional methods. Rhenium sheet and plate can be sheared, but the cold-worked area should be removed during subsequent grinding and polishing.

Vibration radiusing using loose abrasive and frequencies of 23 to 30 Hz can be used to eliminate sharp corners on all refractory metals. The abrasive action rounds edges, eliminating the need for hand filing and polishing. Additional information is available in the article "Machining of Refractory Metals" in *Machining*, Volume 16 of the 9th Edition of *Metals Handbook*.

Forming

Niobium and tantalum sheet are formed using a number of techniques, including conventional form (but usually not shear) spinning, hydroforming, bulge forming, and chemical milling. For the production of a complex part, the designer may need to use several forming and joining operations before the final part shape is achieved. The forming behavior of these metals is similar to that of mild steel, except that they are more prone to galling, seizing, and tearing. In thicknesses from 0.1 to 1.5 mm (0.004 to 0.060 in.), tantalum and niobium can be readily blanked, punched, stamped, or deep drawn at room temperature in steel dies (6% *t* clearance, where *t* is the sheet thickness). Sheet must have a homogeneous, fine grain size (generally ASTM No. 5 or finer) for satisfactory results. Coarse-grain sheet is likely to fail by localized necking during severe forming.

For conventional forming of molybdenum sheet, as well as for blanking, punching, and shearing with heated dies, the following temperature-thickness relationships apply:

Thickness		Temperature	
mm	in.	°C	°F
0.5	0.02	20	70
0.5–1.0	0.02–0.04	95–165	200–325
1.0	0.04	480–540	900–1000

Despite difficulties in fabrication, tungsten is used in more applications than any other refractory metal besides tantalum. Many tungsten parts are die formed or deep drawn. Recrystallization and thermal conductivity data for tungsten appear in Fig. 4 and 5.

Fabricators can shear, draw, or form tungsten by a variety of techniques if they

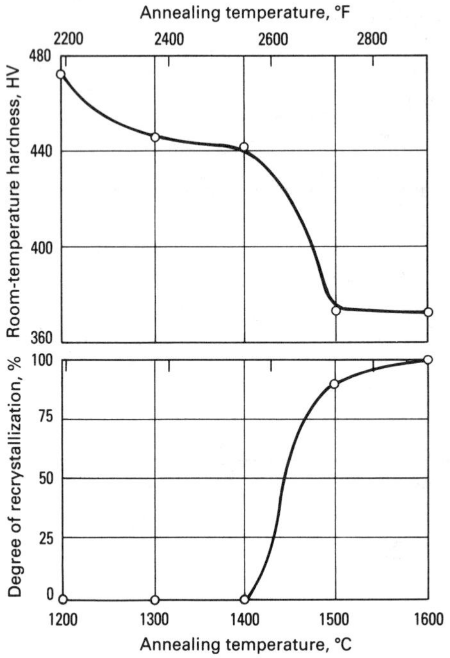

Fig. 4 Recrystallization behavior of undoped tungsten bar

understand its directional and recrystallization properties. Thin sections can be formed into simple shapes from room temperature to about 95 °C (200 °F). Heavier sections, however, require higher forming temperatures:

Thickness		Temperature	
mm	in.	°C	°F
0.25–0.4	0.010–0.016	205–260	400–500
0.4–1.0	0.016–0.039	540	1000
1.0	0.039	1260–1595	2300–2900

Punching and shearing must be done hot in accordance with the same temperature-thickness relationships. Material 1.3 mm (0.050 in.) or greater in thickness should be sheared to within 1.6 to 3.2 mm (1/16 to 1/8 in.) of final dimensions, and parts should be finished by edge grinding.

Table 7 gives blanking pressures, shear strengths, and approximate blanking temperatures for tungsten blanks of various sizes. Shear strengths were derived from strain gage tests on disks about 32 and 50 mm (1¼ and 2 in.) in diameter.

In rolling, open die forming, and closed die forming of tungsten and molybdenum,

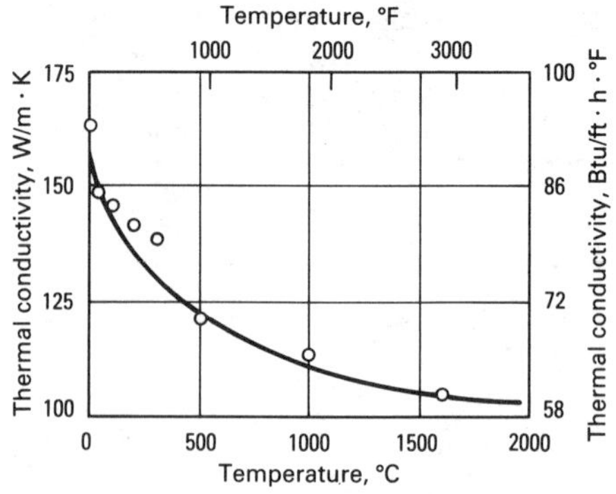

Fig. 5 Thermal conductivity of undoped tungsten

rolls and dies are heated to 425 to 540 °C (800 to 1000 °F). Otherwise, conventional techniques prevail.

The refractory metals can be spun in air if careful attention to temperature and time are maintained. Generally, the metal to be formed is mounted on a heavy spinning lathe and one or (preferably) two rollers. An oxypropane torch is used to heat the workpiece to the following temperatures:

Metal	Temperature °C	Temperature °F
Niobium		
Protected	400–620	750–1150
Unprotected	425 max	800 max
Tantalum		
Protected	480–650	900–1200
Unprotected	480 max	900 max
Molybdenum	480–1065	900–1950
Tungsten	760–1315	1400–2400

Form spinning, shear spinning (flow turning), and extrusion spinning can be used singly or in combination to fabricate refractory metal plates, sheets, or tubular blanks into configurations that are impractical to produce by conventional forming processes. Often, the only alternative is a combination of open die forging and machining, which is comparatively expensive. Spinning involves relatively low tooling and finishing costs and short setup times; it also can produce parts within relatively tight dimensional tolerances.

Small-size tubing of molybdenum, niobium, and tantalum is made by a variety of techniques, depending on the desired quality, quantity, and size. Most heat exchangers require tubing 1.6 to 13 mm (1/16 to ½ in.) in diameter and 0.25 to 0.75 mm (0.010 to

Table 7 Blanking characteristics of tungsten sheet

Blank thickness		Blank diameter		Blanking temperature		Blanking pressure		Shear strength	
mm	in.	mm	in.	°C	°F	kN	lbf	MPa	ksi
1.5	0.060	51	2	1000	1830	128	28 750	480	70
2.3	0.090	51	2	1050	1920	145	32 500	405	59
3.2	0.125	51	2	1100	2010	189	42 500	380	55
1.5	0.060	32	1.250	950	1740	67	15 000	440	64
2.3	0.090	32	1.250	1000	1830	91	20 500	400	58
3.2	0.125	32	1.250	1100	2010	98	22 000	390	57

0.030 in.) in wall thickness. Methods for producing tubes of niobium, molybdenum, and tantalum alloys include gas tungsten arc welding and drawing, extruding a tube shell and reducing (rocking) it to finished size, and cupping and drawing. A low-temperature back-extrusion spinning technique and a floating-mandrel extrusion technique have been used for producing tantalum and molybdenum tube shells. Tantalum tubing for chemical process applications is either welded and drawn or seamless. Tantalum has good drawing properties and can be reduced in area up to 60% without intermediate annealing. Niobium alloy tubing is produced using a similar process.

Tubing of tungsten and its alloys can be made by extrusion. Tungsten tubing has been extruded from billets at ram speeds of 0.13 to 0.2 m/s (0.43 to 0.66 ft/s). Seamless tungsten tubing has been produced experimentally by a filled-billet technique. Tungsten tubing is also made by sinking with a deformable mandrel, by tube drawing with a removable hardened mandrel, and by plug drawing.

Chemical vapor deposition (CVD) has proved useful for fabricating freestanding refractory metal parts. Tungsten tubing produced by CVD can range from 0.025 to 305 mm (0.001 to 12 in.) in inside diameter, 0.10 to 3.18 mm (0.004 to 0.125 in.) in wall thickness, and up to 1.8 m (6 ft.) in length. Other processes such as extrusion cost more and impose greater limitations on size. Tungsten tubing produced by the CVD process has lower mechanical properties than as-worked wrought tubing because of the inherently low ductility of its columnar microstructure. After annealing, however, CVD tungsten displays superior properties. The CVD technique is also used for making tubing of molybdenum, tungsten-rhenium alloys, and other refractory metals in diameters of 3.2 mm (⅛ in.) or less.

Because the crystal structure of rhenium is hexagonal close packed rather than body-centered cubic, it does not exhibit a ductile-to-brittle transition temperature. Annealed rhenium is very ductile and can be bent, coiled, or rolled. Rhenium typically will undergo 20% tensile elongation versus less than 2% for tungsten. However, even though the fabrication of rhenium is possible by conventional methods such as swaging, rolling, and drawing, it does require some special techniques. For the purpose of cold fabrication, the hardness of the metal must be brought below 300 HV, which usually requires 2-h anneals at temperatures above 1700 °C (3090 °F) for material of commercial purity. Traditional hot working generally results in hot shortness caused by the formation and melting of rhenium heptoxide (Re_2O_7) (melting point, T_m, 297 °C, or 567 °F; boiling point, T_b, 363 °C, or 685 °F) at the grain boundaries. Hot swaging and hot extrusion above 1000 °C (1830 °F) are possible if the metal is not exposed to oxygen. Previous cold work helps to prevent hot shortness.

The preferred fabrication technique for sintered compacts involves limiting initial cold reductions to a maximum of approximately 3% per pass between anneals to avoid cracking. In the production of strip, 10 to 20% reductions between anneals can be achieved after an initial 10% reduction pass. Rolling at room temperature is generally performed on combination two-high and four-high mills with small-diameter (19.0 to 44.5 mm, or 0.75 to 1.75 in.) tungsten carbide work rolls. Tubing has been fabricated by electron beam welding a 13 mm (1 in.) wide and 0.20 mm (0.008 in.) thick rhenium strip to form a tube approximately 7.88 mm (0.31 in.) in diameter. In addition to a room-temperature straightening and sizing operation, which achieves about 3% strain, the tubing is put through five alternating cold drawing and 1425 °C (2595 °F) dry hydrogen annealing steps with about a 5 to 7% reduction in area per drawing step to achieve a 7.62 mm (0.30 in.) diameter and a 0.18 mm (0.007 in.) wall final size.

Cleaning

Cleaning is critical throughout the fabrication process, especially for niobium and tantalum. Cleaning should both precede and follow welding, heat treating, or any thermal process. Cleaning for these metals is accomplished in a hot alkaline solution (minimum 10-min exposure), followed by a chemical cleaning in a mixture of hydrofluoric, nitric, and sulfuric acids. A coupon of the same alloy must accompany each lot of hardware to record material removal. The coupon should have a reference point for measurement. Each cleaning operation should remove approximately 0.0025 mm (0.0001 in.) per side. The activity of the acid should be checked before the hardware is placed in the acid. If the oxide is thicker than 0.0025 mm (0.0001 in.) per side, it is likely that the entire component has been contaminated.

After cleaning, the part should be handled with clean, lint-free white gloves, and the edges to be welded should be wrapped in a clean, lint-free material such as plastic. Welding should commence as soon as possible after cleaning; the time between cleaning and welding should never exceed 4 h. All weld tooling should be thoroughly cleaned with methyl ethyl ketone (MEK) or an equivalent residue-free compound; an argon or helium cover gas should be kept flowing at 0.6 to 1.1 m^3/h (20 to 40 ft^3/h) during welding. If copper tooling is used, it must be chromium plated to avoid copper contamination.

Joining

All refractory metals can be joined by electron beam welding, gas tungsten arc welding, or resistance welding. Two major problems are encountered in joining: chemical changes due chiefly to atmospheric contamination and microstructural changes resulting from thermal cycling. The latter changes include grain growth and different stages of precipitation hardening (solution, precipitation, and overaging). Preheating and postheating generally are required to minimize deleterious effects arising from precipitation hardening as well as from the residual stresses normally induced by welding.

Although recrystallization and grain growth are unavoidable in weldments of wrought tungsten and molybdenum, proper choice of welding process and procedure can localize these effects. Electron beam welding has proved effective in achieving full weld penetration with an extremely narrow heat-affected zone. As larger EB chambers become available, size limitations for electron beam welding have become less restrictive. Chambers capable of handling hardware up to 1.5 m (5 ft) in the longest dimension are in commercial use.

Rhenium welds made by inert gas or EB methods are extremely ductile and can be formed further at room temperature. Care must be taken during welding to protect the rhenium against oxidation, however. All other refractory metals suffer losses in ductility and increases in ductile-to-brittle transition temperature when welded, but niobium and tantalum alloys are less affected than are molybdenum and tungsten alloys. Tantalum and niobium alloys generally retain greater than 75% joint efficiency after gas tungsten arc welding. Preheating is not required, but postweld annealing can restore large amounts of ductility and toughness to commercial alloys. Table 8 summarizes recommended postweld annealing treatments for selected refractory metal alloys; Table 9 lists recommended welding conditions.

For niobium and tantalum alloys, joint design is particularly important. The surface to be melted must be twice the thickness of the thickest component; that is, if welding a 1.5 mm (0.060 in.) part to a 1.0 mm (0.040 in.) part, weld height (thickness)

Table 8 Recommended postweld annealing treatments for selected refractory alloys

	Annealing temperature(a)			
	Gas tungsten arc welds		Electron beam welds	
Alloy	°C	°F	°C	°F
C-103	1315	2400	1315	2400
Cb-752	1205	2200	1315	2400
FS-85, C-129Y	1315	2400	1205	2200
T-111, T-222	1315	2400	1315	2400
Ta-2.5W	1260	2300	1260	2300
Ta-10W	Not annealed		Not annealed	

(a) 1 h at temperature

Table 9 Typical conditions for welding 0.9 mm (0.035 in.) refractory metal sheet

Alloy	Gas tungsten-arc welds: Speed, mm/min	in./min	Clamp spacing, mm	in.	Current, A(a)	Arc gap, mm	in.	Electron beam welds: Speed, mm/min	in./min	Clamp spacing, mm	in.	Deflection(b), mm	in.	Voltage, kV	Current, mA
C-103	760	30	9.5	3/8	80	1.5	0.06	1270	50	4.8	3/16	1.3	0.050	150	3.2
C-129Y	760	30	9.5	3/8	110	1.5	0.06	1270	50	13	1/2	1.3	0.050	150	4.1
Cb-752	760	30	9.5	3/8	87	1.5	0.06	380	15	4.8	3/16	1.3	0.050	150	3.3
FS-85	380	15	9.5	3/8	90	1.5	0.06	1270	50	4.8	3/16	1.3	0.050	150	4.4
T-111	380	15	9.5	3/8	115	1.5	0.06	380	15	13	1/2	1.3	0.050	150	3.8
T-222	760	30	6.4	1/4	190	1.5	0.06	380	15	13	1/2	1.3	0.050	150	4.5
Ta-10W	190	7.5	6.4	1/4	118	1.5	0.06	380	15	13	1/2	1.3	0.050	150	3.8

(a) Direct current, straight polarity. (b) Beam deflection at 60 cycles parallel to weld direction

must be 3.0 mm (0.120 in.). Weld tooling should be of hard chromium-plated copper. No copper can contact the refractory metal. Before use, the hard chromium plate should be wire brushed to check adhesion. If the chromium flakes, the tooling must be stripped and replated.

For gas tungsten arc welding, the weld zone should be well flushed with inert cover gas before striking the arc, and the fusion zone should be allowed to cool below 205 °C (400 °F) with gas coverage. Prior to welding, all burrs should be removed by draw filing with a file used only for one particular alloy. All hand welding should be accomplished in chambers that are evacuated prior to backfilling with argon and/or helium.

Although pure niobium shows no evidence of an aging reaction, Nb-1Zr undergoes abrupt losses in strength and ductility when treated at 815 to 980 °C (1500 to 1800 °F) for up to 500 h. Welds are subject to such embrittlement but can be restored to a ductile condition by postweld vacuum annealing at 1040 to 1205 °C (1900 to 2200 °F) for 3 h. This treatment produces overaging, preventing embrittlement on subsequent heating at a lower temperature.

In contrast to welds in niobium and tantalum, which retain good ductility, welds in molybdenum and tungsten are brittle (<50% joint efficiency), and thus these metals are difficult to join. Before welding, molybdenum and tungsten must be preheated above their ductile-to-brittle transition temperatures to prevent fracture. Sections in thicknesses of 0.64 mm (0.025 in.) and less demand special attention in this respect and, at best, present serious cracking problems. Welds in these metals are always brittle, and joint efficiency depends on the reinforcing effect of the weld bead. Resistance welding is feasible, but some problems with electrode sticking can arise. Resistance Welding Manufacturers' Association (RWMA) class I copper electrodes show the least susceptibility to sticking. Projection welding can result in relatively high mechanical properties.

Tungsten is the most difficult refractory metal to join for satisfactory high-temperature service. Welding, especially the EB process, offers the best compromise for joining tungsten for service at high temperatures. Mechanical joints are unsatisfactory unless molybdenum fasteners are used. Diffusion bonding is impractical because of severe tooling problems. Brazing for relatively low-temperature applications is done using precious metals (silver, palladium, and platinum alloys) and transition metals (nickel and manganese alloys) as filler metals.

Table 10 lists typical brazing filler metals and their maximum service temperatures for all refractory metal systems. Molybdenum brazing has received much attention: Brazed molybdenum honeycomb configurations are used for structural and heat shield applications at temperatures from 1370 to 1650 °C (2500 to 3000 °F). Low-temperature brazing processes have been developed for TZM. The high remelt temperatures of the filler metals listed in Table 10 permit relatively high service temperatures.

Niobium and its alloys may be silicide-coated with a chromium- and titanium-containing material before being subjected to temperature-oxidizing environments; the preferred braze alloy is Ti-33Cr because it is compatible with the coating. Foil, 0.13 mm (0.005 in.) thick, is fit metal-to-metal to ensure good filleting of the joint. The foil should be held in place with resistance spot welds. Cleanliness of the mating surfaces will ensure good flow of the alloy. A clean vacuum furnace is heated to 1315 °C (2400 °F) and held for 5 min, then increased to 1480 °C (2700 °F) and held for 8 min. The parts must be furnace cooled to 205 °C (400 °F) before exposure to air. After brazing, the hardware should be pickled and diffusion treated at 1315 °C (2400 °F) in a vacuum for a period of 16 h. When possible, the parts should be wrapped in tantalum foil to minimize contamination.

Diffusion bonding also is used to join refractory metals, primarily niobium and tantalum. The same rationale is required as for brazing. Vanadium foil, 0.05 to 0.08 mm (0.002 to 0.003 in.) thick is placed in the joint and weighted using molybdenum or tungsten tooling. Diffusion bonding with vanadium is considered superior to brazing because a bimetallic system is not necessary, and the joint is microstructurally clean because vanadium forms a continuous solid solution with niobium and tantalum.

Coatings

Surface protection is the most significant obstacle to widespread use of refractory metals in high-temperature oxidizing environments. The existing temperature ceiling of about 1650 °C (3000 °F) is dictated by coating limitations: Coatings have insufficient life at reduced pressures (below ~13 kPa, or 100 torr) and high temperatures (~1370 °C, or 2500 °F) in oxidizing atmospheres and give unreliable protection, particularly at edges and corners. From a practical standpoint, however, niobium is the

Table 10 Typical brazing filler metals and service temperatures

Filler metal	Maximum service temperature, °C	°F
For niobium alloys		
Si-Cr-Ni	980	1800
4Be-48Zr-48Ti	925	1700
Zr-6Be-19Nb	925	1700
Ti-0.5Si	1370	2500
Zr-0.1Be-16Ti-25V	1205	2200
V-35Nb	1205	2200
Ti-50Zr	1650	3000
Titanium	1760	3200
Ti-33Cr	1370	2500
Ti-3Al-11Cr-13V	1650	3000
For tantalum alloys		
Hf-7Mo	2095	3800
V-20Nb-20Ta	1870	3400
Ti-15Ta-25V	1650	3000
Ta-10Hf-(15-70)Nb	2205	4000
Nb-(30-50)Hf	2205	4000
Ta-10Hf	2205	4000
Nb-1.3B	1925	3500
Copper	980	1800
For molybdenum alloys		
Ti-3Be-25Cr	1595	2900
Pt-Mo	1650	3000
Zr-Ti	1230	2250
Cu-Au	815	1500
Ni-Cu	1205	2200
V-35Nb	1205	2200
Ti-30V	1370	2500
Ti-13Ni-25Cr	1760	3200
Co-10Ni-15W-20Cr	1315	2400
For tungsten		
Ag-Mn	870	1600
V-Nb-Ta	1925	3500
V-Ti-Ta	1925	3500
W-25Os	2205	4000
W-3Re-50Mo	2205	4000
Mo-5Os	1925	3500
Niobium	1650	3000
Tantalum	2205	4000
For rhenium		
Ag-Cu	730	1350
Vanadium	1900	3450

only refractory metal for which commercial uses of coated parts have been developed. These coatings are primarily of the silicide type; aluminides are also used, but to a much lesser degree. Silicide coatings have been developed for molybdenum; however, they have seen virtually no commercial use because any failure of the coating at high temperatures causes catastrophic failure by vaporization of MoO_3. Further details on coating systems are described in the section "Niobium" in this article.

Niobium

Sam Gerardi, Fansteel Inc., Precision Sheet Metal Division

COMMERCIALLY PURE EB NIOBIUM is ductile and easy to fabricate at room temperature by conventional forming practices. Niobium alloys are used extensively for aerospace applications because they are relatively light in weight and high in elevated-temperature strength.

Alloy C-103 has been widely used for rocket components that require moderate strength at temperatures of about 1095 to 1370 °C (2000 to 2500 °F). Alloy Nb-1Zr is used in nuclear applications because it has a low thermal neutron absorption cross section, good corrosion resistance, and good resistance to radiation damage. It is used extensively for liquid metal systems operating at temperatures from 980 to 1205 °C (1800 to 2200 °F). The Nb-1Zr alloy combines moderate strength with excellent fabricability. As a result, it is used for parts in sodium vapor or magnesium vapor lamps.

Vapor deposition of Nb-1Zr or niobium on the inside surface of type 316 stainless steel tubing improves the performance of the tubing in many chemical process applications without degrading the mechanical properties of the stainless steel. An intermediate layer of pure niobium under Nb-1Zr improves adherence to the steel substrate.

Alloys C-129Y, FS-85, and Cb-752 have shown higher elevated-temperature tensile and creep strengths than C-103 while maintaining good fabricability, coatability, and thermal stability. They are used for leading edges, nose caps for hypersonic flight vehicles, rocket nozzles, gas turbines, and guidance structures for reentry vehicles. Alloy C-3009 is being evaluated for potential use in fasteners and gun barrels.

An Nb-46.5Ti alloy is being used as a low-temperature superconductor material. Superconducting magnets for magnetic resonance imaging machines employ the alloy in the superconducting composite wire, in which very fine filaments of the alloy (with thicknesses of <1 μm, or 40 μin.) are embedded in a copper matrix (see the article "Niobium-Titanium Superconductors" in this Volume and the section "The Manufacture of Commercial Superconductors" in the article "Wire, Rod, and Tube Drawing" in *Forming and Forging*, Volume 14 of the 9th Edition of *Metals Handbook*). Magnets required by the superconducting super collider may contain nearly 800 Mg (1.8×10^6 lb) of the alloy. The more brittle niobium-tin (Nb_3Sn) superconductor has a higher transition (critical) temperature (T_c) than does niobium-titanium, 18.1 K versus 9 K. Use of the niobium-tin alloy has been limited by fabrication difficulties. An Nb-55Ti alloy is used in significant quantities for fasteners for aerospace structures.

Recently, niobium of very low total interstitial content has been produced for superconducting microwave cavity electron accelerators. Chemical analysis for carbon, oxygen, hydrogen, and nitrogen is unreliable at very low levels; therefore, purity is determined by measuring the residual resistance ratio (RRR) value, which is defined as direct current resistivity of niobium at room temperature to the resistivity of niobium at 4 K in the normal conducting state. An RRR value greater than 250, which represents total carbon, oxygen, hydrogen, and nitrogen contents of less than 25 ppm, can be achieved by electron beam drip melting in a modern high-vacuum furnace at pressures of 1 mPa (10^{-5} torr) at rates of 45 kg/h (100 lb/h) or less.

There is also a substantial market for niobium carbide, primarily in Europe, as an additive in conjunction with tantalum carbide for cemented carbide cutting tools.

Production

Reduction Processes. At present, the major process for the recovery of niobium is the aluminothermic reduction of pyrochlore concentrates to ferroniobium. Niobium metal is purified by a chlorination process wherein volatile $NbCl_5$ is distilled and then hydrolyzed to the oxide. The metal is then recovered by a second aluminothermic reduction:

$$3Nb_2O_5 + 10Al \xrightarrow{\text{accelerator}} 6Nb + 5Al_2O_3 \qquad \text{(Eq 1)}$$

During the exothermic reaction, oxide impurities slag from the molten niobium. Carbothermic reduction has also been practiced.

Another niobium recovery process involves the collection of niobium oxide as a by-product in the processing of tantalum ores. The ore is digested in mixed acids containing hydrofluoric acid and sulfuric acid, and solvent extraction is employed for purification. After aluminothermic reduction of the recovered oxide, the metal is further purified and consolidated by EB melting.

Table 11 Typical composition of niobium and C-103 niobium alloy powder made by the hydride-dehydride process

Element	Analysis, ppm: Niobium	C-103
Niobium	≥99.7(a)	≥87.2(a)
Oxygen	1820	1980
Tantalum	800	2800
Hafnium	<20	9.8(a)
Zirconium	<20	1800
Titanium	20	0.91(a)
Carbon	500	194
Iron	100	200
Aluminum	<20	<20
Nitrogen	197	62
Silicon	30	<20
Copper	<40	<40
Cobalt	<10	<10
Boron	<1	<10
Hydrogen	150	50
Nickel	<20	<20
Molybdenum	<20	100
Tungsten	<50	1100
Other elements(b)	<20	<20

(a) Analysis in %. (b) Other elements include cadmium, chromium, magnesium, manganese, lead, tin, vanadium, and zinc. Source: Fansteel, Inc

Fig. 6 Particle shape of niobium powder made by electron beam melting, hydriding, crushing, and degassing. 250×

Powder Production. Powders are produced from ingot by hydriding, crushing, and dehydriding; in addition, some recent efforts have been directed toward producing complex metastable alloy powders, such as niobium-aluminum and niobium-silicon alloys, by liquid metal atomization and rapid quenching. The particle structure of degassed hydride niobium powder (Fig. 6) is completely analogous to that of a tantalum powder produced in a similar process for capacitors. Typical compositions of niobium and C-103 alloy powder made by this process are compared in Table 11. Niobium powders produced by the hydride-crush-degas process are not normally used for capacitors, and thus milling to a very fine particle size (that is, a high surface area) is not required. Normally, powders are crushed to pass an 80-mesh screen, and a mean particle size of 10 to 15 μm (400 to 600 μin.) is typical. Although thermal agglomer-

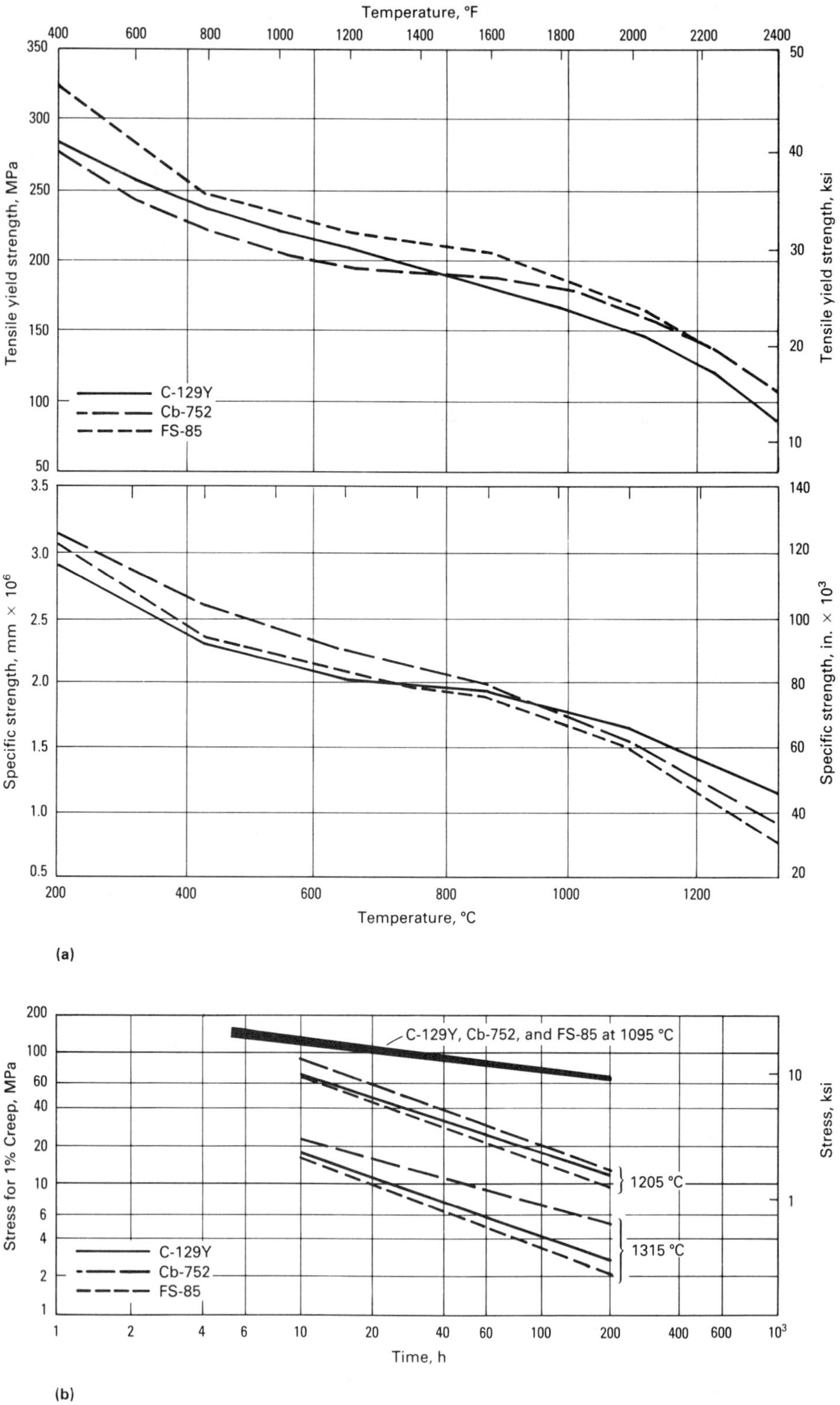

Fig. 7 Effect of temperature on the mechanical properties of three niobium alloys coated with Si-20Cr-20Fe silicide coating. (a) Tensile yield strength and specific strength. (b) Stress. Specific strength is the ratio of tensile yield strength (F_{ty}) (in lb/in.^{2}) to mass density (ρ) (in lb/in.^{3}) and has been converted directly from inches to millimeters. Stress levels are based on material thickness prior to application of coating.

ation is feasible and is commonly employed to form clusters for very fine tantalum powders, it is rarely necessary for niobium.

Niobium powder is frequently used as the starting material to blend with alloying agent powders. The blend is pressed to bars and melted, thereby promoting alloy homogeneity. Niobium alloy scrap, which is reduced to powder by the hydride-dehydride process, can also be incorporated into the alloy blends. One P/M dispersion-strengthened alloy containing 0.5% TiO_2 is claimed to have improved yield strength and high-temperature grain stability.

Coatings

The general fabrication practices for niobium are described in the section "Production of Refractory Metals" in this article. However, niobium and its alloys are the only refractory metals for which large parts are coated to prevent oxidation in high-temperature service (>425 °C, or 800 °F). Early coating development for niobium was centered on pack cementation and chemical vapor deposition. However, experience with these two processes showed that they increased the ductile-to-brittle transition temperature of the metal and caused part distortions. Later, spurred by the needs of the Apollo space program, techniques to apply slurry coatings of complex aluminides and, subsequently, silicides were devised. When properly prepared and applied, these coatings were more reliable, exhibited excellent cyclic performance characteristics without drastic mechanical property deterioration, and caused minimal hardware distortion.

The Si-20Cr-20Fe composition, made using elemental powder suspended in nitrocellulose lacquer with a thermotropic gelling agent, became the mainstay coating. Comparative mechanical properties for three coated niobium alloys (Cb-752, C-129Y, and FS-85) are shown in Fig. 7 to 9. Other variants contain hafnium silicide, which gives the final coating a higher remelt temperature. Methods for applying the slurry include dipping, spraying, and touch-up painting. Following application of approximately 0.08 mm (0.003 in.) of slurry per side, the coating is heated to about 1300 to 1400 °C (2370 to 2550 °F) for reaction bonding and diffusion.

The improvements in high-temperature oxidation resistance with the use of slurry coating technology have not been achieved without side effects. Compared with base metal, coated metal has lower strength and ductility, higher emissivity, and increased weight. A more serious limitation imposed by the presence of coatings is a reduction in design and fabrication options. For aerospace applications, because of the coating requirement, sharp edges must be eliminated. Spot welding and riveting are not recommended for coated hardware fabrication.

Chemical Properties

Niobium forms an oxide coating in most acid environments. This coating provides excellent corrosion resistance, especially to nitric and hydrochloric acids. Strong alkaline solutions and hydrofluoric acid attack niobium severely. Table 12 presents typical data on the corrosion of niobium in

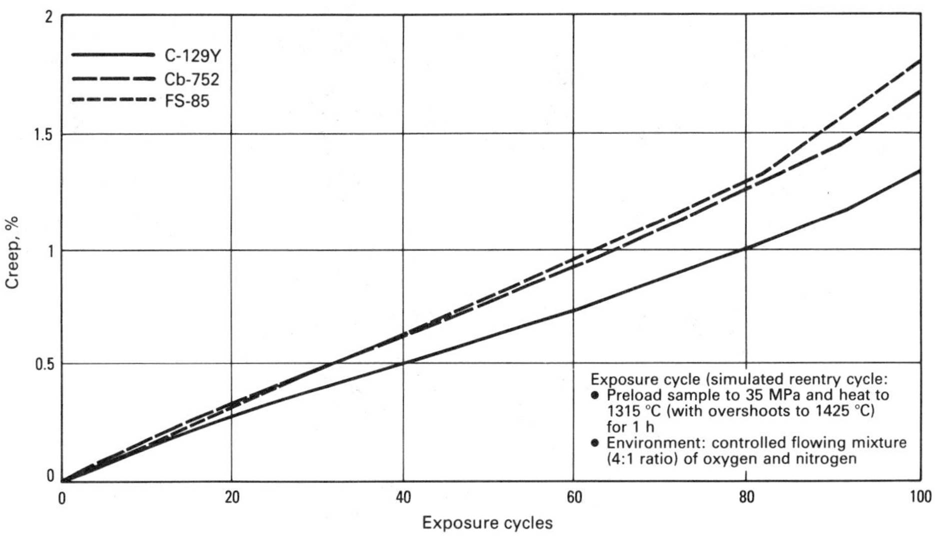

Fig. 8 Cyclic creep in three niobium alloys coated with Si-20Cr-20Fe silicide coating

various acids and alkalis (additional information is available in the article "Corrosion of Niobium and Niobium Alloys" in *Corrosion*, Volume 13 of the 9th edition of *Metals Handbook*). At elevated temperatures, the metal reacts with halogens, oxygen, nitrogen, carbon, hydrogen, and sulfur. It forms high-melting-point compounds with elements such as carbon, boron, silicon, and nitrogen.

Niobium can be used in contact with liquid lithium, sodium, and sodium-potassium eutectic at temperatures well above 800 °C (1470 °F). Addition of 1% Zr increases the resistance of niobium to embrittlement caused by oxygen absorbed from the liquid metal.

Table 12 Corrosion of niobium in aqueous media

Solution	Temperature °C	Temperature °F	Duration of test, days	Loss in weight(a) $mg/m^2 \cdot d$	Loss in weight(a) $oz/ft^2 \cdot yr$	Condition of specimen at end of test
20% HCl	21	70	82	2.5	0.03	No change
Concentrated HCl	21	70	82	6	0.072	Slight etch; not embrittled
	100	212	67	234	2.79	Brittle
Concentrated HNO_3	100	212	67	nil	nil	No change
Aqua regia	22	72	6	nil	nil	No change
H_2SO_4						
20% by volume	21	70	3650	0.2	0.0025	No change
25% by volume	21	70	3650	0.3	0.0036	No change
Concentrated (98%)	21	70	3650	5.6	0.067	Partial embrittlement (18.3% drop in toughness)
	50	122	67	48	0.57	Brittle
	100	212	32	1 131	13.5	Brittle
	150	302	2	12 470	149.3	Brittle
	175	347	1	≥83 200	≥995	Completely dissolved
	100	212	42	464	5.56	Pitted and brittle
85% H_3PO_4	21	70	82	0.7	0.0084	No change
	100	212	31	193	2.32	Brittle
20% tartaric acid	22	72	82	nil	nil	No change
10% oxalic acid	21	70	82	33	0.40	Brittle
NH_4OH	21	70	82	nil	nil	No change
20% Na_2CO_3	100	212	50	74	0.88	Brittle
5% NaOH	21	70	31	66	0.79	Action at surface of liquid
	100	212	5	1 086	13.0	Brittle
5% KOH	21	70	31	442	5.3	Action at surface of liquid
	100	212	5	2 744	32.8	Brittle
30% H_2O_2	21	70	61	11	0.13	Oxide film; not embrittled

(a) Original specimen dimensions: thickness, 0.2 mm (0.008 in.); surface area, 26 cm^2 (4 $in.^2$). 75% of specimen surface was immersed in the liquid.

Table 13 Effect of temperature on the tensile properties of Nb-1Zr

Temperature °C	Temperature °F	Tensile strength MPa	Tensile strength ksi	Yield strength MPa	Yield strength ksi	Elongation, %
20	70	345	50	255	37	15
1095	2000	185	27	165	24	. . .
1650	3000	83	12	69	10	. . .

Mechanical and Physical Properties

Properties of unalloyed niobium are shown in Table 2. Additional information on unalloyed niobium is available in the section "Niobium" in the article "Properties of Pure Metals" in this Volume. Property data for selected niobium alloys are listed below.

Nb-1Zr

Commercial Names

UNS number. Commercial grade, R04261; reactor grade, R04251
Trade name. Wah Chang WC-1Zr, Fansteel 80
Common name. Nb-1Zr

Specifications

ASTM. B 391, B 392, B 393, B 394

Chemical Composition

Composition limits. Commercial grade: 98.5 Nb min, 0.8 to 1.2 Zr, 0.0100 C max, 0.0300 N max, 0.0300 O max, 0.0020 H max, 0.01 Hf max, 0.01 Fe max, 0.005 Mo max, 0.005 Ni max, 0.005 Si max, 0.2 Ta max, 0.05 W max. For reactor grade, Fe is 0.005 max, O is 0.015 max, Ta is 0.1 max, and W is 0.03 max.
Consequence of exceeding impurity limits. Increasing interstitial content decreases ductility of the material.

Mechanical Properties

Tensile properties. Recrystallized: typical tensile strength, 241 MPa (35 ksi); yield strength, 138 MPa (20 ksi); elongation, 20% in 25.4 mm (1 in.). See also Table 13.
Shear strength. See Table 14.
Elastic modulus. Tension, 68.9 GPa (10×10^6 psi)
Impact strength. See Table 15.
Creep-rupture properties. See Fig. 10.

Mass Characteristics

Density. 8.59 g/cm^3 (0.31 $lb/in.^3$)

Thermal Properties

Liquidus temperature. 2407 °C (4365 °F)
Coefficient of linear thermal expansion. 7.54 μm/m · K (4.19 μin./in. · °F) at 20 to 400 °C (68 to 750 °F)
Specific heat. 0.270 kJ/kg · K (0.065 Btu/lb · °F) at 20 °C (68 °F)
Thermal conductivity. 41.9 W/m · K (24.2 Btu/ft · h · °F) at 25 °C (77 °F)

Electrical Properties

Electrical resistivity. 14.7 nΩ · m at 0 °C (30 °F)

Chemical Properties

Resistance to specific corroding agents. Especially resistant to liquid metals

Fabrication Characteristics

Machinability. 80% of C36000 (free-cutting

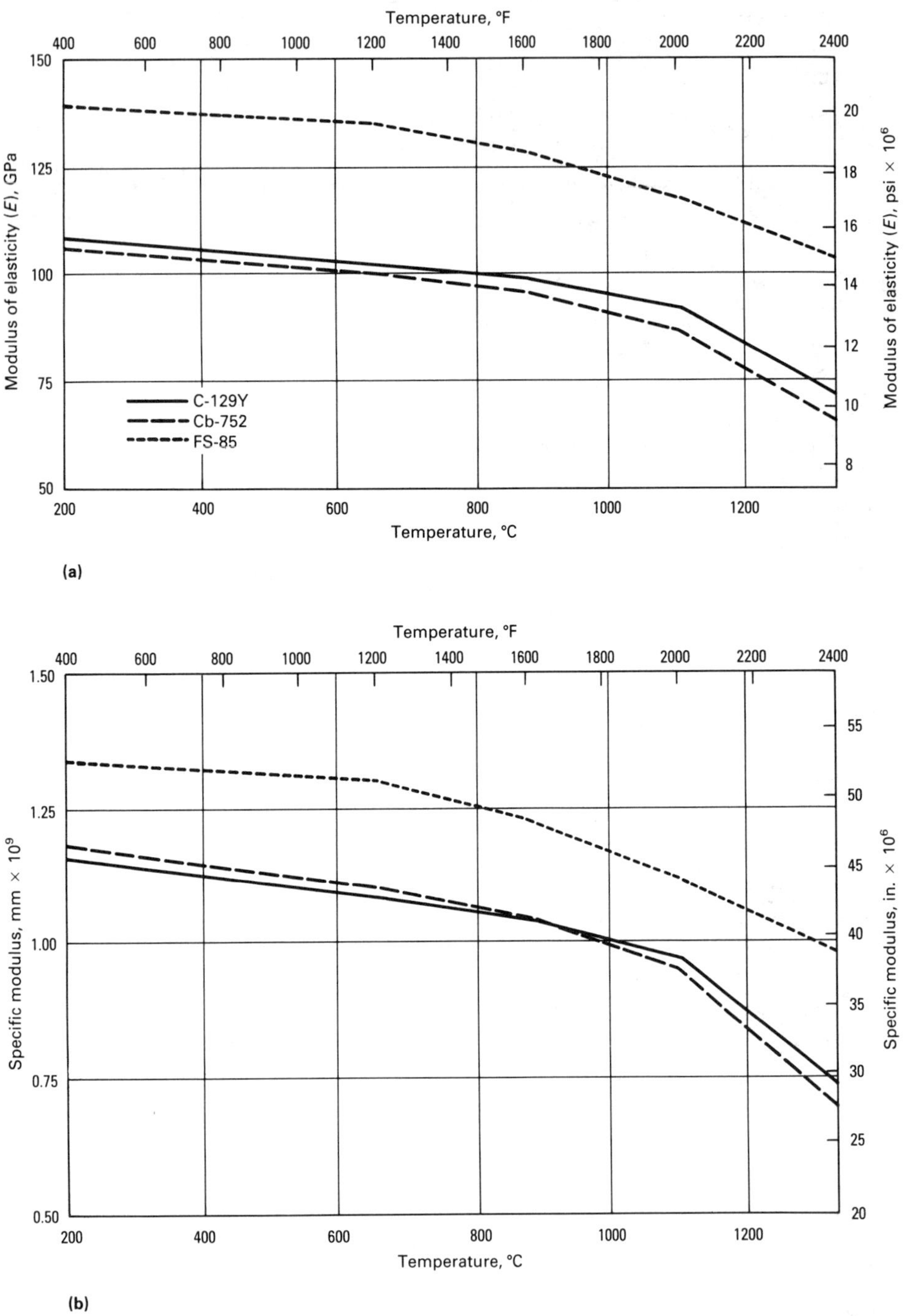

Fig. 9 Effect of temperature on the elastic properties of three niobium alloys coated with Si-20Cr-20Fe silicide coating. (a) Modulus of elasticity. (b) Specific modulus. Specific modulus is the ratio of the modulus of elasticity (E) (in lb/in.^{2}) to mass density (ρ) (in lb/in.^{3}); it has been converted directly from inches to millimeters.

Table 14 Shear strength for Nb-1Zr rivets

Fastener type	Diameter mm	Diameter in.	Shear strength at 20 °C (70 °F) MPa	20 °C (70 °F) ksi	870 °C (1600 °F) MPa	870 °C (1600 °F) ksi
Huck rivet	3.18	0.125	265	38.5	220	32.0
			300	43.5	220	32.0
Deutsch rivet	3.18	0.125	240	35.0	180	26.0
			230	33.0	. . .	. . .
Du Pont explosive	3.18	0.125	185	27.0	90	13.0
			180	26.0	90	13.0

brass)

Forgeability. 75% at 650 to 980 °C (1200 to 1800 °F)

Formability. Extrusion: reduction ratio of 10:1 at 1065 °C (1950 °F). Rolling: 85% reduction at 205 to 315 °C (400 to 600 °F) and at finish. Readily formable by conventional metal-forming processes

Weldability. Can be joined by electron beam welding, resistance welding, and gas tungsten arc welding

Recrystallization temperature. 980 to 1205 °C (1800 to 2200 °F)

Hot-working temperature. 1095 to 1205 °C (2000 to 2200 °F)

Stress-relief temperature. 1 h at 900 to 980 °C (1650 to 1800 °F)

C-103
89Nb-10Hf-1Ti

Commercial Names

Trade name. WC 103, C-103

Chemical Composition

Composition limits. 0.0100 C max, 9 to 11 Hf, 0.7 to 1.3 Ti, 0.7 Zr max, 0.0300 O, 0.0300 N max, 0.0020 H max, 0.5 W max, 0.5 Ta max, bal Nb

Applications

Typical uses. Thrust chambers and radiation skirts for rocket and aircraft engines, guidance structure for glide reentry vehicles, thermal shields, piping or containers for chromic and other acids, piping for liquid alkali metal containment, and sodium-vapor lamp electrodes

Precautions in use. For elevated-temperature applications, aluminide or silicide coatings should be used. Not recommended for use in hydrofluoric acid or strong alkaline solutions

Mechanical Properties

Tensile properties. Typical. Cold rolled: tensile strength, 725 MPa (105 ksi); yield strength, 670 MPa (97 ksi); elongation, 4.5% in 50 mm (2 in.). Recrystallized: tensile strength, 405 MPa (59 ksi); yield strength, 310 MPa (45 ksi); elongation, 26% in 50 mm (2 in.). See also Table 16.

Hardness. 230 HV

Elastic modulus. Tension: 87 GPa (12.6 × 10^6 psi) at 20 °C (68 °F); 43 GPa (6.3 × 10^6 psi) at 1370 °C (2500 °F); 25 GPa (3.6 × 10^6 psi) at 1480 °C (2700 °F); 10 GPa (1.5 × 10^6 psi) at 1650 °C (3000 °F)

Creep-rupture properties. See Fig. 11.

Mass Characteristics

Density. 8.87 g/cm^3 (0.32 lb/in.^{3}) at 25 °C (77 °F)

Thermal Properties

Liquidus temperature. 2350 °C (4260 °F)

Coefficient of linear thermal expansion. 8.10 μm/m · K (4.5 μin./in. · °F) at 20 to

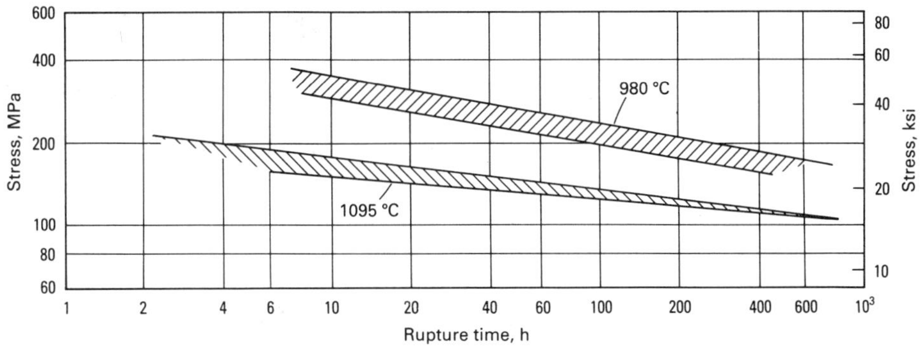

Fig. 10 Stress-rupture properties of Nb-1Zr

Table 15 Charpy impact strength of Nb-1Zr

Condition	Temperature °C	Temperature °F	Impact energy J	Impact energy ft · lbf	Impact fracture
Unnotched specimens					
As-rolled	24	75	210	156	None
	−73	−100	180	133	Partial
Stress relieved 1 h at 900 °C (1650 °F)	24	75	175	129	None
	−73	−100	170	126	None
Recrystallized 1 h at 1205 °C (2200 °F)	24	75	174	128	None
	−73	−100	164	121	None
Notched specimens					
As-rolled	24	75	>81	>60	Partial(a)
	−73	−100	93	69	Partial
Stress relieved 1 h at 900 °C (1650 °F)	24	75	160	119	Partial
	−73	−100	129	95	Partial
Recrystallized 1 h at 1205 °C (1650 °F)	24	75	126	93	Partial
	−73	−100	156	116	None

(a) Specimen stopped hammer, 81 J (60 ft · lbf) range

1205 °C (68 to 2200 °F)
Specific heat. 0.340 kJ/kg · K (0.082 Btu/lb · °F) at 20 °C (70 °F)
Thermal conductivity. 41.9 W/m · K (24.2 Btu/ft · h · °F) at 25 °C (77 °F)

Chemical Properties

General corrosion behavior. A protective oxide forms on C-103 in most acid media, providing excellent corrosion resistance. However, the alloy is severely attacked by hydrofluoric acid and strong alkaline solutions.

Resistance to specific corroding agents. Excellent resistance to nitric acid of all concentrations and to dilute hydrochloric acid

Table 16 Typical tensile properties of arc cast C-103 sheet

Temperature(a) °C	Temperature(a) °F	Direction(b)	Tensile strength MPa	Tensile strength ksi	Yield strength at 0.2% offset MPa	Yield strength at 0.2% offset ksi	Elongation in 25 mm (1 in.), %
0.75 mm (0.03 in.) thick, cold rolled							
RT	RT	L	725	105	660	96	4.5
		T	745	108	640	93	4
1095	2000	L	235	34	160	23	39
		T	215	31	185	27	35
1370	2500	L	90	13	76	11	87
		T	90	13	76	11	80
1 mm (0.04 in.) thick, stress relieved 1 h at 870 °C (1600 °F)							
RT	RT	L	640	93	605	88	9
1095	2000	L	180	26	125	18	63
1370	2500	L	76	11	69	10	>75
1480	2700	L	55	8	48	7	>73
0.75 mm (0.03 in.) thick, recrystallized 1 h at 1315 °C (2400 °F)							
RT	RT	L	405	59	345	50	26
1095	2000	L	185	27	125	18	45
1370	2500	L	83	12	69	10	>70
1480	2700	L	62	9	55	8	>70
1650	3000	L	34	5	28	4	>70

(a) RT, room temperature. (b) L, longitudinal; T, transverse

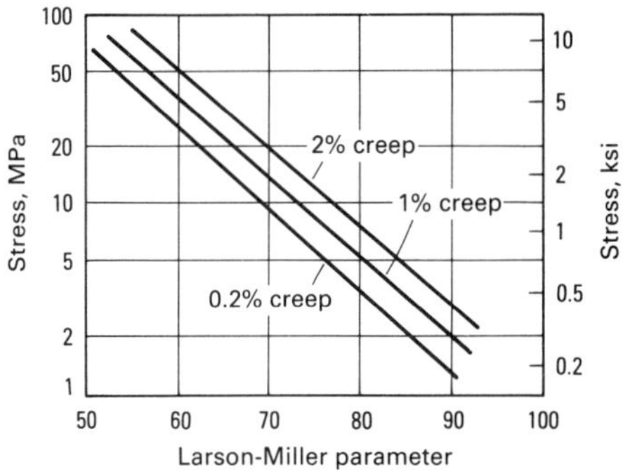

Fig. 11 Larson-Miller parameter plot for recrystallized C-103. Larson-Miller parameter = $T \cdot 10^{-3}$ (20 + log t), where T is the test temperature in K and t is the rupture time in hours

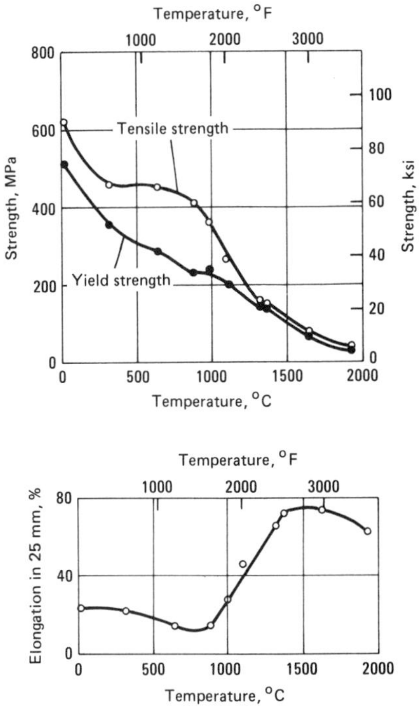

Fig. 12 Tensile properties of C-129Y sheet

Fabrication Characteristics

Forgeability. 60% total reduction at 1205 to 925 °C (2200 to 1700 °F)
Hot formability. Extrusion: reduction ratio is 10:1 at 1205 °C (2200 °F). Rolling: 50% reduction at 425 °C (800 °F); 60 to 80% reduction at finish
Weldability. Good gas tungsten arc weldability
Recrystallization temperature. 1040 to 1315 °C (1900 to 2400 °F)
Stress-relief temperature. 1 h at 870 °C (1600 °F)

C-129Y
80Nb-10W-10Hf-0.1Y

Commercial Name

Trade name. C-129Y

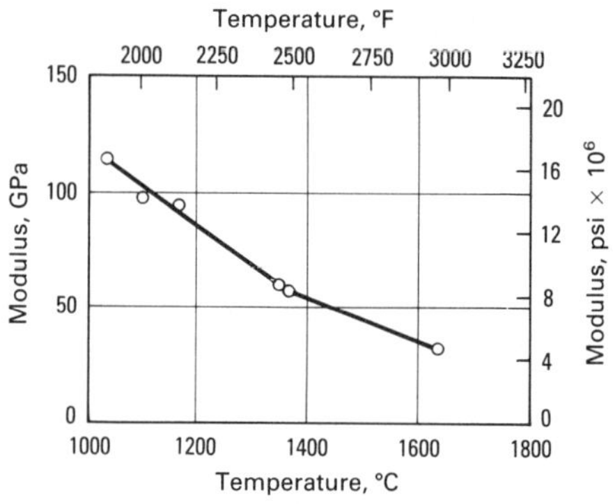

Fig. 13 Static modulus of elasticity for C-129Y

Chemical Composition

Composition limits. 9 to 11 W, 9 to 11 Hf, 0.05 to 0.3 Y, 0.5 Ta, 0.5 Zr max, 0.015 C max, 0.025 O max, 0.015 N max, 0.0015 H max
Consequence of exceeding impurity limits. Increasing interstitial content decreases material ductility.

Applications

Typical uses. For high-temperature applications, space vehicles, missiles; leading edges, nose caps for hypersonic flight vehicles, rocket nozzles; guidance structure for glide reentry vehicles, and so on
Precautions in use. Interstitial contamination should be avoided during welding. A postweld annealing at 1205 to 1315 °C (2200 to 2400 °F) for 1 h is recommended. For elevated-temperature applications, silicide or aluminide coatings are required.

Mechanical Properties

Tensile properties. Tensile strength, 620 MPa (90 ksi); yield strength, 515 MPa (75 ksi); elongation, 25% in 25 mm (1 in.). See also Fig. 12.
Hardness. Recrystallized: 220 HV
Elastic modulus. Tension, 112 GPa (16.2 × 10^6 psi). See Fig. 13.
Creep-rupture properties. See Fig. 14 and 15.

Mass Characteristics

Density. 9.50 g/cm^3 (0.343 lb/in.^{3})

Thermal Properties

Liquidus temperature. 2400 °C (4350 °F)
Coefficient of linear thermal expansion. 6.88 μm/m · K (3.82 μin./in. · °F) at 20 to 1100 °C (70 to 2010 °F)
Specific heat. 0.268 kJ/kg · K (0.064 Btu/lb · °F) at 1095 °C (2000 °F)
Thermal conductivity. 69.6 W/m · K (40 Btu/ft · h · °F)

Chemical Properties

General corrosion behavior. Good elevated-temperature properties combined with heat and oxidation resistance. Excellent corrosion resistance to most acid media. However, C-129Y is severely attacked by hydrofluoric acid and strong alkaline solutions.

Fabrication Characteristics

Machinability. 75% of C36000 (free-cutting brass)
Forgeability. 50% at 930 to 1205 °C (1705 to 2200 °F)
Formability. Extrusion: reduction ratio of 4:1 at 1205 °C (2200 °F). Rolling: reduction of 50% at 430 °C (805 °F), 60 to 70% reduction at finish
Weldability. Weldments exhibit ductility as low as −170 °C (−275 °F) if measures to prevent atmosphere contamination are taken.
Recrystallization temperature. 1315 °C (2400 °F)
Annealing temperature. 980 to 1315 °C (1800 to 2400 °F)
Hot-working temperature. 980 to 1315 °C (1800 to 2400 °F)
Stress-relief temperature. 1 h at 870 °C (1800 °F)

Cb-752 Nb-10W-2.5Zr

Commercial Name

Trade name. Cb-752

Chemical Composition

Composition limits. 9 to 11 W, 2 to 3 Zr, 0.015 C max, 0.02 O max, 0.01 N max, 0.001 H max
Consequence of exceeding impurity limits. Increasing interstitial content decreases alloy ductility.

Applications

Typical uses. Guidance structure for glide reentry vehicles, jet engine structure, thermal radiation, and ducting for space power systems
Precautions in use. Resistance to high-temperature oxidation is poor, and protective coatings are required for elevated-temperature applications.

Mechanical Properties

Tensile properties. Annealed sheet: tensile strength, 540 MPa (78 ksi); yield strength, 400 MPa (58 ksi); elongation, 20% minimum in 50 mm (2 in.). At 1205 °C (2200 °F): tensile strength, 195 MPa (28 ksi); yield

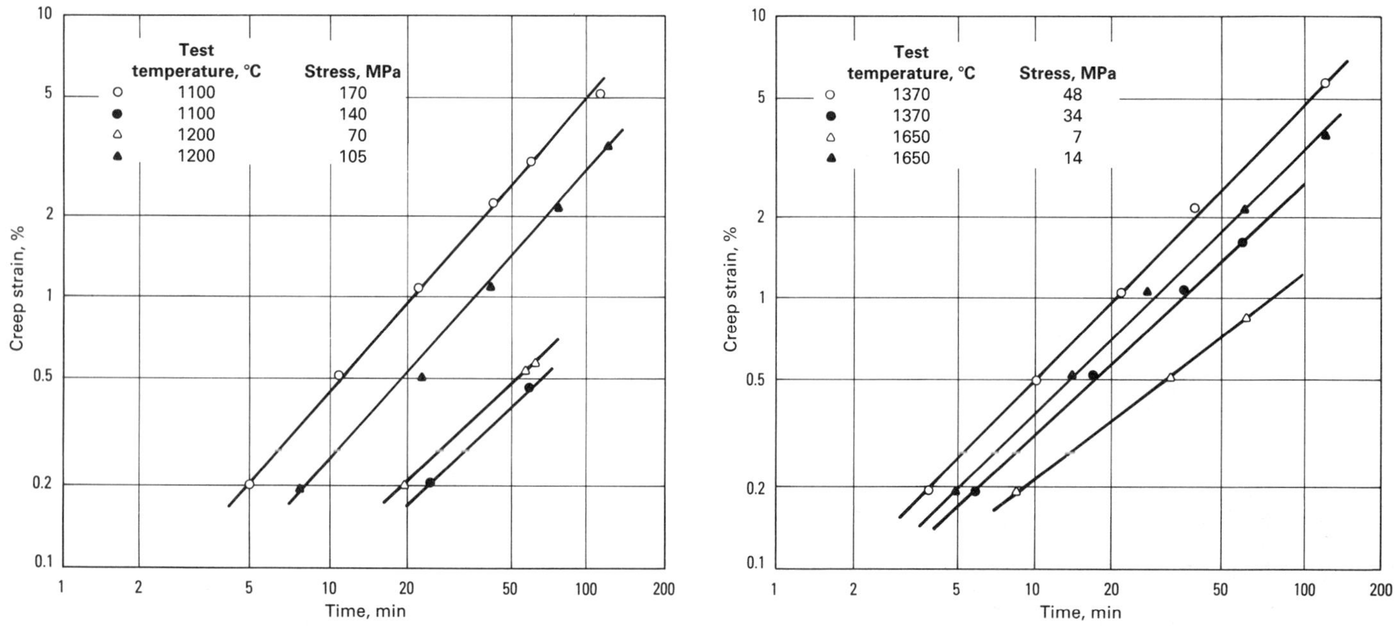

Fig. 14 Creep curves for C-129Y sheet in vacuum. C-129Y sheet, 1 mm (0.04 in.) thick, was annealed 1 h at 1315 °C (2400 °F) and tested in vacuum at 13 mPa (10^{-4} torr).

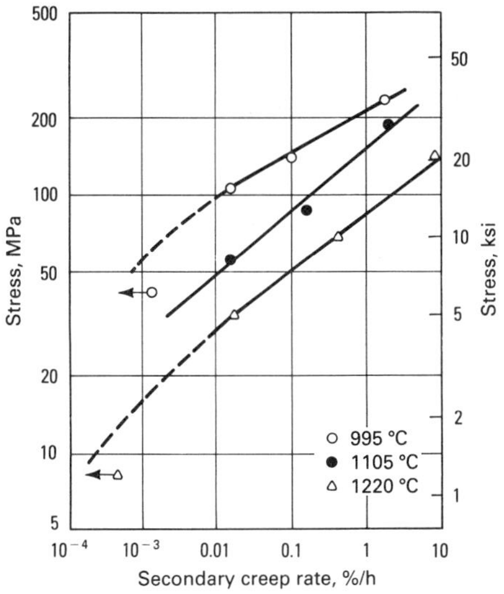

Fig. 15 Secondary creep rate versus stress for C-129Y sheet, cold worked 50%

strength, 150 MPa (22 ksi); elongation, 25% minimum in 50 mm (2 in.). Figure 16 shows elongation in 25 mm (1 in.).
Shear strength. 425 MPa (62 ksi)
Compressive properties. Compressive strength, 351 MPa (50.9 ksi)
Bearing properties. Bearing strength, ultimate, 703 MPa (102 ksi); bearing strength, yield, 625 MPa (91 ksi)
Hardness. 180 HK, annealed 1 h at 1370 °C (2500 °F)
Elastic modulus. Tension, 110 GPa (16 × 10^6 psi)
Fatigue strength. 275 MPa (40 ksi) at 10^6 cycles
Creep-rupture properties. See Fig. 17.

Mass Characteristics

Density. 9.03 g/cm³ (0.326 lb/in.³) at 25 °C (77 °F)

Thermal Properties

Liquidus temperature. 2425 °C (4400 °F)
Coefficient of linear thermal expansion. 7.4 μm/m · K (4.1 μin./in. · °F) at 20 to 1205 °C (68 to 2200 °F)
Specific heat. 0.281 kJ/kg · K (0.067 Btu/lb · °F) at 540 °C (1000 °F); temperature coefficient, 0.0335 per K at 0 to 540 °C (32 to 1000 °F). See also Fig. 18.
Thermal conductivity. 48.7 W/m · K (28 Btu/ft · h · °F) at 760 °C (1400 °F); temperature coefficient, 0.0219 per K at 205 to 760 °C (400 to 1400 °F). See also Fig. 18.

Chemical Properties

General corrosion behavior. Excellent corrosion resistance to most acid media, but severely attacked by hydrofluoric acid and strong alkaline solutions
Resistance to specific corroding agents. Good corrosion resistance to oxygen-free liquid metals, for example, potassium and lithium at elevated temperatures (980 to

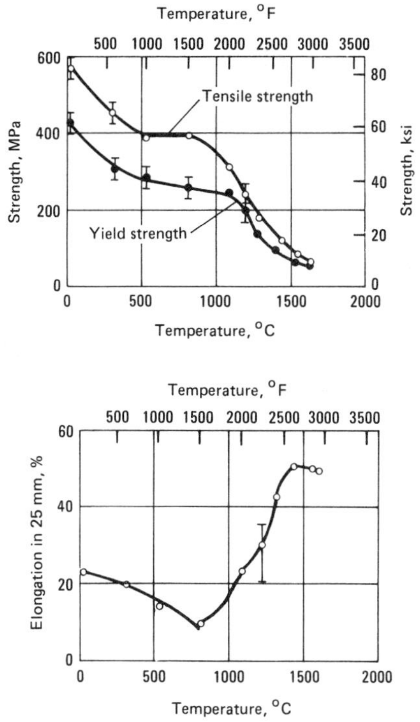

Fig. 16 Effect of test temperature on average tensile properties of duplex-annealed Cb-752 sheet

1205 °C, or 1800 to 2200 °F) for approximately 4000 h

Fabrication Characteristics

Machinability. 80% of C36000 (free-cutting brass)
Forgeability. 50% at 930 to 1205 °C (1705 to 2200 °F)
Formability. The alloy can be formed using most of the conventional methods. Primary ingot breakdown must be done above the recrystallization temperature. Subsequent working may be accomplished in the range from room temperature to 425 °C (800 °F).
Weldability. Fusion welding by gas tungsten arc and electron beam processes can be readily accomplished if measures are taken to prevent atmosphere contamination.
Recrystallization temperature. 1205 to 1315 °C (2200 to 2400 °F)
Annealing temperature. 1205 to 1315 °C (2200 to 2400 °F)
Solution temperature. 1425 to 1540 °C (2600 to 2800 °F)
Aging temperature. 1095 °C (2000 °F)
Hot-working temperature. 980 to 1315 °C (1800 to 2400 °F)
Stress-relief temperature. 1 h at 980 to 1095 °C (1800 to 2000 °F)

FS-85
Cb-28Ta-10W-1Zr

Chemical Composition

Composition limits. 0.01 C max, 26 to 29 Ta, 10 to 12 W, 0.6 to 1.1 Zr, 0.03 O max, 0.015 N max, 0.001 H max

Mechanical Properties

Tensile properties. See Table 17 and Fig. 19.
Creep-rupture properties. See Table 18.
Elastic modulus:

Temperature °C	Temperature °F	Modulus of elasticity GPa	Modulus of elasticity 10^6 psi
RT(a)	RT(a)	140	20
980	1800	125	18
1095	2000	125	18
1205	2200	110	16
1540	2800	105	15
1595	2900	83	12
1650	3000	83	12

(a) RT, room temperature

Mass Characteristics

Density. 10.61 g/cm³ (0.383 lb/in.³)

Thermal Properties

Liquidus temperature. 2590 °C (4695 °F)
Coefficient of linear thermal expansion. 9.0 μm/m · K (5.0 μin./in. · °F) at 20 to 1315 °C (68 to 2400 °F)

Fabrication Characteristics

Forgeability. 50% at 930 to 1290 °C (1705 to 2355 °F)
Formability. Extrusion: reduction ratio of 4:1 at 1205 °C (2200 °F). Rolling: 40% reduction at 205 to 370 °C (400 to 700 °F); 50 to 65% reduction at finish
Recrystallization temperature. 1095 to 1370 °C (2000 to 2500 °F)
Stress-relief temperature. 1 h at 1010 °C (1850 °F)

Tantalum

Charles Pokross, Fansteel Inc.

CURRENTLY, THE LARGEST USE of tantalum is in electrolytic capacitors. Tantalum P/M anodes are used in solid and wet electrolytic capacitors, and precision tantalum foil is used, although to a lesser extent, in foil capacitors. Tantalum is also used in chemical process equipment such as heat exchangers, condensers, thermowells, and

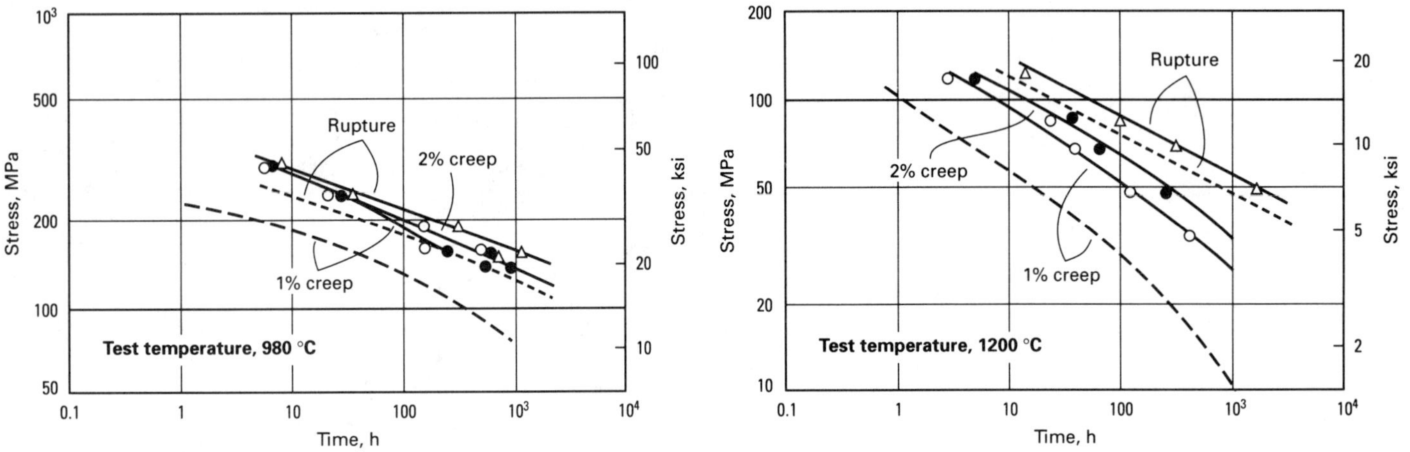

Fig. 17 Total creep curves for Cb-752 sheet. Data points represent material duplex annealed, then aged 1 h at 1595 °C (2900 °F). Dashed lines are for duplex-annealed material that did not undergo aging treatment.

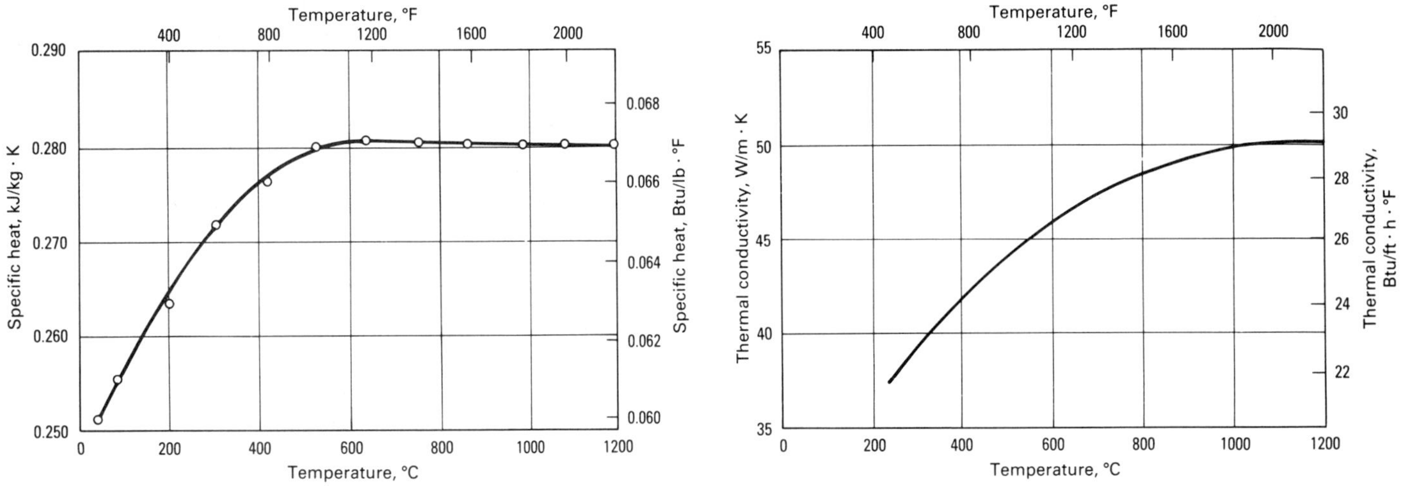

Fig. 18 Thermal properties of Cb-752

lined vessels; most notably, it is used for the condensing, reboiling, preheating, and cooling of nitric acid, hydrochloric acid, sulfuric acid, and combinations of these acids with many other chemicals. Because of its high melting point, tantalum is used for heating elements, heat shields, and other components in high-temperature vacuum furnaces, and it has found some use for trueing of grinding wheels. Tantalum and its alloys have been used in specialized aerospace and nuclear applications and have found increasing use in military components. Because of its corrosion resistance to body fluids, it is used in prosthetic devices and in surgical staples. Tantalum is used as an alloying element in superalloys. Tantalum carbide is an important constituent in complex cemented carbides used in cutting tools.

Production

The production of tantalum metal is accomplished by the extraction of tantalum from either ores or certain tin slags, primarily those from Thailand and Malaysia; further extraction is necessary to separate tantalum from other metals present. The purified extract is recovered by precipitation of $Ta(OH)_5$, which is then calcined to the pentoxide or by crystallization with potassium fluoride to the intermediate salt, potassium fluorotantalate (K_2TaF_7).

Several methods for reducing tantalum compounds to tantalum metal have been developed, but sodium reduction of K_2TaF_7 to produce tantalum metal powder is the most commonly used today. The product of the sodium reduction can then be further refined by melting. The powder may also be pressed and sintered into bar or sold as capacitor-grade powder. By varying the parameters of sodium reduction (for example, time, temperature, sodium feed rate, and diluent), powders of various particle sizes can be manufactured.

A wide range of sodium-reduced capacitor powders are currently available, with unit capacitances ranging from 5000 μF · V/g to greater than 25 000 μF · V/g. Capacitor powders are also manufactured from hydrided, crushed, and degassed EB-melted ingot. These melt-grade powders have higher purity than the sodium-reduced types and have better dielectric properties. However, unit capacitance is usually lower for EB-type powder.

Table 17 Room temperature tensile properties of FS-85 sheet, 0.8 mm (0.030 in.) thick

Strain rate, 2%/min

Condition	Tensile strength MPa	Tensile strength ksi	Yield strength at 0.2% offset MPa	Yield strength at 0.2% offset ksi	Elongation in 25 mm (1 in.), %	Reduction in area, %
Stress relieved	830	120	730	106	11	47
Recrystallized 1 h at 1260 °C (2300 °F)	585	85	475	69	22	54
Recrystallized and Cr-Ti-Si coated	580	84	470	68	18	32

Table 18 Creep and stress-rupture data for FS-85 sheet

Temperature °C	Temperature °F	Stress MPa	Stress ksi	Secondary creep rate, %/h	Rupture time, h
1 mm (0.04 in.) sheet, cold worked 50%					
1095	2000	115	17	0.0321	>158
		135	19.8	0.171	...
		140	20	0.0363	...
		185	27	0.606	...
1205	2200	69	10	<0.0845	...
		83	12	0.119	...
		97	14	0.246	...
		110	16	0.525	...
		125	18	1.06	...
		130	19	1.32	10.42
		140	20	2.20	...
1315	2400	90	13	1.85	9.98
1425	2600	69	10	1.88	9.29
1.5 mm (0.063 in.) sheet, cold worked 94%					
980	1800	240	35	0.930	6.51
1095	2000	140	20	0.462	23.78
		185	27	...	2.33
1205	2200	130	19	5.13	2.72
1315	2400	90	13	3.99	4.93
		90	13	3.24	6.12
		83(a)	12(a)	...	11.21
1425	2600	69	10	4.14	>4.7
		69(a)	10(a)	3.45	5.01

(a) Cold worked 94% and annealed 1 h at 1315 °C (2400 °F)

Tantalum and its alloys are produced in semifinished metallic form by further processing of the sodium-reduced powders. The powder is isostatically pressed into bars, which can then be electron beam melted or sintered at high temperature under vacuum. Tantalum ingots up to 305 mm (12 in.) in diameter can be produced by electron beam melting.

The EB-melting process utilizes evaporation, volatilization of suboxides, and carbon deoxidation as purifying reactions. All of these reactions are more favorable in high vacuum (<130 mPa, or 10^{-3} torr). Because of its high melting point and very low vapor pressure, tantalum can be produced with a purity exceeding 99.95%. In addition, certain metallurgical properties can be imparted to the ingot by the addition of a vacuum arc remelt of the EB-melted ingot. One such property is grain size refinement. A flow chart outlining production of tantalum products from ore is shown in Fig. 20.

Fabrication

Unalloyed tantalum and tantalum alloy ingots can be broken down by either forging or extrusion. Arc cast ingots should be extruded only after upsetting and side forging. Powder metallurgy sintered bars can be rolled directly without any prior breakdown. Tantalum products can subsequently be manufactured by standard cold-working techniques, such as rolling, drawing, tube reducing, and swaging. Typical reductions between anneals are 75 to 80%, but reductions in excess of 95% are not uncommon. Rolled sheet having a controlled, predominant {111} crystallographic texture is being produced because of its superior drawing and forming characteristics. The texture control is achieved by a specific thermomechanical process history.

Powder metallurgy tantalum has superior deep-drawing properties, but it should not be welded because of the porosity that forms in the heat-affected zone. However, EB-melted tantalum can be used for various welded products, including welded and drawn tubing.

Corrosion

Tantalum oxidizes in air at temperatures above 300 °C (570 °F). It is attacked by hydrofluoric acid, fuming sulfuric acid, and strong alkalis. Salts that hydrolyze to form hydrofluoric acid or strong alkalis also attack tantalum. The metal can be embrittled by hydrogen if it is the cathodic member of a galvanic couple exposed in an acid environment or if it is exposed to a hydrogen-containing atmosphere at elevated temperature. Other agents that can attack tantalum include bromine plus methanol, and halogen gases (fluorine at or above room temperature; chlorine at 250 °C, or 480 °F; bromine at 300 °C , or 570 °F; and iodine at somewhat higher temperatures).

Tantalum has excellent resistance to corrosion by most acids, by most aqueous salt solutions, and by organic chemicals. It also has good resistance to many corrosive gases and liquid metals.

Mechanical and Physical Properties

Selected mechanical and physical properties for commercially prepared pure EB-melted and P/M tantalum and tantalum alloys are given in Table 19. Additional properties for these materials appear in the compilations that follow.

Tantalum

Typical chemical impurity limits for both EB-welded and P/M chemically pure tantalum are:

Element	Maximum content, wt% EB-melted tantalum	Maximum content, wt% P/M tantalum
Carbon	0.01	0.01
Oxygen	0.015	0.03
Nitrogen	0.01	0.01
Hydrogen	0.0015	0.0015
Niobium	0.1	0.1
Iron	0.01	0.01
Titanium	0.01	0.01
Tungsten	0.05	0.05
Molybdenum	0.02	0.02
Silicon	0.005	0.005
Nickel	0.01	0.01
Other	0.01	0.01

Additional information on unalloyed tantalum is available in the section "Tantalum" in the article "Properties of Pure Metals" in this Volume.

Ta-2.5W

Commercial Name

Trade name. Tantaloy 63 metal, Cabot-6

Chemical Composition

Composition limits. 2.0 to 3.0 W, 0.5 Nb max

Applications

Typical uses. An EB-melted solid-solution alloy used for heat exchangers, linings for towers, valves, and tubing

Ta-7.5W

Commercial Name

Trade name. 61 metal

Chemical Composition

Composition limits. 7 to 8 W, bal Ta

Applications

Typical uses. A P/M product typically cold drawn into wire for springs, elastic parts for gas chlorinators, and elastic parts for other equipment subjected to severe acid conditions

Ta-10W

Commercial Name

Trade name. Tantaloy 60 metal, Cabot-10

Chemical Composition

Composition limits. 9 to 11 W, bal Ta

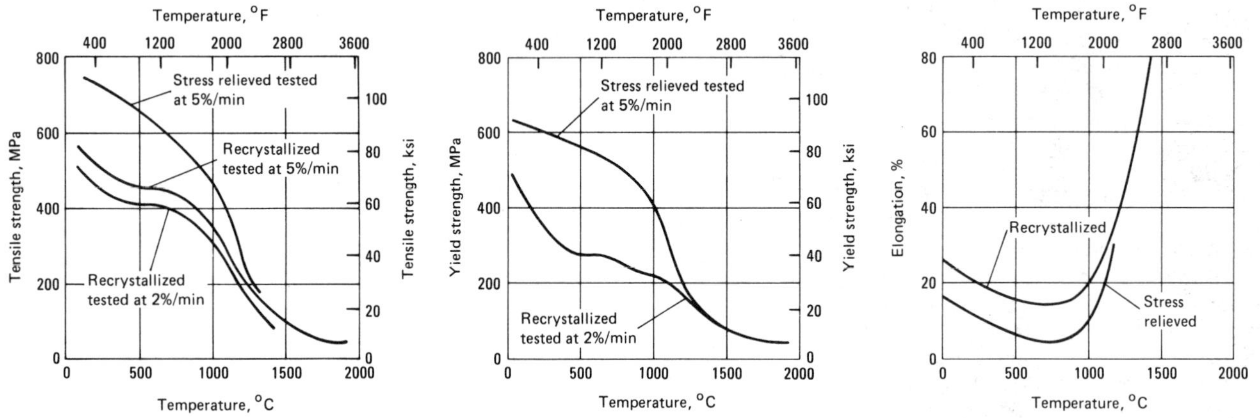

Fig. 19 Effect of temperature on the typical tensile properties of FS-85

Applications

Typical uses. Used at temperatures up to 2480 °C (4500 °F) in aerospace applications, such as hot gas metering valves, rocket engine extension skirts, complex manifold assemblies, and fasteners. Chemical process industry applications include machined solid valves, internal seats and plugs for large valves, liners requiring abrasion and corrosion resistance, and disks used in patching glass-lined steel vessels; also used for tubing in some nuclear applications.

Molybdenum

Walter A. Johnson, Institute of Materials Processing,
Michigan Technological University

IN ITS MOST COMMON application, molybdenum is used as an alloying element in cast irons, steels, heat-resistant alloys, and corrosion-resistant alloys to improve hardenability, toughness, abrasion resistance, corrosion resistance, and strength and creep resistance at elevated temperatures. In its pure form or as a base alloy, molybdenum is used in a wide range of industries in tools (see the articles "Wrought Tool Steels" and "P/M Tool Steels" in *Properties and Selection: Irons, Steels, and High-Performance Alloys*, Volume 1 of the 10th Edition of *Metals Handbook*) and components that can perform satisfactorily at high temperatures or under severe abrasive or corrosive conditions.

In the electrical and electronic industries, molybdenum is used in cathodes, cathode supports for radar devices, current leads for thoria cathodes, magnetron end hats, and mandrels for winding tungsten filaments. Molybdenum is also used as a filler metal for brazing tungsten. Molybdenum resistance heating elements are used in electric furnaces that operate at temperatures up to 2205 °C (4000 °F).

Molybdenum is important in the missile industry, where it is used for high-temperature structural parts such as nozzles, leading edges of control surfaces, support vanes, struts, reentry cones, heat-radiation shields, heat sinks, turbine wheels, and pumps. Molybdenum alloys are particularly well-suited for use in airframes because of their high stiffness, high recrystallization temperature, retention of mechanical properties after thermal cycling, and good creep strength. Alloy Mo-0.5Ti has been used in many aerospace applications, but TZM is preferred where higher hot strength is needed.

In the metalworking industry, molybdenum is used for die casting cores; for hot work tools such as piercer points and extrusion and isothermal forging dies; for boring bars, tool shanks, and chill plates; and for tips on resistance welding electrodes. It is also used for cladding, for equipment for trueing grinding wheels, for molds, and for thermocouples.

Molybdenum has also been useful in the nuclear, chemical, glass, and metallizing industries. Service temperatures for molybdenum alloys in structural applications are limited to a maximum of about 1650 °C (3000 °F). Pure molybdenum has good resistance to hydrochloric acid and is used for acid service in chemical process industries.

Production

Figure 21 is a flow chart for the production of molybdenum mill products. Molybdenum oxides are converted to metallic powders via conventional hydrogen reduction processes. These powders can then be cold pressed and sintered to billet. The P/M billets can be used as arc melting electrodes, or they can undergo subsequent metalworking directly from the P/M billet.

Forging induces strain hardening in the billet, thereby controlling mechanical properties at temperatures below recrystallization. Upset forging breaks up the cast microstructure and improves transverse mechanical properties. Unalloyed molybdenum and TZM can be readily forged with a variety of tools, including steam hammers, drop hammers, and hydraulic forging presses using either open or closed dies.

High-capacity forging equipment is the most desirable because it minimizes workpiece heat loss, thereby extending the working time period and reducing the need for reheating. Minimum reported forging section size is 3.56 mm (0.14 in.), which can be accomplished using standard hot work tool steels. Unalloyed molybdenum and TZM typically are forged in the 870 to 1260 °C (1600 to 2300 °F) temperature range. Billet heating is conducted in commercial gas or oil-fired furnaces. Billets and workpieces will lose weight from volatilization of the oxide at temperatures above 650 °C (1200 °F); however, there is no scale formation. Weight losses of 1 to 5% can be anticipated. Molybdenum and its alloys have high thermal conductivity and low specific heat; therefore, rapid cooling and frequent billet reheats can be anticipated. These materials are typically reheated for 10 to 15 min to a temperature of 1250 °C (2280 °F). Annealing of the forging is accomplished at 1290 °C (2355 °F) for 1 h; stress relief is conducted at 980 °C (1800 °F) for 1 h for pure molybdenum and 1 h at 1200 °C (2190 °F) for TZM.

Molybdenum and its alloys are readily extruded to form a variety of shapes including tubes, round to round bars, round to square bars, and round to rectangular bars.

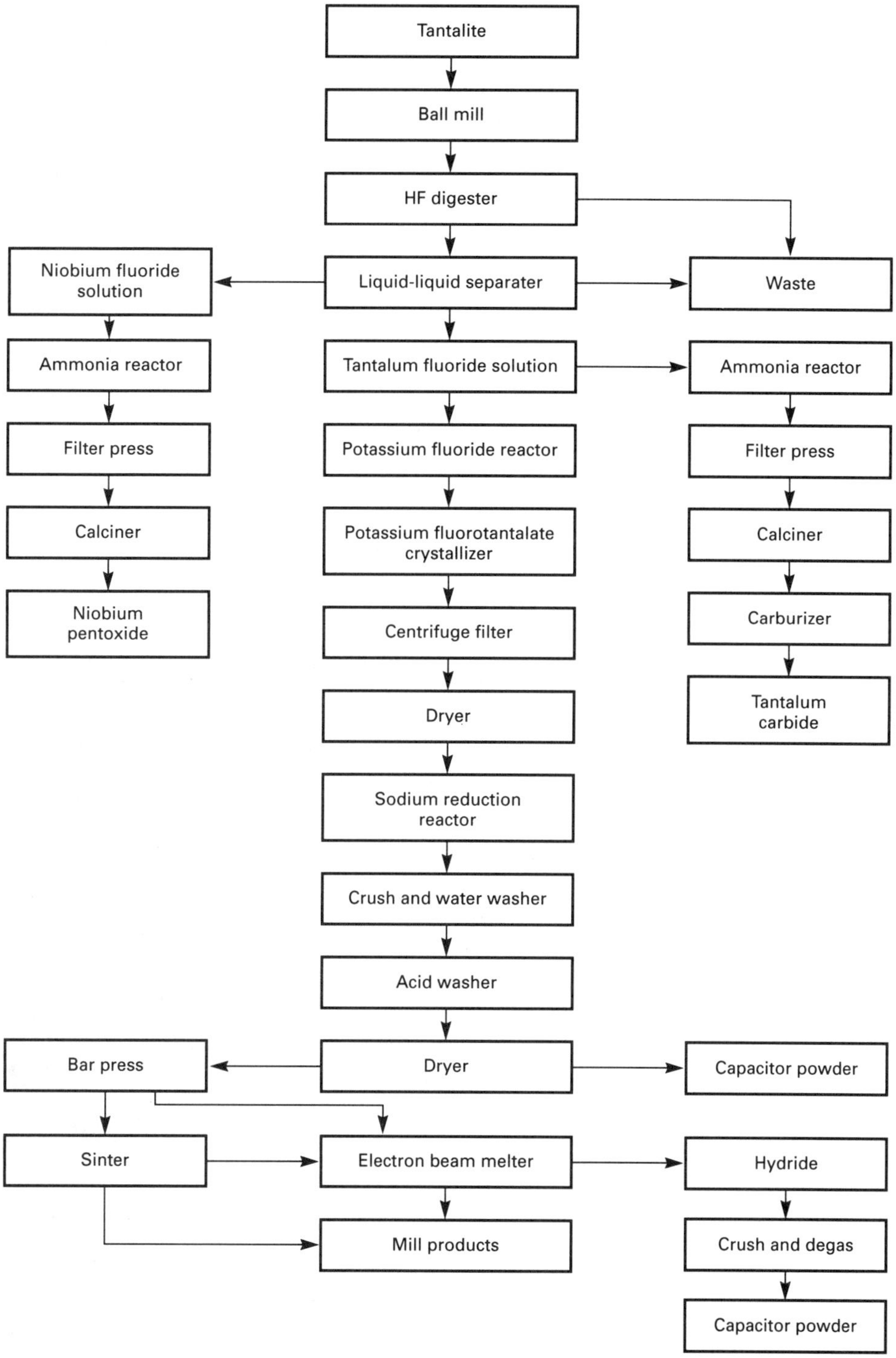

Fig. 20 Processing sequence for tantalum from ore to finished products

Pure molybdenum is typically extruded in the temperature range from 1065 to 1090 °C (1950 to 1995 °F), and TZM is extruded in the temperature range from 1120 to 1150 °C (2050 to 2100 °F). Large tubes and rings are fabricated from back-extruded solid billets. Additional ring-forming operations are undertaken via ring rolling.

Molybdenum and its alloys can be fabricated in sheet form by conventional rolling and cross-rolling processes. Molybdenum and TZM sheet are typically supplied in the annealed condition.

Corrosion Resistance

Molybdenum has particularly good resistance to corrosion by mineral acids, provided that oxidizing agents are not present. The metal is relatively inert in carbon dioxide, hydrogen, ammonia, and nitrogen atmospheres at temperatures up to about 1095 °C (2000 °F); it is also relatively inert in reducing atmospheres containing hydrogen sulfide. Molybdenum has excellent resistance to corrosion by iodine vapor, bromine, and chlorine up to clearly defined temperature limits and good resistance to attack by several liquid metals, including bismuth, lithium, magnesium, potassium, and sodium. In inert atmospheres, it is unaffected at temperatures up to at least 1750 °C (3180 °F) by refractory oxides such as alumina, zirconia, beryllia, magnesia, and thoria. Molybdenum is subject to attack by fused caustic alkalis but not by aqueous caustic solutions. Molten tin, aluminum, iron, and cobalt attack molybdenum severely, as do molten oxidizing salts such as potassium nitrate and potassium carbonate.

Because unprotected molybdenum oxidizes rapidly at temperatures above 500 °C (930 °F) in oxidizing atmospheres, it is not suitable for continued service under such conditions unless it is protected by an adequate coating. Silicide coatings, which provide the best temperature protection, have not been applied commercially to any significant extent.

Mechanical Properties

The mechanical properties of molybdenum and molybdenum alloys greatly depend on the amount of working performed below the recrystallization temperature and on the ductile-to-brittle transition temperature. The minimum recrystallization temperature for molybdenum is 900 °C (1650 °F). Detailed information on the properties of unalloyed molybdenum is available in the section "Molybdenum" in the article "Properties of Pure Metals" in this Volume.

Table 20 summarizes the chemistry and basic mechanical property data for 0.38 mm (0.015 in.) wire fabricated from unalloyed molybdenum and several of the newer molybdenum alloys. Additional property data for molybdenum alloys appear in the compilations that follow.

Mo-0.5Ti
Mo-0.5Ti-0.02C

Commercial Names

UNS number. R03620
ASTM designation. Molybdenum alloy 362

Specifications

ASTM. B 384, B 385, B 386, B 387

Chemical Composition

Composition limits. 0.010 to 0.040 C, 0.010 Fe max, 0.001 N max, 0.005 Ni max, 0.003 O max, 0.010 Si max, 0.40 to 0.55 Ti, bal Mo

Mechanical Properties

Tensile properties. Typical tensile strength: 895 MPa (130 ksi) at 20 °C (70 °F); 415 MPa (60 ksi) at 1095 °C (2000 °F); 76 MPa (11 ksi) at 1650 °C (3000 °F). Typical yield strength:

Table 19 Typical properties of tantalum and tantalum-base alloys

Grade(a)	Hardness HV	Density g/cm³	Density lb/in.³	Melting point °C	Melting point °F	Temperature °C	Temperature °F	Tensile strength MPa	Tensile strength ksi	Yield strength MPa	Yield strength ksi	Elongation, %	Modulus of elasticity GPa	Modulus of elasticity 10^6 psi
Commercially pure tantalum, EB melted	110	16.9	0.609	3000	5430	20	70	205	30	165	24	40	185	27
						200	390	190	27.5	69	10	30	...	...
						750	1380	140	20	41	6	45	160	23
						1000	1830	90	13	34	5	33	...	...
Commercially pure tantalum, P/M	120	16.6	0.600	3000	5430	20	70	310	45	220	32	30	185	27
63 metal, EB melted	130	16.7	0.602	3005(b)	5440(b)	20	70	345	50	230	33	40	195	28
						200	390	315	46	195	28	33	...	...
						750	1380	180	26	83	12	22	...	...
						1000	1830	125	18	69	10	20	...	...
Ta-10W, EB melted	245	16.8	0.608	3030	5490	20	70	550	80	460	67	25	205	30
						200	390	515	75	400	58	...	...	...
						750	1380	380	55	275	40	...	150	22
						1000	1830	305	44	205	30	...	...	...
Ta-7.5W, P/M														
Wire	325	16.8	0.606	3025(b)	5477(b)	20	70	1035	150	1005	146	6	200	29
Sheet	400	16.8	0.606	3025(b)	5477(b)	20	70	1165	169	875	127	7	200	29
Ta-40Nb, EB melted	...	12.1	0.437	2705	4900	...	...	275	40	193	28	25	...	...

(a) EB, electron beam; P/M, powder metallurgy. (b) Estimated

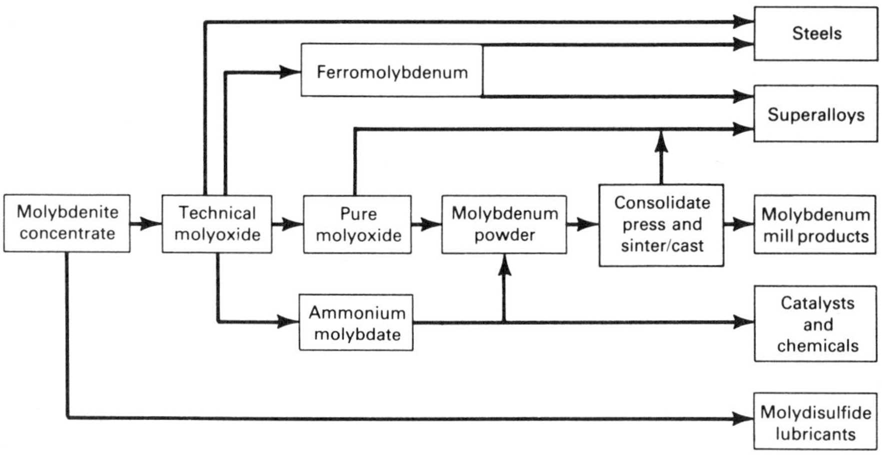

Fig. 21 Processing sequence for molybdenum from ore to finished products

Table 20 Room- and elevated-temperature tensile properties of 380 μm (15 mil) molybdenum wire

Material designation	Composition	Temperature °C	Temperature °F	Ultimate tensile strength MPa	Ultimate tensile strength ksi	Elongation, %
Unalloyed molybdenum	Mo	20	70	1350	196	4.1
		1000	1830	305	44	2.4
		1100	2010	140	20	10.3
		1200	2190	115	17	12.5
MT-104	Mo-0.5Ti-0.08Zr-0.01C	20	70	1565	227	3.1
		1000	1830	1020	148	2.7
		1100	2010	795	115	3.2
		1200	2190	675	98	2.8
Mo + 45 W	Mo-45W	20	70	1980	287	3.6
		1000	1830	1095	159	2.3
		1100	2010	950	138	2.3
		1200	2190	745	108	2.2
HCM	Mo-1.1Hf-0.07C	20	70	1795	260	2.9
		1000	1830	1270	184	3.4
		1100	2010	1185	172	3.3
		1200	2190	1035	150	3.0
HWM-25	Mo-25W-1.0Hf-0.035C	20	70	1935	281	3.2
		1000	1830	1350	196	3.3
		1100	2010	1225	178	3.1
		1200	2190	1075	156	4.6
HWM-45	Mo-45W-0.9Hf-0.03C	20	70	2135	310	3.6
		1000	1830	1460	212	3.3
		1100	2010	1295	188	2.6
		1200	2190	1170	170	2.4

825 MPa (120 ksi) at 20 °C (70 °F); 345 MPa (50 ksi) at 1095 °C (2000 °F); 48 MPa (7 ksi) at 1650 °C (3000 °F)
Elongation. 10% at 20 °C (70 °F)
Elastic modulus. Tension: 315 GPa (46 × 10^6 psi) at 20 °C (70 °F); 180 GPa (26 × 10^6 psi) at 1095 °C (2000 °F)

Mass Characteristics

Density. 10.2 g/cm³ (0.367 lb/in.³)

Thermal Properties

Liquidus temperature. 2610 °C (4730 °F)
Coefficient of linear thermal expansion. 6.1 μm/m · K (3.41 μin./in. · °F) at 20 to 1010 °C (68 to 1850 °F)

Fabrication Characteristics

Recrystallization temperature. 1315 to 1425 °C (2400 to 2600 °F)
Stress-relief temperature. 1 h at 1095 to 1205 °C (2000 to 2200 °F)

TZC
Mo-1Ti-0.3Zr

Chemical Composition

Nominal composition. 1.25 Ti, 0.3 Zr, 0.15 C, bal Mo

Applications

Typical uses. Aerospace equipment and components

Mechanical Properties

Tensile properties. Stress relieved: tensile strength, 995 MPa (144 ksi); yield strength, 725 MPa (105 ksi); elongation, 22% in 50 mm (2 in.); reduction in area, 36%. At 1095 °C (2000 °F): tensile strength, 640 MPa (93 ksi). At 1315 °C (2400 °F): tensile strength, 415 MPa (60 ksi)

TZM
Mo-0.5Ti-0.1Zr

Commercial Names

UNS number. Arc cast, R03630; P/M, R03640

Table 21 Typical tensile properties of TZM

Temperature °C	Temperature °F	Tensile strength MPa	Tensile strength ksi	Yield strength at 0.2% offset MPa	Yield strength at 0.2% offset ksi	Elongation in 50 mm (2 in.), %
Stress-relieved condition						
20	70	965	140	860	125	10
1095	2000	490	71	435	63	...
1650	3000	83	12	62	9	...
Recrystallized material						
20	70	550	80	380	55	20
1095	2000	505	73	...	...	...
1315	2400	369	53.5	...	...	...

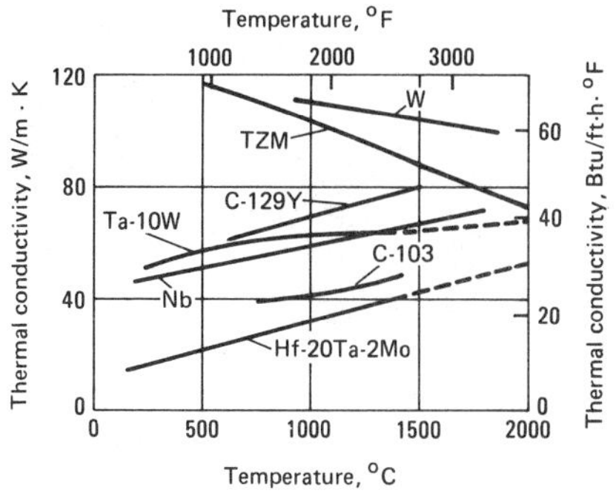

Fig. 22 Thermal conductivities of TZM and selected other refractory metals and alloys

ASTM designation. Arc cast: molybdenum alloy 363. P/M: molybdenum alloy 364

Specifications

ASTM. B 384, B 385, B 386, B 387

Chemical Composition

Composition limits. Arc cast: 0.40 to 0.55 Ti, 0.06 to 0.12 Zr, 0.01 to 0.04 C, 0.010 Fe max, 0.010 Si max, 0.005 Ni max, 0.001 N max, 0.0030 O max, 0.0005 H max. For P/M products: 0.002 N max, 0.030 O max, and 0.005 Si max; all other limits remain the same.

Applications

Typical uses. Used in heat engines, heat exchangers, nuclear reactors, radiation shields, extrusion dies, boring bars

Mechanical Properties

Tensile properties. See Table 21.

Elastic modulus. Tension, 315 GPa (46 × 10^6 psi) at 20 °C (68 °F); 205 GPa (30 × 10^6 psi) at 1095 °C (2000 °F)

Mass Characteristics

Density. 10.16 g/cm^3 (0.367 lb/in.^{3}) at 20 °C (68 °F)

Thermal Properties

Liquidus temperature. 2620 °C (4750 °F)
Coefficient of linear thermal expansion. 4.9 μm/m · K (2.7 μin./in. · °F) at 20 to 40 °C (68 to 100 °F)
Thermal conductivity. See Fig. 22.

Fabrication Characteristics

Recrystallization temperature. 1425 to 1595 °C (2600 to 2900 °F)
Stress-relief temperature. 1 h at 1095 to 1260 °C (2000 to 2300 °F)

Tungsten

Walter A. Johnson, Institute of Materials Processing, Michigan Technological University

TUNGSTEN is consumed in four forms:

- Tungsten carbide
- Alloying additions
- Pure tungsten
- Tungsten-based chemicals

Tungsten carbide accounts for about 65% of tungsten consumption. It is combined with cobalt as a binder to form the so-called cemented carbides, which are used in cutting and wear applications (see the article "Cemented Carbides" in this Volume). Metallic tungsten and tungsten alloy mill products account for about 16% of consumption. Tungsten and tungsten alloys dominate the market in applications for which a high-density material is required, such as kinetic energy penetrators, counterweights, flywheels, and governors. Other applications include radiation shields and x-ray targets. In wire form, tungsten is used extensively for lighting, electronic devices, and thermocouples. Tungsten chemicals make up approximately 3% of the total consumption and are used for organic dyes, pigment phosphors, catalysts, cathode-ray tubes, and x-ray screens.

The high melting point of tungsten makes it an obvious choice for structural applications exposed to very high temperatures. Tungsten is used at lower temperatures for applications that can use its high elastic modulus, density, or shielding characteristics to advantage.

Production

Tungsten and tungsten alloys can be pressed and sintered into bars and subsequently fabricated into wrought bar, sheet, or wire. Many tungsten products are intricate and require machining or molding and sintering to near-net shape and cannot be fabricated from standard mill products.

Shortly before World War II, an easily machinable, relatively ductile family of tungsten-base materials containing a relatively soft and ductile binder phase was developed. These materials, commonly called tungsten heavy metals, are a classic example of the application of liquid-phase sintering to the production of P/M parts. In this case, the basic metal is tungsten, and the liquid phase in which tungsten is partly soluble is primarily nickel. In the original heavy-metal alloys, it was found that the addition of copper was desirable because it lowered the melting temperature of the liquid phase, thereby lowering the sintering temperature. The resulting tungsten-nickel-copper alloy had good mechanical properties, fair ductility, and good machinability. Subsequently, tungsten-nickel-iron alloys that had greater ductility than the tungsten-nickel-copper materials were developed. It was also found that the tungsten-nickel-iron alloys with higher percentages of tungsten could be sintered to near-theoretical density, thereby producing materials of even higher specific gravity.

Tungsten

Tungsten mill products can be divided into three distinct groups on the basis of recrystallization behavior. The first group consists of EB-melted, zone-refined, or arc-melted unalloyed tungsten; other very pure forms of unalloyed tungsten; or tungsten alloyed with rhenium or molybdenum. These materials exhibit equiaxed grain structures upon primary recrystallization. The recrystallization temperature and grain size both decrease with increasing deformation.

The second group, consisting of commercial grade or undoped P/M tungsten, demonstrates the sensitivity of tungsten to purity. Like the first group, these materials exhibit equiaxed grain structures (Fig. 23), but their recrystallization temperatures are higher than those of the first-group materi-

Fig. 23 Recrystallized microstructure of undoped tungsten wire

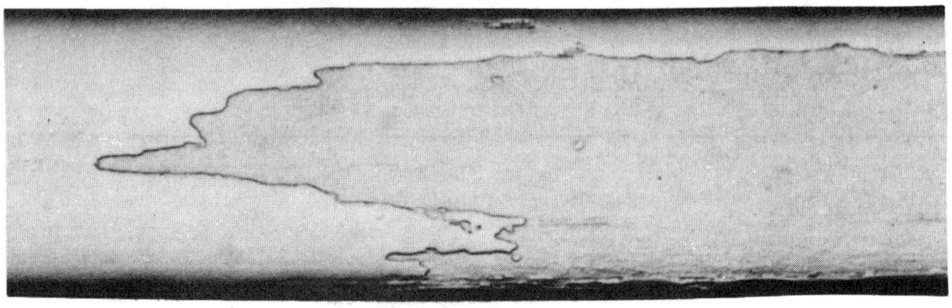

Fig. 24 Recrystallized microstructure of doped tungsten wire

als. Also, these materials do not necessarily exhibit decreases in recrystallization temperature and grain size with increasing deformation. In EB-melted tungsten wire, the recrystallization temperature can be 900 °C (1650 °F) or lower, whereas in commercially pure (undoped) tungsten it can be as high as 1205 to 1400 °C (2200 to 2550 °F).

The third group of materials consists of AKS-doped tungsten (that is, tungsten doped with aluminum-potassium-silicon), doped tungsten alloyed with rhenium, and undoped tungsten alloyed with more than 1% ThO_2. These materials are characterized by higher recrystallization temperatures (>1800 °C, or 3270 °F) and unique recrystallized grain structures (Fig. 24). The structure of heavily drawn wire or rolled sheet consists of very long interlocking grains. This structure is most readily found in AKS-doped tungsten or in doped tungsten alloyed with 1 to 5% Re. The potassium dopant is spread out in the direction of rolling or drawing; when heated, it volatilizes into a linear array of submicron-size bubbles. These bubbles pin grain boundaries in the manner of a dispersion of second-phase particles. As the rows of bubbles become finer and longer with increasing deformation, the recrystallization temperature rises, and the interlocking structure becomes more pronounced. A comparative impurity analysis of the three grades of tungsten is given in Table 22. Higher concentrations of rhenium (7 to 10%) destroy this effect. In W-2ThO_2, the occurrence of this elongated, interlocking structure depends on the thermomechanical treatment

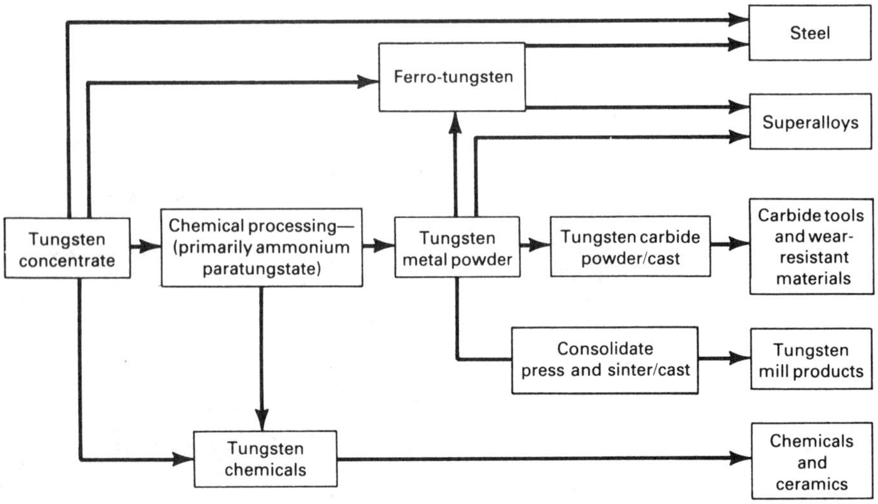

Fig. 25 Processing sequence for tungsten from ore to finished products

Table 22 Typical purity of the three commercial grades of tungsten

Impurity element	Concentration, ppm, in tungsten: Electron beam zone refined	Undoped	Doped
Iron	1	10	11
Nickel	2	5	5
Silicon	5	21	47
Aluminum	<2	<5	15
Potassium	<1	12	91
Oxygen	10	27	36
Carbon	20	31	24

and on the fineness of the thoria dispersion. Addition of 1.5% or more ThO_2 raises the recrystallization temperature of tungsten in much the same way as the potassium dopant raises it, but ThO_2 additions generally result in a much finer grain structure. Rhenium in amounts up to about 5% inhibits recrystallization; in greater amounts, it lowers resistance to recrystallization.

Tungsten Alloys

Three tungsten alloys are produced commercially: tungsten-ThO_2, tungsten-molybdenum, and tungsten-rhenium. The W-ThO_2 alloy contains a dispersed second phase of 1 to 2% thoria. The thoria dispersion enhances thermionic electron emission, which in turn improves the starting characteristics of gas tungsten arc welding electrodes. It also increases the efficiency of electron discharge tubes and imparts creep strength to wire at temperatures above one-half the absolute melting point of tungsten.

A flow diagram outlining the processing of tungsten ore concentrate into major products is shown in Fig. 25. Tungsten mill products, sheet, bar, and wire are all produced via powder metallurgy. These products are available in either commercially pure (undoped) tungsten or commercially doped (AKS-doped) tungsten. These additives improve the recrystallization and creep properties of tungsten, which are especially important when tungsten is used for incandescent lamp filaments. Wrought P/M stock can be zone refined by EB melting to produce single crystals that are higher in purity than the commercially pure product. Electron beam zone-melted tungsten single crystals are of commercial interest for applications requiring single crystals with very high electrical resistance ratios.

Processes for Manufacturing Tungsten Heavy-Metal Alloys. Heavy-metal alloys usually are produced from a mixture of elemental, high-purity, fine-particle-size metal powders. The tungsten powder has an average particle size of about 2 to 3 μm (80 to 120 μin.) and is 99.99% pure. Fine high-purity nickel powder (such as carbonyl nickel), fine electrolytic copper powder, and fine high-purity iron powder (such as carbonyl iron) are used. The powders are blended in a powder blender or ball mill for

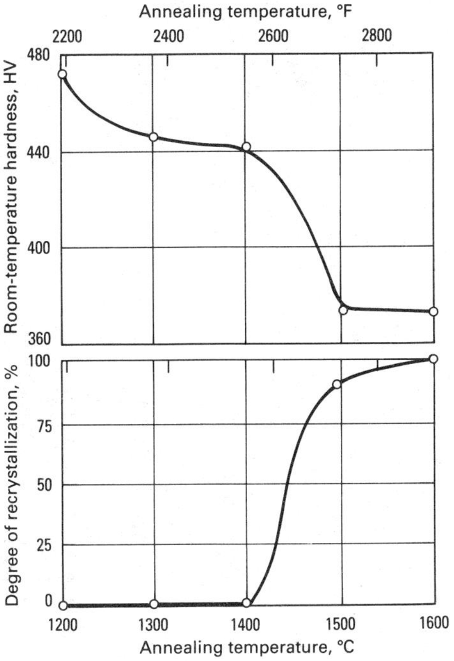

Fig. 26 Recrystallization behavior of undoped tungsten bar

sufficient time to produce a homogeneous mixture and to achieve an apparent density compatible with the molding operation. If molding is by isostatic pressing, no binder is required. If molding is by pressing in a steel or carbide die in a hydraulic or mechanical press, the powder is coated with paraffin or another suitable organic binder. Molding pressures of about 70 to 140 MPa (10 to 20 ksi) are used. The molded compact must be designed to allow for considerable shrinkage during the sintering operation, usually of the order of 20% lineal or more than 50% by volume. Because of the high shrinkage, most parts produced from these alloys require finish machining if close dimensional tolerances are required.

Sintering. The molded parts are usually sintered in box-type electric sintering furnaces by stoking. The furnaces must have molybdenum or tungsten heating elements because sintering temperatures range from about 1425 to 1650 °C (2600 to 3000 °F), depending on the exact composition of the alloy. In some instances, vacuum furnaces are used for sintering these materials, but normally the operation utilizes dry hydrogen or dissociated ammonia for the sintering atmosphere. Sintering times at temperature range from about 20 min for small parts to several hours for large blanks. Part weights can range from a few grams to 20 kg (45 lb) or more.

During sintering, rapid densification of the compact occurs as the fine tungsten particles dissolve in the liquid phase and then reprecipitate on the larger tungsten particles. The compact shrinks in this process, and a very dense structure is produced

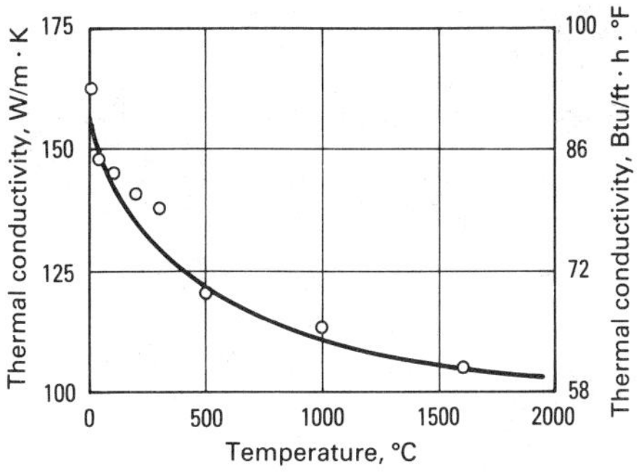

Fig. 27 Thermal conductivity of undoped tungsten

with rounded tungsten-rich grains that are considerably greater in diameter than the original tungsten particles. The blanks are cooled to room temperature in the cooling chamber of the furnace and then removed. Tensile bars and other test blanks usually are sintered from each powder mix and tested for mechanical and physical properties before the mix is approved for production.

Hot Pressing. Some very large parts are produced by hot pressing rather than by cold pressing and sintering. Hot pressing usually is done by leveling the powder mix in a graphite mold and heating the mold in an induction coil while light pressure—sufficient to compact the mix to the required density at temperatures similar to those for sintering—is applied to the assembly. Hot-pressed compacts of this type usually are more brittle and lower in strength than the cold-pressed and sintered materials. Also, the graphite mold may cause a carburized layer to form on the surface of the blank that is difficult to remove in machining.

Coatings

Some promising systems for protecting tungsten from atmospheric exposure at temperatures from 1650 to 2205 °C (3000 to 4000 °F) have been developed, including:

- Roll cladding with tantalum-hafnium alloys
- Slurry-type coatings of iridium-base alloys such as Ir-30Rh
- Duplex and triplex silicide-base coating systems that combine slurry, slip, chemical vapor deposition, and pack cementation processes

Corrosion and Chemical Resistance

At room temperature, tungsten is generally resistant to most chemicals, but it can be easily dissolved with a solution of nitric and hydrofluoric acids. At higher temperatures, tungsten becomes more prone to attack. At about 250 °C (480 °F), it reacts

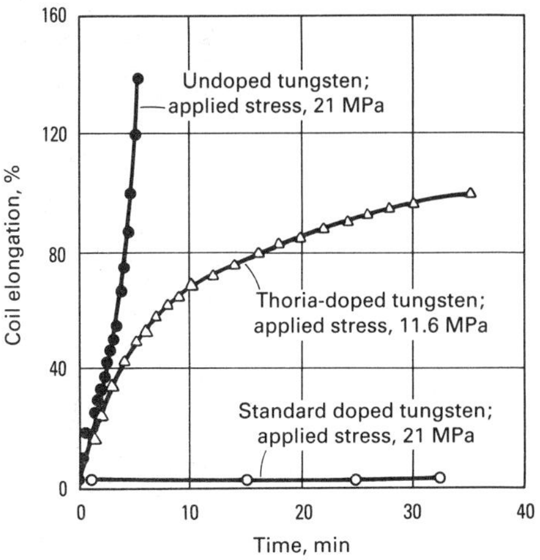

Fig. 28 Creep curves for coiled tungsten wires at 2500 °C (4530 °F)

rapidly with phosphoric acid and chlorine. It begins to oxidize readily at 500 °C (930 °F); at 1000 °C (1830 °F), tungsten reacts with many gases, including water vapor, iodine, bromine, and carbon monoxide. Above 1000 °C (1830 °F), tungsten begins to form compounds with various metals.

Mechanical and Physical Properties

Undoped Tungsten and Tungsten Alloys. Tungsten has high tensile strength and good creep resistance. At temperatures above 2205 °C (4000 °F), tungsten has twice the tensile strength of the strongest tantalum alloys and is only 10% denser. However, its high density, poor low-temperature ductility, and strong reactivity in air limit its usefulness. Maximum service temperatures for tungsten range from 1925 to 2480 °C (3500 to 4500 °F), but surface protection is required for use in air at these temperatures.

Wrought tungsten (as-cold worked) has high strength, strongly directional mechanical properties, and some room-temperature toughness. However, recrystallization occurs rapidly above 1370 °C (2500 °F) and produces a grain structure that is crack sensitive at all temperatures.

Mechanical property data for unalloyed tungsten and tungsten-molybdenum and tungsten-rhenium alloys are shown in Fig. 26 to 31. Additional information on the properties of undoped tungsten is available in the section "Tungsten" in the article "Properties of Pure Metals" in this Volume.

Recrystallized tungsten undergoes a ductile-to-brittle transition above 205 °C (400 °F). Only by heavy warm or cold working is the DBTT lowered to below room temperature (Fig. 32). Annealing raises the DBTT of cold-worked tungsten until it approaches that of recrystallized material.

The exact ductile-to-brittle transition temperature is influenced by many factors,

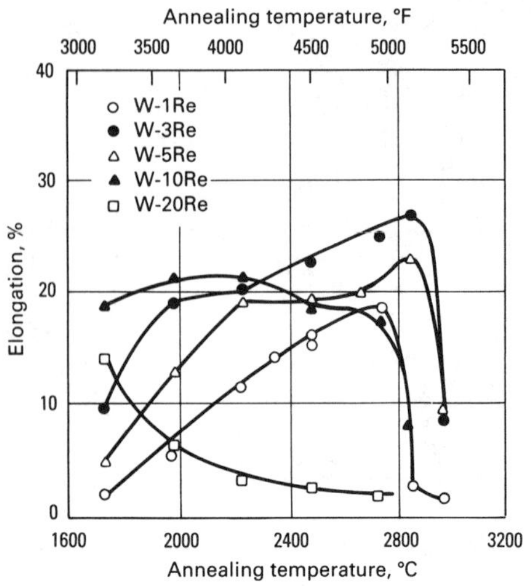

Fig. 29 Room-temperature ductility of annealed wire for five tungsten-rhenium alloys

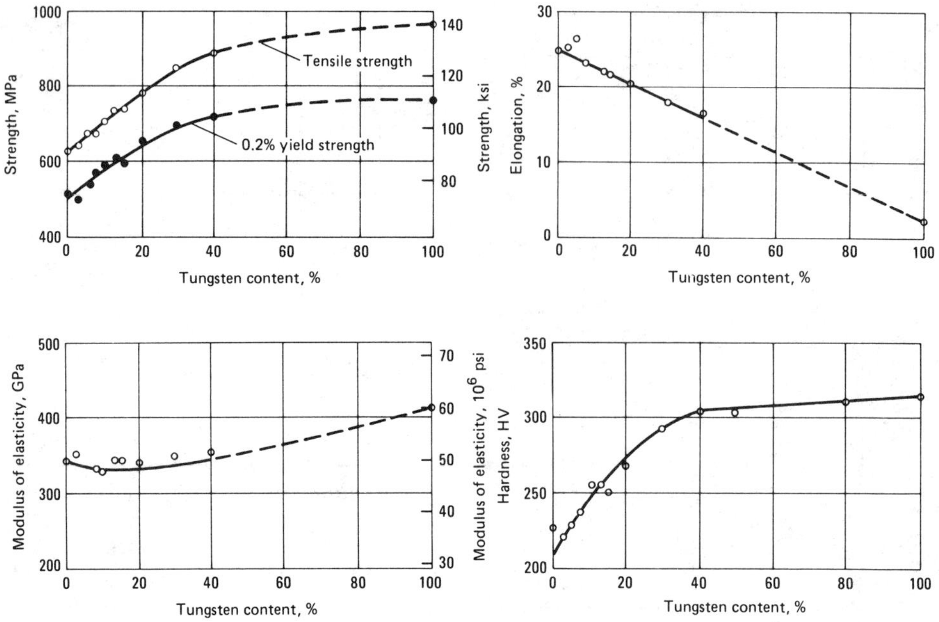

Fig. 30 Effect of tungsten content on the room-temperature mechanical properties of tungsten-molybdenum alloys

including grain size, strain rate, and impurity levels. The DBTT decreases with grain size unless the grains are larger than 1 mm (0.04 in.) in diameter. The DBTT also drops with increases in strain rate, but it climbs rapidly as impurity levels increase. Like all brittle metals, tungsten is very notch sensitive. Therefore, removal of even minute surface flaws by grinding, oxidizing, or electrolytic polishing prior to service improves ductility and lowers the DBTT.

Alloying can have a beneficial effect on the DBTT; the effect of rhenium in producing a ductile alloy is the best-known example. Doping with AKS dopant or alloying with a dispersion of thoria retards recrystallization, thereby improving the ductility of annealed wire. In addition, a fine dispersion of thoria causes a decrease in grain size, which in turn promotes a reduction in the DBTT.

Below the DBTT, recrystallized tungsten fails by a combination of cleavage and grain-boundary fracture. Near the DBTT, fracture by cleavage increases. At higher temperatures, usually above 500 °C (930 °F), grain-boundary and ductile fracture predominate. Generally, grain-boundary fracture predominates in commercially pure tungsten and ductile fracture in AKS-doped tungsten.

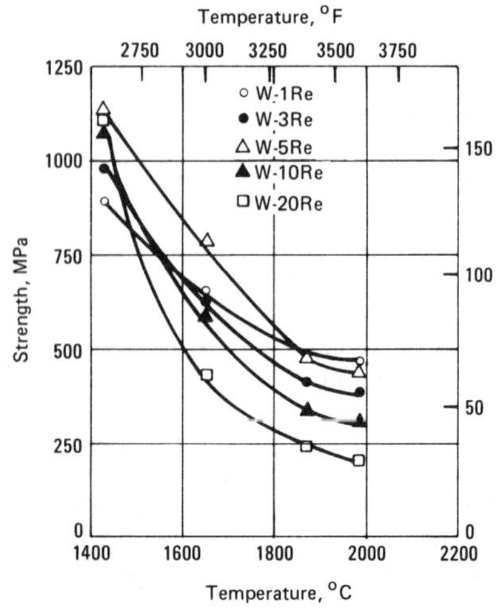

Fig. 31 Short-time tensile strengths of five tungsten-rhenium alloys

Aside from its uses in abrasive and wear-resistant tools and as an alloying element, tungsten finds its primary commercial application in filaments for incandescent lamps. Thoria particles and the potassium bubble dispersion that occurs in AKS-doped tungsten impede the annealing process that progressively eliminates substructure. This allows tungsten to retain hardness and tensile strength at temperatures higher than those at which commercially pure or refined tungsten is softened and weakened. It also improves the creep resistance of tungsten wire at elevated temperatures. Upon recrystallization, a nonsag, interlocking grain structure forms. This structure gives tungsten wire added creep resistance at high temperatures, allowing tungsten filaments in incandescent lamps to burn at high temperatures without sagging.

Alloying with rhenium improves the tensile strength of AKS-doped or undoped tungsten. Although small additions of less than 5% Re cause softening of tungsten-rhenium alloys, hardness increases when solid-solution strengthening becomes the overriding factor. Alloying with molybdenum has a softening effect that is proportional to molybdenum content.

Tungsten is not as anisotropic in elastic behavior as are some other cubic metals, but its stress-strain curve does vary somewhat with crystallographic direction.

Tungsten Heavy-Metal Alloys. Minimum mechanical properties of machinable heavy-metal tungsten alloys are specified at the time of purchase. Three specifications are in general use: MIL-T-21014, ASTM B 459, and AMS 7725. The specifications for machinable high-density tungsten-base alloys usually divide them into four classes based on composition (Table 23) and three types based on tensile properties (Table 24). Tables 25 and 26 give typical mechanical and physical properties of tungsten heavy-metal alloys according to these class and type divisions.

Class 1 alloys are basically tungsten-nickel-copper or tungsten-nickel-iron alloys. The tungsten-nickel-copper alloys of this class typically contain 90% W, 6 to 7% Ni, and 3 to 4% Cu. Minor additions of other metals, such as molybdenum or cobalt, can be added to modify properties such as hardness. Class 1 tungsten-nickel-iron alloys usually contain 90% W, 5 to 7.5% Ni, and 3 to 5.5% Fe.

Class 2, 3, and 4 alloys are usually tungsten-nickel-iron alloys with tungsten contents in the range shown in Table 23. They contain a balance of nickel-iron in a ratio of 4Ni:1Fe (class 2), 7Ni:3Fe (class 3), and 1Ni:1Fe (class 4). Sometimes a portion of the iron may be replaced with copper.

Electrical Properties

The electrical resistivity and temperature coefficient of electrical resistivity properties of tungsten are both strongly affected by purity and deformation. The effects of recovery annealing on these two properties for commercially pure tungsten wire are shown in Table 27. The product of resistivity and temperature coefficient is a nearly

Table 23 Classification of tungsten heavy-metal alloys by composition, density, and hardness

Class	Tungsten content, %	Density g/cm³	Density lb/in.³	Hardness, HRC	Type classification(a)
1	89–91	16.85–17.25	0.609–0.623	30–36	I
1	89–91	16.85–17.25	0.609–0.623	32 max	II, III
2	91–94	17.15–17.85	0.620–0.645	33 max	II, III
3	94–96	17.75–18.35	0.641–0.663	34 max	II, III
4	96–98	18.25–18.85	0.659–0.681	35 max	II, III

(a) See Table 24.

constant value that is independent of the degree of residual cold work. The addition of rhenium, molybdenum, or thoria increases the resistivity of tungsten wire but has no appreciable effect on its temperature coefficient. Some typical electrical resistivity data are shown in Fig. 33 to 35.

Thermocouples in which tungsten is one of the thermoelements are used extensively at very high temperatures. Tungsten-molybdenum thermocouples, for example, can be used at temperatures up to 2205 °C (4000 °F) if maintained in a protective envelope or a reducing atmosphere.

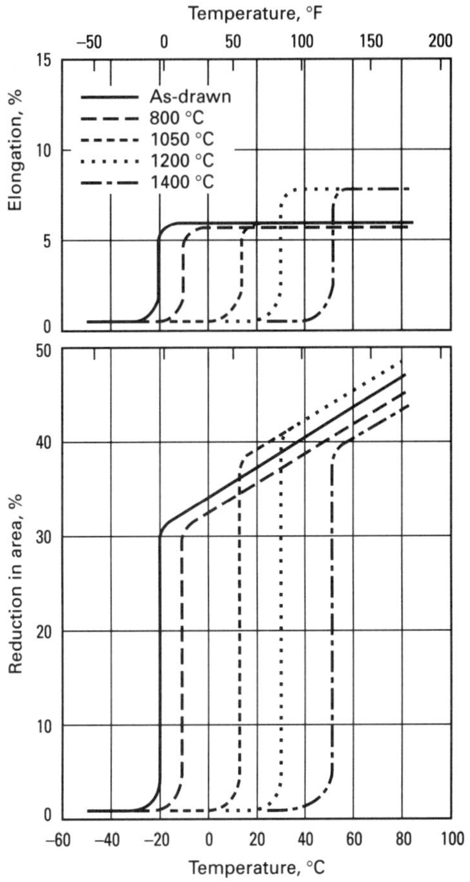

Fig. 32 Variation of DBTT with annealing temperature for undoped tungsten. Data are for 10-min recovery annealing of heavily worked 0.75 mm (0.030 in.) diam wire.

Rhenium

Toni Grobstein, Robert Titran, and Joseph R. Stephens,
NASA Lewis Research Center

PLATINUM-RHENIUM REFORMING CATALYSTS are the major rhenium end-use products and account for about 85% of rhenium consumption. Rhenium catalysts are exceptionally resistant to poisoning from nitrogen, sulfur, and phosphorus. They are used for the hydrogenation of fine chemicals and for hydrocracking, reforming, and the disproportionation of olefins, including increasing the octane rating in the production of lead-free petroleum products. Rhenium is also used in the production of heating elements, x-ray tubes and targets, and metallic coatings. Iridium-coated rhenium nozzles for small chemical rockets and resistojet thrusters are used in space for satellite orientation. Rhenium is a solid-solution-strengthening alloying element in superalloys; in tungsten and molybdenum-based alloys, it markedly increases room-temperature ductility (this increase is known as the rhenium effect).

Rhenium metal is widely used in filaments for mass spectrographs and ion gages because of its high electrical resistivity and low vapor pressures at high temperatures. Rhenium-molybdenum alloys are superconductive at 10 K. Rhenium is used as an electrical contact material because of its wear resistance and its ability to withstand arc erosion. Thermocouples made of rhenium-tungsten are used for measuring temperatures up to 2200 °C (3990 °F), and rhenium wire is used in photoflash lamps for photography.

Table 24 Classification of tungsten heavy-metal alloys by tensile properties

Type	Tensile strength MPa	Tensile strength ksi	0.2% yield strength MPa	0.2% yield strength ksi	Elongation, %
I	900	130	725	105	1.5
II	650	94	520	75	2.0
III	415	60	...	...	1.0

Relatively little development work has been done for rhenium-base alloys as compared with that for other refractory metals. The use of rhenium in aerospace applications has been restricted by its high density; in terrestrial applications, its short supply and consequent high cost have been the limiting factors. For example, the addition of 3% Re to tungsten wire doubles the cost of the wire.

Occurrence and Production

Most rhenium occurs in porphyry copper deposits. Identified sources are estimated to be about 4.5×10^3 Mg (5.0×10^3 tons) in the United States, and approximately 5.9×10^3 Mg (6.5×10^3 tons) in the rest of the world. The United States relies on imports for most of its rhenium supply, with 71% coming from Chile. It is estimated that the United States consumed about 6.35 Mg (7.00 tons) of rhenium in 1989. Rhenium is available as perrhenic acid ($HReO_4$), ammonium perrhenate (NH_4ReO_4), and metal powder. In 1988, the average price of rhenium metal was $1.05/g ($475/lb); the price was $0.66/g ($300/lb) for ammonium perrhenate.

Ammonium perrhenate is converted to metal powder by hydrogen reduction. The reduction is carried out at 380 °C (715 °F) and is followed by a purification and reduction cycle at 700 to 800 °C (1290 to 1470 °F) to remove any residual rhenium oxide. The powder is generally consolidated by cold pressing at about 205 MPa (30 ksi) to a density of 35 to 40% using stearic acid in ether as a lubricant on the punch and the die walls. Subsequent sintering at 1200 °C (2190 °F) for 2 h in vacuum results in little densification but increases the mechanical strength of the compact and burns off volatile impurities. Finally, resistance heating in a vacuum or hydrogen atmosphere at 2700 to 2900 °C (4890 to 5250 °F) produces sintered compacts with densities of more than 90%.

Electron beam remelting is sometimes used to reduce the impurity content of rhenium compacts. Chemical vapor deposition is also a practical fabrication method.

Corrosion Resistance

Rhenium oxidizes catastrophically at temperatures above 600 °C (1110 °F). Oxidation occurs as a result of the formation of

Table 25 Typical mechanical properties of commercial machinable heavy-metal tungsten alloys

Alloy(a)	Density g/cm^3	Density lb/in.^{3}	Tensile strength MPa	Tensile strength ksi	Yield strength at 0.2% offset MPa	Yield strength at 0.2% offset ksi	Elongation in 25 mm (1 in.), %	Hardness, HRC	Proportional limit MPa	Proportional limit ksi	Modulus of elasticity GPa	Modulus of elasticity psi × 10^6	Coefficient of linear thermal expansion µm/m · °C	Coefficient of linear thermal expansion µin./in. · °F	Magnetic properties
Tungsten-nickel-copper															
Class 1	17.0	0.614	785	114	605	88	4	27	205	30	275	40	5.5	3.1	Virtually nonmagnetic
Tungsten-nickel-iron															
Class 1	17.0	0.614	895	130	615	89	16	27	260	38	275	40	5.4	3.0	Slightly magnetic
Class 3	18.0	0.650	925	134	655	95	6	29	350	51	310	45	5.3	2.9	Slightly magnetic
Class 4	18.5	0.667	795	115	690	100	3	32	450	65	345	50	5.0	2.8	Slightly magnetic

(a) For a key to the four classes of tungsten heavy-metal alloys, see Table 23.

Table 26 Additional properties of machinable heavy-metal tungsten alloys

Alloy type(a)	Modulus of rupture (flexure) MPa	Modulus of rupture (flexure) ksi	Proportional limit MPa	Proportional limit ksi	Modulus of elasticity GPa	Modulus of elasticity psi × 10^6	Modulus of rigidity GPa	Modulus of rigidity psi × 10^6	Angle of twist at rupture	Shear strength MPa	Shear strength ksi	Electrical conductivity, %IACS
Type I												
Minimum	1380	200	310	45	205	30	130	19	80°	895	130	13.5
Average	1585	230	425	62	305	44	...	...	100°	...	...	14
Type II												
Minimum	1240	180	...	...	170	25	130	19	160°	550	80	13
Average	1515	220	170	25	275	40	132	19.2	166°	560	81	14
Type III(b)												
Minimum	690	100	...	...	...	...	...	...	...	...	...	...
Average	...	...	...	...	...	...	...	...	...	...	...	...

(a) For a key to the three type divisions of tungsten heavy-metal alloys, see Table 24. (b) This type is used almost exclusively for radiation shielding; data for properties other than modulus of rupture are not available.

rhenium heptoxide (Re_2O_7), which has a melting point of 297 °C (567 °F) and a boiling point of 363 °C (685 °F). The white oxide vapor has been reported to be nonpoisonous. Iridium is currently used as an oxidation-resistant coating for rhenium at high temperatures. Rhenium is unique among the refractory metals in that it does not form a carbide; however, it is similar to the other metals in the group in that it is resistant to liquid lithium metal corrosion. Rhenium is resistant to water cycle corrosion in high-temperature filaments in vacuum. Rhenium has good resistance to sulfuric acid and hydrochloric acid but can be dissolved by nitric acid; it is also resistant to aqua regia at room temperature. In addition, rhenium is resistant to attack by molten tin, zinc, silver, copper, and aluminum.

Mechanical and Physical Properties

Temperature-dependent tensile strength, elastic modulus, and physical property data for rhenium are presented in the introductory section of this article (see Table 2 and Fig. 1 and 2). One of the most outstanding characteristics of rhenium is its very high strain-hardening rate, which is about 3.5 times that of tungsten or molybdenum. The general trend of existing data indicates about a twofold increase in hardness for 25% deformation. This unusually rapid work hardening requires frequent intermediate annealing in inert or reducing atmospheres during fabrication, with low cold reduction levels to avoid cracking. Because impurity levels are critical to fabricability, a vacuum level of 1 to 0.1 mPa (10^{-5} to 10^{-6} torr) or a dry hydrogen atmosphere is used. Hydrogen-nitrogen mixtures, such as dissociated ammonia or annealing hydrogen (H_2 + $7N_2$), have also been successfully used. The ultimate tensile strength of annealed rhenium sheet has been reported to increase from 1158 MPa to 2220 MPa (168 to 322 ksi) as a result of 30.7% cold reduction. Detailed property data for unalloyed rhenium are given in the section "Rhenium" in the article "Properties of Pure Metals" in this Volume.

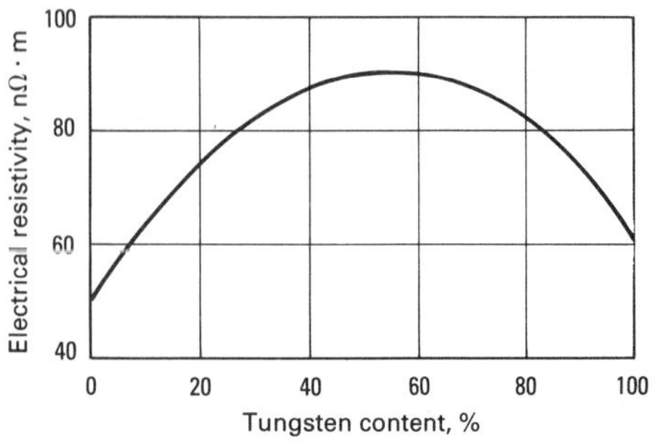

Fig. 33 Effect of tungsten content on the specific electrical resistivity of tungsten-molybdenum alloys

Refractory Metal Fiber-Reinforced Composites

Toni Grobstein and Donald W. Petrasek, NASA Lewis Research Center

REFRACTORY METAL WIRES, in spite of their poor oxidation resistance and high density, have received a great deal of attention as fiber reinforcement materials for use in high-temperature composites. Although the theoretical specific strength potential of refractory alloy fiber-reinforced composites is less than that of ceramic fiber-reinforced composites, the more ductile metal fiber systems are more tolerant of fiber-matrix reactions and thermal expansion mismatches. When refractory metal fibers are used to reinforce a ductile and oxidation-resistant matrix, they are protected from oxidation, and the specific strength of the composite is much higher than that of superalloys at elevated temperatures.

The majority of the studies conducted on this topic have been on refractory wire and superalloy composites that use tungsten or molybdenum wire (available as lamp filament or thermocouple wire) as the reinforcement material. These refractory alloy wires were not designed for use in composites, nor were they developed to achieve optimum mechanical properties in the temperature range of interest for component application, 1000 to 1200 °C (1830 to 2190 °F). The stress-rupture properties of a tungsten lamp filament wire used in early studies

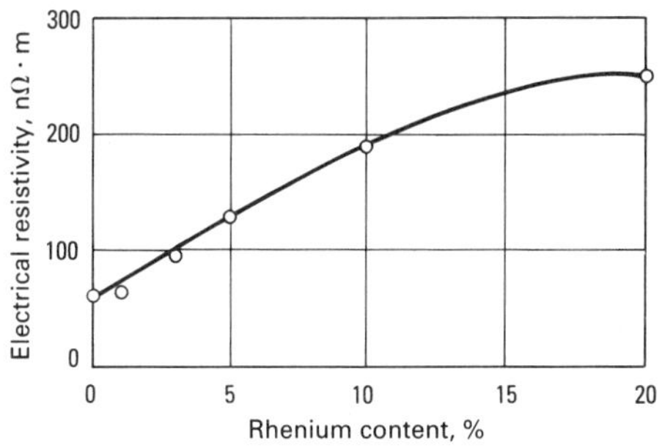

Fig. 34 Specific electrical resistivity of tungsten-rhenium alloys as a function of rhenium content

Table 27 Effect of annealing on the electrical resistivity and temperature coefficient of drawn tungsten wire

Annealing temperature °C	°F	Electrical resistivity, μΩ · m	Temperature coefficient	Matthiessen's rule(a)
As-drawn		617	0.355	219
400	750	591	0.376	222
600	1110	543	0.415	225
800	1470	523	0.433	226
1000	1830	518	0.440	228
1205	2200	500	0.432	216
2500	4530	484	0.481	233

(a) Product of specific resistance and temperature coefficient

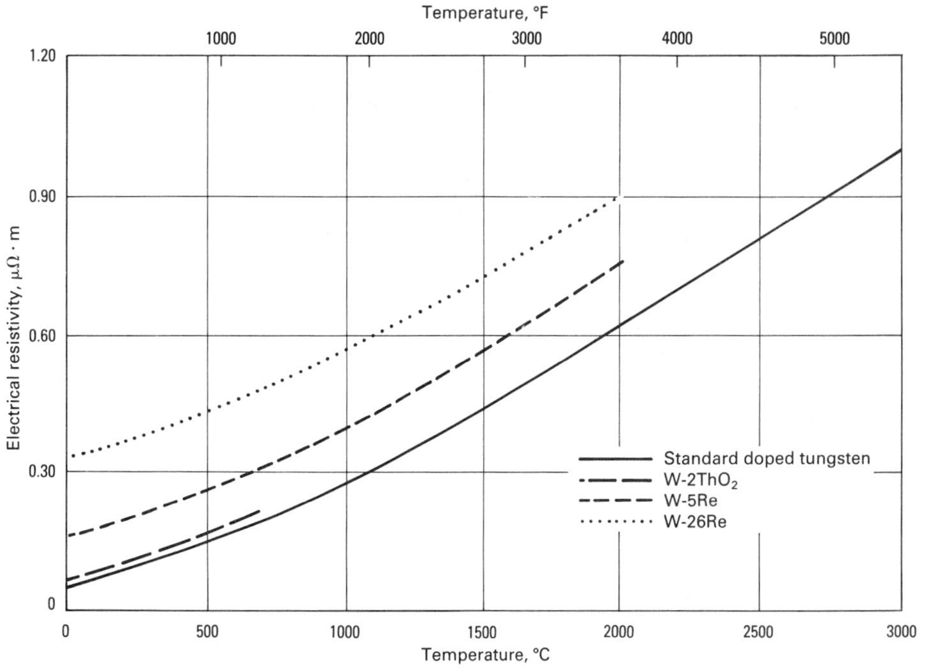

Fig. 35 Effect of temperature on the electrical resistivity of standard doped tungsten and of tungsten-rhenium alloys

were superior to those of rod and bulk forms of tungsten, and this wire showed promise for use as composite reinforcement. After the need for stronger wire was recognized, high-strength tungsten, tantalum, molybdenum, and niobium alloys that were originally used for rod and/or sheet fabrication were drawn into wire.

Excellent progress has been made in providing wires with increased strength. Tungsten alloy wires have been fabricated that have tensile strengths 2.5 times higher than those obtained for potassium-doped tungsten lamp filament wire. The strongest wire fabricated, tungsten-rhenium-hafnium-carbon, has a tensile strength of 2165 MPa (314 ksi) at 1093 °C (2000 °F), which is more than 6 times the strength of the strongest nickel-base or cobalt-base superalloy. Although the ultimate tensile strength values of the tungsten alloy wires were higher than those obtained for molybdenum, tantalum, or niobium wires, their advantage is lessened when the higher density of tungsten is taken into account. Nevertheless, high-strength tungsten alloy wires rank alongside molybdenum wires as offering the most promise for composite applications.

Processing of Composites. The consolidation of matrix and fibers into a composite material with useful properties is one of the most difficult steps in developing composites reinforced with refractory metal wire. Fabrication methods are currently in the laboratory phase of development because satisfactory techniques have not yet been developed for producing large numbers of specimens for extensive property characterization. Fabrication techniques currently being developed can be classified as either liquid-phase or solid-phase methods.

Liquid-phase methods consist of casting the molten matrix using investment casting techniques so that the matrix infiltrates the bundle of fibers. The molten metal must wet the fibers, form a chemical bond, and yet be controlled so as not to degrade the fibers by dissolution, reaction, or recrystallization.

Solid-phase methods generally use processing temperatures much lower than those reached during liquid-phase processing; diffusion rates are therefore much lower, and reaction with the fiber can be less severe. The prerequisite for solid-state processing is that the matrix be in either wire, sheet, foil, or powder form. Cold pressing followed by sintering or hot pressing is used to consolidate the matrix and fiber into a composite component.

Mechanical and Thermal Properties. Refractory fiber-reinforced superalloy composites have demonstrated strengths significantly above those of the strongest superalloys. Tungsten fiber-reinforced superalloy composites, in particular, are potentially useful as high-temperature (1000 to 1200 °C, or 1830 to 2190 °F) materials because of their microstructural stability and superior resistance to stress-rupture and creep deformation, thermal shock, and low- and high-cycle fatigue. Compared with conventional superalloys, refractory metal fiber-reinforced composites have improved ductility, impact damage resistance, and thermal conductivity.

Refractory fiber-reinforced niobium alloy composites have demonstrated a potential for use at temperatures in excess of 1200 °C (2190 °F). The tensile and creep strength properties of these composites have been improved an order of magnitude by adding 50 vol% tungsten fiber to the niobium alloys.

Applications. Refractory metal alloy fiber-reinforced composites are being considered for many different and demanding applications. For example, tungsten fibers are being investigated for use as a reinforcement in copper for strengthening high-conductivity materials in regeneratively cooled rocket nozzles. A volume fraction addition of 10% tungsten fibers in copper has a dramatic effect on the strength of the material without significantly decreasing its thermal conductivity. With the addition of the tungsten fibers, copper can be used at temperatures and stresses that would normally exceed its yield strength.

Tungsten fiber-reinforced superalloy composites are being developed for use in rocket engine turbine blades (Fig. 36). Tungsten fiber-reinforced superalloy composites have a highly attractive combination of properties at temperatures from 870 to 1100 °C (1600 to 2010 °F); these properties make them well suited for advanced rocket engine turbopump blade applications. The composites offer the potential of significantly improved operating life, higher operating temperature capability, and reduced strains induced by transient thermal conditions during engine start and shutdown.

Tungsten fiber-reinforced niobium alloy systems are being investigated for potential

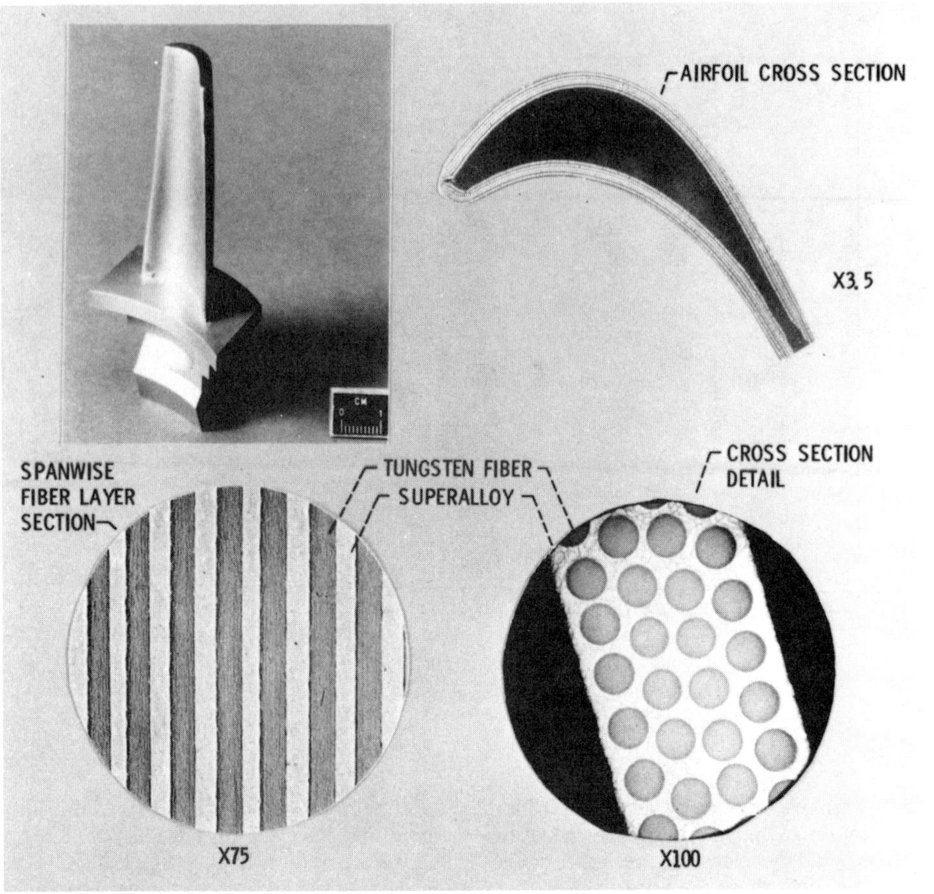

Fig. 36 Location and structure of tungsten fibers in fiber-reinforced superalloy composite turbine blades for rocket engine turbopumps. Courtesy NASA Lewis Research Center

long-term high-temperature applications in space power systems. In addition, molybdenum-base fibers are being proposed for use in intermetallic-matrix composites for aerospace applications.

SELECTED REFERENCES

- *Aerospace Structural Metals Handbook*, Code 5201, 1963; Code 5206, 1966; Code 5208, 1967; Code 5209, 1971; Code 5211, 1973; U.S. Department of Defense, Mechanical Properties Data Center
- Applied Superconductivity Conference, *IEEE Trans. Magn.*, Vol 11 (No. 2), March 1975
- R. Bakish, Ed., *Electron Beam Melting and Refining: State of the Art 1989*, Bakish Materials Corporation
- H. Bildstein and Hugo M. Ortner, Ed., *Eleventh International Plansee Seminar '85*, Metallwerk Plansee GmbH
- J.A. DeMastry, *Corrosion Studies of Tungsten, Molybdenum, and Rhenium in Lithium*, Vol 3, *Nuclear Applications*, 1967, p 127-134
- D.L. Douglass and F.W. Kunz, *Columbium Metallurgy*, Interscience, 1961
- "The Engineering Properties of Tungsten and Tungsten Alloys," DMIC Report 191, Battelle Memorial Institute, 1962
- Paper presented at the First International Symposium on Tantalum, Rothenburg Ob Der Tauber, West Germany, Tantalum Producers International Study Center, May 1978
- S. Foldes, "State of the Science for Tungsten Alloys," General Electric Technical Information Series 62-LMC-107, 1962
- B.W. Gonser, Ed., *Rhenium*, Elsevier, 1962
- M. Hoch, "The High Temperature Specific Heat of Body-Centered Cubic Refractory Metals," GEMP-696, General Electric Company, 1969
- R.I. Jaffee, "Implications of Rhenium Research in the Design of Refractory Alloys," DMIC Memorandum 7, Battelle Memorial Institute, 1959
- KBI Product Data Bulletin 31321-PD2, Kawecki-Berylco Division, Cabot Corporation
- R. Kieffer and H. Braun, *Vanadin-Niob-Tantal*, Springer-Verlag, 1963
- R.D. Larrabee, "The Spectral Emissivity and Optical Properties of Tungsten," Technical Report 328, Massachusetts Institute of Technology, 1957
- K.C. Li and C.Y. Wang, *Tungsten, Its History, Geology, Ore Dressing, Metallurgy, Chemistry, Analysis, Applications and Economics*, Reinhold, 1955
- K. Lo, J. Bevk, and D. Turnbull, Critical Currents in Liquid-Quenched Nb-Al, *J. Appl. Phys.*, Vol 48, 1977
- I. Machlin, R.T. Begley, and E.D. Weisert, Ed., *Refractory Metal Alloys: Metallurgy and Technology*, Plenum Publishing, 1968, p 455-460
- D.R. Mash, D.W. Bauer, and M. Schussler, Fabricating the Refractory Metals, *Met. Prog.*, Vol 99 (No. 2-4), Feb-April 1971
- D.J. Maykuth, "The Availabilities and Properties of Rhenium," DMIC Memorandum 19, Battelle Memorial Institute, 1959
- *Metal Statistics 1988*, American Metal Market, Fairchild Publications
- *Mineral Commodity Summaries 1989*, U.S. Bureau of Mines, 1989
- M.S. Neuberger, D.L. Grigsby, and W.H. Veazie, Jr., *Niobium Alloys and Compounds*, Vol 4, *Handbook of Electronic Materials*, Plenum Publishing, 1972
- "Physical and Mechanical Properties of Tungsten and Tungsten-Base Alloys," DMIC Report 127, Battelle Memorial Institute, 1960
- John H. Port and Joseph M. Pontelandolfo, Fabrication and Properties of Rhenium and Rhenium-Molybdenum Alloys, in *Reactive Metals*, W.R. Clough, Ed., Interscience, p 555-574
- John W. Pugh, Refractory Metals: Tungsten, Tantalum, Columbium, and Rhenium, in *High Temperature Materials*, R.F. Hehemann and G.M. Ault, Ed., John Wiley & Sons, 1959, p 307-318
- G.W.P. Rengstorff, "High Purity Metals," DMIC Report 222, Battelle Memorial Institute, 1966, p 33-37
- G.D. Rieck, *Tungsten and Its Compounds*, Pergamon Press, 1967
- E.M. Savitskii and M.A. Tylkina, Ed., *Study and Use of Rhenium Alloys*, Amerind, 1978
- F.F. Schmidt and H.R. Ogden, "The Engineering Properties of Columbium and Columbium Alloys," DMIC Report 188, Battelle Memorial Institute, 1963
- F.F. Schmidt and H.R. Ogden, "The Engineering Properties of Tantalum and Tantalum Alloys," DMIC Report 189, Battelle Memorial Institute, 1963
- K. Schulze, D. Bach, D. Lupton, and F. Schreiber, Purification of Niobium, in *Niobium: Proceedings of the International Symposium: San Francisco, 1981*, The Metallurgical Society, 1984, p 213–272
- H.G. Sell, G.H. Keith, R.C. Koo, R.H. Schnitzel, and R. Corth, "Physical Metallurgy of Tungsten and Tungsten Base Alloys," Technical Report WADD-TR-60-37, Parts I-III, U.S. Armed Services Technical Information Agency, 1961
- J.G. Sessler and V. Weiss, *Aerospace Structural Metals Handbook*, Vol 11A, *Nonferrous Heat Resistant Alloys*,

AFML-TR-68-115, U.S. Air Force Materials Laboratory, 1969
- K.D. Sheffler, "Generation of Long Time Creep Data on Refractory Alloys at Elevated Temperatures," ER-7442, TRW Inc., Jan 1970
- C.T. Sims *et al*, "Investigations of Rhenium," WADC TR 54-371, Battelle Memorial Institute, 1954
- F.T. Sisco and E. Epremian, *Columbium and Tantalum*, John Wiley & Sons, 1963
- C.J. Smithells, *Tungsten, Its Metallurgy, Properties and Applications*, Chemical Publishing, 1953
- T.E. Tietz and J.W. Wilson, *Behavior and Properties of Refractory Metals*, Stanford University Press, 1965, p 206-222
- R.H. Titran, "Creep Behavior of Tantalum Alloy T-222 at 1365 to 1700°K," NASA Technical Note TN D-7673, June 1974
- *Tungsten*, Metallwerk Plansee AG, 1971
- S.Y.H. Yih and C.T. Wang, *Tungsten: Sources, Metallurgy, Properties and Applications*, Plenum Publishing, 1979

Introduction to Titanium and Titanium Alloys

James D. Destefani*, Bailey Controls Company

TITANIUM has been recognized as an element for 200 years. Only in the last 40 years or so, however, has the metal gained strategic importance. In that time, commercial production of titanium and titanium alloys in the United States has increased from zero to more than 23 million kg/yr (50 million lb/yr).

The catalyst for this remarkable growth was the development by Dr. Wilhelm J. Kroll of a relatively safe, economical method to produce titanium metal in the late 1930s. Kroll's process involved reduction of titanium tetrachloride ($TiCl_4$), first with sodium and calcium, and later with magnesium, under an inert gas atmosphere (Ref 1). Research by Kroll and many others continued through World War II. By the late 1940s, the mechanical properties, physical properties, and alloying characteristics of titanium were defined and the commercial importance of the metal was apparent.

Commercial titanium production soon began in earnest in the United States, and by 1956 U.S. production of titanium mill products was more than 6 million kg/yr (13 million lb/yr) (Ref 2).

Alloy development progressed rapidly. The beneficial effects of aluminum additions were realized early on, and titanium-aluminum alloys were soon commercially available. Two alloys that are still widely used, Ti-6Al-4V and Ti-5Al-2.5Sn, were both developed in the early 1950s. The Ti-6Al-4V alloy, in fact, accounts for more than half of the current U.S. titanium market (Ref 3).

General Metal Characteristics. The rapid growth of the titanium industry is testimony to the metal's high specific strength and corrosion resistance. With density about 55% that of steel, titanium alloys are widely used for highly loaded aerospace components that operate at low to moderately elevated temperatures, including both airframe and jet engine components (see the section "Applications" in this article).

*Formerly with ASM INTERNATIONAL

Titanium's corrosion resistance is based on the formation of a stable, protective oxide layer. This passivating behavior makes the metal useful in applications ranging from chemical processing equipment to surgical implants and prosthetic devices. The corrosion behavior of titanium is discussed in detail in the article "Corrosion of Titanium and Titanium Alloys" in Volume 13 of the 9th Edition of *Metals Handbook*.

Current titanium technology encompasses a variety of products and processes. Some of the latest developments, which are briefly reviewed in the section "New Developments" in this article, include new sponge production and melting practices, titanium-matrix composites, oxide dispersion-strengthened powder metallurgy (P/M) alloys with novel compositions and properties, superplastic forming and diffusion bonding (SPF/DB) of titanium alloy sheet and plate, and titanium-base ordered intermetallic compounds.

This article is intended to provide an overview of contemporary titanium technology. Detailed information on the properties, processing, and application of specific titanium alloys and product forms is available in the articles "Wrought Titanium and Titanium Alloys," "Titanium and Titanium Alloy Castings," "Titanium P/M Products," "Metal-Matrix Composites," and "Ordered Intermetallics" in this Volume.

Alloy Types

Titanium exists in two crystallographic forms. At room temperature, unalloyed (commercially pure) titanium has a hexagonal close-packed (hcp) crystal structure referred to as alpha (α) phase. At 883 °C (1621 °F), this transforms to a body-centered cubic (bcc) structure known as beta (β) phase. The manipulation of these crystallographic variations through alloying additions and thermomechanical processing is the basis for the development of a wide range of alloys and properties. These phases also provide a convenient way to categorize titanium mill products. Based on the phases present, titanium alloys can be classified as either α alloys, β alloys, or $\alpha + \beta$ alloys.

Alpha alloys contain elements such as aluminum and tin. These α-stabilizing elements work by either inhibiting change in the phase transformation temperature or by causing it to increase (Ref 4). Alpha alloys generally have creep resistance superior to β alloys, and are preferred for high-temperature applications. The absence of a ductile-to-brittle transition, a feature of β alloys, makes α alloys suitable for cryogenic applications.

Alpha alloys are characterized by satisfactory strength, toughness, and weldability, but poorer forgeability than β alloys (Ref 5). This latter characteristic results in a greater tendency for forging defects. Smaller reductions and frequent reheating can minimize these problems.

Unlike β alloys, alpha alloys cannot be strengthened by heat treatment. They most often are used in the annealed or recrystallized condition to eliminate residual stresses caused by working.

Alpha + beta alloys have compositions that support a mixture of α and β phases and may contain between 10 and 50% β phase at room temperature. The most common $\alpha + \beta$ alloy is Ti-6Al-4V (Ref 4). Although this particular alloy is relatively difficult to form even in the annealed condition, $\alpha + \beta$ alloys generally have good formability.

The properties of these alloys can be controlled through heat treatment, which is used to adjust the amounts and types of β phase present. Solution treatment followed by aging at 480 to 650 °C (900 to 1200 °F) precipitates α, resulting in a fine mixture of α and β in a matrix of retained or transformed β phase.

Beta alloys contain transition elements such as vanadium, niobium, and molybdenum, which tend to decrease the temperature of the α to β phase transition and thus promote development of the bcc β phase. They have excellent forgeability over a wid-

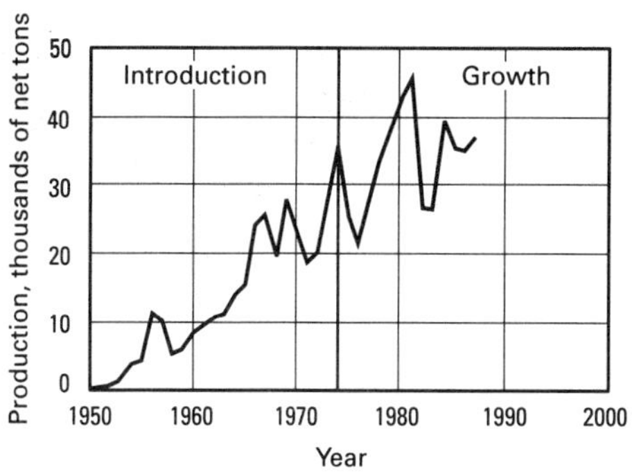

Fig. 1 Growth of U.S. titanium ingot production, 1951 to 1989. Introduction and growth indicate phases of titanium product life cycle. Source: Ref 6

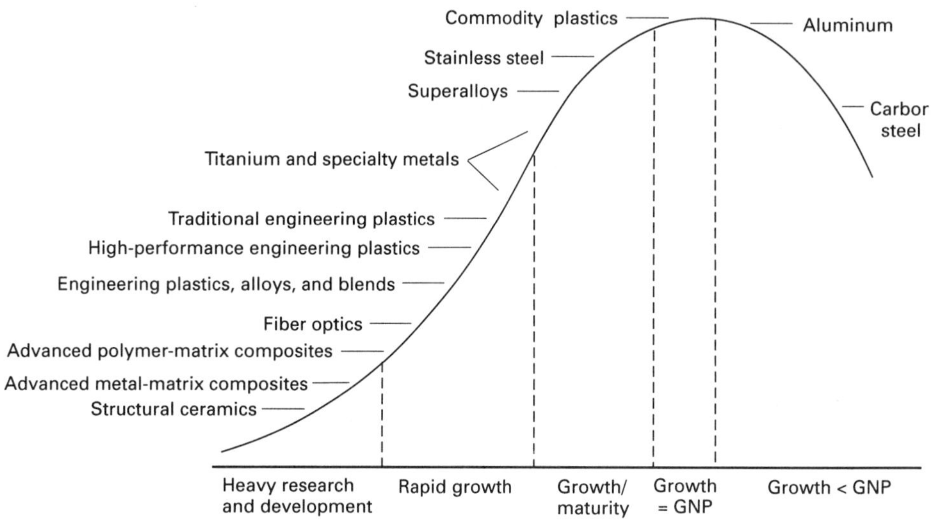

Fig. 2 Schematic product life cycle curve, showing position of various technologies on curve. Depending or market area, titanium ranges from rapid growth to growth/maturing stage. Source: Ref 6

er range of forging temperatures than α alloys, and β alloy sheet is cold formable in the solution-treated condition.

Beta alloys have excellent hardenability, and respond readily to heat treatment. A common thermal treatment involves solution treatment followed by aging at temperatures of 450 to 650 °C (850 to 1200 °F). This treatment results in formation of finely dispersed α particles in the retained β.

Market Development

From its inception, the titanium industry was tied very closely to the market for commercial and military jet aircraft. Dependence on the aerospace industry, which is cyclical in nature, resulted in numerous setbacks. Despite this, growth of the U.S. titanium industry has been relatively steady. Figure 1 illustrates the increase of U.S. titanium ingot production since 1951.

The Product Life Cycle of Titanium (Ref 6). Product life cycle theory has been used for nearly 40 years to analyze the rise and fall of product demand. A typical product life cycle begins with a product's introduction into the marketplace. As shown in Fig. 2, this is followed by several more or less well-defined stages of rapid growth, maturity, and ultimate decline as replacement products enter the marketplace. Using product life-cycle models originally developed for and applied to the U.S. steel and aluminum industries, this theory has recently been applied to relate U.S. titanium demand to industrial economic growth.

Table 1 Past and predicted average annual U.S. demand for titanium and titanium alloys

Product form	Demand, kg × 1000 (lb × 1000) 1984–1988	1989–1993	1994–1999
Ingot	34 635	38 909	44 182
	(76 196)	(85 600)	(97 200)
Mill products	20 630	23 273	26 545
	(45 387)	(51 200)	(58 400)
Castings	392	727	1 455
	(862)	(1 600)	(3 200)

Source: Ref 6

Reference 6 concludes that, despite continued development of new alloys and product forms, titanium has moved rapidly through its product life cycle to maturity in the aircraft industry. The metal is still in the growth stage in applications where corrosion resistance is important, such as the marine and biomedical industries. Other commercial and consumer applications, such as in the automotive industry and in architecture, are only in the developmental stage.

These more varied applications should strengthen and stabilize demand for titanium, making titanium producers less susceptible to fluctuations in any one application area. This diversification should accelerate as the industry continues to mature and titanium makes the transition from a technology product to a commodity product.

Market Trends. Reference 6 also used product life-cycle analysis to forecast future demand for titanium mill products through the end of the century. Table 1 compares past and predicted average annual demands for titanium ingot, castings, and mill products.

Applications

Aerospace applications—including use in both structural (airframe) components and jet engines—still account for the largest share of titanium alloy use. Titanium, in fact, was so successful as an aerospace material that other potential applications were not fully exploited. These have only more recently begun to be explored; some are in development stages, while others are using or starting to use significant quantities of metal. These include:

- Applications where titanium is used for its resistance to corrosion, such as chemical processing, the pulp and paper industry, marine applications, and energy production and storage
- Biomedical applications that take advantage of the metal's inertness in the human body for use in surgical implants and prosthetic devices
- Special applications that exploit unique properties such as superconductivity (alloyed with niobium) and the shape-memory effect (alloyed with nickel)
- New application areas where the metal's high specific strength is important, such as the automotive industry
- Consumer applications ranging from cameras to jewelry, musical instruments, and sports equipment

Table 2 provides a list of many uses for titanium in all of these application areas.

Aerospace Applications

High specific strength, good fatigue resistance and creep life, and good fracture toughness are characteristics that make titanium a preferred metal for aerospace applications. Figure 3 illustrates the rapid increase in use of titanium alloys in both airframe and engine applications for commercial aircraft.

Airframe Components. The earliest production application of titanium was in 1952, for the nacelles and firewalls of the Douglas DC-7 airliner. Since that time titanium and titanium alloys have been used for structural components on aircraft ranging from the Boeing 707, to the supersonic SR-71 Blackbird reconnaissance aircraft, to space satellites and missiles.

Jet Engine Components (Ref 8). Titanium fan disks (Fig. 4), turbine blades and vanes, and structurals are commonly used in aircraft turbine engines. Titanium research is an important aspect of the drive to increase engine efficiencies, and use of titanium in jet engine hot sections is expected to in-

Table 2 Applications for titanium and titanium alloys

Application area	Typical uses
Aerospace	
Airframes	Fittings, bolts, landing gear beams, wing boxes, fuselage frames, flap tracks, slat tracks, brake assemblies, fuselage panels, engine support mountings, undercarriage components, inlet guide vanes, wing pivot lugs, keels, firewalls, fairings, hydraulic tubing, deicing ductings, SPF parts
Engines	Compressor disks and blades, fan disks and blades, casings, afterburner cowlings, flange rings, spacers, bolts, hydraulic tubing, hot-air ducts, helicopter rotor hubs
Satellites, rockets	Rocket engine casings, fuel tanks
Chemical processing	Storage tanks, agitators, pumps, columns, frames, screens, mixers, valves, pressurized reactors, filters, piping and tubing, heat exchangers, electrodes and anode baskets for metal and chlorine-alkali electrolysis
Energy industry	
Power generating plants	Condensers, cooling systems, piping and tubing, turbine blades, generator retaining rings, rotor slot wedges, linings for FGD units, nuclear waste disposal
Geothermal energy	Heat exchangers, evaporators, condensers, tubes
Marine engineering	
Shipbuilding	Heat exchangers, condensers, piping and tubing, propellers, propeller and rudder shafts, data logging equipment, gyrocompasses, thruster pumps, lifeboat parts, radar components, cathodic protection anodes, hydrofoil struts
Diving equipment	Deep-sea pressure hulls, submarines (Soviet Union), submarine ball valves (United States)
Seawater desalination	Vapor heaters, condensers, thin-walled tubing
Offshore installations	Cooling equipment, condensers, heat exchangers, piping and tubing, flanges, deep-drilling riser pipes, flexible risers, desulfurizers, catalytic crackers, sour water strippers, regenerators, structural components
Biomedical engineering	Hip- and knee-joint prostheses, bone plates, screws and nails for fractures, pacemaker housings, heart valves, instruments, dentures, hearing aids, high-speed centrifugal separators for blood, wheelchairs, insulin pumps
Deep drilling	Drill pipes, riser pipes, production tubulars, casing liners, stress joints, instrument cases, wire, probes
Automotive industry	Connecting rods, valves, valve springs and retainers, crankshafts, camshafts, drive shafts, torsion bars, suspension assemblies, coil springs, clutch components, wheel hubs, exhaust systems, ball and socket joints, gears
Machine tools	Flexible tube connections, protective tubing, instrumentation and control equipment
Pulp and paper	Bleaching towers, pumps, piping and tubing
Food processing	Tanks (dairies, beverage industry), heat exchangers, components for packaging machinery
Construction	Facing and roofing, concrete reinforcement, monument refurbishment (Acropolis), anodes for cathodic protection
Superconductors	Wire rod of Ti-Nb alloys for manufacture of powerful electromagnets, rotors for superconductive generators
Fine art	Sculptures, fountain bases, ornaments, doorplates
Consumer products	
Jewelry industry	Jewelry, clocks, watches
Optics	Eyeglass frames, camera shutters
Sports equipment	Bicycle frames, tennis rackets, shafts and heads for golf clubs, mountain climbing equipment (ice screws, hooks), luges, bobsled components, horse shoes, fencing blades, target pistols
Musical instruments	Harmonica reeds, bells
Personal security and safety	Armor (cars, trucks, helicopters, fighter aircraft), helmets, bulletproof vests, protective gloves
Transportation	Driven wheelsets for high-speed trains, wheel tires
Cutting implements	Scissors, knives, pliers
Shape-memory alloys	Nickel-titanium alloys for springs and flanges
Miscellaneous	Pens, nameplates, telephone relay mechanisms, pollution-control equipment, titanium-lined vessels for salt-bath nitriding of steel products

Source: Ref 7

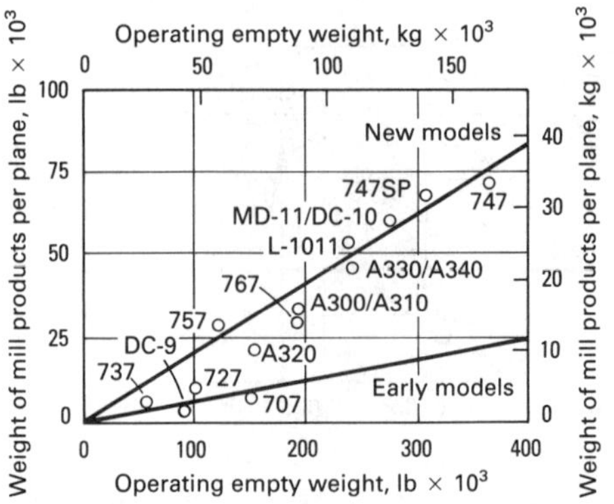

Fig. 3 Increase of titanium consumption on commercial aircraft for both airframe and engine applications. Source: Ref 3

crease as materials capable of withstanding higher temperatures are developed (Ref 9; see also the section "New Developments" in this article).

Titanium-base intermetallic compounds are another class of materials that promise increased engine thrust-to-weight ratios. These are discussed briefly in the section "New Developments" in this article and in the article "Ordered Intermetallic" in this Volume.

Use of precision titanium castings in jet engine applications such as inlet cases and compressor frames is on the rise. The article "Titanium and Titanium Alloy Castings" in this Volume contains more information on titanium casting technology.

Corrosion Applications

Commercially pure titanium is more commonly used than titanium alloys for corrosion applications, especially when high strength is not a requirement. Economics are often the deciding factor in selection of titanium for corrosion resistance. Some of the most common applications where the corrosion resistance of titanium is important are briefly described here. A comprehensive review of the corrosion behavior of titanium materials is available in the article "Corrosion of Titanium and Titanium Alloys" in Volume 13 of the 9th Edition of *Metals Handbook*.

Chemical and Petrochemical Processing. Titanium equipment including vessels, pumps, fractionation columns, and storage tanks is essential in the manufacture of certain chemicals (Ref 10). Figure 5 illustrates two different uses for titanium in the chemical processing industry, and the article "Corrosion in the Chemical Processing Industry" in Volume 13 of the 9th Edition of *Metals Handbook* contains more information on the use of titanium in the industry.

Marine Engineering. Titanium use in ship designs and for offshore oil platforms has increased steadily in the last few years. Applications include propeller and rudder shafts, thruster pumps, lifeboat parts, deep-sea pressure hulls, and submarine components (Ref 7). More information on titanium in marine applications is available in the article "Marine Corrosion" in Volume 13 of the 9th Edition of *Metals Handbook*.

Energy Production and Storage. Titanium plate-type heat exchangers, condensers, and piping and tubing are common in energy facilities using seawater for cooling. In power generating plants, titanium steam-turbine blades and generator retaining rings are used. A critical application is in the main condensers of nuclear power plants, which must remain leak-free (Ref 10). Titanium-

Fig. 4 Forged Ti-6Al-4V jet engine fan disks are 890 mm (35 in.) in diameter and weigh 249 kg (548 lb). Courtesy of Wyman-Gordon Company

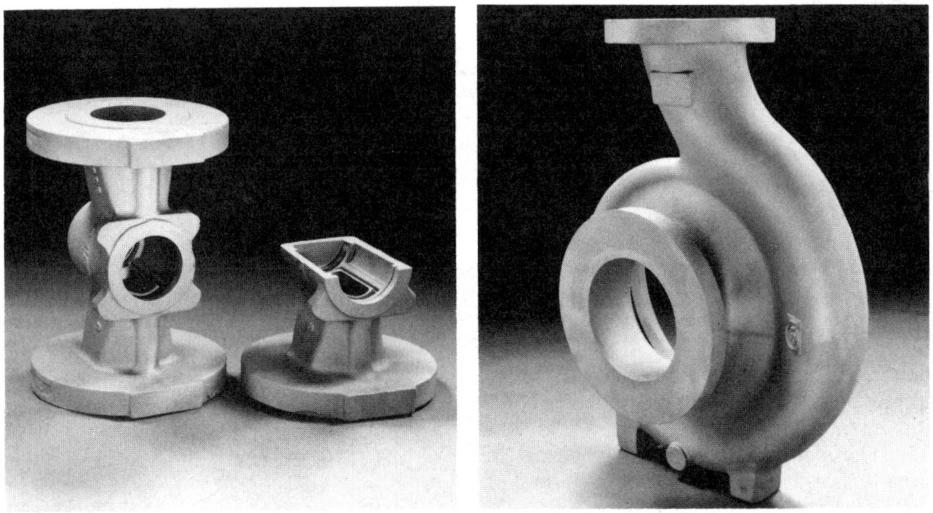

(a) (b)

Fig. 5 Two common corrosion applications for commercially pure titanium components. (a) Valve body. (b) Pump body. Both are used in the chemical processing industry. Courtesy of Oregon Metallurgical Corporation

clad steel produced by roll cladding also is used for condenser and heat-exchanger tubesheets (Ref 10).

Two relatively new uses for titanium alloys are in flue gas desulfurization (FGD) units used to scrub emissions from coal-fired power plants, and as canisters to contain low-level radioactive waste such as spent fuels from nuclear power plants. These applications are discussed in the articles "Corrosion of Emission-Control Equipment" and "Corrosion in the Nuclear Power Industry" in Volume 13 of the 9th Edition of *Metals Handbook*.

Surgical Implants and Prosthetic Devices. The value of titanium in biomedical applications lies in its inertness in the human body, that is, resistance to corrosion by body fluids. Titanium alloys are used in biomedical applications ranging from implantable pumps and components for artificial hearts, to hip and knee implants. Titanium implants with specially prepared porous surfaces promote ingrowth of bone, resulting in stronger and longer-lasting bonds between bone and implant (see the article "Corrosion of Metallic Implants and Prosthetic Devices" in Volume 13 of the 9th Edition of *Metals Handbook*).

A recent biomedical application for titanium alloys is the use of Ti-15Mo-5Zr-3Al wire for sutures and for implant fixation. Using titanium wire eliminates the galvanic corrosion that can occur when titanium implants come in contact with other implant materials such as stainless steels and cobalt-base alloys (Ref 11). Another biomedical application exploits the shape-memory effect seen in nickel-titanium alloys to create compressive stresses that promote knitting of broken bones. Shape-memory Ni-Ti alloys also have been employed experimentally to dilate blood vessels, thus increasing the flow of blood to vital organs (Ref 12).

Other Applications

The unique properties of titanium make it attractive to designers in a variety of industries. Titanium is still relatively expensive compared to steel and aluminum, but increasing use of the metal in the areas discussed in this section is expected to accelerate cost reductions, resulting in still more growth in application diversity (Ref 10).

Automotive Components (Ref 10). At least one automobile maker is investigating the use of titanium in valve systems and suspension springs; however, no manufacturer has yet used titanium on production models. Automotive parts considered to have excellent commercial potential for use of titanium are valves and valve retainers. Racing automobiles have made extensive use of titanium alloys for engine parts (Fig. 6), drive systems, and suspension components for some years, while a titanium alloy connecting rod has been used successfully by a Japanese motorcycle manufacturer. Development of low-cost, durable surface treatments is considered essential to the increased automotive use of titanium.

Architecture. Japanese architects have used titanium as a building material for some time (Ref 10). An example is the roof

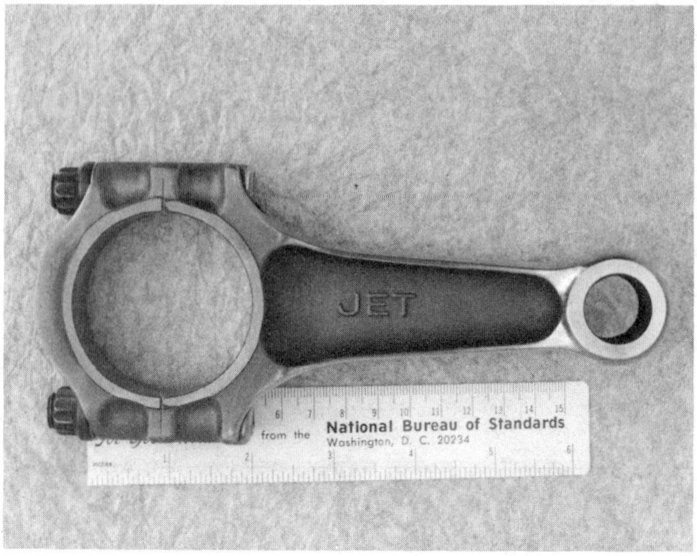

Fig. 6 Forged alloy connecting rod for a racing engine is indicative of increasing automotive applications for titanium. Component, courtesy of Jet Engineering Inc.; photograph by R.T. Kiepura, ASM INTERNATIONAL

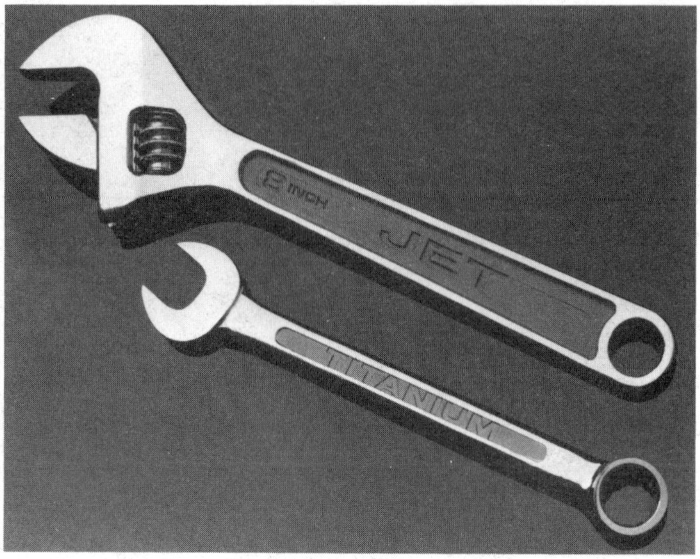

Fig. 7 Lightweight forged titanium alloy wrenches are typical of growing consumer applications for titanium. Courtesy of Jet Engineering Inc.

of the Kobe Municipal Aquarium, which used approximately 11 000 kg (24 000 lb) of titanium. Although more costly than stainless steels, titanium is considered cost-effective in structures erected in the tropics and other areas where buildings are exposed to strong, warm sea winds (Ref 10).

Consumer Goods. Interest in titanium as a material for a wide variety of consumer products is on the rise. Figure 7 shows a consumer application, and Table 2 lists numerous decorative and functional consumer applications for titanium.

New Developments

Several titanium processing and materials technologies currently in various stages of development have the potential to profoundly impact future titanium use, and for this reason they merit explanation. These include development of new sponge production and melting processes, oxide dispersion-strengthened alloys prepared using P/M techniques, titanium-base intermetallic compounds, titanium-matrix composites, SPF/DB of titanium alloys, and increased use of titanium scrap materials.

Sponge Production (Ref 13). Recent work has aimed at not only improving the efficiency of the Kroll process, but also developing new production methods. One of the recently developed methods involves reduction of sodium fluorotitanate (Na_2TiF_5) by an aluminum-zinc alloy to produce a molten titanium-zinc alloy. The zinc is then removed from this by evaporation. Another process uses electrolysis to reduce either $TiCl_4$ or titanium dioxide (TiO_2) to titanium metal.

Melting Practice. Titanium sponge is most commonly double vacuum-arc remelted with recycled scrap material and alloying elements to produce titanium alloy ingot. Electron beam and plasma cold-hearth melting are relatively new melting practices designed to minimize internal ingot defects. Longer dwell times in the liquid pool, longer solution periods, and better mixing prevent nonmetallic inclusions and unmelted refractory metals from being incorporated into the ingot.

Powder Metallurgy Alloys. Powder metallurgy production techniques such as rapid solidification processing and mechanical alloying are being used to produce titanium alloys with novel compositions that would be impossible to achieve through conventional processing (Ref 9, 14). Titanium alloys produced by these methods may contain rare earth elements such as cerium, or large quantities of β stabilizers, which tend to segregate under normal processing conditions. Oxide dispersion-strengthening, an approach widely used to enhance the properties of nickel-base alloys, is also possible using P/M techniques to incorporate dispersion-forming elements such as silicon and boron into the titanium alloy matrix.

Titanium-Base Intermetallic Compounds. Ordered intermetallics with composition near Ti_3Al (actually Ti-24Al-11Nb) have better oxidation resistance, lower density, improved creep resistance, and higher modulus than conventional titanium alloys (Ref 15). These materials have the potential to greatly increase the thrust-to-weight ratio of aircraft engines. Full-scale heats of Ti_3Al have been produced and fabricated using conventional equipment into billet, plate, and sheet. The article "Ordered Intermetallics" in this Volume contains more information on titanium aluminide intermetallics.

Titanium-Matrix Composites. Metal-matrix composites (MMCs) combine the attributes of the base (matrix) metal with those of a reinforcing phase. In the case of titanium-base MMCs, this combination of properties translates to low density with increased high-temperature strength and stiffness (Fig. 8). Titanium-matrix composites have been fabricated using a variety of techniques, including P/M processing (Ref 16, 17). More information on MMCs is available in the article "Metal-Matrix Composites" in this Volume.

Superplastic Forming and Diffusion Bonding. Superplastic forming and concurrent diffusion bonding of titanium alloy sheet

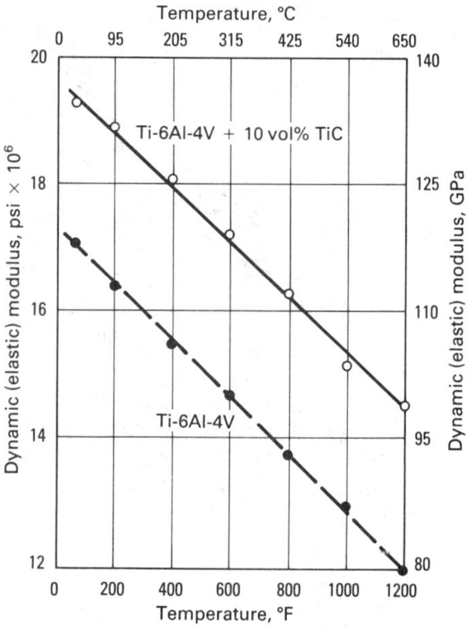

Fig. 8 High-temperature strength and stiffness of a titanium MMC compared to conventional alloy Ti-6Al-4V. Produced using powder metallurgy techniques, the MMC consists of a Ti-6Al-4V matrix reinforced with 10% titanium carbide (TiC) particles. Source: Ref 16

components is a technology that has moved out of the laboratory and into commercial production (Ref 18, 19). The process has the potential to drastically reduce the number of parts and fasteners needed in airframe structures and other complex components. More information on SPF/DB of titanium also is available in the articles "Forming of Titanium and Titanium Alloys" and "Superplastic Sheet Forming" in Volume 14 of the 9th Edition of *Metals Handbook*.

Recycling of Titanium Scrap. As the titanium industry has matured, the use of recycled material has increased. In recent years even machine turnings and chips have been approved for recycling, and U.S. titanium producers used nearly 18 million kg (40 million pounds) of titanium scrap in 1988 (Ref 20). More information on recycling of titanium alloys is available in the article "Recycling of Nonferrous Alloys" in this Volume.

REFERENCES

1. W.J. Kroll, How Commercial Titanium and Zirconium Were Born, *J. Franklin Inst.*, Vol 260, Sept 1955, p 169-192
2. *Titanium: The Industry, Its Future, Its Equities*, F.S. Smithers and Company, 1957, p 7, 33-67
3. H.B. Bomberger, F.H. Froes, and P.H. Morton, Titanium—A Historical Perspective, in *Titanium Technology: Present Status and Future Trends*, F.H. Froes, D. Eylon, and H.B. Bomberger, Ed., Titanium Development Association, 1985, p 3-17
4. E.W. Collings, *The Physical Metallurgy of Titanium Alloys*, American Society for Metals, 1984, p 2
5. M.J. Donachie, Jr., *Titanium: A Technical Guide*, ASM INTERNATIONAL, 1988, p 28
6. O.E. Nelson, The Product Life Cycle of Titanium, paper presented at the Annual Conference of the Titanium Development Association, Tucson, AZ, 13 Oct 1989
7. K.-H. Kramer, Titanium Applications—A Critical Review, in *Proceedings of the Sixth World Conference on Titanium*, P. Lacombe, R. Tricot, and G. Beranger, Ed., Societe Francaise de Metallurgie, 1988, p 521
8. Y. Honnorat, Titanium Alloys Use in Turbojet Engines, in *Proceedings of the Sixth World Conference on Titanium*, P. Lacombe, R. Tricot, and G. Beranger, Ed., Societe Francaise de Metallurgie, 1988, p 365
9. R. Sundaresan, A.G. Jackson, and F.H. Froes, Dispersion Strengthened Titanium Alloys Through Mechanical Alloying, in *Proceedings of the Sixth World Conference on Titanium*, P. Lacombe, R. Tricot, and G. Beranger, Ed., Societe Francaise de Metallurgie, 1988, p 855
10. Y. Fukuhara, Nonaerospace Applications of Titanium, in *Proceedings of the Sixth World Conference on Titanium*, P. Lacombe, R. Tricot, and G. Beranger, Ed., Societe Francaise de Metallurgie, 1988, p 381
11. Y. Ito, Y. Sasaki, and T. Shinke, Beta Titanium Wire for Surgical Implant Uses, in *Proceedings of the Sixth World Conference on Titanium*, P. Lacombe, R. Tricot, and G. Beranger, Ed., Societe Francaise de Metallurgie, 1988, p 405
12. N.I. Koryagin, New Trends in Titanium Application, in *Proceedings of the Sixth World Conference on Titanium*, P. Lacombe, R. Tricot, and G. Beranger, Ed., Societe Francaise de Metallurgie, 1988, p 49
13. T. Tanaka, New Development in Titanium Elaboration—Sponge, Melting and Casting, in *Proceedings of the Sixth World Conference on Titanium*, P. Lacombe, R. Tricot, and G. Beranger, Ed., Societe Francaise de Metallurgie, 1988, p 11
14. R. Sundaresan and F.H. Froes, Development of the Titanium-Magnesium Alloy System Through Mechanical Alloying, in *Proceedings of the Sixth World Conference on Titanium*, P. Lacombe, R. Tricot, and G. Beranger, Ed., Societe Francaise de Metallurgie, 1988, p 931
15. J.D. Destefani, Advances in Intermetallics, *Adv. Mater. Process.*, Feb 1989, p 37-41
16. S. Abkowitz and P. Weihrauch, Trimming the Cost of MMCs, *Adv. Mater. Process.*, July 1989, p 31-34
17. C.M. Cooke, D. Eylon, and F.H. Froes, Development of Rapidly Solidified Titanium Matrix Composites, in *Proceedings of the Sixth World Conference on Titanium*, P. Lacombe, R. Tricot, and G. Beranger, Ed., Societe Francaise de Metallurgie, 1988, p 913
18. P.-J. Winkler, Recent Advances in Superplasticity and Superplastic Forming of Titanium Alloys, in *Proceedings of the Sixth World Conference on Titanium*, P. Lacombe, R. Tricot, and G. Beranger, Ed., Societe Francaise de Metallurgie, 1988, p 1135
19. E. Tuegel, M.O. Pruitt, and L.D. Hefti, SPF/DB Takes Off, *Adv. Mater. Process.*, July 1989, p 36-41
20. *Titanium 1988 Statistical Review*, Titanium Development Association, 1989, p 5

Wrought Titanium and Titanium Alloys

S. Lampman, ASM INTERNATIONAL

THE WROUGHT product forms of titanium and titanium-base alloys, which include forgings and the typical mill products, constitute (on a weight basis) more than 70% of the market in titanium and titanium-alloy production. Various specifications for wrought titanium-base products are listed in Table 1. The wrought products are the most readily available product form of titanium-base materials, although cast and powder metallurgy (P/M) products are also available for applications that require complex shapes or the use of P/M techniques to obtain microstructures not achievable by conventional ingot metallurgy. Powder metallurgy of titanium has not gained wide acceptance and is restricted to space and missile applications. Cast and P/M titanium-base products are discussed in the subsequent articles in this Volume.

The primary reasons for using titanium-base products stem from the outstanding corrosion resistance of titanium and/or its useful combination of low density (≈4.5 g/cm^3, or 0.16 lb/in.^{3}) and high strength (minimum 0.2% yield strengths vary from 480 MPa, or 70 ksi, for some grades of commercial titanium to about 1100 MPa, or 160 ksi, for structural titanium alloy products and over 1725 MPa, or 250 ksi, for special forms such as wires and springs). Some titanium alloys (especially the low-interstitial alpha alloys) are also useful in subzero and cryogenic applications because these alpha alloys do not exhibit a ductile-brittle transition.

Another important characteristic of titanium-base materials is the reversible transformation (or allotropy) of the crystal structure from an alpha (α) (hexagonal close-packed) structure to a beta (β) (body-centered cubic) structure when the temperatures exceed a certain level. This allotropic behavior, which depends on the type and amount of alloy contents, allows complex variations in microstructure and more diverse strengthening opportunities than those of other nonferrous alloys such as copper or aluminum. This diversity of microstructure and properties depends not only on alloy additions but also on thermomechanical processing. By varying thermal or mechanical processing, or both, a broad range of properties can be produced in titanium alloys.

The elevated-temperature strength and creep resistance of titanium-base materials (along with the pickup of interstitial impurities due to the chemical reactivity of titanium) limits the elevated-temperature application of wrought and cast products to about 540 °C (1000 °F) or perhaps 600 °C (1100 °F) in some cases. For higher temperatures, Ti-aluminide products are an active area of research and development (see the article "Ordered Intermetallics" in this Volume).

Commercially Pure Titanium

Pure titanium wrought products, which have minimum titanium contents ranging from about 98.635 to 99.5 wt% (Table 2), are used primarily for corrosion resistance. Titanium products are also useful in applications requiring high ductility for fabrication but relatively low strength (Table 2) in service.

Corrosion Resistance and Chemical Reactivity. Although titanium is a highly reactive metal, titanium also has an extremely high affinity for oxygen and thus forms a very stable and highly adherent protective oxide film on its surface. This oxide film, which forms spontaneously and instantly when fresh metal surfaces are exposed to air and/or moisture, provides the excellent corrosion resistance of titanium. However, anhydrous conditions in the absence of a source of oxygen may result in titanium corrosion, because the protective film may not be regenerated if damaged. This is particularly true of crevice corrosion. Titanium and titanium alloys may be subject to localized attack in tight crevices exposed to hot (>70 °C, or 160 °F) chloride, bromide, iodide, fluoride, or sulfate-containing solutions. Crevices can stem from adhering process stream deposits or scales, metal-to-metal joints (for example, poor weld joint design or tube-to-tubesheet joints), and gasket-to-metal flange and other seal joints. The mechanism for crevice corrosion of titanium is similar to that for stainless steels, in which oxygen-depleted reducing acid conditions develop within tight crevices.

General corrosion rates for unalloyed titanium (99.2 wt% Ti with traces of oxygen) in selected media are given in Table 3. These data should be used only as a guideline for general performance. Rates may vary depending on changes in medium chemistry, temperature, length of exposure, and other factors. Also, total suitability of an alloy cannot be assumed from these values alone, because other forms of corrosion, such as localized attack, may be limiting. These and other factors affecting the corrosion of titanium are discussed in more detail in *Corrosion*, Volume 13 of the 9th Edition of *Metals Handbook*.

Precautions in Use. Hydrogen embrittlement of titanium can occur in pickling solutions (or other hydrogenating solutions) at room temperature and at elevated temperatures during air exposure or in exposures to reducing atmospheres. Nonetheless, titanium alloys are widely used in hydrogen-containing environments and under conditions in which galvanic couples or cathodic charging (impressed current) causes hydrogen to be evolved on metal surfaces. Although hydrogen embrittlement has been observed, traces of moisture or oxygen in hydrogen gas containing environments effectively form the protective oxide film, thus avoiding or limiting hydrogen uptake (Ref 2-5). On the other hand, anhydrous hydrogen gas atmospheres may lead to absorption, particularly as temperatures and pressures increase. Hydrogen embrittlement can also occur at relatively low hydrogen levels due to the hydrogen in the material in the presence of a stress riser under certain conditions.

Elevated temperature atmospheric exposure also results in oxygen and nitrogen contamination that increases in severity with increasing temperature and time of exposure. Violent oxidation reactions can occur between titanium and liquid oxygen or between titanium and red fuming nitric acid. Titanium alloys exhibit good corrosion resistance to white fuming nitric acids.

Crystal Structure. Pure titanium at room temperature has an alpha (hexagonal close-packed) crystal structure, which transforms to a beta (body-centered cubic) structure at

Table 1 Various specifications for wrought products of titanium and titanium alloys

Issuing agency or the name of the standard or specification	Specification or standard for: Plate, sheet, or strip	Forgings	Bar or billet	Rod or wire	Pipe or tubes	Extrusions	Other
American							
Aerospace Material Specifications (AMS) issued by Society of Automobile Engineers (SAE)	4900–4902, 4905–4919	4920, 4921, 4924, 4928, 4930, 4965–4967, 4970, 4971, 4973, 4974, 4976, 4978, 4979, 4981, 4983, 4984, 4986, 4987	4921, 4924, 4926, 4928, 4930, 4965, 4967, 4970–4972, 4974, 4975, 4977–4981, 4995, 4996	Welding wire, 4951, 4953, 4954–4956; Other, 4959 and 4982	4941–4944	Flash-welded ring extrusions, 4933–4936	Bolts and screws, 7640; Spring wire, 4959; and rings in listings for extrusions and bars
American Society for Testing and Materials (ASTM)	B 265	B 381 and F 620	Bar and billet, B 348, F 67	...	B 337 and B 338	...	Nuts, F 467; Bolts, F 468; Surgical implants, F 67, F 136 and F 620
SAE (see also AMS listings)(a)	MAM 2242		MAM 2241, MAM 2245	MAM 2245	...	...	Shapes, MAM 2245
Military	MIL-T-9046	MIL-T-24585, MIL-T-9047, MIL-F-83142 (premium quality forgings)	MIL-T-9047	MIL-T-24585 (rod), MIL-R-81588 (welding rod and wire)	...	MIL-T-81556 (aircraft quality bar and shape extrusions)	MIL-T-40635: high-strength wrought Ti alloys for critical components
Asian							
Japanese Industrial Standards (JIS)	H4600	...	H4650	Rod, H4650; Wire, H4670; Welding wire, H3331	H4630 and H4631	...	...
Japanese Titanium Society industrial standards (TIS)(b)	TIS 7912	TIS 7607	TIS 7915	Rod, TIS 7915; Wire, TIS 7916	TIS 7913, TIS 7914	...	...
South Korean standards	D 5577	...	D5604	Rod, D 5604; Wire, D 5576; Rod and wire, D 5577; Welding wire, D 7030	D 5574, D 5575	...	...
European							
Association Européenne des Constructeurs de Matériel Aérospatial (AECMA)	prEN2517, prEN2525–prEN2528	prEN2520, prEN2522, prEN2524, prEN2531	prEN2518, prEN2519, prEN2521, prEN2530, prEN2532–prEN2534	...	...	...	...
Deutsche Industrie Normen (DIN) standards (Germany)	DIN 17860, V LN 65039, and LN 9293	DIN 17864, V LN 65040	DIN 17862, V LN 65040	Wire, DIN 17863	...	...	Bolts, LN 65047, Joining elements, LN 65072
French	AIR 9182 (sheet)	AIR 9183	AIR 9183	...	...	...	Bolts and screws, AIR 9184;
British Standards Institution	2TA.1, 2TA.2, 2TA.6, 2TA.10, 2TA.21, TA.52, TA.56–TA.59	2TA.4, 2TA.5, 2TA.8, 2TA.9, 2TA.12, TA.13, 2TA.23, TA.24, TA.39, TA.41–TA.44, TA.47, TA.48, TA.50, TA.51, TA.54, TA.55	Bar and section: 2TA.3, 2TA.7, 2TA.11, 2TA.22, TA.38, TA.45, TA.46, TA.49, TA.53	Wire for fasteners, TA.28	...	...	Fasteners, TA.28; Surgical implants, BS 3531 (part 1, 2)

(a) MAM, metric aerospace recommended practice. (b) Except for the TIS 7607 standard for forgings, the listed TIS are for titanium-palladium alloys only. Source: Adapted from Ref 1

a temperature of about 885 °C (1625 °F). This transformation temperature can be raised or lowered depending on the type and amount of impurities or alloying additions. The addition of alloying elements also divides the single temperature for equilibrium transformation into two temperatures—the alpha transus, below which the alloy is all-alpha, and the beta transus, above which the alloy is all-beta. Between these temperatures, both alpha and beta are present. Depending on the level of impurities, the

Table 2 Comparison of various specifications for commercially pure titanium mill products

Designation	Chemical composition, % max: C	H	O	N	Fe	Other	Total others	Tensile properties(a): Ultimate strength, MPa	Ultimate strength, ksi	Yield strength, MPa	Yield strength, ksi	Minimum elongation, %
JIS Class 1	...	0.015	0.15	0.05	0.20	...	...	275–410	40–60	165(b)	24(b)	27
ASTM grade 1 (UNS R50250)	0.10	(c)	0.18	0.03	0.20	...	...	240	35	170–310	25–45	24
DIN 3.7025	0.08	0.013	0.10	0.05	0.20	...	...	295–410	43–60	175	25.5	30
GOST BT1-00	0.05	0.008	0.10	0.04	0.20	...	0.10 max	295	43	...	...	20
BS 19–27t/in.^{2}	...	0.0125	...	...	0.20	...	...	285–410	41–60	195	28	25
JIS Class 2	...	0.015	0.20	0.05	0.25	...	...	343–510	50–74	215(b)	31(b)	23
ASTM grade 2 (UNS R50400)	0.10	(c)	0.25	0.03	0.30	...	...	343	50	275–410	40–60	20
DIN 3.7035	0.08	0.013	0.20	0.06	0.25	...	...	372	54	245	35.5	22
GOST BT1-0	0.07	0.010	0.20	0.04	0.30	...	0.30 max	390–540	57–78	...	...	20
BS 25-35t/in.^{2}	...	0.0125	...	...	0.20	...	...	382–530	55–77	285	41	22
JIS Class 3	...	0.015	0.30	0.07	0.30	...	...	480–617	70–90	343(b)	50(b)	18
ASTM grade 3 (UNS R50500)	0.10	(c)	0.35	0.05	0.30	...	...	440	64	377–520	55–75	18
ASTM grade 4 (UNS R50700)	0.10	(c)	0.40	0.05	0.50	...	...	550	80	480	70	20
DIN 3.7055	0.10	0.013	0.25	0.06	0.30	...	...	460–590	67–85	323	47	18
ASTM grade 7 (UNS R52400)	0.10	(c)	0.25	0.03	0.30	0.12–0.25 Pd	...	343	50	275–410	40–60	20
ASTM grade 11 (UNS R52250)	0.10	(c)	0.18	0.03	0.20	0.12–0.25 Pd	...	240	35	170–310	24.5–45	24
ASTM grade 12 (UNS R53400)	0.10	0.015	0.25	0.03	0.30	0.2–0.4 Mo, 0.6–0.9 Ni	...	480	70	380	55	12

(a) Unless a range is specified, all listed values are minimums. (b) Only for sheet, plate, and coil. (c) Hydrogen limits vary according to product form as follows: 0.015H (sheet), 0.0125H (bar), and 0.0100 H (billet). Source: Adapted from Ref 1

beta transus is about 910 ± 15 °C (1675 ± 25 °F) for commercially pure titanium with 0.25 wt% O_2 max and 945 ± 15 °C (1735 ± 25 °F) with 0.40 wt% O_2 max. For the various ASTM grades of commercially pure titanium, typical transus temperatures (with an uncertainty of about ±15 °C, or ±25 °F) are:

Designation	Typical β transus, °C	Typical β transus, °F	Typical α transus, °C	Typical α transus, °F
ASTM grade 1	888	1630	880	1620
ASTM grade 2	913	1675	890	1635
ASTM grade 3	920	1685	900	1650
ASTM grade 4	950	1740	905	1660
ASTM grade 7	913	1675	890	1635
ASTM grade 12	890	1635	...	...

Typical unit cell parameters for an alpha crystal structure at 25 °C (77 °F) are:

$$a = 0.2950 \text{ nm}$$
$$c = 0.4683 \text{ nm}$$

Impurity elements (commonly oxygen, nitrogen, carbon, and iron) influence unit cell dimensions. The typical unit cell parameter for the beta structure is 0.329 nm at 900 °C (1650 °F).

The microstructure of unalloyed titanium at room temperature is typically a 100% alpha-crystal structure. As amounts of impurity elements increase (primarily iron), small but increasing amounts of beta are observed metallographically, usually at alpha grain boundaries. Annealed unalloyed titanium may have an equiaxed or acicular alpha microstructure. Acicular alpha occurs during beta-to-alpha transformation on cooling through the transformation temperature range. Platelet width decreases with cooling rate. Equiaxed alpha can only be produced by recrystallization of material that has been extensively worked in the alpha phase. The presence of acicular alpha, therefore, is an indication that the material has been heated to a temperature above the beta transus. A beta structure cannot be retained at low temperatures in unalloyed titanium, except in small quantities in materials containing beta stabilizing contaminants such as iron.

Effect of Impurities on Mechanical Properties. Besides the effect on transformation temperatures and lattice parameters, impurities also have important effects on the mechanical properties of titanium. Residual elements such as carbon, nitrogen, silicon, and iron raise the strength and lower the ductility of titanium products. The effect of carbon, oxygen, and nitrogen is shown in Fig. 1.

Basically, oxygen and iron contents determine strength levels of commercially pure titanium. In higher strength grades, oxygen and iron are intentionally added to the residual amounts already in the sponge to provide extra strength. On the other hand, carbon and nitrogen usually are held to minimum residual levels to avoid embrittlement.

When good ductility and toughness are desired, the extra-low interstitial (ELI) grades are used. In ELI grades, carbon, nitrogen, oxygen, and iron must be held to acceptably low levels because they lower the ductility of the final product (see, for example, the effect of carbon, oxygen, and nitrogen in Fig. 1).

The titanium for ingot production may be either titanium sponge or reclaimed scrap. In either case, stringent specifications must be met for control of ingot composition. Most important are the hard, brittle, and refractory titanium oxide, titanium nitride, or complex titanium oxynitride particles that, if retained through subsequent melting operations, could act as crack initiation sites in the final product.

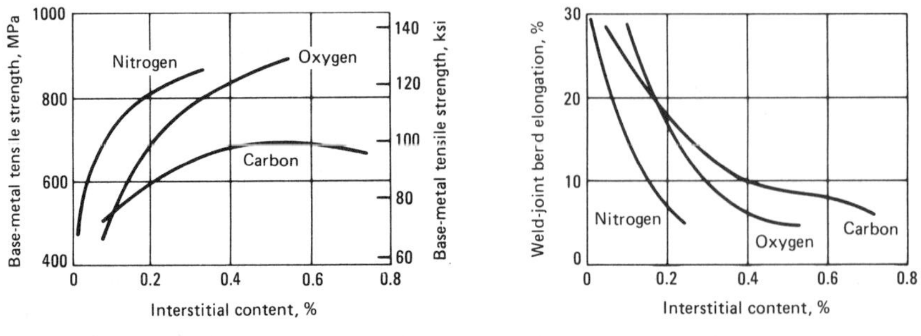

Fig. 1 Effects of interstitial-element content on strength and ductility of unalloyed titanium

Table 3 Corrosion rates for unalloyed titanium (99.2% Ti) in selected media

Corrodent	Concentration, %	Temperature °C	Temperature °F	Corrosion rate μm/yr	Corrosion rate mils/yr
Acetic acid	5, 25, 75	100	212	nil	nil
	50, 99.5	100	212	0.25	0.01
Aluminum chloride, aerated	25	25	77	<2	<0.1
	5, 10	60	140	<2.5	<0.1
	10	100	212	<2.5	<0.1
Ammonium chloride	1, 10, saturated	20–100	68–212	<13	<0.5
Ammonium sulfate	5	25	77	nil	nil
	Saturated + 5% H_2SO_4	25	77	25	1
Aqua regia (3:1)	100	25	77	nil	nil
	100	77	170	890	35
Calcium chloride	28	Boiling		nil	nil
	5, 10, 20	100	212	<25	<1
Calcium hypochlorite	Saturated	25	77	nil	nil
	2, 6	100	212	1.3	0.05
Chlorine					
Saturated with H_2O	...	25	77	125	5
More than 0.013% H_2O	...	79	175	nil	nil
Dry	...	32	90	Rapid	Rapid
Copper nitrate	Saturated	25	77	nil	nil
Cupric chloride	20, 40	Boiling		nil	nil
Ferric chloride	10, 20	25	77	nil	nil
	5	60	140	nil	nil
	10–40	Boiling		nil	nil
	30	93	200	nil	nil
	10–30	100	212	<13	<0.5
	5 + 10% NaCl	100	212	<13	<0.5
Ferric sulfate	10	25	77	nil	nil
Ferrous sulfate	Saturated	25	77	nil	nil
Hydrochloric acid	5	35	95	<50	<2
	10	35	95	1000	40
	20	35	95	4400	175
Hydrochloric acid plus copper sulfate	10 + 0.05	65	150	<50	<2
	10 + 0.1	65	150	<25	<1
	10 + 0.2, 0.25, or 0.5	65	150	nil	nil
	10 + 1	65	150	<25	<1
Hydrogen sulfide	Saturated water	25	77	<125	<5
Lactic acid	10–85	100	212	<125	<5
	10–100	Boiling		<125	<5
Lead acetate	Saturated	25	77	nil	nil
Magnesium chloride	5–40	Boiling		nil	nil
	5–40	100	212	<125	<5
Nitric acid	5	100	212	<25	<1
	10	100	212	<50	<2
	40–50, 69.5	100	212	<25	<1
	65	175	347	<125	<5
	40	200	392	<1250	<50
	70	270	518	<1250	<50
	20	290	554	300	12
Phosphoric acid	5–30	25	77	<50	<2
	35–85	25	77	<1250	<50
	85	38	100	1000	40
	5–35	60	140	<1250	<50
	10	79	175	1250	50
	5	100	212	<1250	<50
Seawater	...	25	77	nil	nil
Silver nitrate	50	25	77	nil	nil
Sulfuric acid	15	25	77	nil	nil
	1	60	140	nil	nil
	3	60	140	1.3	0.05
	5	60	140	730	29
Zinc chloride	Saturated	25	77	nil	nil
	10	Boiling		nil	nil
	20	100	212	<125	<5

Titanium sponge is manufactured by first chlorinating the ore (most commonly rutile or synthetic rutile) and then reducing the resulting $TiCl_4$ with either sodium or magnesium metal. Sodium-reduced sponge is leached with acid to remove the NaCl by-product of reduction. Magnesium-reduced sponge may be leached, inert-gas swept, or vacuum distilled to remove the excess $MgCl_2$ by-product. Vacuum distilling results in lower residual levels of magnesium, hydrogen, and chlorine. Modern melting techniques remove volatile substances from sponge, so that ingot of high quality can be produced regardless of which method is used for production of sponge. Electrolytic methods are being used to produce a very high purity titanium sponge on a pilot-plant scale.

Reclaimed scrap makes production of ingot titanium more economical than production solely from sponge. If properly controlled, addition of scrap (commonly referred to as revert) is fully acceptable and can be used even in materials for critical structural applications, such as rotating components for jet engines.

All forms of scrap can be remelted—machining chips, cut sheet, trim stock, and chunks. To be utilized properly, scrap must be thoroughly cleaned and carefully sorted by alloy and by purity before being remelted. During cleaning, surface scale must be removed, because adding titanium scale to the melt could produce refractory inclusions or excessive porosity in the ingot. Machining chips from fabricators who use carbide tools are acceptable for remelting only if all carbide particles adhering to the chips are removed; otherwise, hard high-density inclusions could result. Improper segregation of alloy revert could produce off-composition alloys and could potentially degrade the properties of the resulting metal.

Melting Practice for Ingot Production. Double melting is considered necessary for all applications to ensure an acceptable degree of homogeneity in the resulting product. Triple melting is used to achieve better uniformity. Triple melting also reduces oxygen-rich or nitrogen-rich inclusions in the microstructure to a very low level by providing an additional melting operation to dissolve them.

Most titanium and titanium alloy ingot is melted twice in an electric-arc furnace under vacuum—a procedure known as the double consumable-electrode vacuum-melting process. In this two-stage process, titanium sponge, revert, and alloy additions are welded together and then are melted to form ingot. Ingots from the first melt are used as the consumable electrodes for second-stage melting. Processes other than consumable-electrode arc melting are used in some instances for first-stage melting of ingot for noncritical applications. Usually, all melting is done under vacuum, but in any event the final stage of melting must be done by the consumable-electrode vacuum-arc process. Newer hearth melting technologies utilizing electron beams or plasma as a heat source are casting commercially pure slabs in ore melting operations.

Segregation and other compositional variations directly affect the final properties of mill products. Melting technique alone does not account for all segregation and compositional variations and thus cannot be correlated with final properties.

Melting in a vacuum reduces the hydrogen content of titanium and essentially removes other volatiles. This tends to result in high purity in the cast ingot. However, anomalous operating factors such as air

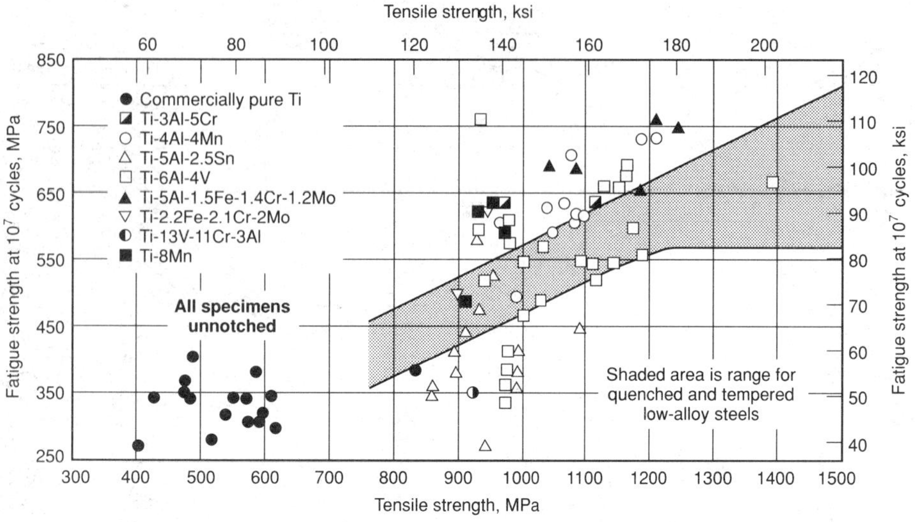

Fig. 2 Fatigue data for titanium and titanium alloys, compared with quenched and tempered low-alloy steels

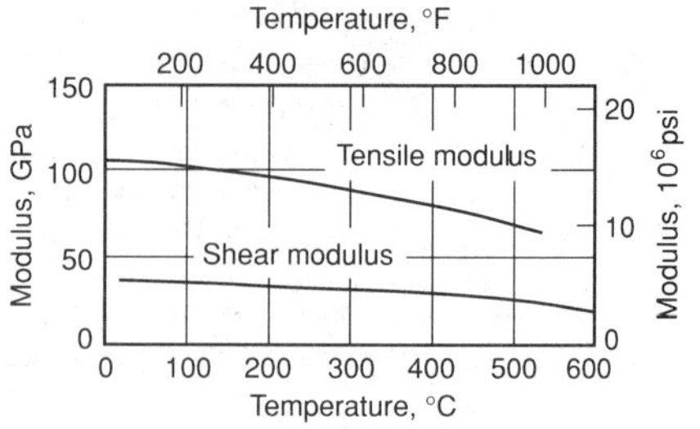

Fig. 3 Variation of elastic modulus with temperature for unalloyed titanium (ASTM grade 4)

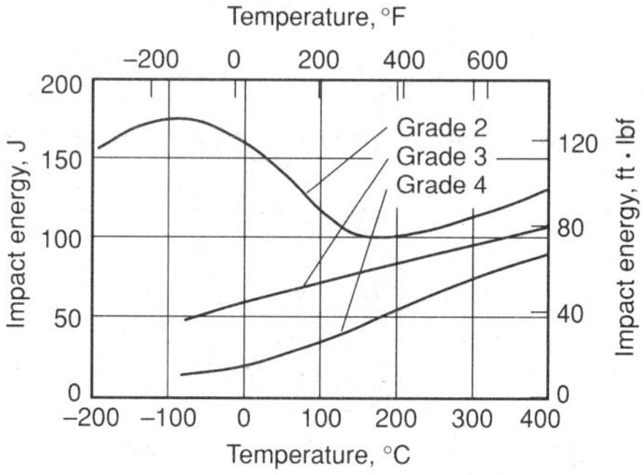

Fig. 4 Charpy V-notch impact strength of unalloyed titanium (ASTM grades 2, 3, and 4) at low temperatures

leaks, water leaks, arc-outs, or large variations in power level affect both soundness and homogeneity of the final product.

Still another factor is ingot size. Normally, ingots are 650 to 900 mm (26 to 36 in.) in diameter and weigh 3600 to 9000 kg (8000 to 20 000 lb). Larger ingots are economically advantageous to use and are important in obtaining refined macrostructures and microstructures in very large sections, such as billets with diameters of 400 mm (16 in.) or greater. Ingots up to 1065 mm (42 in.) in diameter and weighing more than 13.5 Mg (30 000 lb) have been melted successfully, but there appear to be limitations on the improvements that can be achieved by producing large ingots due to increasing tendency for segregation with increasing ingot size.

Segregation in titanium ingot must be controlled because it leads to several different types of imperfections that cannot be readily eliminated by homogenizing heat treatments or combinations of heat treatment and primary mill processing. In aircraft-grade titanium, type I and type II imperfections are not acceptable because they degrade critical design properties.

Type I imperfections, usually called high interstitial defects, are regions of interstitially stabilized alpha phase that have substantially higher hardness and lower ductility than the surrounding material and also exhibit a higher beta transus temperature. They arise from very high nitrogen or oxygen concentrations in sponge, master alloy, or revert. Type I imperfections frequently, but not always, are associated with voids or cracks. Although type I imperfections sometimes are referred to as low-density inclusions (LDI's), they often are of higher density than is normal for the alloy.

Type II imperfections, sometimes called high-aluminum defects, are abnormally stabilized alpha-phase areas that may extend across several beta grains. Type II imperfections are caused by segregation of metallic alpha stabilizers, such as aluminum and contain an excessively high proportion of primary (untransformed) alpha having a microhardness only slightly higher than that of the adjacent matrix. Type II imperfections sometimes are accompanied by adjacent stringers of beta—areas low in both aluminum content and hardness. This condition is generally associated with closed solidification pipe into which alloy constituents of high vapor pressure migrate, only to be incorporated into the microstructure during primary mill fabrication. Stringers normally occur in the top portions of ingots and can be detected by macroetching or anodized blue etching. Material containing stringers usually must undergo metallographic review to ensure that the indications revealed by etching are not artifacts.

Properties of Commercially Pure Titanium. The minimum tensile properties of titanium mill products are listed in Table 2. In terms of tensile and fatigue strength, commercially pure titanium is not as strong as steel or titanium alloys (Fig. 2). Titanium has an intermediate modulus of elasticity (Fig. 3), which can be influenced by texture.

Impact Toughness. Commercially pure titanium has impact strengths comparable to that of quenched and tempered low-alloy steel. Titanium may even exhibit an increase in toughness at low temperatures (Fig. 4), depending on the control of interstitial impurities and brittle refractory constituents.

Creep Behavior. Whether yield strength or creep strength for a given maximum allowable deformation is the significant selection criterion depends on which is lower at the service temperature in question. Between 200 and 315 °C (400 and 600 °F), the deformation of titanium (and some titanium alloys with an alpha or near-alpha crystal structure) loaded to the yield point does not increase with time. Thus, creep strength is seldom a factor in this range. Above 315 °C (600 °F), creep strength becomes an important selection criterion. Typical creep behavior of titanium is shown in Fig. 5.

Physical properties of titanium are described in the article "Properties of Pure Metals" in this Volume. Of the various physical properties shown in Fig. 6, titanium is characterized by a somewhat low thermal conductivity. However, even though titanium has a low thermal conductivity compared to that of other metals, its heat-transfer rate is greater than that of most copper-base alloys. The reason for this has to do with favorable surface film characteristics and lack of corrosion. The major factor in heat transfer relates to material thickness, corrosion resistance, and surface films, not the thermal conductivity of the metal. In this regard, titanium has the following advantages:

- Good strength
- Resistance to erosion and erosion-corrosion
- Very thin, conductive oxide surface film
- Hard, smooth surface that limits adhesion of foreign materials
- Surface promotes dropwise condensation

Consequently, titanium is a useful material in heat exchangers (Fig. 7).

Commercially pure titanium with minor alloy contents include various titanium-pal-

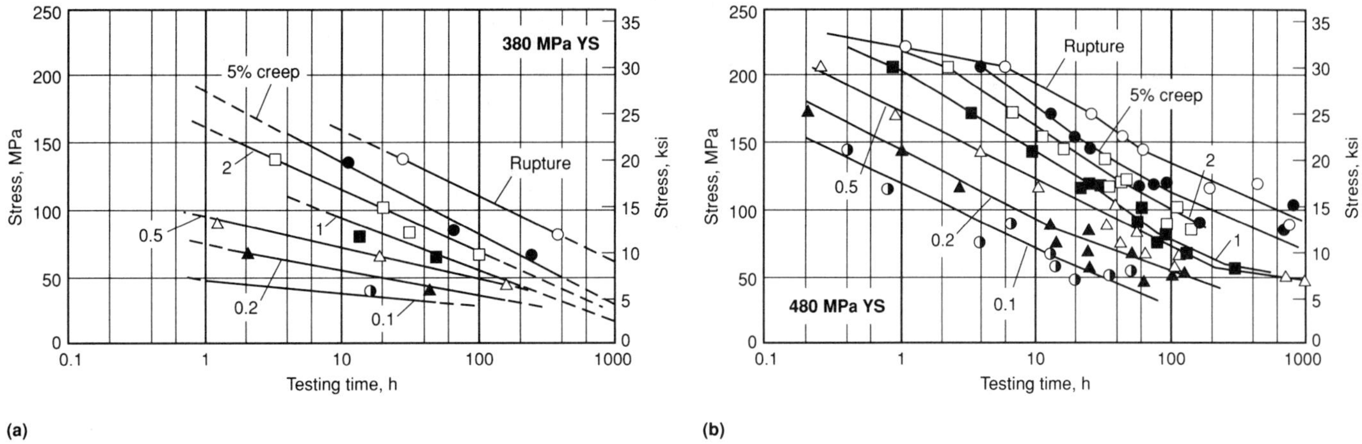

Fig. 5 Creep characteristics at 425 °C (800 °F) for mill-annealed titanium (99.0% Ti) with a 0.2% yield strength (YS) of (a) 380 MPa (55 ksi) and (b) 480 MPa (70 ksi)

ladium grades and alloy Ti-0.3Mo-0.8Ni (ASTM grade 12 or UNS R53400). The alloy contents allow improvements in corrosion resistance and/or strength.

Titanium-palladium alloys with nominal palladium contents of about 0.2% Pd (Table 2) are used in applications requiring excellent corrosion resistance in chemical processing or storage applications where the media is mildly reducing or fluctuates between oxidizing and reducing. The palladium-containing alloys extend the range of titanium application in hydrochloric, phosphoric, and sulfuric acid solutions (Table 4). Characteristics of good fabricability, weldability, and strength level are similar to those of corresponding unalloyed titanium grades.

Palladium additions of less than specified minimums are less effective in promoting an improved corrosion resistance. Excess palladium (above specified range) is not cost effective. Only alpha soluble amounts of palladium are added to make titanium-palladium alloys; therefore, microstructures are essentially the same as for equivalent grades of unalloyed titanium. Titanium-palladium intermetallic compounds formed in this system have not been reported to occur with normal heat treatments.

Alloy Ti-0.3Mo-0.8Ni (UNS R53400, or ASTM grade 12) has applications similar to those for unalloyed titanium but has better strength (Fig. 8) and corrosion resistance (Fig. 9). However, the corrosion resistance of this alloy is not as good as the titanium-palladium alloys. The ASTM grade 12 alloy is particularly resistant to crevice corrosion (Fig. 10) in hot brines (see the section "Corrosion Resistance and Chemical Reactivity" in this article for a brief discussion on crevice corrosion). The microstructure of R53400 is either equiaxed or acicular alpha with minor amounts of beta. Acicular alpha microstructures are found primarily in welds or heat-affected zones.

In a series of crevice corrosion tests, Ti-0.3Mo-0.8Ni was completely resistant in 500-h exposures to the following boiling solutions: saturated $ZnCl_2$ at pH of 3.0; 10% $AlCl_3$; $MgCl_2$ at pH of 4.2; 10% NH_4Cl at pH of 4.1; saturated NaCl, and saturated NaCl + Cl_2, both at pH of 1.0; and 10% Na_2SO_4 at pH of 1.0. In a similar test in boiling 10% $FeCl_3$, crevice corrosion was observed in metal-to-Teflon crevices after 500 h. Ti-0.3Mo-0.8Ni also exhibits the following typical corrosion rates:

Environment	Corrosion rate mm/yr	Corrosion rate mils/yr
Wet Cl_2 gas	0.00089	0.035
5% NaOCl + 2% NaCl + 4% NaOH(a)	0.06	2.4
70% $ZnCl_2$	0.005–0.0075	0.2–0.3
50% citric acid	0.013	0.5
10% sulfamic acid	11.6	455
45% formic acid	nil	nil
88–90% formic acid	0–0.56	0–22
90% formic acid(b)	0.56	2.2
10% oxalic acid	104	4100

(a) No crevice corrosion in metal-to-metal or metal-to-Teflon crevices. (b) Anodized specimens

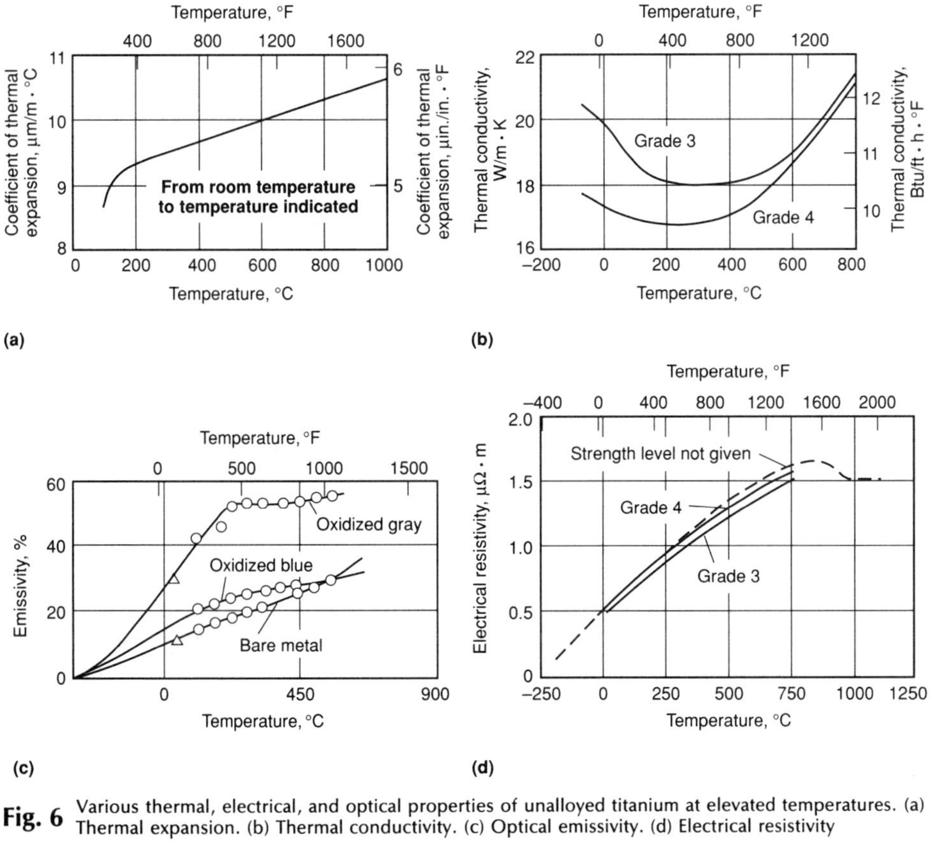

Fig. 6 Various thermal, electrical, and optical properties of unalloyed titanium at elevated temperatures. (a) Thermal expansion. (b) Thermal conductivity. (c) Optical emissivity. (d) Electrical resistivity

Titanium Alloys

Tables 5(a), 5(b), and 5(c) list the compositions of various titanium alloys. Because the allotropic behavior of titanium allows

Fig. 7 Solid titanium heat exchanger using commercially pure ASTM grades 2, 7, and 12. Courtesy of Joseph Oat Corporation

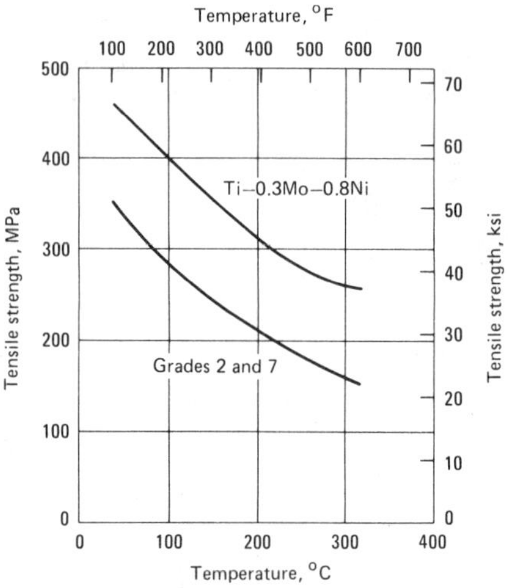

Fig. 8 Minimum tensile strength of low-strength titanium metals

Table 4 Comparative corrosion rates for Ti-Pd, grade 7, and unalloyed titanium, grade 2

Corrodent	Concentration, %	Temperature, °C	Temperature, °F	Corrosion rate: Grade 7, mm/yr	Grade 7, mils/yr	Grade 2, mm/yr	Grade 2, mils/yr
Aluminum chloride	10	100	212	<0.025	<1	<0.025	<1
	25	100	212	0.025	1	50	2020
Chlorine (wet)	. . .	Room		<0.025	<1	<0.025	<1
Citric acid	50	Boiling		<0.025	<1	0.4	17
Hydrochloric acid (HCl) (N_2 saturated)	3	190	374	0.025	1	>28	>1120
	5	190	374	0.1	4	>28	>1120
	10	190	374	8.8	350	>28	>1120
	15	190	374	40	1620	. . .	. . .
HCl (O_2 saturated)	3	190	374	0.13	5	>28	>1120
	5	190	374	0.13	5	>28	>1120
	10	190	374	9.2	368	>28	>1120
Sodium chloride	Brine	93	200	<0.025	<1	. . .	. . .
	10	190	374	<0.025	<1	. . .	. . .
	23(a)	Boiling		. . .	. . .	nil	nil
Sulfuric acid (N_2 saturated)	1	100	212	. . .	. . .	7	282
	1	190	374	0.13	5	. . .	. . .
	5	100	212	. . .	. . .	26.5	1060
	5	190	374	0.13	5	. . .	. . .
	10	190	374	1.5	59	. . .	. . .
Formic acid	50	Boiling		0.075	3	3.6	143
Hydrochloric acid	5	Boiling		0.18	7	>10	>400
Oxalic acid	1	Boiling		1.13	45	45	1800
Phosphoric acid	50	70	158	1.8	71	10	405
	10	Boiling		3.2	127	11	439
Sulfuric acid	5	Boiling		0.5	20	48	1920

(a) Acidified: pH 1.2

diverse changes in microstructures by variations in thermomechanical processing, a broad range of properties and applications can be served with a minimum number of grades. This is especially true of the alloys with a two-phase, $\alpha + \beta$, crystal structure.

The most widely used titanium alloy is the Ti-6Al-4V alpha-beta alloy. This alloy is well understood and is also very forgiving with variations in fabrication operations, despite its relatively poor room-temperature shaping and forming characteristics (compared to steel and aluminum). Alloy Ti-6Al-4V, which has limited section size hardenability, is most commonly used in the annealed condition.

Other titanium alloys are designed for particular application areas. For example:

- *Alloys Ti-5Al-2Sn-2Zr-4Mo-4Cr (commonly called Ti-17) and Ti-6Al-2Sn-4Zr-6Mo* are designed for high strength in heavy sections at elevated (moderate) temperatures
- *Alloys Ti-6242S, IMI 829, and Ti-6242* (Ti-6Al-2Sn-4Zr-2Mo) are designed for creep resistance
- *Alloys Ti-6Al-2Nb-1Ta-1Mo and Ti-6Al-4V-ELI* are designed both to resist stress corrosion in aqueous salt solutions and for high fracture toughness
- *Alloy Ti-5Al-2.5Sn* is designed for weldability, and the ELI grade is used extensively for cryogenic applications
- *Alloys Ti-6Al-6V-2Sn, Ti-6Al-4V, and Ti-10V-2Fe-3Al* are designed for high strength at low-to-moderate temperatures

The typical applications of other titanium alloys are listed in Table 6.

Effects of Alloy Elements

In titanium alloys, the principal effect of an alloying element is its effect on the alpha-to-beta transformation temperature. Some elements stabilize the alpha crystal structure by raising the alpha-to-beta transformation temperature, while other elements stabilize the beta structure by lowering the alpha-to-beta transformation temperature.

Table 7 classifies the common alloying elements as alpha or beta stabilizers. The addition of alloying elements also divides the single temperature for equilibrium transformation into two temperatures—the alpha transus, above which the alpha phase begins transformation to beta, and the beta transus, above which the alloy is all-beta. Between these temperatures, both alpha and beta are present. Transus temperatures

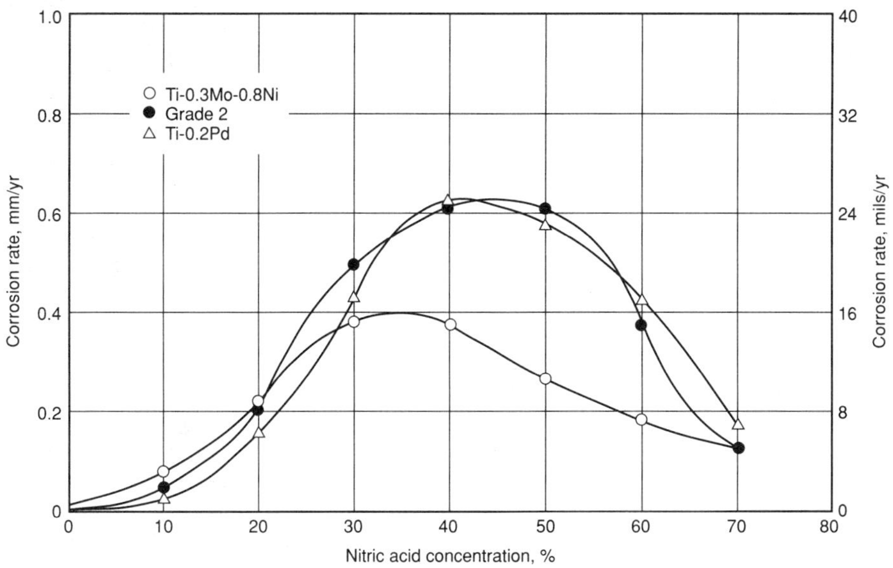

Fig. 9 Corrosion of titanium metals in boiling nitric acid. Solution replaced with fresh solution every 24 h; total exposure time, 480 h

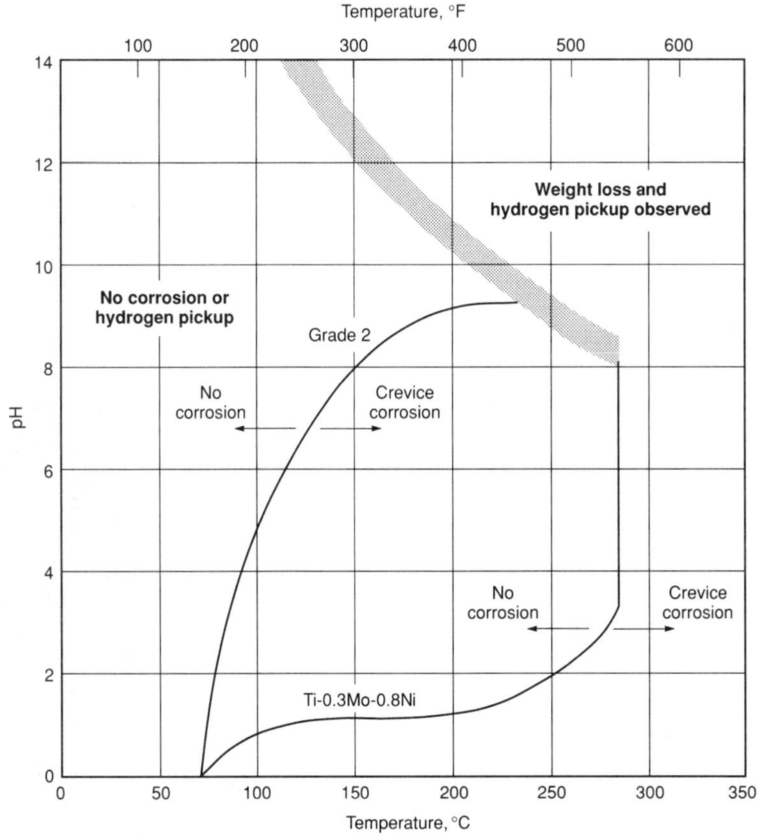

Fig. 10 Crevice corrosion of Ti-0.3Mo-0.8Ni and grade 2 unalloyed Ti in saturated NaCl solution. Shaded band represents transition zone between active and passive behavior.

vary with impurity levels and the uncertainty range of alloy additions.

Alpha Stabilizers. Aluminum is the primary alpha stabilizer in titanium alloys. Other alloying elements that favor the alpha crystal structure and stabilize it by raising the alpha-beta transformation temperatures include gallium, germanium, carbon, oxygen, and nitrogen.

Beta stabilizers are classified into two groups: beta isomorphous and beta eutectoid. Isomorphous alpha phase results from the decomposition of the metastable beta in the first group, whereas in the second group, an intimate eutectoid mixture of alpha and a compound form.

The isomorphous group consists of elements that are completely miscible in the beta phase; included in this group are molybdenum, vanadium, tantalum, and niobium.

The eutectoid-forming group, which has eutectoid temperatures as much as 335 °C (600 °F) below the transformation temperature of unalloyed titanium, includes manganese, iron, chromium, cobalt, nickel, copper, and silicon. Active eutectoid formers (for example, nickel or copper) promote rapid decomposition, and sluggish eutectoid formers (for example, iron or manganese) induce a slower reaction.

Aluminum is a principal alpha stabilizer in titanium alloys that increases tensile strength, creep strength, and the elastic moduli. The maximum solid solution strengthening that can be achieved by aluminum is limited, because above 6% Al promotes ordering and Ti_3Al (α_2) formation, which is associated with embrittlement. Thus, aluminum content of all titanium alloys is typically below 7%. Formation of α_2, which is closely related to O_2 content, can actually occur at lower levels of aluminum.

Tin has extensive solid solubilities in both alpha and beta phases and is often used as a solid solution strengthener in conjunction with aluminum to achieve higher strength without embrittlement. Tin is a less potent alpha stabilizer than aluminum, but does retard the rates of transformation. Tin is used as the main alpha stabilizer in IMI-679 (Table 5b). This alloy has a good combination of strength and temperature capability but higher density and lower modulus than Ti-6Al-2Sn-4Zr-2Mo-0.1Si (Ti-6242S), which uses aluminum as its main alpha stabilizer. Tin will react in concert with aluminum to promote ordering, $Ti_3(Al,Sn)$.

Zirconium forms a continuous solid solution with titanium and increases strength at low and intermediate temperatures. The use of zirconium above 5 to 6% may reduce ductility and creep strength (Ref 6). Zirconium is a weak beta stabilizer (Ref 7), but does retard the rates of transformation.

Molybdenum is an important beta stabilizer that promotes hardenability and short-time elevated-temperature strength. Molybdenum makes welding more difficult (Ref 8) and reduces long-term, elevated-temperature strength.

Niobium is a beta stabilizer that is added primarily to improve oxidation resistance at high temperatures.

Iron is a beta stabilizer that tends to reduce creep strength (see the section "Elevated-Temperature Mechanical Properties" in this article). Reduced iron content is utilized in alloy Ti-1100 (Table 5a) as a way of improving creep strength.

Carbon is an alpha stabilizer that also widens the temperature difference between

Table 5(a) Compositions of various alpha and near-alpha titanium alloys

Product specification	Impurity limits, wt% max						Alloying elements, wt%(a)				
	N	C	H	Fe	O	Max total others or max each	Al	Sn	Zr	Mo	Others
Ti-2.5Cu (AECMA designation, Ti-P11)											
Bars (AECMA standards prEN2523 and 2521)	0.05	0.08	0.01	0.2	0.2	0.4 total others	...	...	...	...	2.0–3.0Cu
Sheet or strip (prEN2128) and forgings (prEN2522 and 2525)	0.05	0.08	0.012	0.2	0.2	0.4 total others	...	...	...	...	2.0–3.0Cu
Ti-5Al-2.5Sn (UNS designation R54520)											
DIN17851 (alloy WL3.7115)	0.05	0.08	0.02	0.5	0.2	...	4.0–6.0	2.0–3.0	...	...	...
AMS 4910 (plate, sheet, strip)	0.05	0.08	0.02	0.5	0.2	0.005Y(b)	4.50–5.75	2.00–3.00	...	...	...
AMS 4926 (bars, rings) and AMS 4966 (forgings)	Impurity limits same as AMS 4910						4.00–6.00	2.00–3.00	...	...	...
ASTM B 265 (plate, sheet, strip)	0.05	0.10	0.02	0.4	0.2	(b)	4.00–6.00	2.00–3.00	...	...	0.12–0.25Pd
ASTM B 348 (bar, billet) and ASTM B 381 (forgings)	0.05	0.10	0.0125	0.4	0.2	(b)	4.00–6.00	2.00–3.00	...	...	...
3620-TA7 (Chinese)	0.05	0.10	0.015	0.3	0.2	0.15Si	4.00–6.00	2.00–3.00	...	...	...
Ti-5Al-2.5Sn-ELI (UNS designation R54521)											
AMS 4909 (plate, sheet, strip)	0.035	0.05	0.0125	0.25	0.12	O + Fe = 0.32, 0.005Y, 0.05 each, 0.3 total	4.50–5.75	2.00–3.00	...	...	...
AMS 4924 (bars, forgings)	0.035	0.05	0.0125	0.25	0.12	O + Fe = 0.32, others(b)	4.70–5.6	2.00–3.00	...	...	...
VT51 (U.S.S.R.)	0.05	0.10	0.015	0.30	0.02	0.15Si	4.00–5.00	2.00–3.00	...	...	...
Ti-8Al-1V-1Mo (UNS R54810)(c)											
AECMA, Ti-P66	Impurity limits not available						8	...	...	1	1V
AMS 4915, 4916, 4933 (rings), 4955 (wire), 4972 (bars, forgings), 4973 (forgings)	0.05	0.08	0.015	0.30	0.12	0.005Y, (b)	7.35–8.35	...	...	0.75–1.25	0.75–1.25V
MIL-R-81588 (ring, wire)	0.015	0.035	0.005	0.20	0.12	0.3 total	7.35–8.35	...	...	0.75–1.25	0.75–1.25V
Ti-6242 (UNS R54620)(c)											
AMS 4919, 4975, 4976	0.05	0.05	0.0125	0.25	0.15	(d), 0.1Si, 0.005Y	5.50–6.50	1.8–2.2	3.6–4.4	1.8–2.2	...
U.S. government (military)	0.04	0.05	0.015	0.25	0.15	0.13Si, 0.3 max others	5.50–6.50	1.8–2.2	3.6–4.4	1.8–2.2	...
Ti-6Al-2Nb-1Ta-0.8 Mo (UNS R56210)											
Typical	0.02	0.03	0.0125	0.12	0.10	...	6	...	...	0.8	2Nb, 1Ta
U.S. government (military)	0.03	0.05	0.0125	0.25	0.10	0.4 total	5.5–6.5	...	...	0.5–1.00	1.5–2.50Nb, 0.5–1.5Ta
Ti-679 (UNS R54790)											
Typical	0.04	0.04	0.008	0.12	0.17	...	2.25 (nom)	11	5	1	0.2Si, nom
AMS 4974 (bars, forgings)	0.04	0.04	0.0125	0.12	0.15	(b), 0.005Y	2.0–2.5	10.5–11.5	4.0–6.0	0.8–1.2	0.15–0.27Si
British TA.18, TA.19, TA.25, and TA.26	...	...	0.0125	0.20	...	...	2.0–2.5	10.5–11.5	4.0–6.0	0.8–1.2	0.1–0.5Si, 78.08 Ti min
British TA.20, TA.27	...	...	0.015	0.20	...	...	2.0–2.5	10.5–11.5	4.0–6.0	0.8–1.2	Same as TA.27
Other near-α alloys											
Ti-6242S(c)(e)	...	...	...	...	...	...	6	2	4	2	0.08Si
Ti-5Al-5Sn-2Zr-2Mo(f)	0.03	0.05	0.0125	0.15	0.13	...	5	5	2	2	0.25Si
Ti-6Al-2Sn-1.5Zr-1Mo	...	...	...	...	...	...	6	2	1.5	1	0.35Bi, 0.1Si
IMI 685	...	...	...	...	...	...	6	...	5	0.5	0.25Si
IMI 829	...	...	...	...	...	...	5.5	3.5	3	0.25	1Nb, 0.3Si
IMI 834	...	...	...	...	...	...	5.5	4.5	4	0.5	0.7Nb, 0.4Si, 0.06C
Ti-1100	...	...	...	0.02	0.07	...	6	2.75	4	0.4	0.45Si

(a) Unless a range is specified, values are nominal quantities. (b) 0.1 max each and 0.4 max total. (c) Depending on heat treatment, these alloys may be considered either near-α or α-β and are also listed in table 5(b) for α-β alloys. (d) 0.1 max each and 0.3 max total. (e) In the United States, alloy Ti-6242S is typically classified as a "superalpha" or "near-α" alloy, although it is closer to being an α-β alloy with its typical heat treatment. (f) Semicommercial alloy with a UNS designation of R54560

the alpha transus and the beta transus. Typically, beta stabilizers cause a widening (or flattening) between the alpha and beta transus temperature. In Fig. 11, for example, the lean beta stabilizer content of alloy IMI 829 produces a near-alpha alloy with a steep beta-transus approach curve. In contrast, an alloy with additional beta stabilizer (in this case Ti-6Al-4V) results in an alpha-beta alloy with a flattened approach curve.

The use of carbon to flatten the approach curve while also stabilizing the alpha phase is the basis for near-alpha alloy IMI 834 (Ref 10). Alloy IMI 834 is heat treated high in the alpha-beta region (Fig. 11) to give about 7.5 to 15 vol% of primary alpha in a fine grain (~0.1 mm) matrix of transformed beta. This combination of equiaxed alpha and transformed beta provides a good combination of creep and fatigue strength (Ref 9). Carbon also improves strength and fatigue performance.

Alloy Classes

Titanium alloys are classified as alpha alloys, alpha-beta alloys, and beta alloys. Alpha alloys have essentially all-alpha microstructures. Beta alloys have largely all-beta microstructures after air cooling from the solution treating temperature above the beta transus. Alpha-beta alloys contain a mixture of alpha and beta phases at room temperature. Within the alpha-beta class,

Table 5(b) Compositions of various alpha-beta titanium alloys

Product specification(s)	Impurity limits, wt% max: N	C	H	Fe	O	Max others, each or total	Alloying elements, wt%(a): Al	Sn	Zr	Mo	Others
Ti-6Al-4V (UNS R56400)											
Typical	0.05	0.10	(b)	0.3	0.2	...	6	...	...	...	4
Alloy Ti-P63 in AECMA standard prEN2530 for bars	0.05	0.08	0.01	0.3	0.2	0.4 total	5.5–6.75	...	...	...	3.5–4.5V
Alloy Ti-P63 in AECMA standard prEN2517 for sheet, strip, plate	0.05	0.08	0.012	0.3	0.2	0.4 total	5.5–6.75	...	...	...	3.5–4.5V
DIN 17851 (alloy WL3.7165)	0.05	0.08	0.015	0.3	0.2	...	5.5–6.75	...	...	...	3.5–4.5V
AMS 4905 (plate)	0.03	0.05	0.0125	0.25	0.12	(c), 0.005Y	5.6–6.3	...	...	...	3.6–4.4V
AMS 4906 (sheet, strip)	0.05	0.08	0.0125	0.30	0.20	0.4 total	5.5–6.75	...	...	...	3.5–4.5V
AMS 4911 (plate, sheet, strip)	0.05	0.08	0.015	0.30	0.20	(c), 0.005Y	5.5–6.75	...	...	...	3.5–4.5V
AMS 4920, 4928, 4934, and 4967 (rings, forgings, wires)	0.05	0.10	0.0125	0.30	0.20	(c), 0.005Y	5.5–6.75	...	...	...	3.5–4.5V
AMS 4954 (wire)	0.03	0.05	0.015	0.30	0.18	(c), 0.005Y	5.5–6.75	...	...	...	3.5–4.5V
ASTM B 265 (plate, sheet)	0.05	0.10	0.015	0.40	0.20	(c)	5.5–6.75	...	...	...	3.5–4.5V, 0.12–0.25Pd
ASTM F 467 (nuts) and F 468 (bolts)	0.05	0.10	0.0125	0.40	0.20	(c)	5.5–6.75	...	...	...	3.5–4.5V
Ti-6Al-4V-ELI (UNS R56401)											
AMS 4907 and 4930	0.05	0.08	0.0125	0.25	0.13	(c), 0.005Y	5.5–6.75	...	...	...	3.5–4.5V
AMS 4996 (billet)	0.04	0.10	0.0125	0.30	0.13–0.19	(d)	5.5–6.75	0.1 max	0.1 max	0.1 max	3.5–4.5V
ASTM F 135 (bar)	0.05	0.08	0.0125	0.25	0.13	...	5.5–6.75	...	...	...	3.5–4.5V
ASTM F 467 (nuts) and F 468 (bolts)	0.05	0.10	0.0125	0.40	0.20	...	5.5–6.75	...	...	...	3.5–4.5V
Ti-6Al-6V-2Sn (UNS R56620)											
Typical	0.04	0.05	0.015	0.35–1.0	0.20	...	6	2	...	...	0.75Cu, 6V
AMS 4918, 4936, 4971, 4978	0.04	0.05	0.015	0.35–1.0	0.20	(c), 0.005Y	5.0–6.0	1.5–2.5	...	...	0.35–1.00Cu, 5.0–6.0V
AMS 4979 (bars, forgings)	0.04	0.05	0.015	0.35–1.0	0.20	(c)	5.0–6.0	1.5–2.5	...	...	Same as above
Other α-β alloys											
UNS 56080 (in AMS 4908)	0.05	0.08	0.015	0.50	0.20	...	...	...	...	...	8.0Mn
UNS 56740 (in AMS 4970)	0.05	0.10	0.013	0.30	0.20	...	7	...	...	4	...
Ti-6246 (UNS R56260)	0.04	0.04	0.0125	0.15	0.15	...	6	2	4	6	...
Ti-17 (see also Table 5c)	0.04	0.05	0.0125	0.30	0.13	...	5	2	2	4	4.0Cr
Ti-6Al-2Sn-2Zr-2Cr-2Mo	0.03	0.05	0.0125	0.25	0.14	...	5.25–6.25	1.75–2.25	1.75–2.25	1.75–2.25	0.20–0.27Si, 1.75–2.25Cr
IMI-551	...	...	...	...	...	...	4	4	...	4	0.5Si
Ti-3Al-2.5V (in AMS 4943)	0.02	0.05	0.015	0.30	0.12	...	2.5–3.5	...	...	...	2.0–3.0V
IMI 550	...	...	...	...	...	...	4	2	...	4	...
IMI 679	...	...	...	...	...	...	2	11	4	1	0.25Si
IMI 700	...	...	...	...	...	...	6	...	5	4	1Cu, 0.2Si
Ti-8Al-1Mo-1V(e)	0.05	0.08	0.015	0.30	0.12	...	8	...	...	1	1V
Ti-6242(e)	0.05	0.05	0.0125(f)	0.25	0.15	0.3 total	5.5–6.5	1.8–2.2	3.6–4.4	1.8–2.2	...
Ti-6242S(e)	...	...	...	...	...	...	6	2	4	2	0.08Si

(a) Unless a range is specified, values are nominal quantities. (b) Typical hydrogen limits of 0.0150H (sheet), 0.0125H (bar), and 0.0100H (billet). (c) 0.1 max each, 0.4 max total. (d) 0.1 max Cu, 0.1 max Mn, 0.001 Y, total others 0.20 max. (e) These alloys are considered either a near-α or an α-β alloy (see Table 5a). (f) 0.0100 max H for bar and billet and 0.0150 max H for sheet and forgings

an alloy that contains much more alpha than beta is often called a near-alpha alloy. The names super-alpha and lean-beta alpha are also used for this type of alpha-beta alloy. For the purposes of this discussion, the near-alpha alloys are grouped with the alpha alloys, even though they may have some microstructural similarities with the alpha-beta alloys.

Alpha alloys (Table 5a) such as Ti-5Al-2.5Sn are slightly less corrosion resistant but higher in strength than unalloyed titanium. Alpha alloys generally are quite ductile, and the ELI grades retain ductility and toughness at cryogenic temperatures. Alpha alloys cannot be strengthened by heat treatment because the alpha structure is a stable phase. The principal microstructural variable of alpha alloys is the grain size. For a fixed composition, short-time strength (yield) and long-time strength (creep rupture) are influenced by grain size and stored energy (if any) of deformation.

The principal alloying element in alpha alloys is aluminum, but certain alpha alloys, and most commercial unalloyed titanium, contain small amounts of beta-stabilizing elements. Alpha alloys that contain small additions of beta stabilizers (Ti-8Al-1Mo-1V or Ti-6Al-2Nb-1Ta-0.8Mo, for example) sometimes have been classed as superalpha or near-alpha alloys. Although they contain some retained beta phase, these alloys consist primarily of alpha and may behave more like conventional alpha alloys in that their response to heat treatment (age hardening) and processing more nearly follows that of the alpha alloys than the conventional alpha-beta alloys.

Because near-alpha alloys contain some beta stabilizers, near-alpha alloys can exhibit microstructural variations (Fig. 12) similar to that of alpha-beta alloys. The microstructures can range from equiaxed alpha (Fig. 12a), when processing is performed in the alpha-beta region, to an acicular structure (Fig. 12c) of transformed beta after processing above the beta transus. Because these microstructural variations are related to different property improvements (Table 8), the processing temperatures of near-alpha alloys generally influence properties in the following way:

Property	β processed	α/β processed
Tensile strength	Moderate	Good
Creep strength	Good	Poor
Fatigue strength	Moderate	Good
Fracture toughness	Good	Poor
Crack growth rate	Good	Moderate
Grain size	Large	Small

Table 5(c) Compositions of various beta titanium alloys

Designation	Specifications	Impurity limits, wt% max: N	C	H	Fe	O	Max others, each or total	Alloying elements, wt%(a): Al	Sn	Zr	Mo	Others
Ti-13V-11Cr-3Al (UNS 58010)	AMS 4917	0.05	0.05	0.025	0.35	0.17	(b)	2.5–3.5	. . .	. . .	. . .	12.5–14.5V, 10.0–12.0Cr
	AMS 4959 (wire)	0.05	0.05	0.030	0.35	0.17	(b), 0.005Y	2.5–3.5	. . .	. . .	. . .	12.5–14.5V, 10.0–12.0Cr
	MIL-T-9046, MIL-R-81588	0.05	0.05	0.025	0.15–0.35	0.17	0.4 total	2.5–3.5	. . .	. . .	. . .	12.5–14.5V, 10.0–12.0Cr
	MIL-T-9047; MIL-F-83142	0.05	0.05	0.025	0.35	0.17	. . .	2.5–3.5	. . .	. . .	. . .	12.5–14.5V, 10.0–12.0Cr
	High-toughness grade	0.015	0.04	0.008	. . .	0.11(max), 0.08(nom)	(c)	2.5–3.5	. . .	. . .	. . .	12.5–14.5V, 10.0–12.0Cr
Ti-8Mo-8V-2Fe-3Al (UNS R58820)	MIL-T-9046, MIL-T-9047, and MIL-F-83142	0.05	0.05	0.015	1.6–2.4	0.16	0.4 total	2.6–3.4	. . .	. . .	7.5–8.5	7.5–8.5V
Beta C (UNS R58640)	Same as above	0.05	0.05	0.015	0.30	0.12	0.4 total	3.0–4.0	. . .	3.5–4.5	3.5–4.5	7.5–8.5V
Beta III	AMS: 4977, 4980 ASTM: B 348, B 265, B 337, and B 338	0.05	0.10	0.020	0.35	0.18	0.4 total	. . .	3.75–5.25	4.5–7.5	10.0–13.0	. . .
Ti-10V-2Fe-3Al	Forging alloy	0.05	0.05	0.015	1.6–2.5	0.13	(c)	2.5–3.5	. . .	. . .	. . .	9.25–10.75V
Ti-15-3	Sheet alloy	0.03	0.03	0.015	0.30	0.13	(c)	2.5–3.5	2.5–3.5	. . .	. . .	14–16V, 2.5–3.5Cr
Ti-17(d)	Engine compressor alloy	0.05	0.05	0.0125	0.25	0.08–0.13	(c)	4.5–5.5	1.6–2.4	1.6–2.4	3.5–4.5	3.5–4.5Cr
Transage 175	High-strength, elevated-temperature	0.05	0.08	0.015	0.20	0.15	(b)(e)	2.2–3.2	6.5–7.5	1.5–2.5	. . .	12.0–14.0V
Transage 134	High-strength alloy	0.05	0.08	0.015	0.20	0.15	(b)(e)	2.0–3.0	1.5–2.5	5.5–6.5	. . .	11.0–13.0V
Transage 129	. . .	. . .	. . .	. . .	. . .	. . .	. . .	2	2	11	. . .	11.5V

(a) Unless a range is specified, values are nominal quantities. (b) 0.1 max each, 0.4 max total. (c) 0.1 max each, 0.3 max total. (d) Alloy Ti-17 is an α-rich near-β alloy that might be classified as an α-β alloy, depending on heat treatment. (e) 0.005 max Y and 0.03 max B

In heat treating titanium alloys above the beta transus, a coarse beta grain size is likely unless adequate precautions are taken in forging and/or heat treatment. In contrast, a beta grain size of ≈0.1 mm can be achieved by processing near-alpha alloys high in the alpha-beta region (that is, near the beta transus) as compared to a typical beta grain size of 0.5 to 1.0 mm for beta-processed alloys. The quench rate also has a significant effect on the transformation product in that slow rates will give aligned alpha plates, which tend to be good for creep but somewhat worse than the faster quenched structures, basket-weave alpha, in fatigue.

Alpha-beta alloys (Table 5b), which contain one or more alpha stabilizers plus one or more beta stabilizers, can be strengthened by heat treatment or thermomechanical processing. Generally, when strengthening is desired, the alloys are rapidly cooled from a temperature high in the alpha-beta range or even above the beta transus. This solution treatment is followed by an intermediate-temperature treatment (aging) to produce an appropriate mixture of alpha and transformed beta products. Response to heat treatment is a function of cooling rate from the solution temperature and therefore may be affected by section size.

Like the near-alpha alloy in Fig. 12, the microstructure of alpha-beta alloys can take on different forms, ranging from equiaxed to acicular or some combination of both. Equiaxed structures are formed by working an alloy in the alpha-beta range and annealing at lower temperatures. Acicular structures (Fig. 13c) are formed by working or heat treating above the beta transus and rapid cooling. Rapid cooling from temperatures high in the alpha-beta range (Fig. 13d and e) will result in equiaxed primary (prior) alpha and acicular alpha from the transformation of beta structures. Generally, there are property advantages and disadvantages for each type of structure. Table 8 compares, on a relative basis, the advantages of each structure.

By a suitable manipulation of forging and heat treatment schedules, a wide range of properties is attainable in alpha-beta alloys. In particular, the alpha-beta alloys are more responsive to aging than the near-alpha alloys. The near-alpha alloys are less responsive to aging because little, if any, change in properties can be expected when phases are in a nearly equilibrium condition prior to aging.

In the alpha-beta alloys, the presence of nonequilibrium phases, such as alpha-prime or metastable beta, results in substantial increases in tensile and yield strengths following the aging treatment. Table 9, for example, shows the response to heat treatment for the widely used Ti-6Al-4V alloy. The tensile data show that no response to aging occurs upon furnace cooling from solution temperatures. Only a slight response occurs upon air cooling (microstructures in Fig. 13b and d), while the greatest response is experienced with water quenching from the solution temperature (microstructures in Fig. 13c and e). Good response to aging takes place upon water quenching from the beta field (Fig. 13c); however, ductilities are quite low (Table 9). The best combination of properties can be produced by solution treating and rapidly quenching from close to but below the beta transus temperature (Fig. 13d or e), followed by an aging treatment (Table 9).

Beta alloys (Table 5c) are sufficiently rich in beta stabilizers (and lean in alpha stabilizers) that the beta phase can be completely retained with appropriate cooling rates. Beta alloys are metastable, and precipitation of alpha phase in the metastable beta is a method used to strengthen the alloys. Beta alloys contain small amounts of alpha-stabilizing elements as strengthening agents.

As a class, beta and near-beta alloys offer increased fracture toughness over alpha-beta alloys at a given strength level, with the advantage of heavy section heat treatment capability. However, beta and near-beta alloys may require close control of processing and fabrication steps to achieve optimal properties, though this is not always the case. In the past, beta alloys had rather limited applications, such as springs and fasteners, where very high strength was required.

In recent years, however, beta alloys have received closer attention because their fracture toughness characteristics respond to the increased need for damage tolerance in aerospace structures. In addition, some beta alloys containing molybdenum have good corrosion characteristics. Beta alloys also exhibit:

Table 6 Typical applications of various titanium-base materials

Nominal contents and common name or specifications	Available mill forms	General description	Typical applications
Commercially pure titanium			
Unalloyed titanium: see Table 2	Bar, billet, extrusions, plate, sheet, strip, wire, rod, pipe, tubing, castings	For corrosion resistance in the chemical and marine industries, and where maximum ease of formability is desired. Weldability: good	Jet engine shrouds, cases, airframe skins, firewalls, and other hot-area equipment for aircraft and missiles; heat-exchangers; corrosion resistant equipment for marine and chemical-processing industries. Other applications requiring good fabricability, weldability, and intermediate strength in service
Ti-0.2Pd: ASTM grades 7 and 11	Bar, billet, extrusions, plate, sheet, strip, wire, pipe, tubing, castings	The Pd-containing alloys extend the range of application in HCl, H_3PO_4, and H_2SO_4 solutions. Characteristics of good fabricability, weldability, and strength level are similar to those of corresponding unalloyed titanium grades.	For corrosion resistance in the chemical industry where media are mildly reducing or vary between oxidizing and reducing
Ti-0.3Mo-0.8Ni: ASTM grade 12	Bar, billet, extrusions, plate, sheet, strip, wire, pipe, tubing, castings	Compared to unalloyed Ti, Ti-0.3Mo-0.8Ni has better corrosion resistance and higher strength. The alloy is particularly resistant to crevice corrosion in hot brines.	For corrosion resistance in the chemical industry where media are mildly reducing or vary between oxidizing and reducing
α alloys			
Ti-2.5 Cu: AECMA Ti-P11, or IMI 230	Bar, billet, rod, wire, plate, sheet, extrusions	Ti-2.5Cu combines the formability and weldability of titanium with improved mechanical properties from precipitation strengthening.	Useful for its improved mechanical properties, particularly up to 350 °C (650 °F). Aging doubles elevated-temperature properties and increases room-temperature strength by 25%.
Ti-5Al-2.5Sn (UNS R54520)	Bar, billet, extrusions, plate, sheet, wire, castings	Air frame and jet engine applications requiring good weldability, stability, and strength at elevated temperatures	Gas turbine engine casings and rings, aerospace structural members in hot spots, and chemical-processing equipment that require good weldability and intermediate strength at service temperatures up to 480 °C (900 °F)
Ti-5Al-2.5Sn-ELI (UNS R54521)	Same as UNS R54520	Reduced level of interstitial impurities improves ductility and toughness.	High-purity grade for pressure vessels for liquefied gases and other applications requiring better ductility and toughness, particularly in hardware for service to cryogenic temperatures
Near-α alloys			
Ti-8Al-1Mo-1V (UNS R54810)	Bar, billets, extrusions, plate, sheet, wire, forgings	Near-α or α-β microstructure (depending on processing) with good combination of creep strength and fatigue strength when processed high in the α-β region (that is, near the β transus)	Fan blades are main use; forgings for jet engine components requiring good creep strength, high strength at elevated temperatures (compressor disks, plates, hubs). Other applications where light, high strength, highly weldable material with low density is required (cargo flooring)
Ti-6Al-2Sn-4Zr-2Mo (Ti-6242, or UNS 54620)	Bar, billet, sheet, strip, wire, forgings	Used for creep strength and elevated-temperature service. Fair weldability	Forgings and flat-rolled products used in gas turbine engine and air-frame applications where high strength and toughness, excellent creep resistance, and stability at temperatures up to 450 °C (840 °F) are required
Ti-6Al-2Sn-4Zr-2Mo-0.1Si (Ti-6242S)	Same as UNS54620 but also castings	Silicon imparts additional creep resistance.	Same as UNS 54620 but maximum-use temperature up to about 520 °C (970 °F)
Ti-6Al-2Nb-1Ta-0.8Mo (UNS R56210)	Plate, sheet, strip, bar, wire, rod	...	Plate for naval shipbuilding applications, submersible hulls, pressure vessels, and other high-toughness applications
Ti-2.25Al-11Sn-5Zr-1Mo (Ti-679, UNS R54790)	Forgings, bar, billet, plate	...	Jet engine blades and wheels, large bulkhead forgings, other applications requiring high-temperature creep strength plus stability and short-time strength
Ti-5Al-5Sn-2Zr-2Mo-0.25Si (Ti-5522S, UNS 54560)	Forged billet and bar, special products available in plate and sheet	Semicommercial; no longer used	Specified in MIL-T-9046 and MIL-T-9047
IMI-685 (Ti-6Al-5Zr-0.5Mo-0.2Si)	Rod, bar, billet, extrusions	Weldable medium-strength alloy	Alloy for elevated-temperature uses up to about 520 °C (970 °F)
IMI-829 (Ti-5.5Al-3.5Sn-3Zr-1Nb-0.3Mo-0.3Si)	Rod, bar, billet, extrusions	Weldable, medium-strength alloy with good thermal stability and high creep resistance up to 600 °C (1110 °F)	Elevated-temperature alloy for service up to about 580 °C (1075 °F)
IMI-834 (Ti-5.8Al-4Sn-3.5Zr-0.7Nb-0.5Mo-0.3Si)	Rod, bar, billet, extrusions	Weldable, high-temperature alloy with improved fatigue performance as compared to IMI 829 and 685	Maximum-use temperature up to about 590 °C (1100 °F)
Ti-1100	...	Elevated-temperature alloy	Maximum-use temperature of 590 °C (1100 °F)

(continued)

Table 6 (continued)

Nominal contents and common name or specifications	Available mill forms	General description	Typical applications
α-β alloys			
Ti-6Al-4V (UNS R56400 and AECMA Ti-P63)	Bar, billet, rod, wire, plate, sheet, strip, extrusions	Ti-6Al-4V is the most widely used titanium alloy. It is processed to provide mill-annealed or β-annealed structures, and is sometimes solution treated and aged. Ti-6Al-4V has useful creep resistance up to 300 °C (570 °F) and excellent fatigue strength. Fair weldability	Ti-6Al-4V is used for aircraft gas turbine disks and blades. It is extensively used, in all mill product forms, for airframe structural components and other applications requiring strength at temperatures up to 315 °C (600 °F); also used for high-strength prosthetic implants and chemical-processing equipment. Heat treatment of fastener stock provides tensile strengths up to 1100 MPa (160 ksi).
Ti-6Al-4V-ELI (UNS R56401)	Same as UNS R56400	Reduced interstitial impurities improve ductility and toughness.	Cryogenic applications and fracture-critical aerospace applications.
Ti-6Al-7Nb (IMI-367)	Rod, bar, billet, extrusions	High-strength alloy with excellent biocompatibility	Surgical implant alloy
Corona 5 (Ti-4.5Al-5Mo-1.5Cr)	Alloy researched for plate, forging, and superplastic forming sheet	Improved fracture toughness over Ti-6Al-4V with less restricted chemistry. Easier to work than Ti-6Al-4V	Once investigated as a possible replacement for Ti-6Al-4V in aircraft, but no longer considered of interest
Ti-6Al-6V-2Sn (UNS R56620)	Bar, billet, extrusions, plate, sheet, wire	In the forms of sheet, light-gage plate, extrusions, and small forgings, this alloy is used for airframe structures where strength higher than that of Ti-6Al-4V is required. Usage is generally limited to secondary structures, because attractiveness of higher strength efficiency is minimized by lower fracture toughness and fatigue properties.	Applications requiring high strength at temperatures up to 315 °C (600 °F). Rocket engine case airframe applications including forgings, fasteners. Limited weldability. Susceptible to embrittlement above 315 °C (600 °F)
Ti-8Mn (UNS R56080)	Sheet, strip, plate	Limited usage	Aircraft sheet and structural parts
Ti-7Al-4Mo (UNS R56740)	Bar and forgings	Limited usage	Jet engine disks, compressor blades and spacers, sonic horns
Ti-6Al-2Sn-4Zr-6Mo (UNS R56260)	Sheet, plate, and bar or billet for forging stock	Should be considered for long-time load-carrying applications at temperatures up to 400 °C (750 °F) and short-time load-carrying applications. Limited weldability	Forgings in intermediate temperature range sections of gas turbine engines, particularly in disk and fan blade components of compressors
Ti-6Al-2Sn-2Zr-2Cr-2Mo-0.25Si	Forgings, sheet	Heavy section forgings requiring high strength, fracture toughness, and high modulus	Forgings and sheet for airframes
Ti-3Al-2.5V (UNS R56320)	Bar, tubing, strip	Normally used in the cold-worked stress-relieved condition	Seamless tubing for aircraft hydraulic and ducting applications; weldable sheet; mechanical fasteners
IMI 550 and 551	Rod, bar, billet, extrusions	High-strength alloys; IMI 551 has increased room-temperature strength due to higher tin contents than IMI 550.	Two high-strength alloys with useful creep resistance up to 400 °C (750 °F)
β alloys			
Ti-13V-11Cr-3Al (UNS R58010)	Sheet, strip, plate, forgings, wire	High-strength alloy with good weldability	High-strength airframe components and missile applications such as solid rocket motor cases where extremely high strengths are required for short periods of time. Springs for airframe applications. Very little use anymore
Ti-8Mo-8V-2Fe-3Al (UNS R58820)	Rod, wire, sheet, strip, forgings	Limited weldability	Rod and wire for fastening applications; sheet, strip, and forgings for aerospace structures
Ti-3Al-8V-6Cr-4Zr-4Mo (Beta C)	Sheet, plate, bar, billet, wire, pipe, extrusions, castings	High-strength alloy with excellent ductility not available in other β alloys. Excellent cold-working characteristics; fair weldability	Airframe high-strength fasteners, rivets, torsion bars, springs, pipe for oil industry and geothermal applications
Ti-11.5Mo-6Zr-4.5Sn (Beta III)	Not being produced anymore	Excellent forgeability and cold workability. Very good weldability	Aircraft fasteners (especially rivets) and sheet metal parts where cold formability and strength potential can be used to greatest advantage. Possible use in plate and forging applications where high-strength capability, deep hardenability, and resistance to stress corrosion are required and somewhat lower aged ductility can be accepted
Ti-10V-2Fe-3Al	Sheet, plate, bar, billet, wire, forgings	The combination of high strength and high toughness available is superior to any other commercial titanium alloy. For applications requiring uniformity of tensile properties at surface and center locations	High-strength airframe components. Applications up to 315 °C (600 °F) where medium to high strength and high toughness are required in bar, plate, or forged sections up to 125 mm (5 in.) thick. Used primarily for forgings
Ti-15V-3Al-3Cr-3Sn (Ti-15-3)	Sheet, strip, plate	Cold formable β alloy designed to reduce processing and fabrication costs. Heat treatable to a tensile strength of 1310 MPa (190 ksi)	High-strength aircraft and aerospace components
Ti-5Al-2Sn-2Zr-4Mo-4Cr (Ti-17)	Forgings	α-rich near-β alloy that is sometimes classified as an α-β alloy. Unlike other β or near-β alloys, Ti-17 offers good creep strength up to 430 °C (800 °F).	Forgings for turbine engine components where deep hardenability, strength, toughness, and fatigue are important. Useful in sections up to 150 mm (6 in.)
Transage alloys	Sheet, plate, bar, forging	Developmental	High-strength (Transage 134) and high-strength elevated-temperature (Transage 175) alloys

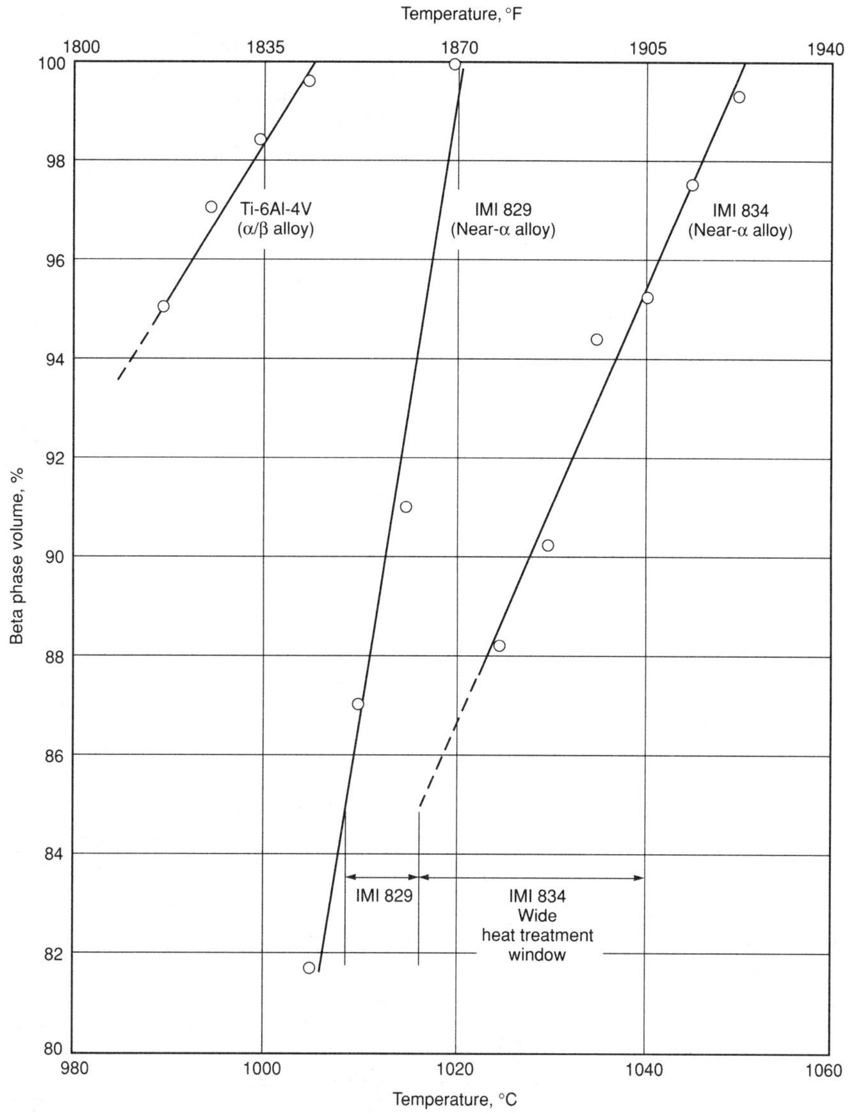

Fig. 11 Beta transus approach curves of IMI 834, IMI 829, and Ti-6Al-4V. Source: Ref 9

- Better room-temperature forming and shaping characteristics than alpha-beta alloys
- Higher strength than alpha-beta alloys at temperatures where yield strength (instead of creep strength) is the applicable criterion
- Better response to heat treatment (solution treatment, quenching, and aging) in heavier sections than the alpha-beta alloys

The use of beta alloys is increasing. Alloy Ti-10V-2Fe-3Al is used for forgings, alloy Ti-15V-3Cr-3Al-3Sn is used for sheet applications, and alloy Ti-3Al-8V-6Cr-4Mo-4Zr is being utilized for springs and extrusions.

Terminology in Classifying Beta Alloys. Although there are a number of ways to define the term beta alloy, T. Duerig and J. Williams (Ref 11) suggest the following operational definition: "A beta-titanium alloy is any titanium composition which allows one to quench a very small volume of material into ice water from above the material's beta-transus temperature without martensitically decomposing the beta phase." The point of such a detailed definition is to exclude all titanium alloys in which martensite can be formed athermally or with the assistance of residual stresses that may arise during the quenching of large pieces. It also excludes the diffusional decomposition of beta, which is section size dependent through cooling rates.

Within this definition, Fig. 14 illustrates the constitution of beta alloys. Within the general class of beta alloys, the solute lean alloys tend to decompose much more readily than do the more stable, solute rich alloys. Therefore, it is useful to divide the general classification of beta alloys into two subclassifications: the lean beta alloys and the rich beta alloys. In Fig. 14, for example, alloys which form the brittle metastable phase, omega (ω) phase, during aging would be defined as lean alloys, and alloys which

Table 7 Ranges and effects of some alloying elements used in titanium

Alloying element	Range (approx), wt%	Effect on structure
Aluminum	2–7	α stabilizer
Tin	2–6	α stabilizer
Vanadium	2–20	β stabilizer
Molybdenum	2–20	β stabilizer
Chromium	2–12	β stabilizer
Copper	2–6	β stabilizer
Zirconium	2–8	α and β strengthener (see text)
Silicon	0.05 to 1	Improves creep resistance

Table 8 Relative advantages of equiaxed and acicular morphologies in near-alpha and alpha-beta alloys

Equiaxed:

Higher ductility and formability
Higher threshold stress for hot-salt stress corrosion
Higher strength (for equivalent heat treatment)
Better low-cycle fatigue (initiation) properties

Acicular:

Superior creep properties
Higher fracture-toughness values
Slight drop in strength (for equivalent heat treatment)
Superior stress-corrosion resistance
Lower crack-propagation rates

are too stable to decompose isothermally to a β + ω mixture would be classified as rich alloys. Alternatives to this definition would be to define the lean alloys as those that deform by either a twinning or a martensitic shearing process when in the solution treated and quenched condition, or to give a processing-oriented definition that would identify the lean alloys as those that can be effectively thermomechanically processed in the α + β phase field (although this definition is certainly the least distinct of the three) (Ref 11). In terms of the most common commercial alloys, these three definitions basically coincide: any alloy classified as lean or rich by one definition would be classed the same way by either of the other definitions, although one could, without doubt, develop compositions that could not be unambiguously defined by all three definitions. Nevertheless, these definitions are more meaningful than the terms metastable beta and near-beta because all commercial beta alloys are metastable and decompose into alpha-beta structures.

Microstructural Constituents

The basis for microstructural manipulation during heat treatment of titanium alloys centers around the β → α transformation that occurs in these alloys during cooling. This transformation can occur by nucleation and growth, or it can occur martensitically, depending on the alloy composition and the cooling rate. The martensitic product is usually hcp and is designated α′. There also is an orthorhombic martensite, designated α″, which forms in alloys that contain higher concentrations of refractory

Table 9 Effect of heat treatment on the tensile properties of Ti-6Al-4V

Treatment(a)	Tensile strength MPa	Tensile strength ksi	Yield strength MPa	Yield strength ksi	Elongation, %	Reduction in area, %
1065 °C (1950 °F)/WQ(b)	1108	160.7	954	138.3	7.7	19.2
After aging	1170	169.7	1057	153.3	8.5	19.2
955 °C (1750 °F)/WQ(b)	1120	162.3	954	138.3	17.0	60.2
After aging	1183	171.6	1069	155.0	16.5	56.4
900 °C (1650 °F)/WQ	1117	162.0	924	134.0	15.2	53.9
After aging	1117	162.0	1014	147.0	15.3	47.5
845 °C (1550 °F)/WQ	1009	146.4	772	112.0	20.0	54.7
After aging	1178	156.3	977	141.7	16.5	48.8
1065 °C (1950 °F)/AC(b)	1060	153.7	944	137.0	7.0	10.3
After aging	1060	153.7	940	136.3	9.8	16.0
955 °C (1750 °F)/AC(b)	955	144.3	846	122.7	17.8	54.1
After aging	1020	148.0	898	130.3	16.1	45.7
900 °C (1650 °F)/AC	1002	145.3	869	126.0	17.5	54.7
After aging	1029	149.3	938	136.0	17.3	50.2
845 °C (1550 °F)/AC	1020	148.0	878	127.3	17.8	47.7
After aging	1036	150.3	931	135.0	16.8	46.9
1065 °C (1950 °F)/FC	1041	151.0	938	136.0	10.5	15.6
After aging	1011	146.6	938	136.0	9.5	15.4
955 °C (1750 °F)/FC	940	136.3	836	121.3	18.8	46.0
After aging	967	140.3	883	128.0	18.2	49.1
900 °C (1650 °F)/FC	963	139.6	855	124.0	16.5	43.3
After aging	963	139.6	876	127.0	16.8	48.3
845 °C (1550 °F)/AC	997	144.6	924	134.0	17.3	48.9
After aging	1060	154.0	954	138.3	17.0	49.6

(a) Aging in all instances: 540 °C (1000 °F) for 4 h; air cool. WQ, water quench; AC, air cool; FC, furnace cool. β transus: 1000 ± 14 °C (1820 ± 25 °F). All specimens are 16 mm (⅝ in.) diameter bars. (b) See Fig. 13 for corresponding microstructures before aging.

elements such as molybdenum, tantalum, or niobium. Literally all thermomechanical processing is conducted above the M_s temperature for either α′ or α″. Alloys that contain enough β-stabilizing elements to depress the M_s temperature below room temperature can be rapidly cooled to retain the metastable β phase. More detailed information of the phase transformations in titanium alloys is given in several of the "Selected References" listed at the end of this article.

Alpha Structures. Equiaxed alpha grains (Fig. 12a) usually are developed by annealing cold-worked alloys above the recrystallization temperature. Elongated alpha grains (Fig. 15b) result from unidirectional working of the metal and are commonly found in longitudinal sections of rolled or extruded alloys. Elongated alpha may be enhanced by the prior presence of blocky and/or grain-boundary alpha.

Primary alpha refers to the alpha phase in a crystallographic structure that is retained from the last high-temperature alpha-beta working or heat treatment. The morphology of alpha is influenced by the prior thermomechanical history.

Transformed Beta. Although some of the areas of alpha phase that appear in micrographs of heat-treated titanium and titanium alloys may have been present before the heat treatment (primary alpha), other areas of alpha have been produced by transformation from beta. The alpha in these latter areas appears in different structures known as serrated, acicular, platelike, Widmanstätten, and alpha prime (martensite). The term transformed beta is used to describe these various alpha structures plus any beta that may remain at room temperature.

Acicular alpha, which is the most common transformation product formed from beta during cooling, is produced by nucleation and growth along one set of preferred crystallographic planes of the prior-beta matrix (Fig. 12a) or along several sets of planes (Fig. 15d); in the latter instance, a basket-weave appearance results that is characteristic of a Widmanstätten structure. Acicular alpha and Widmanstätten alpha are generally interchangeable terms.

Under some conditions, the long grains of alpha that are produced along preferred planes in the beta matrix take on a wide, platelike appearance. Under other conditions, grains of irregular size and with jagged boundaries, called serrated alpha, are produced.

Alpha prime (hexagonal martensite) is a nonequilibrium supersaturated alpha structure produced by diffusionless (martensitic) transformation of beta. The needlelike structure, similar in appearance and in mode of formation to martensite in steel, is often difficult to distinguish from that of acicular alpha, although acicular alpha usually is less well-defined and has curved rather than straight sides.

Alpha double prime (orthorhombic martensite) is a supersaturated nonequilibrium orthorhombic phase formed by a diffusionless transformation of the beta phase in certain alloys.

Alpha-2 (α_2) or Ti_3Al is an ordered phase that can form within the alpha phase in alloys containing more than 6% Al. The reaction is promoted by increased oxygen.

Omega is a nonequilibrium, submicroscopic phase that forms as a nucleation growth product, often thought to be a transition phase during the formation of alpha from beta. It occurs in metastable beta alloys and can lead to severe embrittlement.

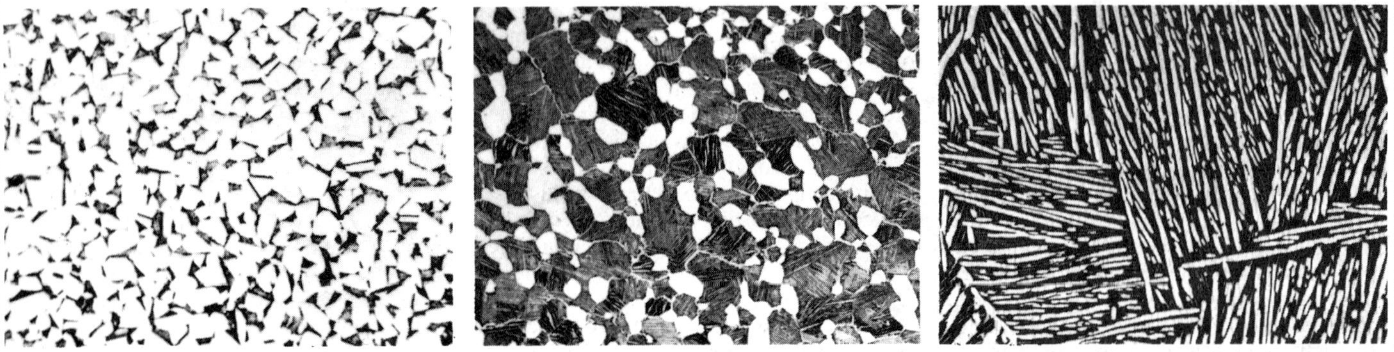

(a) Forged with a starting temperature of 900 °C (1650 °F), which is below the normal temperature range for forging Ti-8Al-1Mo-1V

(b) Forged with a starting temperature of 1005 °C (1840 °F), which is within the normal range, and air cooled

(c) Forged with a starting temperature of 1093 °C (2000 °F), which is above the beta transus temperature, and rapidly air cooled after finish forging

Fig. 12 Microstructures of near-alpha alloy Ti-8Al-1Mo-1V after forging with different starting temperatures. (a) Equiaxed alpha grains (light) in a matrix of alpha and beta (dark). (b) Equiaxed grains of primary alpha (light) in a matrix of transformed beta (dark) containing fine acicular alpha. (c) Transformed beta containing coarse and fine acicular alpha (light). Etchant: Kroll's reagent (192). All micrographs at 250X

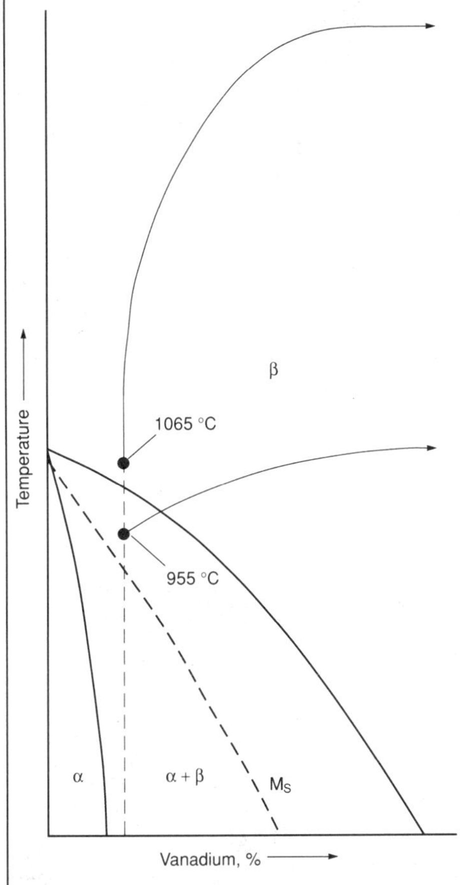

(a)

Air cooled

(b) Acicular alpha (transformed beta) with prior beta grain boundaries

Water quenched

(c) Alpha prime (martensite) matrix with beta (dark) and prior beta grain boundaries

(d) Grains of primary alpha (light) in a matrix of transformed beta containing acicular alpha

(e) Equiaxed primary alpha in a matrix of alpha prime (martensite)

Fig. 13 Microstructures of alloy Ti-6Al-4V after cooling from different areas of the phase field shown in (a). The specimens represented in micrograph (e) provided the best combination of strength and ductility after aging. See the text and Table 9. Etchant: 10 HF, 5 HNO_3, 85 H_2O. All micrographs at 250X

It typically occurs during aging at low temperatures, but can also be induced by high hydrostatic pressures. It can also form athermally upon quenching within beta of certain compositions.

Beta Structures. In alpha-beta and beta alloys, some equilibrium beta is present at room temperature. A nonequilibrium, or metastable, beta phase can be produced in alpha-beta alloys that contain enough beta-stabilizing elements to retain the beta phase at room temperature upon rapid cooling from between the alpha transus and beta transus temperatures. The composition of the alloy must be such that the temperature for the start of martensite formation is depressed to below room temperature. Metastable beta is partially or completely transformed to martensite, alpha, or eutectoid decomposition products with thermal or strain energy activation during processing or service exposure.

Beta flecks are alpha-lean regions in an alpha-beta microstructure. This beta-rich region has a beta transus measurably below that of the matrix. Beta flecks have reduced amounts of primary alpha (or may even be devoid of alpha) and, in alpha-beta alloys, may exhibit a different morphology than the primary alpha in the surrounding matrix. Beta flecks have a higher content of beta stabilizers than the matrix and, through partitioning, probably are lean in alpha stabilizers.

Beta flecks are attributed to microsegregation during solidification of ingots of alloys that contain strong beta stabilizers. They are most often found in products made from large-diameter ingots. Beta flecks also may be found in beta-lean alloys such as Ti-6Al-4V that have been heated to a temperature near the beta transus during processing.

Beta flecks are not considered harmful in alloys lean in beta stabilizers if they are to

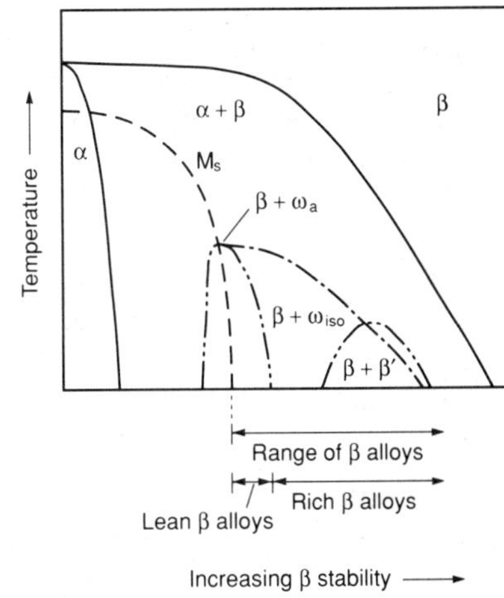

Fig. 14 Schematic phase diagram of a beta-stabilized titanium system, indicating the compositional range that would be considered beta alloys and the subdivision of this range into the lean and rich beta alloys. Source: Ref 11

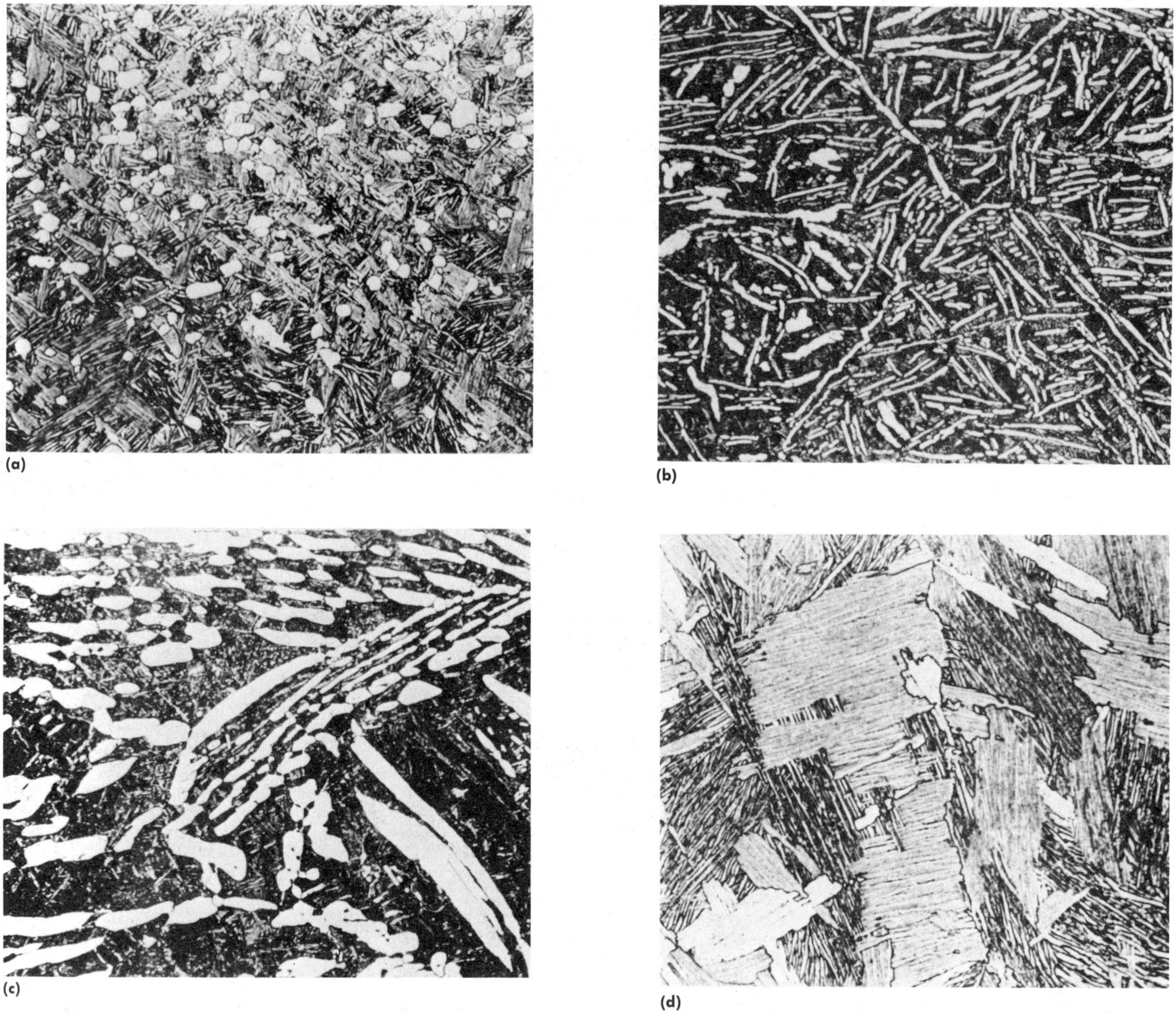

Fig. 15 Microstructures corresponding to different combinations of properties in Ti-6Al-4V forgings. (a) 6% equiaxed primary alpha plus fine platelet alpha in Ti-6Al-4V alpha-beta forged, then annealed 2 h at 705 °C (1300 °F) and air cooled. (b) 23% elongated, partly broken up alpha plus grain-boundary alpha in Ti-6Al-4V, alpha-beta forged and water quenched, then annealed 2 h at 705 °C and air cooled. (c) 25% blocky (spaghetti) alpha plates plus very fine platelet alpha in Ti-6Al-4V alpha-beta forged from a spaghetti-alpha starting structure, then solution treated 1 h at 955 °C (1750 °F) and reannealed 2 h at 705 °C. (d) 92% alpha basket-weave structure in Ti-6Al-4V beta forged and slow cooled, then annealed 2 h at 705 °C. Structures in (a) and (b) produced excellent combinations of tensile properties, fatigue strengths and fracture toughness. Structure in (c) produced very poor combinations of mechanical properties. Structure in (d) produced good fracture toughness, but poor tensile properties and fatigue resistance. Source: Ref 12

be used in the annealed condition. However, they constitute regions that incompletely respond to heat treatment, and thus microstructural standards have been established for allowable limits on beta flecks in various alpha-beta alloys. Beta flecks are more objectionable in beta-rich alpha-beta alloys and beta alloys than in leaner alloys.

Aged Structures. Aging of martensite results in the formation of equilibrium $\alpha + \beta$ but most aged martensite structures cannot be distinguished from unaged martensite by light microscopy. Precipitation of alpha during aging of beta results in some darkening of the aged-beta structure. Aging, or stressing, could change metastable beta to alpha or to eutectoid products.

Wrought Alloy Processing

Because the microstructures of titanium alloys are readily affected by process variables, microstructural control is basic to successful processing of titanium alloys. Undesirable structures (grain-boundary alpha, beta fleck, "spaghetti" or elongated alpha) can interfere with optimal property development (Fig. 15). Titanium ingot structures, discussed in the section "Melting Practice for Ingot Production," can also carry over to the final product.

Several factors are important in the processing of titanium and titanium alloys. Among the most important are:

- Amounts of specific alloying elements and impurities
- Melting process used to make ingot
- Method for mechanically working ingots into mill products
- The final step employed in working, fabrication, or heat treatment

This section focuses on primary fabrication, in which ingots are converted into general

Table 10 Standard forging temperatures for manufacturing titanium billet stock

Alloy	β transus °C	β transus °F	Ingot breakdown °C	Ingot breakdown °F	Intermediate °C	Intermediate °F	Finish °C	Finish °F
Commercially pure titanium								
Grades 1–4	900–955	1650–1750	955–980	1750–1800	900–925	1650–1700	815–900	1500–1650
α and near-α alloys								
Ti-5Al-2.5Sn	1030	1890	1120–1175	2050–2150	1065–1095	1950–2000	1010–1040	1850–1900
Ti-6Al-2Sn-4Zr-2Mo-0.08Si	995	1820	1095–1150	2000–2100	1010–1065	1850–1950	955–980	1750–1800
Ti-8Al-1Mo-1V	1040	1900	1120–1175	2050–2150	1065–1095	1950–2000	1010–1040	1850–1900
α-β alloys								
Ti-8Mn	800	1475	925–980	1700–1800	845–900	1550–1650	815–845	1500–1550
Ti-6Al-4V	995	1820	1095–1150	2000–2100	980–1040	1800–1900	925–980	1700–1800
Ti-6Al-6V-2Sn	945	1735	1040–1095	1900–2000	955–1010	1750–1850	870–940	1600–1725
Ti-7Al-4Mo	1005	1840	1120–1175	2050–2150	1010–1065	1850–1950	955–980	1750–1800
β alloy								
Ti-13V-11Cr-3Al	720	1325	1120–1175	2050–2150	1010–1065	1850–1950	925–980	1700–1800

mill products, and secondary fabrication of finished shapes from mill products. Secondary fabrication refers to manufacturing processes such as die forging, extrusion, hot and cold forming, machining, chemical milling, and joining, all of which are used for producing finished parts from mill products. Each of these processes may strongly influence properties of titanium and its alloys, either alone or by interacting with effects of processes to which the metal has previously been subjected.

Primary Fabrication

Primary fabrication includes all operations that convert ingot into general mill products—billet, bar, plate, sheet, strip, extrusions, tube, and wire. Besides the reduction of section size, the basic objective of primary processing is the refinement of grain size and the production of a uniform microstructure. Primary fabrication is very important in establishing final properties, because many secondary fabrication operations may not involve sufficient reductions for grain refinement by recrystallization. However, some secondary fabrication processes, such as forging and ring rolling, do impart sufficient reduction to play the major role in establishing material properties. In fact, forgings usually recrystallize more uniformly because forging is an efficient method of introducing large amounts of stored energy in the material.

Because titanium alloys utilize many of the same methods (and sometimes the same processing facilities) as other metals, the primary fabrication processes were first designed around the capabilities of steel mill equipment. As the titanium industry matured, special furnace equipment, presses, and mills were developed in response to the different processing requirements of titanium alloys. One of the basic distinctions is the high reactivity of titanium and the possibility of surface contamination (see the section "Heat Treatment" for a discussion on surface contamination). Other major factors affected by thermomechanical processing that appear to be important are: primary alpha morphology, primary alpha volume fraction, and grain boundary alpha. The effects of thermomechanical processing on these microstructural features, which are important in both alpha-beta and beta alloys, are discussed in Ref 13.

In beta alloys, however, thermomechanical processing affects not only the microstructure, but also the decomposition kinetics of the metastable beta phase during aging. The increased dislocation density after working beta alloys leads to extensive heterogeneous nucleation of the equilibrium alpha phase and can thus suppress formation of the brittle omega phase (Ref 13). Strain-rate sensitivity as a function of beta content is also important (see, for example, the section "Superplastic Forming" in this article). Because the flow stress of beta alloys can be affected by their high strain-rate sensitivity, primary processing is frequently accomplished at temperatures higher than those for other titanium alloys with the attendant higher number of reheats and increased wear on forging dies. Beta alloys thus exhibit slightly greater evidence of sensitivity to the thermomechanical processing route in obtaining and retaining uniformly recrys-

Table 11 Variation of typical room-temperature tensile properties with section size for four titanium alloys

Section size(a) mm	Section size(a) in.	Tensile strength MPa	Tensile strength ksi	Yield strength MPa	Yield strength ksi	Elongation(b), %	Reduction in area, %
6Al-4V(c)							
25–50	1–2	1015	147	965	140	14	36
102	4	1000	145	930	135	12	25
205	8	965	140	895	130	11	23
330	13	930	135	860	125	10	20
6Al-4V-ELI(c)							
25–50	1–2	950	138	885	128	14	36
102	4	885	128	827	120	12	28
205	8	885	128	820	119	10	27
330	13	870	126	795	115	10	22
6Al-6V-2Sn(c)							
25–50	1–2	1105	160	1035	150	15	40
102	4	1070	155	965	145	13	35
205	8	1000	145	930	135	12	25
8Al-1Mo-1V							
25–50	1–2(d)	985	143	905	131	15	36
102	4(e)	910	132	840	122	17	35
205	8(f)	1000	145	895	130	12	23
6Al-2Sn-4Zr-2Mo+Si(g)							
25–50	1–2	1000	145	930	135	14	33
102	4	1000	145	930	135	12	30
205	8	1035	150	940	136	12	28
330	13	1000	145	825	120	11	21

(a) Properties are in longitudinal direction for sections 50 mm (2 in.) or less, and in transverse direction for sections 100 mm (4 in.) or more, in section size. (b) In 50 mm (2 in.). (c) Annealed 2 h at 700 °C (1300 °F) and air cooled. (d) Annealed 1 h at 900 °C (1650 °F), air cooled, then heated 8 h at 600 °C (1100 °F) and air cooled. (e) Annealed 1 h at 1010 °C (1850 °F), air cooled, then heated to 566 °C (1050 °F). (f) Annealed 1 h at 1010 °C (1850 °F) and oil quenched. (g) Annealed 1 h at 954 °C (1750 °F), air cooled, then heated 8 h to 600 °C (1100 °F) and air cooled

Table 12 Typical rolling temperatures for several titanium metals

Alloy	Rolling temperatures: Bar °C	Bar °F	Plate °C	Plate °F	Sheet °C	Sheet °F
Commercially pure titanium						
Grades 1–4	760–815	1400–1500	760–790	1400–1450	705–760	1300–1400
α and near-α alloys						
Ti-5Al-2.5Sn	1010–1065	1850–1950	980–1040	1800–1900	980–1010	1800–1850
Ti-6Al-2Sn-4Zr-2Mo	955–1010	1750–1850	955–980	1750–1800	925–980	1700–1800
Ti-8A-1Mo-1V	1010–1040	1850–1900	980–1040	1800–1900	980–1040	1800–1900
α-β alloys						
Ti-8Mn	...	...	705–760	1300–1400	705–760	1300–1400
Ti-4Al-3Mo-1V	925–955	1700–1750	900–925	1650–1700	900–925	1650–1700
Ti-6Al-4V	955–1010	1750–1850	925–980	1700–1800	900–925	1650–1700
Ti-6Al-6V-2Sn	900–955	1650–1750	870–925	1600–1700	870–900	1600–1650
Ti-7Al-4Mo	955–1010	1750–1850	925–955	1700–1750	925–955	1700–1750
β alloy						
Ti-13V-11Cr-3Al	955–1065	1750–1950	980–1040	1800–1900	730–900	1350–1650

Table 13 Tensile properties of unidirectionally rolled Ti-6Al-4V sheet

Gage mm	Gage in.	Tensile strength MPa	Tensile strength ksi	Yield strength MPa	Yield strength ksi	Elongation(a), %	Tensile modulus GPa	Tensile modulus 10^6 psi
Longitudinal direction								
0.737	0.029	945	137	870	126	7.0	100	14.5
1.016	0.040	970	141	855	124	6.5	106	15.4
1.168	0.046	915	133	860	125	6.5	105	15.2
1.524	0.060	985	143	925	134	6.5	104	15.1
1.778	0.070	995	144	915	133	8.0	105	15.3
Transverse direction								
0.737	0.029	1105	160	1061	154	7.5	130	18.8
1.016	0.040	1195	173	1105	160	7.5	145	21.1
1.168	0.046	1225	178	1165	169	7.5	140	20.2
1.524	0.060	1125	163	1090	158	8.0	125	18.2
1.778	0.070	1095	159	1055	153	9.5	135	19.5

(a) In 50 mm (2 in.)

tallized microstructure and less tendency for texturing when compared with α + β or alpha alloys.

Reduction to Billet. Generally, the first breakdown of production ingot is a press cogging operation done in the beta temperature range. Modern processes utilize substantial amounts of working below the beta transus to produce billets with refined structures. These processes are carried out at temperatures high in the alpha region to allow greater reduction and improved grain refinement with a minimum of surface rupturing. Where maximum fracture toughness is required, beta processing (or alpha-beta processing followed by beta heat treatment) is generally preferred. Table 10 gives standard forging-temperature ranges for manufacture of billet stock.

Some billets intended for further forging, rolling, or extrusion go through a grain-refinement process. This technique, developed in the early 1970s, utilizes the fact that titanium recrystallizes when it is heated above the beta transus. However, because grain boundary alpha forms when most (and especially beta-rich) alloys are cooled from above the beta transus, working in the alpha-beta region may be needed to control the formation of grain boundary alpha. Working during continuous cooling through the beta transus is a very effective method of eliminating grain boundary alpha (Ref 13). By starting with grain-refined billet, secondary fabricators may be able to produce forgings that meet strict requirements with respect to macrostructure, microstructure, and mechanical properties without extensive hot working below the beta transus.

Final tensile properties of alpha-beta alloys are strongly influenced by the amount of processing in the alpha-beta field—both below the beta transus temperature and after recrystallization. Such processing increases the strength of high-alpha grades in large section sizes. With modern processing techniques, billet and forged sections readily meet specified tensile properties prior to final forging. Table 11 shows how billet and forging section size affects room-temperature tensile properties of various titanium alloys.

Rolling of Bar, Plate, and Sheet. Roll cogging and hot roll finishing of bar, plate, and sheet are now standard operations, and special rolling and auxiliary equipment have been installed by the larger titanium producers to allow close control of all rolling operations. Rolling processes used by each manufacturer are proprietary and in some respects unique, but because all techniques must produce the same specified structures and mechanical properties, a high degree of similarity exists among the processes of all manufacturers.

A representative range of temperatures used for hot rolling of titanium metals is presented in Table 12. Rolling at these temperatures produces end products with the desired grain structures. When production limitations require the suppression of aging reaction kinetics, the beta phase stability of the beta alloys allows manufacture of hot band and cold-rolled strip product. Because of the body-centered cubic crystal structure of the beta phase, flat-rolled products and even cold-rolled strip are relatively free of in-plane texture (Ref 14). This makes possible strip and plate mill products with very uniform properties.

Bars up to about 100 mm (4 in.) in diameter are unidirectionally rolled, and their properties commonly reflect total reduction in the alpha-beta range. For example, a round bar 50 mm (2 in.) in diameter rolled from a Ti-6Al-4V billet 100 mm (4 in.) square typically is 140 to 170 MPa (20 to 25 ksi) lower in tensile strength than rod 7.8 mm (5/16 in.) in diameter rolled on a rod mill from a billet of the same size at the same rolling temperatures. For bars about 50 to 100 mm (2 to 4 in.) in diameter, strength does not decrease with section size, but transverse ductility and notched stress-rupture strength at room temperature do become lower. In diameters greater than about 75 to 100 mm (3 to 4 in.), annealed Ti-6Al-4V bars usually do not meet prescribed limits for notched stress rupture at room temperature—1170 MPa (170 ksi) minimum to cause rupture of a notched specimen in 5 h—unless the material is given a special duplex anneal. Transverse ductility is lower in bars about 65 to 100 mm (2½ to 4 in.) in diameter because it is not possible to obtain the preferred texture throughout bars of this size.

Plate and sheet commonly exhibit higher tensile properties in the transverse direction relative to the final rolling direction (Table 13). Cross rolling is used to achieve a balance in transverse and longitudinal properties. Unidirectional rolling (Table 13) is not yet used for alpha-beta titanium alloys.

Directionality in properties is observed only as a slight drop in transverse ductility of plate greater than 25 mm (1 in.) thick. Military, AMS, and customer specifications all prescribe lower minimum tensile and yield strengths as plate thickness increases. For forming applications, some customers specify a maximum allowable difference between tensile strengths in the transverse and longitudinal directions.

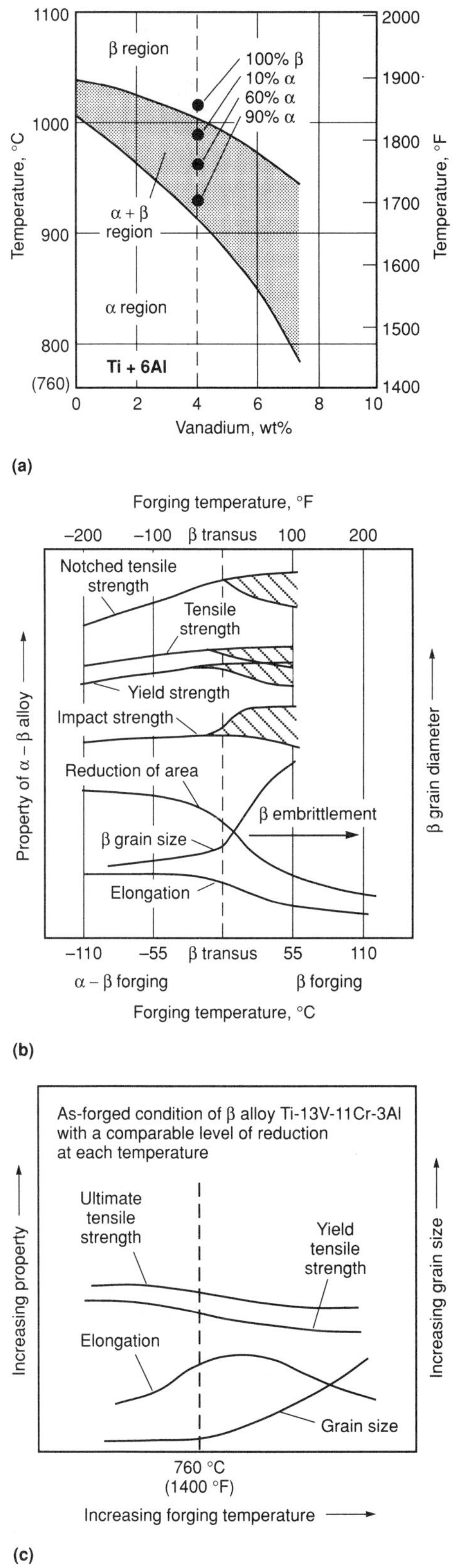

Fig. 16 Effects of forging and heat treating temperatures on properties of titanium alloys. (a) Phase diagram of alpha and beta contents with a base composition of titanium + 6 wt% Al. (b) Generalized effect of processing temperature on beta grain size and room-temperature mechanical properties of an alpha-beta alloy. (c) Generalized effect of processing temperature on as-forged room-temperature mechanical properties of beta alloy Ti-13V-11Cr-3Al

Table 14 Thermochemical schedules for producing various combinations of properties in Ti-6Al-4V forgings

Initial microstructure	Blocker forging temperature range	Finish forging temperature range	Finish forging reduction, %	Cooling after forging	Heat treated condition	Final microstructure
Best combinations of properties						
...	α-β	α-β	...	Air cooled	Annealed	6% equiaxed α plus fine platelet α
Grain-boundary α	α-β	α-β	...	Air cooled	Annealed	26% elongated partly broken up grain-boundary primary α plus fine platelet α
Grain-boundary α	α-β	α-β	...	Water quenched	Annealed	23% elongated partly broken up primary α plus very fine platelet α
...	β	α-β	10	Air cooled	Annealed	63% fine elongated primary α plus fine platelet α
Subnormal properties						
Spaghetti α	α	α	...	Air cooled	STOA(a)	25% blocky primary α plates plus very fine platelet α
...	β	α-β	10	Water quenched	STOA(a)	43% coarse elongated primary α plates plus very fine platelet α
...	β	β	...	Slow cooled	Annealed	92% α basketweave structure

(a) STOA, solution treated and overaged

Forging

Forging is a common method of producing wrought titanium articles, and titanium alloy forgings are produced by all of the forging methods currently available, including open-die (or hand) forging, closed-die forging, upsetting, roll forging, orbital forging, spin forging, mandrel forging, ring rolling, and forward and backward extrusion. Selection of the optimal forging method for a given forging shape is based on the desired forging shape, the sophistication of the design of the forged shape, the cost, and the desired mechanical properties and microstructure. In many cases, two or more forging methods are combined to achieve the desired forging shape, to obtain the desired final part microstructure, and/or to minimize cost. For example, open-die forging frequently precedes closed-die forging to preshape or preform the metal to conform to the subsequent closed dies, to conserve the expensive input metal, and/or to assist in overall microstructural development and grain flow control.

Titanium alloys are forged into a variety of shapes and types of forgings, with a broad range of final part forging design criteria based on the intended application. As a class of materials, however, titanium alloys are considerably more difficult to forge than aluminum alloys and alloy steels, particularly with conventional forging techniques, which use nonisothermal die temperatures of 535 °C (1000 °F) or less and moderate strain rates. Therefore, titanium alloy forgings, particularly closed-dic forgings, are typically produced to less highly refined final forging configurations than are typical of aluminum alloys (although precision forgings in titanium alloys are produced to the same design and tolerance criteria as aluminum alloys; see the article "Forging of Titanium Alloys" in *Forming and Forging*, Volume 14 of the 9th Edition of *Metals Handbook*).

Most titanium alloy forgings are thermally treated after forging, with heat treatment processes ranging from simple stress-relief annealing to multiple-step processes of solution treating, quenching, aging, and/or annealing designed to modify the microstructure of the alloy to meet specific mechanical property criteria. However, the working history and forging parameters used in titanium alloy forging also have a significant impact on the final microstructure (and therefore the resultant mechanical properties) of the forged alloy—perhaps to a greater extent than in any other commonly forged material. Therefore, the forging process in titanium alloys is used not only to create cost-effective forging shapes but also, in combination with thermal treatments, to create unique and/or tailored microstructures to achieve the desired final mechanical properties through thermomechanical processing techniques. In fact, one of the main purposes of die forging is to obtain a combination of mechanical properties that generally does not exist in bar or billet. Tensile strength, creep resistance, fatigue strength, and toughness all may be better in forgings than in bar or other forms.

The hot deformation processes conducted during the forging of the three classes of titanium alloys form an integral part of the overall thermomechanical processing of these alloys to achieve the desired microstructure. By the design of the working process history from ingot to billet to forging, and particularly the selection of metal temperatures and deformation conditions during the forging process, significant changes in the morphology of the allotropic phases of titanium alloys are achieved that in turn dictate the final mechanical properties and characteristics of the alloy.

The key to successful forging and heat treatment is the beta transus temperature. Figure 16 shows the possible locations for temperature of forging and/or heat treatment of a typical alpha-beta alloy such as Ti-6Al-4V. The higher the processing temperature in the $\alpha + \beta$ region, the more beta is available to transform upon cooling. Upon quenching from above the beta transus, a completely transformed, acicular structure arises. The form of the transformed beta structures produced by processing depends on the exact location of the beta transus, which varies from heat to heat of a given alloy, and also on the degree and nature of deformation produced. Section size is important, and the number of working operations can be significant. Conventional forging may require two or three operations, whereas isothermal forging may require only one.

Fundamentally, there are two principal metallurgical approaches to the forging of titanium alloys:

- Forging predominantly below the beta transus (alpha-beta forging)
- Forging predominantly above the beta transus (beta forging)

However, within these fundamental approaches, there are several possible variations that blend these two techniques into processes that are used commercially to achieve controlled microstructures that tailor the final properties of the forging to specification requirements and/or intended service applications. Table 14, for example, summarizes four thermomechanical schedules that produced optimal combinations of properties in Ti-6Al-4V test forgings: excellent tensile strength, good-to-excellent notch fatigue strength, low-cycle fatigue strength, and fracture toughness. Also included in the table are three schedules that produced subnormal properties. The microstructures of Ti-6Al-4V shown in Fig. 15 correspond to two of the schedules that produced good combinations of properties and two that produced inferior combinations. Note the substantial difference in microstructure in the same final product, which, in combination with the resulting properties, demonstrates that control of

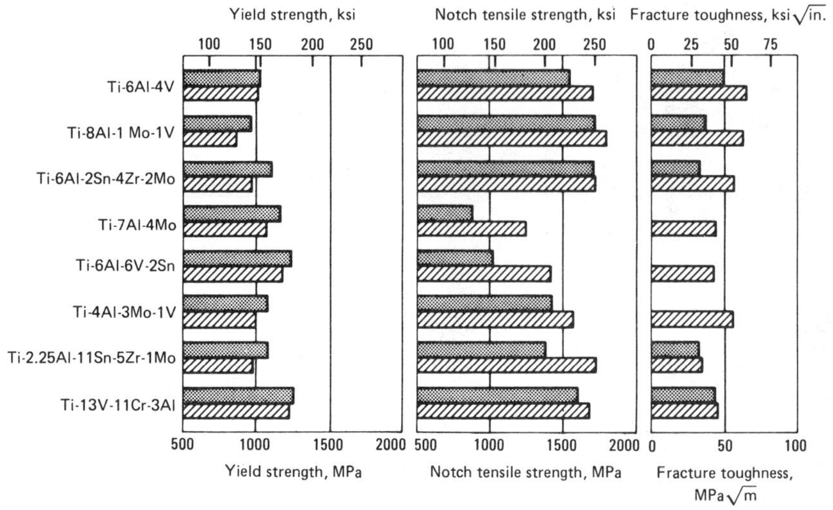

Fig. 17 Comparison of typical mechanical properties of alpha-beta forged and beta forged titanium alloys. Shaded bars represent alpha-beta forged material; striped bars, beta forged material

thermomechanical processing can control the microstructures and corresponding final properties of forgings.

Conventional alpha-beta forging of titanium alloys, in addition to implying the use of die temperatures of 540 °C (1000 °F) or less, is the term used to describe a forging process in which most or all of the forging deformation is conducted at temperatures below the beta transus of the alloy. This forging technique involves working the material at temperatures where both alpha and beta phases are present, with the relative amounts of each phase being dictated by the composition of the alloy and the actual temperature used. With this forging technique, the resultant as-forged microstructure is characterized by deformed or equiaxed primary alpha in a transformed beta matrix; the volume fraction and morphology of primary alpha is dictated by the alloy composition and the actual working history and temperature. Alpha-beta forging is typically used to develop optimal strength/ductility combinations and optimal high/low-cycle fatigue properties. With alpha-beta forging, the effects of working on microstructure, particularly alpha morphology changes, are cumulative; therefore, each successive alpha/beta working operation adds to the structural changes achieved in earlier operations. In the beta alloys, manipulation of the alpha phase during forging is less prevalent; therefore, the beta alloys are typically forged above the beta transus.

Beta forging, as the term implies, is a forging technique for alpha, beta, and alpha-beta alloys in which most or all of the forging work is done at temperatures above the beta transus of the alloy. In commercial practice, beta forging techniques typically involve supertransus forging in the early and/or intermediate stages with controlled amounts of final deformation below the beta transus of the alloy. However, isothermal beta forging is finding use in production of the more creep-resistant components of titanium alloys.

The beta-forged alloys tend to show a transformed beta or acicular microstructure, whereas alpha-beta forged alloys show a more equiaxed structure. Because each structure has unique capabilities (Table 8), tradeoffs are required in developing either an equiaxed or acicular structure. Figure 17, for example, compares alpha-beta forging versus beta forging for several titanium alloys. Although yield strength after beta forging was not always as high as that after alpha-beta forging, values of notch tensile strength and fracture toughness were consistently higher for the beta-forged material.

Consequently, beta forging is typically used to enhance fracture-related properties, such as fracture toughness and fatigue crack propagation resistance, and to enhance the creep resistance of alpha and alpha-beta alloys. In fact, several recently developed alpha alloys (such as IMI 829 and 834) are designed to be beta forged to develop the desired final mechanical properties. There is often a loss in strength and ductility with beta forging as compared to alpha-beta forging.

In beta forging, the working influences on microstructure are not fully cumulative; with each working-cooling-reheating sequence above the beta transus, the effects of the prior working operations are at least partially lost because of recrystallization from the transformation upon heating above the beta transus of the alloy. Beta forging, particularly of alpha and alpha-beta alloys, has the advantages of significant reduction in forging unit pressures and reduced cracking tendency, but it must be done under carefully controlled forging process conditions to avoid nonuniform working, excessive grain growth, and/or poorly worked structures, all of which can result in final

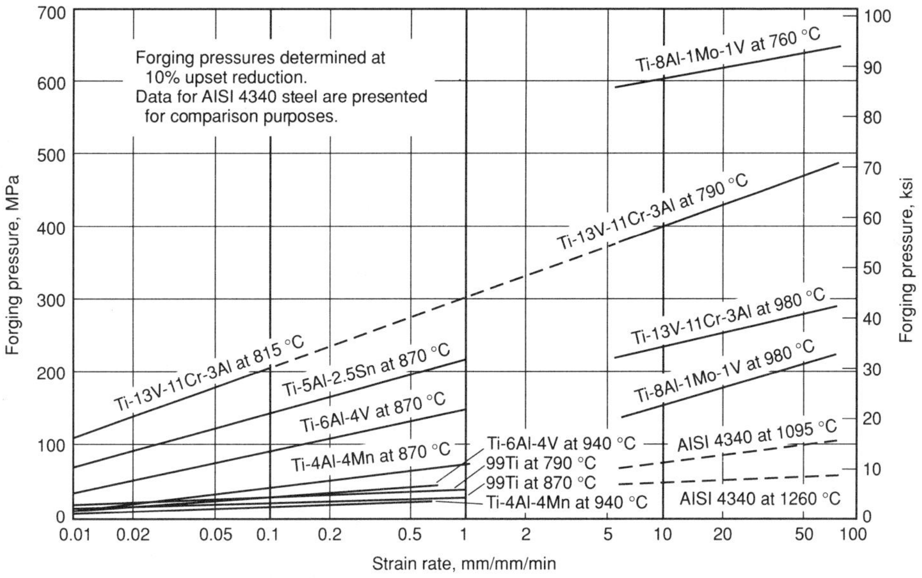

Fig. 18 Effect of strain rate on forging pressures for several titanium alloys at various forging temperatures. Data for AISI 4340 steel are presented for comparison purposes.

forgings with unacceptable or widely variant mechanical properties within a given forging or from lot to lot of the same forging.

Effect of Deformation Rate. Titanium alloys are highly strain-rate sensitive in deformation processes such as forging—considerably more so than aluminum alloys or alloy steels. The strain-rate sensitivity at forging temperatures is much higher for the beta and near-beta alloys, with the result that alloys such as Ti-13V-11Cr-3Al show marked increases in strength or flow stress as the deformation rate is increased. For example, at 788 °C (1450 °F) this alloy requires 50% more energy at a typical hammer velocity of 508 cm/sec (200 in./sec) than at a typical press velocity of 2.8 cm/sec (1.5 in./sec). However, the differences are much less for alpha and $\alpha + \beta$ alloys (Ref 13).

From the known strain-rate sensitivity of titanium alloys, it appears to be advantageous to deform these alloys at relatively slow strain rates in order to reduce the resistance to deformation in forging (Fig. 18); however, under the nonisothermal conditions present in the conventional forging of titanium alloys, the temperature losses encountered by such techniques far outweigh the benefits of forging at slow strain rates. Therefore, in the conventional forging of titanium alloys with relatively cool dies, intermediate strain rates are typically employed as a compromise between strain-rate sensitivity and metal temperature losses in order to obtain the optimal deformation possible with a given alloy. As discussed in the section "Isothermal and Hot-Die Forging" below, major reduction in resistance to deformation of titanium alloys can be achieved by slow strain-rate forging techniques under conditions where metal temperatures losses are minimized through dies heated to temperatures at or close to the metal temperature.

With rapid deformation rate forging techniques, such as the use of hammers and/or mechanical presses, deformation heating during the forging process becomes important. Because titanium alloys have relatively poor coefficients of thermal conductivity, temperature nonuniformity may result, giving rise to nonuniform deformation behavior and/or excursions to temperatures that are undesirable for the alloy and/or final forging mechanical properties. As a result, in the rapid strain-rate forging of titanium alloys, metal temperatures are often adjusted to account for in-process heat-up, or the forging process (sequence of blows, and so on) is controlled to minimize undesirable temperature increases, or both. Therefore, within the forging temperature ranges outlined in Table 15, metal temperatures for optimal titanium alloy forging conditions are based on the type of forging equipment to be used, the strain rate to be employed, and the design of the forging part.

Hot-die and isothermal forging are special categories of forging processes in which the die temperatures are significantly higher than those used in conventional hot-forging processes. This has the advantage of reducing die chill and results in a process capable of producing near-net and/or net shape parts. Therefore, these processes are also referred to as near-net shape forging processes. These processing techniques are primarily used for manufacturing airframe structures and jet-engine components made of titanium and nickel-base alloys, but they have also been used in steel transmission gears and other components.

In the isothermal forging process, the dies are maintained at the same temperature as the forging stock. This eliminates the die chill completely and maintains the stock at a constant temperature throughout the forging cycle. The process permits the use of extremely slow strain rates, thus taking advantage of the strain-rate sensitivity of flow stress for certain alloys. The process is capable of producing net shape forgings that are ready to use without machining or near-net shape forgings that require minimal secondary machining.

The hot-die forging process is characterized by die temperatures higher than those in conventional forging, but lower than those in isothermal forging. Typical die temperatures in hot-die forging are 110 to 225 °C (200 to 400 °F) lower than the temperature of the stock. When compared with isothermal forging, the lowering of die temperature allows wider selection of die materials, but the ability to produce very thin and complex geometries is compromised.

The alloys used for hot-die and isothermal forging include titanium alloys such as Ti-

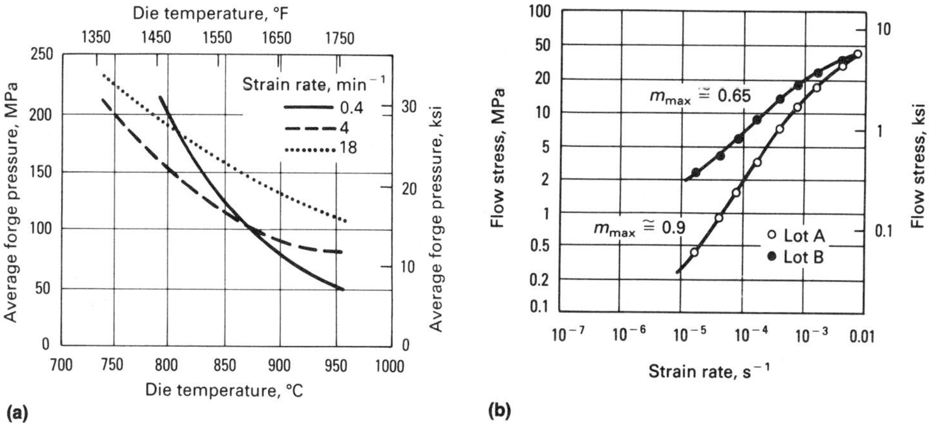

Fig. 19 Forging pressure and flow stress of Ti-6Al-4V. (a) Effect of die temperature at various strain rates. (b) Effect of grain size distribution on flow stress versus strain rate data for Ti-6Al-4V at 927 °C (1700 °F). Lot A, average grain size of 4 μm and grain size range of 1 to 10 μm; lot B, average grain size of 4.6 μm but grain size range of 1 to >20 μm

Table 15 Recommended forging temperature ranges for commonly forged titanium alloys

Alloy	Beta transus (β_t) °C	Beta transus (β_t) °F	Process(a)	Forging temperature(b) °C	Forging temperature(b) °F
α/near-α alloys					
Ti-C.P.(c)	915	1675	C	815–900	1500–1650
Ti-5Al-2.5Sn(c)	1050	1925	C	900–1010	1650–1850
Ti-5Al-6Sn-2Zr-1Mo-0.1Si	1010	1850	C	900–995	1650–1925
Ti-6Al-2Nb-1Ta-0.8Mo	1015	1860	C	940–1050	1725–1825
			B	1040–1120	1900–2050
Ti-6Al-2Sn-4Zr-2Mo(+0.2Si)(d)	990	1815	C	900–975	1650–1790
			B	1010–1065	1850–1950
Ti-8Al-1Mo-1V	1040	1900	C	900–1020	1650–1870
IMI 685 (Ti-6Al-5Zr-0.5Mo-0.25Si)(e)	1030	1885	C/B	980–1050	1795–1925
IMI 829 (Ti-5.5Al-3.5Sn-3Zr-1Nb-0.25Mo-0.3Si)(e)	1015	1860	C/B	980–1050	1795–1925
IMI 834 (Ti-5.5Al-4.5Sn-4Zr-0.7Nb-0.5Mo-0.4Si-0.06C)(e)	1010	1850	C/B	980–1050	1795–1925
α-β alloys					
Ti-6Al-4V(c)	995	1825	C	900–980	1650–1800
			B	1010–1065	1850–1950
Ti-6Al-4V-ELI	975	1790	C	870–950	1600–1740
			B	990–1045	1815–1915
Ti-6Al-6V-2Sn	945	1735	C	845–915	1550–1675
Ti-6Al-2Sn-4Zr-6Mo	940	1720	C	845–915	1550–1675
			B	955–1010	1750–1850
Ti-6Al-2Sn-2Zr-2Mo-2Cr	980	1795	C	870–955	1600–1750
Ti-17 (Ti-5Al-2Sn-2Zr-4Cr-4Mo(f)	885	1625	C	805–865	1480–1590
			B	900–970	1650–1775
Corona 5 (Ti-4.5Al-5Mo-1.5Cr)	925	1700	C	845–915	1550–1675
			B	955–1010	1750–1850
IMI 550 (Ti-4Al-4Mo-2Sn)	990	1810	C	900–970	1650–1775
IMI 679 (Ti-2Al-11Sn-4Zr-1Mo-0.25Si)	945	1730	C	870–925	1600–1700
IMI 700 (Ti-6Al-5Zr-4Mo-1Cu-0.2Si)	1015	1860	C	800–900	1470–1650
β, near-β, and β alloys					
Ti-8Al-8V-2Fe-3Al	775	1425	C/B	705–980	1300–1800
Ti-10V-2Fe-3Al	805	1480	C/B	705–785(g)	1300–1450(g)
			B	815–870	1500–1600
Ti-13V-11Cr-3Al	675	1250	C/B	650–955	1200–1750
Ti-15V-3Cr-3Al-3Sn	770	1415	C/B	705–925	1300–1700
Beta C (Ti-3Al-8V-6Cr-4Mo-4Zr)	795	1460	C/B	705–980	1300–1800
Beta III (Ti-4.5Sn-6Zr-11.5Mo)	745	1375	C/B	705–955	1300–1750
Transage 129 (Ti-2Al-11.5V-2Sn-11Zr)	720	1325	C/B	650–870	1200–1600
Transage 175 (Ti-2.7Al-13V-7Sn-2Zr)	760	1410	C/B	705–925	1300–1700

(a) C, conventional forging processes in which most or all of the forging work is accomplished below the β_t of the alloy for the purposes of desired mechanical property development. This forging method is also referred to as α-β forging. B, β forging processes in which some or all of the forging is conducted above the β_t of the alloy to improve hot workability or to obtain desired mechanical property combinations. C/B, either forging methodology (conventional or β) is employed in the fabrication of forgings or for alloys, such as β alloys, that are predominately forged above their β_t but may be finish forged at subtransus temperatures. (b) These are recommended metal temperature ranges for conventional α-β, or β forging processes for alloys for which the latter techniques are reported to have been employed. The lower limit of the forging temperature range is established for open-die forging operations in which reheating is recommended. (c) Alloys for which there are several compositional variations (primarily oxygen or other interstitial element contents) that may affect both β_t and forging temperature ranges. (d) This alloy is forged and used both with and without the silicon addition; however, the β_t and recommended forging temperatures are essentially the same. (e) Alloys designed to be predominately β forged. (f) Ti-17 has been classified as an α-β and as a near-β titanium alloy. For purposes of this article, it is classified as an α-β alloy. (g) Temperature for finish forging; primary forging performed at about 845 °C (1550 °F).

6Al-4V, Ti-6Al-2Sn-4Zr-2Mo-0.1Si, and Ti-10V-2Fe-3Al. Isothermal forging of alpha-beta alloys is technically feasible, although high process and tooling costs, catastrophic die failures, and other engineering problems associated with very high process temperatures combine to minimize its use on conventional alpha-beta alloys.

Die Temperature. Proper selection of die temperature is one of the critical factors in process design for hot-die and isothermal forging. The effect of die temperature on forging pressure is illustrated in Fig. 19 for Ti-6Al-4V. As shown in Fig. 19, a decrease in die temperature from 955 to 730 °C (1750 to 1350 °F) may result in doubling the forging pressure and may affect the shape capability available.

Die Temperature in Conventional Forging. The dies used in the conventional forging of titanium alloys, unlike some other materials, are heated to facilitate the forging process and to reduce metal temperature losses during the forging process—particularly surface chilling, which may lead to inadequate die filling and/or excessive cracking. Table 16 lists the recommended die temperatures used for several titanium alloy forging processes employing conventional die temperatures. Dies are usually preheated to these temperature ranges using the die heating techniques discussed below. In addition, because the metal temperature of titanium alloys exceeds that of the dies, heat transfer to the dies occurs during conventional forging, frequently requiring that the dies be cooled to avoid die damage. Cooling techniques include wet steam, air blasts, and, in some cases, water.

Table 16 Die temperature ranges for the conventional forging of titanium alloys

Forging process/equipment	Die temperature °C	Die temperature °F
Open-die forging		
Ring rolling	150–260	300–500
	95–260	200–500
Closed-die forging		
Hammers	95–260	200–500
Upsetters	150–260	300–500
Mechanical presses	150–315	300–600
Screw presses	150–315	300–600
Orbital forging	150–315	300–600
Spin forging	95–315	200–600
Roll forging	95–260	200–500
Hydraulic presses	315–480	600–900

Extrusion

Extrusion is used as an alternative to rolling as a mill process in order to make rodlike and seamless pipe products. Properties are affected by processing conditions in much the same way as they are for rolled or forged products. The properties of extruded products, however, are not identical to those of die-forged structures. Titanium extrusions are typically produced in the beta phase (beta extruded). Even where similar microstructures are produced, the thermomechanical working possible in open- and closed-die forging permits much more control over the resultant properties. One of the more unusual applications of extrusion has been in the production of tapered wing spars for a military aircraft.

Forming

Titanium and titanium alloy sheet and plate are strain hardened by cold forming. This normally increases tensile and yield strengths and causes a slight drop in ductility. Beta alloys generally are easier to form than are alpha and alpha-beta alloys. Titanium metals exhibit a high degree of springback in cold forming. To overcome this characteristic, titanium must be overformed or, as is done frequently, hot sized after cold forming.

In all forming operations, titanium and its alloys are susceptible to the Bauschinger effect. This is a drop in compressive yield strength in one loading direction caused by tensile deformation in another direction and vice versa. The Bauschinger effect is most pronounced at room temperature; plastic deformation (1 to 5% tensile elongation) at room temperature always introduces a significant loss in compressive yield strength, regardless of the initial heat treatment or strength of the alloys. At 2% tensile strain, for instance, the compressive yield strengths of Ti-4Al-3Mo-1V and Ti-6Al-4V drop to less than half the values for solution-treated material. Increasing the temperature reduces the Bauschinger effect; subsequent full thermal stress relieving completely removes it.

Table 17 Superplastic characteristics of titanium alloys

Alloy	Test temperature °C	Test temperature °F	Strain rate, s^{-1}	Strain rate sensitivity factor, m	Elongation, %
Commercially pure titanium	850	1560	1.7×10^{-4}	...	115
α-β alloys					
Ti-6Al-4V	840–870	1545–1600	1.3×10^{-4} to 10^{-3}	0.75	750–1170
Ti-6Al-5V	850	1560	8×10^{-4}	0.70	700–1100
Ti-6Al-2Sn-4Zr-2Mo	900	1650	2×10^{-4}	0.67	538
Ti-4.5Al-5Mo-1.5Cr	870	1600	2×10^{-4}	0.63–0.81	>510
Ti-6Al-4V-2Ni	815	1500	2×10^{-4}	0.85	720
Ti-6Al-4V-2Co	815	1500	2×10^{-4}	0.53	670
Ti-6Al-4V-2Fe	815	1500	2×10^{-4}	0.54	650
Ti-5Al-2.5Sn	1000	1830	2×10^{-4}	0.49	420
Near-β and β alloys					
Ti-15V-3Sn-3Cr-3Al	815	1500	2×10^{-4}	0.50	229
Ti-13Cr-11V-3Al	800	1470	...	...	<150
Ti-8Mn	750	1380	...	0.43	150
Ti-15Mo	800	1470	...	0.60	100

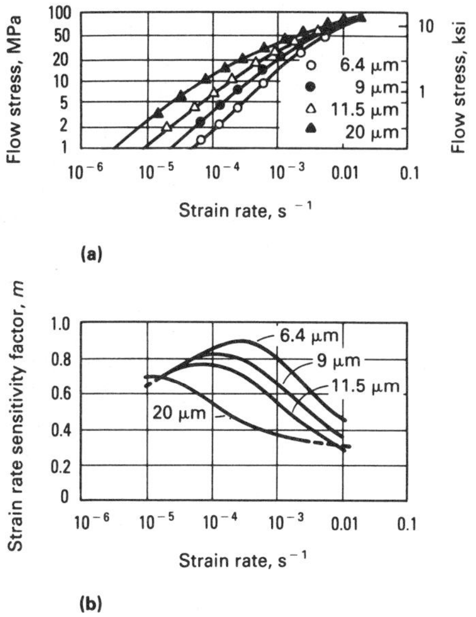

Fig. 20 Flow stress (a) and strain-rate sensitivity factor m (b) versus strain rate for Ti-6Al-4V materials with four different grain sizes. Test temperature: 927 °C (1700 °F)

Temperatures as low as the aging temperature might remove most of the Bauschinger effect in solution-treated titanium alloys. Heating or plastic deformation at temperatures above the normal aging temperature for solution-treated Ti-6Al-4V causes overaging to occur and, as a result, all mechanical properties decrease.

Cold Forming. Commercially pure titanium and most beta titanium alloys, such as Ti-15V-3Sn-3Cr-3Al and Ti-3Al-8V-6Cr-4Zr-4Mo, can be cold formed to a limited extent. Alloy Ti-8Al-1Mo-1V sheet can be cold formed to shallow shapes by standard methods, but the bends must be of larger radii than in hot forming and must have shallower stretch flanges. The cold forming of other alloys generally results in excessive springback, requires stress relieving between operations, and requires more power. Titanium and titanium alloys are commonly stretch formed without being heated, although the die is sometimes warmed to 150 °C (300 °F). For the cold forming of all titanium alloys, formability is best at low forming speeds.

To improve dimensional accuracy, cold forming is generally followed by hot sizing. Hot sizing and stress relieving are ordinarily needed to reduce stress and to avoid delayed cracking and stress corrosion. Stress relief is also needed to restore compressive yield strength after cold forming. Hot sizing is often combined with stress relieving, with the workpiece being held in fixtures or form dies to prevent distortion.

The only true cold-formable titanium alloy is Ti-15V-3Sn-3Cr-3Al, but hot sizing is probably required for all but brake forming. Properties must be developed with an aging treatment (8 h at 540 °C, or 1000 °F, is typical). Because of the high springback rates encountered with this alloy, more elaborate tooling must be used.

Hot forming of titanium alloys at temperatures from 595 to 815 °C (1100 to 1500 °F) increases formability, reduces springback, takes advantage of a lesser variation in yield strength, and allows for maximum deformation with minimum annealing between forming operations. It also eliminates the need for subsequent stress relief. The true net effect in any forming operation depends on total deformation and actual temperature during forming. Titanium metals also tend to creep at elevated temperatures; holding under load at the forming temperature (creep forming) is another alternative for achieving the desired shape without having to compensate for extensive springback. Severe forming must be done in hot dies, generally with preheated stock.

The greatest improvement in the ductility and uniformity of properties for most titanium alloys is at temperatures above 540 °C (1100 °F). At still higher temperatures, some alloys exhibit superplasticity (see the section "Superplastic Forming" below). However, contamination is also more severe at the higher temperatures. Above about 650 °C (1200 °F), forming should be done in vacuum or under a protective atmosphere, such as argon, to minimize oxidation. Coatings can also be used to minimize contamination. Most hot-forming operations are done at temperatures above 540 °C (1000 °F). For applications in which the utmost in ductility is required, temperatures below 315 to 425 °C (600 to 800 °F) are usually avoided.

Temperatures generally must be kept below 815 °C (1500 °F) to avoid marked deterioration in mechanical properties. Superplastic forming, however, is performed at 870 to 925 °C (1600 to 1700 °F) for alloys such as Ti-6Al-4V. At these temperatures, care must be taken not to exceed the beta transus temperature of Ti-6Al-4V. Heating temperature and time at temperature is controlled so that the titanium is hot for the shortest time practical and the metal temperature is in the correct range.

Scaling and Embrittlement. Titanium is scaled and embrittled by oxygen-rich surface layers formed at temperatures higher than 540 °C (1000 °F) commonly referred to as alpha case. The subsequent removal of scale and embrittled surface, or a protective atmosphere, should be considered for any heating above 540 °C (1000 °F). Argon gas is a commonly used atmosphere for superplastic forming.

Aging. Some hot-forming temperatures are high enough to age a titanium alloy. Heat-treatable beta and alpha-beta alloys generally must be reheat treated (solution annealed) after hot forming. Alpha-beta alloys should not be formed above the beta transus temperature.

Because of aging, scaling, and embrittlement, as well as the greater cost of working at elevated temperatures, hot forming is ordinarily done at the lowest temperature that will permit the required deformation. When maximum formability is required, the forming should be done at the highest temperature practical that will retain the mechanical properties and serviceability required of the workpiece.

Tools. Titanium alloys are often formed hot in heated dies in presses that have a slow, controlled motion and that can dwell in the position needed during the press cycle. Hot forming is sometimes done in dies that include heating elements or in dies that are heated by the press platens. Press platens heated to 650 °C (1200 °F) can transmit enough heat to keep the working faces of the die at 425 to 480 °C (800 to 900 °F). Other methods of heating include electrical-resistance heating and the use of quartz lamps and portable furnaces.

Accuracy. Hot forming has the advantage of improved uniformity in yield strength, especially when the forming or sizing temperature is above 540 °C (1000 °F). However, care must be taken to limit the accumulation of dimensional errors resulting from:

- Differences in thermal expansion
- Variations in temperature
- Dimensional changes from scale formation
- Changes in dimensions of tools
- Reduction in thickness from chemical pickling operations

Superplastic Forming. Superplasticity is a term used to indicate the exceptional ductility that certain metals can exhibit when deformed under proper conditions. Although there are several different types of superplasticity, only the micrograin superplasticity is of importance in the fabrication of parts. For micrograin superplasticity, the high ductilities are observed only under certain conditions, and the basic requirements for this type of superplasticity are:

- Very fine grain size material (of the order of 10 μm, or 400 μin.)
- Relatively high temperature (greater than about one-half the absolute melting point)
- A controlled strain rate, usually 0.0001 to 0.01 s^{-1}
- A two-phase structure (alpha and beta in titanium)

Because of these requirements, only a limited number of commercial alloys are superplastic, and these materials are formed using methods and conditions that are different from those used for conventional metals.

However, some of the titanium alloys (Table 17) have been found to be superplastic as conventionally produced, without any alloy modifications nor special mill-processing methods to make them superplastic. The characteristic flow properties of a superplastic metal are exemplified in Fig. 20 for a Ti-6Al-4V alloy tested at 927 °C (1700 °F). It is well known that the primary factor related to this behavior is the rate of change of flow stress with strain rate, usually measured and reported as m, the strain-rate sensitivity exponent:

$$m = \frac{\partial \ln \sigma}{\partial \ln \dot{\epsilon}} \qquad \text{(Eq 1)}$$

where σ is the flow stress and $\dot{\epsilon}$ is the strain rate. The higher the m value of an alloy, the greater its superplasticity.

The metallurgical variables affecting superplastic behavior in titanium alloys include grain size, grain size distribution, alpha morphology grain growth kinetics, diffusivity, phase ratio of alpha and beta, and texture. Alloy composition is also significant and can have a pronounced effect on α-β phase ratio and on diffusivity.

Table 17 shows that the alpha-beta titanium alloys seem to exhibit greater superplasticity than other titanium alloys. The alpha and beta phases are quite different in terms of crystal structure (hexagonal close-packed for alpha, and body-centered cubic for beta) and diffusion kinetics. Beta phase

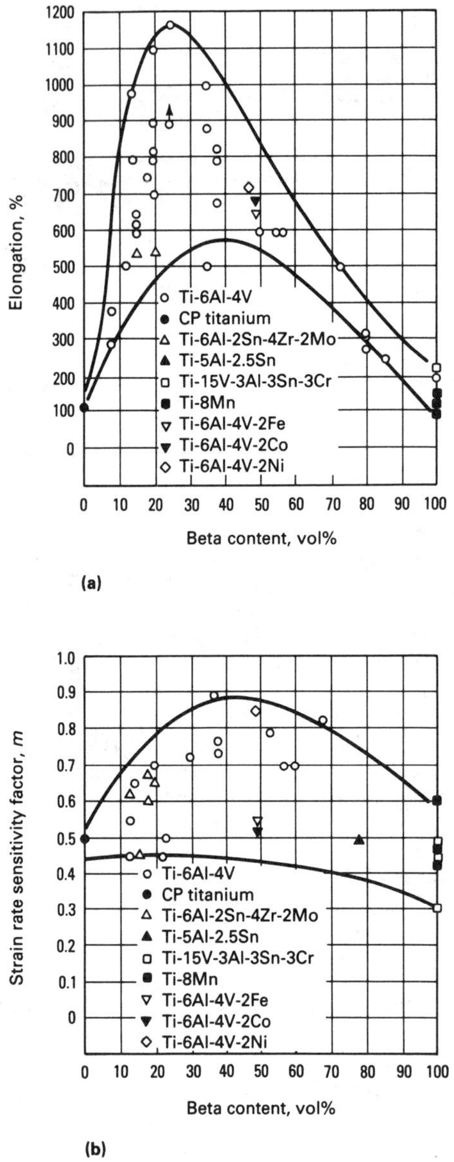

Fig. 21 Elongation (a) and m value (b) as a function of beta-phase content for several titanium alloys. See text for details.

exhibits a diffusivity approximately two orders of magnitude greater than that of alpha phase. For this reason alone it should be expected that the amount of beta phase present in a titanium alloy would have an effect on superplastic behavior.

Figure 21 shows elongations and m values for several titanium alloys as a function of the volume fraction of beta phase present in the alloys. It can be readily seen that elongation values reach a peak at approximately 20 to 30 vol% beta phase (Fig. 21a), while m values peak at beta contents of about 40 to 50 vol% (Fig. 21b). Because m is usually considered to be a good indicator of superplasticity, this discrepancy in the location of maxima of the curves in Fig. 21 may be surprising. It is believed that the difference stems from a grain growth effect during superplastic deformation. Beta phase is known to exhibit more rapid grain coarsening than alpha, and the maximum ductility may be the result of a balance between moderated grain growth (due to the presence of alpha phase) and enhanced diffusivity (due to the presence of beta).

The superplastic forming of titanium alloys is currently being used to fabricate a number of sheet metal components for a range of aircraft and aerospace systems. Hundreds of parts are in production, and significant cost savings are being realized through the use of superplastic forming. Other advantages of superplastic forming over other forming processes include the following:

- Very complex part configurations are readily formed
- Lighter, more efficient structures are possible
- It is performed in a single operation, reducing fabrication time
- Depending on part size, more than one piece can be produced per machine cycle
- The force needed for forming is supplied by a gas, resulting in the application of equal amounts of pressure to all areas of the workpiece

The limitations of the process include:

- Heat-resistant tool materials that contain minimal amounts of nickel are required
- Equipment requirements are extensive
- Long preheat times are necessary to reach the forming temperature
- A protective atmosphere, such as argon, is required

Several processes are used in the superplastic forming of titanium alloys. Among these are blow forming, vacuum forming, thermoforming, deep drawing, and superplastic forming/diffusion bonding. All of these processes are discussed in more detail in *Forming and Forging*, Volume 14 of the 9th Edition of *Metals Handbook*.

Joining

Adhesive bonding, brazing, mechanical fastening, metallurgical bonding, and welding are all used routinely and successfully to join titanium and its alloys. The first three processes do not affect the properties of these metals as long as joints are properly designed. Metallurgical bonding includes all solid-state joining processes in which diffusion or deformation play the major role in bonding the members together.

Because these processes are performed at elevated temperatures, metallurgical effects, either normally caused by heating at that temperature or resulting from contamination, should be anticipated. Except for adhesive bonds, properly processed joints have the same properties as the base metal and, because bonding is carried out at a temperature

Table 18 Typical tensile, bend, and hardness data for as-welded titanium and several titanium alloys

Material condition	Tensile strength		Yield strength		Elongation, %	Minimum bend radius	Hardness	
	MPa	ksi	MPa	ksi			Knoop	Rockwell
Ti Grade 1								
Unwelded sheet	315	46	215	31	50.4	0.7*t*	140	63.5 HRB
Single-bead weld	345	50	255	37	37.5	1.0*t*	140	55.8 HRB
Multiple-bead weld	365	53	270	39	37.7	...	...	...
Transverse weld	325	47(a)	...	...	...	...	...	...
Ti Grade 2								
Unwelded sheet	460	67	325	47	26.2	2.9*t*	165	80.6 HRB
Single-bead weld	505	73	380	55	18.3	2.9*t*	175	83.1 HRB
Multiple-bead weld	510	74	385	56	13.3	...	...	...
Transverse weld	475	69(a)	...	...	...	...	...	...
Ti Grade 3								
Unwelded sheet	545	79	395	57	25.9	1.9*t*	175	94.4 HRB
Single-bead sheet	605	88	475	69	15.5	4.7*t*	220	92.4 HRB
Multiple-bead weld	615	89	480	70	14.7	...	...	...
Transverse weld	560	81(a)	...	...	...	...	...	...
Ti Grade 4								
Unwelded sheet	660	96	530	77	22.3	3.2*t*	215	23.4 HRC
Single-bead weld	695	101	580	84	16.4	5.6*t*	240	21.2 HRC
Multiple-bead weld	710	103	585	85	16.0	...	...	...
Transverse weld	660	96(a)	...	...	...	...	...	...
Ti-5Al-2.5Sn-ELI								
Unwelded sheet	850	123	805	117	15.7	3.8*t*	265	33.2 HRC
Single-bead weld	920	133	770	112	9.8	5.9*t*	310	28.0 HRC
Multiple-bead weld	935	136	820	119	7.5	...	...	...
Transverse weld	850	123(a)	...	...	...	...	...	...
Ti-6Al-2Nb-1Ta-1Mo								
Unwelded sheet	895	130	855	124	9.7	2.8*t*	275	29.6 HRC
Single-bead weld	930	135	800	116	5.9	7.7*t*	300	27.7 HRC
Multiple-bead weld	945	137	815	118	5.7	...	...	...
Transverse weld	890	129(a)	...	...	...	...	...	...
Ti-3Al-2.5V								
Unwelded sheet	705	102	670	97	15.2	4.0*t*	230	23.6 HRC
Single-bead weld	705	102	600	87	12.7	5.4*t*	250	19.6 HRC
Multiple-bead weld	745	108	625	91	11.2	...	...	...
Transverse weld	710	103(a)	...	...	...	...	...	...
Ti-6Al-4V								
Unwelded sheet	1000	145	945	137	11.0	2.6*t*	320	32.2 HRC
Single-bead weld	1060	154	920	133	3.5	10.5*t*	350	35.9 HRC
Multiple-bead weld	1090	158	945	137	3.2	...	...	...
Transverse weld	1015	147(a)	...	...	...	...	...	...
Ti-8Al-1Mo-1V								
Unwelded sheet	1060	154	1020	148	15.0	2.9*t*	325	36.0 HRC
Single-bead weld	1085	157	930	135	5.5	7.0*t*	345	35.2 HRC
Multiple-bead weld	1115	162	960	139	3.2	...	...	...
Transverse weld	1060	154(a)	...	...	...	...	...	...
Ti-6Al-6V-2Sn								
Unwelded sheet	1060	154	1005	146	9.8	2.8*t*	350	34.0 HRC
Single-bead weld	1295	188	1255	182	0.3	25.6*t*	420	46.8 HRC
Multiple-bead weld	1280	186	...	...	0.1	...	...	...
Single-bead weld after furnace cool from 830 °C	1050	152	990	144	3.7	15.5*t*	...	...
Ti-13V-11Cr-3Al								
Unwelded sheet	965	140	910	132	13.9	2.7*t*	300	30.6 HRC
Single-bead weld	950	138	925	134	11.6	2.7*t*	320	30.1 HRC
Multiple-bead weld	925	134	875	127	9.1	...	...	...
Transverse weld	950	138(a)	...	...	...	...	...	...

(a) Fracture occurred in base metal.

high in the alpha-beta field, material properties appear similar to those resulting from high-temperature annealing. With most alloys, a final low-temperature anneal will produce properties characteristic of typical annealed material.

Welding has the greatest potential for affecting material properties. In all types of welds, contamination by interstitial impurities such as oxygen and nitrogen must be minimized to maintain useful ductility in the weldment. Alloy composition, welding procedure, and subsequent heat treatment are highly important in determining the final properties of welded joints. Table 18 reviews mechanical properties for representative alloys and types of welds. The data can be summarized as follows:

- Welding generally increases strength and hardness
- Welding generally decreases tensile and bend ductility
- Welds in unalloyed titanium grades 1, 2,

Table 19 Summary of heat treatments for alpha-beta titanium alloys

Heat treatment designation	Heat treatment cycle	Microstructure
Duplex anneal	Solution treat at 50–75 °C below T_β(a), air cool and age for 2–8 h at 540–675 °C	Primary α, plus Widmanstätten α + β regions
Solution treat and age	Solution treat at ~40 °C below T_β, water quench(b) and age for 2–8 h at 535–675 °C	Primary α, plus tempered α' or a β + α mixture
Beta anneal	Solution treat at ~15 °C above T_β, air cool and stabilize at 650–760 °C for 2 h	Widmanstätten α + β colony microstructure
Beta quench	Solution treat at ~15 °C above T_β, water quench and temper at 650–760 °C for 2 h	Tempered α'
Recrystallization anneal	925 °C for 4 h, cool at 50 °C/h to 760 °C, air cool	Equiaxed α with β at grain-boundary triple points
Mill anneal	α + β hot work + anneal at 705 °C for 30 min to several hours and air cool	Incompletely recrystallized α with a small volume fraction of small β particles

(a) T_β is the β-transus temperature for the particular alloy in question. (b) In more heavily β-stabilized alloys such as Ti-6Al-2Sn-4Zr-6Mo or Ti-6Al-6V-2Sn, solution treatment is followed by air cooling. Subsequent aging causes precipitation of α phase to form an α + β mixture. Source: Ref 13

and 3 do not require postweld treatment unless the material will be highly stressed in a strongly reducing atmosphere. In such event, stress relieving or annealing may prove useful

- Welds in more beta-rich alpha-beta alloys such as Ti-6Al-6V-2Sn have a high likelihood of fracturing with little or no plastic straining. Weld ductility can be improved by postweld heat treatment consisting of slow cooling from a high annealing temperature
- Rich beta-stabilized alloys can be welded, and such welds exhibit good ductility. The aging kinetics of the weld metal may be substantially different than that of the parent metal.

Electron-beam and laser welds are made without filler metal and weld beads have high ratios of depth to width. This combination allows excellent welds to be made in heavy sections, with properties very close to those of the base metal and little distortion.

Welding must be done under strict environmental controls to avoid pickup of interstitials that can embrittle the weld metal. Small and moderate size weldments ordinarily are enclosed within environmentally controlled chambers during welding or by shielding gas that protects the immediate weld zone. Larger weldments are made with the aid of portable chambers that only partly enclose the components, or with the aid of "trailers," both of which maintain a protective atmosphere on both front and back sides of the weld until it has cooled below about 480 °C (1000 °F).

Heat Treatment

Titanium and titanium alloys are heat treated for the following purposes:

- To reduce residual stresses developed during fabrication (stress relieving)
- To produce an optimal combination of ductility, machinability, and dimensional and structural stability (annealing)
- To increase strength (solution treating and aging)
- To optimize special properties such as fracture toughness, fatigue strength, and high-temperature creep strength

These various types of heat-treating cycles are not applicable to all titanium alloys. The alpha and near-alpha titanium alloys can be stress relieved and annealed, but high strength cannot be developed in these alloys by any type of heat treatment. The commercial beta alloys, on the other hand, all contain metastable beta, which thus allows strengthening during aging as the retained beta decomposes. The beta alloys offer great potential for age hardening and frequently utilize the stability of their beta phase to provide large section hardenability. For beta alloys, stress-relieving and aging treatments can be combined, and annealing and solution treating may be identical operations.

Finally, the alpha-beta alloys, as the name suggests, exhibit heat-treatment characteristics between that of the alpha class and the beta class. Alpha-beta alloys can exhibit age hardening from the decomposition of beta, but these alloys do not exhibit the same section size hardenability as the beta alloys due to the lesser amounts of retained beta. Nonetheless, the alpha-beta alloys are the most versatile in that certain microstructures (Table 19) can be enhanced by processing in either the alpha-beta region or the beta-phase region. In general, the beta or alpha-beta processing of alpha-beta alloys has the following effects on properties:

Property	β processed	α/β processed
Tensile strength	Moderate	Good
Creep strength	Good	Poor
Fatigue strength	Moderate	Good
Fracture toughness	Good	Poor
Crack growth rate	Good	Moderate
Grain size	Large	Small

Beta processing of near-alpha alloys for creep strength is useful because the near-alpha characteristic permits them to be worked or heat-treated in the beta-phase field without risk of the loss of room-temperature ductility encountered in other titanium alloys processed in this way. The near-alpha alloys may also be worked high in the alpha-beta to obtain an intermediate microstructure with a mixture of equiaxed and acicular alpha (Fig. 12b). This intermediate type of microstructure, which provides a good combination of fatigue and creep strength, is achieved in the near-alpha IMI 834 alloy by the use of carbon additions and processing high in the alpha-beta region (see the section "Carbon" in this article).

Stress Relieving. Titanium and titanium alloys can be stress relieved without adversely affecting strength or ductility. Stress-relieving treatments decrease the undesirable residual stresses that result from:

- Nonuniform hot forging deformation from cold forming and straightening
- Asymmetric machining of plate (hogouts) or forgings
- Welding and cooling of castings
- Residual thermal stresses generated during the cooling of parts with nonuniform cross sections

Removal of such stresses helps maintain shape stability and eliminates unfavorable conditions, such as the loss of compressive yield strength commonly known as the Bauschinger effect. Titanium mill producers offer a stress-free plate desirable for machining hogouts. This is accomplished by high-temperature creep flattening.

Stress-relieving treatments must be based on the metallurgical response of the alloy involved. Generally, this requires holding at a temperature sufficiently high to relieve stresses without causing an undesirable amount of precipitation or strain aging in alpha-beta and beta alloys, or without producing undesirable recrystallization in single-phase alloys that rely on cold work for strength. The higher temperatures usually are used with shorter times, and the lower temperatures with longer times, for effective stress relief. During stress relief of solution-treated and aged titanium alloys, care should be taken to prevent overaging to lower strength. This usually involves selection of a time-temperature combination that provides partial stress relief.

Uniformity of cooling is critical, particularly in the temperature range from 480 to 315 °C (900 to 600 °F). Oil or water quenching should not be used to accelerate cooling because this can induce residual stresses by unequal cooling. Furnace or air cooling is acceptable.

There are no economical nondestructive testing methods that can measure the efficiency of a stress-relief cycle other than direct measurement of residual stresses by x-ray diffraction. No significant changes in microstructure due to stress-relieving heat treatments can be detected by optical microscopy.

Annealing of titanium and titanium alloys serves primarily to increase fracture toughness, ductility at room temperature, dimensional and thermal stability, and creep resistance. Many titanium alloys are placed in service in the annealed state. Because improvement in one or more properties generally is obtained at the expense of some other property, the annealing cycle should be selected according to the objective of the treatment. Common annealing treatments are:

- Mill annealing
- Duplex annealing
- Triplex annealing
- Recrystallization annealing
- Beta annealing

Mill annealing (Table 19) is a general-purpose treatment given to all mill products. It may not be a full anneal, and may leave traces of cold or warm working in the microstructures of heavily worked products (particularly sheet). Duplex, triplex, and beta annealing alter the shapes, sizes, and distributions of phases to those required for improved creep resistance or fracture toughness. Both recrystallization and beta annealing treatments are used to improve fracture toughness. Beta annealing is done at temperatures above the beta transus of the alloy being annealed.

Straightening, sizing, and flattening may be combined with annealing by use of appropriate fixtures. Straightening of titanium alloys is often necessary in order to meet dimensional requirements. Unlike aluminum alloys, titanium alloys are not easily straightened when cold, because the high yield strength and modulus of elasticity of these alloys result in significant springback.

At annealing/aging temperatures, many titanium alloys have creep resistance low enough to permit straightening during annealing. With proper fixturing, and in some instances judicious weighting, sheet-metal fabrications and thin, complex forgings have been straightened with satisfactory results. However, if the annealing/aging temperature is below about 540 to 650 °C (1000 to 1200 °F), depending on the alloy, the times needed to accomplish the desired creep straightening can be long.

Stabilization Annealing. In alpha-beta titanium alloys, thermal stability is a function of beta-phase transformations. During cooling from the annealing temperature, beta may transform and, under certain conditions and in certain alloys, may form the brittle intermediate phase omega. A stabilization annealing treatment is designed to produce a stable beta phase capable of resisting further transformation when exposed to elevated temperatures in service. Alpha-beta alloys that are lean in beta, such as Ti-6Al-4V, can be air cooled from the annealing temperature without impairing their stability. Furnace (slow) cooling may

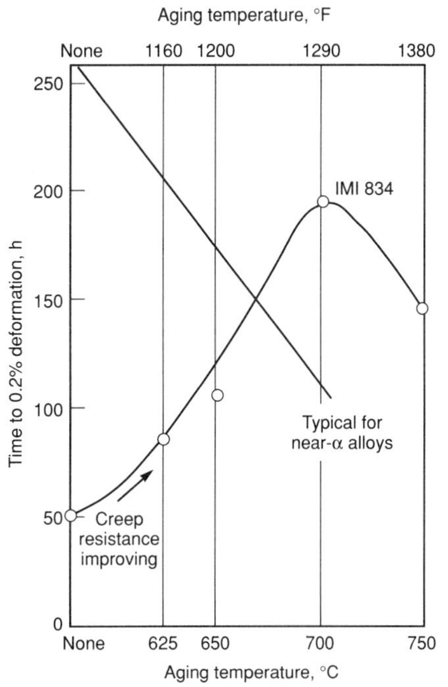

Fig. 22 Effect of aging temperature on creep performance of IMI 834. A higher aging temperature allows more stress relief to be induced, which is important for thick section disks. Source: Ref 9

promote formation of Ti_3Al, an ordering reaction that can degrade resistance to stress corrosion. Slight increases in strength (up to 34 MPa, or 5 ksi) can be gained in Ti-6Al-4V and in Ti-6Al-6V-2Sn by cooling from the annealing temperature to 540 °C (1000 °F) at a rate of 56 °C/h (100 °F/h).

Solution Treating and Aging. A wide range of strength levels can be obtained in alpha-beta or beta alloys by solution treating and aging. Except for the IMI 700 and similar alloys (which depend on age hardening from copper precipitates), the origin of heat treating responses of titanium alloys lies in the instability of the high-temperature beta phase at lower temperatures. Heating an alpha-beta alloy to the solution-treating temperature produces a higher ratio of beta phase. The beta is transformed to beta and martensite by quenching; on subsequent aging, decomposition of the unstable martensite and the small amount of residual beta phase occurs, providing high strength.

To obtain high strength with adequate ductility, it is necessary to solution treat at a temperature high in the alpha-beta field, normally 28 to 83 °C (50 to 150 °F) below the beta transus of the alloy. If high fracture toughness or improved resistance to stress corrosion is required, beta annealing or beta solution treating may be desirable. A change in the solution-treating temperature of alpha-beta alloys alters the amount of beta phase and consequently changes the response to aging (see Table 9). Selection of solution-treating temperature usually is based on practical considerations such as the desired level of tensile properties and the amount of ductility to be obtained after aging.

Because solution treating involves heating to temperatures only slightly below the beta transus, proper control of temperature is essential. If the beta transus is exceeded, tensile properties (especially ductility) are reduced and cannot be fully restored by subsequent thermal treatment. Although the reduction in ductility is not drastic and may be acceptable, the near-alpha and alpha-beta alloys are usually solution treated below the beta transus to obtain an optimum balance of ductility, toughness, and creep strength.

Beta alloys may be obtained from producers in the solution-treated, solution treated and aged, as-forged, or annealed conditions depending on product form, gage, and if forming is to be done. If reheating is required, soak times should be only as long as necessary to obtain complete solutioning. Solution-treated temperatures for beta alloys are above the beta transus; because no second phase is present, grain growth can proceed rapidly.

Aging. The final step in heat treating titanium alloys to high strength normally consists of reheating to an aging temperature between 425 and 650 °C (800 and 1200 °F). In the case of alloy IMI 834, however, an aging time up to approximately 700 °C (1300 °F) optimizes creep strength (Fig. 22).

During aging of some highly beta-stabilized alpha-beta alloys, beta transforms first to a metastable transition phase referred to as omega phase. Retained omega phase, which produces brittleness unacceptable in alloys heat treated for service, can be avoided by severe quenching and rapid reheating to aging temperatures above 425 °C (800 °F). Because a coarse alpha phase forms, however, this treatment might not produce optimal strength properties. An aging practice that ensures that aging time and temperature are adequate to precipitate alpha and revert omega usually is employed. Aging above 425 °C (800 °F) generally is adequate to complete the reaction.

Overaging. Aging at or near the annealing temperature will result in overaging. This condition, called solution treated and overaged, or STOA, is sometimes used to obtain modest increases in strength while maintaining satisfactory toughness and dimensional stability.

Other Special Thermal Treatments. Certain physical properties, such as notch strength, fracture toughness, and fatigue resistance, can be enhanced in some alloys by special thermal treatments. Three such treatments are given below:

- *Solution treating and overaging of Ti-6Al-4V*: Heat 1 h at 955 °C (1750 °F), water quench, then 2 h at 705 °C (1300 °F), air cool. Advantages: improved

Table 20 Typical physical properties of wrought titanium alloys

See Table 15 for transus temperatures.

Nominal composition, %	Coefficient of linear thermal expansion, μm/m · K (μin./in. · °F) At 20–100 °C (70–212 °F)	At 20–205 °C (70–400 °F)	At 20–315 °C (70–600 °F)	At 20–425 °C (70–800 °F)	At 20–540 °C (70–1000 °F)	At 20–650 °C (70–1200 °F)	At 20–815 °C (70–1500 °F)	Electrical resistivity(a), μΩ · m	Thermal conductivity(a), W/m · K	Density(a) g/cm³	Density(a) lb/in.³
Commercially pure titanium											
ASTM grades 1, 2, 3, 4, 7, and 11	8.6 (4.8)	...	9.2 (5.1)	...	9.7 (5.4)	10.1 (5.6)	10.1 (5.6)	0.42–0.52	16	4.51	0.163
α alloys											
5Al-2.5-Sn	9.4 (5.2)	...	9.5 (5.3)	...	9.5 (5.3)	9.7 (5.4)	10.1 (5.6)	1.57	7.4–7.8	4.48	0.162
5Al-2.5Sn (low O_2)	9.4 (5.2)	...	9.5 (5.3)	...	9.7 (5.4)	9.9 (5.5)	10.1 (5.6)	1.80	7.4–7.8	4.48	0.162
Near α											
8Al-1Mo-1V	8.5 (4.7)	...	9.0 (5.0)	...	10.1 (5.6)	10.3 (5.7)	...	1.99	...	4.37	0.158
11Sn-1Mo-2.25Al-5.0Zr-1Mo-0.2 Si	8.5 (4.7)	...	9.2 (5.1)	...	9.4 (5.2)	...	...	1.62	6.9	4.82	0.174
6Al-2Sn-4Zr-2Mo	7.7 (4.3)	...	8.1 (4.5)	...	8.1 (4.5)	...	...	1.9	7.1 at 100 °C	4.54	0.164
5Al-5Sn-2Zr-2Mo-0.25Si	...	...	...	...	...	...	10.3 (5.7)			4.51	0.163
6Al-2Nb-1Ta-1Mo	...	...	...	...	...	9.0 (5.0)	...	...	6.4	4.48	0.162
IMI 685	9.8 (5.4)	9.3 (5.2)	9.5 (5.3)	9.8 (5.4)	10.1 (5.6)	...	...	1.68	4.2	4.45	0.161
IMI 829	...	9.45 (5.3)	...	9.8 (5.4)	...	9.98 (5.5)	...	...	...	4.54	0.164
IMI 834	...	10.6 (5.9)	...	10.9 (6.1)	...	11 (6.1)	...	...	...	4.55	0.164
α-β alloys											
8Mn	8.6 (4.8)	9.2 (5.1)	9.7 (5.4)	10.3 (5.7)	10.8 (6.0)	11.7 (6.5)	12.6 (7.0)	0.92	10.9	4.73	0.171
3Al-2.5V	9.5 (5.3)	...	9.9 (5.5)	...	9.9 (5.5)	...	...	...	...	4.48	0.162
6Al-4V	8.6 (4.8)	9.0 (5.0)	9.2 (5.1)	9.4 (5.2)	9.5 (5.3)	9.7 (5.4)	...	1.71	6.6–6.8	4.43	0.160
6Al-4V (low O_2)	8.6 (4.8)	9.0 (5.0)	9.2 (5.1)	9.4 (5.2)	9.5 (5.3)	9.7 (5.4)	...	1.71	6.6–6.8	4.43	0.160
6Al-6V-2Sn	9.0 (5.0)	...	9.4 (5.2)	...	9.5 (5.3)	...	...	1.57	6.6(b)	4.54	0.164
7Al-4Mo	9.0 (5.0)	9.2 (5.1)	9.4 (5.2)	9.7 (5.4)	10.1 (5.6)	10.4 (5.8)	11.2 (6.2)	1.7	6.1	4.48	0.162
6Al-2Sn-4Zr-6Mo	9.0 (5.0)	9.2 (5.1)	9.4 (5.2)	9.5 (5.3)	9.5 (5.3)	...	...	...	7.7(c)	4.65	0.168
6Al-2Sn-2Zr-2Mo-2Cr-0.25Si	...	...	9.2 (5.1)	...	...	...	...	...	...	4.57	0.165
IMI 550	8.8 (4.9)	9.0 (5)	9.2 (5.1)	9.3 (5.2)	9.7 (5.4)	10.1 (5.6)	...	1.58	7.5	4.60	0.166
IMI 679	8.2 (4.6)	8.9 (4.9)	9.3 (5.2)	9.4 (5.2)	9.6 (5.3)	...	...	...	...	4.84	0.175
β alloys											
13V-11Cr-3Al	9.4 (5.2)	9.9 (5.5)	10 (5.55)	10.1 (5.6)	10.2 (5.7)	10.4 (5.8)	...	...	...	4.82	0.174
8Mo-8V-2Fe-3Al	...	...	...	...	...	...	...			4.84	0.175
3Al-8V-6Cr-4Mo-4Zr	8.7 (4.8)	9 (5)	9.4 (5.2)	9.6 (5.3)	...	...	...			4.82	0.174
11.5Mo-6Zr-4.5Sn	7.6 (4.2)	8.1 (4.5)	8.5 (4.7)	8.7 (4.8)	8.7 (4.8)	...	...	1.56	...	5.06	0.183
15V-3Cr-3Al-3Sn	8.5 (4.7)	8.7–9 (4.8–5)	9.2 (5.1)	9.4 (5.3)	9.7 (5.4)	...	...	1.47	8.08	4.71	0.170
5Al-2Sn-2Zr-4Cr	9 (5)	9.2 (5.1)	9.4 (5.2)	9.5 (5.3)	...	...	...	...	...	...	...

(a) Room temperature. (b) At 93 °C (200 °F). (c) In solution treated and aged condition

notch strength, fracture toughness, and creep strength at strength levels similar to those obtained by regular annealing

- *Recrystallization annealing of Ti-6Al-4V or Ti-6Al-4V-ELI*: Heat 4 h or more at 925 to 955 °C (1700 to 1750 °F), furnace cool to 760 °C (1400 °F) at a rate no higher than 56 °C/h (100 °F/h), cool to 480 °C (900 °F) at a rate no lower than 370 °C/h (670 °F/h), air cool to room temperature. Advantages: improved fracture toughness and fatigue-crack-growth characteristics at somewhat reduced levels of strength. This is usually used with ELI material
- *Beta annealing of Ti-6Al-4V, Ti-6Al-4V-ELI, and Ti-6Al-2Sn-4Zr-2Mo. Ti-6Al-4V or Ti-6Al-4V-ELI*: Heat 5 min to 1 h at 1010 to 1040 °C (1850 to 1900 °F), air cool to 650 °C (1200 °F) at a rate of 85 °C/min (150 °F/min) or higher, then 2 h at 730 to 790 °C (1350 to 1450 °F), air cool. Advantages: improved fracture toughness, high-cycle fatigue strength, creep strength, and resistance to aqueous stress corrosion. Ti-6Al-2Sn-4Zr-2Mo: Heat ½ h at 1020 °C (1870 °F), air cool, then 8 h at 595 °C (1100 °F), air cool. Advantages: improved creep strength at elevated temperatures as well as improved fracture toughness

Post Heat Treating Requirements. Titanium reacts with the oxygen, water, and carbon dioxide normally found in oxidizing heat treating atmospheres and with hydrogen formed by decomposition of water vapor. Unless the heat treatment is performed in a vacuum furnace or in an inert atmosphere, oxygen will react with the titanium at the metal surface and produce an oxygen-enriched layer commonly called "alpha case." This brittle layer must be removed before the component is put into service. It can be removed by machining, but certain machining operations may result in excessive tool wear. Standard practice is to remove alpha case by other mechanical methods or by chemical methods, or by both.

Hydrogen Contamination. Titanium is chemically active at elevated temperatures and will oxidize in air. However, oxidation is not of primary concern. The danger of hydrogen pickup is of greater importance than that of oxidation. This is not normally a problem, but it could be a problem if using a steel heat treating furnace with a reducing atmosphere. Use of these furnaces should only be after complete purging. Current specifications limit hydrogen content to a maximum of 125 to 200 ppm, depending on alloy and mill form. Above these limits, hydrogen embrittles some titanium alloys, thereby reducing impact strength and notch tensile strength and causing delayed cracking. Beta alloys are more susceptible to hydrogen contamination but are also more tolerant of hydrogen.

Heat Treatment Verification. Hardness is not a good measure of the adequacy of the thermomechanical processes accomplished during the forging and heat treatment of titanium alloys, unlike most aluminum alloys and many heat-treatable ferrous alloys. Therefore, hardness measurements are not used to verify the processing of titanium alloys. Instead, mechanical property tests (for example, tensile tests and fracture toughness) and metallographic/microstructural evaluation are used to verify the thermomechanical processing of titanium alloy

Table 21 Minimum and average mechanical properties of wrought titanium alloys at room temperature

Nominal composition, %	Condition	Minimum and average tensile properties(a): Ultimate tensile strength, MPa (ksi)	Minimum and average tensile properties(a): 0.2% yield strength, MPa (ksi)	Minimum and average tensile properties(a): Elongation, %	Minimum and average tensile properties(a): Reduction in area, %	Average or typical properties: Charpy impact strength, J (ft · lbf)	Average or typical properties: Hardness	Average or typical properties: Modulus of elasticity, GPa (10^6 psi)	Average or typical properties: Modulus of rigidity, GPa (10^6 psi)	Average or typical properties: Poisson's ratio	Average or typical properties: Bend radius for thickness (*t*) over 1.8 mm (0.07 in.)
Commercially pure titanium											
99.5 Ti (ASTM grade 1)	Annealed	240–331 (35–48)	170–241 (25–35)	30	55	...	120 HB	102.7 (14.9)	38.6 (5.6)	0.34	2*t*
99.2 Ti (ASTM grade 2)	Annealed	340–434 (50–63)	280–345 (40–50)	28	50	34–54 (25–40)	200 HB	102.7 (14.9)	38.6 (5.6)	0.34	2.5*t*
99.1 Ti (ASTM grade 3)	Annealed	450–517 (65–75)	380–448 (55–65)	25	45	27–54 (20–40)	225 HB	103.4 (15.0)	38.6 (5.6)	0.34	2.5*t*
99.0 Ti (ASTM grade 4)	Annealed	550–662 (80–96)	480–586 (70–85)	20	40	20 (15)	265 HB	104.1 (15.1)	38.6 (5.6)	0.34	3.0*t*
99.2 Ti(b) (ASTM grade 7)	Annealed	340–434 (50–63)	280–345 (40–50)	28	50	43 (32)	200 HB	102.7 (14.9)	38.6 (5.6)	0.34	2.5*t*
98.9 Ti(c) (ASTM grade 12)	Annealed	480–517 (70–75)	380–448 (55–65)	25	42	...	...	...	102.7 (14.9)	...	2.5*t*
α alloys											
5Al-2.5Sn	Annealed	790–862 (115–125)	760–807 (110–117)	16	40	13.5–20 (10–15)	36 HRC	110.3 (16.0)	...	...	4.5*t*
5Al-2.5Sn (low O_2)	Annealed	690–807 (100–117)	620–745 (90–108)	16	...	43 (32)	35 HRC	110.3 (16.0)	...	...	...
Near α											
8Al-1Mo-1V	Duplex annealed	900–1000 (130–145)	830–951 (120–138)	15	28	20–34 (15–25)	35 HRC	124.1 (18.0)	46.9 (6.8)	0.32	4.5*t*
11Sn-1Mo-2.25Al-5.0Zr-1Mo-0.2Si	Duplex annealed	1000–1103 (145–160)	900–993 (130–144)	15	35	...	36 HRC	113.8 (16.5)	...	...	...
6Al-2Sn-4Zr-2Mo	Duplex annealed	900–980 (130–142)	830–895 (120–130)	15	35	...	32 HRC	113.8 (16.5)	...	...	5*t*
5Al-5Sn-2Zr-2Mo-0.25Si	975 °C (1785 °F) (½ h), AC + 595 °C (1100 °F) (2 h), AC	900–1048 (130–152)	830–965 (120–140)	13	...	...	...	113.8 (16.5)	...	0.326	...
6Al-2Nb-1Ta-1Mo	As-rolled 2.5 cm (1 in.) plate	790–855 (115–124)	690–758 (100–110)	13	34	31 (23)	30 HRC	113.8 (17.5)	...	...	...
6Al-2Sn-1.5Zr-1Mo-0.35Bi-0.1Si	β forge + duplex anneal	1014 (147)	945 (137)	11	...	...	...	...	...	...	...
IMI 685 (Ti-6Al-5Zr-0.5Mo-0.25Si)	β heat treated at 1050 °C, OQ, + aged 24h at 550 °C	882–917 (128–133)	758–815 (110–118)	6–11 (on 5*D*)	15–22	43 (32)	...	~125 (~18)	...	...	...
IMI-829 (Ti-5.5Al-3.5Sn-3Zr-1Nb-0.25Mo-0.3Si)	β heat treated at 1050 °C, AC, + aged 2h at 625 °C	930 (min) (35)	820 (min) (119)	9 (min) on 5*D*	15 (min)	...	...	...	...	...	...
IMI-834 (Ti-5.5Al-4.5Sn-4Zr-0.7Nb-0.5Mo-0.4Si-0.06C)	α–β processed	1030 (min) (149)	910 (min) (132)	6 (min) on 5*D*	15 (min)	...	...	...	...	...	...
α–β alloys											
8Mn	Annealed	860–945 (125–137)	760–862 (110–125)	15	32	...	...	113.1 (16.4)	48.3 (7.0)	...	...
3Al-2.5V	Annealed	620–689 (90–100)	520–586 (75–85)	20	...	54 (40)	...	106.9 (15.5)	...	...	...
6Al-4V	Annealed	900–993 (130–144)	830–924 (120–134)	14	30	14–19 (10–14)	36 HRC	113.8 (16.5)	42.1 (6.1)	0.342	5*t*
	Solution + aging	1172 (170)	1103 (160)	10	25	...	41 HRC	...	...	...	
6Al-4V (low O_2)	Annealed	830–896 (120–130)	760–827 (110–120)	15	35	24 (18)	35 HRC	113.8 (16.5)	42.1 (6.1)	0.342	...
6Al-6V-2Sn	Annealed	1030–1069 (150–155)	970–1000 (140–145)	14	30	14–19 (10–14)	38 HRC	110.3 (16.0)	...	...	4.5*t*
	Solution + aging	1276 (185)	1172 (170)	10	20	...	42 HRC	...	...	...	...
7Al-4Mo	Solution + aging	1103 (160)	1034 (150)	16	22	18 (13)	38 HRC	113.8 (16.5)	44.8 (6.5)	...	...
	Annealed	1030 (min) (50)	970 (min) (140)	...	...	...	...	...	...	...	...
6Al-2Sn-4Zr-6Mo	Solution + aging	1269 (189)	1172 (170)	10	23	8–15 (6–11)	36–42 HRC	113.8 (16.5)	...	...	...

(continued)

(a) If a range is given, the lower value is a minimum; all other values are averages. (b) Also contains 0.2 Pd. (c) Also contains 0.8 Ni and 0.3 Mo. AC, air-cooled

Table 21 (continued)

Nominal composition, %	Condition	Minimum and average tensile properties(a): Ultimate tensile strength, MPa (ksi)	0.2% yield strength, MPa (ksi)	Elongation, %	Reduction in area, %	Average or typical properties: Charpy impact strength, J (ft · lbf)	Hardness	Modulus of elasticity, GPa (10^6 psi)	Modulus of rigidity, GPa (10^6 psi)	Poisson's ratio	Bend radius for thickness (*t*) over 1.8 mm (0.07 in.)
α–β alloys (continued)											
6Al-2Sn-2Zr-2Mo-2Cr-0.25Si	Solution + aging	1276 (185)	1138 (165)	11	33	20 (15)	...	122 (17.7)	46.2 (6.7)	0.327	...
	Annealed	1030 (min) (150)	970 (min) (140)	...	...	...	...	...	...	...	...
Corona 5 (Ti-4.5Al-5Mo-1.5Cr)	β annealed plate	910 (132)	817 (118)	...	...	...	...	...	...	...	...
	β worked plate	945 (137)	855 (124)	...	...	...	...	...	...	...	...
	α-β worked	935 (131)	905 (131)	...	...	...	...	...	...	...	...
IMI 550 (Ti-4Al-4Mo-2Sn-0.5Si)	Solution at 900 °C, AC, + aging of 25 mm (1 in.) slice	1100 (160)	940 (136)	7 on 5*D*	15	23 (17)	...	~115 (~17)	...	...	...
β alloys											
13V-11Cr-3Al	Solution + aging	1170–1220 (170–177)	1100–1172 (160–170)	8	...	...	...	101.4 (14.7)	42.7 (6.2)	0.304	...
	Solution + aging	1276 (185)	1207 (175)	8	...	11 (8)	40 HRC	...	...	...	...
8Mo-8V-2Fe-3Al	Solution + aging	1170–1310 (170–190)	1100–1241 (160–180)	8	...	...	40 HRC	106.9 (15.5)	...	...	...
3Al-8V-6Cr-4Mo-4Zr (Beta C)	Solution + aging	1448 (210)	1379 (200)	7	...	10 (7.5)	...	105.5 (15.3)	...	...	...
	Annealed	883 (min) (128 min)	830 (min) (120 (min))	15	...	...	...	...	...	...	...
11.5Mo-6Zr-4.5Sn (Beta III)	Solution + aging	1386 (210)	1317 (191)	11	...	...	...	103 (15)	...	...	...
	Annealed	690 (min) (100 (min))	620 (min) (90 (min))	...	...	...	...	...	...	...	...
10V-2Fe-3Al	Solution + aging	1170–1276 (170–185)	1100–1200 (160–174)	10	19	...	...	111.7 (16.2)	...	...	...
Ti-15V-3Cr-3Al-3Sn (Ti-15-3)	Annealed	785 (114)	773 (112)	22	...	...	...	...	...	...	...
	Aged	1095–1335 (159–194)	985–1245 (143–180)	6–12	...	...	...	...	...	...	...
Ti-5Al-2Sn-2Zr-4Mo-4Cr(Ti-17)	Solution + aging	1105–1240 (160–180)	1305–1075 (150–170)	8–15	20–45	...	...	...	...	...	...
Transage 134 plate	Solution + aging	1055–1380 (153–200)	1000–1310 (145–190)	5–12	10–38	...	...	...	...	...	...
Transage 175 (extruded bar)	Solution + aging	1305 (189)	1250 (180)	10	39						
Transage 175 at 425 °C (800 °F)	Solution + aging	1080 (157)	925 (134)	10	56	...	...	...	...	...	...

(a) If a range is given, the lower value is a minimum; all other values are averages. (b) Also contains 0.2 Pd. (c) Also contains 0.8 Ni and 0.3 Mo. AC, air-cooled

forgings. Mechanical property and microstructural evaluations vary, ranging from the destruction of forgings to the testing of extensions and/or prolongations forged integrally with the parts.

Properties

The titanium alloys, with their high strengths and low densities, can often bridge the properties gap between aluminum and steel alloys, providing many of the desirable properties of each. For example, titanium, like aluminum, is nonmagnetic and has good heat-transfer properties (despite its relatively low thermal conductivity as discussed in the section "Commercially Pure Titanium" in this article). The thermal expansion coefficient of titanium alloys (Table 20), ranging from about 9 to 11 ppm/°C (5 to 6 × 10^{-6} in./in. · °F), is slightly lower than that of most steels and less than half that of aluminum. In addition, titanium is nontoxic and biologically compatible, making it useful for surgical-implant devices.

Other important characteristics of titanium alloys depend on the class of alloy (Fig. 23) and the morphology of the alpha constituents (Table 8). In the near-alpha and alpha-beta alloys, the variations in the alpha morphology are achieved with different heat treatments (Table 19). A fine equiaxed alpha (which is associated with high tensile strength, good ductility, and resistance to

Table 22 Yield strength and plane strain fracture toughness of various titanium alloys

Alloy	α morphology or processing method	Yield strength, MPa	Yield strength, ksi	Plane-strain fracture toughness (K_{Ic}), MPa $\sqrt{m}$	Plane-strain fracture toughness (K_{Ic}), ksi $\sqrt{in.}$
Ti-6Al-4V	Equiaxed	910	130	44–66	40–60
	Transformed	875	125	88–110	80–100
	α–β rolled + mill annealed(a)	1095	159	32	29
Ti-6Al-6V-2Sn	Equiaxed	1085	155	33–55	30–50
	Transformed	980	140	55–77	50–70
Ti-6Al-2Sn-4Zr-6Mo	Equiaxed	1155	165	22–23	20–30
	Transformed	1120	160	33–55	30–50
Ti-6Al-2Sn-4Zr-2Mo forging	α + β forged, solution treated and aged	903	131	81	74
	β forged, solution treated and aged	895	130	84	76
Ti-17	α-β processed	1035–1170	150–170	33–50	30–45
	β processed	1035–1170	150–170	53–88	48–80

(a) Standard oxygen (~0.20 wt%) Source: Adapted from Ref 1, Ref 13, Ref 15

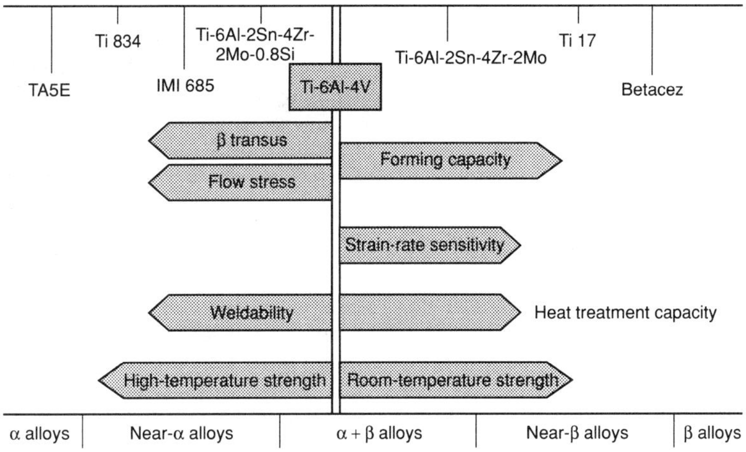

Fig. 23 Main characteristics of the different titanium alloy families

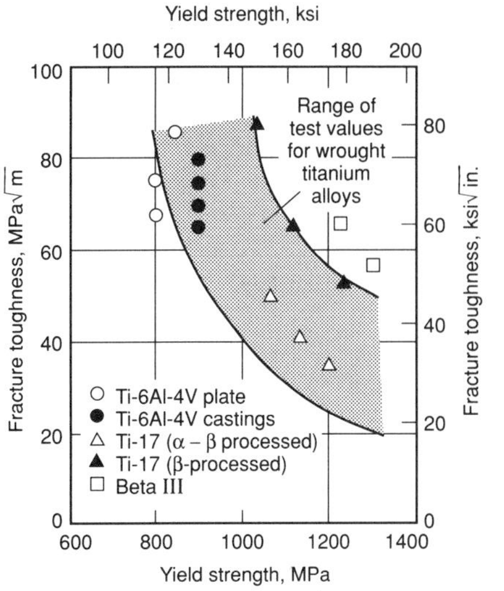

Fig. 24 Fracture toughness of Ti-6Al-4V castings compared to Ti-6Al-4V plate and to other Ti alloys. Sources: Ref 1 and 15

fatigue-crack initiation) occurs when alpha-beta alloys are processed well below the beta transus, while an acicular alpha (which is associated with excellent creep strength, high fracture toughness, and resistance to fatigue-crack propagation) occurs by heating above the beta transus and subsequent beta transformation during cooling and aging. Finally, an intermediate microstructure (Fig. 12b) can be achieved by processing near-alpha alloys close to the beta transus. The objective of an intermediate microstructure with acicular and equiaxed alpha is to provide good creep strength without excessively compromising fatigue strength.

Beta transus temperatures are listed in Table 15. These are typical values that can vary by about ±15 °C (±25 °F) depending on actual composition and impurity levels. Titanium mill producers generally certify the beta transus temperature for each heat supplied, because the beta transus temperature will vary from heat to heat due to small differences in chemistry, particularly oxygen content.

Room-Temperature Tensile Properties. Average tensile properties (and some typical minimum property guarantees) for titanium mill products are listed in Table 21. The effects of alpha morphology and section size are shown in Tables 22 and 23, respectively.

In terms of the principal heat treatments used for titanium, beta annealing decreases strength by 35 to 100 MPa (5 to 15 ksi) depending on prior grain size, average crystallographic texture, and testing direction. Solution treating and aging can be used to enhance strength at the expense of fracture toughness in alloys containing sufficient beta stabilizer (that is, 4 wt%, or more).

Fracture toughness can be varied within a nominal titanium alloy by as much as a multiple of two or three. This may be accomplished by manipulating alloy chemistry, microstructure, and texture. Some trade-offs of other desired properties may be necessary to achieve high-fracture toughness. Strength is often achieved in titanium alloys at the expense of K_{Ic} (Fig. 24).

There are significant differences among titanium alloys in fracture toughness, but there also is appreciable overlap in their properties. Table 22 gives examples of typical plane-strain fracture toughness ranges for alpha-beta titanium alloys. From these data it is apparent that the basic alloy chemistry affects the relationship between strength and toughness. From Table 22 it also is evident that transformed microstructures may greatly enhance toughness while only slightly reducing strength.

Within the permissible range of chemistry for a specific titanium alloy and grade, oxygen is the most important variable insofar as its effect on toughness is concerned. In essence, if high fracture toughness is required, oxygen must be kept low, other things being equal. Reducing nitrogen, as in Ti-6Al-4V-ELI, is also indicated, but the effect is not as strong as it is with oxygen.

Improvements in toughness can be obtained by providing either of two basic types of microstructures:

- Transformed structures, or structures transformed as much as possible, because fractures in such structures must proceed along many faceted paths
- Equiaxed structures composed mainly of regrowth alpha that have both low dislocation-defect densities and low concentrations of nitrogen and oxygen (the so-called "recrystallization annealed" structures)

According to some work, plane-strain fracture toughness is proportional to the fraction of transformed beta in the alloy. The subject is a complex one without clear-cut, empirical rules. Furthermore, the enhancement of fracture toughness at one stage of an operation—for example, a forging billet—does not necessarily carry over to a forged part. Because welds in alloy Ti-6Al-4V contain transformed products, one would expect such welds to be relatively high in toughness.

Fatigue Properties. For a given tensile strength, the fatigue strength of titanium alloys compares favorably with quenched

Table 23 Relation of tensile strength of solution treated and aged titanium alloys to size

	Tensile strength of square bar in section size of:											
	13 mm (½ in.)		25 mm (1 in.)		50 mm (2 in.)		75 mm (3 in.)		100 mm (4 in.)		150 mm (6 in.)	
Alloy	MPa	ksi	MPa	ksi	MPa	ksi	MPa	ksi	MPa	ksi	MPa	ksi
Ti-6Al-4V	1105	160	1070	155	1000	145	930	135	...	...	...	...
Ti-6Al-6V-2Sn (Cu + Fe)	1205	175	1205	175	1070	155	1035	150	...	...	...	...
Ti-6Al-2Sn-4Zr-6Mo	1170	170	1170	170	1170	170	1140	165	1105	160	...	...
Ti-5Al-2Sn-2Zr-4Mo-4Cr (Ti-17)	1170	170	1170	170	1170	170	1105	160	1105	160	1105	160
Ti-10V-2Fe-3Al	1240	180	1240	180	1240	180	1240	180	1170	170	1170	170
Ti-13V-11Cr-3Al	1310	190	1310	190	1310	190	1310	190	1310	190	1310	190
Ti-11.5Mo-6Zr-4.5Sn (Beta III)	1310	190	1310	190	1310	190	1310	190	1310	190	...	...
Ti-3Al-8V-6Cr-4Zr-4Mo (Beta C)	1310	190	1310	190	1240	180	1240	180	1170	170	1170	170

Table 24 Strain control low-cycle fatigue life of Ti-6242S at 480 °C (900 °F)

Test frequency, cycles/min	Total strain range, %	Number of cycles to failure: Acicular structure	Number of cycles to failure: Equiaxed alpha structure
0.4	1.2	1 196	10 500(a)
10	1.2	3 715	31 000(a)
0.4	2.5	273	722
10	2.5	353	1 166

(a) Run out of test. Source: Ref 17

and tempered low-alloy steel (Fig. 2). Fatigue life is usually divided into two regimes: low-cycle fatigue (in which failures occur in 10^4 cycles or less) and high-cycle fatigue (in which failures occur in more than 10^4 cycles). The effects of microstructure on these fatigue regimes are discussed below and in Ref 13 and 16. In general, for low-cycle fatigue and high-cycle fatigue there is a great value in high fracture toughness and low crack-propagation rates.

Low-cycle fatigue (LCF) is very difficult to quantify owing to the wide range of variables and to the limited amount of published data. In general, data are available for both load-controlled and strain-controlled tests. Table 24 gives some strain-controlled LCF data. Results of an LCF study in Ti-6Al-4V are also shown in Fig. 25. In the figure, time to the first crack (at a fixed strain) varies with microstructure. Note that time to crack initiation is optimized with a structure having high amounts of transformed beta, yet still having about 10% of primary alpha. (However, the crack-propagation resistance of the beta-processed structure still exceeds that of alpha-processed material.)

In high-cycle fatigue, surface condition is an important variable affecting fatigue strength. This effect is illustrated in Fig. 26, which shows the effect of different machining operations and shot peening on the fatigue limits of Ti-5Al-2.5Sn. In terms of microstructure, there is general agreement that the Widmanstätten or colony alpha-

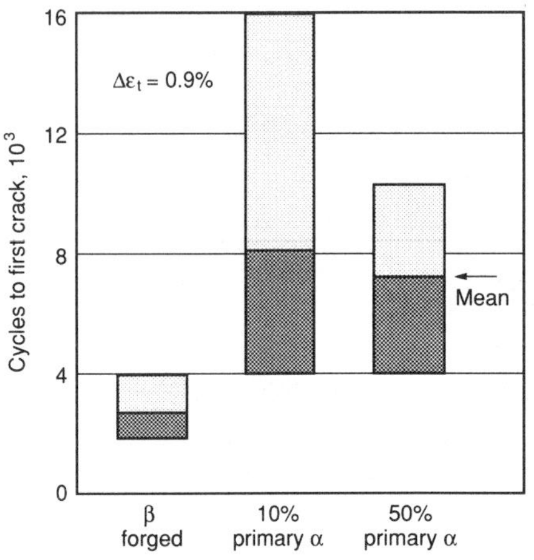

Fig. 25 LCF life of Ti-6Al-4V alloy with different structures: beta forged (100% transformed beta); 10% primary alpha (balance transformed beta); 50% primary alpha

beta microstructure has decidedly poorer fatigue strength. Solution treated and aged material has good fatigue strength, but not as good as fine-grained equiaxed material or beta-quenched material.

Fatigue crack propagation (FCP) is affected by several variables such as strength, microstructure, and texture. In addition, because of the reactive nature of titanium, environmental effects on crack propagation should also be expected. In general, only the more severe environments (such as 3.5% NaCl solution) affect FCP rates by an order of magnitude, or more. The environmental effects are minimized by beta annealing. Gaseous atmospheres also may play a role in affecting FCP rates.

In terms of microstructure, FCP is affected in much the same way as fracture toughness; that is, crack-propagation rates are reduced with a transformed microstructure. Annealing methods are also important. Generally, beta-annealed microstructures have the lowest fatigue-crack growth rates, whereas mill-annealed microstructures yield the highest growth rates. Mill-annealed alloys also exhibit considerable scatter in FCP rates (Fig. 27), because of variations in microstructure, texture, and strength.

Finally, different titanium alloys may have different FCP characteristics just as they have different fracture toughness characteristics. Selected data indicate that fatigue cracks propagate more rapidly in Ti-6Al-2Sn-4Zr-6Mo than in Ti-8Al-1Mo-1V or Ti-6Al-2Sn-4Zr-2Mo under the same test conditions. This may be a simple effect of strength. However, the relative amounts of beta phase may lead to intrinsically different fatigue-crack propagation characteristics. The Ti-6Al-2Sn-4Zr-6Mo alloy is also more easily textured. Figure 27 shows the range of FCP rates in Ti-6Al-4V and a beta alloy, which attains higher strength over greater section sizes than Ti-6Al-4V.

Elevated-Temperature Mechanical Properties. As shown schematically in Fig. 23, high-temperature strength is associated with the alpha and near-alpha alloys. However, when creep strength is not a factor in an elevated-temperature application, the short-time elevated-temperature tensile strengths of beta alloys have a distinct advantage (Fig. 28). Up to about 425 °C (800 °F), beta alloys also have a higher specific strength than H11 die steel (Fig. 28b). The alpha and alpha-beta alloys do not compare as favorably with H11 steels in terms of various specific strengths (Fig. 29).

Nonetheless, near-alpha and alpha/beta titanium alloys have replaced the steels once used in aircraft turbines. Figure 30 compares the strength/density behavior of two titanium alloys with the behavior of three steels used at one time or another in the lower temperature regimes of aircraft gas turbine engines. Compared with steels, titanium alloys are superior up to about 540 °C (1000 °F). This obvious advantage led to initial use of titanium in aircraft engines, first as compressor blades (Pratt & Whitney Aircraft J-57, Rolls-Royce Avon) and then as disks (Pratt & Whitney

Table 25 Temperature range and chemical composition of high-temperature titanium alloys, listed in order of introduction

Alloy designation	Year of introduction	Useful maximum temperature °C	Useful maximum temperature °F	Approximate nominal chemical composition, wt%: Al	Sn	Zr	Mo	Nb	V	Si	Others
Ti-6Al-4V	1954	300	580	6	...	...	...	...	4	...	...
IMI-550	1956	425	795	4	2	...	4	...	...	0.5	...
Ti-811	1961	400	750	8	...	...	1	...	1	...	...
IMI-679	1961	450	840	2	11	5	1	...	...	0.2	...
Ti-6246	1966	450	840	6	2	4	6	...	...	...	...
Ti-6242	1967	450	840	6	2	4	2	...	...	...	...
Hylite 65(a)	1967	520	970	3	6	4	0.5	...	...	0.5	...
IMI-685	1969	520	970	6	...	5	0.5	...	...	0.25	...
Ti-5522S(a)	1972	520	970	5	5	2	2	...	...	0.2	...
Ti-11(a)	1972	540	1000	6	2	1.5	1	...	...	0.1	0.3Bi
Ti-6242S	1974	520	970	6	2	4	2	...	...	0.1	...
Ti-5524S(a)	1976	500	930	5	5	2	4	...	...	0.1	...
IMI-829	1976	580	1080	5.5	3.5	3	0.3	1	...	0.3	...
IMI-834(a)	1984	590	1100	5.5	4	4	0.3	1	...	0.5	0.06C
Ti-1100	...	590	1100	6	2.75	4	0.4	...	...	0.45	0.02 Fe (max)

(a) Not yet used commercially. Sources: Ref 18 and 19

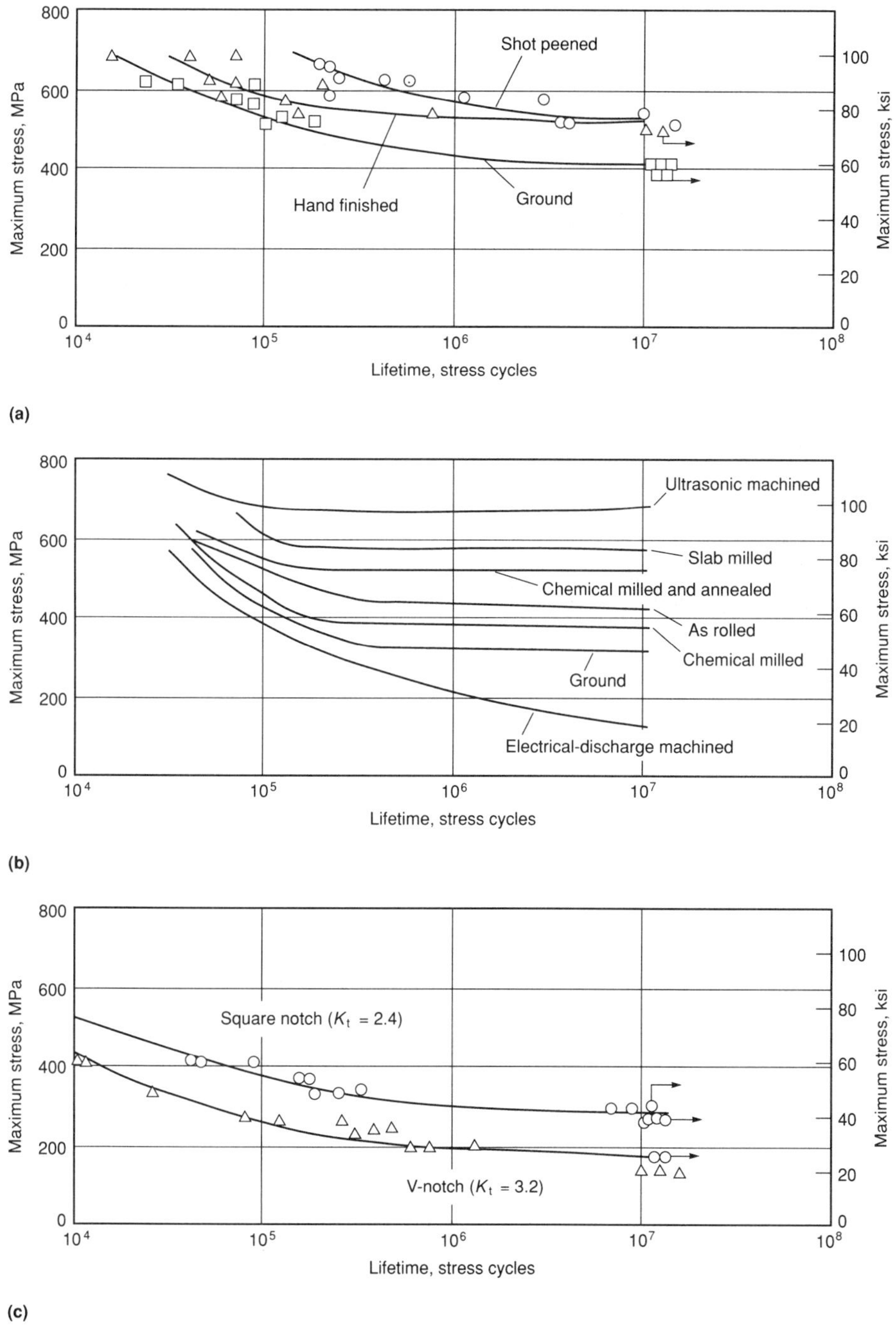

Fig. 26 Rotating-beam fatigue strength of Ti-5Al-2.5Sn. (a) and (b) Fatigue strengths for different types of surface finish. (c) Notch fatigue strength for two different types of notches

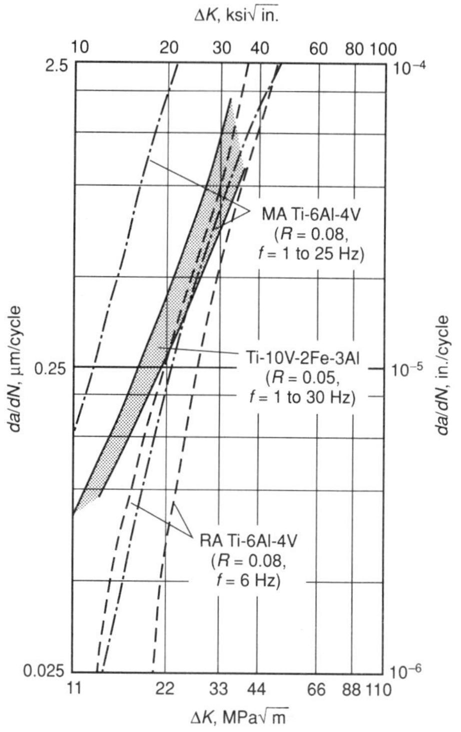

Fig. 27 Comparison of fatigue crack growth rates of beta alloy Ti-10V-2Fe-3Al with mill-annealed (MA) and recrystallization-annealed (RA) Ti-6Al-4V

Aircraft JT-3D). In fact, titanium alloys made possible the fan-type gas turbine engines now in use and have been the subject of ongoing development for elevated-temperature applications (Table 25).

Elevated-Temperature Titanium Alloys. Since 1950, a number of titanium alloys have been developed for elevated-temperature application. The starting alloy was Ti-6Al-4V, which was soon superceded by near-alpha and other alpha/beta alloys with optimization of certain elevated-temperature properties such as short-time strength or long-term creep strength. Alloy Ti-6246 (Ti-6Al-2Sn-4Zr-6Mo), for example, emphasizes short-time strength with its higher beta-stabilizer content, while alloy Ti-6242 (Ti-6Al-2Sn-4Zr-2Mo) provides good long-term creep strength with its content of alpha stabilizers.

Aluminum, tin, zirconium, and oxygen influence strength and ductility in various degrees depending on their amount. After studying the creep and tensile properties of the then known titanium quaternary systems, Rosenberg (Ref 20) arrived at an empirical formula for use in the design of high-temperature titanium alloys:

$$\text{Al} + \tfrac{1}{3}\text{Sn} + \tfrac{1}{6}\text{Zr} + 10 \times \text{O}_2 \leqq 9 \qquad \text{(Eq 2)}$$

The formula gives the maximum combined weight percent of the alloying elements. All commercial alloys presently in service still meet this requirement. Other important alloying elements are molybdenum, silicon, and niobium. Molybdenum enhances hardenability and enhances short-time high-temperature strength or improves strength at lower temperatures. Minor silicon additions improve creep strength (Fig. 31), while niobium is added primarily for oxidation resistance at elevated temperature.

The current temperature limit of titanium alloys is near 590 °C (1100 °F) for IMI-834 and Ti-1100 (Table 25). Alloy IMI-834 has a near-alpha composition with carbon additions (see the section "Carbon"), which is processed high in the alpha-beta region (≅85% beta) so as to reduce beta grain coarsening and achieve a mixture of equiaxed alpha with acicular alpha. This intermediate type of microstructure (shown in Fig. 12b for alloy Ti-8Al-1Mo-1V) provides good creep strength without excessively compromising fatigue strength. Alloy Ti-1100 controls the levels of molybdenum and iron (Fig. 32) to achieve high creep strength (Fig. 33).

The current temperature limit of 590 °C (1100 °F) is due mainly to long-term surface and bulk metallurgical stability problems; creep strength obviously is another factor

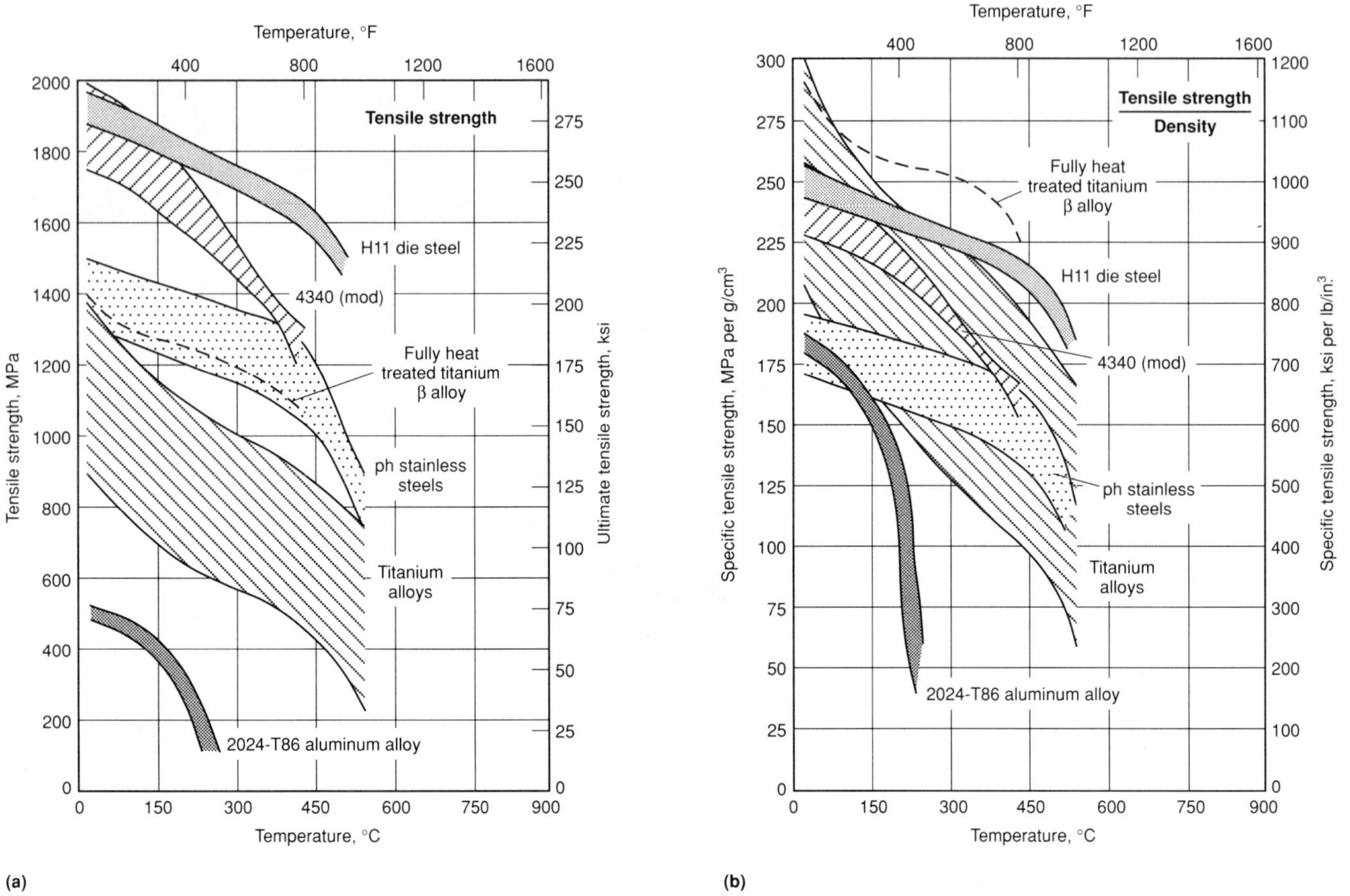

Fig. 28 Comparison of short-time tensile strength and tensile strength/density ratio for titanium alloys, three classes of steel, and 2024-T86 aluminum alloy. Data are not included for annealed alloys with less than 10% elongation or heat-treated alloys with less than 5% elongation.

(depending on deformation limits and stress level). To push the temperature limit higher, there are at least four generic approaches in new alloy system development (Ref 18):

- To develop creep resistance alloys based on a fine dispersion of the α_2 phase (Ti_3Al) in an alpha-titanium matrix (the $\alpha + \alpha_2$ alloy class)
- To develop alloys based on the intermetallic Ti_3Al (α_2)
- To develop alloys using the intermetallic TiAl (γ) as a base
- To strengthen P/M titanium alloys by the incorporation of dispersoids during rapid solidification
- Use of metal-matrix composites

In the $\alpha + \alpha_2$ alloys, the α_2 phase is generally considered to have an embrittling effect on titanium alloys, but it has been demonstrated that niobium (Ref 22) or niobium combined with other beta stabilizing elements (Ref 23) can be used to improve the ductility of $\alpha + \alpha_2$ alloys. Intermetallic Ti_3Al (α_2) and TiAl (γ) matrix materials, which occur at increased aluminum contents, are discussed in the article "Ordered Intermetallics" in this Volume. Titanium P/M products are discussed in the article so-named in this Volume.

Creep Strength. Whether yield strength or creep strength for a given maximum allowable deformation (for example, 0.1% creep strain in 150 h, which is used for aircraft gas-turbine compressor parts) is the significant selection criterion depends on which is lower at the service temperature in question. Between 200 and 315 °C (400 and 600 °F), the deformation of many titanium alloys loaded to the yield point does not increase with time. Thus, creep strength is seldom a factor in this range. Above 315 °C (600 °F), creep strength becomes an important selection criterion. Typical creep strengths are compared in Fig. 33 and 34. Specific creep strengths with some nickel-base alloys are compared in Fig. 35.

In near-alpha and $\alpha + \beta$ titanium alloys, the creep strength is increased by heat treating or processing the material above the beta transus temperature. Upon cooling, this results in an acicular alpha structure that is associated with improved creep strength. However, because an acicular structure also degrades fatigue performance (which is of particular importance in aircraft), an intermediate microstructure (Fig. 12b) may be desired.

Alloy IMI 834 is aged at ~700 °C (1300 °F) for creep strength (Fig. 22). To obtain maximum creep resistance and stability in the near-alpha alloy Ti-8Al-1Mo-1V and Ti-6Al-2Sn-4Zr-2Mo, a duplex annealing treatment is employed. This treatment begins with solution annealing at a temperature high in the alpha-beta range, usually 28 to 56 °C (50 to 100 °F) below the beta transus for Ti-8Al-1Mo-1V and 19 to 56 °C (35 to 50 °F) below the beta transus for Ti-6Al-2Sn-4Zr-2Mo. Forgings are held for 1 h (nominal) and then air or fan cooled depending on section size. This treatment is followed by stabilization annealing for 8 h at 595 °C (1100 °F). Final annealing temperature should be at least 55 °C (100 °F) above the maximum anticipated service temperature. Maximum creep resistance can be developed in Ti-6Al-2Sn-4Zr-2Mo by beta annealing or beta processing.

Creep-Fatigue Interaction. At room temperature and in nonaggressive environments (and except at very high frequencies), the frequency at which loads are applied has little effect on the fatigue strength of most metals. The effects of frequency, however, become much greater as the temperature increases or as the presence of corrosion becomes more significant. At high temperatures, creep becomes more of a factor, and the fatigue strength seems to depend on the total time stress is applied

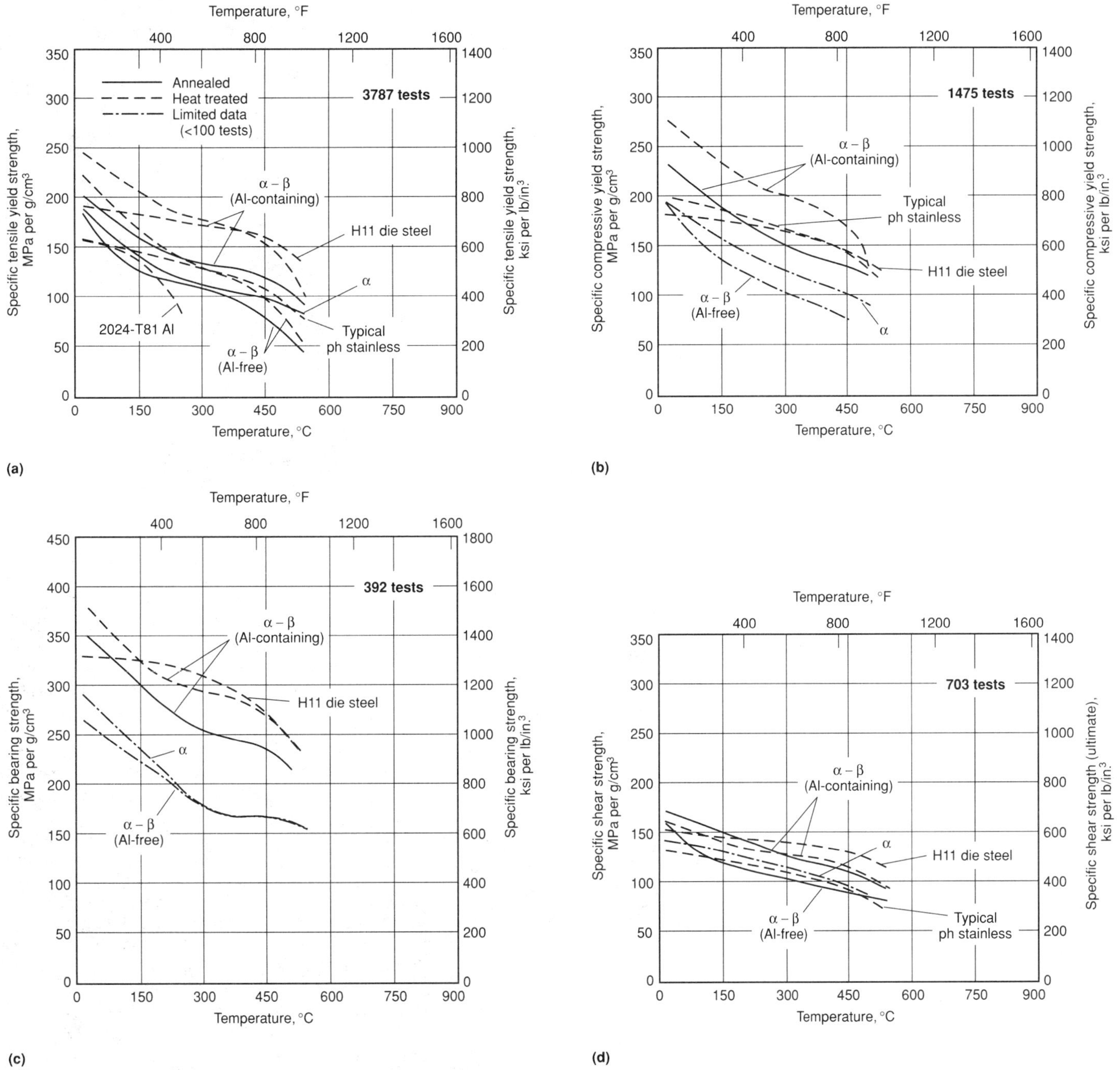

Fig. 29 Specific strengths versus temperature for titanium alloy classes and H11 die steel

rather than solely on the number of cycles. The behavior occurs because the continuous deformation (creep) under load at high temperatures affects the propagation of fatigue cracks. This effect is referred to as creep-fatigue interaction. The quantification of creep-fatigue interaction effects and the application of this information to life prediction procedures constitute the primary objective in time-dependent fatigue tests. Time-dependent fatigue tests are also used to assess the effect of load frequency on corrosion fatigue.

Like other metals, creep-fatigue interaction in titanium alloys can be evaluated in terms of a reduction in low-cycle fatigue strength caused by the introduction of a "hold-time" or dewll at the peak of each stress (or strain) cycle. This effect of creep-fatigue interaction has been a subject of various studies (Ref 24-27), with particular emphasis on the beta-processed near-alpha alloys. In general, creep-fatigue interaction becomes more of a factor with lower rupture ductility (or embrittlement from factors such as internal hydrogen, Ref 25).

Thermal stability is the ability of alloys to retain their original mechanical properties after prolonged service at elevated temperature. An alloy is thermally unstable if it undergoes microstructural changes during use at elevated temperature that affect its properties adversely. Instability may cause either embrittlement or softening, depending on the nature of the microstructural changes. Thermal stability is measured by comparing the properties of an alloy at room temperature before and after exposure (stressed or unstressed) at elevated temperature.

Titanium alloys are generally stable over the temperature ranges where they resist oxidation and retain their useful strength. The alpha alloys are generally stable up to 540 °C (1000 °F) for exposure periods of

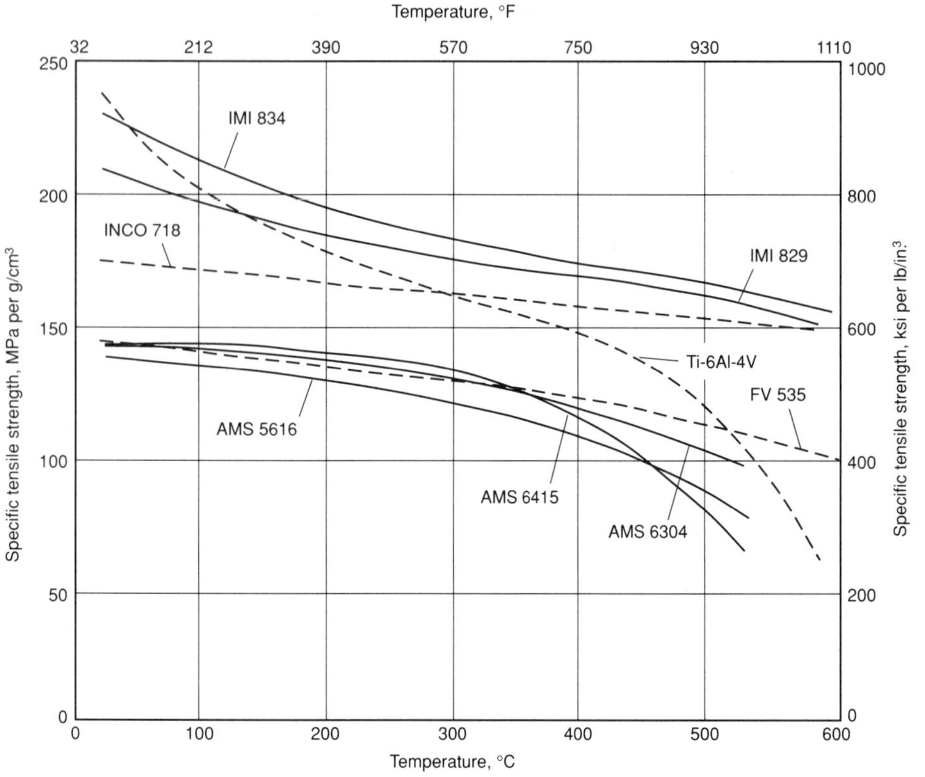

Fig. 30 Specific tensile strength of various titanium alloys compared with steels once used in aircraft turbines

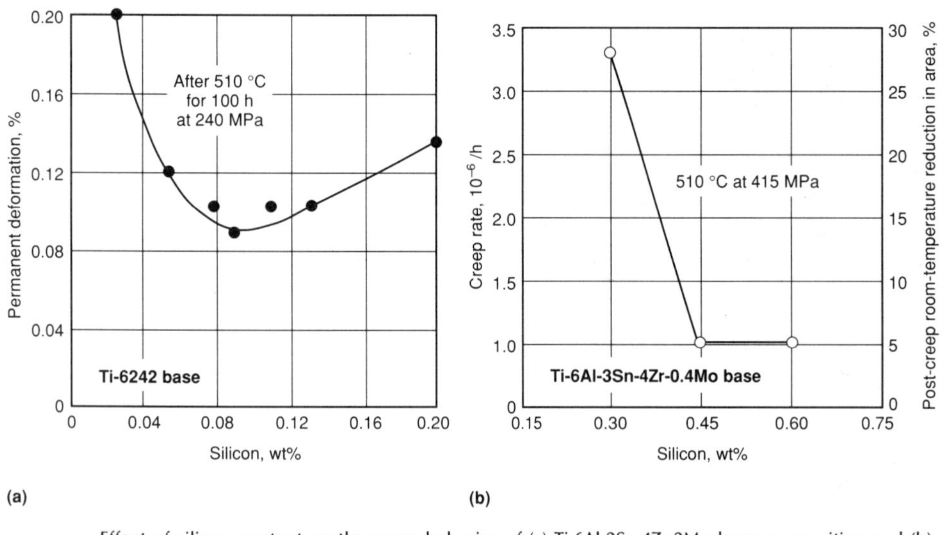

Fig. 31 Effect of silicon content on the creep behavior of (a) Ti-6Al-2Sn-4Zr-2Mo base composition and (b) Ti-6Al-3Sn-4Zr-0.4Mo base composition. Sources: Ref 18 and 19

1000 h or more, except that alloys high in aluminum, such as Ti-8Al-2Nb-1Ta and Ti-8Al-1Mo-1V, will undergo a small, but generally tolerable, amount of hardening and some loss in ductility because of formation of Ti_3Al (α_2) in the microstructure. Elongation and reduction in area, after exposure to the elevated temperature, will be 10% or more.

The stability of commercial alpha-beta alloys depends on composition and heat treatment. In the mill-annealed condition, the alloys may be considered stable up to 315 to 370 °C (600 to 700 °F), although measurable changes in properties will usually accompany exposure to stress and temperature for long times. Properly fabricated and heat treated, these alloys are generally stable up to about 425 °C (800 °F) in the heat treated condition for periods of 1000 h or more.

Properties of titanium alloys may deteriorate during exposure to elevated temperature and stress because of surface cracking. Cracking may result from oxidation or from stress corrosion caused by atmospheres or surface films containing salt or other halides. When conditions are likely to cause surface cracking, tests should be made to determine the susceptibility of the alloys selected to the surface conditions intended.

Low-Temperature Properties. Unalloyed titanium and alpha titanium alloys have hexagonal close-packed (hcp) crystal structures, which accounts for the fact that the properties of these metals do not follow the same trends at subzero temperatures as do the properties of metals with fcc or bcc structures. In particular, alpha titanium may not exhibit a ductile-to-brittle transition in Charpy V-notch data (Fig. 4).

Many of the available titanium alloys have been evaluated at subzero temperatures, but service experience at such temperatures has been gained only for Ti-5Al-2.5Sn and Ti-6Al-4V alloys. These alloys have very high strength-to-weight ratios at cryogenic temperatures and have been the preferred alloys for special applications at temperatures from −195 to −270 °C (−320 to −452 °F). Commercially pure titanium may be used for tubing and other small-scale cryogenic applications that involve only low stresses in service.

The Ti-5Al-2.5Sn alloy usually is used in the mill-annealed condition and has a 100% alpha microstructure. The Ti-6Al-4V alloy may be used in the annealed condition or in the solution treated and aged condition, but for maximum toughness in cryogenic applications the annealed condition usually is preferred. The Ti-6Al-4V alloy is an alpha-beta alloy that has significantly higher yield and ultimate tensile strengths than the all-alpha alloy.

Because interstitial impurities such as iron, oxygen, carbon, nitrogen, and hydrogen tend to reduce the toughness of these alloys at both room and subzero temperatures, ELI grades are specified for critical applications. The composition limits for these alloys are given in Table 26. Note that the iron and oxygen contents of the ELI grades are substantially lower than those of the standard, or normal interstitial (NI), grades. The NI grades are suitable for service to −195 °C (−320 °F); for temperatures below −195 °C, ELI grades generally are specified. For ELI grades, reduced creep strength at room temperature must be considered in design for pressure-vessel service. In Ti-5Al-2.5Sn, stress rupture occurs at stresses below the yield strength.

Precautions. There are two precautions that should be emphasized in considering titanium and titanium alloys for service at cryogenic temperatures: titanium and titanium alloys must not be used for transfer or storage of liquid oxygen, and titanium must not be used where it will be exposed to air while below the temperature at which oxygen will condense on its surfaces. Any abrasion or impact of titanium creating a clean, oxide-free surface that is in contact

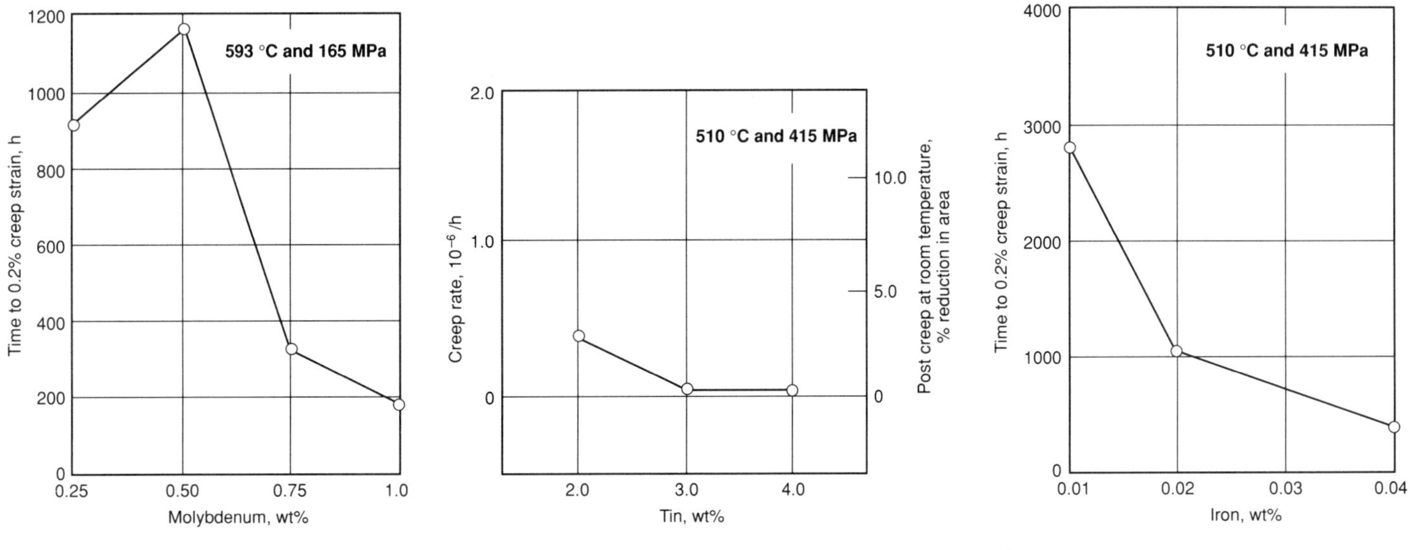

Fig. 32 Chemistry effects on base composition similar to Ti-6Al-3Sn-4Zr-0.4Mo-0.45Si. Source: Ref 19

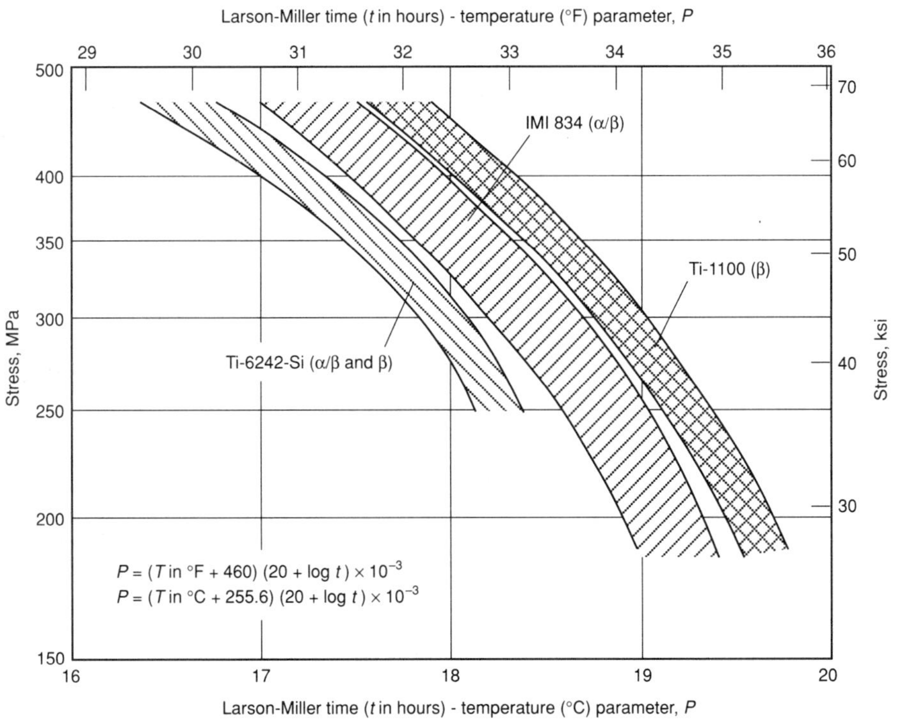

Fig. 33 Creep properties of Ti-6Al-2Sn-4Zr-2Mo(Si), IMI-834, and Ti-1100 alloys. With alpha-beta or beta processing as indicated. Source: Ref 21

Table 26 Compositions of titanium alloys used in cryogenic applications

	Composition, %								
Alloy	Al	Sn	V	Fe max	O max	C max	N max	H max	M max
Ti-75A	...	...	...	...	0.40	0.20	0.07	0.0125	...
Ti-5Al-2.5Sn	4.0–6.0	2.0–3.0	...	0.50	0.20	0.15	0.07	0.020	0.30
Ti-5Al-2.5Sn(ELI)(a)	4.7–5.6	2.0–3.0	...	0.20	0.12	0.08	0.05	0.0175	...
Ti-6Al-4V	5.5–6.75	...	3.5–4.5	...	...	...	...	...	...
Ti-6Al-4V(ELI)(a)	5.5–6.5	...	3.5–4.5	0.15	0.13	0.08	0.05	0.015	...

(a) Extra-low interstitial

with liquid oxygen will cause ignition. Pressure vessels in contact with liquid oxygen in the Apollo launch vehicles were produced from Inconel 718 rather than from Ti-6Al-4V alloy to avoid this problem.

Typical tensile properties of titanium and of titanium alloys Ti-5Al-2.5Sn and Ti-6Al-4V at room temperature and at subzero temperatures are presented in Table 27. Marked increases in yield and tensile strengths are evident for commercial titanium and for titanium alloys as test temperature is reduced from room temperature to −253 °C (−423 °F). In the cryogenic temperature range, these alloys have the highest strength-to-weight ratios of all fusion-weldable alloys that retain nearly the same strength in the weld metal as in the base metal. Yield and tensile strengths of an electron-beam weldment of Ti-5Al-2.5Sn(ELI) sheet are presented in Table 27.

The notch strengths given in Table 27 indicate that these two alloys retain sufficient notch toughness for use to −253 °C (−423 °F). However, the tensile data do not show any substantial improvement in ductility or notch toughness for the ELI grade of Ti-5Al-2.5Sn sheet over the normal interstitial grade except at very low temperatures. The recrystallization annealing treatment used for the Ti-6Al-4V(ELI) forging was developed as a means of improving fracture toughness in large forgings and thick plate.

Values of Young's modulus for titanium alloys increase substantially as test temperature is decreased, as shown in Table 27 and by ultrasonic data (Ref 28), which shows an approximate linear increase in the elastic modulus of Ti-2.5Sn and Ti-6Al-4V from about 112 GPa (16.2 × 10^6 psi) at room temperature to about 122 GPa (17.7 × 10^6

Table 27 Typical tensile properties of titanium and two titanium alloys

Temperature °C	Temperature °F	Tensile strength MPa	Tensile strength ksi	Yield strength MPa	Yield strength ksi	Elongation, %	Reduction in area, %	Notch tensile strength(a) MPa	Notch tensile strength(a) ksi	Young's modulus GPa	Young's modulus 10^6 psi
Ti-75A sheet, annealed, longitudinal orientation											
24	75	580	84.3	465	67.6	25	...	785	114	...	...
−78	−108	750	109	615	89.2	25	...	...	...	...	...
−196	−320	1050	152	940	136	18	...	1100	159	...	...
−253	−423	1280	186	1190	173	8	...	875	127	...	...
Ti-75A sheet, annealed, transverse orientation											
24	75	585	85.1	475	69.0	25	...	800	116	...	...
−78	−108	760	110	645	93.4	20	...	905	131	...	...
−196	−320	1060	153	965	140	14	...	1120	163	...	...
−253	−423	1340	194	1260	182	7	...	880	128	...	...
Ti-5Al-2.5Sn sheet, nominal interstitial annealed, longitudinal orientation											
24	75	850	123	795	115	16	...	1130	164	105	15.4
−78	−108	1080	156	1020	148	13	...	1310	190	115	16.6
−196	−320	1370	199	1300	188	14	...	1630	236	120	17.7
−253	−423	1700	246	1590	231	7	...	1430	208	130	18.5
Ti-5Al-2.5Sn sheet, nominal interstitial annealed, transverse orientation											
24	75	895	130	860	125	14	...	1170	170	...	...
−78	−108	1050	152	1020	148	12	...	1250	181	...	...
−196	−320	1430	208	1370	198	12	...	1630	236	...	...
−253	−423	1670	242	1610	234	6	...	1290	187	...	...
−268	−450	1590	231	...	...	1.5	...				
Ti-5Al-2.5Sn (ELI) sheet, annealed, longitudinal orientation											
24	75	800	116	740	107	16	...	1060	154	115	16.4
−78	−108	960	139	880	128	14	...	1190	173	125	18.0
−196	−320	1300	188	1210	175	16	...	1560	226	130	18.6
−253	−423	1570	228	1450	210	10	...	1670	242	130	19.2
Ti-5Al-2.5Sn (ELI) sheet, annealed, transverse orientation											
24	75	805	117	760	110	14	...	1100	159	110	16.0
−78	−108	950	138	895	130	12	...	1260	182	125	18.1
−196	−320	1300	188	1230	179	14	...	1570	228	130	18.9
−253	−423	1570	228	1480	214	8	...	1530	222	140	20.1
Ti-5Al-2.5Sn (ELI) sheet/weldment, annealed, EB weld											
24	75	815	118	785	114	...	...	...	...	...	...
−196	−320	1300	189	1210	176	...	...	...	...	...	...
−253	−423	1510	219	1380	200	...	...	...	...	...	...
Ti-5Al-2.5Sn (ELI) plate, annealed, longitudinal orientation											
24	75	765	111	705	102	33	43	...	...	...	...
−253	−423	1430	208	1390	202	17	32	...	...	...	...
Ti-5Al-2.5Sn (ELI) forgings, as forged, tangential orientation											
24	75	835	121	760	110	15	36	...	...	...	...
−78	−108	980	142	905	131	12	31	...	...	...	...
−196	−320	1260	182	1100	159	15	30	...	...	...	...
−253	−423	1420	206	1260	182	13	22	...	...	...	...
Ti-6Al-4V (ELI) sheet, annealed, longitudinal orientation											
24	75	960	139	890	129	12	...	1120	162	110	16.2
−78	−108	1160	168	1100	160	9	...	1220	177	115	16.6
−196	−320	1500	217	1420	206	10	...	1460	211	120	17.5
−253	−423	1770	256	1700	246	4	...	1500	217	130	18.6
Ti-6Al-4V (ELI) sheet, annealed, transverse orientation											
24	75	960	139	895	130	12	...	1130	164	110	16.0
−78	−108	1170	169	1100	160	12	...	1260	183	115	16.5
−196	−320	1500	218	1460	212	11	...	1440	209	125	18.2
−253	−423	1750	254	1700	246	4	...	1550	225	130	19.2
Ti-6Al-4V (ELI) plate, annealed, longitudinal orientation											
24	75	890	129	840	122	15	37	...	...	...	...
−253	−423	1640	238	1600	232	...	8	...	...	...	...
Ti-6Al-4V (ELI) forgings, as forged, longitudinal orientation											
24	75	970	141	915	133	14	40	1330	193	...	...
−78	−108	1160	168	1120	163	13	31	1560	226	...	...
−196	−320	1570	227	1480	214	11	31	1900	276	...	...
−253	−423	1650	239	1570	227	11	24	1820	264	...	...
Ti-6Al-4V (ELI) forgings, recrystallization annealed(b)											
24	75	890	129	825	120	14	41	...	...	110	16.1
−196	−320	1430	207	1370	198	10	16	...	...	120	17.5

(a) K_t = 6.3 for all three sheet forms; K_t = 5 to 8 for Ti-6Al-4V (ELI) forgings. (b) Recrystallization annealing treatment: 930 °C (1700 °F) 4 h, furnace cool to 760 °C (1400 °F) in 3 h, cooled to 480 °C (900 °F) in ¾ h, air cool

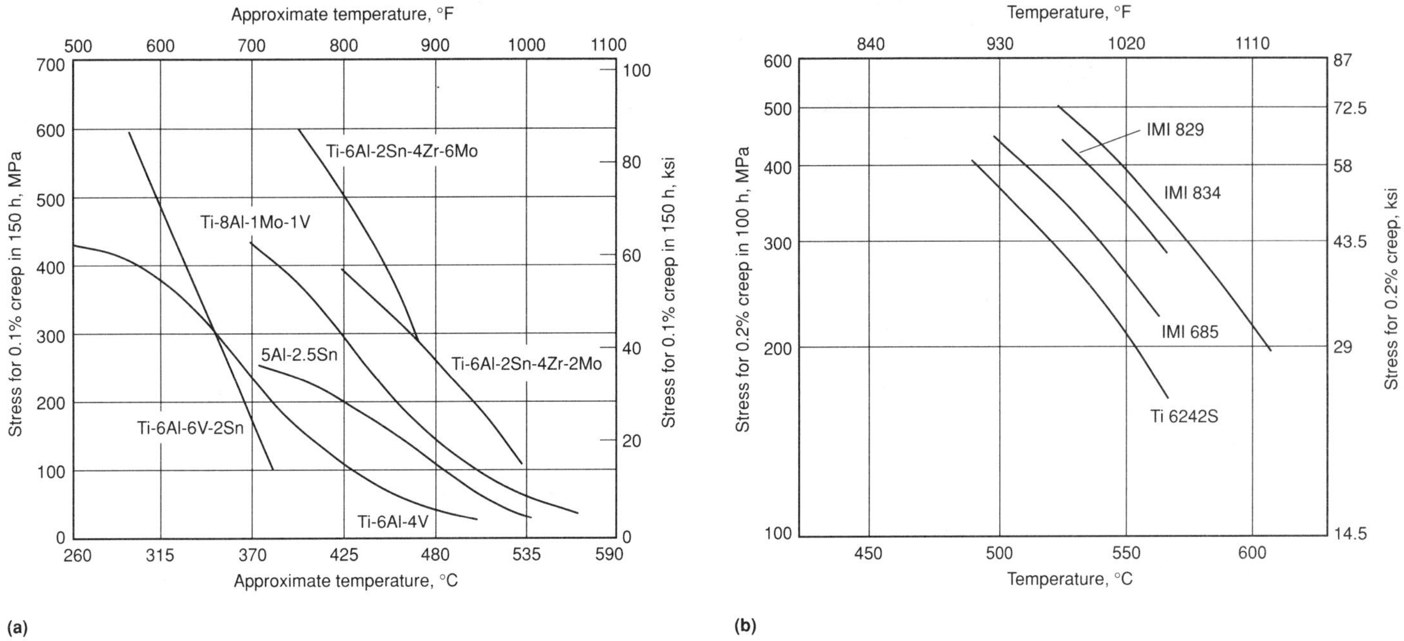

Fig. 34 Comparison of creep strengths for various titanium alloys

psi) and 123 GPa (17.8 × 10^6 psi) at −200 °C (−330 °F) for Ti-6Al-4V and Ti-5Al-2.5Sn, respectively. The same ultrasonic testing (Ref 28) showed a decrease in Poisson's ratio to about 0.31 for Ti-6Al-4V at −200 °C (−330 °F).

Fracture Toughness. Available data on plane-strain fracture toughness (K_{Ic}) at subzero temperatures for alloys Ti-5Al-2.5Sn and Ti-6Al-4V (summarized in Volume 3 of the 9th Edition of *Metals Handbook*), indicate a modest reduction in fracture toughness to about 39 MPa$\sqrt{m}$ (35.5 ksi$\sqrt{in.}$) and 42 MPa$\sqrt{m}$ (38 ksi$\sqrt{in.}$) for Ti-6Al-4V and Ti-5Al-2.5Sn NI grades, respectively, at −195 °C (−320 °F). The ELI grades have better toughness than the corresponding normal interstitial grades at subzero temperatures. The limited data for electron-beam weldments indicate that at −195 °C (−320 °F) there is a slight reduction in toughness in both fusion and heat-affected zones when compared to the base metal in Ti-6Al-4V(ELI) weldments.

Fatigue-Crack-Growth Rates. Data on fatigue-crack-growth rates for Ti-5Al-2.5Sn and Ti-6Al-4V alloys in Volume 3 of the 9th Edition of *Metals Handbook* indicate that low temperature has no effect on the fatigue-crack-growth rates for Ti-5Al-2.5Sn and Ti-6Al-4V(NI). However, over part of the ΔK range, the fatigue-crack-growth rates for Ti-6Al-4V(ELI) are higher at cryogenic temperatures than at room temperature at the same ΔK values.

Fatigue strength at subzero temperatures becomes more sensitive to notches and the presence of welded joints. Therefore, in designing welded structures of titanium alloys that will be subjected to fatigue loading at subzero temperatures, the weld areas usually should be thicker than the remaining areas. Hemispheres for spherical pressure vessels are machined so that the butting sections for the equatorial welds are thicker than the remaining sections, excluding inlet and discharge ports.

ACKNOWLEDGMENT

The author would like to thank Rodney R. Boyer of the Boeing Commercial Aircraft Company and Stan R. Seagle of RMI Titanium Company for their review of the manuscript and their meaningful suggestions for improvement.

REFERENCES

1. M.J. Donachie, Jr., *Titanium: A Technical Guide*, ASM INTERNATIONAL, 1988
2. L.C. Covington and R.W. Schutz, "Corrosion Resistance of Titanium," TIMET Corporation, 1982
3. J.B. Cotton, *Chem. Eng. Prog.*, Vol 66 (No. 10), 1970, p 57
4. L.C. Covington, "Factors Affecting the Hydrogen Embrittlement of Titanium," Paper presented at Corrosion/75 (Toronto, Canada), National Association of Corrosion Engineers, April 1975
5. L.C. Covington, *Corrosion*, Vol 35 (No. 8), Aug 1979, p 378-382
6. V.K. Grigorovich, *Metallurgica*, Toplivo Izvestia, Academy of Sciences, USSR, No. 5, 1960, p 38
7. C.F. Yolton, F.H. Froes, and R.F. Malone, Alloy Element Effects in Metastable Titanium Alloys, *Metall. Trans. A*, Vol 10A, 1979, p 132-134
8. W.A. Baeslack III, D.W. Becker, and F.H. Froes, Advances in Titanium Alloy Welding Metallurgy, *J. Met.*, Vol 36 (No. 5), 1984, p 46-58
9. D.F. Neal, Development and Evaluation of High Temperature Titanium Alloy IMI 834, in *Sixth World Conference on Titanium Proceedings (Part 1)*, Société Française de Métallurgie, 1988, p 253-258
10. D.F. Neal and P.A. Blenkinsop, Titanium Alloy, European Patent 0107419A1, 1984
11. T. Duerig and J. Williams, Overview: Microstructure and Properties of Beta Titanium Alloys, in *Beta Titanium Alloys in the 1980's*, R.R. Boyer and H.W. Rosenberg, Ed., AIME Metallurgical Society, 1984, p 20
12. R.B. Sparks and J.R. Long, "Improved Manufacturing Methods for Producing High Integrity More Reliable Titanium Forgings," AFML TR-73-301, Wyman Gordon Company, Feb 1974
13. J.C. Williams and E.A. Starke, Jr., The Role of Thermomechanical Processing in Tailoring the Properties of Aluminum and Titanium Alloys, in *Deformation, Processing, and Structure*, G. Krauss, Ed., American Society for Metals, 1984
14. G.A. Lenning *et al.*, "Cold Formable Titanium Sheet," Contract F33615-78-C-5116, AFWAL-TR-82-4174, U.S. Air Force Wright Aeronautical Laboratories, 1982
15. R.R. Boyer and H.W. Rosenberg, *Beta Titanium Alloys in the 1980's*, The Metallurgical Society of AIME, 1984, p 407, 438

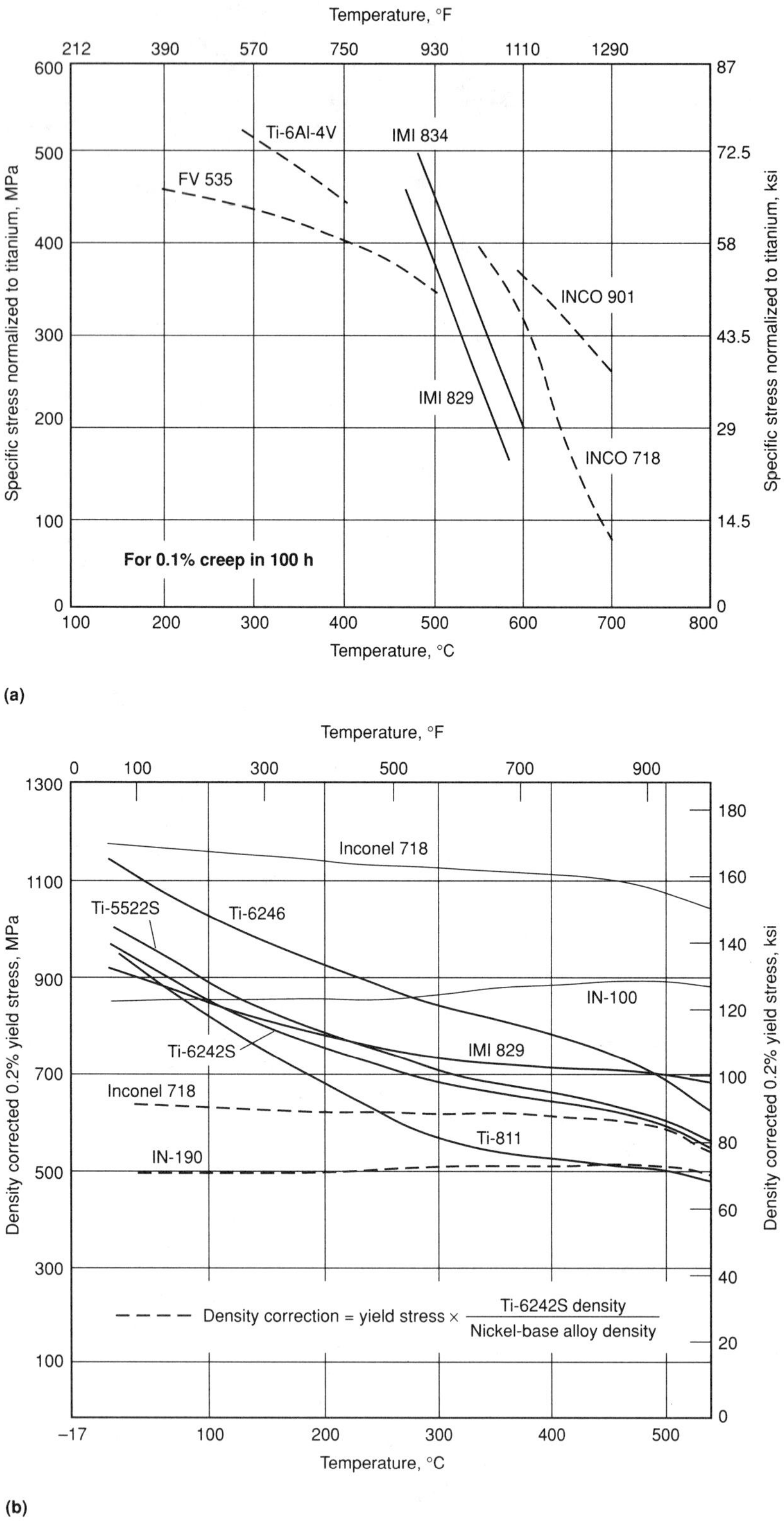

Fig. 35 Specific creep strengths and yield strengths for various titanium-base and nickel-base alloys

16. J.C. Chesnutt, C.G. Rhodes, and J.C. Williams, The Relationship Between Mechanical Properties, Microstructure and Fracture Topography in $\alpha+\beta$ Titanium Alloys, in *STP 600*, American Society for Testing and Materials, 1976
17. D. Eylon, M.E. Rosenblum, and S. Fujishiro, High Temperature Low Cycle Fatigue Behavior of Near Alpha Titanium Alloys, in *Titanium '80, Science and Technology*, H. Kimura and O. Izumi, Ed., TMS-AIME, 1980, p 1845-1854
18. D. Eylon *et al.*, High-Temperature Titanium Alloys—A Review, in *Titanium Technology: Present Status and Future Trends*, Titanium Development Association, 1985
19. P.J. Bania, Ti-1100: A New High-Temperature Titanium Alloy, in *Sixth World Conference on Titanium Proceedings (Part I)*, Société Française de Métallurgie, 1988, p 825-830
20. H.W. Rosenberg, Titanium Alloying in Theory and Practice, *The Science, Technology and Application of Titanium*, R.I. Jaffee and N.E. Promisel, Ed., Pergamon Press, 1970, p 851-859
21. J.S. Park *et al.*, The Effects of Processing on the Properties of Forgings from Two New High Temperature Titanium Alloys, in *Sixth World Conference on Titanium Proceedings*, Société Française de Métallurgie, 1988, p 1283-1288
22. S.M.L. Sastry and H.A. Lipsitt, Ordering Transformation and Mechanical Properties of Ti_3Al and Ti_3Al-Nb Alloys, *Metall. Trans. A*, Vol 8A, 1977, p 1543-1552
23. C.G. Rhodes, C.H. Hamilton, and N.E. Paton, "Titanium Aluminides for Elevated Temperatures Applications," AFML-TR-78-130, U.S. Air Force Materials Laboratory, 1978
24. W.J. Evans and C.R. Gostelow, *Metall. Trans. A*, Vol 10A, 1979, p 1837-1846
25. J.E. Hack and G.R. Leverant, *Metall. Trans. A*, Vol 13A, 1982, p 1729-1737
26. D.F. Neal, Creep Fatigue Interactions in Titanium Alloys, in *Sixth World Conference on Titanium Proceedings*, Société Française de Métallurgie, 1988, p 175-180
27. M.R. Winstone, Effect of Texture on the Dwell Fatigue of a Near-Alpha Titanium Alloy, in *Sixth World Conference on Titanium Proceedings*, Société Française de Métallurgie, 1988, p 169-173
28. C.W. Fowlkes and R.L. Tobler, Fracture Testing and Results for a Ti-6Al-4V Alloy at Liquid Helium Temperature, *Eng. Frac. Mech.*, Vol 8 (No. 3), 1976, p 487-500

SELECTED REFERENCES

- R.R. Boyer and H.W. Rosenberg, Ed., *Beta Titanium Alloys in the 1980's*, The Metallurgical Society of AIME, 1984
- E.W. Collins, *The Physical Metallurgy of Titanium Alloys*, American Society for Metals, 1984
- F.H. Froes, D. Eylon, and H.B. Bomberger, *Titanium Technology: Present Status and Future Trends*, Titanium Development Association, 1985
- R.I. Jaffee and N.E. Promisel, *The Science, Technology and Application of Titanium*, Pergamon Press, 1970

- J.C. Williams and M.J. Blackburn, A Comparison of Phase Transformations in Three Commercial Titanium Alloys, *ASM Quart. Trans.*, Vol 60, 1967, p 373
- J.C. Williams, Phase Transformations in Titanium Alloys: A Review, in *Titanium Science and Technology*, Vol 3, R.I. Jaffee and H.M. Burte, Ed., Plenum Press, 1973, p 1433
- J.C. Williams, Precipitation in Titanium Base Alloys, in *Precipitation Processes in Solids*, K.C. Russell and H.I. Aaronson, Ed., American Institute for Mining, Metallurgical, and Petroleum Engineers, 1976, p 191
- J.C. Williams, Phase Transformations in Ti Alloys—A Review of Recent Developments, *Proc. 3rd Int. Conf. on Titanium* (Moscow, USSR), J.C. Williams and A.F. Belov, Ed., Plenum Press, 1982, p 1477–1498

Titanium and Titanium Alloy Castings

Daniel Eylon, Graduate Materials Engineering, University of Dayton
Jeremy R. Newman and John K. Thorne, TiTech International, Inc.

SINCE THE INTRODUCTION OF TITANIUM and titanium alloys in the early 1950s, these materials have in a relatively short time become backbone materials for the aerospace, energy, and chemical industries (Ref 1). The combination of high strength-to-weight ratio, excellent mechanical properties, and corrosion resistance makes titanium the best material choice for many critical applications. Today, titanium alloys are used for demanding applications such as static and rotating gas turbine engine components. Some of the most critical and highly stressed civilian and military airframe parts are made of these alloys.

The use of titanium has expanded in recent years to include applications in nuclear power plants, food processing plants, oil refinery heat exchangers, marine components, and medical prostheses (Ref 2). However, the high cost of titanium alloy components may limit their use to applications for which lower-cost alloys, such as aluminum and stainless steels, cannot be used. The relatively high cost is often the result of the intrinsic raw material cost of the metal, fabricating costs, and the metal removal costs incurred in obtaining the desired final shape. As a result, in recent years a substantial effort has focused on the development of net shape or near-net shape technologies to make titanium alloy components more competitive (Ref 3). These titanium net shape technologies include powder metallurgy (P/M), superplastic forming (SPF), precision forging, and precision casting. Precision casting is by far the most fully developed and the most widely used titanium net shape technology (for comparison, see the article "Titanium P/M Products" in this Volume).

The annual shipment of titanium castings in the United States increased by 260% between 1979 and 1989 (Fig. 1). With a trend line still strong in the upward direction, this makes titanium casting the fastest growing segment of titanium technology. In fact, the number of sales dollars of castings shipped has grown faster than the number of pounds shipped because of the increasing complexity of configurations being produced, that is, configurations that are closer to net shape, larger in size, and of higher quality for more critical applications.

Even at current levels (approaching 450 Mg, or 9.9×10^5 lb, annually), castings still represent less than 2% of total titanium mill product shipments. This is in sharp contrast to the ferrous and aluminum industries, where foundry output is 9% (Ref 5) and 14% (Ref 6) of total output, respectively. This suggests that the growth trend of titanium castings will continue as users become more aware of industry capability, suitability of cast components in a wide variety of applications, and the net shape cost advantages.

The term castings often connotes products with properties generally inferior to wrought products. This is not true with titanium cast parts. They are generally comparable to wrought products in all respects and quite often superior. Properties associated with crack propagation and creep resistance can be superior to those of wrought products. As a result, titanium castings can be reliably substituted for forged and machined parts in many demanding applications (Ref 7, 8). This is due to several unique properties of titanium alloys. One is the $\alpha + \beta$-to-β allotropic phase transformation at a temperature range of 705 to 1040 °C (1300 to 1900 °F), which is well below the solidification temperature of the alloys. As a result, the cast dendritic β structure is transformed during the solid state cooling to an $\alpha + \beta$ platelet structure (Fig. 2a), which is also typical of β-processed wrought alloy. Furthermore, the convenient allotropic transformation temperature range of most titanium alloys enables the as-cast microstructure to be improved by means of post-cast cooling rate changes and subsequent heat treatment or by hot isostatic pressing (HIP).

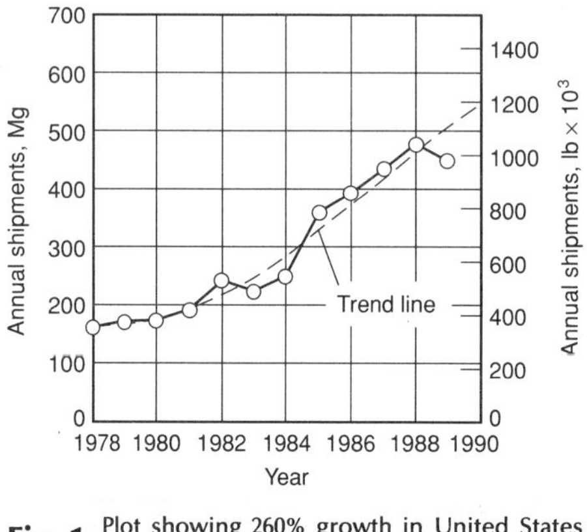

Fig. 1 Plot showing 260% growth in United States titanium casting production in the 10-year period from 1979 to 1989. Source: Ref 4

Another unique property is the high reactivity of titanium at elevated temperatures, leading to an ease of diffusion bonding. As a result, the hot isostatic pressing of titanium castings yields components with no subsurface porosity. At the HIP temperature range of 815 to 980 °C (1500 to 1800 °F), titanium dissolves any microconstituents deposited on internal pore surfaces, leading to complete healing of casting porosity as the pores are collapsed during the pressure and heat cycle. Both the elimination of casting porosity and the promotion of a favorable microstructure improve mechanical properties. However, the very high reactivity of titanium in the molten state presents a challenge to the foundry. Special, and sometimes relatively expensive, methods of melting (Ref 9), moldmaking, and surface cleaning (Ref 7, 8) may be required to maintain product integrity. Additional information on the hot isostatic pressing of castings can be found in Volume 15 of the 9th Edition of *Metals Handbook*.

Historical Perspective of Titanium Casting Technology

Although titanium is the fourth most abundant metallic element in the earth's crust (0.4 to 0.6 wt%) (Ref 9), it has emerged only recently as a technical metal. This is the result of the high reactivity of titanium, which requires complex methods and high energy input to win the metal from the oxide ores. The required energy per ton is 1.7 times that of aluminum and 16 times that of steel (Ref 10). From 1930 to 1947, metallic titanium extracted from the ore as a

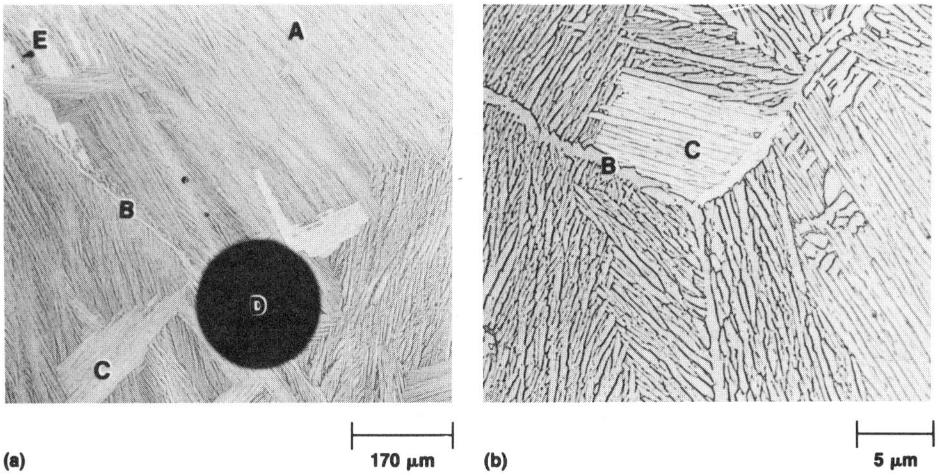

Fig. 2 Comparison of the microstructures of (a) as-cast versus (b) cast + HIP Ti-6Al-4V alloys illustrating lack of porosity in (b). Grain boundary α (B) and α plate colonies (C) are common to both alloys; β grains (A), gas (D), and shrinkage voids (E) are present only in the as-cast alloy.

powder or sponge form was processed into useful shapes by P/M methods to circumvent the high reactivity in the molten form (Ref 11) (see the article "Titanium P/M Products" in this Volume).

Melting Methods. The melting of small quantities of titanium was first experimented with in 1948 using methods such as resistance heating, induction heating, and tungsten arc melting (Ref 12, 13). However, these methods never developed into industrial processes. The development during the early 1950s of the cold crucible, consumable-electrode vacuum arc melting process, or skull melting, by the U.S. Bureau of Mines (Ref 13, 14) made it possible to melt large quantities of contamination-free titanium into ingots or net shapes. Additional information on numerous melting methods is available in the articles "Melting Furnaces" and "Vacuum Melting and Remelting Processes" in Volume 15 of the 9th Edition of *Metals Handbook*.

First Castings. The shape casting of titanium was first demonstrated in the United States in 1954 at the U.S. Bureau of Mines using machined high-density graphite molds (Ref 13, 15). The rammed graphite process developed later, also by the U.S. Bureau of Mines (Ref 16), led to the production of complex shapes. This process, and its derivations, are used today to produce large parts for marine and chemical-plant components (such as the pump and valve components shown in Fig. 3a) because of the rigidity and strength of the mold. Some aerospace components such as the aircraft brake torque tubes, landing arrestor hook, and optic housing shown in Fig. 3(b) have also been produced by this method.

Molding Methods

Rammed Graphite Molding. The traditional rammed graphite molding process uses powdered graphite mixed with organic binders (see the article "Rammed Graphite Molds" in Volume 15 of the 9th Edition of *Metals Handbook*). Patterns typically are made of wood. The mold material is pneumatically rammed around the pattern and cured at high temperature in a reducing atmosphere to convert the organic binders to pure carbon. The molding process and the tooling are essentially the same as those used for cope and drag sand molding in ferrous and nonferrous foundries. In the 1970s, derivations of rammed graphite mold materials were developed using components of more traditional sand foundries, along with inorganic binders. This resulted in more dimensionally stable and less costly molds that were capable of containing molten titanium without undue metal/mold reaction and with easier mold removal from the cast parts.

Lost-Wax Investment Molding. The principal technology that allowed the proliferation of titanium alloy castings in the aerospace industry was the investment casting method, which was introduced in the mid-1960s (see the article "Investment Casting" in Volume 15 of the 9th Edition of *Metals Handbook*). This method, already used at the dawn of the metallurgical age, more than 5000 years ago, for casting copper and bronze tools and ornaments (see the article "History of Casting" in Volume 15 of the 9th Edition of *Metals Handbook*), was later adapted to enable the production of high-quality steel and nickel-base cast parts. The adaptation of this method to titanium casting technology required the development of ceramic slurry materials that had minimum reaction with the extremely reactive molten titanium.

Refractory Oxide Shell Systems. Proprietary lost-wax ceramic shell systems have been developed by the several foundries engaged in titanium casting manufacture. Of necessity, these shell systems must be relatively inert to molten titanium and cannot be made with the conventional foundry ceramics used in the ferrous and nonferrous industries. Usually, the face coats are made with special refractory oxides and appropriate binders. After the initial face coat ceramic is applied to the wax pattern, more traditional refractory systems are used to add shell strength by means of repeated backup ceramic coatings. Regardless of face coat composition, some metal/mold reaction inevitably occurs from titanium reduction of the ceramic oxides. The oxy-

(a) (b)

Fig. 3 Typical titanium parts produced by the rammed graphite process. (a) Pump and valve components for marine and chemical-processing applications. (b) Brake torque tubes, landing arrestor hook, and optic housing components used in aerospace applications

Table 1 Status and capacity of titanium foundries in the United States, Japan, and Western Europe in 1990

Foundry	Maximum pour weight		Approximate maximum envelope size				Melt stock	Use of postcast HIP
			Rammed graphite		Investment casting			
	kg	lb	mm	in.	mm	in.		
Arwood Corp. (CT)	180	400	...	...	1220 diam × 1220	48 diam × 48	Billet	Always
Duriron (OH)	20	50	...	...	510 diam × 760	20 diam × 30	Revert	Often
Howmet Corp. (MI and VA)	730	1600	...	...	1525 diam × 1525	60 diam × 60	Billet	Always
Oremet Corp. (OR)	750	1650	1525 diam × 1830	60 diam × 72	...	...	Billet and revert	Seldom
PCC (OR)	770	1700	...	...	1525 diam × 1220	60 diam × 48	Billet and revert	Always
Rem Products (OR)	180	400	...	...	815 diam × 508	32 diam × 20	Billet	Often
Schlosser Casting Co. (OR)	90	200	...	...	760 diam × 610	30 diam × 24	Billet	Often
Tiline, Inc. (OR)	750	1650	...	...	1370 diam × 610	54 diam × 24	Billet and revert	Always
TiTech International, Inc. (CA)	400	875	915 diam × 610	36 diam × 24	915 diam × 610	36 diam × 24	Billet and revert	Often
PCC France (France)	270	600	...	...	1220 diam × 1220	48 diam × 48	Billet and revert	Always
Tital (West Germany)	180	400	1145 diam × 760	45 diam × 30	1015 diam × 635	40 diam × 25	Billet	Always
Settas (Belgium)	820	1800	1525 diam × 1220	60 diam × 48	610 diam × 610	24 diam × 24	Billet and revert	Often
VMC (Japan)	180	400	1270 diam × 635	50 diam × 25	Research and development		Billet and revert	Seldom

gen-rich surface of the casting stabilizes the α phase. In β and α + β alloys, a metallographically distinct α-case layer on the cast surface is usually formed. This α-case layer may be removed later by means of chemical milling using an acid etchant. It should be noted that this α-case layer is not noticeable in 100% α alloys such as commercially pure (C.P.) titanium or Ti-5Al-2½Sn alloys.

Foundry practices focus on methods to control both the extent of the metal/mold reaction and the subsequent diffusion of reaction products below the cast surface. The diffusion of reaction products into the cast surface is time-at-temperature dependent. The depth of surface contamination can vary from nil on very thin sections to more than 1.5 mm (0.06 in.) on thick sections. On critical aerospace structures, the brittle α case is removed by chemical milling. The depth of surface contamination must be taken into consideration in the initial wax pattern tool design. Hence, the wax pattern and casting are made slightly oversize, and final dimensions are achieved through careful chemical milling. Metal superheat, mold temperature and thermal conductivity, *g* force (if centrifugally cast), and rapid postcast heat removal are other key factors in producing a satisfactory product. These parameters are interchangeable, that is, a high *g* force centrifugal pour into cold molds may achieve the same relative fluidity as a static pour into heated molds.

Other Refractory Shell Systems. The combination of graphite powder, graphite stucco, stucco, and organic binders has also been used as a shell system for the investment casting of titanium. After dewax, the shell is fired in a reducing atmosphere to remove or pyrolyze the binders before casting. This technology has not been promoted as much as the use of refractory oxide shell systems and is presently primarily of historic interest.

Additional Molding Systems. In addition to the rammed graphite and investment molding methods, a poured ceramic mold has been used to produce large parts that require good dimensional accuracy. This method, developed in the late 1970s, was used to a limited extent for several years.

Semipermanent, reusable molds, frequently made from machined graphite, have been used successfully since the earliest U.S. Bureau of Mines work (Ref 13, 15) but only on relatively simple-shape parts that allow metal volumetric shrinkage to occur without restriction. The method is economical only when reasonably high volumes are required, that is, thousands of parts, because of the high cost of the solid mold material.

A titanium sand casting technique based on conventional foundry moldmaking practices has been under development at the U.S. Bureau of Mines (Ref 17). Because the mold materials are less costly and the cast part is easier to remove from the sand mold than from other methods of titanium casting, this development could lower production costs. However, surface quality problems are restricting the use of this method thus far.

Foundries and Capacities. Table 1 lists the major titanium casting foundries in the industrial western world and summarizes the use and capacities of the various titanium casting practices, including the use of hot isostatic pressing.

Alloys

All production titanium castings to date are based on traditional wrought product compositions. As such, the Ti-6Al-4V alloy dominates structural casting applications. This alloy similarly has dominated wrought industry production since its introduction in the early 1950s, becoming the benchmark alloy against which others are compared. However, other wrought alloys have been developed for special applications, with better room-temperature or elevated-temperature strength, creep, or fracture toughness characteristics than those of Ti-6Al-4V. In addition, these same alloys are being cast when net shape casting technology is the most economical method of manufacture. As with Ti-6Al-4V, other cast titanium alloys have properties generally comparable to those of their wrought counterparts.

Chemistry and Demand. Table 2 lists the most prevalent casting alloy chemistries and the most characteristic attribute of each in comparison with Ti-6Al-4V, plus current approximate market share.

Typical Properties. Table 3 is a summary of room-temperature tensile properties for various alloys. These properties, which are typical, vary depending on microstructure as influenced by foundry parameters such as solidification rate and any postcast HIP and heat treatments.

Specifications. Industrywide specifications, listed in Table 4 for reference, give more detail on mechanical property guarantees and process control features. In addition, most major aerospace companies have comparable specifications. *MIL Handbook V, Aerospace Design Specifications* does not presently include titanium alloy castings, but it is expected that such information will be incorporated in the near future. As with wrought products, commercially pure titanium castings are used almost entirely in corrosion-resistant applications. Commercially pure titanium pumps and valves are the principal components made as titanium castings for corrosion-resistant applications. These are used extensively in chemical and petrochemical plants and in numerous marine applications (seawater pumps, for example, are a very important application).

Newer Alloys. As aircraft engine manufacturers seek to use cast titanium at higher operating temperatures, Ti-6Al-2Sn-4Zr-2Mo and Ti-6Al-2Sn-4Zr-6Mo are being specified more frequently (see Tables 2 and 3). Other advanced high-temperature titanium alloys for service up to 595 °C (1100 °F) such as Ti-1100 and IMI-834 are being developed as castings. The alloys mentioned above exhibit the same degree of elevated-temperature superiority as do their wrought counterparts over the more commonly used Ti-6Al-4V (Fig. 4).

Extra low interstitial (ELI) grade Ti-6Al-4V has been used for critical cryogenic

Table 2 Comparison of cast titanium alloys

Alloy	Estimated relative use of castings	Nominal composition, wt% O	N	C	H	Al	Fe	V	Cr	Sn	Mo	Nb	Zr	Si	Special properties(a)(b)
Ti-6Al-4V	85%	0.18	0.015	0.04	0.006	6	0.13	4	...	...	...	...	...	...	General purpose
Ti-6Al-4V ELI(b)	1%	0.11	0.010	0.03	0.006	6	0.10	4	...	...	...	...	...	...	Cryogenic toughness
Commercially pure titanium (grade 2)	6%	0.25	0.015	0.03	0.006	...	0.15	...	...	...	...	...	...	...	Corrosion resistance
Ti-6Al-2Sn-4Zr-2Mo	7%	0.10	0.010	0.03	0.006	6	0.15	...	...	2	2	...	4	...	Elevated-temperature creep
Ti-6Al-2Sn-4Zr-6Mo	<1%	0.10	0.010	0.03	0.006	6	0.15	...	...	2	6	...	4	...	Elevated-temperature strength
Ti-5Al-2.5Sn	<1%	0.16	0.015	0.03	0.006	5	0.2	...	...	2.5	...	...	...	...	Cryogenic toughness
Ti-3Al-8V-6Cr-4Zr-4Mo (Beta-C)	<1%	0.10	0.015	0.03	0.006	3.5	0.2	8.5	6	...	4	...	4	...	RT strength
Ti-15V-3Al-3Cr-3Sn (Ti-15-3)	<1%	0.12	0.015	0.03	0.006	3	0.2	15	3	3	...	...	...	...	RT strength
Ti-1100	<1%	0.07	0.015	0.04	0.006	6.0	0.02	...	...	2.75	0.4	...	4.0	0.45	Elevated-temperature properties
IMI-834	<1%	0.10	0.015	0.06	0.006	5.8	0.02	...	...	4.0	0.5	0.7	3.5	0.35	Elevated-temperature properties
Total	**100%**														

(a) Superior, relative to Ti-6Al-4V. (b) RT, room temperature. (c) ELI, extra low interstitial

space shuttle service where fracture toughness is an important design criteria. The most recent alloy to receive attention in the casting industry is the metastable β alloy Ti-15V-3Al-3Cr-3Sn (Ti-15-3) (see Tables 2 and 3). Originally developed as a highly cold-formable and subsequently age-hardened sheet material, this alloy is highly castable and readily heat treated to a 1275 MPa (185 ksi) tensile strength level, making it a serious candidate for the replacement of high-strength precipitation-hardening (PH) stainless steels such as 17-4 PH. The full density advantage of titanium of about 40% is preserved because strength levels are comparable in both materials. Figure 5 shows typical room-temperature tensile data following several simple aging cycles. The data show excellent 25% elongation at 700 MPa (100 ksi) yield strength in the solution-annealed (as hot isostatically pressed) condition, and subsequent aged tensile strength capability of as much as 1400 MPa (200 ksi) with 3 to 4% elongation.

Titanium-aluminide castings are being developed for application in the compressor sections of aircraft gas turbine engines subjected to the highest temperatures. Compositions based on both the α_2 (Ti_3Al) and γ (TiAl) ordered phases have been cast experimentally, with the former being closer to limited-production status. The low ductility of these alloys at room temperature has been the major producibility challenge. It is anticipated that the service potential for titanium aluminides in the 595 to 925 °C (1100 to 1700 °F) temperature range will eventually be realized. The difficulty in machining shapes in these brittle alloys may increase the advantage of net shape methods such as castings or powder metallurgy (see the article "Titanium P/M Products" in this Volume).

Because Ti-6Al-4V dominates the industry, much more metallurgical and mechanical test data are available on this alloy. These data are discussed in the section "Microstructure of Ti-6Al-4V" below.

Microstructure of Ti-6Al-4V

Cast Microstructure. To understand the relatively high mechanical property levels of titanium alloy castings and the many improvements made in recent years, it is necessary to understand the microstructures of castings and their influence on the mechanical behavior of titanium. The phase transformation from β to α + β leads to the elimination of the dendritic cast structure. The existence of such dendrites during the solidification stage is evident in the surface morphology of shrinkage pores (Fig. 6). The phase transformation, which in the alloy Ti-6Al-4V is typically initiated at 995 °C (1825 °F), results in the microstructural features shown in Fig. 2(a). This microstructure, which will be discussed in detail, is very similar to a β-processed wrought microstructure and therefore has similar properties. Thus, in the study and development of titanium alloy castings, it is possible to draw much information from the vast

Table 3 Typical room-temperature tensile properties of titanium alloy castings (bars machined from castings)

Specification minimums are less than these typical properties.

Alloy(a)(b)	Yield strength MPa	Yield strength ksi	Ultimate strength MPa	Ultimate strength ksi	Elongation, %	Reduction of area, %
Commercially pure (grade 2)	448	65	552	80	18	32
Ti-6Al-4V, annealed	855	124	930	135	12	20
Ti-6Al-4V ELI	758	110	827	120	13	22
Ti-1100, Beta-STA(c)	848	123	938	136	11	20
Ti-6Al-2Sn-4Zr-2Mo, annealed	910	132	1006	146	10	21
IMI-834, Beta-STA(c)	952	138	1069	155	5	8
Ti-6Al-2Sn-4Zr-6Mo, Beta-STA(c)	1269	184	1345	195	1	1
Ti-3Al-8V-6Cr-4Zr-4Mo, Beta-STA(c)	1241	180	1330	193	7	12
Ti-15V-3Al-3Cr-3Sn, Beta-STA(c)	1200	174	1275	185	6	12

(a) Solution-treated and aged (STA) heat treatments may be varied to produce alternate properties. (b) ELI, extra low interstitial. (c) Beta-STA, solution treatment with β-phase field followed by aging

Table 4 Standard industry specifications applicable to titanium castings

MIL-T-81915	Titanium and titanium alloy castings, investment
AMS-4985A	Titanium alloy castings, investment or rammed graphite
AMS-4991	Titanium alloy castings, investment
ASTM B 367	Titanium and titanium alloy castings
MIL-STD-2175	Castings, classification and inspection of
MIL-STD-271	Nondestructive testing requirements for metals
MIL-STD-453	Inspection, radiographic
MIL-Q-9858	Quality program requirement
MIL-I-6866B	Inspection, penetrant method of
MIL-H-81200	Heat treatment of titanium and titanium alloys
ASTM E 155	Reference radiographs for inspection of aluminum and magnesium castings
ASTM E 192	Reference radiographs, investment steel castings
ASTM E 186	Reference radiographs, steel castings 50-102 mm (2-4 in.)
ASTM E 446	Reference radiographs, steel castings up to 50 mm (2 in.)
ASTM E 120	Standard methods for chemical analysis of titanium and titanium alloys
ASTM E 8	Methods of tension testing of metallic materials
AMS-2249B	Chemical-check analysis limits for titanium and titanium alloys
AMS-4954	Titanium alloy welding wire Ti-6Al-4V
AMS-4956	Titanium alloy welding wire Ti-6Al-4V, extra low interstitial

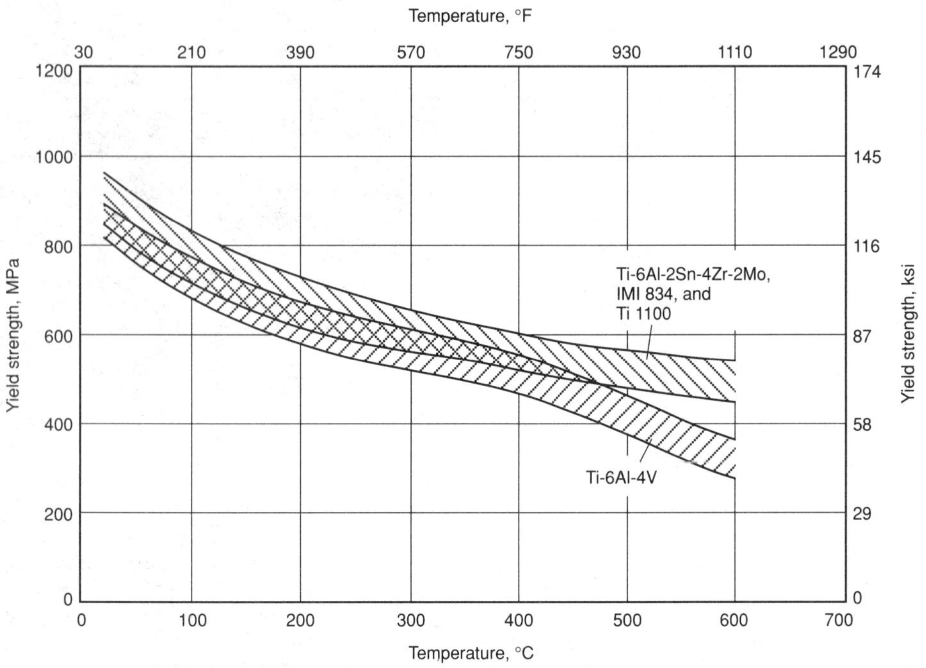

Fig. 4 Plot of yield strength versus temperature to compare elevated-temperature properties of cast Ti-6Al-2Sn-4Zr-2Mo, IMI 834, and Ti 1100 alloys with standard cast Ti-6Al-4V alloy

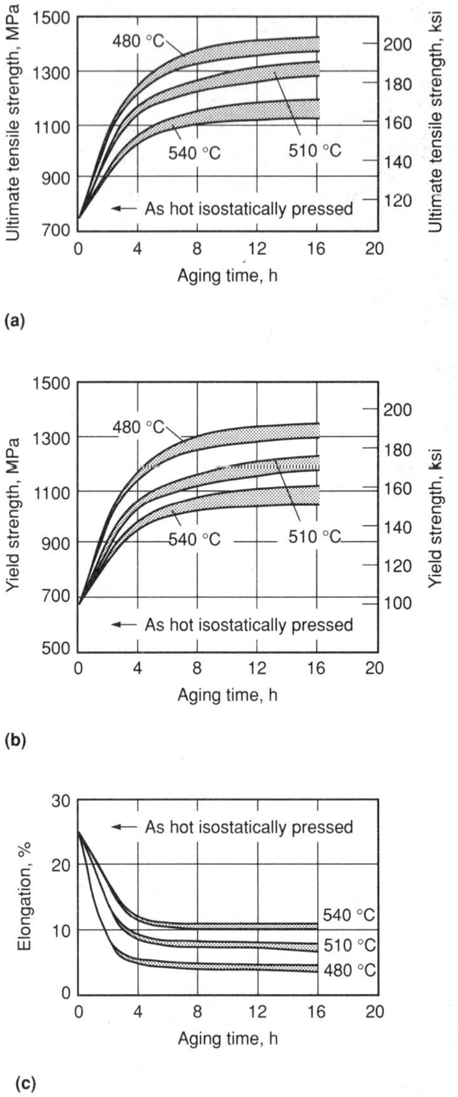

Fig. 5 Aging curves showing typical room-temperature tensile properties of HIP and aged Ti-15V-3Al-3Cr-3Sn (Ti-15-3) castings. (a) Ultimate tensile strength. (b) Yield strength. (c) Elongation

knowledge of conventional titanium ingot metallurgy.

Hot isostatic pressing is now becoming almost a standard practice for all titanium cast parts produced for the aerospace industry (Table 1). As a result, cast + HIP microstructure also needs to be considered (Fig. 2b). Because the HIP temperature is typically well below the β transus temperature, where there is no growth of the β grains, α plates, or their colonies, the as-cast (Fig. 2a) and the cast + HIP (Fig. 2b) microstructures look very much alike, except for the lack of porosity in the latter.

As-Cast and Cast + HIP Microstructures. Because most castings for demanding applications are produced with Ti-6Al-4V alloy (Table 2), only microstructures of this α + β alloy will be reviewed here.

Beta Grain Size. Beta grains (A, in Fig. 2a) develop during the solid-state cooling stage between the solidus/liquidus temperature and the β transus temperature. As a result, large and thick sections, which cool at a slower rate, show larger β grains. The size range of the β grains is from 0.5 to 5 mm (0.02 to 0.2 in.). As will be further discussed, large β grains may lead to large α plate colonies. This is beneficial for fracture toughness, creep resistance, and fatigue crack propagation resistance (Ref 18, 19) and detrimental for low- and high-cycle fatigue strength and tensile elongation (Ref 20, 21).

Grain Boundary α. This α phase (B, in Fig. 2a) is formed along the β grain boundaries when cast material is cooled through the α + β phase field (in Ti-6Al-4V this is typically from 995 °C, or 1825 °F, down to room temperature). This phase is plate shaped and represents the largest α plates in the cast structure. The length of these plates can equal the β grain radius. Because of its long dimension and planar shape, it has been found to be very detrimental to fatigue crack initiation at room temperature (Ref 22, 23) and at elevated temperatures (Ref 23, 24) both in cast and ingot metallurgy (I/M) materials. Many postcast thermal treatments eliminate this phase to improve fatigue life.

Alpha Plate Colonies. Alpha platelets (C, in Fig. 2a) are the transformation products of the β phase when cooled below the β transus temperature. The hexagonal close-packed (hcp) orientation of these plates is related to the parent body-centered cubic (bcc) β phase orientation through one of the 12 possible variants of the Burgers relationship (Ref 25, 26):

$$\{110\}\beta \parallel (0001)\alpha$$

$$\langle 111\rangle\beta \parallel \langle 11\bar{2}0\rangle\alpha$$

When cooling rates are relatively slow, such as in thick-section castings, many adjacent α platelets transform into the same Burgers variant and form a colony of similarly aligned and crystallographically oriented platelets. The large colonies (C, in Fig. 2a) may be associated with early fatigue crack initiation (Ref 21), the result of heterogeneous basal slip across the plates (Ref 27). At the same time, the large colony structure is beneficial for fatigue crack propagation resistance (Ref 28, 29). Because α platelet colonies cannot grow larger than the β grains, titanium castings with large prior β grains typically have large colonies. The individual α platelets are typically 1 to 3 μm (40 to 120 μin.) in thickness and 20 to 100 μm (0.0008 to 0.004 in.) in length (Ref 30, 31). The typical colony size range in Ti-6Al-4V castings is 50 to 500 μm (0.002 to 0.02 in.) (Ref 22, 30, 31).

As a general rule, slower solid-state cooling rates, such as in thick cast sections, result in microstructures with larger β grains, a longer and thicker grain-boundary α phase, thicker α platelets, and larger α platelet colonies.

Porosity. Gas (D, in Fig. 2a) and shrinkage voids (E) are typical phenomena in as-cast titanium products. Hot isostatic pressing, however, closes and heals these pores. This is demonstrated by comparing the as-cast microstructure in Fig. 2(a) with the cast + HIP structure in Fig. 2(b). The reactiveness of the titanium at the HIP

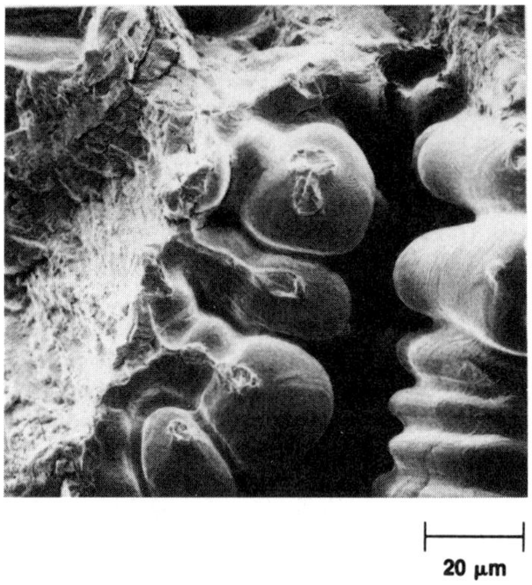

Fig. 6 Dendritic structure present in the surface shrinkage porosity of an as-cast Ti-6Al-4V component

temperature range of 900 to 955 °C (1650 to 1750 °F) leads to dissolution of all microconstituents deposited on the pore surfaces leading to complete healing of casting porosity. Hot isostatic pressing also causes a degree of α plate coarsening. It should be noted that all aerospace-related titanium alloy cast parts are delivered after hot isostatic pressing (see Table 1).

Modification of Microstructure. Most Ti-6Al-4V titanium castings produced commercially today are supplied in the annealed condition. However, much microstructural modification development work has been done recently, and it can be expected that solution-treated and aged or other postcast thermal processing will eventually become specified on cast parts requiring certain property enhancement such as fatigue or tensile strength. The following section reviews several of these developmental procedures and their results.

The modification of microstructure is one of the most versatile tools available in metallurgy for improving the mechanical properties of alloys. This is commonly achieved through a combination of cold or hot working followed by the heat treatment known as thermomechanical processing. Net shapes such as castings or P/M products cannot be worked, which limits the options for controlling microstructures. A substantial amount of work has been done in recent years to improve the microstructures of titanium alloy net shape products, with an emphasis on Ti-6Al-4V material. Most treatment schemes can be successfully applied to both cast parts (Ref 8) and P/M compacts (see the article "Titanium P/M Products" in this Volume) (Ref 32, 33). In the case of titanium alloy castings, the main goal has been to eliminate the grain-boundary α phase, the large α plate colonies, and the individual α plates. This is accomplished either by solution treatments or by a temporary alloying with hydrogen. In some cases, the hydrogen and solution treatments are combined. The details of these methods, including the appropriate references, are listed in Table 5. The typical resulting microstructures of the α-β solution treatment (ABST), β solution treatment (BST), broken-up structure (BUS), and high-temperature hydrogenation (HTH) methods are shown in Fig. 7(a), (b), (c), and (d), respectively. As can be seen from the photomicrographs, these treatments are successful in eliminating the large α plate colonies and the grain-boundary α phase. As discussed below, a substantial improvement of both tensile and fatigue properties is achieved with these processes.

Mechanical Properties of Ti-6Al-4V

Oxygen Influence. Figure 8 is a frequency distribution of tensile properties from separately cast test bars representing hot isostatically pressed and annealed Ti-6Al-4V castings. Oxygen, a carefully controlled alloy addition, is in the 0.16 to 0.20% range, which is common for many aerospace specifications.

Some specifications allow a 0.25% maximum oxygen content. The resultant properties with oxygen in the 0.20 to 0.25% range are typically about 69 to 83 MPa (10 to 12 ksi) higher than those shown in Fig. 8 with slightly lower ductility levels. In this case, it is possible to guarantee 827 MPa (120 ksi) yield strength and 896 MPa (130 ksi) ultimate tensile strength levels with 6% minimum elongation. This strength level is the same minimum guarantee for wrought-annealed Ti-6Al-4V.

Microstructure Influence. Because the microstructure of titanium alloy cast parts is very similar to that of β-processed wrought or I/M material, many properties of hot isostatically pressed castings, such as tensile strength, fracture toughness, fatigue crack propagation, and creep, are at the same levels as with forged and machined parts. Tensile strength and fracture toughness properties of cast, cast + HIP, and cast + HIP + heat-treated material (Table 5) are compared in Table 6 to wrought β-annealed data. To provide a complete review, properties of castings treated by many of the methods listed in Table 5 are also included. At the present time, fracture toughness data are available for only a few of the conditions. As can be seen, some of the treated conditions present properties in

Table 5 Thermal and thermochemical methods for modifying the microstructure of α + β titanium alloy net shape products

Method(a)	Typical solution treatment(b)	Hydrogenation temperature °C	Hydrogenation temperature °F	Intermediate treatment(c) °C	Intermediate treatment(c) °F	Dehydrogenation temperature °C	Dehydrogenation temperature °F	Typical annealing or aging treatment	Applied to product forms(d)	Ref
BUS	1040 °C (1900 °F) for ½ h	...	...	...	...	...	...	845 °C (1550 °F) for 24 h	Cast, P/M, I/M	34–37
GTEC	1050 °C (1925 °F) for ½ h	...	...	...	...	...	...	845 °C (1550 °F) for ½ h and 705 °C (1300 °F) for 2 h	Cast	38
BST	1040 °C (1900 °F) for ½ h and GFC	...	...	...	...	...	...	540 °C (1000 °F) for 8 h	Cast, I/M	39
ABST	955 °C (1750 °F) for 1 h and GFC	...	...	...	...	...	...	540 °C (1000 °F) for 8 h	Cast, I/M	39
HVC (Hydrovac process)	...	650	1200	870(e)	1600(e)	760	1400	...	P/M, I/M	40, 41
TCT	1040 °C (1900 °F) for ½ h	595	1100	Cool to RT		760	1400	...	Cast, P/M, I/M	41–43
CST	...	870	1600	No intermediate step (continuous process)		815	1500	...	Cast	44
HTH	...	900	1650	Cool to RT		705	1300	...	Cast, P/M, I/M	45

(a) Most data apply to Ti-6Al-4V. β transus temperature approximately 995 °C (1825 °F). (b) GFC, gas fan cooled. (c) RT, room temperature. (d) P/M, powder metallurgy; I/M, ingot metallurgy. (e) Glass encapsulated prior to heat treatment

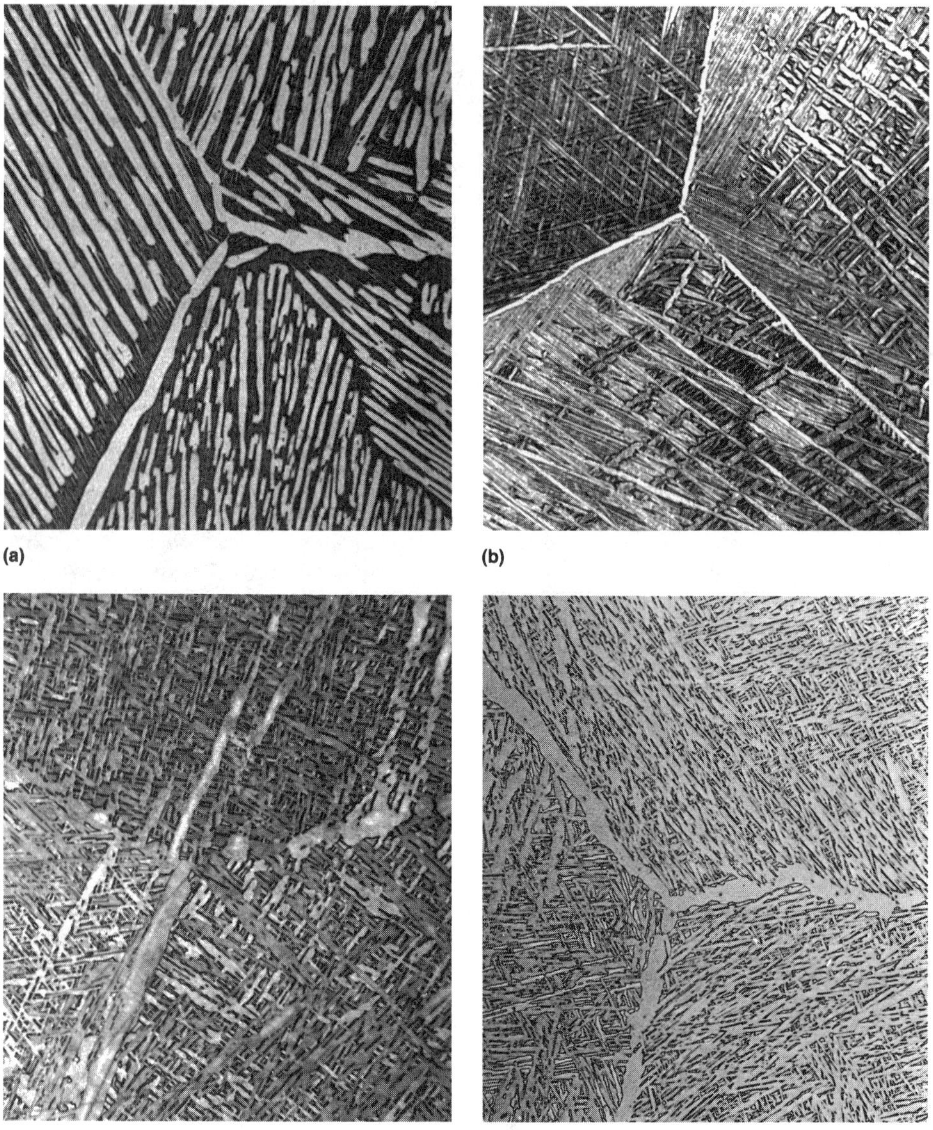

Fig. 7 Photomicrographs of microstructures resulting from a variety of hydrogen and solution heat treatments used to eliminate large α plate colonies and grain boundary α phase in α + β titanium alloys. (a) ABST. (b) BST. (c) BUS. (d) HTH. See Table 5 for details of heat treatments.

excess of I/M β-annealed material. However, it should be noted that tests were done on relatively small cast coupons. Properties of actual cast parts, especially large components, could be somewhat lower, the result of coarser grain structure or slower quench rates. Of special interest are the hydrogen-treated conditions (such as thermochemical treatment, or TCT; constitutional solution treatment, or CST; and HTH, in Table 5) that result in very high tensile strength (as high as 1124 MPa, or 163 ksi) with tensile elongation as high as 8%.

Fatigue and Fatigue Crack Growth Rate. The fatigue crack growth rate (FCGR) behavior of cast Ti-6Al-4V is also, as expected, very similar to that of β-processed wrought Ti-6Al-4V (Ref 50-52). This is demonstrated in Fig. 9 in which the scatterband of the FCGR of cast and cast-HIP alloys is compared to β-processed I/M (Ref 18, 53).

The scatterbands of smooth axial room-temperature fatigue results of cast, cast + HIP (Ref 15, 47, 48, 54-57), and wrought Ti-6Al-4V are shown in Fig. 10. This figure clearly indicates that the HIP process results in a substantially improved fatigue life well into the wrought-annealed region. The fatigue properties of aerospace quality castings have always been an important issue, because in most other alloy systems this is the property that is most degraded, compared to wrought products. However, because of the complete closure and healing of gas (D, in Fig. 2a) and shrinkage (E, in Fig. 2a) pores by HIP and the inherent β-annealed microstructure, it is possible to obtain fatigue life comparable to wrought material in premium investment cast and hot isostatic pressed parts. As indicated previously (Table 5), substantial work has been done in recent years to modify the microstructure of cast parts to produce fatigue properties either equivalent or superior to the best wrought-annealed products. Figure 11(a) compares the smooth fatigue life of Ti-6Al-4V treated by ABST, BST, BUS, CST, Garrett treatment (GTEC), and HTH (Table 5) to wrought material scatterband. As can be seen, all of these treatments were successful in improving fatigue life above average wrought levels. The hydrogen treatments (CST and HTH) resulted in the highest improvement in fatigue strength. However, it should be noted that wrought products subjected to the same treatments result in comparable improvements in fatigue strength.

Another approach to the improvement of fatigue of cast parts is the selection of high-strength cast alloy rather than Ti-6Al-4V. Figure 11(b) compares the fatigue strength of investment cast Ti-3Al-8V-6Cr-4Zr-4Mo (Beta C) and Ti-10V-2Fe-3Al (Ti-10-2-3) in solution-treated and aged (STA) condition to wrought-annealed Ti-6Al-4V (Ref 58). Figure 11(b) shows that fatigue strength in excess of 1000 MPa (145 ksi) can be obtained with high-strength cast alloys.

Casting Design

The best casting design is usually achieved by means of a thorough review by the manufacturer and user when the component is still in the preliminary design stage (see the article "Casting Design" in Volume 15 of the 9th Edition of *Metals Handbook*). Additional features may be incorporated to reduce machining cost, and components may be integrated to eliminate later fabrication. Specifications and tolerances may be reviewed vis-à-vis foundry capabilities, producibility, and pattern tool concepts to achieve the most practical and cost-effective design (see the articles "Dimensional Tolerances and Allowances" and "Patterns and Patternmaking" in Volume 15 of the 9th Edition of *Metals Handbook*). When minimum cast part weight is critical, such as in aerospace components, the capability of the foundry to produce varying wall thicknesses, for example, may be beneficial. Often, cast features that cannot be economically duplicated by any other method may be readily produced.

Titanium castings present the designer with few differences in design criteria, compared with other metals. Ideal designs do not contain isolated heavy sections or uniform heavy walls of large area so that centerline shrinkage cavities and regions with a coarse microstructure may be avoided. From a practical sense, however, ideal tapered walls to promote directional solidification are not usually a reality. The advent of hot isostatic pressing to heal internal as-cast shrinkage cavities has offered the designer much more freedom; however,

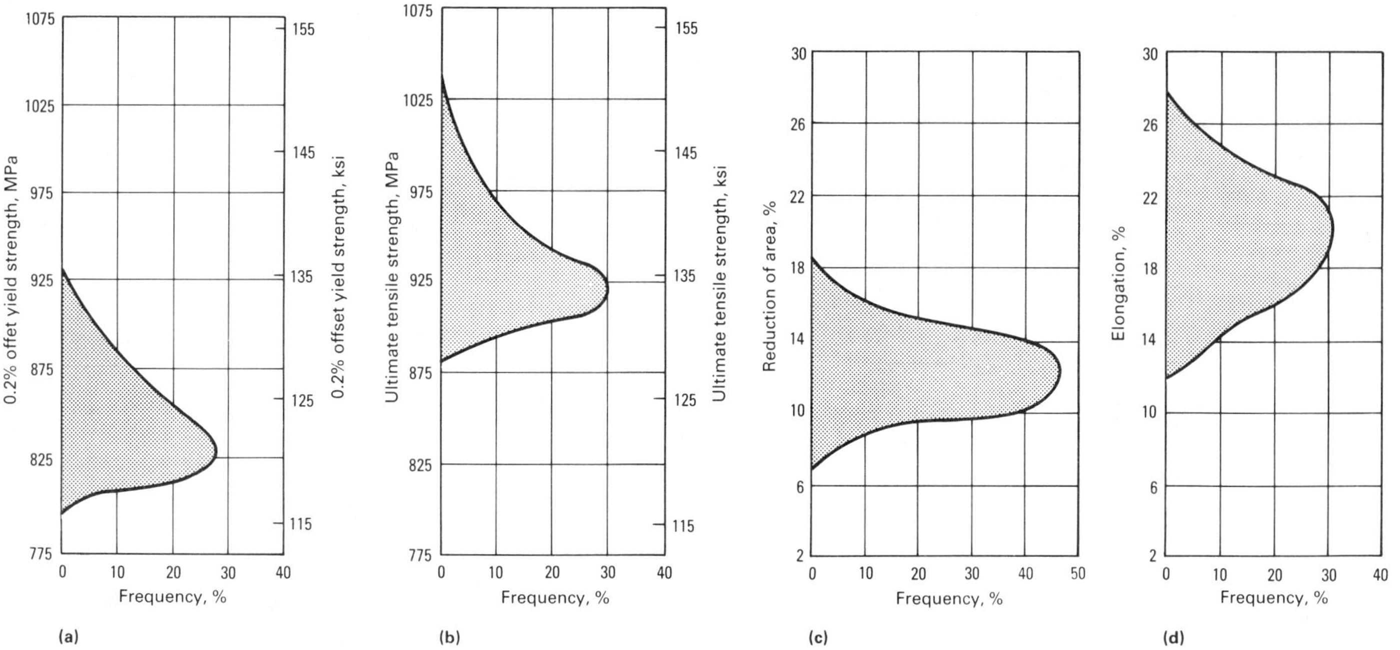

Fig. 8 Frequency distribution of tensile properties of hot isostatically pressed and annealed Ti-6Al-4V casting test bars. Percent frequency is plotted versus (a) 0.2% offset yield strength, (b) ultimate tensile strength, (c) percent reduction of area, and (d) percent elongation. Alloy composition is 0.16 to 0.20% O_2; sample size is 500 heats. Source: Ref 46

there still is a practical limit to the size of an internal cavity that can be healed through hot isostatic pressing without contributing significant surface or structural deformation due to the collapse of internal pores.

The lost-wax investment process provides more design freedom for the foundry to feed a casting properly than does the traditional sand or rammed graphite approach. It is normal practice to use adequate gates and risers to subsequently hot isostatically pressed investment castings to achieve reasonably good as-cast internal x-ray quality so that hot isostatic pressing will not cause extensive surface or structural deformation.

The usual required minimum practical wall thickness for investment castings is 2.0 mm (0.080 in.); however, sections as thin as 1.1 mm (0.045 in.) are routinely made. Even thinner walls may be achieved by chemical milling beyond that required for α-case removal; however, as-cast wall variation is not improved and becomes a larger percentage of the resultant wall thickness. Sand or rammed graphite molded castings have a usual minimum wall thickness of 4.75 mm (0.187 in.), although 3.0 mm (0.12 in.) is not unreasonable for short sections.

Fillet radii should be as generous as possible to minimize the occurrence of hot tears. Although 0.76 mm (0.030 in.) radii are produced, the preferred minimum is 3.0 mm (0.12 in.). A rule of thumb is that a fillet radius should be 0.5 times the sum of the thicknesses of the two adjoining walls.

With proper tool design, zero draft walls are possible. To promote directional solidification, a 3° included draft angle may be preferred. Hot isostatic pressing will close any centerline shrinkage cavities in zero draft walls, making it unnecessary to provide draft. Draft requirements are also dependent on foundry practice, with rammed graphite tooling usually requiring draft, and investment casting typically not requiring draft.

Tolerances. Typically, the major area of concern is the true position of a thin-section surface with respect to a datum. Surface areas of approximately 129 cm² (20 in.²) or greater in sections of less than approximately 5.08 mm (0.200 in.) thickness are susceptible to distortion, depending on adjoining sections. The high strength of titanium compared with that of aluminum and low elastic modulus compared with that of steel present challenges in straightening and in main-

Table 6 Tensile properties and fracture toughness of Ti-6Al-4V cast coupons compared to typical wrought β-annealed material

Material condition(a)	Yield strength MPa	Yield strength ksi	Ultimate tensile strength MPa	Ultimate tensile strength ksi	Elongation, %	Reduction of area, %	K_{Ic} ksi$\sqrt{in.}$	K_{Ic} MPa$\sqrt{m}$	Ref
As-cast	896	130	1000	145	8	16	97	107	37, 47
Cast HIP	869	126	958	139	10	18	99	109	37, 39, 48
BUS(b)	938	136	1041	151	8	12	...	...	37, 39
GTEC(b)	938	136	1027	149	8	11	...	...	38
BST(b)	931	135	1055	153	9	15	...	...	39
ABST(b)	931	135	1020	148	8	12	...	...	39
TCT(b)	1055	153	1124	163	6	9	...	...	41, 49
CST(b)	986	143	1055	153	8	15	...	...	44
HTH(b)	1055	153	1103	160	8	15	...	...	45
Typical wrought β annealed	860	125	955	139	9	21	83	91	18, 19

(a) All conditions (except as-cast) are cast plus HIP. (b) See Table 5 for process details.

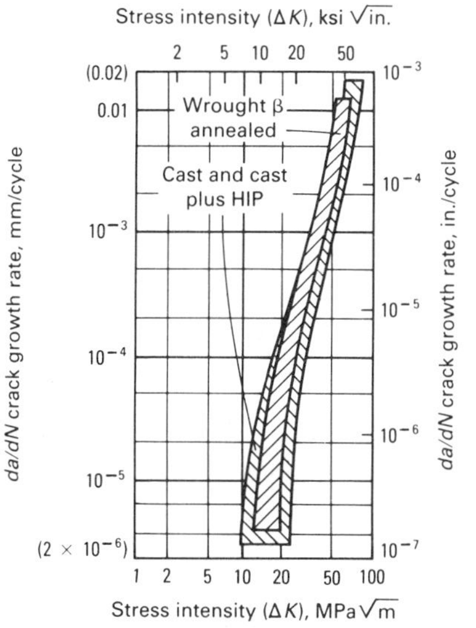

Fig. 9 Scatterband comparison of FCGR behavior of wrought I/M β-annealed Ti-6Al-4V to cast and cast HIP Ti-6Al-4V data

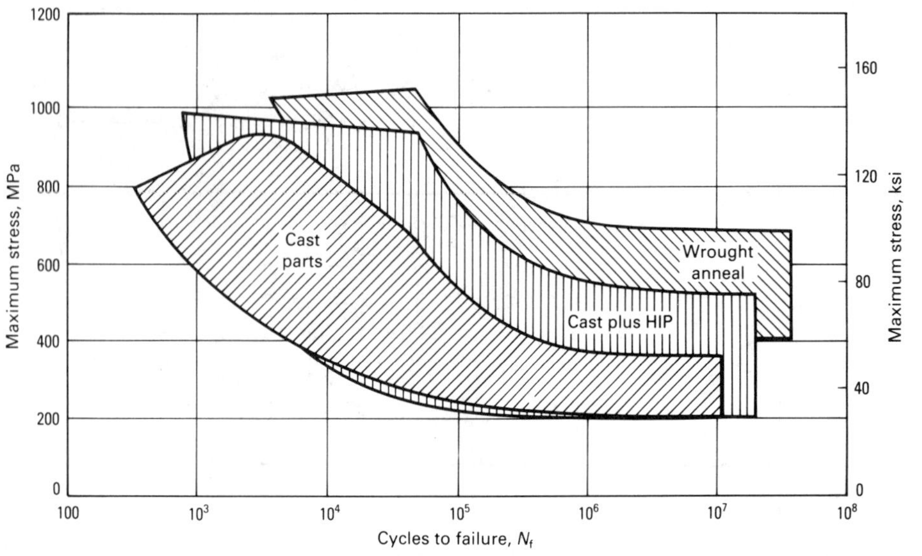

Fig. 10 Comparison of smooth axial room-temperature fatigue rate in cast and wrought Ti-6Al-4V at room temperature with $R = +0.1$

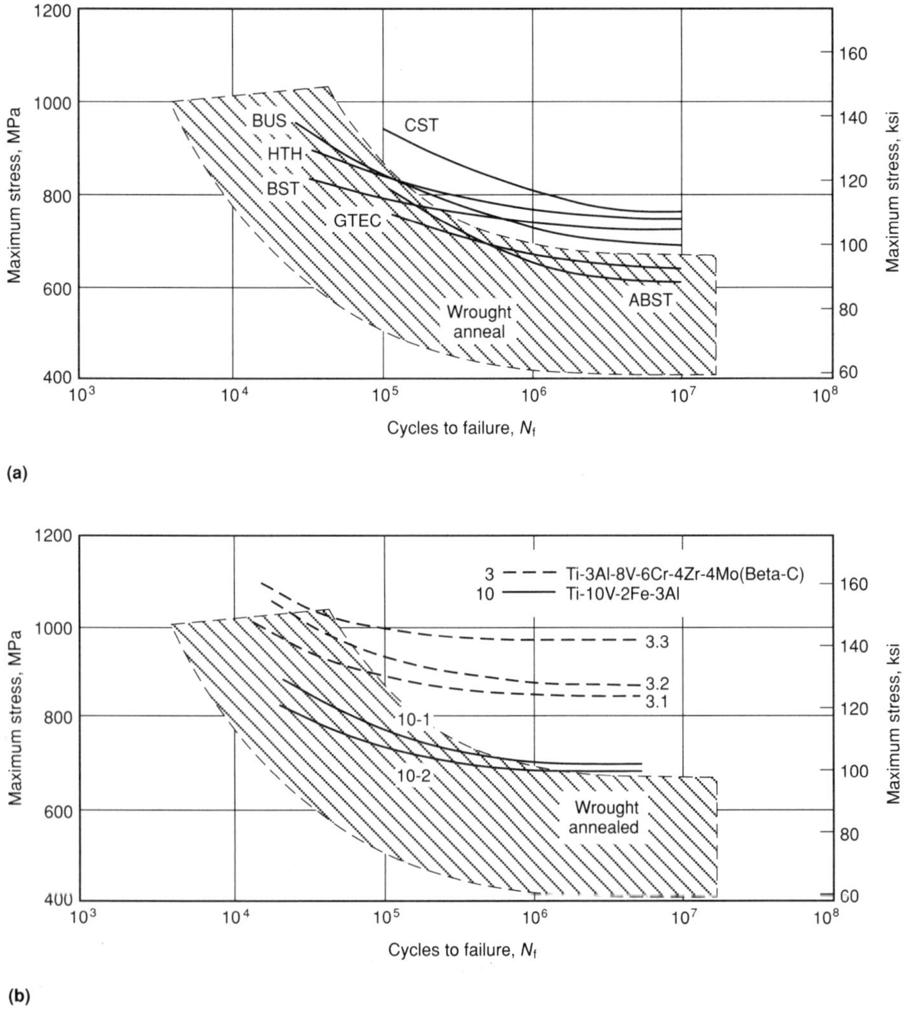

Fig. 11 Comparison of wrought (I/M) annealed Ti-6Al-4V scatterband with (a) Ti-6Al-4V investment castings subjected to various thermal and hydrogen treatments (see Table 5) and (b) heat-treated β titanium alloy castings. For data in (a), smooth axial fatigue measured at room temperature with $R = +0.1$; frequency = 5 Hz using triangular wave form

taining extremely tight, true positions. General tolerance band capabilities for linear dimensions are shown in Table 7.

Hot sizing fixtures have been used increasingly to help control critical casting dimensions. This technique typically involves the use of steel fixtures to "creep" the casting into final tolerances in an anneal or stress-relief heat treatment by the weight of the steel or the use of differential thermal expansion of the steel relative to the titanium.

Standard casting industry thickness tolerances of ±0.76 mm (±0.030 in.) for rammed graphite and ±0.25 mm (±0.010 in.) for investment cast walls are more difficult to maintain with titanium primarily because of the influence of chemical milling. As mentioned earlier, for critical applications it is necessary to mill all surfaces chemically to remove the α case. This operation is subject to variation because of part geometry and bath variables, and because it is usually manually controlled. Standard industry surface finishes are shown in Table 8.

Melting and Pouring Practice

Vacuum Consumable Electrode. The dominant, almost universal, method of melting titanium is with a consumable titanium electrode lowered into a water-cooled copper crucible while confined in a vacuum chamber. This skull melting technique (see the section "Vacuum Arc Skull Melting and Casting" in the article "Vacuum Melting and Remelting Processes" in Volume 15 of the 9th Edition of *Metals Handbook*) prevents the highly reactive liquid titanium from reacting and dissolving the crucible because it is contained in a solid skull frozen against the water-cooled crucible wall. When an adequate melt quantity has been obtained, the residual electrode is quickly retracted, and the crucible is tilted for pouring into the molds. A skull of solid titanium remains in the crucible for reuse in a subsequent pour or for later removal.

Superheating. The consumable electrode practice affords little opportunity for superheating the molten pool because of the cooling effect of the water-cooled crucible. Because of limited superheating, it is common either to pour castings centrifugally, forcing the metal into the mold cavity, or to pour statically into preheated molds to obtain adequate fluidity. Postcast cooling takes place in a vacuum or in an inert gas atmosphere until the molds can be safely removed to air without oxidation of the titanium.

Electrode Composition. Consumable titanium electrodes are either I/M-forged billet, consolidated revert wrought material, selected foundry returns, or a combination of all of these (see Table 1). Casting specifications or user requirements can dictate the composition of revert materials used in electrode construction. Figure 12 shows a typical centrifugal casting furnace arrangement.

Table 7 General linear and diametric tolerance guidelines for titanium castings

Size, mm	Size, in.	Total tolerance band(a), Investment cast	Total tolerance band(a), Rammed graphite process
25 to <102	1 to <4	0.76 mm (0.030 in.) or 1.0%, whichever is greater	1.52 mm (0.060 in.)
102 to <305	4 to <12	1.02 mm (0.040 in.) or 0.7%, whichever is greater	1.78 mm (0.070 in.) or 1.0%, whichever is greater
305 to <610	12 to <24	1.52 mm (0.060 in.) or 0.6%, whichever is greater	1.0%
≥610	≥24	0.5%	1.0%
Examples			
254 mm	10 in.	1.78 mm (0.070 in.) total tolerance band or ±0.89 mm (±0.035 in.)	2.54 mm (0.100 in.) total tolerance band or ±1.27 mm (±0.050 in.)
508 mm	20 in.	3.05 mm (0.120 in.) total tolerance band or ±1.52 mm (±0.060 in.)	5.08 mm (0.200 in.) total tolerance band or ±2.54 mm (±0.100 in.)

(a) Improved tolerances may be possible depending on the specific foundry capabilities and overall part-specific requirements.

Table 8 Surface finish of titanium castings

Process	NAS 823 surface comparator(a)	rms equivalent(b), μm	rms equivalent(b), μin.
Investment			
As-cast	C-12	3.2	125
Occasional areas of...	C-25	6.3	250
Rammed graphite			
As-cast	C-30 to 40	7.5–10	300–400
Occasional areas of...	C-50	12.5	500
Hand finished	C-12 to 25	3.2–6.3	125–250

(a) NAS, National Aerospace Standards. (b) rms, root mean square

Chemical Milling

Residual surface contamination, or α case, is typically removed from as-cast aerospace parts before further processing. This is to eliminate the possibility of the diffusion of these contaminants into the part during subsequent HIP or heat treatment. Chemical milling is normally conducted in solutions based on hydrofluoric and nitric acid mixtures plus additives designed to enhance surface finish and control hydrogen pickup. Hydrogen pickup is more likely the higher the β-phase content of the alloy and is also influenced by etch rate and bath temperature. Subsequent vacuum anneals may be used to remove hydrogen picked up in chemical milling. The general objectives are to remove the entire as-cast surface uniformly to the extent of maximum α-case depth and to retain the dimensional integrity of the part.

Hot Isostatic Pressing

Hot isostatic pressing may be used to ensure the complete elimination of internal gas (D, in Fig. 2a) and shrinkage (E, in Fig. 2a) porosity. The cast part is chemically cleaned and placed inside an autoclave, where it is typically subjected to an argon pressure of 103 MPa (15 ksi) at 900 to 955 °C (1650 to 1750 °F) for a 2 h hold time (Ti-6Al-4V alloy) for void closure and diffusion bonding. Recently, an HIP pressure of 206 MPa (30 ksi) has been employed in the hot isostatic pressing of high-temperature titanium alloys to ensure pore closure in these harder-to-deform materials. This practice has been shown to reduce the scatterband of fatigue property test results and improve fatigue life significantly (Fig. 10). HIP temperature may coarsen the α platelet structure, causing a slight debit in tensile strength, but the benefits of HIP normally exceed this slight decrease in strength, and the practice is widely used for aerospace titanium alloy cast parts.

Weld Repair

The weld repair of titanium castings is an integral step in the manufacturing process and is used to eliminate surface-related defects, such as HIP-induced surface depressions or surface-connected pores that did not close during the HIP cycle. Tungsten inert-gas (TIG) welding practice in argon-filled glove boxes is used with weld filler wire of the same composition as the parent metal. Generally, all weld-repaired castings are stress relief annealed. Excellent-quality weld deposits are routinely obtained in proper practice. Weld deposits may have higher strength but lower ductility than the parent metal because of microstructural differences due to the fast cooling rate of the welding process and some oxygen pickup. Those differences may be eliminated by a postweld solution heat treatment, but standard practice is for stress relief or anneal only. Also, welding rods containing lower oxygen-content alloys are commonly used.

Heat Treatment

Conventional heat treatment of titanium castings is for stress relief anneal after any weld repair. The Ti-6Al-4V alloy is typically heat treated at 730 to 845 °C (1350 to 1550 °F). This is done in a vacuum to ensure the removal of any hydrogen pickup from chemical milling and to protect the titanium chemically milled surface from oxidation. As with HIP and weld repair, castings must be chemically clean prior to heat treatment if diffusion of surface contaminants is to be avoided. Alternate heat treatments for property improvement, such as the solution treating and aging (STA) of Ti-6Al-4V alloy castings, are available. Numerous other

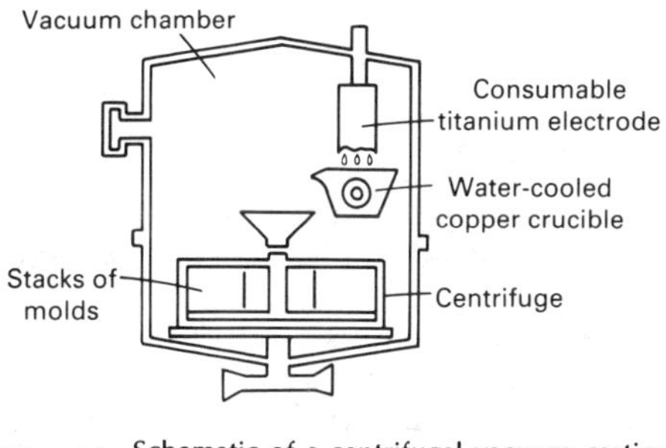

Fig. 12 Schematic of a centrifugal vacuum casting furnace

Fig. 13 Investment cast titanium components for use in corrosive environments

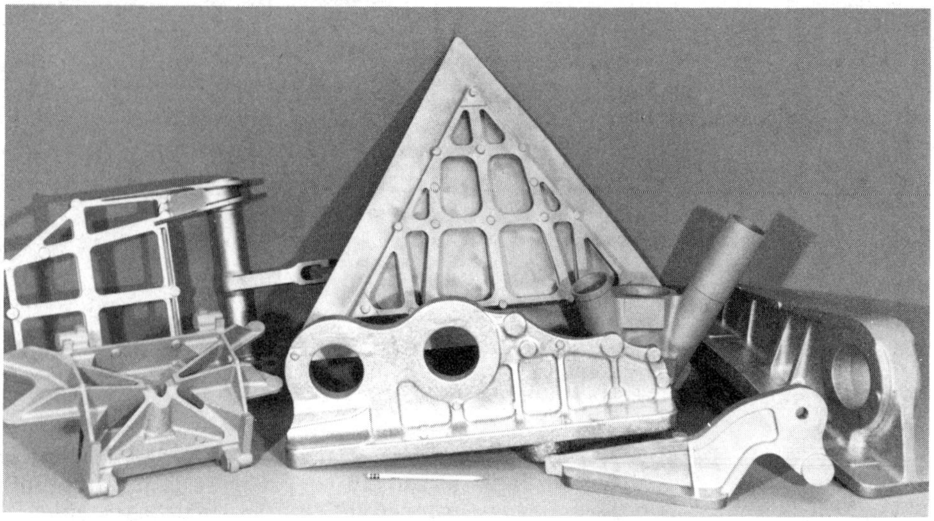

Fig. 14 Investment cast titanium alloy airframe parts

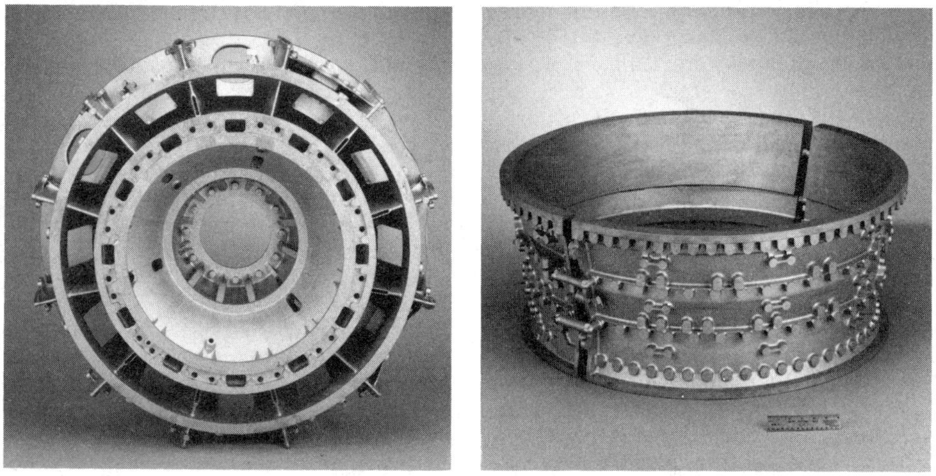

Fig. 15 Typical investment cast titanium alloy components used for gas turbine applications

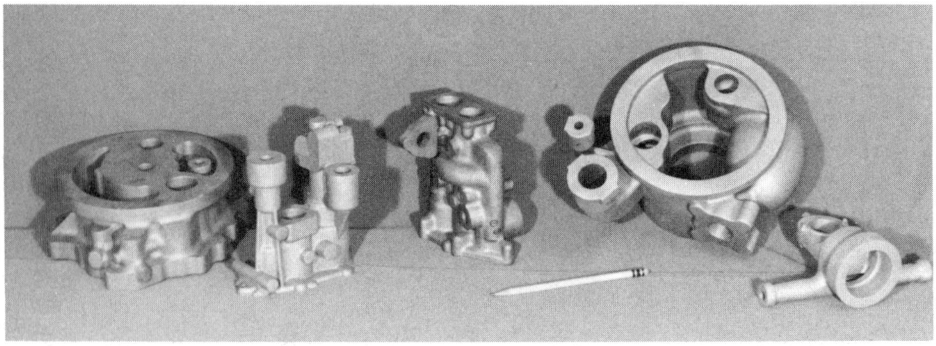

Fig. 16 Titanium hydraulic housings produced by the investment casting process

heat treatments are in various stages of development, as discussed in an earlier section (Table 5).

Final Evaluation and Certification

Titanium castings are produced to numerous quality specifications. Typically, these require some type of x-ray and dye penetrant inspection, in addition to dimensional checks using layout equipment, dimensional inspection fixtures, and coordinate measuring machines. Metallurgical certifications may include HIP and heat treatment run certifications, as well as chemistry, tensile properties, and microstructural examination of representative coupons for the absence of surface contamination.

In the absence of universally accepted x-ray standards, it is common practice to use steel or aluminum reference radiographs (Table 4). Because internal discontinuities in titanium do not necessarily appear the same as they do in other metals, it is necessary to have an expert evaluation of radiographs for proper interpretation. Currently, an industry task force is working on the development of radiographic standards for titanium castings through the American Society for Testing and Materials (ASTM).

Product Applications

The titanium castings industry is relatively young by most foundry standards. The earliest commercial applications, in the 1960s, were for use in pump and valve components requiring corrosion-resistant properties. These applications continue to dominate the rammed graphite production method; however, in more recent years, some users have justified the expense of lost-wax investment tooling for some commercial corrosion-resistant casting applications (see Fig. 13).

Aerospace use of rammed graphite castings became a production reality in the early 1970s for aircraft brake torque tubes, missile wings, and hot gas nozzles. As the more precise investment casting technology developed and the commercial use of HIP became a reality in the mid-1970s, titanium castings quickly expanded into critical airframe (Fig. 14) and gas turbine engine (Fig. 15) applications. The first components were primarily Ti-6Al-4V, the workhorse alloy for wrought aerospace products, and castings were often substituted for forgings, with the addition of some features possible only through net-shape technique. This trend has continued. With continuing experience in manufacturing and specifying titanium castings, applications have expanded from relatively simple, less-critical components for military engines and airframes to large, complex structural shapes for both military and commercial engines and airframes. Today, titanium cast parts are routinely produced for critical structures such as space shuttle attachment fittings, complex airframe structures, engine mounts, compressor cases and frames of many types, missile bodies and wings, and hydraulic housings (Fig. 16). Quality and dimensional capabilities continue to be improved. Titanium castings are used for framework for very sensitive optical equipment because of their relative stiffness, light weight, and the compatibility of the coefficient of thermal expansion of titanium with that of optical glasses (Fig. 17). Applications are evolving for engine airfoil shapes that include individual vanes and integral vane rings for stators, as well as a few

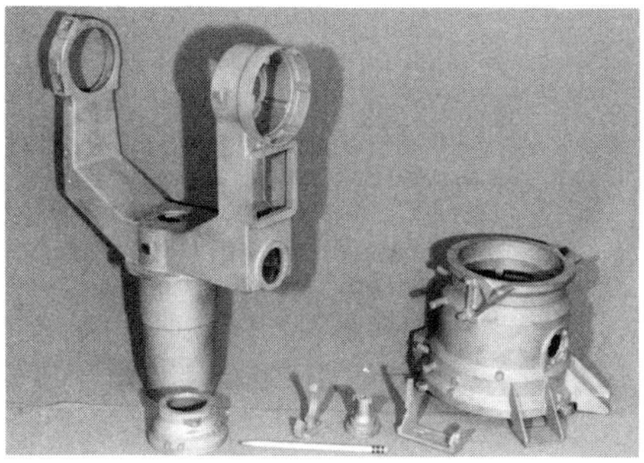

Fig. 17 Titanium housings for aerospace optical applications produced by the investment casting process

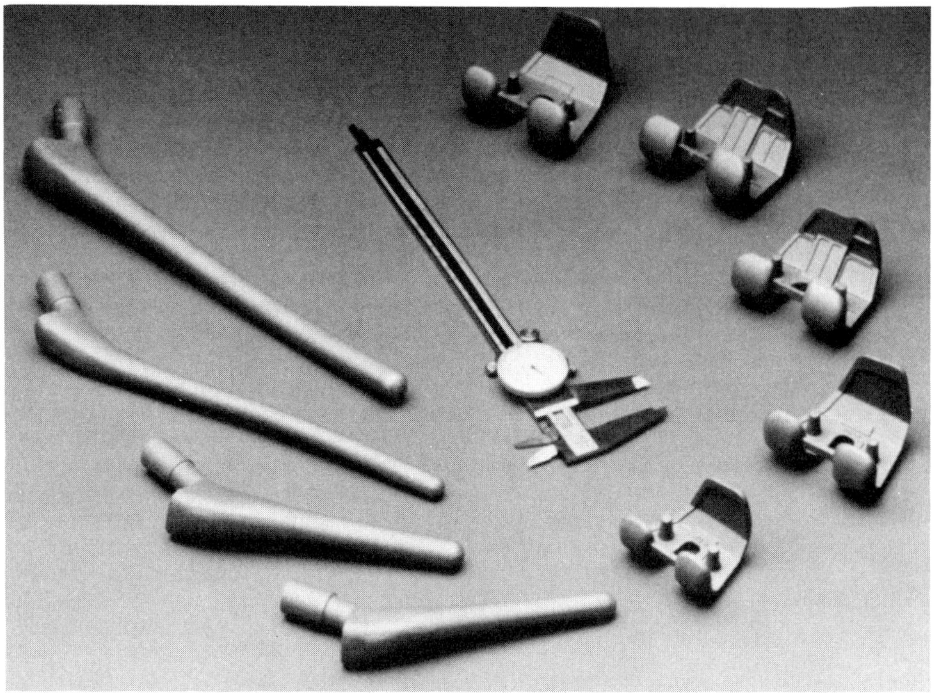

Fig. 18 Titanium surgical knee and hip implant prostheses manufactured by the investment casting process

rotating parts that would otherwise be made from wrought product. Growth will continue as users seek to take advantage of the flexibility of design inherent in the investment casting process and the improvement in the economics of net and near-net shapes. Also, a great deal of work is currently being done in the area of producing rotating cast parts used for fan blades, vanes, and compressor blades for advanced gas-turbine engines.

In spite of the wide acceptance of titanium castings for airframe applications, growth has been somewhat hindered because of the lack of an industrywide data base to establish whether casting factors (derived from early aluminum castings) are, in fact, a necessity for titanium. Such standards are now being considered with the probable elimination of design casting factors (Ref 59).

Foundry size capabilities are expanding to allow the manufacture of larger airframe and static gas turbine engine structures. Widespread routine use of aerospace titanium castings is anticipated as the titanium foundry industry conforms with well-established quality and product standards, and user understanding and confidence continue to be gained from satisfactory product performance.

Concurrent with the above trend, investment cast titanium is increasingly being specified for medical prostheses because of its inertness to body fluids, an elastic modulus approaching that of bone, and the net shape design flexibility of the casting process. Custom-designed knee and hip implant components (Fig. 18) are routinely produced in volume. Some of these are subsequently coated with a diffusion-bonded porous titanium surface to facilitate bone ingrowth or an eventual fixation of the metal implant with the organic bone structure. Of special interest is the use of titanium for a hip joint implant that requires high-fatigue strength properties due to cyclic loading for which Ti-6Al-4V is ideally suited.

REFERENCES

1. H.B. Bomberger, F.H. Froes, and P.H. Morton, Titanium—A Historical Perspective, in *Titanium Technology: Present Status and Future Trends*, F.H. Froes, D. Eylon, and H.B. Bomberger, Ed., Titanium Development Association, 1985, p 3-17
2. *Titanium for Energy and Industrial Applications*, D. Eylon, Ed., The Metallurgical Society, 1981, p 1-403
3. *Titanium Net Shape Technologies*, F.H. Froes and D. Eylon, Ed., The Metallurgical Society, 1984, p 1-299
4. "Titanium 1989, Statistical Review 1979-1988," Annual Report of the Titanium Development Association, 1989
5. American Foundrymen's Society, private communication, 1987
6. Aluminum Association, private communication, 1987
7. D. Eylon, F.H. Froes, and R.W. Gardiner, Developments in Titanium Alloy Casting Technology, *J. Met.*, Vol 35 (No. 2), Feb 1983, p 35-47; *Titanium Technology: Present Status and Future Trends*, F.H. Froes, D. Eylon, and H.B. Bomberger, Ed., Titanium Development Association, 1985, p 35-47
8. D. Eylon and F.H. Froes, "Titanium Casting—A Review," in *Titanium Net Shape Technologies*, F.H. Froes and D. Eylon, Ed., The Metallurgical Society, 1984, p 155-178
9. H.B. Bomberger and F.H. Froes, The Melting of Titanium, *J. Met.*, Vol 36 (No. 12), Dec 1984, p 39-47; *Titanium Technology: Present Status and Future Trends*, F.H. Froes, D. Eylon, and H.B. Bomberger, Ed., Titanium Development Association, 1985, p 25-33
10. E.W. Collings, *Physical Metallurgy of Titanium Alloys*, American Society for Metals, 1984, p 1-261
11. "Titanium: Past, Present and Future," NMAR-392, National Materials Advisory Board, National Academy Press, 1983; PB83-171132, National Technical Information Service
12. W.J. Kroll, C.T. Anderson, and H.L. Gilbert, A New Graphite Resistor Vacuum Furnace and Its Application in Melting Zirconium, *Trans. AIME*, Vol 175, 1948, p 766-773
13. R.A. Beahl, F.W. Wood, J.O. Borg, and H.L. Gilbert, "Production of Titanium Castings," Report 5265, U.S. Bureau of Mines, Aug 1956, p 42
14. A.R. Beahl, J.O. Borg, and F.W. Wood, "A Study of Consumable Electrode Arc Melting," Report 5144, U.S. Bureau of Mines, 1955
15. R.A. Beahl, F.W. Wood, and A.H. Robertson, Large Titanium Castings Produced Successfully, *J. Met.*, Vol 7 (No. 7), July 1955, p 801-804
16. S.L. Ausmus and R.A. Beahl, "Expendable Casting Molds for Reactive Metals," Report 6509, U.S. Bureau of Mines, 1964, p 44
17. R.K. Koch and J.M. Burrus, "Bezonite-Bonded Rammed Olivine and Zircon Molds for Titanium Casting," Report 8587, U.S. Bureau of Mines, 1981
18. G.R. Yoder, L.A. Cooley, and T.W. Crooker, "Fatigue Crack Propagation Resistance of Beta-Annealed Ti-6Al-4V Alloys of Differing Interstitial Oxygen Content," *Metall. Trans. A*, Vol 9A, 1978, p 1413-1420
19. R.R. Boyer and R. Bajoraitis, "Standardization of Ti-6Al-4V Processing Conditions," AFML-TR-78-131, Air Force Materials Laboratory, Boeing Commercial

Airplane Company, Sept 1978
20. D. Eylon, T.L. Bartel, and M.E. Rosenblum, High Temperature Low Cycle Fatigue of Beta-Annealed Titanium Alloy, *Metall. Trans. A*, Vol 11A, 1980, p 1361-1367
21. D. Eylon and J.A. Hall, Fatigue Behavior of Beta-Processed Titanium Alloy IMI-685, *Metall. Trans. A*, Vol 8A, 1977, p 981-990
22. D. Eylon, Fatigue Crack Initiation in Hot Isostatically Pressed Ti-6Al-4V Castings, *J. Mater. Sci.*, Vol 14, 1979, p 1914-1920
23. D. Eylon and W.R. Kerr, The Fractographic and Metallographic Morphology of Fatigue Initiation Sites, in *Fractography in Failure Analysis*, STP 645, American Society for Testing and Materials, 1978, p 235-248
24. D. Eylon and M.E. Rosenblum, Effects of Dwell on High Temperature Low Cycle Fatigue of a Titanium Alloy, *Metall. Trans. A*, Vol 13A, 1982, p 322-324
25. W.G. Burgers, *Physics*, Vol 1, 1934, p 561-586
26. J.C. Williams, Kinetics and Phase Transformation, in *Titanium Science and Technology*, Vol 3, R.I. Jaffee and H.M. Burte, Ed., Plenum Press, 1973, p 1433-1494
27. D. Schechtman and D. Eylon, On the Unstable Shear in Fatigued Beta-Annealed Ti-11 and IMI-685 Alloys, *Metall. Trans. A*, Vol 9A, 1978, p 1273-1279
28. G.R. Yoder and D. Eylon, On the Effect of Colony Size on Fatigue Crack Growth in Widmänstatten Structure Alpha+Beta Alloys, *Metall. Trans. A*, Vol 10A, 1979, p 1808-1810
29. D. Eylon and P.J. Bania, Fatigue Cracking Characteristics of Beta-Annealed Large Colony Ti-11 Alloy, *Metall. Trans. A*, Vol 9A, 1978, p 1273-1279
30. R.J. Smickley and L.P. Bednarz, Processing and Mechanical Properties of Investment Cast Ti-6Al-4V ELI Alloy for Surgical Implants: A Progress Report, in *Titanium Alloys in Surgical Implants*, STP 796, H.A. Luckey and F. Kubli, Ed., American Society for Testing and Materials, 1983, p 16-32
31. R.J. Smickley, Heat Treatment Response of HIP'd Cast Ti-6Al-4V, in *Proceedings of the WesTech Conference*, ASM INTERNATIONAL and Society of Manufacturing Engineers, 1981
32. F.H. Froes, D. Eylon, G.E. Eichelman, and H.M. Burte, Developments in Titanium Powder Metallurgy, *J. Met.*, Vol 32 (No. 2), 1980, p 47-54
33. F.H. Froes and D. Eylon, Powder Metallurgy of Titanium Alloys—A Review, in *Titanium, Science and Technology*, Vol 1, G. Lutjering, U. Zwicker, and W. Bunk, Ed., Deutsche Gesellschaft für Metallkunde, E.V., 1985, p 267-286; *Powder Metall. Int.*, Vol 17 (No. 4), 1985, p 163-167 and continued in Vol 17 (No. 5), 1985, p 235-238; *Titanium Technology: Present Status and Future Trends,* F.H. Froes, D. Eylon, and H.B. Bomberger, Ed., Titanium Development Association, 1985, p 49-59
34. D. Eylon and F.H. Froes, Method for Refining Microstructures of Cast Titanium Articles, U.S. Patent 4,482,398, Nov 1984
35. D. Eylon and F.H. Froes, Method for Refining Microstructures of Prealloyed Powder Metallurgy Titanium Articles, U.S. Patent 4,534,808, Aug 1985
36. D. Eylon and F.H. Froes, Method for Refining Microstructures of Blended Elemental Powder Metallurgy Titanium Articles, U.S. Patent 4,536,234, Aug 1985
37. D. Eylon, F.H. Froes, and L. Levin, Effect of Hot Isostatic Pressing and Heat Treatment on Fatigue Properties of Ti-6Al-4V Castings, in *Titanium, Science and Technology*, Vol 1, G. Lutjering, U. Zwicker, and W. Bunk, Ed., Deutsche Gesellschaft für Metallkunde, E.V., 1985, p 179-186
38. D.L. Ruckle and P.P. Millan, Method for Heat Treating Cast Titanium Articles to Improve Their Mechanical Properties, U.S. Patent 4,631,092, Dec 1986
39. D. Eylon, W.J. Barice, and F.H. Froes, Microstructure Modification of Ti-6Al-4V Castings, in *Overcoming Material Boundaries*, Vol 17, Society for the Advancement of Material and Process Engineering, 1985, p 585-595
40. W.R. Kerr, P.R. Smith, M.E. Rosenblum, F.J. Gurney, Y.R. Mahajan, and L.R. Bidwell, Hydrogen as an Alloying Element in Titanium (Hydrofac), in *Titanium '80, Science and Technology*, H. Kimura and O. Izumi, Ed., The Metallurgical Society, 1980, p 2477-2486
41. R.G. Vogt, F.H. Froes, D. Eylon, and L. Levin, Thermo-Chemical Treatment (TCT) of Titanium Alloy Net Shapes, in *Titanium Net Shape Technologies*, F.H. Froes and D. Eylon, Ed., The Metallurgical Society, 1984, p 145-154
42. L. Levin, R.G. Vogt, D. Eylon, and F.H. Froes, Method for Refining Microstructures of Titanium Alloy Castings, U.S. Patent 4,612,066, Sept 1986
43. L. Levin, R.G. Vogt, D. Eylon, and F.H. Froes, Method for Refining Microstructures of Prealloyed Powder Compacted Articles, U.S. Patent 4,655,855, April 1987
44. R.J. Smickley and L.E. Dardi, Microstructure Refinement of Cast Titanium, U.S. Patent 4,505,764, March 1985
45. C.F. Yolton, D. Eylon, and F.H. Froes, High Temperature Thermo-Chemical Treatment (TCT) of Titanium With Hydrogen, in *Proceedings of the Fall Meeting*, The Metallurgical Society, 1986, p 42
46. TiTech International, Inc., unpublished research
47. F.C. Teifke, N.H. Marshall, D. Eylon, and F.H. Froes, Effect of Processing on Fatigue Life of Ti-6Al-4V Castings, in *Advanced Processing Methods for Titanium*, D. Hasson, Ed., The Metallurgical Society, 1982, p 147-159
48. R.R. Wright, J.K. Thorne, and R.J. Smickley, Howmet Turbine Components Corporation, Ti-Cast Division, private communication, 1982; Technical Bulletin TB 1660, Howmet Corporation
49. L. Levin, R.G. Vogt, D. Eylon, and F.H. Froes, Fatigue Resistance Improvement of Ti-6Al-4V by Thermo-Chemical Treatment, in *Titanium, Science and Technology*, Vol 4, G. Lutjering, U. Zwicker, and W. Bunk, Ed., Deutsche Gesellschaft für Metallkunde, E.V., 1985, p 2107-2114
50. L.J. Maidment and H. Paweltz, An Evaluation of Vacuum Centrifuged Titanium Castings for Helicopter Components, in *Titanium '80, Science and Technology*, H. Kimura and O. Izumi, Ed., The Metallurgical Society, 1980, p 467-475
51. J.-P. Herteman, "Properties d'Emploi de l'Alliage de Titane T.A6V Moule Densifie ou Non par Compaction Isostatic à Chaud," Centre D'essais Aeronautique de Toulouse, Technical Report 30/M/79, July 1979
52. W.H. Ficht, "Centrifugal Cast Titanium Compressor Case," Paper presented at the Manufacturing Technology Advisory Group Meeting, General Electric Company, Aircraft Engine Group, Lynn, MA, 1979
53. D. Eylon, P.R. Smith, S.W. Schwenker, and F.H. Froes, Status of Titanium Powder Metallurgy, in *Industrial Applications of Titanium and Zirconium: Third Conference*, STP 830, R.T. Webster and C.S. Young, Ed., American Society for Testing and Materials, 1984, p 48-65
54. J.K. Kura, "Titanium Casting Today," MCIC-73-16, Metals and Ceramics Information Center, Dec 1973
55. J.R. Humphrey, Report IR-162, REM Metals Corporation, Nov 1973
56. M.J. Wynne, Report TN-4301, British Aircraft Corporation, Nov 1972
57. H.D. Hanes, D.A. Seifert, and C.R. Watts, *Hot Isostatic Processing*, Battelle Press, 1979, p 55
58. D. Eylon, W.J. Barice, R.R. Boyer, L.S. Steel, and F.H. Froes, Castings of High Strength Beta Titanium Alloys, in *Sixth World Conference on Titanium*, Part II, P. Lacomb, R. Tricot, and G. Beranger, Ed., Les Éditions de Physique, 1989, p 655-660
59. R.J. Tisler, Fatigue and Fracture Characteristics of Ti-6Al-4V HIP'ed Investment Castings, in *Proceedings of the International Conference on Titanium*, Titanium Development Association, Oct 1986, p 23-41

Titanium P/M Products

Daniel Eylon, Graduate Materials Engineering, University of Dayton,
F.H. (Sam) Froes, Institute for Materials and Advanced Processes, College of Mines, University of Idaho

TITANIUM, the recently introduced member of the family of major structural metals, is the fourth most abundant structural metal in the crust of the earth after aluminum, iron, and magnesium. The development of its alloys and processing technologies started only in the late 1940s (Ref 1); thus, titanium metallurgy just missed being a factor in the Second World War. The difficulty in extracting titanium from ores, its high reactivity in the molten state, its forging complexity, its machining difficulty, and its sensitivity to segregation and inclusions necessitated the development of special processing techniques. These special techniques have contributed to the high cost of titanium raw materials, alloys, and final products. On the other hand, the low density of titanium alloys provides high structural efficiencies based on a wide range of mechanical properties, coupled with an excellent resistance to aggressive environments. These alloys have contributed to the quality and durability of military high-Mach-number aircraft, light helicopters, and turbofan jet engines as well as the increased reliability of heat exchanger units, and surgical body implants.

Despite the combination of low density, high mechanical performance, and excellent corrosion resistance, the high cost of titanium alloys made them a design choice only when lower-cost alloys could not be used. The drive to develop net-shape technologies, such as casting and powder metallurgy (P/M), has been going on for many years. It has been spurred on by the desire to minimize alloy waste and to reduce or eliminate the cost of machining. This article focuses on the properties and applications of titanium P/M compacts. Titanium casting technology, which represents an alternative production method and is a more widely used net-shape technique, is discussed in the article "Titanium and Titanium Alloy Castings" in this Volume.

Because of difficulties encountered with early melting practices, powder metallurgy was used in the beginning stages of titanium technology to produce alloy ingots (Ref 1). Titanium P/M has been developed as a net-shape technique only in the last 15 years (Ref 2-5). In general, P/M can be divided into two major categories (Ref 6, 7):

- Elemental P/M, in which a blend of elemental powders, along with master alloy or other desired additions, is cold pressed into shape and subsequently sintered to higher density and uniform chemistry
- Prealloyed P/M, which is based on hot consolidation of powder produced from a prealloyed stock

In general, the blended elemental (BE) method produces parts at a low cost, but the parts often are less than fully dense; this technique is typically used for iron (Ref 8), copper (Ref 9), and heavy-metal (Ref 10) alloys. The prealloyed (PA) method is used for making fully dense high-performance components from aerospace alloys such as nickel (Ref 11, 12), aluminum (Ref 13), and beryllium (Ref 14). Titanium P/M has incorporated both methods: BE is used to produce lower-cost parts that are less-than-fully dense, and PA is used for higher-cost, high-performance, full-density compacts (Ref 15). Recent developments in powder selection, compaction techniques, and postcompaction treatments have made it possible to obtain full density in titanium BE P/M. Properties exceeding those of ingot metallurgy (I/M) products have been achieved in BE and PA products. This article highlights the properties and applications of both BE and PA titanium P/M compacts. It includes major recent developments that have led to improved performance in BE and PA products, but it does not cover the developments in titanium rapid solidification alloys that have not yet reached the commercialization stage. Detailed discussions of powder production methods and shape-making techniques are available in *Powder Metallurgy*, Volume 7 of the 9th Edition of *Metals Handbook* (Ref 6, 7). These processes are briefly described here in only enough detail to rationalize properties and applications.

Mechanical Properties of Titanium P/M Products

The mechanical properties of titanium P/M products depend on alloy composition and on the density and final microstructure of the compact. The compact density and microstructure depend on the nature of the powder, on the specific consolidation technique employed, and on postcompaction treatments such as secondary pressing or heat treatment. To date, most components produced by the various P/M methods have been made from Ti-6Al-4V, the most common aerospace titanium alloy. As a result, the majority of the P/M data available in the literature is for this alloy. However, these technologies are also very well suited for other alloys, such as the high-strength β alloys and the high-temperature near-α alloys. This article includes data on a wide variety of alloys to highlight the range of potential applications available for titanium P/M.

Blended Elemental Compacts

The blended elemental method, which is basically a pressing and sintering P/M technique, involves cold pressing or cold isostatic pressing (CIP) a blend of fine elemental titanium and master alloy powders that have been sintered. Titanium sponge fines (−100 mesh) are the most common elemental powder used in this process; these particles are obtained as by-products of the Hunter or Kroll reduction processes (Ref 4). The metallic titanium sponge produced by these processes is vacuum arc melted into ingots. The titanium sponge fines that are too small to be used in the melting process are available at a relatively low cost. This powder has an irregular shape (Fig. 1a), which makes it easy to cold press into green shapes. The powder is sintered at temperatures in the range of 1150 to 1315 °C (2100 to 2400 °F) in a vacuum to prevent gas contamination that can severely degrade compact properties. The high sintering temperature is needed to provide particle bonding and to homogenize the chemistry. It is well above the β transus (that is, the lowest

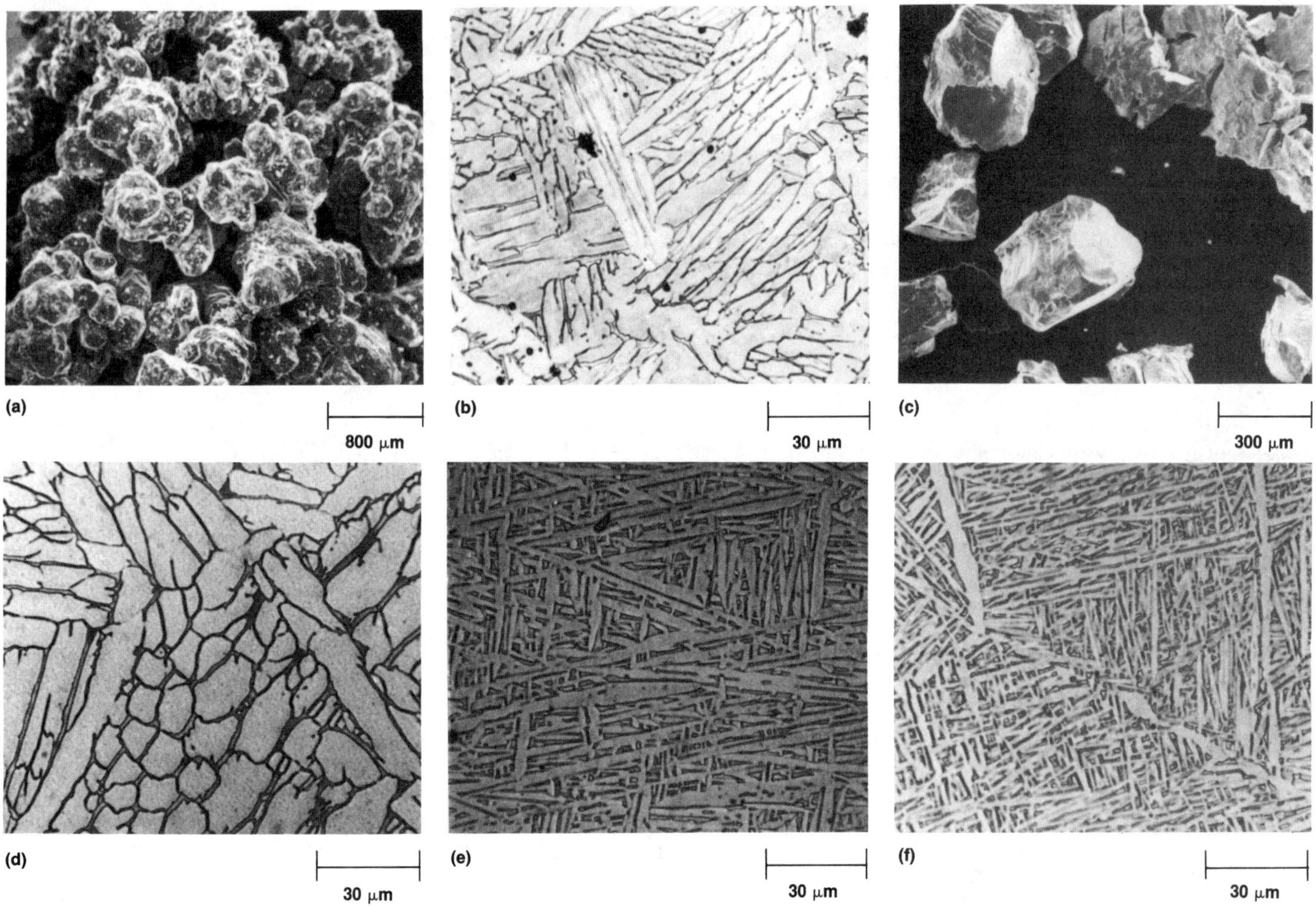

Fig. 1 Photomicrographs of titanium BE materials. (a) −100 mesh titanium sponge fines. (b) Microstructure of a pressed and sintered 99% dense Ti-6Al-4V compact. (c) Crushed hydrogenated-dehydrogenated titanium ingot or machine turnings. (d) Microstructure of a fully dense, pressed and sintered, and hot isostatically pressed Ti-6Al-4V compact. (e) Microstructure of a Ti-6Al-4V compact treated to produce a broken-up structure. (f) Microstructure of a Ti-6Al-4V compact treated with thermochemical processing

equilibrium temperature at which the material is 100% β) of all common titanium alloys, and as a result the compact microstructure in α + β alloys consists of colonies of similarly aligned coarse α plates (Fig. 1b). The plates are about 8 μm (320 μin.) wide and 25 μm (1000 μin.) long. The colonies are about 50 μm (0.002 in.) in diameter. The prior β grains are about 80 μm (0.0032 in.) in diameter. This microstructure is much finer than ingot material treated at the same temperature because of the inherent porosity of the powder compact (Ref 16). The porosity is the result of sodium chloride residues (Kroll process) in the sponge from the reduction process (Ref 4). The sponge fines contain from 0.12 to 0.15% Cl, and, as can be seen in Fig. 1(a), the resulting porosity cannot be entirely closed, even after secondary operations such as hot pressing or hot isostatic pressing (HIP) (Ref 17).

BE compacts have a green density of 85 to 90% after 415 MPa (60 ksi) cold pressing. After vacuum sintering, they have a density of 95 to 99%. Control of particle size and size distribution can produce compacts that are 99% dense. In ferrous, copper, and heavy-metal P/M alloys, 99% is considered to be full density; however, in titanium alloy compacts, such a level of residual porosity (Fig. 1a) will degrade both fatigue and fracture properties. A substantial effort has been made to entirely eliminate the porosity so that BE P/M parts can be used for fatigue-critical aerospace applications. Postsintering HIP densification can lead to 99.8% density and improved properties (Ref 4, 5). However, it is impossible to entirely eliminate the porosity with postsintering hot-pressing operations. During hot pressing, the chlorides present in the compact become volatile and create pockets of insoluble gas. Under the HIP pressure, these relatively large pressurized cavities (Fig. 2a) will break up into a multitude of submicron voids (Fig. 2b) (Ref 16-18) with sodium chloride in the center of the cavity (Fig. 2c). Both macrovoids and microvoids have an approximate hexagonal shape that is associated with basal-plane facets, which are the most energetically stable planes of the hexagonal close-packed structure (Ref 19). During cooling to room temperature, the gas in the voids transforms into cubic chloride crystals, as can be seen in the transmission electron microscopy image in Fig. 2(b).

To obtain pore-free 100% density material such as that produced by ingot metallurgy, the BE method must use chloride-free titanium powder (Ref 20). One source for such powder is commercially pure titanium ingot material or machine turnings embrittled by hydrogenation that are subsequently crushed, and dehydrogenated. This powder is angular (Fig. 1c), and the sintered microstructure is much coarser (Fig. 1d) because of the lack of porosity during sintering.

Tensile Properties and Fracture Toughness. As with data for other titanium technologies, most of the published BE data are on Ti-6Al-4V. Table 1 is a comprehensive listing of tensile properties of Ti-6Al-4V BE compacts processed under a variety of conditions. Table 2 provides the more limited available information on the properties of additional titanium BE alloys.

Blended Elemental Ti-6Al-4V. As indicated in Table 1, most Ti-6Al-4V BE compact

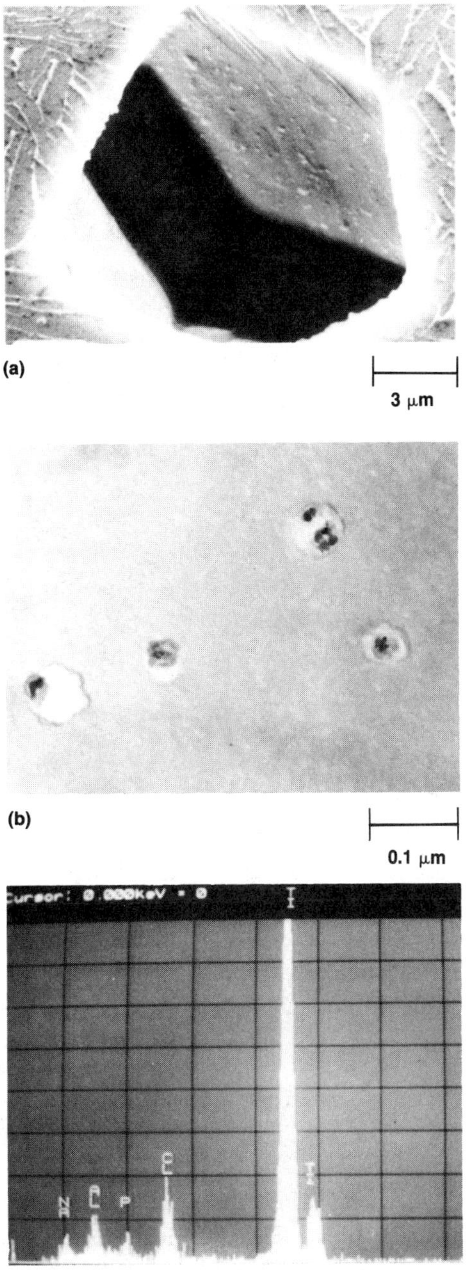

Fig. 2 Chlorine-induced porosity in a titanium BE compact. (a) Scanning electron microscopy photomicrograph of large-size residual porosity in a sectioned Ti-6Al-4V BE compact. (b) Transmission electron microscopy photomicrograph of a Ti-6Al-4V BE compact after postsintering HIP at 925 °C (1700 °F). (c) Chemical analysis showing sodium chloride contaminant at the center of a micropore

conditions exceed minimum MIL-T-9047 specifications. The process details for each condition are in the corresponding references listed in the table. The final shape of the compact can be achieved through a number of process sequences:

- Pressing and sintering
- Pressing and sintering plus HIP
- Pressing and sintering plus rolling

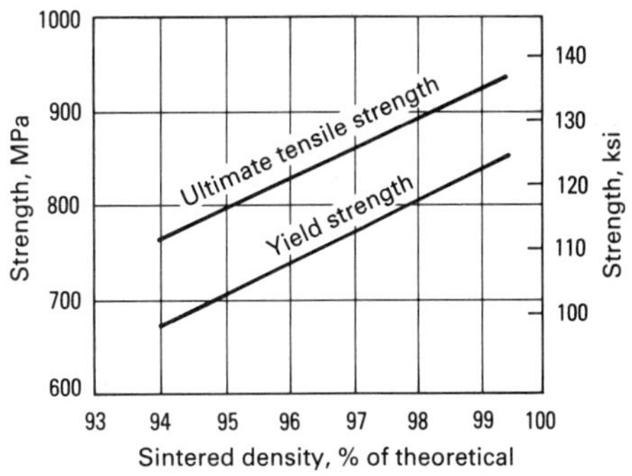

Fig. 3 Effect of sintered density on the yield and tensile strengths of press and sintered Ti-6Al-4V BE compacts. Source: Ref 22

- CIP and sintering
- CIP, sintering, and HIP (collectively designated as cold and hot isostatic pressing, or CHIP)
- CIP, sintering, and rolling
- CIP, sintering, and forging

By controlling the process parameters, it is possible to obtain compacts with densities between 92 and 100% of the theoretical density. The yield and tensile strength of the compacts are proportional to the density (Fig. 3). The fracture toughness also increases with density (Fig. 4). Above 98% density, BE compacts have K_{Ic} values at the level of mill-annealed I/M materials. However, I/M materials with coarse lenticular microstructures similar to those of BE compacts (Fig. 1d) will have much higher K_{Ic} values (70 to 100 MPa$\sqrt{m}$, or 65 to 90 ksi$\sqrt{in.}$). The relatively lower K_{Ic} level of the BE compacts is probably the result of higher oxygen levels (Ref 26) and residual porosity.

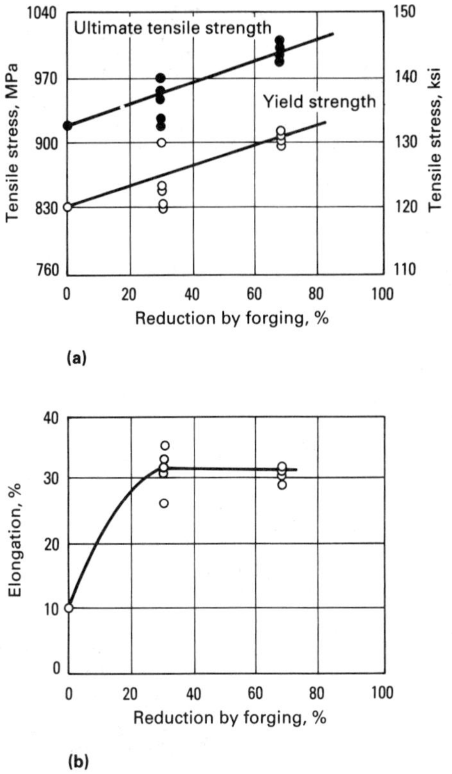

Fig. 5 Effect of forging deformation on Ti-6Al-4V BE compacts (Hunter reduction process sponge fines) isothermally forged at 925 °C (1700 °F). (a) Tensile strength and yield strength. (b) Tensile elongation. Source: Ref 17

Blended elemental compacts can be used as forging preforms (Ref 17). The strong effect of forging deformation on tensile strength and elongation is shown in Fig. 5.

Additional BE Alloys. The limited available mechanical test data for other BE

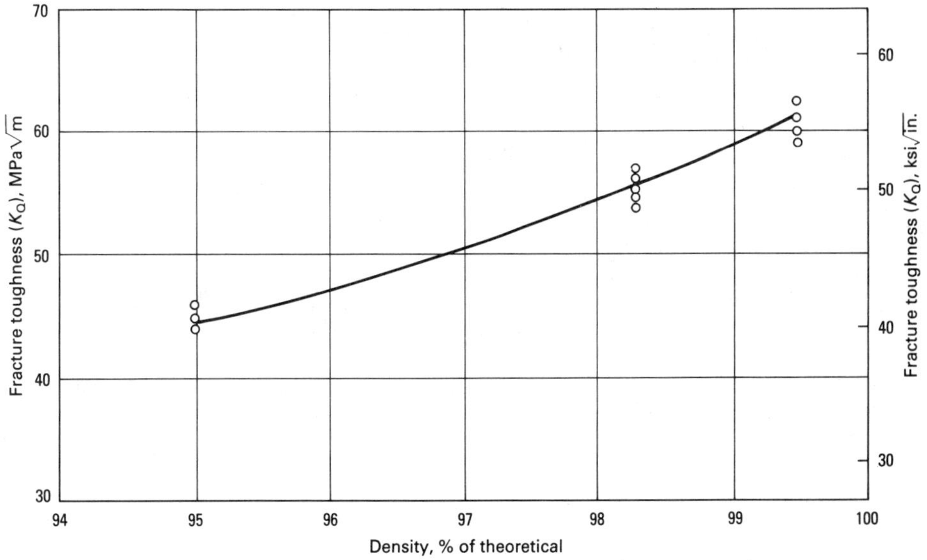

Fig. 4 Effect of density on the fracture toughness of press and sintered Ti-6Al-4V BE compacts. The values are not valid K_{Ic} and thus are labeled as K_Q. Source: Ref 21

Table 1 Tensile and fracture toughness properties of Ti-6Al-4V BE compacts processed under various conditions

Condition(a)	0.2% yield strength MPa	ksi	Ultimate tensile strength MPa	ksi	Elongation, %	Reduction in area, %	K_{Ic} or (K_Q) MPa√m	ksi √in.	Density, %	Chlorine, ppm	O_2, ppm	Ref
Pressed and sintered (96% dense)	758	110	827	120	6	10	...	...	96	1200	...	21
Pressed and sintered (98% dense)	827	120	896	130	12	20	...	...	98	1200	...	21
Pressed and sintered (MR-9 process)(99.2% dense)	847	123	930	135	14	29	38	35	99.2	1200	...	21, 22
Pressed and sintered plus HIP	806	117	875	127	9	17	41	37	≥99	1500	2400	23, 24
CIP and sintered plus HIP	827	120	916	133	13	26	...	...	99.4	1500	2400	24
Pressed and sintered plus α/β forged	841	122	923	134	8	9	...	...	≥99	1500	...	25
Pressed and sintered plus α/β forged	951	138	1027	149	9	24	49	45	99	1200	...	26
Pressed and sintered (92% dense)	827	120	910	132	10	...	...	...	92	1500	2100	17
Plus α/β 30% isothermally forged	841	122	930	135	30	...	...	...	99.7	1500	2100	17
Plus α/β 70% isothermally forged	896	130	999	145	30	...	...	...	99.8	1500	2100	17
CIP and sintered plus HIP (low chlorine)	827	120	923	134	16	34	...	...	99.8	160	...	24
CIP and sintered plus HIP (ELCl)	882	128	985	143	11	36	...	...	100	<10	...	27
Plus BUS treated	951	138	1034	150	7	15	...	...	...	...	...	27
Plus TCP treated	1007	146	1062	154	14	20	...	...	...	...	...	30
Rolled plate, CIP and sintered plus HIP												
Mill annealed (L or TL)	903	131	958	139	10	26	(72)(b)	(65)(b)	≥99	200	1600	28, 29
Mill annealed (T or LT)	923	134	965	140	14	31	(71)(b)	(64)(b)	≥99	200	1600	28, 29
Recrystallization annealed (L or TL)	888	129	916	133	4	8	(75)(b)	(68)(b)	≥99	200	1600	28, 29
Recrystallization annealed (T or LT)	868	126	937	136	5	9	(67)(b)	(61)(b)	≥99	200	1600	28, 29
β annealed (L or TL)	841	122	937	136	10	26	(89)(b)	(81)(b)	≥99	200	1600	28, 29
β annealed (T or LT)	875	127	958	139	7	20	(92)(b)	(84)(b)	≥99	200	1600	28, 29
Minimum properties (MIL-T-9047)	827	120	896	130	10	25	...	...	...	...	...	4

(a) HIP, hot isostatic pressing; CIP, cold isostatic pressing; ELCl, extra-low chlorine powder; BUS, broken-up structure; TCP, thermochemical processing; L, longitudinal; TL, transverse longitudinal; T, transverse; LT, longitudinal transverse (TL and LT per ASTM E 399). (b) Precracked Charpy, K_V. Source: Ref 4, 17, 21–30

alloys, such as Ti-6Al-2Sn-4Zr-6Mo, Ti-5Al-2Cr-1Fe, and Ti-4.5Al-5Mo-1.5Cr, are listed in Table 2. The most detailed work has been done on the Ti-10V-2Fe-3Al alloy, with some results reported at levels close to those for I/M materials (Ref 33). However, more data are needed for these alloys before reliable parameters for property levels and optimum processes can be established. It is interesting to note that in the case of Ti-10V-2Fe-3Al, a tensile strength of 1268 MPa (184 ksi) with 10% elongation can be achieved with BE methods. The CermeTi, listed in Table 2, is essentially a metal-matrix composite with a Ti-6Al-4V base and titanium carbide particulate reinforcement (10 to 15% TiC is typical) that is produced with a BE P/M process.

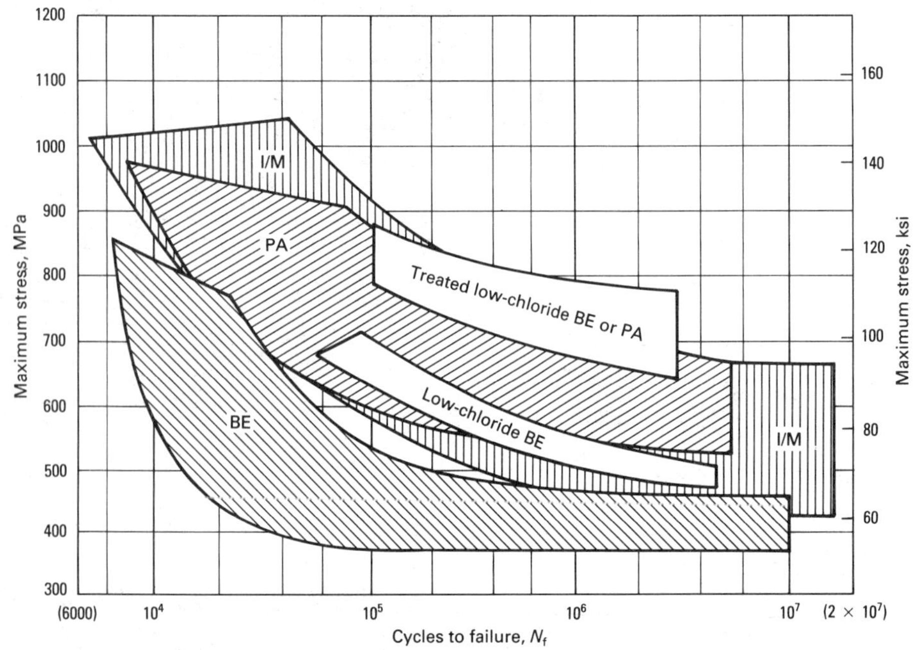

Fig. 6 Comparison of the room-temperature fatigue life scatterbands of BE and PA Ti-6Al-4V compacts to that of a mill-annealed I/M alloy

Fatigue Strength and Crack Propagation. The fatigue life scatterband of chloride-containing Ti-6Al-4V BE compacts is compared in Fig. 6 to a mill-annealed I/M alloy. The effect of low chloride levels and postsintering treatments on fatigue strength is shown in Fig. 7. The effect of compact density on fatigue strength is shown in Fig. 8. The fatigue strength of BE compacts is inherently low when compared to that of mill-annealed I/M products because of the inherent chloride and related porosity of BE materials. This limits the use of the lower-cost pressing and sintering or CIP and sintering processes to applications that are not fatigue critical, such as missile components. By increasing density through secondary pressing operations and through the use of chloride-free titanium powder, it is possible to further improve fatigue strength (Fig. 8). However, this increases the cost of these products, thereby negating one of their primary advantages (Ref 15, 35).

Very limited data is available on the fatigue crack growth rate of Ti-6Al-4V BE compacts. Figure 9 shows that the fatigue crack growth rate of this material is between that of a β-annealed material and that of a mill-annealed I/M material (Ref 26).

Table 2 Tensile and fracture toughness properties of BE titanium alloy compacts processed under various conditions

Alloy and condition(a)	0.2% yield strength MPa	ksi	Ultimate tensile strength MPa	ksi	Elongation, %	Reduction in area, %	K_{Ic} or (K_Q) MPa$\sqrt{m}$	ksi $\sqrt{in.}$	Density, %	Chlorine, ppm	Ref
Ti-5Al-2Cr-1Fe											
Pressed and sintered plus HIP	980	142	1041	151	20	39	...	...	≥99	310	31
Ti-4.5Al-5Mo-1.5Cr (Corona 5)											
Pressed and sintered plus HIP	951	138	1000	145	17	39	(64)	(58)	≥99	310	31
Ti-6Al-2Sn-4Zr-6Mo											
Pressed and sintered, no STA or HIP	1068	155	1109	161	2	1	31	28	99	150	32
Ti-10V-2Fe-3Al											
Pressed and sintered, HIP (1650 °C, or 3000 °F), and STA (775-540 °C, or 1425-1005 °F)	1233	179	1268	184	9	...	30	27	99	1900	33
Pressed and sintered, HIP, and STA (750-550 °C, or 1380-1020 °F)	1102	160	1158	168	10	...	32	29	99	1900	33
Pressed and sintered, no STA or HIP	854	124	930	135	9	12	51	46	98	150	32
Ti-6Al-4V + 10% TiC (CermeTi)											
Pressed and sintered plus HIP(b)	792	115	799	116	1	...	...	...	...	...	34

(a) HIP, hot isostatic pressing; STA, solution treatment and aging. (b) High modulus (Young's modulus of 20 × 10^6 psi, or 140 GPa). Source: Ref 31-34

The BE material tested had a porosity of 1 to 2 vol%, which at this level seems not to adversely influence the fatigue crack growth rate.

Prealloyed Compacts

While BE compacts are produced and used in a wide range of densities, PA P/M parts are acceptable only at 100% density (Ref 2-5, 15, 35). The titanium PA powders are commercially available as spherical particles that have high tap density (65%) and good powder flow and mold fill characteristics. Two main production methods are used for making clean PA powder:

- Gas atomization (Ref 36)
- Plasma rotating-electrode process (PREP) (Ref 37), a modification of the older rotating-electrode process (REP) (Ref 38)

It is also possible to produce PA powders by comminution (Ref 39) and coreduction (Ref 40, 41) methods. However, because of insufficient mechanical property data, these techniques will not be discussed in this article.

Hot isostatic pressing is the primary compaction method for PA powders (Ref 2-5, 15, 35), but vacuum hot pressing (VHP) (Ref 42, 43), extrusion (Ref 44), and rapid omnidirectional compaction (ROC) (Ref 45-48) have also been successfully used. The shape-making step is achieved by containing the powder in a shaped, evacuated mild-steel can. The compaction is typically carried out at a temperature below the β transus to minimize reaction with the can. Processing in the α + β phase field results in a coarse low-aspect-ratio α structure (Fig. 10a). This material is most commonly compared to mill-annealed I/M material because of its microstructure and full density.

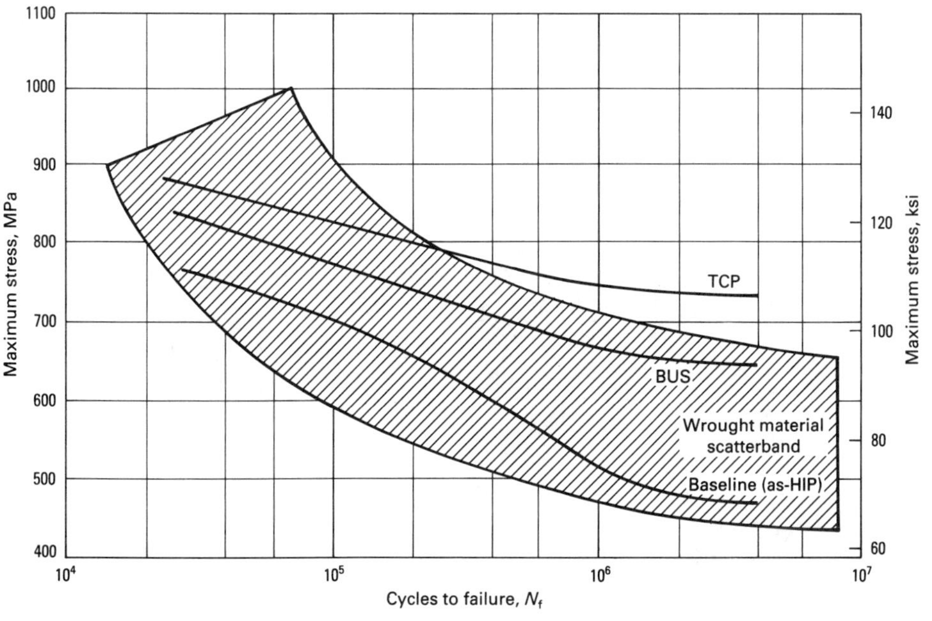

Fig. 7 Comparison of the fatigue strengths of fully dense extra-low chloride Ti-6Al-4V BE compacts with the scatterband for an I/M alloy. The BE compacts were tested in the as-HIP, broken-up structure (BUS), and thermochemically processed (TCP) conditions. Smooth axial fatigue data were obtained at room temperature. Stress ratio (*R*), 0.1; frequency (*f*), 5 Hz with triangular waveform. Source: Ref 27

Powder cleanliness is one of the main factors governing the quality of PA compacts. Because of the full compact density, even a low level of contamination with foreign particles will lead to a substantial loss of inherent properties such as fatigue strength (Ref 49). As a result, only data obtained from clean powders are considered in this article. Also, because fully dense PA compacts are considered for more demanding applications than are the less-dense BE compacts, more mechanical test data have been developed within the aerospace industry on PA P/M than on BE P/M compacts. Only data considered to be typical are reviewed in this article. The majority of PA work has been done on Ti-6Al-4V.

Tensile Properties and Fracture Toughness. Table 3 is a comprehensive listing of the tensile properties of Ti-6Al-4V PA compacts processed under various conditions. Table 4 provides limited information on the properties of additional alloys. When the alloy compacts are produced using HIP (Ref 55), VHP (Ref 42), or ROC (Ref 48) at higher pressures but at lower temperatures, higher strength levels without losses in ductility are achieved. This is the result of the substantial microstructure refinement developed during high-pressure low-temperature powder processing. Similarly, post-compaction hot work, such as rolling (Ref 57) or forging (Ref 58), results in microstructural refinement that improves tensile strength and ductility. Property improvement after postcompaction treatments are discussed in the section "Postcompaction Treatments" in this article.

Additional PA Alloys. Table 4 shows the extent to which other PA titanium alloys have been studied. Almost all major I/M alloys have been evaluated in the PA P/M

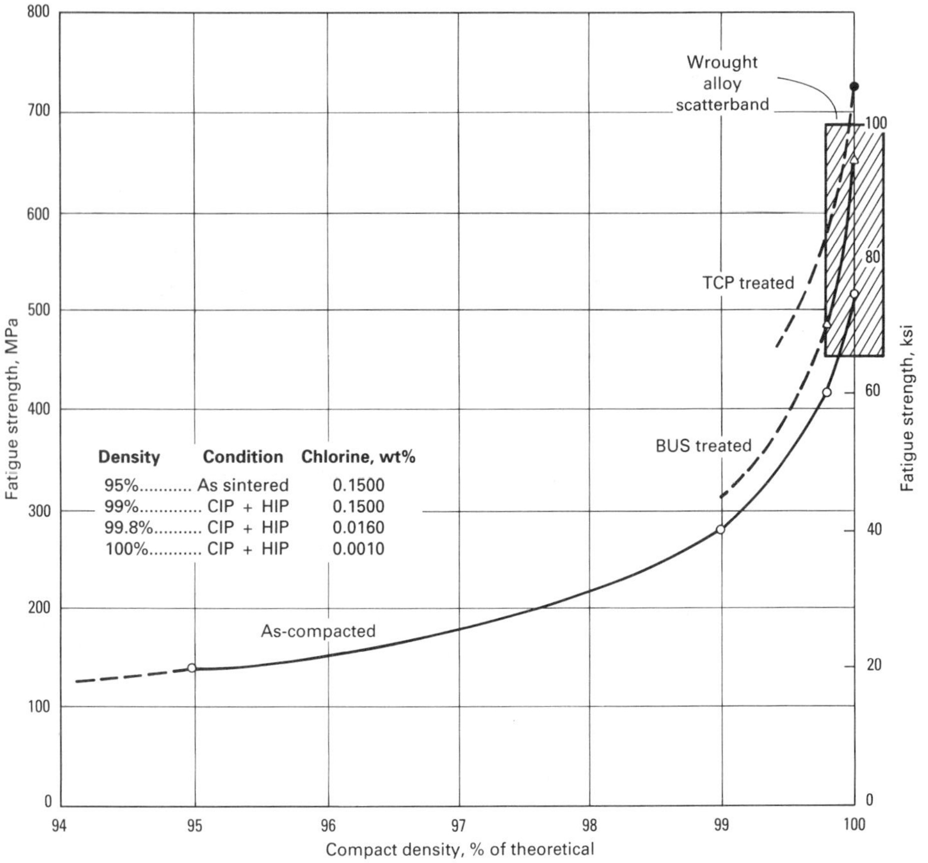

Fig. 8 Effect of compact density on fatigue strength of CIP and sintered Ti-6Al-4V BE compacts. Note that the higher densities are only possible in the low-chloride material. BUS, broken-up structure; TCP, thermochemical processing. Source: Ref 15

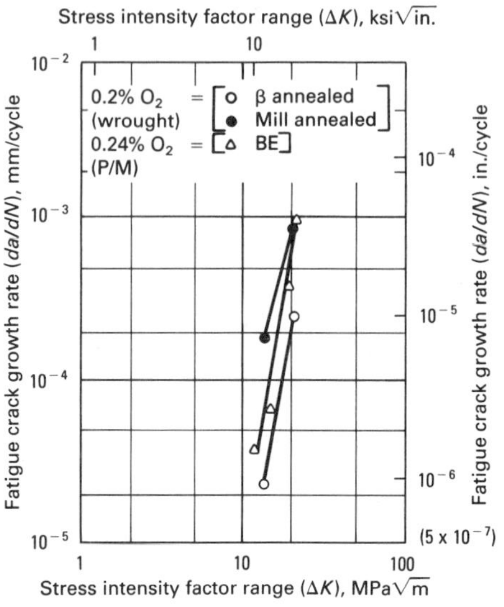

Fig. 9 Comparison of fatigue crack propagation rates of BE and I/M Ti-6Al-4V as a function of the stress intensity factor range at room temperature in air. Stress ratio (R), +0.1 ($R = \sigma_{min}/\sigma_{max}$, where σ_{min} is the minimum stress and σ_{max} is the maximum stress); frequency (f), 5 Hz. Source: Ref 26

form. These include the high-strength metastable β alloys (Ref 33, 52, 65-68), the versatile α + β alloys (Ref 63, 64, 66), the high-temperature near-α alloys (Ref 59, 60), and the ordered titanium aluminide alloys (Ref 70-72). Of special interest is the alloy Ti-1.3Al-8V-5Fe (Ref 67-69): This alloy has a remarkable tensile strength of 1516 MPa (220 ksi) with 8% elongation. With conventional I/M methods, the high iron content of this alloy results in segregation problems. Powder metallurgy, on the other hand, produces fine-grain, homogeneous, and segregation-free products. Such alloys have the potential of expanding the market for titanium P/M technology.

Fatigue Strength and Crack Propagation. The smooth-bar fatigue life scatterband of Ti-6Al-4V PA compacts is compared to that of a mill-annealed I/M alloy in Fig. 6 (Ref 4). The P/M data were obtained by testing high-cleanliness REP and PREP compacts that had undergone hot isostatic pressing; some of the compacts received a postcompaction heat treatment. The data for the PA compacts are at equivalent levels to the best I/M results. Powder contamination must be avoided to maintain a high fatigue strength in these materials. The effect of 50, 150, and 350 μm (0.002, 0.006, and 0.014 in.) diam contaminants on the fatigue strength of Ti-6Al-4V PA compacts is shown in Fig. 11 (Ref 49). Even 50 μm (0.002 in.) contaminant particles are sufficient to noticeably reduce fatigue strength. Figure 12 compares the fatigue characteristics of an actual P/M component to those of an I/M material: Figure 12(a) shows results for smooth-bar high-cycle fatigue, Fig. 12(b) covers notched-specimen high-cycle fatigue, and Fig. 12(c) shows data for strain-controlled low-cycle fatigue. In general, high-cycle fatigue results are in the range of 10^6 to 10^8 cycles to failure, and low-cycle fatigue results fall below 10^5 cycles to failure. For all three tests, the P/M alloy performance was

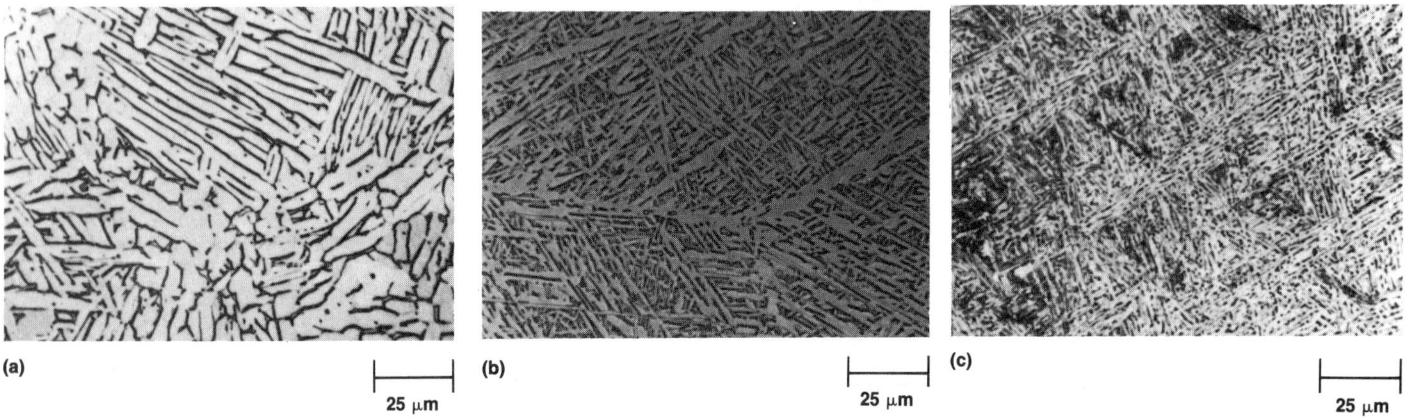

Fig. 10 Microstructure of Ti-6Al-4V PA compacts. (a) As-HIP. (b) Treated to produce a broken-up structure. (c) Thermochemically treated

Table 3 Tensile and fracture toughness properties of Ti-6Al-4V PA compacts processed under various conditions

Condition(a)	0.2% yield strength MPa	0.2% yield strength ksi	Ultimate tensile strength MPa	Ultimate tensile strength ksi	Elongation, %	Reduction in area, %	K_{Ic} or (K_Q) MPa$\sqrt{m}$	K_{Ic} or (K_Q) ksi$\sqrt{in.}$	Titanium PA powder preparation: Powder process	Compaction temperature °C	Compaction temperature °F	Other variables	Ref
HIP	861	125	937	136	17	42	(85)	(77)	PREP	925	1695	...	50
HIP (PSV) and β annealed	1020	148	1095	159	9	21	(67)	(61)	PSV	950	1740	975 °C (1785 °F) anneal	43, 50
HIP and BUS treated	965	140	1048	152	8	17	...	...	PREP	925	1695	...	51
HIP and TCP treated	931	135	1021	148	10	16	...	...	PREP	925	1695	...	30
HIP and annealed (700 °C, or 1290 °F) (REP)	820	119	889	129	14	41	(76)	(69)	REP	955	1750	...	52
HIP, annealed (700 °C, or 1290 °F), and STA (955-480 °C, or 1750-855 °F)	1034	150	1130	164	9	34	...	...	REP	955	1750	...	52
HIP and annealed (700 °C, or 1290 °F) (PREP)	882	128	944	137	15	40	(73)	(67)	PREP	955	1750	...	53
ELI; HIP (as-compacted)	855	124	931	135	15	41	(99)	(90)	REP	955	1750	1300 ppm O_2	54
ELI; HIP and β annealed	896	130	951	138	10	24	93	85	REP	955	1750	1020 °C (1870 °F) anneal	54
HPLT and HIP (as-compacted)	1082	157	1130	164	8	19	...	...	PREP	650	1200	315 MPa (46 ksi)	55
HPLT, HIP, and RA (815 °C, or 1500 °F)	937	136	1013	147	22	38	...	...	PREP	650	1200	315 MPa (46 ksi)	55
HIP and rolled (955 °C, or 1750 °F) (T)	958	139	992	144	12	35	...	...	REP	925	1695	75% rolling reduction	56
HIP, rolled (955 °C, or 1750 °F), and β annealed L or LT	820	119	896	130	13	31	73	66	REP	925	1695	75% rolling reduction	56
T or TL	813	118	896	130	11	23	61	55	REP	925	1695	75% rolling reduction	56
HIP, rolled (950 °C, or 1740 °F), and STA (960-700 °C, or 1760-1290 °F)	924	134	1041	151	15	35	...	...	REP	950	1740	60% rolling reduction	57
HIP, forged (950 °C, or 1740 °F), and STA (960-700 °C, or 1760-1290 °F)	1000	145	1062	154	14	35	...	...	REP	915	1680	56% forging reduction	58
VHP (830 °C, or 1525 °F) (as-compacted)	945	137	993	144	19	38	...	...	REP	830	1525	...	42
VHP (760 °C, or 1400 °F) (as-compacted)	972	141	1014	147	16	38	...	...	REP	760	1400	...	42
ROC (900 °C, or 1650 °F) (as-compacted)	882	128	904	131	14	50	...	...	PREP	900	1650	As-ROC	46
ROC (900 °C, or 1650 °F) and RA (925 °C, or 1695 °F)	827	120	882	128	16	46	...	...	PREP	900	1650	925 °C (1695 °F) RA	46
ROC (650 °C, or 1200 °F) (as compacted)	1131	164	1179	171	10	23	...	...	PREP	600	1110	As ROC	48
ROC (600 °C, or 1100 °F) and RA (815 °C, or 1500 °F)	965	140	1020	148	15	43	...	...	PREP	600	1110	815 °C (1500 °F) RA	48
Minimum properties (MIL-T-9047)	827	120	896	130	10	25	...	...	...	...	...	...	4

(a) HIP, hot isostatic pressing; PSV, *pulverization sous vide* (powder under vacuum), French-made powder (Ref 42); BUS, broken-up structure; TCP, thermochemical processing; REP, rotating-electrode process; STA, solution treated and aged; PREP, plasma rotating-electrode process; ELI, extra-low interstitial; HPLT, high-pressure low-temperature compaction; RA, recrystallization annealed; T, transverse; L, longitudinal; LT, longitudinal-transverse; TL, transverse-longitudinal; VHP, vacuum hot pressing; ROC, rapid omnidirectional compaction. Source: Ref 30, 42, 43, 46, 48, 50-58

comparable to or exceeded that of the I/M material. The component tested is an actual Ti-6Al-4V P/M part used in military airframes (Ref 61).

Figure 13 compares the fatigue crack growth rate of Ti-6Al-4V PA compacts to that of an I/M material with a similar composition and microstructure. Rates are at equivalent levels for both materials, even in PA material with a low level of contamination (Ref 74).

Postcompaction Treatments

Most P/M alloys that are subjected to postcompaction working or to lower-temperature consolidation display improved tensile and fatigue strengths as a result of microstructure refinement. However, in most cases, process economics do not allow subsequent working because it nullifies the objectives of a true net-shape technology. Therefore, only those postcompaction methods leading to microstructure refinement without the use of working will be considered in this article. Two approaches that meet this requirement are heat treatment and thermochemical processing (TCP).

Heat Treatment. In the case of BE Ti-6Al-4V, the only successfully used heat treatment has been the broken-up structure (BUS) treatment in which a β quench is followed by 850 °C (1560 °F) long-term annealing (Ref 27). After such treatment, the microstructure of the alloy is showing broken-up α phase in a matrix of β (Fig. 1e). This microstructure provides a significant improvement in both tensile and fatigue strengths (Fig. 7).

The BUS method is an improvement over standard heat treatments, which typically provide higher tensile properties but no increase in fatigue strength properties.

In the case of Ti-10V-2Fe-3Al BE compacts, β solution treatment and subsequent aging resulted in materials with good combinations of tensile strength and ductility

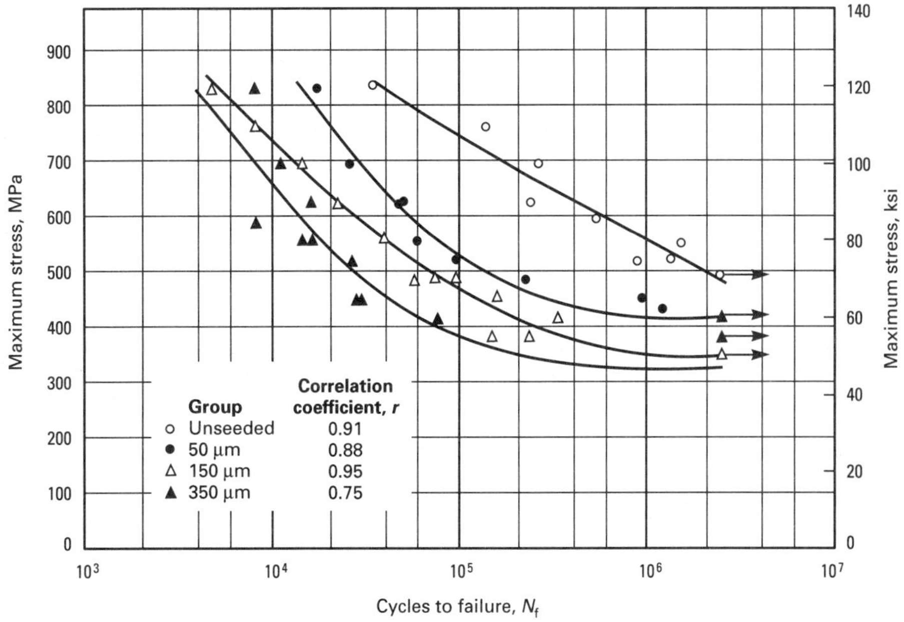

Fig. 11 Effect of contaminant particles on the room-temperature fatigue strength of Ti-6Al-4V PA compacts. Unseeded compacts are compared with SiO_2-seeded PREP compacts. Stress ratio (R), +0.1; frequency (f), triangular waveform load/time cycle at 5 Hz. Source: Ref 49

Table 4 Tensile and fracture toughness properties of PA titanium alloy compacts processed under various conditions

Alloy and condition(a)	0.2% yield strength MPa	0.2% yield strength ksi	Ultimate tensile strength MPa	Ultimate tensile strength ksi	Elongation, %	Reduction in area, %	K_{Ic} or (K_Q) MPa$\sqrt{m}$	K_{Ic} or (K_Q) ksi$\sqrt{in.}$	Titanium PA powder preparation: Powder process(b)	Compaction temperature °C	Compaction temperature °F	Other variables	Ref
Ti-5.5Al-3.5Sn-3Zr-0.25Mo-1Nb-0.25Si (IMI 829)													
HIP and STA (1060-620 °C, or 1940-1150 °F)	951	138	1089	158	18	22	...	...	PREP	1040	1905	...	59
ROC and STA (1060-620 °C, or 1940-1150 °F)	909	132	1034	150	18	20	...	...	PREP	...	...	α + β ROC	59
Ti-6Al-5Zr-0.5Mo-0.25Si (IMI 685)													
HIP and STA (1050-550 °C, or 1920-1020 °F)	970	141	1020	148	11	19	...	...	PREP	950	1740	...	60
Ti-6Al-2Sn-4Zr-2Mo													
HIP and STA (1050-550 °C, or 1920-1020 °F)	924	134	1034	150	17	36	...	...	PREP	910	1670	...	61
Ti-6Al-2Sn-4Zr-6Mo													
HIP, forged (920 °C, or 1690 °F), and annealed (705 °C, or 1300 °F)	1165	169	1296	188	11	37	...	...	REP	900	1650	920 °C (1690 °F), 70% forging reduction	62
Ti-6Al-6V-2Sn													
HIP and annealed (760 °C, or 1400 °F)	1008	146	1055	153	18	37	59	54	PREP	900	1650	...	63, 64
Ti-5Al-2Sn-2Zn-4Cr-4Mo (Ti-17)													
HIP and STA (800-635 °C, or 1470-1175 °F)	1123	163	1192	173	8	11	...	...	REP	915	1680	...	52
Ti-4.5Al-5Mo-1.5Cr (Corona 5)													
HIP and aged (705 °C, or 1300 °F)	944	137	999	145	13	...	(75)	(68)	REP	845	1555	(c)	65
HIP and aged (760 °C, or 1400 °F)	916	133	971	141	14	...	(79)	(72)	REP	845	1555	(c)	65
Ti-10V-2Fe-3Al													
HIP and STA (745-490 °C, or 1375-915 °F)	1213	176	1310	190	9	13	...	...	PREP	775	1425	...	33
HIP, forged, and STA (750-495 °C, or 1380-925 °F)	1286	186	1386	201	7	20	28	25	PREP	775	1425	750 °C (1380 °F), 70% forging reduction	33
HIP, forged, and STA (750-550 °C, or 1380-1020 °F)	1065	155	1138	165	14	41	55	50	PREP	775	1425	750 °C (1380 °F), 70% forging reduction	33
ROC (as-compacted)	965	140	1007	146	16	54	...	...	PREP	650	1200	...	48
ROC and STA (760-510 °C, or 1400-950 °F)	1296	188	1400	203	6	26	...	...	PREP	650	1200	...	
Ti-11.5Mo-6Zr-4.5Sn (Beta III)													
β HIP and STA (745-510 °C, or 1375-950 °F)	1288	187	1378	200	8	18	...	...	PREP	760	1400	...	66
Ti-1.3Al-8V-5Fe													
β extruded and STA (705 °C, or 1300 °F)	1392	202	1482	215	8	7	...	...	PREP	760	1400	...	67
β extruded and STA (770 °C, or 1420 °F)	1461	212	1516	220	8	20	...	...	GA	760	1400	...	68
β HIP and STA (675 °C, or 1245 °F)	1315	191	1414	205	5	10	...	...	GA	725	1335	...	69
Ti-24Al-11Nb													
HIP (1065 °C, or 1950 °F) and STA (1175 °C, or 2145 °F)	510	74	606	88	2	2	...	...	PREP	1065	1950	...	70
HIP (925 °C, or 1700 °F) and STA (1175 °C, or 2145 °F)	696	101	765	111	2	2	...	...	PREP	925	1695	...	71
Ti-25Al-10Nb-3Mo-1V													
ROC (as-compacted)	710	103	854	124	5	6	...	...	PREP	1050	1920	...	72

(a) HIP, hot isostatic pressing; STA, solution treated and aged; ROC, rapid omnidirectional compaction. (b) PREP, plasma rotating-electrode process; REP, rotating-electrode process; GA, gas atomization. (c) Weld study sample. Source: Ref 33, 48, 59-72

(Ref 33). However, the K_{Ic} was found to be too low (Table 2), possibly because of the high chloride levels and the associated porosity. The Ti-6Al-4V PA compacts responded well to the BUS treatment (Ref 51) (Fig. 10b), as well as to solution treatment and aging (Ref 52). Figure 6 shows the improvement in fatigue strength of both BE and PA Ti-6Al-4V compacts as a result of a microstructure refinement brought about by heat treatment.

Thermochemical Processing. The TCP method (Ref 75) involves the use of hydrogen as a temporary alloying element to refine the microstructure of titanium alloys. This method is very suitable for net-shape products because no hot or cold work is needed to refine the microstructure. An example of the refinement obtained in Ti-6Al-4V P/M products can be seen by comparing Fig. 1(d) with Fig. 1(f) and Fig. 10(a) with Fig. 10(c). This microstructural refinement provides slightly higher strength levels than those typically obtained in I/M or conventional P/M materials (Table 3); more significantly, it substantially enhances the fatigue behavior of the P/M products (Fig. 6 to 8).

Applications of Titanium P/M Products

The two distinctively different titanium P/M technologies, the blended element and the prealloyed methods, not only produce compacts with different sets of properties, but also with two different price ranges. The relative low cost of titanium sponge fines and the volume production capability of the pressing and sintering technology allow the production of BE complex-shape aerospace alloy parts at a cost of under $100/kg ($45/lb). Prealloyed powders, on the other hand, require an expensive melt stock, ultraclean handling, and expensive compaction tools. As a result, the higher-performance fully dense PA parts are currently priced above $200/kg ($90/lb), although it is projected that volume production will bring this price down substantially. The differences in density, property, and price target these two technologies to different application markets. Because of their lower cost, more BE components than PA parts are currently in use. The introduction of gas-atomized powder (Ref 36) is expected to lower powder costs and make PA products more cost competitive.

Blended Elemental Products. On the low end of the density scale (20 to 80%), commercially pure (CP) titanium filters are produced for electrochemical and other corrosion-resistant applications (Fig. 14a). Higher-density pressed and sintered CP titanium parts, such as the assorted nuts shown in Fig. 14(b), are made commercially for the chemical-processing industry.

For more demanding applications, Ti-6Al-4V BE components with densities from 98% to close to 100% are produced by the CIP and sintering and by the pressing and sintering methods. Very complex shapes, such as the impeller shown in Fig. 15(a) or the McDonnell-Douglas F-18 pivot fitting shown in Fig. 15(b), can be produced by CIP using elastomeric molds (Ref 7). Part size is currently limited to a maximum length of 610 mm (24 in.) by the availability of CIP equipment. The possible dimensional tolerances for small parts are ±0.5 mm (±0.02 in.). The missile housing (Fig. 15c) and the lens housing (Fig. 15d) are production run parts made by the CIP method. The airframe prototype part (Fig. 15e) is made out of chloride-free powder and is fully dense.

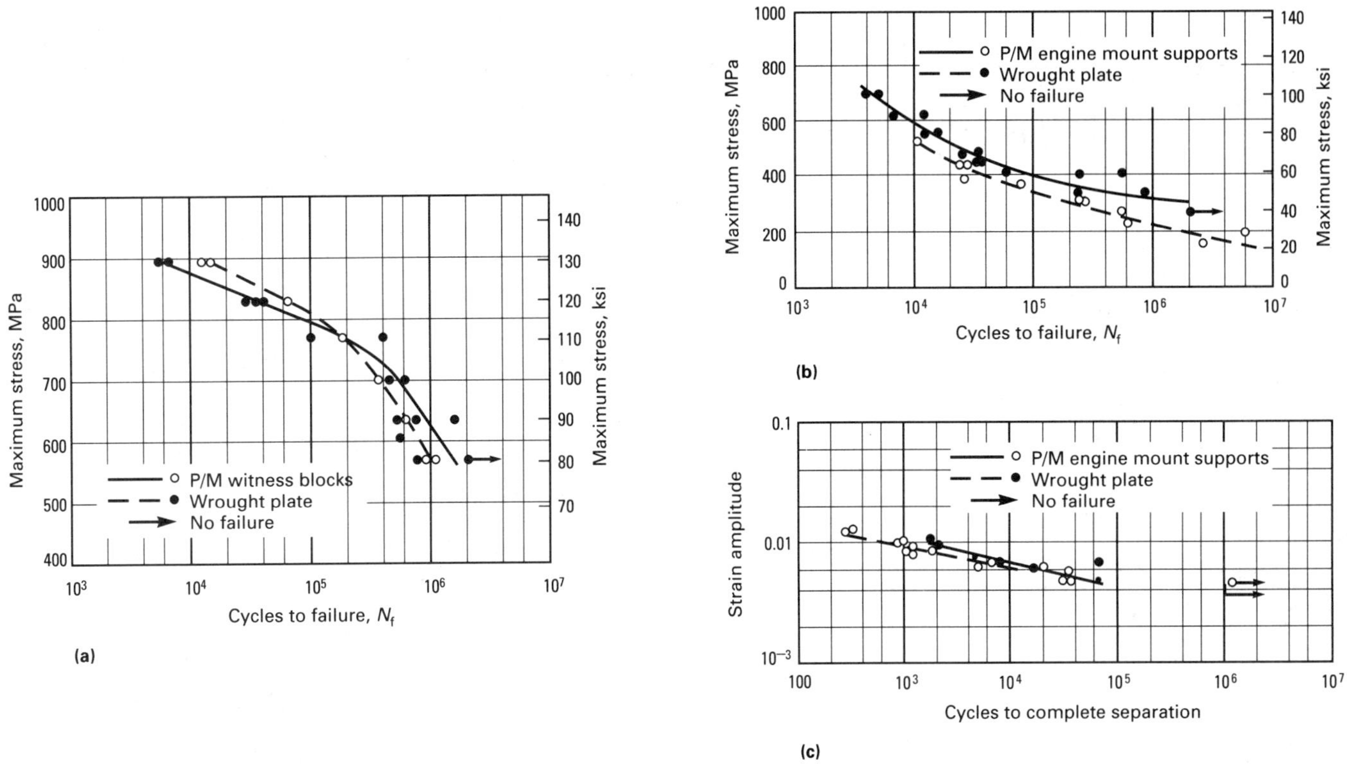

Fig. 12 Comparison of the fatigue strengths at room temperature in air of PREP HIP Ti-6Al-4V P/M components to those of I/M products. (a) Load-controlled smooth-specimen high-cycle fatigue for large bars (13 mm, or ½ in., in diameter). Stress ratio (R), 0.1. (b) Load-controlled notched-specimen high-cycle fatigue. Stress concentration factor (K_t), 3; stress ratio (R), 0.1. (c) Strain-controlled low-cycle fatigue for small specimens (6.4 mm, or ¼ in., in diameter). Stress ratio (R), −1. Source: Ref 73

The pressing and sintering method is more volume oriented. The mirror hub (Fig. 15f) is an example of such a production part. Recently, Ti-6Al-4V BE parts are being considered for use in the automotive industry in an effort to increase performance at a moderate cost; the cylinder in Fig. 15(g) is an example of a BE automotive component.

Prealloyed Products

The relatively high product cost of PA parts has thus far limited the consideration of potential applications of PA technology to, for the most part, the manufacture of critical aerospace components. As previously discussed, property levels have been attained in properly processed PA compacts that match those of high-quality I/M parts. The decision-making process for using PA P/M is based primarily on economic considerations. A number of demonstration parts are now flying in the F-15 (for example, the Ti-6Al-4V keel splice former shown in Fig. 16(a) and in the F-18 (for example, the Ti-6Al-4V engine mount support fitting shown in Fig. 16b) fighter planes. An example of a true net-shape PA rotating engine component is shown in Fig. 16(c).

The largest HIP chamber available to date is 1350 mm (54 in.) in diameter and 2400 mm (96 in.) in height. This limits the compact size, unless welding of PA sections is used as it is for the part shown in Fig. 16(d). The increased demand for the brittle and hard-to-machine titanium aluminides, which are used for higher-temperature applications, is creating a new interest in titanium PA P/M technology. A demonstrator impeller made out of titanium aluminide PREP powder is shown in Fig. 16(e). All of the above-mentioned PA parts were made by the Crucible ceramic mold process (Ref 7).

Future Trends in Titanium P/M Technology

Clearly, sufficient data is now available to allow both prealloyed and blended elemental Ti-6Al-4V P/M compacts to be used with confidence. However, cost remains a major concern: Use of the PA P/M approach is difficult to justify for parts with approximately the same, or perhaps even slightly higher, mechanical property levels. Only a significantly lower cost for PA compacts would enable them to replace reliable I/M materials. The production of a low-cost powder either by a scaled-up gas atomization process or by a direct chemical method would be a significant breakthrough. The BE P/M technique can be cost effective for less critical parts, and its increased use in this area is likely. However, as with the PA material, the production of fully dense chloride-free BE products is likely to be stymied by cost unless a breakthrough occurs.

The trend with other conventional titanium P/M alloys is likely to follow that described above for the Ti-6Al-4V alloy. An exception is the high-strength Ti-1.3Al-8V-5Fe (Ti-185) alloy (Ref 67-69), which cannot be satisfactorily made by the I/M approach because of the segregation of the iron. This alloy should be strictly classified with rapid solidification alloys (discussed below) that require rapid transformation from the liquid to solid states.

The ordered intermetallic titanium aluminides are much more difficult to process and machine than the conventional titanium alloys. In fact, their production characteristics approach those of the superalloys. Thus, the cost benefits to be gained by a net-shape P/M approach are great (particularly for the equiatomic titanium aluminide, TiAl), and these benefits could accelerate the acceptance of titanium P/M methods, particularly with the development of a lower-cost powder.

Research is being conducted to expand the boundaries of conventional titanium P/M technology. Efforts are in progress to evaluate the possibility of using advanced techniques such as rapid solidification (Ref 76, 77), mechanical alloying (Ref 78), and nanostructures (Ref 79) to enhance the behavior of titanium-base materials.

Work on rapid solidification has focused on increasing the temperature capability of both terminal alloys and intermetallic com-

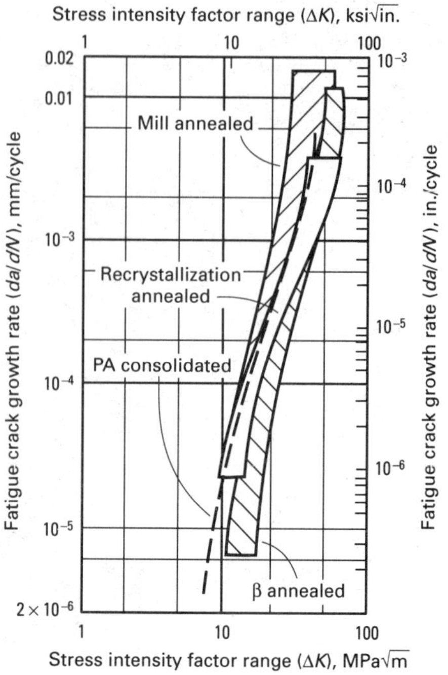

Fig. 13 Comparison of the fatigue crack growth rate at room temperature in air of Ti-6Al-4V PA compacts with that of an I/M alloy material. Stress ratio (*R*), 0.1; frequency (*f*), 5 to 30 Hz (5 Hz for a PA compact). Source: Ref 74

positions by dispersion strengthening. However, while some improvements have been made, they are not considered to be significant enough to warrant the extra cost and concern over product quality assurance associated with a P/M method. The rapid solidification technique has two major drawbacks. The first is that it has been unable to produce more than about 6 vol% of second-phase particles. The second drawback is that rapid solidification results in a β grain size that is much smaller than is desirable and that there is a lack of elongated α phase, which would be formed on cooling into the α-β phase field after a β anneal. Unfortunately, a β anneal, which corrects the second drawback, results in unacceptable coarsening of the dispersoids.

Mechanical alloying of titanium alloys is at a very early stage, but it does exhibit the potential to increase the volume percentage of dispersoids for elevated-temperature applications. In addition, there are indications that the normally immiscible titanium and magnesium can be combined by mechanical alloying to produce a low-density titanium alloy.

Very preliminary results on the mechanical alloying of titanium-magnesium and titanium-eutectoid formers such as nickel and copper suggest that a very fine nanoscale microstructure (~10^{-9} m scale) can be obtained; such a microstructure could have novel physical and mechanical properties.

ACKNOWLEDGMENT

The authors wish to acknowledge the assistance of Cheryl Seitz in the preparation of the manuscript. The help of J. Moll, G. Chanani, and S. Abkowitz in obtaining additional data is highly appreciated.

REFERENCES

1. H.B. Bomberger, F.H. Froes, and P.H. Morton, Titanium—A Historical Perspective, in *Titanium Technology: Present Status and Future Trends*, F.H. Froes, D. Eylon, and H.B. Bomberger, Ed., Titanium Development Association, 1985, p 3-17
2. F.H. Froes, D. Eylon, G.E. Eichelman, and H.M. Burte, Developments in Titanium Powder Metallurgy, *J. Met.*, Vol 32, (No. 2), Feb 1980, p 47-54
3. F.H. Froes and D. Eylon, Titanium Powder Metallurgy—A Review, in *Titanium Net-Shape Technologies*, F.H. Froes and D. Eylon, Ed., The Metallurgical Society of AIME, 1984, p 1-20
4. F.H. Froes and D. Eylon, Powder Metallurgy of Titanium Alloys—A Review, in *Titanium, Science and Technology*, Vol 1, G. Lutjering, U. Zwicker, and W. Bunk, Ed., DGM, 1985, p 267-286; *Powder Metall. Int.*, Vol 17 (No. 4), 1985, p 163-167, continued in Vol 17 (No. 5), 1985, p 235-238; *Titanium Technology: Present Status and Future Trends*, F.H. Froes, D. Eylon, and H.B. Bomberger, Ed., Titanium Development Association, 1985, p 49-59
5. F.H. Froes and D. Eylon, Powder Metallurgy of Titanium Alloys, *Int. Mater. Rev.*, in press
6. F.H. Froes and D. Eylon, Production of Titanium Powder, in *Metals Handbook*, Vol 7, 9th ed., *Powder Metallurgy*, American Society for Metals, 1984, p 164-168
7. F.H. Froes, D. Eylon, and G. Friedman, Titanium P/M Technology, in *Metals Handbook*, Vol 7, 9th ed., *Powder Metallurgy*, American Society for Metals, 1984, p 748-755
8. Automotive Applications, in *Metals Handbook*, Vol 7, 9th ed., *Powder Metallurgy*, American Society for Metals, 1984, p 617-621
9. R.W. Stevenson, P/M Copper-Based Alloys, in *Metals Handbook*, Vol 7, 9th ed., *Powder Metallurgy*, American Society for Metals, 1984, p 733-740
10. T.W. Penrice, Kinetic Energy Penetra-

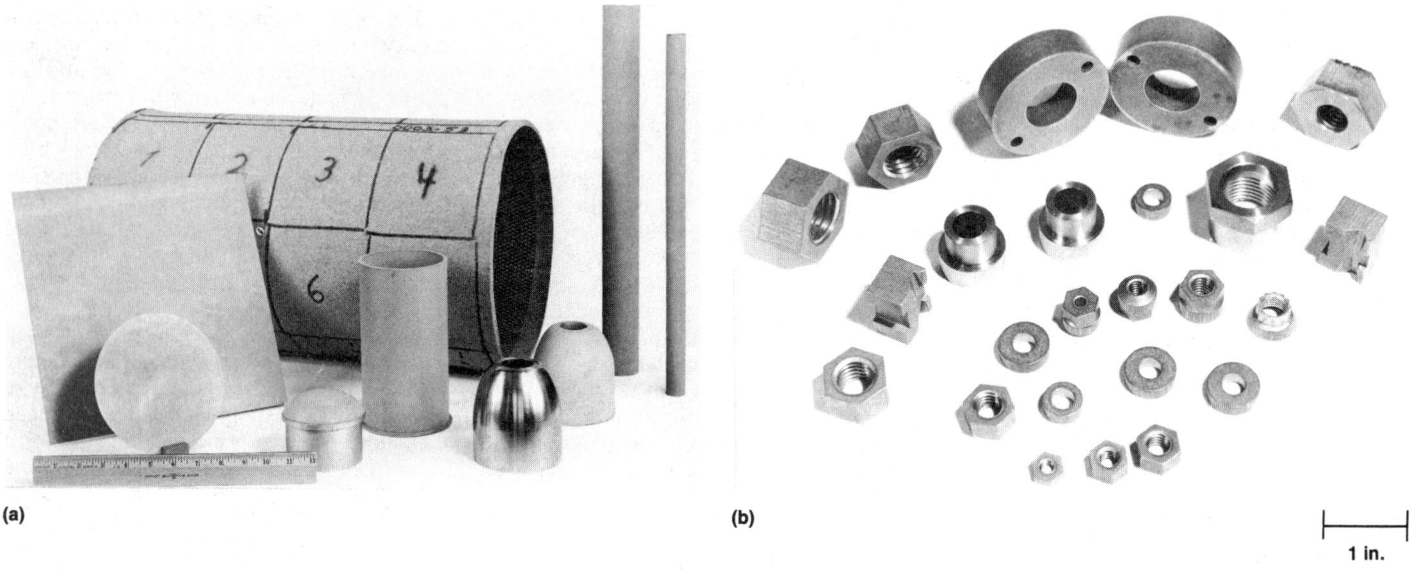

Fig. 14 Commercially pure titanium BE parts. (a) Assortment of porous filters for electrochemical processes. (b) Assortment of parts for the chemical industry. Courtesy of Clevite Industries

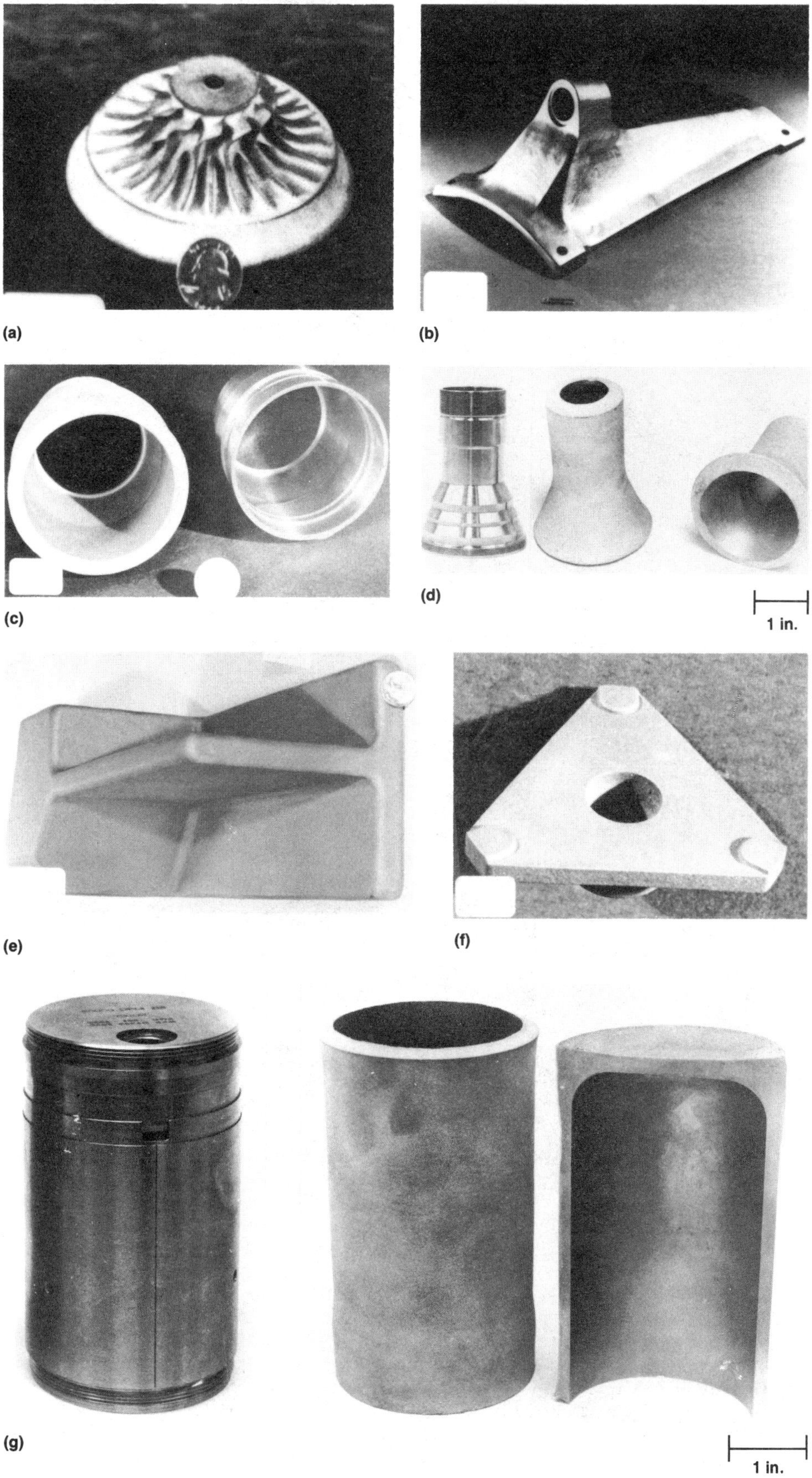

Fig. 15 Aerospace and automotive Ti-6Al-4V components produced by the BE method. (a) Impeller. (b) F-18 fighter plane pivot fitting. (c) Missile housing. (d) Lens housing. (e) Prototype for a 100% dense airframe component. (f) Net-shape 35 mm (1⅜ in.) diam mirror hub. (g) Automotive cylinder. Courtesy of Dynamet Technology (a, c, and e), Metal Powder Industries Federation (b), Clevite Industries (d, g), and Valform (f)

tors, in *Metals Handbook*, Vol 7, 9th ed., *Powder Metallurgy*, American Society for Metals, 1984, p 688-691

11. G.H. Gessinger, *Powder Metallurgy of Superalloys*, Butterworths, 1984
12. B.L. Ferguson, Aerospace Applications, in *Metals Handbook*, Vol 7, 9th ed., *Powder Metallurgy*, American Society for Metals, 1984, p 646-651
13. R.W. Stevenson, Aluminum P/M Technology, in *Metals Handbook*, Vol 7, 9th ed., *Powder Metallurgy*, American Society for Metals, 1984, p 741-748
14. Beryllium P/M Technology, in *Metals Handbook*, Vol 7, 9th ed., *Powder Metallurgy*, American Society for Metals, 1984, p 755-762
15. F.H. Froes and D. Eylon, Titanium Powder Metallurgy—A Review, in *PM Aerospace Materials*, Vol 1, MPR Publishing, 1984, p 39-1 to 39-19
16. F.H. Froes, C.M. Cooke, D. Eylon, and K.C. Russell, Grain Growth in Blended Elemental Ti-6Al-4V Powder Compacts, in *Sixth World Conference on Titanium*, Part III, P. Lacombe, R. Tricot, and G. Beranger, Ed., Les Editions de Physique, 1989, p 1161-1166
17. I. Weiss, D. Eylon, M.W. Toaz, and F.H. Froes, Effect of Isothermal Forging on Microstructure and Fatigue Behavior of Blended Elemental Ti-6Al-4V Powder Compacts, *Metall. Trans. A*, Vol 17A (No. 3), 1986, p 549-559
18. H.I. Aaronson, D. Eylon, and F.H. Froes, Observations of Superledges Formed on Sideplates During Precipitation of Alpha from Beta Ti-6%Al-4%V, *Scr. Metall.*, Vol 21 (No. 11), 1987, p 1421-1425
19. G. Welsch, Y.-T. Lee, P.C. Eloff, D. Eylon, and F.H. Froes, Deformation Behavior of Blended Elemental Ti-6Al-4V Compacts, *Metall. Trans. A*, Vol 14A (No. 4), 1983, p 761-769
20. D. Eylon, R.G. Vogt, and F.H. Froes, Property Improvement of Low Chlorine Titanium Alloy Blended Elemental Powder Compacts by Microstructure Modification, in *Progress in Powder Metallurgy*, Vol 42, compiled by E.A. Carlson and G. Gaines, Metal Powder Industries Federation, 1986, p 625-634
21. P.J. Andersen, V.M. Svoyatytsky, F.H. Froes, Y. Mahajan, and D. Eylon, Fracture Behavior of Blended Elemental P/M Titanium Alloy, in *Modern Developments in Powder Metallurgy*, Vol 13, H.H. Hausner, H.W. Antes, and G.D. Smith, Ed., Metal Powder Industries Federation, 1981, p 537-549
22. J. Park, M.W. Toaz, D.H. Ro, and E.N. Aqua, Blended Elemental Powder Metallurgy of Titanium Alloys, in *Titanium Net Shape Technologies*, F.H. Froes and D. Eylon, Ed., The Metallurgical Society of AIME, 1984, p 95-105
23. S. Abkowitz, Isostatic Pressing of

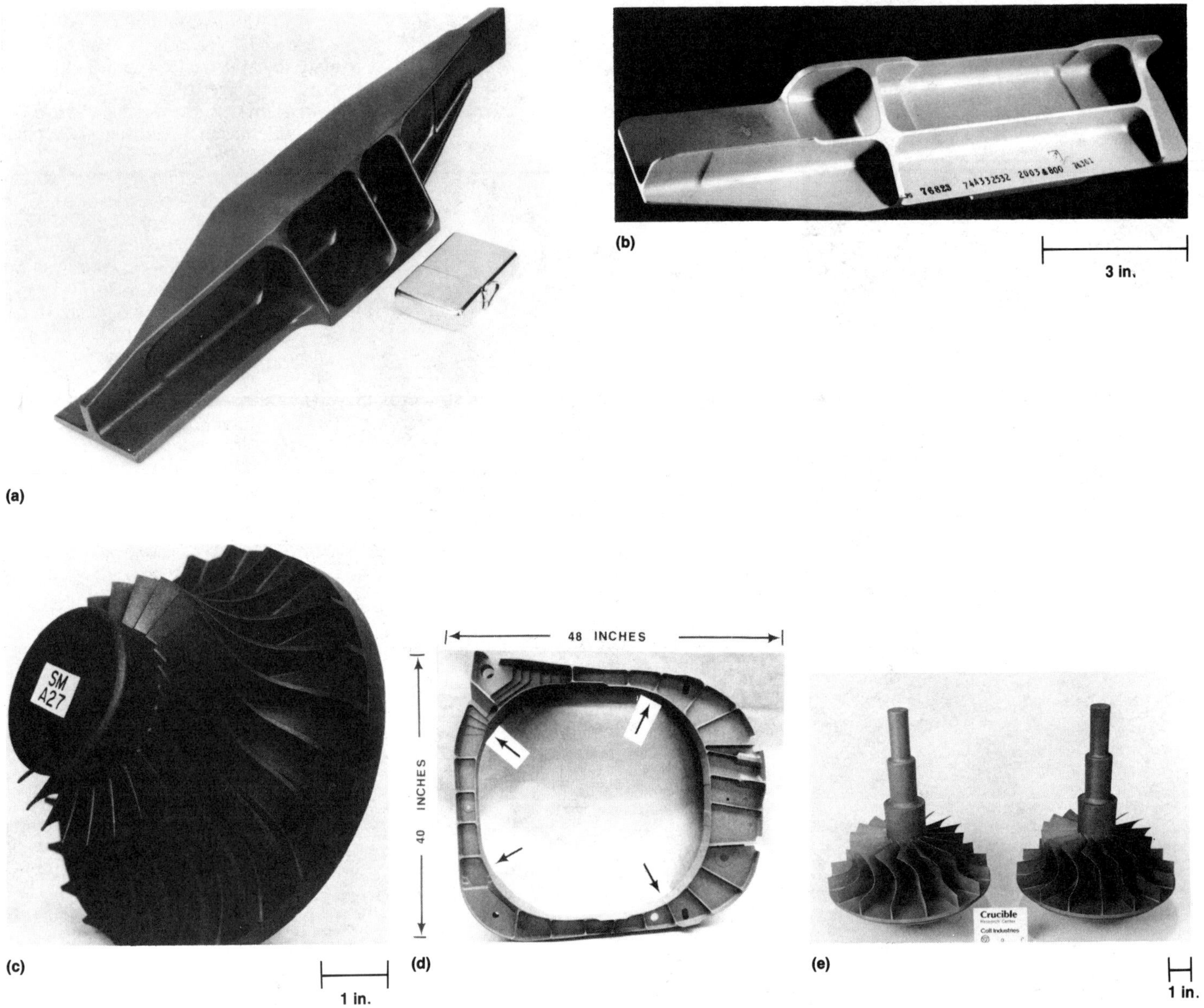

Fig. 16 Prealloyed HIP Ti-6Al-4V aerospace parts produced by the Crucible ceramic mold method. (a) F-14 fighter plane fuselage brace. (b) F-18 fighter plane engine mount support fitting. (c) Cruise missile engine impeller. (d) Four-section welded nacelle frame structure. (e) Titanium aluminide demonstrator impeller. All courtesy of Crucible Research Center

Complex Shapes From Titanium and Titanium Alloys, in *Powder Metallurgy of Titanium Alloys*, F.H. Froes and J.E. Smugeresky, Ed., The Metallurgical Society of AIME, 1980, p 291-302

24. S. Abkowitz, G.J. Kardys, S. Fujishiro, F.H. Froes, and D. Eylon, Titanium Alloy Shapes from Elemental Blend Powder and Tensile and Fatigue Properties of Low Chloride Compositions, in *Titanium Net Shape Technologies*, F.H. Froes and D. Eylon, Ed., The Metallurgical Society of AIME, 1984, p 107-120
25. R.R. Boyer, J.E. Magnuson, and J.W. Tripp, Characterization of Pressed and Sintered Ti-6Al-4V Powders, in *Powder Metallurgy of Titanium Alloys*, F.H. Froes and J.E. Smugeresky, Ed., The Metallurgical Society of AIME, 1980, p 203-216
26. Y. Mahajan, D. Eylon, R. Bacon, and F.H. Froes, Microstructure Property Correlation in Cold Pressed and Sintered Elemental Ti-6Al-4V Powder Compacts, in *Powder Metallurgy of Titanium Alloys*, F.H. Froes and J.E. Smugeresky, Ed., The Metallurgical Society of AIME, 1980, p 189-202
27. D. Eylon, R.G. Vogt, and F.H. Froes, Property Improvement of Low Chlorine Titanium Alloy Blended Elemental Powder Compacts by Microstructure Modification, in *Progress in Powder Metallurgy*, Vol 42, compiled by E.A. Carlson and G. Gaines, Metal Powder Industries Federation, 1986, p 625-634
28. P.R. Smith, C.M. Cooke, A. Patel, and F.H. Froes, in *Progress in Powder Metallurgy*, Vol 38, J.G. Bewley and S.W. McGee, Ed., Metal Powder Industries Federation, 1983, p 339-359
29. P.R. Smith, F.H. Froes, and C.M. Cooke, in *Materials and Processes—Continuing Innovations*, Vol 28, Society for the Advancement of Material and Process Engineering, 1983, p 406-421
30. C.F. Yolton, D. Eylon, and F.H. Froes, Microstructure Modification of Titanium Alloy Products by Temporary Alloying with Hydrogen, in *Sixth World Conference on Titanium*, Part III, P. Lacombe, R. Tricot, and G. Beranger, Ed., Les Editions de Physique, 1989, p

1641-1646

31. M. Hagiwara, Y. Kaieda, and Y. Kawabe, Improvement of Mechanical Properties of Blended Elemental α-β Ti Alloys by Microstructural Modification, in *Titanium 1986, Products and Applications*, Vol II, Titanium Development Association, 1987, p 850-858
32. J.E. Smugeresky and N.R. Moody, Properties of High Strength, Blended Elemental Powder Metallurgy Titanium Alloys, in *Titanium Net-Shape Technologies*, F.H. Froes and D. Eylon, Ed., The Metallurgical Society of AIME, 1984, p 131-143
33. R.R. Boyer, D. Eylon, C.F. Yolton, and F.H. Froes, Powder Metallurgy of Ti-10V-2Fe-3Al, in *Titanium Net-Shape Technologies*, F.H. Froes and D. Eylon, Ed., The Metallurgical Society of AIME, 1984, p 63-78
34. S. Abkowitz and P. Weithrauch, Trimming the Cost of MMC, *Adv. Mater. Proc.*, Vol 136 (No. 1), July 1989, p 31-34
35. F.H. Froes, H.B. Bomberger, D. Eylon, and R.G. Rowe, Potential of Titanium Powder Metallurgy, in *Competitive Advances in Metals and Processes*, Vol 1, R.J. Cunningham and M. Schwartz, Ed., Society for the Advancement of Material and Process Engineering, 1987, p 240-254
36. C.F. Yolton, Gas Atomized Titanium and Titanium Aluminide Alloys, in *Powder Metallurgy in Aerospace and Defense Technologies*, Metal Powder Industries Federation, 1989
37. E.J. Kosinski, The Mechanical Properties of Titanium P/M Parts Produced From Superclean Powders, in *Progress in Powder Metallurgy*, Vol 38, J.G. Bewley and S.W. McGee, Ed., Metal Powder Industries Federation, 1983, p 491-592
38. P.R. Roberts and P. Loewenstein, Titanium Alloy Powders Made by the Rotating Electrode Process, in *Powder Metallurgy of Titanium Alloys*, F.H. Froes and J.E. Smugeresky, Ed., The Metallurgical Society of AIME, 1980, p 21-35
39. J.P. Laughlin and G.J. Dooley III, The Hydride Process for Producing Titanium Alloy Powders, in *Powder Metallurgy of Titanium Alloys*, F.H. Froes and J.E. Smugeresky, Ed., The Metallurgical Society of AIME, 1980, p 37-46
40. J.A. Megy, U.S. Patent 4,127,409, Nov 1978
41. G. Buttner, H.-G. Domazer, and H. Eggert, U.S. Patent 4,373,947, Feb 1983
42. W.H. Kao, D. Eylon, C.F. Yolton, and F.H. Froes, Effect of Temporary Alloying by Hydrogen (Hydrovac) on the Vacuum Hot Pressing and Microstructure of Titanium Alloy Powder Compacts, in *Progress in Powder Metallurgy*, Vol 37, J.M. Capus and D.L. Dyke, Ed., Metal Powder Industries Federation, 1982, p 289-301
43. J. Devillard and J.-P. Herteman, Evaluation of Ti-6Al-4V Powder Compacts Fabricated by the PSV Process, in *Powder Metallurgy of Titanium Alloys*, F.H. Froes and J.E. Smugeresky, Ed., The Metallurgical Society of AIME, 1980, p 59-70
44. I.A. Martorell, Y.R. Mahajan, and D. Eylon, "Property Modification of Ti-10V-2Fe-3Al by Low Temperature Processing," Unpublished report, 1987
45. C.A. Kelto, Rapid Omnidirectional Compaction, in *Metals Handbook*, Vol 7, 9th ed., *Powder Metallurgy*, American Society for Metals, 1984, p 542-546
46. Y.R. Mahajan, D. Eylon, C.A. Kelto, T. Egerer, and F.H. Froes, Modification of Titanium Powder Metallurgy Alloy Microstructures by Strain Energizing and Rapid Omnidirectional Compaction, in *Titanium, Science and Technology*, Vol 1, G. Lutjering, U. Zwicker, and W. Bunk, Ed., DGM, 1985, p 339-346; *Powder Metall. Int.*, Vol 17 (No. 2), 1985, p 75-78; *Titanium Net-Shape Technologies*, F.H. Froes and D. Eylon, Ed., The Metallurgical Society of AIME, 1984, p 39-51
47. Y.R. Mahajan, D. Eylon, C.A. Kelto, and F.H. Froes, Evaluation of Ti-10V-2Fe-3Al Powder Compacts Produced by the ROC Method, in *Progress in Powder Metallurgy*, Vol 41, H.I. Sanderow, W.L. Giebelhausen, and K.M. Kulkarni, Ed., Metal Powder Industries Federation, 1986, p 163-171; *Met. Powder Rep.*, Vol 41 (No. 10), Oct 1986, p 749-752
48. D. Eylon, C.A. Kelto, A.F. Hayes, and F.H. Froes, Low Temperature Compaction of Titanium Alloys by Rapid Omnidirectional Compaction (ROC), in *Progress in Powder Metallurgy*, Vol 43, compiled by C.L. Freeby and H. Hjort, Metal Powder Industries Federation, 1987, p 33-47
49. S.W. Schwenker, D. Eylon, and F.H. Froes, Influence of Foreign Particles on Fatigue Behavior of Ti-6Al-4V Prealloyed Powder Compacts, *Metall. Trans. A*, Vol 17A (No. 2), 1986, p 271-280
50. J.-P. Herteman, D. Eylon, and F.H. Froes, Mechanical Properties of Advanced Titanium Powder Metallurgy Compacts, in *Titanium, Science and Technology*, Vol 1, G. Lutjering, U. Zwicker, and W. Bunk, Ed., DGM, 1985, p 303-310; *Powder Metall. Int.*, Vol 17 (No. 3), 1985, p 116-118
51. L. Levin, R.G. Vogt, D. Eylon, and F.H. Froes, Fatigue Resistance Improvement of Ti-6Al-4V by Thermo-Chemical Treatment, in *Titanium, Science and Technology*, Vol 4, G. Lutjering, U. Zwicker, and W. Bunk, Ed., DGM, 1985, p 2107-2114
52. R.E. Peebles and C.A. Kelto, Investigation of Methods for the Production of High Quality, Low Cost Titanium Alloy Powders, in *Powder Metallurgy of Titanium Alloys*, F.H. Froes and J.E. Smugeresky, Ed., The Metallurgical Society of AIME, 1980, p 47-58
53. R.E. Peebles and L.D. Parsons, Study of Production Methods of Aerospace Quality Titanium Alloy Powder, in *Titanium Net-Shape Technologies*, F.H. Froes and D. Eylon, Ed., The Metallurgical Society of AIME, 1984, p 21-28
54. G.R. Chanani, W.T. Highberger, C.A. Kelto, and V.C. Petersen, Application of Titanium Powder Metallurgy for Manufacture of a Large and Complex Naval Aircraft Component, in *Powder Metallurgy of Titanium Alloys*, F.H. Froes and J.E. Smugeresky, Ed., The Metallurgical Society of AIME, 1980, p 279-290
55. D. Eylon and F.H. Froes, HIP Compaction of Titanium Alloy Powders at High Pressure and Low Temperature (HPLT), *Met. Powder Rep.*, Vol 41 (No. 4), April 1986, p 287-293; *Titanium, Rapid Solidification Technology*, F.H. Froes and D. Eylon, Ed., The Metallurgical Society, 1986, p 273-289
56. R.F. Geisendorfer, Powder Metallurgy Titanium 6Al-4V Plate, in *Powder Metallurgy of Titanium Alloys*, F.H. Froes, and J.E. Smugeresky, Ed., The Metallurgical Society of AIME, 1980, p 151-162
57. R.F. Vaughan and P.A. Blenkinsop, "A Metallurgical Assessment of Ti-6Al-4V Powder," in *Powder Metallurgy of Titanium Alloys*, F.H. Froes and J.E. Smugeresky, Ed., The Metallurgical Society of AIME, 1980, p 83-92
58. D. Eylon, F.H. Froes, D.G. Heggie, P.A. Blenkinsop, and R.W. Gardiner, Influence of Thermomechanical Processing on Low Cycle Fatigue of Ti-6Al-4V Powder Compacts, *Metall. Trans. A*, Vol 14A, 1983, p 2497-2505
59. N.R. Osborne, D. Eylon, and F.H. Froes, Compaction and Net-Shape Forming of Ti-829 Alloy by PM ROC Processing, in *Advances in Powder Metallurgy*, compiled by T.G. Gasbarre and W.F. Jandeska, Metal Powder Industries Federation, 1989
60. B. Borchert, H. Schmid, and J. Wortmann, Microstructure and Strength of PM Ti-685, in *Titanium, Science and Technology*, Vol 1, G. Lutjering, U. Zwicker, and W. Bunk, Ed., DGM, 1985, p 295-302
61. V.K. Chandhok, J.H. Moll, C.F. Yolton, and G.R. McIndoe, Advances in P/M Titanium Shape Technology Using the Ceramic Mold Process, in *Overcoming Material Boundaries*, Vol 17, Soci-

ety for the Advancement of Material and Process Engineering, 1985, p 495-506

62. I. Weiss, F.H. Froes, D. Eylon, and C.C. Chen, Control of Microstructure and Properties of Ti-6Al-2Sn-4Zr-6Mo Powder Forgings, in *Titanium Net-Shape Technologies*, F.H. Froes and D. Eylon, Ed., The Metallurgical Society of AIME, 1984, p 79-94
63. R.H. Witt and I.G. Weaver, Titanium PM Components for Airframes, in *Titanium Net-Shape Technologies*, F.H. Froes and D. Eylon, Ed., The Metallurgical Society of AIME, 1984, p 29-38
64. R.H. Witt and W.T. Highberger, Hot Isostatic Pressing of Near-Net Titanium Structural Parts, in *Powder Metallurgy of Titanium Alloys*, F.H. Froes and J.E. Smugeresky, Ed., The Metallurgical Society of AIME, 1980, p 255-265
65. D.W. Becker, W.A. Baeslack III, and F.H. Froes, Welding of Corona 5 PM Product, in *Powder Metallurgy of Titanium Alloys*, F.H. Froes and J.E. Smugeresky, Ed., The Metallurgical Society of AIME, 1980, p 217-228
66. C.F. Yolton, P/M Beta Titanium Alloys for Landing Gear Applications, in *Progress in Powder Metallurgy*, Vol 42, compiled by E.A. Carlson and G. Gaines, Metal Powder Industries Federation, 1986, p 635-653
67. R.G. Vogt, D. Eylon, and F.H. Froes, Production of High Strength Beta Titanium Alloy Through Powder Metallurgy, in *Titanium, Rapid Solidification Technology*, F.H. Froes and D. Eylon, Ed., The Metallurgical Society, 1986, p 195-199
68. R.R. Boyer, E.R. Barta, C.F. Yolton, and D. Eylon, PM of High Strength Titanium Alloys, in *Powder Metallurgy in Aerospace and Defense Technologies*, Metal Powder Industries Federation, 1989
69. C.F. Yolton and J.H. Moll, Evaluation of a High Strength Rapidly Solidified Titanium Alloy, in *Progress in Powder Metallurgy*, Vol 43, compiled by C.L. Freeby and H. Hjort, Metal Powder Industries Federation, 1987, p 49-63
70. C.F. Yolton, T. Lizzi, V.K. Chandhok, and J.H. Moll, Powder Metallurgy of Titanium Aluminide Components, in *Progress in Powder Metallurgy*, Vol 42, compiled by E.A. Carlson and G. Gaines, Metal Powder Industries Federation, 1986, p 479-488
71. V.S. Moxson and G.I. Friedman, Powder Metallurgy of Titanium Aluminides, in *Progress in Powder Metallurgy*, Vol 42, compiled by E.A. Carlson and G. Gaines, Metal Powder Industries Federation, 1986, p 489-500
72. N.R. Osborne, W.J. Porter, and D. Eylon, Unpublished report, 1989
73. A.S. Sheinker, G.R. Chanani, and J.B. Bohlen, Evaluation and Application of Prealloyed Titanium P/M Parts for Airframe Structures, *Int. J. Powder*, Vol 23 (No. 3), 1987, p 171-176
74. S.W. Schwenker, A.W. Sommer, D. Eylon, and F.H. Froes, Fatigue Crack Growth Rate of Ti-6Al-4V Prealloyed Powder Compacts, *Metall. Trans. A*, Vol 14A (No. 7), July 1983, p 1524-1528
75. F.H. Froes and D. Eylon, Thermochemical Processing (TCP) of Titanium Alloys by Temporary Alloying With Hydrogen, in *Hydrogen Effects on Material Behavior*, A.W. Thompson and N.R. Moody, Ed., The Metallurgical Society, 1990
76. F.H. Froes and R.G. Rowe, Rapidly Solidified Titanium, in *Rapidly Solidified Alloys and Their Mechanical and Magnetic Properties*, Vol 58, B.C. Giessen, D.E. Polk, and A.I. Taub, Ed., Materials Research Society, 1986, p 309-334
77. R.G. Rowe and F.H. Froes, Titanium Rapid Solidification—Alloys and Processes, in *Processing of Structural Metals by Rapid Solidification*, F.H. Froes and S.J. Savage, Ed., ASM INTERNATIONAL, 1987, p 163-173
78. R. Sundaresan and F.H. Froes, Mechanical Alloying, *J. Met.*, Vol 39 (No. 8), Aug 1987, p 22-27
79. F.H. Froes and C. Suryanarayana, Nanocrystalline Metals for Structural Applications, *J. Met.*, June 1989, p 12-17

Zirconium and Hafnium

R. Terrence Webster, Teledyne Wah Chang Albany

ZIRCONIUM was discovered by Klaproth in 1789, but it was not isolated as a metal until 1824, when Berzelius prepared an impure zirconium metal powder. In 1925, VanArkel and DeBoer developed a purified metal using the iodide decomposition process. This process is still used today to purify zirconium and hafnium metal extracted from their ores. In 1947, the magnesium reduction method for extracting the metal from zirconium tetrachloride was developed at the U.S. Bureau of Mines in Albany, OR by W.J. Kroll.

The properties of zirconium established by the U.S. Bureau of Mines indicate that it is ductile and has useful mechanical properties similar to those of titanium and austenitic stainless steel (see the article "Wrought Stainless Steels" in Volume 1 of the 10th Edition of *Metals Handbook*). Zirconium has excellent resistance to many corrosive media, including superheated water, and it is transparent to thermal energy neutrons. These properties prompted the U.S. Navy to use zirconium in water-cooled nuclear reactors as cladding for uranium fuel. In 1958, zirconium became available for industrial use and began to supplant stainless steel as a fuel cladding in commercial power station nuclear reactors. Also, the chemical-processing industries began to use zirconium in several severe corrosion environments.

Today, a high proportion of zirconium is used in water-cooled nuclear reactors; the next largest use is in chemical-processing equipment. Additional uses are in flashbulbs, incendiary ordnance, and gettering contaminating gases in sealed devices such as vacuum tubes.

Hafnium was discovered in 1922 by Coster and DeHevesy. It was shown to occur in the same minerals with zirconium in quantities of 1.5 to 4%. Development programs at Oak Ridge National Laboratory produced a liquid-liquid separation process for separating the hafnium fraction from zirconium.

Hafnium has a high-capture cross section for thermal energy neutrons and has consequently found use in nuclear reactor control rods. Previously, the principal application of hafnium was as a neutron absorption material; currently, however, more hafnium is used in superalloys than in reactor applications. Both zirconium and hafnium are essential alloying metals in other metal systems such as aluminum, copper, magnesium, and titanium alloys and the superalloys.

Metal Processing

Zirconium, hafnium, and titanium are produced from ore that generally is found in a heavy beach sand containing zircon, rutile, and ilmenite. Zircon, which is also a gemstone, is a zirconium-hafnium silicate (Zr-Hf, SiO_4) with a zirconium-to-hafnium ratio of 50 to 1. The zircon is separated from rutile, ilmenite, and other minerals by standard ore dressing methods.

The manufacturing of purified zirconium and hafnium metals is a complicated process (Fig. 1). Extraction is accomplished by mixing the zircon with carbon in the form of coke; the mixture is then chlorinated to produce a high-temperature gas stream containing silicon tetrachloride in addition to zirconium and hafnium tetrachlorides. The zirconium-hafnium tetrachloride is selectively condensed, and the silicon tetrachloride is collected and sold as a by-product.

To produce the industrial grades of zirconium, the crude zirconium-hafnium tetrachloride can be sublimated and reduced to the metal with magnesium. Zirconium for nuclear applications must be separated from hafnium. This is accomplished by one of two processing methods, a liquid-liquid separation process or a distillation process.

Liquid-Liquid Separation Process (Ref 1)

The crude zirconium-hafnium tetrachloride ($ZrCl_4$, $HfCl_4$) is dissolved in water and hydrochloric acid, the zirconium cores are complexed with ammonium thiocyanate (NH_4CNS), and the hafnium is extracted by methyl isobutyl ketone (MIBK). The hafnium content is reduced after many extraction phases to less than 100 ppm. The hafnium is stripped from the solvent and leaves the system in a separate stream from the zirconium stream. The two streams are treated in similar fashion to produce the pure metals (Fig. 1 from step 9 forward).

The aqueous zirconyl chloride is treated first with sulfuric acid and then heated to precipitate basic zirconium sulfate. The basic zirconium sulfate is recovered by filtration and treated with ammonia solution to produce zirconium hydroxide, which is calcined to produce a finely powdered zirconium oxide. The oxide is blended with carbon and then chlorinated to hafnium-free zirconium tetrachloride. This product is further purified by sublimation; it is then reduced to the metal with magnesium by the Kroll process. The resulting porous metal is called sponge metal.

Distillation Separation Process

In the distillation separation process (Ref 2), the crude product obtained after chlorinating the zircon sand is purified by sublimation. The pure tetrachloride vapors are continuously fed into the middle of a separating column containing a solvent made of melted KCl-$AlCl_3$ that is flowing from top to bottom. The chlorides are fed at a rate of about 900 kg/h (2000 lb/h), and the column is maintained as consistently as possible to a temperature of about 350 °C (660 °F) (Fig. 2).

The vapors of the zirconium tetrachloride, generated in a reboiler at about 500 °C (930 °F), rise in a counterflow against the descending solution of potassium chloroaluminate saturated with zirconium and hafnium chlorides. The rising vapor flow is progressively enriched with hafnium chloride, while the stream of liquid going downward progressively loses its hafnium chloride.

The solvent leaving the reboiler at a temperature of about 500 °C (930 °F) still contains a small percentage of zirconium chloride. This remaining content is completely removed with a stripping column, and the solvent can be recycled to the top of the column through the absorber-condenser. The zirconium tetrachloride stripped in the nitrogen stream (flowing at a rate of 50 m^3/h, or 1800 ft^3/h) is cooled and con-

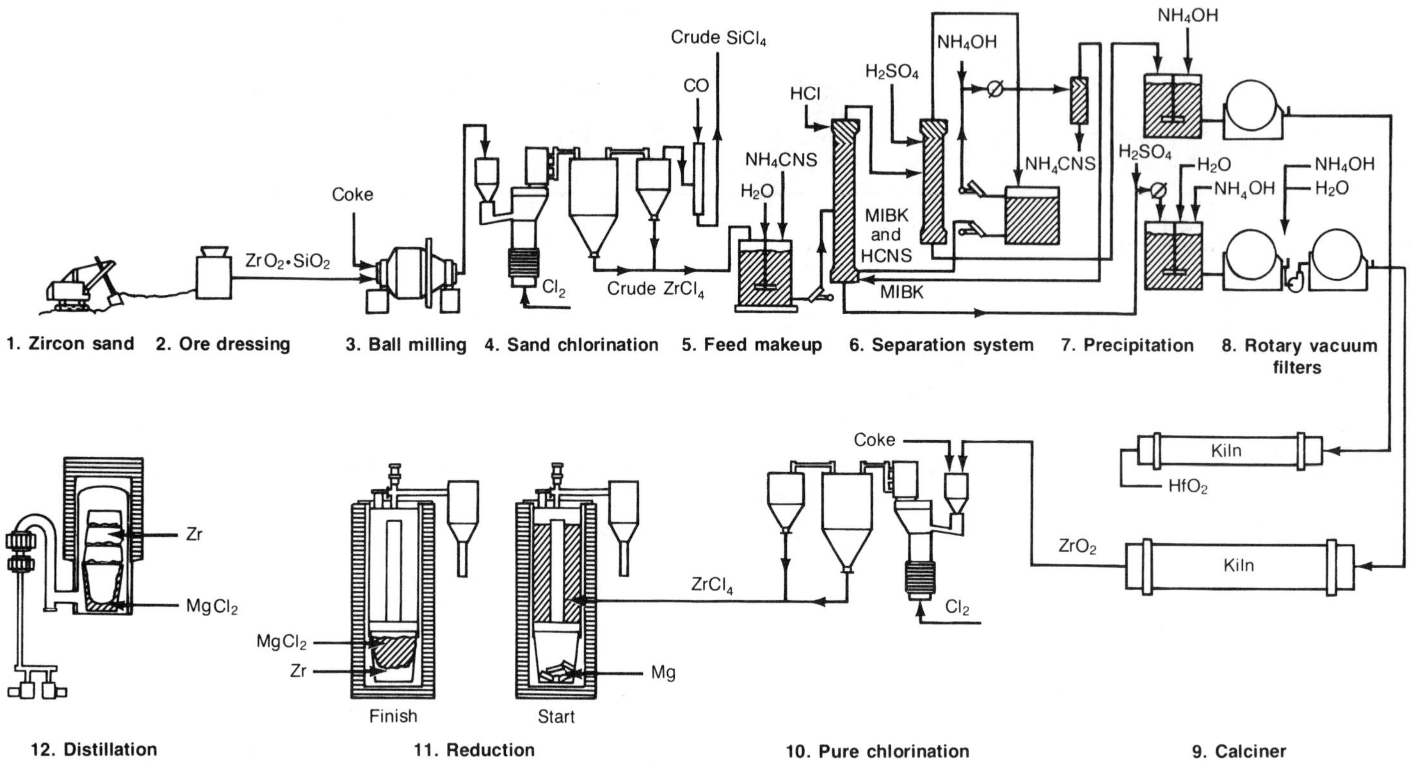

Fig. 1 Flow diagram of the 12 steps required to extract and separate zirconium and hafnium from raw material zircon ore

densed. This tetrachloride is used directly in the Kroll reduction.

The solvent, stripped of the zirconium tetrachloride and pumped to the top of the column, is fed to an absorber-condenser at a temperature of about 350 °C (660 °F). The solvent dissolves many of the vapors coming out of the column. A small amount of these vapors, enriched in hafnium tetrachloride (30 to 50% $HfCl_4$ content), is not dissolved but instead is condensed by cooling (Fig. 2). This small amount of product has to be reprocessed with the same equipment, but with distinct adjustments, in order to obtain pure hafnium tetrachloride.

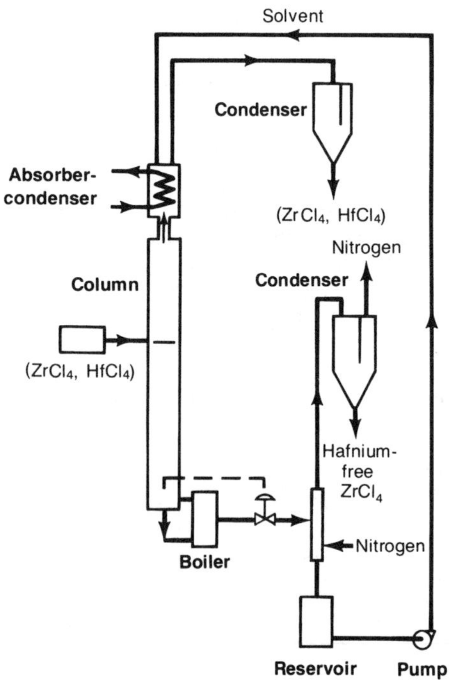

Fig. 2 Flow diagram for hafnium extraction by distillation from molten salt

Refining

Hafnium sponge metal contains too much oxygen to be ductile and consequently needs to be further refined. For a few applications, zirconium also requires additional refining. The most commonly used method is the iodide decomposition process. In this process, the sponge metal is reacted with iodine in an evacuated and sealed vessel; the resulting metal iodide is decomposed on an electrically heated wire, forming large metal crystals. The other refining methods are electron beam melting, molten salt electrorefining, and zone refining.

Melting

Zirconium and hafnium are consolidated by electric arc melting in a vacuum or under an inert gas atmosphere using the consumable electrode, cold mold technique.

Zirconium. Zirconium sponge, recycled material from processing, and alloying elements are combined into an electrode for vacuum arc melting. Zirconium alloys are melted at least twice, with the first ingot becoming the electrode for the second melting. Ingots are about 760 mm (30 in.) in diameter and weigh about 6 tonnes (6.6 tons).

Hafnium ingots are usually melted from a mixture of crystal bar, sponge, and scrap selected to meet the required oxygen content. Hafnium ingots are melted in a similar fashion to that of zirconium but are smaller, generally 230 mm (9 in.) in diameter and 500 kg (1100 lb) in weight.

Primary Fabrication

Forging. Hot forging of zirconium ingots is performed starting at 1050 °C (1920 °F), with reheating if the temperature falls below 550 °C (1022 °F). The billets are usually solution annealed in the β-phase region

Table 1 Variation in allotropic transformation temperature with oxygen content for unalloyed zirconium

Temperature		Phases present at oxygen content of:		
°C	°F	1640 ppm	1370 ppm	970 ppm
955	1750	β	β	β
930	1710	α + β	β	β
925	1700	α + β	β	β
920	1690	α + β	β	β
915	1680	α	α + β	β
910	1670	α	α + β	β
905	1660	α	α + β	β
895	1640	α	α + β	α + β
890	1630	α	α	α + β
885	1625	α	α	α + β
865	1590	α	α	α
855	1575	α	α	α

Fig. 3 Zirconium cooled from 950 °C (1740 °F). Cooling transformation resulted in a Widmanstätten structure. 75×

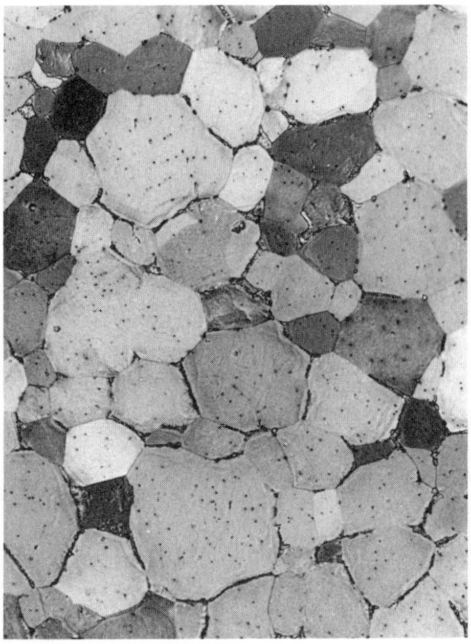

Fig. 4 Zirconium heated at 880 °C (1615 °F) showing impurities deposited in the grain boundaries. 450×

Fig. 5 Recrystallized α zirconium. 300×

above 1000 °C (1830 °F); solution annealing is followed by a water quench. This last step is performed to homogenize the composition and to improve the ductility and corrosion properties of the zirconium wrought products. Hafnium is forged at temperatures between 1100 and 650 °C (2010 and 1200 °F), with frequent reheats and light reductions.

Hot Rolling. Zirconium and hafnium are hot rolled in the 800 to 550 °C (1470 to 1020 °F) temperature range. Hot-rolled products are annealed at 760 °C (1400 °F).

When heated above 650 °C (1200 °F), zirconium and hafnium form a very refractory white or tan oxide that overlays a black, tenacious adherent oxide layer. These oxide layers are mechanically removed by blasting with metal or abrasive grit, followed by pickling with a solution containing 30% nitric acid and 3 to 5% hydrofluoric acid.

Cold rolling requires strong stiff mills to obtain good reductions and maintain good shape. Lubrication is required to produce a smooth surface finish and to avoid the pickup of the metals on the mill rolls. Reductions vary widely with grade, thickness, and mill configuration. Reductions between anneals are generally 25 to 75% for zirconium and slightly less for its alloys. Hafnium, unless very low in oxygen, can be reduced only 15 to 25% between annealing heat treatments.

Annealing is usually done in a vacuum to preserve the cold-rolled surface. Argon or helium atmospheres can also be used. Annealing is conducted at a temperature of about 705 °C (1300 °F) for several hours to ensure recrystallization. The strip must be clean prior to annealing because rolling lubricants and most soils will break down and cause discoloration and contamination of the metal during heat treatment.

Extrusion of Tubing. Zirconium is extruded at temperatures ranging from 800 to 675 °C (1470 to 1245 °F). Hafnium is extruded at temperatures above 960 °C (1760 °F).

Most tubing is seamless and made from extruded hollows. Bare zirconium will adhere to extrusion tooling, running along both the tools and the extrusion when conventional lubricants are used. Zirconium is currently extruded using either a metal cladding, a proprietary solid film lubricant, or the Ugine-Sejournet glass lubrication technique. Extrusion conditions and ratios vary widely for producing tube hollows. Some large pipe sizes are extruded to size, but most tubes are cold reduced from larger extrusions.

The primary method for the cold reduction of zirconium tubing is pilgering, also called rocking or tube reducing. The most common machines employ a set of reciprocating grooved rolls and a tapered mandrel. As the larger tube is fed into this device, it is rolled in small increments to the smaller size. Very large reductions of 70% or more can be made in a single pass. Plug and mandrel die drawing are also possible; however, these processes are no longer in wide use because of lubrication problems and because they produce only small reductions of about 15% per step.

The manufacture of zirconium tubing for nuclear fuel cladding requires careful manipulation of the cold reductions so that the correct crystallographic texture or preferred orientation is developed. It is desirable to have high ratios of wall thickness reduction to tube diameter reduction. This ratio is called the *Q*-factor.

Rod and Wire. Hafnium is not fabricated into tubing. Hafnium and zirconium are hot swaged at 800 to 550 °C (1470 to 1020 °F) and cold drawn to rod and wire.

Casting. Zirconium castings for valves, pumps, and other parts are cast in both rammed and machined graphite molds or in

Table 2 Thermal neutron capture cross section of various materials

Element or alloy	Thermal neutron cross section, b(a)
Beryllium	0.009
Magnesium	0.059
Lead	0.17
Zirconium	0.18
Zircaloy-4	0.22
Aluminum	0.22
Tin	0.65
Niobium	1.1
Iron	2.4
Molybdenum	2.4
300 series stainless steel	3.1
Nickel	4.5
Titanium	5.6
Hafnium	113
Cadmium	2.4×10^3
Gadolinium	44×10^3

(a) b, barns

molds made by the investment technique. The metal is usually melted by consumable arc skull-melting methods. Additional information is available in the article "Zirconium and Zirconium Alloys" in Volume 15 of the 9th Edition of *Metals Handbook*.

Secondary Fabrication

Machining. Three basic machining parameters should be used for all operations involving zirconium and hafnium alloys:

- Slow speeds
- Heavy feeds
- A flood coolant system using a water-soluble oil lubricant

Zirconium and hafnium exhibit a marked tendency to gall and work harden. Therefore, higher-than-normal clearance angles on tools are needed to penetrate the previously work-hardened surface and cut a clean, coarse chip.

Good results can be obtained with both cemented-carbide and high-speed tools. However, cemented carbide usually provides a better finish and higher productivity. Zirconium and hafnium alloys machine to an excellent finish, and the operation requires relatively light horsepower compared with that for alloy steel. Fine chips should not be allowed to accumulate on or near the machining equipment because they can be easily ignited. The chips should be continually removed and stored, preferably under water in remote and isolated areas that are far removed from the production site. Additional information is available in the article "Machining of Reactive Metals" in Volume 16 of the 9th Edition of *Metals Handbook*.

Grinding. The grinding methods used for zirconium and hafnium involve standard grinding machine equipment. The grinding characteristics of zirconium and hafnium alloys are similar to those of other metals, and both wheel and belt grinding can be used. The use of straight grinding oil or oil coolant produces a better finish and higher yields; these substances also prevent ignition of dry grinding swarf. Conventional grinding speeds and feeds can be used. Both silicon carbide and aluminum oxide can be used as abrasives, but silicon carbide generally gives better results. Additional information is available in the article "Machining of Reactive Metals" in Volume 16 of the 9th Edition of *Metals Handbook*.

Tube Bending. The same techniques and equipment used to cold form stainless steels are also used on zirconium tube. Springback caused by work-hardening behavior may be encountered, and provisions for this should be made for any bending operation.

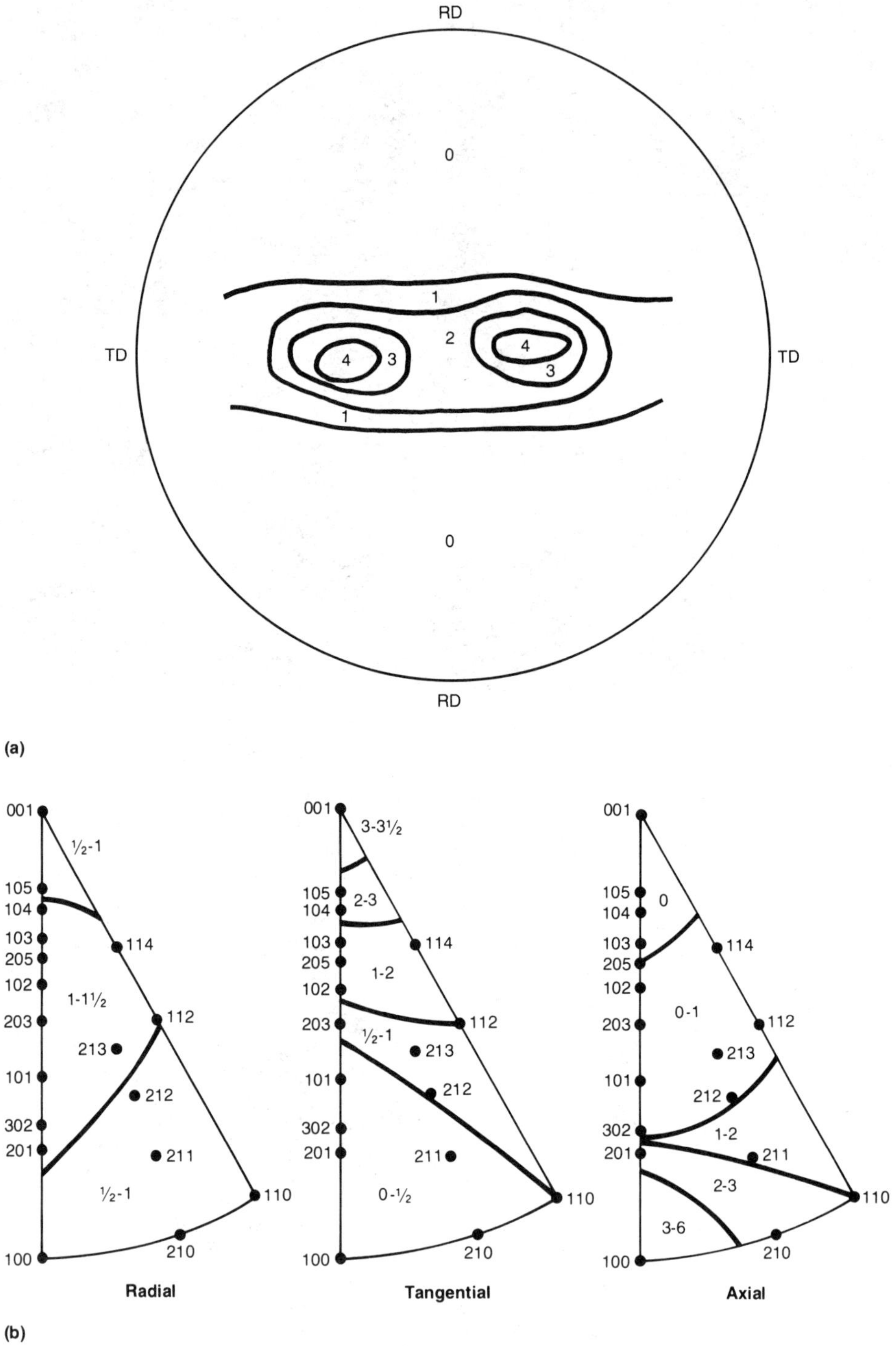

Fig. 6 Typical pole figures for showing anisotropy in zirconium. (a) Stereographic basal pole figure for hot-rolled Zircaloy-2 plate. Numbers indicate the relative densities of the poles in multiples of random occurrence. RD, rolling direction; TD, transverse direction. (b) Inverse pole figure for a typical Zircaloy-2 tubing sample

Table 3 Compositions and tensile properties for nuclear grades of zirconium

Designation		Nominal composition, wt%						Minimum tensile strength		Minimum 0.2% yield strength		Elongation in 50 mm (2 in.), %
Grade	UNS No.	Sn	Fe	Cr	Ni	Nb	O(a)	MPa	ksi	MPa	ksi	
Unalloyed reactor grade	R60001	...	...	...	...	...	0.8	290	42	138	20	25
Zircaloy-2	R60802	1.4	0.1	0.1	0.05	...	0.12	413	60	241	35	20
Zircaloy-4	R60804	1.4	0.2	0.1	...	...	0.12	413	60	241	35	20
Zr-2.5Nb	R60901	...	...	...	...	2.6	0.14	448	65	310	45	20

(a) Typical content

For cold forming, a minimum bend radius of approximately three times the outside diameter is advisable. Hot forming at temperatures from 200 to 425 °C (390 to 795 °F) or the use of special bending techniques is required for bends of smaller radius. To prevent buckling and wall thinning, both the inside and outside tube surfaces at the bend area must be in tension during any bending operation.

Drawing and Spinning. In spite of the work-hardening characteristics of zirconium, it has good hot and cold formability. Designs that eliminate severe or abrupt section changes and that allow generous radii are essential. Dies of nongalling material with tolerances and clearances comparable to those of the dies used for austenitic stainless steels should be employed. As in the case of tube bending, die design should allow for the springback tendency of the material.

Welding. Zirconium and hafnium have a better weldability than some more common construction materials, provided that the proper procedure is followed. Proper shielding from air with inert gases such as argon or helium is very important when welding these metals. Because of the reactivity of zirconium and hafnium to most gases at welding temperatures, welding without proper shielding will allow the absorption of oxygen, hydrogen, and nitrogen from the atmosphere and thus embrittle the weld.

Zirconium and hafnium are most commonly welded by the gas tungsten arc welding (GTAW) technique. Other welding methods used for these materials include gas metal arc welding (GMAW), plasma arc welding, electron beam welding, and resistance welding.

Zirconium and hafnium have low coefficients of thermal expansion and thus experience little distortion during welding. Inclusions are not usually a problem in the welds because these metals have a high solubility for their own oxides, and because no fluxes are used in welding, flux entrapment is eliminated. Zirconium and hafnium both have a low modulus of elasticity; therefore, residual stresses are low in a finished weld. However, stress relief of these welds has been found to be beneficial. A stress-relief temperature of 550 °C (1020 °F) should be used for both zirconium and hafnium.

Zirconium and hafnium are subject to severe embrittlement by relatively minute amounts of impurities, especially nitrogen, oxygen, carbon, and hydrogen. They have a high affinity for these elements at the welding temperature. Because of this high affinity for gaseous elements, zirconium and hafnium must either be welded using arc welding processes with inert shielding gases, such as argon or helium, or be welded in a vacuum. Hafnium ductility is affected to a greater degree by oxygen absorption than is zirconium ductility.

Arc welding in an inert blanket using either a tungsten electrode or consumable electrodes of zirconium gives the best results. The weld puddle, the bead just behind the weld puddle, and the backside of the weld must all be protected from the atmosphere in some manner. The weld puddle and the bead just behind the weld puddle can be protected by secondary shielding, such as a trailing shield.

The most common techniques used for welding zirconium and hafnium are the inert gas GTAW and GMAW methods. This equipment can be set up and used in the manual or automatic welding modes. Alternating current or direct current can be used for gas tungsten arc welding. Straight polarity is preferred for welding with a consumable electrode filler wire because it results in a more stable arc. Additional information is available in the article "Arc Welding of Reactive Metals" in Volume 6 of the 9th Edition of *Metals Handbook.*

Metallurgy of Zirconium and Its Alloys

Zirconium and most of its alloys exhibit strong anisotropy because of two characteristics of the metal: Zirconium has a hexagonal close-packed (hcp) crystal structure at room temperature, and it undergoes allotropic transformation to a body-centered cubic (bcc) structure at about 870 °C (1600 °F). The strong anisotropy profoundly influences the engineering properties of zirconium and its alloys and must be taken into account when selecting and processing a zirconium metal. The most common alloys are rather dilute α alloys with characteristics generally similar to those of unalloyed zirconium.

The Allotropic Transformation

In zirconium, the low-temperature α phase has a hcp crystal structure. This phase transforms to a bcc structure at about 870 °C (1600 °F). Small amounts of impurities, particularly oxygen, strongly affect the transformation temperature (Table 1).

The transformation on cooling generally results in a Widmanstätten structure of α zirconium; β phase cannot be retained even by rapid quenching (Fig. 3). The more rapid the cooling rate, the finer the platelets of the Widmanstätten structure. The phase stability of zirconium is influenced by α- and β-stabilizing elements and by low-solubility intermetallic compound formers.

Alpha-stabilizing elements raise the temperature of the allotropic α-to-β transformation. These elements include aluminum, antimony, tin, beryllium, lead, hafnium, nitrogen, oxygen, and cadmium. Phase diagrams for many of the binary alloy systems formed between these various elements and zirconium exhibit a peritectic or a peritectoid reaction at the zirconium-rich end.

Beta-stabilizing elements lower the α-to-β transformation temperature. Typical β stabilizers include iron, chromium, nickel, molybdenum, copper, niobium, tantalum, vanadium, thorium, uranium, tungsten, titanium, manganese, cobalt, and silver. For binary alloy systems between zirconium and these elements, there usually is a eutectoid reaction, and often a eutectic reaction as well, at the zirconium-rich end of the phase diagram.

Table 4 Physical properties of hafnium

Property	Quantity or crystalline structure
Atomic number	72
Atomic weight	178.5
Density, g/cm^3 (lb/in.^{3})	13.09 (0.47)
Melting point, °C (°F)	2222 (4032)
Boiling point, °C (°F)	3100 (5612)
Allotropic transformation, °C (°F)	1760 (3200)
Crystal structure	
α phase	hcp (<1760 °C, or 3200 °F)
β phase	bcc (>1760 °C, or 3200 °F)
Coefficient of linear thermal expansion per °C (°F) × 10^{-6}	5.9 (10.6)
Thermal conductivity (at 50 °C, or 120 °F), W/m · K (Btu · in./ft^2 · h · °F)	22.3 (155)
Specific heat, J/kg · K (cal$_{IT}$/g · K)	145 (0.035)
Electrical resistivity, μΩ · cm	35.1
Temperature coefficient of resistivity per °C × 10^{-3} at 20 °C (70 °F)	4.4

Table 5 Typical mechanical properties for fully annealed hafnium products

Product form	Test temperature, °C (°F)(a)	Test direction	Ultimate tensile strength, MPa	ksi	Yield strength at 0.02% offset, MPa	ksi	Elongation in 50 mm (2 in.), %
Rod	RT	Longitudinal	485	70	240	35	25
	315 (600)	Longitudinal	310	45	125	18	40
Plate	RT	Longitudinal	470	68	195	28	25
	RT	Transverse	450	65	310	45	25
	315 (600)	Longitudinal	275	40	125	18	45
	315 (600)	Transverse	235	34	165	24	48
Strip	RT	Longitudinal	450	65	170	25	30
	RT	Transverse	450	65	275	40	30
	315 (600)	Longitudinal	275	40	95	14	45
	315 (600)	Transverse	240	35	165	24	50

(a) RT, room temperature

Low-solubility intermetallic compound formers such as carbon, silicon, and phosphorus have very low solubility in zirconium, even at temperatures in excess of 1000 °C (1830 °F). They readily form stable intermetallic compounds that are relatively insensitive to heat treatment.

Impurities such as iron and chromium are soluble in β zirconium but relatively insoluble in α zirconium, where they exist primarily as intermetallic compounds. The size and distribution of these secondary phases are largely governed by reactions that take place during the last transformation from β to α and by subsequent mechanical working at lower temperatures.

Heating at temperatures near the α-β transition, or in the α + β region, causes migration of many impurities to grain boundaries. This migration impairs ductility and corrosion resistance, particularly in zirconium alloys (Fig. 4). However, it is beneficial in some instances for unalloyed zirconium.

Cold Work and Recrystallization

The degree to which unalloyed zirconium can be cold worked depends both on metal purity and on the method of reduction. Zirconium work hardens rapidly, reaching maximum hardness and strength after cold reduction of only about 20%. However, reductions of about 50% are common during cold rolling, and reductions of 80% can be accomplished in some instances. Reductions of 90% or more can be obtained by starting with very soft metal and using machines that feature multiaxial loading (such as cold Pilger machines or Sendzimir rolling mills). Reductions during cold drawing are generally about 15 to 30%. Initially, deformation results in twinning, which reorients the lattice for slip; slip is the primary mechanism for cold working.

Recrystallization is a function of the amount of cold work, temperature, and time, with time playing a relatively small role. In heavily cold-worked material, recrystallization commences at about 510 °C (950 °F). Process annealing of such material is usually conducted at 620 to 790 °C (1150 to 1450 °F). Recrystallization will occur in times as short as 15 min, but much longer times are normally used to ensure that the entire furnace load reaches temperature. Grain growth is nearly nonexistent at the usual annealing temperatures; times of 100 h or more are required to produce grain growth of 2 to 3 ASTM sizes (Fig. 5).

Large grains can be grown by annealing after cold reduction of about 5 to 8%. The most common source of large grains is

Table 6 Typical properties and mechanical properties of zirconium alloys

Property	Reactor grade and grade 702	Zr-2.5Nb, grade 705, and grade 706	Zircaloy-2, Zircaloy-4, and grade 704
Physical			
Density at 20 °C (70 °F), g/cm³	6.50	6.44	6.56
Crystal structure			
α-phase	hcp (<865 °C, or 1590 °F)	...	hcp (<865 °C, or 1590 °F)
β-phase	bcc (>865 °C, or 1590 °F)	bcc (>854 °C, or 1569 °F)	bcc (>865 °C, or 1590 °F)
(α + β) phase	...	hcp + bcc (954 °C, or 1569 °F)	...
Melting point, °C (°F)	1852 (3365)	1840 (3344)	1850 (3362)
Boiling point, °C (°F)	4377 (7910)	4380 (7916)	4375 (7907)
Coefficient of thermal expansion per °C (°F) × 10^{-6} at 25 °C (75 °F)	5.89 (10.6)	6.3 (11.3)	6.0 (10.8)
Thermal conductivity at 300–800 K, W/m · K (Btu · ft/h · ft² · °F)	22 (13)	17.1 (10)	21.5 (12.7)
Specific heat, J/kg · K (cal_{IT}/g · K)	285 (0.068)	285 (0.068)	285 (0.068)
Vapor pressure, kPa (mm Hg)			
At 2000 °C (3630 °F)	1.3 × 10^{-3} (0.01)	...	...
At 3600 °C (6510 °F)	120 (900)	...	...
Electrical resistivity, μΩ · cm at 20 °C (70 °F)	39.7	55.0	74.0
Temperature coefficient of resistivity per °C at 20 °C (68 °F)	0.0044	...	...
Latent heat of fusion, kJ/kg (cal_{IT}/g)	250 (60.4)	...	...
Latent heat of vaporization, MJ/kg (cal_{IT}/g)	6.49 (1550)	...	...
Mechanical			
Modulus of elasticity, GPa (10^6 psi)	99.3 (14.4)	97.9 (14.2)	99.3 (14.4)
Shear modulus, GPa (10^6 psi)	36.2 (5.25)	34.5 (5.0)	36.2 (5.25)
Poisson's ratio at ambient temperature	0.35	0.33	0.37

Table 7 Compositions and mechanical properties of industrial alloy grades of zirconium

Designation: Grade	UNS number	Nominal composition, wt%: Zr + Hf (min)	Hf (max)	Fe + Cr	Sn	O	Minimum tensile strength, MPa	ksi	Minimum 0.2% yield strength, MPa	ksi	Elongation in 50 mm (2 in.), %
702	R60702	99.2	4.5	0.2	...	0.16	379	55	207	30	16
704	R60704	97.5	4.5	0.3	1.5	0.18	413	60	241	35	14
705	R60705	95.5	4.5	0.2	...	0.18	552	80	379	55	16
706	R60706	95.5	4.5	0.2	...	0.16	510	74	345	50	20

reannealing after a straightening or forming operation that imparts only a small amount of cold work.

Anisotropy and Preferred Orientation

The relationships among the preferred orientation of zirconium crystal structure, the working practice that caused it, and the properties that result from it are complex and have been studied in great detail. For most engineering applications, it is important to understand that wrought forms of zirconium and its alloys have different tensile properties in the rolling direction (or longitudinal direction) than they have in the transverse direction. Yield strength is higher in the transverse direction; tensile strength is slightly higher in the rolling direction.

When the crystallographic texture of zirconium is discussed, it is usually presented in the form of a stereographic basal pole figure of the type shown in Fig. 6(a), or as an inverse pole figure of the type shown in Fig. 6(b).

The Role of Oxygen

Oxygen was originally considered a troublesome impurity in zirconium, and considerable effort was devoted to its elimination. But when oxygen levels were finally reduced below 1000 ppm, it was found that required strength levels in Zircaloy could no longer be met. The status of oxygen then changed to one of a controlled solid-solution alloying agent. Early methods for determining oxygen content were crude and relatively imprecise, so hardness (which is roughly related to oxygen content but much easier to measure) became the controlling attribute and is still widely used to express the purity or grade of both unalloyed zirconium and zirconium alloys.

The oxygen content of Kroll process sponge varies from about 500 to 2000 ppm, depending on the number of purification steps and the effectiveness of each step. A hardness of 125 HB indicates soft sponge with an oxygen content of about 800 ppm, whereas 165 HB is considered hard and indicates an oxygen content of about 1600 ppm. Crystal bar zirconium generally has a hardness level of below 100 HB and contains less than 100 ppm oxygen.

Oxygen is a potent strengthener at room temperature, but much of its effectiveness is lost at elevated temperature.

Zirconium Alloys

The most common zirconium alloys, Zircaloy-2 and Zircaloy-4, contain the strong α stabilizers tin and oxygen, plus the β stabilizers iron, chromium, and nickel. There is an extensive $\alpha + \beta$ field from about 790 to 1010 °C (1450 to 1850 °F). Iron, chromium, and nickel form intermetallic compounds, and the distribution of these compound phases is critical to the corrosion resistance of the alloys in steam and hot water. These alloys are generally forged in the β region, then solution treated at about 1065 °C (1950 °F) and water quenched. Subsequent hot working and heat treating is done in the α region (below 790 °C, or 1450 °F) to preserve the fine, uniform distribution of intermetallic compounds that results from solution treating and quenching.

Except for being somewhat stronger and less ductile than unalloyed grades, the Zircaloys are quite similar to unalloyed zirconium in metallurgical behavior.

The only other zirconium alloy that has significant commercial importance is Zr-2.5Nb. In zirconium, niobium is a mild β stabilizer; eutectoid reaction is induced when the niobium content exceeds about 1%. (The eutectoid point occurs at 20% Nb.) The mechanical and physical properties of Zr-2.5Nb are very similar to those of the Zircaloys; however, its corrosion resistance is slightly inferior to that of the Zircaloys.

Applications According to Alloy Grades

Nuclear Applications. A high corrosion resistance to high-temperature water and steam in combination with a transparency to thermal energy neutrons (Table 2) make zirconium a highly desirable material for uranium nuclear fuel cladding and other reactor internal structures. Zirconium alloys are used in pressurized-water reactors and boiling-water reactors in the United States, and in Canadian deuterium uranium (CANDU) reactors.

The high corrosion resistance of zirconium alloys results from the natural formation of a dense stable oxide on the surface of the metal. This film is self healing, it continues to grow slowly at temperatures up to approximately 550 °C (1020 °F), and it remains tightly adherent. Additional information is available in the section "Corrosion of Zircaloy-Clad LWR Fuel Rods" of the article "Corrosion in the Nuclear Power Industry" in Volume 13 of the 9th Edition of *Metals Handbook*. Hafnium has the same high corrosion resistance to high-temperature water and steam as does zirconium, but it also has a very high thermal neutron cross section (see Table 2) and consequently is used as a control rod material, mainly in naval reactors.

Zirconium Alloy Grades. Four reactor grades of zirconium and zirconium alloys are available:

- Reactor grade zirconium (UNS R60001) is highly purified unalloyed zirconium

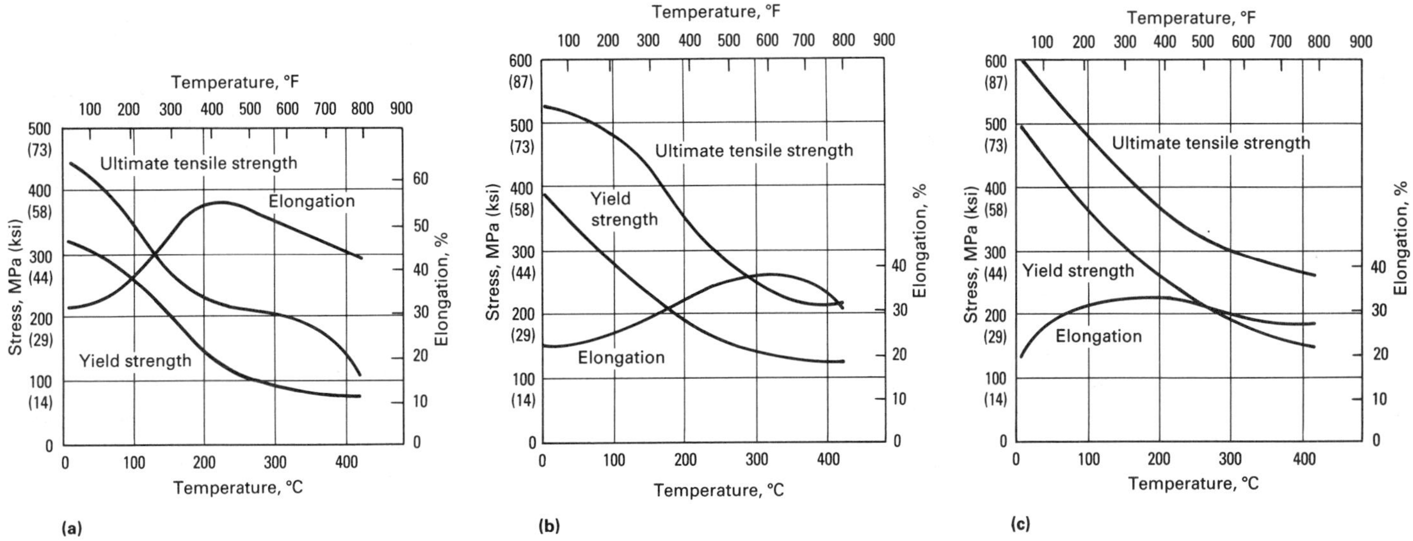

Fig. 7 Typical tensile properties of three industrial-grade zirconium alloys. (a) Grade 702. (b) Grade 704. (c) Grade 705

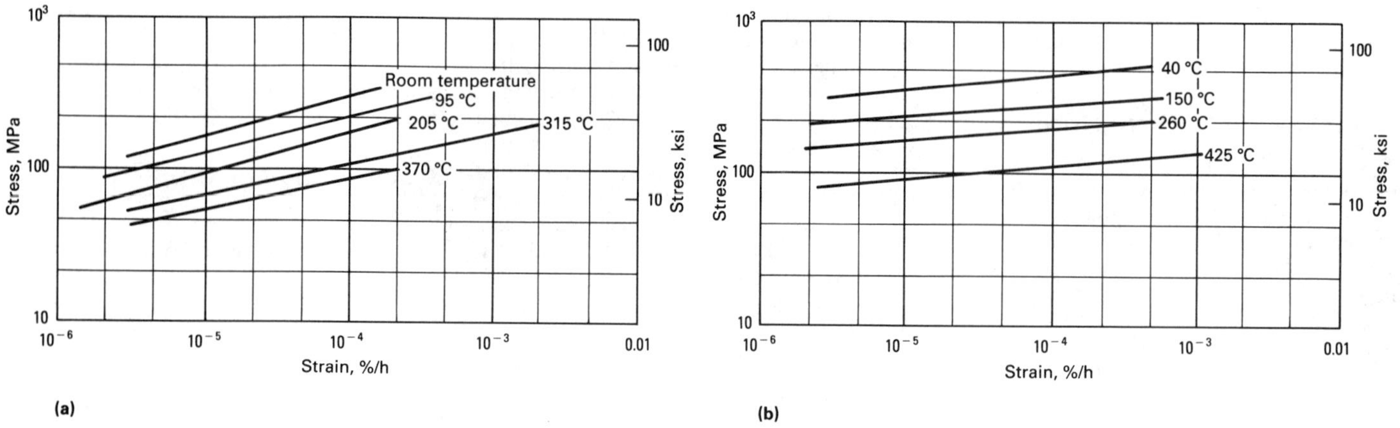

Fig. 8 Plot of minimum creep rate versus stress for two industrial-grade zirconium alloys. (a) Grade 702. (b) Grade 705

currently used as an inner lining of zirconium alloy fuel cladding tubes. Its resistance to corrosion by hot water and steam is highly variable

- Zircaloy-2 (UNS R60802) is a zirconium-tin alloy with small amounts of iron, chrome, and nickel that is highly resistant to hot water and steam. It is widely used in nuclear reactor service
- Zircaloy-4 (UNS R60804) is a variation of Zircaloy-2, but it contains no nickel and has a higher, more closely controlled iron content. This alloy is also used in nuclear service; it absorbs less hydrogen than Zircaloy-2 when exposed to corrosion in water and steam
- Zr-2.5Nb (UNS R60901) has high corrosion resistance to superheated water, and it is heat treatable to a higher level of strength than that which can be achieved with the Zircaloys. It was developed for use in the pressure tubes of heavy water reactors (such as the CANDU reactors)

The compositions and mechanical properties of the nuclear alloys are summarized in Table 3.

Hafnium is used in control rods in the purified unalloyed form. Hafnium products contain approximately 250 ppm O_2 for fabrication ductility. Available product forms are plate, sheet, foil, rod, and wire. The ASTM specifications B 737, for both hot-rolled and cold-finished hafnium rod and wire, and B 776, for hafnium and hafnium alloy flat-rolled products, relate to these applications. The physical properties of hafnium are listed in Table 4; Table 5 gives the mechanical properties of hafnium products.

Industrial Applications. Generally, zirconium alloys are less resistant to chemical solutions than is unalloyed zirconium. Zirconium has excellent corrosion resistance to acids, alkalies, organic compounds, and salt solutions. The few media that will attack zirconium include hydrofluoric acid, ferric or cupric chloride, aqua regia, concentrated sulfuric acid, and moist chlorine gas (see the article "Corrosion of Zirconium and Hafnium" in Volume 13 of the 9th Edition of *Metals Handbook*).

Zirconium industrial alloys are the hafnium-containing commercial grades of zirconium. They are used in nonnuclear applications such as chemical processing equipment, but they are similar in properties to the nuclear-grade alloys (Table 6). The industrial alloy grades of zirconium are:

- Grade 702 (UNS R60702)—commercial pure zirconium that is similar to reactor-grade zirconium
- Grade 704 (UNS R60704)—a Zr-1.5Sn alloy that is similar to Zircaloy-2 and Zircaloy-4
- Grade 705 (UNS R60705)—Zr-2.5Nb, a

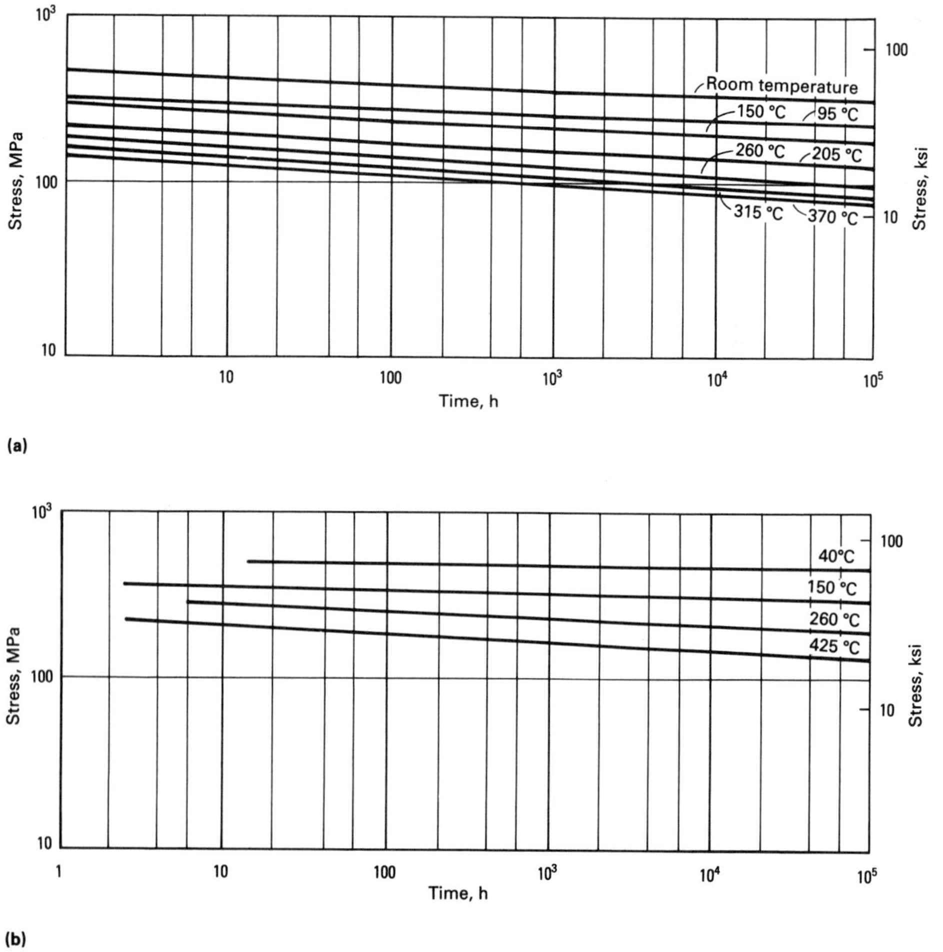

Fig. 9 Stress-rupture curves for two industrial-grade zirconium alloys. (a) Grade 702. (b) Grade 705

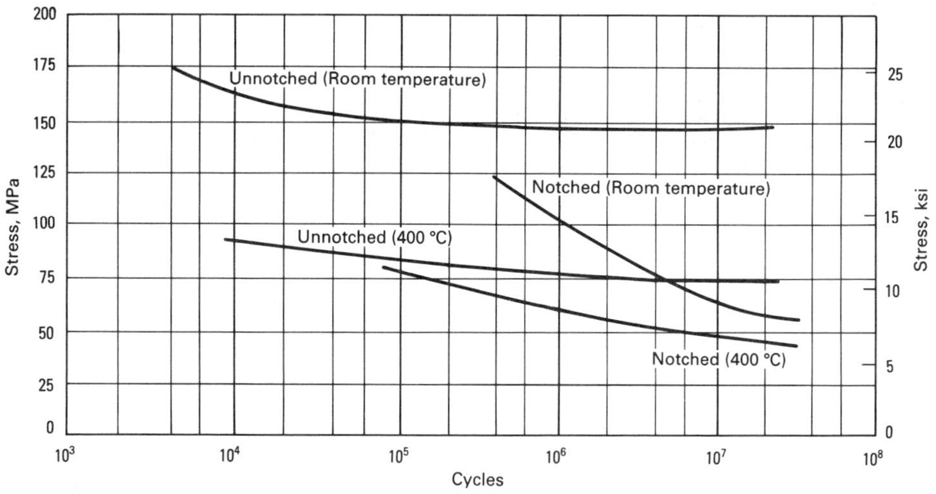

Fig. 10 Flexure fatigue curves for zirconium alloy grade 702

higher-strength alloy that is similar to nuclear-grade R60901

- Grade 706 (UNS R60706—an alloy that differs from grade 705 only in having lower oxygen content, lower tensile and yield strength properties, and greater elongation

Mechanical Properties. The chemical and mechanical properties of these alloys are summarized in Table 7. The ASTM specifications covering zirconium alloys for industrial applications include B 493 (forgings), B 523 (seamless and welded tubing), B 550 (bar and rod), B 551 (flat-rolled products), B 658 (pipe), B 653 (fittings), and B 752 (castings).

Typical tensile properties for grades 702, 704, and 705 are shown in Fig. 7. The minimum creep rates for grades 702 and 705 are shown in Fig. 8. The stress-rupture properties of grades 702 and 705 are shown in Fig. 9. Fatigue strength curves for grade 702 are shown in Fig. 10.

REFERENCES

1. B. Lustman and F. Kevse, Jr., Ed., *Metallurgy of Zirconium*, McGraw-Hill, 1955
2. L. Moulin, P. Thorvenin, and P. Brun, New Process for Zirconium and Hafnium Separation, in *Zirconium in the Nuclear Industry Sixth International Symposium*, STP 824, American Society for Testing and Materials, 1984

SELECTED REFERENCES

- J.H. Schemel, *Manual on Zirconium and Hafnium*, STP 639, American Society for Testing and Materials, 1977
- D.E. Thomas and E.T. Hayes, Ed., *The Metallurgy of Hafnium*, Naval Reactors, Division of Reactor Development, U.S. Atomic Energy Commission, 1960

Uranium and Uranium Alloys

K.H. Eckelmeyer, Sandia National Laboratories

URANIUM is a moderately strong and ductile metal that can be cast, formed, and welded by a variety of standard methods. It is used in non-nuclear applications primarily because of its very high density (19.1 g/cm^3, or 0.690 lb/in.^{3}; 68% greater than lead). Uranium is frequently selected over other very dense metals because it is easier to cast and/or fabricate than the refractory metal tungsten and much less costly than such precious metals as gold and platinum. Typical non-nuclear applications for uranium and uranium alloys include radiation shields, counterweights, and armor-piercing kinetic energy penetrators.

Natural uranium contains approximately 0.7% of the fissionable isotope U-235 and 99.3% U-238. Ore of this isotopic ratio is processed by mineral beneficiation and chemical procedures to produce uranium hexafluoride (UF_6). Isotopic separation is performed at this stage. This produces both enriched UF_6, which contains more than the natural isotopic abundance of U-235 and is subsequently processed and used for nuclear applications, and depleted UF_6, which typically contains 0.2% U-235. Access to enriched UF_6 is tightly controlled, but depleted material can be purchased for industrial applications. The UF_6 is reduced to uranium tetrafluoride (UF_4), commonly called green salt, by chemical reaction with hydrogen. The UF_4 is then reduced with magnesium or calcium in a closed vessel at elevated temperature, producing 150 to 500 kg (330 to 1100 lb) ingots of metallic uranium commonly referred to as derbies. These derbies are typically vacuum induction remelted and cast into the shapes required for engineering components or for subsequent mechanical working. Alloying elements can also be added during this melting step. Ingot breakdown and primary fabrication processes, such as forging, rolling, and extruding, can be readily carried out between 550 and 640 °C (1020 and 1180 °F) or between 800 and 900 °C (1470 and 1650 °F). The 650 to 780 °C (1200 to 1435 °F) range is avoided because cracking commonly occurs at these temperatures. Secondary fabrication processes such as rolling, swaging, and straightening are commonly done between room temperature and 500 °C (930 °F). Following heat treatment, machining to final dimensions can be carried out by most conventional cutting and grinding techniques, but uranium and its alloys are generally considered difficult to machine. Therefore, special tools and conditions must be applied.

This article presents an overview of the processing and properties of uranium and uranium alloys. Each section includes a few key references, but additional information can be found in previously published reviews (Ref 1-8) and reports dealing with specific alloys and/or topics referenced in these reviews.

Environmental, Safety, and Health Considerations

While depleted uranium can be melted, fabricated, and machined using conventional metallurgical practices, its mild radioactivity, chemical toxicity, and pyrophoricity require that special precautions be taken in its processing. This section gives a brief overview of the principal hazards and precautions associated with processing depleted uranium. More complete information is available in Ref 9 to 11, and information on health and safety considerations in processing other metals is available in the article "Toxicity of Metals" in this Volume. Organizations that process depleted uranium should have their facilities and procedures regularly reviewed and approved by an occupational health and safety organization for compliance with current guidelines and statutory regulations. Personnel and work areas should also be tested and inspected regularly.

Radioactivity. Depleted uranium is only mildly radioactive and is listed as a "low specific activity" (LSA) material in shipping regulations. The primary radiological hazards associated with this material are beta (β) and alpha (α) emission. The β-ray dose rate at the surface of a uranium slug is 0.23 rad/h. The dose rate of this modestly penetrating radiation decreases dramatically with distance from the source, due to absorption in the air and geometric effects. As a result, working near depleted uranium and normal handling of this material does not result in excessive exposures to β radiation. Nonetheless, significant β exposures could result from continuous very close contact, so unnecessary contact (such as carrying material in pockets) should be avoided. In addition, it has recently been shown that during melting and casting operations U-238 daughter products such as Th-234 and Pa-234 tend to float to the surface of the molten uranium and are left as residue in the crucible and on the surface of the casting. These daughter products have much higher activities and produce more energetic β rays. They have half-lives of 1.17 min (Pa-234) to 24.1 days (Th-234). Hence, they result in substantially increased β doses in the vicinities of recently melted materials and warrant greater precautions in areas where such materials are being handled.

Alpha radiation is also emitted by depleted uranium, but this nonpenetrating radiation is almost totally absorbed in 10 mm (0.4 in.) of air or in the 0.07 mm (0.003 in.) thick protective layer of skin and therefore presents no external health hazard. If finely divided particles become airborne and are inhaled, however, they can result in damaging α irradiation of delicate lung tissue. Hence, it is important to ensure that airborne uranium concentrations remain below the Occupational Safety and Health Administration standard of 0.25 mg/m^3 of air.

Toxicity. Depleted uranium is about as chemically toxic as other heavy metals, such as lead. Ingestion of excessive amounts of dust or fumes from these metals can cause problems such as kidney damage. Worker ingestion problems can best be avoided by careful control of any finely divided material and by good personal hygiene practices such as washing hands after handling uranium and avoiding eating in areas where uranium is being processed.

Pyrophoricity. Finely divided uranium is pyrophoric; therefore, machining chips and grinding residue must be handled carefully to avoid the danger of fire. Sparking frequently occurs during cutting and grinding

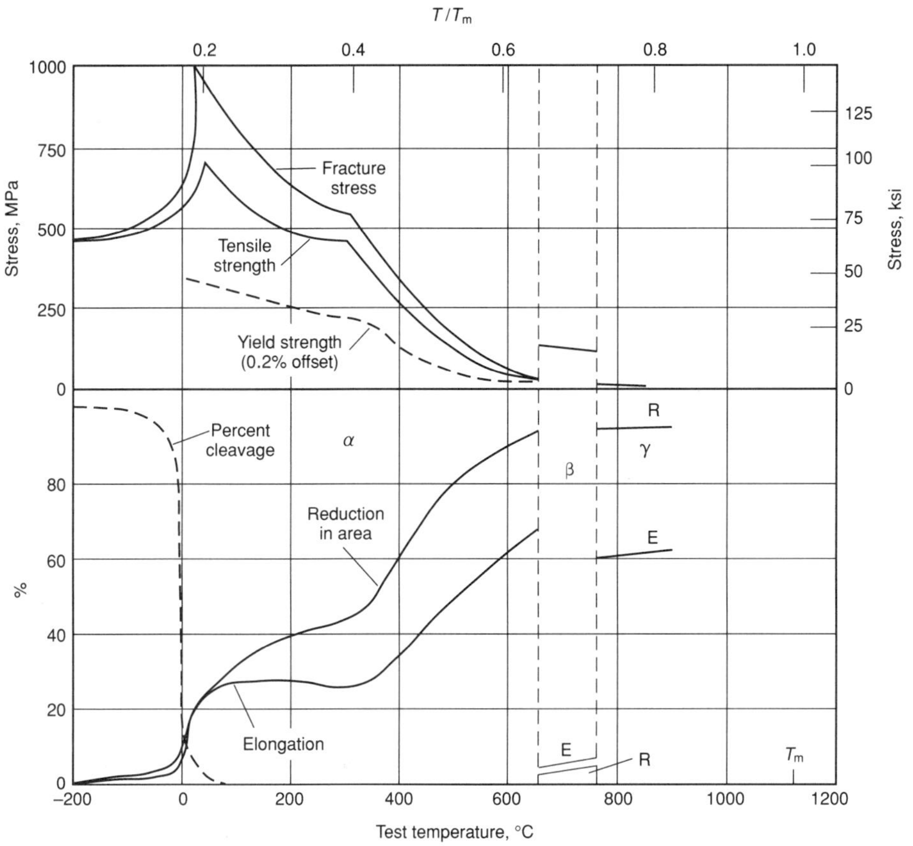

Fig. 1 Effect of temperature on the tensile properties of unalloyed uranium. T_m, melting temperature. Source: Ref 12

operations, and these sparks can ignite even coarse machining chips such as lathe turnings. More finely divided waste, such as grinding residue, can ignite spontaneously due to the heat produced by natural oxidation. These fire hazards can be minimized by using liberal amounts of machining fluid, keeping machining waste submerged in water or oil, removing chips from tools and work areas frequently, and avoiding mixed metal chips. In addition, dry powder fire extinguishers should be readily available in the event that dry chips ignite. Water should not be used on uranium fires because it reacts with the hot metal and generates hydrogen, adding to the combustion.

As can be seen from the previous paragraphs, the danger associated with the radioactivity, toxicity, and pyrophoricity of large pieces of depleted uranium is minimal, but it increases substantially as the material becomes more finely divided. As a result, operations that produce uranium fumes or finely divided particulates must be carefully vented and the exhausts passed through well-monitored filtering systems. In addition, work areas should be checked regularly for the presence of excessive levels of airborne material or residue on floors, furniture, and equipment. Workers should also change footwear and clothing when entering or leaving areas where finely divided residue is likely to be present, such as areas where uranium is being machined or where hot-worked uranium (with its powdery oxide surface) is handled. Finally, personnel who work with uranium should have their radiation exposures monitored by dosimeters and their ingestion levels checked by periodic urinalyses.

Because of uranium's radioactivity and toxicity, disposal of uranium residue and release of uranium-containing effluents are strictly regulated. Facilities that process uranium must be equipped with the appropriate air- and liquid-handling systems and must monitor their effluents regularly to ensure that regulatory standards are being met.

Processing and Properties of Unalloyed Uranium (Ref 1-8, 12-16)

Unalloyed uranium is typically induction melted and poured into molds to produce either engineering components or ingots for subsequent mechanical working. Uranium is a highly reactive metal, so all melting operations must be done in a vacuum. Uranium also reacts with many common crucible materials. To avoid such reactions, vacuum induction furnace crucibles and molds are usually made of graphite with zirconia or yttria washes applied to their interior surfaces to minimize carbon pickup by the molten metal. Unalloyed uranium castings are frequently used in applications such as counterweights and radiation shields, where mechanical properties are not primary design concerns.

Hot and cold fabrication techniques are frequently used to improve ductility and produce forms such as sheets, rods, and hollow cylinders of unalloyed uranium. Solid uranium exhibits three polymorphic forms: gamma (γ) phase (body-centered cubic) above 771 °C (1420 °F), β phase (tetragonal) between 665 and 771 °C (1229 and 1420 °F), and α phase (orthorhombic) below 665 °C (1229 °F). The effects of crystal structure and temperature on mechanical properties are shown in Fig. 1. The high-temperature γ phase is very soft and ductile, and dynamic recrystallization occurs when the metal is worked in this temperature range. Unfortunately, however, oxidation also occurs very rapidly at these temperatures. This results in poor surface quality and a substantial amount of finely divided oxide debris. Hence, relatively little γ-phase metalworking is done except in instances where severe deformation requirements and/or limitations in the tonnage of the available processing equipment demand maximum ductility and/or minimum flow stress. The intermediate-temperature β phase is brittle in polycrystalline form and is unsuitable for metalworking processes. The low-temperature α phase deforms by a number of slip modes, but slip becomes more restricted and increasing amounts of twinning occur as the temperature decreases. The combination of low flow stress, high ductility, and dynamic recrystallization make the high α region (550 to 640 °C; 1020 to 1180 °F) ideal for many primary metalworking operations, such as forging, rolling, and extrusion. The oxidation rate at this temperature is only one-third that in the γ region, and heating in preparation for hot working can be done in molten salt baths so the hot material is exposed to air for only a short time, resulting in substantial reductions in oxidation and improvements in surface quality. The material continues to be quite ductile and workable at temperatures down to 150 to 200 °C (300 to 390 °F), but the absence of recrystallization below about 500 °C (930 °F) results in substantial work hardening when secondary metalworking processes such as rolling and swaging are carried out in the 150 to 500 °C (300 to 930 °F) range. The α phase becomes less ductile below 100 to 150 °C (212 to 300 °F), but secondary fabrication processes such as swaging, drawing, spinning, and straightening can be done at room temperature if only modest deformations are required.

The mechanical properties (particularly ductility) of α uranium near room temperature vary substantially depending on pro-

Table 1 Effect of heat treatment on tensile properties of uranium

Form	Heat treatment	Yield strength, MPa (ksi)	Tensile strength, MPa (ksi)	Elongation, %
Cast	None	205 (30)	450 (65)	6
Cast	500 °C (930 °F), 1 h in salt	215 (31)	460 (66)	8
Cast	650 °C (1200 °F), 2 h in vacuum + 630 °C (1165 °F), 24 h in vacuum(a) (hydrogen outgassed)	185 (27)	560 (81)	13
Cast	720 °C (1330 °F) in vacuum quench, 550 °C (1020 °F), 24 h in vacuum(a) (β quenched and hydrogen outgassed)	295 (43)	700 (101)	22
600 °C (1110 °F) rolled	550 °C (1020 °F), 1 h in salt (α-rolled)	270 (39)	575 (83)	12
600 °C (1110 °F) rolled	630 °C (1165 °F), 2 h in vacuum(a) (α-rolled and hydrogen outgassed)	270 (39)	720 (103)	31
300 °C (570 °F) rolled	630 °C (1165 °F), 2 h in vacuum(a)(b) (warm rolled and hydrogen outgassed)	220 (32)	750 (109)	49

(a) Properties dependent on adequate time for hydrogen outgassing. (b) Several sequences of warm rolling and annealing; final anneal at 550 °C (1020 °F)

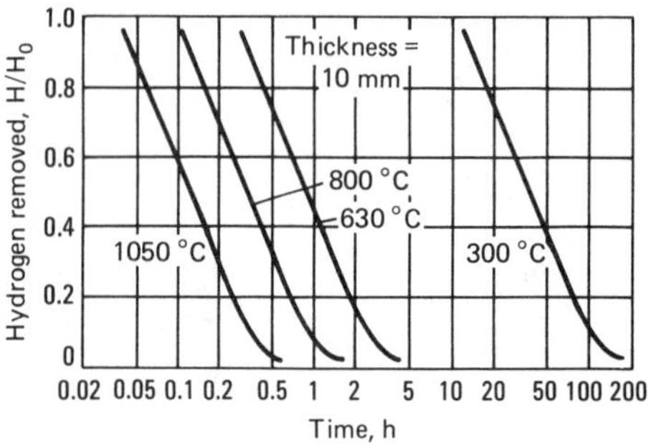

Fig. 2 Time required for hydrogen outgassing at the center of 10 mm thick plate at various temperatures. Times for hydrogen removal vary as thickness is squared; for example, the time required for 90% hydrogen removal (H/H_0 = 0.1) in 3.0 cm thick plate at 630 °C (1165 °F) is $3.0^2 \times 2.5$ h, or 22.5 h. Source: Ref 16

cessing history and impurity content (Table 1). This variability is largely the result of a ductile-to-brittle transition that occurs at about room temperature, as shown in Fig. 1. The ductile-to-brittle transition temperature is sensitive to various metallurgical variables, but in particular can be suppressed to lower temperatures (thus increasing room-temperature ductility) by decreasing hydrogen content and grain size and by leaving some residual warm work in the material. Thus, as-cast material, which typically contains a moderate amount of hydrogen (1 to 3 ppm, by weight) and exhibits very large grains, has low room-temperature ductility (~6% elongation). Ductility can be substantially improved by lowering hydrogen content via vacuum heat treatment (~1 Pa, or 10^{-2} torr) in the high α region (550 to 640 °C, or 1020 to 1180 °F). The time required for hydrogen outgassing depends on temperature and section thickness, as shown in Fig. 2. Hydrogen outgassing can be done more quickly in the γ region (~800 °C, or 1470 °F), but due to the higher solubility of hydrogen in the γ phase, a substantially better vacuum is required (~10 mPa, or 10^{-4} torr). Additional improvements in the ductilities of castings can be obtained by refining the grain size via β phase heat treatment and quenching.

Rolling or other forms of hot working carried out in the high α region (550 to 640 °C, or 1020 to 1180 °F) cause recrystallization to much finer grains than are typical of castings; thus wrought products typically have higher ductilities than do castings. Preheating in salt to the deformation temperature, however, frequently introduces hydrogen into the material, which limits the ductility gains imparted by working and grain size refinement. Vacuum heat treatment in the high α region, after rolling, removes this hydrogen, resulting in additional improvements in ductility. Optimum ductility can be obtained by warm rolling at about 300 °C (570 °F) with vacuum anneals at 630 °C (1165 °F) as required; after the final warm rolling increment a 550 °C (1020 °F) vacuum anneal is performed. This provides a very fine-grained low-hydrogen material with outstanding room-temperature ductility. Eliminating the final annealing step and leaving some warm work in the material decreases this ductility but suppresses the ductile-to-brittle transition temperature by about 40 °C (70 °F), thus resulting in increased low-temperature ductility.

Impurity elements generally have very low solubilities in uranium, so when present they contribute to increased densities of inclusions and second-phase particles. It is generally accepted that these conditions have adverse effects on ductility; however, the extent of many of these effects have not been quantified in careful experiments where the effects of other significant variables, such as hydrogen content, have been controlled or eliminated. Carbon is a particularly notorious impurity because it forms carbide inclusions, which clearly decrease ductility, and because it is introduced to a limited extent during melting in graphite crucibles despite oxide coatings that are applied to the crucibles to minimize this effect. This buildup of carbon is the most serious impediment to recycling uranium scrap. A typical specification for relatively high-purity material is shown below:

Element	Amount (by weight)
Uranium	99.85% min
Carbon	100 ppm max
Iron	75 ppm max
Magnesium	5 ppm max
Nickel	50 ppm max
Silicon	75 ppm max
Manganese	35 ppm max
Aluminum	15 ppm max
Calcium	50 ppm max

Unalloyed uranium is readily weldable provided that it is protected from atmospheric oxygen. Electron beam welding in a vacuum is the most commonly used method, but shielded arc welds have also been successful. Coarse grains in the weld region limit the strengths and ductilities of welded parts to those typical of castings.

The susceptibility of uranium to oxidation and corrosion must be seriously considered, particularly in applications of unalloyed uranium. Unalloyed uranium oxidizes rapidly in air, forming a dark coating of uranium dioxide after only a few hours of exposure. Exposure to water results in rapid corrosion, particularly if the solutions are somewhat acidic or contain even small amounts of chloride (Cl^-) ions. Protective coatings, such as electroplated nickel or ion-plated aluminum, are frequently applied to minimize such environmental effects.

Uranium Alloys

Uranium is frequently alloyed to improve its corrosion resistance and mechanical properties. Alloying results in substantial decreases in density, hence it is desirable to obtain the necessary properties with small amounts of alloying additions. These alloys are produced by vacuum induction or vacuum arc melting and, like unalloyed uranium, can be fabricated hot (gamma and high alpha regions: 800 to 900 °C, or 1470 to 1650 °F; and 500 to 640 °C, or 930 to 1180 °F), warm (150 to 500 °C, or 300 to 930 °F), or cold (room temperature). A wide range of properties can be obtained by post-fabrication heat treatment. More information on the properties and processing of uranium alloys is available in Ref 2 to 7 and 17.

Heat treatment of uranium alloys is based on the fact that common alloying elements (molybdenum, niobium, titanium, and zirconium) are highly soluble in the high-temperature phase but substantially less soluble in the lower-temperature β and α phases (Fig. 3). This results in eutectoid or eutectoid-like phase diagrams, such as that for the U-Ti system shown in Fig. 4. Uranium alloys are generally solution heat treated at approximately 800 °C (1470 °F) to put the alloying additions into solid solution in the γ phase, and then cooled at various rates to room temperature. Slow cooling permits the γ phase to diffusionally decompose, as illus-

Table 2 Mechanical properties of annealed uranium alloys

Alloy	Yield strength, MPa (ksi)	Tensile strength, MPa (ksi)	Elongation, %	Reduction in area, %
Unalloyed U	220 (32)	650 (94)(a)	13–50(a)	12–45(a)
U-0.75 Ti	520 (75)	1070 (155)	10	10
U-2.0 Mo	415 (60)	830 (120)(b)	10(b)	10(b)
U-2.3 Nb	480 (70)	965 (140)	25	31

Note: All with ~50 ppm C and <0.5 ppm hydrogen unless noted. (a) Varies substantially with prior processing and grain size. (b) 210 ppm C; UTS, elongation, and reduction in area are expected to be higher with decreased carbon.

trated in Fig. 5. The product of such diffusional decomposition is analogous to pearlite in steels and consists of alternating platelets of essentially alloy-free α uranium and alloy-enriched second phases. Such two-phase structures exhibit somewhat higher strengths than unalloyed uranium, but because they contain high fractions of alloy-free α-phase uranium, little or no improvement in corrosion resistance is obtained. The properties of several alloys in the slowly cooled condition are shown in Table 2. These types of materials are frequently used in applications such as large, thick-walled cylindrical radiation shields for nuclear material shipping casks, where strengths higher than those of unalloyed uranium are needed but where large size and wall thickness preclude quenching to obtain even higher strengths and ductilities.

Rapid quenching from the high-temperature γ-phase field suppresses diffusional decomposition, resulting in the formation of supersaturated metastable phases, as illustrated in Fig. 5. These phases generally exhibit better combinations of strength and ductility than does unalloyed uranium. In addition, their supersaturation makes them amenable to additional strengthening by age hardening. Finally, the presence of alloying elements in supersaturated solid solution substantially improves their corrosion resistance. The properties of several alloys in various quenched or quenched-and-aged conditions are shown in Table 3. Quenched and aged uranium alloys are commonly used in applications requiring good combinations of strength, ductility, and corrosion

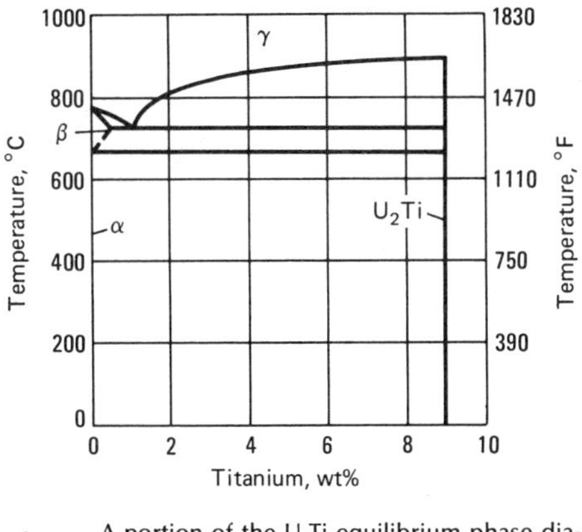

Fig. 4 A portion of the U-Ti equilibrium phase diagram. Source: Ref 17

resistance, such as kinetic energy penetrators.

Solution heat treatment (Ref 16-20) is generally done at about 800 °C (1470 °F) for 1 to 8 h in a vacuum. The temperature must be high enough to ensure that the material is completely transformed to γ phase and that the alloying elements are taken into solid solution. An inert environment is needed to keep the material from rapidly oxidizing at this high temperature. Vacuum heat treatment at about 1 mPa (10^{-4} torr) not only prevents oxidation but removes hydrogen from the material, thus avoiding internal hydrogen embrittlement. The length of the vacuum solution heat treatment is dictated primarily by hydrogen outgassing considerations.

Uranium alloys are most frequently used in either the as-quenched or quenched-and-aged condition. In either case, the material must be cooled from the solution treatment temperature rapidly enough to suppress diffusional decomposition of the γ phase and permit a nonequilibrium supersaturated structure to be obtained. Critical cooling rate varies substantially with alloy composition. Dilute alloys (those containing less than 2 to 3 wt% alloying addition) generally exhibit high critical cooling rates (in the vicinity of 50 to 150 °C/s, or 90 to 270 °F/s). These rates can only be obtained by water-

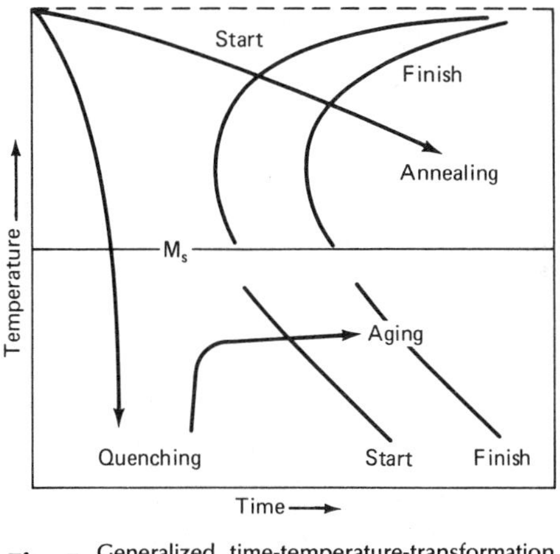

Fig. 5 Generalized time-temperature-transformation diagram showing heat treatments employed with uranium alloys. Slow cooling results in diffusional decomposition of γ phase to coarse dual-phase microstructures. Quenching results in diffusionless transformation of γ phase to supersaturated martensites, which can be subsequently age hardened. M_s, martensite start temperature

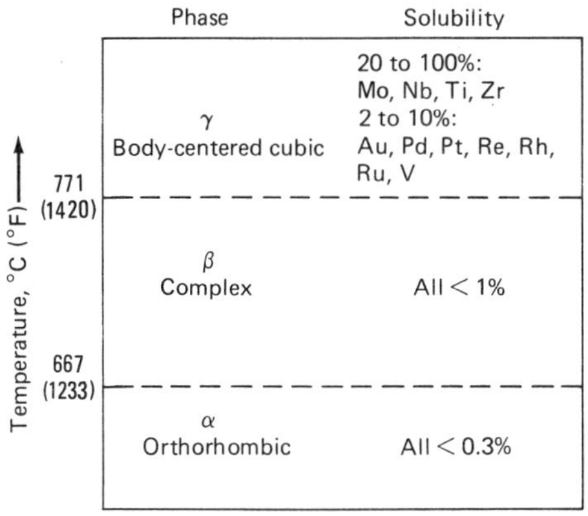

Fig. 3 Solubilities of alloying elements in γ, β, and α polymorphs of uranium. Source: Ref 17

quenching relatively thin sections; thus, quenched or quenched-and-aged dilute alloys cannot be used for heavy section applications. U-0.75Ti, for example, cannot be effectively quenched in sections thicker than about 25 mm (1 in.). More highly alloyed materials generally exhibit lower critical cooling rates (decreased quench rate sensitivity) and can be quenched in thicker sections and/or less severe quenching media, such as oil. Some very highly alloyed materials, such as U-10Mo, have such low critical cooling rates that they do not have to be quenched at all; cooling in an inert atmosphere is sufficient.

Subcritical quenching has a substantial influence on mechanical properties, age hardenability, and corrosion resistance. The effect of quench rate on the tensile properties of U-6Nb is shown in Fig. 6. Quench rates in excess of approximately 10 °C/s (18 °F/s) completely suppress diffusional decomposition and produce a supersaturated variant of the α phase, which is soft and ductile and which exhibits outstanding corrosion resistance. As the cooling rate decreases to below 10 °C/s (18 °F/s), diffusional decomposition occurs to a progressively greater extent. This results in increases in strength and decreases in ductility and corrosion resistance. At 0.2 °C/s (0.4 °F/s), complete diffusional decomposition occurs, resulting in very low ductility and poor corrosion resistance. Further decreases in cooling rate produce coarser two-phase microstructures with lower strengths and somewhat better ductilities but with no improvement in corrosion resistance. Similar trends are observed in other uranium alloys, but with different characteristic quench rates, depending on alloy composition.

Plunge quenching of rods greater than 20 mm (0.8 in.) in diameter into water causes radial heat extraction, which results in sufficiently high triaxial tensile stresses along the centerline and can cause internal fracturing or centerline bursting. This can be avoided by lowering the bar into the quench

Table 3 Properties and applications of heat-treated uranium alloys

Alloy	Density, g/cm^3	Processing	Hardness	Yield strength, MPa (ksi)	Tensile strength, MPa (ksi)	Elongation, %(a)	Reduction in area, %(a)	Corrosion resistance	Used for applications requiring	Precautions, comments
Ualloyed U	19.1	Cast, β-quenched, hydrogen outgassed	93 HRB	295 (43)	700 (101)	22	. . .	Poor	Complex shapes, low strength requirements	. . .
Unalloyed U	19.1	α-rolled, hydrogen outgassed	94 HRB	270 (39)	720 (104)	31	. . .	Poor	Sheet, rod, formed parts, low strength requirements	. . .
U-0.75Ti	18.6	γ-quenched	36 HRC	650 (94)	1310 (190)	31	52	Fair	Moderate strength, high ductility	Section thickness limitations, high residual stresses, low hydrogen required
U-0.75Ti	18.6	γ-quenched, aged 380 °C (715 °F), 6 h	42 HRC	965 (140)	1565 (227)	19	29	Fair	High strength, moderate ductility	Section thickness limitations, high residual stresses, low hydrogen required
U-0.75Ti	18.6	γ-quenched, aged 450 °C (840 °F), 6 h	52 HRC	1215 (176)	1660 (241)	<2	<2	Fair	Maximum hardness, low ductility	Low ductility and toughness, very sensitive to stress-corrosion cracking if stressed
U-2.0Mo	18.5	γ-quenched, aged 550 °C (1020 °F), 5 h	34 HRC	675 (98)	1100 (160)	23	25	Poor	Moderate strength, low residual stresses	. . .
U-2.3Nb	18.5	γ-quenched, aged 600 °C (1110 °F), 5 h	32 HRC	545 (79)	1060 (154)	28	33	Fair	Moderate strength, low residual stresses	. . .
U-4.5Nb	17.9	γ-quenched, aged 260 °C (500 °F), 16 h	42 HRC	900 (130)	1190 (173)	10	8	Good	High strength, moderate corrosion resistance	Reduced sensitivity to quench rate
U-6.0Nb	17.3	γ-quenched	82 HRB	160 (23)	825 (120)	31	34	Excellent	High corrosion resistance, high ductility	Low sensitivity to quench rate
U-10Mo	16.3	γ-quenched	28 HRC	900 (130)	930 (134)	9	30	Excellent	High corrosion resistance, high strength	Very low sensitivity to quench rate, very susceptible to stress-corrosion cracking
U-7.5Nb-2.5Zr	16.4	γ-quenched	20 HRC	540 (78)	850 (123)	23	50	Excellent	High corrosion resistance, moderate strength, high ductility	Very low sensitivity to quench rate, very susceptible to stress-corrosion cracking

(a) All based on high-purity alloys with low hydrogen contents

bath at a controlled rate, typically 5 to 10 mm/s (0.2 to 0.4 in./s). Such controlled quenching results in more longitudinal heat extraction, thus reducing stress triaxiality and preventing centerline bursting.

Microstructures and Properties of Quenched Alloys (Ref 6, 17, 21-23). Quenching from the γ-phase field produces a variety of metastable supersaturated phases whose microstructures, crystal structures, and properties vary in a regular manner with alloy content. These variations are illustrated in Fig. 7, and photomicrographs of the various structures are shown in Ref 6.

Relatively small alloying additions result in progressive increases in hardness and strength. Very dilute alloys exhibit irregular grain structures typical of those produced by massive transformation and similar to those of unalloyed uranium. Somewhat more concentrated alloys exhibit acicular martensitic microstructures. Both of these phases are orthorhombic, in which the lattice parameters (particularly the *b* parameter) are modified by the presence of alloying atoms in solid solution.

Further increases in alloy content cause a transition to a thermoelastic, or banded, martensite. The hardness and strength of thermoelastic martensites decrease with increasing alloy content, apparently due to increasing mobilities of the boundaries of the many fine twins produced during the transformation. Midway in the thermoelastic martensite composition range, the crystal structure changes from orthorhombic to monoclinic, as one lattice angle departs gradually from 90 deg. This change in crystal structure has little apparent effect on mechanical behavior. The martensitic variants of α-uranium are frequently termed α'_a, α'_b, and α''_b. The subscripts a and b denote the acicular and banded morphologies, respectively, and the prime and double prime superscripts denote the orthorhombic and monoclinic crystal structures, respectively.

Additional increases in alloy content produce a transition to γ°, a tetragonal variant of elevated-temperature γ uranium. The γ° phase exhibits an equiaxed grain structure similar to that of γ uranium, but differences in crystal structure can be detected by x-ray diffraction. The hardness and strength of γ° increase with increasing alloy content.

While the microstructures and resulting mechanical properties vary with composition in this complex manner, corrosion resistance is controlled primarily by the amount of alloying addition in solid solution. Hence, corrosion resistance increases progressively with increasing alloy content.

Age Hardening (Ref 24-28). The phases produced by quenching are supersaturated substitutional solid solutions; therefore, they can be further strengthened by age hardening. The as-quenched phases are relatively soft (hardnesses range from 92 HRB to 35 HRC, with yield strengths from 140 to 700 MPa, or 20 to 100 ksi) and ductile (15 to 32% tensile elongation). Aging increases their hardness and strength and decreases their ductility, as shown in Fig. 8 and 9. Age hardening occurs at temperatures below approximately 450 °C (840 °F) due to fine-scale microstructural changes observable only by transmission electron microscopy or other high-resolution techniques. Hardnesses in excess of 50 HRC and yield strengths over 1200 MPa (175 ksi) can be obtained by full aging, but most alloys exhibit little or no ductility in the fully aged condition. Most applications employ partially aged material in order to obtain an acceptable balance of strength and ductility.

Overaging occurs at temperatures in excess of approximately 450 °C (840 °F) by decomposition of the supersaturated metastable microconstituents to the equilibrium phases. This decomposition, which commonly takes place by cellular or discontinuous precipitation, can be easily revealed by optical metallography (Ref 6). Overaging decreases hardness and yield strength and increases ductility (relative to the fully aged condition), but tensile elongations in excess

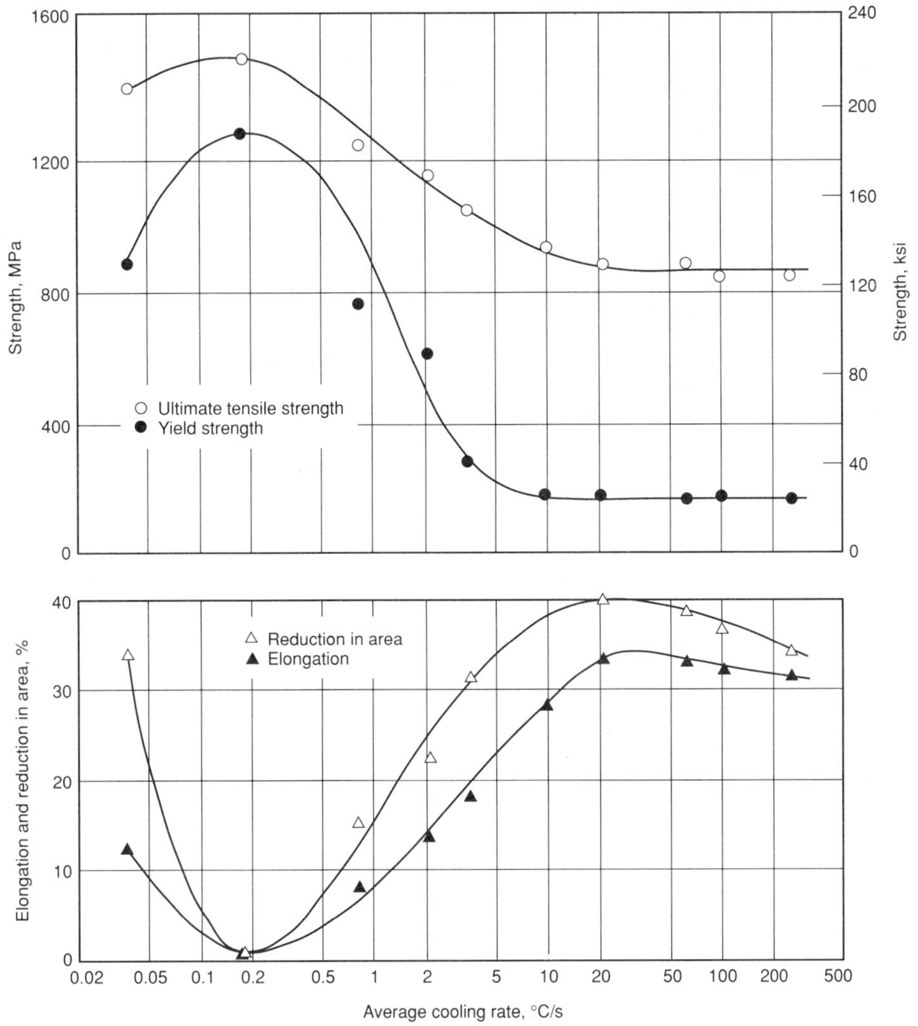

Fig. 6 Effect of cooling rate on tensile properties of U-6.0Nb. Source: Ref 19

of 10% are not usually obtained until yield strength has decreased below 800 MPa (115 ksi). In addition, most alloy-induced increases in corrosion resistance are lost because the α phase no longer contains the alloying additions in supersaturated solid solution. For these reasons, alloys are infrequently used in the overaged condition, except where only moderate strengths (600 to 800 MPa, or 90 to 115 ksi) are required and low residual stresses are strongly desired. Fully overaged dilute alloys such as U-2Mo and U-2.3Nb offer attractive combinations of properties in these cases, as shown in Table 3.

Residual Stresses and Stress Relief (Ref 20, 29). Residual stresses can be introduced into uranium alloys by cold and warm working operations, by the quenching step inherent in solution heat treatment, and by welding operations. These stresses can be large enough to seriously degrade manufacturability and service life. Common consequences of residual stresses include centerline bursting, when bars greater than approximately 20 mm (0.8 in.) in diameter are plunge-quenched into water; dimensional distortion, when parts containing residual stresses are machined; and delayed cracking in welds in high-strength alloys. Surface tensile stresses are particularly dangerous because they can result in stress-corrosion cracking, which is known to occur when uranium alloys with tensile surface stresses are exposed to environments as benign as moist air.

Residual stresses can be relieved either thermally or mechanically. Thermal stress relief occurs in the same temperature range as age hardening; thus a material that is partially aged is also partially stress relieved, but only fully aged and overaged structures are fully stress relieved. As discussed in the previous section, fully overaged alloys that exhibit good ductility, such as U-2Mo and U-2.3Nb, can be used in applications requiring very low residual stresses and only moderate strength. It is important that these materials be slowly cooled from the 550 to 600 °C (1020 to 1110 °F) overaging temperatures, as rapid cooling will reintroduce substantial residual stresses.

Mechanical stress relief can be accomplished by imposing a small amount of plastic deformation on a part after quenching. For example, quenched hollow cylinders, which frequently exhibit large compressive stresses near the outer surfaces and tensile stresses near mid-wall, can be upset forged 1 to 3% at room temperature. This results in plastic deformation of both the tensile and compressively stressed regions, and markedly reduces the magnitudes of the residual stresses (Fig. 10). Mechanical stress relieving is frequently used to minimize run-out problems on quenched parts that must be machined to exacting dimensional tolerances.

Delayed Cracking (Ref 7, 16, 30-32). Uranium alloys are very susceptible to delayed cracking. This problem can result either from small amounts of hydrogen impurities in the metal or from external environments as apparently benign as moist or dry air. Aqueous solutions containing as little as a few ppm Cl^- are a particularly notorious cause of stress-corrosion cracking. In all cases susceptibility to delayed cracking increases the increasing strength of the material. Therefore, soft materials such as unalloyed uranium, as-quenched U-6Nb, and fully overaged alloys, are relatively immune, whereas materials that have been quenched and aged to high strength are more prone to delayed cracking.

Susceptibility to hydrogen, either as internally dissolved hydrogen or in the form of atmospheric moisture, is most severe in dilute alloys. The alloy U-0.75Ti, for example, can be severely embrittled by testing at low strain rates in humid air or by the presence of as little as 0.2 ppm of dissolved hydrogen. More concentrated alloys, such as U-6Nb, however, are relatively immune to atmospheric moisture and can tolerate greater than 10 ppm of dissolved internal hydrogen without ductility losses.

Sensitivity to atmospheric oxygen, on the other hand, is most severe in concentrated alloys, such as U-10Mo and U-7.5Nb-2.5Zr. These alloys fail by intergranular fracture when they are highly stressed in oxygen-containing environments such as dry air.

Aqueous solutions containing even low concentrations of Cl^- cause severe cracking in all high-strength uranium alloys. In moderately concentrated solutions (500 ppm Cl^-), cracking has been observed in the absence of an applied stress due to the wedging effect of the lower-density corrosion product.

Because of the sensitivity of high-strength uranium alloys to delayed cracking, it is desirable to minimize the magnitudes of residual stresses. This can be done by mechanical stress relief after quenching in alloys that will be aged to high-strength levels, or by complete overaging in cases where decreased corrosion resistance and only moderate strengths can be tolerated. It

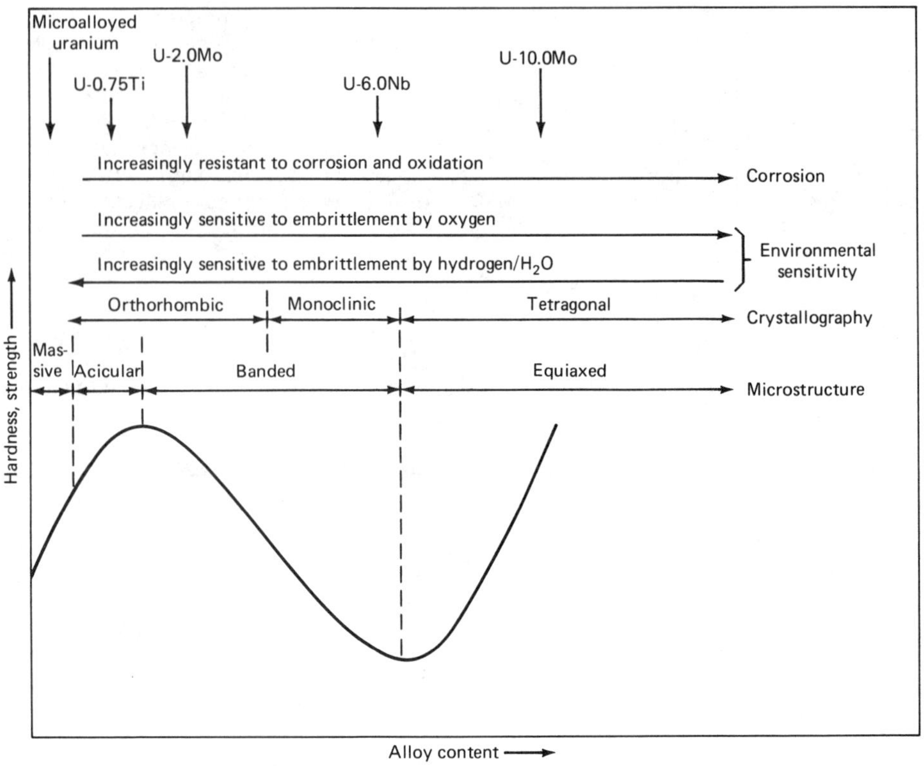

Fig. 7 Effects of alloy composition on microstructure, crystal structure, and properties of quenched uranium alloys

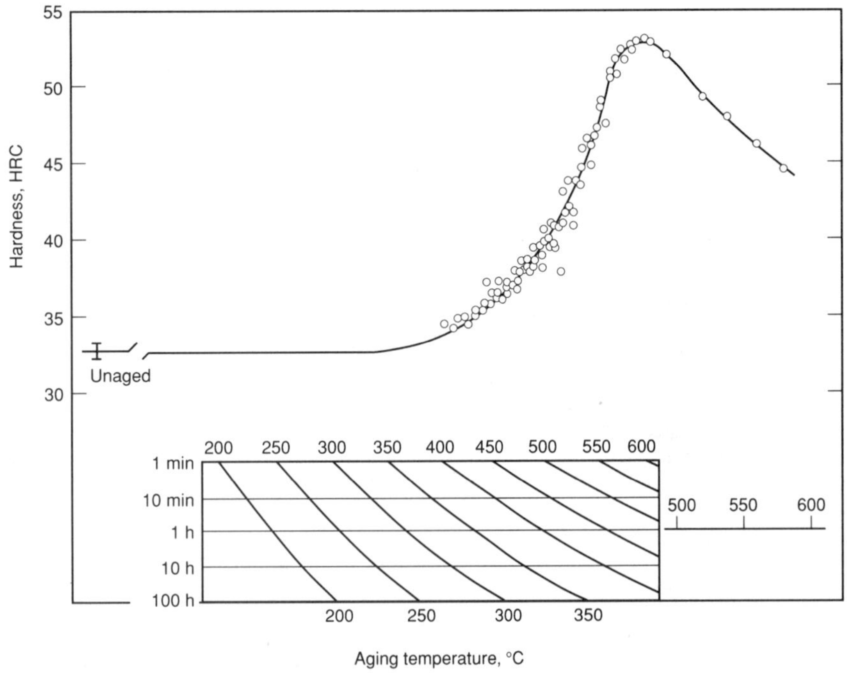

Fig. 8 Effects of aging temperature and time on hardness of U-0.75Ti

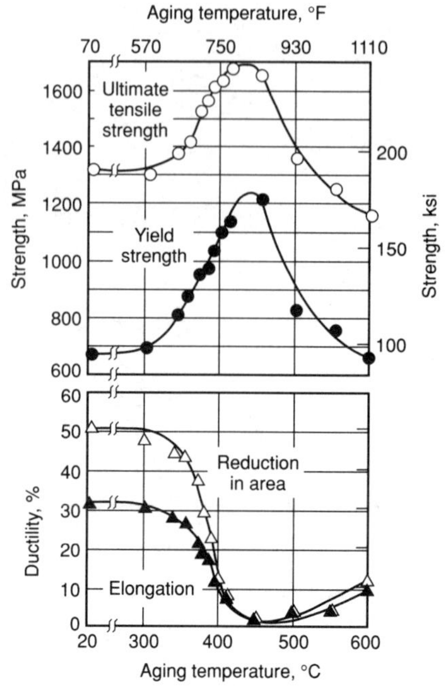

Fig. 9 Effect of aging temperature on tensile properties of U-0.75Ti aged 6 h

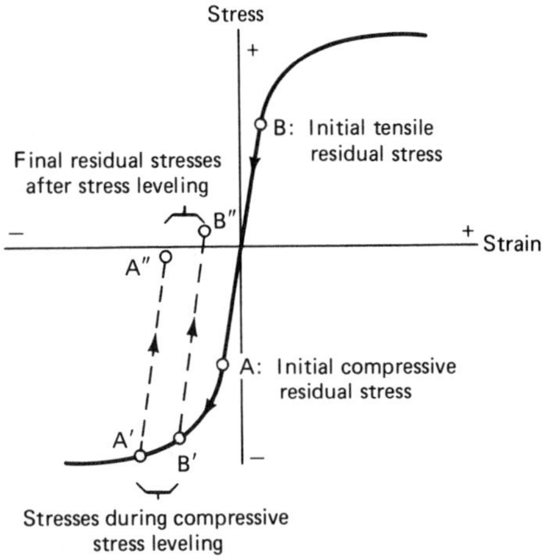

Fig. 10 Stress-strain curve of quenched uranium alloy illustrating initial residual stresses at the surface (A) and interior (B) and how compressive mechanical stress relief reduces residual stress magnitudes

is particularly critical to avoid tensile residual stresses at free surfaces in components that are heat treated to high strengths.

Welding (Ref 33, 34). Welding of low-strength uranium alloys in inert environments is relatively straightforward, but delayed cracking frequently occurs in high-strength alloys due to the environmental effects described in the previous section. The relationship between delayed cracking susceptibility and strength is believed to be associated with the magnitudes of the stresses that develop upon cooling the weld metal and heat-affected zone. The magnitudes of these stresses are controlled (limited) by the yield strengths of the material. In low-strength materials, the elastic stresses are lower than those required to cause cracking, so sound welds are obtained. In high-strength materials, much larger residual stresses can be supported, so cracking occurs after the welded structure has been removed from the inert welding environment. Caution must be exercised when welding even normally soft materials, however, because the cooling rates associated with the welding process might not result in the intended microstructure. If hard microconstituents are formed in the weld region due to inadequate quenching, serious cracking problems can result. For example, U-

Table 4 Effect of composition on tensile properties of β-quenched microalloyed uranium

Composition, ppm(a)	Yield strength, MPa (ksi)	Tensile strength, MPa (ksi)	Elongation, %	Reduction in area, %
Unalloyed U	200 (29)	750 (109)	33	37
240 Al	220 (32)	780 (113)	25	21
270 Fe	310 (45)	870 (126)	35	34
570 Fe	325 (47)	955 (138)	31	32
440 Si	305 (44)	860 (125)	31	28
1000 Si, 420 Fe	495 (72)	1005 (145)	18	14
1000 Si, 480 Fe, 150 C	455 (66)	975 (141)	13	11

Note: Heat treatment, 730 °C (1345 °F) for 4 h in vacuum, water quenched. (a) All materials contain ~50 ppm each of C, Fe, Al, Si unless otherwise noted. Source: Ref 36

Table 5 Effect of heat treatment on tensile properties of microalloyed uranium

Composition, ppm	Heat treatment(a)	Yield strength, MPa (ksi)	Tensile strength, MPa (ksi)	Elongation, %	Reduction in area, %
1000Si-420Fe-50C	γ-quenched	485 (70)	940 (136)	13	9
1000Si-420Fe-50C	β-quenched	500 (73)	1000 (145)	18	14
1000Si-420Fe-50C	α-annealed (550 °C, or 1020 °F, 4 h)	380 (55)	955 (139)	31	42
1000Si-480Fe-150C	γ-quenched	485 (70)	780 (113)	4	3
1000Si-480Fe-150C	β-quenched	455 (66)	975 (141)	13	11
1000Si-480Fe-150C	α-annealed (550 °C, or 1020 °F, 4 h)	340 (49)	1020 (148)	21	27

(a) All heat treatments in vacuum. Source: Ref 36

6Nb can be successfully joined by electron beam welding because heat sinking by the adjacent cold metal provides a sufficient quench rate to produce the soft α_b'' microstructure. Cracking could result, however, from the use of alternate welding processes that cause more heating of adjacent material, thus slowing the quench rate and resulting in the formation of harder microconstituents in the weld or heat-affected regions.

Specific Alloys and Classes of Alloys

Microalloyed Uranium (Ref 35, 36). Very dilute alloys (containing less than 0.3 wt% total alloying additions) can be used in applications that require slightly higher yield strength than that available in unalloyed uranium but no improvements in corrosion resistance. Common alloying elements employed include aluminum, iron, molybdenum, silicon, and vanadium, and are added during the vacuum induction melting process. Small amounts of these elements are soluble in the γ and β phases but not in the α phase. Quenching enables these elements to be retained in supersaturated solid solution in the α phase, where they have the greatest effect on strength, as shown in Table 4. The microstructures of these quenched alloys consist of irregular "massive" grains, identical in appearance to unalloyed uranium. Unlike these intentional alloying additions, carbon does not go into solution to any significant extent. Instead, the presence of tramp carbon results in the formation of carbide inclusions that decrease ductility.

If these microalloyed materials are slowly cooled (or annealed in the high α region) second phases form, apparently during the β → α transformation. These result in some strengthening, but not as much as when the alloying elements are retained in solid solution. Table 5 compares the properties of γ-quenched, β-quenched, and α-annealed microalloys. As shown in that table, annealing results in decreased strength but improved ductility. Carbon promotes the formation of carbide inclusions that reduce ductility regardless of heat treatment condition.

Vanadium additions have been reported to decrease the grain size of cast uranium, resulting in increases in both strength and ductility. As shown in Fig. 11, this effect is maximized at 0.2 wt% V.

Little work has been done on topics such as hydrogen cracking, stress-corrosion cracking, and welding of microalloyed uranium. Since these alloys exhibit relatively low strengths and cannot sustain high residual stresses, it seems likely that they would be relatively free from delayed cracking problems. It is probable, however, that the presence of small quantities of dissolved hydrogen and/or low strain rate testing in humid air would cause substantial decreases in tensile ductility.

Uranium-Titanium Alloys (Ref 15-18, 20, 26, 28-32). The alloy U-0.75Ti is used for applications that require the highest combinations of strength and ductility. Like unalloyed uranium, this alloy is produced by vacuum induction melting. During the melting process, some of the titanium reacts with tramp carbon to form solid titanium carbides, which float to the surface of the melt and can be removed, rather than uranium and niobium carbides, which form during the solidification process and are dispersed throughout the ingot or casting. Good carbon control and excellent microcleanliness are thus possible with U-Ti alloys.

U-0.75Ti is vacuum-solution heat treated at 800 °C (1470 °F) to remove hydrogen and put the titanium into solution in the γ phase, and then water quenched to room temperature to produce supersaturated α_a' martensite. The quenching process introduces high residual stresses, particularly when the quenched parts are thick. The magnitudes of these stresses are frequently reduced by mechanical stress leveling in order to avoid delayed cracking and/or reduce the amount of distortion that occurs during subsequent machining. The α_a' martensite exhibits a yield strength of 650 MPa (95 ksi) due to the solid solution effect of titanium. Further strengthening can be obtained by age hardening, in the 330 to 450 °C (625 to 840 °F) range, to form fine coherent precipitates of U_2Ti in the α_a' martensite. Overaging occurs at temperatures exceeding 450 °C (840 °F), due to cellular decomposition of the α_a' martensite to the equilibrium α and U_2Ti phases. More complete descriptions of the microstructural changes that occur during heat treatment are given in Ref 3, 5, and 6.

The mechanical properties of quenched and aged U-0.75Ti are summarized in Fig. 8 and 9. Aging results in substantial increases in hardness and strength but decreases in ductility. Overaging reduces strength but produces only modest increases in ductility because a semicontinuous film of brittle U_2Ti forms along the martensite plate boundaries during the overaging process. Because of this, the material is virtually never used in the overaged condition. Most applications that require a balance of strength and ductility specify aging at about 380 °C (715 °F) for 4 to 6 h.

The ductility of U-0.75Ti can be severely decreased by the presence of small amounts of hydrogen, as shown in Fig. 12. This hydrogen embrittlement phenomenon is strongly strain-rate dependent and is most

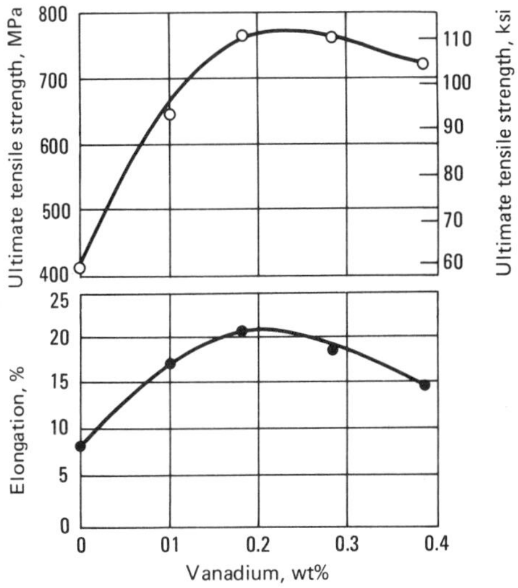

Fig. 11 Tensile properties of cast uranium-vanadium alloys. Source: Ref 35

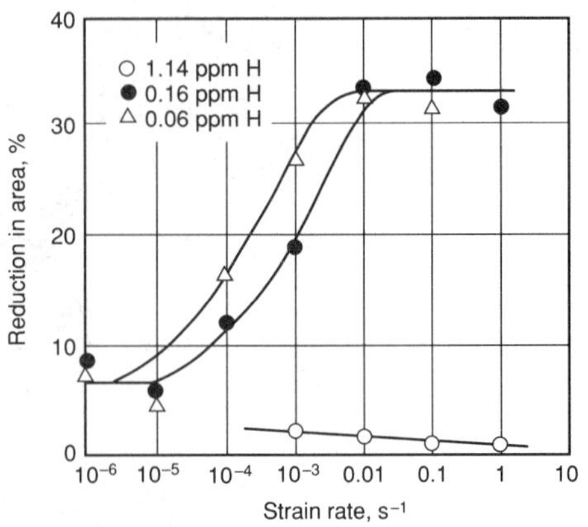

Fig. 12 Effects of hydrogen content and strain rate on ductility of U-0.75Ti with yield strength of 965 MPa. Source: Ref 32

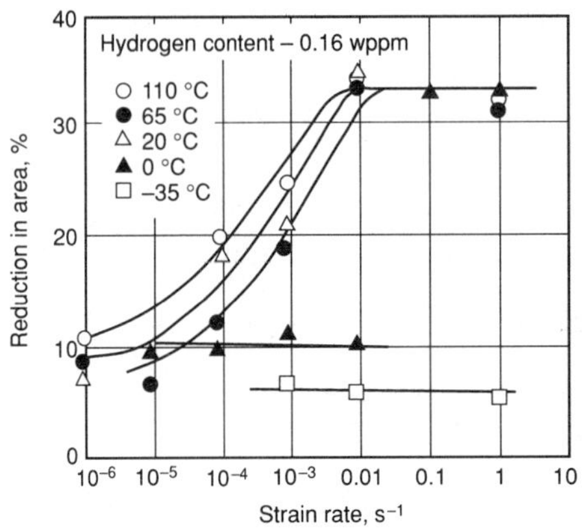

Fig. 13 Effect of temperature and strain rate on ductility of U-0.75Ti with yield strength of 965 MPa. The alloy contained 0.16 ppm hydrogen. Source: Ref 32

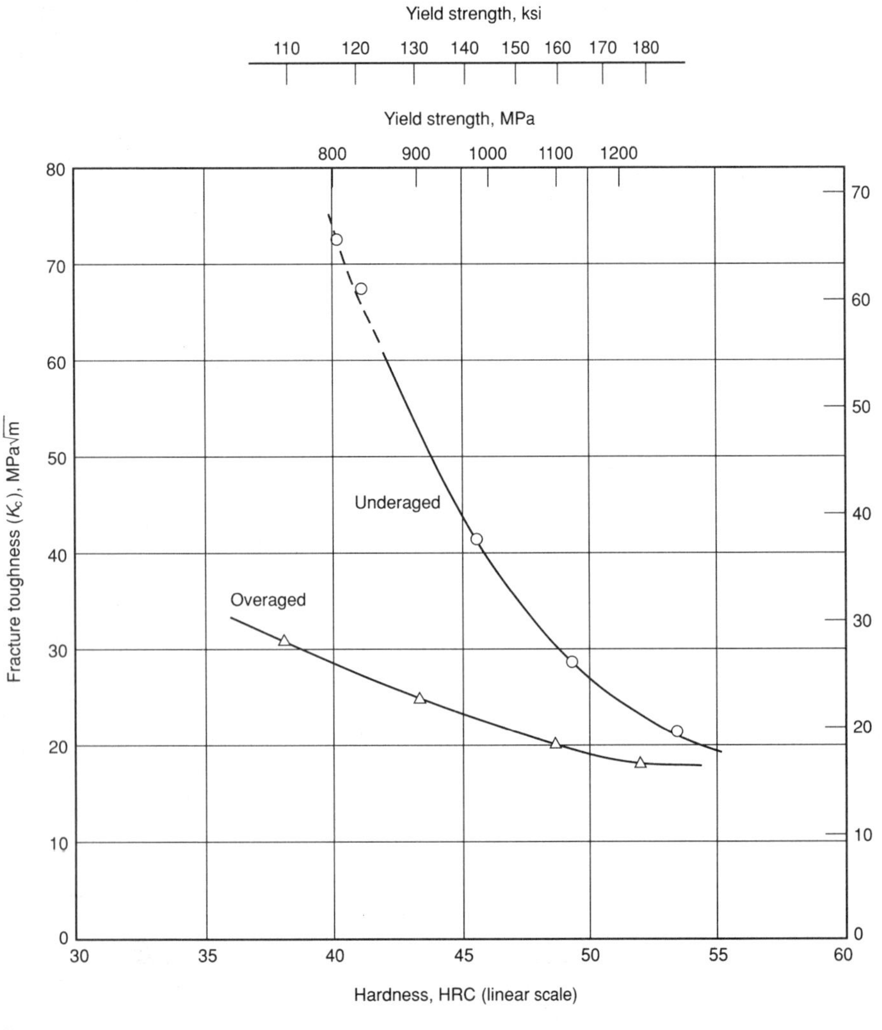

Fig. 14 Effect of aging on fracture toughness of U-0.75Ti. Source: Ref 17

Table 6 K_{ISCC} values for U-0.75Ti

Environment	K_{ISCC}, MPa$\sqrt{m}$ (ksi$\sqrt{in.}$): Aged 380 °C (715 °F), 6 h (Yield strength 985 MPa, or 143 ksi)	Aged 450 °C (840 °F), 6 h (Yield strength 1200 MPa, or 174 ksi)
Dry air	42 (38)	29 (26)
Air (100% relative humidity)	28 (25)	15 (14)
50 ppm Cl⁻ solution	18 (16)	...
3.5% NaCl solution	18 (16)	10 (9)

Source: Ref 30

severe at low strain rates. This problem can be avoided by solution heat treatment in a vacuum of higher than 1 mPa (10^{-4} torr) for periods long enough to reduce the hydrogen content to a low value. The times required for effective hydrogen removal are shown in Fig. 2. Atmospheric moisture has a similar embrittling effect on this alloy, apparently because water vapor reacts with the surface of the metal to produce UO_2 plus atomic hydrogen. Because of this, tensile testing is frequently carried out in a dry environment to avoid scatter in the ductility data due to variations in relative humidity. The tensile properties of U-0.75Ti are also very dependent on test temperature (Fig. 13). Low ductilities are obtained at temperatures below approximately 0 °C (32 °F) regardless of strain rate, hydrogen content, and testing environment.

The fracture toughness behavior of U-0.75Ti parallels its ductility, as shown in Fig. 14. For the commonly used heat treatment approach (380 °C, or 715 °F, for 6 h), the yield strength and fracture toughness values are approximately 950 MPa (140 ksi) and 45 MPa$\sqrt{m}$ (41 ksi$\sqrt{in.}$), respectively. This results in a critical flaw size in the vicinity of 0.7 mm (0.03 in.), which is detectable by nondestructive inspection techniques.

Unfortunately, subcritical flaw growth is possible in U-0.75Ti in environments as benign as humid air. Values of K_{ISCC} in various environments are given in Table 6. This susceptibility to stress-corrosion cracking increases the importance of quench-induced residual stresses. Fortunately, the surface stresses in most parts of simple shape are compressive, but tensile surface stresses can result from machining complex features into such parts. Residual stress magnitudes can be reduced by both mechanical stress leveling and aging, but aging is of limited utility in this regard because it also reduces K_{ISCC}.

U-0.75Ti exhibits substantial quench rate sensitivity. Quench rates of approximately 200 °C/s (360 °F/s) are required to obtain 100% martensite. This can only be achieved by water quenching of plates thinner than about 10 mm (0.4 in.), as shown in Fig. 15. The lower cooling rates that occur during water quenching of thicker parts are insufficient to completely prevent diffusional decomposition of the γ phase. The diffusionally formed transformation products characteristic of these lower cooling rates exhibit substantially lower ductilities than the α_a' martensite, as shown in Fig. 16. In addition, since they are not supersaturated with titanium, they cannot be subsequently age hardened. For most applications, however, acceptable performance can be obtained if approximately 50% martensite is present at mid-thickness. This enables sat-

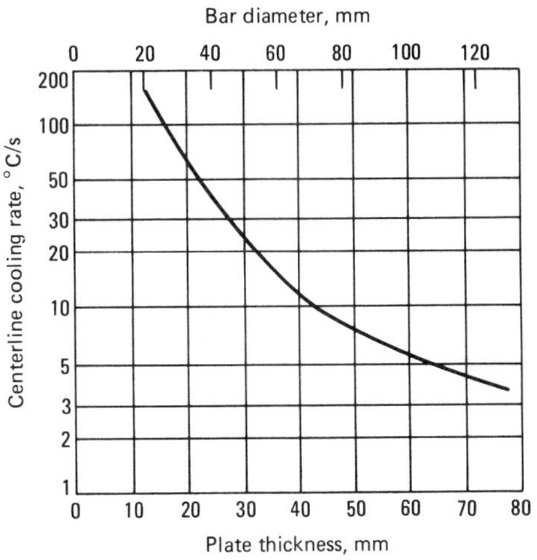

Fig. 15 Centerline cooling rates in plates and bars quenched into 20 °C water

isfactory processing of U-0.75Ti plates up to about 28 mm (1.1 in.) thick and rods up to about 46 mm (1.8 in.) in diameter.

These quench-rate-sensitivity limitations in section size can be reduced somewhat by modifying alloy composition. Decreasing the titanium content increases the kinetics of diffusional decomposition of the γ phase, but it also increases the martensite start temperature. These effects interact in a way that reduces the critical quench rate for martensite formation with decreasing titanium content. The effects of titanium content on the sizes of parts that can be effectively heat treated is shown in Table 7. As shown, reducing the titanium content to 0.6 wt% enables larger parts to be effectively heat treated. Such a reduction in titanium content, however, also decreases the yield strength corresponding to any given aging treatment by about 60 MPa (9 ksi). This strength reduction can be overcome by aging at a slightly higher temperature, but this results in decreased ductility. Optimum properties thus are obtained by keeping titanium content as high as can be tolerated without encountering excessive quench-rate-sensitivity problems. Below about 0.5 wt% titanium, the γ phase no longer transforms to martensite on quenching but undergoes a massive transformation to a variant of α phase that exhibits very little age hardenability. For this reason 0.5 wt% Ti is the lower limit for useful U-Ti alloys. Increases in titanium content, conversely, result in superior combinations of strength and ductility, but increases in the critical quench rate for martensite formation make these alloys difficult to process except in very thin sections.

Table 7 Critical cooling rates and section sizes for 50% martensite at centerline in U-Ti alloys

Titanium content, %	Critical cooling rate, °C/s (°F/s)	Plate thickness, mm (in.)	Bar diameter, mm (in.)
0.60	10 (18)	42 (1.7)	72 (2.8)
0.75	28 (50)	28 (1.1)	46 (1.8)
1.0	40 (72)	24 (0.95)	40 (1.6)
1.5	>100 (>180)	<15 (<0.6)	<25 (<1)

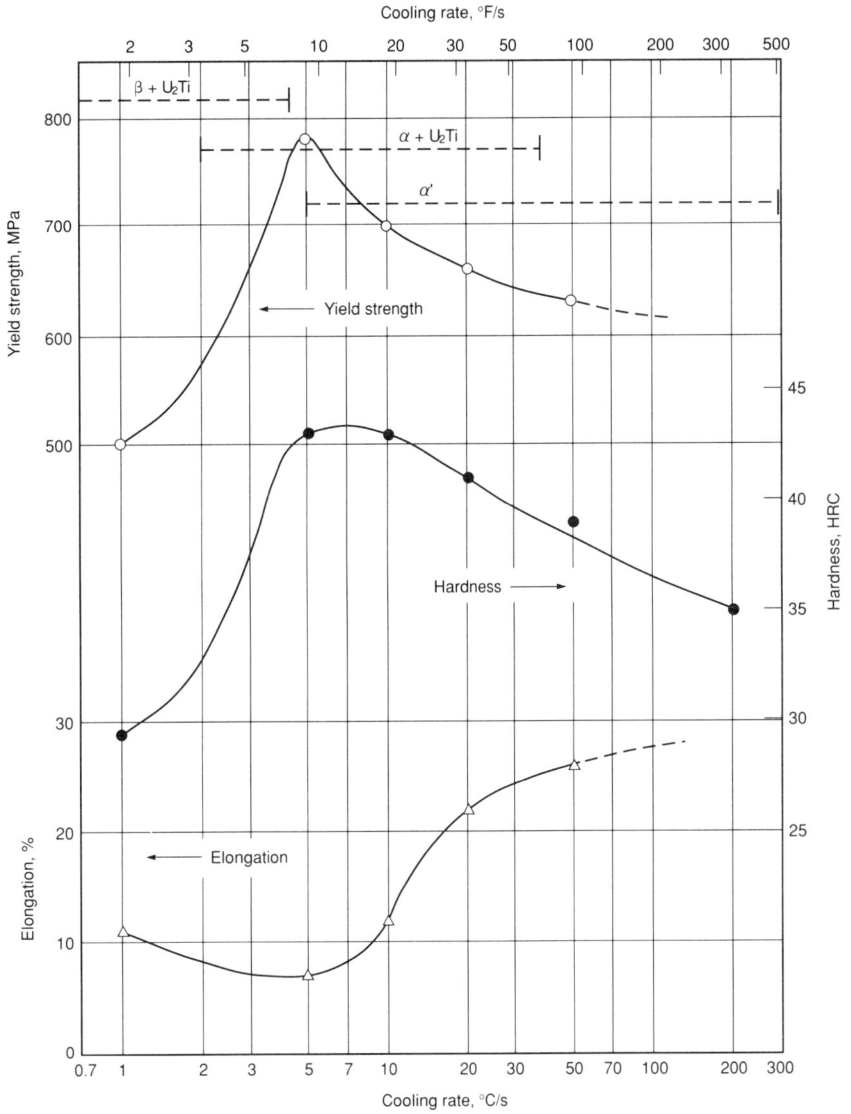

Fig. 16 Effect of quench rate on properties of U-0.75Ti. Dashed lines show approximate microstructural regions. Source: Ref 18

U-Ti alloys are difficult to weld because their high yield strengths permit residual stresses in the vicinity of 650 MPa (95 ksi) to develop in the weld and heat-affected regions. These stresses result in cold cracking when the material is removed from the protective welding environment and exposed to moist air. Because of this, mechanical joints must be made between U-Ti alloy components.

The titanium in solid solution makes U-0.75Ti somewhat less susceptible to oxidation and corrosion than unalloyed uranium. Nevertheless, protective coatings, such as electroplated nickel and ion-plated aluminum, are used in many applications.

Uranium-Niobium Alloys (Ref 16, 17, 19, 23, 25, 27, 30, 31, 37, 38). The alloy U-6Nb is used for applications requiring good corrosion resistance and ductility. This alloy is vacuum arc melted because of its high melting temperature and tendency to segregate during solidification. After forming, the material is vacuum-solution heat treated at 800 °C (1470 °F), to put the niobium into solid solution and to remove hydrogen, and is then quenched to room temperature to produce supersaturated α_b'' martensite. The as-quenched material exhibits a yield strength of only 160 MPa (25 ksi) much lower than that of U-0.75Ti. Because of this low yield strength, U-6Nb cannot sustain high residual stresses, hence stress relief is not required. This low yield strength results from the ease with which the many twin boundaries in the α_a'' martensite move. This twin boundary movement, coupled with the $\alpha_b'' \leftrightarrow \gamma°$ martensitic transformation that occurs

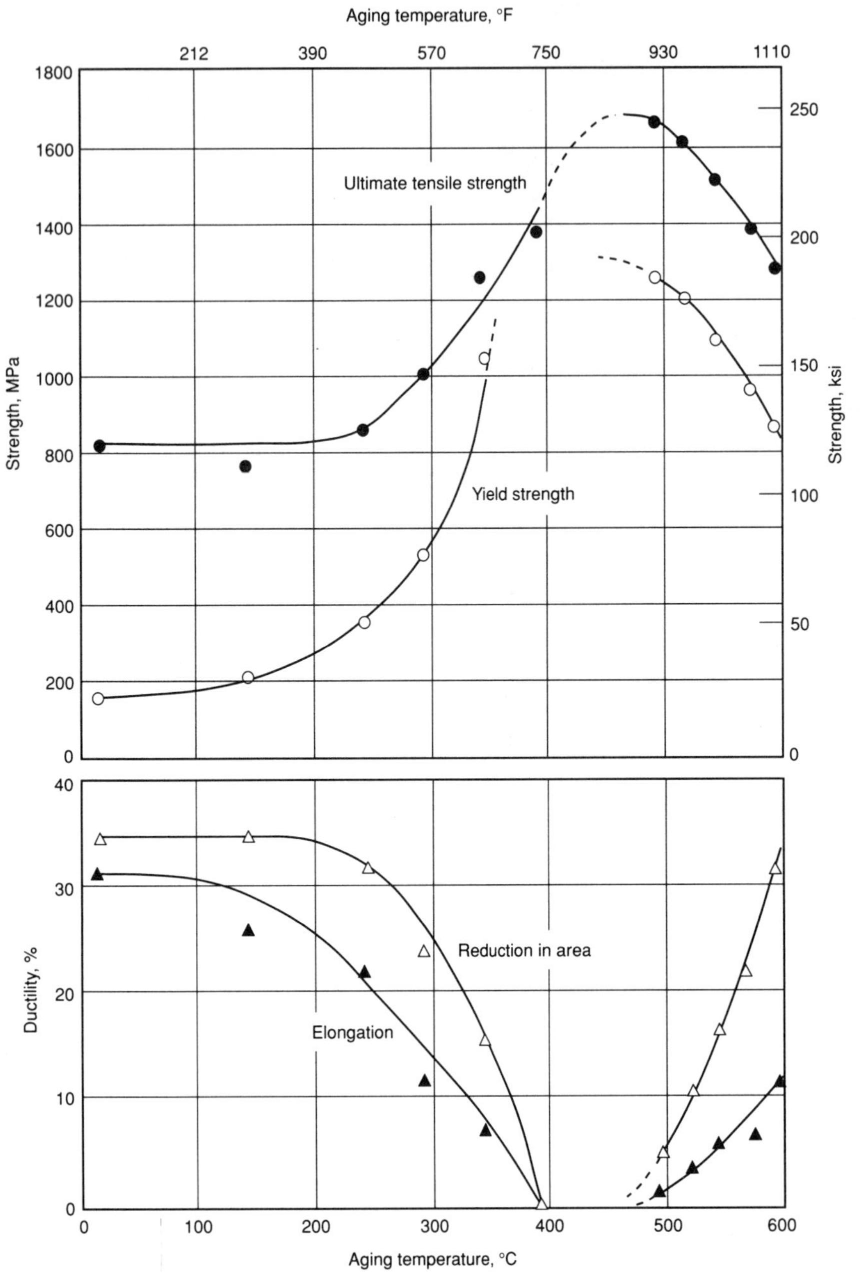

Fig. 17 Effect of aging temperature on tensile properties of U-6.3Nb aged 1 h. Source: Ref 19

slightly above room temperature, results in a shape memory effect that can cause temperature-dependent dimensional instabilities and make it difficult to machine parts to close tolerances. These dimensional instabilities can be overcome by age hardening the martensite slightly, usually at 150 °C (300 °F) for 2 h. This aging treatment increases the yield strength somewhat, but has little effect on ductility.

U-6Nb is most frequently used in either the as-quenched or slightly aged (dimensionally stabilized) condition, where it exhibits good corrosion resistance and high ductility. Aging in the 150 to 400 °C (300 to 750 °F) range produces additional increases in strength and decreases in ductility, as shown in Fig. 17, but the combinations of strength and ductility obtainable in this alloy are not as good as those in U-0.75Ti. Overaging occurs at temperatures in excess of 400 °C (750 °F).

U-6Nb is substantially less sensitive to quench rate than U-0.75Ti. Single-phase supersaturated microstructures can be obtained at cooling rates as low as 10 °C/s (18 °F/s), as shown in Fig. 6. This permits plates as thick as 40 mm (1.6 in.) and bars up to 70 mm (2.75 in.) in diameter to be processed by water quenching to produce 100% martensitic microstructures. Thinner parts can be quenched in less severe media, such as oil.

The properties of U-6Nb are far less influenced by variations in hydrogen content, atmospheric moisture, test temperature, and strain rate than those of U-0.75Ti. In addition, this alloy is less susceptible to stress-corrosion cracking than many other alloys. To some extent, this is due to the materials generally being used in a low-strength condition; when aged to high strength it becomes increasingly prone to stress-corrosion cracking. It is also readily weldable and, since high residual stresses cannot be supported by this low-strength material, welds are generally not prone to delayed cracking. U-6Nb exhibits good resistance to corrosion, so it does not typically require protective coatings.

Other U-Nb alloys have been investigated, with niobium contents ranging from 2 to 12 wt%. Alloys with 2 to 4 wt% Nb exhibit better combinations of strength and ductility when aged to higher strengths, but not as good as U-0.75Ti. These leaner alloys also require more drastic quenches than does U-6Nb, and are more prone to hydrogen cracking when aged to high-strength conditions, particularly the U-2.3Nb alloy. Fully overaged U-2.3Nb, however, exhibits good ductility and resistance to delayed cracking at moderate strength levels, as shown in Table 3. Alloys with 8 to 12 wt% Nb exhibit excellent corrosion resistance but are increasingly susceptible to stress-corrosion cracking due to their higher strengths.

Uranium-molybdenum alloys (Ref 22, 24, 25, 28, 39) have been extensively investigated but are not widely used in high-performance applications. U-Mo alloys are similar to U-Nb alloys except they are easier to melt (can be vacuum induction melted), are less resistant to corrosion, and are more susceptible to stress-corrosion cracking. U-2Mo is the most common of these alloys and is used primarily in the as-cast condition for shielding in radioactive material shipping casks. U-2Mo can be solution treated, quenched, and aged, but the resulting strength-ductility combinations are not as good as those obtainable in U-0.75Ti. When slow cooled from 800 °C (1470 °F), this alloy exhibits about twice the yield strength of unalloyed uranium (Table 2). This, plus ease of melting and casting, makes it well suited for thick-walled castings, such as those used in radioactive shipping containers. It can also be used in the quenched and fully overaged condition for applications requiring moderate strengths and very low residual stresses (Table 3).

U-10Mo exhibits good corrosion resistance and is very quench rate insensitive, but its relatively high yield strength (~900 MPa, or 130 ksi), permits it to sustain high residual stresses and it stress-corrosion cracks severely. For these reasons, it is rarely used.

Other Alloys (Ref 27, 40, 41). Many ternary, quaternary, and higher-order alloys have also been investigated. In most cases these were relatively highly alloyed materials (α_b'' or γ° microstructures). Little evidence has been presented indicating that these alloys exhibit better strength-ductility combinations, reduced quench rate sensitivity, or improved resistance to hydrogen or stress-corrosion cracking than their simpler binary counterparts. For this reason, these higher-order alloys have found limited application.

ACKNOWLEDGMENTS

The author wishes to thank A.D. Romig Jr., B.C. Odegard, and T.N. Simmons of Sandia National Laboratories, G.M. Ludtha of Martin Marietta Energy Systems, and G.B. Dudder of Battelle Pacific Northwest Laboratory for their review of this manuscript and helpful suggestions.

REFERENCES

1. A.N. Holden, *Physical Metallurgy of Uranium*, Addison-Wesley, 1958
2. W.D. Wilkinson, *Uranium Metallurgy*, Vol 1, 2, Interscience, 1962
3. *Physical Metallurgy of Uranium Alloys*, J.J. Burke *et al.*, Ed., Brook Hill, 1976
4. P. Loewenstein, Industrial Uses of Depleted Uranium, in *Properties and Selection: Stainless Steels, Tool Materials and Special-Purpose Metals*, Vol 3, *Metals Handbook*, 9th ed., American Society for Metals, 1980, p 773
5. *Metallurgical Technology of Uranium and Uranium Alloys*, Vol 1, 2, 3, American Society for Metals, 1982
6. K.H. Eckelmeyer, Metallography of Uranium and Uranium Alloys, in *Metallography and Microstructures*, Vol 9, *Metals Handbook*, 9th ed., American Society for Metals, 1985, p 476
7. L.J. Weirick, Corrosion of Uranium and Uranium Alloys, in *Corrosion*, Vol 13, *Metals Handbook*, 9th ed., American Society for Metals, 1987, p 813
8. J.A. Aris, Machining of Uranium and Uranium Alloys, in *Machining*, Vol 16, *Metals Handbook*, 9th ed., ASM INTERNATIONAL, 1989, p 874
9. "Occupational Health Guideline for Uranium and Insoluble Compounds," U.S. Department of Health and Human Services, 1978
10. Health Physics Manual of Good Practice for Uranium Facilities, EG&G 2530, UC-41, NTIS DE88-013620, June 1988
11. M.D. Henderson, "Evaluation of Radiation Exposure in Metal Preparation Depleted Uranium Process Areas," Y/DQ-5, Martin Marietta Energy Systems, Inc., 1989
12. D.M.R. Taplin, The Tensile Properties and Fracture of Uranium Between −200 °C and 900 °C, *J. Aust. Inst. Met.*, Vol 12, 1967, p 32
13. J.S. Daniel, B. Lesage, and P. Lacombe, The Influence of Temperature on Slip and Twinning in Uranium, *Acta Metall.*, Vol 19, 1971, p 163
14. H. Inouye and A.C. Schaffhauser, "Low Temperature Ductility and Hydrogen Embrittlement of Uranium—A Literature Survey," ORNL-TM-2563, Oak Ridge National Laboratory, 1969
15. D.M.R. Taplin and J.W. Martin, The Effects of Grain Size and Cold Work on the Tensile Properties of Alpha Uranium, *J. Less Common Met.*, Vol 7, 1964, p 89
16. G.L. Powell, Internal Hydrogen Embrittlement in Uranium Alloys, in *Metallurgical Technology of Uranium and Uranium Alloys*, Vol 3, American Society for Metals, 1982, p 877
17. K.H. Eckelmeyer, Diffusional Transformations, Strengthening Mechanisms and Mechanical Behavior, in *Metallurgical Technology of Uranium and Uranium Alloys*, Vol 1, American Society for Metals, 1982, p 129
18. K.H. Eckelmeyer and F.J. Zanner, Quench Rate Sensitivity in U-0.75 wt%Ti, *J. Nucl. Mater.*, Vol 67, 1967, p 33
19. K.H. Eckelmeyer, A.D. Romig Jr., and L.J. Weirick, The Effect of Quench Rate on the Microstructure, Mechanical Properties, and Corrosion Behavior of U-6 Wt.Pct.Nb, *Met. Trans.*, Vol 15A, 1984, p 1319
20. B.H. Llewellyn, G.A. Aramayo, G.A. Ludtka, J.E. Park, M. Siman-Tov, and K.F. Wu, "Comparisons of Analytical and Experimental Results in Immersion Quenching of U-0.75%Ti Cylinders," Y/DV-560, Martin Marietta Energy Systems Y-12 Plant, 1986
21. J. Lehmann and R.F. Hills, Proposed Nomenclature for Phases in Uranium Alloys, *J. Nucl. Mater.*, Vol 2, 1960, p 261
22. R.F. Hills, B.R. Butcher, and B.W. Howlett, The Mechanical Properties of Quenched U-Mo Alloys, Tensile Tests on Polycrystalline Specimens, Part 1, *J. Nucl. Mater.*, Vol II, 1964, p 149
23. K. Tangri and D.K. Chaudhuri, Metastable Phases—Uranium Alloys With High Solute Solubility in the BCC Gamma Phase, The System U-Nb, Part I, *J. Nucl. Mater.*, Vol 15, 1965, p 278
24. G.H. May, The Annealing of a Quenched Uranium-5 at.% Molybdenum Alloy, *J. Nucl. Mater.*, Vol 7, 1962, p 72
25. K.H. Eckelmeyer, Aging Phenomena in Dilute Uranium Alloys, in *Physical Metallurgy of Uranium Alloys*, J.J. Burke *et al.*, Ed., Brook Hill, 1976, p 463
26. A.M. Ammons, Precipitation Hardening in Uranium Rich Uranium-Titanium Alloys, in *Physical Metallurgy of Uranium Alloys*, J.J. Burke *et al.*, Ed., Brook Hill, 1976, p 511
27. R.J. Jackson, Elastic, Plastic, and Strength Properties of U-Nb and U-Nb-Zr Alloys, in *Physical Metallurgy of Uranium Alloys*, J.J. Burke *et al.*, Ed., Brook Hill, 1976, p 611
28. K.H. Eckelmeyer and F.J. Zanner, The Effect of Aging on the Mechanical Behavior of U-0.75 wt% Ti and U-2.0 wt% Mo, *J. Nucl. Mater.*, Vol 62, 1976, p 37
29. K.H. Eckelmeyer, "Residual Stresses in Uranium and Uranium Alloys," SAND 85-1427, Sandia National Laboratories, 1985
30. N.J. Magnani, Stress Corrosion Cracking of Uranium Alloys, in *Physical Metallurgy of Uranium Alloys*, J.J. Burke *et al.*, Ed., Brook Hill, 1976, p 935
31. J.W. Koger, Overview of Corrosion, Corrosion Protection, and Stress Corrosion Cracking of Uranium and Uranium Alloys, in *Metallurgical Technology of Uranium and Uranium Alloys*, Vol 3, American Society for Metals, 1982, p 751
32. C. Odegard, K.H. Eckelmeyer, and J.J. Dillon, "The Embrittlement of U-0.8% Ti by Absorbed Hydrogen," in *Proc. 4th Int. Conf. Hydrog. Eff. Mater. Behav.*, to be published
33. P.W. Turner and L.D. Johnson, Joining of Uranium Alloys, in *Physical Metallurgy of Uranium Alloys*, J.J. Burke *et al.*, Ed., Brook Hill, 1976, p 145
34. G.L. Mara and J.L. Murphy, Welding of Uranium and Uranium Alloys, in *Metallurgical Technology of Uranium and Uranium Alloys*, Vol 3, American Society for Metals, 1982, p 633
35. C. Collot and R. Reisse, A Study of the Structure and Mechanical Properties of Uranium-Low Vanadium Alloys, *Mem. Sci. Rev. Met.*, Vol 68, 1971
36. R.L. Ludwig, "The Effect of Small Additions of Silicon, Iron, and Aluminum on the Room Temperature Tensile Properties of High Purity Uranium," Y-2286, Union Carbide Corporation Y-12 Plant, 1983
37. R.J. Jackson and D.V. Miley, "Tensile Properties of Gamma Quenched and Aged Uranium-Based Niobium Alloys," *Trans. ASM*, Vol 61, 1968, p 336
38. R.A. Vandermeer, J.C. Ogle, and W.G. Northcutt Jr., A Phenomenological Study of the Shape Memory Effect in Polycrystalline Uranium-Niobium Alloys, *Metall. Trans.*, Vol 12A, 1981, p 733
39. A.M. Nomine, D. Bedere, and D. Miannay, The Influence of Physio-Chemical Parameters on the Mechanical Properties of Some Isotropic Uranium Alloys, in *Physical Metallurgy of Uranium Alloys*, J.J. Burke *et al.*, Ed.,

Brook Hill, 1976, p 657
40. J. Greenspan, D.A. Colling, and F.J. Rizzitano, Polynary Uranium Alloys, in *Physical Metallurgy of Uranium Alloys*, J.J. Burke *et al.*, Ed., Brook Hill, 1976, p 701
41. K.H. Eckelmeyer, The Effects of Heat Treatment on the Microstructure and Mechanical Behavior of U-0.75 wt% Mo-0.75 wt% Nb-0.75 wt% Zr-0.50 wt% Ti, *J. Nucl. Mater.*, Vol 68, 1977, p 92

Beryllium

A. James Stonehouse and James M. Marder, Brush Wellman Inc.

BERYLLIUM is a metal with an unusual combination of physical and mechanical properties that make it particularly effective in optical components, precision instruments, and specialized aerospace applications. In each of these three general application areas, beryllium is selected because of its combination of low weight, high stiffness, and specific mechanical properties such as a precise elastic limit. It is also useful because it is transparent to x-rays and other high-energy electromagnetic radiation. In addition, beryllium is used as a small additive in some copper-base and nickel-base alloys (see the article "Beryllium-Copper and Other Beryllium-Containing Alloys" in this Volume).

Unalloyed beryllium is readily joined by brazing. Fusion welding is not advisable in most situations, although beryllium can be fusion welded with aluminum filler metals when extreme care is exercised. Beryllium can be extruded into bar, rod, and tubing or rolled into sheet. The surface of beryllium can be polished to a very reflective mirror finish, and this finish is particularly effective at infrared wavelengths. Beryllium can be plated with nickel, silver, gold, and aluminum, and the surface also can be anodized or chromate conversion coated to provide a measure of corrosion resistance. Beryllium can be machined to extremely close tolerances; this attribute, in combination with its excellent dimensional stability, allows beryllium to be used for the manufacture of extraordinarily precise and stable components.

Almost all of the beryllium in use is a powder metallurgy (P/M) product. Powder processing is required for a number of reasons. Castings of beryllium generally have porosity and other casting defects that make them unsuitable for use in critical applications. This stems from the high melting point, the high melt viscosity, and the narrow solid-liquid range of beryllium. The high melting point (1283 °C, or 2341 °F) promotes reaction of the molten metal with potential casting mold materials; the high melt viscosity and the narrow solid-liquid range prevent the easy filling of complex castings. If high superheat temperatures are used to reduce viscosity, mold reaction limits the integrity of the component. In addition to the limitation on structural integrity imposed by casting, the grain size of as-cast beryllium is quite coarse (>50 μm, or 0.002 in.). The ductility and strength of beryllium depend primarily on grain size and obey the Hall-Petch relationship; these properties require grain sizes of less than about 15 μm (600 μin.) in structural components. Powder metallurgy processing allows grain sizes as low as 1 to 10 μm to be achieved when required. Consolidation by vacuum hot pressing, hot isostatic pressing, pressing and sintering, or other processes can produce parts with density values in excess of 99.5% of the theoretical value of 1.8477 g/cm^3 (0.067 lb/in.^{3}).

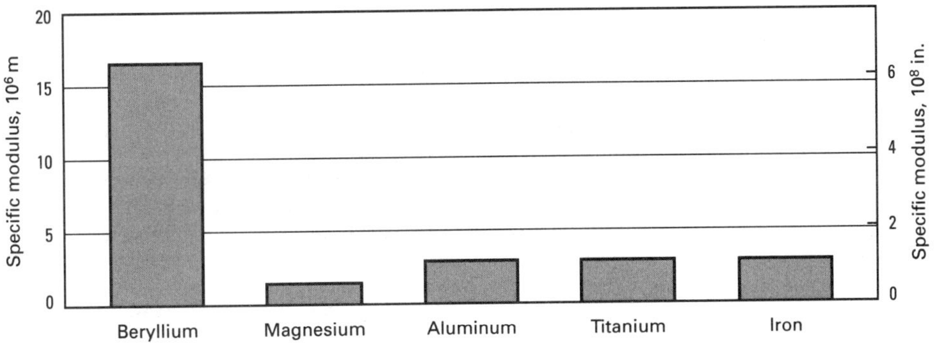

Fig. 1 Specific modulus of lightweight materials

Table 1 Selected physical properties of beryllium

Property	Amount
Elastic modulus, GPa (10^6 psi)	303 (44)
Density, g/cm^3 (lb/in.^{3})	1.8477 (0.067)
Thermal conductivity, W/m · K (Btu/h · ft · °F)	210 (121)
Coefficient of thermal expansion, 10^{-6}/°C (10^{-6}/°F)	11.5 (6.4)
Specific heat at room temperature, kJ/kg · K (Btu/lb · °F)	2.17 (0.52)
Melting point, °C (°F)	1283 (2341)
Mass absorption coefficient (Cu K-alpha), cm^2/g	1.007
Specific modulus, m (in.)(a)	16.7 × 10^6 (6.56 × 10^8)

(a) The specific modulus is defined (in inches) from the ratio of the elastic modulus (in psi) and the density (in lb/in.^{3}).

The application of P/M processing to beryllium does impose its own limits on the characteristics of the product. For example, the use of excessively fine powders increases the oxide content because of the increase in specific surface area. The oxide content of beryllium is one of the determinants of physical and mechanical properties; therefore, knowledge and control of the powder process are important to the proper fabrication of a beryllium component. Also, powder shape affects the anisotropy of mechanical properties.

Properties of Importance

Of the mechanical and physical properties given for beryllium in the article "Properties of Pure Metals" in this Volume, the most noteworthy are its density, elastic modulus, mass absorption coefficient, and the other physical properties presented in Table 1. The specific modulus, which is the ratio of the modulus of elasticity to the density, is of particular interest: The specific modulus for beryllium is substantially greater than that of other aerospace structural materials such as steel, aluminum, titanium, or magnesium (Fig. 1). Also, beryllium has the highest heat capacity (1820 J/kg · °C, or 0.435 Btu/lb · °F) among metals and a thermal conductivity comparable (210 W/m · K, or 121 Btu/ft · h · °F) to that of aluminum (230 W/m · K, or 135 Btu/ft · h · °F). Beryllium thus may prove to be an efficient substrate material for conducting waste heat away from active solid-state electronic components, particularly in aerospace applications.

The microyield strength of beryllium is of particular concern in guidance system components for ships, aircraft, and missiles. In a gyroscopic system, permanent errors can be introduced by yielding of the guidance components. A permanent set of 1 part in 10^6 (microyield), as opposed to the common yield criterion of 2 parts in 10^4, is the yield strength value used in evaluating gyroscope materials. Instrument grades of beryllium thus have a microyield strength acceptance criterion.

Infrared Reflectivity. The second important area of application for instrument-grade beryllium is in infrared optics. At long-wave infrared (LWIR) wavelengths, beryllium has a reflectivity in excess of 99% of the incident intensity. This enables beryllium to be used in LWIR surveillance and in deep-space observatories. The mirrors for satellites such as the Infrared Astronomical Satellite often are beryllium because the metal combines infrared reflectivity with light weight and high stiffness.

The low mass absorption coefficient of beryllium makes it practically transparent to x-rays and other high-energy electromagnetic radiation. It is therefore used as a window material in x-ray tubes and detectors such as those used in energy dispersive analysis of x-ray equipment.

Beryllium Mining and Refining

While a number of beryllium-containing minerals have been identified, only beryl ($3BeO\text{-}Al_2O_3\text{-}6SiO_2$) and bertrandite ($4BeO\text{-}2SiO_2\text{-}H_2O$) have been commercially significant to date. Beryl occurs in isolated pockets in pegmatites and, with the possible exception of operations in the Soviet Union and China, is normally recovered as a hand-sorted by-product of mining operations for other minerals such as feldspar, spodumene, or mica. Commercial beryl normally contains about 10% BeO (3.6% Be). Until Brush Wellman began commercial extraction from a bertrandite-bearing deposit near Delta, UT in 1969, beryl was the only source of beryllium. The Utah deposit consists of mineralization in a water-laid tuff and is mined by open-pit methods. This ore, which averages 0.6% BeO, is the principal source of beryllium for the subsequent extraction process (Fig. 2), although Brush Wellman processes both bertrandite ore and beryl at its extraction facility in Delta, UT. The beryl is purchased from foreign sources as a means of husbanding the available domestic reserves of beryllium-containing ore. The mine and mill at Delta, UT, contain an estimated 50 years of proven reserves of bertrandite ore.

Beryl is treated by arc melting, water quenching, and elevated-temperature sulfating to make the beryllium accessible to the extraction process. The bertrandite mineral is soluble in sulfuric acid without this treatment; this factor, along with the extensive nature of the deposit, provides the advantage of Utah bertrandite over imported beryl. The product of both ore inputs is a crude solution of beryllium sulfate. A purified beryllium hydroxide is obtained by a solvent extraction procedure followed by hydroxide precipitation. For the production of beryllium metal, the hydroxide is dissolved in ammonium bifluoride, purified, and then crystallized from aqueous solution as ammonium fluoroberyllate [$(NH_4)_2BeF_4$]. This salt is thermally decomposed to the anhydrous fluoride (BeF_2) and NH_4F gas. The fluoride is reacted with magnesium to yield beryllium and magnesium fluoride (MgF_2). The primary beryllium is in the form of pea-to-marble-sized pebbles.

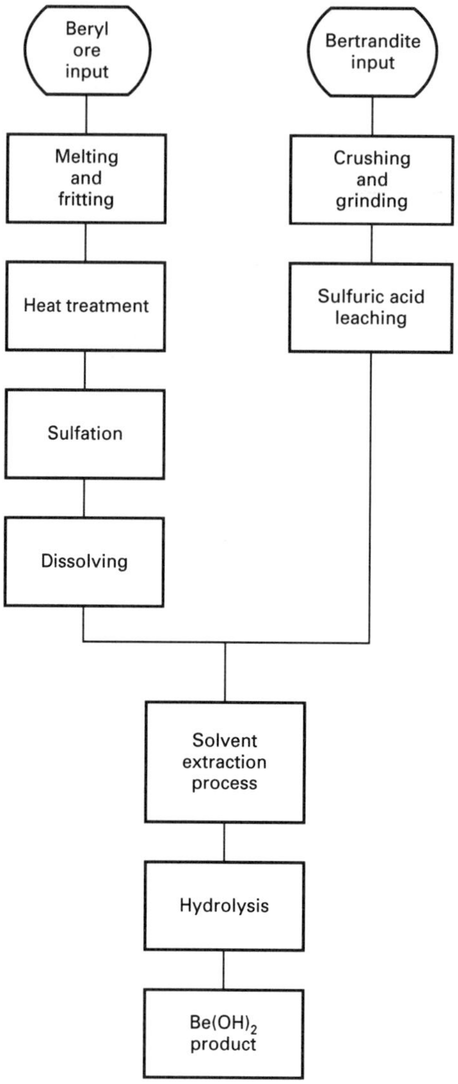

Fig. 2 Beryllium extraction processes used at the Delta, UT, extraction facility

The electrowinning of beryllium by fused-salt electrolysis of beryllium chloride ($BeCl_2$) from a variety of low-melting-point baths is possible, as is the electrorefining of beryllium using a bath of $KCl\text{-}LiCl\text{-}BeCl_2$. However, these procedures are not known to be in commercial operation at this time.

Beryllium pebbles from magnesium reduction are vacuum cast to remove residual reduction slags and any excess magnesium. A similar step is necessary with the electrolytic materials to remove the trapped chlorides from the fused-salt bath. The present commercial practice is to melt the metal in magnesium oxide crucibles and then pour it into graphite molds yielding about 180 kg (400 lb). The vacuum cast ingot is the input for beryllium powder manufacture.

Beryllium Powder Production Operations

Current Industrial Practices. Although ingot castings are not useful as commercial products themselves because of the inherent limitations of cast beryllium (as previously described), the ingot is quite acceptable from the chemical analysis point of view and provides a starting point for subsequent P/M operations. The ingot is converted into chips using a lathe and a multihead cutting tool. The chips are ground to powder using one of several mechanical methods: ball milling, attritioning, or impact grinding. These processes produce powders with varying characteristics, particularly with regard to particle shape.

Ball milling and attritioning are relatively slow processes, and they activate slip and fracture primarily upon the basal planes of the hexagonal close-packed beryllium crystal. This fracture mode gives rise to a flat-plate particle morphology. When the particles manufactured by this process are loaded into a die for consolidation, the flat surfaces align and give rise to areas of preferred orientation.

The impact-grinding process, which relies upon a high-velocity gas stream to accelerate beryllium particles and drive them against a beryllium target, activates fracture upon additional crystal planes other than the basal plane, resulting in a blocky particle. The blocky particles exhibit less tendency to align preferentially during powder loading, thereby reducing the tendency for preferred orientation in the final consolidated product. The reduction in preferred orientation leads to improved overall ductility, a property particularly sensitive to orientation, in all directions in the consolidated component. For this reason, impact grinding has largely replaced attritioning and ball milling as the major powder production technique.

Atomization has recently been introduced as a powder production technique. Atomization has several potential advantages over mechanical comminution methods. Atomization typically produces a spherical particle, and the use of such a particle is logically the most effective means of eliminating property anisotropy resulting from the shape of pow-

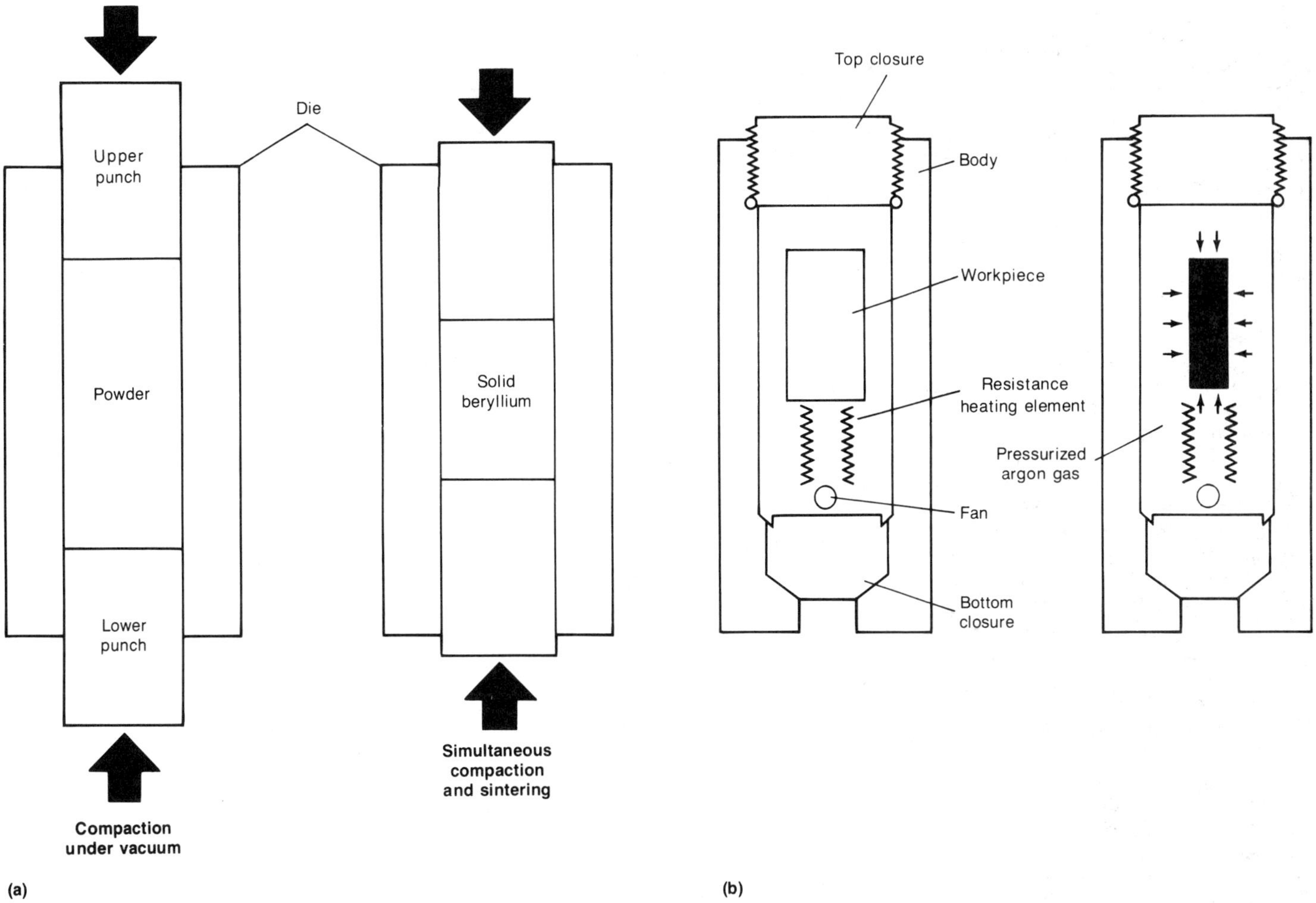

Fig. 3 Schematic diagrams of two powder consolidation methods. (a) Vacuum hot pressing. In this method, a column of loose beryllium powders is compacted under vacuum by the pressure of opposed upper and lower punches (left). The billet is then brought to final density by simultaneous compaction and sintering in the final stages of pressing (right). (b) Hot isostatic pressing. In this process, the powder is simultaneously compacted and sintered to full density inside a pressure vessel within a resistance-heated furnace (left). The powder is placed in a container, which collapses when pressure is exerted evenly on it by pressurized argon gas (right).

der particles. There are also economic factors that favor atomization. The inert-gas atomization process is capable of very high production rates, and the effectiveness of such a process could conceivably reduce production costs significantly.

Three atomization techniques are currently in various stages of developmental investigation. The inert-gas atomization process is furthest along in the development cycle and has shown the capability of producing clean fine-grain powders with excellent flow and packing characteristics. The properties of consolidated billets made from this material have been equal or superior to those made with comparable mechanically comminuted material. Centrifugal atomization, as represented by the plasma rotating-electrode process and the rapid solidification technique is in a less-advanced stage of application development.

Powder Consolidation Methods

Once powder is made, it must be consolidated. This step in the production of beryllium components has effects on the properties of the consolidated material that are as profound as the effects of the size and chemistry of the powder. Until the mid-1980s the predominant consolidation method was vacuum hot pressing of powder into right circular cylinders. This remains the technique with the highest production volume, but net-shape technology is transforming the industry because of the cost advantages associated with greater material utilization.

Vacuum Hot Pressing. Schematic diagrams of the vacuum hot pressing (VHP) and the hot isostatic pressing (HIP) processes are shown in Fig. 3. In the VHP process, the powder is vibratory loaded into a die, and the die is placed into a vacuum furnace. Hydraulically driven punches are inserted into the die cavity, vacuum is established, and the heating and pressing cycle is initiated. Consolidation of beryllium powder is performed in vacuum because of the reactive nature of the powder. Densities in excess of 99% of theoretical are typically achieved. The sizes of vacuum hot presses range from 200 to 1800 mm (8 to 72 in.) in diameter. Pressures up to 8 MPa (1200 psi) and temperatures up to 1100 °C (2000 °F) are used in the consolidation process. The consolidated billets are then machined into components.

Near-net shape processes are being developed that take advantage of various combinations of HIP, cold isostatic pressing (CIP), and sintering. The use of a variety of processes allows parts to be manufactured that meet all design requirements at the lowest manufacturing cost. The use of near-net shape processing is not an entirely new development. Cold pressing of powder followed by vacuum sintering to achieve final density has been used for a number of years to make parts such as brakes for F-14, C-5A, and S-3A aircraft. In recent years, however, the implementation of HIP technology has fostered a more aggressive development effort.

Sintering of beryllium to achieve high densities requires temperatures in excess of 1200 °C (2200 °F); these elevated temperatures cause grain growth and a resultant

Table 2 Chemistry of commercial grades of beryllium

Beryllium grade	Beryllium components, % Be, min	BeO, max	Maximum impurities, ppm Al	C	Fe	Mg	Si	Other, each
Structural grades								
S-65B	99.0	0.7	600	1000	800	600	600	400
S-200F and S-200FH	98.5	1.5	1000	1500	1300	800	600	400
Instrument grades								
I-70A	99.0	0.7	700	700	1000	700	700	400
O-50	99.0	0.5	700	700	1000	700	700	400
I-220B	98.0	2.2	1000	1500	1500	800	800	400
I-400B	94.0	4.25 min	1600	2500	2500	800	800	400

Table 3 Mechanical property requirements for structural grades of beryllium

Grade	0.2% yield strength MPa	ksi	Ultimate strength MPa	ksi	Elongation, %
S-65B	207	30	290	42	3.0
S-200F, S-200FH	240	35	325	47	2.0

reduction in mechanical properties. Hot isostatic pressing allows powders to be consolidated at temperatures as low as 650 to 725 °C (1200 to 1340 °F), although 1000 °C (1830 °F) is a more common consolidation temperature. Very high strength is achieved with the lower-temperature cycle, but with a loss of ductility. A temperature of 1000 °C (1830 °F) gives an excellent combination of strength and ductility. In addition, if HIP is employed as a final densification step, 100% theoretical density can be achieved. This density level is not possible with either sintering or the VHP process.

There are several variations on the net-shape processes that rely on HIP to provide final densification. The most straightforward of these is the direct-HIP process, in which powder is loaded into a shaped mild-steel can. The can is then degassed in a vacuum at about 600 to 700 °C (1100 to 1300 °F) in order to remove both air and the gasses that are adsorbed onto the powder surface. The can is then hermetically sealed and isostatically pressed. After HIP consolidation, the can is removed by either machining or acid leaching. If the can is properly designed, the shape of the beryllium part can be used with a minimum of machining or secondary processing. The direct-HIP process yields the greatest control over grain size—and therefore over properties—because it permits consolidation with the greatest temperature latitude.

Another variant of the net-shape process combines cold isostatic pressing with hot isostatic pressing. In this method, powder is loaded into an elastic (for example, neoprene or latex) bag and isostatically pressed by means of a liquid at room temperature to produce a green compact of about 80% theoretical density. The compact can then be vacuum sintered to achieve a density of 98% or more. At this density level, a can is not necessary for further densification by HIP because the porosity is isolated rather than interconnected. The expense of fabricating a complex disposable can is thus eliminated, and the bag that is used in place of the can has a relatively low cost and is reusable. The trade-off is that the high sintering temperature used in the bag method results in grain growth. However, this is offset somewhat by the elimination of residual porosity. Also, the CIP process permits the production of parts with complex, three-dimensional complexity that cannot be made by the uniaxial cold-pressing route.

Beryllium Grades and Their Designations

The distinctions among grades of beryllium are generally made according to mechanical properties, which vary with grain size and specimen orientation. Physical properties generally do not vary greatly among grades. In view of these considerations, it is obvious that the specification of a grade of beryllium must identify the consolidation technique as well as powder characteristics.

Beryllium grades, as opposed to alloys, are nominally beryllium with beryllium oxide as the only other major component. Although oxide inclusions are regarded as undesirable contaminants in many systems, the oxide in beryllium is desirable as a grain-boundary pinning agent. The finer the powder, the greater the oxide content. When very fine-grain sizes are required, the high oxide content of the powder acts to stabilize the grain size during consolidation. Beryllium oxide contents of commercial grades vary from an allowed maximum of 0.5% in O-50, an optical grade, to a required minimum of 4.25% in I-400, an instrument grade. Table 2 presents chemical compositions of currently available commercial grades of beryllium.

Structural grades of beryllium are indicated by the prefix S in their designations. Property requirements for commercially available structural grades are presented in Table 3. In general, these grades are produced to meet ductility and minimum strength requirements. S-200F, an impact-ground powder grade, is the most commonly used grade of beryllium. This grade evolved from S-200E, which used attritioned powder as the input material. Recently, S-200FH has been introduced to the marketplace; the suffix H designates consolidation by HIP. The properties of S-200F include a minimum ductility of 2% elongation in all directions within the vacuum-hot-pressed billet, an increase of 1% over its predecessor, S-200E. In addition, S-200F has yield and ultimate strengths that are somewhat higher than those of S-200E, reflecting the general improvement in powder-processing techniques and consolidation methods.

Grade S-65 also is an impact-ground powder product. This grade was formulated to meet the damage tolerance requirements for use in the Space Shuttle. Grade S-65 sacrifices some strength for improved ductility. The 3% minimum ductility requirement is achieved by using impact-ground powder in combination with tailored heat treatments. The heat treatments produce a desirable morphology of iron-aluminum-beryllium-base precipitates.

Low levels of iron and aluminum are present in commercial grades of beryllium (see Table 2). Although these elements cannot be eliminated economically, they can be balanced, and heat treatments can be applied to form discrete grain-boundary precipitates of $AlFeBe_4$. This minimizes the iron in solid solution or in the compound $FeBe_{11}$, either of which is embrittling. The precipitates also eliminate aluminum from the grain boundaries, thus precluding hot shortness at elevated temperatures.

At moderate temperatures, beryllium develops substantial ductility. At 800 °C (1470 °F), for example, the elongation of S-200F is in excess of 30%. Yield strength and ultimate strength decrease with increasing temperature, but usable strength and modulus are maintained up to approximately 600 to 650 °C (1100 to 1200 °F). The changes in strength and ductility of S-200F with temperature are shown in Fig. 4.

Instrument grades of beryllium, which are designated by the prefix I, were developed to meet the specific needs of a variety of precision instruments. These instruments generally are used in inertial-guidance systems where high geometrical precision and resistance to plastic deformation on a part-per-million scale are required. The resistance to deformation at this level is measured by microyield strength.

In addition to those grades developed to meet the needs of inertial guidance systems, grade I-70 was developed specifically for optical components in satellite imaging systems. Because the large mirrors used in aerospace optics must retain precise geometry throughout complex loading spectra, no differentiation was drawn between

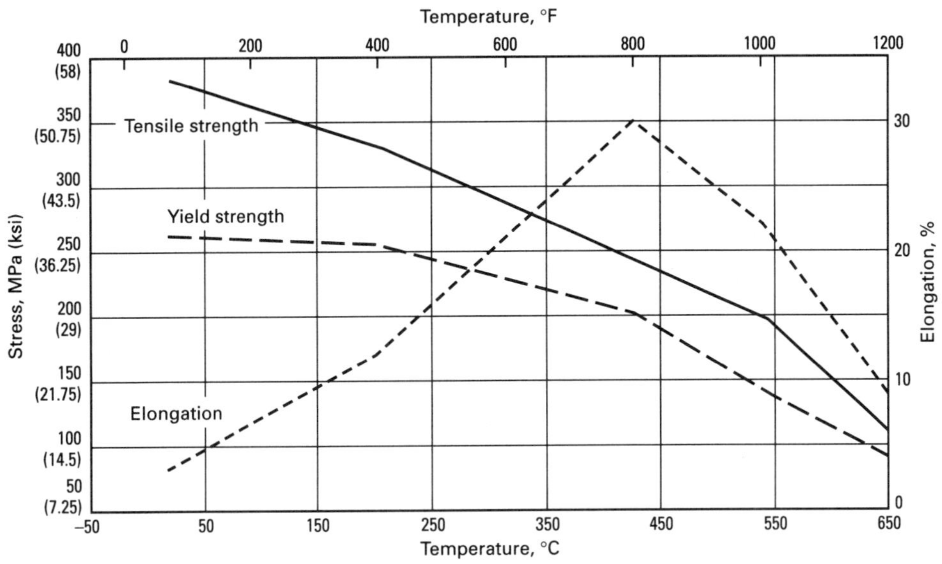

Fig. 4 Tensile properties of S-200F beryllium at elevated temperatures

Table 4 Mechanical property requirements for instrument grades of beryllium

Instrument grade	0.2% yield strength MPa	0.2% yield strength ksi	Tensile strength MPa	Tensile strength ksi	Elongation, %	Microyield strength MPa	Microyield strength ksi
I-70A	172	25	240	35	2.0	...	...
O-50	172	25	240	35	2.0	...	...
I-220B	275	40	380	55	2.0	34.5	5
I-400A	...	...	345	50	...	62	9

grades used for optical instruments and those used for inertial-guidance instruments. Recently, however, O-50, a grade developed specifically for the qualities of infrared reflectivity and low scatter has initiated the use of the prefix O to indicate an optical grade. The property requirements for commercially available instrument grades are given in Table 4.

The property requirement that differentiates instrument grades from structural grades is microyield strength. In I-400, ductility has practically been ignored in order to attain microyield strength values of 60 to 70 MPa (9 to 10 ksi). Grade I-220 strikes a balance between ductility and microyield strength, with a 2% elongation requirement and a microyield strength of 35 to 40 MPa (5 to 6 ksi). As greater insight into the relationship between properties and powder processing, composition, and heat treatment is obtained, further property improvements should be made possible.

Wrought Products and Fabrication. Consolidated beryllium block can be rolled into plate, sheet, and foil, and it can be extruded into shapes or tubing at elevated temperatures. At present, working operations typically warm work the material, thereby avoiding recrystallization. The strength of warm-worked products increases significantly as the degree of working increases. The in-plane properties of sheet and plate (rolled from a P/M material) increase as the gage decreases, as shown in Table 5. As with most hexagonal close-packed materials, it develops substantial texture as a result of these working operations. The texture in sheet, for example, generally results in excellent in-plane strength and ductility, with almost no ductility in the short-transverse (out-of-plane) direction. Similar property trade-offs resulting from the anisotropy caused by warm working are inherent in extruded tube and rod. Sheet can be formed at moderate temperatures by standard forming methods. For some special applications, there has been interest in the improved formability of ingot-derived rolling stock as opposed to material rolled from a block of P/M materials. The ingot-derived stock has improved weldability and formability, primarily because of its reduced oxide content.

In many instances, complex components must be built up from shapes made from sheet and foil. For example, beryllium honeycomb structures have been made by brazing formed sheet, as have complex satellite structural components. Beryllium tubing made by extrusion has also been used in satellite components. Space probes such as the Galileo Jupiter explorer have used extruded beryllium tubing to provide stiff lightweight booms for precision antenna and solar array structures.

Table 5 In-plane properties of beryllium products rolled from a P/M source block

Thickness mm	Thickness in.	0.2% yield strength MPa	0.2% yield strength ksi	Ultimate strength MPa	Ultimate strength ksi	Elongation, %
11–15	0.45–0.60	275	40	413	60	3
6–11	0.25–0.45	310	45	448	65	4
<6	<0.25	345	50	482	70	10

Health and Safety Considerations

Beryllium has been commercially produced for more than 50 years, and its toxicity has been recognized and successfully controlled for the last 30 years. Information on the toxicity of beryllium is contained in the article "Toxicity of Metals" in this Volume.

The main concern associated with the handling of beryllium is the effect on the lungs when excessive amounts of respirable beryllium powder or dust are inhaled. Two forms of lung disease are associated with beryllium: acute berylliosis and chronic berylliosis. The acute form, which can have an abrupt onset, resembles pneumonia or bronchitis. Acute berylliosis is now rare because of the improved protective measures that have been enacted to reduce exposure levels.

Chronic berylliosis has a very slow onset. It still occurs in industry and seems to result from the allergic reaction of an individual to beryllium. At present, there is no way of predetermining those who might be hypersensitive. Sensitive individuals exposed to airborne beryllium may develop the lung condition associated with chronic berylliosis.

Exposure Limits. Two in-plant exposure limits have been set by the Occupational Safety and Health Administration to prevent beryllium disease. The first is a maximum atmospheric concentration of 2 $\mu g/m^3$ of air averaged over an 8-h day. The second is a short-exposure limit of 25 $\mu g/m^3$ of air for a duration of less than 30 min. The U.S. Environmental Protection Agency (EPA) has set a nonoccupational limit of 0.01 $\mu g/m^3$ of air averaged over a 1-month period outside of a beryllium facility. The EPA limits the emission of beryllium into the environment to 10 g in any 24-h period. To meet these requirements, beryllium producers must adhere to industrial-hygiene standards and use air pollution control measures when dusts, mists, and fumes might be created. Historically, these control measures appear to have been effective in preventing chronic beryllium disease.

Precious Metals

Precious Metals and Their Uses

A.R. Robertson, Englehard Corporation

THE EIGHT PRECIOUS METALS, listed in order of their atomic number as found in periods 5 and 6 (groups VIII and Ib) of the periodic table, are ruthenium, rhodium, palladium, silver, osmium, iridium, platinum, and gold. Atomic, structural, and physical properties of the precious metals, which are also referred to as the noble metals, are listed in Table 1. Additional property data can be found in the articles "Properties of Precious Metals" and "Properties of Pure Metals" in this Volume.

Precious metals are of inestimable value to modern civilization. Their functions in jewelry, coins, and bullion, and as catalysts in devices to control auto exhaust emissions are widely understood. But in certain other applications, their functions are not as spectacular and, although vital to the application, are largely unknown except to the users. Many facets of daily life are influenced by precious metals and their alloys. For example, precious metals are used in dental restorations and dental fillings (see the section "Precious Metals in Dentistry" in this article). Precious metal solders are used in dentistry and in the jewelry and electronics industries. Thin precious metal films are used to form the electronic circuits. Much of our clothing today is produced with the aid of precious metals that are used in spinnerettes for producing synthetic fibers. Precious metals perform as catalysts in various processes; for example, widely used agricultural fertilizers are produced with the aid of a platinum-rhodium alloy catalyst woven in the form of gauze, and auto emissions are reduced through the use of platinum-group alloy catalysts. Electrical contacts containing palladium are essential to telephone communications. Certain organometallic compounds containing plati-

Table 1 Selected properties of precious metals

Property	Value for indicated metal							
	Platinum	Palladium	Iridium	Rhodium	Osmium	Ruthenium	Gold	Silver
Atomic number	78	46	77	45	76	44	79	47
Atomic weight, amu	195.09	106.4	192.2	102.905	190.2	101.07	196.967	107.87
Crystal structure(a)	fcc	fcc	fcc	fcc	hcp	hcp	fcc	fcc
Electronic configuration (ground state)	$5d^9 6s$	$4d^{10}$	$5d^7 6s^2$	$4d^8 5s$	$5d^6 6s^2$	$4d^7 5s$	$5d^{10} 6s$	$4d^9 5s^2$
Chemical valence	2,4	2,4	3,4	3	4,6,8	3,4,6,8	1,3	1,2,3
Density at 20 °C (70 °F), g/cm³ (lb/in.³)	21.45 (0.774)	12.02 (0.434)	22.65 (0.818)	12.41 (0.448)	22.61 (0.816)	12.45 (0.449)	19.32 (0.697)	10.49 (0.378)
Melting point, °C (°F)	1769 (3216)	1554 (2829)	2447 (4437)	1963 (3565)	3045 (5513)	2310 (4190)	1064.4 (1948)	961.9 1763.4
Boiling point, °C (°F)	3800 (6870)	2900 (5250)	4500 (8130)	3700 (6690)	5020 ± 100 (9070 ± 180)	4080 ± 100 (7375 ± 180)	2808 (5086)	2210 (4010)
Electrical resistivity at 0 °C (32 °F), μΩ · cm	9.85	9.93	4.71	4.33	8.12	6.80	2.06	1.59
Linear coefficient of thermal expansion, μin./in./°C	9.1	11.1	6.8	8.3	6.1	9.1	14.16	19.68
Electromotive force versus Pt-67 electrode at 1000 °C (1830 °F), mV	. . .	−11.457	12.736	14.10	. . .	9.744	12.34(b)	10.70(c)
Tensile strength, MPa (ksi)								
As-worked wire	207–241 (30–35)	324–414 (47–60)	2070–2480 (300–360)(d)	1379–1586 (200–230)(d)	. . .	496 (72)(d)	207–221 (30–32)	290 (42)
Annealed wire	124–165 (18–24)	145–228 (21–33)	1103–1241 (160–180)	827–896 (120–130)	. . .	. . .	124–138 (18–20)	125–186 (18.2–27)
Elongation in 50 mm (2 in.), %								
As-worked wire	1–3	1.5–2.5	15–18(d)	2	. . .	3(d)	4	3–5
Annealed wire	30–40	29–34	20–22	30–35	. . .	. . .	39–45	43–50
Hardness, HV								
As-worked wire	90–95	105–110	600–700(d)	. . .	. . .	. . .	55–60	. . .
Annealed wire	37–42	37–44	200–240	120–140	300–670	200–350	25–27	25–30
As-cast	43	44	210–240	. . .	800	170–450	33–35	. . .
Young's modulus at 20 °C (70 °F), GPa (10⁶ psi)								
Static	171 (24.8)	115 (16.7)	517 (75)	319 (46.5)	558 (81)	414 (60)	77 (11.2)	74 (10.8)
Dynamic	169 (24.5)	121 (17.6)	527 (76.5)	378 (54.8)	. . .	476 (69)	. . .	. . .
Poisson's ratio	0.39	0.39	0.26	0.26	. . .	. . .	0.42	0.37(e)

(a) fcc, face-centered cubic; hcp, hexagonal close packed. (b) At 800 °C (1470 °F). (c) At 700 °C (1290 °F). (d) Hot worked. (e) Annealed. Source: Engelhard Industries Division, Engelhard Corporation num are significant drugs for cancer chemotherapy.

Resources and Consumption

Metal specialists at the U.S. Bureau of Mines continually survey the market in silver, gold, and platinum-group metals to determine present availability and usage and to forecast future trends. *Mineral Commodity Profiles* and *Mineral Industry Surveys* covering these metals are issued periodically.

Much of the information in this section was obtained from Ref 1 to 8. For the latest available information concerning these metals, it is recommended that the most recent issues of these publications be consulted.

Silver. In recent years, the United States has been a net importer of silver. Imports, including unrefined silver, supplied about 89 million troy ounces of silver to the U.S. supply in 1988. Domestic mine production added 53 million troy ounces to this total, and refining of old scrap increased the U.S. silver level to a total of approximately 217 million troy ounces during this same period. The U.S. supply is obtained from primary and secondary sources. About 25% of primary silver is obtained from predominantly silver ores; the remaining primary silver is a by-product of the refining of copper, lead, zinc, and other metals. In addition, significant quantities of silver are derived as a by-product of gold mining. The top states for mine production are Idaho, Nevada, Montana, Arizona, and Utah.

Five smelting and refining companies produce the major portion of domestic primary silver. The smelters and refineries treat ores, concentrates, residues, and precipitates from company mines and plants in addition to materials purchased from other sources. Silver scrap is recycled by several primary smelters and a considerable number of small secondary refineries. In addition, secondary silver is recovered by several trading and fabricating companies and is recycled by end-product manufacturers.

In recent years, government regulations relating to the environment and to control of emissions of hazardous compounds have limited the operation of some base metal smelters that recover silver as a by-product. Silver in its ionic form, as in waste discharged from electroplating plants, is considered a potential source of pollution and a health hazard. Because of increasing concerns about environmental matters, governmental agencies can be expected to step up their efforts to minimize discharges from processing plants.

The following table presents primary statistics for U.S. silver demand in 1987. A general indication of the annual demand pattern for silver can be drawn from this data:

End use	Demand, troy ounces × 10^6
Electroplated ware	2.5
Sterling ware	3.8
Jewelry and arts	4.2
Photography	60.2
Dental and medical supplies	1.3
Brazing alloys and solders	5.6
Mirrors	1.0
Batteries	2.5
Contacts and conductors	22.7
Bearings	0.3
Coin, medallions, and commemorative objects	4.2
Catalysts	2.5
Other	4.5
Total U.S. demand	**115.3**

Several factors can affect the supply and demand of silver. One factor is the appreciable amount of silver required for monetary purposes; another is the speculative or investor market in refined silver bars and sacks of domestic coins. In addition, there has been some interest in the collection of commemorative medallions and limited-use objects fabricated from silver. Whether the latter end use will continue to grow or will decline in the future is a matter of considerable interest. In any event, depending on prices, a large potential secondary supply of silver is available in the form of coins, silverware, jewelry, and commemorative objects.

Gold. Because of its aesthetic beauty and enduring physical properties, gold is important not only to industry and the arts, but also as a commodity having long-term value. In the past, gold was considered to be mainly a monetary metal. However, starting in the late 1950s, more gold was used by manufacturers and investors than was used for monetary purposes. Since 1968, gold has become, to a considerable extent, a free-market commodity, with prices free to adjust to supply and demand. Despite this open market, almost half of the total world supply of gold, estimated at 2.4 billion troy ounces, is in various government vaults tied down by agreements among large industrial nations.

About 1.7 million troy ounces of gold per year are mined and reclaimed from old scrap in the United States (see, for example, the section "Recycling of Electronic Scrap" in the article "Recycling of Nonferrous Alloys" in this Volume). This amount falls far short of the amount required by U.S. industry. The requirements for bullion and coins are almost equal to those of industry. Because of this general demand for gold, the net inflow of this metal from foreign sources is large. In 1988, for example, 3.0 million troy ounces were imported, mostly in the form of refined metal.

About 50% of U.S. refinery production comes from gold ores, and the remainder from by-products of the refining of copper and other base metals. Refinery production in the United States includes gold from domestic mines, from imported ores and base bullion, and from domestic and foreign scrap. In recent years about 5 to 10% of U.S. refinery production has been derived from foreign ores, base bullion, and scrap.

About 5 to 10% of the total U.S. supply of gold comes from old scrap, which is defined as metal discarded after use. New scrap, which is generated during manufacturing, is usually reclaimed by the fabricator and is not considered part of the market supply.

The United States has appreciable gold resources, some of which are marginally profitable to recover. The price is now high enough to encourage growth of production at a modest rate, but environmental restraints on placer mining and the high cost of developing lode deposits currently dictate that the United States will continue to import most of its gold.

The largest foreign producer of gold, the Republic of South Africa, produced about 20 million troy ounces of the estimated 1988 total world production of 59 million troy ounces. Other important gold producers are the Soviet Union, Canada, South America, Asia, and Oceania. According to the U.S. Bureau of Mines, world resources are adequate to meet the forecast demand for this metal to the year 2000.

Data for U.S. gold demand in 1988 illustrates the general pattern of consumption of gold in fabricated products:

End use	Demand, troy ounces × 10^3
Jewelry and arts	1774.0
Dental supplies	247.0
Industrial products	1176.0
Total U.S. demand	**3197.0**

Source: U.S. Bureau of Mines

The use of gold for jewelry accounts for approximately 55% of the gold consumed. Dental uses generally amount to 7 to 8% of annual demand. Industrial requirements are generally centered in the electronics industry. However, even though the total number of end uses for gold in electronics continues to be high, considerable emphasis has been placed on reducing the use of gold in present applications because of its increasing cost. Bars, medallions, coins, and related products amounted to less than ¼ of 1% of the total U.S. gold consumption in 1988.

Platinum-Group Metals. The six closely related metals in the platinum group commonly occur together in nature. These transition elements are near neighbors in periods 5 and 6 of group VIII in the periodic table. Ruthenium, rhodium, and palladium each have a density of approximately 12 g/cm^3; osmium, iridium, and platinum each have a density of about 22 g/cm^3 (see Table 1).

The platinum-group metals are among the scarcest of metallic elements, and for this reason their cost is high. They occur as native alloys or in mineral compounds in placer deposits, sometimes together with gold; they also occur in lode deposits in basic or ultrabasic rocks, where they may be found together with nickel and copper. Most of the world supply of platinum-group metals is currently extracted from lode deposits in the Republic of South Africa, the Soviet Union, and Canada.

Another major source of platinum-group metals is old scrap obtained from obsolete equipment, spent catalyst, and discarded jewelry. This source is increasing in importance with the broadening industrial applications of these metals. Material of each type can be concentrated by a number of different methods, then refined by any of several chemical processes that conclude with a heating step to convert precipitates to metal in a porous, somewhat powdery form called sponge. Sponge is the most common form in commercial metal transactions, although ingots and shot are also traded.

In lode deposits, platinum-group metals are often associated with nickel and copper sulfide, which may be the principal products of mining (as in Canada and the Soviet Union) or important coproducts (as in the Republic of South Africa). In the ores, the proportions of the six platinum-group metals vary from one lode deposit to another. Canadian deposits in the Sudbury District contain approximately equal amounts of platinum and palladium; South African deposits contain more than twice as much platinum as palladium. Generally, platinum and palladium together account for about 80 to 85% of the platinum-group metals present in any given ore, followed by (in order of decreasing presence) ruthenium, rhodium, iridium, and osmium.

The composition of placer deposits differs somewhat from that of lode deposits. Placer deposits are characterized by the nearly complete absence of palladium and the common presence of gold. It appears that palladium and, to a certain extent, platinum, rhodium, and ruthenium are dissolved away during placer formation; in the well-established placer deposit in Witwatersrand, South Africa, only osmium and iridium are present (as the alloy osmiridium). At present, the only economically important placers are found in Colombia, South Africa, and the Soviet Union; together they account for about 2% of total world production.

The United States depends almost entirely on foreign sources for platinum-group metals. In 1988, net import reliance as a percentage of apparent U.S. consumption was approximately 93%. Apparent U.S. consumption during 1988 is estimated at 2.7 million troy ounces. The sources of U.S. supply between 1984 and 1987 were the Republic of South Africa, 44%; the United Kingdom, 16%; the Soviet Union, 9%; all others, 31%.

One company in Texas recovers platinum-group metals as by-products of the copper-refining process. In addition, about 24 smaller refineries located mostly on the East and West Coasts, handle or in some way process domestic scrap. However, most of them treat only platinum and palladium, and only three or four refine all six metals.

At present, U.S. mine production and reserves of platinum-group metals are small; untapped domestic resources appear to be large but are not well explored. The heavy dependence the United States places on foreign sources for these critical metals has strategic implications as well as a substantial impact on the U.S. balance of payments. The need for exploration and development of U.S. resources is apparent.

The U.S. demand for platinum-group metals is projected to grow at an annual rate of about 2.5%, with an estimated 1990 demand of 3.3 million troy ounces. Demand in the rest of the world is forecast to grow more slowly than in the U.S.; it is expected to reach a level of about 9.7 million troy ounces in the year 2000. World reserves and resources appear to be more than adequate to meet this demand.

The U.S. demand patterns in 1988 for three of the most-used platinum-group metals were:

End use	1988 Demand, troy ounces × 10^3 Platinum	Palladium	Rhodium
Automotive	609 000	160 000	65 000
Chemical	61 976	81 343	3 091
Dental and medical	10 871	227 747	142
Electrical	108 660	386 710	3 508
Glass	19 896	350	2 748
Jewelry and decorative	11 932	7 356	5 254
Petroleum	36 730	26 111	45
Miscellaneous	76 394	135 581	22 787
Total U.S. demand	**935 459**	**1 025 198**	**102 575**

The United States and Japan currently use about 60% of the platinum-group metals produced. Western Europe and the Soviet Union essentially divide the remaining 40%. Automotive emission requirements (which require the use of a platinum-palladium catalyst) began in 1974 in the United States and presently are substantial (this application alone accounted for 43% of total consumption in 1987). In Japan, about three-fourths of the platinum goes into jewelry, whereas in the United States and Western Europe, about 5 to 15% of this metal is used in jewelry. Substantial increases in the use of platinum-group metals can be expected in the European Economic Community with the gradual implementation of restrictions on automobile emissions from 1988 through 1993.

Trade Practices

Precious metals are bought and sold in troy ounces or, in markets where the metric system is used, in kilograms. One kilogram is equal to 32.15 troy ounces. The troy system of weights is based on the troy ounce of 480 grains, 31.1 grams, or 20 pennyweight. One troy ounce is equal to 1.097 avoirdupois ounces.

Silver and Gold. The term fineness refers to the weight portion of silver or gold in an alloy, expressed in parts per thousand. For example, 1000 fine silver (also called fine silver) is pure silver, or 100% silver, and 1000 fine gold is 100% pure gold. Gold bullion that is commercially traded is at least 995 fine or higher. Sterling silver is 925 fine, or 925 parts (also 92.5%) silver and 75 parts (or 7.5%) copper. Until 1964 the U.S. coin silver was an alloy of 90% Ag (900 fine) and 10% Cu. Silver bullion that is traded has a silver content ranging from 999 fine to 999.9 fine. Gold and copper are silver impurities in any fineness of silver bullion.

Another way of indicating gold purity is by the karat, which is a unit of fineness equal to the 1/24th part of pure gold. In this system, 24 karat (24 k) gold is 1000 fine or pure gold. The most popular jewelry golds in the United States are:

Karat designation	Gold content
24 k	100% Au (99.95% min)
18 k	18/24ths, or 75% Au
14 k	14/24ths, or 58.33% Au
10 k	10/24ths, or 41.67% Au

Each category differs from the others in number, type, and proportions of the base metal additions. Gold alloys used in jewelry are always specified by karats, whereas those used in dentistry and for electronic purposes are designated by percentage.

Jewelry golds range from light yellow through deep yellow to reds and greens, and also include a family of whites. Each alloying element has a different effect on the color of gold:

- *Silver*: As the proportion of silver increases, gold changes in hue from yellow to greenish-yellow to white
- *Copper*: As copper content increases, gold becomes redder in appearance
- *Nickel*: Nickel has the effect of whitening gold. The so-called white golds substitute nickel for silver
- *Zinc*: Zinc is considered a decolorizer. Some red golds (copper-containing alloys) are converted to a substitution of zinc for some of the copper and silver

Trade practice rules for the jewelry industry in the United States, set by the Federal Trade Commission, require that any article labeled gold contain at least 10 k gold, with a tolerance of 1/2 k. In gold cladding (gold adhered to base metal stock), the ratio of the weight of the material is indicated, together with the karat of the cladding. For example, one-tenth 12 k gold filled stock is a base metal surfaced with a layer or layers of 12 k gold alloy that make up 10% of the weight of the composite article. Such an article, if assayed into, would be found to contain 5% gold.

The designation gold filled is limited by stamping regulations to articles in which the weight of the coating is at least 1/20th of the total. Lower ratios may be stamped rolled gold plate. The quality mark cannot be applied to articles surfaced with an alloy of less than 10 k. In the cases of the gold-filled and rolled gold plate materials, the karat gold is bonded to the base metal substrate by soldering, brazing, welding, or mechanical means.

Platinum-Group Metals. Platinum and palladium are traded on the New York Mercantile Exchange in respective lots of 50 and 100 troy ounces; on the Chicago Mercantile Exchange, they are traded in units of 100 troy ounces. On the New York Exchange, the metals can be in the form of bar or sheet.

The required purity of platinum may vary according to the end use or application. Although commercial-grade platinum must be at least 99.8% pure, platinum with a purity of at least 99.9% is required for alloying, laboratory ware, and contacts. Platinum of even higher purity, sometimes with controlled impurities, is used for other specialized applications such as thermocouples and resistance thermometers. The present U.S. thermometric standard platinum, designated Pt 67, is 99.999% pure.

Federal regulations stipulate that an article of trade can be marked platinum only if it contains at least 98.5% platinum-group metals, of which no more than 5% may consist of platinum-group metals other than platinum; that is, the material must contain a minimum of 93.5% Pt. Special stamping provisions cover some jewelry alloys. All platinum jewelry sold in the United Kingdom must be hallmarked.

Alloys used for dental purposes are rather complex in composition, and metallurgical considerations are the dominant factors in their design. Various specifications have been established by the American Dental Association and by the federal government, but these do not cover all of the dental alloys (see the section "Precious Metals in Dentistry" in this article).

Special Properties

The precious metals have unusual combinations of properties that often are superior to those of other materials. In some cases, these property combinations make them the only materials that can meet the specialized requirements of an advanced technology or industrial application. The initial investment in these metals or their alloys may be high, but it is offset by long, reliable service and by ease of refining. Also, the refining process is marked by a high recovery rate because the precious metals are virtually indestructable. This high rate of recovery can make their use, for many applications, economical as well as efficient.

The precious metals share a number of properties that distinguish them from other metals or alloys, including corrosion resistance, good electrical conductivity, catalytic activity, and excellent reflectivity. However, each of these metals also has distinctive individual characteristics.

Silver, a bright white metal, is very soft and malleable in the annealed condition. It does not oxidize at room temperature, but it is attacked by sulfur. Nitric, hydrochloric, and sulfuric acids attack silver, but the metal is resistant to many organic acids and to sodium and potassium hydroxide.

In commercial applications, the special chemical properties, superior thermal and electrical conductivity, high reflectivity, malleability, ductility, and/or corrosion resistance of silver justify its high initial cost. In addition, uses have been established in photography, brazing, batteries, medicines, dentistry, mirror backings (silver-backed mirrors may become a more significant use as solar energy technology is developed), bearings, catalysts, coinage, and nuclear control rods.

The use of silver in photography is based on the ability of exposed silver halide salts to undergo a secondary image amplification process called development. In silver solders, the controlling factor is the rather low melting temperature of the alloys and their ability to wet various base metals at temperatures below the melting points of the metals to be joined. Such alloys do not dissolve or attack steel in normal usage, are ductile, have sufficient strength over a wide range of temperatures, and are capable of joining a wide variety of materials. Silver alloys are finding more and more use as replacements for the lead-tin solders traditionally used in residential plumbing.

Silver that contains varying amounts of dispersed cadmium oxide (≦20% CdO) is used in medium- and heavy-duty electrical contacts (see the article "Electrical Contact Materials" in this Volume). In this composite material, silver imparts its good electrical and thermal conductivity as well as its low surface contact resistance, and the dispersed cadmium oxide improves resistance to sticking and welding and provides good resistance to arc erosion (good arc quenching). The susceptibility of fine silver contacts to sulfidation precludes their use in low-current, low-voltage, and low-contact-force applications. In general, they should not be used below 10 V (except at high currents) or in situations where a voltage drop of 0.2 V will be troublesome; in addition, they are not suitable for application in low-level audio circuits because of the electrical noise they would introduce.

Silver is used in engine bearings because it has good lubrication properties as well as moderate hardness, good thermal conductivity, and low solubility in iron.

The good mechanical properties of certain silver-tin-mercury and silver-tin-copper-mercury alloys, and the small dimensional changes that occur during setting of these alloys, are the basis for the extended use of silver in dental amalgams (see the discussion of dental amalgams in the article "Properties of Precious Metals" that follows).

Sterling silver (silver-copper alloy) retains its long-established position in uses where elegant appearance is of paramount importance. For jewelry and tableware, high reflectivity makes silver particularly attractive. Much work has been done in developing a nontarnishing sterling silver, but no such alloy has yet been produced. Various thin protective coatings, such as rhodium, have been used on silver objects that are not likely to scratch.

Silver-clad copper, brass, nickel, and iron are produced for a variety of uses, ranging from electrical conductors and contacts to components for chemical equipment. Silver is also used in various chemical processes, including catalytic applications such as the production of formaldehyde or the oxidation of ethylene.

Silver coatings are applied to glass and ceramics by spreading a special silver paste on the material and then warming it to red heat. These coatings are widely used in electronic devices and automotive applications. Chemical methods for applying conductive coatings to plastics and glass are also used as the base for electroplate. Organometallic solutions containing silver are applied and fired in the production of conductors, electrical grounds and shields, resistance heaters, electrode terminals, and conductive bases for electroplating.

The rapid diffusion of oxygen through silver at elevated temperatures can be an advantage or disadvantage depending on the application. This phenomenon has been used to advantage in the internal oxidation of base metal alloying constituents (such as cadmium, rare earths, cerium, or calcium) in silver alloys. The resulting silver composite containing fine, well-dispersed oxide particles has been used in electrical contact applications (see the article "Electrical Contact Materials" in this Volume).

Electrodeposited silver is used widely for electrical, electronic, industrial, and decorative applications. Heavy electrodeposits can be used for surfacing chemical equipment and for bearings.

Gold is a bright, yellow, soft, and very malleable metal. Its special properties include corrosion resistance, good reflectance, resistance to sulfidation and oxidation, freedom from ionic migration, ease of alloying with other metals to develop special properties, and high electrical and thermal conductivity. Because gold is easy to fashion, has a bright pleasing color, is nonallergenic, and remains tarnish free indefinitely, it is used extensively in jewelry. For much the same reasons, it has long been used in dentistry in inlays, crowns, bridges, and orthodontic appliances (see the section "Precious Metals in Dentistry" in this article).

Gold is used to a considerable extent in electronic devices, particularly in printed circuit boards, connectors, keyboard contactors, and miniaturized circuitry (see *Packaging*, Volume 1 of the *Electronic Materials Handbook* published by ASM INTERNATIONAL). Because electronic devices employ low voltages and currents, it is important that the coated components remain completely free from tarnish films and that they remain chemically and metallurgically stable for the life of the equipment.

Gold is a good reflector of infrared radiation; for this reason, gold films are used in radiant heating and drying devices as well as in thermal barrier windows for large buildings. A much-publicized use of gold reflective coatings has been for protecting space vehicle components and space suits from excessive solar radiation that could raise temperatures substantially.

Fired-on gold organometallic compounds are used to decorate porcelain and glassware. Chemically inert gold rupture discs are used in chemical process equipment. Because of its good resistance to corrosion and wear, the gold alloy 70Au-30Pt has been used in the perforated spinnerettes through which cellulose acetate fibers are extruded. Gold has also been used in other industrial applications, such as sliding electrical contacts, fine-wire gold connectors for the semiconductor industry, vacuum and sputter-deposited films or coatings for interconnecting links in thin-film integrated circuits, gold brazing alloys for joining jet engine components, and gold alloys in thermocouples for both cryogenic service down to liquid helium temperature and high-temperature use up to 1300 °C (2370 °F).

Among the gold alloys, the Au-Ag-Cu-Pt-Pd alloys are used in dentistry because of their good mechanical properties, response to age-hardening treatments, nobility, and moderate melting points. The Au-Ag-Cu (yellow) golds and Au-Ni-Cu-Zn (white and suntan) golds have a relatively good resistance to tarnishing and corrosion, and adequate mechanical properties; these attributes, along with social custom and the available colors of the materials, account for their use for jewelry, eyeglass frames, and rings. These alloys are also used for certain rubbing contacts in small electrical devices. Gold-silver alloys containing about 70% Au, generally with a few percent platinum, resist both oxidation and sulfidation, and have other properties useful for low-current electrical contacts.

Pure gold is readily electrodeposited and it, as well as rhodium and palladium, is used for surfacing certain high-frequency conductors for service in environments where silver corrodes. A substantial quantity of electrodeposited gold is used for surfacing plug-type electrical connectors. Platinum metals may be required at higher temperatures to minimize diffusion and adhesion (sticking). Gold alloys are also electrodeposited on jewelry and other items where appearance is important. For some pieces, several layers of gold and other metals are deposited successively, and the article is subsequently heated to produce an alloy by diffusion.

Pure gold has high reflectivity in the red and infrared spectral ranges and therefore is sometimes used for surfacing infrared reflectors. Although pure gold resists nitric, sulfuric, and hydrochloric acids as well as many other corrosives, its applications are limited because of its susceptibility to attack by halogens, its softness and relatively low melting point, and, to some extent, its cost. However, gold sometimes is used as a lining for small calorimeter bombs and as a corrosion-resistant solder. The hard 70Au-30Pt alloy has been used for rayon spinnerettes, but it generally has been replaced in this application by Pt-10Rh.

Platinum-Group Metals. All six platinum-group metals are closely related and commonly occur together in nature. Their most distinctive trait in the metallic form is their exceptional resistance to corrosion. Of the six metals, platinum has the most outstanding properties and is the most used. Second in industrial importance is palladium, which is the lightest metal of the group.

Rhodium occasionally is fabricated in the unalloyed form, but it is more commonly used as an alloying element with platinum, and to a lesser extent, with palladium. In the unalloyed form, iridium is fabricated into large crucibles that are used in the production of single crystals of yttrium-aluminum garnet and gadolinium-gallium garnet, a substrate for bubble memory devices. It also finds considerable use as an alloying element for platinum and rhodium. Ruthenium is mainly used as an alloying element for platinum and palladium. When alloyed with platinum and palladium, rhodium, iridium, and ruthenium (in order of increasing effectiveness) act as hardening agents. Osmium forms a toxic oxide at ambient temperature and is therefore a difficult metal to utilize. A naturally occurring alloy of osmium and iridium called osmiridium is very hard and has been used for fountain pen tips and phonograph needles. The important properties of the platinum-group metals are outlined below.

Platinum is a white, very ductile metal that remains bright in air at all temperatures up to its melting point. Platinum has the following engineering characteristics:

- High melting point (1769 °C, or 3216 °F)
- Readily strengthened by alloying with compatible precious metals
- Can be electroplated
- Virtually nonoxidizable
- Resists molten glass and molten salts in oxidizing atmospheres
- Low vapor pressure
- Low electrical resistivity and, conversely, a high temperature coefficient of electrical resistivity; this combination makes it eminently suited for measuring elements in resistance thermometers
- Stable electrical contact resistance
- Stable thermoelectric behavior (the Pt-10Rh versus Pt thermocouple is the defining instrument on the International Practical Temperature Scale of 1968)
- High thermionic work function
- Special magnetic properties when alloyed with cobalt
- High thermal conductivity
- High resistance to spark erosion (hence its use in spark plugs)
- Excellent catalytic activity
- Coefficient of thermal expansion matching that of common glass

Platinum resists practically all chemical reagents and is soluble only in acids that generate free chlorine, such as aqua regia.

Palladium is a white, very ductile metal with properties similar in many respects to those of platinum. Palladium has the following engineering characteristics:

- A density of 12.02 g/cm^3, which is approximately 56% that of platinum and 63% that of gold. It can be used in place of lower-cost gold alloys without sacrificing the good corrosion resistance of gold
- High melting point (1554 °C, or 2829 °F)
- Excellent ductility
- Easily cold worked
- Outstanding ability to form extensive ductile solid solutions with other metals
- Can be electroplated, electroformed, and deposited via electroless methods
- Effective whitener for gold
- Good catalytic activity

Palladium resists tarnishing in ordinary atmospheres, but it does tarnish slightly upon outdoor exposure to sulfur-contaminated environments. When palladium is heated in air to 400 to 800 °C (750 to 1475 °F), a thin oxide film is formed; this film decomposes at higher temperatures, leaving the metal with a bright appearance. Hydrochloric acid

Table 2 Industrial applications of precious metals

Application	Special requirements	Metal or alloy
Electrical and electronic devices		
Spark plug electrodes	Resistance to corrosion and erosion	Thoriated Pt-4W, Ir, ODS Pt, Pd-Au
Jet engine glow plugs	Relight on flameout	Rh-Pt
Leads for thermistors	Freedom from oxidation	Pt and Ag plus binder
Transistor junctions	Doping contact	Au and doping alloy
	Nondoping contact	Ir-Pt
Resistors and potentiometers	High resistivity, low temperature coefficient, and low contact resistance	8W-Pt, 5Mo-Pt, 10Ru-Pt, Au-Pd-Fe, dental-type alloys
Resistance wire and resistance film	High resistivity, low temperature coefficient, and low contact resistance	Au-Pd-Pt
Electrodes for ceramic condensers	Applicability, nonoxidizing, solderability	Ag or Pt, with bonding agent
Electrodes for air condensers	Corrosion resistance	Ag and Au
Conductors in printed circuits	Corrosion resistance, solderability, wear resistance (Rh)	Ag, Au, Rh, Pd (Ag may lead to ionic shorting)
Connectors (such as terminals, lugs, and tabs)	Low contact resistance, solderability	Ag, Au; Pd electro- or electroless plate
High-temperature wiring	Conductivity, oxidation resistance, low contact resistance	Pt-clad base metal, solid Ag, Ag-Mg-Ni
Fuses	High conductivity and oxidation resistance	Ag-Au
Solid leads in mercury contact devices	Negligible solubility, freedom from oxidation	Pt where wetting required; also 10Ir-Pt. Ir where no wetting desired. Rh-plated steel for collector rings
Bonding in vacuum devices requiring vacuum-tight low-vapor-pressure seals	Desired melting point and low vapor pressure	28Cu-72Ag, 20Cu-80Au, 40Ni-60Pd, Au-Pd
Brazing alloys for tungsten	Ductility, high melting point, vapor pressure	Platinum
Instrument applications		
Sensing elements for resistance thermometers	Stable and known resistance, high temperature coefficient	Ultrapure Pt
Thermocouples	Stable temperature relation	10Rh-Pt vs Pt, 6Rh-Pt vs 30Rh-Pt, 13Rh-Pt vs Pt, 5Rh-Pt vs 20Rh-Pt, Au-Pd vs Rh-Pt, Au-Pd vs Ir-Pt
	For sensing ultrahigh temperature in oxygen-free atmosphere	Ir-Rh vs Ir
	High electromotive force	Au-Pd vs Rh-Pt, Au-Pd vs Au-Pd-Pt
Thermocouple connectors	Low-resistance joints with base metal wires	Platinum plate
Galvanometer suspensions	Corrosion resistance, strength, and conductivity	40Cu-60Pd (slow cooled), 14 k Au, Ag-Cu
Galvanometer pivots	Hardness and corrosion resistance	60 Os-Ru alloy
Contact parts in low level switches	Low electrical contact resistance, good wear resistance	Rh electroplate; 69Au-6Pt, 25Ag; Pt, Pd, and hard dental alloys
Slip rings, brushes for selsyns	Low contact resistance, good wear resistance, and minimum friction	18 k Au, dental alloys, 60Pd-40Cu, Ag, Au electroplate, Rh electroplate
Sensing elements for gas analyzers	Catalytic action proportional to gas content	Pd-Pt, platinum metal
Glass and ceramics industries		
Tanks and crucibles for optical glass	Insolubility, high melting point, noncontaminating	Pure platinum
Glass and ceramics industries (continued)		
Bushings and valves for fiberglass	Insolubility, high strength	10Rh-Pt, 20Rh-Pt, ODS Pt-Rh
Crucibles for continuous melting glass frit	Noncontaminating	Platinum
Crucibles for melting optical salt crystals	Insolubility, high melting point, noncontaminating	Platinum
Metallized glass and ceramics, metal film bonded to ceramic by heat	Nonoxidizing, desired color	Liquid-bright Au and Pt pastes
Metallized glass and ceramics, metal film produced by vacuum sublimation	Desired properties	Au, Pd, Rh, Ag, and alloys
Heater windings for glass, ceramic, and ferrite research	Nonoxidizing, high melting point, low vapor pressure	Pt, 20Rh-Pt, and 40Rh-Pt
Chemical industry		
Septum in a hydrogen purification system	Selective transmission	Pd, 60Pd-40Ag
Catalyst for removal of oxygen from H_2	Activity at low temperature	Pd on alumina
Septum in an oxygen purification system	Selective transmission	Pure silver
Catalyst for production of nitrogen or nitrogen-hydrogen heat-treating atmosphere from ammonia	Activity and long life	Platinum metal
Catalyst for production of formaldehyde from methanol	Activity	Silver
Catalyst for production of ethylene oxide from ethylene	Activity	Silver
Catalyst for destruction of odoriferous or hazardous contaminants	Activity	Platinum metal
Catalyst for ammonia plus air to yield HNO_3	Long life, high efficiency	Rh-Pt
Catalyst for ammonia, air, and methane to yield HCN	Long life, high efficiency	Rh-Pt
Rayon spinnerettes	Corrosion resistance, strength, ductility	Rh-Pt, Pt-Au
High-temperature HCl containers	Corrosion resistance	Platinum
Electrochemical applications		
Insoluble anode for electrolytic protection	Non-film-forming, high corrosion resistance	Platinum, 20Pd-Pt, and 50Pd-Pt
Insoluble anode for production of persulfates and perchlorates, and for electroplating	Corrosion resistance in chlorides, sulfates; proper anodic reaction	Platinum and 5Ir-Pt
Positive plates in primary and secondary batteries	Corrosion resistance, conductivity, and depolarization	$Ag-Ag_2O_2$
Fuel cell electrodes	Catalytic activity, corrosion resistance	Platinum metals
Container for tantalum capacitors	Corrosion resistance, high conductivity	Silver
Aerospace applications		
Brazing alloy in stainless steel systems for handling rocket fuels and oxidizers	Corrosion resistance, compatibility	Au-Cu-Ni, Au-Ni-Cr
Special uses		
Crucible for molten lead	Insolubility and high melting point	Ir under oxygen-free atmosphere
Crucible for molten bismuth	Insolubility and high melting point	Ru under oxygen-free atmosphere
Crucible for molten NaOH	High corrosion resistance	Silver
Container for high-temperature sulfur and sulfur gases	High corrosion resistance	Gold

(continued)

Table 2 (continued)

Application	Special requirements	Metal or alloy	Application	Special requirements	Metal or alloy
Special uses (continued)			**Reflectors**		
Container for high-temperature SO_2	Corrosion resistance, ductility	Pure Pt, pure Au, Au-Pt alloy	Visible and infrared reflecting surface	High efficiency	Ag where protected; Rh where exposed
Container for high-temperature (1000 °C, or 1830 °F) H_2S	Corrosion resistance, ductility	Gold, platinum	Ultraviolet and infrared reflecting surface	High and uniform reflectivity	Rhodium
Container for S and H_2S (<1000 °C, or <1830 °F)	Corrosion resistance, ductility	Gold	Red and infrared reflecting surface	High long-wave reflectivity	Gold
Neutron absorber	High absorption cross section	Iridium	**Safety devices**		
Intense gamma ray source	Radiation energy; moderate half-life	Iridium	Over-pressure protector (frangible disk)	Reproducible tensile properties, corrosion resistance	0.6Ir-Pt, Ag, Au
Magnet	Highest known energy product and corrosion resistance, ductility	23Co-Pt	Fuse wire for temperature-limiting fuse	Required and constant melting point oxidation resistance	Gold
Laboratory ware	Corrosion and heat resistance	Platinum, 0.6Ir-Pt, 3.5Rh-Pt			

and sulfuric acid attack palladium slightly; nitric acid, ferric chloride, and most halogens attack it readily. Palladium absorbs hydrogen, which will diffuse at a relatively rapid rate when the metal is heated. This reaction is the basis for laboratory apparatus for purifying hydrogen. Palladium has found increasing use in dental porcelain fused to metal alloys (for example, Au-Pt-Pd-Ag alloys).

Rhodium is a hard white metal. It is fairly ductile when hot. Rhodium is the whitest platinum-group metal and remains bright under all atmospheric conditions at ordinary temperatures. It resists hot aqua regia. High oxidation resistance and a high melting point (1963 °C, or 3565 °F) permit rhodium to be used for fabricating items for use at high temperature. Rhodium has high specular reflectivity and the highest electrical and thermal conductivities of any platinum-group metal.

Iridium is a white metal that has limited malleability at room temperature; however, it can be worked at elevated temperatures. Iridium oxidizes visibly when heated in air (to temperatures of 600 to 1000 °C (1100 to 1850 °F), but it remains bright at higher temperatures. Acids or aqua regia do not attack it, but molten salts do. Iridium has exceptional corrosion resistance, and this property coupled with a high-temperature (≦1650 °C, or 3000 °F) strength comparable to that of tungsten and a high melting point (2447 °C, or 4437 °F) permit its use in crucibles for melting nonmetallic substances at temperatures as high as 2100 °C (3800 °F). Iridium has a high modulus of elasticity (517 GPa, or 75 × 10^6 psi). It is the only known metal that can be used for short periods of time at temperatures up to 2000 °C (3650 °F) in air without undergoing catastrophic failure. Iridium is catalytically active and is the heaviest of all metals (22.65 g/cm^3).

Ruthenium is a very hard white metal that cannot be worked cold. It can be worked after being heated to a fairly high temperature, but only with extreme difficulty. Ruthenium resists common acids, including aqua regia, at temperatures up to 100 °C (212 °F). Ruthenium has a high resistance to contamination by lead. Like iridium, it is principally used as a hardener for platinum and palladium. Ruthenium has a high melting point (2310 °C, or 4190 °F), is exceptionally hard, and has a high elastic modulus (414 GPa, or 60 × 10^6 psi). In the absence of oxygen, ruthenium exhibits good resistance to attack by molten lithium, sodium, potassium, copper, silver, and gold. It has low electrical contact resistance at temperatures up to 600 °C (1100 °F) and resists any tendency of the contacts to weld together at these temperatures.

Osmium is a white hard metal that is not malleable at room or elevated temperatures. It forms a toxic oxide at ambient temperatures. Osmium is used as an alloying element to provide other precious metals with extreme hardness and resistance to corrosion. Osmium has the highest melting point of all the platinum-group metals (3045 °C, or 5513 °F) and the second-highest density (22.61 g/cm^3).

Commercial Forms and Uses

Semifinished Products. Silver, gold, platinum, palladium, and rhodium can be drawn to rod and wire as small as 25 μm (0.001 in.) in diameter. Iridium can be drawn to diameters as small as 75 μm (0.003 in.). Some of the platinum alloys containing iridium or rhodium can be drawn to diameters of 7.5 μm (0.0003 in.).

Sheet, strip, ribbon, and foil in a broad range of alloys, sizes, and thicknesses can be produced. Silver, gold, platinum, and some of its alloys can be rolled to thicknesses as small as 2.5 μm (0.0001 in.), but tolerances cannot be guaranteed. Clad materials can be obtained as wire, sheet, strip, and formed parts, with a great variety of substrate materials.

Tube is manufactured in a wide range of sizes and in round, half-round, and square sections. Seamless tube made of platinum, palladium, gold, and most alloys of these metals is manufactured in sizes ranging from 0.4 mm (0.016 in.) outside diameter × 0.1 mm (0.004 in.) wall thickness up to 44 mm (1.750 in.) outside diameter × a 5 mm (0.200 in.) wall thickness. Tube in larger sizes or made of less ductile materials such as platinum alloyed with rhodium (≧25%) or iridium (>25%) is manufactured only as seamed tube with a 3 to 75 mm (⅛ to 3 in.) inside diameter × a 0.25 to 2.5 mm (0.010 to 0.100 in.) wall thickness. Pure rhodium and pure iridium are usually furnished as seamed tube with a 3 to 40 mm (⅛ to 1½ in.) inside diameter × a 0.25 to 0.6 mm (0.010 to 0.025 in.) wall thickness and as single lengths about 150 mm (6 in.) long. Base metal tube is available with an outer cladding or an inner lining of platinum, gold, silver, or any of the commercial precious metal alloys.

Precious metal powders are produced for a wide range of electronic and industrial applications. Electronic powders are chemically precipitated to produce particle sizes of less than 10 μm and tend to be high in surface area. They are used as inks in hybrid circuits. Flake powders tend to produce shinier, smooth films; spherical particles more often result in dull-appearing surfaces. Platinum, palladium, 40Pt-20Pd-40Au, 10Pt-20Pd-70Au, 7.5Pt-22.5Pd-70Au, and 75Au-25Pd powders have been used for electronic purposes. Trials are usually necessary to determine the most suitable powder from the standpoint of both cost and performance.

Powders intended for industrial uses are composed of mixtures of particles that range in size from about 2 to 3 μm to as large as 840 μm (20 mesh), depending on the size required. These powders are suitable for use in powder metallurgy parts, as protective coatings against hostile industrial environments, as raw materials in alloy manufacture, and for various other uses.

Industrial Uses. Requirements and materials for more than 65 industrial applications of precious metals are cited in Table 2. Additional information on selection and application of precious metal contacts is included in the article "Electrical Contact Materials" in this Volume.

Coatings. Several cladding or coating processes are used to produce composite articles with precious metal surfaces. Table 3 lists the most important coating processes for precious metals, with characteristics, common thickness ranges, and typical applications of each.

Jewelry. Gold, the first jewelry metal, still is the most popular. The popularity of gold is maintained by tradition, its distinctive color, and the karat mark. Color and karat are the primary factors to be considered in the selection of a particular gold. Yellow is the most popular color, but red, green, and white karat golds are also available. The 14 k golds are the most popular in the United States, although significant quantities of all kinds of jewelry are made of 18 k gold. At the same time, there is significant use of 10 k gold, especially for rings set with synthetic colored stones. Gold-plated jewelry generally is produced for mass market jewelry lines rather than for fine jewelry lines.

Hand crafting is usual where only a few exclusive creations are made, but where many duplicates will be required, die forming or casting is appropriate. For simple rings, mechanical forming methods are justified where more than a thousand units are required; casting is more cost effective for smaller quantities. When complex shapes such as watchcases are produced from clad or filled stock, intricate and expensive dies are required to maintain uniformity of the cladding.

Platinum is frequently used to make settings for the finest jewelry. In addition to its high intrinsic worth, the workability and strength of platinum ensure reliable retention of jewels, and its white color enhances the brilliance of diamonds. Manufacturers use Pt-10Ir for either wrought or cast items; in some instances, Pt-5Ru may be used. The 15 and 20% Ir alloys are preferred for some of the more delicate pieces, such as small chain.

Where any excess weight is objectionable, as in earrings, palladium is preferred to platinum because of its lower density. Palladium has platinumlike characteristics that have led to its increased use, particularly in quality jewelry.

All of the white metals—platinum, palladium, and white gold—are frequently finished with rhodium plate for whiteness and wear resistance. Sterling is the standard silver jewelry alloy in spite of its tendency to tarnish.

Table 3 Precious metal coatings

Method	Characteristics	Thickness range	Examples of applications
Mechanical and thermal bonding (cladding)			
Brazing, hot pressing, hot and cold rolling, puddling, casting	100% density, good adhesion, high wear resistance, uniform thickness	≥2.5 μm (≥0.1 mil)	Precious-metal-clad base metals for jewelry, electrical contacts, chemical apparatus, or other industrial uses; applicable for all malleable precious metals and alloys
Vacuum coating			
Vacuum metallizing	Fairly uniform coating, transparent layers, good adhesion	0.025–12.5 μm (1–500 μin.)	For decorative purposes, reflectors (rhodium on glass), condensers for electronic devices (mostly metals on paper, plastic, or lacquered surfaces); applicable for Ag and Au. Nucleation with Ag required prior to applying Zn on plastic condensers
Cathode sputtering	Very even coating, good adhesion, high density	1.2–125 μm (0.05–5 mils)	For improved corrosion resistance, silver in surgical gauzes, gold on thin Al alloy foils, diaphragms, mirrors
Electrochemical and chemical coating			
Electroplating	Reasonably dense and usually well-adhering deposits; mechanical and physical properties depend greatly on plating conditions	0.15–125 μm (6–5000 μin.)	Decorative uses, improved corrosion and wear resistance, electrical contacts; applicable to a wide range of elemental precious metals and some of their alloys
Fired-on films			
Formulated organometallic solutions, thermal decomposition	Thin, well-adhering film	0.05–0.25 μm (2–10 μin.)	Ceramic and electronic uses, printed circuits, decorations; applicable to bright Au, Ir, Pt, Pd, and Ag, mostly on nonmetallic surfaces
Resins containing very fine suspended metal particles with a low-melting inorganic glass flux	Thick, adhering films	12–40 μm (0.5–1.5 mils)	Electronic applications
Chemical decomposition coating	Thin, well-adhering film	Usually very thin	Mirrors

Precious Metals in Dentistry

THE CHANGES IN THE USE of precious metals in dentistry over the last two decades has been dramatic. The use of gold has declined substantially; it has been replaced in many dental applications by palladium or by nonmetallic dental materials. Reference 9 provides information on the materials science specific to the dental industry. Reference 10 is a comprehensive review of the multidiscipline approach involved in considering materials for the oral environment.

This section is a brief review of the dentistry materials presently in use, with an emphasis on the applications of precious metals. The corrosion characteristics of dental alloys are thoroughly reviewed in the article "Tarnish and Corrosion of Dental Alloys" in *Corrosion*, Volume 13 of the 9th Edition of *Metals Handbook*, and reference is made to this article in several tables.

Classification of Dental Alloys

A variety of alloys are available for dental applications:

- Direct filling alloys
- Crown and bridge alloys
- Partial denture alloys
- Porcelain fused to metal (PFM) alloys
- Wrought wire alloys
- Implant alloys
- Soldering alloys

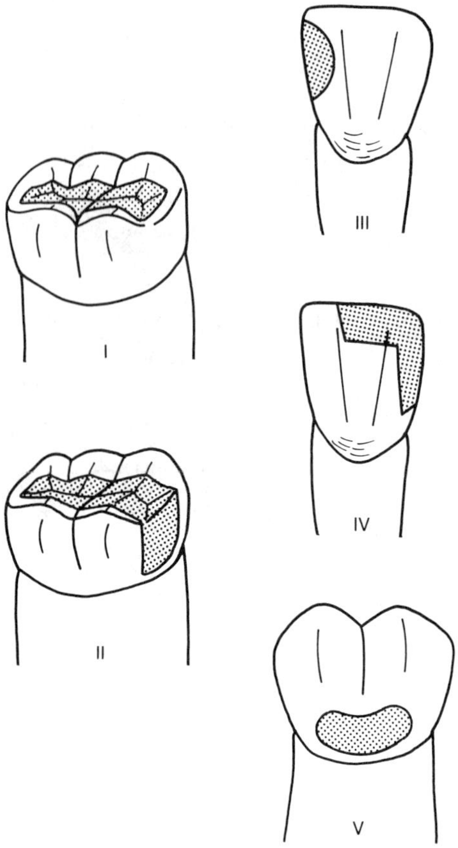

Fig. 1 Different classes of inlay preparation. Classes I and II involve one or two surfaces of a posterior tooth. The restorations can be made of a soft gold alloy (80Au-10Ag-9Cu-1Pd), but are usually made of silver amalgam. Alternate materials are either composite resin or dental porcelain bonded onto the remaining tooth. Classes III, IV, and V inlays are generally made of composite resin. Source: J.F. Jelenko & Company

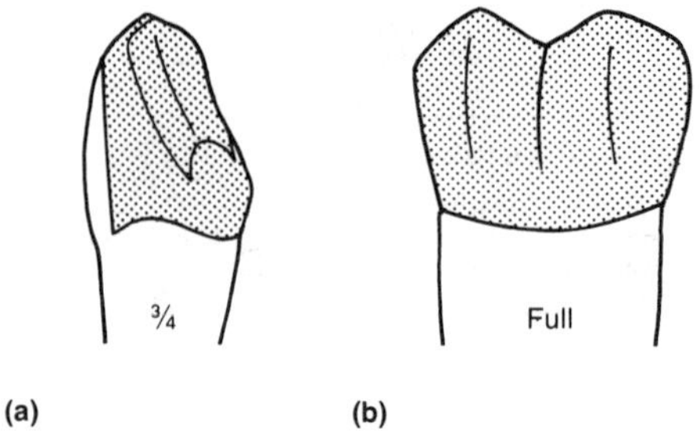

Fig. 2 Gold alloy dental crowns. (a) Three-quarter crown, which covers three surfaces of a tooth. (b) Full crown, which covers the entire tooth. These types of restorations as well as bridgework (multiple crowns) are made from gold alloys containing 40 to 78 wt% Au. Source: J.F. Jelenko & Company

Of these categories, precious metals are used in direct filling alloys, crown and bridge alloys, PFM alloys, and soldering alloys. Figures 1 and 2 show a number of typical restorations fabricated from some of these alloys.

Direct Filling Alloys. Compositions for direct filling restorations usually consist of silver-tin-copper-zinc alloy amalgams. Pure gold in the form of cohesive foil, mat, or powder is used only in very limited applications.

Amalgams are produced by combining mercury with alloy particles by a process referred to as trituration. About 42 to 50% Hg is initially triturated with the high-copper types; increased quantities of mercury are used with the low-copper types. High-speed mechanical amalgamators mix the materials in a matter of seconds. The plastic amalgam mass after trituration is inserted in the cavity by a condensation process. This is accomplished by pressing small amalgam increments together until the entire filling is formed. For amalgams using excess mercury during trituration, the excess mercury is condensed to the top of the setting amalgam mass and scraped away. Table 4 presents compositions for a number of different amalgam alloys.

Crown and Bridge Alloys. Alloys for all-alloy cast crown and bridge restorations are usually gold-, silver-, or nickel-base compositions, although iron-base and other alloys have also been used. The gold-base alloys contain silver and copper as principal alloying elements, with smaller additions of palladium, platinum, zinc, indium, and other precious metals such as ruthenium and iridium as grain refiners.

Table 5 lists compositions of precious metal crown and bridge alloys. Physical and mechanical properties for dental alloys that contain precious metals are provided in Table 6.

Partial Denture Alloys. This type of restoration consists of an alloy substructure upon which a pink acrylic resin and plastic teeth are placed. A cobalt-chromium alloy is almost always used in place of gold.

Porcelain Fused to Metal Alloys. Alloys for porcelain fused to alloy restorations are gold-, palladium-, or nickel-base compositions. The gold-base alloys are divided into gold-platinum-palladium, gold-palladium-silver, and gold-palladium types. The palladium-base alloys are palladium-silver alloys or palladium-gallium alloys with additions of either copper or cobalt. The nickel-base alloys are alloyed primarily with chromium and with minor additions of molybdenum and other elements (about 40% of PFM alloys are made from Ni-15Cr-3Mo-2Be). In contrast to alloys for crown and bridge use, alloys fused to porcelain contain low concentrations of oxidizable elements such as tin, indium, iron, and gallium for the precious-metal-containing alloys; and aluminum, vanadium, and others for the nickel- and cobalt-base alloys. During the heating cycle, these elements form oxides on the surface of the alloy and combine with the porcelain at the firing temperatures to promote chemical bonding.

Table 4 Compositions of selected dental amalgam alloys

Alloy(a)	Composition, wt% Ag	Sn	Cu	Zn
Low-copper amalgam				
1	75.0	24.6	0.1	0.3
2	72.0	26.0	1.0	1.0
3	72.8	26.2	2.4	1.0
4	69.0	26.6	3.5	0.9
5	68.0	26.0	5.1	0.9
High-copper amalgam				
6	69.8	20.0	9.7	0.5
7	69.3	18.1	11.6	1.0
8	69.8	16.2	13.5	0.5
9	62.0	18.5	18.5	1.0
10	63.5	16.9	19.5	0.2
11	59.5	27.6	12.2	0
12	60.0	22.0	13.0	(5.0 In)
13	49.5	30.0	20.0	(0.5 Pd)
14	41.0	32.5	26.5	0
15	40.0	31.2	28.8	0

(a) Numbers are provided for reference purposes only; they are not alloy designations. Source: *Corrosion*, Volume 13 of the 9th Edition of *Metals Handbook*

Stringent demands are placed on the alloy system meant to be used as a substrate for the baking on or firing of a porcelain veneer. The thermal expansion coefficients of alloy and porcelain must be matched so that the porcelain will not crack and break away from the alloy as the material is cooled from the firing temperature to room temperature. Thermal expansion coefficients of porcelains are in the range of 14×10^{-6} to 15×10^{-6} in./in. · °C. Selection of an alloy with a slightly larger coefficient by about 0.05% is recommended so that the alloy will be under slight compression.

The alloy must have a high melting point so that it can withstand the firing temperatures involved with the porcelain. However, excessively high temperatures will preclude the use of conventional dental equipment; therefore, a temperature of 1300 to 1350 °C (2370 to 2460 °F) is about maximum. The porcelain firing procedures require an alloy with high hardness, strength, and modulus so that thin sections of the alloy substrate can support the porcelain, especially at the firing temperatures. Compositions and properties of precious metal PFM alloys are given in Tables 6 and 7.

Wrought orthodontic wires are composed of stainless steel, cobalt-chromium-nickel, nickel-titanium, and β titanium alloys. The superelastic nickel-titanium alloys are described in the article "Shape Memory Alloys" in this Volume.

Implant Alloys. Alloys that have found applications in support structures implanted in the lower or upper jaws are composed of cobalt-chromium, nickel-chromium, stainless steel, and titanium and its alloys.

Soldering Alloys. Gold-base and silver-base solder alloys (Table 8) are used for the joining of separate alloy components. Base

Table 5 Compositions of crown and bridge alloys that contain precious metals

Alloy(a)	Composition, wt% Au	Pt	Pd	Ag	Cu + Zn
1.	91.7	. . .	. . .	5.6	2.7
2.	81.0	. . .	4.0	12.0	3.0
3.	78.0	. . .	2.0	13.0	7.0
4.	74.0	. . .	4.0	12.0	10.0
5.	69.0	3.0	4.0	12.0	12.0
6.	64.0	2.0	2.0	19.0	13.0
7.	62.0	. . .	3.0	26.0	9.0
8.	60.0	. . .	4.0	27.0	9.0
9.	59.5	. . .	4.0	25.0	11.5
10.	59.0	. . .	4.0	23.0	14.0
11.	48.5	. . .	3.5	35.0	13.0
12.	50.0	. . .	4.0	25.0	21.0
13.	20.0	. . .	20.0	40.0	20 In + Zn
14.	39.0	1.0	6.0	41.0	13.0
15.	15.0	1.0	23.0	44.0	17.0
16.	40.0	. . .	5.9	40.5	13.6
17.	42.0	2.0	8.0	9.0	39.0
18.	. . .	. . .	25.0	70.0	5.0
19.	41.0	1.0	4.0	9.0	45.0
20.	. . .	. . .	25.0	59.0	16.0
21.	26.0	. . .	10.0	. . .	64.0
22.	. . .	. . .	8.0	70.0	22.0 In

(a) Numbers are provided for reference purposes only; they are not alloy designations. Source: *Corrosion*, Volume 13 of the 9th Edition of *Metals Handbook*

alloy solders with high fusing temperatures are also used for the joining of nickel-chromium and other alloys. In many cases, the term brazing would be more appropriate than joining, but the former term is seldom used in dentistry. The gold-containing solders are used almost exclusively in bridgework because of their superior tarnish and corrosion resistance. The use of silver-base solders is mainly limited to the joining of stainless steel and cobalt-chromium wires in orthodontic appliances.

Gold-base solders are largely gold-silver-copper alloys to which amounts of zinc, tin, indium, and other elements have been added to control melting temperatures and flow during melting. The silver-base solders are basically silver-copper-zinc alloys to which smaller amounts of tin have been added. The higher-fusing solders to be used with the high-fusing alloys are usually specially formulated for a particular alloy composition because not all alloys have good soldering characteristics.

Table 6 Properties of precious-metal-containing dental alloys

Alloy(a)	Precious metal content, % Au	Pt	Pd	Ag	Casting temperature °C	°F	Melting range °C	°F	Density, g/cm^3	Hardness(b) HB	HV	Ultimate tensile strength(b) MPa	ksi	Yield strength(b) MPa	ksi	Elongation, %(b)	Coefficient of thermal expansion, 10^{-6}/°C at 600 °C
Porcelain fused to metal alloys																	
1	87.5	10.0	1.0	. . .	1260	2300	1040–1140	1900–2085	19.2	150	165	483	70	414	60	5	14.7
2	87.5	4.5	6.0	1.0	1260	2300	1150–1175	2100–2150	18.3	165	182	500	72.5	450	65.3	5	14.7
3	86.0	10.0	2.0	. . .	1260	2300	1070–1190	1960–2170	19.2	170	190	586	85	517	75	5	14.7
4	75.0	. . .	18.0	1.0	1300	2372	1085–1185	1990–2165	17.0	210	230	620	90	517	75	5	14.7
5	69.0	. . .	18.5	9.0	1290	2350	1165–1250	2130–2280	16.7	190	210	662	96	517	75	6	14.7
6	52.5	. . .	27.0	16.0	1315	2400	1205–1260	2200–2300	13.8	200	220	690	100	552	80	10	14.7
7	51.5	. . .	38.5	. . .	1345	2450	1270–1305	2320–2380	13.5	200	220	793	115	572	83	20	14.1
8	45.0	. . .	40.0	5.0	1345	2450	1135–1290	2080–2350	13.5	200	220	793	115	572	83	20	14.2
9	35.0	. . .	57.0	. . .	1345	2450	1130–1300	2065–2370	13.0	220	245	814	118	558	81	20	14.0
10	. . .	. . .	88.0	. . .	1345	2450	1160–1285	2120–2340	11.0	210	235	793	115	572	83	25	14.2
11	2.0	. . .	85.0	1.0	1345	2450	1105–1290	2020–2360	11.0	. . .	270	862	125	658	95.5	20	14.2
12	2.0	. . .	79.0	. . .	1345	2450	1135–1245	2075–2275	11.0	240	265	690	100	552	80	20	14.3
13	2.0	. . .	76.0	. . .	1345	2450	1105–1250	2020–2280	10.9	321	340	1145	166	796	115.5	20	14.0
14	2.0	. . .	60.0	26.0	1370	2500	1230–1305	2250–2380	10.7	172	189	655	95	462	67	20	14.7
15	. . .	. . .	60.0	28.0	1370	2500	1230–1305	2250–2380	10.7	172	189	655	95	462	67	20	14.8
16	. . .	. . .	54.0	38.5	1315	2400	1160–1285	2115–2340	10.7	170	187	724	105	462	67	25	15.3
Soft inlays and medium-hard inlays and crowns																	
17	83	. . .	1.0	10.0	1015	1860	945–960	1730–1760	16.6	Q73	Q80	Q345	Q50	Q103	Q15	Q35	. . .
18	77	. . .	1.0	14.0	1025	1880	925–960	1695–1760	15.9	Q92	Q101	Q400	Q58	Q186	Q27	Q38	. . .
Hard inlays, crowns, and fixed bridgework																	
19	74.5	. . .	3.5	11.0	1030	1890	930–960	1710–1760	15.5	Q110/H165	Q121/H182	Q434/H531	Q63/H77	Q207/H276	Q30/H40	Q39/H19	. . .
20	66.0	. . .	4.0	20.0	1030	1890	925–950	1700–1740	14.7	Q120/H165	Q132/H182	Q441/H662	Q64/H96	Q283/H524	Q41/H76	Q38/H17	. . .
21	62.0	. . .	3.0	25.0	1020	1870	880–950	1615–1740	14.3	Q130/H190	Q145/H210	Q490/H662	Q71/H96	Q290/H483	Q42/H70	Q35/H15	. . .
22	60.0	. . .	4.0	27.0	1020	1870	915–965	1675–1770	14.2	Q103/H175	Q113/H193	Q407/H593	Q59/H86	Q203/H358	Q29.5/H52	Q34/H11	. . .
23	58.0	. . .	3.5	27.0	1020	1870	915–965	1675–1770	14.0	Q130/H190	Q143/H210	Q427/H621	Q62/H90	Q290/H552	Q42/H80	Q28/H10	. . .
24	55.0	. . .	5.0	28.5	1010	1850	880–945	1615–1735	13.7	Q150/H205	Q165/H225	Q517/H703	Q75/H102	Q324/H586	Q47/H85	Q34/H13	. . .
25	50.0	. . .	4.0	35.0	980	1800	860–910	1580–1760	13.2	Q120/H210	Q136/H236	Q496/H690	Q72/H100	Q303/H579	Q44/H84	Q35/H10	. . .
26	46.0	. . .	6.0	39.5	980	1800	845–915	1550–1680	12.8	Q125/H210	Q138/H231	Q448/H690	Q65/H100	Q241/H586	Q35/H85	Q30/H13	. . .
27	20.0	. . .	20.0	38.0	1150	2100	820–1020	1510–1870	11.9	135	150	552	80	252	36.5	8	. . .
28	2.0	. . .	27.0	61.0	1120	2050	960–1055	1760–1930	10.8	Q150/H155	Q165/H170	Q586/H620	Q85/H90	Q241/H448	Q35/H65	Q10/H10	. . .
29	. . .	. . .	25.0	70.0	1175	2150	1020–1100	1870–2010	10.6	Q130/H140	Q143/H154	Q434/H469	Q63/H68	Q262/H324	Q38/H47	Q10/H8	. . .
Extra-hard inlays, thin crowns, fixed bridgework, and partial dentures																	
30	69.0	3.0	3.5	12.5	1030	1890	920–945	1690–1730	15.2	Q135/H240	Q149/H264	Q490/H776	Q71/H112.5	Q276/H493	Q40/H71.5	Q35/H7	. . .
31	66.5	. . .	3.5	14.5	1020	1870	890–915	1635–1675	14.5	Q150/H215	Q165/H237	Q455/H765	Q66/H111	Q296/H603	Q43/H87.5	Q35/H4	. . .
32	60.0	. . .	4.0	22.0	970	1780	890–905	1630–1660	14.1	Q135/H225	Q149/H248	Q483/H883	Q70/H128	Q300/H672	Q43.5/H97.5	Q34/H3	. . .
33	56.0	. . .	4.0	25.0	980	1800	870–930	1600–1710	13.6	Q169/H231	Q186/H254	Q503/H745	Q73/H108	Q372/H720	Q54/H104.5	Q38/H2.5	. . .
34	42.0	. . .	9.0	26.0	980	1800	845–970	1555–1780	12.6	Q175/H265	Q193/H292	Q586/H883	Q85/H128	Q448/H841	Q65/H122	Q18/H3	. . .
35	15.0	. . .	25.0	45.0	1120	2050	930–1020	1705–1870	11.3	Q165/H245	Q180/H270	Q576/H696	Q83.5/H101	Q434/H586	Q63/H85	Q10/H6	. . .
36	3.0	. . .	30.0	50.0	1100	2010	965–1030	1770–1890	10.5	Q150/H230	Q170/H255	Q607/H862	Q88/H125	Q427/H724	Q62/H105	Q20/H4.5	. . .

(a) Numbers are provided for reference purposes only; they are not alloy designations. (b) Q, quenched or softened condition; H, hardened condition. Source: J.F. Jelenko & Company

Table 7 Compositions of precious metal porcelain fused to metal alloys

Alloy(a)	Composition, wt%				
	Au	Pt	Pd	Ag	Other
1.	87.5	4.2	6.7	0.9	0.3 Fe, 0.4 Sn
2.	84.8	7.9	4.6	1.3	1.3 In, 0.1 Ir
3.	54.2	...	25.4	15.7	4.6 Sn
4.	51.4	...	29.5	12.1	6.8 In
5.	59.4	...	36.4	...	4.0 Ga
6.	19.9	0.9	39.0	35.9	3 Ni, 1.2 Ga
7.	...	...	60.5	32.0	7.5 In
8.	1.8	...	77.8	...	10.4 Ga, 10 Cu
9.	...	...	78.0	...	In, Sn, Cu, Ga, Co

(a) Numbers are provided for reference purposes only; they are not alloy designations. Source: *Corrosion*, Volume 13 of the 9th Edition of *Metals Handbook*

Table 8 Compositions of selected silver- and gold-base dental solders

Type	Composition, wt%						
	Au	Pd	Ag	Cu	Sn	In	Zn
Silver	...	...	52.6	22.2	7.1	...	14.1
Gold	45.0	...	20.6	28.4	4.3	...	2.9
	63.0	2.7	19.0	8.6	...	6.5	...

Source: *Corrosion*, Volume 13 of the 9th Edition of *Metals Handbook*

Properties

The diversity in available alloys exists so that alloys with specific properties can be used when needed. For example, the mechanical property requirements of alloys used for crown and bridge applications are different from the requirements of alloys used for porcelain fused to alloy restorations. Even though crown and bridge alloys must possess sufficient hardness and rigidity when used in stress-bearing restorations, excessively high strength makes grinding, polishing, and burnishing difficult, and it is likely to lead to excessive wear of the occluding teeth. Alloys used with PFM restorations are used as substrates for the overlaying porcelain. In this case, the high strength and rigidity of the alloys more closely match the properties of the porcelain. Also, the higher sag resistance of the alloy at the temperatures used for firing the porcelain results in less distortion and lower levels of retained residual stresses.

Similarly, alloys used for partial denture and implant applications must possess increased mechanical properties for resistance to failures. However, clasps contained within removable partial denture devices are often fabricated from a more ductile alloy, such as a gold-base alloy, than from cobalt-chromium or nickel-chromium alloys. This ensures that the clasps possess sufficient ductility for adjustments without breakage from brittle fractures.

Other property requirements in specific systems include:

- Matching the thermal expansion coefficients between porcelain and the substrate alloy for PFM restorations
- Negligible setting contractions for direct filling amalgams
- Specific modulus to yield strength ratios for orthodontic wires

Alloy color is often a consideration because high gold prices have led to the use of alloys with lower gold contents. Lighter and pale-yellow gold alloys, as well as white gold alloys, are currently most prevalent.

Table 6 presents some typical mechanical properties for a number of different alloy systems used in dentistry. Additional information on the compositions, properties, and applications of dental alloys can be found in Ref 9 and 10, and in the article "Tarnish and Corrosion of Dental Alloys" in *Corrosion*, Volume 13 of the 9th Edition of *Metals Handbook*.

REFERENCES

1. R.G. Reese, Jr., Silver, in *Mineral Commodity Profiles*, U.S. Bureau of Mines, 1983
2. R.G. Reese, Jr., Silver, in *Mineral Facts and Problems*, Bulletin 675, U.S. Bureau of Mines, 1985
3. J.M. Lucas, Gold, in *Mineral Facts and Problems*, Bulletin 675, U.S. Bureau of Mines, 1985
4. Platinum-Group Metals in the Third Quarter 1988, *Miner. Ind. Surv.*, 1988
5. Platinum-Group Metals in the Fourth Quarter 1988, *Miner. Ind. Surv.*, 1988
6. Gold and Silver in June 1989, *Miner. Ind. Surv.*, Aug 1989
7. Metal Statistics 1988, 81st Annual Ed., *Am. Met. Mark.*, 1988
8. *Eng. Min. J.*, Vol 190 (No. 3), March 1989
9. R.W. Phillips, *Skinner's Science of Dental Materials*, 8th ed., W.B. Saunders Company, 1982
10. B.R. Lang, M.E. Razzaoug, and H.F. Morris, Ed., *International Workshop on Biocompatibility, Toxicity and Hypersensitivity to Alloy Systems Used in Dentistry*, University of Michigan School of Dentistry, 1986

Properties of Precious Metals

Silver and Silver Alloys

Commercially Pure Silver

Compiled by C.D. Coxe (deceased), A.S. McDonald, and G.H. Sistare, Jr. (deceased), Handy & Harman
Reviewed for this Volume by A.M. Reti, Handy & Harman

Applications

Typical uses. The largest single use for commercially pure silver is for photographic emulsions. The second largest use is in the electrical and electronic industries, for electrical contacts in the medium to high current and voltage categories for conductors, and in primary batteries. Silver is deposited on glass to form mirrors, and particles of metallic silver about 1 to 5 μm (40 to 200 μin.) in diameter are used in pastes for metallizing other nonconducting materials. Silver sputtering targets are used for application of transparent thin films to architectural glass, automotive windshields, and microelectronic components. Silver anodes of various shapes are used in the electroplating industry.

In the chemical industry, silver is used as a catalyst for the dehydrogenation of methanol to make formaldehyde, and in the oxidation of ethylene to ethylene oxide. Silver may also be used to line reactors and vessels, particularly caustic evaporators or crystallizers. Silver also is used as a liner in heavy-duty journal bearings.

Mechanical Properties

Tensile properties. Typical: tensile strength, 125 MPa (18 ksi) for 5 mm (0.2 in.) diam wire annealed at 565 °C (1050 °F); yield strength (divider method), 54 MPa (7.9 ksi). See also Fig. 1.

Hardness. Research on the effect of oxygen on the hardness of annealed silver of various purities indicates that oxidation of impurities during oxidizing anneals generally causes a substantial increase in surface hardness and restrains grain growth, effects that are absent in spectroscopically pure silver. Very pure silver had a hardness of 25 HV after a hydrogen anneal at 650 °C (1200 °F), and 27 HV after annealing in air at 650 °C (1200 °F).

Poisson's ratio. 0.37 for annealed material; 0.39 for hard-drawn material

Elastic modulus. Tension, 71 GPa (10.3×10^6 psi)

Mass Characteristics

Density. 10.49 g/cm^3 (0.379 lb/in.^{3}) or 5.527 troy ounces/in.^{3} at 20 °C (68 °F); density is lowered by cold work and probably by oxygen.

Thermal Properties

Melting point. For oxygen-free silver, 961.93 °C (1763.5 °F)

Coefficient of linear thermal expansion. 19.68 μm/m · K (10.93 μin./in. · °F) at 0 to 100 °C (32 to 212 °F); 20.61 μm/m · K (11.45 μm/in. · °F) at 0 to 500 °C (32 to 930 °F)

Specific heat. 0.234 kJ/kg · K (0.056 Btu/lb · °F) at 0 °C (32 °F), 0.237 kJ/kg · K (0.0568 Btu/lb · °F) at 100 °C (212 °F)

Thermal conductivity. 418.68 W/m · K (2902 Btu · in./ft^2 · h · °F) at 0 °C (32 °F)

Electrical Properties

Electrical conductivity. Effect of percentage reduction for extremely pure 2.3 mm (0.091 in.) diam wire at 20 °C (68 °F):

Reduction, %	%IACS
Annealed	102.8
10.2	102.2
20.0	101.0
37.0	99.7
48.6	99.5
60.0	99.4
68.5	98.4
74.0	98.1

The electrical conductivity of commercial drawn wire may be much lower than 98 to 99%.

Electrical resistivity. 1.59 μΩ · cm at 0 °C (32 °F); 177 μΩ · cm at 20 °C (68 °F) for annealed 2.3 mm (0.091 in.) diam wire; temperature coefficient, 4.1×10^{-3}/°C (6.5×10^{-3} μΩ · cm/°C) from 0 to 100 °C (32 to 212 °F)

Chemical Properties

General corrosion behavior. Silver does not appear to oxidize at room temperature in air and thus differs from copper, but it is attacked and blackened by ozone. Silver oxide, however, does exist and has extremely high resistivity. Sulfur attacks silver rapidly, as it does copper, and the rate of tarnishing of silver in indoor atmospheres is determined by the supply of sulfur atoms, because the coating is nonprotective. This sulfide decreases the reflectivity of silver and also increases the electrical contact resistance, particularly at low currents, because it is nonohmic in character. The rate of sulfidation of silver indoors in a large city is of the order of 7 mg/m^2 · d. Much work has been done in searching for a tarnish-resistant high-silver alloy, but it appears that substantial additions of noble metals are required to achieve this goal, about 50% Pd or 70% Au being needed for complete resistance. Various protective plates have been used to protect silver from tarnishing. Of these, rhodium plate applied over a very thin nickel plate is the most successful and maintains a pleasing appearance but is little used.

Resistance to specific corroding agents. Silver is resistant to acetic acid and has been used for condensers handling this acid. It is also resistant to various other organic acids and foods that are free from sulfur. It shows good resistance to phenol and to hydrofluoric and phosphoric acids, provided that these also are substantially free from sulfur.

Silver is attacked by all the low-melting molten metals, such as mercury, sodium and potassium and their mixtures, lead, tin, indium, and bismuth; consequently, the use of silver in heat exchangers and other devices that employ liquid-metal heat-transfer mediums should be avoided.

Silver is resistant to sodium and potassium hydroxides, is used in the laboratory for caustic fusions, and has also been considered for large equipment. However, silver creeps at the fusion temperatures of these hydroxides, and its use for large equipment would require supporting vessels. It is attacked by moist bromine, iodine, and chlorine, and vigorously by HCl, HI, and HBr. Alkaline cyanides, in the presence of air or other oxidizing agents, dissolve silver rapidly. Nitric acid that contains traces of nitrous acid attacks silver vigorously, as does hot concentrated sulfuric acid. Hot dilute sulfuric acid also attacks silver.

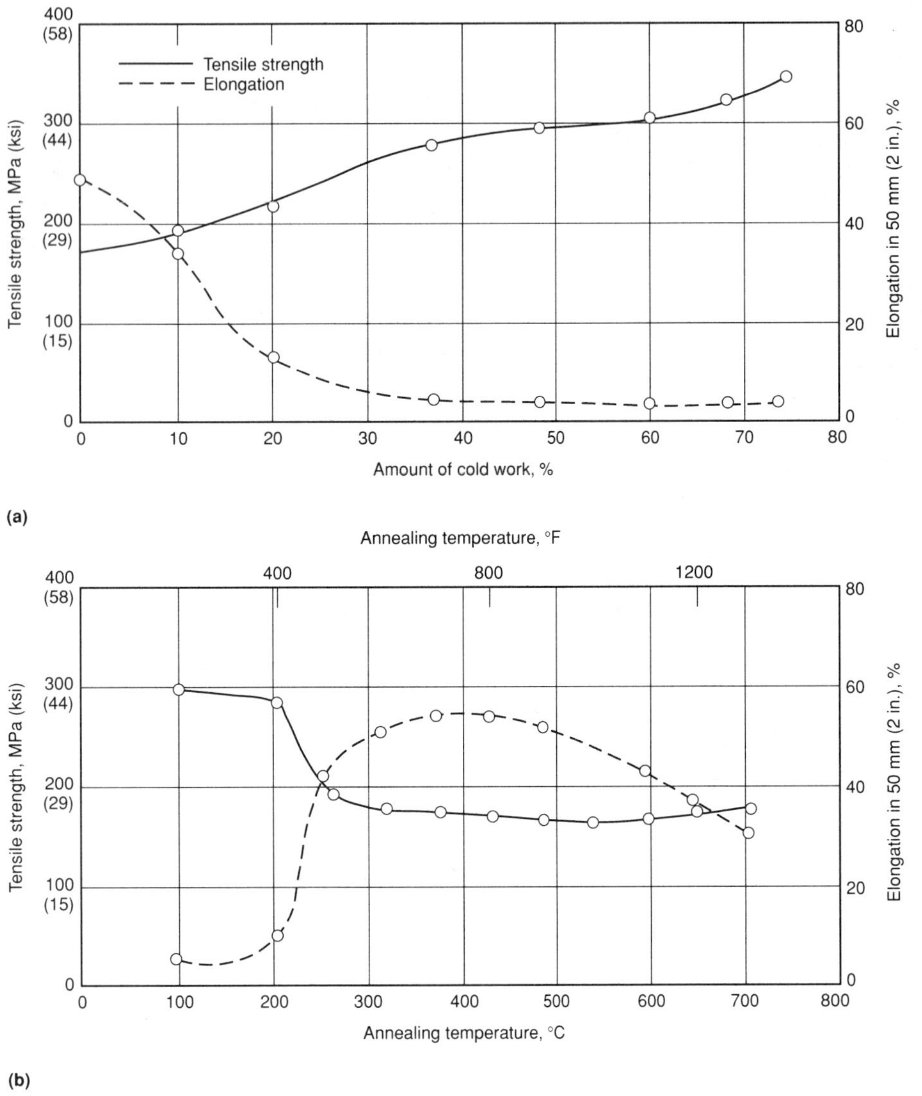

Fig. 1 Tensile properties of commercial fine silver, 2.3 mm (0.091 in.) diam wire. (a) Cold drawn after annealing. (b) Cold drawn 49% before annealing

Fabrication Characteristics

Recrystallization temperature. 20 to 200 °C (68 to 390 °F), depending on purity

Silver-Copper Alloys

Compiled by C.D. Coxe (deceased), A.S. McDonald, and G.H. Sistare, Jr. (deceased), Handy & Harman
Reviewed for this Volume by A.M. Reti, Handy & Harman

Commercial Names

Common names. Sterling silver (92.5 min Ag), coin silver (90Ag-10Cu)

Chemical Composition

Composition limits. Sterling silver must contain at least 92.5% Ag. The remainder is unrestricted but is normally copper because, in general, other metals have proved less desirable and are less-effective hardeners. Coin silver is 90% Ag and 10% Cu. The eutectic alloy contains 28.1% Cu.

Applications

Typical uses. Silver-copper alloys have been used for thousands of years. Copper is effective in hardening silver, but lowers the melting point considerably and lowers the electrical and thermal conductivities appreciably. Sterling silver is used for flat and hollow tableware and various items of jewelry. Coin silver with 10% Cu and 90% Ag was used for U.S. silver coins and is used for electrical contacts operating under service conditions where pure silver is considered too soft and is more likely to pit. The 28% Cu eutectic alloy finds some use as a brazing or soldering alloy. With heavy cold work, it is quite strong and is used for spring-type electrical contacts.

Mechanical Properties

Tensile properties. See Fig. 2 and 3.
Hardness. See Fig. 3.

Electrical Properties

Electrical conductivity. See Fig. 2.
Electrical resistivity. The addition of copper to silver raises the resistivity to a greater extent if the copper is held in solution by quenching and to a lesser extent if it is precipitated by aging or slow cooling. In Fig. 3, the curve labeled "Commercial wire, annealed" shows resistivity typical of commercial phosphorus-deoxidized wire, annealed between 480 and 540 °C (900 and 1000 °F) and cooled to room temperature in 1 h. The decrease in resistivity between 20 and 28% Cu is not significant; deviations of this amount can be expected from lot to lot of the same composition.

Chemical Properties

General corrosion behavior. At ordinary temperatures, the presence of copper in solid solution in silver will have little effect on the resistance of the metal to corrosion. The presence of small areas of the slightly less noble copper-rich phase might be expected to cause difficulty because of electrolytic effects, but apparently the difference between the potentials is small enough for the duplex alloys to behave satisfactorily in their usual applications. In seawater, however, and in similar electrolytes, some selective attack may be anticipated. At slightly elevated temperatures, copper oxidizes selectively. This behavior is of some consequence in electrical contacts, since it necessitates higher contact pressure. At approximately 595 °C (1100 °F), oxidizing atmospheres will cause rapid oxidation of the copper, and oxygen will diffuse to a considerable depth, forming a substance called fire. One hour of exposure to air at this temperature will oxidize the 7.5% Cu alloy to a depth of 0.08 mm (0.003 in.). This was formerly very troublesome, but with the production of well-deoxidized alloys and the use of nonoxidizing atmospheres, effects from the oxidation of copper have been minimized.
Resistance to specific corroding agents. Because sulfur tarnishes the silver-copper alloys in about the same way it tarnishes silver, sulfur must be excluded, or the silver protected by appropriate coatings or wrappings, if appearance is to be maintained without polishing.

Fabrication Characteristics

Processing. In melting silver-copper alloys, oxygen content should be brought to a low level before pouring at 1050 to 1095 °C (1920 to 2000 °F). Where the electrical conductivity is not important, final deoxidation with 0.025% P is convenient. Cadmium has been used as a partial deoxidizer; 0.5% or more is required. Lithium also is being used with success. Melting the material under a cover of broken graphite and pouring the alloy through a reducing flame during casting also gives good results. The alloy may be re-

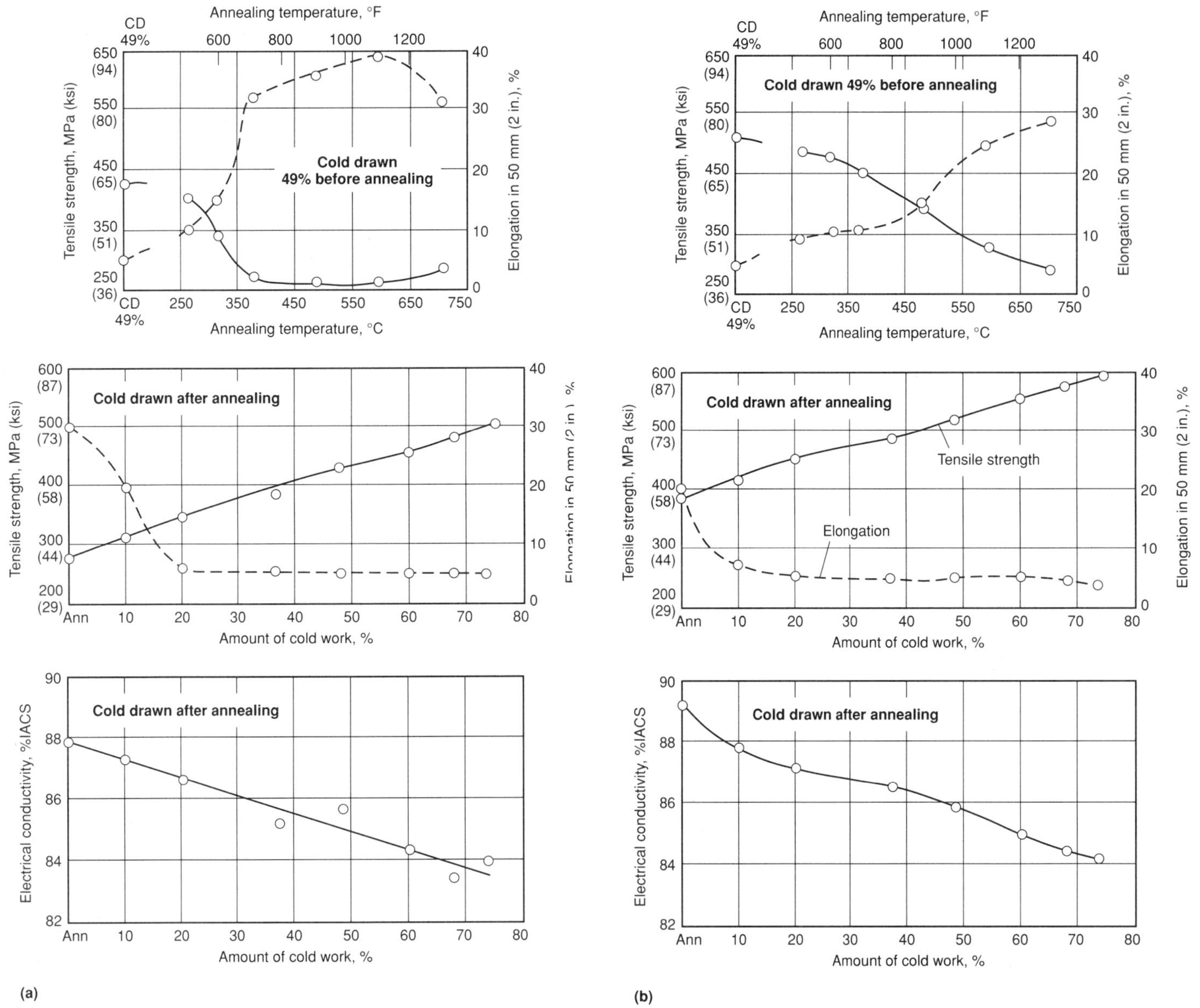

Fig. 2 Tensile properties and electrical conductivity of silver-copper alloys. (a) Sterling silver (92.5Ag-7.5Cu). (b) Eutectic alloy (72Ag-28Cu). Samples are cold-drawn 2.3 mm (0.091 in.) diam wire. CD, cold drawn; Ann, annealed

duced approximately 60% in rolling, and short anneals are suitable at 540 to 675 °C (1000 to 1250 °F) in a steam atmosphere or a salt bath. Where the higher temperature is used, quenching is required for producing full softness. Alternation of oxidizing and reducing atmospheres is very damaging. Where light oxidation has occurred, pickling in a hot sulfuric-acid solution (5 to 10%) is suitable. Heavy oxidation, or fire, can be eliminated only by removing considerable metal, either mechanically or chemically.

Heat treatment. Sterling silver can be age hardened without difficulty, and part of the merit of this composition in providing acceptable properties after miscellaneous treatments results from some hardening on cooling in air. The solubility of copper in silver at 650 °C (1200 °F) is about 4%, and at 730 °C (1350 °F) about 6%, so sterling silver processed at these temperatures is duplex with small amounts of the copper-rich phase scattered through the silver-rich matrix. Aging treatments cause precipitation of the copper-rich phase, and if prolonged, increase the electrical conductivity considerably. Coin silver will remain duplex after any annealing treatment and ages in much the same manner as the 7.5% Cu alloy. Both alloys respond to an aging treatment of 2 h at about 280 °C (535 °F) or 1 h at 300 °C (575 °F). The mechanical properties of coin silver are virtually the same as those of sterling silver after the usual annealing treatments at about 650 °C (1200 °F), because the composition of the silver-rich phase will be the same. Alloys containing 20 to 30% Cu have much more of the copper-rich phase and show less age hardening. In practice, relatively little deliberate use is made of the precipitation-hardening phenomenon in the silver-copper alloys. The solution temperature 705 to 730 °C (1300 to 1350 °F) is rather close to the solidus temperature 780 °C (1435 °F) and requires better temperature control than is available to many artisans who work with these alloys. In alloys heavily deoxidized with phosphorus, incipient melting in the grain boundaries may occur at 705 to 730 °C (1300 to 1350 °F), and when this happens the piece is likely to crack during quenching. Furthermore, these alloys are extremely soft at the solution temperature and are easily damaged. On the other hand, when soldering is done, the metal surrounding the joint may be heated to the solution temperature, and air cooling

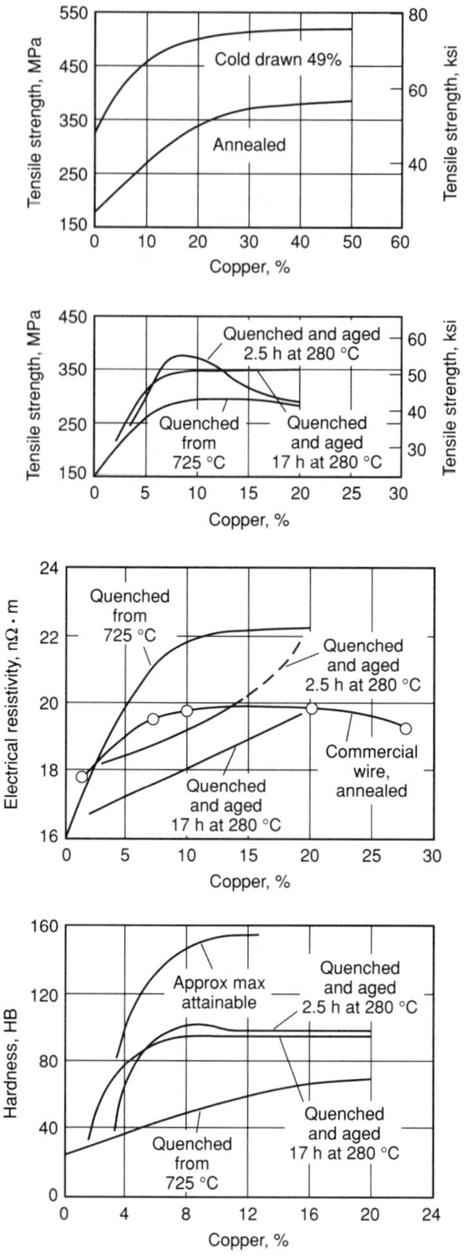

Fig. 3 Effect of copper content on properties of silver-copper alloys

will cause some hardening. This counteracts the softening that would otherwise result from the soldering of work-hardened metal.

Silver-Base Brazing Filler Metals

Compiled by C.D. Coxe (deceased), A.S. McDonald, and G.H. Sistare, Jr. (deceased), Handy and Harman
Revised by C.W. Philp, Handy & Harman

Commercial Names

Common name. Silver brazing filler metals
Former names. Silver solders, hard solders, silver brazing alloys

Specifications

ANSI/AWS. A5.8

Chemical Composition

Composition limits. See Table 1.

Applications

Typical uses. Filler metal for brazing copper, nickel, and cobalt alloys, tool steels, stainless steels, and precious metals. Filler metal for brazing carbide tips onto cutting tools, or wear-resisting tips made of tungsten or molybdenum onto copper-alloy resistance welding electrodes

Mechanical Properties

Tensile properties. Typical: tensile strength (approximate range), 275 to 415 MPa (40 to 60 ksi). The strength of silver brazing filler metals declines rapidly at elevated temperatures. Short-time tests on the filler metal 50Ag-15.5Cu-16.5Zn-18Cd indicate that the loss in strength at 205 °C (400 °F) will approximate 20 to 30% of the strength at room temperature and 50% at 260 °C (500 °F). The filler metals that contain 5% or less of zinc and cadmium have better elevated-temperature strengths than those high in zinc and cadmium.

Thermal Properties

Liquidus temperature. See Table 1.
Solidus temperature. See Table 1.
Brazing temperature range. See Table 1.

Electrical Properties

Electrical conductivity. The electrical conductivity of silver brazing filler metals varies from about 10 to 80% IACS. Filler metals with higher silver content and lower zinc content have the highest conductivity. The silver-copper eutectic (72Ag-28Cu) has conductivity of approximately 77% IACS. The high-silver, low-zinc Ag-Cu-Zn filler metals also find some use in electrical contacts.

Chemical Properties

General corrosion behavior. The corrosion resistance of silver-base brazing filler metals is better than that of most of the nonferrous base metal alloys with which they are used.

Fabrication Characteristics

Formability. Silver-base brazing filler metals are malleable and ductile and can be fabricated into sheet and wire with 50% or greater reductions between anneals.

Silver-Magnesium-Nickel Alloys

Compiled by G.M. Wityak, Handy & Harman

Chemical Composition

Composition limits. 0.25 Mg max, 0.25 Ni, bal Ag

Applications

Typical uses. The unique oxidation-hardenable characteristics of this material allow it to excel in applications where stable, high thermal and electrical conductivity are required. This attribute is particularly useful in electromechanical load-carrying devices. This material is generally formed prior to oxidation, and complex shapes in the form of relay springs, sliding contact arms, and severely deformed contact faces can be fabricated without cracking. Once oxidized, the shapes are retained, but for all practical purposes can no longer be plastically deformed. These parts can be easily brazed, cleaned, and plated without a change in hardness since there is no heat-affected zone (HAZ) to be concerned with as is the case with cold-worked material.
Precautions in use. After oxidation, the ductility of this material is limited, particularly in bending. Care should be taken in the design of the device such that service loads and deflections do not exceed the elastic limit of the material. In almost all cases, failure will be in the form of brittle fracture.

Mechanical Properties

Tensile properties. See Table 2.

Fabrication Characteristics

Heat treatment. This silver-magnesium-nickel alloy is unique in that it can be dispersion strengthened via internal oxidation. This is accomplished by heating the alloy in an air or oxygen atmosphere. The magnesium present in solid solution will precipitate to form submicroscopic magnesium oxide particles. The fine dispersion of these hard refractory compounds in the soft, fine silver matrix imparts great strength. The nickel, because of its limited solubility in silver, acts as a grain refiner. This is particularly important during the oxidation operation since the grain boundaries slow the diffusion rate of magnesium outward.

In almost all applications, the oxidation operation is undertaken after forming. Prior to oxidation, this alloy is very ductile, exhibiting forming characteristics very similar to commercial fine silver. After oxidation the strength of the material is about twice that of the unoxidized alloy, and ductility is very limited.

After oxidation this material cannot be restored to its prior ductile condition. Sustained exposure to a reducing atmosphere at a high temperature will soften the material slightly.

It is important to understand that internal oxidation is a diffusion-controlled process. Material thickness, oxygen concentration, temperature, and time are all interrelated variables. Figure 4 shows the oxidation times required as a function of thickness and temperature.

Table 1 Nominal composition and solidification temperatures for silver-base brazing filler metals

AWS designation(a)	UNS No.	Composition, wt%: Ag	Cu	Zn	Cd	Ni	Sn	Li	Mn	Other elements, total(b)	Solidification temperatures: Solidus °C	Solidus °F	Liquidus °C	Liquidus °F	Brazing temperature range °C	Brazing temperature range °F
BAg-1	P07450	44.0–46.0	14.0–16.0	14.0–18.0	23.0–25.0	...	...	...	...	0.15	607	1125	618	1145	618–760	1145–1400
BAg-1a	P07500	49.0–51.0	14.5–16.5	14.5–18.5	17.0–19.0	...	...	...	...	0.15	627	1160	635	1175	635–760	1175–1400
BAg-2	P07350	34.0–36.0	25.0–27.0	19.0–23.0	17.0–19.0	...	...	...	...	0.15	607	1125	702	1295	702–843	1295–1550
BAg-2a	P07300	29.0–31.0	26.0–28.0	21.0–25.0	19.0–21.0	...	...	...	...	0.15	607	1125	710	1310	710–843	1310–1550
BAg-3	P07501	49.0–51.0	14.5–16.5	13.5–17.5	15.0–17.0	2.5–3.5	...	...	...	0.15	632	1170	688	1270	688–816	1270–1500
BAg-4	P07400	39.0–41.0	29.0–31.0	26.0–30.0	...	1.5–2.5	...	...	...	0.15	671	1240	779	1435	779–899	1435–1650
BAg-5	P07453	44.0–46.0	29.0–31.0	23.0–27.0	...	...	...	...	...	0.15	663	1225	743	1370	743–843	1370–1550
BAg-6	P07503	49.0–51.0	33.0–35.0	14.0–18.0	...	...	...	...	...	0.15	688	1270	774	1425	774–871	1425–1600
BAg-7	P07563	55.0–57.0	21.0–23.0	15.0–19.0	...	...	4.5–5.5	...	...	0.15	618	1145	652	1205	652–760	1205–1400
BAg-8	P07720	71.0–73.0	bal	...	...	...	...	...	...	0.15	779	1435	779	1435	779–899	1435–1650
BAg-8a	P07723	71.0–73.0	bal	...	...	...	...	0.25–0.50	...	0.15	766	1410	766	1410	766–871	1410–1600
BAg-9	P07650	64.0–66.0	19.0–21.0	13.0–17.0	...	...	...	...	...	0.15	671	1240	718	1325	718–843	1325–1550
BAg-10	P07700	69.0–71.0	19.0–21.0	8.0–12.0	...	...	...	...	...	0.15	691	1275	738	1360	738–843	1360–1550
BAg-13	P07540	53.0–55.0	bal	4.0–6.0	...	0.5–1.5	...	...	...	0.15	718	1325	857	1575	857–968	1575–1775
BAg-13a	P07560	55.0–57.0	bal	...	...	1.5–2.5	...	...	...	0.15	771	1420	893	1640	871–982	1600–1800
BAg-18	P07600	59.0–61.0	bal	...	...	...	9.5–10.5	...	...	0.15	602	1115	718	1325	718–843	1325–1550
BAg-19	P07925	92.0–93.0	bal	...	...	...	...	0.15–0.30	...	0.15	760	1400	891	1635	877–982	1610–1800
BAg-20	P07301	29.0–31.0	37.0–34.0	30.0–34.0	...	...	...	...	...	0.15	677	1250	766	1410	766–871	1410–1600
BAg-21	P07630	62.0–64.0	27.5–29.5	...	...	2.0–3.0	5.0–7.0	...	...	0.15	691	1275	802	1475	802–899	1475–1650
BAg-22	P07490	48.0–50.0	15.0–17.0	21.0–25.0	...	4.0–5.0	...	...	7.0–8.0	0.15	680	1260	699	1290	699–830	1290–1525
BAg-23	P07850	84.0–86.0	...	...	...	...	...	...	bal	0.15	960	1760	970	1780	970–1038	1780–1900
BAg-24	P07505	49.0–51.0	19.0–21.0	26.0–30.0	...	1.5–2.5	...	...	...	0.15	660	1220	705	1305	705–843	1305–1550
BAg-26	P07250	24.0–26.0	37.0–39.0	31.0–35.0	...	1.5–2.5	...	...	1.5–2.5	0.15	705	1305	800	1475	800–870	1475–1600
BAg-27	P07251	24.0–26.0	34.0–36.0	24.5–28.5	12.5–14.5	...	...	...	...	0.15	605	1125	745	1375	745–860	1375–1575
BAg-28	P07401	39.0–41.0	29.0–31.0	26.0–30.0	...	...	1.5–2.5	...	...	0.15	650	1200	710	1310	710–843	1310–1550
BAg-33	P07252	24.0–26.0	29.0–31.0	26.5–28.5	16.5–18.5	...	...	...	...	0.15	607	1125	682	1260	682–760	1260–1400
BAg-34	P07380	37.0–39.0	31.0–33.0	26.0–30.0	...	...	1.5–2.5	...	...	0.15	650	1200	721	1330	721–843	1330–1550

(a) AWS, American Welding Society. (b) The brazing alloy shall be analyzed for the specific elements for which values are shown in this table. If the presence of other elements is indicated in the course of this work, the amount of those elements shall be determined to ensure that their total does not exceed the limit specified for other elements.

Dental Amalgam

Compiled by R.M. Waterstrat and N.W. Rupp
American Dental Association Health Foundation
Paffenbarger Research Center
National Institute of Standards and Technology

Dental amalgam is a silver-mercury alloy used for restoring lost tooth structure. In this application it is commonly referred to as a silver filling. It is essentially a metallic composite consisting mainly of the intermetallic compounds $\gamma(Ag_3Sn)$ and $\gamma_1(Ag_2Hg_3)$, with smaller amounts of either $\gamma_2(Sn_7Hg)$ or $\eta'(Cu_6Sn_5)$, depending on the copper content. It is prepared in the dental operatory by grinding or milling silver-tin alloy particles with liquid mercury. Small amounts of moisture from the hand or from mixing equipment, if added during mixing, may cause excessive expansion of alloys containing zinc. The alloy mixture remains in a plastic condition for several minutes following its preparation and may be compacted, shaped, or carved after insertion in a tooth cavity, for example. The mixture hardens by a diffusion reaction in which liquid mercury is replaced by solid mercury compounds such as Ag_2Hg_3 and Sn_7Hg. Excess mercury in the alloy after packing causes the alloy to expand and flow or creep excessively. Excessive working of the mixture during amalgamation or packing reduces or eliminates desired setting expansion. Improper compaction increases the rate of corrosion and decreases physical properties.

Dental amalgam typically contains 40 to 50% Hg, 20 to 35% Ag, 12 to 15% Sn, 2 to 15% Cu, and under 1% Zn. The applicable specifications, however, apply only to the silver-tin alloy particles and liquid mercury used in preparing the amalgam. These are ANSI-MD 156.1 and ISO R1559 for the silver-tin alloy, and ANSI-MD 1556.6 and ISO R1560 for the mercury. Composition limits for the alloy particles are 65% Ag min, 29% Sn max, 6% Cu max, 3% Hg max, 2% Zn max. These limits may be exceeded or other elements added, but if so, the manufacturer must submit the composition and the results of adequate biological and clinical tests to show that the resulting amalgam is safe and effective for use in the mouth, as directed by the manufacturer. Alloys having higher copper content, for example, have been shown to produce improved creep resistance, corrosion resistance, and durability, without producing undesirable biological reactions.

Mechanical Properties

Tensile properties. Tensile strength, 10 to 17% of compressive strength

Compressive properties. Compressive strength, 275 to 345 MPa (40 to 50 ksi) after

Table 2 Tensile properties of unoxidized and oxidized silver-magnesium-nickel alloy strips

Alloy designation	Thickness, mm	Thickness, in.	Temper	Ultimate tensile strength, MPa	Ultimate tensile strength, ksi	Yield strength (0.2% offset), MPa	Yield strength (0.2% offset), ksi	Elongation, % (50 mm, or 2 in., gage length)	Reduction, %
Unoxidized	...	...	Soft	250	36	130	19	28	0
	...	...	½ hard	325	47	315	46	4	21
	...	...	¾ hard	325	47	315	46	3	29
	...	...	Hard	330	48	285	41	2	37
Oxidized(b)	0.15	0.006	...	425	62	380	55	9	...
	0.18–0.25	0.007–0.010	...	415	60	360	52	12	...
	0.30–0.45	0.012–0.018	...	400	58	340	49	13	...
	0.48–0.61	0.019–0.024	...	385	56	325	47	14	...

Tensile properties(a). (a) Unoxidized tensile data is typical; oxidized tensile data is minimum. (b) Modulus of elasticity, 83 GPa (12×10^6 psi)

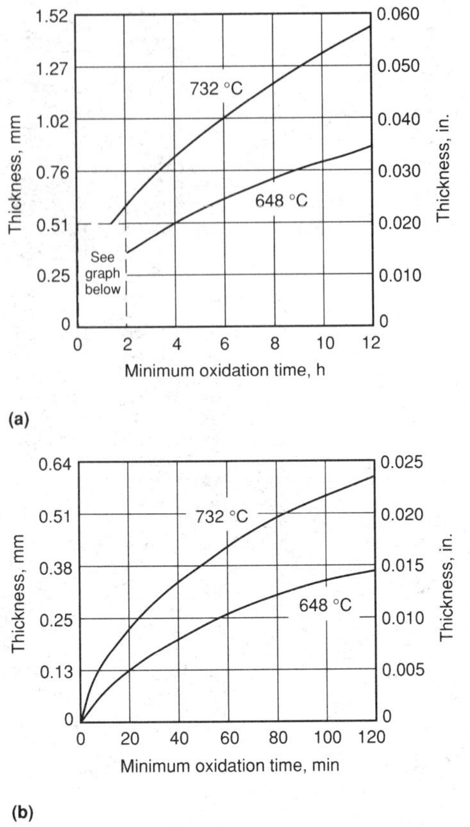

Fig. 4 Effect of strip thickness and temperature on the oxidation times of silver-magnesium-nickel alloy. (a) 0 to 2-h interval. (b) 2 to 12-h interval. Oxidation times given are minimum values obtained by testing in air. Oxidation times for alloys oxidized in pure oxygen are 60% of values shown in figure.

5 days
Hardness. 90 HK
Elastic modulus. Tension, 60 GPa (8.7 × 10^6 psi)
Creep-rupture characteristics. All amalgams flow or creep when subjected to loading. Loads of 10% of the compressive strength may result in a creep rate of 2 μm/m · s (80 μin./in. · s) at room temperature.

Mass Characteristics

Density. 11 g/cm³ (0.397 lb/in.³)
Solidification shrinkage. Expands or contracts 0 ± 0.2% during hardening by a diffusion reaction at room temperature

Thermal Properties

Thermal properties depend to some extent on the copper content. The amalgams containing $\gamma_2(Sn_7Hg)$ compound begin to break down and sweat mercury at about 75 °C (167 °F), whereas amalgams containing $\eta'(Cu_6Sn_5)$ compound are stable up to 120 °C (248 °F).

Coefficient of linear thermal expansion. 22 to 28 μm/m · K (12 to 16 μin./in. · °F) near body temperature

Electrical Properties

Standard electrode potential. The standard electrode potential of dental amalgam is −0.5 V versus gold in normal sodium-chloride solution.

Chemical Properties

Corrosion resistance. The color of dental amalgam is silvery white. Slight tarnishing or corrosion may occur in the oral environment. Electrolytic corrosion and pitting may result from contact with other metals. Dental amalgam is attacked readily by inorganic acids.

Gold and Gold Alloys

Commercial Fine Gold

Compiled by J.A. Bard
Matthey Bishop, Inc.
Reviewed for this Volume by
James Klinzing, Johnson Matthey, Inc.

Commercial Names

Common name. Called proof gold if more than 99.99% Au

Applications

Typical uses. The usual grade of refined gold contains from 99.95 to 99.98% Au and is suitable for most purposes, including dental and jewelry alloys. Metal that contains 99.5% Au is acceptable for international exchange and by the U.S. Mint without a refining penalty. Coin gold containing only copper as a hardener with 89.9 to 91.7% Au may also be acceptable.

Gold of high purity is employed for decorative and dental uses, for surfacing china and glass, as a thin film on glass for selective light filters stable over a wide range of temperature, for thermal limit fuses to protect electric furnaces, as a target in x-ray apparatus, as a freezing-point standard, as a high-melting solder to produce vacuum-tight pressure welds, for the lining of chemical equipment, and clad on phosphor bronze or nickel silver for contact springs in radio-frequency circuits.

Besides decoration and for infrared reflectors, electroplated gold has wide electrical application: in waveguides to provide a coating resistant to corrosion and tarnishing, on grid wires to suppress secondary emission, on variable-resistor terminals to give low-noise internal contact, for adhesion and flexibility of coating on vibrating and flexing components; on contacts for low and stable contact resistance, low cathode-glow discharge and capacitive-current weight loss, and low-rms noise voltage. It is also used as a stop-off in electroplating. Gold is evaporated or sputtered onto selected areas of solid-state electronic devices such as silicon transistors and integrated-circuit chips to provide electrical terminals for these devices, and onto which small-diameter fine gold wires may be thermocompression bonded for electrical connection to the lead frames or other external circuits. Gold and silicon, or germanium, make low-melting eutectics, and this may be done *in situ* simply by heating pure gold in contact with silicon to produce a solder that bonds the semiconductor to its base or other terminals. The low-melting gold-tin eutectic (prealloyed) is also used for similar purposes. Other solders may include gold with antimony for *n*-type semiconductors and indium for *p*-type semiconductors.
Precautions in use. As of January 1, 1975, a license is not required to buy or sell gold, but future transactions should take into account any changes in federal regulations. In melting, avoid contamination with base metals, particularly lead, bismuth, and the like. Keep atmosphere oxidizing during melting. Avoid contact with hydrochloric acid containing free chlorine; aqua regia; concentrated sulfuric acid containing oxidizing agents; arsenic and phosphoric acids; and alkali cyanides, particularly in the presence of oxygen.

Mechanical Properties

Tensile properties. See Table 3.
Hardness. See Table 3.
Poisson's ratio. 0.42 (form not known)
Elastic modulus. See Table 3.
Fatigue strength. 31.7 MPa (4.6 ksi) at 10^7 cycles of reversed bending

Mass Characteristics

Density. 19.32 g/cm³ (0.698 lb/in.³) at 20 °C (68 °F)

Table 3 Mechanical properties of proof gold (≥99.99% Au)

Condition	Tensile strength MPa	Tensile strength ksi	Yield strength (0.2% offset) MPa	Yield strength (0.2% offset) ksi	Elongation(a), %	Hardness, HB	Modulus of elasticity GPa	Modulus of elasticity 10^6 psi
Cast	125	18	...	...	30	33	74.5	10.8
Wrought, annealed	130	19	nil	nil	45	25	79.9	11.6
60% reduction	220	32	205	30	4	58	79.3	11.5

(a) In 50 mm (2 in.)

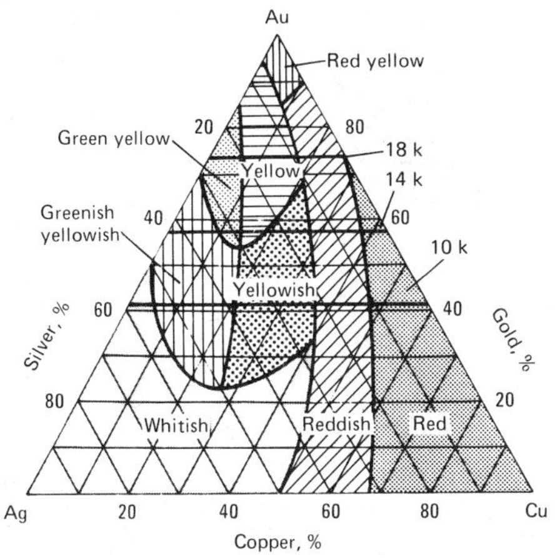

Fig. 5 Color chart for gold-silver-copper alloys for jewelry and dental applications

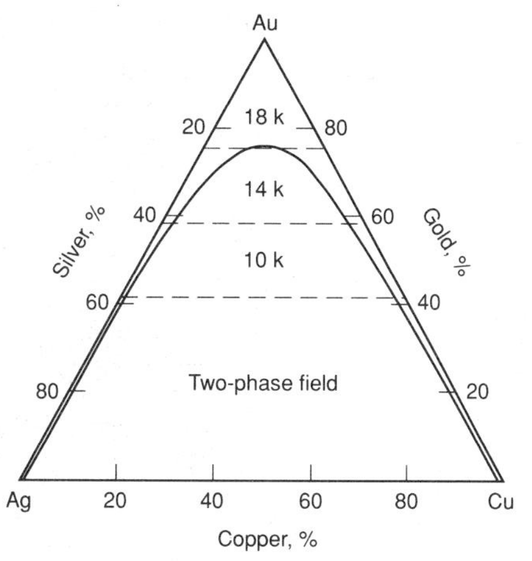

Fig. 6 Isothermal section of the gold-silver-copper ternary phase diagram at 370 °C (700 °F)

Thermal Properties

Melting point. 1064 °C (1948 °F)
Coefficient of linear thermal expansion. 14.2 μm/m · K (7.9 μin./in. · °F) at 20 °C (68 °F)
Specific heat. 0.130 kJ/kg · K (0.0312 Btu/lb · °F) at 18 °C (64 °F)
Thermal conductivity. 300 W/m · K (2100 Btu · in./ft^2 · h · °F) at 0 °C (32 °F)

Electrical Properties

Electrical conductivity. Volumetric, 73.4% IACS at 20 °C (68 °F)
Electrical resistivity. 20 to 22 nΩ · m at 0 °C (32 °F), 23.5 nΩ · m at 20 °C (68 °F)
Relative attenuation. 1.19 (copper = 1); in waveguide 10.16 × 22.86 mm (0.400 × 0.900 in.) ID (λ = 32 mm), 0.139 dB/m; in waveguide 4.32 × 10.67 mm (0.170 × 0.420 in.) ID (λ = 12.5 mm), 0.6 dB/m. Skin depth: λ = 10 mm (0.4 in.), 0.45 μm (18 μin.); λ = 100 mm (4 in.), 1.43 μm (57 μin.); λ = 1 m (40 in.), 4.53 μm (180 μin.). Noise voltage: 0.6 μV rms at 0.5 to 200 Hz (gold ring, graphite brush; pressure, 1.08 kPa (0.157 psi); speed, 0.35 m/s, or 1.2 ft/s). Contact erosion: cathode glow discharge in air: weight loss, 0.886 (platinum = 1). Capacitive current: weight loss, 1.14 (platinum = 1)

Fabrication Characteristics

Formability. Suited to forming by all methods
Weldability. Torch braze with silver solder, no flux, any flame; oxyacetylene weld with gold, no flux, any flame; resistance weld by any method
Annealing temperature. 300 °C (575 °F), but usually no annealing is required
Hot-working temperature. Can be worked at any temperature below the melting point
Casting temperature. 1095 to 1300 °C (2000 to 2370 °F)

Gold-Silver-Copper Alloys

Compiled by G.H. Sistare, Jr. (deceased), and A.S. McDonald, Handy & Harman
Reviewed for this Volume by A.M. Reti, Handy & Harman

Commercial Names

Common names. Green, yellow, and red golds

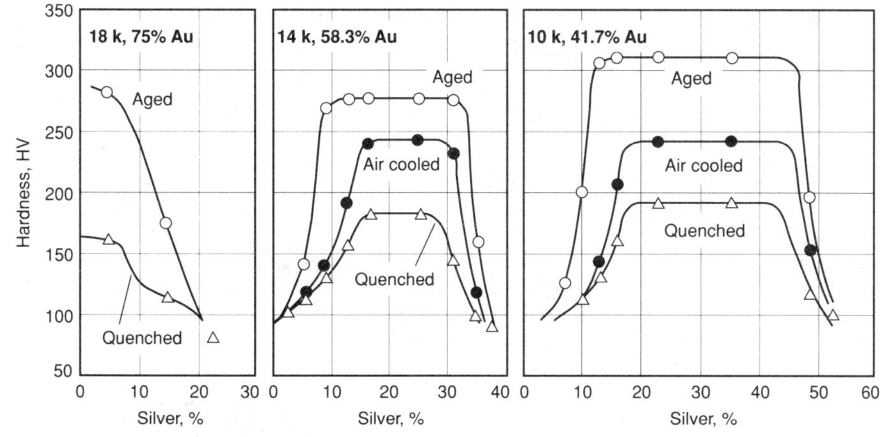

Fig. 7 Variation of hardness with silver content for gold-silver-copper alloys

Chemical Composition

Alloy types. Most of the commercially important colored alloys for jewelry and dental applications are based on the gold-silver-copper system (see Fig. 5)—frequently modified by the addition of zinc, and sometimes of nickel, for jewelry alloys, and of palladium and platinum for dental alloys.

In the ternary phase diagram, the two-phase field of the silver-copper system extends well in toward the gold corner of the diagram (see Fig. 6). Alloys in the single-phase solid-solution area on both the silver-rich and copper-rich sides are generally soft and not hardenable, except for the order-hardening gold-copper alloys containing approximately 75% Au. Two-phase alloys near the single-phase limit at 370 °C (700 °F) are quite soft when annealed and may be precipitation hardened by solution annealing, quenching, and aging at 260 to 315 °C (500 to 600 °F). Alloys lying farther into the two-phase region are harder in the annealed condition.

Figure 7 shows the effect of composition on hardness of gold-silver-copper alloys at three karat levels. The colors indicated in the triangular diagram may be modified by additions of other metallic elements. Zinc is frequently added to gold-silver-copper alloys:

- As a deoxidizer
- To lighten the color (it makes reddish alloys more yellow)
- To lessen the hardening that may occur on air cooling
- To lower the melting temperature for gold solder

Where it is desirable to reduce the grain size of cast gold-silver-copper alloys, fractional percentages of iridium or rhodium, plus ruthenium, have been used, particularly in the dental field, to refine the structure. In the wrought jewelry alloys, additions are occasionally desired to reduce the rate of grain growth, and a very small amount of cobalt, or less desirably, iron can be used to accomplish this; nickel has some effect in this direction but its solubility is relatively high, particularly in the low-silver alloys. The addition of considerable percentages of nickel lightens the color and increases the solid-solution hardness. Iron may cause inclusions, and cobalt and nickel will form low-solubility phases with some deoxidizers.

The 18 k gold-silver alloy has a good green color but is too soft for general use, except as a finishing plate, whereas the red 18 k gold-copper alloy is troublesome to work because of ordering transformation in the solid state. The 18 k gold alloys of the gold-silver-copper type are yellow in color. A wider range of colors is available in the 14 and 10 k alloys.

Because the properties of 10 and 14 k alloys are controlled largely by the ratio of

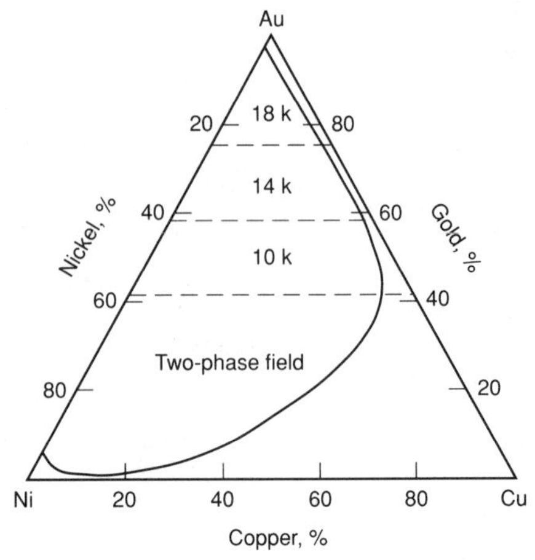

Fig. 8 Isothermal section of the gold-nickel-copper ternary phase diagram at 315 °C (600 °F)

Table 4 Typical compositions and firecracking tendencies of gold-nickel-copper-zinc white golds

Alloy	Historical sequence	Composition, wt% Au	Ni	Cu	Zn	Fire-cracking tendency
18 k	Early	75	17.30	2.23	5.47	Marked
14 k	Early	58.33	15.17	18.04	8.46	Marked
14 k	Intermediate	58.33	10.82	22.08	8.77	Moderate
14 k	Modern	58.33	12.21	23.47	5.99	Slight
10 k	Early	41.67	21.24	25.25	11.84	Marked
10 k	Intermediate	41.67	15.12	30.96	12.25	Moderate
10 k	Modern	41.67	17.08	32.85	8.40	Slight

Table 5 Typical mechanical properties of gold-platinum alloy (70Au-30Pt)

Condition	Proportional limit MPa	ksi	Yield strength MPa	ksi	Hardness(a), HB
Annealed(b)	200	28.8	245	35.4	130
Annealed(c)	...	...	...	...	114
Hard(d)	447	64.9	570	82.5	169

(a) 500 kg load. (b) 1095 °C (2000 °F), air cooled. (c) 1000 °C (1830 °F), quenched. (d) 66% reduction

silver to copper, regardless of the gold content, alloys can be converted from one gold content to another gold content having similar properties, by addition of pure gold to low-karat alloys or by addition of standard base alloy to high-karat golds.

The metallurgy of gold-silver-copper alloys has been treated in detail (Ref 1) and further contributions were made on the age-hardening and ordering transformation within this ternary system (Ref 2, 3).

Applications

Typical uses. These alloys are used mostly in jewelry but sometimes are used for slip rings and brushes on electrical instruments. Parts may be cast to shape, made from rolled or drawn stock, or made from clad material comprising a layer of gold on one or both sides of a core made of nickel silver, pure nickel, brass, or bronze. Certain of these alloys can be electrodeposited, and sometimes coatings are produced by electrodeposition followed by heating to cause diffusion with the underlying metal.

Gold-Nickel-Copper Alloys

Compiled by G.H. Sistare, Jr. (deceased), and A.S. McDonald, Handy & Harman
Reviewed for this Volume by A.M. Reti, Handy & Harman

Commercial Names

Common name. White gold

Chemical Composition

Alloy types. Historically, white golds developed from a patented gold-nickel-zinc alloy containing about 80% of gold that was offered as a substitute for platinum jewelry. In extending the concept to conventional karat levels, it was immediately discovered that some copper was essential for workability at 18 k. Subsequently, it was found that sizeable amounts of copper were essential if 14 and 10 k alloys were to be worked at all. Modern white golds are based on alloys in the gold-nickel-copper system to which 5 to 12% Zn is added.

The gold-nickel-copper system (discussed in detail in Ref 4) is similar to the gold-silver-copper system in that both are dominated by a two-phase immiscibility gap in the solid-state ternary field. In the gold-nickel-copper system, the immiscibility gap in the gold-nickel binary system extends into the ternary field. In the gold-silver-copper system, the silver-copper binary eutectic generates an immiscibility gap that extends into the ternary. Note however, that in one case (gold-nickel-copper), the two-phase field is opposite the copper corner, while in the other (gold-silver-copper), the two-phase field is opposite the gold corner. Compare the gold-nickel-copper isothermal section in Fig. 8 with the gold-silver-copper isothermal section in Fig. 6. It follows that in constant karat pseudobinary sections (that is, sections at fixed gold content) the two-phase field is asymetrical in the gold-nickel-copper system and symmetrical in the gold-silver-copper system. Thus, white golds based on fixed nickel-copper ratios are dissimilar at different karat levels, whereas yellow golds based on fixed silver-copper ratios, by and large, are readily converted from one karat to another.

Gold-nickel-copper-base white golds work harden faster and are harder after annealing than gold-silver-copper-base yellow golds. All gold-nickel-copper-base white golds lie within the two-phase region at room temperature. The 18 and 14 k alloys can be homogenized at elevated temperatures. The 10 k alloys do not homogenize with any practical heat treatment. The 18 k alloys can be age hardened, but the 14 or 10 k alloys are not age-hardenable modified with an addition of cobalt. It takes about 0.75% Co at 14 k and 1.75% Co at 10 k. In both instances, the cobalt is substituted for copper. These modifications are mostly used as casting alloys and usually can be aged as-cast without first solution annealing.

Gold-nickel-copper-base white golds are very susceptible to firecracking. They firecrack when they are given a full anneal after light cold working (reductions of less than 50%). Expedients such as stress relieving prior to annealing are ineffective in preventing firecracking. White-gold compositions are the result of a compromise between color and firecracking tendency. Increasing the copper content reduces the firecracking tendency but offsets the whitening effect of nickel. This can be compensated for with the addition of zinc, which tends to decolorize the copper and enhance the whitening effect of nickel. Unfortunately, it also enhances the firecracking tendency. This is illustrated in Table 4, which lists typical compositions of gold-nickel-copper-zinc white golds in order of decreasing karat. At 18 k, little can be done to reduce firecracking tendency; with 75% gold, any significant replacement of nickel with copper would result in an unacceptable color. At 14 k, considerable amounts of copper can be added, and at 10 k, even greater amounts of copper are tolerable. Note in the historical sequence in Table 4 how the increase in copper and decrease in zinc contents have progressively led to a decrease in firecracking tendency, while maintaining an acceptable color.

Gold-Platinum Alloy 70Au-30Pt

Compiled by D.J. Accinno
Engelhard Industries, Inc.
Reviewed for this Volume by
James Klinzing, Johnson Matthey, Inc.

Applications

Typical uses. Because of its high corrosion resistance, this alloy is used for spinnerettes in the numerous methods of rayon produc-

Table 6 Nominal composition and solidification temperatures of gold-base brazing filler metals

AWS designation(a)	UNS No.	Composition, wt%				Other elements, total(b)	Solidification temperatures					
		Au	Cu	Pd	Ni		Solidus °C	Solidus °F	Liquidus °C	Liquidus °F	Brazing temperature range °C	Brazing temperature range °F
BAu-1	P00375	37.0–38.0	bal	. . .	. . .	0.15	991	1815	1016	1860	1016–1093	1860–2000
BAu-2	P00800	79.5–80.5	bal	. . .	. . .	0.15	891	1635	891	1635	891–1010	1635–1850
BAu-3	P00350	34.5–35.5	bal	. . .	2.5–3.5	0.15	974	1785	1029	1885	1029–1091	1885–1995
BAu-4	P00820	81.5–82.5	. . .	. . .	bal	0.15	949	1740	949	1740	949–1004	1740–1840
BAu-5	P00300	29.5–30.5	. . .	33.5–34.5	35.5–36.5	0.15	1135	2075	1166	2130	1166–1232	2130–2250
BAu-6	P00700	69.5–70.5	. . .	7.5–8.5	21.5–22.5	0.15	1007	1845	1046	1915	1046–1121	1915–2050

(a) AWS, American Welding Society. (b) The brazing filler metal will be analyzed for those specific elements for which values are shown in this table. If the presence of other elements is indicated in the course of this work, the amount of those elements will be determined to ensure that their total does not exceed the limit specified.

tion that employ high corrosive chemicals. The alloy is also used as a high-melting-point platinum solder.

Mechanical Properties

Tensile properties. Typical: tensile strength, 639 MPa (92.7 ksi); reduction in area, 50%. See also Table 5.
Hardness. See Table 5.
Elastic modulus. Tension, 113.8 GPa (16.51 × 10^6 psi)

Mass Characteristics

Density. Annealed, 19.92 g/cm^3 (0.720 lb/$in.^3$) at 20 °C (68 °F)

Thermal Properties

Liquidus temperature. 1450 °C (2642 °F)
Solidus temperature. 1228 °C (2242 °F)

Electrical Properties

Electrical resistivity. 340 nΩ · m (quenched); 220 nΩ · m (aged) at 20 °C (68 °F). Temperature coefficient, 0.0059 nΩ · m/K at 0 to 1200 °C (32 to 2190 °F)

Fabrication Characteristics

Joining. Braze with gold solder; no flux; any flame
Annealing temperature. 1095 °C (2000 °F) in air

Gold-Base Brazing Filler Metals

Compiled by C.E. Fuerstenau
Lucas-Milhaupt, Inc.

Commercial Names

Common name. Gold brazing filler metals
Former names. Hard solder, gold brazing alloys

Specifications

ANSI/AWS. A5.8 (Ref 5)

Chemical Composition

Composition limits. See Table 6.

Applications

Typical uses. Filler metal for brazing stainless steel, tungsten, molybdenum, nickel, and cobalt-base alloys. Filler metal for brazing thin sections due to low rate of interaction with the base metals. Commonly used in applications where strength and corrosion resistance are needed at elevated temperatures, such as in some jet engine components

Mechanical Properties

Tensile properties. Typical: tensile strength (approximate range), 415 to 550 MPa (60 to 80 ksi). The strength of gold-base filler metals is good at elevated temperatures, but it does decline. Short-term tests on filler metal 82Au-18Ni indicate that the loss in strength at 425 °C (800 °F) will approximate 20% of the strength at room temperature and 35% at 650 °C (1200 °F). Those filler metals containing nickel will exhibit higher strength at room and elevated temperatures.

Thermal Properties

Liquidus temperature. See Table 6.
Solidus temperature. See Table 6.
Typical brazing temperature. See Table 6.

Electrical Properties

Electrical conductivity. The electrical conductivity of these filler metals is approximately 5% IACS and up for the various compositions. Those filler metals containing nickel exhibit the lowest conductivity.

Chemical Properties

General corrosion behavior. The gold-base filler metals exhibit excellent oxidation resistance, even at elevated temperature. They also resist attack by water and salt water.

Fabrication Characteristics

Formability. Gold-base filler metals are relatively malleable and ductile. They can be fabricated into sheet and wire form easily.

Platinum and Platinum Alloys

Commercially Pure Platinum 99.95% Pt

Compiled by Edward D. Zysk (deceased)
Engelhard Minerals & Chemicals Corp.
Reviewed for this Volume by
James Klinzing and Lisa Dodson,
Johnson Matthey, Inc.

Applications

Typical uses. Of the platinum group metals, platinum is the least rare, and it is the most widely used because of its general corrosion resistance, high melting point, appearance, and ductility. Platinum of the highest purity is required for use in resistance thermometers and thermocouples. Various alloying elements such as rhodium, ruthenium, and iridium, and for special purposes, other hardeners are employed to develop higher mechanical properties or to protect against special corrosion conditions. Platinum or its alloys are used for the cathodic protection of ship hulls, for electrical contacts, brushes, precision potentiometer wire, chemical production, laboratory ware, spinnerettes for synthetic fibers, anodes in both solid and clad form, and for jewelry. It is also used as a crucible liner for producing high-purity optical glass or as a bushing in the extrusion of fiberglass. A more recently developed application for platinum and its alloys is in cardiac pacemakers and other biomedical specialty items. Platinum is an outstanding catalyst for oxidation, as in the production of H_2SO_4 and HNO_3; for hydrogenation as in the production of vitamins and other chemicals; and in the petroleum reforming process as in the production of high-octane gasolines. Certain organometallic compounds containing platinum have significant antitumor activity.

Mechanical Properties

Tensile properties. Typical: annealed at 700 °C (1290 °F): tensile strength, 125 to 165 MPa (18 to 24 ksi); proportional limit, <13.8 MPa (<2 ksi); elongation, 30 to 40% in 50

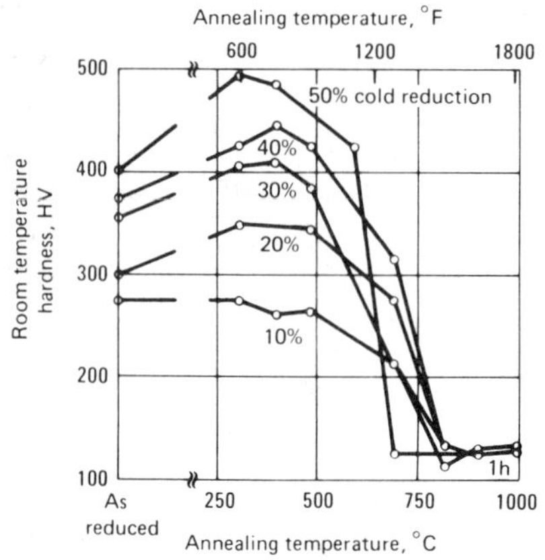

Fig. 9 Room-temperature hardness of commercially pure (99.9%+) platinum after warm rolling and annealing

mm (2 in.). Hard drawn, 50% cold worked: tensile strength, 205 to 240 MPa (30 to 35 ksi); elongation, 1 to 3% in 50 mm (2 in.)
Effect of low temperature. Coarse grain material: 125 MPa (18.2 ksi) at room temperature; 283 MPa (41.0 ksi) at −195 °C (−317 °F); 565 MPa (82.0 ksi) at −253 °C (−425 °F). Fine grain material: 124 MPa (18.0 ksi) at 21 °C (70 °F); 448 MPa (65.0 ksi) at −195 °C (−317 °F) (Ref 6)
Effect of elevated temperature. Annealed thermocouple quality (about 99.99% pure): 143 MPa (20.7 ksi) at room temperature; 90 MPa (13.0 ksi) at 400 °C (750 °F); 55 MPa (8.0 ksi) at 800 °C (1470 °F); 34 MPa (5.0 ksi) at 1000 °C (1830 °F); 21 MPa (3.0 ksi) at 1200 °C (2190 °F). Tensile strength of 99.98% pure material of two grain sizes is given in Ref 6 for −253 to 827 °C (−423 to 1521 °F).
Hardness. Annealed at 700 °C (1290 °F): 37 to 42 HV; hard drawn, 50% cold work: 90 to 95 HV. Electrodeposited: approximately 600 HV. Effect of rolling and annealing, see Fig. 9 and Ref 7. Effect of alloying, see Fig. 10.
Poisson's ratio. 0.39 (Ref 8)
Elastic modulus. At 20 °C (68 °F), annealed at 700 °C (1290 °F). Tension: static, 171 GPa (24.8 × 10^6 psi); dynamic, 169 GPa (24.5 × 10^6 psi). Hard drawn, 50% cold work, tension: static, 156 GPa (22.6 × 10^6 psi)
Creep-rupture characteristics. For platinum and platinum-palladium alloys, see Ref 9.

Mass Characteristics

Density. 21.46 g/cm³ (0.775 lb/in.³) at 25 °C (77 °F)

Thermal Properties

Melting point. 1769 °C (3217 °F) (Ref 10)
Coefficient of linear thermal expansion. 9.1 μm/m · K (5.1 μin./in. · °F) from 20 to 100 °C (68 to 212 °F) (Ref 11)
Specific heat. 0.132 kJ/kg · K (0.0314 Btu/lb · °F) at 0 °C (32 °F) (Ref 12)
Latent heat of fusion. 113 kJ/kg
Thermal conductivity. 71.1 W/m · K (493 Btu · in./ft² · h · °F) at 0 °C (32 °F) (Ref 13)

Electrical Properties

Electrical resistivity. 98.5 nΩ · m at 0 °C (32 °F); 106 nΩ · m at 20 °C (68 °F). Temperature coefficient: 0.0039/K from 0 to 100 °C (32 to 212 °F). Effect of alloying, see Fig. 11

Optical Properties

Color. Silver white
Spectral reflectance. Bulk: 70.1% at 589 nm. Electrodeposited: 58.4% at 441 nm; 59.1% at 589 nm; 59.4% at 668 nm (Ref 15, 16, 17)

Chemical Properties

General corrosion behavior. See the article "Corrosion of the Noble Metals" in Volume 13, *Corrosion*, of the 9th Edition of *Metals Handbook*. See also Ref 18.
Resistance to specific corroding agents. Resistant to reducing or oxidizing acids at room temperature; attacked by aqua regia (a mixture of nitric and hydrochloric acids); attacked slowly by hydrochloric acid plus other oxidizing agents. Resistant to ferric chloride at room temperature; hydrobromic acid plus bromine attacks at room temperature. All of the free halogens attack at elevated temperatures; hydrochloric acid in the absence of oxidizing agents does not attack, and platinum is useful against this normally active gas up to 1095 °C (2000 °F). Sulfur dioxide does not attack even at 1095 °C (2000 °F) (Ref 19).

As an anode, platinum is outstanding and is used commercially in sulfuric and persulfuric acids, various sulfate-chloride plating electrolytes, and in chlorates with very little corrosion. If electrolyzed with alternating current, chlorides may attack, a characteristic exploited in etching platinum and platinum alloys.

Platinum is highly resistant to acid potassium sulfate, sodium carbonate, potassium nitrate at moderate temperatures, and to sodium carbonate at 800 to 900 °C (1475 to 1650 °F) under nonoxidizing conditions. Although attacked vigorously by molten alkali cyanides

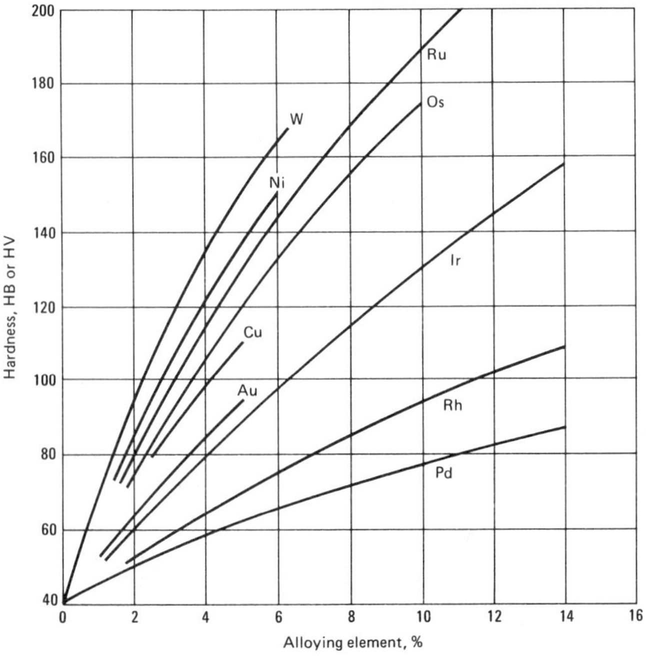

Fig. 10 Effect of various alloying additions on the hardness of annealed platinum

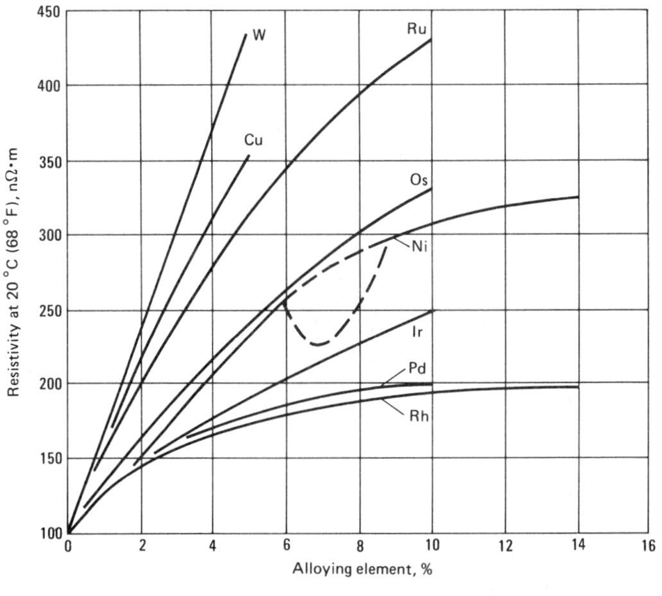

Fig. 11 Effect of various alloying additions on the electrical resistivity of platinum. Source: Ref 14

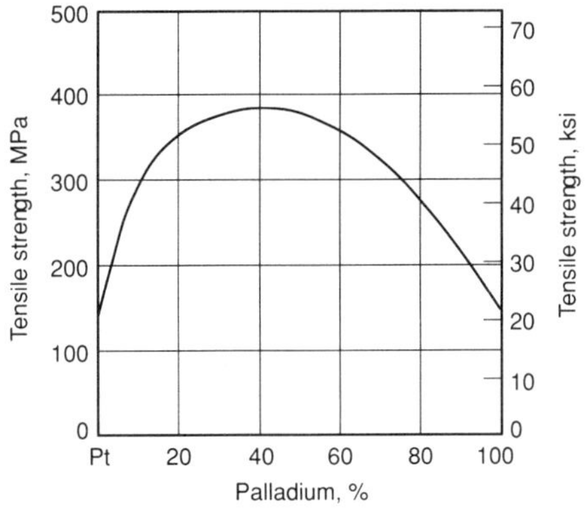

Fig. 12 Tensile strength of annealed platinum-palladium alloys as a function of palladium content

and polysulfides, it is quite resistant to the normal sulfides plus alkali. Certain phosphates attack at high temperatures and care must be taken to avoid reducing conditions, particularly when compounds of arsenic, phosphorus, tin, lead, or iron are present. It is resistant to molten glasses, especially to those low in lead and arsenic. Platinum, even in the form of thin leaf, is resistant to corrosion and tarnishing on exposure to the atmosphere, including urban sulfur.

Fabrication Characteristics

Annealing. Annealing temperature depends on the purity of the material and amount of prior cold work. Figure 9 shows the effect of reduction during rolling of 99.99% pure platinum; grain size after annealing platinum of this purity is determined almost entirely by prior reduction; virtually no grain growth occurs on the usual short anneals. It is probable, however, that platinum free from oxygen in solution will show grain growth after recrystallization and may have a still lower annealing temperature.

Air is the preferable atmosphere for annealing platinum; hot-reducing atmospheres, particularly where silica, iron, or easily reduced oxides are nearby, are almost certain to result in contamination. Annealing at too frequent intervals can result in substantial growth, causing the orange peel effect during subsequent working or polishing. To prevent the formation of orange peel, the reduction in cross-sectional area before annealing should not be less than 30%. Annealing for too long a time as well as at too high a temperature can result in thermal etching (grains of metal become clearly visible).

Precautions in working. The maintenance of oxidizing conditions throughout processing is essential to avoid contamination. To remove iron, pickling in hot hydrochloric acid after rolling and before annealing is essential for high-purity wire and sheet.

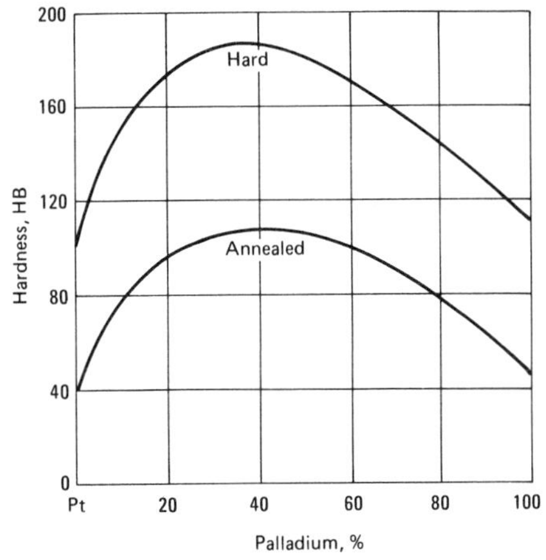

Fig. 13 Hardness of platinum-palladium alloys as a function of palladium content

Platinum-Palladium Alloys

Compiled by J. Hafner and R. Volterra
Metals and Controls Division, Texas Instruments, Inc.
Reviewed for this Volume by
James Klinzing and Lisa Dodson,
Johnson Matthey, Inc.

Applications

Typical uses. Platinum-palladium alloys are used in place of pure platinum for jewelry in Europe; in the United States, stamping laws do not provide for this type of alloy. Platinum-palladium alloys, with or without additions of other metals, are used for electrical contacts. Platinum with up to 20% Pd is used as an insoluble anode in seawater, and low-palladium platinum alloys are being used in the glass industry.

Mechanical Properties

Tensile properties. See Fig. 12.
Hardness. See Fig. 13.

Structure

Microstructure. Platinum and palladium form a continuous series of solid solutions. Solidus and liquidus curves are close together; the maximum interval between them is about 60 °C (about 110 °F) (50Pt-50Pd). No transformations in the solid state have been reported.

Electrical Properties

Electrical resistivity. See Fig. 14.
Temperature coefficient of resistivity. See Fig. 14.

Chemical Properties

General corrosion resistance. The alloys containing less than 25% Pd perform the same as pure platinum in most chemical mediums. The resistance to nitric acid decreases as the palladium content increases, but an alloy with only 2% Pt is as resistant to this reagent as a 14 k gold alloy. The platinum-rich alloys do not discolor when heated in air, but the palladium-rich alloys darken between 400 and 750 °C (750 and 1380 °F), because of the formation of palladium oxide, which is stable in that interval of temperature but which decomposes at higher temperature. Prolonged heating at high temperature causes slight weight loss by volatilization; however, at 900 °C (1650 °F) in oxygen, the loss in weight is less than that of pure platinum. Presumably, this is caused by the adsorption of oxygen in palladium. The solubility of hydrogen in palladium is reduced significantly by the addition of platinum. At over 34% Pt, only adsorption can be observed. The 90Pt-10Pd alloy has the lowest corrosion rate in seawater.

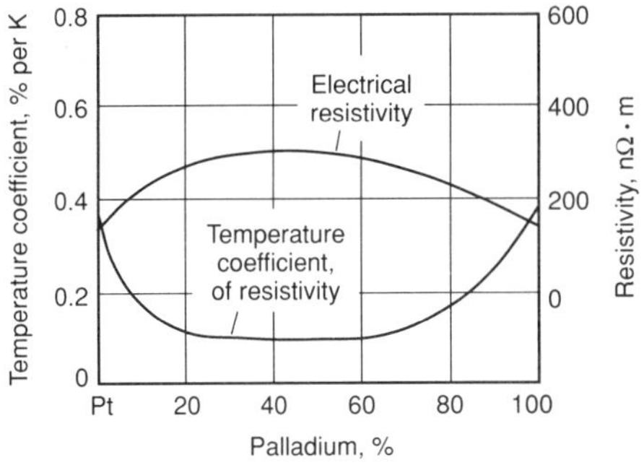

Fig. 14 Electrical resistivity of platinum-palladium alloys as a function of palladium content

Fabrication Characteristics

Workability. All platinum-palladium alloys can be cold worked. Those with high palladium content should be annealed in inert or nitrogen atmospheres. Recommended annealing temperature is 950 °C (1740 °F).

Platinum-Iridium Alloys

Compiled by J. Hafner and R. Volterra
Metals and Controls Division
Texas Instruments, Inc.
Reviewed for this Volume by
James Klinzing and Lisa Dodson,
Johnson Matthey, Inc.

Applications

Typical uses. Platinum-iridium alloys are used in the electrical, electrochemical, chemical, medical, and jewelry fields. The Pt-10% Ir alloys are also used for standards of length and weight because of their permanence. Some applications of different alloys are:

Application	Iridium, %
Laboratory ware	0.4–0.6
Jewelry	5–15
Medical	10
Electrical contacts	10–25
Electrodes for electrochemical processes	10
Tubing for pens, hypodermic needles, spring elements	25–30

Mechanical Properties

Tensile properties. See Table 7 and Fig. 15.
Hardness. See Table 7 and Fig. 16.

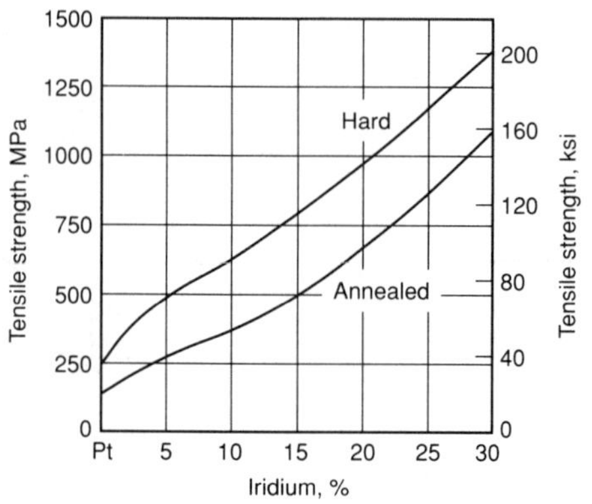

Fig. 15 Tensile strength of platinum-iridium alloys as a function of iridium content

Structure

Microstructure. The platinum-iridium alloys solidify in a complete series of solid solutions. Below about 995 °C (1825 °F), a very sluggish separation into two solid solutions has been reported. The duplex region ranges from 7 to 99% Ir at 700 °C (1290 °F).

Mass Characteristics

Density. See Table 7.

Electrical Properties

Electrical resistivity. See Table 7 and Fig. 17.

Chemical Properties

General corrosion resistance. The resistance to corrosion and tarnishing of the platinum-iridium alloys is excellent. The resistance to aqueous solutions of halogens and aqua regia increases with iridium content. Because iridium forms a volatile oxide at high temperatures, there is a noticeable loss by evaporation when these alloys are annealed in air above 900 °C (1650 °F). The loss by evaporation increases with temperature and iridium content.

Fabrication Characteristics

Workability. The workability of the platinum-iridium alloys decreases with increasing iridium content. The practical limit of workability for cast alloys is about 40% Ir. Alloys with about 25 to 30% Ir can be hot or cold worked. Cold reductions to 75% can be used for alloys with 20% Ir or less; permissible reductions for alloys of higher iridium content are lower. Annealing temperatures between 1000 and 1200 °C (1830 and 2190 °F) are suitable for alloys to 10% Ir. For alloys of higher iridium content, temperatures near 1400 °C (2550 °F) are recommended. Platinum alloys with 10 to 90% Ir have been reported to be susceptible to precipitation hardening. Noticeable changes in physical properties can be obtained by heating at approximately 800 °C (1475 °F) after quenching from about 1700 °C (3090 °F).

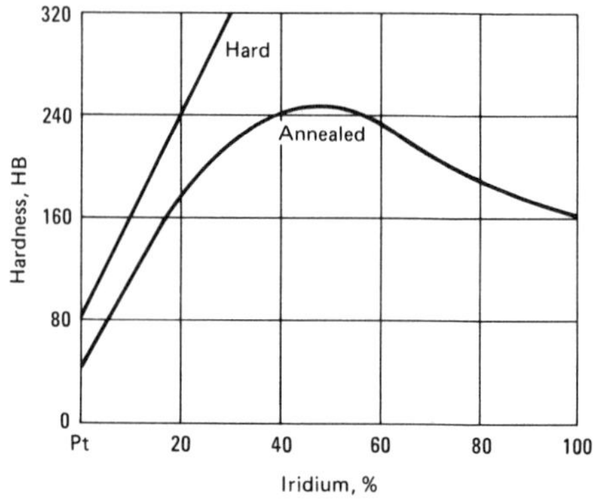

Fig. 16 Hardness of platinum-iridium alloys as a function of iridium content

Platinum-Rhodium Alloys (3.5 to 40% Rh)

Compiled by R.B. Green
Radio Corporation of America
Reviewed for this Volume by
James Klinzing and Lisa Dodson,
Johnson Matthey, Inc.

Applications

Typical uses. Rhodium is the preferred addition to platinum for most applications at high temperatures under oxidizing conditions, because unlike most other hardeners, rhodium is not selectively volatilized. The 10% Rh alloy is used more than any of the other alloys in this series. In the production of nitric acid, it is used as a catalyst for the oxidation of ammonia by air. This alloy, with its composition controlled closely, is the positive element in the standard thermocouples (Pt-10Rh versus platinum) that are used to define the International Practical Temperature Scale of 1968 in the range from 630.74 °C (1167.3 °F) to the gold point (1064.43 °C, or 1948.0 °F). Some use is made of the Pt-13Rh versus platinum thermocouple in instruments that are calibrated for it.

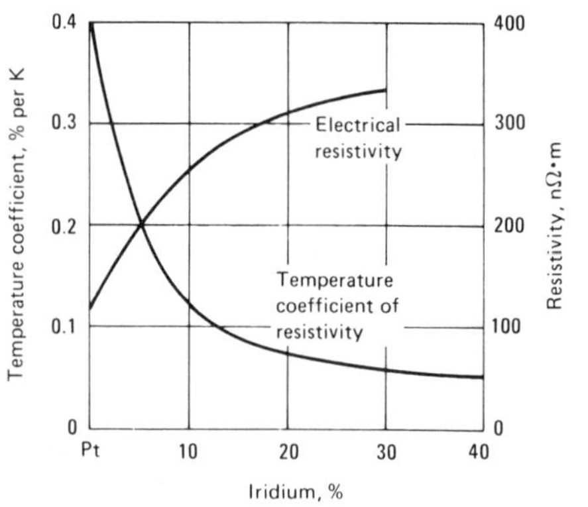

Fig. 17 Electrical resistivity of platinum-iridium alloys as a function of iridium content

Table 7 Typical properties of platinum-iridium alloys

% Iridium and temper	Tensile strength MPa	Tensile strength ksi	Hardness, HB	Density g/cm^3	Density lb/in.^{3}	Electrical resistivity, nΩ · m	Temperature coefficient(a), per °C
5% annealed	275	40	90	21.49	0.777	190	0.00188
5% hard	485	70	140				
10% annealed	380	55	130	21.53	0.778	250	0.00126
10% hard	620	90	185				
15% annealed	515	75	160	21.57	0.780	285(b)	0.00102
15% hard	825	120	230				
20% annealed	690	100	200	21.61	0.781	310	0.00081
20% hard	1000	145	265				
25% annealed	860	125	240	21.66	0.783	330	0.00066
25% hard	1170	170	310				
30% annealed	1105	160	280	21.70	0.784	350	0.00058
30% hard	1380	200	360				
35% annealed	...	...	...	21.79	0.787	360	0.00058
35% hard	...	...	...				

(a) Of electrical resistivity at 0 to 160 °C (32 to 320 °F). (b) By interpolation

Table 8 Typical properties of platinum-rhodium alloys

% Rhodium and temper	Tensile strength MPa	Tensile strength ksi	Elongation(a), %	Hardness, HB	Density g/cm^3	Density lb/in.^{3}	Electrical resistivity(b), nΩ · m	Temperature coefficient(c), per °C
3.5% annealed	170	25	35	60	20.90	0.755	166	0.0022
3.5% hard(d)	415	60	...	120				
5.0% annealed	205	30	35	70	20.65	0.746	175	0.0020
5.0% hard(d)	485	70	...	130				
10% annealed	310	45	35	90	19.97	0.722	192	0.0017
10% hard(d)	620	90	2	165				
20% annealed	485	70	33	120	18.74	0.677	208	0.0014
20% hard(d)	895	130	2	210				
30% annealed	540	78	30	132	17.62	0.637	194	0.0013
30% hard(d)	1060	154	0.5	238				
40% annealed	565	82	30	150	16.63	0.601	175	0.0014
40% hard(d)	1255	182	0.5	290				

(a) In 50 mm (2 in.). (b) At 20 °C (68 °F). (c) Of electrical resistivity at 20 to 100 °C (68 to 212 °F). (d) Hard, as cold worked, 75% reduction

The following thermocouples are accepted internationally and standard temperature/electromotive tables are available (for further information see NBS Monograph 125 and latest addition of ASTM Monograph 565):

- Pt-10Rh versus Pt (Type S)
- Pt-13Rh versus Pt (Type R)
- Pt-30Rh versus Pt-6Rh (Type B)

Types R and S thermocouples are generally used to 1400 °C (2552 °F) for extended service, whereas the Type B couple may be used to 1600 °C (2910 °F). The preferred atmosphere for the use of these thermocouples is air. For emf stability, high-purity alumina insulators and protection tubes should be used.

Pt-3.5Rh alloy is used as a crucible and shows very little loss in weight at high temperatures. Platinum-rhodium alloys are used in the glass industry; they stand up well on contact with molten glass. The Pt-10Rh alloy is used for feeder dies and in handling glasses of high melting point. The Pt-10Rh alloy also is used for rayon spinnerettes. Pt-10Rh and Pt-20Rh alloys are used as windings in high-temperature furnaces that operate under oxidizing atmospheres. The 40% Rh alloy has been used as a winding in furnaces operating between 1500 and 1800 °C (2730 and 3275 °F).

Mechanical Properties

Tensile properties. See Table 8.
Hardness. See Table 8.

Mass Characteristics

Density. See Table 8.

Electrical Properties

Electrical resistivity. See Table 8.
Temperature coefficient of electrical resistivity. See Table 8.

Fabrication Characteristics

Workability. These alloys can be worked hot or cold. Alloys containing up to 20% Rh have readily been cold worked. Alloys with a higher rhodium content may require more hot work before they can be cold worked successfully. The hot-working temperature range is between 900 and 1200 °C (1650 and 2190 °F), and the annealing temperature range is between 900 and 1000 °C (1650 and 1830 °F). If proper melting procedures are followed, reductions up to 90% between anneals are possible.

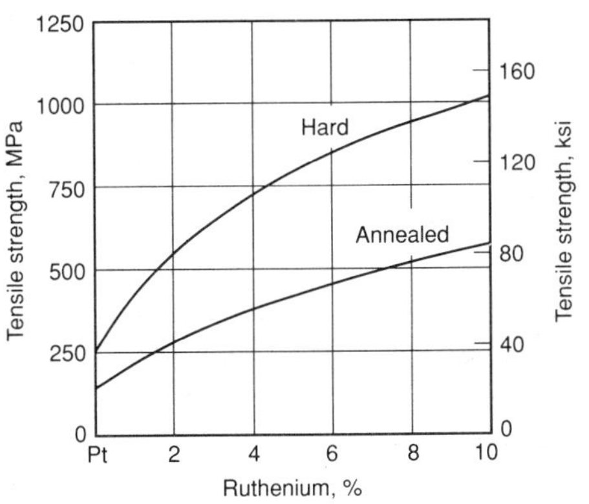

Fig. 18 Tensile strength of platinum-ruthenium alloys as a function of ruthenium content. Initially reduced by 75%, then annealed 15 min

Platinum-Ruthenium Alloys

Compiled by F.E. Carter
Engelhard Industries, Inc.
Minerals and Chemicals Corp.
Reviewed for this Volume by
James Klinzing and Lisa Dodson,
Johnson Matthey, Inc.

Applications

Typical uses. The alloy that contains 5% Ru is used in jewelry and has properties essentially the same as the 10% Ir alloy—the so-called hard platinum of the jewelry trade. The same alloy is also used for laboratory electrode stems and for certain other chemical equipment, but it is not completely suitable for service at high temperature under strongly oxidizing conditions. Platinum-ruthenium alloys are frequently employed as electrical contacts—the 5% Ru alloy in the medium-duty field, the 10% alloy in aircraft magnetos, and the 14% alloy for heavy-duty contacts. Complex platinum-base alloys that contain 4 to 5% Ru are being used to some extent for spark plug electrodes in aircraft. The 10 and 11% Ru platinum alloys are about equally in demand, for electrical contacts and hypodermic needles.

Mechanical Properties

Tensile properties. Tensile strength: 5% Ru: annealed, 415 MPa (60 ksi); hard, 795 MPa (115 ksi). 10% Ru: annealed, 570 MPa (83 ksi); hard, 1035 MPa (150 ksi). Elongation in 50 mm (2 in.): 5% Ru: annealed, 34%; hard, 2%; 10% Ru: annealed, 31%; hard, 2%. See also Fig. 18.

Hardness. 5% Ru: annealed, HB 130, HRB 70; hard, HB 210, HRB 94. 10% Ru: annealed, HB 190, HRB 86; hard, HB 280. 14% Ru, annealed, HV 240

Mass Characteristics

Density. 5% Ru, 20.67 g/cm^3 (0.747 lb/in.^{3}); 10% Ru, 19.94 g/cm^3 (0.720 lb/in.^{3})

Electrical Properties

Electrical resistivity. 5% Ru, 315 nΩ · m; 10% Ru, 430 nΩ · m; 14% Ru, 460 nΩ · m. See also Fig. 19.

Temperature coefficient of resistivity. 5% Ru, 0.0009/°C at 0 to 1000 °C (32 to 1830 °F); 10% Ru, 0.0008/°C at 0 to 1000 °C (32 to 1830 °F); 14% Ru, 0.00036/°C at 0 to 100 °C (32 to 212 °F)

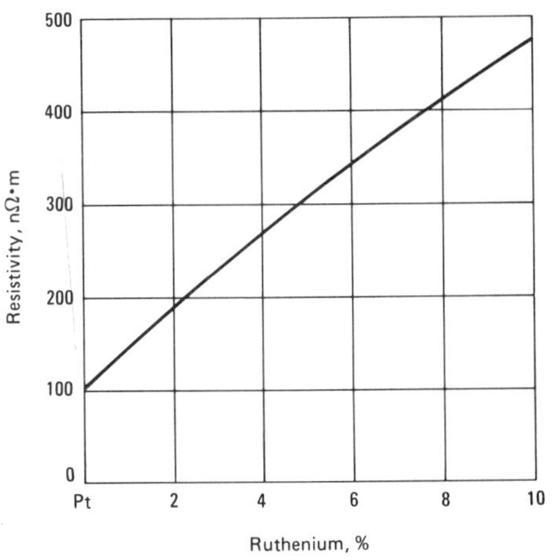

Fig. 19 Electrical resistance of platinum-ruthenium alloys as a function of ruthenium content

Fabrication Characteristics

Workability. The 5% Ru alloy is hot worked between 900 and 1200 °C (1650 and 2200 °F); and the 10% Ru alloy, between 995 and 1300 °C (1825 and 2375 °F). Annealing temperature for the 5% alloy is about 1000 °C (1825 °F); and for the 10% alloy, about 1095 °C (2000 °F). Maximum reduction between anneals should not exceed 90% for the 5% Ru alloy, or 75% for the 10% alloy. The 14% Ru alloy approaches the practical limit of workability. The atmosphere for high-temperature anneals should be only slightly oxidizing, since excessively oxidizing atmospheres cause loss of ruthenium in the same manner as with iridium alloys.

79Pt-15Rh-6Ru

Compiled by J.D. Mitilineos
Sigmund Cohn Corp.
Reviewed for this Volume by
James Cohn, Sigmund Cohn Corp.

Commercial Names

Trade name. Alloy No. 851

Chemical Composition

Composition limits. 78.9 to 80.1 Pt, 14.9 to 15.1 Rh, 5.0 to 6.1 Ru

Applications

Typical uses. This alloy has remarkably high tensile strength and hardness, combined with excellent corrosion resistance, weldability, stability, and shelf life. In solid-solution form, it is used for wire-wound potentiometers, galvanometers, suspension strips, bridgewires, contacts, catalytic glow plugs, and corona wire for copiers.

Mechanical Properties

Tensile properties. Typical: tensile strength, 2070 MPa (300 ksi); yield strength, 1515 MPa (220 ksi); elongation, 2% in 254 mm (10 in.)

Table 9 Mechanical properties of platinum-tungsten alloys

Condition	Tungsten, %	Tensile strength MPa	Tensile strength ksi	Hardness, HV
Annealed at 1200 °C (2190 °F)..............	2	570	82	100
	4	770	112	133
	6	860	125	158
	8	895	130	180
Hard, 99.8% reduced.....	2	1345	195	...
	4	1690	245	...
	6	1930	280	...
	8	2070	300	...
Hard, 50% reduced	2	...	...	170
	4	...	...	220
	6	...	...	260
	8	...	...	300

Hardness. 371 HK
Elastic modulus. Tension, 205 GPa (30×10^6 psi)

Mass Characteristics

Density. 18.6 g/cm^3 (0.67 lb/in.^{3})

Thermal Properties

Solidus temperature. 1880 °C (3415 °F)
Coefficient of linear thermal expansion. 15.6 μm/m · K (8.69 μin./in. · °F)
Thermal electromotive force versus Cu. 3.24 m V/°C

Electrical Properties

Electrical resistivity. 308 nΩ · m at 0 °C (32 °F); temperature coefficient, 0.06%/°C (0.185 nΩ · m/°C at 0 to 100 °C (32 to 212 °F)

Chemical Properties

General corrosion behavior. Excellent resistance to corrosion
Resistance to specific corroding agents. Very slowly attacked even by aqua regia. Can only be put into solution with a caustic fusion

Fabrication Characteristics

Weldability. Excellent

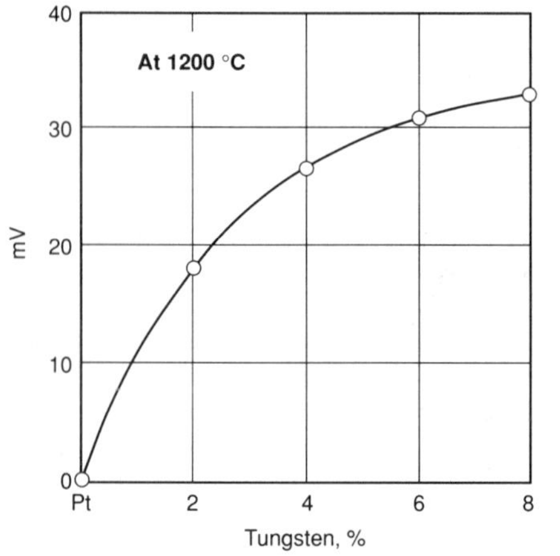

Fig. 22 Thermal electromotive force of platinum-tungsten alloys versus platinum as a function of tungsten content

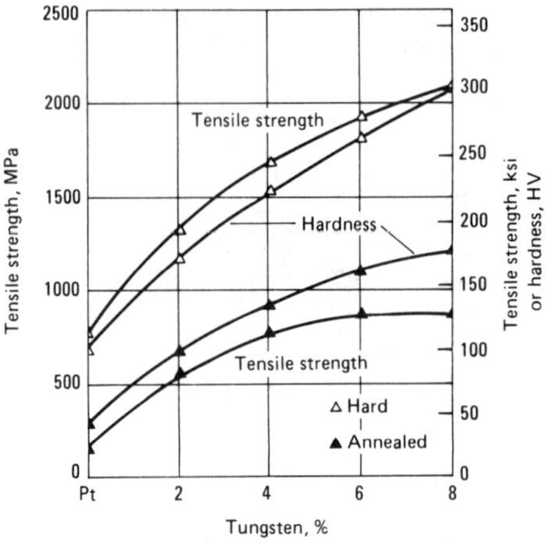

Fig. 20 Mechanical properties of platinum-tungsten alloys as a function of tungsten content

Platinum-Tungsten Alloys

Compiled by J.D. Mitilineos
Sigmund Cohn Corp.
Reviewed for this Volume by
James Cohn, Sigmund Cohn Corp.

Chemical Composition

Nominal compositions. Four platinum-tungsten alloys are commonly used: 2%, 4%, 6%, and 8% W.

Applications

Typical uses. The 4% W platinum alloy was originally developed for spark plug electrodes in aircraft engines. Its resistance to lead contamination is superior in the hard-drawn condition. It was also used for grids in radar tubes because of reduced electron emission. These uses have diminished over the years.

The Pt-8% W alloy has been used for potentiometer wire because it has excellent wear resistance and low electrical noise characteristics. Pt-8% W has been used as a bridgewire and heating element. As with many strong, high-platinum-content alloys, wide use of this alloy is made in the medical community due to its high degree of compatibility with human tissue and excellent fatigue resistance.

Precautions in use. The Pt-8% W alloy is not recommended for use at high temperature under oxidizing conditions because of the selective oxidation of tungsten.

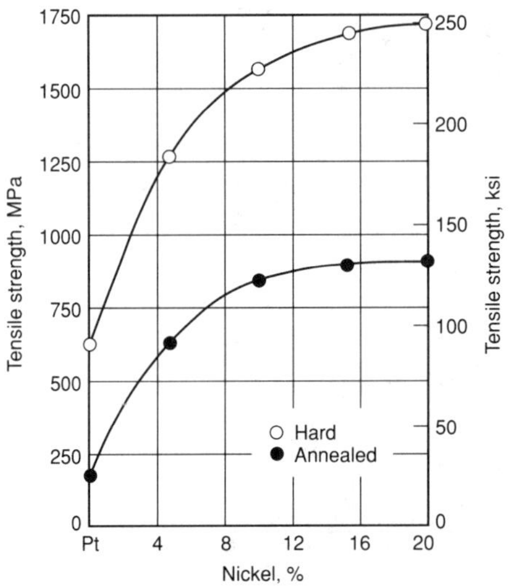

Fig. 23 Tensile strength of platinum-nickel alloys as a function of nickel content

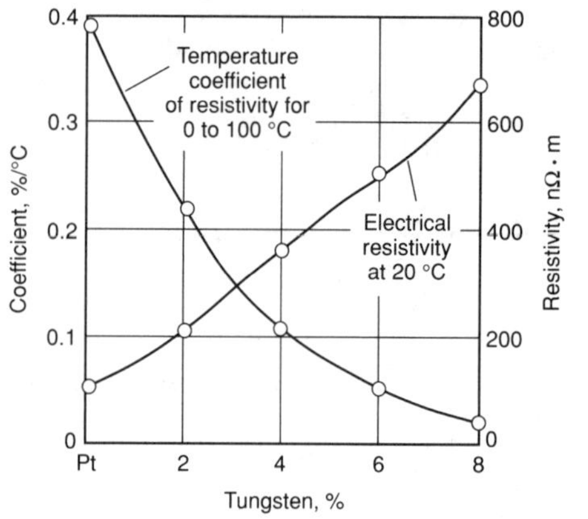

Fig. 21 Electrical resistivity of platinum-tungsten alloys as a function of tungsten content

Mechanical Properties

See Table 9 and Fig. 20.
Hardness. Sheet: See Table 9 and Fig. 20.

Thermal Properties

Liquidus temperature. 8% W, 1910 °C (3470 °F)
Solidus temperature. 8% W, 1870 °C (3400 °F)

Electrical Properties

Electrical resistivity. At 20 °C (68 °F) annealed: 2% W, 215 nΩ · m; 4% W, 360 nΩ · m; 6% W, 530 nΩ · m; 8% W, 665 nΩ · m. See also Fig. 21.

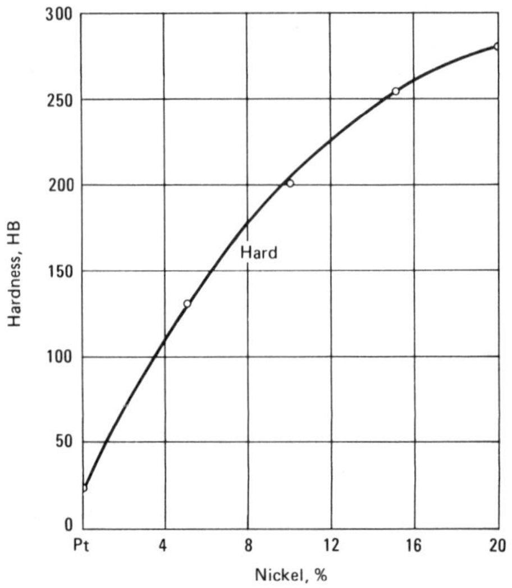

Fig. 24 Hardness of platinum-nickel alloys as a function of nickel content

Table 10 Electrical properties of platinum-nickel alloys

Condition	Nickel, %	Resistivity, nΩ · m	Coefficient, Ω/Ω per K
Annealed	5	236	0.00179
	10	298	0.00135
	15	330	0.00114
	20	350	0.00102
Hard	5	244	0.00170
	10	304	0.00125
	15	440	0.00105
	20	360	0.00094

Temperature coefficient of electrical resistivity. Annealed 0 to 100 °C (32 to 212 °F): 2% W, 0.0022/K; 4% W, 0.0011/K; 6% W, 0.0006/K; 8% W, 0.00025/K. See also Fig. 21.
Thermal electromotive force versus Pt. At 1200 °C (2190 °F), cold junction at 0 °C (32 °F): 2% W, 19 mV; 4% W, 26.5 mV; 6% W, 31.5 mV; 8% W, 34 mV. See also Fig. 22.

Platinum-Nickel Alloys

Compiled by J.D. Mitilineos
Sigmund Cohn Corp.
Reviewed for this Volume by
James Cohn, Sigmund Cohn Corp.

Chemical Composition

Nominal compositions. Platinum-nickel alloys range in composition from 0 to 20% N.

Applications

Typical uses. These alloys have long been used for their strength at high temperatures. Taut band strips for electrical meters are made from 10% Ni-Pt alloys. In addition to high tensile strength, this alloy is remarkably free of hysteresis.
Precautions in use. Selective oxidation of nickel limits the use of platinum-nickel alloys at high temperature under oxidizing conditions.

Mechanical Properties

Tensile properties. Tensile strength:

Nickel, %	Tensile strength MPa	ksi
Annealed		
5	640	93
10	815	120
15	910	130
20	910	130
Hard, 90% reduced		
5	1240	180
10	1550	225
15	1690	245
20	1725	250

See also Fig. 23.
Hardness. 5% Ni, 130 HB; 10% Ni, 200 HB; 15% Ni, 255 HB; 20% Ni, 280 HB. See also Fig. 24.
Elastic modulus. Tension for 10% Ni-Pt alloy: 170 GPa (25 × 10⁶ psi)

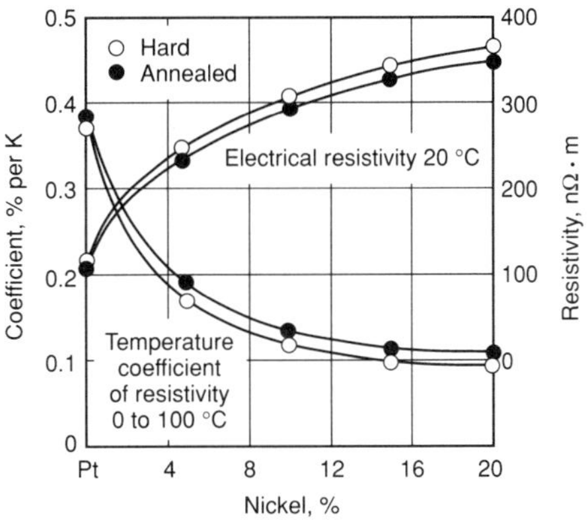

Fig. 25 Electrical resistivity of platinum-nickel alloys as a function of nickel content

Electrical Properties

Electrical resistivity. At 20 °C (68 °F): See Table 10 and Fig. 25.
Temperature coefficient of electrical resistivity. At 0 to 100 °C (32 to 212 °F): See Table 10 and Fig. 25.

Platinum-Cobalt Permanent Magnet Alloys

Compiled by A.R. Robertson
Engelhard Corporation

W.J. Jellinghaus in 1936 discovered that certain platinum-cobalt alloys had an unusually high coercive force, H_c (Ref 20). To date, coercivities up to 540 kA · m^{-1} (6.8 kOe) have been observed in these alloys at room temperature. The most valuable data on these alloys have been published in a series of papers by J.B. Newkirk, R. Smoluchowski, A.H. Geisler, and D.L. Martin; the last of the series is cited below (Ref 21). The platinum-cobalt alloys used for permanent magnets are near the 50 at.% composition (23.3 wt% Co). In this region the alloys are disordered face-centered cubic (fcc) at high temperature and ordered face-centered tetragonal (fct) at low temperature. Alloys near 30 at.% Co form a superlattice similar to Cu_3Au, fcc with cobalt atoms on the corner sites and platinum atoms on the face centers. The strain induced by dimensional change in a rigid alloy when it is ordered results in considerable hardening.

Applications

Typical uses. When an alloy of platinum and cobalt is properly processed, it exhibits extremely high-energy products. For this reason, this material is used in areas where the length-to-thickness ratio of most other magnetic materials would be unfavorable. In many critical applications that require minimum space or weight and stable temperature performance without fear of attrition, platinum-cobalt magnets may be the best choice. Typical uses include focusing magnets, hearing-aid magnets, magnetic phonograph cartridges, electric watch magnets, rotors in miniature motors, and gyro bearings. Other applications include platinum-cobalt films for use in digital magnetooptic recording, medical implants for atrial stimulation and sensing, as well as in a system for delivery of magnetic emboli via a guided catheter to specific cerebral arteries.

Properties and Fabrication

The magnetic characteristics of the platinum-cobalt alloys can be varied by adjusting composition within the range from 40 to 60 at.% Co, and by varying the heat treatment. Curves of magnetic properties as functions of time and temperature of aging are given in papers by D.L. Martin (Ref 22, 23). According to Martin:

- The disordered phase has a higher saturation $(B_i)_P$ and residual induction B_r than the ordered alloy
- The coercive force, H_c, and the intrinsic coercive force, H_{ci}, and the related maximum energy product $(BH)_{max}$ increase to a maximum and then decrease with additional aging time
- Peak values of H_c, H_{ci}, and $(BH)_{max}$ were obtained in alloys aged at 600 °C (1110 °F). The coercive force reaches its maximum before ordering is complete
- The effect of cobalt on the magnetic properties was not established precisely. However, alloys with 49 to 50 at.% Co have the highest coercive force and the highest $(BH)_{max}$
- Magnetic properties depend not only on time and temperature of aging but also on the temperature at which the alloys are previously heated for disordering and the rate at which they are cooled from the disordering temperature
- The platinum-cobalt alloys can be prepared either by melting or by powder metallurgy; with proper techniques, they can be worked hot or cold, but sulfur content must be controlled to prevent hot shortness. The disordered phase is softer and more ductile than the ordered phase and can be retained at room temperature by quenching

Additional information is available in the article "Permanent Magnet Materials" in this Volume.

Zirconia-Grain-Stabilized Platinum and Platinum Alloys

Compiled by James Klinzing and
Lisa Dodson
Johnson Matthey Inc.

The use of platinum and its alloys (up to 25% Rh-Pt) is well established in the automotive, chemical, glass, electrical, and dental industries. However, these applications often mean high-temperature operation of these materials. At high temperatures these materials are weak and subject to creep.

Table 11 Mechanical and electrical properties of zirconia-grain-stabilized (ZGS) platinum alloys compared with conventional platinum and platinum alloys

Alloy	Density		Specific resistance, μΩ · cm	Temperature coefficient of resistance, per °C	Hardness, HV	Ultimate tensile strength at elevated temperatures													
						20 °C (70 °F) annealed		1000 °C (1830 °F)		1100 °C (2010 °F)		1200 °C (2090 °F)		1300 °C (2370 °F)		1400 °C (2550 °F)		1500 °C (2730 °F)	
	g/cm³	lb/in.³				MPa	ksi	MPa	ksi	MPa	ksi	MPa	ksi	MPa	ksi	MPa	ksi	MPa	ksi
100% Pt	21.4	0.773	10.6	0.0039	40	124	18.0	23	3.4	17	2.4	12.8	1.85	7.86	1.14	3.9	0.57	...	
Pt-10Rh	20.0	0.723	18.4	0.0017	90	331	48.0	82.0	11.9	61	8.8	47	6.8	38	5.5	30	4.4	23	3.4
Pt-20Rh	18.8	0.679	20.0	0.0017	115	483	70.0	230	33.4	162	23.5	99.3	14.4	68.6	9.95	49	7.1	38	5.5
ZGS Pt	21.4	0.773	11.12	0.0031	60	183	26.5	51.0	7.4	45	6.5	37	5.4	35	5.1	28	4.1	23	3.4
ZGS Pt-10 Rh	19.8	0.715	21.2	0.0016	110	355	51.5	163	23.7	140	20.3	125	18.2	92.4	13.4	83.4	12.1	70.3	10.2

They are also subject to contamination failure.

Zirconia-grain-stabilized (ZGS) platinum materials are produced by incorporating a fine, insoluble phase dispersed uniformly throughout the platinum metal matrix, a process called dispersion strengthening (see the article "Dispersion-Strengthened Nickel-Base and Iron-Base Alloys" in this Volume).

Grain stabilization provides greater resistance to grain growth, dislocation movement, and grain-boundary sliding. These problems normally occur in conventional platinum at high temperatures causing sagging, bulging, and cracking.

Applications

Typical uses. Product applications where high temperature causes creep, distortion, and ultimately failure of unsupported conventional platinum and its alloys. ZGS platinum-rhodium bushings used for the production of continuous filament glass fiber resist creep-induced sagging and eliminate the need for costly platinum supports.

Possible weight reduction in product design is also an advantage of using ZGS platinum. For instance, glass-carrying apparatus can be designed with thinner wall sections and have 50% greater useful life; lightweight thermocouple pockets may be fabricated with walls one-half the conventional thickness; and in stirrers for specialty glasses, ZGS platinum often makes possible the elimination of molybdenum and ceramic cores, doubling of service life, and reduction of running problems.

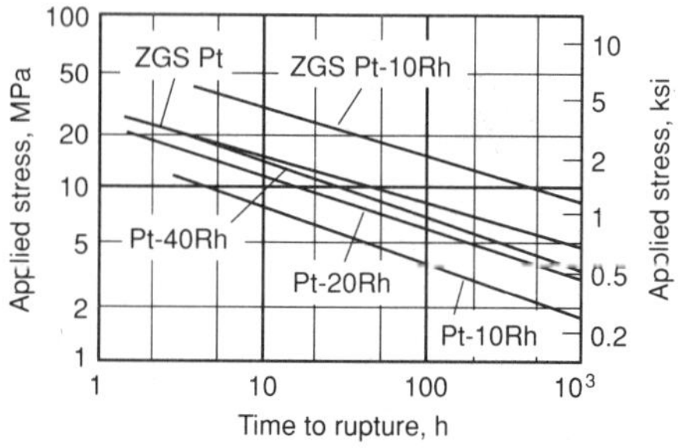

Fig. 26 Stress-rupture properties of ZGS platinum, ZGS Pt-10Rh, and the commercially important conventional alloys. Samples were 1.5 mm (0.060 in.) thick sheets tested in air at 1400 °C (2550 °F).

Physical, Electrical, and Tensile Properties

Table 11 shows the typical properties of ZGS platinum and its alloys at room temperature, and compares the tensile strength of ZGS materials at various temperatures with conventional platinum and its alloys. Figure 26 illustrates the improvement in high-temperature life achieved by grain stabilization.

The time to failure at 1400 °C (2550 °F) under a stress of 9.83 MPa (1.425 ksi) shows that ZGS platinum lasts two to three times as long as the best conventional platinum-rhodium alloy, and ZGS Pt-10Rh lasts more than 10 times as long as ZGS platinum.

Fabrication Characteristics

Machinability. The machining characteristics of ZGS platinum and its alloys are, in general, similar to but somewhat better than those of pure platinum, platinum-rhodium alloys, or pure nickel. ZGS materials gall readily, therefore tools must be kept sharp to minimize galling and pickup on the machining tools. Sulfur-free oils normally are recommended for machining.

Workability. ZGS platinum and its alloys can be hot forged or hot rolled readily in the temperature range of 1200 to 1500 °C (2190 to 2730 °F). It is essential to maintain oxidizing conditions throughout processing to prevent contamination. Before annealing, iron should be removed from the hot-worked surface by pickling in hot hydrochloric acid.

ZGS platinum and its alloys can be cold worked easily by rolling, drawing, and extruding. Wire can be cold drawn with a reduction of 98% or more without intermediate annealing. The material can be easily formed, deep drawn, pierced, punched, spun, and so on.

ZGS platinum work hardens at a rate intermediate between pure platinum and 10% Rh-Pt, making ZGS platinum much easier to fabricate than conventionally melted platinum-rhodium alloys. At the same time, ZGS 10% Rh-Pt work hardens at a rate intermediate between the 10% Rh-Pt alloy and the 20% Rh-Pt alloy. This makes ZGS 10% Rh-Pt easier to fabricate than the 20% Rh-Pt alloy. Cold-worked ZGS platinum and its alloys can be softened by recrystallization at a temperature only slightly higher than that used for pure platinum and its alloys.

Corrosion Resistance

ZGS platinum and its alloys are similar to pure platinum and its alloys in their corrosion-resistance characteristics. It is resistant to corrosion by single acids, alkalies, aqueous solutions of simple salts, and organic materials. Even at elevated temperatures, it is resistant to dry hydrogen chloride and to sulfurous gases. Aqua regia and hydrochloric acid containing an oxidizing agent will attack ZGS materials as will free halogens to some degree at elevated temperatures.

ZGS platinum and its alloys are outstanding in their resistance to oxidation, remaining untarnished on heating in air at all temperatures. It is also essentially inert to many molten salts and it resists the action of fused glasses if oxidizing conditions are maintained. A number of low-melting metals, including lead, tin, antimony, zinc, and arsenic will readily alloy with and attack ZGS materials at their melting temperatures.

Palladium and Palladium Alloys

Commercially Pure Palladium 99.85% Pd

Compiled by E.M. Wise and R.F. Vines
The International Nickel Company, Inc.
Reviewed for this Volume by
James Klinzing, Johnson Matthey, Inc.

Applications

Typical uses. Palladium resembles platinum in appearance, ductility, and strength. Its nobility and melting point are somewhat lower than those of platinum, but its lower cost-per-unit weight and lower density pro-

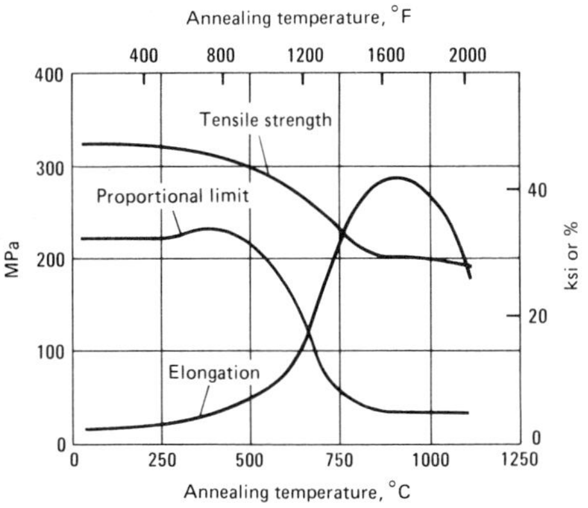

Fig. 27 Tensile properties of cold-drawn deoxidized palladium as a function of annealing temperature. Annealing time was 5 min.

vide economic advantages. The major use of palladium is for contacts in light-duty electrical relays, where its freedom from tarnishing provides extreme reliability and noise-free transmission, required in voice circuits. Its effectiveness as a catalyst accounts for its use in the removal of oxygen from atmospheres used for heat treatment, the recombination of hydrogen and oxygen, the hydrogenation of terpines, and also the manufacture of organics such as vitamins. Hydrogen will diffuse selectively through a palladium septum, yielding pure gas. However, the gas must be initially free from sulfur. Palladium hardened with ruthenium provides an all-precious-metal white jewelry alloy that sets off diamonds to advantage. Generally a solid solution former, palladium is employed as the major or auxiliary element in dental, electrical contact, and special resistance alloys. Palladium is an important constituent in the high-temperature solders that have low vapor pressures, excellent wettability, and minimum penetration into austenitic alloys. Palladium with a minimum purity of 99.8% (UNS P03980) meeting ASTM B 589, and MIL-E-46065 (MR) is available commercially.

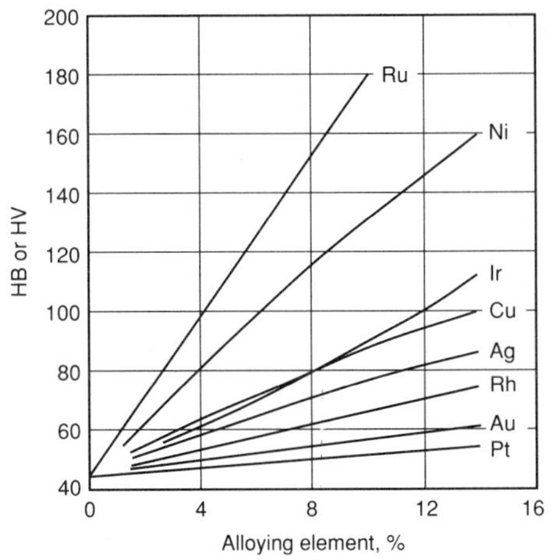

Fig. 30 Effect of various alloying additions on the hardness of annealed palladium

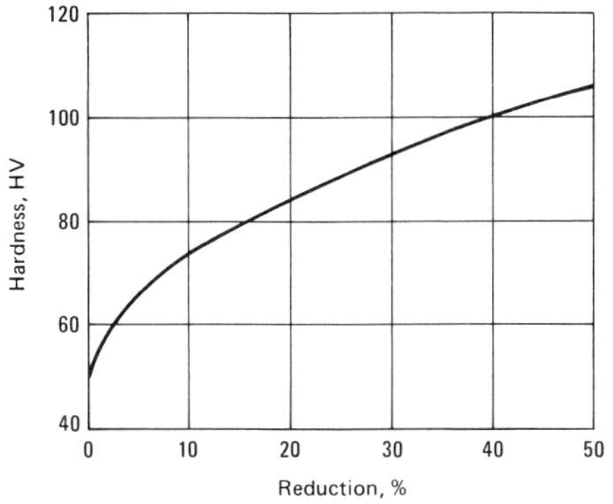

Fig. 28 Hardness of cold rolled palladium as a function of reduction during rolling

Precautions in use. Contamination with low-melting-point metals causes embrittlement, with base metals causes hardening and decreased corrosion resistance, and with silicon causes loss of hot strength.

Mechanical Properties

Tensile properties. Typical: 1.3 mm (0.05 in.) wire. After annealing at high temperature: tensile strength, 145 MPa (21 ksi); elongation, 24%. After annealing at 800 °C (1470 °F): tensile strength, 172 MPa (25 ksi); elongation, 30%. Tensile strength at various temperatures, material annealed at 1095 °C (2000 °F): at room temperature, 193 MPa (28.0 ksi); at 400 °C (750 °F), 125 MPa (18.1 ksi); at 800 °C (1470 °F), 57 MPa (8.3 ksi); at 1000 °C (1830 °F), 26 MPa (3.8 ksi)

The tensile properties of palladium are mildly sensitive to the kind and amount of residual deoxidizing agents present. Palladium containing such residuals may have a tensile strength of 172 to 207 MPa (25 to 30 ksi) as annealed and 324 MPa (47 ksi) after cold drawing 50%. The tensile properties of deoxidized palladium, initially reduced 50% by cold drawing and annealed for 5 min at various temperatures, are shown in Fig. 27. The optimum anneal for this material is about 800 °C (1470 °F).

Hardness. Rolled and annealed: 37 to 42 HV. Effect of cold rolling, see Fig. 28. Effect of annealing, see Fig. 29. Electrodeposited palladium is much harder than wrought material, ranging from 190 HV for metal from the

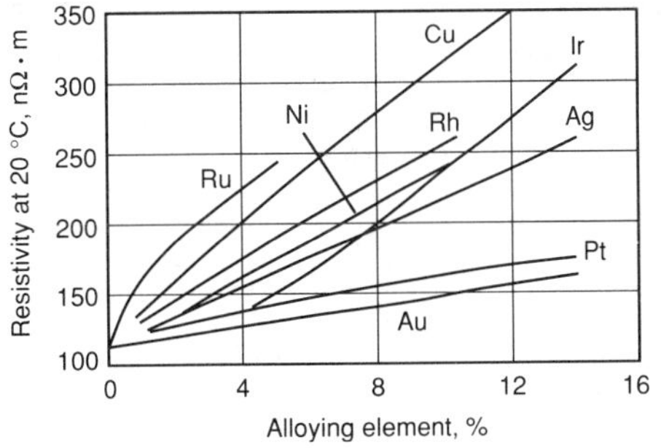

Fig. 31 Effect of various alloying additions on the electrical resistivity of palladium

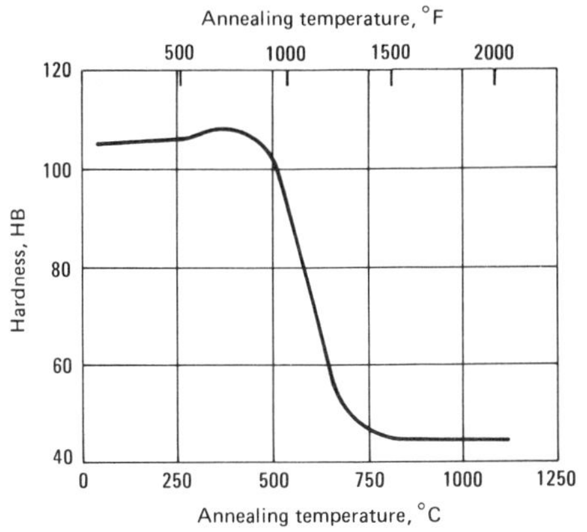

Fig. 29 Hardness of palladium as a function of annealing temperature

chloride bath, to about 400 HV for deposits from the complex nitrite baths. For jewelry and other applications where a stronger material is desired, palladium usually is hardened with ruthenium. The relative effects of the various additions of the hardness of annealed material are given in Fig. 30.

Elastic modulus. Tension: 112 GPa (16.3 × 10^6 psi)

Mass Characteristics

Density. 12.02 g/cm^3 (0.434 lb/in.^{3}) at 20 °C (68 °F)

Thermal Properties

Melting point. 1552 °C (2826 °F)

Coefficient of linear thermal expansion. 11.76 μm/m · K (6.53 μin./in. · °F) at 20 °C (68 °F)

Specific heat. 0.245 kJ/kg · K (0.0584 Btu/lb · °F) at 0 °C (32 °F)

Thermal conductivity. 76 W/m · K (526 Btu · in./ft^2 · h · °F) at 18 °C (64 °F)

Electrical Properties

Electrical conductivity. 16% IACS at 20 °C (68 °F)

Electrical resistivity. 108 nΩ · m at 20 °C (68 °F) and 100 nΩ · m at 0 °C (32 °F). Effect of alloying, see Fig. 31

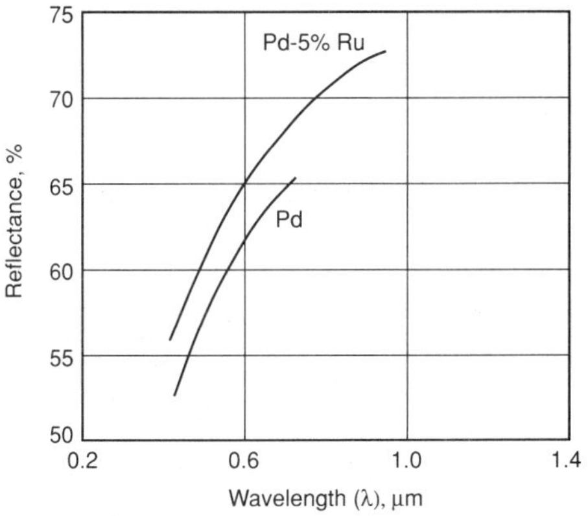

Fig. 32 Reflectance of palladium and Pd-5%Ru as a function of wavelength

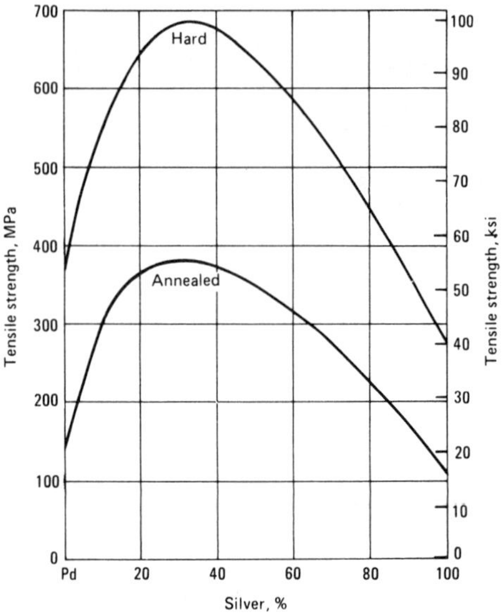

Fig. 33 Tensile strength of palladium-silver alloys as a function of silver content

Optical Properties

Reflectance. 62.8% in white light. Increases slightly in going from red to blue (see Fig. 32)

Chemical Properties

General corrosion behavior. Generally, palladium is less corrosion resistant than platinum, but more corrosion resistant than silver. In ordinary atmospheres palladium is resistant to tarnish, but some discoloration may occur during exposure to moist industrial atmospheres that contain sulfur dioxide. Adding palladium to gold or silver alloys improves the tarnish resistance.

Resistance to specific corroding agents. At room temperature, palladium is resistant to corrosion by hydrofluoric, perchloric, phosphoric, and acetic acids. It is attacked slightly by sulfuric, hydrochloric, and hydrobromic acids, especially in the presence of air; and it is attacked readily by nitric acid, ferric chloride, hypochlorites, and moist chlorine, bromine, and iodine. Palladium is resistant to molten sodium or potassium nitrate, but not to sodium peroxide, sodium hydrate, or sodium carbonate.

Fabrication Characteristics

Precautions in melting. Torch melting of small lots of palladium alloys is common in jewelry and dental applications. For larger melts, palladium is best inductively melted under an argon or lean hydrogen-nitrogen gas cover with care to prevent contamination with silicon, which causes hot shortness. The melt is deoxidized with 0.05% Al or calcium boride just before pouring.

Hot-working temperature. 760 to 1095 °C (1400 to 2000 °F)

Recrystallization temperature. 595 °C (1100 °F)

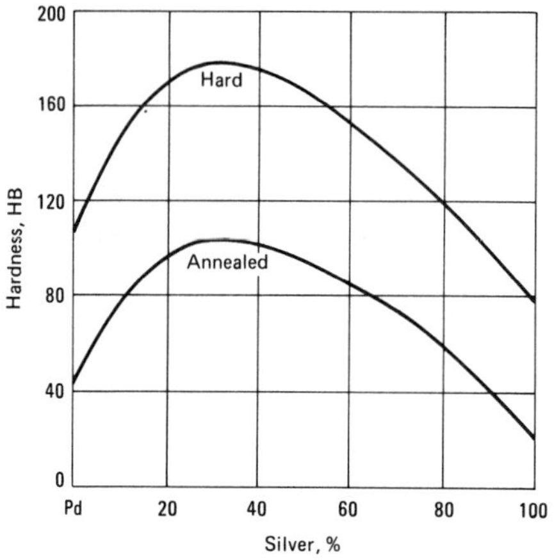

Fig. 34 Hardness of palladium-silver alloys as a function of silver content

Annealing. Nitrogen-hydrogen mixtures, nitrogen, argon, or steam provide suitable annealing atmospheres. The pure metal may be annealed at about 800 °C (1470 °F); 1000 to 1100 °C (1830 to 2010 °F) is required for some of the harder alloys. Slow cooling of palladium from 815 °C (1500 °F) to 425 °C (800 °F) will cause a blue oxide coating to form. To avoid this, the metal should be quenched in water or cooled in a nitrogen atmosphere. Cooling in hydrogen will cause a phase change to occur, with accompanying distortion.

Joining. Palladium can be melted with an oxyhydrogen torch or welded with plasma or gas-tungsten arc welding (GTAW) equipment. An oxidizing oxyacetylene flame is desirable for soldering palladium with platinum solders melting from 1095 to 1300 °C (2000 to 2375 °F). A gas-air torch and lower-melting white gold solders are employed for soldering palladium jewelry.

Palladium-Silver Alloys

Compiled by J. Hafner and R. Volterra, Texas Instruments, Inc.
Revised by Robert S. Mroczkowski,* AMP Incorporated

Applications

Typical uses. Alloys with 1, 3, 10, 40, 50, and 60% Pd are employed for electrical contacts, the lower-palladium alloys showing less transfer than silver, and the higher-palladium alloys providing surety of contact, due to the absence of a tarnish film.

The 60Pd-40Ag alloy, which has electrical resistivity of 42 μΩ · cm (252 Ω · circ mil/ft) at 20 °C (68 °F) and a temperature coefficient of resistivity of only 0.00003/°C (0.00002/°F) between 0 and 100 °C (32 and 212 °F), is used for precision resistance wires.

60Pd-40Ag alloy is also being used as a contact finish for electronic connectors

*Mr. Mroczkowski would like to acknowledge the support of A. Epstein and P. Lees (Technical Materials, Inc.), who assisted in this revision.

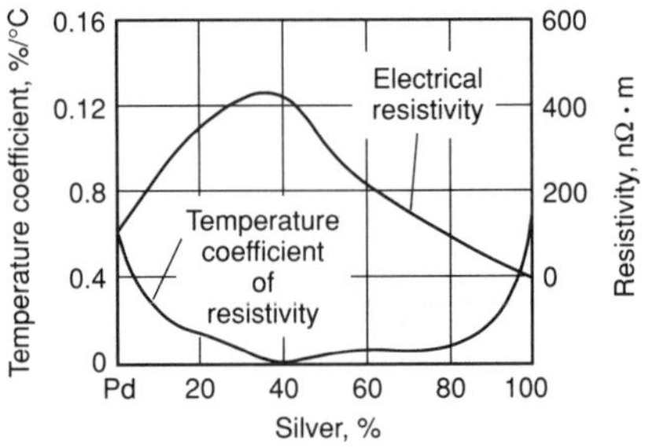

Fig. 35 Electrical resistivity of palladium-silver alloys as a function of silver content

used in low-level dry-circuit applications. A surface layer of gold is used for optimum performance in these applications.

Palladium-silver alloys are used for brazing stainless steel, Inconel, and other heat-resistant alloys; the 90Ag-10Pd alloy has a flow point of 1065 °C (1950 °F) and is much less likely to dissolve or penetrate the base metal than are nickel-base brazing alloys.

Mechanical Properties

Tensile properties. See Fig. 33.

Hardness. See Fig. 34.

Modulus of elasticity. Maximum at about 35% Pd content

Structure

Microstructure. Palladium and silver form a continuous series of solid solutions. Solidus and liquidus curves are close, with a maximum separation of about 60 °C (about 110 °F) for the 30% Pd alloy. No transformations in the solid state have been determined.

Mass Characteristics

Density. Can be calculated from the ratio of the components and the values of the same properties of the unalloyed metals

Thermal Properties

Coefficient of thermal expansion. Can be calculated from the ratio of the components and the values of the same properties of the unalloyed metals

Electrical Properties

Electrical resistivity. See Fig. 35.

Optical Properties

Color. The color of the alloys varies with palladium content; alloys with as little as 15 to 20% Pd have the color of palladium.

Chemical Properties

General corrosion resistance. Alloys with more than about 50% Pd resist tarnishing. Nitric acid dissolves all the alloys, but they are quite resistant to hydrochloric acid, except in the presence of oxidizing agents. Cyanides attack all the alloys, particularly those rich in silver. Metallographic etching can be performed with the Jewett-Wise etch (10% KCN, 10% $NH_4S_2O_8$). The high-temperature corrosion in oxygen decreases with increasing palladium content; the 25% Pd alloy has only half as much weight loss as fine silver at 900 °C (1650 °F). Addition of silver to palladium increases the solubility of hydrogen up to 30 to 40% Ag, decreasing it to zero at about 75% Ag. The 60% Pd-40% Ag alloy has been suggested as a selective diffusion septum for the separation of hydrogen from other gases. The separation of high-purity hydrogen also has employed a 75Pd-25Ag alloy (U.S. Patent 2,773,561).

60Pd-40Ag contact systems subject to vibration will catalyze organic vapors resulting in a surface polymer film that degrades electrical contact performance. The 60Pd-40Ag alloy minimizes such film formation. The presence of the silver, however, increases susceptibility to sulfur- and chloride-bearing environments.

Fabrication Characteristics

Workability. All the alloys can be cold worked. Annealing, preferably at 850 °C (1560 °F), must be in inert or nitrogen atmospheres, to prevent oxidation.

In electronic connector applications, the 60Pd-40Ag alloy is primarily used in inlay form.

60Pd-40Cu

Compiled by H.T. Reeve
Bell Telephone Laboratories
Reviewed for this Volume by
A.R. Robertson, Engelhard Corporation

Applications

Typical uses. 60Pd-40Cu alloy was devised for electrical contacts in the milliampere range, in circuits containing sufficient capacity so that a considerable rush of current occurs on closure. It is also used where a hard material is required for slip rings running against brushes of the same material, and is generally given a final heat treatment to convert it to the ordered, highly conducting condition. This conversion is accelerated greatly by cold work or by the presence of hydrogen in solution.

Mechanical Properties

Tensile properties. Typical: tensile strength: annealed, 515 MPa (75 ksi); hard drawn, 1330 MPa (193 ksi)

Mass Characteristics

Density. 10.6 g/cm^3 (0.383 lb/in.^{3}) at 20 °C (68 °F)

Thermal Properties

Liquidus temperature. Approximately 1224 °C (2235 °F)
Solidus temperature. Approximately 1196 °C (2185 °F)

Electrical Properties

Electrical resistivity. Annealed and quenched, 350 nΩ · m; 35 nΩ · m ordered,

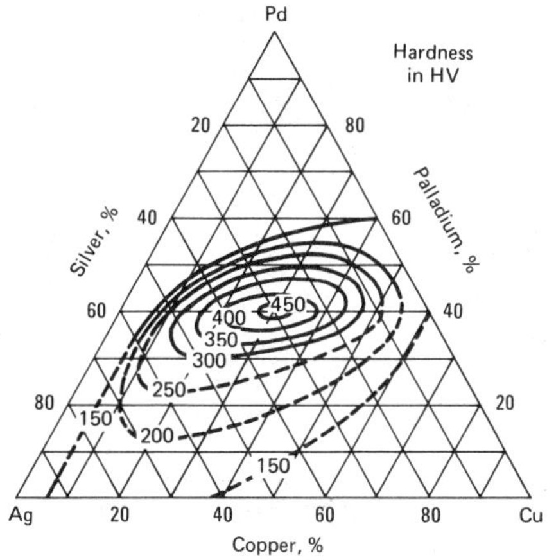

Fig. 36 Maximum hardness of aged palladium-silver-copper alloys

which is best produced by heating cold-worked material at 300 °C (570 °F)
Temperature coefficient of electrical resistivity. Annealed, 0.00032/K (0.00018 °F); ordered, 0.00224/K (0.00125/°F) at 20 to 100 °C (68 to 212 °F)

Palladium-Silver-Copper Alloys

Compiled by P.J. Cascone
J.F. Jelenko & Company

Applications

Typical uses. Alloys that contain about 45% Pd with sufficient copper to make them age hardenable to the desired level are used in dentistry, as are variants that contain a small percentage of platinum or gold. These alloys are used also for electrical contacts subjected to sliding wear or for applications that require good spring properties. About 1% Zn may be present, and this, as well as platinum, appears to accelerate age hardening. The alloys containing 10 to 25% Pd are used for high-strength brazed joints.

Mechanical Properties

Hardness. The hardening response of this system is quite good, as shown in Fig. 36. Although the alloy 40Pd-30Ag-30Cu can be readily worked in the quenched condition, upon age hardening it attains a hardness in excess of 450 HV.

Thermal Properties

The liquidus and solidus features of the palladium-silver-copper system follow the general pattern of the gold-silver-copper system. Palladium increases the liquidus and solidus temperatures of the silver-copper alloys much more rapidly than gold. The silver-copper eutectic persists into the ternary liquidus diagram, terminating at about 30Pd-45Ag-25Cu. The eutectic decomposition on the silver-copper side degenerates into a dome-shape two-phase region in the ternary diagram enabling most of the alloys to be age hardenable (Ref 24). Below 600 °C (1110 °F) the ordered phases $PdCu_5$ and Pd_3Cu_5 appear (Ref 25). These phases extend across the ternary diagram to the silver-rich side of the immiscibility field. Additional sources for phase equilibrium data are available in Ref 26.

Chemical Properties

Resistance to tarnishing. The silver-palladium alloys that contain 50 to 60 wt% or at.% Pd, have good resistance to tarnishing, but that of the corresponding copper-palladium alloys is not quite so good. If substantial age hardening is required, the palladium content should not exceed about 45%, but small amounts of gold or platinum can be added without impairing the hardening, and may actually increase it along with the resistance to tarnishing. As a result, a whole series of useful quaternary alloys exists between the Cu-Ag-Pd and Cu-Ag-Au systems.

Fabrication Characteristics

Melting and working. At high temperatures, most of the ternary alloys are solid solutions and all are workable after quenching. The alloys must be melted in such a manner that oxygen, silicon, and sulfur are low in the metal when it is cast. Final deoxidation with a few hundredths percent of calcium or calcium boride often is useful. Annealing in nitrogen at about 800 °C (1475 °F) is suitable for most of the alloys, and the age-hardenable ones should be cooled fairly rapidly for full softness. If much zinc is present, oil quenching may be required, to prevent partial hardening during cooling. Treatment for ¼ to 5 h between 400 and 455 °C (750 and 850 °F) is an effective method for hardening many alloys.

Palladium-Silver-Gold Alloys 40Ag-30Pd-30Au

Compiled by P.J. Cascone
J.F. Jelenko & Company

Applications

Typical uses. The palladium-silver-gold alloys that can be very easily clad to other metals are used when high resistance to chemical corrosion is needed and when other material (for example, tantalum) may present fabrication difficulties. The addition of indium or tin, within solid solubility limits, results in a series of useful dental alloys to which porcelain is fused. The palladium-gold-silver alloys that are made susceptible to precipitation hardening by additions of small amounts of other metals, such as copper, are the base of a group of dental alloys. However, these are heterogeneous two-phase alloys and may be less corrosion-resistant than the solid-solution alloys, which have no such additions.

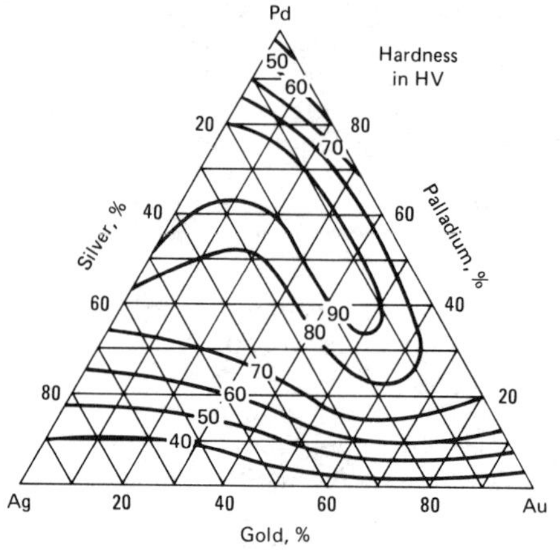

Fig. 37 Hardness of annealed palladium-silver-gold alloys

Mechanical Properties

Tensile properties. 30Pd-40Ag-30Au alloy: typical tensile strength, 640 MPa (92.5 ksi)
Hardness. 30Pd-40Ag-30Au alloy: 130 HB after cold rolling. See also Fig. 37.

Thermal Properties

This single-phase system exhibits a narrow melting range throughout. There are no solid-state transformations. Various properties of this system are given in Ref 24.

Chemical Properties

General corrosion behavior. In binary alloys, the addition of 20% Au to palladium or about 60% Au to silver results in substantial resistance to nitric acid, and the ternary alloys from these points to the gold corner of the ternary diagram are very resistant to nitric acid. These alloys are particularly useful where the presence of halogen acids precludes the use of even the best of the corrosion-resistant base metal alloys. The alloy 40Pd-30Au-30Ag shows a loss of 0.06 g/m^2 in the atmosphere formed by boiling 20% hydrochloric acid and air. The alloys containing 50 to 60% Pd or about 70% Au are tarnish-resistant. At high temperatures, the alloys do not oxidize in air but they dissolve oxygen, and any subsequent treatment in a reducing atmosphere will produce surface imperfections.

Fabrication Characteristics

Workability. These alloys can be cold worked without difficulty but must be annealed in an inert atmosphere (for example, nitrogen) with a low dew point to avoid palladium oxidation.
Annealing temperature. 850 °C (1560 °F)

Table 12 Typical mechanical properties of 95.5Pd-4.5Ru

Condition	Proportional limit MPa	Proportional limit ksi	Yield strength MPa	Yield strength ksi	Hardness(a), HV
Annealed(b)	270	39	350	51	152
Annealed(c)	···	···	···	···	126
Hard (60% red)	···	···	···	···	286

(a) 5 kg load. (b) 1000 °C (1830 °F), air cooled. (c) 1000 °C (1830 °F), quenched

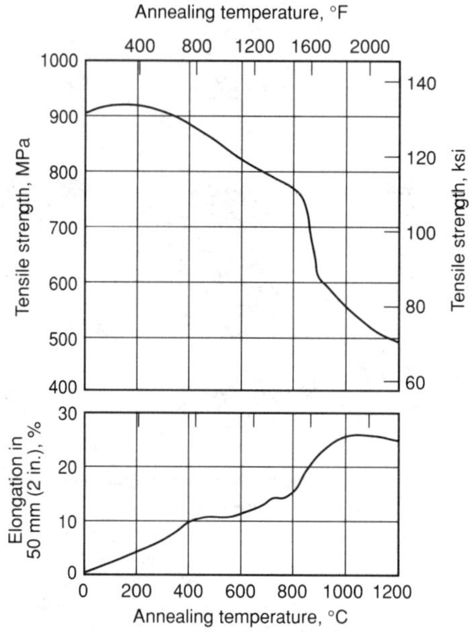

Fig. 38 Effect of annealing temperature on the tensile properties of 95.5Pd-4.5Ru

95.5Pd-4.5Ru

Compiled by D.J. Accinno
Engelhard Corporation

Applications

Typical uses. Jewelry, electrical contacts, reduction of nitric oxide by monolithic-supported palladium-ruthenium alloys. This alloy is standard for palladium jewelry in the United States; both wrought and cast forms used. This alloy, including richer ruthenium alloys (containing up to 12% Ru) is used for electrical contacts. The latter percentage is normally accepted as the limit of commercial workability.

Mechanical Properties

Tensile properties. Typical. Annealed: tensile strength, 380 MPa (55 ksi); elongation, 30% in 50 mm (2 in.). Cold drawn: tensile strength, 560 MPa (81 ksi); elongation, 3% in 50 mm (2 in.). See also Fig. 38 and Table 12.
Hardness. See Table 12.
Elastic modulus. 141 GPa (20.4 × 10^6 psi)

Mass Characteristics

Density. Annealed: 12.07 g/cm^3 (0.436 lb/in.^{3}); as-cast: 11.62 g/cm^3 (0.420 lb/in.^{3}) at 20 °C (68 °F)

Electrical Properties

Electrical resistivity. 242 nΩ · m at 20 °C (68 °F)
Temperature coefficient of electrical resistivity. 0.0013/°C (0.0007/°F) at 0 to 100 °C (32 to 212 °F)
Thermal electromotive force versus Pt:

Temperature °C	Temperature °F	mV
200	390	+2.58
400	750	+5.73
600	1110	+9.00
800	1470	+12.12
1000	1830	+15.30
1200	2190	+18.36

Fabrication Characteristics

Melting and casting. Melting with city gas-oxygen torch and the use of MgO or high-purity Al_2O_3-lined crucibles is common practice, although the use of a zirconia crucible yields superior results and is preferred. To obtain sound castings, deoxidation by 0.05 to 0.1% Al just prior to casting is advised. When using high-frequency induction to melt the charge, magnesia or preferably zirconia-lined crucibles with an argon covering atmosphere are the best approach. Final deoxidation of this melt with a small amount of calcium boride is generally recommended.
Annealing temperature. 900 °C (1650 °F) is suitable for annealing, based on elongation and tensile strength (see Fig. 38). Annealing in nitrogen plus 3 to 7% hydrogen is desirable for high-quality jewelry. Torch annealing should be done in an oxidizing flame.
Joining. Soldering and melting can be readily accomplished with an oxyacetylene torch (oxidizing flame). To avoid hot tearing of cast jewelry, care must be exercised to avoid possible contamination with oxygen or sulfur and particularly silicon and phosphorus.

REFERENCES

1. A.S. McDonald and G.H. Sistare, *Gold Bull.*, Vol 11 (No. 3), 1978, p 66-73
2. K. Yasuda, *Gold Bull.*, Vol 20 (No. 4), 1987, p 90-103
3. M. Nakagawa and K. Yasuda, *J. Less-Common Met.*, Vol 138 (No. 1), 1988, p 95-106
4. A.S. McDonald and G.H. Sistare, *Gold Bull.*, Vol 11 (No. 4), 1978, p 128-131
5. "Filler Metals for Brazing," ANSI/AWS A5.8-89, American Welding Society, 1989
6. R.P. Carreker, Jr., Report 55-RL-1413, General Electric Company Research Lab, 1955
7. E. Gruneisen, *Ann. Phys.*, Vol 25, 1908, p 825
8. W. Koster and J. Scherb, *Z. Metallkd.*, Vol 49, 1958, p 501
9. E.P. Sadowski, H.J. Albert, D.J. Accinno, and J.S. Hill, Stress Rupture Properties of Some Platinum and Palladium Alloys, AIME Metallurgical Society Conference, *Refractory Metals and Alloys*, Vol II, M. Semchysen and J.J.

Harwood, Ed., Interscience, 1961
10. The International Practical Temperature Scale of 1968 Amended Edition of 1975, *Metrologia*, Vol 12, 1976, p 7-17
11. P. Hidnert and W. Sander, NBS Circular 486, U.S. Department of Commerce, National Bureau of Standards, 1950
12. F.N. Jaeger and E. Rosenbohm, *Physics*, Vol 6, 1939, p 1123
13. R.W. Powell, R.P. Tye, and M.J. Woodman, *Platinum Met. Rev.*, Vol 6, 1962, p 138
14. R.F. Vines and E.M. Wise, *Platinum Metals and Their Alloys*, International Nickel Company, Inc., 1941
15. P. Drude, *Ann. Phys.*, Vol 39, 1890, p 481
16. W. Meier, *Ann. Phys.*, Vol 31, 1910, p 1017
17. G. Hass and L. Hadley, Optical Properties of Metals, in *American Institute of Physics Handbook*, 2nd ed., 1965, p 6-107 to 6-118
18. *Corrosion Handbook*, John Wiley & Sons, 1948
19. E.M. Wise and J.T. Eash, *Trans. AIME*, Vol 128, 1938, p 282
20. W.J. Jellinghaus, *Z. Tech. Phys.*, Vol 17, 1936, p 33
21. J.B. Newkirk and R.J. Smoluchowski, *Appl. Phys.*, Vol 22, 1951, p 290
22. D.L. Martin, Effects of Temperature on Remanence Magnetics, in *Proceedings of the Conference on Magnetism and Magnetic Materials*, American Institute of Electrical Engineers, 1957, p 188
23. D.L. Martin, Processing and Properties of Cobalt Platinum Permanent Magnet Alloys, *Trans. Metall. Soc. AIME*, Vol 212, Aug 1958, p 478-485
24. *Konstitution der Ternären Metallischer Systeme* (No. 11), W.M. Guertler, Ed., Rotadruck Ernst Jaster, 1960
25. *Constitution of Binary Alloys, First Supplement*, R.P. Elliott, Ed., McGraw-Hill, 1965, p 378
26. *Multicomponent Alloy Constitution Bibliography 1955-1973*, A. Prince, Ed., The Metals Society, 1978

SELECTED REFERENCES

- *Source Book on Brazing and Brazing Technology*, M.M. Schwartz, Ed., American Society for Metals, 1980
- Technical Data Sheet D-36, Handy & Harman

Rare Earth Metals

K.A. Gschneidner, Jr., B.J. Beaudry, and J. Capellen, Iowa State University*

THE RARE EARTH ELEMENTS comprise about one-fifth of the naturally occurring elements of the periodic table. The rare earths include the Group IIIA elements scandium, yttrium, and the lanthanide elements (lanthanum, cerium, praseodymium, neodymium, promethium, samarium, europium, gadolinium, terbium, dysprosium, holmium, erbium, thulium, ytterbium, and lutetium) in the periodic table of elements (see the periodic table in the article "Properties of Pure Metals" in this Volume). This definition will be used throughout this article, although many scientists and engineers use the term rare earths to mean the 15 lanthanide elements and do not consider scandium (Sc) and yttrium (Y) to be such.

Introductory textbooks view the rare earth elements as being so chemically similar to one another that collectively they can be considered as one element. To a certain extent this is correct—many applications are based on this close similarity—but a closer examination reveals vast differences in their behaviors and properties. For example, the melting points of the lanthanide elements vary by a factor of almost two between lanthanum (918 °C) and lutetium (1663 °C), the end members of the trivalent lanthanide series. This difference is much larger than that found in many of the groups of the periodic table.

In addition to the normal trivalent state exhibited by most of the rare earth metals, two of the lanthanides (europium and ytterbium) are divalent. This also accounts for vast differences in properties (for example, the vapor pressure of lanthanum at 1000 °C, or 1830 °F, is only *one billionth* that of ytterbium).

The term rare implies that these elements are scarce, but in fact, the rare earths are quite abundant and exist in many viable deposits throughout the world. Of the 83 naturally occurring elements, the 16 naturally occurring rare earths as a group lie in the 50th percentile of the elemental abundances. Cerium (Ce), the most abundant, ranks 28th and thulium (Tm), the least abundant, ranks 63rd.

Research Grade Versus Commercial Grade

The rare earth metals are extremely reactive elements forming stable oxides, sulfides, hydrides, and other compounds. Thus they are difficult to prepare in a high-purity form without a great deal of effort and cost. For research, generally high-purity materials are necessary to determine the intrinsic properties and behaviors of the metals and alloys. However, for many applications, especially metallurgical ones, such high purities are not needed; therefore, industry makes no effort to prepare high-purity metals, alloys, or compounds.

The major impurities in the rare earth metals (either research grade or commercial grade) are the interstitial impurities hydrogen, carbon, nitrogen, and oxygen (Ref 1). The other rare earth impurity concentrations in a given rare earth metal, and usually most nonrare earth metallic impurities, are low in both grades of metals relative to their interstitial concentrations. Research-grade metals are usually ≧99.8 at% pure, although ≧99.95 at% metals can be prepared at great effort. The hydrogen and oxygen contents of the ≧99.8 at% pure metals are usually of the order of 200 to 400 ppm atomic, whereas the carbon content is <100 ppm atomic and the nitrogen is <10 ppm atomic. The highest metallic impurities are about 10 ppm atomic, but most are <1 ppm atomic.

Commercial-grade rare earth metals are about 98 at% pure but occasionally can be as low as 95 at% pure (Ref 1). The major impurities are usually oxygen (1 to 2 at%), followed by hydrogen (0.5 to 1 at%), and the container material used to melt the rare earth metals (molybdenum, tantalum, or carbon) (0.5 at%). These impurity levels can have pronounced effects on some of the physical properties (Ref 1), and thus the data, information, and discussions that follow concerning the physical and chemical properties of the rare earths are based on research-grade materials. The last section of this article on applications is concerned with commercial-grade material. In some commercial applications, such as permanent magnets, impurities are helpful because they prevent domain wall motion, thus improving the magnetic strength of the magnet material. In other cases, the amount of rare earth added to an alloy is less than 1 wt%, and the impurities in the rare earth metal are diluted to an insignificant concentration having no effect.

Preparation and Purification

Rare earth elements are found in nature intimately mixed in varying proportions depending on the ore. Separation into pure-component rare earths is done on a large scale by liquid-liquid extraction and by ion exchange on a smaller scale. For more information on the separation techniques, the reader is referred to the review by Powell (Ref 2). The preparation of the pure metal from the separated oxide and its subsequent purification is discussed in this section.

The procedures established at the Ames Laboratory (Ref 3) use the rule "Make pure reactants, keep them pure." Twelve of the rare earth metals are made by the calcium reduction of the fluoride in tantalum crucibles under an argon atmosphere. To obtain the high-purity reactants, the pure oxide is reacted with a dynamic anhydrous hydrogen fluoride (HF)-argon atmosphere in a platinum-lined Inconel tube furnace at 650 °C (1200 °F). To achieve higher purity, this fluoride is melted in a platinum crucible under a dynamic HF-argon atmosphere. Calcium is purified by distillation under a partial pressure of helium. The distilled calcium and melted fluoride are handled in a purified helium atmosphere to keep the calcium pure. Stoichiometric quantities of calcium and the rare earth fluoride, RF_3, are wcighcd, mixcd, and packcd into a tantalum crucible in a helium-filled glove box. The filled crucible is heated in an induction furnace under an argon atmosphere to above the melting points of all of the reac-

*Dr. Gschneidner is affiliated with the Ames Laboratory, DOE, the Department of Materials Science and Engineering, and the Rare-earth Information Center; Mr. Beaudry is associated with the Ames Laboratory, DOE; and Mr. Capellen is on the staff of the Rare-earth Information Center.

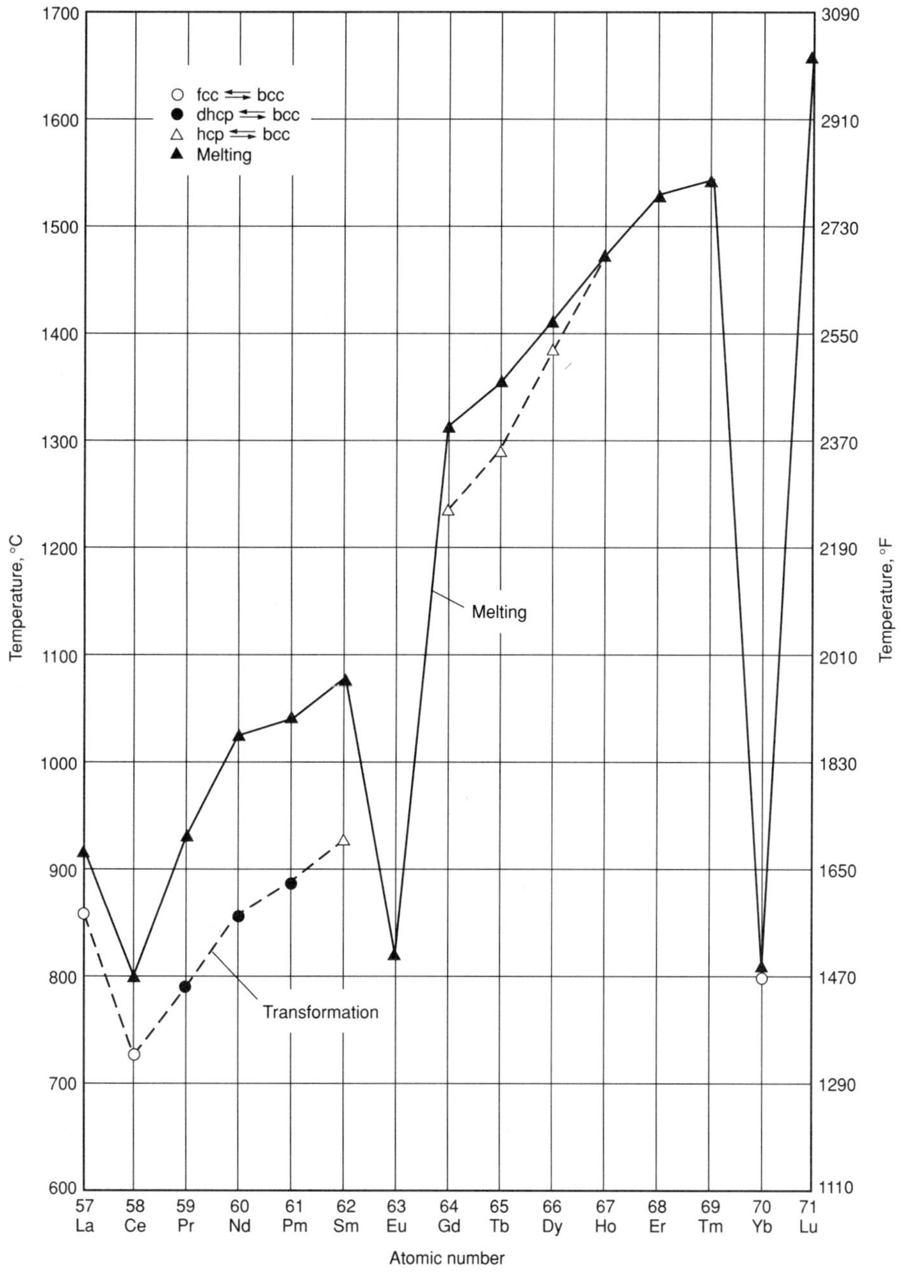

Fig. 1 The melting and close-packed (fcc, dhcp, hcp) to bcc transformation temperatures versus the lanthanide atomic number

tants and products to achieve reaction and separation of the liquid metal and slag. After the reduction step, the slag is removed and residual volatile impurities (Ca, CaF_2, RF_3) are removed by vacuum melting.

In establishing procedures to vacuum melt and/or distill the calcium-reduced metals, the variation in their melting points (Fig. 1) and vapor pressures (or boiling points as shown in Fig. 2) provides a natural grouping. Metals with low melting points and high boiling points (La, Ce, Pr, and Nd) prepared by the calcium reduction of the fluorides are vacuum melted at ~1800 °C (3300 °F) then cooled slowly and held molten just above their melting point to permit tantalum dissolved at high temperatures to precipitate out of solution and settle to the bottom of the crucible. The metals with high melting points and high boiling points (Gd, Tb, Y, and Lu) dissolve too much tantalum at their melting points (0.09 to 1.4 at%), and they are distilled from the reduction crucible into a condenser (an inverted tantalum crucible), leaving the tantalum in the distilland. Impurities such as oxygen and carbon are found in the distillate in these metals with high boiling points. The four remaining metals (Dy, Ho, Er, and Sc), which are also prepared by calcium reduction of their fluorides, have high melting points and relatively low boiling points and can be vaporized below their melting points (sublimed). In the sublimation process they are purified with respect to oxygen, nitrogen, and carbon, as well as tantalum.

Four of the rare earth metals (Sm, Eu, Tm, and Yb) have low boiling points (Fig. 2) and are prepared directly from their oxide by reaction with lanthanum, cerium, or mischmetal chips. The equilibrium established between the reductant and oxide is driven to completion by the vaporization of the volatile metal. The as-reduced/distilled metals (Sm, Eu, Tm, or Yb) are readily purified by vacuum sublimation. Multiple sublimations can be used to further purify these four metals as well as Dy, Ho, Er, and Sc.

Solid-state electrolysis has been used to obtain small quantities of ultrahigh-purity rare earth metals. In this method, a high direct current is passed through a rod of the metal in ultrahigh vacuum (10^{-10} Pa, or 10^{-12} torr). The current heats the metal and transports the impurities to the anode or cathode. Zone refining was combined with solid-state electrolysis to obtain the highest purity thus far achieved in gadolinium and neodymium (Ref 4). The lowest-melting metals (La and Ce) and the low-boiling-point metals (Sm, Eu, Tm, and Yb) are not readily purified by solid-state electrolysis. Float zone melting of lanthanum and cerium under ultrahigh vacuum and multiple sublimations of Sm, Eu, Tm, and Yb under ultrahigh vacuum yields the purest metals.

Physical Properties

Because impurities have pronounced effects on the properties of rare earth elements, the following discussions and information on properties are based on research-grade materials. Additional information on the properties of the rare earth elements is given in the article "Properties of Pure Metals" in this Volume.

The Electronic Configurations. The number of 4*f* electrons in a given lanthanide element depends on the state of matter and also on its chemical environment. This is illustrated in Table 1, where the electronic configurations of the rare earth (R) elements are given for the free gaseous atom (the ground state), the metallic form, and the R^{2+}, R^{3+}, and R^{4+} ionic species. The free-atom configurations are of indirect interest to the material scientist or engineer, but they are important in:

- Certain thermochemical cycle calculations, in particular those that involve the heat of sublimation, that is, R (metal) → R (gas) at 298 K
- Physical processes that involve vaporization (see the section "Boiling Points and Sublimation Energies" in this article)

Considering the number of 4*f* electrons in the R^{3+} state as normal, then the lanthanide series as a whole has $4f^n$ electrons, where $n = 0$ for La, $n = 1$ for Ce, ... to $n = 14$ for

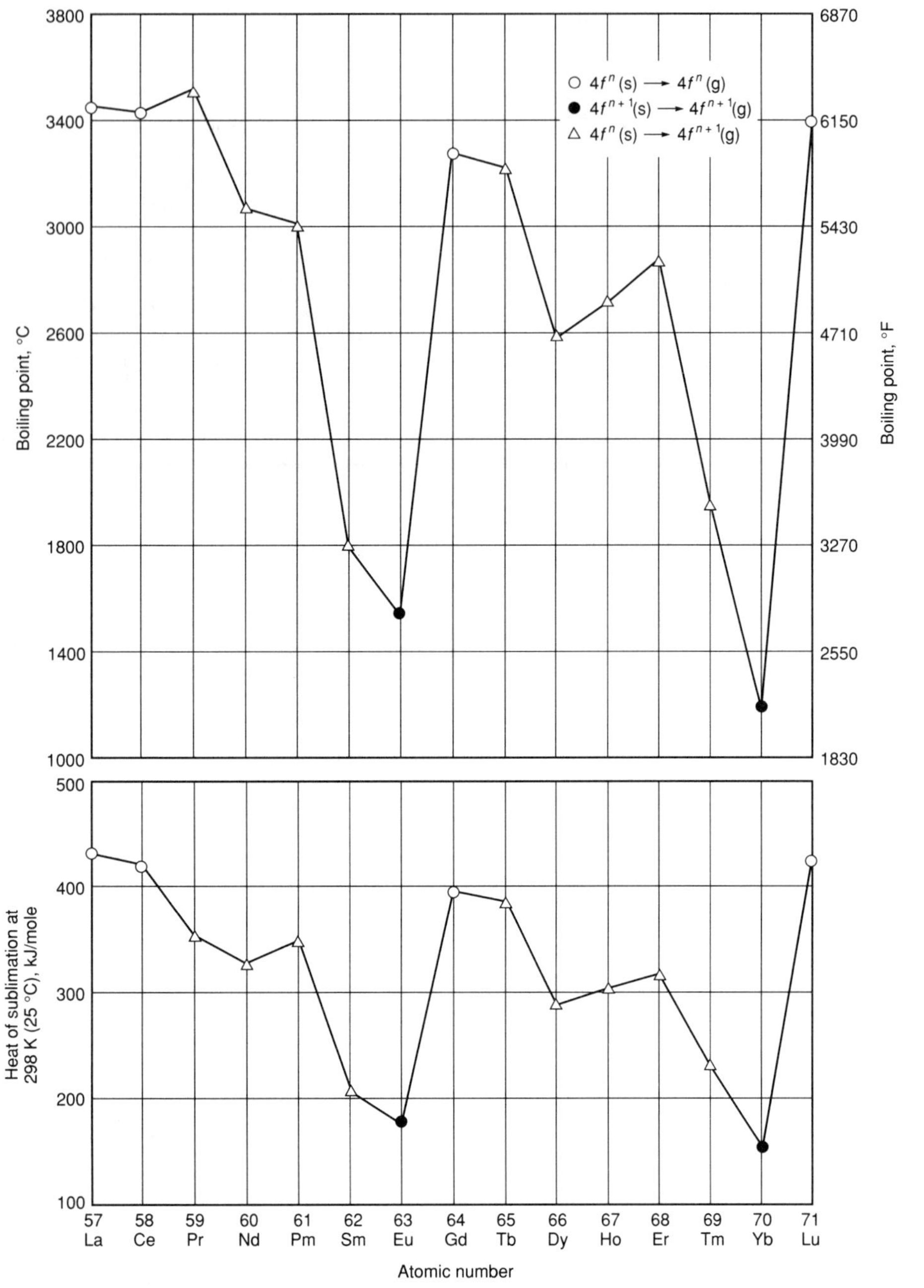

Fig. 2 The boiling points and heats of sublimation versus the lanthanide atomic number. s, solid; g, gas

Lu. However, in the gaseous state the most common configuration is $4f^{n+1}6s^2$, whereas only La, Ce, Gd, and Lu have the normal $4f^n$ configuration ($4f^n5d^16s^2$). The normal configuration for the metallic state is $4f^n$ $(5d6s)^3$ (that is, trivalent) with only europium (Eu) and ytterbium (Yb) having the divalent $4f^{n+1}$ $(5d6s)^2$ configuration.

As seen in Table 1, the R^{3+} valence state is common to all of these elements, although a few of the lanthanides exhibit other valence states, such as R^{4+} for Ce, Pr, and Tb, and R^{2+} for Sm, Eu, and Yb. The Ce^{4+}, Eu^{2+}, and Yb^{2+} states are quite common in nature and are used by industry to separate these three elements from the remaining rare earths by relatively cheap chemical methods. Some important uses also depend upon these nontrivalent states.

The overall chemical and metallurgical properties of the rare earth elements are due to their outer electrons $5d6s$ ($3d4s$ for Sc, and $4d5s$ for Y). There is some variation in the chemical properties of the lanthanides due to the lanthanide contraction (discussed below) and hybridization of the $4f$ electrons with the valence electrons. For many uses it is not cost effective to separate the lanthanides in order to use the element that gives the best properties, and so they are used as mixed rare earths.

For properties that depend upon the number of $4f$ electrons (that is, magnetic and optical properties), separated, and sometimes quite pure, individual rare earth elements are generally required.

Structure, Metallic Radius, Atomic Volume, and Density. The rare earth metals crystallize in all of the common metallic structures and several unique ones. The room-temperature structures as one proceeds along the lanthanide series are:

- A double c hexagonal close-packed (dhcp) structure for lanthanum
- A face-centered cubic (fcc) structure for cerium
- A dhcp structure for praseodymium, neodymium, and promethium
- A unique nine-layer hexagonal structure for samarium (Sm-type structure)
- A body-centered cubic (bcc) structure for europium
- A hexagonal close-packed (hcp) structure for gadolinium through thulium (that is, Gd, Tb, Dy, Ho, Er, and Tm)
- An fcc structure for ytterbium
- An hcp structure for lutetium

Both scandium and yttrium have hcp structures at room temperature (Ref 3, 5). At high temperatures, many of the metals (La through Sm, Gd, Tb, Dy, Sc, and Y) transform to a bcc phase before melting. The remaining four hcp metals (Ho, Er, Tm, and Lu) are monotropic as is bcc europium. At intermediate temperatures, lanthanum crystallizes in the fcc structure, and samarium crystallizes in the hcp structure. Below room temperature cerium transforms to a dhcp, and upon further cooling an fcc structure that is about 15 vol% smaller than the room-temperature fcc phase. This large-volume contraction is due to an apparent valence increase of ~⅔ of an electron per atom. Both terbium (Tb) and dysprosium (Dy) undergo an hcp to an orthorhombic distortion due to magnetoelastic effects when these metals order magnetically. Ytterbium (Yb) upon cooling just below room temperature becomes hcp.

The metallic radii for a coordination number of twelve (CN = 12) are shown in Fig. 3, where it is seen that the radii of the trivalent lanthanide metals fall on a smooth curve from lanthanum (La) to lutetium (Lu) with a cusp at gadolinium. The tendency of cerium (Ce) toward tetravalency is obvious, in that its metallic radius lies below that established by the normal trivalent metals. The divalent character of europium (Eu) and of ytterbium (Yb) is also obvious with their large metallic radii. This divalent character is evident in other physical properties and plays an important role in their preparation, purification, alloying behavior, and compound formation. The radius of yttrium (Y) is essentially identical to that of gadolinium (Gd), whereas that of scandium (Sc) is 0.16406 nm, which is significantly smaller

Table 1 The electronic structures of the rare earth elements in various states of matter

Element	Neutral atom configuration		4f configuration of known oxidation states: M^{2+}	M^{3+}	M^{4+}	Metallic state number of electrons: Valence	4f
Sc		$3d4s^2$	...	0	...	3	0
Y		$4d5s^2$	...	0	...	3	0
La		$5d6s^2$	...	0	...	3	0
Ce	$4f$	$5d6s^2$	...	1	0	3	1
Pr	$4f^3$	$6s^2$	...	2	1	3	2
Nd	$4f^4$	$6s^2$	...	3	...	3	3
Pm	$4f^5$	$6s^2$	...	4	...	3	4
Sm	$4f^6$	$6s^2$	6	5	...	3	5
Eu	$4f^7$	$6s^2$	7	6	...	2	7
Gd	$4f^7$	$5d6s^2$	...	7	...	3	7
Tb	$4f^9$	$6s^2$	...	8	7	3	8
Dy	$4f^{10}$	$6s^2$	...	9	...	3	9
Ho	$4f^{11}$	$6s^2$	...	10	...	3	10
Er	$4f^{12}$	$6s^2$	...	11	...	3	11
Tm	$4f^{13}$	$6s^2$	...	12	...	3	12
Yb	$4f^{14}$	$6s^2$	14	13	...	2	14
Lu	$4f^{14}$	$5d6s^2$	...	14	...	3	14

than that of 0.17349 nm for lutetium (the smallest lanthanide metal).

The atomic volumes of the lanthanides, as expected from the geometrical relationship between the radius and volume, vary in an identical manner to the function of atomic number as the radius (Fig. 3). The density shows a reciprocal behavior, with lanthanum (La) the least dense of the trivalent lanthanides and lutetium (Lu) the most dense. Furthermore, the densities of europium (Eu) and ytterbium (Yb) are significantly less than those of their immediate respective neighbors. Because of their small masses scandium (Sc) and yttrium (Y) are significantly less dense than any of the lanthanides.

Melting and Transformation Temperatures. The melting points for the lanthanide metals are shown in Fig. 1. It is noted that the melting points rise rapidly from 918 °C (1684 °F) for lanthanum (La) up to 1663 °C (3025 °F) for lutetium (Lu). The divalent character of europium (Eu) and ytterbium (Yb) is evident in their low melting points, which are comparable to those of the alkaline earth metals. The melting points of scandium (Sc) and yttrium (Y) are relatively high, 1541 °C (2805 °F) and 1522 °C (2772 °F), respectively, close to those for erbium (Er), 1529 °C (2784 °F), and thulium (Tm), 1545 °C (2813 °F).

The close-packed to bcc-phase transformations are also shown in Fig. 1 and they tend to follow the melting-point trend. But as seen, the transformation temperature from gadolinium to holmium (Gd, Tb, Dy, Ho) increases more rapidly as a function of atomic number than the melting point, and, thus, the bcc phase becomes metastable with respect to liquid formation at holmium and for the remaining three trivalent lanthanide metals. Scandium and yttrium also have an hcp-to-bcc transformation at 1337 °C (2439 °F) and 1478 °C (2692 °F), respectively.

Boiling Points and Sublimation Energies. The boiling points of the lanthanide metals vary as a function of atomic number in a sawtoothlike manner (Fig. 2), which contrasts to the smooth variation observed for the radii (Fig. 3) and melting points (Fig. 1) of the lanthanide metals (other than the divalent europium and ytterbium metals). The boiling points vary from extremely high temperatures in several lanthanide metals (La, Ce, Pr, Gd, Tb, and Lu) and also yttrium (which is not shown in Fig. 2), to quite low temperatures for other lanthanides (Sm, Eu, Tm, and Yb). The remaining five lanthanides (Nd, Pm, Dy, Ho, and Er) and scandium (Sc) have intermediate boiling-point temperatures between these extremes. The large difference in boiling points (vapor pressures) and melting temperatures means that there is considerable variation in the methods used to prepare and purify the rare earths, as discussed in the section "Preparation and Purification" in this article.

The large variation in the boiling points of the lanthanide metals occurs because the electronic configurations of both the solid and the gas phases are involved in the process. This is best seen in the heats of sublimation, which are related to the boiling

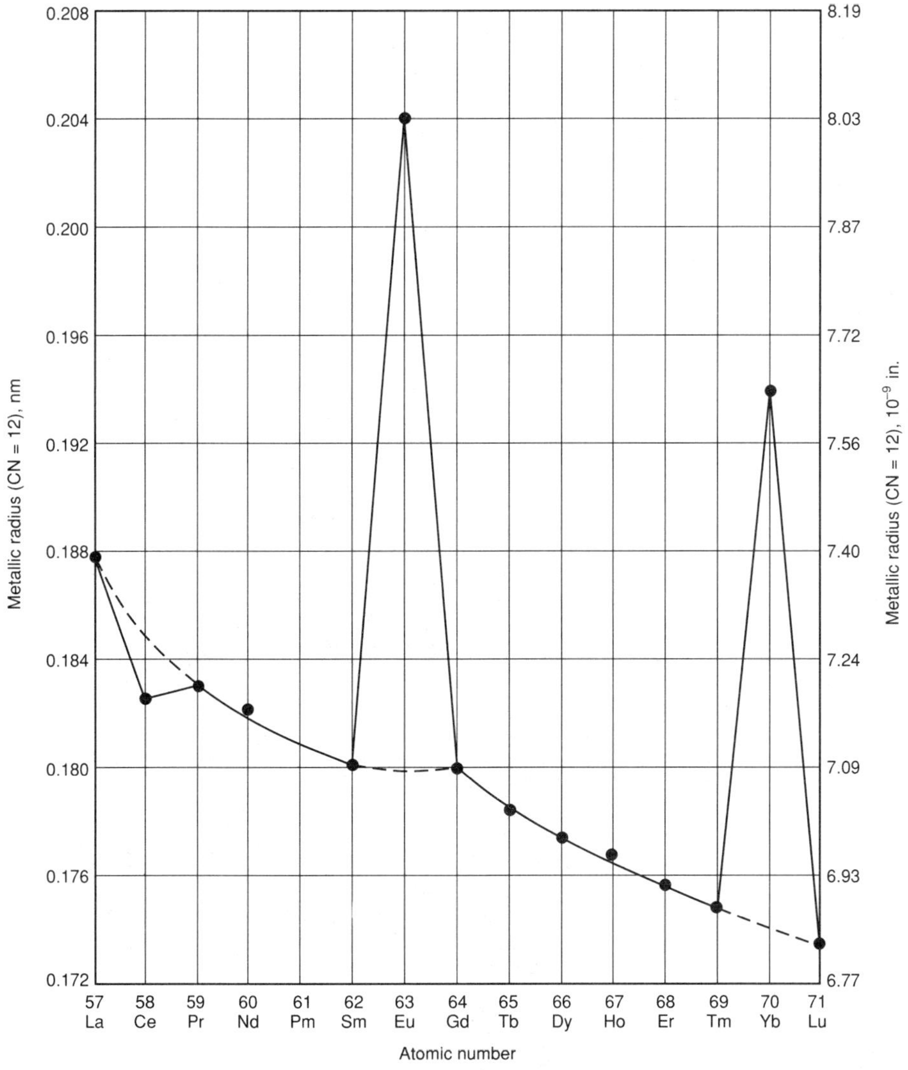

Fig. 3 The metallic radius for a coordination number (CN) of twelve versus the lanthanide atomic number

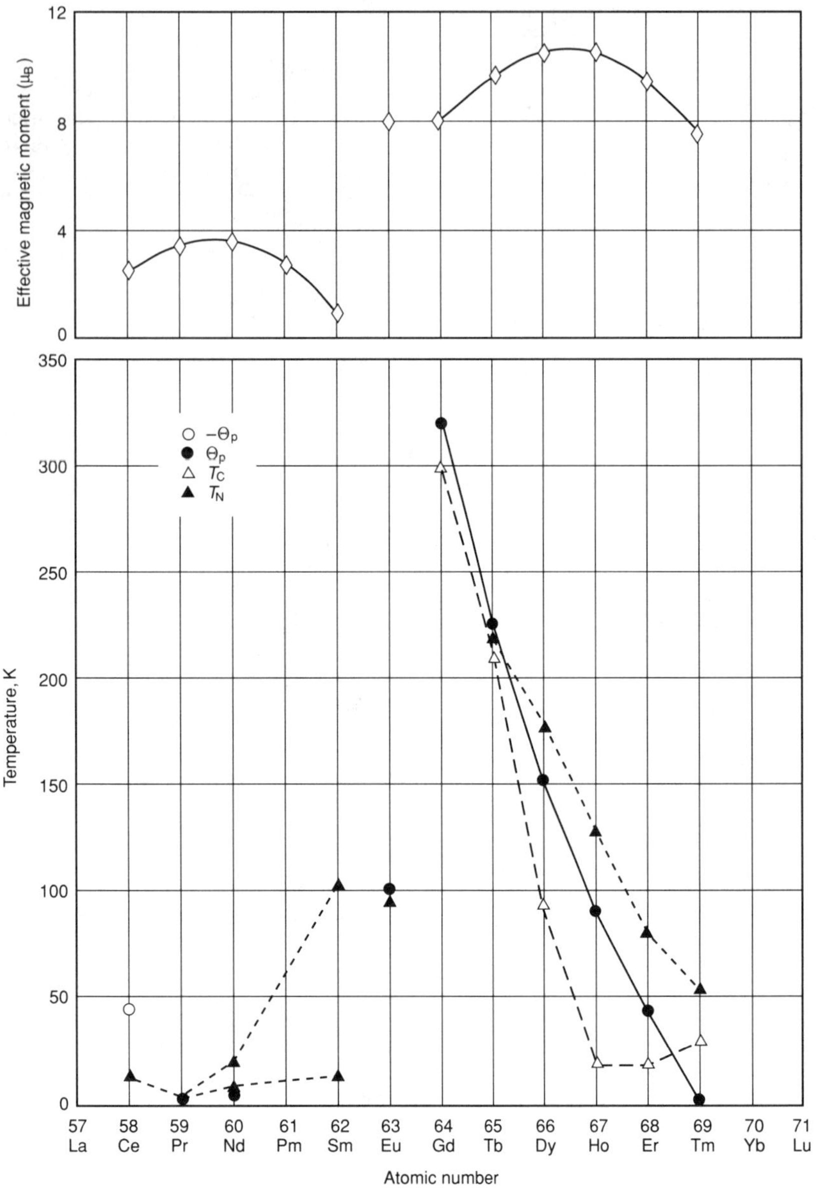

Fig. 4 The various magnetic temperatures (lower portion) and effective magnetic moments (upper portion) of the lanthanide metals. The paramagnetic ordering temperatures, θ_p, which can be negative (Ce) or positive (the other lanthanides) is obtained at high temperatures above the magnetic ordering temperatures, T_N or T_C. The symbol T_C represents the Curie or ferromagnetic ordering temperature, and T_N represents the Néel or antiferromagnetic ordering temperature. The lines are not connected to europium (Eu) because it is divalent and the other magnetic lanthanides are trivalent.

points via Trouton's rule and plotted at the bottom of Fig. 2. The heat of sublimation is the energy required to remove one gram atom from the solid at 25 °C (77 °F). In Fig. 2 the open circles represent the metals that are trivalent in the solid state ($4f^n$) and vaporize to a trivalent gaseous atom ($4f^n$) (see Table 1). This vaporization process requires ~420 kJ/mole. The solid circles in Fig. 2 denote the two divalent metals ($4f^{n+1}$) that vaporize to a divalent gaseous atom ($4f^{n+1}$). But in this case these elements (europium and ytterbium) only require ~164 kJ/mole to complete the process. The elements indicated by open triangles are trivalent in the solid ($4f^n$), vaporize to a divalent gaseous atom ($4f^{n+1}$), and have intermediate sublimation energies. Thus, the electronic configurations play an important role in the vaporization process, and in turn on the preparation, purification, and physical and chemical metallurgy of the lanthanide metals.

Magnetic properties. The second most extensively studied property of rare earth metals—second only to crystal structure determinations—is their magnetic behavior. The reason for this is the measurement of the magnetic properties yields information about the *f* electrons of the lanthanide element in the substance being examined. These studies have led to many interesting and exciting scientific discoveries and to one of the major uses of two rare earths in the metallic form, namely the use of neodymium in Nd-Fe-B permanent magnets and samarium in Sm-Co permanent magnets.

The 4*f* electrons are energetically and radially buried in an atom, do not enter into the bonding, and are only slightly influenced by the external environment around the lanthanide atom. Unpaired 4*f* electrons have a magnetic moment that gives rise to larger magnetic susceptibilities, two to four orders of magnitude larger than those of normal metals. Because the 4*f* electrons are radially buried, they do not directly overlap (as do the 3*d* electrons of manganese, iron, cobalt, and nickel, which accounts for the magnetic behaviors of the 3*d* transition metals). The 4*f* electrons on one lanthanide atom communicate with those on another lanthanide via the valence (primarily the *s*) electrons. As the free valence electron moves through the solid it is polarized by the first 4*f* electron (its spin is aligned parallel with the 4*f* spin), and as it passes by the second atom the 4*f* electron of the second atom is in turn polarized by the valence electron. This indirect exchange, called the RKKY interaction, is weaker than the direct overlap found in the 3*d* metals, but it is sufficiently strong for the 4*f* electrons to align magnetically at about room temperature (for Gd) or below (for the other lanthanides). Many unusual magnetic structures have been found, not only in the metals, but also in their compounds (Ref 6). In the metals, the 4*f* electrons of the first members of the lanthanide series (called the light lanthanides) align antiparallel to each other (antiferromagnets). Their ordering temperatures, called the Néel temperatures (T_N), are shown in Fig. 4. In these metals, two ordering temperatures are found (even in cerium, which are only 1 K apart). The higher ordering temperature is due to magnetic alignment of 4*f* spins on the hexagonal sites in the dhcp or Sm-type structures, whereas the lower one is due to ordering on the cubic sites. In the case of the last half of the lanthanide series (called the heavy lanthanides) they order antiferromagnetically at high temperatures, and then upon further cooling they order ferromagnetically or ferrimagnetically (see Fig. 4), with the exception of gadolinium (Gd), which is a ferromagnet at all temperatures (Ref 7).

The magnetic ordering temperatures are a maximum at gadolinium (Gd), which has seven unpaired 4*f* electrons (the maximum possible) and then fall off to each side as the number of unpaired 4*f* electrons is reduced by one with each succeeding lanthanide. The magnetic moment, however, depends not only on the number of unpaired 4*f* electrons but also on their orbital motion. This gives rise to the peak observed in the magnetic moment of the light lanthanides

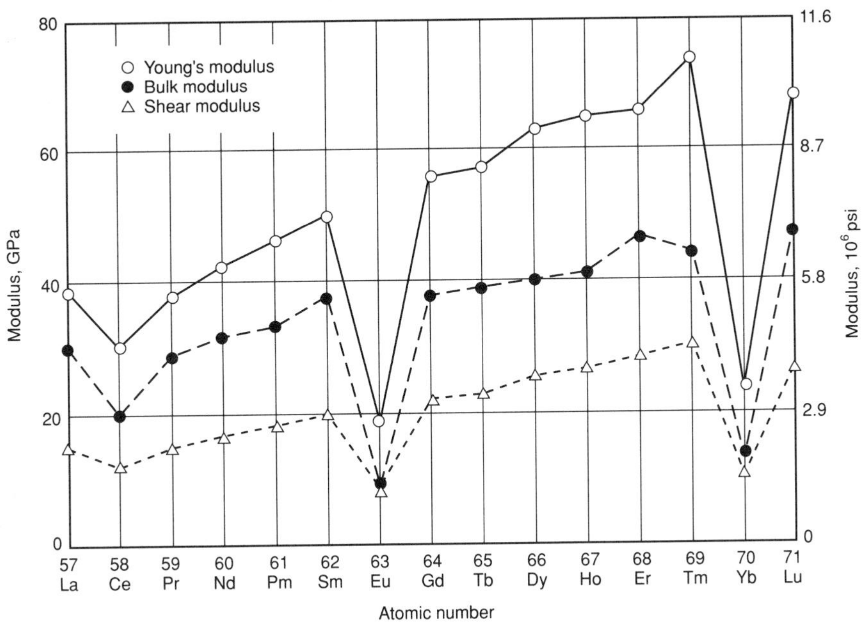

Fig. 5 The bulk elastic constants of the lanthanide metals versus the atomic number

(at Pr-Nd) and a peak in the heavy lanthanides (at Dy-Ho) as shown in the top part of Fig. 4.

The combination of the light lanthanides with the iron group metals, in the appropriate intermetallic compounds, results in the highest-strength permanent magnets known today, the Sm-Co and Nd-Fe-B families. These magnetic materials are discussed in the section "Magnetic Materials" of this article and in the article "Permanent Magnet Materials" in this Volume.

Elastic and Mechanical Properties. As reported by Scott (Ref 8) in his review of elastic and mechanical properties of rare earth metals, the values reported in literature vary considerably because of the wide range of impurity levels in the metals; variations by a factor of 10 are not uncommon, especially for the mechanical properties. The values given here are thought to be the most reliable and the closest to the intrinsic property of the given metal.

In general, the elastic properties increase with increasing purity. The chosen values for the lanthanide metals are plotted in Fig. 5 as a function of the atomic number. The anomalies are clearly evident at cerium (Ce) (premonition of the $\gamma \rightarrow \alpha$ transformation) and divalent europium (Eu) and ytterbium (Yb). Furthermore, there is an increase in the elastic moduli as a function of increasing atomic number until a maximum is reached at thulium (Tm). As expected the single-crystal c_{ij} values exhibit the same trend as observed in Fig. 5 for the bulk elastic constants. The elastic constants are similar in magnitude to those of aluminum, zinc, cadmium, and lead.

The hardness and strength values for the lanthanides follow the same periodic trend as observed for the elastic moduli shown in Fig. 5, but the experimental scatter of the data is larger. Cerium, europium, and ytterbium have anomalous low values for the same reason as discussed above for the elastic constants. The increase of the mechanical properties from the light lanthanides to the heavy lanthanides also seems to occur, and is similar to that of aluminum at the low range of the reported lanthanide values, and falls between those of aluminum and titanium for the upper range of values. Most of the rare earths do not neck down before fracture, and therefore, the ultimate tensile strength and fracture strength are nearly the same.

Chemical Properties

Chemical Reactivity. The rare earths are extremely reactive metals, especially with respect to the normal atmospheric gases (Ref 5). The light trivalent lanthanides will oxidize upon exposure to air at room temperature and therefore should be stored in vacuum or under helium or argon in sealed containers (Ref 3). The heavy lanthanides, and scandium (Sc) and yttrium (Y) do not oxidize at room temperature; they form a protective oxide coating just as aluminum, which prevents further oxidation. The rate of oxidation depends on several variables and is higher when:

- The impurity level (of most common impurities) is high
- The relative humidity is high (Ref 5)
- The temperature is high (Ref 5)
- The atomic number of the lanthanide is low (Ref 5)

The presence of impurities such as carbon, iron, calcium, and many of the *p*-group elements such as zinc, gallium, germanium, and their congeners greatly increases the rate of oxidation. For the *p*-group elements the rate of oxidation increases as one goes down the periodic table in a given group. The presence of iron is quite important in the major application of cerium lighter flints.

This chemical reactivity is due to the large negative free energy of formation of the oxides—among the most negative of all the elements in the periodic table (see Fig. 6). This chemical reactivity is responsible for some uses of the rare earth metals, such as getters in vacuum tubes, lighter flints, incendiary devices, and getters in metals and alloys.

Divalent europium (Eu) oxidizes much more readily than any of the trivalent rare earth metals, and special precautions are necessary when handling this metal (Ref 3, 5). Divalent ytterbium (Yb), however, is relatively inert and can be handled in air without any difficulties.

The rare earth metals (R) react slowly with N_2 and high temperatures are required to observe any appreciable reaction. Furthermore, the formation of RN on the surface greatly reduces any further nitridation (Ref 5). The free energies of formation of the nitrides are shown in Fig. 7. These data show that the rare earth nitrides are among the most stable nitrides, exceeded only by TiN and ZrN in stability.

The metals will easily hydride at elevated temperatures (400 to 600 °C, or 750 to 1100 °F). Unless special care is taken, when the metal is hydrided up to and beyond RH_2 in hydrogen content, the solid material fragments (Ref 5).

The rare earth metals will react exothermically with sulfur, selenium, and phosphorus. If heated to the appropriate temperature the reaction will take off and could seriously damage the crucible, furnace, vacuum enclosures, and so forth. However, at low temperatures some of the rare earths will hardly react (for example, the heavy lanthanides with sulfur). The free energies of formation of some of the rare earth sulfides and oxysulfides are shown in Fig. 8.

Metallography and Surface Passivation. The recommended metallographic polishing method is to electropolish the metal in 6% perchloric acid dissolved in absolute methanol at dry-ice temperatures, which leaves the metal with a shiny silvery color and a passivated surface (Ref 3, 5). This electropolishing procedure is also used to clean the metal surface after mechanical fabrication and/or heat-treating operations. However, when cleaning or polishing cerium or ytterbium, the low-temperature phases β-Ce or α-Yb form at the dry-ice temperature. The room-temperature forms may be maintained by chemically polishing or cleaning with Roman's solution, a complex solution of organic and mineral acids.

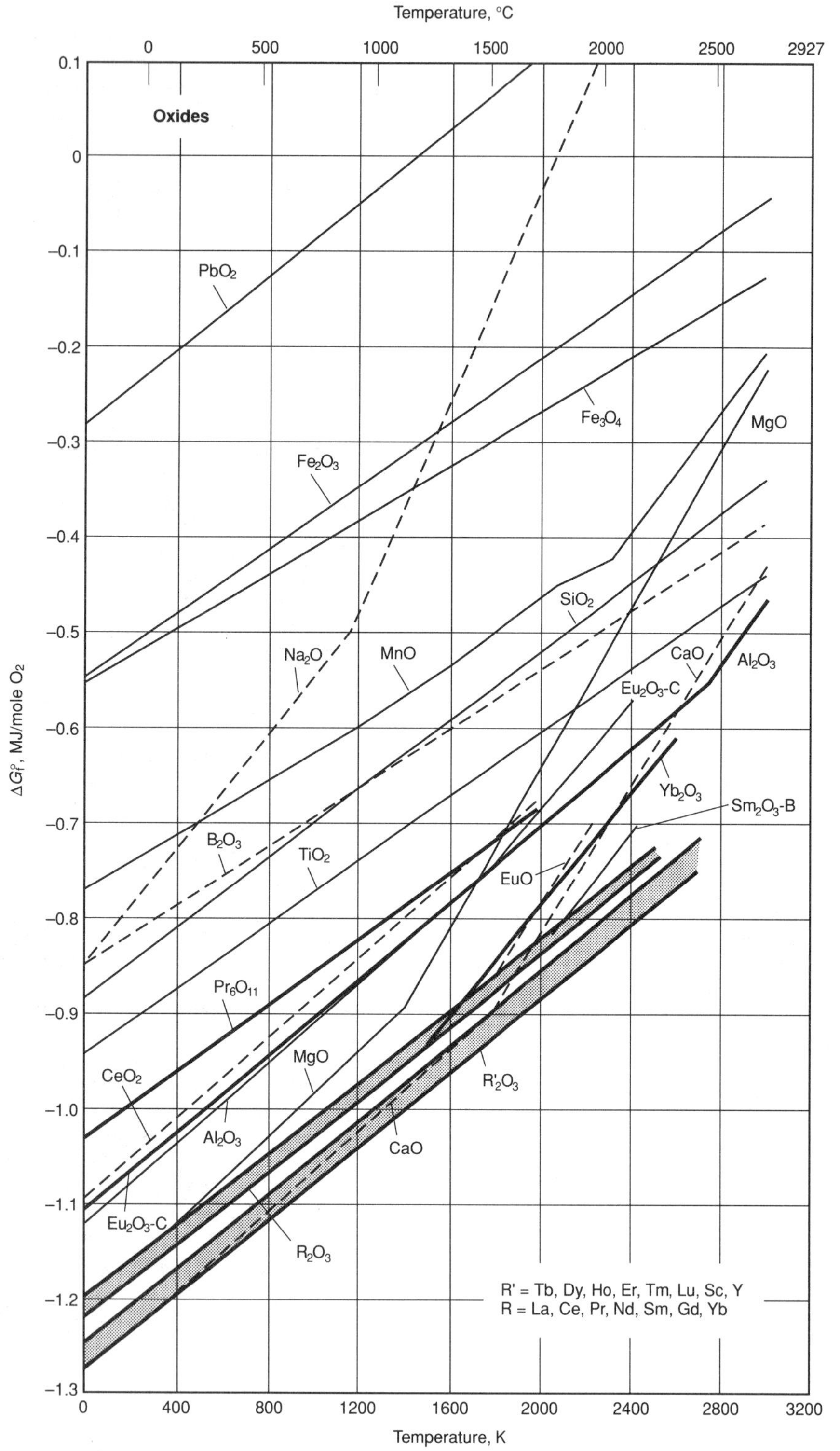

Fig. 6 The standard free energies of formation of the rare earth and some selected nonrare earth oxides. Because the values of the light lanthanide metal (R) sesquioxides lie close to one another (also the heavy rare earth metal [R′] sesquioxides), the free energies are drawn in a broad band for the two groups, except where departures become evident. Broken lines are used for clarity.

Alloy Formation

The rare earth metals are fairly large, electropositive elements (Fig. 9). Because of their large sizes they are not readily dissolved in the solid state of most of the common metals, and because they are considerably more electropositive than most metals, the rare earths tend to form compounds with them (Ref 5).

Solid Solution Alloys. The rare earth metals form extensive solid solutions with each other and with zirconium and thorium. The divalent metals (magnesium, zinc, cadmium, and mercury) form extensive solid solutions in the high-temperature bcc forms of the rare earth metals, but the rare earths are essentially insoluble in these divalent metals. The bcc phases of the rare earth metals can be retained metastably at room temperature by rapid quenching of alloys containing magnesium and cadmium. The rare earths form extensive solid solutions in silver and gold, but these two metals do not dissolve to any extent in the rare earths (Ref 5).

Compound Formation. The rare earth metals form compounds with the elements to the right of the group VIA elements in the periodic table, except the rare gases, but not with the elements to the left of the group VIIA elements, except hydrogen, beryllium, and magnesium. The light lanthanides generally form fewer compounds than the heavy lanthanides and yttrium. It has been estimated that more than 3000 binary compounds are formed by the rare earth metals with the other elements in the periodic table. The melting points of the intermetallic compounds from the VIIA through the IIB groups are comparable to those of the component metals and rarely exceed the melting point of the highest-melting pure metal, and if so by less than 100 °C (210 °F). From the IIIB through VIB groups the melting points for at least one of the compounds are much higher (several hundreds to more than one thousand °C) than that of the highest-melting component element (generally the rare earth element). Boron and carbon are exceptions because of their elemental high melting points, but most of the rare earth borides and carbides are high-melting compounds. The crystal chemistry of the more than 3000 rare earth binary compounds has been extensively studied; but because of space limitations, the reader is referred to Gschneidner and Daane (Ref 5) and references cited therein for further information.

Liquid Immiscibility. The rare earth metals form immiscible liquids with the alkali, alkaline earth, the group VA metals, and uranium. For chromium and molybdenum, only the light lanthanides, form immiscible liquids, whereas the heavy lanthanides scandium (Sc) and yttrium (Y) form simple eutectics. In the case of manganese, only lanthanum and cerium form immiscible liquids and no intermetallic phases; the remaining rare earths form one or more intermetallic compounds with manganese and presumably no immiscible liquids. Divalent europium (Eu) and ytterbium (Yb) form immiscible liquids with the trivalent rare earth metals (R). The width of the immiscibility gap appears to be larger in the Eu-R systems than in the corresponding Yb-R system. Furthermore, the immiscibility-gap width becomes increasingly larger with increasing atomic number of R for

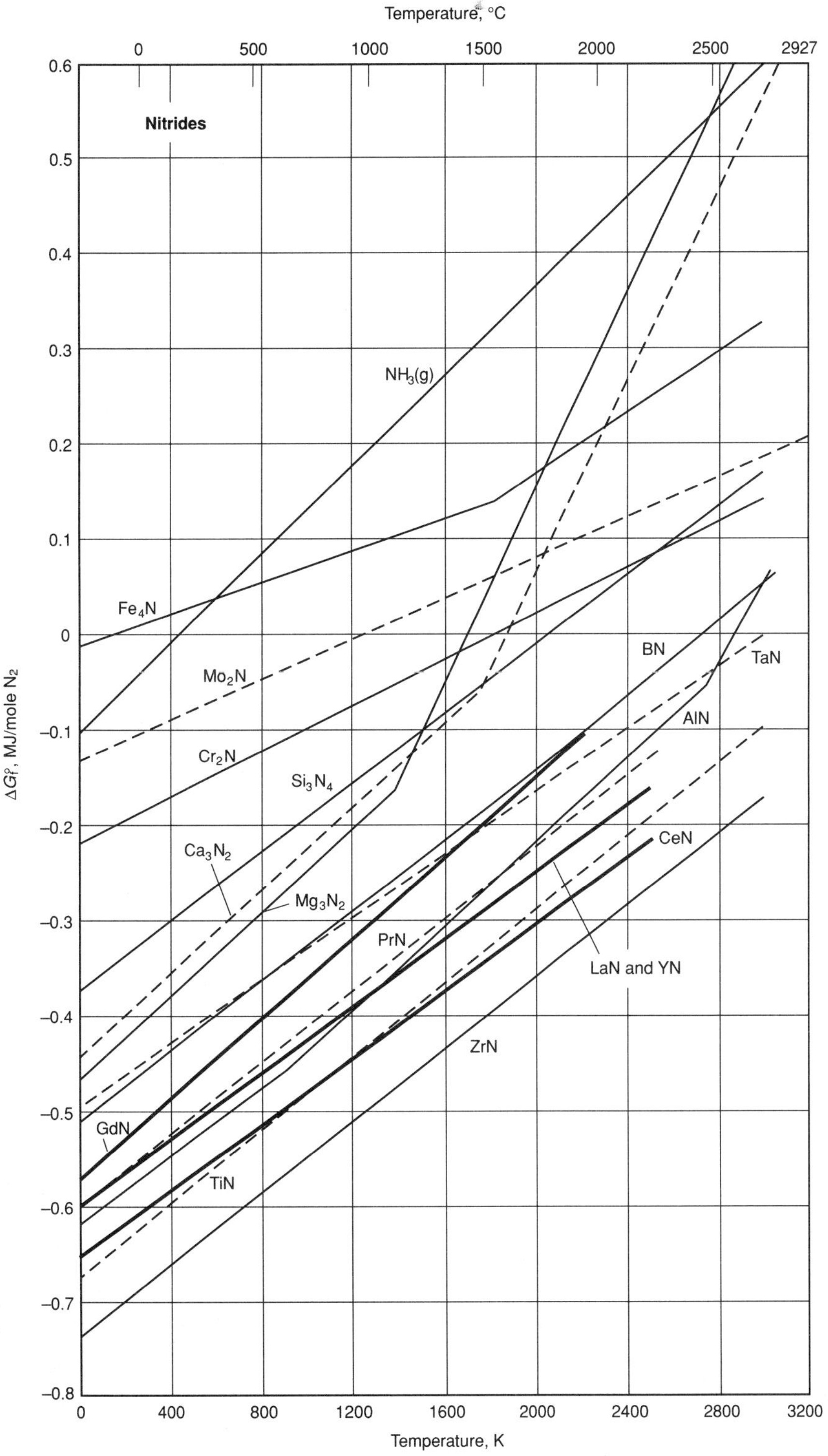

Fig. 7 The standard free energies of formation of the rare earth and some selected nonrare earth nitrides as a function of temperature

a series of R-Yb alloys, which is consistent with the sizes of the elements (Fig. 3).

Applications

This section describes various applications of commercial-grade rare earth elements and commercial alloys, which include rare earth elements as additives. A number of commercial alloys with rare earth constituents are given in Tables 2 and 3. Table 2 lists alloys where rare earth elements (usually Y, La, or Ce) are added as metal alloying components. Table 3 lists materials in which rare earths (primarily Y) are added as finely dispersed oxides.

Alloy Additives. The application of rare earths as alloy additives in metallurgy depends on one or more of the following properties:

- A high chemical affinity for carbon, nitrogen, oxygen, sulfur, and other tramp elements
- Metallic size
- Low vapor pressure
- Alloy formation properties

In many of these applications the rare earths are added as the naturally occurring mixture of elements as reduced from monazite or bastnasite ore. This material is called mischmetal and has the approximate rare earth distribution of 50% Ce, 30% La, 15% Nd, and 5% Pr.

Ductile Iron (Ref 9). Rare earths, which are added as mischmetal, cerium, lanthanum, or yttrium, are used to control the microstructure and chemical form of the excess carbon in cast irons. With no additive the carbon forms graphite flakes and results in gray iron—a cast iron that is easily machinable, and has high thermal conductivity, but low tensile strength. The rare earths (R) modify the carbon morphology in cast irons by removing free oxygen and sulfur from the melt through formation of stable compounds such as oxysulfides (R_2O_2S), which act as nuclei for the growth of spheroidal graphite. This ductile or nodular iron has improved tensile strength, is more ductile, but has a lower thermal conductivity than gray iron. In addition, the rare earths tie up undesirable trace elements, such as lead and antimony, as intermetallic compounds.

Steels (Ref 10). In the late 1960s it was found that rare earth additions, as mischmetal or as a mixed rare earth silicide, would reduce the sulfur content to extremely low levels and control the morphology of the sulfide inclusion. This sulfide shape control greatly improved transverse shelf energy of the high-strength low-alloy (HSLA) steels, which made the forming of the steel into intricate parts and shapes much more efficient and economical. This application lasted into the early 1980s in the United States, Europe, and Japan, but now has been replaced almost exclusively by calcium. The People's Republic of China, however, still uses the rare earths in its steels. Several of the commercial alloys are listed in Table 2.

Superalloys. The rare earth elements are sometimes added to superalloys, which are a broad class of heat-resistant alloys based on iron, cobalt, or nickel. These alloys are essential for use in the gas turbine engines that power present-day jets, electrical generators, and so on. They are also used in environments where good corrosion resistance is required. Less than 1 wt% of rare

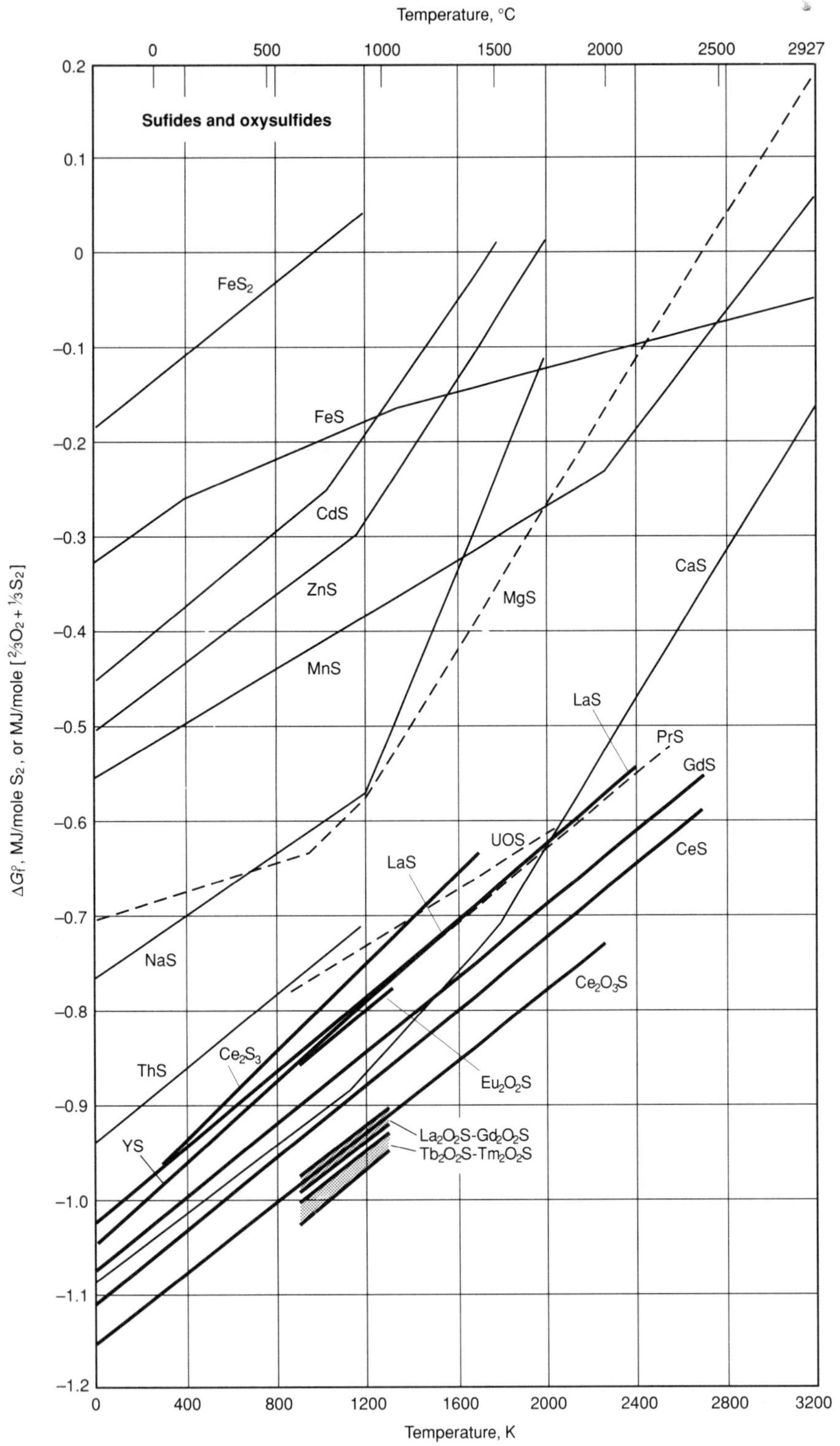

Fig. 8 The standard free energies of formation of the rare earth and some selected nonrare earth sulfides and oxysulfides as a function of temperature. The broad bands marked La_2O_2S-Gd_2O_2S and Tb_2O_2S-Tm_2O_2S contain the free energy versus temperature curves for R_2O_2S with R = La, Pr, Nd, Sm, and Gd, and R′ = Tb, Dy, Ho, Er, and Tm, respectively.

earths, which are used as individual metals, dramatically improves the performance of these alloys. For example, lanthanum raises the operating temperature of nickel-base Hastelloy X from about 950 °C (1750 °F) to about 1100 °C (2000 °F). An extensive list of commercial superalloys that contain rare earths is shown in Table 2.

Magnesium Alloys (Ref 11). The alloy behavior of magnesium is notable for the variety of elements with which it forms solid solutions, including the rare earths. Alloys developed early used mischmetal (MM) to reduce microporosity in wrought alloys such as Mg-1.25Zn-0.17MM (Ref 11). This alloy was difficult to cast, but other alloys containing various amounts of zinc, zirconium, and mischmetal were developed with good castability. Rare earth additions are especially effective in improving the creep resistance of magnesium-base alloys. The rare earths also refine the grain size and improve the strength, ductility, toughness, weldability, machinability, and corrosion resistance. Recently developed alloys have contained separate rare earths. Didymium (an Nd-Pr mixture) is the most effective, followed by cerium-free mischmetal, mischmetal, cerium, and lanthanum, in order of decreasing effectiveness. An Mg-Al-Zn-Nd alloy has good corrosion resistance in an aqueous saline solution. Also an Mg-Y-Nd-Zr alloy was shown to have good corrosion resistance, good castability, and stability to 300 °C (570 °F). Some of the commercial alloys are listed in Table 2.

Recently, just as with aluminum (see below), it has been discovered that a melt-spun amorphous magnesium alloy containing 10 at% Ce and 10 at% Ni has a tensile fracture strength more than twice as large as conventional, optimum age-hardened alloys. It also has good ductility.

Aluminum Alloys. The addition of mischmetal to aluminum-base alloys used for high-tension transmission lines improves tensile strength, heat resistance, vibration resistance, corrosion resistance, and extrudability. Two aluminum-base alloys used in the automobile industry contain 22Si-1MM and 2.5Cu-1.5Ni-0.8Mg-1.2Fe-1.2Si-0.15MM (all wt%). These alloys are used for making cast parts with good high-temperature properties and fatigue strength. They are also used in the aircraft, small engine, and other fields.

A new development involving the use of rare earths in aluminum is the low-density glassy alloys containing about 90 at% Al, 5 to 9 at% transition metals, and about 5 at% rare earths (Ref 12). The transition metals studied include iron, cobalt, nickel, and rhodium, and the rare earths studied include cerium, neodymium, and yttrium. The melt-spun produced materials have extremely high tensile strengths, about twice that of the best crystalline commercial alloy. The alloys begin to crystallize between 250 to 300 °C (480 to 570 °F). They are also quite ductile and because of their low density are of interest to the aerospace industry.

An alloy of aluminum containing 8 wt% Fe and 4 wt% Ce is made by rapid solidification of the melt and processed by powder metallurgy techniques. Quick quenching of the alloy allows large amounts of insoluble metallic elements to be finely dispersed within the aluminum matrix and produces a dispersion-strengthened alloy. The alloy has creep resistance, elevated-temperature

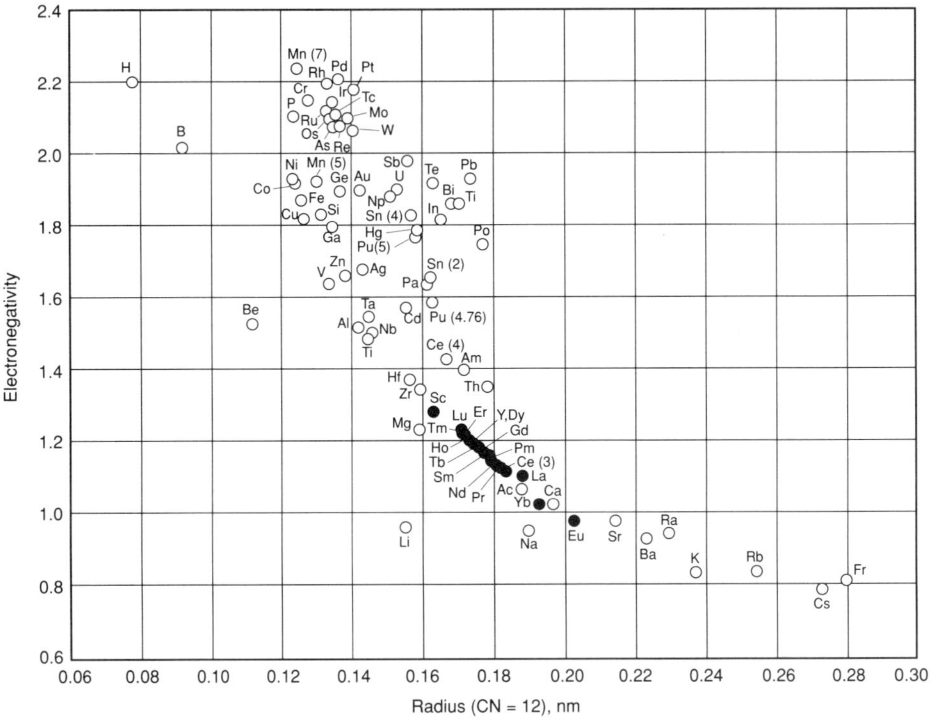

Fig. 9 A plot of the electronegativity (in Pauling's units) versus the metallic radius for a coordination number (CN) of 12 of the elements. The rare earth elements are indicated by the solid points.

tensile strength, thermal stability, and corrosion resistance.

Another related area of application is the use of 1 to 3 wt% mischmetal in aluminum-carbon composites to improve the wetting of the carbon and thus the incorporation of the graphite dispersoid into the metal matrix. The ultimate tensile strength is improved by about 20%, and much more carbon can be incorporated (Ref 13).

Titanium Alloys. The rare earths in titanium alloys are usually present as dispersoids but are also present in solution. The dispersoids are normally oxides but can be sulfides, oxycarbides, or oxysulfides. The dispersoids produced by rapid solidification are ultrafine and serve as barriers to the movement of dislocations, and thus affect the behavior of the alloys in a number of ways. The finer and more stable the dispersoid at high temperatures, the better the properties of the alloys. The smallest and most closely spaced dispersoids are obtained with erbium and yttrium. The dispersoids also maintain their small size upon annealing to about 1000 °C (1830 °F). Alloys with most of the other rare earths and mischmetal have been investigated with various degrees of success. Among the alloys that have found practical applications are those with erbium (Er) or yttrium (Y), added as metals or oxides. Some of these are Ti-6Al-2Sn-4Zr-2Er, Ti-6Al-2Sn-4Zr-2Mo-0.1Si-2Er, Ti-8Al-4Y, Ti-3Er, and Ti-(0.74-1.84)Y (all wt%).

Copper. Mischmetal or yttrium additions to oxygen-free high-conductivity (OFHC) copper improve the oxidation resistance, with little or no adverse effect on the electrical conductivity. At 0.1 wt% Y the oxidation resistance is nearly doubled at 600 °C (1100 °F). Mischmetal (0.1 wt%) greatly improves the hot workability and deep drawing characteristics of bronzes containing less than 1 wt% lead. The mischmetal (MM) forms $MMPb_3$ and prevents the liquefaction of the lead at grain boundaries. In leaded bearing bronzes containing 11 to 50 wt% lead, MM additions guarantee a high uniformity in the lead distribution, promote favorable dendritic solidification, produce a better appearance, and improve the mechanical properties. Mischmetal at 0.2 wt% added to a leaded bearing bronze reduced the coefficient of friction by a factor of four.

Zinc. The main use of rare earths in zinc is in the form of Galfan, a Zn-5Al-0.05MM (in wt%) alloy used in galvanizing baths (Ref 14). The alloy was developed by the International Lead Zinc Research Organization. Galfan was found to be superior in corrosion resistance and formability and equal in weldability and paintability when compared to the normal galvanizing alloys.

Oxide-Dispersion-Strengthened (ODS) Alloys. Yttrium oxide (Y_2O_3) is widely used as a dispersed oxide in superalloys (Ref 15, 16). The oxide is introduced into the alloy by mechanical alloying, which is a high-energy ball milling process that permits solid-state processing and gives the biggest improvement in properties. The milled powders are hot extruded or hot isostatically pressed, which produces a fine dispersion of yttrium oxide in a segregation-free matrix. These compacted alloys have a capacity for secondary recrystallization that results in oriented coarse grains with improved high-temperature creep resistance, strength, and oxidation resistance (Ref 17). Application of these high-strength, high-temperature alloys has been primarily in gas turbine blades, vanes, and combustors. However, any application that requires high-temperature strength creep resistance and corrosion resistance can use oxide-dispersion-strength superalloys. Table 3 lists a number of commercial ODS alloys.

Lighter Flints. A 50 to 75 wt% mischmetal-iron alloy containing a few other alloying additives is used as a lighter flint. This alloy, due to the pyrophoric nature of cerium, ignites or sparks when sharply struck. A large free energy of formation of the rare earth oxides (see Fig. 6) is the main basis for this application. Lighter flints were first sold in 1908, making this one of the oldest uses of the rare earths. They are still used today, although the volume is smaller than it was 20 years ago. Furthermore, lighter flints are the only metallurgical market in which the rare earths (considering the group as one component) are the major constituent.

Magnetic Materials. This section briefly reviews the use of rare earth elements in magnetic materials. Additional information on magnetic materials containing earth elements is given in the article "Permanent Magnet Materials" in this Volume.

Samarium-Cobalt Permanent Magnets. Although the magnetic properties of the rare earth materials had been studied extensively in the 1950s and early 1960s, it was not until 1966 that Strnat and Hoffer noted that YCo_5 had an extremely large magnetocrystalline anisotropy and an unheard of theoretical energy product, which suggested that it would make an excellent permanent magnet (Ref 5). Within a few years, first $SmCo_5$ and then Sm_2Co_{17} were found to be the best permanent magnets ever produced. Most of the strength of the magnet comes from the cobalt atoms, but the role of the samarium (or other rare earths such as Pr, Ce, and Y) is crucial for the permanent magnet properties. The magnetic moments of the 3*d* electrons of cobalt and the 4*f* electrons of samarium couple parallel to each other (which further increases the magnetic strength), but because the 4*f* moments in the hexagonal crystalline environment are difficult to rotate in an applied magnetic field, the cobalt moments are locked in, giving rise to the superior permanent magnet properties. About 15 years later these magnets were, for the most part, superseded by the even more powerful Nd-Fe-B permanent magnets (see below). However, the high magnetic-ordering (Curie) temperatures of the Sm-Co alloys (700 to

Table 2 Commercial alloys containing rare earth metals

Designation	Alloy type	Rare earth	Composition, wt%	Remarks
AiResist 13	Co superalloy	Y	0.1	High-temperature parts
AiResist 213	Co superalloy	Y	0.1	Hot corrosion resistance
AiResist 215	Co superalloy	Y	0.17	Hot corrosion resistance
FSX 418	Co superalloy	Y	0.15	Oxidation resistance
FSX 430	Co superalloy	Y	0.03–0.1	Oxidation and hot corrosion resistance
Haynes 188	Co superalloy	La	0.05	Oxidation resistance, strength
Haynes 1002	Co superalloy	La	0.05	. . .
Melco 2	Co superalloy	Y	0.15	. . .
Melco 9	Co superalloy	Y	0.13	. . .
Melco 10	Co superalloy	Y	0.10	. . .
Melco 14	Co superalloy	Y	0.18	. . .
C-207	Cr	Y	0.15	. . .
CI-41	Cr	Y + La	0.1 (total)	. . .
253	Fe superalloy	Ce	0.055	. . .
GE 1541	Fe superalloy	Y	1.0	. . .
GE 2541	Fe superalloy	Y	1.0	. . .
Haynes 556	Fe superalloy	La	0.02	High temperature, up to 1095 °C
ICF 42	High-strength steel	R	. . .	(a)
ICF 45	High-strength steel	R	. . .	(a)
ICF 50	High-strength steel	R	. . .	(a)
VAN 50	High-strength steel	Ce	. . .	(a)
VAN 60	High-strength steel	Ce	. . .	(a)
VAN 70	High-strength steel	Ce	. . .	(a)
VAN 80	High-strength steel	Ce	. . .	(a)
EK 30A	Mg (Zr, Zn)	R	3.0	Creep resistance
EK 41A	Mg (Zr, Zn)	R	4.0	Creep resistance
EZ 33A	Mg (Zr, Zn)	R	3.0	Creep resistance
QE 22A	Mg (Zr, Ag)	R	1.2–3.0	Creep resistance
QE 222A	Mg (Zr, Ag)	Dm(b)	2	Creep resistance
WE 54	Mg (Zr)	Y + R	5.25 + 3.5	High strength, weldability
ZE 10A	Mg (Zn)	R	0.17	Creep resistance
ZE 41A	Mg (Zr, Zn)	R	1.2	. . .
ZE 63A	Mg (Zr, Zn)	R	2–3	Creep resistance
ZE 63B	Mg (Zr, Zn, Ag)	R	2–3	Creep resistance
C129Y	Nb	Y	0.1	. . .
Hastelloy N	Ni superalloy	Y	0.26	. . .
Hastelloy S	Ni superalloy	La	0.05	High stability
Hastelloy T	Ni superalloy	La	0.02	Low thermal expansion
Haynes 214	Ni superalloy	Y	0.02	Oxidation resistance
Haynes 230	Ni superalloy	La	0.5	High-temperature strength
Melni 19	Ni superalloy	La	0.17	. . .
Melni 22	Ni superalloy	La	0.16	. . .
René Y	Ni superalloy	La	0.05–0.3	. . .
Udimet 500 + Ce	Ni superalloy	Ce	. . .	. . .
Unimet 700 + Ce	Ni superalloy	Ce	0.2–0.5	. . .

(a) Rare earth (R) or cerium added for inclusion shape control. (b) Dm, Didymium, alloy of 80Nd-20Pr

Table 3 Oxide dispersion-strengthened alloys

Designation	Alloy type	Rare earth oxide	Amount, wt%	Remarks
MA 956	Fe superalloy	Y_2O_3	0.5	. . .
Haynes 8077	Ni superalloy	Y_2O_3	1.0	Developmental alloy
IN 853	Ni superalloy	Y_2O_3	1.2	Corrosion resistance
MA 753	Ni superalloy	Y_2O_3	1.3	
MA 754	Ni superalloy	Y_2O_3	0.6	High-temperature alloy
MA 758	Ni superalloy	Y_2O_3	0.6	Resistant to molten glass
MA 953	Ni superalloy	La_2O_3	0.9	. . .
MA 957	Ni superalloy	Y_2O_3	0.25	Intermediate-temperature alloy
MA 6000	Ni superalloy	Y_2O_3	1.1	Creep resistance

900 °C, or 1300 to 1650 °F) gives them a distinct advantage over the Nd-Fe-B alloy (Curie temperature of ~300 °C, or 570 °F) for high-temperature (>100 °C) applications.

Nd-Fe-B Permanent Magnets (Ref 5). More recently, a new family of rare earth permanent magnets was discovered in 1981, and their superior permanent magnet properties were realized by 1983. The major component in these magnets is the tetragonal $Nd_2Fe_{14}B$. In this case the major contribution to the magnetic strength comes from the iron, plus some from the neodymium. The role of the neodymium atom (as it is for the samarium atom in the Sm-Co alloys) is to lock in the magnetic moments of the iron and to prevent them from rotating in an applied magnetic field. The Nd-Fe-B alloys have a higher magnetic energy product than the Sm-Co alloys, but the major advantage of the former is that neodymium and iron are cheaper than the samarium and cobalt, respectively. The main disadvantage of the $Nd_2Fe_{14}B$ compound (as noted above in the previous subsection) is its low Curie temperature. However, the Curie point can be improved by the substitution of cobalt for iron and dysprosium for neodymium. Dysprosium additions also substantially increase the intrinsic coercivity and reduce the reversible temperature coefficient and remanence. Currently this application is the largest market for an individual rare earth *metal* (neodymium), and it is currently growing at a rate better than 25% per year.

Terfenol (Ref 5). The ternary intermetallic compound $[(Tb_{0.3}Dy_{0.7})Fe_2]$, which is known as Terfenol, exhibits giant magnetostrictions in an applied field, ~100 times larger than in nickel. That is, when a magnetic field is applied to a magnetostrictive material, it will expand or contract. Conversely, when stress is applied to the material, a magnetic pulse is generated. Its magnetostrictive properties were discovered in 1971 and commercial production began about 15 years later. Some of the uses of terfenol include sonar devices, micropositioners, and liquid-control valves.

Magnetic Refrigerants. The intermetallic compound $PrNi_5$, which has the hexagonal $CaCu_5$ structure, is used to obtain extremely low temperatures in conjunction with the nuclear magnetic cooling of copper. The $PrNi_5$ is used as the first stage and copper as the second stage in a two-stage adiabatic demagnetization unit. The entire unit is cooled down in the presence of two magnetic fields (one around the $PrNi_5$ stage and the second around the copper) by a dilution refrigerator to ~25 mK. When the ~6 T field is slowly removed from the $PrNi_5$ stage, the entire unit is cooled down to ~5 mK. When the field is reduced around the copper in the second stage, a temperature of ~30 μK can then be reached. In 1983 Japanese scientists set a new record of 27 μK for *the lowest working temperature at which useful experiments could be performed on materials other than the refrigerant itself* (Ref 5).

Other lanthanides, especially gadolinium compounds, have been used as magnetic refrigerants for cooling gases or systems to various temperatures as low as 4 K. In this case the magnetic refrigerant rotates through a magnetic field and in so doing warms and cools itself. As the refrigerant enters the magnetic field, it warms up and thus allows heat to be removed and vented by exchange gas (just as heat is removed during the compression cycle of an air conditioner or refrigerator). When the refrigerant leaves the magnetic field, it cools down to cool the system or gas (just as cooling occurs during the expansion of the coolant in a refrigerator). The gadolinium refrigerant is rotated at a speed of ~5 rpm. The main advantages of the magnetic refrigerator are that it is quite compact and has a large refrigeration power per unit volume. It also is quite reliable, has a long lifetime, and is vibration free. Magnetic refrigerators are claimed to be more efficient than most cryogenic cooling systems, especially below the temperature of liquid nitrogen (77 K) (Ref 18).

Magnetooptical Materials (Ref 19). A fairly recent commercial development is the use of amorphous rare earth-transition metal alloys for information storage as magnetooptic discs.* In this application amorphous $Tb_{25}(Fe_{0.9}Co_{0.1})_{75}$ thin films (500 to 2000 Å thick) are RF or DC sputtered on a substrate and coated by a transparent ceramic film. A laser light is used to write, read, or erase the information on the amorphous alloy by making use of the Kerr rotation. Storage densities of ~10^8/cm^2 have been achieved, which is 15 to 50 times larger than the densities found in a conventional magnetic hard disk.* Gadolinium is sometimes used instead of terbium, and neodymium additions are used to increase the Kerr rotation. The initial application of these magnetooptic storage devices is for personal computers and work stations.

Hydrogen Storage Alloys. A large number of compounds formed between iron, cobalt, and nickel with the rare earth metals have the ability to absorb large quantities of hydrogen. For example, $LaNi_5$ will form $LaNi_5H_6$ under a few atmospheres of H_2 pressure. This compound has a larger number of hydrogen atoms per cubic volume than liquid H_2. Furthermore, the hysteresis of the absorption/desorption cycle is small, and a large heat is associated with this reaction. Some of the applications using these properties include hydrogen storage, heat pumps, heat engines, isotope separation, hydrogen gas separation and purification, energy storage, and catalysis (Ref 20). A rechargeable battery based on the rare earth-nickel-hydride cell could develop into the most important technological application of these materials (Ref 21).

Miscellaneous Applications. Rare earth elements have a variety of specialized applications in instruments and materials. Some miscellaneous applications are described below.

Electron Emitter-LaB_6 (Ref 5). The superior thermionic properties of LaB_6 have been known for nearly 40 years. LaB_6 is composed of B_6 clusters and lanthanum atoms in a CsCl-like arrangement. This compound has a melting point >2500 °C (4500 °F) and a low vapor pressure that when combined with its low work function and excellent thermionic emission current make LaB_6 a better electron emitter than tungsten. The thermionic emission can be improved by using ⟨111⟩ oriented single crystals. LaB_6 is used in electron guns of electron microscopes where high intensities are highly desirable, if not essential.

High-Pressure Gage. At high pressure ytterbium becomes a semiconductor at ~2 GPa (290 ksi) and room temperature, and as a result the electrical resistance increases by an order of magnitude from its ambient pressure value. A further increase of pressure results in the room-temperature fcc polymorph transforming to the high- (temperature) pressure bcc form at ~4 GPa (580 ksi), with a resultant sharp drop of the resistivity to a value comparable to the 1 atmosphere value (Ref 5). Scientists and engineers make use of this large resistivity increase and drop as a high-pressure gage. The ytterbium is usually used in the form of a thin wire.

Getters. The reactivity of the rare earth metals with O_2, N_2, CO_2, H_2O, and so forth, and the large, negative free energies of formation for the oxides, nitrides, and hydrides (among the most negative in the periodic table, see Fig. 6 and 7) account for their past use as getters in vacuum tubes such as television and cathode-ray tubes. The rare earth metals, usually a mischmetal alloy, react with the residual gases in the tube, thus increasing the lifetime of the filament. Rare earth metals are not used in this way anymore.

Corrosion Protection of Metals. In the past few years scientists, especially in Australia, have found that the rare earths can act as inhibitors in aqueous corrosion and as coatings for corrosion protection. The rare earths, as ions in solution, are nearly as effective as chromates in inhibiting corrosion of aluminum alloys, mild steel, and zinc in aqueous solutions. Cerium hydroxide peroxide coatings, ~100 nm thick, have been applied on aluminum alloys, zinc, cadmium, magnesium, and steel to provide corrosion protection that is as good as the common coatings used (that is, zinc, cadmium, and chromates). The main impetus for using the rare earths in corrosion protection is to replace many of the environmentally unacceptable materials currently being used by nontoxic materials and still afford good corrosion protection. The rare earths are nontoxic and seem to provide the necessary protection.

*The conventional spelling is "disc" in the optical technologies and "disk" in the magnetic technology.

REFERENCES

1. K.A. Gschneidner, Jr., Preparation and Purification of Rare Earth Metals and Effect of Impurities on Their Properties, in *Science and Technology of Rare Earth Materials*, E.C. Subbarao and W.E. Wallace, Ed., Academic Press, 1980, p 25-47
2. J.E. Powell, Separation Chemistry, in *Handbook on the Physics and Chemistry of Rare Earths*, Vol 3, K.A. Gschneidner, Jr. and L. Eyring, Ed., North-Holland, 1979, p 81-109
3. B.J. Beaudry and K.A. Gschneidner, Jr., Preparation and Basic Properties of the Rare Earth Metals, in *Handbook on the Physics and Chemistry of Rare Earths*, Vol 1, K.A. Gschneidner, Jr. and L. Eyring, Ed., North-Holland, 1978, p 173-232
4. D. Fort, B.J. Beaudry, and K.A. Gschneidner, Jr., Ultrapurification of Rare Earth Metals: Gadolinium and Neodymium, *J. Less-Common Met.*, Vol 134, 1987, p 27-44
5. K.A. Gschneidner, Jr. and A.H. Daane, Physical Metallurgy, in *Handbook on the Physics and Chemistry of Rare Earths*, Vol 11, K.A. Gschneidner, Jr. and L. Eyring, Ed., North-Holland, 1988, p 409-484
6. S.K. Sinha, Magnetic Structures and Inelastic Neutron Scattering: Metals, Alloys and Compounds, in *Handbook on the Physics and Chemistry of Rare Earths*, Vol 1, K.A Gschneidner, Jr. and L. Eyring, Ed., North-Holland, 1978, p 489-589
7. K.A. McEwen, Magnetic and Transport Properties of the Rare Earths, in *Handbook on the Physics and Chemistry of Rare Earths*, Vol 1, K.A. Gschneidner, Jr. and L. Eyring, Ed., North-Holland, 1978, p 411-488
8. T.E. Scott, Elastic and Mechanical Properties, in *Handbook on the Physics and Chemistry of Rare Earths*, Vol 1, K.A. Gschneidner, Jr. and L. Eyring, Ed., North-Holland, 1978, p 591-705
9. H.F. Linebarger and T.K. McCluhan, The Role of the Rare Earth Elements in the Production of Nodular Iron, in *Industrial Applications of Rare Earth Elements*, K.A. Gschneidner, Jr., Ed., ACS Symposium Series 164, American Chemical Society, 1981, p 19-42
10. L.A. Luyckx, The Rare Earth Metals in Steel, in *Industrial Applications of Rare Earth Elements*, K.A. Gschneidner, Jr., Ed., ACS Symposium Series 164, American Chemical Society, 1981, p 43-78
11. I.S. Hirschhorn, Metallurgical Applications of the Rare Earth Metals, *Mod. Cast.*, Vol 55 (No. 6), 1969, p 94-96
12. R.W. Cahn, Aluminum-Based Glassy Alloys, *Science*, Vol 341, 1989, p 183-184
13. R. Upadhyaya, B.C. Pai, K.G. Satyanarayana, and A.D. Damodaran, Studies on the Additions of Mischmetal to Al-Alloy Matrix Composites, in *Rare Earths. Extraction, Preparation and Applications*, R.G. Bautista and M.M. Wong, Ed., The Minerals, Metals, Materials Society, 1988, p 261-268
14. S.F. Radtke and D.C. Herrschaft, Role

of Misch Metal in Galvanizing With a Zn-5% Al Alloy, *J. Less-Common Met.*, Vol 93, 1983, p 253-259

15. J.D. Whittenberger, Elevated Temperature Mechanical Properties and Residual Tensile Properties of Two Cast Superalloys and Several Nickel-Base Oxide Dispersion Strengthened Alloys, *Met. Trans.*, Vol 12A, 1981, p 193-206
16. H.E. Chandler, Superalloy Update, *Met. Prog.*, Vol 123 (No. 7), 1983, p 21-28
17. E. Grundy and W.H. Patton, Properties and Applications of Hot Formed O.D.S. Alloys, in *High Temperature Alloys, Their Exploitable Potential*, J.B. Marriott, M. Merz, J. Nihoul, and J. Ward, Ed., Elsevier, 1988, p 327-335
18. Magnetic Refrigeration, *Supercond. Ind.*, Vol 2, Spring 1989, p 34-41
19. K.H.J. Buschow, Magneto-Optical Properties of Alloys and Intermetallic Compounds, in *Ferromagnetic Materials*, Vol 4, E.P. Wohlfarth and K.H.J. Buschow, Ed., Elsevier, 1988, p 493-595
20. K.H.J. Buschow, Hydrogen Absorption in Intermetallic Compounds, in *Handbook on the Physics and Chemistry of Rare Earths*, Vol 6, K.A. Gschneidner, Jr. and L. Eyring, Ed., North-Holland, 1984, p 1-111
21. J.J.G. Willems and K.H.J. Buschow, Permanent Magnets to Rechargeable Hydride Electrodes, *J. Less-Common Met.*, Vol 129, 1987, p 13-39

Germanium and Germanium Compounds

J.H. Adams, Eagle-Picher Industries, Inc.

GERMANIUM (Ge) is a semiconducting metalloid element found in Group IV A and period 4 of the periodic table. Although it looks like a metal, it is fragile like glass. Its electrical resistivity is about midway between that of metallic conductors and that of good electrical insulators. Although it was first isolated by Winkler in 1886, no commercial application was found for it until the early 1940s, when it was found to have interesting electrical properties. Its first significant use was in solid-state electronics, and with it the transistor was invented. Indeed, the entire modern field of semiconductors owes its development to the early successful use of germanium. Germanium is still used in the field of electronics, but its use in the field of infrared optics surpassed its electronic applications in the 1970s. Germanium has also found widespread use in the fields of gamma ray spectroscopy, catalysis, and fiber optics. The physical, thermal, and electronic properties of germanium metal are given in Table 1. Table 2 lists the optical properties of germanium.

Sources

The crust of the earth is estimated to contain 1.5 to 7 g of germanium per ton. Germanium usually occurs widely dispersed in minerals such as sphalerite; it rarely occurs in concentrated form. Almost all germanium production has been from zinc smelters. Copper smelters are the second largest source. There are only a few actual minerals of germanium, some with germanium concentrations up to about 8%. Most of these have occurred in Africa, with the highest concentration near Tsumeb, Southwest Africa (now Namibia). Reference 1 reviews 19 germanium minerals found near Tsumeb. Several papers on the geochemistry of germanium are collected in Ref 2.

Germanium also occurs in significant concentrations in many coals around the world. The concentration of germanium within many coal veins varies from top to bottom, with the highest concentration occurring in the upper and lower few centimeters. It is assumed that this distribution within the vein indicates the deposition of germanium from solution after the vein was formed. When coal is burned in power-generating or coking plants, the germanium tends to concentrate in the fly ash or flue dust produced. Any recovery of germanium from coal would most likely be from such ash or dust. Significant germanium recovery from coal in Britain was reported in the 1950s (Ref 3), and smaller amounts have been reported from other countries since then.

Chemical Properties

This section focuses on the chemical properties of various germanium compounds. The physical, thermal, electronic, and optical properties of germanium metal are summarized in Tables 1 and 2.

Germanium Metal. Germanium is quite stable in air up to 400 °C (750 °F) where slow oxidation begins. Oxidation becomes noticeably more rapid above 600 °C (1100 °F). The metal resists concentrated hydrochloric acid, concentrated hydrofluoric acid, and concentrated sodium hydroxide solutions, even at their boiling points. It is not attacked by cold sulfuric acid but does react slowly with hot sulfuric acid. Nitric acid attacks germanium more readily at all temperatures than does sulfuric acid. Germanium reacts readily with mixtures of nitric and hydrofluoric acids and with molten alkalies; it reacts more slowly with aqua regia. The principal reaction route for the mixed acids is the oxidation of the germanium with one constituent followed by the dissolution of the oxide by the other constituent. The reaction with fused alkalies is a direct oxidation with the release of hydrogen. Germanium also reacts readily with the halogens to form the respective tetrahalides.

In compounds, germanium can have a valence of either 2 or 4. Although the divalent compounds tend to be less stable than the tetravalent ones, most can be stored at room temperature for years with no change in composition. At higher temperatures, most of the divalent compounds decompose. Reference 4 reviews the syntheses and properties of many germanium compounds (including divalent ones) and provides a good discussion of the properties of germanium bonds. An excellent earlier review of inorganic germanium compounds is given in Ref 5.

Germanium Halides. Germanium tetrachloride ($GeCl_4$) is made by the reaction of hydrochloric acid on germanium concentrates containing oxides and/or germanates. It can also be made by the reaction of chlorine on heated metallic germanium. The properties of $GeCl_4$ are shown in Table 3.

Germanium tetrachloride is soluble in solvents such as acetone, absolute ethanol, benzene, carbon disulfide, carbon tetrachloride, chloroform, and diethyl ether. It is only slightly soluble in concentrated hydrochloric acid, with the solubility dropping with acid normality and reaching a minimum at about 5 *N*. At HCl concentrations below about 5 *N*, the tetrachloride begins to hydrolyze to GeO_2 (see the section "Germanium Oxides" below). Germanium tetrachloride is insoluble in concentrated sulfuric acid and does not react with it. The solubility of free chlorine in $GeCl_4$ can reach as high as 4 wt%, especially at low temperatures.

Germanium tetrabromide ($GeBr_4$) and tetraiodide (GeI_4) can be easily prepared by the reaction of the respective halogen with germanium metal and by the reaction of GeO_2 with HBr and HI solutions, respectively. The preparation of germanium tetrafluoride (GeF_4) is not so straightforward, and pure GeF_4 is usually made by decomposing barium hexafluorogermanate at about 700 °C (1300 °F).

Germanium Oxides. Germanium dioxide (GeO_2) is usually made by the hydrolysis of $GeCl_4$ with water. It is also made by the ignition of germanium disulfide. Solid GeO_2

Table 1 Physical, thermal, and electronic properties of germanium

Property	Value
Physical	
Atomic number	32
Atomic weight	72.59
Crystal structure	Diamond cubic
Density at 25 °C (77 °F), g/cm³ (lb/in.³)	5.323 (0.1924)
Atomic density at 25 °C (77 °F), atoms/cm³	4.416×10^{22}
Lattice constant at 25 °C (77 °F), nm	0.565754
Liquid surface tension at melting point, N/m (lbf/ft)	0.650 (0.0445)
Modulus of rupture, MPa (ksi)	110 (16)
Mohs hardness	6.3
Poisson's ratio at 125–375 K (−235 to 215 °F)	0.278
Natural isotopic abundance, %	
mass number 70	20.4
mass number 72	27.4
mass number 73	7.8
mass number 74	36.6
mass number 76	7.8
Thermal	
Melting point, °C (°F)	937.4 (1719)
Boiling point, °C (°F)	2830 (5126)
Heat capacity at 25 °C (77 °F), J/kg · K (Btu/lb · °F)	322 (0.07696)
Latent heat of fusion, J/g (Btu/lb)	466.5 (200.7)
Latent heat of vaporization, J/g (Btu/lb)	4602 (1980)
Heat of combustion, J/g (Btu/lb)	7380 (3175)
Heat of formation, J/g (Btu/lb)	4006 (1723)
Vapor pressure, kPa (psi)	
At 2080 °C (3775 °F)	1.33 (0.193)
At 2440 °C (4425 °F)	13.3 (1.93)
At 2710 °C (4910 °F)	53.3 (7.73)
At 2830 °C (5125 °F)	101.3 (14.69)
Coefficient of linear expansion, 10^{-6}/K (10^{-6}/°F)	
At 100 K (−280 °F)	2.3 (1.3)
At 200 K (−100 °F)	5.0 (2.8)
At 300 K (80 °F)	6.0 (3.3)
Thermal conductivity, W/m · K	
At 100 K (−280 °F)	232
At 200 K (−100 °F)	96.8
At 300 K (80 °F)	59.9
At 400 K (260 °F)	43.2
Electronic	
Intrinsic resistivity at 25 °C (77 °F), Ω · cm	53
Intrinsic conductivity type	N (negative)
Intrinsic electron drift mobility at 25 °C (77 °F), cm²/V · s	3800
Intrinsic hole drift mobility at 25 °C (77 °F), cm²/V · s	1850
Band gap, direct, minimum at 25 °C (77 °F), eV	0.67
Band gap, direct, minimum at 0 K, eV	0.744
Number of intrinsic electrons at 25 °C (77 °F), cm⁻³	2.12×10^{13}

exists in soluble, insoluble, and vitreous forms. The properties of these three forms are given in Table 4. The soluble form is the usual product of hydrolyzing $GeCl_4$. The insoluble form can be prepared by heating soluble oxide at 300 to 900 °C (570 to 1650 °F), especially in the presence of about 0.5 wt% alkali halides. The glassy, or vitreous, form is prepared by melting either of the other forms and then cooling the melt.

Table 2 Optical properties of germanium

Wavelength, µm	Refractive index at 300 K (80 °F)	Absorption coefficient, cm⁻¹	Through 1 cm thickness (uncoated) Reflection, %	Absorption, %	Transmission, %
1.8	4.134	7.0	37.3	62.7	0.0
1.9	4.120	0.68	41.0	38.2	20.8
2.0	4.108	0.010	53.6	1.0	45.4
4.0	4.0255	0.0047	53.0	0.5	46.5
6.0	4.0122	0.0068	52.8	0.7	46.5
8.0	4.0074	0.0150	52.4	1.5	46.1
10.0	4.0052	0.0215	52.2	2.1	45.7
10.6	4.0048	0.0270	52.0	2.6	45.4
11.0	4.0045	0.0295	51.8	2.9	45.3
11.9	4.0040	0.200	46.9	16.4	36.7
12.0	4.0039	0.170	47.6	14.4	38.0
13.0	4.0035	0.160	47.8	13.7	38.5
14.0	4.0032	0.149	48.2	12.8	39.0
15.0	4.0029	0.385	43.3	27.1	29.6
16.0	4.0026	0.530	41.4	33.4	25.2
18.0	4.0022	2.00	36.3	58.2	5.5
20.0	4.0018	2.15	36.2	59.0	4.8

Germanium monoxide (GeO) can best be prepared in pure form by heating a mixture of germanium and GeO_2 in the absence of oxygen. At temperatures above 710 °C (1310 °F), GeO sublimes from the mixture and condenses as a glassy deposit in the cooler part of the reaction vessel. Germanium monoxide is stable at room temperature.

Germanates are usually prepared by the fusion of GeO_2 with alkali oxides or carbonates in platinum crucibles. Sodium heptagermanate ($Na_3HGe_7O_{16}$) is precipitated by the neutralization of a sodium hydroxide solution of GeO_2 with hydrochloric acid to a pH above 7.

Germanides can be formed by melting other metals with germanium in the proper stoichiometric concentrations and then freezing the melt. They can also be prepared by vacuum sintering the two metals together; sintering is usually followed by long annealing. Other procedures include the thermal dissociation of one germanide into another and the electrolysis of fused mixed salts. The preparation and properties of about 200 germanides have been tabulated and reviewed (Ref 6). One of the germanides that has been prepared most often is magnesium germanide (Mg_2Ge).

Germanes, or germanium hydrides, are commonly prepared by the reaction of a germanide, such as Mg_2Ge, with hydrochloric acid. They can also be produced by the reduction of $GeCl_4$ with lithium aluminum hydride and by the reduction of GeO_2 by sodium borohydride in water solution. The preparation and properties of the germanes are reviewed in Ref 7 and 8.

Miscellaneous Inorganic Compounds. Germanium nitride (Ge_3N_4) is about as inert as tetragonal GeO_2. It is prepared most easily from germanium powder and ammonia at 700 to 850 °C (1300 to 1560 °F). The nitride does not react with most mineral acids, aqua regia, or caustic solutions, even when hot. Germanium disulfide (GeS_2) is an unusual and useful compound because it is insoluble in strong acids such as 6 *N* HCl and 12 *N* H_2SO_4. This insolubility permits the recovery of germanium from acid solutions by gassing with H_2S. The disulfide can also be made by the reaction of GeO_2 with sulfur.

Organogermanium Compounds. The field of organogermanium chemistry has drawn widespread interest for many years. Organogermanium compounds are generally characterized as having low chemical reactivity and relatively high thermal stability. The synthesis of many begins with a Grignard reaction. Many excellent reviews of the organogermanium literature have been published (Ref 4, 8, 9-19). During the 1980s, several organogermanium compounds were produced in commercial quantities. These included spirogermanium (which is 2-aza-8-germanspiro decane-2-propamine-8,8-diethyl-N,N-dimethyl dihydrochloride) and carboxyethyl germanium sesquioxide. These compounds have been studied extensively for their anticancer and blood pressure effects.

Table 3 Properties of germanium tetrachloride

Property	Value
Molecular weight	214.40
Color	Colorless
Density at 25 °C (77 °F), g/cm³ (lb/in.³)	1.874 (0.0677)
Melting point, °C (°F)	−49.5 (−57.1)
Boiling point, °C (°F)	83.1 (181.58)
Refractive index at 25 °C (77 °F), 0.5893 µm	1.464
Heat capacity (constant pressure) of vapor at 25 °C (77 °F), J/kg · K (Btu/lb · °F)	449 (0.1073)
Heat of vaporization at boiling point, J/g (Btu/lb)	137 (58.94)
Heat of formation at 25 °C (77 °F), J/g (Btu/lb)	−3318 (−1427)
Vapor pressure, Pa (psi)	
At 225 K (−55 °F)	10^2 (0.0145)
At 253 K (−5 °F)	10^3 (0.145)
At 294 K (69 °F)	10^4 (1.45)
At 356 K (181 °F)	10^5 (14.5)
At 462 K (372 °F)	10^6 (145)
At 550 K (530 °F) (T_c)	3.850×10^6 (558)

Table 4 Properties of solid forms of germanium dioxide

Property	Soluble	Insoluble	Vitreous
Structure	Hexagonal	Tetragonal	Amorphous
Density at 25 °C (77 °F), g/cm^3	4.228	6.239	3.637
Melting point, °C (°F)	1116 (2040)	1086 (1987)	. . .
Solubility in water at 25 °C (77 °F), g/L	4.53	Insoluble	5.18
Solubility in water at 100 °C (212 °F), g/L	13	Insoluble	. . .
Solubility in HCl, HF, and NaOH solutions	Soluble	Insoluble	Soluble

Manufacturing and Processing

Ore Processing. None of the minerals mentioned earlier in the section "Sources" is mined solely for its germanium content. Almost all of the germanium recovered worldwide is a by-product of other metals, primarily zinc, copper, and lead. The enriched copper-lead concentrates from Tsumeb, Namibia, have been treated in a vertical retort from which germanium sulfide is sublimed and separated (Ref 3). The copper-zinc ores of Katanga, Zaire, have been treated by roasting with H_2SO_4, followed by leaching and selective precipitation of the germanium with MgO (Ref 3). In the United States, zinc concentrates have been roasted and then sintered for germanium recovery. In this process, the sinter fume is chemically leached, and the germanium is selectively precipitated from the leach solution by fractional neutralization; it is then sent to the germanium refinery (Ref 20). Because of the low solubility of germanium sulfide and tannate in acid solutions, germanium has been recovered by precipitating it from acid solutions with H_2S or tannic acid. Sulfide precipitates are usually oxidized with sodium chlorate or permanganate, followed by, or concurrent with, dissolution in concentrated HCl and distillation of the resulting $GeCl_4$. Tannic acid precipitates are usually upgraded by igniting the precipitate to the oxide and dissolving the oxide in concentrated HCl, with subsequent distillation of the $GeCl_4$. Germanium can also be recovered from the still residue in the distillation of zinc metal (Ref 21).

From June 1986 to September 1987, germanium was recovered from the Apex Mine near St. George, UT. This was an unusual operation in that germanium was the major product from the mine, with gallium a valuable by-product. Copper was also recovered because the Apex Mine was an old abandoned copper mine. The process involved dissolution of the screened ore in sulfuric acid, followed by cementation of the copper, solvent extraction of the gallium, and precipitation of the germanium with H_2S. The GeS_2 precipitate was oxidized with sodium chlorate and dissolved in HCl. The $GeCl_4$ formed was distilled and then hydrolyzed to a crude GeO_2. Unfortunately, the operation was not a financial success, and the producer was forced into bankruptcy. The assets of the producer were purchased in 1989 by the Hecla Mining Company, which made major changes in the processing circuit, including the addition of a solvent extraction system for germanium recovery. Start-up of the revised operation was scheduled for March 1990.

In electrolytic zinc plants, which have become increasingly important for environmental reasons, germanium is precipitated, usually along with iron, during the purification of the $ZnSO_4$ electrolyte prior to electrolysis. Germanium is one of several impurities that have an adverse effect on zinc electrolysis. If the germanium concentration is high enough in the separated solids, economic recovery of germanium is possible. During recent years, several solvent extraction and ion exchange processes have been developed (Ref 22-28) that provide better germanium separations, primarily from $ZnSO_4$ electrolytes.

Purification. Regardless of the source of the germanium, all germanium concentrates are purified by similar techniques. The ease with which concentrated germanium oxides and germanates react with concentrated hydrochloric acid, and the convenient boiling point of the resulting $GeCl_4$ (83.1 °C, or 181.6 °F), make chlorination a standard refining step. An oxidizing agent is often added to the primary distillation or to the subsequent fractionation, or both, to suppress the volatility of arsenic (Ref 29). Other purification steps are used to separate certain other objectionable impurities, if present. The fractionation is usually done in glass or quartz because most subsequent uses of $GeCl_4$ require impurity levels of no more than about 1 mg/kg (1 ppm).

The purified $GeCl_4$ is hydrolyzed with deionized water to produce GeO_2, which is removed by filtration and dried. The dried GeO_2 is reduced with hydrogen at about 760 °C (1400 °F) to germanium metal powder, which is subsequently melted and cast into so-called first-reduction, or as-reduced, bars. These bars are then subjected to zone refining to produce intrinsic or electronic-grade germanium metal. Zone refining is ideally suited for the refining of germanium; in fact, the original development of this procedure was prompted by the need for ultrapure germanium (Ref 30). Zone refining results in polycrystalline germanium, usually containing less than 100 ng total impurities per gram of germanium and less than 0.5 ng of electrically active impurities per gram of germanium. For extremely high-purity applications, such as for uncompensated gamma ray detector crystals, germanium has been refined to impurity concentrations of less than 0.0003 ng/g (0.0003 ppb). However, zone refining is not equally efficient in removing all impurities. Boron and silicon are not removed easily with this method and must be removed from germanium before the zone-refining step.

For use as a semiconductor, refined germanium is grown into single crystals from the melt. The electronic properties are controlled by the addition of selected impurities (dopants) to the melt before crystal growth begins. The grown crystal is sliced and fabricated into devices, and the germanium scrap generated is recycled through the refinery into intrinsic germanium.

For use in infrared optics, zone-refined germanium is recast or grown into forms suitable for lens and window manufacture. After the germanium is annealed, it is cut and ground into lens or window blanks, which are then polished, coated, and assembled into an infrared system.

Environmentally, the production of germanium is quite harmless. Among the products produced, the chemicals consumed, and the by-products made, only arsenic would normally be considered a problem. The bulk of the arsenic is separated from the germanium at the smelter, and the small amounts entering the germanium-refining plant are easily controlled. The acids, bases, and chlorine used in processing and in gas scrubbing can be neutralized and held in permanent containment ponds or disposed of in hazardous-waste landfills. No hazardous substances are known to be discharged into surface waters or municipal treatment facilities from germanium-refining plants.

Economic Aspects

World reserves of germanium have been estimated at 4000 Mg (4400 tons), but it is impossible to discuss reserves without considering price. In many applications the cost of germanium is a very small part of the overall cost, and increasing the germanium price substantially would have little impact on its use in such applications. However, substantial increases in its price would expand germanium reserves significantly.

Because germanium is almost always recovered as a by-product, its price and availability over the long range are subject to supply and demand considerations for its host products, usually zinc and copper, as well as for itself. This is not the case over the short term (6 to 24 months) because producers often recover germanium from stockpiles of smelter residues that can last for years. Therefore, short-term pricing is largely controlled only by demand. Figure 1 plots the price levels of intrinsic germanium in the United States over the past 35 years; prices are adjusted for inflation. The No-

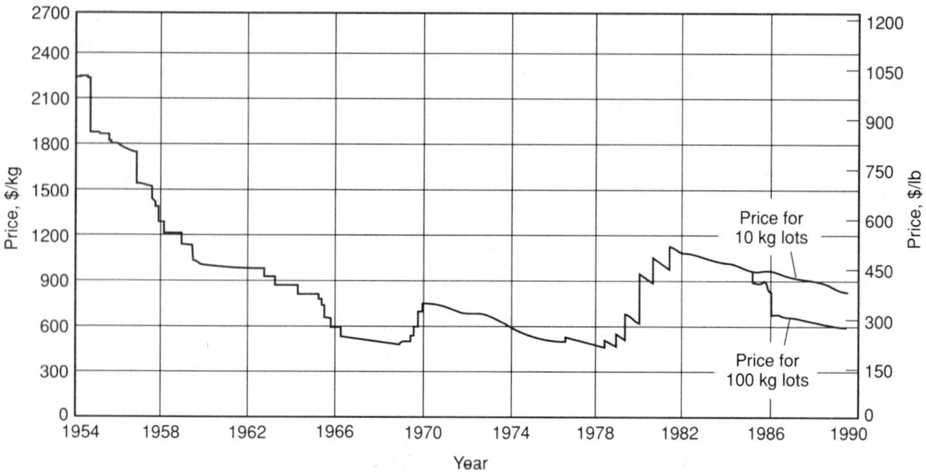

Fig. 1 Historical price of intrinsic germanium in the United States in constant 1983 dollars

vember 1978 adjusted price of \$469/kg was an all-time low. During 1985 a two-tiered pricing structure was implemented: The higher price is for 10 kg lots, and the lower price applies to 100 kg lots. The prices in December 1989 were \$1060 (adjusted price \$841) for 10 kg lots and \$750 (\$595) for 100 kg lots.

Information concerning world production rates of germanium is difficult to obtain. This is due in part to the fact that there are very few producers. It is known that world production exceeded 100 Mg/yr (110 tons/yr) during the period of peak demand for semiconductors. Production has varied between 30 and 80 Mg/yr (33 and 88 tons/yr) during the 1970s and between 80 and 110 Mg/yr (88 and 120 tons/yr) in the 1980s.

An annual review of germanium is published by the U.S. Bureau of Mines (Ref 31) with a broader survey published every 5 years (Ref 32). The *Engineering and Mining Journal* also published an annual germanium review through 1986 (Ref 33). The economics of germanium, with special emphasis on world trade, have been reviewed by Roskill Information Services (Ref 34-40).

Specifications

Most electronic-grade GeO_2 is 99.9999% minimum purity (6N) as measured spectrographically and has 0.1 to 0.2% maximum volatile content, a bulk density of 1.5 to 2.0 g/cm^3 as specified by the customer, a metal billet resistivity of at least 5 ohm-cm average on the last third of the billet to freeze, and an N-type billet conductivity. The GeO_2 intended for chemical use, such as for use as a catalyst, is at least 5N pure spectrographically, at least 98% finer than 325 mesh, and at least 99.9% soluble in water and/or ethylene glycol.

Most intrinsic germanium metal sold is N-type with a resistivity of at least 40 ohm-cm at 25 °C (77 °F) or 50 ohm-cm at 20 °C (68 °F). Germanium metal prepared for use in infrared optics is usually specified to be N-type with a resistivity of 5 to 40 ohm-cm, to be stress-free and fine annealed, and to have certain minimum transmission (or maximum absorption) characteristics in the 3 to 5 or 8 to 12 μm wavelength ranges. Either polycrystalline or single-crystal material will be specified.

Germanium single crystals intended for electronic applications will usually be specified according to conductivity type, dopant, resistivity, orientation, and maximum dislocation density. They will be specified to be lineage-free unless the specified resistivity is below about 0.5 ohm-cm. Minority carrier lifetime and majority carrier mobility are occasionally specified.

The $GeCl_4$ refined for use in making optical fibers is usually specified to contain less than 1 to 5 ppb of each of seven impurities: chromium, manganese, iron, cobalt, nickel, copper, and zinc. Limits are sometimes specified for a few other elements. Also of concern are hydrogen-bearing impurities; therefore, maximum limits of 10 to 20 ppm are usually placed on HCl, OH, CH_2, and CH_3 contents.

Analytical and Test Methods

The analysis of ores for germanium is usually done with an emission spectrograph but can be done in the field with the phenylfluorone method (Ref 41). Analysis of germanium refinery samples is usually done after fusion of the sample with KOH or NaOH in nickel crucibles. Following distillation of the $GeCl_4$ from HCl solution of the fusion, the germanium can be determined in one of three ways:

- Gravimetrically, usually by precipitation of GeS_2 from acid solution and by ignition to GeO_2
- Titrimetrically, usually by reduction with sodium hypophosphite and titration with KIO_3 solution
- Spectrally, with an atomic absorption spectrophotometer

The last procedure is not considered as accurate as the first two. Excellent reviews of the analytical chemistry of germanium are given in Ref 42 and 43.

Analysis of refined germanium products is done in a wide variety of ways, including several methods that have become ASTM standards (Ref 44). Electronic grade GeO_2 is analyzed using an emission spectrograph to determine its spectrographic purity. Its volatile content is measured in accord with ASTM F 5 and its bulk density with F 6. Other ASTM standards cover the preparation of a metal billet from a sample of the oxide (F 27), and the determination of the conductivity type (F 42) and resistivity (F 43) of the billet.

The type and resistivity of all grades of germanium metal are also measured in accord with F 42 and F 43. The transmission characteristics of optical grade germanium are determined with an infrared spectrophotometer, and the measurement of the interstitial oxygen content of the metal is covered in F 120 and F 122. Germanium single crystals can be further evaluated in accord with ASTM F 26, F 28, F 76, F 334, F 389, and F 398.

The overall spectrographic purity of $GeCl_4$ can be determined by using an emission spectrograph to examine a sample of GeO_2 that has been produced from the $GeCl_4$ by hydrolysis with deionized water. The trace metal impurities of concern to fiber optic producers are determined by flameless atomic absorption, and the hydrogen-bearing impurity concentrations are measured by infrared absorption of the $GeCl_4$.

Toxicology

Germanium compounds generally have a low order of toxicity (Ref 45). Only germanium hydride is considered toxic, with a maximum time-weighted average 8-h safe exposure limit of only 0.2 ppm (Ref 46). The lethal dose median for GeO_2 is 750 mg/kg, and that of germanium is 586 mg/kg (Ref 47). The toxicity of specific germanium compounds usually must be considered more from the standpoint of the other part of the compound than from the germanium content. The biological activity of germanium is reviewed in Ref 8, 11, and 13.

Uses

The main use for germanium remained in the field of semiconductor electronics into the 1970s. Its earliest use in 1941 was as a solid-state diode that performed as a detector in radar systems. This device was adapted to radio circuits as a radio frequency detector, and the germanium transistor was invented in 1947. This invention revolutionized the electronics industry and caused a sharp increase in the demand for germanium. The use of germanium in these conventional semiconductor roles reached a peak in the early 1960s, with world consumption averaging over 100 Mg/yr (110 tons/yr). Except for a brief upsurge in this demand during 1969, there has been a general decline in this application of germanium since the early 1960s.

The use of germanium as a semiconductor substrate deserves special mention. In this application, single-crystal wafers of germanium are used as substrates for the epitaxial deposition of gallium arsenide (GaAs) or gallium arsenide phosphide (GaAsP) for use as light-emitting diodes or solar cells. These substrates take the place of more expensive gallium arsenide wafers. Many metric tons of germanium were consumed in the mid-1970s for the production of GaAsP-Ge light-emitting diodes for calculators and watches (Ref 48). The large-scale production of GaAs-Ge solar cells did not begin until 1989, and production is expected to increase in the 1990s. Although the use of germanium as a substrate for GaAs solar cells provides stronger, lighter, and cheaper cells while maintaining the high conversion efficiency of gallium arsenide, one study (Ref 49) suggests that the widespread use of germanium for this application would require several times the present world output of germanium. However, there are large, untapped sources of germanium, and it is not inconceivable that, with sufficient incentive, such demands could be met in time.

The largest use of germanium is in the field of infrared optics. In this application, the transparency of germanium to infrared wavelengths longer than 2 μm and its high refractive index are utilized rather than its electrical properties. Other advantageous properties of germanium for this use are its low dispersion; easy machinability; reasonable strength; low price compared to other infrared materials; good resistance to atmospheric oxidation, to moisture, and to chemical attack; and availability in large sizes. It had been estimated that the world demand for germanium for infrared devices would increase to 55 to 70 Mg/yr (60 to 77 tons/yr) during the 1980s (Ref 50). It is now estimated that 55 to 65 Mg/yr (60 to 72 tons/yr) was indeed consumed in this application from 1987 to 1989.

Infrared devices are principally used for military applications. Infrared systems with germanium usually operate in the 8 to 12 μm range and usually contain several germanium lenses, a germanium window, and a color-correcting lens made from a Ge-Sb-Se glass, a Ge-As-Se glass, or ZnSe. Some of the viewers also utilize a germanium infrared detector.

The U.S. government was so impressed with the importance of germanium for infrared applications that a decision was made in April 1987 to add intrinsic germanium metal to its strategic stockpile. From May 1987 through September 1989, stockpile orders were placed for 69 Mg (76 tons) of germanium. Deliveries were scheduled from July 1987 through January 1991. Some additional germanium is expected to be ordered for the stockpile.

Nonmilitary infrared applications for germanium include CO_2 lasers, intrusion alarms, and police and border patrol surveillance devices. Germanium is used as a thin film coating for infrared materials to decrease reflection losses and/or to provide heavy filtering action below 2 μm. Germanium metal is also used in specially prepared germanium single crystals for gamma ray detectors. Both the older lithium-drifted detectors and the purer, more expensive intrinsic detectors, which do not have to be stored in liquid nitrogen, do an excellent job of spectral analysis of gamma radiation and are important analytical tools.

The primary application of germanium dioxide is in the preparation of germanium metal. However, there are several other uses that provide significant markets for the oxide. The largest of these is its use in place of antimony oxide (Sb_2O_3) as a catalyst in the reaction of ethylene glycol with terephthalic acid in the production of polyester fibers and polyethylene terephthalate (PET) resins. Although more expensive than Sb_2O_3, GeO_2 produces a polyester fiber that does not yellow with age, which is especially attractive to makers of white shirts and other white fabrics. The PET resins are used almost entirely in making beverage bottles such as 2- and 3-liter soft drink bottles. Germanium dioxide produces a stronger, clearer bottle than those made with Sb_2O_3, but the change to GeO_2 is also being driven by the concern over the long-term effects of trace contamination of the bottle contents with antimony. The use of GeO_2 in this application is more widespread in the Far East than in other parts of the world.

Another significant use of GeO_2 is the production of bismuth germanium oxide crystals ($Bi_4Ge_3O_{12}$). These scintillation crystals are used primarily in positron emission tomography scanners, which are expected to find increased use as suitable radioisotopes become less expensive. Germanium dioxide is also used in significant quantities in the production of spirogermanium and carboxyethyl germanium sesquioxide (Ref 45), both of which have medical applications. In addition, GeO_2 is included in a few special glass formulations, primarily to increase the refractive index of the glass.

The only significant application of $GeCl_4$, besides its use in the production of GeO_2, is in optical fibers (Ref 51). In this application, $GeCl_4$ is converted to GeO_2, which is deposited, along with SiO_2 and sometimes B_2O_3 and/or P_2O_5, on the inside of a pure quartz tube and subsequently collapsed to form a solid rod or preform. The preform is then drawn into fine fibers that can be used as optical waveguides, primarily in the 0.8 to 1.6 μm wavelength region. Such fibers have proved very useful, especially in long-distance telephone lines. Phone signals have been transmitted over 100 km (65 miles) on such fibers without amplification. Germanium dioxide provides the higher refractive index fiber core, which prevents signal loss; also, it produces an extremely clear glass that provides very low signal absorption at the selected wavelengths at which infrared emitters are available.

Other uses of germanium include the use of magnesium germanate as a phosphor, the addition of lead germanate to barium titanate capacitors, the use of germanium-gold alloys in dental fillings and precision castings, and the use of germanium single crystals as x-ray monochromators for high-energy physics applications.

REFERENCES

1. W.E. Wilson, Ed., *Tsumeb! The Worlds Greatest Mineral Locality*, The Mineralogical Record, 1977
2. J.N. Weber, Ed., *Geochemistry of Germanium*, Dowden, Hutchinson, & Ross, 1973
3. *Eng. Min. J.*, Vol 157, 1956, p 75
4. F. Glocking, *The Chemistry of Germanium*, Academic Press, 1969
5. O.H. Johnson, *Chem. Rev.*, Vol 51, 1952, p 431
6. G.V. Samsonov and V.N. Bondarev, *Germanides*, A. Wald, trans., Primary Sources, 1970
7. F.G.A. Stone, *Hydrogen Compounds of the Group IV Elements*, Prentice-Hall, 1962, p 63-76
8. E.G. Rochow, Germanium, in *Comprehensive Inorganic Chemistry*, Vol 2, J.C. Bailar, Jr., H.J. Emeleus, R. Nyholm, and A.F. Trotman-Dickenson, Ed., Pergamon Press, 1973, p 1-41
9. O.H. Johnson, *Chem. Rev.*, Vol 48, 1951, p 259
10. E.G. Rochow, D.T. Hurd, and R.N. Lewis, *The Chemistry of Organometallic Compounds*, John Wiley & Sons, 1957
11. F. Rijkens, *Organogermanium Compounds*, Germanium Research Committee, 1960
12. D. Quane and R.S. Bottei, *Chem. Rev.*, Vol 63, 1963, p 403
13. F. Rijkens and G.J.M. Van der Kerk, *Investigations in the Field of Organogermanium Chemistry,* Germanium Research Committee, 1964
14. F. Glocking, *Quart. Rev. Chem. Soc.*, Vol 20, 1966, p 45

15. M. Dub, *Organometallic Compounds*, Vol 2, 2nd ed., Springer-Verlag, 1967
16. K.A. Hooton, Organogermanium Compounds, in *Preparative Inorganic Reactions*, Vol 4, W.L. Jolly, Ed., John Wiley & Sons, 1968, p 85-176
17. N. Hagihara, Ed., *Handbook of Organometallic Compounds*, W.A. Benjamin, 1968, p 449-467
18. M. Lesbre, P. Mazerolles, and J. Satgé, *The Organic Compounds of Germanium*, John Wiley & Sons, 1971
19. R.C. Weast, Ed., *CRC Handbook of Chemistry and Physics*, 59th ed., CRC Press, 1978, p C-692 to C-697
20. J.A. O'Connor, *Chem. Eng.*, Vol 59, 1952, p 158
21. A. Lebleu, P. Fossi, and J. Demarthe, U.S. Patent 4,090,871, May 1978
22. A. DeSchepper and A. Van Peteghem, U.S. Patent 3,883,634, May 1975
23. D. Rouillard, *et al.*, U.S. Patent 4,389,379, June 1983
24. A. DeSchepper *et al.*, U.S. Patent 4,432,951, Feb 1984
25. A. DeSchepper *et al.*, U.S. Patent 4,432,952, Feb 1984
26. D. Boateng *et al.*, U.S. Patent 4,525,332, June 1985
27. G. Cote *et al.*, U.S. Patent 4,568,526, Feb 1986
28. W. Krajewski and K. Hanusch, U.S. Patent 4,666,686, May 1987
29. H.R. Harner, *Rare Metals Handbook*, 2nd ed., C.A. Hampel, Ed., Reinhold, 1961, p 188-197
30. W.G. Pfann, *Zone Melting*, 2nd ed., John Wiley & Sons, 1966, p 134
31. T.O. Llewllyn, Minor Metals, in *Bureau of Mines Minerals Yearbook*, Preprint, U.S. Department of the Interior, 1988
32. P.A. Plunkert, Germanium, in *Mineral Facts and Problems*, U.S. Department of the Interior, 1985
33. J.H. Adams, *Eng. Min. J.*, Vol 187, May 1986, p 46
34. *Germanium: World Survey of Production, Consumption and Prices With Special Reference to Future Trends*, Roskill Information Services, 1974
35. *Germanium: World Survey of Production, Consumption and Prices With Special Reference to Future Trends*, Statistical Supplement, Roskill Information Services, 1976
36. *Germanium: World Survey of Production, Consumption and Prices With Special Reference to Future Trends*, 2nd ed., Roskill Information Services, 1977
37. *Germanium: World Survey of Production, Consumption and Prices With Special Reference to Future Trends*, 2nd ed., Statistical Supplement, Roskill Information Services, 1979
38. *The Economics of Germanium*, 3rd ed., Roskill Information Services, 1981
39. *The Economics of Germanium*, 4th ed., Roskill Information Services, 1984
40. *The Economics of Germanium*, 5th ed., Roskill Information Services, 1988
41. H.J. Cluley, *Analyst*, Vol 76, 1951, p 523
42. J.R. Musgrave, Germanium, in *Treatise on Analytical Chemistry*, Part II, Vol 2, I.M. Kolthoff, P.J. Elving, and E.B. Sandell, Ed., John Wiley & Sons, 1962, p 208-245
43. V.A. Nazarenko, *Analytical Chemistry of Germanium*, N. Mandel, trans., John Wiley & Sons, 1974
44. *Electronics*, Vol 10.05, *Annual Book of ASTM Standards*, American Society for Testing and Materials
45. A. Furst, Biological Testing of Germanium, in *Toxicology and Industrial Health*, Vol 3, 1987, p 167-204
46. Threshold Limit Values for Chemical Substances and Physical Agents with Intended Changes for 1978, in *Proceedings of the American Conference of Governmental Industrial Hygienists*, 1978, p 19
47. H.E. Christensen, *Toxic Substances List*, National Institute for Occupational Safety and Health, 1972
48. J.H. Adams, *Eng. Min. J.*, Vol 178, March 1977, p 181
49. J.W. Litchfield, R.L. Watts, W.E. Gurwell, J.N. Hartley, and C.H. Bloomster, *A Methodology for Identifying Materials Constraints to Implementation of Solar Energy Technologies*, Battelle Memorial Institute, 1978
50. J.R. Piedmont and R.J. Riordan, The Supply of Germanium for Future World Demands, in *Proceedings of the Fourth European Electro-Optics Conference,* Vol 164, Society of Photo-Optical Instrumentation Engineers, 1978, p 216-222
51. D.B. Keck, P.C. Schultz, and F. Zimar, U.S. Patent 3,737,292, June 1973

Gallium and Gallium Compounds

Deborah A. Kramer, Division of Mineral Commodities, U.S. Bureau of Mines

GALLIUM-BASE COMPONENTS can be found in a variety of products ranging from compact disk players to advanced military electronic warfare systems. Compared with components made of silicon, a material gallium arsenide (GaAs) has replaced in some of these applications, components made of GaAs can emit light, have greater resistance to radiation, and operate at faster speeds and higher temperatures. However, GaAs components are more costly and more difficult to fabricate than those of silicon, so they are used only in applications where the advantage of their properties significantly outweighs their cost disadvantage.

Gallium occurs in very low concentrations in the crust of the Earth and virtually all primary gallium is recovered as a by-product, principally from the processing of bauxite to alumina. Most gallium applications require very high purity levels, and the metal must be refined before use until it contains no more than 1 ppm total impurities. Most gallium metal recovery and refining facilities are in Europe.

Complex processing techniques are used to produce single crystals of gallium and GaAs (Fig. 1); complex techniques also are required for the fabrication of gallium and GaAs optoelectronic devices and integrated circuits (ICs). Japan and the United States lead the world in GaAs crystal and device fabrication. Considerable investments are being made to increase processing efficiency, develop new devices, and increase the applications of current GaAs-base components.

Fig. 1 High-purity gallium single crystal. Approximate size, 75 to 100 mm (3 to 4 in.) high. Courtesy of INGAL International Gallium GmbH

Uses

Gallium has limited commercial applications in its metallic form. Its principal use is in the manufacture of semiconducting compounds, mainly GaAs and gallium phosphide (GaP). Over 90% of the gallium consumed in the United States is used for optoelectronic devices and ICs. Optoelectronic devices—light-emitting diodes (LEDs), laser diodes, photodiodes, and solar (photovoltaic) cells—take advantage of the ability of GaAs to convert electrical energy into optical energy and vice versa. The principal market for optoelectronic devices is in nonmilitary applications, including communications systems and consumer electronic goods. Gallium-arsenide-base integrated circuits are used primarily in defense applications, although developments in recent years have increased their use in the commercial sector. Gallium-arsenide-base integrated circuits are important, particularly in defense applications, because they can send information about five times faster, withstand more radiation, and operate at higher temperatures than comparable silicon-base integrated circuits.

Optoelectronic Devices

An LED is a semiconductor that emits light when an electric current is passed through it. Light-emitting diodes have been in commercial use for many years. The first commercial applications for LED technology were in displays for hand-held calculators and digital watches. Today, LEDs are used in visual displays in automobiles, calculators, appliances, consumer electronic equipment, and a wide variety of industrial equipment. Light-emitting diodes are also used as a light source in short-distance fiber-optic communications systems.

Light-emitting diodes consist of layers of an epitaxially grown material on a substrate. These epitaxial layers are normally gallium aluminum arsenide (GaAlAs), gallium arsenide phosphide (GaAsP), or indium gallium arsenide phosphide (InGaAsP); the substrate material is either GaAs or GaP. The materials used to fabricate LEDs determine the color of light that is emitted. With GaP substrates, the wavelength of light can cover the spectrum from 555 nm (pure green) to about 700 nm (red). With GaAs substrates, light emitted from the LED is limited to wavelengths at the red and infrared end of the spectrum.

Laser diodes operate on the same principle as LEDs, but they convert electrical energy to a coherent light output. Laser diodes, also called semiconductor lasers or injection laser diodes, principally consist of an epitaxial layer of GaAs, GaAlAs, or InGaAsP on a GaAs substrate. The two most commonly used laser diodes are GaAlAs and InGaAsP diodes.

Gallium aluminum arsenide laser diodes operate at about 780 to 900 nm and are used in a wide variety of consumer products and in communications systems. Consumer product applications for GaAlAs diodes include uses in compact disk players, nonimpact laser printers, and optical video disk players. Communications applications include short-range fiber-optic communications systems, satellite communications, radar transmission, and local cable television transmission systems. In 1988, GaAlAs laser diodes operating in the visible end of the spectrum (670 to 680 nm) were introduced to compete with helium-neon gas lasers in several applications. The principal application for GaAlAs visible laser diodes is in bar code scanners, although they can also be used in laser printers, pointers, and a variety of specialized applications (Ref 1).

Indium gallium arsenide phosphide laser diodes operate at longer wavelengths (1300 to 1500 nm). They are primarily used for transmitting high-frequency long-distance signals in fiber-optic communication systems and in cable television supertrunks.

Photodiodes, or detectors, are used to detect a light impulse generated by a source, such as an LED or laser diode, and convert it to an electrical impulse. Photodiodes are fabricated from the same materials as LEDs and are used primarily as light detectors in fiber-optics systems. There are two types of gallium-base photodiodes: Photodiodes of GaAlAs epitaxially grown on a GaAs substrate are used to detect light at short wavelengths, and photodiodes of InGaAsP on an indium phosphide (InP) substrate are used to detect light at longer wavelengths.

Because of its ability to convert light to electrical energy, GaAs is an excellent material for solar cells. Although solar cells are not in widespread use, they have been used to power communications satellites. The advantage GaAs has in this application is its electrical conversion efficiency; GaAs solar cells have been demonstrated to convert 22% of the available sunlight to electricity, compared with about 16% for silicon solar cells. Because of their higher energy efficiency, GaAs solar cells can be smaller than those constructed of silicon and still provide the same power to the satellite. Consequently, a satellite can carry a greater payload when GaAs is used as the solar cell material. Gallium arsenide is also more resistant to radiation than silicon; consequently, GaAs solar cells have a longer life in space environments (Ref 2).

One defense application of optoelectronic GaAs is in night vision equipment. The GaAs component converts infrared radiation to visible light, enabling soldiers to see at night. Four layers of GaAlAs are epitaxially deposited on a GaAs substrate. The substrate and two of the layers are removed, yielding a thin GaAlAs film. Fabrication of these devices is closely controlled to prevent defects in the crystal structure. Even small imperfections in the night vision device cannot be tolerated.

Substitute materials are available for GaAs-base devices in many of the optoelectronic applications. Liquid crystal displays (LCDs), organic compounds that change their light reflection and refraction properties when a current is applied, are the most common substitute for LEDs. For example, LEDs have virtually been replaced by LCDs in one of their original applications—digital watches.

The principal competition for gallium-base laser diodes is from InP devices; because InP devices emit light at a longer wavelength than GaAs devices, they are suitable for fiber-optic communications applications. However, InP technology is at an earlier stage of development than GaAs technology; thus, InP devices are more costly. Germanium- or silicon-base devices are the primary substitutes for gallium-base photodiodes. Competing materials for solar cells are silicon, copper indium diselenide, and cadmium telluride. Although these materials are not as efficient in thin-film solar cells as GaAs, they are generally less costly.

Integrated Circuits

Although ICs currently represent a smaller share of the GaAs market than optoelectronic devices, they are considered to have potential for greater growth. Two types of ICs are produced commercially: analog and digital. Analog ICs are designed to process signals generated by radar and military electronic warfare systems, as well as those generated by satellite communications systems. Digital ICs essentially function as memory and logic elements of computers.

Analog or microwave ICs are used principally in defense applications. Although silicon technology is preferred for signals at frequencies of 3 GHz or less, such as those in television, radios, and computers, silicon operates too slowly at higher frequencies. For these higher frequencies, up to 30 GHz, GaAs microwave ICs are used. One type of GaAs IC, the monolithic microwave integrated circuit (MMIC), combines several discrete components on one chip and can perform functions that used to require bulky circuits consisting of vacuum tubes and waveguides.

One application of GaAs MMICs is in phased-array radar systems. With the development of the GaAs MMIC, the size of radar components can be decreased significantly, with improved signal-to-noise ratios. In phased-array radar, the antenna elements are fixed in a matrix in a single plane rather than in a rotating dish, and they are steered electronically to allow objects in the sky to be individually identified. Because GaAs MMICs are small, the size of the antennas can be reduced significantly, perhaps enough to eventually enable phased-array radar systems to fit on an airplane. Current phased-array radar systems are too large for this application.

Another defense application of GaAs MMICs is in expendable decoys designed to provide fighter aircraft with protection against radar-directed antiaircraft missiles. These decoys contain a small radar transmitter and receiver that, when ejected from an aircraft, begin transmitting the same frequency of radar energy as that reflected from the aircraft, although at a higher strength. Thus, a radar-directed missile would home in on the decoy rather than the aircraft. Because of the small size of GaAs MMIC components in the radar, these decoys are only about 150 mm (6 in.) long and 25 mm (1 in.) in diameter.

Gallium arsenide MMICs are also a component of solid-state phased-array jammers. These jammers can be mounted on aircraft to receive and jam radar signals. With GaAs MMIC technology, the need for jamming pods is eliminated, providing space to carry additional weapons.

Gallium arsenide technology is in its infancy, and many defense applications for MMICs are still being developed. These military applications include space-based radar, missile seekers, "smart" munitions, and other electronic warfare devices, as well as navigation and communications equipment. Gallium arsenide MMIC technology may spread to the commercial sector; potential applications include direct-broadcast satellite receivers and business communications equipment such as cellular telephones.

Digital ICs. The first digital GaAs-base ICs were introduced into the U.S. market as off-the-shelf products in 1984. Consequently, their use in computer systems has thus far been limited. Most GaAs components have been used in high-speed supercomputers that are still being developed. Because of the high cost of GaAs as compared with that of silicon, GaAs digital ICs are not expected to replace silicon ICs in high-volume commercial applications such as personal computers. The use of GaAs digital ICs will instead be confined to operations in which large quantities of data must be interpreted in a very short time, such as in weather-forecasting and surveillance satellites. Gallium arsenide digital ICs also may be used in space-borne signal-processing applications, such as those required for implementation of the U.S. Strategic Defense Initiative.

Although digital GaAs components are not expected to replace silicon in most high-volume applications, two ICs introduced in 1988 were designed to be exact, yet higher-speed, replacements for silicon components. One device is a static random-access memory IC that can be used in supercomputers and as cache storage in less-powerful systems. The other is a programmable logic device for use in personal computers, workstations, network servers, and other applications (Ref 3).

Other Applications

Gallium is used in applications other than those in which its semiconductor properties are important. Gallium oxide is used in making single-crystal garnets for special applications. As used in the electronics industry, the term garnet refers to compounds of mixed M_2O_3 metal oxides. Gallium gadolinium garnet (GGG) is used as the substrate for a bubble memory device. The single crystals are produced by conventional means, and an epitaxial layer is added that contains rare earth oxides. These rare earth oxides provide magnetic domains, or bubbles, that can be oriented to store information and moved by an electric field for information readout.

Memory devices can be made with silicon materials, bubbles, or magnetic tape or disks. These three technologies compete on the bases of cost and size. Silicon devices are low in cost and small, but they require power to retain information; otherwise the data are volatile and subject to loss. Although bubbles are costlier and disks are bulkier than comparable silicon devices, they are nonvolatile, which makes them attractive for certain applications. Although not commonly used in most computer applications, GGG bubble memories are suited to dirty environments and environments that are subject to wide fluctuations in temperature. Commercial applications for GGG bubble memories include petrochemical data collection and plant machine control.

Gallium, scandium, and gadolinium oxides are used in another mixed-oxide single-crystal garnet. Gallium scandium gadolinium garnet (GSGG) has demonstrated efficiency as a laser host with potential inertial fusion energy applications.

Small quantities of metallic gallium are used for low-melting-point alloys, for dental alloys, and as components in some magnesium, cadmium, and titanium alloys. Gallium is also used in high-temperature thermometers and as a substitute for mercury in switches; it is suitable for these applications because it has the longest liquid range of any element. Gallium has additional uses in glasses and mirror coatings.

Properties and Grades

Gallium arsenide has several properties that give it advantages over silicon in many applications. These advantages are particularly prominent in optoelectronic applications. When stimulated by an electric current, GaAs gives off either visible or infrared light; silicon only gives off energy in the form of infrared radiation or heat. This makes GaAs a useful material for fabricating LEDs and laser diodes, applications for which silicon cannot be used. Both GaAs and silicon can convert light to electrical energy, which makes them useful for photodiodes and solar cells, but GaAs can convert more of the available light to electrical energy, making it more energy efficient.

In IC applications, the properties of GaAs make it especially suitable for defense applications. Gallium arsenide is about ten times more resistant to radiation than silicon. This resistance is essential in satellite operations in space, where components are exposed to damaging radiation from the sun. Gallium arsenide circuits also can operate at higher temperatures than can those of silicon: The maximum temperature is approximately 350 °C (660 °F) for GaAs and 275 °C (525 °F) for silicon. Therefore, the use of GaAs circuits reduces the need for bulky cooling equipment. Also, electrons move up to five or six times faster through GaAs than through silicon, making operation of GaAs-base circuits much faster than that of silicon circuits. High-speed circuit technology is growing in importance because defense applications are becoming increasingly sophisticated and require split-second decision-making capabilities. Typical physical properties of GaAs are provided in Table 1.

Table 1 Selected physical properties of GaAs

Property	Amount
Molecular weight	144.6
Melting point, K	1511
Density, g/cm^3	
At 300 K (solid)	5.3165 ± 0.0015
At 1511 K (solid)	5.2
At 1511 K (liquid)	5.7
Lattice constant, nm	0.5654
Adiabatic bulk modulus, dyne · cm^{-2}	7.55×10^{11}
Thermal expansion, K^{-1}	
At 300 K	6.05×10^{-6}
At 1511 K	7.97×10^{-6}
Specific heat, J · g^{-1} · K^{-1}	
At 300 K	0.325
At 1511 K	0.42
Thermal diffusivity at 300 K, cm^2 · s^{-1}	0.27
Latent heat, J · cm^{-3}	3290
Band gap, eV	1.44
Refractive index at 10 μm	3.309
Dielectric constant	
Static	12.85
Infrared	10.88
Electron mobility, cm^2 · V^{-1} · s^{-1}	
At 77 K	205 000
At 300 K	8500
Hole mobility at 300 K	400
Intrinsic resistivity at 300 K, Ω · cm	3.7×10^8

Source: Ref 4

All of these properties make GaAs attractive, but GaAs has several drawbacks that limit its use only to those applications where its properties are crucial. The first of these drawbacks is its high cost. Another drawback is that GaAs fabrication is a much more intricate process than silicon fabrication. It is much more difficult to grow a single-crystal ingot from two elements than it is from one, especially when the arsenic tends to diffuse out of the melt at temperatures lower than those required for GaAs crystal growth. Consequently, GaAs wafers have more imperfections in the crystal structure than silicon wafers, and these imperfections may adversely affect the electronic properties of a device constructed on the wafer. Gallium arsenide also has lower production yields than silicon. From ingot to usable wafers, GaAs has an effective yield of about 15%. Gallium arsenide wafers are brittle and subject to breakage during device fabrication, decreasing the effective yield still further.

Purity requirements for the raw materials to produce GaAs are stringent. For optoelectronic devices, the gallium and arsenic must be at least 99.9999% pure; for ICs, a purity of 99.99999% is required. These two purity levels are referred to by several names: 99.9999% pure gallium is often called 6 nines (6N) or optoelectronic grade, and 99.99999% pure gallium is called 7 nines (7N), semiinsulating, or IC grade. For 7N gallium, the total of the impurities must be less than 100 ppb. In addition to the difficulty of consistently producing material with such high purity, it is difficult to analyze for the small quantity of impurities. Certain impurities can cause more problems during GaAs production than others. The impurities of most concern are calcium, carbon, copper, iron, magnesium, manganese, nickel, selenium, silicon, sulfur, tellurium, and tin. Generally these elements should be present in concentrations of less than 1 ppb in both the gallium and arsenic. Lead, mercury, and zinc should be present in concentrations of less than 5 ppb. Although aluminum, chlorine, and sodium are often present, each of their concentrations should be less than 10 ppb.

Resources

Although gallium is as abundant in the crust of the Earth as lead, it is widely disseminated and is rarely found in concentrations greater than 0.1%. Consequently, gallium is nearly always recovered as a by-product during processing of ores or other materials to recover other metals. The principal materials in which gallium is found are bauxite, coal, phosphate ores, and sphalerite (zinc ore). Of these, gallium is currently commercially recovered during the processing of bauxite to alumina and the processing of sphalerite to zinc.

Bauxite generally is considered the best source of by-product gallium because gallium occurs in virtually all bauxites and is somewhat concentrated during the extraction of alumina from bauxite via the Bayer process. The gallium content of bauxite varies depending on the individual deposit, averaging about 50 ppm for the world. Bauxites containing high quantities of gallium, 70 to 80 ppm, are found in India, Suriname, and the United States. Figure 2 shows an estimate of the world gallium reserves available from bauxite, based on bauxite reserves, and the average gallium content of the bauxite in each country (Ref 5).

Although world gallium reserves of over 1 million metric tonnes (or megagrams, Mg) are available from bauxite, much of this bauxite will not be mined for many decades, and only about 40% of the available gallium is recoverable with current technology. According to U.S. Bureau of Mines projections, Australia, Jamaica, the Soviet Union, and the United States have the most potential for gallium recovery from bauxite.

Because gallium is not recovered at many of the alumina plants throughout the world,

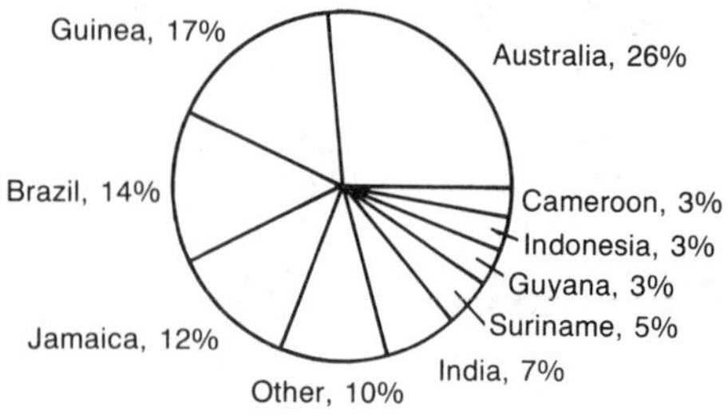

Fig. 2 World gallium reserves in bauxite. Total is 400 000 metric tonnes, based on a 40% recovery.

and because most of the gallium originally contained in the bauxite does not dissolve during the alumina extraction process, large quantities of gallium are discarded in the red mud residue. Although not currently considered a gallium resource material, the red mud residue represents a large potential gallium resource. Much of the residue is contained in tailings ponds near the alumina refineries and would be an easily accessible resource.

Zinc ores also represent a significant source of gallium, although not all zinc ores contain gallium. Sphalerite, a zinc sulfide mineral, generally contains detectable quantities of gallium, but little quantitative information is available to present an accurate assessment of the gallium potential of these ores. Based on the assumption that the average gallium content of sphalerite is 50 ppm, domestic sphalerite reserves of 21 $\times 10^6$ Mg contain 1050 Mg of gallium. Total world reserves of 147 $\times 10^6$ Mg of sphalerite may contain as much as 7350 Mg of gallium. The countries with the largest sphalerite reserves are Canada, the United States, and Australia. As with bauxite, much of this ore will not be mined for many decades and represents a long-term source of gallium.

Coal fly ash and phosphate flue dusts also contain gallium, but because of the availability of gallium from bauxite and sphalerite, it is unlikely that these materials would be used as principal sources of gallium. However, technology for the recovery of gallium from these materials has been developed.

Recovery Technology

Gallium Recovery From Bauxite. Throughout the world, alumina is recovered from bauxite by the Bayer process. In this process, alumina is extracted from bauxite through digestion with a hot caustic solution. After the slurry is cooled and solid residue is separated from the aluminum-containing liquor, the solution is seeded with alumina trihydrate crystals to precipitate the dissolved aluminum as alumina trihydrate. Alumina trihydrate is separated from the solution and calcined to produce alumina, and the caustic solution is recycled to the bauxite digestion step.

Because gallium is chemically similar to aluminum, it tends to remain with aluminum during processing. When the aluminum is extracted during digestion, gallium is also extracted. Gallium is not removed from the solution during subsequent processing steps, and because the solution is recycled, gallium builds up to an equilibrium concentration of 100 to 125 ppm. When gallium recovery is desired, a bleed stream is separated from the caustic solution before it is recycled to the digestion step. Crude gallium metal, 97.0 to 99.9% pure (3N), is recovered from the bleed solution by two principal processes: the Beja process and the de la Bretèque process. Simplified flowsheets for these processes are shown in Fig. 3 and 4.

In the Beja process, carbon dioxide is injected into the bleed solution to precipitate aluminum not recovered in the Bayer process as alumina trihydrate. The trihydrate is separated from the solution, and the gallium-containing solution is carbonated again. In the second carbonation, a gallium precipitate containing between 0.3 and 1% Ga is recovered. Both of the carbonation steps are carefully controlled so that about 90% of the aluminum is removed during the first carbonation and 90% of the gallium is precipitated during the second carbonation. After the gallium precipitate is separated from the solution, which is recycled to the Bayer process, the precipitate is dissolved in a caustic solution to increase the gallium-to-aluminum ratio. This solution is electrolyzed to recover crude gallium as a liquid.

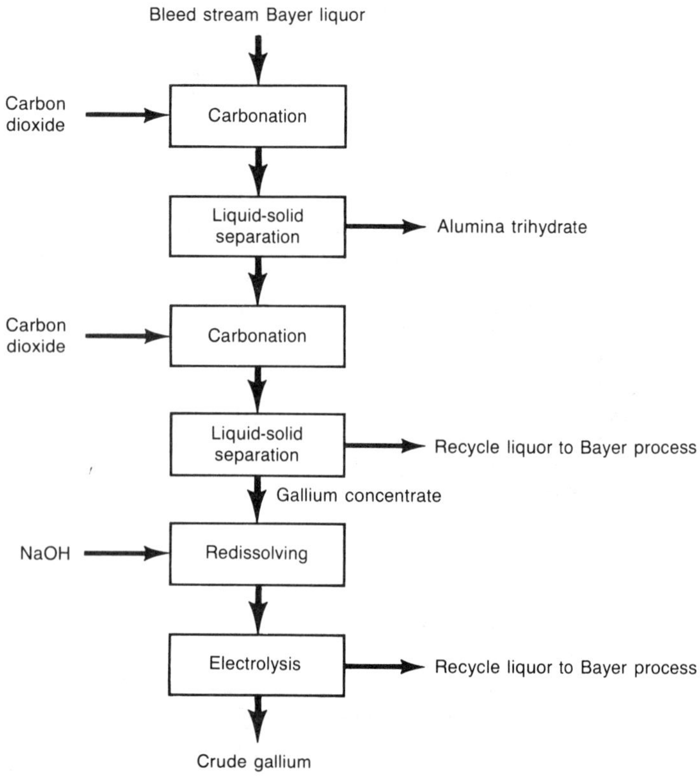

Fig. 3 Beja process for recovering crude gallium from Bayer liquors

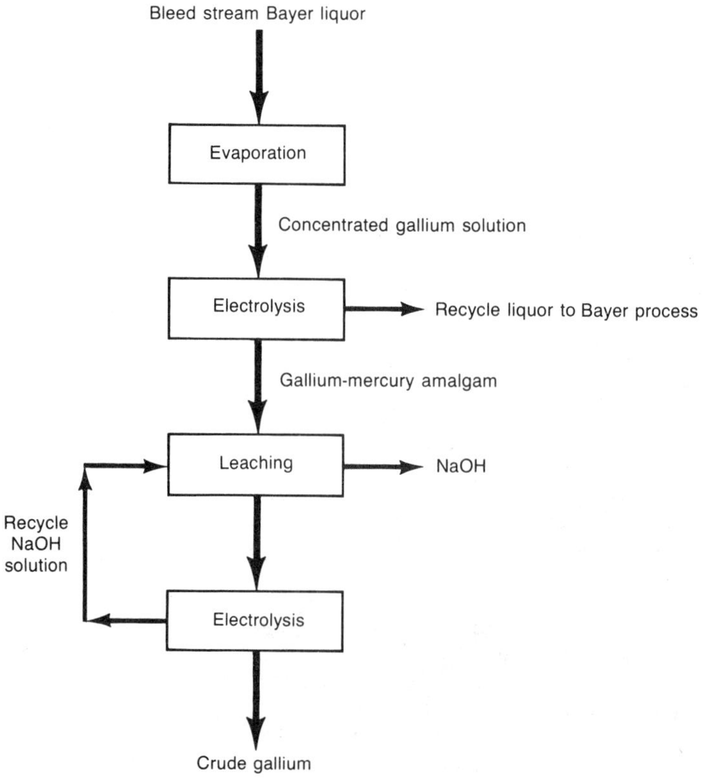

Fig. 4 The de la Bretèque process for recovering crude gallium from Bayer liquors

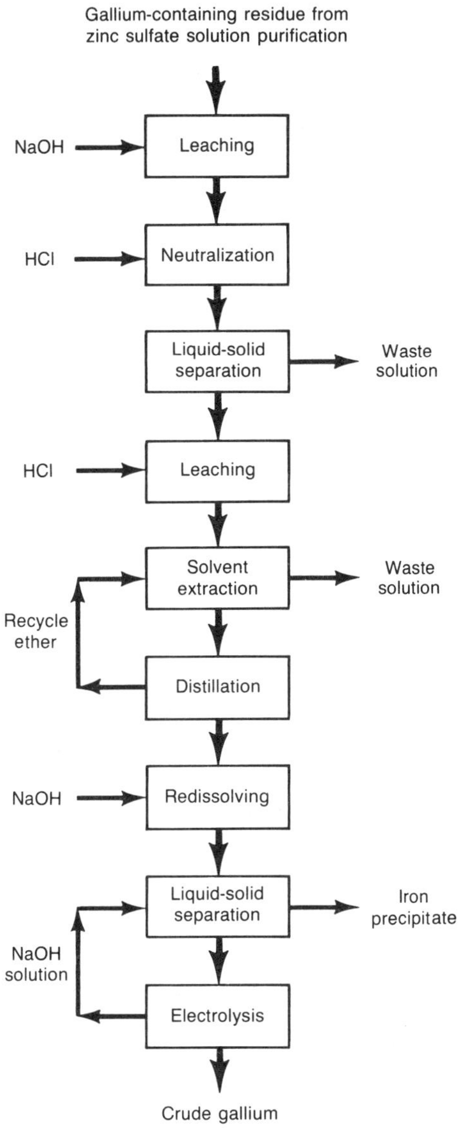

Fig. 5 Gallium recovery from zinc ore

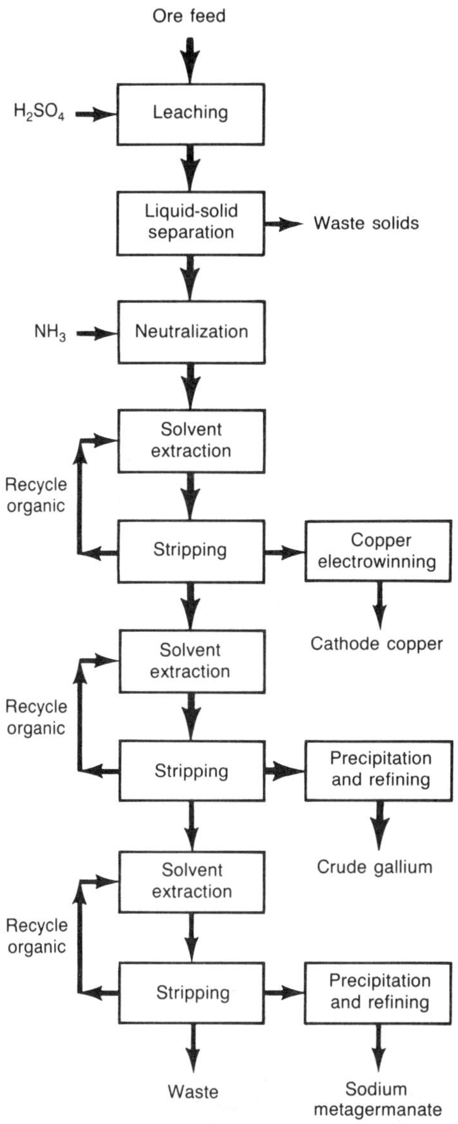

Fig. 6 Gallium recovery from the Hecla Mining Company mine near St. George, UT

The spent solution is recycled to the Bayer process (Ref 6).

In the de la Bretèque process, the bleed stream from the Bayer process is concentrated by evaporation to increase the gallium concentration. The concentrated solution is directly electrolyzed using a highly agitated mercury cathode. The agitation allows the gallium to form an amalgam with the mercury. When the gallium concentration reaches about 1% in the amalgam, it is drawn off and leached with a caustic solution. This yields a concentrated gallium solution from which crude gallium can be recovered by electrolysis (Ref 7).

Because mercury losses are significant owing to the high level of cathode agitation in the de la Bretèque process, a modification to the process was developed by Vereinigte Aluminium Werke AG (VAW) of the Federal Republic of Germany. Instead of a gallium-mercury amalgam prepared by electrolysis, this process uses a sodium-mercury amalgam prepared by electrolyzing the caustic solution with a mercury cathode. Gallium is then extracted by a cementation process as the gallium in solution replaces the sodium in the amalgam. Subsequent gallium recovery follows the same steps as in the de la Bretèque process.

Although the de la Bretèque process and the VAW modification are the most commonly used processes, several companies have developed proprietary recovery techniques that they claim are less costly than conventional processes. Rhône-Poulenc S.A. uses a liquid-liquid extraction technique at its plants in France and Australia to recover gallium from Bayer liquors. The Sumitomo Chemical Company of Japan uses an unidentified absorbent to extract gallium directly from Bayer liquors.

Gallium Recovery From Zinc Ore. Dowa Mining Company, the only company that currently recovers gallium from zinc ore, uses an electrolytic method for recovering zinc. In this method, a roasted zinc concentrate is leached with sulfuric acid to produce a zinc sulfate solution, which is neutralized to remove impurities. Impurities that precipitate from the zinc sulfate solution include gallium, aluminum, and iron. Leaching this residue with a caustic solution extracts the gallium, along with the aluminum and iron impurities. After the remaining residue is separated from the gallium-containing solution, the solution is neutralized to precipitate the metal hydroxides. The hydroxide solids are leached with hydrochloric acid to dissolve gallium and aluminum, and the gallium is separated from the aluminum in solution by solvent extraction with ether. Distillation of the ether solution yields a gallium-rich residue that still contains some iron, which is removed by treating the residue with a strong caustic solution that extracts the gallium and leaves the iron as a solid. Iron residue is filtered from the gallium-containing solution, and crude gallium is recovered by electrolysis. A simplified flowsheet for this process is shown in Fig. 5.

Gallium Recovery From Other Sources. Other sources of gallium that have been investigated include phosphate flue dust, coal fly ash, aluminum smelter flue dusts, and iron oxide minerals found in Utah. The only sources that have been commercially treated to recover gallium are the minerals in Utah and aluminum smelter flue dusts. In 1986, St. George Mining Corporation, a subsidiary of Musto Explorations Ltd., began recovery of gallium from an abandoned copper mine near St. George, UT. Although much of the copper had been mined, the remaining iron oxide minerals contained an average of 0.042% Ga. St. George Mining ceased operation in 1987 and filed for bankruptcy. Hecla Mining Company purchased the property in 1989 and restarted operations in 1990, after completing process development.

In the Hecla process (Fig. 6), copper, germanium, and gallium are separated from the ore by a sulfuric acid leach. After neutralization of the leach liquor with ammonia, separation of the three metals is accomplished by selective solvent extraction. Cathode copper, 4N gallium metal, and a sodium metagermanate concentrate are recovered with this process.

In July 1987, Elkem A/S of Norway began producing crude gallium by using aluminum smelter flue dust as a source material. Dusts generated at two smelters in Mosjoen and Tyssedal are blended to yield material with average concentrations of:

Element	Concentration, %
Carbon	33
Fluorine	17
Oxygen	17
Aluminum	13
Sodium	9
Iron	6
Sulfur	3
Calcium	1.5
Gallium	0.5

Leaching the flue dust with hydrochloric acid extracts the gallium. Solids are filtered from the liquid phase and mixed with portland cement before disposal. The liquid phase undergoes a series of solvent extraction stages to separate gallium from dissolved impurities. After cleaning and stripping, crude gallium is recovered by electrolysis of the water phase. A simplified flowsheet for this process is shown in Fig. 7.

Gallium Purification. Most applications for gallium require metal with a purity of either 6N or 7N. Crude gallium is purified in essentially two steps. The first step produces 99.99% pure (4N) gallium, and the second produces metal with a purity of 6N to 7N.

Many of the impurities in crude gallium occur in the surface oxide or as finely dispersed phases in the metal. Liquid gallium filtration and heating under vacuum remove these types of impurities. Metallic impurities can be reduced to less than 0.01%, producing 4N gallium, by sequential washing with hydrochloric acid. Another method that can be used is electrolytic refining, which involves anodic dissolution of gallium in an alkaline solution and then deposition at a liquid gallium cathode.

The principal method used to produce 6N and 7N metal is gradual crystallization of molten gallium. In this process, impurities remain in the liquid phase and do not contaminate the gallium crystal. Crystallization is repeated until gallium of the desired purity is obtained. Another method of producing high-purity gallium is to convert the gallium to a halide compound, such as gallium trichloride, which is then zone refined. High-purity gallium is recovered by electrolysis of the halide compound.

Fabrication of GaAs Crystals

Gallium arsenide single crystals are more difficult to fabricate (grow) than those of silicon. With silicon, only one material needs to be controlled, whereas with GaAs, a one-to-one ratio of gallium atoms to arsenic atoms must be maintained. At the same time, arsenic volatilizes at the temperatures needed to grow crystals. To prevent a loss of arsenic, which would result in the formation of an undesirable gallium-rich crystal, GaAs ingots are grown in an enclosed environment.

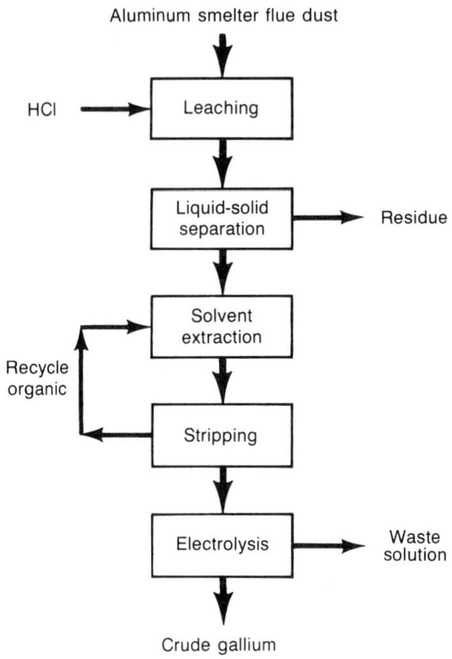

Fig. 7 Elkem A/S process for recovering gallium from aluminum smelter flue dust

Crystal Growth Methods. Two basic methods are used to fabricate GaAs single-crystal ingots. One is the boat-growth horizontal Bridgeman (HB) technique, which is also known as the gradient freeze technique; the other is the liquid-encapsulated Czochralski (LEC) technique. Ingots produced by the HB method are D-shaped and have a typical cross-sectional area of about 13 cm^2 (2 in.^{2}). Single-crystal ingots grown by the LEC method are round and are generally 75 mm (3 in.) in diameter, with a cross-sectional area of about 45 cm^2 (7 in.^{2}). Some 100 mm (4 in.) diam GaAs ingots have been produced by the LEC method, but they are not yet the industry standard.

In HB growth, gallium and arsenic in the proper ratio are placed in one end of a silicon dioxide (quartz) or pyrolytic boron nitride boat. A seed GaAs crystal is contained at the other end of the boat. The boat is placed in a sealed quartz tube, which is evacuated to a very low pressure. The tube is placed in a multiple-zone furnace, where the gallium and arsenic react to form GaAs. The compound is heated to 1240 °C (2265 °F), the melting point of GaAs. The GaAs melt is slowly cooled from the seed end, resulting in single-crystal growth.

In the LEC method of crystal growth, carefully weighed pieces of gallium and arsenic are melted in a pressurized vessel (crystal puller). The GaAs melt is contained in a crucible constructed of either high-purity quartz or pyrolytic boron nitride. The melt is covered with a layer of boric oxide, which retards arsenic loss from the melt by sublimation. A seed crystal is lowered through the boric oxide into the melt and slowly withdrawn as both the seed and crucible are rotating.

HB Versus LEC Ingots. Each of the two methods produces GaAs ingots with particular advantages and disadvantages. Crystals formed by the boat-growth HB method are particularly suitable for optoelectronic applications because their structure is highly perfect with respect to dislocations. Optoelectronic devices also require crystals with a high doping concentration, which boat-grown crystals readily provide because the silicon dissolved from the quartz boat contributes to crystal doping. However, the high doping concentration becomes a problem if semiinsulating GaAs crystals for ICs are being produced. The silicon impurities are called shallow donors, or *N*-type dopants. To compensate for these impurities, either a controlled quantity of gallium oxide can be added to the melt, or chromium (a deep acceptor, or *P*-type dopant) can be added. Also, crystal growth can be accomplished in a boron nitride container, which eliminates any contact with silicon-containing material during growth.

The shape of the HB-grown ingot makes it inconvenient for subsequent wafer processing because automated wafer-processing systems are designed to handle round wafers. Ingots grown by LEC generally contain more crystal structure defects, that is, they have higher dislocation densities, than HB-grown ingots. This affects the electronic properties of the device constructed on the GaAs. Dislocation densities in HB wafers normally run between 500 and 20 000/cm^2 (3000 and 130 000/in.^{2}); those in LEC wafers can be as large as 100 000/cm^2 (645 000/in.^{2}). Because chips are batch processed wafer by wafer, the larger the wafer, the more chips per wafer and the lower the cost per chip. A 75 mm (3 in.) diam LEC-grown wafer can yield more of the same-size chips than can a 50 × 40 mm (2 × 1.5 in.) HB wafer. Conversely, the capital costs for an HB system are significantly less than those for an LEC system.

Wafer Processing and Doping. After the ingots are grown, the ends are cut off, and the ingots are shaped by grinding the edges. Ingots are then sliced into wafers (Fig. 8). Wafers go through several stages of surface preparation, polishing, and testing before they are ready for device manufacture or epitaxial growth. Wafer preparation steps are done in a clean room and with minimal contact to avoid introducing surface contaminants. In LEC growth, the effective yield from starting material to finished wafers is currently less than 15%.

Pure GaAs is semiinsulating, which means that it is not a conductor of electricity. In order for GaAs to conduct electricity, a small number of atoms of another element must be incorporated into the GaAs crystal structure; the incorporation of these atoms is called doping. The atoms act as

Fig. 8 LEC-grown GaAs ingot and wafers. Courtesy of Morgan Semiconductor Division of Ethyl Corporation

electron donors or electron acceptors. Electron donor atoms have one more electron than the atoms that they are replacing, and this extra electron is free to move within the crystal as an electrical charge carrier. Electron acceptors have one less electron than the atoms they are replacing and behave as positively charged particles to serve as electrical charge carriers (Ref 8).

Before devices can be manufactured from GaAs wafers, the wafers must be doped with another metal or metals. Normally, this is accomplished either by ion implantation or by some type of epitaxial growth. Because GaAs is a semiinsulating substrate, no special isolation areas are required to separate each device fabricated on the chip. The lack of these isolation areas results in more-compact higher-density circuits, and adds to the speed advantage of GaAs.

In ion implantation, ions of another material are implanted into specific areas of the semiinsulating GaAs to make those areas electronically active. Areas of the chip that are to remain semiinsulating are covered with a photoresist mask before ion implantation. The process of ion implantation may be repeated several times with different metals on different areas of the chip, depending on the type and complexity of the device being manufactured. After ion implantation, the GaAs must be annealed at about 850 °C (1560 °F) to activate the implanted dopants and remove crystal damage incurred during implantation. When annealing, as in crystal growth, several techniques, including encapsulation of the chip, are used to prevent arsenic losses at the elevated temperature. After doping, optoelectronic device or IC manufacture can be completed through deposition of layers of metals and insulators by various techniques.

A similar technique, called ion cluster beam, is not used as frequently as ion implantation. In this technique, ions are grouped together and implanted into the wafer at lower speeds than those used in ion implantation. This process is reported to result in less damage to the crystal structure.

The deposition of an epitaxial layer is another means of creating electronically active regions on the GaAs substrate. There are four principal methods for growing epitaxial layers: liquid-phase epitaxy (LPE), vapor-phase epitaxy (VPE), metal-organic chemical vapor deposition (MOCVD), and molecular beam epitaxy (MBE). Liquid-phase epitaxy is generally not considered suitable for complex semiconductor production because it cannot be as precisely controlled as the other three techniques. In LPE, the substrate wafer is contained in a graphite boat within a quartz furnace tube, where it is contacted with solutions containing the metals to be deposited. Cooling the solution causes the metals to precipitate on the substrate. LPE produces relatively thick epitaxial layers, and the boundaries between layers are gradual rather than sharply defined.

Two methods of VPE are used to grow epitaxial layers on a GaAs substrate: the hydride method and the chloride method. In VPE, GaAs substrates are mounted in a reactor. To make GaAsP epitaxial layers, two gaseous streams are introduced into the reactor. In the hydride process, one gas stream combines arsine (AsH_3) and phosphine (PH_3) with a hydrogen carrier gas; the other gas stream is a hydrochloric acid gas that has been passed over a gallium reservoir to form gallium trichloride, and that also is mixed with a hydrogen carrier gas. Dopants are added to the gas streams if necessary. Gallium trichloride reacts with the AsH_3 and PH_3 gases to deposit a GaAsP layer on the substrate. In the chloride process, arsenic trichloride and phosphorus trichloride gases are substituted for AsH_3 and PH_3. Vapor-phase epitaxy technology can coat multiple wafers at the same time, and the layer thickness, molecular composition, and dopant concentration can be more closely controlled than with LPE.

In MOCVD, wafers are placed in a quartz reactor, which is maintained at atmospheric or slightly reduced pressure and at a temperature between 650 and 750 °C (1200 and 1380 °F). Metals to be deposited are in the forms of gases that chemically combine on the heated substrate. For example, to prepare a GaAlAs layer, gallium and aluminum are present in the form of organic gases, generally trimethyl or triethyl gallium and aluminum [$(CH_3)_3Ga$ or $(C_2H_5)_3Ga$ and $(CH_3)_3Al$ or $(C_2H_5)_3Al$] in a hydrogen carrier gas. Arsenic is in the form of AsH_3 in the hydrogen carrier gas. Dopants may also be added. The flow rates of these gases are carefully controlled. As the gases mix in the reactor and contact the hot wafers, they react to form GaAlAs and methane or ethane, and the GaAlAs deposits on the substrate wafers.

With MBE, the GaAs substrate is mounted on a heating block in a reactor maintained under a vacuum, along with effusion cells containing the elements to be deposited. For a GaAlAs layer, the effusion cells would contain gallium, aluminum, arsenic, and dopants. The elements are heated to temperatures that cause them to evaporate. By precise opening and closing of mechanical shutters in front of the effusion cells, the concentration of each element as it deposits can be carefully controlled.

With both MOCVD and MBE, the process can be repeated to build many thin layers of materials with differing compositions. After the epitaxial layers are deposited, device manufacture can be completed through deposition of metallic and insulating layers.

As with crystal growth methods, both MOCVD and MBE have advantages and disadvantages. Metal-organic chemical vapor deposition can coat multiple wafers at a time, whereas MBE systems can coat only one. Molecular beam epitaxy requires a vacuum, whereas MOCVD can be performed at atmospheric pressure. The cost of MOCVD equipment is approximately one-third the cost of MBE equipment. Molecular beam epitaxy provides the most precise control over the composition and thickness of the epitaxial layers, and it also provides the greatest reproducibility. In addition, MOCVD uses AsH_3 gas and therefore must be performed in a room equipped with safety equipment to prevent the toxic gas from escaping.

Secondary Recovery

Sources of Scrap. Because of the low yield in processing gallium to optoelectronic de-

Table 2 Estimated world primary gallium production

Country	Annual production, kg								
	1980	1981	1982	1983	1984	1985	1986	1987	1988
China	3 000	3 400	2 600	5 100	3 500	5 000	6 000	6 000	6 000
Czechoslovakia	500	1 650	1 700	2 000	2 500	3 300	3 000	3 200	2 000
France	4 300	4 600	3 700	7 000	8 500	9 500	15 500	14 000	12 000
Germany, Federal Republic of	2 300	3 000	4 000	5 300	6 000	5 500	7 000	7 000	7 000
Hungary	1 500	1 500	2 000	3 000	3 000	2 800	3 200	3 000	2 500
India	...	...	...	...	...	...	...	...	100
Japan	3 000	3 000	3 000	3 000	10 000	10 000	10 000	10 000	6 000
Norway	...	...	...	...	...	...	...	500	1 500
United States	3 000	1 500	1 560	...	...	...	750(a)	(b)	...
Total	**17 600**	**18 650**	**18 560**	**25 400**	**33 500**	**36 100**	**45 450**	**43 700(c)**	**37 100**

(a) Reported figure. (b) Withheld to avoid disclosing individual company proprietary data. (c) Excluding U.S. production

Table 3 Estimated world secondary gallium production

Country	Annual production, kg(a)								
	1980	1981	1982	1983	1984	1985	1986	1987	1988
Canada	...	...	...	...	...	...	...	5 000	5 000
Germany, Federal Republic of	500	500	700	1000	2 100	1500	1 500	1 500	2 000
Japan	3000	5000	4000	5000	7 000	4000	9 000	7 000	8 000
United Kingdom	200	300	800	1000	1 000	1000	1 500	1 500	1 000
United States	...	...	...	...	...	...	...	2 400	2 500
Total	**3700**	**5800**	**5500**	**7000**	**10 100**	**6500**	**12 000**	**17 400**	**18 500**

(a) New scrap only

vices or ICs, substantial quantities of new scrap are generated during the various processing stages. These wastes have varying gallium and impurity contents, depending upon the processing step from which they result. Gallium-arsenide-based scrap, rather than metallic gallium, represents the bulk of the scrap that is recycled.

During the processing of gallium metal into a GaAs device, waste is generated during the GaAs ingot formation. If the ingot formed does not exhibit single-crystal structure or if it contains excessive quantities of impurities, it is considered to be scrap. Also, some GaAs remains in the reactor after the ingot is produced and can be recycled.

During the wafer preparation and polishing stage, significant quantities of wastes are generated. Before wafers are sliced from the ingot, both ends of the ingot are cut off and discarded because impurities are concentrated at the tail end of the ingot and crystal imperfections occur at the seed end. These ends represent up to 25% of the weight of the ingot. As the crystal is sliced into wafers, two types of wastes are generated: saw kerf, which is essentially GaAs sawdust, and broken wafers. When the wafers are polished with an abrasive lapping compound, a low-grade waste is generated.

During the epitaxial growth process, various wastes are produced, depending on the growth method used. In LPE, metallic gallium contaminated with arsenic and dopant metals results, and in VPE, exhaust gases containing GaAs are produced. Because GaAs is a brittle material, wafers can break during the fabrication of electrical circuitry on their surfaces. These broken wafers can also be recycled.

The gallium content of these waste materials ranges from less than 1 to 99.99%. Wastes from LPE normally have the highest gallium content, 98 to 99.99%. Ingot ends and wafers broken during processing generally contain 39 to 48% Ga, VPE exhaust gases contain 6 to 15% Ga, saw kerf contains up to 30% Ga (wet basis), and lapping compound wastes contain less than 1% Ga. These wastes are contaminated with small quantities of many impurities, the most common being aluminum oxide, copper, chromium, germanium, indium, silicon, silicon carbide, tin, and zinc. Wafers broken during the fabrication of electrical circuitry also contain gold and silver impurities. In addition to metallic impurities, the scrap may be contaminated with materials introduced during processing, such as water, silicone oils, waxes, plastics, and glass.

In the processing of GaAs scrap, the material is crushed, if necessary, and then dissolved in a hot acidic solution. This acid solution is neutralized with a caustic solution to precipitate the gallium as gallium hydroxide, which is filtered from the solution and washed. The gallium hydroxide filter cake is redissolved in a caustic solution and electrolyzed to recover 3N to 4N gallium metal. This metal can be refined to 6N or 7N gallium by conventional purification techniques if desired.

Some GaAs manufacturers recycle their own scrap, or scrap may be sold to metal traders, to a company that specializes in recycling GaAs, or to the GaAs manufacturer's gallium supplier, who can recover the gallium and return it to the customer. In general, the prices commanded by GaAs scrap parallel the price fluctuations of 4N gallium metal. Also, prices are dependent on the type and gallium content of the scrap; saw kerf sells for a lower price than ingot scrap, which in turn sells for a lower price than metallic (LPE) scrap.

Although GaAs scrap is an important component in the flow of gallium materials throughout the world, it cannot be considered an additional long-term source of world gallium supply. Gallium arsenide scrap that is recycled is new scrap, which means that it has not reached the consumer as an end product and is present only in the closed-loop operations among the companies that recover gallium from GaAs scrap and the wafer and device manufacturers. Because this closed loop occasionally crosses international boundaries, it is difficult to distinguish between gallium recovered from scrap and virgin gallium when evaluating the gallium supply of an individual country.

World Supply and Demand

Little information is published detailing gallium production and trade data. The United States and Japan are the only countries for which detailed data are available. In many cases, no distinctions are made in published figures among virgin, recycled, and purified gallium. As an example, the United States ships some GaAs scrap to the Federal Republic of Germany for gallium recovery, and the recovered gallium is returned to the United States. This gallium may be counted twice as a part of the domestic supply. Or, one country recovers virgin gallium and ships it to a second country for refining to 7N gallium. Each country may count this as production, thus doubling the quantity of gallium that appears to be available. Consequently, the figures determined for gallium supply and demand are subject to significant interpretation.

Gallium Production. Tables 2 and 3 show estimates of both primary and secondary gallium production. These figures were derived from U.S. production data, published by the U.S. Bureau of Mines; U.S. import data, supplied by the U.S. Commerce Department; and production and import data for Japan, published in *Roskill's Letter From Japan*. Because most of the world's gallium demand is centered in Japan and the United States, these sources are believed to provide data on about 85% of the gallium produced in the world.

High-Purity Arsenic Production. Arsenic is recovered as arsenic trioxide in about 20 countries from the smelting or roasting of nonferrous metal ores or concentrates. Arsenic metal, which accounts for only about 3% of the world demand for arsenic, is produced by the reduction of arsenic trioxide. Commercial-grade arsenic metal, 99% pure arsenic, is produced in only a few countries, and this grade accounts for the majority of arsenic metal production. High-purity arsenic, 4N purity or greater, is used in the semiconductor industry.

Gallium Arsenide Ingot, Wafer, and Device Manufacturers. Table 4 lists companies involved in various phases of GaAs wafer and device manufacture. As is evident from the number of companies listed for these countries, most of the advanced GaAs manufacturing occurs in the United States and Japan. Some companies are fully integrated from GaAs ingot manufacture through device manufacture, whereas others make either wafers or devices.

Research and Development

Considerable research is being done concerning all phases of gallium extraction, GaAs material properties, and GaAs-based device manufacturing. Because GaAs IC manufacture is still in the developmental stage, much of the research activity centers on designing and manufacturing devices.

The U.S. Department of Defense sponsors a great deal of gallium research through Defense Advanced Research Projects Agency (DARPA) and the National Aeronautics and Space Administration (NASA), as well as through the laboratories of the service branches. Over the past few years, the focus of DARPA in funding projects has been to increase the efficiency of processing GaAs devices. Although a variety of microwave and digital ICs have been fabricated from GaAs, many of these are prototype devices. Projects funded through DARPA have been principally designed to increase the limited production of the prototype devices to full-scale manufacturing. Improving the manufacturing process may allow more complex ICs to be developed with increased radiation resistance and faster speed. The principal focus of NASA-funded research, on the other hand, has been the investigation of optoelectronic devices, particularly solar cells. The main thrust of this research is to increase the energy efficiency and reduce the cost of GaAs-base solar cells.

In 1986, the Department of Defense began a $135 million program to develop MMICs for military electronic applications. The program is called MIMIC, for microwave/millimeter wave monolithic integrated circuit. MIMIC would provide funds for companies that are already involved in GaAs research to accelerate their activities.

Most of the companies that are involved in the commercial GaAs market, both in optoelectronic devices and ICs, are involved in the development of devices that optimize the properties of GaAs. Among the new devices being developed are the high-electron-mobility transistor (HEMT), the heterojunction bipolar transistor (HBT), the ballistic transistor, and the quantum-well laser. HEMT consists of an undoped GaAs substrate with a thin epitaxial layer of silicon-doped GaAlAs on top. When an electric current is passed through the HEMT, electrons from the impurity atoms in the GaAlAs layer fall into the GaAs layer, where they move very fast. The HBT operates in essentially the same manner, but the GaAlAs layer is more highly doped. Both HEMTs and HBTs could increase the signal processing speed in MMICs and digital ICs.

A ballistic transistor is basically a sandwich structure with two GaAlAs layers on both sides of an ultrathin GaAs layer. As in an HEMT device, electrons from the GaAlAs layer fall into the GaAs layer and pick up speed. However, because the GaAs layer is so thin, electrons pass through the GaAs layer and into the second GaAlAs layer without slowing down. This enhanced electron movement could increase the speed of digital ICs and would allow MMICs to operate at high frequencies.

The quantum-well laser is fabricated in the same way and with the same material as the ballistic transistor, but instead of passing through the GaAs layer, electrons are trapped in this layer. By confining the charge carriers to this very small area, the chance is increased that they will recombine to emit light. Consequently, this structure increases the amount of light generated for a specific electrical signal (Ref 9). Development of these new devices has been made possible with the advent of MOCVD and MBE, which are capable of depositing ultrathin layers on a substrate.

With increased emphasis on developing new devices, demands have been placed on the GaAs substrate manufacturers to supply better-quality and more-uniform substrates. Consequently, GaAs wafer manufacturers have been refining their crystal growth techniques to produce material with fewer defects, to improve the yield from gallium and arsenic metals to GaAs wafers, and to increase the scale of production. At the same time, wafer manufacturers are trying to produce large-diameter wafers that ultimately could increase the yield from wafer to device.

Companies involved in epitaxial growth are also working to improve properties such as the uniformity in the thickness and composition of the epitaxial layers. Recently, metal-organic molecular beam epitaxy (MOMBE), also referred to as chemical beam epitaxy, has been developed to combine the advantages of MOCVD and MBE. These advantages include superior epitaxial layer thickness and uniformity, defect-free surfaces, the ability to grow layers on more than one wafer at a time, and the ability to introduce and control phosphorus atoms for optoelectronic device fabrication. The MOMBE technology was introduced in early 1987.

Work is also being done on combining GaAs with other materials to take advantage of the best qualities in each material. Prototypes of GaAs epitaxial layers grown on silicon substrates have been produced, and sample quantities have been shipped to customers for testing. By using GaAs layers on a silicon wafer, the superior structural properties of silicon can be combined with the electrical and optical properties of GaAs. Larger, more durable wafers can be produced with light-emitting properties and increased radiation resistance.

Two methods can be used for producing the combination GaAs-silicon wafers. Gallium arsenide can be deposited by MOCVD or MBE over the entire silicon wafer, a method called blanket epitaxy, or islands of GaAs can be epitaxially deposited on the silicon wafer, a method called selective epitaxy. Wafers produced by blanket epitaxy could replace bulk GaAs wafers for GaAs MMICs and digital ICs. Wafers produced by selective epitaxy can combine silicon ICs with GaAs optoelectronic devices, GaAs MMICs, or GaAs digital ICs. Blanket epitaxial wafers would require less gallium than that consumed in the fabrication of bulk GaAs wafers, and selective epitaxial wafers would allow GaAs to be used in areas in which its use is not currently feasible.

In solar cells, where GaAs has not supplanted silicon to any great degree, epitaxially deposited GaAs layers on germanium substrates may represent a hybrid substitute material for silicon. Gallium arsenide is fragile and can only be deposited in thick layers on a GaAs substrate. This puts GaAs

Table 4 Gallium arsenide ingot, wafer, and device manufacturers

Company	Ingot and wafer manufacture(a)		Epitaxy(b)				Device manufacture		
	LEC	HB	LPE	VPE	MOCVD	MBE	Optoelectronic	Analog	Digital
Canada									
Cominco Electronic Materials Ltd.	X								
France									
Picogiga						X			
The Philips Group				X					
Thomson CSF					X	X			
Germany, Federal Republic of									
Wacker Chemitronic AG	X	X							
Japan									
Dowa Mining Company	X								
Fujitsu Ltd.							X	X	
Furukawa Company Ltd.	X								
Hitachi Cable Ltd.	X	X	X	X					
Hitachi Manufacturing Company Ltd.							X	X	
Iwaki Company Ltd.	X								
Japan Victor Corporation							X		
Matsushita Electric Corporation							X	X	
Mitsubishi Electric Corporation						X		X	
Mitsubishi Metal Corporation	X	X		X					
Mitsubishi Monsanto Chemical Company Ltd.	X	X	X	X					
NEC Corporation		X	X						
Nippon Mining Company Ltd.	X								
Oki Electric Industry Company Ltd.		X							
Sanyo Electric Company Ltd.		X							
Sharp Corporation		X							
Shin-Etsu Semiconductor Corporation	X								
Showa Denko K.K.	X								
Stanley Electric Company Ltd.		X							
Sumitomo Electric Industries Ltd.	X	X	X	X					
Sumitomo Metal Mining Company Ltd	X	X							
Toshiba Corporation	X					X			
Sweden									
Semitronics AB	X								
United Kingdom									
General Electric Company (U.K.)			X		X	X	X	X	
MCP Electronic Materials Ltd.		X							
United States									
Airtron Division of Litton Industries	X								
Anadigics Inc.					X			X	X
Applied Solar Energy Corporation					X		X		
AT&T Bell Laboratories					X	X	X		X
Bertram Laboratories	X								
Crystal Specialties, Inc.		X	X	X	X				
Epitronics Corporation			X		X				
Ford Microelectronics Division of Ford Motor Company	X	X							
General Electric Company		X	X	X	X				
General Instrument Corporation	X			X					
GigaBit Logic					X	X			
Harris Microwave Semiconductor Corporation	X	X	X					X	
Hewlett Packard Company	X	X	X	X			X	X	
Honeywell Inc.					X	X	X		X
Hughes Aircraft Company	X						X	X	X
IBM Corporation								X	X
ITT Corporation					X		X	X	X
Kopin Corporation					X		X		
Laser Diode, Inc.		X					X	X	
M/A-Com Inc.	X	X	X						
McDonnell Douglas Corporation									X
Motorola Inc.						X	X		
Pacific Monolithics Inc.							X	X	
Rockwell International Corporation	X			X	X		X	X	
Siemens Corporation		X				X			
Spire Corporation				X		X			
Texas Instruments Inc.	X	X				X	X	X	
Triquint Semiconductor Inc.							X	X	
TRW Inc.		X			X	X	X		
Varo, Inc.				X		X			
Vitesse Semiconductor Corporation								X	
Westinghouse Electric Company	X		X	X			X		

(a) LEC, liquid-encapsulated Czochralski technique; HB, horizontal Bridgeman technique. (b) LPE, liquid-phase epitaxy; VPE, vapor-phase epitaxy; MOCVD, metal-organic chemical vapor deposition; MBE, molecular beam epitaxy at a disadvantage in comparison with silicon, which is sturdy and can be epitaxially grown in thinner layers. Germanium substrates are stronger and less costly than GaAs substrates, and GaAs epitaxial layers can be grown thinner using MOCVD. Consequently, the increased energy efficiency and radiation resistance of GaAs solar cells can be exploited while reducing the total weight of GaAs-based solar cells. While providing the same amount of power as silicon solar cells, GaAs-on-germanium solar cells can be made smaller, thereby allowing a satellite to carry a larger payload.

By continuing to push the limits of GaAs technology, researchers have also developed the optical equivalent of the transistor, a GaAs-base IC that controls light in the same manner a transistor controls electrical current. Thousands of alternating layers of GaAs and GaAlAs, each 40 atoms thick, are used in the construction of the IC. When a voltage is applied, the material becomes transparent, allowing a laser beam to shine through. A second, less-powerful laser beam concentrates the electrical voltage in certain layers, which become opaque. Thus the second laser beam controls the transmission of the first laser beam. The outgoing light beam from one device can then be used as an input for a second device. Development of these devices could be a step toward developing an optical computing device that would use light rather than electrical power to transmit information.

Basic research is being performed on the extraction of gallium from nontraditional source materials. The U.S. Bureau of Mines has investigated the extraction of gallium from phosphorus flue dust and low-grade domestic resources (Ref 10, 11). Work is also being done by private firms on the recovery of gallium from coal fly ash and phosphorus flue dust.

Strategic Factors

Despite the fact that gallium is currently being used in some sophisticated military and satellite systems, and is planned to be incorporated into additional systems, it has not been designated as a material to be added to the U.S. National Defense Stockpile. In 1986, government and private agencies assessed the need to stockpile gallium, but it was determined that in the event of a national emergency, gallium supplies would be adequate. If consumption increases dramatically, it is likely that this assessment would be reevaluated.

Because the United States produces only small quantities of gallium metal, imports must supply the bulk of the U.S. demand. This import dependence is likely to continue. A gallium extraction plant is scheduled to open in Utah in 1990, but this plant will not have the capability to produce 6N and 7N metal. Consequently, most of the high-purity metal will continue to be imported from Europe; small quantities may be purified in the United States.

With rapid technological progress, especially in GaAs IC development, the status of world supply and demand is changing dramatically. The status of GaAs has advanced in the past decade or so from that of a laboratory curiosity to that of a material with distinct applications and almost no effective substitutes at present. The development of fiber-optic telecommunications systems, the advent of sophisticated electronic military warfare, the widespread use of consumer electronics, and the need to process vast quantities of data in the shortest time possible have provided the impetus for implementing a large number of GaAs research and development programs. By continuing to push the limits of GaAs technology, its applications have expanded. At the same time, continuing research into developing other alternate materials, such as InP, superconductors, and organic polymer semiconductors, may yield materials with properties superior to those of GaAs. Development of these potential substitutes could radically alter the future of GaAs technology.

ACKNOWLEDGMENT

This article was reprinted with permission from ''Gallium and Gallium Arsenide: Supply, Technology, and Uses,'' Report IC 9208, U.S. Bureau of Mines, 1988.

REFERENCES

1. J. Dreyfuss, Visible-Wavelength Laser Diodes, *Lasers & Optronics*, Vol 7 (No. 8), Aug 1988, p 53-58
2. K. Zwiebel, Photovoltaic Cells, *Chem. Eng. News*, Vol 64 (No. 27), 7 July 1986, p 34-48
3. B.C. Cole, Special Report: This Time, GaAs Is For Real, *Electronics*, Vol 61 (No. 12), June 1988, p 65-81
4. J. Mun, *GaAs Integrated Circuits, Design and Technology*, Macmillan, 1988, p 6
5. F.E. Katrak and J.C. Agarwal, Gallium: Long-Run Supply, *J. Met.*, Vol 33 (No. 9), Sept 1981, p 33-36
6. M. Beja, Method of Extracting Gallium Oxide From Aluminous Substances, U.S. Patent 2,574,008, Nov 1951
7. P. de la Bretèque, Method of Recovering Gallium From an Alkali Aluminate Lye, U.S. Patent 2,793,179, May 1957
8. W.R. Frensley, Gallium Arsenide Transistors, *Sci. Am.*, Vol 257 (No. 2), Aug 1987, p 80-87
9. H. Brody, Ultrafast Chips at the Gate, *High Technol.*, Vol 6 (No. 3), March 1986, p 28-35
10. J.C. Judd, M.P. Wardell, and C.F. Davidson, Extraction of Gallium and Germanium From Domestic Resources, in *Light Metals 1988*, The Metallurgical Society of AIME, 1987, p 857-862
11. D.L. Neylan, C.P. Walters, and B.W. Haynes, Gallium Extraction From Phosphorus Flue Dust by a Sodium Carbonate Fusion-Water Leach Process, in *Recycle and Recovery of Secondary Metals*, The Metallurgical Society of AIME, 1986, p 727-733

SELECTED REFERENCES

- M.H. Brodsky, Progress in Gallium Arsenide Semiconductors, *Sci. Am.*, Feb 1990, p 68-75
- Leroy L. Chang and Klaus Ploog, Ed., *Molecular Beam Epitaxy and Heterostructures*, Kluwer, 1985
- *IEEE Gallium Arsenide Integrated Circuit Symposium*, Institute of Electrical and Electronics Engineers, 1988
- *Packaging*, Vol 1, *Electronic Materials Handbook*, ASM INTERNATIONAL, 1989
- S.M. Sze, *Semiconductor Devices, Physics and Technology*, John Wiley & Sons, 1985

Indium and Bismuth

Laurence G. Stevens and C.E.T. White, Indium Corporation of America

INDIUM and BISMUTH, although distinct in several properties and areas of application, have relatively low melting temperatures and some common areas of application. One common application area, for example, is in low-melting-temperature solders (that is, solders with a melting point or range below the tin-lead eutectic temperature of 183 °C, or 360 °F). These solders fall into the general group of either indium-base solders or bismuth-base solders.

Another common and significant use of bismuth and indium is in fusible alloys. Most fusible alloys contain large percentages of bismuth and occasionally some indium. However, the largest application of bismuth is in chemical products such as cosmetics, pharmaceuticals, and industrial or laboratory chemicals. Indium is primarily used in solders and fusible alloys, although it has found increasing application in other technologies such as semiconductors and solar cells.

Indium

Indium was discovered in 1863 by F. Reich and H.T. Richter at the Freiburg School of Mines in Germany while they were checking local zinc ores for thallium by spectrograph. The new element was named indium from the characteristic indigo blue lines of its spectrum. Subsequent work on flue dusts from the zinc works in Goslar, Germany yielded the pure metal, which was first exhibited in 1867. Indium, however, remained a laboratory curiosity until the 1920s.

Occurrence. Indium does not occur naturally in concentrated deposits. It is widely distributed in nature in the form of minerals, although generally in very low concentrations. The crust of the earth has been estimated to contain 0.1 ppm of indium, which means that the element has about the same relative abundance as silver.

The geochemical properties of indium are such that it tends to occur in nature with base metals of groups II-B and IV-A of the periodic table. Historically, indium has been recovered almost exclusively as a by-product of zinc produced from sphaleritic and marmatitic zinc ores. It has, however, been reported in other base metal ores, particularly those of copper, lead, and tin. The association of indium with zinc ores tends to be in the high-temperature sulfide deposits typically found in the western United States. The lower-temperature strataband deposits of the Mississippi Valley area are, in general, barren of indium. Indium has been found in ores from many countries besides the United States and Germany, including Australia, Bolivia, Canada, Finland, Italy, Japan, Peru, Sweden, and the Soviet Union. Most ores contain less than 0.001% In, and many contain less than 0.0001% In. Like many of the rarer metals, indium becomes concentrated in by-products during the recovery of the major metals. Because indium is most frequently associated with zinc, commercial production comes from zinc residues, slag, flue dusts, and metallic intermediates in zinc smelting and associated lead smelting.

Recovery. A variety of methods are employed in the recovery of indium, depending on the source material and its indium content. Among the more common methods are leaching the indium-containing by-product in sulfuric or hydrochloric acid, purifying the leach solution with indium strips, and sponging the crude indium on zinc or aluminum sheets. Solvent extraction with diphosphoric (2-ethylhexyl) acid or tributyl phosphate is effective in removing indium from dilute solutions. Another recovery procedure involves precipitating indium phosphate selectively from slightly acidic solutions, converting the phosphate to the oxide by leaching in a strong caustic soda solution, and then reducing the oxide to metal. Indium that is distilled with zinc in zinc retort smelting processes concentrates in the zinc-lead bottom metal during the first-stage evaporation and reflux purification of zinc, and can be separated as a high-grade slag. Indium is recovered from the slag by leaching and sponging on zinc or aluminum. The decline in production of zinc by zinc retort smelting processes has reduced the importance of this procedure in the recovery of indium. Other techniques have been developed for more-complex feed materials.

Sponge indium generally is from 99.0 to 99.5% pure and requires upgrading for most uses, particularly for those in the semiconductor industry. Refining techniques involve soluble-anode electrolysis. Suitable addition agents are required to obtain a satisfactory deposit. For the highest-purity grades used for compound semiconductors, refining is supplemented by other methods. Purities of 99.97, 99.99, 99.999, and 99.9999% are obtained.

Production. The first reported production of indium was in 1867, when H.T. Richter, one of its discoverers, exhibited 500 g (1.1 lb) that had been extracted from the flue dust of the Goslar, Germany zinc works. Only laboratory quantities were produced during the next 50 years, and it was not until 1926 that any appreciable amount of indium metal was produced. In that year, Dr. William S. Murray, who later founded the Indium Corporation of America, produced 7¾ kg (250 troy oz) of metal from a complex sulfide ore at Kingman, AZ. Production from this general area totalled about 1 metric ton (32 000 troy oz) in the period from 1926 to 1934. The Anaconda Company began production of indium from its Montana properties in 1934, with a peak production rate of 2.5 metric tons (82 000 troy oz) in 1944.

Additional uses developed in the period after the Second World War, and total production had climbed to 5.0 metric tons (160 000 troy oz) per year by 1950 with the advent of production from the American Smelting and Refining Company (ASARCO). In 1945, the then Cerro De Pasco Corporation began producing indium in Peru and reached a production level of 1.2 metric tons (40 000 troy oz) per year by 1948. Canadian production commenced in 1942 in research quantities and reached substantial levels in 1955 when Cominco Limited completed a new plant with a capacity in excess of 15.5 metric tons (500 000 troy oz) per year. Subsequently, production of indium was commenced on a commercial scale in West Germany (Preussag AG), Belgium (Métallurgie Hoboken-Overpelt S.A.-N.V.), France and Italy (Ste Miniere et Métallurgique de Penarroya), Japan (Nippon Mining Company, Dowa Mining Company, Mitsui Mining & Smelting Company, and Sumitomo Metal Mining Company), East Germany (V.E.B. Bunt Metal), the Netherlands (Billiton B.V.), the United Kingdom (Capper Pass & Sons Ltd.), and the Soviet Union (various government combines). Production from the People's Republic of China has recently appeared on the market. New and/or increased production in Canada by Cominco Ltd. and the Falconbridge Ltd./Indium Corporation has been announced and will commence in the early 1990s.

Few production figures are reported by producers, and total world reported production includes estimated outputs for various countries. In 1973, estimated worldwide indium production reached a peak up to that point of 60.4 metric tons (1 941 000 troy oz). Subsequent production levels have fluctuated from a low of 40 metric tons (1 300 000 troy oz) in 1982 to the 1989 estimate (Ref 1) of close to 100 metric tons (3 200 000 troy oz).

Pricing History. The price of indium was first quoted in the "Metal and Mineral Markets" section of the Engineering and Mining Journal in September 1930; the price was set at $15 per gram. The price gradually decreased as demand developed and additional supplies became available. The quoted price was $72 per kg ($2.25/troy oz) in 1945. Prices for the basic commercial grade (99.97% pure metal) held in the range of $48 to $72 per kg ($1.50 to $2.25/troy oz) until 1972. After 1972, prices began an upward trend and peaked at close to $320 per kg ($10.00/troy oz). Recently, however, prices have fallen because of increasing supplies. The most recent quotations set the price at $230 to $250 per kg ($7.15 to $7.80/troy oz). The higher-purity grades command premiums depending on the final purity desired.

Properties. Indium (atomic number 49) is in subgroup III-A of the periodic table and is a silvery-white metal with a brilliant metallic luster. Indium is softer than lead (it can be scratched with the fingernail, for example) and is highly malleable and ductile. The highly plastic nature of indium permits almost indefinite deformation under compression. Its elongation is abnormally low because indium does not work harden. Indium retains its plasticity and ductility even under cryogenic conditions.

Among the other interesting properties of indium is the wide spread between its melting point (156.6 °C, or 313.9 °F) and its boiling point (2080 °C, or 3775 °F) and its ability to wet glass, quartz, and many ceramics. As an additive, indium tends to harden and strengthen tin- and lead-based solders; its most marked effect is on lead-base solders. Another feature of indium is the improved thermal fatigue resistance obtained in the binary lead-indium system and the ternary lead-silver-indium system as compared to that of the lead-tin and lead-tin-silver systems.

The chemical properties of indium are largely determined by its position in group III-A of the periodic table. The properties of elements in this subgroup are determined by the behavior of the incomplete outer electronic shell, which consists of 2 S electrons and 1 P electron; thus, principal valences of 1 and 3 may be anticipated. The increasing stability of the S electrons in the higher atomic numbers of this subgroup point to characteristic valences of 1 for the higher atomic numbers and 3 for the lower numbers. Indium, which is in an intermediate position in the subgroup, displays both valences, but its most common valence is 3.

Typical mechanical, thermal, and electrical properties of pure indium are given in the article "Properties of Pure Metals" in this Volume. In the electromotive series, indium falls between thallium and cadmium with a standard potential of 0.34 V. It is readily attacked by hot mineral acids but not by water or alkalies.

Initial Areas of Application. The first commercial use of indium was in the production of dental alloys. This was initiated in 1933, when the Williams Gold Company of Buffalo, NY, in conjunction with Dr. William S. Murray, developed a series of gold-base dental alloys containing indium. Indium acts as an oxygen scavenger, resulting in alloys with improved tensile strength, ductility, and resistance to discoloration. Indium is still used in dental alloys.

Dr. Murray originally was interested in using indium as an addition to silver-plated flatware. During the period from 1926 to 1934, he and his associates developed a series of indium plating baths. In 1934, the Indium Corporation of America was incorporated to carry on the work. The original indium plating baths were of the alkaline cyanide type, but most plating at the present time is done from a sulfamate type bath that has fewer environmental restrictions. Other baths are occasionally used, including fluoroborate and sulfate baths.

The flatware application did not prove successful, but other applications of indium plating, in particular the plating of bearings, became important. Indium was applied in the form of an electroplate on the lead layer of steel-backed silver-lead bearings; the plate was then diffused into the lead layer. The indium addition gave the bearing improved strength and hardness, increased corrosion resistance, and improved antiseizure properties. This type of bearing found extensive use in aircraft piston engines during World War II, but its use has declined since the advent of jet engines. Nevertheless, indium still finds use in high-performance engine components such as the crankshaft bearing shown in Fig. 1.

It was noted as early as 1935 that the addition of indium to low-melting alloys such as Wood's and Lipowitz's metals caused the melting point to drop 1.45 °C (2.6 °F) for each 1% of indium added, with a minimum melting point of 47 °C (117 °F) reached at an indium level of 19.1%. Based on these initial observations, a series of alloys has been developed that contain bismuth as the major component to which indium additions are made. Melting points of these alloys extend from 47 to 146 °C (117 to 295 °F). These alloys are used in lens blocking and in temperature overload devices such as safety links, fuses, and sprinkler plugs. These are now classified as members of the family of fusible alloys, which are described in more detail in the section "Bismuth" in this article.

Low-Melting-Temperature Indium-Base Solders. Besides being used as a strengthening agent in lead-base solders, indium is also used as a base material in low-melting-temperature solders. Many indium-containing solders have been developed with enhanced properties such as reduced gold scavenging and resistance to thermal fatigue and alkaline corrosion.

Table 1 lists various indium-base solder alloys. These alloys are, in general, easy to fabricate. They are available as preforms (discs, squares, rectangles, washers, and spheres) and as wire, ribbon, and foil. They are also available in solder creams of either dispensing or screen-printing quality.

The 52% In solder has a low melting point (118 °C, or 244 °F), making it useful for the assembly of devices that are susceptible to temperature damage if conventional solders are used. It will also wet glass, quartz, and many ceramics, which makes it valuable for glass-to-metal seals. The series of solders containing 19 to 80% In (Table 1) have melting points ranging from 147 to 280 °C (117 to 535 °F). In comparison with conventional tin-lead solders, these indium-lead solders have improved thermal fatigue characteristics, and they greatly reduce the scavenging of gold surfaces. The temperature range is wide enough to permit two- or even three-step soldering. The 5% In solder is a high-temperature silver-bearing solder with good thermal fatigue properties. It is used extensively in the assembly of diodes and rectifiers.

The application techniques for indium-base alloys are similar to those used for the conventional tin-lead solders. In the case of preforms, oven heating is used for short runs, and conveyor-type furnaces are used for large runs. In special cases, the use of induction heating, heat guns, or reducing atmospheres is recommended. Vapor-phase soldering with indium-base alloys continues to gain in importance, particularly for the joining of back-plane connector pins to printed boards. Wave soldering has been performed satisfactorily with these alloys, but indium-base solders tend to dross slightly more than do tin-lead alloys.

Indium-base solders are generally considered to be specialty solders. They possess

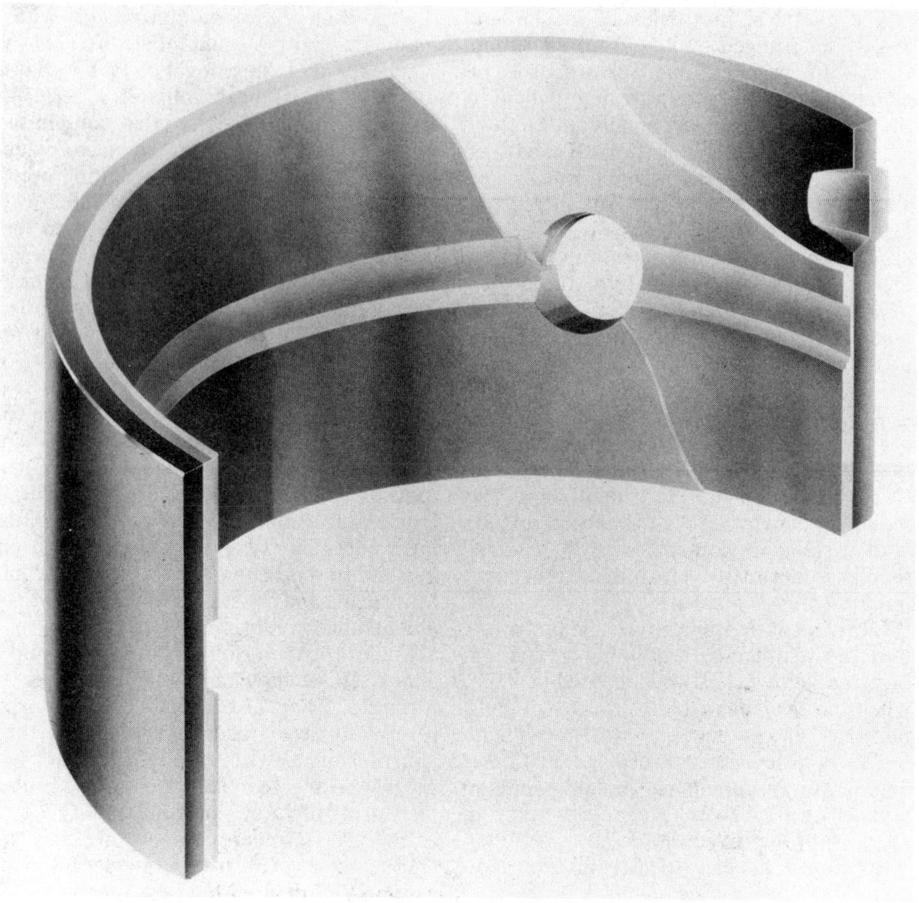

Fig. 1 Indium-plated crankshaft bearing for a high-performance reciprocating engine. The indium is applied as an electroplate on a lead-bronze shell. Courtesy of Vandervel America, Inc.

special properties that make them valuable for specific applications such as those described below.

Glass-to-Metal Seals. Pure indium, the 52In-48Sn alloy, and the 97In-3Ag alloy will wet glass, quartz, and many ceramics. Therefore, they find use in glass-to-metal seals; also, because of their low vapor pressure, they are useful as seals in vacuum systems. They retain their plasticity down to liquid-helium temperatures and thus can be used for sealing cryogenic systems.

Resistance to Thermal Fatigue. The indium-lead and indium-lead-silver alloys listed in Table 1 have a much greater resistance to thermal fatigue than do the conventional lead-tin solders. This advantage, coupled with the marked reduction in the scavenging and leaching of gold surfaces associated with these alloys, has led to their use in electronic assemblies.

Silver-Palladium Compatibility. Conventional tin-lead solders cannot be used to solder in silver-palladium metallizations when high service temperatures are required because of the formation of the palladium-tin (PdSn) intermetallic. When tin-lead solders are used, a significant reduction in adhesion strength can occur at a temperature of 150 °C (300 °F) in a few hundred hours, leading to ultimate failure of the solder bond. Lead-silver-indium solders are compatible with silver-palladium conductors at service temperatures up to 200 °C (400 °F).

Large Temperature Differentials. The lead-indium composition is a solid-solution system (noneutectic) with an available temperature range from the melting point of indium (156.6 °C, or 313.9 °F) to the melting point of lead (327 °C, or 621 °F). This wide range permits a choice of alloys with temperature differentials large enough for step soldering.

Corrosion Resistance. Indium-base solders have good resistance to alkaline corrosion. However, corrosion resistance in the presence of traces of halide ions is not satisfactory, necessitating the use of hermetic seals or conformal coatings.

Other Applications. Besides solder and fusible alloys, indium is used in a variety of other applications. Nuclear reactor control rods containing 80% Ag, 15% In, and 5% Cd were developed in the 1950s and have been used in the majority of pressurized water reactors built since that time. Other important applications are described below.

Sodium Lamps. A major application of indium in Europe has been in the manufacture of low-pressure sodium lamps, which are mainly used for outdoor lighting. The indium is applied as an oxide coating on the inside of the glass cylinder that forms the outer envelope of the lamp. The coating reflects the infrared waves emitted by the lamp while letting the visible light pass through, thus permitting the lamp to operate at a higher temperature. This yields an improved lumens-per-watt efficiency in comparison with that of conventional incandescent lamps. The lamp emits an orange light and has not found favor in the United States, where the blue-white light of the mercury vapor lamp is preferred.

Conductive films of indium oxide and indium-tin-oxide on glass find many applications. These include conductive patterns for liquid crystal displays (LCDs) and windshield defoggers and deicers as well as transparent electrodes for flat-panel displays. The high transparency of these coatings to visible light makes them well suited for these applications. Because these coatings reflect infrared radiation while passing visible light, they have the potential to be used for residential and commercial building glass.

Seals and Gaskets. The softness and plasticity of indium make it an excellent material not only for solder applications but also for gaskets and seals. Indium has the ability to work into the oxide skin of other metals and thus can improve the electrical and thermal conductivity of junctions of these metals while acting as a metallic seal against corrosion.

Semiconductors. With the invention of the germanium transistor in 1946, a major new market for indium developed in the production of alloy junction transistors. In these transistors, indium is used in the P-N junction, which is formed by alloying discs or spheres of indium into a wafer of N-doped germanium. This use of indium peaked around 1969 and 1970, but it has declined since then because germanium has been replaced by silicon in most semiconductor applications. Germanium semiconductors are now produced only for replacement purposes.

Intermetallic semiconductors formed with indium and group V elements such as antimony, arsenic, and phosphorus have received considerable attention in recent years. Indium antimonide has been used for infrared detectors but has been limited to military applications because it does not develop optimum parameters until cooled to liquid-nitrogen temperatures (77 K). Indium phosphide has more potential for commercial applications, such as a quaternary laser diode containing indium-gallium-arsenic-phosphorus. Gunn effect diodes, IMPATT (impact avalanche transit time) diodes, and millimeter wave oscillator circuits are under development.

Solar Cells. Indium phosphide and indium-copper-diselenide/cadmium sul-

Table 1 Indium-base and indium-alloyed solders

Alloy designation in ASTM B 774(a)	Nominal composition, wt%(a) In	Pb	Sn	Other	Liquidus temperature °C	°F	Solidus temperature °C	°F	Electrical conductivity, % of copper	Thermal conductivity at 85 °C (185 °F), W/m · °C	Thermal coefficient of expansion at 20 °C (68 °F), ppm/°C	Tensile strength MPa	ksi	Bond holding strength MPa	ksi	Applications
Indium-base solders																
···	100	···	···	···	157	313	157	314	24.0	0.78	29	4	0.575	6	0.89	Nonmetallic, special joining
290	97	···	···	3 Ag	143	290	143	290	23.0	0.73	22	5.5	0.80	···	···	Nonmetallic, microwire, special joining
···	95	···	···	5 Bi	150	302	125	257	···	···	···	···	···	···	···	···
···	90	···	···	10 Ag	237	459	141	285	22.1	0.67	15	11.4	1.65	11	1.6	Nonmetallic, special joining
300–302	80	15	···	5 Ag	150	302	143	290	13	0.43	10	17.6	2.55	14.8	2.15	General and multipurpose, microwire
···	74	···	···	26 Cd	123	253	123	253	···	···	···	···	···	···	···	···
320–345	70	30	···	···	174	345	160	320	8.8	0.38	28	23.8	3.45	···	···	General and multipurpose, microwire, microcream
···	70	9.6	15	5.4 Cd	125(b)	257(b)	···	···	···	0.39	27	10	1.47	13.8	2	···
···	60	40	···	···	185	365	174	345	7.0	0.29	27	28.6	4.15	···	···	Multipurpose, microwire, microcream
244	52	···	48	···	118	244	118	244	11.7	0.34	20	11.9	1.72	11.2	1.63	Nonmetallic, multipurpose, special joining
···	50	···	50	···	125	257	118	244	11.7	0.34	20	11.9	1.72	11.2	1.63	Nonmetallic, general purpose
···	50	50	···	···	209	408	180	356	6.0	0.22	27	32.2	4.67	18.5	2.68	General and multipurpose, microwire, microcream
···	44	···	42	14 Cd	93	200	93	200	···	0.36	24	18.1	2.63	···	···	Special joining, general purpose
···	40	20	40	···	130	266	120	250	···	···	···	···	···	···	···	···
Indium-alloyed solders																
···	40	60	···	···	225	437	195	383	5.2	0.19	26	35	5	···	···	Multipurpose, microwire, microcream
174	26	···	17	57 Bi	79	174	79	174	···	···	···	···	···	···	···	Low-temperature eutectic
···	25	75	···	···	264	508	250	482	4.6	0.18	26	37.6	5.45	24.3	3.52	General and multipurpose, microwire, microcream
···	25	37.5	37.5	···	181	358	134	274	7.8	0.23	23	36.3	5.26	30	4.3	Multipurpose
136	21	18	12	49 Bi	58	136	58	136	2.43	···	12.8	43.4	6.3	···	···	Special joining
···	20	26	54	···	144	291	135	275	···	···	···	···	···	···	···	···
···	19	81	···	···	280	536	270	518	4.5	0.17	27	38.3	5.55	···	···	General and multipurpose, microwire
117	19	22.6	8.3	44.7 Bi	47	117	47	117	···	···	···	···	···	···	···	Low-temperature eutectic
307–323	12	18	70	···	162(b)	324(b)	···	···	12.2	0.45	24	36.7	5.32	29	4.19	General purpose
···	5	95	···	···	314	598	292	558	5.1	0.21	29	30	4.33	22.2	3.22	···
···	5	90	···	5 Ag	310	590	290	554	5.6	0.25	27	39.5	5.73	22	3.18	···
···	5	92.5	···	2.5 Ag	300(b)	572(b)	···	···	5.5	0.25	25	31.4	4.56	19.5	2.83	General purpose, special joining, microwire

(a) For alloys specified in ASTM B 774, impurity limits are 0.001% Ag and 0.01% Bi max (when nominals are not specified), 0.08% Cu max, 0.1% Sb max, 0.08% Zn max. (b) Melting point

fide are under active investigation as materials for solar cells. Among photovoltaic materials, copper-indium-diselenide is one of the best at absorbing light.

Bismuth

Unlike indium, bismuth has been known by mankind for many centuries. It was probably not recognized as a specific metal by the early Orientals, Greeks, and Romans, but Europeans, in particular, were becoming aware of its properties by the Middle Ages. In the 15th century, Basil Valentine referred to it as wismut. This was Latinized to bisemutum by the early metallurgist Georgus Agricola in the 16th century. By the middle of the 18th century, work by several metallurgists in Europe had resulted in the widespread recognition of bismuth as a specific metal.

Occurrence. The concentration of bismuth in the crust of the earth has been estimated to be 0.2 ppm, the same order of magnitude as that of silver. Some natural bismuth is found in veins associated with silver, lead, zinc, and tin in areas of Bolivia, Canada, and Germany. Specific bismuth ores include bismite (Bi_2O_3), bismuthinite (Bi_2S_3), and bismutite ($Bi_2O_3 \cdot CO_2 \cdot H_2O$). However, the majority of the bismuth produced is recovered as a by-product during the smelting and refining of lead, copper, tin, silver, and gold ores; lead and copper ores supply the vast majority of the metal. Mine production has been reported from Australia, Bolivia, Canada, the People's Republic of China, the Federal Republic of Germany, Japan, Korea, Mexico, Peru, Romania, the United States, the Soviet Union, and Yugoslavia. Production reported in Belgium, Italy, and the United Kingdom is from imported ores and by-products (Ref 2).

Recovery. Bismuth is recovered primarily during the smelting of copper and lead ores. In copper smelting, a portion of the bismuth is volatilized in the copper converter and caught along with such elements as lead, arsenic, and antimony as a dust in a baghouse or Cottrell system. This dust is transferred to a lead-smelting operation. A major part of the bismuth remains with the metallic copper. During electrolytic refining of the copper, the bismuth accumulates in the anode slime along with such other impurities as lead, selenium, tellurium, and the precious metals. The procedure for handling the anode slimes is such that the bismuth is collected in the lead.

Bismuth is found in most lead ores and accompanies the lead through the smelting and refining operations. In the furnace kettle process for refining lead bullion, the bismuth is not removed unless it exceeds 0.05%, the specified limit for bismuth in commercial lead. The two most important methods for removing bismuth from lead are the Betterton-Kroll process and the Betts process. The Betterton-Kroll process is based on the formation of high-melting compounds such as Ca_2Bi_2 and Mg_3Bi_2 that

Table 2 World mine and refinery production of bismuth by country

Country(a)	Mine output (metal content), 10^3 lb(b) 1984	1985	1986	1987(c)	1988(d)	Refined metal, 10^3 lb(b) 1984	1985	1986	1987(c)	1988(d)
Australia(d)(e)	2980	3090	2200	772	880	...	...	...	...	...
Belgium(d)	...	...	...	...	...	850	1350	2200	1900	1750
Bolivia	7	351	95	2	40	...	...	...	...	...
Canada(f)	366	443	337	364	430	330(d)	395(d)	310(d)	330(d)	385
China(d)	570	570	570	570	600	570	570	570	570	600
Federal Republic of Germany	...	...	...	...	...	880(d)	880(d)	880(d)	880(d)	...
France	174	154	209	200(d)	200	...	...	...	...	...
Italy	...	...	...	...	...	57	119	146	97	100
Japan(g)	375(d)	430(d)	420(d)	365(d)	355	1241	1415	1411	1204	1155(h)
Mexico(g)	1005(d)	2140(d)	1740(d)	2350(d)	2160	955	2039	1651	2231	2059(h)
Peru	1433	1731	1334	908	730	1111	1627	1254	853	680
Republic of Korea(g)	278(d)	298(d)	300(d)	320(d)	310	278	298	300	320	310
Romania(d)	180	180	180	170	145	180	180	180	170	145
Union of Soviet Socialist Republics(d)	180	185	185	190	190	180	185	185	190	190
United Kingdom(d)	...	...	...	...	...	330	330	330	400	300
United States	(i)	(i)	(i)	(i)	(i)	(i)	(i)	(i)	(i)	(i)
Yugoslavia(g)	66(d)	150(d)	46(d)	161(d)	66	66	150	46	161	66
Total	**7614**	**9722**	**7616**	**6372**	**6106**	**7028**	**9538**	**9463**	**9306**	**7740**

(a) In addition to the countries listed, Brazil, Bulgaria, the German Democratic Republic, Greece, Mozambique, and Namibia are believed to have produced bismuth, but available information is inadequate for formulation of reliable estimates of output levels. (b) Data were originally compiled in units of 10^3 lb; to convert to metric tons, multiply by 0.4536. (c) Preliminary data. (d) Estimated data. (e) It is believed that bismuth-rich residues were stockpiled at the mine head during the period 1983–1985 and released into the world market in subsequent years. (f) Figures listed under mine output are reported in Canadian sources as production of refined metal and bullion plus recoverable bismuth content of exported concentrate. (g) Mine output figures have been estimated based on reported metal output figures. (h) Reported figure. (i) Withheld to avoid disclosing company proprietary data; excluded from total. Source: U.S. Bureau of Mines

separate from the molten lead bullion bath and can be skimmed off as dross. In the actual process, magnesium and calcium-lead are stirred into the molten bullion, the charge is cooled, and the bismuth dross is skimmed off. The bismuth dross is then melted in small kettles, and the entrapped lead is removed by liquation. The dross is subsequently treated with lead chloride or chlorine to remove the calcium and magnesium; further chlorination is used to remove the lead. A final treatment with caustic soda produces 99.95% Bi.

In the Betts process, lead bullion is electrolyzed in a solution of lead fluosilicate and free fluosilicic acid with pure-lead cathodes. The impurities, including bismuth, are collected in the anode slimes, which are filtered, dried, and smelted. The resulting metal is cupelled, driving the bismuth into the slag and litharge. The slag is reduced to metal containing 20 to 25% Bi, which is then refined to a purity of 99.95% with the same procedure used in the Betterton-Kroll process. Final refining involves repeated treatments with caustic soda and niter followed by a finishing treatment with caustic to yield bismuth that is 99.995% pure.

Bismuth occurring as an oxide or carbonate can be recovered by leaching with hydrochloric acid followed by precipitation and separation of the oxychloride. Removal of copper from the filtrate is accomplished with scrap iron. Repeated precipitations are made, and the purified oxychloride is smelted with lime and charcoal to yield the pure metal.

World production of bismuth per year has varied from 3190 metric tons (3515 tons) in 1984 to 4326 metric tons (4770 tons) in 1985. Mine output ran from 2770 to 4410 metric tons (3053 to 4860 tons) during this period. Actual production capacity is 6940 metric tons (7650 tons). A summary of bismuth production by country is given in Table 2.

Pricing History. Bismuth prices have generally held in the \$4.40 to \$5.50 per kg (\$2.00 to \$2.50/lb) range since the 1950s; occasional excursions above this level have been brought about by special conditions. The price trend firmed in 1984 and peaked at \$14.44 per kg (\$6.55/lb) in January 1985. Subsequent prices have varied between \$6.60 and \$13.20 per kg (\$3.00 and \$6.00/lb). The quoted price in December 1989 was \$9.25 to \$10.12 per kg (\$4.20 to \$4.60/lb).

Properties. Bismuth (atomic number 83) is found in subgroup V-A of the periodic table. It is a brittle crystalline metal with a high metallic lustre and a pinkish tinge. It is one of the two metals (gallium is the other) that expand upon freezing; for bismuth, this expansion is 3.32%. Bismuth has the lowest thermal conductivity of any metal with the exception of mercury. It is the most diamagnetic of all metals, with a mass susceptibility of -1.35×10^6. When influenced by a magnetic field, bismuth displays the greatest increase in resistivity of all the metals. Its thermal conductivity, however, decreases in a magnetic field. Unlike most metals, bismuth has an electrical resistance that is greater in the solid state than in the liquid state by a ratio of approximately 2.

Bismuth, like other members of its subgroup, forms two sets of compounds in which it is trivalent or pentavalent; the trivalent compounds are the more common. Bismuth is soluble in mineral acids but not in water or alkalis. It has the tendency to form oxysalts, particularly with the chloride and nitrate. Mechanical and thermal properties of bismuth are given in the article "Properties of Pure Metals" in this Volume.

Uses. A summary of bismuth consumption in the United States by category is given in Table 3. The chemicals category in Table 3 includes bismuth compounds used as catalysts in the manufacture of plastics and in the synthesis of methanol. The largest use of bismuth in this category is in the manufacture of acrylonitrile using bismuth phosphomolybdate. Bismuth oxychloride is included in cosmetics to give pearlescence to such products as lipstick, eye shadow, and nail polish. Bismuth appears in pharmaceutical products primarily as oxysalts in various indigestion remedies.

Bismuth is used as an additive in steel and aluminum to improve machinability; In

Table 3 U.S. consumption of bismuth metal by category

Use	Bismuth metal consumed, 10^3 lb(a) 1984	1985	1986	1987	1988
Chemicals(b)	1573	1325	1462	1650	1497
Fusible alloys	609	610	639	736	733
Metallurgical additives	424	668	772	1088	1086
Other alloys	20	21	28	24	26
Other(c)	22	20	18	23	34
Total	**2648**	**2644**	**2919**	**3521**	**3376**

(a) Original data were compiled in units of 10^3 lb; to convert to metric tons, multiply by 0.4536. (b) Includes industrial and laboratory chemicals, cosmetics, and pharmaceuticals. (c) Includes experimental uses. Source: U.S. Bureau of Mines

Table 4 Suitability of selected fusible alloys for various applications

Alloy compositions and melting temperatures are given in Tables 5 and 6.

Application	Suitable alloys
Matrix metal for	
Large bearings	X
Punch and die assemblies	Y
Small bearings	Y
Anchoring for	
Rods and tubular members	Y
Bushings in drill jigs, plates, and fixtures	C, Y, Z
Inserts and parts in ceramics, plastics, metal powders, and wood	G, Z
Locator members in aircraft and automotive assembly drilling, inspection, and welding fixtures	C, H, Z
Hold-down bolts in floors	G, Z
Magnets in chucks, instruments, and work-holding devices	H, Z
Needles in lace and textile machinery	G, H, T
Patterns in foundry match plates	G, H, T
Precision parts for testing and inspection	A, B, C, G, Z
Reamers for axial and concentric alignment in turret tool holders	C, H
Shafts in permanent-magnet rotors	H, Y
Chucking for	
Lens buffing and grinding	A, C
Gem cutting	G, Z
Metal spinning of reentrant and bottleneck shapes	C, G
Fusible cores for	
Electroforming	G, H, Z
Founding	C, G
Holding and forming fiberglass and plastic laminates	C, G
Compound wax patterns	B, C
Punch and die applications	
Light sheet metal embossing dies	C, G, H, Y, Z
Form blocks for bending, forming, and joggling of extruded shapes, sheet metal, and tubing	H, Y
Soft-metal dies for lost-wax patterns	H, Z
Pilots for die casting and plastic trim dies	H
Repairing broken dies	Y
Stretch press form blocks	G, H
Punch and die applications (continued)	
Stripper plates in stamping dies	Y
Guerin process dies	C, G, H, Y
Mold applications	
Duplicating plaster or plastic patterns	G, H, Z
Casting plastics, phenolics, elastomers, epoxies, and polyesters	G, H, Z
Forming sheet and tubular polyethylene plastics by vacuum or air pressure	H, Z
Plastic teeth (dental)	H
Prosthetic development work	A, H, Z
Encapsulating avionic, electrical, and electronic components	H, Z
Pattern applications	
Dental models (special compositions)	A
Duplicate patterns for foundry match plates, ceramics, plastics, and pottery	C, H, Z
Lining and sealing core boxes	H
Model airplane, railroad, and ship parts	G, H, T
Spray-metallizing, altering, and repairing patterns and core boxes	T, Z
Tracer models for pantograph engraving and duplicating	H, Z
Miscellaneous applications	
Filler for tube and mold bending	C, G
Low-melting solders	...
Heat-transfer mediums in constant-temperature baths	...
Seals for bright-annealing and nitriding furnaces	...
Seals for adjustment screws on torque wrenches and instruments	...
Hold-down clamp pads	...
Ammunition composites	...
Fire detection apparatus and alarm systems	...
Safety plugs for tanks and cylinders for compressed gas	...
Fusible elements in automatic sprinkler heads and fire door release links	...
Automatic shutoffs for gas and electric water-heating systems	...
Selenium rectifiers such as the counterelectrode	...
Lead and bismuth additions to aluminum and other metals and alloys to obtain free-cutting materials	...

wrought iron it reduces the formation of graphite on freezing. Another application of bismuth is in thermoelectric devices containing intermetallic compounds of bismuth with selenium and/or tellurium; these devices make use of the Peltier effect for refrigeration. Some recently developed high-temperature superconductors also contain bismuth compounds, in particular those in the bismuth-strontium-calcium-copper-oxygen system.

Fusible alloys include a group of binary, ternary, quaternary, and quinary alloys containing bismuth, lead, tin, cadmium, and indium. The term fusible alloy refers to any of the more than 100 white-metal alloys that melt at relatively low temperatures, that is, below the melting point of tin-lead eutectic solder (183 °C, or 360 °F). The melting points of these alloys range as low as 47 °C (116 °F). Fusible alloys are used for lens blocking and tube bending, for anchoring chucks and fixtures, and for mounting thin sections such as gas turbine blades for machining. The eutectic fusible alloys, which can be tailored to give a specific melting point, find application in temperature control devices and in fire protection devices such as sprinkler heads. Applications of selected fusible alloys are listed in Table 4.

Many of the fusible alloys used in industrial applications are based on eutectic compositions (see Table 5). Under ambient temperature, such an alloy has sufficient strength to hold parts together; at a specific elevated temperature, however, the fusible alloy link will melt, thus disconnecting the parts. In fire sprinklers, the links melt when dangerous temperatures are reached, releasing water from piping systems and extinguishing the fire. Boiler plugs and furnace controls react similarly because an increase in temperature beyond the safety limits of the furnace or boiler operation will melt the plug. When the fusible alloy link melts, pressure or heat in the boiler can be dissipated, or feeding of the fuel supply can be ceased, thereby reducing operation to a safe level.

In addition to eutectic alloys, each of which melts at a specific temperature, there are numerous noneutectic fusible alloys, which melt over a range of temperatures.

Table 5 Compositions and melting temperatures of selected eutectic fusible alloys

	Melting temperature		Composition, %				
Alloy(a)	°C	°F	Bi	Pb	Sn	Cd	Other
A	47	117	44.70	22.60	8.30	5.30	19.10 In
B	58	136	49.00	18.00	12.00	...	21.00 In
C	70	158	50.00	26.70	13.30	10.00	...
D	91.5	197	51.60	40.20	...	8.20	...
E	95	203	52.50	32.00	15.50	...	...
F	102.5	217	54.00	...	26.00	20.00	...
G	124	255	55.50	44.50	...	...	...
H	138.5	281	58.00	...	42.00	...	...
I	142	288	...	30.60	51.20	18.20	...
J	144	291	60.00	...	...	40.00	...
K	177	351	...	...	67.75	32.25	...
L	183	362	...	38.14	61.86	...	...
M	199	390	...	...	91.00	...	9.00 Zn
N	221	430	...	...	96.00	...	3.50 Ag
O	236	457	...	79.7	...	17.7	2.60 Sb
P	247	477	...	87.0	...	...	13.00 Sb

(a) Letter designations are intended only for identification of alloys in Tables 4, 7, and 8.

Table 6 Compositions, yield temperatures, and melting temperature ranges of selected noneutectic fusible alloys

Alloy(a)	Yield temperature °C	Yield temperature °F	Melting temperature range °C	Melting temperature range °F	Composition, % Bi	Pb	Sn	Cd
Q	70.5	159	70–73	158–163	50.50	27.8	12.40	9.30
R	72	162	70–79	158–174	50.00	34.5	9.30	6.20
S	72.5	163	70–84	158–183	50.72	30.91	14.97	3.40
T	72.5	163	70–90	158–194	42.50	37.70	11.30	8.50
U	75	167	70–101	158–214	35.10	36.40	19.06	9.44
V	96	205	95–104	203–219	56.00	22.00	22.00	...
W	96	205	95–149	203–300	67.00	16.00	17.00	...
X	101	214	101–143	214–289	33.33	33.34	33.33	...
Y(b)	116	241	103–227	217–440	48.00	28.50	14.50	...
Z	150	302	138–170	281–338	40.00	...	60.00	...
ZZ(c)	...	...	61–65	142–149	48.0	25.63	12.77	9.60

(a) Letter designations are intended only for identification of alloys in Tables 4, 7, and 8. (b) Also contains 9.00% Sb. (c) Also contains 4.0% In

Table 7 Physical properties of selected fusible alloys

Alloy compositions are given in Tables 5 and 6.

Alloy	Density g/cm³	Density lb/in.³	Tensile strength MPa	Tensile strength ksi	Hardness, HB	Maximum 30-s load MPa	Maximum 30-s load ksi	Safe sustained load MPa	Safe sustained load ksi	Electrical conductivity, % compared with copper
Eutectic alloys (Table 5)										
A	8.86	0.32	37	5.4	12	...	...		...	3.34
B	8.58	0.31	43	6.3	14	...	...		...	2.43
C	9.38	0.339	41	6.0	9.2	70	10	2	0.3	4.17
G	10.5	0.380	44	6.4	10.2	55	8	3	0.5	1.75
H	8.72	0.315	55	8.0	22	100	15	3	0.5	5.00
Noneutectic alloys (Table 6)										
T	9.44	0.341	37	5.4	9	62	9.0	2	0.3	4.27
Y	9.5	0.343	90	13	22	110	16	2	0.3	2.57
Z	8.2	0.296	55	8.0	22	103	15	3	0.5	7.77
ZZ	9.5	0.342	34	4.95	11	...	...	...	...	3.27

Table 8 Cumulative growth and shrinkage data for selected fusible alloys

Alloy compositions are given in Tables 5 and 6.

Alloy	Cumulative growth or shrinkage in mm/mm for a test bar(a) at the specified time after casting: 2 min	6 min	30 min	1 h	2 h	5 h	500 h
Eutectic alloys (Table 5)							
A	+0.0005	+0.0002	0.0000	−0.0001	−0.0002	−0.0002	−0.0002
B	+0.0003	+0.0002	+0.0001	0.0000	−0.0001	−0.0002	−0.0002
C	+0.0025	+0.0027	+0.0045	+0.0051	+0.0051	+0.0051	+0.0057
G	−0.0008	−0.0011	−0.0010	−0.0008	−0.0004	0.0000	+0.0022
H	+0.0007	+0.0007	+0.0006	+0.0006	+0.0006	+0.0005	+0.0005
Noneutectic alloys (Table 6)							
T	−0.0004	−0.0007	−0.0009	0.0025	+0.0016	+0.0018	+0.0025
Y	+0.0008	+0.0014	+0.0047	+0.0048	+0.0048	+0.0049	+0.0061
Z	−0.0001	−0.0001	−0.0001	−0.0001	−0.0001	−0.0001	−0.0001
ZZ	+0.0020	+0.0022	+0.0040	+0.0046	+0.0046	+0.0046	+0.0052

(a) Data are in mm/mm (in./in.) compared to cold mold dimensions of a 13 × 13 × 250 mm (½ × ½ × 10 in.) test bar with a weight of about 0.45 kg (1 lb).

Selected eutectic alloy compositions are listed in Table 6.

A fusible alloy with a long melting range is useful in staking rods and tubing in assemblies because the alloy is distributed around part surfaces while still molten and provides a firm anchorage after it solidifies.

Properties of Fusible Alloys. Table 7 lists various physical properties of selected fusible alloys. Fusible alloys are ageable, and thus their mechanical properties often depend on the period of time that has elapsed since casting, as well as on the casting conditions and the solidification rate. Test conditions also affect mechanical property values. For example, many fusible alloys can appear brittle when subjected to sudden shock but exhibit high ductility under slow rates of strain.

In certain alloys, normal thermal contraction due to cooling after solidification can be partly, completely, or more than compensated for by expansion due to aging. For example, bismuth alloys containing 33 to 66% Pb exhibit net expansion after solidification and during subsequent aging. Some fusible alloys show no contraction (shrinkage) and expand rapidly while still warm; others show slight shrinkage during the first few minutes after solidification and then begin to expand; in still others, expansion does not commence until some time after the fusible alloy casting has cooled to room temperature. Cumulative growth and shrinkage data for selected fusible alloys are given in Table 8.

Each of the three characteristics—net expansion, net contraction, and little or no volume change—can provide specific advantages, depending on the application. For example, a wood pattern used for making molds must be of somewhat greater dimensions than those desired in the casting to compensate for shrinkage of the casting on solidification and during cooling to room temperature. Where metal patterns are cast from a master wood pattern, two such allowances will have to be made unless the alloy used for the metal patterns possesses zero shrinkage. Fusible alloys with eutectic compositions are often used for casting metal patterns from wood masters because they undergo definite growth that is sufficient to allow for cleaning of production castings without reducing dimensions below required values. The growth characteristics of fusible alloys are often used to advantage when a metal part is to be firmly anchored in a lathe chuck. After the part is machined, the fusible alloy is melted away.

In general, the load-bearing capacity of fusible alloys is good, although some deformation will occur under prolonged stress. In addition, hardness and other mechanical properties of many fusible alloys change gradually with time, probably because of the same microstructural changes caused by aging that affect growth or shrinkage.

Bismuth-Base Solders. Alloys rich in bismuth generally are not considered to be good solders. When using the traditional halide-activated fluxes at the low soldering temperatures required with these alloys, the flux often does not reach its activation temperature. Consequently, fluxing activity is inefficient, and wetting can be poor. In these instances, fluxes with other activation mechanisms should be used to improve soldering quality (for example, fluxes with organic acids and/or chelation agents).

Although alloys rich in bismuth can be difficult to use as solders, bismuth-base solder alloys (Table 9) do find use in such applications as:

- Soldering heat-treated surfaces when a higher soldering temperature would result in a softening of the part
- Soldering joints where the adjacent material is heat sensitive and deterioration would occur at a higher soldering temperature
- Step soldering operations where a low temperature is necessary to protect a nearby solder joint

Table 9 Bismuth-base solder alloys

Composition, wt%					Melting temperatures			
					Solidus		Liquidus	
Bi	Pb	Sn	Cd	In	°C	°F	°C	°F
44.7	22.6	8.3	5.3	19.1	47	117	47	117
49	18	12	...	21	58	136	58	136
32.5	...	16.5	...	51.0	60	140	60	140
48	25.63	12.77	9.6	4.0	61	142	65	149
50	26.7	13.3	10.0	...	70	158	70	158
57	...	17	...	26	79	174	79	174
42.5	37.7	11.3	8.5	...	71	160	88	190
52.5	32.0	15.5	...	...	95	203	95	203
46	20	34	...	...	100	212	100	212
67	...	...	...	33	109	229	109	229
55.5	44.5	...	...	...	124	255	124	255
57.42	1.00	41.58	...	...	135	275	135	275
58	...	42	...	...	138	281	138	281
14	43	43	...	...	144	291	163	325
40	...	60	...	...	138	281	170	338
12.6	47.47	39.93	...	...	146	294	176	349

- Construction of temperature-overload devices such as safety links, fuses, and plugs where positive-pressure contact is too variable or inconsistent for assembly operation
- Soldering low-temperature alloys such as pewter
- Machine-soldering operations for through-hole soldering of very thick multilayer printed circuit boards
- Assembly operations, such as surface mounting, where the integrated circuit packages would be vulnerable to thermal damage at the temperatures required for conventional tin-lead soldering
- Assembly operations using injection-molded circuit boards where the glass transition temperature is too low for the use of tin-lead alloys

In addition, several bismuth-base alloys (particularly the 42Sn-58Bi eutectic alloy) are being used for wave techniques, during which surface-mounted devices are directly exposed to molten solder. This direct exposure has provided an incentive to lower solder temperatures to the range of 150 to 170 °C (300 to 340 °F) by using bismuth-base solders. Conventional tin-lead solders require temperatures between 240 and 250 °C (460 and 480 °F). In some respects such as fatigue strength and dissolution of copper, bismuth-base joints from wave soldering are superior to those made by conventional tin-lead soldering; however, their ductility is substantially lower.

Table 9 lists various bismuth-base solders. These alloys tend to be more difficult to fabricate than indium-base solders, and thus the range of available preforms is not as large. They can be supplied as wire, rod, sheet, and ingots. Solder creams of the 42Sn-58Bi eutectic alloy, the 43Sn-43Pb-14Bi composition, and the 57.42Bi-41.58Sn-1.00Pb alloy are available. Most of the other listed alloys could also possibly be used for solder creams; however, in some cases they would have to be compounded just prior to use because of the aggressive fluxes required.

The 58% Bi and 14% Bi solder alloys are commonly used for low-melting-temperature soldering. Joint strength is comparable to that produced with tin-lead solders, although the ductility is somewhat lower. Leaching of copper and beryllium-copper is substantially reduced when these alloys are used. The 46% Bi alloy has a melting point at the boiling point of water and characteristics similar to those of the 58% Bi and 14% Bi alloys.

REFERENCES

1. *Mineral Commodity Summaries*, U.S. Bureau of Mines, 1989, p 76-77
2. *Minerals Yearbook–Bismuth*, U.S. Bureau of Mines, 1988, p 1-5

SELECTED REFERENCES

- S.C. Carapella and H.E. Howe, Bismuth and Bismuth Alloys, in *Encyclopedia of Chemical Technology*, Vol 3, M. Grayson, Ed., John Wiley & Sons, 1978, p 912-921
- H.E. Howe, Bismuth, in *Rare Metals Handbook*, C.A. Hampel, Ed., Rheinhold, 1961, p 58-68
- H.E. Howe, Bismuth, in *Encyclopedia of the Chemical Elements*, C.A. Hampel, Ed., Rheinhold, 1968, p 56-65
- S.C. Liang, C.E.T. White, and R.A. King, Indium, in *Encyclopedia of the Chemical Elements*, C.A. Hampel, Ed., Rheinhold, 1968, p 283-290
- J.R. Mills, R.A. King, and C.E.T. White, Indium, in *Rare Metals Handbook*, C.A. Hampel, Ed., Rheinhold, 1961, p 220-238
- E.F. Milner and C.E.T. White, Indium, in *Encyclopedia of Chemical Technology*, Vol 13, M. Grayson, Ed., John Wiley & Sons, 1980, p 207-212

Special-Purpose Materials

Magnetically Soft Materials

Douglas W. Dietrich, Carpenter Technology Corporation

MAGNETIC MATERIALS are broadly classified into two groups with either hard or soft magnetic characteristics. Hard magnetic materials are characterized by retaining a large amount of residual magnetism after exposure to a strong magnetic field. These materials typically have coercive force, H_c, values of several hundred to several thousand oersteds (Oe) and are considered to be permanent magnets. The coercive force is a measure of the magnetizing force required to reduce the magnetic induction to zero after the material has been magnetized. In contrast, soft magnetic materials become magnetized by relatively low-strength magnetic fields, and when the applied field is removed, they return to a state of relatively low residual magnetism. Soft magnetic materials typically exhibit coercive force values of approximately 400 $A \cdot m^{-1}$ (5 Oe) to as low as 0.16 $A \cdot m^{-1}$ (0.002 Oe). Soft magnetic behavior is essential in any application involving changing electromagnetic induction such as solenoids, relays, motors, generators, transformers, magnetic shielding, and so on.

Important characteristics of magnetically soft materials also include:

- High permeability
- High saturation induction
- Low hysteresis-energy loss
- Low eddy-current loss in alternating flux applications
- In specialized cases, constant permeability at low field strengths and/or a minimum or definite change in permeability with temperature

Cost, availability, strength, corrosion resistance, and ease of processing are several other factors that influence the final selection of a soft magnetic material.

Magnetically soft materials manufactured in large quantities include high-purity iron, low-carbon irons, silicon steels, iron-nickel alloys, iron-cobalt alloys, ferritic iron-chrome alloys, and ferrites. Soft magnetic amorphous materials are being produced commercially; however, their characteristics are covered in the article "Metallic Glasses" in this Volume.

Ferromagnetic Properties

In crystalline materials, the basis for ferromagnetism lies in the alignment of magnetic moments from noncompensated electron spins in the 3*d* shell of the transition series elements such as iron, nickel, and cobalt. In ferromagnetic materials that are below their Curie temperature, the magnetic moments of adjacent atoms are coupled parallel to each other. For a small volume of material, all of the individual magnetic moments are aligned in one direction. This small volume is magnetized to saturation and is known as a magnetic domain. An adjacent volume of material, also magnetized to saturation, may have the summation of its magnetic moments point in another direction. Where two such volumes meet (with differing alignments), a domain boundary wall must exist. The total magnetization of a sample of material is the net vector summation of all the individual component domain magnetization vectors. In the demagnetized state, the net summation of all domains approaches zero. The net magnetization of a material can be changed by domain wall movement and/or by rotation of the individual domain magnetization vectors. The energies required to cause domain rotation and wall motion are associated with the materials' crystalline structure, grain size, residual stress, impurities, and so on.

In crystalline ferromagnetic materials, magnetization occurs spontaneously in preferred easy directions. The easy directions are those in which the crystalline anisotropy is a minimum. In iron the easy direction is the cube edge $\langle 100 \rangle$. In nickel, the easy magnetization direction is the $\langle 111 \rangle$ cube diagonal (Fig. 1). The crystalline anisotropy constant (K_1) is a measure of the energy required to turn the spontaneous magnetization vector from the preferred direction into the direction of the applied magnetizing field. If K_1 approaches zero, it is relatively easy to turn the magnetization in any direction, and the permeability is likely to be high.

As an external magnetic field is applied to a ferromagnetic material, the magnetic domains that happen to coincide with the applied field grow at the expense of less favorable domains. Upon further increase of the external field the domains rotate into a parallel direction to this field. Particularly during domain rotation, the crystalline lattice spacing may be altered so that the material expands or contracts slightly. This change in dimensions when magnetized is known as magnetostriction. Positive magnetostriction is an elongation of the material in the direction of the applied field. The change is extremely small; for example, pure nickel has a saturation magnetostriction coefficient (λ_1) of approximately -38×10^{-6} $\Delta l/l$. In very high permeability alloys, λ_1 approaches zero. Conversely, applying external stress to ferromagnetic material causes the magnetic hysteresis loop to change.

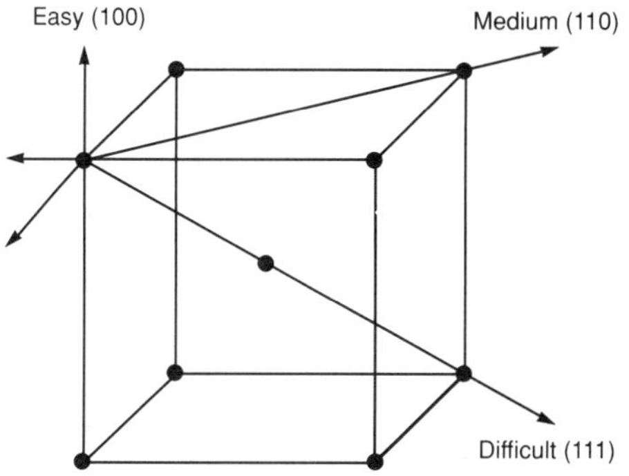

Fig. 1 Crystallographic orientation of iron showing ease of magnetization in the three principal directions

The ferromagnetic and electrical properties of materials can be divided into two general categories: those that are structure sensitive and those that are structure insensitive. Structure insensitive refers to those properties not markedly affected by small changes in gross composition, small amounts of certain impurities, heat treatment, or plastic deformation. Several generally accepted structure-insensitive properties are the saturation induction (B_s), resistivity (ρ), and Curie temperature (T_c). These properties are largely dependent upon the composition of the alloy selected

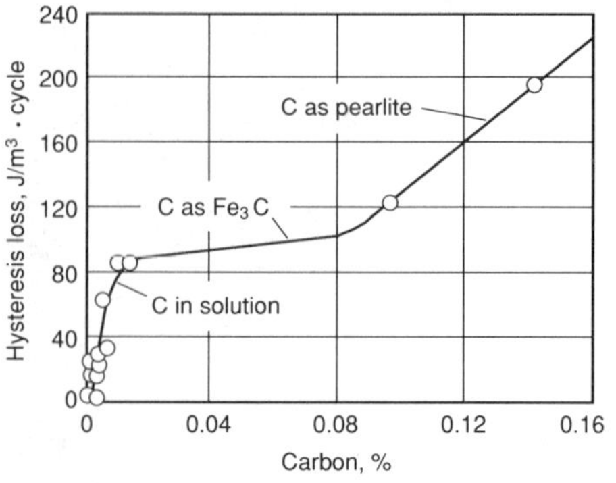

Fig. 2 Relationship between carbon content and hysteresis loss for unalloyed iron. Induction $B = 1$ T (10 kG).

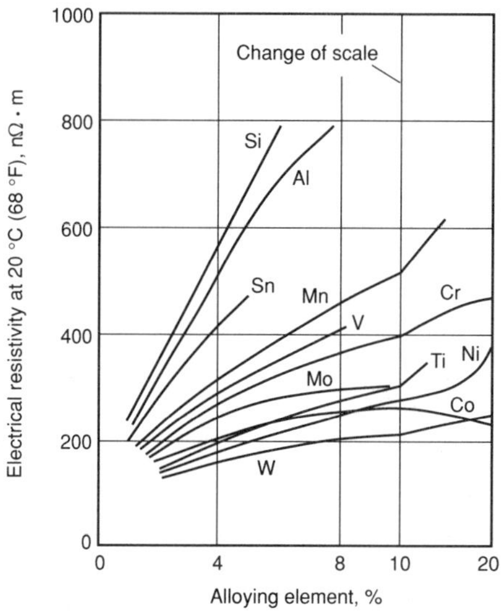

Fig. 3 Effect of alloying elements on electrical resistivity of iron

and are not changed substantially in the process of manufacturing a component from the alloy.

Structure-sensitive properties are those drastically affected by impurities: residual strain, grain size, and so on. Permeability (μ), coercive force (H_c), hysteresis losses (W_h), residual induction (B_r), and magnetic stability are all considered to be structure sensitive. A means of controlling structure-sensitive properties is through manufacturing processing of the alloy and/or by the proper use of a final annealing heat treatment.

Effect of Impurities on Magnetic Properties

Elements such as carbon, oxygen, nitrogen, and sulfur are commonly found as impurities in all alloys (see the articles "Preparation and Characterization of Pure Metals" in this Volume). Even in very low concentrations these elements tend to locate at interstitial sites in the crystalline lattice; thus, the lattice can be severely strained. Very minor concentrations may interfere with the easy movement of magnetic domains and impair soft magnetic properties. Figure 2 shows the approximate relationship between carbon content and the hysteresis loss of iron. Hysteresis losses are similarly related to sulfur and oxygen content. Furthermore, if carbon and/or nitrogen remains in the alloy uncombined, or if these elements exceed their respective solubility limits near room temperature, they may migrate in time and precipitate in a form of fine particles that can pin the magnetic domain walls. This causes a hardening of the magnetic properties known as aging.

Steel producers utilize raw materials and melting methods that provide impurity levels for those alloys guaranteed to provide a certain level of magnetic performance. In certain cases, particularly with fully processed silicon steels, the producer then utilizes a decarburization heat treatment to further reduce the carbon content of the as-supplied strip product. This process is not economically or physically possible for all soft magnetic alloys and for heavier strip or bar product forms. Thus, it is often desirable and necessary that the consumer anneal the parts in a strongly reducing, nonoxidizing atmosphere as part of the component manufacturing process. Annealing of the final part further reduces impurities, particularly carbon and sulfur, below the levels that can be achieved by melting control alone. In iron alloys and silicon steels, the content of the finished part should be less than 0.003% C to optimize soft magnetic properties and minimize aging.

Effect of Alloying Additions on Magnetic Properties

The major constituents of most soft magnetic alloys are one or more of the common ferromagnetic elements: iron, nickel, or cobalt. Most useful combinations of these elements and the typical additional alloying additions made to soft magnetic alloys are fully substitutional. They contribute to the control of crystalline lattice structure to promote high permeability, low coercive force, and low hysteresis loss.

Certain alloying additions may also increase electrical resistivity that helps to reduce eddy-current losses in alternating current (ac) devices. For example, pure iron can exhibit good soft magnetic properties and has a high saturation induction. It is used extensively in direct current (dc) applications and small fractional horsepower motors; however, its low electrical resistivity results in high eddy-current losses in ac applications. Figure 3 shows the changes in resistivity that result from additions of various elements to iron.

Silicon. As a result, iron alloys containing 1 to 4% silicon are commonly used in ac applications. Well-annealed pure iron is very soft, typically ranging from 20 to 40 HRB. The addition of silicon also strengthens the annealed alloy. Iron containing ~2.5% Si exhibits an annealed hardness of ~90 HRB.

Silicon additions greater than approximately 2.5% to pure iron can eliminate the transformation from α to γ phase found in pure iron. Consequently, higher silicon-content alloys can be annealed to promote grain growth at high temperatures without passing through a phase transformation. The lack of a phase transformation also facilitates the development of preferentially oriented (cube on edge) grain structure in silicon steels. The oriented silicon steels typically contain 3.15% Si.

Cobalt. Most alloying additions made to iron lower its saturation induction (B_s) as shown in Fig. 4. However, the addition of cobalt results in increased saturation induction up to approximately 2.46 T (24.6 kG) at approximately 35% Co.

Vanadium. The addition of vanadium to 50Co-50Fe alloys can improve processing by allowing quenching to obtain ductility for cold rolling of strip products.

Phosphorus may be added to pure irons and silicon irons to enhance stampability and machinability and to aid the sintering of powdered irons.

Chromium is added to iron to produce ferritic stainless steels with suitable soft magnetic characteristics for certain applications.

Additional Elements. Additions of molybdenum, copper, and/or chrome can be made to ~80Ni-Fe alloys to optimize crystallographic parameters to achieve very high permeability. The significance of each of these alloying additions is reviewed in greater detail in the section "Classification of Alloys" in this article.

Effect of Heat Treatment

In certain special cases, particularly fully processed lamination iron and silicon steels, the as-supplied coil product can possibly be stamped and the laminations placed into service without a reheat treatment. Fully processed lamination materials represent a large volume of the soft magnetic alloy market but they are not typical of most magnetic alloys. Nearly all of the other materials discussed in this article should be annealed as a finished (or nearly finished) component to develop their optimum soft magnetic properties. Even fully processed grades of silicon steel may require a low-temperature stress-relief heat treatment for removal of fabrication stresses.

Most alloys supplied for the stamping of laminations are provided in a cold-rolled unannealed condition. This condition provides best stamping characteristics, acceptable flatness, and minimum burr. Even when thin or heavy sheet and strip products

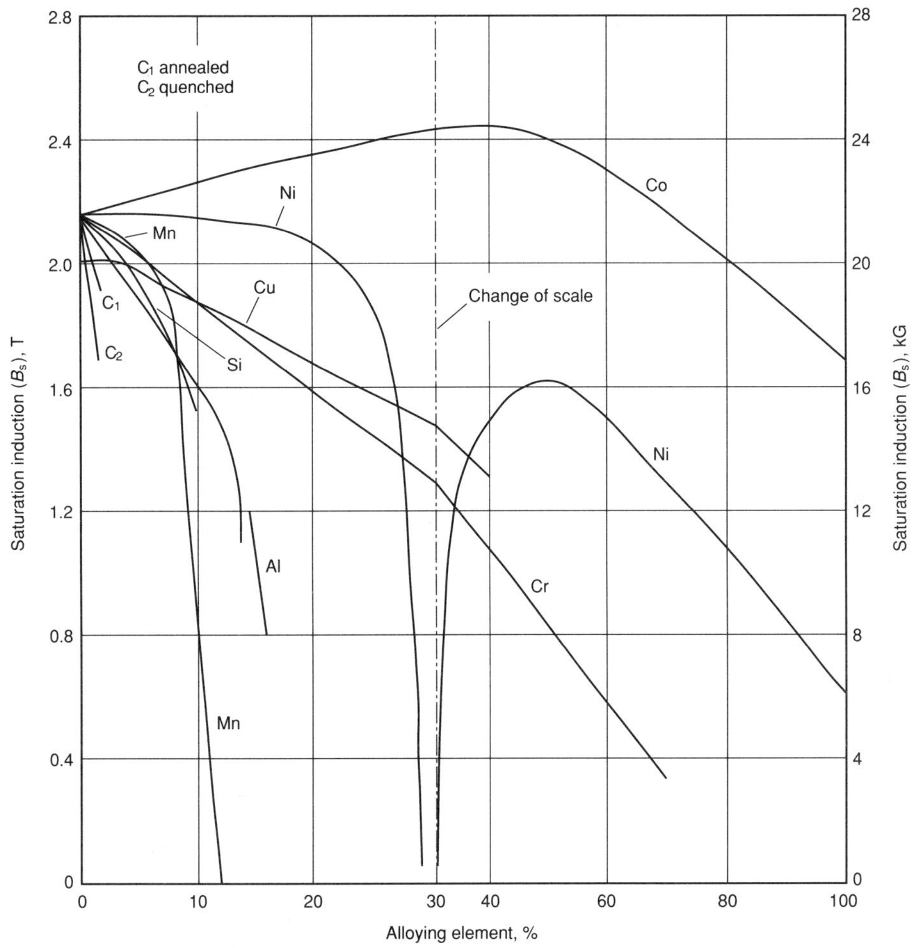

Fig. 4 Effect of alloying elements on room-temperature saturation induction of iron

are produced in a mill-annealed condition for forming, bending, or deep drawing (see the article "Carbon and Low-Alloy Sheet and Strip" in Vol 1 of the 10th Edition of *Metals Handbook*), the mill anneal is intended to provide mechanical properties suitable for the fabrication operations. These products usually exhibit a fine grain structure to prevent cracking and orange peel during forming. The mill anneal is not intended to provide the soft magnetic properties in the as-shipped product. Similarly, bar products are generally produced to provide optimum machinability, and wire products are made for formability by bending or perhaps cold heading. Parts made from bar or wire product forms also require annealing to obtain optimum soft magnetic properties.

Three major objectives of annealing are to:

- Eliminate stresses
- Promote/control grain growth
- Further reduce impurities, particularly carbon, nitrogen, and sulfur

Annealing may also control a particular phase transformation or provide a critical degree of ordering in some high-permeability alloys.

Minimizing Residual Stress. The structure-sensitive properties of soft magnetic materials are very strongly influenced by residual stress, both remaining from the manufacturing of the material as well as from the stresses introduced by fabrication of the part. In oriented silicon steels, a compressive stress of as little as 3.45 to 6.9 MPa (0.5 to 1.0 ksi) can increase core loss by 50 to 100%. Very high permeability nickel-iron tape cores can be damaged just by squeezing them by hand. Thus, great care must be taken to remove residual stresses by annealing and to minimize new stresses introduced during assembly. Machined parts should be annealed as near to finished dimensions as possible. If finishing operations are required for close tolerances after annealing, they should be limited to light machining or grinding passes.

Maximizing Grain Size. For most applications of nonoriented magnetic materials, grain size should be as large as possible. A fine grain structure has a large grain-boundary surface area per unit volume. Grain boundaries represent a physical discontinuity of the crystalline structure that impedes the movement of magnetic domains. A coarse grain structure provides less grain-boundary surface area per unit volume and generally results in softer dc magnetic properties (Fig. 5a). In ac applications, as the frequency is increased, there may be an optimum grain-size range that provides best magnetic performance. Figure 5(b) presents data for oriented silicon steel for which an optimum grain-size range exists.

Reducing Impurity Levels. The exposure of stamped laminations or machined components to annealing temperatures suitable for the alloy will result in stress relief and the desirable grain growth. These two objectives of annealing can be achieved in a vacuum furnace or inert (protective) atmosphere furnace. Adequate soft magnetic properties for the particular application may result. However, for optimum properties, most soft magnetic alloys should also undergo decarburization during the annealing treatment. Suitable annealing atmospheres for the reduction of carbon include forming gas (5 to 10% H_2 and 90 to 95% N_2), dissociated ammonia, or pure hydrogen. Stronger reducing effects are achieved with greater hydrogen content. A moist atmosphere (10 °C, or 50 °F dew point) may also promote rapid decarburization in low-carbon irons and some silicon irons, but the atmosphere must remain nonoxidizing to the iron and silicon present in the alloy. Thus, moist annealing is usually used only on irons and low-silicon-content steels and at temperatures of ~845 °C (1550 °F) or lower. If the section size of the part is large, decarburization may be ineffective. However, smaller parts and strip or foil products may benefit greatly by decarburization.

The thermal cycles and furnace atmospheres used to heat treat soft magnetic alloys vary greatly depending upon the alloy system, processing history, and properties desired. Each of the subsequent sections provides general information about the heat treatment of various alloys. However, it is strongly recommended that the producers of the alloys be contacted for specific information regarding annealing of their products.

Alloy Classifications and Magnetic Testing Methods

The comparison of magnetic test data among different materials, or even different forms of the same alloy, can be very misleading if the data are not developed by similar methods. Factors such as annealing temperature, annealing atmosphere, cooling rate, specimen configuration, and test frequency all have a profound effect on the magnetic test data. Various standards organizations have developed both material and testing method standards for magnetic materials. Throughout this article, reference will be made to American Society for Testing and Materials (ASTM) test methods and materials standards. Test method standards describe appropriate test equipment, elec-

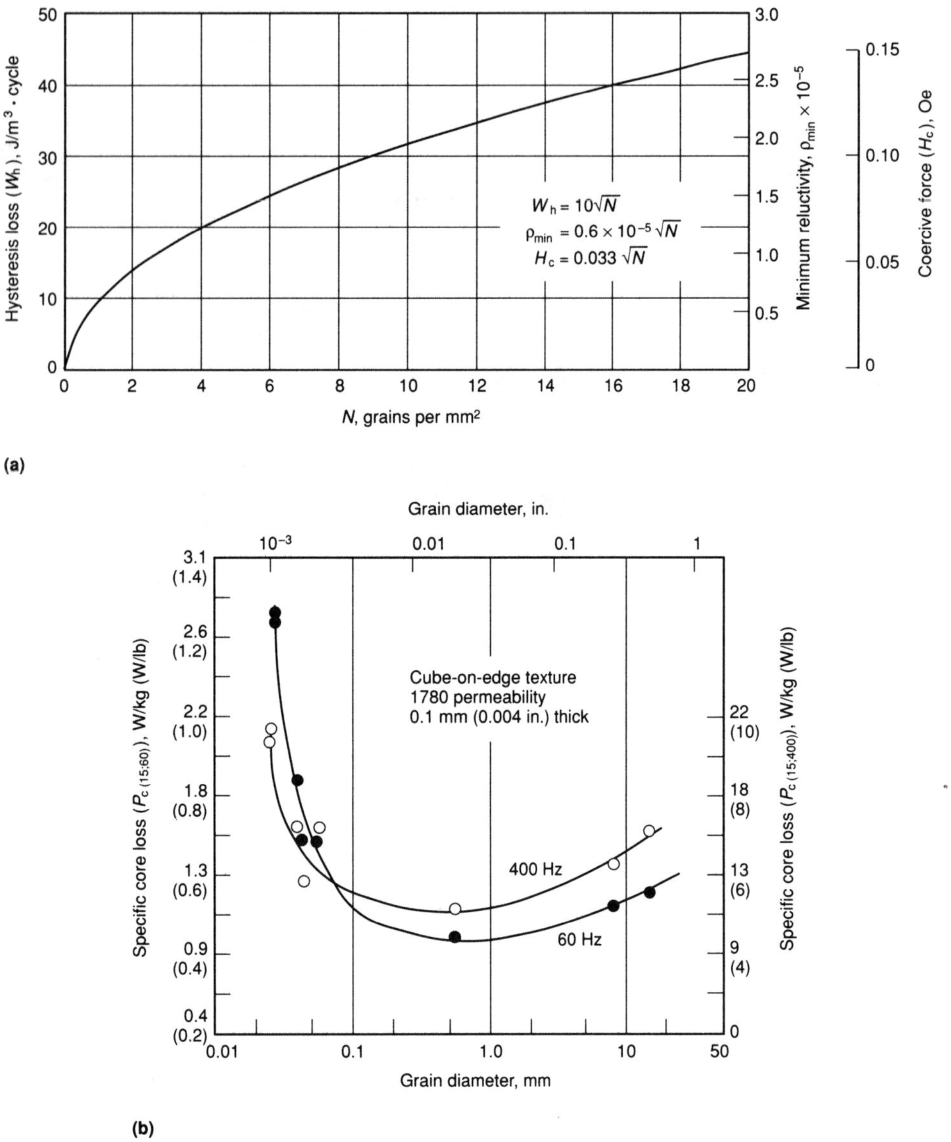

Fig. 5 Effect of grain size on magnetic properties of pure iron and silicon iron. (a) Relationship between grain size and hysteresis loss for high-purity iron at B = 1 T (10 kG). (b) Variation of core loss with grain size for samples of 3.15 Si-Fe having similar cube-on-edge textures and chemical purity. $P_{c(15;60)}$ is the measured magnetic core loss at 15 kG and 60 Hz; $P_{c(15;400)}$ is the measured magnetic core loss at 15 kG and 400 Hz.

trical circuits, and specimen configurations to provide meaningful reproducible test results. Material standards have been prepared for many of the more commonly used soft magnetic alloys. They generally provide typical magnetic data and magnetic test capability limits using appropriate test methods. The material standards often provide grade designation, physical and mechanical properties, chemical analyses, and appropriate heat-treating information. ASTM A 340-87 provides symbols and definitions of terms related to magnetics.

High-Purity Iron

For many years, extremely high-purity iron has been produced for researching its magnetic characteristics. Those impurities that have the strongest detrimental effect on its magnetic properties are carbon, sulfur, and nitrogen. These elements can all be reduced to levels well below their room-temperature solubility in iron by annealing at 1300 to 1500 °C (2370 to 2730 °F) in hydrogen for several hours. It is necessary to cool slowly from the high temperature through the γ to α transformation to produce excellent soft magnetic properties. Heat treatment of iron in this temperature range is not a normal commercial practice but has been used for research into the capabilities of extremely pure iron. Bozorth reported a maximum dc permeability of greater than 10^6 for purified iron in the mid-1930s.

The saturation induction of iron based upon a density of 7.878 g/cm³ (0.2846 lb/in.³) is reported as 2.158 T (21.58 kG). The electrical resistivity is 9.8 μΩ · cm (59 Ω · circ mil/ft) at 20 °C (68 °F), and the temperature coefficient of resistivity is 0.0065/°C (0.0036/°F).

Commercially Pure Irons—Vacuum Induction Melted. Commercial high-purity irons are available from specialty steel manufacturers with a purity of approximately 99.8% Fe. This product is produced by vacuum induction melting (VIM), sometimes followed by vacuum consumable electrode remelting. It is manufactured in billet, bar, wire, or strip forms suitable for fabrication by forging, forming, or machining. Because of its high saturation induction and low electrical resistivity (10.7 μΩ · cm), high-purity irons have been used primarily in dc applications as flux carriers, electromagnetic lenses, and pole pieces or pole caps. The high-purity irons have been used in high-vacuum systems with low outgassing demands. Following fabrication of the parts, they must be annealed to develop the desired soft magnetic properties. Maximum dc permeabilities of approximately 1.7×10^4 and coercivity of approximately 20 $A \cdot m^{-1}$ (0.25 Oe) can be attained in bar product forms. Annealing is typically performed in a strongly reducing atmosphere, such as hydrogen, at temperatures between 815 to 980 °C (1500 to 1800 °F), for 4 h, followed by furnace cooling. The magnetic-property capabilities of this product are guaranteed by the producer at levels dependent upon the product form, method of testing, and annealing process employed.

Commercially Pure Irons—Air-Furnace Melted. Commercial soft magnetic irons with a purity of approximately 99.1 to 99.8% iron are also available in billet, bar, wire, or strip forms. Historically, Armco Electromagnetic Iron generally fits into this category. Iron of this purity level can be produced by the electric-arc melting process. These irons have a saturation induction of about 2.15 T (21.5 kG), a specific gravity of 7.86, and an electrical resistivity of approximately 13 μΩ · cm at 20 °C (68 °F). The as-supplied carbon content of low-carbon magnetic iron is below 0.025%, typically 0.010% or less. The machinability of low-carbon iron bars is somewhat difficult due to the physical softness of the product. A variation of this product contains approximately 0.15% P, which strengthens the ferritic structure and enhances its machinability. The phosphorus content is not detrimental to the soft magnetic capability of the iron. These grades of low-carbon magnetic iron are covered by ASTM specification A 848-87 and are sold to guaranteed magnetic-property capability limits.

After fabrication, the parts must be annealed to develop the desired magnetic characteristics. A typical annealing cycle consists of heating to approximately 815 to 845 °C (1500 to 1550 °F) in a reducing atmosphere (forming gas $90N_2$-$10H_2$, dissociated NH_3, or pure hydrogen) for 1 to 4 h at temperature, followed by furnace cooling. Bar products will typically exhibit a dc maximum permeability of 5×10^3 and a coercivity below 80 A $\cdot m^{-1}$ (1.0 Oe). These grades of low-carbon irons are produced to minimize magnetic aging following proper annealing. A typical guarantee would be 5% maximum increase in coercive force after aging at 100 °C (212 °F) for 200 h. Producers should be contacted for specific information concerning their products. Typical dc magnetic properties for annealed bar and heavy strip product forms of low-carbon magnetic iron and silicon irons are given in Table 1 and Table 2.

Low-Carbon Steels

For applications that require less than superior magnetic properties, low-carbon steels such as 1008, 1010, 12L14, and so on, are sometimes used (see the article "Classification and Designation of Carbon and Low-Alloy Steels" in Volume 1 of the 10th Edition of *Metals Handbook*). Such steels are not sold to magnetic quality specifications and may show considerable variation in quality, depending upon melting methods and the physical condition of the as-supplied product. Parts made from these steels will generally show improved soft magnetic characteristics if annealed as normally recommended for higher-quality, soft magnetic low-carbon iron. The degree of improvement is not guaranteed, however, and the parts may be subject to considerable magnetic aging over time. Due to their low cost, availability, and machinability, they have been used as pole pieces in electromagnets, magnetic clutches, and other noncritically designed flux carriers.

Compressed Powdered Iron. For applications in which complicated magnetic parts would otherwise require considerable machining, it may be helpful to press iron powder in a mold and sinter the part in vacuum or in a reducing atmosphere. Figure 6 shows anticipated effects of sintering temperature and sintering time on magnetic permeability of powdered iron, expressed as a percentage of the permeability of annealed, low-carbon, soft magnetic irons, for a constant magnetizing force.

Depending upon the powder metallurgy (P/M) techniques, starting powder quality, and final part configuration, the density of finished P/M iron parts can range from 6.2 g/cm³ (0.224 lb/in.³) to virtually full density of 7.8 g/cm³ (0.282 lb/in.³). It has been shown that the magnetic properties, as well as electrical resistivity, are a function of the density of the component. Higher saturation induction, residual induction, and maximum permeability result from increased density, whereas coercive force and resistivity are lowered.

There are presently two ASTM standards prepared that specifically cover magnetic properties of soft magnetic iron P/M parts (ASTM A 811-83) and iron powders containing 0.45 to 0.80% P intended for use in P/M-produced soft magnetic components (ASTM A 839-87).

The use of P/M techniques to make magnetically soft components may eliminate all machining operations and save a substantial amount of the total cost, compared to conventional manufacturing. If stress-inducing secondary machining operations are required to finish the sintered P/M part, then final stress-relief annealing may be required to restore its magnetic characteristics.

Low-Carbon Lamination Steel. Low-carbon sheet and strip products are available in a variety of thicknesses, commonly called

Table 1 Typical dc magnetic properties of low-carbon irons and silicon-iron bar products and heavy strip (>1.0 mm, or >0.040 in.) after annealing compared

Metal or alloy	Product form	ASTM test method(a)	Maximum permeability, $\mu_m \times 10^3$	From 1.5 T (15 kG) B_r, T (kG)	H_c, $A \cdot m^{-1}$ (Oe)	Saturation induction, B_i T	kG	Resistivity μΩ · cm
Low-carbon irons:								
Vacuum-melted high-purity iron	Bar	A 341	8	0.11–0.58 (1.1–5.8)	32 (0.40)	2.15	21.5	10
	Bar	A 596	17	1.00 (10)	20 (0.25)	2.15	21.5	10
	Strip	A 596	17	1.00 (10)	20 (0.25)	2.15	21.5	10
Air-melted magnetic iron(b)	Bar	A 341	7	0.25–1.20 (2.5–12)	60 (0.75)	2.15	21.5	13
	Bar	A 596	8	1.42 (14.2)	68 (0.85)	2.15	21.5	13
	Strip	A 596	11	1.38 (13.8)	48 (0.60)	2.15	21.5	13
Low-carbon steel:								
1010 steel	...	...	3.8	0.90 (9)	80–160 (1.0–2.0)	2.15	21.5	13
Silicon irons(c):								
1 Si-Fe	Strip	A 341	7.7	0.80–1.10 (8–11)	44 (0.55)	2.10	21.0	25
	Strip	A 596	14.8	1.31 (13.1)	32 (0.40)	2.10	21.0	25
1 Si-Fe (FM)(d)	Bar	A 341	9	0.20–1.10 (2–11)	40 (0.50)	2.10	21.0	25
2.5 Si-Fe	Strip	A 596	18	1.30 (13.0)	28 (0.35)	2.05	20.5	40
2.5 Si-Fe (FM)(d)	Bar	A 341	10	0.25–1.20 (2.5–12.0)	32 (0.40)	2.05	20.5	40
	Bar	A 596	14	1.37 (13.7)	24 (0.30)	2.05	20.5	40
4 Si-Fe	Bar	A 596	18.5	1.08 (10.8)	24 (0.30)	1.95	19.5	58

(a) ASTM A 341, straight length test samples using permeameter; ASTM A 596, ring test samples, machined or stamped. (b) ASTM Material Standard A 848-87. (c) All silicon irons shown are included in ASTM Material Standard A 867-86. (d) FM, free machining

Table 2 Nominal compositions, typical annealing cycles, and typical mechanical properties of low-carbon iron and silicon-iron bar products and heavy strip (>1.0 mm, or 0.040 in.)

Metal or alloy	Nominal composition C	Si	Al(a)	P	Fe	Annealing cycle used to develop Table 1 data(b) Temperature	Time, h	Atmosphere	Cooling(c)	Annealed mechanical properties: Ultimate tensile strength MPa	ksi	Yield strength MPa	ksi	Elongation, %	Hardness, HRB
Low-carbon irons:															
Vacuum-melted high-purity iron	<0.01	0.02	...	...	~99.9	955 °C (1750 °F)	4	Wet H_2	Furn	235	34	124	18	55	<20
Air-melted magnetic iron	0.01	0.1	...	...	~99.5	845 °C (1550 °F)	4	Wet H_2	Furn	283	41	165	24	48	38
Low-carbon steel:															
1010 steel	0.10	...	...	...	~99	...	...	...	...	310	45	172	25	32	42
Silicon irons:															
1 Si-Fe	0.02	1.1	0.5	...	bal	845 °C (1550 °F)	4	Wet H_2	Furn	345	50	172	25	35	50
1 Si-Fe (FM)(d)	0.02	1.1	0.5	0.15	bal	845 °C (1550 °F)	4	Wet H_2	Furn	434	63	262	38	40	70
2.5 Si-Fe	0.02	2.5	0.5	...	bal	1065 °C (1950 °F)	4	Dry H_2	Furn	517	75	414	60	35	85
2.5 Si-Fe (FM)(d)	0.02	2.4	0.5	0.15	bal	1065 °C (1950 °F)	4	Dry H_2	Furn	551	80	448	65	35	88
4 Si-Fe	0.02	4.0	0.5	...	bal	1065 °C (1950 °F)	4	Dry H_2	Furn	655	95	517	75	30	95

(a) Up to 0.5% Al optional—usually substituted for some of the silicon content. (b) Annealing cycles given only as an example—other cycles may be more desirable depending upon equipment available and specific application. (c) Furn, furnace cooled nominally 55 °C/h (100 °F/h) to 110 °C/h (200 °F/h). (d) FM, free machining

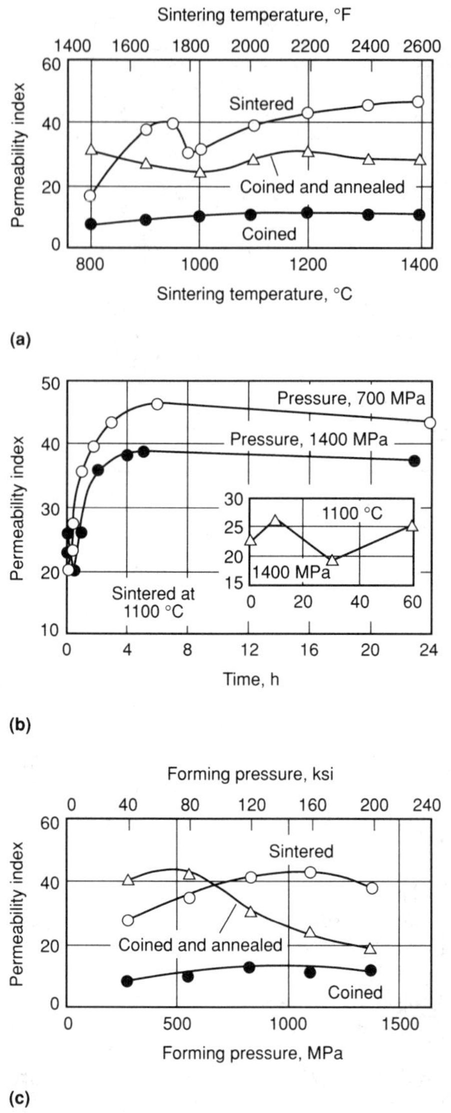

Fig. 6 Variations in the permeability index of P/M iron as a function of (a) sintering temperature, (b) duration of sintering, and (c) forming pressure. The magnetic permeability (for a constant magnetizing force) is shown as a percentage of the permeability of annealed, hot-rolled, low-carbon steel in (a) and (c).

motor lamination steel. These steels have low carbon content, typically less than 0.06%, and may have phosphorus and manganese added to increase resistivity and improve punchability. Motor lamination steel is usually intended for applications wherein the laminations will be annealed by the purchaser to develop the desired magnetic properties. The magnetic-property capability is not guaranteed by the producer except for special classes of this product. ASTM standard A 726-85 covers this category of product.

Similar low-carbon iron, but melted to tighter analysis requirements, is produced as fully processed cold-rolled lamination steel as described in ASTM A 840-85. Fully processed product refers to lamination steel manufactured by hot rolling, pickling, cold rolling to finish thickness, and continuous annealing to achieve core loss values below the maximum guaranteed limits. It is intended for applications where the lamination can be used in the as-stamped condition, without the need for further heat treatment. Although low-carbon steels exhibit power losses higher than those of silicon steels, they have better permeability at high flux density. This combination of magnetic properties, coupled with low price, makes low-carbon steels especially suitable for applications such as fractional-horsepower motors, which are used intermittently.

Silicon Steels (Flat-Rolled Products)

The beneficial effects of silicon additions to iron include:

- Increase of electrical resistivity
- Suppression of the γ loop enabling desirable grain growth
- Development of preferred orientation grain structure

The addition of silicon also reduces magnetocrystalline anisotropy energy, and at ~6.5% Si content reduces the magnetostriction constants to nearly zero. High-permeability and low hysteresis losses can therefore be attained at the 6.5Si-Fe composition. On the negative side, the addition of silicon to iron lowers magnetic saturation, lowers Curie temperature, and seriously decreases mechanical ductility. At silicon levels above ~4%, the alloy becomes brittle and difficult to process by cold-rolling methods; thus, few commercial steels contain more than ~3.5% Si.

The commercial grades of silicon steel in common use are made mostly in electric or basic oxygen furnaces. Continuous casting and/or vacuum degassing (V-D) may be employed. Flat-rolled silicon-iron sheet and strip has low sulfur content, typically below 0.025%, with better grades below 0.01%. Manganese may be present up to approximately 0.70%. Residual elements such as chromium, molybdenum, nickel, copper, and phosphorus may also be present. The major alloying addition is silicon plus up to 0.6% Al (optional). These alloys are not generally sold on the basis of their composition, but rather are sold based upon controlled magnetic properties, particularly ac core losses.

Table 3 gives examples of properties specified by ASTM and American Iron and Steel Institute (AISI) for standard grades of electrical steel. The AISI designations were adopted in 1946 to eliminate the wide variety in nomenclature formerly used. When originally adopted, the AISI designation number approximated ten times the maximum core loss, in watts per pound, exhibited by 29 gage (0.36 mm, or 0.014 in.) samples when tested at a flux density of 1.5 T (15 kG) and a magnetic circuit frequency of 60 Hz. Note that fully processed M-36 tested as 0.36 mm (0.014 in.) strip now has a maximum allowable core loss of 4.2 W/kg (1.9 W/lb), not an approximate level of 7.9 W/kg (3.6 W/lb).

The AISI designations are still in common use, but the newer ASTM designations provide more specific information regarding the grade identified. A typical ASTM designation is 47S200. The first two digits of the ASTM designation indicate the thickness in mm (×100). Following these digits is a letter (C, D, F, S, G, H, or P) that indicates the material type and the respective magnetic test conditions. The last three digits provide an indication of the maximum allowable core loss in units of either (watts/kg) × 100, or (watts/lb) × 100. If the core-loss value is expressed in watts/kg, the grade designation takes the suffix M, indicating an ASTM metric standard. Several ASTM flat-rolled products specifications are written in English and metric versions, such as A 677-84 and its companion metric specification A 677M-83. A general summary of the ASTM code letter designation is shown in Table 4. Refer to ASTM standard A 664-87 for a complete explanation of the identification practice and the conditions that apply to the test parameters.

The typical relative peak permeability at 60 Hz and 1.5 T (15 kG) and typical 60 Hz rms excitation to produce 1.5 T (15 kG) induction are shown, because these are useful design parameters for the application of these materials. Best permeability at high induction is obtained in steels with lower silicon contents. Low core loss is obtained with higher silicon contents, larger grains, lower impurity levels, and thinner gages. Nonoriented silicon irons are preferred for motor laminations.

Nonoriented Silicon Steels. Nonoriented (isotropic) flat-rolled products are available in semiprocessed and fully processed conditions and contain 0.5 to 3.5% Si. The vast majority of finished nonoriented silicon steel is sold in either full-width coils (860 to 1220 mm, or 34 to 48 in. widths) or slit-width coils, but some are sold as sheared sheets. Fully processed electrical steel frequently is coated with organic or inorganic materials after mill annealing to reduce eddy currents in lamination stacks.

Semiprocessed Grades. The carbon contents of semiprocessed grades are relatively low, usually below 0.030%. However, semiprocessed product is not sufficiently decarburized for general use as supplied; therefore, decarburization and annealing to develop potential magnetic quality and to avoid magnetic aging must be done by the user. Anneals of this type are typically performed at temperatures between 790 and 845 °C (1450 and 1550 °F) for approximately 1 h with a suitable decarburization atmosphere. The atmosphere must contain sufficient moisture to promote decarburization without excessive oxidation of the metal.

Table 3 Magnetic and thermal properties of selected conventional and high permeability flat-rolled electrical steels

AISI type (approximate equivalent)	Nominal (Si + Al) content, %	Thickness		ASTM designation	Maximum core loss at 60 Hz and B = 1.5 T (15 kG)		Typical relative peak permeability, μ_p		Typical 60 Hz rms excitation to produce 1.5T (15 kG) $A \cdot cm^{-1}$	Saturation induction, b_s		Thermal conductivity	
		mm	in.		W/kg	W/lb	at 60 Hz and 1.5 T (15 kG)	at H_p of 0.8 kA · m^{-1} (10 Oe)		T	kG	W/m · K	Btu/ft · h · °F
Nonoriented													
Semiprocessed (ASTM A 683)(a)													
M-47	1.10	0.64	0.025	64S350	7.71	3.50	2000	...	3.50	...	...	...	...
	1.10	0.47	0.019	47S300	6.61	3.00	1900	...	3.50	...	...	...	...
M-45	1.70	0.64	0.025	64S280	6.17	2.80	1900	...	3.50	...	...	...	...
	1.70	0.47	0.019	47S250	5.51	2.50	1750	...	4.00	...	...	...	...
M-43	2.00	0.64	0.025	64S260	5.73	2.60	1850	...	3.50	...	...	...	...
	2.00	0.47	0.019	47S230	5.07	2.30	1700	...	4.00	...	...	...	...
M-36	2.40	0.64	0.025	64S230	5.07	2.30	1750	...	4.00	...	...	...	...
	2.40	0.47	0.019	47S200	4.41	2.00	1600	...	4.50	...	...	...	...
M-27	2.70	0.64	0.025	64S213	4.69	2.13	1450	...	4.50	...	...	...	...
	2.70	0.47	0.019	47S188	4.14	1.88	1300	...	5.50	...	...	...	...
...	3.00	0.64	0.025	64S194	4.28	1.94	1300	...	5.50	...	...	...	...
...	3.00	0.47	0.019	47S178	3.92	1.78	1000	...	6.50	...	...	...	...
Fully processed (ASTM A 677)(b)													
...	0.50	0.64	0.025	64F600	13.22	6.00	1600	...	3.5	...	...	...	...
...	0.80	0.47	0.019	47F450	9.92	4.50	1450	...	4.0	...	...	...	...
M-47	1.05	0.64	0.025	64F470	10.36	4.70	1500	...	4.5	2.11	21.1	37.7	21.8
	1.05	0.47	0.019	47F380	8.38	3.80	1300	...	4.0	2.11	21.1	37.7	21.8
M-45	1.85	0.64	0.025	64F340	7.49	3.40	1300	...	5.0	2.07	20.7	25.1	14.5
	1.85	0.47	0.019	47F290	6.39	2.90	1250	...	5.0	2.07	20.7	25.1	14.5
M-43	2.35	0.64	0.025	64F270	5.95	2.70	1200	...	5.5	2.04	20.4	20.9	12.1
	2.35	0.47	0.019	47F230	5.07	2.30	1100	...	5.5	2.04	20.4	20.9	12.1
M-36	2.65	0.64	0.025	64F240	5.29	2.40	1000	...	6.5	2.02	20.2	18.8	10.9
	2.65	0.47	0.019	47F205	4.52	2.05	930	...	6.5	2.02	20.2	18.8	10.9
	2.65	0.36	0.014	36F190	4.19	1.90	800	...	7.0	2.02	20.2	18.8	10.9
M-27	2.8	0.64	0.025	64F225	4.96	2.25	950	...	7.0	2.02	20.2	18.8	10.9
	2.8	0.47	0.019	47F190	4.19	1.90	870	...	7.5	2.02	20.2	18.8	10.9
	2.8	0.36	0.014	36F180	3.97	1.80	760	...	8.0	2.02	20.2	18.8	10.9
M-22	3.2	0.64	0.025	64F218	4.80	2.18	870	...	8.0	2.00	20.0	18.8	10.9
	3.2	0.47	0.019	47F185	4.08	1.85	750	...	8.5	2.00	20.0	18.8	10.9
	3.2	0.36	0.014	36F168	3.70	1.68	690	...	9.0	2.00	20.0	18.8	10.9
M-19	3.3	0.64	0.025	64F208	4.58	2.08	800	...	8.5	1.99	19.9	16.7	9.7
	3.3	0.47	0.019	47F174	3.83	1.74	660	...	9.0	1.99	19.9	16.7	9.7
	3.3	0.36	0.014	36F158	3.48	1.58	500	...	10.0	1.99	19.9	16.7	9.7
M-15	3.5	0.47	0.019	47F168	3.70	1.68	375	...	10.0	1.98	19.8	16.7	9.7
	3.5	0.36	0.014	36F145	3.20	1.45	300	...	12.0	1.98	19.8	16.7	9.7
Oriented													
Fully processed (ASTM A 876)(c)													
M-6	3.15	0.35	0.014	35G066	1.45	0.66	...	>1800	...	2.00	20.0	16.7	9.7
	3.15	0.35	0.014	35H094	2.07(d)	0.94	...	>1800	...	2.00	20.0	16.7	9.7
M-5	3.15	0.30	0.012	30G058	1.28	0.58	...	>1800	...	2.00	20.0	16.7	9.7
	3.15	0.30	0.012	30H083	1.83(d)	0.83	...	>1800	...	2.00	20.0	16.7	9.7
M-4	3.15	0.27	0.011	27G051	1.12	0.51	...	>1800	...	2.00	20.0	16.7	9.7
	3.15	0.27	0.011	27H074	1.63(d)	0.74	...	>1800	...	2.00	20.0	16.7	9.7
...	3.15	0.23	0.009	23G046	1.01	0.46	...	>1800	...	2.00	20.0	16.7	9.7
...	3.15	0.23	0.009	23H071	1.56(d)	0.71	...	>1800	...	2.00	20.0	16.7	9.7
...	3.15	0.27	0.011	27P066	1.45(d)	0.66	...	>1880	...	2.00	20.0	16.7	9.7
...	3.15	0.30	0.012	30P070	1.54(d)	0.70	...	>1880	...	2.00	20.0	16.7	9.7

(a) Refer to ASTM A 683-84 and companion specification A 683M-84 (metric) for detailed information. (b) Refer to ASTM A 677-84 and companion specification A 677M-83 (metric) for detailed information. (c) Refer to ASTM A 876-87 and companion specification A 876M-87 for detailed information. (d) B = 1.7 T (17 kG)

An atmosphere of 20% H_2, 80% N_2, and a dew point of 15 °C (55 °F) often meets these requirements. However, it is necessary to contact the manufacturer for specific recommendations and procedures for safe and proper annealing of these alloys.

Semiprocessed electrical steels are usually supplied without a surface insulation coating or with only a thin, tightly adhering oxide to provide insulation resistance.

Fully processed grades are process annealed by the manufacturer in moist hydrogen at about 825 °C (1520 °F) to reduce carbon. The final annealing operation is carried out by the manufacturer at a higher temperature (up to 1100 °C, or 2010 °F for continuous strip) to promote grain growth and development of magnetic properties. The desirable magnetic characteristics are thus produced during the manufacturing so additional heat treatment by the purchaser is generally not required. All coils are sampled and tested in accordance with ASTM specifications A 343, A 347, and A 804 method of test, and graded as to quality. These products are primarily intended for commercial power frequency (50 to 60 Hz) applications and are sold to maximum core-loss limits at a particular induction, typically 1.5 T (15 kG).

Oriented Silicon Steels. Grain size is important in silicon steel with regard to core losses and low flux-density permeability. However, for high-flux density permeability, crystallographic orientation is a major controlling factor. Like iron, silicon steels are more easily magnetized in the direction of the cube edge, {100}, as shown in Fig. 7.

As mentioned previously, when the silicon content in pure iron exceeds approximately 2½%, the allotropic transformation of iron from α to γ is suppressed. The absence of this transformation allows the higher silicon-iron alloy to be fully ferritic up to the melting point. This behavior permits the manufacturer of these strip products to apply special cold-rolling and heat-treating techniques to promote secondary recrystallization in the final anneal. The processing results in a well-developed crystallographic texture with the cube edge parallel to the rolling direction {110}{001}, often referred to as the Goss or cube-on-edge orientation. Conventional-

Table 4 ASTM letter code designations for electrical steel and lamination steel grades (ASTM standard A 664-87)

Letter code	Grade	Condition	Core loss test parameters with cyclic frequency, *f*, at 60 Hz: Induction, *B*, T	kG	Test specimen	Specimen condition
C	Low-carbon lamination steel	Fully processed	1.5	15	50/50 grain Epstein	As sheared
D	Low-carbon lamination steel	Semiprocessed	1.5	15	50/50 grain Epstein	Quality development annealed(a)
F	Nonoriented electrical steel	Fully processed	1.5	15	50/50 grain Epstein	As sheared
S	Nonoriented electrical steel	Semiprocessed	1.5	15	50/50 grain Epstein	Quality development annealed(a)
G	Grain-oriented electrical steel	Fully processed	1.5	15	Parallel grain Epstein	Stress-relief annealed(b)
H	Grain-oriented electrical steel	Fully processed	1.7	17	Parallel grain Epstein	Stress-relief annealed(b)
P	Grain-oriented electrical steel, high permeability	Fully processed	1.7	17	Parallel grain Epstein	Stress-relief annealed(b)

Note: Refer to ASTM standards for detailed information regarding test methods, heat treatments, and so on. (a) Quality development anneal—either 790 °C (1450 °F) or 845 °C (1550 °F) soak 1 h. Temperature depends on particular grade. (b) Stress relief anneal—usually in the range from 790 °C (1450 °F) to 845 °C (1550 °F) for 1 h

oriented (anisotropic) grades contain about 3.15% Si.

Around 1970, improved {110}⟨001⟩ crystallographic texture was developed by modification of composition and processing. The improved high-permeability material usually contains about 2.9 to 3.2% Si. Conventional grain-oriented 3.15% Si steel has grains about 3 mm (0.12 in.) in diameter. The high-permeability silicon steel tends to have grains about 8 mm (0.31 in.) or larger in diameter. Ideally, grain diameter should be less than 3 mm (0.12 in.) to minimize excess eddy-current effects from domain-wall motion. Special coatings provide electrical insulation and induced tensile stresses in the steel substrate. In this case, the induced stresses lower core loss and minimize noise in transformers.

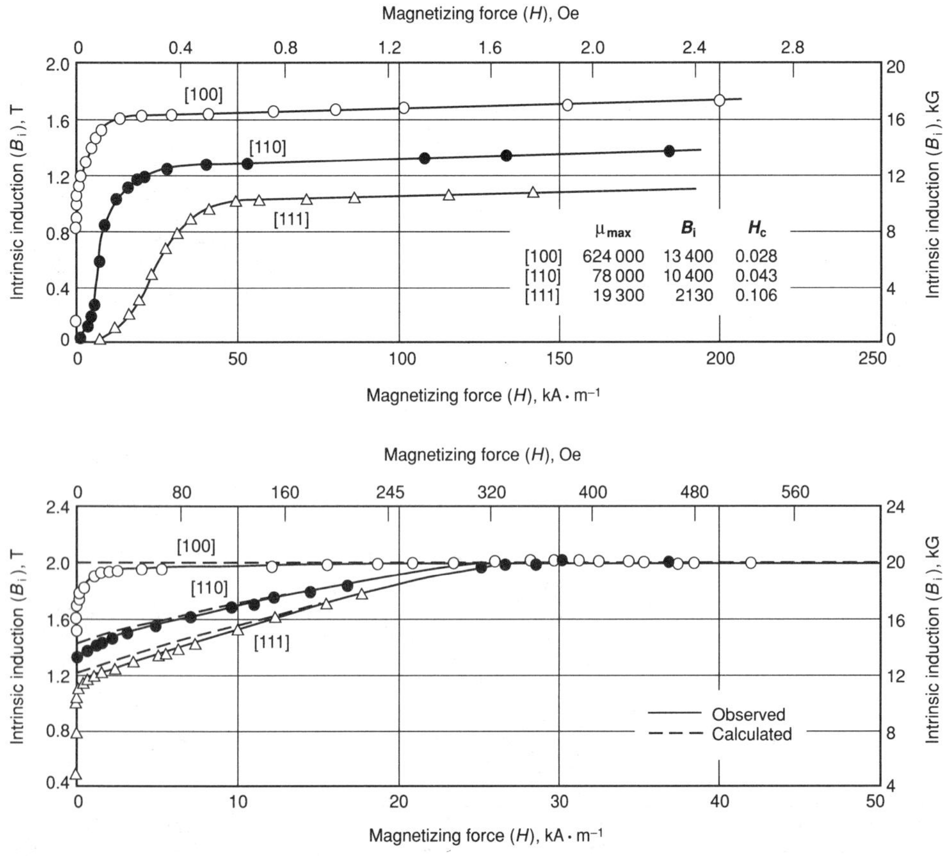

Fig. 7 Observed and calculated *B-H* curves for [100], [110], and [111] directions in single crystals of 3 to 3.5% Si steel

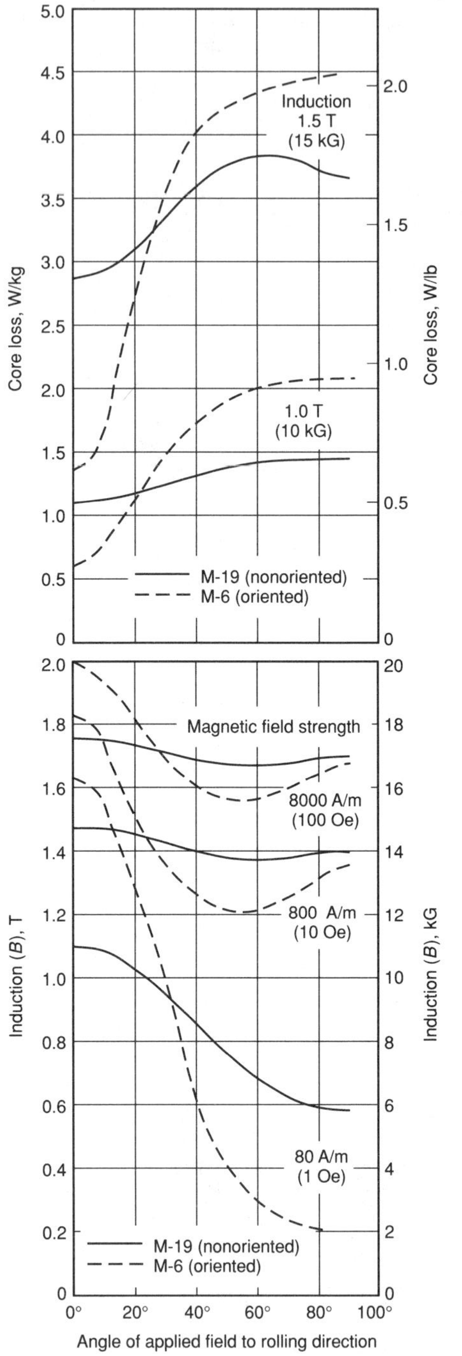

Fig. 8 Comparative flux densities and core losses for nonoriented M-19 and oriented M-6 electrical steels as a function of the direction of applied field. Steel thickness is 0.36 mm (0.014 in.).

Oriented Silicon Steel Versus Nonoriented Silicon Steel. Figure 8 compares the variation in flux density and core loss, with respect to direction of rolling, for oriented and nonoriented silicon steels. These curves indicate the advantage of using oriented steels in a manner such that the critical flux path of the application is parallel to the sheet or strip cold-rolling direction. Corresponding *B-H* curves and half hysteresis loops are given in Fig. 9.

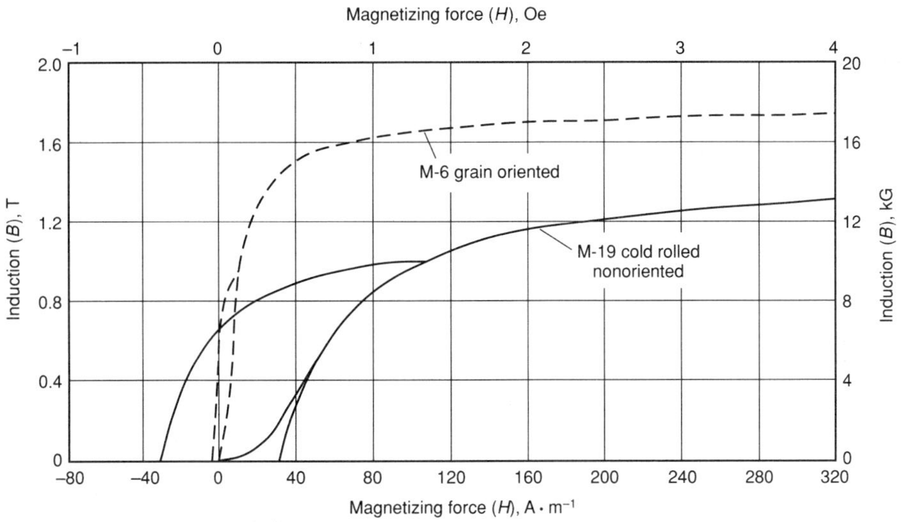

Fig. 9 Half hysteresis loops and dc magnetization curves for grain-oriented M-6 and cold-rolled nonoriented M-19 steels. Steel thickness is 0.36 mm (0.014 in.).

Table 5 Silicon contents, mass densities, and applications of electrical steel sheet and strip

ASTM specification	AISI type	Nominal (Si + Al) content, %	Assumed density, g/cm^3	Characteristics and applications
Lamination steel				
A 726 or A 840	...	0	7.85	High magnetic saturation; magnetic properties may not be guaranteed; intermittent-duty small motors
Nonoriented electrical steels				
A 677 or A 677M (fully processed) and A 683 or A 683M (semiprocessed)	M-47	1.05	7.80	Ductile, good stamping properties, good permeability at high inductions; small motors, ballasts, relays
	M-45	1.85	7.75	Good stamping properties, good permeability at moderate and high inductions, good core loss; small generators, high-efficiency continuous-duty rotating machines, ac and dc
	M-43	2.35	7.70	
	M-36	2.65	7.70	Good permeability at low and moderate inductions, low core loss; high reactance cores, generators, stators of high-efficiency rotating machines
	M-27	2.80	7.70	
	M-22(a)	3.20	7.65	Excellent permeability at low inductions, lowest core loss; small power transformers, high-efficiency rotating machines
	M-19(a)	3.30	7.65	
	M-15(a)	3.50	7.65	
Oriented electrical steels				
A 876 or A 876 M	M-6	3.15	7.65	Grain-oriented steel has highly directional magnetic properties with lowest core loss and highest permeability when flux path is parallel to rolling direction; heavier thicknesses used in power transformers, thinner thicknesses generally used in distribution transformers. Energy savings improve with lower core loss.
	M-5	3.15	7.65	
	M-4	3.15	7.65	
	M-3	3.15	7.65	
High-permeability oriented steel				
	...	2.9–3.15	7.65	Low core loss at high operating inductions

(a) ASTM A 677 only

Aging. High-grade, fully processed, silicon electrical steel does not age as received from the mill because its carbon content has been reduced to about 0.003% or less. With higher carbon contents, however, core loss can increase with time because of carbide precipitation. Also, silicon steel may age appreciably if not correctly heat treated in a manner that completely stabilizes its physical structure. Table 5 lists typical applications of electrical steel sheet and strip.

Silicon-Iron Bar and Heavy Strip

Silicon irons containing between approximately 1 and 4% Si (and optionally up to approximately 0.5% Al) have been used extensively for the manufacture of ac and dc relay cores, printer hammers, flux-path components, pole pieces, and rapidly activated solenoids such as automotive fuel injectors, and so on. For high-volume screw-machine applications, free-machining grades containing approximately 0.15% P are produced. At this level, phosphorus has no significant effect on the magnetic characteristics. These grades have been commonly known as relay steels or silicon core irons.

These products are typically supplied in a condition suitable for forging, machining, stamping, or forming. After fabrication, the parts should be annealed to remove stress, to increase grain size, and to reduce their carbon content. ASTM standard A 867-86 describes the characteristics of silicon-iron bar products and provides dc magnetic-property capability, which can be guaranteed by the alloy manufacturers. Tables 1 and 2 present dc magnetic properties for silicon-iron bar and heavy strip products.

Silicon irons containing approximately 1% Si exhibit the phase transformation discussed for low-carbon irons. As a result, it is common practice to anneal these alloys at 815 to 870 °C (1500 to 1600 °F), which is below the Ac_1 transformation temperature. Alloys containing ~2.5% Si can be annealed at higher temperatures and show continued improvement in permeability with increasing annealing temperature. Figure 10 shows the dc normal induction curves for 2.5Si-Fe annealed at four temperatures from 730 to 1065 °C (1350 to 1950 °F). The maximum permeabilities ranged from a low of approximately 1.7×10^3 to a high of approximately 1.4×10^4 in the example, thus illustrating the importance of proper annealing. The manufacturer of these alloys should be contacted for specific annealing information.

Iron-Aluminum Alloys

Although aluminum and silicon have similar effects on electrical resistivity and some magnetic properties of iron, aluminum is seldom substituted for silicon because of the resulting difficulties in fabrication. Aluminum is used most commonly as small (<0.5%) additions to the better grades of nonoriented silicon steel to increase electrical resistivity and thereby reduce eddy currents without impairing cold workability. Alloys of 12% Al or 16% Al and iron have high resistivity and can provide high permeability. At low flux densities, the magnetic properties of these alloys can be made to approach those of some of the low nickel-content nickel-iron alloys (see the section "Nickel-Iron Alloys" in this article). High-aluminum alloys are very difficult to pro-

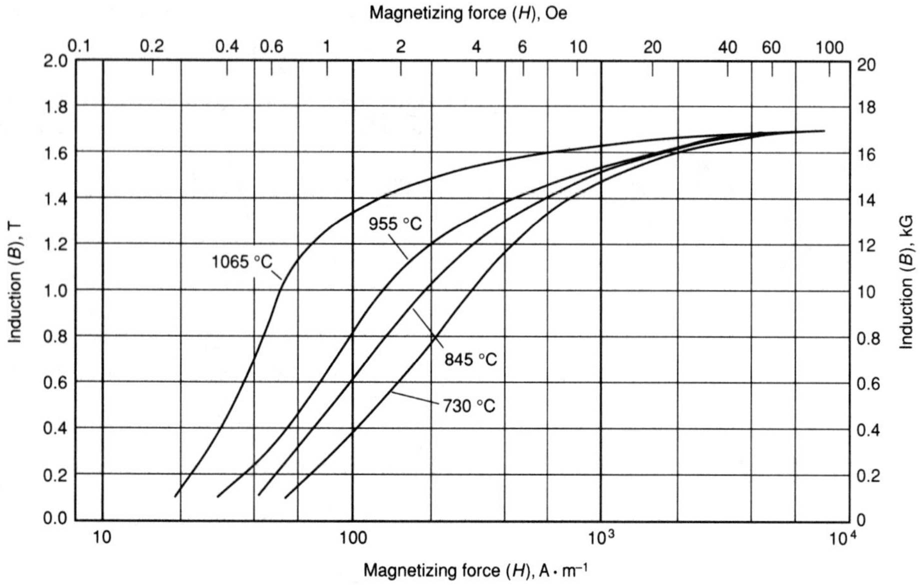

Fig. 10 Direct-current normal induction curves for 2.5% Si-Fe heavy strip, that was annealed at various temperatures. Ring test specimens are from 1.52 mm (0.060 in.) thick strip, annealed in pure dry hydrogen, 2 h at heat, then furnace cooled. Tested per ASTM standard A 596

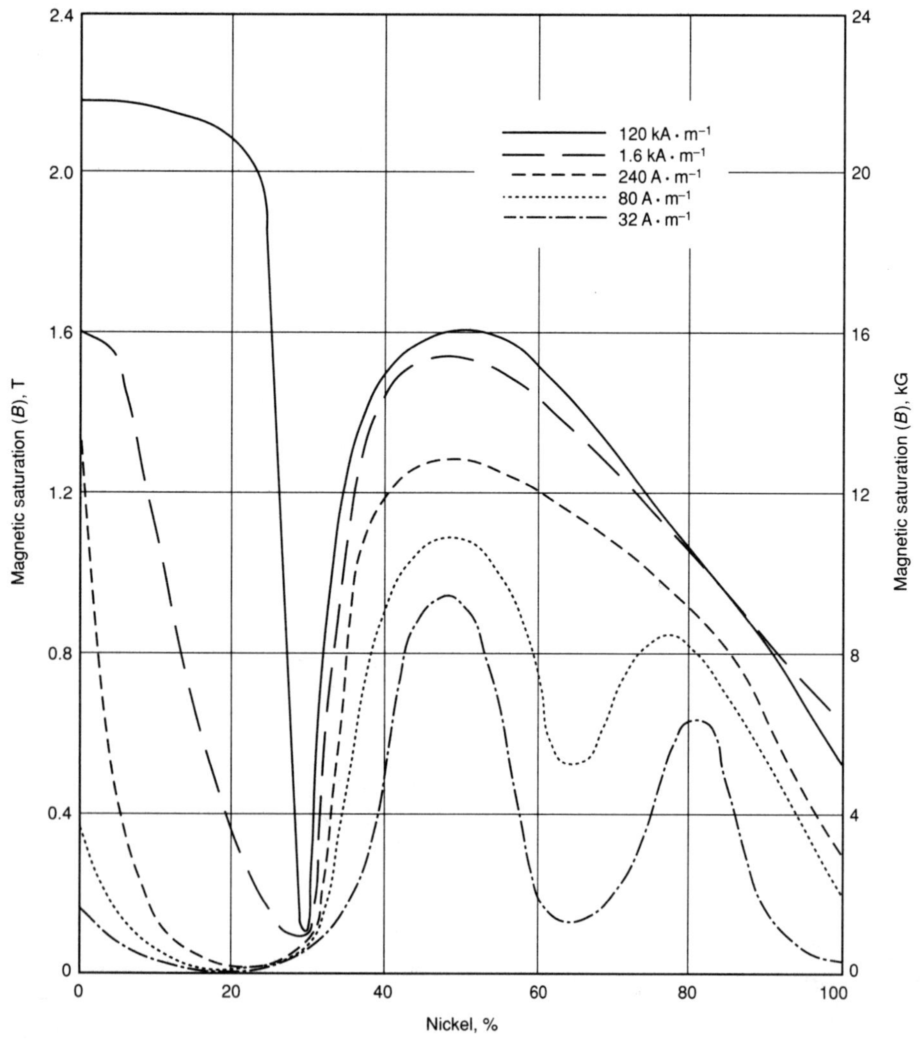

Fig. 11 Magnetic saturation of binary nickel-iron alloys at various field strengths. All samples were annealed at 1000 °C (1830 °F) and cooled in the furnace.

cess and have not been readily available in large quantities.

Nickel-Iron Alloys

The effect of nickel content in nickel-iron alloys on saturation induction (B_s) and on initial permeability (μ_0) after annealing are illustrated in Fig. 11 and Fig. 12. Below ~28% Ni, the crystalline structure is bcc low-carbon martensite if cooled rapidly and ferrite and austenite if cooled slowly, and these alloys are not considered useful for soft magnetic applications. Above ~28% Ni, the structure is fcc austenite. The Curie temperature in this system is approximately room temperature at ~28% Ni and increases rapidly up to ~610 °C (1130 °F) at 68% Ni. Thus, these austenitic alloys are ferromagnetic. The magnetic properties are controlled by saturation magnetization and the magnetic anisotropy energies, particularly magnetocrystalline (K_1) and magnetostrictive (λ_1) anisotropies.

Two broad classes of commercial alloys have been developed in the nickel-iron system. The high-nickel alloys (about 79% Ni) have high initial and maximum permeabilities and very low hysteresis losses, but they also have a saturation induction of only ~0.8 T (8 kG). The low-nickel alloys (about 45 to 50% Ni) are lower in initial and maximum permeability than the 79% Ni alloys, but are still much higher than silicon irons. The low-nickel alloys have a saturation induction of about 1.5 T (15 kG). Values of initial permeability (at *B* of 4 mT, or 40 G) above 1.2×10^4 are typically obtained in low-nickel alloys, and values above 6.5×10^4 are obtained for 79Ni-4Mo-Fe alloys at 60 Hz using 0.36 mm (0.014 in.) thick laminations. Maximum dc permeabilities of 1.4×10^5 for low-nickel alloys and 3.75×10^5 for high-nickel alloys are routinely attained.

Figure 13 contains data from early laboratory studies to illustrate the effect of both composition and heat treatment on initial permeability. To obtain very high magnetic permeability, both the magnetocrystalline anisotropy (K_1) and the magnetostrictive anisotropy (λ_1) must be minimized. The magnetostrictive anisotropy is highly dependent upon the alloy composition. Generally, there is little that the purchaser can do to the alloy that will change this characteristic. However, in the high-nickel alloys, the magnetocrystalline anisotropy can be altered by the appropriate annealing cycle cooling rate. In these alloys, short-range atomic ordering occurs as the alloy is cooled from ~760 °C (1400 °F) to ~400 °C (750 °F). The degree of ordering has a profound effect on K_1, and, therefore, each composition will have an optimum cooling rate that minimizes the net anisotropy energies and can result in very high permeability.

High-Nickel Alloys. Figure 13 illustrates that in ~78.5Ni-Fe the initial permeability was low after either furnace cooling or

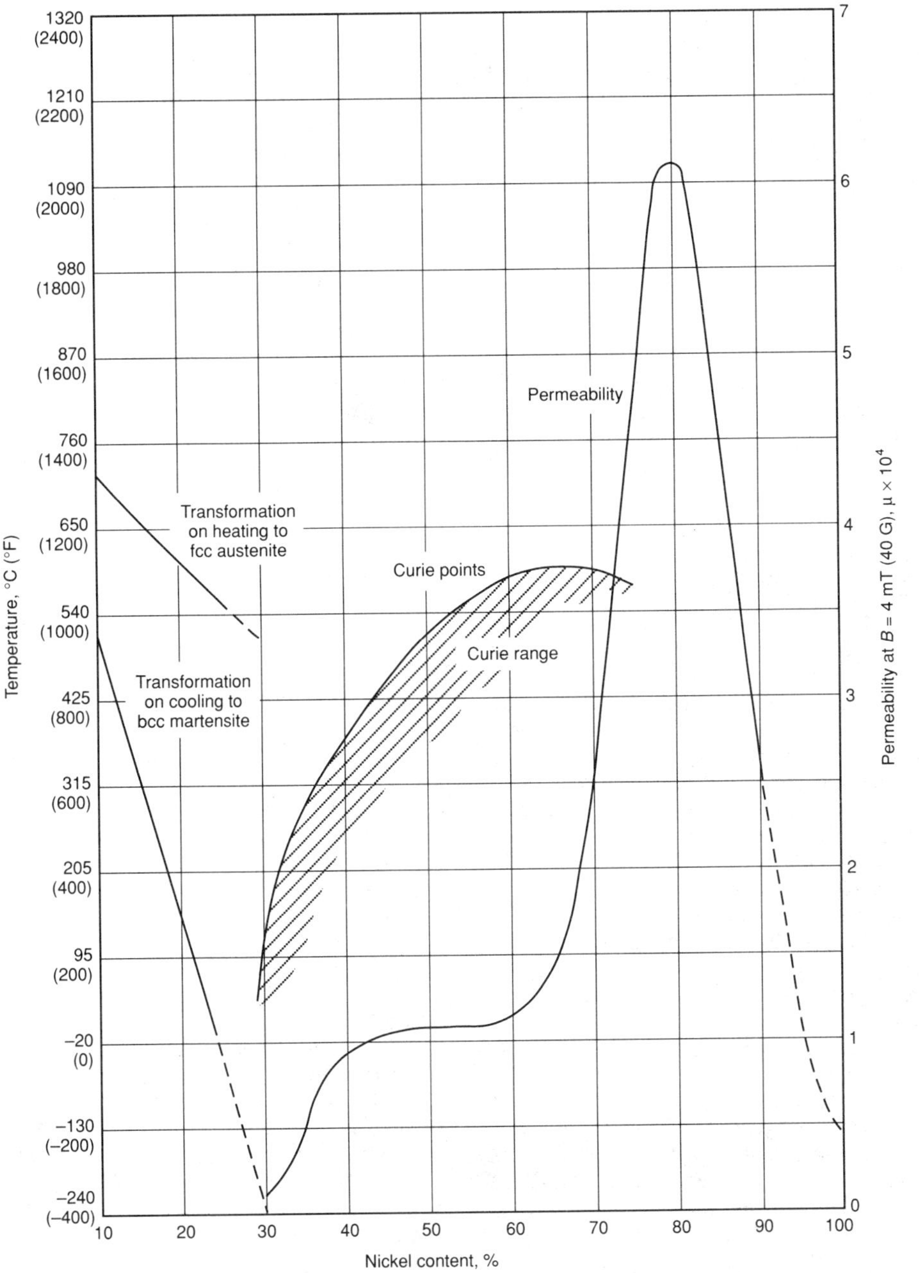

Fig. 12 Effect of nickel content on initial permeability, Curie temperature, and transformation in nickel-iron alloys

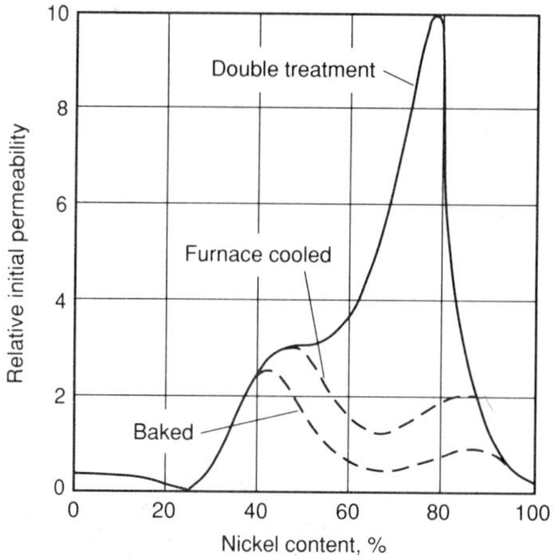

Fig. 13 Relative initial permeability at 2 mT (20 G) for Ni-Fe alloys given various heat treatments. Treatments were as follows: furnace cooled—1 h at 900 to 950 °C (1650 to 1740 °F), cooled at 100 °C/h (180 °F/h); baked—furnace cooled plus 20 h at 450 °C (840 °F); double treatment—furnace cooled plus 1 h at 600 °C (1110 °F) and cooled at 1500 °C/min (2700 °F/min).

baking at 450 °C (840 °F). However, if the same alloy was rapidly cooled from 600 °C (1110 °F), the initial permeability was increased dramatically. High-purity 78.5Ni-Fe can exhibit an initial dc permeability of 5 $\times 10^4$ and a maximum permeability of 3 $\times 10^5$. These properties are obtained on ring laminations annealed in dry hydrogen at 1175 °C (2150 °F), rapid furnace cooled to room temperature, then reheated to 600 °C (1110 °F), and oil quenched. This alloy has limited commercial use because the complex heat treatment is not easily performed on parts. Also, its electrical resistivity is only 16 μΩ · cm, which allows large eddy-current losses in ac applications.

Alloying additions of 4 to 5% Mo, or of copper and chromium to ~79Ni-Fe, alter the kinetics of ordering and the magnetostrictive anisotropy energy. Alloying also increases the electrical resistivity; however, saturation induction is reduced to the 0.8 T (8 kG) level. Specialty steel manufacturers produce a variety of carefully designed very high-permeability alloys that permit the use of practical commercial annealing procedures. Popular alloys include the Moly-Permalloys (typically 80Ni-4 to 5Mo-bal Fe) and MuMetals (typically 77Ni-5Cu-2Cr-bal Fe).

High-permeability alloys must also be of high commercial purity. Air- and vacuum-melting practices are both used to produce low-nickel alloys, but nearly all of the high-nickel alloys are produced by VIM. Figure 14 shows a historical perspective of the change in initial permeability of 80Ni-4Mo-Fe alloys when VIM became widely used around 1960. Interstitial impurities such as carbon, sulfur, oxygen, and nitrogen must be minimized by special melting procedures and by careful final annealing of laminations and other core configurations. Sulfur contents higher than several ppm and carbon in excess of 20 ppm are detrimental to final-annealed magnetic properties in high-nickel alloys.

Laminations or parts made from these high-nickel alloys are usually commercially annealed in pure dry hydrogen (dew point less than −50 °C, or −58 °F, at ~1000 to 1205 °C, or 1830 to 2200 °F, for several hours to eliminate stresses, to increase grain size, and to provide for alloy purification). They are cooled at any practical rate down to the critical ordering temperature range. The rate of cooling through the ordering range is typically 55 °C/h (100 °F/h) to 350 °C/h (630 °F/h), depending upon the alloy being heat treated. Although the cooling rate below the ordering range is not critical, stresses due to rapid quenching must be avoided.

Vacuum furnaces may be used to anneal some high-nickel soft magnetic alloys if the application does not demand the optimum magnetic properties. However, dry hydrogen is strongly recommended for annealing nickel-iron alloys. Parts must always be thoroughly degreased to remove oils (particularly sulfur-bearing oils) prior to annealing.

Low-Nickel Alloys. In alloys containing approximately 45 to 50% Ni, the effect of cooling rate on initial permeability is not

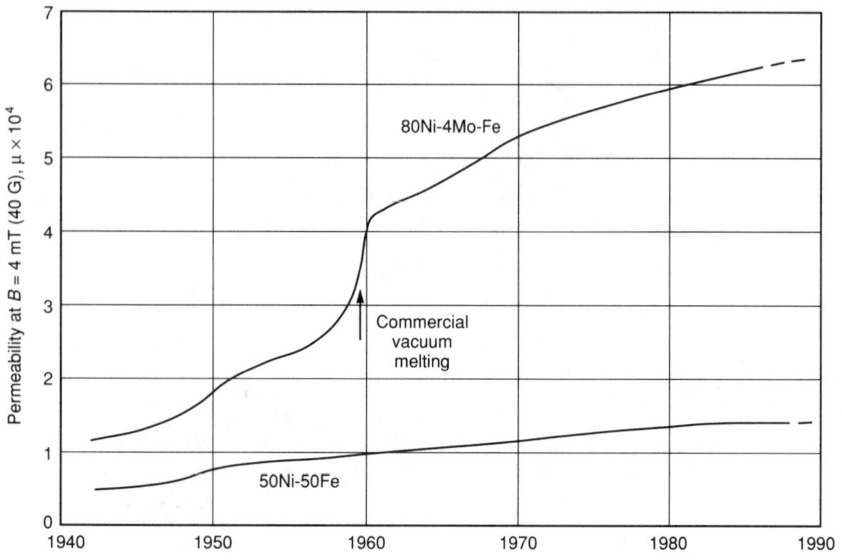

Fig. 14 Progress in initial permeability values of commercial-grade nickel-iron alloys since early 1940s. Frequency, *f*, is 60 Hz. Thickness of annealed laminations was 0.36 mm (0.014 in.).

great, as evidenced in Fig. 13. The typical annealing cycle to develop high permeability for these low-nickel alloys is similar to the high-nickel cycle, except that any cooling rate between ~55 °C/h (100 °F/h) and ~140 °C/h (252 °F/h) is usually suggested. A dry hydrogen atmosphere is also recommended for annealing low-nickel alloys.

Magnetic Properties of Nickel-Iron Alloys. The nickel-iron alloys are generally manufactured as strip or sheet product; however, billet, bar, and wire can be produced as needed. Strip products are usually supplied in a cold-rolled condition for stamping laminations or as thin foil for winding of tape toroidal cores. Strip and sheet products may also be supplied in a low-temperature, mill-annealed, fine-grain condition suitable for forming and deep drawing. ASTM standard A 753 describes the as-supplied condition and the magnetic property capabilities of many of the higher-volume-usage nickel-iron alloys. Tables 6, 7, and 8 provide typical dc and ac magnetic characteristics and some mechanical properties of nickel-iron alloys.

Nickel-iron alloys can be processed and heat treated to develop a wide range of properties for a variety of applications. The scope of these applications is too extensive to report in detail in this article, thus only a brief listing of several applications can be presented (Table 9). The high-nickel alloys have been used extensively for magnetic shielding, high-quality, low-noise audio frequency transformers, ground fault interrupter cores, antishoplifting devices, tape recorder head laminations, magnetometer bobbin cores, and so on. Domestic producers of the specialty alloys listed in Table 9 are shown in Table 10.

Various articles present extremely high permeability figures for Supermalloy, which was originally developed as a high-purity vacuum-induction-melted 79Ni-5Mo-Fe alloy at a time when most nickel-iron alloys were still being air melted. Furthermore, Supermalloy data generally refers to thin foil (0.1 mm, or 0.004 in., and less) tape toroidal cores, annealed in pure dry hydrogen at ~1290 °C (2350 °F), held for several hours, and very carefully cooled at the optimum rate. In some cases, the cooling may be performed under the influence of a magnetic field. It should, therefore, be apparent that Supermalloy properties represent a specialized case of handling a high-quality 79Ni-5Mo-Fe alloy. Supermalloy properties cannot be obtained in product forms such as bars or heavy strip, even if the same basic alloy was used to manufacture the heavier products.

Most applications of the 50Ni-Fe alloys are based on the requirements of moderately high saturation and permeabilities greater than those available on silicon irons. Heavy strip of 50Ni-Fe has been used for the manufacture of sensitive relays and safety shutoff valves for gas-fired devices. Lamination thicknesses 0.50 to 0.1 mm (0.020 to 0.004 in.) can be produced as a semioriented product that provides its best magnetic properties, after annealing, in the direction of cold rolling. Referred to as the transformer grade, it has been used primarily in audio frequency transformers. Lamination thicknesses can also be manufactured in a nonoriented condition used for rotating components such as resolvers, servosynchros, and rotor laminations. This condition is generally known as motor or rotor grade.

In 50% Ni alloys, the magnetocrystalline anisotropy is not zero. By proper melt control, manufacturing practice, and careful annealing, a strong cube texture can be developed, particularly in tape toroidal cores. These cores will exhibit a square hysteresis loop. The ratio of remanence to saturation induction (B_r/B_s) will typically exceed 0.98. These products have been known as Deltamax (Allegheny Ludlum Corporation), Hipernik V (Westinghouse Electric Corporation), HyRa 49 (Carpenter Technology Corporation), and Orthonol (Spang & Company, Specialty Metals Division).

Table 6 Typical dc magnetic properties of annealed high-permeability nickel-iron alloys. Data for 0.30 to 1.52 mm (0.012 to 0.060 in.) thickness strip; ring laminations annealed in dry hydrogen at 1175 °C (2150 °F) (unless otherwise noted), 2 to 4 h at temperature. ASTM A 596

Alloy	Permeability		Approximate induction at maximum permeability, μ_m		Residual induction B_r		Coercive force, H_c		Saturation induction, (B_s)		Resistivity, $\mu\Omega \cdot cm$
	Initial × 10^3	Maximum, μ_m × 10^3	T	kG	T	kG	$A \cdot m^{-1}$	Oe	T	kG	
Low nickel											
45Ni-Fe	7(c)	90	0.6	6	0.68	6.8	4	0.05	1.58	15.8	50
49Ni-Fe(a)	6.1(c)	64	0.8	8	0.96	9.6	8	0.10	1.55	15.5	47
49Ni-Fe(b)	14(c)	140	0.78	7.8	0.97	9.7	4	0.05	1.55	15.5	47
49Ni-Fe	17(c)	180	0.75	7.5	0.90	9.0	2.4	0.03	1.55	15.5	47
45Ni-3Mo-Fe	6(c)	60	0.62	6.2	0.89	8.9	4.8	0.06	1.45	14.5	65
High nickel											
78.5Ni-Fe	50(d)	300	0.35	3.5	0.50	5.0	1.0	0.013	1.05	10.5	16
79Ni-4Mo-Fe	90(d)	400	0.28	2.8	0.35	3.5	0.3	0.004	0.79	7.9	59
75Ni-5Cu-2Cr-Fe	85(d)	375	0.25	2.5	0.34	3.4	0.4	0.005	0.77	7.7	56

(a) Annealed at 955 °C (1750 °F). (b) Annealed at 1065 °C (1950 °F). (c) Measured at B = 10 mT (100 G). (d) Measured at B = 4 mT (40 G)

Table 7 Typical ac magnetic properties of annealed high-permeability nickel-iron alloys

Nominal composition	Thickness, mm (in.)	Cyclic frequency, Hz	Impedance permeability, $\mu_z \times 10^3$, at indicated induction, B(a): B = 4 mT (40 G)	B = 20 mT (200 G)	B = 0.2 T (2 kG)	B = 0.4 T (4 kG)	B = 0.8 T (8 kG)
49Ni-Fe(c)	0.51 (0.020)	60	10.2	16.5	31.3	40.1	. . .
	0.36 (0.014)	60	12	19.4	37.3	48.2	54.7
	0.25 (0.010)	60	12	20.5	42.5	54.9	68.9
	0.15 (0.006)	60	12	21	47	63.5	85.3
	0.51 (0.020)	400	4.7	5.9	11.7	11.3	. . .
	0.36 (0.014)	400	6.1	7.9	14.4	17.7	13.3
	0.15 (0.006)	400	8.8	12.6	21.8	28.6	35
79Ni-4Mo-Fe	0.36 (0.014)	60	68	77	100	. . .	. . .
79Ni-5Mo-Fe	0.15 (0.006)	60	90	110	170	. . .	. . .
	0.10 (0.004)	60	110	135	230	. . .	. . .
	0.03 (0.001)	60	100	120	180	. . .	. . .
	0.36 (0.014)	400	23.2	25.4	30.5	. . .	. . .
	0.15 (0.006)	400	49.7	52.4	64.5	. . .	. . .
	0.03 (0.001)	400	89.6	105.2	180.4	. . .	. . .
49Ni-Fe(d)	0.36 (0.014)	60	. . .	. . .	. . .	. . .	. . .
	0.15 (0.006)	60	. . .	. . .	. . .	. . .	. . .
	0.36 (0.014)	400	. . .	. . .	. . .	. . .	. . .
	0.15 (0.006)	400	. . .	. . .	. . .	. . .	. . .

Nominal composition	Inductance permeability, $\mu_L \times 10^3$, DU laminations(b): B = 4 mT (40 G)	B = 20 mT (200 G)	B = 0.2 T (2 kG)	B = 0.4 T (4 kG)	B = 0.8 T (8 kG)
49Ni-Fe(c)	. . .	. . .	. . .	. . .	. . .
	. . .	. . .	. . .	. . .	. . .
	. . .	. . .	. . .	. . .	. . .
	. . .	. . .	. . .	. . .	. . .
	. . .	. . .	. . .	. . .	. . .
	. . .	. . .	. . .	. . .	. . .
79Ni-4Mo-Fe	. . .	. . .	. . .	. . .	. . .
79Ni-5Mo-Fe	. . .	. . .	. . .	. . .	. . .
	. . .	. . .	. . .	. . .	. . .
	. . .	. . .	. . .	. . .	. . .
	. . .	. . .	. . .	. . .	. . .
	. . .	. . .	. . .	. . .	. . .
	. . .	. . .	. . .	. . .	. . .
	. . .	. . .	. . .	. . .	. . .
49Ni-Fe(d)	18.6	35.8	78	110	135
	19.6	39.2	98.5	142	215
	11.8	17.6	36.4	55	30
	12.2	18.5	48.3	95	164

Nominal composition	Core loss in mW/kg (mW/lb) at indicated induction B: B = 4 mT (40 G)	B = 20 mT (200 G)	B = 0.2 T (2 kG)	B = 0.4 T (4 kG)	B = 0.8 T (8 kG)
49Ni-Fe(c)	. . .	. . .	. . .	. . .	. . .
	0.011 (0.005)	0.21 (0.097)	15 (6.7)	48 (21.7)	160 (73)
	. . .	. . .	. . .	. . .	. . .
	0.009 (0.004)	0.21 (0.094)	13 (5.8)	44 (19.9)	135 (62)
	. . .	. . .	. . .	. . .	. . .
	0.21 (0.094)	4.34 (1.97)	282 (128)	905 (410)	3880 (1760)
	0.15 (0.069)	3.20 (1.45)	238 (108)	705 (320)	2310 (1050)
79Ni-4Mo-Fe	. . .	0.099 (0.045)	6.50 (2.95)	. . .	. . .
79Ni-5Mo-Fe	. . .	0.051 (0.023)	3.00 (1.36)	. . .	. . .
	. . .	0.024 (0.011)	1.60 (0.73)	. . .	. . .
	. . .	. . .	. . .	. . .	. . .
	0.11 (0.050)	2.20 (1.00)	160 (72.5)	. . .	. . .
	0.044 (0.020)	0.99 (0.45)	65.9 (29.9)	. . .	. . .
	. . .	. . .	. . .	. . .	. . .
49Ni-Fe(d)	0.011 (0.005)	0.22 (0.10)	15 (6.6)	51 (23)	185 (83)
	0.007 (0.003)	0.13 (0.06)	8.6 (3.9)	31 (14)	105 (47)
	0.20 (0.091)	4.4 (2.00)	306 (139)	1010 (460)	4800 (2200)
	0.11 (0.052)	2.38 (1.08)	172 (78.0)	550 (250)	1700 (790)

(a) Tested per ASTM A 772 method: thicknesses >0.13 mm (0.005 in.) tested using ring specimens; <0.13 mm (0.005 in.) tested via tape toroid specimens. (b) Per ASTM A 346 method; DU, interleaved U-shape transformer. (c) Nonoriented rotor or motor grade. (d) Transformer semioriented grade

The relatively low Curie temperature of nickel-iron alloys in the range from 28% Ni to ~52% Ni provides an anomalous thermal-expansion behavior. Thus, these alloys are used as low or controlled expansion alloys and glass-to-metal sealing alloys. Compositions, particularly around 51Ni-Fe, have been used in wire form to manufacture dry reed switches. This application requires thermal-expansion and glass-sealing characteristics matching those of the glass envelope, as well as soft magnetic behavior.

Constant Permeability With Changing Temperature. In all magnetic materials, magnetic properties change with temperature. The change in flux density with temperature for iron tested at four different values of magnetizing force is plotted in Fig. 15. Operation of a device at a flux density of 1.5 T (15 kG) would be only slightly affected by variations in operating temperatures near ambient. There is a similar minimized temperature effect for all materials, except that the flux density for optimum operations depends on the materials, given the proper flux density and temperature range. Large changes occur for all ferromagnetic alloys at temperatures approaching the Curie temperature of the alloy. Approximate Curie temperatures for several soft magnetic alloys are listed below:

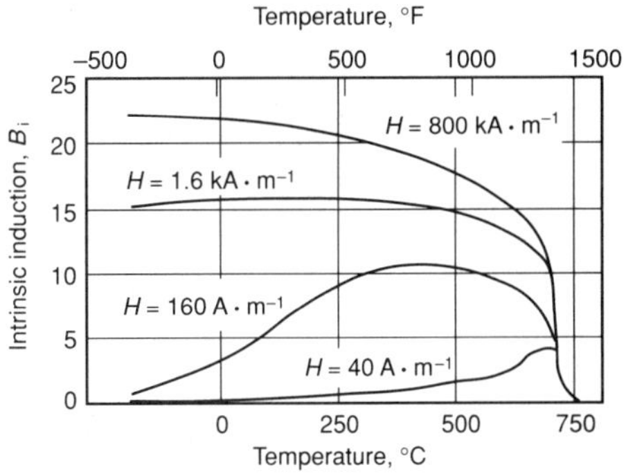

Fig. 15 Variation of induction with temperature for iron, at four different values of magnetizing force

Nominal alloy composition	Curie temperature, T_c °C	°F
29–32Ni-Fe	20 to 150	68 to 300
36 Ni-Fe	275	530
77Ni-5Cu-2Cr-Fe	405	760
79Ni-4Mo-Fe	455	850
50Ni-Fe	500	930
3Si-Fe	730	1350
0.5Si-Fe	755	1390
Pure iron	770	1420
27Co-0.6Cr-Fe	925	1700
2V-49Co-49Fe	970	1775

It is not always possible to operate a material at the best flux density for temperature stability. One way of obtaining better temperature stability is to use magnetic materials in insulated powder form, such as pressed Permalloy powder cores, which have good temperature stability due to the presence of many built-in air gaps. Another method involves use of nickel-iron alloys, such as Isoperm or Conpernik (Westinghouse Electric Corporation), that have been drastically cold rolled and then underannealed to produce a partly strained alloy less sensitive to temperature changes.

However, each of these methods sacrifices the higher permeability of the basic alloys that could be used in the magnetic circuit. The air gap effect of the powdered cores shears the hysteresis loop, and the residual stress remaining in alloys such as Conpernik and Isoperm has the same effect. (It is not known if Conpernik or Isoperm are in commercial production.)

Alloys for Magnetic Temperature Compensation. Many measuring instruments

Table 8 Typical heat treatments and physical properties of nickel-iron alloys

Alloy nominal composition	ASTM standard	Annealing treatment(a)	Hardness	Yield strength MPa	Yield strength ksi	Ultimate tensile strength MPa	Ultimate tensile strength ksi	Elongation, %	Specific gravity
45Ni-Fe	A 753 type 1	Dry hydrogen, 1120 to 1175 °C (2050 to 2150 °F), 2 to 4 h, cool at nominally 85 °C/h (150 °F/h)	48 HRB	165	24	441	64	35	8.17
49Ni-Fe	A 753 type 2	Same as 45Ni-Fe	48 HRB	165	24	441	64	35	8.25
45Ni-3Mo-Fe	...	Same as 45Ni-Fe	...	...	...	...	...	...	8.27
78.5Ni-Fe	...	Dry hydrogen, 1175 °C (2150 °F), 4 h rapid cool to RT(b), reheat to 600 °C (1110 °F), 1 h, oil quench to RT	50 HRB	159	23	455	66	35	8.60
80Ni-4Mo-Fe	A 753 type 4	Dry hydrogen, 1120 to 1175 °C (2050 to 2150 °F), 2 to 4 h cool thru critical ordering temperature range, ~760 to 400 °C (1400 to 750 °F) at a rate specified for the particular alloy, typically 55 °C/h (100 °F/h) up to ~390 °C/h (700 °F/h)	58 HRB	172	25	545	79	37	8.74
80Ni-5Mo-Fe	A 753 type 4	Same as 80Ni-4Mo-Fe	58 HRB	172	25	545	79	37	8.75
77Ni-5Cu-2Cr-Fe	A 753 type 3	Same as 80Ni-4Mo-Fe	50 HRB	125	18	441	64	27	8.50

(a) All nickel-iron soft magnetic alloys should be annealed in a dry (−50 °C, or −58 °F) hydrogen atmosphere, typically for 2 to 4 h; cool as recommended by producer. Vacuum annealing generally provides lower properties, which may be acceptable depending upon specific application. (b) RT, room temperature

Table 9 Application of nickel-iron and iron-cobalt magnetically soft alloys

Application	Specialty alloy(a)	Special property
Instrument transformer	79Ni-4Mo-Fe, 77Ni-5Cu-2Cr-Fe, 49Ni-Fe	High permeability, low noise and losses
Audio transformer	79Ni-4Mo-Fe, 49Ni-Fe, 45Ni-Fe, 45Ni-3Mo-Fe	High permeability, low noise and losses, transformer grade
Hearing aid transformers	79Ni-4Mo-Fe	High initial permeability, low losses
Radar pulse transformers	2V-49Co-49Fe, oriented 49Ni-Fe, 79Ni-4Mo-Fe, 45Ni-3Mo-Fe	Processed for square hysteresis loop, tape toroidal cores
Magnetic amplifiers	Oriented 49Ni-Fe, 79Ni-4Mo-Fe	Processed for square hysteresis loop, tape toroidal cores
Transducers	2V-49Co-49Fe, 45-50Ni-Fe	High saturation magnetostriction
Shielding	79Ni-4Mo-Fe, 77Ni-5Cu-2Cr-Fe, 49Ni-Fe	High permeability at low induction levels
Ground fault (GFI) interruptor core	79Ni-4Mo-Fe	High permeability, temperature stability
Sensitive dc relays	45 to 49Ni-Fe, 78.5Ni-Fe	High permeability, low losses, low coercive force
Electromagnet pole tips	2V-49Co-49Fe, 27Co-0.6Cr-Fe	High saturation induction
Tape recorder head laminations	79Ni-5Mo-Fe	High permeability, low losses (0.05 to 0.03 mm (0.002 to 0.001 in.)
Telephone diaphragm armature	2V-49Co-49Fe	High incremental permeability
Temperature compensator	29 to 36Ni-Fe	Low Curie temperature
High-output power generators	2V-49Co-49Fe, 27Co-0.6Cr-Fe	High saturation
Dry reed magnetic switches	51Ni-Fe	Controlled expansion glass/metal sealing
Chart recorder (instrument) motors, synchronous motors	49Ni-Fe	Moderate saturation, low losses, nonoriented grade
Loading coils	81-2 brittle Moly-Permalloy	Constant permeability with changing temperature

(a) See Table 10 for domestic producers of alloys listed.

and other devices depend upon maintaining constant flux, produced by a permanent magnet, across an air gap. Unfortunately, as a permanent magnet is warmed, it loses strength and the air gap flux density changes. To compensate for such changes, a certain amount of the magnetic flux can be shunted around the air gap of the instrument by using an alloy with high negative magnetic temperature coefficient in the temperature range of interest. The amount of shunted flux, therefore, decreases with increasing ambient temperature, forcing more flux through the gap than would normally occur. Nearly complete compensation for temperature changes can be made by correct design of parts. Watt-hour meters and automobile speedometers are examples in which a temperature compensator shunt has been used to compensate for the change in pole strength of the permanent magnet (and the change in electrical resistivity of the aluminum drag disk) over their designed working-temperature range.

Nickel-iron alloys containing between ~28% and 36% Ni are frequently used commercially for this purpose because the Curie temperature for these alloys range from about room temperature up to ~275 °C (530 °F). Table 11 and Fig. 16 show how some typical nickel-iron alloys vary in permeability with temperature at H of 3.7 kA · m^{-1} (46 Oe). Temperature-compensation alloys are produced commercially by several producers, among them Carpenter Technology Corporation. The alloys are generally manufactured as annealed strip products.

Iron-Cobalt Alloys

Pure iron has a saturation induction of 2.158 T (21.58 kG). Higher saturation values can be achieved only in alloys of iron and cobalt. The highest known value is approximately 2.46 T (24.6 kG), which occurs at a cobalt content of ~35%.

In Fig. 17, the dc magnetic induction response at several magnetizing forces from 0.25 to 1400 kA · m^{-1} (0.003 to 17 kOe) for binary iron-cobalt compositions is shown. The composition corresponding to ~50Co-50Fe is of particular interest due to its peaked response at low magnetizing force (that is, high permeability). The room-temperature crystallographic structure of iron-cobalt alloys containing up to approximately 70% Co is ferritic (α-Fe), except that the 50Co-50Fe compositional range orders very rapidly to α′, a brittle cesium-chloride type structure, upon cooling below ~725 °C (1340 °F). The binary 50:50 alloy was originally known as Permindur (Western Electric Company), but was never widely manufactured due to its extreme brittleness. In the 1930s, Bell Laboratories Inc. developed a modified alloy by adding ~2% V, which slowed the embrittling reaction sufficiently so that hot-rolled strip (thickness less than 3.04 mm, or 0.120 in.) could be rapidly cooled, retaining the ductile structure. The vanadium addition also provides increased resistivity (to 43 μΩ · cm), but does lower saturation induction slightly. Vanadium additions greater than ~2% shift the Fe-Co-V phase boundary limits, producing semihard and hard magnetic materials such as Remendur 27, P6, and Vicalloy—which are covered in the article "Permanent Magnet Materials" in this Volume. The fully heat-treated coercive force values for these alloys are ~2.2, 4.4, and 24 kA · m^{-1} (27, 55, and 300 Oe), respectively.

Alloy 2V-49Co-49Fe. The soft magnetic alloy consisting of essentially 2V-49Co-49Fe has been known as Vanadium Permendur (Allegheny Ludlum Corporation), 2V-Permendur, Supermendur, and is currently domestically produced as Hiperco

Table 10 Domestic producers of nickel-iron and iron-cobalt magnetically soft alloys. See Table 9 for applications

Alloy	Trade name: Carpenter Technology Corporation	Allegheny Ludlum Corporation	AlTech	Spang & Company, Specialty Metals Div.
45 to 49Ni-Fe	High permeability 49	4750	4750	Alloy 48
Square loop 49Ni-Fe	HyRa 49	Delta Max	...	Orthonol
45Ni-3Mo-Fe	...	Monimax	...	...
77Ni-5Cu-2Cr-Fe	HyMu 77	MuMetal	...	MuMetal
79Ni-4Mo-Fe	HyMu 80, HyMu 800, HyMu 80 Mark II Hipernom	4–79 Permalloy	...	Permalloy 80
2V-49Co-49Fe	Hiperco 50 and 50A	2V Permendur	...	Permendur V
27Co-0.6Cr-Fe	Hiperco 27	...	...	...
30 to 32Ni-Fe	Temperature compensator	...	...	

Table 11 Permeability of nickel-iron magnetic alloys used for temperature compensation

Alloy(a)	Permeability(b), for H = 3.7 kA · m^{-1} (46 Oe) at: 45 °C (−50 °F)	−20 °C (0 °F)	10 °C (50 °F)	25 °C (77 °F)	40 °C (100 °F)	65 °C (150 °F)	95 °C (200 °F)	120 °C (250 °F)
29.0% Ni	92	70	25	3	2	...	...	...
29.5% Ni	120	102	74	46	27	4	...	...
31.0% Ni	156	140	120	110	98	73	45	15
32.5% Ni	212	202	191	180	170	145	120	85

(a) Remainder is iron in all alloys. (b) Temperature-permeability properties can be varied by manufacturing procedure and thermal treatment at finished size. Above values are average and representative of temperature-permeability properties available in a commercial product.

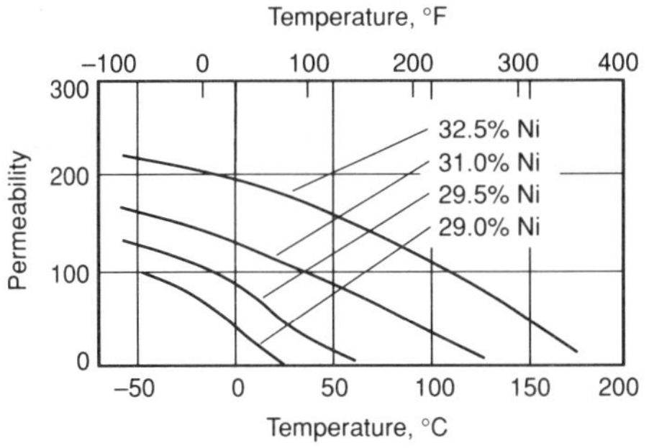

Fig. 16 Effect of nickel content on the permeability-temperature characteristics of annealed nickel-iron temperature-compensator alloys at H of 3.7 kA · m^{-1} (46 Oe)

50, Hiperco 50A (Carpenter Technology Corporation), or Permendur V (Spang & Company, Specialty Metals Division). With specialized mill processing, this alloy can be cold rolled to nearly any gage (0.025 mm, or 0.001 in., thickness, or less). In typical lamination thicknesses (0.64 to 0.15 mm, or 0.025 to 0.006 in.), the cold-rolled unannealed strip can be stamped (30 to 39 HRC) into various rotor/stator/transformer lamination configurations. Thinner strip is generally wound into tape toroidal cores. This alloy is also available in round bars and can be processed in certain wire sizes. Bar products cannot be provided in a disordered condition because it is physically impossible to cool the bars at a sufficiently rapid rate. Machining, handling, and the application of bar-product forms must be performed carefully due to limited ductility.

Alloy 27Co-0.6Cr-Fe. Another iron-cobalt alloy composition, containing 27Co-0.6Cr-bal Fe, is commercially available, designated as Hiperco 27 (Carpenter Technology Corporation). It is primarily produced as a lamination strip product and as a bar product. The 27% Co content provides saturation levels similar to the 2V-49Co-49Fe composition, and does not require the same degree of specialized manufacturing. The addition of 0.6% Cr increases electrical resistivity to 19 μΩ · cm. Its lower cobalt content makes the alloy less expensive, but it is not as magnetically soft as the 2V-49Co-49Fe alloy. In dc field applications (pole pieces, flux return members, and so on), energy losses due to low resistivity and relatively low permeability (hysteresis losses) may not be of concern. Therefore, the 27Co-Fe alloy may be preferred due to its high saturation and lower price. In ac applications, particularly as frequency is increased, the 2V-49Co-49Fe alloy is generally selected.

Magnetic Properties of Iron-Cobalt Alloys. The iron-cobalt alloys exhibit a high positive magnetostrictive coefficient, which has made them useful in transducers (sonar) and in extremely accurate positioning devices. The magnetostrictive coefficients at saturation are approximately 36 × 10^{-6} and 60 × 10^{-6} $\Delta l/l$ for the 27Co-Fe and the 2V-49Co-49Fe alloys, respectively.

All iron-cobalt alloys must be heat treated as finished laminations or as machined components. The heat treatments vary depending upon the intended application. The 2V-49Co-49Fe alloys and the 27Co-0.6Cr-Fe alloys have both been used extensively as laminations in generator and motor applications for aircraft, submarines, tanks, and so on, where the high flux density of these alloys permits the design of equipment with less weight and bulk. For high-speed rotating components, relatively low annealing temperatures such as 760 to 790 °C (1400 to 1450 °F) may be employed to maintain strength, but at some sacrifice in soft magnetic characteristics. For applications where strength is not important, higher annealing temperatures are used. Both alloys undergo a sluggish transformation back to the α phase when heated above their respective Ac_1 temperatures and cooled to room temperature. For 2V-49Co-49Fe of high purity, this temperature is approximately 900 °C (1650 °F). The Ac_1 temperature for 27Co-Fe is approximately 955 °C (1750 °F). These alloys should not be annealed above their respective Ac_1 temperature unless extremely slow cooling (less than 11 °C/h, or 20 °F/h) is employed to allow complete transformation from the γ phase to the α phase. A typical annealing cycle for the 2V-49Co-49Fe alloy to develop good soft magnetic properties is: dry hydrogen atmosphere 845 to 870 °C (1550 to 1600 °F), 4 h at soak, and cool nominally at 110 °C/h (200 °F/h). If annealed below the Ac_1 temperature, the cooling rate is not critical.

Iron-cobalt alloys tend to form a thin, tight, blue oxide film during annealing. This can be prevented in extremely low dew-point furnaces or hard vacuum furnaces but may not always be preventable in larger commercial annealing facilities. The thin blue oxide layer is generally not harmful to magnetic properties. In fact, oxide is sometimes deliberately developed on laminations by holding in a moist atmosphere at ~540 °C (1000 °F) to provide interlamination resistance to reduce eddy-current losses.

In specialized applications, usually tape toroids, high-purity 2V-49Co-49Fe cores may be final annealed under the influence of an applied magnetic field. Magnetic-field annealing can be beneficial to providing a particularly square magnetic hysteresis loop with low core loss. Supermendur (Western Electric Company) is a high-purity variation of the 2V-49Co-49Fe alloy, which is specially processed and domain oriented by heat treatment in a magnetic field. Data presented for Supermendur generally applies only to thin-strip (0.1 mm, or 0.004 in., and less) processed and often magnetic-field annealed. Properties reported for Supermendur include a maximum permeability of 7 × 10^4 occurring at 2.0 T (20 kG), coercive force of 16 A · m^{-1} (0.20 Oe), and B_r of 2.14 T (21.4 kG) from B_s of 2.4 T (24 kG). Core-loss values of ~18 W/kg (8 W/lb) at 400 Hz and 2.0 T (20 kG) have been reported for 0.1 mm (0.004 in.) thick strip. It is not possible to reproduce the high permeability values or loop-squareness data reported for Supermendur in bar, heavy strip, or even lamination stock product forms. When comparing the properties of various soft magnetic alloys, it is always necessary to know

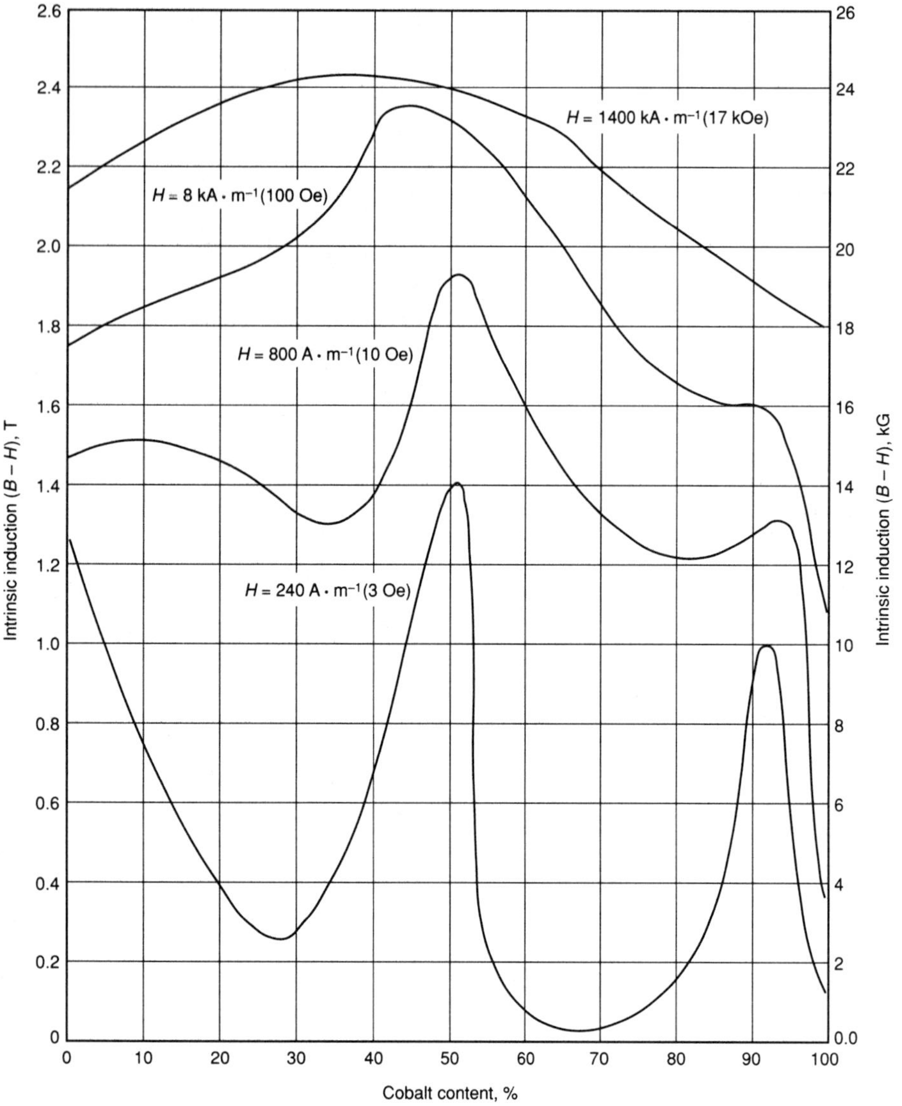

Fig. 17 Intrinsic induction of iron-cobalt alloys at various magnetizing levels

all the details of the product form, test method, and annealing practices employed.

Some typical magnetic and mechanical properties of several iron-cobalt alloys are shown in Tables 12 and 13. Several applications of Co-Fe alloys are listed in Table 9. Both the 27Co-Fe and 2V-49Co-49Fe alloys are included in ASTM standard A 801-86. Figure 18 provides a comparison of the typical dc normal induction properties of iron-cobalt alloys as well as nickel-iron, low-carbon iron, and ferritic (430F) stainless products. It is strongly suggested that the material producers be contacted for more detailed information regarding variations of these alloys manufactured for specific purposes, and for the proper selection and processing methods.

Ferrites

Ferrites for high-frequency applications are ceramics with characteristic spinel-magnetic structures ($M \cdot Fe_2O_4$, where M is a metal) and usually comprise solid solutions of iron oxide and one or more oxides of other metals such as manganese, zinc, magnesium, copper, nickel, and cobalt. They are unique among magnetic materials in their outstanding magnetic properties at high frequencies, which result from very high resistivities ranging from about $10^8\ \Omega \cdot$ cm to as high as $10^{14}\ \Omega \cdot$ cm. Hence, at frequencies where eddy-current losses for metals become excessive, ferrites make ideal soft magnetic materials. Because ferrites have inherently high corrosion resistance, parts made of these materials normally do not require protective finishing.

Disadvantages of ferrites include low magnetic saturation, low Curie temperature, and relatively poor mechanical properties compared with those of metals. Ferrites are produced from powdered raw materials by mixing, calcining, ball milling, pressing to shape, and firing to the desired magnetic properties. The final product is hard, brittle, and unmachinable, and thus close dimensional tolerances must be obtained by grinding.

Types of Ferrite. Many different ferrites are available for magnetic use. They can be classified into three general types:

- Square-loop ferrites for computer memories
- Linear ferrites for transformers and for inductors in filters
- Microwave ferrites for microwave devices

In recent years, due to increasing use of semiconductors for computer memories, square-loop ferrites have decreased in importance.

Microstructure and composition have much stronger influences on the magnetic properties of ferrites than on those of metals. Hence, properties of finished ferrite parts can vary drastically with purity and structure of raw materials, with the nature of binders used, and with the ceramic-processing technique employed. In general, lithium ferrites, Mn-Mg-Zn ferrites, and Mn-Mg-Di ferrites are used for computer memories. (Di is the symbol for didymium, a mixture of the rare earth elements praseodymium and neodymium.) Lithium ferrite is higher in Curie temperature and saturation magnetization, but lower in switching speed, than Mn-Mg-Zn and Mn-Mg-Di ferrites. Linear ferrites comprise Mn-Zn and Ni-Zn ferrites. Mn-Zn ferrite is higher in saturation magnetization, but lower in resistivity, than Ni-Zn ferrite. Mn-Zn ferrite is preferred for frequencies up to about 1 MHz. For microwave applications, Ni-Zn, Mg-Mn-Al, and Mg-Mn-Cu ferrites are used, as well as garnets of the type $M_{3+x}Fe_{5-x}O_{12}$ (where M = Y + Al, or M = Y + Gd + Al).

Stainless Steels

The magnetic behavior of stainless steels varies considerably, ranging from paramagnetic (nonmagnetic) in fully austenitic grades, to hard or permanent magnetic behavior in the hardened martensitic grades, to soft magnetic properties in ferritic stainless steels.

Austenitic Stainless Steels. All austenitic stainless steels are paramagnetic (nonmagnetic) in the annealed, fully austenitic condition. The dc magnetic permeabilities range from ~1.003 to ~1.005 when measured at magnetizing forces of 16 $kA \cdot m^{-1}$ (200 Oe). The permeability increases with cold work due to deformation-induced martensite, a ferromagnetic structure. For certain grades such as types 302 and 304, the increase in magnetic permeability can be appreciable, resulting in these grades being weakly ferromagnetic in the heavily cold-worked condition. This phenomenon is il-

Table 12 Typical dc and ac magnetic properties of annealed iron-cobalt alloys in the form of 0.15 to 0.5 mm (0.006 to 0.020 in.) thick lamination strip products

Alloy nominal composition	Annealing temperature °C	Annealing temperature °F	Thickness mm	Thickness in.	dc induction in T (kG) at indicated H(a): H=160 A · m^{-1} (2 Oe)	H=0.8 kA · $^{-1}$ (10 Oe)	H=4 kA · m^{-1} (50 Oe)	H=8 kA · m^{-1} (100 Oe)	H=20 kA · m^{-1} (250 Oe)
2V-49Co-49Fe ...	845	1550	All	All	1.35 (13.5)	2.16 (21.6)	2.30 (23.0)	2.36 (23.6)	2.41 (24.1)
27Co-0.6Cr-Fe ...	845	1550	All	All	0.5 (5)	1.35 (13.5)	1.92 (19.2)	2.12 (21.2)	2.30 (23.0)
2V-49Co-49Fe ...	875	1610	0.51	0.020	...	...	...	...	...
			0.36	0.014	...	...	...	...	...
			0.25	0.010	...	...	...	...	...
			0.20	0.008	...	...	...	...	...
			0.15	0.006	...	...	...	...	...
			0.51	0.020	...	...	...	...	...
			0.36	0.014	...	...	...	...	...
			0.25	0.010	...	...	...	...	...
			0.20	0.008	...	...	...	...	...
			0.15	0.006	...	...	...	...	...
27Co-0.6Cr-Fe ...	845	1550	0.36	0.014	...	...	...	...	...
			0.36	0.014	...	...	...	...	...

Alloy nominal composition	Frequency, Hz	ac total core loss, P_c, in W/kg (W/lb) at indicated peak induction, B(b): B = 1 T (10 kG)	B=1.5 T (15 kG)	B = 2 T (20 kG)	Saturation induction, B_s, T	B_s, kG	From H= 8 kA · m^{-1} (100 Oe): H_c, A · m^{-1}	H_c, Oe	B_r, T	B_r, kG
2V-49Co-49Fe.....	...	...	...	...	2.42	24.2	72	0.9	1.6	16
27Co-0.6Cr-Fe	...	...	...	...	2.43	24.3	130	1.6	1.4	14
2V-49Co-49Fe.....	60	1.5 (0.67)	2.89 (1.31)	4.76 (2.16)	...		...		...	
	60	1.4 (0.65)	2.69 (1.22)	4.17 (1.89)	...		...		...	
	60	1.3 (0.57)	2.31 (1.05)	3.57 (1.62)	...		...		...	
	60	1.2 (0.56)	2.29 (1.04)	3.48 (1.58)	...		...		...	
	60	1.2 (0.53)	2.1 (0.95)	3.08 (1.40)	...		...		...	
	400	21 (9.5)	53.8 (24.4)	112 (50.8)	...		...		...	
	400	17 (7.5)	36.8 (16.7)	67.4 (30.6)	...		...		...	
	400	13 (6.0)	27.3 (12.4)	46.9 (21.3)	...		...		...	
	400	12 (5.4)	23.6 (10.7)	38.3 (17.4)	...		...		...	
	400	10 (4.7)	20 (9.0)	31.5 (14.3)	...		...		...	
27Co-0.6Cr-Fe	60	3.13 (1.42)	5.55 (2.52)	8.00 (3.63)	...		...		...	
	400	36.8 (16.7)	73.2 (33.2)	110 (50.0)	...		...		...	

(a) Ring laminations tested per ASTM A 596 method. (b) Ring laminations tested per ASTM A 697 method

Table 13 Typical mechanical and physical properties of iron-cobalt alloy 0.15 to 0.38 mm (0.006 to 0.015 in.) thick lamination strips

Alloy nominal composition	Anneal condition	Tensile properties: 0.2% yield strength MPa	0.2% yield strength ksi	Ultimate tensile strength MPa	Ultimate tensile strength ksi	Elongation, %	Hardness, HRC	Resistivity, μΩ · cm	Specific gravity
2V-49Co-49Fe(a)	As-cold rolled	1295	188	1336	194	1	38	43	8.12
	760 °C (1400 °F), 2-h cool at ~100 °C/h (180 °F/h)	427	62	799	116	8	...	43	8.12
	845 °C (1550 °F), 2-h cool at ~100 °C/h (180 °F/h)	365	53	696	101	7	...	43	8.12
27Co-0.6Cr-Fe(b).....	As-cold rolled	1137	165	1143	166	6	34	19	7.95
	760 °C (1400 °F), 4-h cool at ~165 °C/h (330 °F/h)	434	63	709	103	14	...	19	7.95
	845 °C (1550 °F), 4-h cool at ~165 °C/h (330 °F/h)	310	45	551	80	15	...	19	7.95

(a) Hiperco 50 (Carpenter Technology Corporation), Permendur V (Spang & Company, Specialty Metals Division). (b) Hiperco 27 (Carpenter Technology Corporation)

lustrated graphically in Fig. 19 for nine austenitic stainless steels.

The differing performance among grades is a reflection of their composition. In particular, nickel increases austenite stability, thereby decreasing the work-hardening rate and the rate of increase of magnetic permeability. Consequently, the higher-nickel grades exhibit lower magnetic permeabilities than the lower-nickel grades when cold worked in equivalent amounts.

The magnetic permeabilities achievable in austenitic stainless steels are very low when compared to conventional magnetic materials. Consequently, it is their nonmagnetic behavior that is of more concern. Certain applications, such as housings and components for magnetic detection equipment used for security, measuring, and control purposes, require that the steel be nonmagnetic, since the presence of even weakly ferromagnetic parts can adversely affect performance. If the magnetic permeability of an austenitic stainless steel is of particular concern, it can be measured by relatively simple means, as described in ASTM standard A 342-Method No. 6. The equipment described is commercially available at relatively low cost.

Ferritic stainless steels are ferromagnetic and have been used as soft magnetic components in products such as solenoid housings, cores, and pole pieces. Although their magnetic properties are not generally as good as conventional soft magnetic alloys, they have been successfully used for magnetic components that must withstand corrosive environments. As such, they offer a cost-effective alternative to plated iron and silicon-iron components. In addition, the relatively high electrical resistivity of ferritic stainless steels has resulted in superior ac performance.

Special restricted analyses of AISI type 430F are produced for use in solenoid valve components. The ASTM A 838-85 specification provides typical properties for these alloys. Alloy type 1 is 430F containing approximately 0.4% Si and exhibiting an electrical resistivity of 60 μΩ · cm. When fully mill annealed, it has a hardness of approximately 78 HRB. Its maximum dc permeability is approximately 2×10^3, with a coercivity of approximately 160 A · m^{-1} (2 Oe). Alloy type 2 is a higher-silicon version of 430F, with an electrical resistivity of 76 μΩ · cm and a fully annealed hardness of 82 HRB. Despite its higher hardness, alloy type 2 typically exhibits a dc permeability of 2.6×10^3 and a coercivity of 130 A · m^{-1} (1.6 Oe). Both alloys are available in round centerless ground (C.G.) bar form fully processed, so that in many applications they are suitable for high-volume screw machining of parts, passivation, and placement into service without annealing. Hex bars and other special-shape products may only be available in a cold-drawn condition suitable for machining, but they may require annealing of the parts to develop soft magnetic properties.

Magnetic properties of selected ferritic stainless steels are listed in Table 14.

Martensitic and Precipitation-Hardenable Stainless Steels. All martensitic and most precipitation-hardenable stainless steels are ferromagnetic. Due to the stresses induced by hardening, these grades exhibit permanent magnetic properties in the hardened condition. For a given grade, the coercive force tends to increase with increasing hardness, rendering these alloys more difficult to demagnetize. If the hardenable martensitic stainless steels are used in the annealed condition, they suffer from:

- Poorer magnetic properties due to the presence of a significant volume of chrome-carbides, which contribute to pinning domain-wall movement

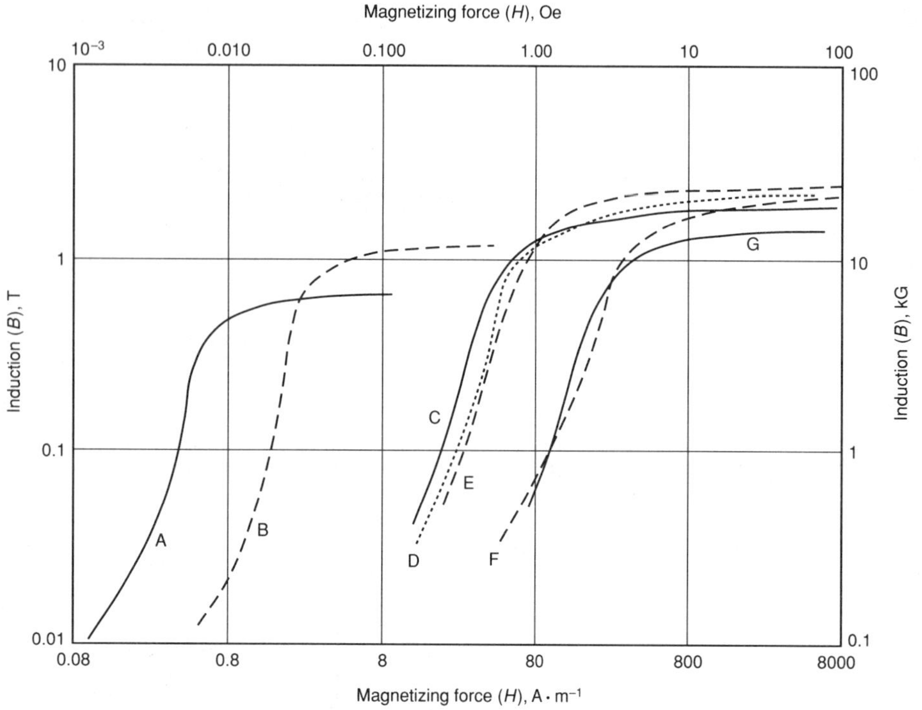

Fig. 18 Direct current normal induction characteristics of several soft magnetic materials annealed at indicated temperature: A, 79Ni-4Mo-Fe (1175 °C, or 2150 °F); B, 49Ni-Fe (1175 °C, or 2150 °F); C, 2.5Si-Fe (1065 °C, or 1950 °F); D, Air melt iron (845 °C, or 1550 °F); E, 2V-49Co-49Fe (875 °C, or 1610 °F); F, 27Co-0.6Cr-Fe (845 °C, or 1550 °F); and G, 430F (as shipped)

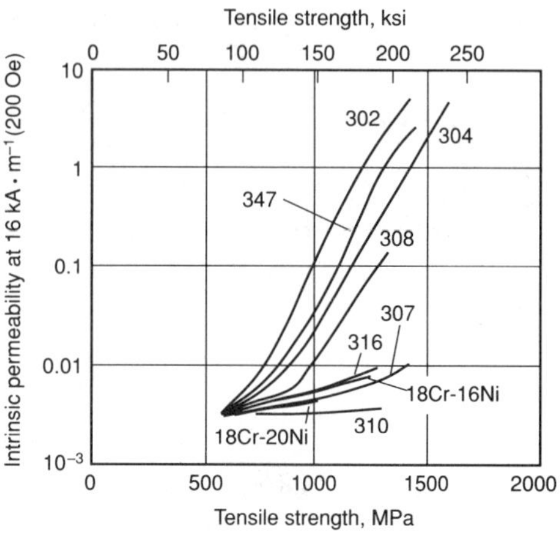

Fig. 19 Correlation of increased tensile strength from cold working and the permeability of cold-worked austenitic stainless steels. Annealed hot-rolled strips 2.4 to 3.2 mm (0.095 to 0.125 in.) thick before cold reduction. For normal permeability values, add unity to the numbers given on vertical scale.

- Reduced corrosion resistance due to matrix depletion of chromium

The ferritic nonhardenable 430 or 430F grades are preferred for soft magnetic applications for these reasons. Magnetic properties of selected martensitic stainless steels are shown in Table 14.

Additional information is available in the article "Wrought Stainless Steels" in *Properties and Selection: Irons, Steels, and High-Performance Alloys*, Volume 1 of the 10th Edition of *Metals Handbook*.

Corrosion Resistance of Magnetically Soft Materials

In specific applications, corrosion resistance may limit the choices of magnetically soft materials. Pure irons and silicon irons corrode readily in mild atmospheres. Consequently, these materials may require protection by painting, plating, potting, or molding.

Iron Alloys and Silicon Steels. If parts machined from iron or silicon-iron alloys are to be plated, they are annealed to develop soft magnetic properties first. Annealing should generally be performed in a dry atmosphere to prevent the formation of iron/silicon oxides that may impair plating adherence. If oxides are formed during annealing, they must usually be removed by a mechanical means. Acid-bath cleaning of these materials usually results in a rough-pitted surface and undersize dimensions.

Nickel-Iron Alloys. Alloys of ~50Ni-Fe have only mild resistance to corrosion and will rust in industrial environments. For example, annealed 50Ni-Fe alloys have shown approximately 15 to 25% of their surface area rusted after 200-h exposure to neutral salt spray at 35 °C (95 °F). High-nickel soft magnetic alloys possess fair corrosion resistance, but they will tarnish with time in industrial atmospheres. Plating, painting, or an epoxy coating can generally be used on these materials if required.

Ferritic Stainless Steels. Where the base material must be corrosion resistant, the ferritic stainless steels may be used. Allowance must be made, however, for lower permeability, high coercive force, and lower saturation than available in most other soft magnetic grades. The nonhardening types such as 430 or 430F solenoid varieties provide the best soft magnetic properties of the stainless grades.

Table 14 Magnetic properties of selected ferritic and martensitic stainless steels. Data determined on round bars 9.53 to 15.88 mm (0.375 to 0.625 in.) in diameter per ASTM A 341 using Fahy permeameter.

Grade	ASTM A 838	Condition(a)	Rockwell hardness	Maximum permeability, μ_m	Coercive force, H_c A · m⁻¹	Coercive force, H_c Oe	Resistivity, μΩ · cm
Martensitic:							
Type 410	...	A	B85	750	480	6	57
		H	C41	95	2900	36	
Type 416	...	A	B85	750	480	6	57
		H	C41	95	2900	36	
Type 420	...	A	B90	950	800	10	55
		H	C50	40	3600	45	
Type 440B	...	H	C55	62	5100	64	60
Ferritic:							
Type 430F (solenoid quality)	Alloy 1	A	B78	2000	160	2.0	60
Type 430FR (solenoid quality)(b)	Alloy 2	A	B82	2600	128	1.6	76
Type 446	...	A	B85	1000	360	4.5	67

(a) A, fully annealed; H, heat treated for maximum hardness. (b) Carpenter Technology Corporation

Selection of Alloys for Power Generation Applications

The tonnage of magnetically soft materials such as nickel-iron or cobalt-iron alloy used for specialized applications is small compared to the volume of materials used for fractional horsepower motors, heavy rotating equipment, and power frequency transformers. Materials used in these industries range from commercial low-carbon steel sheet and strip to high-silicon and grain-oriented silicon steels.

Table 15 Soft magnetic materials used for motors and generators. Where more than one grade of silicon sheet is shown, they are listed in the order of increasing cost and efficiency as a result of electrical properties.

Type of motor	Material thickness mm	Material thickness in.	Material
Starting motors			
Automotive	0.35	0.014	1008
	0.46	0.0185	1008
	0.63	0.025	1008
Medium, 0.75 to 75 kW (1 to 100 hp)	0.35	0.014	M-36
	0.46	0.0185	M-43
	0.63	0.025	1008
Large, 75 kW (100 hp) min	0.35	0.014	M-19
	0.46	0.0185	M-36, M-27
	0.63	0.025	M-43, M-36
Motors and generators for intermittent operation(a)			
Miniature	0.46	0.0185	M-50, M-43
	0.63	0.025	1008, M-50
Gyros	0.35	0.014	M-15
	0.46	0.0185	M-15
Selsyns	0.35	0.014	M-15, 45 to 50% Ni iron
Fractional, 0.19 kW (¼ hp)	0.63	0.025	1008
Fractional, 0.37 kW (½ hp)	0.63	0.025	M-43
Fractional, 0.56 kW (¾ hp)	0.63	0.025	M-36
Medium and large	0.46	0.0185	M-43, M-36, M-27
	0.63	0.025	M-43, M-36, M-27
Motors and generators for continuous operation			
Fractional, 0.19 kW (¼ hp)	0.63	0.025	1008
Fractional, 0.37 kW (½ hp)	0.63	0.025	M-43
Fractional, 0.56 kW (¾ hp)	0.63	0.025	M-36
Medium, 0.75 to 75 kW (1 to 100 hp)	0.35	0.014	M-22, M-19
	0.46	0.0185	M-36, M-27
Large, 75 to 3800 kW (100 to 5000 hp)	0.35	0.014	M-19, M-15
	0.46	0.0185	M-27, M-19, M-15
	0.63	0.025	M-27, M-19
Large, >3800 kW (5000 hp)	0.35	0.014	M-15, M-6
	0.46	0.0185	M-19, M-15

Frequency	Material thickness μm	Material thickness mils	Material
High-frequency motors			
To 400 Hz	180	7	3% Si steel(b)
	380	15	M-19, M-15
800 to 1200 Hz	125	5	3% Si steel(b)
	180	7	3% Si steel(b)
Servo motors	125	5	3% Si steel(b)
Synchronous motors	100–355	4–14	45 to 50% Ni iron

(a) 1008 steel is used in 0.76 mm (0.030 in.) for all applications in this category. (b) Cold rolled, nonoriented

Tables 15 and 16 can be used as guides to compositions, grades, and gages previously used for several applications; these tables also take into account factors such as cost, availability, punchability, temperature, and corrosion resistance. Several different alloys may be suitable for a given set of conditions.

There is a gradual increase in price per pound of silicon irons as the alloy core losses improve (that is, decreasing M number). In considering the relationship between electrical properties and gage, it may appear practical to downgrade to a less expensive, lower-efficiency alloy, but to use a thinner gage. For example, a stack of 0.36 mm (0.014 in.) M-43 laminations (29 gage) and the same size stack of 0.47 mm (0.0185 in.) M-36 laminations (26 gage) may be similar in electrical properties, even though M-36 alloy is more efficient and slightly more expensive. However, selection on this basis does not consider all aspects of the cost.

Most important of these is the cost of making laminations from the strip or sheet. A punch press running at constant speed will produce the same number of laminations regardless of material thickness. Hence, it would require 31% more machine time to produce a stack of a given size from 29 gage (0.36 mm, or 0.014 in.) than from 26 gage (0.47 mm, or 0.0185 in.) sheet. Annealing, stacking, and handling costs also increase as the gage becomes thinner. Therefore, where electrical efficiency is critical and thin-gage material is used, the major cost is in making the stack with thinner laminations. It is false economy to downgrade the material and sacrifice electrical properties for only slight savings in material cost.

Motors and Generators

Table 15 lists information on alloys used for motors and generators.

Starting Motors. For small starting motors that operate infrequently and for short periods of time, there is little emphasis on electrical efficiency. For this reason, the least expensive core material has generally been used, regardless of gage. The most common gage for such applications is 24 gage (0.63 mm, or 0.025 in.). For heavier starting motors (up to 75 kW, or 100 hp), 29 gage (0.36 mm, or 0.014 in.) M-36 is more efficient. For the largest starting motors, efficiency is of greater importance, and upgrading of material is justified.

Intermittent Service Devices. Table 15 also lists materials and thicknesses used for ac and dc motors and generators for intermittent service. For small motors used in highly competitive, light-duty consumer items, electrical efficiency had been of secondary consideration. To some extent, higher-quality silicon irons are now being selected for these applications today.

For gyros and selsyns, which are specialized applications demanding more efficiency, higher-grade materials in thinner gages are recommended. Large industrial motors, even in intermittent operation, require more efficiency than small motors, and it is for this reason that both intermediate gages and intermediate materials are recommended for large motors.

Continuous Service Devices. Compositions and gages of sheets for motors and generators for continuous operation are also presented in Table 15. Upgrading of material for the more rigorous service is recommended but the same principles of selection should be followed. In high-speed rotating machinery, yield strength of the magnetic rotor material may be the decisive factor in alloy selection.

Airborne power generation requires compact, high-output equipment that places further demands on the material. Generally, the high-cobalt alloy, 2V-49Ni-49Fe in 0.15 to 0.38 mm (0.006 to 0.015 in.) lamination thickness, has been selected primarily due to its high saturation induction and low hysteresis losses. The 27Co-0.6Cr-Fe alloy has also been used for this application.

Transformers

Table 16 lists examples of magnetically soft silicon irons for nonrotating equipment, the highest-volume use of which are power transformers. In 50 to 60 Hz power transformers, an important consideration is weight. The volume of a transformer is closely proportional to that of the core, and the weight of core and coil are usually about equal. For minimum weight, material with the highest possible operating flux density must be used. Over-voltage requirements must be considered. Regular oriented

Table 16 Soft magnetic materials used for transformers

Type	Material thickness, mm	Material thickness, in.	Material
Continuous duty(a)			
Distribution	0.27	0.011	M-3, M-4
	0.30	0.012	M-5
	0.35	0.014	M-6
Power	0.30	0.012	M-5
	0.35	0.014	M-6
Voltage regulator	0.30	0.012	M-5
	0.35	0.014	M-15
	0.63	0.025	M-22
Welding transformer	0.30	0.012	M-5
	0.35	0.014	M-6
	0.63	0.025	M-43, M-36, M-27

Application	Standard electrical steels	Other alloys
Special application transformers		
Instrument	M-15, M-6 0.30–0.63 mm (0.012–0.025 in.)	2V-49Co-49Fe, 79Ni-4Mo-Fe, 45–50% Ni iron
Radio, power	M-27, M-22, M-19 0.30–0.35 mm (0.012–0.014 in.)	2V-49Co-49Fe, 79Ni-4Mo-Fe
Radio, audio	M-19, M-17, M-15, M-6, M-5 0.35–0.46 mm (0.014–0.0185 in.)	45–50% Ni iron
Radar pulse transformers	25 to 100 μm (1–4 mil): oriented 3% Si steel; 125–180 μm (5–7 mil): nonoriented 3% Si steel	Oriented 45–50% Ni iron, 79Ni-4Mo-Fe, 79Ni-5Mo-Fe, 2V-49Co-49Fe, 45Ni-3Mo-Fe (13 to 100 μm, or 0.5–4 mil)
Chokes, power	M-22, M-19, M15, M6	
Chokes, radio	. . .	Carbonyl irons, ferrites
Ballasts	M-27, M-22 0.46–0.63 mm (0.0185–0.025 in.)	. . .
Miscellaneous bell ringing and toy	1008	. . .

(a) For core laminations for welding transformers, M-27 and M-22 are recommended in 0.46 mm (0.0185 in.) sheet.

grades M-6 to M-4 exhibit flux densities of about 1.8 to 1.84 T (18 to 18.4 kG) in a field of 800 A · m^{-1} (10 Oe). This limits design flux density to about 1.7 T (17 kG). Somewhat higher flux density is attainable with the high-permeability grades designated P by ASTM nomenclature in Table 3.

Although core weight is important, values of energy losses for transformers are receiving more attention as costs of generating electrical power escalate. This factor is placing emphasis on the importance of reducing core loss by lower operating inductions or through the use of higher-quality grades of core material, which are thinner, have higher permeability, and lower hysteresis losses.

Noise produced in transformers results ultimately from magnetostriction of the core material, which varies with operating flux density, silicon content, operating strain, and type of surface insulation. Flux density is important in magnetostriction and joint noise. For typical designs of power transformers using grain-oriented material, a 10% reduction in flux density will reduce noise about 3 decibels (dB).

Design and Fabrication of Magnetic Cores

The purpose of the core metal in a motor, generator, or transformer is to offer the best path for the magnetic lines of flux, and its success in this respect is measured by its permeability. Cores are usually composed of a large number of thin metal laminations that are fabricated by punching from thin sheets of metal, and after being enameled are assembled to form a core. The enamel forms an insulation between laminations that reduces the eddy currents induced in the metal of the core by transformer action. Normal oxidation scale is frequently sufficient insulation for this.

Interlaminar insulation is necessary for high electrical efficiency in the magnetic core, whether the application is static or rotating. For small cores used in fractional-horsepower motors, an oxide surface on the laminations may insulate the core adequately. Insulations of AISI types C-1, C-2, C-3, C-4, and C-5 are used for more rigorous requirements.

Organic-Type Insulation. Types C-1 and C-3 are organic and cannot be successfully applied to laminations before annealing. They are unsuitable for electrical equipment operated at high temperatures or for power transformers with certain types of coolants. However, they improve the punchability of the sheet steel.

Inorganic-Type Insulation. Inorganic types C-4 and C-5 are used when insulation requirements are severe and when annealing temperatures up to 790 °C (1450 °F) must be withstood. Typical values of interlaminar resistance for these two types are between 3 and 100 Ω · cm/lamination under a pressure of 2070 kPa (300 psi). These coatings also can be made to impart residual tensile stresses in the steel substrate, which can improve magnetic properties.

Core insulation must be sufficiently thin and uniform so as to have no more than 2.0% effect on the lamination factor (solidity of the core). To calculate the required insulation for most operations at power frequency, the square of the resistivity, in ohm-centimeters per lamination, should at least equal the square of the width of the magnetic path, in inches. This usually ensures a negligible interlaminar loss that is less than 1.0% of the core loss.

Core Selection Based on Ease of Fabrication. Laminations usually are fabricated by punching the required shape from flat or coiled sheet or from coiled strip. Selection of grade and gage is based primarily on electrical requirements. Differences in punchability may be considered in final selection. Studies of electrical sheet indicate that punchability decreases as silicon content increases, but test results are not conclusive. High-silicon alloys are inherently more abrasive, but they are also more brittle and can be punched with less roll and drag at the edges than can the more ductile low-silicon grades.

Optimum punchability also depends on factors other than steel composition or method of manufacture. If it is difficult to punch a fully processed sheet, it may be advisable to use a semiprocessed grade with magnetic properties not fully developed by the supplier. In this situation, the laminations must be annealed to produce the desired properties.

The advantage of the better punchability of semiprocessed electrical sheet is offset somewhat by the need for better control of annealing conditions during fabrication and by adhesion of laminations during annealing. It is impractical to grade steel for electrical properties after laminations have been fabricated. Therefore, the supplier must grade the product by annealing samples cut from the sheet before it is shipped. The lower cost of the semiprocessed material may influence the designer's decision.

SELECTED REFERENCES

- R. Ball, *Soft Magnetic Materials*, Heyden & Son Ltd., 1979
- C.W. Chen, *Magnetism and Metallurgy of Soft Magnetic Materials*, North-Holland, 1977
- *Ferromagnetic Materials—A Handbook on the Properties of Magnetically Ordered Substances*, E.P. Wohlfarth, Ed., Elsevier, 1980
- Hard and Soft Magnetic Materials With Applications Including Superconductivity, in *Proceedings of the Conference on Hard and Soft Magnetic Materials*, J.A. Salsgiver, K.S.V.L. Narasimhan, P.K. Rastogi, A.R. Traut, D.J. Bailey, and

H.R. Sheppard, Ed., ASM INTERNATIONAL, Oct 1987

- C. Heck, *Magnetic Materials and Their Applications*, Crane, Russak & Company, Inc., 1974
- *Magnetic Properties; Metallic Materials for Thermostats, Electrical Resistance, Heating, Contacts*, Vol 03.04, *Annual Book of ASTM Standards*, American Society for Testing and Materials
- *Soft and Hard Magnetic Materials With Applications*, J.A. Salsgiver, K.S.V.L. Narasimhan, P.K. Rastogi, H.R. Sheppard, and C.M. Maucione, Ed., Proceedings of a symposium held in conjunction with American Society for Metals (Lake Buena Vista, FL), Oct 1986
- "Steel Products Manual—Flat-Rolled Electrical Steel," American Iron and Steel Institute, March 1978

Permanent Magnet Materials

Revised by J.W. Fiepke, Crucible Magnetics, a Division of Crucible Materials Corporation

PERMANENT MAGNET is the term used to describe solid materials that have sufficiently high resistance to demagnetizing fields and sufficiently high magnetic flux output to provide useful and stable magnetic fields. Permanent magnets are normally used in a single magnetic state. This implies insensitivity to temperature effects, mechanical shock, and demagnetizing fields. This article does not consider magnetic memory or recording materials in which the magnetic state is altered during use. It does include, however, hysteresis alloys used in motors.

Permanent magnet materials include a variety of alloys, intermetallics, and ceramics. Commonly included are certain steels, Alnico, Cunife, iron-cobalt alloys containing vanadium or molybdenum, platinum-cobalt, hard ferrites, and rare-earth alloys. Each type of magnet material possesses unique magnetic and mechanical properties, corrosion resistance, temperature sensitivity, fabrication limitations, and cost. These factors provide designers with a wide range of options in designing magnetic parts.

Permanent magnet materials are based on the cooperation of atomic and molecular moments within a magnet body to produce a high magnetic induction. This induced magnetization is retained because of a strong resistance to demagnetization. These materials are classified ferromagnetic or ferrimagnetic and do not include diamagnetic or paramagnetic materials. The natural ferromagnetic elements are iron, nickel, and cobalt. Other elements, such as manganese or chromium, can be made ferromagnetic by alloying to induce proper atomic spacing. Ferromagnetic metals combine with other metals or with oxides to form ferrimagnetic substances; ceramic magnets are of this type. Although scientific literature lists many magnetic substances, relatively few have gained commercial acceptance because of the commercial requirement for low cost and high efficiency.

Permanent magnet materials are marketed under a variety of trade names and designations throughout the world. The United States designations will be used here; other designations are listed in Table 1.

Permanent magnet materials are developed for their chief magnetic characteristics: high induction, high resistance to demagnetization, and maximum energy content. Magnetic induction is limited by composition; the highest saturation induction is found in binary iron-cobalt alloys. Resistance to demagnetization is conditioned less by composition than by shape or crystal anisotropies and the mechanisms that subdivide materials into microscopic regions. Precipitations, strains and other material imperfections, and fine particle technology are all used to obtain a characteristic resistance to demagnetization.

Maximum energy content is most important because permanent magnets are used primarily to produce a magnetic flux field (which is a form of potential energy). Maximum energy content and certain other characteristics of materials used for magnets, are best described by its hysteresis loop. Hysteresis is measured by successively applying magnetizing and demagnetizing fields to a sample and observing the related magnetic induction.

Fundamentals of Magnetism

For understanding a permanent magnet, Faraday's concept of representing a magnetic flux field by lines of force is very useful. The lines of force radiate outward from a north pole and return at a south pole. The lines of force can be revealed by a powder pattern made by sprinkling iron powder on a paper placed above a bar magnet. The number of lines per unit area is the magnetic induction and is designated B. Induction in the magnet consists of lines of force due to the magnetic field and lines of magnetization due to the ferromagnetism of the magnet:

$$B = H + B_i \qquad \text{(Eq 1)}$$

where H is the magnetic field strength and B_i is the intrinsic induction.*

*International Electrotechnical Commission specification TC 68 has recommended J, the symbol for magnetic polarization, be used in place of B_i because intrinsic induction is also known as magnetic polarization.

Magnetic Hysteresis. A hysteresis loop is a common method of characterizing a permanent magnet. The intrinsic induction is measured as the magnetizing field is changed (see Fig. 1). Starting with a virgin state of the material at the origin 0, induction increases along curve I to the point marked +S as the field is increased from zero to maximum. The point +S is the point at which induction no longer increases with higher magnetizing field, and is known as the saturation induction. When the magnetizing field is reduced to zero in permanent magnets, most of the induction is retained. In Fig. 1, when the field is reduced through zero and reversed to −S, the induction decreases from +S to B_r to −S. At zero field, there is a remanent magnetization in the sample, defined as B_r; the value of B_r approaches the saturation induction in well prepared permanent magnet materials. This point on the hysteresis loop is called residual induction.

If the field is increased again in the positive direction, the induction passes through $-B_r$ to +S as shown, and not through the origin. Thus, there is a hysteresis effect, and this plot is called the hysteresis loop. The two halves of the loop are generally symmetrical and form a major loop, which represents the maximum energy content, or

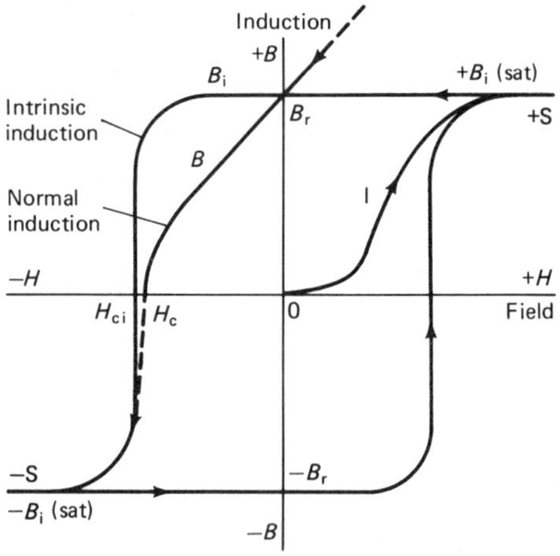

Fig. 1 Major hysteresis loop for a permanent magnet material. B_i (sat) is the saturation induction.

Table 1 Principal magnet designations and their suppliers

Designation	Magnet type	Country	Company
Alnico	Alnico	U.S.	Arnold, CMSC Corp., Crucible, Hitachi Permanent Magnet Co., Thomas & Skinner
		England	Magnetic Materials Group, Mullard, S G Magnets, Swift Levicks
		France	Giffey-Pretre, Ugimag
		Germany	Baermann, Krupp, Magnetfabrik-Bonn, Thyssen
		India	Elpro
		Italy	Centro, Elett. Lombarda Sampas
		Japan	Daido, Hitachi, Mitsubishi, Sumitomo, Tokin, Toshiba, Warabi
		Netherlands	Philips
		Spain	Aceros Hamsa
Alnicomax	Anisotropic Alnico	Italy	Elett. Lombarda
Arnemax	NdFeB	U.S.	Arnold
Arnox	Hard ferrite	U.S.	Arnold
Bonded magnets	Bonded ferrite	U.S.	B.F. Goodrich, Dynacast, Electrodyne, Electro-Kinesis, Gen-Corp., Magnets, Inc., Pantasote, Stackpole, Tengam, 3M, Xolox
		France	Ugimag
		Germany	Baermann, Krupp, Thyssen
		Italy	Industria Ossido
		Japan	Cosmo-TME, Dai Ichi, Fuji, Hitachi, MG Co., Sumitomo, TDK
		Netherlands	Philips
	Bonded NdFeB	U.S.	Delco, Dynacast, Electrodyne, IG Technologies, Neomag, Stackpole, Tengam, 3M, Xolox
		England	Magnetic Materials Group
		France	Ugimag
		Germany	Baermann, Krupp, Magnetfabrik-Bonn
		Japan	Cosmo-TME, Dai Ichi, Fuji, Hitachi, MG Co., Mitsubishi, Suwa-Seikosha
	Bonded Co rare-earth	U.S.	Dynacast, 3M, Xolox
		England	Magnetic Materials Group
		Germany	Baermann
		Japan	Cosmo-TME, Dai Ichi, Fuji, MG Co., Mitsubishi, Suwa-Seikosha
Ceramagnet	Hard ferrite	U.S.	Stackpole
Coalni	Isotropic Alnico	Italy	Centro
Coalnimax	Anisotropic Alnico	Italy	Centro
Cobalt platinum	Co-Pt	U.S.	Engelhard
Coercimax	Anisotropic Alnico	Italy	Centro
Columax	Anisotropic Alnico	U.S.	Thomas & Skinner
Coramag	Co rare-earth	France	Ugimag
Cormax	Co rare-earth	Japan	Sumitomo
Crucore	Co rare-earth	U.S.	Crucible
Crumax	NdFeB	U.S.	Crucible
CS1	CuNiFe	Japan	Sumitomo
CS2	CuNiCo	Japan	Sumitomo
CS3	Vicalloy	Japan	Sumitomo
CuNiFe	CuNiFe	U.S.	IG Technologies
F340	Isotropic ferrite	U.S.	D.M. Steward
F730, 830	Anisotropic ferrite	U.S.	D.M. Steward
FB	Hard ferrite	U.S.	Allen-Bradley/TDK
FB	Hard ferrite	Japan	TDK
Ferriflex	Bonded ferrite	France	Ugimag
Ferrimag	Hard ferrite	U.S.	Crucible
Ferrinet	Hard ferrite	Japan	Tokin
Ferrite magnets	Hard ferrite	U.S.	Allen-Bradley/TDK, Arnold, Crucible, Delco, D.M. Steward, General Magnetic, Hitachi, Magno-Ceram, National Magnetics, Stackpole
		England	Magnetic Materials Group, Mullard, Swift Levicks
		France	Ugimag
		Germany	Baermann, Bosch, Krupp, Magnetfabrik-Bonn, Magnetfabrik-Schramberg, Thyssen
		Italy	Centro, Industria Ossido
		Japan	Daido, Fuji, Hitachi, Hokko Denshi, Nihon-Ugimag, Sumitomo, Taiyo Yuden, TDK, Tokin, Tokyo Ferrite, Tokyo Magnet Chemical, Yokohama, Sumitoku
		Netherlands	Philips
		Spain	Aceros Hamsa
Ferromax	Hard ferrite	Japan	Daido
Ferroxdure	Hard ferrite	France	Giffey-Pretre
		Italy	Centro, Sampas
		Netherlands	Philips
FXD	Hard ferrite	Japan	Sumitomo
Genox	Hard ferrite	U.S.	General Magnetic
Gumox	Bonded ferrite	Germany	Magnetfabrik-Bonn
Hicorex	Co rare-earth	U.S., Japan	Hitachi
Hicorex-Nd	NdFeB	U.S., Japan	Hitachi
HLRA	Co rare-earth	England	Magnetic Polymers
Incor	Co rare-earth	U.S.	IG Technologies
Indalloy	CoMoFe	U.S.	IG Technologies
Koerox	Hard ferrite	Germany	Krupp
Koerzit	Alnico	Germany	Krupp
Koerzit T, H	Vicalloy	Germany	Krupp
Koroseal	Bonded ferrite	U.S.	B.F. Goodrich
Lanthanet	Co rare-earth	Japan	Tokin
M	Hard ferrite	U.S.	Allen-Bradley/TDK
Magnadur	Hard ferrite	England	Mullard
Magnalox	Bonded ferrite	U.S.	Xolox
Maxalco	Alnico	Italy	Sampas
MK	Alnico	Japan	Mitsubishi
MQ	NdFeB	U.S.	Delco
MRC	Co rare-earth	Japan	Mitsubishi
MRP	Bonded Co rare-earth	Germany	Baermann
MVC	FeCoV	Japan	Mitsubishi
NeIGT	NdFeB	U.S.	IG Technologies
NEO	NdFeB	U.S.	Electron Energy
Neobond	Bonded NdFeB	U.S.	Xolox
Neodymium-iron	NdFeB	U.S.	Crucible, Delco, Electron Energy, Hitachi, IG Technologies, Thomas & Skinner
		England	Magnetic Materials Group
		France	Ugimag
		Germany	Baermann, Krupp, Magnetfabrik-Schramberg, Thyssen, Vacuumschmelze
		Japan	Hitachi, Sumitomo, Shin-Etsu, TDK
		Netherlands	Philips
Neomax	NdFeB	Japan	Sumitomo
Neorec	NdFeB	Japan	TDK
NFW	Alnico	Japan	Daido
NiAlCo	Isotropic Alnico	France	Giffey-Pretre, Ugimag
NKS	Alnico	Japan	Sumitomo
Oerstit	Alnico	Germany	Thyssen
Ox	Hard ferrite	Germany	Magnetfabrik
Oxilit	Bonded ferrite	Germany	Thyssen
Oxit	Hard ferrite	Germany	Thyssen
Placo	Co-Pt	Germany	Magnetfabrik
Plastalloy	Bonded ferrite	U.S.	3M
Plasto ferrite	Bonded ferrite	France	Ugimag
Prac	Bonded Alnico	Germany	Magnetfabrik
Rarebond	Bonded Co rare-earth	Japan	Fuji
Rarebond-N	Bonded NdFeB	Japan	Fuji
Rarenet	Co rare-earth	Japan	Shin-Etsu
Rec	Co rare-earth	Japan	TDK
Reco	Isotropic Alnico	England	Mullard
		Netherlands	Philips
Recoma	Co rare-earth	France	Ugimag
		Netherlands	Philips
Refema	NdFeB	France	Ugimag
Remco	Co rare-earth	U.S.	Electron Energy
RM, RN	Bonded ferrite	Japan	Sumitomo
Safe-Nialco	Isotropic Alnico	Spain	Aceros Hamsa
Safe-Supernialco	Anisotropic Alnico	Spain	Aceros Hamsa
SAM	Bonded Co rare-earth	Japan	Suwa Seikosha

(continued)

Table 1 (continued)

Designation	Magnet type	Country	Company
Samarium cobalt	Co rare-earth	U.S.	Crucible, Electron Energy, Hitachi, IG Technologies, Permanent Magnet Co., Recoma, Thomas & Skinner
		England	Magnetic Materials Group, S G Magnets, Swift Levicks
		France	Ugimag
		Germany	Baermann, Krupp, Magnetfabrik-Bonn, Magnetfabrik-Schramberg, Thyssen, Vacuumschmelze
		Japan	Daido, Hitachi, Sumitomo Shin-Etsu, TDK, Tokin, Toshiba
		Netherlands	Philips
		Spain	Imanes
		Switzerland	Brown, Boveri & Co.
Samlet	Bonded Co rare-earth	Japan	Suwa-Seikosha
Secolit	Co rare-earth	Germany	Thyssen
Serem	Co rare-earth	Japan	Shin-Etsu
Spinal	Isotropic ferrite	France	Ugimag
Spinalor	Anisotropic ferrite	France	Ugimag
Sprox	Bonded ferrite	Germany	Magnetfabrik
Stabon	Bonded ferrite	U.S.	Stackpole
Supermagloy	Co rare-earth	England	Swift Levicks
Tascore	Co rare-earth	U.S.	Thomas & Skinner
Ticonal	Anisotropic Alnico	England	Mullard
		France	Giffey-Pretre, Ugimag
		Netherlands	Philips
TMK	Alnico	Japan	Tokin
Tromalit	Bonded Alnico	France	Ugimag
Vacodym	NdFeB	Germany	Vacuumschmelze
Vacomax	Co rare-earth	Germany	Vacuumschmelze
YBM	Hard ferrite	Japan	Hitachi
YCM	Alnico	Japan	Hitachi
YRM	Bonded ferrite	Japan	Hitachi

the amount of magnetic energy that can be stored in the material. Innumerable minor loops can be measured within the major loop, measurements being made to show the effects of lesser fields on magnets under operating conditions.

Demagnetization. The particular value of the demagnetizing field needed to reduce B_i (or J) to zero is called the intrinsic coercive force H_{ci}. Figure 1 includes the normal demagnetization curve derived from the intrinsic curve. The field required to reduce induction B to zero is the normal coercive force H_c. The important practical features of the curve for application to permanent magnet materials are the numerical values of B_r, H_c, H_{ci}, and the area within the hysteresis loop.

Because a permanent magnet most often is used to provide a flux field in a space outside itself, the material rests within its own field, which is a self-demagnetizing field. Therefore, for practical applications, a magnet designer is interested primarily in the second quadrant of the hysteresis loop, called the demagnetization curve (see Fig. 2). This curve represents the resistance to demagnetization and, in an affirmative sense, the ability of a material to establish a magnetic field in an air gap or adjoining magnetic material.

Magnetic Energy. The maximum magnetic energy available for use outside the magnet body is proportional to the largest rectangle that fits inside the normal demagnetization curve. It is indicated by the product $(B_dH_d)_{max}$ and is usually cited as the figure of merit for determining the quality of permanent magnet materials.

A characteristic useful in selecting permanent magnet materials subjected to varying demagnetizing conditions is the permeability (that is, the ratio of the induction to the corresponding magnetizing force) at the operating point:

$$\mu = B_d/H_d \qquad \text{(Eq 2)}$$

For example, a straight-line demagnetization curve where $B_r = H_c$ would have the ideal permeability of 1.0; a magnet of such a material would recover spontaneously all flux when a partial demagnetizing field is removed. The corresponding intrinsic curve would be flat out to the knee, and the material would retain maximum energy. Rare-earth alloy and high coercivity hard ferrite magnets come closest to ideal permanent magnet behavior.

Figure 3 is the product curve of B and H at each point along the demagnetization curve, plotted against B. On the demagnetization curve, each value of B or H involves the other as a coordinate variable. The maximum value of their product—$(B_dH_d)_{max}$—represents the maximum magnetic energy that a unit volume of the material can produce in an air gap. Often, the most efficient design for a magnet is that which employs the magnet at the flux density corresponding to the $(B_dH_d)_{max}$ value.

The amount of total external magnetic flux available from a magnet operating in an open-circuit condition (that is, some flux both in air or nonmagnetic substance) depends on its shape. This relation is shown in Fig. 4 for one specific shape. The permeability μ is the ratio of the total external permeance B_d to that of the permeance of the space occupied by the magnet, H_d, and is equal to the slope of the demagnetization curve.

An enlarged plot of the first and second quadrants of the intrinsic induction curve is given in Fig. 5.

The intrinsic demagnetization curve is of interest to both the materials scientist and the applications engineer. Material scientists are concerned about the effect of composition and processing on the various intrinsic parameters of the material: B_{is}, B_r, and H_{ci}. Applications engineers are concerned about the flux density in an air gap due to both B_i and H. Accordingly, they are interested in the normal induction curve and in the values of B_r, H_c, and $(BH)_{max}$. Design engineers use the intrinsic induction curve to predict performance while the magnet is under the influence of temperature, armature reaction, or other demagnetizing forces.

Lines of constant energy product (B_dH_d) usually are plotted in the second quadrant area of a hysteresis loop. As illustrated in Fig. 6 to 9, they appear as a series of hyperbolic curves superimposed on the rectangular *B-H* grid of the demagnetization curves. The maximum values of external energy are therefore readily available in relation to the demagnetization curve. In this form, the grid constitutes an efficient guide for the design engineer. In practice, a magnet with a fixed air gap would have one fixed B_d/H_d operating point on the demagnetization curve corresponding to the material being used. For variable air gaps, such as are produced by relative movement between the armature and field poles of electrical machinery, the external energy available at the air gap changes continuously, resulting in a so-called minor loop with minimum and maximum values. In practice, the minor loop is plotted on the demagnetization curve to determine location of the loop on the curve, and to evaluate the extent of flux variation within the minor loop cycle. Efficient design of equipment using permanent magnets, such as magnetos, small generators and motors, requires that the minor loop operate near the $(B_dH_d)_{max}$ point.

Magnetically soft materials differ from permanent magnet materials not only in their higher permeabilities, but also, and more significantly, in their much lower resistance to demagnetization. The best magnetically soft materials have H_c values of virtually zero. The hysteresis loop of such a material retraces itself through or near the origin point with each cycle.

Conversely, permanent magnet materials have wide hysteresis loops, characterized by high values of H_{ci}, which range from about 8 to $>1.6 \times 10^3$ kA · m^{-1} (100 Oe to >20 kOe).

Commercial Permanent Magnet Materials

Table 2 lists most of the permanent magnet materials commercially available in the

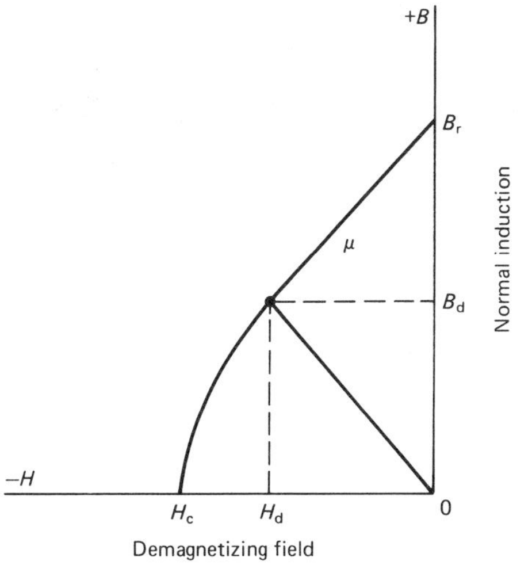

Fig. 2 Normal demagnetization curve for a permanent magnet material

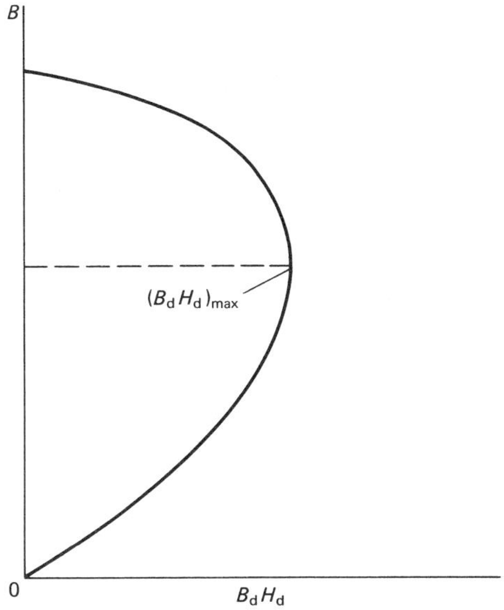

Fig. 3 Typical energy-product curve for a permanent magnet material

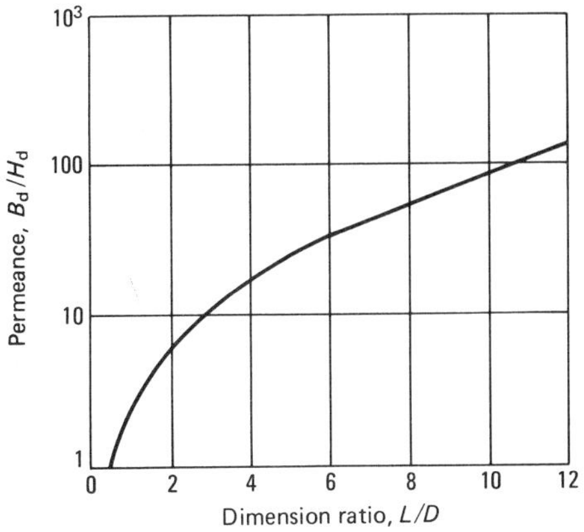

Fig. 4 Relation between magnetic properties and dimensions of straight bar magnets of circular cross section. L is the length of the bar, and D is the bar diameter.

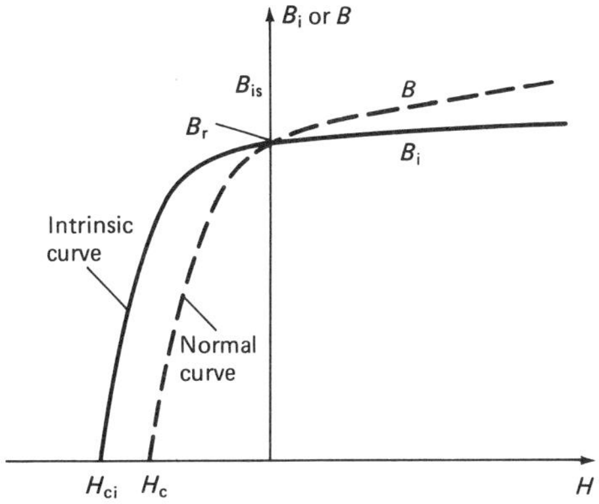

Fig. 5 Intrinsic magnetization curve, B_i, in the first and second quadrants compared with the curve for B

United States and their nominal compositions. Magnetic properties are given in Table 3. Figures 6 to 9 present demagnetization curves associated with the materials listed in Table 2. Physical and mechanical properties are summarized in Table 4. The production of permanent magnet materials is controlled to achieve magnetic characteristics and other properties are allowed to vary according to the manufacturing process used. The selection of materials and the design of permanent magnets for particular applications is a well-defined engineering art; design assistance is available from most producers.

Magnet Steels

Until about 1930, all the commercial permanent magnet materials were quench-hardening steels. Up to about 1910, plain carbon steels containing about 1.5% C were the principal magnet alloys. Alloy steels with up to 6% W were then developed, and later, high-carbon steels with 1 to 6% Cr came into use. Coercive forces for this group of alloys ranged from 3.2 to 5.6 kA · m^{-1} (40 to 70 Oe) (Fig. 6a). The most significant improvement in the quench-hardening steels came in 1917, when the Japanese introduced a cobalt steel containing 36% Co and having a coercive force as high as 20 kA · m^{-1} (250 Oe). The use of magnet steels has declined and today these steels are considered obsolete.

Magnet Alloys

The first advance away from magnet steels came in 1931 with the development of a series of ternary alloys of iron and cobalt plus molybdenum or tungsten. These alloys became known as Remalloy, Cunife, Cunico, and Vicalloy. With the exception of Cunife, these alloys are now obsolete. The magnetic properties of the materials are shown in Fig. 6.

Cunife. Commercial Cunife contains approximately 20% Fe, 20% Ni, and 60% Cu. This composition is in the two-phase region of the phase diagram. The material is quenched from about 1000 °C (1830 °F) to give a homogeneous face-centered cubic (fcc) structure. The quenched specimens are already fully magnetic and contain iron-nickel rich clusters, 5 to 10 nm (50 to 100 Å) in cluster size, in a copper-rich matrix. It is then aged at 650 °C (1200 °F) to develop the amounts of both phases to optimum proportions. It is usually cold worked in stages to maximize directional magnetic properties in the final shape. The nature of the phase diagram, the periodicity of the microstructure and x-ray diffraction effects all support the view that the magnetic structure develops by spinodal decomposition. The coercive force can be accounted for on the basis of shape anisotropy of iron-nickel-rich magnetic regions in a copper-rich nonmagnetic matrix. The material is extremely anisotropic, with the superior magnetic properties in the direction of rolling, a factor that must be considered in magnet design.

Cold working produces a crystallographic texture which is developed strongly by subsequent annealing. The spinodal decomposes along definite crystallographic planes, and the texture is developed by the shape of the particulates. An additional contribution to magnetic texture may arise from straightforward deformation of the partially decomposed spinodal phase. Aging in a magnetic field has no influence on the magnetic properties because the Curie temperature is too far below the decomposition temperature.

The mechanical softness in this alloy system permits easy cold reduction and working, thus leading to many applications in the form of wire or tape. Optimum properties are obtained only after 95% or more cold reduction.

Alnico Alloys

Alnico alloys are one of the major classes of permanent magnet materials. The Alnicos vary widely in composition and in preparation, to give a broad spectrum of properties, costs, and workability. Alnico alloys are sold under a variety of names throughout the world (see Table 1). As a group, Alnico alloys are brittle and hard and can be machined only by surface grinding, electrical discharge machining, or electrochemical milling. They resist atmospheric corrosion well up to 500 °C (930 °F). Magnetic properties are negligibly affected by vibration or shock. Generally, Alnico is superior to other permanent magnet materials in resisting temperature effects on magnetic performance. Typical compositions and properties are summarized in Tables 2, 3, and 4; demagnetization curves are shown in Fig. 7.

Alnicos 1 through 4 are isotropic and generally lower in cost than the other Alnico alloys but are much lower in magnetic characteristics. The remainder of the alloys

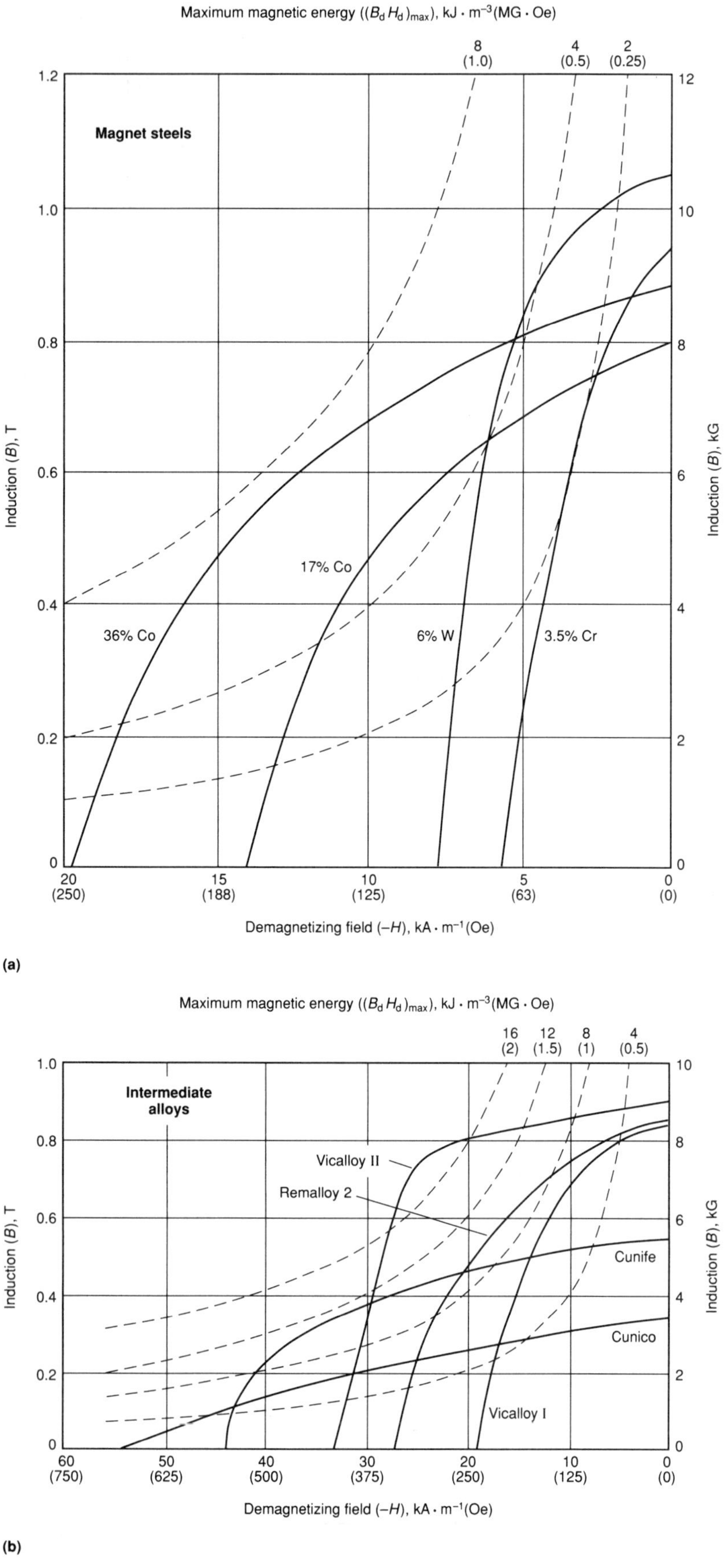

Fig. 6 Demagnetization curves for obsolete permanent magnet materials. (a) Magnet steels. (b) Intermediate alloys. Among intermediate alloys, only Cunife is still used.

are generally anisotropic. Maximum properties, and therefore greatest economy, are obtained when the device is designed to make use of oriented material. Optimum properties are achieved by casting and heat treating.

Magnets made by sintering Alnico powders or by bonding are used where small or intricate shapes to precise tolerances are required. Sintered Alnico is produced by blending powders and then pressing and sintering just below the melting temperature in an oxygen-free atmosphere. The sintered alloys have mechanical properties superior to those of cast Alnicos, but the magnetic properties generally are slightly lower. Sintered magnets are given the same heat treatments as cast.

In general, optimum properties are developed by solution treating at 1100 °C (2010 °F), where the alloy is in equilibrium as a body-centered cubic (bcc) phase, followed by cooling at a rapid, but critical, rate. Between 900 and 800 °C (1650 and 1470 °F), the alloy separates into two nonequilibrium bcc phases. One is almost pure iron and the other, a weakly magnetic phase, is roughly FeNiAl. The resulting microstructure is typical of a structure resulting from spinodal decomposition. Various combinations of H_c and B_r can be obtained.

Various other elements are often added to Alnicos. For example, to lessen the deleterious effect of carbon, carbide stabilizers such as titanium or niobium may be added. Titanium and copper increase H_c at the expense of B_r, but niobium increases H_c without decreasing B_r. Increasing the basic cobalt content by about 20% or more results in a major improvement in magnetic performance, because such large amounts of cobalt make it possible to develop a preferred orientation by heat treating the material in a magnetic field. Less striking improvements in H_c and B_r are obtained in nonoriented samples.

The ability of the magnetic field to influence the orientation (and thus the anisotropy) of the decomposing phase originates in the mechanism of decomposition. In processing the most commonly used alloy in this family, Alnico 5, for instance, the molten alloy is cast, then solution treated above 1250 °C (2280 °F). If zirconium and/or silicon is present, the alloy is solution treated at 900 to 925 °C (1650 to 1700 °F). The alloy is then placed in a magnetic field and cooled at a controlled rate from the solution treating temperature. Finally, the alloy is aged at 600 to 500 °C (1110 to 930 °F). An Alnico casting ordinarily can be shaped only by grinding or electrolytic machining, although hot working and certain other very specialized processing techniques are possible. Final finishing generally is done by grinding.

The critical phenomenon in this process is the spinodal decomposition of the high-

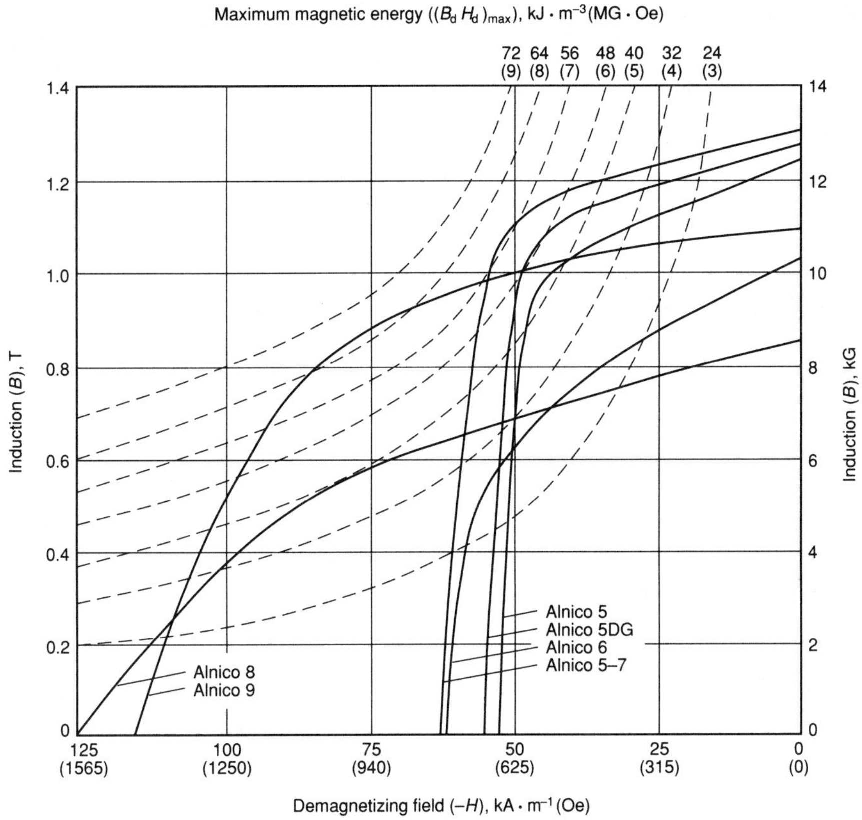

Fig. 7(a) Magnetization curves for anisotropic cast Alnico permanent magnet materials

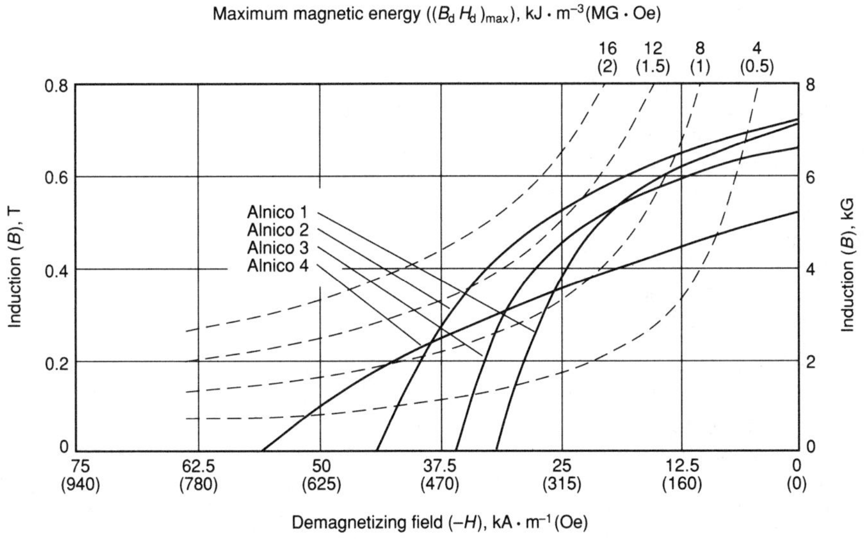

Fig. 7(b) Magnetization curves for isotropic cast Alnico permanent magnet materials

temperature α-phase into an Fe-Co-rich α-phase and a Ni-Al-rich α′-phase. In the high-cobalt Alnico alloys, heat treating in a magnetic field appears to favor decomposition of parallel compositional waves and to suppress transverse waves. This effect is strongest if aging is carried out just below the intersection, on the phase diagram, of the Curie temperature and the temperature where the spinodal instability sets in. This condition limits the number of alloys that respond to magnetic aging. The addition of cobalt to FeNiAl promotes magnetic aging by moving the Curie temperature and decomposition temperature closer together.

Because the magnetic structure is influenced by crystallographic orientation during its formation, careful development of the proper ⟨100⟩ crystal texture is required to achieve the best properties in the Alnicos. In commercial alloys the magnetic properties of Alnico 5DG, which is a partly oriented material (DG indicates directed grain), are significantly better than those of standard Alnico 5; those of Alnico 5-7, which has almost perfect orientation, are even better.

In the construction of magnets from crystal oriented Alnico, designs are limited by the possible shapes in which a properly oriented magnetic field can be established during heat treatment. (Magnetic fields are easy to create, for instance, in the shapes of straight lines, circles, or arcs of circles.) This is not a major limitation on component design, however, because it can generally be overcome by using segmented magnets or by secondary fabrication of simple cast shapes, or both.

Platinum-Cobalt Alloys

Although platinum-cobalt magnets are expensive, they are useful in certain applications. Platinum-cobalt is isotropic, ductile, easily machined, resistant to corrosion and high temperatures, and has magnetic properties superior to all except the rare earth/cobalt alloys. Best magnetic properties are obtained at an atomic ratio of 50Pt: 50Co. Above about 820 °C (1510 °F) the alloy has a disordered fcc structure. Below this temperature, ordering develops a slightly tetragonal structure. To process this alloy to develop a $(BH)_{max}$ of 72 kJ · m^{-3} (9 MG · Oe), the following treatment is used:

- Heat to 1000 °C (1830 °F) to fully disorder
- Cool at a controlled rate to room temperature
- Age at 600 °C (1110 °F) for about 5 h

The final structure is only partly ordered; its physical structure is without distinction until overaged to the completely ordered structure. Values of $(BH)_{max}$ of more than 80 kJ · m^{-3} (10 MG · Oe), have been achieved by fabricating parts using powder metallurgical techniques and heat treating with essentially the same treatment as that used for cast parts. Typical properties are given in Tables 3 and 4 and the demagnetization curve is shown in Fig. 8. Platinum-cobalt magnets have been replaced by rare-earth materials and are seldom used.

Cobalt and Rare-Earth Alloys

Permanent magnet materials based on combinations of cobalt and the lighter rare-earth (lanthanide) metals are the materials of choice for most small, high-performance devices operating between 175 to 350 °C (345 to 660 °F). These materials are manufactured by powder metallurgy methods and have low-temperature coefficients which can be altered by additions of a heavy rare-earth, such as gadolinium or holmium. Samarium or praseodymium are the best choices for the rare-earth component in

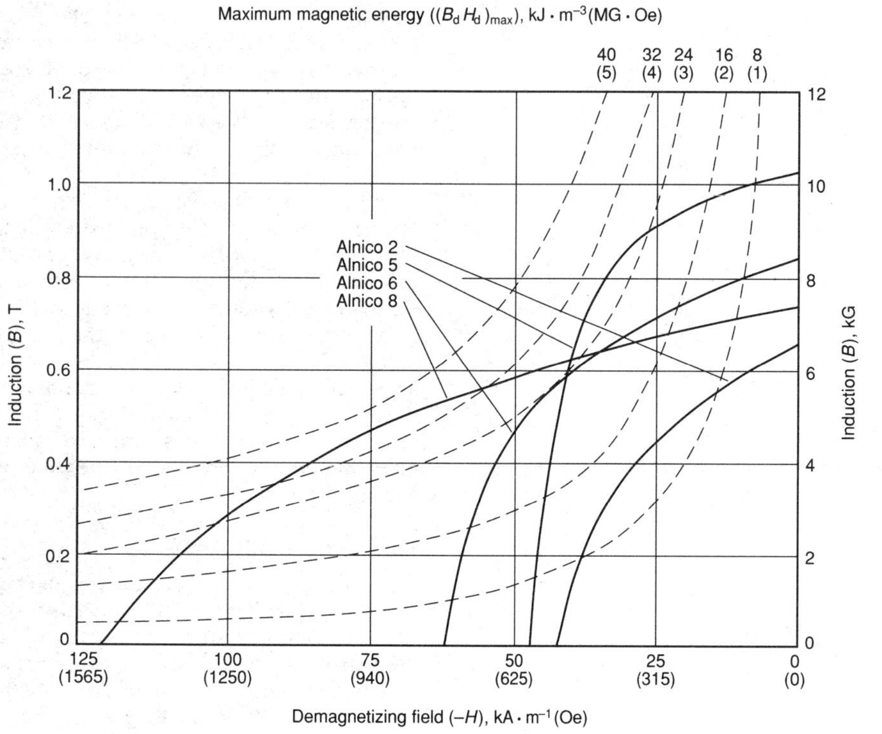

Fig. 7(c) Magnetization curves for sintered Alnico permanent magnet materials

commercial rare-earth cobalt permanent magnet materials.

An alloy need not contain only a single lanthanide metal; mixtures are often used. Likewise, a portion of the cobalt can be replaced with copper and iron to obtain desired magnetic characteristics. In some alloys containing copper and iron, some of the samarium is replaced with cerium for cost savings. These are of the $(RE)_2(TM)_{17}$ type, where RE is rare earth and TM is mostly cobalt with some substitution of iron and copper as mentioned.

Sintered $SmCo_5$, $Co_5(RE)$, $(RE)_2(TM)_{17}$ materials with minor elemental additions produce some of the highest magnetic quality permanent magnets. Typical properties of commercially available materials are given in Tables 3 and 4, and demagnetization curves are presented in Fig. 8.

Because of the variety of materials, few standard designations have been developed. The four compositions listed in Table 2 (arbitrarily numbered 1 to 4) represent the range of magnetic quality available from the many producers of these materials.

Bonded permanent magnets utilizing $SmCo_5$ or Sm_2Co_{17} powders likewise are available from a number of magnet producers. Data are presented on selected grades in Tables 3 and 4 and on demagnetization curves in Fig. 6(c). Compression molding is usually favored as the molding method, although injection molding is possible.

It was the discovery of the large magnetocrystalline anisotropy of Co_5Y that initiated the interest in $Co_5(RE)$ alloys as permanent magnet materials. An energy product of nearly 225 kJ · m^{-3} (28 MG · Oe) was achieved in a single particle of Co_5Y. However, no investigator has reported the achievement of a usefully high coercive force in a $Co_5(RE)$ permanent magnet containing a substantial amount of yttrium.

Cerium is by far the most abundant, and potentially the least expensive, of all rare-earth elements. Procedures for preparing large quantities of Co_5Ce have been described in the literature, but this alloy does not make good magnets because of low saturation magnetization and stability problems. However, because cerium-rich misch metal (MM) is at present less expensive than cerium metal, and because $Co_5(MM)$ has some permanent magnet properties superior to those of Co_5Ce, very little effort has been made to develop permanent magnets with Co_5Ce as the basic magnetic phase. Most development efforts involving cerium metal have been conducted with precipitation-hardening alloys of the 2:17 type, where some of the samarium is replaced with cerium.

Magnetic Properties. Among the binary $Co_5(RE)$ phases, Co_5Pr has the highest potential energy product. Moreover, praseodymium is more than eight times as plentiful as samarium in bastnasite, the major source of rare-earth metals in the United States. The crystal anisotropy of Co_5Pr is much lower than that of samarium cobalt, leading to significantly lower values of H_{ci}. However, several producers substitute praseodymium for up to 50% of the samarium in $SmCo_5$ alloys to get maximum values of residual induction with acceptable levels of H_{ci}.

Sintered cobalt and rare-earth materials are hard and brittle, very much like the Alnicos. Magnets made of these materials often are pressed to final shape to eliminate machining; magnets are also produced by cutting or slicing them from blocks or other simple shapes.

Cobalt and rare-earth materials have coercive forces much higher than those of other permanent magnet materials. Nevertheless, they can be satisfactorily magnetized in fields lower than those necessary to achieve saturation induction. Virgin magnets—that is, magnets that have never been exposed to a magnetizing field after final heat treatment—can be magnetized in fields of about 1.2 MA · m^{-1} (15 kOe).

To obtain anisotropic properties, tooling and dies are designed to compact powders in orienting fields in a manner similar to that used for ferrites. Bonded magnets are produced by simply compacting aligned powder mixed with plastic or soft metal binders. Magnetic quality of bonded magnets is lower than that of sintered magnets.

Applications. In general, cobalt and rare-earth magnets cost more than any other type, except platinum-cobalt magnets. Cobalt and rare-earth magnets have replaced most platinum-cobalt types, particularly in microwave applications, where not only the lower price but also the higher coercivity of cobalt and rare-earth magnets are advantageous. The unique combination of properties in these magnets [high H_c, high $(BH)_{max}$, recoil permeability of one, and temperature stability] has led to a variety of new designs in traveling-wave tubes, other charged particle-beam focusing devices, watches, motors and generators, couplers, and magnetic bearings.

Hard Ferrites

Hard ferrites, also known as ceramic permanent magnet materials, are predominantly complex oxides. The ferrite most commonly used for magnets is $SrO \cdot 6Fe_2O_3$. Prior to the mid-1970s, $BaO \cdot 6Fe_2O_3$ was produced in major quantities. Ferrites are commonly listed with the chemical composition $MO \cdot 6Fe_2O_3$, where M represents barium, strontium, or the combination of the two. Various additives such as SiO_2 or Al_2O_3 are beneficial in increasing coercivity and aid in sintering. Hard ferrites have coercive forces three to eight times the coercive forces obtained for most Alnicos. Both sintered and bonded ferrite magnets are used extensively.

A typical sintered ferrite magnet is made by mixing SrO with Fe_2O_3. This mixture is calcined above 1095 °C (2000 °F) to form the complex ferrite oxide compound. After

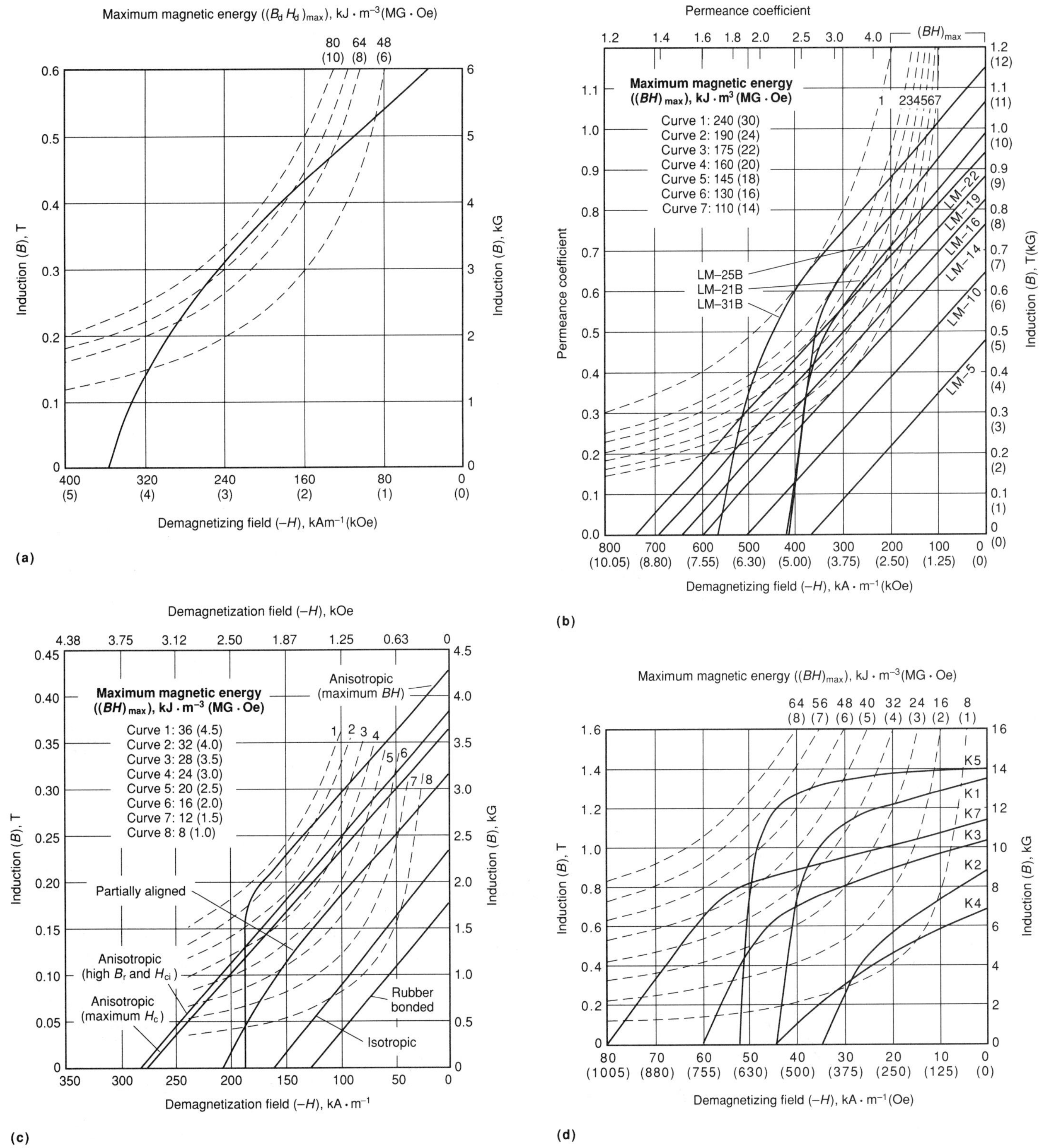

Fig. 8 Demagnetization curves for permanent magnet materials. (a) Platinum-cobalt alloys. (b) Cobalt and rare-earth alloys. (c) Strontium-ferrite alloys. (d) Iron-chromium-cobalt alloys

milling into fine particles, the ferrite compound is compacted in a die and sintered at 1200 to 1300 °C (2190 to 2370 °F). Sintering develops a typical ceramic structure machinable only by diamond wheel grinders and slicing tools. Shrinkage during sintering is higher than for sintered metal magnets. About 15% shrinkage occurs, yielding a sintered magnet with a density about 95% of theoretical.

Nonoriented ferrites are almost isotropic. To obtain anisotropic magnets with a resultant twofold to threefold improvement in magnetic properties, the original powdered mixture is then ground to a powder of about 1 μm (40 μin.) particle size. This powder is compacted in a die within an orienting field and finally sintered at 1200 to 1300 °C (2190

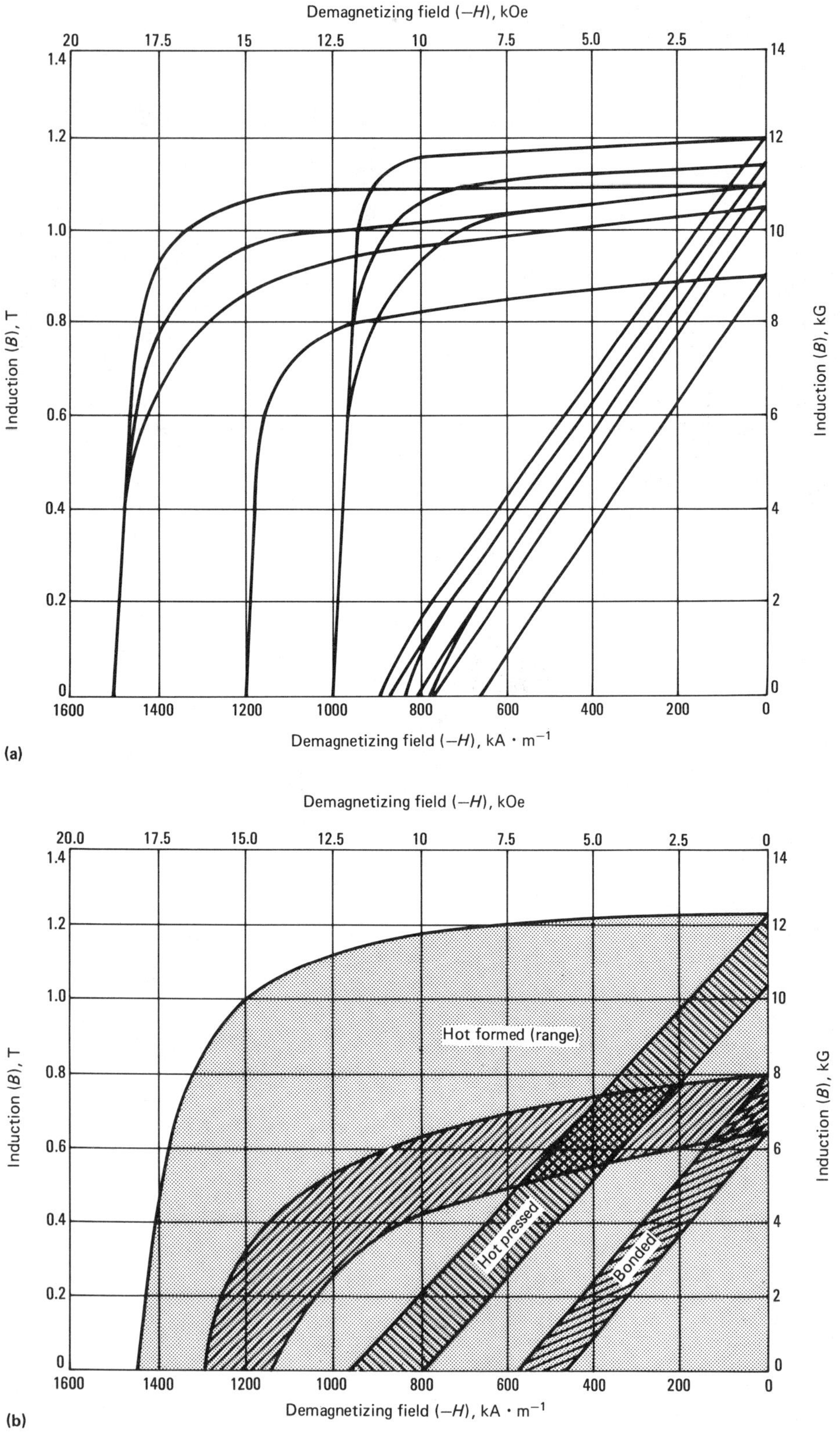

Fig. 9 Demagnetization curves for neodymium-iron-boron alloy magnets. (a) Sintered. (b) Prepared from rapidly solidified ribbon

to 2370 °F). Similar powder is used to prepare bonded magnets by mixing it with an appropriate proportion of a polymer and pressing, injection molding, or extruding to final shape. Magnetic properties generally are inferior to sintered magnets but dimensional tolerances can be held close and processing costs kept low.

Ceramic permanent magnet materials have high electrical resistivities and are poor conductors of heat. Although ferrites are not affected by high temperatures or atmospheric corrosion, the magnetic properties are more temperature-dependent than they are with other permanent magnet materials. Coercive forces decrease with lowering temperatures; flux density B_d decreases at a rate of 0.19%/°C (0.11%/°F) with rising temperature. The ferrites lose a portion of their magnetic flux at low subzero temperatures, the amount of loss depending on the *L*/*D* ratio. For example, the irreversible loss at −57 °C (−70 °F) is 3% when the *L*/*D* ratio is 0.09; however, when the *L*/*D* ratio is 0.50, no loss occurs. Typical properties are given in Tables 3 and 4 and in the demagnetization curves in Fig. 6(c). Although the $(BH)_{max}$ of the ferrites is relatively low, their low cost and high coercive force make them attractive for applications such as separators, magnetos, motors, speakers, and so on.

Tooling to produce a ferrite magnet to a specific size/configuration is expensive (that is, up to $50 000). Normally the tooling has a large number of cavities and produces a number of pieces in each press cycle. This means the application must use a large quantity of pieces to justify the tooling expense and obtain the lowest cost.

Iron-Chromium-Cobalt Alloys

Anisotropic iron-chromium-cobalt permanent magnet alloys have magnetic properties that are somewhat comparable to Alnico 5. The magnetic hardening of these alloys is performed by tempering after solution treatment. Prior to the tempering cycle the material is ductile and can be machined. The isotropic grades of this alloy have replaced the Remalloy and Vicalloy alloys in many applications. Typical properties are summarized in Tables 3 and 4; demagnetization curves are shown in Fig. 8.

Neodymium-Iron-Boron Alloys

Introduced in 1983, neodymium-iron-boron alloy permanent magnet materials have become the material of choice for a wide range of permanent magnet devices and applications. This newest family of permanent magnet materials is based on combinations of neodymium, iron, and boron. Processing of these magnets has been accomplished by two different techniques:

- Conventional powder metal sintering
- Consolidation of rapidly solidified materials

Praseodymium has been substituted for neodymium with favorable results. Likewise, additions of aluminum, dysprosium, gallium, and other elements are made to obtain desired magnetic characteristics.

Conventional Powder Metallurgy Processing. The powder metallurgy process for

Table 2 Nominal compositions, Curie temperatures, and magnetic orientations of selected permanent magnet materials

Designation	Nominal composition	Approximate Curie temperature °C	°F	Magnetic orientation(a)
3½% Cr steel	Fe-3.5Cr-1C	745	1370	No
6% W steel	Fe-6W-0.5Cr-0.7C	760	1400	No
17% Co steel	Fe-17Co-8.25W-2.5Cr-0.7C	...	...	No
36% Co steel	Fe-36Co-3.75W-5.75Cr-0.8C	890	1630	No
Cast Alnico 1	Fe-12Al-21Ni-5Co-3Cu	780	1440	No
Cast Alnico 2	Fe-10Al-19Ni-13Co-3Cu	810	1490	No
Cast Alnico 3	Fe-12Al-25Ni-3Cu	760	1400	No
Cast Alnico 4	Fe-12Al-27Ni-5Co	800	1475	No
Cast Alnico 5	Fe-8.5Al-14.5Ni-24Co-3Cu	900	1650	Y, H
Cast Alnico 5DG	Fe-8.5Al-14.5Ni-24Co-3Cu	900	1650	Y, H, C
Cast Alnico 5-7	Fe-8.5Al-14.5Ni-24Co-3Cu	900	1650	Y, H, C
Cast Alnico 6	Fe-8Al-16Ni-24Co-3Cu-2Ti	860	1580	Y, H
Cast Alnico 7	Fe-8Al-18Ni-24Co-4Cu-5Ti	840	1540	Y, H
Cast Alnico 8	Fe-7Al-15Ni-35Co-4Cu-5Ti	860	1580	Y, H
Cast Alnico 9	Fe-7Al-15Ni-35Co-4Cu-5Ti	...	...	Y, H, C
Cast Alnico 12	Fe-6Al-18Ni-35Co-8Ti	...	...	No
Sintered Alnico 2	Fe-10Al-17Ni-12.5Co-6Cu	610	1490	No
Sintered Alnico 4	Fe-12Al-28Ni-5Co	800	1475	No
Sintered Alnico 5	Fe-8.5Al-14.5Ni-24Co-3Cu	900	1650	Y, H
Sintered Alnico 6	Fe-8Al-16Ni-24Co-3Cu-2Ti	860	1580	Y, H
Sintered Alnico 8	Fe-7Al-15Ni-35Co-4Cu-5Ti	860	1580	Y, H
Cunife	20Fe-20Ni-60Cu	410	770	Y, R
Bonded ferrite A	$BaO\text{-}6Fe_2O_3$ + organics	450	...	No, P
Bonded ferrite B	$BaO\text{-}6Fe_2O_3$ + organics	450	...	No
Sintered ferrite 1	$BaO\text{-}6Fe_2O_3$	450	840	No,P
Sintered ferrite 2	$BaO\text{-}6Fe_2O_3$	450	840	Y, A
Sintered ferrite 3	$BaO\text{-}6Fe_2O_3$	450	840	Y, A
Sintered ferrite 4	$SrO\text{-}6Fe_2O_3$	460	860	Yes
Sintered ferrite 5	$SrO\text{-}6Fe_2O_3$	460	860	Yes
Bonded neodymium	NdFeB + organics	...	...	Y, P, E
Hot-formed neodymium	...	...	...	Y, R
Hot-pressed neodymium	NdFeB + organics	...	...	Y, R
Sintered neodymium	NdFeB	310	590	Y, A
FeCrCo	...	640	1185	Y, R
Platinum cobalt	76.7Pt-23.3Co	480	900	No
Cobalt rare earth 1	$SmCo_5$	725	1340	Y, A
Cobalt rare earth 2	$SmCo_5$	725	1340	Y, A
Cobalt rare earth 3	$SmCo_5$	725	1340	Y, A
Cobalt rare earth 4	Sm_2Co_{17}	800	1475	Y, A
Bonded Co rare earth	...	...	...	Y, P, E

(a) Y, yes; H, orientation developed during heat treatment; C, columnar crystal structure developed; P or E, some orientation developed during pressing or extrusion; R, orientation developed by rolling or other mechanical working; A, orientation developed predominantly by magnetic alignment of powder prior to compacting but alignment influenced by pressing forces also

producing permanent magnets consists of alloy preparation, premilling, milling, particle alignment/pressing, sintering/heat treatment, grinding, coating, and magnetizing. The high reactivity of neodymium or other rare earths and their alloys requires the suppression of contamination during alloy preparation and processing. In particular, oxidation by O_2 or H_2O or both must be kept to a minimum through all fine powder handling and sintering stages. Any oxidation of the alloy occurring during processing depletes the alloy of the rare-earth components and shifts the composition of the rich side of the phase diagram. This usually results in the production of an alloy having unfavorable magnetic qualities.

To obtain powder compacts with maximum magnetization, the powder is magnetically aligned and pressed. The powder compaction is performed by die pressing or by isostatic pressing. A maximum energy product of 360 $kJ \cdot m^{-3}$ (45 MG · Oe) has been reported (Ref 1). When die pressing, the aligning field is established in the cavity with its magnetic axis either in the direction of pressing or at right angles. Multicavity tooling is generally used.

Isostatic pressing is normally carried out on powders prealigned in a pulsed field capable of producing much higher fields than used in the die pressing method. This improves the degree of particle alignment and results in higher unit magnetic properties than those obtained for die-pressed pieces.

Because of their high iron content, neodymium-iron-boron magnets are susceptible to corrosion and require a corrosion-resistant coating for the many applications. Presently, electrophoretic coating by a cathodic process is widely used. This is a multistage procedure consisting of surface cleaning and pretreatment. An epoxy coating approximately 25 μm (1000 μin.) thick is typical of this process more suitable for larger pieces, where the cost of holding the individual piece during the coating process is not prohibitive. Other organic and metallic coatings are also used.

Data are presented on selected grades of commercially available materials in Tables 3 and 4 and on demagnetization curves in Fig. 9. A graph depicting the full range of sintered neodymium-iron-boron available is shown in Fig. 10. It is amazing to see such a proliferation of materials in only a 6-year period. Powders made by this method will not produce a usable bonded magnet.

Rapidly Solidified Material Consolidation. One major American automobile manufacturer utilizes a patented process to produce neodymium-iron-boron magnets. The process involves the use of rapidly solidified ribbons having a uniform, finely crystalline microstructure consisting primarily of the $Nd_2Fe_{14}B$ phase.

Table 3 Nominal magnetic properties of selected permanent magnet materials

For nominal compositions, see Table 2; for mechanical and physical properties, see Table 4.

Designation	H_c		H_{ci}		B_r		B_{is}		$(BH)_{max}$		B_d		H_d		Required magnetizing field		Permeance coefficient at $(BH)_{max}$	Average recoil permeability, G/Oe
	kA · m^{-1}	Oe	kA · m^{-1}	Oe	T	kG	T	kG	kJ · m^{-3}	MG · Oe	T	kG	kA · m^{-1}	Oe	kA · m^{-1}	kOe		
3½% Cr steel	5.3	66	...	...	0.95	9.5	...	...	2.3	0.29	...	...	...	...	...	...	...	...
6% W steel	5.9	74	...	...	0.95	9.5	...	...	2.6	0.33	...	...	...	...	...	...	...	...
17% Co steel	14	170	...	...	0.95	9.5	...	...	5.2	0.65	...	...	...	...	...	...	...	...
36% Co steel	19	240	...	...	0.975	9.75	...	...	7.4	0.93	...	...	...	...	...	...	...	...
Cast Alnico 1	35	440	36	455	0.71	7.1	1.05	10.5	11	1.4	0.45	4.5	24	305	160	2.0	14	6.8
Cast Alnico 2	44	550	46	580	0.725	7.25	1.09	10.9	13	1.6	0.45	4.5	28	350	200	2.5	12	6.4
Cast Alnico 3	38	470	39	485	0.70	7.0	1.00	10.0	11	1.4	0.43	4.3	26	320	200	2.5	13	6.5
Cast Alnico 4	58	730	62	770	0.535	5.35	0.86	8.6	10	1.3	0.30	3.0	34	420	280	3.5	8.0	4.1
Cast Alnico 5	50	620	50	625	1.25	12.5	1.35	13.5	42	5.25	1.02	10.2	42	525	240	3.0	18	4.3
Cast Alnico 5DG	52	650	52	655	1.29	12.9	1.40	14.0	49	6.1	1.05	10.5	46	580	280	3.5	17	4.0
Cast Alnico 5-7	58	730	59	735	1.32	13.2	1.40	14.0	59	7.4	1.15	11.5	51	640	280	3.5	17	3.8
Cast Alnico 6	60	750	...	...	1.05	10.5	1.30	13.0	30	3.7	0.71	7.1	42	525	320	4.0	13	5.3
Cast Alnico 7	84	1 050	...	...	0.857	8.57	0.945	9.45	30	3.7	...	...	...	...	400	5.0	8.2	...
Cast Alnico 8	130	1 600	138	1 720	0.83	8.3	1.05	10.5	40	5.0	0.506	5.06	76	950	640	8.0	5.0	3.0
Cast Alnico 9	115	1 450	...	...	1.05	10.5	...	...	68	8.5	...	...	...	...	560	7.0	7.0	...
Cast Alnico 12	76	950	...	...	0.60	6.0	...	...	14	1.7	0.315	3.15	43	540	400	5.0	5.6	...
Sintered Alnico 2	42	525	44	545	0.67	6.7	1.10	11.0	12	1.5	0.43	4.3	28	345	200	2.5	12	6.4
Sintered Alnico 4	56	700	61	760	0.52	5.2	...	...	10	1.2	0.30	3.0	32	400	280	3.5	...	7.5
Sintered Alnico 5	48	600	48	605	1.04	10.4	1.205	12.05	29	3.60	0.785	7.85	37	465	240	3.0	18	4.0
Sintered Alnico 6	61	760	63	790	0.88	8.8	1.15	11.5	22	2.75	0.55	5.5	40	500	320	4.0	12	4.5
Sintered Alnico 8	125	1 550	134	1 675	0.76	7.6	0.94	9.4	36	4.5	0.46	4.6	80	1 000	640	8.0	5.0	2.1
Cunife	44	550	44	555	0.54	5.4	0.59	5.9	12	1.5	0.40	4.0	26	325	200	2.5	12	3.7
Bonded ferrite A	155	1 940	...	...	0.214	2.14	...	...	8	1.0	0.116	1.16	...	...	960	12.0	1.3	1.1
Bonded ferrite B	92	1 150	...	...	0.14	1.4	...	...	3	0.4	...	...	...	...	640	8.0	1.2	1.1
Sintered ferrite 1	145	1 800	276	3 450	0.22	2.2	...	...	8	1.0	0.11	1.1	72	900	800	10.0	1.2	1.2
Sintered ferrite 2	175	2 200	185	2 300	0.38	3.8	...	...	27	3.4	0.185	1.85	132	1 650	800	10.0	1.1	1.1
Sintered ferrite 3	240	3 000	292	3 650	0.32	3.2	...	...	20	2.5	0.16	1.6	130	1 600	800	10.0	1.1	1.1
Sintered ferrite 4	175	2 200	185	2 300	0.40	4.0	...	...	30	3.7	0.215	2.15	135	1 700	960	12.0	1.2	1.05
Sintered ferrite 5	250	3 150	287	3 590	0.355	3.55	...	...	24	3.0	0.173	1.73	138	1 730	1 200	15.0	1.0	1.05
NdFeB (sintered)	848	10 600	>1 350	>17 000	1.16	11.6	...	...	255	32	0.60	6.0	425	5 300	>2 000	>25.0	1.13	...
Bonded NdFeB	430	5 400	720	9 000	0.69	6.9	...	...	76	9.5	0.315	3.15	240	3 000	...	...	1.05	...
Hot-pressed NdFeB	560	7 000	1 280	16 000	0.80	8.0	...	...	110	13.7	0.38	3.8	295	3 700	...	...	1.05	...
Hot-formed NdFeB	880	11 000	1 200	15 000	1.20	12.0	...	...	274	34.2	0.59	5.9	465	5 800	...	...	1.05	...
Platinum cobalt	355	4 450	430	5 400	0.645	6.45	...	...	74	9.2	0.35	3.5	215	2 700	1 600	20.0	1.2	1.2
Cobalt rare earth 1	720	9 000	1 600	20 000	0.92	9.2	0.98	9.8	170	21	...	...	...	...	2 400	30.0	...	...
Cobalt rare earth 2	640	8 000	>2 000	>25 000	0.86	8.6	...	...	145	18	0.44	4.4	330	4 100	2 400	30.0	...	1.05
Cobalt rare earth 3	535	6 700	>1 200	>15 000	0.80	8.0	...	...	120	15	0.40	4.0	295	3 700	2 400	30.0	...	1.1
Cobalt rare earth 4	640	8 000	>640	>8 000	1.13	11.3	...	...	240	30	0.60	6.0	400	5 000	>1 600	>20.0	1.2	...

Isotropic magnets can be made by cold compaction of the rapidly solidified powder with a resin binder or by hot pressing to full density. Because of the extremely fine grain size, field alignment and compaction of milled isotropic ribbons is extremely difficult and not practical. Anisotropy can be achieved by hot deformation. Energy products of 320 kJ · m^{-3} (40 MG · Oe) have been achieved by this method. Data on epoxy-bonded, hot-pressed, and hot-formed magnets produced from rapidly solidified neodymium-iron-boron alloys are found in Tables 3 and 4 and on demagnetization curves in Fig. 9.

All permanent magnet materials are temperature dependent. A change in temperature causes the working point on the operating slope (see Fig. 11) to change. As long as the working point stays within the linear region of the demagnetization curve, the changes in flux density are reversible. In all other cases, any change in flux density is irreversible and can only be regained by remagnetization. To avoid irreversible changes in the flux density through temperature variations, the working point must remain within the linear section of the demagnetization curve. The large temperature coefficient of coercivity of neodymium-iron-boron materials causes the linear region to shrink, or a knee begins to appear on the demagnetization curve. Material scientists have not been successful in altering the temperature coefficients of the alloy. The solution to keeping the demagnetization curve linear has been to develop alloys with high intrinsic coercive forces. The design engineer must consider the operating point, operating temperature, and any external demagnetizing force when selecting a specific material grade.

Selection and Application

Permanent magnets are superior to electromagnets for many uses because they maintain their fields without an expenditure of electrical power and without the generation of heat.

Tables 3 and 4 and Fig. 6 to 9 give nominal properties only. Even under the most carefully controlled manufacturing conditions, some variation from these nominal values must be expected and considered in practical application. Figures 12 and 13 are examples of variations in energy values for two magnet materials.

Economic Considerations

Cost per pound is seldom considered in selecting a magnet material. Cost per unit of magnetic energy is more significant and is more often the basis of comparison. Alnico

Table 4 Nominal mechanical and physical properties of selected permanent magnet materials

See Table 2 for composition, Curie temperatures and magnetic orientations; see Table 3 for nominal magnetic properties.

Designation	Density, g/cm³	Tensile strength		Transverse modulus of rupture		Hard-ness, HRC	Coefficient of linear expansion		Electrical resistivity, nΩ · m	Maximum service temperature	
		MPa	ksi	MPa	ksi		μm/m · K	μin./in. · °F		°C	°F
3½% Cr steel	7.77	...	...	...	...	60–65	12.6	7.01	290	...	...
6% W steel	8.12	...	...	...	...	60–65	14.5	8.06	300	...	...
17% Co steel	8.35	...	...	...	...	60–65	15.9	8.84	280	...	...
36% Co steel	8.18	...	...	...	...	60–65	17.2	9.56	270	...	...
Cast Alnico 1	6.9	28	4.1	96	14	45	12.6	7.01	750	540	1004
Cast Alnico 2(a)	7.1	21	3.1	52	7.5	45	12.4	6.89	650	540	1004
Cast Alnico 3	6.9	83	12	157	23	45	13.0	7.23	600	480	896
Cast Alnico 4	7.0	63	9.1	167	24	45	13.1	7.28	750	590	1094
Cast Alnico 5(a)(b)	7.3	37	5.4	73	11	50	11.4	6.34	470	540	1004
Cast Alnico 5DG	7.3	36	5.2	62	9.0	50	11.4	6.34	470	...	...
Cast Alnico 5-7	7.3	34	4.9	55	8.0	50	11.4	6.34	470	540	1004
Cast Alnico 6(a)	7.4	157	23	314	46	50	11.4	6.34	500	540	1004
Cast Alnico 7	7.3	108	16	...	...	60	11.4	6.34	580	...	...
Cast Alnico 8	7.3	64	9.3	...	...	56	11.0	6.12	500	540	1004
Cast Alnico 9	7.3	48	6.9	55	8.0	56	11.0	6.12	...	...	...
Cast Alnico 12	7.4	275	40	343	50	58	11.0	6.12	620	480	896
Sintered Alnico 2	6.8	451	65	480	70	43	12.4	6.89	680	480	896
Sintered Alnico 4	6.9	412	60	588	85	...	13.1	7.28	680	590	1094
Sintered Alnico 5	7.0	343	50	392	57	44	11.3	6.28	500	540	1004
Sintered Alnico 6	6.9	382	55	755	110	44	11.3	6.28	530	540	1004
Sintered Alnico 8	7.0	...	...	382	55	43	...	...	...	...	...
Bonded ferrite A(c)	3.7	4.4	0.63	...	...	...	94	52	~10^{13}	95	203
Sintered ferrite 1(d)	4.8	49	7.1	...	...	...	10	6	~10^{13}	400	752
Sintered ferrite 2	5.0	...	...	...	...	...	10	6	~10^{13}	400	752
Sintered ferrite 3	4.5	...	...	...	...	...	18	10	~10^{13}	400	752
Sintered ferrite 4	4.8	...	...	...	...	...	...	...	10^{13}	400	752
Sintered ferrite 5	4.5	...	...	...	...	...	...	...	10^{13}	...	...
Sintered NdFeB	7.5	830	120	...	...	58	5	3	1600	150	300
Bonded NdFeB	5.8	13.5	2	...	...	36	...	...	10^9	130	265
Hot-pressed NdFeB	7.5	...	...	...	...	58	...	...	1600	150	300
Hot-formed NdFeB	7.4	...	...	...	...	58	...	...	1600	150	300
Cunife	8.6	686	99	...	...	95 HRB	12	6.7	180	350	662
Platinum cobalt	15.5	1370	199	1 570	230	26	11	6.1	280	350	662
Cobalt rare earth(e)	8.2	3430	498	13 730	1 990	50	511; 131	284; 72.8	500	250	482

(a) Specific heat: 460 J/kg · K (0.11 Btu/lb · °F). (b) Thermal conductivity: 25 W/m · K (170 Btu · in./ft² · h · °F) at room temperature. (c) Thermal conductivity: 0.62 W/m · K (4.3 Btu · in./ft² · h · °F). (d) Thermal conductivity: 5.5 W/m · K (38 Btu · in./ft² · h · °F). (e) Specific heat: 380 J/kg · K (0.09 Btu/lb · °F). Thermal conductivity: 15 W/m · K (104 Btu · in./ft² · h · °F)

5 is one of the more expensive of the Alnico group by the pound, but because of its superior magnetic properties is the most economical Alnico alloy for many applications. Alnico 6 and Alnico 5DG are both slightly, but not significantly, more costly by the pound than Alnico 5. Alnico 2 costs about 35% less per pound than Alnico 5, but is significantly more costly on the basis of magnetic energy.

Magnet prices for all alloys are strongly influenced by size, shape, quantity purchased, and method of manufacture (cast, sintered, or wrought). A significant increase in price may also be associated with a stringent tolerance requirement.

All of the magnet parameters are important and need to be considered when designing a specific magnetic circuit. The overriding consideration is to make the trade-offs necessary to obtain the best overall design for the system in which the magnetic circuit operates. This means that for some systems the neodymium-iron-boron magnet may be the material of choice and prove to be the most economical one.

Optimum Alloy Usage

Table 5 classifies various permanent magnet materials in relation to the properties relevant to the specific application. The number and range of applications utilizing permanent magnets has increased dramatically, making a full listing impractical. Table 6 lists the more predominant applications.

Alnico 5 was for many years the material recommended for most applications. The development of the hard ferrite material provided the design engineers with a lower cost option for many existing applications and made possible new ones. Permanent magnet dc-motor development became active, and large magnetic separators arrived. Ferrite separators were critical to the success of a new age in low-grade iron ore mining. Then engineers began replacing the Alnico in magnetos and loud speakers. Some problems arose along the way, but hard ferrites became the dominant permanent magnet material and still hold that position.

When samarium cobalt was introduced, many doubted that such a high-price material would ever be used in any volume production. New applications and conversion from existing alloys to the samarium-cobalt type eliminated these doubts. For a specific group of applications, these alloys are the only viable choice.

Neodymium-iron-boron permanent magnet materials have already been established as the material to use in voice coil motors and other devices. Figure 14 projects the 1988 world market for permanent magnet materials.

Hysteresis Applications. In most applications, permanent magnets are used to supply flux in an air gap and operate only in the second quadrant of the hysteresis loop. In a few specialized applications, notably hysteresis torque devices, it is necessary to consider the entire hysteresis loop in evaluating a magnetic material.

In hysteresis torque devices, the driving force is provided by a rotating magnetic field. The rotor is usually a thin-wall cylinder magnet that is not premagnetized but that can be magnetized by the rotating field. Immediately after starting, the induced poles rotate within the rotor, subjecting the material to repeated transversals of the hysteresis loop. Hysteresis causes the induced rotor poles to lag behind those of the applied field, producing an accelerating torque. The torque developed is proportional to the area of the hysteresis loop through which the magnet is driven.

If torque were the only requirement, the hysteresis material chosen would be a magnet with the largest possible hysteresis loop (and therefore the largest hysteresis loss). However, two other factors must be consid-

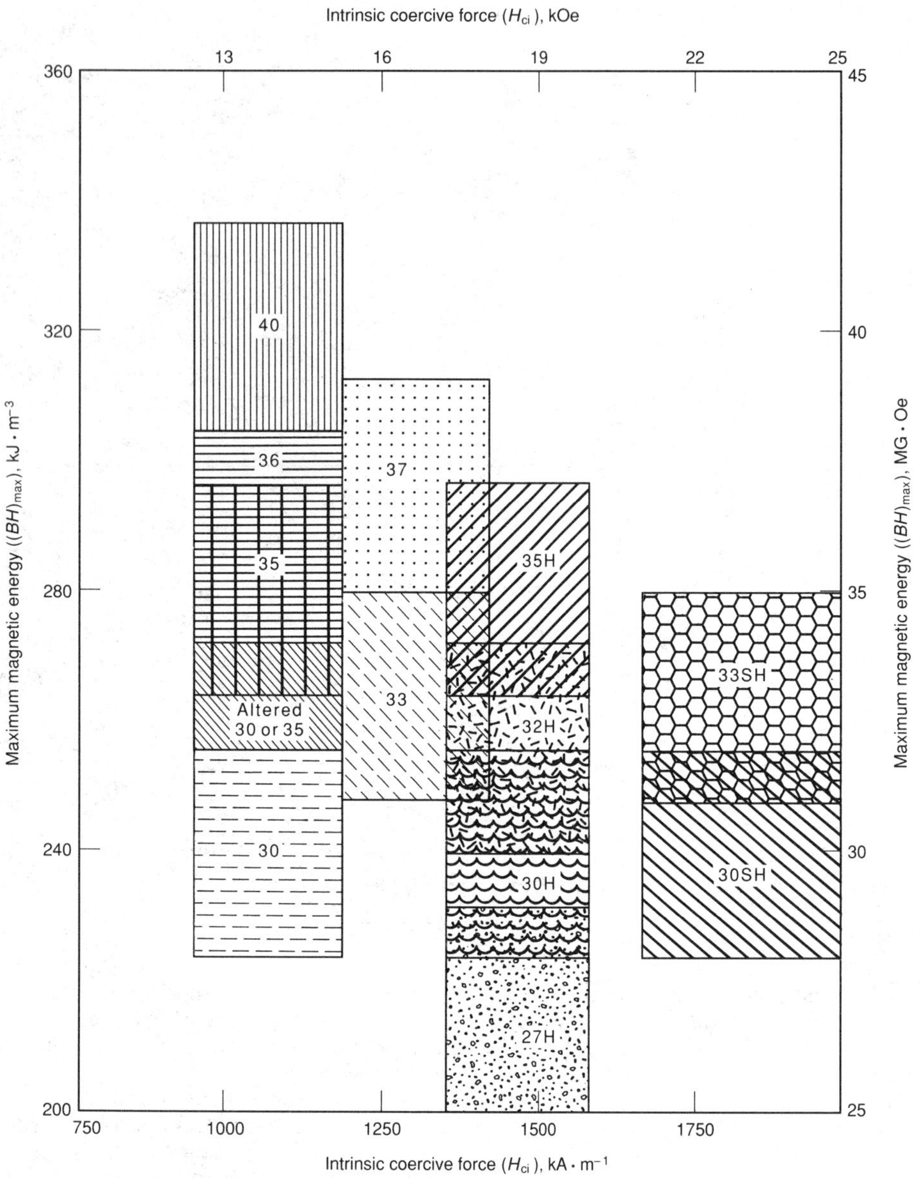

Fig. 10 Graphic presentation of the range of various sintered neodymium-iron-boron materials. Coercive force designations: H, high; SH, super high. Source: Sumitomo Special Metals Company

ered. Because the applied field must be capable of magnetizing the rotor, there is a practical limit to the coercive force that can be used.

Materials that are very high in coercive force seldom are used for these applications. For example, Alnico 6 develops the largest hysteresis loss, as indicated in Fig. 15, which shows hysteresis loss as a function of magnetizing force for various permanent magnet materials. Therefore, Alnico 6 would produce the greatest amount of torque. However, to produce this torque would require a magnetizing force of more than 80 kA · m^{-1} (1 kOe), a magnetizing force that is not practical for most applications.

Because the efficiency of the motor is determined by the magnetic characteristics of the rotor, the shape of the hysteresis loop is important. A convenient measure of this shape is the energy factor, η. The energy factor is defined as the ratio of the area of the hysteresis loop to the area of a rectangle drawn through the extremes (peak B and H) of the curve and ranges from 0.5 to 0.75. Two materials may have the same intercepts in terms of induction at a given magnetizing force, yet have different energy factors.

In operation, materials are not magnetized to saturation, but operate within a minor hysteresis loop. The actual loop is determined not only by the material, but also by the intensity of the magnetizing field. If only a small field is available, greater torque is developed by a low-energy material that is operated near its saturation value than by a higher-energy material that is little affected by the small applied field.

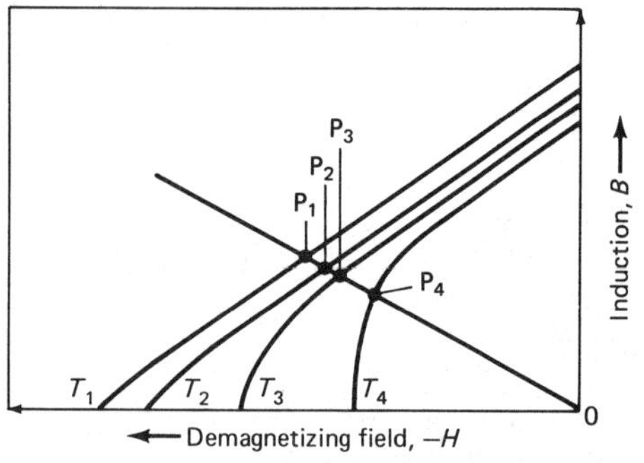

Fig. 11 Effect of increasing temperature on working point of neodymium-iron-boron permanent magnet materials

There is one value of magnetizing force for each material at which it operates most efficiently—that is, at which it produces maximum torque for a given magnetizing force. In Fig. 15, this point is marked with a dot on the curve for each material. The efficiency of a hysteresis material is the ratio of hysteresis loss to magnetomotive force required. Peak efficiencies of commonly used materials are:

Alloy	Efficiency
P-6 alloy	0.330
Cast Alnico 5	0.304
Cast Alnico 6	0.202
Vicalloy	0.197
17% Co steel	0.158
36% Co steel	0.142
Cast Alnico 2	0.124
3.5% Cr steel	0.117

The materials referenced above are in limited use and thus may be difficult to purchase. The material currently being utilized in hysteresis applications is isotropic iron-chromium-cobalt (Fig. 16).

Stabilization and Stability. There is an important group of permanent magnet applications where the accuracy or performance of the device is drastically affected by very small changes (1% or less) in the strength of the magnet. These applications include braking magnets for watt-hour meters, magnetron magnets, special torque motor magnets, and most dc panel and switchboard instrument magnets. Operation of these devices requires extreme accuracy over a moderate range of conditions, or moderate accuracy over an extreme range of conditions.

If the nature and magnitude of the conditions are known, it often is possible to predict the flux change. It also may be possible, by exposing the magnet to certain influences in advance, to render the magnet insensitive to subsequent changes in service. For many years, permanent magnets in instruments have exhibited long-term stability of the order of one part per thousand

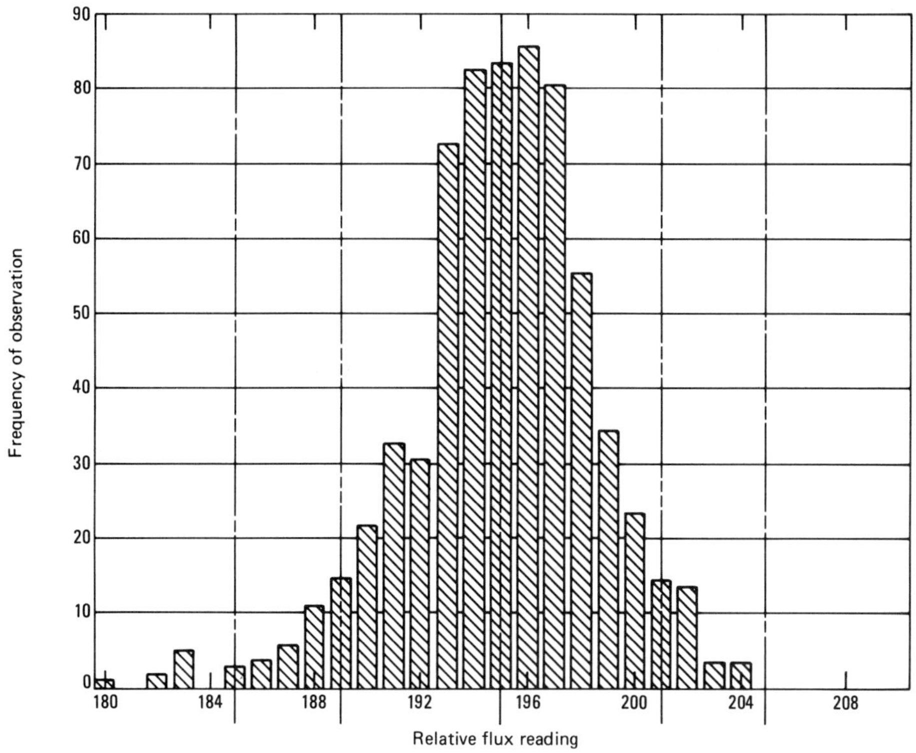

Fig. 12 Distribution of flux for 14 lots of neodymium-iron-boron magnets

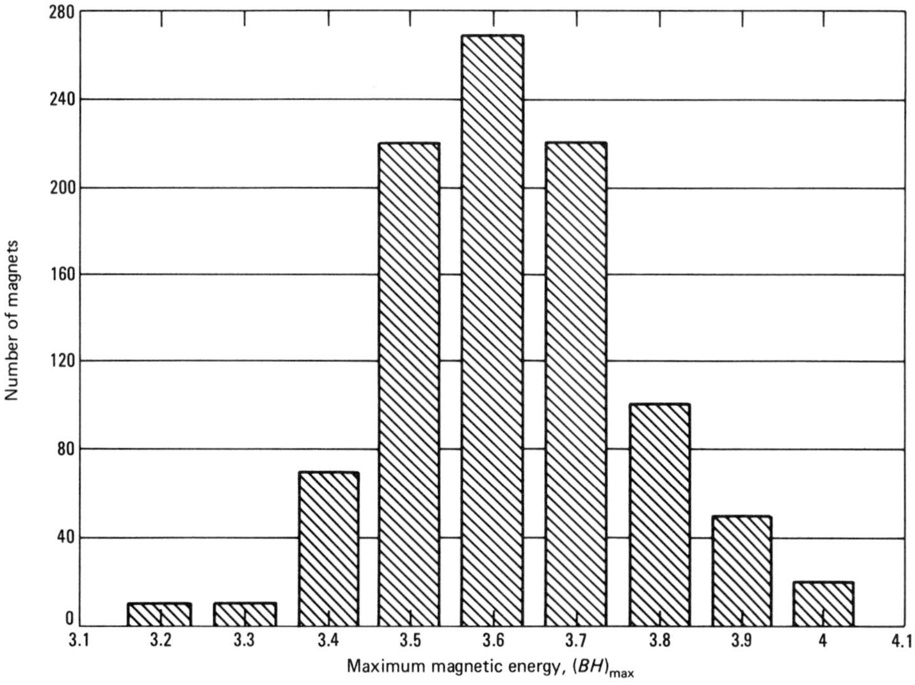

Fig. 13 Distribution of maximum magnetic energy for hard ferrite magnets

(0.1%). More recently, investigations in conjunction with inertial guidance systems for space vehicles have shown that long-term stability of the order of one to 10 ppm (0.0001 to 0.001%) can be achieved. This incredible stability of a magnetic field achieved with modern permanent magnets contrasts sharply with the instability of very early (steel) permanent magnets, in which both structural and magnetic changes caused a significant loss of magnetization with time.

Irreversible Changes. Losses in magnetization with time can be classified as either reversible or irreversible. Irreversible changes are defined as changes where the affected properties remain altered after the influence responsible for the change has been removed. For example, if a magnet loses field strength under the influence of elevated temperature and if the flux does not return to its original value when the magnet is cooled to room temperature, the change is considered irreversible. Full flux is restored by remagnetizing unless metallurgical changes permanently altered the hysteresis loop.

Changes in Metallurgical State. Irreversible changes begin to occur at different temperatures for different alloys. These changes usually depend on both time and temperature, and thus short exposures above the recommended temperatures may be tolerated. These changes may take the form of growth of the precipitate phase, such as in Alnico and Cunife; precipitation of another phase, such as γ precipitation in Alnico; an increase in the amount of an ordered phase, such as in platinum-cobalt; an increase in grain size, as in $SrO{\cdot}6Fe_2O_3$; oxidation, as occurs with metals, or reduction, as occurs with oxides; radiation damage; cracking; or changes in dimensions.

Typical permanent changes in H_c due to changes in metallurgical structure are shown in Fig. 17. Results for the steels and Alnico 5 were obtained on samples that initially were in the optimum magnetic state. The remainder of the materials were aged, starting from their quenched, as-cast, or as-prepared state, as indicated in the figure. The temperature at which changes in properties first become noticeable corresponds to the beginning of metallurgical changes; this temperature corresponds closely to the maximum temperature to which each material can be exposed, even after aging to the optimum magnetic state.

Irreversible metallurgical changes often can be counteracted, and original properties restored, by a suitably chosen thermal treatment. For example, if Alnico 5 has become degraded by exposure to 700 °C (1290 °F), it may be solution treated at 1300 °C (2370 °F), cooled in a magnetic field, and aged at 600 °C (1110 °F) to reattain the optimum metallurgical structure.

A nuclear environment is known to cause changes in metallurgical structure and thus may cause changes in magnetic properties. Permanent magnet materials tested were not affected by neutron (n) irradiation at levels below about 3×10^{17} n/cm^2 (2×10^{18} n/in.^{2}). Results of later work at levels up to 10^{20} n/cm^2 (6×10^{20} n/in.^{2}) showed some degradation. The Alnicos are not affected by radiation up to 5×10^{20} n/cm^2 (3×10^{21} n/in.^{2}) at neutron energies greater than 0.4 eV, and up to 2×10^{19} n/cm^2 (1×10^{20} n/in.^{2}) for neutron energies greater than 2.9 MeV. Radiation effects were found to be independent of temperature, but high tem-

Table 5 Classification of permanent magnetic materials on the basis of application-relevant properties

Property	Performance: High ⟵					⟶ Low
Energy (density)	NdFeB	(RE)Co	Alnico, FeCrCo	Ferrites	FeCoVCr	AlNi
$(BH)_{max}$, kJ · m^{-3} (MG · Oe)	320 (40)	. . .	. . .	. . .	. . .	8 (1)
Stability against demagnetization	(RE)Co	NdFeB		Ferrites	Alnico, FeCrCo	FeCoVCr
H_{ci}, kA · m^{-1} (Oe)	2000 (25 000)	. . .		. . .	. . .	8 (100)
μ_{rec}	1.0	. . .		. . .	. . .	10
Hysteresis loss	Alnico, FeCoVCr, FeCrCo					
Reversible temperature variation	Alnico, FeCoVCr	FeCrCo		(RE)Co	NdFeB	Ferrites
$\alpha(\phi)$, %/K	−0.01	. . .	. . .	. . .	. . .	−0.2
Curie temperature,	Alnico, FeCrCo		(RE)Co, FeCoVCr	Ferrites		NdFeB
°C (°F)	900 (1650)		. . .	. . .		300 (570)
Stability with high-temperature operation	Ferrites		Alnico, FeCoVCr, FeCrCo	(RE)Co		NdFeB
Possibility for choosing a preferred direction (PD)	Alnico, FeCrCo		(a) Ferrites; (RE)Co; NdFeB			FeCoVCr, FeCrCo,
PD cause	Magnetic field with heat treatment		Magnetic field with shaping by pressing or injection molding			Mechanical deformation
Elasticity	FeCoVCr, CrVCo					All others
Physical strength (stability with handling)	FeCoVCr, FeCrCo	NdFeB	Alnico, AlNi	Ferrites		(RE)Co
Economy, relative cost per unit of magnetic energy, $	Ferrites		Alnico, FeCrCo	NdFeB	(RE)Co	FeCoVCr
	1		. . .	5	. . .	20

peratures tended to counteract radiation effects.

Tests have been done on $SmCo_5$, Sm_2Co_{17}, and neodymium-iron-boron magnets irradiated at up to 8 × 10^{19} n/cm^2 (5 × 10^{20} n/in.^{2}) by several national laboratories. These tests showed that Sm_2Co_{17} was found to be less sensitive by a factor of 10 to neutron irradiation than $SmCo_5$ material. Neodymium-iron-boron proved to be extremely sensitive to ionizing radiation, losing over 50% of its flux at 4 × 10^4 Gy (4 × 10^6 rad). The lower Curie temperature of neodymium-iron-boron probably accounts for its sensitivity to radiation.

Temperature Effects. The properties of a magnet vary with temperature in a man-

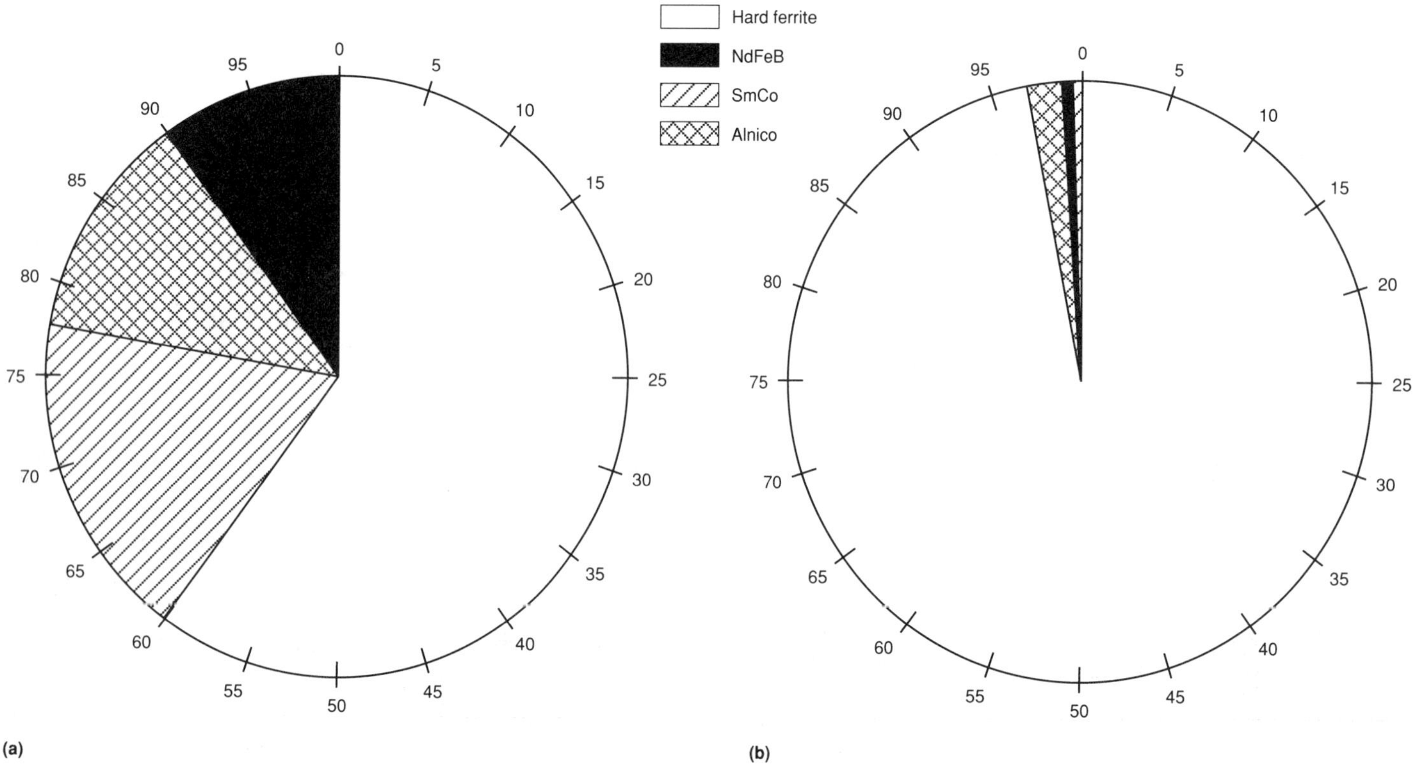

Fig. 14 Breakdown of global permanent magnet market in terms of monetary value (a) and product weight (b). Source: Ref 2

Table 6 Applications of permanent magnet materials

Application	Recommended material	Primary reason for selection	Alternative material	Condition or reason favoring selection of alternative material
Aircraft magnetos, military or civilian	SmCo	Maximum energy per unit volume	Cast Alnico 5	Availability or cost restraint
Alternators	SmCo	Compactness and reliability	Ferrite Alnico	Where space is available for a larger volume of material of lower magnetic energy and cost
Magnetos for lawn mowers, garden tractors, and outboard engines	Ferrite	Adequate magnetic energy at lower cost than Alnico	Alnico NdFeB	Higher energy material is required
Small dc motors	Bonded ferrite	Shape favors fabrication; adequate magnetic energy at lower cost	Bonded NdFeB Sintered ferrite	Higher magnetic energy is required
Large dc motors	SmCo	Maximum energy per unit volume	NdFeB	Where lower cost is required, operating temperature is low
Automotive dc motors	Ferrite	Adequate magnetic energy at lower cost than alternate materials	Bonded NdFeB	Higher magnetic energy and less weight
Automotive cranking motors	Ferrite	Adequate magnetic energy at lower cost than alternate materials	Bonded NdFeB	Higher magnetic energy and less weight
Voice coil motors (computers)	NdFeB	High energy	SmCo	Availability
Acoustic transducers	Ferrite	Low cost	NdFeB	Higher magnetic energy allows smaller size and weight
Magnetic couplings (small gap)	Ferrite	Adequate magnetic energy at lower cost	Bonded NdFeB	Higher torque is required
Magnetic couplings (large gap)	NdFeB	High energy	SmCo	High operating temperature
Transport systems	NdFeB	High energy	SmCo	Availability
Separators	Ferrite	Adequate magnetic energy at lower cost	NdFeB	High magnetic energy required
Magnetic resonance imaging	NdFeB	High energy	Ferrite	Where space is available for a larger volume of material of lower energy
Magnetic focusing systems	NdFeB	High energy	SmCo	High operating temperatures or low-temperature coefficient is required
Synchronous hysteresis motors	Isotropic FeCrCo	Shape favors fabrication from wrought material	Cobalt steel	Availability
Holding devices	Ferrite	Adequate magnetic energy at low cost	Alnico	Where holding force versus temperature must not vary over wide ranges
Ammeters and voltmeters	Alnico	Low temperature coefficient	Not available	
Watt-hour meters	Alnico 5 or 6	Low temperature coefficient	Not available	

ner that usually can be predicted. The variation of B_{is} with temperature can be calculated from theory, provided detailed knowledge of the crystallographic and magnetic structure of the magnetic phase is available. In many other instances, such information is not yet available, but direct measurements of B_{is} versus T have been made. Examples of such curves for many permanent magnet materials are shown in Fig. 18. The curves in Fig. 18 include some that represent changes in B_r with T, which depend extensively on properties of the system, such as particle size, sample shape, demagnetizing fields, and domain structures. Changes with temperature must be determined experimentally for the specific magnet and configuration of interest. Typical reversible and irreversible changes are given in Tables 7 and 8 as a function of the length-to-diameter ratio for a number of permanent magnet materials.

Changes in H_{ci} with temperature can be predicted from the changes with temperature of anisotropy and magnetization. This assumes knowledge of the physical origin of all anisotropies contributing to H_{ci}. Experimental results are shown for many magnets in Fig. 19. For a case where uniaxial anisotropy predominates, as in $SrO{\cdot}6Fe_2O_3$, good agreement between calculated and experimental results is obtained. In a case where shape anisotropy is dominant, calculated and experimental results also are in good agreement, especially when the small crystal anisotropy contributions are considered. In the case of Alnico, crystal anisotropy is more in evidence. In addition, there is greater uncertainty as to the effect of the so-called nonmagnetic phase, especially at lower temperatures where the nonmagnetic phase may contribute appreciable magnetization. In the case of steels, the temperature dependence based on the inclusion mechanism is difficult to predict.

Demagnetization curves may change in both shape and peak values with changes in temperature. Families of demagnetization curves at various temperatures are shown in Fig. 20 for Alnico and hard ferrite.

Changes in magnetic state may be caused by temperature effects, such as ambient temperature changes or statistical local temperature fluctuations within the material; mechanical effects, such as mechanical shock or acoustical noise; or magnetic field effects, such as external fields, circuit reluctance changes, or magnetic surface contacts. In all of these situations, the loss of magnetization may be restored by remagnetizing.

Mechanical shock and vibration add energy to a permanent magnet, and decrease the magnetization in the same manner as discussed for the case of thermal energy. The only difference is that energy imparted thermally to the magnet is precisely kT, where k is the Boltzmann constant and T is the temperature, whereas the energy imparted mechanically usually is not known. Thus, repetitive shocks or continual vibration should decrease the magnetization by the same logarithmic relations as for thermal effects, but where time is replaced, for example, by number of impacts.

Little work has been done regarding stabilization to minimize mechanical effects because it is seldom found necessary. It is generally conceded that modern permanent magnet materials are not affected by mechanical shock.

Reversible Changes. A loss of magnetization caused by a disturbing influence, such as temperature or an external magnetic field, is considered reversible if the original

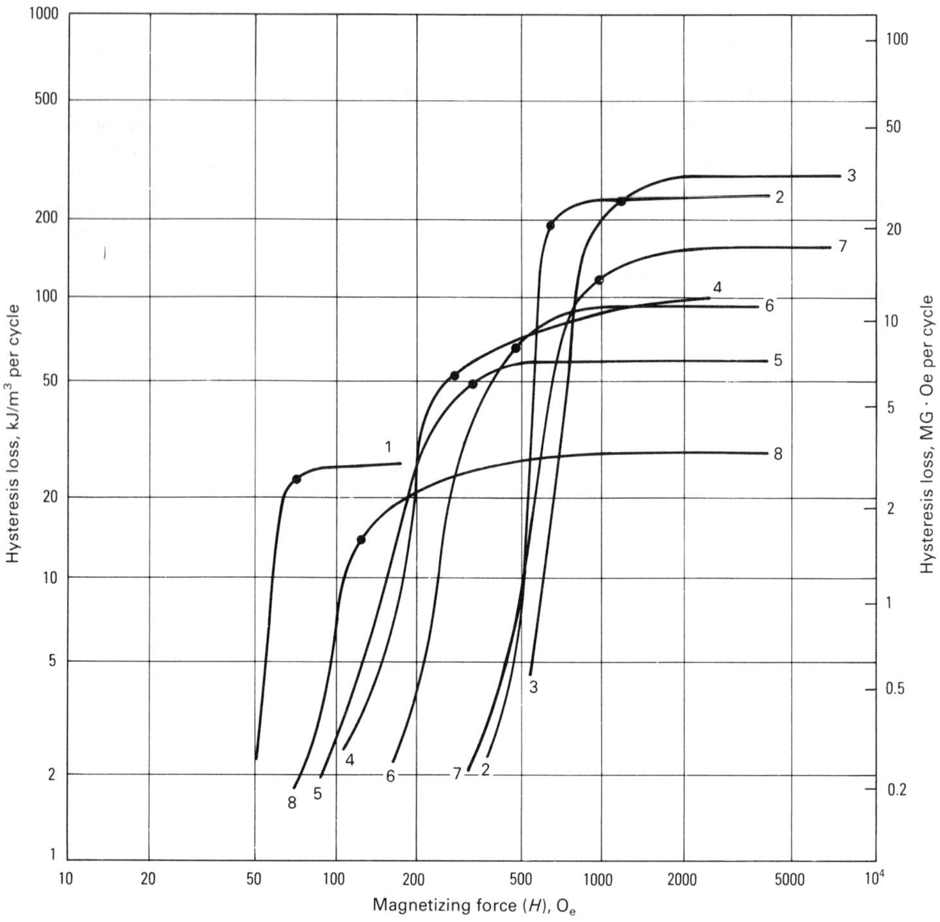

Fig. 15 Hysteresis loss versus magnetizing force for various permanent magnet materials. Data points indicate maximum efficiency. 1, P-6 alloy; 2, cast Alnico 5; 3, cast Alnico 6; 4, Vicalloy; 5, 17% Co steel; 6, 36% Co steel; 7, cast Alnico 2; 8, 3½% Cr steel

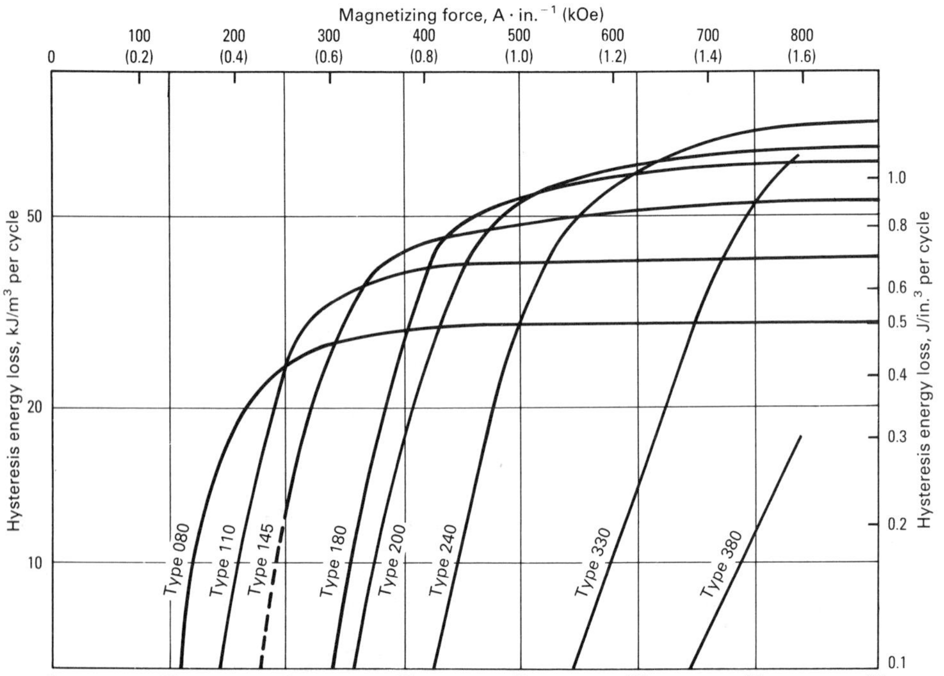

Fig. 16 Hysteresis energy loss versus magnetizing force of isotropic iron-chromium-cobalt alloys

properties of the magnet return when the disturbing influence is removed.

Time Effects at Constant Temperature. In ferromagnetic materials, the intensity of magnetization does not instantly attain its equilibrium value when the applied field is suddenly changed. This time dependence may be due to eddy current effects or to reversible or irreversible magnetic viscosity. In general, eddy current effects are important only for a very short time—normally, less than a second after a change in the applied field. Such effects are not considered here. Reversible magnetic viscosity has been shown to be due to ionic diffusion in the crystal lattice and thus has a time-temperature dependence characteristic of diffusion processes. The time constant is:

$$\tau = \tau_\infty \exp (E/RT) \qquad \text{(Eq 3)}$$

where τ_∞ is the time constant at infinite temperature, R is the reluctance (the reciprocal of permeance) and E is the activation energy, normally 0.1 to 1 eV. The time constant appears to be important only in magnetically soft materials, and only at high frequencies.

Irreversible magnetic viscosity is important to the stability of permanent magnets. Irreversible magnetic viscosity is due to the influence of thermal fluctuations on magnetization or the domain process responsible for magnetization. The effect of thermal agitation has been considered in terms of the energy required to activate irreversible domain processes. The time-temperature dependence of magnetization was shown to be given by:

$$M(t) = S \ln t \qquad \text{(Eq 4)}$$

where $S = \lambda N M_s kT$. Here, N is the number of blocks, or regions of magnetization M_s per unit volume; and λ is the constant probability density of energy E of all these blocks. Because these factors are all relatively independent of temperature (except near the Curie temperature), S is nearly directly proportional to T. The results of experiments are in agreement with this equation, as shown in Fig. 21. Aging at room temperature results in losses in magnetization for many materials, as shown in Fig. 22.

Effects of Temperature Variations. Various permanent magnet materials undergo changes in magnetization as the temperature is cycled above and below room temperature (see Fig. 23). For a long bar operating above $(BH)_{max}$ (that is, B/H = 35), as in Fig. 23(a), the change in M is reversible. For a shorter bar operating below $(BH)_{max}$, as in Fig. 23(b), the first cooling cycle results in a substantial loss in magnetization. After the initial low-temperature exposure, the changes in M are reversible, but at a level below the initial magnetization. Results on an even shorter bar are shown in

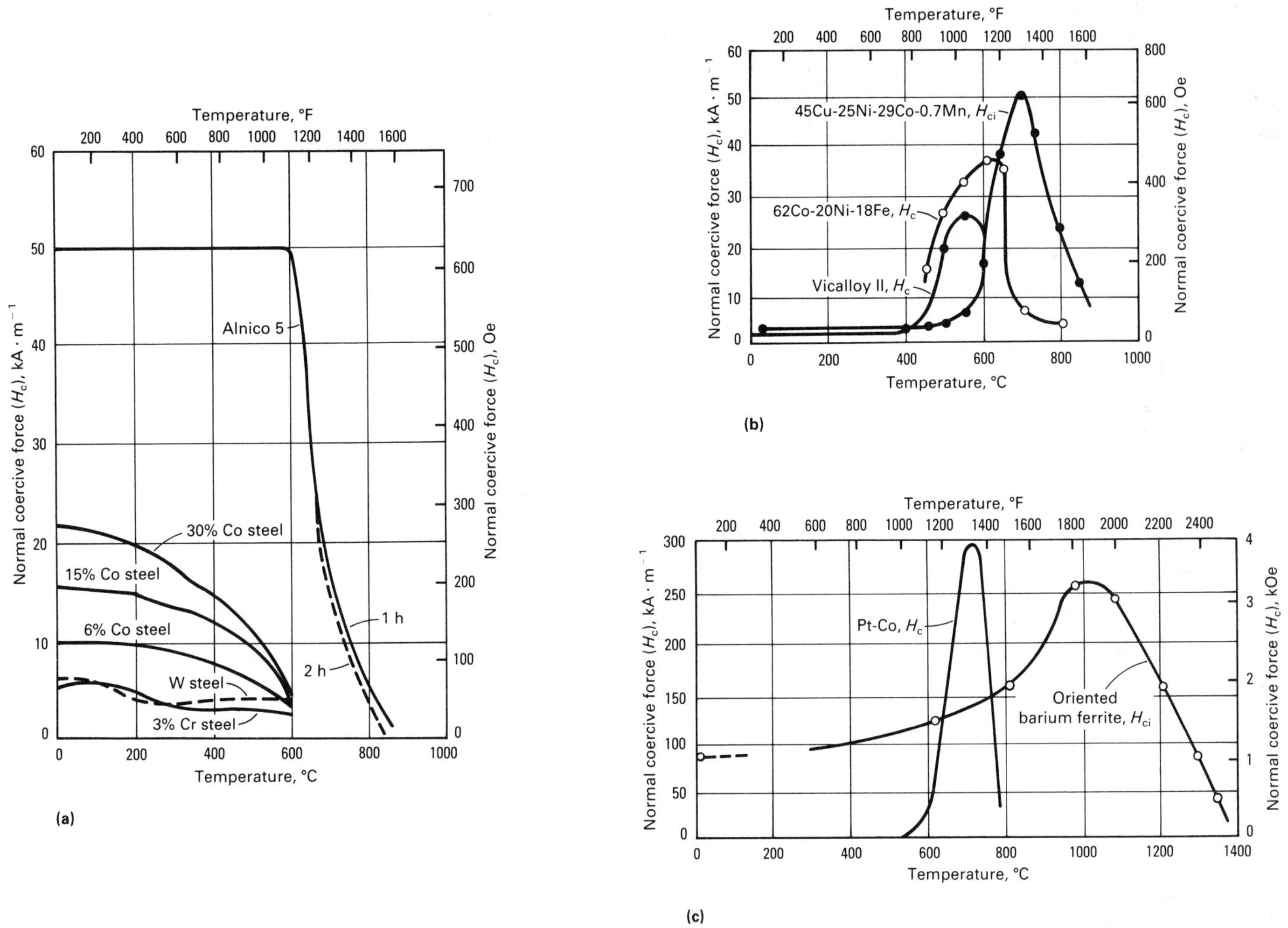

Fig. 17 Irreversible changes in H_c and H_{ci} for various permanent magnet materials

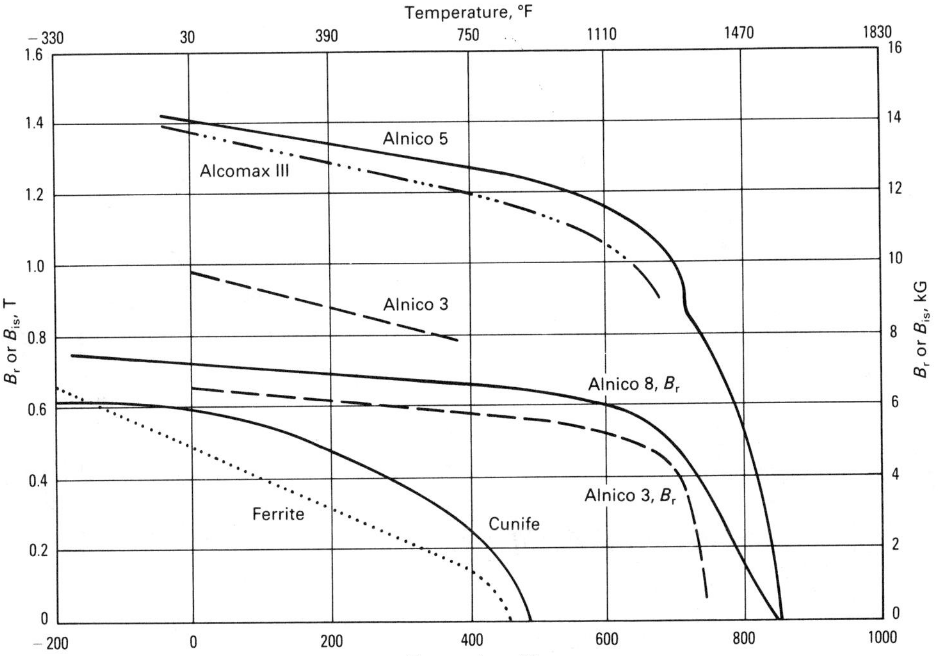

Fig. 18 Temperature dependence of saturation magnetization, B_{is}, or remanence B_r, for various permanent magnet materials

Fig. 23(c). These data suggest that by proper choice of dimensions, a reversible coefficient of approximately zero could be achieved over a limited range of temperature.

Design Considerations. Stability can have a significant influence on choice of magnet material, as well as on component shape and magnetic circuit arrangement. For example, the rather drastic change in coercive force of oriented hard ferrite with temperature (Fig. 19) requires special considerations in design. Here, the lowest permeance coefficient (B/H) that can be used is established by stability considerations rather than by magnetic circuit analysis.

For the more widely used permanent magnet materials, reversible changes in magnetization are encountered by cooling below room temperature (Table 7). Because the reversible remanence changes are closely approximated by a straight line, a reversible temperature coefficient is listed. The values of the temperature coefficient vary with the material. When the values of the coefficient are very small, they may be of a different sign for different magnet shapes.

Table 7 Magnetization changes on cooling below room temperature (20 °C, or 70 °F)

Material	Dimensional ratio, L/D	% irreversible loss at room temperature after exposure to −190 °C (−310 °F)	−60 °C (−75 °F)	Reversible temperature coefficient, % remanence change per °C
Alnico 2	5.29	0	0	−0.025
	3.69	0	0	−0.021
	2.66	0	0	−0.018
	1.77	0	0	−0.009
	0.94	0	0	−0.014
Alnico 5	8.00	0	0	−0.022
	5.36	4.6	1.4	−0.012
	3.63	9.0	2.5	−0.002
	2.72	6.2	3.6	+0.010
	1.84	7.9	2.1	+0.016
	0.94	8.5	3.4	+0.007
Alnico 6	8.00	0	0	−0.045
	6.03	1.8	0.4	−0.020
	3.57	8.5	1.3	−0.007
	2.70	10.1	4.1	+0.007
	1.78	10.5	4.2	+0.022
	0.89	7.9	3.1	+0.046
Alnico 8	5.62	0	0	−0.013
	2.85	0.5	0.1	+0.003
	1.91	0.7	0.3	+0.015
	1.01	1.3	0.5	+0.033
Barium ferrite (isotropic)	0.50	4.0	0(a)	−0.19
	0.28	...	1.3	−0.19
	0.09	...	2.4	−0.19
Barium ferrite ($[BH]_{max}$ anisotropic)	1.20	...	0(a)	−0.19
	0.50	...	...	−0.19
Barium ferrite ($[H_c]_{max}$ anisotropic)	0.40	...	0(a)	−0.19
Platinum cobalt	31.67	0	0	−0.015
	16.02	0.3	0	−0.015
	10.63	0.2	0	−0.015
	5.56	0.2	0	−0.015
Rare earth cobalt 1, 2, and 3	10	0	0	−0.030 to −0.045
	2	0	0	−0.030 to −0.045
	1	0	0	−0.030 to −0.045
	0.5	0	0	−0.030 to −0.045
NdFeB	4	Not available	0	−0.12
	2	Not available	0	−0.12
	1	Not available	0	−0.12
	0.5	Not available	0	−0.12

(a) In the case of the low-temperature irreversible loss occurring in oriented barium ferrite, only the smallest dimension ratio resulting in no irreversible loss at −60 °C (−75 °F) is shown. This is the recommendation of a major producer to avoid catastrophic loss. The minimum dimensional ratio will depend upon the lowest temperature to be encountered.

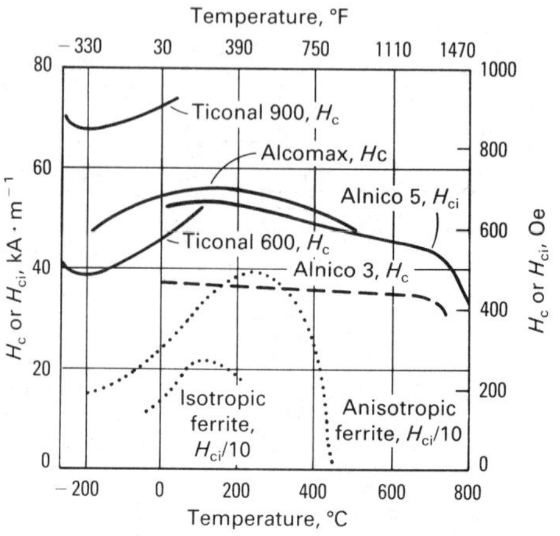

Fig. 19 Temperature dependence of normal coercive force H_c, and intrinsic coercive force, H_{ci}, for various permanent magnet materials

Consequently, it is often possible to carefully design magnet shape to yield very small variations in remanence with temperature. Similar changes may result upon heating above room temperature (Table 8). It is important to distinguish between irreversible losses and reversible changes. It is common practice prior to use to cycle a magnet between the temperature extremes to be encountered in service. Nearly all of the irreversible loss is encountered in one temperature cycle, but in some instances four or five cycles may be necessary.

In applications that are extremely sensitive to magnetization changes, it is very common to use a temperature compensating circuit to counteract reversible changes

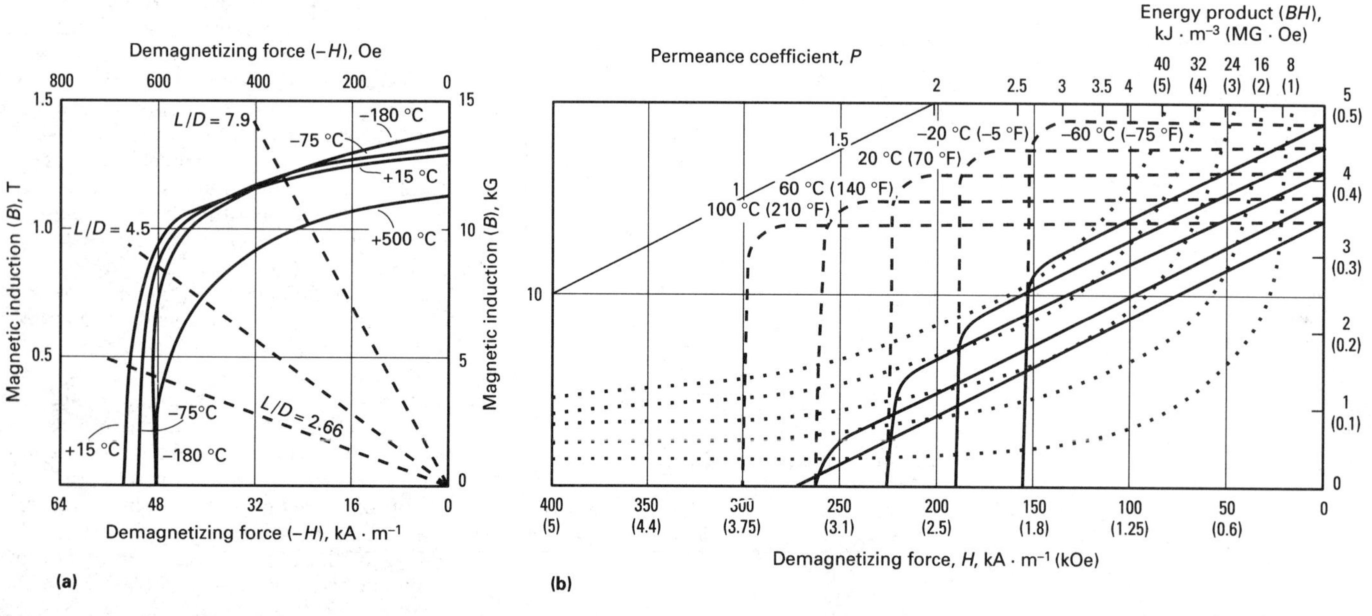

Fig. 20 Demagnetization curves for Alcomax (a) and for oriented ferrite (b) at various temperatures

Table 8 Magnetization losses on heating above room temperature (20 °C, or 70 °F)

Column I: Percent irreversible remanence loss at room temperature after heating to indicated temperature
Column II: Percent of initial room-temperature remanence found stable at indicated temperature

Material	Dimensional ratio (L/D)	Temperature, °C (°F): 100 (212) I	100 (212) II	200 (390) I	200 (390) II	300 (570) I	300 (570) II	400 (750) I	400 (750) II	500 (930) I	500 (930) II
Alnico 2	8.00	2.0	98	3.1	94	4.2	90	6.1	86	8.2	80
	3.62	3.1	98	4.0	92	6.9	88	8.6	84	12.0	78
	2.00	3.5	97	4.7	91	7.4	89	10.7	85	13.1	81
Alnico 5	8.00	0.1	99.9	0.2	96	0.4	93.6	0.7	91.2	1.2	88.0
	4.68	0.4	99.6	0.8	96.3	1.1	93.8	1.7	91.1	2.0	88.2
	2.00	0.5	99.4	1.7	96.6	2.1	94.1	2.6	92.2	3.0	88.6
Alnico 6	20.00	0.1	98.2	0.2	95.6	0.4	93.0	0.8	89.7	1.8	86.5
	4.12	0.5	98.7	0.9	95.6	1.2	92.7	2.0	89.4	3.0	85.2
	2.00	0.7	99.1	1.2	97.2	1.5	94.2	2.1	90.5	3.3	86.0
Alnico 8	5.62	0.7	98.8	...	...	...	...	...	...	...	...
	2.85	0.7	99.0	...	...	...	...	...	...	...	...
	1.91	0.9	99.4	...	...	...	...	...	...	...	...
	1.01	1.0	99.8	...	...	...	...	...	...	...	...
Barium ferrite (all grades)	All	0	85	0	68	0	50	...	...	...	...
Platinum cobalt	31.67	0	97.9	...	...	...	...	...	...	...	...
	16.02	0	98.5	...	...	...	...	...	...	...	...
	10.63	0	98.8	...	...	...	...	...	...	...	...
	5.56	0	97.9	...	...	...	...	...	...	...	...
Rare earth cobalt	0.5	0.1	96.5	0.5	91.2	...	...	...	...	...	...
	0.4	0.3	96.3	2.0	89.7	...	...	...	...	...	...
	0.3	0.6	95.9	2.5	89.2	...	...	...	...	...	...
	0.2	1.0	95.5	3.5	88.2	...	...	...	...	...	...
	0.1	1.8	94.7	5.5	86.2	...	...	...	...	...	...
NdFeB(a)	1.7	0.6(b)	...	...	...	...	...	...	...	...	...
	1.04	1.3(b)	...	...	...	...	...	...	...	...	...
	0.2	6.5(b)	...	...	...	...	...	...	...	...	...

(a) H_{ci} = 1440 kA · m^{-1} = 18 kOe. (b) At 140 °C (285 °F)

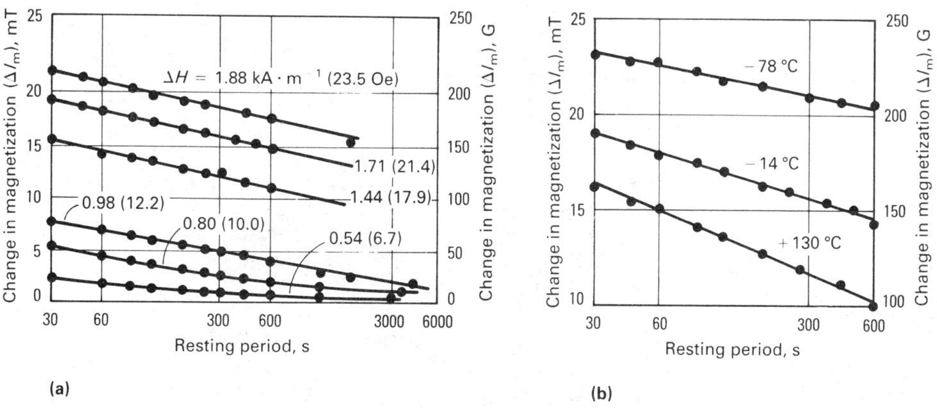

Fig. 21 Changes in magnetization, ΔH, with time for Alnico magnets. (a) At −14 °C (7 °F). (b) For ΔH = 1.71 kA · m^{-1} (21.40 Oe)

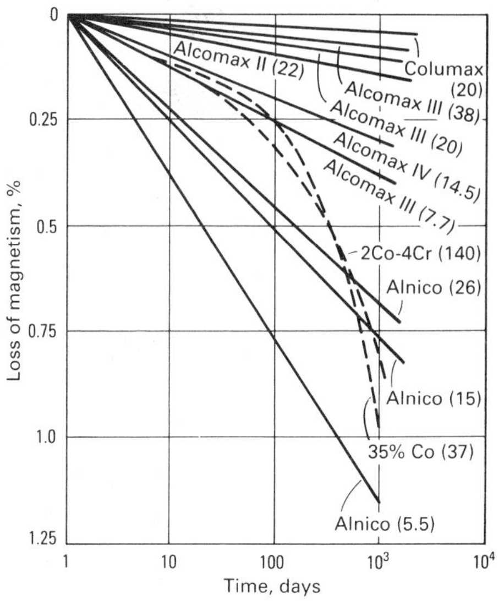

Fig. 22 Changes in magnetization with time at room temperature for various permanent magnet materials. Numbers in parentheses are working points, *B/H*.

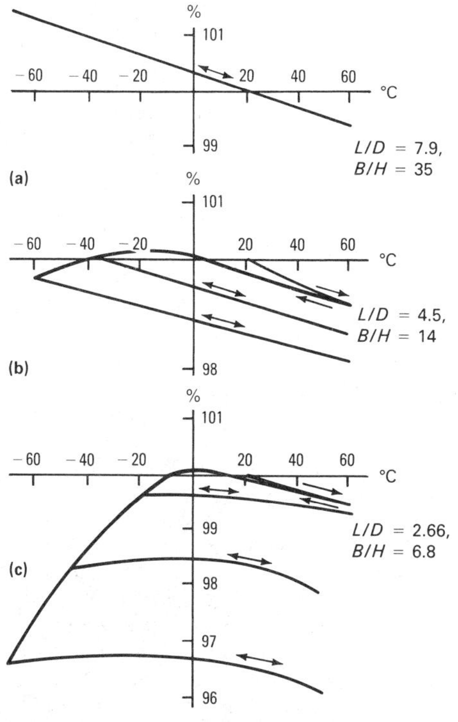

Fig. 23 Temperature variation for Alcomax III bars of various length-to-diameter ratios and working points

over the operating temperature range. Temperature-sensitive iron-nickel alloys are used as magnetic shunts for this purpose. A shunt is mounted beside the permanent magnet and simply diverts flux from the air gap as the temperature decreases. Temperature compensation by shunting requires overdesign of the magnet to allow for the loss in flux through the shunt at low operating temperatures.

Exposure at Very High Temperatures. There is considerable interest in using permanent magnets at temperatures approaching the Curie temperature of the permanent magnet material. The anisotropic Alnico 5 and Alnico 6 have been considered for use at 500 to 700 °C (930 to 1290 °F). At these temperatures, metallurgical effects as well as irreversible and reversible temperature effects are present. Alnico 5 exposed to 700 °C (1290 °F) for 20 h resulted in the reduction of $(BH)_{max}$ and H_c to approximately one half of their initial values. It is possible to program such changes into equipment and devices to allow permanent magnets to function for a limited time at extreme temperatures.

Stress Effects. Some magnets subjected to tension or compression show large changes in properties. This is especially true of Vicalloy, as shown in Fig. 24, and Cunife, as shown in Fig. 25. The changes are reversible, often even after considerable deformation has occurred. The changes are due to the contribution that stress makes to the total anisotropy of the system.

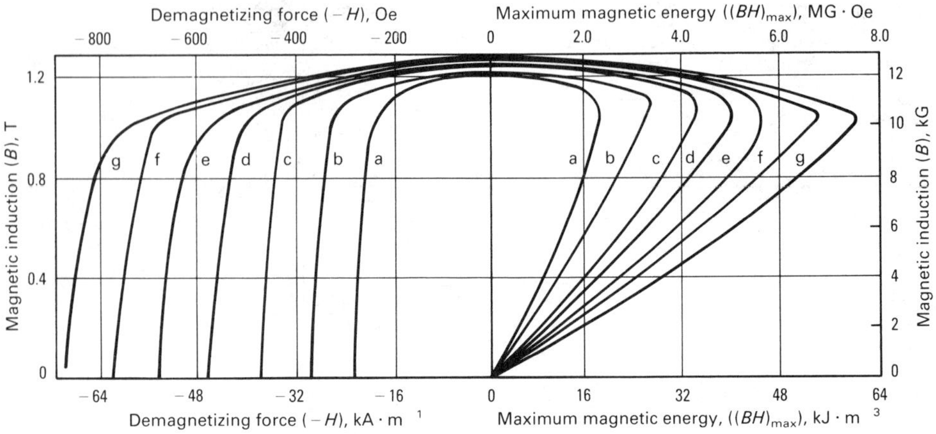

Fig. 24 Effect of stress on the magnetization curves for Vicalloy. Applied stresses: a, 0; b, 500 MPa (73 ksi); c, 990 MPa (144 ksi); d, 1490 MPa (216 ksi); e, 1990 MPa (289 ksi); f, 2490 MPa (361 ksi); g, 2990 MPa (434 ksi)

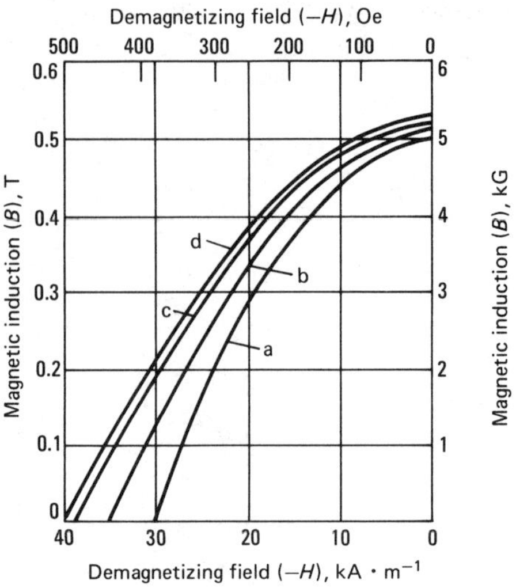

Fig. 25 Effect of stress on the magnetization curves for Cunife. Applied stresses: a, 0; b, 385 MPa (56 ksi); c, 570 MPa (83 ksi); d, 635 MPa (92 ksi)

Table 9 Recommended magnetizing fields, demagnetizing methods, and maximum service temperature

Permanent magnet material	Magnetizing field		Demagnetizing method	Maximum service temperature	
	$kA \cdot m^{-1}$	kOe		°C	°F
Steels, Cunife, Vicalloy, Remalloy	80	1	ac field	100	210
Alnico 3 to 6	240	3	ac field	500	930
Alnico 8 and 9	480	6	ac field	500	930
Ferrite 1 to 8	800	10	Curie temperature	300	570
Pt-Co	1600	20	Curie temperature	325	620
$SmCo_5$	1200–4000(a)	15–50(a)	Heat treatment	300	570
Sm_2Co_{17}	1600–3200(a)	20–40(a)	Heat treatment	350	660
Sintered NdFeB	1600–2800(a)	20–35(a)	Curie temperature	100–200(b)	210–390(b)
Bonded NdFeB	1600–3600(a)	20–45	(c)	125	255
Hot-pressed NdFeB	3600	45	Curie temperature	150	300
Hot-formed NdFeB	2000	25	Curie temperature	100–150(b)	210–300(b)
FeCrCo	240	3	ac field	500	930

(a) Depending on previous magnetic history. (b) Depends on alloy, H_{ci}, and operating slope (c) Not available

Magnetization Prior to Use. Magnets are magnetized in applied fields supplied by dc or pulsed-current electromagnets. Where practical, saturating magnetizing fields are recommended to gain full use of magnetic potential energy (see Table 9). Most magnets are demagnetized by heating to the Curie temperature or by applying an ac or dc field to reduce the measured induction to zero. Materials such as Alnico and samarium cobalt cannot be demagnetized by exposure to their Curie temperature because the metallurgical changes alter the permanent magnet properties.

Partial demagnetization may be needed to reduce the flux density B_d to some calibrated level, or to prestabilize against anticipated magnetic losses. These losses can occur due to external demagnetizing forces, such as in electric motors, or by temperature cycles. Partial demagnetization is accomplished by initial exposure to the operating environment or by applying an ac field equivalent to about twice the amount of knockdown anticipated.

Calibration of rare-earth magnets can introduce other difficulties. The process of exposing the magnet to the field levels required to demagnetize the magnet to a specific level is established. The difficulty is due to changes in the flux distribution of the magnet. Such changes may have an adverse effect on flux stability.

Handling of all permanent magnets is best done in the nonmagnetized condition. There is less risk of attracting dirt and chips, of snapping magnetic objects onto the magnet with possible injury, and of partial demagnetization due to mechanical shock. The extremely strong magnetic fields and high mechanical forces developed by rare-earth magnets may create hazards to personnel through chipping, shattering, or pinching on impact. Always make sure that magnetized rare-earth magnets are handled under control when they come near each other or other ferromagnetic materials. Metallic chips and fine particles are susceptible to rapid oxidation (burning) and spontaneous (pyrophoric) ignition.

REFERENCES

1. K.S.V.L. Narasimhan, Iron Based Rare Earth Magnets, *J. Appl. Phys.*, Vol 57, 1985
2. J. Ormerod, The Processing and Application of Sintered NdFeB Based Permanent Magnets, in *Proceedings of the Ninth International Workshop on Rare-Earth Magnets*, 1987

SELECTED REFERENCES

- J.R. Cost, R.D. Brown, A.L. Giorgi, and J.T. Stanley, "Radiation Effects in Rare Earth Permanent Magnets," Report LA-UR 87-1455, Los Alamos National Laboratory, 1987
- R.D. Cullity, *Introduction to Magnetic Materials*, Addison-Wesley, 1972
- D.I. Gordon, Environmental Evaluation of Magnetic Materials, *Electro-Technology*, Vol 67 (No. 1), Jan 1961, p 118-125
- *Hard and Soft Magnetic Materials With Applications Including Superconductivity*, Proceedings of a conference on hard and soft materials held in conjunction with ASM INTERNATIONAL's Materials Week '87, D.J. Bailey, K.S.V.L. Narasimhan, P.K. Rastogi, J.A. Salsgiver, and A.R. Traut, Ed., 1987
- R.I. Joseph, Ballistic Demagnetizing Factor in Uniformly Magnetized Cylinders, *J. Appl. Phys.*, Vol 37, 1966, p 4639
- M. McCaig and A.E. Clegg, *Permanent Magnets in Theory and Practice*, Halsted Press, 2nd ed., John Wiley & Sons, 1987
- L. Moskowitz, *Permanent Magnet Design and Application Handbook*, Cahners, 1976
- R.J. Parker and R.J. Studders, *Permanent Magnets and Their Applications*, John Wiley & Sons, 1962
- "Permanent Magnet Guidelines," PMG-88, Magnetic Materials Producers Association, 1988
- *Permanent Magnets and Magnetism*, D. Hadfield, Ed., Niffe Books and John Wiley & Sons, 1962

- *Proceedings of Second International Workshop on Rare Earth Cobalt Permanent Magnets*, K.J. Strnat, Ed., 1976
- K. Schuler and K. Brinkman, *Dauermagnete-Werkstoffe and Anwendungen*, Springer-Verlag, 1970
- R.S. Sery, R.H. Lundsten, and D.I. Gordon, *Radiation Damage Thresholds for Permanent Magnets*, Naval Ordnance Laboratory, 18 May 1961
- *Soft and Hard Magnetic Materials With Applications*, Proceedings of a symposium held in conjunction with ASM INTERNATIONAL's Materials Week '86, C.M. Maucione, K.S.V.L. Narasimhan, P.K. Rastogi, J.A. Salsgiver, and H.R. Sheppard, Ed., ASM INTERNATIONAL, Oct 1986
- F.C. Spreadbury, *Permanent Magnets*, Pitman and Sons, 1949
- "Standard Specifications for Permanent Magnet Materials," 0100-87, Magnetic Materials Producers Association, 1987
- W.L. Zingery *et al.*, Evaluation of Long-Term Magnet Stability, *J. Appl. Phys.*, Vol 1, March 1966

Metallic Glasses

W.L. Johnson, Keck Laboratory of Engineering, California Institute of Technology

METALLIC GLASSES can be prepared by solidification of liquid alloys at cooling rates sufficient to suppress the nucleation and growth of competing crystalline phases. Their discovery in 1960 by Pol Duwez and his colleagues was made possible by the innovation of rapid quenching methods. This article presents a historical survey of the study of metallic glasses and other amorphous metals and alloys from their inception to the present. This includes a discussion of synthesis and processing methods, structure and morphology, and a description of the electronic, magnetic, chemical, and mechanical properties of metallic glasses. In addition, the development of metallic glasses as materials for technical applications will be described.

Historical Introduction and Background

Traditionally, a glass is considered to be a vitrified liquid. This is a liquid that is cooled below its thermodynamic melting point and fails to crystallize, but solidifies nevertheless. In a rather well-defined temperature range below the melting point, an undercooled liquid undergoes configurational freezing. The viscosity (η) of the liquid rises rapidly with falling temperature in this temperature range to values normally associated with the solid state. An example is shown in Fig. 1(a). For a typical liquid metal, viscosities are measured in units of centipoise. Over a temperature range of tens of degrees, this value rises to 10^{16} poise, a value generally taken to indicate a solid. The heat capacity of the liquid shows an anomaly of the form indicated in Fig. 1(b). The peak in the heat capacity (or the temperature at which the rate of change of viscosity is maximum) is traditionally used to define the glass transition temperature of the undercooled liquid. To experimentally achieve vitrification, crystallization of the undercooled liquid must be avoided.

Many naturally occurring minerals, such as volcanic obsidian, exhibit a glassy structure. The earliest evidence of glass synthesis by man was recorded by the historian Pliny in the first century (Ref 1). He writes

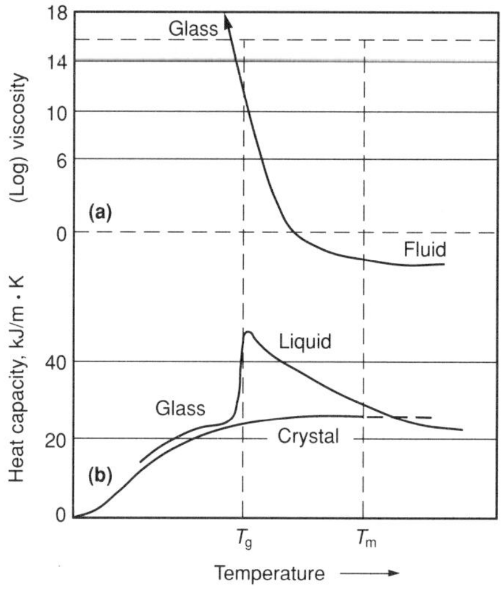

Fig. 1 (a) Temperature dependence of the viscosity of an undercooled melt. (b) Heat capacity of an undercooled melt as a function of temperature. Also shown is the typical heat capacity of the corresponding crystalline solid (at the same composition).

of a band of Phoenician sailors who about the year 5000 B.C. built a fire over blocks of soda from a ship's cargo. As the fire died, the fused soda sank into the sand, forming a shiny glass rivulet. Remains from Middle Eastern civilizations dating back 5000 years provide direct evidence of glassmaking.

One may naturally ask why the discovery of metallic glasses did not occur until the twentieth century. The answer lies in the ease with which metallic melts undergo crystallization. In contrast to the obsidian and silicate glasses mentioned above, metallic glasses can only be produced when the melt is cooled at high rates. The time required to nucleate and grow crystals in an undercooled melt varies enormously with the nature of the atomic and molecular units that make up the liquid. These units must organize themselves into a crystalline nucleus of critical size in order to initiate the process of crystallization. In metallic melts, where the fundamental units can be viewed as roughly spherical individual atoms, the formation of a crystalline nucleus occurs with relative ease. It is unimpeded by kinetic hindrances that arise when complex, covalently bonded molecular units are organized to form a nucleus (as in materials such as silica, polymers, and so forth). As such, the rate of crystallization in the undercooled melt tends to be relatively high in metallic liquids. The theory of nucleation and growth of crystals in undercooled metals was developed over 40 years ago by Turnbull and others (Ref 2). It was shown that the rate of crystal nucleation and growth is a sensitive function of the degree of undercooling $\Delta T = T_m - T$, where T_m is the thermodynamic melting point. The rate of crystal nucleation at a given ΔT is a rapidly increasing function of ΔT. In metals, this typically permits undercoolings of ~100 K to be achieved with no observable crystallization in laboratory time scales. At greater undercoolings, the crystallization rate rises very rapidly with falling temperature, and copious nucleation and growth of crystals is observed. The rise in the nucleation rate is finally halted when the configurational freezing temperature is approached. Using classical nucleation theory, Turnbull developed a model for the rate of crystal nucleation ($\dot{N}$) (measured as the number of crystalline nuclei/cm^3 · s^{-1}) in metallic melts. He proposed a formula that can be written as:

$$\dot{N} = \text{A} \exp[-\text{B}/(T - T_g)]\exp(-\Delta G^N/k_B T) \quad \text{(Eq 1)}$$

where A and B are constants, T_g is the glass transition temperature, ΔG^N is the free-energy barrier that opposes the formation of a crystalline nucleus, and k_B is the Boltzmann constant. ΔG^N decreases with increasing undercooling. It roughly varies as:

$$\Delta G^N = C\sigma^3/[\Delta S_F(T_m - T)]^2 \quad \text{(Eq 2)}$$

where C is a geometric constant of order unity, σ is the interfacial energy per unit area between a liquid and a crystal, and ΔS_F is the entropy of fusion per unit volume of liquid. Improvements on this simple formula have been developed (Ref 3), but the version in Eq 2 is adequate for this discussion. These equations lead to a nucleation rate that depends on temperature in a manner illustrated in Fig. 2 for a typical choice of the constants (A, B, C, σ, ΔS_F) and

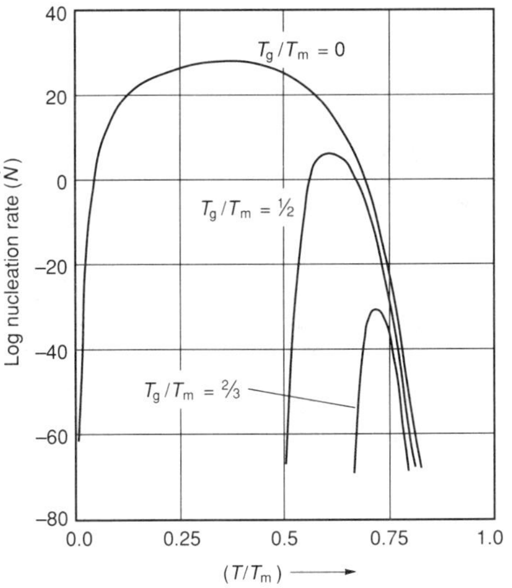

Fig. 2 Temperature dependence of nucleation rate ($\dot{N}$) in an undercooled melt. The nucleation rate is given in units of nuclei/cm^3 · s^{-1}.

several values of the ratio T_g/T_m. A sharp peak in the nucleation rate as a function of undercooling is associated with the rising thermodynamic driving force for crystallization (Δg_{XL}), which is measured as the free energy drop per unit volume on crystallization:

$$\Delta g_{XL} = \Delta S_F(T_m - T) \qquad \text{(Eq 3)}$$

This competes against a falling fluidity (the inverse of viscosity):

$$\eta^{-1} \sim \exp[-B/(T - T_g)] \qquad \text{(Eq 4)}$$

The maximum rate of nucleation determines the rate at which cooling must be carried out in order to suppress crystallization. As seen in Fig. 2, this maximum rate depends sensitively on the value of T_g compared with the melting point T_m. When T_g/T_m is relatively large (for example, $T_g/T_m = 2/3$ in Fig. 2), the maximum rate of nucleation $\dot{N}$ $\sim 10^{-30}$ nuclei/cm^3 · s^{-1} occurs when T/T_m ~ 0.7. When T_g/T_m is only somewhat smaller (for example, $T_g/T_m = 1/2$ in Fig. 2), a maximum rate of $\dot{N} = 10^8$ nuclei/cm^3 · s^{-1} is predicted. The ratio T_g/T_m thus plays a key role in determining whether or not crystallization can be avoided during the quenching of the melt.

In binary alloys where the liquidus and solidus curves fall to relatively low temperatures compared with the melting points of the elemental constituents, the ratio T_g/T_m (here T_m is interpreted as the liquidus/solidus temperature) is larger than in pure metals. Deep-eutectic features in phase diagrams tend to satisfy this condition. As such, compositions lying near deep-eutectic points tend to be good glass-forming compositions. Turnbull pointed out this deep-eutectic criterion during the early 1960s.

Once the nucleation rate is known, the critical cooling rate required for glass formation can be determined. In pure metals, this cooling rate has been estimated to be of the order of 10^{13} K/s. Clearly, this is very difficult to achieve in practical situations. Fortunately, in certain metallic alloys (for example, deep-eutectic alloys), the required rates are more modest, of the order of 10^4 to 10^5 K/s. Although not normally achieved in cooling bulk metallic melts, such rates are feasible under special circumstances. The techniques used to achieve these cooling rates are discussed in detail in the section "Synthesis and Processing Methods" in this article.

The brief discussion above explains why metallic glass formation is normally not observed in everyday experimental situations and answers the question of why metallic glasses were not discovered until the twentieth century. The following section traces the development of special quenching techniques capable of producing metallic glasses. Further, it will be shown that the final glassy state produced by quenching a melt can also be achieved by a variety of other methods in which the parent phase is not a liquid.

It is traditional to refer to materials produced by methods other than melt quenching as amorphous materials. The term amorphous also applies to melt-quenched glasses and is therefore a more general name applied to any material having a structure and physical characteristics similar to those of a configurationally frozen liquid. Specifically which properties of a particular material qualify it as amorphous will be addressed in the sections on structure and properties. The terms amorphous and glassy have become more or less interchangeable. This partially reflects the tendency to view materials in terms of structure and properties rather than in terms of how they are prepared. In fact, the structure and properties of amorphous solids often depend (albeit in a subtle manner) on the synthesis method used to produce them. As such, the distinction between "glass" and the more general term "amorphous solid" remains somewhat useful.

Synthesis and Processing Methods

Rapid Quenching From the Melt. As mentioned in the previous section, the production of metallic glasses by liquid quenching requires rather high cooling rates. To achieve such rates, heat must be extracted from the melt along a temperature gradient. A molten sample of typical dimension R and initial temperature T_m (the melting point of the material) will require a total cooling time τ (to ambient temperature) of the order of:

$$\tau \sim (R^2/\kappa) \qquad \text{(Eq 5)}$$

where κ is the thermal diffusivity of the metal. It is given by $\kappa = K/C$ where K is the thermal conductivity and C is the heat capacity per unit volume. The cooling rate achieved ($\dot{T}$) will be of the order of:

$$\dot{T} = dT/dt = T_m/\tau = (KT_m/CR^2) \qquad \text{(Eq 6)}$$

Taking $T_m \sim 1000$ K ~ 1 W/cm · s^{-1} · deg^{-1} (typical of a molten metal) and $C \sim 2$ J/cm^3 · deg^{-1} (also typical of a metallic liquid), gives

$$\dot{T} = 500/R^2 \qquad \text{(Eq 7)}$$

To achieve cooling rates of 10^6 K · s^{-1}, the sample dimension must be of the order of 100 μm or less. In addition, the achievement of such cooling rates presupposes that the liquid sample can be suddenly brought into intimate thermal contact with a large, thermally conductive mass initially at ambient temperature. This restrictive combination of conditions requires one to produce a thin molten layer and to bring it suddenly into contact with a cold, highly conductive surface.

In 1959, Pol Duwez and his colleagues devised a clever method to accomplish this (Ref 4). A liquid droplet is melted in a nonreactive crucible under an inert atmosphere. The droplet is suddenly subjected to an acoustical shock wave that both atomizes and accelerates the atomized droplets against a cold copper surface at ambient temperature. The individual atomized droplets are of micron dimensions. The device is shown schematically in Fig. 3. This technique is referred to as gun quenching. The droplets impact on the substrate, spread as they slide along, and freeze by heat conduction to the underlying copper strip. This method produces a rather irregular sample, but it has been shown to be capable of achieving cooling rates of 10^6 to 10^8 K · s^{-1}. It was an ideal tool for studying solidification of metals at high cooling rates. With it Duwez *et al.* (Ref 5) produced the first metallic glasses in the gold-silicon binary system.

A somewhat refined version of the method, used later by the Duwez group, is illustrated in Fig. 4 and is called the piston and anvil technique. In this method, the droplet is melted and allowed to fall between the faces of an anvil and a pneumatically accelerated piston. The droplet is struck by the face of the piston, carried onto the anvil, and spread by the momentum of the moving piston into a thin layer. The layer subsequently solidifies by conduction of heat onto the piston and anvil surfaces. Cooling rates of the order of 10^5 to 10^6 K · s^{-1} are achieved. The sample produced is a thin foil (the size of a nickel) of thickness ranging from 30 to 50 μm. Owing to their rather uniform thickness, such samples were ideal for scientific studies.

Both the gun method and piston and anvil method were used by the Duwez group and by the Giessen and Grant group at M.I.T. during the 1960s to produce samples of

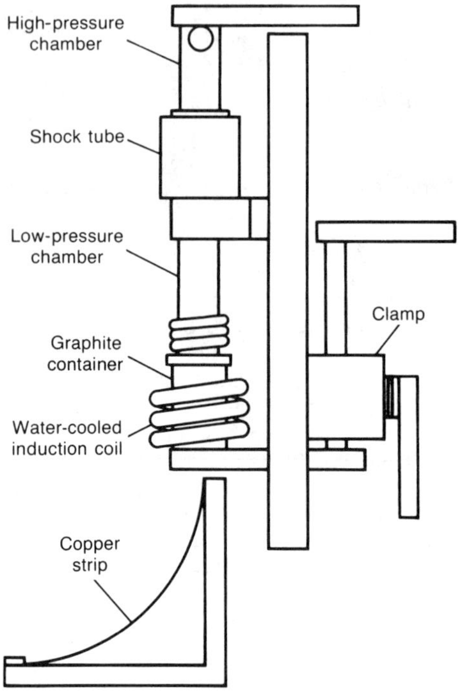

Fig. 3 Schematic showing the gun quenching apparatus used by Duwez *et al.* to carry out rapid solidification experiments. Source: Ref 4

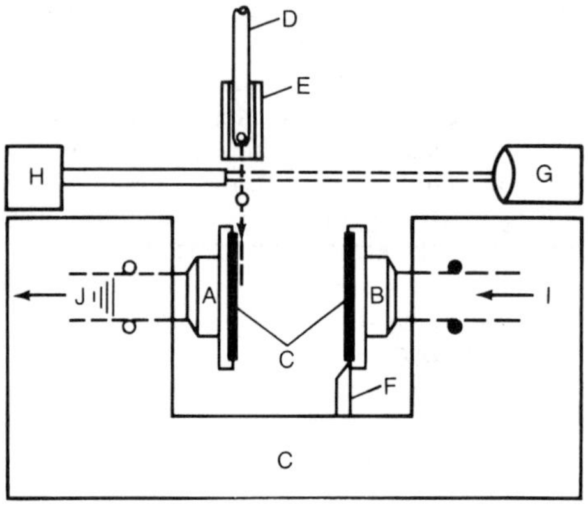

Fig. 4 Schematic drawing of the piston and anvil device used for rapid solidification of liquid drops at cooling rates of 10^4 to 10^6 K/s. The device was developed to produce relatively uniform foils of metallic glass by the Duwez group. The droplet is melted, ejected from the crucible, detected during fall by the photocell, and then quenched by the pneumatically driven piston onto the anvil. A, anvil; B, piston; C, chassis; D, crucible containing the sample droplet; E, heating element; F, latch (releases piston); G, light source; H, photocell and timing circuits; I, pneumatic drive system for piston; J, pneumatic cushion for anvil. Source: Ref 4

Fig. 5 Sheet of metallic glass prepared using the planar-flow casting method. Such sheets are used to wind power-distribution transformer cores.

glassy metallic alloys for scientific study. Using the rapid solidification methods, the Duwez group found a generic class of metallic glass-forming alloys now referred to as the metal-metalloid systems. These alloys contain about 80 at.% of a transition metal (for example, iron, nickel, cobalt, and so on) and 20 at.% of a metalloid element (for example, silicon, phosphorus, boron, carbon, and so on). The Giessen group discovered a second class of metallic liquids that form glasses at rapid solidification rates. These are referred to as the metal-metal systems. These alloys contain an early transition metal or rare earth element (for example, zirconium, titanium, niobium, tantalum, and so on) alloyed with a late transition element (for example, nickel, cobalt, iron, palladium, and so on). The composition ranges of these glass-forming systems were found to be more varied than those of the metal-metalloid glasses. Both classes of alloys yielded glasses that were stable at room temperature and above. Some of these alloys could be heated to temperatures of the order of 500 °C (930 °F) or more without crystallization on laboratory time scales.

By 1970, it became increasingly apparent that if metallic glasses were to be more than a laboratory curiosity, then a method of continuous fabrication capable of producing larger sample quantities was necessary. Several researchers developed the continuous casting, or melt spinning, processes in response to this need (Ref 6, 7). In these processes, a continuously flowing jet of liquid metal is forced onto the exterior surface of a rapidly rotating substrate wheel. Upon contact with the wheel, the jet spreads onto the surface and solidifies. A solid continuous ribbon is thrown from the wheel in a continuous manner by centripetal force. Multiple-jet casters, in which adjoining jets fuse together on the wheel, were initially developed to produce a wider product.

A major breakthrough in this method was the planar-flow casting technique. Patented by Narasimhan at the Allied Corporation (now Allied-Signal Inc.) (Ref 8), this method involves a broad continuous planar-melt jet produced by a slotted nozzle that flows continuously onto the surface of a rotating drum. In its commercial realization, it leads to the production of uniform sheets of metallic glass having a thickness of ~15 to 150 μm, widths up to ~1 m (40 in.), and essentially unlimited length. The cooling rates achieved by this method are of the same order as those achieved in the original methods of Duwez. The broad sheets of metallic glass produced are ideal for a number of commercial applications. A photograph of such sheets is shown in Fig. 5. The uses of such samples will be discussed further in the section dealing with applications.

Vapor Quenching and Electrodeposition. As mentioned earlier, there are other methods of producing materials that have atomic structures and physical properties similar to those of a configurationally frozen liquid. It can be argued that these amorphous solids are in essentially the same metallurgical state as a melt-quenched glass. As such, it is appropriate to review these methods.

The first involves the atom-by-atom deposition of the material onto a substrate. The supply of atoms is provided in several different ways. These include production of metallic vapor by thermal evaporation (such as using electron beam guns), atom-by-atom erosion of a solid target by energetic ions (sputtering), or deposition of metallic ions from an electrolytic solution (electrodeposition and electroless deposition). There are also other variations (such as chemical vapor deposition). All of these methods have in common the fact that the parent material consists of individual atoms (or small atomic clusters) that are deposited atomic layer by atomic layer onto a relatively cold substrate.

The production of amorphous metals by this method was pioneered by Buckel and Hilsch (Ref 9) during the 1950s at the University of Gottingen in Germany. Using thermal evaporation of simple metals (tin, lead, and so forth) and alloys of simple metals (tin-copper, lead-copper, and so forth) they showed that quenching of the metallic vapor onto a cryogenically cooled substrate ($T < 10$ K) could lead to the growth of an amorphous film. Buckel and Hilsch devised methods such as *in situ* electron diffraction to demonstrate that the structure of these films was similar to that of a liquid. These experiments involved some of the earliest reflection electron diffraction studies of thin films in a high-vacuum chamber. In addition, they developed *in situ* methods for measuring the electrical resistivity of films. Using these methods, they found that the amorphous films became superconducting at temperatures higher than those of the corresponding equilibrium phases. In addition, they were able to show that the amorphous films crystallized at relatively low temperatures. In the case of pure metals (for example, tin, lead, bismuth, and so on), crystallization was observed when the films were heated above ~20 K.

Similar studies of films of pure metals (copper, gold, and so on) and alloy films (copper-gold) formed by quenching on a cryogenic substrate were carried out somewhat later by Mader *et al.* (Ref 10), who pointed out that cryoquenched films of pure noble metals were not amorphous, whereas alloys of noble metals did become amorphous. They attributed this to the fact that alloys of elements having different atomic sizes crystallize with greater difficulty under cryoquenching conditions. They used computer simulation of the deposition process to verify this hypothesis.

Studies of cryoquenched pure metals were extended to the transition elements by Collver and Hammond (Ref 11). They deposited thin films of the 4*d* and 5*d* transition elements onto cryogenically cooled substrates and found that many of the films produced were amorphous. Like Buckel and Hilsch, they observed that amorphous films of pure metals tended to crystallize at temperatures far below room temperature. On the other hand, they discovered that certain amorphous alloy films (for example, molybdenum-rhenium alloys) exhibited far greater stability against crystallization. Some of these films could be heated to room temperature and above (~400 °C, or 750 °F) without crystallization.

Since these early studies, many groups have studied vapor-deposited amorphous alloys (Ref 12). For many alloys, it is unnecessary to cool the substrate in order to obtain an amorphous film. Furthermore, amorphous films so produced often remain amorphous up to rather elevated temperatures (~700 °C, or 1290 °F). Vapor deposition has become a common method of producing amorphous films for a variety of technical applications. Various methods of producing metallic vapor or atomic scale clusters are currently used. Among these are sputtering, chemical vapor deposition, electron beam evaporation, and so forth. Sputtering utilizes a dc or radio frequency (rf) generated plasma discharge to accelerate inert gas ions into a metal target. The ion impact on the target energizes secondary metal atoms (or ions), which are ejected from the target. These metal atoms are subsequently collected on a substrate and a metal film is grown. The original target is often a metal alloy.

In other cases, two or more elemental targets are simultaneously bombarded (cosputtering) to produce an alloy film. The sputtering method has been perfected with the development of commercially available sputter guns. These guns use magnetic fields to concentrate the inert gas plasma in the vicinity of the target. As such they are often very efficient in producing secondary metal atoms. This variation of sputtering is called magnetron sputtering. In commercially available units, rather high thin-film deposition rates (10 to 100 nm/s) are achievable, and thus reasonably thick films can be deposited in practical time scales.

Chemical vapor deposition employs volatile gases (for example, metal fluorides, metal chlorides, metal carbonyls, and so forth). These gases are heated in a reaction chamber and caused to decompose into a volatile component (chlorine, for example) and pure metal vapor. The volatile component is swept away in the reaction chamber while the metal vapor is deposited onto a substrate. Simultaneous decomposition of two reactants can be used to deposit alloy films. This technique has been used to form amorphous alloy coatings on complex-shape substrates and irregular surfaces.

Electrodeposition is another atom-by-atom deposition method used to produce amorphous alloy films and coatings. It is much like chemical vapor deposition, except metal ions instead of atoms are obtained by dissolution of metal salts in an aqueous solution. When a current is passed through the electrolytic cell, these metal ions are collected on an anode. A film is built up on the anode as the process proceeds. This method has long been used in commercial plating. For example, electrolytically deposited nickel films have long been used in commercial applications. These films, in fact, often contain (in addition to nickel) substantial amounts of other elements (for example, phosphorus). For sufficiently high concentrations of these other elements, the films are amorphous. It is likely that the earliest amorphous alloy films were made by this process over 60 years ago. Unfortunately, at that time no systematic study of the structure of these films was made.

Solid-State Methods. We have thus far discussed the production of amorphous metals and metallic glasses by quenching of liquid metals or quenching of metallic vapors. An obvious question is whether it is possible to produce an amorphous metal or metallic glass by inducing a crystalline solid to transform to the amorphous state. Such a transformation could be called a crystal-to-glass transformation. As mentioned earlier, the amorphous phase is not an equilibrium phase. Without exception, it is believed that the lowest free-energy state of a solid is a crystalline phase or a mixture of crystalline phases. How, then, can a crystalline metallic solid be induced to transform to the glassy state?

Clearly, the crystalline system must initially be driven away from equilibrium. An excess free energy must be stored in the crystalline system that is sufficient to render it thermodynamically less stable than a competing glassy phase. Figure 6 illustrates the basic principle involved. A nonequilibrium crystalline state is initially created by suitable means. The Gibbs free energy of this state is greater than that of the amorphous state. On the other hand, the equilibrium crystalline state has still lower Gibbs free energy. In order for the nonequilibrium crystalline state to transform to the amorphous phase, kinetics must favor this transformation.

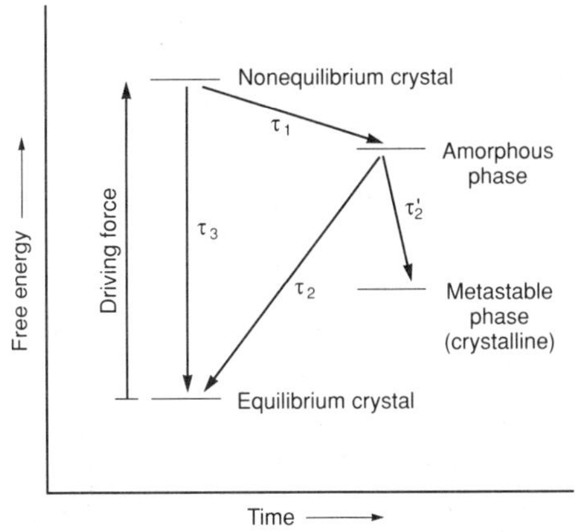

Fig. 6 Illustration of the free-energy relationship and transformation time scales (τ_1, τ_2, τ_3) for a nonequilibrium crystalline solid that undergoes a crystal-to-glass transformation

In Fig. 6, three time constants (τ_1, τ_2, and τ_3) are indicated. These time constants give the characteristic time scale for the respective transformations. In order that the amorphous phase be formed and retained, the transformation from the nonequilibrium crystalline state to the amorphous state must be kinetically preferred over the two other competing transformations:

$$\tau_1 < \tau_2 \text{ and } \tau_1 < \tau_3 \quad \text{(Eq 8)}$$

The first condition ensures that the amorphous state formed will not crystallize to the equilibrium state during the time scale of the experiment; the second condition ensures that the nonequilibrium crystal will not directly transform to the equilibrium crystalline state. In the second condition, the amorphous phase must be metastable during the time scale of the experiment. It is well known that amorphous phases crystallize by a process of nucleation and growth of crystals. The theory of crystallization of an undercooled melt can be applied to the crystallization of the glass. As such, by a suitable choice of kinetic conditions (sufficiently low temperature), one can always ensure that the second condition of Eq 8 is satisfied. Essentially, solid-state amorphization must be carried out below that temperature where the glassy product would normally crystallize. Practically speaking, this requires that the experiments be carried out below the T_g of the amorphous alloy to be produced. In fact, both conditions in Eq 8 can be simultaneously satisfied in a number of actual situations. A few of these situations are dis-

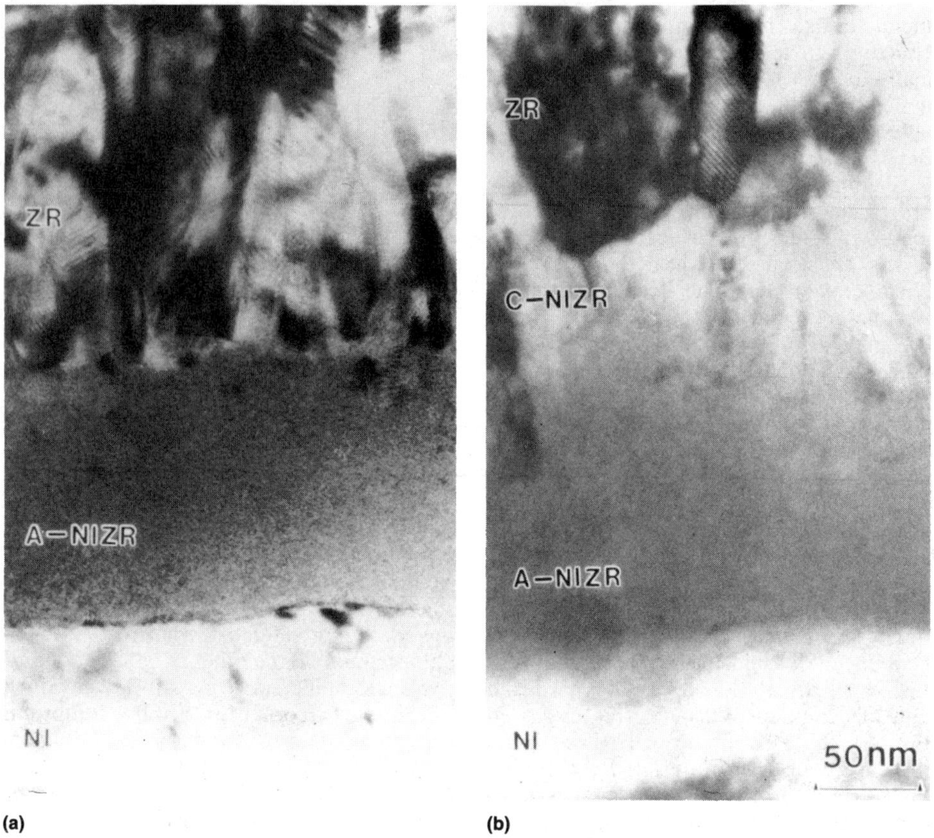

Fig. 7 Electron micrograph showing a cross section of a thin-film diffusion couple consisting of layers of nickel and zirconium. (a) The diffusion couple has been reacted at 300 °C (570 °F) for 6 h. An amorphous interlayer having a thickness of about 80 nm has formed and grown at the nickel-zirconium interface. (b) The diffusion couple has been reacted for 18 h at 300 °C (570 °F). Formation and growth of the amorphous interlayer has now been succeeded by later growth of the crystalline intermetallic compound NiZr.

cussed in this article; more extensive discussions are available in Ref 10 and 11.

One example of solid-state amorphization is the reaction of two crystalline metals in a diffusion couple to form an amorphous phase. This phenomenon has been observed in numerous binary diffusion couples (Ref 13, 14) and was first reported by Schwarz and Johnson (Ref 15). The example described here is the nickel-zirconium binary couple. Figure 7 is an electron micrograph showing a nickel-zirconium diffusion couple in cross section. The couple was reacted for 6 h at a temperature of 300 °C (570 °F). Between the nickel and zirconium layers, a featureless and uniform gray layer having a thickness of the order of 80 nm has formed. This zone is amorphous. It can be shown that this amorphous phase initially nucleates at the nickel-zirconium interface at grain boundaries of zirconium grains. Apparently, the nucleation barrier for amorphous phase formation is lower than that for formation of the equilibrium intermetallic compounds of the nickel-zirconium binary system. Figure 8 shows an enthalpy of mixing diagram at T = 300 °C (570 °F) for the nickel-zirconium system. Because the enthalpies of mixing in this system are far larger than TS_{mix} (where S_{mix} is the entropy of mixing), the free energy of mixing can be approximated by the enthalpy of mixing.

In equilibrium, a diffusion couple should ultimately form one or a mixture of two of the terminal solid solutions or equilibrium compounds. These are the lowest free-energy states. On the other hand, the amorphous phase has lower free energy than a physical mixture of the pure metals or their solid solutions. This is precisely the situation described in Fig. 6. The initial free energy of the system is greater than that of the amorphous phase. The free energy of the amorphous phase is greater than that of the ultimate equilibrium phase, which consists of one or a mixture of two intermetallic compounds. In this case one can associate τ_1 with the time required to nucleate and grow the amorphous layer to a certain thickness (for example, 80 nm). The time constant τ_2 is associated with the time required for the amorphous phase to crystallize into one or more of the equilibrium intermetallic compounds, whereas the time constant τ_3 is the time required to nucleate and grow an intermetallic compound in the original diffusion couple. At the temperature where this solid-state reaction is carried out, it is clear that the conditions described by Eq 8 are satisfied.

A second case of solid-state amorphization involves the use of an external driving force to raise the enthalpy (and free energy) of an initially single-phase crystalline material. Irradiation of a crystal by high-energy particles (ions, electrons, and neutrons) has been shown to be an effective means of accomplishing this. The crystal is held at a relatively low temperature, and irradiation induces damage in the crystal in the form of point defects (vacancies, interstitial atoms, and chemical disorder). If the ambient temperature is sufficiently low, recovery of the crystal can be suppressed. The recovery is mediated by the mobility of the defects. Thus the temperature chosen must be sufficiently low to suppress defect mobility. Defect mobility is characterized by a time scale τ_d, which can be identified with τ_3 in Eq 8. If τ_d is sufficiently long compared with the time scale of the experiment, then the defects formed will be trapped in the crystal. Under these conditions, the defect concentration can be driven far outside of equilibrium limits. The excess enthalpy stored in the crystal in the form of defects then provides the thermodynamic driving force for amorphization. This raises the free energy of the crystal above that of the amorphous phase. If a suitable nucleation site for an amorphous phase exists at the ambient temperature, then a crystal-to-glass transition occurs. Often, the amorphous phase nucleates most easily at extended defect sites in the original crystal (for example, grain boundaries, dislocations, a free surface, or other extended defect). For example, during electron irradiation of the intermetallic compound Ti_3Cu_4, Meshi *et al.* (Ref 16) observed nucleation of the amorphous phase along dislocations. In the absence of extended defects, it may be possible to observe homogenous nucleation of the amorphous phase. Apparently, for a sufficiently damaged crystal, the nucleation barrier for amorphous phase formation becomes very small, or perhaps even vanishes. The amorphous phase then forms without the need for thermal activation. The crystal-to-glass transformation can occur even at cryogenic temperatures under these circumstances. It would appear that the crystal lattice becomes unstable against amorphization for sufficiently high defect concentrations. The mechanism of instability has been a subject of considerable debate. Fecht and the author (Ref 17) have argued that such instability may be relative to an entropy catastrophe. Experimentally, it has been observed that the shear modulus of the crystal is substantially reduced as instability is approached. Figure 9 shows the results of an irradiation experiment by Okamoto *et al.* (Ref 18) in which the shear sound velocity in a Zr_3Al Cu_3Au-type alloy is observed to dramatically decrease during irradiation at cryogenic temperatures. The irradiation ultimately results in amorphization of the crystal.

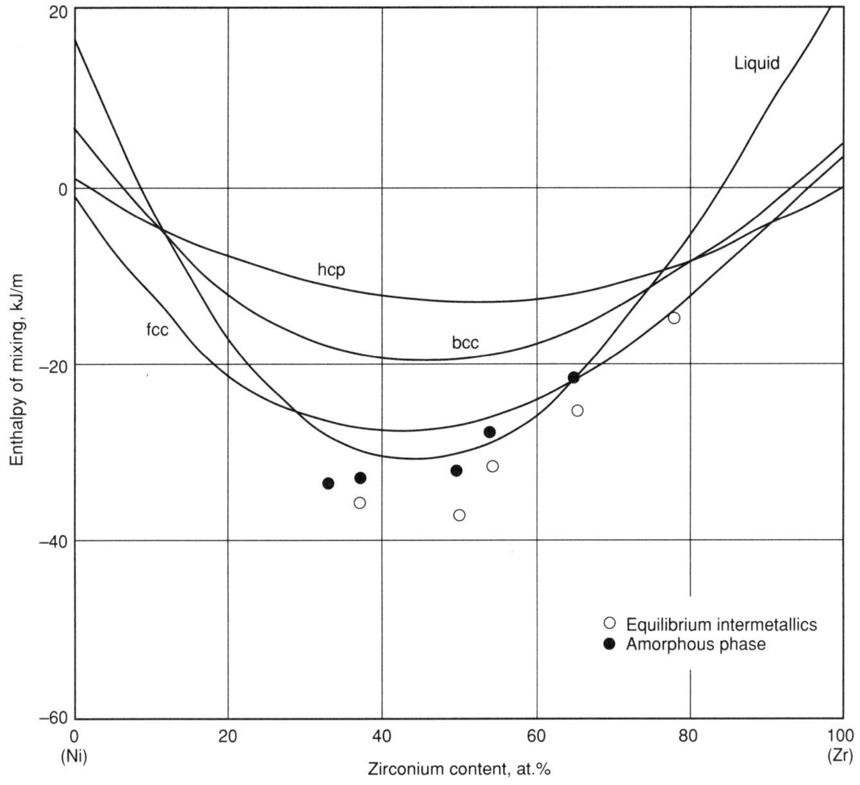

Fig. 8 Enthalpy of mixing versus composition for the nickel/zirconium binary system at a fixed temperature of 300 °C (570 °F). The composition dependence of the enthalpy of mixing of various solid-solution phases (hcp, bcc, and fcc) and the undercooled liquid/amorphous phase are shown. The enthalpy of mixing of the intermetallic compounds is indicated by data points.

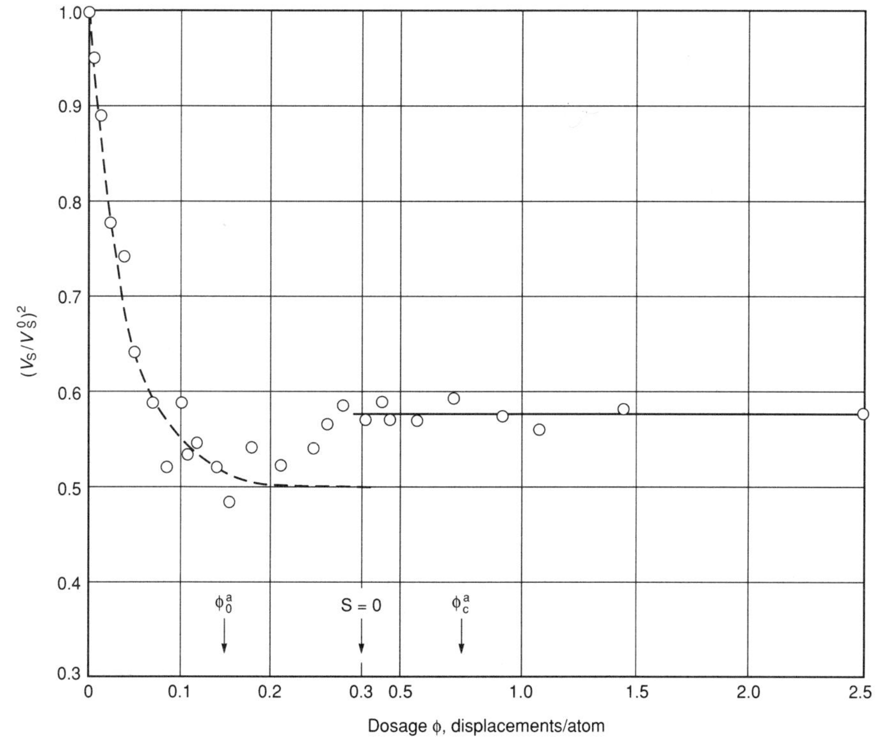

Fig. 9 Relative change in shear sound velocity in crystalline Zr_3Al as a function of irradiation dose by 1 MeV krypton ions. Dosage has been converted to conventional units of irradiation-induced displacements per atom of sample. ϕ_0^a and ϕ_c^a refer to the doses at which the onset and completion of the crystal-to-amorphous transformation occur as observed by electron microscopy. The ratio V_s/V_s^0 is that of the shear sound velocity in the irradiated/unirradiated sample. Source: Ref 18

Severe mechanical deformation of crystalline intermetallic compounds has also been found to result in solid-state amorphization. In this case, repeated deformation of a crystal at relatively high strain rates is carried out in a high-energy ball mill. Dislocations, dislocation walls, grain boundaries (formed by collapse of dislocation arrays), and antiphase boundaries (formed by motion of partial dislocations) are trapped in the crystal in concentrations far in excess of those ordinarily found. This has been observed to result in amorphization of the crystal (Ref 19). This amorphization by mechanical attrition has been found to be an effective means of producing large quantities of amorphous alloy powders.

There are other methods of solid-state amorphization too numerous to detail in the space available here. Amorphization during absorption of hydrogen by a crystal (Ref 20), and mechanical alloying of physical mixtures of metals (Ref 21) are examples. All of these techniques have in common the underlying description implied by Eq 8 and Fig. 6. Equation 8 expresses the kinetic constraints required, and Fig. 6 expresses the fact that a thermodynamic driving force is required if such transformations are to occur.

Structure of Metallic Glasses and Amorphous Metals

Structural Models, Diffraction Experiments, and Crystallization. Because metallic glasses were initially prepared by rapid solidification of a liquid, it is natural to assume that their atomic scale structure is related to that of the liquid phase. On the other hand, glasses are in fact solids in which the configuration of atoms is frozen in much the same manner as it is frozen in a crystalline solid. At temperatures sufficiently low to suppress atomic diffusion, each atom is confined to move in the potential well created by its neighbors. It is often presumed that the local atomic arrangements in glasses resemble those of the corresponding crystal. From these notions, two types of models have historically been put forward to describe the atomic arrangements in metallic glasses. These are loosely called continuous random network and microcrystalline models.

The continuous random network models trace their roots to the work of Bernal and Finney (Ref 22) in the early 1960s. Bernal studied mechanical models of hard spheres packed randomly in an elastic bladder and characterized the atomic arrangements that occur in such hard-sphere packings. For example, he determined the pair correlation function obtained when large samples of "dense randomly packed spheres" were analyzed. The pair correlation function, $\rho(r) - \rho_0$ (ρ_0 = average atomic density), is basically the deviation of the atomic density

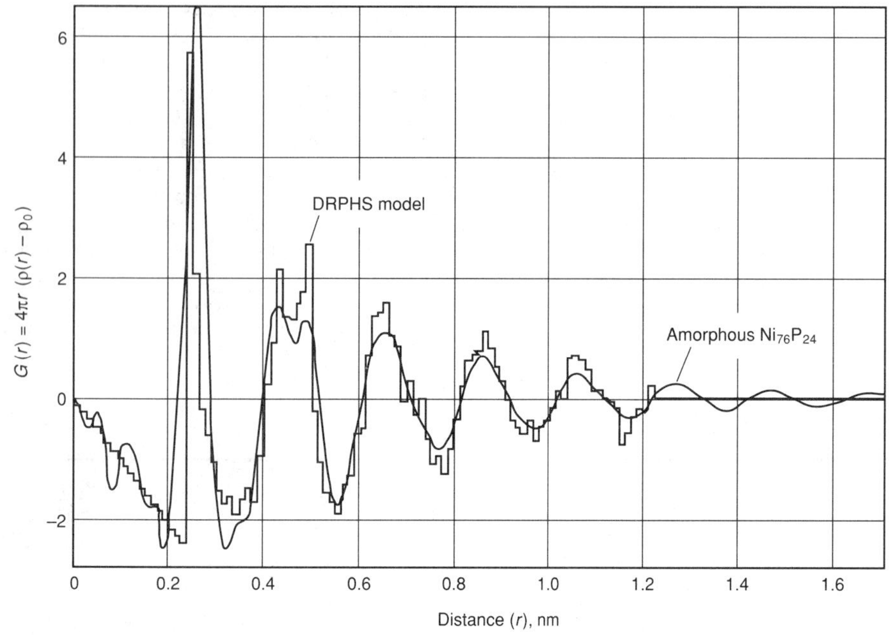

Fig. 10 The reduced radial distribution function for a dense random packing of hard spheres compared with that for the metallic glass $Ni_{76}P_{24}$. Source: Ref 22, 23

from its average value as one moves out radially from a typical atom in the packing. Oftentimes it is useful to plot another function, $G(r) = 4\pi r[\rho(r) - \rho_0]$. This function is called the reduced radial distribution function (see the article "Radial Distribution Function Analysis," in *Materials Characterization*, Volume 10 of the 9th Edition of *Metals Handbook*).

For a dense random packing of hard spheres (DRPHS), Bernal and Finney obtained a result that is illustrated in Fig. 10. At the time of their work, x-ray diffraction studies of pure liquid metals had also been carried out, and pair correlation functions of liquid metals were known with reasonable accuracy. The Bernal-Finney DRPHS model was found to give a pair correlation function that agreed remarkably well with that obtained on real liquid metals. The DRPHS model became an accepted model for the structure of liquid metals.

In the early 1970s, Cargill (Ref 23) measured the pair correlation functions of several binary amorphous alloys using x-ray diffraction techniques. He attempted to compare these correlation functions to those of Bernal. In addition, he computed the pair correlation function that would be obtained if an assembly of small microcrystals (~1.5 nm in dimension) were brought together. He concluded that the DRPHS pair correlation function provided a far better description of an amorphous $Ni_{76}P_{24}$ alloy than did the correlation function computed for an assembly of microcrystals. Incidentally, he considered microcrystals of phases that appear in the equilibrium phase diagram of the binary nickel-phosphorus system. On this basis, he hypothesized that the structure of metallic glasses was better described as liquidlike than microcrystalline.

Following Cargill's early work, several investigators undertook to improve upon Bernal's DRPHS model by replacing the hard spheres with soft spheres interacting through pair potentials. They used computers to construct and relax large models (containing of the order of 1000 atoms) of randomly packed atoms to obtain more realistic random atomic packing models. Generally speaking, these elaborate computer models were found to give somewhat better agreement with experimental correlation functions of amorphous metals than the simple Bernal hard-sphere model.

This work was extended to binary alloys containing A and B atoms. The pair potentials for AA, BB, and AB interactions were allowed to vary in the computer simulations, and the partial pair correlation functions for AA pairs, BB pairs, and AB pairs could be separately evaluated. It was found possible to reproduce the detailed features of experimental pair correlation functions of binary amorphous alloys (for example, see Ref 24). Using neutron scattering and isotope substitution techniques, it became possible to experimentally separate the partial pair correlations for the AA, BB, and AB pairs in binary alloys (Ref 25). These separate partial pair correlations could be compared with those obtained in the computer simulations. It was found that not only the total pair distribution but even the partial (AA, BB, and AB) pair correlations in actual alloys could be described by computer simulation provided a suitable choice of pair potentials (AA, BB, and AB) was made. These binary dense random packing models became the standard for describing the structure of metallic glasses.

From the study of partial pair correlation functions it became apparent that binary amorphous alloys frequently exhibit short-range chemical ordering. For example, metal-metalloid glasses containing a transition metal A (for example, gold, nickel, iron, and so forth) and a metalloid element (for example, silicon, germanium, boron, phosphorus, and so forth) and having a concentration near ~20 at.% of the metalloid element were found to possess a common type of chemical ordering. Diffraction studies of partial pair correlation functions showed that the transition metal atoms possessed nearest-neighbor atoms of both the metal and metalloid type, whereas the metalloid atoms had only transition metal atoms as nearest-neighbors. Simply put, nearest neighboring metalloid atoms are excluded in the glassy structure. Such short-range order was inconsistent with the notion that metallic glasses are chemically random.

This discovery led some investigators to explore models for amorphous structure based on assemblies of ordered atomic clusters. For example, it was argued that in $Ni_{80}B_{20}$, $Pd_{80}Si_{20}$, and other similar metal-metalloid glasses, the glass structure could be viewed as being built of trigonal prismatic clusters of nickel atoms surrounding a boron atom, each cluster containing 6 to 9 atoms (Ref 26). It was further argued that these molecular clusters were similar to those observed in crystalline phases of the same composition (for example, Ni_3B). These models might be classified as intermediate between dense random packings and microcrystalline structures. It should be noted, however, that the molecular units that comprise the model contain fewer than, or roughly the same, number of atoms as would be present in one unit cell of the corresponding crystalline phase. As such, the units can hardly be called microcrystals. On the other hand, the ability of these models to reproduce experimental pair correlation functions with high accuracy certainly suggests that the structure of metallic glasses is far from random.

In recent years, the development of ultrahigh-resolution imaging techniques in electron microscopes has further revealed the extent of short-range ordering in metallic glasses. Current electron microscopes are capable of direct imaging with spatial resolution of 0.2 nm or less. This permits direct imaging of lattice planes in crystals. When examined with such instruments, metallic glasses often exhibit remarkable structure on the nanometer scale.

For example, the well-studied metal-metalloid glasses often locally exhibit microcrystal-like structures. An example is shown in Fig. 11 for a $Pd_{75}Si_{25}$ glass. Local features that resemble lattice planes are

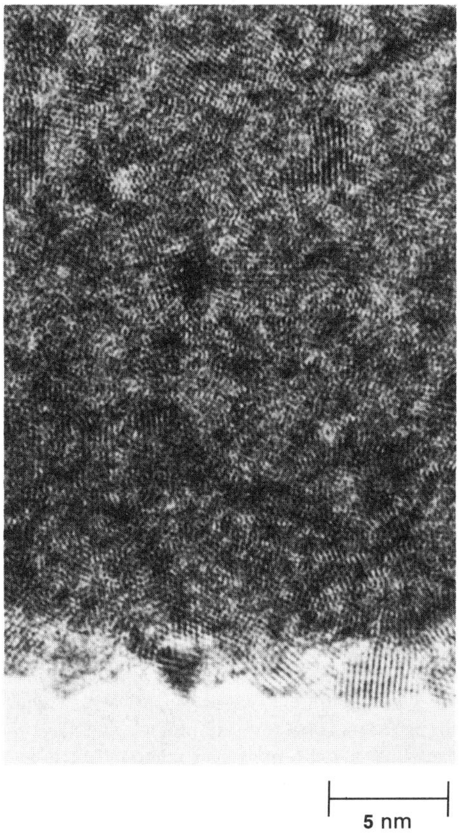

Fig. 11 High-resolution transmission electron micrograph of a thin-film sample of amorphous $Pd_{75}Si_{25}$. Notice the textured appearance on the scale of ~2 nm, as well as the apparent lattice fringes on the same scale.

seen to extend over distances of the order of 2 nm. It is tempting to interpret such features as microcrystallines. Nevertheless, these zones extend over distances of only a few times the lattice constants of the corresponding crystal structures. As such, the zones apparently do not represent a thermodynamically stable crystalline nucleus. The regions do not grow spontaneously upon heating. This distinction is important. If the zones are smaller than the dimension of a critical crystalline nucleus, then they will not spontaneously coarsen on heating. In a true microcrystalline alloy, spontaneous coarsening of the grain size would occur upon heating.

Spaepen (Ref 27) has used this idea to develop kinetic criteria for distinguishing a glass from a microcrystalline alloy. The nucleation and growth of a crystallite in a liquid or glass is described by nucleation theory. On the other hand, grain coarsening is described by a different kinetic model (Ref 28). By observing the crystallization of a metallic glass during heating at a constant rate or by isothermal annealing, it should be possible to distinguish nucleation and growth of crystals from coarsening of microcrystals. In most actual experiments of this type, the results favor the nucleation and growth kinetics. The microcrystalline-like zones seen in Fig. 11 do not therefore appear to constitute thermodynamically stable crystalline nuclei.

The formation of a stable crystalline nucleus is a thermally activated process that involves surmounting a nucleation barrier. As a consequence, the glass is said to be metastable and distinct from the crystalline state. This type of thermodynamic criterion would seem to provide a clear means of distinguishing glassy structures from microcrystalline structures. The argument, however, presupposes that a nucleation barrier for crystallization is an essential property of amorphous materials.

From Eq 2, it can be seen that the existence of a nucleation barrier is a consequence of the finite interfacial energy (σ) of the glass/crystal interface. Because melting (and crystallization) are normally assumed to be first-order phase transitions, this would seem to follow. On the other hand, if melting were to be a second-order or continuous phase transition, then a finite interfacial energy might not be required (Ref 17). Under such circumstances, the glass might crystallize without a nucleation barrier and the Spaepen criteria might fail to distinguish a glass from a microcrystalline alloy.

Finally, it should be mentioned that extended atomic ordering in metallic glasses need not necessarily be related to crystallographic ordering. Recent computer studies of undercooled liquids have suggested that icosahedral short-range order may develop in an undercooled liquid metal (Ref 29, 30). Such work suggests that long-range orientational ordering with icosahedral symmetry may even develop without translational symmetry. Such ordering is similar to that observed in the recently discovered quasicrystals (Ref 31). It is not consistent with ordinary translational symmetry and therefore is unlike the order found in any real crystal. As such, the atomic ordering seen in metallic glasses at the nanometer scale need not be of the same type as that observed in the corresponding equilibrium crystalline phases.

Dependence of Stucture on Synthesis Method and Thermal History. In the section "Synthesis and Processing Methods," a variety of methods for preparing amorphous metals were discussed. As mentioned, it is by no means obvious that the structure of a glass prepared by undercooling a liquid is the same as that of a sputter-deposited film or electrodeposited layer of similar composition. This raises a natural question: Does the amorphous phase have a unique structure in the sense that crystals have a unique structure? In answer to this question, it should be noted that if an attempt is made to measure the entropy of a glass by measuring the heat capacity curve of a liquid in the undercooled regime and then measuring the heat capacity of the glass below T_g, an

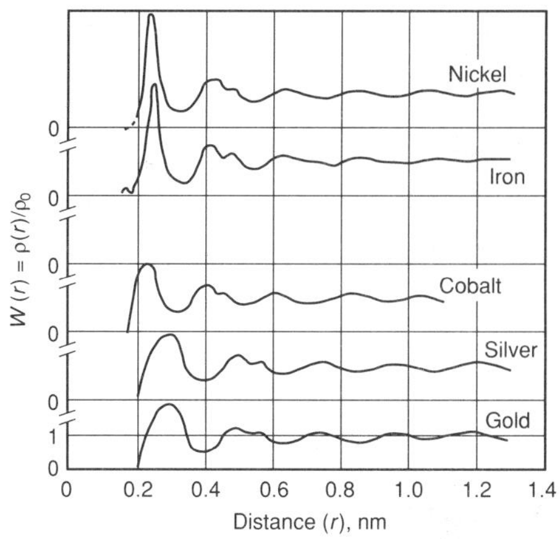

Fig. 12 The pair distribution function (W), of several pure amorphous metals prepared by vapor deposition onto a cryogenically cooled substrate. Notice the similarity with the pair correlation function of Bernal shown in Fig. 10. Source: Ref 32, 33

apparent residual entropy for the glass at $T = 0$ K will be found. This would suggest that a glass can have many equivalent configurations. It would follow that there is no unique glass structure. It might then be asked whether various glass structures are similar. Does the structure depend on the method of synthesis or the thermal history of the glass?

It is found experimentally that the pair correlation functions of alloys prepared by different methods exhibit the same qualitative features. For example, Fig. 12 shows the pair distribution function of several vapor-quenched amorphous pure metal films. The films were quenched onto cryogenically cooled substrates (Ref 32). A comparison with Fig. 10 reveals that these vapor-quenched pure amorphous metal films exhibit all of the essential features of the DRPHS model, the same model that works so well in describing the structure of both liquid metals and amorphous metal-metalloid alloys (Ref 24).

Comparative studies of sputtered, vapor-quenched, electrodeposited, and liquid-quenched alloys (Ref 33) have generally shown that alloys of the same composition prepared by different methods have similar pair correlation functions. It would seem that the short-range atomic ordering of amorphous alloys is at least qualitatively well defined and rather independent of the method of preparation of the sample. On the other hand, there do seem to be systematic differences. For example, the maxima and minima in the pair correlation functions appear to be somewhat less pronounced in alloys prepared by vapor quenching than those of alloys prepared by liquid quenching. This suggests that sputtered or vapor-quenched alloys are somewhat more disordered. The density of vapor-quenched alloys is often found to be somewhat less (~1 to 3%) than that of the corresponding liquid-quenched alloys. All of these facts suggest subtle differences in various amorphous structures.

In addition to structural differences related to synthesis, it is also found that the pair correlation function of a single amorphous alloy sample changes on thermal annealing at temperatures below T_g. This structural relaxation phenomenon in amorphous alloys is of importance and has been studied in detail for liquid-quenched metallic glasses by numerous investigators (Ref 34). Structural changes on annealing have been found to be both irreversible and reversible. A sample quenched from the liquid state by rapid solidification exhibits subtle but irreversible changes in structure on reheating to temperatures below T_g. These structural changes have been found to be accompanied by a slight densification of the order of a few tenths of 1%. When the annealing is repeatedly and alternately carried out at two different temperatures below T_g, smaller but reversible structural change can also be observed. The effect of these reversible and irreversible structural changes on the physical properties of metallic glasses is discussed in more detail in the following sections.

Finally, it should be pointed out that when thin films and sputtered samples are annealed below T_g, the structural changes that occur tend to be in the direction of achieving a structure more similar to the liquid-quenched structure. The same films tend to densify on annealing, achieving an ultimate density closer to that of liquid-quenched samples. We could summarize these observations by saying that amorphous samples prepared by different methods have a tendency to relax toward a common structure upon annealing at temperatures near T_g.

Thermodynamic Properties

The Glass Transition and Crystallization. As mentioned previously, the glass transition is a phenomenon that occurs when a liquid is undercooled and undergoes configurational freezing to a solid. This is accompanied by a dramatic increase in viscosity and a dramatic drop in atomic mobility (Fig. 1a). The heat capacity of the liquid is observed experimentally to exhibit a maximum (Fig. 1b). Below this maximum, the glassy solid exhibits a falling heat capacity much like that of an ordinary crystalline solid.

In general, the ease with which crystallization occurs in metallic melts makes direct observation of the glass transition rather difficult. Chen and Turnbull were the first to directly observe the glass transition in liquid gold-silicon glass-forming alloys (Ref 35). They pointed out the importance of this observation in establishing that the liquid-quenched alloys of Duwez were indeed like conventional glasses.

Table 1 Glass transition temperature (T_g) and crystallization temperature (T_x) of metal-metal and metal-metalloid metallic glasses

Alloy	T_g, K	T_x, K
$Au_{81}Si_{19}$	292	320
$Au_{55}Pb_{22.5}Sb_{22.5}$(a)	312.9	337.3
$Ni_{80}P_{20}$	622	640
$Pd_{80}P_{20}$	610	630
$Pd_{77.5}Cu_{6.0}Si_{16.5}$(b)	645	...
$Pd_{77.5}Cu_{6.0}Si_{16.5}$(c)	666	...
$Pd_{77.5}Cu_{6.0}Si_{16.5}$(d)	690	...
$Fe_{80}P_{13}C_7$	705	730
$Fe_{80}B_{20}$ (Metglas 2605)	>713	713
$Zr_{50}Cu_{50}$	707	755.5
$Zr_{35}Cu_{65}$	781	815
$Zr_{72}Ni_{28}$	642	671
$Zr_{60}Ni_{40}$	713	751
$Zr_{36}Ni_{64}$	834	864
$Ta_{80}Si_{10}B_{10}$	...	1225
$(Ta_{0.3}W_{0.7})Si_{10}B_{10}$	...	1450
$W_{40}Re_{40}B_{20}$	...	1300

Note: Unless otherwise indicated, heating rate was 10 to 30 K/s. (a) Heating rate was 20 K/s. (b) Heating rate was 10 K/s. (c) Heating rate was 10^3 K/s. (d) Heating rate was 10^5 K/s. Data compiled from a large number of sources

The ability to observe the glass transition upon heating of metallic glasses generally depends upon the heating rate employed. The necessity of avoiding nucleation and growth of crystals upon heating generally requires that a rather high heating rate be used. The experimentally observed glass-transition temperature T_g (as indicated, for example, by the heat capacity maximum) is itself found to depend slightly on the heating rate. Table 1 lists the glass-transition temperatures of several common metallic glasses. The values are typically obtained by locating the maximum in the heat capacity during heating in a differential scanning calorimeter. At higher heating rates, T_g is observed to be slightly displaced (by a few degrees) to higher temperatures. Because the glass transition is a kinetic phenomenon, this is not surprising.

The values listed in Table 1 correspond to typical heating rates (10 to 30 $K \cdot s^{-1}$) employed in differential scanning calorimetry. It has been argued that an ideal glass transition would be observed in the limit where the heating rate vanishes. This has led to speculation regarding the possible existence of an underlying thermodynamic phase transition. Theories of the ideal glass transition have been developed. The earliest of these is the Gibbs Dimarzio theory (Ref 36), which was based on the concept of the disappearance of free volume with lowering temperature in the undercooled liquid. Cohen and Turnbull have further developed this theory (Ref 37).

More recently, Cohen and Grest (Ref 38) have developed a theory that combines the free-volume concept with percolation theory. In this rather sophisticated theory, the glass transition corresponds to the percolation of liquidlike atomic cells throughout the solid structure. The liquidlike cells consist of atoms with a shell of nearest neighbors that defines a local volume exceeding some critical volume. When these cells percolate throughout the structure, the solid develops fluidity. Because theories of the glass transition refer to a transition observed in the limit of very slow cooling rates, they cannot in general be compared with experimental data.

Generally speaking, the viscosity, atomic diffusion constant (see below), and other phenomena that involve atomic rearrangements all exhibit the same temperature dependence in the vicinity of the glass transition. This is not surprising because the same types of atomic jumps are required for all of these processes. In particular, the nucleation and growth of crystals in the glass are controlled by the same types of atomic jumps. Near T_g, the rates of these atomic jumps increase precipitously (see Fig. 1a). In metallic glasses, crystallization rates tend to become significant at temperatures in the vicinity of T_g. Often, a crystallization temperature (T_x) is defined for metallic glasses. In reality, the crystallization temperature depends on the time scale of an experiment and cannot be precisely defined. T_x is generally higher when experiments are carried out at high heating rates (or short times) than when experiments are carried out at low heating rates (long times). In any case, T_x is nevertheless fairly well defined when practical time scales are considered. The rate of nucleation of crystals in a metallic glass is described by the same classical nucleation theory used to describe nucleation of crystals in the undercooled melt (see Fig. 2). It clearly is a very rapidly increasing function of temperature in the vicinity of T_g. Table 1 contains a list of observed crystallization temperatures for several common metallic glasses of both the metal-metal and metal-metalloid type. Notice that only in cases where $T_x > T_g$ is it possible to measure both. For the metallic glasses in Table 1, T_x is generally rather close to T_g. This reflects the fact that when atomic mobility becomes appreciable (near T_g), crystallization occurs with relative ease.

Heat Capacity and Two-Level Systems. The specific heat at constant pressure (c_p) of metallic glasses has many features in common with that of crystalline metals. Both electronic and lattice contributions are observed. The heat capacity vanishes at very low temperatures, increases over an intermediate temperature range, and saturates at a value near the Dulong-Petit value (3R/mol where R is the gas constant) at higher temperatures (below but near T_g). Near T_g, an anomalous increase is observed as previously discussed. As in crystalline metals, the electronic contribution to the heat capacity is linear with temperature:

$$c_p^e = \gamma T \quad \text{(Eq 9)}$$

and can be used to determine the electronic density of states at the Fermi level (Ref 39). At low temperatures, it can be clearly separated from a lattice heat capacity of the form:

$$c_p^l = \beta T^3 \quad \text{(Eq 10)}$$

which can be associated with the usual Debye contribution. Measurement of the coefficient β can, as in crystalline materials, be used to determine a Debye temperature for the amorphous phase (Ref 40). In general, the Debye temperature (θ_D) is found to be somewhat lower in amorphous materials than in the corresponding crystalline state. This is thought to reflect a decrease in the shear sound velocity in glasses compared with crystals. Direct sound velocity measurements in metallic glasses (Ref 41) show a reduction of about 15% in the shear sound velocity of the glass compared with that of the crystal at the same composition.

Prior to 1972, this was thought to be a rather complete picture of the thermal excitations in metallic glasses. At this time, Anderson *et al.* (Ref 42) and Phillips (Ref 43) independently proposed a model to explain an anomalous low-temperature contribution to the heat capacity of nonmetallic glasses. In insulating glasses, this small anomalous contribution takes a linear form:

$$c_p^a = \gamma^a T \quad \text{(Eq 11)}$$

where the superscript "a" stands for anomalous. Because insulating glasses have no conduction electrons, this contribution to the heat capacity cannot be electronic in origin. Anderson and Phillips proposed that it arises from atomic configurations called two-level systems. They proposed that these two-level systems would contribute an excess linear contribution to the lattice heat capacity at low temperatures.

In fact, experiments show that γ^a is rather small. It is typically between one and two orders of magnitude smaller than the γ that arises from electronic excitations in metals. In the case of amorphous metals, this anomalous contribution is obscured by the electronic heat capacity. As such, it was originally thought that this contribution could not be measured in amorphous metals.

This problem was overcome by Graebner (Ref 44), who recognized that in a superconducting metallic glass far below T_c (the superconducting transition temperature) the ordinary electronic heat capacity is exponentially reduced. As such, he was able to demonstrate an anomalous and linear nonelectronic contribution arising from two-level systems. It is now widely recognized that the anomalous linear heat capacity at low temperatures arising from two-level systems is a general feature of glasses that distinguishes them from crystalline solids. Two-level systems also influence the low-temperature behavior of the thermal conductivity and the attenuation of sound at low temperatures. These effects are characteristic of amorphous metals (Ref 45).

Thermal Transport. The thermal conductivity of metallic glasses at room temperature is generally of the same order of magnitude as crystalline alloys. It is typically dominated by electronic transport. The high degree of atomic disorder results in substantial electron scattering and a short electron mean free path. As such, the magnitude of the thermal conductivity at ambient temperatures tends to be similar to that observed in very disordered crystalline alloys. On the other hand, at cryogenic temperatures the thermal transport properties of metallic glasses tend to exhibit unusual features that are associated with the aforementioned two-level systems (Ref 45). The two-level excitations scatter long wavelength phonons, leading to a characteristic T^2 dependence of the thermal conductivity at very low temperatures (<1 K).

As in the case of the heat capacity, this lattice contribution to the thermal conductivity can be isolated from the electronic contribution by using a superconducting amorphous metal at temperatures well below the superconducting transition temperature (Ref 45).

Atomic Diffusion. Below the glass transition temperature, atomic diffusion constants of metallic glasses are presumably small because the atomic diffusion that is characteristic of a liquid metal is rapidly frozen out as the glass transition sets in. Several investigators have studied atomic diffusion in metallic glasses at temperatures below T_g (Ref 46, 47). In general, it has been found that the diffusion constant *D* follows an Arrhenius law:

$$D = D_0 e^{-Q/T} \quad \text{(Eq 12)}$$

over the limited ranges of temperatures investigated. Values of activation energy *Q* are found to be similar to those found for self-diffusion in crystalline metals. The values of *D* tend to range from 10^{-14} to 10^{-20} $cm^2 \cdot s^{-1}$ over temperatures ranging from T_g down to temperatures at which *D* is not measurable by available experimental techniques (typically ~100 °C, or 180 °F, below T_g).

Deviations from Arrhenius behavior have been observed in a few experiments carried out over larger temperature ranges. This is thought to indicate that the atomic jumps observed in diffusion in glasses involve a distribution of activation energies (a distribution of *Q*) rather than a single, well-defined activation energy, as would be characteristic of vacancy jumping in crystals. Further, the activation energy to create a vacancylike defect in a glass would vary from one atomic site to another. As such, the apparent activation energy would be expected to decrease with decreasing temperature, leading to deviations from a simple Arrhenius law. Attempts to develop a theory of atomic self-diffusion in metallic glasses have been based on the concepts of vacancylike defects (Ref 46). In the case of impurity diffusion by small atoms like hydrogen, the diffusion is believed to proceed by an interstitial mechanism in much the same manner as the interstitial diffusion of hydrogen carbon, nitrogen, and so on, in crystalline metals (Ref 48).

Mechanical Properties

Several good reviews of the mechanical properties of metallic glasses can be found in the literature (see, for example, Ref 49 to 52). Here, only a brief description of some of the important features will be given.

Deformation Mechanisms: Homogeneous and Inhomogeneous Flow. Metallic glasses undergo deformation under an applied stress τ via two different mechanisms, homogeneous and inhomogeneous deformation. Homogeneous deformation occurs at low stresses (typically $\tau < \mu/100$, where μ is the yield stress of the material) and at relatively low temperatures compared to T_g. At sufficiently low stresses, the flow is essentially Newtonian viscous flow. At higher temperatures (near T_g) or higher stress levels ($\tau > \mu/50$), deformation occurs inhomogeneously through the formation of localized shear bands.

The overall behavior has been described by Spaepen and Taub (Ref 50) using an Ashby deformation map (Fig. 13). As can be seen in Fig. 13, the mode of deformation also depends on the rate of deformation. Higher strain rates are associated with inhomogeneous deformation. At sufficiently high strain rates, deformation becomes inhomogeneous at all temperatures below T_g.

Inhomogeneous deformation occurs in localized shear bands. The individual shear bands are typically 20 to 40 nm in dimension. Total plastic strains of the order of 10 or more may occur within an individual shear band. The overall plastic strain of the specimen depends on the density of shear bands. The deformation mechanism within an individual shear band is believed to be fluidlike. The transition to fluid behavior within the band is believed to be triggered by the stress-driven creation of free volume, adiabatic heating, or some combination of the two.

Yield Strength, Hardness, and Elastic Constants. The elastic constants of metallic glasses are quite similar to those observed in the corresponding crystalline materials. The bulk modulus is typically reduced by a few percent, whereas the shear modulus is typically 15% smaller than in the corresponding crystalline material. Young's modulus is typically somewhat smaller than that of the corresponding crystalline material.

By contrast, values of yield strength (σ_y) and hardness (H) are typically much greater than those of ductile crystalline solids. In fact, the yield strengths of metallic glasses compared with their elastic constants place them among the strongest of known solids.

A typical measure of the relative strength of a solid is the ratio of yield strength to

Young's modulus (σ_y/E). This ratio is typically 0.02 for metallic glasses, which yield plastically in a tensile test. By comparison, iron whiskers have a value of 0.05, and aluminum nitride (AlN) fibers exhibit a value of 0.02. These values approach the theoretical maximum-attainable yield strength. The hardness of metallic glasses corresponds to the yield strength :

Glass composition	Vickers hardness, kg/mm^2	Yield strength, MPa (ksi)	Young's modulus, GPa (psi × 10^6)
$Pd_{77.5}Cu_6Si_{16.5}$	500	1570 (227)	88.2 (12.8)
$Fe_{80}B_{20}$	1100	3626 (526)	166.6 (24.2)
$W_{40}Re_{40}B_{20}$.......	2400	>7840 (>1136)	...

Generally, it is found that $H/\sigma_y \sim 3$ for metallic glasses. This suggests that refractory glasses such as $W_{40}Re_{40}B_{20}$ should have yield strengths of the order of >7840 MPa (1136 ksi). Unfortunately, because metallic glasses are very susceptible to brittle fracture, their practical applications are limited.

Failure, Fracture Toughness, and Embrittlement. Metallic glass ribbons that are free of imperfections fail in tension in an essentially plastic manner. Failure tends to coincide with yielding, and it typically occurs along a plane oriented at 45 ° to the axis along which the stress is applied. This is characteristic of a ductile failure mechanism. The coincidence of failure with yielding shows that metallic glasses do not work harden. When the fracture surface of a failed ribbon is examined, a characteristic veinlike pattern is typically observed (Ref 50). This pattern is thought to be characteristic of a fluidlike instability. A similar pattern is produced when a layer of viscous fluid is sandwiched between two solid surfaces that are subsequently pulled apart. Taub and Spaepen (Ref 50), for example, have discussed the nature of this fluidlike instability and compared it with those originally described by Saffman and Taylor (Ref 53). This failure mechanism is an essentially plastic mechanism. The failure occurs by plastic deformation.

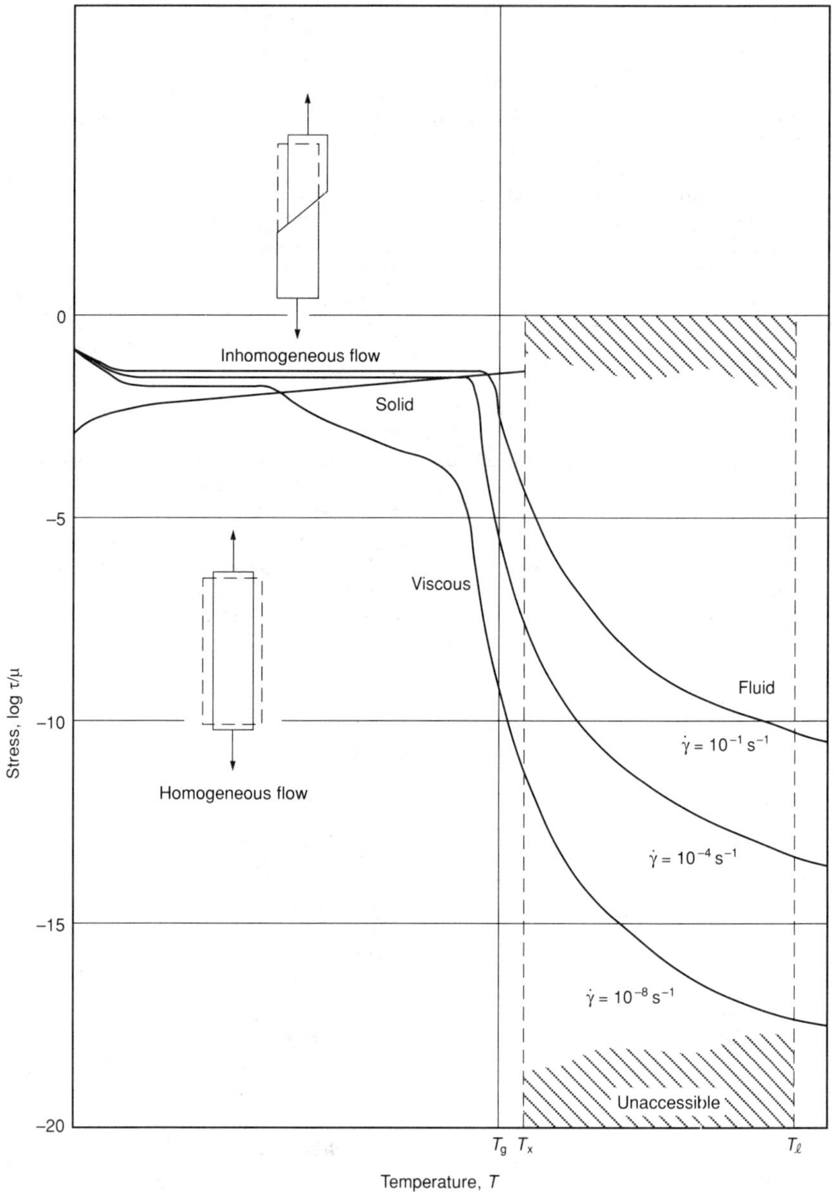

Fig. 13 Ashby flow map for a metallic glass showing the temperatures and stress levels (applied stress τ versus yield stress μ) that result in homogeneous/inhomogeneous flow. The strain rate is denoted by $\dot{\gamma}$; the liquidus temperature is T_ℓ.

Imperfections in real metallic glasses often cause failure to occur by brittle fracture (Ref 50, 52). Brittle fracture occurs when the tensile stress at a stress concentrator, such as a microcrack, reaches the theoretical fracture stress of the material (σ_{th}) before shear-induced plastic flow can blunt the crack and relieve the stress. In metallic glasses containing defects such as crystallites (which act as stress concentrators), brittle fracture is frequently observed to occur. At very low temperatures, where the viscosity is very high, brittle failure is also a more prominent mechanism of failure.

Many metallic glasses also exhibit an annealing embrittlement effect when subjected to annealing temperatures below T_g and below that which would be required to initiate crystallization. This has been attributed to various mechanisms by different authors. Densification and loss of free volume upon such annealing causes an accompanying rise in viscosity. This encourages embrittlement. Other authors have attributed the annealing embrittlement to microsegregation or clustering of metalloid elements (in metal-metalloid glasses). Metallic glasses containing high concentrations of metalloid elements are especially prone to annealing embrittlement (Ref 54). Finally, it is interesting that the embrittlement effect in metal-metalloid glasses can be reversed by neutron irradiation (Ref 55). In the same experiment, neutron irradiation was observed to result in lowered density of the glass. This suggests that the loss of free volume is related to the embrittlement.

Metallic glasses generally exhibit rather low fracture toughness. This is presumably related to their high strength (Ref 52). It is also clear that samples containing imperfections such as crystallites exhibit lower values of fracture toughness than samples free of such imperfections. For defect-free samples, the fracture toughness of iron-base metallic glasses has been found to be consistent with that observed in steels when the higher yield strength of metallic glasses is taken into account. Figure 14 compares the plane-strain fracture toughness values of an 18% Ni maraging steel and an AISI 4340 alloy steel as a function of yield strength with fracture toughness values for two commercial iron-base metallic glasses produced in ribbon form.

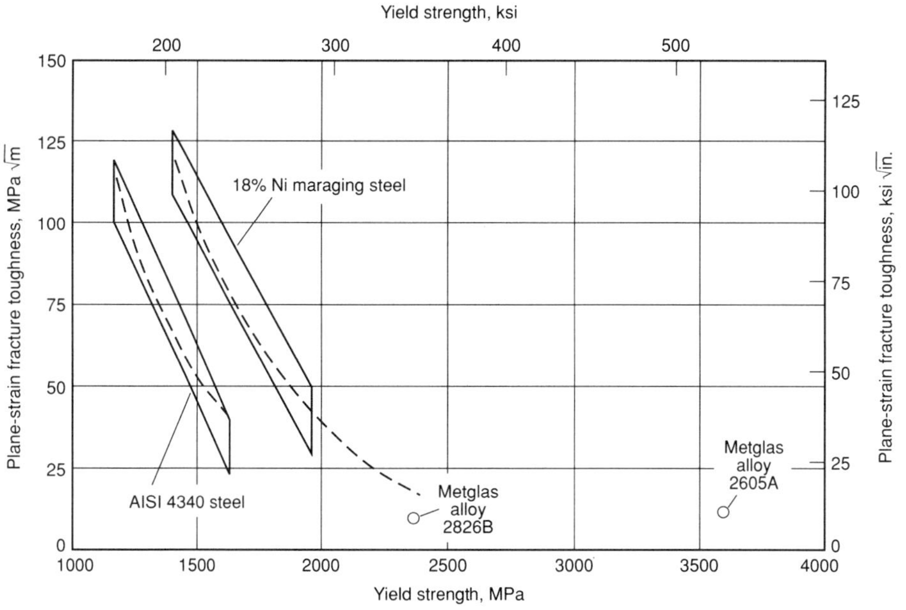

Fig. 14 The plane-strain fracture toughness of two ferrous metallic glasses compared with that of two steels. The lower fracture toughness of the metallic glasses is consistent with their higher yield strength. Source: Ref 52

Electronic and Magnetic Properties

Electrical Transport Properties. Metallic glasses exhibit electrical transport properties that are characteristic of metals (Ref 56, 57). The electrical resistivity (ρ) at ambient temperature ranges from about 50 μΩ · cm to about 250 μΩ · cm. In contrast to crystalline metals and alloys, where ρ decreases rapidly with decreasing temperature, ρ varies little with temperature for metallic glasses. Furthermore, ρ increases with temperature in some metallic glasses and decreases with temperature in others. This behavior is generally quite similar to that found in the corresponding liquid alloy. Extrapolation of metallic glass resistivities to higher temperatures is in fact found to join rather smoothly to the resistivity curve of the liquid alloy. The resistivity values for a number of metallic glasses are shown in Fig. 15. The temperature dependence of the resistivity can be generally characterized by a temperature coefficient of resistivity (α), defined as $\alpha = \rho^{-1}(d\rho/dT)$, where ρ and $d\rho/dT$ are conventionally measured at ambient temperature. This coefficient takes on values ranging from -2×10^{-4} to 2×10^{-4}.

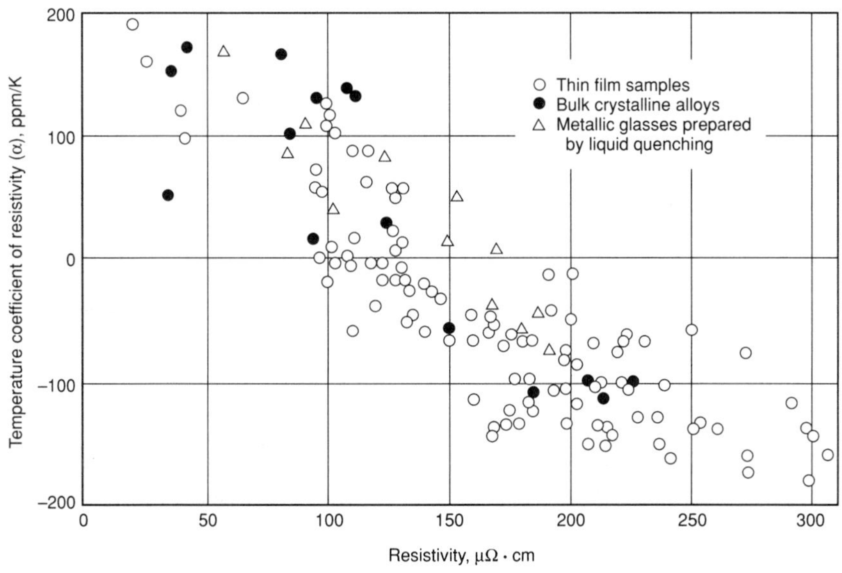

Fig. 15 The Mooij correlation relating the temperature coefficient of resistivity (α) to the absolute value of resistivity at ambient temperature for a large number of metallic samples. Metallic glasses with absolute resistivity less than ~150 μΩ · cm have positive values of α, whereas those with resistivity greater than this tend to have negative values of α.

An interesting correlation has been observed between α and the absolute value of ρ; it is referred to as the Mooij correlation (Ref 58). This correlation is also illustrated in Fig. 15. The Mooij correlation is in fact found to apply generally to all metallic solids and is related to the behavior of electrical resistivity in the limit where the electron mean free path approaches atomic distances. In metallic glasses, where atomic disorder leads to strong electron scattering, resistivities are typically high and the change in the sign of α is common.

Theories of electrical conductivity in metallic glasses are basically an extension of earlier theories developed to describe electron scattering in liquid metals. The reader is referred to the above-mentioned review articles for more details.

Other electronic transport properties of metallic glasses that have been studied in detail are the Hall coefficient, thermopower, and magnetoresistivity. The Hall coefficient is found to have both positive and negative signs in metallic glasses. As in the case of crystalline metals, this has been interpreted to indicate electronlike and holelike conduction mechanisms.

The thermopower of metallic glasses is essentially linear with temperature, although low-temperature anomalies have been observed that are related to electron-photon scattering and to electron localization effects. Magnetoresistivity is generally positive (resistivity increases with the application of a magnetic field).

Magnetic Properties. Many metallic glasses contain atoms that carry a magnetic moment. Metal-metalloid glasses containing iron, nickel, and cobalt are examples. In the early 1960s, shortly after the discovery of metallic glasses by Duwez, the question naturally arose as to whether such materials could undergo a transition to the ferromagnetic state as temperature was lowered. At the time, it was not clear whether the development of ferromagnetic order of atomic spins in solids required an underlying crystalline lattice. The disordered atomic structure of metallic glasses implies that the exchange interactions between neighboring spins should vary according to the local atomic environment. It is not clear whether such disorder would suppress ferromagnetism. The answer to this question came in the course of studies of the glass $Fe_{78}P_{12}C_{10}$ by the Duwez group (Ref 59). Magnetic measurements on this glass clearly revealed that it undergoes a ferromagnetic transition at temperatures near 400 °C (750 °F) and develops a spontaneous magnetization. Since this initial discovery, ferromagnetism has been found in a large variety of metallic glasses (Ref 60-62).

The first systematic studies of the ferromagnetic properties of metallic glasses con-

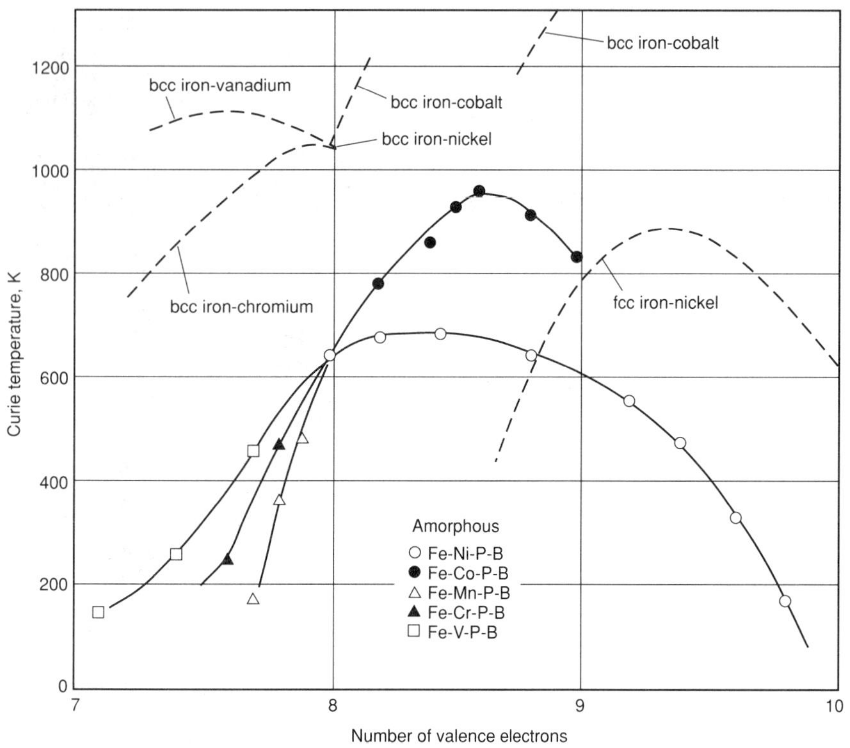

Fig. 16 Ferromagnetic Curie temperatures of several ferrous-group metallic glasses as a function of the total valence of the metallic component. All the alloys have fixed metalloid concentrations of 10 at.% P and 10 at.% B. Also shown are trends in the Curie temperature for related crystalline solid solutions of two transition metals. A similar trend in Curie temperatures is observed in corresponding crystalline solutions of the ferrous-group metals. Source: Ref 63

taining ferrous-group metals were carried out by Mizogouchi *et al.* (Ref 63). Figure 16 is taken from their work and shows the variation of the Curie temperature of several series of ferrous-group metallic glasses of overall composition $TM_{80}P_{10}B_{10}$, where TM stands for a ferrous transition metal (chromium, manganese, iron, cobalt, or nickel) or a combination of two such metals. The horizontal axis in the figure represents the average number of valence electrons for the transition metal component (for example, 8 for Fe, 9 for Co, and so on). It can be seen, for example, that iron-cobalt-base glasses have the highest Curie temperatures and that the Curie temperature of the glasses is generally somewhat lower than that of crystalline transition metals or solid solutions of transition metals that contain no metalloids. This depression in Curie temperature can at least in part be explained by the dilution effect of the metalloid elements. The atomic disorder in metallic glasses may also play a role. In general, there exists no first-principles theory of the Curie temperature.

The saturation magnetic moment of ferromagnetic metallic glasses varies in a systematic way with the valency of the transition metal component, reaching a maximum of about 2 Bohr magnetons per transition metal atom in iron-cobalt-base metal-metalloid glasses. Again, these saturation magnetizations are somewhat lower than those of corresponding crystalline alloys even when the dilution effect of metalloid elements is taken into account. This reduction in saturation magnetization again seems to be associated at least in part with the disorder in the local atomic environments in metallic glasses. Variations of the saturation moment with metalloid element concentration and type have also been studied. It is believed that electron charge transfer effects between transition metal and metalloid atoms can account for many of these systematic variations.

The ferrous-group metal-metalloid elements have been found to exhibit a very low intrinsic magnetic anisotropy. The intrinsic anisotropy is related to the local atomic structure, and unlike crystal field anisotropies, it is random in direction. This random atomic scale anisotropy varies on a scale of about 1 nm, and in the ferrous-group metal-metalloid glasses its magnitude is exceedingly small. On scales larger than this, these glassy magnets are intrinsically very homogeneous. This leads to a very low intrinsic coercive force (H_c). In the absence of surface defects (which couple to magnetic domain walls via the demagnetizing field), internal stresses (which lead to magnetoelastic coupling of the magnetization to the stress field), second-phase precipitates (crystallites), or induced anisotropies, the coercive force of these amorphous magnets is among the smallest found in any magnetic material. For ferrous-group metallic glasses, typical values of H_c in high-quality stress-free ribbons are often found to lie in the range of a few millioersteds (mOe). This property has made metallic glasses very attractive for applications requiring a soft magnetic material.

When a soft magnetic material is subjected to an ac magnetic field, the energy dissipated by the induced changes in the magnetization is related not only to the coercive force but also to eddy current loss arising from the changing magnetization within the sample. Eddy current loss is proportional to the conductivity of the material. Because amorphous metals have relatively low electrical conductivity, these ac losses are expected to be smaller than those in corresponding crystalline materials. The combination of low coercive force and low conductivity is ideal for applications in ac transformers (see the section "Technology and Applications" in this article).

Amorphous alloys typically exhibit magnetostriction effects comparable in magnitude to those observed in similar crystalline materials. These magnetoelastic effects are related to the coupling of stress fields in the material to the magnetization. Internal stress fields raise the coercive force of the material. In melt-quenched ribbons, quenched-in internal stresses couple to the magnetization. These stresses can typically be eliminated by an appropriate annealing treatment (Ref 61), resulting in restoration of the intrinsic coercive force. Annealing in a magnetic field or cold working of the material are both found to produce induced magnetic anisotropies. Using these techniques, special features (such as a large remanance magnetization) can be produced.

Amorphous Superconductors. Superconductivity in amorphous metals was first reported by Buckel and Hilsh in their work on thin films of simple metals (for example, tin, lead, and bismuth) quenched onto cryogenically cooled substrates (Ref 9). They found, for example, that the semimetal bismuth became a superconducting metal when quenched onto a liquid-helium-cooled substrate to form an amorphous phase. The critical superconducting transition temperature (T_c) of amorphous metallic bismuth was reported to be 6 K. These amorphous bismuth films were observed to crystallize on heating to ~20 K. This absence of thermal stability against crystallization made further study very difficult. Later studies by Collver and Hammond (Ref 11) extended these results to cryoquenched films of transition metals. They found, for example, that amorphous molybdenum films became superconducting at nearly 7 K, whereas the superconducting transition temperature of crystalline molybdenum was less than 1 K.

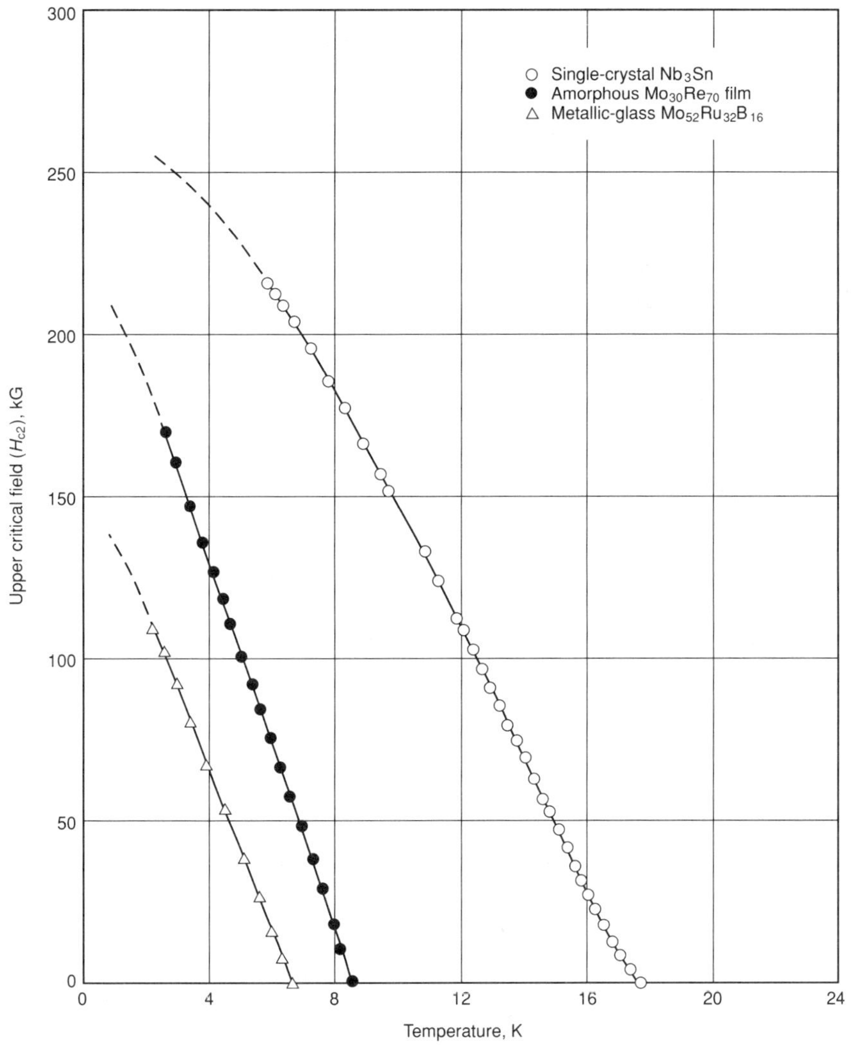

Fig. 17 The upper critical field (H_{c2}) as a function of temperature for the amorphous superconductors $Mo_{52}Ru_{32}B_{16}$ and $Mo_{30}Re_{70}$ compared with that of crystalline Nb_3Sn, a commercially used high-field superconductor

By contrast, amorphus niobium had transition temperatures of only about 6 K, whereas crystalline niobium had a transition temperature of 9.3 K. Both amorphous molybdenum and niobium films were found to crystallize near 40 to 50 K. Certain superconducting transition metal alloys (for example, $Mo_{50}Re_{50}$, with $T_c \sim 8$ K) were found to be stable against crystallization upon heating to room temperature.

Superconductivity in bulk metallic glasses prepared by rapid quenching methods was first reported by the author and colleagues (Ref 64) for the alloy $La_{80}Au_{20}$ with T_c of 3.6 K. Later work showed that a variety of metallic glasses exhibit superconductivity. Metal-metalloid glasses based on molybdenum, ruthenium, rhenium, and niobium were found to exhibit superconductivity. One well-studied series of alloys, $Mo_{80-x}Ru_xB_{20}$ (with $20 < x < 60$), exhibits T_c ranging from 5 K to 7 K. The superconducting properties of these metallic glasses were studied in detail.

Amorphous superconductors have many special features that arise from the high degree of atomic disorder present in the amorphous phase (Ref 65, 66). As mentioned earlier, this disorder leads to strong electron scattering and a short electron mean free path (of the order of interatomic distances). For superconductors, this leads to a rather small superconducting coherence length (ξ). For typical amorphous superconductors, ξ ranges from about 4 nm to about 10 nm. By contrast, crystalline metals have coherence lengths of about 50 to 100 nm.

On the other hand, the London penetration depth (λ) for amorphous superconductors is typically quite large. The short coherence length and large penetration depth influence many of the properties of amorphous superconductors. For example, amorphous superconductors are extreme type II superconducting materials with very small lower critical fields (H_{c1}) and very large upper critical fields (H_{c2}).

Superconductivity in the mixed state persists to very high magnetic fields. The upper critical fields of the superconducting metallic glass $Mo_{52}Ru_{32}B_{16}$ and an amorphous $Mo_{30}Re_{70}$ thin film are shown in Fig. 17, where it is compared with the high-field crystalline superconductor Nb_3Sn. The latter material is used to construct high-field superconducting magnets. Even though they possess a lower transition temperature, the amorphous superconductors have a comparable critical field at low temperatures.

Crystalline inclusions and other inhomogeneities in amorphous superconductors result in pinning of magnetic flux vortices in the mixed state (between H_{c1} and H_{c2}). As a consequence, such inclusions can enhance the critical current density (J_c) of the superconductor in the presence of an applied magnetic field. Using this phenomenon, amorphous superconductors can be engineered to carry current densities of the order of 10^5 A/cm^2 (6.4×10^5 A/in.^{2}) in magnetic fields of the order of 50 to 100 kOe. These parameters have prompted consideration of metallic glass superconducting ribbons as high-field superconducting materials. The relatively modest transition temperatures of amorphous superconductors (<9 K) have thus far prevented their application. More detailed information on superconductivity can be found in the Section "Superconducting Materials" in this Volume.

Chemical Properties

As previously discussed, metallic glasses and other amorphous metals have an atomic scale structure similar to that of liquid metals. This structure has been described earlier in terms of continuous random packings. The continuous nature of the structure implies the absence of spatially extended defects such as grain boundaries or dislocations, which commonly occur in crystalline materials. Computer simulation studies in which dislocationlike defects were introduced into a model metallic glass structure showed that these defects undergo spontaneous relaxation and are subsequently difficult to recognize (Ref 67). In crystalline or polycrystalline metals, such extended defects oftentimes serve as chemically distinct sites at which surface reactions such as oxidation preferentially occur. The absence of well-defined extended defects in amorphous metals results in a chemically more uniform surface, which in turn leads to interesting chemical properties. For example, certain metallic glasses have been found to have excellent corrosion resistance in a variety of chemically hostile environments.

Ferrous metal-metalloid glasses containing modest amounts of chromium and molybdenum (for example, $Fe_{72}Cr_8P_{13}C_7$ and $Fe_{75}Mo_{15}P_{13}C_7$) have been found to undergo spontaneous passivation and exhibit high resistance to corrosion in strongly corrod-

Fig. 18 Various parts that use metallic glasses. Most prominently featured are two spools of as-cast amorphous alloy for high-frequency and antitheft applications. Also shown (right side) are four wound magnetic cores made from amorphous alloy ribbon. Several high-frequency epoxy-encapsulated small cores are shown in the foreground. Such cores are used in switch-mode power supplies, magnetic amplifiers, chokes, and other electronic devices to be incorporated in electronic assemblies such as the one shown. At the lower right are bar code markers containing a strip of metallic glass. In the background is a woven blanket of metallic glass used in magnetic shielding. Courtesy of H.H. Liebermann, Allied-Signal Inc.

ing solutions such as 12 *M* hydrochloric acid (Ref 68). Such alloys are found to have corrosion resistance significantly better than that of stainless steels containing similar amounts of chromium. This is attributed to the chemical and structural homogeneity of the metallic glass on the atomic scale.

Despite the absence of extended defects, amorphous metals do possess structural fluctuations on the scale of atomic distances. Each local atomic environment is distinct and different. As such, the local electronic orbital configurations of surface atoms tend to vary over the surface, and the surface exhibits a distribution of chemically active sites. This feature leads to interesting catalytic properties. For example, amorphous iron-nickel-metalloid alloys have been used for catalytic synthesis of hydrocarbons by hydrogenation of carbon monoxide (Ref 69).

Technology and Applications

As will be described in this section, metallic glasses are being used in an increasingly wide variety of applications. Figure 18 shows a number of components that utilize amorphous metals.

Soft Magnetic Materials. The homogeneity, low coercive force, and relatively high permeability of ferrous amorphous alloys are attractive features in many magnetic applications. The innovation of the planar flow casting method for producing uniform sheets of metallic glass in widths ranging up to 1 m (40 in.), thicknesses of 30 to 50 μm, and essentially unlimited length has resulted in the commercialization of metallic glasses as a core material for power-distribution transformers. Distribution transformers are used by the utility companies in the final voltage step-down in the network that supplies power to residences, stores, offices, and small industries.

These transformers are put into service for a period of 25 to 40 years, and during that time the primary coil is continuously energized. The significance of this duty pattern lies in the fact that the core material in the transformer is cycled at a frequency of 50 or 60 Hz continuously during the

Table 2 Core losses and coil losses in a conventional silicon steel and several metallic glass power-distribution transformers

Type/company	Capacity, kVa	Core loss, W	Coil loss at full load, W
Conventional grain-oriented silicon steel	25	85	240
Alloy 2605S (Allied-Signal)	25	16	235
General Electric	25	18	330
Osaka Transformers	30	30	390
Takaoka Electrical	20	18.9	348

Source: Ref 70, 71

lifetime of the transformer. During each cycle, energy is lost to the magnetic core material. In addition, the alternating field produces eddy currents and accompanying losses within the transformer. Core losses are essentially determined by the coercive force, whereas the eddy current losses are proportional to the electrical conductivity of the material and inversely proportional to the thickness of the individual sheets of magnetic material from which the core is wound.

The low electrical conductivity of metallic glasses compared with conventional crystalline core materials (grain-oriented silicon steels) makes amorphous metals extremely attractive for reducing eddy current losses. The low coercive forces compared with competing crystalline materials reduce core loss. Taken together with the fact that the continuous planar flow casting process leads to low fabrication costs, these features have led to the introduction of metallic glasses into the commercial transformer core market (Ref 70).

The use of amorphous metals in this application can reduce energy losses, thereby reducing costs. For a single transformer of typical size (a power output of 25 kVA), a conventional iron-silicon transformer can be designed to 98.7% efficiency at full load but generate 85 W of core loss independent of the loading. Over a lifetime of 35 years, this loss amounts to 2.6×10^4 kWh. At a cost of $0.06/kWh, this amounts to over $1500. Table 2 illustrates one analysis of losses for several types of transformers. The reduction in core losses arising from the incorporation of metallic glass as the core material is substantial.

It should be noted that the coil losses in Table 2 refer to a fully loaded transformer. During periods of no load, these losses are essentially eliminated, whereas the core losses occur continuously and are weighted far more heavily in a total cost analysis. The innovation of metallic glass transformers has provided new impetus for the development of more efficient conventional iron-silicon transformers. A complex set of economic factors will ultimately determine how widespread the use of metallic glass transformers will become.

The use of metallic glasses as soft magnetic materials is increasing in a number of other applications. These include inductive components such as saturable reactors, inverter transformers, and chokes for magnetic switches (used in power supplies for pulsed high-power lasers). Other applications in the areas of magnetic sensors and transducers are also currently under development. Many of these applications exploit the low coercivity combined with induced magnetic anisotropy effects. Induced anisotropies (by magnetic annealing) allow tailoring to achieve saturation at very low fields and other characteristics. These features are important in many magnetic applications.

Brazing Materials. Metallic glasses essentially melt to a fluid at temperatures near T_g. In typical metal-metalloid glasses, compositions are chosen that lie near deep eutectic features in the equilibrium phase diagram. As such, the liquidus and solidus curves of the equilibrium alloys often plunge to relatively low temperatures at glass-forming compositions. For example, in iron-base and nickel-base metal-metalloid glasses, the equilibrium solidus curves generally lie in the temperature range of 700 to 1100 °C (1300 to 2000 °F) at the glass-forming compositions. On the other hand, the T_g of the glass typically lies at still lower temperatures, ranging from 300 to 500 °C (570 to 930 °F). The fact that these alloys melt at such low temperatures and that some conventional brazing alloys have glass-forming compositions has been exploited in their use as brazing materials. For example, ribbons and sheets of nickel-base metal-metalloid glasses having T_g near 300 °C (570 °F) are used for specialized low-temperature brazing. The availability of thin sheets that can be cut to a variety of shaped preforms to fit the joining surfaces has made this a successful commercial product for making brazed joints with excellent strength and integrity.

Coatings. The high yield strength and hardness of many metallic glasses and amorphous metals and the observed corrosion resistance of those containing chromium, molybdenum, and other transition elements suggest possible applications as protective wear and corrosion-resistant coatings. Some preliminary studies in this area have been encouraging. For example, sputter deposition of molybdenum-boron-ruthenium and molybdenum-iron-chromium-boron-phosphorus amorphous alloys onto steel substrates to thicknesses of only a few microns was found to lead to substantial improvement of both the wear and corrosion resistance of the steel (Ref 72). Losses in sliding wear under specified loading conditions were reduced by several orders of magnitude (compared with the uncoated surface), and corrosion resistance in various hostile chemical environments was also significantly improved. A key factor in corrosion resistance is the avoidance of microcracks and other defects in the amorphous coating, which act as sites for preferential chemical attack. With regard to wear resistance, it was also noticed in the above studies (Ref 72) that the sliding friction coefficient of the amorphous coating was exceedingly small. When compared with a crystalline coating of comparable hardness, the amorphous coating exhibits substantially reduced wear. The wear rate divided by the hardness at a particular load (during, for example, sliding wear) is a good indication of the intrinsic wear characteristics of a surface. Based on the limited data available, it appears that low sliding friction is associated with intrinsically lower wear rates for amorphous coatings when compared with crystalline coatings of comparable hardness.

Recently, it has been observed that certain glass-forming alloys in their equilibrium crystalline form, when subjected to sliding wear under sufficient load, undergo a crystal-to-amorphous transformation in a thin surface layer (of the order ~1 μm in thickness). When this transformation occurs, a substantial reduction in the sliding friction is observed along with an accompanying reduction in wear rate (Ref 73). The mechanism by which the surface layer in the wear scar transforms to the amorphous state is likely related to the observation of amorphization induced by mechanical attrition and mechanical alloying. These phenomena appear to be a fertile ground for future research.

Reinforcing Fibers. The high yield strength of metallic glasses has led to a number of applications as reinforcing fibers in composite materials. A simple example is in the area of reinforcing concrete. Metallic glass fiber can be easily and cheaply produced by a variation of the melt-spinning process in which the melt stream is steadily broken up. The incorporation of only a few volume percent of metallic glass fibers in concrete has been observed to increase the overall work required to cause fracture by 100 times. This substantial improvement in fracture toughness is achieved at a modest increase in cost. Other attempts have been made to introduce metallic glass fibers and ribbons in metal matrices (Ref 74). Others have considered the use of metallic glass ribbons in plastic composites, as belting in radial tires, and so forth. Some refractory metal-metalloid glasses have tensile yield strengths in excess of 7 GPa (10^6 psi); therefore, fibers of these materials are expected to compete with those of the most refractory materials currently used in fabricating composites.

Bulk Metallic Glasses. The high yield strength, hardness, and other properties of some metallic glasses have also led to interest in producing bulk amorphous materials for certain applications. Because rapid solidification techniques require that one dimension of the sample be relatively small in order to achieve the high cooling rates required to bypass crystallization, the production of bulk three-dimensional materials can only be accomplished by consolidation of prequenched powders or ribbon materials. Production of metallic glass powders in large quantities has been demonstrated using variations of the rapid quenching method whereby the liquid is atomized into small droplets and cooled, for example, by thermal conduction in a fluid medium. Spray atomization, high-velocity gas jet atomization, and other methods have been employed successfully for this purpose. When droplet sizes below ~50 μm are achieved, cooling rates adequate for glass formation can be obtained in many alloys. Another simple method of powder production is to crush metallic glass ribbons by grinding or milling. Low temperatures (liquid nitrogen cooling) and hydrogenation of the ribbons are often used during the milling process in order to promote brittle fracture of the ribbon material. Several companies have used this method to produce commercial quantities of powder-flake material. Mechanical alloying is an ideal and relatively inexpensive method of producing metallic glass powder materials in large quantities; this method has only recently been exploited. All of these powder production methods lead to precursor materials that can be subsequently used as the input material for consolidation techniques.

Several consolidation methods have been successfully applied to metallic glass powders. These include shock consolidation, explosive forming, sintering at temperatures below T_g, hot extrusion near T_g, and hot rolling near T_g. In the shock consolidation method (Ref 75), an intense shock wave is produced by a supersonic flyer plate that impacts onto a green powder compact. As the shock front moves through the green compact, densification occurs and energy is preferentially dissipated on the particle boundaries. If the shock intensity and duration are properly chosen, interparticle melt layers are adiabatically formed as densification occurs. This is followed by a quench of the molten zones by heat conduction to the still-cold particle interiors. This rapid quench is sufficient to retain the glassy structure. Successful consolidation requires proper choices of powder grain size, shock intensity (typical peak pressures of the order of 1 Mbar are required), and duration (shock front durations of tens of microseconds). Bulk metallic glasses have been achieved by this method with densities approaching 99% or more of the intrinsic density. Such specimens exhibit bulk mechanical properties that approach the intrinsic values found for high-quality ribbons.

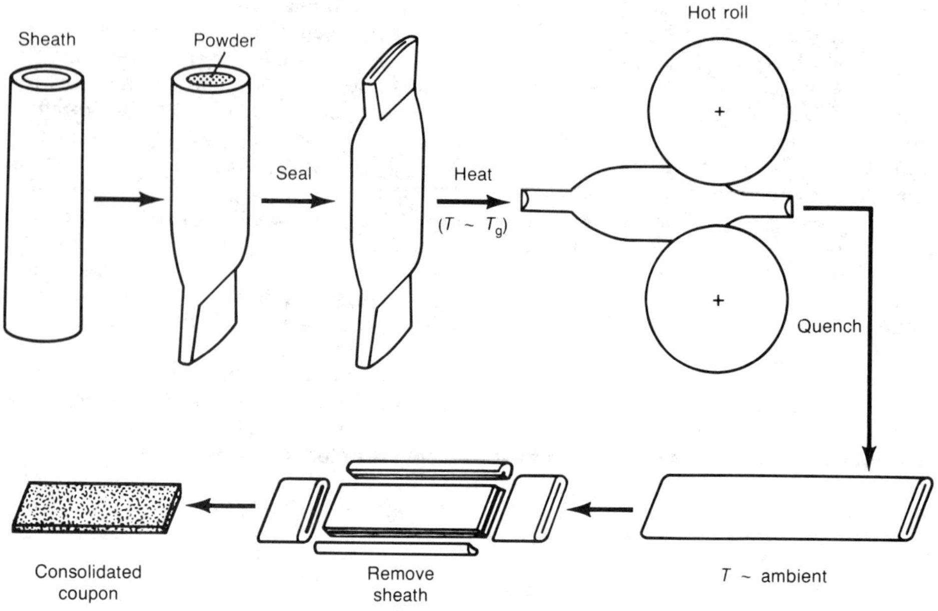

Fig. 19 Schematic illustration of the method used by Shingu to consolidate metallic glass powders into bulk materials. The sheathed sample is heated to temperatures near its glass transition by immersion in a hot salt bath. The heated sample is rolled and then quenched in water. The final consolidated specimen is removed from the sheath. Source: Ref 77

Explosive forming is similar to shock consolidation. The primary difference lies in the fact that a controlled and shaped shock front is generated. This is accomplished by surrounding the green compact by a suitably shaped explosive charge that burns in a temporal sequence, producing a spatially and temporally controlled shock front. Cline and Hopper (Ref 76) were the first to demonstrate successful consolidation of rods, cylinders, and other shapes using this method.

Recently, Shingu (Ref 77) has exploited the metastability of metallic glasses against crystallization at temperatures near T_g to develop hot extrusion and rolling methods of consolidation that use the homogeneous fluidlike flow of metallic glasses at these elevated temperatures. His hot-rolling method is illustrated schematically in Fig. 19. In this method, the sample is enclosed in a jacket and heated to a temperature near T_g, where fluidlike viscosity is achieved. It is then rolled, formed, and quickly quenched to low temperature. Under favorable circumstances, nucleation and growth of crystals do not occur during processing. A fully consolidated glass is then achieved with density, strength, and so on, approaching the intrinsic values. This method of consolidation has been applied to a number of metal-metalloid and metal-metal glasses.

Controlled Crystalline Microstructures. Amorphous metals and alloys are interesting precursor materials for the production of crystalline materials with controlled microstructure. The nucleation and growth of crystalline phases from an amorphous matrix at temperatures near T_g can be controlled with high precision. Starting with an amorphous precursor material, microcrystalline alloys can be produced with controlled grain sizes and with phase distributions not achievable in conventionally processed alloys. For example, this technique has been used to produce ultrafine-grain iron-neodymium-boron alloys with grain sizes that are optimized to achieve hard magnetic properties (Ref 78, 79). Optimum permanent magnet materials are actually achieved by starting with a rapidly quenched material that already possesses an ultrafine grain structure due to nucleation and growth of crystals during the rapid quench.

Future Developments. As the properties of metallic glasses and amorphous metals become better characterized, and as economical synthesis and processing methods become available, it is certain that other applications will be developed. The ultimate commercial future of metallic glasses and amorphous alloys will depend in part on economic factors. On the other hand, the unique structural, mechanical, electronic, magnetic, and chemical properties of these materials are certain to lead to as yet unexplored areas of technical application. Metallic glasses have a relatively short (30-year) history compared with the majority of currently used engineering materials. It is reasonable to believe that the evolution of this class of engineering materials is still in its relatively early stages.

REFERENCES

1. H. Logan, *How Much Do You Know About Glass?*, Dodd Publishing, 1951
2. D. Turnbull, *Solid State Physics*, Vol 3, Academic Press, 1956, p 225-3063
3. F. Spaepen and D. Turnbull, *Rapidly Quenched Metals*, Vol II, N.J. Grant and B.C. Giessen, Ed., MIT Press, 1976, p 205-230
4. P. Duwez, *Trans. ASM*, Vol 60, 1967, p 607
5. P. Duwez, R. Willens, and Clement, *Nature*, Vol 187, 1960, p 809
6. H.H. Liebermann and C.D. Graham, Jr., *IEEE Trans. Mag.*, Vol 12 (No. 6), 1976, p 921
7. D. Polk, U.S. Patent 3,881,542, 1975; S. Kavesh, U.S. Patent 3,881,540, 1975
8. M.C. Narasimhan, U.S. Patent 4,142,571, 1979
9. W. Buckel and R. Hilsch, *Z. Phys.*, Vol 138, 1954, p 109; also, *Z. Phys.*, Vol 146, 1956, p 27
10. S. Mader and A.S. Nowick, *Appl. Phys. Lett.*, Vol 7, 1965, p 57
11. M.M. Collver and R.H. Hammond, *Phys. Rev. Lett.*, Vol 30, 1973, p 92
12. G. Bergmann, *Phys. Rep.*, Vol 27C, 1976, p 161
13. W.L. Johnson, *Prog. Mater. Sci.*, Vol 30, 1986, p 81
14. K. Samwer, *Phys. Rep.*, Vol 161, 1988, p 1
15. R.B. Schwarz and W.L. Johnson, *Phys. Rev. Lett.*, Vol 51, 1983, p 415
16. D.E. Luzzi and M. Meshi, *Res Mech.*, Vol 21, 1987, p 207
17. H. Fecht and W.L. Johnson, *Nature*, Vol 334, 1988, p 50
18. P.R. Okamoto, L.E. Rehn, J. Pearson, R. Bhadra, and M. Grimsditch, *J. Less-Common Met.*, Vol 140, 1988, p 231
19. L. Schultz, *Mater. Sci. Eng.*, Vol 97, 1988, p 15
20. X.L. Yeh, W.L. Johnson, and K. Samwer, *Appl. Phys. Lett.*, Vol 42, 1983, p 242
21. C.C. Koch, O.B. Cavin, C.G. McKamey, and J.O. Scarbrough, *Appl. Phys. Lett.*, Vol 43, 1983, p 1017
22. J.D. Bernal, *Proc. R. Soc.*, Vol 37, 1959, p 355; also, J.L. Finney, *Proc. R. Soc.*, Ser A319, 1970, p 479
23. G.S. Cargill III, *Solid State Physics*, Vol 30, F. Seitz, D. Turnbull, and H. Ehrenreich, Ed., Academic Press, 1975, p 227; *J. Appl. Phys.*, Vol 42, 1970, p 12
24. D.S. Boudreaux and J.M. Gregor, *J. Appl. Phys.*, Vol 48, 1977, p 152
25. J. Bletry and J.F. Sadoc, *J. Phys. F, Met. Phys.*, Vol 5, 1975, p L110; H. Ruppersberg, D. Lee, and C.N.J. Wagner, *J. Phys. F, Met. Phys.*, Vol 10, 1980, p 1645
26. P. Gaskell, in *Glassy Metals II*, H.J. Güntherodt and H. Beck, Springer-Verlag, 1983, p 5-47
27. F. Spaepen, *Mater. Res. Soc. Symp.*, Vol 132, 1989, p 127
28. C.V. Thompson, H.J. Frost, and F.

Spaepen, *Acta Metall.*, Vol 35, 1987, p 887
29. P.J. Steinhardt, D.R. Nelson, and M. Ronchetti, *Phys. Rev. Lett.*, Vol 47, 1981, p 1297
30. S. Sachdev and D.R. Nelson, *Phys. Rev. Lett.*, Vol 53, 1984, p 1947
31. D. Schectman, I. Blech, D. Gratias, and J.W. Cahn, *Phys. Rev. Lett.*, Vol 53, 1984, p 1951
32. T. Ichikawa, *Phys. Status Solidi*, Vol A19, 1973, p 707
33. L.B. Davies and P.J. Grundy, *Phys. Status Solidi*, Vol A8, 1971, p 189; also, *J. Non-Cryst. Solids*, Vol 11, 1972, p 179
34. T. Egami, *Rep. Prog. Phys.*, Vol 47, 1984, p 1601
35. H.S. Chen and D. Turnbull, *Appl. Phys. Lett.*, Vol 10, 1967, p 284
36. J.H. Gibbs and E.A. Di Marzio, *J. Chem. Phys.*, Vol 28, 1958, p 373
37. M.H. Cohen and D. Turnbull, *J. Chem. Phys.*, Vol 31, 1959, p 1164
38. G.S. Grest and M.H. Cohen, *Phys. Rev. B*, Vol 21, 1980, p 4113; M.H. Cohen and G.S. Grest, *Phys. Rev. B*, Vol 20, 1979, p 1077
39. W.L. Johnson and M. Tenhover, in *Glassy Metals: Magnetic, Chemical, and Structural Properties*, R. Hasegawa, Ed., CRC Press, 1983, p 65-105
40. D. Weaire and P.C. Taylor, in *Dynamical Properties of Solids*, Vol 4, G.K. Horton and A.A. Maradudin, Ed., North-Holland, 1980, p 1-61; J.B. Suck and H. Rudin, in *Glassy Metals I*, H.J. Güntherodt and H. Beck, Ed., Springer-Verlag, 1983, p 217-260
41. B. Golding, B.G. Bagley, and F.S.L. Hsu, *Phys. Rev. Lett.*, Vol 29, 1972, p 69
42. P.W. Anderson, B.I. Halperin, and C.M. Varma, *Philos. Mag.*, Vol 25, 1972, p 1
43. W.A. Phillips, *J. Low Temp. Phys.*, Vol 7, 1972, p 351
44. J.E. Graebner, B. Golding, R.J. Schutz, F.S.L. Hsu, and H.S. Chen, *Phys. Rev. Lett.*, Vol 39, 1977, p 1480
45. J.L. Black, in *Glassy Metals I*, H.J. Güntherodt and H. Beck, Ed., Springer-Verlag, 1983, p 167-190
46. H. Kronmuller and W. Frank, *Radiation Effects and Defects in Solids*, Vol 108, 1989, p 81
47. B. Cantor and R.W. Cahn, in *Amorphous Metallic Solids*, F.E. Luborsky, Ed., Butterworths, 1983, p 487
48. B.S. Berry and W.C. Pritchet, *Phys. Rev. B*, Vol 24, 1981, p 2299; R.C. Bowman, Jr., A.J. Maeland, and W.K. Rhim, *Phys. Rev. B*, Vol 26, 1982, p 6362
49. H. Kimura and T. Masumoto, in *Amorphous Metallic Alloys*, F.E. Luborsky, Ed., Butterworths, 1983, p 187
50. F. Spaepen and A. Taub, in *Amorphous Metallic Alloys*, F.E. Luborsky, Ed., Butterworths, 1983, p 231
51. J.J. Gilman, *J. Appl. Phys.*, Vol 46, 1975, p 1625
52. L. Davis, in *Metallic Glasses*, J.J. Gilman and H. Leamy, Ed., American Society for Metals, 1978, p 190
53. P.G. Saffman and G.I. Taylor, *Proc. R. Soc.*, Vol A235, 1958, p 312
54. F.E. Luborsky and J.L. Walter, *J. Appl. Phys.*, Vol 47, 1976, p 3648
55. E.A. Kramer, W.L. Johnson, and C. Cline, *Appl. Phys. Lett.*, Vol 35, 1979, p 815
56. P.J. Cote and L.V. Miesel, in *Glassy Metals I*, H.J. Güntherodt and H. Beck, Ed., Springer-Verlag, 1981, p 141
57. H.J. Güntherodt and H.U. Kunzi, in *Metallic Glasses*, J.J. Gilman and H.J. Leamy, Ed., American Society for Metals, 1976, p 247
58. J.H. Mooij, *Phys. Status. Solidi*, Vol A17, 1973, p 521
59. P. Duwez and S.C.H. Lin, *J. Appl. Phys.*, Vol 38, 1967, p 4096
60. J. Durand, in *Glassy Metals II*, H. Güntherodt and H. Beck, Springer-Verlag, 1983, p 343
61. R.C. O'Handley, in *Amorphous Metallic Alloys*, F.E. Luborsky, Ed., Butterworths, 1983, p 257
62. F.E. Luborsky, in *Amorphous Metallic Alloys*, F.E. Luborsky, Ed., Butterworths, 1983, p 360
63. T. Mizoguchi, K. Yamauchi, and H. Miyajima, in *Amorphous Magnetism*, H.O. Hooper and A.M. de Graaf, Ed., Plenum Press, 1973, p 325
64. W.L. Johnson, S.J. Poon, and P. Duwez, *Phys. Rev. B*, Vol 11, 1975, p 150
65. W.L. Johnson, in *Glassy Metals I*, H.J. Güntherodt and H. Beck, Ed., Springer-Verlag, 1983, p 191; W.L. Johnson and M. Tenhover, in *Glassy Metals: Magnetic, Chemical, and Structural Properties*, R. Hasegawa, Ed., CRC Press, 1983, p 65
66. G. Bergmann, *Phys. Rep.*, Vol 27C, 1976, p 161
67. P. Chaudhari, F. Spaepen, and P. Steinhardt, in *Glassy Metals I*, H.J. Güntherodt and H. Beck, Ed., Springer-Verlag, 1983, p 127
68. K. Kobayashi, K. Asami, and K. Hashimoto, *Proc. 4th Int. Conf. Rapidly Quenched Met.*, T. Masumoto and K. Suzuki, Ed., Japan Institute of Metals (Sendai, Japan), 1982, p 1443
69. A. Yokoyama *et al.*, *Scr. Metall.*, Vol 15, 1981, p 365
70. D. Raskin and C.H. Smith, in *Amorphous Metallic Alloys*, F.E. Luborsky, Ed., Butterworths, 1983, p 381
71. R. Schulz *et al.*, *Mater. Sci. Eng.*, Vol 99, 1988, p 19
72. A.P. Thakoor, J.L. Lamb, S.K. Khanna, M. Hehra, and W.L. Johnson, *J. Appl. Phys.*, Vol 58, 1985, p 3409
73. D.S. Scruggs, U.S. Patent No. 4,725,512, 1988
74. S.J. Cytron, *J. Mater. Sci. Eng.*, Vol 1, 1982, p 211
75. R.B. Schwarz, P. Kasiraj, T. Vreeland, Jr., and T. Ahrens, *Acta Metall.*, Vol 32, 1984, p 1243
76. C. Cline and R. Hopper, *Scr. Metall.*, Vol 11, 1977, p 1137
77. P. Shingu, *Mater. Sci. Eng.*, Vol 97, 1988, p 137
78. General Motors Corporation, U.S. Patent Application 544728
79. G.E. Carr *et al.*, *Mater. Sci. Eng.*, Vol 99, 1988, p 147

Electrical Resistance Alloys

Revised by Robert A. Watson, The Kanthal Corporation, Bo Jönsson, Kanthal AB, George A. Fielding, Harrison Alloys, Inc., Donald V. Cunningham, Emerson Electric, Wiegand Division, C. Dean Starr, C. Dean Starr, Inc.

ELECTRICAL RESISTANCE ALLOYS include those types used in instruments and control equipment, heating elements, and devices that convert heat generated to mechanical energy. In this article they are classified as Resistance Alloys, Heating Alloys, and Thermostat Metals. The primary focus will be on metals used in these three classes, although some mention is also made of nonmetallic materials used in similar applications.

Resistance Alloys

The primary requirements for resistance alloys are uniform resistivity, stable resistance (no time-dependent aging effects), reproducible temperature coefficient of resistance, and low thermoelectric potential versus copper. Properties of secondary importance are coefficient of expansion, mechanical strength, ductility, corrosion resistance, and ability to be joined to other metals by soldering, brazing, or welding. Availability and cost are also factors.

Nominal compositions and physical properties of metals and alloys used to make resistors for instruments and controls are listed in Table 1.

Resistance alloys must be ductile enough so that they can be drawn into wire as fine as 0.01 mm (0.0004 in.) in diameter or rolled into narrow ribbon from 0.4 to 50 mm (1/64 to 2 in.) wide and from 0.025 to 3.8 mm (0.001 to 0.15 in.) thick.

Alloys must be strong enough to withstand fabrication operations, and it must be easy to procure an alloy that has consistently reproducible properties. For instance, successive batches of wire must have closely similar electrical characteristics: if properties vary from lot to lot, resistors made of wire from different batches may cause a given model of instrument to exhibit widely varying performance under identically reproduced conditions, or may cause large errors in a given instrument when a resistor from one batch is used as a replacement part for a resistor from another batch.

Coefficients of expansion of both the resistor and the insulator on which it is wound must be considered because stresses can be established that will cause changes in both resistance and temperature coefficient of resistance. It is equally important that consideration be given to the choice between single-layer and multiple-layer wound resistors, because of the difference in rate of heat dissipation between the two styles.

In design of primary electrical standards of very high accuracy, cost of resistance material is not a consideration. For ordinary production components, however, cost may be the deciding factor in material selection.

Resistors

Resistors for electrical and electronic devices may be divided into two arbitrary classifications on the basis of permissible error: those employed in precision instruments in which over-all error is considerably less than 1%, and those employed where less precision is needed. The choice of alloy for a specific resistor application depends on the variation in properties that can be tolerated.

In many electronic devices, resistors whose error in resistance value is 5 to 10% are entirely satisfactory. Most resistors for this classification are made of carbon. Carbon resistors are not discussed in this article. Here, we are concerned chiefly with metallic resistors such as wirewound precision resistors and potentiometers, resistance thermometers, and ballast resistors.

Some applications of resistance materials require devices with large temperature coefficients of resistance, either positive or negative. A device of this type is called a thermistor. Thermistors are made almost exclusively of ceramic semiconductor materials.

Precision resistors (those with less than 1% error) require careful material selection. The ideal material for a precision resistor should have a temperature coefficient of resistance equal to zero for the temperature range over which the resistor will operate. In addition, to ensure freedom from thermoelectric effects, it should have a small or negligible thermoelectric potential versus copper, which is the material normally used for the connecting conductor. Temperature differentials may exist among various junctions between a resistance wire and a connecting wire, resulting in a network of thermocouples that can cause parasitic electromotive forces in the circuit; this effect is especially critical in precise dc circuits. In an apparatus where extreme precision is required, it is advisable to make the connecting wires of the same material as the resistors or to design the apparatus so that all dissimilar-metal junctions are at the same operating temperature.

Selection of material for, and specific dimensions of, a precision resistor must include consideration of equipment size and heat-dissipation characteristics. Temperature excursions from the ambient or from a specified operating temperature may be undesirable, because they may cause net changes in resistance that will affect the stability or accuracy of the instrument. The magnitude of the change in resistance can be calculated using the temperature coefficient of resistance. For example, a resistor made of a low-resistivity material could be several times larger than one made of a higher-resistivity material and yet achieve the same total resistance. The large resistor would have a much greater surface area and therefore could dissipate much more heat, and thus, despite its low resistivity, would attain a lower steady-state temperature than would be possible for a small, high-resistivity resistor operating under the same conditions. Alloys used for precision resistors generally have resistivities ranging from 0.5 to 1.35 $\Omega mm^2/m$ (300 to 800 $\Omega \cdot$ circ mil/ft).

Resistance thermometers are commonly made of copper, nickel, or platinum; these devices are precision resistors whose resistance change with temperature is stable and reproducible over specified ranges of temperature. For resistance thermometers, the larger the temperature coefficient of resistance of the material, the greater the accuracy and ease of measurement. Temperature coefficients of relatively pure metals are greatly affected by small amounts of impurities. In fact, one of the most sensitive

Table 1 Typical properties of electrical resistance alloys

Basic composition, %	Resistivity(a), nΩ · m(b)	TCR, ppm/°C(c)	Thermoelectric potential versus Cu, μV/°C	Coefficient of thermal expansion(d), μm/m · °C	Tensile strength(a)		Density(a)	
					MPa	ksi	g/cm^3	lb/in.^{3}
Radio alloys								
98Cu-2Ni	50	1400 (25–105 °C)	−13 (25–105 °C)	16.5	205–410	30–60	8.9	0.32
94Cu-6Ni	100	700 (25–105 °C)	−13 (25–105 °C)	16.3	240–585	35–85	8.9	0.32
89Cu-11Ni	150	450 (25–105 °C)	−25 (25–105 °C)	16.1	240–515	35–75	8.9	0.32
78Cu-22Ni	300	180 (25–105 °C)	−36 (0–75 °C)	15.9	345–690	50–100	8.9	0.32
Manganins								
87Cu-13Mn	480	±15 (15–35 °C)	1 (0–50 °C)	18.7	275–620	40–90	8.2	0.30
83Cu-13Mn-4Ni	480	±15 (15–35 °C)	−1 (0–50 °C)	18.7	275–620	40–90	8.4	0.31
85Cu-10Mn-4Ni(e)	380	±10 (40–60 °C)	−1.5 (0–50 °C)	18.7	345–690	50–100	8.4	0.31
Constantans								
57Cu-43Ni	500	±20 (25–105 °C)	−43 (25–105 °C)	14.9	410–930	60–135	8.9	0.32
55Cu-45Ni	500	±40 (−55–105 °C)	−42 (0–75 °C)	14.9	455–860	66–125	8.9	0.32
53Cu-44Ni-3Mn	525	±70 (−55–105 °C)	−38 (0–100 °C)	14.9	410–930	60–135	8.9	0.32
Nickel-chromium-aluminum alloys								
75Ni-20Cr-3Al-2(Cu, Fe, or Mn)	1333	±20 (−55–105 °C)	1.0 (25–105 °C)	12.6	825–1380	120–200	8.1	0.29
72Ni-20Cr-3Al-5Mn	1375	±20 (−55–105 °C)	1.0 (25–105 °C)	13	690–1380	100–200	7.1	0.26
Nickel-base alloys								
78.5Ni-20Cr-1.5Si	1080	80 (25–105 °C)	3.9 (25–105 °C)	13.5	790–1380	115–200	8.3	0.30
76Ni-17Cr-4Si-3Mn	1330	±20 (−55–105 °C)	−1 (20–100 °C)	15	900–1380	130–200	7.8	0.28
71Ni-29Fe	208	4300 (25–105 °C)	−40 (25–105 °C)	15	480–1035	70–150	8.4	0.31
68.5Ni-30Cr-1.5Si	1187	90 (25–105 °C)	−1.2 (25–105 °C)	12.2	825–1380	120–200	8.1	0.29
60Ni-16Cr-22.5Fe-1.5Si	1125	150 (25–105 °C)	0.9 (25–105 °C)	13.5	725–1345	105–195	8.4	0.30
37Ni-21Cr-40Fe-2Si	1080	300 (20–100 °C)	. . .	16.0	585–1135	85–165	7.96	0.288
35Ni-20Cr-43.5Fe-1.5Si	1000	400 (25–105 °C)	−1.1 (25–105 °C)	15.6	585–1135	85–165	8.1	0.29
Iron-chromium-aluminum alloys								
73.5Fe-22Cr-4.5Al	1350	60 (25–105 °C)	−3.0 (0–100 °C)	11	690–965	100–140	7.25	0.262
73Fe-22Cr-5Al	1390	40 (25–105 °C)	−2.8 (0–100 °C)	11	690–965	100–140	7.15	0.258
72.5Fe-22Cr-5.5Al	1450	20 (25–105 °C)	−2.6 (0–100 °C)	11	690–965	100–140	7.1	0.256
81Fe-15Cr-4Al	1250	±50 (25–105 °C)	−1.2 (0–100 °C)	11	620–900	90–130	7.43	0.268
Pure metals								
Aluminum (99.99+)	26.55	4290(a)	−3.4 (0–50 °C)	23.9(a)	50–110	7–16	2.70	0.098
Copper (99.99)	16.73	4270 (0–50 °C)	0	16.5(a)	115–130	17–19	8.96	0.324
Gold (99.999+)	23.50	4000 (0–100 °C)	0.2 (0–100 °C)	14.2(a)	130	19	19.32	0.698
Iron (99.94)	970	5000(a)	12.2 (0–100 °C)	11.7(a)	180–220	26–32	7.87	0.284
Molybdenum (99.9)	52	3300(a)	6.9 (0–100 °C)	4.9	690–2140	100–310	10.22	0.369
Nickel (99.8)	80	6000 (20–35 °C)	−22 (0–75 °C)	15	345–760	50–110	8.90	0.322
Platinum (99.99+)	105	3920 (0–100 °C)	7.6 (0–100 °C)	8.9(a)	125	18	21.45	0.775
Silver (99.99)	16	4100(a)	−0.2 (0–100 °C)	19.7	125	18	10.49	0.379
Tantalum (99.96)	125	3820 (0–100 °C)	−4.3 (0–100 °C)	6.5(a)	690–1240	100–180	16.6	0.600
Tungsten (99.9)	55	4500(a)	3.6 (0–100 °C)	4.3(a)	1825–4050	265–590	19.25	0.695

(a) At 20 °C (68 °F). (b) To convert to Ω · circ mil/ft, multiply by 0.6015. (c) Temperature coefficient of resistance is $(R - R_0)/R_0\ (t - t_0)$, where R is resistance at t °C and R_0 is resistance at the reference temperature t_0 °C. (d) At 25 to 105 °C. (e) Shunt manganin

tests of the purity of a metal is measurement of its temperature coefficient of resistance, which decreases sharply with increasing impurity or alloy content.

Ballast resistors are used extensively in industrial circuits to maintain constant currents over long periods of time. In such an application, a ballast resistor must be able to dissipate energy in such a way as to control current over a wide range of voltages. Wires with the proper temperature coefficient of resistance can be made to change resistance rapidly with changes in current, due to self heating, in such a manner that the current in the circuit will remain nearly constant even when there are fluctuations in voltage across the circuit. Because ballast resistors operate at elevated temperatures, mechanical properties are important also. Typical materials used in ballast resistors are pure iron, pure nickel, and nickel-iron alloys such as 71Ni-29Fe (see Table 1).

Reference resistors and virtually all other applications of resistance alloys demand temperature coefficients of resistance lower than ±20 ppm/°C (±20 μΩ/Ω · °C). This requirement stems from the fact that, for these applications, resistance errors resulting from the small changes in ambient temperature that are continually taking place cannot be tolerated. In the most demanding of these applications, resistors often are mounted in thermally insulated containers and are carefully maintained at a temperature slightly above the maximum anticipated ambient temperature.

The most important requirement of a resistor used as a reference standard is that its value be predictable within narrow limits over long periods of time. Many reference resistors exhibit a nearly linear change in resistance with time. Hence, resistance between dates of calibration can be determined by interpolation; resistance at future points in time can be determined by extrapolation, but undue reliance should not be placed on extrapolated values. Figure 1 shows the change in resistance with time for a 10-kΩ resistor made of a Ni-Cr-Al-Cu alloy.

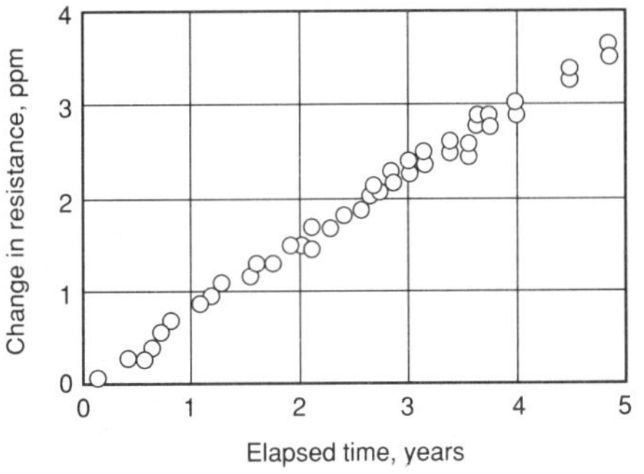

Fig. 1 Change in resistance of a 10-kΩ resistor with time

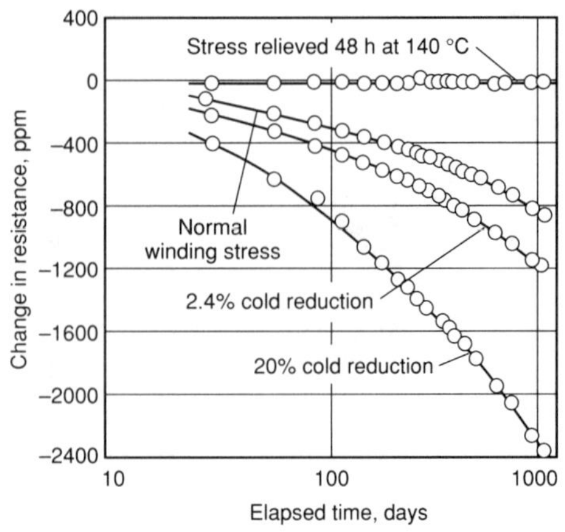

Fig. 2 Change in resistance of manganin resistors upon aging at room temperature

Stability, or the ability to maintain a specific value of resistance within narrow limits over a long period of time, is an important requirement of materials for precision resistors and reference resistors. Principal sources of instability are:

- Relief of residual stresses during service
- Time-dependent or time-temperature-dependent metallurgical changes, such as precipitation, of a second phase
- Corrosion or oxidation
- Humidity

Effect of Stress Relief. Residual stresses often are relieved at room temperature over long periods of time through a process known as stress relaxation. Stress relaxation alters the resistance of a coil at a rate of change that increases with the original level of residual stress. For this reason, only carefully preannealed wires are used for precision resistors. Stresses induced during winding, weaving, or other operations in fabrication of resistors from preannealed wire must be kept to a minimum. Thorough annealing of finished resistors is not always possible because the wires may be enameled or may be coated with a textile insulation of only moderate resistance to heat. Textiles limit the maximum temperature to about 140 °C (285 °F), and the highest rated enamels have a 220 °C (430 °F) maximum temperature that can be used for stress relieving finished resistors.

Figure 2 shows the effect of residual stress on the stability of a manganin alloy subjected to different amounts of cold work. The top curve illustrates that a low-temperature stress-relieving treatment substantially eliminates the stresses that would, in time, have been eliminated due to natural relaxation at room temperature.

Resistors represented by the top curve were stress relieved at 140 °C (285 °F) for 48 h to stabilize their resistance within about 20 ppm of the nominal value. Resistors not stress relieved, as represented by the other curves, continue to change in resistance almost indefinitely. For most modern, hermetically sealed precision resistors annealed at 140 °C (285 °F), the change in resistance does not exceed 10 ppm/year, and for many it does not exceed 5 ppm/year. However, resistors made of manganin that are used as reference standards require greater stability, and stress relief at 140 °C (285 °F) is not adequate. One-ohm resistors of the best grade (the double-wall type) are treated as follows. A coil of wire is wound on a steel mandrel and annealed at about 500 °C (930 °F) in a protective atmosphere for 6 h or more. The coil is removed and slipped over an insulated tube of the same diameter as the mandrel, and then is hermetically sealed using a second tube slightly greater in diameter. In most resistors of this type, the change in resistance does not exceed 1 ppm/year.

Effect of Metallurgical Stability. The second factor affecting stability of precision resistors is the metallurgical stability of the alloy being used as the resistance element; any metallurgical change will be detrimental. All resistance alloys are single-phase solid-solution alloys; thus, the changes in resistance that occur are relatively small but not insignificant. Changes in resistance are caused by internal changes such as long-range order-disorder reactions in 71Ni-29Fe alloys, short-range order or clustering in quaternary nickel-chromium alloys, and even minor ordering in manganin alloys. Accordingly, resistance of these alloys is affected by heat treatment and by rates of cooling from heat-treating temperatures. Power resistors that can operate as high as 300 °C (570 °F) can in effect be heat treated during service. The net effect during service can be an increase in resistance for nickel-chromium alloys, a decrease for manganin, and either an increase or decrease for nickel-iron alloys.

Effect of Corrosion. The third factor affecting stability of resistors is corrosion and/or oxidation. Corrosion of the resistance element will decrease its effective cross section, resulting in a corresponding increase in resistance. If the corrosive attack is selective, changes will occur in temperature coefficient of resistance and thermal emf, as well as in resistivity. These corrosive effects may be minimized by protecting the wire with an enamel or plastic coating. One relatively common source of corrosive attack, but one that is often overlooked, is flux residue at soldered or brazed joints. Another less obvious cause of instability is the presence of tin-containing solder. Intergranular stress corrosion, believed to originate during thermal stress-relieving treatments, may cause open circuits.

Effect of Humidity. The fourth factor affecting the stability of enameled wire-wound resistors is humidity. The change in resistance is dependent on enamel thickness, wire diameter, and change in the amount of moisture to which the resistor is exposed. Increasing the moisture for a given resistor will cause a positive change in resistance, whereas decreasing the moisture causes a negative shift. This factor can be eliminated where it is physically possible to hermetically seal the resistor.

Combinations of these four factors—residual stresses, metallurgical instability, corrosion or oxidation, and humidity—account for the complex changes in resistance that often occur in resistors.

Solderability or Joining. The ease with which alloys can be soldered, brazed, or welded is an important consideration in selection of materials for precision resistors. Improperly brazed or soldered joints frequently cause resistance instability in the circuit. Metals to be soldered must be cleaned prior to tinning so that solder can completely wet the surfaces and maintain electrical continuity. For copper-nickel alloys this is relatively simple because protective oxide coatings are not formed on these alloys. Nickel-chromium alloys must be tinned immediately after cleaning and before an inherent protective oxide forms.

Pressure Coefficients of Resistance. The resistance value of a resistor may change if the hydrostatic pressure on the resistance element is changed; for manganin this change is about 23 pΩ/Ω · Pa (0.16 pΩ/Ω · psi). Sealed resistors may also be affected by changes in external pressure. In a double-wall one-ohm resistor, for example, a change in pressure on the inner tube will cause a change in tube diameter, thus altering the length of wire wound on the tube. The magnitude of the resistance change depends in part on the thickness of the wall, and for commercial resistors is typically less than the hydrostatic pressure coefficient (PCR) of manganin. Unsealed resistors wound on mica cards containing air bubbles may have pressure coefficients several times greater than that predicted from the hydrostatic pressure coefficient of the alloy. This effect is important only if there is a large change in pressure, which would be most likely if there were a large change in elevation above sea level.

Types of Resistance Alloys

Copper-nickel resistance alloys, generally referred to as *radio alloys*, have very low resistivities and moderate temperature coefficients of resistance (TCR), as shown in the first four listings in Table 1. Resistivity of radio alloys increases, and TCR decreases, as nickel content increases. Thermal emf is negative with respect to copper, the magnitude being directly proportional to nickel content. All radio alloys can be readily soldered or brazed. Those with 12 and 22% nickel have high enough resistance to permit welding. Because of their high copper contents, radio alloys have low resistance to oxidation and thus are restricted to applications involving low operating temperatures. They are used chiefly for resistors that carry relatively high currents, and for this reason rapid dissipation of heat from the surface of the resistor is desirable. In this application, resistor temperature may vary over a wide range, but temperature changes are relatively unimportant.

Copper-manganese-nickel resistance alloys, generally referred to as manganins, have been adopted almost universally for precision resistors, slide wires and other resistive components with values of 1 kΩ or less, and are also used for components with values up to 100 kΩ.

Originally, manganin was the name of a specific alloy, but the term is now generic and covers several different compositions (see Table 1). All manganins are moderate in resistivity (from 380 to 480 nΩ · m, or 230 to 290 Ω · circ mil/ft) and low in TCR (less than ±15 ppm/°C).

Manganins are stable solid-solution alloys. The electrical stability of these alloys, verified by several decades of experience, is such that their resistance values change no more than about 1 ppm per year when the material is properly heat treated and protected. Manganin-type alloys are characterized by rather steep, parabolic relations between resistance and temperature (see Fig. 3). This severely restricts the range of temperature over which resistance is stable, thus limiting the use of manganins to devices for which operating temperatures are both stable and predictable. For some applications, the maximum of the parabola (peak, or peak temperature) is kept near room temperature by controlling composition, minimizing the effects of small changes in ambient temperature. The temperature coefficient of commercial manganin is usually less than ±10 ppm/°C for an interval of 10 °C (18 °F) on either side of the peak.

When instruments are designed for operation above ambient temperature, the chemical composition of the manganin is chosen so that the peak will occur in the operating temperature range. So-called "shunt manganin," which carries high currents and consequently gets hot in use, usually has a peak temperature from 45 to 65 °C (115 to 150 °F).

Manganins are susceptible to selective oxidation or preferential corrosive attack. This may occur during heat treatment, wire manufacture, or coil fabrication. Selective oxidation results in formation of a copper-rich (manganese-depleted) zone on the wire. This copper-rich sheath has the effect of greatly increasing the temperature coefficient of resistance and raising the peak temperature well beyond the range where any precision resistor would ordinarily be used.

The resistivity of manganin—roughly 500 nΩ · m (300 Ω · circ mil/ft) at 25 °C (77 °F)—is adequate for most instrumentation purposes. The thermoelectric potential versus copper is very low, usually less than −2 μV/°C from 0 to 100 °C (32 to 212 °F).

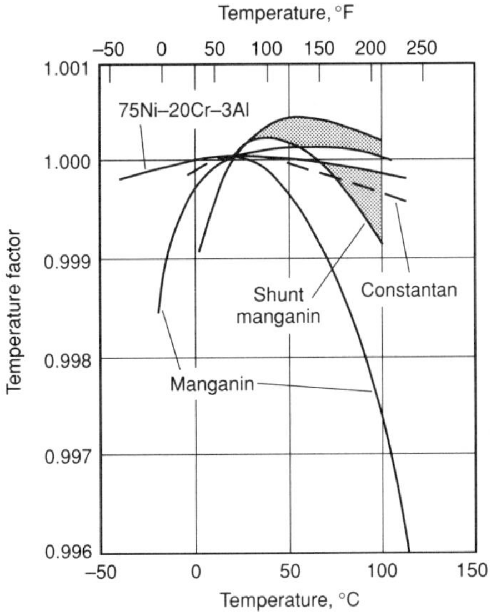

Fig. 3 Variation of resistance with temperature for four precision resistor alloys. To calculate resistance at temperature, multiply resistance at room temperature by the temperature factor.

Constantan, like manganin, has become a generic term for a series of alloys that have moderate resistivities and low temperature coefficients of resistance. Nominally, constantans are 55Cu-45Ni alloys, but specific compositions vary from about 50Cu-50Ni to about 65Cu-35Ni. The temperature coefficient of conventional constantan can be held within ±20 ppm/°C of ambient temperature. However, the difference in TCR between the low (−55 to 25 °C, or −67 to 77 °F) and high-temperature ranges (25 to 105 °C, or 77 to 220 °F) is about 20 ppm. Thus, the specification is ±20 ppm/°C over one temperature range or ±40 ppm/°C over both ranges. A variation of constantan with 3% Mn improves the flatness of the resistance temperature curve and provides a TCR of ±20 ppm/°C from −55 to 105 °C (−67 to 220 °F). All constantans contain iron and cobalt in addition to manganese.

The temperature coefficient of resistance of constantan is very low and parabolic like that of manganin, but remains flat over a much wider range (Fig. 3). Other properties are given in Table 1; specific property values vary somewhat with composition. Constantans are considerably more resistant to corrosion than manganins.

Use of constantans as electrical resistance alloys is restricted largely to ac circuits, because thermoelectric potential versus copper is quite high for these materials (about 40 μV/°C at room temperature). However, if the circuit voltage is high enough to overshadow thermoelectric effects, constantans may be used in dc circuits as well.

Nickel-Chromium-Aluminum Resistance Alloys. Nickel-chromium alloys containing small amounts of other metals—usually aluminum plus either copper, manganese, or iron—have resistivities about 2½ to 3½ times that of manganin. Ni-Cr-Al resistance alloys have been adopted almost universally for the construction of wire-wound precision resistors having resistance values of about 100 kΩ, and are also used for resistors with values as low as about 100 Ω. The temperature coefficients of resistance of these alloys are vastly superior to those of manganin and constantan, being less than ±20 ppm/°C between −55 and 105 °C (−67 and 220 °F). The difference in TCR between the hot region (25 to 105 °C) and the cold region (−55 to 25 °C) is about 20 ppm/°C for constantan (and about 10 ppm/°C for newer constantan), but only about 5 ppm/°C for the original quaternary Ni-Cr-Al alloys, and only 1 ppm/°C for the new quaternary alloys. The high resistivity and low TCR of Ni-Cr-Al alloys are obtained by an order-disorder type of heat treatment at approximately 540 °C (1000 °F). Therefore, if desired, the temperature coefficient can be decreased without resorting to melt selection, which is required for alloys that do not respond to heat treatment. The availability of smaller temperature coefficient ranges is dependent on wire size and alloy composition. Table 2 gives the available commercial ranges. Electrical stability of quaternary Ni-Cr-Al alloys is excellent—1 to 10 ppm/year or less. Their thermoelectric potential versus copper is also excellent—about 1 μV/°C at temperatures from 0 to 100 °C (32 to 212 °F).

As indicated in Table 1, the mechanical properties of Ni-Cr-Al alloys are higher than those of manganin and constantan. Wires made of Ni-Cr-Al alloys are available in diameters as small as 0.01 mm (0.0004 in.), whereas wires of copper-base alloys such as constantan are seldom produced in diameters smaller than 0.025 mm (0.001 in.). Because the resistance of a wire varies inversely with the square of its diameter, it is possible with small-diameter Ni-Cr-Al wires to produce miniature resistors that are exceedingly high in resistance.

The Ni-Cr-Al alloys resist oxidation better than other commercial electrical resistance alloys. This is an advantage in resistors that are not covered with enamel, teflon, or other coatings. It is a disadvantage for making acceptable soldered or brazed joints, because it necessitates greater care in joint preparation. However, suitable soldered or silver-brazed connections can be made readily using appropriate fluxes.

Other Precision Resistance Materials. In high-resistance precision resistors, where TCR limits are less stringent, 80Ni-20Cr alloys may be used. 80Ni-20Cr alloys have temperature coefficients from four to six-

Table 2 Temperature coefficient of resistance of nickel-chromium-aluminum base alloys as a function of wire diameter

Wire diameter, mm	Wire diameter, in.	Alloy I(a)	Alloy II(b)
		Both ranges (−55 to 25 °C, and 25 to 105 °C)	
>0.287	>0.0113	±20	±10
0.25–0.0787	0.010–0.0031	±10	±8
0.07–0.013	0.00275–0.0005	±5	±3
		One range (−55 to 25 °C, or 25 to 105 °C)	
>0.287	>0.0113	±15	±10
0.25–0.0787	0.010–0.0031	±7	±5
0.07–0.013	0.00275–0.0005	±4	±3

(a)Alloy I, 75Ni-20Cr-3Al-2(Cu, Fe, or Mn). (b) Alloy II, 72Ni-20Cr-3Al-5Mn

Table 3 TCR values for 71Ni-29Fe and pure nickel determined using various reference temperatures

Reference temperature, °C	Temperature range, °C	TCR, ppm/°C
71Ni-29Fe alloy		
20	20–100	4500
25	25–100	4300
Nickel (99.9% purity) (270)		
0	0–100	6730
20	20–100	6150
25	25–100	6000
Nickel (99% purity) (205)		
0	0–100	5250
20	20–100	4750
25	25–100	4620

teen times the nominal value for Ni-Cr-Al alloys. Other similar alloys, such as Ni-Cr-Fe alloys, are used primarily to accomplish specific design objectives because they permit designers to vary wire diameter and wire coatings in order to accommodate design constraints, such as severe space limitations. All of these precision resistance alloys have low thermoelectric potentials versus copper at temperatures from 0 to 100 °C.

Aside from electrical resistance and temperature coefficient, several other design factors must be taken into consideration in selection of precision resistance metals for specific applications. For example, pure nickel (99.8%) is used in precision instruments that require a high positive temperature coefficient; pure copper is used for compensating resistors in precision instruments where the temperature coefficient of moving coils must be matched; and pure platinum is used for heating elements in thermocouple-type ac-to-dc converters as well as for precision resistance bulbs. Aluminum is seldom used as a resistor material, but because of its favorable ratio of weight to resistance it is often used in windings for the moving coils of permanent-magnet electrical instruments.

Alloys such as 71Ni-29Fe and pure metals such as nickel and platinum have low resistivities and high-temperature coefficients. For these metals, it is important that operating temperature be specified carefully, because the temperature coefficient varies quite sharply with temperature (see Table 3).

Although their high-temperature coefficients eliminate them from use in ordinary precision resistors, pure metals are useful in other applications such as temperature-measurement instruments and ballast devices.

Semiprecision Resistance Alloys. The alloys in Table 1 other than those previously discussed are used chiefly for rheostats and potentiometers. These semiprecision resistance alloys are not made to the rigid specifications that apply to manganins, constantans, or quaternary Ni-Cr-Al alloys. For these alloys, long-term stability generally is not critical, and such factors as thermal emf versus copper and temperature coefficient of resistance are relatively unimportant.

Semiprecision resistance alloy wires are produced in diameters ranging from a lower limit of 0.013 mm (0.0005 in.)—a lower limit of 0.02 mm (0.0008 in.) for certain Cu-Ni alloys—to an approximate maximum of 6.35 mm (¼ in.). Above 0.078 mm (0.0031 in.), wires are made only in standard B&S gages; below 0.078 mm, both standard B&S gages and intermediate sizes are available. Also, ribbon and strip usually are available in thicknesses from 0.013 to 6.35 mm (0.0005 to ¼ in.). Ribbon and strip in thicknesses near the lower end of the foregoing range usually are available only as cold rolled stock, and those in thicknesses near the extreme upper end of the range usually are available only as hot rolled stock.

Dimensional tolerances are seldom specified for applications in which resistivity and electrical resistance per unit length, rather than physical dimensions, are of primary importance. Typical tolerances on resistance per unit length are:

- For hot rolled ribbon rods, ±8
- For cold rolled ribbon, ±5%
- For cold drawn round wire finer than 0.05 mm (0.002 in.), ±10%
- For cold drawn wire 0.05 to 0.1 mm (0.002 to 0.004 in.), ±8%
- For cold drawn wire 0.1 mm (0.004 in.) and heavier, ±5%

Closer tolerances are available.

Thermostat Metals

A thermostat metal is a composite material (usually in the form of sheet or strip) that consists of two or more materials bonded together, of which one may be a nonmetal. Because the materials bonded together to form the composite differ in thermal expansion, the curvature of the composite is altered by changes in temperature; this is the fundamental characteristic of any thermostat metal. A thermostat metal is, therefore, a complete, self-contained transducing system capable of transforming heat directly into mechanical energy for control, indicating, or monitoring purposes.

In applications such as circuit breakers, thermal relays, motor overload protectors, and flashers, the change in temperature necessary for operation of the element is produced by the passage of current through the element itself—in other words, the change is produced by I^2R heating. In certain other applications, any increase in the temperature of the thermostat element caused by I^2R heating is objectionable, and a thermostat metal with low electrical resistivity is required.

For circuit breakers and similar devices, there are thermostat metals that differ in electrical resistivity but are similar in other properties. This allows a manufacturer to design a complete series of circuit breakers of different ratings in which the thermostat elements are all of the same size but have different electrical resistances. Resistivity is varied by incorporating a layer of a low-resistivity metal between outer layers of two other metals that have high resistivities and that differ widely in expansion coefficient.

In one series of commercial thermostat metals with resistivities ranging from 165 to 780 nΩ · m (100 to 470 Ω · circ mil/ft) at 24 °C (75 °F), high-purity nickel is used for the intermediate layer. In a series with resistivities from 33 to 165 nΩ · m (20 to 100 Ω · circ mil/ft), high-conductivity copper alloys are employed for the intermediate layer.

The use of a manganese-copper-nickel alloy having a resistivity of 1745 nΩ · m (1050 Ω · circ mil/ft) for one of the outer layers has extended the practical upper resistivity limit of thermostat metals to 1620 nΩ · m (975 Ω · circ mil/ft) at 24 °C.

Tolerances on resistivity at a standard temperature vary from ±3 to ±10%, depending on the type of thermostat metal and its resistivity.

About 30 different alloys are used to make over 50 different thermostat metals. Most of these 30 alloys are nickel-iron, nickel-chromi-

Table 4 Properties of thermostat metals frequently selected for some common service temperatures

Temperature range of maximum sensitivity		Composition		Resistivity at 24 °C (75 °F)		Flexivity(a)	
°C	°F	High-expanding side	Low-expanding side	nΩ · m	Ω · circ mil/ft	μm/m · °C	μin./in. · °F
−20 to 150	0–300	75Fe-22Ni-3Cr	64Fe-36Ni	780	470	26.3	14.6
−20 to 200	0–400	75Fe-22Ni-3Cr	Pure Ni	160	95	8.3	4.6
		72Mn-18Cu-10Ni	64Fe-36Ni	1120	675	38.5	21.4
120–290	250–550	67Ni-30Cu-1.4Fe-1Mn	60Fe-40Ni	565	340	16.6	9.2
150–450	300–850	66.5Fe-22Ni-8.5Cr	50Fe-50Ni	580	350	11.2	6.2

(a)At 40-150 °C (100-300 °F). See ASTM B 106 for standard test method for determining flexivity of thermostat metals.

um-iron, chromium-iron, high-copper, and high-manganese alloys.

Thermostat metals are available as strip or sheet in thicknesses ranging from 0.13 to 3.2 mm (0.005 to 0.125 in.) and widths from 0.5 to 300 mm (0.020 to 12 in.). They are easily formed into the required shapes. Thermostat metals usually are selected on the basis of the temperature range in which they are required to operate. They are available for various operating ranges between −185 and 540 °C (−300 and 1000 °F). Properties and typical bimetal combinations for several temperature ranges are given in Table 4.

Heating Alloys

Resistance heating alloys are used in many varied applications—from small household appliances to large industrial process heating systems and furnaces. In appliances or industrial process heating, the heating elements are usually either open helical coils of resistance wire mounted with ceramic bushings in a suitable metal frame, or enclosed metal-sheathed elements consisting of a smaller-diameter helical coil of resistance wire electrically insulated from the metal sheath by compacted refractory insulation. In industrial furnaces, elements often must operate continuously at temperatures as high as 1300 °C (2350 °F) for furnaces used in metal-treating industries, 1700 °C (3100 °F) for kilns used for firing ceramics, and occasionally 2000 °C (3600 °F) or higher for special applications.

The primary requirements of materials used for heating elements are high melting point, high electrical resistivity, reproducible temperature coefficient of resistance, good oxidation resistance, absence of volatile components, and resistance to contamination. Other desirable properties are good elevated-temperature creep strength, high emissivity, low thermal expansion, and low modulus (both of which help minimize thermal fatigue), good resistance to thermal shock, and good strength and ductility at fabrication temperatures.

Table 5 gives physical and mechanical properties, and Table 6 presents recommended maximum operating temperatures for resistance heating materials for furnace applications. Of the four groups of materials listed in these tables, the first group (Ni-Cr and Ni-Cr-Fe alloys) serves by far the greatest number of applications.

The ductile wrought alloys in the first group have properties that enable them to be used at both low and high temperatures in a wide variety of environments. The Fe-Cr-Al compositions (second group) are also ductile alloys. They play an important role in heaters for the higher temperature ranges, which are constructed to provide more effective mechanical support for the element. The pure metals that comprise the third group have much higher melting points. All of them except platinum are readily oxidized and are restricted to use in nonoxidizing environments. They are valuable for a limited range of application, primarily for service above 1370 °C (2500 °F). The cost of platinum prohibits its use except in small, special furnaces.

The fourth group, nonmetallic heating-element materials, are used at still higher

Table 5 Typical properties of resistance heating materials

Basic composition	Resistivity(a), Ω · mm²/m(b)	Average change in resistance(c), %, from 20 °C to: 260 °C	540 °C	815 °C	1095 °C	Thermal expansion, μm/m · °C, from 20 °C to: 100 °C	540 °C	815 °C	Tensile strength MPa	ksi	Density g/cm³	lb/in.³
Nickel-chromium and nickel-chromium-iron alloys												
78.5Ni-20Cr-1.5Si (80-20)	1.080	4.5	7.0	6.3	7.6	13.5	15.1	17.6	655–1380	95–200	8.41	0.30
77.5Ni-20Cr-1.5Si-1Nb	1.080	4.6	7.0	6.4	7.8	13.5	15.1	17.6	655–1380	95–200	8.41	0.30
68.5Ni-30Cr-1.5Si (70-30)	1.180	2.1	4.8	7.6	9.8	12.2	. . .	. . .	825–1380	120–200	8.12	0.29
68Ni-20Cr-8.5Fe-2Si	1.165	3.9	6.7	6.0	7.1	. . .	12.6	. . .	895–1240	130–180	8.33	0.30
60Ni-16Cr-22Fe-1.5Si	1.120	3.6	6.5	7.6	10.2	13.5	15.1	17.6	655–1205	95–175	8.25	0.30
37Ni-21Cr-40Fe-2Si	1.08	7.0	15.0	20.0	23.0	14.4	16.5	18.6	585–1135	85–165	7.96	0.288
35Ni-20Cr-43Fe-1.5Si	1.00	8.0	15.4	20.6	23.5	15.7	15.7	. . .	550–1205	80–175	7.95	0.287
35Ni-20Cr-42.5Fe-1.5Si-1Nb	1.00	8.0	15.4	20.6	23.5	15.7	15.7	. . .	550–1205	80–175	7.95	0.287
Iron-chromium-aluminum alloys												
83.5Fe-13Cr-3.25Al	1.120	7.0	15.5	. . .	. . .	10.6	. . .	. . .	620–1035	90–150	7.30	0.26
81Fe-14.5Cr-4.25Al	1.25	3.0	9.7	16.5	. . .	10.8	11.5	12.2	620–1170	90–170	7.28	0.26
73.5Fe-22Cr-4.5Al	1.35	0.3	2.9	4.3	4.9	10.8	12.6	13.1	620–1035	90–150	7.15	0.26
72.5Fe-22Cr-5.5Al	1.45	0.2	1.0	2.8	4.0	11.3	12.8	14.0	620–1035	90–150	7.10	0.26
Pure metals												
Molybdenum	0.052	110	238	366	508	4.8	5.8	. . .	690–2160	100–313	10.2	0.369
Platinum	0.105	85	175	257	305	9.0	9.7	10.1	345	50	21.5	0.775
Tantalum	0.125	82	169	243	317	6.5	6.6	. . .	345–1240	50–180	16.6	0.600
Tungsten	0.055	91	244	396	550	4.3	4.6	4.6	3380–6480	490–940	19.3	0.697
Nonmetallic heating-element materials												
Silicon carbide	0.995–1.995	−33	−33	−28	−13	4.7	. . .	. . .	28	4	3.2	0.114
Molybdenum disilicide	0.370	105	222	375	523	9.2	. . .	. . .	185	27	6.24	0.225
$MoSi_2$ + 10% ceramic additives	0.270	167	370	597	853	13.1	14.2	14.8	. . .	. . .	5.6	0.202
Graphite	9.100	−16	−18	−13	−8	1.3	. . .	. . .	1.8	0.26	1.6	0.057

(a)At 20 °C (68 °F). (b) To convert to Ω·circ mil/ft, multiply by 601.53. (c) Changes in resistance may vary somewhat, depending on cooling rate.

Table 6 Recommended maximum furnace operating temperatures for resistance heating materials

Basic composition, %	Approximate melting point °C	Approximate melting point °F	Maximum furnace operating temperature in air °C	Maximum furnace operating temperature in air °F
Nickel-chromium and nickel-chromium-iron alloys				
78.5Ni-20Cr-1.5Si (80-20)	1400	2550	1150	2100
77.5Ni-20Cr-1.5Si-1Nb	1390	2540		
68.5Ni-30Cr-1.5Si (70-30)	1380	2520	1200	2200
68Ni-20Cr-8.5Fe-2Si	1390	2540	1150	2100
60Ni-16Cr-22Fe-1.5Si	1350	2460	1000	1850
35Ni-30Cr-33.5Fe-1.5Si	1400	2550		
35Ni-20Cr-43Fe-1.5Si	1380	2515	925	1700
35Ni-20Cr-42.5Fe-1.5Si-1Nb	1380	2515		
Iron-chromium-aluminum alloys				
83.5Fe-13Cr-3.25Al	1510	2750	1050	1920
81Fe-14.5Cr-4.25Al	1510	2750		
79.5Fe-15Cr-5.2Al	1510	2750	1260	2300
73.5Fe-22Cr-4.5Al	1510	2750	1280	2335
72.5Fe-22Cr-5.5Al	1510	2750	1375	2505
Pure metals				
Molybdenum	2610	4730	400(a)	750(a)
Platinum	1770	3216	1500	2750
Tantalum	3000	5400	500(a)	930(a)
Tungsten	3400	6150	300(a)	570(a)
Nonmetallic heating-element materials				
Silicon carbide	2410	4370	1600	2900
Molybdenum disilicide	(b)	(b)	1700–1800	3100–3270
$MoSi_2$ + 10% ceramic additives	(b)	(b)	1900	3450
Graphite	3650–3700(b)	6610–6690(c)	400(d)	750(d)

(a) Recommended atmospheres for these metals are a vacuum of 10^{-4} to 10^{-5} mm Hg, pure hydrogen, and partly combusted city gas dried to a dew point of 4 °C (40 °F). In these atmospheres the recommended temperatures would be:

Element	Vacuum	Pure H_2	City gas
Mo	1650 °C (3000 °F)	1760 °C (3200 °F)	1700 °C (3100 °F)
Ta	2480 °C (4500 °F)	Not recommended	Not recommended
W	1650 °C (3000 °F)	2480 °C (4500 °F)	1700 °C (3100 °F)

(b) See the property data on molybdenum disilicide at the end of this article. (c) Graphite volatilizes without melting at 3650 to 3700 °C (6610 to 6690 °F). (d) At approximately 400 °C (750 °F) (threshold oxidation temperature), graphite undergoes a weight loss of 1% in 24 h in air. Graphite elements can be operated at surface temperatures up to 2205 °C (4000 °F) in inert atmospheres.

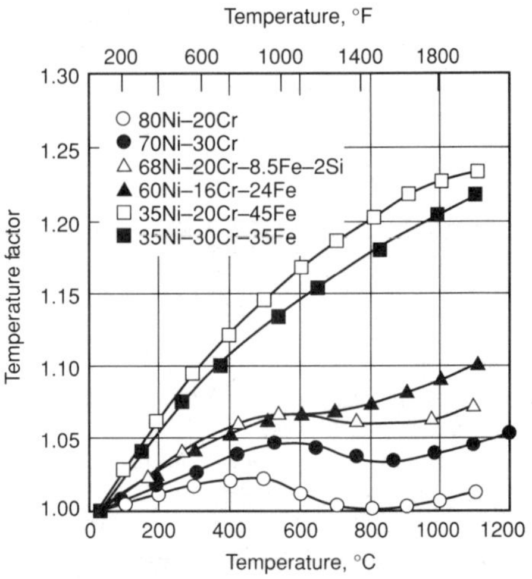

Fig. 4 Variation of resistance with temperature for six Ni-Cr and Ni-Cr-Fe alloys. To calculate resistance at temperature, multiply resistance at room temperature by the temperature factor.

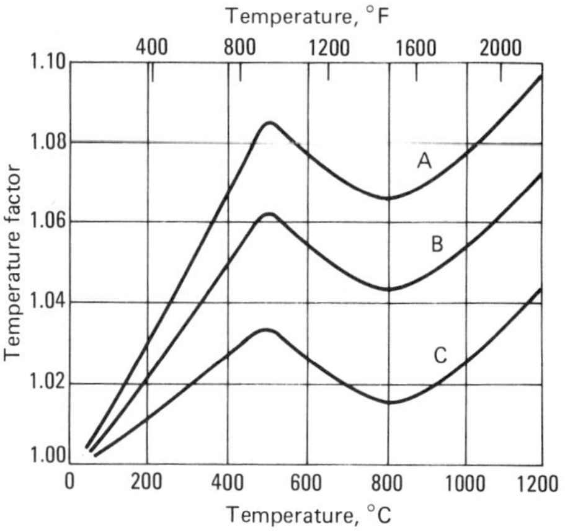

Fig. 5 Variation of resistance with temperature for 80Ni-20Cr heating alloy. Curve A is for a specimen cooled rapidly after the last production heat treatment. Curve C is for a specimen cooled slowly after the last production heat treatment. Curve B represents the average value for material as delivered by the producer. To calculate resistance at temperature, multiply the resistance at room temperature by the temperature factor.

temperatures. Silicon carbide can be used in oxidizing atmospheres at temperatures up to 1650 °C (3000 °F); three varieties of molybdenum disilicide are effective up to maximum temperatures of 1700, 1800, and 1900 °C (3100, 3270, and 3450 °F) in air. Molybdenum disilicide heating elements are gaining increased acceptance for use in industrial and laboratory furnaces. Among the desirable properties of molybdenum disilicide elements are excellent oxidation resistance, long life, constant electrical resistance, self-healing ability, and resistance to thermal shock. Nonmetallic heating elements described are considerably more fragile as compared to metal heating alloys.

Nickel-Chromium and Nickel-Chromium-Iron Alloys. The resistivities of Ni-Cr and Ni-Cr-Fe alloys are high, ranging from 1000 to 1187 nΩ · m (600 to 714 Ω · circ mil/ft) at 25 °C. Figure 4 shows that the resistance changes more rapidly with temperature for 35Ni-20Cr-45Fe than for any other alloy in this group. The curve for 35Ni-30Cr-35Fe (which is no longer produced) is similar, but slightly lower. The other four curves, which are for alloys with substantially higher nickel contents, reflect relatively low changes in resistance with temperature. For these alloys, rate of change reaches a peak near 540 °C (1000 °F), goes through a minimum at about 760 to 870 °C (1400 to 1600 °F) and then increases again. For Ni-Cr alloys, the change in resistance with temperature depends on section size and cooling rate. Figure 5 presents values for a typical 80Ni-20Cr alloy. The maximum change (curve A) occurs with small sections, which cool rapidly from the last production heat treatment. The smallest change occurs for heavy sections, which cool slowly. The average curve (curve B) is characteristic of medium-size sections.

Iron-Chromium-Aluminum Alloys. Fe-Cr-Al heating alloys are higher in electrical resistivity and lower in density than Ni-Cr and Ni-Cr-Fe alloys. Resistivity of Fe-Cr-Al alloys depends on both aluminum and chromium contents, with aluminum being predominant (see Fig. 6).

These alloys have excellent resistance to oxidation at elevated temperatures because reaction with atmospheric oxygen forms a protective layer of relatively pure alumina. At about 1200 °C (2200 °F), this oxide consists of nearly pure Al_2O_3. This gray-white protective skin has extremely high dielectric strength. The electrical resistivity of aluminum oxide is 10^{12} Ω · m at room temperature, and at about 1100 °C (2000 °F) it is still 10^4 Ω · m. Under normal operating conditions, deterioration of the oxide surface layer, and the resulting aluminum depletion, are fairly slow provided that there is no contact with certain refractories at temperatures above 980 °C (1800 °F). The time required for a 10% change in resistance varies from 75 to 100% of heater life (time to burnout), depending on the particular melt, the size of the heater, and the operating temperature.

Tensile strength of Fe-Cr-Al alloys is relatively low, as shown in Table 7. Because of this low strength, the weight of the lower terminal in straight-wire testing

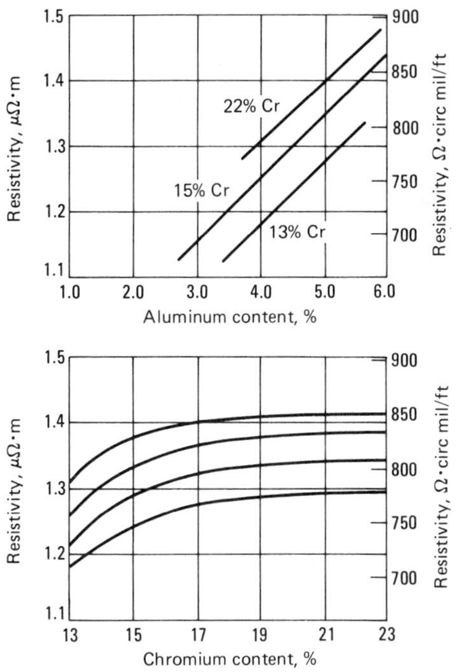

Fig. 6 Effects of aluminum and chromium on resistivity of Fe-Cr-Al heating alloys

causes the wire to stretch; consequently, life tests on Fe-Cr-Al alloys are conducted using U-shape specimens. Life of heaters made from Fe-Cr-Al alloys is also influenced by the fact that these alloys exhibit large increases in resistance with time as a result of their grain-growth characteristics.

Iron-chromium-aluminum alloys undergo a metallurgical change that causes brittleness after cyclic exposure to high temperatures. As a result, when heaters made of these alloys fail, repair is more difficult than with nickel-chromium alloys. However, new Fe-Cr-Al material has been developed that overcomes several of the disadvantages of the traditional Fe-Cr-Al materials. This new Fe-Cr-Al material is produced with powder metallurgical techniques and results in an alloy that has higher hot strength, as well as having a hot resistance change with time, no worse than nickel-chromium.

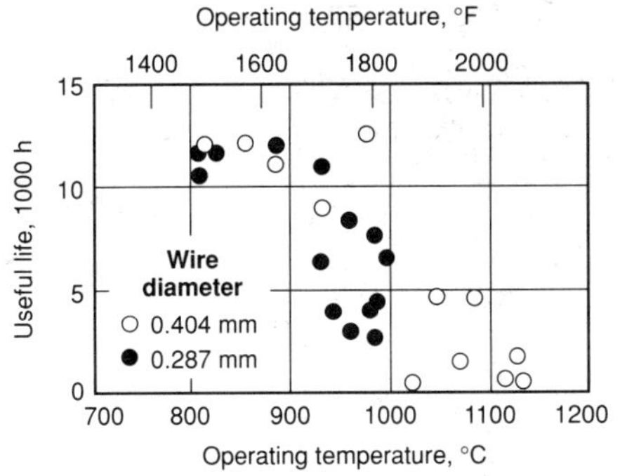

Fig. 7 Total life versus temperature for two sizes of 74.5Fe-20Cr-5Al-0.5Co wire. The alloy included 0.5% Co, which served as a means of identification years ago by one manufacturer.

Figure 7 presents data on life of 74.5Fe-20Cr-5Al-0.5Co wire 0.40 and 0.29 mm (0.0159 and 0.0113 in.) in diameter at a series of temperatures. With Fe-Cr-Al alloys, as with Ni-Cr alloys, life of a heater of given size decreases as temperature increases.

Pure metals used as heating alloys (see Table 5) have very low resistivities and very high-temperature coefficients of resistance. The tensile strengths of molybdenum and tungsten are quite high, even at elevated temperatures, and that of tantalum is medium (see Table 7). Above 800 °C (1500 °F), all three of these refractory metals are substantially stronger than Ni-Cr, Ni-Cr-Fe, and Fe-Cr-Al alloys at any given temperature.

Platinum can be used for temperatures up to 1500 °C (2750 °F) because it has excellent resistance to oxidation in air. Molybdenum, tantalum, and tungsten must be kept below 400, 500, and 300 °C (750, 930, and 570 °F), respectively, because they are subject to catastrophic oxidation in air at moderate and elevated temperatures.

Nonmetallic Materials. Of the nonmetallic heating materials, graphite and silicon carbide are much higher in resistivity than wrought metallic heating materials. The temperature coefficient of resistance of these materials is negative. Molybdenum disilicide elements have relatively low resistivity at room temperature and a very high positive temperature coefficient of resistance.

All of these materials have low tensile strengths. However, because the resistivity of silicon carbide is so high, silicon carbide elements are made in large cross sections in order to reduce resistance to reasonable values, and as a result these elements can withstand relatively high mechanical loads. Graphite has poor oxidation resistance and should not be used above 400 °C (750 °F). Silicon carbide, however, has excellent oxidation resistance and can be used at temperatures up to 1600 or 1650 °C (2900 or 3000 °F).

Two materials composed of about 90% molybdenum disilicide and 10% refractory oxides have been produced. One has a maximum operating temperature of 1700 °C (3100 °F), while the other serves at temperatures up to 1800 °C (3270 °F).

Pure molybdenum disilicide is too brittle for practical use as an electric heating element. Addition of metallic and ceramic binding agents reduces this brittleness to a practical level. Sintering in the presence of a liquid phase then produces a material that is essentially free of porosity. The high resistance to thermal shock of $MoSi_2$ has enabled $MoSi_2$ elements to undergo, without damage, a test in which they were cycled from room temperature to 1650 °C (3000 °F) through 20 000 cycles.

The excellent high-temperature performance of molybdenum disilicide elements is brought about by the chemical reaction that takes place on the element surface.

Above 980 °C (1800 °F), the material reacts with oxygen to form a silicon dioxide (quartz glass) coating, which protects the base material against chemical attack including further oxidation. This film is "self-healing:" surface cracks developed by mechanical damage are covered by a new coating of quartz glass when the element is again heated above 980 °C (1800 °F) in air.

Design of Open Resistance Heaters

Regardless of which heating alloy is selected, design of the heating element is important. One of the most important rules is to allow for unhindered expansion and contraction so as to avoid concentration of stresses as the temperature changes.

For service at lower temperatures, particularly from 400 to 600 °C (750 to 1100 °F), formed heating elements are used in ovens. In this construction, a heater support is made of two high-alloy rods spaced approximately 300 mm (12 in.) apart in a frame made of angle sections. The rods, whose length is determined by rated electrical input, contain spool insulators around which is wound a ribbon element made of a heating

Table 7 Elevated-temperature tensile strength of selected resistance heating materials

	Tensile strength at:							
	425 °C (800 °F)		650 °C (1200 °F)		870 °C (1600 °F)		1100 °C (2000 °F)	
Heating material	MPa	ksi	MPa	ksi	MPa	ksi	MPa	ksi
Nickel-chromium and nickel-chromium-iron alloys								
68.5Ni-30Cr-1.5Si	735	107	675	98	205	30	75	11
78.5Ni-20Cr-1.5Si	715	104	620	90	170	25	75	11
68Ni-20Cr-8.5Fe-1.5Si	760	110	655	95	195	28	75	11
Iron-chromium-aluminum alloys								
79.5Fe-15Cr-5.2Al	480	70	205	30	48	7	...	...
73.5Fe-22Cr-4.5Al	525	76	165	24	14	2	...	...
72.5Fe-22Cr-5.5Al	550	80	345	50	52	7.5	26	3.8
Pure metals								
Tungsten	560	81	525	76	395	57	295	43
Molybdenum	620	90	585	85	365	53	235	34
Tantalum	315	46	315	46	280	41	195	28

Table 8 Typical length and spacing of loops in Ni-Cr, Ni-Cr-Fe, and Fe-Cr-Al heating elements

Maximum operating temperature		Loop spacing(a) for ribbon width of: Up to 19 mm (3/4 in.)		Over 19 mm (3/4 in.)		Maximum loop length(b)	
°C	°F	mm	in.	mm	in.	mm	in.
Sidewall heaters							
540–760	1000–1400	50–75	2–3	65–75	2.5–3	450	18
760–1100	1400–2000	50–75	2–3	65–75	2.5–3	450	18
1100–1175	2000–2150	65–75	2.5–3	75	3	300	12
Roof heaters							
540–760	1000–1400	50–75	2–3	50–75	2–3	300	12
760–1100	1400–2000	50–75	2–3	50–75	2–3	300	12
1100–1175	2000–2150	50–75	2–3	50–75	2–3	300	12
Floor heaters							
540–760	1000–1400	40 min	1.5 min	55 min	2.25 min	450	18
760–1100	1400–2000	40 min	1.5 min	55 min	2.25 min	450	18
1100–1175	2000–2150	55 min	2.25 min	55 min	2.25 min	450	18

(a) Loop spacing is the lineal distance between centers for two adjacent bends on the same side of the element. (b) Loop length is the overall lineal distance from one side of the element to the other. For elements over 300 mm (12 in.) in loop length, the loops must be separated or supported by additional insulators. Two end supports or one hanger and a bottom guide separator are sufficient for elements under 300 mm (12 in.) in length.

Table 9 Typical ribbon size and electrical capacity of Ni-Cr, Fe-Cr-Al, and Ni-Cr-Fe heating elements

Maximum operating temperature		Size of ribbon: Min thickness		Width		Power density	
°C	°F	mm	in.	mm	in.	kW/m^2	$W/in.^2$
540–760	1000–1400	0.75	0.030	Any	Any	21.5	14
760–925	1400–1700	1.8	0.070	13–40	½–1½	18.5	12
925–1100	1700–2000	2.3	0.090	20–40	¾–1½	15.5	10
1100–1175	2000–2150	2.5	0.100	20–40	¾–1½	12.5	8

alloy. In a similar alternative construction, the ribbon element is replaced by a continuous helical coil of 5-gage or smaller wire.

Ribbon sizes for oven heaters range from 0.09 to 0.20 mm (0.0035 to 0.008 in.) thick, and from 9.5 to 16 mm (⅜ to ⅝ in.) wide. Oven heaters are rated to give maximum output at a watt density of approximately 8 kW/m^2 (5 $W/in.^2$). (Watt density is obtained by dividing total power input to the elements by total surface area of the heater.) For 120- or 240-V oven heaters operating under normal conditions, expected life of Ni-Cr elements in air is three to five years (depending on temperature, atmosphere, and cyclic conditions).

For furnace temperatures up to 1175 °C (2150 °F), sinuous loop elements generally are formed from ribbon having a width-to-thickness ratio of about 12 to 1 and dimensions varying from 0.76 to 3.2 mm (0.030 to 0.125 in.) in thickness and from 13 to 38 mm (½ to 1½ in.) in width. The dimensions for various temperature ranges are listed in Table 8. Round rods of resistance material also may be formed into elements. Rod-type elements have been used by several furnace manufacturers.

Dimensional relationships of loops are important in achieving the desired combination of uniform furnace temperature and long heater life. Recommended loop dimensions for various locations within a furnace are shown in Table 8 for several different ranges of operating temperature. Ribbon size and watt density are correlated with operating temperature in Table 9. When designed in accordance with these recommendations, Ni-Cr elements have life expectancies of up to seven years at temperatures of 540 to 925 °C (1000 to 1700 °F), two to five years at 980 to 1100 °C (1800 to 2000 °F), and one to two years at 1100 to 1175 °C (2000 to 2150 °F).

Iron-chromium-aluminum heating elements used for temperatures up to 1300 °C (2350 °F) also may be made from ribbon. However, Fe-Cr-Al ribbon must be formed into short, sinuous loops having a loop spacing of 50 mm (2 in.) or less, and requires better loop support than Ni-Cr ribbon. In the design of heating elements, for maximum life the designer should consider factors such as the lowest practical voltage, maximum cross-sectional area, and lowest watt density consistent with design and reasonable cost.

Refractory Supports. Ceramic refractories that come in contact with heating elements may influence selection of heating alloys. Below 1000 °C (1825 °F), protective oxides that form on the surfaces of Ni-Cr, Ni-Cr-Fe, and Fe-Cr-Al alloys do not react with ceramic oxides, including refractory grades of SiO_2, Al_2O_3, CaO, Na_2O, MgO, K_2O, and ZrO_2. Above 1000 °C, pure MgO, Al_2O_3, and ZrO_2 are recommended. Many ordinary refractory-grade materials become conductive at such temperatures; the sodium and potassium contents of the refractory material should be low to prevent this. Sulfur-containing refractories should not be used with Ni-Cr, Ni-Cr-Fe, or Fe-Cr-Al alloys. Refractories also must be as low as possible in ferric oxide, if they are to be used with Fe-Cr-Al resistance elements.

Use of molybdenum, tungsten, or platinum heating elements at temperatures above 1200 °C (2200 °F) necessitates use of pure oxide refractories. High-purity alumina (99%) and magnesia are the most satisfactory. Zirconia becomes conductive above 1300 °C (2350 °F); silica decomposes and embrittles platinum at about 1200 °C (2200 °F). Consequently, neither zirconia nor silica can be used at or above these temperatures.

In cyclic temperature applications, some alloys elongate continuously and at the same time continuously decrease in cross-sectional area. Iron-chromium-aluminum alloys do this even in the absence of applied external force. A popular but unproved explanation of this growth is as follows: first, the alloy oxidizes at elevated temperature; on cooling, the high compressive strength of the oxide layer forces the weaker core to elongate; on reheating, the oxide is weak in tension so that it cracks, and reoxidation takes place; thus growth continues in a cyclic manner. Growth is detrimental because it causes large changes in resistance, so suitable steps must be taken in the design of high-temperature elements to minimize cyclic growth.

Iron-chromium-aluminum heating alloys must be adequately supported for use at temperatures near 1300 °C (2350 °F).

Nickel-chromium alloys, which exhibit little or no growth, do not require special support at their maximum operating temperatures, which are 200 °C (360 °F) lower than those of Fe-Cr-Al alloys (1400 °C versus 1200 °C). Although all electrical heating alloys have low tensile properties at high temperatures, Ni-Cr and Ni-Cr-Fe alloys have higher strength at elevated temperature than do Fe-Cr-Al alloys. For example, 80Ni-20Cr tested at 1100 °C (2000 °F), and a strain rate of 3.3 mm/m · s (0.2 in./in. · min) exhibited yield and tensile strengths of 41 and 48 MPa (6 and 7 ksi), respectively. An Fe-Cr-Al alloy tested under identical conditions exhibited values of only 12 and 16 MPa (1.8 and 2.3 ksi). The new P/M Fe-Cr-Al alloy has values of about 20 MPa (2.9 ksi).

One way of establishing allowable stresses in heating elements is to define allowable stress as the load that will produce 0.1% creep in 1000 h or 1% creep in 10 000 h. For 80Ni-20Cr, this load is 1.4 MPa (200 psi) at 1100 °C (2000 °F). For the P/M formed Fe-Cr-Al product, the load is 0.8 MPa (116 psi) at 1100 °C (2000 °F).

Fabrication of Open Resistance Heaters

Annealed wire, rod, or ribbon of any common nickel-chromium heating alloy can be bent at room temperature around a mandrel whose diameter equals the diameter or thickness of the stock. Some difficulty has been experienced in making such bends in 80Ni-20Cr alloys, because these alloys tend to strain age during forming if they have been heated in the range 100 to 200 °C (200 to 400 °F) between forming stages.

Iron-chromium-aluminum alloys are harder at room temperature, and somewhat more difficult to form, than Ni-Cr alloys. They can be shaped into heating elements by techniques much like those used for making elements from Ni-Cr alloys, but heavy-gage material may require preheating to 150 °C (300 °F) to facilitate forming of sharp radii.

Joining. Nickel-chromium and nickel-chromium-iron alloys are readily joined by welding. Preferred processes include resistance welding, gas-shielded metal-arc welding, and oxyfuel gas welding. Filler metal of essentially the same composition as the base alloy should be used. These alloys can be brazed, but heating elements are seldom joined in this way because brazing introduces a zone whose melting temperature is lower than that of the base metal.

Iron-chromium-aluminum alloys are weldable, but they should be welded as quickly as possible to prevent grain growth. Postheating, which sometimes is done for stress relief, must be done at a temperature low enough to prevent grain growth. A temperature range of 700 to 800 °C (1290 to 1470 °F) is recommended.

Sheathed Heaters

Nickel-chromium (80Ni-20Cr), nickel-chromium-iron (60Ni-16Cr-22.5Fe-1.5Si), and iron-chromium-aluminum alloys are extensively used as heating elements in sheathed heaters, where compacted granular magnesium oxide provides electrical insulation between a wire element and a metallic sheath. This construction permits operation at high watt density without rapid degradation. Various grades of magnesium oxide are available for use at different temperature levels.

Insulation resistance and heater life are affected by factors such as chemical and physical characteristics of the insulation, operating temperature, and atmospheric conditions within the heater sheath. In a sheathed heater helix, strain aging can lead to nonuniform stretching during forming of the helix, which in turn can lead to hot spots in service. Penetration of oxidation products into the insulating layer of magnesium oxide may occur, causing a reduction in insulation resistance with prolonged cyclic testing or use. This can result in excessive electrical leakage to the metallic sheath and failure. In the preparation of element wire and fabrication of heaters, care must be taken to ensure uniformity of cross section and to avoid surface damage that could shorten service life.

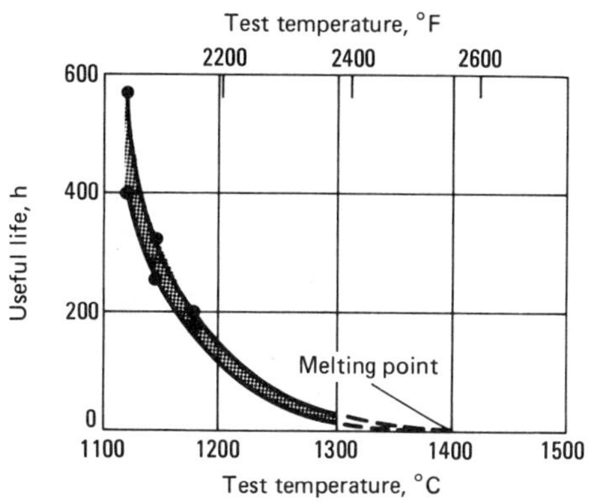

Fig. 8 Variation in useful life with temperature in cyclic testing of 80Ni-20Cr heating elements. Life is defined as time to burnout or to 10% change in electrical resistance, whichever occurs first. Tests were conducted in accordance with ASTM B 76; specimens were at temperature for 2 min out of every 4-min cycle.

In the design of sheathed heaters, it must be recognized that the operating temperature of the internal heating element is considerably higher than that of the external sheath. Selection of heater material is based on expected operating temperature and environment. Electrical-grade, fused magnesium oxide is used as insulation in most sheathed heaters. The MgO is approximately 94% pure; however, higher purity can be obtained if required. The melting point of MgO is approximately 2750 °C (5000 °F), which is far above the melting point of metal resistor alloys or sheath materials. Electrical insulating properties of MgO decrease with increase in temperature. Consideration must be given to the dielectric strength of MgO at operating temperature when designing metal-sheathed heaters. Iron-chromium-aluminum alloys should only be used either when tube temperature is below 700 °C (1290 °F), or when the wire temperature is below 850 °C (1560 °F).

Service Life of Heating Elements

Maximum service temperature is one of the most important factors governing service life of heating elements. Whether the temperature is constant or intermittent also has a marked effect. Data from accelerated laboratory testing (Fig. 8) illustrate the effects of temperature on 80Ni-20Cr heating elements. The graph shows a rapid decrease in element life as temperature increases above 1120 °C (2050 °F).

The life of a heating element is dependent upon many factors. These include method of melting, composition, environment, and other design parameters. Manufacturers of heating elements use an ASTM life test to control their product. Even under simplified and closely controlled test conditions, there is still a variation in element life, as shown by the data in Fig. 9 for 56 identical specimens of 80Ni-20Cr tested at 1175 °C (2150 °F) in air. In actual service, the construction of a resistor, environment, and the operation per se can vary so greatly that direct correlation with the ASTM life test is not possible. Thus, manufacturers of wire and ribbon use the ASTM test as an internal control tests, whereas manufacturers of devices or appliances use tests appropriate to the unit being evaluated.

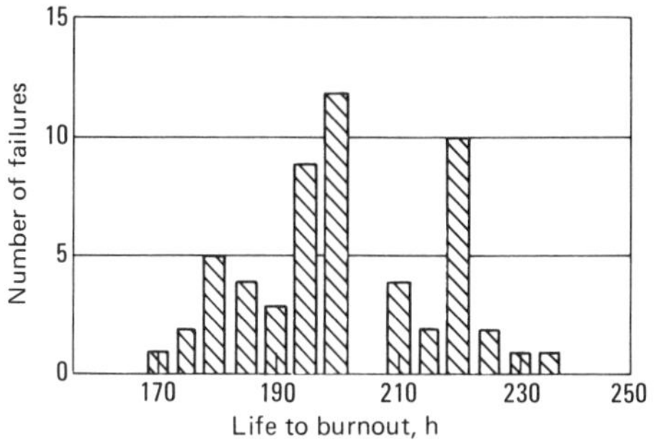

Fig. 9 Distribution of life to burnout in 56 identical tests of 80Ni-20Cr elements at 1175 °C (2150 °F)

The life of a heating element increases with ribbon thickness or wire diameter, as shown in Fig. 10. In some applications, heaters may be required to operate for 10 to 15 years without element failure. Predictions, based on data from accelerated life tests, that elements will achieve such extended service lives are not reliable. Often, maximum heater temperatures must be lowered considerably from those normally given in data sheets, in order to ensure exceptionally long element life.

Oxidation resistance of alloys used for heating elements is critical. In addition to the inherent oxidation resistance necessary for any alloy used at elevated temperatures, heating alloys also must have adherent oxides that resist spalling during temperature cycling. Because elements are heated electrically to attain temperature, oxidation and spalling in a localized area result in a local increase in resistance with consequent increase in local temperature, which creates a hot spot and shortens the life of the element. Because localized oxidation or spalling increases total resistance only slightly, the current through the element remains essentially constant, and I^2R heating causes an increase in temperature only in the region of increased resistance.

The effect of composition on the life of heating-element alloys is evaluated by both static and cyclic testing. In static testing, the element is heated to an elevated temperature and held for a prescribed time, and the weight gain or loss due to oxidation is measured. In cyclic testing, the element is alternately:

- Heated to an elevated temperature and held for a prescribed time
- Cooled to ambient temperature and held for an equal time

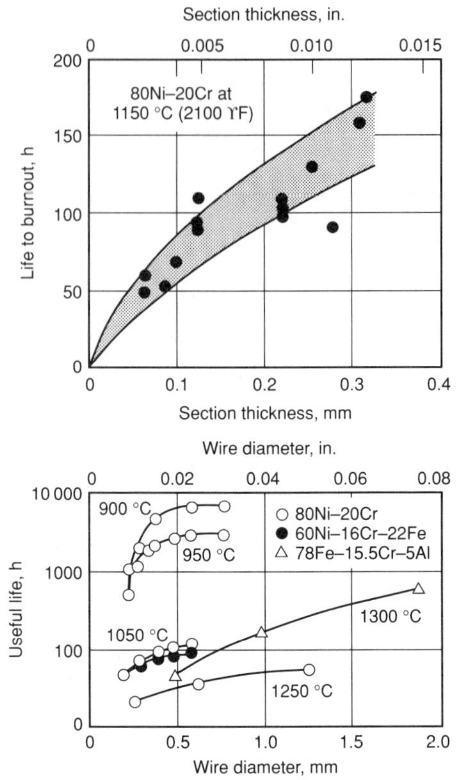

Fig. 10 Variation of heater life with section thickness or wire diameter. Life is defined as time to burnout or to 10% change in resistance, whichever occurs first.

The weight loss in percent, or total testing time in hours (total life), is used as a measure of the quality of the alloy. Either low weight loss or long life is desired. For heating elements used with a fixed voltage, life is defined as the time in hours to burnout or to a 10% increase in total resistance, whichever occurs first. Any increase in total resistance results in a decrease in power and consequently a decrease in temperature.

Figure 11(a) compares two alloys that were continuously oxidized in air at 1175 °C (2150 °F). Cylindrically coiled strips 0.13 by 9.5 by 250 mm (0.005 by 0.375 by 10 in.) were used. Oxidation was evaluated by determining the thickness of the oxide layer. Figure 11(b) presents test results for the same two alloys oxidized under similar conditions, except that heating was intermittent.

Chemical composition of heating-element alloys in the first two groups in Table 5 is carefully controlled to achieve resistance to intermittent oxidation in oxidizing environments. Figure 12 presents data on weight loss of Ni-Cr-Fe alloys as nickel-plus-chromium content is increased from just over 20% to almost 100%. In practice, nickel-plus-chromium content is maintained above 50% to inhibit the oxidation and spalling that lead to large weight changes in high-iron compositions. Figure 13 shows the weight gain in Ni-Cr alloys after continuous oxidation for 100 h at 1100 °C (2000 °F). Increasing the chromium content of Ni-Cr alloys causes a substantial decrease in oxidation rate. Minimum weight gain is obtained at about 30% Cr.

Minor chemical constituents are important in governing life at elevated temperatures. Close control of minor elements, especially silicon, has substantially increased life of Ni-Cr heating alloys. An example of the effect of silicon on the life of 80Ni-20Cr alloys is shown in Fig. 14.

The five graphs in Fig. 15 show the effects of manganese and silicon contents on rate of oxidation of 80Ni-20Cr alloys under the conditions of time and temperature indicated. Specimens for these tests were first polished with emery papers through No. 0000. They were tested in oxygen at a pressure of 10 kPa (76 mm Hg), which constitutes accelerated testing. Such tests performed under controlled conditions give some useful information, but they cannot necessarily be correlated with element life

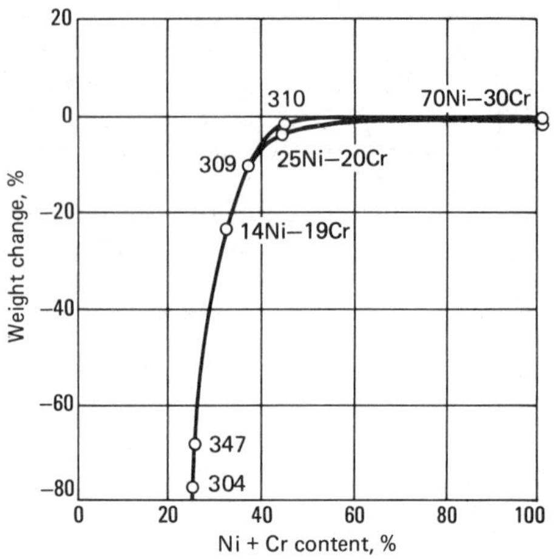

Fig. 12 Intermittent oxidation of Ni-Cr, Ni-Cr-Fe, and stainless-steel heating elements in air. Weight change was determined after 400 h in a cyclic test at 980 °C (1800 °F) where the power was cycled 15 min on and 15 min off.

for any specific application. Element life is related to oxide stability and adherence under actual conditions of use; life is affected by furnace atmosphere, temperature cycling, and contact with other materials (such as refractories).

Addition of 1% Nb to 80Ni-20Cr increases resistance to preferential oxidation of chromium, commonly called "green rot." Green rot occurs when Ni-Cr alloys are exposed to environments with low partial pressures of oxygen at temperatures from 870 to 1040 °C (1600 to 1900 °F); maximum oxidation rate occurs at 955 °C (1750 °F). Green rot is most common in 90Ni-10Cr alloys used in thermocouples. However, it can occur in 80Ni-20Cr alloys exposed for long periods of time. Although green rot in 70Ni-30Cr alloys is theoretically possible, it has not occurred during testing in which such alloys have been exposed for over fifteen years.

Addition of niobium to a 35Ni-20Cr-45Fe alloy stabilizes the carbon by forming nio-

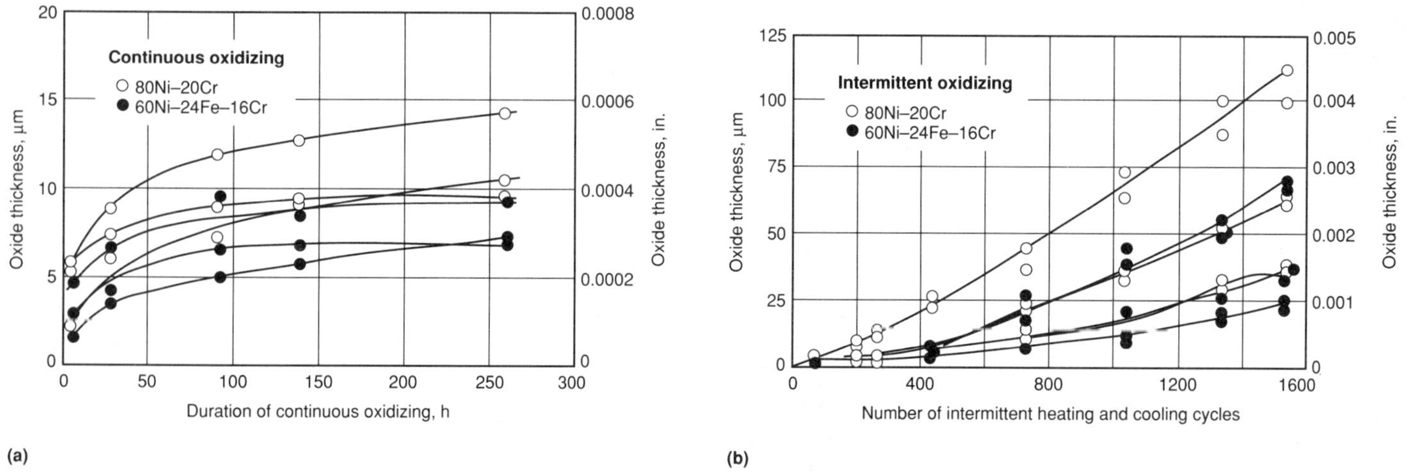

Fig. 11 Typical rates of oxidation at 1175 °C (2150 °F) for two common heating alloys. In both continuous tests (a) and intermittent tests (b), cylindrically coiled strips 0.13 by 9.5 by 250 mm (0.005 by ⅜ by 10 in.) were heated in air. In the intermittent oxidizing tests, power was cycled 7.5 min on and 7.5 min off.

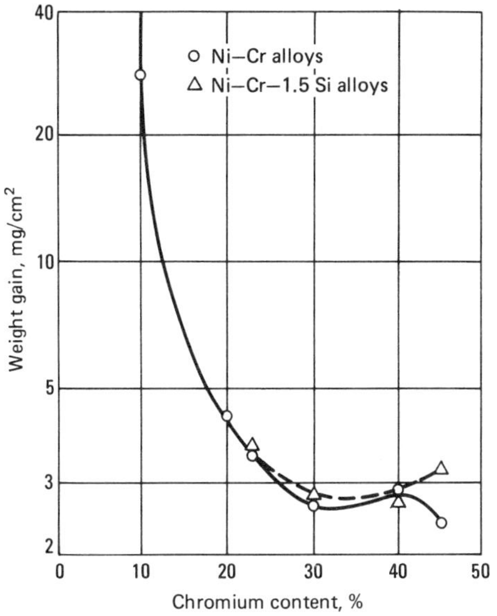

Fig. 13 Continuous oxidation of Ni-Cr heating alloys held 100 h at 1100 °C (2000 °F)

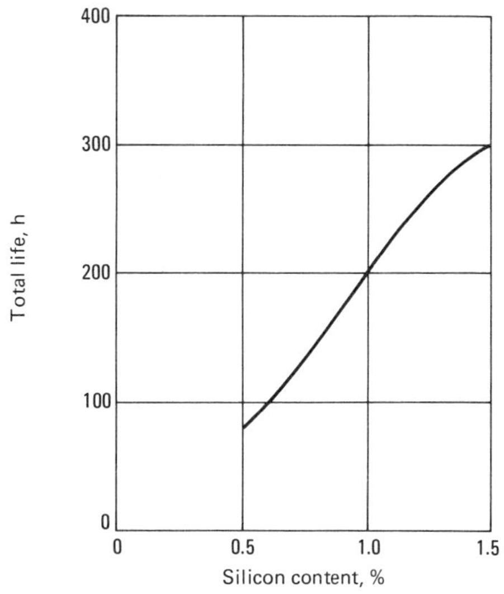

Fig. 14 Total life versus silicon content for 80Ni-20Cr alloys. Total life was determined for 0.64 mm (0.025 in.) diam wire in intermittent tests in air at 1175 °C (2150 °F).

bium carbides. If carbon is tied up in this manner, heating elements remain ductile when stressed during service.

Atmospheres

Based on element temperature, Table 10 rates serviceabilities of various heating-element materials as good, fair, or not recommended for the temperatures and atmospheres indicated. Element temperatures are always higher than furnace control temperatures; the difference depends on watt-density loading on the element surface. Thus, when furnaces are operated near maximum element temperature in the more active atmospheres, watt-density loading should be lower and element cross-sectional area should be higher.

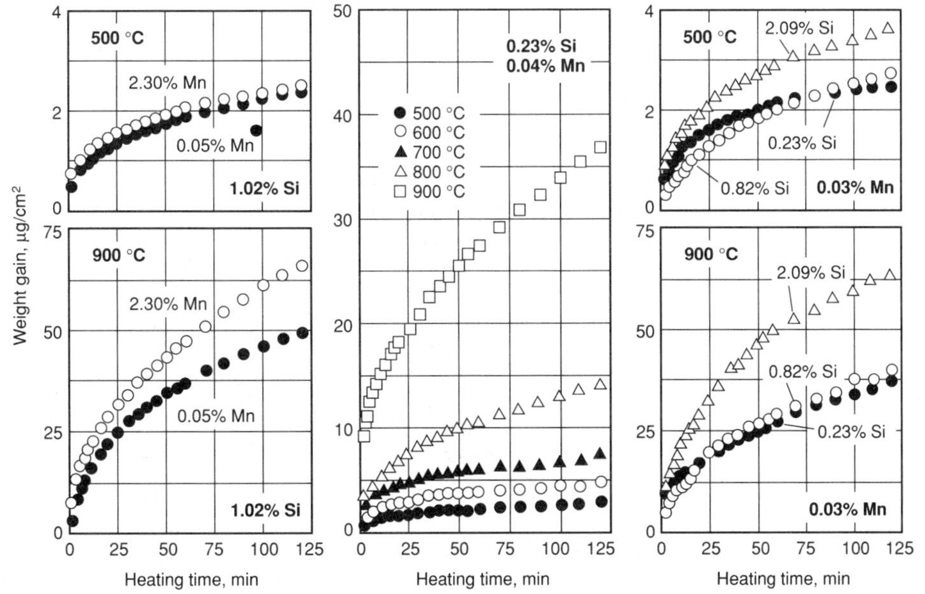

Mn	Si	Cr	Ni	Fe	Al	C	Useful life, h
2.30	1.02	20.00	bal	0.25	0.13	0.06	105
0.05	1.02	20.00	bal	0.25	0.15	0.06	106
0.04	0.23	20.00	bal	0.25	0.20	0.06	63
0.03	2.09	20.00	bal	0.25	0.17	0.06	124
0.03	0.82	20.00	bal	0.25	0.15	0.06	178

Fig. 15 Effects of silicon and manganese contents on continuous oxidation of 80Ni-20Cr alloys. Alloys can be identified in the table and in the graphs, by manganese and silicon contents. Specimens were first polished with emery papers through No. 0000 and then tested in pure oxygen at 10 kPa (0.1 atm, technical), which constitutes accelerated testing.

With the exception of molybdenum, tantalum, tungsten, and graphite, commonly used resistor materials have satisfactory life in air and in most other oxidizing atmospheres.

Oxidizing Atmospheres. Nickel-chromium and nickel-chromium-iron alloys are the most widely used heating materials in electric heat-treating furnaces. The 80Ni-20Cr alloys are more commonly used than the 60Ni-16Cr-20Fe or the 35Ni-20Cr-45Fe types. In fact, most electric-furnace manufacturers provide 80Ni-20Cr elements as standard, both because they permit a wider range of furnace temperatures and because it is usually more economical to stock only a limited number of heater materials. The 80Ni-20Cr alloys permit a wider range of operating temperatures because they have the greatest resistance to oxidation, and therefore can be used at higher temperatures than other Ni-Cr and Ni-Cr-Fe alloys. Heating elements of lower nickel content are required for certain special applications, such as where an oxidizing atmosphere contaminated with sulfur, lead, or zinc is present.

The iron-chromium-aluminum alloys are widely used in furnaces operating at 800 to 1300 °C (1500 to 2350 °F). In general, Ni-Cr heating elements are unsuitable above 1150 °C (2100 °F) because the oxidation rate in air is too great and the operating temperature is too close to the melting point of the alloy, although some Ni-Cr elements have been used at element temperatures up to 1200 °C (2200 °F). The Fe-Cr-Al elements historically have been recommended for operation in air. Recent use in many reducing atmospheres indicates that the performance of Fe-Cr-Al elements is comparable to Ni-Cr resistance elements in most environments. In addition, Fe-Cr-Al elements can generally be used at higher temperatures than Ni-Cr elements.

For temperatures above 1300 °C (2350 °F), silicon carbide or molybdenum disilicide elements are employed in industrial furnaces. Here again, maximum life of heating elements is obtained in air. These nonmetallic materials give fair service life in slightly reducing atmospheres at temperatures up to 1300 °C (2350 °F) for SiC and 1500 °C (2750 °F) for $MoSi_2$. They can be used in both oxidizing and slightly reducing atmospheres more commonly than Fe-Cr-Al elements, which are recommended only for service in air or inert atmospheres.

Platinum has been used in some small laboratory furnaces up to 1480 °C (2700 °F) in air. Because of the high cost of platinum, it is used only in special applications where silicon carbide cannot be worked into the furnace design. Platinum is restricted to service in air and cannot be used in reducing atmospheres. Although the initial cost of platinum is high, it has a high salvage value.

Table 10 Comparative life of heating-element materials in various furnace atmospheres

See Table 11 for atmosphere compositions.

Element material	Relative life and maximum operating temperature in: Oxidizing (air)	Reducing: dry H_2 or type 501	Reducing: type 102 or 202	Reducing: type 301 or 402	Carburizing: type 307 or 309	Reducing or oxidizing, with sulfur	Reducing, with lead or zinc	Vacuum
Nickel-chromium and nickel-chromium-iron alloys								
80Ni-20Cr	Good to 1150 °C	Good to 1175 °C	Fair to 1150 °C	Fair to 1000 °C	Not recommended(a)	Not recommended	Not recommended	Good to 1150 °C
60Ni-16Cr-22Fe	Good to 1000 °C	Good to 1000 °C	Good to fair to 1000 °C	Fair to poor to 925 °C	Not recommended	Not recommended	Not recommended	...
35Ni-20Cr-43Fe	Good to 925 °C	Good to 925 °C	Good to fair to 925 °C	Fair to poor to 870 °C	Not recommended	Fair to 925 °C	Fair to 925 °C	...
Iron-chromium-aluminum alloys								
Fe, 22Cr, 5.8Al, 1Co	Good to 1400 °C	Fair to poor to 1150 °C(b)	Good to 1150 °C(b)	Fair to 1050 °C(b)	Not recommended	Fair	Not recommended	Good to 1150 °C(b)
22Cr, 5.3Al, bal Fe	Good to 1400 °C	Fair to poor to 1050 °C(b)	Good to 1050 °C(b)	Fair to 950 °C(b)	Not recommended	Fair	Not recommended	Good to 1050 °C(b)
Pure metals								
Molybdenum	Not recommended(c)	Good to 1650 °C	Not recommended	Not recommended	Not recommended	Not recommended	Not recommended	Good to 1650 °C
Platinum	Good to 1400 °C	Not recommended	Not recommended	Not recommended	Not recommended	Not recommended	Not recommended	...
Tantalum	Not recommended	Not recommended	Not recommended	Not recommended	Not recommended	Not recommended	Not recommended	Good to 2500 °C
Tungsten	Not recommended	Good to 2500 °C(d)	Not recommended	Not recommended	Not recommended	Not recommended	Not recommended	Good to 1650 °C
Nonmetallic heating element materials								
Silicon carbide	Good to 1600 °C	Fair to poor to 1200 °C	Fair to 1375 °C	Fair to 1375 °C	Not recommended	Good to 1375 °C	Good to 1375 °C	Not recommended
Graphite	Not recommended	Fair to 2500 °C	Not recommended	Fair to 2500 °C	Fair to poor to 2500 °C	Fair to 2500 °C in reducing	Fair to 2500 °C	...
Molybdenum disilicide	Good to 1850 °C	1350 °C	1600 °C	1400 °C	1350 °C	...	...	...

Note: Inert atmosphere of argon or helium can be used with all materials. Nitrogen is recommended only for the nickel-chromium group. Temperatures listed are element temperatures, not furnace temperatures. (a) Special 80Ni-20Cr elements with ceramic protective coatings designated for low voltage (8 to 16 V) can be used. (b) Must be oxidized first. (c) Special molybdenum heating elements with $MoSi_2$ coating can be used in oxidizing atmospheres. (d) Good with pure H_2 only

Elements made of 90% molybdenum disilicide and 10% refractory oxide mixtures perform well at continuous temperatures of 1700 and 1800 °C (3100 and 3270 °F) (depending on type) in air and in other oxidizing or inert atmospheres.

Carburizing Atmospheres. Unpurified exothermic (type 102) and purified exothermic (type 202) atmospheres are less harmful than endothermic (type 301) or charcoal (type 402) atmospheres, which are higher in carbon potential. The higher-carbon-potential atmospheres have a tendency to carburize Ni-Cr alloys, especially at higher temperatures. Chromium is a strong carbide former and may pick up enough carbon to lower the melting point of the alloy, causing localized fusion in the heating element. For this reason, in reducing atmospheres of high-carbon potential, it is safer to limit the operating temperature of 80Ni-20Cr to about 1000 °C (1850 °F).

Unprotected heating elements made of Ni-Cr alloys are not recommended for use at more than 30 volts in enriched endothermic carburizing or carbonitriding (type 309) atmospheres. Short life of heating elements in these types of atmospheres may be caused by carbon deposits on the element or refractory and by carburization of the element alloy. Carbon deposits may also short out extension wires and terminals, and cause them to melt. In recent years, a coated Ni-Cr heating element has been developed for use in carburizing atmospheres; in this element, the alloy is protected with a high-temperature ceramic coating that resists carburization. The element is designed to operate at low voltage (8 to 10 volts) to prevent arcing at the terminals in carbon-impregnated brickwork.

In recent times, when more and more carburizing furnaces are being converted from fossil fuels to electricity, it has been found that molybdenum disilicide can operate safely in both carburizing and reducing atmospheres at element temperatures up to 1500 °C (2700 °F). This makes it possible to eliminate the radiant tubes often used to protect metallic heating elements, thereby greatly increasing the efficiency of these furnaces by allowing faster recovery when a cold charge is placed in the furnace. Radiant tubes form a thermal barrier that slows heat transfer from the heating elements to the charge.

Reducing Atmospheres. In conventional heat-treating terminology, a reducing atmosphere is one that will reduce iron oxide on steel. Reducing atmospheres are of several types, as shown in Table 11.

With the exception of dry hydrogen and dissociated ammonia (type 501), all atmospheres listed as reducing in Table 10 are oxidizing to Ni-Cr and Fe-Cr-Al alloys. Even hydrogen or dissociated ammonia will selectively oxidize chromium in a Ni-Cr alloy unless the gas is extremely dry. The type of oxide produced in "reducing" atmospheres is entirely different from that produced in air. The oxide produced in air is a green-to-black, impervious type that retards further oxidation of the underlying metal. It is usually a combination of Cr_2O_3 and $NiO{\cdot}Cr_2O_3$. The oxide produced on Ni-Cr elements in reducing atmospheres is green and porous and allows the atmosphere to internally oxidize the base metal. This type of attack, frequently referred to as green rot, takes place over a limited temperature range—870 to 1040 °C (1600 to 1900 °F)—in any atmosphere that is oxidizing to chromium and reducing to nickel, and occurs as particles or stringers of Cr_2O_3 surrounding metallic nickel.

Among the listed reducing atmospheres, type 501 has the smallest effect on Ni-Cr heating elements. At temperatures above 1100 °C (2000 °F), a Ni-Cr element will have better life in dry hydrogen than in air, because oxidation in air occurs more rapidly at elevated temperatures. Wet hydrogen,

Table 11 Types and compositions of standard furnace atmospheres

See Table 10 for comparative life of heating elements in these atmospheres.

Type	Description	Composition, vol%					Typical dew point	
		N_2	CO	CO_2	H_2	CH_4	°C	°F
Reducing atmospheres								
102(a)	Exothermic unpurified	71.5	10.5	5.0	12.5	0.5	27	80
202	Exothermic purified	75.3	11.0	...	13.0	0.5	−40	−40
301	Endothermic	45.1	19.6	0.4	34.6	0.3	10	50
502	Charcoal	64.1	34.7	...	1.2	...	−29	−20
501	Dissociated ammonia	25	...	...	75	...	−51	−60
Carburizing atmospheres								
307	Endothermic + hydrocarbon	No standard composition					...	...
309	Endothermic + hydrocarbon + ammonia	No standard composition					...	...

(a) This atmosphere, refrigerated to obtain a dew point of 4 °C (40 °F), is widely used.

on the other hand, will cause preferential oxidation, and, around 950 °C (1750 °F), green rot will occur.

Graphite heating elements have been used for laboratory applications at temperatures near 1370 °C (2500 °F) in atmospheres free from O_2, CO_2, and H_2O. Silicon carbide elements give fair life in some reducing atmospheres; however, the maximum operating temperature is lower than for operation in air.

The poorest life for silicon carbide elements is obtained in hydrogen or dissociated ammonia. All atmospheres, including air, must be relatively dry; wet atmospheres shorten the life of silicon carbide elements. Silicon carbide is not recommended for use in carburizing atmospheres because it absorbs carbon, thus reducing electrical resistance and overloading the power supply.

Molybdenum disilicide heating elements can be safely used in carbon monoxide environments at 1500 °C (2730 °F), in dry hydrogen at 1350 °C (2460 °F), and in moist hydrogen at 1460 °C (2660 °F). As a rule of thumb, any combination of temperature and atmosphere that does not attack silica glass is compatible with molybdenum disilicide.

Atmosphere Contamination. Sulfur, if present, will appear as hydrogen sulfide in reducing atmospheres and as sulfur dioxide in oxidizing atmospheres. Sulfur contamination usually comes from one or more of the following sources: high-sulfur fuel gas used to generate the protective atmosphere; residues of sulfur-base cutting oil on the metal being processed; high-sulfur refractories, clays, or cements used for sealing carburizing boxes; and the metal being processed in the furnace. Sulfur is destructive to Ni-Cr and Ni-Cr-Fe heating elements. Pitting and blistering of the alloy occur in oxidizing atmospheres, and a Ni-S eutectic that melts at 645 °C (1190 °F) may form in any type of atmosphere. The higher the nickel content, the greater the attack. Therefore, if sulfur is present and cannot be eliminated, Fe-Cr-Al elements are preferred over those made of nickel-base alloys.

Lead and zinc contamination of a furnace atmosphere may come from the work being processed. This is a common occurrence in sintering furnaces for processing powder metallurgy parts. In the presence of a reducing atmosphere, lead will vaporize from leaded bronze powders (such as those used to make sintered bronze bushings) and attack the heating elements, forming lead chromate. Metallic lead vapors are even more harmful than sulfur to Ni-Cr alloys, and will cause severe damage to a heating element in a matter of hours if unfavorable conditions of concentration and temperature exist. Higher-nickel alloys are affected more than lower-nickel alloys. Elements made of 35Ni-20Cr-43.5Fe-1.5Si give satisfactory life for sintering lead-bearing bronze powders at 845 °C (1550 °F) in reducing atmospheres; 80Ni-20Cr elements give poor life in this application.

Zinc contamination results from zinc stearate used as a lubricant and binder when P/M compacts are pressed. The zinc stearate volatilizes when the compacts are heated and may carburize the heating element. (Brazing of nickel silvers, which contain at least 18% Zn, also results in a high concentration of zinc vapors in the furnace atmosphere.) Zinc vapors, which alloy with Ni-Cr heating elements and result in poor life, may be eliminated at the higher sintering temperatures by using a separate burn-off furnace at 650 °C (1200 °F), with heating elements protected by full muffle, by sheathing, or by a high-temperature ceramic protective coating. If these precautions are not feasible, silicon carbide elements (which are not affected by sulfur, lead, or zinc contamination) should be used at both low and high temperatures when contamination is anticipated.

The Ni-Cr, Ni-Cr-Fe, Fe-Cr-Al, and molybdenum disilicide heating elements should not be used in the presence of uncombined chlorine or other halogens.

Vacuum Service. For vacuum heating, 80Ni-20Cr elements have been used at temperatures up to 1150 °C (2100 °F). The 80Ni-20Cr alloys generally are not satisfactory much above 1150 °C (2100 °F), because the vapor pressure of chromium is high enough for chromium to vaporize from the elements, resulting in poor life, contamination of the material being processed, and loss of vacuum. Because of this, watt density must be kept low, especially at higher temperatures.

In vacuum heating, the estimated maximum operating temperature at which weight loss by evaporation from refractory-metal heating elements will not exceed 1% in 100 h is:

Metal	Temperature	
	°C	°F
Tungsten	2550	4620
Tantalum	2400	4350
Molybdenum	1900	3470
Platinum	1600	2910

Molybdenum disilicide heating elements are not suitable for use in high vacuum.

Properties of Electrical Resistance Alloys

80Ni-20Cr

Commercial Name

Common name. 80-20 alloy
UNS designation. N06003

Specifications

ASTM. B 344 and B 267
DIN. 17470 (Material number 2.4869)

Chemical Composition

Composition limits. 19 to 21 Cr, 1 Fe max, 1.0 Mn max, 0.15 C max, 0.75 to 1.75 Si, 0.01 S max, bal Ni

Applications

Typical uses. Electric heating elements for household appliances, industrial process heating, and industrial furnaces; fine-wire resistors for electronic applications
Precautions in use. Avoid sulfur-bearing and reducing atmospheres at high temperatures and furnace cements containing phosphoric acid or with high sulfur content.

Table 12 Variation of tensile properties of 80Ni-20Cr wire with temperature

Annealed wire 0.72 mm (0.0285 in.) in diameter

Temperature		Tensile strength		Elongation,
°C	°F	MPa	ksi	%
20	68	860	125	30
425	800	725	105	22
650	1200	620	90	20
870	1600	205	30	17
1040	1900	97	14	15

Mechanical Properties

Tensile properties. Annealed: tensile strength, 790 MPa (115 ksi); elongation, 25 to 35% in 50 mm or 2 in.; reduction in area, 55%. Hard: tensile strength, 1380 MPa (200 ksi); elongation, 0 to 1% in 50 mm or 2 in.; reduction in area, 0 to 1%. See also Table 12.
Hardness. Annealed: 85 to 90 HRB. Hard: 100 to 105 HRB
Elastic modulus. Tension, 215 GPa (31 × 10^6 psi)

Structure

Crystal structure. fcc

Mass Characteristics

Density. 8.4 g/cm^3 (0.30 lb/in.^{3}) at room temperature

Thermal Properties

Liquidus temperature. 1390 °C (2530 °F)
Coefficient of linear thermal expansion. 17.3 μm/m · K (9.6 μin./in. · °F) at 20 to 1000 °C (68 to 1830 °F)
Specific heat. 450 J/kg · K (0.107 Btu/lb · °F) at room temperature
Thermal conductivity. 13.4 W/m · K (7.7 Btu/ft · h · °F) at 100 °C (212 °F)

Electrical Properties

Electrical conductivity. Volumetric, 1.6% IACS at 25 °C (77 °F)
Electrical resistivity. 1.080 Ω · mm^2/m (650 Ω · circular mil/ft) at 20 °C (68 °F); temperature coefficient, 80 μΩ/Ω · K at −65 to 150 °C (−85 to 300 °F). See also Fig. 16.

Chemical Properties

General corrosion behavior. Highly resistant to oxidizing atmospheres up to 1200 °C (2200 °F). It is subject to corrosion in sulfur-bearing atmospheres at elevated temperatures and in certain reducing atmospheres at around 925 °C (1700 °F).

Fabrication Characteristics

Weldability. Can be soldered, brazed, or welded using oxyfuel-gas, carbon-arc, or resistance methods
Heat treatment. Annealing only
Annealing temperature. 870 to 1040 °C (1600 to 1900 °F)
Reduction between anneals. 65%
Hot-working temperature. 1205 °C (2200 °F)

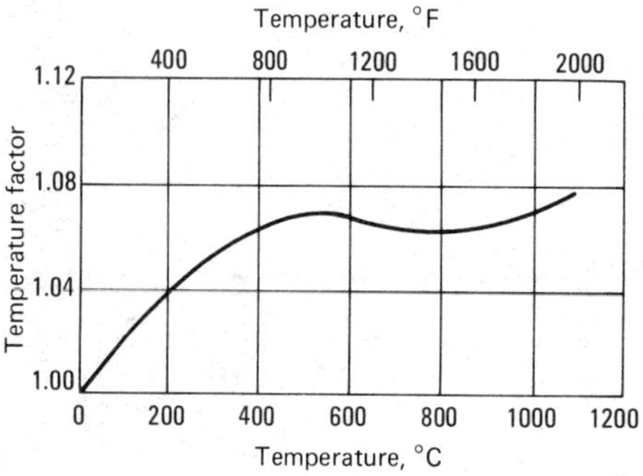

Fig. 16 Electrical resistance versus temperature for 80Ni-20Cr. To find resistance at any given temperature, multiply resistance at room temperature by the temperature factor.

70Ni-30Cr

Commercial Name

Common name. 70-30 NiCr
UNS designation. N06008

Specification

DIN. 17470 (Material number 2.4658)

Chemical Composition

Composition limits. 29 to 31 Cr, 1 Fe max, 1.0 Mn max, 0.15 C max, 0.75 to 1.75 Si, 0.01 S max, bal Ni

Applications

Typical uses. Elements for industrial furnaces. Particularly good in reducing atmospheres. Not subject to "green rot"
Precautions in use. Avoid sulfur-bearing atmospheres and using furnace cements containing phosphoric acid or with high sulfur contents. Do not use in pure N_2.
Tensile properties. Annealed: tensile strength, 825 MPa (120 ksi); elongation, 36 to 40% in 50 mm or 2 in.; reduction in area, 50%; Hard: tensile strength, 1380 MPa (200 ksi); elongation, 0 to 1% in 50 mm or 2 in.

Structure

Crystal structure. fcc

Mass Characteristics

Density. 8.11 g/cm^3 (0.29 lb/in.^{3}) at room temperature

Thermal Properties

Liquidus temperature. 1365 °C (2490 °F)
Coefficient of linear thermal expansion. 17 μm/mm · K (9.4 μin./in. · °F) at 20 to 1000 °C (68 to 1830 °F)
Specific heat. 460 J/kg · K (0.110 Btu/lb · °F) at room temperature
Thermal conductivity. 13.7 W/m · K (7.9 Btu/ft · h · °F) at 100 °C (212 °F)

Electrical properties

Electrical conductivity. Volumetric, 1.5% IACS at 25 °C (77 °F)
Electrical resistivity. 1.18 Ω · mm^2/m (710 Ω · circular mil/ft) at 25 °C (77 °F); temperature coefficient, 90 μΩ/Ω · K at 20 to 100 °C (68 to 212 °F)

Chemical Properties

General corrosion behavior. Highly resistant to oxidizing atmospheres up to 1260 °C (2300 °F). Not subject to preferential oxidation ("green rot") in reducing atmospheres. It is subject to corrosion in sulfur-bearing atmospheres at elevated temperatures.

Fabrication Characteristics

Weldability. Can be soldered, brazed, or welded using oxyfuel-gas, carbon arc, or resistance methods
Heat treatment. Annealing only. Annealing temperature 870 to 1120 °C (1600 to 2050 °F)
Reduction between anneals. 60%

60Ni-22Fe-16Cr

Commercial Name

Common name. 60Ni-16Cr alloy
UNS designation. N06004

Specifications

ASTM. B 344 and B 267
DIN. 17470 (Material number 2.4867)

Chemical Composition

Composition limits. 57 Ni min + Co, 14 to 18 Cr, 1.0 Mn max, 0.75 to 1.75 Si, 0.15 C max, 0.01 S max, bal Fe

Applications

Typical uses. Heat-resisting elements for heating devices such as toasters, percolators, waffle irons, flat irons, ironing machines, heater pads, hair driers, permanent wave equipment, and hot water heaters. Electrical usage in high-resistance rheostats for electronic equipment, oxidized wire with high-resistance coating for close-wound rheostats, potentiometers, and thermocouples. Corrosion-resisting usage in dipping baskets for acid pickling and cyanide hardening, automatic pickling machine parts, filters, enameling racks, containers for molten salts

Mechanical Properties

Tensile properties. Sand cast: tensile strength, 450 MPa (65 ksi); elongation, 2% in 50 mm or 2 in.; reduction in area, 2%. Annealed: tensile strength, 690 MPa (100 ksi); elongation, 30% in 50 mm or 2 in.; reduction in area, 47%. Hard: tensile strength, 1345 MPa (195 ksi); elongation, 0 to 2% in 50 mm (2 in.). See also Table 13.
Hardness. Sand cast: 92 HRB. Annealed: 83 HRB

Structure

Microstructure. fcc
Metallography. Rough polish with emery paper through 000; use levigated alumina on polishing cloth to finish. Etching for grain size: Marble's reagent preferably or aqua

Table 13 Nominal tensile properties of 60Ni-22Fe-16Cr at elevated temperatures

Temperature		Cast material			Wrought material		
		Tensile strength		Elongation,	Tensile strength		Elongation,
°C	°F	MPa	ksi	%	MPa	ksi	%
20	68	450	65	2.0	725	105	30
540	1000	325	47	2.1	365	53	12
650	1200	295	43	2.3	260	38	13
760	1400	255	37	2.5	200	29	23
870	1600	150	22	3.7	130	19	32
980	1800	105	15	6.0	62	9	45

regia (3 HCl, 1 HNO_3); sometimes less HCl is used and sometimes it is diluted to 50% with water, depending on the worked or annealed condition of the alloy. Use aqua regia for microstructure. Upon completion of polishing, this alloy often exhibits a cold-worked film obscuring the true structure, and hence must be repolished on the last stage and re-etched. This is common to other stainless austenitic materials. Usual method of reporting grain size—grains per sq mm

Mass Characteristics

Density. 8.25 g/cm^3 (0.298 lb/in.^{3}) at 20 °C (68 °F)
Patternmaker's shrinkage. 2.0%

Thermal Properties

Liquidus temperature. 1375 °C (2510 °F)
Coefficient of linear thermal expansion. 17.0 μm/m · K (9.4 μin./in. · °F) at 20 to 1000 °C (68 to 1830 °F)
Specific heat. 450 J/kg · K (0.107 Btu/lb · °F) at 20 °C (68 °F)
Thermal conductivity. 13.4 W/m · K (7.7 Btu/ft · h · °F) at 100 °C (212 °F)

Electrical Properties

Electrical conductivity. Volumetric, 1.5% IACS at 25 °C (77 °F)
Electrical resistivity. 1.125 Ω · mm^2/m (675 Ω · circular mil/ft) at 25 °C (77 °F); temperature coefficient, 150 μΩ/Ω · K at 20 to 100 °C (68 to 212 °F). See also Fig. 17.

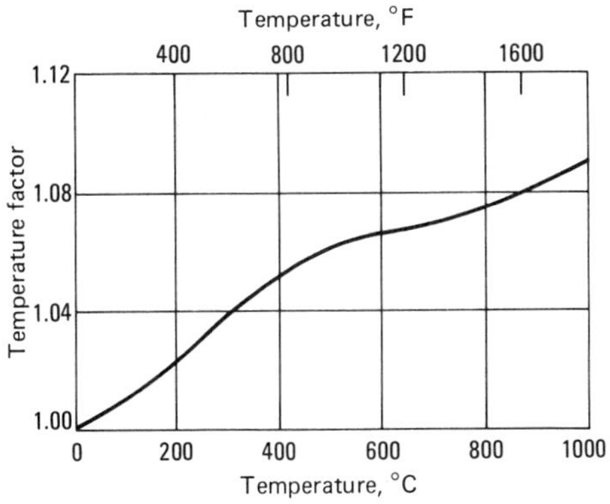

Fig. 17 Electrical resistance versus temperature for 60Ni-22Fe-16Cr. To find resistance at any given temperature, multiply resistance at room temperature by the temperature factor.

Chemical Properties

Corrosion testing. Strauss test, using changes in both weight and electrical resistance as criteria for evaluation
Resistance to specific corroding agents:

Corrosive agent	Resistance
Acid, acetic	Resistant
Acid, hydrochloric	Moderate general attack
Acid, lactic	Resistant
Acid, nitric	Resistant
Acid, phosphoric	Moderate general attack when hot
Acid, sulfuric	Resistant
Air	Resistant
Alcohol, ethyl	Resistant
Alcohol, methyl	Resistant
Ammonia	Resistant
Ammonium nitrate	Resistant
Carbon dioxide	Resistant
Carbon tetrachloride	Resistant
Copper sulfate	Moderate intergranular attack
Foodstuffs	Resistant
Fruit products	Resistant
Hydrogen sulfide	Severe general attack when hot
Lead and its compounds	Severe attack when hot
Milk	Resistant
Mineral oils	Resistant
Motor fuel	Resistant
Oxygen	Resistant
Petroleum products	Resistant when cold
Photographic solutions	Resistant
Potassium hydroxide	Resistant
Potassium nitrate	Resistant
Sodium carbonate	Resistant
Sodium hydroxide	Resistant
Sodium nitrate	Resistant
Sugar	Resistant
Sulfur	Severe general attack when hot
Sulfur dioxide	Severe general attack when hot
Water, distilled	Resistant
Water, rain	Resistant

Fabrication Characteristics

Formability. Suited to forming by hot and cold rolling, forging, drawing, pressing, and bending
Weldability. Soft solder with 50Pb-50Sn, using HCl plus Zn flux. Silver braze (silver solder) with low-melting-point filler metals, using borax flux and a neutral flame. Braze with low-brass filler metal, using borax flux. Oxyfuel-gas weld with 60Ni-24Fe-16Cr filler metal, using no flux and a neutral or carburizing flame
Casting temperature. Sand and ingot. 1510 to 1540 °C (2750 to 2800 °F)
Annealing temperature. 760 to 1090 °C (1400 to 2000 °F)
Reduction between anneals. 75% max in area
Standard finishes. Castings are sandblasted, forgings pickled, and wrought mill products given various finishes.

37Ni-21Cr-2Si-40Fe

Commercial Name

Common name. 37-21 alloy

Chemical Composition

37 Ni, 20-21 Cr, 1.0 Mn max, 2.0 Si, bal Fe

Application

Typical uses. This alloy has a lower resistivity than 60Ni-16Cr at room temperature, but it has a higher temperature coefficient. It finds use as heating elements in appliances.

Mechanical Properties

Tensile properties. Annealed: tensile strength, 585 MPa (85 ksi); elongation, 25 to 35% in 50 mm or 2 in.; Hard: tensile strength, 1135 MPa (165 ksi); elongation, 0 to 1% in 50 mm or 2 in.

Structure

Microstructure. fcc
Metallography. Same as 60Ni-22Fe-16Cr

Mass Characteristics

Density. 7.95 g/cm^3 (0.29 lb/in.^{3}) at room temperature

Thermal Properties

Liquidus temperature. 1380 °C (2515 °F)
Coefficient of linear thermal expansion. 19.0 μm/m · K (10.6 μin./in. · °F) at 20 to 1000 °C (68 to ±830 °F)
Specific heat. 460 J/kg · K (0.110 Btu/lb · °F) at 20 °C (68 °F)
Thermal conductivity. 13.0 W/m · K (7.5 Btu/ft · h · °F) at 100 °C (212 °F)

Electrical Properties

Electrical conductivity. Volumetric, 1.6% IACS at 25 °C (77 °F)
Electrical resistivity. 1.08 Ω · mm^2/m (650 Ω · circular mil/ft) at 25 °C (77 °F); temperature coefficient, 240 μΩ/Ω · K at −65 to 150 °C (−85 to 300 °F)

Chemical Properties

Corrosion testing. Same as 60Ni-22Fe-16Cr

Fabrication Characteristics

Formability. Suited to forming by hot and cold rolling, forging, drawing, pressing, and bending
Weldability. Soft solder, silver braze, and braze with same methods as 60Ni-24Fe-15Cr alloy; Oxyfuel-gas weld with 37Ni-42Fe-21Cr filler metal, using no flux and a neutral or carburizing flame
Casting temperature. 1525 to 1550 °C (2775 to 2825 °F)

Table 14 Nominal tensile properties of 35Ni-43Fe-20Cr at elevated temperatures

Temperature		Cast material			Wrought material		
		Tensile strength		Elongation,	Tensile strength		Elongation,
°C	°F	MPa	ksi	%	MPa	ksi	%
20	68	425	62	2	705	102	32
540	1000	295	43	4	345	50	22
650	1200	250	36	6	250	36	20
760	1400	235	34	8.5	185	27	26
870	1600	130	19	20	125	18	28
980	1800	83	12	25	62	9	30

Annealing temperature. 760 to 1175 °C (1400 to 2150 °F)
Reduction between anneals. 75% max
Hot-working temperature. 980 to 1260 °C (1800 to 2300 °F)
Standard finishes. See 60Ni-22Fe-16Cr.

35Ni-43Fe-20Cr

Commercial Name

Common name. 35Ni-20Cr alloy

Specifications

ASTM. B 344

Chemical Composition

Composition limits. 34 to 37 Ni, 18 to 21 Cr, 1.0 Mn max, 0.25 C max, 3.0 Si max, 0.03 S max, bal Fe

Applications

Typical uses. Although 35Ni-43Fe-20Cr has lower electrical resistivity than 60Ni-22Fe-16Cr near room temperature, it has a higher temperature coefficient. It is normally used at temperatures up to 815 °C (1500 °F) for heavy-duty rheostats, low-priced electrical appliances, and resistors operating in cracked gas atmospheres. It is not subject to "green rot" and is used as heating elements in industrial electric furnaces operating below 1040 °C (1900 °F).

Mechanical Properties

Tensile properties. Sand cast: tensile strength, 425 MPa (62 ksi); elongation, 2% in 50 mm or 2 in.; reduction in area, 2%. Annealed: tensile strength, 585 MPa (85 ksi); elongation, 30% in 50 mm or 2 in.; reduction in area, 45%. Hard: tensile strength, 1205 MPa (175 ksi); elongation, 1 to 2% in 50 mm (2 in.). See also Table 14.

Structure

Microstructure. fcc
Metallography. Same as 60Ni-22Fe-16Cr

Mass Characteristics

Density. 7.95 g/cm^3 (0.287 lb/in.^{3}) at 20 °C (68 °F)
Patternmaker's shrinkage. 2.0%

Thermal Properties

Liquidus temperature. 1390 °C (2540 °F)
Coefficient of linear thermal expansion. 28.4 μm/m · K (15.8 μin./in. · °F) at 20 to 500 °C (68 to 930 °F)
Specific heat. 460 J/kg · K (0.110 Btu/lb · °F) at 20 °C (68 °F)
Thermal conductivity. 13 W/m · K (7.5 Btu/ft · h · °F) at 100 °C (212 °F)

Electrical Properties

Electrical conductivity. Volumetric, 1.7% IACS at 25 °C (77 °F)
Electrical resistivity. 1.0 Ω · mm^2/m (610 Ω · circular mil/ft) at 25 °C (77 °F); temperature coefficient, 310 μΩ/Ω · K at 20 to 500 °C (68 to 930 °F). See also Fig. 18.

Chemical Properties

Corrosion testing. Same as 60Ni-22Fe-16Cr

Fabrication Characteristics

Formability. Suited to forming by hot and cold rolling, forging, drawing, pressing, and bending
Weldability. Soft solder, silver braze, and braze with the same methods as for 60Ni-22Fe-16Cr alloy. Oxyfuel-gas weld with 35Ni-43Fe-20Cr filler metal, using no flux, and a neutral or carburizing flame
Casting temperature. 1525 to 1550 °C (2775 to 2825 °F)
Annealing temperature. 760 to 1090 °C (1400 to 2000 °F)
Reduction between anneals. 75% max
Hot-working temperature. 980 to 1260 °C (1800 to 2300 °F)
Standard finishes. See 60Ni-22Fe-16Cr.

Constantan 45Ni-55Cu

Chemical Composition

Composition limits. Variable; several specific compositions—all having an approximate nominal composition of 42 to 45 Ni, bal Cu—are commonly supplied. Composition limits of both major and minor constituents are varied to suit the specific application.

Applications

Typical uses. This alloy has about the highest electrical resistivity, the lowest temperature coefficient of resistance, and the highest thermal emf against platinum of any of the copper-nickel alloys. Because of the first two of these properties, it is used for electrical resistors, and because of the latter property, for thermocouples.
Precautions in use. Maximum temperature for resistor use is 500 °C (930 °F); maximum temperature for thermocouple use is 900 °C (1650 °F).

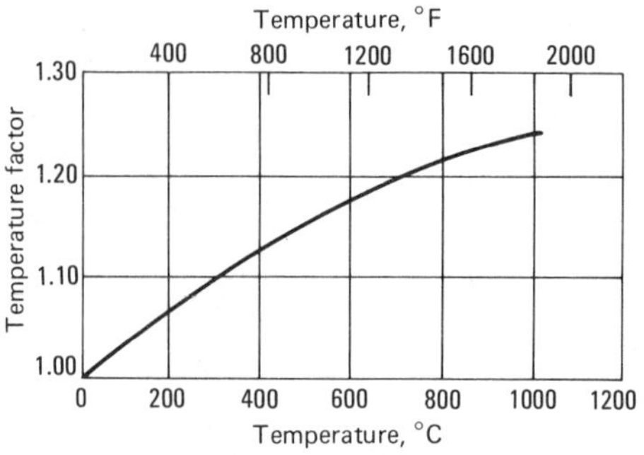

Fig. 18 Electrical resistance versus temperature for 35Ni-43Fe-20Cr. To find resistance at any given temperature, multiply resistance at room temperature by the temperature factor.

Mechanical Properties

Tensile properties. Wrought: tensile strength, annealed, 415 MPa (60 ksi); cold worked, 930 MPa (135 ksi). Cast: tensile strength, 380 MPa (55 ksi); yield strength, 145 MPa (21 ksi); elongation, 32% in 50 mm or 2 in.; reduction in area, 4%
Shear strength. 295 MPa (43 ksi)
Hardness. 75 to 85 HB, 48 to 54 HRB
Impact resistance. Charpy: 41 J (30 ft · lbf)

Mass Characteristics

Density. Wrought: 8.9 g/cm^3 (0.32 lb/in.^{3}) at 20 °C (68 °F). Cast: 8.6 g/cm^3 (0.31 lb/in.^{3})
Patternmaker's shrinkage. 2.1%

Thermal Properties

Liquidus temperature. 1280 °C (2330 °F)
Solidus temperature. 1220 °C (2225 °F)
Coefficient of linear thermal expansion. 14.9 μm/m · K (8.3 μin./in. · °F) at 20 to 100 °C (68 to 212 °F); 16.3 μm/m · K (9.0 μin./in. · °F) at 20 to 500 °C (68 to 930 °F); 18.8 μm/m · K (10.4 μin./in. · °F) at 20 to 1000 °C (68 to 1830 °F)
Specific heat. 395 J/kg · K (0.094 Btu/lb · °F) at 20 °C (68 °F)
Thermal conductivity. 21 W/m · K (12.3 Btu/ft · h · °F) at 20 °C (68 °F)

Electrical Properties

Electrical resistivity. 0.5 to 0.525 Ω · mm^2/m (300 to 315 Ω · circular mil/ft) at 20 °C (68 °F); temperature coefficient, ±20 μΩ/Ω · K at 25 to 150 °C (77 to 300 °F). The coefficient may be either positive or negative, depending on small variations in composition and on variations in the amount of cold work. In any event, the value of the coefficient is small.
Thermoelectric potential. The basic alloy is modified by additions of manganese and iron to give slightly different emf characteristics in thermocouple service, as specified by different pyrometer manufacturers. For more detailed information on this subject, see the article in this Volume entitled "Thermocouple Materials."

Fabrication Characteristics

Formability. Suited to forming by hot and cold rolling, forging, drawing, pressing, and bending
Weldability. Can be welded, brazed, and soldered by conventional methods
Casting temperature. Minimum: 1350 °C (2460 °F)
Annealing temperature. 870 to 980 °C (1600 to 1800 °F)
Reduction between anneals. 80% max
Hot-working temperature. 870 to 1120 °C (1600 to 2050 °F)

22Cr-5.3Al-Fe Balance

Chemical Composition

Typical 22Cr-5.3Al-balance Fe. Nominal aluminum content will vary from 4.5 to 5.8%. Values given below for 5.3% Al

Applications

Typical uses. Electric resistance heating elements in industrial and laboratory furnaces as well as in household appliances
Precautions in use. Keep the material as clean as possible prior to the first firing, which forms the protective oxide. Choose suitable support materials for use at the highest temperatures.

Mechanical Properties

Tensile properties. Annealed: tensile strength, 690 MPa (100 ksi); elongation, 15 to 25% in 50 mm or 2 in.; Hard: tensile strength, 1035 MPa (150 ksi); elongation, 0 to 1%
Curie point. 600 °C (1100 °F)

Mass Characteristics

Density. 7.15 g/cm^3 (0.259 lb/in.^{3}) at room temperature

Thermal Properties

Liquidus temperature. 1500 °C (2730 °F)
Coefficient of linear thermal expansion. 15 μm/m · K (8.3 μin./in. · °F)
Specific heat. 460 J/kg · K (0.11 Btu/lb · °F) at room temperature
Thermal conductivity. 16 W/m · K (9.25 Btu/ft · h · °F) at 20 °C (68 °F)

Electrical Properties

Electrical resistivity. 1.39 μΩ · m (835 Ω · circular mil/ft) at 20 °C (68 °F)

Chemical Properties

General corrosion behavior. Highly resistant to oxidizing atmospheres up to 1400 °C (2550 °F). Affinity for oxygen and subsequent oxide formation is so great that only a small partial pressure of oxygen is necessary for oxide to form. Pre-oxidation of element can assist in obtaining maximum life of the heating element. Does not react as severely as nickel to sulfur attack. Oxide can provide sufficient dielectric for use of close-wound coils in small appliances.

Table 15 Maximum service temperatures for $MoSi_2$ heating elements

Atmosphere	Temperature °C	Temperature °F
Air	1700(a)	3100(a)
Nitrogen	1590	2900
Argon, helium	1450	2640
Dry hydrogen	1350	2460
Moist hydrogen(b)	1460	2660
Carbon dioxide	1590	2900
Carbon monoxide	1450	2730
Sulfur dioxide	1590	2900
Cracked, partly burnt ammonia(c)	1400	2550
Methane	1350	2460

(a) For "1700 °C" material; for "1800 °C" material, maximum service temperature is 1800 °C (3270 °F). (b) Dew point 15 °C (60 °F). (c) Approximately 8% H_2

Fabrication Characteristics

Weldability. Can be welded using resistance welding. TIG or carbon arc recommended for larger elements
Heat treatment. Annealing only
Annealing temperature. 850 °C (1560 °F)
Minimum reduction between annealing. 35%

Molybdenum Disilicide $MoSi_2$, $MoSi_2$ + 10% Ceramic Additives

Chemical Composition

Composition limits. 90% $MoSi_2$, 10% metal and ceramic additives

Applications

Typical uses. Electric resistance heating elements in industrial and laboratory furnaces. Normally supplied in hairpin shape consisting of terminals twice the diameter of the hot zone. Most common hot zone diameters are 3, 6, and 9 mm (0.12, 0.24, 0.35 in.). Will operate in oxidizing atmospheres at temperatures of 1700 °C (3100 °F) for $MoSi_2$, and 1800 °C (3275 °F) for $MoSi_2$ + 10% ceramic additives. Also usable at lower temperatures in reducing atmospheres. For maximum recommended service temperatures, see Table 15.
Precautions in use. Avoid elemental chlorine, other halogens, and high vacuum environments.

Mechanical Properties

Tensile strength. 195 MPa (28 ksi) at room temperature
Compressive strength. 2350 MPa (340 ksi) at room temperature
Hardness. Knoop: 1280 at 20 °C (68 °F), decreasing to 640 at 700 °C (1290 °F)
Elastic modulus. Tension, 410 GPa (59 × 10^6 psi) at room temperature
Bending strength. 345 MPa (50 ksi) at 20 °C (68 °F), 135 MPa (19 ksi) at 1100 °C (2010 °F)

Structure

Crystal structure. Pure $MoSi_2$: body-centered tetragonal

Mass Characteristics

Density. 5.6 g/cm^3 (0.202 lb/in.^{3}) at room temperature

Thermal Properties

Incipient melting temperature. 2050 °C (3720 °F). Decomposes before melting at approximately 1740 °C (3165 °F) for $MoSi_2$, and 1825 °C (3315 °F) for $MoSi_2$ + 10% ceramic additives
Coefficient of linear thermal expansion. 7.1 to 8.8 μm/m · K (3.9 to 4.9 μin./in. · °F) at 20 to 1500 °C (68 to 2730 °F)
Specific heat. 420 J/kg · K (0.10 Btu/lb · °F) at 20 °C (68 °F)

Electrical Properties

Electrical resistivity. For $MoSi_2$ + 10% ceramics:

Temperature °C	Temperature °F	1700 °C material μΩ · m	1800 °C material μΩ · m
20	68	465	440
1370	2500	4800	4570
1650	3000	6060	5750

Thermoelectric potential. In microvolts versus Pt: $5.13\ T + 1.88 \times 10^{-2}\ T^2 - 5.22 \times 10^{-6}\ T^3$
Temperature of superconductivity. 1.3 K
Emissivity. 0.34 at 370 °C, 0.60 at 1370 °C, 0.75 to 0.80 at 1500 °C in air due to formation of SiO_2 coating

Chemical Properties

General corrosion behavior. In normal use, material has a SiO_2 glass coating, and any combination of temperature and atmosphere that is harmful to quartz glass will deteriorate the elements.

Fabrication Characteristics

Sintering temperature. 1620 °C (2950 °F)
Hot-working temperature. $MoSi_2$, 1450 to 1700 °C (2640 to 3090 °F); $MoSi_2$ + 10% additives, 1500 to 1800 °C (2730 to 3270 °F)

Electrical Contact Materials

Revised by Yuan-Shou Shen, Engelhard Corporation
Pat Lattari and Jeffrey Gardner, Texas Instruments, Inc.
Harold Wiegard, J.M. Ney Company

ELECTRICAL CONTACTS are metal devices that make and break electrical circuits. Contacts are made of either elemental metals, composites, or alloys that are made by the melt-cast method or manufactured by powder metallurgy (P/M) processes. Powder metallurgy facilitates combinations of metals which ordinarily cannot be achieved by alloying.

A majority of contact applications in the electrical industry utilize silver-type contacts, which include the pure metal, alloys, and powder metal combinations. Silver, which has the highest electrical and thermal conductivity of all metals, is also used as a plated, brazed, or mechanically bonded overlay on other contact materials—notably, copper and copper-base materials. Other types of contacts used include the platinum group metals, tungsten, molybdenum, copper, copper alloys, and mercury. Aluminum is generally a poor contact material because it oxidizes readily, but is used in some contact applications because of its good electrical and mechanical properties and its availability and cost.

Selection Criteria

Electrical contact materials are used in diverse service conditions, and no metal has all the desired properties required to accomplish the objectives of different contact applications. The usefulness of an electrical contact material also depends on a variety of electrical and mechanical properties, service life, load conditions, and economics. The choice of contact materials in particular applications is discussed in the section "Recommended Contact Materials" in this article.

The desirable properties of electrical contact materials include such characteristics as high electrical conductivity to minimize the heat generated during passage of current; high thermal conductivity to dissipate both the resistive and arc heat developed; high resistance to chemical reactions in all application environments so as to avoid formation of insulating oxides, sulfides, and other compounds; and immunity to arcing damage on the making and breaking of electrical contact. The melting point of electrical contact materials should also be high enough to limit arc erosion, metal transfer, and welding or sticking, but low enough to increase resistance to reignition in switching. (When the melting point is high, contacts continue to heat gas in the contact gap after the current drops to zero, thus facilitating reignition.) Other important characteristics of electrical contact materials include:

- Vapor pressure should be low to minimize arc erosion and metal transfer
- Hardness should be high to provide good wear resistance, yet ductility should also be high enough to ensure ease of fabrication
- Purity of the material should be maintainable at a level that ensures consistent performance. Finally, the material should be available at low cost and should not present an environmental hazard in use or in the necessary fabrication process

Circuit Characteristics. Proper choice of electrical contact materials depends on whether the current is alternating or direct, whether the circuit voltage or amperage is high or low, and whether the voltage at contacts during interruption of the circuit is high or low. Whether the load is inductive, capacitive or resistive, or is a motor load, is also important. Allowance should be made for overload, where such a condition can be expected, and attention should be given to the method of arc suppression. Consideration must also be given to potential hazards during service life of the electrical contact. For example, failure of a railroad signal to open might cause a fatal accident, although failure to close would only be troublesome. For such applications, carbon-to-metal contacts are employed because they cannot become welded together, even under the most adverse conditions.

Mechanical Factors. A basic factor in the selection of a contact material is the force required to close a contact of a given material. For example, platinum, palladium, and gold are used where reliable closure with low force is required.

Other mechanical factors affecting the contact force or the selection of a contact material include:

- The nature of the contact force
- The frequency of operation
- The speed of opening and closing of the contact
- The manner in which the contact is made (wipe, slice, or butt)
- The degree of chatter on opening or closing
- The size of contact gap when the contact is fully opened
- The method of operation—whether by cam, simple lever or push button, electromagnet, or bimetal thermostatic control

Environmental factors influencing selection of a contact material are type of atmosphere, ambient temperature, and whether contacts are exposed to ambient atmosphere or enclosed in a hermetically sealed, artificial environment such as might be specified for use in outer space. Atmospheric pressure and minor contaminants that are present affect contact life and, in some instances, may cause a sharp decrease in service life. Gases may cause tarnishing, as does the presence of humidity, airborne salts, dust, and organic vapors.

Service Life. The required service life for contacts varies according to the application. This may range from a few operations, on missiles and detection systems, to 100 million cycles in automotive vibrators or 40 years in telephone relays. Likewise, the dormant time between successive openings or closings may vary widely—from milliseconds or microseconds in vibrators to months or years in alarms and similar safety devices.

Failure Modes of Make-Break Contacts

In an electric make-break switching device, there are two different types of contacts: arcing contacts and sliding contacts. The arcing contact points, which usually consist of a pair of thin shaped slabs, perform the actual duty of making, carrying, and breaking the current. Arcing contacts differ from sliding contacts in that the moving member of the switching device travels perpendicular to the contact surfaces. As a result, arcs generated during opening and closing actions always strike and consequently damage the conducting surfaces.

Arcing, except in a circuit with an extremely low potential or low current, is a major factor—if not the main factor—causing failure of contact points.

When a pair of contacts opens in a live circuit, an arc is often generated between the contact pair, which remains until they are separated by a certain gap. Relatively less severe arcing occurs when the contacts close. Arcing also occurs when the moving contact bounces away from the stationary contact during closing. The arc causes contact erosion by blowing away the molten metal droplets, vaporizing the material, and transforming the metal to ion jets. Sometimes the material vaporizes from one contact and then condenses onto the other contact, thereby altering the surface configuration of both contacts. This is known as material transfer.

Welding. When a pair of contacts close, the arcs generated during closing and bouncing of a moving contact melt a small portion of both contacts. On reclosure, solidification of the molten material welds the contact pair in the same manner as fusion welding.

Another type of welding occurs after the contacts are made. To make a pair of contacts more conductive, a mechanical load is always applied on the contact pairs. Theoretically, the load could make two rigid contact surfaces touch at no more than three points. However, the touching points at both surfaces yield either elastically or plastically, resulting in larger areas of contact. These constricted regions carry the current through the contacts and form regions of high current density. Heat is generated in these areas and, if the temperature becomes high enough, the two contact points eventually are welded together.

Occasionally, the strength of the weld exceeds the opening force of the switching device, resulting in catastrophic failure of the entire electrical system because the contacts fail to open on command.

Bridge Formation. When a pair of contacts opens, the contact area gradually decreases because of the gradual lessening of contact pressure. The continuous opening action causes the contact areas to reach a stage at which the current density of the constricted areas is so great that it melts the material in these regions. Continuous separation of the contact points now pulls the molten metal, forming a current-carrying bridge. The temperature of the molten bridge continues to rise as the contact points pull apart. It may become high enough to evaporate the material and finally break the circuit. This "bridge" phenomenon during the opening of a pair of contacts slightly damages the surfaces of the contacts and evaporates some of the bridge material. This generally results in pitting of one contact surface and buildup of material on the other; an uneven continuous transfer may eventually erode one of the contacts. Furthermore, the surface asperities from the continuous bridge formation may interlock the contact pair and interfere with their mechanical separation.

Oxidation of the contact surfaces, which may be accelerated by the heat from arcing, is a serious problem because most metallic oxide films are nonconductive or semiconductive. The oxide film may easily increase contact resistance. In high-current circuits, this may cause excessive contact heating. In low-voltage and low-current circuits, the oxide films can grow so thick that they completely insulate the contact surfaces before the contact bodies erode. This happens more frequently when a pair of contacts operates in a hostile environment such as a polluted industrial atmosphere. Condensed organic polymers also play a role in precious metal contacts at light loads. These polymers come from monomers which evaporate from resins and are polymerized on the active catalytic metal surfaces of contacts.

Property Requirements for Make-Break Arcing Contacts

The four failure modes discussed above determine the requirements of materials for arcing contacts. The most important requirements are listed below. In selecting a material, it is often necessary to reach a compromise that provides adequate properties without jeopardizing essential qualities of the component as a whole, such as reliability, life, and cost.

- *Electrical conductivity*: Because the conduction of electricity between the pair of contacts depends on only a few constricted spots, the higher the electrical conductivity, the less the amount of heat that will be generated by high current density in these spots
- *Thermal properties*: High melting and boiling points decrease evaporation loss caused by high arcing heat. High thermal conductivity disperses the heat rapidly and quenches the arc
- *Chemical properties*: Contact materials should be corrosion resistant so that insulating films (either oxides or other compounds) do not form easily when the contacts operate in a hostile environment
- *Mechanical properties*: The major loads applied to a contact pair are the closing force and the impact between movable and stationary contact points during closing. An induced relative movement between two contact surfaces always exists when closing. In some devices, such as certain types of relays, a wiping motion is purposely designed into the device to destroy any oxide films that form. However, friction between wiping surfaces produces wear of the contacts upon repeated opening and closing. Generally, hard materials are more resistant to wear. However, hard materials often have high contact resistances and low thermal conductivities, both of which contribute to a greater tendency to contact welding. Hard materials also have high tensile strengths, which may or may not be advantageous in electrical contact applications
- *Fabrication properties*: Contact materials should have the capability of being welded, brazed, or otherwise joined to backing materials. In addition, they should have sufficient malleability to enable them to be shaped, or they should be capable of being formed by P/M techniques

None of the elementary metallic elements used for arcing contacts meets all of the criteria for an ideal contact material. For instance, silver has the best electric and thermal conductivity and good oxidation resistance, but its resistance to arcing and mechanical wear is low. Tungsten resists arcing and withstands mechanical wear, but it has poor conductivity and poor oxidation resistance. Properties of a contact material can usually be improved by combining metals, either by alloying or by powder metallurgy, but improvement often is achieved at the expense of other properties. For example, the alloying element that increases the hardness of silver also decreases its conductivity.

Sliding Contacts

The applications of sliding contacts are usually quite different from those of arcing contacts. Friction, contact temperature, mechanical considerations, and wear also are different.

The fundamental difference between arcing contacts and sliding contacts is that sliding contacts require films on the contact faces to facilitate sliding without seizure or galling; shear must take place within this film with only minor disturbance of both materials. A lubricant of some kind is always necessary. This can be provided by graphite if there is moisture present—such as in an environment having a dew point of about −20 °C (−4 °F) or higher. Alternatively, lubrication can be provided by very thin oil films, although excessive oil vapor causes over-filming. It can also be provided by molybdenum disulfide, and other chalcogenides of molybdenum, tungsten, and niobium. Oxygen, sulfur, and other contaminants cause increased filming.

In applications in air, a drop in voltage can result from an equilibrium between oxidation and filming (which tend to increase the drop) and fretting or film breakdown and cleaning action (which tend to decrease the drop). In the absence of lubricants, fretting and oxidation are most important. In inert or reducing gases, oxidation is largely eliminated, and the voltage drop decreases until counteracted by mechanical factors. Noble metals that are properly lubricated also minimize voltage drops in air.

Brush contacts generally contain an appreciable amount of metal if they are intended for use in low-voltage (<24 V) applications. Large quantities of brush contacts are used in automotive and related industries as starter brushes and auxiliary motor brushes; copper-graphite is the principal material. Silver-graphite brushes are used primarily in instruments and in outer space applications. Some silver-graphite brushes are used in seam welders and similar equipment.

Oxidation of sliding contacts is similar to that of arcing contacts, except that the surface disturbed by friction oxidizes more rapidly. In most applications in air, the metal surface generates a film that is a complex mixture of graphite, oxide, sulfide, and water, which tends to decrease the conducting area.

The surfaces generated on metal-graphite brushes as they wear are effective cleaning agents in that they abrade films and keep larger areas available for conduction. Even so, it is sometimes advisable to have additional abrasive material in the brushes to prevent over-filming in critical atmospheres.

Because the major factor in friction is the shear strength of any film that is present, the composition of this film, as affected by atmospheric contaminants, is important. Table 1 lists common materials that affect friction in sliding contacts and the mechanism by which they affect contact friction.

Brush Materials. Considering the range of commercially available metal powders, graphites, other lubricants, and processing variables, there are unlimited possibilities for development of suitable brush materials. However, only a limited number of commercial grades have been developed; a list of grades compiled by one supplier appears in Table 2.

More brush contacts are made from copper and its alloys than from any other class of material. In applications where copper metals undergo substantial oxidation, silver metals may be used. Tungsten or, more rarely, molybdenum is used where a high melting point is required. Platinum, palladium, and gold are used where reliable closure with low force is required. Brushes clad or electroplated with precious metals, and brushes made of sintered alloys, are important for general applications in power-switching relays.

Table 1 Effects of atmospheric contaminants on friction between carbon brushes and copper

Contaminant	Effect
CCl_4	Friction becomes high and erratic
Cl_2	Small amounts, friction decreases; large amounts, friction increases
SO_2	Friction increases
H_2S	Small amounts, friction decreases; large amounts (detectable by smell), friction increases
Silicones	With current, friction increases; without current, friction decreases
Steam, H_2O	Friction may increase or decrease
Tobacco smoke	Friction increases
Oil vapor	Small amounts, friction decreases; large amounts, friction increases

Interdependence Factors. When contacts are attached to a carrier, which is usually a copper alloy, the properties of the carrier material and the properties of the interface between contact and carrier (that is, the area of bond and the conductivity across the interface) are critical to ultimate performance. The contact carrier serves as a heat sink as well as a structural member and electrical conductor. The overall efficiency of the system depends on the contact, the contact carrier, and the method of attachment, all of which affect the size of the contact required for a specific application. To conserve precious metal, the contact materials, carrier material, and method of attachment must be optimized. Some high-strength, high-conductivity copper alloys are used for carriers because of their structural properties and resistance to softening at brazing temperatures.

The attachment method that provides minimum interface alloying, minimum softening of the carrier, and maximum bond area generally produces the best combination of properties for the contact system as a whole. Common attachment methods include brazing, resistance welding, percussion welding, and resistance induction torch welding. Welding methods provide the most localized heat input and therefore minimize softening of the carrier, which may occur during brazing methods. A more weldable backing such as steel, nickel, or monel is often clad to the precious metal to provide optimum welding compatibility with the carrier. Projections can also be coined or rolled into the contact backing to assist in welding.

Percussion welding does not require a special backing for attachment. However, percussion welding has fallen out of favor due to relatively high cost in comparison with brazing, welding, and mechanical attachment (riveting). The overwhelming majority of contacts in the United States and Europe are brazed, either in strip or part form.

Copper Contact Alloys

High electrical and thermal conductivities, low cost, and ease of fabrication ac-

Table 2 Properties of metal-graphite electrical contact materials

Composition, % Metal	Composition, % Graphite	Approximate density, g/cm³	Electrical conductivity, % IACS	Applications
Copper-graphite materials				
30	70	2.60	0.11	Alternators; small auxiliary motors; low-metal, long-life brushes
30	70	2.50	2	Automotive auxiliary and appliance motors
36	64	2.75	3	Automotive heaters and blower motors
40	60	2.75	4	Automotive and other small auxiliary starting motors
40	60	2.75	2.5	Automotive heaters and ac motors
50	50	3.05	0.73	Automotive alternators
50	50	2.97	6	Automotive auxiliary and appliance motors
50	50	3.18	0.83	Industrial truck motors
62	38	3.65	3	Automotive starters. Excellent grade for low-humidity applications; excellent filming properties
65	35	3.15	3	Starters
75	25	3.25	0.51	ac wound motors and rotary converters
95	5	6.30	34	Collector roll brushes
92	8	7.30	41	High-current-carrying brush material for grounding applications
96	4	7.75	42	Automotive starters
Silver-graphite materials				
90	10	7.5	42	Instruments, fuel pumps
50	50	3.4	2	Welding machines, motors
85	3(a)	7.8	45	Antenna motors(b)
75	20(c)	5.2	12	Antenna motors and generators(b)

(a) Plus $12MoS_2$. (b) With altitude protection. (c) Plus $5MoS_2$

Table 3 Properties of copper metals used for electrical contacts

UNS number	Solidus temperature °C	Solidus temperature °F	Electrical conductivity, % IACS	Hardness OS035 temper	Hardness H02 temper	Tensile strength OS035 temper MPa	Tensile strength OS035 temper ksi	Tensile strength H02 temper MPa	Tensile strength H02 temper ksi
C11000	1065	1950	100	40 HRF	40 HRB	220	32	290	42
C16200	1030	1886	90	54 HRF	64 HRB(a)	240	35	415(a)	60(a)
C17200	865	1590	15–33(b)	60 HRB(c)	93 HRB(d)	495(c)	72(c)	655(d)	95(d)
C23000	990	1810	37	63 HRF	65 HRB	285	41	395	57
C24000	965	1770	32	66 HRF	70 HRB	315	46	420	61
C27000	905	1660	27	68 HRF	70 HRB	340	49	420	61
C50500	1035	1900	48	60 HRF	59 HRB	276	40	365	53
C51000	975	1785	20	28 HRB	78 HRB	340	49	470	68
C52100	880	1620	13	80 HRF	84 HRB	400	58	525	76

(a) H04 temper. (b) Depends on heat treatment. (c) TB00 temper. (d) TD02 temper

count for the wide use of copper alloys in electrical contacts. The main disadvantage of copper contacts is low resistance to oxidation and corrosion. In many applications, the voltage drop resulting from the film developed by normal oxidation and corrosion is acceptable. In some circuit breaker applications, the contacts are immersed in oil to prevent oxidation. In other applications, such as in drum controllers, sufficient wiping occurs to maintain fairly clean surfaces, thus providing a circuit of low resistance. In some applications, such as knife switches, plugs, and bolted connectors, contact surfaces are protected with grease or coatings of silver, nickel, or tin. In power circuits, where oxidation of copper is troublesome, contacts frequently are coated with silver. Vacuum-sealed circuit breakers use oxygen-free copper contacts (wrought or powder metal) for optimum electrical properties.

In air, copper does not provide high resistance to arcing, welding, or sticking. Where these characteristics are important, copper-tungsten or copper-graphite mixtures are used. However, when used in a helium atmosphere, a Cu-CdO contact performs similarly to an Ag-CdO contact. Copper alloys are used for high currents in vacuum interrupters.

Pure copper is relatively soft, anneals at low temperatures, and lacks the spring properties sometimes desired. Some copper alloys, harder than pure copper and having much better spring properties, are listed in Table 3. The annealing temperature and the elevated-temperature properties of copper can be increased by additions of 0.25% Zn, 0.5% Cr, 0.03 to 0.06% Ag (10 to 20 oz per ton) or small amounts of finely dispersed metal oxides, such as Al_2O_3, with little loss of conductivity. On the other hand, improved mechanical properties are obtained only at the expense of electrical conductivity. Precipitation-hardened alloys, dispersion-hardened alloys, and powder metal mixtures can provide a wide range of mechanical and electrical properties.

Applications. Copper-base metals are commonly used in plugs, jacks, sockets, connectors, and sliding contacts. Because of tarnish films, the contact force and amount of slide must be kept high to avoid excessive contact resistance and high levels of electrical noise. Yellow brass (C27000) is preferred for plugs and terminals because of its machinability. Phosphor bronze (C50500 or C51000) is preferred for thin socket and connector springs and for wiper-switch blades because of its strength and wear resistance. Nickel silver is sometimes preferred over yellow brass for relay and jack springs because of its high modulus of elasticity and strength, resistance to tarnishing, and better appearance. Sometimes, copper alloy parts are nickel plated to improve surface hardness, reduce corrosion, and improve appearance. However, nickel carries a thin but hard oxide film that has high contact resistance; very high contact force and long slide are necessary to rupture the film. To maintain low levels of resistance and noise, copper metals should be plated or clad with a precious metal.

Silver Contact Alloys

Silver, which has the highest electrical and thermal conductivities of all metals, is the most widely used material in pure or alloyed form for a considerable range of arcing contacts (1 to 600 A). Because of its electrical and thermal conductivity, silver is widely used in contacts that remain closed for long periods of time and, in the form of electroplate, as a coating for connection plugs and sockets. It is also used on contacts subject to occasional sliding, such as in rotary switches, and to a limited extent for low-resistance sliding contacts, such as slip rings.

The various types of silver-base contact materials include:

- Unalloyed silver or silver overlays
- Binary, ternary, and multicomponent silver alloys such as silver-cadmium, silver-copper, silver-copper-nickel, or silver-cadmium-nickel contact materials
- Composite silver-base materials with refractory constituents such as cadmium oxide, magnesium oxide, tin oxide, graphite, tungsten, tungsten carbide, or molybdenum

The composite silver-bases are often made using P/M processes, which allow the combination of constituents that ordinarily cannot be alloyed. However, some silver-base composites with cadmium oxide or magnesium oxide can also be made by preparing binary alloys of silver and cadmium or of silver and magnesium, then converting the cadmium or magnesium into an oxide by internal oxidation. Contact materials with refractory or oxide constituents are described in the section "Composite Materials" in this article.

This section describes the properties of silver and the silver alloys used as contact materials. Mechanical properties and hardness of pure silver are improved by alloying, but its thermal and electrical conductivities are adversely affected. Figure 1 shows the effect of different alloying elements on the hardness and electrical resistivity of silver. Properties of the principal silver metals used for electrical contacts are given in Table 4.

Electrical and Thermal Conductivity. Because silver has the highest electrical and thermal conductivities of all metals at room temperature, it can carry high currents without excessive heating, even when dimensions of the contacts are only moderate. Although good thermal conductivity is desired once the contact is in service, such conductivity increases the difficulty of assembly welding unless a higher resistance layer is added by cladding.

In component assemblies, migration of silver through and around electrical insulation may cause failure of the insulation. When in contact with certain materials, such as phenol fiber, and when under electric potential, silver migrates ionically through or across the insulating material, producing thread-like connections that lower the resistance across the insulation. This reduces insulating qualities, and the reduction is even greater if moisture is present in the atmosphere. Insulators must be designed with care to avoid this hazard.

Oxidation Resistance. Silver is used instead of copper chiefly because of its resistance to oxidation in air. In general, silver oxide is not a problem on silver contacts, whether or not the contacts make and break

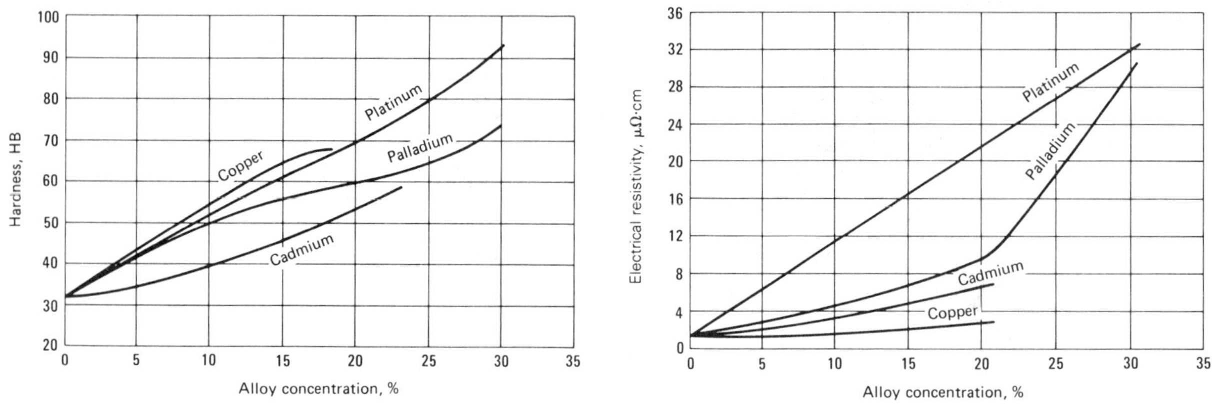

Fig. 1 Hardness and electrical resistivity versus alloy content for silver alloy contacts

the circuit. However, silver oxide can be produced by exposure to ozone, as well as by other methods. This oxide has high resistivity, is decomposed slowly on heating at about 175 °C (350 °F), is decomposed rapidly at about 350 °C (650 °F) and is removed by arcing. This "self-cleaning" characteristic is most unusual among metals and is a chief reason for the attraction of silver as a contact material.

Silver is vulnerable to attack by sulfur or sulfide gases in the presence of moisture. The resulting sulfide film may produce significant contact resistance, particularly where contact force, voltage, or current is low. Direct current brings silver ions from the matrix into the sulfide where they form connecting bridges. Therefore, particularly at high direct current, the film becomes somewhat conducting. The resistance of a silver sulfide film decreases as temperature increases—Ag_2S decomposes slowly at 360 °C (680 °F) and more rapidly at higher temperatures. In addition, the film may increase erosion and entrap dust.

Limitations of Silver Contacts. Silver will provide a fairly long contact life for make-break contacts and will handle up to 600 A. In pure silver contacts, difficulties sometimes arise from transfer of metal from one electrode to the other, which leads to the formation of buildups on one contact surface and holes in the other. When used in dc circuits, silver contacts are subject to ultimate failure by mechanical sticking as a direct result of metal transfer. The direction of transfer is generally from the positive contact to the negative, but under the influence of arcing, the direction may be reversed. With high currents or inductive loads, it may be desirable to shunt the load with a resistance-capacitance protection network to reduce erosion.

When arcing produces a glow discharge in air, the rate of erosion of silver is unusually high because of a chemical interaction with air to form $AgNO_2$.

For low resistance and low noise levels, the design of the contact device must provide sufficient force and slide to break through any silver sulfide film and maintain film-free metal-to-metal contact at the interface. Connectors should have high slide force and several newtons normal force. Rotary switches that have up to 490 mN (0.11 lbf) normal force and considerable slide should have a protective coating of grease to reduce sulfiding and to remove abrasive particles. In low-noise transmission circuits, silver should not be used on relay and other butting contacts that have less than 195 mN (0.044 lbf) force; other precious-metal coatings, such as gold or palladium, should be used instead of silver.

A silver sulfide film has a characteristic voltage drop of several tenths of a volt. Where this drop is tolerable, silver contacts will provide reliable contact closure. Failure to close, however, may be greater than with other precious metal contacts because of impacted dirt, with a sulfide film acting as a dirt catcher.

For many applications, silver is too soft to give acceptable mechanical wear. Alloying additions of copper, cadmium, platinum, palladium, gold, and other elements are effective in increasing the hardness and modifying the contact behavior of silver. These additions do, however, lower both the electrical conductivity and the oxidation resistance relative to pure silver.

Fine silver (99.9 Ag) has the highest electrical and thermal conductivities of all metals. It has a high current capacity, which limits heat generation, and a high thermal conductivity, which allows contacts to readily dissipate the heat generated by arc-

Table 4 Properties of silver metals used for electrical contacts

Alloy	Solidus temperature		Electrical conductivity, % IACS	Hardness, HR15T		Tensile strength				Density, g/cm³	Elongation in 50 mm or 2 in., %	
						Annealed		Cold worked				
	°C	°F		Annealed	Cold worked	MPa	ksi	MPa	ksi		Annealed	Cold worked
99.9Ag	960	1760	104	30	75	170	25	310	45	10.51	55	5
99.55Ag-0.25Mg-0.2Ni	...	...	70	61	77	207	30	345	50	10.34	35	6
99.47Ag-0.18Mg-0.2Ni-0.15Cu	...	...	75	64	84	...	...	...	...	10.38	...	...
99Ag-1Pd	...	...	79	44	76	180	26	324	47	10.14	42	3
97Ag-3Pd	977	1790	58	45	77	186	27	331	48	10.53	37	3
97Ag-3Pt	982	1800	45	45	77	172	25	324	47	10.17	37	3
92.5Ag-7.5Cu	821	1510	88	65	81	269	39	455	66	10.34	35	5
90Ag-10Au	971	1780	40	57	76	200	29	317	46	11.03	28	3
90Ag-10Cu	775	1430	85	70	83	276	40	517	75	10.31	32	4
90Ag-10Pd	1000	1830	27	63	80	234	34	365	53	10.57	31	3
86.8Ag-5.5Cd-0.2Ni-7.5Cu	...	...	43	72	85	276	40	517	75	10.10	43	3
85Ag-15Cd	877	1610	35	51	83	193	28	400	58	10.17	55	5
77Ag-22.6Cd-0.4Ni	...	...	31	50	85	241	35	469	68	10.31	55	4
75Ag-24.5Cu-0.5Ni	...	...	75	78	85	310	45	552	80	10.00	32	4
72Ag-28Cu	775	1430	84	79	85	365	53	552	80	9.95	20	5
60Ag-23Pd-12Cu-5Ni	...	...	11	86	93	517	78	758	110	10.51	22	3

Table 5 Properties of gold metals used for electrical contacts

Alloy	Solidus temperature °C	°F	Electrical conductivity, % IACS	Hardness, HR15T Annealed	Cold worked	Tensile strength Annealed MPa	ksi	Cold worked MPa	ksi	Density, g/cm³
99Au	1085	1985	74	40	65	...	...	...	...	19.36
90Au-10Cu	932	1710	16	76	91	400	58	705	102	17.18
75Au-25Ag	1029	1885	17	50	77	...	...	...	...	15.96
71.5Au-14.5Cu-8.5Pt-4Ag-1Zn	925	1700	11	88	96	...	...	...	...	15.9
72.5Au-14Cu-8.5Pt-4Ag-1Zn	954	1750	10	88	96	...	...	...	...	16.11
72Au-26.2Ag-1.8Ni	...	...	14	61	81	230	33	345	50	15.56
71Au-5Ag-9Pt-15Cu	...	...	8	88.5	75(a)	700	101	1170	170	16.02
69Au-25Ag-6Pt	1029	1885	10	70	84	275	40	415	60	15.92
50Au-50Ag	...	...	...	...	...	...	...	...	...	13.59

(a) Rockwell 15N

ing. Silver also has good oxidation resistance, and therefore a low-resistance and low-voltage drop across the contact interface can be maintained.

The low boiling and melting points of silver are disadvantages. Fine silver contacts tend to weld easily, and usually have high erosion loss (see "Silver-Cadmium Alloys" below). The low hardness and low mechanical strength of fine silver result in high rates of mechanical wear.

Fine silver contacts are used in low-current (<20 A) applications such as switches and relays in appliances and automotive products. Because fine silver is very ductile, it can be fabricated into many designs, including contacts in the form of solid, tubular, and composite rivets and solid buttons.

Silver-Copper Alloys. Copper additions improve the hardness of silver appreciably and slightly decrease its conductivity. However, copper decreases the tarnish resistance of silver; hence, the oxidized film increases contact resistance. Switching devices that have silver-copper contacts should have a high closing force and large wiping action to break down the oxide films.

Silver-copper alloys are used in place of fine silver where electrical, mechanical, and atmospheric conditions are compatible. For the same application, silver-copper alloys usually cost less than fine silver.

Addition of a small amount of nickel to a silver-copper alloy (as in Ag-24.5Cu-0.5Ni, for example) makes the oxide film brittle, so that switching devices can use less closing force.

Silver-Cadmium Alloys. Cadmium improves the arc-quenching ability of silver, and also increases its resistivity and mechanical strength. Silver-cadmium alloys are more resistant to arc erosion and welding than fine silver and silver-copper alloys.

Ag-22.6Cd-0.4Ni Alloy. Because of the nickel addition, the oxidation film of this alloy is also brittle. Ag-22.6Cd-0.4Ni is used in electrical gages and automotive voltage regulators, where the closing force is light and where a stable resistance and low transfer rate are required. It is also used to make positive contact and retard material transfer when paired with Ag-3Pd alloy in polarized low-voltage circuits.

Ag-15Cd alloy, which typically undergoes internal oxidation to form a Ag-CdO composite, is a widely used composition. The Ag-CdO composite (discussed in the section "Composite Materials" in this article) exhibits low welding tendencies at the contact interface and is the material most commonly used to switch light or medium current in ac or dc circuits such as line starters, solenoid relays for automotive starters, and other devices subjected to high-surge current.

Ag-5.5Cd-0.2Ni-7.5Cu alloy has excellent resistance to corrosion and good spring properties. It is used to make current-carrying spring contacts in television tuners, collector rings, and rf switches.

Silver-Platinum Alloys. A small addition of platinum increases the hardness, wear resistance, and corrosion resistance of silver, but concurrently decreases its electrical conductivity. Silver-platinum alloys are used in switching devices having low closing force where cost is not the main concern.

Silver-Palladium Alloys. Palladium improves the wear resistance of silver, but also decreases its conductivity. Silver-palladium alloys are less susceptible to oxidation than fine silver. Ag-3Pd alloy is used as the negative contact paired with Ag-22.6Cd-0.4Ni in low-voltage dc circuits. Silver and palladium form a complete solid solution, and their alloys have very good fabricability.

Silver-Gold Alloys. Gold increases hardness and improves oxidation resistance of silver. The tarnish films on contact surfaces are more stable than those of any other alloy. Ag-10Au is primarily used in ac and dc relays with current capacities less than 0.5 A where high reliability is essential. This alloy is very ductile and can be fabricated in the same manner as fine silver.

Multi-Component Alloys. Ag-0.25Mg-0.20Ni and Ag-0.18Mg-0.20Ni-0.15Cu have similar properties. In low-current dc applications (voltage regulators, thermal gages, and relays), these materials provide low transfer characteristics. Mechanical properties can be improved by internally oxidizing the alloying elements into oxides.

Ag-23Pd-12Cu-5Ni has high hardness (good resistance to wear), better tarnish resistance, and a higher melting point than fine silver. It is limited to light current applications. Because of its high hardness, the alloy is used as brush contacts in potentiometers and other sliding applications. This alloy is also made into disks for composite rivets.

Gold Contact Alloys

Pure gold has unsurpassed resistance to oxidation and sulfidation, but a low melting point and susceptibility to erosion limit its use in electrical contacts to situations where the current is not more than 0.5 A. Although oxide and sulfide films do not form on gold, a carbonaceous deposit is sometimes formed when a gold contact is operated in the presence of organic vapors. The resistance of this film may be several ohms.

When gold is used in contact with palladium or rhodium, very low contact resistances have been reported.

The low hardness of gold can be increased by alloying with copper, silver, palladium, and platinum, but usage is necessarily restricted to low-current applications because of the low melting point.

Properties of gold and its alloys are listed in Table 5. If low tarnish rates and low contact resistance are to be preserved, the gold content should not be less than about 70%.

Fine Gold. The unique property of fine gold (99.9 Au) as a contact material is its superb tarnish resistance in air. Only platinum is more tarnish resistant, but fine gold is less costly than platinum. Pure gold is very soft and susceptible to mechanical wear, metal transfer, and welding. Pure gold electroplated or roll bonded over a base metal substrate is used in dry circuit connectors and relays to improve reliability. It is widely used in computers and telecommunications equipment where reliability is a major concern.

Au-26.2Ag-1.8Ni and Au-27Ag-3Ni Alloys. Silver and nickel increase the hardness of gold, thereby increasing resistance to mechanical wear and deformation. Au-26.2Ag-1.8Ni resists welding and transfer better than pure gold and is used in devices that carry less than 0.5 A current where high reliability is required. The alloy is ductile and has good fabricability. Alloy Au-27Ag-3Ni is more widely used than Au-26.2Ag-1.8Ni.

Au-25Ag-6Pt. Both silver and platinum increase the hardness of gold. Au-25Ag-6Pt is employed in low-current and low-closing-force relays such as those used in telecommunication systems, where high reliability is required. Under conditions of erosion, contacts have long life if the current is limited to 0.4 A. The alloy is also highly satisfactory for use in sliding contacts, such as in rotary switches or low-pressure slip rings, because it has good wear resistance and maintains low contact resistance. It is less susceptible to polymer formation than palladium and, where this is important, its greater cost may be justified.

Au-25Ag and Au-50 Ag. Silver increases the hardness of gold, but decreases its tarnish resistance. Gold-silver alloys are used where a higher degree of reliability is required than can be obtained with silver-base alloys.

Au-5Ag-9Pt-15Cu provides good tarnish resistance as well as high hardness and strength. It is used as a contact where a large wipe is required. It is also used as brush contacts against slip rings made of Au-26.2Ag-1.8Ni.

Au-10Cu. Copper increases the hardness of gold with only a small sacrifice in corrosion resistance. Au-10Cu is used in low-voltage dc devices such as alternators or voltage regulators, and as a positive contact paired with a platinum-iridium negative contact. Under light closing forces, this combination provides a low transfer rate and good anti-welding characteristics.

Au-14.5Cu-8.5Pt-4.5Ag-1Zn. This multiple-component alloy is age hardenable. Compared to other high-gold-content alloys, it can be two to three times harder. The high gold, silver, and copper content make it a low producer of frictional polymers; thus, an excellent material for applications with voltages too low to electrically puncture these films. The combined strength and nobility of this alloy make it the ideal material for sliding contacts, slip rings, and resistance wire. This alloy is the preferred mating material for Pd-30Ag-14Cu-10Au-10Pt-1Zn alloy. Applications also include self-contained cantilever beam contacts.

Precious Metals of the Platinum Group

Platinum and palladium are the two most important metals of the platinum group. These metals have a high resistance to tarnishing, and therefore provide reliable contact closure for relays and other devices having contact forces of less than 490 mN (0.110 lbf). Their high melting points, low vapor pressure, and resistance to arcing make them suitable for contacts that close and open the load, particularly in the range up to 1 A. The low electrical and thermal conductivities of these metals, as well as their cost, generally exclude them from use at currents above about 5 A.

Palladium has an arcing limit only slightly less than that of platinum and gives comparable performance in relays for telephones and similar services handling 1 A or less. Palladium is a satisfactory substitute for platinum in these applications.

Chemical Properties. Platinum has a high resistance to corrosion, including resistance to oxidation, sulfidation, and salt water. It will not form a stable oxide at any temperature.

Palladium is resistant to oxidation at ordinary temperatures. If heated above 350 °C (660 °F), it will oxidize slowly to form an oxide that is stable at room temperature. However, the oxide is decomposed promptly on heating to 800 °C (1470 °F) or by arcing. The oxide is not considered to be a significant factor in the reliability of closure of telephone-type relays.

The presence of organic vapors in the contact area can seriously influence the life and reliability of electrical contacts, particularly the low-force precious metal contacts universally employed in high-reliability low-noise circuits. The damaging organic vapors may arise from coil forms, wire coatings, insulation, soldering flux, potting and sealing compounds, and other organics in associated electrical equipment, as well as from external sources.

Organic contamination may produce two distinctly different forms of contact damage: activation and polymer formation.

Activation is the development of a carbon deposit on the contacting surfaces, formed by the decomposition of the organic contaminant in the arc. This deposit markedly increases arc erosion. The carbon deposits decrease the current needed to sustain an arc and prolong the arcing time. A 95% reduction in contact life may result from activation brought about by the presence of organic vapors. Activation can be reduced or eliminated by using insulating materials that are not sources of organic vapors, by adequate ventilation and, perhaps, by absorbing the organic vapors in a getter such as active carbon.

Polymer formation is the development of a polymer-like insulating brown powder on contacts in dry circuits (those not carrying current on make or break), and may lead to transient open circuits. The insulating brown powder is believed to result from the adsorption of the organic vapor on the contact surface, followed by its polymerization by the friction associated with contact operation. The sliding motion both forms the polymer and pushes it outside the slide area, where it builds up as a brown powder. A transient open circuit occurs when some of the built-up powder falls into the contact area.

Controlled experiments have shown that the type of contact metal influences the amount of polymer formed. The greatest amount of polymer is formed on the platinum metals; lesser amounts are formed on gold and some base metals; polymer does not form on silver.

Elimination of materials that give rise to organic vapors is a possible solution to polymer formation, but one that is difficult to carry out. From a practical standpoint, the problem has been solved in telephone circuits by cladding one of a mating pair of palladium contacts with a very thin layer of gold. In dry circuits, the one gold surface significantly reduces polymer formation, although in working circuits, the gold soon wears off and exposes the palladium base.

Erosion and Sticking. The arcing current limit for platinum group metals is about 1 A, and contact life is long if the current is kept below this value. With currents higher than 1 A or with inductive loads, it may be desirable to shunt the load or contact with a resistance-capacitance network to reduce erosion, and to reduce failures caused by snagging of pits and build-ups. In general, for equal volumes of contact metal, the life of platinum or palladium contacts is about ten times the life of silver contacts.

Resistance and Noise. Palladium is used almost universally on relays and relay-type switch contacts in telephone systems within the United States for talking circuit transmission. In this service, palladium is essentially noise-free and is used in preference to platinum or gold alloys because it is more economical.

In a few isolated instances, where the palladium talking circuit contacts have been subject to vibration in service, noise troubles have developed because of polymer formation. In these few instances, the difficulty has been met by the use of gold alloys, which greatly reduces the production of polymer.

Fine Platinum. Platinum has higher melting and boiling points (1799 °C, 3270 °F, and 4530 °C, 8186 °F, respectively) than those of gold (1063 °C, 1945 °F and 2970 °C, 5380 °F), and has excellent corrosion resistance.

Fine platinum (99.9 Pt) provides a very low and consistent surface resistance at a wide range of temperatures. Contacts made with platinum remain clear in hostile environments. It is usually used in light-force relays with a current capacity of up to 2 A when reliability is the most important factor.

Properties of platinum and its alloys are listed in Table 6. The effect of alloying on the hardness and electrical resistivity of platinum is shown in Fig. 2.

Platinum-Iridium Alloys. Platinum and iridium form a complete solid solution. The

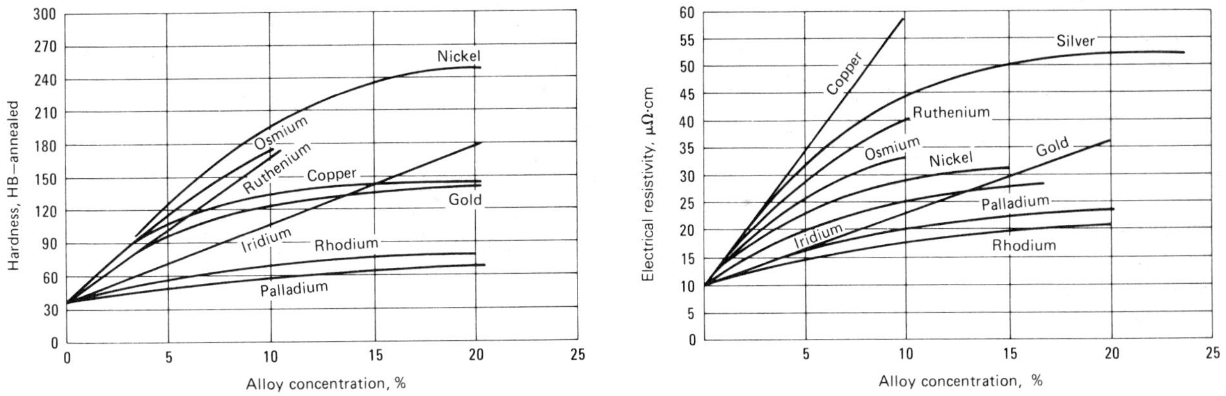

Fig. 2 Hardness and electrical resistivity versus alloy content for platinum contacts

physical properties, melting point, hardness, and mechanical strength of Pt-Ir alloys increase almost linearly with the amount of iridium in platinum, without affecting the corrosion resistance of the latter. This group of alloys is used in low-current ac and dc circuits when the mechanical forces are high and high wear resistance and strength are required.

Fabrication of platinum-iridium alloys becomes difficult when the iridium content is high. These alloys are usually used in the form of disks to make composite rivet faces.

Platinum-Ruthenium Alloys. Ruthenium forms a solid solution with platinum up to 79% Ru. Ruthenium increases the hardness, strength, and wear resistance of platinum, but high ruthenium content makes the alloy brittle. To ensure good fabricability, the ruthenium content should not exceed 15%.

Platinum-ruthenium alloys have good tarnish resistance and cost less than platinum-iridium alloys. Pt-Ru alloys are used as positive contacts in low-voltage dc circuits, almost always paired with negative contacts made of tungsten. Pt-Ru alloys can be made into rivets or disks for composite rivets.

Pt-18.4Pd-8.2Ru. This ternary alloy has properties similar to those of platinum-ruthenium and platinum-iridium alloys, but costs less. It is used in low-voltage dc circuits as the positive contact and paired with tungsten as the negative contact. This alloy has poor fabricability and can be used in the form of disks for composite rivets.

Fine Palladium. The oxidation resistance of fine palladium (99.9 Pd) is second only to that of platinum. The boiling point (3950 °C, 7142 °F) and melting point (1552 °C, 2825 °F) are slightly lower than those of platinum. However, palladium costs about one-fourth as much as platinum, and usually replaces platinum when cost is the main concern. Table 6 gives properties of palladium metals. Figure 3 shows the effect of alloying elements on palladium.

Palladium is used in light-closing-force and low-current applications. It has very good fabricability and can be easily made into rivets or composite rivets.

Palladium-Ruthenium Alloys. Palladium forms only a limited solid solution with ruthenium. Ruthenium increases the hardness of palladium without sacrificing its corrosion resistance. Palladium-ruthenium alloys are used in low-current relays where closing forces are high. These alloys are used mostly as positive contacts paired with tungsten in dc circuits. In some applications, they are used to replace platinum-ruthenium alloys when cost is important. Practically all of the palladium-ruthenium alloys are used in the form of rivets because of their poor fabricability.

Pd-40Cu. Corrosion resistance of Pd-40Cu is good, but it is inferior to that of other precious metals. The alloy has high hardness and is used primarily as brush contacts and in instruments and gages. Because of its very poor headability, it can be used only in disk form to make composite rivets.

Pd-40Ag. Silver improves the hardness of palladium. Pd-40Ag costs less than fine palladium but has the same corrosion resis-

Table 6 Properties of platinum and palladium metals used for electrical contacts

Alloy	Solidus temperature, °C	Solidus temperature, °F	Electrical conductivity(a), % IACS	Hardness, HR15T, Annealed	Hardness, HR15T, Cold worked	Tensile strength, Annealed, MPa	Tensile strength, Annealed, ksi	Tensile strength, Cold worked, MPa	Tensile strength, Cold worked, ksi	Density, g/cm^3	Elongation(a) in 50 mm (2 in.), %
99.9Pt	1770	3220	15	60	73	138	20	241	35	21.45	35
95Pt-5Ru	1775	3230	5	84	89	414	60	793	115	20.57	18
92Pt-8Ru	...	...	4	86	91	483	70	896	130	20.27	15
90Pt-10Ir	1780	3240	7	87	92	379	55	620	90	21.52	12
89Pt-11Ru	1815	3300	4	91	96	586	85	1034	150	19.96	12
86Pt-14Ru	1843	3350	3	93	99	655	95	1172	170	19.06	10
85Pt-15Ir	1787	3250	6	90	95	517	75	827	120	21.52	12
80Pt-20Ir	1810	3290	5	93	97	689	100	1000	145	21.63	12
75Pt-25Ir	1819	3310	5	95	98	862	125	1172	170	21.68	10
73.4Pt-18.4Pd-8.2Ru	...	...	4	90	92	517	75	862	125	17.77	12
65Pt-35Ir	1899	3450	4	97	99	965	140	1344	195	21.80	8
99.9Pd	1554	2830	16	62	78	193	28	324	47	12.17	28
95Pd-5Ru	1593	2900	8	79	89	372	54	517	75	12.00	15
89Pd-11Ru	1650	3000	6	85	92	483	70	689	100	12.03	13
72Pd-26Ag-2Ni	1382	2520	4	82	90	469	68	689	100	11.52	13
60Pd-40Ag	1338	2440	4	65	91	372	54	689	100	11.30	28
60Pd-40Cu	1200	2190	8	82	92	565	82	1331	193	10.67	20
35Pd-9.5Pt-9Au-14Cu-32.5Ag	1085	1985	5	90	94	689	100	1034	150	11.63	18
35Pd-10Pt-10Au-14Cu-30Ag	1015	1860	5	90	98	827	120	1240	180	11.8	20
44Pd-38Ag-16Cu-1Pt-1Zn	1032	1890	7	91	96	758	110	1205	175	10.8	15

(a) For material in annealed condition

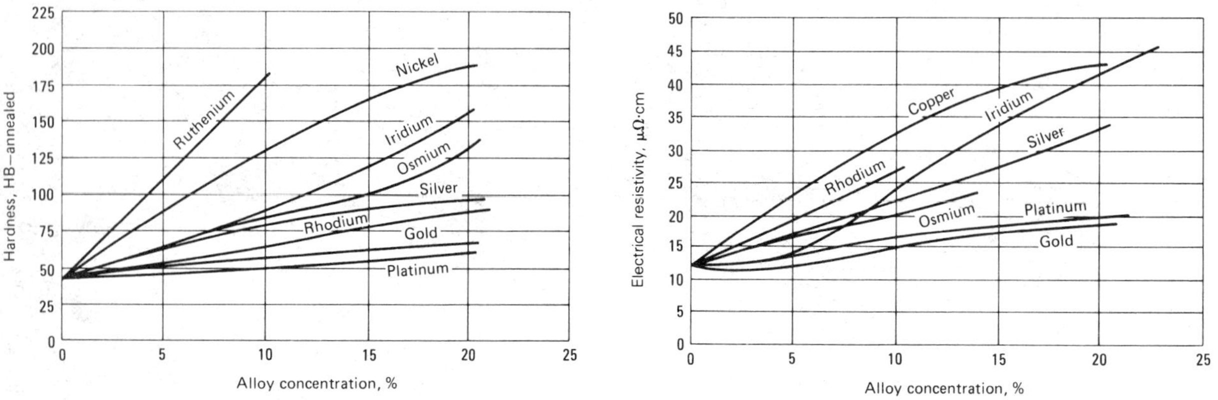

Fig. 3 Hardness and electrical resistivity versus alloy content for palladium contacts

tance against a sulfiding atmosphere. It is used in high-closing-force contacts with less than 1 A current. It has a fair headability, but is used mostly in disk form for composite rivets.

The Pd-40Ag alloy is often clad to base metals for electronic connector applications for increased wear resistance compared to gold alloys. A particular structure of gold diffused into the Pd-40Ag alloy has been found to offer superior properties for electronic connectors used in telecommunications. When clad to copper-base metals this system offers corrosion resistance and wear resistance at lowest cost.

Multiple-Component Alloys. Table 6 lists multiple-component alloys that are designed to increase mechanical properties and decrease cost, with some sacrifice of corrosion resistance.

Pd-9.5Pt-9.0Au-32.5Ag alloy is used for brushes and slide contacts. It has a modulus of elasticity of 115 GPa (17 × 10^6 psi) and a proportional limit of 930 MPa (135 ksi), which are the highest for precious-metal contacts.

Pd-26Ag-2Ni is used in ac or dc contact devices where operation frequency is high, such as in business machines and computers. In dc circuits, it is also used as a positive contact paired with tungsten.

Pd-30Ag-14Cu-10Au-10Pt-1Zn has one of the highest strength and nobility combinations of all other precious-metal contact materials. This alloy has excellent ductility in the annealed condition, which makes it extremely well suited for forming, drawing, and other deformation processes. After the forming operation, the material is age hardened to achieve desired mechanical properties.

Because this alloy is age hardenable, the mechanical properties can be altered to provide the maximum flexibility to the designer. The age-hardening characteristics also provide resistance to stress relaxation at elevated temperatures, which allows for the use of very low contact forces at elevated temperatures without loss of contact pressure. Before hardening, this alloy has good workability owing to its ductility in the annealed condition. After hardening, it also has superior wear properties as a result of hardness values that approach 400 HV. Applications include potentiometers, brushes, make-break contacts, spring arms, and probes for integrated and printed circuits.

Pd-38Ag-16Cu-1Pt-1Zn is another age-hardenable alloy. It is more economical than Pd-30Ag-14Cu-10Au-10Pt-1Zn because it does not have gold and has less platinum. It has high strength and hardness in the age-hardened condition, providing wear resistance and making it ideal for self-contained cantilever beam contacts. The combined 45% noble and 38% semi-noble content give this alloy good resistance to tarnish and corrosion.

Precious Metal Overlays

Silver, gold, rhodium and, to a lesser extent, platinum and palladium are employed in clad and electroplated contacts. Electrodeposition and cladding compete for many of the same applications. Clad overlays are favored because of their lower porosity, but electrodeposits are slightly less expensive for those applications requiring a very thin layer of precious metal. Electrodeposits frequently have higher hardness than the annealed wrought material; rhodium, which has a deposit hardness in excess of 600 HV in the annealed condition, is an outstanding example. High hardness accounts for the superior wear resistance of electroplated rhodium where rubbing or wiping occurs. Even at a thickness of 0.13 to 0.50 μm (5 to 20 μin.) over silver or base-metal contacts, rhodium improves wear resistance and minimizes tarnishing.

Electroplated gold is employed on silver contacts to minimize tarnishing. Nickel underlayers (barrier coats) are used to prevent migration of silver through the gold plate. Recent studies indicate that migration of silver along nickel grain boundaries is rapid at high temperatures. Hence, a nickel barrier coat is questionable for high-temperature applications. Other work in this area has disclosed that palladium can be substituted for gold as a protective coating for extension of shelf life. Electroplated gold also is used on palladium contacts to minimize polymer formation in dry circuits. However, gold electrodeposits on both silver and palladium contacts soon wear off if the contacts wipe, rub, or arc.

Electroplated silver is sometimes applied to copper-base materials to make less expensive components. Electroplated silver is slightly harder than annealed wrought silver. Palladium and platinum electroplated on silver have improved tarnish resistance. Platinum, palladium, and gold electroplates are used on silver to prevent the development of conducting filaments in insulating supports.

Clad overlays are used extensively in applications requiring precious-metal thicknesses from 0.025 to 1.3 mm (0.001 to 0.050 in.). These thicknesses cannot be obtained by electroplating. These applications include electromechanical devices operating in current ranges where arcing and subsequent erosion are likely to occur. In these devices, sufficient precious metal must be available to survive the erosion that will occur over the required device cycle life.

Tungsten and Molybdenum

Most tungsten and molybdenum contacts are made in the form of composites with silver or copper as the other principal component. Tungsten, which was one of the earliest metals other than copper and silver adopted for electrical contact applications, has the highest boiling point (5930 °C, or 10 700 °F) and melting point (3110 °C, or 5625 °F) of all metals; it also has very high hardness at both room and elevated temperatures. Therefore, as a contact material, it offers excellent resistance to mechanical wear and electrical erosion. Its main disadvantages are low corrosion resistance and high electrical resistance. After a short period of operation, an oxidized film will build up on tungsten contacts, resulting in very high contact resistance. Considerable force is required to break through the film, but high pressure and considerable impact cause little damage to the underlying metal because of its high hardness. Tungsten contacts are used in switching devices with closing forces of more than 20 N (4.5 lbf) and in circuits with high voltages and currents not more than 5 A, such as automotive ignitions, vibrators, horns, voltage regulators, magnetos, and electric razors. In low-voltage dc devices, tungsten is always used as the negative contact, and is paired with a positive contact made of precious metal.

Tungsten rods or strips that are consolidated by swaging or rolling from sintered powder compacts have very poor ductility. They cannot be cold worked, in contrast to other contact materials. Tungsten disks are usually cut from rods or punched from strips and then brazed directly to functional parts such as breaker arms, brackets, or springs.

Properties such as grain size, grain configuration, and the degree of fibrous structure, which affect contact behavior, are controlled by using special swaging methods and annealing cycles. Tungsten disks usually are supplied with a ground finish, but they can also be electrochemically polished to obtain high-luster surfaces.

The high boiling and melting points of molybdenum—5560 °C (10 040 °F) and 2610 °C (4730 °F), respectively—are second only to those of tungsten and rhenium. Molybdenum is not used as widely as tungsten because it oxidizes more readily and erodes faster on arcing than tungsten. Nevertheless, because the density of molybdenum (10.2 Mg/m^3 or 0.369 $lb/in.^3$) is about half that of tungsten (19.3 Mg/m^3 or 0.697 $lb/in.^3$), use of molybdenum is advantageous where mass is important. Its cost by volume is also lower.

In addition to its use in make-break contacts, molybdenum is widely used for mercury switches because it is not attacked, but only wetted, by mercury.

Like tungsten, molybdenum strips and sheets are made by swaging or rolling sintered powder compacts. Disks made from rods or sheets are brazed to blanks or other structural components. Table 7 lists the properties of tungsten and molybdenum, and Fig. 4 and 5 show the effect of temperature or diameter on various properties.

Aluminum

As a contact metal, aluminum is generally poor because it oxidizes readily. Where aluminum is used in contacting joints, it should be plated or clad with copper, silver, or tin. Aluminum should never be used for

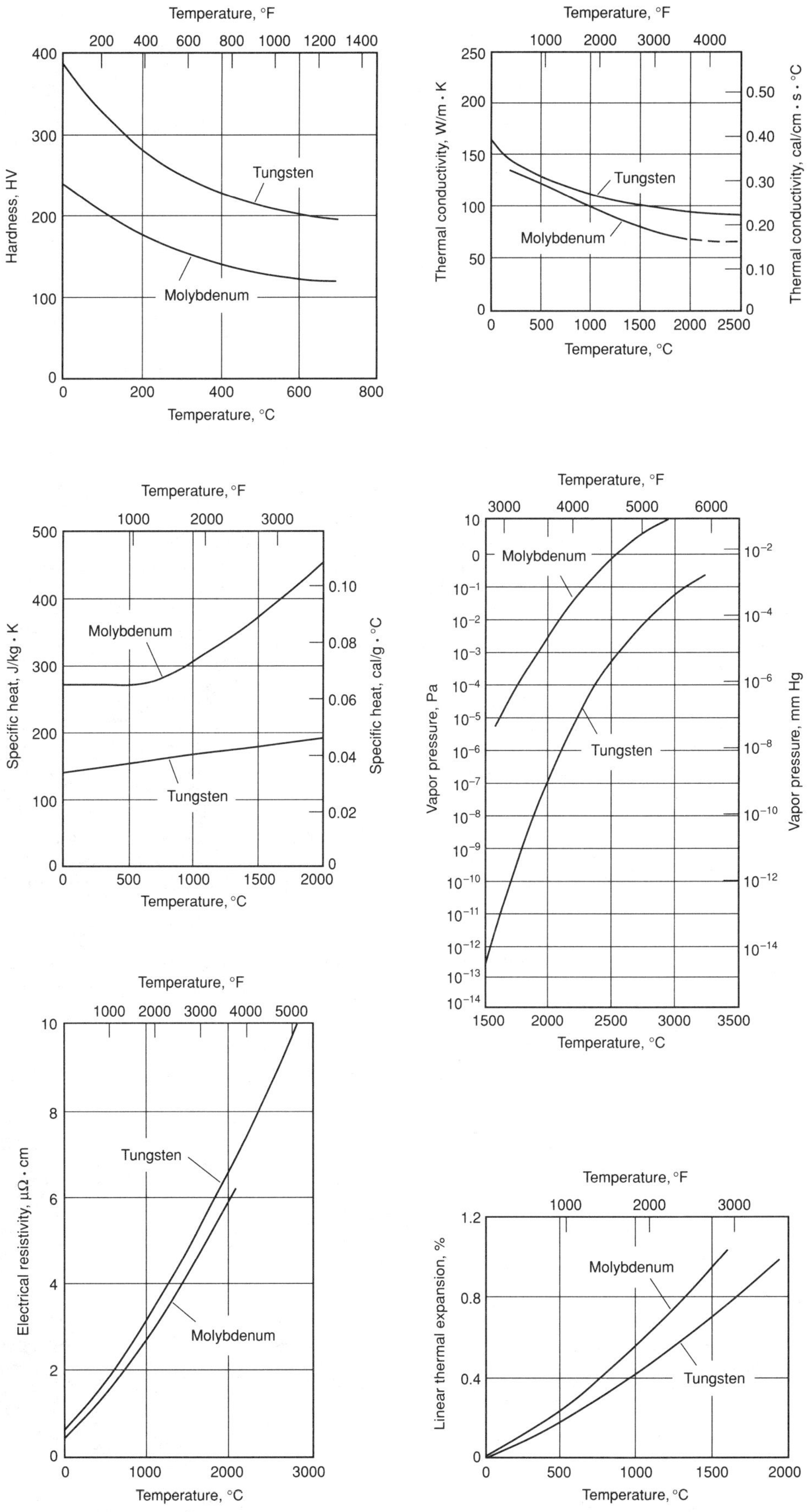

Fig. 4 Variation of properties with temperature for tungsten and molybdenum

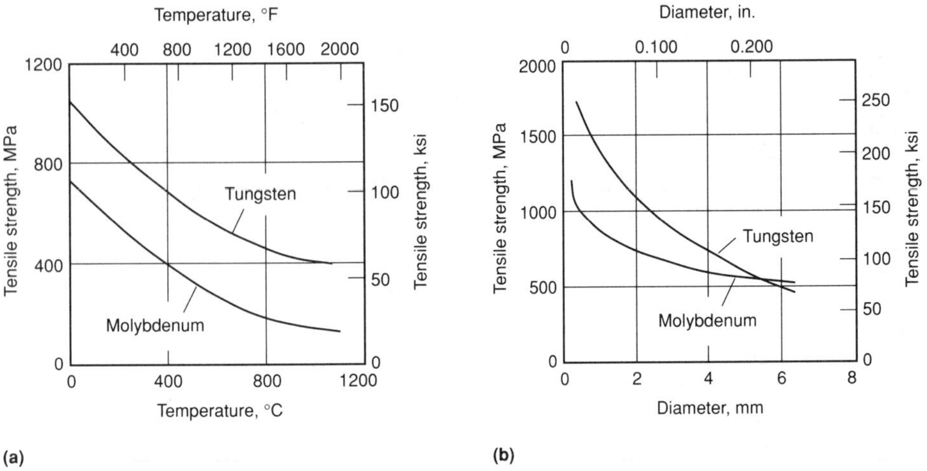

Fig. 5 Tensile strength of tungsten and molybdenum. (a) Variation with temperature. (b) Variation with diameter for tungsten and molybdenum rod

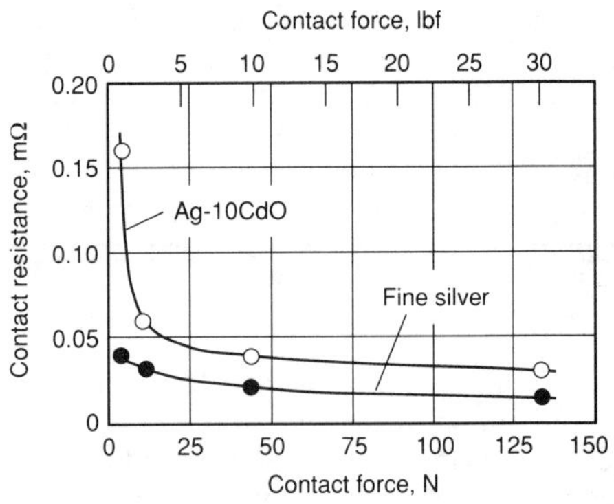

Fig. 6 Contact resistance versus force for fine silver and Ag-CdO contacts. Unarced contacts were 12.7 mm (½ in.) in diameter with a 38 mm (1½ in.) spherical radius. Resistance measurements were made with ac current at 50 A and 60 Hz.

power applications where arcing is present. For instance, if aluminum contacts were substituted for silver in a motor starter, an explosion due to noninterruption of current on motor-starter de-energization would probably occur on load interruption.

Composite Materials

Composite electrical contacts are made from three categories of materials:

- Those that contain refractory constituents such as tungsten or molybdenum carbide
- Those that contain semirefractory constituents such as cadmium oxide, magnesium oxide, and tin oxide
- Those that contain elements (such as silver and nickel) that do not conventionally alloy but which are formed by P/M processes to produce contact materials with unique properties

The various types of composite contact materials (Table 8) generally have a base material of silver, copper, or refractory metals and their carbides. The refractory-base and silver-base contacts are used in switching devices operated in air. Copper-base composite contacts are used in vacuum and oil switching devices.

Table 8 presents the compositions and properties of various composite contact materials. Because manufacturing methods affect the properties of materials with the same composition, the manufacturing methods are also given in Table 8. The most common methods of producing composite electrical contact materials are described in the section "Composite Manufacturing Methods" in this article.

Data published by contact manufacturers usually include density, hardness, and electrical conductivity (Table 8). These data provide designers of electrical devices with the basic properties of a composite contact. Other properties, such as contact resistance, may depend on operational parameters such as force (Fig. 6).

Characteristics that relate directly to failure modes such as arc erosion or material transfer are usually described in a qualitative manner. Very few quantitative data pertaining to these characteristics have been published because these properties depend on several test parameters. For instance, the arc erosion rate is affected by various mechanical factors:

- Opening force and opening speed
- Closing force and closing speed
- Bouncing of the movable contact
- Wiping distance
- Gap between opposing contacts

or electrical factors:

- Current—both amperage and whether ac or dc
- Voltage
- Power factor (inductive/capacitive)

Because each variable can greatly affect the arc erosion rate of a composite, it is virtually impossible to define a universal test to evaluate erosion rate.

Published data on erosion rate and welding frequency usually are collected under very specific conditions. They are valid only for qualitative description in a specific set of circumstances and cannot be extrapolated to suit other applications. The only means of learning how a composite will perform in a specific application is to test it extensively in the device in which it will be

Table 7 Typical properties of tungsten and molybdenum

Some of the physical properties of tungsten and molybdenum vary considerably with cross-sectional area and grain structure.

Tungsten	
Hardness, HRA (HV)	70 (385)
Modulus of elasticity, GPa (10^6 psi)	
At 20 °C (68 °F)	405 (59)
At 1000 °C (1830 °F)	325 (47)
Density, g/cm^3	19.3
Melting point, °C (°F)	3410 (6170)
Boiling point, °C (°F)	5900 (10 650)
Specific heat (Fig. 4), J/kg (Btu/lb · °F), at 20 °C (68 °F)	140 (0.033)
Thermal conductivity (Fig. 4), W/m · K (Btu/ft · h · °F), at 20 °C (68 °F)	130 (75)
Coefficient of linear thermal expansion (Fig. 4), μm/m · K, at 20 °C (68 °F)	4.43
Specific resistance (Fig. 4), nΩ · m at 20 °C (68 °F)	5.5
Electrical conductivity, % IACS, at 20 °C (68 °F)	31
Molybdenum	
Hardness, HRA (HV)	58 (210)
Modulus of elasticity, GPa (10^6 psi)	
At 20 °C (68 °F)	325 (47)
At 1000 °C (1830 °F)	270 (39)
Density, g/cm^3	10.22
Melting point, °C (°F)	2622 (4750)
Boiling point, °C (°F)	4800 (8672)
Specific heat (Fig. 4), J/kg (Btu/lb · °F) at 20 °C (68 °F)	270 (0.065)
Thermal conductivity (Fig. 4), W/m · K (Btu/ft · h · °F) at 20 °C (68 °F)	155 (89)
Coefficient of linear thermal expansion (Fig. 4), μm/m · K at 20 °C (68 °F)	5.53
Specific resistance (Fig. 4), nΩ · m at 20 °C (68 °F)	5.2
Electrical conductivity, % IACS at 20 °C (68 °F)	33

Table 8 Properties of composites for electrical make-break contacts

Nominal composition, %	Manufacturing method(a)	Density, g/cm³ Calculated	Density, g/cm³ Typical	Electrical conductivity, % IACS	Hardness	Tensile strength MPa	Tensile strength ksi	Modulus of rupture MPa	Modulus of rupture ksi	Data source(b)	Application examples
Molybdenum-silver											
90Ag-10Mo	PSR	10.47	10.38	65–68	35–40 HRB	...	...	...	...	A	Air conditioner controls
80Ag-20Mo	PSR	10.44	10.36	59–62	38–42 HRB	...	...	...	...	A	Light and medium duty applications, automotive circuit breakers
75Ag-25Mo	PSR	10.42	10.33	58–61	44–47 HRB	...	...	...	...	A	
70Ag-30Mo	PSR	10.41	10.31	56–60	46–48 HRB	...	...	...	...	A	
65Ag-35Mo	PSR	10.39	10.30	55–64	49–55 HRB	...	...	...	...	A	Automatic circuit protectors, starting switches
60Ag-40Mo	PSR	10.38	10.28	55–62	55–62 HRB	...	...	...	...	A	
50Ag-50Mo	INF	10.35	10.10–10.24	45–52	70–80 HRB	...	...	758	110	C,A	Air and oil-circuit breakers, arcing tips, traffic signal relays, home circuit breakers
	PSR	10.35	10.14	50	65 HRB	...	...	552	80	C,A	
45Ag-55Mo	INF	10.33	10.10–10.32	44–58	75–82 HRB	...	...	...	...	A	
40Ag-60Mo	INF	10.32	10.10–10.22	42–49	80–90 HRB	...	...	...	...	C,A	Aircraft switches, breaker arcing tips, electric razors, air and oil-circuit breakers
	PSR	10.32	10.12	45	50–68 HRB(c)	...	...	676	98	C,A	
35Ag-65Mo	INF	10.30	10.00–10.08	40–45	82–92 HRB	...	...	...	...	A	
30Ag-70Mo	INF	10.29	10.00–10.31	35–45	85–95 HRB	414	60	931	135	C,A	Air circuit breakers, low-erosion arcing tips
25Ag-75Mo	INF	10.27	10.27	31–34	93–97 HRB	414	60	958	139	C,A	
20Ag-80Mo	INF	10.26	10.23–10.26	28–32	96–98 HRB	407	59	965	140	C,A	Arcing contacts, heavy duty electrical applications
15Ag-85Mo	INF	10.24	10.18	28–31	97–102 HRB	...	...	...	...	G	
10Ag-90Mo	INF	10.23	10.13	27–30	97–102 HRB	...	...	...	...	G	Semiconducting material
Silver/cadmium oxide											
97.5Ag-2.5CdO	PSR	10.42	10.21	85	22 HRF(c)	110(c)	16(c)	...	...	C	Aircraft circuit breakers, aircraft relays, automotive relays, truck controls, snap switches, contactors, motor controllers, circuit breakers, governor relays
	PSE	10.42	10.42	95	37 HRF(c) 60 HRF(d)	131(c) 172(d)	19(c) 25(d)	...	...	E,C	
95Ag-5CdO	PSR	10.35	9.50–10.14	80–90	32 HRF(c)	110(c)	16(c)	...	...	C,A	
	PSE	10.35	10.35	92	40 HRF(c) 70 HRF(d)	131(c) 172(d)	19(c) 25(d)	...	...	E,C	
	IO	10.35	10.35	80	40 HRF(c) 75 HRF(d)	186(c) 241(d)	27(c) 35(d)	...	...	E,C	
	PPSE	10.35	10.35	85	70 HRF(c) 90 HRF(d)	207(c) 248(d)	30(c) 36(d)	...	...	E,C	
90Ag-10CdO	PSR	10.21	9.30–9.80	72–85	42 HRF(c)	103(c)	15(c)	...	...	C,A	
	PSE	10.21	10.21	84–87	46 HRF(c) 80 HRF(d)	172(c) 228(d)	25(c) 33(d)	...	...	W,E,A,C,T	
	IO	10.21	10.21	75	45 HRF(c) 81 HRF(d)	186(c) 262(d)	27(c) 38(d)	...	...	W,E,C,M	
	SF	...	...	...	...	...	...	...	...	E,W	
	PPSE	10.21	10.21	82	71 HRF(c) 90 HRF(d)	269(c) 317(d)	39(c) 46(d)	...	...	E,C	
88Ag-12CdO	PSE	10.3	10.2	81.0	60 HV(c) 95 HV(d)	...	...	...	...	E,M,W	
87Ag-13CdO	...	...	9.20	43	56 HRF(c)	...	...	...	...	A	
86.7Ag-13.3CdO	IO	10.11	10.11	68	48 HRF(c) 84 HRF(d)	200(c) 262(d)	29(c) 38(d)	...	...	E,C	
86.5Ag-13.5CdO	PPSE	10.11	10.11	75	70 HRF(c) 90 HRF(d)	276(c) 324(d)	40(c) 47(d)	...	...	E,C	
85Ag-15CdO	PSR	10.06	8.60–9.58	55–75	35 HRF(c)	83(c)	12(c)	...	...	E,C,A,M	
	PSE	10.06	9.90–10.06	55–75	57 HRF(c) 80 HRF(d)	193(c) 241(d)	28(c) 35(d)	...	...	E,T,C,A,M,W	
	IO	10.06	10.06	65	50 HRF(c) 85 HRF(d)	207(c) 269(d)	30(c) 39(d)	...	...	W,C,E	Pressure and temperature controls
	SF	...	...	...	...	...	...	...	...	E,W	
	PPSE	10.06	10.06	72	70 HRF(c) 90 HRF(d)	276(c) 331(d)	40(c) 48(d)	...	...	C	
83Ag-17CdO	IO	10.01	10.01	62	52 HRF(c) 88 HRF(d)	214(c) 276(d)	31(c) 40(d)	...	...	C,E	Aircraft circuit breakers, aircraft relays, truck controls, contactors, circuit breakers, governor relays
	PPSE	10.01	10.01	70	70 HRF(c) 90 HRF(d)	276(c) 352(d)	40(c) 51(d)	...	...	C,E	
80Ag-20CdO	PPSE	9.93	9.93	68	70 HRF(c) 90 HRF(d)	276(c) 345(d)	40(c) 50(d)	...	...	C,E	
75Ag-25CdO	PPSE	9.79	9.79	60	...	...	...	...	...	C,E	

(continued)

(a) PSR, press-sinter-re-press; INF, press-sinter-infiltrate; PS, press-sinter; PSE, press-sinter-extrude; IO, internal oxidation; PPSE, preoxidize-press-sinter-extrude; SF, oxidized from one direction. (b) A: Advance Metallurgy, Inc., McKeesport, PA. C: Contacts, Materials, Welds, Inc., Indianapolis, IN. E: Engelhard Industries, Plainville, MA. G: Gibson Electric Inc., Delmont, PA. S: Stackpole Carbon Co., St. Marys, PA. T: Texas Instruments Inc., Attleboro, MA. M: Metz Degussa, South Plainville, NJ. W: Art Wire-Duduco, Cedar Knolls, NJ. (c) Annealed. (d) Cold worked

Table 8 (continued)

Nominal composition, %	Manufacturing method(a)	Density, g/cm³ Calculated	Density, g/cm³ Typical	Electrical conductivity, % IACS	Hardness	Tensile strength MPa	Tensile strength ksi	Modulus of rupture MPa	Modulus of rupture ksi	Data source(b)	Application examples
Silver-graphite											
99.75Ag-0.25C	PSR	10.41	9.70–10.40	95–103	33–45 HRF(c) 70–73 HRF(d)	172(c) 255(d)	27(c) 37(d)	...	...	C,A,S	Automotive regulators, low voltage make-break contacts, sliding contacts
99.5Ag-0.5C	PSR	10.31	9.60–10.30	92–102	26–44 HRF(c) 69–72 HRF(d)	169(c) 252(d)	24.5(c) 36.5(d)	...	...	C,A,S	
99.25Ag-0.75C	PSR	10.22	10.21	90–100	39 HRF(c) 70 HRF(d)	165(c) 247(d)	24(c) 35.8(d)	...	...	C	
99Ag-1C	PSR	10.13	9.40–10.12	87–99	24–36 HRF(c) 68–69 HRF(d)	162(c) 241(d)	23.5(c) 35(d)	...	...	C,A,S	
98.5Ag-1.5C	PSR	9.96	10.04	97	33 HRF(c) 66 HRF(d)	152(c) 231(d)	22(c) 33.5(d)	...	...	C,A	
98Ag-2C	PSR	9.79	9.15–9.57	82–90	22 HRF(c) 65 HRF(d)	...	...	...	...	C,A,S	Mate with other contact materials in circuit breakers
97Ag-3C	PSR	9.46	8.80	55–62	20 HRF(c) 60 HRF(d)	...	...	...	...	A,S	
97Ag-3C	PSE	9.10	8.90	86	42 HV	...	...	...	...	W	
96Ag-4C	PSE	9.15	8.8	79	41 HV	...	...	...	...	W	
95Ag-5C	PSR	8.88	8.30–8.68	55–62	25 HRF(d)	...	...	...	...	C,A,S	
	PSE	8.88	8.84	75	40 HRF(d)	...	...	...	...	W,C	
93Ag-7C	PSR	8.37	7.80	50–57	15 HRF(c) 45 HRF(d)	...	...	...	...	C,A,S	
90Ag-10C	PSR	7.69	6.30–7.20	43–53	13 HRF(c) 30 HRF(d)	...	...	...	...	W,C,A,S	
Silver-iron											
90Ag-10Fe	PSR	10.16	9.60–10.25	87–92	48 HRF(c) 81 HRF(d)	214(c) 272(d)	31(c) 39.5(d)	...	...	C,A	Wall switches, thermostat controls
Silver-nickel											
99.7Ag-0.3Ni	...	10.49	...	100	53 HR15T(c) 79 HR15T(d)	...	...	...	...	T	...
95Ag-5Ni	PSR	10.41	9.80–10.41	80–95	32 HRF(c) 84 HRF(d)	165(c)	24(c)	...	...	C,A,S	Appliance switches
90Ag-10Ni	PSR	10.31	9.70–10.32	75–90	35 HRF(c) 89 HRF(d)	172(c)	25(c)	...	...	W,C,S,A,E	Low rating line starters
85Ag-15Ni	PSR	10.22	9.50–10.02	66–80	40 HRF(c) 93 HRF(d)	186(c)	27(c)	...	...	W,C,A,S, A,E	Circuit breakers
80Ag-20Ni	PSR	10.13	9.30–9.50	63–75	52–59 HRF(c) 80 HRF(d)	...	...	...	...	W,E,A,S	Circuit breakers, disconnect switches
75Ag-25Ni	PSR	10.05	9.20	59	61 HRF(c)	...	...	...	...	S	
70Ag-30Ni	PSR	9.96	9.40–9.53	55–56	42 HRF(c) 87 HR(d)	...	...	...	...	W,C,S,A	
65Ag-35Ni	PSR	9.88	9.00	49	26 HR30T(c)	...	...	...	...	S	
60Ag-40Ni	PSR	9.80	8.90–9.60	44–47	40 HR30T(c) 92 HR30T(d)	241(c) 414(d)	35(c) 60(d)	...	...	W,C,S,A	
	PSE	9.80	9.60	60	46 HR30T(c)	...	...	...	...	S	
55Ag-45Ni	PSR	9.71	8.80	41	25 HR30T(c)	...	...	...	...	S	
50Ag-50Ni	PSR	9.63	9.00	38	50 HR30T(c)	...	...	...	...	S	
45Ag-55Ni	PSR	9.56	8.50	35	30 HR30T(c)	...	...	...	...	S	
40Ag-60Ni	PSR	9.48	8.80	32	35 HR30T(c) 97 HR(d)	...	...	...	...	S	Circuit breakers
	PSE	9.48	9.30	40	68 HR30T(c)	...	...	...	...	S	Transformer protectors, contactors, relays
35Ag-65Ni	PSR	9.40	8.60	30	40 HR30T(c)	...	...	...	...	S	
30Ag-70Ni	PSR	9.32	8.50	27	40 HR30T(c)	...	...	...	...	S	
25Ag-75Ni	PSR	9.25	8.20	24	40 HR30T(c)	...	...	...	...	S	
20Ag-80Ni	PSR	9.17	8.00	21	35 HR30T(c)	...	...	...	...	S	
Silver/tin oxide											
92Ag-8SnO_2	PSE	10.08	10.00	88	58 HV(c) 92 HV(d)	205–230	30–33.5	...	...	E,M,W	Light switches, relays, motor vehicle switches
90Ag-10SnO_2	PSE	9.98	9.97	82	64 HV(c) 98 HV(d)	215(c)	31(c)	...	...	E,M	Light switches, relays motor vehicle switches
88Ag-12SnO_2	PSE	9.70	9.68	72	72 HV(c) 105 HV(d)	...	...	...	...	E,M,W	Low-voltage motor contactors and switches rated to 10A. Low-voltage circuit breakers rated to 100A
Silver/zinc oxide											
92Ag-8ZnO	PSE	9.81	9.80	77	60–65 HV(c)	...	...	...	...	E,M,W	Low-voltage circuit breakers rated to 200A

(continued)

(a) PSR, press-sinter-re-press; INF, press-sinter-infiltrate; PS, press-sinter; PSE, press-sinter-extrude; IO, internal oxidation; PPSE, preoxidize-press-sinter-extrude; SF, oxidized from one direction. (b) A: Advance Metallurgy, Inc., McKeesport, PA. C: Contacts, Materials, Welds, Inc., Indianapolis, IN. E: Engelhard Industries, Plainville, MA. G: Gibson Electric Inc., Delmont, PA. S: Stackpole Carbon Co., St. Marys, PA. T: Texas Instruments Inc., Attleboro, MA. M: Metz Degussa, South Plainville, NJ. W: Art Wire-Duduco, Cedar Knolls, NJ. (c) Annealed. (d) Cold worked

Table 8 (continued)

Nominal composition, %	Manufacturing method(a)	Density, g/cm³ Calculated	Density, g/cm³ Typical	Electrical conductivity, % IACS	Hardness	Tensile strength MPa	Tensile strength ksi	Modulus of rupture MPa	Modulus of rupture ksi	Data source(b)	Application examples
Tungsten carbide-silver											
65Ag-35WC	INF	11.86	11.53–11.85	55–60	50–65 HRB	272	39.5	483	70	C,A	Aircraft contactors, lighting relays, low-voltage switches, circuit breakers
	PSR	11.86	11.10–11.80	50–60	50–62 HRB	...	...	...	...	C,A	
60Ag-40WC	PSR	12.09	11.40–11.92	46–55	60–70 HRB	...	...	...	...	A	
58Ag-42WC	PSR	12.17	11.86–11.97	50–55	75–85 HRB	...	...	...	...	C	
50Ag-50WC	INF	12.56	12.12–12.50	43–52	75–85 HRB	276	40	793	115	C,A	
40Ag-60WC	INF	13.07	12.70–12.92	40–47	90–100 HRB	379	55	827	120	C,A	Heavy-duty circuit breakers
38Ag-62WC	INF	13.18	12.92–13.29	35–38	90–100 HRB	552	80	...	...	C	
35Ag-65WC	INF	13.35	12.90–13.18	30–37	95–105 HRB	...	...	...	...	A	Semiconducting material
20Ag-80WC	PSR	14.23	13.2	19	400 HV(c) 470 HV(d)	...	...	...	...	M	Suppression of tungstate formation in low-voltage and high-voltage circuit breakers
Tungsten-silver											
90Ag-10W	PSR	11.00	10.30–11.20	90–95	20–33 HRB	...	...	...	...	C,A	Controls, automatic circuit protectors, wall switches
85Ag-15W	PSR	11.27	10.60–11.30	85–90	25–38 HRB	...	...	...	...	A	Current-carrying contacts in circuit breakers, light-duty contactors
80Ag-20W	PSR	11.55	10.90–11.70	80–85	30–43 HRB	...	...	...	...	A	
70Ag-30W	PSR	12.16	12.00	72–80	40–47 HRB	...	...	...	...	A	
65Ag-35W	PSR	12.48	12.1	68	80 HV(c) 90 HV(d)	...	...	...	...	M	
60Ag-40W	PSR	12.84	12.10–12.60	60–65	50–60 HRB	...	...	...	...	A	
35Ag-65W	INF	14.92	14.20–14.77	45–53	80–93 HRB	...	...	827	120	C,M	Automotive starting switches, circuit breakers
	PS	14.92	13.90–14.20	47–50	85–87 HRB	...	...	...	...	C	
	PSR	14.92	14.65–14.74	47–50	55–65 HRB(c)	...	...	572	83	C	
30Ag-70W	INF	15.42	15.02	40–50	85–93 HRB	...	...	...	...	A	Motor starters, aircraft equipment, circuit breakers, contactors, computers, arcing tips
27.5Ag-72.5W	INF	...	15.56	49	90 HRB	483	70	896	130	C	
	PSR	...	15.44	...	58–68 HRB(c)	...	...	586	85	C	
25Ag-75W	INF	15.96	15.25–15.40	40–50	85–95 HRB	...	...	...	...	A	
20Ag-80W	INF	16.53	16.18	35–40	91–100 HRB	...	...	...	...	A	
15Ag-85W	INF	17.14	16.60–17.05	32–41	90–100 HRB	448	65	758	110	A,C	Motor governors
10Ag-90W	PSR	17.81	17.25	29–35	95–105 HRB	379	55	758	110	A	Semiconducting material
Tungsten carbide-copper											
50Cu	INF	11.39	11.00–11.27	42–47	90–100 HRF	...	...	1103	160	C,A	Arcing contacts in oil switches, wiping shoes in power transformers
44Cu	INF	11.77	11.64	43	99 HRF	...	...	1241	180	C	
30Cu	INF	12.78	12.65	30	38 HRC	...	...	...	...	...	
Tungsten-copper											
75Cu-25W	PSR	10.37	9.45–10.00	50–79	35–60 HRB	...	...	414	60	C,A	Current-carrying contacts
70Cu-30W	...	10.70	10.45	76	59–66 HRB	...	...	...	...	A	
65Cu-35W	...	11.06	11.40	72	63–69 HRB	...	...	...	...	A	Vacuum interrupter
60Cu-40W	...	11.45	11.76	68	69–75 HRB	...	...	...	...	A	Oil-circuit breakers, arcing tips
50Cu-50W	INF	12.30	11.90–11.96	45–63	60–81 HRB	...	...	...	...	A	
44Cu-56W	INF	12.87	12.76	55	79 HRB	434	63	827	120	C	
40Cu-60W	INF	13.29	12.80–12.95	42–57	75–86 HRB	...	...	...	...	A	Oil-circuit breakers, reclosing devices, arcing tips, tap change arcing tips, contactors
35Cu-65W	INF	13.85	13.35	54	83–93 HRB	...	...	...	...	A	
32Cu-68W	INF	14.20	13.95	50	90 HRB	...	...	896	130	C	
30Cu-70W	INF	14.45	13.85–14.18	36–51	86–96 HRB	...	...	1000	145	C,A	Circuit breaker runners, arcing tips, tap change arcing tips
26Cu-74W	INF	14.97	14.70	46	98 HRB	621	90	1034	150	...	
25Cu-75W	INF	15.11	14.50	33–48	90–100 HRB	...	...	...	...	A	Vacuum switches, arcing tips, oil-circuit breakers
20Cu-80W	INF	15.84	15.20	30–40	95–105 HRB	758	110	...	...	C	
15Cu-85W	PSR	16.45	16.0	20	190 HV(c) 260 HV(d)	...	...	...	...	M	
13.4Cu-86.6W	INF	16.71	16.71	33	20 HRC	621	90	1034	150	C	
10.4Cu-89.6W	INF	17.22	17.22	30	30 HRC	765	111	1138	165	C	
Tungsten-graphite-silver											
48Ag-51.75W-0.25C	PSR	13.21	13.38	65	55 HRB	...	...	552	80	C	Circuit breakers, arcing tips
46Ag-53W-1C	PSR	13.58	12.85	55	85 HRB	...	...	...	...	C	
45Ag-50W-5C	PSR	11.00	10.60	37–43	45–55 HRB	...	...	621	90	A	
Complex composite contacts											
88Ag-10Ni-2C	PSR	9.63	9.37	70	26 HRF(c) 64 HRF(d)	...	...	...	...	C,A	Sliding contacts
25Ag-50Fe-25Cu	PSR	8.67	8.52	21	84 HRF(c) 94 HRF(d)	...	...	...	...	C	Circuit breakers

(a) PSR, press-sinter-re-press; INF, press-sinter-infiltrate; PS, press-sinter; PSE, press-sinter-extrude; IO, internal oxidation; PPSE, preoxidize-press-sinter-extrude; SF, oxidized from one direction. (b) A: Advance Metallurgy, Inc., McKeesport, PA. C: Contacts, Materials, Welds, Inc., Indianapolis, IN. E: Engelhard Industries, Plainville, MA. G: Gibson Electric Inc., Delmont, PA. S: Stackpole Carbon Co., St. Marys, PA. T: Texas Instruments Inc., Attleboro, MA. M: Metz Degussa, South Plainville, NJ. W: Art Wire-Duduco, Cedar Knolls, NJ. (c) Annealed. (d) Cold worked

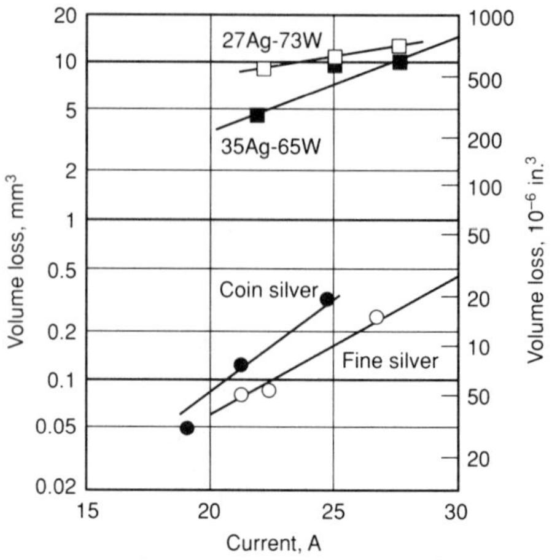

Fig. 7 Contact erosion characteristics of silver and silver-tungsten contacts. Test conditions were 115 V, 60 Hz, and 1.0 power factor for 100 000 operations at 60 operations per minute. Closing and opening speeds were 38 mm/s (1½ in./s). Closing force was 980 mN (0.22 lbf) and opening force, 735 mN (0.165 lbf).

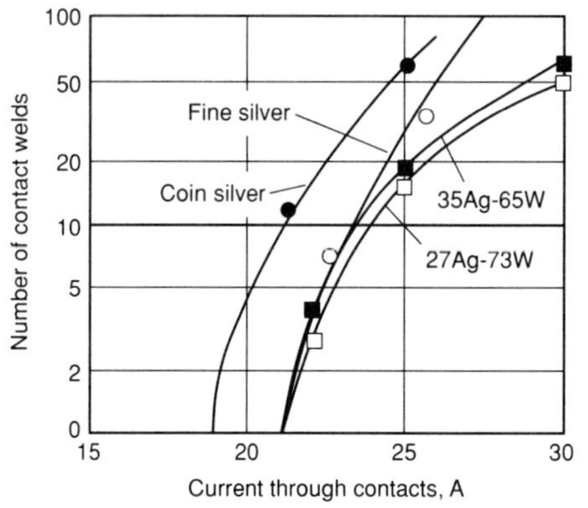

Fig. 8 Contact welding characteristics of silver and silver-tungsten contacts. Operation characteristics are the same as for Fig. 7.

used. Examples of test data are given in Fig. 6 to 9 and in the section "Life Tests" in this article.

Refractory Metal and Carbide-Base Composites

Refractory metals and their carbides are distinguished by high melting and boiling points, and high hardness, but poor electrical and thermal conductivities and poor oxidation resistance. In pure elemental form, refractory metals perform well only under low-current conditions.

Forming a composite can compensate for these drawbacks. For example, the development of composite contact materials involving silver or copper with tungsten or molybdenum or their carbides has resulted in materials that can withstand higher currents and more arcing than other contact materials, without experiencing sticking or rapid erosion. The refractory metal content may vary from 10 to 90%, although 40 to 80% usually is used in air- and oil-immersed circuit breaker devices. Refractory metals offer good mechanical wear resistance and resistance to arcing. The silver and copper provide the good electrical and thermal conductivities.

Because silver or copper does not alloy with tungsten, molybdenum, or their carbides, P/M processes are required in fabrication. Depending on the composition, refractory metals containing silver or copper contact materials are made either by pressing and sintering or by the press-sinter-infiltrate method. When infiltration is used, either all refractory metal powder is compacted to shape, or a small amount of silver or copper powder is blended with the refractory metal, compacted, and sintered in a reducing atmosphere. The sintered compact is then returned to the furnace; silver or copper is added to act as the infiltrant.

Most infiltrated composite contacts use silver as the infiltrant because of its excellent thermal and electrical conductivities, as well as its superb oxidation resistance. Copper infiltrant, which costs less but has very poor corrosion resistance, is used for composites that operate in noncorrosive environments such as oil, vacuum, or inert atmospheres. At temperatures above the melting point of the infiltrant, the liquid metal penetrates and fills the interconnecting voids of the pressed and sintered compact. Densities of 96 to 99% of theoretical can be achieved by this process. Infiltrated contact materials find use as current-carrying contacts in air- and oil-immersed circuit breakers, heavy-duty relays, automotive starters, and switches. Lower properties can be obtained by pressing and sintering.

In a material made by infiltration, the function of the infiltrant (silver or copper) is twofold. First, because silver or copper does not alloy with tungsten, molybdenum, or carbides, the conductivity of the composite depends strictly on the volume percentage of infiltrant. Second, during arcing, the high temperature melts the infiltrant; consequently, the heat of fusion absorbs (quenches) a portion of the heat generated by the arc. Theoretically, the skeleton, which is made of a high-melting element, will not begin to melt until all the low-melting component evaporates. The refractory skeleton also prevents molten infiltrant from flowing by capillary action. Because of this, erosion loss of the contact is low. Properties (such as the erosion data in Fig. 7 and 9) of the contact vary with the composition of the composite. A composite with high skeletal composition has high hardness and better wear resistance, but lower current-carrying capacity. On the other hand, a high-silver composite possesses high electrical and thermal conductivities, and undergoes lower temperature rise, but is softer.

Compositions of Refractory Component. There is a lower limit for the composition of the skeleton material. Generally, when the amount of refractory or carbide is less than about 30 vol%, it is difficult to form a sound and uniform skeleton to accommodate the amount of silver. For practical purposes, the skeleton material should amount to a minimum of 50 wt% for tungsten and molybdenum, and 35 wt% for tungsten carbide. Any composite containing lesser amounts than these limits should be made by the press-sinter-re-press method, and should be considered a silver-base composite in which the function of the refractory material is to reinforce the silver matrix. For compounds with 60% or less tungsten, the classical method of mixing the powders, pressing, sintering (generally below the copper melting point), and re-pressing might also be used. Materials with 60 to 80% W are generally produced by infiltration, either of loose tungsten powder or of a pressed and sintered tungsten compact.

Tungsten, tungsten carbide, and molybdenum powders are the most commonly used materials for making skeletons for infiltrated contacts. Composites with tungsten skeletons have the best arc-interrupting and arc-resisting characteristics and the best arc-erosion resistance. Their antiwelding properties are moderate (Fig. 8). High-

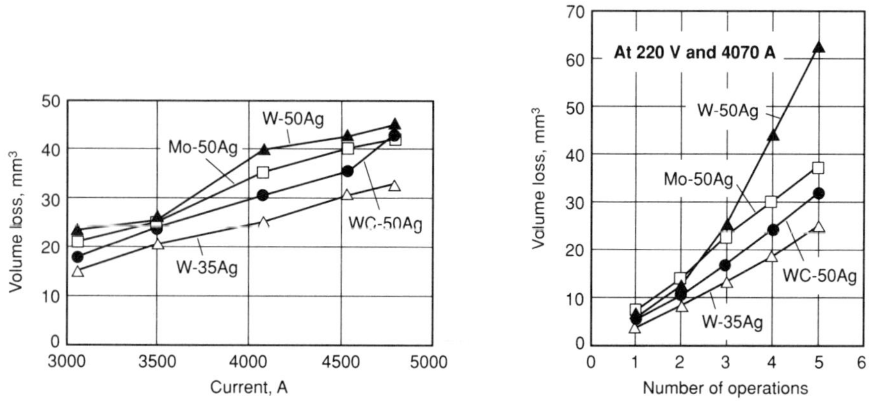

Fig. 9 Results of short-circuit tests on Ag-W, Ag-Mo, and Ag-WC

energy devices usually use silver-infiltrated composites having a tungsten skeleton.

Composites with tungsten carbide skeletons have better resistance to welding, better anticorrosion properties, and more stable contact resistance compared with other infiltrated composites. Devices that handle switching arcs usually use composites based on tungsten carbide skeletons.

For a combination of properties, or sometimes for a special requirement, a skeleton made of a mixture of tungsten and tungsten carbide is used. The blended powder contains either the mixture of tungsten and tungsten carbide or a mixture of tungsten and graphite. In the latter case, the graphite and part of the tungsten react to form tungsten carbide during sintering.

Composites with molybdenum skeletons have relatively low contact resistance and behave well in circuit-interrupting devices. For the same current-carrying capacity, a molybdenum-base composite costs less than the other two, but the antiwelding and anticorrosion properties of molybdenum-base composites are inferior to those with tungsten or tungsten carbide skeletons. Figure 9 compares the erosion characteristics of a molybdenum composite with tungsten and tungsten carbide composites.

Silver-Base Composites

The main advantage of a silver composite over a silver alloy is that the bulk conductivity of a silver composite depends generally on the percentage of silver by volume. An alloying element in solution greatly decreases the conductivity of silver. For instance, the volume of silver in Ag-15CdO composite is less than that in Ag-15Cd alloy, yet the electrical conductivity of the former (65% IACS) is much greater than that of the latter (35% IACS).

In silver composites, the second phase forms discrete particles that are dispersed in the silver matrix. The dispersed phase improves the matrix in two ways. First, it increases the hardness of the composite material in a manner similar to dispersion hardening. Second, in the region where two mating contacts touch upon closure, the second phase particles reduce the surface area of silver-to-silver contact. This greatly reduces the tendency to stick or weld. In cases where the contacts do weld, the second-phase oxide particles (which are weaker and more brittle than silver) behave as slag inclusions and reduce the strength of the weld, allowing the device contact-separating force to pull the contacts apart.

Silver-base composites can be divided into two types: type 1 uses a pure element or carbide as the dispersed phase; type 2 uses oxides as the dispersed phase. In both types, the hardness increases and the conductivities decrease as the volume fraction of dispersed phase increases, and vice versa.

Silver-Base Composites With a Pure Element or Carbide. In type 1, the dispersed phase functions as a hardener and improves the mechanical properties of the silver matrix. The dispersed phase also promotes improved electrical performance such as antiwelding properties. Elements used include tungsten, tungsten carbide, molybdenum, nickel, iron, graphite, and mixtures of these materials.

Silver-Tungsten and Silver-Molybdenum Composites. Silver composites (made by the press-sinter-re-press method using tungsten, tungsten carbide, and molybdenum as the dispersed phases) show electrical conductivities similar to those of infiltrated composites of the same components. However, their mechanical properties are inferior because the dispersed phases do not form a refractory skeleton.

Silver-Nickel Composites. One of the elements typically combined with silver by P/M processes is nickel. Nickel is more effective as a hardening agent than copper; consequently, silver nickel is considerably harder than coin silver. At the same time, nickel does not increase contact resistance appreciably, particularly in combinations that include 15 wt% Ni or less. Silver nickel is combined in proportions ranging to about 40 wt%.

Composites with nickel as the dispersed phase resist mechanical deformation or peening under impact and possess good antiwelding properties. Silver-nickel composite contacts can be used as both members of a contact pair. Sometimes, a silver-nickel composite is used as the moving contact operating against a stationary contact of a different composite such as silver-graphite.

The combinations most widely used are 60Ag-40Ni and 85Ag-15Ni. These materials are very ductile and can be formed in all of the shapes in which silver contacts are used, including very thin sheets for facing large contact areas. This material is ideal for use under heavy sliding pressures. It does not gall like fine silver and coin silver, but instead takes on a smooth polish. It is therefore suitable for sliding contact purposes, as well as for make-break contacts. Silver nickel can handle much higher currents than fine silver before it begins to weld. It has a tendency to weld when operated against itself. Therefore, it is frequently used against silver graphite.

The 60Ag-40Ni composite is the hardest material in the silver-nickel series. It is the most suitable for sliding contact in which pressure is high. This alloy also has the lowest rate of wear under sliding action. It is less ductile than silver-nickel materials containing less nickel, but it is still sufficiently ductile for all conventional manufacturing processes.

The 85Ag-15Ni composite is the most widely used material in the silver-nickel series. Because of its ideal mechanical properties, 85Ag-15Ni is an ideal material for motor-starting contactors and is superior in this type of application to fine silver, coin silver, and copper. It is also suitable as a general-purpose contact for various types of relays and switches.

The contact resistance of clean 85Ag-15Ni contacts that have not operated under load tend to be slightly lower for fine silver. However, in make-break circuits, silver tends to gradually increase contact resistance. This increase is not necessarily permanent, as contact resistance varies with the effects of arcing on the contacts. Generally, average resistance is higher than the initial resistance before the contacts operate. The contact resistance of 85Ag-15Ni is similar, except that it usually varies within a narrower range. Exhibiting nearly constant contact resistance is more important than possessing low contact resistance.

85Ag-15Ni exhibits a lower contact resistance and is also harder than coin silver.

Another advantage of 85Ag-15Ni is its low flammability; that is, it makes a smaller arc than other materials. In testing of more than 40 contact materials, 85Ag-15Ni exhibited the lowest arc energy. Low arc energy is important in that the ability to break a circuit with as little flame as possible is desirable. This characteristic was primarily responsible for the adoption of 85Ag-15Ni for relays in aircraft electrical systems.

Silver-Graphite Composites. Graphite is also combined with silver by P/M techniques. Graphite in silver-base composites serves as a good lubricant, reducing the damage caused by frictional forces. Silver-graphite composites are used chiefly as sliding or brush contacts. These materials have high resistance to welding and are also used as make-break contacts. In circuit breakers, they are usually paired with silver-nickel composites.

The most frequently used composition is 95Ag-5C, although graphite compositions ranging from 0.25 to 90% with the remainder silver have been used. This material was developed as a circuit breaker contact material. The addition of graphite prevents welding. Frequently, 95Ag-5C is used in combination with silver nickel or silver tungsten contacts. It is also used in combination with pure nickel contacts and with fine silver contacts. Silver graphite is soft compared to other types of contact materials, and electrical and mechanical erosion is more rapid.

95Ag-5C has been widely used as a material for contacts in molded-case circuit breakers, sliding contacts, and contact brushes. This material is only moderately ductile and can be rolled into sheets and punched into contacts of various shapes. However, it cannot be headed to make solid rivets or bent to any great extent without cracking. It can be coined to a moderate extent. 95Ag-5C contacts can be individually molded. Depending on size, shape, and quantity, contacts of this material are either punched from rolled slabs, extruded, or individually molded from powders. Copper is combined with graphite as a substitute for silver in certain applications.

A modified form of silver graphite is silver-nickel-graphite. Typical compositions are 88Ag-10Ni-2C and 77Ag-20Ni-3C. These materials are substantially harder than 95Ag-5C and exhibit superior wear resistance, but offer less protection against welding. Like 95Ag-5C, they can be manufactured from slabs or by molding individually.

Composites of silver-iron exhibit good antiwelding and good wear characteristics when used in creep-type thermostat devices. These materials have poor corrosion resistance.

Silver-Base Composites With Dispersed Oxides. Type 2 silver-base composites use semirefractory oxides as the dispersed phase. These silver-base composites are produced by a variety of methods such as internal oxidation, pre-oxidation, and conventional P/M processes (see the section "Composite Manufacturing Methods" in this article).

The semirefractory component of Type 2 silver-base composites includes metal oxides such as CdO, SnO_2, or ZnO. In general, the semirefractory constituents promote nonsticking qualities or provide increased resistance to wear.

The silver-cadmium oxide group of electrical contact materials is the most widely used of all the silver semirefractory contact materials. The addition of 5 to 15% cadmium oxide to silver imparts excellent nonsticking and arc quenching qualities.

Because of its resistance against arc erosion and its low contact resistance which does not increase even after switching, Ag-CdO has proved to be a universally good contact material for many switching devices. Silver-cadmium oxide contact materials are well suited for contactors and motor starters, but are also used in circuit breakers, relays, and switches with medium to low currents.

Ag-CdO material has antiwelding and antierosion properties united with constant resistance, examples of its main advantage of well-combined properties. Another favorable quality is that it has good workability. It can be fabricated by either the internal oxidation (least costly), preoxidation, or P/M methods. The Ag-CdO material can also be cold reduced or rolled quite easily. For instance, Ag-15CdO material can endure more than 70% cold reduction.

Silver-tin oxide, which is used widely in Europe as a contact material, is a class of composite materials that has the potential to replace silver-cadmium oxide composites in many electrical contact applications. However, general comparisons of silver-tin-oxide ($AgSnO_2$) contacts with Ag-CdO contacts are difficult because results may depend on the specific conditions of testing. Previous concerns on the toxicity of CdO, which was one of the motivations for using Ag-SnO_2 contacts, have also been relaxed in Japan and Europe. The toxicity of CdO must be distinguished from the highly toxic nature of cadmium.

Like Ag-CdO contacts, Ag-SnO_2 contacts can be produced by internal oxidation or P/M techniques. One drawback of the Ag-SnO_2 composite is that a third element (such as indium) must be added to achieve internal oxidation when the silver alloys contain more than 4% Sn. The oxidized material also does not allow a high level of cold reduction because of its brittleness. Therefore, a press-sinter-re-press method or extruded method is the most feasible way to fabricate silver-tin oxide, although extruded products are more brittle than extruded Ag-CdO powder of similar compositions. For example, extended Ag-10SnO_2 can be subject to a maximum of 30% cold reduction compared to over 60% for Ag-12CdO.

Another drawback is the higher temperature rise of Ag-SnO_2 contacts (as compared to Ag-CdO) after arcing. This troublesome characteristic has, however, been eliminated with Ag-SnO_2 materials made by P/M methods.

Table 8 lists three grades of commercially available silver-tin oxide composite contact materials. Ag-SnO_2 contact materials cannot be easily brazed or welded. To be able to braze silver-tin oxide contacts, they are made with at least two layers, the contact layer and brazable or weldable fine silver layer. The brazing alloy can be applied separately in the shape of paste, wire, or foil, or it is already clad onto the semifinished product.

Silver-zinc oxide (Ag-ZnO) is another composite material that has been tested and marketed for contact applications. Silver-zinc oxide, like silver-tin oxide composite, cannot take high cold reduction because of the brittleness of the oxidized material. When internal oxidization is used, the maximum zinc content cannot exceed 6% for good oxidation. Typical applications of a commercially available Ag-ZnO composite are listed in Table 8.

Multiple-Component Composites. There is no ideal material to meet all conditions for contact applications. If required by manufacturers of switching devices, contact manufacturers can offer composite materials consisting of as many as four or five components. Most of these composites serve only special purposes. They are not universally accepted and generally cost more. Two common three-component composites are listed in Table 8.

Composite Manufacturing Methods

The methods used to manufacture composite contact materials can be classified into three major categories:

- Standard P/M processes, for producing composites from materials that cannot be conventionally alloyed
- Internal oxidation processes, for producing silver-base composites with dispersed oxides
- Hybrid consolidation, which is a combination of the internal oxidation and P/M consolidation processes

P/M Methods

Infiltration is used exclusively for making refractory metal and carbide-base composite contact materials. Metal powder or carbide powder is first blended to the desired composition with or without a small amount of binder to impart green strength, then is pressed and sintered into a skeleton of the required shape. Silver or copper is then infiltrated into the pores of the skeleton. This method produces the most densified composites, generally 97% or more of theoretical density. Complete densification is not possible because of the presence of some closed pores in the sintered skeleton. After infiltration, the contact is sometimes chemically or electrochemically etched so that only pure silver appears on the surface. The contact thus treated has better corrosion resistance and performs better in the early stages of use.

Press-Sinter. For small refractory-metal contacts (not exceeding about 25 mm, or 1 in., in diameter), a high-density material can be obtained by pressing a blended powder of exact final composition into shape and then sintering it at the melting temperature of the low-melting-point component (liquid-phase sintering). In some cases, an activating agent such as nickel, cobalt, or iron is added to improve the sintering effect on the refractory metal particles. For this process, powders of much finer particle size are required so that more bonding surface exists. However, the skeleton formed by this process is weaker than that formed by the infiltration process. Formation of the skeleton usually shrinks the apparent volume of the refractory portion of the composition, thus bleeding out the molten component onto the surface of the finished contact.

Press-Sinter-Re-press. The press-sinter-re-press process is used for all categories of contact materials, especially those in the silver-base category. Blended powders of the correct composition are compacted to the required shape and then sintered. Afterward, the material is further densified by a second pressing (re-pressing). Sometimes the properties can be modified by a second sintering or annealing. The versatility of this

Table 9 Comparison of Ag-CdO material made by different methods

Properties	Press-sinter-re-press	Press-sinter-extrude	Internal oxidation	Preoxidize-press-sinter-extrude
Performance characteristics				
Resistance to arc erosion	3	2	1	1
Resistance to sticking and welding	1	1	2	2
Low contact resistance and temperature rise	1	1	1	1
Arc interruption	3	2	1	1
Resistance to corrosion	1	1	1	1
Material characteristics				
High mechanical properties	3	2	2	1
Resistance to annealing	3	2	2	1
Electrical and thermal conductivity	2	1	1	1
Flexibility of composition	2	2	2	1
Uniform cadmium oxide distribution	1	1	3	1

Note: 1 indicates that under most conditions this is the preferred material; 2 indicates that under most conditions the material is preferable to 3, but not as good as 1; 3 indicates that the material may be acceptable, but under typical operating conditions it is not as good as 1 or 2.

process makes it applicable for contacts of any configuration and of any material. However, it is difficult to obtain material with as high a density as is obtained with other processes. Material thus produced also may have weak bonding between particles.

Press-Sinter-Extrude. Blended powder of final composition is pressed into an ingot and sintered. The ingot is then extruded into wires, slabs, or other desired shapes. The extruded material may be subsequently worked by rolling, swaging, or drawing. Material made by this method is usually fully dense.

The press-sinter-extrude process is used mostly for silver-base composites. Other processes used for manufacturing silver-base composite contacts are direct extrusion or direct rolling of loose powder. Although they appear to be uncommon, they are economically feasible if the equipment is properly designed and built.

Internal Oxidation

Silver-base composites with dispersed metal oxides can be produced by internal oxidation. In this process, a silver alloy (such as a silver-cadmium alloy) is first cast into ingots, which are rolled into strips or fabricated further into the finished product form. The silver-alloy material is then heated in air or oxygen, so that the oxygen diffuses into the alloy and forms metal oxide particles (such as CdO in the case of a Ag-Cd alloy) dispersed in the silver matrix.

Internal oxidation is used in the production of a substantial portion of Ag-CdO composites. The initial Ag-Cd alloy can be internally oxidized either in strip or finished product form. The silver-cadmium alloy is heated between 800 to 900 °C (1470 to 1650 °F) in a furnace with air, oxygen-enriched air, or pressurized oxygen. Under this condition, the oxygen species diffuse into the silver-cadmium alloy and oxidize the cadmium species. Upon the completion of the oxidation, the cross section of the material will display a microstructure of cadmium oxide particles embedded in a silver matrix. Contact parts are punched from the strip and then coined into required shapes.

The size of the CdO particles and the uniformity of their dispersion are dependent on the temperature and the partial pressure of oxygen. Reduced temperature decreases coalescence of cadmium prior to being oxidized and thereby causes a finer dispersion of CdO. Increasing the partial pressure of oxygen in the furnace increases the diffusion rate of oxygen into the silver. This also causes a finer CdO dispersion by reducing the time available for the cadmium to coalesce.

During the internal oxidation of a silver-cadmium alloy, the cadmium species become depleted in zones when the oxygen front moves into the silver-cadmium alloy. The cadmium atoms before the oxygen front immediately diffuse into the zone against the oxygen front. As the oxidation front moves from the surfaces of the strip toward the center, the concentration of the cadmium species becomes increasingly dilute as compared to the original composition. Hence, after the oxidation is completed, the cross section will display a significant oxide-deficient or -depleted zone in the center of the contact body.

For some applications the presence of the depletion zone is detrimental, requiring its removal or displacement from the center. There are two common methods to achieve this result. In the first, an oxidation barrier, such as ceramic glaze, is applied to one surface so that the oxidation can proceed from only one side. The second method is to laminate two Ag-Cd sheets of the same size and to form a package by welding along the four edges. After oxidation, the sheets are separated. The oxide-deficient zone will appear on one side (the inner side of the package) of each sheet.

Package rolling is another technique for reducing the size of the depletion region. In this method, very thin Ag-Cd sheets are first oxidized. Then a number of sheets (for example, 16 sheets) are stacked together and hot-bond rolled into one slab. The cross section of the final product displays very thin depleted zones equal to the number of sheets.

Hybrid Consolidation

Various hybrid techniques utilize a combination of internal oxidation and P/M methods. These methods are used to produce a finer average oxide size and/or a more uniform distribution of cadium oxides in the matrix of a Ag-CdO composite. Hybrid consolidation methods include:

- Preoxidize-press-sinter-extrude
- Coprecipitation

Table 9 compares Ag-CdO composites manufactured by different methods.

The preoxidized-press-sinter-extrude process combines the oxidation process and the press-sinter-extrude process. Commercially, it is called "preoxidized process." The purpose of this method is to redistribute the oxide-deficient center of Ag-CdO composites.

The preoxidized process is used exclusively for making silver-cadmium oxide (Ag-CdO) material. Alloys are reduced to small particles in the shape of flakes, slugs, or shredded foil. These particles are oxidized and then consolidated with the press-sinter-extrude process. Material made by this method is more uniform than the same material made by conventional internal oxidation. Mechanical properties are superior to those of the same material made by the press-sinter-re-press method.

The Ag-CdO particulates are made by one of four methods and then are pressed into ingots, sintered, and extruded according to standard metallurgical method. There are four processes to prepare the particulates:

- *Granulated Wire*: Ag-Cd alloy is first made into wire and oxidized. The oxidized wire is then chopped into granules with a length of about 3 mm (⅛ in.)
- *Low-Pressure Water Atomization*: The molten Ag-Cd alloy is atomized by water at a pressure of 100 to 200 kPa (15 to 30 psi). The approximately quarter-inch particulates are in the form of thin twisted flakes. Then the flakes are oxidized for consolidation
- *High-Pressure Water Atomization*: The molten Ag-Cd alloy is atomized with high water pressure, usually higher than 2750 kPa (400 psi). The powder sizes range between 40 mesh (420 μm) and 270 mesh (53 μm). Then the alloy powders are oxidized before consolidation

Coprecipitation. Conventional blending or mechanical mixing of silver and cadmium oxide powders begins by dissolving the proper amounts of silver and cadmium metals in nitric acid. Compounds of silver and cadmium coprecipitate from the solution when the pH value of the solution is changed by adding either hydroxide or carbonate solutions. During subsequent calcination at about 500 °C (930 °F), the compound mixture decomposes to form a mixture of silver and cadmium oxide. Alkali-metal content can be controlled in the ppm range by adequate washing. Controlled amounts of sodium, potassium, and lithium may enhance electrical life. Excessive amounts of these elements can lead to rapid erosion, restrike, and generally poor electrical life. Depending on device design, the range may be from 10 to 300 ppm. Contacts are consolidated from this mixture by conventional P/M methods. The microstructure of contacts made by this method displays a finer particle size and a more uniformly dispersed CdO phase than material made by conventional blending.

Life Tests

Example 1. Wear of Telephone Plugs (Fig. 10). A simulated service test, representing the service of telephone plug bodies, has provided a comparison of the wear properties of 2011-T3 aluminum and half-hard brass rubbing against nickel-silver.

The outside diameter of each sleeve was measured at intervals of 60° around the circumference. After the original diameter of approximately 6.32 mm (0.2490 in.) was measured, the plug was mounted on a test fixture designed to simulate normal mating of the plug and corresponding jack. Upon entering the jack, the sleeve rubbed against a nickel-silver spring contact. The test lasted for one million operations at a rate of about 30 per minute. To get more uniform wear, the plugs were rotated about 35° after each 100 000 operations. There was considerably less wear on the aluminum alloy plugs than on the brass, as shown in Fig. 10.

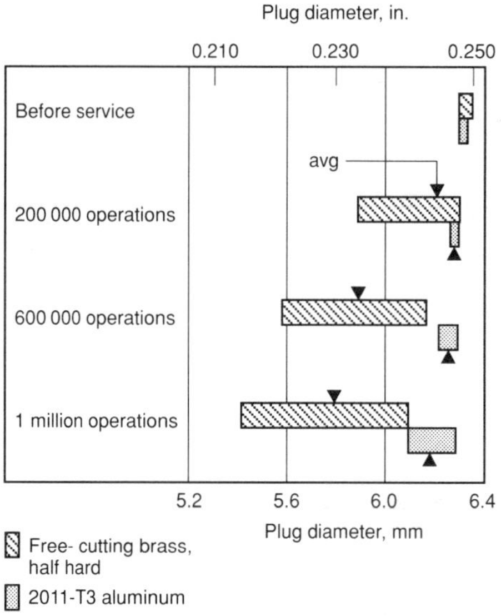

Fig. 10 Wear of aluminum and brass telephone plugs in simulated service. Test plugs were cycled at 30 operations per minute and rotated 35° after each 100 000 operations. Measurements were taken after each 200 000 operations.

Example 2. Life Tests Using an ASTM Microcontact Tester (Fig. 11). Effect of voltage on contact resistance and contact area for 100 000 operations, using 0.38 mm (0.015 in.) diameter 80Pt-20Ir contacts with a load of 100 mA, is shown in Fig. 11. Tests were made using an ASTM microcontact tester that allowed selection of any make-break contact force up to 49 mN (0.011 lbf). Contacts were protected from dust by a glass cover during testing. In these tests, the make force was 10 mN (0.002 lbf) and the break force was 5 mN (0.001 lbf). Ten readings were taken at each interval, resulting in the spread shown in Fig. 11.

Low noninductive voltages had little effect on contact operation. There appeared to be more consistency in the 3 and 6 V readings, and they finished at lower resistance.

Example 3. Life Tests Using a Movable-Coil Relay. An accelerated life test was conducted on five contact materials mounted in a movable-coil type of relay. These contact materials were solid silver, solid 80Pt-20Ir, and gold, ruthenium, and rhodium plated on the platinum-iridium alloy. The moving contacts were two 20 mm (0.80 in.) diam flat disks mounted one on each side of a phosphor bronze strip. The stationary contacts were pointed wire having a 0.075 mm (0.003 in.) radius at the point, mounted in adjusting screws, one on each side of the moving contact. The contacts were operated with a force of about 980 μN (2.2×10^{-4} lbf) on one side and no mechanical load on the other. The electrical load was purely resistive: 1 mA at 9 V dc. Contact resistances were measured using an ohmmeter that operated on 1.5 V, 30 mA, and a force of 29 μN (6.5×10^{-6} lbf). Results of this test support these conclusions:

- If sticking were not a problem, silver would be the best material, having low contact resistance up to 12 million operations. However, silver began to stick after 800 000 operations
- Gold-plated platinum-iridium had low contact resistance, and there was no sticking until about 3.5 million operations
- The three platinum-family metals were similar, each having higher contact resistance than silver or gold from the start. However, platinum-iridium did not increase in resistance as much as the other two and finished with the lowest contact resistance of the three platinum metals

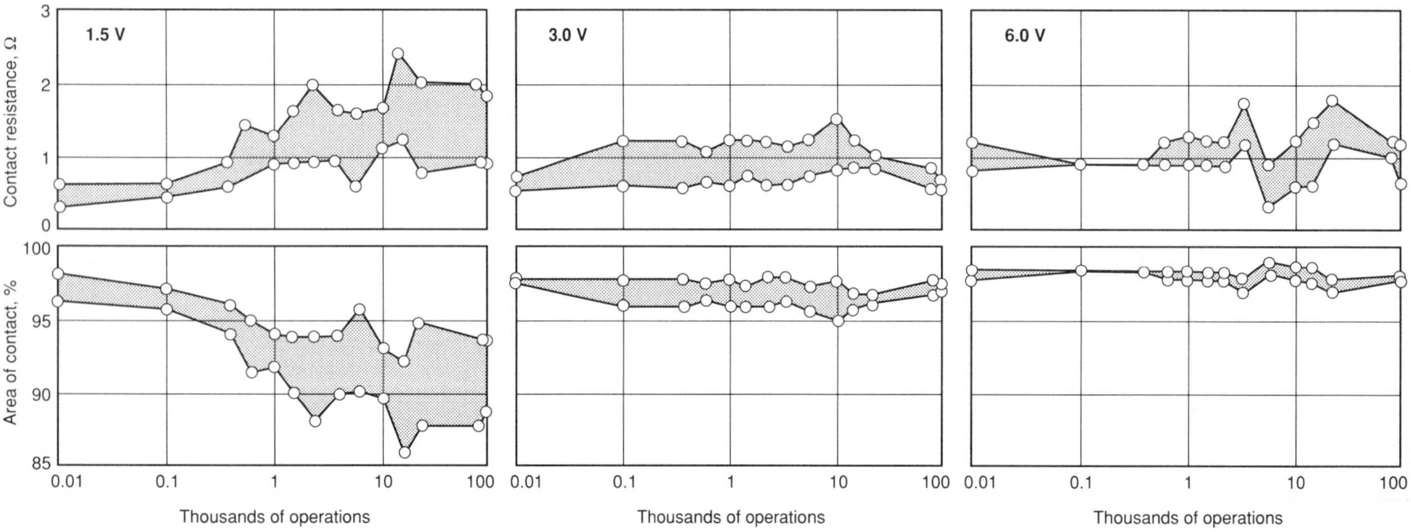

Fig. 11 Effect of voltage on contact resistance and contact area for 80Pt-20Ir contacts 0.4 mm (0.015 in.) thick. (Upper row of graphs) Effect of voltage on contact resistance at a load of 100 mA. Contacts were held in a clamp-type holder and tested in a closed jar. Shaded area represents spread for ten readings. (Lower row of graphs) Effect of voltage on contact area for tests same as above. Initial contact area was 96 to 98%.

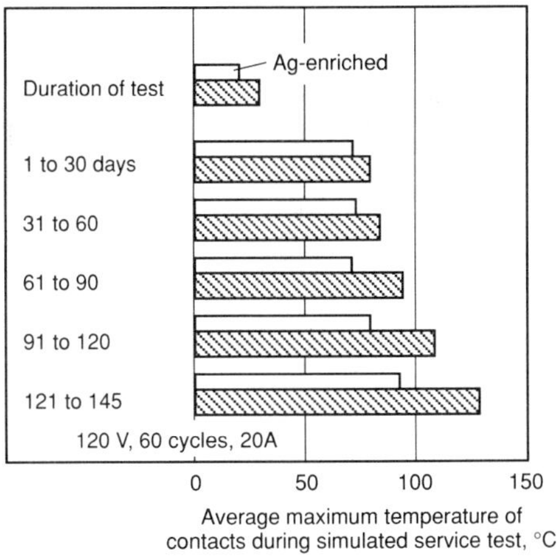

Fig. 12 Effect of time on maximum temperature of Ag-W contacts, compared with similar contacts having silver-enriched surfaces. See text for description of test.

An examination of the contacts at the end of the test revealed that there had been no arcing and that the load was not sufficient to erode the contacts. Thus, failure (high contact resistance) occurred solely because of wear.

Although silver and gold were superior for applications involving frequent make-break, platinum-iridium would still be chosen for use where the relay would be idle for long periods of time, as in a burglar alarm.

Example 4. Life Tests in Circuit Breakers (Fig. 12). Ag-W contacts and the same kind of contacts having silver-enriched surfaces were tested for about five months in circuit breakers to determine how the average maximum temperature of the contacts changed during this simulated service. The current through the contact was 20 A at 120 V, 60 Hz, ac. The breakers were mounted in an enclosure where the temperature was 40 °C (105 °F); outside the enclosure it was 25 °C (77 °F). The breakers were operated in the closed position for 8½ h per day from Monday through Friday, but remained open overnight and on weekends (one make and one break per day). The temperature values shown in Fig. 12 are averages of six readings.

The maximum temperatures for the Ag-W contacts changed almost linearly from 78 to 127 °C (172 to 260 °F) during the period. For the first three months, there was no appreciable change for the silver-enriched contacts. The temperature increases significantly during the fourth and fifth months, reaching about 95 °C (203 °F) at the end of the test. This is 23 °C (42 °F) higher than for the first period, but 32 °C (57 °F) lower than the temperature developed in the same period for the contacts that were not silver enriched.

Effect of Atmosphere. Life tests involving butt contacts have been made for several materials in atmospheres of helium, hydro-

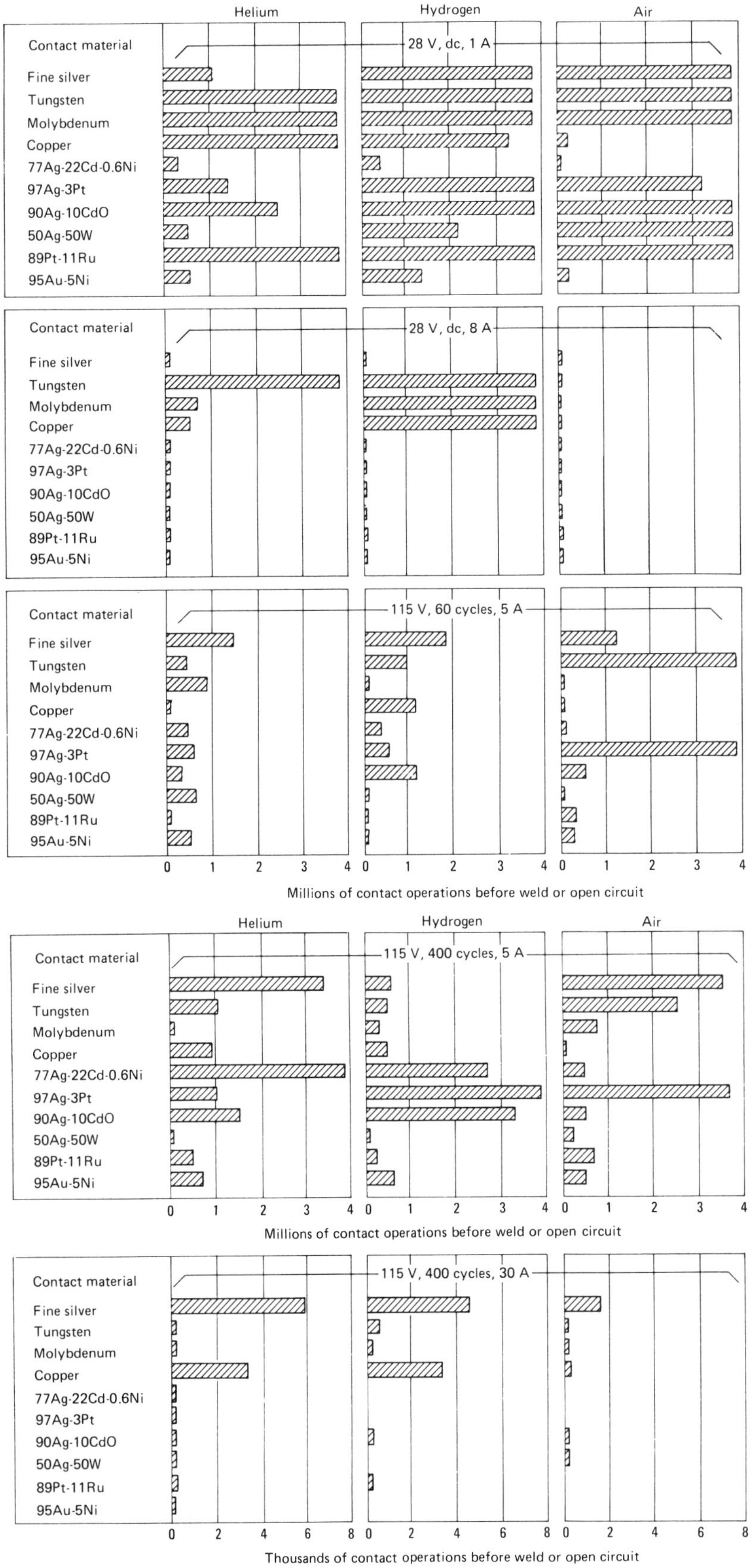

Fig. 13 Effect of atmosphere and current characteristics on life of contact materials. Tests were performed with a purely resistive load and a rate of five operations per second. Action was entirely butting, the contact being closed magnetically under a force of 20 g (0.2 N, or 0.045 lbf) and opened by a spring to 0.8 mm (1⁄32 in.) width of gap. Welding or open circuit was the criterion of failure. Tests were discontinued after 3.88 million operations.

gen, and air under different operating conditions. Extensive data are presented in Fig. 13. The contacts were subjected to five operations per second, the closing being made magnetically to a closed force of 195 mN (0.044 lbf); the opening was accomplished by a spring to a gap of 0.8 mm (1/32 in.).

The criterion of failure was a weld or an open circuit, whichever developed first. If failure did not occur in 3.88 million operations, the test was discontinued.

Not only can the effect of the different atmospheres be determined for any one of the five conditions used, but comparisons can be made of direct and alternating current, variations in current, and variations in frequency. The current conditions were:

Volts	Amperes	Frequency, Hz
28 dc	1	...
28 dc	8	...
115 ac	5	60
115 ac	5	400
115 ac	30	400

For the direct-current test, none of the ten materials gave long life in air when the current was 8 A. Six materials lasted for the full 3.88 million operations in air when the current was 1 A: fine silver, tungsten, molybdenum, 90Ag-10CdO, 50Ag-50W, and 89Pt-11Ru. Tungsten was the only material to last the full test period when helium was used as the test medium and the direct current was 8 A. Copper, molybdenum, and tungsten lasted the full test period when hydrogen was the test medium and the direct current was 8 A.

The alloy containing 97Ag-3Pt had long life when air and alternating current at 5 A were used, regardless of frequency. The lives of 77Ag-22.4Cd-0.6Ni, 97Ag-3Pt and 90Ag-10CdO, in a hydrogen atmosphere, were increased considerably by increasing the frequency from 60 to 400 Hz. In general, the life of a material was decreased by increasing the current. Fine silver showed a definite superiority over all other materials at 30 A. Copper was a fairly strong rival at 30 A, except in air.

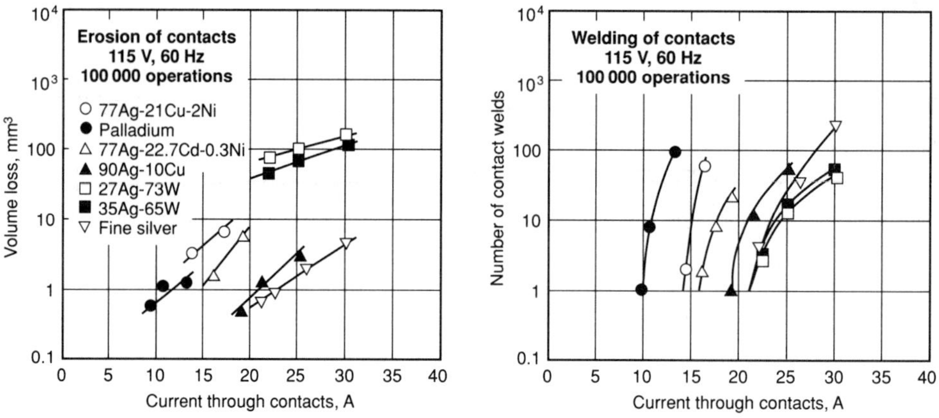

Fig. 14 Comparison of erosion and welding characteristics of selected electrical contact materials. All contacts were butting, with a closing force of 980 mN (0.22 lbf), an opening force of 735 mN (0.165 lbf) and closing and opening speeds of 38 mm/s (1½ in./s). Contacts operated at the rate of 60 per minute.

Erosion and Wear. Three simulated tests, the results of which are shown in Fig. 14 and 15, are related to erosion and wear. The following contact materials are dealt with in Fig. 14: fine silver, 90Ag-10Cu, 77Ag-21Cu-2Ni, 77Ag-22.7Cd-0.3Ni, palladium, 35Ag-65W, and 27Ag-73W.

Alternating current at 115 V and 60 Hz was employed, with the test lasting 100 000 operations at the rate of 60 per minute. The contacts were butt-type with closing and opening speeds of 38 mm (1½ in.) per second. The closing force was 980 mN (0.22 lbf), and the opening force 735 mN (0.165 lbf).

The number of contact welds was far less for fine silver and for the silver-tungsten sintered products than for palladium. The first two are about the same, but palladium can carry only about 40% as much current through the contact for the same number of contact welds.

The volume loss for fine silver at 25 A is about the same as that of palladium at 12 A. The volume loss for 27Ag-73W and 35Ag-65W is about 100 times that of fine silver at the same amperage.

In Fig. 15, comparison is made of copper, tungsten, 40Cu-60W, and 32Cu-68W, when tested for loss in volume for 30 operations in 10C transformer oil at 240 V, 60 Hz, 1400 A. The volume loss for 40Cu-60W was about two-thirds that for 32Cu-68W, and both were far below that for tungsten or copper.

In the second test, silver, Ag-CdO, and silver-tungsten were used to determine the number of operations required for failure in short circuit in a circuit breaker operating at 12 V dc, 7.5 A, interrupting a 300-A short circuit. A substantial increase in life is realized with a proper polarized combination: Ag-W positive and Ag-Fe negative.

Life of Polarized Contacts. Some information on life of polarized contacts has already been given in connection with erosion and wear (Fig. 15). It was shown that the polarized combination of 20Ag-80W or 35Ag-65W positive and 90Ag-10Fe negative, operating at 12 V dc, 7.5 A, with short-circuit interruption of 300 A, had a longer life than a similar combination where the tungsten content of the sintered contact was as low as 50%.

Test results are shown in Fig. 16 for a device operating with various contact materials in a dc circuit, operating at 12 V, 1.04 A, where polarized contact materials were used, with different materials as positive and negative terminals. The effect of the polarity of the contact materials is clearly

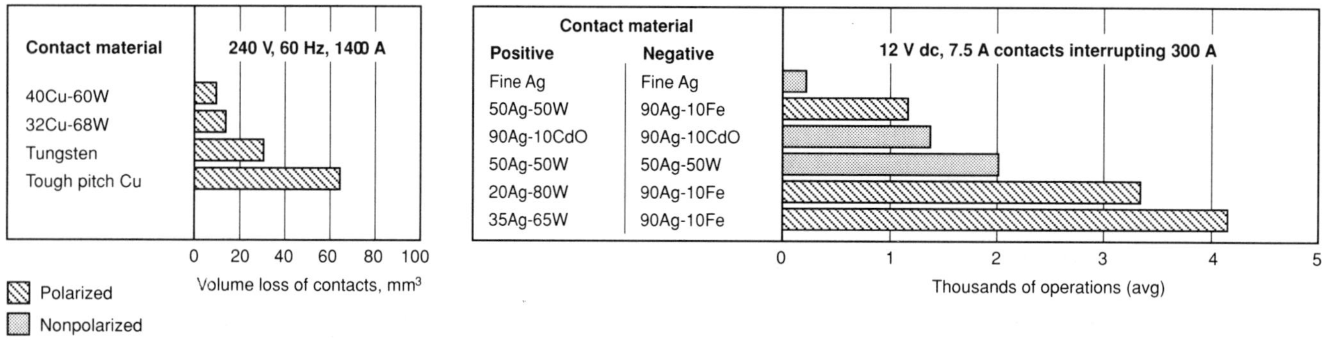

Fig. 15 Contact materials compared for erosion and wear characteristics. (Left) Four contact materials compared for volume loss after 30 operations in 10C transformer oil at a power factor of 70%, maximum arc time of ½ cycle, one operation per minute, using a closing force of 13 N (3 lbf) and an average opening speed of 2.4 m/s (8 ft/s). (Right) Life of polarized contact pairs and nonpolarized pairs based on the number of short circuits observed.

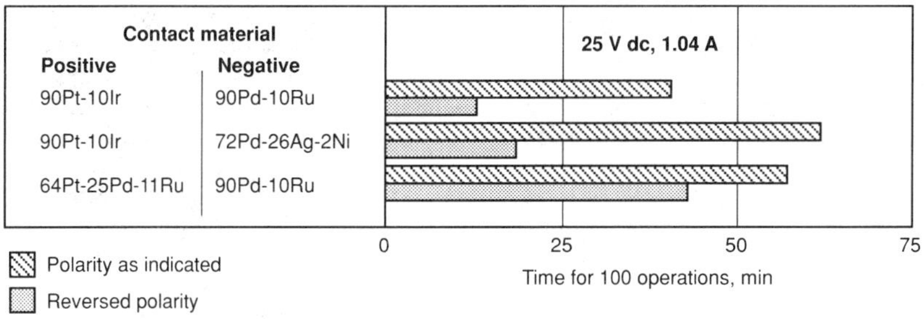

Fig. 16 Effect of composition and polarity on life of some platinum and palladium alloy contacts actuated by a bimetal element. Sustained arcing affects the bimetal, thus indicating effectiveness of contact combinations by time for 100 operations.

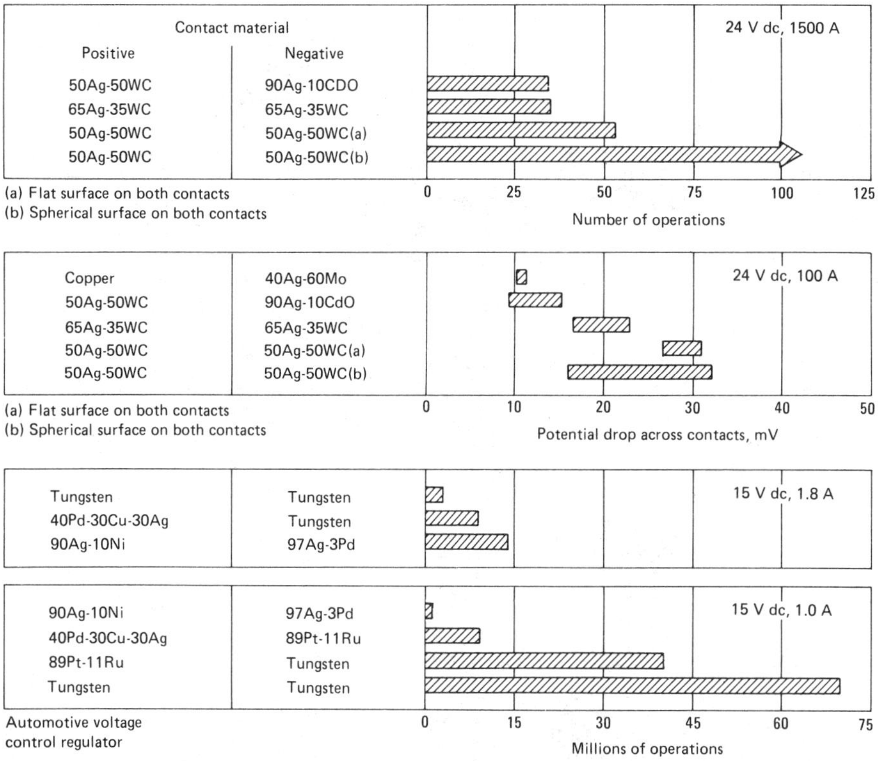

Fig. 17 Effect of composition, combination, and electrical characteristics on contact life of various materials. 24 V dc, 1500 A: A comparison of several polarized contact material combinations for their susceptibility to welding when subjected to an overload of 1500 A at 24 V. The 50Ag-50WC combination with a spherical surface shows considerably longer life than some other combinations of silver sintered products; it did not weld in 100 operations. 24 V dc, 100 A: Shown is the scatter of contact potential after 200 000 operations for several samples of different alloy combinations. Under conditions of relatively high current, the contact force used in these tests was approximately 2.2 to 2.8 N. 15 V dc, 1.8 A: A comparison of the life of three contact material combinations intended for use in automotive voltage regulators. At this level of amperage, tungsten with tungsten failed because of insulation resulting from oxide formation after 3 million operations. Palladium-copper-silver with tungsten failed as a result of sticking. Silver-nickel with silver-palladium failed from metal transfer. 15 V dc, 1.0 A: Life comparison for four contact material combinations intended for use in automotive voltage regulators. Silver-nickel with silver-palladium made the poorest showing and failed from sticking. Palladium-copper-silver with platinum-ruthenium, and platinum-ruthenium with tungsten failed from oxide insulation. Tungsten with tungsten, which failed from oxide insulation at 1.8 A, survived about 70 million operations.

demonstrated. Instead of plotting the number of operations obtained with each polarized contact combination, the time required to complete 100 cycles is used. Because sustained arcing affects the temperature rise of the contacts, and also the bimetal element used to make and break the circuit, the time required to complete 100 cycles is a good indication of the effectiveness of different contact combinations, as well as the effect of polarity on a given combination.

Of the materials and combinations tested, 90Pd-10Ru positive and 90Pt-10Ir negative gave the shortest time (12 min) for the completion of 100 operations; 90Pt-10Ir positive and 72Pd-26Ag-2Ni negative required the longest time (62 min). It takes less time for 100 cycles when the palladium alloy is positive and the platinum alloy is negative than for the reversed polarity. The time for 100 cycles for the combination 90Pd-10Ru positive and 90Pt-10Ir negative is about one-third that for the same combination with reversed polarity.

Figure 17 shows that polarized contacts, using 50Ag-50WC at both terminals for 24 V dc and 1500 A, have considerably longer life than some other combinations of sintered silver products. Results are also given in this same figure for automotive voltage regulators for 15 V dc, 1.0 A and 15 V dc, 1.8 A. Tungsten-tungsten had a very long life of about 70 million operations when the current was 1.0 A. However, when the current was increased to 1.8 A, the life was reduced to 3 million operations. Positive 90Ag-10Ni and 97Ag-3Pd negative was the best combination tested for the higher amperage but the poorest of all combinations for the lower amperage.

Recommended Contact Materials

Fixed and stationary contacts for operations at low frequency are made of less expensive materials such as copper metals and aluminum metals. The ordinary two-prong domestic appliance plug is this type in its simplest form. Plug connectors are manually connected and disconnected. There is little trouble from arcing, pitting, and welding because the current is controlled at some other contact where it will be interrupted before the plug is engaged or removed. The life of the spring is the life of the contact; it is the spring that usually fails. Therefore, spring bronze is used for plugs frequently connected and disconnected that do not need the highest conductivity. Where higher conductivity is needed, silver on copper is better.

For bolted connectors, low contact resistance is important. Silver-coated copper is ideal in a noncorrosive environment. For wet-cell battery terminals, where corrosion is severe, lead against lead is used. Many combinations of metals are used for bolted connectors, with factors such as temperature, corrosivity, conductivity, and cost affecting selection.

Power Circuits. Materials selected for contacts in power circuits (Tables 10 to 12) usually have high electrical and thermal conductivity, and high resistance to arc erosion and to welding or sticking. Because these contacts are relatively large, precious metals like platinum are seldom used except in extremely corrosive atmospheres. Bare copper, copper faced or plated with silver, and aluminum faced or plated with silver are commonly used for stationary contacts carrying high current. Carbon, occasionally mixed with copper or silver in a sintered product, is commonly used for brushes against copper commutators or slip rings for sliding contacts. Occasionally, silver or silver-plated copper is used. Sintered products consisting of tungsten, tungsten carbide, or molybdenum mixed with copper or silver are used for spark gaps and for mov-

Table 10 Recommended materials for fixed or stationary contacts for power circuits

Materials are listed in order of decreasing preference.

Alloy	Advantages(a)
Plug connectors (1 to 100 A) 1 to 10 000 operations	
Brass	(b, c)
Bronze	(c, d, e)
Sn on Cu	(f, g, h)
Plug connectors (100 to 100 000 A) 1 to 10 000 operations	
Cu	(b, f)
Ag	(f, g, h, i)
Ag on Cu	(b, f, g, h, i)
Ag on Al	(b, g, h, i)
Cr-Cu	(d, e)
Blade connectors (10 to 100 000 A) 10 to 10 000 operations	
Cu	(b, f)
Ag on Cu	(f, g, h)
Bolted connectors (100 to 1000 A) 1 to 100 operations	
Brass	(b, c)
Bronze	(e, j)
Sn on Cu	(f, g, h)
Bolted connectors (1000 to 1 000 000 A) 1 to 100 operations	
Cu	(b, f)
Ag on Cu	(f, g, h)
Sn on Cu	(f, g)
Ag on Al	(b, g, h)

(a) With fixed or stationary contacts, failure is ultimately caused by deterioration of the contact surface. (b) Low-cost material. (c) Material easy to fabricate. (d) Wear resistance. (e) Material provides lower contact resistance. (f) Electrical conductivity. (g) Surface oxidation resistance. (h) Material provides lower contact resistance. (i) Material permits use of a low contact force. (j) Material has higher strength.

ing contacts that are required to interrupt high currents at high voltage with heavy arcing.

Liquid metals are seldom used in power circuits, but mercury and the sodium-potassium eutectic are used against solid contacts where a circuit must be completed between a stationary part and one involving an unusual type of motion, or where currents of more than 100 000 A must be "collected" from rotating equipment, as is accomplished with the liquid eutectic.

Brushes. A cursory examination of Table 11 shows that sliding power contacts have a wide range of application. Although the list is not complete, it covers machines with current ranges from 1 to 10 000 A. Even with this broad application range, there is one condition that must be satisfied in any successful sliding contact. Somewhere between the two surfaces that are moving relative to each other, there must be a film or region in shear that prevents seizing or welding of the clean surfaces. Trouble ensues when the film disappears in plating generators because of low humidity, or when the film can no longer be formed because of lack of oxygen and moisture, as in aircraft flying at high altitudes. Under normal conditions, this film is inherent.

Lubricants added to the brush material also help in preventing cold welds. In certain power stations operating at room temperature in the wintertime, the relative humidity must be kept above 25% to prevent rapid wear and dusting of brushes. In aircraft applications, adjuvants are helpful in maintaining this film. The identity of such films has not been clearly established. Graphite is most effective, but it requires the presence of considerable moisture and oxygen.

There are no simple rules for selecting materials for brushes, but the usual practice is to start with a material in the correct resistance range. If this is unsuitable, the reasons for failure, which might be poor commutation, overfilming, high wear, or arcing, must be considered and brush properties improved.

For many functions, it is imperative to select contact materials that do not contribute unduly to high-frequency radio noise, as may be generated by sliding contacts and brushes. Such noise may be minimized by having low or uniform contact resistance between brushes and rings, a condition that can be effected through the use of graphite, silver-graphite, gold-graphite, or other noble metal brushes against alloys of silver, gold, or platinum, or against graphite. A good expedient is to use two or more brushes against a single ring to have parallel circuits, the ultimate being fiber brushes, molybdenum, or some other metal that gives the effect of many separately supported parallel contacts. In most applications, silver-graphite brushes containing

Table 11 Recommended materials for sliding contacts for power circuits

Materials are listed in order of decreasing preference.

Alloy	Advantages	Cause of failure
Power brushes (10 to 10 000 A) continuous slide		
Electrographite	(a, b, c)	(d)
Carbon graphite	(a, b, c)	(d)
Fractional horsepower brushes (1 to 10 A) continuous slide		
Electrographite	(a, b, c)	(e)
Carbon graphite	(a, b, c)	(e)
Resin bonded	(a, b, c)	(e)
Aviation brushes (10 to 1000 A) continuous slide		
Electrographite	(a, b, c, f)	(g)
Carbon graphite plus BaF, CdI, MoS_2 for altitude performance	(a, b, c, f)	(g)
Automotive starter brushes (10 to 1000 A) continuous slide		
Copper graphite	(a, b, h, i)	(j)
Automotive generator brushes (10 to 100 A) continuous slide		
Electrographite	(a, b, c)	(k)
Carbon graphite	(a, b, c)	(k)
Automotive auxiliary brushes (1 to 10 A) continuous slide		
Copper graphite	(a, b, h, i)	(l)
Plating generator brushes (100 to 10 000 A) continuous slide		
Copper graphite	(a, b, h, i)	(m)
Alternating-current and slip ring brushes (1 to 1000 A) continuous slide		
Electrographite	(a, b)	(n)
Carbon graphite	(a, b)	(n)
Copper graphite	(a, b, h, i)	(n)
Commutators (10 to 10 000 A) continuous slide		
Cu	(a, h)	(o)
Ag on Cu	(h, i, p)	(o)
Silver-bearing Cu	(b, h, q)	(o)
Zirconium-bearing Cu	(b, h, q, r)	(o)
Slip rings (10 to 10 000 A) continuous slide		
Stainless steel	(b, q)	(o)
Silver copper	(h, i, p)	(o)
Bronze	(a, b, s)	(o)
Tool steel	(a, b, q)	(o)
Wire against slider or trolley wheel (10 to 1000 A) continuous slide or roll		
Bronze wire against bronze wheels	(a, b)	(t)
Ag-Cu against Ag-W	(b, h, i, p, u)	(t)
Cd-Cu wire against Cu-C sliders	(h, p)	(t)
Cd-Sn-Cu wire against hard Cu	(h, o)	(t)
Steel against cast iron	(a, b)	(t)
Liquid to collector assembly (10 to 10 000 A) continuous dip		
Mo against Hg	(b, h, u)	(v)
Steel against Hg	(a, b, q)	(v)
Bearings and swivels (10 to 10 000 A) intermittent slide		
Brass	(a)	(o)
Steel	(p)	(o)
Cu	(h)	(o)
Ag-graphite	(b, w)	(o)
Ag	(h, i, p)	(o)
Bronze	(b, q)	(o)
Graphite	(u)	(o)

(a) Low-cost material. (b) Wear resistance. (c) High contact resistance for commutation. (d) Wear, poor commutation. (e) Wear, arcing. (f) Suitable for operation in dry air or at altitude. (g) Wear, dusting, poor commutation. (h) Electrical conductivity. (i) Lower contact resistance. (j) Wear, high resistance. (k) Wear, arcing, poor performance. (l) Wear, poor performance. (m) Wear, dusting, grooving of commutators. (n) Wear, sparking, grooving. (o) Wear, arc erosion. (p) Surface oxidation resistance. (q) Higher strength. (r) Higher annealing temperature. (s) Ease of fabrication. (t) Wear. (u) Arc-erosion resistance. (v) Liquids practically never, solids by arc erosion. (w) Less sticking and welding tendency from 5 to 50% graphite are used against silver rings. At high altitude, silver-graphite must have a protecting adjuvant to prevent the rapid wear that may ensue in dry rarefied air.

Circuit Breakers. Contact materials recommended for use in air in circuit breakers having a maximum current rating of 800 A and a maximum voltage rating of 600 V are:

Stationary contact	Moving contact
Silver-tungsten	Silver-tungsten
Silver-WC	Silver-WC
Silver-molybdenum	Silver-WC
Silver-nickel	Silver-molybdenum
Fine silver	Silver-tungsten
Silver-molybdenum	Silver-tungsten

Vibrators have severe requirements as contact materials because of the high localized temperature caused by the rapidly repeated making and breaking of contact. As a result, only a small amount of current at low voltage can be handled if the operation is continuous and reasonable life is expected. Contact materials recommended for vibrators are listed in the first group of Table 13.

Automobile horns usually operate for only a short length of time at each operation. They carry 10 to 20 A with a closing force of 25 to 30 N (6 to 7 lbf). Tungsten is almost universally used as the contact material because of the combination of good conductivity and high melting point. The high pressure is needed to break down an oxide film that develops.

Tungsten is also used for contacts in radio vibrators that operate in a protected environment with low current and pressure for comparatively long periods of time. Silver and silver alloy points are used in buzzers that operate with very low current for short periods of time. Platinum or gold is used in critical applications, such as fire alarms, where reliability is essential.

Voltage Regulators. A wide range of contact materials is used in automotive voltage regulators because even minor differences in alternator field circuits or in suppression devices can have a marked effect on contact life. Generally, regulator contacts are polarized, the substance with the higher melting point being specified for the side that ordinarily loses material. A most successful contact combination is a tungsten negative contact against platinum-ruthenium positive contact. Other commonly used materials are tungsten against tungsten at higher voltage with good environmental protection; silver alloys against palladium alloys where cost is important; and gold against platinum-iridium where high current and low induced voltage prevail. In these combinations, the material listed first is used for the positive contact.

Most regulator contacts fail through development of an open circuit because of oxides or other contaminants that come from the contact arc, hydrocarbon vapors,

Table 12 Recommended materials for make-break contacts for power circuits

Alloy	Advantages	Cause of failure
Tap changers (10 to 100 A) no-load make-break, 100 to 100 000 V, 10 000 max operations		
Bronze wiper against brass	(a)	(b)
Cu-Cd against Cu-Cd	(c, d, e)	(b)
Brass or bronze against brass or bronze	(a)	(b)
Tap changers (100 to 100 000 A) no-load make-break 100 to 100 000 V, 10 000 max operations		
Cu	(f)	(b)
Cu-Cd	(f, h, i)	(b)
Ag on Cu	(c, h, i)	(b)
Ag-Ni	(f, h)	(b)
Ag-Cu-Ni	(f, h)	(b)
Cr-Cu	(c, f, i)	(b)
Tap changers (10 to 10 000 A) load make-break, 100 to 100 000 V, 10 000 max operations		
Cu-W	(e, j, k)	(l)
Cu-WC	(e, j, k)	(l)
Ag-graphite	(c, d, g, m)	(l)
Ag-W	(c, d, e, g, j, k)	(l)
Ag-Ni	(a, f, h, i)	(l)
Contactors and motor starters (10 to 10 000 A) alternating current, 10 to 100 V, 10 000 000 max operations		
Ag	(c, d, g, j, p)	(l)
Ag-CdO	(e, f, h, k, m)	(l)
Ag-Cd	(e, g, j)	(l)
Cu	(a, c, j)	(l)
Cu-W, Ag-W, Ag-Ni	(e, f, h, k, m)	(l)
Contactors and motor starters (10 to 10 000 A) direct current, 1 to 1000 V, 10 000 000 max operations		
Ag-CdO	(e, f, h, j, k)	(l)
Ag-Cd	(e, i, j)	(l)
Ag	(c, d, g, i)	(l)
Ag-Ni	(e, h)	(l)
Ag-W or Ag-WC	(e, f, h, j, k)	(l)
Air circuit breakers (10 to 100 A) current-carrying and arcing, 10 to 1000 V, 1 000 000 max operations		
Ag-Cu-Ni	(a, c)	(n)
Ag-Ni	(e, h, j)	(n)
Ag-graphite	(c, d, g)	(n)
Ag-W	(e, f, h, j, k)	(n)
Ag-CdO	(c, d, e, g, j, k, m)	(n)
Air circuit breakers (100 to 1000 A) current-carrying and arcing, 10 to 10 000 V, 100 000 max operations		
Ag-W	(e, h, j, k)	(n)
Ag-CdO	(c, d, e, j, g, m)	(n)
Ag-WC	(c, g, h, j)	(n)
Air circuit breakers (100 to 100 000 A) current-carrying only, 100 to 100 000 V, 10 000 max operations		
Cu	(a, c)	(n)
Ag on Cu	(c, d, g, m)	(n)
Ag-Ni	(e, h)	(n)
Ag-graphite	(c, d, g, h, m)	(n)
Air circuit breakers (1000 to 1 000 000 A) arcing tip only, 100 to 100 000 V, 10 000 max operations		
Cu-W	(a)	(n)
Ag-W, Ag-WC, Ag-Mo	(h, j, k)	(n)
Magnet steel	(a)	(n)
Oil circuit breakers (10 to 10 000 A) current-carrying and arcing, 1000 to 100 000 V, 100 000 max operations		
Cu-W	(a)	(n)
Ag-W	(d, e, g)	(n)
Ag-Mo	(c, g, m)	(n)
Oil circuit breakers (100 to 100 000 A) current-carrying only, 1000 to 1 000 000 V, 10 000 max operations		
Cu	(a, f)	(n)
Ag on Cu	(c, d, g)	(n)
Ag-Ni	(f, h)	(n)
Oil circuit breakers (100 to 1 000 000 A) arcing tip only, 1000 to 1 000 000 V, 10 000 max operations		
Cu-W	(a)	(n)
Cu-WC	(k)	(n)
Ag-W	(c, d, g)	(n)
Ag-Mo	(c, d, e, g)	(n)
Spark gaps (10 to 1 000 000 A) 100 to 1 000 000 V, 1 000 000 max operations		
W	(j)	(n)

(a) Low-cost material. (b) Surface deterioration, wear. (c) Electrical conductivity. (d) Lower contact resistance. (e) Less sticking and welding tendency. (f) Higher strength. (g) Surface oxidation resistance. (h) Wear resistance. (i) Ease of fabrication. (j) Arc-erosion resistance. (k) Resistance to transfer and pitting. (l) Sticking, arc erosion. (m) Allows low contact force. (n) Arc-erosion, material transfer and pitting, or overheating

Table 13 Recommended materials for light power and engineering contacts of current range of 0.1 to 30 A

Materials are listed in order of decreasing preference.

Alloy	Advantages	Cause of failure	Applications
Electromagnetic vibrators (0.1 to 30 A)(a)			
W-0.5Mo	(b, c, d, e, f)	(g)	Automotive voltage regulators, bells, buzzers, horns, radio vibrators
Pd-Ag-Ni-W			
Pt-Ru-W			
Ag-Mo			
Au-Pt-Ir			
Thermomechanical thermostats (0.1 to 30 A)(h)			
Ag	(f, i, j, k, l, n)	(m)	Household heating and cooling, cooking, electric blankets
Ag-Pd	(e, j, k, l)		
Ag-Cd	(b, c)		
Hg-Pt	(j)		
Hg-Mo	(b, c, e, j)		
Ag-CdO	(b, c, e)		
Pt-Ir	(d, j, k, n)		
Au	(j, k, n)		
Manual or electromagnetic snap-action switches (0.1 to 30 A)(o)			
Bronze	(f, l, p)	(q)	Lighting, appliances, engineering equipment
Ag-CdO	(b, c, e)		
Ag	(f, i, j, k, l, n)		
Pt	(d, j, k, n)		
Au	(j, k, n)		
Electromagnetic relays (0.1 to 30 A)(r)			
Ag-CdO	(b, c, e)	(s)	Control systems, engineering equipment, lighting, appliances
Ag	(i, j, k, n)		
Ag-Cu	(b, d, f)		
Ag-Cu-Ni	(d, f)		
Pd	(j, k, n)		

(a) Fast action with force of 0.137 to 4.45 N (½ to 16 oz). Contact, butting plus slight wipe. 1 to 110 V; 1 billion max operations. (b) Less sticking and welding tendency. (c) Arc-erosion resistance. (d) Wear resistance. (e) Resistance to transfer and pitting. (f) Low-cost material. (g) Wear, welding transfer, and arcing (open). (h) Slow or fast action with 2.7 N (10 oz) max force. Contact, butting to considerable wipe. 300 V max; 100 000 max operations. (i) Electrical conductivity. (j) Lower contact resistance. (k) Surface oxidation resistance. (l) Ease of fabrication. (m) Arcing, welding and contamination by dust. (n) Allows low contact force. (o) Fast action with 0.2 to 4.45 N (1 to 16 oz) force. Contact, butting to wipe. 300 V max; 1 million max operations. (p) Spring properties. (q) Arcing (open), and welding. (r) Fast action with 0.2 to 4.45 N (1 to 16 oz) force. Contact, butting to considerable wipe. 300 V max; 1 million max operations. (s) Wear (open), some welding (closed)

or dust particles. Some failures for the same equipment at a different level of field current are caused by welding. Hundreds of different contact material combinations have been used in voltage regulators.

Switches. Two classes of snap-action switches must be considered. The manual type is found in the walls of homes and offices and on electric ranges, ovens, and other similar appliances. Lower-cost contact materials are used for manual switches because substantial slide or wipe of contact surfaces and high contact force can be tolerated.

The other type is a precision snap-action switch that may be operated electromagnetically or mechanically by precision rotating or sliding cams with little movement or low operating forces. This type of switch is used on equipment such as machine tools, precision controls, and thermostats.

Bronze is unsuitable for precision snap-action switches because it corrodes too readily. In circuits that are sensitive to resistance and in which the voltage is low (1 V or less), either silver, gold, or platinum alloys are used. In more than 80% of heat thermostats, mercury-molybdenum or mercury-platinum switches are used; the remainder are of materials such as fine silver, 90Ag-10CdO, 90Pt-10Ir, tungsten, 35Pd-30Ag-14Cu-10Pt-10Au-1Zn, 90Ag-10Fe, and 69Au-25Ag-6Pt.

Mercury switches are well suited for use in thermostats because they promise absolute reliability in making and breaking of contact. However, they must be kept in the desired position.

Fine silver is used extensively if the voltage is as high as 20 V and the current is 1 A or higher. Where a current of 10 to 30 A is controlled and the voltage is higher than 110 V, Ag-CdO is used, especially if the making and breaking of electrical contact is slow.

The precious metals, particularly gold alloys, are recommended where low voltage and so-called dry circuits are involved. It is good practice to have multiple contacts when there is a need for high reliability in making electrical contact.

It has been demonstrated experimentally that reliability can be increased as much as 2700% by using two contacts rather than one, and with the pressure force on each contact being only half of the force on the single contact.

Telecommunications Equipment. In electrical contacts for use in telecommunications equipment, the contact resistance must be low enough to ensure satisfactory circuit operation and to prevent excessive transmission losses. The contact resistance also must be constant to avoid noise modulation in the transmitted signal. Contact separation should not become unreliable because of welding, snagging, or excessive surface roughness due to arcing. The contacts should not become excessively worn due to mechanical action or eroded due to arcing.

The types of contacts most frequently used in telecommunications equipment are:

- Connector contacts that close with considerable slide and high force but are not required to be changed often—for example, a multiwire cable plug and its socket
- Sliding contacts that operate with considerable slide but with low force to make operation easier and to reduce mechanical wear from frequent operation. The contact in a rotary switch is an example
- Butting contacts that generally close with light force and without much slide. The low force and slide minimize power requirements and wear, thereby permitting operation at a high rate. Telephone relays feature this kind of contact

Table 14 lists materials recommended for telecommunications equipment. The most frequently used metals are platinum, palladium, iridium, ruthenium, gold, silver, copper, nickel, and tungsten. The precious metals, because of their tarnish resistance, provide greater reliability of closure and sometimes greater resistance to arc erosion. The base metals, where they can be used, are more economical and provide greater freedom from welding and snagging of pits and buildups arising from arcing.

Microcontacts are considered to be those contacts having a closure force of 50 mN (0.01 lbf) or less and carrying currents measured in milliamperes at voltages of 120 V or less. Materials recommended for microcontacts are given in Table 15.

Silver metals are generally used as microcontacts that have more than 10 mN (0.002 lbf) force and preferably some wipe. Other metals plated with silver are also used, as in magnetically operated microcontacts. In such devices, the fixed contact is a magnet that must be capable of being plated with a metal that can carry the current. The moving contact is iron or a magnetically soft alloy, which also must be plated. These contacts are used in meter-movement relays that have an actuating current of as little as 2 mA. When the soft iron on the pointer is deflected so that it enters the field of the magnetic contact, magnetic attraction causes the contacts to close with a force of 18 to 40 mN (0.004 to 0.009 lbf), and with some impact and wipe. Silver is used in armature-type sensitive relays less frequently than palladium, which is better than silver in such low-force low-current applications. Palladium is used, for example, for brushes in micropotentiometers.

Table 14 Recommended contact materials for telecommunication equipment

Voltage, V	Current, A	Closed force, g	Make and break with V and A	Expected life, max No. of operations	Materials used(a)	Advantages	Usual cause of failure
Connectors							
Plug and jack							
...	...	500	No	10^5	Brass against Ni-brass springs	(b, c, d)	(e)
					Ni-plated brass against Ag-plated brass springs	(f, g)	(e)
Vacuum tube plug and socket							
...	...	High	No	10^2	Ni-plated brass against phosphor bronze springs	(b, c, d)	...
...	...	High	No	10^2	Ni-plated brass against Ag-plated phosphor bronze springs	(f, g)	...
Multicontact connector							
...	...	150	No	10^3	Brass against phosphor bronze springs	(b, c, d)	(h)
...	...	150	No	10^3	Au-plated phosphor bronze against phosphor bronze springs	(f, g, i)	(h)
...	...	150	No	10^3	Au-plated copper against phosphor bronze springs	(c, f, g)	(h)
Sliding contacts							
Electromechanical rotary switches (telephone type)							
50 max	0.1 max	20–65	Not usually	10^6	Phosphor bronze wiper against brass or phosphor bronze terminal	(b, d)	(j)
50 max	0.1 max	20–65	Not usually	10^6	69Au-25Ag-6Pt overlay on phosphor bronze terminal	(f, g, i)	(j)
50 max	0.1 max	20–65	Not usually	10^6	69Au-25Ag-6Pt overlay on phosphor bronze or brass for both wiper and terminal	(f, g, i)	(j)
Manual rotary switches							
50 max	0.1 max	20–65	Not usually	10^6	Ag-plated brass	(b)	(k)
50 max	0.1 max	20–65	Not usually	10^6	90Ag-10Cu rotor blades against 90Ag-5Cu-5Zn or against 90Ag-5Cu-5Cd clips	(i)	(k)
50 max	0.1 max	20–65	Not usually	10^6	69Au-25Ag-6Pt overlay on brass for both blades and clips	(f, g, i)	(k)
Slip rings and brushes							
25 max	0.01 max	25	No	...	Ag	(b)	(l)
25 max	0.01 max	25	No	...	90Ag-10Cu	(i)	(l)
25 max	0.01 max	25	No	...	70Au-30Ag	(f, g, i)	(l)
25 max	0.01 max	25	No	...	Pd-Ag-Cu	(f, g, m)	(l)
25 max	0.01 max	25	No	...	Coin silver	(b, c, f, m)	(l)
25 max	0.01 max	25	No	...	Au-Ag-Pt	(f, g, m)	(l)
25 max	0.01 max	25	No	...	Au-Pt	(f, g, m, n)	(l)
Butting contacts							
Sensitive relays							
50 max	0–1	1–5	No	10^9	Pd	(b)	(o)
50 max	0–1	1–5	No	10^9	69Au-25Ag-6Pt	(p)	(o)
50 max	0.4 max	1–5	Yes	10^7	Pd	(b)	(o)
50 max	0.4 max	1–5	Yes	10^7	69Au-25Ag-6Pt	(p)	(o)
50 max	0.4–1	1–5	Yes	10^7	Pd	(b)	(o)
50 max	0.4–1	1–5	Yes	10^7	Pt	(f, g)	(o)
Telegraph relays							
±130	0.060	1–5	Yes	10^8	W against 60Pd-40Cu	(q)	Sticky
General-purpose relays							
50 max	1 max	5–50	No	10^9	Pd	(b)	(o)
50 max	1 max	5–50	No	10^9	69Au-25Ag-6Pt	(p)	(o)
50 max	0.4 max	5–50	Yes	10^8	Pd	(b)	(o)
50 max	0.4 max	5–50	Yes	10^8	69Au-25Ag-6Pt	(p)	(o)
50 max	0.4–1	5–50	Yes	10^8	Pd	(b)	(o)
50 max	0.4–1	5–50	Yes	10^8	Pt	(f, g)	(o)
10–50	1 max	20–50	No	10^9	Ag	(b)	(o)
10–50	0.4 max	20–50	Yes	10^7	Ag	(b)	(o)
Switches							
50 max	0.4 max	50–250	Yes	10^6	69Au-25Ag-6Pt	(i)	Dust
50 max	0.4 max	50–250	Yes	10^6	Pt-Ru or Pt-Ir	(i)	Dust
50 max	1 max	5–50	Yes	10^6	Pd	(b)	Dust
50 max	1 max	5–50	Yes	10^6	Pt	(f, g)	Dust

(a) Materials are listed in the order of decreasing preference. (b) Low-cost material. (c) Ease of fabrication. (d) Spring properties. (e) Wear and surface deterioration. (f) Surface oxidation resistance. (g) Lower contact resistance. (h) Surface deterioration. (i) Wear resistance. (j) Surface deterioration, resistance, and noise. (k) Resistance and noise. (l) Wear, resistance, and noise. (m) Allows low contact force. (n) Electrical conductivity. (o) Dust, polymer, carbonaceous deposits, and erosion sticking. (p) Reduced polymer formation. (q) Material provides improved resistance to metal transfer and pitting.

Table 15 Recommended materials for microcontacts

Current, microamperes	Voltage, V	Closed force, g	Slide or wipe	Expected operations, millions	Materials used(a)	Advantages	Usual cause of failure
Meter-movement relays							
Magnetic contacts							
100 max	120 max, ac or dc	2–4	Small	2–5	Plated bright Au	(b, c, d, e)	(g)
					Plated bright Ag	(b, c, e, f)	(g)
Load-current-aiding type							
5–25	75–120, dc	½–1	Small	2–5	Pt-Ir	(b, c, d, e, h)	(g)
Sensitive contacts							
200 max	6 max, dc	0.001–0.1	Small	2–20	Pt-Ir	(b, c, d, e, i)	(g)
Armature-type sensitive relays							
3000 max	120 max, ac or dc	1–5	Small	0.1–1.0	Pd	(b, c, d, i)	(g)
					Au-plated Pd	(b, c, d, e, f)	
					Au	(b, c, d, e, f)	
Micropotentiometers							
Brushes							
100 max	120 max, ac or dc	0.1–5	Intermittent		71.5Au-14.5Cu-8.5Pt-4.5Ag-1Zn	(b, c, d, e, h)	(l)
					35Pd-30Ag-14Cu-10Au-10Pt-1Zn	(b, c, d, h)	
					44Pd-38Ag-16Cu-1Pt-1Zn	(b, c, d, h)	
					69Au-25Ag-6Pt	(b, c, d)	
					60Pd-40Ag	(h, k)	
Miniature slip rings							
Wire							
1 max	120 max, ac or dc	1–5	Continuous		71.5Au-14.5Cu-8.5Pt-4.5Ag-1Zn	(b, c, d, e, h)	(m)
					35Pd-30Ag-14Cu-10Au-10Pt-1Zn	(b, c, d, h)	
					44Pd-38Ag-16Cu-1Pt-1Zn	(b, c, d, h)	
					90Pt-10Rh	(b, c, d)	
Brush rings							
1 max	120 max, ac or dc	1–5	Continuous		71.5Au-14.5Cu-8.5Pt-4.5Ag-1Zn	(h, d)	(m)
					35Pd-30Ag-14Cu-10Au-10Pt-1Zn	(h)	
					69Au-25Ag-6Pt	(b, c, d)	
					Hard Au plate	(b, c, d, e)	
					Rh plate	(h)	(l)

(a) Materials are listed in order of decreasing preference. (b) Surface oxidation resistance. (c) Lower contact resistance. (d) Low contact force. (e) Reduced polymer formation. (f) Electrical conductivity. (g) Frictional polymer and wear. (h) Wear resistance. (i) Less sticking and welding. (j) Arc erosion resistance. (k) Low-cost material. (l) Wear, noise, and polymer. (m) Wear and noise

Platinum-iridium (usually 80Pt-20Ir) is the material most commonly used where the contact force is less than 10 mN (0.002 lbf). Contacts made from 80Pt-20Ir are used in meter-movement relays, both load-current-contact-aiding (LCCA) and sensitive types. In both types, the contacts must close with extremely low force. With the LCCA type, the load current through the contact also passes through an extra winding on the moving coil and thereby increases the closure force. Initial contact often is made with less than 10 μN (2×10^{-6} lbf) of force. Although hardness in a contact material is an advantage, it can also be a disadvantage. For instance, 70Pt-30Ir and 65Pt-35Ir are too hard for most microcontact applications.

Gold has become more widely used in microcontacts since the development of sealed relays. It is a less potent catalyst than members of the platinum family and therefore does not collect as much polymer on contacting surfaces. Like silver, it is easily plated and is used in this form on magnetic contacts. Gold is also plated over palladium in sealed relays to decrease the polymer effect. In miniature slip rings and micropotentiometers, gold alloys are used as brushes, rings, and wire.

Slip ring-brush assemblies are the mechanical devices that transfer electrical information in applications such as gyroscopes, strain gages, and video tape heads by means of contact surfaces that move relative to one another. The smaller of the two moving parts is referred to as the brush, wiper, or contact. By nature these applications require low electrical noise, low friction, long life, and high reliability. Various combinations of metals are used to meet the above demands.

The alloys to be used in slip ring-brush assemblies are selected on an empirical basis, with consideration for environmental conditions, performance requirements, shelf life, and cost. The advantages of wrought precious metals are their low and predictable oxidation rates, long shelf life, and low electrical noise. When compromises are economically necessary, the ring material is changed because it represents a large volume of material compared with the brush.

In general the brush material should be harder than the ring material because the wear of the brush takes place over a relatively small area as compared to the ring. Although the brush should be harder than the ring, it should not be so hard as to act as a cutting tool against the ring material.

A lower-cost alternative to solid wrought precious metals is the electrodeposition of a noble metal, usually a hardened gold. Some use is made of rhodium-plated surfaces, but unless a thin overplate of gold is used, rhodium is not suitable for low noise requirements because of the formation of insulating polymers. Several potential problems with plated surfaces can significantly affect performance and reliability. Porosity, entrapped plating salts, and changes in chemistry through the plating thickness can cause inconsistent performance.

Availability

Silver, gold, platinum, palladium, and most of their ductile alloys are available as stamped contacts. Except for material from which contact disks are produced, a variety of stock sizes is not maintained because, in

Table 16 Commercially available forms of electrical contact materials

Alloy	Product form: Solid rivet	Wire	Strip	Tape	Disks	Attached(a)	Composite weld disks(b)	Clad(c)	Rings	Brushes	Mig method: Melting	P/M
100 Ag	x	x	x	x	x	x	x	x	x	...	x	...
100 Pd	x	x	x	x	x	x	x	x	x	...	x	...
100 Au	x	x	x	x	x	...	...	x	...	...	x	...
100 Ru	...	...	x	...	x	x	...	...	...	...	x	x
100 Ir	...	x	x	...	x	x	...	...	...	...	x	x
100 Pt	x	x	x	...	x	x	x	x	...	x	x	x
100 Os	...	...	...	...	x	x	...	...	...	...	...	x
100 Rh	...	x	x	...	x	...	...	...	Plate	...	x	...
92.5Ag-7.5Cu	x	x	x	...	x	x	x	x	...	...	x	...
90Ag-10Cu	x	x	x	x	x	x	x	x	x	...	x	...
75Ag-24.5Cu-0.5Ni	x	x	x	x	x	x	x	x	x	...	x	...
72Ag-28Cu	...	x	x	...	...	...	...	x	...	x	x	...
99Ag-1Pd	x	x	x	x	x	x	x	x	...	x	x	...
97Ag-3Pd	x	x	x	x	x	x	x	x	...	x	x	...
90Ag-10Pd	x	x	x	x	x	x	x	x	...	...	x	...
90Ag-10Au	x	x	x	...	x	x	...	...	...	...	x	...
97Ag-3Pt	x	x	x	...	x	x	...	...	x	...	x	...
85Ag-15Cd	x	x	x	...	x	x	x	x	...	...	x	...
95Ag-5CdO	x	x	x	x	x	x	x	x	...	...	x	x
90Ag-10CdO	x	x	x	x	x	x	x	x	...	...	x	x
85Ag-15CdO	x	x	x	x	x	x	x	x	...	...	x	x
90Ag-10Fe	x	...	x	...	x	x	x	...	...	...	...	x
90Ag-10W	x	...	...	...	x	x	x	...	...	...	...	x
50Ag-50WC	...	...	...	...	x	x	x	...	...	...	...	x
65Ag-35WC	...	...	...	...	x	x	x	...	...	...	...	x
75Ag-25Zn	...	...	x	...	x	...	...	...	...	...	x	...
85Ag-15Ni	...	...	...	...	x	x	x	...	x	...	...	x
70Ag-30Ni	...	...	...	...	x	x	x	...	...	...	...	x
70Ag-30Mo	...	...	...	...	x	x	x	...	...	...	...	x
97Ag-3 graphite	...	...	...	...	x	x	x	...	...	...	...	x
95Ag-5 graphite	...	...	...	...	x	x	x	...	...	x	...	x
60Pd-40Ag	x	x	x	...	x	x	x	x	...	...	x	...
60Pd-40Cu	x	x	x	...	x	x	x	...	...	x	x	...
95Pd-5Ru	x	x	x	...	x	x	x	...	...	...	x	...
75Au-25Ag	x	x	x	...	x	x	x	x	...	...	x	...
90Au-10Cu(Coin)	x	x	x	...	x	x	x	x	x	...	x	...
95Pt-5Ru	...	...	x	...	x	x	x	...	...	...	x	...
90Pt-10Ru	...	...	x	...	x	x	x	...	...	...	x	...
90Pt-10Rh	...	...	x	...	x	x	x	...	...	...	x	...
90Pt-10Ir	...	x	x	...	x	x	x	...	...	...	x	...
85Pt-15Ir	...	x	x	...	x	x	x	...	...	...	x	...
80Pt-20Ir	...	x	x	...	x	x	x	...	...	...	x	...
75Pt-25Ir	...	x	x	...	x	x	x	...	...	...	x	...
65Pt-35Os	...	...	...	...	x	x	x	...	...	...	...	x
69Au-25Ag-6Pt	x	x	x	...	x	x	x	...	x	x	x	...
60Ru-35Ir-5Pt	...	...	...	...	x	x	x	...	...	...	x	...

(a) Contact disks attached to screws, rivets, blades and bars. (b) Composite welding-type buttons produced for resistance welding attachment. Backings of nickel, Monel, and steel. (c) Clad materials including overlay, throughlay, edgelay, strip of precious metal on (or in) base metal

general, no two applications are identical. For disks, material in strip form, varying in thickness from 0.25 mm to 1 mm (0.010 to 0.040 in.) is available.

Except for tungsten and molybdenum, contact materials are ductile enough so that they can be produced in all contact forms. Tungsten and molybdenum and some of the P/M materials have lower ductility and are available in fewer forms. The commercially available forms of common electrical contact materials are listed in Table 16.

Silver, gold, platinum, palladium, and nearly all the alloys of these metals, as well as tungsten, molybdenum, and the various sintered products of silver and the refractory metals, can be used to produce steel-back contacts. Steel-back contacts have been made in the form of screws, rivets, or buttons for projection welding.

Powder metallurgy materials are available with final densities up to 99% of theoretical and with high-conductivity surfaces as inserts or overlays. Both P/M and wrought materials may be attached to appropriate carriers by brazing, welding, or diffusion bonding, even though they contain cadmium oxide. For percussion welding of Ag-CdO contacts, backing is not required. For resistance welding, a fine silver backing is needed.

P/M contacts are available with a silver matrix for air and oil applications, and with a copper matrix for oil applications only. The second phase may be tungsten, molybdenum, graphite, tungsten carbide, cadmium oxide, or zinc oxide.

Materials with a high content of refractory metal (50% or more) are usually made by infiltration. In this process, the refractory metal powder is pressed into the desired size and shape with a controlled porosity. The compact is sintered at high temperature in a reducing atmosphere, and then molten silver or copper is infiltrated into the porous sintered compact.

Compositions of a lower refractory content are made by blending powders, pressing them to a desired size and shape, sintering the pressed compact at a high temperature, and re-pressing to size the parts and to increase the density of the compact.

Although parts usually are molded to final shape, they are sometimes finished by machining or grinding to obtain special shapes or unusually close tolerances. When only a few parts are needed, they may be machined from bars to save the cost of expensive dies for pressing a P/M compact.

Sintered materials are available in sizes ranging from rectangles or disks about 0.8 mm (1/32 in.) thick by 3 mm (1/8 in.) square or 3 mm (1/8 in.) in diameter to bars 200 mm (8 in.) long. Most are available in widths up to 75 mm (3 in.) and thicknesses up to 12.5 mm (1/2 in.). For some materials, these dimensions may be exceeded.

In small sizes (up to about 12.5 mm, or 1/2 in., square), the higher refractory compositions are frequently made with serrated surfaces coated with excess silver or silver-base brazing alloy so that they can be attached easily to a backing by welding or by resistance brazing in a welding machine. Larger sizes are generally attached to backing by silver-alloy brazing. The materials are often pre-tinned with silver-base brazing alloy in a controlled atmosphere for good wetting of the contact material.

Many of the low-refractory contact materials are fabricated as disks, rectangles, and special-contour facings that also are attached to backing by silver-alloy brazing. Most materials with 90% or more silver are ductile enough to allow fabrication as rivets, and to allow assembly by staking or spinning. Among silver-graphite materials, those containing more than 0.5% graphite cannot be satisfactorily cold headed into rivets.

Ag-CdO mixtures are available as round, rectangular, or special-shape facings, and also as rivets.

Cost

Copper is the least expensive elemental material used as electrical contacts, followed by silver, palladium, gold, and platinum in that order. Copper is the most frequently used backing material to reduce cost. Sometimes nickel- or steel-backed contacts are used. Silver, gold, platinum, palladium, and their alloys all have been used to produce nickel-steel-back contacts in the form of screws, rivets, or buttons for projection welding. Nickel will not rust during fabrication or use.

ACKNOWLEDGMENT

The editors would like to express their thanks to Hendrick Slaats of Engelhard Corporation and Ernest M. Jost of Chemet Corporation for their helpful suggestions during the review of this article.

Thermocouple Materials

T.P. Wang, Thermo Electric Company, Inc.

ACCURATE MEASUREMENT of temperature is one of the most common and vital requirements in science, engineering, and industry. Measurement of temperature is generally thought to be one of the simplest and most accurate measurements that can be made. This is a misconception. Unless proper techniques are employed, highly inaccurate readings can occur, and either useless data can be generated or materials can be misprocessed. Also, under certain conditions, it may be difficult or impossible to obtain accurate temperature measurements regardless of whether proper techniques are employed.

Nine types of instruments, under appropriate conditions and within specific operating ranges, may be used for measurement of temperature: thermocouple thermometers, radiation thermometers, resistance thermometers, liquid-in-glass thermometers, filled-system thermometers, gas thermometers, optical fiber thermometers, Johnson noise thermometers, and bimetal thermometers. The success of any temperature-measuring system depends not only on the capacity of the system but also on how well the user understands the principles, advantages, and limitations of its application.

The thermocouple thermometer is by far the most widely used device for measurement of temperature. Its favorable characteristics include good accuracy, suitability over a wide temperature range, fast thermal response, ruggedness, high reliability, low cost, and great versatility of application.

Essentially, a thermocouple thermometer is a system consisting of a temperature-sensing element called a thermocouple, which produces an electromotive force (emf) that varies with temperature, a device for sensing emf, which may include a printed scale for converting emf to equivalent temperature units, and an electrical conductor (extension wires) for connecting the thermocouple to the sensing device. Although any combination of two dissimilar metals and/or alloys will generate a thermal emf, only seven thermocouples are in common industrial use today. These seven have been chosen on the basis of such factors as mechanical and chemical properties, stability of emf, reproducibility, and cost. They will be discussed individually after a review of the principles and practice of measuring temperature by use of thermocouples.

Principles of Thermocouple Thermometers

The principle on which thermocouples depend was discovered in 1821 by Seebeck, who found that when two dissimilar metals are joined in a closed circuit an electromotive force is generated if the two junctions are maintained at different temperatures. This *thermal emf* induces an electric current to flow continuously through the circuit and is termed *Seebeck emf* in honor of its discoverer.

Figure 1(a) is a schematic diagram of two electrical conductors, A and B, whose two junctions are exposed to different temperatures, T_1 and T_2. The thermal emf generated in this circuit, E_{AB}, is expressed:

$$E_{AB} = f\,[A, B, (T_2 - T_1)] \qquad \text{(Eq 1)}$$

The thermal emf E_{AB} is a vector quantity. Its magnitude and direction depend on the material characteristics of A and B as well as the temperature difference between the hot and cold junctions, $T_2 - T_1$, providing that A and B are homogeneous in composition.

A circuit diagram describing a thermocouple in a voltmeter circuit is shown in Fig. 1(b). Thermoelement A is represented by a battery E_A and resistance R_A. E_A is the emf output of thermoelement A with reference to a certain standard, and R_A is the resistance of thermoelement A. Similarly, E_B is the emf output of thermoelement B with reference to the same standard and R_B is the resistance of thermoelement B.

The voltage drop between terminals A and B is given by the following equation:

$$V_{AB} = E_A - E_B - I(R_A + R_B + R_S) \qquad \text{(Eq 2)}$$

where R_S is the resistance of a large resistor in series with the thermoelements to minimize the effect of the resistance of the thermoelements. If $E_A - (E_A + V_{AB})$ is positive, the thermoelectric current I will flow continuously from A to B at the cold junction. In this case, A is termed the *positive thermoelement* and B the *negative thermoelement* of the thermocouple.

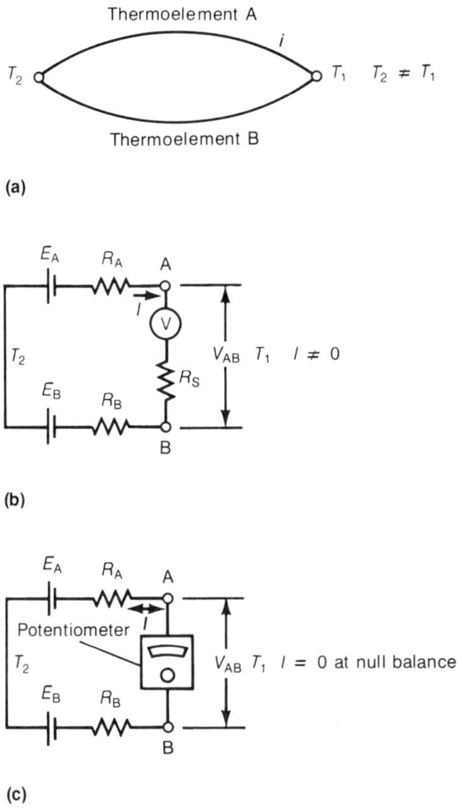

Fig. 1 Schematic (a) and equivalent circuit diagrams (b) and (c) of a typical thermocouple system. (b) Thermocouple in a voltmeter circuit. (c) Thermocouple in a potentiometer circuit. Adapted from Ref 1

In Fig. 1(c), a potentiometer is connected across the terminals in place of the voltmeter. A bucking voltage is applied at the potentiometer until it is equal in magnitude and opposite in direction to the thermoelectric voltage E_{AB}. At null balance, there is no current flow. All the *IR* terms in Eq 2 become zero. Under this condition:

$$V_{AB} = E_{AB} = E_A - E_B \qquad \text{(Eq 3)}$$

Then, the measured emf at the potentiometer V_{AB} is the thermal emf of the thermo-

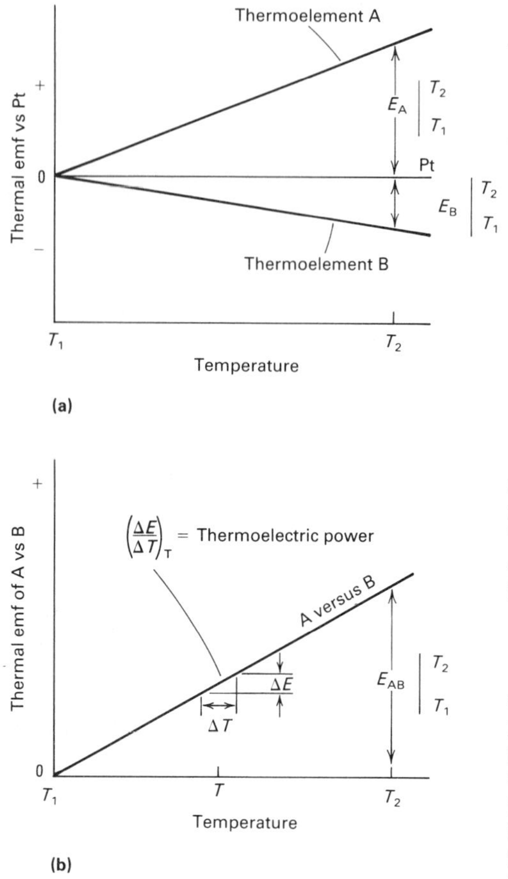

Fig. 2 Thermal emf of two thermoelements with respect to platinum and to each other. (a) Thermal emf of thermoelements A and B versus Pt. (b) Thermal emf of thermocouple A—B

couple AB. It may be observed from Eq 3 that thermal emf is a bulk property. It is independent of the resistance, and hence of the diameter, of the wire.

The thermoelectric property of an electrical conductor is usually expressed in millivolts against a common reference. Platinum 67, because of its purity and excellent oxidation resistance at elevated temperatures, is the United States thermometric reference standard. This standard can be obtained from the National Institute of Standards and Technology (formerly NBS).

The change in emf with temperature for two electrical conductors A and B, which are, respectively, positive and negative with reference to platinum, is shown in Fig. 2(a). The emf values of both conductors change linearly with temperature. If A and B are joined as a couple and the two ends are exposed to temperatures T_1 and T_2, the emf of the couple AB at temperatures between T_1 and T_2 is equal to the algebraic difference in emf between the positive thermoelement A versus Pt and the negative thermoelement B versus Pt at temperatures in the same range. This is shown in Fig. 2(b) and expressed in the equation:

$$E_{AB}\Big|_{T_1}^{T_2} = E_A\Big|_{T_1}^{T_2} - E_B\Big|_{T_1}^{T_2} \qquad \text{(Eq 4)}$$

where $E_{AB}\Big|_{T_1}^{T_2}$ is the emf of thermocouple AB between T_1 and T_2, in millivolts,

$E_A\Big|_{T_1}^{T_2}$ is the emf of thermoelement A vs Pt between T_1 and T_2, in millivolts, and

$E_B\Big|_{T_1}^{T_2}$ is the emf of thermoelement B vs Pt between T_1 and T_2, in millivolts.

Thermoelectric power at a given temperature T is defined as the rate of change of thermal emf with respect to temperature. The thermoelectric power of the thermocouple AB at temperature T is the slope of its emf/temperature curve,

$$\frac{\Delta E}{\Delta T}$$

as shown in Fig. 2(b). The thermoelectric power of a thermoelement at a given temperature T is the rate of change of its emf, referenced to Pt, with respect to temperature. In using a thermocouple for temperature measurement, it is essential that the thermoelectric power of the thermocouple be fairly large and uniform within the applicable temperature range.

In case the emf/temperature relationship of the thermocouple AB is well established we may determine the temperature difference between T_2 and T_1 by measuring with a potentiometer the generated thermal emf E_{AB}. Note that a thermocouple really does not measure the temperature of the hot junction T_2 but measures the temperature difference $T_2 - T_1$.

The ice point (0 °C, or 32 °F) is set universally as the reference cold junction (T_1) for all established emf tables. If we use the ice point as our reference cold junction T_1, then the measured emf does correspond to the temperature of the hot junction T_2.

Measurement of Temperature by a Thermocouple. A setup for measurement of temperature by use of a thermocouple is illustrated schematically in Fig. 3. The welded junction of thermocouple PN is inserted into an electric furnace, the temperature of which is to be measured. The ice-point cold junction is provided by two mercury U-tubes embedded in a Dewar flask packed with shaved ice. The legs of the thermocouple are inserted into the mercury U-tubes and connected to the positive and negative terminals of a potentiometer by insulated copper wires. The temperature of the furnace then can be obtained by measuring the emf generated by the thermocouple and referenced to the established emf table for that particular thermocouple. Commercially

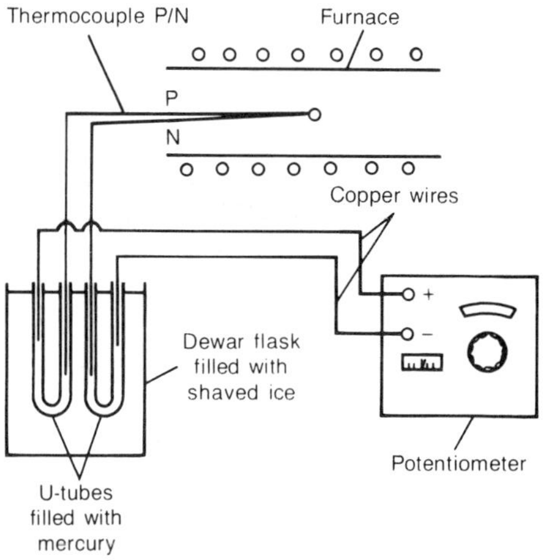

Fig. 3 Schematic diagram of the experimental setup for measuring temperature using a thermocouple and an ice-point reference junction. Adapted from Ref 2

available automatic compensating cold junctions can be used in place of the abovementioned mercury U-tubes to achieve a 0 °C reference junction. These may be built into an indicating or recording instrument used to measure the emf developed by the thermocouple or external of the measuring instrument.

In the absence of an ice junction, the thermocouple wires may be connected directly to the terminals of the potentiometer. The ambient temperature of the terminals is measured by a thermometer and converted to emf in millivolts from the emf table. The total emf generated by the thermocouple between the hot junction and the ice point is the sum of the emf thus measured by the potentiometer and this ambient-temperature correction factor. The temperature of the hot junction can be obtained by referring this total emf to the established table.

For additional information, see ASTM E 563, "Standard Recommended Practice for Preparation and Use of Freezing Point Reference Baths."

Preparation of the Measuring Junction. The two dissimilar thermoelements must be joined at the temperature-measuring junction to form the thermocouple. The joint must have good thermal and electrical conductivity without adversely affecting the mechanical and electrical properties of the thermocouple wires at this joint.

Prior to being joined, the thermoelements are straightened to facilitate insertion into hard-fired ceramic insulators. In this operation, care should be taken to avoid excessive cold working of the wires, which has a deleterious effect on the emf of the couple. After being cut to the desired length, the thermocouple wires are cleaned carefully (to remove lubricant residue, fingerprints, and other contaminants) with a suitable solvent such as methyl ethyl ketone, Freon TF, or isopropyl alcohol, prior to joining.

Table 1 Properties of standard thermocouples

Type	Thermoelements	Base composition	Melting point, °C	Resistivity, nΩ · m	Recommended service	Max temperature °C	Max temperature °F
J	JP	Fe	1450	100	Oxidizing or reducing	760	1400
	JN	44Ni-55Cu	1210	500			
K	KP	90Ni-9Cr	1350	700	Oxidizing	1260	2300
	KN	94Ni-Al, Mn, Fe, Si, Co	1400	320			
N	NP	84Ni-14Cr-1.4Si	1410	930	Oxidizing	1260	2300
	NN	95Ni-4.4Si-0.15Mg	1400	370			
T	TP	OFHC Cu	1083	17	Oxidizing or reducing	370	700
	TN	44Ni-55Cu	1210	500			
E	EP	90Ni-9Cr	1350	700	Oxidizing	870	1600
	EN	44Ni-55Cu	1210	500			
R	RP	87Pt-13Rh	1860	196	Oxidizing or inert	1480	2700
	RN	Pt	1769	104			
S	SP	90Pt-10Rh	1850	189	Oxidizing or inert	1480	2700
	SN	Pt	1769	104			
B	BP	70Pt-30Rh	1927	190	Oxidizing, vacuum or inert	1700	3100
	BN	94Pt-6Rh	1826	175			

For applications below about 500 °C (about 1000 °F), base-metal thermocouple wires may be silver brazed using borax as flux. Above this temperature, thermocouple junctions usually are prepared by welding. Noble-metal thermocouples should always be joined by welding. Thermocouples are usually welded using gas, electric-arc, resistance, tungsten-inert gas, and plasma-arc processes. In gas welding, a neutral flame is required (preferably oxidizing for noble metals). Prior to gas or arc welding, the ends of types E, J, K, and T thermocouple wires are first twisted one and a half turns.

Effecting a hot junction in a sheathed thermocouple requires a higher degree of skill, special equipment, and considerable care. After the sheath has been stripped away, joining usually is done by gas-tungsten-arc or plasma-arc welding. A clean, dry, and well-lighted work area is required to produce a finished element of good integrity. An oven capable of continuous operation at 90 °C (200 °F) should be available for storage of unsealed sheathed thermocouples during unavoidable delays in forming of junctions. Use of such an oven will minimize a pickup of airborne moisture and other contaminants.

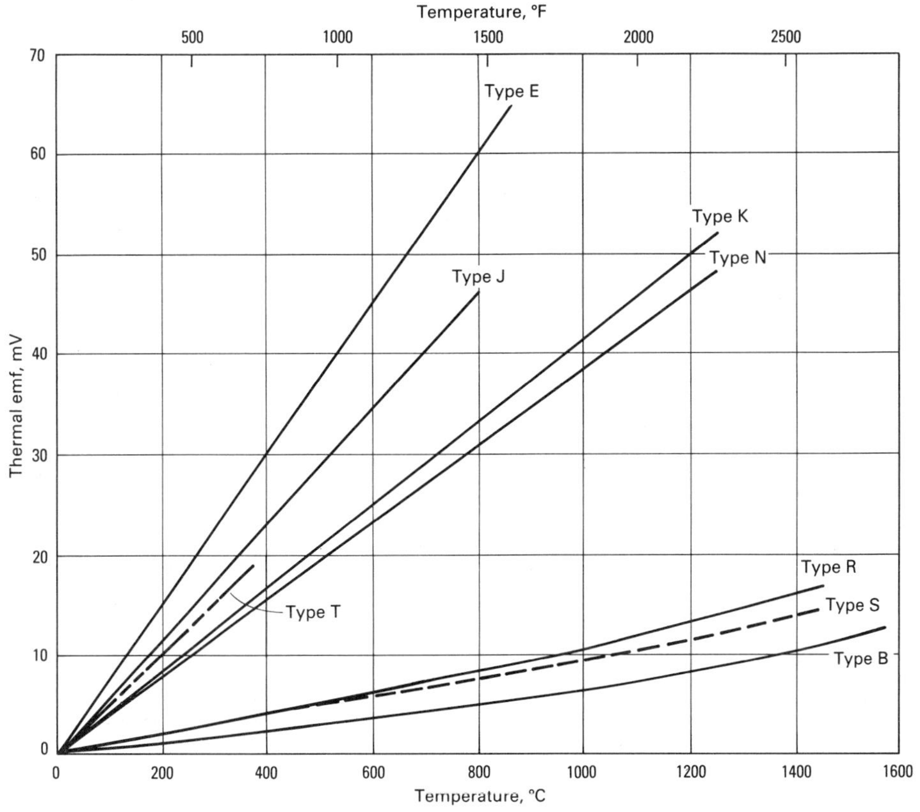

Fig. 5 Thermal emf curves for ISA standard thermocouples. Thermal emf plots are based on IPTS-68 (1974).

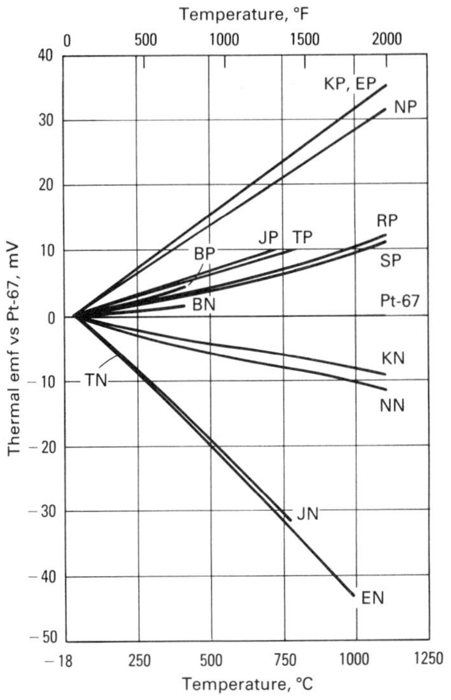

Fig. 4 Thermal emf of standard thermoelements. Adapted from Ref 1

Thermocouple Materials

Commercially available thermocouples are grouped according to material characteristics (base metal or noble metal) and standardization. At present, five base-metal thermocouples and three noble-metal thermocouples have been standardized and given letter designations by ANSI (American National Standards Institute), ASTM (American Society for Testing and Materials), and ISA (Instrument Society of America). Among the remaining thermocouples in use, some have not been assigned letter designations because of limited usage, and some are being considered for standardization.

Standard Thermocouples

The base compositions, melting points, and electrical resistivities of the individual thermoelements of the seven standard thermocouples are presented in Table 1. Maximum operating temperatures and limiting factors in environmental conditions are listed also.

The relationships between emf and temperature for the individual thermoelements with reference to Platinum 67 and for the seven standard thermocouples are shown in Fig. 4 and 5, respectively. Tolerances for initial calibration of standard thermocouples (those meeting established tables within a specified tolerance) are listed in Table 2.

Type J. The type J thermocouple is widely used, primarily because of its versatility and low cost. In this couple, the positive ther-

Table 2 Initial calibration tolerances for thermocouples when the reference junction is at 0 °C (32 °F)

Thermocouple type	Temperature range, °C	Initial calibration tolerance: Standard (whichever is greater)	Special (whichever is greater)
T	0 to 350	±1 °C or ±0.75%	±0.5 °C or 0.4%
J	0 to 750	±2.2 °C or ±0.75%	±1.1 °C or 0.4%
E	0 to 900	±1.7 °C or ±0.5%	±1 °C or ±0.4%
K	0 to 1250	±2.2 °C or ±0.75%	±1.1 °C or ±0.4%
N	0 to 1250	±2.2 °C or ±0.75%	±1.1 °C or ±0.4%
R or S	0 to 1450	±1.5 °C or ±0.25%	±0.6 °C or ±0.1%
B	800 to 1700	±0.5%	...
T(a)	−200 to 0 °C	±1 °C or ±1.5%	(b)
E(a)	−200 to 0 °C	±1.7 °C or ±1%	(b)
K(a)	−200 to 0 °C	±2.2 °C or ±2%	(b)

(a) Thermocouples and thermocouple materials are normally supplied to meet the limits of error specified in the table for temperatures above 0 °C. The same materials, however, may not fall within the subzero limits of error given in the second section of the table. If materials are required to meet the subzero limits, the purchase order must so state. Selection of materials usually will be required. (b) Little information is available to justify establishment of special limits of error for subzero temperatures. Limited experience suggests the following limits for types E and T thermocouples: Type E, −200 to 0 °C ± 1 °C or ±0.5%; Type T, 200 to 0 °C ± 0.5 °C or ±0.8%. These limits are given only as a guide for discussion between purchaser and supplier. Due to the characteristics of the materials, subzero limits of error for type J thermocouples and special subzero limits for type K thermocouples are not listed. Source: Adapted from ANSI MC96.1

moelement is iron and the negative thermoelement is constantan, a 44Ni-55Cu alloy. As shown in Fig. 4, the emf of iron is positive with reference to platinum, but the emf of constantan is the most negative with respect to platinum among all thermoelements. The thermoelectric power of the type J couple as a whole is about 55 μV/°C (30 μV/°F) over the temperature range from 0 to 750 °C (32 to 1380 °F), a value higher than that of any other standard couple except the type E couple.

The commercial grade of iron used as the positive leg of the type J thermocouple contains small amounts of carbon, cobalt, manganese, and silicon. Therefore, its emf can vary significantly from one heat to another. Accordingly, the emf of this iron is shown in Fig. 4 as a shaded band instead of a single line. However, constantan of a slightly different composition can be obtained commercially to match the iron wire on hand so that the thermocouple as a whole meets established emf/temperature requirements. As shown in Table 2, the initial calibration tolerance (to established table values) for standard-grade type J couples is ±2.2 °C (±4 °F) or ±3/4% of temperature, whichever is greater over the range from 0 to 750 °C (32 to 1380 °F).

Type J couples can be used in both oxidizing and reducing atmospheres at temperatures up to about 760 °C (1400 °F). They find extensive use in heat-treating applications in which they are exposed directly to the furnace atmosphere. The No. 8 gage (3.25 mm, or 0.128 in., diam) type J couple can have a useful service life of about 1000 h at 760 °C (1400 °F) in an oxidizing atmosphere. The same size couple can be used at higher temperatures in a reducing atmosphere. Smaller-gage type J couples are also available for laboratory use or applications where quicker heat-sensing response is desired. As thermocouple wire diameter decreases, the recommended upper temperature limit also decreases, as shown in Table 3.

Type K thermocouples, like type J couples, are also widely used in industrial applications. The positive thermoelement is a 90Ni-9Cr alloy; its thermal emf versus platinum is the most positive. The negative thermoelement is a 94% Ni alloy containing silicon, manganese, aluminum, iron, and cobalt as alloying constituents; its emf is negative with respect to platinum. The thermoelectric power of the type K couple is close to 40 μV/°C (22 μV/°F) over the extended temperature range from 0 to 1100 °C (32 to 2000 °F); at temperatures up to 1250 °C (2300 °F), it is still close to 35 μV/°C (19 μV/°F). Commercial type K couples are Chromel-Alumel (trademark of Hoskins Manufacturing Company), T_1-T_2 (trademarks of Harrison Alloys, Inc.), and Tophel-Nial (trademark of Carpenter Technology Corporation).

Type K thermocouples can be used up to 1250 °C (2280 °F) in oxidizing atmospheres. As shown in Table 2, initial calibration of standard-grade type K couples should be within ±2.2 °C (±4 °F) or ±3/4% of table values (whichever is greater) up to 1260 °C (2300 °F). The maximum operating temperature of No. 8 gage wire (3.25 mm, or 0.128 in., diam) is 1260 °C (2300 °F). For smaller-diameter wire, recommended upper temperature limits are lower, as shown in Table 3.

Type K couples should not be used in elevated-temperature service in reducing atmospheres or in environments containing sulfur, hydrogen, or carbon monoxide. At elevated temperatures in oxidizing atmospheres, uniform oxidation takes place, and the oxide formed on the surface of the positive (90Ni-10Cr) thermoelement is a spinel, $NiO\text{-}Cr_2O_3$. However, in reducing atmospheres, preferential oxidation of chromium takes place, forming only Cr_2O_3. The presence of this greenish oxide (commonly known as "green rot") depletes the chromium content, causing a very large negative shift of emf (up to −2 MV) and rapid deterioration of the thermoelement.

The type K positive thermoelement is susceptible to short-range ordering (changes from a random to an ordered atomic structure in localized regions) on aging at about 500 °C (930 °F). This causes a change in emf of about 0.2 MV, which is equivalent to a change of 5 °C (9 °F). This change can be essentially eliminated by preaging the thermoelement at 500 °C (930 °F). However, the short-range ordering is a reversible process. The type K positive thermoelement will change back to its initial disordered condition if heated to 800 °C (1470 °F) or above, followed by rapid cooling.

Besides the above considerations, type K couples are quite versatile. They are the only standard base-metal thermocouples that can be used for sensing temperatures from 900 to 1260 °C (1650 to 2300 °F).

Type T thermocouples are used extensively for cryogenic measurements. The positive thermoelement is copper, the thermal emf of which is positive with respect to

Table 3 Recommended upper temperature limits for protected thermocouples of various wire sizes

Type of thermocouple	No. 8 AWG, 3.25 mm (0.128 in.) °C	°F	No. 14 AWG, 1.63 mm (0.064 in.) °C	°F	No. 20 AWG, 0.81 mm (0.032 in.) °C	°F	No. 24 AWG, 0.51 mm (0.020 in.) °C	°F	No. 28 AWG, 0.33 mm (0.013 in.) °C	°F
T	...	...	370	700	260	500	200	400	200	400
J	760	1400	590	1100	480	900	370	700	370	700
E	870	1600	650	1200	540	1000	430	800	430	800
K	1260	2300	1090	2000	980	1800	870	1600	870	1600
N	1260	2300	1090	2000	980	1800	870	1600	870	1600
R, S	...	...	...	...	...	...	1480	2700	...	...
B	...	...	...	...	...	...	1700	3100	...	...

Upper temperature limit. Source: Adapted from ANSI MC96.1

platinum. Commercial-grade oxygen-free high conductivity (OFHC) copper (C10100), unlike commercial grades of iron, is of high purity and is very homogeneous. Its emf is quite uniform from lot to lot. The constantan (44Ni-55Cu) used for the negative thermoelement of the type T couple has the same base composition as that of the constantan used in the type J couple, but is slightly different in minor alloying constituents. Because of this, the two types of constantan have different emf characteristics. As manufactured, the copper-constantan couple has an emf output that conforms to the emf table for type T thermocouples to within ±1 °C (±2 °F) at temperatures from 0 to 350 °C (32 to 660 °F). From −200 to 0 °C (−330 to 32 °F), the tolerance on the initial calibration is ±1% of temperature. Type T couples can be used in either oxidizing or reducing atmospheres. They should not be used above 370 °C (700 °F) because of the poor oxidation resistance of copper.

Type E. The positive thermoelement of the type E thermocouple is 90Ni-9Cr, the same as that of the type K thermocouple; the negative element is 44Ni-55Cu, the same as that of the type T couple. Among all standard thermoelements, these two are the most positive and most negative with respect to Platinum 67. Therefore, the thermoelectric power of the type E couple is the highest among all standard couples. Type E couples are used primarily for power-generation applications such as thermopiles.

The recommended maximum operating temperature for type E thermocouples is 870 °C (1600 °F). Like type K thermocouples, type E couples should be used only in oxidizing atmospheres, because their use in reducing atmospheres results in preferential oxidation of chromium (green rot).

Type N. The Nicrosil/Nisil thermocouple was developed for oxidation resistance and emf stability superior to those of type K thermocouples at elevated temperatures. The positive thermoelement is Nicrosil (nominal composition, 14 Cr, 1.4 Si, 0.1 Mg, bal Ni), and the negative thermoelement is Nisil (nominal composition, 4.4 Si, 0.1 Mg, bal Ni). Reference 3 presents emf values and various physical properties of Nicrosil/Nisil thermocouples. These couples have been shown (Ref 4-6) to have longer life and better emf stability than type K thermocouples at elevated temperatures in air, both in the laboratory and in several industrial applications (see Fig. 6). The Nicrosil/Nisil thermocouple was standardized as the type N thermocouple by ASTM. It was established as a standard thermocouple material by ISA and other technical societies in the late 1980s.

Type S. The type S thermocouple served as the interpolating instrument for defining the International Practical Temperature Scale of 1968 (amended in 1975) from the freezing point of antimony (630.74 °C, or

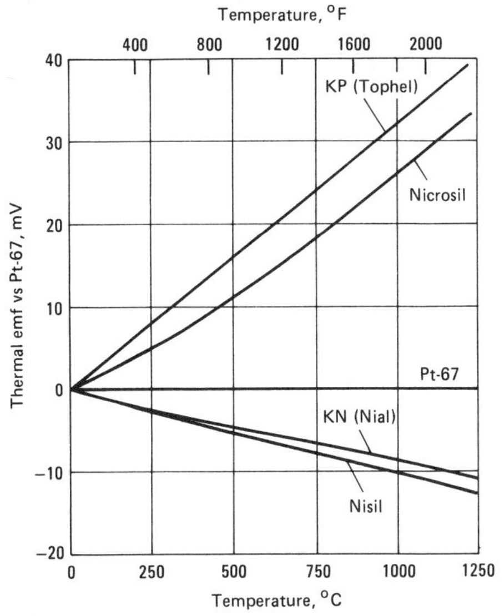

Fig. 6 Thermal emf of Nicrosil, Nisil, and type K thermoelements versus Pt-67. Adapted from Ref 7

1167.33 °F) to the freezing point of gold (1064.43 °C, or 1947.97 °F). It is characterized by a high degree of chemical inertness and stability at high temperatures in oxidizing atmospheres. The materials used in the legs of this thermocouple, Pt-10Rh and platinum, both are ductile and can be drawn into fine wire (as small as 0.025 mm, or 0.001 in., in diameter for special applications). For general-purpose use, wire 0.51 mm (0.020 in.) in diameter is commonly used.

The thermoelectric output of the type S thermocouple is about 6 μV/°C (3.3 μV/°F) at temperatures from 0 to 100 °C (32 to 212 °F) and about 11.5 μV/°C (6.4 μV/°F) at 1000 °C (1830 °F). Type S couples can be used in intermittent service up to 1750 °C (3180 °F) (the melting point of platinum is 1769 °C, or 3216 °F) and can be used continuously up to 1500 °C (2730 °F) if properly protected. Because of its low emf output, this thermocouple is not used for measuring subzero temperatures. Type S couples that match standard emf/temperature values within ±1.5 °C (±2.7 °F) or ±0.25% (whichever is greater), and under special conditions within ±0.6 °C (±1 °F) or ±0.1%, can be obtained from reliable sources.

The type S couple is widely used in industrial laboratories as a standard for calibration of base-metal thermocouples and other temperature-sensing instruments. It is commonly used for controlling processing of steel, glass, and many refractory materials. It should be used in air or in oxidizing or inert atmospheres. It should not be used unprotected in reducing atmospheres in the presence of easily reduced oxides, atmospheres containing metallic vapors such as lead or zinc, or atmospheres containing nonmetallic vapors such as arsenic, phosphorus, or sulfur. It should not be inserted directly into metallic protection tubes and is not recommended for service in vacuum at high temperatures except for short periods of time. Because the negative leg of this couple is fabricated from high-purity platinum (approximately 99.99% for commercial couples and 99.995%+ for special grades), special care should be taken to protect the couple from contamination by the insulators used as well as by the operating environment.

Type R. The type R thermocouple (Pt-13Rh/Pt) has characteristics similar to those of the type S couple. In 1922 it was found that the British Pt-10Rh alloy had a higher emf than the U.S. version but was unstable due to the presence of 0.34% Fe. In order to produce an alloy free from iron (and therefore stable) but with an emf that met the calibration of existing instruments (in other words, having an emf output equivalent to that of the couple using impure Pt-10Rh element), it was necessary to increase the rhodium content to 13%. The emf output of the type R thermocouple is slightly higher than that of the type S couple. End-use applications for type S couples also apply to type R.

Type B thermocouples (Pt-30Rh/Pt-6Rh) may be used in still air or inert atmospheres for extended periods at temperatures up to 1700 °C (3100 °F) and intermittently up to 1760 °C (3200 °F) (Pt-6Rh leg melts at approximately 1826 °C, or 3319 °F). Because both of its legs are platinum-rhodium alloys, the type B couple is less sensitive than type R or type S to pickup of trace impurities from insulators or from the operating environment. Under corresponding conditions of temperature and environment, type B thermocouples exhibit less grain growth and less drift in calibration than type R or type S thermocouples.

The type B couple also is suitable for short-term use in vacuum at temperatures up to about 1700 °C (3100 °F); its emf stability varies with temperature, time at temperature, and degree of vacuum. It should not be used in reducing atmospheres, or in those containing metallic or nonmetallic vapors, unless suitably protected with ceramic protection tubes. It should never be inserted directly into a metallic primary protection tube.

This couple has a very small emf in the normal reference range from 0 to about 100 °C (32 to about 212 °F), and particularly from 0 to 50 °C (32 to 120 °F). Errors arising because of uncertainties in the temperature of the reference junction, or as a result of that temperature being ignored, are relatively small for measurements of high temperature (over 1000 °C, or 1830 °F). The thermoelectric power of the type B thermocouple is 9.1 μV/°C (5.1 μV/°F) at 1000 °C (1830 °F) and 11.3 μV/°C (6.3 μV/°F) at 1400 °C (2550 °F).

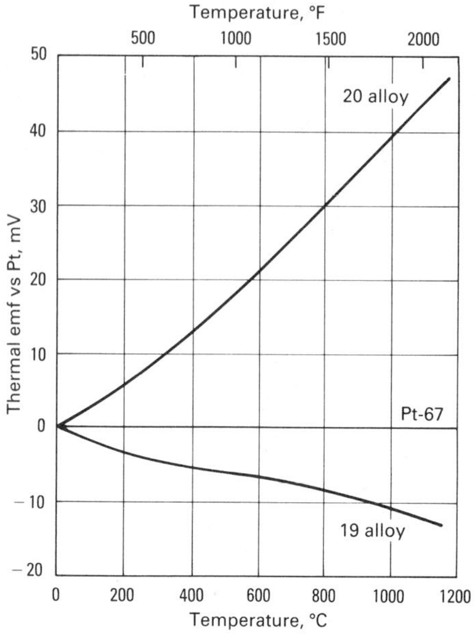

Fig. 7 Thermal emf of 19 alloy and 20 alloy versus Pt-67

Table 4 Properties of two thermocouples: 19 alloy/20 alloy and Nicrosil-Nisil

Property	19 alloy	20 alloy	Nicrosil	Nisil
Nominal composition	Ni-1Co	Ni-18Mo	Ni-14Cr-1.4Si	Ni-4.4Si-0.1Mg
Melting point, °C (°F)	1450 (2640)	1425 (2600)	1410 (2570)	1400 (2550)
Specific gravity	8.9	9.1	8.52	8.70
Thermal conductivity, W/m · K at 20 °C	50	15	130	230
Coefficient of thermal expansion, μm/m · °C (20 to 100 °C)	13.6	11.9	13.3	12.1
Magnetic susceptibility	Magnetic	Magnetic	Nonmagnetic	Nonmagnetic
Resistivity, nΩ · m at 20 °C	80	1650	930	370
Temperature coefficient of resistance, μΩ/Ω · °C (20 to 100 °C)	3050	290	100	900
Tensile strength, MPa (ksi)	415 (60)	895 (130)	760 (110)	655 (95)
Yield strength, MPa (ksi)	170 (25)	515 (75)	415 (60)	380 (55)
Elongation, %	35	35	30	35

Nonstandard Thermocouples

19 Alloy/20 Alloy. The 19 alloy/20 alloy thermocouple was developed for temperature-sensing and control applications at elevated temperatures in hydrogen or in reducing atmospheres. The positive thermoelement is the 20 alloy, which has a nominal composition of 82Ni-18Mo. The negative thermoelement is the 19 alloy, the nominal composition of which is 99Ni-1Co. Values of emf versus platinum for 19 alloy and 20 alloy are given in Fig. 7. The emf of the 19 alloy/20 alloy thermocouple is somewhat larger than that of the type K couple. Physical, electrical, and mechanical properties of the 19 alloy/20 alloy thermocouple are listed in Table 4.

19 alloy/20 alloy thermocouples can be used in hydrogen or in reducing atmospheres over the entire range from 0 to 1260 °C (32 to 2300 °F) with excellent performance. Hotchkiss (Ref 8) showed that, after exposure at about 950 °C (1750 °F), a type K couple was out of calibration by about 2 mV (about 50 °C, or 90 °F) in the negative direction whereas a 19 alloy/20 alloy thermocouple remained essentially in calibration. The oxidation resistance of 19 alloy/20 alloy is not good when compared with that of the type K couple. 19 alloy/20 alloy thermocouples should not be used in oxidizing atmospheres above about 650 °C (1200 °F).

Iridium-Rhodium. Three iridium-rhodium thermocouples are commercially available: 60Ir-40Rh/Ir, 50Ir-50Rh/Ir, and 40Ir-60Rh/Ir. Of these three combinations, 60Ir-40Rh/Ir appears to be preferred at this time. Properties of iridium-rhodium couples are given in Table 5 and Fig. 8.

Iridium-rhodium couples are suitable for use for limited periods of time in air or other oxygen-carrying atmospheres at temperatures up to about 2000 °C (3600 °F), and generally are used for such service at temperatures above the range in which types R, S, and B thermocouples are employed. They can be used in inert atmospheres and in vacuum, but not in reducing atmospheres (easily reduced oxides in contact with iridium or with Ir-Rh alloys are sources of contamination). These couples have been used for short periods of time at temperatures up to only 60 °C (110 °F) below the melting point of the alloy leg—that is, up to 2180 °C (3960 °F) for 60Ir-40Rh, up to 2140 °C (3880 °F) for 50Ir-50Rh, and up to 2090 °C (3790 °F) for 40Ir-60Rh.

After being hot worked, iridium thermoelements have a fibrous structure and are reasonably ductile. However, annealing causes the structure to become equiaxed, with resultant decreases in ductility and handleability. This should be considered if any preinstallation fabrication is anticipated.

Compensating extension wires are available for iridium-rhodium thermocouples—copper for the positive leg and stainless steel for the negative leg.

Platinum-Molybdenum. The Pt-5Mo/Pt-0.1Mo thermocouple is used for measuring temperatures from 1100 to about 1500 °C (2000 to about 2700 °F) under neutron radiation (type K couples are employed for temperatures up to 1100 °C). Basic characteristics of Pt-5Mo/Pt-0.1Mo couples are given in Table 6. Platinum alloys containing

Table 5 Properties of iridium-rhodium thermocouples

Property	60Ir-40Rh versus Ir	50Ir-50Rh versus Ir	40Ir-60Rh versus Ir
Nominal operating temperature range, °C (°F), in:			
Wet hydrogen	(a)	(a)	(a)
Dry hydrogen	(a)	(a)	(a)
Inert atmosphere	2100 (3812)	2050 (3722)	2000 (3632)
Oxidizing atmosphere	(a)	(a)	(a)
Vacuum	2100 (3812)	2050 (3722)	2000 (3632)
Approximate microvolts per °C (per °F)			
Mean over nominal operating range	5.3 (2.9)	5.7 (3.2)	5.2 (2.9)
At top temperature of normal range	5.6 (3.1)	6.2 (3.5)	5.0 (2.8)
Melting temperature, °C (°F), nominal:			
Positive thermoelement	2250 (4082)	2202 (3996)	2153 (3907)
Negative thermoelement	2443 (4429)	2443 (4429)	2443 (4429)
Stability with thermal cycling	Fair	Fair	Fair
Ductility (of more brittle thermoelement) after use	Poor	Poor	Poor

(a) Not recommended. Source: Adapted from Ref 9

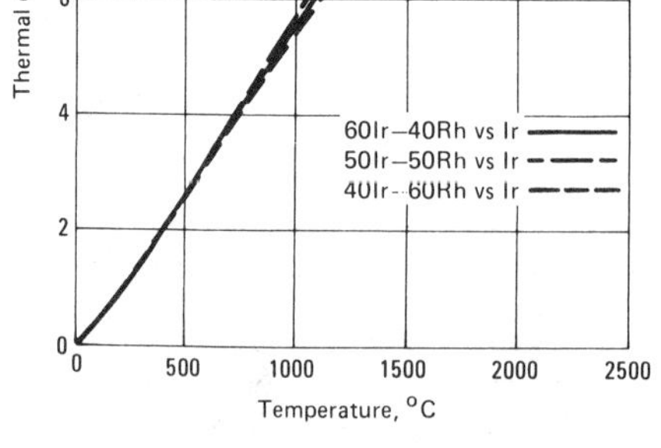

Fig. 8 Thermal emf of iridium-rhodium/iridium thermocouples. Adapted from Ref 9

Table 6 Properties of platinum molybdenum thermocouples

Property	Pt-5Mo versus Pt-0.1Mo
Nominal operating temperature range, °C (°F), in:	
Reducing atmosphere (nonhydrogen)	(a)
Wet hydrogen	(a)
Dry hydrogen	(a)
Inert atmosphere (helium)	1400 (2552)
Oxidizing atmosphere	(a)
Vacuum	(a)
Maximum short-time temperature, °C (°F)	1550 (2822)
Approximate microvolts per °C (°F):	
Mean, over nominal operating range	29 (16)
At top temperature of normal range	30 (17)
Melting temperature, °C (°F), nominal:	
Positive thermoelement	1788 (3250)
Negative thermoelement	1770 (3218)
Stability with thermal cycling	Good
High-temperature tensile properties	Fair
Stability under mechanical working	Good
Ductility (of most brittle thermoelement) after use	Good
Resistance to handling contamination	Fair
Recommended extension wire 70 °C (158 °F) maximum:	
Positive conductor	Cu
Negative conductor	Cu-1.6Ni

(a) Not recommended. Source: Adapted from Ref 9

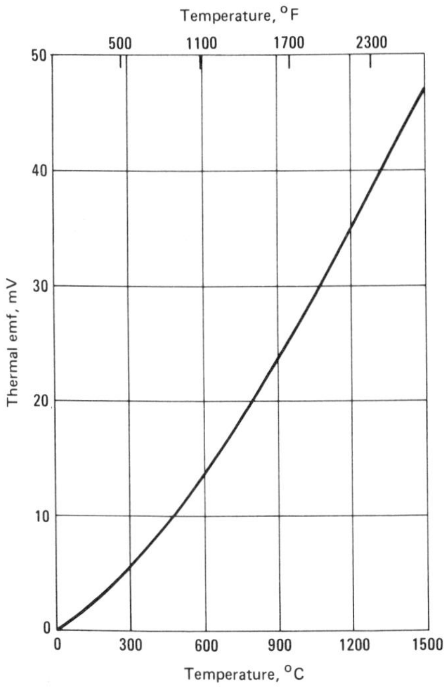

Fig. 9 Thermal emf of Pt-5Mo/Pt-0.1Mo thermocouples. Adapted from Ref 9

Table 7 Properties of Platinel thermocouples

Property	Platinel II	Platinel I
Nominal operating temperature range, °C (°F), in:		
Reducing atmosphere (nonhydrogen)	(a)	(a)
Wet hydrogen	(a)	(a)
Dry hydrogen(b)	1010 (1850)	1010 (1850)
Inert atmosphere	1260 (2300)	1260 (2300)
Oxidizing atmosphere	1260 (2300)	1260 (2300)
Vacuum	(a)	(a)
Maximum short-time temperature (<1 h), °C (°F)	1360 (2480)	1360 (2480)
Approximate microvolts per °C (°F):		
Mean, over nominal operating range (100 to 1000 °C)	42.5 (23.5)	41.9 (23.3)
At top temperature of normal range (1000 to 1300 °C)	35.5 (19.6)	33.1 (18.4)
Melting temperature, °C (°F), nominal:		
Positive thermoelement—solidus	1500 (2732)	1580 (2876)
Negative thermoelement—solidus	1426 (2599)	1426 (2599)
Stability with thermal cycling	Good	Good
High-temperature tensile properties	Fair	Fair
Ductility (of most brittle thermoelement) after use	Good	Good
Recommended extension wire at approximately 800 °C (1472 °F):		
Positive conductor	Type KP	Type KP
Negative conductor	Type KN	Type KN

(a) Not recommended. (b) High-purity alumina insulators are recommended. Source: Adapted from Ref 9

rhodium are not suitable for use under neutron radiation, because the rhodium is slowly transmuted to palladium.

The output of the Pt-5Mo/Pt-0.1Mo couple is high and increases with temperature in a uniform manner (see Fig. 9). Detailed emf/temperature tables (at 1 °C intervals) are available from suppliers. Compensating lead wires, good from 0 to 70 °C (32 to 160 °F), are available for Pt-5Mo/Pt-0.1Mo thermocouples—copper for the positive leg and Cu-1.6Ni for the negative leg.

Platinel. The Platinel thermocouple (trademark of Englehard Corporation), an all-noble-metal combination, was metallurgically designed to approximate the emf/temperature characteristics of the type K couple. Actually, two combinations have been produced: Platinel I and Platinel II (see Table 7). Both have negative legs of 65Au-35Pd. The positive leg of the Platinel I couple is 83Pd-14Pt-3Au, and the positive leg of the Platinel II couple consists of 55Pd-31Pt-14Au. The Platinel II couple has superior high-temperature fatigue properties and appears to be preferred over Platinel I.

Figure 10 compares thermal emf of Platinel thermocouples with that of the type K couple. The emf match with type K is good at elevated temperatures, but some departure occurs at lower temperatures. In an application involving measurement of turbine inlet temperatures in an aircraft engine, the connection between the thermocouple and type K extension wire is effected at 800 °C (1470 °F) where the emf match is excellent. In this application, only about 13 mm (½ in.) of Platinel II wire is used, and the remainder is type K. Other base-metal extension wires capable of matching the emf of the Platinels very closely over the range from 0 to 160 °C (32 to 320 °F) are also available.

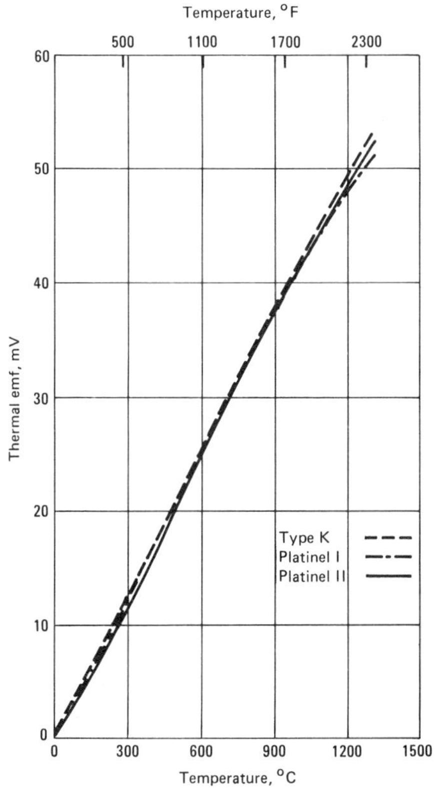

Fig. 10 Thermal emf of Platinel thermocouples compared with that of type K thermocouple. Adapted from Ref 9

Platinel couples can be used unprotected (insulators only) in air to 1200 °C (2190 °F) for extended periods of time and to 1300 °C (2370 °F) for shorter periods. Platinel II aged in commercial hydrogen for 1000 h at

1000 °C (1830 °F) showed reasonably good stability. Drift did not exceed 0.75%. It is recommended that precautions usually followed with platinum-rhodium thermocouples also be observed with Platinels. In particular, it should be noted that phosphorus, sulfur, and silicon have deleterious effects on the life of Platinel thermocouples.

Tungsten-Rhenium. Three tungsten-rhenium thermocouples are commercially available: W/W-26Re, doped W-3Re/W-25Re, and W-5Re/W-26Re. All three couples have been used at temperatures up to 2760 °C (5000 °F), but they usually are employed only below 2315 °C (4200 °F) due to temperature limitations of ceramic insulators.

Early use of tungsten-rhenium thermocouples, particularly W/W-26Re couples, indicated that these couples might be capable of measuring high temperatures with reasonable accuracy; however, one serious drawback was immediately observed. The tungsten leg, when heated to its recrystallization temperature of 1200 °C (2200 °F), became embrittled, an effect that was not experienced with the opposite leg (W-25Re or W-26Re). Early research showed that addition of 10% rhenium to the tungsten element did much to retain ductility after recrystallization. However, although large additions of rhenium to tungsten solved this problem, they also lowered the emf of the thermocouple. Consequently, other techniques intended to retain room-temperature ductility were employed, including special processing and doping combined with addition of 5% or less rhenium to the tungsten element.

The "dope," in the form of potassium, silicon, and aluminum compounds, is added during preparation of the tungsten powder. With the exception of potassium, these additives are volatile and are almost eliminated during processing. Doping, however, assists in formation of a microstructure characterized by large, elongated grains whose boundaries make relatively small angles with the wire axis. This structure is similar to that found in the well-known "nonsag" tungsten filament wire. Within recent years, microvoids or "bubbles" containing potassium "plated" on the inside surface were found decorating the boundaries of these elongated grains. It is now accepted that these voids promote formation of the desired elongated grains.

The thermal emf values of the three W-Re combinations are compared in Fig. 11, and other pertinent properties are shown in Table 8. All three thermocouple combinations are supplied as matched pairs guaranteed to meet the emf outputs given in producer-developed tables within ±1% (see ASTM E 452 for calibration procedure). Compensating extension wires are available for each combination.

Important factors controlling the use of W-Re thermocouples at high temperatures include:

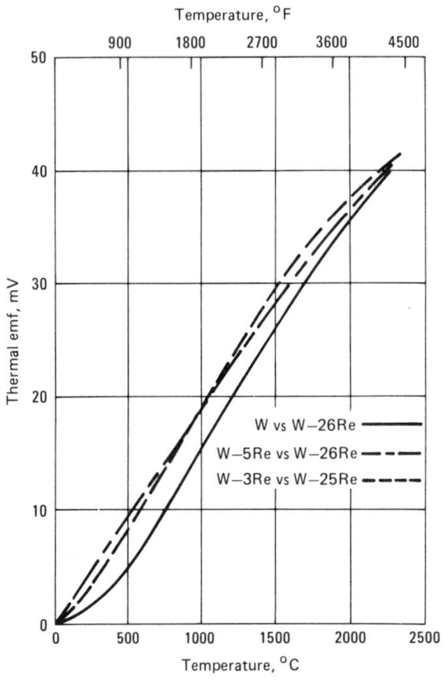

Fig. 11 Thermal emf of tungsten-rhenium thermocouples. Adapted from Ref 9

- Insulation, sheaths, and protection tubes (choice of insulation, sheaths, and protection tubes depends on operating temperature and environment)
- Diameter of thermoelements (larger diameters for higher temperatures)
- Atmosphere (vacuum, high-purity hydrogen, high-purity inert atmospheres required)

There is evidence that selective vaporization of rhenium occurs at temperatures of the order of 1900 °C (3450 °F) and higher when bare, unprotected W-Re couples are used in vacuum. This is not a problem, however, when these couples are protected with suitable refractory-metal sheaths.

For swaged-type thermocouples, maximum service temperature is affected by the diameter of the thermocouple wire as well as by the thickness of the ceramic insulating material (wire-to-wire and wire-to-sheath resistance). The problem here is mainly "shunt" error. At high temperatures, the resistivity of ceramic insulating materials decreases exponentially with temperature. Therefore, at a sufficiently high temperature, the insulation shunt resistance between thermocouple wires becomes comparable with wire resistance, and shunting results. The error in thermocouple reading may be positive or negative. Increasing the insulation thickness will also help to increase maximum allowable operating temperature.

Thermocouple Extension Wires

Thermocouple extension wires, also known as extension wires or lead wires, are electrical conductors used for connecting the thermocouple wires to the temperature measuring and control instrument. Extension wires usually are supplied in cable form, with positive and negative wires electrically insulated from each other. The chief reasons for using extension wires are economy and mechanical flexibility.

- *Economy*. Base-metal thermoelements, which cost less than $10 per pound in 1980, are always used as extension wires for the noble-metal thermocouple wires, which in 1980 cost about $700 per troy ounce. For base-metal thermocouples, use of extension wires permits periodic replacement of the thermocouple, which is exposed to elevated temperatures, without replacing the insulated extension-wire cables
- *Mechanical Flexibility*. Insulated solid or stranded wires in sizes from 14 to 20 gage are used as extension wires. This lends mechanical sturdiness and flexibility to the thermocouple circuitry while permitting the use of larger-diameter (usually 3.2 mm, or ⅛ in.) base-metal thermocouples for improved oxidation resistance and service life, or smaller-diameter (usually 0.5 mm, or 0.020 in.) noble-metal thermocouple wire to save cost

Circuitry of Thermocouple Wires and Extension Wires. A schematic diagram and circuitry of thermocouple and extension wires are shown in Fig. 12. One end of the positive extension wire PX and one end of the negative extension wire NX are joined to the positive thermoelement P and the negative thermoelement N, respectively, at the head junction. The temperature at the head junction is usually less than 205 °C (400 °F). The other ends of PX and NX are connected to the positive and the negative terminals, respectively, of the measuring or control instrument.

By substituting in the equation for Kirchoff's law and rearranging terms, the emf of the thermocouple and extension wire assembly between hot- and cold-junction temperatures T_2 and T_1 can be shown to be equal to the sum of the emf of the thermocouple PN between T_2 and T_H (the head-junction temperature) and the emf of extension wire PX-NX between T_H and T_1 (see Eq A, in Fig. 12).

The emf of the thermocouple PN between T_2 and T_1, without the use of any extension wire, can be expressed as the sum of the emf of the couple PN between T_2 and T_H and the emf of the couple between T_H and T_1 (see Eq B, in Fig. 12).

In order that the thermocouple and extension wire assembly generates the same emf as that of the thermocouple PN between the hot-junction temperature T_2 and the cold-junction temperature T_1, the following condition must be met (Eq C, in Fig. 12):

Table 8 Properties of tungsten-rhenium thermocouples

Property	W versus W-26Re	W-3Re versus W-25Re	W-5Re versus W-26Re
Nominal operating temperature range, °C (°F), in:			
Dry hydrogen	2760 (5000)	2760 (5000)	2760 (5000)
Inert atmosphere	2760 (5000)	2760 (5000)	2760 (5000)
Vacuum(a)	2760 (5000)	2760 (5000)	2760 (5000)
Maximum short-time temperature, °C (°F)	3000 (5430)	3000 (5430)	3000 (5430)
Approximate microvolts per °C (°F):			
Mean, over nominal operating range 0 to 2316 °C (32 to 4200 °F)	16.7 (9.3)	17.1 (9.5)	16.0 (8.9)
At top temperature of normal range 2316 °C (4200 °F)	12.1 (6.7)	9.9 (5.5)	8.8 (4.9)
Melting temperature, °C (°F), nominal:			
Positive thermoelement	3410 (6170)	3360 (6080)	3350 (6062)
Negative thermoelement	3120 (5648)	3120 (5648)	3120 (5648)
Stability with thermal cycling	Good	Good	Good
High-temperature tensile properties	Good	Good	Good
Stability under mechanical working	Fair	Fair	Fair
Ductility (of most brittle thermoelement) after use	Poor	Poor to good depending on atmosphere or degree of vacuum	Poor to good depending on atmosphere or degree of vacuum
Resistance to handling contamination	Good	Good	Good
Extension wire	Available	Available	Available

(a) Preferential vaporization of rhenium may occur when bare (unsheathed) couple is used at high temperatures and high vacuum. Check vapor pressure of rhenium at operating temperature and vacuum before using bare couple. Source: Adapted from Ref 9

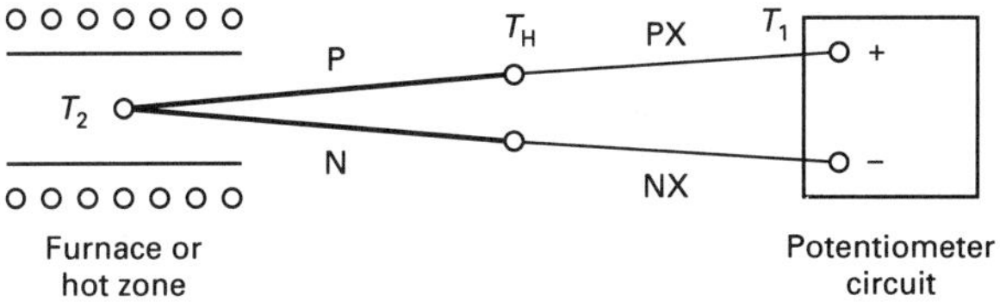

$$E_{assembly}\Big|_{T_1}^{T_2} = E_{PX}\Big|_{T_1}^{T_H} + E_P\Big|_{T_H}^{T_2} + E_N\Big|_{T_2}^{T_H} + E_{NX}\Big|_{T_H}^{T_1} = (E_P - E_N)\Big|_{T_H}^{T_2} + (E_{PX} - E_{NX})\Big|_{T_1}^{T_H} \quad \text{(Eq A)}$$

$$E_{PN}\Big|_{T_1}^{T_2} = (E_P - E_N)\Big|_{T_1}^{T_2} = (E_P - E_N)\Big|_{T_H}^{T_2} + (E_P - E_N)\Big|_{T_1}^{T_H} \quad \cdots \text{(Eq B)}$$

$$(E_P - E_N)\Big|_{T_1}^{T_H} = (E_{PX} - E_{NX})\Big|_{T_1}^{T_H} \quad \text{(Eq C)}$$

For the extension wire: $E_P = E_{PX}\Big|_{T_1}^{T_H}$ and $E_N = E_{NX}\Big|_{T_1}^{T_H}$

For an alternative extension wire: $E_P \neq E_{PX}\Big|_{T_1}^{T_H}$ and $E_N \neq E_{NX}\Big|_{T_1}^{T_H}$

Fig. 12 Computation of thermal emf compensated to 0 °C (32 °F) for a normal industrial thermocouple setup. In this example, the emf of the assembly is calculated from Eq A; the remaining equations define terms used in Eq A. In the sketch and equations, P and N designate elements in the thermocouple wire, and PX and NX designate elements in the extension wire. T_2 = hot-junction temperature; T_1 = cold-junction temperature; and T_H = head-junction temperature (205 °C, or 400 °F). Adapted from Ref 3

$$(E_P - E_N)\Big|_{T_1}^{T_H} = (E_{PX} - E_{NX})\Big|_{T_1}^{T_H} \quad \text{(Eq 5)}$$

where E_P is the emf of the positive thermoelement versus platinum, E_N is the emf of the negative thermoelement versus platinum, E_{PX} is the emf of the positive extension wire versus platinum, E_{NX} is the emf of the negative extension wire versus platinum, T_1 is the cold-junction temperature, and T_H is the head junction temperature. This expression means that the emf of the thermocouple PN within the temperature range T_1 to T_H must be the same as the emf of the extension-wire couple PX-NX within the same temperature range. It is not necessary, however, for the emf of the extension-wire couple to match that of the thermocouple over the entire range from T_1 to T_2. Extension wires of the same composition as that of the thermocouple wire are termed *extension wires*. Wires of different alloys used as extension wires to develop the same emf as that of the thermocouple wire from 0 to 200 °C (32 to 400 °F) are called *alternate* or *compensating extension wires*.

Extension wires are used for base-metal thermocouples. Alternate extension wires are used for noble-metal thermocouples. Base compositions, physical properties, ranges of application, and initial calibration tolerances of the extension wires for standardized thermocouples are listed in Table 9.

Error Analysis. The error introduced by incorporation of extension wire in the thermocouple circuitry can be expressed:

$$\Delta E_{assembly}\Big|_{T_1}^{T_2} = \Delta E_{PX-NX}\Big|_{T_1}^{T_H} - \Delta E_{PN}\Big|_{T_1}^{T_H} + \Delta E_{PN}\Big|_{T_1}^{T_2} \quad \text{(Eq 6)}$$

where

$\Delta E_{assembly}\Big|_{T_1}^{T_2}$ is the deviation (of initial calibration) of the thermocouple and extension wire assembly from established emf table between temperatures T_1 and T_2,

$\Delta E_{PX-NX}\Big|_{T_1}^{T_H}$ is the deviation of the extension wire between T_1 and T_H,

$\Delta E_{PN}\Big|_{T_1}^{T_H}$ is the deviation of the thermocouple between T_1 and T_H, and

$\Delta E_{PN}\Big|_{T_1}^{T_2}$ is the deviation of the thermocouple between T_1 and T_2.

Equation 6 shows that, besides the initial calibration error of the thermocouple over the temperature range T_1 to T_2, an addition-

Table 9 Properties of thermocouple extension wires

Thermocouple type	Extension wire	Base composition	Resistivity, nΩ · m	Melting point °C	Melting point °F	Temperature range, °C (°F)	Initial calibration tolerance
J	JPX	Fe	100	1450	2640	0 to 200	±2.2 °C
	JNX	45Ni-55Cu	500	1210	2210	(32 to 390)	(±4 °F)
K	KPX	90Ni-10Cr	700	1350	2460	0 to 200	±2.2 °C
	KNX	95Ni-AlSiMg	320	1210	2210	(32 to 390)	(±4 °F)
N	NPX	Ni-14Cr-1.4Si	930	1410	2570	0 to 200	±2.2 °C
	NNX	Ni-4.4Si-0.1Mg	370	1400	2550	(32 to 390)	(±4 °F)
T	TPX	Cu	17	1083	1981	−60 to 100	±1.0 °C
	TNX	45Ni-55Cu	500	1210	2210	(−75 to 212)	(±1.8 °F)
E	EPX	90Ni-10Cr	700	1450	2642	0 to 200	±1.7 °C
	ENX	45Ni-55Cu	500	1450	2642	(32 to 390)	(±3 °F)
R or S	SPX	Cu	17	1083	1981	0 to 200	±5 °C
	SNX	Cu-1Ni-0.3Mn	45	1100	2010	(32 to 390)	(±57 μV)
B(a)	BPX(b)	Cu-2Mn	150	1100	2010	0 to 200	±33 μV
	BNX	Cu	17	1083	1981	(32 to 390)	

(a) Cu/Cu extension wire can be used if head-junction temperature is 100 °C or less. (b) Proprietary alloy. Can be used up to 300 °C (570 °F) with initial calibration of +50 μV at this temperature. Source: Adapted from ANSI MC96.1

al term is introduced when extension is used. This term is equal to the difference of initial calibration of the extension-wire couple and the thermocouple between the cold-junction temperature T_1 and head-junction temperature T_H. This additional term can be minimized by judiciously choosing a pair of extension wires or alternate extension wires, the initial emf calibration of which closely matches that of the thermocouple wire between T_1 and T_H.

Color Coding of Thermocouple Wires and Extension Wires

For many years ISA has coordinated an effort to standardize color coding of thermocouple and extension wires in the United States. The main objective has been to establish uniformity in designation of various types of thermocouples and extension wires to provide, by means of insulation color, identification of wires by type or composition as well as by polarity when used as part of a thermocouple system (Ref 9). The present U.S. color designations, as indicated in ANSI MC96.1 (1982), are given in Tables 10 and 11. Foreign and international color codes are given in Table 12. Color coding is not uniform throughout the world. United Kingdom, France, Germany, Japan, U.S.S.R., and China have their own color codes. At present, the International Electrotechnical Commission is adopting a new international color code in an attempt at world standardization.

Thermocouple Calibration

The temperature/emf relationship for a specific thermocouple combination is a definite physical property and thus does not depend on details of the apparatus or method used for determining this relationship. Consequently, thermocouples can be calibrated by any of several methods, the choice of which depends on type of thermocouple, temperature range, accuracy required, size of wires, apparatus available, and personal preference.

Calibration of a thermocouple is achieved through determination of its electromotive force (emf) at a series of known temperatures, which when coupled with a standardized means of interpolation will give values of emf over the entire temperature range in which it will be used. A standard thermometer that indicates temperatures on a universally acceptable scale is required, as well as a means of measuring the emf of the thermocouple and a controlled heat source wherein the thermocouple and the standard can be brought to the same temperature.

Only the basic points of calibration techniques will be described in this review; the reader is directed to other, more detailed sources of information, such as Ref 9 and 10.

Temperature Scales (Ref 9 and 11). Meaningful measurement of temperature requires a scale with appropriate units, just as measurement of length requires a yardstick or meter stick with all of its subdivisions. The ideal temperature scale is known as the thermodynamic scale. However, measurement of temperature on this scale (using a gas thermometer) is extremely difficult even under laboratory conditions. For many years prior to 1927, the need for a more practical temperature scale had been apparent.

In 1927, such a scale, named the International Temperature Scale (ITS 27), was adopted by the Seventh General Conference on Weights and Measures. Among other advantages, this scale served to unify the existing national temperature scales (Germany, Britain, United States, and so forth). The scale was revised in 1948, and in a 1960 modification, the word "Practical" was inserted in the name of the scale, which now became the International Practical Temperature Scale. The scale was revised again in 1968, and was amended in 1975 (Ref 11).

The International Practical Temperature Scale of 1968 (amended in 1975), or IPTS 68 (amended 1975), was designed in such a

Table 10 Color coding of duplex insulated thermocouple wire

Thermocouple Type	Positive wire	Negative wire	Color of insulation Overall(a)	Positive(a)	Negative
T	TP	TN	Brown	Blue	Red
J	JP	JN	Brown	White	Red
E	EP	EN	Brown	Purple	Red
K	KP	KN	Brown	Yellow	Red
N	NP	NN	Brown	Orange	Red

(a) A tracer color of the positive wire code color may be used in the overall braid. Source: Adapted from ANSI MC96.1

Table 11 Color coding of single conductor and duplex insulated thermocouple extension wires

Extension wire type: Type	Positive	Negative	Color of insulation for duplex extension wires: Overall	Positive	Negative(a)	Color of insulation for single conductor extension wires(b): Positive	Negative(c)
T	TPX	TNX	Blue	Blue	Red	Blue	Red-blue trace
J	JPX	JNX	Black	White	Red	White	Red-white trace
E	EPX	ENX	Purple	Purple	Red	Purple	Red-purple trace
K	KPX	KNX	Yellow	Yellow	Red	Yellow	Red-yellow trace
N	NPX	NNX	Orange	Orange	Red	Orange	Red-orange trace
R or S	SPX	SNX	Green	Black	Red	Black	Red-black trace
B	BPX	BNX	Gray	Gray	Red	Gray	Red-gray trace

(a) A tracer having the color corresponding to the positive wire code color may be used on the negative wire color code. (b) *NOTE OF CAUTION:* In the procurement of random lengths of single conductor insulated extension wire, it must be recognized that such wire is commercially combined in matching pairs to conform to established calibration curves. Therefore, it is imperative that all single conductor insulated extension wire be procured in pairs, at the same time, and from the same source. (c) The color identified as a trace may be applied as a tracer, braid, or by any other readily identifiable means. Source: Adapted from ANSI MC96.1

Table 12 Foreign and international color codes of thermocouple extension wire cable

Letter code	Conductor	United Kingdom, BS1843	Germany, DIN43714	Japan, JIS1610	France, NFC42-323	IEC, 584-3 (1989)
T	Positive	White	Red	Red	Yellow	Brown
	Negative	Blue	Brown	White	Blue	White
	Overall	Blue	Brown	Brown	Blue	Brown
J	Positive	Yellow	Red	Red	Yellow	Black
	Negative	Blue	Blue	White	Black	White
	Overall	Black	Blue	Yellow	Black	Black
E	Positive	Brown	Red	Red	Yellow	Purple
	Negative	Blue	Black	White	Purple	White
	Overall	Brown	Black	Purple	Purple	Purple
K	Positive	Brown	Red	Red	Yellow	Green
	Negative	Blue	Green	White	Purple	White
	Overall	Red	Green	Blue	Yellow	Green
N	...	...	...	...	...	(a)
R	Positive	White	Red	Red	Yellow	Orange
	Negative	Blue	White	White	Green	White
	Overall	Green	White	Black	Green	Orange
S	Positive	White	Red	Red	Yellow	Orange
	Negative	Blue	White	White	Green	White
	Overall	Green	White	Black	Green	Orange
B	Positive	No standard	Red	Red	No standard	...
	Negative	Use copper	Gray	Gray	Use copper	...
	Overall	wire	Gray	Gray	wire	...

(a) Pink was proposed but not yet approved for type N thermocouples for IEC 584-3.

way that the temperature measured on it closely approximates the thermodynamic temperature; the difference is within the limits of the present accuracy of measurement.

The IPTS 68 (amended 1975) is based on the assigned values of the temperatures of 13 reproducible equilibrium states (defining fixed points) and on standard instruments calibrated at these temperatures (Ref 11). Interpolation is provided by formulas used to establish the relations between indications on standard instruments and values of International Practical Temperature.

The IPTS 68 uses both International Practical Kelvin Temperature, symbol T_{68}, and International Practical Celsius Temperature, t_{68}. The relation between T_{68} and t_{68} is the same as that between T and t on the Thermodynamic Scale—that is, $t_{68} = T_{68} - 273.15$ K. The units of T_{68} and t_{68} are the kelvin symbol, K, and the degree Celsius symbol, °C, as in the case of thermodynamic temperature T and Celsius temperature t. The standard instruments used are:

- Platinum resistance thermometer 13.81 to 903.89 K (−434.81 to 1167.33 °F)
- Pt-10Rh/Pt thermocouple 630.74 °C to 1064.43 °C (gold point)
- Above 1064.43 °C (1947.97 °F), defined in terms of the Planck radiation law using 1064.43 as a reference temperature (Optical Pyrometer)

The International Temperature Scale of 1990 (ITS-90) is a new temperature scale, which became effective worldwide on 1 Jan 1990. The ITS-90 was developed to replace current temperature scale IPTS-68. This was done to overcome the deficiencies in accuracy and reproducibility of the existing scale and to incorporate the advances made in the last 20 years in thermometry. A new set of defining fixed points, such as the triple point of water (0.01 °C, or 273.16 K) and freeze points of high-purity metals was established. Platinum thermocouples are no longer used as the interpolating instrument for ITS-90. Resistance thermometers (primarily platinum resistance thermometers) are used instead. Above the gold point (1064.43 °C, or 1947.97 °F), radiation thermometers are used. As a result, temperatures on the ITS-90 scales are in much better agreement with thermodynamic values than are those on IPTS-68 and its subsequent revision in 1975.

The magnitude of the change in temperature from IPTS-68 to ITS-90 is within ±0.4 °C (±0.7 °F) from 0 to 1000 °C (32 to 1830 °F). Therefore, the existing tables for thermocouples and resistance temperature detectors (RTDs) need to be corrected to reflect these changes. The National Institute of Science and Technology (NIST) is preparing new tables of thermocouples and RTDs for ITS-90. For further details, see Ref 12.

Changes have also occurred in voltage and electrical resistance scales. The change in the Volt scale will be 9.3 ppm in the United States. (The changes in other parts of the world are slightly different.) The change in the Ohm scale will be 1.69 ppm. Instruments having this sensitivity range should be corrected to reflect the changes in volt and ohm scales.

Methods of Thermocouple Calibration. Initial calibration of a thermocouple can be done by any of the following methods:

- Freezing-point calibration
- Direct thermoelement emf measurement versus platinum
- Thermoelement comparison method
- Calibration of thermocouples by comparison methods

In the freezing-point method of calibration, the emf output of the thermocouple as a whole is measured during the cooling cycle of molten pure metals. In the second and third methods, the emf of both the positive and negative thermoelements are individually measured versus platinum or another calibrated standard.

Freezing-Point Calibration. In the calibration of a thermocouple at freezing points, the thermocouple (properly protected) is slowly immersed in the molten metal. The metal is brought essentially to a uniform temperature at the beginning of freezing by holding its temperature constant at about 10 °C (18 °F) above the freezing point for several minutes and then cooling slowly, or by agitating the metal with the thermocouple protection tube just before freezing begins. The emf of the thermocouple is observed at regular intervals of time. These values are plotted, and the emf corresponding to the flat portion of the cooling curve is the emf at the freezing point of the metal.

Metals of sufficient purity that may be used in freezing-point calibrations are:

- Tin with a freezing point of 231.928 °C (449.470 °F)
- Indium with a freezing point of 156.5985 °C (313.8773 °F)
- Zinc with a freezing point of 419.527 °C (787.149 °F)
- Aluminum with a freezing point of 660.323 °C (1220.581 °F)
- Silver with a freezing point of 961.78 °C (1763.20 °F)
- Gold with a freezing point of 1064.18 °C (1947.52 °F)
- Copper with a freezing point of 1084.62 °C (1984.32 °F)
- Nickel with a freezing point of 1455 °C (2651 °F)
- Palladium with a freezing point of 1554 °C (2829 °F)
- Platinum with a freezing point of 1769 °C (3216 °F)

The triple point of water and the first seven freeze points of the above metals are the primary reference points in ITS-90 up to 1064.18 °C (1947.52 °F). The freezing points of the other metals listed above are secondary reference points. Of all these metals, antimony and tin have marked tendencies to undercool before freezing, but such undercooling will not be excessive if the liquid metal is stirred.

A Pt-10Rh/Pt thermocouple may be calibrated using values obtained at the freezing point of aluminum (660.323 °C), the freezing point of silver (961.78 °C), and the freezing point of gold (1064.18 °C).

The emf developed by a homogeneous thermocouple at the freezing point of a metal is constant and reproducible if all of the following conditions are fulfilled:

- The thermocouple is protected from contamination
- The thermocouple is immersed in the freezing-point sample sufficiently far to eliminate heating or cooling of the junction by heat flow along the wires and protection tube
- The reference junctions are maintained at a constant and reproducible temperature
- The freezing-point sample is of sufficient purity
- The metal is maintained at an essentially uniform temperature during freezing

Freezing points can be reproduced under industrial conditions within 0.1 to 5 °C (0.2 to 9 °F) for calibrations between the ice point and the melting point of platinum. Because of difficulty in testing, fixed points at temperatures above the freezing point of copper (1084.62 °C) usually are expressed as melting points rather than freezing points. Complete units including freezing-point sample, crucible, and heating source are available commercially.

See Ref 10 and 13 to 15 for additional information on the freezing-point method of calibrating thermocouples.

Direct emf Measurement Versus Platinum. The method used for direct measurement of thermoelement emf versus platinum at a fixed temperature is illustrated in Fig. 13. The thermocouple wire specimens, a primary platinum standard and a calibrated reference-grade type R couple (Pt-13Rh/Pt) are welded together at one end. The multiple couple is in turn welded into a heat sink, and the whole assembly is inserted halfway into a 2 m (6 ft) horizontal electric furnace to a depth of approximately 1 m (3 ft).

Precision potentiometers are used so that measurement of emf versus platinum for a test specimen, and measurement of temperature by the calibrated platinum thermocouple, can be made simultaneously. The test specimens, the calibrated platinum couple, and the platinum standard are inserted into mercury U-tubes embedded in Dewar flasks packed with shaved ice and are electrically connected to two precision potentiometers as shown in Fig. 13. Simultaneous measurements can be made as soon as the furnace temperature and the cold-junction temperature reach equilibrium.

Consider the case of emf measurements of Tophel, a type K positive thermoelement, at 980 °C (1800 °F). It is not necessary

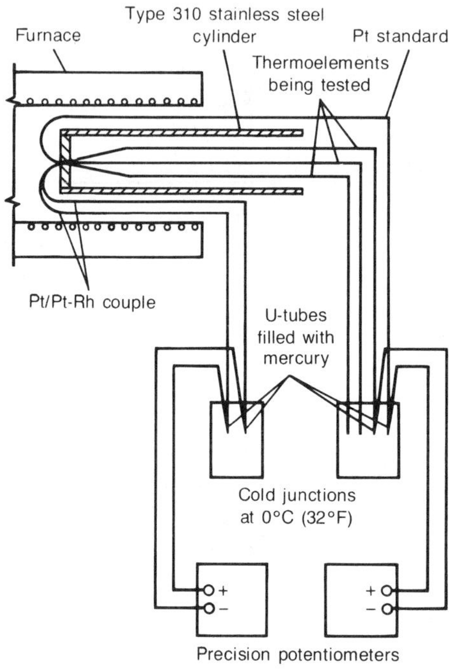

Fig. 13 Experimental setup for direct measurement of the emf of thermoelements versus platinum. Adapted from Ref 16

that the furnace temperature as measured by the calibrated Pt/Rh couple be exactly 980 °C (1796 °F). In actual practice, all that is needed is that the furnace temperature be within ±2.8 °C (±5 °F) of the desired temperature. The emf of the thermoelement at the desired test temperature can be computed with the following equation:

$$E_D = E_M - (T_M - T_D)\zeta \qquad \text{(Eq 7)}$$

where E_D is the corrected specimen emf versus Pt at the desired temperature T_D, E_M is the measured emf at the measuring temperature T_M, and ζ is the thermoelectric power of the thermoelement versus Pt at T_D. As an example of the use of Eq 7, the following data were generated for a sample of Tophel:

Emf of Pt/Rh couple, mV	10.247
	10.248
	10.248
Corresponding temperature, T_M, °C (°F)	983 (1801.4)
Measured emf, E_M, mV	31.999
	32.000
	32.000
Thermoelectric power, mV/°C (mV/°F)	0.0306 (0.017)

Note:
E_D = Corrected emf of sample vs Pt at 1800 °F
= 32.000 mV − (1801.4 − 1800) × 0.017 mV/°F
= 32.000 mV − 0.024 mV
= 31.976 mV

This value is in excellent agreement with National Bureau of Standards (NBS) calibrations on both ends of a single coil of Tophel: 31.970 mV and 31.980 mV. (The example shown above is in IPTS-68 scale.)

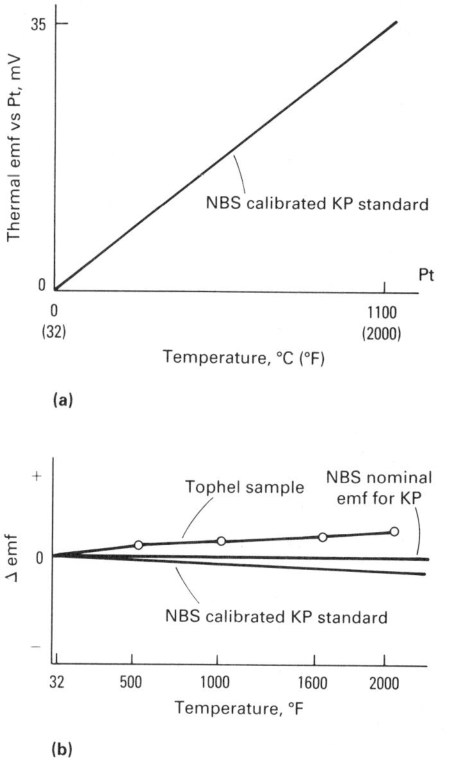

Fig. 14 Thermal emf plots for KP thermoelements illustrating the comparison method of emf measurement. (a) emf of KP standard versus platinum. (b) Δemf of Tophel sample and KP standard thermoelements versus NBS nominal emf for type KP. Adapted from Ref 2

Comparison Method. In industrial practice, a thermoelement is calibrated at several fixed temperatures against a thermocouple standard of the same alloy calibrated by NIST. The procedure in ASTM E 207 describes the preferred standard method. The emf of the test specimen can be obtained by the following general equation:

$$\text{emf vs Pt} = \Delta\text{emf}_{\text{specimen vs std}} + \text{emf}_{\text{std vs Pt}} \qquad \text{(Eq 8)}$$

Figure 14(a) shows the emf versus Pt curve for a Tophel standard, a type K positive thermoelement. The type K positive thermoelement has a large thermoelectric power versus Pt. For example, its thermoelectric power versus Pt at 980 °C (1800 °F) is 31 μV/°C (17 μV/°F). An error of 34 μV would have been introduced by the temperature measurement error of only 1.1 °C (2 °F) in the case of direct measurement of emf versus Pt. However, this is not so in the comparison method, as can be readily observed from Fig. 14(b).

Figure 14(b) shows values of Δemf versus NBS nominal emf as ordinates against temperature as abscissa. The emf deviations of the NIST-calibrated Tophel standard from NIST nominal emf values for the type K positive thermoelement are plotted with a heavy line. The deviation of the test specimen from NIST nominal emf at any test temperature can be obtained graphically by plotting measured Δemf on the chart or by use of the following equation:

$$\Delta emf_{specimen\ vs\ NBS\ nominal\ emf} = \Delta emf_{specimen\ vs\ std} + (emf_{std\ vs\ Pt} - NIST\ nominal\ emf\ vs\ Pt) \quad \text{(Eq 9)}$$

Equation 9 is a general equation applicable to both the positive and negative thermoelements. If we calculate Δemf for both thermoelements and substitute these values in Eq 4, then:

$$\Delta emf_{couple} = \Delta E_P - \Delta E_N \quad \text{(Eq 10)}$$

Calibration of Thermocouples by Comparison Methods. Calibration of a thermocouple by comparing it to a working standard is sufficiently accurate for most purposes and can be done conveniently in most industrial and technical laboratories. The emf of the thermocouple being calibrated is measured at selected calibration points, the temperature of each point being measured by a standard thermocouple (usually one calibrated by NIST) or other standard thermometer. Test points are selected on the basis of thermocouple type, temperature range to be covered, accuracy required, and end use.

The accuracy obtained with this technique depends on the ability of the observer to bring the junction of the thermocouple to the same temperature as that of the sensing portion of the standard used, such as the measuring junction of a standard thermocouple or the sensitive portion of a resistance or liquid-in-glass thermometer. The accuracy obtained is further limited by the accuracy of the standard. The method of bringing both measuring junctions to the same temperature depends on type of thermocouple, type of standard, and method of heating.

Potentiometric instruments or high-impedance electronic instruments are used to measure emf, thus eliminating instrument loading as a contributor of significant error.

Additional information relative to this calibration method and attainable accuracies may by found in ASTM E 220, "Standard Method for Calibration of Thermocouples By Comparison Techniques," and in ANSI MC96.1.

Reference Tables for Thermocouples

Practical use of thermocouples requires that the selected thermocouple meet an established or standardized temperature/emf relationship within acceptable tolerance limits. Because the thermocouple in a thermoelectric thermometer system is replaced periodically due to drift, failure, or other reasons, conformance to an established temperature/emf relationship is necessary in order to permit interchangeability when commercially available readout equipment is used. Such widely acceptable reference tables have S and T thermocouples and are available in NBS Monographs 124 (Cryogenic) and 125 (Standard Couples), ANSI MC96.1 and ASTM E 230. Less detailed versions of these tables (at intervals of 10 °C, or 18 °F) usually may be obtained from producers or distributors of these thermocouples.

For other nonstandard thermocouples, including those that do not have letter designations, tables usually are developed by producers and are available from either producers or suppliers. Additionally, temperature/emf values for three W/Re combinations have been published "for information" by ASTM in the standards book containing standards related to thermocouples, and one combination (W-3Re/W-25Re) has values published in ASTM E 696.

All tables, in order to gain wide acceptance, must conform to an internationally recognized temperature scale. At this time, the scale is IPTS 68, and the latest published tables should conform to it. However, a large quantity of control or measurement instruments still in use are in compliance with IPTS 48, and replacement thermocouples for these instruments are purchased on this scale. The difference in the two scales may or may not be significant depending on the application. In this regard, particular attention should be paid to types S, R, and B thermocouples for use above 1000 °C (1830 °F). See Ref 11 for differences arising from use of either scale.

Initial Calibration Tolerances. Table 2 lists manufacturers' tolerances for initial calibration of all standardized thermocouples. For example, a brand new type K couple could be in error by as much as ±4.2 °C (±7.5 °F) when used for temperature measurement at 540 °C (1000 °F). The deviation in emf of a thermocouple from the standard table value is equal to the algebraic difference of the individual emf deviations of the thermoelements, as shown in the following equation:

$$\Delta E_{couple} = \Delta E_P - \Delta E_N \quad \text{(Eq 11)}$$

where ΔE_{couple} is the emf deviation of the couple from the table value, in millivolts; ΔE_P is the emf deviation of the positive thermoelement from NBS nominal value, in millivolts; and ΔE_N is the emf deviation of the negative thermoelement from NBS nominal value, in millivolts.

The deviations in initial calibration of a typical type K couple from NBS table values are illustrated in Fig. 15. The corresponding deviation of initial calibration expressed in temperature is obtained as follows:

$$\Delta T = \frac{\Delta E_{couple}}{\text{Th.p.}} \quad \text{(Eq 12)}$$

where ΔE is the emf deviation of the couple at a certain temperature and Th.p. is the thermoelectric power of the couple at the same temperature.

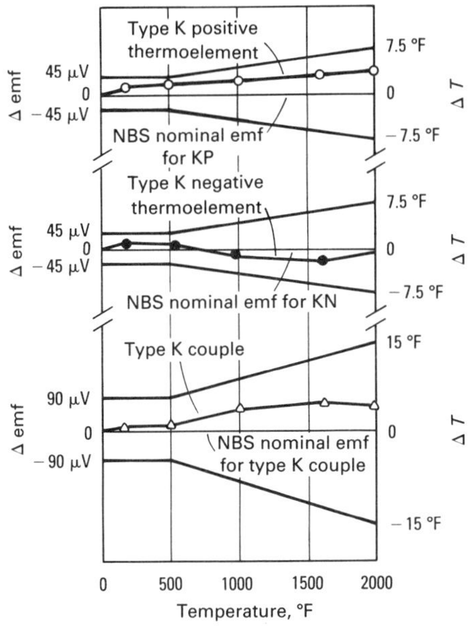

Fig. 15 Thermal emf plots for a type K thermocouple illustrating the method of evaluating emf deviation. Adapted from Ref 2

Change of Calibration During Service

Any thermocouple can be subject to failure (of a type that creates an open circuit) during service. Failure can be caused by localized melting of the thermoelements as a result of overheating, by vibration resulting in fatigue failure, or by gradual reduction of wire diameter through high-temperature oxidation. Prior to failure, the emf calibration of a thermocouple will change, primarily as a result of the individual or combined changes in chemical composition, homogeneity, and structure that take place in the thermoelements. The magnitudes and directions of these changes are dependent on temperature, time, wire diameter, and environmental conditions.

Effect of Environment on Base-Metal Thermocouples. The change in emf, as a function of test temperature, of a 3.25 mm diam (0.128 in. diam) type K thermocouple on exposure to air up to 1100 °C (2000 °F) is shown in Fig. 16. After 10 h of exposure, the emf of the couple had changed about 3 °C (5 °F) at 1100 °C (2000 °F) in the positive direction. The change had increased to +10 °C (+18 °F) after exposure for 1000 h. At lower test temperatures, the magnitude of the change is smaller. A decrease in silicon and chromium contents of the positive thermoelement through preferential oxidation causes a net change in emf. Similarly, the change in emf of the negative thermoelement is attributed to preferential oxidation of its alloy constituents Si, Mn, Al, and Fe (Ref 16 and 17).

The oxidation resistance and emf stability of type J couples are inferior to those of

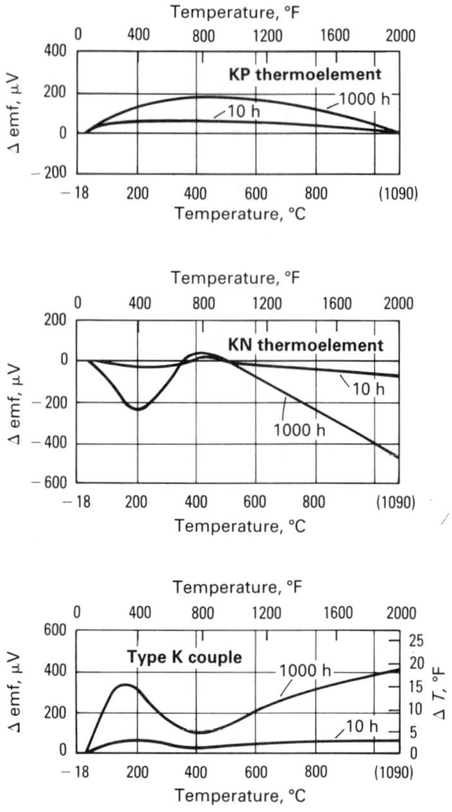

Fig. 16 Changes in thermal emf of a type K thermocouple resulting from long-time exposure in air at temperatures up to 1100 °C (2000 °F)

type K couples, and type J couples should not be used above 760 °C (1400 °F).

Type K couples are recommended for use in inert atmospheres at elevated temperatures only for short intervals. Type J couples, on the other hand, are stable and can perform better in inert environments than in air.

Type K thermocouples are not recommended for use in reducing or hydrogen-bearing atmospheres. Type J couples are stable and can be expected to perform well at temperatures up to 760 °C (1400 °F),

Effect of Environment on Bare Pt-Rh Thermocouples. Pt-10Rh/Pt, Pt-13Rh/Pt, and Pt-30Rh/Pt-6Rh thermocouples can be used with very good results continuously in air or in oxidizing atmospheres to 1500 °C, or 2730 °F, for types S and R; to 1700 °C, or 3090 °F, for type B) and intermittently to temperatures approaching the melting point of platinum (1769 °C, or 3216 °F) for types S and R and to 1780 °C (3235 °F) for type B. For these couples in these atmospheres, life is governed by the temperature of operation, partial pressure of oxygen, rate of change of the atmosphere in the vicinity of the hot junction, and method of mounting of the thermocouple.

It is generally agreed that volatile oxides of platinum and rhodium are formed when these metals are heated at high temperatures and are the principal cause of the loss of metal. Experience indicates slightly more rhodium than platinum volatilizes in air, which results in a negative drift of the couple after long periods of operation. A negative drift of 6 to 9 °C (11 to 16 °F) has been reported for a thermocouple that was in continuous use in air at 1290 °C (2350 °F) for over three years.

Bare Pt-Rh thermocouples can be used in inert atmospheres such as argon, helium, or nitrogen with very good results.

As far as can be ascertained, reducing gases such as carbon monoxide and hydrogen do not have adverse effects on types S, R, and B couples directly, but it is suspected that these gases reduce impurity oxides such as silica, which is usually present in alumina. The silicon reduced from the silica is known to unite with platinum to form a low-melting eutectic (830 °C, or 1530 °F). Close contact of these couples by easily reduced oxides of any metal should not be permitted.

Type S, type R, and type B thermocouples have been used for short periods of time in vacuum. Long-time exposure to vacuum is not recommended.

It has been reported that unstable hydrocarbons crack in contact with hot platinum-group metals, causing damage to these metals in the form of a fine intergranular precipitate of carbon.

Halogen gases have harmful effects on platinum-group metals at high temperatures.

Direct contact between bare couples and compounds of easily reduced metals such as lead, bismuth, and antimony should be avoided at high temperatures, because such contact results in formation of low-fusing platinum alloys.

Unprotected platinum thermocouples are attacked by phosphorus, arsenic, sulfur, and vapors of metals such as zinc and lead. This attack generally results in brittleness and hot shortness.

All contact between bare couples and caustic alkalis, nitrates, cyanides, alkaline earths, and the hydroxides of barium and lithium should be avoided, because these substances attack platinum at red heat.

Judicious use of insulators and protection tubes will eliminate problems with many of the contaminants listed above. A variety of ceramics, some of which are gastight, are available, but it should be kept in mind that no single ceramic will suit all applications.

Certain precautions should be followed when platinum-group metals are in contact with ceramics in reducing atmospheres. It is generally known that when platinum or platinum alloys in contact with silica are heated in reducing atmospheres above 1200 °C (2200 °F), platinum silicides with low melting points (as low as 830 °C, or 1530 °F) are formed at the grain boundaries, which results in embrittlement of the wire. It has been shown that this attack also occurs at and above 1100 °C (2000 °F) when platinum and platinum alloys are adjacent to but not in contact with silica-bearing materials.

In experiments conducted at about 1100 °C (2000 °F), in which silica was present in the alumina insulation, and carbon and sulfur were also present (residue of drawing compound on wire), thermocouple wires failed due to melting. Based on failure analysis, it was hypothesized that a volatile compound of SiS_2 is first formed, which serves to transport silicon present in the insulator or protection tube to the platinum or platinum alloy. This compound decomposes in contact with the hot platinum, and the liberated sulfur recombines with additional silicon in the refractory.

It is quite obvious that all traces of lubricating oils, drawing compounds, or other sulfur-bearing compounds should be removed from the thermocouple assembly, because silica is present in varying amounts in all commercial refractories (particularly mullite and sillimanite). Fractures in wires, caused by platinum silicide, generally present a melted appearance, and the fracture surface contains a number of glazed areas.

Insulation and Protection

To operate properly, thermocouple wires must be electrically insulated from one another at all points other than the measuring junction and must be protected from the operating environment.

Thermocouple Wire Insulation. For cryogenic applications (below 0 °C and as low as about 4 K) varnish or varnish-type coatings are used to insulate thermoelements from one another. The coating usually is selected on the basis of good electrical resistance, ease of application, and ability to withstand flexing at the very low temperatures. Formvar, polyurethane, teflon, Pyre-ML-Polyimide (E.I. Du Pont de Nemours and Company, Inc.), and GE 7031 (General Electric Company) have been used for this purpose. In particular, the polyimide coating has not only good electrical resistance but also excellent flexing strength at very low temperatures.

A variety of material, polyvinyl chloride, thermoplastic elastomer, fluoroethylene, synthetic polyimide fiber, polyimide film, fiberglass fibers, high-temperature fiberglass fibers, vitrified silica fibers, and ceramic fibers are available for insulating thermocouple from ambient to 1370 °C (2500 °F). The maximum service temperatures on these thermocouple insulation materials are listed in Table 13. The material is listed essentially in the order of dielectric strength at increasing operating temperatures. Besides dielectric strength, which dictates the maximum operating temperature of the insulation, the resistance to chemicals, mois-

Table 13 Maximum service temperatures, advantages and limitations of thermocouple wire insulation

Code	Material	Maximum temperature °C	°F	Advantages and limitations
P	Polyvinyl chloride	105	221	Resistance to chemicals and moisture
R(a)	Thermoplastic elastomer	125	257	Application to −55 °C (−65 °F). Flame resistant
N	Nylon(b)	150	300	Resistant to chemicals. Flammable
TZ	Tefzel(b)	150	300	Resistant to chemicals. Nonflammable
TEX	Teflon(b)	200	400	Resistant to chemicals. Nonflammable
PFA, TF	Teflon(b)	260	500	Resistant to chemicals. Nonflammable
B(a)	Polyamide fiber	260	500	Replacement for asbestos. Nonflammable. Good abrasion resistance
K	Kapton(b)	260	500	Resistance to chemicals and abrasion
G	Fiberglass	500	932	Nonflammable, resists oils
Q	High-temperature fiberglass	700	1300	Nonflammable, resists oils
HG	Refrasil(c), vitrified silica	1000	1832	Excellent dielectric at high temperature. Poor abrasion resistance
Cefir(a)	Nextel 312(c), ceramic fiber	1200	2200	Excellent dielectric at high temperature. Good abrasion resistance, moisture resistant
...	Nextel 440(d), ceramic fiber	1370	2500	Excellent dielectric at high temperature. Good abrasion resistance, moisture resistance

(a) Trademark of Thermo Electric Company, Inc. (b) Trademark of E. I. Du Pont de Nemours & Company, Inc. (c) Trademark of Thompson Company. (d) Trademark of 3M Company

Table 14 Maximum service temperatures for protection tubes

Materials	Maximum service temperature °C	°F
Carbon steel	540	1000
Wrought iron	700	1300
Cast iron	700	1300
304 stainless steel	870	1600
316 stainless steel	870	1600
Chrome iron (446)	980	1800
Nickel	980	1800
Inconel	1150	2100
Porcelain	1650(a)	3000(a)
Silicon carbide	1650	3000
Sillimanite	1650(a)	3000(a)
Aluminum oxide	1760(a)	3200(a)

(a) Horizontal tubes should receive additional support above 1480 °C (2700 °F). Source: Adapted from ANSI MC96.1

ture, flame, and abrasion are also listed for comparison.

Each type has its own advantages and limitations, and a knowledge of these advantages and limitations is essential if accurate and reliable measurements are to be made. It is important that these types of insulation be selected only after consideration of exposure temperatures, heating rates, number of temperature cycles, mechanical handling, moisture, routing of wires, and chemical deterioration.

Ceramic Insulation. At temperatures above approximately 300 °C (570 °F), hard-fired ceramic insulators are used on most bare thermocouple elements. Such insulators are available with single, double, or multiple bores, and in a variety of shapes, diameters, and lengths. The thermocouple supplier should be consulted on the type or types of insulation available for each specific application. The hard-fired ceramic insulators that are used with base-metal thermocouples are mullite, aluminum oxide, and steatite. Steatite is the most commonly used material for fish-spline insulators.

Platinum-rhodium thermocouples (types R, S, and B) for use below 1000 °C (1830 °F) may be insulated with quartz, mullite, sillimanite, or porcelain. Mullite and sillimanite have been used in industrial applications involving oxidizing atmospheres and temperatures from 1000 to 1400 °C (1830 to 2550 °F), but 99% Al_2O_3 is preferable for such service. Because both of these materials contain silica in various proportions, care should be taken to prevent promotion of a reducing atmosphere via carbonaceous impurities (such as residual lubricant on thermoelements). For all laboratory uses, for industrial uses in slightly reducing atmospheres (above 1000 °C, or 1830 °F), for critical applications, and for all uses of type B couples to around 1750 °C (3180 °F), pure, sintered, dense alumina (99.5% min Al_2O_3) is recommended. This insulation should be of one-piece, full-length construction to provide maximum protection from contamination.

For iridium-rhodium and tungsten-rhenium thermocouples, choice of insulation depends on temperature of use as well as environment. Hard-fired insulators of high-purity alumina may be used to approximately 1800 °C (3270 °F). From 1800 °C (3270 °F) to approximately 2300 °C (4170 °F), beryllium oxide (melting point: 2565 °C, or 4650 °F) should be considered. However, when beryllium oxide is used, certain safety precautions are necessary.

When hard-fired, dense, beryllia insulators are used, dimensional changes should be considered in design of the temperature-measuring system if it is to be used at or above approximately 2150 °C (3900 °F). At this temperature, beryllia undergoes a phase change. The problem is not serious in swaged thermocouples when crushable beryllia is used.

Thoria, which has a melting point higher than that of beryllia, has been used at temperatures up to about 2500 °C (4500 °F). However, the low electrical resistivity of this ceramic material limits its applications at very high temperatures. Hafnia has been used on an experimental basis with some success.

In addition to conventional thermocouple assemblies with hard-fired ceramic insulators, sheathed, compacted, ceramic-insulated thermocouples are in common use. Magnesium oxide generally is used as the insulating material. A more detailed discussion of this type of construction is presented in a later section on metal-sheathed thermocouples.

Protection. Closed-end tubes made of metal, porcelain, mullite, sillimanite, quartz, or pyrex-glass may be used to prevent contamination of thermocouple sensing elements by the environment and to provide mechanical protection and support. Such tubes are called protection tubes. In some instances, two concentric tubes are employed. A protection tube must be large enough in inside diameter to accommodate an insulated matched couple (positive and negative thermoelements joined at the hot end). However, larger-diameter tubes may be used for strength, to permit insertion of a checking thermocouple alongside the service thermocouple, and to provide an adequate diameter-to-length ratio. Metallic protection tubes are generally available in pipe sizes of ½ in., ¾ in., and 1 in.

Bare, insulated, base-metal thermocouples may be inserted directly into base-metal protection tubes. For noble-metal thermocouples, however, a ceramic protection tube generally is employed between the couple and the base-metal protection tube. For severe operating environments at elevated temperatures, platinum or platinum-rhodium protection tubes may be used. Bare but insulated noble-metal thermocouples may be inserted directly into these tubes. In any event, protection tubes must be internally clean and free of sulfur-bearing compounds, oils, and easily reduced oxides.

A wide range of metal and ceramic protection tubes is available commercially (see for example Table 14). This allows for the selection of a particular protection tube for a specific application.

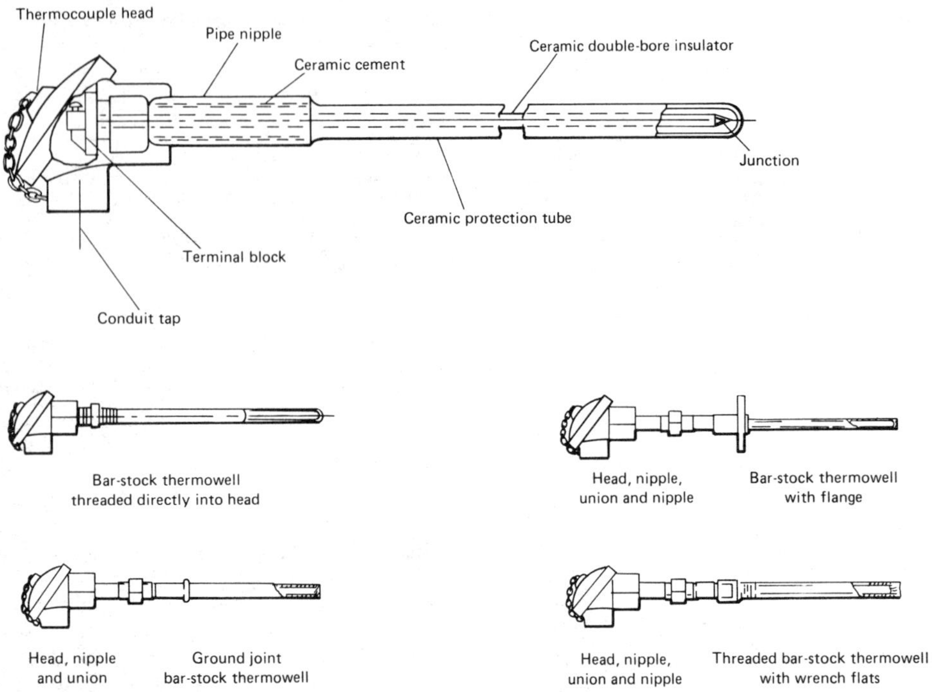

Fig. 17 Typical industrial thermocouples insulated with hard-fired ceramics. Adapted from Ref 9

Steel protection tubes may be used at temperatures up to about 500 °C (930 °F). Stainless steels of the 18-8 variety may be used at up to 800 °C (1470 °F), and stainless steels of higher alloy content at up to about 1000 °C (1830 °F). The 80Ni-20Cr alloys and certain Ni-Cr-Fe alloys may be used to around 1100 °C (2000 °F), with the latter having better resistance to sulfur. It should be remembered that high-nickel alloys should not be used in sulfur-containing atmospheres at temperatures above 400 °C (750 °F).

Protection tubes for platinum-rhodium thermocouples (types R, S, and B) have been made of quartz for service at temperatures up to about 1000 °C (1830 °F), and of mullite for service up to around 1650 °C (3000 °F). Both materials have good resistance to thermal shock. However, in order to ensure long life and emf stability, fused alumina tubes or insulators are preferable for such couples at temperatures above 1200 °C (2200 °F). Fused alumina tubes are more expensive than mullite tubes and have lower resistance to thermal shock. Double ceramic tubes, comprising fused-alumina primary tubes and mullite secondary tubes, are used in certain applications.

Metal protection tubes made of iridium, tantalum, tungsten, and molybdenum, and of Ir-Rh, Nb-1Zr, W-26Re, and Mo-50Re alloys, have been used to protect tungsten-rhenium thermocouples at high temperature. The noble metal iridium is the only known metal that may be used in air unprotected for short periods of time at temperatures up to approximately 2100 °C (3800 °F) without undergoing catastrophic failure. Iridium-rhodium alloys may be used under similar circumstances up to 2000 °C (3600 °F). Experience has shown that, within their recommended temperature ranges, Ir-Rh alloys have better oxidation resistance than that of iridium.

The refractory metal tubes noted above must always be used in inert atmospheres or in a good vacuum. Of these, the tantalum and Nb-1Zr tubes, because of their excellent cold workability, have found extensive use in swaged-type W-Re thermocouples. They are presently being used at temperatures up to approximately 2100 °C (3800 °F).

The Mo-50Re alloy has some interesting possibilities as a material for protective sheaths. This alloy has some cold workability and, more important, is still ductile at room temperature after exposure to temperatures above its recrystallization temperature. At these high temperatures, cleanness of both wire and tubing (thermowell or swaged sheath) is very important. It has been found that carbon present in the tubing (possibly lubricant residue) can react with beryllium oxide insulators at temperatures below 2000 °C (3600 °F).

Protecting wells are employed for thermocouples used in liquids and gasses at high pressure. These wells are made of metal, and they may be turned and drilled from bar stock or built up by welding. Materials such as stainless steels (18-8), carbon steel, and 14% chromium iron are used to fabricate these wells depending on end use.

The foregoing paragraphs describe the procedures used to insulate and protect conventional thermocouples (''bare-wire'' thermocouples). Sheathed thermocouple elements (''swaged-type'' thermocouples) are fabricated from commercially available sheathed thermocouple wires. Fabricating such thermocouples successfully requires special equipment, special precautions, and more skill than is usually required for fabricating conventional bare-wire thermocouple assemblies. This type of thermocouple assembly is described briefly in the following section.

Thermocouple Assemblies

Conventional Thermocouples. Some typical thermocouple assemblies employed in industrial applications are shown in Fig. 17. In the assembly shown at the top of this figure, a closed-end pipe protection tube may be substituted for the nipple and ceramic protection tube in the base-metal thermocouple applications. For additional details, see also ANSI MC96.1 ''Temperature Measurement Thermocouples,'' and suppliers' literature.

Metal-Sheathed Thermocouples. In metal-sheathed couples, the wires are insulated from each other and from the sheath by means of compressed pure refractory oxide powder. The resulting assembly (thermocouple wires, oxide powder, and integral sheath) is flexible enough to be formed around a diameter equal to four times that of the assembly, without damage.

Fabrication of a metal-sheathed thermocouple is simple and begins with matched thermocouple wires surrounded by a partly sintered ceramic material held within a metal tube. By swaging, drawing, or any other mechanical-reduction process, the assembly is reduced in diameter. As a result of this working, the insulation is first broken into powder and then is compacted around the wires while the assembly is elongated.

An assembly produced in this fashion should have a minimum insulation resistance of 100 MΩ at 500 V dc for sizes larger than 1.6 mm (1/16 in.) in outside diameter. This requires some care during fabrication and use of dry, uncontaminated compacted ceramic. Because of the hygroscopic nature of powdered ceramics—especially MgO—moisture can be absorbed through the exposed ends of the sheath by capillary action. For this reason, the metal-sheathed couple or cable should be purchased with the ends closed by welding or suitably sealed in some other manner. Under certain circumstances, organic seals may not be suitable for this purpose. It has been reported that cable sealed with an organic material leaked when shipped by air freight, with a resulting decalibration of the thermocouple. The following precautions should be exercised when handling compacted ceramic-insulated thermocouples, in order to preserve the integrity of the insulation:

- Never leave an end of a sheathed couple exposed for more than 2 or 3 min; seal ends immediately. Use appropriate seal, depending on method of shipping
- Expose ends only in areas of low relative humidity
- Store sheathed assemblies in an area that is warm (above 38 °C, or 100 °F) and dry (relative humidity less than 25%)

Sheaths are selected to suit specific end-use requirements. The materials that have been used for this purpose include: types 304, 310, 316, 321, 347, and 440; stainless steels; platinum alloys; Hastelloy X; copper; aluminum; Inconel 600; Inconel 702; and tantalum and niobium alloys.

Depending on temperature and application, magnesia, alumina, beryllia, or thoria may be used for the insulation. Grounded (to sheath) or ungrounded junctions may be supplied as required, the former having faster response time in temperature sensing.

Among the advantages of compacted sheathed thermocouple construction are small dimensions (as small as 0.5 mm, or 0.020 in., OD) and flexibility. In addition, the assembly is completely resistant to thermal shock, to which more conventional assemblies comprising hard-fired insulators and outer refractory ceramic sheaths are prone.

Criteria for Selection of Thermocouples for Industrial Applications

No thermocouple meets all requirements of temperature measurements over the entire range from cryogenic through 2700 °C (4900 °F). However, each of the previously discussed standard or nonstandard thermocouples possesses characteristics most desirable for a particular application. The following criteria should be given careful consideration during selection and design of thermocouple systems:

Performance requirements

- Accuracy in temperature measurement
- EMF stability
- Service life

Operating environment

- Temperature range
- Time at temperature
- Temperature gradient
- Thermal cycling
- Effect of pressure or vacuum
- Nuclear radiation
- Chemical composition

Cost and availability

- Initial and replacement cost of thermocouple (parts and labor)
- Initial and replacement cost of thermocouple extension wire (parts and labor)
- Initial and replacement cost of thermocouple accessories (parts and labor)
- Downtime
- Delivery time (immediate or extended)

Design selection

- Thermocouple and extension wires (types)
- Temperature/emf relationship and temperature range
- Sensitivity of couple
- Available wire diameter
- Insulation and protection (types)
- Bare-wire versus sheathed construction, with proper end sealing for sheathed construction
- Assembly configuration and type of measuring junction
- Chemical, physical, and mechanical properties (electrical resistance, temperature coefficient of resistivity, coefficient of expansion, thermal conductivity, density, specific heat)

Cryogenic Applications. With the exception of types J, E, T, N, and K, standard thermocouples developed for use at moderate or high temperatures are too low in sensitivity at cryogenic temperatures to be of any practical value in cryogenic applications. Of these five, type E, and to a lesser extent type T, are suitable for general low-temperature service down to −200 °C (−330 °F).

Advocated for applications at still lower temperatures (below 50 K, and possibly as low as 4 K) is a thermocouple consisting of gold plus a trace amount of iron versus either the KP or the EP thermoelement (90Ni-9Cr; see Table 1). Actually, three different Au-Fe thermoelements have been used: Au-0.02 at.% Fe, Au-0.03 at.% Fe, and Au-0.07 at.% Fe. Of these three, the latter may have wider application.

To ensure a high emf in the cryogenic range as well as reproducibility from lot to lot, the Au-Fe alloys must be carefully prepared. High-purity (99.999%) gold is used, and trace amounts of iron are added. For example, only 57 ppm (by weight) of iron are added to the gold in making Au-0.02 at.% Fe, 85 ppm in making Au-0.03 at.% Fe, and 200 ppm in producing Au-0.07 at.% Fe.

If standard thermocouples (types E and T) are intended for cryogenic use (to −200 °C, or −325 °F), the supplier should be notified of this intention in order to facilitate selection of materials. If a Au-Fe/KP or Au-Fe/EP couple is being considered, it should be kept in mind that the gold-bearing thermoelement must be made to special order because it is not a stocked item. Wires used in cryogenic applications generally are fine-gage wires (0.10 to 0.15 mm, or 0.004 to 0.006 in., in diameter).

Good Thermocouple Practice

After the proper thermocouple and extension wire have been selected, care must be taken in installation of the thermocouple system to ensure that errors are not introduced that can affect service. Following are some precautions that should be observed:

- Avoid cold working of thermocouple and extension wire. Excessive deformation can adversely affect accuracy of the thermocouple. Severe bending, flexing, or hammering of the thermocouple wire should be avoided. If cold working does occur, heat treatment should be considered to remove its effects. The degree of cold work that may be tolerated without annealing will depend on the end use and accuracy required
- Extraneous junctions should be monitored—as when connecting lead wires in a thermocouple circuit and when connecting the circuit to a recorder. The solution is to maintain a uniform ambient temperature in the vicinity of these junctions and at the measuring device
- Provide adequate protection. Generally, thermocouples must be equipped with suitable protection in the form of wells or protection tubes to guard the immersed portion against physical damage or contamination

A protecting tube is a tube designed to enclose a temperature-sensing device and to protect it from the deleterious effects of the environment. It may provide for attachment to a connection head, but it is not primarily designed for pressure-tight attachment. A thermowell is a pressure-tight receptacle adapted to receive a temperature-sensing element and is provided with external threads or other means for pressure-tight attachment to a vessel. There are many varieties of these tubes and wells, in various metals, alloys, and refractory materials, available on the market today to meet special requirements. Adequate protection may also be obtained through the use of metal sheathed, mineral insulated thermocouples. In the latter case, the sheath is an integral part of the thermocouple assembly. On the debit side, it should be stated that the protecting tube interferes with ideal temperature measurement and control. It decreases the sensitivity of measurement (speed of response) and increases installation space and cost.

- Select largest practical wire size for a particular end use. The largest practical wire size should be used, consistent with end use requirements such as rapid response, flexibility, and available space. Heavy gage thermocouples have greater long-term stability at high temperatures than thermocouples of a lighter gage but also have a slower response. Speed of response, or rate at which the thermocouple detects temperature changes, may be of vital concern in many applications; particularly where these changes occur rapidly. However, many factors of heat transfer affect the speed of response of a thermocouple, and the mass of the couple is only one of them

- Thermocouples should be located properly to achieve maximum benefits. The thermocouple should be placed so that the measured temperature is representative of the equipment or medium that is being studied. For example, stagnant areas (not at representative temperature) or exposure to direct flame impingement (unless desired) could result in erroneous readings. If the thermocouple is immersed in a fluid (liquid or gas), the depth of immersion should be sufficient to minimize heat transfer away from the measuring junction. For many applications, a minimum immersion depth of ten times the outside diameter of the protection tube is considered adequate to prevent serious temperature errors (readings on low side). Also to be considered is the probability of radiation heat transfer to a bare thermocouple junction from the environment. In this case, radiation shields should be used
- When measuring high temperatures, install the thermocouple vertically whenever possible to prevent sagging of the protection tube. However, care should be taken to properly support the thermocouple within the tube. This is particularly true for the noble metal thermocouples when used at elevated temperatures greater than 600 °C (1100 °F). In this case, the thermocouple assembly may be supported by resting the bead of the thermocouple and the insulator on the bottom of the tube (on high purity alumina powder when a metal tube is used)
- Make sure that the protecting tube or well extends far enough beyond the outer surface of the vessel or heat source to bring the connecting head to approximately ambient temperature (particularly with type K using alternate extension wire and types R or S)
- When making thermocouples, clean the free ends of the wire well before fastening them to the connecting head, and be sure that they are inserted with proper polarity as identified on the terminal block

Maintenance of Thermocouples

Scheduled maintenance can be beneficial. The life and reliability of a thermocouple measuring system can be improved, and the likelihood of catastrophic failure is reduced if recommended calibration and maintenance procedures are followed. In addition, any gradual aging or drift can be determined by periodic observation and recording of thermocouple behavior. Based on information obtained in the maintenance program, scheduled replacement of the thermocouple probe can be made before it has deteriorated beyond acceptable limits or before failure. The portion of the Metal Treating Institute specification MTI2000, "Quality Assurance Specifications for Performance of Heat Treating Processes," relating to temperature measurement is one example of a planned maintenance approach.

The personnel employed in the maintenance program should be familiar with the operating procedures upon which the system is based. Of primary importance, the equipment used for maintenance should be in good working order.

The following items are generally considered a good basis for planned maintenance:

- Thermocouples should be checked regularly at intervals determined by experience. For example, checking base metal thermocouples once a month may be sufficient for some applications but not for others. Exceptions could vary greatly
- If a thermocouple must be removed for examination, carefully reinsert it so as not to change the depth of immersion. Most important of all, do not decrease immersion
- It is preferable to check thermocouples in place. However, rather than checking thermocouples, it may be preferable to replace thermocouples after a predetermined average life has been achieved. This would ensure nearly perfect operation without the periodic problem of checking thermocouples
- A type K thermocouple should not be exposed to temperatures greater than 760 °C (1400 °F) if it is to be used for accurate temperature measurement below 540 °C (1000 °F)
- "Burned out" protection tubes should not be used, otherwise the thermocouple may be damaged or ruined. In particular, old, previously utilized, protection tubes and insulators should not be used with new noble metal thermocouples because of possible contamination (especially true for types R and S thermocouples)
- Contacts must be kept clean if switches are used in the thermocouple circuit
- When recording or indicating potentiometers are connected in parallel for operation from a single thermocouple, the circuit should be analyzed carefully to determine possible effects of one instrument on the other

Troubleshooting. The prime requisite in troubleshooting is a good knowledge and understanding of how the temperature-measuring system operates and why it fails to operate. Familiarity with the previous history of operation of a particular system can be important in quickly determining and correcting a problem.

In general, troubleshooting entails systematic checking of the system one section at a time. Readings should be taken with independent instruments, or one can substitute components that were previously found to be in good working order. This is continued until the problem has been isolated. If an independent source of heat is used to stimulate the system, it may not be possible to produce the same temperature gradients found in normal operation, and the emf results obtained with aged wire will not correspond with those obtained in actual operation.

Because it would be difficult, if not impossible, to discuss all potential operating problems, several common ones have been selected to get an idea of what may be expected in actual practice:

- First make sure that the correct extension or compensating extension wire has been used. This is particularly true for installations that may have a number of different types of thermocouples. In this case, it is not unusual to find extension wire designed for one type of thermocouple to be used with another. Also, it is not unusual to find that the initially purchased extension wire or thermocouples were wrong. A common mistake is to use compensating extension wire intended for the type R or type S thermocouple with the type B couple
- The positive extension wire must be connected to the positive thermoelement and the negative extension wire must be connected to the negative thermoelement. A very large temperature error will be introduced if the polarity is reversed. Figure 18 shows that a temperature error of about 180 °C (355 °F) is observed when a type K couple is used in measuring a temperature of 980 °C (1800 °F) with its extension wire in a reversed position
- Check the polarity of spliced extension wire and make certain that the splice will not be subject to intermittent or high-resistance contact. The system may be installed by electricians or mechanics who may not appreciate the seriousness of polarity reversal
- Make sure that the thermocouple is suitable for the instrument used and that the connections at the terminal block are tight
- If the extension wire color coding differs from the standard U.S. color coding shown in Tables 10, 11, and 12 (see also ANSI Standard MC96.1, Temperature Measurement Thermocouples), particularly in new installations, check to see that this particular extension wire is correct for this thermocouple
- When there is doubt about the type of a thermocouple used, there are several ways by which this can be determined quickly. This can be done visually, as for example distinguishing the noble metal thermocouples (types R, S, and B) from the base metal ones (types J, K, E, and T); copper leg of the type T thermocou-

$$E_t = E_{PN}\Big|_{T_H}^{T_2} + E_{PXNX}\Big|_{T_1}^{T_H}$$

Case 1: correct polarity

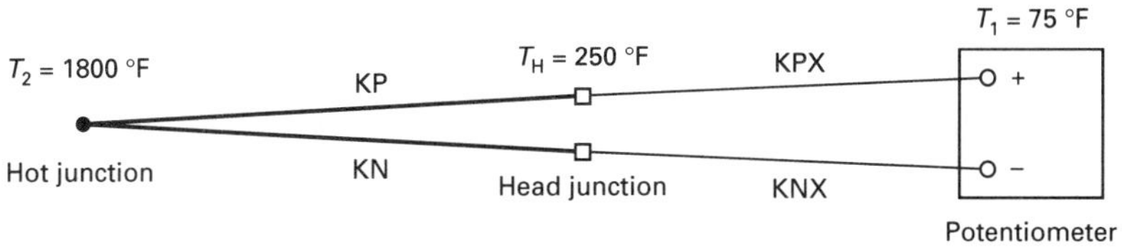

$$E_t = E_{KPKN}\Big|_{250\ °F}^{1800\ °F} + E_{KPXKNX}\Big|_{75\ °F}^{250\ °F} + E_{KPXKNX}\Big|_{32\ °F}^{75\ °F}$$

E_t = (40.62 − 4.97) + (4.97 − 1.00) + 1.00 mV

E_t = 40.62 mV; equivalent to 1800 °F

Case 2: reversed polarity

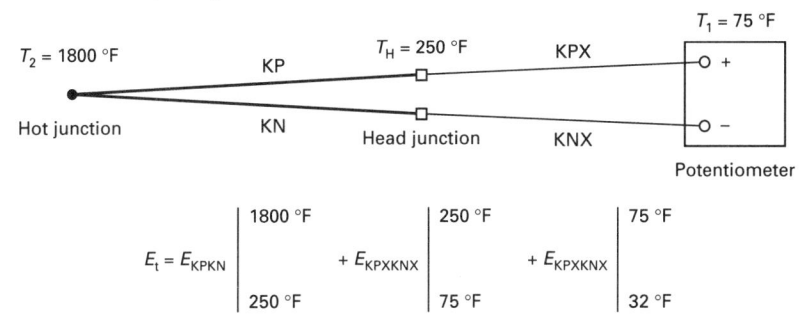

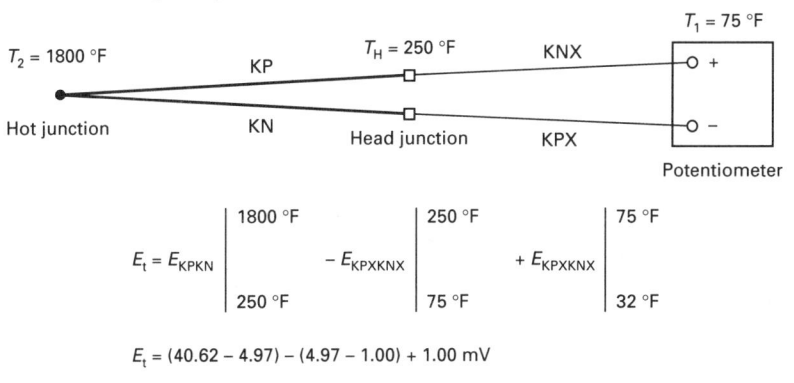

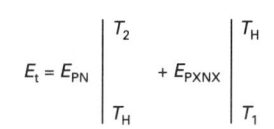

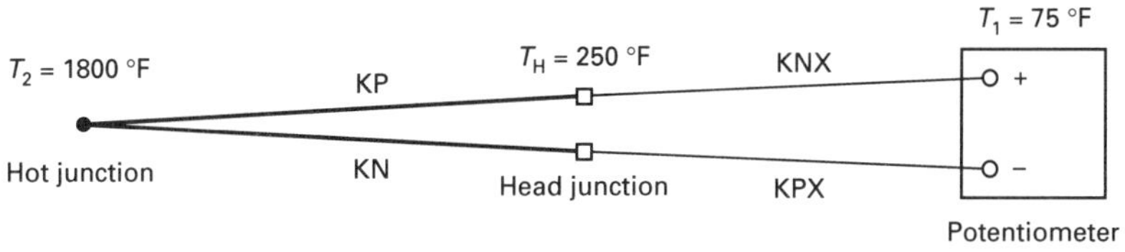

$$E_t = E_{KPKN}\Big|_{250\ °F}^{1800\ °F} - E_{KPXKNX}\Big|_{75\ °F}^{250\ °F} + E_{KPXKNX}\Big|_{32\ °F}^{75\ °F}$$

E_t = (40.62 − 4.97) − (4.97 − 1.00) + 1.00 mV

E_t = 32.68 mV; equivalent to 1445 °F

Fig. 18 Effect of reversed polarity at the head junction on emf output of a type K thermocouple assembly. Adapted from Ref 2

ple. With the use of a magnet, the positive legs of the types E, J, and K couples can be distinguished (magnetic). Other checks consist of making up a thermocouple of the lead wire and checking the output at a fixed temperature; this is also true for distinguishing whether the couple is type R or S. Checks on lead wire may not be necessary where standard color coding is clearly distinguishable but may be useful in old installations where the color coding has become faded

- Checking the resistance of a thermocouple circuit will indicate immediately whether it is in good condition. Low resistance may be equated to the probability of good performance. High resistance may be an indication that the thermocouple is nearing the end of its useful life or that there is a loose connection. In particular, this is a good test in installations where a large number of couples are connected to a readout device through switches

REFERENCES

1. T.P. Wang, Temperature Sensors, *Instrument and Control Systems*, Vol 40, 1967, p 100
2. T.P. Wang, "EMF Measurements," Technical Paper MF77-958, Society of Manufacturing Engineers, 1977
3. T.P. Wang and C.D. Starr, The HI BX, A New Type B Thermocouple Extension Wire, *ISA Trans.*, Vol 16 (No. 3), 1977, p 85
4. T.P. Wang and C.D. Starr, Nicrosil-Nisil Thermocouples in Production Furnaces in the 538 °C (1000 °F) to 1177 °C (2150 °F) Range, *ISA Trans.*, Vol 18 (No. 4), 1979, p 83
5. T.P. Wang and C.D. Starr, Electromotive Force Stability of Nicrosil-Nisil, *ASTM J. Test. Eval.*, Vol 8 (No. 4), July 1980, p 192
6. C.D. Starr and T.P. Wang, A New Stable Nickel-Base Thermocouple, *ASTM J. Test. Eval.*, Vol 21, 1976, p 42
7. N.A. Burley, R.L. Powell, and G.W. Burns, "The Nicrosil Versus Nisil Thermocouple Properties and Thermoelectric Reference Data," NBS Monograph 161, National Bureau of Standards, 1978
8. A.G. Hotchkiss and H.M. Webber, *Protective Atmospheres*, Wiley, 1953, p 1295
9. *Manual on the Use of Thermocouples in Temperature Measurement*, ASTM STP 470B (revised 1980), American Society for Testing and Materials, 1980
10. E.H. McClaren, The Freezing Points of High Purity Metals as Precision Temperature Standards, in *Temperature, Its Measurement and Control in Science and Industry*, Vol 3, Part 1, Reinhold, 1962, p 185
11. The International Practical Temperature Scale of 1968, Amended Edition of 1975, *Metrologia*, Vol 12, 1976
12. B.W. Mangum and G.T. Furukawa, "Guidelines for Realizing the International Temperature Scale of 1990 (ITS-90)," NIST Technical Note 1265, and "Thermocouple Temperature and emf Tables for the ITS-90," NIST Technical Note 175, National Institute of Science and Technology, to be published
13. W.F. Roeser and S.T. Lomberger, "Methods of Testing Thermocouple Materials," NBS Circular 590, National Bureau of Standards, 1958
14. W.G. Trabolt, in *Temperature, Its Measurement and Control in Science and Industry*, Vol 3, Part 2, Reinhold, 1962, p 45
15. "Thermocouple Reference Table Based on IPTS 68," NBS Monograph 125, National Bureau of Standards, 1973
16. C.D. Starr and T.P. Wang, Effect of Oxidation on Stability of Thermocouples, *Trans. ASTM*, Vol 63, 1963
17. T.P. Wang, A.J. Gottlieb, and C.D. Starr, "The EMF Stability of Type K Thermocouple Alloys," Society of Automotive Engineers, 1969

SELECTED REFERENCES

- B.W. Mangum and G.T. Furukawa, "New Temperature Scale (ITS-90)," NIST Technical Note 1265, National Institute of Science and Technology, to be published
- "Precision Measurement and Calibration Temperature," NBS Special Publication, Vol II, National Bureau of Standards, 1968
- "Thermocouple Temperature and emf Tables for the ITS-90," NIST Technical

Note 175, National Institute of Science and Technology, to be published
- "Thermocouple Thermometers," Scientific Apparatus Makers Association, PMC Standard No. 8-10-1963, Process Measurement and Control Section, 1963
- E.D. Zysk, Platinum Metal Thermocouples, in *Temperature, Its Measurement and Control in Science and Industry*, Vol 3, Part 2, Reinhold, 1962
- E.D. Zysk, "Noble Metals in Thermometry," Engelhard Industries Technical Bulletin, Vol 5 (No. 3), Dec 1964
- E.D. Zysk and D.A. Toenshoff, "Calibration of Refractory Metal Thermocouples," Paper 12, Presented at a conference of Instrument Society of America, Oct 1966
- E.D. Zysk and A.R. Robertson, Newer Thermocouple Materials, in *Temperature, Its Measurement and Control in Science and Industry*, Vol 4, Part 3, Instrument Society of America, 1967

Low-Expansion Alloys

Revised by Earl L. Frantz, Carpenter Technology Corporation

LOW-EXPANSION ALLOYS include various binary iron-nickel alloys and several ternary alloys of iron combined with nickel-chromium, nickel-cobalt, or cobalt-chromium alloying. Many of the low-expansion alloys are identified by trade names:

- *Invar*, which is a 64%Fe-36%Ni alloy with the lowest thermal expansion coefficient of iron-nickel alloys
- *Kovar*, which is a 54%Fe-29%Ni-17%Co alloy with coefficients of expansion closely matching those of standard types of hard (borosilicate) glass
- *Elinvar*, which is a 52%Fe-36%Ni-12%Cr alloy with a zero thermoelastic coefficient (that is, an invariable modulus of elasticity over a wide temperature range)
- *Super Invar*, which is a 63%Fe-32%Ni-5%Co alloy with an expansion coefficient smaller than Invar but over a narrower temperature range

Besides these common trade names, alloy compositions are also selected to have appropriate expansion characteristics for a particular application. Low-expansion alloys are used in applications such as:

- Rods and tapes for geodetic surveying
- Compensating pendulums and balance wheels for clocks and watches
- Moving parts that require control of expansion, such as pistons for some internal-combustion engines
- Bimetal strip
- Glass-to-metal seals
- Thermostatic strip
- Vessels and piping for storage and transportation of liquefied natural gas
- Superconducting systems in power transmissions
- Integrated-circuit lead frames
- Components for radios and other electronic devices
- Structural components in optical and laser measuring systems

Low-expansion alloys are also used with high-expansion alloys (65%Fe-27%Ni-5%Mo, or 53%Fe-42%Ni-5%Mo) to produce movements in thermoswitches and other temperature-regulating devices.

Iron-Nickel Alloys

Alloys of iron and nickel have coefficients of linear expansion ranging from a small negative value (−0.5 ppm/°C) to a large positive (20 ppm/°C) value. Figure 1 shows the effect of nickel content on the linear expansion of iron-nickel alloys at room temperature. In the range of 30 to 60% Ni, it is possible to select alloys with appropriate expansion characteristics. The alloy containing 36% nickel (with small quantities of manganese, silicon, and carbon amounting to a total of less than 1%) has a coefficient of expansion so low that its length is almost invariable for ordinary changes in temperature. This alloy is known as Invar, which is a trade name (of Imphy, S.A.) meaning invariable.

After the discovery of Invar, an intensive study was made of the thermal and elastic properties of several similar alloys. Iron-nickel alloys that have nickel contents higher than that of Invar retain to some extent the expansion characteristics of Invar. Alloys that contain less than 36% nickel have much higher coefficients of expansion than alloys containing 36% or more nickel. Further information on iron-nickel alloys besides Invar is given in the section "Iron-Nickel Alloys Other Than Invar" in this article.

Invar

Invar (UNS number K93601) and related alloys have low coefficients of expansion over only a rather narrow range of temperature (see Fig. 2). At low temperatures in the region from A to B, the coefficient of expansion is high. In the interval between B and C, the coefficient decreases, reaching a minimum in the region from C to D. With increasing temperature, the coefficient begins again to increase from D to E, and thereafter (from E to F), the expansion curve follows a trend similar to that of the nickel or iron of which the alloy is composed. The minimum expansivity prevails only in the range from C to D.

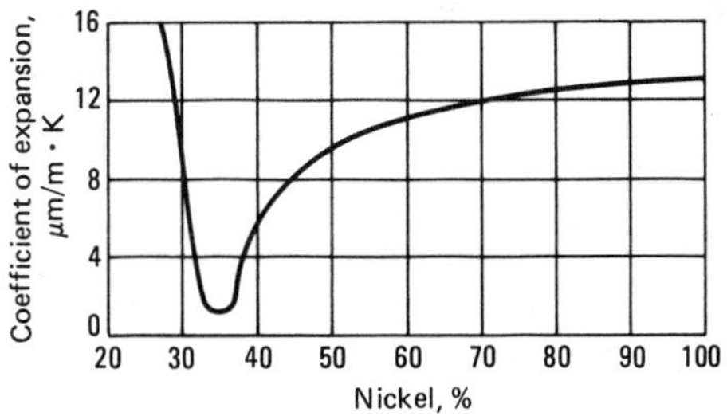

Fig. 1 Coefficient of linear expansion at 20 °C versus Ni content for Fe-Ni alloys containing 0.4% Mn and 0.1% C

In the region between D and E in Fig. 2, the coefficient is changing rapidly to a higher value. The temperature limits for a well-annealed 36% Ni iron are 162 and 271 °C (324 and 520 °F). These temperatures correspond to the initial and final losses of magnetism in the material (that is, the Curie temperature). The slope of the curve between C and D is then a measure of the coefficient of expansion over a limited range of temperature.

Table 1 gives coefficients of linear expansion of iron-nickel alloys between 0 and 38 °C (32 and 100 °F). The expansion behavior of several iron-nickel alloys over wider ranges of temperature is represented by curves 1 to 5 in Fig. 3. For comparison, Fig. 3 also includes the similar expansion obtained for ordinary steel.

Effects of Composition on Expansion Coefficient. The effect of variation in nickel content on linear expansivity is shown in Fig. 1. Minimum expansivity occurs at about 36% Ni, and small additions of other metals have considerable influences on the position of this minimum. Because further additions of nickel raise the temperature at which the inherent magnetism of the alloy disappears, the inflection temperature in the

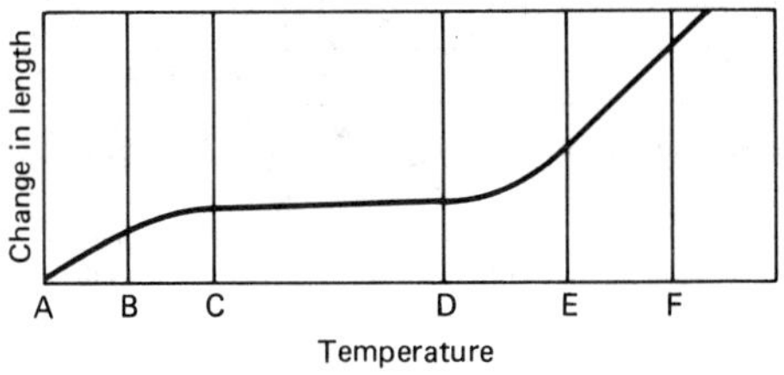

Fig. 2 Change in length of a typical Invar over different ranges of temperature

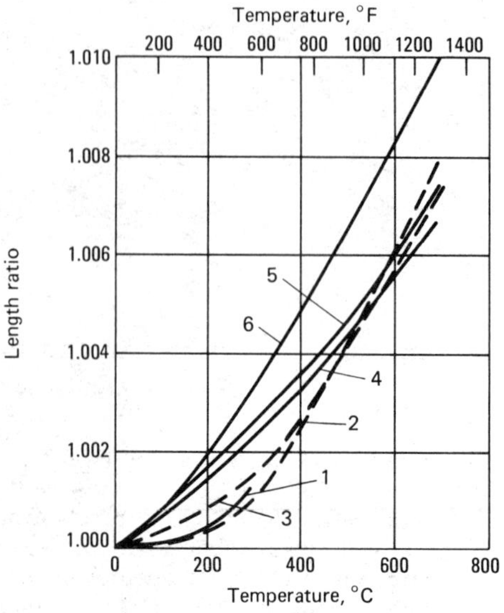

Fig. 3 Thermal expansion of Fe-Ni alloys. Curve 1, 64Fe-31Ni-5Co; curve 2, 64Fe-36Ni (Invar); curve 3, 58Fe-42Ni; curve 4, 53Fe-47Ni; curve 5, 48Fe-52Ni; curve 6, carbon steel (0.25% C)

expansion curve (Fig. 2) also rises with increasing nickel content.

The addition of third and fourth elements to Fe-Ni provides useful changes of desired properties (mechanical and physical), but significantly changes thermal expansion characteristics. Minimum expansivity shifts toward higher nickel contents when manganese or chromium is added, and toward lower nickel contents when copper, cobalt, or carbon is added. Except for the ternary alloys with nickel-iron-cobalt compositions (Super-Invars), the value of the minimum expansivity for any of these ternary alloys is, in general, greater than that of a typical Invar alloy.

The effects of additions of manganese, chromium, copper, and carbon are shown in Fig. 4. Additions of silicon, tungsten, and molybdenum produce effects similar to those caused by additions of manganese and chromium; the composition of minimum expansivity shifts towards higher contents of nickel. Addition of carbon is said to produce instability in Invar, which is attributed to the changing solubility of carbon in the austenitic matrix during heat treatment.

Table 1 Thermal expansion of Fe-Ni alloys between 0 and 38 °C

Ni, %	Mean coefficient, μm/m · K
31.4	3.395 + 0.00885 *t*
34.6	1.373 + 0.00237 *t*
35.6	0.877 + 0.00127 *t*
37.3	3.457 − 0.00647 *t*
39.4	5.357 − 0.00448 *t*
43.6	7.992 − 0.00273 *t*
44.4	8.508 − 0.00251 *t*
48.7	9.901 − 0.00067 *t*
50.7	9.984 + 0.00243 *t*
53.2	10.045 + 0.00031 *t*

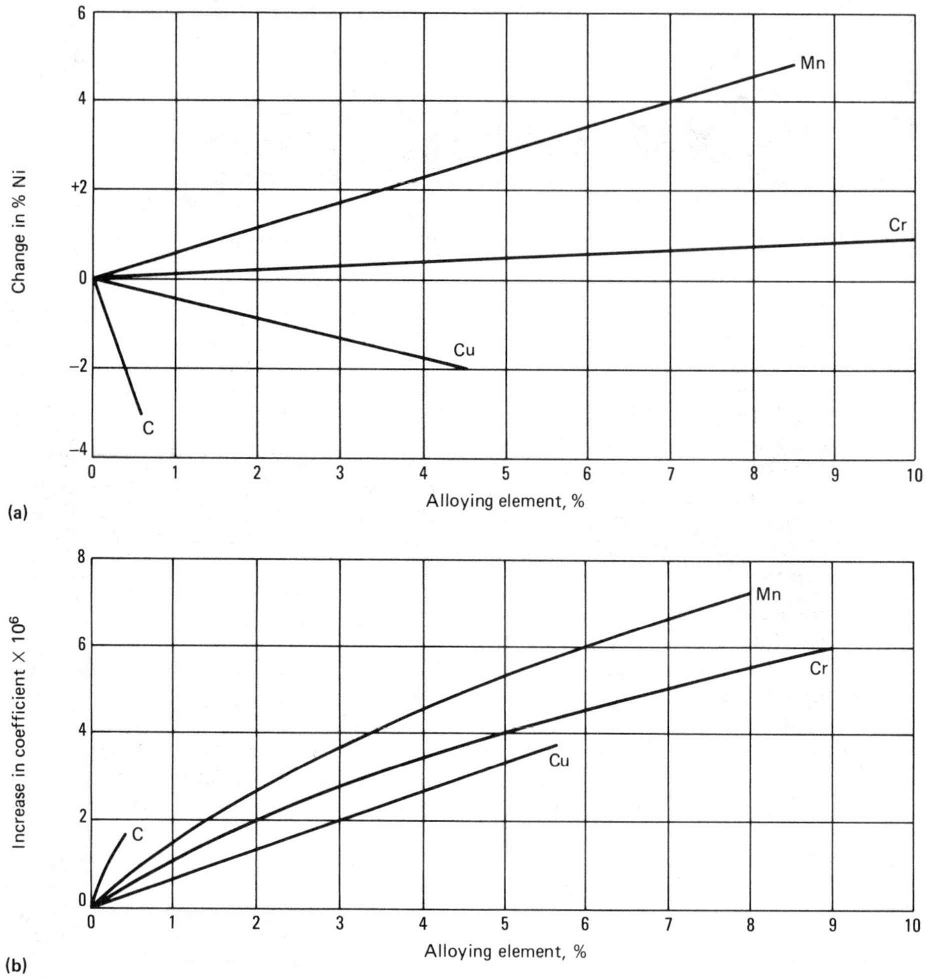

Fig. 4 Effect of alloying elements on expansion characteristics of Fe-Ni alloys. (a) Displacement of nickel content caused by additions of manganese, chromium, copper, and carbon to alloy of minimum expansivity. (b) Change in value of minimum coefficient of expansion caused by additions of manganese, chromium, copper, and carbon

Effects of Processing. Heat treatment and cold work change the expansivity of Invar alloys considerably. The effect of heat treatment for a 36% Ni Invar alloy is shown in Table 2. The expansivity is greatest in well-annealed material and least in quenched material.

Hot workability is enhanced by very close control of deoxidation and degassing during the melt process. Considerable care must be used in hot working of iron-nickel alloys because at hot-working temperature they have a tendency to check and break up when carelessly handled. Invar and related alloys should be annealed in a reducing atmosphere. Because they are susceptible to intercrystalline oxidation during annealing, they should be processed in an atmosphere that contains a large percentage of a neutral gas (such as nitrogen) and a small percentage of a reducing gas. Cold rolling and drawing of iron-nickel alloys are quite similar to corresponding processing procedures for nickel.

Heat Treatment. The iron-nickel binary alloys are not hardenable by heat treatment. Annealing practice should be adjusted to be consistent with requirements of the intended application. Exposure to temperatures and times that promote excessive grain growth will limit further fabricating steps that require extreme bending, forming, deep drawing, chemical etching, and so forth.

Annealing is done at 750 to 850 °C (1380 to 1560 °F). When the alloy is quenched in

Table 2 Effect of heat treatment on coefficient of thermal expansion of Invar

Condition	Mean coefficient, μm/m · K
As forged	
At 17–100 °C (63–212 °F)	1.66
At 17–250 °C (63–480 °F)	3.11
Quenched from 830 °C (1530 °F)	
At 18–100 °C (65–212 °F)	0.64
At 18–250 °C (65–480 °F)	2.53
Quenched from 830 °C and tempered	
At 16–100 °C (60–212 °F)	1.02
At 16–250 °C (60–480 °F)	2.43
Cooled from 830 °C to room temperature in 19 h	
At 16–100 °C (60–212 °F)	2.01
At 16–250 °C (60–480 °F)	2.89

water from these temperatures, expansivity is decreased, but instability is induced both in actual length and in coefficient of expansion. To overcome these deficiencies and to stabilize the material, it is common practice to stress relieve approximately at 315 to 425 °C (600 to 800 °F) and to age at a low temperature 90 °C (200 °F) for 24 to 48 hours.

Cold drawing also decreases the thermal expansion coefficient of Invar alloys. The values for the coefficients in the following table are from experiments on two heats of Invar:

Material condition	Expansivity, ppm/°C
Direct from hot mill	1.4 (heat 1)
	1.4 (heat 2)
Annealed and quenched	0.5 (heat 1)
	0.8 (heat 2)
Quenched and cold drawn (>70% reduction with a diameter of 3.2 to 6.4 mm, or 0.125 to 0.250 in.)	0.14 (heat 1)
	0.3 (heat 2)

By cold working after quenching, it is possible to produce material with a zero, or even a negative, coefficient of expansion. A negative coefficient may be increased to zero by careful annealing at a low temperature. However, these artificial methods of securing an exceptionally low coefficient may produce instability in the material. With lapse of time and variation in temperature, exceptionally low coefficients usually revert to normal values. For special applications (geodetic tapes, for example), it is essential to stabilize the material by cooling it slowly from 100 to 20 °C (212 to 68 °F) over a period of many months, followed by prolonged aging at room temperature. However, unless the material is to be used within the limits of normal atmospheric variation in temperature, such stabilization is of no value. Although these variations in heat-treating practice are important in special applications, they are of little significance for ordinary uses.

Magnetic Properties. Invar and all similar iron-nickel alloys are ferromagnetic at room temperature and become paramagnetic at higher temperatures. Because additions in nickel contents raise the temperature at which the inherent magnetism of the alloy disappears, the inflection temperature in the expansion curve rises with increasing nickel content. The loss of magnetism in a well-annealed sample of a true Invar begins at 162 °C (324 °F) and ends at 271 °C (520 °F). In a quenched sample, the loss begins at 205 °C (400 °F) and ends at 271 °C (520 °F). Figure 5 shows how the Curie temperature changes with nickel content in iron.

Table 3 Physical and mechanical properties of Invar

Solidus temperature, °C (°F)	1425 (2600)
Density, g/cm^3	8.1
Tensile strength, MPa (ksi)	450–585 (65–85)
Yield strength, MPa (ksi)	275–415 (40–60)
Elastic limit, MPa (ksi)	140–205 (20–30)
Elongation, %	30–45
Reduction in area, %	55–70
Scleroscope hardness	19
Brinell hardness	160
Modulus of elasticity, GPa (10^6 psi)	150 (21.4)
Thermoelastic coefficient, μm/m · K	500
Specific heat, at 25–100 °C (78–212 °F), J/kg · °C (Btu/lb · °F)	515 (0.123)
Thermal conductivity, at 20–100 °C (68–212 °F), W/m · K (Btu/ft · h · °F)	11 (6.4)
Thermoelectric potential (against copper), at −96 °C (−140 °F), μV/K	9.8

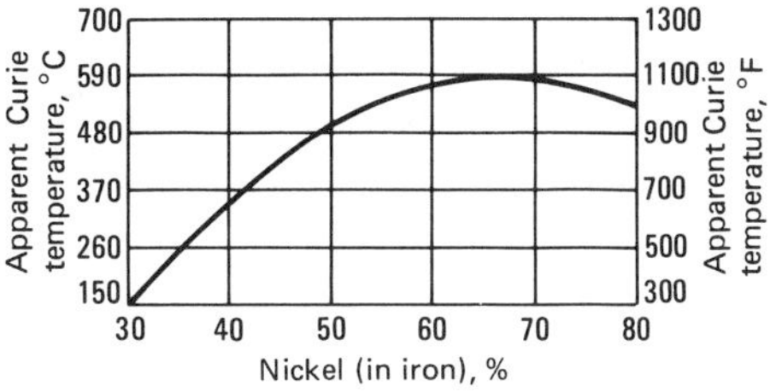

Fig. 5 Effect of nickel content on the Curie temperature of iron-nickel alloys

The thermoelastic coefficient, which describes the changes in the modulus of elasticity as a function of temperature, varies according to the nickel content of iron-nickel low-expansion alloys. Invar has the highest thermoelastic coefficient of all low-expansion iron-nickel alloys, while two alloys with 29 and 45% nickel have a zero thermoelastic coefficient (that is, the modulus of elasticity does not change with temperature). However, because small variations in nickel content produce large variations in the thermoelastic coefficient, commercial application of these two iron-nickel alloys with a zero thermoelastic coefficient is not practical. Instead, the iron-nickel-chromium Elinvar alloy provides a practical way of achieving a zero thermoelastic coefficient.

Electrical Properties. The electrical resistivity of 36Ni-Fe Invar is between 750 and 850 nΩ · m at ordinary temperatures. The temperature coefficient of electrical resistivity is about 1.2 mΩ/Ω · K over the range of low expansivity. As nickel content increases above 36%, the electrical resistivity decreases to ~165 nΩ · M at ~80% NiFe. This is illustrated in Fig. 6.

Other Physical and Mechanical Properties. Table 3 presents data on miscellaneous properties of Invar in the hot-rolled and forged conditions. The effects of temperature on mechanical properties of forged 66Fe-34Ni are illustrated in Fig. 7.

The binary iron-nickel alloys are not hardenable by heat treatment. Significant increases in strength can be obtained by cold working some product forms such as wire, strip, and small-diameter bar. Table 4 shows tensile and hardness data for both 36% and 50% nickel-iron alloys after cold working various percent cross-section reductions.

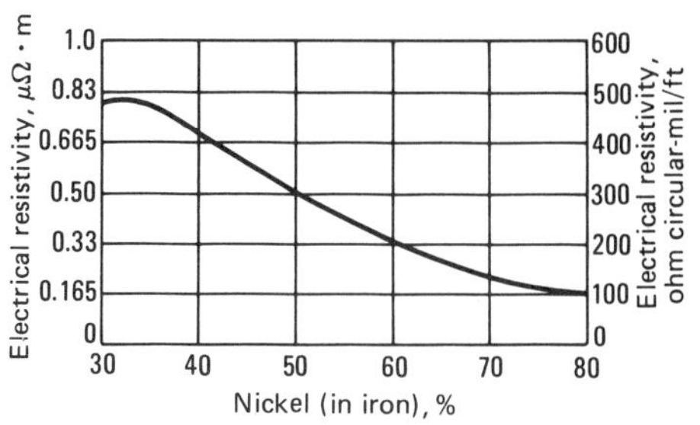

Fig. 6 Effect of nickel content on electrical resistivity of nickel-iron alloys

Mechanical properties such as tensile strength and hardness decrease rapidly with increasing service temperatures. Selected elevated-temperature data for iron-nickel alloys are shown in Table 5.

Corrosion Resistance. The iron-nickel low-expansion alloys are not corrosion resistant, and applications in even relatively mild corrosive environments must consider their propensity to corrode. A comparison to corrosion of iron, in both high humidity and salt spray environments, is shown in Fig. 8 and Table 6. Rust initiation occurs in approximately 24 hours for nickel contents less than ~40% in high-humidity tests. Severe corrosion occurs after 200 hours exposure to a neutral salt spray at 35 °C (95 °F).

Machinability. The iron-nickel alloys can be machined using speeds or feeds that are

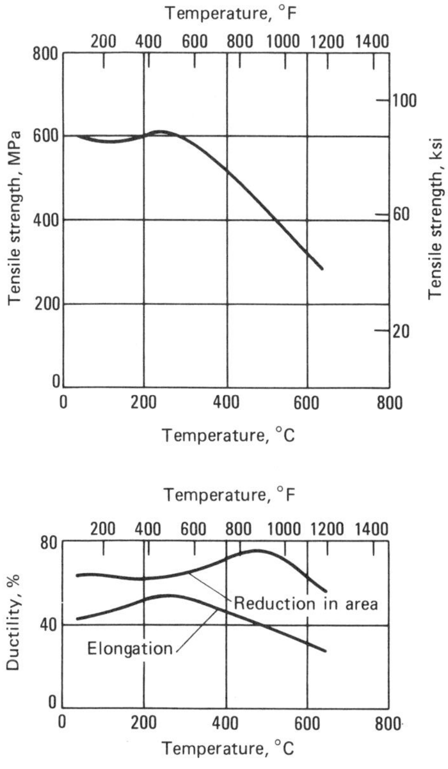

Fig. 7 Mechanical properties of a forged 34% Ni alloy. Alloy composition: 0.25 C, 0.55 Mn, 0.27 Si, 33.9 Ni, balance Fe. Heat treatment: annealed at 800 °C (1475 °F) and furnace cooled

Table 4 Mechanical properties of Invar and a 52% Ni-48% Fe glass-sealing alloy

UNS number (alloy name)	0.2% yield strength MPa	ksi	Ultimate tensile strength MPa	ksi	Elongation, %	Approximate equivalent hardness, HRB
K93601 (Invar 36% Ni)						
As annealed	260	38	470	68	37	75
10% cold worked	370	54	565	82	23	86
30% cold worked	550	80	675	98	10	95
50% cold worked	640	93	725	105	5	96
70% cold worked	703	102	730	106	3	97
K14052 (glass-sealing alloy 52% Ni)						
As annealed	235	34	538	78	32	83
10% cold worked	525	76	640	93	19	92
30% cold worked	715	104	750	109	6	99
50% cold worked	770	112	814	118	3	100
70% cold worked	800	116	834	121	2	26 HRC

Table 5 Typical tensile properties at elevated temperatures for some low-expansion nickel-iron alloys

UNS number (alloy name)	Test temperature °C	°F	0.2% yield strength MPa	ksi	Ultimate tensile strength MPa	ksi	Elongation, %	Reduction of area, %
Invar 36% Ni (K93601)	24	75	265	38.5	483	70	44	81.5
	150	300	139	20.2	405	59	44.5	77.5
	315	600	95	14	420	61	50	73
	480	900	90	13	275	40	63	73
42% Ni low-expansion alloy (K94100)	24	75	295	43	550	80	43.7	73.5
	150	300	225	32.5	510	74	45.6	67.1
	315	600	188	27.3	495	71.8	52.8	67.1
	480	900	157	22.8	370	54	43	58.4
49% Ni low expansion alloy	24	75	300	43.3	538	78	46.2	79.3
	150	300	243	35.3	483	70	43.2	75.6
	315	600	223	32.3	462	67	42	73.5
	480	900	217	31.5	385	55.8	35.5	51.9

Table 6 Effects of relative humidity on selected nickel-iron low-expansion alloys

Specimens exposed to 95% relative humidity for 200 h at 35 °C (95 °F)

Alloy type (UNS number)	Condition	Portion of surface rusted (average), %(a)	First rust (three specimens), h
Electrical iron	Annealed	70	1, 1, 1
	Cold rolled	50	1, 1, 2
30% Ni temperature-compensator alloy(b)	Annealed	5	24, 24, 24
	Cold rolled	>5	24, 24, 24
Invar 36% Ni (K93601)	Annealed	Few rust spots	48, 48, 96
	Cold rolled	Few rust spots	96, 96, 96
42% Ni low-expansion alloy (K94100)	Annealed	0	...
	Cold rolled	0	...
49% Ni low-expansion alloy	Annealed	0	...
	Cold rolled	0	...
52% Ni glass-sealing alloy (K14052)	Annealed	0	...
	Cold rolled	0	...
80% Ni-4.5% Mo(c)	Annealed	0	...
	Cold rolled	0	...

(a) Visual estimate of the percentage of surface rusted. (b) Provided under the tradename of "Temperature Compensation 30" by Carpenter Technology Corporation. (c) Provided under the tradenames of "MolyPermalloy" by Allegheny Ludlum and "HyMu80" by Carpenter Technology Corporation

modified to accommodate their gummy and stringy characteristics. As a general comparison to other austenitic alloys, they are similar to 316 stainless steel. Using single point turning as a measure of machinability, the iron-nickel low-expansion alloys exhibit a 25% machinability rating compared to resulfurized carbon steel (such as B 1112). Some general machining parameters for iron-nickel alloys are shown in Table 7. There are "free-cut" varieties of Invar-type alloys available. These require minor additions of other elements (such as selenium) which when combined with moderate adjustments to other residual elements (such as manganese) will produce a twofold improvement in the machinability characteristic of these alloys. Some increase in thermal expansion characteristic results from the modified compositions.

Welding. Invar can be successfully welded using most standard arc-welding processes. In general, preparation for welding should be similar to stainless steels and should include proper cleaning and handling. Joint designs should allow easy access to the weld because of poor weld pool fluidity but also should limit total weld volume to reduce shrinkage problems. Preheating and postheating are not required

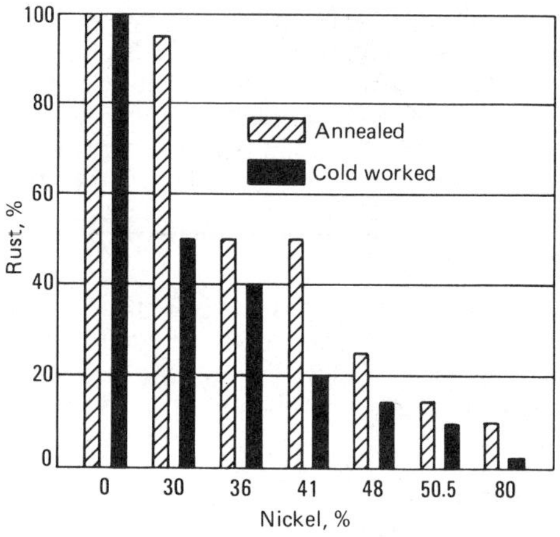

Fig. 8 Rust versus nickel content from 200 h neutral salt spray at 35 °C (95 °F)

Table 7 Examples of various machining parameters for iron-nickel low-expansion alloys

Turning (single-point and box tools)	
Roughing	
Depth of cut, mm (in.)	2.5 (0.1)
Speed, m/min (ft/min)	9 (30)
Feed, mm/rev (in./rev)	0.25 (0.010)
Finishing	
Depth of cut, mm (in.)	0.5 (0.020)
Speed, m/min (ft/min)	6 (20)
Feed, mm/rev (in./rev)	0.05 (0.002)
Turning (cutoff and form tools)	
Speed, m/min (ft/min)	6 (20)
Feed, mm/rev (in./rev) with a tool width of:	
3.2 mm (0.125 in.)	0.025 (0.001)
6.4 mm (0.250 in.)	0.025 (0.001)
13 mm (0.50 in.)	0.038 (0.0015)
25 mm (1.0 in.)	0.025 (0.001)
50 mm (2.0 in.)	0.018 (0.0007)
Drilling	
Speed, m/min (ft/min)	10 (35)
Feed, mm/rev (in./rev) for a drill diameter of:	
1.6 mm (1/16 in.)	0.025 (0.001)
3.2 mm (1/8 in.)	0.075 (0.003)
6.4 mm (1/4 in.)	0.10 (0.004)
13 mm (1/2 in.)	0.20 (0.008)
20 mm (3/4 in.)	0.25 (0.010)
25 mm (1 in.)	0.30 (0.012)
38 mm (1 1/2 in.)	0.38 (0.015)
50 mm (2 in.)	0.45 (0.018)
Tapping speed, m/min (ft/min)	
≤7 threads per 25 mm (1 in.)	1.8 (6)
8–15 threads per 25 mm (1 in.)	2.4 (8)
6–24 threads per 25 mm (1 in.)	3.65 (12)
>24 threads per 25 mm (1 in.)	4.5 (15)
End milling parameters	
With 0.5 mm (0.020 in.) radial depth of cut:	
Speed, m/min (ft/min)	20 (65)
Feed, mm/tooth (in./tooth), with 13 mm (1/2 in.) cutter diam	0.05 (0.002)
Feed, mm/tooth (in./tooth), with 25–50 mm (1–2 in.) cutter diam	0.10 (0.004)
With 1.5 mm (0.06 in.) radial depth of cut:	
Speed m/min (ft/min)	15 (50)
Feed, mm/tooth (in./tooth), with 13 mm (1/2 in.) cutter diam	0.075 (0.003)
Feed, mm/tooth (in./tooth), with 25–50 mm (1–2 in.) cutter diam	0.125 (0.005)

Table 8 Expansion characteristics of Fe-Ni alloys

Composition, % Mn	Si	Ni	Inflection temperature °C	°F	Mean coefficient of expansion, from 20 °C to inflection temperature, μm/m · K
0.11	0.02	30.14	155	310	9.2
0.15	0.33	35.65	215	420	1.54
0.12	0.07	38.70	340	645	2.50
0.24	0.03	41.88	375	710	4.85
...	...	42.31	380	715	5.07
...	...	43.01	410	770	5.71
...	...	45.16	425	800	7.25
0.35	...	45.22	425	800	6.75
0.24	0.11	46.00	465	870	7.61
...	...	47.37	465	870	8.04
0.09	0.03	48.10	497	925	8.79
0.75	0.00	49.90	500	930	8.84
...	...	50.00	515	960	9.18
0.25	0.20	50.05	527	980	9.46
0.01	0.18	51.70	545	1015	9.61
0.03	0.16	52.10	550	1020	10.28
0.35	0.04	52.25	550	1020	10.09
0.05	0.03	53.40	580	1075	10.63
0.12	0.07	55.20	590	1095	11.36
0.25	0.05	57.81	None		12.24
0.22	0.07	60.60	None		12.78
0.18	0.04	64.87	None		13.62
0.00	0.05	67.98	None		14.37

and should be avoided. A low interpass temperature (150 °C, or 300 °F max) should be maintained.

Welding is most commonly performed using the gas-tungsten-arc or gas-metal-arc processes. Gas-tungsten-arc welding can be accomplished with argon and/or helium shielding gases. Welding is best performed with a freshly ground thoriated tungsten electrode. Gas-metal-arc welding can be successfully performed in all metal transfer modes, depending primarily on base metal thickness. Shielding gases should be argon or argon-helium mixtures. Other nonarc welding processes (such as resistance welding) may also be used.

When a filler metal is needed, a matching composition will provide the best match in thermal expansion properties. Invarod weld filler metal (a 36Ni-Fe alloy containing ~1% Ti and 2.5% Mn) has been successfully used for matching expansion characteristics. If a matching composition is not available, a high-nickel filler metal conforming to AWS A5.14 ERNi-1 or ERNiCrFe-5 can be used. These materials will result in a weld with different thermal expansion properties.

Iron-Nickel Alloys Other Than Invar

Although iron-nickel alloys other than Invar have higher coefficients of thermal expansion, there are applications where it is advantageous to have nickel contents above or below the 36% level of Invar. The alloy containing 39% Ni, for example, has a coefficient of expansion corresponding to that of low-expansion glasses.

Alloys that contain less than 36% Ni have much higher coefficients of expansion than alloys with a higher percentage. Alloys containing less than 36% Ni include temperature-compensator alloys (30 to 34% Ni). These exhibit linear changes in magnetic characteristics with temperature change. They are used as compensating shunts in metering devices and speedometers.

Iron-nickel alloys that have nickel contents higher than that of Invar retain to some extent the expansion characteristics of Invar. Because further additions of nickel raise the temperature at which the inherent magnetism of the alloy disappears, the inflection temperature in the expansion curve (Fig. 2) rises with increasing nickel content. Although this increase in range is an advantage in some circumstances, it is accompanied by an increase in coefficient of expansion. Table 8 and Fig. 9 present additional information on the coefficients of expansion of nickel-iron alloys at temperatures up to the inflection temperature. They also give data on alloys with up to 68% Ni.

Of significant commercial interest are those alloys containing approximately 40% to 50% nickel-iron alloys. Typical compositions and thermal expansions for some of these alloys are given in Table 9.

The 42% Ni-irons are widely used in applications for their low-expansion characteristics. These include semiconductor packaging components, thermostat bimetals, incandescent light bulb glass seal leads (copper clad), and seal beam lamps.

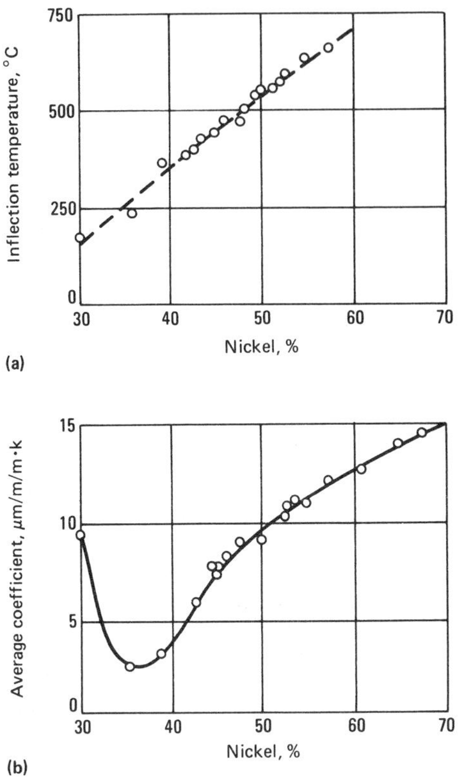

Fig. 9 Effect of nickel content on expansion of Fe-Ni alloys. (a) Variation of inflection temperature. (b) Variation of average coefficient of expansion between room temperature and inflection temperature

Table 9 Composition and typical thermal expansion coefficients for common iron-nickel low-expansion alloys

Alloy	ASTM specification	Composition(a), % C(max)	Mn(max)	Si(max)	Ni(nom)
42 Ni-Iron	F 30	0.02	0.5	0.25	41
46 Ni-Iron	F 30	0.02	0.5	0.25	46
48 Ni-Iron	F 30	0.02	0.5	0.25	48
52 Ni-Iron	F 30	0.02	0.5	0.25	51
42 Ni-Iron (Dumet)	F 29	0.05	1.0	0.25	42
42 Ni-Iron (Thermostat)	B 753	0.10	0.4	0.25	42

Alloy	Typical thermal expansion coefficients from room temperature to: 300 °C (570 °F) ppm/°C	ppm/°F	400 °C (750 °F) ppm/°C	ppm/°F	500 °C (930 °F) ppm/°C	ppm/°F
42 Ni-Iron	4.4	2.4	6.0	3.3	7.9	4.4
46 Ni-Iron	7.5	4.2	7.5	4.2	8.5	4.7
48 Ni-Iron	8.8	4.9	8.7	4.8	9.4	5.2
52 Ni-Iron	10.1	5.6	9.9	5.5	9.9	5.5
42 Ni-Iron (Dumet)	...	...	6.6	3.7	...	...
42 Ni-Iron (Thermostat)	5.8(b)	3.2(b)	5.6(c)	3.1(c)	5.7(d)	3.15(d)

(a) Balance of iron with residual impurity limits of 0.25% max Si, 0.015% max P, 0.01% max S, 0.25% max Cr, and 0.5% max Co. (b) From room temperature to 90 °C (200 °F). (c) From room temperature to 150 °C (300 °F). (d) From room temperature to 370 °C (700 °F)

Table 10 Type, composition, and typical thermal expansion for some iron-nickel-chromium glass-seal alloys

Alloy type	ASTM specifications	Composition(a), % Mn(max)	Si(max)	Cr(nom)	Ni(nom)
42-6	F 31	0.25	0.25	5.75	42.5
45-5	...	0.25	0.30	6.00	45.0
48-5	...	0.30	0.20	6.00	47.5

Alloy type	Average thermal expansion coefficients from room temperature to: 200 °C (390 °F) ppm/°C	ppm/°F	300 °C (570 °F) ppm/°C	ppm/°F	400 °C (750 °F) ppm/°C	ppm/°F	500 °C (930 °F) ppm/°C	ppm/°F
42-6	7.1	3.9	8.3	4.6	10.0	5.55	11.5	6.4
45-5	8.2	4.55	8.7	4.8	10.0	5.55	11.2	6.2
48-5	...	...	9.4	5.2	10.3	5.7	...	...

(a) Balance of iron with 0.05% max C, 0.015% max P, 0.015% max S, and 0.50% max Co

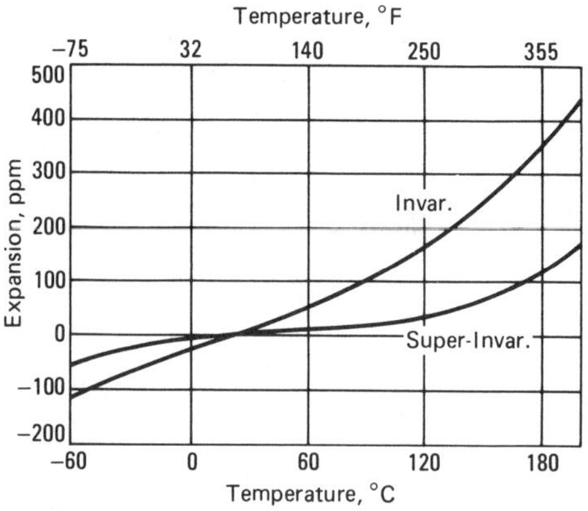

Fig. 10 Comparison of thermal expansion for Super-Invar (63% Fe, 32% Ni, 5% Co) and Invar (64% Fe, 36% Ni) alloys

Dumet wire is an alloy containing 42% Ni. It is clad with copper to provide improved electrical conductivity and to prevent gassing at the seal. It can replace platinum as the seal-in wire in incandescent lamps and vacuum tubes.

The 43 to 47% Ni-iron alloys are commonly used for glass seal leads, grommets, and filament supports. This group of alloys includes Platinate (36% Ni to 64% Fe), which has a coefficient of thermal expansion equivalent to that of platinum (9.0 ppm/°C).

Iron-Nickel-Chromium Alloys

Elinvar is a low-expansion iron-nickel-chromium alloy with a thermoelastic coefficient of zero over a wide temperature range. It is more practical than the straight iron-nickel alloys with a zero thermoelastic coefficient, because its thermoelastic coefficient is less susceptible to variations in nickel content expected in commercial melting.

Elinvar is used for such articles as hairsprings and balance wheels for clocks and watches and for tuning forks used in radio synchronization. Particularly beneficial where an invariable modulus of elasticity is required, it has the further advantage of being comparatively rustproof.

The composition of Elinvar has been modified somewhat from its original specification of 36% Ni and 12% Cr. The limits now used are 33 to 35 Ni, 61 to 53 Fe, 4 to 5 Cr, 1 to 3 W, 0.5 to 2 Mn, 0.5 to 2 Si, and 0.5 to 2 C. Elinvar, as created by Guillaume and Chevenard, contains 32% Ni, 10% Cr, 3.5% W, and 0.7% C.

Other iron-nickel-chromium alloys with 40 to 48% Ni and 2 to 8% Cr are useful as glass-sealing alloys because the chromium promotes improved glass-to-metal bonding as a result of its oxide-forming characteristics. The most common of these contain approximately 42 to 48% nickel with chromium of 4 to 6%. Although chromium additions increase the minimum thermal expansion and lower inflection points (Curie temperature), they have a beneficial effect on the glass-sealing behavior of these alloys. The chromium promotes formation of a surface chromium oxide that improves wetting at the metal/glass interface. Some of this metal oxide is absorbed by the glass during the actual glass seal and promotes a higher-strength metal/glass bond (graded seals). Compositions and thermal expansions for some Fe-Ni-Cr alloys are shown in Table 10.

Iron-Nickel-Cobalt Alloys

Replacement of some of the nickel by cobalt in an alloy of the Invar composition lowers the thermal expansion coefficient and makes the alloy's expansion characteristics less susceptible to variations in heat treatment. These iron-nickel-cobalt alloys (known as Super-Invars), however, have a more restrictive temperature range of useful application. In its restricted temperature range, the expansion coefficient of a Super-Invar alloy is lower than that of Invar (unless the Invar is in the cold-worked condition).

Super-Invar. Substitution of ~5% Co for some of the nickel content in the 36% Ni (Invar) alloy provides an alloy with an expansion coefficient even lower than Invar. A Super-Invar alloy with a nominal 32% Ni and 4 to 5% Co will exhibit a thermal expansion coefficient close to zero, over a relatively narrow temperature range. Figure 10 compares thermal expansion for 32% Ni-5% Co Super-Invar with that of an Invar alloy.

Cobalt has been added to other Fe-Ni alloys in amounts as high as 40%. Such additions increase the coefficient of expansion at room temperature. However, because they also raise the inflection temperature, they produce an alloy with a moderately low coefficient of expansion over a wider range of temperature. If θ is inflection temperature in °C, X is nickel content, Y is cobalt content, and Z is manganese content. The inflection temperature of any low-expansion Fe-Ni-Co alloy is approximated by $\theta = 19.5\,(X + Y) - 22Z - 465$. Carbon content does not significantly affect the inflection temperature.

For practical applications, these Fe-Ni-Co alloys require that Ni+Co content be sufficient to lower the martensite start temperature (M_s) to well below room tempera-

Table 11 Minimum coefficient of expansion in low-expansion Fe-Ni alloys containing titanium

Ti, %	Optimum Ni, %	Minimum coefficient of expansion, μm/m · K
0	36.5	1.4
2	40.0	2.9
3	42.5	3.6

Table 12 Mechanical properties of low-expansion Fe-Ni alloys containing 2.4 Ti and 0.06 C

Condition	Tensile strength MPa	ksi	Yield strength MPa	ksi	Elongation(a), %	Hardness, HB
42Ni-55.5Fe-2.4Ti-0.06C(b)						
Solution treated	620	90	275	40	32	140
Solution treated and age hardened	1140	165	825	120	14	330
Solution treated, cold rolled 50% and age hardened	1345	195	1140	165	5	385
52Ni-45.5Fe-2.4Ti-0.06C(c)						
Solution treated	585	85	240	35	27	125
Solution treated and age hardened	825	120	655	95	17	305

(a) In 50 mm (2 in.). (b) Inflection temperature, 220 °C (430 °F); minimum coefficient of expansion, 3.2 μm/m · K. (c) Inflection temperature, 440 °C (824 °F); minimum coefficient of expansion, 9.5 μm/m · K

Table 13 Thermoelastic coefficients of constant modulus Fe-Ni-Cr-Ti alloys

Composition, % Ni	Cr	C	Ti	Thermoelastic coefficient, annealed condition, μm/m · K	Range of possible coefficients(a), μm/m · K
42	5.4	0.06	2.4	0	18 to −23
42	6.0	0.06	2.4	36	54 to 13
42	6.3	0.06	2.4	−36	−18 to −60

(a) Any value in this range can be obtained by varying the heat treatment.

Table 14 Mechanical properties of constant-modulus alloy 50Fe-42Ni-5.4Cr-2.4Ti

Condition	Tensile strength MPa	Tensile strength ksi	Yield strength MPa	Yield strength ksi	Elongation(a), %	Hardness, HB	Modulus of elasticity GPa	Modulus of elasticity 10^6 psi
Solution treated	620	90	240	35	40	145	165	24
Solution treated and aged 3 h at 730 °C (1345 °F)	1240	180	795	115	18	345	185	26.5
Solution treated and cold worked 50%	930	135	895	130	6	275	175	25.5
Solution treated, cold worked 50% and aged 1 h at 730 °C (1345 °F)	1380	200	1240	180	7	395	185	27

(a) In 50 mm (2 in.)

Table 15 Composition and thermal expansion coefficients of high-strength controlled-expansion alloys

Alloy designation	Composition, %	260 °C (500 °F) ppm/°C	260 °C (500 °F) ppm/°F	370 °C (700 °F) ppm/°C	370 °C (700 °F) ppm/°F	415 °C (780 °F) ppm/°C	415 °C (780 °F) ppm/°F	Inflection temperature °C	Inflection temperature °F
Incoloy 903 and Pyromet CTX-1	0.03 C, 0.20 Si, 37.7 Ni, 16.0 Co, 1.75 Ti, 3.0 (Nb + Ta), 1.0 Al, 0.0075 B, bal Fe	7.51	4.17	7.47	4.15	7.45	4.14	440	820
Incoloy 907 and Pyromet CTX-3	0.06 C max, 0.5 Si, 38.0 Ni, 13.0 Co, 1.5 Ti, 4.8 (Nb + Ta), 0.35 Al max, 0.012 B max, bal Fe	7.65	4.25	7.50	4.15	7.55	4.20	415	780
Incoloy 909 and Pyromet CTX-909	0.06 C max, 0.40 Si, 38.0 Ni, 14.0 Co, 1.6 Ti, 4.9 (Nb + Ta), 0.15 Al max, 0.012 B max, bal Fe	7.75	4.30	7.55	4.20	7.75	4.30	415	780

(Coefficient of thermal expansion, from room temperature to: 260 °C, 370 °C, 415 °C)

Table 16 Typical tensile properties of high-strength, controlled-expansion alloys

Alloy designation	Test temperature °C	Test temperature °F	Ultimate tensile strength MPa	Ultimate tensile strength ksi	0.2% yield strength MPa	0.2% yield strength ksi	Elongation, %	Reduction in area, %
Incoloy 903 and Pyromet CTX-1	Room temperature		1480	215	1310	190	15	45
	540	1000	1310	190	1035	150	15	45
Incoloy 907 and Pyromet CTX-3	Room temperature		1170	170	825	120	15	25
	540	1000	1035	150	690	100	15	40
Incoloy 909 and Pyromet CTX-909	Room temperature		1310	190	1070	155	10	20
	540	1000	1135	165	895	130	15	50

ture. Nickel-cobalt contents for M_s temperatures of about −100 °C (−150 °F) can be approximated by:

$$Y = 0.0795\ \theta + 4.82 + 19W - 18.1$$

$$X = 41.9 - 0.0282\ \theta - 37Z - 19W$$

where W is carbon content.

Kovar is a nominal 29%Ni-17%Co-54%Fe alloy that is a well-known glass-sealing alloy suitable for sealing to hard (borosilicate) glasses. Kovar has a nominal expansion coefficient of approximately 5 ppm/°C and inflection temperature of ~450 °C (840 °F) with an M_s temperature less than −80 °C (−110 °F). The Dilver-P alloy produced by Imphy, S.A., is a competitive grade with the Kovar alloy of Carpenter Steel.

Special Alloys

Iron-Cobalt-Chromium Low-Expansion Alloys. An alloy containing 36.5 to 37% Fe, 53 to 54.5% Co, and 9 to 10% Cr has an exceedingly low, and at times, negative (over the range from 0 to 100 °C, or 32 to 212 °F) coefficient of expansion. This alloy has good corrosion resistance compared to low-expansion alloys without chromium. Consequently, it has been referred to as "Stainless Invar." Fernichrome, a similar alloy containing 37% Fe, 30% Ni, 25% Co, and 8% Cr, has been used for seal-in wires for electronic components sealed in special glasses.

Hardenable Low-Expansion Alloys. Alloys that have low coefficients of expansion, and alloys with constant modulus of elasticity, can be made age hardenable by adding titanium. In low-expansion alloys, nickel content must be increased when titanium is added. The higher nickel content is required because any titanium that has not combined with the carbon in the alloy will neutralize more than twice its own weight in nickel by forming an intermetallic compound during the hardening operation.

As shown in Table 11, addition of titanium raises the lowest attainable rate of expansion and raises the nickel content at which the minimum expansion occurs. Titanium also lowers the inflection temperature. Mechanical properties of alloys containing 2.4% titanium and 0.06% carbon are given in Table 12.

In alloys of the constant-modulus type containing chromium, addition of titanium allows the thermoelastic coefficients to be varied by adjustment of heat-treating schedules. The alloys in Table 13 are the three most widely used compositions. The recommended solution treatment for the alloys that contain 2.4% Ti is 950 to 1000 °C (1740 to 1830 °F) for 20 to 90 min., depending on section size. Recommended duration of aging varies from 48 h at 600 °C (1110 °F) to 3 h at 730 °C (1345 °F) for solution-treated material.

For material that has been solution treated and subsequently cold worked 50%, aging time varies from 4 h at 600 °C (1100 °F) to 1 h at 730 °C (1350 °F). Table 14 gives mechanical properties of a constant-modulus alloy containing 42% Ni, 5.4% Cr, and 2.4% Ti. Heat treatment and cold work markedly affect these properties.

High-Strength, Controlled-Expansion Alloys. There is a family of Fe-Ni-Co alloys strengthened by the addition of niobium and titanium that show the strength of precipitation-hardened superalloys while maintaining low coefficients of thermal expansion typical of certain alloys from the Fe-Ni-Co system. Compositions of the alloys are shown in Table 15; typical mechanical properties are presented in Table 16. The combination of exceptional strength and low coefficient of expansion makes this family useful for applications requiring close operating tolerances over a range of temperatures. Several components for gas turbine engines are produced from these alloys. Further information on low-expansion superalloys is contained in the article "Nickel and Nickel Alloys" in this Volume.

Table 17 Tradenames of various low-expansion alloys

Nominal composition, %	UNS number	Tradename and producing company
Iron-nickel alloys		
36% Ni, bal Fe	K93601	Invar (INCO and Imphy, S.A.) Invar M63 (Imphy, S.A.) AL-36 (Allegheny Ludlum) Invar "36" (Carpenter Steel)
39% Ni, bal Fe	···	Low expansion "39" (Carpenter Steel)
42% Ni, bal Fe	K94100	Low expansion "42" (Carpenter Steel) AL-42 (Allegheny Ludlum) N42 (Imphy, S.A.)
46% Ni, bal Fe	···	Platinate (same expansion coefficient as platinum) Glass Sealing "46" (Carpenter Steel)
47–48% Ni, bal Fe	···	N47, N48 (Imphy, S.A.)
49% Ni, bal Fe	···	AL-4750 (Allegheny Ludlum) Low expansion "49" (Carpenter Steel)
52% Ni, bal Fe	K14052	Glass Sealing "52" (Carpenter Steel) AL-52 (Allegheny Ludlum) N52 (Imphy, S.A.)
Iron-nickel-cobalt alloys		
32% Ni, 5% Co, bal Fe	···	Super Invar (INCO) Super Invar 32-5 (Carpenter Steel)
Iron-nickel-chromium alloys		
42% Ni, 6% Cr, bal Fe	K94760	Sealmet 4 (Allegheny Ludlum) Glass Sealing "42-6" (Carpenter Steel) N426 (Imphy, S.A.) SNC-K (Toshiba)

Engineering Applications

Use of alloys with low coefficients of expansion has been confined mainly to such applications as geodetic tape, bimetal strip, glass-to-metal seals, and electronic and radio components. Almost all variable condensers are made of Invar. Struts on jet engines are made of Invar to ensure rigidity with temperature changes. Close control of residuals (such as sulfur, phosphorus, aluminum, and nitrogen) has resulted in a readily weldable alloy (Invar M63 in Table 17) which has been extensively used for making tanks of liquid natural gas ships.

More recent applications of the iron-nickel "low" expansion alloys include structural components for optical and laser measurement systems, and lay-up tooling for graphite/epoxy composite components. Significant quantities of these alloys find application in substrates and housings for hermetic packaging of semiconductors where ceramic components require some matching of thermal expansion.

There is increasing use of Invar-type alloys for shadow masks in color television picture tubes. The low thermal expansion of Invar prevents excessive distortion of this shadow mask as internal temperatures increase during operation of the picture tube.

SELECTED REFERENCE

- *Physical Metallurgy of Controlled Expansion Invar-Type Alloys*, The Metallurgical Society, 1989

Shape Memory Alloys

Darel E. Hodgson, Shape Memory Applications, Inc., Ming H. Wu, Memry Corporation,
and Robert J. Biermann, Harrison Alloys, Inc.

THE TERM SHAPE MEMORY ALLOYS (SMA) is applied to that group of metallic materials that demonstrate the ability to return to some previously defined shape or size when subjected to the appropriate thermal procedure. Generally, these materials can be plastically deformed at some relatively low temperature, and upon exposure to some higher temperature will return to their shape prior to the deformation. Materials that exhibit shape memory only upon heating are referred to as having a one-way shape memory. Some materials also undergo a change in shape upon recooling. These materials have a two-way shape memory.

Although a relatively wide variety of alloys are known to exhibit the shape memory effect, only those that can recover substantial amounts of strain or that generate significant force upon changing shape are of commercial interest. To date, this has been the nickel-titanium alloys and copper-base alloys such as Cu-Zn-Al and Cu-Al-Ni.

A shape memory alloy may be further defined as one that yields a thermoelastic martensite. In this case, the alloy undergoes a martensitic transformation of a type that allows the alloy to be deformed by a twinning mechanism below the transformation temperature. The deformation is then reversed when the twinned structure reverts upon heating to the parent phase.

History

The first recorded observation of the shape memory transformation was by Chang and Read in 1932 (Ref 1). They noted the reversibility of the transformation in AuCd by metallographic observations and resistivity changes, and in 1951 the shape memory effect (SME) was observed in a bent bar of AuCd. In 1938, the transformation was seen in brass (copper-zinc). However, it was not until 1962, when Buehler and co-workers (Ref 2) discovered the effect in equiatomic nickel-titanium (Ni-Ti), that research into both the metallurgy and potential practical uses began in earnest. Within 10 years, a number of commercial products were on the market, and understanding of the effect was much advanced. Study of shape memory alloys has continued at an increasing pace since then, and more products using these materials are coming to the market each year (Ref 3, 4).

As the shape memory effect became better understood, a number of other alloy systems that exhibited shape memory were investigated. Table 1 lists a number of these systems (Ref 5) with some details of each system. Of all these systems, the Ni-Ti alloys and a few of the copper-base alloys have received the most development effort and commercial exploitation. These will be the focus of the balance of this article.

General Characteristics

The martensitic transformation that occurs in the shape memory alloys yields a thermoelastic martensite and develops from a high-temperature austenite phase with long-range order. The martensite typically occurs as alternately sheared platelets, which are seen as a herringbone structure when viewed metallographically. The transformation, although a first-order phase change, does not occur at a single temperature but over a range of temperatures that varies with each alloy system. The usual way of characterizing the transformation and naming each point in the cycle is shown in Fig. 1. Most of the transformation occurs over a relatively narrow temperature range, although the beginning and end of the transformation during heating or cooling actually extends over a much larger temperature range. The transformation also exhibits hysteresis in that the transformation on heating and on cooling does not overlap (Fig. 1). This transformation hysteresis (shown as T in Fig. 1) varies with the alloy system (Table 1).

Table 1 Alloys having a shape memory effect

Alloy	Composition	Transformation-temperature range °C	Transformation-temperature range °F	Transformation hysteresis Δ°C	Transformation hysteresis Δ°F
Ag-Cd	44/49 at.% Cd	−190 to −50	−310 to −60	≈15	≈25
Au-Cd	46.5/50 at.% Cd	30 to 100	85 to 212	≈15	≈25
Cu-Al-Ni	14/14.5 wt% Al 3/4.5 wt% Ni	−140 to 100	−220 to 212	≈35	≈65
Cu-Sn	≈15 at.% Sn	−120 to 30	−185 to 85		
Cu-Zn	38.5/41.5 wt% Zn	−180 to −10	−290 to 15	≈10	≈20
Cu-Zn-X (X = Si, Sn, Al)	a few wt% of X	−180 to 200	−290 to 390	≈10	≈20
In-Ti	18/23 at.% Ti	60 to 100	140 to 212	≈4	≈7
Ni-Al	36/38 at.% Al	−180 to 100	−290 to 212	≈10	≈20
Ni-Ti	49/51 at.% Ni	−50 to 110	−60 to 230	≈30	≈55
Fe-Pt	≈25 at.% Pt	≈−130	≈−200	≈4	≈7
Mn-Cu	5/35 at.% Cu	−250 to 180	−420 to 355	≈25	≈45
Fe-Mn-Si	32 wt% Mn, 6 wt% Si	−200 to 150	−330 to 300	≈100	≈180

Source: Ref 5

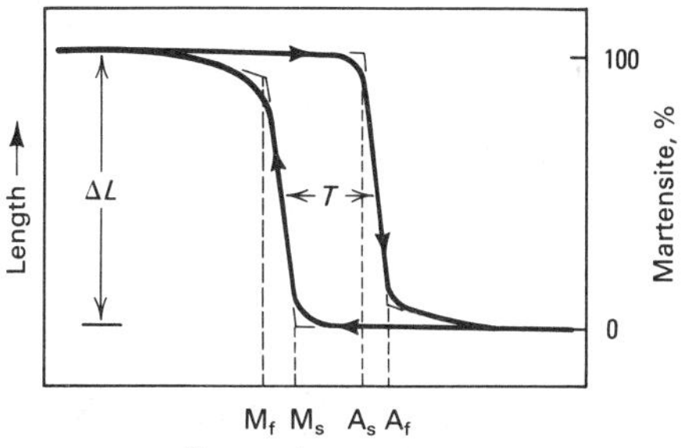

Fig. 1 Typical transformation versus temperature curve for a specimen under constant load (stress) as it is cooled and heated. T, transformation hysteresis. M_s, martensite start; M_f, martensite finish; A_s, austenite start; A_f, austenite finish

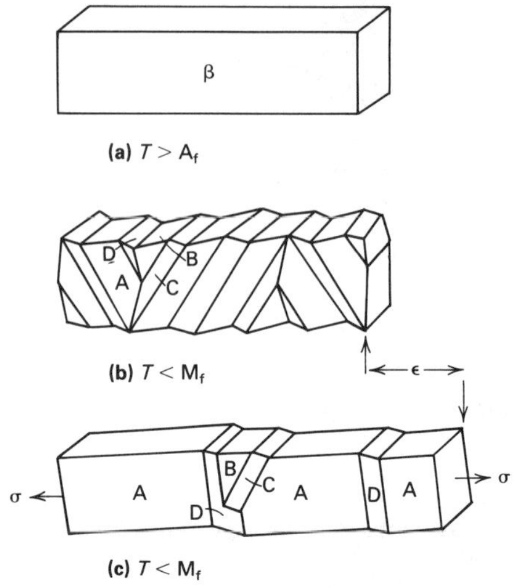

Fig. 2 (a) A β phase crystal. (b) Self-accommodating, twin-related variants A, B, C, and D, after cooling and transformation to martensite. (c) Variant A becomes dominant when stress is applied.

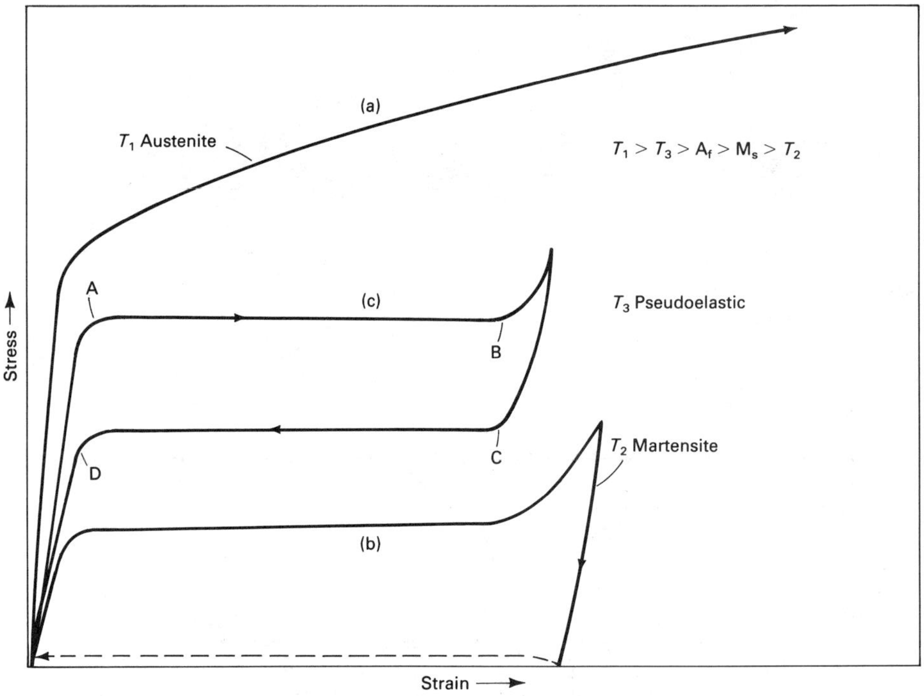

Fig. 3 Typical stress-strain curves at different temperatures relative to the transformation, showing (a) Austenite. (b) Martensite. (c) Pseudoelastic behavior

Crystallography of Shape Memory Alloys

Thermoelastic martensites are characterized by their low energy and glissile interfaces, which can be driven by small temperature or stress changes. As a consequence of this, and of the constraint due to the loss of symmetry during transformation, thermoelastic martensites are crystallographically reversible.

The herringbone structure of athermal martensites essentially consists of twin-related, self-accommodating variants (Fig. 2b). The shape change among the variants tends to cause them to eliminate each other. As a result, little macroscopic strain is generated. In the case of stress-induced martensites, or when stressing a self-accommodating structure, the variant that can transform and yield the greatest shape change in the direction of the applied stress is stabilized and becomes dominant in the configuration (Fig. 2c). This process creates a macroscopic strain, which is recoverable as the crystal structure reverts to austenite during reverse transformation.

Thermomechanical Behavior

The mechanical properties of shape memory alloys vary greatly over the temperature range spanning their transformation. This is seen in Fig. 3, where simple stress-strain curves are shown for a nickel-titanium alloy that was tested in tension below, in the middle of, and above its transformation-temperature range. The martensite is easily deformed to several percent strain at quite a low stress, whereas the austenite (high-temperature phase) has much higher yield and flow stresses. The dashed line on the martensite curve indicates that upon heating after removing the stress, the sample "remembered" its unstrained shape and reverted to it as the material transformed to austenite. No such shape recovery is found in the austenite phase upon straining and heating, because no phase change occurs.

An interesting feature of the stress-strain behavior is seen in Fig. 3(c), where the material is tested slightly above its transformation temperature. At this temperature, martensite can be stress induced. It then immediately strains and exhibits the increasing strain at constant stress behavior, seen in AB. Upon unloading, though, the material reverts to austenite at a lower stress, as seen in line CD, and shape recovery occurs, not upon the application of heat but upon a reduction of stress. This effect, which causes the material to be extremely elastic, is known as pseudoelasticity. Pseudoelasticity is nonlinear. The Young's modulus is therefore difficult to define in this temperature range as it exhibits both temperature and strain dependence.

In most cases, the memory effect is one way. That is, upon cooling, a shape memory alloy does not undergo any shape change, even though the structure changes to martensite. When the martensite is strained up to several percent, however, that strain is retained until the material is heated, at which time shape recovery occurs. Upon recooling, the material does not spontaneously change shape, but must be deliberately strained if shape recovery is again desired.

It is possible in some of the shape memory alloys to cause two-way shape memory. That is, shape change occurs upon both heating and cooling. The amount of this shape change is always significantly less than obtained with one-way memory, and very little stress can be exerted by the alloy as it tries to assume its low-temperature shape. The heating shape change can still exert very high forces, as with the one-way memory.

A number of heat-treatment and mechanical training methods have been proposed to create the two-way shape memory effect (Ref 6, 7). All rely on the introduction of microstructural stress concentrations, which cause the martensite plates to initiate in particular directions when they form upon cooling, resulting in an overall net-shape change in the desired direction.

Characterization Methods

There are four major methods of characterizing the transformation in SMAs and a large number of minor methods that are only rarely used and will not be discussed.

The most direct method is by differential scanning calorimeter (DSC). This technique measures the heat absorbed or given off by a small sample of the material as it is heated and cooled through the transformation-temperature range. The sample can be very small, such as a few milligrams, and because the sample is unstressed this is not a factor in the measurement. The endotherm and exotherm peaks, as the sample absorbs or gives off energy due to the transformation, are easily measured for the beginning, peak, and end of the phase change in each direction.

The second method often used is to measure the resistivity of the sample as it is heated and cooled. The alloys exhibit interesting changes and peaks in the resistivity (by up to 20%) over the transformation-temperature range; however, correlating these changes with measured phase changes or mechanical properties has not always been very successful. Also, there are often large changes in the resistivity curves after cycling samples through the transformation a number of times. Thus, resistivity is often measured as a phenomenon in its own right, but is rarely used to definitely characterize one alloy versus another.

The most direct method of characterizing an alloy mechanically is to prepare an appropriate sample, then apply a constant stress to the sample and cycle it through the transformation while measuring the strain that occurs during the transformation in both directions. The curve shown in Fig. 1 is the direct information one obtains from this test. The values obtained for the transformation points, such as M_s and A_f, from this method are offset to slightly higher temperatures from the values obtained from DSC testing. This happens because the DSC test occurs at no applied stress, and the transformation is not stress induced; therefore, increasing test stress will lead to increasing transformation-temperature results. This test is directly indicative of the property one can expect in a mechanical device used to perform some function using shape memory. Its disadvantages are that specimens are often difficult to make, and results are quite susceptible to the way the test is conducted.

Finally, the stress-strain properties can be measured in a standard tensile test at a number of temperatures across the transformation-temperature range, and from the change in properties the approximate transformation-temperature values can be interpolated. This is very imprecise, though, and is much better applied as a measure of the change in properties of each phase, due to such things as work hardening or different heat treatments.

Commercial SME Alloys

The only two alloy systems that have achieved any level of commercial exploitation are the Ni-Ti alloys and the copper-base alloys. Properties of the two systems are quite different. The Ni-Ti alloys have greater shape memory strain (up to 8% versus 4 to 5% for the copper-base alloys), tend to be much more thermally stable, have excellent corrosion resistance compared to the copper-base alloys' medium corrosion resistance and susceptibility to stress-corrosion cracking, and have much higher ductility. On the other hand, the copper-base alloys are much less expensive, can be melted and extruded in air with ease, and have a wider range of potential transformation temperatures. The two alloy systems thus have advantages and disadvantages that must be considered in a particular application.

Nickel-Titanium Alloys. The basis of the nickel-titanium system of alloys is the binary, equiatomic intermetallic compound of Ni-Ti. This intermetallic compound is extraordinary because it has a moderate solubility range for excess nickel or titanium, as well as most other metallic elements, and it also exhibits a ductility comparable to most ordinary alloys. This solubility allows alloying with many of the elements to modify both the mechanical properties and the transformation properties of the system. Excess nickel, in amounts up to about 1%, is the most common alloying addition. Excess nickel strongly depresses the transformation temperature and increases the yield strength of the austenite. Other frequently used elements are iron and chromium (to lower the transformation temperature), and copper (to decrease the hysteresis and lower the deformation stress of the martensite). Because common contaminants such as oxygen and carbon can also shift the transformation temperature and degrade the mechanical properties, it is also desirable to minimize the amount of these elements.

The major physical properties of the basic binary Ni-Ti system and some of the mechanical properties of the alloy in the annealed condition are shown in Table 2. Note that this is for the equiatomic alloy with an A_f value of about 110 °C (230 °F). Selective work hardening, which can exceed 50% reduction in some cases, and proper heat treatment can greatly improve the ease with which the martensite is deformed, give an austenite with much greater strength, and create material that spontaneously moves itself both on heating and on cooling (two-way shape memory). One of the biggest challenges in using this family of alloys is in developing the proper processing procedures to yield the properties desired.

Because of the reactivity of the titanium in these alloys, all melting of them must be done in a vacuum or an inert atmosphere. Methods such as plasma-arc melting, electron-beam melting, and vacuum-induction melting are all used commercially. After ingots are melted, standard hot-forming processes such as forging, bar rolling, and extrusion can be used for initial breakdown. The alloys react slowly with air, so hot working in air is quite successful. Most cold-working processes can also be applied to these alloys, but they work harden extremely rapidly, and frequent annealing is required. Wire drawing is probably the most widely used of the techniques, and excellent surface properties and sizes as small as 0.05 mm (0.002 in.) are made routinely.

Fabrication of articles from the Ni-Ti alloys can usually be done with care, but some of the normal processes are difficult. Machining by turning or milling is very difficult except with special tools and practices. Welding, brazing, or soldering the alloys is generally difficult. The materials do respond well to abrasive removal, such as grinding, and shearing or punching can be done if thicknesses are kept small.

Table 2 Properties of binary Ni-Ti shape memory alloys

Properties	Property value
Melting temperatures, °C (°F)	1300 (2370)
Density, g/cm^3 (lb/in.^{3})	6.45 (0.233)
Resistivity, μΩ · cm	
Austenite	≈100
Martensite	≈70
Thermal conductivity, W/m · °C (Btu/ft · h · °F)	
Austenite	18 (10)
Martensite	8.5 (4.9)
Corrosion resistance	Similar to 300 series stainless steel or titanium alloys
Young's modulus, GPa (10^6 psi)	
Austenite	≈83 (≈12)
Martensite	≈28–41 (≈4–6)
Yield strength, MPa (ksi)	
Austenite	195–690 (28–100)
Martensite	70–140 (10–20)
Ultimate tensile strength, MPa (ksi)	895 (130)
Transformation temperatures, °C (°F)	−200 to 110 (−325 to 230)
Latent heat of transformation, kJ/kg · atom (cal/g · atom)	167 (40)
Shape memory strain	8.5% maximum

Heat treating to impart the desired memory shape is often done at 500 to 800 °C (950 to 1450 °F), but it can be done as low as 300 to 350 °C (600 to 650 °F) if sufficient time is allowed. The SMA component may need to be restrained in the desired memory shape during the heat treatment; otherwise, it may not remain there.

Commercial copper-base shape memory alloys are available in ternary Cu-Zn-Al and Cu-Al-Ni alloys, or in their quaternary modifications containing manganese. Elements such as boron, cerium, cobalt, iron, titanium, vanadium, and zirconium are also added for grain refinement.

The major alloy properties are listed in Table 3. The martensite-start (M_s) temperatures and the compositions of Cu-Zn-Al alloys are plotted in Fig. 4. Compositions of Cu-Al-Ni alloys usually fall in the range of 11 to 14.5 wt% Al and 3 to 5 wt% Ni. The martensitic transformation temperatures can be adjusted by varying chemical composition. Figure 4 and the following empirical relationships are useful in obtaining a first estimate:

- Cu-Zn-Al: M_s(°C) = 2212 − 66.9 (at.% Zn) − 90.65 (at.% Al) (Ref 8)
- Cu-Al-Ni: M_s(°C) = 2020 − 134 (wt% Al) − 45 (wt% Ni) (Ref 9)

Table 3 Properties of copper-base shape memory alloys

Property	Property value: Cu-Zn-Al	Property value: Cu-Al-Ni
Thermal properties		
Melting temperature, °C (°F)	950–1020 (1740–1870)	1000–1050 (1830–1920)
Density, g/cm³ (lb/in.³)	7.64 (0.276)	7.12 (0.257)
Resistivity, μΩ · cm	8.5–9.7	11–13
Thermal conductivity, W/m · °C (Btu/ft · h · °F)	120 (69)	30–43 (17–25)
Heat capacity, J/kg · °C (Btu/lb · °F)	400 (0.096)	373–574 (0.089–0.138)
Mechanical properties		
Young's modulus, GPa (10^6 psi)(a)		
β phase	72 (10.4)(a)	85 (12.3)(a)
Martensite	70 (10.2)(a)	80 (11.6)(a)
Yield strength, MPa (ksi)		
β phase	350 (51)	400 (58)
Martensite	80 (11.5)	130 (19)
Ultimate tensile strength, MPa (ksi)	600 (87)	500–800 (73–116)
Shape memory properties		
Transformation temperatures, °C (°F)	<120 (250)	<200 (390)
Recoverable strain, %	4	4
Hysteresis, Δ°C (Δ°F)	15–25 (30–45)	15–20 (30–35)

(a) The Young's modulus of shape memory alloys becomes difficult to define between the M_s and the A_s transformation temperatures. At these temperatures, the alloys exhibit nonlinear elasticity, and the modulus is both temperature- and strain-dependent.

The melting of Cu-base shape memory alloys is similar to that of aluminum bronzes. Most commercial alloys are induction melted. Protective flux on the melt and the use of nitrogen or inert-gas shielding during pouring are necessary to prevent zinc evaporation and aluminum oxidation. Powder metallurgy and rapid solidification processing are also used to produce fine-grain alloys without grain-refining additives.

Copper-base alloys can be readily hot worked in air. With low aluminum content (<6 wt%), Cu-Zn-Al alloys can be cold finished with interpass annealing. Alloys with higher aluminum content are not as easily cold workable. Cu-Al-Ni alloys, on the other hand, are quite brittle at low temperatures and can only be hot finished.

Manganese depresses transformation temperatures of both Cu-Zn-Al and Cu-Al-Ni alloys and shifts the eutectoid to higher aluminum content (Ref 10). It often replaces aluminum for better ductility.

Because copper-base shape memory alloys are metastable in nature, solution heat treatment in the parent β-phase region and subsequent controlled cooling are necessary to retain β phase for shape memory effects. Prolonged solution heat treatment causes zinc evaporation and grain growth and should be avoided. Water quench is widely used as a quenching process, but air cooling may be sufficient for some high-aluminum content Cu-Zn-Al alloys and Cu-Al-Ni alloys. The as-quenched transformation temperature is usually unstable. Postquench aging at temperatures above the nominal A_f temperature is generally needed to establish stable transformation temperatures.

Cu-Zn-Al alloys, when quenched rapidly and directly into the martensitic phase, are susceptible to the martensite stabilization effect (Ref 11). This effect causes the reverse transformation to shift toward higher temperatures. It therefore delays and may completely inhibit the shape recovery. For alloys with M_s temperatures above the ambient, slow cooling or step quenching with intermediate aging in the parent β-phase state should be adopted.

The thermal stability of copper-base alloys is ultimately limited by the decomposition kinetics. For this reason, prolonged exposure of Cu-Zn-Al and Cu-Al-Ni alloys at temperatures above 150 °C (300 °F) and 200 °C (390 °F) respectively, should be avoided. Aging at lower temperatures may also shift the transformation temperatures. In case of aging in the β phase, this results from the change in long-range order (Ref 12). When aged in the martensitic state, the alloys exhibit an aging-induced martensite stabilization effect (Ref 11). For high-temperature stability, Cu-Al-Ni is generally a better alloy system than Cu-Zn-Al. However, even for moderate temperature applications, which demand tight control of transformation temperatures, these effects need to be evaluated.

Applications

There is a wide variety of uses for the shape memory alloys. The following will illustrate one or two products in several categories of application.

Free recovery is illustrated when an SMA component is deformed while martensitic, and the only function required of the shape memory is that the component return to its previous shape (while doing minimal work) upon heating. A prime application of this is the blood-clot filter developed by M. Simon (Ref 13). The Ni-Ti wire is shaped to anchor itself in a vein and catch passing clots. The part is chilled so it can be collapsed and inserted into the vein, then body heat is sufficient to turn the part to its functional shape.

Constrained Recovery. The most successful example of this type of product is undoubtedly the Cryofit hydraulic couplings made by Raychem Corporation (Ref 14). These fittings are manufactured as cylindrical sleeves slightly smaller than the metal tubing they are to join. Their diameters are then expanded while martensitic, and, upon warming to austenite, they shrink in diameter and strongly hold the tube ends. The tubes prevent the coupling from fully recovering its manufactured shape, and the stresses created as the coupling attempts to do so are great enough to create a joint that, in many ways, is superior to a weld.

Similar to the Cryofit coupling, the Betalloy coupling (Ref 15) is a Cu-Zn-Al coupling also designed and marketed by Raychem Corporation for copper and aluminum tubing. In this application, the Cu-Zn-Al shape memory cylinder shrinks on heating and acts as a driver to squeeze a tubular liner onto the tubes being joined. The joint strength is enhanced by a sealant coating on the liner.

Force Actuators. In some applications, the shape memory component is designed to exert force over a considerable range of motion, often for many cycles. Such an application is the circuit-board edge connector made by Beta Phase Inc. (Ref 16). In this electrical connector system, the SMA component is used to force open a spring when the connector is heated. This allows force-free insertion or withdrawal of a circuit board in the connector. Upon cooling, the Ni-Ti actuator becomes weaker and the spring easily deforms the actuator while it closes tightly on the circuit board and forms the connections.

Based on the same principle, Cu-Zn-Al shape memory alloys have found several applications in this area. One such example is a fire safety valve, which incorporates a Cu-Zn-Al actuator designed to shut off toxic or flammable gas flow when fire occurs (Ref 17).

Proportional Control. It is possible to use only a part of the shape recovery to accurately position a mechanism by using only a selected portion of the recovery because the transformation occurs over a range of temperatures rather than at a single temperature. A device has been developed by Beta Phase Inc. (Ref 18) in which a valve controls the rate of fluid flow by carefully heating a shape-memory-alloy component just enough to close the valve the desired amount. Repeatable positioning within 0.25 μm (10^{-5} in.) is possible with this technique.

Superelastic Applications. A number of products have been brought to market that use the pseudoelastic (or superelastic) property of these alloys. Eyeglass frames that use superelastic Ni-Ti to absorb large deformations without damaging the frames are

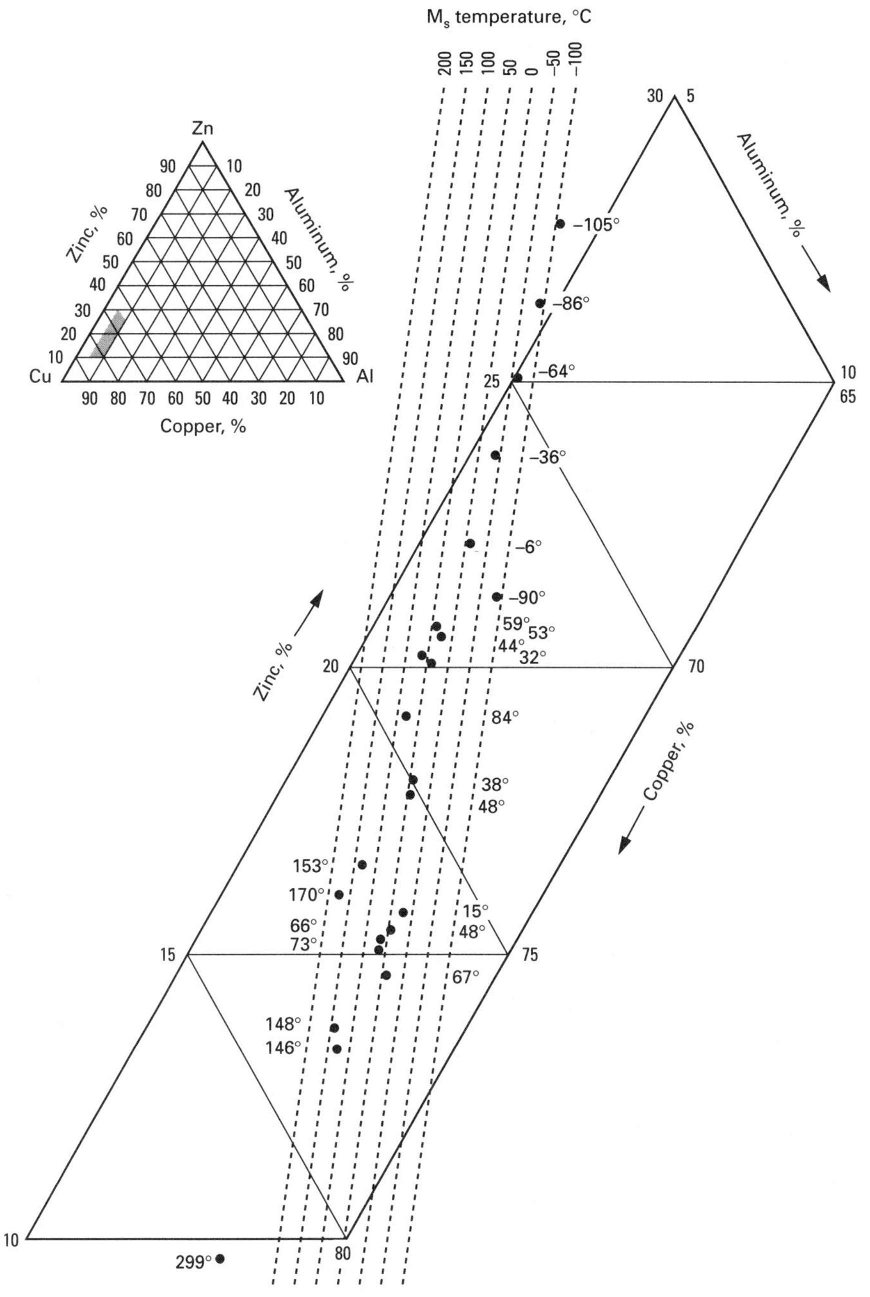

Fig. 4 M_s temperatures and compositions of Cu-Zn-Al shape memory alloys

now marketed, and guide wires for steering catheters into vessels in the body have been developed using Ni-Ti wire, which resists permanent deformation if bent severely. Arch wires for orthodontic correction using Ni-Ti have been used for many years to give large rapid movement of teeth.

The properties of the Ni-Ti alloys, particularly, indicate their probable greater use in biomedical applications. The material is extremely corrosion resistant, demonstrates excellent biocompatibility, can be fabricated into the very small sizes often required, and has properties of elasticity and force delivery that allow uses not possible any other way.

Future Prospects

Although specific products that might use the Ni-Ti alloys in the future cannot be foretold, some directions are obvious. The cost of these alloys has slowly decreased as use has increased, so uses that require lower-cost alloys to be viable are being explored. Alloy development has yielded several ternary compositions with properties improved over those obtained with binary material, and alloys tailored to specific product needs are likely to multiply. The medical industry has developed a number of products using Ni-Ti alloys because of their excellent biocompatibility and large pseudoelasticity, and many more of these applications are likely. Finally, the availability of small wire that is stable, is easily heated by a small electrical current, and gives a large repeatable stroke should lead to a new family of actuator devices (Ref 19). These devices can be inexpensive, are reliable for thousands of cycles, and are expected to move Ni-Ti into the high-volume consumer marketplace.

Recent interest in the development of iron-base shape memory alloys has challenged the concept that long-range order and thermoelastic martensitic transformation are necessary conditions for shape memory effect. Among the alloys, Fe-Pt (Ref 20), Fe-Pd (Ref 21), and Fe-Ni-Co-Ti (Ref 22) can be heat treated to exhibit thermoelastic martensitic transformation, and, therefore, shape memory effect. However, alloys such as Fe-Ni-C (Ref 23), Fe-Mn-Si (Ref 24), and Fe-Mn-Si-Cr-Ni (Ref 25) are not ordered and undergo nonthermoelastic transformation, and yet exhibit good shape memory effect. These alloys are characteristically different from conventional shape memory alloys in that they rely on stress-induced martensite for shape memory effect, exhibit fairly large transformation hysteresis, and, in general, have less than 4% recoverable strain. The commercial potential of these alloys has yet to be determined, but the effort has opened up new classes of alloys for exploration as shape memory alloys. These new classes include β-Ti alloys and iron-base alloys.

REFERENCES

1. L.C. Chang and T.A. Read, *Trans. AIME*, Vol 191, 1951, p 47
2. W.J. Buehler, J.V. Gilfrich, and R.C. Wiley, *J. Appl. Phys.*, Vol 34, 1963, p 1475
3. *Proceedings of Engineering Aspects of Shape Memory Alloys* (Lansing, MI), 1988
4. D.E. Hodgson, *Using Shape Memory Alloys*, Shape Memory Applications, 1988
5. K. Shimizu and T. Tadaki, *Shape Memory Alloys*, H. Funakubo, Ed., Gordon and Breach Science Publishers, 1987
6. J.R. Willson, *et al.*, U.S. Patent 3,625,969, 1972
7. A.D. Johnson, U.S. Patent 4,435,229, 1972
8. L. Delaey, M. Chandrasekaran, W. DeJonghe, W. Rapacioli, and A. Deruyttere, INCRA Research Report 238, International Copper Research Association
9. K. Sugimoto, *Bull. Jpn. Inst. Met.*, Vol 24, 1985, p 45
10. P.L. Brook, U.S. Patent 4,166,739, Sept 1979
11. M. Ahlers, *Proceedings of International Conference on Martensitic Transformations* (Nara, Japan), 1986, p 786

12. D. Schofield and A.P. Miodownik, *Met. Technol.*, Vol 7, 1980, p 167
13. M. Simon, *et al.*, *Radiology*, Vol 172, 1989, p 99-103
14. J.D. Harrison and D.E. Hodgson, *Shape Memory Effects in Alloys*, J. Perkins, Ed., Plenum Press, 1975, p 517
15. Product Brochure, Raychem Corporation, Menlo Park, CA
16. J.F. Krumme, *Connect. Technol.*, Vol 3 (No. 4), April 1987, p 41
17. E. Waldbusser, *Semicond. Saf. Assoc. J.*, Aug 1987, p 34
18. D.E. Hodgson, *Proceedings of Engineering Aspects of Shape Memory Alloys* (Lansing, MI), 1988
19. Product brochure, Dynalloy Inc., Irvine, CA
20. M. Foos, C. Frantz, and M. Gantois, *Shape Memory Effects in Alloys*, J. Perkins, Ed., Plenum Press, 1975, p 407
21. T. Sohmura, R. Oshima, and F.E. Fujita, *Scr. Metall.*, Vol 14, 1980, p 855
22. T. Maki, K. Kobayashi, M. Minato, and I. Tamura, *Scr. Metall.*, Vol 18, 1984, p 1105
23. S. Kajiwara, *Trans. Jpn. Inst. Met.*, Vol 26, 1985, p 595
24. A. Sato, K. Soma, E. Chishima, and T. Mori, in *Proceedings*, International Conference on Martensitic Transformations (Louvain, Belgium), 1982, p C4-797
25. H. Otsuka, H. Yamada, H. Tanahashi, and T. Maruyama, in *Proceedings*, International Conference on Martensitic Transformations (Sydney, Australia), 1989

Metal-Matrix Composites

John V. Foltz, Metallic Materials Branch, Naval Surface Warfare Center, White Oak Laboratory
Charles M. Blackmon, Applied Materials Technology Branch, Naval Surface Warfare Center, Dahlgren Laboratory

METAL-MATRIX COMPOSITES (MMCs) are a class of materials with potential for a wide variety of structural and thermal management applications. Metal-matrix composites are capable of providing higher-temperature operating limits than their base metal counterparts, and they can be tailored to give improved strength, stiffness, thermal conductivity, abrasion resistance, creep resistance, or dimensional stability. Unlike resin-matrix composites, they are nonflammable, do not outgas in a vacuum, and suffer minimal attack by organic fluids such as fuels and solvents.

The principle of incorporating a high-performance second phase into a conventional engineering material to produce a combination with features not obtainable from the individual constituents is well known. In a MMC, the continuous, or matrix, phase is a monolithic alloy, and the reinforcement consists of high-performance carbon, metallic, or ceramic additions. Reinforced intermetallic compounds such as the aluminides of titanium, nickel, and iron are also discussed in this article (for more information on intermetallic compounds, see the article "Ordered Intermetallics" in this Volume).

Reinforcements, characterized as either continuous or discontinuous, may constitute from 10 to 60 vol% of the composite. Continuous fiber or filament reinforcements include graphite (Gr), silicon carbide (SiC), boron, aluminum oxide (Al_2O_3), and refractory metals. Discontinuous reinforcements consist mainly of SiC in whisker (w) form, particulate (p) types of SiC, Al_2O_3, or titanium diboride (TiB_2), and short or chopped fibers of Al_2O_3 or graphite. Figure 1 shows cross sections of typical continuous and discontinuous reinforcement MMCs.

The salient characteristics of metals as matrices are manifested in a variety of ways; in particular, a metal matrix imparts a metallic nature to the composite in terms of thermal and electrical conductivity, manufacturing operations, and interaction with the environment. Matrix-dominated mechanical properties, such as the transverse elastic modulus and strength of unidirectionally reinforced composites, are sufficiently high in some MMCs to permit use of the unidirectional lay-up in engineering structures.

This article will give an overview of the current status of MMCs, including information on physical and mechanical properties, processing methods, distinctive features, and the various types of continuously and discontinuously reinforced MMCs. More information on the processing and properties of MMCs is available in the Section "Metal, Carbon/Graphite, and Ceramic Matrix Composites" in *Composites*, Volume 1 of the *Engineered Materials Handbook* published by ASM INTERNATIONAL.

Property Prediction

Property predictions of MMCs can be obtained from mathematical models, which require as input a knowledge of the properties and geometry of the constituents. For metals reinforced by straight, parallel continuous fibers, three properties that are frequently of interest are the elastic modulus, the coefficient of thermal expansion, and the thermal conductivity in the fiber direction. Reasonable values can be obtained from rule-of-mixture expressions for Young's modulus (Ref 1):

$$E_c = E_f v_f + E_m v_m \qquad \text{(Eq 1)}$$

coefficient of thermal expansion (Ref 2):

$$\alpha_c = \frac{\alpha_f v_f E_f + \alpha_m v_m E_m}{E_f v_f + E_m v_m} \qquad \text{(Eq 2)}$$

and thermal conductivity (Ref 3):

$$k_c = k_f v_f + k_m v_m \qquad \text{(Eq 3)}$$

where v is volume fraction, and E, α, and k are the modulus, coefficient of thermal expansion, and thermal conductivity in the fiber direction, respectively. The subscripts c, f, and m refer to composite, fiber, and matrix, respectively.

Processing Methods

Processing methods for MMCs are divided into primary and secondary categories. Primary processing is the operation by which the composite is synthesized from its raw materials. It involves introducing the reinforcement into the matrix in the appropriate amount and location, and achieving proper bonding of the constituents. Secondary processing consists of all the additional steps needed to make the primary composite into a finished hardware component.

Many reinforcement and matrix materials are not inherently compatible, and such materials cannot be processed into a composite without tailoring the properties of an interface between them. In some composites the coupling between the reinforcing agent and the metal is poor and must be enhanced. For MMCs made from reactive constituents, the challenge is to avoid excessive chemical activity at the interface, which would degrade the properties of the material. These problems are usually resolved either by applying a surface treatment or coating to the reinforcement or by modifying the composition of the matrix alloy.

Solidification processing (Ref 4, 5), solid-state bonding, and matrix deposition techniques have been used to fabricate MMCs. Solidification processing offers a near-net-shape manufacturing capability, which is economically attractive. Developers have explored various liquid metal techniques that use multifilament yarns, chopped fibers, or particulates as the reinforcement. A castable ceramic/aluminum MMC is now commercially available (Ref 6); cast components of this composite are shown in Fig. 2. Solid-state methods use lower fabrication temperatures with potentially better control of the interface thermodynamics and kinetics. The two principal categories of solid-state fabrication are diffusion bonding of materials in thin sheet form (Ref 7) and powder metallurgy techniques (Ref 8). Matrix deposition processes, in which the matrix is deposited on the fiber, include electrochemical plating, plasma spraying, and physical vapor deposition (Ref 7). A new method, metal spray deposition, is currently being investigated (Ref 9). After deposition process-

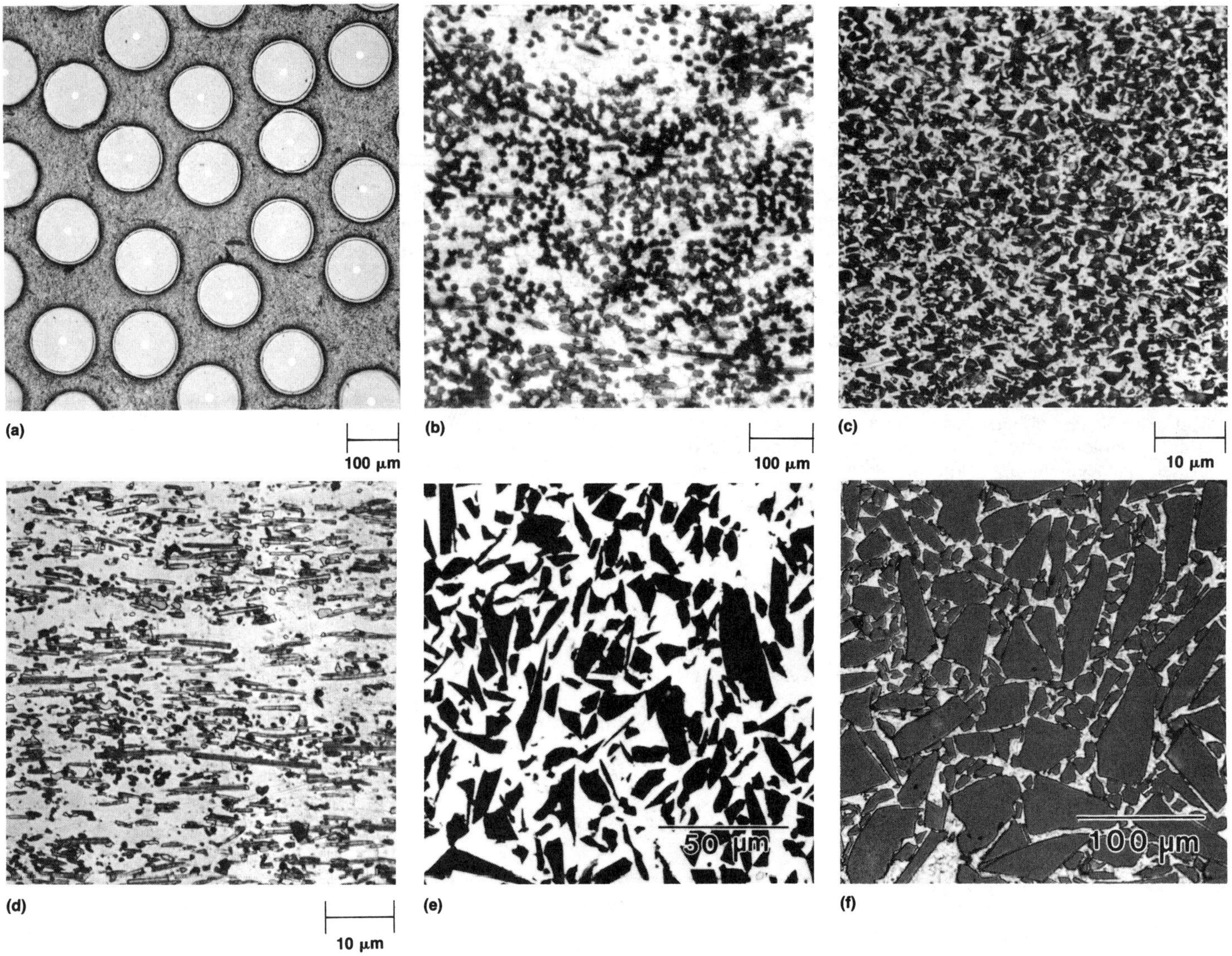

Fig. 1 Cross sections of typical fiber-reinforced MMCs. (a) Continuous-fiber-reinforced boron/aluminum composite. Shown here are 142 μm diam boron filaments coated with B_4C in a 6061 aluminum alloy matrix. (b) Discontinuous graphite/aluminum composite. Cross section shows 10 μm diam chopped graphite fibers (40 vol%) in a 2014 aluminum alloy matrix. (c) A 6061 aluminum alloy matrix reinforced with 40 vol% SiC particles. (d) Whisker-reinforced (20 vol% SiC) aluminum MMC. (e) and (f) MMCs manufactured using the PRIMEX^{TM} pressureless metal infiltration process. (e) An Al_2O_3-reinforced (60 vol%) aluminum MMC. (f) A highly reinforced (81 vol%) MMC consisting of SiC particles in an aluminum alloy matrix. The black specks in the matrix are particles of an inorganic preform binder and do not indicate porosity. (a) and (b) Courtesy of DWA Composite Specialties, Inc. (c) and (d) Courtesy of Advanced Composite Materials Corporation. (e) and (f) Courtesy of Lanxide Corporation

ing, a secondary consolidation step such as diffusion bonding often is needed to produce a component.

Which secondary processes are appropriate for a given MMC depends largely on whether the reinforcement is continuous or discontinuous. Discontinuously reinforced MMCs are amenable to many common metal forming operations, including extrusion, forging, and rolling. Because a high percentage of the materials used to reinforce discontinuous MMCs are hard oxides or carbides, machining can be difficult, and methods such as diamond sawing, electrical discharge machining (Ref 10), and abrasive waterjet cutting (Ref 11) are sometimes utilized (see *Machining*, Volume 16 of the 9th Edition of *Metals Handbook* for more information about these machining methods).

Aluminum-Matrix Composites

Most of the commercial work on MMCs has focused on aluminum as the matrix metal. The combination of light weight, environmental resistance, and useful mechanical properties has made aluminum alloys very popular; these properties also make aluminum well suited for use as a matrix metal. The melting point of aluminum is high enough to satisfy many application requirements, yet low enough to render composite processing reasonably convenient. Also, aluminum can accommodate a variety of reinforcing agents, including continuous boron, Al_2O_3, SiC, and graphite fibers, and various particles, short fibers, and whiskers (Ref 12). The microstructures of various aluminum matrix MMCs are shown in Fig. 1.

Continuous Fiber Aluminum MMC. Boron/aluminum is a technologically mature continuous fiber MMC (Fig. 1a). Applications for this composite include tubular truss members in the midfuselage structure of the Space Shuttle orbiter and cold plates in electronic microchip carrier multilayer boards. Fabrication processes for B/Al composites are based on hot-press diffusion bonding or plasma spraying methods (Ref 13). Selected properties of a B/Al composite are given in Table 1.

Continuous SiC fibers (SiC_c) are now commercially available; these fibers are candidate replacements for boron fibers because they have similar properties and offer a potential cost advantage. One such SiC fiber is SCS, which can be manufactured with any of several surface chemistries to

Fig. 2 Discontinuous silicon carbide/aluminum castings. Pictured are a sand cast automotive disk brake rotor and upper control arm, a permanent mold cast piston, a high-pressure die cast bicycle sprocket, an investment cast aircraft hydraulic manifold, and three investment cast engine cylinder inserts. Courtesy of Dural Aluminum Composites Corporation

Table 1 Room-temperature properties of unidirectional continuous fiber aluminum-matrix composites

Property	B/6061 Al	SCS-2/6061 Al	P100 Gr/6061 Al	FP/Al-2Li(a)
Fiber content, vol%	48	47	43.5	55
Longitudinal modulus, GPa (10^6 psi)	214 (31)	204 (29.6)	301 (43.6)	207 (30)
Transverse modulus, GPa (10^6 psi)	···	118 (17.1)	48 (7.0)	144 (20.9)
Longitudinal strength, MPa (ksi)	1520 (220)	1462 (212)	543 (79)	552 (80)
Transverse strength, MPa (ksi)	···	86 (12.5)	13 (2)	172 (25)

(a) FP is the proprietary designation for an alpha alumina (α-Al_2O_3) fiber developed by E.I. Du Pont de Nemours & Company, Inc. Source: Ref 14–16

enhance bonding with a particular matrix, such as aluminum or titanium (Ref 14). The SCS-2 fiber, tailored for aluminum, has a 1 μm (0.04 mil) thick carbon rich coating that increases in silicon content toward its outer surface.

Silicon carbide/aluminum MMCs exhibit increased strength and stiffness as compared with unreinforced aluminum, and with no weight penalty. Selected properties of SCS-2/Al are given in Table 1. In contrast to the base metal, the composite retains its room-temperature tensile strength at temperatures up to 260 °C (500 °F) (Fig. 3). This material is the focus of development programs for a variety of applications; an example of an advanced aerospace application for an SCS/Al MMC is shown in Fig. 4.

Graphite/aluminum (Gr/Al) MMC development was initially prompted by the commercial appearance of strong and stiff carbon fibers in the 1960s. As shown in Fig. 5, carbon fibers offer a range of properties, including an elastic modulus up to 966 GPa (140 psi $\times$ 10^6) and a negative coefficient of thermal expansion down to -1.62×10^{-6}/°C (-0.9×10^{-6}/°F). However, carbon and aluminum in combination are difficult materials to process into a composite. A deleterious reaction between carbon and aluminum, poor wetting of carbon by molten aluminum, and oxidation of the carbon are significant technical barriers to the production of these composites (Ref 19). Three processes are currently used for making commercial Gr/Al MMCs: liquid metal infiltration of fiber tows (Ref 20), vacuum vapor deposition of the matrix on spread tows (Ref 21, 22), and hot press bonding of spread tows sandwiched between sheets of aluminum (Ref 19). With both precursor wires and metal-coated fibers, secondary processing such as diffusion bonding or pultrusion is needed to make structural elements. Squeeze casting also is feasible for the fabrication of this composite (Ref 23).

Precision aerospace structures with strict tolerances on dimensional stability need stiff, lightweight materials that exhibit low thermal distortion. Graphite/aluminum MMCs have the potential to meet these requirements. Unidirectional P100 Gr/6061 Al pultruded tube (Ref 15) exhibits an elastic modulus in the fiber direction significantly greater than that of steel, and it has a density approximately one-third that of steel (Table 1). Reference 24 contains additional data for P100 Gr/Al.

In theory, Gr/Al angle-plied laminates can be designed to provide a coefficient of thermal expansion (CTE) of exactly zero by selecting the appropriate ply-stacking arrangement and fiber content. In practice, a near-zero CTE has been realized, but expansion behavior is complicated by hysteresis attributed to plastic deformation occur-

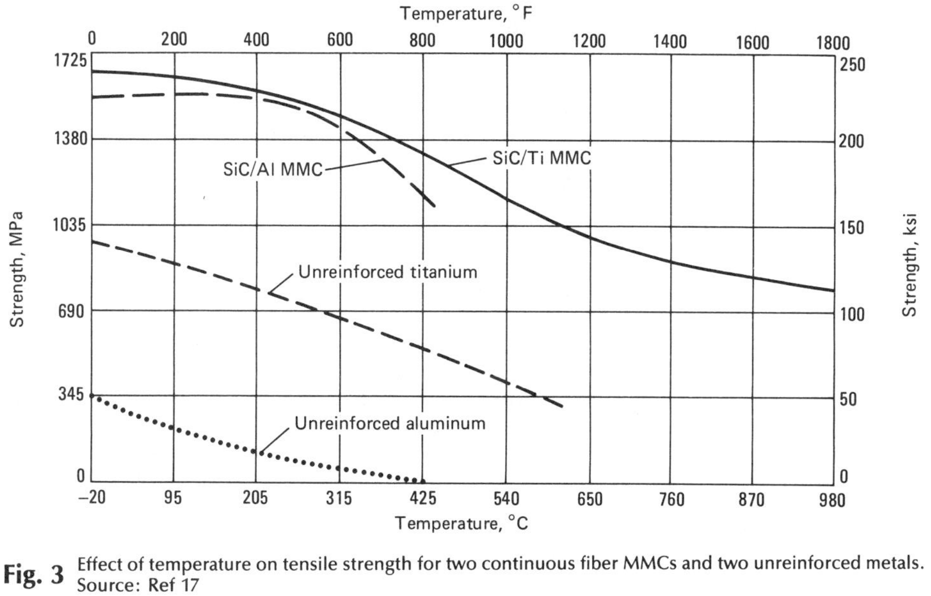

Fig. 3 Effect of temperature on tensile strength for two continuous fiber MMCs and two unreinforced metals. Source: Ref 17

Fig. 4 Advanced aircraft stabilator spar made from an SCS/Al MMC. Courtesy of Textron Specialty Materials

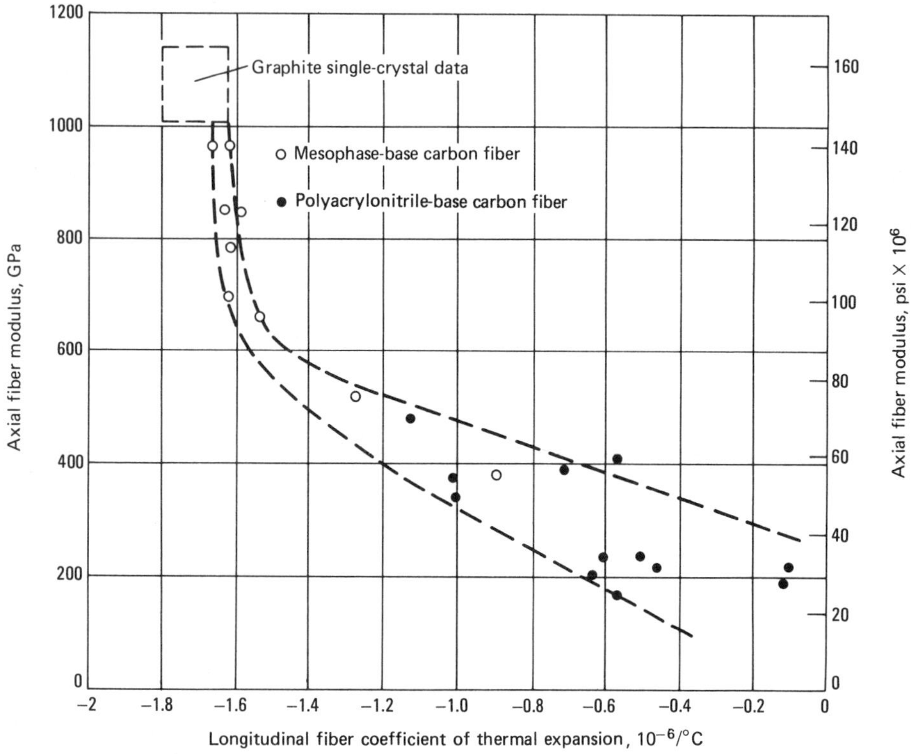

Fig. 5 Carbon fiber axial modulus versus axial coefficient of thermal expansion for mesophase (pitch-base) and polyacrylonitride-base (pan-base) graphite fibers. Source: Ref 18

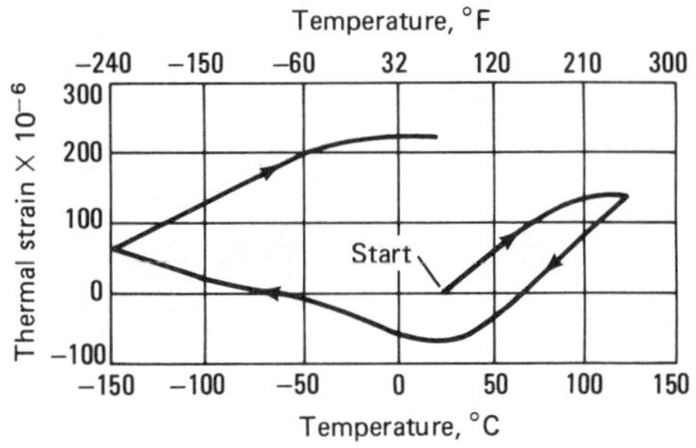

Fig. 6 Thermal expansion in the fiber direction of a P100 Gr/6061 Al single-ply unidirectional composite laminate. Source: Ref 25

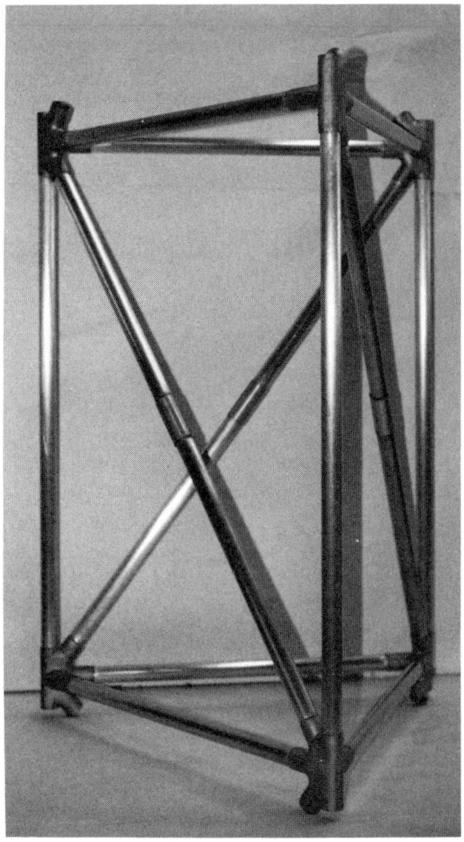

Fig. 7 Space truss made of ±12° lay-up graphite/aluminum tubes. Axial coefficient of thermal expansion is −0.072 × 10^{-6}/°C per bay. Courtesy of DWA Composite Specialties, Inc.

ring in the matrix during thermal excursions (Fig. 6). Full-scale segments of a Gr/Al space truss (Fig. 7) have been fabricated and successfully tested. The advent of pitch-based graphite fibers with three times the thermal conductivity of copper (Ref 26) suggests that a high-conductivity low-CTE version of Gr/Al can be developed for electronic heat sinks and space thermal radiators.

Aluminum oxide/aluminum (Al_2O_3/Al) MMCs can be fabricated by a number of methods, but liquid or semisolid-state processing techniques are commonly used. Certain of the oxide ceramic fibers used as reinforcements are inexpensive and provide the composite with improved properties as compared with those of unreinforced aluminum alloys. For example, the composite has an improved resistance to wear and thermal fatigue deformation and a reduced coefficient of thermal expansion. Continuous fiber Al_2O_3/Al MMCs are fabricated by arranging Al_2O_3 tapes in a desired orientation to make a preform, inserting the preform into a mold, and infiltrating the preform with molten aluminum via a vacuum assist (Ref 27). Reinforcement-to-matrix bonding is achieved by small additions of lithium to the melt. The room-temperature properties of a unidirectional Al_2O_3/Al (FP/Al-2Li) are given in Table 1; additional mechanical properties of continuous Al_2O_3/Al are given in Ref 28.

Discontinuous Aluminum MMCs. Discontinuous silicon carbide/aluminum (SiC_d/Al) is a designation that encompasses materials with SiC particles, whiskers, nodules, flakes, platelets, or short fibers in an aluminum matrix (see Fig. 1). Several companies are currently involved in the development of powder metallurgy SiC_d/Al, using either particles or whiskers as the reinforcement phase (Ref 8). A casting technology exists for this type of MMC, and melt-produced ingots can be procured in whatever form is needed—extrusion billets, ingots, or rolling blanks—for further processing (Ref 29). Arsenault and Wu (Ref 30) compared powder metallurgy and melt-produced discontinuous SiC/Al composites to determine if a correlation exists between strength and processing type. They found that if the size, volume fraction, distribution of the reinforcement, and bonding with the matrix are the same, then the strengths of the powder metallurgy and melt-produced MMCs are the same.

Whiskers in discontinuously reinforced MMCs can be oriented in processing to provide directional properties. McDanels (Ref 31) evaluated the effects of reinforcement type, matrix alloy, reinforcement content, and orientation on the tensile behavior of SiC_d/Al composites made by powder metallurgy techniques. He concluded that these composites offer a 50 to 100% increase in elastic modulus as compared with unreinforced aluminum (Fig. 8). He also found that these materials have a stiffness approximately equivalent to that of titanium but with one-third less density. Tensile and yield strengths of SiC_d/Al composites are up to 60% greater than those of the unreinforced matrix alloy. Selected properties of SiC_d/Al MMCs are given in Table 2. Studies of the elevated-temperature mechanical properties of SiC_d/Al with either 20% whisker or 25% particulate reinforcement indicate that SiC_d/Al can be used effectively for long-time exposures to temperatures of at least 200 °C (400 °F) and for short exposures at 260 °C (500 °F) (Ref 31, 33).

Discontinuous silicon carbide/aluminum MMCs are being developed by the aerospace industry for use as airplane skins, intercostal ribs, and electrical equipment

Table 2 Properties of discontinuous silicon carbide/aluminum composites

Property	SiC_p/Al-4Cu-1.5Mg(a)	SiC_w/Al-4Cu-1.5Mg(b)
Reinforcement content, vol%	20	15
Longitudinal modulus, GPa (10^6 psi)	110 (16)	108 (15.7)
Transverse modulus, GPa (10^6 psi)	105 (15)	90 (13)
Longitudinal tensile strength, MPa (ksi)	648 (94)	683 (99)
Transverse tensile strength, MPa (ksi)	641 (93)	545 (79)
Longitudinal strain to failure, %	5	4.3
Transverse strain to failure, %	5	7.4

(a) 12.7 mm (0.5 in.) plate. (b) 1.8–3.2 mm (0.070–0.125 in.) sheet. Source: Advanced Composite Materials Corporation

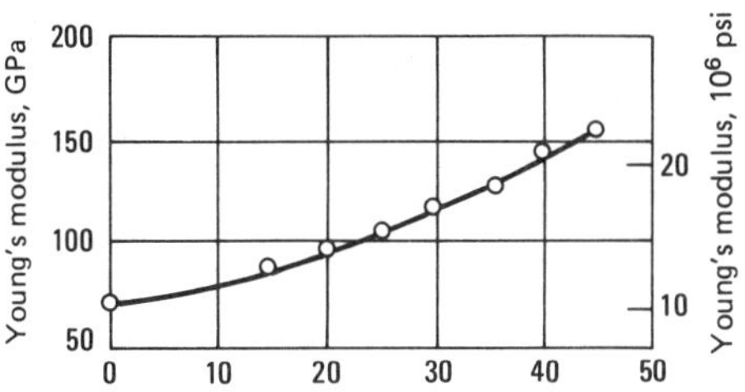

Fig. 8 Effect of reinforcement content on the Young's modulus of a particulate-reinforced SiC/2124-T6 Al MMC. Source: Ref 32

racks (Fig. 9). These composites can be tailored to exhibit dimensional stability, that is, resistance to microcreep, which is important for precision mirror optics and inertial measurement units (Fig. 10). In the electronics industry, metals such as iron-nickel alloys that are now used for packaging materials and heat sinks are candidates for replacement by SiC_d/Al. The composite has lower density, better thermal conductivity (≥160 W/m · K), and can be made to have a low coefficient of thermal expansion (Fig. 11). Hybrid electronic packages made from SiC_d/Al are shown in Fig. 12.

Most Gr/Al MMC development work has focused on using continuous fibers as reinforcement. However, various solidification processing techniques have been investigated for use in the production of cast particle Gr/Al for applications needing an inexpensive antifriction material (Ref 35).

Discontinuous Al_2O_3/Al MMCs are made using short fibers, particles, or compacted staple fiber preforms as reinforcements. The addition of chopped Al_2O_3 fibers to an agitated, partially solid aluminum alloy slurry has been used to produce a castable discontinuous MMC (Ref 36). Squeeze casting has attracted much attention because the process minimizes material and energy use, produces net shape components, and offers a selective reinforcement capability (Ref 37, 38).

A recent development in MMC fabrication technology is the proprietary Lanxide PRIMEX™ process, which involves pressureless metal infiltration into a ceramic preform. This process has been used to produce an Al_2O_3/Al composite by the infiltration of a bed of alumina particles with a molten alloy that was exposed to an oxidizing atmosphere. The matrix material of the resultant composite is composed of a mixture of the oxidation reaction product and unreacted aluminum alloy (Ref 39). The Lanxide process offers net shape capability (Fig. 13), and the properties of composites produced by this method can be tailored to fit specific applications.

Aluminum oxide/aluminum MMCs are candidate materials for moving parts of automotive engines, such as pistons (Ref 40), connecting rods (Ref 16), piston pins, and various components in the cylinder head and valve train (Ref 41). Examples of automotive parts fabricated from MMCs are shown in Fig. 14. These components were fabricated from aluminum-base composites with reinforcements typically of silicon carbide or alumina in volume loadings ranging from 5 to 25%. A variety of processing techniques can be used to fabricate such parts. For example, production of the combustion bowl area of the diesel piston involved squeeze casting a ceramic preform with metal; the cylinder liner was sand mold cast; and the connecting rod was made using novel forming processes specifically adapted for composites. Other possible automotive applications for this class of materials include brake rotors, brake calipers, and drive shafts.

A Toyota diesel engine piston selectively reinforced with an aluminosilicate ceramic compact is currently in production (Ref 42). Selective reinforcement of the all-aluminum piston with a ceramic fiber preform provides wear resistance equal to that of a piston with an iron insert, and the thermal transport is only marginally lower than that of unreinforced aluminum. With the elimination of the iron insert, piston weight is reduced, and high-temperature strength and thermal stability are enhanced.

Magnesium-Matrix Composites

Magnesium composites are being developed to exploit essentially the same properties as those provided by aluminum MMCs: high stiffness, light weight, and low CTE. In practice, the choice between aluminum and magnesium as a matrix is usually made on the basis of weight versus corrosion resistance. Magnesium is approximately two-thirds as dense as aluminum, but it is more active in a corrosive environment. Magnesium has a lower thermal conductivity, which is sometimes a factor in its selection. Three types of magnesium MMCs are currently under development: continuous fiber Gr/Mg for space structures (Ref 43), short staple fiber Al_2O_3/Mg for automotive engine components (Ref 44), and discontinuous SiC or B_4C/Mg for engine components (Ref

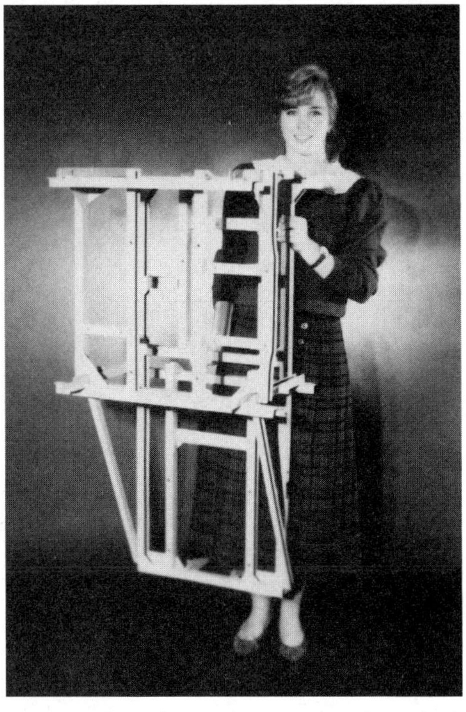

Fig. 9 Lightweight aircraft equipment racks made of particulate SiC/Al. Courtesy of DWA Composite Specialties, Inc. and Lockheed ASD

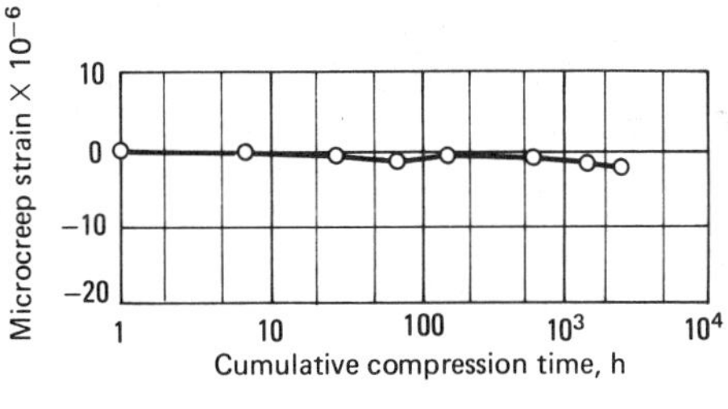

Fig. 10 Microcreep behavior of 2124-T6 aluminum reinforced with 30 vol% SiC particulate. Performance of composite indicates long-term dimensional stability. Source: Ref 32

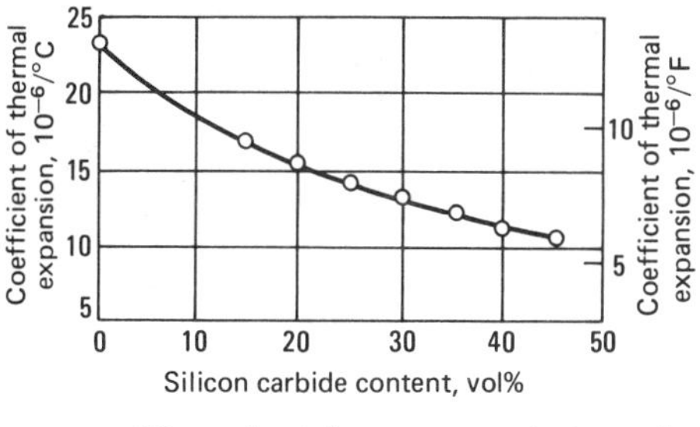

Fig. 11 Effect of reinforcement content on the room-temperature coefficient of thermal expansion for a SiC_p/2124-T6 Al MMC. Source: Ref 34

Fig. 12 Electronic packages made from SiC_d/Al (60 vol% SiC) MMCs. Courtesy of Lanxide Corporation

45) and low-expansion electronic packaging materials (Ref 46). Processing methods for all three types parallel those used for their aluminum MMC counterparts.

The production of the continuous-fiber Gr/Mg composite involves the titanium-boron coating method of making composite wires, physical vapor deposition of the matrix on fibers, or diffusion bonding of fiber-thin sheet sandwiches to make panels. A casting technology exists for Gr/Mg that involves the deposition of an air-stable silicon dioxide coating on the fibers from an organometallic precursor solution (Ref 47). Magnesium wets the coating, permitting incorporation of the matrix by near-net-shape casting procedures (Ref 48). Testing of a unidirectionally reinforced Gr/Mg MMC in the fiber direction recorded modulus values in agreement with Eq 1 and a tensile strength of 572 MPa (83 ksi) (Ref 43). A P100 Gr/AZ91C Mg unidirectional laminate was shown to have a lower CTE and smaller residual strain than those of a P100 Gr/6061 Al MMC after both composites had undergone thermal cycling between −155 °C and 120 °C (−250 and 250 °F) (Ref 25).

Titanium-Matrix Composites

Titanium was selected for use as a matrix metal because of its good specific strength at both room and moderately elevated temperatures and its excellent corrosion resistance. In comparison with aluminum, titanium retains its strength at higher temperatures; it has increasingly been used as a replacement for aluminum in aircraft and missile structures as the operating speeds of these items have increased from subsonic to supersonic. Efforts to develop titanium MMCs were hampered for years by processing problems stemming from the high reactivity of titanium with many reinforcing materials. Reference 49 is a review of titanium MMC technology. Silicon carbide is now the accepted reinforcement; the SCS-6 fiber is an example of one commercially available type. The SCS-6 fiber has a 140 μm (5.6 mil) diameter, a 33 μm (1.3 mil) carbon core, and a carbon-rich surface (Fig. 15).

Although a number of processing techniques have been evaluated for titanium MMCs, only high-temperature/short-time roll bonding, hot isostatic pressing, and vacuum hot pressing have been used to any substantial degree. Plasma spraying also is employed to deposit a titanium matrix onto the fibers (Ref 50). Properties for a representative unidirectional SiC/Ti laminate are given in Table 3. The elevated-temperature strength of the SiC/Ti composite is significantly greater than that of unreinforced titanium (Fig. 3). Potential applications for continuous titanium MMCs lie primarily in the aerospace industry and include major aircraft structural components (Ref 51) and fan and compressor blades for advanced turbine engines.

Table 3 Room-temperature properties of a unidirectional SiC_c/Ti MMC

Property	SCS-6/Ti-6Al-4V
Fiber content, vol%	37
Longitudinal modulus, GPa (10^6 psi)	221 (32)
Transverse modulus, GPa (10^6 psi)	165 (24)
Longitudinal strength, MPa (ksi)	1447 (210)
Transverse strength, MPa (ksi)	413 (60)

Source: Ref 14

Titanium MMCs with discontinuous reinforcements are in the early stages of development (Ref 52). This type of composite has a moderate stiffness and elevated-temperature strength advantage over monolithic titanium alloys. It also offers a near-net-shape manufacturing capability with the use of powder metallurgy techniques; therefore, it may be more economical to fabricate than continuous fiber titanium MMCs.

Copper-Matrix Composites

Copper appears to have potential as a matrix metal for composites that require thermal conductivity and high-temperature strength properties superior to those of aluminum MMCs. Copper MMCs with continuous and discontinuous reinforcements are being evaluated.

Continuous tungsten fiber reinforced copper composites were first fabricated in the late 1950s as research models for studying stress-strain behavior, stress-rupture and creep phenomena, and impact strength and conductivity in MMCs (Ref 53). The composites were made by liquid-phase infiltration. On the basis of their high strength at temperatures up to 925 °C (1700 °F), W/Cu MMCs are now being considered for use as

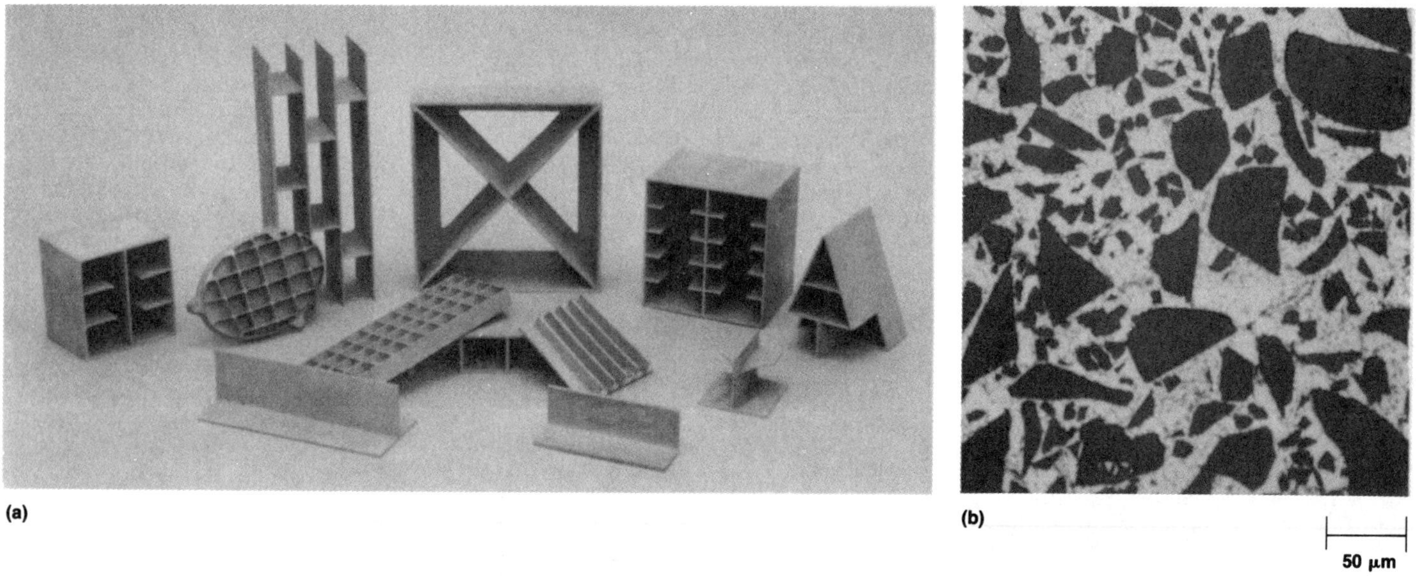

Fig. 13 Discontinuous silicon carbide/aluminum MMC (60 vol% SiC) produced by the PRIMEX™ process. (a) Near-net-shape components fabricated from the composite. (b) Composite microstructure. Courtesy of Lanxide Corporation

Fig. 14 Automotive components fabricated from MMCs. Clockwise from left: experimental piston for a gasoline engine, experimental cylinder liner, production piston for a heavy-duty diesel truck engine, and experimental connecting rod. Courtesy of Ford Motor Company

liner materials for the combustion chamber walls of advanced rocket engines (Ref 54).

Continuous Gr/Cu MMCs. Interest in continuous Gr/Cu MMCs gained impetus from the development of advanced graphite fibers. Copper has good thermal conductivity, but it is heavy and has poor elevated-temperature mechanical properties. Pitch-base graphite fibers have been developed that have room-temperature axial thermal conductivity properties better than those of copper (Ref 26). The addition of these fibers to copper reduces density, increases stiffness, raises the service temperature, and provides a mechanism for tailoring the coefficient of thermal expansion. One approach to the fabrication of Gr/Cu MMCs uses a plating process to envelop each graphite fiber with a pure copper coating, yielding MMC fibers flexible enough to be woven into fabric (Ref 55). The copper-coated fibers must be hot pressed to produce a consolidated component. Table 4 compares the thermal properties of aluminum and copper MMCs with those of unreinforced aluminum and copper. Graphite/copper MMCs have the potential to be used for thermal management of electronic components (Ref 55), satellite radiator panels (Ref 56), and advanced airplane structures (Ref 57).

***In situ* Composites.** Discontinuous MMCs formed by the working of mixtures of individual metal phases exhibit strengths as much as 50% higher than those predicted in theory from the strength of the individual constituents (Ref 8). These materials are called *in situ* composites because the elongated ribbon morphology of the reinforcing phase is developed in place by heavy mechanical working, which can consist of extrusion, drawing, or rolling. This approach has been applied to the fabrication of discontinuous refractory metal/copper composites, with niobium/copper serving as the prototype. Niobium/copper maintains high strength at temperatures up to 400 °C (750 °F), and it remains stronger than high-temperature copper alloys and dispersion-hardened copper up to 600 °C (1110 °F) (Ref 58). These composites are candidates for applications such as electrical contacts that require good strength plus conductivity at moderate temperatures.

Table 4 Thermal properties of unreinforced and reinforced aluminum and copper

Material	Reinforcement content, vol%	Density		Axial thermal conductivity		Axial coefficient of thermal expansion	
		g/cm^3	lb/ft^3	W/m · °C	Btu/ft · h · °F	10^{-6}/°C	10^{-6}/°F
Aluminum	0	2.71	169	221	128	23.6	13.1
Copper	0	8.94	558	391	226	17.6	9.7
SiC_p/Al	40	2.91	182	128	74	12.6	7
P120 Gr/Al	60	2.41	150	419	242	−0.32	−0.17
P120 Gr/Cu	60	4.90	306	522	302	−0.07	−0.04

Source: Ref 34

Superalloy-Matrix Composites

Superalloys are commonly used for turbine engine hardware and, therefore, superalloy-matrix composites were among the first candidate materials considered for upgrading turbine performance by raising component operating temperatures. Superalloy MMCs were developed to their present state over a period of years, starting from the early 1960s. The following summary is drawn from the review in Ref 59.

High-temperature strength in superalloy MMCs has been achieved only through the use of refractory metal reinforcements (tungsten, molybdenum, tantalum, and niobium fibers with compositions specially modified for this purpose). The strongest fiber developed, a tungsten alloy, exhibited a strength of more than 2070 MPa (300 ksi) at 1095 °C (2000 °F), or more than six times the strength of the superalloy now used in the Space Shuttle main engine.

Much of the early work on superalloy MMCs consisted of fiber-matrix compatibility studies, which ultimately led to the use of matrix alloys that exhibit limited reaction with the fibers. Tungsten fibers, for example, are least reactive in iron-base matrices, and they can endure short exposures at temperatures up to 1195 °C (2190 °F) with no detectable reaction.

Fabrication of superalloy MMCs is accomplished via solid-phase, liquid-phase, or deposition processing. The methods include investment casting, the use of matrix metals in thin sheet form, the use of matrix metals in powder sheet form made by rolling powders with an organic binder, powder metallurgy techniques, slip casting of metal alloy powders, and arc spraying. Iron-, nickel-, and cobalt-base MMCs have been made, and a wide range of properties have been achieved with these MMCs, including elevated-temperature tensile strength, stress-rupture strength, creep resistance, low- and high-cycle fatigue strength, impact strength, oxidation resistance, and thermal conductivity (Ref 59). The feasibility of making a component with a complex shape was shown using a first-stage convection-cooled turbine blade as a model from which a W/FeCrAlY hollow composite blade was designed and fabricated. Additional information on superalloy MMCs reinforced with refractory metals can be found in the article "Refractory Metals and Alloys" in this Volume.

Intermetallic-Matrix Composites

One disadvantage of superalloy MMCs is their high density, which limits the potential minimum weight of parts made from these materials. High melting points and relatively low densities make intermetallic-matrix composites (IMCs) viable candidates for lighter turbine engine materials

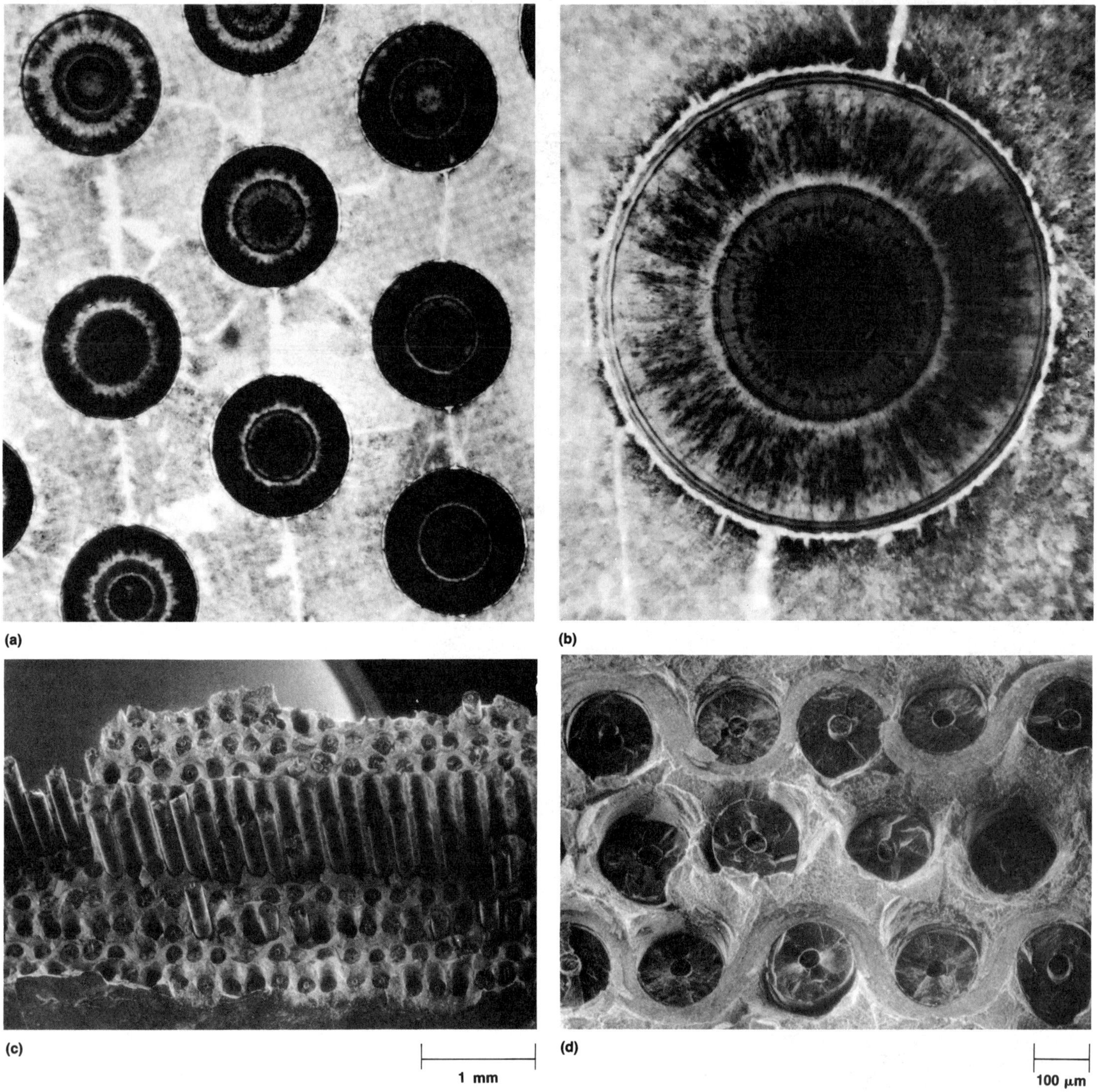

Fig. 15 Continuous-fiber-reinforced titanium-matrix MMCs. (a) Hot-pressed SiC fibers (SCS-6, 35 vol%) in a Ti-6Al-4V matrix. Fiber thickness, 140 μm; density, 3.86 g/cm^3. (b) Chemical vapor deposited SiC fiber (SCS-6) showing the central carbon monofilament substrate and the carbon-rich surface. Fiber properties: thickness, 140 μm; tensile strength, 3450 MPa (500 ksi); modulus of elasticity, 400 GPa (58 × 10^6 psi); density, 3.0 g/cm^3. (c) Fracture surface of a hot-pressed SCS-6 SiC/titanium MMC plate. (d) Close-up view of fractured SCS-6 fibers. Courtesy of Textron Specialty Materials, a subsidiary of Textron, Inc.

(Ref 60). An intermetallic compound differs from an alloy in that the former has a fixed compositional range, a long-range order to the arrangement of atoms within the lattice, and a limited number of slip systems available for plastic deformation. At present, the IMC technology is in its infancy, and many critical issues remain to be addressed.

Aluminides of nickel, titanium, and iron have received most of the early attention as potential matrices for IMCs. Work on aluminide IMCs is concerned with developing methods to fabricate reproducible specimens with useful properties; work is also being done on characterizing the interface chemical reactions of fiber/matrix combinations. Candidate reinforcements for commercially available intermetallic materials are SiC and Al_2O_3 fibers, refractory metal fibers, and particulates such as titanium carbide (TiC) and titanium diboride (TiB_2). Research is being done to find methods for growing advanced single-crystal fibers and using refractory metal aluminides and silicides as matrices (Ref 61). Key factors in selecting a reinforcement/matrix combination are chemical compatibility at the processing temperature and an approximate match of thermal expansion coefficients between the material pair to minimize residual fabrication stresses.

Reference 62 is an overview of the development of nickel aluminide IMCs, and it describes the various processing techniques used to make this composite. These techniques include hot pressing, diffusion bonding, hot extrusion, reactive sintering, and liquid infiltration. Reference 63 presents evidence that silicon carbide cannot serve as a reinforcement for nickel aluminide IMCs without the use of a diffusion barrier coating. A gas pressure liquid infiltration technique has been used to produce continuous fiber Al_2O_3/NiAl (Ref 64). Reference 65 describes a powder cloth method for the fabrication of a 40 vol% continuous fiber SiC/Ti_3Al + Nb IMC. Data on IMC properties are very limited.

The XD composites are a proprietary class of discontinuous reinforcement *in situ* composites. The XD technology uses a casting process to produce a fine, closely spaced, and uniform distribution of second-phase particles (Ref 66). The dispersoids are formed and grown *in situ* instead of being mechanically mixed as a separate additive. This approach to making ceramic-stiffened composites has been demonstrated for a number of metals as well as for titanium and nickel aluminides (Ref 67). Strength levels of greater than 690 MPa (100 ksi) were measured at 20 °C (70 °F) and at 800 °C (1470 °F) for a two-phase lamellar Ti-45 at.% Al alloy reinforced with equiaxed TiB_2 ceramic particulates (Ref 66).

REFERENCES

1. Z. Hashin and B.W. Rosen, The Elastic Moduli of Fiber-Reinforced Materials, *J. Appl. Mech. (Trans. ASME)*, June 1964, p 223
2. D.E. Bowles and S.S. Tompkins, Prediction of Coefficients of Thermal Expansion for Unidirectional Composites, *J. Compos. Mater.*, Vol 23, April 1989, p 370
3. G.S. Springer and S.W. Tsai, Thermal Conductivities of Unidirectional Materials, *J. Compos. Mater.*, Vol 1, 1967, p 166
4. A. Mortensen, J.A. Cornie, and M.C. Flemings, Solidification Processing of Metal-Matrix Composites, *J. Met.*, Feb 1988, p 12
5. P.K. Rohatgi, R. Asthana, and S. Das, Solidification, Structures, and Properties of Cast Metal-Ceramic Particle Composites, *Int. Met. Rev.*, Vol 31 (No. 3), 1986, p 115
6. D.E. Hammond, Foundry Practice for the First Castable Aluminum/Ceramic Composite Material, *Mod. Cast.*, Aug 1989, p 29
7. T.W. Chou, A. Kelly, and A. Okura, Fibre-Reinforced Metal-Matrix Composites, *Composites*, Vol 16 (No. 3), July 1985, p 187
8. D.L. Erich, Metal-Matrix Composites: Problems, Applications, and Potential in the P/M Industry, *Int. J. Powder Metall.*, Vol 23 (No. 1), 1987, p 45
9. J. White, T.C. Willis, I.R. Hughes, and R.M. Jordan, Metal Matrix Composites Produced by Spray Deposition, in *Dispersion Strengthened Aluminum Alloys*, Y.W. Kim and W.M. Griffith, Ed., The Minerals, Metals and Materials Society, 1988, p 693
10. M. Ramula and M. Taya, EDM Machinability of SiC_w/Al Composites, *J. Mater. Sci.*, Vol 24, 1989, p 1103
11. P.K. Rohatgi, N.B. Dahotre, S.C. Gopinathan, D. Alberts, and K.F. Neusen, Micromechanism of High Speed Abrasive Waterjet Cutting of Cast Metal Matrix Composites, in *Cast Reinforced Metal Composites*, S.G. Fishman and A.K. Dhingra, Ed., ASM INTERNATIONAL, 1988, p 391
12. F.A. Girot, J.M. Quenisset, and R. Naslain, Discontinuously-Reinforced Aluminum Matrix Composites, *Compos. Sci. Technol.*, Vol 30, 1987, p 155
13. K.G. Kreider and K.M. Prewo, Boron-Reinforced Aluminum, in *Metallic Matrix Composites*, Vol 4, K.G. Kreider, Ed., *Composite Materials*, Academic Press, 1974, p 400
14. "Silicon Carbide Composite Materials," Data Sheet, Textron Specialty Materials, 1989
15. R.B. Francini, Characterization of Thin-Wall Graphite/Metal Pultruded Tubing, in *Testing Technology of Metal Matrix Composites*, STP 964, P.R. DiGiovanni and N.R. Adsit, Ed., American Society for Testing and Materials, 1988, p 396
16. F. Folgar, Fiber FP/Metal Matrix Composite Connecting Rods: Design, Fabrication and Performance, *Ceram. Eng. Sci. Proc.*, Vol 9 (No. 7-8), 1988, p 561
17. D. Hughes, Textron Unit Makes Reinforced Titanium, Aluminum Parts, *Aviat. Week Space Technol.*, 28 Nov 1988
18. E.G. Wolff, Stiffness-Thermal Expansion Relationships in High Modulus Carbon Fibers, *J. Compos. Mater.*, Vol 21, Jan 1987, p 81
19. M.U. Islam and W. Wallace, Carbon Fibre Reinforced Aluminium Matrix Composites. A Critical Review, *Adv. Mater. Manuf. Process.*, Vol 3 (No. 1), 1988, p 1
20. M.F. Amateau, Progress in the Development of Graphite-Aluminum Composites Using Liquid Infiltration Technology, *J. Compos. Mater.*, Vol 10, Oct 1976, p 279
21. M. Yoshida, S. Ikegami, T. Ohsaki, and T. Ohkita, Studies on Ion-Plating Process for Making Carbon Fiber Reinforced Aluminum and Properties of the Composites, in *Proceedings of the 24th National SAMPE Symposium*, Vol 24, Society for the Advancement of Material and Process Engineering, 1979, p 1417
22. D.J. Bak, Vapor Deposition Improves Metal Matrix Composites, *Des. News*, 16 June 1986
23. R.J. Sample, R.B. Bhagat, and M.F. Amateau, High Pressure Squeeze Casting of Unidirectional Graphite Fiber Reinforced Aluminum Matrix Composites, in *Cast Reinforced Metal Composites*, S.G. Fishman and A.K. Dhingra, Ed., ASM INTERNATIONAL, 1988, p 179
24. L. Rubin, "Data Base Development for P100 Graphite Aluminum Metal Matrix Composites," Aerospace Report TOR-0089 (4661-02)-1, Aerospace Corporation, Sept 1989
25. S.S. Tompkins and G.A. Dries, Thermal Expansion Measurements of Metal Matrix Composites, in *Testing Technology of Metal Matrix Composites*, STP 964, P.R. DiGiovanni and N.R. Adsit, Ed., American Society for Testing and Materials, 1988, p 248
26. L.M. Sheppard, Challenges Facing the Carbon Industry, *Ceram. Bull.*, Vol 67 (No. 12), 1988, p 1897
27. A.K. Dhingra, Metal Matrix Composites Reinforced with Fibre FP(α-Al_2O_3), *Philos. Trans. R. Soc. (London) A*, Vol 294, 1980, p 559
28. H.R. Shetty and Tsu-Wei Chou, Mechanical Properties and Failure Characteristics of FP/Aluminum and W/Aluminum Composites, *Metall. Trans. A*, Vol 16A, May 1985, p 853
29. D.M. Schuster, M. Skibo, and F. Yep, SiC Particle Reinforced Aluminum by Casting, *J. Met.*, Nov 1987, p 60
30. R.J. Arsenault and S.B. Wu, A Comparison of PM Vs. Melted SiC/Al Composites, *Scr. Metall.*, Vol 22, 1988, p 767
31. D.L. McDanels, Analysis of Stress-Strain, Fracture, and Ductility Behavior of Aluminum Matrix Composites Containing Discontinuous Silicon Carbide Reinforcement, *Metall. Trans. A*, Vol 16A, June 1985, p 1105
32. W.R. Mohn and D. Vukobratovich, Engineered Metal Matrix Composites for Precision Optical Systems, *SAMPE J.*, Jan-Feb 1988, p 26
33. P.L. Boland, P.R. DiGiovanni, and L. Franceschi, Short-Term High-Temperature Properties of Reinforced Metal Matrix Composites, in *Testing Technology of Metal Matrix Composites*, STP 964, P.R. DiGiovanni and N.R. Adsit, Ed., American Society for Testing and Materials, 1988, p 346
34. C. Thaw, R. Minet, J. Zemany, and C.

Zweben, Metal Matrix Composite Microwave Packaging Components, *SAMPE J.*, Nov-Dec 1987, p 40
35. P.K. Rohatgi, S. Das, T.K. Dan, Cast Aluminum-Graphite Particle Composites—A Potential Engineering Material, *J. Inst. Eng. (India)*, Vol 67, Mar 1987, p 77
36. C.G. Levi, G.J. Abbaschian, and R. Mehrabian, Interface Interactions During Fabrication of Aluminum Alloy-Alumina Fiber Composites, *Metall. Trans. A*, Vol 9A, May 1978, p 697
37. M.W. Toaz, Squeeze Cast Composites, in *Proceedings of AFS's International Conference on Permanent Mold Casting of Aluminum* (Detroit, MI), American Foundrymen's Society, 1989
38. H. Fukunaga, Squeeze Casting Processes for Fiber Reinforced Metals and Their Mechanical Properties, in *Cast Reinforced Metal Composites*, S.G. Fishman and A.K. Dhingra, Ed., ASM INTERNATIONAL, 1988, p 101
39. M.S. Newkirk, A.W. Urquhart, H.R. Zwicker, and E. Breval, Formation of Lanxide™ Ceramic Composite Materials, *J. Mater. Res.*, Vol 1 (No. 1), Jan-Feb 1986, p 81
40. M.D. Smalc, The Mechanical Properties of Squeeze Cast Diesel Pistons, in *Engine Components—New Materials and Manufacturing Processes*, J.M. Bailey, Ed., ICE Vol 1, American Society of Mechanical Engineers, 1985, p 29
41. J. Dinwoodie, "Automotive Applications for MMC's Based on Short Staple Alumina Fibres," SAE Paper 870437, Society of Automotive Engineers, 1987
42. T. Donomoto, K. Funatani, N. Miura, and N. Miyake, "Ceramic Fiber Reinforced Piston for High Performance Diesel Engines," SAE Paper 830252, Society of Automotive Engineers, 1984
43. B.J. Maclean and M.S. Misra, Thermal-Mechanical Behavior of Graphite/Magnesium Composites, in *Mechanical Behavior of Metal-Matrix Composites*, J.E. Hack and M.F. Amateau, Ed., The Metallurgical Society of AIME, 1982, p 195
44. J. Dinwoodie and I. Horsfall, New Development With Short Staple Alumina Fibres In Metal Matrix Composites, in *Proceedings of The Sixth International Conference on Composite Materials (ICCM-VI)*, American Institute of Mining, Metallurgical, and Petroleum Engineers, 1987, p 2.390
45. B.A. Mikucki, Ceramic Fibers Boost Magnesium's Potential, *Mod. Cast.*, July 1989, p 49
46. A.L. Geiger and M. Jackson, Low-Expansion MMCs Boost Avionics, *Adv. Mater. Proc. inc. Met. Prog.*, July 1989, p 23
47. H.A. Katzman, Fibre Coatings for the Fabrication of Graphite-Reinforced Magnesium Composites, *J. Mater. Sci.*, Vol 22, 1987, p 144
48. D.M. Goddard, Report on Graphite/Magnesium Castings, *Met. Prog.*, April 1984, p 49
49. P.R. Smith and F.H. Froes, Developments in Titanium Metal Matrix Composites, *J. Met.*, March 1984, p 19
50. N. Newman and R. Pinckert, Materials for the NASP, *Aerosp. Am.*, May 1989, p 24
51. S.D. Forness and S. Pollock, SPF/DB Ti F-15 Horizontal Stabilator With B_4CB/Ti 15-3-3-3 TMC Skins, in *Proceedings of Department of Defense Eighth Metal Matrix Composites (MMC) Technology Conference*, Vol 1, Metal Matrix Composites Information Analysis Center, June 1989
52. S. Abkowitz and P. Weihrauch, Trimming the Cost of MMCs, *Adv. Mater. Proc. inc. Met. Prog.*, July 1989, p 31
53. D.L. McDanels, "Tungsten Fiber Reinforced Copper Matrix Composites. A Review," NASA Technical Paper 2924, National Aeronautics and Space Administration, Sept 1989
54. L.J. Westfall and D.W. Petrasek, "Fabrication and Preliminary Evaluation of Tungsten Fiber Reinforced Copper Composite Combustion Chamber Liners," NASA Technical Memorandum 100845, National Aeronautics and Space Administration, May 1988
55. D.A. Foster, Electronic Thermal Management Using Copper Coated Graphite Fibers, *SAMPE Q.*, Oct 1989, p 58
56. D.L. McDanels and J.O. Diaz, "Exploratory Feasibility Studies of Graphite Fiber Reinforced Copper Matrix Composites for Space Power Radiator Panels," NASA TM-102328, National Aeronautics and Space Administration, Sept 1989
57. T.M.F. Ronald, Advanced Materials to Fly High in NASP, *Adv. Mater. Proc. inc. Met. Prog.*, May 1989, p 29
58. J.D. Verhoeven, W.A. Spitzig, F.A. Schmidt, and C.L. Trybus, Deformation Processed Cu-Refractory Metal Composites, *Mater. Manuf. Proc.*, Vol 4 (No. 2), 1989, p 197
59. D.W. Petrasek, R.A. Signorelli, T. Caulfield, and J.K. Tien, Fiber Reinforced Superalloys, in *Superalloys, Supercomposites and Superceramics*, Academic Press, 1989, p 625
60. J.R. Stephens and M.V. Nathal, Status and Prognosis for Alternative Engine Materials, in *Superalloys 1988*, S. Reichman, D.N. Duhl, G. Maurer, S. Antolovich, and C. Lund, Ed., The Metallurgical Society, 1988, p 183
61. R. Bowman and R. Noebe, Up-and-Coming IMCs, *Adv. Mater. Proc. inc. Met. Prog.*, Aug 1989, p 35
62. J.M. Yang, W.H. Kao, and C.T. Liu, Development of Nickel Aluminide Matrix Composites, *Mater. Sci. Eng.*, Vol A107, 1989, p 81
63. J.M. Yang, W.H. Kao, and C.T. Liu, Reinforcement/Matrix Interaction in SiC Fiber-Reinforced Ni_3Al Matrix Composites, in *Proceedings of the High-Temperature Ordered Intermetallic Alloys III Symposium*, C.T. Liu, A.I. Taub, N.S. Stoloff, and C.C. Koch, Ed., *Materials Research Society Symposium Proceedings*, Vol 133, 1989, p 453
64. S. Nourbakhsh, F.L. Liang, and H. Margolin, Fabrication of a Ni_3Al/Al_2O_3 Unidirectional Composite by Pressure Casting, *Adv. Mater. Manuf. Proc.*, Vol 3 (No. 1), 1988, p 57
65. P.K. Brindley, P.A. Bartolotta, and S.J. Klima, "Investigation of a SiC/Ti-24Al-11Nb Composite," NASA Technical Memorandum 100956, National Aeronautics and Space Administration, 1988
66. L. Christodoulou, P.A. Parrish, and C.R. Crowe, XD™ Titanium Aluminide Composites, in *Proceedings of the Symposium on High Temperature/High Performance Composites*, F.D. Lemkey, S.G. Fishman, A.G. Evans, and J.R. Strife, Ed., *Materials Research Society Symposium Proceedings*, Vol 120, 1988, p 29
67. A.R.C. Westwood and S.R. Winzer, Advanced Ceramics, in *Advancing Materials Research*, P.A. Psaras and H.D. Langford, Ed., National Academy Press, 1987, p 225

Ordered Intermetallics

C.T. Liu and J.O. Stiegler, Metals and Ceramics Division, Oak Ridge National Laboratory
F.H. (Sam) Froes, Institute for Materials and Advanced Processes, Colleges of Mines, University of Idaho

ORDERED INTERMETALLIC compounds constitute a unique class of metallic materials that form long-range ordered crystal structures (Fig. 1) below a critical temperature, generally referred to as the critical ordering temperature (T_c). These ordered intermetallics usually exist in relatively narrow compositional ranges around simple stoichiometric ratios.

Ordered intermetallic alloys with relatively low critical ordering temperatures (<700 °C, or 1290 °F) were studied quite extensively in the 1950s and 1960s, following the discovery of unusual dislocation structures and mechanical behavior associated with ordered lattices (Ref 1-4). Deformation in ordered alloys is controlled by the glide of superlattice or paired dislocations, as illustrated in Fig. 2 for a two-dimensional ordered lattice having an AB composition. The first, or leading, dislocation creates a layer of antiphase domain (which can be thought of simply as a layer of wrong bonding), and the second, or following, dislocation restores the order. The relatively low mobility of superlattice dislocations at higher temperatures gives rise to anomalous yield behavior; that is, yield strength increases rather than decreases with increasing test temperature (Ref 5-12). The anomalous yielding has been observed in many ordered intermetallics, such as Ni_3Al (Ref 5-7) and Cu_3Au (Ref 8) alloys. The results obtained by these studies are summarized in Ref 1.

The interest in ordered intermetallics subsided in the latter part of the 1960s because of severe embrittlement problems encountered with the compounds. Most strongly ordered intermetallics are so brittle that they simply cannot be fabricated into useful structural components (Ref 1-4). Even when fabricated, these compounds have a low fracture toughness that severely limits their use as engineering materials. However, in the latter part of the 1970s, some prominent results were reported that showed that the ductility and fabricability of ordered intermetallics could be dramatically improved by alloy design efforts using physical metallurgy principles. The ductility of Co_3V was substantially improved by macroalloying with iron additions that reduced the average electron concentration and changed the ordered crystal structure from hexagonal to cubic (Ref 13-17). The alloys $(Fe, Co, Ni)_3V$ with the cubic $L1_2$ ordered structure exhibited more than 40% ductility at room temperature (Ref 14). The ductility of polycrystalline Ni_3Al was dramatically increased by microalloying with boron additions (Ref 18), which segregated to grain boundaries and suppressed brittle intergranular fracture (Ref 19-21). Both cases have demonstrated the feasibility of achieving high tensile ductility in strongly ordered intermetallic alloys.

The recent search for new high-temperature structural materials has stimulated further interest in ordered intermetallics (Ref 22-24). These compounds generally exhibit promising high-temperature properties be-

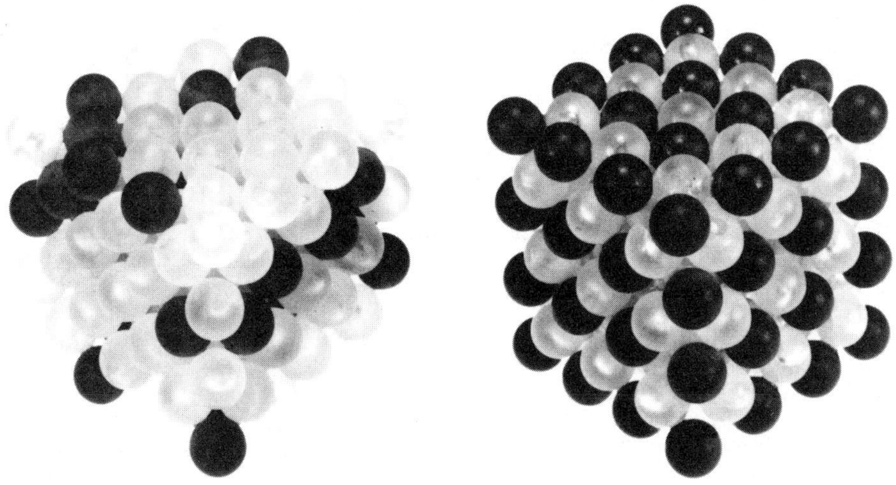

(a) (b)

Fig. 1 Atomic arrangements of conventional alloys and ordered intermetallic compounds. (a) Disordered crystal structure of a conventional alloy. (b) Long-range ordered crystal structure of an ordered intermetallic compound

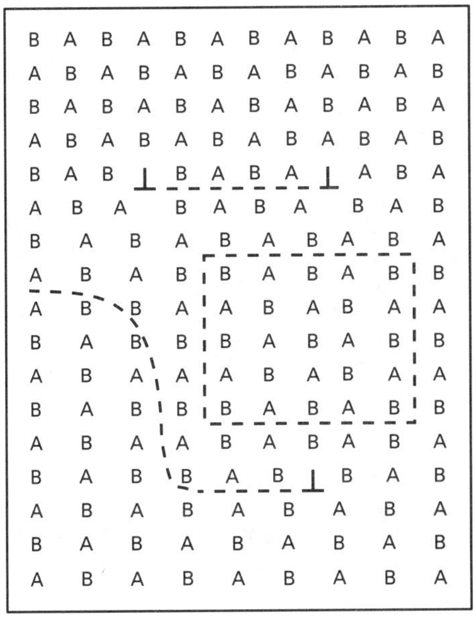

Fig. 2 Schematic representation of a superlattice dislocation in a two-dimensional simple cubic lattice, along with two thermally produced antiphase boundaries, one of which terminated on an ordinary dislocation. Source: Ref 1

Table 1 Properties of nickel, iron, and titanium aluminides

Alloy	Crystal structure(a)	Critical ordering temperature (T_c) °C	°F	Melting point (T_m) °C	°F	Material density g/cm³	Young's modulus GPa	10^6 psi
Ni_3Al	$L1_2$ (ordered fcc)	1390	2535	1390	2535	7.50	179	25.9
NiAl	B2 (ordered bcc)	1640	2985	1640	2985	5.86	294	42.7
Fe_3Al	$D0_3$ (ordered bcc)	540	1000	1540	2805	6.72	141	20.4
	B2 (ordered bcc)	760	1400	1540	2805	...	...	...
FeAl	B2 (ordered bcc)	1250	2280	1250	2280	5.56	261	37.8
Ti_3Al	$D0_{19}$ (ordered hcp)	1100	2010	1600	2910	4.2	145	21.0
TiAl	$L1_0$ (ordered tetragonal)	1460	2660	1460	2660	3.91	176	25.5
$TiAl_3$	$D0_{22}$ (ordered tetragonal)	1350	2460	1350	2460	3.4	...	...

(a) fcc, face-centered cubic; bcc, body-centered cubic; hcp, hexagonal close packed

cause the long-range ordered superlattice lowers dislocation mobility and diffusion processes at elevated temperatures (Ref 1-4, 22-24). However, because of the brittleness problem, the intermetallics have been used mainly as strengthening constituents in structural materials. For example, high-temperature nickel-base superalloys owe their outstanding strength properties to a fine dispersion of precipitated particles of the ordered γ′ phase (Ni_3Al) embedded in a ductile disordered matrix.

Recent research has focused on understanding the brittle fracture and low ductility in ordered intermetallics (Ref 1-4, 22-29). Possible causes for brittleness include:

- Insufficient number of deformation modes
- High yield strength or hardness caused by difficulty in the generation and glide of dislocations
- Poor cleavage strength or low surface energy
- Planar slip and localized deformation
- High strain rate sensitivity (which promotes brittle crack propagation at crack tips)
- Grain boundary weakness
- Environmental embrittlement

In some cases, the brittleness results from strong resistance to the motion of dislocations, to the point that cleavage or intergranular fracture may be favored. In many cases, however, the dislocations are relatively mobile. Brittleness results either from low-symmetry crystal structures that do not possess enough independent slip systems to permit arbitrary deformation or from the presence of grain boundaries that are too weak to resist the propagation of cracks. Recently, it has been found that quite a number of ordered intermetallics, such as iron aluminides (Ref 28-29), exhibit environmental embrittlement at ambient temperatures. The embrittlement involves the reaction of water vapor in air with reactive elements (aluminum, for example) in intermetallics to form atomic hydrogen, which drives into the metal and causes premature fracture.

In recent years, alloying and processing have been employed to control the ordered crystal structure, microstructural features, and grain-boundary structure and composition to overcome the brittleness problem of ordered intermetallics (Ref 22-24). Success in this work has inspired parallel efforts aimed at improving strength properties. The results have led to the development of a number of attractive intermetallic alloys having useful ductility and strength.

Alloy design work has been centered primarily on aluminides of nickel, iron, and titanium (Ref 22-24). These materials possess a number of attributes that make them attractive for high-temperature applications. They contain sufficient amounts of aluminum to form, in oxidizing environments, thin films of alumina (Al_2O_3) that often are compact and protective (Ref 30). These materials have low densities, relatively high melting points, (Table 1) and good high-temperature strength properties.

Crystal structures showing the ordered arrangements of atoms in several of these aluminides are illustrated in Fig. 3. For most of the aluminides listed in Table 1, the critical ordering temperature is equal to the melting temperature. Others disorder at somewhat lower temperatures, and Fe_3Al passes through two ordered structures ($D0_3$ and $B2$) before becoming disordered. Deviations from stoichiometry are accommodated either by the incorporation of vacancies in the lattice (for example, NiAl) (Ref 31-34) or by the location of antisite atoms in one of the sublattices. Many of the aluminides exist over a range of compositions, but the degree of order decreases as the deviation from stoichiometry increases. Additional elements can also be incorporated without losing the ordered structure. For example, in Ni_3Al, silicon atoms are located on aluminum sites, cobalt atoms on nickel sites, and iron atoms on either (Ref 35). In many instances, the so-called intermetallic compounds can be used as bases for alloy development to improve or optimize properties for specific applications.

All of the aluminides discussed in this article suffer from low-temperature (room-temperature) embrittlement, which has been severe enough to preclude their use as structural materials. In several cases, however, metallurgical solutions have been discovered that offer the possibility of engineering applications (Ref 22-24). This article summarizes research and development of nickel aluminides based on Ni_3Al and NiAl, iron aluminides based on Fe_3Al and FeAl, titanium aluminides based on Ti_3Al and TiAl, and other aluminides and intermetallics such as silicides; this article also provides a brief summary on applications of intermetallics.

This article focuses almost exclusively on compounds under development as structural materials. Information on other intermetallics, such as samarium-cobalt materials for permanent magnets and niobium-base superconductive materials, is available in the articles "Rare Earth Metals," "Permanent Magnet Materials," and "A15 Superconductors" in this Volume.

Nickel Aluminides

The nickel-aluminum phase diagram shows two stable intermetallic compounds, Ni_3Al and NiAl, formed on the nickel-rich end (Ref 36). The compound Ni_3Al has an $L1_2$ crystal structure, a derivative of the face-centered cubic (fcc) crystal structure; NiAl has a $B2$ structure, a derivative of the body-centered cubic (bcc) crystal structure (see Fig. 3). Because of the different crystal structures, the two nickel aluminides have quite different physical and mechanical properties.

Ni_3Al Aluminides

Intergranular Fracture and Alloying Effects. The aluminide Ni_3Al is of interest because of its excellent strength and oxidation resistance at elevated temperatures. As mentioned earlier, Ni_3Al is the most important strengthening constituent in nickel-base superalloys. Single crystals of Ni_3Al are ductile at ambient temperatures, but polycrystalline materials fail by brittle grain-boundary fracture with very little plasticity (Ref 19, 37, 38). This effect persists even in very high-purity materials where no grain-boundary segregation of impurities can be detected, suggesting that the brittleness is an intrinsic feature (Ref 19, 20, 39, 40). The observation of this characteristic turned attention toward a search for segregants that might act in a beneficial way.

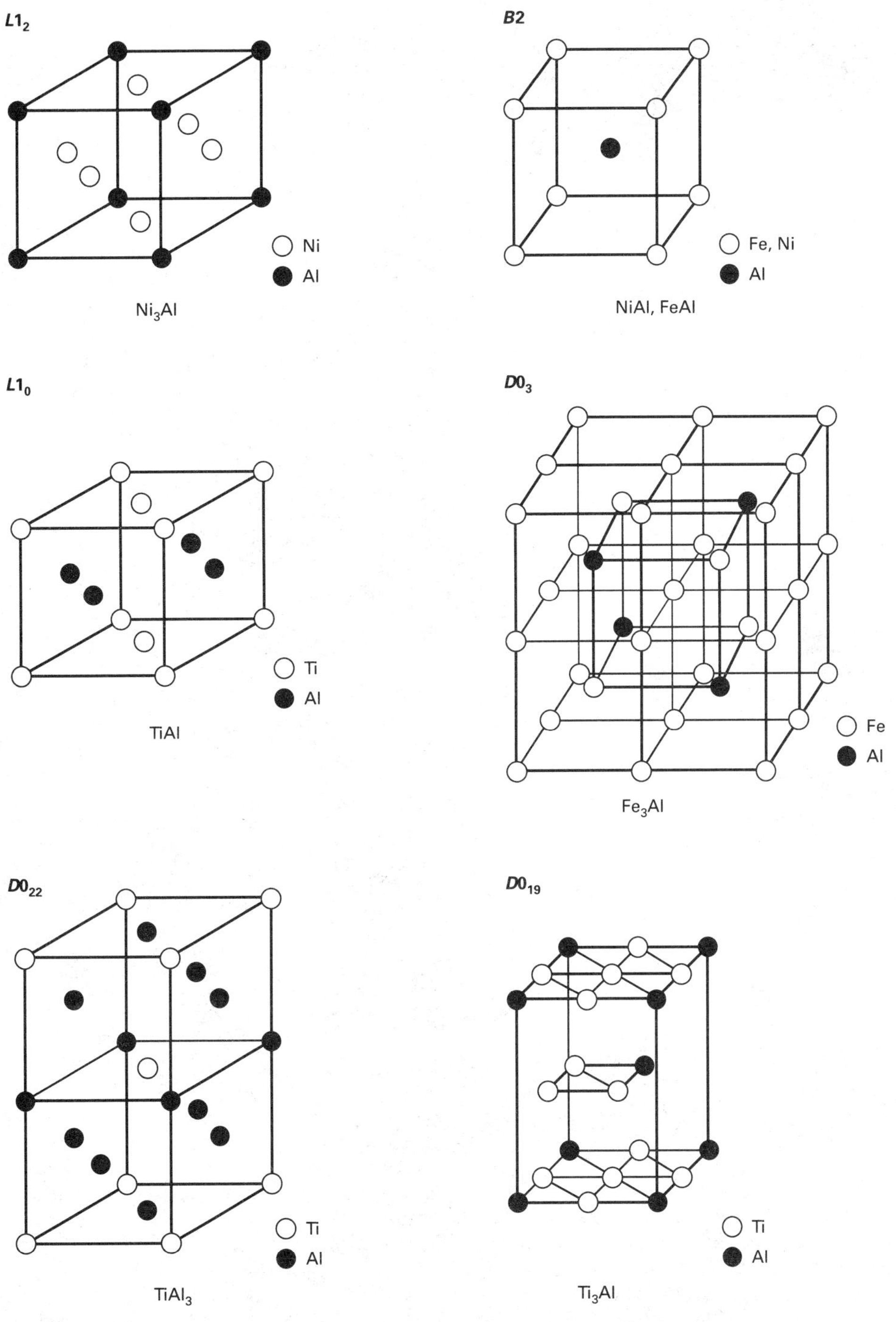

Fig. 3 Crystal structures of nickel, iron, and titanium aluminides

Studies of segregants led to the startling discovery (Ref 18-21) that small (~0.1 wt%) boron additions not only eliminated the brittle behavior of Ni_3Al but converted the material to a highly malleable form exhibiting tensile ductility as high as 50% at room temperature (Fig. 4) (Ref 19, 20). The beneficial effect of boron is, however, dependent on stoichiometry, and boron is effective in increasing the ductility of Ni_3Al only in alloys containing less than 25 at.% aluminum (Ref 20, 41). Both Auger spectroscopy (Ref 20) and imaging atom probe (Ref 42-45) studies have demonstrated the strong segregation of boron to grain boundaries, although the extent of segregation varies along boundaries and even at different points along the same boundary. Segregation is less strong in aluminum-rich alloys (Ref 20). The beneficial effect of boron has been attributed to an increase in the intrinsic strength or cohesion of the grain boundary (Ref 20, 46-50) and enhancement of dislocation generation and facilitation of slip transmission across grain boundaries (Ref 51-54). Both suggested mechanisms can be rationalized from boron-induced atomic disordering in the grain-boundary region (Ref 54-58).

The strong effect of boron on ductility is unique. For example, carbon, which is chemically similar to boron, has no comparable effect (Ref 59). A limited but still useful amount of ductility can be developed by much larger additions of substitutional elements such as iron (Ref 60, 61), manganese (Ref 60), chromium (Ref 60), or beryllium (Ref 62). In these cases, the beneficial effect has been suggested to be related to a reduction in the electronegativity difference between average nickel and aluminum atoms and the formation of more homogeneous atomic bonding (Ref 60, 63-66) across grain boundaries in Ni_3Al.

Environmental Embrittlement at Elevated Temperatures. Ductility at elevated temperatures depends on the test environment (Ref 67-70). In vacuum, the ductility of boron-doped nickel-rich alloys (<23% Al) remains high at all test temperatures, although a moderate minimum appears in the vicinity of 800 °C (1470 °F) (see Fig. 5 and Ref 67, 68). In tests conducted in environments containing oxygen, the ductility minimum is much deeper. In fact, ductility approaches zero with full intergranular fracture in tests conducted in air at 760 °C (1400 °F). The loss in ductility is due to a dynamic effect that requires the simultaneous application of a tensile stress and the presence of oxygen (Ref 67-70). It appears that the formation of protective Al_2O_3 films is too slow to deter rapid intergranular crack propagation. Chromium additions in the range of 6 to 10% restore the intermediate temperature ductility (Ref 71-73), possibly because of the more rapid formation of protective chromia (Cr_2O_3) films. Ductility at intermediate temperatures can also be improved by the production of an elongated grain structure in Ni_3Al (Ref 74).

Anomalous Dependence of Yield Strength on Temperature. Ni_3Al is one of a number of intermetallic alloys that exhibit an engineering yield strength (0.2% offset) that increases with increasing temperature (Ref 6, 7, 9). This is shown in Fig. 6, which is a plot of yield stress as a function of test temperature. The anomalous yielding effect, which is lower at lower strains, occurs because of extremely rapid work hardening. The work hardening is caused by the cross slip of screw dislocation segments from the primary {111} slip planes to {100} planes, where they become pinned and much less mobile (Fig. 7). Driving forces for cross slip include anisotropy of the energy of antiphase boundaries formed between the superdislocation pairs required for deformation in the ordered lattice (Ref 9, 10, 76, 77) and the torque exerted between the screw dislocation pairs arising from elastic anisotropy

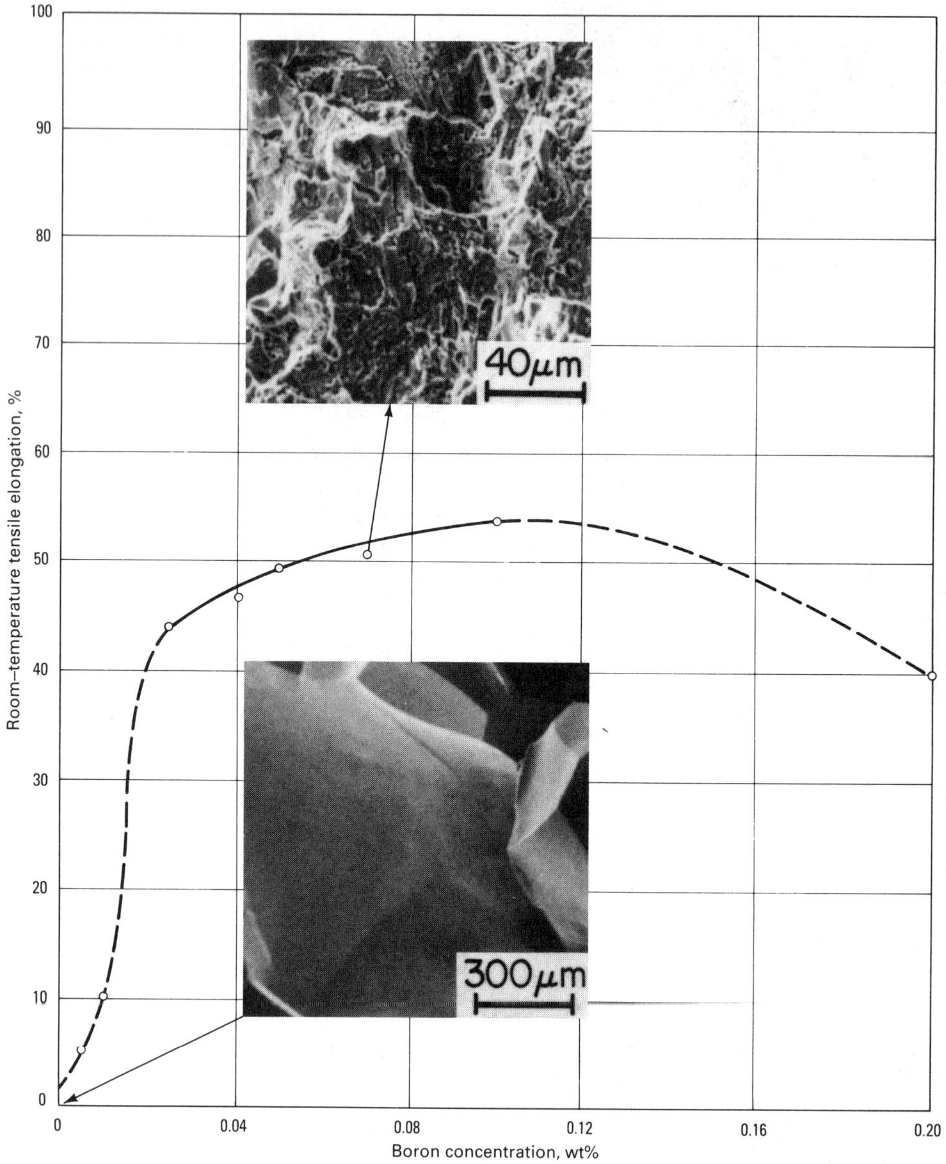

Fig. 4 Effect of boron additions on the room-temperature tensile elongation and fracture behavior of Ni_3Al (24 at.% Al)

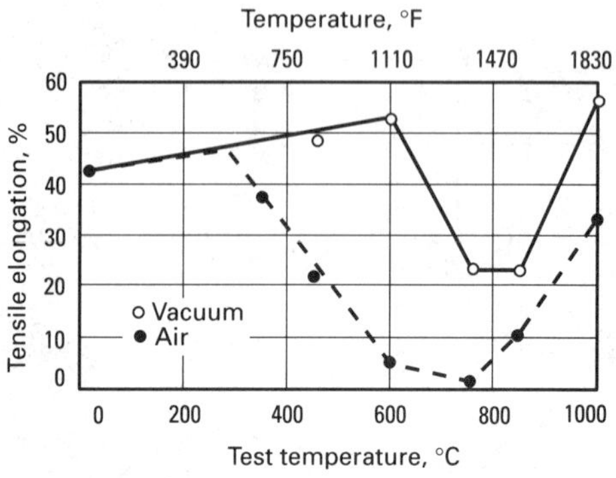

Fig. 5 Tensile elongation of alloy IC-145 (Ni-21.5Al-0.5Hf-0.1B at.%) in vacuum and in air

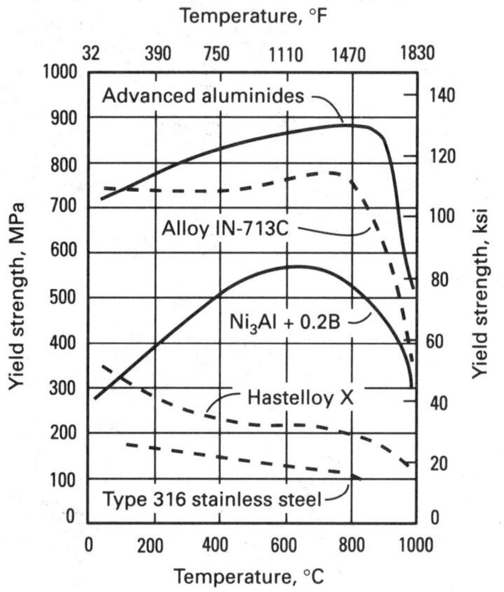

Fig. 6 Yield strength versus test temperature for Ni_3Al alloys, two superalloys, and type 316 stainless steel. Source: Ref 75

(Ref 11, 12). In either case, the cross slip pinning process is thermally activated, which leads to the positive temperature dependence of the yield strength shown in Fig. 6. The reduction of yield strength at high temperatures occurs because of enhanced dislocation mobility on {100} planes, which reduces the effectiveness of the pinning centers formed by the cross slip process. The anomalous yielding behavior makes Ni_3Al stronger than many commercial solid-solution alloys (such as type 316 stainless steel and Hastelloy alloy X) at elevated temperatures (Fig. 6).

Solid-Solution Hardening. Ni_3Al doped with boron serves as a design base for ductile and strong materials for structural uses. The aluminide is capable of being hardened by solid-solution effects because it can dissolve substantial alloying additions without losing the advantage of long-range order. One study constructed the solubility lobes of ternary Ni_3Al phase ($L1_2$) at 1000 °C (1830 °F) for various alloying elements (Ref 35). The elements that dissolve substantially in Ni_3Al can be divided into three groups (Fig. 8). The first group of elements, including silicon, germanium, titanium, vanadium, and hafnium, substitutes almost exclusively on aluminum sublattice sites. The second group, consisting of copper, cobalt, and platinum, substitutes on nickel sublattice sites. The third group, which includes elements such as iron, manganese, and chromium, substitutes on both sublattice sites. Guard and Westbrook (Ref 5) first suggested that the electronic structure (that is, the position of elements in the periodic table) rather than the atom size factor plays a dominant role on the substitution behavior. The extent of solid solution in Ni_3Al, however, is controlled by the atomic size misfit and the difference in the heats of formation between Ni_3Al and Ni_3X.

The room-temperature solid-solution hardening of Ni_3Al depends on the substitutional behavior of alloying elements, atomic size misfit, and the degree of nonstoichiometry of the alloy. A review of the mechanical properties of Ni_3Al indicated that it could be hardened more effectively by the elements substituting on aluminum sites and, to a lesser degree, by the elements substituting on nickel or on both nickel and aluminum sites (Ref 78). In addition, the strengthening is most pronounced for stoichiometric alloys and aluminum-rich alloys; it is much less pronounced for nickel-rich alloys. The solid-solution hardening is quite complex at elevated temperatures (Ref 79, 80). The hardening effects depend strongly on crystal orientation and test temperature, which cannot be explained by the classic solid-solution models developed based on the elastic interaction between dislocations and symmetric defects (Ref

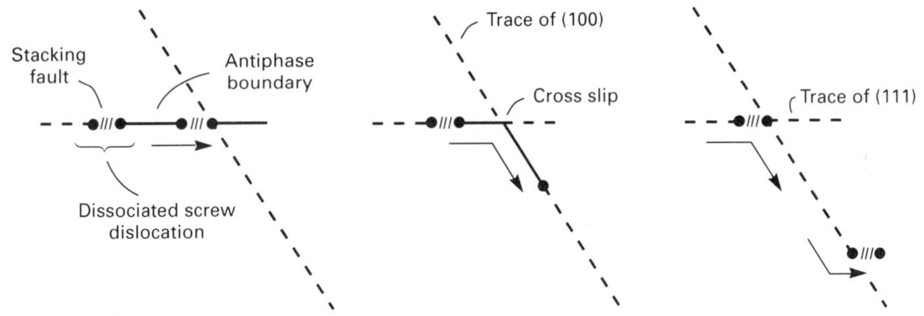

Fig. 7 Mechanism of cross-slip pinning as proposed in Ref 76 (after Ref 77)

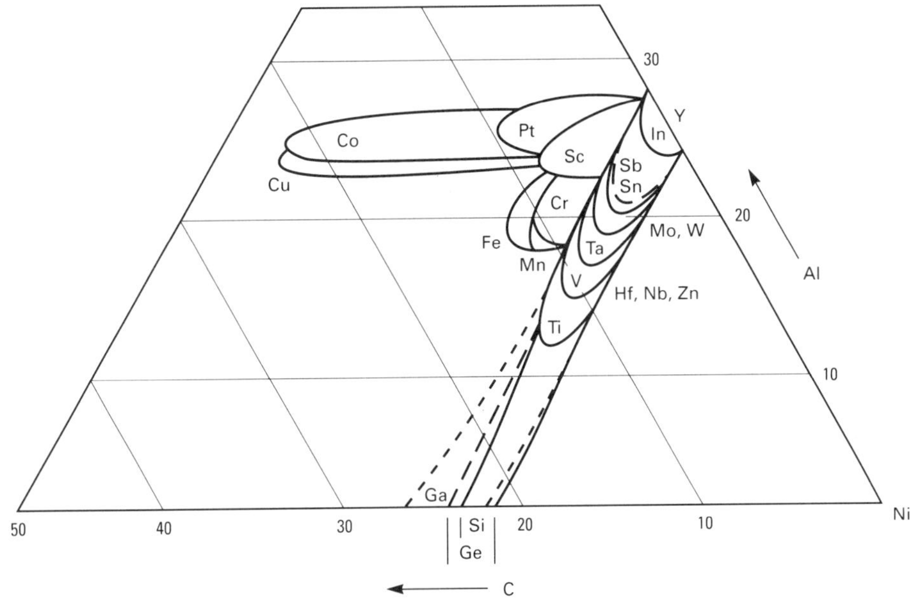

Fig. 8 Semischematic depiction of the solubility lobes of ternary Ni_3Al phase around 1000 °C (1830 °F). Source: Ref 35

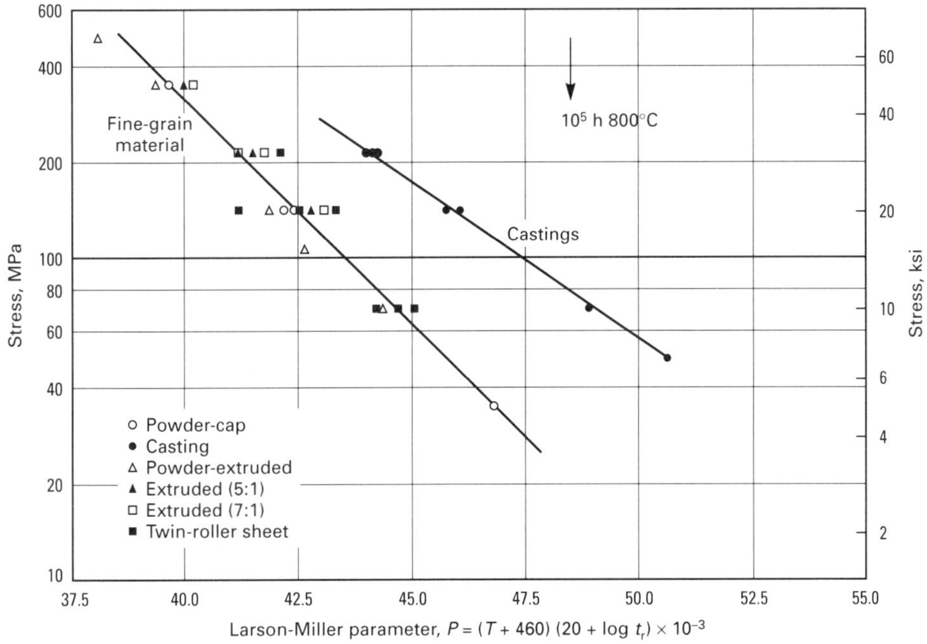

Fig. 9 Larson-Miller parameter (P) plot showing the effect of processing on the creep-rupture properties of IC-221 (Ni-16.1Al-8Cr-1Zr-0.8B, at.%). Tests were conducted in the temperature range of 650 to 870 °C (1200 to 1600 °F) for times ranging from 10 to 12 464 h. Source: Ref 84

81). For example, hafnium is found to be more effective in hardening at 850 °C (1560 °F) than at room temperature (Ref 75). These observations suggest that the unusual hardening is related to how solutes affect the cross slip pinning processes occurring at elevated temperatures. Hafnium and zirconium are most effective in strengthening Ni_3Al at elevated temperatures (Ref 75, 82). Solid-solution hardening is the subject of current studies (Ref 83).

Mechanical Properties. The study of ductility and strength of Ni_3Al has led to the development of ductile nickel aluminide alloys for structural applications (Ref 71, 84, 85). The alloys generally contain hafnium, zirconium, tantalum, and molybdenum at levels up to 5 at.% for improving strength at elevated temperatures; they contain up to 10 at.% Cr for enhancing ductility at intermediate temperatures (400 to 900 °C, or 750 to 1650 °F). Boron at levels less than 500 ppm is added for strengthening grain boundaries and increasing ductility at ambient temperature. Figure 6 shows the yield strength of a cast Ni_3Al alloy, which is stronger than nickel-base alloy IN-713C at elevated temperatures. The Ni_3Al alloys generally possess ductilities of 25 to 40% at temperatures up to 700 °C (1290 °F), and 15 to 30% at up to 1000 °C (1830 °F) in air. The chromium-containing Ni_3Al alloys generally contain 5 to 15% of disordered (γ) phase, the amount of which depends on the aluminum concentration.

Creep properties of Ni_3Al alloys have been characterized as functions of stress, temperature, and composition. Hafnium and zirconium additions are most effective in improving the creep resistance of Ni_3Al (Ref 71). Figure 9 shows creep data for the polycrystalline nickel aluminide alloy IC-221 (Ni-16.1Al-8.0Cr-1.0Zr-0.08B, at.%). The creep properties of Ni_3Al alloys, like those of nickel-base superalloys, are sensitive to grain size, but the Ni_3Al alloys have better creep resistance for coarse-grain materials (for example, cast materials). For applications where creep resistance is important, coarse-grain material is more desirable at temperatures greater than 700 °C (1290 °F). Creep properties of single-crystal Ni_3Al alloys containing refractory elements such as tantalum have been studied at temperatures up to 1000 °C (1830 °F) (Ref 85). In general, the creep resistance of Ni_3Al is comparable to that of most of the nickel-base superalloys, but it is not as good as that of some advanced single-crystal nickel-base superalloys used for jet engine turbine blades.

Fatigue and fatigue crack growth are substantially better in Ni_3Al alloys than in nickel-base superalloys in tests below the range of the ductility minimum (Ref 86-88); see Fig. 10 for room-temperature fatigue crack growth. The good fatigue resistance of Ni_3Al and other ordered intermetallic

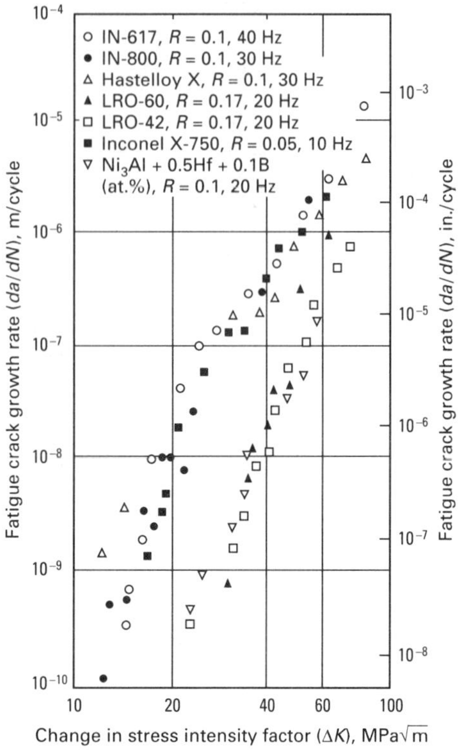

Fig. 10 Crack growth rates of nickel aluminide (Ni-23.5Al-0.5Hf-0.1B, at.%), LRO alloys [$(Fe,Ni)_3(V,Ti)$], and several high-temperature alloys tested in air at 25 °C (80 °F). Source: Ref 87

Fig. 11 Direct cast sheets of IC-50 (Ni-23.5Al-0.5Zr-0.08B, at.%) and IC-218 (Ni-16.7Al-8.0Cr-0.4Zr-0.08B, at.%) Ni_3Al alloys. Source: Ref 84

alloys has been attributed to fine planar slip and superlattice dislocation structure. Dynamic embrittlement in oxidizing environments severely reduces the fatigue resistance of Ni_3Al at temperatures above 500 °C (930 °F) (Ref 86); however, this problem has been alleviated by adding moderate amounts (for example, 8 at.%) of chromium to Ni_3Al (Ref 86-88). Fatigue/creep interactions in single crystals and directionally solidified Ni_3Al alloys have been characterized for temperatures up to 800 °C (1470 °F) (Ref 89-90). Limited results indicate that the performance of single-crystal Ni_3Al alloyed with hafnium and boron is superior to that of Udimet 115 at 760 °C (1400 °F).

Processing and Fabrication. Alloys based on Ni_3Al are susceptible to weld cracking; however, if welding is done with care, sound welds can be made in most of the alloys (Ref 91-93). Welding speed should be reduced, and the boron level has to be limited to about 0.1 at.% to avoid hot cracking. Oxygen is particularly detrimental: Oxide scale on alloy surfaces should be removed prior to welding, and the atmosphere must be controlled to reduce oxygen during welding. Certain alloying additions (iron, for example) have been found to promote weldability.

The unique strength and ductility characteristics of Ni_3Al present correspondingly unique challenges and opportunities in the processing of the aluminide alloys (Ref 71, 84, 94). The alloys generally show excellent ductility at ambient temperatures; however, their hot ductility is sensitive to test temperature, grain size, and alloy composition. Conventional fabrication techniques (such as hot rolling for large ingots) are ineffective because regions near the surface cool to the range of the ductility minimum, which leads to the formation of large intergranular surface cracks. Isothermal forging offers excellent possibilities for fabrication because the alloys exhibit superplastic behavior above about 1000 °C (1830 °F). Conventional hot forging is feasible for fine-grain alloys containing less than 0.3 at.% Zr or Hf. Cold fabrication is effective if the materials can be cast into sheet or rod forms that can be cold formed further without the need for repeated recrystallization treatments. Figure 11 is a photograph of direct cast sheets of alloy IC-218 (Ni-16.7Al-8.0Cr-0.4Zr-0.08B, at.%). The as-cast sheet has excellent ductility and is ready to be fabricated into wrought material by cold rolling. Hot extrusion of powder metallurgy materials is another effective fabrication method for Ni_3Al alloys.

Structural Applications. Although the properties of alloys based on Ni_3Al approach those of established superalloys, the Ni_3Al alloys are unlikely to displace superalloys in aircraft engine applications. The opportunity exists, however, for enhancing the properties of Ni_3Al alloys further through the incorporation of second phases. In addition, alloys based on Ni_3Al could provide an attractive matrix for composite development (Ref 95-99). Monolithic aluminide alloys are likely to find near-term use in applications that take advantage of some of their unique or unusual properties (Ref 84). The potential applications (and the properties they would exploit) include:

- Gas, water, and steam turbines (the excellent cavitation, erosion, and oxidation resistance of the alloys)
- Aircraft fasteners (low density and ease of achieving the desired strength)
- Automotive turbochargers (high fatigue resistance and low density)
- Pistons and valves (wear resistance and capability of developing a thermal barrier by high-temperature oxidation treatment)
- Bellows for expansion joints to be used in corrosive environments (good aqueous corrosion resistance)
- Tooling (high-temperature strength and wear resistance developed through preoxidation)
- Permanent molds (the ability to develop a thermal barrier coating by high-temperature oxidation)

NiAl Aluminides

Nickel-aluminum containing more than about 40 at.% Ni starts to form a single-phase B2-type ordered crystal structure based on the bcc lattice (Ref 36). In terms of physical properties, *B*2 NiAl offers more potential for high-temperature applications than $L1_2$ Ni_3Al. It has a higher melting point (1638 °C, or 2980 °F), a substantially lower density (5.86 g/cm^3 for NiAl versus 7.50 g/cm^3 for Ni_3Al), and a higher Young's modulus (294 GPa, or 42.7 × 10^6 psi, versus 179

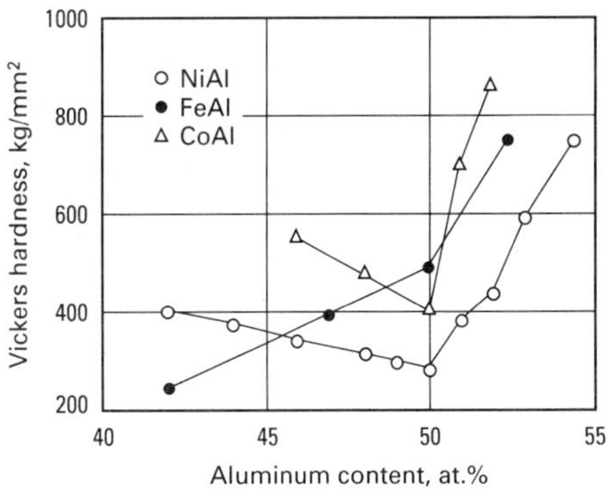

Fig. 12 Vickers hardness of CoAl, FeAl, and NiAl as a function of aluminum content. Source: Ref 107

GPa, or 25.9 × 10^6 psi). In addition, NiAl offers excellent oxidation resistance at high temperatures (Ref 30, 100). In the 1950s and 1960s, NiAl alloys were employed as coating materials for hot components in corrosive environments. The oxidation resistance of NiAl can be further improved by alloying with yttrium and other refractory elements such as hafnium and zirconium (Ref 101, 102).

Structure and Property Relationships. The use of NiAl in structural members suffers two major drawbacks: poor ductility at ambient temperatures and low strength and creep resistance at elevated temperatures. Single crystals of NiAl are quite ductile in compression, but both single and polycrystalline NiAl appear to be brittle in tension at room temperature. The nickel aluminide exhibits mainly {100} slip, rather than {111} slip as commonly observed for bcc materials (Ref 103-105). The lack of sufficient slip systems has been regarded as the major cause of low ductility in NiAl. The aluminide shows a sharp increase in ductility above 400 °C (750 °F) and becomes very ductile above 600 °C (1110 °F) (Ref 38, 106); therefore, fabrication of NiAl at high temperatures presents no major problems.

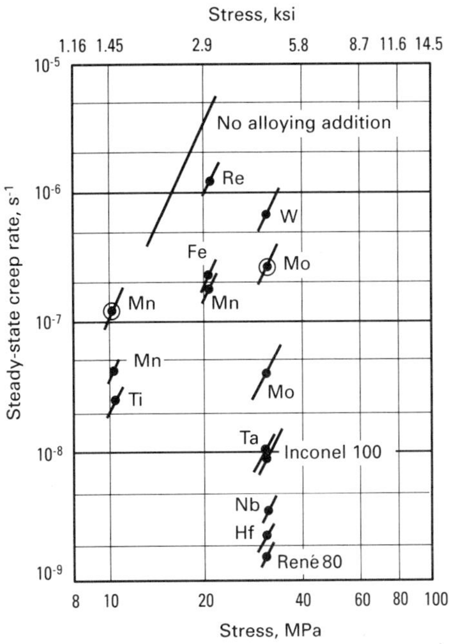

Fig. 13 Compressive creep data for NiAl alloys containing various alloying elements tested at 1300 K. Data for selected conventional superalloys are provided for comparison. Lines drawn through data points are expected slopes. Source: Ref 109

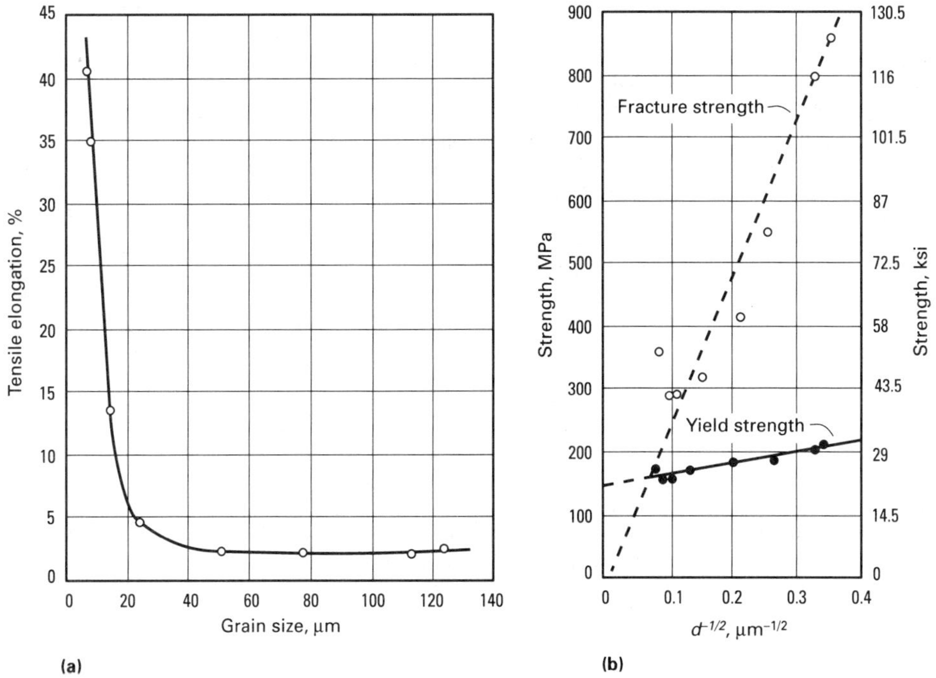

Fig. 14 Effect of grain size (d) on properties of NiAl at 673 K. (a) Tensile elongation. (b) Yield strength and fracture strength. Source: Ref 112

The ordered $B2$ structure exists in NiAl over a solubility range of about 15 at.%. Deviations from stoichiometry are accommodated by the incorporation of vacancies in aluminum-rich alloys and by the formation of antisite defects in nickel-rich alloys (Ref 31-34). The presence of lattice defects has a strong effect on low-temperature strength, with minimum strength occurring at the stoichiometric composition (Fig. 12). The minimum becomes less pronounced with increasing temperature and is essentially eliminated at 600 °C (1110 °F). At all compositions, the yield strength decreases with increasing temperature (Ref 38, 106). Abrupt drops in strength in the range of 400 to 600 °C (750 to 1110 °F) are accompanied by a sharp increase in ductility. The alloys are highly ductile but extremely weak at higher temperatures. For example, Ni-50Al (at.%) showed a yield strength of 35 MPa (5 ksi) and a tensile ductility of greater than 50% at 1000 °C (1830 °F) (Ref 38, 108).

NiAl is quite weak in creep at elevated temperatures (Ref 108-111). However, its creep properties can be substantially improved by alloy additions (Ref 109). Figure 13 shows the compressive creep rate of NiAl alloyed with up to 5 at.% ternary additions. The strength at 1300 K of alloys containing tantalum, niobium, and hafnium is comparable to or even greater than that of the superalloy IN-100. These alloying elements showed very low solubility in NiAl, and the improvement apparently comes from the precipitation of fine second-phase particles that impede the motion of dislocations. It has recently been reported that alloying with 15 at.% Fe to replace nickel lowers diffusion rates and thus reduces the creep rate of NiAl (Ref 110).

Effect of Grain Size and Alloy Stoichiometry on Ductility. Considerable efforts have been devoted to improving the ductility of NiAl at ambient temperatures by controlling microstructure and alloy additions. Schulson and Bakrer (Ref 112) studied ductile-to-brittle transition as a function of grain size and temperature for NiAl and other aluminides and silicides. NiAl specimens produced by hot extrusion of powder metallurgy (P/M) material, even those with extremely fine grain sizes, were brittle at ambient temperatures. They found that grain refinement alone did not improve the room-temperature ductility of NiAl. On the other hand, NiAl (49 at.% Al) showed a sharp brittle-to-ductile transition at a critical grain size at elevated temperatures. For example, NiAl exhibited a sharp increase in ductility at 400 °C (750 °F) for grain sizes less than 20 μm (Fig. 14). A tensile elongation of 40% was achieved at 400 °C (750 °F) for NiAl with a grain size of 3 μm.

Hahn and Vedula (Ref 106) recently investigated tensile elongation and fracture behavior as functions of aluminum concentration in NiAl prepared by hot extrusion of

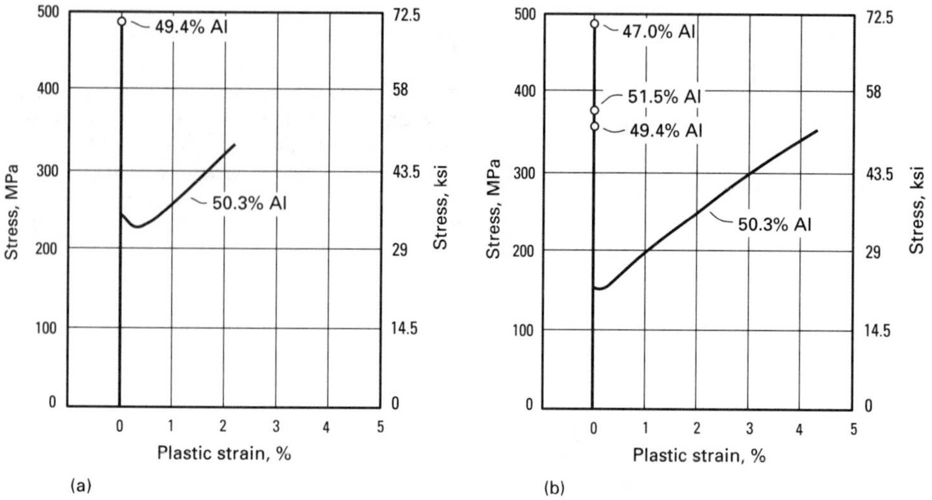

Fig. 15 Effect of stoichiometry on tensile properties of cast and extruded binary NiAl alloys. Nominal strain rate, 1.41 × 10^{-3}/s. (a) Tested at room temperature. (b) Tested at 473 K. Source: Ref 106

cast ingots. They found that NiAl alloys with off-stoichiometric compositions fractured with no appreciable plastic deformation, whereas an alloy with a near-stoichiometric composition exhibited significant tensile elongation (~2%) at both room temperature and 200 °C (390 °F) (Fig. 15). Their findings essentially confirm previous results reported in Ref 113. Although the stoichiometric effect is not well understood at the present time, it is believed that yield strength plays a dominant role in the ductility observed for near-stoichiometric NiAl. As indicated in Fig. 15, the yield strength for the near-stoichiometric composition (Ni-50.3Al) is distinctly lower than that for off-stoichiometric compositions. The low yield strength apparently prevents the initiation and propagation of brittle fracture until a high stress level is reached by strain hardening through plastic deformation.

Alloying Effect and Ductility Improvement. Because NiAl with on- and off-stoichiometric compositions shows mainly brittle intergranular fracture (Ref 106), it is possible to improve its ductility by the control of grain-boundary composition through microalloying. Auger analyses have revealed that grain boundaries in NiAl (50% Al) are clean and free of any segregated impurities (Fig. 16), indicating that they are intrinsically brittle (Ref 114). Boron added to NiAl has a strong tendency to segregate to the grain boundaries and suppress intergranular fracture (Fig. 16). However, there is no attendant improvement in tensile ductility because boron is an extremely potent solid-solution strengthener in NiAl. Unlike boron, both carbon and beryllium are ineffective in suppressing intergranular fracture in NiAl. Beryllium slightly improves the room-temperature tensile ductility of NiAl. In these microalloyed alloys, the nickel and aluminum contents of the grain boundaries are not significantly different from the bulk levels, and no evidence of strong boron-nickel cosegregation has been found (Ref 114).

Attempts have also been made to improve room-temperature ductility by macroalloying NiAl with alloy additions that might change its deformation behavior in bulk material. Additions of chromium, manganese, and vanadium were reported to promote ⟨111⟩ slip vectors in NiAl; however, no improvement in ductility was observed (Ref 115, 116). On the other hand, limited tensile ductility was obtained in NiAl alloyed with cobalt (Ref 117), which promotes additional deformation modes through possibly martensitic transformation. Sufficient iron additions to nickel-rich NiAl, such as Ni-30Al-20Fe (at.%) with a B2 structure, also result in approximately 2% plastic elongation (Ref 118). The same alloy with a fine-grain structure shows 5% elongation when produced by a rapid solidification technique (Ref 119). The alloy Ni-20Al-30Fe, which has a two-phase structure (NiAl + Ni_3Al), exhibited a tensile ductility of 22% when produced by hot extrusion (Ref 120) and a lower ductility (10 to 17%) when produced by rapid solidification (Ref 119, 121).

Potential and Future Work. Although significant progress has been made, NiAl has not yet developed into an engineering material for structural use. Further efforts must be devoted to improving both low-temperature ductility and high-temperature strength. The design options offered by NiAl are certainly attractive enough to motivate additional research into the development of NiAl alloys. At present, major NiAl development efforts are going on in a number of laboratories. NiAl and its alloys are also attractive as matrix materials for intermetallic composites, and a great deal of work has been conducted in this area (Ref 122).

Iron Aluminides

Phase Stability and Potential for Structural Use. Iron aluminides form bcc ordered crystal structures over the composition range of 25 to 50 at.% (Ref 36). The aluminide Fe_3Al exists in the ordered $D0_3$ structure up to 540 °C (1000 °F) and in the $B2$ structure between 540 and 760 °C (1000 and 1400 °F); it has a disordered structure above 760 °C (1400 °F). The $D0_3 \rightarrow B2$ transition temperature decreases and the $B2$ ordered temperature increases with an increase in aluminum concentration above 25%. Only the $B2$ structure is stable at aluminum levels above 36%, and the single-phase field extends to approximately 50 at.% Al (FeAl).

The iron aluminides based on Fe_3Al and FeAl possess unique properties and have development potential as new materials for structural use. This potential is based on the capability of the aluminides to form protective aluminum oxide scales in oxidizing and sulfidizing environments at elevated temperatures (Ref 123-126). In addition to excellent corrosion resistance, the aluminides offer low material cost, low density, and conservation of strategic elements. However, the major drawbacks of the aluminides are their low ductility and fracture toughness at ambient temperature and their poor strength at temperatures above 600 °C (1110 °F). Recently, considerable efforts have been devoted to understanding and improving their mechanical properties through control of grain structure, alloy additions, and material processing.

Mechanical Behavior of Fe_3Al and FeAl. The aluminides show low ductility and brittle fracture at room temperature and their fracture mode depends on aluminum concentration. The aluminides containing less than 40 at.% Al exhibit mainly transgranular cleavage fracture, whereas those with more than 40% Al show essentially brittle intergranular fracture (Ref 127-131). The fracture mode is also sensitive to other parameters such as grain size and impurities. Fe_3Al (25 at.% Al) has been reported to fracture intergranularly when it contains excess carbon (Ref 132). Also, some FeAl alloys with less than 40% Al show grain-boundary fracture when prepared by P/M techniques and contaminated with oxygen (Ref 128).

Mechanical properties of the iron aluminides have been characterized as functions of test temperature and alloy composition in a number of studies (Ref 127-131). In general, yield strength is not sensitive to temperature below 600 to 650 °C (1110 to 1200 °F); above that temperature range, strength shows a sharp drop with temperature. For intermediate and coarse-grain materials, the aluminides with less than about 40% Al generally show a small increase in yield strength with temperature, with strength reaching a peak at temperatures of

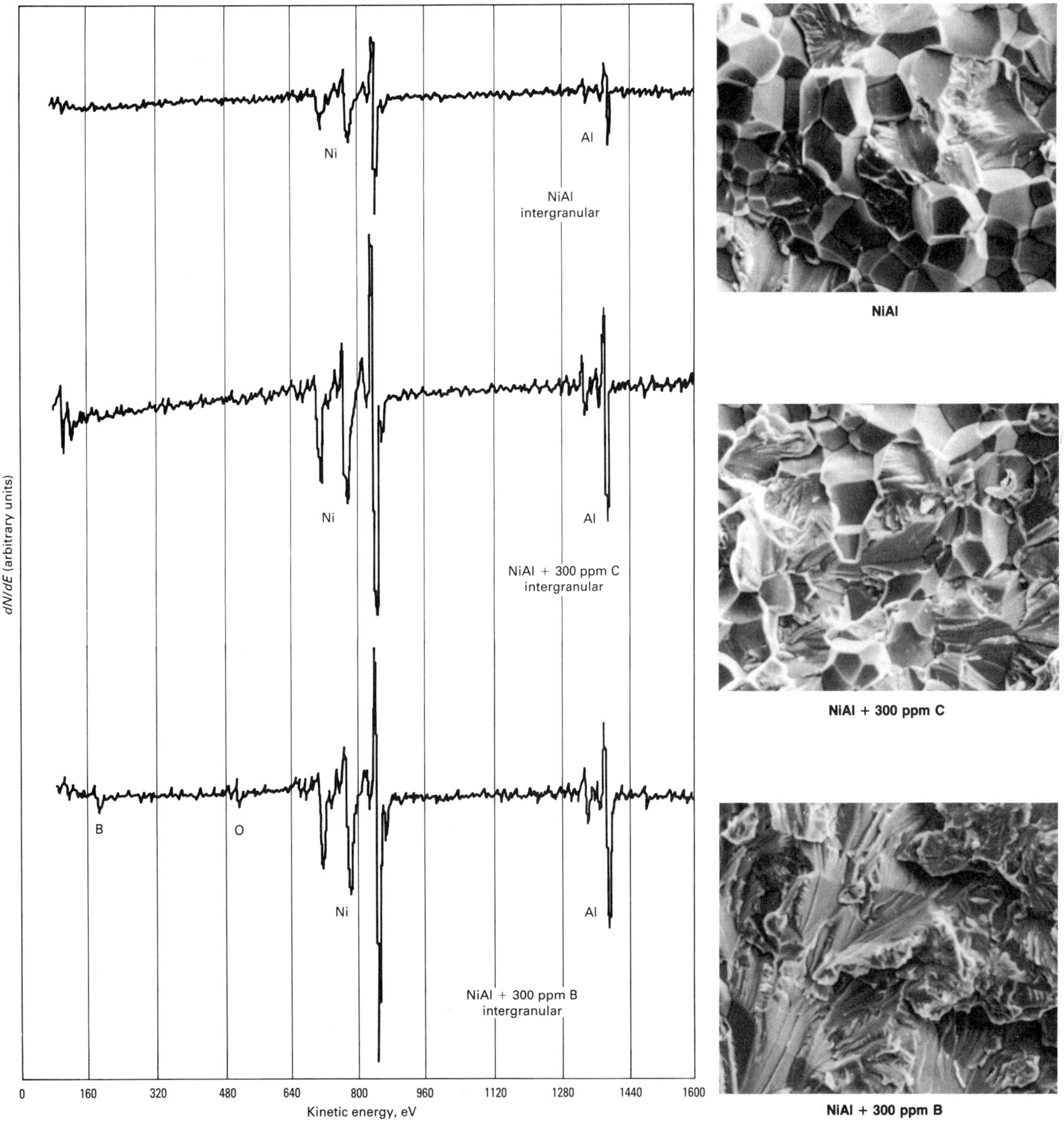

Fig. 16 Results of Auger electron analysis showing the effect of microalloying on the grain-boundary composition and the room-temperature fracture mode of NiAl. Top to bottom: unalloyed NiAl (mainly grain-boundary fracture), NiAl doped with 300 ppm C, and NiAl doped with 300 ppm B. All photomicrographs 400×. Source: Ref 114

about 550 to 650 °C (1020 to 1200 °F). For the aluminides with higher levels of aluminum, the increase in yield strength is suppressed, possibly because of grain-boundary sliding at elevated temperatures (Ref 127, 128, 131).

The room-temperature yield strength of Fe_3Al that contains approximately 25 at.% Al is quite high (550 to 700 MPa, or 80 to 100 ksi) because of the low mobility of single dislocations and particle strengthening due to the presence of a disordered bcc phase (Ref 131, 133). Strength decreases with increasing aluminum concentration and reaches a minimum at an aluminum content of about 30%. With further increase in aluminum concentration, strength shows a moderate increase. Unlike NiAl, the FeAl

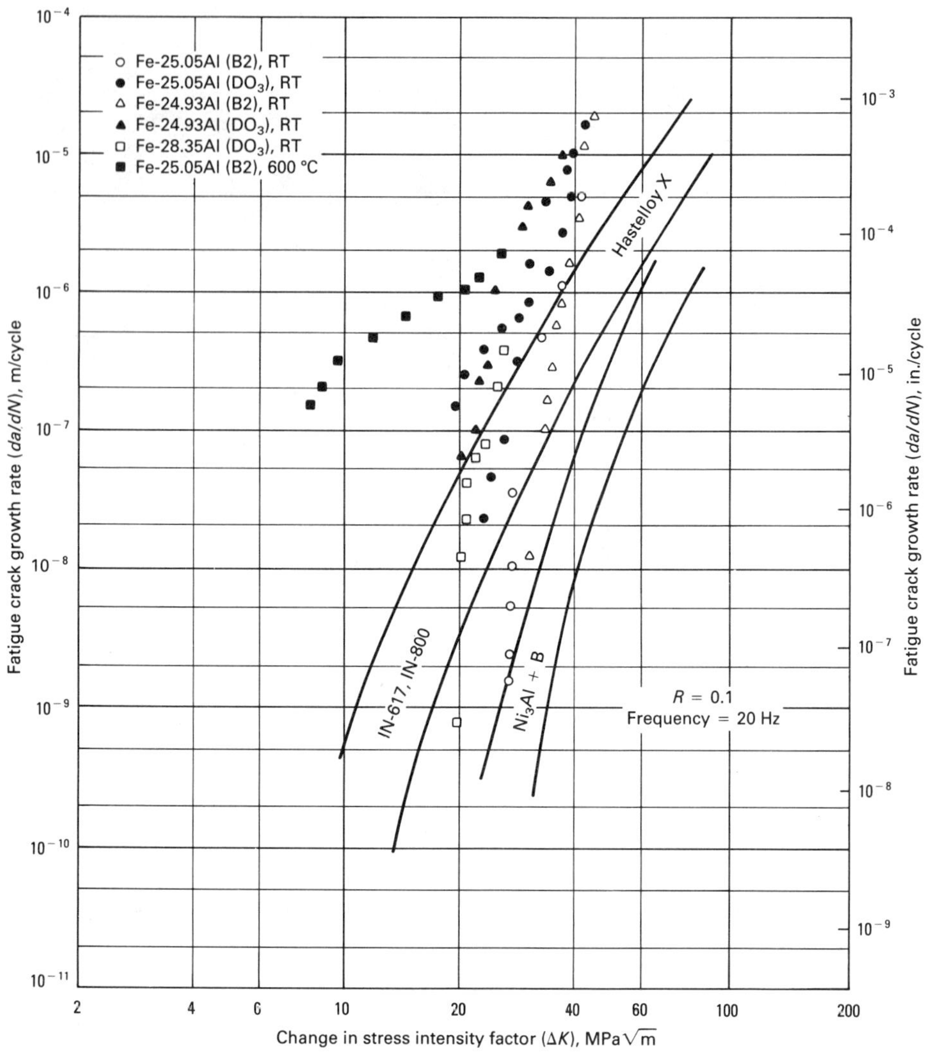

Fig. 17 Effects of aluminum content, crystal structure, and temperature on fatigue crack growth in Fe_3Al. Curves for nickel-base superalloys and Ni_3Al are shown for comparison. Stress ratio (R), 0.1; frequency, 20 Hz. RT, room temperature. Source: Ref 87

aluminide shows a substantial increase in strength and hardness when it approaches a stoichiometric composition (Fig. 12). The cause for this different behavior is not well understood at the present time.

The fatigue and crack growth behavior of Fe_3Al alloys have been studied as functions of aluminum concentration (23.7 to 28.7 at.% Al) and test temperature (20 to 600 °C, or 70 to 1110 °F) (Ref 87). At room temperature, the hyperstoichiometric alloy with 28.7% Al has better fatigue resistance than the hypostoichiometric alloy with 23.7% Al. The better fatigue properties of the higher-aluminum alloys have been attributed to the presence of superdislocations, which cause crack initiation to be more difficult than in the hypostoichiometric alloy. However, the trend is reversed at 500 °C (930 °F). Figure 17 compares the crack growth behavior of Fe_3Al with that of other materials. At room temperature and 500 °C (930 °F), Fe_3Al aluminides are generally intermediate in crack growth rates between nickel-base superalloys and Ni_3Al+B or $(Fe,Ni)_3V$ alloys at low changes in stress level (ΔK); at high ΔK, the crack growth is most rapid in Fe_3Al. At 600 °C (1110 °F), the crack growth rate of Fe_3Al is unusually high as compared with that of the other materials. The reason for this sharp increase in crack growth rate is not well known, but it may be related to the transition of $D0_3$ to $B2$ in Fe_3Al.

Ductility and Slip Behavior. In terms of ductility, the aluminides with less than 40% Al have a room-temperature tensile elongation of about 2 to 4% for coarse-grain materials (that is, materials with grain sizes of 150 to 200 μm) (Ref 28). Elongation increases to 6 to 8% when the grain structure is refined (Ref 134, 135), indicating the effect of grain size on the ductility. A recent study found that the ductility of FeAl with 40% Al was substantially improved by mechanical alloying (Ref 136); the mechanical alloying probably reduced the grain size, thereby reducing the tendency toward brittle intergranular fracture. The ductility of aluminides with up to 30% Al increases with test temperature and reaches more than 40% at 600 °C (1110 °F). Aluminides with more than 35% Al show a decrease in ductility above 600 °C (1110 °F) and reach a minimum ductility around 750 °C (1380 °F). The decrease in ductility is believed to be caused by cavitation in the grain-boundary region when the aluminides are tested under tension (Ref 127, 128).

The aluminides, like iron and steels, slip by {111} dislocations at ambient temperatures. However, the slip changes to {100} type at elevated temperatures. The transition depends on the aluminum concentration (Ref 137, 138), with a general trend of decreasing transition temperature with increasing aluminum level. For example, a transition temperature of about 1000 °C (1830 °F) was reported for the 35% Al alloy; the transition temperature was below 400 °C (750 °F) for the 50% Al alloy (Ref 137). There is no sharp change in ductility around the transition temperature.

Environmental Embrittlement. For more than 40 years, the iron aluminides have been known to be brittle at ambient temperatures; however, the major cause of the brittleness has only recently been identified. Researchers have found that the aluminides are intrinsically quite ductile and that the poor ductility commonly observed in air tests is due to an extrinsic effect—environmental embrittlement (Ref 28, 29). The data in Fig. 18 and Table 2 indicate the effect of test environment on the room-temperature tensile properties of FeAl (36.5% Al) and Fe_3Al (28% Al). Yield strength is not sensitive to environment, but ultimate tensile strength is generally correlated with tensile ductility, which depends strongly on test environment. Aluminides tested in air had a ductility of 2 to 4%. FeAl tested in dry oxygen had a ductility of 17.6%, and Fe_3Al tested in vacuum and dry oxygen had a ductility of 12 to 13%. The water vapor test confirmed the low ductility found in the air tests, indicating that moisture in air is the embrittling agent.

Embrittlement is expected to involve the following chemical reaction at metal surfaces:

$$2Al + 3H_2O \rightarrow Al_2O_3 + 6H^+ + 6e^-$$

The reaction of water vapor with aluminum atoms at crack tips results in the formation of atomic hydrogen that drives into the metal and causes crack propagation. The fact that yield strength (Table 2) is insensitive to ductility and test environment is consistent with the mechanisms of hydrogen embrittlement observed in other ordered intermetallic alloys (Ref 139-147). Molecular hydrogen causes much less embrittlement in the aluminides, possibly because of its lower activity as compared with that of the atomic hydrogen produced from the water vapor reaction.

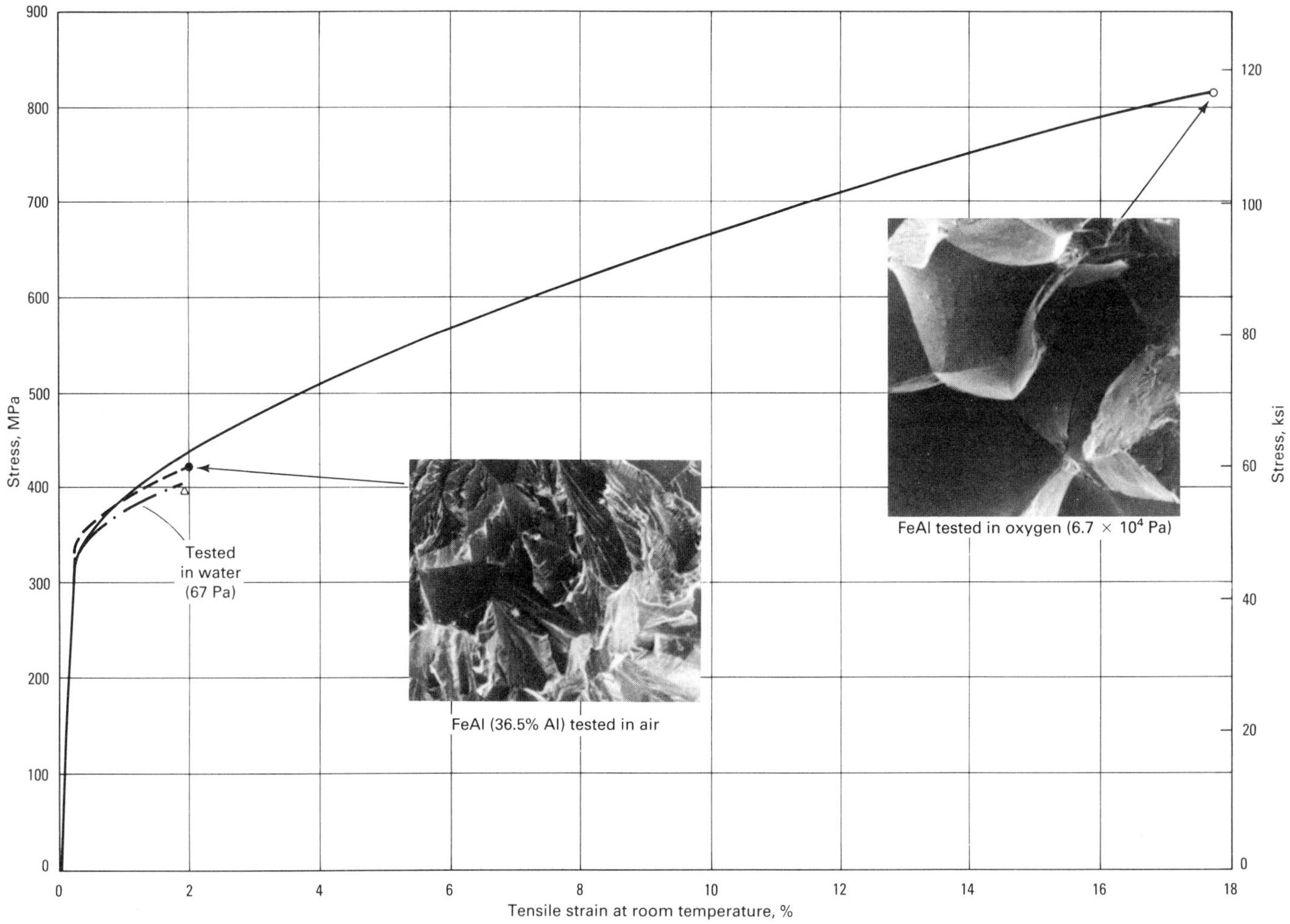

Fig. 18 Effect of test environment on the room-temperature ductility and fracture behavior of FeAl (36.5% Al)

Alloying Effects in Fe_3Al Aluminides. Grain structure refinement by material processing and alloy additions has been shown to be useful in increasing ductility in Fe_3Al aluminides (Ref 148, 149). Additions of titanium diboride (TiB_2) to Fe_3Al powders are very effective in reducing grain size, and they increase the tensile ductility of recrystallized materials from 2% to 5 to 7%. Stress relief following hot working of the same materials results in ductilities as high as 18%. The presence of TiB_2 particles increases the recrystallization temperature from 650 to 1100 °C (1200 to 2010 °F), which means that wrought materials will retain room-temperature ductility even after exposure to temperatures as high as 1000 °C (1830 °F). For these materials, ductility is very high at temperatures above 600 °C (1110 °F), and conventional hot fabrication techniques can be employed without difficulty.

Strength properties of these aluminides are also sensitive to microstructure and the level of aluminum (Ref 125, 148-150) (Fig. 19). Room-temperature yield strength drops sharply with an increase of aluminum above 25%. This drop is, as mentioned before, a result of both the increase in mobility of paired dislocations and the elimination of particle strengthening from the disordered bcc phase. Additions of TiB_2, which reduce the grain size of recrystallized material and stabilize the wrought structure, increase the strength significantly and cause it to be retained to higher temperatures.

Substantial efforts have been made toward improving the elevated-temperature properties of Fe_3Al aluminides by alloy additions of such elements as titanium, molybdenum, silicon, chromium, nickel, manganese, niobium, and tantalum (Ref 148, 149, 151-156). Among these alloying elements, titanium, molybdenum, and silicon

Table 2 Effect of selected test environments on room-temperature tensile properties of iron aluminides

Test environment (gas pressure)	Elongation, %	Yield strength MPa	Yield strength ksi	Ultimate tensile strength MPa	Ultimate tensile strength ksi
Fe_3Al (28% Al)(a)					
Air	4.1	387	56	559	81
Vacuum (~1 × 10^{-4} Pa)	12.8	387	56	851	123
Argon + 4% H_2 (6.7 × 10^4 Pa)	8.4	385	55.8	731	106
Oxygen (6.7 × 10^4 Pa)	12.0	392	56.8	867	126
Water vapor (1.3 × 10^3 Pa)	2.1	387	56	475	69
FeAl (36.5% Al)(a)					
Air	2.2	360	52.2	412	60
Vacuum (<1 × 10^{-4} Pa)	5.4	352	51	501	73
Argon + 4% H_2 (6.7 × 10^4 Pa)	6.2	379	55	579	84
Oxygen (6.7× 10^4 Pa)	17.6	360	52.2	805	117
Water vapor (67 Pa)	2.4	368	53.4	430	62

(a) All specimens were annealed 1 h at 900 °C (1650 °F) + 2 h at 700 °C (1290 °F).

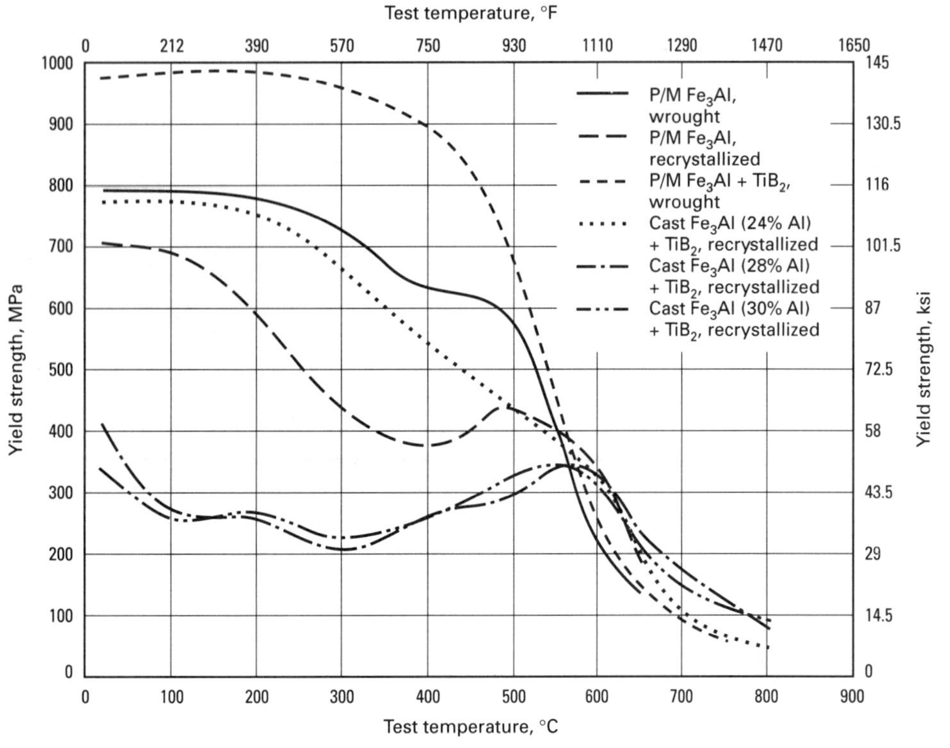

Fig. 19 Yield strength versus test temperature for various Fe_3Al materials

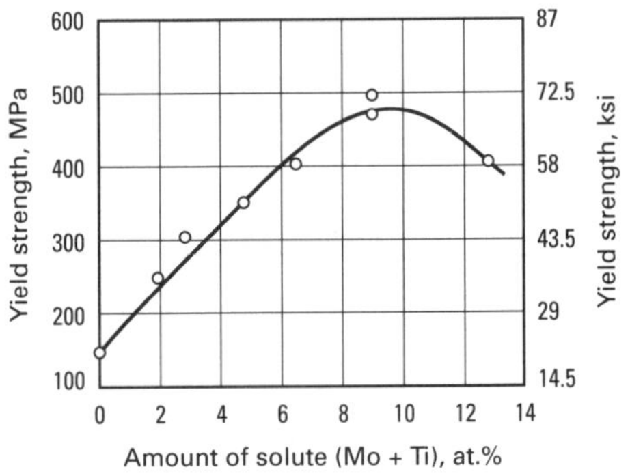

Fig. 20 Yield strength at 650 °C (1200 °F) as a function of the total solute content of molybdenum and titanium in FeAl (30 at.% Al). Source: Ref 151

are most effective in strengthening the aluminides through a solid-solution hardening effect (Ref 151). Figure 20 shows a substantial increase in the 650 °C (1200 °F) yield strength of a 30% Al aluminide brought about by alloying with molybdenum and titanium. In fact, the yield strength is tripled when the combined molybdenum and titanium content reaches 9%. The increase in strength has been related to an increase in the $D0_3 \rightarrow B2$ transition temperature and associated changes in the nature of the dislocations involved in the deformation processes. In terms of creep properties, alloying with molybdenum and additions of TiB_2 particles increases the stress for 100-h rupture life of a P/M aluminide from 28 to 193 MPa (4 to 28 ksi) at 650 °C (1200 °F) (Ref 148, 149). The solubility of niobium, tantalum, zirconium, and hafnium is low (<1%) in iron aluminides, and alloying with 1 or 2% of these elements substantially improves the room- and elevated-temperature strengths of iron aluminides through a precipitation-hardening effect (for example, precipitation of $L2_1$ particles in Fe_3Al containing 2% Nb) (Ref 153, 156).

A recent alloy design of Fe_3Al showed that the ductility of the aluminide prepared by melting and casting and fabricated by hot rolling can be substantially improved by increasing the aluminum content from 25 to 28 or 30 at.% and by adding chromium at a level of 2 to 6% (Ref 154, 155). The increase in the aluminum concentration sharply decreases the yield strength of the aluminide. The beneficial effect of chromium may come from modifying the surface composition and reducing the water vapor and aluminum atom reaction, that is, reducing environmental embrittlement. The mechanical properties of the chromium-modified Fe_3Al alloys can be further improved by thermomechanical treatment and alloy additions of molybdenum and niobium (Ref 157). Some of these alloys show a tensile ductility of more than 15% at room temperature and a yield strength of close to 500 MPa (72.5 ksi) at 600 °C (1110 °F). These ductile Fe_3Al alloys are much stronger than austenitic and ferritic steels such as type 314 stainless steel and Fe-9Cr-1Mo steel. The refractory elements also substantially enhance the creep properties of the Fe_3Al alloys.

Alloying Effect in FeAl Aluminides. FeAl aluminides containing 40% or more aluminum fail at room temperature by intergranular fracture with little tensile ductility (Ref 127, 128). Small additions of boron (0.05 to 0.2%) suppress grain-boundary fracture and allow a small increase in ductility (~3%) of Fe-40Al, but not of Fe-50Al (Ref 158). The beneficial effect of boron is not nearly as dramatic in FeAl as it is in Ni_3Al, but it is nevertheless significant. The ductility of boron-doped FeAl aluminides remains low because the alloys are still embrittled by the test environment (air). It has been found recently that boron-doped FeAl (40% Al) exhibits a high ductility (18%) when tested in dry oxygen to avoid environmental embrittlement (Ref 108).

Boron additions also increase the elevated-temperature strength of FeAl, especially in combination with niobium and zirconium. For example, the creep rate can be lowered by an order of magnitude at 825 °C (1520 °F) by the combination of 0.1% Zr and 0.2% B (Ref 159). Measurements of the activation energy for creep indicate that additions act by slowing diffusional processes rather than through precipitation reactions. Partial replacement of iron with nickel improves the creep properties of FeAl at high temperatures (Ref 110).

Weldability and Corrosion Resistance. Other properties of iron aluminides, including weldability (Ref 160) and corrosion resistance (Ref 126, 157), have been characterized to a limited extent. Fe_3Al is weldable with careful control of welding parameters and minor alloy additions. Additions of TiB_2 promote hot cracking and are detrimental to the weldability of Fe_3Al aluminides. Sound weldments have been achieved in Fe_3Al alloys using both electron beam and gas tungsten arc welding processes.

The iron aluminides are highly resistant to oxidation and sulfidation at elevated temperatures (Ref 123-126). This resistance stems from the ability of the aluminides to form highly protective Al_2O_3 scales. The oxidation resistance generally increases with increasing aluminum content; the major products are α-Al_2O_3 and trace amounts of iron oxides when the aluminides are oxidized at temperatures above 900 °C (1650 °F) (Ref 123). Cyclic oxidation of Fe-40Al alloyed with up to 1 at.% Hf, Zr, and B produced little degradation at temperatures up to 1000 °C (1830 °F) (Ref 124). Aluminide specimens tested at 700 and 870 °C (1290 and 1600 °F) showed no indication of attack in sulfidizing environments, except for the formation of a thin layer of oxides with a thickness in an interference color range (Ref 157). As shown in Fig. 21, the iron aluminide alloys exhibited corrosion rates lower than those of the best existing iron-base alloys (including coating material) by a couple of orders of magnitude when tested in a

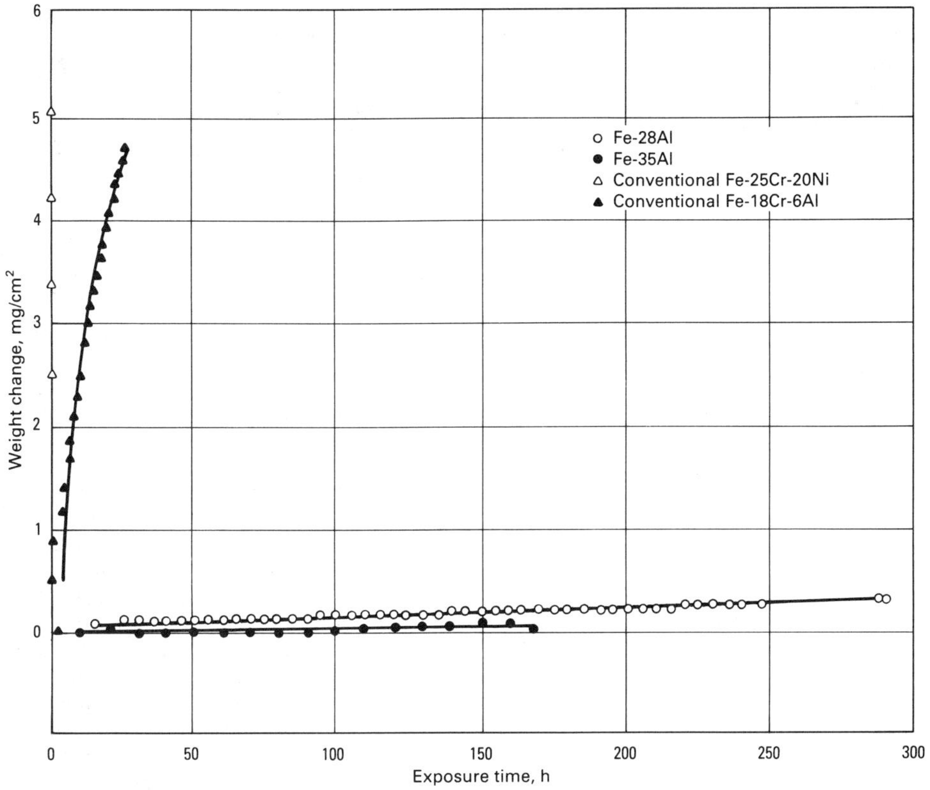

Fig. 21 Comparison of corrosion behavior of iron aluminides with that of conventional iron-base alloys Fe-18Cr-6Al (coating material) and Fe-25Cr-20Ni. All materials were exposed to a severe sulfidizing environment at 800 °C (1470 °F). Source: Ref 126

severe sulfidizing environment at 800 °C (1470 °F). In addition, the aluminides with more than 30% Al are very resistant to corrosion in molten nitrate salt environments at 650 °C (1200 °F).

Potential Applications. Iron aluminides were previously excluded from the realm of structural materials because of their brittleness at ambient temperatures and poor strength at elevated temperatures. Recent research and development activities have demonstrated that adequate engineering ductility (10 to 15%) can be achieved in the aluminides through the control of microstructure and alloy additions. Both tensile and creep strengths of the aluminides are substantially improved by alloying with refractory elements, which results in solution hardening and particle strengthening. The recently developed aluminide alloys are stronger than austenitic steels and ferritic low-alloy steels at ambient and elevated temperatures (Ref 157). Adequate ductility and strength combined with low cost, excellent oxidation and corrosion resistance, low density, and good fabricability make the aluminide alloys promising for structural use at temperatures up to 700 to 800 °C (1290 to 1470 °F). Potential applications include molten salt systems for chemical air separation, automotive exhaust systems, immersion heaters, heat exchangers, catalytic conversion vessels, chemical production systems, coal conversion systems, and so on. Several industrial companies have started to prepare large heats as a first step in the commercialization of iron aluminide alloys. Further research is required to develop a data base (including information on tensile and creep properties, fracture toughness, low- and high-cycle fatigue, crack growth, and elastic modulus) on some promising aluminides for specific engineering applications.

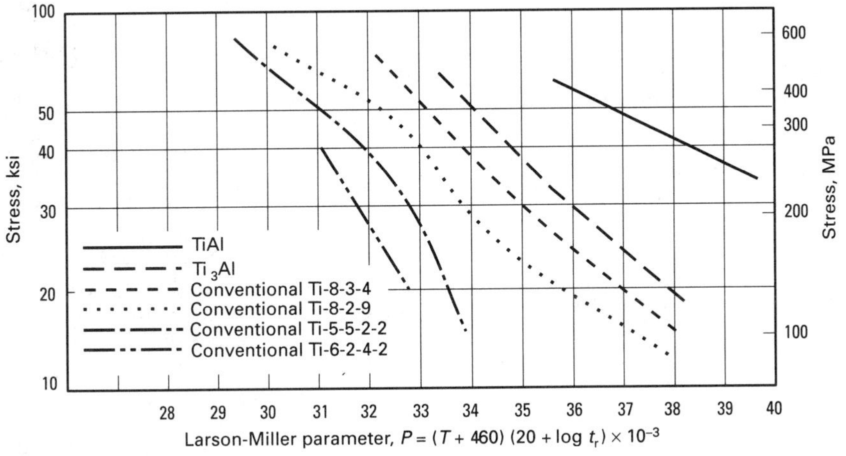

Fig. 22 Comparison of the creep behavior of conventional titanium alloys and titanium aluminide intermetallics. Source: Ref 167

Titanium Aluminides

Because of their low density, titanium aluminides based on Ti_3Al and TiAl are attractive candidates for applications in advanced aerospace engine and airframe components (Ref 161-166). The characteristics of titanium aluminides are presented alongside those of other aluminides in Table 1; the creep behavior of titanium aluminides is compared with that of conventional titanium alloys in Fig. 22. The present materials mix in an advanced jet engine is shown in Fig. 23, and a possible future engine material mix is given in Table 3. Despite a lack of fracture resistance (low ductility, fracture toughness, and fatigue crack growth rate), the titanium aluminides Ti_3Al (α-2) and TiAl (γ) have great potential for enhanced performance. Properties of these aluminides are compared with those of conventional titanium alloys and superalloys in Table 4. Because they have slower diffusion rates than conventional titanium alloys, the titanium aluminides feature enhanced high-temperature properties such as strength retention, creep and stress rupture, and fatigue resistance (Ref 170).

Another negative feature of titanium aluminides, in addition to their low ductility at ambient temperatures, is their oxidation resistance, which is lower than desirable at elevated temperatures (Ref 171-173). The titanium aluminides are characterized by a strong tendency to form TiO_2, rather than the protective Al_2O_3, at high temperatures. Because of this tendency, a key factor in increasing the maximum-use temperatures of these aluminides is enhancing their oxidation resistance while maintaining adequate levels of creep and strength retention at elevated temperatures.

Composite concepts using the titanium aluminides as a matrix are also being actively pursued (Ref 174), specifically to increase "forgiveness," modulus, and elevated-temperature performance; however, these materials will not be considered in this article. A recent detailed review of monolithic tita-

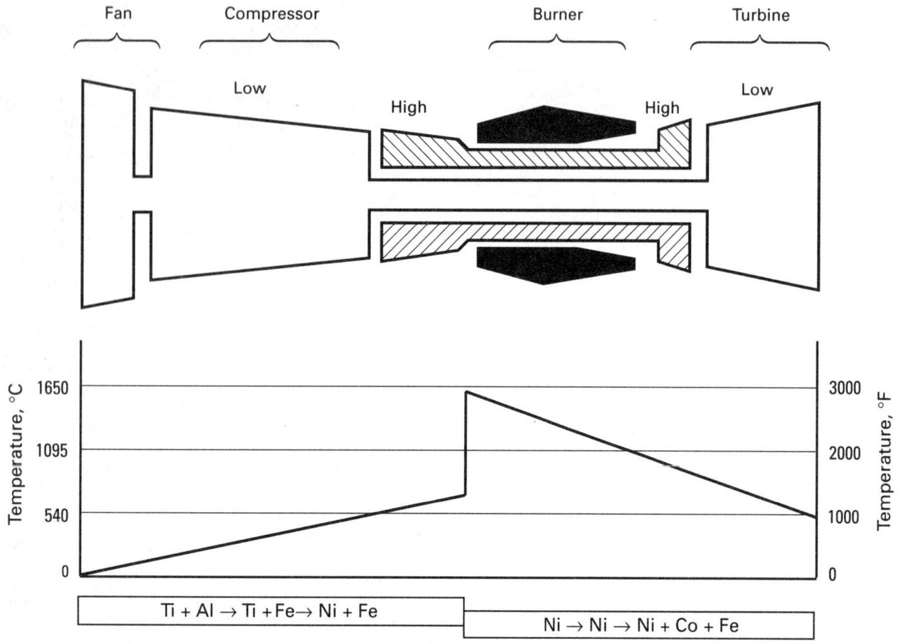

Fig. 23 Material mix in a current advanced aircraft turbofan engine. Source: Ref 168

Table 3 Possible material mix in a 2010 jet turbine engine

Material	Engine use, %
Metal-matrix composites	30
Ceramics (aluminides) and ceramic composites	30
Superalloys	13
Resin (polymer) composites	10
Titanium alloys	10
Steel	7
Aluminum	0
Magnesium	0

Source: Ref 168

nium aluminides is contained in Ref 166, which expands on many of the points made in this article.

Alpha-2 Alloys

Crystal Structure and Deformation Behavior. Ti_3Al, which has an ordered $D0_{19}$ structure, contains three linearly independent slip systems that account for dislocation motion on the basal {0001}, prism {1010}, and pyramidal {0221} planes (Fig. 24) (Ref 175, 176). Prism slip requires only a single dislocation without creating a near-neighbor antiphase boundary, and additional slip requires movement of two dislocations (Ref 26). In addition, two independent slip systems involving $\langle c + a \rangle$ slip occur to satisfy the Von Mises criterion for uniform deformation.

The semicommercial and experimental α-2 alloys developed up to the present time are two phase (α-2 + β/$B2$), with contents of 23 to 25 at.% Al and 11 to 18 at.% Nb. Alloy compositions with current engineering significance are Ti-24Al-11Nb (Ref 177, 178), Ti-25Al-10Nb-3V-1Mo (Ref 165), Ti-25Al-17Nb-1Mo (Ref 179) and modified alloy compositions such as Ti-24.5Al-6Nb-6(Ta,Mo,Cr,V). Increasing the niobium content generally enhances most material properties, although excessive niobium can degrade creep performance. Niobium can be replaced by specific elements for improved strength (molybdenum, tantalum, or chromium), creep resistance (molybdenum), and oxidation resistance (tantalum, molybdenum). However, for full optimization of mechanical properties, control of the microstructure must be maintained, particularly for tensile, fatigue, and creep performance.

The α-2 (Ti_3Al) alloy has a wide range of compositional stability, with aluminum contents of 22 to 39 at.%. The compound is congruently disordered at a temperature of 1180 °C (2155 °F) and an aluminum content of 32 at.%. The stoichiometric composition, Ti-25Al, is stable up to about 1090 °C (1995 °F) (Ref 36). Ternary phase diagrams centered around the α-2 phase are not yet well developed (Ref 180, 181), and even specifics of the Ti_3Al-Nb pseudobinary system are still being debated (Ref 182, 183). At niobium contents of less than 5 at.%, a martensitic transformation to the α phase occurs during rapid cooling from the high-temperature β phase (Ref 182, 184, 185). Increased niobium contents suppress the martensitic transformation, and β or $B2$ can be retained at room temperature (Ref 182, 184, 186). This quenched-in $B2$ phase is metastable, and it contains transitional microstructures and phases such as antiphase boundaries (Ref 184, 185), "tweed" microstructures (Ref 182, 184, 187, 188), the 0 phase, and fine ω phase particles (Ref 187); the ω transforms to α-2 upon heat treatment in a similar manner to the ω → α transformation that occurs in conventional terminal titanium alloys (Ref 161).

Material Processing. Microstructural features that can be varied by thermomechanical processing include primary α-2 grain size and volume fraction, secondary α-2 plate morphology and thickness, and the presence of secondary β grains (Ref 166). Beta processing generally results in elongated Widmanstätten α-2 in large primary β grains in a manner similar to that in conventional titanium alloys (Ref 161).

Up to 4 wt% H can be dissolved in titanium alloys at elevated temperatures. This hydrogen can then be used to improve processibility, and final mechanical properties are enhanced after its removal; removal of the hydrogen can be easily achieved by vacuum annealing (Ref 189, 190). This thermochemical processing technique allows titanium aluminides to be processed at reduced temperatures (Ref 189-192) and results in a finer microstructure (Ref 175, 189, 190, 192).

Mechanical and Metallurgical Properties. Typical mechanical properties for a number of α-2 alloys are listed in Table 5. Production of two-phase alloys by alloying Ti_3Al with β-stabilizing elements results in up to a doubling of strength. Interface strengthening of the two-phase mixture appears to be predominantly responsible for the increased strength, but other strengthening factors, such as long-range order, solid solution, and texture effects, also contribute (Ref 166, 193).

A fine Widmanstätten microstructure with a small amount of primary α-2 grains

Table 4 Properties of titanium aluminides, titanium-base conventional alloys, and nickel-base superalloys

Property	Conventional titanium alloys	Ti_3Al	TiAl	Nickel-base superalloys
Density, g/cm³	4.5	4.1–4.7	3.7–3.9	8.3
Modulus, GPa (10^6 psi)	96–100 (14–14.5)	100–145 (14.5–21)	160–176 (23.2–25.5)	206 (30)
Yield strength, MPa (ksi)(a)	380–1150 (55–167)	700–990 (101–144)	400–650 (58–94)	...
Tensile strength, MPa (ksi)(a)	480–1200 (70–174)	800–1140 (116–165)	450–800 (65–116)	...
Creep limit, °C (°F)	600 (1110)	760 (1400)	1000 (1830)	1090 (1995)
Oxidation limit, °C (°F)	600 (1110)	650 (1200)	900 (1650)	1090 (1995)
Ductility at room temperature, %	20	2–10	1–4	3–5
Ductility at high temperature, %	High	10–20	10–60	10–20
Structure	hcp/bcc	$D0_{19}$	$L1_0$	fcc/L_2

(a) At room temperature. Source: Ref 162–169

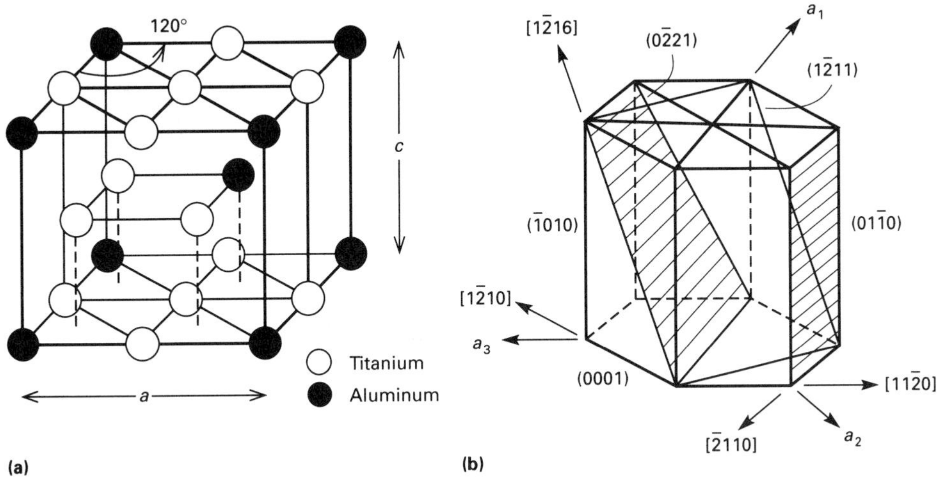

Fig. 24 Crystal structure of Ti_3Al. (a) $D0_{19}$ hexagonal superlattice structure of Ti_3Al with lattice constants of c = 0.420 nm and a = 0.577 nm. (b) Possible slip planes and slip vectors in the structure

exhibits better ductility than microstructures with a coarse Widmanstätten microstructure or an aligned acicular α-2 morphology (Ref 166). The fatigue properties of titanium alloys are strongly influenced by microstructure and work on conventional titanium alloys (Ref 161) suggests that high-ductility alloys perform best under low-cycle fatigue (LCF) conditions (Ref 194). The low ductility exhibited in material with Widmanstätten α plates in α + β alloys is responsible for low high-temperature LCF strength. Early data (Ref 195) suggest that fatigue crack growth rate is relatively insensitive to microstructure, although the coarse Widmanstätten microstructure exhibits the slowest fatigue crack growth rate at low stress intensities. Fracture toughness appears to depend on microstructures as well as alloy composition, but the precise relationship is yet to be defined (Ref 166). A recent detailed investigation into the effect of microstructure on creep behavior in Ti-25Al-10Nb-3V-1Mo has shown that the colony-type microstructure shows better creep resistance than other microstructures (Ref 196). Creep resistance of Ti-25-10-3-1 is raised by a factor of ten in the steady-state regime over that of conventional alloy Ti-1100 (Ti-6Al-3Sn-4Zr-0.4Mo-0.45Si) and two orders of magnitude over that of Ti-6Al-2Sn-4Zr-2Mo-0.1Si (Ref 196). However, 0.4% creep strain in Ti-25-10-3-1 is reached within 2 h.

Additions of silicon and zirconium appear to improve creep resistance (Ref 197), but the most significant improvement is attained by increasing the aluminum content to 25 at.% and limiting β-stabilizing elements to about 12 at.% (Ref 178, 198). However, the Ti-24.5Al-17Nb-1Mo alloy exhibits a rupture life superior to that of other α-2 alloys (Ref 166).

Gamma Alloys

Crystal Structure and Deformation Behavior. The γ-TiAl phase has an $L1_0$ ordered face-centered tetragonal structure (Ref 199-201), which has a wide range (49 to 66 at.% Al) of temperature-dependent stability (Ref 36, 199). At the equiatomic TiAl composition, the c/a ratio is 1.02; tetragonality increases up to c/a = 1.03 with increasing aluminum concentration (Ref 202-204). Within the compositional range specified at off-stoichiometric compositions, excess titanium or aluminum atoms occupy antisites without creating constitutional vacancies (Ref 205). The γ-TiAl phase apparently remains ordered up to its melting point of approximately 1450 °C (2640 °F) (Ref 36).

The layered arrangement of titanium and aluminum atoms on successive (002) planes and the slight tetragonality of c/a = 1.02 (Fig. 25) gives rise to two types of dislocations with $\frac{1}{2}\langle 110\rangle$-type Burgers vectors on {111} in γ-TiAl: ordinary dislocations $\frac{1}{2}\langle 110\rangle$ and superdislocations $\langle 011\rangle = \frac{1}{2}\langle 011\rangle + \frac{1}{2}\langle 011\rangle$ that will leave the superlattice undisturbed (Fig. 25). Another superdislocation, with a Burgers vector of $\frac{1}{2}\langle 112\rangle$, has also been suggested (Ref 206, 207). The superdislocation core can dissociate further into other complex partial dislocations, which are energetically more favorable, involving planar defects such as stacking faults and antiphase boundaries (Ref 206–208). The slip systems and partial dislocations are shown in Fig. 25. The $\frac{1}{6}\langle\bar{1}12\rangle$ on $\{1\bar{1}1\}$ partials form twin dislocations, but $\frac{1}{6}\langle 2\bar{1}1\rangle$ partials are forbidden as twinning dislocations in the $L1_0$ structure (Ref 166, 209, 210).

In single-phase γ alloys containing 52 to 54 at.% Al, deformation at room temperature occurs by motion of both ordinary and superdislocations; however, the superdislocations [011] and [101] are largely immobile because segments of the trailing $\frac{1}{6}[112]$-type superpartials form faulted dipoles that must be extended as deformation progresses (Ref 206-208, 210, 211). Increasing temperature and decreasing aluminum content increase the $\frac{1}{2}\langle 110\rangle$ slip activity as the faulted dipoles disappear and twinning dominates (Ref 166, 211, 212). In two-phase Ti-48Al, the deformation modes of primary γ grains are twinning with ⟨112⟩ twin dislocations and slip by $\frac{1}{2}[110]$-type dislocations.

The extremely low ductility values at ambient temperature and the increased ductility with increasing temperatures strongly influence the observed fracture mode. Tensile and fatigue specimens indicate that the predominant fracture modes are cleavage at

Table 5 Properties of α-2 Ti_3Al alloys with various microstructures

Alloy	Microstructure(a)	Yield strength MPa	Yield strength ksi	Ultimate tensile strength MPa	Ultimate tensile strength ksi	Elongation, %	Plane-strain fracture toughness (K_{Ic}) MPa$\sqrt{m}$	Plane-strain fracture toughness ksi $\sqrt{in.}$	Creep rupture(b)
Ti-25Al	E	538	78	538	78	0.3	...	...	...
Ti-24Al-11Nb	W	787	114	824	119	0.7	...	...	44.7
	FW	761	110	967	140	4.8	...	...	...
Ti-24Al-14Nb	W	831	120	977	142	2.1	...	...	59.5
Ti-25Al-10Nb-3V-1Mo	W	825	119	1042	151	2.2	13.5	12.3	>360
	FW	823	119	950	138	0.8	...	...	...
	C+P	745	108	907	132	1.1	...	...	...
	W+P	759	110	963	140	2.6	...	...	...
	FW+P	942	137	1097	159	2.7	...	...	...
Ti-24.5Al-17Nb	W	952	138	1010	146	5.8	28.3	25.7	62
	W+P	705	102	940	136	10.0	...	...	...
Ti-25Al-17Nb-1Mo	FW	989	143	1133	164	3.4	20.9	19.0	476

(a) E, equiaxed α-2; W, Widmanstätten; FW, fine Widmanstätten; C, colony structure; P, primary α-2 grains. (b) Time to rupture, h, at 650 °C (1200 °F) and 380 MPa (55 ksi). Source: Ref 166

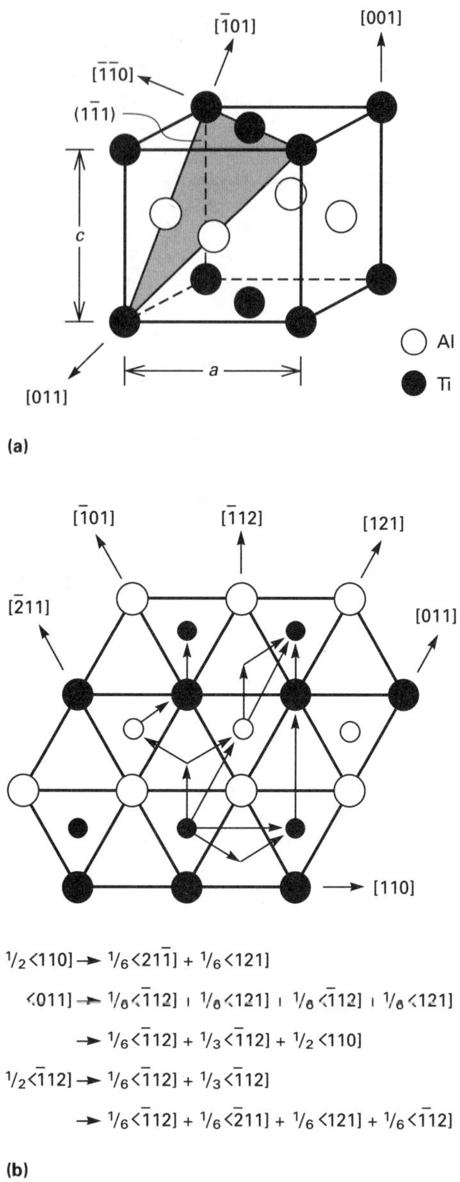

Fig. 25 Crystal structure of γ-TiAl alloys. (a) Ordered face-centered tetragonal ($L1_0$) TiAl structure. Shaded area represents the (111) plane. (b) Slip dislocations on (111) plane, ordinary dislocations ½⟨110⟩, superdislocations ⟨011⟩ and ½⟨112⟩, and twin dislocations ⅙⟨112⟩ with possible dissociations. Source: Ref 202

low temperatures due to dislocation pile-ups and intergranular fracture at temperatures above the brittle-ductile transition (Ref 211-215).

The γ alloys introduced up to the time of publication contain approximately 46 to 52 at.% Al and 1 to 10 at.% M, with M being at least one of the following: vanadium, chromium, manganese, niobium, tantalum, tungsten, and molybdenum (Ref 164, 201, 216-220). These alloys can be divided into two categories: single-phase (γ) alloys and two-phase (γ + α-2) materials (Ref 201). The (α-2 + γ)/γ phase boundary at 1000 °C (1830 °F) occurs at an aluminum content of approximately 49 at.%, depending on the type and level of solute M. Single-phase γ alloys contain third alloying elements such as niobium or tantalum that promote strengthening and further enhance oxidation resistance (Ref 221, 222). Third alloying elements in two-phase alloys can raise ductility (vanadium, chromium, and manganese) (Ref 164, 201, 218-220), increase oxidation resistance (niobium and tantalum) (Ref 221 and 222), or enhance combined properties (Ref 201).

Material Processing. The microstructure of the nominally γ alloys can be single-phase γ or, in slightly leaner compositions, two-phase γ + α-2. By appropriate thermomechanical processing, the morphology of the phases can be adjusted to produce either lamellar or equiaxed morphologies, or a mixture of the two (Ref 166, 201).

The lamellar structure can lead to refinement of the microstructure, improved ductility (Ref 218, 220), and a decreased microstructure scale by recrystallizing the fine γ grains (Ref 223). Optimum ductility occurs at a content of about 10 vol% α-2; when the α-2 phase content exceeds 20 vol%, ductility can be degraded (Ref 219). This ductility behavior is consistent with the fact that α-2 becomes increasingly brittle with increasing aluminum content over 25 at.% (Ref 177). The α-2 plates contain approximately 35 at.% Al.

Control of the microstructure in single-phase γ alloys requires the optimization of grain size and morphology. In two-phase alloys, the volume ratio of lamellar to equiaxed gamma (LG/γG) must also be controlled (Ref 165, 166, 201, 218, 219). A lamellar volume fraction of about 30% gives rise to the optimum combination of properties, with a desirable high-temperature creep resistance and acceptable levels of tensile strength and ductility (Ref 165). Heat treatment temperature and time strongly affect the LG/γG volume ratio. Thermomechanical processing (TMP) refines the microstructure when processing is conducted in such a way that both the α and γ grains are recrystallized in the (α + γ) phase field. Grain morphology varies considerably depending on composition, solution treatment temperature and time, cooling rate, and stabilization temperature and time (Ref 201). Grain size decreases with reduced aluminum content and with additions of vanadium, manganese, and chromium (Ref 219, 220). The number of annealing twins in the γ phase increases as aluminum content decreases or when manganese or vanadium levels are increased (Ref 219). Chromium additions increase the volume fraction of the lamellar structure (Ref 220).

Mechanical and Metallurgical Properties. The strength and ductility of γ alloys are strongly dependent on alloy composition and TMP conditions (Ref 201). Figure 26 shows this variation in binary γ alloys after

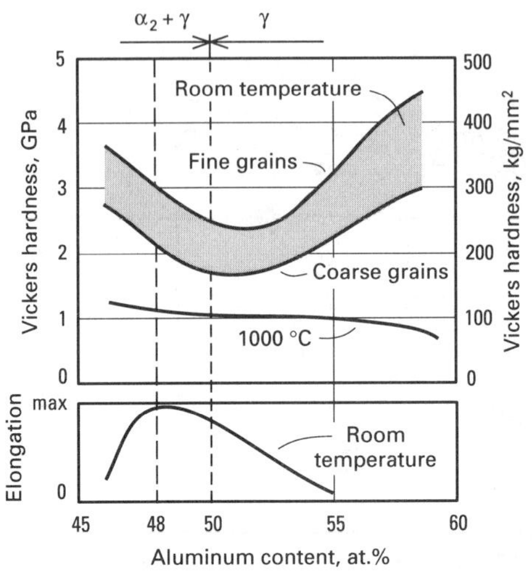

Fig. 26 Effect of aluminum content on room-temperature tensile elongation and hardness of binary γ titanium aluminide alloys. Hardness values at 1000 °C (1830 °F) are also shown. Note the single-phase γ region and the two-phase (α_2 + γ) region. Source: Ref 201

a number of TMP treatments. However, the Ti-52Al alloy demonstrates the lowest hardness value at room temperature, regardless of the TMP treatment (Ref 202, 224-227). At 1000 °C (1830 °F), however, strength tends to decrease gradually with increasing aluminum levels (Ref 201). Tensile strength and hardness vary in the same fashion with variations in aluminum content (Ref 201). Room-temperature tensile elongation is maximum at a composition of approximately Ti-48Al.

Ternary alloys of composition Ti-48Al with approximately 1 to 3% of vanadium, manganese, or chromium exhibit enhanced ductility, but Ti-48Al alloys with approximately 1 to 3% of niobium, zirconium, hafnium, tantalum, or tungsten shows lower ductility than binary Ti-48Al (Ref 201). The brittle-ductile transition (BDT) occurs at 700 °C (1290 °F) in Ti-56Al and it occurs at lower temperatures with decreasing aluminum levels. Increased room-temperature ductility generally results in a reduced BDT temperature. Above the BDT temperature, ductility increases rapidly with temperature, approaching 100% at 1000 °C (1830 °F) for the most ductile γ alloy compositions. The trend bands for variations in yield strength and tensile ductility with test temperature are shown in Fig. 27. The elastic moduli of γ alloys range from 160 GPa to 176 GPa (23 × 10^6 to 25.5 × 10^6 psi) and decrease slowly with temperature (Ref 166, 201).

Low-cycle fatigue experiments (Ref 165) suggest that fine grain sizes increase fatigue life at temperatures below 800 °C (1470 °F). Fatigue crack growth rates for γ alloys are more rapid than those for superalloys, even when density is normalized (Ref 217). Both fracture toughness and impact resistance

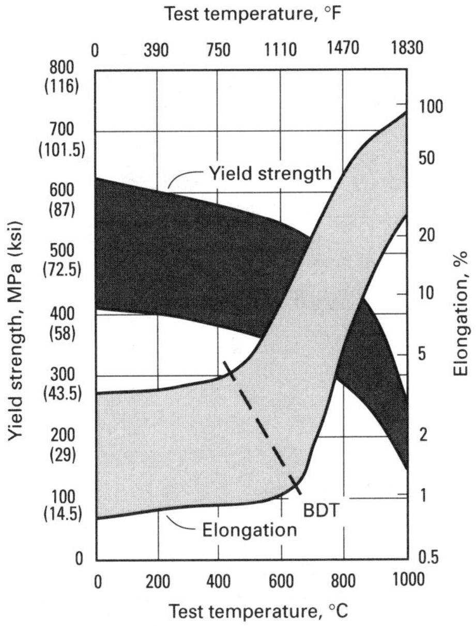

Fig. 27 Ranges of yield strength and tensile elongation as functions of test temperature for γ-TiAl alloys. BDT, brittle-ductile transition. Source: Ref 166, 201

are low at ambient temperatures, but fracture toughness increases with temperature; for example, the plane-strain fracture toughness (K_{Ic}) for Ti-48Al-1V-0.1C is 24 MPa$\sqrt{m}$ (21.8 ksi$\sqrt{in.}$) at room temperature (Ref 165). Fracture toughness is strongly dependent on the volume fraction of the lamellar phase. In a two-phase quaternary γ alloy, a fracture toughness of 12 MPa$\sqrt{m}$ (10.9 ksi$\sqrt{in.}$) is observed for a fine structure that is almost entirely γ; K_{Ic} is greater than 20 MPa$\sqrt{m}$ (18.2 ksi$\sqrt{in.}$) when a large volume fraction of lamellar grains are present (Ref 166, 201). Creep properties of γ alloys, when normalized by density, are better than those of superalloys, but they are strongly influenced by alloy chemistry and TMP. Increased aluminum content and additions of tungsten (Ref 228) or carbon (Ref 165) increase creep resistance. Increasing the volume fraction of the lamellar structure enhances creep properties (Ref 165) but lowers ductility. The level of creep strain from elongation upon initial loading and primary creep is of concern because it can exceed projected design levels for maximum creep strain in the part.

Future Directions and Applications. Because of their low density, the titanium aluminides could replace superalloys in many elevated-temperature applications in airframes, engines, and missiles. The great importance attached to structural integrity in advanced engine and airframe designs means that reliability and reproducibility are of major concern, and the low levels of "forgiveness" in the titanium aluminides are problematic. At present, no flying applications exist for the titanium aluminides, although many components have been fabricated from both the α-2 and γ alloys and have performed quite satisfactorily in ground-based tests.

Other Intermetallics and Development

Research on nickel, iron, and titanium aluminides has recently extended to other aluminides and intermetallics such as trialuminides and silicides. Low ductility and brittle fracture remain the major concerns for structural use of these materials (Ref 22-24). This article will consider only the intermetallic systems that have the potential for structural applications or that contribute to a general understanding of physical metallurgy and mechanical behavior of this class of materials as a whole. The ordered alloys with low T_c, such as Cu_3Au, are not included in this discussion.

Co_3V and Co_3Ti Alloys

Bulk materials of many ordered intermetallics are brittle because of their low-symmetry crystal structures, which have a limited number of deformation modes. The ductility of these alloys can be substantially improved by controlling the ordered crystal structures—in other words, changing the crystal structure from one of low symmetry (an ordered complex hexagonal structure) to one of high symmetry (an ordered cubic structure) through macroalloying. A prominent example is Co_3V alloyed with iron additions (Ref 13-17). Iron acts to lower the average electron concentration, thereby controlling the ordered crystal structure in this transition-metal ordered alloy.

The ordered crystal structure in the pseudobinary Ni_3V-Co_3V-Fe_3V alloy systems can be correlated with electron concentration (*e*/*a*), which is defined as the average number of valence electrons per atom (Ref 14-17, 229-231). With an increase in *e*/*a* in A_3B alloys containing refractory elements, the ordered structure changes systematically from predominantly cubic to predominantly hexagonal in character (Ref 231). In some cases, a further increase in *e*/*a* causes a change in structure from hexagonal to tetragonal.

The alloy Co_3V forms a complex hexagonal ordered structure in which the unit cell contains six close-packed ordered layers (*abcacb*) and has a hexagonality of 33.3%. The alloy is brittle because of insufficient slip systems available in its low-symmetry ordered structure (Ref 17). The electron concentration of Co_3V can be increased by partial replacement of cobalt (*e*/*a* = 9) with nickel (*e*/*a* = 10) to form $(Ni,Co)_3V$. With an increase of *e*/*a*, the hexagonality of the ordered structure increases systematically from 33.3 to 100%. A further increase in *e*/*a* above 8.54 produces a change in structure from hexagonal to tetragonal, similar to Ni_3V ($D0_{22}$). On the other hand, *e*/*a* in Co_3V can be decreased by the partial substitution of iron (*e*/*a* = 8) for cobalt, producing $(Co,Fe)_3V$. These alloys have an *e*/*a* below 7.89, which stabilizes the $L1_2$ ordered cubic structure with the stacking sequence *abc*. By control of *e*/*a*, the $L1_2$-type cubic ordered structure can be stabilized in $(Ni,Co,Fe)_3V$ and $(Ni,Fe)_3V$.

The importance of the ordered cubic structure in $(Ni,Co,Fe)_3V$ alloys is shown in Fig. 28. The cubic ordered alloys with the compositions $(Fe,Co)_3V$, $(Fe,Co,Ni)_3V$, and $(Fe,Ni)_3V$ are all ductile, with tensile elongations of 40% or higher. On the other hand, the hexagonally ordered alloys Co_3V and $(Ni,Co)_3V$ are brittle, with less than 1% elongation at room temperature. The deformation behavior of ordered cubic alloys is similar to that of fcc materials having 12 slip systems. The brittleness of the ordered hexagonal alloys is attributed to their limited deformation modes in the hexagonal ordered structures. The hexagonal ordered alloys have ductilities too low to permit easy fabrication, whereas the cubic ordered alloys have excellent fabricability at both room and elevated temperatures. The results shown in Fig. 28 demonstrate the feasibility of dramatically improving tensile ductility by control of ordered crystal structures.

The $L1_2$-ordered $(Fe,Co,Ni)_3V$ alloys, like many other $L1_2$ alloys, show a positive temperature dependence for yield strength (Fig. 29). Yield strength increases with temperature and reaches a maximum around T_c. Because of this increase, $(Fe,Co,Ni)_3V$ alloys become stronger than conventional disordered solid-solution alloys at elevated temperatures. The strength of $(Fe,Co,Ni)_3V$ alloys decreases sharply above T_c because of the loss of the long-range ordered crystal structure. These alloys have excellent creep and fatigue properties in the ordered state; however, a lack of oxidation resistance limits their use in hostile environments.

The ordered crystal structure in Cu_3Ti-Ni_3Ti-Co_3Ti-Fe_3Ti, like that in Ni_3V-Co_3V-Fe_3V, is controlled by electron concentration (Ref 229). The cubic ordered structure ($L1_2$) has been reported to form in Co_3Ti and $(Co,Ni)_3Ti$ alloys. The alloys with the $L1_2$ structure are very ductile, with more than 40% tensile elongation obtained at room temperature in vacuum (Ref 141). However, Co_3Ti shows some degree of environmental embrittlement when tested in air at ambient temperatures.

Structure Maps for Ordered Intermetallics

The foregoing section has shown the importance of controlling ordered crystal structures in improving the mechanical properties of ordered intermetallics. The stability of ordered intermetallic phases is controlled by four alloy variables: valency

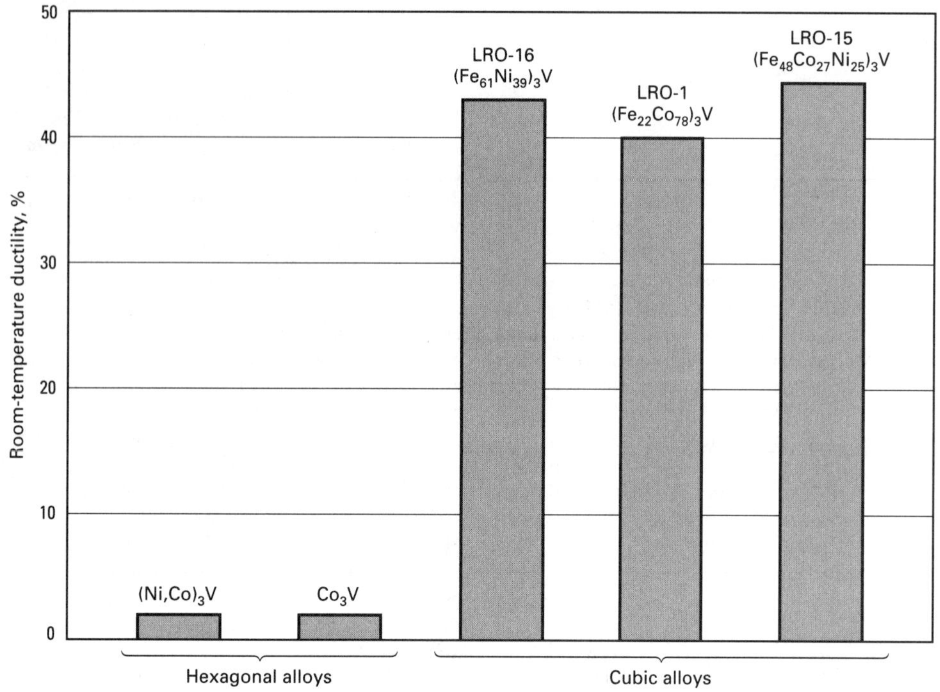

Fig. 28 Comparison of room-temperature tensile elongations of cubic and hexagonal alloys. Source: Ref 17

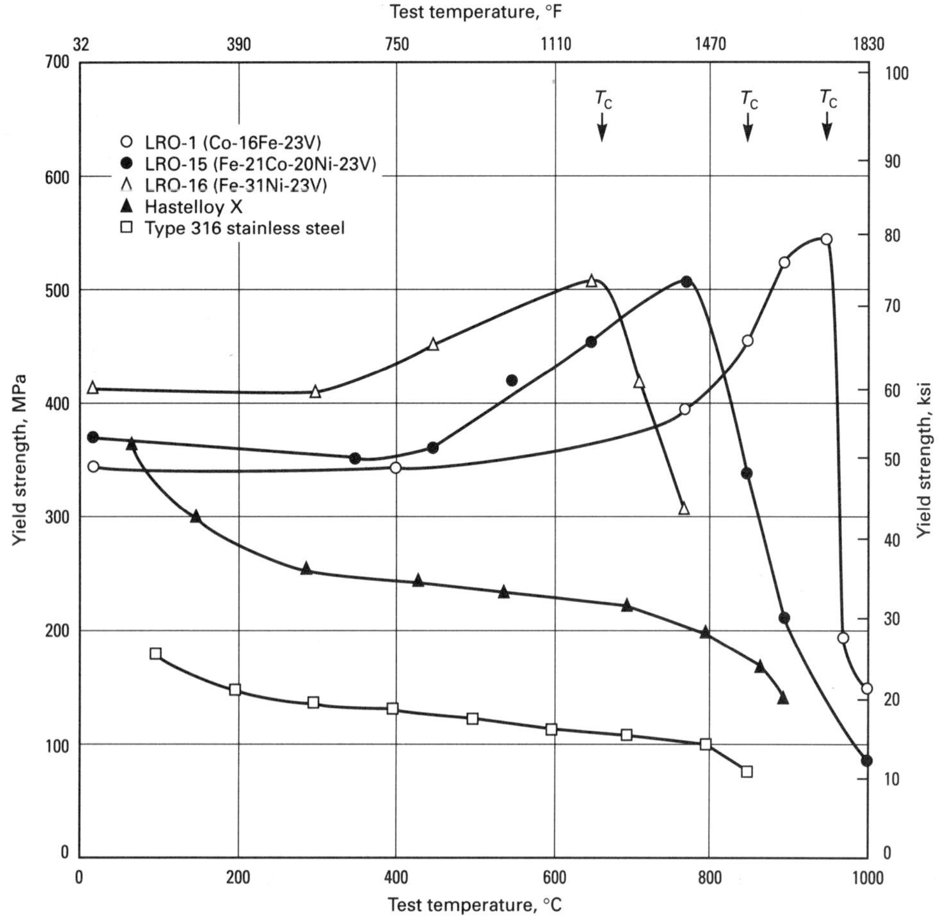

Fig. 29 Variation of yield strength with test temperature for cubic ordered alloys (LRO-1, LRO-15, and LRO-16) and commercial solid-solution alloys Hastelloy X and type 316 stainless steel. T_c, critical ordering temperature. Source: Ref 17

difference (ΔZ), atomic size difference (ΔR), average number of valence electrons per atom (e/a), and angular dependence of the valence orbitals (Ref 232-237). The last variable is related to the quantum character of electrons, which is generally neglected by classical approaches (Ref 235). Based on the first three variables, a number of schemes have been developed for representing intermetallic phases as functions of these variables. Each of these has demonstrated some merit in predicting phase stability in a given class of intermetallic phases.

Recently, Pettifor (Ref 237-239) constructed structure maps for binary intermetallic phases using a single phenomenological parameter called the Mendeleev number (M). Mendeleev numbers are assigned to each alloying element based on its position in a modified periodic table; these numbers, in principle, include all alloying information from both classical and quantum mechanical considerations. The advantage of the Pettifor scheme is that structure maps for intermetallic phases can be represented by simple two-dimensional plots. Examples of the structure maps and detailed explanations of their use are available in Ref 235, 238, and 239. The maps successfully group different structure types into separate domains, and they can represent all kinds of intermetallic phases, instead of just one kind as in other structure schemes.

The Pettifor structure maps are useful tools for controlling the ordered crystal structures in multicomponent intermetallics. For example, the structure maps predict that the ordered phases adjacent to their domain boundary are relatively less stable with respect to the structure type in the adjacent domain. In other words, these phases can be easily altered from one structure type to the other through control of an average Mendeleev number. Pettifor has recently shown that many ordered crystal phases observed in pseudobinary and ternary intermetallic alloys fit well into the domains of structural stability for binary intermetallic phases (Ref 235). His work has demonstrated that the average Mendeleev number serves as a useful parameter for the control of phase stability in intermetallic phases.

Trialuminides and Cleavage Fracture

Trialuminides are materials of composition Al_3X, where X stands for titanium, zirconium, niobium, vanadium, and so on, and they form tetragonal ordered crystal structures ($D0_{22}$ and $D0_{23}$). These aluminides are of technological interest because of their high melting points (≧1350 °C, or 2460 °F), good oxidation resistance, and extremely low density (~4.0 g/cm^3). The trialuminides are, however, extremely brittle, and their brittleness has been attributed to the tetragonal crystal structures and the associated limited slip systems. Recently, considerable effort has been devoted to

Table 6 Mechanical properties of trialuminides

Alloy	Vickers hardness	Yield strength MPa	Yield strength ksi	Fracture toughness MPa $\sqrt{m}$	Fracture toughness ksi $\sqrt{in.}$	Young's modulus (E) GPa	Young's modulus (E) 10^6 psi	Shear modulus (G) GPa	Shear modulus (G) 10^6 psi	Poisson's ratio (ν)	Bulk modulus (K) GPa	Bulk modulus (K) 10^6 psi	K/G
Al-25Sc	142	105	15	3.1	2.8	166	24	68	10	0.22	99	14	1.5
Al-25Zr-6Fe	200(a)	175(a)	25(a)	2.2(a)	2.0(a)	166	24	68	10	0.23	103	15	1.5
Al-23Ti-6Fe-5V	200(b)	270	39	2.1	1.9	...	...	...	...	...	...	...	...
Al-25Ti-8Fe	...	...	...	...	...	192	28	84	12	0.14	89	13	1.1

(a) Source: Ref 239. (b) Source: Ref 238. Other data from Ref 245

improving the mechanical properties of trialuminides by controlling their microstructures and changing the ordered crystal structures from tetragonal $D0_{22}$ and $D0_{23}$ to cubic $L1_2$ (Ref 240-247). Consistent with the Pettifor AB_3 structure map, the cubic ordered structure can be stabilized by lowering the average Mendeleev number, which is accomplished by alloying with moderate amounts (5 to 12 at.%) of chromium, manganese, iron, cobalt, nickel, and copper; these elements are mainly used to substitute for aluminum in $TiAl_3$ and $ZrAl_3$ alloys. These $L1_2$ trialuminide alloys have much better compressive ductility and toughness than Al_3Ti, which has a tetragonal $D0_{22}$ structure, and Al_3Zr, which has a $D0_{23}$ structure; however, all three alloys remain very brittle in tension (Ref 240-247). Recently, a measurable tensile ductility (~0.5%) based on four-point bend tests has been reported for $Al_{66}Cr_9Ti_{25}$ and $Al_{66}Mn_6Ti_{23}V_5$ (Ref 248). Successful achievement of tensile ductility is expected to spur more interest in the development of ductile trialuminide alloys. No stable cubic ordered structure has been reported for the trialuminide alloys based on Al_3V and Al_3Nb (Ref 246).

A possible way to improve the ductility of the $L1_2$ trialuminide alloys is to lower their hardness by controlling their microstructure and alloy composition. Studies (Ref 241, 243, 249) have found that the hardness of $L1_2$ Al_3Ti and Al_3Zr alloys can be substantially reduced from 350 to 200 HV by alloying additions of, for example, vanadium. However, the vanadium-modified aluminide alloys, even with such low hardness levels, exhibit brittle cleavage fracture with virtually no tensile ductility in a manner no different from that of the higher-hardness $L1_2$ trialuminides. Selected-area electron channeling pattern analyses revealed that the {110}-type planes are the predominant cleavage planes, with {100} and ⟨111⟩ as minor planes. This observation is different from that in Ni_3Al, where the {111}-type planes are the major fracture planes (Ref 249).

Brittle cleavage is generally associated with high hardness and high yield strength. The study of $L1_2$ trialuminides has demonstrated an interesting case where brittle cleavage can take place in soft materials. Table 6 shows the room-temperature mechanical properties of three $L1_2$ trialuminides: Al_3Sc, used as a model material to study cleavage fracture in $L1_2$ trialuminides; $Al_{69}Zr_{25}Fe_6$; and $Al_{66}Ti_{23}Fe_6V_5$. All of these aluminides are soft (≦200 HV) and have yield strengths as low as 105 MPa (15 ksi), indicating no difficulty in the generation of dislocations in these materials. The unusual brittle cleavage fracture has been attributed to intrinsic low cleavage strength and the difficulty in emission of dislocations from crack tips. Total-energy calculations (Ref 250) suggest the formation of directional scandium-aluminum or titanium-aluminum bonds as a possible cause for the observed (110) cleavage in the trialuminides. In addition, the calculated cleavage strength of Al_3Sc is substantially lower than that of Ni_3Al.

Ni_3X Alloys and Intergranular Fracture

The study of grain-boundary fracture in Ni_3Al has been extended to other $L1_2$ intermetallics, particularly Ni_3X alloys, where X stands for iron, manganese, aluminum, gallium, silicon, or germanium. In 1985, Takasugi and Izumi (Ref 62, 63) initiated a systematic study of the effect of metallurgical, mechanical, and chemical factors on grain-boundary cohesion in $L1_2$ ordered A_3B alloys. They found that the valency difference (ΔZ) between A and B atoms is the dominant factor controlling grain-boundary cohesive strength, and that the tendency for grain-boundary fracture increases with increasing ΔZ. They also considered the importance of the atomic size difference and postulated that a better correlation can be obtained by a combined consideration of both valency and atomic size differences. Their correlation appears to correctly rank the grain-boundary cohesive strength of $L1_2$ ordered nickel-base alloys in the order $Ni_3Fe > Ni_3Mn > Ni_3Al > Ni_3Ga > Ni_3Si > Ni_3Ge$, which is in agreement with the experimental data listed in Table 7.

Taub *et al.* (Ref 64-66), on the other hand, studied grain-boundary fracture in boron-doped and undoped binary and pseudobinary intermetallic alloys based on Ni_3X (where X stands for aluminum, gallium, silicon, or germanium) that were prepared by melt spinning. They found that both bend ductility and fracture behavior can be better correlated with the electronegativity consideration rather than with the valency difference as proposed by Takasugi and Izumi (Ref 60). In addition to requiring only a single parameter to correlate the data successfully, the electronegativity consideration provides a better understanding of atomic bonding. The electronegativity difference can be generally regarded as a standard scale for measuring a transfer of electrons between atoms. Compared with aluminum, gallium, and silicon atoms, germanium atoms are more electronegative with respective to nickel atoms; consequently, germanium has a greater tendency to pull electron charge from nickel/nickel bonds, thereby further reducing cohesive strength and promoting intergranular fracture in Ni_3Ge. Grain boundaries in Ni_3Ge are weaker than those in Ni_3Al, and boron is considered to be ineffective in increasing the ductility of Ni_3Ge and $Ni_3(Al,Ge)$ alloys containing high levels of germanium.

Takasugi and Izumi and Taub *et al.* have correlated intergranular fracture with the average electron character of the boundary (Ref 63-66). From this correlation, it is possible to manipulate grain-boundary cohesion by macroalloying, that is, replacing constituent atoms with alloy additions. They showed that a partial replacement of aluminum with iron or manganese in Ni_3Al reduces the average valency and electronegativity differences between the nickel and

Table 7 Valency/size effect/electronegativity correlation with ductility in the $L1_2$ Ni_3X alloys

X species	Valency difference (ΔZ)(a)	Lattice dilation ($a - a_{Ni}/a_{Ni}$)	Electronegativity difference (Pauling)	Type of fracture(b) Undoped alloy	Type of fracture(b) Boron-doped alloy
Iron	0.2	+1.0%	−0.08	T	...
Manganese	0.9	+2.2%	−0.36	T	...
Aluminum	3.0	+1.5%	−0.30	I	T
Gallium	3.0	+1.6%	−0.10	I	T
Silicon	4.0	−0.04%	−0.01	I	M
Germanium	4.0	+1.5%	+0.10	I	I

(a) Source: Ref 63. (b) T, transgranular; I, intergranular; M, mixed mode. Unless otherwise noted, data are from Ref 65.

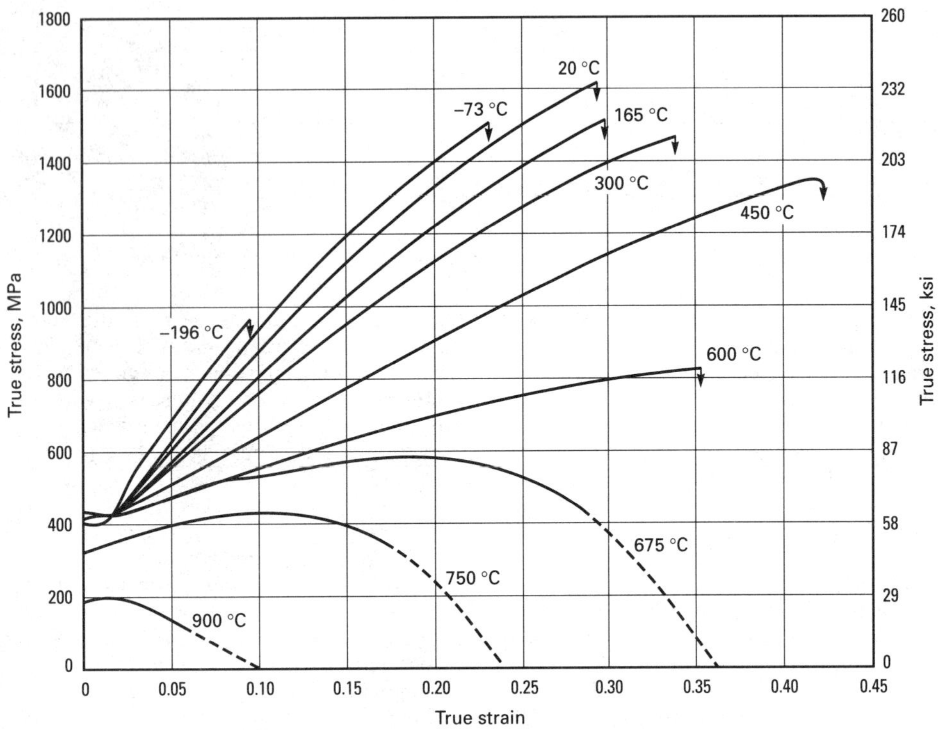

Fig. 30 True stress versus true strain as a function of temperature for Zr_3Al with a grain size of 5 μm. Testing conducted at a strain rate of $2.7 \times 10^{-4}\ s^{-1}$. Source: Ref 254

"aluminum" atoms, thereby improving room-temperature ductility and lowering the propensity for grain-boundary fracture in Ni_3Al (Ref 60). However, the increase in ductility is not as dramatic as that from microalloying with boron, which occupies interstitial sites in Ni_3Al. A combination of both microalloying and macroalloying has also proved to be very effective in increasing the ductility of Ni_3Al alloyed with boron and iron (Ref 61). Beryllium, which was recently verified to occupy the substitutional sites (Ref 251), has only a moderate effect on ductility improvement (Ref 62).

Zr_3Al

Because of its desirable nuclear properties (low-absorption cross section for thermal neutrons), Zr_3Al was studied extensively in the 1970s for potential use as a cladding material for water-cooled nuclear power reactors (Ref 252). Zr_3Al is a line compound formed by the peritectoid reaction at 975 °C (1790 °F) (Ref 36):

$$Zr_2Al + Zr \rightarrow Zr_3Al$$

The aluminide has good oxidation resistance and can be easily fabricated into strip, rod, and tubing (Ref 252). The fabrication can best be done by hot working at a temperature within the βZr + Zr_2Al two-phase field, followed by annealing below the peritectoid temperature to produce near-single-phase Zr_3Al. Working of Zr_3Al is limited to a reduction in area of about 30% (Ref 253).

Deformation and fracture in Zr_3Al have been studied extensively (Ref 252). The yield strength of the aluminide is sensitive to grain size and obeys the Hall-Petch relationship. Figure 30 shows curves for true stress versus true strain as a function of temperature for Zr_3Al with a grain size of 5 μm. The aluminide is quite ductile at low temperatures, with a ductility of 30% at room temperature for specimens prepared by electropolishing. The ductility at ambient temperatures drops when surfaces are damaged by machining or abrading with SiC paper to 1.5 and 15%, respectively (Ref 255, 256), and the fracture mode changes from transgranular to intergranular. This result clearly indicates that the aluminide is very notch sensitive. Yield strength is insensitive to temperatures up to about 600 °C (1110 °F); above that temperature, strength decreases substantially. The aluminide is stronger than zirconium-base alloys such as Zircaloy-2 and Zr-2.5Nb at intermediate temperatures (300 °C, or 570 °F) (Ref 252).

The structure and properties of Zr_2Al have also been characterized in irradiation conditions (Ref 252). The aluminide shows significant swelling upon irradiation. Susceptibility to notch sensitivity is suppressed by fast-neutron irradiation to relatively low exposures; however, the alloy undergoes an embrittling crystalline-to-amorphous phase transformation when irradiated to higher doses at temperatures up to 400 °C (750 °F). The aluminide will remain an experimental material until the problem of radiation damage can be alleviated by some metallurgical means.

$L2_1$ Heusler Alloys

As mentioned before, the aluminide NiAl is relatively weak at elevated temperatures and has, in particular, poor creep resistance. An effective way to improve the creep properties of NiAl is to introduce a Heusler phase ($L2_1$) in it (Ref 257). All the elements of the IVB and VB groups in the periodic table are known to form Heusler-type ternary compounds in the form of Ni_2AlX, where X stands for titanium, tantalum, hafnium, and so on. The unit cell of Heusler alloys is composed of eight $B2$ unit cells in which nickel atoms occupy the corner sublattices and aluminum and titanium atoms form an ordered array on the body-centered sublattices. The Heusler alloys are generally hard and brittle at ambient temperatures, and their hardness can be related to the size difference between X and aluminum atoms and also to internal strains resulting from the complex crystal structure itself. For example, room-temperature hardness is approximately 5 GPa (510 HV) for Ni_2AlTi and approximately 8 GPa (815 HV) for Ni_2AlHf (Ref 258). The fracture mode is basically transgranular cleavage in Ni_2AlHf at temperatures below 600 °C (1110 °F). Above that temperature, the hardness drops and the compression ductility increases abruptly because of the onset of thermally activated processes.

Compared with NiAl for structural applications, Heusler alloys have two major disadvantages: higher brittleness at low temperatures (<600 °C, or 1110 °F) and relatively low melting temperatures. The Heusler alloys are much harder and more brittle than NiAl (Ref 258, 259). They show limited room-temperature ductility (<3%) in compression tests and have been reported to have nil ductility in tension. The melting point of all known Heusler phases (≦1400 °C, or 2550 °F) (Ref 258, 260) is lower than that of NiAl by 250 °C (450 °F). The advantage of the Heusler alloys is that they are stronger and more creep resistant than NiAl at elevated temperatures. Single-phase Ni_2AlTa showed a yield strength of 540 MPa (78 ksi) when tested at 970 °C (1780 °F) at a low strain rate of $1 \times 10^{-5}\ s^{-1}$ (Ref 259). In comparison, NiAl exhibited a yield strength of less than 30 MPa (4 ksi) under the same test conditions.

Some efforts have been devoted to developing two-phase alloys based on NiAl(β) and Ni_2AlX(β′) phases (Ref 257-263); these attempts are analogous to the development of γ/γ′ superalloys. The lattice misfit between β and β′ phases varies with the X elements; misfit is near zero for NiAl/Ni_2AlTa (Ref 259), approximately 1% for NiAl/Ni_2AlTi (Ref 257), and approximately 5% for NiAl/Ni_2AlHf (Ref 258). Consequently, coherent precipitates exist in NiAl/Ni_2AlTa, and no coherency exists between NiAl and Ni_2AlHf. Strutt *et al.* (Ref 261, 262) reported the excellent creep resistance

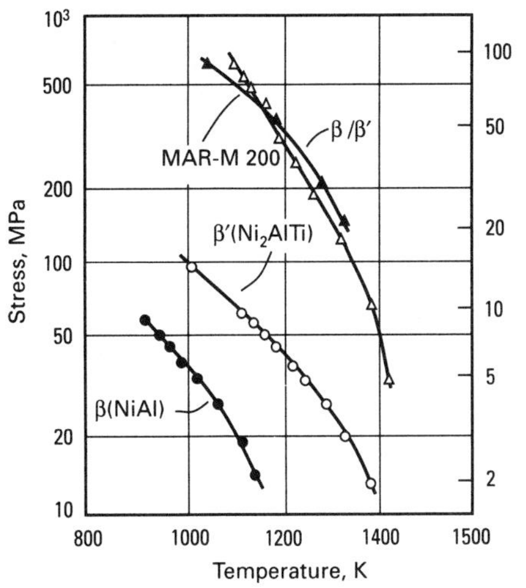

Fig. 31 Creep strength, defined as the stress to maintain a creep rate of 10^{-7} s^{-1} versus temperature for β, β′, β/β′, and nickel-base alloys MAR-M200. Source: Ref 257

of Ni_2AlTi and β/β′ alloys at high temperatures. As shown in Fig. 31, the creep strength of β′ is better than β(NiAl), and the β/β′ alloy is as strong as the advanced directionally solidified superalloy MAR-M 200 (Ref 257). The excellent creep resistance, superior oxidation resistance, and low density of β/β′ alloys certainly warrants their further development for high-temperature structural applications.

Silicides

Silicides have been used as commercial materials because of their excellent oxidation and corrosion resistance in hostile environments. For example, $MoSi_2$ is currently used for heating elements at temperatures to 1800 °C (3270 °F). Ni_3Si is the major constituent of the commercial alloy Hastelloy D, a corrosion-resistant alloy with the unique ability to resist attack by sulfuric acid solutions. This section briefly describes the recent work on a number of silicides, including Ni_3Si, Fe_3Si, $MoSi_2$, Nb_3Si, and Ti_5Si_3. Other potential uses for silicides include high-temperature structural materials and magnetic materials.

Ni_3Si Alloys. The silicide Ni_3Si has an $L1_2$ ordered crystal structure existing below the peritectic temperature of approximately 1035 °C (1900 °F) (Ref 36). It is of commercial interest because of its excellent corrosion resistance in acid environments, particularly sulfuric acid solutions. The engineering use of Ni_3Si is limited by its poor ductility at ambient temperatures and lack of fabricability at high temperatures (Ref 64, 66, 264-267). Because of these problems, the commercial Ni_3Si alloy, Hastelloy D, has to be used in the cast condition with relatively poor mechanical properties.

Grain boundaries in Ni_3Si, like those in Ni_3Al, are intrinsically brittle, as evidenced by Auger spectroscopic analyses (Ref 264-266). As shown in Fig. 32, the ductility of Ni_3Si can be effectively improved by reducing the silicon concentration below 20 at.% or by microalloying with boron, carbon, or beryllium. A dramatic improvement in room-temperature ductility was obtained by adding boron to Ni-18.9Si-3.2Ti (at.%). The increase in ductility from 3% for Ni-22.5Si to 30% for Ni-18.9Si-3.2Ti-0.1%B is accompanied by a change in fracture mode from brittle grain-boundary fracture to ductile dimple failure. Auger analyses showed that boron strongly segregates to grain boundaries, thereby suppressing intergranular fracture.

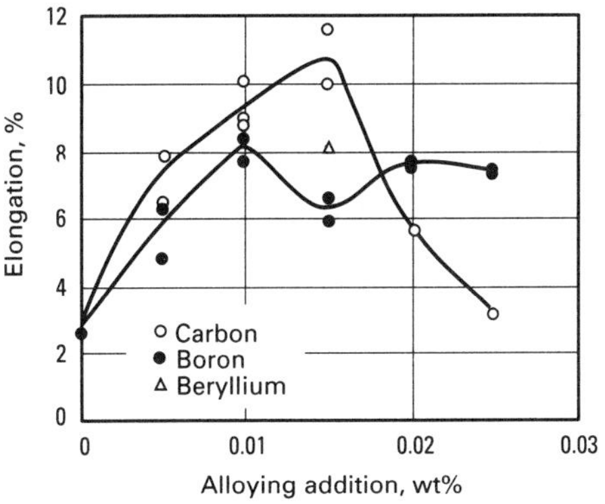

Fig. 32 Effect of microalloying with boron, carbon, and beryllium on the room-temperature ductility of Ni-22.5Si. Source: Ref 267

Just like Ni_3Al, ductile Ni_3Si alloys show a severe reduction in ductility at intermediate temperatures (400 to 800 °C, or 750 to 1470 °F) when tested in tension in air (Ref 267). The loss in ductility is due to dynamic embrittlement involving oxygen in air. The embrittlement can be reduced and the ductility improved by alloying with moderate amounts (4 to 6%) of chromium.

In the early 1970s, it was reported that additions of titanium greatly improved the as-cast properties of Ni_3Si alloys but reduced their hot fabricability (Ref 268, 269). A recently conducted systematic study (Ref 265, 267) of alloying effects for the development of wrought Ni_3Si alloys found that macroalloying with niobium and vanadium is as effective as that with titanium in improving the room-temperature ductility of Ni_3Si alloys. On the other hand, the alloying elements molybdenum, iron, and chromium are beneficial to hot fabricability. Some Ni_3Si alloys containing these three elements showed a superplastic behavior (approximate elongation of 500 to 600%) when tested at 1020 to 1100 °C (1870 to 2010 °F) in air. The strength of Ni_3Si at temperatures up to 700 °C (1290 °F) can be significantly improved by the addition of up to 1% Hf. Some alloys, such as Ni-18.9Si-3.2Cr-0.6Hf-0.15B (at.%) are, in fact, as strong as nickel-base superalloy IN-718 at room and intermediate temperatures. These ductile, strong Ni_3Si alloys have the potential to be used as structural materials for chemical and petrochemical applications.

Fe_3Si Alloys. The iron silicide, Fe_3Si, forms the $D0_3$ structure with a high T_c (1240 °C, or 2265 °F), whereas the iron aluminide, Fe_3Al, has the same structure with a much lower T_c (540 °C, or 1000 °F) (Ref 36). A commercial alloy named Sendust based on Fe_3Si and Fe_3Al has been developed for magnetic applications (Ref 270-276); its typical composition is Fe-9.6Si-5.4Al (wt%). Sendust has been used as a magnetic head core material because of its superior magnetic properties and resistance to wear and corrosion. The alloy is hard and brittle at ambient and elevated temperatures; consequently, it is difficult to fabricate into useful forms.

Single crystals of Sendust deform by ⟨111⟩{110} and ⟨111⟩{112} slip systems, depending on the crystal orientation (Ref 271). Polycrystalline Sendust is very strong at room temperature, with a yield strength of approximately 1240 MPa (180 ksi) and an ultimate tensile strength of approximately 1900 MPa (275 ksi) in compression tests (Ref 272). The yield strength decreases moderately at temperatures below 600 °C (1110 °F), and it decreases sharply above that temperature. The corresponding fracture mode changes from predominantly transgranular cleavage to a mixed mode of cleavage and grain-boundary separation. The cleavage planes are of the {100} type.

Steady-state deformation of Sendust at high temperatures is governed by grain-boundary sliding, migration, and dynamic recrystallization. A careful control of test temperature and strain rate results in extensive deformation and ductile fracture (Ref 273). Figure 33 defines the critical condition for the ductile-to-brittle transition observed in Sendust. Extrapolation of the critical line in Fig. 33 to higher temperatures gives optimum conditions for hot forging and rolling of polycrystalline Sendust; these extrapolated conditions have been supported by experimental data. Hot-rolled materials exhibited better magnetic properties because the segregation of alloying elements introduced during solidification was eliminated by hot rolling (Ref 274). The development of the hot fabrication scheme substantially reduced the production cost of Sendust for magnetic head core applications.

$MoSi_2$ and Refractory Silicides. Molybdenum disilicide ($MoSi_2$) is a line compound with a very high melting point (T_m = 2020 °C, or 3670 °F) (Ref 36). It has an ordered tetragonal structure ($C11_b$) in which the unit cell contains three bcc lattices. $MoSi_2$ is attractive because of its high electrical and thermal conductivities and excellent oxidation resistance at high temperatures. The silicide is capable of forming a thin adhesive self-healing protective layer of silica glass on its surfaces when exposed to oxidizing atmospheres at temperatures up to 1800 °C

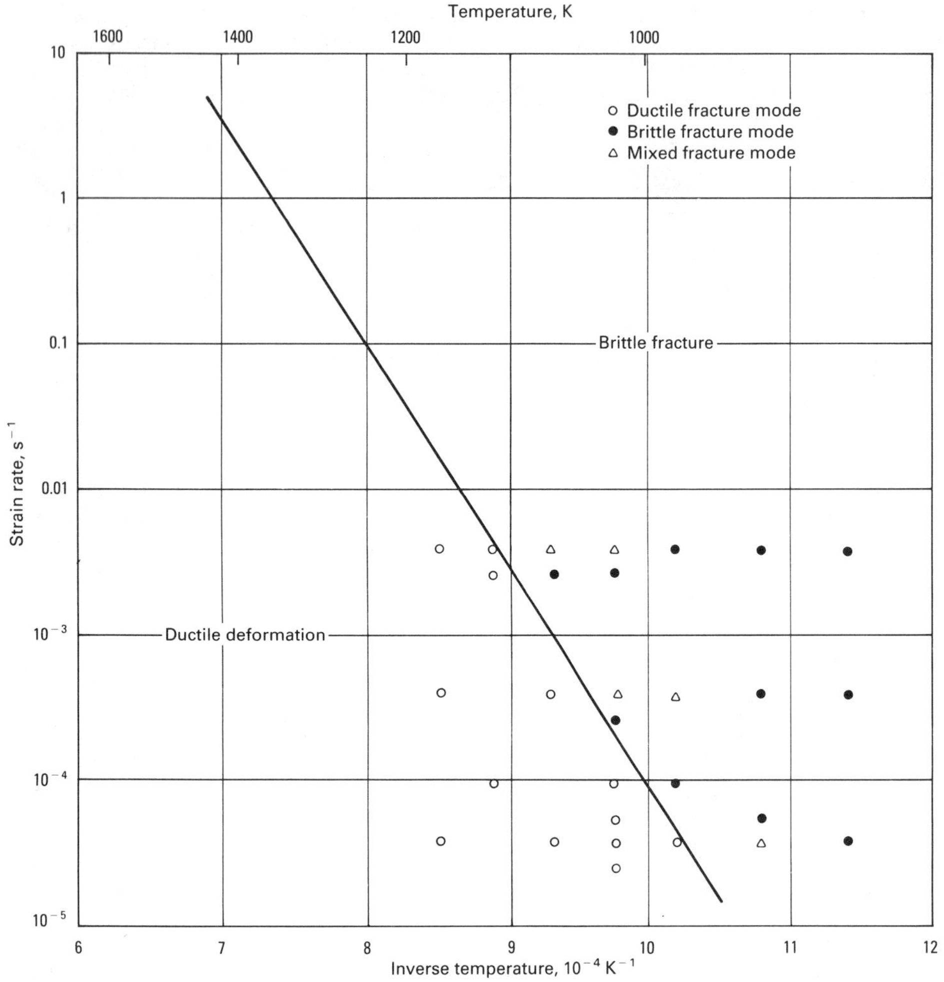

Fig. 33 Optimum conditions for ductile deformation of polycrystalline Sendust alloy (Fe-9.6Si-5.4Al). Source: Ref 273

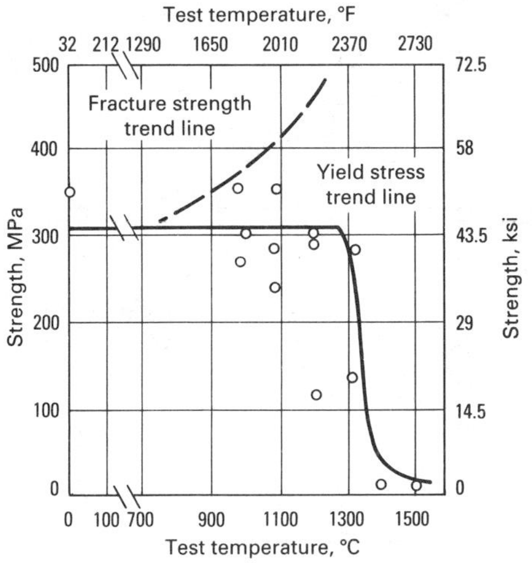

Fig. 34 Strength of $MoSi_2$ as a function of temperature. Data points determined by various investigators and compiled in Ref 281.

(3270 °F). $MoSi_2$ has been used commercially for electrical heating elements in high-temperature furnaces under the trade name of Kanthal (Ref 277). The Kanthal materials, which contain roughly 80% $MoSi_2$ and 20% glass ceramic compounds, are prepared by P/M processing. The ceramic component is added to improve the ductility of these materials at high temperatures and also to increase their electrical resistance. The lifetime of Kanthal heating elements can be as long as 5 years. However, $MoSi_2$ is prone to fast intergranular oxidation with severe material damage when exposed to air at 500 to 800 °C (930 to 1470 °F) (Ref 278). This problem can be alleviated by rapid heating to above 800 °C (1470 °F), or preheating to high temperatures for the formation of protective silica films.

Recently, considerable attention (Ref 278-281) has been given to the development of $MoSi_2$ as a high-temperature structural material for use at temperatures up to approximately 0.8 T_m (1550 °C, or 2820 °F). $MoSi_2$ is very brittle and hard (750 HV) at ambient temperatures. Figure 34 shows strength as a function of test temperature for $MoSi_2$. From 20 to 1250 °C (70 to 2280 °F), the yield strength is around 320 MPa (46 ksi) and is independent of temperature; however, it decreases abruptly above 1300 °C (2370 °F). The silicide exhibits a brittle-to-ductile transition around 925 °C (1700 °F), and above that temperature fracture strength increases sharply, as indicated by the line above the yield stress trend line in Fig. 34. Single crystals of $MoSi_2$ and WSi_2 have been successfully grown by floating-zone techniques. Studies of deformation in these crystals at 900 to 1500 °C (1650 to 2730 °F) revealed that slip occurs mainly on {110} and {013} planes (Ref 279, 280).

As shown in Fig. 34, $MoSi_2$ is not strong at high temperatures, particularly above 1300 °C (2370 °F). Consequently, efforts have been initiated to develop composite materials based on $MoSi_2$ (Ref 281, 282). Both ceramic fibers (or particulates) and ductile metal wires have been selected as reinforcement components. The ceramics, which include silicon carbide (SiC), titanium carbide (TiC), and zirconium diboride (ZrB_2), are compatible with $MoSi_2$, whereas the metal wires, which include niobium, tantalum, and tungsten, react with the silicide to form protective layers of Mo_5Si_3 that slow down interdiffusional processes (Ref 281). Some limited data have demonstrated pronounced improvements in high-temperature strength and room-temperature fracture toughness from these reinforcements. Further studies are required to optimize processing parameters.

Other refractory silicides, including Nb_5Si_3 (Ref 281, 283) and Ti_5Si_3 (Ref 284, 285), have also been studied recently because of their extremely high melting points (>2000 °C, or 3630 °F) and their potential for good oxidation resistance. These silicides form complex ordered crystal structures and are very hard and brittle at ambient temperatures. For example, Ti_5Si_3 showed essentially brittle cleavage fracture with a hardness as high as 1000 HV (Ref 285). The mechanical properties of the monolithic silicides do not appear to be significantly improved by material processing and alloy additions, and current efforts are thus mainly devoted to the development of two-phase structures with silicides as the precipitate phase (Ref 283, 284).

Figure 35 shows the existence of a pseudobinary eutectic between Ti_5Si_3 and Ti_3Al. The eutectic line determined allows control of microstructural variables including morphology, volume fraction, and the size of individual microconstituents in the titanium-silicon-aluminum ternary alloys. Coarse platelets observed in hypereutectic compositions severely embrittle the alloys. On the other hand, the fine two-phase mixture formed at the eutectic compositions provides better mechanical properties and warrants further study. Fine precipitation of Nb_5Si_3 in a niobium matrix results in a pronounced improvement in fracture toughness at ambient temperatures and in bend strength at temperatures up to 1400 °C (2550 °F) (Ref 283).

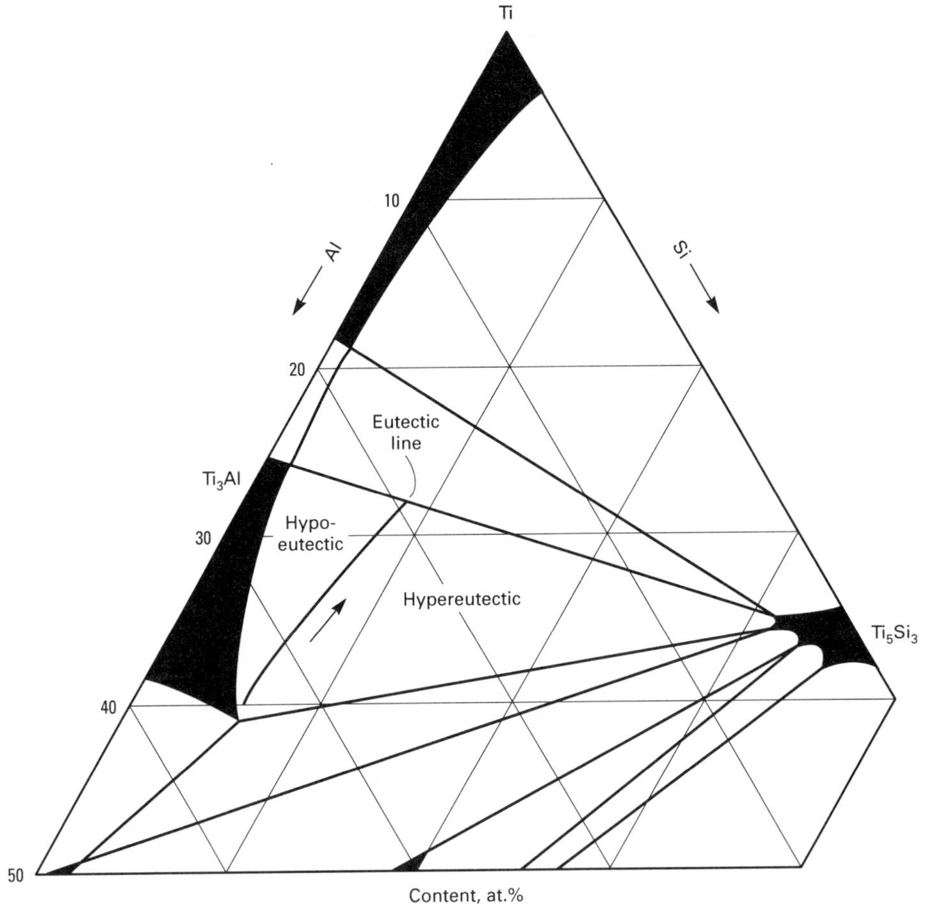

Fig. 35 Projection of the eutectic lines on the 1200 °C (2190 °F) isothermal section of the titanium-aluminum-silicon system. Source: Ref 284

Specific Gravity Versus Melting Point Diagrams for High-Temperature Intermetallics

The search for high-temperature low-density intermetallic compounds for aircraft and space applications has prompted the compilation (Ref 286-288) of structure-insensitive properties of 293 binary intermetallics that melt at temperatures of 1500 °C (2730 °F) and above. The properties include specific gravity, melting point, and elastic modulus; these properties are basically insensitive to processing history, heat treatment, impurities, and resultant microstructures. Figure 36 shows locations of the systems on a specific gravity (ρ) versus melting point (T_m) diagram. The data points are essentially distributed on the upper left side of the diagram and are bounded by an envelope line. Data points located near the envelope line represent the most promising materials for aerospace applications.

The low ρ and high T_m envelopes derived for various types of intermetallic compounds are located separately when plotted in Fig. 36 (Ref 286). The envelopes for closely packed ordered structures like $L1_2$ and $D0_{22}$ are located near the upper left corner, and the envelopes for refractory silicides with $D8_8$-type structures and beryllides with C_{14} and C_{15} ordered structures are close to the overall envelope line shown in Fig. 36. Because of their low density, high melting point, and excellent oxidation resistance, beryllides are promising materials for aerospace applications, and development activities in this area have increased recently (Ref 289-292).

A compilation and representation of data in terms of specific gravity, melting point, elastic modulus, and crystal structures provides a useful guide for preliminary ranking of intermetallics for various uses, particularly aerospace applications. However, other characteristics, including such structure-sensitive properties as low-temperature toughness and high-temperature strength and creep resistance, should also be considered when developing and selecting materials for specific applications.

Summary

Ordered intermetallic compounds based on aluminides and silicides constitute a unique class of metallic materials that have promising physical and mechanical properties for structural applications at elevated temperatures. The attractive properties of these compounds include excellent high-temperature strength, superior resistance to oxidation and corrosion, high melting points, and relatively low material density. However, major drawbacks of ordered intermetallics are their poor ductility and low fracture resistance at ambient temperatures.

For about the past ten years, substantial efforts have been devoted to research and development of intermetallics, and significant progress has been made in understanding their susceptibility to brittle fracture and in improving their ductility and toughness at both low and high temperatures. In a number of cases, significant tensile ductility has been achieved at ambient temperatures by controlling ordered crystal structures, increasing deformation modes, enhancing bulk and grain-boundary cohesive strengths, and controlling surface compositions and test environments. Success in these areas has inspired parallel efforts aimed at improving strength properties.

The alloy design work has been centered primarily on aluminides of nickel, iron, and titanium, and this work has resulted in substantial improvements in the mechanical and metallurgical properties of these materials at ambient and elevated temperatures. At the present time, the ductile aluminide alloys based on Ni_3Al, Fe_3Al, FeAl, and Ti_3Al compositions have been developed to the stage of being ready or close to ready for structural applications. Further research is required to develop a data base that includes information on tensile, creep, fracture toughness, low- and high-cycle fatigue, crack growth, and elastic modulus properties of these aluminides in specific applications.

The research and development work on nickel, iron, and titanium aluminides has recently extended to other aluminides and intermetallics such as Al_3X trialuminides and refractory silicides. Their susceptibility to brittle fracture and poor toughness remains the major factor limiting the potential structural uses of these materials. However, remarkable progress has been made in improving their low-temperature toughness and high-temperature strength. The recent development of new physical metallurgy tools for ordered intermetallics such as structural maps (Ref 225-239) and specific gravity versus melting point diagrams (Ref 286-288) is expected to accelerate the alloy development of these materials.

Ordered intermetallics with improved ductility and toughness have been considered for elevated-temperature structural applications that require high-temperature strength, low material density, and corrosion resistance. Titanium aluminides (Ti_3Al and TiAl) with high specific strengths have been developed for jet engine, aircraft, and related structural applications. Nickel aluminides based on Ni_3Al that have a combi-

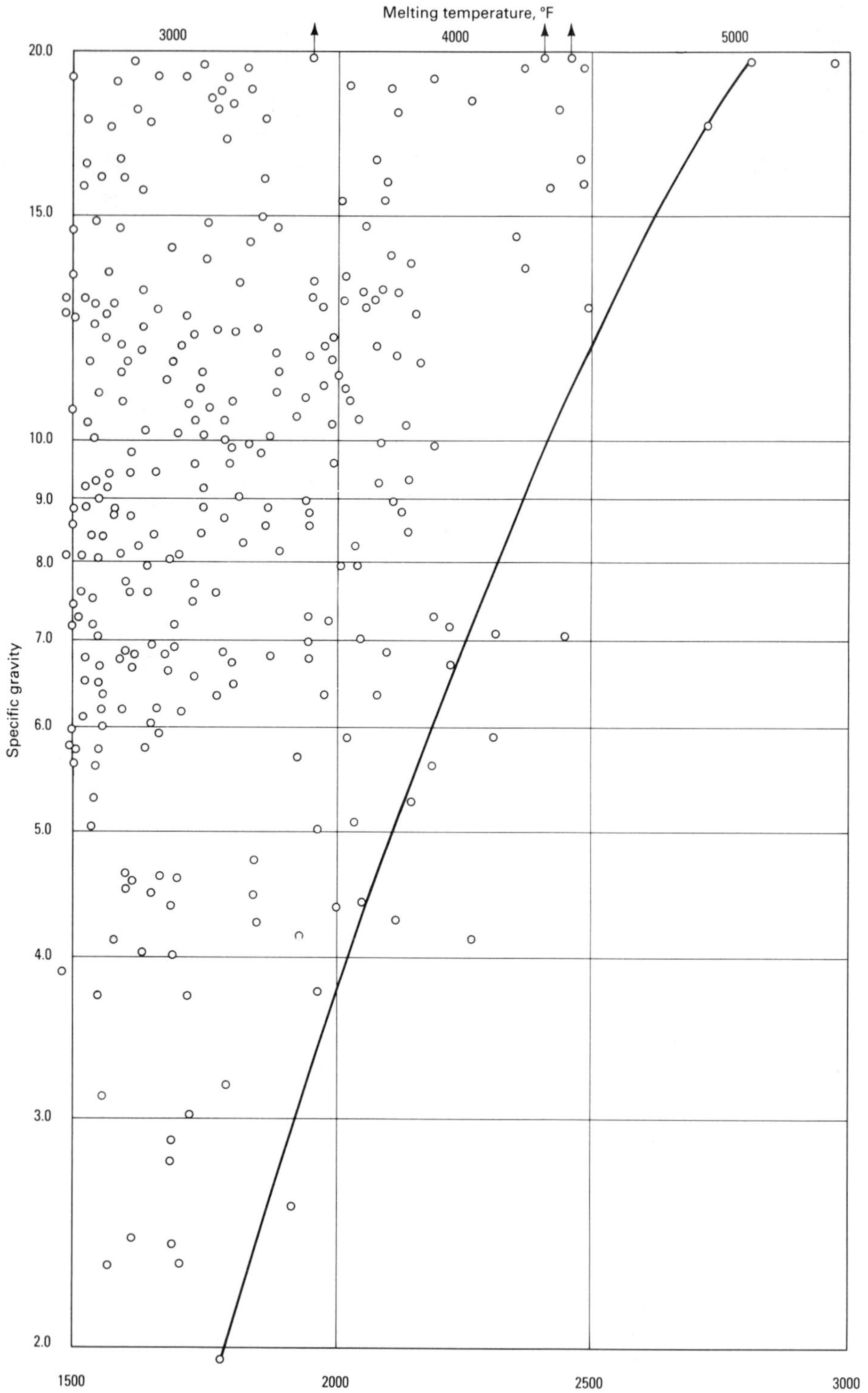

Fig. 36 Melting temperature versus specific gravity for 293 binary intermetallic compounds. The solid line is an empirical approximate envelope to the data. Source: Ref 286

nation of good strength, ductility, and oxidation resistance are currently being developed for use as high-temperature dies, heating elements, and hot components in heat engines and energy conversion systems. It has been found that Fe_3Al and FeAl aluminides possess excellent corrosion resistance in oxidizing, sulfidizing, and molten salt environments. As shown in Fig. 21, the iron aluminide alloys exhibit corrosion rates lower than those of the best existing iron-base alloys (including existing coating materials) by more than two orders of magnitude when tested in a severe sulfidizing environment at 800 °C (1470 °F) (Ref 126). A combination of low material cost and density with adequate ductility and fabricability makes iron aluminide alloys extremely attractive for structural applications in corrosive environments. Because of the superior corrosion resistance of aluminides and silicides, they should make a major impact on next-generation corrosion-resistant materials.

Ordered intermetallics have been employed in many areas besides structural applications. Molybdenum disilicide has been used commercially for electrical heating elements in high-temperature furnaces since 1956 (Ref 277). The $Fe_3(Si,Al)$ alloy with the trade name of Sendust has been developed for magnetic applications because of its superior magnetic properties as well as its wear and corrosion resistance (Ref 270-276). The NiTi alloy called Nitinol (Ref 293, 294) is currently the major material used as a shape memory alloy for systems control in the building, automobile, and automation industries (see the article "Shape Memory Alloys" in this Volume). Considerable efforts are now being devoted to the development of new shape memory alloys based on intermetallics for use at temperatures above ambient (>70 °C, or 160 °F).

ACKNOWLEDGMENT

The authors wish to thank E.P. George, C.G. McKamey, and M.H. Yoo for paper review and valuable discussions. The authors are grateful to Faye Christie, Connie Dowker, Susan Goetz, and Paula Bauer for manuscript preparation. This work is supported in part by the Division of Materials Sciences and Advanced Industrial Materials Program, U.S. Department of Energy, under contract DE-AC05-84OR21400 with Martin Marietta Energy Systems, Inc.

REFERENCES

1. N.S. Stoloff and R.G. Davies, *Prog. Mater. Sci.*, Vol 13 (No. 1), 1966, p 1
2. J.H. Westbrook, Ed., *Mechanical Properties of Intermetallic Compounds*, Wiley, 1959
3. J.H. Westbrook, Ed., *Intermetallic Compounds*, Wiley, 1967
4. B.H. Kear, C.T. Sims, N.S. Stoloff, and J.H. Westbrook, Ed., *Ordered Alloys—Structural Applications and Physical Metallurgy*, Claitor's Publishing, 1970
5. R.W. Guard and J.H. Westbrook, *Trans. AIME*, Vol 215, 1959, p 807-814
6. S.M. Copley and B.H. Kear, *Trans. Metall. Soc. AIME*, Vol 239, 1967, p 977
7. P.H. Thornton, R.G. Davies, and T.L. Johnston, *Metall. Trans.*, Vol 1, 1970,

p 207

8. D.P. Pope, *Philos. Mag.*, Vol 25, 1972, p 917
9. S. Takeuchi and E. Kuramoto, *Acta Metall.*, Vol 21, 1973, p 415
10. V. Paidar, D.P. Pope, and V. Vitek, *Acta Metall.*, Vol 32, 1984, p 435
11. M.H. Yoo, *Scr. Metall.*, Vol 20, 1986, p 915
12. M.H. Yoo, J.A. Horton, and C.T. Liu, *Acta Metall.*, Vol 36, 1988, p 2935
13. C.T. Liu, *Metall. Trans.*, Vol 4, 1973, p 1743
14. C.T. Liu and H. Inouye, *Metall. Trans. A*, Vol 10A, 1979, p 1515
15. C.T. Liu, *J. Nucl. Mater.*, Vol 85/86, 1979, p 907
16. C.T. Liu, *J. Nucl. Mater.*, Vol 104, 1982, p 1205
17. C.T. Liu, *Int. Metall. Rev.*, Vol 29, 1984, p 168
18. A. Aoki and O. Izumi, *Nippon Kinzoku Gakkaishi*, Vol 43, 1979, p 1190
19. C.T. Liu and C.C. Koch, *Trends in Critical Materials Requirements for Steels of the Future: Conservation and Substitution Technology for Chromium*, NBSIR-83-2679-2, National Bureau of Standards, 1983
20. C.T. Liu, C.L. White, and J.A. Horton, *Acta Metall.*, Vol 33, 1985, p 213-219
21. A.I. Taub, S.C. Huang, and K.M. Chang, *Metall. Trans. A*, Vol 15A, 1984, p 399
22. *High-Temperature Ordered Intermetallic Alloys*, Materials Research Society Symposia Proceedings, Vol 39, C.C. Koch, C.T. Liu, and N.S. Stoloff, Ed., Materials Research Society, 1985
23. *High-Temperature Ordered Intermetallic Alloys II*, Materials Research Society Symposia Proceedings, Vol 81, N.S. Stoloff, C.C. Koch, C.T. Liu, and O. Izumi, Ed., Materials Research Society, 1987
24. *High-Temperature Ordered Intermetallic Alloys III*, Materials Research Society Symposia Proceedings, Vol 133, C.T. Liu, A.I. Taub, N.S. Stoloff, and C.C. Koch, Ed., Materials Research Society, 1989
25. L.E. Tanner *et al.*, "Mechanical Behavior of Intermetallic Compounds," Report AST-TDR62-1087, Manlabs, Inc., 1963-1964, parts 1-3
26. H.A. Lipsitt, D. Schechtman, and R.E. Schafrik, *Metall. Trans. A*, Vol 11A, 1980, p 1369
27. K. Aoki and O. Izumi, *Acta Metall.*, Vol 27, 1979, p 807
28. C.T. Liu, E.H. Lee, and C.G. McKamey, *Scr. Metall.*, Vol 23, 1989, p 875
29. C.T. Liu, C.G. McKamey, and E.H. Lee, *Scr. Metall.*, in press
30. E.A. Aitken, *Intermetallic Compounds*, J.H. Westbrook, Ed., Wiley, 1967, p 491-516
31. A.J. Bradley and A. Taylor, *Proc. R. Soc. (London) A*, Vol 136, 1932, p 210
32. A.J. Bradley and A. Taylor, *Proc. R. Soc. (London) A*, Vol 159, 1937, p 56
33. N. Ridley, *J. Inst. Met.*, Vol 94, 1966, p 255
34. M.J. Cooper, *Philos. Mag.*, Vol 8, 1963, p 805
35. S. Ohiai, Y. Oya, and T. Suzuki, *Acta Metall.*, Vol 32, 1984, p 289-298
36. T.B. Massalski, Ed., *Binary Alloy Phase Diagrams*, Vol 1 and 2, American Society for Metals, 1986
37. K. Aoki and O. Izumi, *Trans. Jpn. Inst. Met.*, Vol 19, 1978, p 203
38. E.M. Grala, in *Mechanical Properties of Intermetallic Compounds*, J.H. Westbrook, Ed., Wiley, 1960, p 358-404
39. T. Takasugi, E.P. George, D.P. Pope, and O. Izumi, *Scr. Metall.*, Vol 19, 1985, p 551-556
40. T. Ogura, S. Hanada, T. Masumoto, and O. Izumi, *Metall. Trans. A*, Vol 16A, 1985, p 441-443
41. A.I. Taub, S.C. Huang, and K.M. Chang, Stoichiometry Effects on the Strengthening and Ductilization of Ni_3Al by Boron Modification and Rapid Solidification, in *Failure Mechanisms in High Performance Materials*, J.G. Early, T.R. Shives, and J.H. Smith, Ed., Cambridge University Press, 1985, p 57-65
42. J.A. Horton and M.K. Miller, *Acta Metall.*, Vol 35, 1987, p 133
43. M.K. Miller and J.A. Horton, *J. Phys.*, Vol C7, 1986, p 263
44. J.A. Horton and M.K. Miller, in *High-Temperature Ordered Intermetallic Alloys II*, Materials Research Society Symposia Proceedings, Vol 81, N.S. Stoloff, C.C. Koch, C.T. Liu, and O. Izumi, Ed., Materials Research Society, 1987, p 105-110
45. D.D. Sieloff, S.S. Brenner, and M.G. Burke, in *High-Temperature Ordered Intermetallic Alloys II*, Materials Research Society Symposia Proceedings, Vol 81, N.S. Stoloff, C.C. Koch, C.T. Liu, and O. Izumi, Ed., Materials Research Society, 1987, p 87-97
46. C.L. White, R.A. Padgett, C.T. Liu, and S.M. Yalisove, *Scr. Metall.*, Vol 18, 1984, p 1417-1420
47. G.S. Painter and F.W. Averill, *Phys. Rev. Lett.*, Vol 58, 1987, p 234
48. M.E. Eberhart and D.D. Vvedinsky, *Phys. Rev. Lett.*, Vol 58, 1987, p 61
49. S.P. Chen, A.F. Voter, and D.J. Srolovitz, *J. Phys.*, submitted for publication
50. G.M. Bond, I.M. Robertson, and H.K. Birnbaum, *J. Mater. Res.*, Vol 2, 1987, p 436-440
51. E.M. Schulson, T.P. Weihs, D.V. Viens, and I. Baker, *Acta Metall.*, Vol 33, 1985, p 1587
52. P.S. Khadkikar, K. Vedula, and B.S. Shale, *Metall. Trans. A*, Vol 18A, 1987, p 425
53. I. Baker, E.M. Schulson, and J.A. Horton, *Acta Metall.*, Vol 35, 1987, p 1533-1541
54. A.H. King and M.H. Yoo, in *High-Temperature Ordered Intermetallic Alloys II*, Materials Research Society Symposia Proceedings, Vol 81, N.S. Stoloff, C.C. Koch, C.T. Liu, and O. Izumi, Ed., Materials Research Society, 1987, p 99-104
55. I. Baker, E.M. Schulson, and J.R. Michael, *Philos. Mag.*, Vol B57, 1988, p 379
56. D.N. Sieloff, S.S. Brenner, and Hua Ming-Jian, in *High-Temperature Ordered Intermetallic Alloys III*, Materials Research Society Symposia Proceedings, Vol 133, C.T. Liu, A.I. Taub, N.S. Stoloff, and C.C. Koch, Materials Research Society, 1989, p 155-160
57. E.P. George, C.T. Liu, and R.A. Padgett, *Scr. Metall.*, Vol 23, 1989, p 979-982
58. R.A.D. Mackenzie and S.L. Sass, *Scr. Metall.*, Vol 22, 1988, p 1807
59. S.C. Huang, C.L. Briant, K.M. Chang, A.J. Taub, and E.L. Hall, *J. Mater. Res.*, Vol 1, 1986, p 60-67
60. T. Takasugi, O. Izumi, and N. Masahashi, *Acta Metall.*, Vol 33, 1985, p 1259
61. J.A. Horton, C.T. Liu, and M.L. Santella, *Metall. Trans. A*, Vol 18A, 1987, p 1265-1277
62. T. Takasugi, N. Masahashi, and O. Izumi, *Scr. Metall.*, Vol 20, 1986, p 1317
63. T. Takasugi and O. Izumi, *Acta Metall.*, Vol 33, 1985, p 1247-1258
64. A.I. Taub, C.L. Briant, S.C. Huang, K.M. Chang, and M.R. Jackson, *Scr. Metall.*, Vol 20, 1986, p 129-134
65. A.I. Taub and C.L. Briant, in *High-Temperature Ordered Intermetallic Alloys II*, Materials Research Society Symposia Proceedings, Vol 81, N.S. Stoloff, C.C. Koch, C.T. Liu, and O. Izumi, Ed., Materials Research Society, 1987, p 343-353
66. A.I. Taub and C.L. Briant, *Acta Metall.*, Vol 35, 1987, p 1597-1603
67. C.T. Liu, C.L. White, and E.H. Lee, *Scr. Metall.*, Vol 19, 1985, p 1247-1250
68. C.T. Liu and C.L. White, *Acta Metall.*, Vol 35, 1987, p 643
69. A.I. Taub, K.M. Chang, and C.T. Liu, *Scr. Metall.*, Vol 20, 1986, p 1613
70. C.A. Hippsley and J.H. DeVan, *Acta Metall.*, Vol 37, 1989, p 1485-1496
71. C.T. Liu and V.K. Sikka, *J. Met.*, Vol 38, 1986, p 19-21
72. C.T. Liu, in *Micon 86*, American So-

ciety for Testing and Materials, 1988, p 222-237

73. J.A. Horton, J.V. Cathcart, and C.T. Liu, *Oxid. Met.*, Vol 29, 1988, p 347-365
74. C.T. Liu and B.F. Oliver, *J. Mater. Res.*, Vol 4, 1989, p 294-299
75. C.T. Liu and C.L. White, in *High-Temperature Ordered Intermetallic Alloys*, Materials Research Society Symposia Proceedings, Vol 39, C.C. Koch, C.T. Liu, and N.S. Stoloff, Ed., Materials Research Society, 1985, p 365-380
76. B.H. Kear and H.G.F. Wilsdorf, *Trans. AIME*, Vol 224, 1962, p 383
77. B.H. Kear, *Acta Metall.*, Vol 12, 1964, p 555
78. R.D. Rawlings and A. Staton-BeVan, *J. Mater. Sci.*, Vol 10, 1975, p 505-514
79. L.R. Curwick, Ph.D. dissertation, University of Minnesota, 1972
80. D.P. Pope and C.T. Liu, in *Superalloys, Supercomposites and Superceramics*, J.K. Tien and T. Caulfield, Ed., Academic Press, 1989, p 584-624
81. R.L. Fleischer, *Acta Metall.*, Vol 11, 1963, p 203
82. Y. Mishima, S. Ochiai, and T. Suzuki, *Acta Metall.*, Vol 33, 1985, p 1161
83. D.M. Dimiduk, Ph.D. dissertation, Carnegie-Mellon University, 1989
84. C.T. Liu, V.K. Sikka, J.A. Horton, and E.H. Lee, "Alloy Development and Mechanical Properties of Nickel Aluminides (Ni_3Al) Alloys," ORNL-6483, Oak Ridge National Laboratory, 1988
85. D.L. Anton, D.D. Pearson, and D.B. Snow, in *High-Temperature Ordered Intermetallic Alloys II*, Materials Research Society Symposia Proceedings, Vol 81, N.S. Stoloff, C.C. Koch, C.T. Liu, and O. Izumi, Ed., Materials Research Society, 1987, p 287-295
86. G.E. Fuchs, A.K. Kuruvilla, and N.S. Stoloff, private communication, 1989
87. N.S. Stoloff, G.E. Fuchs, A.K. Kuruvilla and S.J. Choe, in *Mechanical Properties of Intermetallic Compounds*, J.H. Westbrook, Ed., Wiley, 1959, p 247-260
88. G.M. Camus, D.J. Duquette, and N.S. Stoloff, in *High-Temperature Ordered Intermetallic Alloys III*, Materials Research Society Symposia Proceedings, Vol 133, C.T. Liu, A.I. Taub, N.S. Stoloff, and C.C. Koch, Ed., Materials Research Society, 1989, p 579-584
89. R.S. Bellows, E.A. Schwarkopf, and J.K. Tien, *Metall. Trans. A*, Vol 19A, 1988, p 479-486
90. R.S. Bellows and J.K. Tien, *Scr. Metall.*, Vol 21, 1987, p 1659-1662
91. M.L. Santella, S.A. David, and C.L. White, in *High-Temperature Ordered Intermetallic Alloys*, Materials Research Society Symposia Proceedings, Vol 39, C.C. Koch, C.T. Liu, and N.S. Stoloff, Ed., Materials Research Society, 1985, p 495-503
92. S.A. David, W.A. Jemian, C.T. Liu, and J.A. Horton, *Weld. Res. Suppl.*, Jan 1985, p 22s-28s
93. M.L. Santella and S.A. David, *Weld. Res. Suppl.*, May 1986, p 129s-137s
94. V.K. Sikka, in *High-Temperature Ordered Intermetallic Alloys III*, Materials Research Society Symposia Proceedings, Vol 133, C.T. Liu, A.I. Taub, N.S. Stoloff, and C.C. Koch, Ed., Materials Research Society, 1989, p 487-492
95. G.L. Povirk, J.A. Horton, C.G. McKamey, T.N. Tiegs, and S.R. Nutt, *J. Mater. Sci.*, Vol 23, 1988, p 3945-3950
96. J.M. Yang, W.H. Kao, and C.T. Liu, *Metall. Trans. A*, Vol 20A, 1989, p 2459-2469
97. G.E. Fuchs, in *High-Temperature Ordered Intermetallic Alloys III*, Materials Research Society Symposia Proceedings, Vol 133, C.T. Liu, A.I. Taub, N.S. Stoloff, and C.C. Koch, Ed., Materials Research Society, 1989, p 615-620
98. S. Nourbakhsh, F.L. Liang, and H. Margolin, *J. Phys. E, Sci. Instrum.*, Vol 21, 1988, p 898
99. S. Nourbakhsh, F.L. Liang, and H. Margolin, in *High-Temperature Ordered Intermetallic Alloys III*, Materials Research Society Symposia Proceedings, Vol 133, C.T. Liu, A.I. Taub, N.S. Stoloff, and C.C. Koch, Ed., Materials Research Society, 1989, p 459-464
100. J.L. Smialek, *Metall. Trans. A.*, Vol 9A, 1978, p 309
101. J. Jedlinski and S. Miowic, *Mater. Sci. Eng.*, Vol 87, 1987, p 281
102. C.A. Barrett, *Oxid. Met.*, Vol 30, 1988, p 361
103. A. Ball and R.E. Smallman, *Acta Metall.*, Vol 14, 1966, p 1517
104. N.J. Zaluzec and H.L. Fraser, *Scr. Metall.*, Vol 8, 1974, p 1049
105. I. Baker and E.M. Schulson, *Metall. Trans. A*, Vol 15A, 1984, p 1129
106. K.H. Hahn and K. Vedula, *Scr. Metall.*, Vol 23, 1989, p 7
107. J.H. Westbrook, *J. Electrochem. Soc.*, Vol 103, 1956, p 54
108. C.T. Liu, Oak Ridge National Laboratory, unpublished research, 1989
109. K. Vedula, V. Pathare, I. Aslamidis, and R.H. Titran, in *High-Temperature Ordered Intermetallic Alloys*, Materials Research Society Symposia Proceedings, Vol 39, C.C. Koch, C.T. Liu, and N.S. Stoloff, Ed., Materials Research Society, 1985, p 411-421
110. I. Jung, M. Rudy, and G. Sauthoff, in *High-Temperature Ordered Intermetallic Alloys II*, Materials Research Society Symposia Proceedings, Vol 81, N.S. Stoloff, C.C. Koch, C.T. Liu, and O. Izumi, Ed., Materials Research Society, 1987, p 263-274
111. P.R. Strutt and B.H. Kear, in *High-Temperature Ordered Intermetallic Alloys*, Materials Research Society Symposia Proceedings, Vol 39, C.C. Koch, C.T. Liu, and N.S. Stoloff, Ed., Materials Research Society, 1985, p 279-292
112. E.M. Schulson and D.R. Barker, *Scr. Metall.*, Vol 17, 1983, p 519
113. A.G. Rozner and R.J. Wasilewski, *J. Inst. Met.*, Vol 94, 1966, p 169
114. E.P. George and C.T. Liu, *J. Mater. Res.*, Vol 5, 1990, p 754
115. D.B. Miracle, S. Russell, and C.C. Law, in *High-Temperature Ordered Intermetallic Alloys III*, Materials Research Society Symposia Proceedings, Vol 133, C.T. Liu, A.I. Taub, N.S. Stoloff, and C.C. Koch, Ed., Materials Research Society, 1989, p 225-230
116. R. Darolia, D.F. Lahrman, R.D. Field, and A.J. Freeman, in *High-Temperature Ordered Intermetallic Alloys III*, Materials Research Society Symposia Proceedings, Vol 133, C.T. Liu, A.I. Taub, N.S. Stoloff, and C.C. Koch, Ed., Materials Research Society, 1989, p 113-118
117. S.M. Russell, C.C. Law, and M.J. Blackburn, in *High-Temperature Ordered Intermetallic Alloys III*, Materials Research Society Symposia Proceedings, Vol 133, C.T. Liu, A.I. Taub, N.S. Stoloff, and C.C. Koch, Ed., Materials Research Society, 1989, p 627-632
118. S. Guha, P. Munroe, and I. Baker, *Scr. Metall.*, Vol 23, 1989, p 897-900
119. A. Inoue, T. Masumoto, and H. Tomioka, *J. Mater. Sci.*, Vol 19, 1984, p 3097
120. S. Guha, P.R. Munroe, and I. Baker, in *High-Temperature Ordered Intermetallic Alloys III*, Materials Research Society Symposia Proceedings, Vol 133, C.T. Liu, A.I. Taub, N.S. Stoloff, and C.C. Koch, Ed., Materials Research Society, 1989, p 633-638
121. R.D. Field, D.D. Krueger, and S.C. Huang, in *High-Temperature Ordered Intermetallic Alloys III*, Materials Research Society Symposia Proceedings, Vol 133, C.T. Liu, A.I. Taub, N.S. Stoloff, and C.C. Koch, Ed., Materials Research Society, 1989, p 567-572
122. A.R.C. Westwood, *Metall. Trans. A*, Vol 19A, 1988, p 749
123. B. Schmidt, P. Nagpal, and I. Baker, in *High-Temperature Ordered Intermetallic Alloys III*, Materials Research Society Symposia Proceedings, Vol 133, C.T. Liu, A.I. Taub, N.S. Stoloff, and C.C. Koch, Ed., Materials Research Society, 1989, p 755-760

124. J.L. Smialek, J. Doychak, and D.J. Gaydosh, "Oxidation Behavior of FeAl + Hf,Zr,B," NASA TM-101402, NASA Lewis Research Center, 1988
125. C.G. McKamey *et al.*, "Evaluation of Mechanical and Metallurgical Properties of Fe_3Al-Based Aluminides," ORNL/TM-10125, Oak Ridge National Laboratory, Sept 1986
126. J.H. DeVan, in *Oxidation of High-Temperature Intermetallics*, T. Grobstain and J. Doythak, Ed., TMS, 1989
127. I. Baker and D.J. Gaydosh, *Mater. Sci. Eng.*, Vol 96, 1987, p 147
128. M.G. Mendiratta, S.K. Ehlers, and D.K. Chatterjee, in *Rapid Solidification Processing: Principles and Technologies*, National Bureau of Standards, 1983, p 420
129. J.A. Horton, C.T. Liu, and C.C. Koch, in *High-Temperature Alloys: Theory and Design*, J.O. Stiegler, Ed., American Institute of Mining, Metallurgical, and Petroleum Engineers, 1984, p 309-321
130. D.J. Gaydosh, S.L. Draper, and M.V. Nathal, *Metall. Trans. A*, Vol 20A, 1989, p 1701
131. C.G. McKamey, J.A. Horton, and C.T. Liu, in *High-Temperature Ordered Intermetallic Alloys II*, Materials Research Society Symposia Proceedings, Vol 81, N.S. Stoloff, C.C. Koch, C.T. Liu, and O. Izumi, Ed., Materials Research Society, 1987, p 321-327
132. W.R. Kerr, *Metall. Trans. A*, Vol 17A, 1986, p 2298
133. H. Inouye, in *High-Temperature Ordered Intermetallic Alloys*, Materials Research Society Symposia Proceedings, Vol 39, C.C. Koch, C.T. Liu, and N.S. Stoloff, Ed., Materials Research Society, 1985, p 255-261
134. G. Sainfort, P. Mouturat, P. Pepin, J. Petit, G. Cabane, and M. Salesse, *Mem. Étud. Sci. Rev. Métall.*, Vol 60, 1963, p 125
135. P. Morgnand, P. Mouturat, and G. Sainfort, *Acta Metall.*, Vol 16, 1968, p 807
136. S. Strothers and K. Vendula, in *Proceedings of the Powder Metallurgy Conference*, Vol 43, Metal Powder Industries Federation, 1987, p 597
137. M.G. Mendiratta, H.K. Kim, and H.A. Lipsitt, *Metall. Trans. A*, Vol 15A, 1984, p 395
138. Y. Umakoshi and M. Yamaguchi, *Philos. Mag. A*, Vol 44, 1981, p 711
139. T. Takasugi and O. Izumi, *Acta Metall.*, Vol 34, 1986, p 607
140. N. Masahashi, T. Takasugi, and O. Izumi, *Metall Trans. A*, Vol 19A, 1988, p 353
141. O. Izumi and T. Takasugi, *J. Mater. Res.*, Vol 3, 1988, p 426
142. T. Takasugi, N. Masahashi, and O. Izumi, *Scr. Metall.*, Vol 20, 1986, p 1317
143. N. Masahashi, T. Takasugi, and O. Izumi, *Acta Metall.*, Vol 36, 1988, p 1823-1836
144. T. Takasugi and O. Izumi, *Scr. Metall.*, Vol 19, 1985, p 903-907
145. A.K. Kuruvilla, S. Ashok, and N.S. Stoloff, in *Proceedings of the Third International Congress on Hydrogen in Metals*, Vol 2, Pergamon Press, 1982, p 629
146. A.K. Kuruvilla and N.S. Stoloff, *Scr. Metall.*, Vol 19, 1985, p 83
147. G.M. Camus, N.S. Stoloff, and D.J. Duquette, *Acta Metall.*, Vol 37, 1989, p 1497-1501
148. R.G. Bordeau, "Development of Iron Aluminides," AFWAL-TR-87-4009, United Technologies Corporation, Pratt and Whitney, 1987
149. M.G. Mendiratta, Tai-Il Mah, and S.K. Ehlers, "Mechanisms of Ductility and Fracture in Complex High-Temperature Materials," AFWAL-TR-85-4061, Materials Laboratory, U.S. Air Force Wright Aeronautic Laboratories, Airforce Systems Command, July 1985
150. J.O. Stiegler and C.T. Liu, in *Advances in Materials Science and Engineering*, R.W. Cahn, Ed., Pergamon Press, 1988, p 3-9
151. R.S. Diehm and D.E. Mikkola, in *High-Temperature Ordered Intermetallic Alloys II*, Materials Research Society Symposia Proceedings, Vol 81, N.S. Stoloff, C.C. Koch, C.T. Liu, and O. Izumi, Ed., Materials Research Society, 1987, p 329-334
152. C.G. McKamey and J.A. Horton, *Metall. Trans. A*, Vol 20A, 1989, p 751-757
153. D.M. Dimiduk, M.G. Mendiratta, D. Banerjee, and H.A. Lipsitt, *Acta Metall.*, Vol 36, 1988, p 2947-2958
154. C.G. McKamey, J.A. Horton, and C.T. Liu, *Scr. Metall.*, Vol 22, 1988, p 1679
155. C.G. McKamey, J.A. Horton, and C.T. Liu, *J. Mater. Res.*, Vol 4, 1989, p 1156-1163
156. M.G. Mendiratta, S.K. Ehlers, D.M. Dimiduk, W.R. Kerr, S. Mazdiyasni, and H.R. Lipsitt, in *High-Temperature Ordered Intermetallic Alloys II*, Materials Research Society Symposia Proceedings, Vol 81, N.S. Stoloff, C.C. Koch, C.T. Liu, and O. Izumi, Ed., Materials Research Society, 1987, p 393-404
157. C.G. McKamey *et al.*, "Development of Iron Aluminides for Gasification Systems," ORNL-TM-10793, Oak Ridge National Laboratory, July 1988
158. M.A. Crimp and K. Vedula, *J. Mater. Sci.*, Vol 78, 1986, p 193
159. K. Vedula and J.R. Stephens, in *High-Temperature Ordered Intermetallic Alloys II*, Materials Research Society Symposia Proceedings, Vol 81, N.S. Stoloff, C.C. Koch, C.T. Liu, and O. Izumi, Ed., Materials Research Society, 1987, p 381-391
160. S.A. David *et al.*, *Weld. J.*, Vol 68, 1989, p 372s-381s
161. F.H. Froes, D. Eylon, and H.B. Bomberger, Ed., *Titanium Technology: Present Status and Future Trends*, Titanium Development Association, 1985
162. P.J. Bania, An Advanced Alloy for Elevated Temperatures, *J. Met.*, Vol 40 (No. 3), 1988, p 20-22
163. H.H. Lipsitt, in *High-Temperature Ordered Intermetallic Alloys*, Materials Research Society Symposia Proceedings, Vol 39, C.C. Koch, C.T. Liu, and N.S. Stoloff, Ed., Materials Research Society, 1985, p 351-364
164. M.J. Blackburn and M.P. Smith, "Research to Conduct an Exploratory Experimental and Analytical Investigation of Alloys," Technical Report AFWAL-TR-80-4175, U.S. Air Force Wright Aeronautical Laboratories, 1980
165. M.J. Blackburn and M.P. Smith, "R&D on Composition and Processing of Titanium Aluminide Alloys for Turbine Engine," Technical Report AFWAL-TR-82-4086, U.S. Air Force Wright Aeronautical Laboratories, 1982
166. Y.-W. Kim and F.H. Froes, in *Proceedings of the Symposium on High-Temperature Aluminides and Intermetallics*, TMS, in press
167. F.H. Froes, *Mater. Edge*, No. 5, May 1988
168. Eli F. Bradley, "The Potential Structural Use of Aluminides in Jet Engines," Paper presented at the Gorham Advanced Materials Institute Conference on Investment, Licensing and Strategic Partnering Opportunities, Emerging Technology, Applications, and Markets for Aluminides, Iron, Nickel and Titanium (Monterrey, CA), Nov 1990
169. R.E. Schafrik, Dynamic Elastic Moduli of the Titanium Aluminides, *Metall. Trans. A*, Vol 8A, 1977, p 1003-1006
170. H.A. Lipsitt, in *Advanced High Temperature Alloys: Processing and Properties*, S.S. Allen, R.M. Pellous, and R. Widmer, Ed., American Society for Metals, 1986
171. N.S. Choudhury, H.C. Graham, and J.W. Hinze, in *Properties of High Temperature Alloys With Emphasis on Environmental Effects*, Electrochemical Society, 1976, p 668-680
172. M. Khobaib and F.W. Vahldiek, in *Space Age Metals Technology*, Vol 2, F.H. Froes and R.A. Cull, Ed., Soci-

ety for the Advancement of Material and Process Engineering, 1988, p 262-270

173. J. Subrahmanyam, Cyclic Oxidation of Aluminated Ti-14Al-24Nb Alloy, *J. Mater. Sci.*, Vol 23, 1988, p 1906-1910
174. F.H. Froes, *Mater. Edge*, May/June 1989, p 17
175. W.J.S. Yang, Observations of Superdislocation Networks in Ti_3Al-Nb, *J. Mater. Sci. Lett.*, Vol 1, 1982, p 199-202
176. W.J.S. Yang, "C" Component Dislocations in Deformed Ti_3Al, *Metall. Trans. A*, Vol 13A, 1982, p 324
177. M.J. Blackburn, D.L. Ruckle, and C.E. Bevau, "Research to Conduct an Exploratory Experimental and Analytical Investigation of Alloys," Technical Report AFML-TR-78-18, U.S. Air Force Materials Laboratory, 1978
178. M.J. Blackburn and M.P. Smith, "Research to Conduct an Exploratory Experimental and Analytical Investigation of Alloys," Technical Report AFML-TR-81-4046, U.S. Air Force Wright Aeronautical Laboratories, 1981
179. M.J. Blackburn and M.P. Smith, "Development of Improved Toughness Alloys Based on Titanium Aluminides," Interim Technical Report FR-19139, United Technologies, 1988
180. H. Bohm and K. Lohberg, Uber eine Uberstrukturphase vom CsCl-Typ im System Titan-Molybdan-Aluminum, *Z. Metallk.*, Vol 49, 1958, p 173-178
181. T.J. Jewett *et al.*, in *High-Temperature Ordered Intermetallic Alloys III*, Materials Research Society Symposia Proceedings, Vol 133, C.T. Liu, A.I. Taub, N.S. Stoloff, and C.C. Koch, Ed., Materials Research Society, 1989, p 69-74
182. M.J. Kaufman *et al.*, in *Sixth World Conference on Titanium*, Part II, P. Lacombe *et al.*, Ed., Les Editions de Physique, 1989, p 985-990
183. R.G. Rowe, *High Temperature Aluminides and Intermetallics*, S.H. Whang, C.T. Liu, and D. Pope, Ed., TMS, 1990
184. R. Strychor, J.C. Williams, and W.A. Soffa, Phase Transformations and Modulated Microstructures in Ti-Al-Nb Alloys, *Metall. Trans. A*, Vol 19A (No. 2), 1988, p 225-234
185. S.M.L. Sastry and H.A. Lipsitt, Ordering Transformations and Mechanical Properties of Ti_3Al and Ti_3Al-Nb Alloys, *Metall. Trans. A*, Vol 8A, 1977, p 1543
186. W.A. Baeslack III, M.J. Cieslak, and T.J. Headley, Structure, Properties and Fracture of Pulsed Nd:YAG Laser Welded Ti-14.8 wt% Al-21.3 wt% Nb Titanium Aluminide, *Scr. Metall.*, Vol 22, 1988, p 1155-1160
187. J.C. Williams, in *Titanium Technology: Present Status and Future Trends*, F.H. Froes, D. Eylon, and H.B. Bomberger, Titanium Development Association, 1985, p 75-86
188. A.G. Jackson, K. Teal, and F.H. Froes, in *High-Temperature Ordered Intermetallic Alloys II*, Materials Research Society Symposia Proceedings, Vol 81, N.S. Stoloff, C.C. Koch, C.T. Liu, and O. Izumi, Ed., Materials Research Society, 1987, p 143-149
189. F.H. Froes and D. Eylon, *Hydrogen Effects on Materials Behavior*, A.W. Thompson and N.R. Moody, Ed., TMS, 1990
190. F.H. Froes, D. Eylon, and C. Suryanarayana, *J. Met.*, March 1990
191. W.H. Kao *et al.*, in *Progress in Powder Metallurgy*, Vol 37, Metal Powder Industries Federation, 1982, p 289-301
192. C.H. Ward *et al.*, in *Sixth World Conference on Titanium*, Part II, P. Lacombe, R. Tricot, and G. Beranger, Ed., Les Editions de Physique, 1989, p 1009-1014
193. C.H. Ward *et al.*, in *Sixth World Conference on Titanium*, Part II, P. Lacombe, R. Tricot, and G. Beranger, Ed., Les Editions de Physique, 1989, p 1103-1108
194. R.W. Hertzberg, *Deformation and Fracture Mechanics of Engineering Materials*, 2nd ed., John Wiley & Sons, 1983
195. M.A. Stucke and H.A. Lipsitt, in *Titanium Rapid Solidification Technology*, F.H. Froes and D. Eylon, Ed., TMS, 1986, p 255-262
196. W. Cho, "Effect of Microstructure on Deformation and Creep Behavior of Ti-25Al-10Nb-3V-1Mo," Technical Report, U.S. Air Force Office of Scientific Research, Oct 1988
197. C.G. Rhodes, in *Sixth World Conference on Titanium*, Part I, P. Lacombe, R. Tricot, and G. Beranger, Ed., Les Editions de Physique, 1989, p 119-204
198. M.G. Mendiratta and H.A. Lipsitt, Steady-State Creep Behavior of Ti_3Al-Base Intermetallics, *J. Mater. Sci.*, Vol 15, 1980, p 2985-2990
199. H.R. Ogden *et al.*, Constitution of Titanium-Aluminum Alloys, *Trans. AIME*, Vol 191, 1951, p 1150-1155
200. D. Clark, K.S. Kepson, and G.I. Lewis, A Study of the Titanium-Aluminum System up to 40 at.% Aluminum, *J. Inst. Met.*, Vol 91, 1962-1963, p 197
201. Y.-W. Kim, Intermetallic Alloys Based on Gamma Titanium Aluminide, *J. Met.*, Vol 41 (No. 7), 1989, p 24-30
202. E.S. Bumps, H.D. Kessler, and M. Hansen, Titanium-Aluminum System, *Trans. AIME*, Vol 194, 1952, p 609-614
203. P. Duwez and J.L. Taylor, Crystal Structure of TiAl, *J. Met.*, 1952, p 70
204. S.C. Huang, E.L. Hall, and M.F.X. Gigliotti, in *High-Temperature Ordered Intermetallic Alloys II*, Materials Research Society Symposia Proceedings, Vol 81, N.S. Stoloff, C.C. Koch, C.T. Liu, and O. Izumi, Ed., Materials Research Society, 1987, p 481-486
205. R.P. Elliott and W. Rostoker, The Influence of Aluminum on the Occupation of Lattice Sites in the TiAl Phase, *Acta Metall.*, Vol 2, 1954, p 884-885
206. G. Hug, A. Loiseau, and A. Lasalmonie, Nature and Dissociation of the Dislocations in TiAl Deformed at Room Temperature, *Philos. Mag. A*, Vol 54 (No. 1), 1986, p 47-65
207. T. Kawabata and O. Izumi, Dislocation Structures in TiAl Single Crystals Deformed at 77K, *Scr. Metall.*, Vol 21, 1987, p 433-434
208. G. Hug, A. Loiseau, and P. Veyssiere, Weak-Beam Observation of a Dissociation Transition in TiAl, *Philos. Mag. A*, Vol 57 (No. 3), 1988, p 499-523
209. D.W. Pashley, J.L. Robertson, and M.J. Stowell, The Deformation of Cu Au I, *Philos. Mag. A*, 8th series, Vol 19, 1969, p 83
210. D. Schechtman, M.J. Blackburn, and H.A. Lipsitt, The Plastic Deformation of TiAl, *Metall. Trans.*, Vol 5, 1974, p 1373
211. H.A. Lipsitt, D. Schechtman, and R.E. Schafrik, The Deformation and Fracture of TiAl at Elevated Temperatures, *Metall. Trans. A*, Vol 6A, 1975, p 1991
212. E.L. Hall and S.-C. Huang, in *High-Temperature Ordered Intermetallic Alloys III*, Materials Research Society Symposia Proceedings, Vol 133, C.T. Liu, A.I. Taub, N.S. Stoloff, and C.C. Koch, Ed., Materials Research Society, 1989, p 693-698
213. T. Kawabata and O. Izumi, Dislocation Reactions and Fracture Mechanism in TiAl $L1_0$ Type Intermetallic Compound, *Scr. Metall.*, Vol 21, 1987, p 435-440
214. T. Kawabata *et al.*, Bend Tests and Fracture Mechanisms of TiAl Single Crystals at 293-1083 K, *Acta Metall.*, Vol 36 (No. 4), 1988, p 963-975
215. S.M.L. Sastry and H.A. Lipsitt, Fatigue Deformation of TiAl Base Alloys, *Metall. Trans. A*, Vol 8A, 1977, p 299
216. M.J. Blackburn and M.P. Smith, Titanium Alloys of the TiAl Type, U.S. Patent 4,294,615, 1981
217. M.J. Blackburn, J.T. Hill, and M.P. Smith, "R&D on Composition and Processing of Titanium Aluminide Alloys for Turbine Engines," Technical Report AFWAL-TR-84-4078, U.S. Air

Force Wright Aeronautical Laboratories, 1984
218. S.-C. Huang and E.L. Hall, in *High-Temperature Ordered Intermetallic Alloys III*, Materials Research Society Symposia Proceedings, Vol 133, C.T. Liu, A.I. Taub, N.S. Stoloff, and C.C. Koch, Ed., Materials Research Society, 1989, p 373-383
219. T. Tsujimoto and K. Hashimoto, in *High-Temperature Ordered Intermetallic Alloys III*, Materials Research Society Symposia Proceedings, Vol 133, C.T. Liu, A.I. Taub, N.S. Stoloff, and C.C. Koch, Ed., Materials Research Society, 1989, p 391-396
220. T. Kawabata, T. Tamura, and O. Izumi, in *High-Temperature Ordered Intermetallic Alloys III*, Materials Research Society Symposia Proceedings, Vol 133, C.T. Liu, A.I. Taub, N.S. Stoloff, and C.C. Koch, Ed., Materials Research Society, 1989, p 329-334
221. D.J. Maykuth, "Effects of Alloying Elements in Titanium," DMIC Report 136B, Battelle Memorial Institute, May 1961
222. I.A. Zelonkov and Y.N. Martynchik, Oxidation Resistance of Alloys of Compound TiAl with Niobium at 800 and 1000C, *Metallofiz., Nauk. Dumka*, Vol 42, 1972, p 63-66
223. C.R. Feng, D.J. Michel, and C.R. Crowe, in *High-Temperature Ordered Intermetallic Alloys III*, Materials Research Society Symposia Proceedings, Vol 133, C.T. Liu, A.I. Taub, N.S. Stoloff, and C.C. Koch, Ed., Materials Research Society, 1989, p 669-674
224. H.R. Ogden *et al.*, Mechanical Properties of High Purity Ti-Al Alloys, *J. Met.*, Feb 1952
225. M.J. Blackburn and M.P. Smith, "The Understanding and Exploitation of Alloys Based on the Compound TiAl (Gamma Phase)," Technical Report AFML-TR-79-4056, U.S. Air Force Materials Laboratory, 1979
226. T. Tsujimoto *et al.*, Structures and Properties of an Intermetallic Compound TiAl Based Alloys Containing Silver, *Trans. Jpn. Inst. Met.*, Vol 27 (No. 5), 1986, p 341-350
227. S.-C. Huang, E.L. Hall, and M.F.X. Gigliotti, in *Sixth World Conference on Titanium*, Part II, P. Lacombe, R. Tricot, and G. Beranger, Ed., Les Editions de Physique, 1989, p 1109-1114
228. S.M. Barinov *et al.*, Temperature Dependence of Strength and Ductility of the Decomposition of Titanium Aluminide, *Izv. Akad. Nauk SSSR*, Vol 5, 1983, p 170-174
229. A.K. Sinha, *Trans. Metall. Soc. AIME*, Vol 245, 1969, p 911
230. J.H.N. VanVucht, *J. Less-Common Met.*, Vol 11, 1966, p 308
231. P.A. Beck, *Adv. X-Ray Anal.*, Vol 12, 1969, p 1
232. P. Villars, *J. Less-Common Met.*, Vol 92, 1983, p 215-238
233. P. Villars, *J. Less-Common Met.*, Vol 99, 1984, p 33-43
234. P. Villars, *J. Less-Common Met.*, Vol 102, 1985, p 199-211
235. D.G. Pettifor, *Mater. Sci. Technol.*, Vol 4, 1988, p 675
236. E. Mooser and W.B. Pearson, *Acta Crystallogr.*, Vol 12, 1959, p 1015-1022
237. D.G. Pettifor, *New Sci.*, Vol 110 (No. 1510), 1986, p 48-53
238. D.G. Pettifor, *J. Phys. C, Solid State Phys.*, Vol 19, 1986, p 285-313
239. D.G. Pettifor and R. Podloucky, *Phys. Rev. Lett.*, Vol 55, 1985, p 261
240. K.S. Kumar and J.R. Pickens, *Scr. Metall.*, Vol 22, 1988, p 1015
241. E.P. George, W.D. Porter, H.M. Henson, W.C. Oliver, and B.F. Oliver, *J. Mater. Res.*, Vol 4, 1989, p 78
242. E.P. George, W.D. Porter, and D.C. Joy, in *High-Temperature Ordered Intermetallic Alloys III*, Materials Research Society Symposia Proceedings, Vol 133, C.T. Liu, A.I. Taub, N.S. Stoloff, and C.C. Koch, Ed., Materials Research Society, 1989, p 311-315
243. J.H. Schneibel and W.D. Porter, in *High-Temperature Ordered Intermetallic Alloys III*, Materials Research Society Symposia Proceedings, Vol 133, C.T. Liu, A.I. Taub, N.S. Stoloff, and C.C. Koch, Ed., Materials Research Society, 1989, p 335-340
244. C.D. Turner, W.O. Powers, and J.A. Wert, *Acta Metall.*, Vol 37, 1989, p 2635-2644
245. J.H. Schneibel, P.F. Becher, and J.A. Horton, *J. Mater. Res.*, Vol 3, 1988, p 1272
246. P.R. Subramanian, J.P. Simons, M.G. Mendiratta, and D.M. Dimiduk, in *High-Temperature Ordered Intermetallic Alloys III*, Materials Research Society Symposia Proceedings, Vol 133, C.T. Liu, A.I. Taub, N.S. Stoloff, and C.C. Koch, Ed., Materials Research Society, 1989, p 51-56
247. K.S. Kumar, "Review: Ternary Intermetallics in Al-Refractory Metal-X (X = V, Cr, Mn, Fe, Co, Ni, Cu, Zn) Systems," MML-JL89-46, Martin Marietta Labs, April 1989
248. S. Zhang, J.P. Nic, and D.E. Mikkola, *Scr. Metall.*, to be published
249. E.P. George *et al.*, *J. Mater. Res.*, to be published
250. C.L. Fu, *J. Mater. Res.*, to be published
251. N. Masahashi, T. Takasugi, and O. Izumi, *Acta Metall.*, Vol 36, 1988, p 1815-1822
252. E.M. Schulson, *Int. Met. Rev.*, Vol 29, 1984, p 195
253. E.M. Schulson and M.J. Stewart, *Metall. Trans. B*, Vol 7B, 1976, p 363-368
254. E.M. Schulson and J.A. Roy, *Acta Metall.*, Vol 26, 1978, p 15-28
255. E.M. Schulson and J.A. Roy, *J. Nucl. Mater.*, Vol 71, 1977, p 124-133
256. E.M. Schulson and J.A. Roy, *J. Nucl. Mater.*, Vol 60, 1976, p 234-236
257. P.R. Strutt and B.H. Kear, in *High-Temperature Ordered Intermetallic Alloys*, Materials Research Society Symposia Proceedings, Vol 39, C.C. Koch, C.T. Liu, and N.S. Stoloff, Ed., Materials Research Society, 1985, p 279-292
258. M. Takeyama and C.T. Liu, *J. Mater. Res.*, to be published
259. H.R. Pak *et al.*, *Mater. Sci. Eng.*, to be published
260. R.D. Field, R. Darolia, and D.F. Lahrman, *Scr. Metall.*, Vol 23, 1989, p 1469-1474
261. P.R. Strutt, R.A. Dodd, and G.M. Rowe, in *2nd International Conference on the Strength of Metals and Alloys*, American Society for Metals, 1970
262. P.R. Strutt, R.S. Polvani, and J.C. Ingram, *Metall. Trans. A*, Vol 7A, 1976, p 23
263. J.D. Wittenberger, *J. Mater. Res.*, Vol 4, 1989, p 1164
264. T. Takasugi, E.P. George, D.P. Pope, and O. Izumi, *Scr. Metall.*, Vol 19, 1985, p 551-556
265. W.C. Oliver and C.L. White, in *High-Temperature Ordered Intermetallic Alloys II*, Materials Research Society Symposia Proceedings, Vol 81, N.S. Stoloff, C.C. Koch, C.T. Liu, and O. Izumi, Ed., Materials Research Society, 1987, p 241-246
266. I. Baker, R.A. Padgett, and E.M. Schulson, *Scr. Metall.*, Vol 23, 1989, p 1969-1974
267. W.C. Oliver, in *High-Temperature Ordered Intermetallic Alloys III*, Materials Research Society Symposia Proceedings, Vol 133, C.T. Liu, A.I. Taub, N.S. Stoloff, and C.C. Koch, Ed., Materials Research Society, 1989, p 397-402
268. K.J. Williams, *J. Inst. Met.*, Vol 97, 1969, p 112
269. K.J. Williams, *J. Inst. Met.*, Vol 99, 1971, p 310
270. S. Hanada, S. Watanabe, T. Sato, and O. Izumi, *Trans. Jpn. Inst. Met.*, Vol 22, 1981, p 873-881
271. S. Hanada, S. Watanabe, T. Sato, and O. Izumi, *J. Jpn. Inst. Met.*, Vol 45, 1981, p 1279-1284
272. S. Hanada, T. Sato, S. Watanabe, and O. Izumi, *J. Jpn. Inst. Met.*, Vol 45, 1981, p 1285-1292
273. S. Hanada, T. Sato, S. Watanabe, and O. Izumi, *J. Jpn. Inst. Met.*, Vol 45, 1981, p 1293-1299
274. S. Watanabe, T. Sato, and S. Hanada,

J. Jpn. Inst. Met., Vol 47, 1983, p 329-335

275. S. Hanada *et al.*, *Trans. Jpn. Inst. Met.*, Vol 25, 1984, p 348-355
276. S. Watanabe *et al.*, *Trans. Jpn. Inst. Met.*, Vol 25, 1984, p 477-486
277. *Kanthal Super Handbook*, Kanthal Furnace Products, 1986
278. J. Schlichting, *High Temp.—High Press.*, Vol 10, 1978, p 241-269
279. Y. Ymakoshi, T. Hirano, T. Sakagami, and T. Yamane, *Scr. Metall.*, Vol 23, 1989, p 87-90
280. K. Kimura, M. Nakamura, and T. Hirano, *J. Mater. Sci.*, to be published
281. P. Meschter and D.S. Schwartz, *J. Met.*, Vol 41, 1989, p 52-55
282. J.-M. Yang, W. Kai, and S.M. Jeng, *Scr. Metall.*, Vol 23, 1989, p 1953-1958
283. M.G. Mendiratta and D.M. Dimiduk, in *High-Temperature Ordered Intermetallic Alloys III*, Materials Research Society Symposia Proceedings, Vol 133, C.T. Liu, A.I. Taub, N.S. Stoloff, and C.C. Koch, Ed., Materials Research Society, 1989, p 441-446
284. J.S. Wu, P.A. Beaven, R. Wagner, C. Hartig, and J. Seeger, in *High-Temperature Ordered Intermetallic Alloys III*, Materials Research Society Symposia Proceedings, Vol 133, C.T. Liu, A.I. Taub, N.S. Stoloff, and C.C. Koch, Ed., Materials Research Society, 1989, p 761-766
285. C.T. Liu, E.H. Lee, and T.J. Henson, "Initial Development of High-Temperature Titanium Silicide Alloys," ORNL-6435, Oak Ridge National Laboratory, Jan 1988
286. R.L. Fleischer, *J. Mater. Sci.*, Vol 22, 1987, p 2281-2288
287. R.L. Fleischer, R.S. Gilmore, and R.J. Zabala, *J. Appl. Phys.*, Vol 64, 1988, p 2964
288. A.I. Taub and R.L. Fleischer, *Science*, Vol 243, 1989, p 616
289. R.M. Paine, A.J. Stonehouse, and W.W. Beaver, *Corrosion*, Vol 20, 1963, p 307-313
290. R.L. Fleischer and R.J. Zabala, *Metall. Trans. A*, Vol 20A, 1989, p 1279
291. A.J. Carbone *et al.*, *Scr. Metall.*, Vol 22, 1988, p 1903-1906
292. T.G. Nieh, J. Wadsworth, and C.T. Liu, *J. Mater. Res.*, Vol 4, 1989, p 1347
293. W.J. Buehler and F.I. Wang, *Ocean Eng.*, Vol 1, 1968, p 105-120
294. I.M. Schetky, *Sci. Am.*, Vol 241, 1979, p 74-82

Dispersion-Strengthened Nickel-Base and Iron-Base Alloys

J.J. deBarbadillo and J.J. Fischer, Inco Alloys International, Inc.

MECHANICAL ALLOYING (MA) was originally developed for the manufacture of nickel-base superalloys strengthened by both an oxide dispersion and γ' precipitate. Now in its third decade of advancement, mechanical alloying provides a means for producing powder metallurgy (P/M) dispersion-strengthened alloys of widely varying compositions with a unique set of properties. At present, commercial quantities of material are available in the nickel-, iron-, and aluminum-base alloy systems.

It has been known for some time that the strength of metals at high temperature could be increased by the addition of a dispersion of fine refractory oxides. While many methods can produce such dispersions in simple metal systems, these techniques are not applicable to the production of the more highly alloyed materials required for gas turbines and critical industrial applications. Conventional P/M techniques, for example, either do not produce an adequate dispersion or do not permit the use of reactive elements such as aluminum and chromium. These elements confer beneficial characteristics, including corrosion resistance and intermediate temperature strength. In contrast, the mechanical alloying process was developed to introduce a fine inert oxide dispersion into superalloy matrices that contain reactive alloying elements.

Mechanical Alloying Alloy Applications

MA ODS (oxide dispersion-strengthened) alloys were used first in aircraft gas-turbine engines and later in industrial turbines. Components include vane airfoils and platforms, blades, nozzles, and combustor/augmentor assemblies. As experience was gained with production, fabrication, and use of the alloys, this knowledge was applied to the manufacture of component parts in numerous industries. These include diesel-engine glow plugs, heat-treatment fixtures (including shields, baskets, trays and mesh belts, and skid rails for steel plate and billet heating furnaces), burner hardware for coal- and oil-fired power stations, gas sampling tubes, thermocouple tubes, and a wide variety of components used in the production or handling of molten glass.

Mechanical Alloying Process

The mechanical alloying process may be defined as a method for producing composite metal powders with a controlled microstructure. The process involves repeated fracturing and rewelding of a mixture of powder particles in a highly energetic ball charge. On a commercial scale, the process is carried out in vertical attritors or horizontal ball mills.

During each collision of the grinding balls, many powder particles are trapped and plastically deformed. The process is illustrated schematically in Fig. 1. Sufficient deformation occurs to rupture any absorbed surface-contaminant film and expose clean metal surfaces. Cold welds are formed where metal particles overlay, producing composite metal particles. At the same time, other powder particles are fractured. Figure 1 shows two metallic constituents as indicated by light and cross-hatched particles, although in a commercial alloy there may be several constituents.

As the process progresses, most of the particles become microcomposites similar to the one produced in the collision of Fig. 1. The cold welding, which tends to increase the size of the particles involved, and the fracturing (of the particles), which tends to reduce particle size, reach a steady-state balance. This leads to a relatively coarse and stable overall particle size. The internal structure of the particles, however, is continually refined by the repeated plastic deformation.

Production of ODS Superalloy Powders. A uniform distribution of submicron refractory oxide particles must be developed in a highly alloyed matrix for the production of ODS alloys. This requires a powder mixture more varied in composition and particle size than indicated schematically in Fig. 1. A typical powder mixture may consist of fine (4 to 7 μm) nickel powder, −150 μm chromium, and −150 μm master alloy. The master alloy may contain a wide range of elements selected for their roles as alloying constituents or for gettering of contaminants. About 2 vol% of very fine yttria, Y_2O_3 (25 nm, or 250 Å) is added to form the dispersoid. The yttria becomes entrapped along the weld interfaces between fragments in the composite metal powders. After completion of the powder milling a uniform interparticle spacing of about 0.5 μm (20 μin.) is achieved.

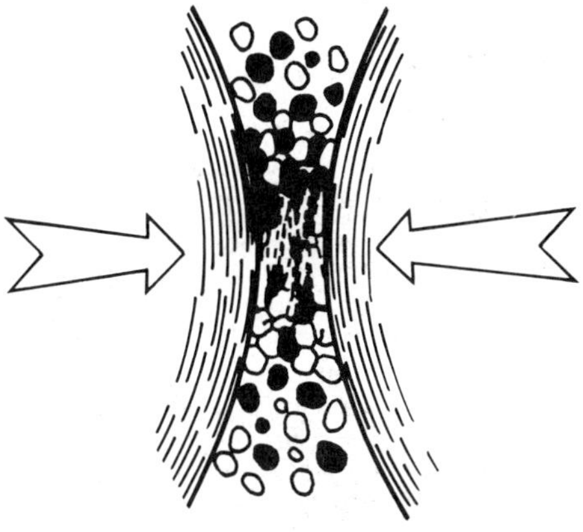

Fig. 1 Schematic depicting the formation of composite powder particles at an early stage in the mechanical alloying process

Consolidation and Property Development. The production of powder containing a uniform dispersion of fine refractory oxide particles in a superalloy matrix is only the first step in achieving the full potential of this type of alloy. These powders must be consolidated and worked under conditions that develop coarse grains during a secondary recrystallization heat treatment. A schematic representation of key operations in the production sequence of a selected product is shown in Fig. 2. It must be emphasized that while the powder-making process is unique to mechanical alloying, all suc-

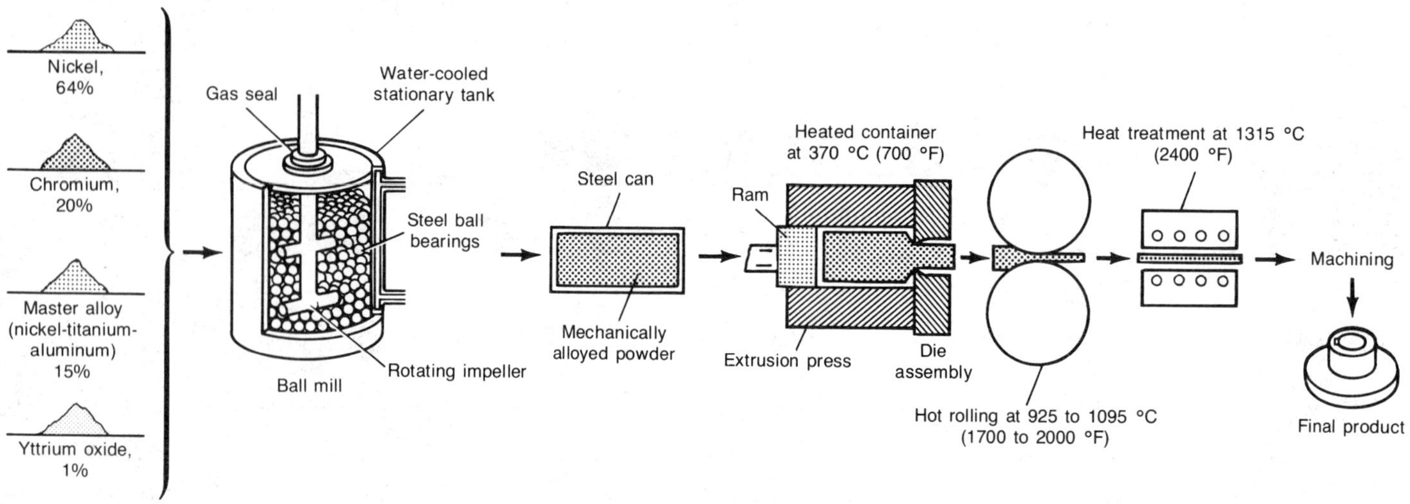

Fig. 2 Schematic showing typical process operations used in the production of MA ODS products

ceeding operations are done on standard mill equipment used to produce wrought high-performance alloys. MA ODS alloys are now commercially available as bar, plate, sheet, tube, wire, shapes, and forgings. However, not all alloys are available in all product forms.

As noted previously, the final properties of MA ODS alloys are dependent on the grain structure as well as the presence of the fine dispersoid. In most products, a grain-coarsening anneal at about 1315 °C (2400 °F) is provided after fabrication. In bar stock, for example, relatively coarse grains are formed that are elongated in the direction of extrusion and working. This elongated structure is necessary for achievement of maximum elevated-temperature properties. Grain aspect ratio, the average grain dimension parallel to the applied stress divided by the average grain dimension perpendicular to the applied stress, has a strong effect on elevated-temperature stress-rupture properties. For many bar products, a high grain aspect ratio is desirable. Plate and sheet products tend to exhibit pancake-shape grains. Through careful control of processing conditions, these grains can be made to be equiaxed in the plane of the sheet, thus providing nearly isotropic properties in the plane of the sheet. Structures for tubing and forgings may be more complex.

Additional information is available in the Section "Nickel- and Iron-Based Dispersion-Strengthened Alloys" in the article "Dispersion-Strengthened Materials" in *Powder Metallurgy*, Volume 7 of the 9th Edition of *Metals Handbook*.

Commercial Alloys

The most common mechanically alloyed ODS alloys include MA 754, MA 758, MA 956, MA 6000, and MA 760.

Alloy MA 754

Alloy MA 754 was the first mechanically alloyed ODS superalloy to be produced on a large scale. This material is basically a Ni-20Cr alloy strengthened by about 1 vol% Y_2O_3 (see Table 1). It is comparable to TD NiCr (an earlier ODS material strengthened by thoria, ThO_2) but has a nonradioactive dispersoid, and because of its higher strength, has been extensively used for aircraft gas-turbine vanes and high-temperature test fixtures.

Microstructure. The microstructure of a commercially produced rectangular bar shows the elongation of the grains along the direction of working. Grain width in the long transverse direction is somewhat greater than the grain thickness. The details of the grain structure in the longitudinal and transverse sections are shown in Fig. 3. The longitudinal view shows the maximum and minimum grain dimensions, whereas the transverse view shows the extreme irregularity of grain boundaries typical of ODS materials. Although it is not obvious from the photomicrograph, this alloy possesses a

Table 1 Nominal composition of selected mechanically alloyed materials

Alloy designation	Ni	Fe	Cr	Al	Ti	W	Mo	Ta	Y_2O_3	C	B	Zr
MA 754	bal	...	20	0.3	0.5	...	...	...	0.6	0.05	...	...
MA 758	bal	...	30	0.3	0.5	...	...	...	0.6	0.05	...	...
MA 760	bal	...	20	6.0	...	3.5	2.0	...	0.95	0.05	0.01	0.15
MA 6000	bal	...	15	4.5	2.5	4.0	2.0	2.0	1.1	0.05	0.01	0.15
MA 956	...	bal	20	4.5	0.5	...	...	...	0.5	0.05	...	...

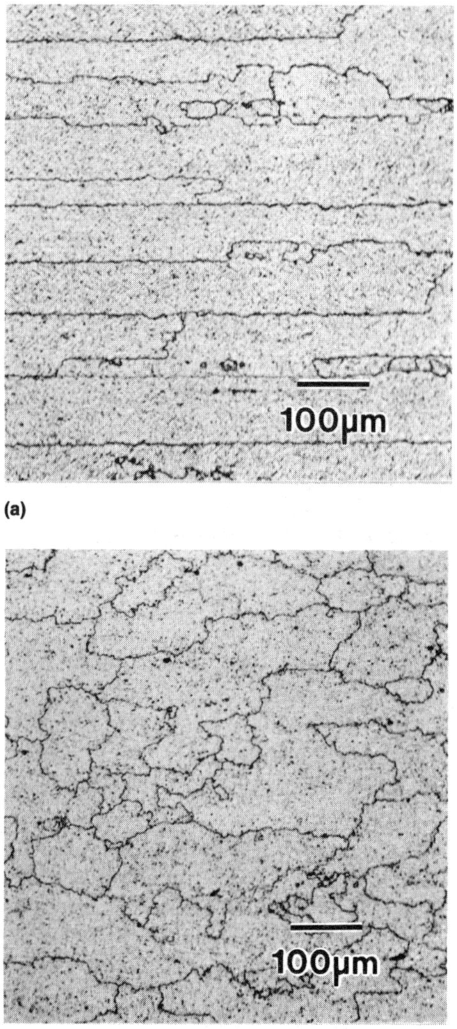

Fig. 3 Alloy MA 754 microstructure shown from two different views. (a) Longitudinal. (b) Transverse. Note high grain aspect ratio shown in longitudinal section (a).

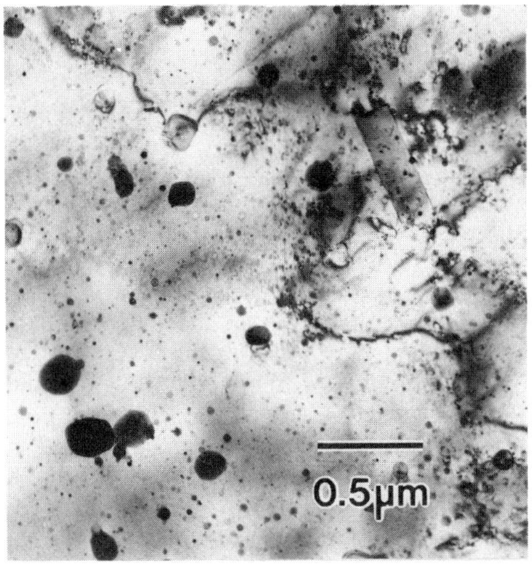

Fig. 4 Transmission electron microscopy (TEM) photomicrograph of alloy MA 754 microstructure showing uniform distribution of fine oxides and scattered coarser carbonitrides

strong (100) crystallographic texture in the longitudinal direction. This texture has been associated with optimum thermal fatigue resistance.

The oxide dispersoid distribution in MA 754 is shown in Fig. 4. The very fine, dark particles are the uniform dispersion of stable yttrium aluminates formed by the reaction between the added yttria, excess oxygen in the powder, and aluminum added to the getter oxygen. The larger dark particles are titanium carbonitrides.

Elevated-Temperature Strength. The tensile properties of MA 754 bar are shown in Fig. 5. The properties shown are for the longitudinal direction. Long transverse strength is similar, but ductility is considerably lower.

In Fig. 6, the 1095 °C (2000 °F) longitudinal stress-rupture properties of MA 754 bar are compared to those of TD NiCr, thoriated nickel bar, alloy MAR-M 509 (a cast cobalt-base alloy), and alloy 80A, a conventional nickel-base alloy having a composition similar to the matrix of MA 754. MA 754, like other ODS materials, has a very flat log stress-log rupture life slope compared to conventional alloys.

The elevated-temperature stress-rupture properties of MA 754 bar are dependent on testing direction, as indicated in Table 2. The rupture-stress capability in the longitudinal direction is consistently higher than that in the long transverse direction, reflecting the differences in grain aspect ratio in the two directions. When MA 754 is produced as cross-rolled plate with coarse equiaxed pancake grains, equal longitudinal and transverse stress-rupture properties are observed. In this form, the rupture strength is about 80% that of the longitudinal bar.

Physical Properties. Important physical properties of alloy MA 754 are given in Table 3. The relatively high melting point, 1400 °C (2550 °F), and low room-temperature modulus of elasticity in the longitudinal direction 151 GPa (22 × 10[6] psi) are especially important. The low modulus, indicating a strong (100) crystallographic texture in the direction of the long grain dimension, has been shown to give superior thermal fatigue resistance.

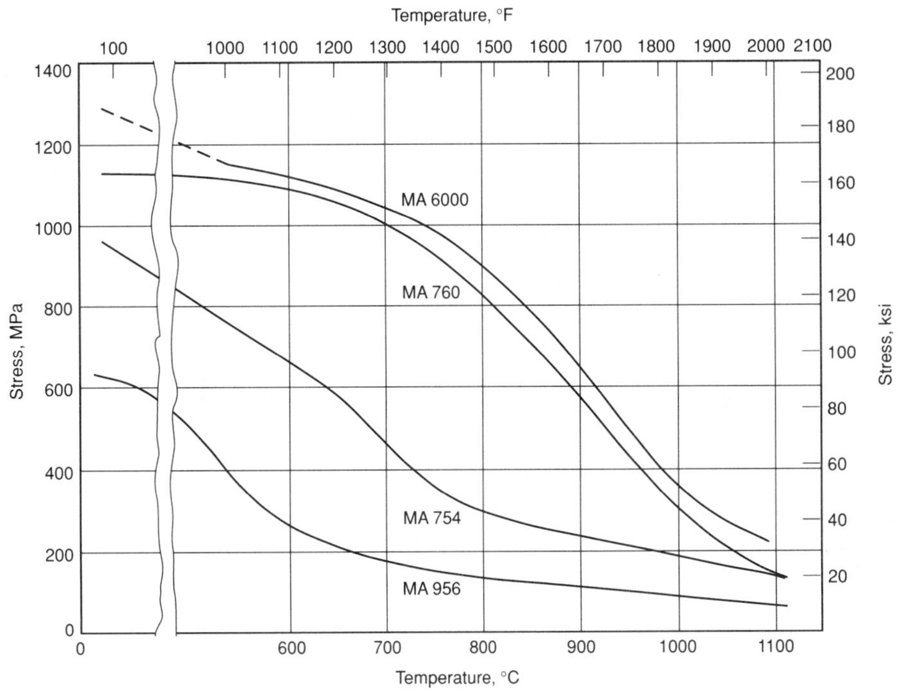

Fig. 5(a) Effect of temperature on the tensile strength of selected MA ODS alloys. Data is for longitudinal direction.

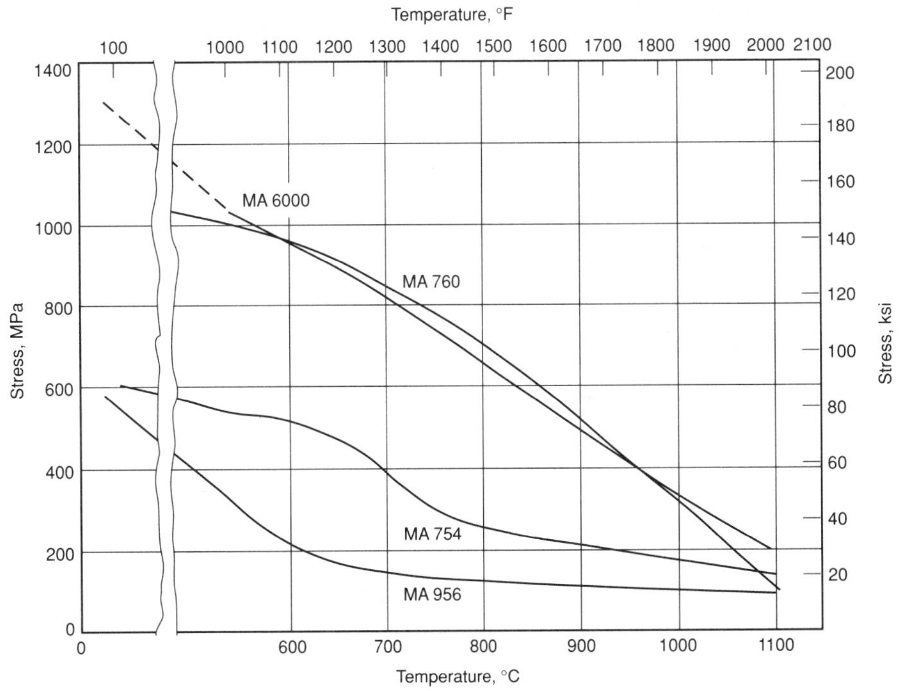

Fig. 5(b) Effect of temperature on the yield strength (0.2% offset) of selected MA ODS alloys. Data is for longitudinal direction.

Alloy MA 758

Alloy MA 758 is a higher-chromium version of MA 754 (see Table 1). This alloy was developed for applications in which the higher chromium content is needed for greater oxidation resistance. The mechanical properties of this alloy are similar to those of MA 754 when identical product

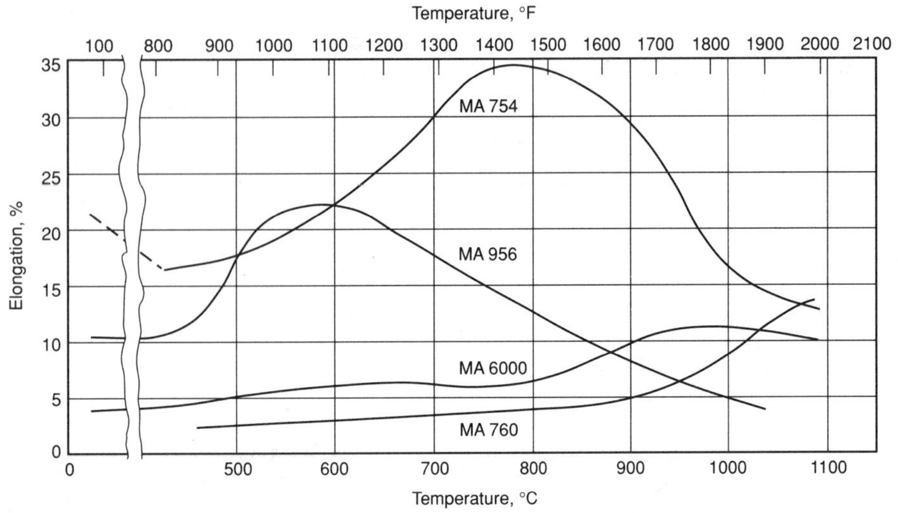

Fig. 5(c) Effect of temperature on the elongation of selected MA ODS alloys. Data is for longitudinal direction.

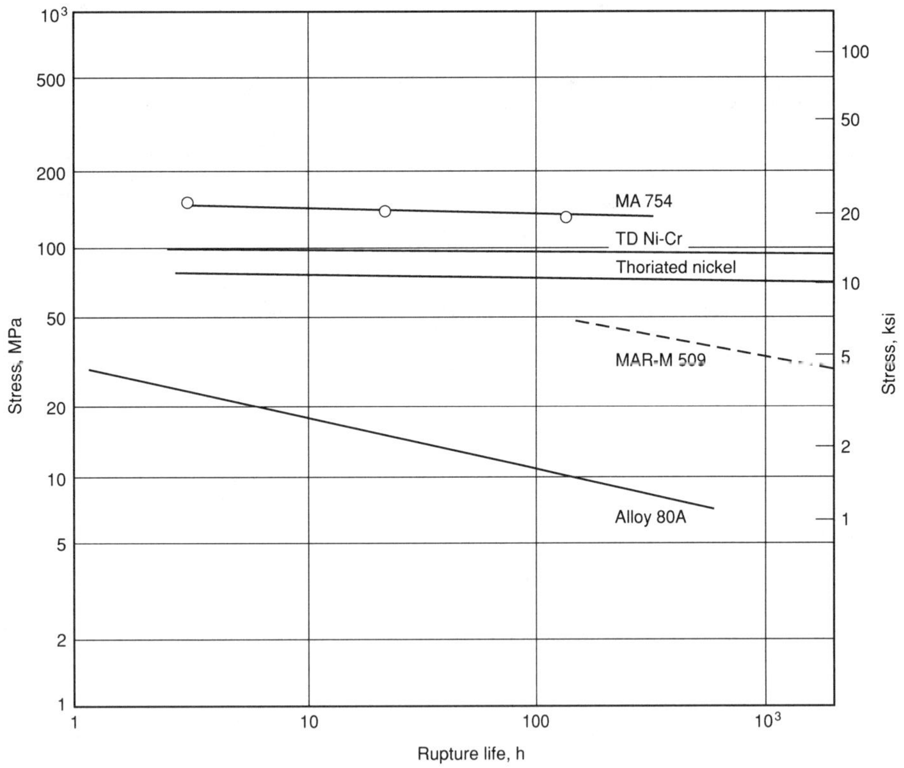

Fig. 6 Comparison of stress-rupture properties of alloy MA 754 bar to other alloy material bars at 1095 °C (2000 °F)

Table 2 Stress-rupture properties of Alloy MA 754 bars

Temperature		Longitudinal — Stress to produce rupture in				Long transverse — Stress to produce rupture in			
		100 h		1000 h		100 h		1000 h	
°C	°F	MPa	ksi	MPa	ksi	MPa	ksi	MPa	ksi
650	1200	284	41.2	256	37.2	241	35.0	208	30.2
760	1400	214	31.1	199	28.8	172	25.0	149	21.6
870	1600	170	24.7	158	22.9	108	15.6	91	13.2
980	1800	136	19.7	129	18.7	63	9.1	46	6.6
1095	2000	102	14.8	94	13.6	38	5.5	24	3.5
1150	2100	90	13.1	78	11.3	23	3.4	17	2.4

forms and grain structures are compared. This alloy has found applications in the thermal processing industry and is also used in the glass-processing industry.

Alloy MA 956

The production of alloy MA 956 demonstrates the ability to add large amounts of metallic aluminum by mechanical alloying (see Table 1). This material is a ferritic iron-chromium-aluminum alloy, dispersion-strengthened with yttrium aluminates formed by the addition of about 1 vol% of yttria. Because of its generally good hot and cold fabricability, MA 956 has been produced in the widest range of product forms of any MA ODS alloy. In sheet form, this alloy is produced by a sequence of hot and cold working, which yields large pancake-shape grains following heat treatment. This grain structure ensures excellent isotropic properties in the plane of the sheet. MA 956 is used in the heat-treatment industry for furnace fixturing, racks, baskets, and burner nozzles. It also is used in advanced aerospace sheet and bar components, where good oxidation and sulfidation resistance are required in addition to high-temperature strength properties.

Mechanical Properties. The tensile properties of MA 956 are shown in Fig. 5. The tensile strength of this alloy is quite a bit lower than that of the other MA materials at low temperatures. However, the strength-versus-temperature curve is extremely flat so that the strength of this alloy exceeds that of all non-ODS sheet materials at approximately 1095 °C (2000 °F).

The stress-rupture properties of MA 956, at elevated temperatures in both the longitudinal and transverse directions, are given in Table 4.

Physical Properties. Alloy MA 956 has a very high melting point (1480 °C, or 2700 °F), a relatively low density (7.2 g/cm^3, or 0.26 lb/in.^{3}) compared to competitive materials, and a relatively low thermal expansion coefficient (see Table 3). This combination of properties makes the alloy well suited for sheet applications such as gas-turbine combustion chambers.

Alloy MA 6000

Alloy MA 6000 has a composition based on an alloy-development philosophy similar to that of the more sophisticated cast and wrought superalloys. This is because it contains a critical balance of elements to produce strength at intermediate and elevated temperatures, along with oxidation and hot-corrosion resistance. Alloy MA 6000 combines γ' hardening from its aluminum, titanium, and tantalum content for intermediate strength, with oxide dispersion-strengthening from the yttria addition for strength and stability at very high temperatures. Oxidation resistance comes from its aluminum and chromium contents, while titanium, tantalum, chromium, and tungsten act in concert to provide sulfi-

Table 3 Physical properties of selected mechanical alloying ODS materials

Alloy	Melting point, °C	Melting point, °F	Modulus of elasticity, GPa	Modulus of elasticity, psi × 10^6	Mass density, g/cm^3	Mass density, lb/in.^{3}	Coefficient of expansion, at 20 to 980 °C (70 to 1800 °F), μm/m · K	Coefficient of expansion, μin./in. · °F
MA 754	1400	2550	151	22	8.3	0.30	16.9	9.41
MA 956	1480	2700	269	39.0	7.2	0.26	14.8	8.22
MA 6000	1296–1375	2365–2507	203	29.4	8.11	0.29	16.7	9.3

dation resistance. The tungsten and molybdenum also act as solid solution strengtheners in this alloy. MA 6000 is an ideal alloy for gas-turbine vanes and blades where exceptional high-temperature strength is required.

Microstructure. Alloy MA 6000 has a highly elongated coarse grain structure that results from the thermomechanical processing (TMP) followed by high-temperature annealing. It has proved useful to utilize zone annealing to achieve the optimum grain aspect ratio for this alloy. As a result, the only product forms presently available are bar or small forgings.

Zone annealing is performed by slowly passing a heating element down the axis of the bar. In practice, either resistance or induction heating can be used. A typical zone annealing speed is 100 mm/h (4 in./h).

The microstructure of MA 6000 is shown in Fig. 7. Note the high-volume fraction of γ′ (45 to 50 vol%), and the very fine dispersoid particles present in both the γ′ (dark irregular particles) and lighter matrix.

Mechanical Properties. The elevated-temperature properties of alloy MA 6000, in terms of the specific rupture strength for 1000-h life as a function of temperature, are compared with those of directionally solidified alloy DS MAR-M 200 containing hafnium and a thoriated nickel alloy bar in Fig. 8. This diagram clearly shows the effect of the two strengthening mechanisms in alloy MA 6000. At intermediate temperatures, around 815 °C (1500 °F), the strength of MA 6000 approaches that of the complex, highly alloyed alloy DS MAR-M 200 containing hafnium and is almost four times that of an unalloyed ODS metal such as thoriated nickel (TD nickel). At high temperatures (~1095 °C, or 2000 °F), where the alloy DS MAR-M 200 containing hafnium has lost most of its strength due to growth and dissolution of its γ′ precipitate, MA 6000 has useful strength due to the presence of the oxide dispersion. At temperatures between these extremes, the strength of MA 6000 is superior to both the cast nickel-base superalloy and the ODS metal because the two strengthening mechanisms supplement one another.

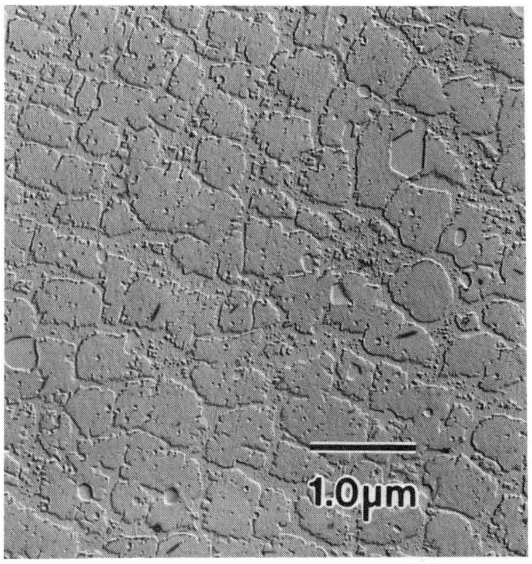

Fig. 7 Microstructure of heat-treated alloy MA 6000, showing high-volume fraction γ′ and dispersoid phases

Alloy MA 760

Alloy MA 760 is an age-hardened nickel-base alloy with a composition designed to provide a balance of high-temperature strength, long-term structural stability, and oxidation resistance. Its primary use is expected to be for industrial gas turbines. The composition of this alloy is shown in Table 1. It is similar to MA 6000 in that its strength is supplemented by γ′ age hardening. Its properties also benefit from zone annealing to give coarse elongated grains. The stress-rupture properties of alloy MA 760 exceed those of MA 754 but are exceeded by those of MA 6000 (see Fig. 5 and 8).

Oxidation and Hot-Corrosion Properties

Because the MA ODS alloys are normally used uncoated at very high temperatures in hostile environments, the resistance of the alloys to oxidation, carburization, sulfidation, and oxide fluxing is important. While all of the alloys are resistant to the effects of these deleterious chemical processes, the relative resistance varies with specific alloy and environment.

Figure 9 shows the resistance of the MA ODS alloys to cyclic oxidation at various temperatures. All of the alloys form protective scales at 1000 °C (1830 °F), and thus are highly resistant to oxidation. At higher temperatures, the relatively low-chromium MA 6000 shows increased weight loss, whereas the other alloys are highly stable. Under the most severe conditions, MA 6000 would require coating for long-time exposure. Coatings and procedures suitable for this alloy have been reported in the technical literature. The oxidation resistance of MA 956 is unsurpassed by any existing commercial sheet alloy. However, accelerated oxidation may occur in this alloy during long-time exposure at temperatures above 1200 °C (2190 °F) depending on the environment.

Evaluation of oxidation-sulfidation resistance of gas-turbine alloys is frequently done in a burner-rig test. Representative data for selected alloys tested at 927 °C (1700 °F) is shown in Fig. 10. MA 956 exhibits extremely high resistance to this form of attack. MA 6000, though less resistant than MA 956, was comparable to the cast alloy IN-738 in this test.

The MA ODS materials have also been evaluated in a wide range of specialized environments. MA 956 has proved to be especially resistant to carburization, as illustrated in Table 5.

These alloys also show excellent resistance to attack by molten glass. C glass and lime glass are two glasses whose effect on mechanically alloying materials has been evaluated. These glasses have the following compositions:

Compound	Composition, wt%, C glass	Composition, wt%, Lime glass
SiO_2	65	73
Al_2O_3	4	1.7
Na_2O	8.5	16.3
CaO	14	4.7
MgO	3	3.1
B_2O_3	5	...
H_2O	...	0.4
Li_2O	...	0.15

Based on a 5-day immersion test, MA 754 and MA 758 demonstrate high corrosion resistance to molten C glass:

Alloy	Metal loss, mm	Metal loss, mil
MA 754	0.04	1.6
MA 758	0.03	1.2

Based on a 240-h immersion test in lime glass at 1150 °C (2100 °F), MA 754 has corrosion-resistance properties intermediate between those of MA 956 and MA 758:

Alloy	Mass change, mg · cm^{-2}	Mass change, lb · in.$^{-2}$ × 10^{-4}
MA 956	4	0.57
MA 754	28	4.0
MA 758	42	6.0

It is well known that the relative performance of alloys in glass is dependent upon glass composition, temperature, impurity level, velocity, and other factors. Consequently, the tabular information shown in this section should be considered only as illustrative of the generally high resistance of MA ODS alloys in these molten glass environments.

Fabrication of MA ODS Alloys

The mill product forms available vary from alloy to alloy, depending on factors

Table 4 Stress-rupture properties of Alloy MA 956 sheet

Temperature		Longitudinal — Stress to produce rupture in						Transverse — Stress to produce rupture in					
		10 h		100 h		1000 h		10 h		100 h		1000 h	
°C	°F	MPa	ksi	MPa	ksi	MPa	ksi	MPa	ksi	MPa	ksi	MPa	ksi
980	1800	84	12.2	75	10.9	67	9.7	72	10.4	70	10.2	63	9.1
1100	2010	64	9.3	57	8.3	51	7.4	64	9.3	57	8.3	...	...
1150	2100	50	7.3	39	5.7	...	...	31	4.5	24	3.5	...	...

such as ease of fabrication and applicable forming methods.

Bars. All of the alloys are available as bars, and much of the data reported in the literature refer to bar properties. All of the bar products can be precision forged, and MA 754 forged airfoils have been in commercial use for years. The high-temperature properties of forgings can be equivalent to those of annealed bar, provided that care is taken in the design of the part and the thermomechanical processing is controlled to produce the desired grain structure and orientation. Forgings of MA 6000 and MA 760 with optimal properties in the airfoil axis can be obtained by zone annealing after forging. Both seamless and flat-butt-welded rings with desired properties in the hoop direction have been made from MA 754 and MA 758.

Plate products are available for MA 754, MA 758, and MA 956. Equiaxed properties can be obtained through control of rolling conditions. Plate is readily amenable to a

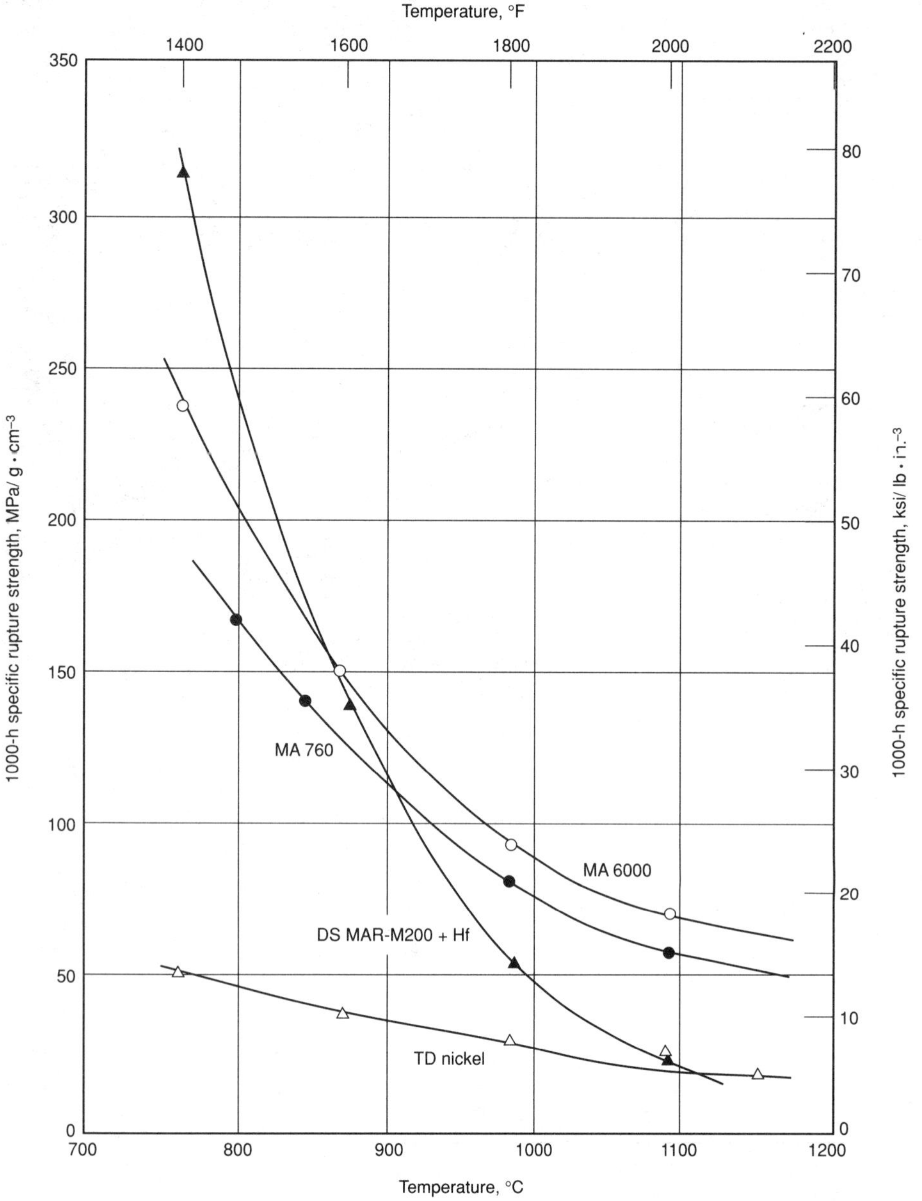

Fig. 8 Effect of temperature on the 1000-h specific rupture strength of MA 760, MA 6000, DS MAR-M 200, and TD nickel

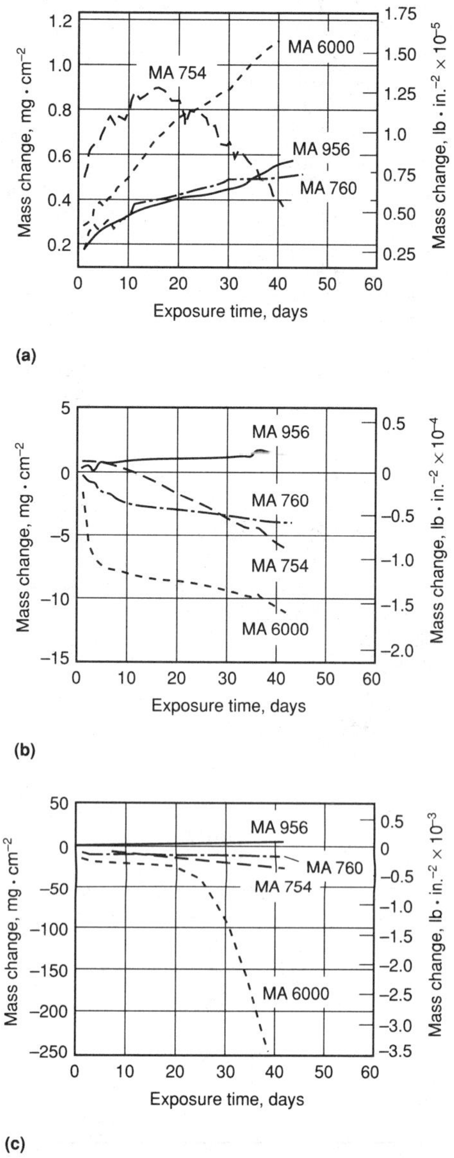

Fig. 9 Effect of temperature on mass change for four mechanically alloyed materials exposed to air containing 5% H_2O vapor. (a) 1000 °C (1830 °F). (b) 1100 °C (2010 °F). (c) 1200 °C (2190 °F)

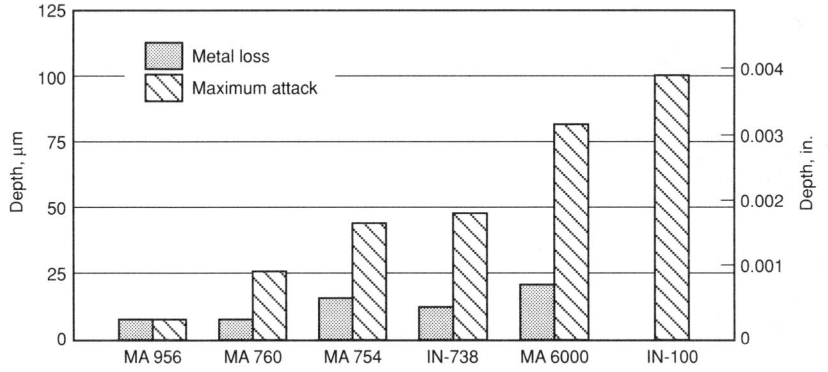

Fig. 10 Comparison of the oxidation-sulfidation resistance of MA ODS alloys with that of superalloys IN-738 and IN-100. Tested in a burner rig for 500 h at 925 °C (1700 °F) using an air-to-fuel ratio that varied from 27:1 to 21:1. JP-5 fuel contained 0.3% S. Temperature test cycle consisted of the alloy held at temperature for 1 h and then cooled for 3 min. No metal loss data for IN-100 because sample was destroyed in 50 h. Metal loss is defined as loss of diameter due to oxide and sulfide scale formation. Maximum attack is defined as loss of diameter due to internal oxidation and sulfidation.

Table 5 Comparison of the carburization resistance of Alloy MA 956 with Alloy 800

Test duration was 100 h at 1095 °C (2000 °F) temperature in a H_2 + 2% CH_4 atmosphere

Alloy	Weight change: Undescaled, mg/cm^3 (lb/in.^{3} × 10^{-6})	Weight change: Descaled, mg/cm^3 (lb/in.^{3} × 10^{-6})	Metal loss µm	Metal loss in.	Maximum attack µm	Maximum attack in.
MA 956	0.07 (2.5)	−0.42 (−15)	10	0.0004	10	0.0004
Alloy 800	33.74 (1225)	29.89 (1085)	132	0.00528	7615	0.3046

variety of hot-forming operations, including hot shear spinning. Optimal formability and minimum flow stress is obtained when the plate is in the fine-grain (unrecrystallized) condition. The standard grain-coarsening anneal is then applied to the formed component.

Sheet. The only alloy currently available in sheet form is MA 956. This material, which is readily cold rolled to standard sheet tolerance, is commercially available in gages down to thicknesses of 0.25 mm (0.010 in.) and widths up to 610 mm (24 in.). A wide variety of components have been cold formed from MA sheet by standard metal-forming operations. Experience has shown that warming to about 95 °C (200 °F) is necessary to prevent cracking because this alloy undergoes a ductile-to-brittle transition in the vicinity of room temperature.

Additional Product Forms. MA 956 has also been produced in a number of other forms for special applications. These forms include pipe, thin-wall tube, and fine wire.

Joining of MD ODS Alloys

Many applications for MA ODS alloys require some method of joining. Procedures that involve fusion of the base metal destroy the unique microstructure that is responsible for the high-temperature strength of these alloys. Accordingly, fusion welds that are needed for attachment or positioning for brazing should be located in areas of relatively low stress. Procedures such as gas-tungsten-arc welding (GTAW), electron-beam welding (EBW), and pulsed laser-beam welding (LBW) have all been used successfully on a limited scale. MA 956 sheet assemblies have also been made using resistance spot welding (RSW).

As might be expected, nonfusion processes are required in order to obtain tensile and stress-rupture properties approaching those of the parent metal. Vacuum diffusion bonding (DB) and diffusion brazing (DFB) are now used extensively for assembly of aircraft engine components. Riveting operations using similar alloy rivets have also been applied for nonaircraft applications.

SELECTED REFERENCES

- J.S. Benjamin, Dispersion Strengthened Superalloys by Mechanical Alloying, *Met. Trans.*, Vol 1, 1970, p 2943-2951
- R.C. Benn, P. Deb, and D.H. Boone, *Proceedings of the High Temperature Coatings Symposium* (Orlando, FL), M. Khobaib, Ed., AIME/ASM Conference, Oct 1986
- R.C. Benn and G.M. McColvin, The Development of ODS Superalloys for Industrial Gas Turbines, in *Superalloys 1988*, S. Reichman, D.N. Duhl, G. Maurer, S. Antolovich, and C. Lund, Ed., The Metallurgical Society, 1988, p 75-80
- J.J. Fischer, I. Astley, and J.P. Morse, *Proceedings of the Third International Symposium on Superalloys: Metallurgy and Manufacture* (Seven Springs, PA), Sept 1976, p 361-371
- J.J. Fischer and R.M. Haeberle, Commercial Status of Mechanically Alloyed Materials, *Mod. Devel. Powder Metall.*, Vol 18-21, 1988, p 461-477
- "Frontiers of High Temperature Materials," in Proceedings of an International Conference on Oxide Dispersion Strengthened Superalloys by Mechanical Alloying, Inco Alloy Products, 1981
- "Frontiers of High Temperature Materials II," in Proceedings of an International Conference on Oxide Dispersion Strengthened Superalloys by Mechanical Alloying, Inco Alloy Products, 1983
- E. Grundy, Other Applications of Superalloys, *Mater. Sci. Technol.*, Vol 3, Sept 1987, p 782-790
- T.J. Kelly, Welding of Mechanically Alloyed ODS Materials, in *Trends in Welding Research in the United States*, S.A. David, Ed., American Society for Metals, 1982, p 471-488
- *New Materials by Mechanical Alloying Techniques*, E. Arzt and L. Schultz, Ed., Deutsche Gesellschaft fur Metallkunde, Oberursel, West Germany, 1989
- G.D. Smith and J.J. Fischer, High Temperature Corrosion Resistance of Mechanically Alloyed Products in Gas Turbine Environments, *Proceedings of the 1990 ASME Turbo Exposition* (Brussels, Belgium), American Society of Mechanical Engineers, to be published
- Solid State Powder Processing, in *Proceedings of International Symposium* (Warrendale, PA), J.J. deBarbadillo and A.H. Clauer, Ed., 1990
- J.H. Weber, High Temperature Oxide Dispersion Strengthened Alloys, The 1980s—Pay Off Decade for Advanced Materials, *SAMPE J.*, Vol 25, 1980

Cemented Carbides

A.T. Santhanam, P. Tierney, and J.L. Hunt, Kennametal Inc.

CEMENTED CARBIDES belong to a class of hard, wear-resistant, refractory materials in which the hard carbide particles are bound together, or cemented, by a soft and ductile metal binder. These materials were first developed in Germany in the early 1920s in response to demands for a die material having sufficient wear resistance for drawing tungsten incandescent filament wires to replace the expensive diamond dies then in use. The first cemented carbide to be produced was tungsten carbide (WC) with a cobalt binder. Although the term cemented carbide is widely used in the United States, these materials are better known internationally as hard metals.

Tungsten carbide was first synthesized by the French chemist Henri Moissan in the 1890s (Ref 1). There are two types of tungsten carbide: WC, which directly decomposes at 2800 °C (5070 °F), and W_2C, which melts at 2750 °C (4980 °F) (Ref 2, 3). Early attempts to produce drawing dies from a eutectic alloy of WC and W_2C were unsuccessful, because the material had many flaws and fractured easily. The use of powder metallurgy techniques by Schroeter in 1923 paved the way for obtaining a fully consolidated product (Ref 4). Schroeter blended fine WC powders with a small amount of iron, nickel, or cobalt powders and pressed the powders into compacts, which were then sintered at approximately 1300 °C (2400 °F). Cobalt was soon found to be the best bonding material. Over the years, the basic WC-Co material has been modified to produce a variety of cemented carbides, which are used in a wide range of applications, including metal cutting, mining, construction, rock drilling, metal forming, structural components, and wear parts. Approximately 50% of all carbide production is used for metal cutting applications.

This article discusses the manufacture and composition of cemented carbides and their microstructure, classifications, physical and mechanical properties, and applications. New tool geometries, tailored substrates, and the application of thin, hard coatings to cemented carbides by chemical vapor deposition and physical vapor deposition are examined for metal cutting applications. The current status of cemented carbides in nonmetal cutting applications will also be covered. This article is limited to tungsten carbide cobalt-base materials. Information on metal-bonded titanium carbide materials and steel-bonded tungsten carbide is given in the article "Cermets" in this Volume. Extensive reviews of the scientific and industrial aspects of cemented carbides are available in Ref 5 to 8.

Manufacture of Cemented Carbides

Cemented carbides are manufactured by a powder metallurgy process consisting of a sequence of steps in which each step must be carefully controlled to obtain a final product with the desired properties, microstructure, and performance. The steps include:

- Processing of the ore and the preparation of the tungsten carbide powder
- Preparation of the other carbide powders
- Production of the grade powders
- Compacting or powder consolidation
- Sintering
- Postsinter forming

The sintered product can be directly used or can be ground, polished, and coated to suit a given application.

Preparation of Tungsten Carbide Powder. There are two methods by which tungsten carbide powders are produced from the tungsten-bearing ores. Traditionally, tungsten ore is chemically processed to ammonium paratungstate and tungsten oxides. These compounds are then hydrogen-reduced to tungsten metal powder. The fine tungsten powders are blended with carbon and heated in a hydrogen atmosphere between 1400 and 1500 °C (2500 and 2700 °F) to produce tungsten carbide particles with sizes varying from 0.5 to 30 μm (Fig. 1). Each particle is composed of numerous tungsten carbide crystals. Small amounts of vanadium, chromium, or tantalum are sometimes added to tungsten and carbon powders before carburization to produce very fine (<1 μm) WC powders.

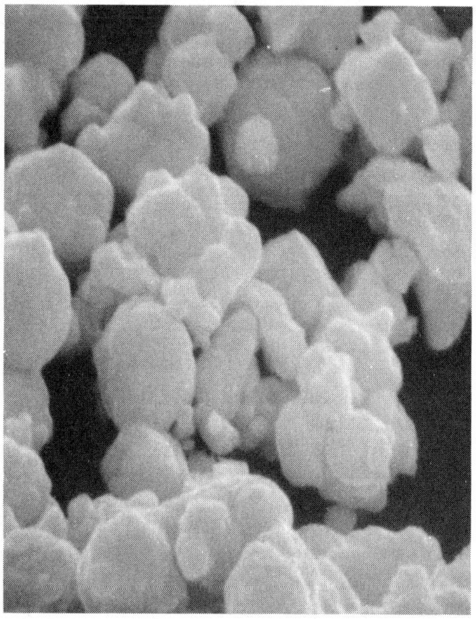

Fig. 1 Tungsten carbide particles produced by the carburization of tungsten and carbon. 10 000×

In a more recently developed and patented process, tungsten carbide is produced in the form of single crystals through the direct reduction of tungsten ore (sheelite) (Ref 9). The ore is mixed with iron oxide, aluminum, carbon, and calcium carbide. A high-temperature exothermic reaction ($2Al + 3FeO \leftrightarrow Al_2O_3 + 3Fe$) at about 2500 °C (4500 °F) produces a molten mass that, when cooled, consists of tungsten carbide crystals dispersed in iron, and a slag containing impurities. The crystalline WC (Fig. 2) is then chemically separated from the iron matrix.

Tungsten-titanium-tantalum (niobium) carbides are used in steel-cutting grades to resist cratering or chemical wear and are produced from metal oxides of titanium, tantalum, and niobium. These oxides are mixed with metallic tungsten powder and carbon. The mixture is heated under a hydrogen atmosphere or vacuum to reduce the oxides and form solid-solution carbides such as WC-TiC, WC-TiC-TaC, or WC-TiC-(Ta, Nb)C. The menstruum method can

Fig. 2 Tungsten carbide single crystals produced by the direct reduction of tungsten ore. 200×

be used to produce WC-TiC solid solution. In this method, the individual carbides are dissolved in liquid nickel. Solid-solution carbides are then precipitated during cooling (Ref 10).

Production of Grade Powders. Cemented carbide grade powders may consist of WC mixed with a finely divided metallic binder (cobalt, nickel, or iron) or with additions of other cubic carbides, such as TiC, TaC, and NbC, depending on the required properties and application of the tool. Intensive milling is necessary to break up the initial carbide crystallites and to blend the various components such that every carbide particle is coated with binder material. This is accomplished in ball mills, vibratory mills, or attritors that use carbide balls. The mills are usually lined with carbide sleeves, although mills lined with low-carbon steel or stainless steel are also used.

Milling is performed under an organic liquid such as heptane or acetone to minimize heating of the powder and to prevent its oxidation. The liquid is distilled off after the milling operation. A solid lubricant such as paraffin wax is added to the powder blend in the final stages of the milling process or later in a blender. The lubricant provides a protective coating to the carbide particles and prevents or greatly reduces the oxidation of the powder. The lubricant also imparts strength to the pressed or consolidated powder mix.

After milling, the organic liquid is removed by drying. In a spray-drying process commonly used in the cemented carbide industry, a hot inert gas such as nitrogen impinges on a stream of carbide particles. This produces free-flowing spherical powder aggregates.

Powder Consolidation. A wide variety of techniques are used to compact the cemented carbide grade powders to the desired shape. Carbide tools for mining and construction applications are pill pressed (pressure applied in one direction) in semiautomatic or automatic presses. Metal cutting inserts are also pill pressed, but may require additional shaping after sintering. Cold isostatic pressing, in which the powder is subjected to equal pressure from all directions, followed by green forming, is also a common practice for wear and metal forming tools. Rods and wires are formed by the extrusion process.

Unlike most other metal powders, cemented carbide powders do not deform during the compacting process. Generally, they cannot be compressed to much above 65% of the theoretical upper limit for density. Despite this low green density, carbide manufacturers have developed the technology for achieving good dimensional tolerances in the sintered product.

Sintering and Postsintering Operations. The first step in the sintering process is the removal of the lubricant (dewaxing) from the powder compact. The pressed compacts are normally set on graphite trays coated with a graphite paint. The compacts are first heated to about 500 °C (900 °F) in a hydrogen atmosphere or vacuum using either semicontinuous or batch-type graphite furnaces.

After lubricant removal, the compacts are heated in a vacuum (0.1 Pa, or 10^{-3} torr) to a final sintering temperature ranging from 1350 to 1600 °C (2460 to 2900 °F), depending on the amount of the cobalt binder and the desired microstructure. The dewaxing and sintering operations can also be performed in a single vacuum cycle using furnaces equipped to condense the lubricant and remove it from the heating chamber.

During the final sintering operation, the cobalt melts and draws the carbide particles together. Shrinkage of the compact ranges from 17 to 25% on a linear scale, producing a virtually pore-free, fully dense product.

In the 1970s, the cemented carbide industry took advantage of hot isostatic pressing (HIP) technology, in which vacuum-sintered material is heated again under a gaseous (argon or helium) pressure of 100 to 150 MPa (15 to 20 ksi) (Ref 11). The temperatures of this additional process are 25 to 50 °C (45 to 90 °F) below the sintering temperature. The high temperatures and pressures employed in the HIP furnace remove any residual internal porosity, pits, or flaws and produce a nearly perfect cemented carbide.

The latest advancement in sintering technology is the sinter-HIP process, which was developed in the early 1980s (Ref 12). In this process, low-pressure hot isostatic pressing (up to about 7 MPa, or 1 ksi) is combined with vacuum sintering, and the pressure is applied at the sintering temperature when the metallic binder is still molten. With this process, void-free products can be produced at costs only slightly higher than those of vacuum sintering.

Postsinter Forming. A large number of cemented carbide products are shaped after sintering because of surface finish, tolerance, and geometry requirements. This forming operation is both time consuming and expensive. The sintered material is formed with metal-bonded diamond or silicon carbide wheels, turned with a single-point diamond tool, or lapped with diamond-containing slurries.

Cemented Carbides for Machining Applications

The performance of cemented carbide as a cutting tool lies between that of tool steel and cermets. Compared to tool steels, cemented carbides are harder and more wear resistant, but also exhibit lower fracture resistance and thermal conductivities than tool steel. Cermets, on the other hand, are more wear resistant than cemented carbides, but may not be as tough. Any comparison of cermets and cemented carbides, however, depends on the percent of binder material and the type and size of carbide grains. Cermets are described in more detail in the article "Cermets" in this Volume and in Volume 16, *Machining*, of the 9th Edition of *Metals Handbook*.

Compositions and Microstructures

The performance of carbide cutting tools is strongly dependent on composition and microstructure, and the properties of cemented carbide tools depend not only on the type and amount of carbide but also on the carbide grain size and the amount of binder metal. The basic physical and mechanical properties of refractory metal carbides used in the production of cemented carbide tools are given in Table 1. Tungsten carbide and molybdenum carbide have hexagonal crystal structures, whereas the carbides of titanium, tantalum, niobium, vanadium, hafnium, and zirconium are cubic. They undergo no structural changes up to their melting points.

Tungsten Carbide-Cobalt Alloys. The first commercially available cemented carbides consisted of tungsten carbide particles bonded with cobalt. These are commonly referred to as straight grades. These alloys exhibit excellent resistance to simple abrasive wear and thus have many applications in metal cutting. Table 2 lists the representative properties of several straight WC-Co alloys.

The commercially significant alloys contain cobalt in the range of 3 to 25 wt%. For machining purposes, alloys with 3 to 12% Co and carbide grain sizes from 0.5 to more than 5 μm are commonly used.

Table 1 Properties of refractory metal carbides

Carbide	Hardness, HV (50 kg)	Crystal structure	Melting point °C	Melting point °F	Theoretical density, g/cm³	Modulus of elasticity GPa	Modulus of elasticity 10^6 psi	Coefficient of thermal expansion, μm/m · K
TiC	3000	Cubic	3100	5600	4.94	451	65.4	7.7
VC	2900	Cubic	2700	4900	5.71	422	61.2	7.2
HfC	2600	Cubic	3900	7050	12.76	352	51.1	6.6
ZrC	2700	Cubic	3400	6150	6.56	348	50.5	6.7
NbC	2000	Cubic	3600	6500	7.80	338	49.0	6.7
Cr_3C_2	1400	Orthorhombic	1800(a)	3250	6.66	373	54.1	10.3
WC	(0001) 2200 (10$\bar{1}$0) 1300	Hexagonal	~2800(a)	5050	15.7	696	101	(0001) 5.2 (10$\bar{1}$0) 7.3
Mo_2C	1500	Hexagonal	2500	4550	9.18	533	77.3	7.8
TaC	1800	Cubic	3800	6850	14.50	285	41.3	6.3

(a) Not congruently melting, dissociation temperature. Source: Ref 8

The ideal microstructure of WC-Co alloys should exhibit only two phases: angular WC grains and cobalt binder phase. Representative microstructures of several straight WC-Co alloys are shown in Fig. 3. The carbon content must be controlled within narrow limits. Too high a carbon content results in the presence of free and finely divided graphite (Fig. 4), which in small amounts has no adverse effects in machining applications. Deficiency in carbon, however, results in the formation of a series of double carbides (for example, Co_3W_3C or Co_6W_6C), commonly known as η phase, which causes severe embrittlement. Because the formation of η phase involves the dissolution of the original carbides into the cobalt binder, η phase appears as an irregularly shaped phase in the microstructure (Fig. 5).

Submicron Tungsten Carbide-Cobalt Alloys. In recent years, WC-Co alloys with submicron carbide grain sizes (Fig. 6) have been developed for applications requiring more toughness or edge strength. Typical applications include indexable inserts and a wide variety of solid carbide drilling and milling tools. Grain refinement in these alloys is obtained by small additions (0.25 to 3.0 wt%) of tantalum carbide, niobium carbide, vanadium carbide, or chromium carbide. Additions can be made before carburization of the tungsten or later in the powder blend. Vanadium carbide is the most effective grain growth inhibitor. Chromium carbide, in addition to being an efficient grain growth inhibitor, imparts excellent mechanical properties. Tantalum carbide is not as effective as vanadium carbide or chromium carbide in grain refinement.

Alloys Containing Tungsten Carbide, Titanium Carbide, and Cobalt. The tungsten carbide-cobalt alloys, developed in the early 1920s, were successful in the machining of cast iron and nonferrous alloys at much higher speeds than were possible with high-speed steel tools, but were subject to chemical attack or diffusion wear when cutting steel. As a result, the tools failed rapidly at speeds not much higher than those used with high-speed steel. This led to the development of WC-TiC-Co alloys.

Tungsten carbide diffuses readily into the steel chip surface, but the solid solution of tungsten carbide and titanium carbide resists this type of chemical attack. Unfortunately, titanium carbide and WC-TiC solid solutions are more brittle and less abrasion resistant than tungsten carbide. The amount of titanium carbide added to tungsten carbide-cobalt alloys is therefore kept to a minimum, typically no greater than 15 wt%. The carbon content is less critical in WC-TiC-Co alloys than in WC-Co alloys, and the η phase does not appear in the microstructure unless carbon is grossly inadequate. In addition, free graphite rarely occurs in these alloys.

Steel-Cutting Grades of Cemented Carbide Alloys. The WC-TiC-Co alloys have given way to alloys of tungsten carbide, cobalt, titanium carbide, tantalum carbide, and niobium carbide. The tungsten carbide-cobalt alloys containing TiC, TaC, and NbC are called complex grades, multigrades, or steel-cutting grades. Adding TaC to WC-TiC-Co alloys partially overcomes the deleterious effect of TiC on the strength of WC-Co alloys. Tantalum carbide also resists cratering and improves thermal shock resistance. The latter property is particularly useful in applications involving interrupted cuts. Tantalum carbide is often added as (Ta,Nb)C because the chemical similarity between TaC and NbC makes their separation expensive. Fortunately, NbC has an effect similar to TaC in most cases. The relative concentrations of tantalum carbide and niobium carbide in these alloys are dependent on the raw material used, the desired composition, the properties, and the microstructure.

Unlike the WC-Co alloys, the microstructure of WC-TiC-(Ta,Nb)C-Co alloys shows three phases: angular WC grains, rounded WC-TiC-(Ta,Nb)C solid-solution grains, and cobalt binder. The solid-solution carbide phase often exhibits a cored structure, indicating incomplete diffusion during the sintering process. Representative microstructures of several WC-TiC-(Ta,Nb)C alloys are shown in Fig. 7. The size and

Table 2 Properties of representative cobalt-bonded cemented carbides

Nominal composition	Grain size	Hardness, HRA	Density g/cm³	Density oz/in.³	Transverse strength MPa	Transverse strength ksi	Compressive strength MPa	Compressive strength ksi	Modulus of elasticity GPa	Modulus of elasticity 10^6 psi	Relative abrasion resistance(a)	Coefficient of thermal expansion, μm/m · K at 200 °C (390 °F)	Coefficient of thermal expansion, μm/m · K at 1000 °C (1830 °F)	Thermal conductivity, W/m · K
97WC-3Co	Medium	92.5–93.2	15.3	8.85	1590	230	5860	850	641	93	100	4.0	...	121
94WC-6Co	Fine	92.5–93.1	15.0	8.67	1790	260	5930	860	614	89	100	4.3	5.9	...
	Medium	91.7–92.2	15.0	8.67	2000	290	5450	790	648	94	58	4.3	5.4	100
	Coarse	90.5–91.5	15.0	8.67	2210	320	5170	750	641	93	25	4.3	5.6	121
90WC-10Co	Fine	90.7–91.3	14.6	8.44	3100	450	5170	750	620	90	22	...	...	...
	Coarse	87.4–88.2	14.5	8.38	2760	400	4000	580	552	80	7	5.2	...	112
84WC-16Co	Fine	89	13.9	8.04	3380	490	4070	590	524	76	5	...	...	...
	Coarse	86.0–87.5	13.9	8.04	2900	420	3860	560	524	76	5	5.8	7.0	88
75WC-25Co	Medium	83–85	13.0	7.52	2550	370	3100	450	483	70	3	6.3	...	71
71WC-12.5TiC-12TaC-4.5Co	Medium	92.1–92.8	12.0	6.94	1380	200	5790	840	565	82	11	5.2	6.5	35
72WC-8TiC-11.5TaC-8.5Co	Medium	90.7–91.5	12.6	7.29	1720	250	5170	750	558	81	13	5.8	6.8	50

(a) Based on a value of 100 for the most abrasion-resistant material

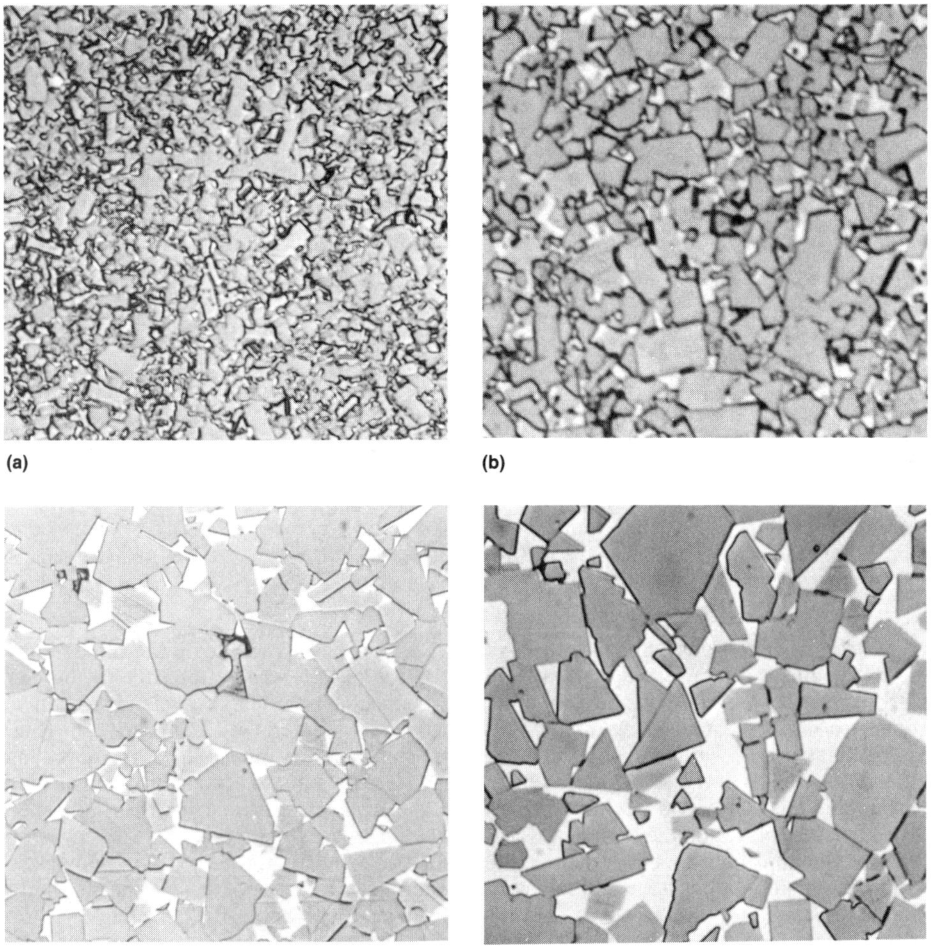

Fig. 3 Microstructures of straight WC-Co alloys. (a) 97WC-3Co alloy, medium grain size. (b) 94WC-6Co alloy, medium grain. (c) 94WC-6Co alloy, coarse grain. (d) 85WC-15Co alloy, coarse grain. All etched with Murakami's reagent for 2 min. 1500×

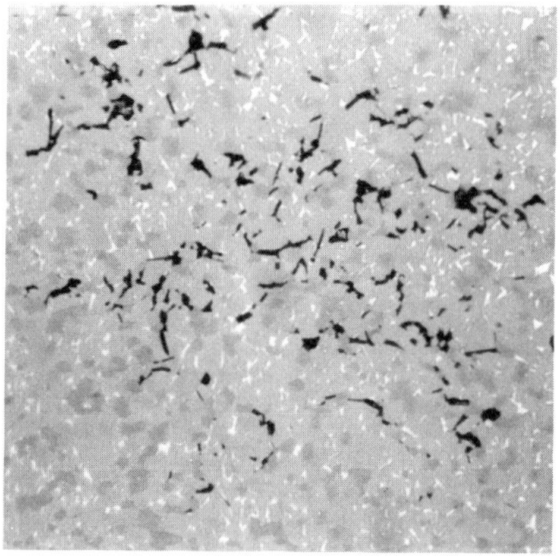

Fig. 4 Free graphite in a tungsten carbide alloy. Black areas contain graphite and are an example of C-type porosity. Polished 86WC-8(Ta,Ti,Nb)C-6Co alloy. 1005×

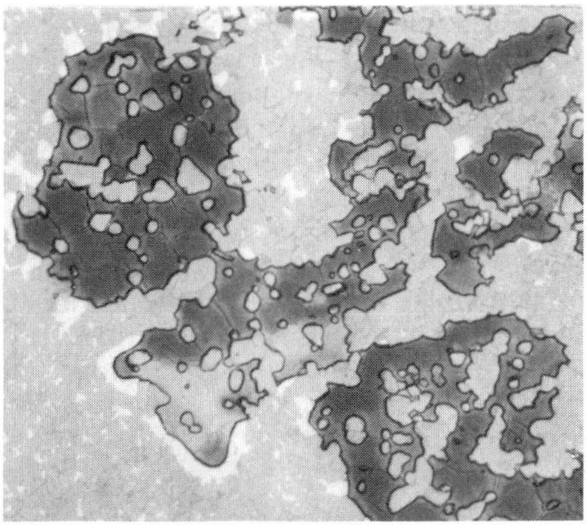

Fig. 5 η phase microstructure. Micrograph shows a $(Co_3W_3)C$-η phase in detail. η phase appears as various shades of gray with clearly defined grain boundaries. Light gray WC particles surrounded by η phase are rounded because of the solubility of tungsten carbide in the binder. 85WC-8(Ta,Ti,Nb)C-7Co alloy etched with Murakami's reagent for 3 s. 900×

Fig. 6 Submicron carbide grain size. 94WC-6Co alloy. Etched with Murakami's reagent for 2 min. 1500×

distribution of the phases vary widely, depending on the amounts and grain sizes of the raw materials employed and on the method of manufacture. Similarly, the properties of these complex alloys also vary widely, as indicated in Table 2 for a few representative steel-cutting grades.

Classification of Cemented Carbides

There is no universally accepted system for classifying cemented carbides. The systems most often employed by producers and users are discussed below. Each system has inherent strengths and weaknesses in describing specific materials, and for this reason close cooperation between user and producer is the best means of selecting the proper grade for a given application.

C-Grade System. The U.S. carbide industry uses an application-oriented system of classification to assist in the selection of proper grades of cemented carbides. This C-grade system does not require the use of trade names for identifying specific carbide grades (Table 3). Although this classification simplifies tool application, it does not reflect the material properties that significantly influence selection of the proper carbide grade. Additionally, the definitions of work materials involved in this classification scheme are imprecise. There is also no universal agreement on the meanings of the terms used to describe the various application categories. Despite these limitations, the C-grade classification has been successfully used by the manufacturing industry since 1942.

ISO Classification. In 1964, the International Organization of Standardization (ISO) issued ISO Recommendation R513 "Application of Carbides for Machining by Chip Removal." The basis for the ISO classification of carbides is summarized in Table 4.

In the ISO system, all machining grades are divided into three color-coded groups:

- Highly alloyed tungsten carbide grades (letter P, blue color) for machining steel
- Alloyed tungsten carbide grades (letter M, yellow color, generally with less TiC than the corresponding P series) for multipurpose use, such as steels, nickel-base superalloys, and ductile cast irons

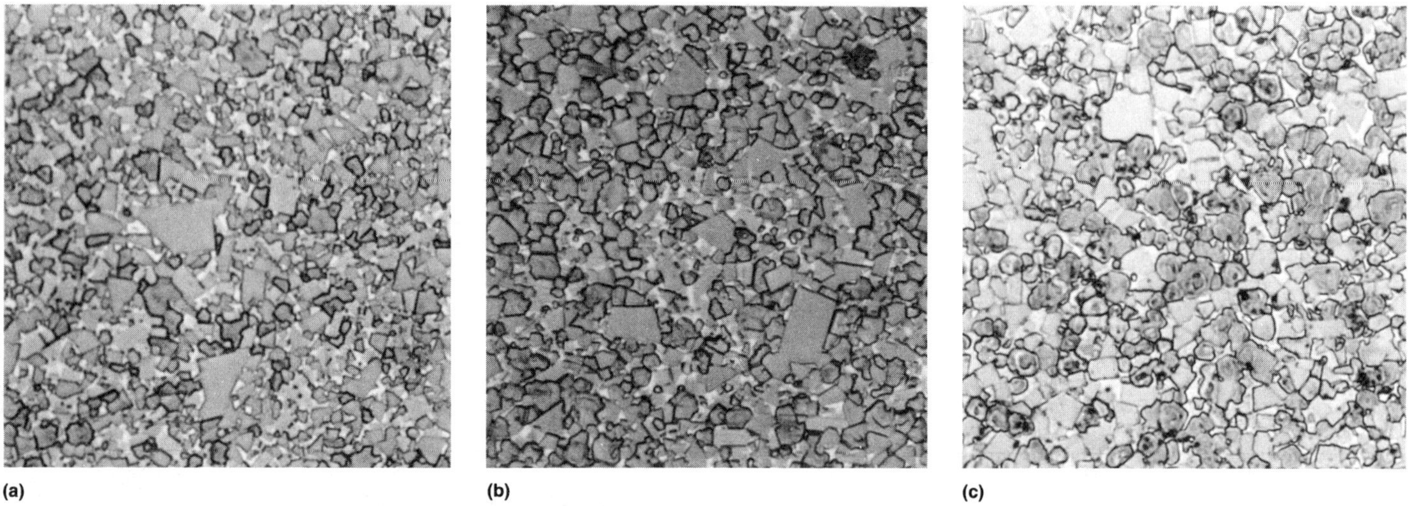

Fig. 7 Representative microstructures of steel-cutting grades of cemented tungsten carbide. (a) 85WC-9(Ta,Ti,Nb)C-6Co alloy, medium grain size. (b) 78WC-15(Ta,Ti,Nb)C-7Co alloy, medium grain. (c) 73WC-19(Ta,Ti,Nb)C-8Co alloy, medium grain. The gray, angular particles are WC, and the dark gray, rounded particles are solid-solution carbides. The white areas are cobalt binder. All etched with Murakami's reagent for 2 min. 1500×

- Straight tungsten carbide grades (letter K, red color) for cutting gray cast iron, nonferrous metals, and nonmetallic materials

Each grade within a group is assigned a number to represent its position from maximum hardness to maximum toughness. P-grades are rated from 01 to 50, M-grades from 10 to 40, and K-grades from 01 to 40. Typical applications are described for grades at more or less regular numerical intervals. Although the coated grades had not been developed at the time the ISO classification system was prepared, one should be able to classify them as easily as the uncoated grades.

Tool Wear Mechanisms

The cutting of metals involves extensive plastic deformation of the workpiece ahead of the tool tip, high temperatures, and severe frictional conditions at the interfaces of the tool, chip, and workpiece. Most of the work of plastic deformation and friction is converted into heat. In cutting, about 80% of this heat leaves with the chip, but the other 20% remains at the tool tip, producing high temperatures (≧1000 °C, or 1800 °F) (Ref 13). The stresses on the tool tip are also high; the actual values are dependent on the workpiece material and the machining conditions. In addition, the tool may experience repeated impact loads during interrupted cuts, and the freshly produced chips may chemically interact with the tool material. The cutting tool is thus subjected to a variety of hostile conditions.

The performance of a tool material is dictated by its response to the above conditions existing at the tool tip. High temperatures and stresses can cause blunting from the plastic deformation of the tool tip, and high stresses may lead to catastrophic fracture. In addition to plastic deformation and fracture, the service life of cutting tools is determined by a number of wear processes, such as crater wear, attrition wear, flank or abrasive wear, thermal fatigue, and depth-of-cut notching.

Crater wear (Fig. 8a and b) occurs on the rake face, where the tool temperatures are higher. Crater wear is caused by a chemical interaction between the rake face of the insert and the hot metal chip flowing over the tool. This interaction may involve diffusion or dissolution of the tool material into the chip. The chemical inertness of the tool material (or its coating) relative to the workpiece material is a requisite property for crater resistance.

Attrition Wear and Built-Up Edge. If machining is done at relatively low speeds, and if the tool tip temperature is not high enough for crater wear or deformation to be significant, attrition may become the dominant wear process. The attrition process may occur when there is an intermittent flow of workpiece material, and this condition is usually associated with a built-up edge (Fig. 9). Built-up edge occurs at low metal cutting speeds and is not a serious problem as long as the edge remains intact on the tool. When machining gray cast iron, built-up edges do not break away from the tool. However, when machining steel at low speeds, built-up edges break off easily. This may result in attrition wear if small fragments of tool material are carried away as the built-up edge breaks off. In some cases, large chunks of tool material may even be carried away. Built-up edge and attrition wear can be minimized by increasing the metal cutting speed, by selecting fine-grain WC-Co alloys, and/or by using positive-rake tools with smooth surface finishes.

Flank or abrasive wear (Fig. 8a and c) is often observed on the flank face and is related to the hardness of the tool material or coating. Harder materials provide greater flank and abrasive wear resistance.

Cemented carbide tools may be subjected to abrasive wear when abrasive particles (such as sand on the surface of castings) or hard carbide or alumina inclusions are present in the workpiece materials. Tools with lower binder contents and/or finer carbide grain sizes can resist abrasive wear.

Thermal Fatigue. Cemented carbide tools sometimes exhibit a series of cracks perpendicular to the tool edge when applied to interrupted cutting operations such as milling (Fig. 10). These thermal cracks are caused by the alternating expansion and contraction of the tool surface as it is heated during the cut and cooled outside the cut. The cracks initiate on the rake face, then spread across the edge and down the flank face of the tool. With prolonged intermittent cutting, lateral cracks appear parallel with,

Table 3 C-grade classification of cemented carbides

C-grade	Application category
Machining of cast iron, nonferrous, and nonmetallic materials	
C-1	Roughing
C-2	General-purpose machining
C-3	Finishing
C-4	Precision finishing
Machining of carbon and alloy steels	
C-5	Roughing
C-6	General-purpose machining
C-7	Finishing
C-8	Precision finishing
Nonmachining applications	
C-9	Wear surface, no shock
C-10	Wear surface, light shock
C-11	Wear surface, heavy shock
C-12	Impact, light
C-13	Impact, medium
C-14	Impact, heavy

Table 4 ISO R513 classification of carbides according to use for machining

Designation(a)	Groups of application: Material to be machined	Groups of application: Use and working conditions
P 01	Steel, steel castings	Finish turning and boring; high cutting speeds, small chip section, accuracy of dimensions and fine finish, vibration-free operation
P 10	Steel, steel castings	Turning, copying, threading, and milling; high cutting speeds, small or medium chip sections
P 20	Steel, steel castings, malleable cast iron with long chips	Turning, copying, milling, medium cutting speeds and chip sections; planing with small chip sections
P 30	Steel, steel castings, malleable cast iron with long chips	Turning, milling, planing, medium or low cutting speeds, medium or large chip sections, and machining in unfavorable conditions(b)
P 40	Steel, steel castings with sand inclusion and cavities	Turning, planing, slotting, low cutting speeds, large chip sections with the possibility of large cutting angles for machining in unfavorable conditions(b) and work on automatic machines
P 50	Steel, steel castings of medium or low tensile strength, with sand inclusion and cavities	For operations demanding very tough carbide: turning, planing, slotting, low cutting speeds, large chip sections, with the possibility of large cutting angles for machining in unfavorable conditions(b) and work on automatic machines
M 10	Steel, steel castings, manganese steel, gray cast iron, alloy cast iron	Turning, medium or high cutting speeds; small or medium chip sections
M 20	Steel, steel castings, austenitic or manganese steel, gray cast iron	Turning, milling; medium cutting speeds and chip sections
M 30	Steel, steel castings, austenitic steel, gray cast iron, high-temperature resistant alloys	Turning, milling, planing; medium cutting speeds, medium or large chip sections
M 40	Mild free-cutting steel, low-tensile steel, nonferrous metals, and light alloys	Turning, parting off, particularly on automatic machines
K 01	Very hard gray cast iron, chilled castings of over 85 scleroscope hardness, high-silicon aluminum alloys, hardened steel, highly abrasive plastics, hard cardboard, ceramics	Turning, finish turning, boring, milling, scraping
K 10	Gray cast iron over 220 HB, malleable cast iron with short chips, hardened steel, silicon aluminum alloys, copper alloys, plastics, glass, hard rubber, hard cardboard, porcelain, stone	Turning, milling, drilling, boring, broaching, scraping
K 20	Gray cast iron up to 220 HB, nonferrous metals: copper, brass, aluminum	Turning, milling, planing, boring, broaching, demanding very tough carbide
K 30	Low-hardness gray cast iron, low-tensile steel, compressed wood	Turning, milling, planing, slotting, for machining in unfavorable conditions(b) and with the possibility of large cutting angles
K 40	Softwood or hardwood, nonferrous metals	Turning, milling, planing, slotting, for machining in unfavorable conditions(b) and with the possibility of large cutting angles

(a) In each letter category, low designation numbers are for high speeds and light feeds; higher numbers are for slower speeds and/or heavier feeds. Also, increasing designation numbers imply increasing toughness and decreasing wear resistance of the cemented carbide materials. (b) Unfavorable conditions include shapes that are awkward to machine; material having a casting or forging skin; material having variable hardness; and machining that involves variable depth of cut, interrupted cut, or moderate to severe vibrations.

and close to, the cutting edge. The thermal and lateral cracks may join together and cause small fragments of tool material to break away. The resistance of WC-Co tools to thermal fatigue can generally be improved by TaC additions.

Depth-of-cut notching (Fig. 8a and d) consists of a high degree of localized wear on both the rake face and the flank face at the depth of the cut line. Notching is common when machining materials that work harden, such as austenitic stainless steels or high-temperature alloys. This type of wear is attributed to the chemical reaction of the tool material with the atmosphere or to abrasion by the hard, sawtooth outer edge of the chip. Depth-of-cut notching may lead to tool fracture; it can be minimized by:

- Increasing the fracture toughness of the tool material
- Increasing the lead angle (for round inserts) or the side cutting angle (for other insert shapes)
- Chamfering the tool edge
- Varying the depth of cut if multiple passes are made

Cemented Carbide Properties

Evaluation of the physical and mechanical properties of tool materials is an important prerequisite to the selection of grades for a given metal cutting application and for tool material development. A number of industry, national, and ISO standards have been developed for determining the selected properties of cemented carbides (Table 5).

Hardness determines the resistance of a material to abrasion and wear. It is affected not only by composition but also by the level of porosity and microstructure. For straight WC-Co alloys of comparable WC grain size, hardness and abrasion resistance decrease with increasing cobalt content (Fig. 11a and b). However, because both composition and microstructure affect hardness, cobalt content and grain size must be considered. At a given cobalt level, hardness improves with decreasing WC grain size (Fig. 11a).

In cemented carbides, hardness is measured by the Rockwell A-scale diamond cone indentation test (HRA) or by the Vickers diamond pyramid indentation test (HV). Both tests are performed on a finely ground, lapped or polished planar surface placed at right angles to the indentor axis. The Rockwell A test employs a load of 60 kg, whereas a range of loads can be used in the Vickers test. For cemented carbides used in machining applications, hardness values range from 88 to 94 HRA and from 1100 to 2000 HV.

Although the Rockwell scale has been used for decades as a measure of hardness, a true indication of the resistance to plastic deformation in metal cutting operations can be obtained only by measuring hardness at elevated temperatures. Measurements of hardness over a wide range of temperatures are therefore valuable for tool selection.

Hardness testers with high-temperature capability (up to 1200 °C, or 2200 °F) are commercially available and are being increasingly used by the cemented carbide industry. Figure 12 shows hot hardness data for a number of cemented carbides. The hardness of these materials decreases monotonically with increasing temperatures.

Compressive Properties. One of the unique properties of cemented carbides is their high compressive strength. Uniaxial compression tests can be performed on straight cylindrical samples or on cylinders having reduced diameters in the middle to localize fracture. The compressive strengths of cemented carbides are greater than those of most other materials. Typical values of compressive strength range from 3.5 to 7.0 GPa (0.5 to 1.0 × 10^6 psi).

The ductility of cemented carbides is generally low at room temperature, so there is little difference between their yield strength and fracture strength. At higher temperatures, however, these materials exhibit a small but finite amount of ductility. Measurement of yield strength is therefore more appropriate at elevated temperatures. High-temperature compressive yield strength is typically measured at 0.2% offset strain. Compression tests are performed in a high-temperature furnace (typically with resistance heating) under a vacuum or in an inert atmosphere. Figure 13 shows yield strength

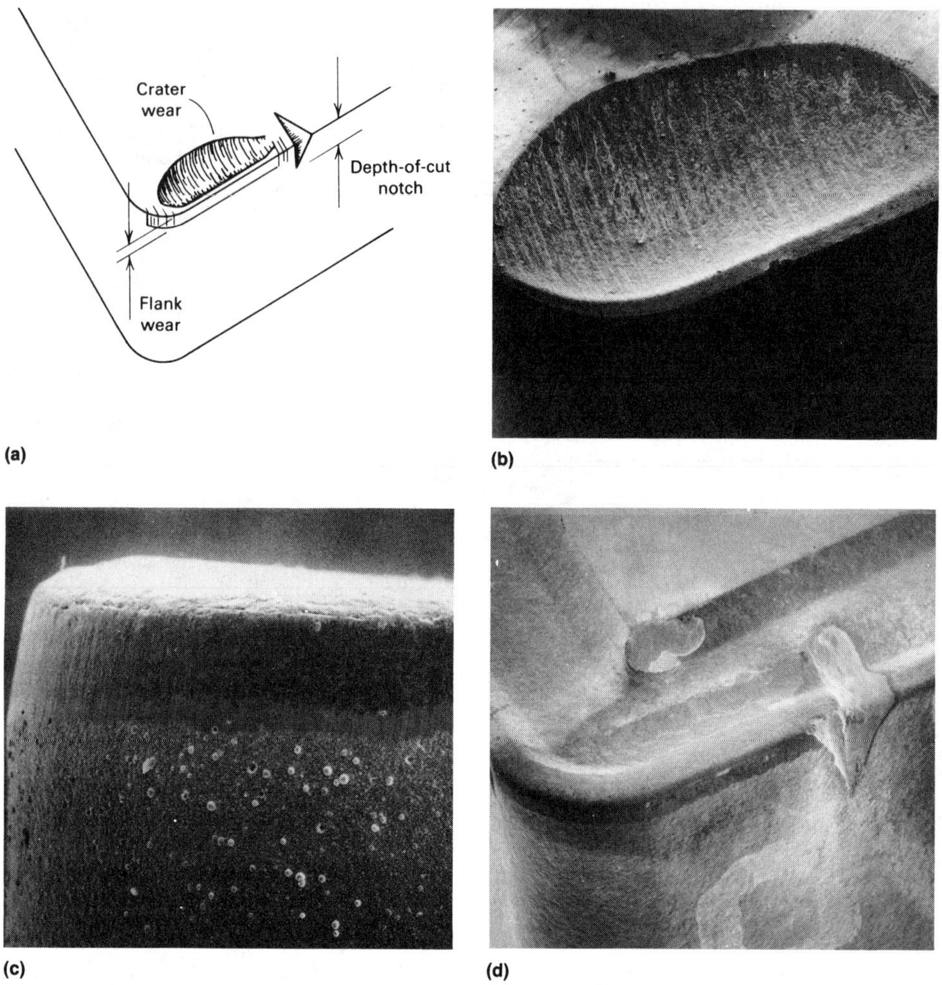

Fig. 8 Crater wear, flank wear, and depth-of-cut notch wear processes. (a) Schematic of wear mechanisms. (b) Crater wear on a cemented carbide tool produced during the machining of plain carbon steel. 15×. (c) Abrasive wear on the flank face of a cemented carbide tool produced during the machining of gray cast iron. 75×. (d) Depth-of-cut notching on a cemented carbide tool produced during the machining of a nickel-base superalloy. 15×

Table 5 Test methods for determining the properties of cemented carbides

Property	Test method: ASTM/ANSI	CCPA(a)	ISO
Abrasive wear resistance	B 611	P112	···
Apparent grain size	B 390	M203	···
Apparent porosity	B 276	M201	4505
Coercive force	···	···	3326
Compressive strength	E 9	P104	4506
Density	B 311	P101	3369
Fracture toughness	···	···	···
Hardness, HRA	B 294	P103	3738
Hardness, HV	E 92	···	3878
Linear thermal expansion	B 95	P108	···
Magnetic permeability	A 342	P109	···
Microstructure	B 657	M202	4499
Poisson's ratio	E 132	P105	···
Transverse rupture strength	B 406	P102	3327
Young's modulus	E 111	P106	3312

(a) Cemented Carbides Producers Association

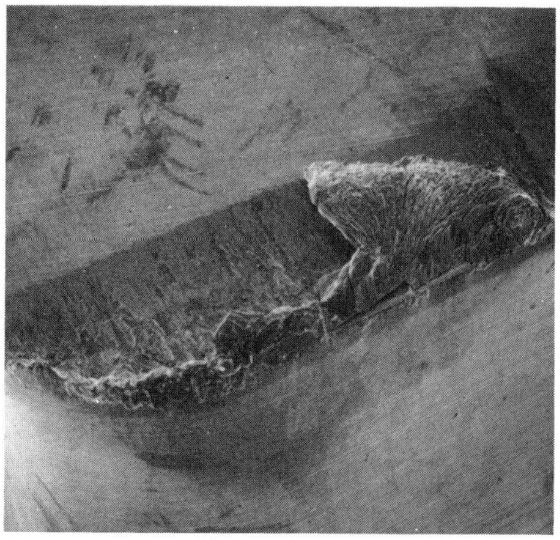

Fig. 9 Built-up edge on a cemented carbide tool. The built-up edge was produced during the low-speed machining of a nickel-base alloy. 20×

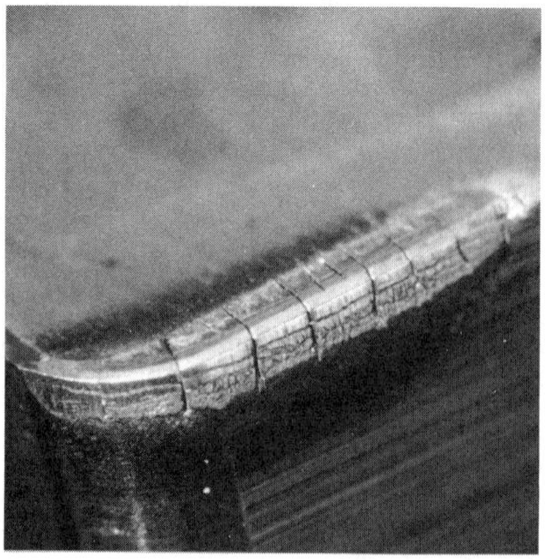

Fig. 10 Thermal cracks in a cemented carbide insert. The thermal cracks are perpendicular to the cutting edge, and the mechanical cracks are parallel to the cutting edge. 15×

data for selected straight and alloyed WC-Co grades. Like hardness, the compressive yield strengths of cemented carbides decrease monotonically with increasing temperature; the rate of decrease depends on the composition and the microstructure. As in metallic materials, fine-grain alloys tend to lose their yield strengths more rapidly with increasing temperature than coarse-grain grades, although at room temperature the former can exhibit high yield strengths.

Transverse Rupture Strength. The most common method of determining the fracture strength of cemented carbides is the transverse rupture test. In this test, a rectangular test bar is placed across two sintered carbide support cylinders, and a gradually increasing load is applied by a third carbide cylinder at the midpoint between the supports. Transverse rupture strength is determined from the dimensions of the test bar, the distance between the supports, and the fracture load. A disadvantage of this test is the large scatter in the experimental data resulting from surface defects introduced into the test specimens during processing. Nevertheless, it is an excellent quality control test, and it is particularly useful for large carbide components.

Figure 11(e) shows the variation in transverse rupture strength with cobalt content. In metal cutting applications, no clear relationship has been established between transverse rupture strength and turning performance. However, there appears to be a good correlation between transverse rupture strength and milling performance (Ref 14). During milling, the tool is subjected to tensile stresses as it leaves the cut, and a material with high transverse rupture strength should be able to resist fracture under these conditions.

Fracture toughness is less sensitive than transverse rupture strength to such extrinsic factors as specimen size, geometry, and surface finish. Fracture toughness is measured by the critical stress intensity factor K_{Ic} (Ref 15-17). This parameter indicates the resistance of a material to fracture in the presence of a sharp crack and thus provides a better measure of the intrinsic strength of the cemented carbide than transverse rupture strength. A variety of specimen geom-

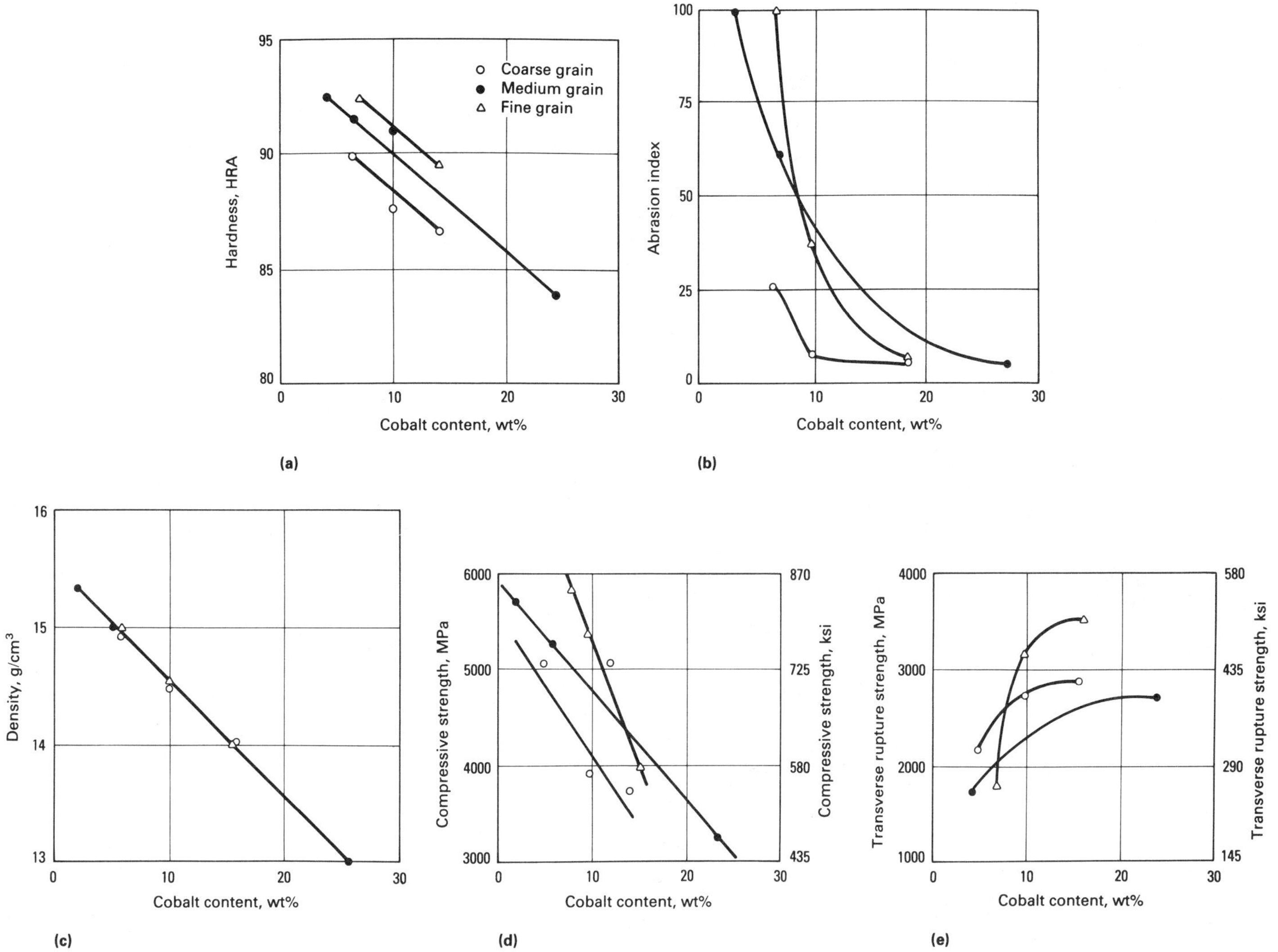

Fig. 11 Variation in properties with cobalt content and grain size for straight WC-Co alloys. (a) Variation in hardness. (b) Variation in abrasion resistance. (c) Variation in density. (d) Variation in compressive strength. (e) Variation in transverse rupture strength

etries have been used in this test, including the single-edge notched beam, the double cantilever beam, the compact tension specimen, and the double torsion specimen. The carbide industry in the United States generally uses commercial equipment for fracture toughness evaluation. The fracture toughness of cemented carbides increases with cobalt content and with WC grain size (Fig. 14). On the other hand, cubic carbide additions lessen the fracture toughness of WC-Co alloys.

As with the other mechanical properties, attention is focused on the development of test techniques to evaluate fracture toughness at elevated temperatures. Figure 15 shows K_{Ic} data for a number of cemented carbides from room temperature to about 1000 °C (1800 °F). Depending on the composition of the cemented carbide, the K_{Ic} parameter is insensitive to temperature, up to about 600 °C (1100 °F), but increases rapidly at higher temperatures. This behavior is reminiscent of the ductile-to-brittle transition observed in quenched-and-tempered steels.

The density, or specific gravity, of cemented carbides is very sensitive to composition and porosity in the sample and is widely used as a quality control test. Density values of cemented carbides range from 15 g/cm^3 for low-cobalt straight WC-Co alloys to about 10 or 12 g/cm^3 for highly alloyed carbide grades (Fig. 11c).

Magnetic Properties. Tungsten carbide-cobalt alloys lend themselves to the analysis of magnetic properties because cobalt is ferromagnetic. The properties measured are magnetic saturation and coercive force. Both free cobalt and solid solutions of cobalt and tungsten contribute to magnetization. The magnetic saturation of pure cobalt is 201×10^{-6} T m^3/kg. With additions of tungsten to cobalt, the magnetic saturation decreases steadily from 201×10^{-6} to about 151×10^{-6} T m^3/kg. In this range, the cemented carbide is characterized by two or three phases (WC and Co, or WC, a solid-solution carbide, and Co). Values below 151 $\times 10^{-6}$ T m^3/kg indicate the presence of η phase. The solubility of tungsten in cobalt is inversely proportional to carbon content. Lower-carbon alloys have more tungsten dissolved in cobalt and are characterized by lower magnetization. Magnetic saturation thus provides an accurate measure of the changes in carbon content in the cemented carbide alloy and is widely used as a quality control test.

The coercive force varies considerably with increasing sintering temperature and indicates the structural changes that take place during sintering. The coercive force of WC-Co alloys reaches a maximum at the optimum sintering temperature and decreases at higher temperatures because of grain growth. Therefore, measurement of the coercive force permits control of the sintering process. The factors influencing

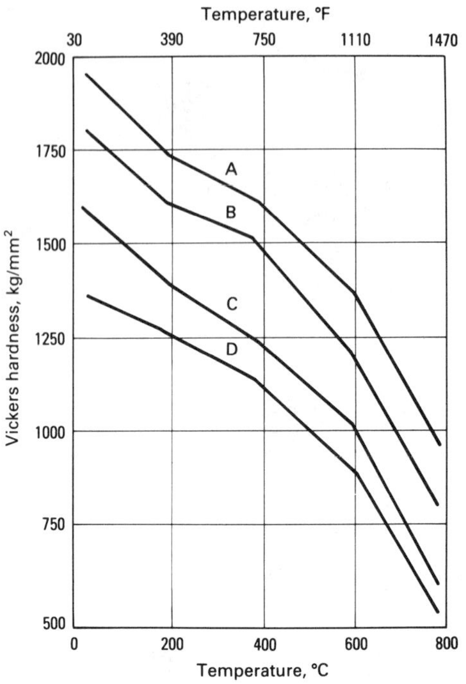

Fig. 12 Variation in microhardness with temperature. Microhardness is based on a 1 kg load, and all alloys are of medium WC grain size. A, 97WC-3Co alloy; B, 94WC-6Co; C, 80WC-12(Ti,Ta,Nb)C-8Co; D, 86WC-2TaC-12Co

the coercive force are complex, varied, and interactive. For a given cobalt and carbon content, the coercive force provides a measure of the degree of distribution of the carbide phase in the microstructure. Commercial units are available for the rapid measurement of magnetic saturation and coercive force in cemented carbides.

Porosity. The properties of a cemented carbide are dependent on its density, which in turn is critically dependent on composition and porosity. Porosity is evaluated on the as-polished material. The American Society for Testing and Materials (ASTM) has established a standard procedure (B 276) that rates three types of porosity:

- Type A, covering pore diameters less than 10 μm
- Type B, covering pore diameters between 10 and 25 μm
- Type C (Fig. 4), covering porosity developed by the presence of free carbon

Type A porosity is rated at a magnification of 200×, while types B and C porosity are rated at 100×. The degree of porosity is given by four numbers ranging in value from 02 to 08. The number provides a measure of pore volume as a percentage of total volume of the sample.

Thermal Shock Resistance. As discussed earlier in this article, cutting tool materials are subjected to thermal shocks during interrupted cutting operations such as milling. Resistance to thermal shock is therefore an important property that determines tool performance in milling. No laboratory test has yet been developed that can consistently predict the resistance to thermal shock of a tool. However, empirical parameters have been suggested that can be used to evaluate tool materials for their probable resistance to thermal shock (Ref 18). A commonly used parameter is $\sigma k/E\alpha$, where σ is the transverse rupture strength, k is the thermal conductivity, E is Young's modulus, and α is the coefficient of thermal expansion. Table 6 lists representative values of this parameter for a number of WC-Co alloys. In general, the higher the value of $\sigma k/E\alpha$, the better the thermal shock resistance.

Abrasive Wear Resistance. Most producers of cemented carbides use a wet-sand abrasion test to measure abrasion resistance. In this test, a sample is held against a rotating wheel for a fixed number of revolutions while the sample and wheel are immersed in a water slurry containing aluminum oxide particles. Comparative rankings are reported, usually on the basis of a wear rating based on the reciprocal of volume loss. Although standard test procedures are available, carbide producers have not agreed on a single test method, and so the values of abrasion resistance cited in the literature vary widely. Because of this variance, it is almost impossible to make valid comparisons among test results reported by different producers. It is also fallacious to use abrasion resistance as a measure of the wear resistance of cemented carbide materials when they are used for cutting steel or other materials; abrasion resistance in a

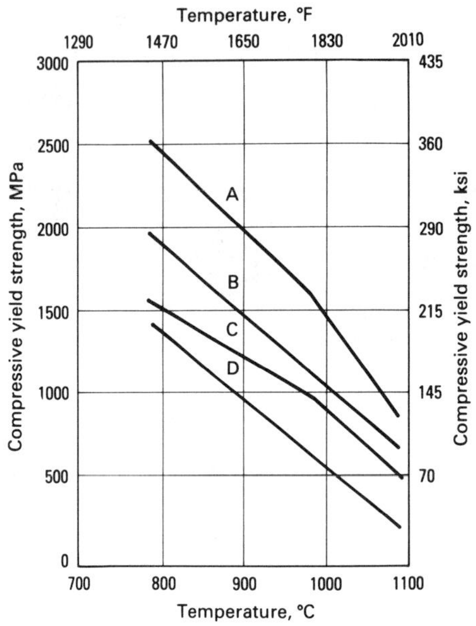

Fig. 13 Variation in compressive yield strength with temperature. Measured at 0.2% offset strain; all alloys characterized by medium grain size. A, 73WC-22(Ti,Ta,Nb)C-5Co alloy; B, 80WC-12(Ti,Ta,Nb)C-8Co; C, 86WC-8(Ti,Ta,Nb)C-6Co; D, 86WC-2TaC-12Co

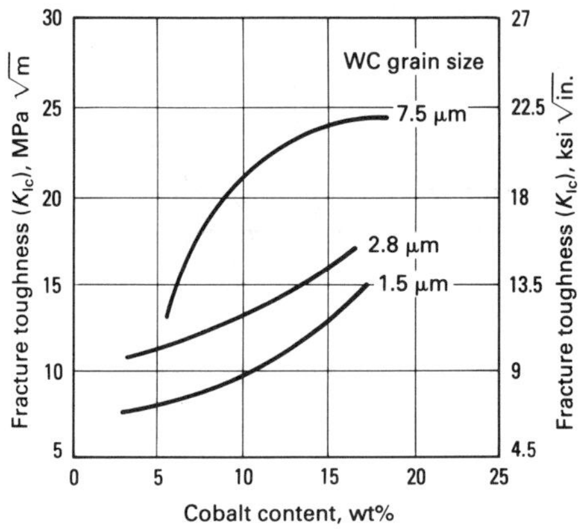

Fig. 14 Variation in fracture toughness (K_{Ic}) with cobalt content for WC-Co alloys with different WC grain sizes. Source: Ref 15

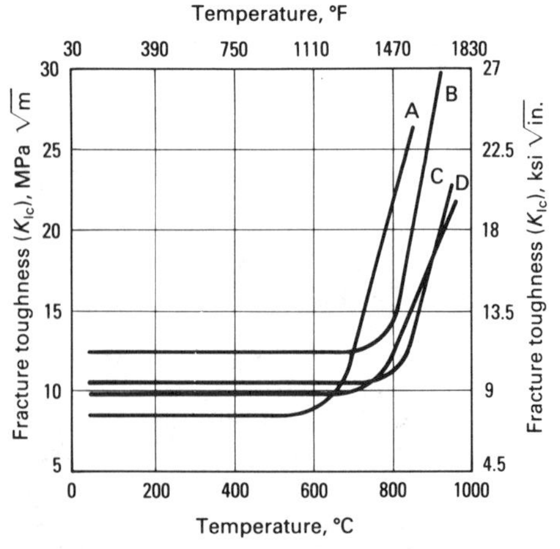

Fig. 15 Variation in fracture toughness (K_{Ic}) with temperature for a number of WC-Co base alloys. A, 86WC-2TaC-12Co; B, 85WC-9(Ti,Ta,Nb)C-6Co; C, 80WC-12(Ti,Ta,Nb)C-8Co; D, 96WC-4Co

Table 6 Thermal shock resistance parameters for cemented carbides

Composition, wt%	WC grain size	Transverse rupture strength, σ, MPa	Transverse rupture strength, σ, ksi	Thermal conductivity (k), W/m · K	Young's modulus, E, GPa	Young's modulus, E, 10^6 psi	Coefficient of thermal expansion (α), μm/m · K	Thermal shock resistance (σk/Eα), kW/m
97WC-3Co	Medium	1590	230	121	641	93	5.0	60
94WC-6Co	Medium	2000	290	100	648	94	5.4	57
90WC-10Co	Fine	3100	450	80	620	90	6.0	67
71WC-12.5TiC-12TaC-4.5Co	Medium	1380	200	35	565	82	6.5	13
72WC-8TiC-11.5TaC-8.5Co	Medium	1720	250	50	558	81	6.8	23

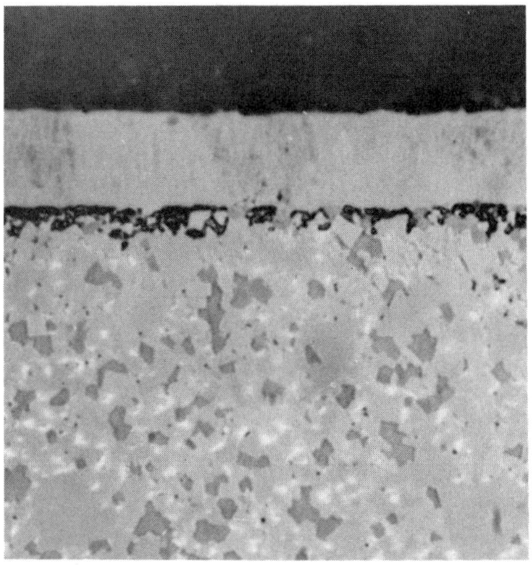

Fig. 16 Decarburization of a TiC coating. Micrograph shows the η phase at the coating/substrate interface of an 85WC-9(Ti,Ta,Nb)C-6Co alloy with an 8 μm (315 μin.) TiC coating. Etched with Murakami's reagent for 3 s. 1500×

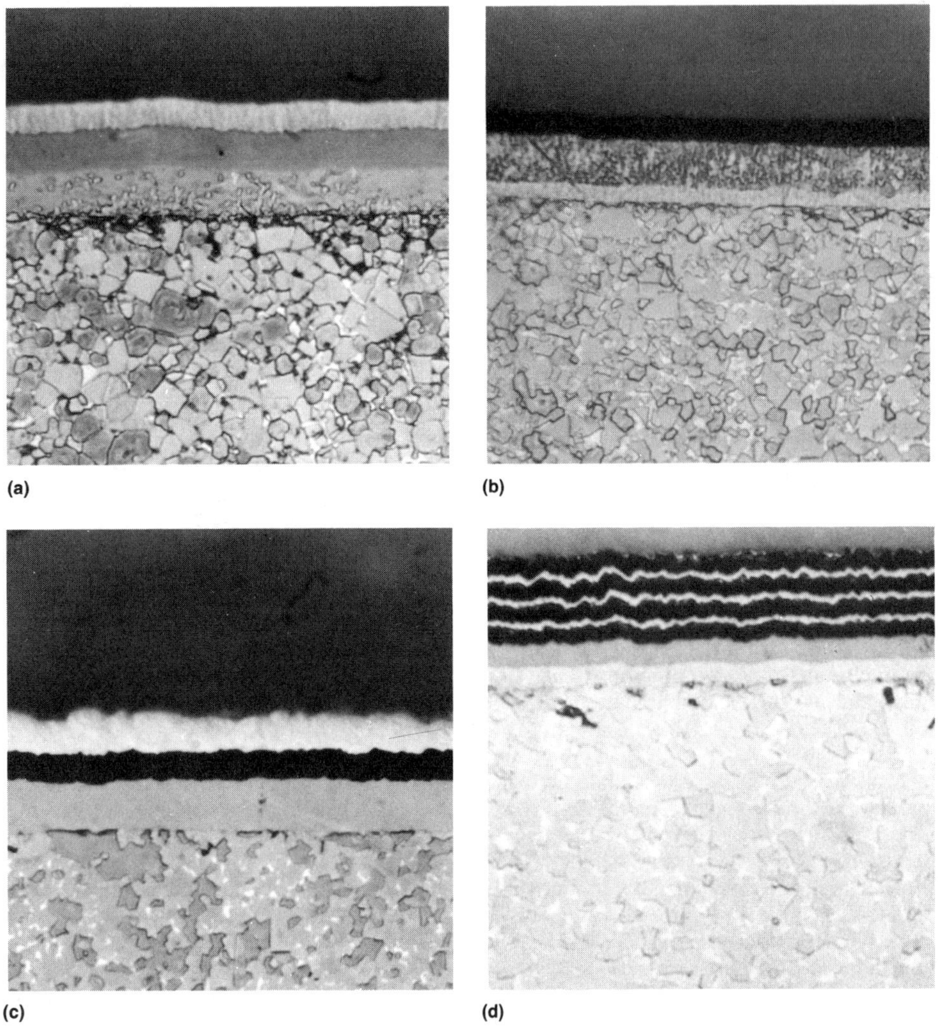

Fig. 17 Multilayer coatings of carbide substrates. (a) 73WC-19(Ti,Ta,Nb)C-8Co alloy with a TiC/TiCN/TiN coating of about 10 μm (400 μin.) in total thickness. (b) 85WC-9(Ti,Ta,Nb)C-6Co with a TiC/Al_2O_3 coating about 9 μm (350 μin.) thick. (c) 85WC-9(Ti,Ta,Nb)C-6Co with a TiC/Al_2O_3/TiN coating about 10 μm (400 μin.) thick. (d) 88WC-7(Ti,Ta,Nb)C-5Co with TiC/TiCN coating supporting multiple alternating coating layers of Al_2O_3 and TiC. All etched with Murakami's reagent for 3 s. 1500×

standard test does not correspond directly to wear resistance in machining operations.

Generally, the abrasion resistance of cemented carbides decreases as cobalt content or grain size is increased (Fig. 11b). Abrasion resistance is also lower for complex carbides than for straight WC grades having the same cobalt content.

Coated Carbide Tools

One of the challenges in the design of cemented carbide tools is the optimization of toughness associated with straight WC-Co alloys with the superior crater wear resistance of alloyed carbides containing high levels of titanium carbide. This challenge has led to the development of coated carbide tools.

Coated carbides account for a major portion of all commercial metal cutting inserts sold in the United States. The success of coated carbides is based on their proven ability to extend tool life on steels and cast irons by a factor of at least two to three. This is accomplished by a reduction in wear processes, especially at higher cutting speeds.

Laminated Coatings. An important development in the production of coated carbide tools occurred in the 1960s, when laminated tips consisting of a base of WC-Co alloy with a sintered layer of high TiC composition were produced. This development not only enabled higher cutting speeds in steel machining but also reduced crater wear on the tool. Although metal cutting productivity increased with the use of these laminated tools, the thermal expansion mismatch between the substrate and the surface layer caused thermal stresses during metal cutting, and the laminate tended to spall during use.

Chemical Vapor-Deposited Coatings. Further development of laminated tools was superseded in 1969 by the application of a thin layer (~5 μm, or 200 μin.) of hard TiC coating to the cemented carbide tool by chemical vapor deposition (CVD) (Ref 19). The impetus for this development came from the Swiss Watch Research Institute, where vapor-deposited TiC coatings had been used on steel watch parts and cases to combat wear on these components.

The CVD coating process consists of heating the tools in a sealed reactor with gaseous hydrogen at atmospheric or lower pressure; volatile compounds are added to the hydrogen to supply the metallic and nonmetallic constituents of the coating. For example, TiC coatings are produced by reacting $TiCl_4$ vapors with methane (CH_4) and hydrogen (H_2) at 900 to 1100 °C (1650 to 2000 °F). The reaction is:

$$TiCl_4(g) + CH_4(g) + H_2(g) \rightleftharpoons TiC(s) + 4HCl(g) + H_2(g)$$

During the TiC deposition process, a secondary reaction often occurs in which carbon is taken from the cemented carbide substrate:

$$TiCl_4(g) + C(s) + 2H_2(g) \rightleftharpoons TiC(s) + 4HCl(g)$$

The resulting surface decarburization leads to the formation of a brittle η phase at the coating/substrate interface (Fig. 16) and sometimes to premature tool failure due to excessive chipping and insufficient edge strength (Ref 20, 21). These problems are particularly prevalent in severe machining operations involving interrupted cuts.

Performance inconsistencies have been largely eliminated by a number of metallurgical and processing innovations. These include improvements in CVD coating technology, which have resulted in coatings with greater uniformity of thickness, more adherence, and more consistent morphology and microstructure with minimum interfacial η phase and associated porosity (Ref 22).

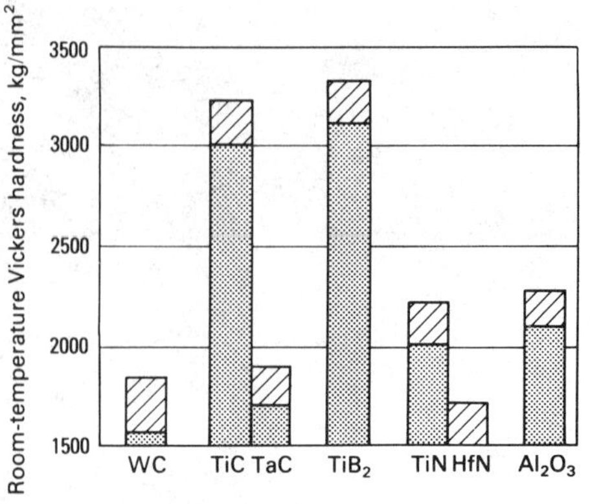

Fig. 18 Room-temperature microhardness of hard coating materials. The hatched area indicates the range of hardness normally observed in these materials. Source: Ref 24

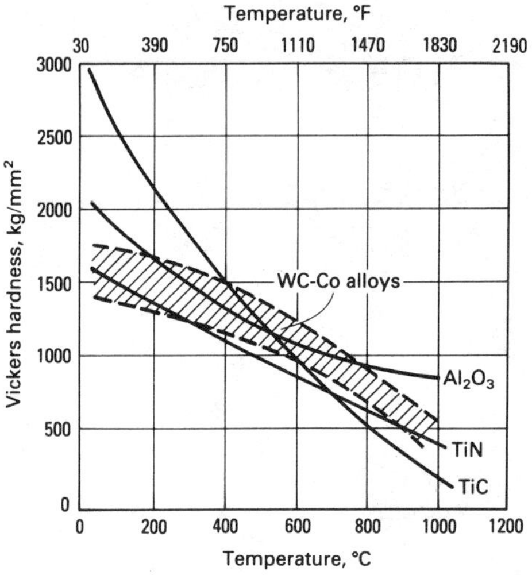

Fig. 19 Temperature dependence of hardness of TiC, Al_2O_3, and TiN. The range of hardness of WC-Co alloys is also shown.

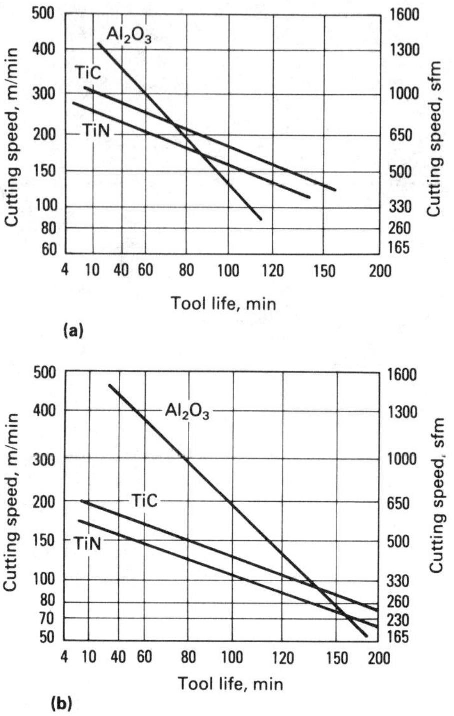

Fig. 20 Tool life diagrams of coated inserts. Tool life is based on a 0.25 mm (0.01 in.) flank wear criterion. (a) Turning 1045 steel with a 2.5 mm (0.1 in.) depth of cut and a 0.40 mm/rev (0.016 in./rev) feed rate. (b) Turning SAE G4000 gray cast iron with a 2.5 mm (0.1 in.) depth of cut and a 0.25 mm/rev (0.01 in./rev) feed rate. Source: Ref 25

Compositions of CVD coatings have also evolved from single-layer TiC coatings with narrow application ranges to multilayer hard coatings. Multilayer coatings have a nominal total thickness of about 10 μm (400 μin.) and use various combinations of TiC, TiCN, TiN, Al_2O_3, and occasionally HfN (Fig. 17).

Multilayer coatings are intended to suppress both crater wear and flank wear. There is also a trend toward the use of multiple alternating coating layers (Fig. 17d), which are believed to produce finer grain sizes and to minimize chipping. These improvements increase tool life and extend the range of application of the coated tool (Ref 23).

Hardness and Tool Life. Figure 18 shows the room-temperature Vickers microhardness of various hard coatings deposited on cemented carbide substrates. The range of hardness for WC-Co alloys is shown for reference. The compounds TiC and TiB_2 have the highest relative hardness of about 3000 kg/mm^2 and offer the best protection against abrasive wear. With increases in temperature, however, TiC loses hardness rapidly, whereas Al_2O_3 retains its hardness to higher temperatures.

At 1000 °C (1800 °F), which is the temperature typically reached on the rake face of the tool during high-speed machining (Ref 13), Al_2O_3 has the highest hardness, followed by TiN and then TiC (Fig. 19). For comparison, the hardness of WC-Co substrates is also shown in Fig. 19. The hot hardness data indicate that the TiC coating would be more effective at the flank face, which rarely exceeds 600 °C (1100 °F) during machining. At lower speeds, TiC also offers adequate rake face protection from abrasive wear. On the other hand, Al_2O_3 would provide better abrasion resistance at higher speeds, as judged from its high hardness at elevated temperatures.

These concepts are illustrated in tool life plots (Fig. 20) based on flank wear criterion for TiC, TiN, and Al_2O_3 coated cemented carbide inserts in turning 1045 steel and gray cast iron workpiece materials. At lower speeds, the TiC coating provides the longest tool life, followed by TiN and Al_2O_3. At higher speeds, the ranking is altered. These rankings reflect the varying temperature dependence of their microhardness shown in Fig. 19.

Hard coatings also reduce frictional forces at the chip/tool interface, which in turn reduces the heat generated in the tool and thus results in lower tool tip temperatures. The compound TiN is particularly noted for this lubricity effect (Ref 26).

Diffusion Wear. The standard free energy of formation is given in Fig. 21 for WC, TiN, TiC, and Al_2O_3 from room temperature to about 2000 °C (3600 °F). This thermodynamic property gives an indication of the extent to which these materials will undergo diffusion wear. The most stable of these materials at all temperatures is Al_2O_3.

The dissolution rate of coating materials into the workpiece can also determine the rate of diffusive wear on the tool (Ref 24). The relative dissolution rates of various coating materials into steel at different temperatures are given in Table 7. The dissolution rate of Al_2O_3 in steel is an order of magnitude lower than the dissolution rates of other refractory compounds. Therefore, Al_2O_3 is the most crater-resistant material and is a very effective coating for cemented carbide tools used in the high-speed machining of steel.

Thermal Expansion and Coating Adhesion. The high temperatures employed for CVD coating generally ensure good bonding between the substrate and the coating. However, coating adhesion can be adversely affected by stresses caused by the thermal expansion mismatch between the substrate and the coating. Table 8 lists the thermal properties of various coating materials as well as WC-Co substrates. The thermal expansion mismatch is lowest for TiC and highest for TiN, whereas the thermal expansion coefficients of the coating materials listed are higher than those of the substrates. As a result, hard coatings on cemented carbide substrates are in residual tension at room temperature. Because the stresses are most severe at tool corners, the CVD-coated tools must be honed before coating. Another reason for honing is to minimize the formation of η phase, which tends to develop to a greater extent at sharp tool edges.

Cobalt Enrichment. Although the early coated tools substantially improved metal

Table 7 Relative dissolution rates of coating materials into iron

Rates are relative to TiC.

Material	Dissolution rate at: 100 °C (212 °F)	500 °C (930 °F)	1100 °C (2000 °F)
WC	1.1×10^{10}	5.4×10^{4}	3.2×10^{2}
TiC	1.0	1.0	1.0
TaC	2.3	1.2	8.0×10^{-1}
TiB_2	9.9×10^{1}	8.5	2.8
TiN	1.0×10^{-8}	1.8×10^{-3}	2.2×10^{-1}
HfN	2.5×10^{-12}	3.8×10^{-5}	2.5×10^{-2}
Al_2O_3	1.1×10^{-24}	8.9×10^{-11}	4.1×10^{-5}

Source: Ref 24

Table 8 Tool material thermal properties

Material	Melting point °C	Melting point °F	Coefficient of thermal expansion, μm/m · K	Thermal conductivity, W/m · K 100 °C (212 °F)	500 °C (930 °F)	1000 °C (2000 °F)
WC-Co	...	...	5–6	38–80	...	...
Co	1492	2717	12.3	70	...	...
WC	~2800	5070	~5	120	...	...
TiC	3100	5612	7.7	33	37	41
TiN	2950	5342	9.4	21	23	26
Al_2O_3	2050	3722	8.4	28	13	6
$M_{12}C$	...	...	7–10	...	...	...

cutting productivity, they were prone to catastrophic fracture when applied at high feed rates or in intermittent cutting operations. One solution to the problem of coating-related tool fracture is to improve the fracture toughness of the substrate by increasing its cobalt content and/or by increasing the binder mean free path (mean thickness of the binder between WC grains). Unfortunately, this approach results in decreased deformation resistance, which can cause tool tip blunting.

A major breakthrough in resolving this conflict between fracture toughness and deformation resistance occurred in the late 1970s, when a TiC/TiCN/TiN-coated tool was developed with a peripheral cobalt-enriched zone (10 to 40 μm, or 400 to 1600 μin., thick), which provided superior edge strength while maintaining the edge and crater wear resistance of the coating layers (Ref 27). The cobalt-enriched zone contained essentially a straight WC-Co composition with nearly three times the nominal cobalt level, while the bulk of the tool insert, containing higher levels of solid-solution cubic carbide and less cobalt, provided the necessary deformation resistance. Thus, a single tool combined the fracture resistance of a high-cobalt alloy with the deformation resistance of a lower-cobalt alloy.

The ability of the cobalt-enriched tool to resist edge chipping and catastrophic fracture is illustrated in Fig. 22, which compares a cobalt-enriched insert with a nonenriched tool after an edge strength test. Tool failures such as that shown in Fig. 22(b) can ruin the workpiece and result in excessive production downtime. Such tool failures can be particularly damaging in flexible machining systems and in untended machining centers.

The development of the cobalt-enriched tool permitted the use of heavy interrupted cuts, such as those encountered in scaled forgings and castings. However, optimum performance was obtained only at lower speeds. Further refinements to the cobalt enrichment concept have expanded the application range of this type of tool to higher speeds (Ref 28). Cutting speeds for cobalt-enriched tools range from 60 to 300 m/min (200 to 1000 sfm), with feeds ranging from 0.1 to 1 mm/rev (0.005 to 0.050 in./rev). These capabilities are suitable for most metal removal applications. Two typical cobalt-enriched microstructures that can cover such a range of machining speeds and feeds are illustrated in Fig. 23 and 24.

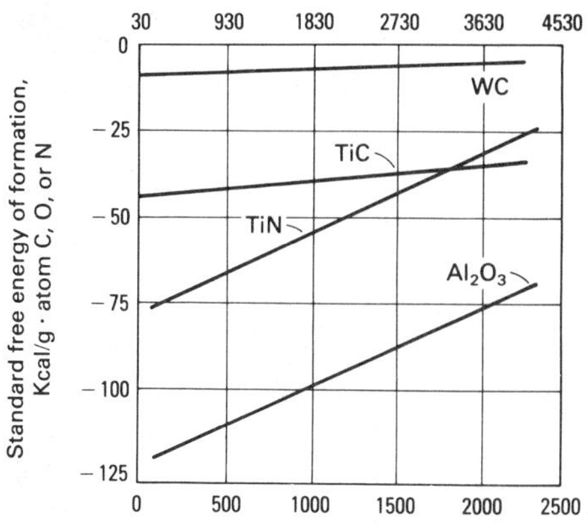

Fig. 21 Variation in the standard free energy of formation of WC, TiN, TiC, and Al_2O_3 with temperature. This parameter gives an indication of the extent to which these materials will undergo diffusion wear.

Physical vapor deposition (PVD) has recently emerged as a commercially viable process for applying hard TiN coatings onto high-speed steel tools (Ref 29). Physical vapor deposition typically employs lower temperatures (~500 °C, or 900 °F) than chemical vapor deposition. Therefore, the PVD process is attractive for use with cemented carbide tools because the lower deposition temperature prevents η phase formation and provides for refinement of the grain size of the coating layer.

It has been shown by standard three-point bend tests that CVD coatings reduce the transverse rupture strength of cemented carbide tools by as much as 30% because of the presence of interfacial η phase and/or tensile residual stress within the coating (Fig. 25). On the other hand, PVD coatings

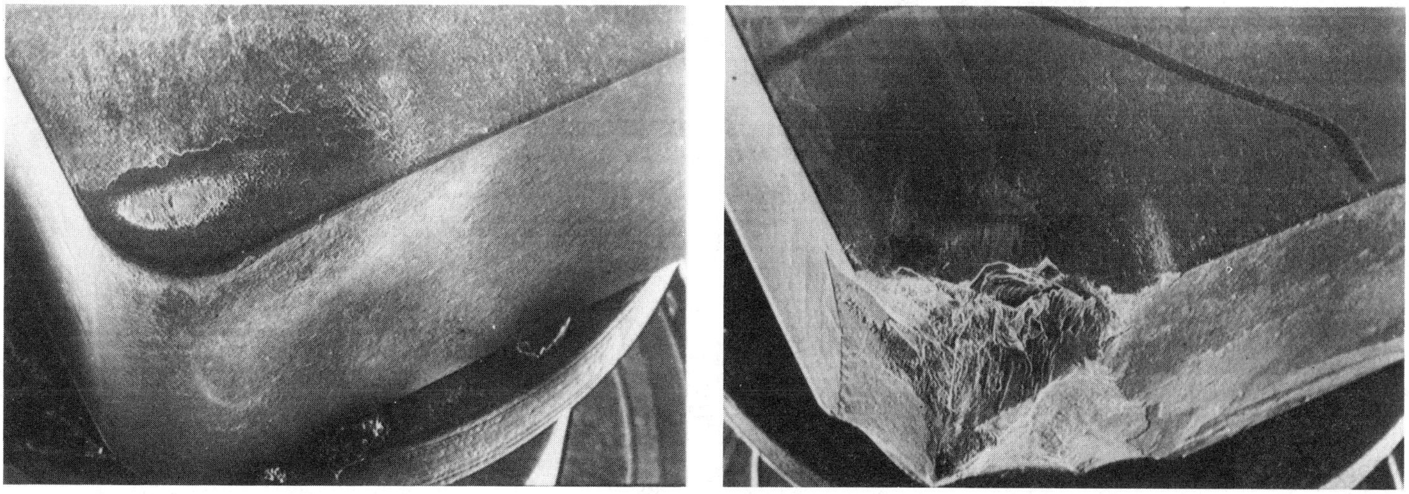

(a) (b)

Fig. 22 Scanning electron macrographs of cobalt-enriched and nonenriched tools after a slotted-bar strength test. Machining parameters: depth of cut, 2.5 mm (0.1 in.); speed, 107 m/min (350 sfm); feed, 0.50 mm/rev (0.02 in./rev). The workpiece was AISI 41L50 steel; average hardness: 26 HRC. (a) Cobalt-enriched tool after 1200 impacts. 10×. (b) Nonenriched tool after 220 impacts. 11×

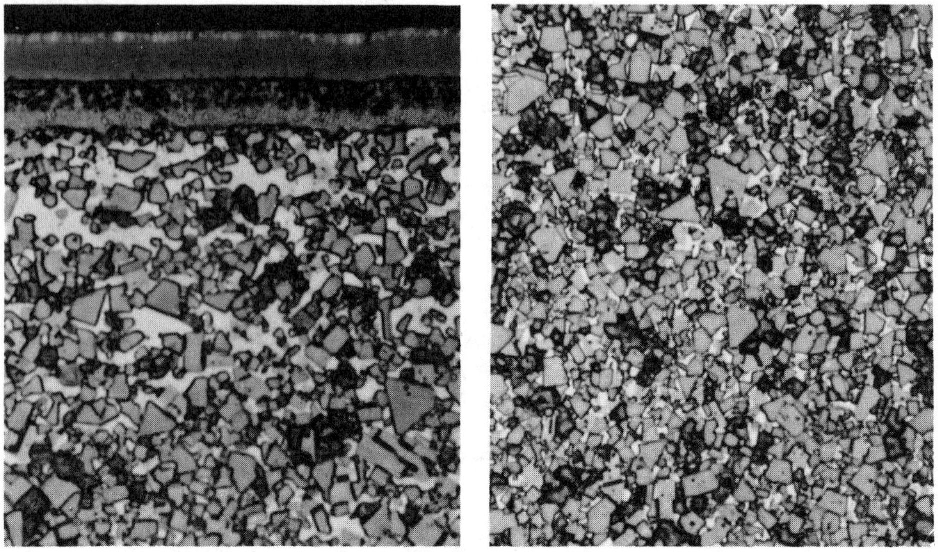

(a) (b)

Fig. 23 Microstructure of a cobalt-enriched coating. 86WC-8(Ti,Ta,Nb)C-6Co tool with a TiC/TiCN/TiN coating. (a) Cobalt-enriched periphery (beneath the coating). (b) Bulk microstructure. Both etched with Murakami's reagent for 2 min. 1500×

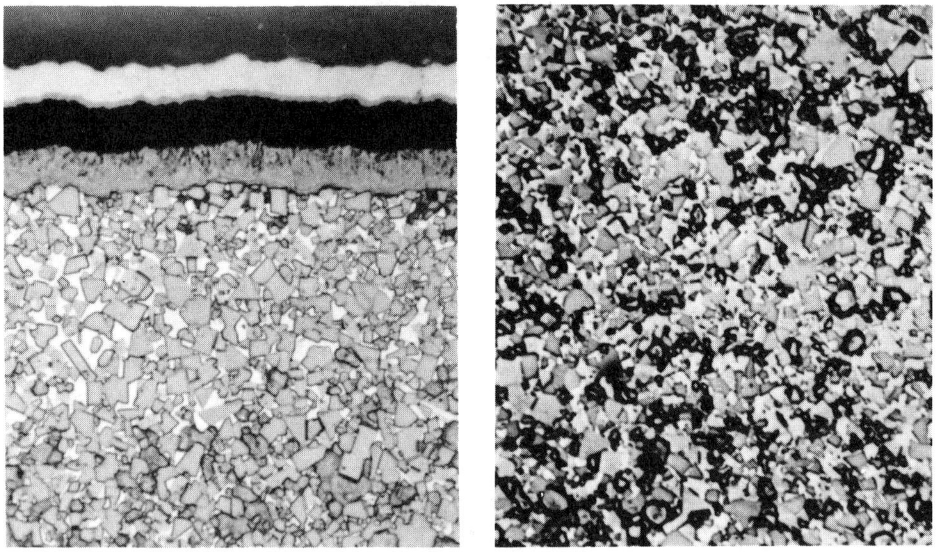

(a) (b)

Fig. 24 Microstructure of a second-generation cobalt-enriched coated tool. 85WC-9(Ti,Ta,Nb)C-6Co tool with a TiC/Al_2O_3/TiN coating. (a) Cobalt-enriched periphery (beneath the coating). (b) Bulk microstructure. Both etched with Murakami's reagent for 2 min. 1500×

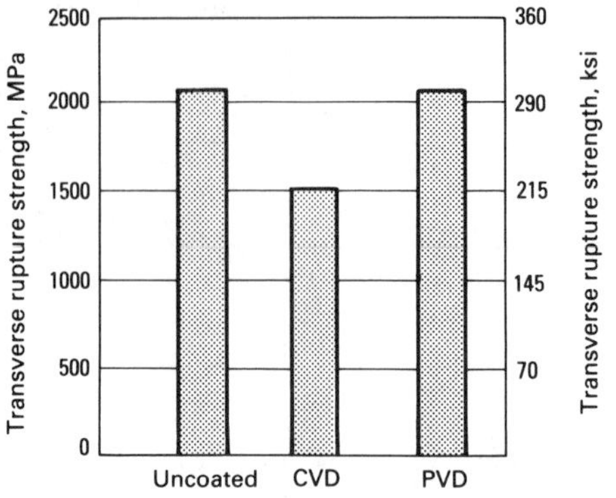

Fig. 25 Comparison of the transverse rupture strength of uncoated and coated carbide tools. Measured by a three-point bend test on 5 × 5 × 19 mm (0.2 × 0.2 × 0.75 in.) specimens of 73WC-19(Ti,Ta,Nb)C-8Co

Fig. 26 An example of PVD TiN coating on a sharp cemented carbide tool. Etched with Murakami's reagent for 3 s. 1140×

do not produce η phase. In addition, PVD coatings may induce a compressive residual stress, depending on the deposition technique. Therefore, PVD coatings do not degrade the transverse rupture strength of the carbide tools (Ref 30, 31).

Another advantage of the PVD process is its ability to coat uniformly over sharp cutting edges (Fig. 26). A sharp edge is desirable in a cutting tool because it leads to lower cutting forces, reduced tool tip temperatures, and finer finishes. In many cases, CVD coatings cannot be applied over sharp cutting edges without heavy η phase formation and/or coating buildup, both of which can lead to rapid edge failures. The PVD coating offers the benefit of the abrasive wear resistance of a hard coating layer without the degradation of edge strength.

The above advantages of PVD coatings are especially beneficial in operations such as milling. As noted previously, milling operations can produce severe thermal and mechanical cracks in cutting tools, and these cracks can lead to edge chipping. The CVD coatings often aggravate this situation through stress raisers, such as thermally induced cracks within the coating as well as interfacial η phase. These problems can be minimized by honing the cutting edge, by reducing the coating thickness, and by controlling η phase formation, but they can be eliminated by the use of PVD coatings. As a result, PVD-coated milling tools have recently become a commercial reality (Ref 32).

It should be noted that PVD coatings will not replace CVD coatings to any great extent in the near future. Currently, PVD has certain limitations. For example, Al_2O_3 coating is not feasible, and multilayer coatings have not yet been commercially developed. Additionally, methods have not yet been developed for coating complex tool geometries.

Tools and Toolholding

Early carbide metal cutting tools consisted of carbide blanks brazed to steel holders or milling cutters. Tools that became dull were resharpened by grinding. Clearance angles, cutting point radii, and other features could also be ground into the tools to suit particular cutting situations. Special chip-breaker grooves designed to curl and

Fig. 27 Indexable insert secured by pin

Fig. 28 Pin-and-clamp method of securing an indexable insert to a steel toolholder

Fig. 29 Screw-on method of securing an indexable insert to a toolholder

break the chips generated in metal cutting were also ground into the early tools.

Although these early carbide metal cutting tools provided significant increases in metal cutting productivity, certain disadvantages have become apparent. Regrinding changes the size of the tool; therefore, the cutting tool/workpiece relationship must be readjusted each time the tool is resharpened. Maintaining consistent geometry is difficult with reground tools, and part quality can suffer. Further, because the braze joint can withstand only a limited range of

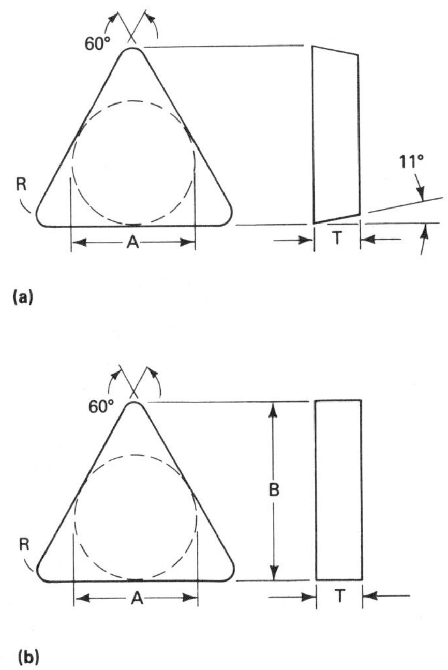

Fig. 30 Positive- and negative-rake geometries. (a) Triangle positive-rake insert. (b) Triangle negative-rake insert

temperature, the selection of usable carbide compositions is restricted, and CVD coating technologies cannot be applied. Nevertheless, brazed tools are still used in applications such as circular saws and small-diameter drills where mechanical clamping is impractical.

Indexable Carbide Inserts. The now-familiar prismatically shaped indexable inserts were introduced in the 1950s. These so-called throwaway inserts resembled brazed tools except that the carbide was secured in the holder pocket by a clamp rather than a braze. When a cutting edge wore, a fresh edge was simply rotated or indexed into place. Several other holding methods were subsequently developed.

The most popular holding method employs a pin that passes through a hole in the insert and forces it into the holder pocket (Fig. 27). Clamps are still widely used, often in conjunction with a holding pin (Fig. 28). Another common holding style employs a screw with a tapered head that fits a conical hole in the insert and thus holds the insert securely (Fig. 29).

Consistency and ease of replacement are the main advantages of indexable insert tooling. Consistent positioning of the cutting edge from index to index simplifies machine tool setup and helps ensure uniform product quality. The use of indexable inserts also eliminates labor costs for regrinding and permits CVD-coated tools and a wide variety of carbide compositions to be utilized.

Indexable carbide inserts are available in both positive and negative geometries (Fig. 30). Negative-rake inserts have excellent resistance to breakage and are well suited to difficult operations involving interrupted cuts. Negative-rake inserts can also be used on both sides, effectively doubling the number of cutting edges available per insert. Although positive-rake inserts can be used on only one side, they cut with lower force, reduce the possibility of distorting the workpiece, and help produce better surface finishes. Both negative- and positive-rake inserts are available in a range of tolerances.

Chipbreaking. In metal cutting operations the relative motion of the tool and workpiece produces a localized shear deformation in the work material immediately ahead of the tool's cutting edge. Workpiece material is thereby removed in the form of chips. The shape and length of the chips depend on machining conditions, type of workpiece material, and cutting tool geometry (Fig. 31). Chip forms not interfering with the cutting operation and easily disposed of are considered acceptable. Unacceptable chip forms include long, stringy chips that can tangle and be hazardous to the machine tool, operator, and workpiece finish, or excessively tight chips which can break cutting tools and scar the workpiece. Each manufacturing plant or shop typically sets standards for acceptable chips based on safety, machine setup, part finish, chip disposal systems, and other factors.

Indexable inserts often feature chipbreaker grooves to control chip formation (Fig. 32). In addition to providing chip control, these grooves often produce lower cutting forces. In fact, advanced chip-control geometries can give negative rake inserts the force-reducing capabilities of positive-rake designs. Customized chip-control geometries are available for special workpiece materials and operations.

Figure 33 illustrates the variety of chip forms produced by a given workpiece ma-

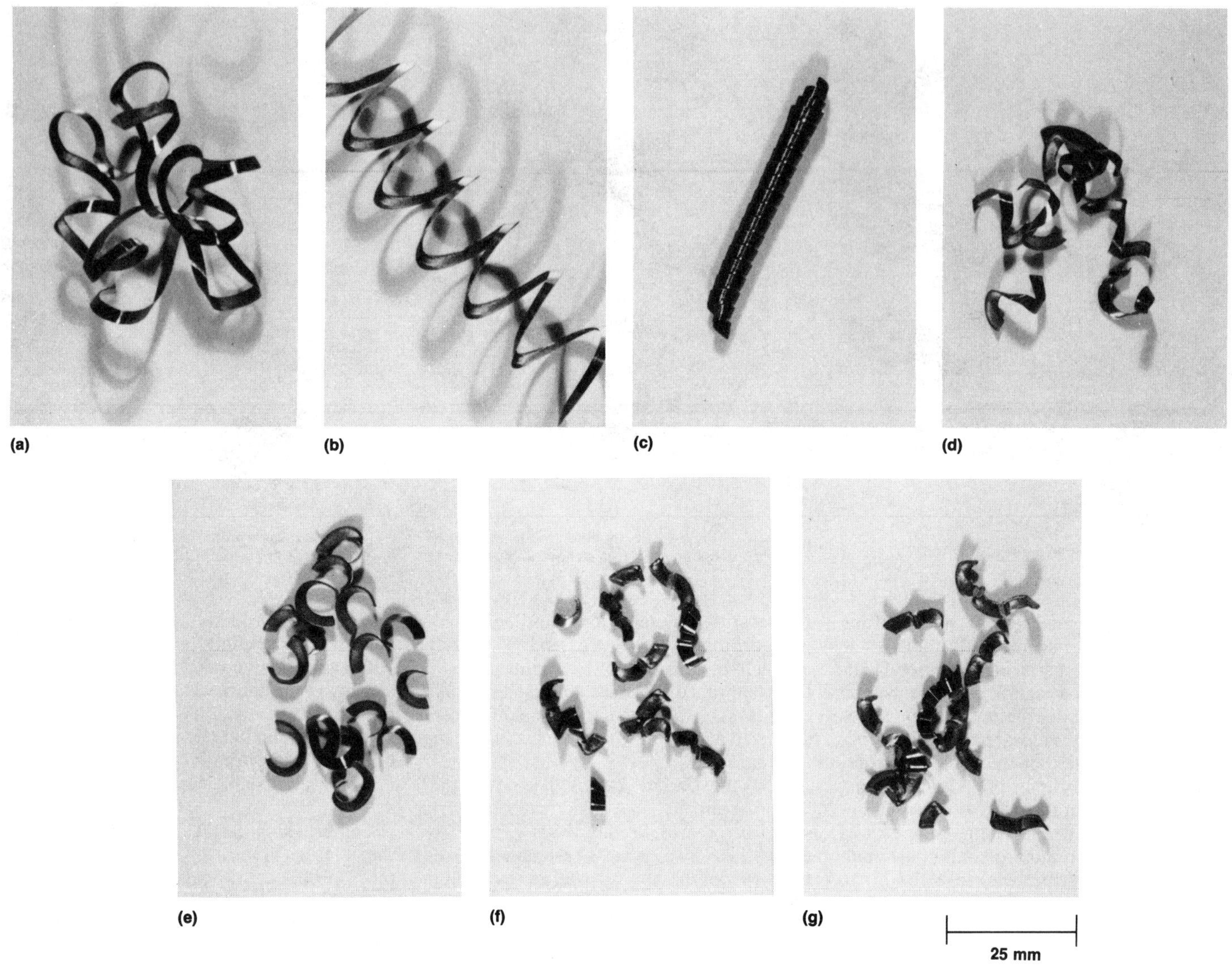

Fig. 31 Types of chips produced in metal cutting operations. (a) Uncontrolled chip, unacceptable. (b) Coil over 75 mm (3 in.) long (loose or tight), unacceptable. (c) Coil less than 75 mm (3 in.) long (loose or tight), acceptable. (d) Short coils; single C-shaped chips, acceptable. (e) Single C- or 6-shaped chips, acceptable. (f) Single with some double C-shaped chips, acceptable. (g) Multiple C-shaped chips, unacceptable

terial, insert geometry, and cutting speed as a function of feed rate and depth of cut. Note the range of feed rate and depth of cut combinations that produce acceptable chip forms in this particular situation.

Tool manufacturers usually provide diagrams that map out application ranges in terms of feed rate and depth of cut for the various chip-control geometries they offer. After cutting speed and insert grade are chosen for a certain workpiece material, these charts enable users to choose geometries to match desired feed rate and depth of cut. Figure 34 presents a simplified version of such a diagram. For high depth of cut and feed rates, roughing geometries would be recommended; for medium depths of cut and feed rates, general purpose geometries are preferred; and for shallow depths of cut and fine feeds, finishing geometries would be chosen. In actual operation, chip control can vary with workpiece material changes, insert nose radii and lead angle, and tool wear; even so, feed rate and geometry are the predominant factors in consistent control of chips.

Edge preparation refers to the practice of modifying the cutting edge itself after the correct overall geometry has been produced on the tool. Edge preparations are applied for two reasons: to prevent the chipping and premature failure of a too-sharp and therefore weak cutting edge, or to provide a slightly rounded (honed) edge that will optimize the effect of CVD coating. The edge preparations most commonly used are hones or chamfers (Fig. 35).

The edge preparation of brazed tools is most often applied by hand with a stone. Because the radius desired is very small (commonly 0.025 to 0.075 mm, or 0.001 to 0.003 in.), it is very difficult to achieve consistent performance. Too little hone can lead to microchipping and subsequent rapid edge wear; too much hone is very much like a preworn edge and also results in shorter tool life. Indexable carbide inserts, especially coated inserts, are usually equipped with a machine-applied hone of the size the manufacturer considers proper for the insert size, style, grade, and application. In general, only enough hone to prevent chipping should be used.

Cemented carbide tools intended for use in heavy-duty machining applications are often chamfered to provide maximum resistance to edge chipping. Although this may result in the loss of some tool life, the trade-off is reasonable if the mode of failure with honed inserts has been breakage rather than wear. Inserts intended for the machining of nonferrous and some aerospace alloys are generally supplied with sharp edges to reduce cutting forces and thus improve tool life.

Drills and end mills are available in both solid cemented carbide and indexable insert

Fig. 32 Indexable inserts with chip-breaker styles

versions. The increased production benefits realized from carbide tools can be as high as 6 to 1 compared to high-speed steel products. Solid carbide tools are available in the smallest sizes, ranging from circuit board drills of 0.13 mm (0.005 in.) in diameter to end mills with diameters up to 38 mm (1.5 in.). Drills and end mills with indexable inserts are generally available in sizes ranging from 16 mm (5⁄8 in.) in diameter up to about 75 mm (3 in.) (Fig. 36 and 37). Solid carbide tools have the advantages of greater rigidity, more cutting edges, and greater precision than comparable tools with indexable inserts. Indexable inserts offer the advantage of repeatability, require no resharpening, and permit the use of a wider variety of grades.

Threading. A number of methods are available for producing thread forms. These can be divided into two major categories: forming the thread, and machining (or cutting) the thread. A common thread forming process is thread rolling, which utilizes two diametrically opposed dies to cold form a thread into a workpiece. A frequent (and sometimes desirable) by-product of this process in certain materials is work hardening at the root of the thread.

Threads are typically machined by one of two methods: tapping, or single-point machining utilizing an indexable insert/toolholder system. Tapping is used where the diameter of the hole is too small for an indexable carbide/toolholder system and/or the machining speed is too slow for carbide. The three popular insert/toolholder systems used for threading are the laydown triangle, the stand-up (on edge) triangle, and several versions of a proprietary stand-up 55° parallelogram design (Fig. 38).

An undesirable by-product of thread machining is the V-shaped chip removed from the workpiece. Indexable inserts with proprietary chip control geometries molded into the top rake surface of the cutting edge are becoming available, and they control and break the chips with varying levels of success. In the more common thread forms, multitooth thread-chasing inserts (Fig. 39) are available as a means of reducing the number of passes required to complete a thread, thus improving productivity. An increasingly popular option available for many thread forms is the cresting insert, which machines the full thread form. Noncresting inserts machine the root and flanks but not the crest of the thread.

Thread milling, a thread machining method that is useful when turning is not possible, is performed on multiaxis computer numerical control machines capable of helical interpolation (Fig. 40). A disadvantage of thread milling is that the thread form it produces is slightly imperfect because of the inability of the cutting tool to clear the helical angle of the thread form as it exits the part. However, the threads are sufficiently accurate for all but the most demanding applications.

Grooving. There are three different grooving insert styles in common use:

- 90° V-bottom (Fig. 41)
- Proprietary stand-up 55° parallelogram (Fig. 38)
- Stand-up (on-edge) triangle

The V-bottom system is the most suitable for deep grooving because the cutting edge of the insert is wider than the body and is directly supported by the toolholder. The compact design and proprietary clamping method of the 55° parallelogram system maximize rigidity in shallow grooving. The on-edge triangle system offers three cutting edges on each insert, as opposed to two in the other common grooving systems. Chip control is a major concern in grooving, and products are becoming available that control and/or break chips with varying levels of success.

Cutoff. The early carbide cutoff tools consisted of carbide inserts brazed onto steel shanks. As in the case of carbide turning inserts, efforts to eliminate tool resharpening costs and to improve performance led to the development of mechanically held replaceable cutoff inserts. These inserts are available in a variety of styles, but most have a vee shape in the top or bottom surface, which is gripped by the steel holder for rigidity. Most cutoff inserts have a single cutting edge and are held either by clamping or by wedging directly into the holder. Chip control is available in either molded or ground geometries (Fig. 42).

Machining Applications

This section will focus on the relationship between workpiece material, cutting tool, and operating parameters. Advances in cemented carbides discussed earlier in this article have produced a wide selection of tool materials. Suggestions are given here to simplify the choice of cutting tool for a given machining application.

Workpiece Materials. Selection of appropriate cutting-tool grades and machining parameters depends initially on a number of workpiece material parameters: chemical composition, microstructure, and hardness. The effects of composition on machinability are often complicated by the synergistic effects of the elements comprising the workpiece alloy. In steels, for example, only generalized observations can be made on the role of various alloying elements on machinability. Workpiece microstructure is a result of the fabrication technique and heat treatment employed to provide desired properties. Variations in microstructure can have a profound effect on machinability. Workpiece hardness is the easiest parameter to document and is therefore widely used as a factor in rating machinability of a material.

Once the workpiece material is understood, the next step is to choose the right cutting-tool grade and machining parameters. Speed is critical in achieving long tool

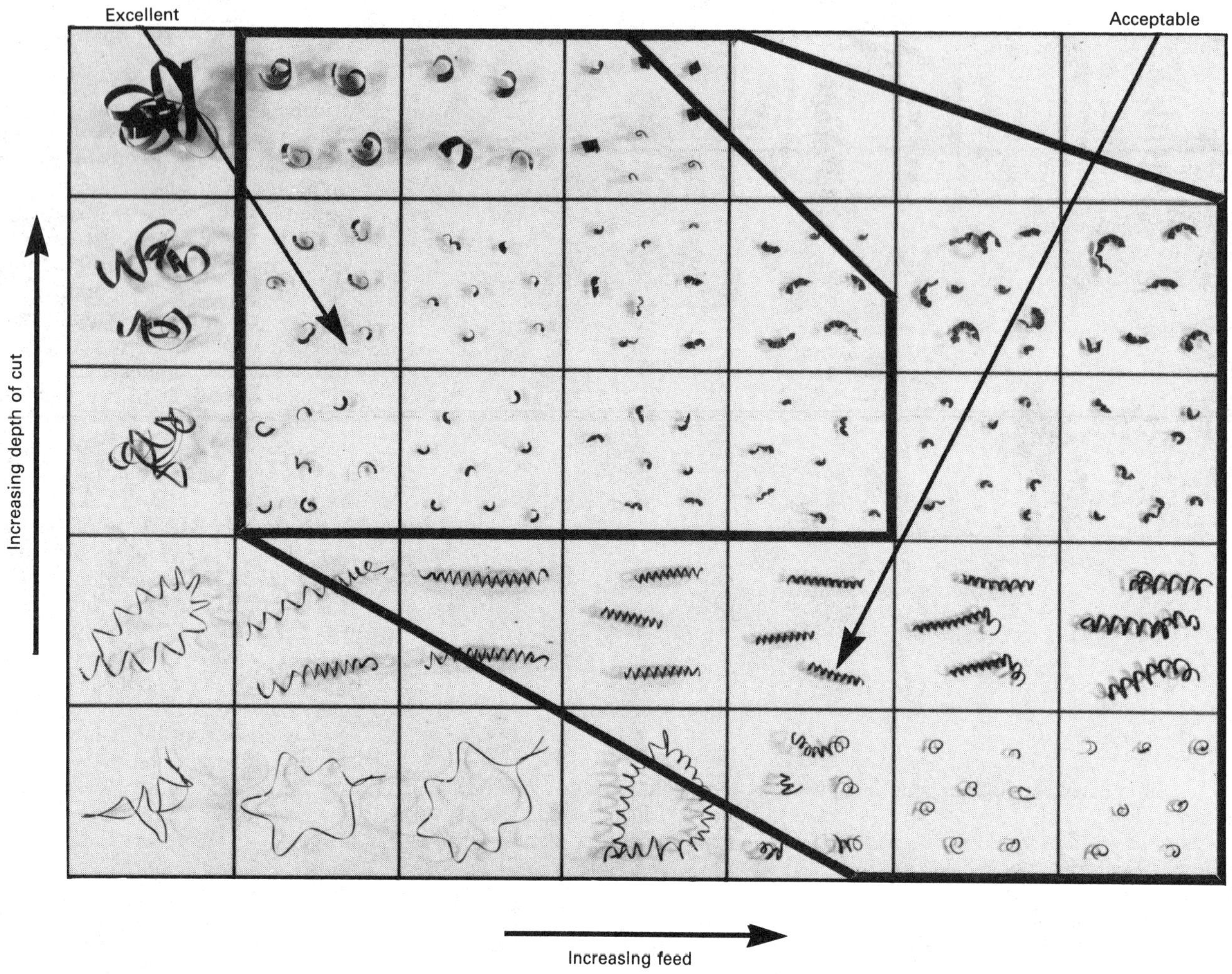

Fig. 33 Types of chips formed as a function of feed rate and depth of cut for a given tool geometry and cutting speed

life, whereas feed rate is important in minimizing the cutting time. The combination of grade, speed, and feed rate determines machining productivity.

Gray Cast Iron. Among the various classifications of cast iron, ferritic and pearlitic gray irons are easiest to machine. The graphite in these materials acts as a lubricant during the cut and produces discontinuous chips. Cast iron is easily machined with negative-rake inserts. Coolants are usually not required but sometimes are used to control dust.

Ductile Nodular Iron. Irons with nodular graphite are stronger than the gray irons and tend to produce curled chips like those from a steel workpiece, rather than the flake-type chip produced in machining gray iron. A C5 carbide grade with a groove-type chip-control geometry is often an appropriate choice in machining these materials.

Austenitic stainless steel is characterized by low thermal conductivity, high work-hardening rate, high ductility and toughness, and high tensile strength. This gummy material produces high temperatures at the chip/tool interface, leading to built-up-edge and tool failure. Tools should be sharp-edged, with positive rakes and adequate chip-control capability. Coolant should be used wherever possible.

Nickel-Base Alloys. These materials are widely used in aerospace industry where their heat and corrosion-resistance properties are fully utilized. They are very abrasive and are susceptible to work hardening. Generally, tools with negative rakes are used in roughing applications, whereas positive-rake tools are employed for finishing. No matter which rake is used, the key is to keep the cutting edge sharp.

Titanium alloys are characterized by low elastic modulus, high strength, high reactivity, and a tendency to work harden. In machining these alloys, rigidity of the tool and a high tool-clearance angle are essential. The cutting edge should be sharp, the speeds slow, and the coolant generous, but the tool should not dwell in the cut because it may work harden the surface of the workpiece. Uncoated fine-grain carbide (C2) is usually the best cutting-tool material for titanium; PVD-TiN-coated tools are also showing promise.

Aluminum is relatively easy to machine, and cutting speeds can be very high, but chip disposal can be a problem. Sharp cutting edges and coolants are helpful when fine workpiece surface finish is desired. On high-silicon aluminum, diamond tools offer the best performance.

Free-machining steels include additives that dramatically increase their machinability. The common additives are lead, sulfur, phosphorus, tellurium, bismuth, and boron. These steels can be machined at dramatically higher speeds than other steels, and good surface finishes can be obtained with negative or positive tools. Chip control is usually not a problem. For a given speed, much longer tool life can be obtained on the free-machining steels than on plain carbon steels.

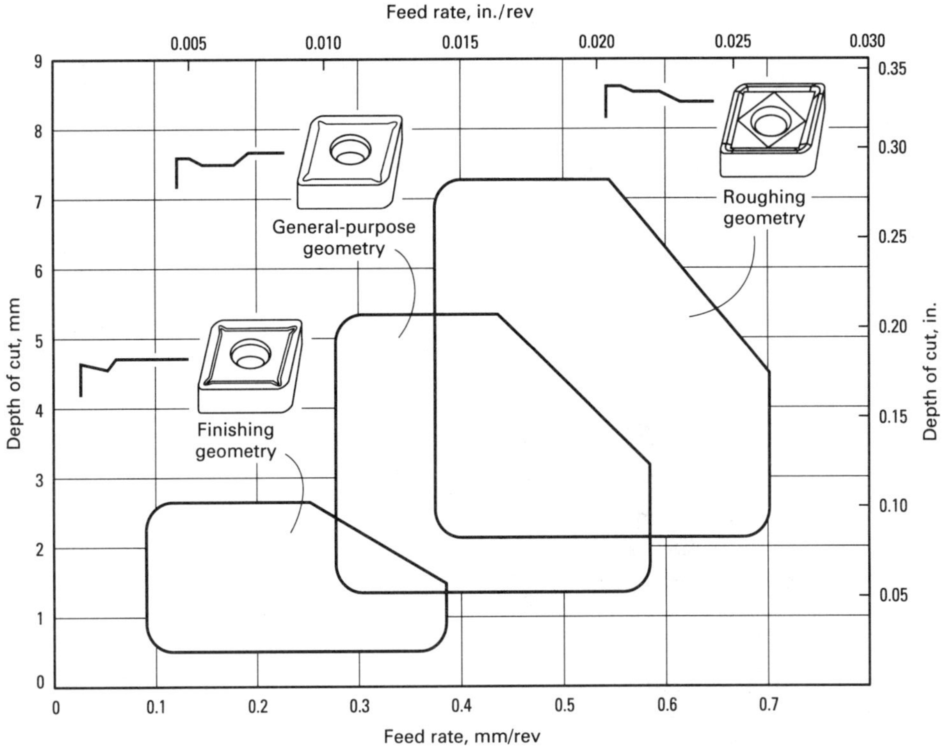

Fig. 34 Simplified diagram showing the application range of different insert geometries in terms of feed and depth of cut

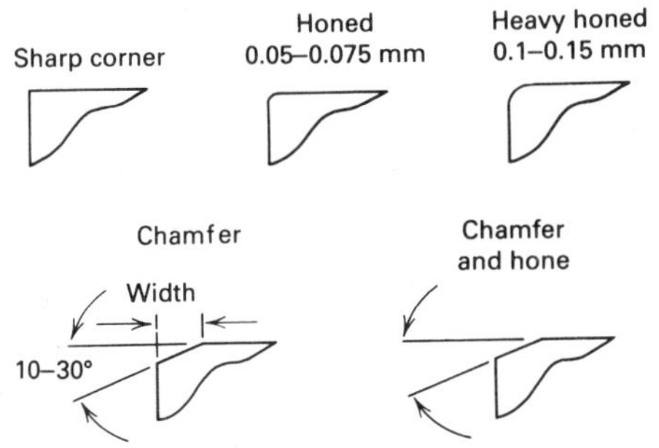

Fig. 35 Various edge preparations for metal cutting tools

Plain Carbon Steels. The machinability of plain carbon steels depends on workpiece carbon content. Plain carbon steels, with carbon content <0.25%, are soft and gummy and require positive-rake tools with sharp cutting edges if a good surface finish is desired. Tool failure while machining these materials is usually caused by buildup on the cutting edge and microchipping of the insert. As the carbon content increases, crater wear becomes the predominant tool-failure mechanism. At lower carbon contents, crater wear is less of a concern, but chip control becomes more difficult.

Alloy steels have higher hardness levels, higher yield strengths, and higher tensile strengths than plain carbon steels. Alloy steels are more difficult to machine than plain carbon steels; decreasing the carbon content of alloy steels improves machinability. Flank wear is the nominal mode of tool failure in alloy steel machining, but crater wear becomes more dominant as machining speed and/or carbon content increases. Machinability can also be a problem with chromium-molybdenum steels, which tend to be abrasive to the tool material. Higher nickel-alloy steels present another machinability challenge because they show a tendency to work harden. Like austenitic stainless steels, higher-nickel-alloy steels require positive-rake tools for productive machining.

Martensitic and Ferritic Stainless Steel. Ferritic stainless steels are not difficult to machine, although some produce stringy chips. Martensitic stainless steels follow the more traditional machining responses to carbon level: higher carbon levels increase the hardness and promote formation of chromium carbide. This decreases machinability and causes abrasive wear to the cutting tool. Sharp-edged positive-rake tools with chip control should be used for both martensitic and ferritic stainless steels.

Cutting tool grades of cemented carbide include coated carbides, uncoated alloyed carbides, and uncoated straight carbides. Selection is discussed below.

Uncoated Straight WC-Co Grades. Despite the advent of coated cemented carbide tools in the late 1960s, uncoated straight WC-Co grades still find a place in many machining operations. Grades in the K30 and K20 categories (C2 and C3) are typically used in machining gray cast iron, high-temperature alloys, austenitic stainless steels, nonferrous alloys (aluminum), and nonmetals. The higher cobalt K40 (C1) grades are also often used on difficult-to-machine workpieces such as chilled cast iron and heat-treated steels, where cutting tool strength and shock resistance have increased importance.

Although cutting-tool materials such as SiAlONs and whisker-reinforced ceramics have provided increases in machining productivity on nickel-base alloys, similar improvements have not occurred in the machining of titanium alloys. Submicron carbide grades (fine-grained carbide) K40 to K20 (C1 to C3) have shown the capacity to enhance machining productivity in titanium alloys, nickel-base materials, and other high-temperature alloys.

Uncoated alloyed carbide grades (P40 to P10, or C5 to C8) are primarily used in machining carbon steels, free-machining steels, tool steels, alloy steels, ferritic stainless steels, and malleable irons. They are widely used when machines do not have sufficient horsepower to utilize the high metal-removal capabilities of advanced coated tools. Uncoated alloyed carbides also find application in brazed form tools, which are typically made with highly specialized geometries mirroring the part being machined. Uncoated alloyed carbide grades are often employed in high positive, sharp-edged geometries for machining special part configurations with thin wall sections and tight tolerances. These parts cannot be subjected to high forces during machining. The CVD-coated tools, which are honed prior to coating, are not as effective in these applications. The recently developed PVD-TiN-coated carbides and cermets with sharp or near-sharp edges are also proving effective in such applications.

Coated Carbide Grades. More than 60% of the metal cutting inserts currently sold in the United States are CVD coated. Coated carbide tools provide abrasion resistance, crater resistance, and edge-buildup resistance, while permitting the use of higher machining speeds. This is illustrated in the tool-life diagram shown in Fig. 43, which compares the performance of a P40 (C5) grade with and without TiC-TiCN-TiN coating and a typical Al_2O_3-coated grade in turning SAE 1045 steel.

The TiC-TiCN-TiN-coated tools are generally used at cutting speeds higher than those used with uncoated tools, but lower than those employed with Al_2O_3-coated tools. The coatings provide good wear characteristics for machining both irons and steels. Combined with specially designed cobalt-enriched substrates, these coatings can also add new dimensions to tool life improvement, especially in interrupted cuts. In these applications the impact strength of the cobalt-enriched substrate

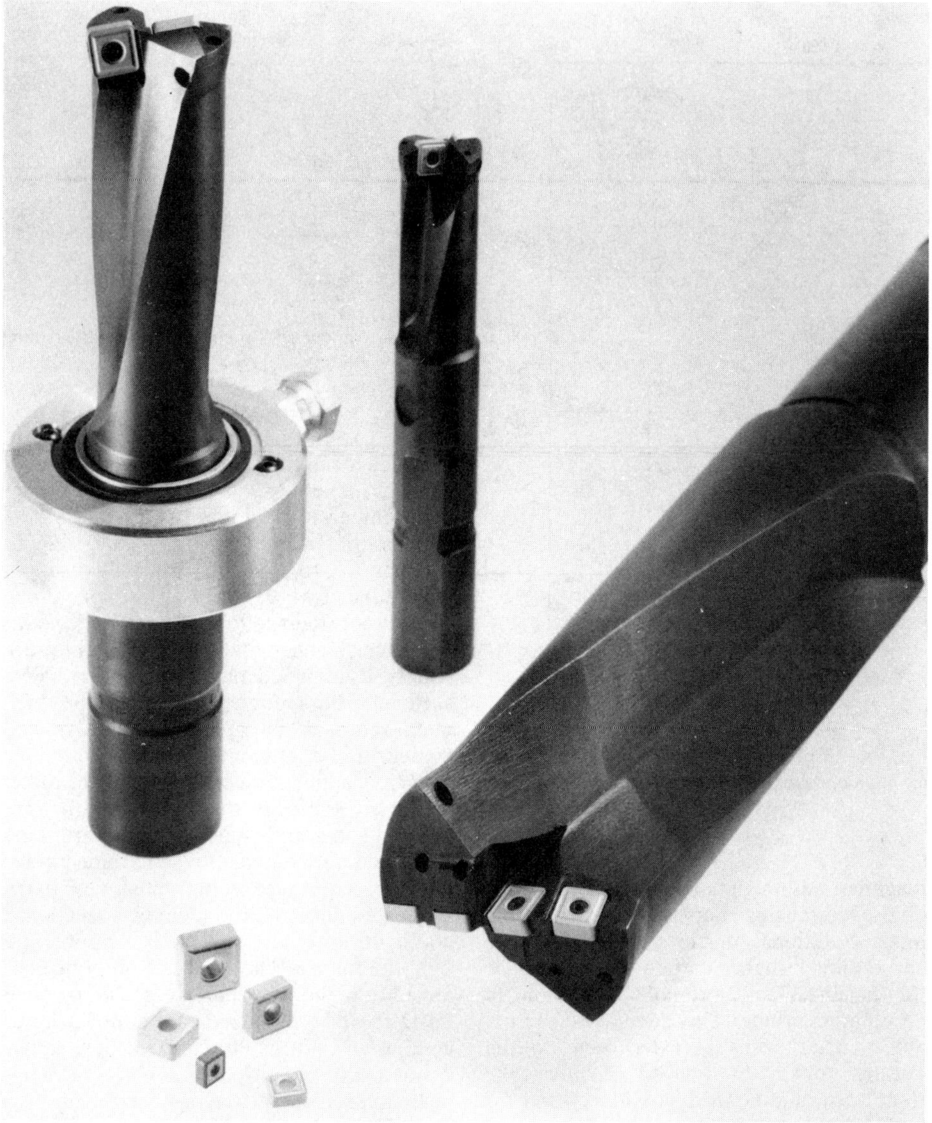

Fig. 36 Metal cutting drills with indexable carbide inserts

works as a partner with wear-resistant coatings.

The TiC-Al_2O_3- and TiC-Al_2O_3-TiN-coated tools are generally employed at higher speeds than the TiC-TiCN-TiN-coated tools. When machining steels at high speeds, the alumina layers provide excellent crater resistance. These grades work well on both irons and steels but are not recommended for materials such as aluminum or titanium. The new alternating multiple-layer coatings (Fig. 17d) are proving to be very effective in the machining of steels and irons.

PVD coatings can be applied to sharp or near-sharp insert edges without the deleterious effect of η phase formation at the coating-substrate interface. The sharp, tough PVD-coated tools are particularly well suited to milling, drilling, grooving, and threading applications. They have also been found to perform well on difficult-to-machine materials such as high-temperature alloys and austenitic stainless steels. Another application of PVD-coated tools is low-speed cutting, where the lubricious effect of the fine-grain TiN coatings helps improve tool life.

Machining Parameters. Recommended speed and feed ranges and starting points for uncoated and coated carbide grades for many common workpiece materials are shown in Table 9. More detailed operating parameters can be obtained from cutting tool suppliers or from independent sources, such as the Metcut Machinability Data Base.

Carbides for Nonmachining Applications

Almost 50% of the total production of cemented carbides is now used for nonmetal cutting applications such as metal and nonmetallic mining, oil and gas drilling, transportation and construction, metalforming, structural and fluid-handling components, and forestry tools. New applications are constantly being identified for carbides, largely because of their excellent combination of properties, including abrasion resistance, mechanical impact strength, compressive strength, high elastic modulus, thermal shock resistance, and corrosion resistance. This section addresses the current status of cemented carbides in nonmachining applications and highlights the specific property requirements and compositions employed. More extensive treatment of carbides used in nonmachining applications is available in Ref 33 to 38.

Compositions and Classification of Carbides

Compositions. The majority of cemented carbides used in mining, construction, oil and gas drilling, and metalforming applications is comprised of straight tungsten carbide-cobalt grades. Alloyed carbides are used only in special applications. A remarkable feature of these metal-bonded carbide alloys is that they can be tailored to provide different combinations of abrasion resistance and toughness by controlling the amount of cobalt and WC grain size. In general, cobalt contents vary from 5 to 30 wt% and WC grain sizes range from <1 to >8 μm and sometimes even up to 30 μm. The selection of a proper grade for a given application depends on an understanding of the complete process and the dominant failure mechanisms observed in the tool material.

Classification of Carbides. The C-grade classification for metal cutting carbide grades adopted by the U.S. carbide industry was discussed earlier in this article. This system has been expanded to include an array of nonmachining applications, as shown in Table 3.

Metalforming Applications

Cemented carbides are employed in metalforming applications because of their combination of high compressive strength, good abrasion resistance, high elastic modulus, good impact and shock resistance, and ability to take and retain excellent surface finish. Typical applications in this category include drawing dies, hot and cold rolling of strips and bars, cold heading dies, forward and back extrusion punches, swaging hammers and mandrels, and can body punches and dies. Table 10 lists nominal composition and properties of representative carbides and their applications. In applications that require high-impact strength, grades with 11 to 25 wt% Co are used. The higher-cobalt grades can survive more severe impacts. If impact strength is not a consideration, alloys with 6 to 10 wt% Co are selected. When wear resistance is of paramount im-

Fig. 37 End mills with indexable carbide inserts

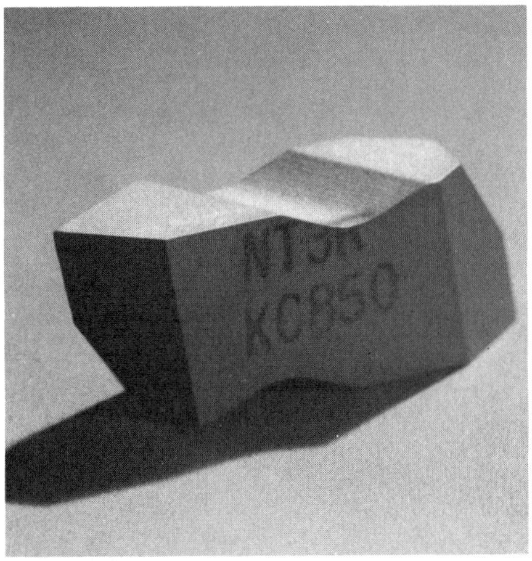

Fig. 38 Proprietary 55° parallelogram carbide insert

Fig. 39 A multitooth thread-chasing insert

portance, grades with lower cobalt contents and finer grain sizes are suitable choices. Cemented carbide compositions with higher cobalt contents (up to 30 wt%) and very coarse grain sizes (up to 30 μm) are employed in hot metalforming applications that demand high toughness and thermal shock resistance. When gall resistance (resistance to metal "pickup" on the tool) is needed, alloy carbides with tungsten-titanium carbide and tantalum-niobium carbides are used. For corrosion resistance applications, grades with finer WC grain sizes and lower cobalt contents are preferred because corrosion in general attacks the cobalt binder, leaving the carbide particles uncemented and making them susceptible to abrasive wear. Tungsten carbides with either nickel binder or combinations of nickel, cobalt, and chromium are also used in applications that require corrosion resistance.

Drawing Dies. As noted earlier in this article, the impetus for the synthesis of WC and subsequent development of cemented carbides came from the wiredrawing industry, where they are employed even today. The critical properties needed for this application are high compressive strength, mctal-to-metal sliding wear resistance, and good surface finish. Toughness is not a primary consideration in this application because no impact is involved. The dies are often supported in steel cases to withstand the tensile hoop stresses developed during drawing. The most commonly used grade is WC-6Co with medium grain size (1 to 2 μm) and hardness of 92 HRA. For drawing very small diameter wires (~0.1 mm, or 0.004 in.), a lower-cobalt alloy (~5 wt%) with finer WC grain size (≦1 μm) and a hardness of 94 HRA is used. Cemented carbide with such a low cobalt content must, however, be hot isostatically pressed to produce a sound, pore-free product. A 9 wt% Co grade is used for drawing larger diameter wires.

Cemented carbides are also employed for drawing tubes, rods, bars, and complex sections. The working stresses in some of these applications are more complex than those in wiredrawing dies, requiring carbides with higher toughness (higher cobalt contents).

Rod Mill Rolls. Carbide rolls are being increasingly used in Morgan, Demag, Ashlow, Krupp, Hille, Moeller, and Neumann mills to provide improved productivity and efficiency over steel rolls for the production of hot-rolled steel rods. The comparative advantages are closer dimensional control, truer roundness, improved rod finish, nongalling tendency, and increased delivery speeds up to 4500 m/min (15 000 ft/min). These performance advantages result from the high compressive strength, hardness and wear resistance, dimensional stability, heat resistance, and rigidity of cemented carbides compared with that of steel rolls. The compositions currently used for hot rolling have 15 to 30 wt% Co with very coarse WC grain sizes (up to 20 μm). Occasionally, compositions with cobalt-nickel-chromium binders are used.

Sendzimir Mill Rolls. Carbide rolls are well suited for the cold reduction and finishing of strip products in Sendzimir mills.

Fig. 40 Thread milling with indexable inserts

Rigidity and dimensional stability are particularly important in this application. The rolls are produced by hot isostatic pressing that provides the pit-free, smooth surface necessary for reducing, sizing, and flattening steel stock where surface finish is extremely critical. The rolls maintain true stock thickness across their entire width to deliver consistently high dimensional accuracy. In addition, improved wear life and an ability to operate at higher speeds than steel rolls provide greater productivity. The compositions used in these applications have medium levels of cobalt (~5.5 wt%) with medium WC grain size (1 to 2 μm).

Wire Flattening Rolls. Carbide rolls are successfully employed to flatten alloy steel wires into strips. The rigidity of the carbide produces uniform thickness across the width of the strip, and the high polish on the rolls imparts a very smooth surface to the steel strip. In addition, roll life is increased many times over that of most tool steel rolls due to the superior wear resistance of tungsten carbide alloy. A high-cobalt alloy (~12 wt%) is employed in this application.

Rebar Rolls. Successful application of carbides in rod mill rolls led to their use in high-speed twist-free mills to produce concrete reinforcing rods. In this application, the rod has ribbed patterns embossed on the surface that are formed by appropriate negative depressions on the roll. The high hardness, wear resistance, and compressive strength of carbide can provide an order of magnitude improvement in tool life over that of steel or cast iron rolls. Typically, a 25 to 30 wt% Co grade is employed in this application.

Slitter Knives. The high abrasive resistance and edge strength of carbides make them suitable for use as slitter knives for trimming steel cans and stainless and carbon steel strips and for cutting abrasive materials in the paper, cellophane, and plastic industries. The carbides are also used in slitting magnetic tapes for audio, video, and computer applications. The grades currently used in these applications are of medium to submicron grain sizes with 6 to 10 wt% Co. The fine grain size offers a sharp cutting edge, good surface finish, and high edge strength, which ensures high edge reliability.

Cold-Forming Applications. Cemented carbides are useful materials in cold-forming equipment such as punches and dies for extrusion or heading punches and dies. The high compressive strength and deformation resistance of cemented carbide make it practical to form a variety of parts not economically produced with steel punches and dies. Some examples of parts produced with cemented carbide punches and dies are wrist pins, bearing races, valve tappets, spark plug shells, bearing retainer cups, and propeller shaft ends. The punches range in size up to 100 mm (4 in.) in diameter and 500 mm (20 in.) in length and can produce 50 000 to 500 000 pieces between reworking, which represents nearly a tenfold increase in life over steel punches.

Back extrusion punches have to withstand heavy shock as well as deformation associated with the process. Generally, a WC-12Co alloy is used. When the compressive load on the punch becomes high, a lower-cobalt alloy (~11 wt%) is a more appropriate choice. For forward and backward extrusion dies, a WC-16Co grade is

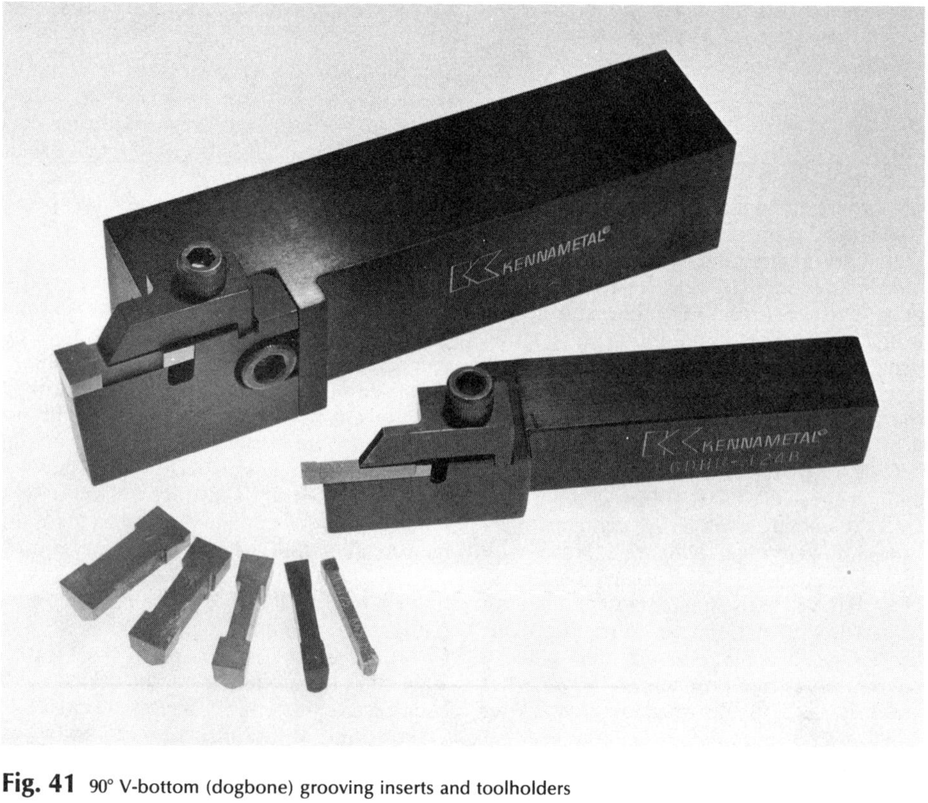

Fig. 41 90° V-bottom (dogbone) grooving inserts and toolholders

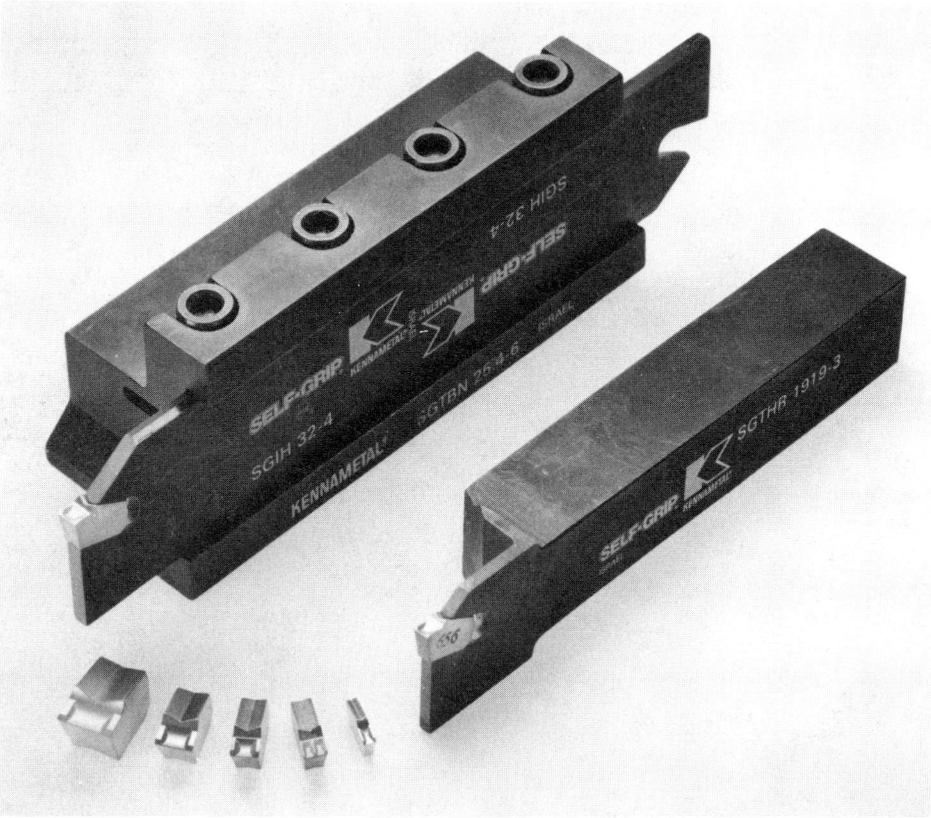

Fig. 42 Cutoff inserts and toolholders

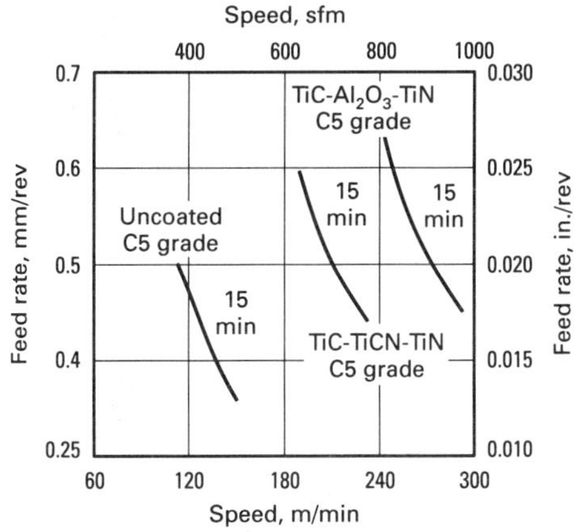

Fig. 43 Tool life comparison of a coated and an uncoated carbide tool. Constant tool life (15 min) plot for an uncoated P40 (C5) carbide and coated P40 (C5) carbides in turning SAE 1045 steel. The depth of cut was 2.5 mm (0.100 in.).

recommended. Applications with more severe impact may require compositions with higher cobalt contents (~20 wt%). By the same token, a lower-cobalt grade (~12 wt% Co) is selected for less severe impact situations. Submicron carbides with hardness in the range of 92 to 93 HRA may also be used for punches. By a judicious choice of cobalt content, it is possible to obtain improved wear resistance while retaining adequate compressive strength in the fine-grain materials.

In cold-heading applications involving the manufacture of nuts, bolts, screws, and other components with formed heads, the dies have to withstand considerable stress and repeated impacts and must therefore possess good fatigue strength. Alloys of 20 to 30 wt% Co with medium grain sizes (2 to 3 μm) and hardnesses of 84 to 85 HRA are selected.

Can Making. The can-making industry employs cemented carbides for blanking, piercing, and drawing sheet metal into the required shapes. The two-piece cans that are popular today are produced by a draw-iron method which employs carbide punches and thin-walled carbide ironing dies. The carbides provide the fine surface finish and high polish required in this application, as well as high resistance to "pickup" due to cold welding or galling. The alloys used for the punches generally have medium grain sizes with 11 to 12 wt% Co and hardness values in the range 89 to 90 HRA. For can-ironing dies, similar grades are used, sometimes with TiC and TaC additions for galling resistance. Submicron-grain carbides with improved wear resistance are also employed for can-ironing dies.

Stamping Punches and Dies. The high elastic modulus of carbides combined with their ability to incorporate fine details makes them ideal tool materials for stamping punches and dies. As in many other metalforming applications, fine-grain carbides are chosen for punches because of their edge retention capability and higher abrasion resistance.

Structural Components

The physical and mechanical properties of cemented tungsten carbides (Table 2) make them appropriate materials for a wide range of structural components, including plungers, boring bars, powder compacting dies and punches, high-pressure dies and punches, pulverizing hammers, carbide feed rolls and chuck jaws, and many others. The predominant wear factors in most applications are high abrasion, attrition, and erosion.

Boring Bars and Plungers. The high elastic modulus of carbides combined with their high compressive strength and wear resistance makes them ideal candidates for use in boring bars, long shafts, and plungers, where reduction in deflection, chatter, and vibration are concerns. In boring machines, for example, the static stiffness of a carbide boring bar is three times that of an alloy steel bar because of the difference in modulus of elasticity of the two materials. With its greater static stiffness, a carbide bar can make three times as heavy a cut with the same deflection, or it can make an equal cut with one-third as much deflection and consequently with greater accuracy. The following data permit comparison of the performance of a steel and a carbide bar:

Item	Steel bar	Carbide bar
Rough boring		
Feed, mm/rev (in./rev)	Oil 0.1 (0.005)	0.2 (0.007)
Speed, rev/min	180	350
Depth of cut, mm (in.)	4.8–6.4 (3/16–1/4)	4.8–6.4 (3/16–1/4)
Finish boring		
Number of cuts	2	1
Feed, mm/rev (in./rev)	0.1 (0.005)	0.1 (0.005)
Speed, rev/min	150	400
Depth of cut, mm (in.)	0.8 (0.030)	0.2 (0.007)
Finish, μm (μin.)	3.8 (150)	2.3 (90)

Powder Compacting Dies and Punches. Tungsten carbide punches and dies are successfully employed in compacting metal, ceramic, and carbide powders prior to sintering. The dies are generally made of 6 wt% Co grade with a medium grain size and hardness of 92 HRA. Powder compacting punches, rams, and core rods employ a higher cobalt grade (~11 wt% Co) and ~90 HRA. They can be made of solid carbide or as a composite that uses tungsten carbide in the wear areas. More recently, fine-grain carbides with 10 wt% Co are also employed in these applications.

Table 9 Recommended speed and feed ranges for uncoated and coated carbide grades

Material	Hardness, HB	Grade(a)	Speed range(b)		Starting feed(c)		Starting speed	
			m/min	sfm	mm/rev	in./rev	m/min	sfm
Gray cast iron (Class 30)	190–220	UC (C2)	61–152	200–500	0.38	0.015	107	350
		PVD	61–183	200–600	0.38	0.015	122	400
		TRI	91–213	300–700	0.38	0.015	137	450
		AOC	168–457	550–1500	0.38	0.015	305	1000
Nodular/ductile cast iron	190–260	UC (C2, C5)	46–122	150–400	0.38	0.015	91	300
		PVD	61–152	200–500	0.38	0.015	107	350
		TRI	91–183	300–600	0.38	0.015	122	400
		AOC	122–259	400–850	0.38	0.015	183	600
Austenitic stainless steel (304 Series)	175–230	UC (C2)	46–122	150–400	0.25	0.010	91	300
		PVD	91–190	300–625	0.25	0.010	137	450
		TRI	91–198	300–650	0.25	0.010	137	450
Nickel-base alloys (Inconel 718)	200–260	UC (fine grain)	21–61	70–200	0.25	0.010	38	125
		PVD	27–61	90–200	0.25	0.010	46	150
		TRI	27–61	90–200	0.25	0.010	49	160
Titanium	250–350	UC (fine grain)	30–76	100–250	0.25	0.010	46	150
		PVD	30–91	100–300	0.25	0.010	53	175
Aluminum	80–120	UC (C2)	152–1219	500–4000	0.38	0.015	305	1000
		PVD	213–1829	700–6000	0.38	0.015	457	1500
Free-machining steels (1100 Series)	175–250	UC (C5)	76–152	250–500	0.38	0.015	122	400
		PVD	91–183	300–600	0.38	0.015	152	500
		TRI	107–244	350–800	0.38	0.015	183	600
		AOC	152–366	500–1200	0.38	0.015	244	800
Plain carbon steels (1000 Series)	200–300	UC (C5)	61–122	200–400	0.38	0.015	91	300
		PVD	61–183	200–600	0.38	0.015	122	400
		TRI	91–213	300–700	0.38	0.015	152	500
		AOC	122–305	400–1000	0.38	0.015	198	650
Alloy steels (4000 Series)	175–250	UC (C5)	61–122	200–400	0.38	0.015	91	300
		PVD	61–137	200–450	0.38	0.015	122	400
		TRI	91–168	300–550	0.38	0.015	137	450
		AOC	122–259	400–850	0.38	0.015	213	700
Martensitic and ferritic stainless steels (400, 500 Series)	23	UC (C5)	61–122	200–400	0.25	0.010	91	300
		PVD	61–152	200–500	0.25	0.010	122	400
		TRI	91–168	300–550	0.25	0.010	137	450
		AOC	122–198	400–650	0.25	0.010	152	500

(a) UC, Uncoated Carbide C1-C8 (K, P, M ISO); TRI, CVD-coated (TiC, TiCN, TiN, HfN combination); AOC, CVD-coated (Al_2O_3 component within coating); PVD, PVD-coated (TiN). (b) This chart is for general-purpose use only. See manufacturing handbooks for specific information. (c) Based on general-purpose machining (turning); vary for roughing to finishing, or vary to obtain acceptable chip control

High-Pressure Dies and Punches. Another successful application area for carbides as dies and pistons is in the manufacture of synthetic diamonds. The high compressive strength of carbides allows them to withstand the ultrahigh pressures (>7 GPa, or 1 × 10^6 psi) generated within the reaction chamber. The carbides are held by large steel retaining rings to keep them in compression. The pistons use carbides with ~6 wt% Co to take advantage of their compressive strength and abrasion resistance. The dies, on the other hand, use carbides with higher cobalt levels (10 to 12 wt%) and medium grain sizes (2 to 3 μm), because, in addition to high compressive strength, the dies require adequate fracture toughness.

Fluid-Handling Components

The rigidity, hardness, and dimensional stability of cemented carbide, coupled with its resistance to abrasion, corrosion, and extreme temperatures, provide superior performance in fluid-handling applications. Several typical examples are discussed below.

Seal Rings. As operating environments have become more demanding, use of cemented carbides for axial mechanical seal rings has become more common. In situations involving corrosion, abrasion, high temperatures and pressures, and high speeds of rotation, carbides provide reliable service. Although often higher in initial cost, these alloys provide impressive savings in long-range costs because they reduce the frequency of maintenance downtime. Because of their dimensional stability and resistance to deformation, carbide seal rings retain their flatness and fine finish and are therefore used over a wide range of temperatures, from sealing hot gases to cryogenic liquids.

For seal rings facing extreme corrosion, a nearly binderless carbide of tantalum and tungsten is employed. For moderate corrosion, alloys with cobalt-chromium binder provide longer service life. A WC-Ni alloy is preferred where other carbides are not sufficiently corrosion resistant.

Table 10 Nominal composition and properties of representative cemented carbides and their applications

Typical application	Binder content, wt%	Grain size	Hardness, HRA
Heavy blanking punches and dies, cold heading dies	20 to 30	Medium	85
Heading dies (severe impact), hot forming dies, swaging dies	11 to 25	Medium to coarse	84
Back extrusion punches, hot forming punches	11 to 15	Medium	88
Back extrusion punches, blanking punches and dies for high shear strength steel	10 to 12	Fine to medium	89
Powder compacting dies, Sendzimir rolls, strip flattening rolls, wire flattening rolls	6	Fine	92
Extrusion dies (low impact), light blanking dies	10 to 12	Fine to medium	90
Extrusion dies (medium impact), blanking dies, slitters	12 to 16	Medium	88
Corrosion resistant grades, valves and nozzles, rotary seals, bearings	6 to 12	Fine to medium	92
Corrosion resistant grade with good impact resistance for valves and nozzles, rotary seals and bearings	6 to 10 Ni	Medium	90
Deep draw dies (nongalling), tube sizing mandrels	10 Co with TiC and TaC	Medium	91

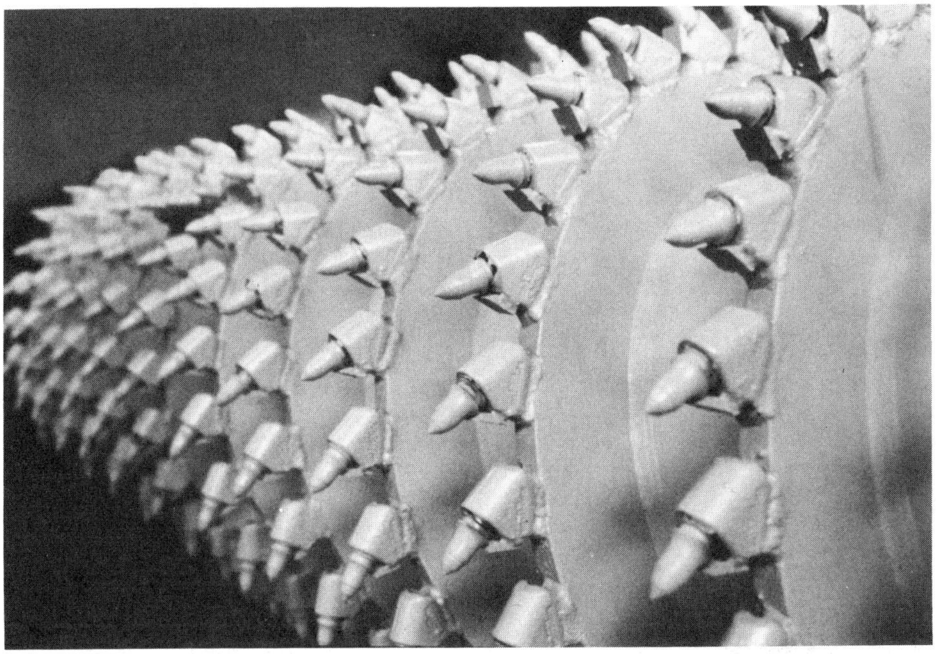

Fig. 44 Carbide-tipped tools mounted on a rotating drum of a road planing machine

Bearings, Valve Stems, Valve Seats. One of the common uses of tungsten carbide bearings is in shaft sleeves and bushings in centrifugal pumps. A variety of fluids, from water to highly abrasive slurries and corrosive liquids such as sulfuric acid, flow through such pumps. Carbide bearings can be used in direct contact with these fluids, thereby eliminating the need for bearing seals. In waste disposal plants, cemented carbide valve stems and seats have successfully replaced stainless steel components in combating the highly abrasive and corrosive slurry of sludge, sand, and water.

The high hardness of carbides minimizes wear by preventing abrasive particles from embedding themselves in the surface and continuing to abrade the mating surface. With most other bearing materials, it is necessary to isolate the bearing from the abrasive particles to avoid frequent downtime for bearing changes.

Nozzles are used to control and direct the spray of powders and liquids in, for example, dusting chemicals on crops, directing coolants onto hot steel, and spraying latex paints, and are also used in sand and grit blasting equipment. All these applications involve highly corrosive and abrasive materials. Cemented carbide is an excellent choice for these nozzles because it can outwear steel 100 to 1 and will thereby maintain the spray pattern and quantity of flow for a longer period of time, extending the service life of the nozzle. Each nozzle application dictates its own configuration. Many applications can use a small carbide nozzle insert held to other base materials by epoxy, braze, shrink fit, or taper fit. This permits the use of carbide without major redesign of a nozzle assembly or the need to manufacture a complex shape from solid carbide.

Transportation and Construction Applications

Among the diverse applications of cemented carbides is a wide range of tools and components for the transportation and construction industries. Examples include tools for road planing, soil stabilization, asphalt reclamation, vertical and horizontal drilling, trenching, dredging, tunnel boring, and forestry, as well as components such as snowplow blades, tire studs, and street sweeper skids.

Road Planing. Carbide-tipped tools provide a productive method for planing pavement surfaces. Mounted on a rotating drum (Fig. 44) of a road planer, carbide tools strip away concrete or asphalt pavement surfaces to any depth up to 250 mm (10 in.). The rotatable carbide-tipped road planing bit is secured in a steel block by means of a retainer sleeve, which may have a flange to improve bit rotation and provide easy bit changing. The bit rotation results in a self-sharpening effect and contributes to long tool life.

The composition of the carbide used in road planing depends on the type of pavement surface. Concrete pavements comprising coarse aggregates are inherently harder than asphalt and require a more impact-resistant grade. In addition to impact resistance, adequate abrasion resistance and fracture toughness are also required. Alloys with ~9.5 wt% Co and coarse WC grain size (10 to 20 μm) can be successfully used for planing concrete pavement. Asphalt pavements require a more abrasion-resistant grade (for example, WC-6Co). In some asphalt planing, the steel bit body holding the carbide undergoes more wear than the carbide because of the undercutting of the steel body by the abrasive asphalt fragments.

Trenching Tools. Tungsten carbide cutter bits for asphalt/concrete trenchers are being increasingly used to make slotting cuts for utility maintenance and have developed in parallel with road planers. Bucket wheel machines equipped with carbide-tipped tools have broadened the application range of trenchers. For example, bucket wheel trenchers have successfully excavated such difficult grounds as Arctic permafrost and Florida coral.

Soil stabilization refers to the mixing and pulverizing of hard soil as a preliminary step in new highway construction or as preparation for resurfacing. The requirements of the tool material are abrasion resistance combined with impact resistance (no heat is generated in the process). Drum mixers for soil stabilization operations are "laced" with cemented carbide-faced tines, or blades. Uniformity of mixing over long distances and avoidance of frequent tool changes are important advantages. The carbides must withstand high-stress abrasion and impact loading. Alloys commonly used have a nominal cobalt content of 10 wt% and medium grain size.

Asphalt Reclamation. Carbide-tipped tools are also employed in continuous asphalt reclamation systems, integrated with road planers to recycle old petroleum-base road surfacing material. The tools are either rotatable or nonrotatable and are designed for making deep cuts in soft to medium-hard abrasive material. A 9 to 10 wt% Co alloy with coarse tungsten carbide grain size performs well in this application.

Skids for Street Sweepers. Because of their superior abrasion resistance, carbide skids are outlasting steel or rubber skids on street sweepers by a factor of 30 to 1 and are eliminating downtime resulting from skid replacement. The long-wearing qualities of cemented carbide skids also help the sweepers ride higher and avoid hang-ups on manholes, railroad tracks, and other obstacles.

Snowplow Blades. Carbide-tipped blades improve snowplowing efficiency by providing a wear life up to 20 times longer than steel blades and by eliminating frequent blade changes. Because abrasion resistance and mechanical impact resistance are essential in this application, alloys with 10 to 12 wt% Co and medium to coarse WC grain size are used. The carbide inserts are either brazed or mechanically bolted into the cutting edge. The blades come in different length segments which can be combined to equip most plows.

As a companion to snowplow blades, carbide snowplow shoes increase blade life through better weight distribution of the blades and the snowplow frame. The carbide shoes also prevent blade wear caused by digging when plowing shoulders. Because the wear on the blade and shoe are similar, no adjustment is needed during operation.

Tire Studs and Golf Shoe Spikes. Cemented carbide tire studs provide excellent traction on icy highways. The wear resistance of tungsten-carbide alloys ensures long wear life for the tire studs, but has also resulted in government regulation of their use because of concerns about pavement damage. Carbide studs have also been used to replace steel spikes in golf shoes. Carbide studs last the lifetime of the shoe, and their taper design minimizes the problem of grass and mud buildup.

Vertical Drilling. Large down-hole construction tools, up to about 380 mm (15 in.) in diameter, are designed with a fixed array of conical cemented carbide cutter bits and a single-blade pilot bit, for water well drilling, pile driving operations, foundation work, and a host of other construction applications. The materials penetrated with these tools are alluviums, clay and sand strata, till, shales, and other soft sedimentary rocks.

Larger bore diameters, from 400 to 600 mm (16 to 24 in.), are drilled with vertical auger bolt-on tooling, for which rotary bit-body arms are designed to accept arrays of carbide conical cutter bits and a center pilot bit assembly (Fig. 45). The carbide-tipped auger has eliminated the need for blasting and air hammers in drilling rocky or frozen ground. Alloys with ~10 wt% Co and medium grain size are preferred in this application. As in most other wear parts, carbide-tipped cutters outlast steel teeth, thereby minimizing tooling costs and tool replacements. Because the cutting action is smooth, the wear and tear on augers, connectors, and drive motors is minimized.

Horizontal Drilling. In horizontal drilling projects such as laying of oil, gas, and water pipelines, carbide-tipped auger heads are dramatically improving the boring productivity. They are particularly helpful in drilling in unconsolidated terrains and soft to medium rock formations. The conical carbide bits in the auger heads are self-sharpening and are easily replaced. Larger auger heads are designed with wing cutters to facilitate the withdrawal of the auger from the drilled hole.

Chain Ditchers and Ditching Saws. Cemented carbides have successfully replaced steel tools in ditching operations, where carbide-tipped chain links can increase productivity in rocky or frozen ground. The tough, self-sharpening carbide bits can cut through sandstone shale up to 760 mm (30 in.) thick at 30 m/h (100 ft/h) with minimal bit changing.

Ditching saws tooled with conical cemented carbide bits are used to cut through

Fig. 45 Vertical auger with carbide-tipped tools used in heavy construction

asphalt and reinforced concrete. The carbides must withstand impact loading and high-stress abrasion. Generally, alloys with 9 to 10 wt% Co and medium to coarse WC grain size are used.

Tunnel and Shaft Boring. Carbide-tipped cutters can help speed production on tunnel-boring projects. They are ideal for use on moles and other tunnel-boring equipment to penetrate boulders, hard-packed glacial till, and solid rock.

Forestry Tools. Cemented carbide cutting tools are well established in forest product and woodworking industries and account for much higher levels of productivity, closer dimensional tolerances, better surface finishes, and lower tool maintenance costs. Examples are circular saws, debarking tools, edging and planing tools, and miscellaneous woodworking tools for both commercial and home use.

Heavy-duty tungsten carbide-tipped circular saws for log end cuts and ripping are subjected to a variety of wear mechanisms: abrasive wear, corrosive wear of cobalt binder by organic acids, and mechanical impact from tramp metal and rock. The saw-tip alloys have 10 to 12 wt% Co with moderately fine WC grain size (1 to 2 μm) to provide good edge strength and impact resistance. To increase corrosion resistance, nickel is sometimes substituted, either partially or wholly, for cobalt as a binder.

Heavy-duty log-debarking tungsten carbide tools are subjected to the same type of wear processes as the circular saw tips, although the main emphasis here is on high fracture toughness.

Cemented carbide-tipped circular saws and edging and planing tools for the dimension lumber, plywood, particle board, and composite industries are exposed to less impact and corrosion but greater abrasive wear conditions. The carbide alloys are correspondingly lower in binder content (~6 wt% Co), relatively finer grained, and higher in hardness.

Tungsten carbide saws, knives, and cutter heads of many designs enable high dimensional accuracy and precision joinery in commercial woodworking shops and in home-use equipment. Because high abrasion resistance and high edge strength are essential in these tools, the carbide alloys have very fine grain structures (≦1 μm), high hardnesses (93 RA), and lower cobalt contents (~6 wt%).

Mining and Oil and Gas Drilling

Recovery of natural resources from the surface of the earth is an important economic activity in which cemented carbides play a crucial role. Included in this category are recovery of metallic ores and nonmetals by underground or open pit mining practices, recovery of minerals such as coal, potash, and trona, and drilling for oil and gas. The method of excavation and the types of tools employed in each of the above application areas depend on the type of geostrata which is involved. The applications can be broadly classified into three types:

- Rotary drilling
- Roto-percussive drilling
- Flat-seam underground mining

Rotary Drilling. There are a number of rotary drilling methods that employ carbide-tipped bits. Two-pronged bits (Fig. 46a) are used for smaller holes. Conical bits can be used for core drilling when mounted on cylinders or for tunnel excavation when attached to large drilling heads. For small boring bits, speeds of rotation usually vary from 100 rev/min (for rock) to 1500 rev/min (for coal). Rotary drilling is efficient and gives high penetration rates in softer rocks; however, in hard abrasive rock, tool wear and fracture are severe.

In harder rock formations, a rolling conical cutter rock bit, called a "tricone" bit, with embedded hemispherical carbide button heads, is employed (Fig. 46b). Rotary drilling may be used for drilling either small or large holes and for raise hole boring.

Roto-percussive drilling is employed on harder rocks. In this operation, a blunt chisel-like cemented carbide insert is impacted against the rock, which is locally pulverized by impact. The insert is indexed around by rotation and again impacted by

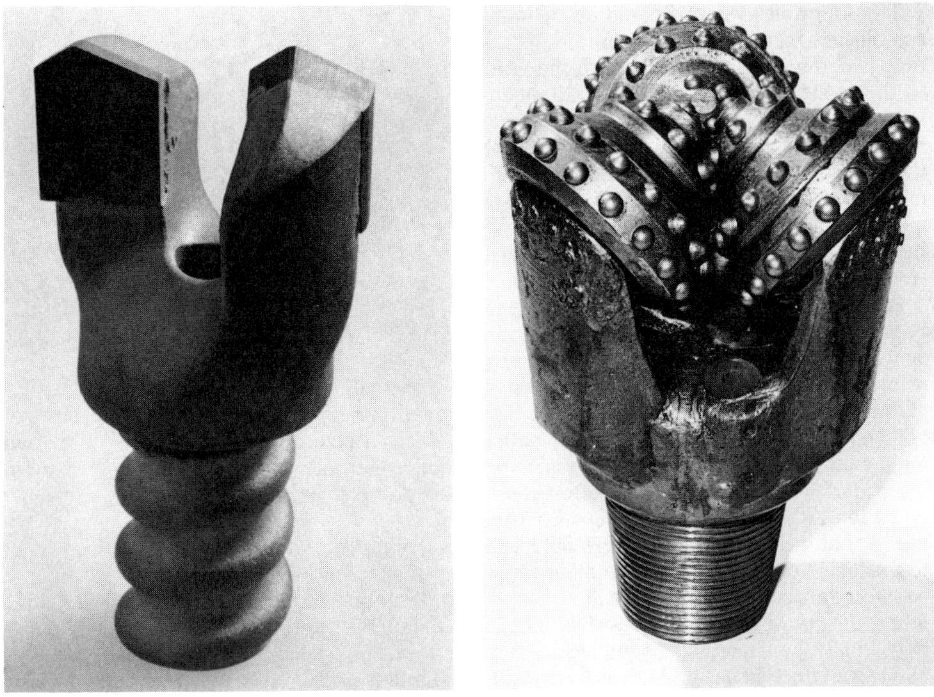

(a) (b)

Fig. 46 Two examples of rotary gas drilling bits. (a) Two-pronged carbide bit used in rotary drilling of small holes. (b) Tricone rotary roller bit with embedded carbide button heads

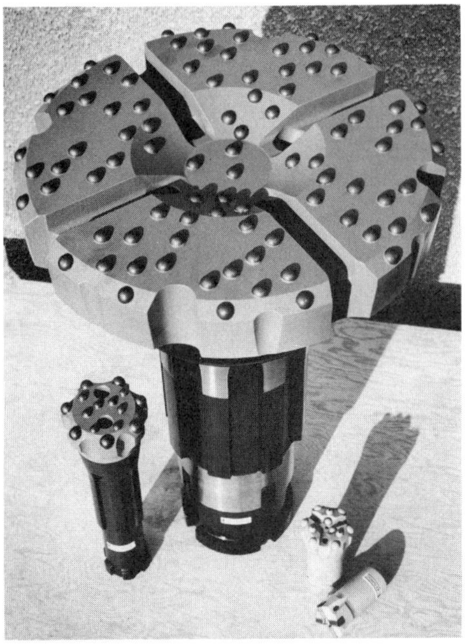

Fig. 47 Button-type carbide percussive drill bits. A carbide cross bit is in the foreground.

pneumatic means. Although the hardest rocks can be drilled in this manner, the penetration rate is significantly slower than that obtained by rotary drilling in softer rocks. The demands on the carbide insert are also severe, requiring good impact strength.

A major development in percussive tools is the use of a number of hemispherical carbide button bits (Fig. 47) instead of the chisel-shape inserts. There are several advantages to this design. First, the bits are self-sharpening so that deeper holes can be drilled before regrinding. Secondly, button bits are more reliable because they are held by an interference fit that can be accurately controlled by precise machining. The button bit design is more suited to drilling harder rocks than softer and tougher rocks. For example, on limestones the chisel design is still favored, whereas for drilling larger holes in hard rocks the button bit is preferred.

Flat-Seam Underground Mining. Excavation of coal, potash, trona, and so forth, from a relatively level bed is carried out by flat-seam mining using brazed carbide tools. These formations usually contain layers of soft and hard material. As a result, both the carbide and the braze may be subjected to thermal stresses.

Underground Mining of Metallic Ores. Recovery of metallic ores from rock veins is carried out by roto-percussive drilling using either chisel-type tungsten carbide inserts brazed to a steel tool holder or domed carbide compacts in button rock bits. In the case of brazed carbide bits, the inserts wear by severe abrasion and chipping. Button bits, on the other hand, show mainly abrasive wear. In button bits, improved steel support for the cutting elements permits use of alloys with lower cobalt contents and associated high abrasive wear resistance, without compromising chipping resistance. Cobalt contents generally range from 5 to 7 wt% with medium-fine tungsten carbide grain structure.

Open Pit Mining. Rotary blast-hole bits are used for open-pit mining for both metals and nonmetals. The bits, mainly of tricone style with tungsten carbide compacts set in alloy steel cones, are used for blast holes up to about 430 mm (17 in.) in diameter. Chipping of the compact is the more important failure mechanism in this application, and carbide alloys with higher cobalt contents, usually in the 7 to 10 wt% range, are commonly employed. Additionally, blast-hole drills are often fitted with drill stabilizer bars featuring flush tungsten carbide buttons to reduce gage wear and increase the cutting efficiency of the bit (Fig. 48).

Coal Mining Tools. Continuous drum mining equipment is often fitted with conical carbide cutter bits (Fig. 49a) designed to penetrate the coal face with a cleaving action and remove little, if any, of the adjacent rock. The bits are subjected to abrasive wear and high-impact loading; however, the tungsten carbide inserts are well supported in pockets in steel tool shanks, allowing long bit life. The most frequently used alloys contain about 7 to 9 wt% Co with a relatively coarse WC grain size (~3 μm).

Fig. 48 Rotary blast-hole tricone bit with embedded carbides and fitted with carbide stabilizer bars

Flat cutter bits (Fig. 49b) with tungsten carbide inserts butt-brazed to the tool shank are also used in continuous drum miners. Alloys with ~9 wt% Co and moderately coarse grain size (2 to 3 μm) are used for these flat bits. Similar alloys are also used

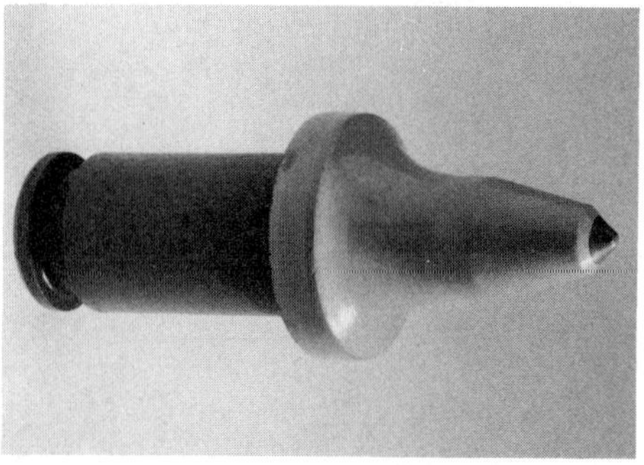

(a)

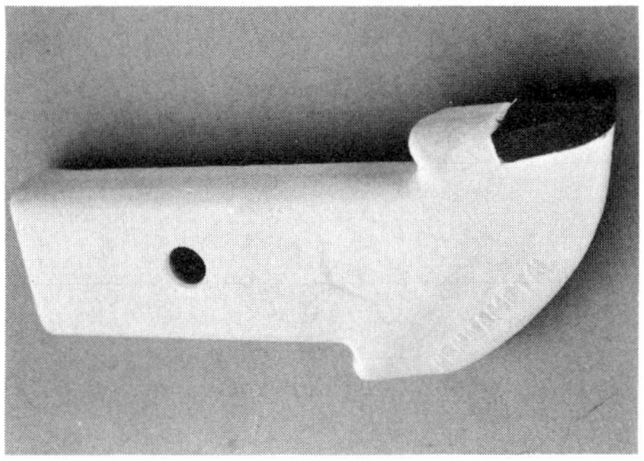

(b)

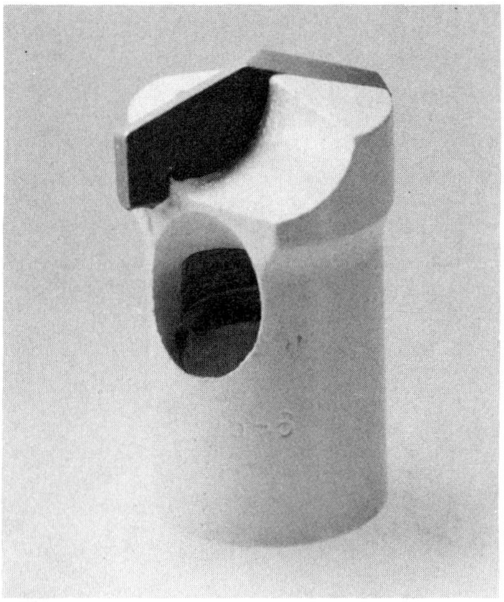

(c)

Fig. 49 Examples of some carbide bits used in coal mining. (a) Conical carbide bit. (b) Flat mining bit with brazed tungsten carbide insert. (c) Tungsten carbide roof-drilling bit with provision for center vacuum

for two-prong rotary-bit auger mining systems.

Tungsten carbide roof-drilling bits (Fig. 49c) are widely employed for anchor bolt drilling in underground mining operations. This application is especially severe, involving heavy abrasion and high mechanical and thermal stresses. Alloys of moderate to low binder contents (6 to 7 wt% Co) and medium to coarse tungsten carbide grain sizes are successfully employed.

The longwall system of mining, a long-established feature of British mining practice, is coming into wider usage in the U.S. In this system, conical, radial, and tangential cutter bits are used on coal shearing drums. Thermal cracking of the carbide due to high temperatures developed during mining and impact damage are dominant failure mechanisms, calling for relatively coarse microstructures and higher cobalt levels, frequently ~11 wt%, in the carbide bit.

Potash and Trona Mining Tools. Mining of potash salts in deep strata in Saskatchewan and some trona in Wyoming is carried out by rotors laced with brazed, flat-type tungsten carbide cutter bits. High frictional forces involved in the operation often raise cutter temperatures, causing incipient salt fusion on insert surfaces. The fused salts often cause corrosion of WC grains and thereby contribute to tool wear. Thermal braze failure and tool damage during machine retooling cause additional problems in this application. Sometimes, the above-mentioned wear factors can be partially counteracted by increasing the mass in cutter bit design and using a high-temperature braze.

Oil and Gas Drilling. Three classes of drill bits dominate oil and gas drilling:

- *Steel tricone bits* for relatively shallow bore holes in soft to medium-hard sedimentary rocks
- *Tungsten carbide tricones* for deeper bores penetrating metamorphic and crystalline rocks
- *Diamond oil bits* for deep bore holes in hard, crystalline rocks

The middle class, tricone bits with arrays of tungsten carbide buttons, is the dominant drilling tool for both exploration and production holes. Heavy-impact loading and high-stress abrasion require larger button sizes and special button shapes. The carbide used is moderately coarse grained with cobalt levels ranging from 10 to 14 wt%. As in blast-hole bits, gage wear is controlled by flush-set tungsten carbide buttons on bit-body clearance surfaces.

Drilling tool design has become extremely competitive in the oil and gas drilling industry. Many new designs have emerged in recent years. One of the most notable is the polycrystalline diamond cutter supported by a base of tungsten carbide; individual cutters are set in an array on the bit crown surface. Penetration is comparatively rapid and cuttings flow well in this design. The tungsten carbide base for each cutter is designed for high impact resistance. The polycrystalline diamond drill bit operates at a depth in between that of steel and tungsten carbide tricone drill bits. Polycrystalline diamond is discussed in the article "Superabrasives and Ultrahard Tool Materials" in this Volume.

Use of Tungsten Carbide in Diamond Cutting Tools. Diamond tools are available in a wide variety of product forms, some of which utilize tungsten carbide. The two major methods of using tungsten carbide in diamond cutting tools are:

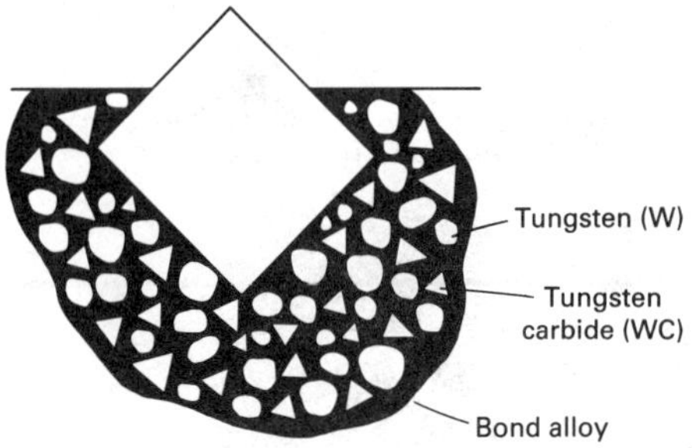

Fig. 50 Schematic of a coarse diamond particle embedded in a matrix alloy of a "surface-set" diamond drill bit

- Polycrystalline diamond bonded to a cemented carbide substrate
- Diamond particles embedded in a matrix alloy with tungsten carbide

The first method of using cemented carbide as a substrate in diamond tools is discussed in the article "Superabrasives and Ultrahard Tool Materials" in this Volume.

In the second method, a matrix alloy is embedded with diamond, which may consist of relatively coarse diamond firmly held on the surface by a matrix alloy (Fig. 50) or smaller diamonds dispersed within the matrix. In diamond drill bits, the use of diamonds set on the surface (Fig. 50) is referred to as a surface-set bit. A bit with diamond dispersed in the alloy matrix is referred to as an impregnated bit.

Diamond tools are employed in a variety of industries: mineral exploration and development, oil and gas exploration and production, and concrete, asphalt and dimension stone cutting. Most diamond drill bits for mineral exploration are designed to recover core samples. Oil bits, on the other hand, are generally used to bore full holes up to about 610 mm (24 in.). All diamond drill bits are designed for rotary drilling machines, with fluid flushing systems for cuttings recovery and drill bit cooling.

The matrix alloys used in diamond tools must support the diamonds firmly and must themselves wear at rates designed to keep the diamonds exposed for maximum cutting efficiency. Alloys that are too hard or wear resistant will not be removed rapidly enough to keep fresh diamond edges exposed. Alloys that are too soft will allow excessive exposure and consequent loss of the diamonds. Tungsten has emerged over the years as the ideal metal for bonding diamonds due to its affinity for carbon in the diamond and consequent bonding that occurs without damage to the diamond. In conjunction with tungsten, WC is used in the matrix formulations to protect diamonds against erosion and certain types of abrasion.

Powder metallurgy techniques are employed in the manufacture of diamond tools. There are many types of tungsten and tungsten carbide powders but only a few perform effectively in diamond tool matrix compositions. Conventional WC powders, produced by solid-state diffusion carburization of tungsten metal powder, are irregular in shape (Fig. 1) and exhibit poor flow and packing characteristics. The chill-cast, eutectic tungsten carbide, WC/W_2C, is carbon deficient and easily forms brittle double carbides (eta phase) during the manufacturing process. However, macrocrystalline tungsten carbide (Fig. 2) and tungsten metal powders derived from it meet the matrix powder requirements in terms of particle sizes and morphologies and are therefore widely used in matrix-powder formulations.

The functions of the tungsten and tungsten carbide in diamond tooling can be achieved only if they are effectively bonded together with appropriate auxiliary metals. The metals most commonly used to form the matrix bond are cobalt, nickel, copper, iron, and various compositions of copper/nickel/zinc, copper/zinc, and copper/manganese.

There are three basic methods by which diamond drill bits are manufactured. The most commonly employed is the infiltration process. This method is ideal for the production of surface-set bits because it does not dislocate the diamonds from their positions in the molds. The hot press method is a single-stage process that assures good quality control. The cold press-sinter process permits use of rapid mechanical pressing methods and neutral or reducing furnace atmospheres. The latter assures an oxygen-free environment which is important to the protection of diamonds, especially in matrix compositions that require high sintering temperatures. The hot press and cold press-sinter processes are used primarily in the manufacture of impregnated diamond tools.

Matrix alloys made by powder metallurgy methods also play an essential part in the manufacture of circular diamond saws for concrete and asphalt highway repair, general concrete cutting in the construction industries, and in the dimension stone industries. They are also essential for the manufacture of thin-wall diamond coring bits for boring conduit openings in cured concrete structures.

ACKNOWLEDGMENT

The authors gratefully acknowledge the contributions to this article made by Gary D. Stephens, William M. Stoll, Dave C. Vale, and Don L. Himler.

REFERENCES

1. H. Moissan, *The Electrical Furnace*, V. Lenher, Trans., Chemical Publishing Company, 1904
2. E.K. Storms, *The Refractory Carbides*, Academic Press, 1978
3. M. Hansen and K. Anderko, *Constitution of Binary Alloys*, McGraw-Hill, 1958
4. K. Schroeter, U.S. Patent 1,549,615, 1925
5. E.M. Trent, Cutting Tool Materials, *Metall. Rev.*, Vol 13 (No. 127), 1948, p 129-144
6. K.J.A. Brookes, *World Directory and Handbook of Hardmetals*, 4th ed., International Carbide Data, 1987
7. E. Lardner, *Powder Metall.*, Vol 21, 1978, p 65
8. H.E. Exner, *Int. Met. Rev.*, Vol 24 (No. 4), 1979, p 149-173
9. P.M. McKenna, U.S. Patent 3,379,503, 1968
10. P.M. McKenna, Tool Materials—Cemented Carbides, in *Powder Metallurgy*, J. Wulff, Ed., 1942, p 454-469
11. H.D. Hanes, D.A. Seifert, and C.R. Watts, *Hot Isostatic Processing*, Battelle Press, 1979, p 20-24
12. R.C. Lueth, Advances in Hardmetal Production, in *Proceedings of the Metal Powder Report Conference* (Luzern), Vol 2, MPR Publishing Services Ltd., 1983
13. P.A. Dearnley, *Met. Technol.*, Vol 10, 1983
14. H. Tanaka, Relationship Between the Thermal, Mechanical Properties and Cutting Performance of TiN-TiC Cermet, in *Cutting Tool Materials*, Conference Proceedings, American Society for Metals, 1981, p 349-361
15. R.C. Lueth, *Fracture Mechanics of Ceramics*, R.C. Bradt *et al.*, Ed., Plenum Press, 1974, p 791-806
16. J.L. Chermant and F. Osterstock, *J. Mater. Sci.*, Vol 11, 1976, p 1939-1951
17. J.R. Pickens and J. Gurland, *Mater. Sci. Eng.*, Vol 33, 1978, p 135-142
18. W.D. Kingery, H.K. Bowen, and D.R. Uhlmann, *Introduction to Ceramics*, 2nd ed., John Wiley & Sons, 1960, p 828
19. C.S. Ekmar, German Patent 2,007,427
20. W. Schintlmeister, O. Pacher, and K. Pfaffinger, in *Chemical Vapor Deposition, Fifth International Conference*, J.M. Blocker, Jr. *et al.*, Ed., The Electrochemical Society Softbound Symposium Series, Electrochemical Society, 1975, p 523
21. W. Schintlmeister, O. Pacher, K. Pfaffinger, and T. Raine, *J. Electrochem. Soc.*, Vol 123, 1976, p 924-929
22. V.K. Sarin and J.N. Lindstorm, *J. Electrochem. Soc.*, Vol 126, 1979, p 1281-1287
23. W. Schintlmeister, W. Wallgram, J. Ganz, and K. Gigl, *Wear*, Vol 100, 1984, p 153-169
24. B.N. Kramer and P.K. Judd, *J. Vac. Sci. Technol.*, Vol A3 (No. 6), 1985, p 2439-2444
25. T.E. Hale, Paper presented at the International Machine Tool Show Technical Conference, National Machine Tools Builders Association, 1982
26. H.E. Hintermann, *Wear*, Vol 100, 1984, p 381-397
27. B.J. Nemeth, A.T. Santhanam, and G.P. Grab, in *Proceedings of the Tenth Plansee Seminar* (Reutte/Tyrol), Metallwerk Plansee A.G., 1981, p 613-627
28. A.T. Santhanam, G.P. Grab, G.A. Rolka, and P. Tierney, An Advanced Cobalt-Enriched Grade Designed to Enhance Machining Productivity, in *High Productivity Machining—Materials and Processes*, Conference Proceedings, American Society for Metals, 1985, p 113-121
29. R.F. Bunshah and A.C. Raghuram, *J. Vac. Sci. Technol.*, Vol 9, 1972, p 1385
30. G.J. Wolfe, C.J. Petrosky, and D.T. Quinto, *J. Vac. Sci. Technol.*, Vol A4 (No. 6), 1986, p 2747-2754
31. D.T. Quinto, G.J. Wolfe, and P.C. Jindal, *Thin Solid Films*, *The International Conference on Metallic Coatings*, Vol 153, 1987, p 19-36
32. D.T. Quinto, C.J. Petrosky, and J.L. Hunt, *Cutting Tool Eng.*, Vol 39, 1987, p 46-52
33. G. Schneider, Jr., *Principles of Tungsten Carbide Engineering*, 2nd ed., Society of Carbide and Tool Engineers, 1989
34. G.A. Wood, in *Proceedings of the 25th Machine Tool Design and Research Conference*, S.A. Tobias, Ed., 1985, p 253-259
35. G.E. Spriggs and D.J. Bettle, *Powder Metall.*, Vol 18 (No. 35), 1975, p 53-70
36. E. Lardner, *Powder Metall.*, Vol 21 (No. 2), 1978, p 65-80
37. J. Larsen-Basse, *Powder Metall.*, Vol 16 (No. 31), 1973, p 1-32
38. W.E. Jamison, in *Wear Control Handbook*, M.B. Peterson and W.O. Winer, Ed., American Society of Mechanical Engineers, 1980, p 859-998

Cermets

John L. Ellis, Consultant, and Claus G. Goetzel, Consultant and Lecturer

CERMET is an acronym that is used worldwide to designate "a heterogeneous combination of metal(s) or alloy(s) with one or more ceramic phases in which the latter constitutes approximately 15 to 85% by volume and in which there is relatively little solubility between metallic and ceramic phases at the preparation temperature" (Ref 1, 2). A good definition of the term ceramic can be found in the *Ceramic Glossary* (Ref 3): "Any of a class of inorganic, nonmetallic products which are subject to a high temperature during manufacture or use. Typically, but not exclusively, a ceramic is a metallic oxide, boride, carbide, or a mixture or compound of such materials; that is, they include anions that play important roles in atomic structures and properties." With particular reference to cermets, this definition of the ceramic component could be broadened to include nitrides, carbonitrides, and silicides.

In a broad sense, cermets are akin to the hard and refractory particulate kind of materials within the general class of metal-matrix composites (see the article "Metal-Matrix Composites" in this Volume). There is a good deal of overlap in the literature, especially in the range of comparable volume fractions of the respective particulate and metallic components. In contrast to composite laminates, the combination of metal and nonmetal in cermets occurs on a microscale. The nonmetallic phase is usually not fibrous, but consists of more or less equiaxed fine grains that are well dispersed in and bonded to the metal matrix. If either the ceramic or the metallic component is predominantly fibrous, the material should be designated as a fiber composite. The bond between the nonmetallic phase and the metal matrix makes important contributions to the cermet; it is strongly affected by the phase relations, solubilities, and wetting properties that exist in the relationship between the ceramic and metallic components.

The size of the ceramic component varies, depending on the system and application. It can be as coarse as 50 to 100 μm, as in some types of cermets based on uranium dioxide (UO_2) that are used for nuclear reactor fuel elements, or as fine as 1 to 2 μm, as in the micrograin type of cemented carbides. If the ceramic component is even finer and is present in small amounts, the material can be considered to belong to the class of dispersion-strengthened alloys and therefore fall outside the accepted definition of cermets.

The basic objective of combining metal and ceramic on an intimate scale is to incorporate the desirable qualities and suppress the undesirable properties of both materials. The most outstanding example of the desirable properties obtained from combining metal and ceramic materials involves the hard-metal types made from cemented carbides (see the article "Cemented Carbides" in this Volume). Cemented carbides have enjoyed a steady expansion over the past six decades. Over that time, the development of the hard-metal/cermet tool materials has moved away from the early tungsten-base carbides to carbide- and nitride-base compositions of increasing complexity (Table 1).

Cermets originally were used for cutting tool applications. Some 45 years ago (Ref 2, 5), they began to be considered for use in more taxing applications, such as propulsion systems. The expectations were that the refractory behavior, strength, and corrosion resistance of the ceramic phase could be mated advantageously on a proportional basis with the high ductility and thermal conductivity of the metallic phase, and that some superior new materials would become available for a multitude of high-temperature applications.

Unfortunately, these goals were not fulfilled, despite a major effort in the United States and Europe during the 1950s. The degree of ductility and toughness imparted by the metallic binder phase remains inadequate for most critical applications, such as turbojet and stationary gas turbine blades or nozzle vanes. In other areas, however, cermets have proven their value as engineering materials, notably in tools based on titanium carbide (TiC) or titanium carbonitride (TiC,N), and in some types of nuclear fuel elements. Cermets based on uranium dioxide, as well as those based on uranium carbide (UC), offer potential for advanced fuel elements. Cermets based on zirconium boride (ZrB_2) or silicon carbide (SiC), and others containing aluminum oxide (Al_2O_3), silicon dioxide (SiO_2), boron carbide (B_4C), or refractory compounds combined with diamonds, possess unique properties. Several are used commercially in a wide range of applications that includes hot-machining tools; shaft seals; valve components and wear parts; ultrahigh-temperature exposed ducts, nozzles, and other rocket engine

Table 1 History of cermet product development and marketing

Year	Composition	Trademark	Manufacturer
1930–1931	WC-Co	G1	Krupp-Widia
1930	TiC-Mo_2C-(Ni, Mo, Cr)	Titanit S	Metallwerk Plansee
1930	TaC-Ni	Ramet	Fansteel Corporation
1933	TiC-TaC-Ni	. . .	Siemens AG
1938–1945	TiC-VC-(Fe, Ni, Co)	. . .	Metallwerk Plansee
1949–1955	TiC-(NbC)-(Ni, Co, Cr, Mo, Al)	WZ	Metallwerk Plansee
	TiC-(Nb, Ta, Ti)C-(Ni, Mo, Co)	Kentanium	Kennametal
1952–1954	TiC-(steel, Mo)	Ferro-TiC	Sintercast (Chromalloy)
1960	TiC-(Ni, Mo)	. . .	Ford Motor Company
1970	Ti(C, N)-(Ni, Mo)	Experimental alloys	Technical University Vienna
1974	(Ti, Mo) (C, N)-(Ni, Mo)	Spinodal Alloy	Teledyne Firth Sterling
1975	TiC-TiN-WC-Mo_2C-VC-(Ni, Co)	KC-3	Kyocera
1977–1980	TiC-Mo_2C-(Ni, Mo, Al)	. . .	Ford Motor Company, Mitsubishi
1980–1983	(Ti, Mo, W) (C, N)–(Ni, Mo, Al)	. . .	Mitsubishi
1988	(Ti, Ta, Nb, V, Mo, W) (C, N)-(Ni, Co)-Ti_2AlN	TTI, TTI 15	Krupp-Widia

Source: Ref 4 and Kennametal, Inc.

components; furnace fixtures and hearth elements; grinding wheels; and diamond-containing drill heads and saw teeth.

A significant application of cermets involves cutting tool materials that utilize titanium carbide or titanium carbonitrides as the hard refractory phase. Frequently, molybdenum carbide (Mo_2C) and other carbides are also built into these cermet formulations. The cratering and flank wear resistance properties of the titanium carbide and titanium carbonitride cermet tool materials are better than those of the conventional cemented-carbide (that is, cobalt-bond tungsten carbide) tool. In comparison to ceramic cutting tools, these cermets permit heavier cuts, which, at high speed, results in a greater amount of metal removal at a comparable level of tool life. Cermets clearly possess characteristics of a cutting tool material that is capable of filling the gap between conventional cemented carbides and ceramics (see *Machining*, Volume 16 of the 9th Edition of *Metals Handbook*).

Classification

Cermets can be classified according to their hard refractory component. In this system, the principal categories of cermets are determined by the presence of six components: carbides, carbonitrides, nitrides, oxides, borides, and miscellaneous carbonaceous substances.

The metallic binder phase can consist of a variety of elements, alone or in combination, such as nickel, cobalt, iron, chromium, molybdenum, and tungsten; it can also contain other metals, such as stainless steel, superalloys, titanium, zirconium, or some of the lower-melting copper or aluminum alloys. The volume fraction of the binder phase depends entirely upon the intended properties and end use of the material. It can range anywhere from 15 to 85%, but for cutting tool applications it is generally kept at the lower half of the scale (for example, 10 to 15 wt%) (Ref 4).

The metallic bond for each cermet is selected in order to produce the desired structure and properties for the specific application. The iron group metals and their alloys dominate as in the cemented tungsten carbide class of hard metals; nickel and, to a lesser extent, cobalt and iron possess a desirable combination of relatively high hardness and good ductility. However, the binder for a cermet can also be chosen from the group of more reactive metals, such as titanium or zirconium, or it can be selected from a series of refractory metals that includes chromium, niobium, molybdenum, and tungsten. Lower-melting metals and alloys, primarily those based on copper and aluminum, round out the list of binders at the bottom of the temperature scale. Aluminum, however, is more commonly associated with metal-matrix composites.

Carbide-base cermets are by far the largest category of cermets, even if the term is used in its narrower sense and excludes the broad field of cemented-carbide cutting tools and wear parts based on tungsten carbide (WC). Since the inception of cermet technology, the dominant concept has been that of a material based on TiC as the primary hard and refractory constituent, with the bonding provided by any of a variety of lower-melting ductile metals or alloys (much the same as those used for cemented tungsten carbides). The TiC cermets have found use in tool and wear resistance applications; in selected high-stress, high-temperature systems; and in corrosive environments. Cermets based on SiC and B_4C, which generally are classified as metal-matrix composites, have gained considerable industrial significance in wear and corrosion resistance, or antifriction, applications; they are also used in nuclear reactor applications. Cermets with a chromium carbide (Cr_3C_2) base have been used for a variety of corrosion resistance applications and as gage blocks; however, they have apparently lost much of their industrial usage.

Carbonitride-base cermets can be produced with or without additions of various other carbides (of which Mo_2C is the most important); they are bonded with the common cemented-carbide binders. At present, these materials are the primary cermets for tool applications. Their enhanced strength, which makes them suitable for high-speed cutting tools, is based on a greatly improved bond between the hard carbide grains and the binder metal. The improved bond is a consequence of a miscibility gap in the quaternary TiC, TiN, MoC, and MoN system that results in a so-called spinodal decomposition into two isostructural phases (Ref 6) with inherently better wettability to the binder (Ref 4).

Nitride-base cermets constitute a special class of tool materials. Titanium nitride (TiN) and especially cubic boron nitride (CBN) produce excellent cutting materials if they are combined with a hard binder metal. Titanium nitride and zirconium nitride (ZrN) bonded with their respective metallic elements have been developed for special heat- and corrosion-resistant purposes.

Oxide-base cermets constitute a category that includes UO_2 or thorium dioxide (ThO_2), which are used for a major fission component in nuclear reactor fuel elements; Al_2O_3 or other highly refractory oxides, used for components in liquid-metal manipulation (for example, pouring spouts) and general furnace parts; and SiO_2, used for a minor constituent in friction elements. Combinations of Al_2O_3 with TiC are suitable for hot-machining tools.

Boride-base cermets have a boride of one of the transition metals as the dominant phase. These cermets provide excellent high-temperature corrosion resistance to attack by active metals, such as aluminum, in the molten or vapor state. A combination of ZrB_2 and SiC is resistant to erosion from the propulsion gases of chemical rockets.

Carbon-containing cermets are materials that contain graphite in varying proportions. They are used for electrical brushes and contacts or as minor constituents to provide some lubrication in friction elements. Also included in this category are diamond particles within metal matrices that are used in special tools.

Fabrication Techniques

The methods used for powder preparation, forming, firing or sintering, and post-treatments of cermets generally are similar to conventional ceramic and powder metallurgy (P/M) processing techniques. Figure 1 is a flow chart of the various P/M techniques applicable to cermets. Table 2 summarizes the relative characteristics of the major forming methods that are practiced in producing carbide-base and most other types of cermets. The principal processes are cold forming and sintering, pressure sintering, and infiltration.

The cold-pressing process includes static uniaxial and isostatic multiaxial compaction. The powder mixtures are compacted at pressures of 35 to 100 MPa (5 to 14.5 ksi). The predominant method involves pressing dry wax-lubricated powder in hardened steel dies with double-action opposing punches. For long rods or tubes of uniform cross section, these dies are used for the extrusion of a paste in which the powder particles are embedded in a suitable organic binder or wax. To form complex or large shapes, the dry powder is placed in a pliable mold and compacted from all sides by hydrostatic pressure inside a sealed, reinforced steel cylinder.

Powder Preparation. The first step in the cermet production process is the mixing and milling of the ingredient powders. The mixtures, consisting of the hard-phase substance in powder form and the pure metal or metals in the proper proportions required for the composition of the binder alloy, are milled in ball mills. The balls are made of tungsten carbide or, more frequently, of a highly sintered cermet. The mill can be lined with the same type of cermet material to reduce the possibility of mixture contamination. In addition to conventional ball mills, high-energy vibratory ball mills and attrition mills are used. With the latter type of mills, substantial savings in milling time, energy, and floor space can be achieved. During the milling process, the hard phase particles are comminuted and thoroughly coated with binder metal. Organic liquids such as hexane are used in the process to minimize the rise in temperature and pre-

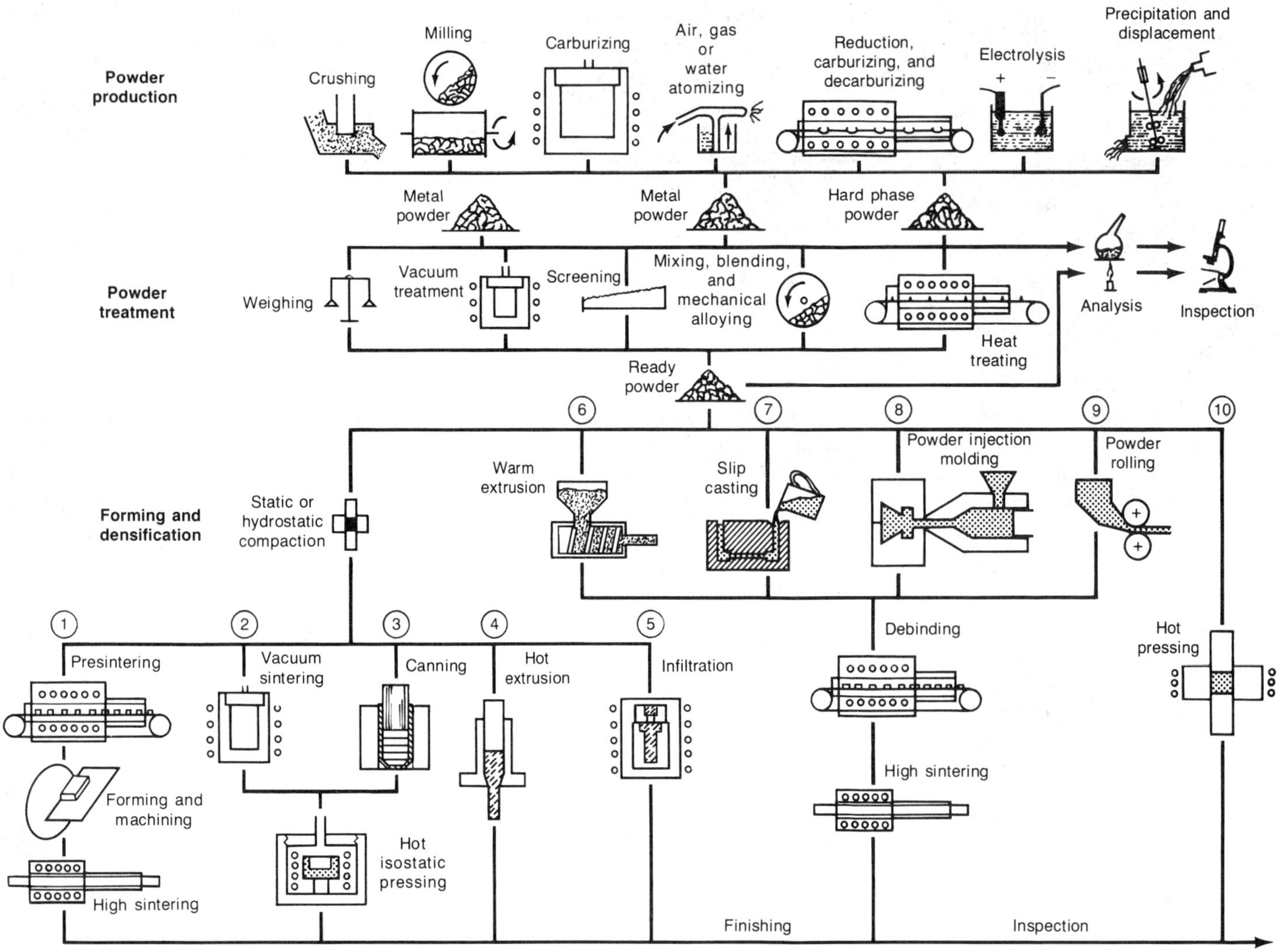

Production method	Products
1. Presintering	Cemented-carbide parts and cermets
2. Vacuum sintering	Steel-bonded carbides (standard pieces) and cermets
3. Canning	Steel-bonded carbides (special pieces)
4. Hot extrusion	Aluminum cermets with moderate amounts of hard-phase additions
5. Infiltration	TiC parts with nickel- or cobalt-base infiltrants and other cermets with about 55–85 vol% hard phase
6. Warm extrusion	Cemented-carbide rods or other slender cermet parts
7. Slip casting	Cermets with high proportions of hard phase
8. Powder injection molding	A wide variety of cermet compositions
9. Powder rolling	Aluminum, copper, and other nonferrous metals with moderate additions of hard-phase components
10. Hot pressing	A wide variety of cermet compositions

Fig. 1 Powder metallurgy production methods for cermet and cemented-carbide products

vent oxidation. After the powders have been milled to a particle size of 325 mesh or finer, the mixtures are dried before further processing and use. A lubricant is then added, and those powder mixtures destined for compaction in automatic presses are agglomerated so that they can flow freely from the hopper to the compacting die.

Static Cold Pressing. Cold-pressing methods for cermet powder mixtures generally follow the well-known powder-compacting techniques used in conventional powder metallurgy. Small cermet parts needed in reasonably large quantities are compacted in special hard-metal dies by automatic presses with double-action opposing punches (Fig. 2). Whether solid or segmented, the dies are shrunk into a strong, tough, heat-treated steel retainer.

The automatic compaction cycle consists of filling powder in the die, compacting the powder, ejecting the compact, and removing it. Two methods are used for ejecting the compact from the die. In the first method, the lower punch moves upward and pushes the bottom of the pressed piece to the level of the die table. In the second method, the die is forced down (withdrawn) over the lower punch until the bottom of the compact is on level with the top of the die (Fig. 3). The second method is gaining favor with many specialists because it allows building shorter, less expensive tools, and because it provides better support for the fragile compact during ejection.

Another interesting variation, particularly for the compaction of fragile cermets, is provided by the so-called anvil press (Fig. 2b). In this method, the powder is pressed against the anvil by the upward motion of a single lower punch. The anvil is then laterally removed to allow ejection of the com-

Table 2 Cermet forming techniques

Technique	Size capability	Shape capability	Mold or die requirements	Production Rate	Production Labor
Static cold pressing	Limited by press capacity	Prismatic shapes without undercuts	Hardened steel or carbide dies	High	Low to moderate
Hydrostatic cold pressing	Limited by capacity of pressure vessel	Simple or complex shapes	Rubber molds	Low	High
Powder rolling	Limited width, long length	Flat and thin	...	High	Low
Warm extrusion	Limited by equipment size	Long pieces with uniform cross section	Hardened steel or carbide dies	High	Low
Powder injection molding	Small pieces	Complex shapes with undercuts	Hardened steel or carbide dies	High	Low
Static hot pressing	Limited by press capacity	Prismatic shapes without undercuts	Graphite or ceramic molds	Low	High
Hot isostatic pressing	Limited by capacity of pressure vessel	Simple or complex shapes	...	Medium	Low to high
Hot extrusion	Depending on press capacity	Long pieces with uniform cross sections	Alloy steel dies	High	Low
Infiltration	Depending on equipment	Intricate shapes feasible	Graphite and ceramic molds	Low	High

pact by the continued upward motion of the punch. Because this cold-pressing technique is a single-action pressing method, an anvil press is acceptable only for relatively thin single-level pieces. Yet, where applicable, the anvil press saves tool costs; in addition, it reduces the ejection path to a minimum, which is an important factor for fragile cermets.

Medium-to-large rectangular pieces are compacted on hydraulic presses in multiple-section dies held together by powerful steel frames. These frames are capable of counteracting the large internal forces that are exerted on the powder mass and transmitted radially to the die sections. Often, at the completion of the cycle, the dies are opened for the careful removal of the fragile cermet compact.

For round pieces it is preferable to use double-action hydraulic presses with a built-in ejection action. These presses can use either one of the two ejection methods:

- If the die is mounted on a hydraulically activated floating platen, a single-action press will adequately provide the effect of a double-action press. In this case ejection is accomplished by the withdrawal method
- When a full-power double-action hydraulic press is available (obviously a heavier and more expensive machine than the single-action press), the die can be mounted on the stationary platen. The compact is ejected by raising the lower punch to the level of the die table at the completion of the pressing cycle

Cold compaction of simple cermet shapes gives generally good results when using adequately lubricated powder mixtures, well-designed tooling, and sturdy presses. Even though the compacts are fragile, they should nevertheless be firm, have adequate green strength and well-formed edges, and be free of laminations or other internal defects. Depending on the composition of the cermet, compaction problems can arise that require modifications in the equipment or the process. Generally, the higher the proportion of the hard phase, the greater the difficulties that arise in compaction; these difficulties may not be evident until after sintering. Also, more problems can be expected with iron-, nickel-, or cobalt-base cermets than with the softer, more malleable aluminum-base compositions. Larger pieces, as measured both in diameter and height, intensify the problems associated with composition. Air entrapment, bridging, laminations, intermittent voids, and variable density throughout the compact are just a few of the problems encountered in the cold compaction of cermets (Ref 7).

Some of these problems can be overcome by adding more lubricant to the mixture, by increasing the taper of the compacting die, or by slowing down the compaction process. Die wall lubrication between pressings often helps to eliminate compaction or ejection problems. Adequate preloading of the die and die lapping in the pressing direction are other precautions that help to overcome problems.

Even for the simplest forms, such as cylindrical or rectangular single-level blanks, not all problems encountered can be solved by the conventional static cold-pressing technique. This is one reason why P/M specialists depend heavily on other, more sophisticated forming processes for the fabrication of a wide variety of products.

Cold Hydrostatic Pressing. High-quality cermet compacts require uniform densification throughout. This can be ideally accomplished by cold hydrostatic pressing. In this method, pressure is applied simultaneously and uniformly from all sides toward the center of gravity of the powder mass while all friction of powders against the die wall is completely eliminated. In order to compact simple or even relatively complex shapes by this method, dry powders are filled into a pliable mold. The powders are settled, and air is removed on a vibrating table; the mold is then sealed and placed into a reinforced steel cylindrical vessel filled with a fluid. After the vessel is closed, hydraulic pressure is built up, thereby compressing the

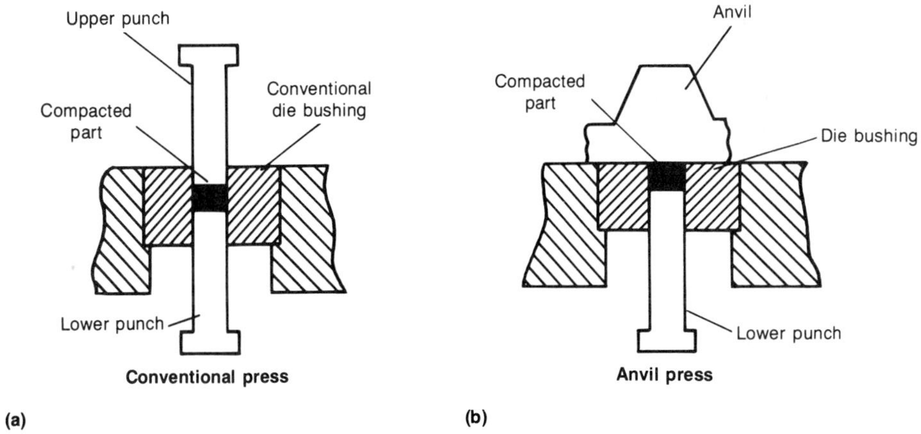

Fig. 2 Static cold pressing with (a) a conventional press and (b) an anvil press. The anvil press has no upper punches; therefore, the misalignment, breakage, and wear problems associated with those punches are eliminated. Courtesy of PTX-Pentronix, Inc.

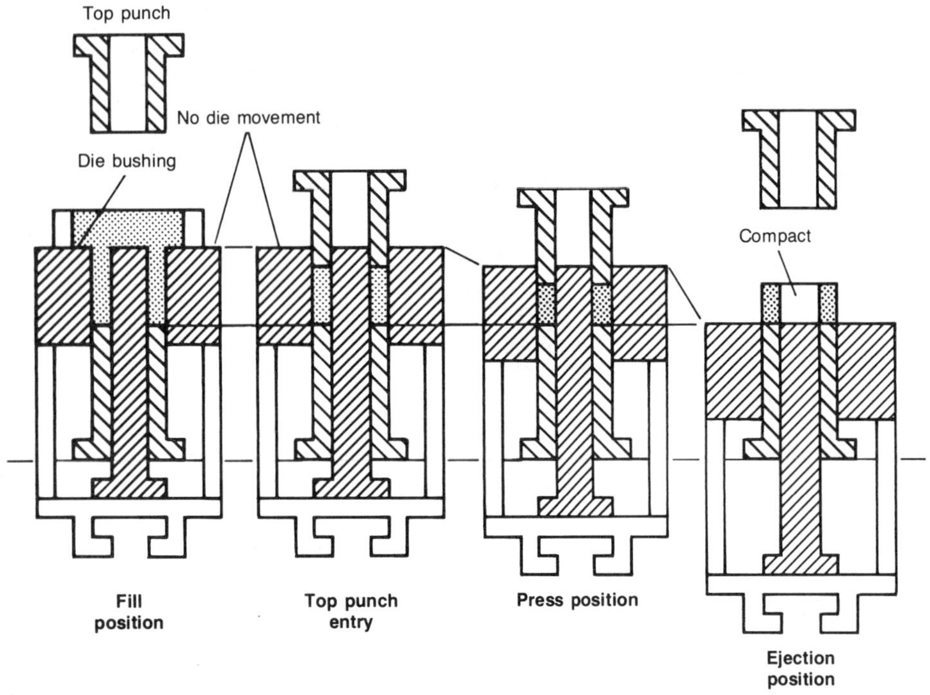

Fig. 3 Withdrawal press cycle with controlled die motion (top and bottom pressure). Courtesy of Dorst America

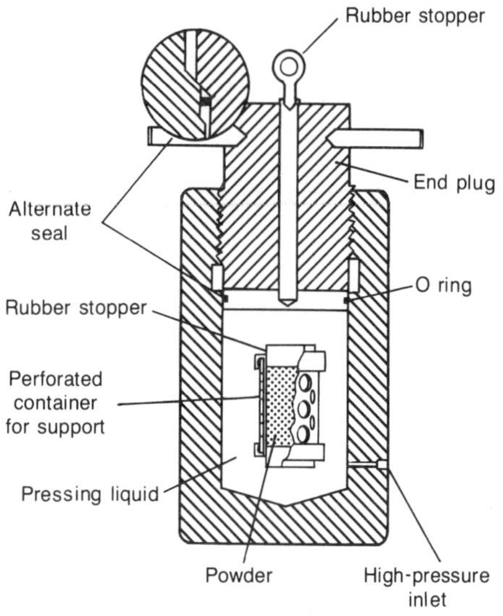

Fig. 4 Schematic of a cold hydrostatic pressing vessel with a wet-bag powder mold. Source: Ref 7

powder contained in the mold. The two hydrostatic pressing methods frequently used for pressing cermets are the wet-bag method and the dry-bag method.

The wet-bag method involves placing one or more powder-filled molds inside a hydrostatic pressing vessel (Fig. 4). The powder-filled mold is pliable and is placed within a perforated container for support. Inside the vessel, the powder-containing mold is completely surrounded by hydraulic fluid. Depending on the size of the vessel and the individual mold, often a number of molds can be placed in the vessel and compacted simultaneously. The entire process—loading the vessel with one or several molds, building up and holding the pressure, releasing the pressure, and reopening and unloading the vessel—is relatively slow. Moreover, filling the mold, assembling the mold, loading the molds into the pressure vessel, and unloading and removing the pressed piece after completion are slow, manual processes that require meticulous attention to detail.

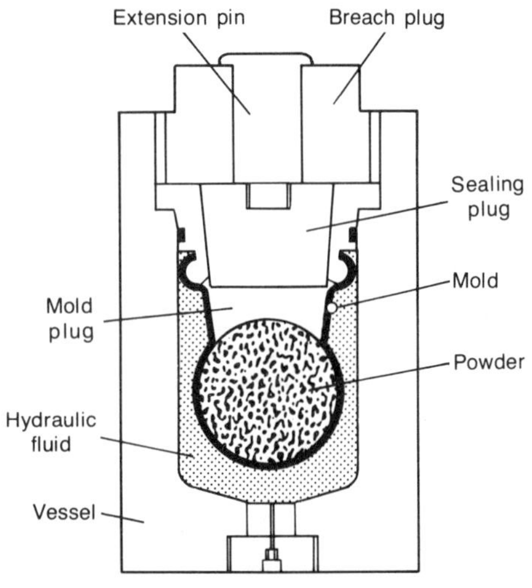

Fig. 5 Schematic of dry-bag hydrostatic pressing equipment. Courtesy of Olin Energy Systems

Dry-bag pressing uses a flexible mold that is permanently sealed in the pressure vessel (Fig. 5). After the mold cavity has been filled with a controlled quantity of powder, the cover plate is closed, and hydraulic pressure is applied. After the pressure is released, the molded piece is removed, and a new cycle commences. Dry-bag pressing is a much faster production process than the wet-bag method and lends itself to automation. Much of this technology has been developed for producing near-net shape ceramic pieces, for example, automotive spark plug bodies. Each dry-mold setup requires special engineering and development (Ref 8).

Advantages and Disadvantages. Hydrostatic pressing offers the following advantages (Ref 7):

- Pressed cermet pieces have a uniform density regardless of size and shape
- Wet-bag method is well suited for large pieces and often is the only practical method for pressing such pieces
- Slender pieces with high ratios of length to cross section are feasible
- Mold cost is low compared to that of rigid compacting dies. Low production quantities can thus be economically produced, especially by the wet-bag method
- Undercuts and varying cross sections are feasible with either the dry-bag or wet-bag method
- Little or no lubricant is required
- Process is well suited for research and development work

The disadvantages of hydrostatic pressing are:

- Dimensional control of compacts is limited. Mold design must accommodate the radial and axial shrinkage caused by hydrostatic pressing as well as the shrinkage that occurs during subsequent sintering
- Surfaces of compacts are less smooth than those of die-pressed pieces
- A high liquid-phase sintering step or encapsulation is necessary before hydrostatically pressed cermet pieces can be densified by hot isostatic pressing
- Equipment cost is high, and equipment utilization can be low
- Labor cost is relatively high

For difficult-to-press cermet compositions with high loading of the hard phase and/or relatively hard metal and alloy binders, cold hydrostatic pressing often is a convenient production method; sometimes it is the only reliable method for working with certain compositions.

Warm Extrusion of Cermet Powder Mixtures. The process of warm extruding cemented ultrafine carbide powder with an admixture of plasticizers has been known for many years. It is successfully used for cermets as well as for forming simple prismatic shapes that have a high ratio of length to cross sections. Cylindrical and triangular shapes and other cross sections can be readily extruded; even tubes are feasible (Ref 9).

(a)

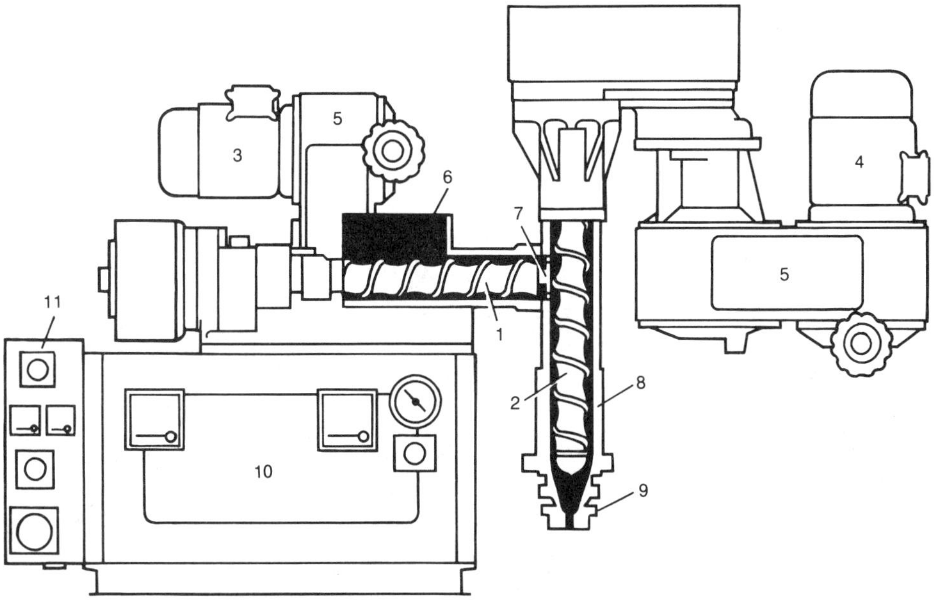

(b)

Fig. 6 Machinery for the warm extrusion of cermet powder mixtures. (a) Extrusion head. (b) Vacuum extrusion press. (1) Feed worm. (2) Compression worm. (3) Feed worm drive. (4) Compression worm drive. (5) Variable belt drive. (6) Feeding hopper. (7) Perforated plate. (8) Coolable compression cylinder. (9) Die support. (10) Vacuum unit. (11) Control box. Courtesy of Dorst America

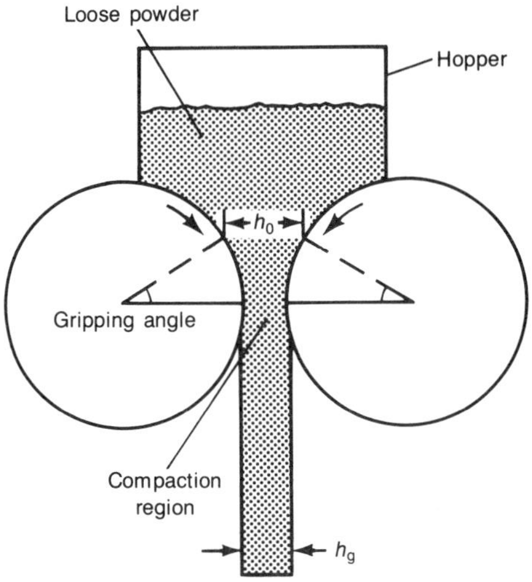

Fig. 7 Schematic of powder rolling with saturated feed and horizontal roll arrangement. Compression ratio, h_0/h_g. Source: Ref 7

Depending on the plasticizer used (for example, polystyrene with an admixture of diphenyl and diphenyl-ether), extrusion requires a temperature somewhere between 160 and 175 °C (320 and 350 °F). Slow and complete debinding under vacuum prior to high sintering is critical in order to avoid distortion, cracking, or microporosity. Screw extruders similar to those used in the plastics industry are adapted to this process (Fig. 6). For the production of a high-quality product, hot isostatic pressing is recommended.

Powder rolling (roll compacting) is a well-known forming process in conventional powder metallurgy that may find application in cermet production. In this process, cermet powder mixtures are fed from a hopper into the gap of a rolling mill and emerge as a continuous strip or sheet. While the horizontal arrangement of the rolls is most convenient for feeding the powder from the hopper into the roll gap (Fig. 7), the vertical arrangement is preferable for feeding the emerging fragile strip horizontally into a series of subsequent operations. The vertical arrangement requires more carefully engineered devices for feeding the powder uniformly into the roll gap (Ref 7).

In contrast to the starting material in slab rolling, the loose powder used in powder rolling has no strength before entering the roll gap and must flow freely or be forced into the gap. During the roll compacting process, the density and physical properties of the powder mixture change. For cermet powder compositions, the feasibility of roll forming a strip of sufficient density and strength depends upon a number of factors, including, but not limited to, the roll diameter and speed, the degree of loading of the cermet mixture with hard-phase substances, the ductility of the metallic phase, and the amount of plasticizer added to the mixture. The presence of the hard component adds to the friction of the powders against the roll and to the internal friction of the powder mixture during the compacting step. This is a favorable characteristic of cermet powders for roll forming; however, it is offset to some extent by the inherently low green strength of the resultant sheet or strip.

The sheet thickness that can be compacted with a given diameter roll is quite limited. A ratio of roll diameter to strip thickness between 600 to 1 and 100 to 1 seems to be the range for various metal powders (Ref 7). It is reasonable to assume that the middle-to-lower range applies to cermets. Special devices are required to prevent the powders from flowing laterally out of the roll gap. A uniform flow of powders over the entire width of the roll is essential for obtaining uniform density in the roll-formed strip. Edge cracking can occur, particularly with

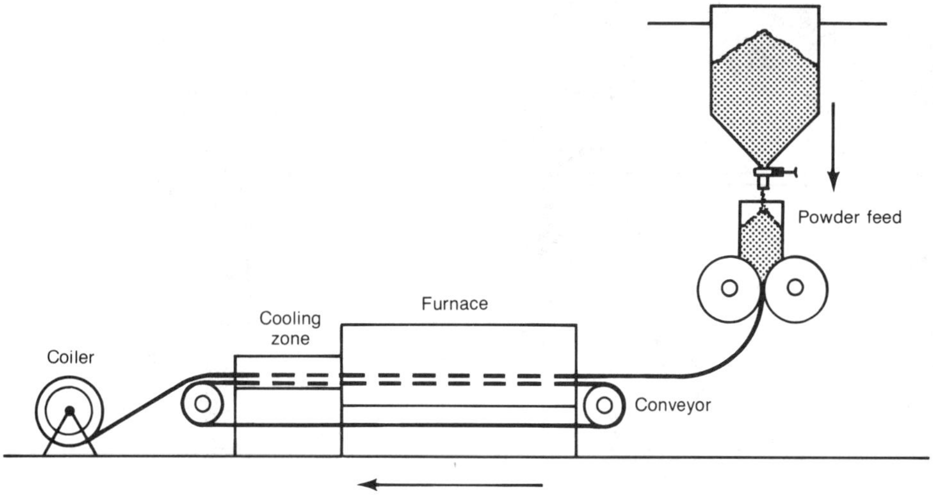

Fig. 8 Powder rolling process with strip reeled into individual rolls after first sintering treatment. Source: Ref 7

heavier strips. An optimum strip thickness has to be established experimentally. Thicker strips are too stiff to be coiled, and thinner ones are too fragile.

Rolling speed is another variable that can only be optimized through experimentation with a given cermet powder composition. Pure metal powders without hard phase have been roll compacted at speeds of 30 m/h (100 ft/h). It remains to be seen whether an output anywhere near this order of magnitude can be obtained with cermets.

A complete powder-rolling line for continuous operation includes debinding and sintering furnaces, rerolling stands, and, if necessary, one or more reannealing furnaces. Up-coiling equipment is needed at the end of the line. This equipment constitutes a major capital investment that is warranted only by a large and continuous demand for the product. Although labor costs for such an operation are low, it may be some time before this line production method finds applications in cermets. A simpler arrangement (Fig. 8) is feasible if, after a debinding step (not shown) and continuous atmosphere sintering, a product emerges of sufficient strength and ductility to permit upcoiling. Roll-compacting arrangements have been proposed for producing a sandwich-type strip consisting of two layers of different compositions.

Slip casting, a method for forming metal powders into a desired shape, follows a technique that has been used for ceramics for a long time. This method uses an aqueous suspension of cermet powders (the slip) that is poured into a porous plastic mold. The liquid is absorbed by the mold, and the powder is deposited on the mold wall. In the case of hollow shapes, the excess slip is drained off after the deposit reaches the required wall thickness; for solid parts the slip must remain and slowly dry.

The water-base slip has low viscosity in order to facilitate pouring, yet it should be stable during standing in order to avoid demixing. Demixing can be a serious problem with cermet powders, particularly those having a substantial difference of specific weight between the hard phase and the binder metal; it can lead to differences in composition and properties from one end of a cermet part to another. Variations in composition can lead to cracks during drying or subsequent sintering. In order to control the viscosity of the slip at the optimum level, it is generally necessary to use a deflocculant and to control the pH.

A typical slip casting mold that has the negative of the form to be cast is shown in Fig. 9. After slow drying, the slip cast part needs a debinding step followed by high sintering.

The resultant part has a higher density than the tap density of its original powder mixture. The fine powders that frequently are used to facilitate slip casting can lead to superior properties in the sintered part (Ref 7).

Slip casting and mold making together are more art than technology. They require knowledge of parameters such as slip viscosity and suspension stability, wetting agents and deflocculants, and slip-mold interaction and mold release. Other important parameters are wall-building rate and casting crack formation. Slip casting requires only a small investment; however, it is labor intensive and is not well suited for mass production. At the present state of technology, cermet parts of a certain complexity are more likely to be suitable for injection metal molding than for slip casting. The former process is more capital intensive, but it is better suited for a medium-to-large production volume.

The P/M injection molding (MIM) process has evoked a great deal of interest since it was first developed in the early 1970s. Commercialization has been slow, mostly because of the long cycle that is required from concept to the point of shipping acceptable parts to a customer. Intensive research and application engineering continues in many laboratories, and more rapid growth is expected in the future (Ref 10, 11).

On a laboratory basis, cermet parts have been made by this process, and its commercialization is underway, particularly in the field of cemented carbides. However, the bulk of current MIM experience is in the area of structural ferrous and nonferrous parts.

The powder injection molding process for cermets (Fig. 10) involves mixing and blending the ingredient metal and hard-phase powders with a suitable polymer binder and then granulating the mixture. The granulated product is heated and injection molded under pressure. The polymer imparts viscous flow characteristics to the mixture to aid in forming, mold filling, and uniform packing. After demolding, the binder is removed, and the remaining cermet structure is densified by sintering and, perhaps, by hot isostatic pressing (Ref 10).

Binder compositions and debinding techniques are the main differences among the various MIM processes. There is no universal binder. A primary requirement of the binder is that it allow flow and packing into the mold cavity. It must wet the powder, and it should be designed to minimize de-

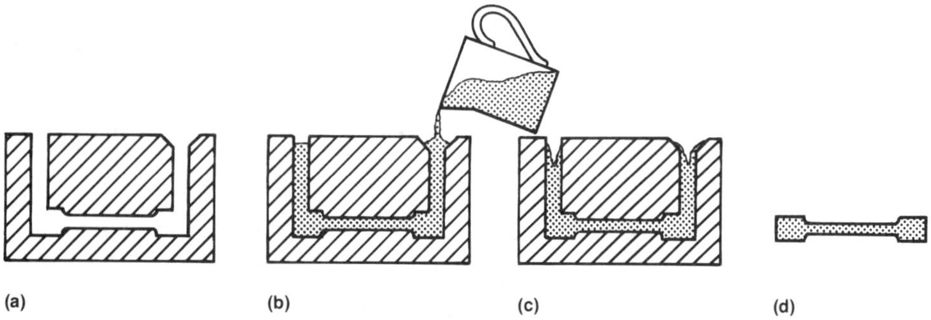

Fig. 9 Schematic of metal powder slip casting. (a) Assembled mold. (b) Filling the mold. (c) Absorbing water from the slip. (d) Finished piece, removed from the mold and trimmed. Source: Ref 7

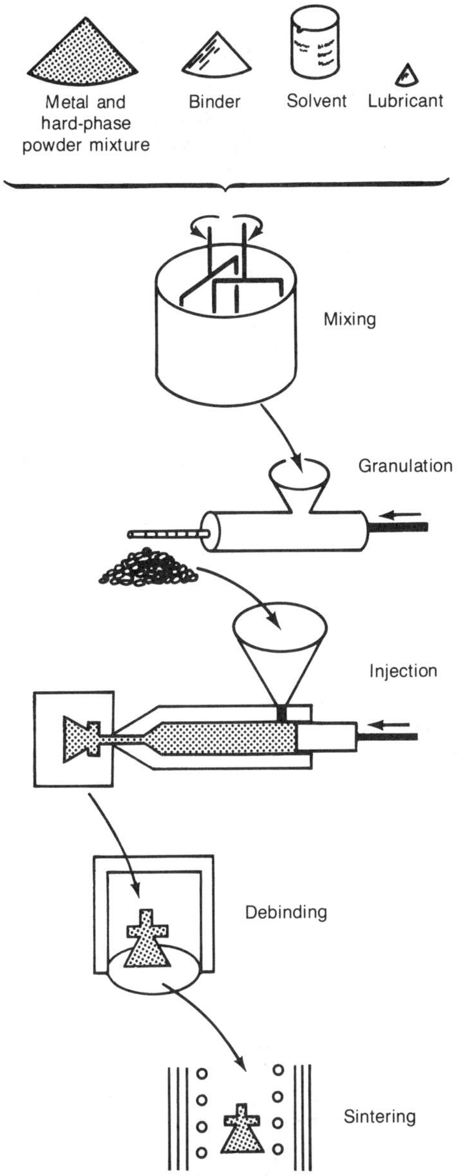

Fig. 10 Schematic of MIM process for cermets. Source: Ref 10

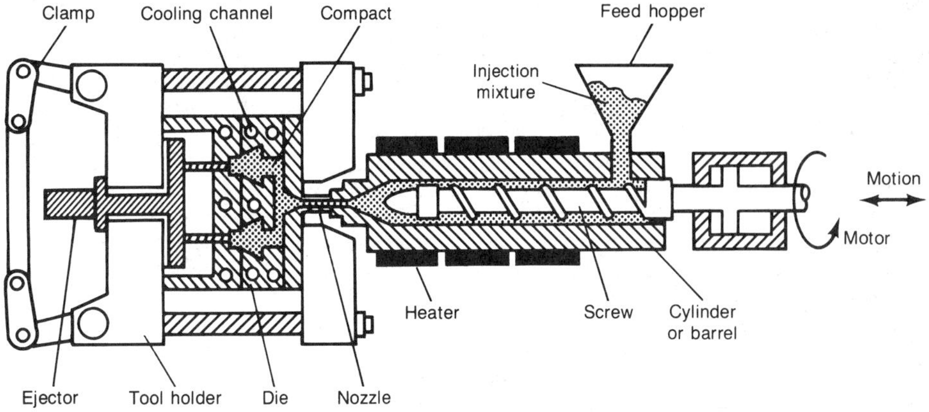

Fig. 11 Mold and injection mechanism for the MIM process. Source: Ref 10

binding time and defects. A multiple-component binder that is not chemically intersoluble allows for progressive extraction in debinding: As one compound is removed and the pores partially opened, the remaining binder holds the particles in place and maintains the shape of the compact. The remainder then vaporizes through the open pores without generating an internal vapor pressure that might cause compact failure. Waxes with additives are most frequently used as binders. The phases of the molding operation are:

- Clamping and filling of the mold
- Maintaining pressure while the compact becomes solid
- Retraction of filling mechanism
- Opening of the mold and ejection of the compact

Mold filling depends on the viscous flow of the feedstock into the mold cavity. The viscosity depends on temperature, shear rate, binder chemistry, powder interfacial chemistry, and loading (Ref 10).

Thermal debinding is the most frequently used technique, but capillary wicking and solvent extraction can also be considered as an alternative method. Complete debinding is required before commencing the sintering cycle. Most cermets require a liquid-phase sintering cycle to achieve complete densification of the compact. A modern furnace that combines debinding, high-vacuum sintering, and a final pressure-sintering cycle can accomplish all of these steps economically.

Applications and Advantages of the MIM Process for Cermets. In recent years, major progress has been made in using the MIM process for the production of heat engine components, military hardware, computers, and aerospace and automotive components. The powder injection molding process offers new opportunities in advanced materials manufacturing (Ref 11), and it offers potential advantages for use in cermet manufacturing technology. General aspects of applying the MIM process in cermet manufacturing include:

- In principle, the MIM process is applicable to cermets without modification of the injection molding machines or the typical mold designs (Fig. 11)
- The production of small- to medium-size complex shapes by the MIM process is feasible, provided that the geometry of the shapes allows for demolding. When this requirement is met, multiple levels, reentrant angles, and undercuts can be accommodated
- On small parts, tolerances of ±3 μm/mm (0.003 in./in.) after sintering can be obtained on conventional P/M parts with the MIM process. Larger tolerances would be needed on cermets to allow for shrinkage when liquid-phase sintering is needed for complete densification
- Small runs (of as few as 2000 parts) are feasible for conventional P/M parts. Because of the higher price level of cermets and the relatively high cost of competitive forming techniques, runs of similar or even smaller size could be economically attractive for cermet parts produced by the MIM process
- With proper debinding techniques and a liquid-phase sintering cycle (perhaps followed by hot isostatic pressing), high-quality parts with good physical properties could be produced by the MIM process
- Excessive mold wear caused by the hard phase during the injection molding of a cermet composition, particularly one with high loading, does not seem to be a serious problem in the MIM process. Future experience will demonstrate if such a mold wear problem exists and to what extent it affects the economics of using the MIM process for cermets

Sintering. Not all cermets require liquid-phase sintering, but the majority use this process to convert the green compacts into solid, strong, and dense products. Sintering temperatures depend entirely on the ceramic-metal system involved and on the choice between solid- and liquid-phase sintering. Typical temperatures range from 850 to 1050 °C (1560 to 1920 °F) for products that contain a bronze, silver, or copper metal matrix; 1300 to 1500 °C (2370 to 2730 °F) for cemented carbides and borides; and 1700 to 2200 °C (3100 to 4000 °F), or even higher, for certain ceramic oxide-base cermets.

For applications requiring fine machining and grinding, as in many cemented-carbide parts and tools, presintering is performed at 1000 to 1100 °C (1830 to 2010 °F) to bond the metallic contact points and give enough green strength to the body so that it can withstand rough machining. Allowance is made for the substantial shrinkage that occurs during subsequent sintering.

Depending on the green density, cermet compacts can shrink during liquid-phase sintering by as much as 18 to 26% linear (45 to 60% by volume). In systems with good sinterability, virtually all porosity is eliminated (Ref 12).

During all sintering processes, particularly during the liquid-phase process, many complicated metallurgical phenomena take place that depend on temperature, furnace atmosphere (hydrogen, inert gas, or vacuum), and the dynamics of the particular ceramic-metal system. For example, metals change into alloys; the hard phase partially dissolves in the liquid phase and changes the composition of the latter phase; portions of the liquid phase can diffuse into the hard phase; and reprecipitation of some elements dissolved in the liquid phase can take place during the cooling portion of the cycle. Also, if carbon is present in the furnace atmosphere (perhaps from the furnace furniture), it will react with oxygen or other elements. The phenomena that occur during the sintering of WC-cobalt systems have been investigated very thoroughly over a long period of time. Ample literature on the basic system and many of its alloy variations is available (Ref 12).

Mechanism of Liquid-Phase Sintering. While not strictly a cermet, the liquid-phase sintering of heavy alloys consisting of tungsten-nickel-copper has interesting ramifications that are applicable to the cermet field (Ref 7). This liquid-phase sintering process includes these principal features:

- The hard phase is partially soluble in the liquid phase during sintering. At temperature, the liquid phase is limited so that the compacts keep their shape
- The sintering temperature must be high enough so that an appreciable amount of liquid phase is present. When these conditions are met, densification takes place
- The finer the particle size of the hard phase, the more rapid and complete the densification of the compact
- Final density is independent of the compacting pressure. To reach theoretical density, compacts pressed at low pressure will shrink correspondingly more than those pressed at high pressure
- The microstructure will show grain growth when compared to the particle size of the original hard-phase powder. This grain growth can be appreciable and is dependent on the sintering time and temperature

When the original hard-phase particles are angular (for example, titanium carbide), they can become rounded during the sintering process. However, this is not always the case because some angular hard substances (for example, tungsten carbide) seem to possess shape memory. During the reprecipitation of dissolved elements from the liquid phase during cooling, angular contours reappear on the hard-phase particles of these substances.

Furnaces. Continuous high-temperature sintering furnaces have been used in the cemented-carbide and refractory metal industries for many years. They are equipped with a hydrogen or protective atmosphere with a low dew point to reduce residual oxygen in the compacts and to prevent oxidation. Continuous pusher-type furnaces equipped with silicon carbide or molybdenum heating elements have been particularly successful and are used for sintering a large volume of small parts. Batch-type vacuum furnaces have become very popular in the last 30 to 40 years. When there is a choice between pusher-type continuous furnaces and batch-type furnaces, the equipment and operating costs favor the former.

In a typical operation, the parts are laid out without packing on graphite plates that are stacked with spacers within the furnace. When direct contact between the graphite and the compacts is undesirable, the plates are lined with an inert ceramic. It is essential to use vacuum equipment when highly reactive powder mixtures are sintered. The optimum level of vacuum to be reached during liquid-phase sintering varies greatly depending on the hard-phase binder system being treated. Several advanced furnace designs provide for initial operation using a hydrogen atmosphere, with a switch to vacuum at a later stage in the sintering cycle. Others provide a pulsating cycle of hydrogen pressure alternating with vacuum.

Static Hot Pressing. Hot pressing is a cermet production method in which the pressure and temperature are applied simultaneously. The powder mixtures are either compacted directly in the hot press mold or prepressed cold in dies and then transferred to the hot press tools of the pressure-sintering furnace. Pressures are considerably lower than for the cold press method. They can range from deadweight loads up to 3 MPa (500 psi) for pressure sintering (of friction elements, for example), or from 10 to 35 MPa (1500 to 5000 psi) for hot pressing; the lower end of the hot-pressing range applies to liquid-phase systems.

Sintering temperatures are reached by induction or resistance heating of the mold, or by direct induction or resistance heating of the powder compact. In the former case, the mold material consists of graphite. This is the more practical process because usually no controlled atmosphere supply is required. The latter method requires ceramic molds, which are sensitive to thermal shock, break easily on product removal, and are costly to produce accurately to the dimensions of the mold opening. The advantage of direct compact heating—that the tooling and surrounding area can remain cool—can be offset by a temperature gradient and the resulting microstructure segregation effects in the product. For most systems with readily oxidizing metal matrices, a controlled atmosphere is required.

The densification effect of conventional static hot pressing is more pronounced than the effect that can be achieved by cold pressing and subsequent sintering. Heating the cermet powder mixture increases its plasticity and produces larger areas of interparticle contact. The surface shearing action that occurs during the process mechanically disrupts surface oxide films and generates clean bonding surfaces (Ref 13). Shape limitations are similar to those for static cold pressure. Prismatic single-level pieces that have no undercuts or reentrant angles are preferred; however, shallow details in the punch faces are acceptable. Very large pieces are well suited for this process.

A typical graphite induction-heated vacuum furnace (Fig. 12) with graphite tooling is capable of double-action hot compacting or repressing of a 125 mm (5 in.) diam billet at pressures up to 90 Mg (100 tons) and temperatures up to 2300 °C (4200 °F) in a high-vacuum or controlled atmosphere. The fully pressed compact is ejected from the die while still hot to reduce cooling time, minimize sticking, and prolong mold life. Items produced in this type of furnace include WC-Co draw dies and friction elements, Al-B_4C billets, stable oxide components, and some boride-base cermets.

Among the various cermet production processes, static hot pressing in such a furnace is the only reliable single-step method for producing a fully dense, high-quality, near-net shape compact from a cermet mixture. However, graphite die life is limited, and cermet compositions that do not react with graphite are preferred. In order to avoid a cermet-graphite reaction, ceramic molds can be used for hot pressing, although they are more fragile and costly than cermet molds. The complete hot press setup (including a vacuum system or atmosphere generator power system, hydraulic system, controls, and instrumentation) is expensive. In addition, static hot pressing is labor intensive because products are pressed one at a time. Therefore, hot isostatic pressing may be more appropriate for many sensitive high-temperature consolidation tasks involving small- and medium-size pieces. For large and very large pieces, vacuum hot pressing is often used because large equipment is readily available (Ref 14). For example, Fig. 12 shows a 225 Mg (250 ton) vacuum hot press with double-action bottom rams and a 1070 mm (42 in.) square platen; the press has a maximum operating temperature of 1315 °C (2400 °F).

Hot isostatic pressing (HIP) has become increasingly popular as a means for producing carbide-base and other cermets of very high and uniform density. Internal flaws and micro- or macroporosity are virtually eliminated in the resultant product. Isostatic pressing is a batch process accomplished in water-cooled pressure vessels capable of withstanding internal pressures of up to 210

Fig. 12 Production-scale 225 Mg (250 ton) vacuum hot press. Courtesy of Vacuum Industries, Inc.

MPa (30 ksi). Heating up to a temperature of 1600 °C (3000 °F) is achieved with a high-frequency or resistance furnace mounted inside the pressure vessel. The pressure medium is an inert gas, usually argon. The pressure medium can also be helium (Ref 8), which at the pressure employed (100 to 150 MPa, or 15 to 20 ksi) has a density close to that of water. More detailed information on HIP equipment is contained in *Powder Metallurgy*, Volume 7 of the 9th Edition of *Metals Handbook*.

Hot isostatic pressing was originally developed for use in gas pressure-assisted diffusion bonding processes such as the encapsulation of nuclear fuel elements (uranium oxide, for example) in a zircaloy sheath. This was soon followed by applications such as powder consolidation and densification of difficult-to-sinter substances and cermet composition (Ref 13). Hot isostatic pressing is most successfully applied in the cemented-carbide industry and in the manufacture of steel-bonded titanium carbide. Notwithstanding the ease with which these cermets sinter to high density, they often have slight localized porosity in the range up to 50 μm; on occasion they have voids in the range from 0.25 to 2.5 mm (0.01 to 0.1 in.) caused by random or accidental contamination.

Hot isostatic pressing is an improvement over static hot pressing in that it eliminates the need for costly and highly perishable molds. However, before compacts are submitted to isostatic pressing, they need to have a sufficiently dense structure (at least at the skin) to inhibit gas penetration. Compacts with lower density and interconnecting pores require a gas-tight encapsulation of some sort before being treated. Three methods are in common use to accomplish this encapsulation (Ref 13). In the first method, compacts are formed and sintered to 95% or more of theoretical density, and the resultant continuous, dense surface structure acts as an impenetrable envelope to the high-pressure gas. Alternatively, the density is raised to a sufficiently high level that no interconnecting porosity remains within the compact. This method is used primarily for small- and medium-size pieces.

The second method involves a steel can that is prepared in accordance with the desired form of the compact. The can is filled with a powder mixture that is densely packed by vibration or pressing. (A cold-compacted or hydrostatically pressed piece could be encapsulated in the steel can instead of the powder mixture.) After loading, the can is closed tightly by welding and then evacuated. This method is most commonly used for medium- and large-size pieces, in situations where the additional cost of a perishable can is economically bearable.

The third method is the Ugine-Sejournet process, in which vitrified glass is used to encapsulate the compact. This method may be more economical than the steel can method.

None of the methods for preventing the high-pressure gas from penetrating the compact are inexpensive. Fortunately, the HIP process is flexible enough to allow for the simultaneous hot isostatic pressing of a number of freestanding or encapsulated compacts. The process cost can be apportioned according to the volume occupied by each piece in the available furnace space.

The cemented-carbide and P/M products that undergo HIP processing are of substantially higher quality than those produced by any other process. The higher quality is a result of the near theoretical densities produced by HIP processing. Pieces with near theoretical densities have high strength levels and reliable physical properties. Hot isostatic pressing is a capital-intensive batch process; when the encapsulation method is used, it is labor intensive as well.

Hot extrusion of cermet billets is unique among the various cermet processes because it is essentially a solid-state process. All of the other previously discussed densification processes involve a liquid phase.

During hot extrusion of powdered material, large hydrostatic compression forces occur. A unidirectional force component first compresses the powder material to full density and then forces the material through the die. Depending on the configuration of the front surface of the extrusion die, a large shear component can absorb as much as one-half of the total energy needed for extrusion. The total amount of one-step deformation (ratio of the cross section of the billet to the cross section of the die opening) is much larger than in any other cermet hot-working process.

Methods. The three basic methods for the hot extrusion of powder mixtures (including some cermets) are shown in Fig. 13. The first method (Fig. 13a) is applicable to loose, coarse magnesium powder or pellets (see the article "Hot Extrusion" in *Powder Metallurgy*, Volume 7 of the 9th Edition of *Metals Handbook*). The material is poured into the hot extrusion container where it heats up to extrusion temperature within 30 s. The rapid advance of the extrusion tool compacts the heated powder and extrudes it though the die. In the second method (Fig. 13b), aluminum-base cermet powder mixtures are transformed into extrusion billets by the cold pressing and hot densification of individual compacts. These billets are then extruded in a conventional manner (Ref 15).

A third widely used technique (Fig. 13c) consists of filling cold powder mixtures into steel cans and compacting the mixtures by means of a penetrating punch while the can is supported in a packing die. (Prepressed compacts can be used instead of the powder mixtures, thus eliminating the compacting step.) A closure is welded over the open end of the can and the air is evacuated from the can. The can is then heated and extruded. As an alternative, a penetrating punch can force the open can and the powder mixture through the extrusion die. In both versions of this method, the extruded product will be sheathed by a thin layer of the material used

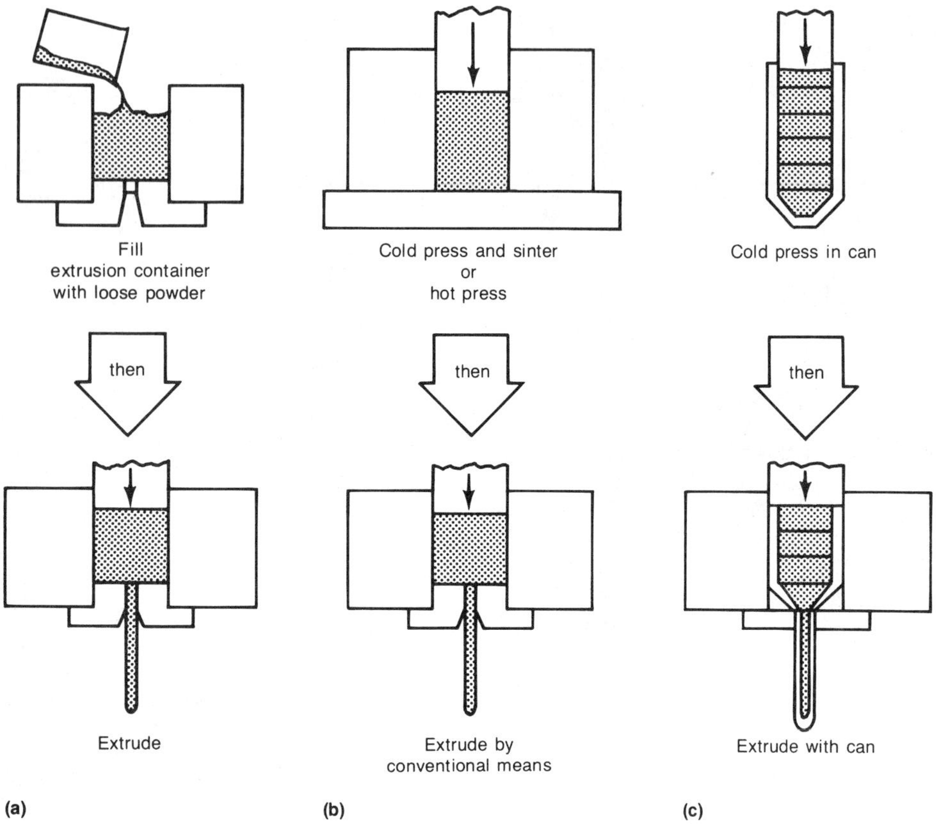

Fig. 13 Three methods for the hot extrusion of powder mixtures. (a) Loose-powder method. (b) Billet method. (c) Steel can method. Source: Ref 7

for canning. This layer must be removed by mechanical stripping or etching.

Application. Hot extrusion is an attractive forming and densification process for cermets. Unfortunately, its application in this broad field has serious limitations caused by the loading of the nonmetallic material, the choice of the metallic component, and the degree of interaction between the two substances. Precise limitations on the volumetric amount of nonmetallic substance have not been established, but when this substance exceeds about 18 to 25 vol%, the composite material exhibits hot shortness to such an extent that the problems of edge cracking, internal lamination, and distortion of the extruded product become intolerable. When the nonmetallic phase occurs in the form of very fine powders, the hot shortness problems are aggravated; they are also aggravated when the nonmetallic substance interacts with the metallic component at the extrusion temperature to form new phases or eutectoids. How to deal with these complex problems for each particular pairing of cermet components is beyond the scope of this review. A careful study of the constitution diagram of the binder metal and the hard phase before undertaking any serious work is recommended.

Like metal alloy billets, cermet billets are much easier to extrude when aluminum or aluminum alloys are used for the metallic components. Billet densification and extrusion occur at lower temperatures and pressures. The expensive canning and decanning process can generally be avoided, as can the use of a protective atmosphere. Moreover, straightening and finishing of the extruded product often can be performed at room temperatures. Defects due to hot shortness in extruded aluminum-base products occur only at higher levels of loading with hard-phase substances.

Compared with cermets containing aluminum or aluminum alloys, cermets with iron, nickel, cobalt, or alloys of these metals as the metallic phase are more technically demanding and more costly to extrude. For example, billet heating and hot compacting require a controlled atmosphere, billet canning is practically unavoidable, and a glass process is required for lubrication and reduction of die wear. Postextrusion finishing, such as the removal of can material, straightening, finish rolling, and so on, also requires more costly processes. The extruded cermet product can be expected to be harder, stiffer, less malleable, and more brittle than the metal binder component. Because of the higher extrusion temperature, metallurgical interaction between the matrix metal or alloy and the hard-phase cermet component is far more likely to occur, particularly if the latter is a nonmetallic compound of the carbide or boride group. Oxide-base hard-phase components are less likely to interact with the metallic component.

The undesirable interaction between the hard substance and the metallic component of the cermet is easier to control with the solid-state extrusion process than with any densification process involving a liquid phase. Also, in spite of the aforementioned limitations and problems, hot extrusion continues to attract the attention of product development engineers as a possible production method for certain cermets, particularly those with an aluminum base. It has the potential to be a relatively low labor cost, yet capital-intensive, mass-production process for high-technology rod or strip material.

Combination Sintering-Compacting. Combination debinding, sintering, and pressure consolidation furnaces have been developed in an attempt to simplify the manufacturing process for cermets and similar products. As stated before, a debinding step is essential before sintering green products that contain admixed lubricants, organic binders, or plasticizers. These additives are needed in varying proportions for static cold pressing, warm extrusion, powder roll compacting, slip casting, and injection metal molding of cermets. After debinding and during liquid-phase sintering, the green compacts shrink to nearly complete density. When densification progresses to the point that pressurized gas can no longer penetrate into the compact, hot isostatic compacting occurs. Gas compacting the already-sintered dense cermet at high gas pressures and at a temperature near that of liquid-phase formation improves the product quality by eliminating all residual porosity, internal flaws, and defects.

Recent experience with WC-Co compacts has shown that using a lower isostatic pressure of only 2.7 MPa (390 psig) can produce compacts with strength and densification nearly equal to those of compacts produced by the high-pressure HIP process (Table 3). Based on these findings, multimode single-chamber pressure furnaces have been developed that are capable of operating in vacuum, with partial pressure, and with positive gas pressure up to 10 MPa (1500 psig); the furnaces operate at temperatures between 1450 and 2200 °C (2640 and 3390 °F) (Ref 16).

This new furnace concept of combining three operations in one cycle offers several advantages over separate sintering and hot isostatic pressing operations:

- Debinding, sintering, and densification under pressure take place in one cycle and in a single vessel
- During the various stages of the process, the compacts do not come in contact with air

Table 3 Comparison of transverse rupture strength for various cemented carbides after hot isostatic pressing and pressure sintering

Processing method	Density, g/cm³	Density, lb/in.³	Hardness, HRA	Transverse rupture strength, MPa	Transverse rupture strength, ksi
6% cobalt					
Vacuum sintering	14.87	0.537	91.6	2180	316
Vacuum sintering and HIP	14.90	0.538	91.9	2645	384
Pressure sintering	14.89	0.538	92.0	2480	360
9% cobalt					
Vacuum sintering	14.59	0.527	91.2	2170	315
Vacuum sintering and HIP	14.58	0.527	91.2	2380	345
Pressure sintering	14.63	0.529	91.2	2843	412
12% cobalt					
Vacuum sintering	14.09	0.509	89.9	2140	310
Vacuum sintering and HIP	14.07	0.508	90.0	2515	365
Pressure sintering	14.11	0.510	90.8	2565	372

Source: Ref 16

- Controlling the cycle with an electronic microprocessor ensures automatic operation and a high degree of program reproducibility (Fig. 14)
- Transfer of parts from one process step to another is avoided, saving labor cost and process time
- Combining processes saves energy

Infiltration is a process that is similar to liquid-phase sintering, except that the solid phase is first formed into a porous skeleton body, and the liquid-metal phase is introduced during sintering from the outside and allowed to penetrate the pore system. Excessive shrinkage associated with *in situ* liquid-phase sintering is avoided, and dimensional stability of the product is obtained, except for about 1% growth that is due to a thin surface film formed by the liquid metal.

This technique is used for systems of two or more components that have widely differing melting temperatures. Aside from hot pressing, infiltration is the only powder processing method that can obtain essentially full density of a near-net shape. All of the other densification processes involve substantial shrinkage and thus destroy shape and dimensional accuracy. By machining, hydrostatic pressing, or powder injection molding the skeletal preforms prior to infiltration, complexities in part design, such as undercuts, reentrant angles, and multiple levels, can be realized to an extent not possible in parts of comparable high density that are made by extrusion or hot pressing. The other unique feature of infiltration is that—under suitable conditions of low contact angles and limited solubilities between the high- and low-melting phases—systems of completely intertwined continuous networks can be obtained. This is of considerable importance for making products that must combine high thermal or electrical conductivity with acceptable levels of strength and abrasion or erosion resistance.

The procedure used for TiC cermets involves two steps (Ref 18). First, an approximately 60% dense carbide skeleton body of near-net shape is formed by mixing the TiC powder with a small percentage of nickel binder and wax, cold pressing the mixture at about 35 MPa (5000 psi) into a slab, vacuum sintering the slab at about 1300 °C (2370 °F), and then machining the contour (for example, a turbine blade). The second step consists of inserting the skeleton shape into a mold assembly that contains the metal in a ceramic tundish on top and provides for the gravity feeding of the liquid to the skeleton at the preferred contact faces. An infiltration arrangement of this type is shown in Fig. 15.

The mold assembly is made of graphite, and its cavity is lined with a refractory ceramic in powder form that interfaces with the TiC skeleton. The ceramic liner is chosen so that it does not react with the titanium carbide up to infiltration temperature and also so that it shrinks at a controlled rate, permitting the formation of a uniform gap all around. The mold assembly is heated in a vacuum furnace to about 1400 to 1500 °C (2550 to 2730 °F); that is, well above the melting temperature of infiltrating alloys, such as 80Ni-20Cr and 70Co-24Cr-6Mo. During infiltration, the liquid metal first fills the gap between the liner and the skeleton

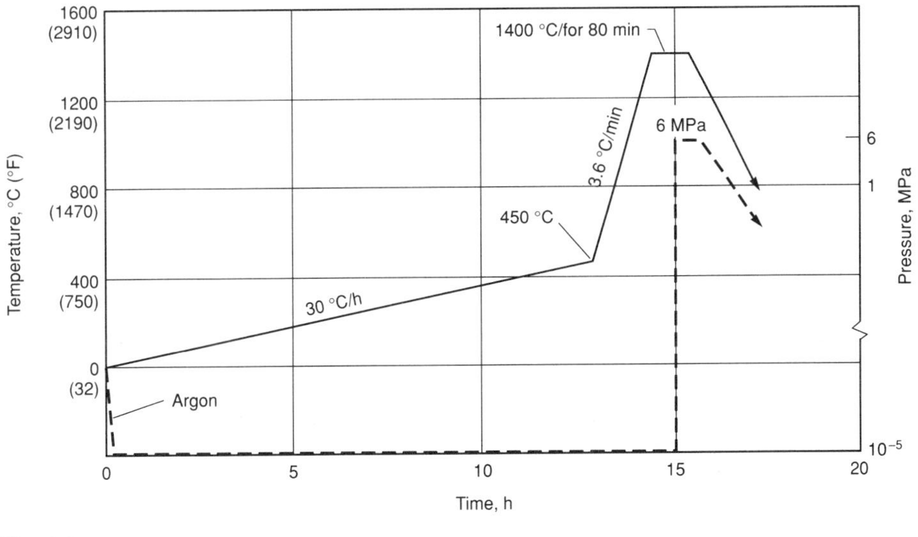

Fig. 14 Schematic cycle diagram for low-pressure dewaxing and overpressure sintering. Source: Ref 17

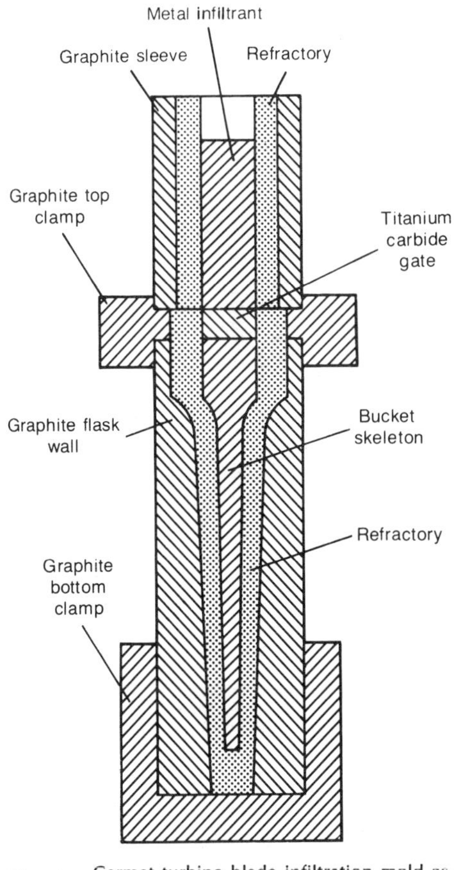

Fig. 15 Cermet turbine blade infiltration mold assembly. Source: Ref 19

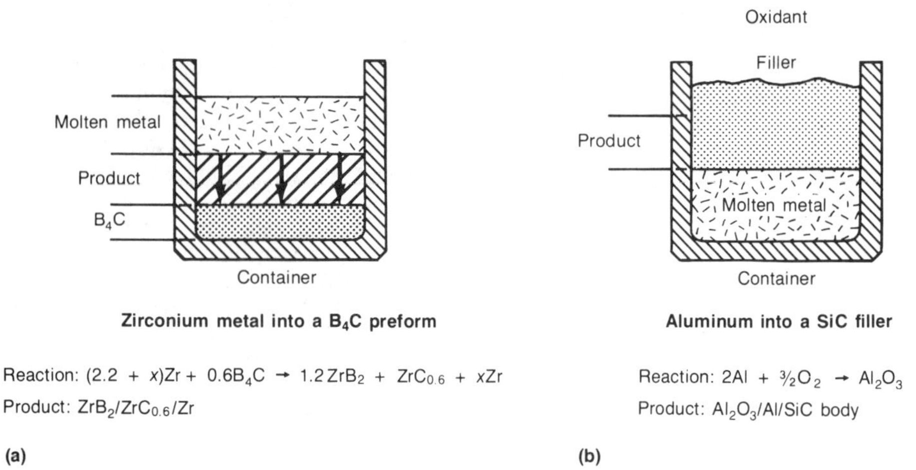

Fig. 16 Schemes for the formation of platelet reinforcements in reaction-infiltrated cermets. (a) Zirconium infiltrant. (b) Aluminum infiltrant. Source: Ref 20

exterior by capillary forces and then penetrates the interior of the porous TiC part. After furnace cooling, the fully infiltrated product can be readily extracted by fragmenting the sintered ceramic liner without degrading the graphite mold assembly, which can be reused.

Graded cermet parts can be produced by varying the density of the TiC skeleton through the use of a special die filling and multiple-step pressing. For example, a turbine blade can be made that has a high concentration of titanium carbide and, therefore, high strength at the center of the foil and in the transition to the root. The turbine blade also has a metal-rich jacket around the foil and especially at the mechanical shock-sensitive blade edges, as well as around the serrated root needed for blade attachment to the turbine disk.

The infiltration process has been successfully applied to other cermet systems; it has been especially successful when used with interfacial reactions. An example of such an application is the production of complex ceramics that are reinforced by microscopic-size platelets of another compound and bonded by a third species (Ref 20). The preformed ceramic is a metalloidal carbide, the binder is a relatively high-melting reactive metal, and the platelets are the reaction product of the carbide and binder. The unique composite microstructure of such a platelet-reinforced ceramic is obtained by gravity infiltration of the molten metal into a porous preform or bed of the carbide. In the case of a B_4C ceramic and a zirconium metal infiltrant, controlled oxidation at the contact faces produces a new phase, ZrB_2, that precipitates abundantly in the form of platelets that reinforce the ceramic (Fig. 16a). A similar result is achieved with SiC preforms or fillers that are infiltrated with molten aluminum metal under oxidizing conditions; the reaction product in this case is Al_2O_3 (Fig. 16b).

The container or mold used in the infiltration process is made from graphite, and it is shaped in accordance with the configurations of the desired product; the process is conducted in an argon atmosphere. The infiltration and reaction temperature depends on the melting point and liquidity of the metal. It can be as high as 2000 °C (3630 °F) in the system involving zirconium. The time to complete penetration and reaction is in the 1 to 2 h range, and the end product typically contains 5 to 15% residual binder metal. The platelet-reinforced infiltrated ceramic system of the type $ZrB_2/ZrC_x/Zr$ exhibits a good combination of high strength, high fracture toughness, and high thermal conductivity. This combination makes it an interesting candidate material for rocket engine components and wear parts (Ref 21). Other systems that have been successfully produced or are potentially workable by the infiltration process include TiB_2 ceramics combined with nickel as second phase (Ref 22), TiC with steel (Ref 23), WC with cobalt (Ref 24), AlN with aluminum (Ref 25), and Al_2O_3 with aluminum (Ref 26).

Infiltration processing of boron carbide and boride-reactive metal cermets has also been used successfully in the development of high-strength, hard, and lightweight products that offer an interesting combination of toughness with high thermal and electrical conductivity (Ref 27). The process involves the infiltration of molten reactive metals, particularly aluminum, into chemically treated boron carbide, or it can use metal-boride starting constituents, such as powders or low aspect ratio fibers, that have been consolidated into a porous ceramic precursor sponge. This process is an alternative to the infiltration of the molten aluminum into thermally modified precursor sponges. Conventional or colloidal chemistry is used in the chemical reaction-controlled casting and infiltration procedures. The potential also exists for consolidating the precursor sponges by injection molding in either a single-step or a two-step process. The key to the process lies in controlling the surface chemistry of the starting constituents. In the two-step process, the first step is the production of highly configured geometries that can be molded by using chemically pretreated binders. In the second step, the binder is volatilized from the precursor, leaving it as a skeleton ready for infiltration.

Bonding and Microstructure

The physiochemical aspects of the bond between the dissimilar phases, and the size of the ceramic grains embedded in the metallic matrix are vitally important to the properties and performance of cermets. Reference 28 contains a discussion of the fundamentals involved in the bonding mechanism, particularly the TiC-metal systems.

Bonding. In general, because of the basic difference in the nature of the ceramic and metallic components in cermets, none of the known solid-state bonds applies by itself. Instead, combinations are formed among ionic bonds, covalent bonds, and metallic bonds. The first type predominates in cermets based on oxide ceramics, and little force beyond simple adhesion exists to hold onto the metallic phase. The second bond type applies chiefly to systems involving silicon and carbon, such as graphite, diamond, and SiC. The strength of this bond is also rather limited. The cermet gains substantially in its mechanical cohesiveness only where metallic bonding combines with covalent bonding. The metal-carbide and metal-boride systems are examples of cermet types in which this combination occurs.

Solubility. The high bond strength between metal and ceramic in cemented carbides and borides is further enhanced by mutual or partial solubility. During sintering, the active surfaces of the carbide or boride particles are dissolved in the liquid phase, and the carbon, boron, and transition metal atoms are reprecipitated on the solid particles during cooling. Depending on the system, these elements remain dissolved in the binder phase in levels that range from trace amounts to several percent.

Even in other cermet types, this partial solubility mechanism is beneficial because it generates a metalloid interface. Metals are known to bond more readily to metalloids, such as silicides and borides, than to oxides. For example, in the $Cr\text{-}Al_2O_3$ system, a surface layer of chromium oxide (Cr_2O_3) on the chromium particles forms a solid solution with Al_2O_3 during sintering in a closely controlled, mildly oxidizing atmosphere. The result is greatly enhanced bonding (Ref 29, 30). Similar metal-metalloid-oxide transition-type bond enhance-

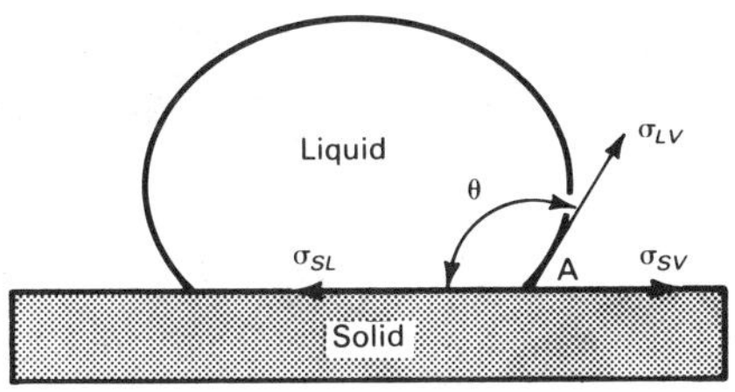

Fig. 17 Surface forces acting at the point of intersection of a liquid resting on a solid. See text for explanation of symbols. Source: Ref 28

ment can be obtained with an intermediate layer of copper oxide (CuO) in the Cu-Al_2O_3 system, or with a layer of titanium nitride (TiN) in the nickel-magnesium oxide (Ni-MgO) system (Ref 28).

Wetting. Another important aspect of the bonding mechanism is the wettability of the solid phase by the liquid metal component. This is controlled by the surface energies of the system during liquid-phase sintering (Ref 28). In Fig. 17, the wetting ability is indicated by the contact angle (θ) that is formed by a liquid drop resting on a solid substrate. The relationships among the surface force vectors are given by the equation:

$$\sigma_{SV} - \sigma_{SL} = \sigma_{LV} \cos \theta$$

where σ_{SV}, σ_{SL}, and σ_{LV} are the surface energies of the solid-vapor interface, solid-liquid interface, and liquid-vapor interface, respectively. The contact angle is a parameter that can be measured with precision. In metal-ceramic oxide systems, the surface energy of the liquid-vapor interface of the metal is greater than the surface energy of the solid-vapor interface of the oxide, and the contact angle is much larger than 90°. Consequently, during liquid-phase sintering or infiltration in a neutral atmosphere, the liquid metal is not retained in the pores of the solid, but tends to sweat out. If the contact angle is less than 90°, however, the liquid metal phase is retained in the pore system of the ceramic; as the contact angle approaches 0, the bond becomes stronger. This is the case with the cemented carbides that have cobalt binders.

Microstructure. The nature of the bond in cermets is very closely related to the microstructure. This is especially significant for the carbide cermets, which have properties that are greatly affected by variables such as the shape, size, and dispersion of carbide grains; the amount of carbide grains in the metal matrix; the composition and structure of the matrix; and, of course, the degree of bonding of the two phases (Ref 31). Although these variables act in conjunction with one another, they can be singled out for specific effects.

A very fine carbide grain size tends to increase strength and hardness, but a somewhat coarser size of about 2.2 μm can provide an increase in the fracture toughness of cemented carbides (Ref 32). Sharp corners prevail in WC grains and cause only minimal harm to the strength of low-binder cemented-carbide grades; in cermets with higher metal contents, however, they affect the localized stress raisers in the ductile metal matrix.

A good dispersion of the carbide grains in the metal matrix provides isolation of the grains, which limits the tendency of a crack initiated in one grain to propagate to others that are coalesced with the first. The same reasoning applies to any secondary hard phases that might form during sintering and bridge the original grains. Chromium carbide and nickel aluminide (Ni_3Al) are examples of reaction products from the matrix that can deposit on the carbide grains of TiC cermets and thereby contribute to the continuity of the brittle hard-phase structure (Ref 30).

The volume fraction of carbide in the cermet has a major influence on the mechanical and physical properties of the end product. Because of the mutual solubility and enhanced bond between the disparate phases, the rule of mixture does not strictly apply to the TiC systems. Generally, strength and hardness increase, and the coefficient of thermal expansion decreases with carbide content. Ductility improves as the metal matrix becomes the continuous phase and as it increases sufficiently in volume to avoid a triaxial state of stress.

The composition of the matrix is important in several respects. High ductility and toughness are essential to relieve the stresses caused by the hard phase and to provide a modicum of safety against catastrophic failure in service. Alloy selection must entail consideration of the matrix as a possible source for the brittle reaction products that can coalesce with the carbide grains into a continuous hard phase. The composition of the metal matrix also influences such properties as oxidation and corrosion resistance, machinability, and weldability.

Good bonding between the carbide and metallic phase is essential because the bond must translate the stresses from one phase to the other. Therefore, any gaps between the surface of a carbide particle and the matrix are detrimental. The importance of these bonds is emphasized by the example of bond improvement in Ni-TiC by the addition of molybdenum. The contact angle of liquid nickel on the TiC in hydrogen is 17°, but the angle changes to almost 0 with the molybdenum additions. As can be seen from micrographs of the two compositions (Fig. 18), the molybdenum-free version suffers excessive carbide grain growth, whereas the Ni-Mo-TiC displays a fine-grain, well-dispersed carbide phase in a continuous metal matrix (Ref 28). Cermets of this type contain molybdenum with nickel in the binder alloy rather than in a solid solution of the carbide (Ti, Mo)C_{1-x}. During sintering,

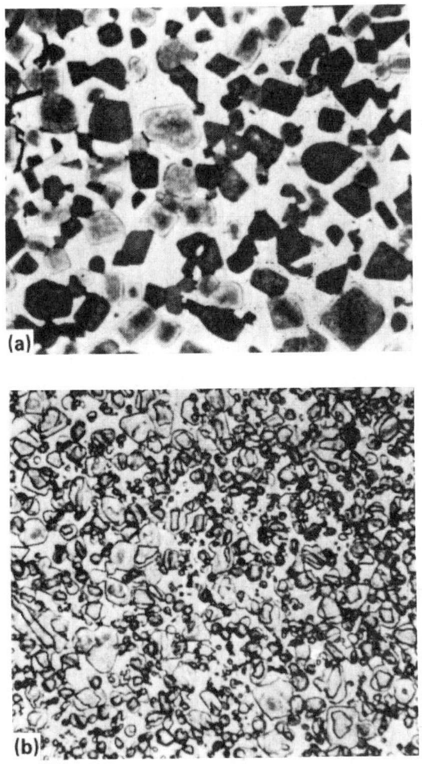

Fig. 18 Microstructure of titanium carbide cermets sintered 1 h in vacuum at 1400 °C (2550 °F) on graphite. (a) 50 wt% TiC and 50 wt% Ni. 1000×. (b) 50 wt% TiC, 37.5 wt% Ni, and 12.5 wt% Mo. 1000×. Source: Ref 28

the molybdenum reacts with the TiC particles to form a case of (Ti, Mo)C_{1-x} that surrounds the TiC core of each particle. This mechanism tends to enhance the wettability of the carbide phase by the binder. The result is strength enhancement, but basic brittleness and, in particular, chip and notch sensitivity are not relieved.

Complexing of Cermet Compositions to Improve Deformation Resistance and Fracture Toughness. Major improvements in the deformation resistance of TiC cermets and concomitant reductions in brittleness are possible by complexing either the binder alloy or the carbide phase, or both (Ref 33). Complexing the binder alloy involves the addition of aluminum, which can produce substantial solid-solution strengthening if correctly applied (that is, in an optimized binder composition such as 22.5% Ni, 10% Mo, and (~7 at.% (Al + Ti). An improvement in the compressive yield strength of the carbide phase is achieved by forming a solid solution of the TiC with 10 wt% VC. The addition of approximately 10 wt% TiN greatly enhances the deformation resistance of the cermet. This enhancement is believed to be at least partially the result both of a grain refinement effect and of the solid-solution hardening of the carbide phase.

If the ratio of titanium nitride to titanium carbide is increased, the resulting carbonitride cermet undergoes a change in microstructure that, under controlled conditions, can greatly improve the strength and fracture toughness of the cermet. In the early 1970s, it was discovered that, in the quaternary systems Ti-Mo-C-N and Ti-W-C-N, a miscibility gap exists in the otherwise complete solid solutions among TiC, TiN, MoC, and MoN (Ref 6). Under controlled processing conditions, the homogeneous single-phase solid-solution (Ti, Mo) (C, N) decomposes spontaneously or spinodally into two isostructural phases (α' and α'') with nearly identical lattice parameters but differing chemical compositions. The α' is essentially titanium carbonitride that contains virtually all of the nitrogen of the original mixture. The α'' phase contains only small amounts of nitrogen but nearly all of the molybdenum or tungsten. The microstructure of vacuum-sintered cermets containing the carbonitride with a nickel-molybdenum binder consists of hard particles; it has a nitrogen-rich titanium carbonitride α' core encased in a molybdenum-rich α'' rim. This α'' phase exhibits better wettability by the binder alloy during liquid-phase sintering under vacuum and produces a cermet of greatly enhanced strength.

Incremental improvements in deformation resistance and fracture toughness are achieved by further complexing the cermet compositions without changing the essential structural features (Ref 4). Present optimization trends encompass the hard component as well as the binder. In the former, the titanium carbonitride solid solution is diluted with Mo_2C, NbC, TaC, VC, and WC, singly or in combination, up to about 40 wt%. The binder is a solid solution of nickel and cobalt, in varying proportions, and generally amounts to 10 to 15 wt% of the entire cermet composition. The alloy is strengthened by titanium and molybdenum absorbed by diffusion from the hard particles during liquid-phase sintering. Aluminum added initially (for example, by nickel-coated aluminum particles) contributes to a further increase in strength, especially at temperatures encountered in tool-cutting operations (Ref 33).

Cermets exhibiting an entirely different microstructure are also possible (Ref 21). Hexagonal platelets of micrometer size can be dispersed inside spheroidal or otherwise equiaxed grains, and both hard components are bonded by softer metal. These microstructural features are the result either of a direct chemical reaction in the starting ingredients during sintering or of a chemical reaction with an extraneous element. The boron-carbide and zirconium system is an example of the former process; the reaction results in an intimate mixture of ZrB_2 platelets and rounded ZrC_x grains evenly dispersed in the unalloyed zirconium matrix. In reactions involving oxygen in gaseous or solid form and various combinations of hard compounds, a variety of particle geometries can be produced in the binder matrices (Ref 26). Typical systems are AlN-Al, TiN-Ti, ZrN-Zr, and SiO-Al. Oxygen can enter the system directly or from an adjacent source such as Al_2O_3 or $BaTiO_3$ contact faces.

Oxide Cermets

This class of material combines oxide ceramic and metallic constituents on a microscopic scale. Thus, it fits the term cermet in the true sense of the word. More than most other mechanically mixed combinations of interstitial compounds and metallic phases, oxide cermets are negatively affected by poor thermal shock resistance and inadequate fracture toughness that limit their usefulness in a great many situations involving high temperatures and dynamic stresses. Some of these materials, however, possess excellent resistance to oxidation or corrosion at high temperatures, and others exhibit unique physical properties, such as nuclear fission. Generally, oxide cermets can be produced to withstand high-temperature stresses greater than those tolerated by most nonmetallic oxide ceramics.

About a half dozen different oxide-ceramic-metal cermets have been developed; several are used industrially. Generally, they differ from oxide dispersion-strengthened alloys by having a ceramic component that is coarser by several orders of magnitude. Also, in most of these cermets, the volume fraction of the oxide is considerably larger than that in oxide dispersion-strengthened materials.

Silicon Oxide Cermets. The classic combination of ceramic and metal can be found in the metallic friction materials, in which the ceramic produces the hard phase. Industrial machinery clutches and heavy-duty brakes, including those for airplanes, are the major fields of application. The ceramic phase is a relatively coarse (for example, 200 mesh) granular SiO_2 to which Al_2O_3 sometimes is added; it amounts to about 2 to 7 vol% of the material. The metallic matrix consists of brass or bronze compositions, and it also can contain iron and lead. All materials have graphite dispersions to provide some degree of lubrication. Conventional P/M techniques, as well as pressure sintering, are employed to produce the friction materials in the form of disks that fit into special attachment cups or plates and strips that are bonded directly to the structural steel support.

Aluminum Oxide Cermets. In this type of cermet, the ceramic is the dominant phase and the metal serves only as a binder. Aluminum oxide cermets are used in cutting tool bits for very high-speed machining with light chip removal (Ref 34). The oxide is milled to great fineness (usually only 1 to 3 μm), then mixed and milled together with nickel powder. Because the binder phase rarely exceeds 5 to 10 vol%, the cermet is very brittle after pressing and sintering, and press lubricants and organic binders are required to facilitate handling. Sintering is carried out in dry hydrogen, in dry nitrogen, or, preferably, in vacuum at temperatures of about 1450 to 1550 °C (2640 to 2820 °F). Finishing is a delicate operation.

A different type of aluminum oxide cermet has been used in the past for high-temperature, heat-resistant applications, such as the furnace components, jet flame holders, pouring spouts, flame protection rods, and seals. These applications met with only limited commercial success over the years. These items were made from complex compositions with a small mass percentage of TiO_2 in the ceramic phase and with molybdenum replacing up to one-fifth of the metallic chromium as the bonding matrix.

However, a similar but less complex composition is used to produce thermocouple protection tubes, which enjoy a solid and expanding market (Ref 35). The tubes are made of a binary 77% Cr and 23% Al_2O_3 cermet. The standard tubular product has an outside diameter of 22 mm (0.876 in.) and an inside diameter of 16 mm (0.625 in.); it is closed at one end of the 910 mm (36 in.) long unit. Other fabricated tubes up to 75 mm (3 in.) in diameter and 600 mm (24 in.) long are on the market.

In the manufacture of these tubes, the powder mixture is ground to a particle size of about 10 μm. Consolidation is achieved by slip casting, cold pressing, or hydrostatic pressing, followed by high-temperature sintering at 1560 to 1700 °C (2840 to 3090 °F). The furnace atmosphere is high-purity hydrogen that contains controlled amounts of water vapor to cause surface oxidation of the chromium particles. The chromium oxide diffuses into the alumina, forming a solid solution at the contact areas that results in strong bonds between the grains (Ref 36). Some typical properties of this type of cermet are listed in Table 4. Figure 19 shows the effect of temperature on the transverse rupture strength, the tensile strength, and the stress-rupture strength of aluminum-oxide cermets.

The chromium content has a significant bearing on the creep resistance of these cermets in the 1380 to 1530 °C (2515 to 2785 °F) temperature range (Ref 37). With up to about 25 vol% Cr, the Al_2O_3 forms a coherent matrix, and the chromium occurs mainly as a statistically distributed phase. For higher chromium concentrations, a network of the metal forms that is fairly continuous at 50 vol%. Consequently, the dominance of the creep strength of the Al_2O_3 is lost as the formation of the metallic network becomes complete.

Many other metals have been mated with Al_2O_3 on an experimental basis (Ref 38, 39, 40) with the objective of developing serviceable, high-temperature cermet materials

Table 4 Composition and properties of aluminum oxide cermets

Property	Cermet: $Cr-Al_2O_3$	Cermet: $Cr-Al_2O_3$
Composition, wt%		
Chromium	72	77
Al_2O_3	28	23
Density, g/cm³ (lb/in.³)	5.9 (0.21)	5.9 (0.21)
Electrical resistivity at 25 °C (75 °F), μΩ · cm	. . .	87
Mean coefficient of thermal expansion, μm/m · °C (μin./in. · °F)		
At 25–800 °C (75–1470 °F)	8.64 (4.80)	. . .
At 25–1000 °C (75–1830 °F)	. . .	8.93 (4.96)
At 25–1315 °C (75–2400 °F)	10.35 (5.75)	. . .
Thermal conductivity at 260 °C (500 °F) avg, W/m · K (Btu · in./ft² · h · °F)	. . .	50.2 (348)
Specific heat, J/kg · K (Btu/lb · °F)	. . .	669 (0.16)
Hardness, HV	. . .	365
Modulus of elasticity, GPa (10⁶ psi)		
At 25 °C (75 °F)	324 (47.0)	259 (37.5)
At 1000 °C (1830 °F)	. . .	225 (32.6)
Transverse rupture strength at 25 °C (75 °F), MPa (ksi)	550 (80)	310 (45)
Tensile strength at 25 °C (75 °F), MPa (ksi)	270 (39)	145 (21)
Compressive strength at 25 °C (75 °F), MPa (ksi)	. . .	760 (110)
Shear modulus at 25 °C (75 °F), GPa (10⁶ psi)	. . .	117 (17)
Shear strength at 25 °C (75 °F), MPa (ksi)	. . .	276 (40)
Poisson's ratio in flexure at 25 °C (75 °F)	. . .	0.20–0.22
Microcharpy unnotched impact resistance at 25 °C (75 °F), J (in. · lb)	1.35 (<12)	. . .
Thermal shock resistance, max temp °C (°F)	1040 (1900)	. . .
Long-time oxidation resistance, max temp °C (°F)	1200 (2200)	1200 (2200)

Source: Ref 35, 36

that have acceptable engineering properties. Metals used in these studies include nickel, cobalt, iron, molybdenum, tungsten, copper, and silver; the main effort was directed at a better understanding of the bonding mechanism. None of these combinations, however, has achieved commercial realization.

Lately, considerable interest has been generated in aluminum as a matrix metal for Al_2O_3, and some ingenious fabrication methods, such as solid-state bonding and reaction infiltration (Ref 41), have been used to produce isometric and complex configurations as well as sheet structures. Because aluminum is the dominant phase, and the Al_2O_3 serves primarily as a reinforcement, these materials should be classified as metal-matrix composites.

Magnesium Oxide Cermets. Chromium also has been used as the metallic phase in magnesia-base cermets (Ref 36). Results of experiments with various metal-ceramic ratios have been reported in the literature, with the magnesium oxide (MgO) fraction ranging from 50 vol% (Ref 38) to as low as 6 vol% (Ref 42). None of these compositions has exhibited a combination of properties that is superior to that of the Al_2O_3-Cr cermets. However, in the MgO-Cr system, an intermediate reaction product (a $MgO{\cdot}Cr_2O_3$ spinel) was observed between the ceramic and metal phases (Ref 36).

The material containing 6% MgO is extrudable and exhibits elongation of 10% and more at room temperature after sintering of the extruded powder mixture (Ref 42). Yield and tensile strengths are about 200 and 350 MPa (30 and 50 ksi) at temperatures up to 600 °C (1100 °F), but they taper off at higher temperatures. These strength properties can be maintained at the higher level up to about 1000 °C (1800 °F) by alloying the chromium with a small amount (for example, 1%) of niobium; however, this degrades the ductility. A measurable degree of room-temperature elongation was also observed in a 30 vol% Cr material that was made by hydrostatically compressing a coarse powder mixture and sintering it at 1600 °C (2900 °F). Unfortunately, nitrides form in this highly refractory material if it is heated in air above 1100 to 1200 °C (2000 to 2200 °F); this causes the cermet to rapidly lose its ductility.

Nickel, iron, cobalt, and alloys of these metals with chromium have also been investigated for use in MgO-base cermets (Ref 38, 40). The MgO-Co cermets, in particular, exhibit interesting mechanical and electrical properties over a wide composition range (Ref 40). For example, the stress-rupture strength at 850 °C (1560 °F) for 100 h of a cermet with 50 wt% Co reaches a peak of 77 MPa (11.2 ksi). In spite of its almost continuous metal phase (~30 vol%), this cermet is an insulator. There is no abrupt change in strength as the material transits from high to low electrical resistance.

Beryllium Oxide Cermets. According to Ryshkewitch (Ref 43), beryllia cermets bonded with tungsten possess better thermal shock resistance and soften at a higher temperature than most of the chromium-alumina materials. They have been used successfully as crucibles, and at one time they were proposed for use in rocket nozzle throat inserts. Ryshkewitch has also proposed the use of combinations of beryllia with up to 50 vol% beryllium metal for high-temperature thermal insulators and nose cones for reentry bodies, despite the extreme brittleness and toxicity of these materials.

Zirconium Oxide Cermets. Zirconia is another ceramic that can be bonded with metal to give useful refractory products. Even when combined with only small amounts of metal, such as 5 to 15 at.% Ti, strong and thermal shock-resistant materials can be produced. These materials are suitable for such applications as crucibles for the melting of rare and reactive metals (Ref 44, 45). If the oxide is combined with molybdenum, the resulting cermet exhibits excellent corrosion resistance against molten steel, good high-temperature strength, and limited sensitivity to thermal shock (Ref 46), especially when the metal content is approximately 50 vol%. Applications include thermocouple sheaths for temperature measurements of metallic melts and extrusion dies used for forming nonferrous metals. Zirconium oxide cermets with a somewhat higher ceramic content, such as 60 vol%, are suitable for use in wear-resistant parts (Ref 46).

Thorium Oxide Cermets. Cronin (Ref 47, 48) discusses metal-ceramic materials in which finely divided thoria is combined with molybdenum or tungsten to form a number of products used in the electronics industry. The principal P/M operations performed on the oxide and metal powders include screening through 325 mesh screens, weighing, dry blending, compacting, sintering in a reducing atmosphere (at 2000 °C, or 3630 °F, for the molybdenum material; at a somewhat higher temperature for tungsten), and finish machining to specified sizes and tolerances.

These products take the form of cylinders and sleeves in high-power pulse magnetrons that can deliver up to several million watts. They are formed into simple disk shapes for use in evacuated electron beam tubes (klystrons), traveling wave tubes, and special-purpose guns. In some high-voltage operations, the thermionic emission cathodes operate over a wide temperature range (1000 to 1700 °C, or 1830 to 3100 °F), but the range is narrower (~1300 to 1500 °C, or 2370 to 2730 °F) for cathodes in the average tubes. Because the ThO_2 is present in the refractory metal as a well-dispersed, fine-particulate minor phase that rarely exceeds 4 to 5 vol%, the materials are usually referred to as a dispersion-type alloy rather than as a cermet in the literature (Ref 47, 48).

Uranium Oxide Cermets. These cermets are used in the fuel elements of nuclear reactor cores. They consist of a dispersion of the fissionable UO_2 in a sintered matrix of aluminum (Ref 49), stainless steel, or tungsten (Ref 50). Compared to plain oxide fuel, these cermets have better retention of fission products and an increased thermal

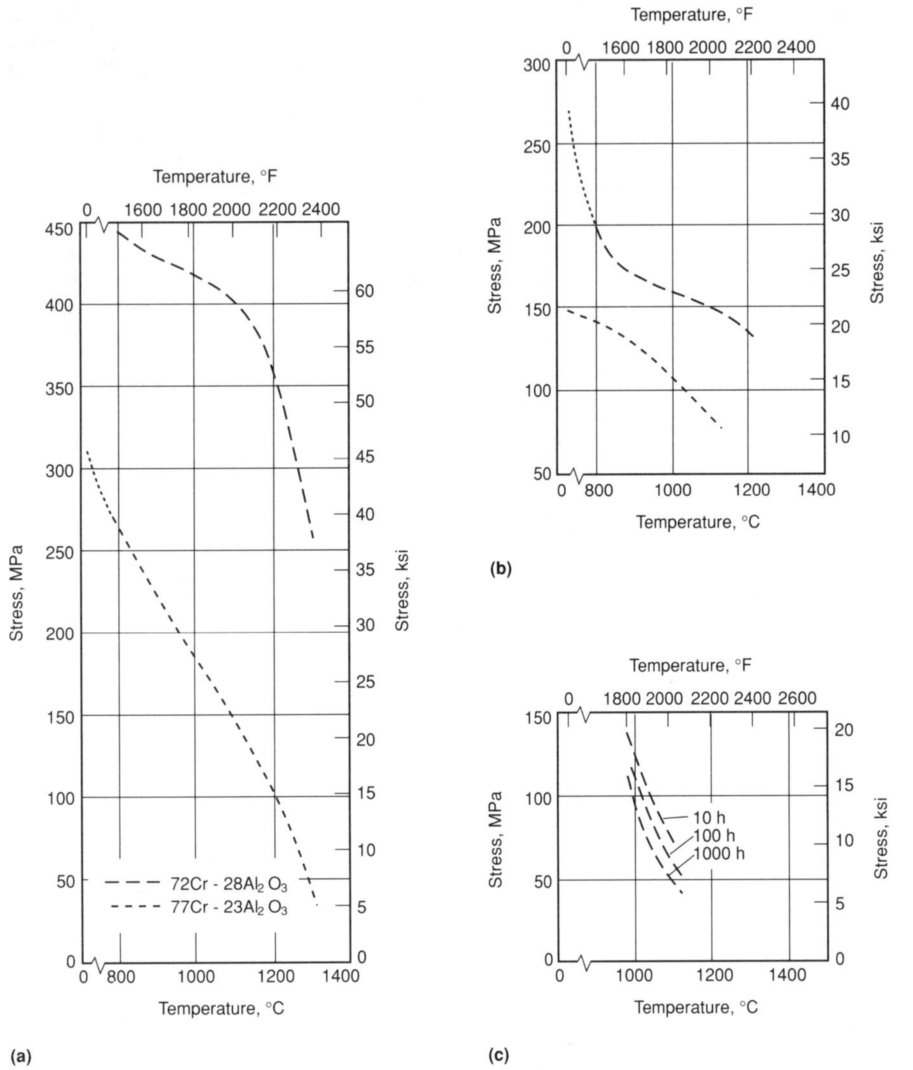

Fig. 19 Effect of temperature on the strength properties of aluminum oxide-chromium cermets. (a) Transverse rupture strength. (b) Tensile strength. (c) Stress-rupture strength. Source: Ref 35, 36

conductivity, which inhibits melting at high operating temperatures. Usually, to ensure a continuous and coherent metal matrix and to limit the radiation damage caused by the fissionable UO_2, the ceramic component is kept below 35 vol% of the cermet. The cermet is contained inside structural stainless steel supports, such as casings or frames.

Powder metallurgy fabrication details for uranium oxide cermets are given in Ref 51. The UO_2 may vary in purity, depending on its processing, and should be of stoichiometric composition. The ceramic particles are fairly coarse, and they must be strong enough to withstand subsequent working without fracturing. Typical particle sizes are 44 μm minimum (+325 mesh) for bicrystals, 35 to 44 μm for monocrystals, and 40 to 50 μm for the diameter of agglomerates of very fine (0.1 to 1 μm) irregular crystals. These UO_2 powders are fired at 1600 to 1700 °C (2900 to 3100 °F) in hydrogen to increase particle size, strength, and density (in the case of the agglomerates). Small additions of TiO_2 increase the sintering rate. Procedures used conventionally for mixing oxide and metal powder must be modified, that is, glovebox or other environmentally controlled operations must be used because of the strong radioactivity of the UO_2. Where large differences in density between the ceramic and metal exist, tumbling is inadequate to prevent segregation, and ball milling is required.

Table 5 lists the density and some physical and nuclear properties of the UO_2 and the various matrix metals in cermet fuels. Standard techniques are used to process the metal-ceramic powder mixture into consolidated cermet fuel. Cold pressing achieves higher densities when pressing lubricants are employed. Because high-temperature sintering usually is inadequate for meeting the high-density and dimensional specifications of the fuel element, either cold working and sizing or machining is required. Some oxide particle fragmentation and waste is unavoidable in this case. Alternate methods of consolidation that yield higher densities and minimize fragmentation, but increase production cost, include hot pressing and hot-working processes such as extrusion, swaging, rolling, and drawing.

Cermets containing 50 vol% each of UO_2 and tungsten have been fabricated into fuel elements for gas-cooled reactor cores that have coolant temperatures of 1500 °C (2730 °F) and higher. This fabrication has been done by such methods as high-energy compaction, isostatic or vacuum hot pressing, powder rolling, and coextrusion (Ref 50).

Other Oxide-Containing Cermets. Recent high-technology developments have increased the demand for new materials with properties tailored to fit a particular application. A typical case in point is the need for cost-effective electronic packaging materials such as semiconductor substrates or heat sinks. These materials require unique combinations of electrical and thermal conductivities, adequate strength for assembly, and a strictly controlled thermal expansion coefficient that allows mating with silicon or other semiconductor substances. Particulate composites or cermets can best fulfill the requirements of the particular system because they offer flexibility in combining the properties of the respective metallic and ceramic components.

Table 5 Properties of uranium dioxide and candidate matrix metals in nuclear reactor fuel cermets

Material	Thermal absorption cross section, barns/atom	Density		Melting point	
		g/cm^3	lb/in.^{3}	°C	°F
Uranium dioxide (0.53 uranium volume ratio)	0.002	10.96	0.397	2500	4530
Beryllium	0.01	1.848	0.067	1277	2330
Magnesium	0.06	1.738	0.063	650	1202
Zirconium	0.18	6.489	0.235	1852	3366
Aluminum	0.23	2.699	0.098	660	1220
Niobium	1.1	8.57	0.31	2468	4474
Molybdenum	2.4	10.22	0.37	2610	4730
Iron	2.4	7.87	0.285	1537	2799
Stainless steel, type 304	2.9	~7.90	~0.286	~1400	~2550
Chromium	3.1	7.19	0.26	1875	3407
Nickel	4.5	8.90	0.322	1453	2647

Source: Ref 51

For example, a composite of iron and cordierite ($2Al_2O_3$-$2MgO$-$5SiO_2$) was developed to achieve a specific and controlled thermal expansion (Ref 52). When processed from powders by static cold pressing in a die and then sintering with the aid of a 0.2 vol% B addition, a fully dense cermet body results that exhibits good interfacial bonding. In compositions of up to 40 vol% cordierite, the ceramic phase dispersed in the iron matrix controls the thermal-expansion coefficient.

Metal-Matrix High-Temperature Superconductor. Another oxide-containing cermet has been developed for a high-temperature superconductor part. In this cermet, a ceramic-copper oxide complex is combined with an easily formable metal for the purpose of rendering the composite fabricative. Copper is a good choice for the metallic phase because it has good strength, ductility, and work-hardening characteristics combined with favorable electrical and thermal conductivities. These properties make copper an outstanding forming aid for the ceramic superconductor as a matrix, as encapsulation, or as both. For the severe working reductions required for bar and wire fabrication, intimate interfacial bonding between the ceramic and the metal is essential. One method of accomplishing this is grinding the $YBa_2Cu_3O_x$ and copper into ultrafine powders, mixing them intensively, and then subjecting the mix to shock wave consolidation inside copper tubing (Ref 53).

Carbide and Carbonitride Cermets

Metal-bonded carbide or carbonitride materials are probably the most important group of cermets at the present time. Logically, all metal-bonded tungsten carbide and titanium carbide materials should fall into the category of cermets. However, it has been customary in the industry to designate all cobalt-bonded tungsten carbide compositions as cemented carbides. This important class of cermets is discussed in the article "Cemented Carbides" in this Volume. This section focuses on the other categories of metal-bonded carbide or carbonitride materials, such as:

- Titanium carbide cermets bonded with nickel or steel
- Titanium carbonitride cermets
- Steel-bonded tungsten carbide cermets
- Chromium carbide cermets
- Metalloid carbide-base cermets

Nickel-Bonded Titanium Carbide Cermets

This class of cermets has received a great deal of attention in recent years. The development of high-thrust turbojet engines for military aircraft, and the advent of commercial jet transports shortly thereafter, exposed the need for better materials for certain critical stationary and, in particular, moving parts in the power plants of these aircraft. A major effort has been mounted to develop TiC cermets for use in these applications.

The underlying motivation for the large development effort conducted in the United States and Europe throughout the 1950s was the desire to combine, on a microscale, the high strength, the relatively good oxidation resistance at elevated temperatures, and the low specific gravity of a ceramic with a metallic alloy phase that imparts good resistance to mechanical and thermal shock. In these TiC cermets, the metallic phase was varied over a broad range, that is, from about 30 to 72 wt%. The principal alloys were Ni-Mo, Ni-Mo-Al, Ni-Cr, and Ni-Co-Cr types (Ref 54, 55, 56). Some more-complex alloys similar to commercial superalloys were also used to bond the TiC. Where high-temperature oxidation resistance of the binder metal was inadequate, such as in Ni-Mo, this property was enhanced in the cermet by complexing the ceramic phases through prealloying a small amount of carbon and niobium, tantalum, or titanium in the solid-solution (Nb, Ta, Ti)C with the TiC (Ref 55).

Titanium carbide powder is produced industrially by a reaction that uses TiO_2 and carbon powders as starting materials. Carburization occurs through the gas phase (carbon monoxide) at a reaction temperature of 1600 to 1700 °C (2900 to 3100 °F). The reaction is terminated only when the free-carbon content of the product is below 0.8%. High-quality powder has only 0.1 to 0.2% free carbon and a minimum of 80.0% Ti.

Two manufacturing routes have been used to produce these cermets, conventional cemented-carbide production practices and infiltration. The first, still in use today, employs uniaxial or isostatic compaction of the ceramic and metal powder mixtures, presintering, shape generation by machining, high-temperature vacuum sintering, and finishing. With fastidious manipulations, particularly during the early processing steps, very complex shapes have been produced to a high degree of dimensional accuracy with this technique. Blades with different airfoil twists, and even completely integrated turbine rotor units with the individual blade foils hogged out of the disk have been made by these cemented-carbide practices.

If high stress-rupture strength and low creep strength at operating temperatures in the 1000 to 1100 °C (1830 to 2000 °F) range are to be derived from the TiC phase alone, the binder content must be low (comparable to that in cemented tungsten carbides). However, TiC cermets with low binder contents have insufficient toughness. Increasing the metal content improves toughness only slightly, and it degrades the strength properties of the material.

The brittle nature of the cemented TiC cermets affected their performance as gas turbine blades. In engine tests reported in Ref 57, blade failures occurred in the base of the fir tree roots, due to insufficient notch toughness; at the tips of the airfoil, due to low resistance to impact by hard carbon particles; and across the airfoil, due to an incapacity to accommodate dynamic interference with protrusions or bows of the stationary shroud by some degree of plastic deformation. Low fatigue resistance near the tip of the deeply scalloped foils was another shortcoming of the material.

Considerable progress was made toward production of an acceptable TiC cermet for these critical turbojet engine applications by increasing the amounts of cermet material and simultaneously strengthening the binder phase through alloying. The composition was varied in accordance with the strength and ductility requirements for different parts of the blades; this was accomplished by using the infiltration process for making graded products. Alloy accumulations were produced at the root and at the edges and tip of the airfoil (Ref 18, 31, 58, 59).

The goal of developing serviceable engine components made of infiltrated carbide cermets was stopped short by simultaneous developments in industrial vacuum metallurgy that produced advanced precipitation-strengthened superalloys capable of operating reliably and for sustained periods at increasingly higher gas temperatures. However, titanium carbide cermets have low density, and they have better high-temperature oxidation resistance than that of cobalt-cemented tungsten carbide; thus, they are still used in several less-sensitive applications, especially those in which such properties are advantageous. Examples are seals and bearings operating at elevated temperatures, sliding contacts, and, especially, wear-resistant parts.

A typical nickel-bonded titanium carbide grade for seal ring applications contains 25 wt% Ni, 8 wt% Mo, 6 wt% NbC, 3 wt% WC, and a balance of TiC. The physical and mechanical properties of this grade are:

Property	Amount
Density, g/cm³ (lb/in.³)	6.1 (0.221)
Thermal conductivity, W/m · K (Btu · in./ft² · h · °F)	17.99 (124.8)
Mean coefficient of thermal expansion, μm/m · °C (μin./in. · °F)	8.4 (4.67)
Hardness, HRA	89.0
Transverse rupture strength, GPa (ksi)	1.7 (247)
Ultimate compressive strength, GPa (ksi)	4.2 (610)
Compressive yield strength, GPa (ksi)	3.8 (550)
Compressive modulus of elasticity, GPa (10⁶ psi)	394 (57.2)
Shear modulus, GPa (10⁶ psi)	156 (22.6)
Poisson's ratio	0.264
Fracture toughness (K_{IC}), MPa · $\sqrt{m}$ (ksi · $\sqrt{in.}$)	10.6 (9.64)

Source: Kennametal, Inc.

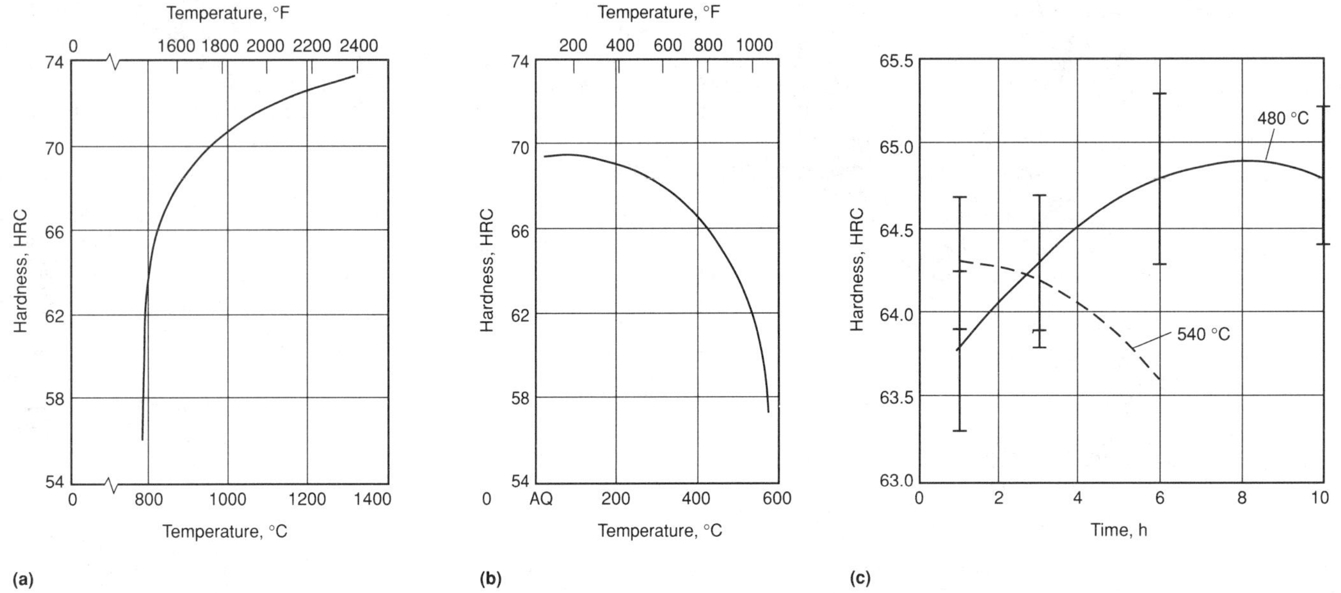

Fig. 20 Room-temperature hardness of heat-treated titanium carbide cermets with ferrous metal binders. (a) Effect of austenitizing temperature on a quench-hardenable material. (b) Effect of tempering temperature on a quench-hardened material. AQ, as-quenched. (c) Effect of aging time at temperature on two materials precipitation hardened at different temperatures

Steel-Bonded Titanium Carbide Cermets

Titanium carbide cermets with steel binders are a direct outgrowth of extensive development work on infiltrated TiC cermets with nickel-chromium and cobalt-molybdenum alloy binders. Development work in the 1950s was directed at solving the severe material problems encountered in the hot end of jet engines and gas turbines. Subsequent TiC-base cermet development efforts focused on the broad areas of tooling and wear.

A new cermet had to fulfill three main requirements to compete effectively in a crowded materials market for highly wear-resistant components and long-lasting tools. New cermets were expected to:

- Be machinable in the annealed condition with conventional cutting tools
- Be hardenable with conventional equipment, without decarburization and without experiencing an undue change in size
- Wear well after hardening on tough applications, giving a performance equivalent or superior to that of conventional cemented tungsten carbide

Within a composition range of about 25 to 50 vol% TiC, a steel-bonded titanium carbide cermet fulfills all three of these requirements. Generally, the addition of titanium carbide by P/M methods alters the properties of a given steel in the direction of higher hardness, higher wear resistance, and higher modulus of elasticity; however, the addition is detrimental to elongation, impact, and fatigue properties.

The upper limits for TiC additions are reached when the cermet is no longer machinable or hardenable without quench cracking. The lower limits are not as well defined. Liquid-phase sintering becomes impractical when a compact with low titanium carbide content loses shape in the process. Figure 20 shows hardness as a function of heat-treating temperature in quench-hardened TiC cermets, and as a function of aging time in precipitation-hardened TiC cermets.

Depending on the choice of steel binders, some grades of steel-bonded titanium carbide have attractive oxidation-, corrosion-, and heat-resistant properties. The proportion of hard phase in the alloy governs the changes in physical and metallurgical properties, following the rule of mixtures.

Comparison of Heat-Treatable Steel-Bonded Carbides With Cobalt-Bonded Tungsten Carbide (Ref 60). In the general field of cemented carbides, the cobalt-bonded grades of tungsten carbide are the ones that have been developed to the greatest extent during the past three decades. The compositions, physical properties, methods of manufacture, and applications of this group of materials are reviewed in the article "Cemented Carbides" in this Volume.

In many applications, cobalt-bonded tungsten carbide has certain disadvantages. The properties of a steel-bonded titanium carbide can be compared with those of cobalt-bonded tungsten carbide:

- Steel-bonded titanium carbides respond to heat treatment and are machinable by conventional means when the binder is in the annealed condition. Cobalt-bonded tungsten carbide, on the other hand, does not respond to heat treatment and is not readily machinable by conventional means
- Fully hardened steel-bonded carbide can be tempered at varying temperatures, thereby obtaining greater toughness than the cobalt-bonded tungsten carbide. However, this gain in toughness is accompanied by some sacrifice in hardness
- Cemented tungsten carbides are high-modulus materials. Steel-bonded titanium carbides have moduli that are not much higher than those of steel
- The coefficients of thermal expansion of the steel-bonded carbides are nearer to those of steel than are those of the cemented tungsten carbides
- Both the tungsten carbide and titanium carbide products can be brazed. In the case of the latter, the braze must have a melting point higher than the austenitizing temperature. Hardening is performed after the brazing operation
- Steel-bonded carbides consisting of 45 vol% TiC have about one-half the density of the available grades of tungsten carbides. This is an important consideration when designing wear-resistant components for astronautical or aeronautical vehicles or for high-velocity rotating equipment
- The steel-bonded carbide compositions contain no cobalt. This is an advantage because cobalt has an exceedingly long half-life, and materials containing cobalt therefore retain radioactivity after exposure to nuclear radiation. Once irradiated, they are difficult to handle for long periods of time
- Most of the steel-bonded titanium carbides and all of the cobalt-bonded tung-

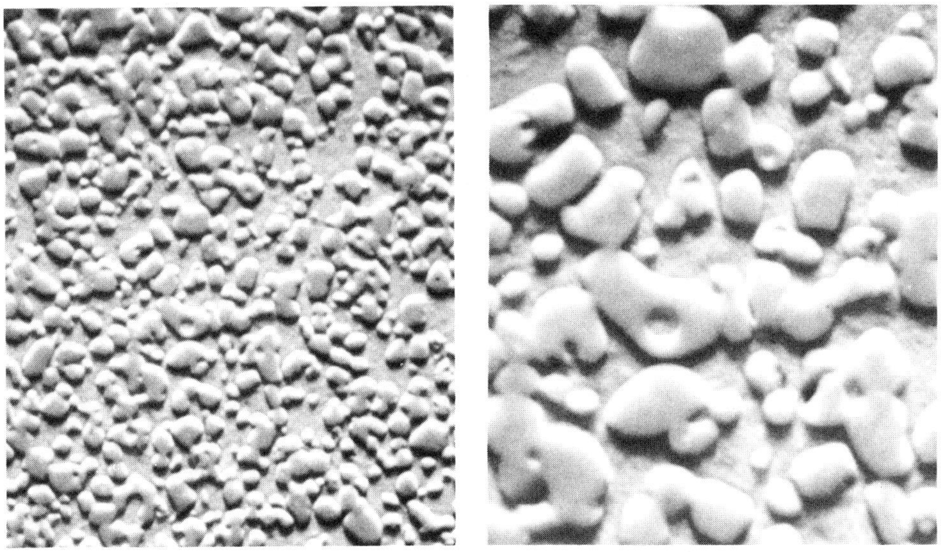

(a) (b)

Fig. 21 Rounded shape of titanium carbide particles in a steel-bonded cermet. (a) 750×. (b) 2000×. Courtesy of Alloy Technology International, Inc.

sten carbides are magnetic. The steel-bonded materials, however, have a much higher percentage of binder, which is generally ferromagnetic. Thus, the steel-bonded materials can be held in a magnetic chuck during such operations as surface grinding

- Raw materials for steel-bonded titanium carbide are inexpensive and abundant in the United States. Tungsten and cobalt, on the other hand, are relatively high on the list of strategic materials
- Unlike the angular shape of tungsten carbide, the shape of titanium carbide particles, when sintered in a steel matrix, is rounded (Fig. 21). When these very hard and rounded titanium carbide particles are exposed at the surface, they resist cold welding and provide inherent lubricity and a low coefficient of friction
- Certain tungsten carbide compositions give better results in important applications involving severe abrasive wear and impact, such as coal mining bits, oil well drilling tools, and cold-heading dies. Cobalt-bonded tungsten carbide, with minor additions of titanium, tantalum, and niobium carbides, also dominates the large field of cutting tools

Comparison of Steel-Bonded Titanium Carbide Cermets With Other Wear-Resistant Materials. The wear resistance of steel-bonded carbides is far greater than that of even the highly wear-resistant machinable tool steels. In applications where tool steel blanking and forming tools have been replaced by steel-bonded carbide tools, 10- to 20-fold improvements in performance have been observed.

When compared with cast cobalt wear-resistant materials, the steel-bonded carbides have the advantage of machinability, hardenability, as well as wear resistance (Ref 60-63).

Titanium Carbide Cermets With Various Steel Binder Compositions. The wetting and solubility characteristics of titanium carbide make it compatible with a great many alloy steels for formulating steel-bonded carbide (SBC) cermets. From the many possible combinations, a few significant ones have been developed into actual cermet grades. Table 6 gives the compositions, properties, and heat treatments of several quench- and precipitation-hardening SBC grades.

Grade C is a general-purpose cermet with a low-chromium, low-molybdenum binder steel composition and 45 vol% TiC. It is comparatively tough and readily machinable in the annealed condition, and it quench hardens to a level of 70 HRC. It is well suited for tool and wear applications in which operating temperatures do not exceed 190 °C (375 °F). Beyond this temperature, the alloy steel binder will overtemper, with a resulting loss of hardness and wear resistance.

Grade CM has a high-chromium, low-molybdenum steel binder composition and contains 45 vol% TiC. This cermet is more heat resistant than grade C and has slightly lower toughness, good machinability, and reliable hardenability from 1080 °C (1975 °F). It will withstand maximum working temperatures of 525 °C (975 °F).

Other SBC grades have lower TiC contents in order to enhance their toughness and thermal shock resistance. Some grades with age-hardening characteristics allow for higher maximum operating temperatures and greater resistance to oxidation and corrosion. Table 7 gives a list of proven applications for TiC cermets.

Manufacturing Steel-Bonded Titanium Carbide Cermets. Two principal manufacturing processes are used for producing standard or special annealed blanks of steel-bonded cermets. The first process, shown in Fig. 1, includes these steps:

- Preparation of powder mixtures by ball milling titanium carbide, iron, carbon, and elemental alloying metals in powder form in the proportions needed for obtaining a specific alloy steel binder
- Static or hydrostatic cold compaction
- High sintering under vacuum in the presence of a liquid phase
- Hot isostatic re-pressing and annealing

The second process, which is used particularly for large pieces or special forms, includes these steps:

- Static or hydrostatic cold compaction
- Encapsulation in a steel can
- Hot isostatic pressing and annealing
- Decanning

Either process yields a product that is largely free from porosity, flaws, or other internal defects. The latter process is also used for bonding cermet pieces to a steel backup support or extension.

Hardening of steel-bonded cermets, such as those listed in Table 6, is accomplished by several processes that are usually selected in accordance with the availability of suitable equipment in the fabrication plant. Oxidation and decarburization should be avoided. Cracking, distortions, and size changes are minimal because only about one-half of the mass is transformed to martensite and the titanium carbide is permanently hard and does not participate in the process. Often, finish grinding after hardening can be avoided because the amount of size change is minimal.

Machining and Grinding. By purchasing suitable standard annealed blanks, a reasonably well-equipped machine or tool shop can produce complex tool or wear parts without nontraditional machining methods such as electric discharge machining. In the annealed condition, steel-bonded TiC cermets, such as those listed in Table 6, can be machined by conventional methods. Most machining operations work better dry than with lubricant. Taps, even when new, should be degreased before use.

Grinding the cermet in the annealed condition achieves a fine surface finish with rapid stock removal. For high-precision parts or tool components, finish grinding is usually accomplished with medium-grain aluminum oxide wheels operating at a relatively high transverse table speed without coolant. Heavy grinding equipment gives good results. Form grinding with form-dressed wheels in the annealed condition and regrinding, if necessary, after hardening together constitute a relatively inexpensive

Table 6 Properties of steel-bonded titanium carbide cermets

Grade	Carbide content, vol%	Matrix alloy type	Heat-treating cycle, °C (°F)/h	Tempering cycle, °C (°F)/h	Hardness, HRC		Relative machinability(a)	Maximum working temperature		Density	
					Annealed	Hardened		°C	°F	g/cm^3	$lb/in.^3$
C	45	Medium-alloy tool steel	955 (1750)/1	190 (375)/1	44	70	1	190	375	6.60	0.239
CM	45	High-chromium tool steel	1080 (1975)/1	525 (975)/1 + 510 (950)/1	48	69	2	525	975	6.45	0.233
CM-25	25	High-chromium tool steel	1080 (1975)/1	485 (900)/1 + 470 (875)/1	32	66	2	540	1000	7.00	0.253
CHW-45	45	Tool steel	1040 (1900)/1	540 (1000)/1 + 540 (1000)/1	45	64	2	540	1000	6.45	0.233
CHW-25	25	Tool steel	985 (1800)/1	540 (1000)/1 + 525 (975)/1	30	61	2	540	1000	7.00	0.253
SK	35	Impact-resistant tool steel	1025 (1875)/1	425 (800)/1 + 425 (800)/1	38	62	1	540	1000	6.80	0.246
CS-40	45	Martensitic stainless steel	1060 (1940)/1	150 (300)/1	50	68	3	370	700	6.45	0.233
PK	42	Maraging steel	485 (900)/3	...	50	61	3	450	840	6.60	0.239
MS-5A	41	Age-hardening martensitic stainless steel	485 (900)/10	...	48	61	1	450	840	6.55	0.237
HT-6A	40	Age-hardening nickel-base steel	870 (1600)/ 8 + 760 (1400)/4	...	46	52	4	985	1800	6.80	0.246
HT-2A	40	Age-hardening nickel-iron	790 (1450)/8	...	46	53	4	760	1400	6.60	0.239

Grade	Carbide content, %	Matrix alloy type	Expansion test range, from 20 °C (70 °F) to:		Expansion coefficient		Transverse rupture strength		Compressive strength		Impact strength(b)		Thermal shock, number of cycles(c)	Linear size change through heat treatment, %
			°C	°F	μm/m · °C	μin./in. · °F	MPa	ksi	MPa	ksi	J/cm^2	$in. \cdot lb/in.^2$		
C	45	Medium-alloy tool steel	190	375	3.53	1.96	1490	216	3585	520	5.66	323	5	+0.048
CM	45	High-chromium tool steel	525	975	5.54	3.08	1275	185	3323	482	3.69	211	2	−0.011
CM-25	25	High-chromium tool steel	540	1000	10.13	5.63	1744	253	3226	468	6.58	376	103	+0.058
CHW-45	45	Tool steel	540	1000	5.72	3.18	1165	169	2206	320	3.14	179	6	+0.019
CHW-25	25	Tool steel	540	1000	6.71	3.73	1979	287	2813	408	6.27	358	106(d)	+0.039
SK	35	Impact-resistant tool steel	540	1000	9.47	5.26	1551	225	2627	381	7.39	422	100	+0.034
CS-40	45	Martensitic stainless steel	370	700	4.41	2.45	1027	149	3123	453	2.59	148	1	+0.016
PK	42	Maraging steel	450	840	3.80	2.11	1379	200	2875	417	7.37	421	84	−0.029
MS-5A	41	Age-hardening martensitic stainless steel	450	840	3.55	1.97	1765	256	2861	415	6.00	343	9	−0.009
HT-6A	40	Age-hardening nickel-base steel	985	1800	8.87	4.93	1317	191	1965	285	5.36	306	11	−0.014
HT-2A	40	Age-hardening nickel-iron	760	1400	13.99	7.77	1241	180	2206	320	5.95	340	18	−0.037

(a) A rating of 1 indicates the greatest ease of machining. (b) Unnotched specimen. (c) Tested by heating the specimen to 1000 °C (1830 °F) and oil quenching; the cycle is repeated until a crack appears. (d) Specimen did not fail. Source: Alloy Technology International

process for creating complex forms in die sections or forming rolls.

Titanium Carbonitride Cermets

Titanium carbonitride cermets, which are widely used as cutting tool materials, evolved from the development of titanium carbide cutting tools in the 1950s. The early cermet cutting tool materials contained 70% TiC, 12% Ni, and 18% Mo_2C. These sintered compositions had densities of 6.08 g/cm^2, Rockwell hardnesses of 92 HRA, and transverse rupture strengths of about 860 MPa (125 ksi). Because they combined high hardness, substantial strength, and low thermal conductivity, they were considered especially suitable for high-speed light-cut machining operations (Ref 4).

Subsequent development work in major industrial countries led to new complex cermet cutting tool compositions based on the titanium carbide and nickel-molybdenum cermet tool material. The titanium carbide cermets with nickel-molybdenum binders were promising high-speed tool materials, but they were inadequate in such areas as toughness and thermal shock resistance. This motivated research efforts such as the two-pronged approach to enhancing cutting performance by strengthening the binder phase and improving the carbide phase (Ref 33). For the former, aluminum seemed to be the most promising binder metal alloying element. Using 0.075 mm (0.003 in.) maximum nose deformation as the tool wear criterion, optimum cutting speeds were achieved with additions of about 7% Al to the matrix (Fig. 22). The maximum strength of the alloy appeared to peak with a content of about 7% (Al + Ti).

After reaching the optimum strengthening effect of the matrix by using an empirical approach to the problem, the research effort used the same approach in its investigation of ways to strengthen the carbide phase. Tool nose deformation resistance showed a clear increase with vanadium carbide (VC) additions in solid solution with the TiC hard phase. At a level of 5% VC, the cutting speed for equivalent deformation was much higher than that of material with no addition. Higher VC additions had a negative effect (Fig. 23). However, further substantial improvements in deformation resistance were achieved by increasing TiN contents, probably because of the grain size refinement and solid-solution hardening effect of this material. Titanium nitride additions to a cermet alloy containing aluminum and vanadium carbide yielded greater improvements in deformation resistance than when each constituent was added individually. Along with increased deformation resistance, the TiN-modified cutting alloys also exhibited greater resistance to thermal cracking, an important factor for applications involving interrupted cuts, such as those which occur in milling operations.

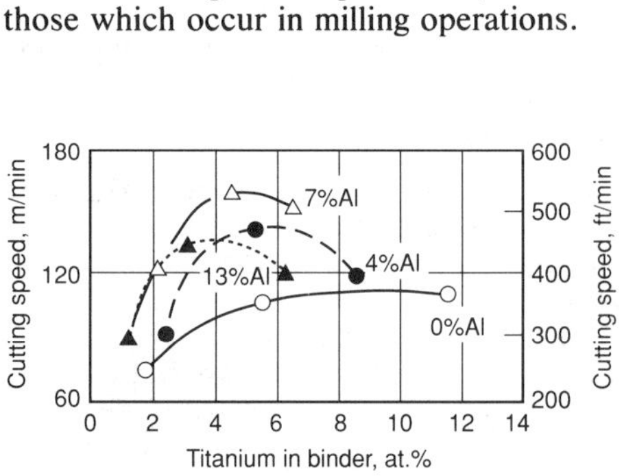

Fig. 22 Cutting speeds for 0.075 mm (0.003 in.) nose deformation versus atomic percent of binder titanium for materials containing four different levels of aluminum. Source: Ref 33

Table 7 Applications of steel-bonded titanium carbide cermets

Application
Wear-resistant parts in valves, metal-stamping and metal-punching equipment
Hone guides
Gages
Core rods
Forming dies
Draw dies
Wiping dies
Guide rolls
Extrusion barrel liners
Molds for plastic injection molding
Paper punches
Screw segments
Tablet dies
Forming dies and mandrels
Pelletizer knives
Pelletizer die faces
Cold- and hot-heading dies
Textile fiber rolls
Fiber venturis
Sizing dies
Furnace furniture and rails
Filament mandrels
Cryogenic pistons
Bed knives
Mixing cups and pintles
Aerospace material

Source: Alloy Technology International Inc.

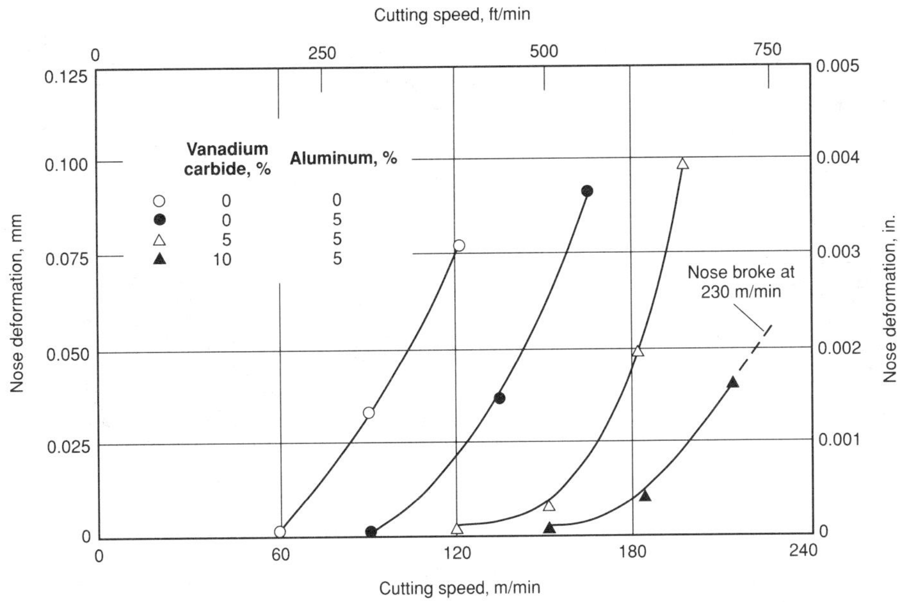

Fig. 23 Tool nose deformation versus vanadium carbide content of cutting tool materials containing 0 or 5% Al in the binder. Material cut was 4340 steel with a hardness of 300 HB. Source: Ref 33

It is interesting to note that in the final analysis, aluminum was no longer required for any of the improvements obtained by adding various strengthening elements to the original TiC-N-Mo composition. Compositions containing TiC-VC-TiN-Ni-Mo were optimized by adjusting the amount of each constituent to obtain the required properties for specific machining applications. With these complex cermets, harder workpieces can be machined at higher cutting speeds, even in applications with intermittent cutting, such as milling.

A somewhat more theoretical approach to the problem of improving on the original TiC-Ni-Mo cutting cermet was used to develop important alternative compositions of titanium carbonitride cermets. These materials were patented in 1976. According to the patent abstract, "The carbonitride alloys are based on selected compositions located within the spinodal range of the systems [that] [have] titanium and group VI metals M as [their] base metals and [have] a gross composition falling within the area ABDE of [Fig. 24]. The binder is selected from metals of the iron group and metals of the group VI refractory transition metals and comprises between 5 and 45 weight percent of the composition" (Ref 64).

Figure 24 shows the complex relationship between, and compositional limitations of, the TiN and CN factions that need to be accommodated to achieve desirable spinodal alloys with superior cutting properties. According to the patent abstract, this figure "defines the gross composition of the carbonitride solid solution $(Ti_x M_y)(C_u N_v)_z$ used as input material in the fabrication of the alloy composition" covered by the patent.

Properties. The microstructure of a typical spinodal titanium carbonitride at 1500× (Fig. 25) shows typical isolated angular carbides of much larger size than the surrounding carbides; they are more or less uniformly distributed throughout the field. Each larger carbide is separated from the next carbide of similar size by a distance of about 1 to 3× its size. Figure 26 shows schematically the α_1 core phase of the cermet, consisting of titanium carbonitride, and the α_2 rim phase, consisting of α_1 enriched with molybdenum carbide. The latter is the transition phase, which is responsible for the strength of the cermet. A typical composition has a hardness of 93 HRA, a density of 6.02 g/cm^3, and a transverse rupture strength of about 1550 MPa (225 ksi). A tougher grade of similar composition (produced by Teledyne-Firth-Sterling) has a hardness of 91.8 HRA, a density of 6.30 g/cm^3, and a transverse rupture strength of 2070 MPa (300 ksi).

Based on general experience with cemented carbides, the transverse rupture strength of the titanium carbonitride cermet would be expected to vary with the proportion of binder metal in the overall composition. Figure 27 shows this relationship in the range of 10 to 17.5% Ni and the dependency of this property on the weight fraction of molybdenum in nickel (Ref 64).

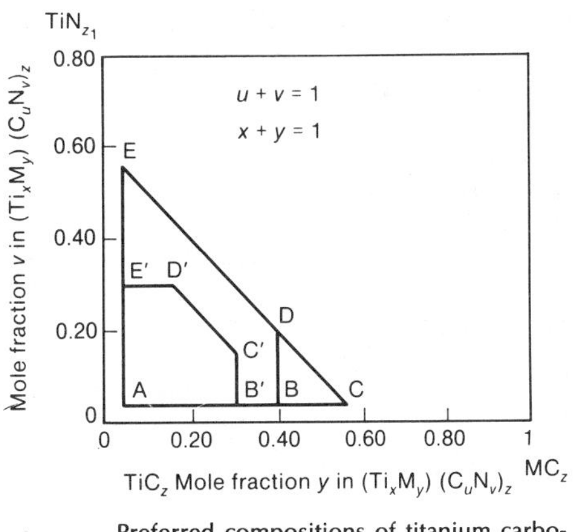

Fig. 24 Preferred compositions of titanium carbonitride cermets. M, molybdenum and/or tungsten; z, number of moles carbon and nitrogen divided by the number of moles titanium and M; z is variable between the limits 0.80 and 1.07. Source: Ref 64

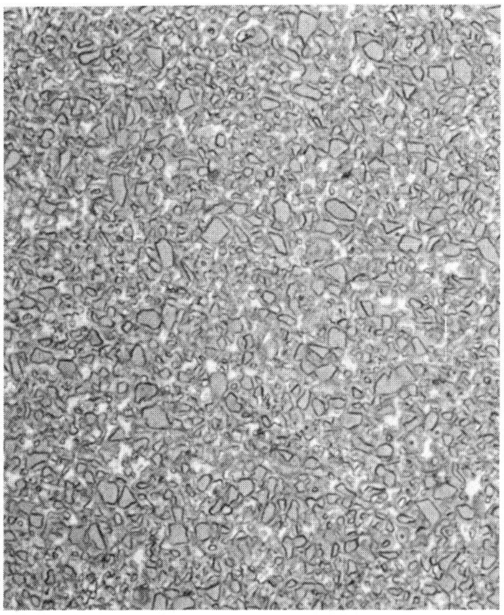

Fig. 25 Microstructure (at 1500×) of a typical spinodal titanium carbonitride cermet. The manufacturing process, called spinodal decomposition, minimizes grain growth and consistently produces nonporous material. Courtesy of Teledyne-Firth-Sterling

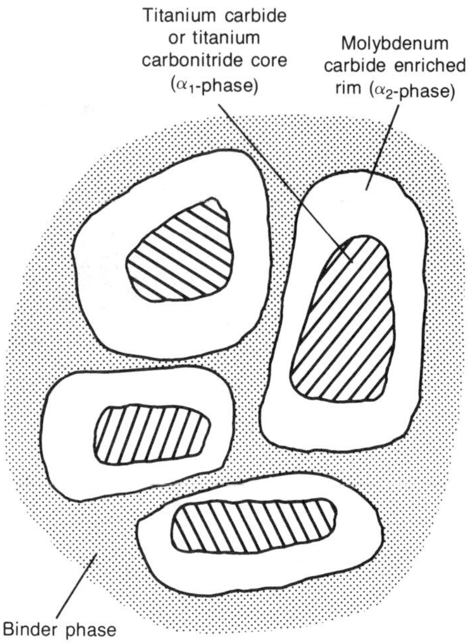

Fig. 26 Schematic of the microstructure of a titanium carbonitride cermet

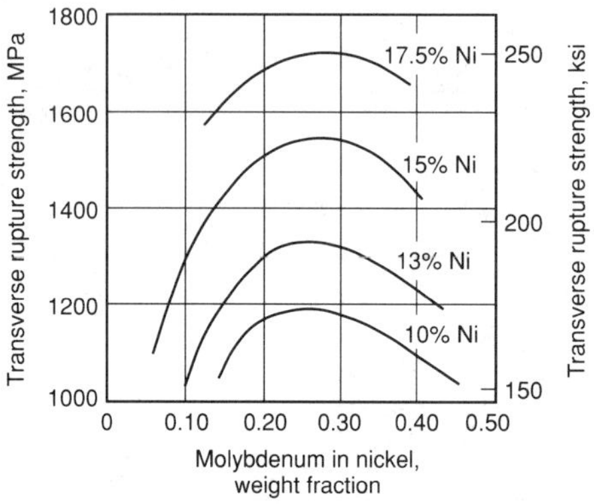

Fig. 27 Effect of binder metal composition on the transverse rupture strength of a titanium carbonitride cermet. Source: Ref 64

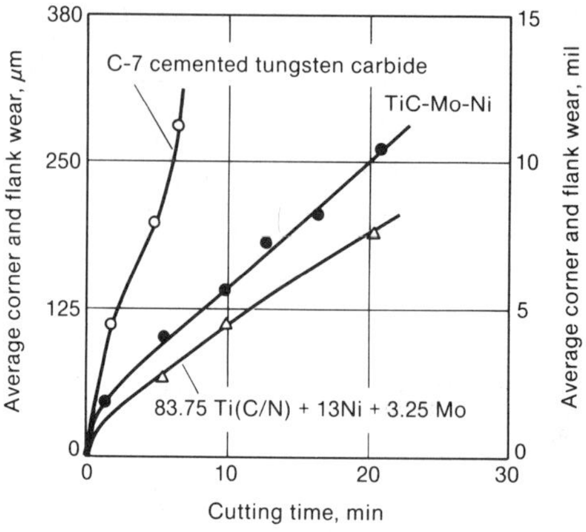

Fig. 28 Comparison of flank wear for two cermets and a cemented carbide when turning 4340 steel. Source: Ref 64

The hardnesses at room and elevated temperatures of these materials are comparable to those of conventional cemented carbides. Yet, the strength and toughness levels of the titanium carbonitride composition are somewhat lower than those of conventional cemented carbides; this would limit the feed rate and depth of cut in heavy roughing applications. On the other hand, Fig. 28 shows the low flank wear on a titanium carbonitride tool in comparison to that of a simple TiC-Mo-Ni cermet and that of a cemented tungsten carbide when turning 4340 steel at various speeds. The wear resistance of the cutting edge is related to the cutting temperature: Pressure welding results in a built-up edge at low temperatures, whereas diffusion and oxidation processes occur at high cutting temperatures. The relatively high free-formation enthalpy of titanium carbonitride increases its resistance to built-up edges, scaling, and crater formation. Favorable flank wear when cutting a tough steel at a relatively high cutting speed is the property of this cermet that enables it to prolong tool life and increase total chip removal between tool changes.

Applications. Titanium carbonitride cermet cutting tools are used for the high-speed milling, roughing, and semifinishing of carbon, alloy, and stainless steels. The resistance to cratering and flank wear exhibited by this material tends to preserve cutting edges; as a result, excellent surface finishes and close tolerances are obtained on longer production runs, even on superalloys and other difficult-to-machine materials. More detailed information on titanium carbonitride tool materials is given in the article "Cermets" in *Machining*, Volume 16 of the 9th Edition of *Metals Handbook*.

Steel-Bonded Tungsten Carbide Cermets

In an attempt to improve on the toughness of steel-bonded TiC cermets, Chinese research metallurgists directed their attention to tungsten carbide as a hard phase. The resulting steel-bonded tungsten carbide cermet tool retained the machinability and hardenability for which this class of materials became known (Ref 65). The higher toughness of the WC-base cermet was attributed to "a small wetting angle with steel under high temperature; even [when] sintering [this composition] in normal hydrogen, dense compacts can be obtained. The solubility of WC in iron group elements is much greater than that of TiC particle in the structure of the alloy maintained" (Ref 65). The physical properties of this new alloy are given in Table 8. The impact strength value of 7.35 J/cm^2 (35 ft · lbf/in.^{2}) is very high not only for steel-bonded cemented carbides but for any cemented-carbide material.

This cermet is used for heavy-impact applications such as cold upsetting or extrusion, heavy punching, cold forging, and ball heading. For example, on ball-heading tools, the performance of the WC steel-bonded cermet was 10 to 100 times superior to that of a previously used heading die.

Procedures for machining in the annealed condition and finish grinding in the hardened condition for these materials are similar to those prescribed for steel-bonded TiC cermets. The tungsten carbide cermets are supplied in the form of standard or special annealed blanks. Any reasonably well-equipped tool shop can fabricate and heat treat special parts in accordance with its needs.

Chromium Carbide Cermets

Cermets that contain chromium carbide as the major constituent possess some unique properties that make them useful for certain applications in the tool and chemical industries (Ref 66, 67). This class of material is essentially a cemented chromium carbide of the Cr_3C_2 modification that is bonded with nickel or a nickel-tungsten alloy. The Cr_3C_2 powder is produced by reacting Cr_2O_3 with carbon at a temperature of about 1600 °C (2900 °F). Minor additions of carbides of lower chromium content are added to control the carbon balance and keep the free-carbon content low. Standard cemented-carbide manufacturing practice is applied to the production of these cermets. Some interesting properties and specific applications of these Cr_3C_2 cermets are:

- Very low density, which makes the material useful in applications such as the production of valve balls in oil well valves
- Relatively high coefficient of thermal expansion, which permits direct brazing to steel, provided that boron-containing fluxes are used
- Bright and durable surfaces of high reflectivity, which permit finishing to optical flatness. These surfaces, together with the thermal expansion characteristics, make the material suitable for gage blocks, micrometer tips, and other measuring tools
- Virtually nonmagnetic nature, which eases measuring tasks, in spite of the nickel binder
- Excellent wear and corrosion resistance, for example, against salt water attack at

Table 8 Properties of steel-bonded tungsten carbide

Property	Amount
Density, g/cm^3 (lb/in.^{3})	~10.2 (~0.37)
Hardness, HRC	
Annealed	33–40
Hardened and tempered	67–68
Transverse rupture strength, MPa (ksi)	2315 (336)
Impact toughness, J/cm^2 (ft · lbf/in.^{2})	7.35 (35)
Modulus of elasticity, GPa (psi × 10^6)	295 (43)

Source: National Machining Carbide, Inc.

Table 9 Properties of typical chromium carbide cermets

Property	Type A	Type B
Composition, wt%		
Cr_3C_2	83	88
Nickel	15	12
Tungsten	2	. . .
Density, g/cm^3 (lb/in.^{3})	7.0 (0.253)	6.9 (0.250)
Electrical resistivity at 25 °C (75 °F), μΩ · cm	84	70
Electrical conductivity,		
Mhos(cm), at 25 °C (75 °F)	0.012	. . .
%IACS	2.1	. . .
Thermal conductivity, at 50 °C (120 °F), W/m · K (Btu · in./ft^2 · h · °F)	10.88 (75.5)	12.55 (87.1)
Mean coefficient of thermal expansion at 25–595 °C (75–1100 °F), μm/m · °C (μin./in. · °F)	10.71 (5.95)	11.10 (6.17)
Hardness at 25 °C (75 °F), HRA	88.3	. . .
Hardness, HV		
At 25 °C (75 °F)	. . .	1300
At 800 °C (1470 °F)	. . .	900
Modulus of elasticity in compression at 25 °C (75 °F), GPa (10^6 psi)	345 (50)	333 (48)
Elastic limit in compression at 25 °C (75 °F), MPa (ksi)	900 (130)	. . .
Compressive strength at 25 °C (75 °F), MPa (ksi)	3450 (500)	3725 (540)
Transverse rupture strength at 25 °C (75 °F), MPa (ksi)	780 (113)	735 (107)
Ductility in compression at 25 °C (75 °F), %	1.1	. . .
Poisson's ratio	0.28	. . .
Izod unnotched impact resistance at 25 °C (75 °F), J (in. · lbf)	0.158 (1.4)	. . .
Resistance to oxidation, max temp. °C (°F)		
Short time	1100 (2000)	. . .
Long time	1000 (1830)	. . .
Resistance to corrosion, 10–24 h immersion, wt loss, g/m^2/day		
50% H_2SO_4	3.1	. . .
35% HNO_3	1.9	. . .
50% NaOH	0.1	. . .
5% lactic acid	0.15	. . .
1*N* solution H_2SO_4	. . .	5
1*N* solution HNO_3	. . .	>10

Source: Ref 66, 67

temperatures up to 85 °C (185 °F), which makes the cermet suitable as a bearing and seal material or for fishing rod guide rings

- Abrasion resistance that is greatly superior to that of any normal corrosion-resistant alloy (Ref 12)
- Outstanding high-temperature erosion resistance at temperatures up to at least 1000 °C (1830 °F)

Applications and Properties. Applications for the chromium carbide cermets include high-temperature bearings and seals; various valve components, nozzles, and guides operating at elevated temperatures; and a multitude of gaging components. Typical physical, mechanical, and chemical properties of representative chromium carbide cermets are given in Table 9. Figure 29 shows the effect of temperature on the thermal conductivity, the expansion coefficient, and the transverse rupture strength of chromium carbide cermets (Ref 66, 67).

Although it has found moderate use in a number of the applications mentioned above, chromium carbide has recently fallen into disfavor and is no longer offered as a standard grade by the cemented-carbide industry. It should be noted, however, that chromium carbide has the potential of opening up an entirely new field of application for cermets in the area of coatings. For example, a coating developed for gas bearing journals blends chromium carbide for wear resistance, nickel alloy for bonding, silver for lubrication up to 500 °C (930 °F), and calcium fluoride-barium fluoride eutectic for lubrication in the 500 to 900 °C (930 to 1650 °F) temperature range (Ref 68). The cermet coatings are plasma sprayed from powder blends and finished by diamond grinding. A typical composition of such a blend is 32% Ni alloy, 10% Ag, 10% BaF_2-CaF_2 eutectic, with the balance consisting of Cr_3C_2.

Other Carbide-Base Cermets

Carbides of other refractory metals, such as zirconium carbide (ZrC), hafnium carbide (HfC), tantalum carbide (TaC), and niobium carbide (NbC), have been produced experimentally and investigated for

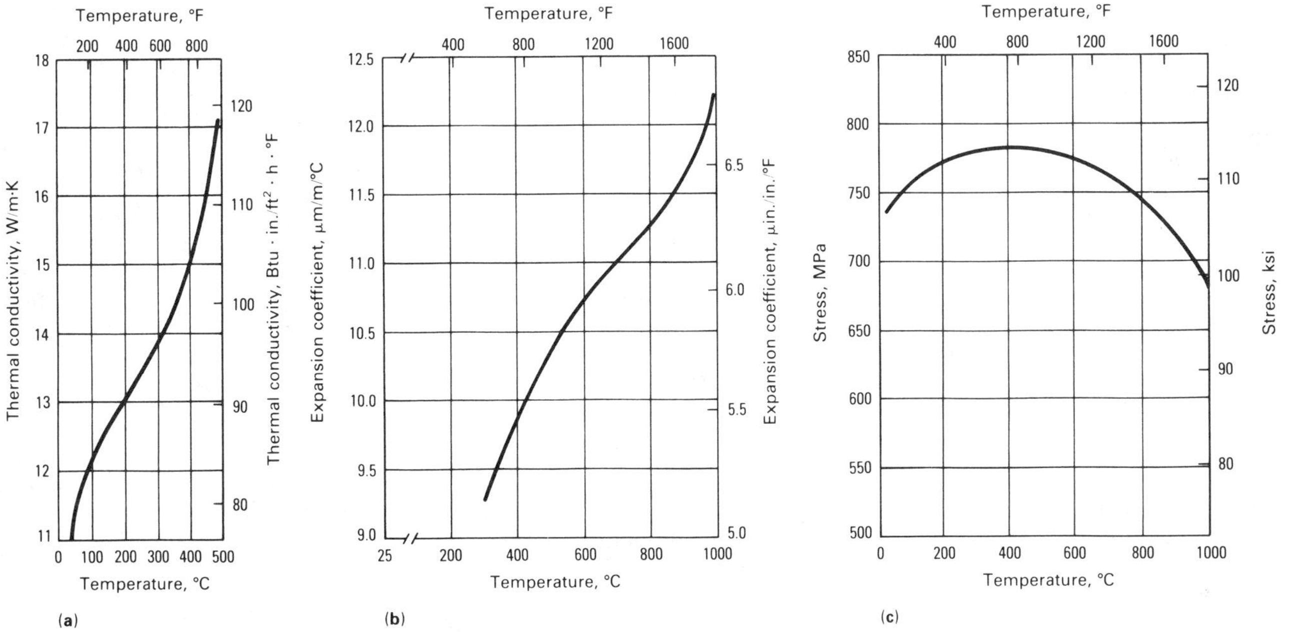

Fig. 29 Effect of temperature on thermal properties and strength of chromium carbide cermets. (a) Thermal conductivity ambient of $83Cr_3C_2$-15Ni-2W. (b) Mean coefficient of thermal expansion from ambient to temperature indicated on scale for $83Cr_3C_2$-15Ni-2W. (c) Transverse rupture strength of $88Cr_3C_2$-12Ni. Source: Ref 66, 67

Table 10 Properties of selected neutron-absorbing materials

Material	Melting temperature °C	°F	Density g/cm^3	lb/in.^{3}	Neutron absorption cross-section, Barns
Boron-10	2000	3630	2.3	0.083	4 000
Boron	2000	3630	2.3	0.083	715
Cadmium	321	610	8.6	0.31	2 500
Dysprosium	1425	2595	8.6	0.31	1 200
Europium	900	1650	5.2	0.19	6 000
Gadolinium	1300	2370	7.9	0.285	40 000
Hafnium	2130	3865	13.4	0.484	115
Samarium	1070	1958	6.9	0.25	8 900

Source: Ref 15

high-temperature applications. They have the highest melting points of all compounds known (Ref 69). Hafnium carbide melts at 3890 °C (7030 °F), tantalum carbide at 3800 °C (6870 °F), zirconium carbide at 3530 °C (6380 °F), and niobium carbide at 3500 °C (6330 °F). All of these carbides exhibit poor oxidation resistance at high temperatures and are extremely brittle. Cementing these carbides with ductile binder metals does not improve these properties sufficiently to make the resulting cermets competitive with the industrial carbides available for high-temperature structural and tool applications (Ref 70). The use of small amounts of tantalum and niobium carbides as additions to titanium carbide to produce cermets of enhanced high-temperature oxidation resistance has been cited before.

Uranium carbide cermets are of some interest in nuclear reactor technology (Ref 51). Because carbon has a low neutron cross section, uranium carbide would be desirable as a fuel element for neutron economy. The compound has higher thermal conductivity than uranium dioxide; also, it has a high melting point (2300 °C, or 4170 °F) and is creep resistant to 1000 °C (1830 °F). The major disadvantages of uranium carbide are brittleness, poor thermal shock resistance, and susceptibility to corrosion in aqueous environments at elevated temperatures. Binder matrix metals chosen for their low thermal-absorption cross section, such as beryllium, zirconium, niobium, molybdenum, or iron, do not alleviate the disadvantages of the UC ceramic phase. Therefore, this type of cermet has not found industrial applications beyond experimental reactor technology in the United States and abroad.

The carbides of the metalloids boron and silicon, B_4C and SiC, are of considerable industrial significance and enjoy such diverse applications as superhard tools and electrical resistor heating elements. These compounds often are processed and used without metallic binder phases, which results in a product outside the material classification for cermets. An important exception, however, involves cermets in which the metalloidal carbide is combined with a metallic matrix phase of major proportion. Some of these have been of long-standing importance to aerospace and nuclear reactor technology; others have more recently come to the fore in avionics and automotive engine manufacturing. The materials are commonly categorized as ceramic-reinforced metal-matrix composites, but those containing a relatively high volume fraction of particulate metalloidal carbide and a discontinuous aluminum matrix are relevant to this article and are discussed below.

Aluminum-Silicon Carbide Cermets. The recently publicized Lanxide process (Ref 26, 41), whereby ceramic-reinforced aluminum-matrix composites are produced by a chemically controlled capillary infiltration process, has produced some very interesting property data (Ref 71). For a cermet containing as reinforcement about 45 vol% of SiC particles ranging in size from about 8 to 25 μm, the following approximate data were established:

Property	Amount
Bending strength, MPa (ksi)	475 (70)
Tensile strength, MPa (ksi)	350 (50)
Modulus of elasticity in tension, GPa (10^6 psi)	180 (26)
Fracture toughness (K_{IC}), MPa$\sqrt{m}$ (ksi$\sqrt{in.}$)	13 (11.8)
Thermal expansion coefficient, μm/m · °C (μin./in. · °F)	14 (7.8)

As the silicon carbide to aluminum ratio is varied either up or down in metal-matrix volume fraction, the property values change accordingly. The material is believed to be of particular interest for structures subjected to sliding abrasion and similar wear conditions.

Aluminum-Boron Carbide Cermets. Natural boron contains 18.8% of an isotope B-10, which has a high neutron cross section (that is, it shows a high capacity for absorbing neutrons). Thus, boron and boron-containing alloys or intermetallic compounds are useful for controlling a nuclear reactor. This makes boron carbide (B_4C) a desirable neutron absorber because of its commercial availability in powder form, its high purity, and its consistent quality. Low density and high chemical stability are other favorable characteristics of this powder. Boron carbide has been included in most aluminum cermets used for neutron absorption elements in nuclear reactors. In some cases, oxides of europium, dysprosium, and samarium are also considered desirable (Table 10).

Cermet components of the Al-B_4C composition have found repeated and diversified applications in certain sectors of the nuclear industry, particularly for those water-cooled reactors that operate within the useful temperature range of aluminum. Some typical components are flat plates with dimensions of 2.5 × 20 × 1370 mm (0.1 × ¾ × 54 in.). Other more complicated shapes include 11 m (36 ft) long reactor control rods with a 43 mm (1.7 in.) outside diameter and a 0.5 mm (0.020 in.) wall thickness. Special P/M processes are required to produce components of this type in accordance with rigid dimensional tolerances and a high degree of critical chemical consistency from end to end.

Successful processes for the manufacture of Al-B_4C cermets include these steps (see Fig. 1):

- Thorough mixing of ingredient powders
- Cold compaction of powders into billets
- Sintering
- Hot consolidation of billets
- Hot extrusion
- Cold rolling

Special multiple-step blending techniques are needed when producing mixtures with very low concentration of the neutron absorber (that is, less than 1%). It has been found that uniform powder blends invariably yield control rods with highly uniform distributions of the neutron absorber over their entire length. The P/M process ensures higher levels of compositional accuracy from end to end than can be verified by chemical analysis. To avoid undesirable contamination, extreme cleanliness is required when switching mixing equipment from one composition to another.

Hot billet densification prior to extrusion is a necessary step. This may involve the expensive step of encapsulation (that is, extruding with a can as shown earlier in Fig. 13). When the hard phase in the cermet exceeds about 15 to 20 vol%, the composition is no longer malleable and hot extrusion becomes more difficult.

Direct roll compaction of Al-B_4C mixtures in the range of 15 to 20 vol% hard phase at room temperature has not produced acceptable results. It is interesting to note, however, that other groups have been successful in producing nuclear control rods by the direct powder-rolling technique. Copper-boron-carbide strip, for example, has been produced by a continuous direct powder-rolling process. The compositions contained as much as 20 wt% (about 50 vol%) boron carbide. The precompacted strip was sintered and brought to full density by a sequence of rerolling and annealing steps (Ref 72). The processing and microstructural characterization of Al-B_4C cer-

Table 11 Properties of metal borides and boride-base cermets

Property	Cermet grades						
	TiB_2(a)	ZrB_2(b)	ZrB_2-B(c)	CrB(d)	CrB-Ni(e)	CrB-Cr-Mo(f)	Mo_2NiB_2(g)
Melting point or range, °C (°F)	2980 (5400)	3040 (5500)	2955–3010 (5350–5450)	2050 (3720)	1650–1760 (3000–3200)	1930–1980 (3500–3600)	1430 (2600)
Density, g/cm³ (lb/in.³)	4.5 (0.163)	6.1 (0.221)	4.97–5.27 (0.180–0.191)	6.15 (0.222)	6.16–6.27 (0.223–0.225)	6.77–7.27 (0.245–0.263)	8.40 (0.305)
Electrical resistivity at 25 °C (75 °F), μΩ · cm	15.3	16	17–23	20	38–58	37–54	66–71
Mean coefficient of thermal expansion, μm/m · °C (μin./in. · °F)	6.39 (3.55)(h)	7.50 (4.17)(h)	5.76 (3.20)(i)	...	9.81 (5.45)(j)	9.90 (5.50)(j)	...
Thermal conductivity at 200 °C (500 °F), W/m · K (Btu · in./ft² · h · °F)	25.9 (180)	23.0 (160)	...	...	...	...	...
Hardness							
HK	3370	2300	...	2140	...	...	...
HRA	...	...	88–90	...	75–86	77–88	88–90
Modulus of elasticity, GPa (10⁶ psi)	365 (53)	441 (64)	...	...	...	...	...
Transverse-rupture strength, MPa (ksi)							
At 20 °C (70 °F)	130 (19)	200 (29)	...	...	813 (118)	...	≤690 (≤100)
At 980 °C (1800 °F)	...	...	434 (63)	...	550–950 (80–138)	620–965 (90–140)	...
Tensile strength, MPa (ksi)	127 (18.4)	196 (28.5)	...	...	...	...	...
Stress-to-rupture strength at 980 °C (1800 °F), MPa (ksi)(k)	...	...	128 (18.5)	...	83–137 (12–20)	96–103 (14–15)	≤82 (≤12)

(a) 100 wt% TiB_2. (b) 100 wt% ZrB_2. (c) 95 wt% ZrB_2, 5 wt% B. (d) 100 wt% CrB. (e) 85 wt% CrB, 15 wt% Ni. (f) 80 wt% CrB, 16 wt% Cr, 4 wt% Mo. (g) 100 wt% Mo_2NiB_2. (h) At 20–760 °C (70–1400 °F). (i) At 20–1205 °C (70–2200 °F). (j) At 20–980 °C (70–1800 °F). (k) 100 h. Source: Ref 79–81

mets and their fabrication into a variety of structural elements has been described (Ref 27, 73) and patented (Ref 74) by Halverson and co-workers.

Boride Cermets

Because metal borides generally are more refractory than titanium carbide, cermets based on borides are of interest for applications that require a material of extreme heat and corrosion resistance, such as materials in contact with reactive hot gases or molten metals. The diborides of the transition metals hafnium, tantalum, zirconium, and titanium have extremely high melting points, descending in the order given from 3250 to 2800 °C (5880 to 5070 °F). Molybdenum boride (MoB) and chromium boride (CrB) melt at considerably lower temperatures (2180 and 1550 °C, or 3960 and 2820 °F, respectively). The oxidation resistance of the transition metal diborides above a temperature of 1100 °C (2000 °F) is considerably better than that of TiC and roughly follows the descending order of the melting point (Ref 75). The oxidation resistance and strength properties at high temperatures can be further enhanced by reacting the boride crystals with small amounts of other thermally stable compounds, such as SiC or molybdenum disilicide ($MoSi_2$), prior to processing the powder into solid bodies.

Because these metal borides have relatively high thermal conductivity and high-temperature stability, they do not depend on a supportive metallic binder matrix for thermal shock resistance and strength as do the Ni-Cr alloys in TiC cermets. The boride phases alone in their highly purified state are extremely hard and abrasive; however, their consolidated bodies pose problems in fabrication to useful products, as well as in service, especially in environments involving dynamic gas or liquid metal flow. This shortcoming can be alleviated in some instances by a metallic binder phase. For thermodynamic reasons, this binder phase generally is limited to 2 to 5 at.% up to a maximum of 10 at.%.

The principal candidate metals for cementing the boride grains are iron, nickel, cobalt, chromium, molybdenum, tungsten, and boron, or some of the alloys of these metals. Low-melting eutectics in the systems boron-iron (1161 °C, or 2122 °F), boron-cobalt (1102 °C, or 2015 °F), and especially boron-nickel (990 °C, or 1814 °F) restrict the amount of the respective binder metals to a small percentage. The effectiveness of iron and cobalt, especially as binders for titanium diboride (TiB_2) and zirconium diboride (ZrB_2), is further diminished by the formation of very brittle intermetallic compounds; whereas chromium and boron, singly or combined, produce tougher, higher-melting eutectics with these borides (Ref 76). Additions of up to 5 wt% B and 10 wt% Mo or W can be successfully used as a binder (of ZrB_2, for example) without forming low-melting phases (Ref 77, 78).

The transition metal borides are produced as pure crystals by such processes as the solid-state reaction of metal boron, the reaction of the metal or its oxide with boron carbide, the reduction of boron and metal oxides with carbon or reactive metals, or fused-salt electrolysis. Mixtures of these borides and binder metal powder are processed into cermet products by ceramic or P/M techniques, such as hydrostatic pressing or slip casting followed by vacuum sintering, or by hot uniaxial or isostatic pressing. The high costs of producing the borides and handling the brittle products with the necessary care have limited applications to those cases in which the unusual properties of these materials are an essential requirement. In addition to the high cost, these materials generally have poor mechanical properties, and thus few boride-base cermets have been able to hold on to practical uses in industry. Table 11 lists the physical and mechanical properties of metal borides and boride-base cermets.

Zirconium Boride Cermets. A comprehensive study of the properties of this transition metal boride was reported some 25 years ago (Ref 81). The very high melting point and good high-temperature mechanical properties, as well as a noticeable reduction in brittleness with rising temperature, make it one of the few borides that have attracted considerable attention. The addition of 2 to 5 wt% B binder to ZrB_2 renders the material suitable for extremely high-temperature applications, including high-performance burner, rocket, and jet reaction systems (Ref 77).

The oxidation resistance of zirconium boride can be further enhanced by reacting it with up to 15% SiC, and the consolidated cermet bodies can successfully withstand oxidizing environments in the 1900 to 2500 °C (3450 to 4530 °F) temperature range (Ref 82, 83). These materials have been the object of an extensive investigation for use as nozzle throat inserts for liquid propellant rockets (Ref 84). Probably the most outstanding characteristics of zirconium boride are its high-temperature corrosion resistance and nonwetting properties when in contact with molten aluminum, brass, zinc, and lead. As a result, applications for this cermet opened up in systems handling molten metals. Typical examples include impellers and bearings in pumps for liquid die casting alloys, spray nozzles for atomizing metal powders, and furnace parts that come in contact with molten reactive metals or vapors.

Titanium Boride Cermets. The physical and mechanical properties at low and high temperatures of titanium boride do not vary greatly from those of zirconium boride (Ref 81). As a single boride or in solid solution with chromium boride (CrB_2), titanium boride has been considered by some to be the most promising of the transition metal borides (Ref 76). Successful applications have included evaporation vessels for reactive metals, electrodes for aluminum refining, and, in general, parts that are exposed to molten zinc and brass. The addition of TiB_2 to TiC in a composite structure has been successfully used for cutting tools, and complex cermets of TiC-TiB_2 with a Co-Si alloy binder (Ref 85, 86) or a TiB-MoSi with a graphite binder (Ref 84) have been experimented with for use as nozzles.

A gradient-type cermet of titanium boride and copper has been produced recently in the form of intermediate ceramic-metal layers that link TiB_2 to pure copper (Ref 87). The composite is the product of a self-propagating high-temperature synthesis. Careful determination of the respective blending ratios of the boride and copper powders for each layer ensures that the reaction proceeds from one layer to the next.

Chromium Boride Cermets. Of the generally excellent corrosion- and oxidation-resistant boride materials, chromium boride was one of the first to be investigated for its high-temperature potential. The compound can be successfully bonded with cobalt, nickel, nickel-chromium, and nickel-copper (Ref 88). A composition containing as much as 15 wt% Ni can be hot pressed without exuding much liquid phase. The cermet is oxidation resistant up to 950 °C (1740 °F) and has a high hot hardness and a transverse rupture strength of about 890 MPa (130 ksi). These properties can be further improved by using Cr_2B crystals and cementing them with up to 10 wt% of an 80Cr-20Mo alloy (Ref 76, 89). These cermets have good stress-to-rupture properties and mechanical shock resistance and at one time were considered as candidate materials for steam and gas turbine blades, valve seats and inserts for internal combustion engines, and exhaust nozzles and tubes for jet engines. Although compositions much higher in nickel (for example, cermets containing the compound Cr_2NiB_4) suffer from the low-melting eutectics, this is used as an advantage for wear- and erosion-resistant overlay coatings and hardfacing applications.

Molybdenum Boride Cermets. The molybdenum borides MoB and Mo_2B have less thermal stability than the previously discussed metal borides, but their electrical properties, hardness, and wear resistance are very good. When cemented with nickel, these cermets have excellent corrosion resistance, for example, to dilute sulfuric acid (Ref 79). Nickel-bonded molybdenum boride exhibits interesting behavior in two areas: First, if the composition corresponds to the compound molybdenum-nickel boride (Mo_2NiB_2), if the cermet contains Mo_2B in addition to Mo_2NiB_2, or if a low-melting, intermetallic binder containing chromium boride and nickel is used, cutting tool materials can be produced from the composition that are comparable to commercial WC tool tips for machining brass, aluminum, and cast iron (Ref 90, 91). Second, the Mo_2NiB_2-type composition has thermal expansion characteristics that closely match those of the refractory metals and a favorable melting temperature; these properties make it ideal for use as a high-temperature braze for molybdenum and tungsten, without risk of excessive grain growth or embrittlement of the primary metal structure (Ref 79, 92). When used in rod form with shielded arc welding equipment, this cermet is suitable for brazing electronic components in applications such as vacuum tubes and magnetrons.

Recently, a molybdenum boride cemented with an iron-base binder phase alloyed with nickel and chromium has shown promise as a cutting tool material (Ref 93). This cermet exhibits good mechanical properties coupled with excellent wear and corrosion resistance. In specific tool applications, such as extrusion dies for hot copper and tools for can making, this boride cermet has performed better than cemented carbides. The role of nickel in the Mo_2FeB_2 cermet and the effect of varying its content up to 10 wt% in the Fe-5B-44.4Mo composition have also been investigated, mainly as part of a study of the corrosion resistance potential of the material (Ref 94). The nickel enters only into the iron-base binder phase, which changes with increasing nickel content from ferritic to martensitic to austenitic. The martensitic binder phase at 2.5% Ni gives the cermet a transverse rupture strength of 2.24 GPa (325 ksi) and a hardness of 86.9 HRA.

Other Refractory Cermets

The nitrides, carbonitrides, and silicides of certain transition metals have gained importance for specific uses in operations involving high temperatures. The main mode of application for these refractory cermets, however, is in the form of coatings, such as TiN and TiC-TiN in various ratios for high-speed cutting tools or $MoSi_2$ for surface protection of molybdenum against high-temperature oxidation. In a very few cases, these compounds are used as solids, either in the pure state or cemented with a lower-melting metallic phase.

Carbonitride- and Nitride-Based Cermets. Titanium nitrides and titanium carbonitrides have been found suitable for use as the hard phase for tool materials (Ref 95). The best binder is an alloy of 70Ni-30Mo, and optimum hardness, in the 1000 to 2000 HV range, is obtained with 10 wt% binder. The hardness increases progressively with the TiC component of the solid solution. The same trend prevails for the hardness of a cermet containing 14 wt% binder: The values increase from about 1400 to 1900 HV for the straight cemented TiC composition. Transverse rupture strength does not follow any trend; the best values reach about 1300 MPa (188 ksi) for a 10 wt% binder composition with a 72-to-18 TiN-TiC ratio and a 14 wt% binder material with a 69-to-17 TiN-TiC ratio. This compares with 1070 and 1275 MPa (155 ksi and 185 ksi), respectively, for the straight TiC cermets with 10 and 14 wt% binder. The hardness of titanium nitride alone cemented with 10% of the 70Ni-30Mo alloy has a hardness level of about 1050 HV and a transverse rupture strength of about 785 MPa (115 ksi). Titanium carbonitride cermets for tool applications are discussed in greater detail in the section "Titanium Carbonitride Cermets" in this article.

Combinations of nitrides and borides, with or without metallic binder, can also be fabricated into tools. A mixture of 60 wt% tantalum nitride (TaN) and 40 wt% ZrB_2 has been hot pressed into tool bits that have performed very well at very high cutting speeds (Ref 96).

Nitride products based on the metalloids boron and silicon, like their carbide counterparts, have gained some significant commercial uses since their early development in the 1950s and 1970s, respectively. The normal hexagonal crystal lattice of boron nitride (BN) can be converted to a cubic crystal form by reacting boron powder with nitrogen at a minimum temperature of 1650 °C (3000 °F) while simultaneously applying pressure in excess of 7000 MPa (1000 ksi) with the aid of special press tools adopted from the manufacture of synthetic diamond. The product is extremely hard and is considered to be one of the best electrical insulators known, especially at high temperatures up to about two-thirds of its melting point, that is, in the vicinity of 2730 °C (4950 °F) (Ref 45, 97).

Cermets exhibiting excellent cutting performance have been achieved by bonding carefully graded particles of the superhard cubic boron nitride with cobalt or similar hard metal binders. Hot pressing is the preferred method of powder consolidation, and tool bits made in this manner outperform tungsten carbide tips by a factor of two-to-one and better (Ref 98).

The nitride of silicon and its combination with different oxides, notably Al_2O_3 (known as the SiAlONs), as well as the different silicon ceramics based on silicon carbide, belong to the increasingly important new class of refractory materials known as structural ceramics. Additives of these cermets are nonmetallic and serve mainly to control the sintering mechanism. They do not contribute to a strengthening of the hard particle structure in the sense of a metallic binder. In fact, they cause a weakening of the grain-boundary network at high temperature in many systems. Therefore, these silicon ceramics are considered to lie outside the material classification for cermets.

Silicide Cermets. The metallic silicides have found commercial use only in isolated instances. This is due chiefly to the extreme brittleness of these compounds and to the concomitant problems encountered when they are fabricated into solid objects. Because of its outstanding high-temperature oxidation resistance, and its favorable coefficients of thermal expansion and electrical resistance, molybdenum disilicide ($MoSi_2$) is an important material for heating elements. Poor resistance to mechanical and thermal shock is the major deficiency of molybdenum disilicide and limits the applications of this material to simple cylindrical or rectangular shapes. Additions of metallic elements to remedy this handicap have been only partially successful, and $MoSi_2$ cermets with nickel, cobalt, and platinum binder metals are still too brittle for fabrication into complex shapes (Ref 99). High-temperature bearings have been made experimentally by infiltrating molten silver into hard matrices containing $MoSi_2$, tungsten disilicide (WSi_2), or vanadium disilicide (VSi_2); these bearings have shown good antifriction behavior against steels at elevated temperatures (Ref 100).

Graphite- and Diamond-Containing Cermets. Materials that contain a combination of carbon in the form of graphite or diamond with metals constitute a border region for cermets and are usually not designated as such. However, because the carbon and metallic components are most often intimately mixed and uniformly distributed in the microstructure, they are pertinent to this discussion.

Graphite-metal combinations for electrical contact applications basically fall into two types of materials. For metallic brushes used in motors and generators, the metallic phase consists of copper or bronze; in the case of sliding contacts involving relatively low rubbing speeds and light contact pressure, the metallic phase is silver. In brushes, the graphite particle content may spread over a wider range, from 5 to 70 wt%. A typical binary composition contains 70% Cu and 30% graphite. To improve wear and bearing properties, many brushes also contain up to 10% Sn and/or Pb and up to 12% Zn (Ref 101). The graphite content in the silver contact composition generally ranges between 2 and 50 wt%.

Graphite-containing metallic friction materials for brake linings and clutch facings have a predominantly metallic matrix to utilize a high thermal conductivity. This property permits rapid energy absorption, making this type of material suitable for service under a more severe wear and temperature environment than that which is possible for organic, resin-bonded asbestos friction elements. The most important contribution of a cermet-type lining material in aircraft brakes probably has been an increased energy capacity without additional weight or the use of a larger unit (Ref 102). The friction coefficient of these cermets is tailored to the requirements of the particular application, principally by varying the ratio of a friction-producing ceramic to the graphite, which acts as a solid lubricant. The metallic matrix phase is essentially a bearing alloy containing 60 to 75 wt% Cu and 5 to 10% each of tin, lead, zinc, and/or iron. Graphite content falls within the 5 to 10% range, and the ceramic, mainly SiO_2 with the possibility of some Al_2O_3 additions, amounts to 2 to 7% (Ref 103).

Cermets composed of diamond, varying in size from coarse splinters to fine dust inside a metal matrix, are used for grinding, lapping, sawing, cutting, dressing, and trueing tools. The size of the diamond is important for the efficiency of the tool; although finish improves as the grain or grit size becomes finer, the cutting speed is slower. For dressing tools, 5 to 35 diamond splinters are embedded per carat with a size of approximately 1 to 2.5 mm (0.04 to 0.1 in.). For rough grinding, the grit size is in the range of 0.15 to 0.5 mm (0.006 to 0.02 in.); for fine polishing, it falls between 0.05 and 0.15 mm (0.002 and 0.006 in.). Even finer diamond powder is used in combination with tungsten carbide for specialized applications such as polishing plane surfaces of hard metal tools or finishing the rolls for Sendzimir-type mills. Typical compositions of these tools contain 12 to 16 wt% diamond dust embedded in a tungsten carbide matrix cemented with 13% Co (Ref 104).

Other metallic bonding substances are based on copper, iron, nickel, molybdenum, or tungsten. Examples for copper matrices are bronzes with 10 to 20% Sn or 2 to 4% Be, which can be strengthened by precipitation hardening, and a 47Cu-47Ag-6Co alloy. Bonding metals suitable for somewhat higher-temperature service include iron-nickel, iron-nickel-chromium, and iron-tin-antimony-lead alloys; Permalloy; and nickel alloys containing 2 to 8% Be. Refractory metal-base matrices are alloys of the molybdenum-copper, molybdenum-cobalt, or tungsten-nickel-copper types and tungsten-nickel-iron heavy alloys (Ref 104). In general, the bond materials must be selected with consideration of lowest possible processing temperatures to avoid the possible transformation of the diamond to graphite.

REFERENCES

1. ASTM Committee C-21, "Report of Task Group B on Cermets," American Society for Testing and Materials, 1955
2. J.R. Tinklepaugh and W.B. Crandall, Chapter 1 in *Cermets*, Reinhold, 1960
3. E.C. Van Schoick, Ed., *Ceramic Glossary*, The Ceramic Society, 1963
4. P. Ettmayer and W. Lengauer, The Story of Cermets, *Powder Metall. Int.*, Vol 21 (No. 2), 1989, p 37-38
5. R. Kieffer and F. Benesovsky, *Hartmetalle*, Springer-Verlag, 1965, p 437-489
6. E. Rudy, Boundary Phase Stability and Critical Phenomena in Higher Order Solid Solution Systems, *J. Less-Common Met.*, Vol 33, 1973, p 43-70
7. F.V. Lenel, *Powder Metallurgy, Principles and Applications*, Metal Powder Industry Federation, 1980
8. P. Popper, *Isostatic Pressing*, British Ceramic Research Association, Heyden & Sons Ltd., 1976
9. R. Kieffer and P. Schwarzkopf, *Hartstoffe and Hartmetalle*, Springer-Verlag, 1953
10. R.M. German, Molding Metal Injection, in *Powder Injection Molding*, Metal Powder Industries Federation, 1989
11. L.F. Pease III, Present Status in PM Injection Molding (MIM): An Overview, in *Progress in Powder Metallurgy*, Vol 43, Metal Powder Industries Federation, 1987
12. K.J.A. Brookes, *World Directory and Handbook of Hardmetals*, 2nd ed., Engineer's Digest Publications, 1979
13. E. Lardner, Metallurgical Applications of Isostatic Hot Pressing, Chapter 10 in *High Pressure Technology*, Marcel Dekker, 1977
14. Vacuum Hot Press Furnaces for Powder Compaction, *Met. Powder Rep.*, Vol 37 (No. 11), 1982
15. J.L. Ellis, Forming of Dispersion Type Aluminum Base Powder Metallurgy Nuclear Products, in *Progress in Powder Metallurgy*, Vol 18, Metal Powder Industries Federation, 1962
16. S.W. Kennedy, "Development in Combination Debinder/Pressure Consolidation Furnace," Technical Note, Vacuum Industries Inc., 1989
17. R.E. Bauer, Sinter-HIP Furnaces Sintering and Compacting in a Combined Cycle, in *Modern Developments in Powder Metallurgy*, Metal Powder Industries Federation, 1988
18. C.G. Goetzel, Infiltration Process, in *Cermets*, Reinhold, 1960, p 73-81
19. H.W. Lavendel and C.G. Goetzel, Recent Advances in Infiltrated Titanium Carbides, in *High Temperature Materials*, R.F. Heheman and G.M. Ault, Ed., John Wiley & Sons, 1959, p 140-154
20. W.B. Johnson, T.D. Claar, and G.H. Schiroky, Preparation and Processing

of Platelet Reinforced Ceramics by the Directed Reaction of Zirconium With Boron Carbide, *Ceram. Eng. Sci. Proc.*, Vol 10 (No. 7/8), 1989

21. T.D. Claar, W.B. Johnson, C.A. Anderson, and G.H. Schiroky, Microstructure and Properties of Platelet Reinforced Ceramics Formed by the Directed Reaction of Zirconium With Boron Carbide, *Ceram. Eng. Sci. Proc.*, Vol 10 (No. 7/8), 1989
22. V.J. Tennery, C.B. Finch, C.S. Yust, and G.W. Clark, Structure-Property Correlations for TiB_2-Based Ceramics Densified Using Active Liquid Metals, in *Proceedings of the International Conference on the Science of Hard Materials*, Plenum, 1983
23. C.G. Goetzel and L.P. Skolnick, Some Properties of a Recently Developed Hard Metal Produced by Infiltration, in *Sintered High-Temperature and Corrosion-Resistant Materials*, F. Benesovsky, Ed., Pergamon Press, 1956, p 92-98
24. R. Kieffer and F. Benesovsky, The Production and Properties of Novel Sintered Alloys (Infiltrated Alloys), *Berg Hüttenmänn. Monatsh.*, Vol 94 (No. 8/9), 1949, p 284-294
25. D.K. Creber, S.D. Poste, M.K. Aghajanian, and T.D. Claar, AlN Composite Growth by Nitridation of Aluminum Alloys, *Ceram. Eng. Sci. Proc.*, Vol 9 (No. 7/8), 1988, p 975
26. M.S. Newkirk, H.D. Lesher, D.R. White, C.R. Kennedy, A.W. Urquhart, and T.D. Claar, Preparation of Lanxide Ceramic Matrix Composites: Matrix Formation by the Directed Oxidation of Molten Metals, *Ceram. Eng. Sci. Proc.*, Vol 8 (No. 7/8), 1987, p 879-882
27. D.C. Halverson, A.J. Pyzik, I.A. Aksay, and W.E. Snowden, Processing of Boron Carbide-Aluminum Composites, in *Advanced Ceramic Materials*, Preprint UCRL-93862, Lawrence Livermore National Laboratory, 1986
28. M. Humenik, Jr. and T.J. Whalen, Physiochemical Aspects of Cermets, in *Cermets*, Reinhold, 1960, p 6-49
29. A.R. Blackburn and T.S. Shevlin, Fundamental Study and Equipment for Sintering and Testing Cermet Bodies: V. Fabrication, Testing and Properties of 30 Chromium-70 Alumina Cermets, *J. Am. Ceram. Soc.*, Vol 34 (No. 11), 1951, p 327-331
30. C.A. Hauck, J.C. Donley, and T.S. Shevlin, "Fundamental Study and Equipment for Sintering and Testing of Cermet Bodies," Report WADC-TR-173, U.S. Air Force, March 1956
31. C.G. Goetzel, Titanium Carbide-Metal Infiltrated Cermets, in *Cermets*, Reinhold, 1960, p 130-146
32. J.L. Chermant and F. Osterstock, Fracture Toughness and Fracture of WC-Co Composites, *J. Mater. Sci.*, Vol 11, 1976, p 1939-1951
33. D. Moskowitz and M. Humenik, Jr., Cemented TiC Base Tools With Improved Deformation Resistance, in *Modern Developments in Powder Metallurgy*, Vol 14, Metal Powder Industries Federation, 1980, p 307-320
34. R.L. Hatschek, Take a New Look at Ceramics/Cermets, Special Report 733, *Am. Mach.*, Vol 125 (No. 5), 1981, p 165-176
35. "UCAR Metal-Ceramic Thermocouple Protecting Tubes Grade LT-1," Catalog section H-8738, Union Carbide Corporation, Carbon Products Division, March 1981
36. T.S. Shevlin, Oxide-Base Cermets, in *Cermets*, Reinhold, 1960, p 97-109
37. G. Engelhardt and F. Thümmler, Creep Deformation of Al_2O_3-Cr Cermets With Cr-Content up to 50 vol%, in *Modern Developments in Powder Metallurgy*, Vol 8, Metal Powder Industries Federation, 1974, p 605-626
38. A.E.S. White *et al.*, Metal-Ceramic Bodies, in *Symposium on Powder Metallurgy 1954*, Special Report No. 58, The Iron and Steel Institute, 1956, p 311-314
39. J.R. Baxter and A.L. Roberts, Development of Metal-Ceramics from Metal Oxide Systems, in *Symposium on Powder Metallurgy 1954*, Special Report No. 58, The Iron and Steel Institute, 1956, p 315-324
40. G.T. Harris and H.C. Child, The Rupture Strength of Some Metal-Bonded Refractory Oxides, in *Symposium on Powder Metallurgy 1954*, Special Report No. 58, The Iron and Steel Institute, 1956, p 325-330
41. M.K. Aghajanian, J.T. Burke, D.R. White, and A.S. Nagelberg, A New Infiltration Process for the Fabrication of Metal Matrix Composites, *SAMPE Q.*, Vol 20 (No. 4), 1989, p 817-823
42. R.V. Watkins, G.C. Reed, and W.L. Schalliol, Hot-Extruded Chromium Composite Powder, in *Progress in Powder Metallurgy*, Vol 20, Metal Powder Industries Federation, 1964, p 149-158
43. E. Ryshkewitch, Oxide-Metal Compound Ceramics, in *Metals for the Space Age*, Springer-Verlag, 1965, p 823-830
44. B.C. Weber and M.A. Schwartz, Metal-Modified Oxides, in *Cermets*, Reinhold, 1960, p 119-121
45. B.C. Weber and M.A. Schwartz, Container Materials for Melting Reactive Metals, in *Cermets*, Reinhold, 1960, p 154-158
46. F. Heitzinger, Molybdenum + Zirconia, A New Metalceramic Material for New Applications, in *Modern Developments in Powder Metallurgy*, Vol 8, Metal Powder Industries Federation, 1974, p 371-390
47. L.J. Cronin, Refractory Cermets, *Am. Ceram. Soc. Bull.*, Vol 30, 1951, p 234-238
48. L.J. Cronin, Electronic Refractory Cermets, in *Cermets*, Reinhold, 1960, p 158-166
49. C.E. Weber and H.H. Hirsch, Dispersion Type Fuel Elements, in *Proceedings of the First International Conference on Peaceful Uses of Atomic Energy*, Vol 9, 1953, p 196-202
50. A.N. Holden, *Dispersion Fuel Elements*, Gordon & Breach, 1967, p 80-91, 152-167
51. P. Loewenstein, P.D. Corzine, and J. Wong, *Nuclear Reactor Fuel Elements*, Interscience, 1962, p 393-394, 396-398
52. L.J.S. Klein and R.M. German, Controlled Thermal Expansion Metal-Ceramic Composites by Co-Sintering, *Int. J. Powder Metall.*, Vol 24 (No. 1), 1988, p 39-46
53. L.E. Murr, A.W. Hare, and N.G. Eror, Metal-Matrix High-Temperature Superconductor, *Adv. Mater. Process.*, Vol 132 (No. 4), Oct 1987, p 36-44
54. J. Wambold and J.C. Redmond, Recent Developments in Sintered Titanium Carbide Compositions, in *High Temperature Materials*, John Wiley, 1959, p 125-139
55. J. Wambold, Properties of Titanium Carbide-Metal Compositions, in *Cermets*, Reinhold, 1960, p 122-129
56. W.L. Havekotte, Titanium Carbide-Base Cermets, in *Sintered High-Temperature and Corrosion-Resistant Materials*, Pergamon Press, 1956, p 111-129
57. G.C. Deutsch, The Use of Cermets as Gas-Turbine Blading, in *High-Temperature Materials*, John Wiley, 1959, p 190-204
58. H.W. Lavendel and C.G. Goetzel, "A Study of Graded Cermet Components for High Temperature Turbine Applications," U.S. Air Force, WADC-TR-57-135, ASTIA Document AD 131031, May 1957
59. H.W. Lavendal and C.G. Goetzel, Recent Advances in Infiltrated Titanium Carbides, in *High Temperature Materials*, John Wiley, 1959, p 140-154
60. J.L. Ellis, E. Gregory, and M. Epner, Heat Treatable Steel-Bonded Carbides—New Construction Materials for Tools and Wear Resistant Components, in *Progress in Powder Metallurgy*, Vol 16, Metal Powder Industries Federation, 1960, p 76-83
61. M. Epner and E. Gregory, Titanium Carbide-Steel Cermets, in *Cermets*, Reinhold, 1960, p 146-149

62. J.L. Ellis, A Machinable, Heat Treatable, and Weldable Cemented Carbide for Tooling Purposes, *Tool Eng.*, Vol 38 (No. 4), 1957, p 103-105
63. S.E. Tarkan and M.K. Mal, Hardening Steel Bonded Carbides, *Met. Prog.*, Vol 105 (No. 5), 1974, p 99, 100, 102
64. E. Rudy, Spinodal Carbonitride Alloys for Tool and Wear Applications, U.S. Patent 3,971,656, 1976
65. Z.-H. Wan, Machinable and Heat-Treatable Steel-Bonded Tungsten Carbide, in *Modern Developments in Powder Metallurgy*, Vol 17, Metal Powder Industries Federation, 1984, p 193-219
66. J. Hinnüber and O. Rüdiger, Chromium Carbide in Hard-Metal Alloys, in *Symposium on Powder Metallurgy 1954*, Special Report No. 58, The Iron and Steel Institute, 1956, p 305-310
67. R.F. Pozzo and J.V. West, Chromium Carbide Applications, in *Cermets*, Reinhold, 1960, p 150-153
68. Harold Sliney, NASA's Inventor of the Year, *NASA Tech. Briefs*, Vol 13 (No. 3), March 1989, p 10-11
69. R.L. Hammer, W.D. Manly, and W.H. Bridges, Carbides and Cermets, in *Reactor Handbook*, Vol 1, 2nd ed., C.R. Tipton, Jr., Ed., Interscience, 1960, p 508
70. R. Kieffer and F. Benesovsky, *Hartmetalle*, Springer-Verlag, 1965, p 469-470
71. J.T. Burke, M.K. Aghajanian, and M.A. Rocazella, Microstructures and Properties of Discontinuous Metal Matrix Composites Formed by a Unique Low Cost Pressureless Infiltration Technique, Paper presented at the International SAMPE Symposium and Exhibition (Reno, NV), May 1989, p 2440-2454
72. E.J. Bradbury, C.R. Sutton, and D.K. Worn, Composite Neutron-Absorbing Materials for Control Rod and Screening Applications, in *Series V Metallurgy and Fuels*, Vol 4, *Metallurgy of Nuclear Reactor Components*, Pergamon Press, 1961
73. D.C. Halverson, A.J. Pysik, and I.A. Aksay, Processing and Microstructural Characterization of B_4C/Al Cermets, Paper presented at the ACS Ninth Annual Conference on Composites and Advanced Ceramic Materials (Cocoa Beach, FL), Jan 1985, Preprint UCRL-91305, p 1-12
74. D.C. Halverson, A.J. Pysik, and I.A. Aksay, Boron-Carbide-Aluminum and Boron-Carbide-Reactive Metal Cermets, U.S. Patent 4,605,440, 1986
75. L. Kaufman, E.C. Clougherty, and J.B. Berkowitz-Mattuck, Oxidation Characteristics of Hafnium and Zirconium Diboride, *Trans. AIME*, Vol 239 (No. 4), 1967, p 458-466
76. R. Kieffer and F. Benesovsky, *Hartmetalle*, Springer-Verlag, 1965, p 475-479
77. R. Steinitz, Borides—Part B: Fabrication, Properties and Applications, in *Modern Materials*, Vol 2, Academic Press, 1960, p 191-224
78. C.E. Halcombe, Jr., "Slip Casting of Zirconium Diboride," Report Y-1819, U.S. Atomic Energy Commission, 28 Feb 1972
79. J.L. Everhart, New Refractory Hard Metals, *Mater. Methods*, Vol 40 (No. 2), Aug 1954, p 90-92
80. J.D. Latva, Selection and Fabrication of Ceramics and Intermetallics, *Met. Prog.*, Vol 82 (No. 4), Oct 1962, p 139-144, 180, 186
81. L. Kaufman and E.V. Clougherty, Investigation of Boride Compounds for High Temperature Applications, *Metals for the Space Age*, Springer-Verlag, 1965, p 722-758
82. L. Kaufman and E.V. Clougherty, "Investigation of Boride Compounds for Very High Temperature Applications," Report RTD-TDR-63-4096, Part 1, U.S. Air Force Materials Laboratory, Dec 1963
83. E.V. Clougherty, R.L. Pober, and L. Kaufman, Synthesis of Oxidation Resistant Metal Diboride Composites, *Trans. AIME*, Vol 242 (No. 6), 1968, p 1077-1082
84. E.V. Clougherty *et al.*, "Research and Development of Refractory Oxidation Resistant Diborides," Report AFSC-ML-TR-68-190, U.S. Air Force Materials Laboratory, Part 1, Oct 1968; Part 2, Vol 1-7, Nov 1969-June 1970; Part 3, May 1970
85. H.M. Greenhouse, R.F. Stoops, and T.S. Shevlin, A New Carbide-Base Cermet Containing TiC, TiB_2 and CoSi, *J. Am. Ceram. Soc.*, Vol 37 (No. 5), 1954, p 203-206
86. E.T. Montgomery *et al.*, "Preliminary Microscopic Studies of Cermets at High Temperatures," U.S. Air Force Report WADC-TR-54-33, Part 1, April 1955, Part 2, Feb 1956
87. Gradient Ceramic/Metals Made by Advanced Methods, *Adv. Mater. Proc.*, Vol 132 (No. 4), Oct 1987, p 20
88. S.J. Sindeband, Properties of Chromium Boride and Sintered Chromium Boride, *Trans. AIME*, Vol 185, Feb 1949, p 198-202
89. I. Binder and D. Moskowitz, "Cemented Borides," PB 121346, Office of Technical Services, U.S. Department of Commerce, 1954-1955
90. R. Steinitz and I. Binder, New Ternary Boride Compounds, *Powder Metall. Bull.*, Vol 6 (No. 4), Feb 1953, p 123-125
91. I. Binder and A. Roth, An Evaluation of Molybdenum Borides as Cutting Tools, *Powder Metall. Bull.*, Vol 6 (No. 5), May 1953, p 154-162
92. A. Blum and W. Ivanick, Recent Developments in the Application of Transition Metal Borides, *Powder Metall. Bull.*, Vol 7 (No. 3-6), April 1956
93. K. Takagi, S. Ohira, T. Ide, T. Watanabe, and Y. Kondo, New P/M Iron-Containing Multiple Boride Base Hard Alloy, in *Modern Developments in Powder Metallurgy*, Vol 16, Metal Powder Industries Federation, 1985, p 153-166
94. K. Takagi, M. Komai, T. Ide, T. Watanabe, and Y. Kondo, Effect of Ni on the Mechanical Properties of Fe, Mo Boride Hard Alloys, *Int. J. Powder Metall.*, Vol 23 (No. 3), 1987, p 157-161
95. R. Kieffer, P. Ettmayer, and M. Freudhofmeier, About Nitrides and Carbonitrides and Nitride-Based Cemented Hard Alloys, in *Modern Developments in Powder Metallurgy*, Vol 5, Plenum Press, 1971, p 201-214
96. F.C. Holtz and N.M. Parikh, Developments in Cutting Tool Materials, *Eng. Dig.*, Vol 28 (No. 1), 1967, p 73, 75, 99
97. Borazon—Man Made Material Is Hard as Diamond, *Mater. Methods*, Vol 45 (No. 5), 1957, p 194, 196
98. N.J. Pipkin, D.C. Roberts, and W.I. Wilson, Amborite—A Remarkable New Cutting Material from De Beers, *Ind. Diamond Rev.*, June 1980, p 203-206
99. R. Kieffer and F. Benesovsky, *Hartmetalle*, Springer-Verlag, 1965, p 487-489
100. R.H. Baskey, An Investigation of Seal Materials for High Temperature Applications, *Trans. Am. Soc. Lub. Eng.*, Vol 3 (No. 1), 1960, p 116-123
101. F.V. Lenel, *Powder Metallurgy*, Metal Powder Industries Federation, 1980, p 556
102. R.H. Heron, Friction Materials—A New Field for Ceramics and Cermets, *Ceram. Bull.*, Vol 34 (No. 12), 1955, p 295-298
103. F.V. Lenel, *Powder Metallurgy*, Metal Powder Industries Federation, 1980, p 485
104. C.G. Goetzel, *Treatise on Powder Metallurgy*, Vol 2, Interscience, 1950, p 171-174

Superabrasives and Ultrahard Tool Materials

T.J. Clark, G.E. Superabrasives; and R.C. DeVries, G.E. Corporate Research and Development Center (Retired)

THE PRINCIPAL superhard materials are found as phases in the boron-carbon-nitrogen-silicon family of elements (Fig. 1). Of these, the superhard materials of commercial interest include silicon nitride (Si_3N_4), silicon carbide (SiC), boron carbide (B_4C), diamond, and cubic boron nitride (CBN). Silicon nitride provides the base composition for the important category of SiAlON ceramics, which are used in structural applications (see the article "Structural Ceramics" in this Volume) and as high-speed cutting tool materials (see the article "Ceramics" in *Machining*, Volume 16 of the 9th Edition of *Metals Handbook*).

The carbides of the metalloids boron and silicon (B_4C and SiC in Fig. 1) are also of considerable industrial significance and enjoy such diverse applications as superhard tools and electrical resistor heating elements. These compounds are processed and used both with or without metallic binder phases. When these two metalloid carbides are used without a metallic binder phase, the resultant material most likely falls within the material group of ceramics. If silicon carbide (SiC) and boron carbide (B_4C) are used with a metallic binder phase, then the resultant material is considered a cermet (see the article "Cermets" in this Volume).

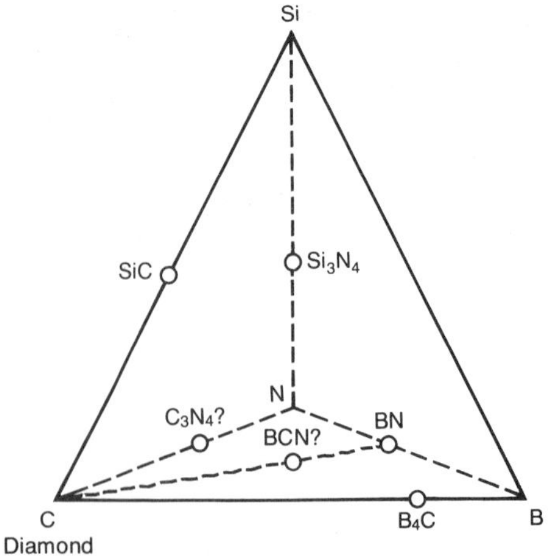

Fig. 1 The carbon-boron-nitrogen-silicon composition tetrahedron showing the principal known superhard materials: the diamond form of carbon, cubic BN, SiC, and B_4C. Polycrystalline aggregates of diamond and SiC as well as Si_3N_4 are also commercially available.

This article focuses exclusively on the superhard materials consisting of either diamond or CBN. The other commercially significant materials in Fig. 1 are discussed in the above-mentioned articles of *Metals Handbook*. Additional information on the superhard nitrides and carbides can be found in Ref 1 and 2. Information on possible new hard materials is available in Ref 3.

The focus of this article is further restricted to synthesized diamond and CBN. The latter does not occur in nature, and the former commands 90% of the industrial diamond market. These materials will be treated in terms of the forms in common use: diamond or CBN grains (loose or bonded) and sintered polycrystalline diamond or CBN tools.

Synthesis of Diamond and Cubic Boron Nitride

The basic objective in the synthesis of diamond and CBN is to transform a crystal structure from a soft hexagonal form to a hard cubic form. In the case of carbon, for example, hexagonal carbon (graphite) would be transformed into cubic carbon (diamond). Synthetic CBN and diamond are produced either as crystalline grains or as sintered polycrystalline products.

The synthesis of CBN or diamond grit can be achieved by static high-pressure high-temperature (HPHT) processing or by dynamic (explosive) techniques. The HPHT method, despite high equipment investment costs, is the predominant technique for producing synthetic diamond and CBN. In addition, diamond is also synthesized under metastable conditions (see the section "Low-Pressure Synthesis of Superhard Coatings" in this article).

High-Pressure High-Temperature Synthesis. The bulk of synthetic CBN and diamond is made by subjecting hexagonal carbon or boron nitride to high temperatures and high pressures with large special-purpose presses or with the commonly used mechanical device known as the uniaxial belt (Ref 4). By the simultaneous application of heat and pressure, hexagonal carbon or boron nitride can be transformed into a hard cubic form. This requires strenuous pressures and temperatures, as illustrated in the graphite-diamond and hexagonal BN-cubic boron nitride equilibrium diagrams (Fig. 2, 3).

It is possible to directly convert graphite to diamond, but very high pressures are required, and the properties of the resultant product are difficult to control. In commercial practice, the required conditions for diamond synthesis can be reduced by the use of solvent/catalysts such as nickel, iron, cobalt, and manganese or alloys of these metals (Ref 5, 6). Figure 4 shows an example of a metal-carbon system at 5.7 GPa (57 kb), where a stable diamond plus liquid region exists. Even with solvent/catalysts it is necessary to simultaneously sustain a pressure of about 5 GPa (50 kb) and a temperature of about 1500 °C (2700 °F), for periods ranging from minutes to hours, to make the variety of products in common use today.

Conditions are similar for the synthesis of CBN, but the reactants are usually alkali, alkaline earth metals, or compounds. Cubic boron nitride can be grown from a variety of solvent/catalysts, including metal systems similar to those used for diamond synthesis (Ref 8). Because the pressure-temperature conditions for the conversion of hexagonal boron nitride (HBN) to CBN are less severe than those for the conversion of graphite to diamond, some sintered polycrystalline products are synthesized by the direct process under static conditions; however, most commercial monocrystalline CBN is made by a solvent/catalyst process.

Explosive Shock Synthesis. The direct conversion of graphite to diamond, or HBN

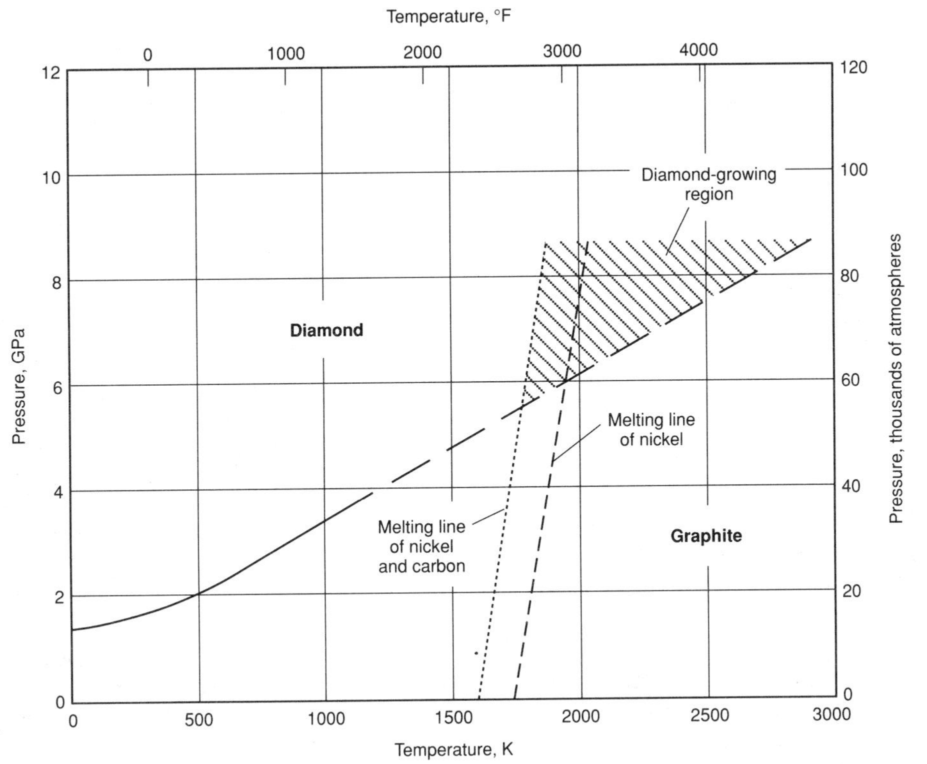

Fig. 2 Pressure-temperature diagram showing the stability regions of diamond and graphite and the role of the solvent/catalyst in lowering the synthesis conditions

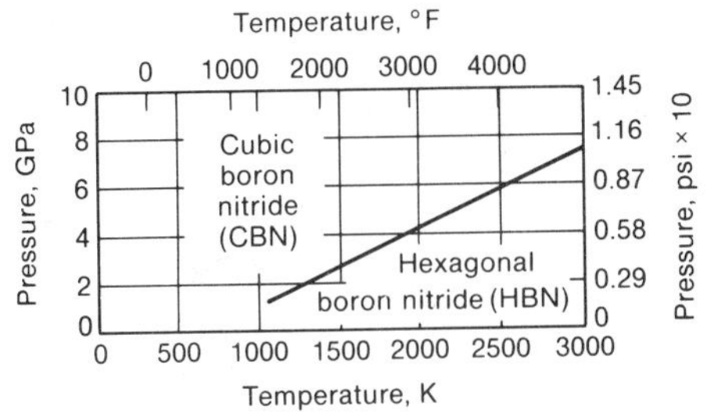

Fig. 3 Equilibrium diagram for HBN and CBN

to CBN, can be done on a commercial scale using explosive shock techniques (Ref 9). The process is relatively simple but produces only fine-grain materials, which are principally used as polishing powders or as possible source materials for sintering into polycrystalline products.

Low-Pressure Synthesis of Superhard Coatings. The history of diamond synthesis under metastable conditions (plasma-assisted, chemical vapor deposition, or physical vapor deposition coating processes) goes back at least to the late 1950s and perhaps even earlier. The efforts of Russian (Ref 10) and Japanese (Ref 11) scientists in the period from 1975 to 1985 made this technique feasible for limited commercial applications. The potential exists to make films or sheets of polycrystalline and single-crystal diamond at temperatures of about 900 °C (1650 °F) and at pressures of less than 1 atmosphere (0.1 MPa). A limited amount of information exists (Ref 12) on grinding or machining applications of these materials. Some films have been made for x-ray windows, speaker diaphragms, and wear surfaces.

Synthesis of Polycrystalline Diamond and Polycrystalline Cubic Boron Nitride. It is possible also to produce polycrystalline diamond (PCD) or polycrystalline cubic boron nitride (PCBN) by sintering (or binding) many individual crystals of diamond or CBN together to produce a larger polycrystalline mass. It is commercial practice to enhance the rate of sintering by the addition of a metal second phase (Ref 13). In addition, the whole mass must again be maintained in the cubic region of the respective temperature-pressure phase diagram to prevent the hard cubic crystals from reverting to the soft hexagonal form. By such high-temperature high-pressure sintering techniques, it is possible to obtain a mass of diamond or CBN in which randomly oriented crystals are combined to produce a large isotropic mass.

An immense range of polycrystalline products can be made of diamond or CBN. Changes in grain size, the second phase employed, the degree of sintering, the particle size distribution, and the presence or absence of inert ceramic, metallic, or nonmetallic fillers are examples of factors that have profound effects on the mechanical, physical, and thermal properties of the final product. By careful formulation it is possible to tailor material properties for particular applications.

Properties of Diamond

The crystal structure of diamond and the lattice structure of graphite are shown in Fig. 5. The conversion from graphite to diamond is accompanied by a 26% decrease in volume. For diamond, all the lattice sites

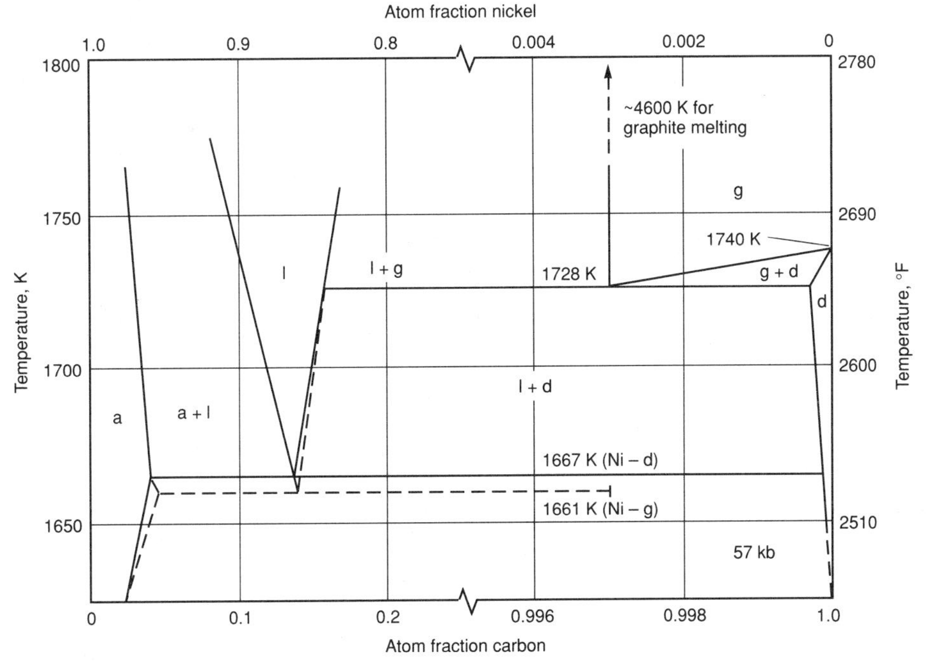

Fig. 4 Nickel-carbon system at 5.7 GPa (57 kb) showing the stability regions of diamond (d) and graphite (g) in equilibrium with liquid (l + d, l + g). a, austenite. Source: Ref 7

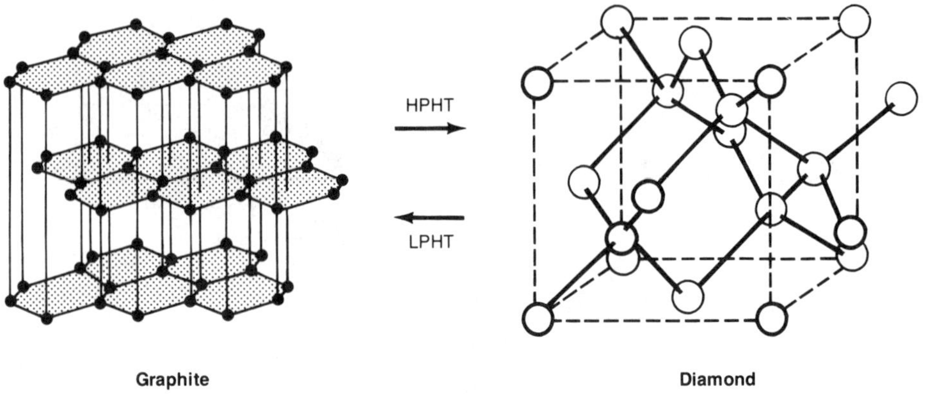

Fig. 5 Arrangement of carbon atoms in diamond and graphite. The arrows indicate the transformation of graphite to diamond at HPHT conditions and the reverse transformation at low pressures and high temperatures (LPHT).

are occupied nominally by carbon, but boron and nitrogen can be substituted for carbon in amounts in the parts-per-million range. Synthesized diamond usually has metal, metal carbide, and graphite inclusions; however, some of the metal may be on defect or interstitial sites and thus may not be visible.

Diamond oxidizes in air above about 600 °C (1100 °F) and back converts into a poorly graphitized form (as indicated by the reverse arrow in Fig. 5) upon heating in the absence of air. The reaction rate for the transformation back into graphite is dependent on conditions, but it is a significant factor at temperatures about 750 °C (1400 °F) in many practical applications. These phenomena impose critical limitations on the use and fabrication of bonded-abrasive tools.

Diamond is chemically inert to inorganic acids, but upon heating it reacts readily with carbide-forming elements such as iron, nickel, cobalt, tantalum, tungsten, titanium, vanadium, boron, chromium, zirconium, and hafnium. Controlled reactivity is important in forming metal bonds, but that same reactivity can limit the use of diamonds in cutting and grinding applications.

The thermal conductivity of some near-perfect diamond crystals can be as high as 5× that of copper at room temperature. Less-perfect materials still have a high conductivity, and this has to be taken into consideration before use in many applications. In terms of electrical conductivity, diamond is an electrical insulator unless doped with boron or, as in some commercial materials, mixed with a metal phase.

Diamond is the hardest practical material known (Table 1). The hardness of single-crystal diamond varies as a function of orientation, but this is important only in single-point tools and in the polishing of gemstones, diamond microtome blades, and diamond surgical knives.

Although hard, diamond is a brittle material and breaks on impact, primarily by cleavage on the four (111) planes. Toughness or friability can be varied considerably for synthesized grains. Thus, it is possible to make a very friable material for some grinding operations and a very tough material for stonecutting. The presence of defects and second phases can be manipulated to influence the fracture properties of synthesized diamond. This control is not available from most natural stones.

Most natural diamonds are essentially octahedral in shape as grown. Irregularly shaped fragments can be obtained by crushing and selection. Synthesized diamond can be reproducibly grown as cubes, cubooctahedrons, and octahedrons (Fig. 6). The cubooctahedral shapes are generally preferred for stone sawing, but they are not always appropriate for grinding.

Synthesized diamond is available in the size range from submicron to about one centimeter. The latter are specialty items for single-point tools, heat sinks, microtomes, surgical blades, and other applications. Table 2 shows the most popular sizes for many applications. For still larger sizes it is more practical to use sintered polycrystalline materials.

Properties of Cubic Boron Nitride

The crystal structure of CBN can be derived from that of diamond (Fig. 5) by substituting boron and nitrogen for carbon on alternate sites in the diamond lattice. The resulting zinc blende structure differs from diamond in having no center of symmetry and a different cleavage plane (110). The soft hexagonal form of boron nitride (HBN) has the same relationship to graphite that CBN has to diamond. Cubic boron nitride may also exhibit a back conversion to a hexagonal structure that is analogous to the back conversion of diamond into graphite.

Cubic boron nitride is nominally boron nitride (that is, B:N = 1:1), with a band gap of about 6.6 eV, and thus should be colorless. However, it is usually amber in color and behaves like an extrinsic semiconductor. Cubic boron nitride can be doped as both p- and n-type. It is most likely to be boron-rich when obtained from conventional processes. A black form is also commercially available. Cubic boron nitride can include solvent/catalyst materials and HBN from the synthesis process. An extremely tough microcrystalline form is also available that is useful in metal and vitreous bonds for heavy-duty applications.

Cubic boron nitride is more resistant both to oxidation and to back conversion into a graphitelike form than is diamond. It can be heated to 1300 to 1400 °C (2350 to 2550 °F) before its protective oxide layer no longer prevents further degradation. Back conversion is not significant until temperatures reach about 1700 °C (3100 °F).

Because the reactivity of CBN with iron-, cobalt-, and nickel-base alloys is much less than that of diamond, CBN fills an important gap in the use of ultrahard materials for the machining of these metals. Cubic boron nitride reacts with strong nitride and boride formers such as titanium, tantalum, zirconium, hafnium, chromium, tungsten, silicon, and aluminum. Under controlled conditions some of these elements can be used as bonding materials. The reactivity of the oxidized surface of boron nitride with alkalis and alkaline earths can be used in making vitreous bonds, but it also can lead to degradation by borate formation in the presence of these reactants.

Theoretically, the thermal conductivity of CBN is slightly more than half that of diamond. Cubic boron nitride also is about half as hard as diamond, but it is about twice as hard as any other material. In contrast to

Table 1 Properties of selected hard materials

	Density			Compressive strength		Coefficient of thermal expansion		Thermal conductivity	
	g/cm^3	$lb/in.^3$	Hardness, HK	GPa	10^6 psi	mm/mm/°C × 10^{-6}	in./in./°F × 10^{-6}	W/m · K	cal/°C · cm · s
Diamond (C)	3.52	0.127	7000–10000	10	1.5	4.8	2.7	2100	5.0
Cubic boron nitride	3.48	0.126	4500	7	1	5.6	3.1	1400	3.3
Silicon carbide (SiC)	3.21	0.116	2700	1.3	0.19	4.5	2.5	42	0.10
Alumina oxide (Al_2O_3)	3.92	0.142	2100	3	0.435	8.6	4.8	33	0.08
Tungsten carbide (WC-Co, 6%)	15.0	0.542	1700	5.4	0.78	4.5	2.5	105	0.25

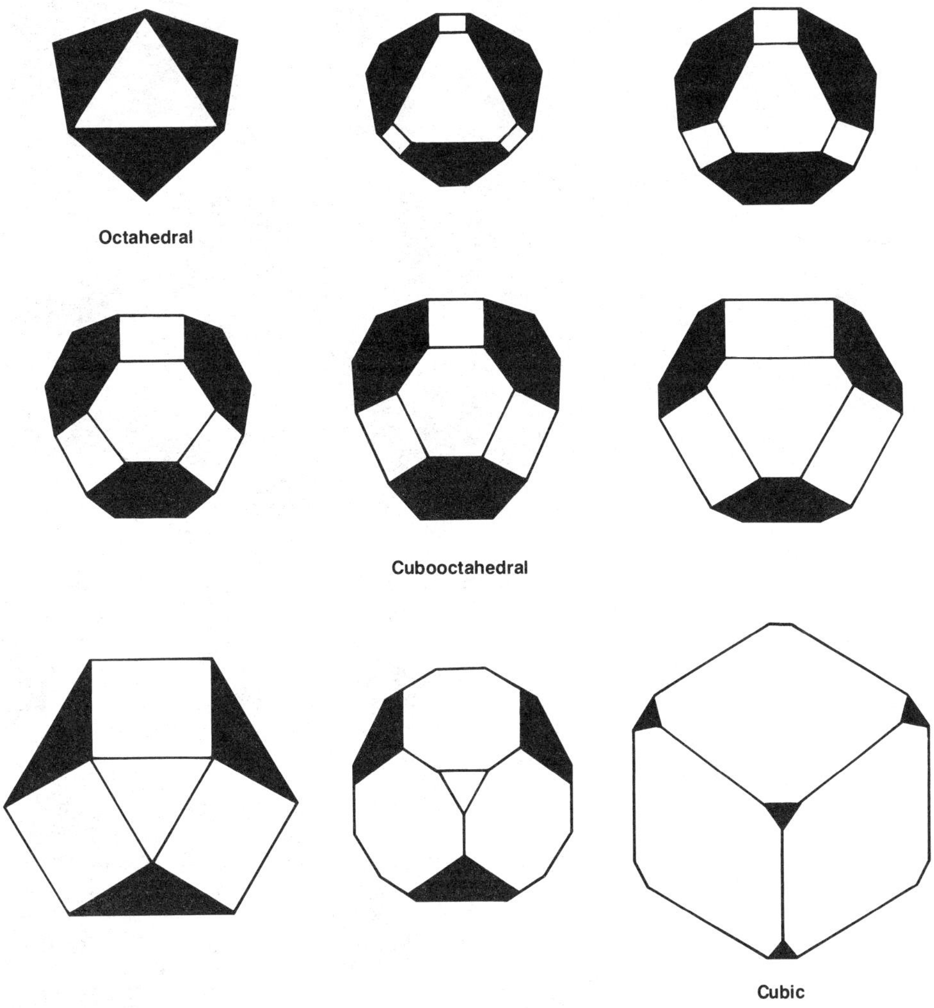

Fig. 6 Shapes of synthesized diamond abrasive grains. Varying proportions of cube (100) and octahedral faces (111) predominate, but (110) and (113) shapes are often present as well.

the four (111) cleavage planes of diamond, CBN cleaves on six (110) planes and therefore is intrinsically more friable than diamond on a single-grain basis.

The preferred growth form for CBN from most solvent systems is a (111) truncated tetrahedron, with some development of cube forms. The size range for CBN grains is from the submicron level to about ½ mm (0.02 in.). Grains larger than 1 mm (0.04 in.) are not usually grown.

Properties of Sintered Polycrystalline Diamond

Sintered PCD, which was developed in 1970 (Ref 13), is a unique material produced by liquid-phase sintering at HPHT conditions. It is characterized by diamond-to-diamond bonding. This sintering process made possible the production of pieces much larger than 1 mm (0.04 in.) with isotropic properties. The commercial product is useful for cutting tools, drill bits, wear surfaces, and wire dies.

Compared to a single crystal, a sintered polycrystalline material is essentially isotropic with respect to wear and cleavage; therefore, its practical toughness in use is improved. Whereas a single crystal is fragile with respect to catastrophic failure by cleavage, a polycrystalline material may chip locally but has no extensive cleavage plane. Wire-drawing dies of sintered PCD outlast single crystals because they do not cleave in hoop tension and because they maintain roundness and dimensions.

Sintered polycrystalline diamond blanks are made *in situ* in an HPHT apparatus, which imparts some limitations to size and shape. Round tool shapes are most common, but squares, triangles, and other shapes are also available. The sintered diamond blanks are produced in both supported and unsupported configurations. In supported structures, cemented tungsten carbide (WC-Co) provides additional strength and a brazeable surface for tool fabrication. A combination of chemical and mechanical bonding exists between the diamond and the substrate by virtue of the transport of cobalt through the diamond layer during HPHT production. Unsupported polycrystalline diamond can be mounted in tools by more conventional diamond-bonding techniques. Subsequent finishing operations are involved in all cases to make tools with tight dimensional and angular tolerances from the blanks. Finishing operations, depending on the tools, can include laser cutting, electrodischarge machine cutting, grinding, lapping, and polishing steps.

Polycrystalline diamond with a metallic second phase has a microstructure of diamond grains with the metallic phase mostly at the grain boundaries (Fig. 7). Both phases are continuous, and the metal phase can be removed chemically. Within the grains it is very common to see deformation twin bands that were produced during the HPHT sintering. These bands are visible in a polished section because they are more wear resistant than the surrounding material.

The sintered diamond contains about 5 to 10 vol% of metal phase and, when made of synthetic diamonds, also may include metals and graphite from the original crystal growth process. Because diamond predominates, the chemical reactivity and stability for oxidation, back conversion, and wetting/bonding will be similar to those for synthetic diamond alone. The metal phase is an added complication, however, with respect to thermal stability. It can enhance graphitization (back conversion), and it can also contribute to degradation above about 700 °C (1300 °F) by thermal stresses that are due to the large differences in the thermal expansion coefficients of diamond and metals. It is advisable not to overheat these tools during bonding and brazing. Without the metal phase, the material is more thermally stable. Some sintered material is made with better thermal expansion matching of the bonding phase (such as silicon

Table 2 Sizes of diamond and CBN grains for grinding applications

		U.S. mesh size ranges													
Superabrasive	Bond	20–30	30–40	40–50	50–60	60–80	80–100	100–120	120–140	140–170	170–200	200–230	230–270	270–325	325–400
Diamond	Resin/vitreous						X	X	X	X	X	X	X	X	X
Diamond	Metal						X	X	X	X	X	X	X	X	X
CBN	Resin/vitreous						X	X	X	X	X	X	X	X	X
CBN (microcrystalline)	Metal/vitreous	X	X	X	X	X	X	X	X	X	X	X	X	X	X

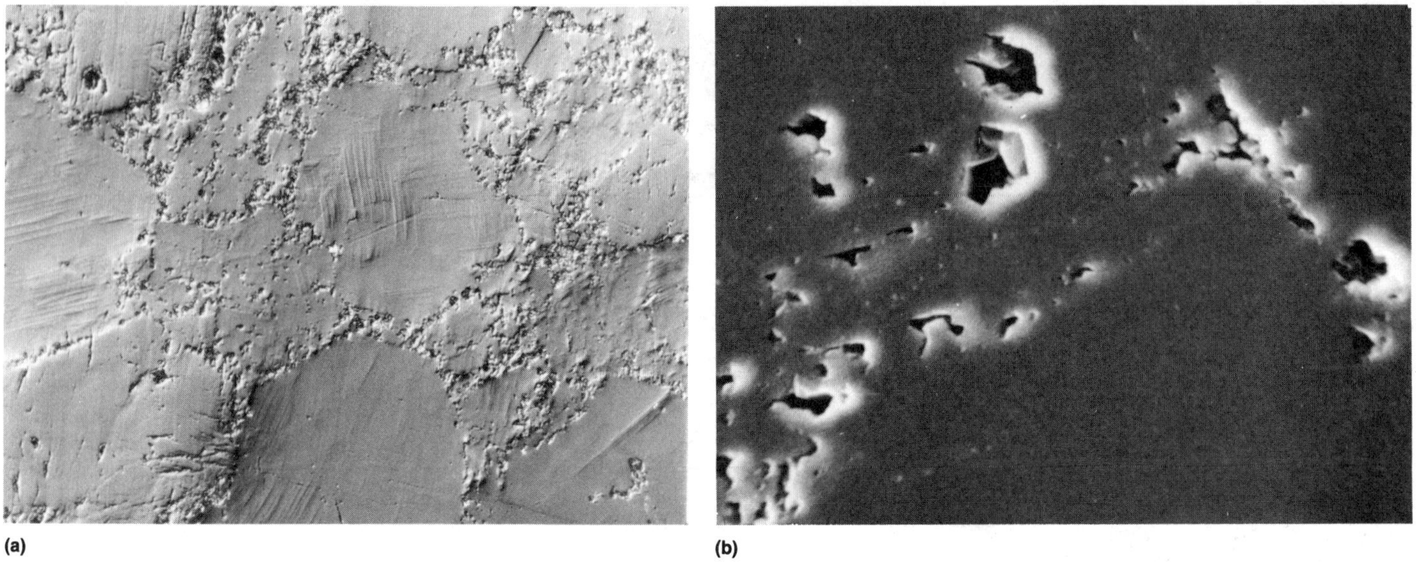

Fig. 7 Microstructure of sintered polycrystalline diamond. (a) Diamond with second phase at the grain boundaries. 225×. (b) Detailed structure of diamond-to-diamond bonding at grain boundary

carbide) with diamond to minimize thermal degradation at the expense of strength.

Properties of Sintered Polycrystalline Cubic Boron Nitride

The microstructure of sintered PCBN is shown in Fig. 8. The major phase is CBN with a metallic second phase from the liquid sintering process. With respect to chemical composition and chemical reactivity (oxidation, back conversion, and wetting/bonding), CBN is the predominant material.

Sintered PCBN can be heated to at least 700 °C (1300 °F) before thermal degradation occurs. The maximum thermal conductivity of sintered PCBN lies in the range of 2.5 to 9.0 W/cm · °C, depending on the processing conditions used to make the samples (Ref 14).

Sintered PCBN is less tough than sintered polycrystalline diamond. As with the diamond version, it has the advantage of isotropic wear and hardness rather than the catastrophic cracking and cleavage that characterize the single crystals in heavy-duty applications. The sizes and shapes available are similar to those for diamond.

Superabrasive Grains

Superabrasive grains are commercially available in a range of sizes, shapes, and qualities (Table 2). Diamond or CBN grains can be used as loose abrasives, as bonded abrasives in grinding wheels and hones, and as bonded abrasives in single-point applications such as turning tools, dressers, and scribes.

Loose Abrasive Grains

Lapping and polishing constitute two major applications of both natural and synthetic loose abrasive grains. Of the synthetic abrasive powders, alumina and silicon carbide are the most widely used in lapping and polishing operations. Silicon carbide is harder than alumina (Table 1) and fractures more easily, thereby providing new cutting edges and extending the useful life of the abrasive.

Synthetic Lapping Abrasives. Silicon carbide, which has sharp edges for cutting, is used for lapping hardened steel or cast iron, particularly when an appreciable amount of stock is to be removed. Fused alumina is also sharp, but it is tougher than silicon carbide and breaks down less readily. Fused alumina is generally more suitable than silicon carbide for lapping soft steels or nonferrous metals.

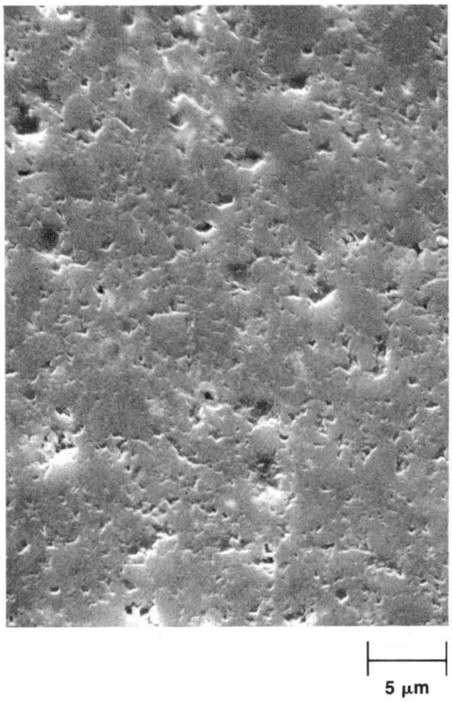

Fig. 8 Microstructure of sintered PCBN

Boron carbide (B_4C), one of the superhard materials shown in Fig. 1, has a hardness of about 2800 HV and is an excellent abrasive for lapping. However, because it costs 10 to 25 times as much as silicon carbide or fused alumina, boron carbide is usually used only for lapping dies and gages, which is often done by hand and in small quantities, using little abrasive. An example of such a use of boron carbide is in the production of synthetic sapphire for electronic applications. The raw material cost is expensive, justifying a high abrasive-processing cost.

Diamond is also used as an abrasive for lapping metals. It is available as a paste or a slurry. Table 3 lists typical sizes of powders used for lapping applications. Fine-mesh diamond and diamond micron powders are the abrasives most often used in lapping. Depending on the needs of the purchaser, these powders can be provided in several types that differ in aggressiveness (the sharpness of cutting points and edges), shape, and toughness.

In a typical lapping machine, the abrasive is applied in a slurry with a water-soluble glycol solution to the surface of a cast iron lap. A continuous feed system ensures that fresh abrasive is always available. To provide uniform lapping on each tool, the workpieces are held under pressure on the lap face and are rotated in a fixture while the lap is turning. When the desired surface is achieved, the tools are replaced with the next set, allowing an efficient semicontinuous operation.

Polishing. As with lapping, aluminum oxide and silicon carbide are widely used synthetic abrasives for polishing. They are

Table 3 Size ranges of micron diamond powders for grinding and polishing

Size ranges, μm							
1/10	0–1	1–3	4–8	10–15	15–20	22–36	40–60
0–1/4	0–2	2–4	5–10	10–20	15–25	30–40	54–80
0–1/2	1–2	3–6	6–12	12–22	20–30	36–54	60–100

harder, more uniform, longer lasting, and easier to control than most natural abrasives. Aluminum oxide grains are very angular and are particularly useful in polishing tougher metals, such as alloy steels, high-speed steels, and malleable and wrought iron. Silicon carbide is usually used in polishing low-strength metals, such as aluminum and copper. It is also applied in polishing hard, brittle materials, such as carbide tools, high-strength steels, and chilled and gray irons. Polishing with loose-grain diamond is more common for nonmetallic workpieces (like granite) than it is for metals. When done on metals, however, the same fine mesh size diamond and diamond micron powders that are used for lapping are employed.

Bonded-Abrasive Grains

The principal use of bonded-abrasive grains is in grinding wheels. The primary metals commonly ground with diamond or CBN are shown in Table 4. Of these applications, the most important worldwide is the grinding of cemented tungsten carbide for producing machine tools, wear surfaces, and dimensioned parts and for resharpening tool blanks. Resin-bonded diamond grinding wheels have become the accepted standard for this application. Cemented tungsten carbide, made by sintering compacted mixtures of tungsten carbide particles with cobalt powders, is a hard, tough, wear-resistant material suitable for use in metal-cutting tools. These same characteristics make it difficult to grind but an ideal workpiece for resin-bonded diamond grinding wheels. Generally, a more friable diamond is best suited for these applications because friable diamond is capable of regenerating cutting edges and points.

One limitation to the economical use of diamond as the superabrasive of choice is its solubility in iron, nickel, cobalt, and alloys based on these metals. Cubic boron nitride is preferred for the grinding of these metals (Table 4), and CBN grains for grinding iron, nickel, cobalt, and their alloys are generally in the 60 to 400 mesh size range (250 to 38 μm).

Grinding wheels are available in a wide variety of sizes and shapes. Selection of the proper wheel for a given application is critical. The grinding wheel manufacturers have years of experience and can provide help as needed. Krar and Ratterman (Ref 15) also give guidelines that can be of help. They indicate that one should first choose the best bond for the application, then specify, in order, wheel diameters and widths, superabrasive mesh sizes, and concentrations. If properly done, this procedure will ensure good wheel life, good material removal rates, and the required workpiece surface finish.

Once the grinding wheels have been fabricated, it is good practice to true them to establish the desired shape; they should then be dressed to ensure good protrusion of the abrasive grains. These operations are covered in detail in Ref 16 and in the article "Superabrasives" in Volume 16 of the 9th Edition of *Metals Handbook*.

Table 4 Metals typically ground or machined with superabrasives and ultrahard tool materials

Metal types	Hardness	Examples of designations or applications	Principal alloying elements	Grind with CBN	Grind with Diamond	Machine with PCBN	Machine with PCD
Hardened steel							
Tool, die, and high-speed steels	>50 HRC	A2, D2, M2, M4, O5, T15	Co, Cr, Mo, V, W	Yes	No	Yes	No
Alloy steels	>50 HRC	4130, 4340, 5150, 52100, 8620, 9260	Cr, Mo, Ni, V	Yes	No	Yes	No
Carbon steels	>50 HRC	1050, 1095	Mn, Si	Yes	No	Yes	No
Stainless steels	>50 HRC						
Austenitic	...	301, 302	Cr, Ni, Mn	(a)	No	Yes	No
Martensitic	>50 HRC	410, 440A	Cr	Yes	No	Yes	No
Cast iron							
Gray iron	>180 HB	Engine blocks, flywheels, crankshafts	C, Si	Yes	No	Yes	No
White iron	>450 HB	Ni-Hard (rolls)	C, Ni, Si, Cr	Yes	No	Yes	No
Ductile iron	>200 HB	Crankshafts, exhaust manifolds	C, Si	(a)	No	(a)	No
Superalloys							
Nickel-base superalloys	>35 HRC	Inconel, René, Waspalloy	Cr, Co, Mo, W, Ti	Yes	No	Yes	No
Cobalt-base superalloys	>35 HRC	Stellite, AiResist, Haynes	Cr, W	Yes	No	Yes	No
Iron-base superalloys	>35 HRC	A-286, Incoloy	Cr, Ni, Mo	Yes	No	Yes	No
Hardfacing materials							
Carbide/Oxide-base materials	>35 HRC	UCAR LA-2, LC-4	Al_2O_3, Cr_2O_3 WC	No	Yes	No	Yes
Metal-base materials	>35 HRC	Stellite, Hastelloy	Mo, Ni, Cr, Co, Fe	Yes	No	Yes	No
Aluminum alloys							
Sand or permanent cast alloys							
Mold cast alloys	40–145 HB	A356, A390	Si, Cu, Mg	No	No	No	Yes
Die cast alloys	65–125 HB	A360, 380, 390	Si, Cu, Zn	No	No	No	Yes
Wrought alloys	40–150 HB	2218, 7049	Cu, Zn, Mg	No	No	No	Yes
Cemented tungsten carbide							
All tool and die grades	84–95 HRA	...	TaC, TiC, Co	No	Yes	No	No
All presintered tool and die grades	...	...	TaC, TiC, Co	No	Yes	No	Yes
Sintered die grades	<90 HRA	...	>6% Co	No	Yes	No	Yes

(a) Can be machined or ground if the equipment and operating conditions are suitable for superabrasives

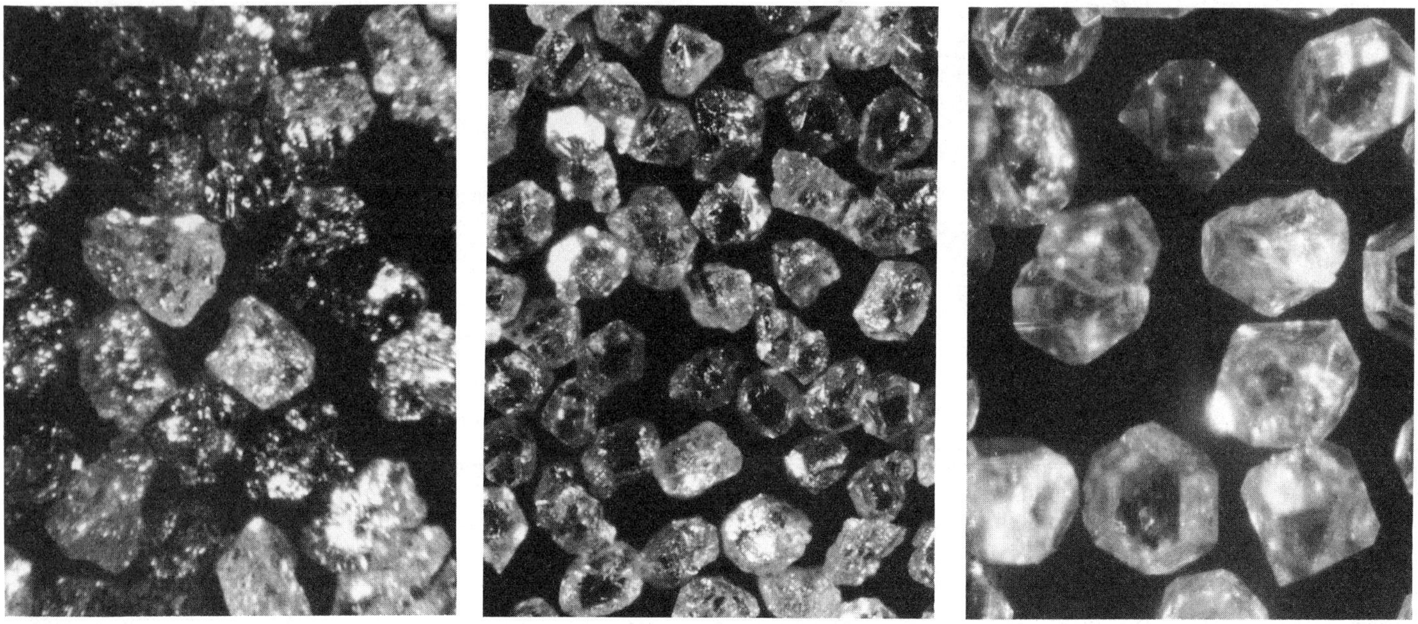

(a) (b) (c)

Fig. 9 Commercially available diamond grains used in various applications. (a) Friable diamond grains especially tailored for resin bond grinding wheels. (b) Diamond grains tailored for use in metal bond grinding wheels. These grains are typically in the 80 to 400 mesh size (350 to 38 μm) range. (c) Synthesized diamond grains for use in diamond saw blade applications, such as for the sawing of marble, granite, and concrete. These grains are in the 20 to 60 mesh size (850 to 250 μm) range.

Bonds. Several commercial bonds are available for grinding wheels. The most common are resin systems, vitreous systems, metal systems, and electroplated systems. All are suitable for use with superabrasives.

Resin Bonds. The diamond types synthesized for use in resin bond grinding wheels are shown in Fig. 9(a). This type of diamond is friable and thus is capable of regenerating cutting edges and points. It is also well suited to the scratching action required for material removal in the grinding of hard materials such as cemented tungsten carbide. Depending on the application, diamond concentrations can range from 50 to 150 (12.5 to 37.5 vol% of superabrasive). Most commercially available wheels have a 75 to 100 concentration (18.75 to 25 vol% of superabrasive).

Nearly all of the common resin bonds are thermosetting resins, and most of the thermosetting resins are phenolic resins. Resin powders and a solvent, such as furfural, are mixed with superabrasive particles and a filler, such as silicon carbide, and then placed in a mold containing the metal core. The resin mixture is cured in the hot press mold at pressures of 35 to 105 MPa (5 to 15 ksi) and temperatures of at least 150 °C (300 °F) for times ranging from 30 min to 2 h. Before use, the wheels are trued to eliminate chatter and to ensure the proper form. After trueing, it is necessary to dress the wheels to ensure abrasive protrusion for free cutting. Resin bond wheels are commercially available in a large range of sizes. The wheels can be used wet or dry and are free cutting, but they have relatively short life and poor form-holding characteristics.

Phenol-Aralkyl Bonds. Recently, work has been made public concerning new phenol-aralkyl bond systems for diamond abrasives (Ref 17). This class of resins can be used for making grinding wheels in the same equipment used for ordinary phenolic resin wheels. The new formulations are claimed to provide significantly improved wear life, cooler cutting, and a superior workpiece surface finish.

Thermoplastic resins, such as polyimides, are of interest as bond systems for heavy-duty grinding wheels. They are characterized by higher temperature stability limits than those of the phenolic resins. However, these resins do soften at high temperature; this can allow the superabrasive particles to move within the softened bond, and grains may be lost prematurely from the wheel. Special rough coatings have been devised to anchor the abrasive grains in such bonds (Fig. 10), and these coatings have proved effective.

Vitreous bond systems are generally tailored from glass or ceramic formulations. Vitreous bonds are finding increasing application with CBN abrasive grains; they are also useful for diamond grain wheels. Most vitreous bond systems are proprietary materials used for the production grinding of steel, cast iron, and superalloys. To be suitable for use with superabrasives, the bonds must have the proper wear characteristics, be formable at moderate temperatures and pressures, and be chemically compatible with the superabrasive grains. Some vitreous bonds meet these criteria with diamond but are too reactive with CBN; these applications require the use of protective metal coatings on the CBN. A significant reaction between the bond and the CBN abrasive can produce gaseous by-products that cause excessive porosity in the bond; this can lead to a loss of abrasive particle material and a weakening of the bond. Properly made vitreous bonds

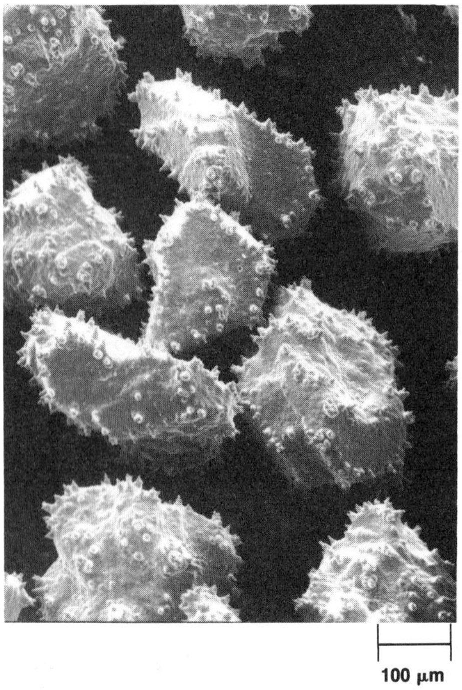

Fig. 10 Diamond of the type shown in Fig. 9(a), but with a special spiked nickel coating. Cubic boron nitride can be coated in a similar fashion.

have several advantages: ease of conditioning, free-cutting characteristics, reduced frictional heat, excellent surface finish capabilities, consistently accurate geometry, and long wheel life (Ref 16).

Metal bond systems are used with superabrasive grinding wheels in applications such as glass and ceramic grinding. Figure 9(b) shows typical diamond grains used in metal bonds for grinding. The grains are stronger than those used in resin bonds (Fig. 9a), but they are not as strong as the grains used in saw blade applications for stone and concrete (Fig. 9c). It is common practice to use softer metals such as bronze for metal bond grinding wheels. These metals wear away during use at a rate that ensures both crystal protrusion at the wear surface and free-cutting action. The two basic processes for the fabrication of metal bond wheels are hot pressing and cold pressing followed by sintering. Processing temperatures range from 600 °C (1100 °F) to greater than 1100 °C (2000 °F), pressures from about 14 to 140 MPa (2 to 20 ksi), and times at temperature from about 15 min to over 1 h. Superabrasive concentrations generally vary from 50 to 100 (12.5 to 25 vol% of superabrasive). These bonds are relatively tough, and they have long life and good form-holding characteristics. For glass and ceramic grinding, the mesh size ranges from 60 to 400 (250 to 38 μm).

Electroplated bond systems are available for grinding wheel fabrication. Superabrasive grains are bonded to wheel cores by electrodeposition of nickel or a nickel alloy. Normally, the layer of superabrasive is tacked down by immersing the core as a cathode into a bed of the superabrasive crystals in a plating solution; the wheel is then removed to a fresh bath for final plating. The final product has a single layer of superabrasive crystals with good particle exposure. Such wheels can be fabricated into complex forms and will hold those forms well for the life of the wheel. The wheels are free cutting but have a relatively short life because they possess only a monolayer of crystals.

Coatings. Superabrasive grains are often coated before being incorporated into the bond systems. The coatings are generally of metals, specifically nickel, cobalt, copper, and titanium. The coatings serve several purposes, depending on the superabrasive and the bond. Many of the resin bonds wet metals better than they wet superabrasives. A good example of this is a phenolic bond with nickel-coated synthetic diamond as compared with the same bond with uncoated synthetic diamond. The bond with the nickel-coated diamond is stronger, aiding retention of the protruding grains. In addition, metal coatings can slow the transfer of heat from the cutting points of the grains to the resin bond, delaying the onset of charring and degradation of the bond and extending the life of the grinding wheel. Coatings can also act as barriers to chemical reactions, such as those that occur between some vitreous bonds and CBN. Detrimental reactions can be eliminated by thin coatings of titanium on the superabrasive surfaces.

While a number of processes can be used for coating superabrasives (for example, chemical vapor deposition, physical vapor deposition, plasma spraying, and sputtering), most commercial coatings are prepared by electroplating techniques. Electrolytic coatings can be applied using a standard or modified Watts bath (Ref 18). Autocatalytic (electroless) coatings are also common and can be applied with baths that require no passage of electric current from external power sources (Ref 19). Autocatalytic coatings can generally be distinguished by the presence of phosphorus from the hypophosphites used as reducing agents. This phosphorus can slightly embrittle the coating, which often improves its performance.

Copper coatings have been designed for superabrasives in resin bond wheels that are used for dry grinding, and the copper-coated wheels are more effective for these applications than those with nickel coatings or uncoated crystals. Copper coatings are normally applied at a 60 wt% concentration. Nickel coatings are more effective in wet grinding with resin bond grinding wheels; they are commercially available at 30 and 56 wt% concentrations. To simplify inventories, some shops prefer to use nickel coatings for all applications, wet or dry. The dry grinding performance of wheels with nickel-coated superabrasives is definitely not as good as that of wheels with copper coatings, but it may be acceptable. The reverse situation, that is, using copper-coated grains in wet grinding applications, gives poor results and is not recommended.

Coated grains are not commonly found in metal bond grinding wheels or in electroplated wheels. There is nothing to restrict their use for special applications, however.

Ultrahard Tool Materials

Ultrahard tool materials of sintered polycrystalline diamond (PCD) or PCBN are commercially available in many shapes, sizes, and compositions. Depending on their type, they are used for cutting, drilling, milling, dressing, and as wear surfaces.

The PCD or PCBN in ultrahard tool blanks often is bonded to a cemented carbide substrate (Fig. 11), which allows brazing to tool shanks or to indexable inserts for use in standard toolholders. Solid PCD and PCBN can also be used as inserts.

The types of metals typically machined with ultrahard tool materials are summarized in Table 4. More detailed information on tool fabrication and applications is available in the article "Ultrahard Tool Materials" in Volume 16 of the 9th Edition of *Metals Handbook*.

Fig. 11 Polycrystalline diamond with substrates. (a) Typical fully round PCD tool blank. This type of blank is brazed or mechanically clamped to extend the usable cutting edge. (b) Typical square PCD tool blank with a long straight cutting edge that is ideal for many applications. (c) Typical triangular PCD tool blank that is useful in single-point turning applications, either as tools or as brazed-in tips on carbide tools

Polycrystalline diamond tool blanks are useful in the machining of nonferrous and nonmetallic materials (Tables 4 and 5) and are commercially available in a variety of shapes and sizes (Fig. 12). An important

Table 5 PCD tool applications

Application
Nonferrous materials
Silicon aluminum
Hypereutectic alloys
Hypoeutectic alloys
Copper alloys
Tungsten carbide
Nonmetallic materials
Woodworking
Fiberboard
Medium-density fiberboard
Chipboard
Hardboard
Composites
Graphite-epoxy
Carbon-carbon fiber
Fiberglass plastic
Ceramics

variable for the end user is the average grain size, which is separated into three grades in Fig. 13: fine (average diamond grain size, 4 μm), medium (5 μm), and coarse (25 μm). As shown in Fig. 13, differences in grain size can cause variations in abrasion resistance, grindability, and workpiece surface finish. As a result of these differences, the areas of preferred application are different for the three grades:

PCD grade	Application
Fine grain	Applications requiring good surface finishes; woodworking
Medium grain	General purpose applications (aluminum alloys with <16% Si)
Coarse grain	Heavy-duty applications; milling applications, interrupted cuts (aluminum alloys with >16% Si)

Table 6 General starting conditions for PCD cutting tools

Workpiece material	Speed, m/min	Speed, ft/min	Feed, mm/rev	Feed, in./rev
Aluminum alloys				
4–8% Si	1280–1980	4200–6500	0.1–0.65	0.004–0.025
9–14% Si	1005–1585	3300–5200	0.1–0.5	0.004–0.020
16–18% Si	305–700	1000–2300	0.1–0.4	0.004–0.015
Copper alloys	610–1005	2000–3300	0.05–0.2	0.002–0.008
Plastics and composites	300–1005	1000–3300	0.1–0.3	0.004–0.012
Sintered tungsten carbide	20–40	65–130	0.15–0.25	0.006–0.010
Manufactured wood	1005–3000	3300–9800	0.1–0.4	0.004–0.015

The techniques for using PCD tool blanks are not very different from those for using conventional ceramic blanks. Where possible, these guidelines should be followed:

- Use a positive rake
- Maintain a sharp cutting edge
- Use the largest nose radius possible
- Use a rigid machine set-up
- Minimize tool overhang
- Use a flood coolant whenever possible

When first machining a new material, the starting conditions suggested in Table 6 can be used. Slight modifications may give improved results, depending on the particular configuration. Polycrystalline diamond tool blanks can be used dry; the high thermal conductivity of the diamond layer removes and distributes heat generated at the cutting edge. However, tool performance is generally improved by the use of coolants (Ref 15). Water-soluble oil emulsions, such as those used in conventional machining with cemented tungsten carbide tools, are adequate if properly applied to the rake surface. They reduce frictional heating and the formation of built-up edges while providing good chip flow.

Fig. 12 Various configurations and sizes of PCD tool blanks. Round shapes are available in standard sizes up to about 34 mm (1.34 in.) in diameter.

Polycrystalline cubic boron nitride (PCBN) tool blanks are useful in the machining of iron, steel, and cobalt- and nickel-base alloys (Tables 4 and 7). This makes them complementary to rather than competitive with the PCD tool blanks. They are generally not recommended for use with superalloys or steels that have hardness of less than 35 and 45 HRC, respectively. The causes and solutions of common problems encountered when using PCBN tools are listed in Table 8.

Polycrystalline CBN blanks are available in a variety of shapes and sizes similar to those of polycrystalline diamond (Fig. 12). Polycrystalline CBN tool blanks can consist of a basic CBN layer (either solid or on a cemented carbide substrate) or a composite abrasive layer (about half CBN and half ceramic). The composite blanks have excellent thermal and wear resistance but lower impact resistance; therefore, they are less applicable to interrupted-cut machining. For milling applications, basic PCBN blanks are preferred.

Tool Geometry. The general guidelines for the use of PCBN tools (Ref 15) are similar but not identical to those for PCD tools. Negative-rake PCBN tools should be used wherever possible because they can withstand high cutting forces.

The lead or side cutting-edge angle should be as large as possible when using PCBN

Table 7 PCBN tool applications

Application
Hard cast iron
Ni-Hard
Alloy cast iron
Chilled cast iron
Nodular cast iron
Soft cast iron
Gray cast iron
Sintered iron
Powder metallurgy products
Hardened steel
Tool steels
Die steels
Case-hardened steels
A, D, and M series steels
Bearing steels
Superalloys
Inconel 600, 718, and 901
René 77 and 95
Hastelloy
Waspaloy
Stellite

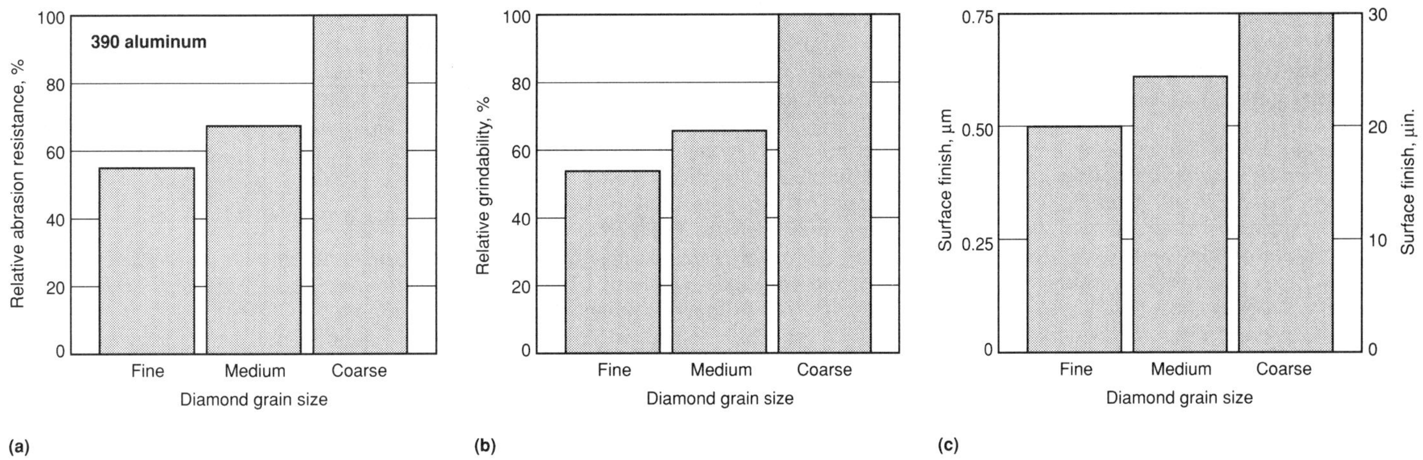

Fig. 13 Variation in tool performance with average grain size in PCD tool blanks. (a) Abrasion resistance. (b) Grindability. (c) Surface finish. Surface finish is also dependent on other factors such as feed rate, tool geometry, and workpiece condition.

tools; it should only rarely be less than 15°. A large lead angle spreads the cut over a wide section of the cutting edge, resulting in a thinner chip, which in turn reduces loading on the tool blank or insert. The reduced loading allows the feed per revolution to be increased without increasing the chances of cutting edge chippage. In addition, a large lead angle helps reduce notching at the depth-of-cut line; notching can occur in an overly hard workpiece, or as the presence of scale on a workpiece.

Sharp corners on cutting tools concentrate stresses and can cause premature load failures. Honing a radius on the edge and chamfering the cutting edge are two available methods for overcoming this problem. A chamfer of 15° with a width of 0.2 mm (0.008 in.) is recommended for most roughing operations. Honing the edge slightly is suggested for finishing operations (Ref 15).

Starting Feeds and Speeds. Good starting conditions are listed in Table 9 for several materials commonly machined with PCBN tools. While feeds and speeds are dependent on workpiece properties, the conditions given in Table 9 generally produce satisfactory results. In the speed ranges shown, the higher speeds are for finishing operations. It is recommended that cutting fluids be used whenever possible.

Table 8 Common problems encountered with PCBN tool blanks

Problem	Cause	Solution
Edge chippage	Improper edge preparation	Chamfer the cutting edge by 15° and 0.2 mm (0.008 in.); ensure a rigid toolholding system
Rapid tool flank wear	Cutting speed too slow (insufficient to soften ahead of tool); or cutting speed too fast (excessive heat generated)	Change speed to recommended rates: For hardened ferrous materials (>45 HRC), 70–130 m/min (230–430 ft/min) For soft gray cast iron (200 HB), 450–915 m/min (1500–3000 ft/min)
	Feed rate too light (thin chip cannot dissipate heat; tool rubbing)	Use a minimum feed rate of 0.1 mm/rev (0.004 in./rev)
	Depth of cut too light (excessive tool rubbing)	Use a minimum depth of cut of 0.125 mm (0.005 in.)
Rapid tool crater wear	Soft tool steel; cutting speed too high (excessive heat)	Use only for steels with a minimum hardness of 45 HRC (see above for speed recommendations)
	Insufficient coolant	Use coolant if possible.

Table 9 General starting conditions for PCBN cutting tools

Classification	Material type	Speed m/min	Speed ft/min	Feed mm/rev	Feed in./rev
Hardened ferrous materials (>45 HRC)	Hardened steels (4340, 8620, M2, T15); hard cast irons (chilled iron, Ni-Hard)	70–130	230–430	0.1–0.5	0.004–0.020
Superalloys (>35 HRC)	Nickel- and cobalt-base alloys (Inconel, René, Stellite, Colmonoy)	200–245	650–800	0.1–0.25	0.004–0.010
Soft cast irons (typically 180–240 HB)	Pearlitic gray iron, Ni-Resist	460–915	1500–3000	0.1–0.65	0.004–0.025
Flame-sprayed materials	Hardfacing materials	60–105	200–350	0.1–0.3	0.004–0.012
Cold-sprayed materials	Hardfacing materials	105–150	350–500	0.1–0.33	0.004–0.013

REFERENCES

1. *Ceram. Bull.*, Vol 67 (No. 6), 1988
2. P. Schwarzkopf and R. Kieffer, *Refractory Hard Materials*, Macmillan, 1953
3. A.Y. Liu and M.L. Cohen, Prediction of New Low Compressibility Solids, *Science*, Vol 245, 1989, p 841-842
4. H.T. Hall, Ultra-High-Pressure, High-Temperature Apparatus: The "Belt," *Rev. Sci. Instrum.*, Vol 31 (No. 2), 1980, p 125-131
5. H.P. Bovenkerk, F.P. Bundy, H.T. Hall, H.M. Strong, and R.H. Wentorf, Jr., Preparation of Diamond, *Nature*, Vol 184, 1959, p 1094-1098
6. R.J. Wedlake, Technology of Diamond Growth, in *The Properties of Diamond*, J. Field, Ed., Academic Press, 1979
7. H.M. Strong and R.E. Hanneman, Crystallization of Diamond and Graphite, *J. Chem. Phys.*, Vol 46, 1967, p 3668-3676
8. R.C. DeVries and J.F. Fleischer, Phase Equilibria Pertinent to the Growth of Cubic Boron Nitride, *J. Cryst. Growth*, Vol 13/14, 1972, p 88-92
9. P.S. DeCarli, Method of Making Diamond, U.S. Patent 3,238,019, March 1966; and P.S. DeCarli and J.C. Jamieson, Formation of Diamond by Explosive Shock, *Science*, Vol 133, 1966, p 1821-1822
10. B.V. Spitsyn, L.L. Bouilov, and B.V. Derjaguin, Vapor Growth of Diamond on Diamond and Other Surfaces, *J. Cryst. Growth*, Vol 52, 1981, p 219-226

11. S. Matsumoto, Y. Sato, M. Tsutsumi, and N. Setaka, Growth of Diamond Particles from Methane-Hydrogen Gas, *J. Mater. Sci.*, Vol 17, 1982, p 3106-3112
12. B. Lux and R. Haubner, Low Pressure Synthesis of Superhard Coatings, *Int. J. Refract. Met. Hard Mater.*, Vol 9, 1989, p 158-174
13. R.H. Wentorf, Jr. and W.A. Rocco, Diamond Tools for Machining, U.S. Patent 3,745,623, July 1973
14. F.R. Corrigan, Thermal Conductivity of Polycrystalline Cubic Boron Nitride in Compacts, *High Pressure Science and Technology*, Vol 1, Plenum Publishing, 1979, p 994-999
15. S.F. Krar and E. Ratterman, *Superabrasives—Grinding and Machining With CBN and Diamond*, McGraw-Hill, 1990
16. B. Nailor, "Trueing Parameters for Conditioning Vitrified Bond CBN Wheels," Paper presented at Advancements in Abrasives, The 27th International Abrasive Engineering Conference, Bloomingdale, IL, Sept 1989
17. G.I. Harris, "Phenol-aralkyl Resin Bonded Wheels," Paper presented at the Industrial Diamond Association Ultra-Hard Materials Seminar, Toronto, Sept 1989
18. N.V. Parthasaradhy, *Practical Electroplating Handbook*, Prentice-Hall, 1989, p 183-186
19. F.A. Lowenheim, *Electroplating—Fundamentals of Surface Finishing*, McGraw-Hill, 1978, p 391-400

SELECTED REFERENCES

- R. Komanduri and Desai, Tool Materials, in *Encyclopedia of Chemical Technology*, Vol 23, 1983, p 273
- R. Komanduri and M.C. Shaw, Surface Morphology of Synthetic Diamonds and Cubic Boron Nitride, *Int. J. Mach. Tool Des. Res.*, Vol 14, 1983, p 63-84

Structural Ceramics

Gerald L. DePoorter, Colorado Center for Advanced Ceramics, Department of Metallurgical and Materials Engineering, Colorado School of Mines
Terrence K. Brog and Michael J. Readey, Coors Ceramics Company

CERAMICS are nonmetallic, inorganic engineering materials processed at a high temperature. The general term "structural ceramics" refers to a large family of ceramic materials used in an extensive range of applications. Included are both monolithic ceramics and ceramic-ceramic composites. Chemically, structural ceramics include oxides, nitrides, borides, and carbides. Many processing routes are possible for structural ceramics and are important because the microstructure, and therefore the properties, are developed during processing.

General properties and uses of structural ceramics are reviewed first. Ceramic processing is described and the relationship of processing, microstructure, and properties presented. Specific structural ceramic materials, including composites, are presented. This article concludes with a discussion of future direction and problems with structural ceramics.

Uses and General Properties of Structural Ceramics

Industrial uses, required properties, and examples of specific applications for structural ceramics are summarized in Table 1. These applications take advantage of the temperature resistance, corrosion resistance, hardness, chemical inertness, thermal and electrical insulating properties, wear resistance, and mechanical properties of the structural ceramic materials. Combinations of properties for specific applications are summarized in Table 1. Ceramics offer advantages for structural applications because their density is about one-half the density of steel, and they provide very high stiffness-to-weight ratios over a broad temperature range. The high hardness of structural ceramics can be utilized in applications where mechanical abrasion or erosion is encountered. The ability to maintain mechanical strength and dimensional tolerances at high temperature makes them suitable for high-temperature use. For electrical applications, ceramics have high resistivity, low dielectric constant, and low loss factors that when combined with their mechanical strength and high-temperature stability make them suitable for extreme electrical insulating applications.

Specific properties of ceramics compared with other materials are discussed in the Section "Properties and Applications of Structural Ceramics" in this article. The text by Kingery (Ref 1) should be consulted for a general discussion of the properties as related to composition and microstructure.

Processing of Structural Ceramics

The processing steps for producing structural ceramics are shown in the flow chart given in Fig. 1. These steps can be grouped into four general categories:

- Raw material preparation
- Forming and fabrication
- Thermal processing
- Finishing

These categories are also indicated on Fig. 1. Only a brief overview of ceramic processing can be included here. For specific details see the text by Reed (Ref 2).

Raw material preparation includes material selection, ceramic-body preparation, mixing and milling, and the addition of processing additives such as binders. Mate-

Table 1 Industry, use, properties, and applications for structural ceramics

Industry	Use	Property	Application
Fluid handling	Transport and control of aggressive fluids	Resistance to corrosion, mechanical erosion, and abrasion	Mechanical seal faces, meter bearings, faucet valve plates, spray nozzles, micro-filtration membranes
Mineral processing power generation	Handling ores, slurries, pulverized coal, cement clinker, and flue gas neutralizing compounds	Hardness, corrosion resistance, and electrical insulation	Pipe linings, cyclone linings, grinding media, pump components, electrostatic precipitator insulators
Wire manufacturing	Wear applications and surface finish	Hardness, toughness	Capstans and draw blocks, pulleys and sheaves, guides, rolls, dies
Pulp and paper	High-speed paper manufacturing	Abrasion and corrosion resistance	Slitting and sizing knives, stock-preparation equipment
Machine tool and process tooling	Machine components and process tooling	Hardness, high stiffness-to-weight ratio, low inertial mass, and low thermal expansion	Bearings and bushings, close tolerance fittings, extrusion and forming dies, spindles, metal-forming rolls and tools, coordinate-measuring machine structures
Thermal processing	Heat recovery, hot-gas cleanup, general thermal processing	Thermal stress resistance, corrosion resistance, and dimensional stability at extreme temperatures	Compact heat exchangers, heat exchanger tubes, radiant tubes, furnace components, insulators, thermocouple protection tubes, kiln furniture
Internal combustion engine components	Engine components	High-temperature resistance, wear resistance, and corrosion resistance	Exhaust port liners, valve guides, head faceplates, wear surface inserts, piston caps, bearings, bushings, intake manifold liners
Medical and scientific products	Medical devices	Inertness in aggressive environments	Blood centrifuge, pacemaker components, surgical instruments, implant components, lab ware

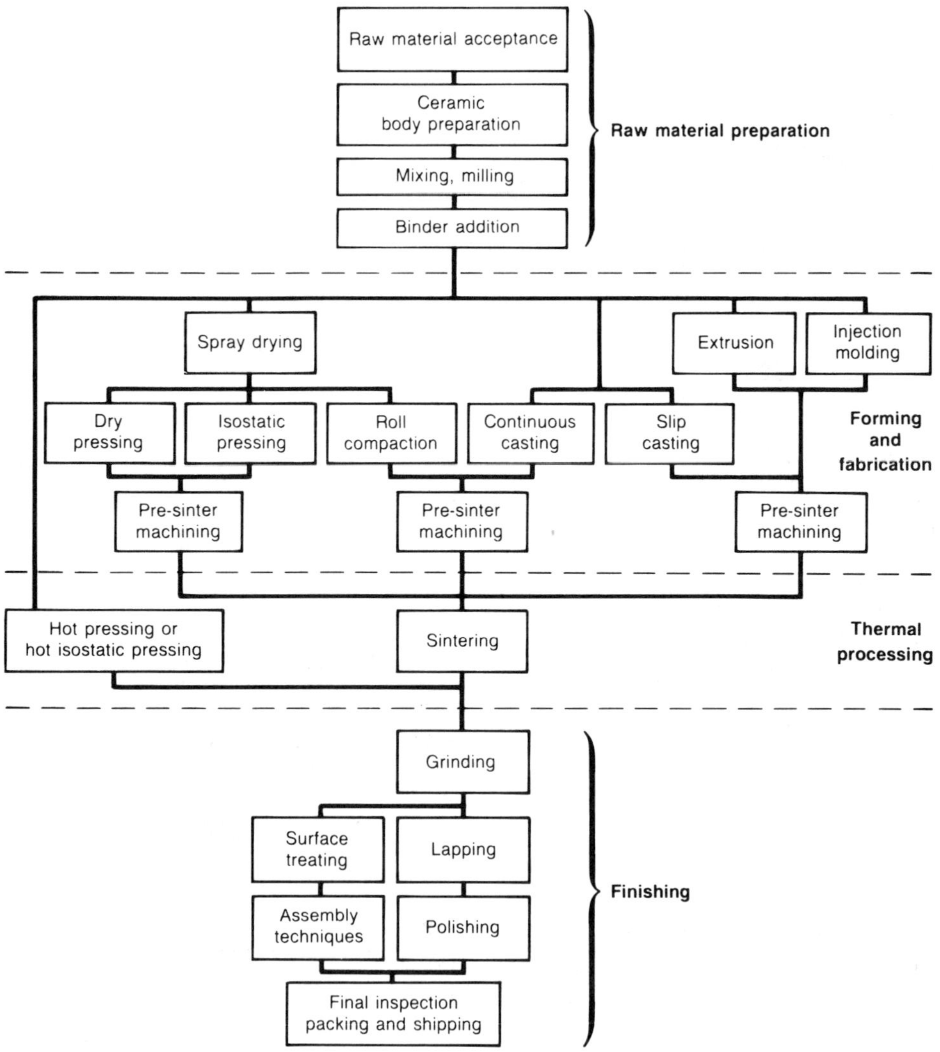

Fig. 1 Flow chart for ceramic processing

rial selection is important because structural ceramics require high-quality starting materials that can be described as industrial inorganic chemicals. For example, the Al_2O_3 powder for alumina ceramics is usually obtained as calcined alumina from the Bayer process, which uses bauxite as the starting material. Zirconia is obtained from industrial sources that process zircon ($ZrSiO_4$) to produce ZrO_2 of 99% purity. Silicon carbide, SiC, is produced by the Acheson process in which silica, SiO_2, and coke are placed in an arc furnace and reacted at 2200 to 2500 °C (4000 to 4500 °F). Silicon nitride, Si_3N_4, is produced by reacting silicon with nitrogen. Chemical techniques are used to produce powders where extremely high purity and very fine particle sizes are required. A detailed description of material selection for structural ceramics is included in the reference text edited by Somiya (Ref 3). The assurance of final product quality starts with well-defined and strict material acceptance criteria.

Ceramic body preparation consists of combining the collection of materials necessary for the final body composition. For oxide systems, the starting materials are generally mixed in aqueous systems and milled to obtain the specified particle-size distribution for the body. If necessary, organic binders are added after the milling and mixing. This results in a slurry or slip, which is the starting material for forming and fabrication of the component.

Forming and Fabrication. Structural ceramics are formed from either powders, stiff pastes, or slurries. The slurry from the preparation procedures is converted to an agglomerated flowable powder by spray drying or to a stiff paste by filter pressing. Structural components are formed by pressing of powders, extrusion of stiff pastes, or by slip casting of slurries. In some cases, pre-sinter machining (green machining) is required.

Forming Structural Ceramics From Powders. Pressing operations are used to make consolidated ceramics starting from a powder. Complex shapes are made to net shape in large volumes by dry pressing in uniaxial double-acting presses using specific tooling made for each part. No further shaping of these components is usually required prior to the thermal processing step. Faucet valve plates, pipe linings, and grinding media are made by pressing.

Isostatic pressing is used for larger components and where extensive pre-sinter machining is required. Powders are placed in flexible tooling and pressure is hydraulically applied in all directions forming a part that is machined to net shape prior to sintering. Electrostatic precipitator insulators, sodium-vapor lamp tubes, and spark plugs are examples of products formed by isopressing.

Forming Structural Ceramics From Stiff Pastes. Stiff pastes are used to form structural ceramics by extrusion. In extrusion the plastic mass is forced through a die at high pressure, which determines the shape of the component. Rods and tubes are usually formed by extrusion.

Forming Structural Ceramics From Slurries or Slips. Slurries, or slips, are used to form structural components by slip casting. In slip casting a porous mold, usually plaster, is filled with slip. Capillary action draws the water from the slip into the mold, which forms a solid layer at the slip-mold interface. When the required wall thickness is reached, the remaining slip is poured from the mold. Thermocouple protection tubes are made by slip casting.

Pre-Sinter Machining. Structural components, when required, are machined to final unfired dimensions after the forming operations described above. Conventional machining techniques such as turning, milling, and drilling are used, and in many cases the machining is done on numerically controlled machines.

Thermal processing for structural ceramics is done either at ambient pressure or with added pressure in the case of hot pressing or hot isostatic pressing (HIP). The final microstructure is developed during thermal processing by sintering, vitrification, or reaction bonding.

Sintering takes place by volume, surface, or grain-boundary diffusion and is a solid-state process. During sintering the pores are removed, the piece is densified, and grain growth occurs if desired for the particular ceramic being processed. Sintering is used for high-purity oxide systems.

Vitrification involves the presence of a liquid phase during thermal processing. The liquid phase provides faster diffusion paths and holds the piece together by capillary action during processing. This results in an amorphous or glassy phase being present in the final microstructure. The final microstructure is created by vitrification for systems with less than 99% pure oxide, porcelains, and Si_3Ni_4 with sintering additives.

Table 2 Properties of various alumina ceramics

Alumina content, %	Bulk density, g/cm^3	Flexure strength, MPa (ksi)	Fracture toughness, MPa$\sqrt{m}$ (ksi$\sqrt{in.}$)	Hardness, GPa (10^6 psi)	Elastic modulus, GPa (10^6 psi)	Thermal conductivity, W/m · K (Btu/ft · h · °F)	Linear coefficient of thermal expansion, ppm/°C (ppm/°F)
85	3.41	317 (46)	3–4 (2.8–3.7)	9 (1.3)	221 (32)	16.0 (9.24)	7.2 (4)
90	3.60	338 (49)	3–4 (2.8–3.7)	10 (1.5)	276 (40)	16.7 (9.65)	8.1 (4.5)
94	3.70	352 (51)	3–4 (2.8–3.7)	12 (1.7)	296 (43)	22.4 (12.9)	8.2 (4.6)
96	3.72	358 (52)	3–4 (2.8–3.7)	11 (1.6)	303 (44)	24.7 (14.3)	8.2 (4.6)
99.5	3.89	379 (55)	3–4 (2.8–3.7)	14 (2.0)	372 (54)	35.6 (20.6)	8.0 (4.4)
99.9	3.96	552 (80)	3–4 (2.8–3.7)	15 (2.2)	386 (56)	38.9 (22.5)	8.0 (4.4)

Source: Coors Ceramic Company

In some cases the thermal processing is aided by adding external pressure during sintering. The pressure can be applied uniaxially in hot pressing or isostatically in hot isostatic pressing. Covalent materials such as silicon carbide and silicon nitride, and composite systems usually undergo hot pressing. Pressure can also be used to suppress the decomposition of materials (such as in the gas-pressure sintering of Si_3Ni_4). Hot pressing is also used in the processing of spinels.

The microstructure is developed by reaction bonding for some covalent structural ceramics such as silicon carbide and silicon nitride. For example, silicon carbide components are formed by mixing together very fine SiC coated with fine carbon, which is exposed to silicon above its melting point. The molten silicon and the carbon react to form silicon carbide in place which bonds the SiC grains together.

Finishing. Additional processing is required where tolerances are tighter than can be achieved by sintering or where a surface must be extremely flat or polished. Diamond grinding is used to provide tight dimensional tolerances. Lapping using abrasive slurries, extremely flat surfaces, and polishing by slurry abrasion will achieve a fine surface finish.

Properties and Applications of Structural Ceramics

Alumina Ceramics. Aluminum oxide, Al_2O_3 (often referred to as alumina), is perhaps the material most commonly used in the production of technical ceramics. The reasons for its wide acceptance are many; alumina has a high hardness, excellent wear and corrosion resistance, and low electrical conductivity. It is also fairly economical to manufacture, involving low-cost alumina powders.

Alumina ceramics actually include a family of materials, typically having alumina contents from 85 to ≧99% Al_2O_3, the remainder being a grain-boundary phase. The different varieties of alumina stem from diverse application requirements. For example, 85% alumina ceramics such as milling media are used in applications requiring high hardness, yet they are economical. Aluminas having purities in the 90 to 97% range are often found in electronic applications as substrate materials, due to the low electrical conductivity. The grain-boundary phase in these materials also allows for a strong bond between the ceramic and the metal conduction paths for integrated circuits. High-purity alumina (>99%) is often used in the production of translucent envelopes for sodium-vapor lamps.

The microstructure and resulting properties of alumina ceramics greatly depend on the percentage of alumina present. For example, high-purity aluminas typically have a fairly simple microstructure of equiaxed alumina grains (Fig. 2), whereas a 96% alumina ceramic will have a more complicated microstructure consisting of alumina grains (often elongated in shape) surrounded by a grain-boundary phase (Fig. 3). Depending on processing, this grain-boundary phase may be amorphous, crystalline, or both. The properties of this family of materials vary widely, as shown in Table 2.

Aluminum titanate, Al_2TiO_5, is a ceramic material that has recently received much attention because of its good thermal shock resistance. Aluminum titanate has an orthorhombic crystal structure, which results in a very anisotropic thermal expansion. The coefficient of thermal expansion (CTE) normal to the *c*-axis of the orthorhombic crystal is -2.6×10^{-6}/°C (-1.4×10^{-6}/°F) whereas the CTE parallel to the *c*-axis is about 11×10^{-6}/°C (6.1×10^{-6}/°F). The resulting thermal expansion coefficient for a polycrystalline material is very low (0.7×10^{-6}/°C, or 0.4×10^{-6}/°F) as shown in Table 3.

The excellent thermal shock resistance of aluminum titanate derives from this considerable thermal expansion anisotropy. During cooling from the densification temperature, the aluminum titanate grains shrink more in one direction than the other, which results in small microcracks developing in the microstructure as the grains actually pull away from each other. Subsequent thermal stresses (either by fast cooling or heating) are thereby dissipated by the opening and closing of the microcracks. Unfortunately, a consequence of the microcracks is that aluminum titanate does not have particularly high strength (25 MPa, or 3 ksi). However, the microcracks do impart very low thermal conductivity, making it an excellent candidate for thermal insulation devices.

The excellent thermal shock resistance of aluminum titanate offers the potential for

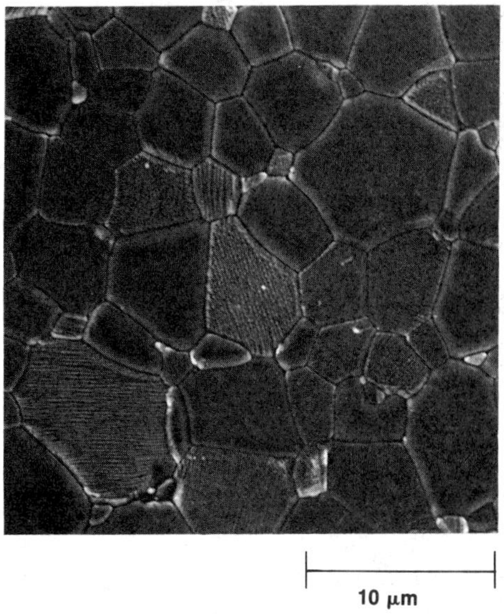

Fig. 2 Scanning electron micrograph of a high-purity Al_2O_3. The sample has been thermally etched to reveal the grain boundaries. Note the equiaxed grain morphology and lack of any intergranular phase.

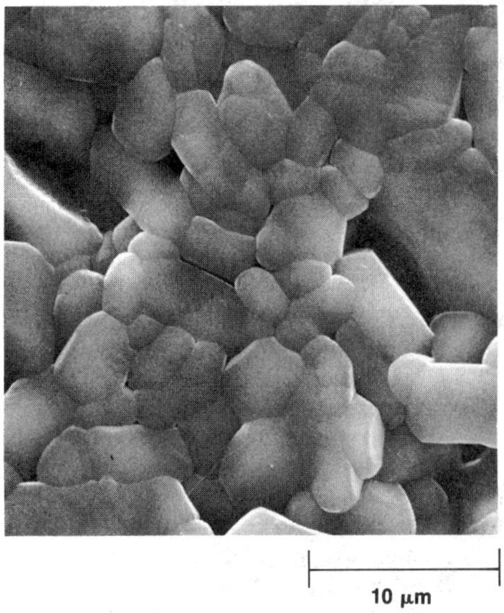

Fig. 3 Scanning electron micrograph of a typical 96% Al_2O_3 ceramic. The sample has been thermally etched to reveal the grain boundaries. The intergranular phase was also removed during etching. Note the tabular morphology of some of the alumina grains.

Table 3 Physical properties of various ceramics

Material	Bulk density, g/cm³	Flexure strength, MPa (ksi)	Fracture toughness, $MPa\sqrt{m}$ ($ksi\sqrt{in.}$)	Hardness, GPa (10^6 psi)	Elastic modulus, GPa	Thermal conductivity, W/m · K (Btu/ft · h · °F)	Linear coefficient of thermal expansion, ppm/°C (ppm/°F)
Aluminum titanate	3.10	25	. . .	. . .	5	1.0	0.7
		(3.6)	. . .	. . .	(0.7)	(0.6)	(0.4)
Sintered SiC	3.10	550	4	29	400	110.0	4.4
		(80)	(3.6)	(4.2)	(58)	(63.6)	(2.4)
Reaction-bond SiC	3.10	462	3–4	25	393	125.0	4.3
		(67)	(2.7–3.6)	(3.6)	(57)	(72.2)	(2.4)
Silicon nitride	3.31	906	6	15	311	15.0	3.0
		(131)	(5.5)	(2.2)	(45)	(8.7)	(1.7)
Boron carbide	2.50	350	3–4	29	350	. . .	. . .
		(51)	(2.7–3.6)	(4.2)	(51)	. . .	. . .

many applications. For example, aluminum titanate has found uses as funnels and ladles in the foundry industry (aluminum, magnesium, zinc, and iron do not wet aluminum titanate). The automotive industry is also investigating aluminum titanate for exhaust port liners and exhaust manifolds.

Silicon carbide, SiC, is ceramic material that has been in existence for decades but has recently found many applications in advanced ceramics. There are actually two families of silicon carbide, one known as direct-sintered SiC, and the other known as reaction-bonded SiC (also referred to as siliconized SiC). In direct-sintered SiC, submicrometer SiC powder is compacted and sintered at temperatures in excess of 2000 °C (3600 °F), resulting in a high-purity product. Reaction-bonded SiC, on the other hand, is processed by forming a porous shape comprised of SiC and carbon-powder particles. The shape is then infiltrated with silicon metal; the silicon metal acts to bond the SiC particles.

The properties of the two families of SiC are similar in some ways and quite different in others. Both materials have very high hardnesses (27 GPa, or 3.9×10^6 psi), high thermal conductivities (typically 110 W/m · K), and high strengths (500 MPa, or 73 ksi). However, the fracture toughness of both materials is generally low, of the order of 3 to 4 $MPa\sqrt{m}$ (2.7 to 3.6 $ksi\sqrt{in.}$). The major differences are found in wear and corrosion resistance. While both are very good in each category, direct-sintered SiC has a greater ability to withstand severely corrosive and erosive environments (the limiting factor for reaction-bonded SiC is the silicon metal).

Applications for SiC ceramics are typically in the areas where wear and corrosion are problems. For example, SiC is often found as pump seal rings and automotive water-pump seals. Silicon carbide's high thermal conductivity also allows them to be used as radiant heating tubes in metallurgical heat-treatment furnaces.

Silicon Nitride. An intense interest in silicon nitride (Si_3N_4) ceramics has emerged over the past few decades. The motivation for such interest lies in the automotive industry, where use of ceramic components in engines would greatly improve operating efficiency. Silicon nitride offers great potential in these applications because of its excellent high-temperature strength of 900 MPa (130 ksi) at 1000 °C (1830 °F), high fracture toughness of 6 to 10 $MPa\sqrt{m}$ (5.5 to 9 $ksi\sqrt{in.}$), and good thermal shock resistance. It also has very good oxidation resistance, a property of particular importance in automotive applications.

The automotive components of interest are turbocharger rotors, pistons, piston liners, and valves. The greatest application of Si_3N_4, however, is as a cutting-tool material in metal-machining applications, where machining rates can be dramatically increased due to the high-temperature strength of Si_3N_4.

Boron carbide, B_4C, is another material that is just now finding applications. The chief advantages of B_4C are its exceptionally high hardness (29 GPa, or 4.2×10^6 psi) and low density (2.50 g/cm³, or 0.09 lb/in.³). However, manufacturing B_4C is difficult because of the high temperatures necessary to effect densification (≫2000 °C, or 3600 °F). Thus in most cases B_4C is densified with pressure, as in hot pressing. This limits the complexity of shapes possible without excessive grinding and machining.

A disadvantage of B_4C is the high cost of the powders and subsequent processing. As such, B_4C has found use only in applications that demand the unique properties of B_4C, namely military armor.

SiAlON is an acronym for silicon-aluminum-oxynitride. SiAlON is fabricated in several ways, but is typically made by reacting Si_3N_4 with Al_2O_3 and AlN at high temperatures. SiAlON is a generic term for the family of compositions that can be obtained by varying the quantities of the original constituents. The advantages of SiAlONs are their low thermal expansion coefficient (2 to 3 × 10^{-6}/°C, or 1 to 1.7 × 10^{-6}/°F) and good oxidation resistance.

The array of potential applications is similar to that of Si_3N_4, namely automotive components and machine tool bits. However, the chemistry of SiAlON is complex, and reproducibility is a major hurdle to becoming more commercially successful. The processing of SiAlONs and their use as cutting-tool materials are discussed in more detail in Volume 16 of the 9th Edition of *Metals Handbook*.

Zirconia. Pure zirconia cannot be fabricated into a fully dense ceramic body using existing conventional processing techniques. The 3 to 5% volume increase associated with the tetragonal-to-monoclinic phase transformation causes any pure ZrO_2 body to completely destruct upon cooling from the sintering temperature. Additives such as calcia (CaO), magnesia (MgO), yttria (Y_2O_3), or ceria (CeO_2) must be mixed with ZrO_2 to stabilize the material in either the tetragonal or cubic phase. Applications for cubic-stabilized ZrO_2 (CSZ) include various oxygen-sensor devices (cubic ZrO_2 has excellent ionic conductivity), induction heating elements for the production of optical fibers, resistance heating elements in new high-temperature oxidizing kilns, and inexpensive diamond-like gemstones. Partially-stabilized or tetragonal-stabilized ZrO_2 systems will be discussed below.

Toughened Ceramics. Decades ago, ceramics were characterized as hard, high-strength materials with excellent corrosion and electrical resistance in addition to high-temperature capability. However, low fracture toughness limited its use in structural applications. The birth of toughened ceramics coincided with industrial applications requiring high-temperature capability, high strength, and an improvement in fracture resistance over existing ceramic materials. The primary driving force toward developing toughened ceramics was the promise of an all-ceramic engine. Several of the materials discussed in this section were or are being considered as ceramic-engine component materials.

Zirconia-toughened alumina (ZTA) is the generic term applied to alumina-zirconia systems where alumina is considered the primary or continuous (70 to 95%) phase. Zirconia particulate additions (either as pure ZrO_2 or as stabilized ZrO_2) from 5 to 30% represent the second phase (Fig. 4). The solubility of ZrO_2 in Al_2O_3 and Al_2O_3 in ZrO_2 is negligible. The ZrO_2 is present either in the tetragonal or monoclinic symmetry. ZTA is a material of interest primarily because it has a significantly higher strength and fracture toughness than alumina.

The microstructure and subsequent mechanical properties can be tailored to specific applications. Higher ZrO_2 contents lead to increased fracture toughness and strength values, with little reduction in hardness and elastic modulus, provided most of the ZrO_2 can be retained in the tetragonal phase. Strengths up to 1050 MPa (152 ksi) and fracture toughness values as high as 7.5 $MPa\sqrt{m}$ (6.8 $ksi\sqrt{in.}$) have been measured (Table 4). Wear properties in

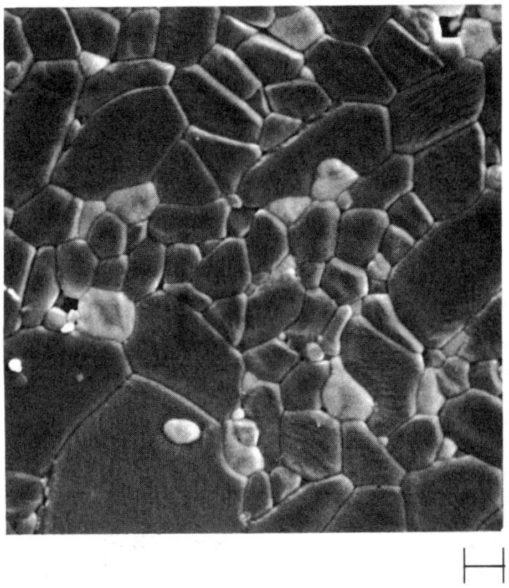

Fig. 4 Scanning electron micrograph of high-purity, zirconia-toughened alumina showing dispersed zirconia phase (white) within an alumina matrix

some applications may also improve due to mechanical property enhancement compared to alumina. These types of ZTA compositions have been used in some cutting-tool applications.

Zirconia-toughened alumina has also seen some use in thermal shock applications. Extensive use of monoclinic ZrO_2 can result in a severely microcracked ceramic body. This microstructure allows thermal stresses to be distributed throughout a network of microcracks where energy is expended opening and/or extending microcracks, leaving the bulk ceramic body intact.

Zirconia-toughened alumina was invented almost 15 years ago. However, commercial success has been limited, partly due to the failure of industry to produce a low-cost ZTA with improved properties and its failure to identify markets allowing immediate penetration. One exception has been the use of ZTA in some cutting-tool applications.

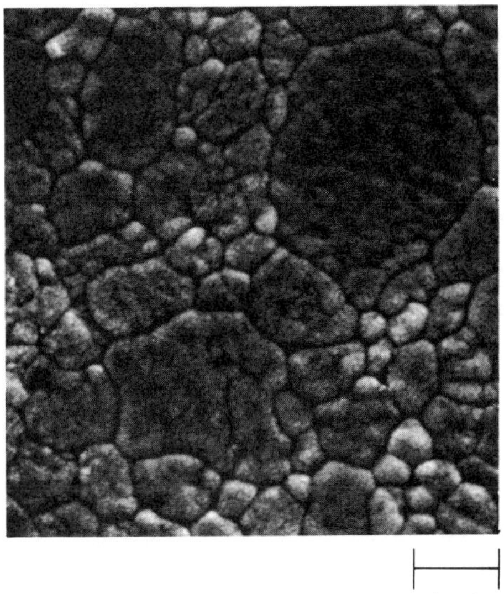

Fig. 5 Scanning electron micrograph of a Y-TZP sample. The larger 3 to 5 μm grains are cubic (~5%); the smaller 1 to 2 μm grains are tetragonal (~95%).

Table 4 Typical physical properties of various ceramics

Material	Bulk density, g/cm³	Flexure strength MPa	Flexure strength ksi	Fracture toughness MPa√m	Fracture toughness ksi√in.	Hardness, GPa	Hardness, 10^6 psi	Elastic modulus GPa	Elastic modulus 10^6 psi
ZTA	4.1–4.3	600–700	87–101	5–8	4.6–7.3	15–16	2–2.3	330–360	48–52
Mg-PSZ	5.7–5.8	600–700	87–101	11–14	10–13	12	1.7	210	30
Y-TZP	6.1	900–1200	130–174	8–9	7.3–8.2	12	1.7	210	30
Alumina-SiC	3.7–3.9	600–700	87–101	5–8	4.6–7.3	15–16	2–2.3	430–380	62–55
Silicon nitride-SiC	3.2–3.3	800–1000	116–145	6–8	5.5–7.3	15–16	2–2.3	300–380	43–55

Transformation-toughened zirconia is a generic term applied to stabilized zirconia systems in which the tetragonal symmetry is retained as the primary zirconia phase. The four most popular tetragonal phase stabilizers are CeO_2, Y_2O_3, CaO, and MgO. The use of these four additives results in two distinct microstructures. MgO- and CaO-stabilized ZrO_2 consist of 0.1 to 0.25 μm tetragonal precipitates within 50 to 100 μm cubic grains. Firing usually occurs within the single cubic-phase field, and phase assemblage is controlled during cooling.

Interest in CaO-stabilized ZrO_2 has waned in recent years. MgO-stabilized ZrO_2 (Mg-PSZ), on the other hand, has enjoyed immense commercial success. Its combination of moderate-high strength of 600 to 700 MPa (87 to 100 ksi), high fracture toughness of 11 to 14 $MPa\sqrt{m}$ (10 to 13 $ksi\sqrt{in.}$), and flaw tolerance enables the use of Mg-PSZ in the most demanding structural ceramic applications. The elastic modulus is approximately 210 GPa (30 × 10^6 psi), and the hardness is approximately 12 to 13 GPa (1.7 to 1.9 × 10^6 psi). Among the applications for this material are extrusion nozzles in steel production, wire-drawing cap stands, foils for the paper-making industry, and compacting dies. Among the toughened or high-technology ceramic materials, Mg-PSZ exhibits the best combination of mechanical properties and cost, for room- and moderate-temperature structural applications.

Yttria-stabilized ZrO_2 (Y-TZP) is a fine-grain, high-strength, and moderate-high fracture toughness material. High-strength Y-TZPs are manufactured by sintering at relatively low sintering temperatures (1400 °C, or 2550 °F). Nearly 100% of the zirconia is in the tetragonal symmetry and the average grain size is approximately 0.6 to 0.8 μm. The tetragonal phase in this microstructure is very stable. Higher firing temperatures (1550 °C, or 2800 °F) result in a high-strength (1000 MPa, or 145 ksi), high fracture toughness (8.5 $MPa\sqrt{m}$, or 7.7 $ksi\sqrt{in.}$), fine-grain material with excellent wear resistance. The microstructure (Fig. 5) consists of a mixture of 1 to 2 μm tetragonal grains (90 to 95%) and 4 to 8 μm cubic grains (5 to 10%). The tetragonal phase in this microstructure is more readily transformable than above due to the larger tetragonal grain size and a lower yttria content in the tetragonal phase, resulting in a tougher material.

Among the applications for Y-TZP are ferrules for fiber-optic assemblies. Material requirements include a very fine-grain microstructure, grain-size control, dimensional control, excellent wear properties, and high strength. The fine-grain microstructure and good mechanical properties lend the Y-TZP as a candidate material for knife-edge applications, including scissors, slitter blades, knife blades, scalpels, and so forth. However, compared to Mg-PSZ, Y-TZP is more expensive, has a lower fracture toughness, and is not nearly as flaw tolerant.

There are some temperature limitations in these materials. Mechanical strength of both Mg-PSZ and Y-TZP may start to deteriorate at temperatures as low as 500 °C (930 °F). Also, the Y-TZP ceramic is susceptible to severe degradation at temperatures between 200 to 300 °C (400 to 570 °F).

Composite Ceramics. The early success of ZTA and partially-stabilized zirconia systems provided the impetus to include toughened ceramics as a candidate for structural applications. However, due to the limited maximum-temperature use of these materials, intense research was generated to determine other toughening mechanisms (besides transformation toughening and dispersed-phase toughening) and alternative toughened-ceramic systems.

Silicon carbide whisker (SiC_w)-reinforced alumina surfaced in the last decade as a potential ceramic-engine component material. Composed of fine equiaxed alumina grains and needlelike SiC whiskers, this material exhibited promising fracture toughness (6.5 $MPa\sqrt{m}$, or 5.9 $ksi\sqrt{in.}$) and strength (600 MPa, or 87 ksi) properties. Al_2O_3-SiC_w composites have been used quite successfully in cutting-tool applications. These composites may also overcome the severe obstacles that currently prevent the use of ceramic materials in some aluminum can tooling applications.

Conventional processing methods can be employed provided the whisker loading is less than approximately 8 vol%. Composites with higher whisker loadings must be hot pressed, or sufficient liquid-glass-phase sintering must occur to fabricate fully dense bodies. The former limits the fired billet size and requires extensive grinding after sintering. The latter limits its high-temperature use.

Silicon Nitride Matrix Composites. High-temperature degradation of the mechanical properties of Al_2O_3-SiC_w composites and the excellent high-temperature strength, oxidation resistance, thermal shock resistance, and fracture toughness of Si_3N_4 caused a recent thrust of interest in fabricating SiC_w-reinforced Si_3N_4. The major phase, Si_3N_4, offers many favorable properties, and the SiC whiskers provide significant improvement in the fracture toughness of the composite. Whisker-reinforced Si_3N_4 is now being touted as the material of choice for hot-section ceramic-engine components, although production is currently limited to laboratory or pilot plant-size fabrication. Processing difficulties, health issues, and raw material costs of all the SiC whisker-reinforced composites have lessened the industrial impact of these materials and may prevent widespread acceptance and use in the near future.

Future Directions and Problems

One of the primary disadvantages of ceramic materials is their brittle nature, characterized by a low fracture toughness. Although significant improvements have been made to increase the fracture toughness, brittleness continues to keep ceramics from more widespread use.

Ceramic Composites. One direction that shows promise is that of composite materials. For example, silicon carbide whiskers have been incorporated into an aluminum oxide matrix, resulting in a composite with greatly improved toughness. The toughening mechanism is probably a combination of whisker pullout and crack bridging, whereby the SiC whisker effectively resists crack propagation. Other types of ceramic-ceramic composites would include adding a transforming phase such as zirconia to a host matrix, allowing transformation toughening to improve the fracture toughness of ceramics.

Metal-Ceramic Composites (Cermets). Another class of composites is metal-ceramic composites (cermets). In this case, a ductile metal phase is incorporated into the brittle ceramic. In the event of a propagating crack, the crack interacts with the metal phase, and the metal then begins to plastically deform. This deformation absorbs energy, acting to increase the toughness of the composite. The development and commercial use of various metal-ceramic composites are the subject of the article "Cermets" in this Volume.

Processing. Another area of importance is the science and technology of ceramic processing, both from an economic and performance sense. Currently manufacturing ceramics is a labor- and capital-intensive industry, where products are often custom-made for customers. Manufacturers are continually striving to increase productivity and reduce costs, very often through intense process engineering and optimization.

Improved processing techniques should also enhance the performance of structural ceramic components, particularly with respect to reliability. Currently ceramics tend to be very flaw sensitive, in that the strength depends on the size of flaw in the microstructure. The flaw size in turn is usually determined by processing conditions. In most ceramics, conventional processing results in a fairly broad flaw size distribution, which yields a broad strength distribution. Since design engineers often need to know the average strength and strength deviation, a large standard deviation will limit the design strength of a component. Therefore, improved processing techniques should reduce the spread in strengths and allow greater opportunities for ceramics in structural applications.

REFERENCES

1. W.D. Kingery, H.K. Bowen, and D.R. Uhlmann, *Introduction to Ceramics*, 2nd ed., John Wiley & Sons, 1976
2. J.S. Reed, *Introduction to the Principles of Ceramic Processing*, John Wiley & Sons, 1988
3. *Advanced Technical Ceramics*, S. Somiya, Ed., Academic Press, 1989

Superconducting Materials

Introduction

D.C. Larbalestier, L.V. Shubnikov Professor of Superconducting Materials in the Applied Superconductivity Center and the Department of Materials Science and Engineering, University of Wisconsin-Madison

SINCE THE DISCOVERY of high-temperature superconductivity in 1986, pictures of the levitation of a magnet above a superconducting sheet have been widely published in both scientific and popular journals. Owing to the widespread distribution of levitation kits to high schools, many students have been able to play with this almost magical property of superconductors. In the six articles that follow, some of the details of how the superconducting state manifests itself in important classes of superconducting materials are described. Following a contribution entitled "Principles of Superconductivity," the manufacture, properties, and applications of various superconducting materials are addressed in the following articles:

- "Niobium-Titanium Superconductors" (the most widely used superconductor)
- "A15 Superconductors" (in which class the important material Nb_3Sn lies)
- "Ternary Molybdenum Chalcogenides (Chevrel Phases)"
- "Thin-Film Materials"
- "High-Temperature Superconductors for Wires and Tapes"

Even with this broad view, however, only a brief flavor of the breadth of the superconducting state and its applications can be given here.

At the beginning of the 1990s, the science and applications of superconductivity find themselves in an interesting state. A vigorous industry has grown up around the applications of low-temperature niobium-base superconductors. This includes a superconducting electronics industry and a substantial industry producing superconducting magnets. Few large laboratories are now without a superconducting magnet, whether used for physical property measurements, nuclear magnetic resonance (NMR) and other resonance experiments, or for investigations of superconductivity itself. Magnetic resonance imaging (MRI) magnets for the NMR imaging of the whole human body are installed in thousands of hospitals worldwide, and enormous magnet assemblies for particle accelerators and plasma fusion experiments have been built.

At present the largest superconducting device is the Tevatron, the 4.8 km (3 mile) circumference 1000 GeV proton accelerator at Fermilab near Chicago. This consists of about 1000 6 m (20 ft) long superconducting magnets. The proven success of this device was vital to the decision to construct the Superconducting Super Collider (SSC). The SSC, now beginning its construction phase near Dallas, will be about 80 km (50 miles) in circumference and contain about 10 000 20 m (65 ft) long superconducting magnets.

The great vitality of the superconducting community has been enormously enhanced by the amazing and very unexpected discovery in early 1986 by Bednorz and Muller of high-temperature superconductivity in the rare earth cuprates (Ref 1). A rapid phase of new discovery quickly produced several new classes of high-temperature superconductors (Ref 2). Enormously important issues of basic physics are posed by the existence of superconductivity at temperatures as high as 125 K and magnetic fields of greater than 50 T (500 kG). At the same time, the potential for applications is enormous. Before considering these issues further, however, it is instructive to go back to the beginning of superconductivity and trace the development of its technology. This overview will provide a foundation for understanding the basic science and potential applications of superconductivity.

Historical Development

The superconducting state was an unexpected outcome of the low-temperature researches of a group led by Kamerlingh Onnes at the University of Leiden (Holland) in 1911. Onnes discovered that mercury lost all resistance when cooled to about 4 K. Two years later, he came to Chicago to report to the third International Conference of Refrigeration (1913). At this time he reviewed the recent research of the Leiden group (Ref 3). This article is quite astonishing, and only extensive quotations can convey the breadth of Onnes's conception of the possibilities of the superconducting state. Onnes commences by describing his initial 1911 experiments on mercury and then proceeds to rapidly sketch whole segments of the technology of superconducting magnets:

> Mercury has passed into a new state, which on account of its extraordinary electrical properties may be called the superconductive state. . . . The behavior of metals in this state gives rise to new fundamental questions as to the mechanism of electrical conductivity.
>
> It is therefore of great importance that tin and lead were found to become superconductive also. Tin has its step-down point at 3.8° K, a somewhat lower temperature than that of the vanishing point of mercury. The vanishing point of lead may be put at 6° K. Tin and lead being easily workable metals, we can now contemplate all kinds of electrical experiments with apparatus without resistance. . . .
>
> The extraordinary character of this state can be well elucidated by its bearing on the problem of producing intense magnetic fields with the aid of coils without iron cores. . . . Theoretically it will be possible to obtain a field as intense as we wish by arranging a sufficient number of amperewindings round the space where the field has to be established. This is the idea of Perrin, who made the suggestion of a field of 100,000 gauss being produced over a fairly large space in this way. He pointed out that by cooling the coil by liquid air the resistance of the coil and therefore the electric work to maintain the field could be diminished. . . . In order to get a field of 100,000 gauss in a coil with an internal space of 1 cm radius, with copper as metal, and cooled by liquid air 100 kilowatt would be necessary. . . . The electric supply, as Fabry remarks, would give no real difficulty, but it would arise from the development of Joule-heat in the small volume of coil, the dimensions of which are measured by centimeters, to the amount of 25 kilogram-calories per second, which in order to be carried off by evaporation of liquid air would require about 0.4 liter of liquid air per second, let us say about 1500 liters of liquid air per hour. . . .

But the greatest difficulty, as Fabry points out, resides in the impossibility of making the small coil give off the relatively enormous quantity of Joule-heat to the liquefied gas. The dimensions of the coil to make the cooling possible must be much larger, by which at the same time the electric work and the amount of liquefied gas required becomes greater in the same proportion. The cost of carrying out Perrin's plan even with liquid air might be about comparable to that of building a cruiser. . . .

We should no more get a solution by cooling with liquid helium as long as the coil does not become superconductive.

The problem which seems hopeless in this way enters a quite new phase when a superconductive wire can be used. Joule-heat comes not more into play, not even at very high current densities, and an exceedingly great number of amperewindings can be located in a very small space without in such a coil heat being developed. A current of 1000 amp/mm^2 density was sent through a mercury wire, and of 460 amp/mm^2 density through a lead wire, without appreciable heat being developed in either. . . .

There remains of course the possibility that a resistance is developed in the superconductor by the magnetic field. If this were the case, the Joule-heat depending on this resistance would have been withdrawn. One of the first things to be investigated as soon as the appliances, which are arranged for making the projected researches on magnetism at helium-temperatures, will be ready, will be this magnetic resistance. We shall see that this plays no role for fields below say 1000 gauss.

The insulation of the wire was obtained by putting silk between the windings, which being soaked by the liquid helium brought the windings as much as possible into contact with the bath. The coil proved to bear a current of 0.8 ampere without losing its superconductivity. There may have been bad places in the wire, where heat was developed which could not be withdrawn and which locally warmed the wire above the vanishing point of resistance. . . .

I think it will be possible to come to a higher current density . . . if we secure a better heat conduction from the bad places in the wire to the liquid helium . . . in a coil of bare lead wire wound on a copper tube the current will take its way, when the whole is cooled to 1.5° K, practically exclusively through the windings of the superconductor. If the projected contrivance succeeds and the current through the coil can be brought to 8 amperes . . . we shall approach to a field of 10,000 gauss. The solution of the problem of obtaining a field of 100,000 gauss could then be obtained by a coil of say 30 centimeters in diameter and the cooling with helium would require a plant which could be realized in Leiden with a relatively modest financial support. . . . When all outstanding questions will have been studied and all difficulties overcome, the miniature coil referred to may prove to be the prototype of magnetic coils without iron, by which in future much stronger and at the same time much more extensive fields may be realized then are at present reached in the interferrum of the strongest electromagnets. As we may trust in an accelerated development of experimental science this future ought not to be far away.

What a description! Many of the points essential to the development of a proper magnet technology were sketched out by Onnes already in 1913. His vision of powerful magnets, the problem of heat removal from compact windings, the attractive economic feasibility of superconducting as opposed to resistive magnets, operation at current densities of 1000 A/mm^2 and temperatures down to 1.5 K—all of these are crucial aspects of our present superconducting magnet technology. Elsewhere in the same article he describes the melting of superconducting wires following an abrupt transition from the superconducting to the normal state and perhaps prefigures modern composite conductor manufacture by considering the properties of a resistive constantan wire coated with a superconducting layer of tin.

Returning to Onnes's own words, it is only the last sentence that strikes a false note. An accelerated progress to applications was reasonable to dream about—but it did not happen. The reason is clearly delineated in a footnote to Onnes's paper: The passage of the electric current through the superconducting wire easily produced a magnetic field of about 0.05 T (500 G), which though weak was sufficiently strong to quench the superconducting state of a type I superconductor such as lead, tin, or mercury. Sadly, more than 20 years passed before there was much understanding of this issue.

In the early 1930s the thermodynamic aspects of the superconducting-to-normal transition were established by Meissner and Ochsenfeld and by Gorter and Casimir (Ref 4). As we know now, the crucial step to applications would have been to identify how to make the transition from a (low-field) type I superconductor to a (high-field) type II superconductor (see the article "Principles of Superconductivity" in this section for an explanation of these terms). This work was in fact underway.

The systematic effects of alloying lead with indium, tellurium, and similar solutes were carried out by Shubnikov's group in Kharkov (Ukrainian Republic) in the period 1935 to 1937, and the basic thermodynamic aspects of the transition—the appearance of a lower (H_{c1}) and an upper critical field (H_{c2}) in the alloys in place of the single small critical field (H_c) of pure lead—were all identified. These were crucial observations. Tragically, Shubnikov's work remained unappreciated by the scientific community as a whole. In 1937 Shubnikov was falsely denounced and sent to a labor camp, dying in prison in 1945. As a political prisoner his work could not be cited by his fellow Soviet scientists, and the scanty accounts of his early researches that had appeared in the Western literature were ignored. An alternative erroneous hypothesis—the filamentary sponge model, which inherently regarded the high-field superconducting properties as being associated with microscopic metallurgical inhomogeneities—was then used to explain the occasional reports of high-field superconductivity (Ref 4).

Further advances had to wait until the 1950s when Soviet theoreticians Ginzburg and Landau addressed the phenomenology of the superconducting-to-normal transition in a magnetic field. It is very striking to recall that even then, Ginzburg and Landau rejected as unphysical those solutions that predicted type II superconductivity. It was left to a persistent student of Landau's (Abrikosov) to explore the "unphysical" type II state. This work was published only in 1957 (Ref 5).

It took the serendipitous experiments of Kunzler *et al.* in late 1960 (Ref 6) to convince the experimentalists that type II superconductivity could indeed realize Onnes's 1913 dreams. Kunzler's experiment showed that a prototype wire of Nb_3Sn could carry a supercurrent of more than 10^5 A/cm^2 in a field of 8.8 T (88.8 kG). Compared to copper, which might operate (resistively) at 10^3 A/cm^2, the advantages of superconductors for high-field magnets became widely appreciated. A rapid advance to applications proceeded during the 1960s, culminating in a wide range of applications in both high-magnetic-field and electronic devices (Ref 7-9).

During the 1960s, 1970s, and 1980s there was a continued interest in the search for new superconductors. However, higher T_c materials were hard to find. The A15 compound Nb_3Sn held the record of 18 K in 1960, and no advance beyond the 23 K of Nb_3Ge was obtained after 1973. In extensive reviews of the field in 1986 (75 years after the discovery of Onnes), virtually no attention was paid to the prospect of developing materials having higher T_c (Ref 7-9).The community had run out of collective ideas.

Fortunately, however, one group at least, that of Muller in Switzerland, was still pursuing higher T_c materials. After several years of unsuccessful efforts, their researches were crowned with success. A mixed-phase ceramic of La-Ba-Cu-O exhibited a T_c onset of about 40 K (Ref 1). Within a very short time T_c had been raised to 92 K ($YBa_2Cu_3O_{7-x}$), 110 K ($Bi_2Sr_2Ca_2Cu_3O_x$), and 125 K ($Tl_2Ba_2Ca_2Cu_3O_x$) (Ref 2). Because major expectations for exciting new physics and applications lie with these materials, we confidently expect that the next edition of *Metals Handbook* will require a comprehensive rewrite of the present introduction.

REFERENCES

1. J.G. Bednorz and K.A. Muller, *Z. Phys.*, Vol B64, 1986, p 189
2. J.C. Philips, *Physics of High T_c Superconductors*, Academic Press, 1989
3. H. Kamerlingh Onnes, *Comm. Physical Lab Leiden Suppl.,* No. 34b, 1913
4. T. Berlincourt, Type II Superconductivity: Quest for Understanding, *IEEE Trans. on Magn.*, Vol 23 (No. 2), March 1987, p 403-412
5. A.A. Abrikosov, *Sov. Phys. JETP*, Vol 5, 1957, p 1174
6. G. Kunzler, Recollection of Events Associated With the Discovery of High Field-High Current Superconductivity, *IEEE Trans. Magn.*, Vol 23 (No. 2), March 1987, p 396-402
7. *Phys. Today, Spec. Issue: Supercond.*, Vol 39 (No. 3), March 1986, p 22-80
8. Kamerlingh Onnes Symposium on the Origins of Applied Superconductivity—75th Anniversary of the Discovery of Superconductivity, *IEEE Trans. on Magn.*, Vol 23 (No. 2), March 1987, p 354-415
9. *Superconducting Devices*, S. Ruggiero and D. Rudman, Ed., Academic Press, 1990

Principles of Superconductivity

William H. Warnes, Department of Mechanical Engineering, Center for Advanced Materials Research, Oregon State University

SINCE ITS DISCOVERY in the early 1900s, superconductivity has been found in a wide range of materials, including pure metals, alloys, compounds, oxides, and organic materials (see Fig. 1 and Table 1). Superconductivity is by no means a rare phenomenon, as there are several hundred superconducting materials known today (Ref 1). The following sections will provide a basic introduction to the principles of superconductivity. Due to the necessarily limited nature of this article, the reader is referred for more information to the large number of excellent texts available in the field:

- References 2 and 3 for a general overview of superconductivity and its applications
- Reference 4 for a midlevel introduction to the theory of superconductivity
- Reference 5 for a comprehensive survey of filamentary superconductors in magnet applications
- References 6 and 7 for technical information on superconducting materials and applications
- References 8 and 9 for an advanced treatment of superconductivity theory

The breadth of this article is further restricted to focus primarily on the principles of superconductivity as they relate to applications. As a result, details of the quantum theory and thermodynamics of superconductivity will be largely left to the references. The few equations that are described use the International System of Units (SI).

The primary physical property of the superconducting state is the complete disappearance of electrical resistance (see Fig. 2) on lowering the temperature below a critical temperature (T_c) (see Table 1). For all superconductors presently known, the critical temperatures are well below room temperature, and they are usually attained by cooling with liquified gases, either at or below atmospheric pressure. The two most common of these coolants are liquid helium and liquid nitrogen (see Table 2).

That the resistance of superconducting materials is (within experimental resolution) zero has been shown by measurements of electrical currents flowing in superconducting loops (Ref 11). Sensitive measurements of the continuously circulating electrical currents after periods of several weeks have shown no measurable decay of the supercurrents, yielding a resistive decay time scale of greater than 10^5 years.

Zero electrical resistance is not the only hallmark of superconductivity. A superconducting material must also exhibit perfect diamagnetism, that is, complete exclusion of an applied magnetic field from the bulk of the superconductor (see Fig. 3). The Meissner effect (also known as Meissner-Ochsenfeld effect) (Ref 12) occurs because circulating supercurrents are induced to flow in a thin sheath at the surface of the superconductor. These currents generate a magnetic field opposing the external field and summing to zero field inside the superconductor. Because these surface currents do not have infinite current density, the external field penetrates the superconductor over the thickness of the sheath. This characteristic distance is called the magnetic penetration depth, $\lambda(T)$ (see Fig. 4), and is a function of temperature. Values of $\lambda(T)$ for several materials at 4.2 K are given in Table 1.

The perfect diamagnetism of superconducting materials implies that the superconducting state will cease to be thermodynamically stable when the magnetic field is large enough. The thermodynamic critical field (H_c) is therefore defined by the difference in the volumetric Gibbs free energies of the normal and superconducting states. This difference is called the condensation energy, and it equals the energy density of the excluded field:

$$G_n(T) - G_s(T) = \mu_0[H_c(T)]^2/2 \qquad \text{(Eq 1)}$$

Application of a magnetic field larger than H_c will destroy the superconducting state.

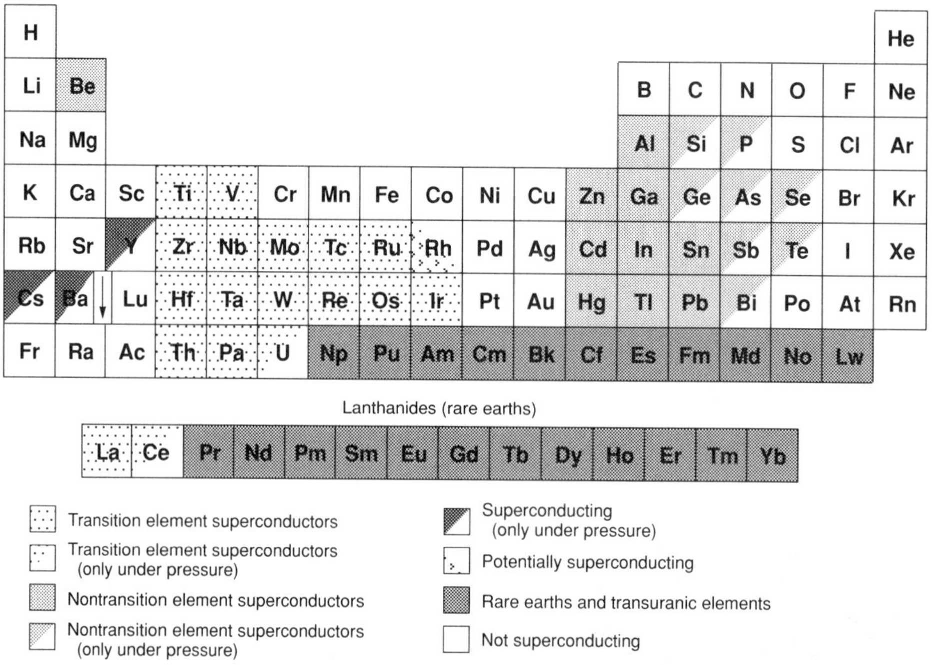

Fig. 1 Periodic table of the elements showing the large number of elements known to have superconducting transitions

Table 1 Approximate superconducting properties of selected superconducting materials

Material	Type	Critical temperature, T_c at 0 T	Parameters at 4.2 K: Thermodynamic critical field, T, at $\mu_0 H_c$	$\mu_0 H_{c1}$	$\mu_0 H_{c2}$	Magnetic penetration depth (λ), nm	Coherence length (ξ), nm	Critical current density (J_c), kA · mm^{-2}
Pb	I	7.3	0.0803(a)	...	...	40	83	...
Nb	II	9.3	0.37	0.25	0.41	30	40	...
Nb45-50-Ti	II	8.9–9.3	0.16	0.009	10.5–11.0	500	10	3 (at 5 T)
Nb_3Sn	II	18	0.46	0.034	19–25	200	6	10 (at 5 T)
Nb_3Ge	II	23	0.16	0.004	36–41	650	4	10 (at 5 T)
NbN	II	16–18	0.16	0.004	20–35	600	5	10 (at 0 T)
$PbMo_6S_8$	II	14–15	0.4	0.005	40–55	240	4	0.8 (at 5 T)
$YBa_2Cu_3O_7$	II	92	0.5	0.05(b)	60(b)	150(b)	15(b)	1 (at 77 K, 0 T)(d)
			0.03	0.01(c)	>200(c)	1000(c)	2–3(c)	

(a) Thermodynamic critical field at 0 K. (b) Measured with field parallel to the *c*-axis. (c) Measured with field parallel to the *a-b* plane. (d) Epitaxial thin film, current in the *a-b* plane

Theoretical Background of Superconductivity

In 1950, Ginzburg and Landau developed a phenomenological theory of superconductivity (Ref 13) invoking a macroscopic quantum mechanical wave function or order parameter (ψ) for which $|\psi(x)|^2 = n_s$, where n_s is the density of superconducting electrons. The minimum distance over which n_s may change significantly (for example, at a normal metal/superconductor boundary, n_s changes from 0 to 1) defines the temperature-dependent coherence length, $\xi(T)$ (see Table 1). In addition, the spatial extent of ψ suggests that superconductivity is a cooperative phenomenon between the conduction electrons extending over significant distances, comparable to the sample dimensions. The long-range ordering of electrons is responsible for the Josephson effects discussed below.

Theoretical work by Bardeen, Cooper, and Schrieffer (BCS) in 1957 (Ref 14) showed that superconductivity could be well described by pairs of conduction electrons of opposite momenta coupling together through a weak attractive interaction. The pair interaction produces a gap in the energy levels of the electrons and allows a

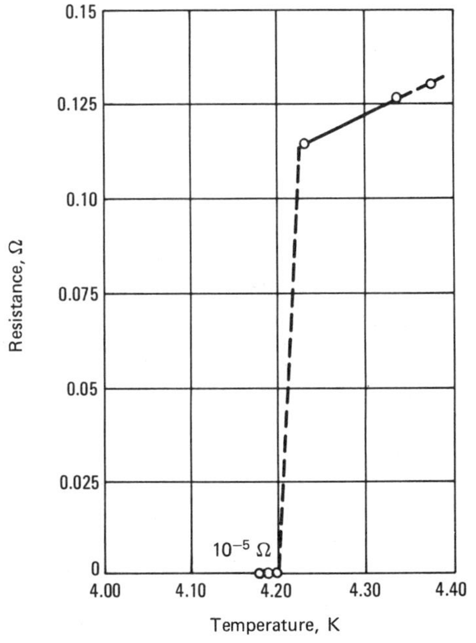

Fig. 2(a) Electrical resistance as a function of temperature for superconductivity discovered in mercury by Kamerling Onnes in 1911. Source: Ref 10

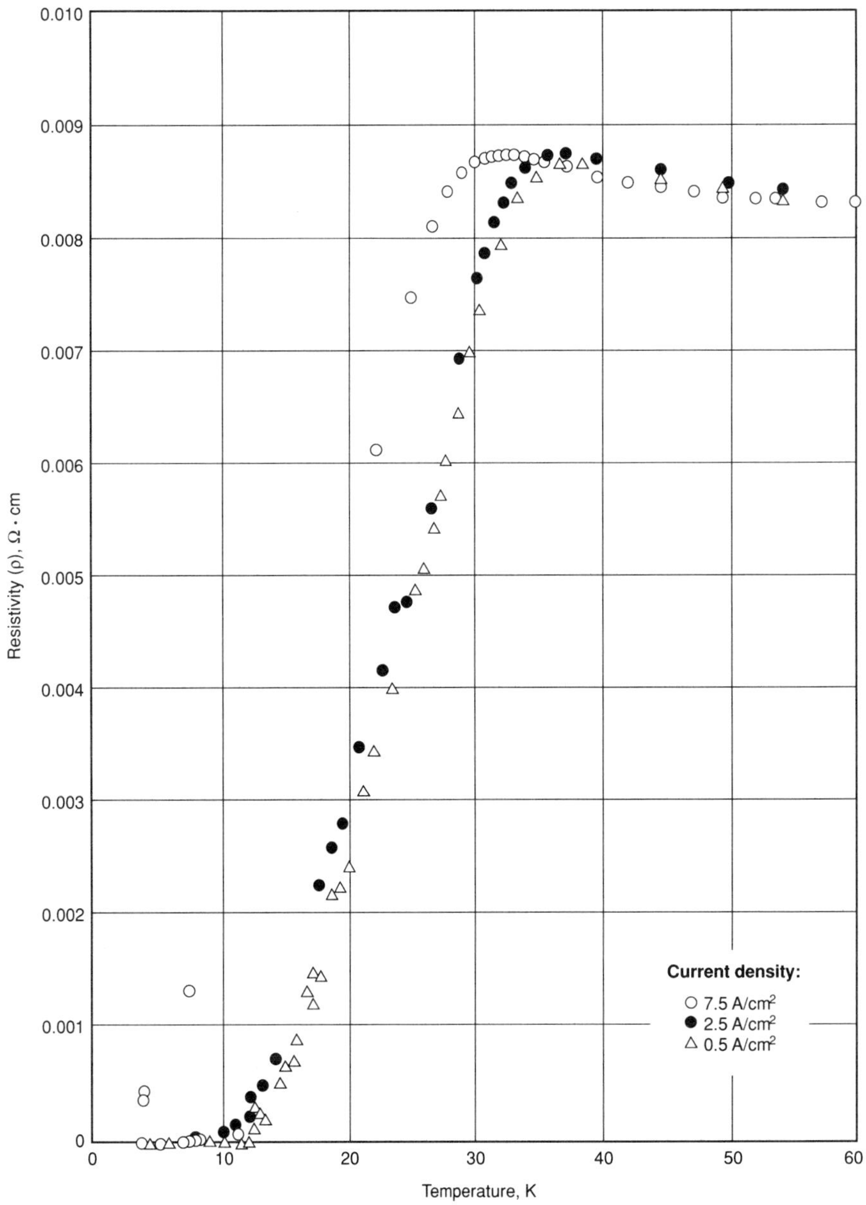

Fig. 2(b) Electrical resistance as a function of temperature for the first high-temperature ceramic (oxide-containing barium) superconductors discovered by Bednorz and Muller in 1986. Source: Ref 10

Table 2 Properties of selected cryogenic cooling fluids

Fluid	T_b at 760 mm Hg, K	Heat of vaporization, J/L × 10^3	Enthalpy at various temperatures, J/L, at 4.2 K	20.3 K	77.4 K	273 K	Cost, ($/L)
Helium	4.215	2.5	260	252	253	253	3.5–5.0
Hydrogen	20.39	31.5	...	959	422	346	0.8–2.0
Nitrogen	77.36	160.6	...	...	356	354	0.4–1.0

net reduction in the free energy of the superconductor by forming electron pairs. The Frohlich electron-phonon interaction (Ref 15) provides a mechanism for attractive interaction between the electrons by coupling them through the exchange of virtual phonons. BCS theory and the electron-phonon mechanism have been very successful in describing many of the experimental results of superconductivity, including the size of the energy gap (Ref 16) and the isotope effect (Ref 17, 18).

Type I and Type II Superconductors. Abrikosov, also in 1957 (Ref 19), showed that the Ginzburg-Landau theory predicted two distinct behaviors for superconductors in an applied magnetic field (see Fig. 5), depending on the value of the dimensionless Ginzburg-Landau parameter, $\kappa(T)$. The $\kappa(T)$ is defined as the ratio of the two characteristic lengths, $\lambda(T)/\xi(T)$, and is only weakly dependent on temperature (see Table 1). When $\kappa(T)$ is small ($<1/\sqrt{2}$), there is a positive surface energy between regions of superconducting and normal phases. This positive surface energy is responsible for the exclusion of magnetic flux from the bulk superconductor because bulk penetration of the magnetic field would produce a normal phase region containing the magnetic field

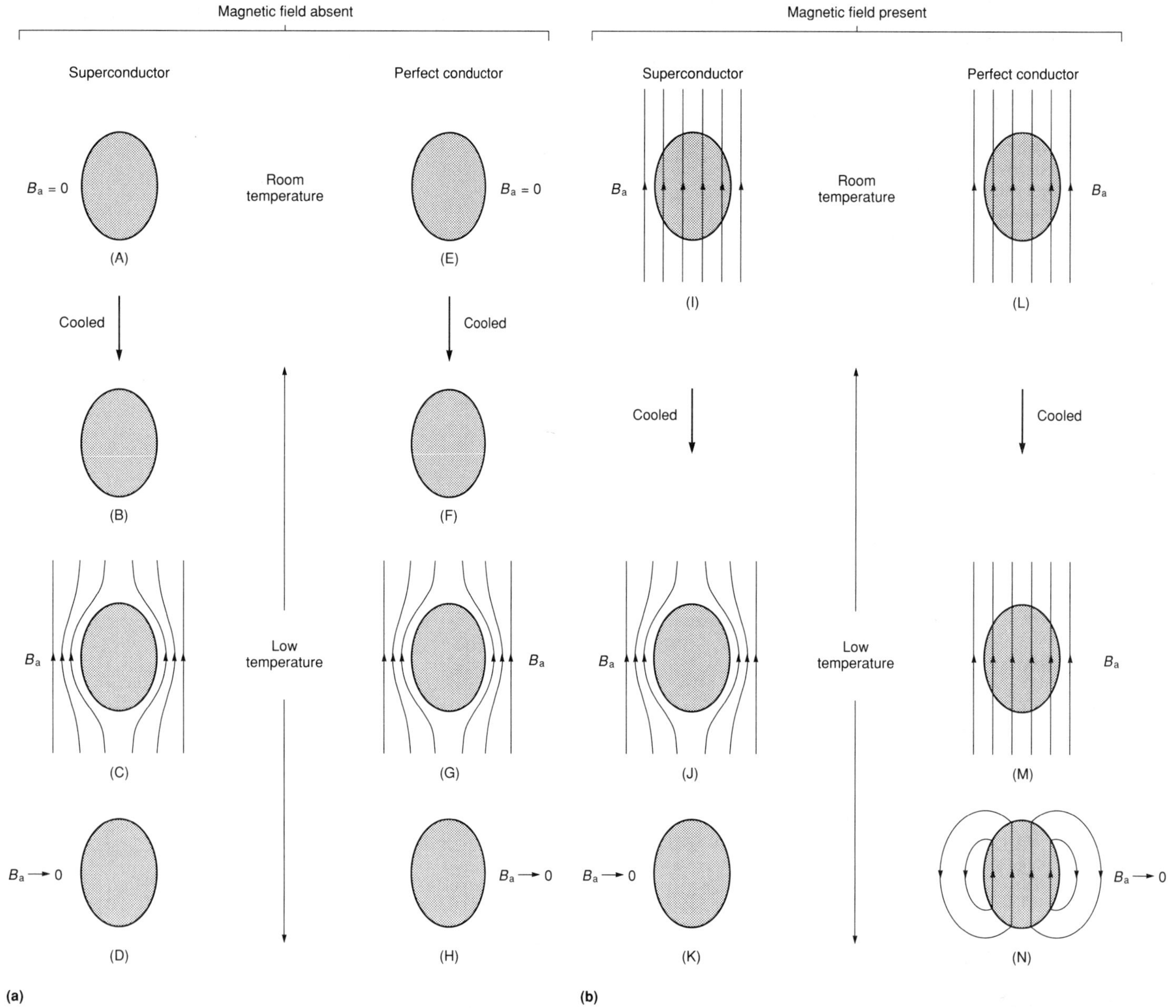

Fig. 3 Comparison of the magnetic behavior of a superconductor to that of a perfect conductor in the presence or absence of an external magnetic field (B_a) when cooled to below the transition temperature. (a) When cooled without being subjected to the magnetic field (A and B) and (E and F), both conductors exhibit exclusion of an applied magnetic field (C and D) and (G and H). (b) When cooled in the presence of a magnetic field (I) and (L), the superconductor excludes the magnetic field, called the Meissner effect (J and K), whereas the perfect conductor traps the field (M and N).

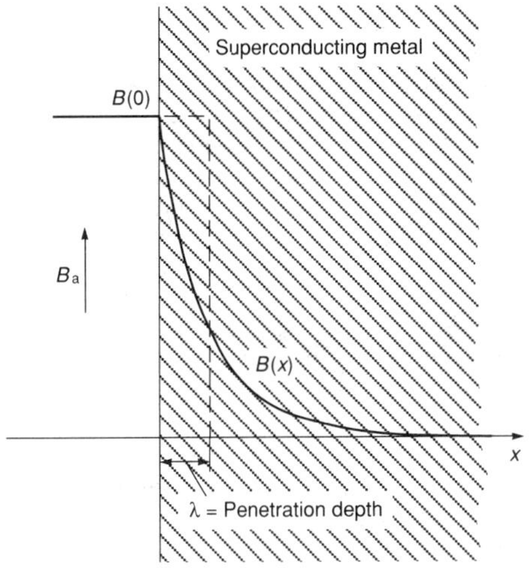

Fig. 4 Currents flowing within a thin sheath at the surface of a superconductor preventing the external applied magnetic field from entering the bulk. The thickness of the current sheath, and the distance over which the magnetic field decays is called the penetration length (λ).

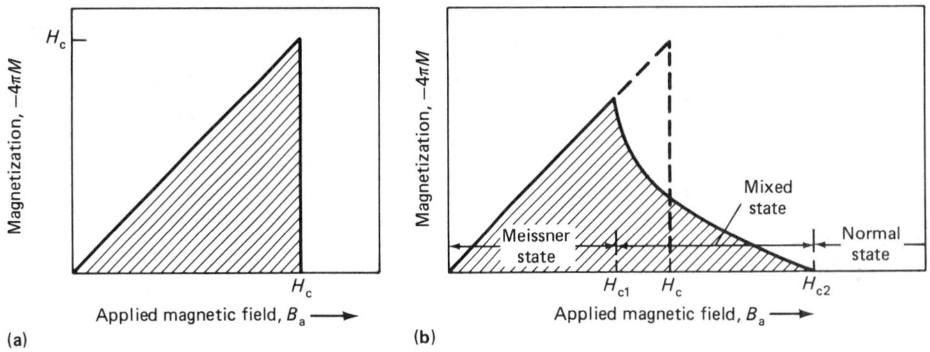

Fig. 5 Plot of magnetization versus applied magnetic field for two classifications of bulk superconductors. (a) Type I. This type exhibits a complete Meissner effect (perfect diamagnetism). The internal field (given by $B = H - 4\pi M$) is zero. Above H_c the material is a normal conductor, and the magnetization is too small to be seen on this scale. (b) Type II. The applied field begins to enter the sample at a field H_{c1} that is lower than H_c. Superconductivity persists in the mixed state up to a high field of H_{c2}, above which the material is a normal conductor. For a given value of H_c, the area under the magnetization curves is the same for both conductors.

within the superconductor. This would require an increase in the free energy equal to the surface energy of the normal-superconducting boundary and therefore is not thermodynamically stable. Materials with $\kappa(T) < 1/\sqrt{2}$, comprising most of the elemental superconductors, are called type I superconductors.

Type II superconductors, for which $\kappa(T) > 1/\sqrt{2}$, have a negative surface energy between normal and superconducting phases. Type II superconductors, consisting of the alloy superconductors, exhibit perfect diamagnetism up to a lower critical magnetic field (H_{c1}), which is smaller than H_c. Below H_{c1}, type II materials show identical magnetic behavior to type I materials. As the magnetic field is increased above H_{c1}, the overall free energy is reduced by creating superconducting/normal phase boundaries, allowing the magnetic field to enter the bulk of the superconductor. The field enters as flux quanta (Φ_0), the smallest unit of magnetic flux, creating a large superconducting/normal phase boundary area (see Fig. 6). The overall reduction in free energy allows type II superconductors to maintain the superconducting state to much larger values of the applied magnetic field before the free-energy balance favors the normal state. Stability of the superconducting state in magnetic fields up to the upper critical magnetic field (H_{c2}) allows type II materials to be exploited for high-magnetic-field applications.

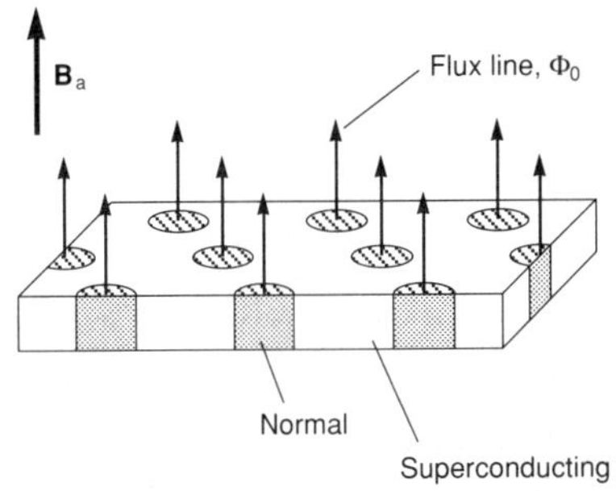

Fig. 6 The magnetic flux line lattice predicted by Abrikosov for type II superconductors in the mixed state. The field enters as individual units of magnetic flux (the flux quantum, Φ_0) in a triangular array. The areal density of the flux lines is equal to the internal magnetic field.

When the applied magnetic field is between H_{c1} and H_{c2}, the superconductor is said to be in the mixed state. The number density of magnetic flux quanta within the superconductor is then determined by the internal field, $B_i = n\Phi_0$, where n is the number per unit area.

Critical Parameters of Superconductivity

Superconductivity can be destroyed not only by large magnetic fields or high temperatures, but also by passing an electric current through the superconductor that is larger per unit cross sectional area than the critical current density (J_c). The J_c is measured in units of A/mm^2. A current density less than J_c will flow in the superconductor with no resistance and thereby result in no power loss or ohmic heating. Current densities larger than J_c produce a voltage loss in the superconductor, generating heat and eventually raising the temperature above T_c. The critical temperature (T_c) and the critical magnetic fields (H_c, H_{c1}, and H_{c2}) are material properties for a given material or composition; they are not affected to any large extent by changes in processing or microstructure. However, within a single material the J_c may vary over several orders of magnitude, and it is very strongly affected by metallurgical microstructure and defect distribution. This provides an opportunity for control of the J_c through appropriate materials processing (Ref 20, 21).

Applications of superconductivity may be broken into two categories: high-magnetic-field and low-magnetic-field applications (Ref 22). High-field applications require that superconductors carry large critical current densities. This is especially true for superconducting magnets and generators (Ref 23). The applied research effort in high-field superconductivity is therefore primarily focused on increasing the J_c. Low-field applications include flux shields, transmission lines, Josephson devices, and resonant cavities; these are primarily limited by T_c and H_c.

The three critical parameters of temperature, magnetic field, and current density are closely interdependent. For example, the H_{c2} decreases with increasing temperature or current. These three parameters define a three-dimensional thermodynamic phase field, within which the superconducting state is stable (see Fig. 7). Design and operation of superconducting devices must keep in mind the overall shape of the phase surface, in order to provide margins of safety in all three parameters. Operating conditions for superconducting magnets are often at temperatures of $T \leqq 0.5\ T_c$.

Many high-field materials have been found to obey a scaling law behavior (Ref 24), with:

$$J_c(T,H) = C[H_{c2}(T)]^n f(h) \quad \text{(Eq 2)}$$

where C is a materials dependent constant and $h = H_a/H_{c2}$ (see Fig. 8). The parameter n varies between 1.5 and 2.5 at low temperatures, while the function $f(h)$ is found to be approximately $h(1 - h)$ for Nb-Ti materials (Ref 25, 26), and $h^{1/2}(1 - h)^2$ for Nb_3Sn (Ref 27).

Measurements of T_c, H_{c2}, and J_c can be made by either resistive or magnetic methods (Ref 28). A typical resistive measurement involves passing a small measuring current through the superconductor and recording the voltage along the superconductor as a function of temperature, magnetic field, or current density. The transition

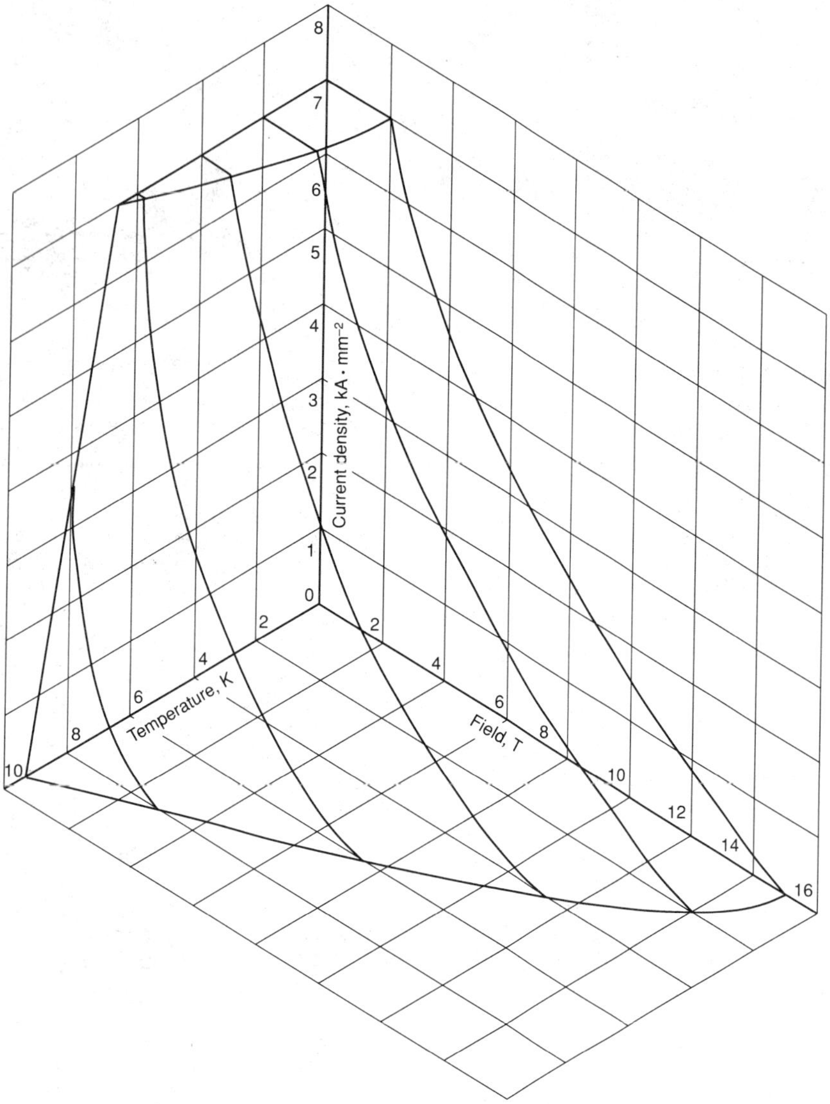

Fig. 7 The critical surface for a niobium-titanium alloy. As long as the state of the superconductor remains within the critical surface, it will be superconducting. The strong interdependence of the three critical parameters (T_c, H_{c2}, and J_c) is clearly seen.

from the normal to the superconducting state can occur over a very small range of temperatures, fields, or current densities in ideal samples (for example, $\Delta T_c < 10^{-5}$ K for a carefully prepared niobium standard (Ref 29). However, in many type II materials the phase transitions are much broader, and thus the location of the superconducting phase boundary is not precise (see Fig. 2 and 9).

Flux Pinning

Abrikosov (Ref 19) was the first to show that the Ginzburg-Landau theory predicted type II superconductivity. He showed that the flux quanta would be arranged in a periodic triangular lattice when the applied magnetic field is $H_{c1} \leqq H_a \leqq H_{c2}$. This periodic magnetic structure is called the Abrikosov, or flux line, lattice (see Fig. 10).

An isolated flux line can be modeled as a cylindrical core of normal-phase material of radius ξ containing a single unit of magnetic flux (Φ_0) surrounded by a circulating shielding supercurrent of extent λ (see Fig. 11).

The passage of an electric transport current through the superconductor produces a Lorentz force between the current and the flux lines:

$$\mathbf{F}_L = \mathbf{J} \times \mathbf{B} \qquad \text{(Eq 3)}$$

(see Fig. 12). In an ideal homogeneous superconductor, the flux lines will move in the direction of the Lorentz force, causing a dissipation of energy due to their viscous flow. The dissipation appears as a voltage in the direction of the current flow, producing a power loss in the superconductor and an accompanying heating. This flux motion is called flux flow, and the resistivity measured during flux flow is found to be proportional to $\rho_n H_a/H_{c2}$, where ρ_n is the normal state resistivity and H_a is the applied field (Ref 32).

By introducing microstructural inhomogeneities to the superconductor (for example, second-phase precipitates, inclusions, voids, dislocation tangles, or grain boundaries), the flux lines can be effectively pinned against the Lorentz force (Ref 33). The basic interaction force between a single flux line and a single pinning center can be viewed as follows: Although the negative surface energy of the superconducting/normal boundary allows the flux line to enter the superconductor, there is still an energy penalty paid to create a flux line equal to the condensation energy times the flux line volume. This increased energy is needed to convert the core of the flux line to the normal state.

If the flux line were positioned on a nonsuperconducting volume defect, such as a void or normal precipitate, the energy necessary to turn the core normal would be saved, and a lower free energy would result. This type of flux pinning, called the core interaction, is the primary source of pinning in two-phase alloys such as niobium-titanium (Ref 34). Grain-boundary pinning described in the model by Zerweck (Ref 35) is thought to be responsible for the high critical currents in Nb_3Sn and other single-phase type II superconductors. Other basic pinning forces include pinning by minority superconducting phases, by magnetic interactions, and by elastic interactions with the strain fields surrounding inclusions and precipitates (Ref 36).

By including pinning centers in the microstructure, the superconductor is capable of carrying substantial currents in an applied magnetic field with no voltage loss or power dissipation. The critical current density is given by equating the maximum Lorentz force to the maximum pinning force:

$$\mathbf{F}_{L_{max}} = \mathbf{J}_c \times \mathbf{B} \qquad \text{(Eq 4)}$$

Flux pinning theory provides an insight into the scale of inhomogeneities required to produce large critical current densities. Volume pinning centers should be about 2ξ in diameter to optimally match the flux line size. In addition, there should be a large number density of pinning centers, spaced by approximately the flux line lattice spacing:

$$a_0 = 1.075(\Phi_0/B)^{1/2} \qquad \text{(Eq 5)}$$

Within an order of magnitude, these dimensions are obtainable and have been observed in high-current-density materials (see Fig. 13).

Flux line motion of any kind produces resistive-type losses in the superconductor. In addition to flux flow losses, which occur at high Lorentz forces, flux lines can be thermally activated to move out of the pinning potential of the pinning centers at current densities appreciably below J_c. This thermally activated flux motion is called

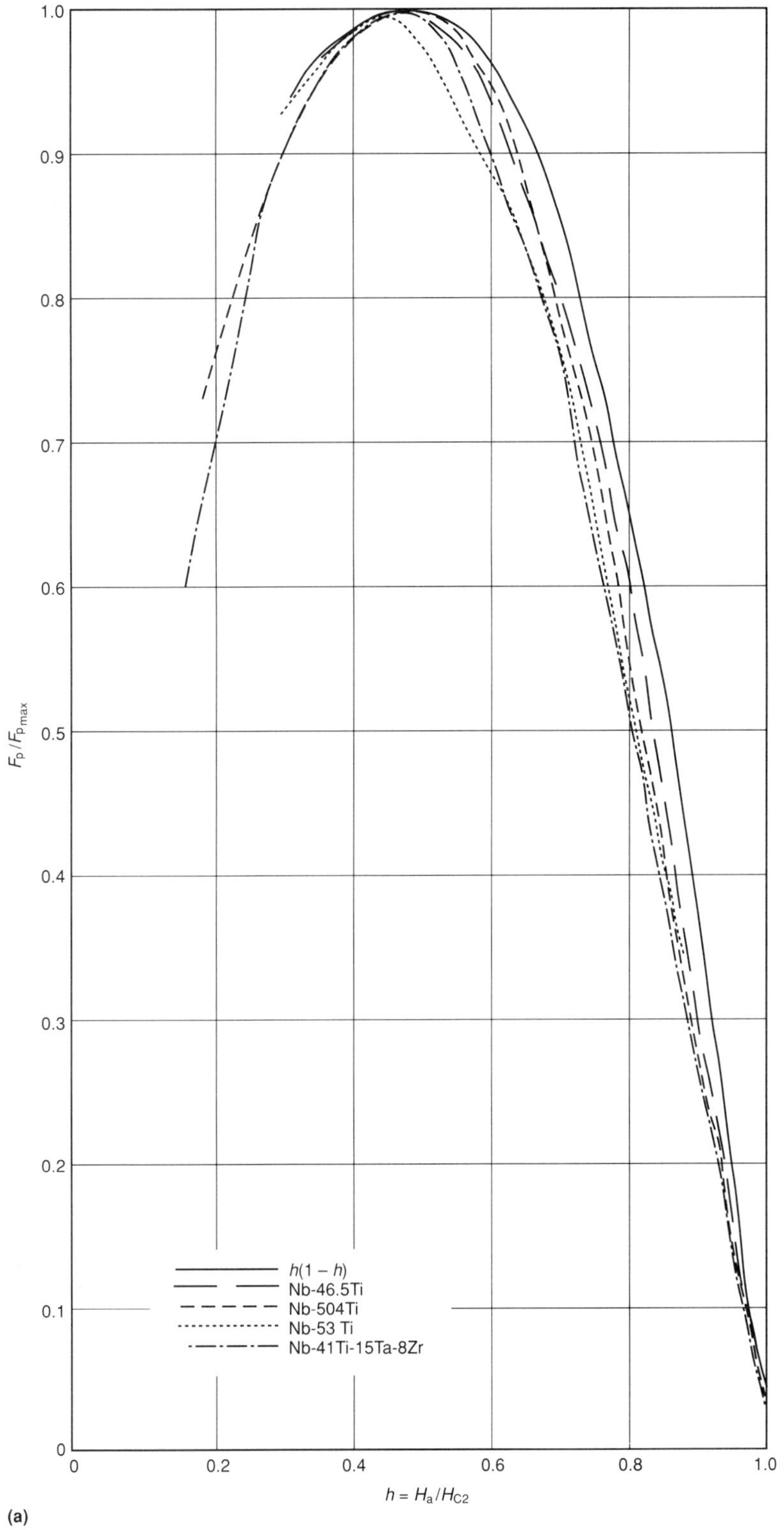

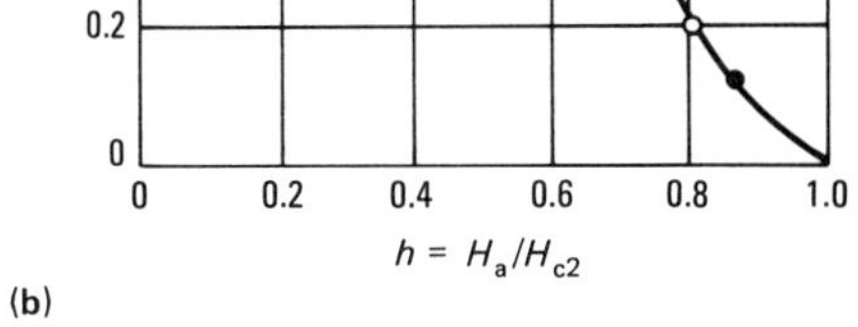

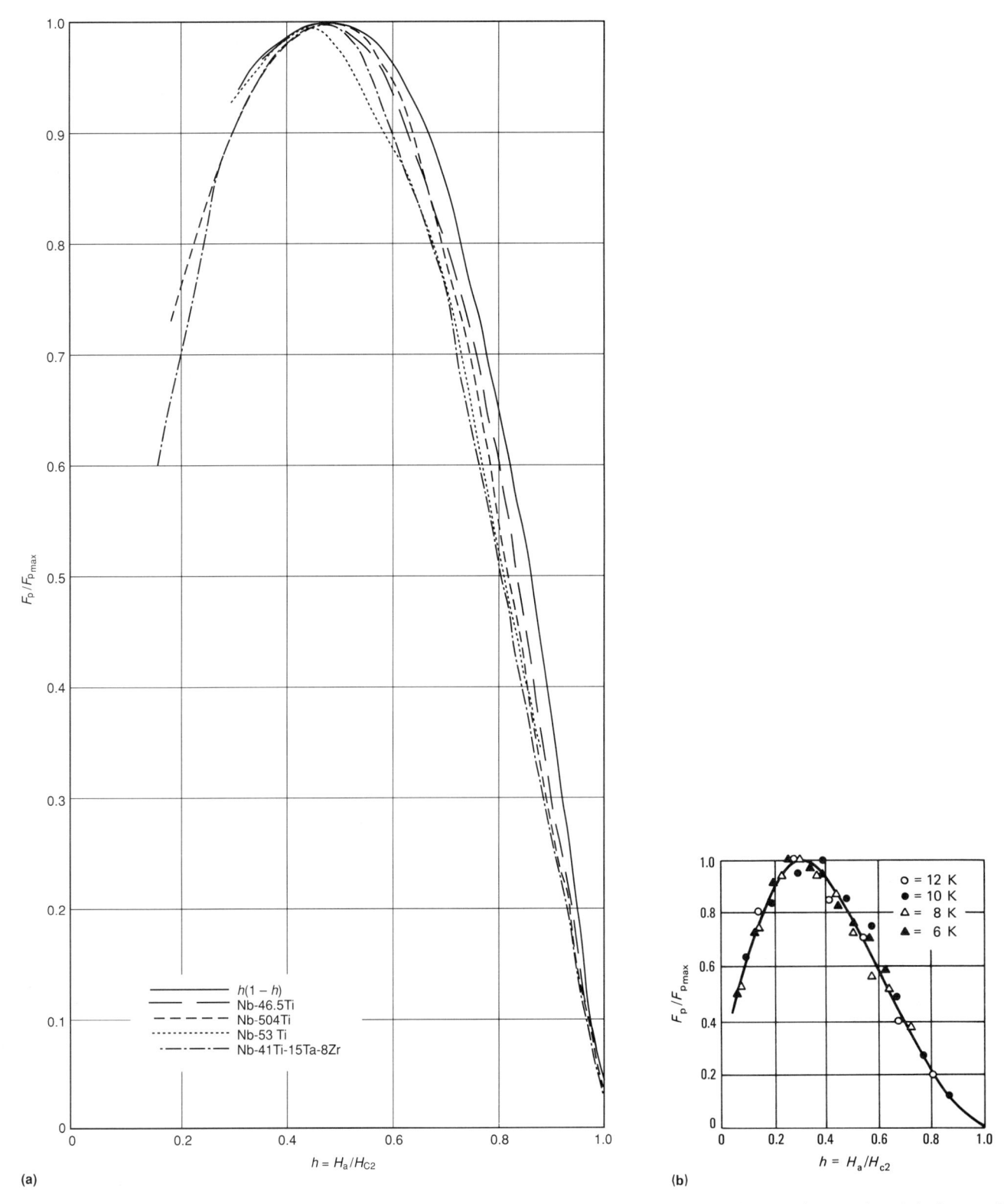

Fig. 8 Scaling law behavior of the critical current density (J_c) for (a) several niobium-titanium alloys (Ref 25) and (b) a Nb_3Sn conductor (Ref 26). In both cases, $F_p = J_cB$ is plotted, scaled by the maximum value versus the reduced applied magnetic field, $h = H_a/H_{c2}$. The niobium-titanium alloys show an $h(1 - h)$ dependence, whereas the Nb_3Sn exhibits an $h^{1/2}(1 - h)^2$ dependence.

flux creep; it was originally proposed by Anderson (Ref 37). For high-field materials at low temperatures, flux creep is not generally a significant problem. However, as the T_c increases (for example, in high-temperature ceramic materials), the thermal energy available to promote flux creep becomes important, producing resistive losses at current densities well below J_c (Ref 38).

Magnetic Properties

The field region below H_c for type I and below H_{c1} for type II materials is called the Meissner state and exists because of the sur-

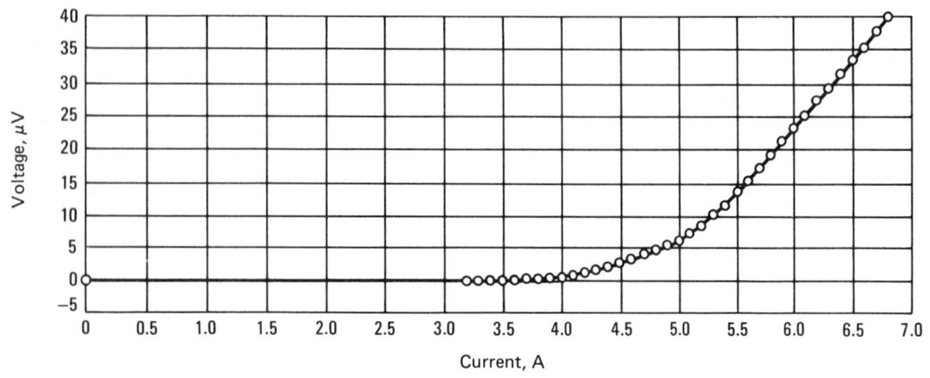

Fig. 9 Broadened critical current transition measured resistively for a niobium-titanium wire. Source: Ref 30

face-energy barrier to flux entry. The exclusion of magnetic flux from the superconductor (except within λ of the surface) in this field range suggests the possibility of using superconductors as flux shields to provide magnetic-field free volumes. Both type I (Ref 39) and type II (Ref 40) superconductors have been used for this purpose. These applications are somewhat limited, however, owing to the generally low values of H_c and H_{c1}.

In type II materials, the critical current density plays an important role in determining the magnetization behavior. The reversible magnetization curve of Fig. 5 is seldom approached in real materials; a hysteretic behavior is more generally found (Fig. 14). Irreversibility of the magnetization as a function of the field is caused by metallurgical defects pinning the flux lines and restricting their movement in and out of the sample. The magnetic hysteresis (ΔM) is therefore proportional to the J_c (Ref 41). The relationship between J_c and magnetization hysteresis may be given as:

$$\Delta M(H) = \mu_0 J_c(H) d \quad \text{(Eq 6)}$$

where d is the diameter of the superconductor. The magnetization of type II superconductors is analogous to the magnetization behavior of magnetic materials such as iron, with the exception that in superconductors it is a diamagnetic rather than a paramagnetic effect. Measurements of ΔM have been used extensively as a means of determining the J_c (Ref 42), especially with samples with a size or shape that renders transport measurements difficult.

Performing a field cycle of 0-H-0 will leave a magnetic field trapped inside type II superconductors that depends on the value of H, d, and the J_c. The magnetic field caused by the magnetization of the superconductor produces a distortion in the transport-current-generated magnetic fields

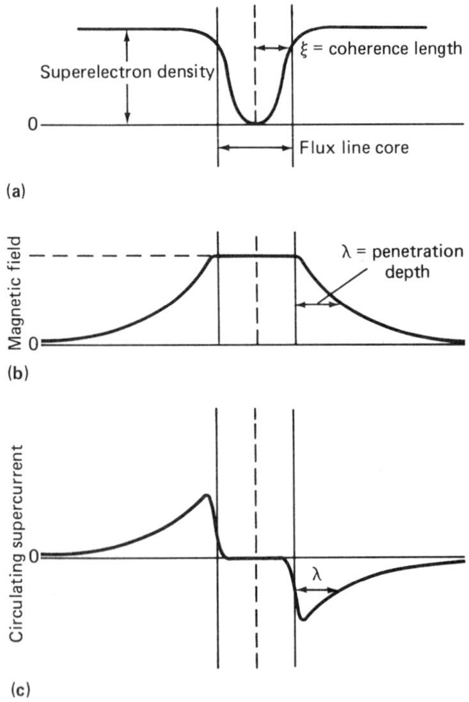

Fig. 11 Model of a single flux line considered as a single unit of magnetic flux, $\Phi_0 = 2 \times 10^{-15}$ Wb, filling a cylindrical volume of radius ξ, the coherence length. (a) The superelectron density rises to its maximum value within about ξ of the core of the flux line. (b) The magnetic field falls off over the distance of the penetration depth (λ). (c) The magnetic field in the core ($\Phi_0/\pi\xi^2$) is generated by circulating supercurrents flowing within λ of the core.

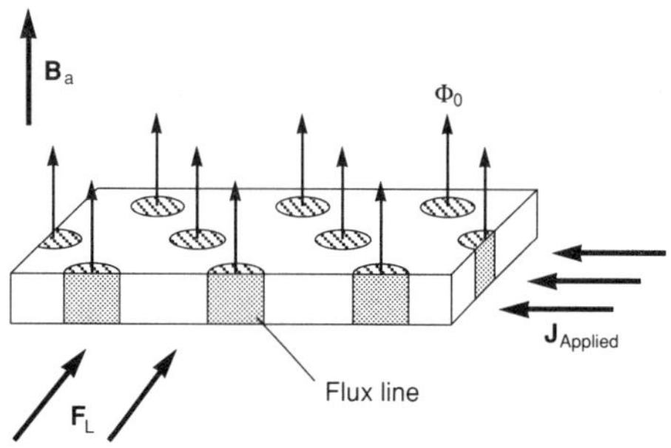

Fig. 12 Transport current density (**J**) flowing through the superconductor. The flux lines experience a reactive force given by the Lorentz force equation, $\mathbf{F}_L = \mathbf{J} \times \mathbf{B}$. In the absence of flux pinning, the Lorentz force will cause the flux lines to flow in a direction perpendicular to both the transport current and the applied field, creating a voltage dissipation in the direction of the current.

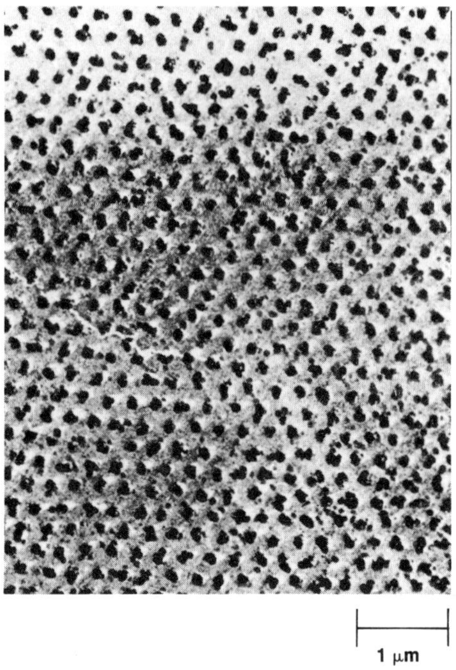

Fig. 10 Triangular flux line lattice in a lead-indium alloy type II superconductor. Small ferromagnetic particles are attracted to the points of high-field density in the core of the flux lines. The flux line positions are seen using a replica in the transmission electron microscopy (TEM). Source: Ref 31

and is a significant problem in designing superconducting magnets of high-field quality, for example, accelerator and magnetic resonance imaging (MRI) magnets (Ref 43).

Since the discovery of high-temperature superconductors in 1986 (Ref 10), the demonstration of magnetic flux exclusion causing a small magnet to levitate above a liquid nitrogen superconductor has become commonplace (see Fig. 15). This experiment was originally performed at 4.2 K using superconducting lead (Ref 44), but is now more commonly performed at liquid nitrogen temperatures with ceramic superconductors. The flux expulsion responsible for levitation with the liquid nitrogen superconductors is not only due to the Meissner effect (because the material is in the mixed state), but is also due to the effect of the flux pinning and the critical current density (Ref 45). Any application involving diamagnetic flux expulsion for levitation (for example, levitated bearing surfaces) will require materials with large J_c values.

Stabilization

A major consideration in operating superconducting devices near their critical surface is stability of the superconducting state to small disturbances. The primary problem is that at the low temperatures necessary for superconductivity, the specific heat of materials is quite small (see Table 3). Even a small energy input will therefore cause a large increase in the temperature of the

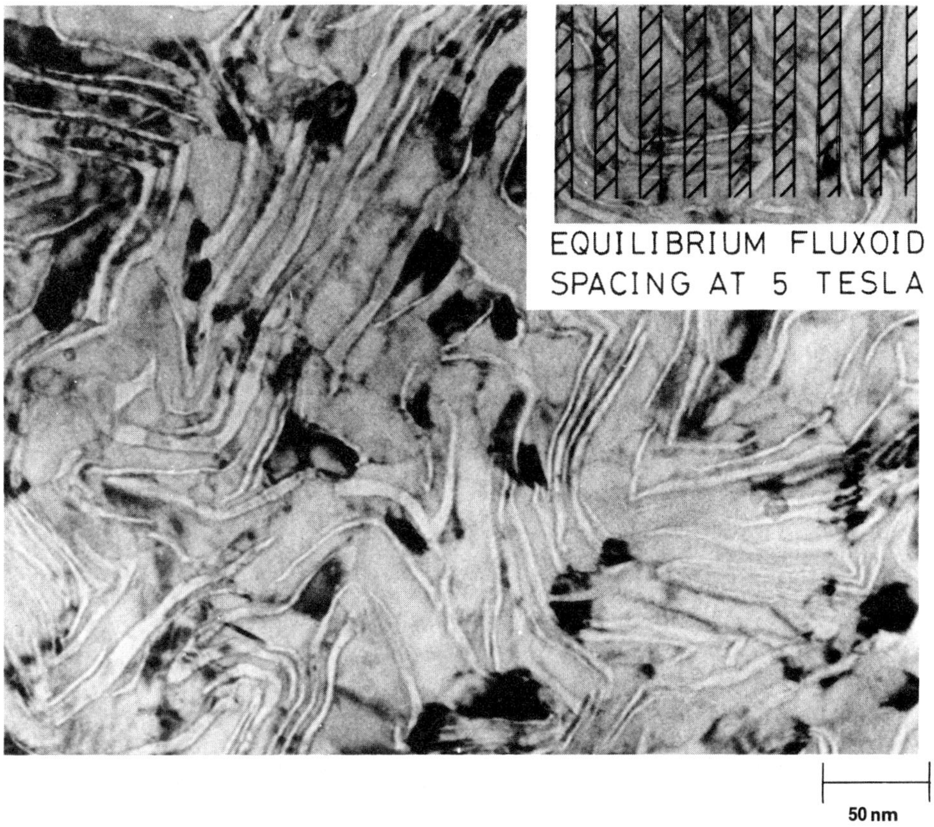

Fig. 13 Transverse cross section TEM photomicrograph of a portion of one filament of a Nb-46.5Ti composite wire. The light streaks are the α-Ti precipitates that are responsible for flux pinning through the core interaction. This wire has a large pinning force, with J_c = 3150 A/mm^2 at 5 T and 4.2 K. The scale of the α-Ti precipitates can be compared to the flux line lattice size and spacing in the inset. Courtesy of Peter Lee, University of Wisconsin, Applied Superconductivity Center

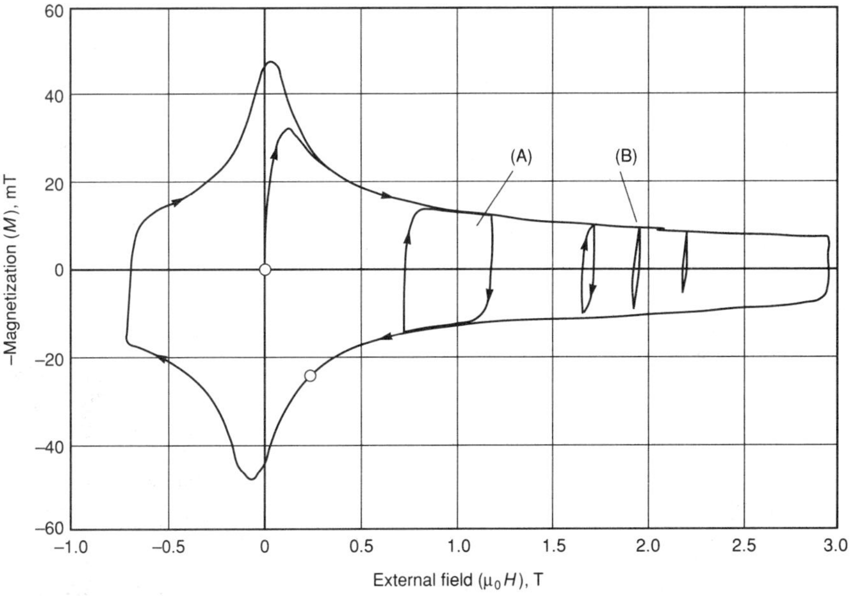

Fig. 14 Hysteretic magnetization of a multifilament niobium-titanium composite wire due to the trapping of magnetic flux by flux pinning centers. At low fields (A) where the J_c is highest, the hysteresis loops are larger than at high fields (B).

superconductor. The temperature increase lowers the critical current density, which changes the magnetic field profile in the superconductor. The flux motion provides a heat input, leading to a further temperature increase and a run-away transition to the

Fig. 15 Levitation of a high-field permanent magnet above a high-T_c superconductor at liquid nitrogen temperatures. The exclusion of magnetic flux by the superconductor due to flux pinning defects creates a magnetic pressure between the magnet and the superconductor that opposes the gravitational force.

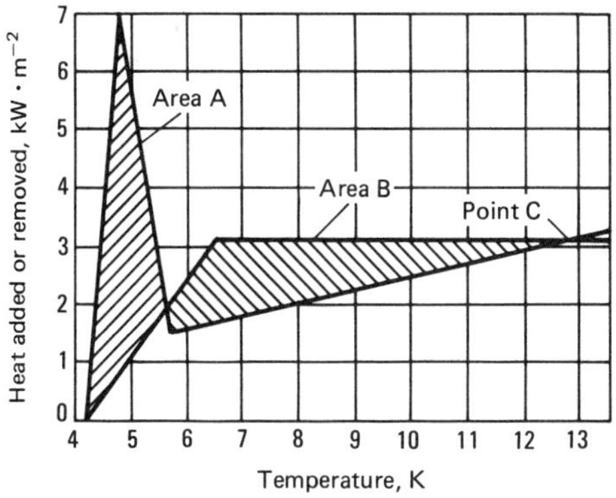

Fig. 16 Equal-area condition for cryogenic stability of a superconductor. (A) is the heat transfer from the superconductor to liquid helium, and (B) is the heat generated in the superconductor by a local disturbance. As long as the area under the cooling curve (A) is greater than the area under the generation curve (B), the superconductor will recover. For this case, the superconductor will be stable for disturbances producing temperature increases as high as Point C, 13 K above ambient temperature.

normal state. Two major sources of transient-energy input in high-field magnets are mechanical disturbances (for example, due to wire movement under the magnetic hoop stresses) (Ref 46), and flux jumping (Ref 47).

The goal in stabilization of superconductors is to prevent a localized disturbance from growing. This can be accomplished in several ways. The simplest, and least satisfactory, is operating the device with enough margin to avoid crossing the critical surface. In this case, the superconductor will be stable if the heat is conducted away from the localized disturbance more quickly than it is generated.

Cryogenic Stability. For larger disturbances, which generate larger heat inputs, it becomes necessary to put the coolant in intimate contact with the superconductor. The criterion for cryogenic stability was determined by Stekly and Zar (Ref 48) and, simply stated, requires the cooling power available in the cryogen to be larger than the

Table 3 Specific heats of selected materials at various temperatures for an applied field of 0 T

Material	Specific heat, J/g · K, at 4.2 K	77 K	300 K
Nb-Ti	6.3×10^{-4}	0.14	...
Cu	1.1×10^{-4}	0.19	0.38
Al	3.0×10^{-4}	0.34	0.90
Steel (AISI 305)	1.9×10^{-4}	0.20	0.42

heat generated by the disturbance (see Fig. 16). To assist in transporting the heat to the cryogen, the superconductor is generally surrounded by a high-thermal-conductivity normal metal such as copper or aluminum.

In addition to higher thermal conductivity, the normal metal matrices provide a high-electrical-conductivity parallel current path. Because the normal-state resistivity (ρ_n) of the superconductor is quite high, the resistive heat input during the disturbance is greatly reduced, allowing recovery from a larger disturbance.

The primary drawback to designing devices that are cryogenically stable is that surrounding the superconductor with coolant significantly dilutes the block current density, thereby reducing the achievable magnetic field. Most large-scale magnets rely heavily on cryogenic stability, however, and designs incorporating normal liquid helium (Ref 49), superfluid helium (Ref 50), and forced-flow liquid helium (Ref 51) have been tested.

Adiabatic Stability. A different stability issue is raised with the problem of flux jumps. A flux jump is the sudden movement of magnetic flux in the superconductor, causing a voltage and generating a local heating. The flux motion can come about in many ways. A slight temperature increase (due to mechanical motion, for example), or an increase in the applied magnetic field causes the local J_c to be reduced. Because the J_c is related to the flux gradient in the superconductor by Maxwell's equations ($\partial B/\partial x = \mu_0 J_c$), the flux must redistribute to match the new gradient. The flux movement generates heat, increasing the temperature, which in turn reduces the J_c still further. The temperature continues to rise, quickly leading to a normal zone in the superconductor.

The heat generated locally by a flux jump is given roughly by $\mu_0 J_c^2 a^2$ where a is the half thickness of the conductor. The condition for adiabatic stability against flux jumps is established by equating the heat generated during the flux jump to the heat capacity of the superconductor. If the superconductor size (a) can be made small enough (typically <50 μm, or 0.002 in.), the temperature will not rise above T_c and the superconductor will recover.

Dynamic Stability. The condition for stability against flux jumping is eased by allowing the heat to be carried away by the coolant. However, in the superconductor the thermal-diffusion time constant (τ_t) is much larger than the magnetic-diffusion time constant (τ_m) and the heat cannot be extracted from the superconductor during the flux jump. For copper and aluminum, τ_m is much larger than τ_t and the flux motion is slowed enough that, for superconductors of small size, the heat can be conducted away safely. This dynamic stability criterion yields a maximum size for the superconductor similar to that determined by the adiabatic criterion, although they arise from quite different mechanisms. In both cases, the stability is ensured if the superconductor is made small enough. This is one of the reasons for producing multifilamentary wires with fine superconducting filaments as discussed in the sections "Alternating Current Losses and RF Effects" and "Josephson Effects" in this article.

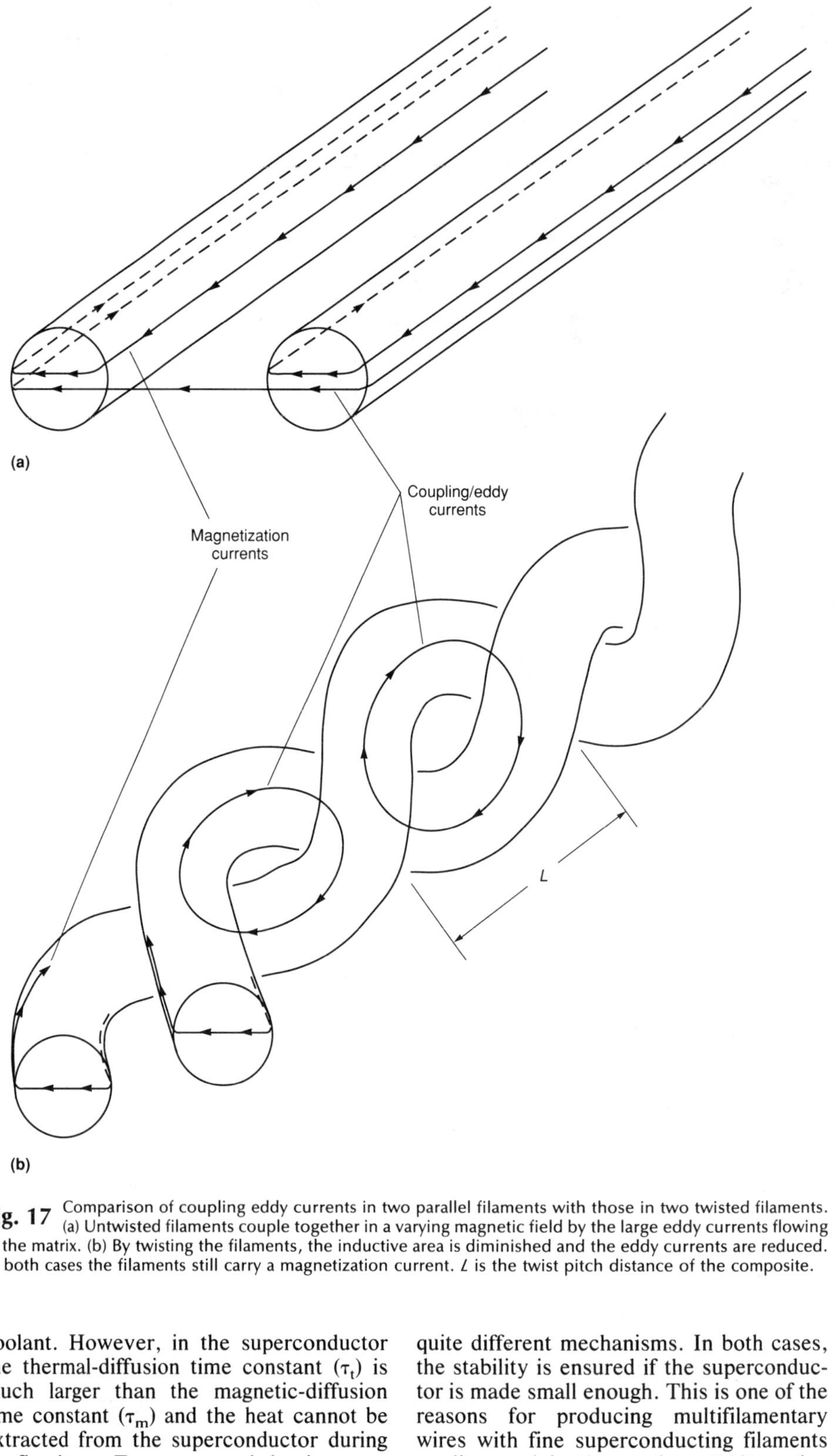

Fig. 17 Comparison of coupling eddy currents in two parallel filaments with those in two twisted filaments. (a) Untwisted filaments couple together in a varying magnetic field by the large eddy currents flowing in the matrix. (b) By twisting the filaments, the inductive area is diminished and the eddy currents are reduced. In both cases the filaments still carry a magnetization current. L is the twist pitch distance of the composite.

A third aspect of superconductor stability relates to the interfilament coupling of multifilamentary composites. Because the filaments are very close together in the composite (commonly <5 μm, or 200 μin.), a

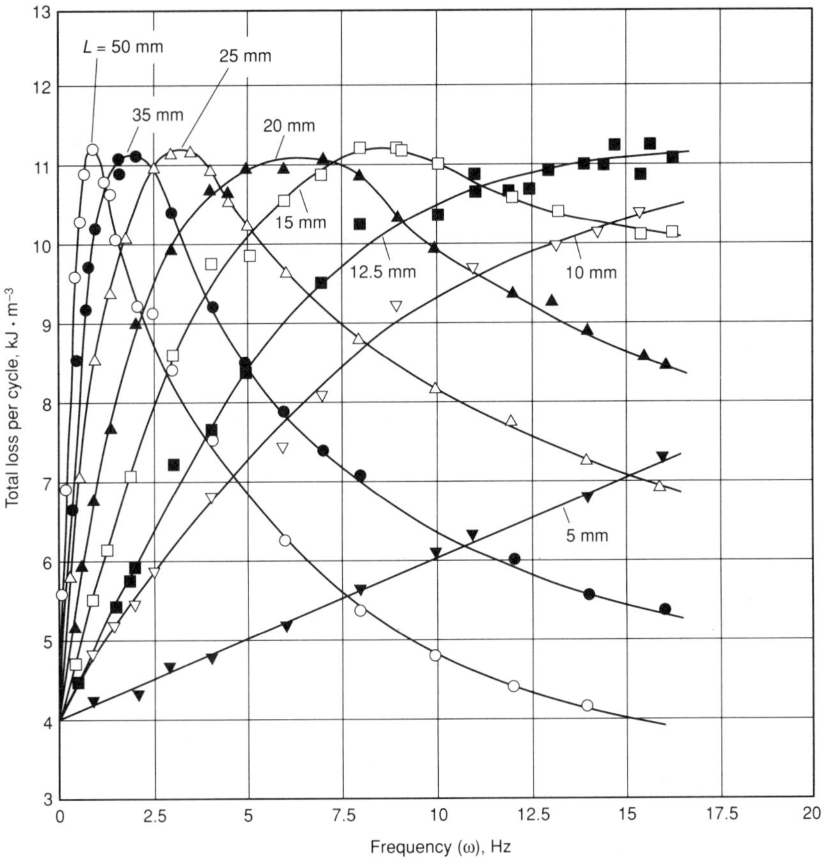

Fig. 18 Measurements of the ac loss in a twisted multifilamentary composite in a dc magnetic field with a small ac ripple. As L decreases, the magnitude of the ac losses decreases dramatically. At low frequencies, the only loss is due to the hysteresis loss. As the frequency increases, the eddy current losses go through a maximum. At high frequencies, the penetration losses become most important.

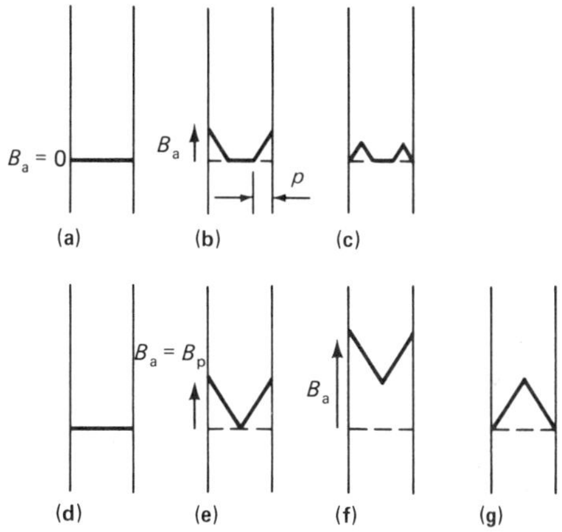

Fig. 19 The critical state model of flux penetration into a superconducting slab. As the applied field is raised from zero (a and b), the field penetrates the surface of the superconductor to a depth p. The gradient of the field ($\partial B/\partial x$), is equal to the critical current density (J_c). When the field is reduced to zero (c), magnetic flux remains trapped inside the superconductor. When a large magnetic field is applied from zero (d and f), the field fully penetrates the sample at the field $B_p = \mu_0 J_c d$. As the field is raised above B_p, the field profile changes across the entire sample (f). When the field is removed (g), the maximum trapped flux is left in the sample.

varying magnetic field will induce eddy currents to flow across the resistive matrix (see Fig. 17). The coupling causes the individual filaments to behave as a single large filament with a diameter that is nearly as large as that of the wire. The increased effective diameter exceeds the flux jump stability criteria, leading to degraded performance of the composite wire. By introducing a twist-pitch to the filament bundle during wire processing, the induced eddy currents can be made to cancel one another, effectively decoupling the filaments and restoring flux jump stability.

These stability criteria were originally developed for superconductors at liquid helium temperatures, but the basic phenomenon is not expected to be different in superconductors at liquid nitrogen temperatures. The primary difference will be in the beneficial effect of the larger specific heats of the device components (see Table 3) and the increased cooling capacity of liquid nitrogen. On the other hand, the increased resistivity and slower thermal diffusion of the normal metals at these higher temperatures will lower the stability margins. The probable effect will be an increase in the stability margins over those of superconductors at liquid helium temperatures (Ref 52).

Alternating Current Losses and RF Effects

Superconductors subjected to time-varying magnetic fields can experience significant losses and therefore power dissipation. Even magnets constructed for constant-field use must be ramped up and down during operation and are thus subject to ac losses. The time-varying magnetic field generates an electromotive force (emf) in the superconductor and therefore results in a resistive loss.

Alternating Current Losses

The power loss associated with time-varying fields can be broken into three components (see Fig. 18 and Ref 23):

- Hysteresis
- Penetration
- Eddy currents

Hysteresis losses occur because the movement of the applied magnetic field into and out of the superconducting filament must overcome the flux pinning, and the loss is therefore dependent on J_c. A change in the applied field of approximately $B_p = \mu_0 J_c d$ will penetrate fully to the center of a superconductor of a diameter d (see Fig. 19). For magnetic field changes less than B_p, the flux penetrates only an outer layer of the superconductor of thickness p, and the bulk of the superconductor is shielded from the changing field. Because flux motion takes place in only a small volume near the surface, the losses per cycle are small. As the applied field change increases, the volume swept by the flux in each cycle increases to a maximum at B_p, at which $p = d/2$. For field changes much larger than B_p, the flux fully penetrates the superconductor. Because the J_c falls as the applied field increases, the size of the magnetization hysteresis is reduced, and the losses are again small.

The picture is somewhat more complicated when a transport current is introduced into the superconductor, but the behavior for field changes larger than B_p is not very different. The hysteresis losses are independent of the frequency of the field variation and are proportional to $B_m J_c d$, where B_m is the amplitude of the field change. This suggests that hysteresis losses can be minimized by reducing the diameter of the superconductor. The partitioning of the superconductor into many small filaments in composite wires has been discussed above for the purpose of stability. Reduction of ac losses is another reason to design composites in this way.

Penetration losses are similar to hysteresis losses, in that they come from the restricted movement of flux, in this case, into and out of the composite as a whole. When the J_c is large, the outer filaments of a conductor may shield the inner filaments from the field change. The outer filaments

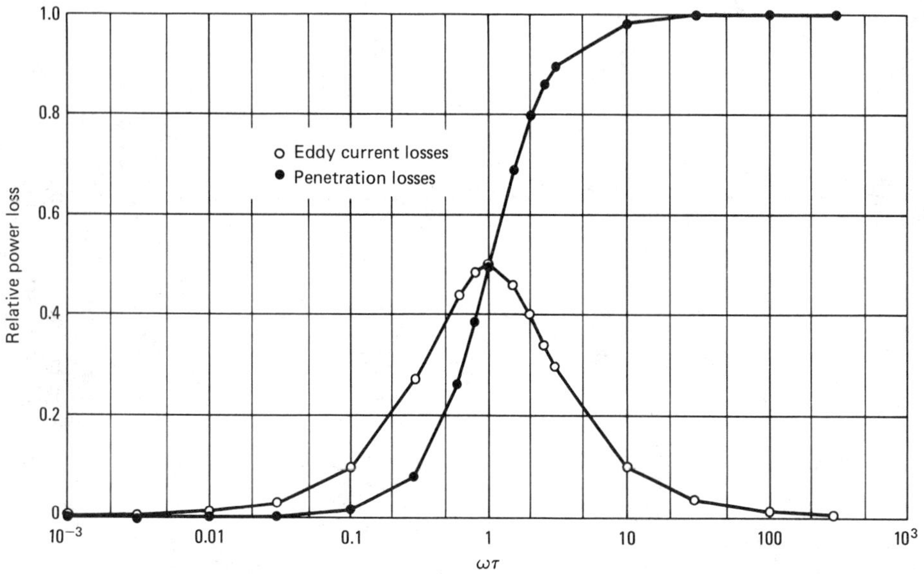

Fig. 20 The behavior of the penetration and eddy current losses as a function of $\omega\tau$, where ω is the frequency of the ac magnetic field and τ is the natural decay time constant of the induced eddy currents

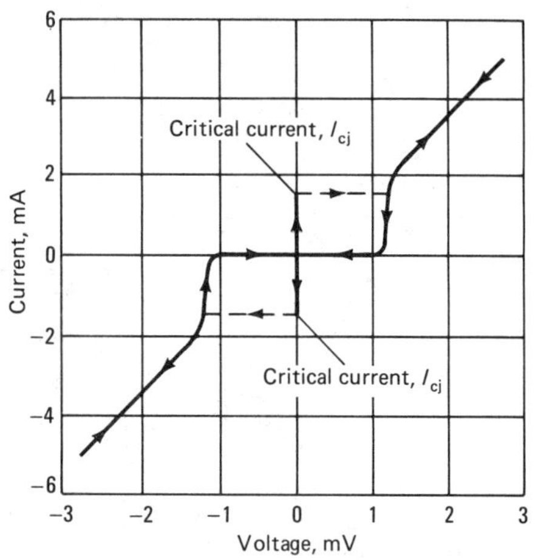

Fig. 21 Current-voltage characteristic produced by the dc Josephson effect. As long as the current through the junction is less than the critical junction current (I_{cj}), the voltage is zero. Note the hysteretic behavior on cycling the current.

couple together to act as a single filament with a diameter equal to that of the wire diameter; these losses become more important as the spacing between filaments is reduced. Unlike the hysteresis losses, the penetration losses are frequency dependent, similar to the dependence of skin depth on frequency in normal conductors. The penetration losses per cycle are proportional to $B_m^2(\omega^2\tau^2)/\omega^2\tau^2 + 1)$, where τ is the natural decay time for the induced currents; these losses become important only for large values of $\omega\tau$ (that is, high frequency) (see Fig. 20 and Ref 23).

Eddy current losses in multifilamentary composites can significantly increase the ac losses because currents are induced to flow in the resistive matrix surrounding the superconducting filaments. The induced eddy current will be proportional to the rate of change of the magnetic field, the resistivity of the matrix, and the effective size of the inductive loop formed by the superconducting filaments. The eddy current losses can be reduced by introducing a twist pitch to the filament bundle, which reduces the area of the loop coupling with the changing magnetic field. Once the filaments have been twisted, the eddy current loss per cycle is roughly $B_m^2\omega\tau/(\omega^2\tau^2 + 1)$. The only effective way to further reduce the eddy current losses at a given frequency is to reduce τ by increasing the matrix resistivity. This has been done for many ac applications (for example, by using copper-nickel alloys for the matrix material) (Ref 53).

Radio Frequency Losses

For high-frequency applications ($\omega > 10$ MHz), superconductors are attractive because of the very low values of surface resistance that they provide (Ref 54). The surface resistance of superconductors was studied by London (Ref 55), who developed the two-fluid model in which the total current at high frequencies is made up of a normal-electron current and a superelectron current. The normal current is concentrated within the classical skin depth of the surface, whereas the supercurrent flows within the penetration depth (λ). At low temperatures, the normal-electron mean free path can become quite large and the surface resistance becomes limited by the residual surface resistance, which is caused by the presence of impurities and defects in the superconductor. The superconducting surface resistance varies as ω^2, whereas the residual surface resistance varies as ω^n, where n varies between 1.5 and 2.0.

The exact source of all the residual resistance losses is still under investigation. However, the surface resistance of superconducting materials at frequencies of 1 GHz is a factor of 10^6 lower than conventional copper at room temperature. The small value of the surface resistance is what makes superconductors attractive for high-frequency applications. In particular, the Q value of a resonant circuit, which depends inversely on the resistance, can be extremely high for a superconducting circuit. Superconductors have found many applications in stable oscillators (with frequency stability of better than 1 part in 10^{12}), microwave generators and detectors, resonant cavities for particle accelerators, filters, tuners, and high-frequency transmission lines and field guides (Ref 56).

Josephson Effects

As mentioned above, the superconducting state can be described as a macroscopic order parameter, or wave function of electron pairs:

$$\Psi(x,t) = |\Psi(x,t)| \exp[i\phi(x,t)] \qquad \text{(Eq 7)}$$

where $|\Psi|^2 = n_s$, the density of superconducting electron pairs, $i = \sqrt{-1}$, and $\phi(x,t)$ is the phase of the wave function, which is dependent on position and time. Because superconductivity is a cooperative phenomenon involving many of the conduction electrons in the superconductor, the phase $\phi(x,t)$ is coherent everywhere. That is, if at an instant in time the value of $\phi(x,t)$ is known at one position and the wavelength of the wavefunction is known, then the phase is known everywhere in the sample. The wavelength of the wave function is determined by the momentum of the superelectron pairs, and it is therefore dependent on the current density and the applied magnetic field.

The wave function, $\psi(x,t)$, can only have one value at each position in the sample at any given time. This means that for a current flowing in a superconducting loop, the phase $\phi(x,t)$ will have to change by $2\pi n$ (where n is an integer) on going once around the loop. The consequence of this is that the magnetic flux contained in the loop will be quantized in units of the flux quantum (Φ_0) (Ref 57).

When a thick nonsuperconducting barrier interrupts the current path, the supercurrent through the barrier is zero and the phases on either side of the barrier act independently. If the barrier thickness is reduced to allow some superconducting electron pairs to tunnel through the barrier, then their wave functions overlap and begin to couple together, producing a fixed phase difference across the gap (Ref 58). The phase difference between the two superconductors ($\Delta\phi$) is given by:

$$\sin(\Delta\phi) = I/I_{cj} \qquad \text{(Eq 8)}$$

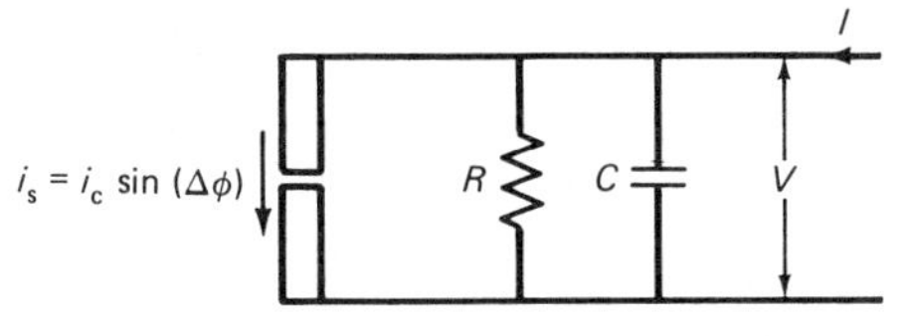

Fig. 22 The effective circuit of a Josephson junction, including a self-capacitance and a normal resistance parallel with the junction. This effective circuit is used to develop $I(\Delta\phi)$ (see Eq 10).

where I is the transport current through the junction, I_{cj} is the critical current of the junction; this current is much smaller than the critical current of the bulk material. This is the dc Josephson effect (Fig. 21).

If the current through the junction is made to vary as a function of time, then the phase difference $\Delta\phi$ will also be a function of time. The changing phase produces a voltage across the junction, which is given by:

$$2\,eV = (h/2\pi)d(\Delta\phi)/dt \qquad \text{(Eq 9)}$$

These two equations, which relate the phase differences across a weak link or tunneling junction to the junction voltage and current, are descriptive of most of the properties of a Josephson junction (Ref 59).

In a junction in which the supercurrent through the junction (I_s) is changing as a function of time, a voltage appears across the junction and normal electrons may tunnel through the weak link, producing a resistive current (I_r). Because the junction is made of two pieces of metal close together, the junction has a capacitance (C) (see Fig. 22). The total current through the junction, consisting of both the supercurrent and the normal current, can now be written as:

$$I = \frac{ch}{4\pi e}\frac{d^2(\Delta\phi)}{dt^2} + \frac{h}{4\pi eR}\frac{d(\Delta\phi)}{dt} + I_c\sin(\Delta\phi) \qquad \text{(Eq 10)}$$

Examination of Eq 10 reveals that for small currents the voltage across the junction will be zero and $\Delta\phi$ will increase as the current increases. There is, however, a critical current above which a dc voltage appears across the junction. Associated with the appearance of the dc voltage, the superelectron current begins flowing back and forth across the junction with a frequency of $\nu = 2eV_{dc}/h$, where e is the charge of an electron. The superelectron current produces an ac ripple voltage on top of the dc voltage. This is known as the ac Josephson effect, and it is the basis for applications involving tunable oscillators and radiation bolometers (Ref 60).

If a magnetic field is applied to the junction, the change in $\Delta\phi$ due to the field is negligible, but there is a large change in the critical current of the junction. By applying a very small magnetic field to a junction carrying a bias current, the junction will develop a resistive voltage that persists until the current is lowered below the critical current. The switching of the junction on (resistive) and off (superconducting) can be extremely fast (on the order of a few picoseconds), and is one of the motivations behind using Josephson junctions as elements in computer devices (Ref 61).

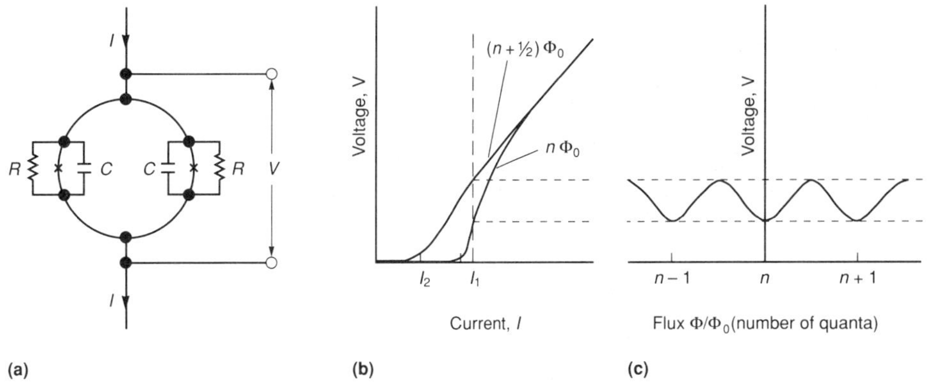

Fig. 23 Effect of applied magnetic field on the critical current of the total loop in a two-junction superconductor. (a) The dc SQUID consists of two junctions carrying a bias current (I). The voltage (V) is measured as a function of the magnetic field. (b) In the presence of a magnetic field, the critical currents of the junctions are lowered from I_1 to I_2. (c) The change in critical current produces a change in the voltage across the SQUID that oscillates with the frequency of the flux quantum (Φ_0).

When a loop of superconductor consisting of two junctions carrying a current is put in a magnetic field, the applied field lowers the I_c of the junctions (see Fig. 23). Because the total phase change of the wave function traveling once around the loop must equal $n2\pi$, the change of phase across the two junctions becomes a strong function of the applied magnetic field. The critical current of the total loop will oscillate as a function of the applied field with a period equal to the flux quantum (Φ_0). This oscillation is due to the interference of the wave functions at the two junctions, and this device is called a dc superconducting quantum interference device (SQUID). By measuring the voltage across the loop, the number of flux quanta contained in the loop can be accurately determined, and the SQUID can therefore be used as a very sensitive magnetic flux detector (Ref 60).

REFERENCES

1. B.W. Roberts, in *Intermetallic Compounds*, J.H. Westbrook, Ed., John Wiley & Sons, 1967
2. *Phys. Today*, March 1986; *Mech. Eng.*, June 1988
3. *Phys. Today*, Aug 1971
4. A.C. Rose-Innes, F.H. Rhoderick, *Introduction to Superconductivity*, Pergamon Press, 1969
5. M.N. Wilson, *Superconducting Magnets*, Oxford University Press, 1983
6. E.W. Collings, *Applied Superconductivity, Metallurgy and Physics of Titanium Alloys*, Vol I and II, Plenum Press, 1986
7. *Superconductor Materials Science: Metallurgy, Fabrication and Applications*, S. Foner and B.B. Schwartz, Ed., Plenum Press, 1981
8. M. Tinkham, *Introduction to Superconductivity*, McGraw-Hill, 1975
9. *Superconductivity*, R.D. Parks, Ed., Vol I and II, Marcel Dekker, 1969
10. J.G. Bednorz and K.A. Muller, *Z. Phys. B*, Vol 64, 1986, p 189
11. J. File and R.G. Mills, *Phys. Rev. Lett.*, Vol 10 (No. 3), 1963, p 93
12. W. Meissner and R. Ochsenfeld, *Naturwissenschaften*, Vol 21, 1933, p 787
13. V.L. Ginzburg, L.D. Landau, *Zh. Eksp. Teor. Fiz.*, Vol 20, 1950, p 1064
14. J. Bardeen, L.N. Cooper, and J.R. Schrieffer, *Phys. Rev.*, Vol 108, 1957, p 1175
15. H. Frohlich, *Phys. Rev.*, Vol 79, 1950, p 845
16. D.M. Ginsberg, *Amer. J. Phys.*, Vol 30 (No. 6), 1962, p 433
17. E. Maxwell, *Phys. Rev.*, Vol 78, 1950, p 477
18. C.A. Reynolds, B. Serin, W.H. Wright, and L.B. Nesbitt, *Phys. Rev.*, Vol 78, 1950, p 487
19. A.A. Abrikosov, *Sov. Phys. JETP*, Vol 5, 1957, p 1174
20. P.J. Lee and D.C. Larbalestier, *Acta Metall.*, Vol 35 (No. 10), 1987, p 2523
21. *Filamentary A15 Superconductors*, M. Suenaga and A.F. Clark, Ed., Plenum Press, 1980
22. *Phys. Today*, March 1986
23. M.N. Wilson, *Superconducting Magnets*, Oxford University Press, 1983
24. W.A. Fietz and W.W. Webb, *Phys. Rev.*, Vol 178 (No. 2), 1969, p 657
25. D.G. Hawksworth and D.C. Larbalestier, *Proceedings of the 8th Symposium on Engineering Problems of Fusion Research*, No. 1, 1979, p 245
26. R.J. Hampshire and M.T. Taylor, *J. Phys. F*, Vol 2, 1972, p 89
27. E.J. Kramer, *J. Appl. Phys.*, Vol 44

(No. 3), 1973, p 1360
28. L.F. Goodrich and F.R. Fickett, *Cryogenics*, May 1982, p 225; F.R. Fickett, *J. Res. Natl. Bur. Stand.*, Vol 90 (No. 2), 1985, p 95
29. F.R. Fickett and A.F. Clark, "Development of Standards for Superconductors," NBSIR 80-1629, National Bureau of Standards, 1979
30. W.H. Warnes, *J. Appl. Phys.*, Vol 6 (No. 5), 1988, p 1651
31. U. Essman and H. Trauble, *Phys. Lett.*, Vol 24A, 1967, p 526
32. Y.B. Kim, C.F. Hempstead, and A.R. Strnad, *Phys. Rev.*, Vol 139 (No. 4A), 1965, p 1163
33. A.M. Campbell and J.E. Evetts, *Adv. Phys.*, Vol 21, 1972, p 199
34. D. Dew-Hughes, *Philos. Mag.*, 1974, p 293
35. G. Zerweck, *J. Low Temp. Phys.*, Vol 42 (No. 1), 1981, p 1
36. H. Ullmaier, *Irreversible Properties of Type II Superconductors*, Springer-Verlag, 1975
37. P.W. Anderson, *Phys. Rev. Lett.*, Vol 9, 1962, p 309
38. C. Giovannella, P. Roualt, A. Campbell, and G. Collin, *J. Appl. Phys.*, Part 2B, Vol 63 (No. 8), 1988, p 4173
39. L.L. vant-Hull and J.E. Mercereau, *Rev. Sci. Instrum.*, Vol 34 (No. 11), 1963, p 1238
40. A.K. Chizhou, *Sov. Phys. Tech. Phys.*, Vol 18 (No. 11), 1974, p 1499
41. C.P. Bean, *Phys. Rev. Lett.*, Vol 8, 1962, p 250
42. A.K. Ghosh, M. Suenaga, T. Asano, A.R. Moodenbaugh, and R.L. Sabatini, *Adv. Cryog. Eng.*, Vol 34, 1988, p 607
43. M.A. Green and R.M. Talman, *IEEE Trans. Mag.*, Part 1, Vol 24 (No. 2), 1988, p 823
44. V. Arkadiev, *Nature*, Vol 160, 1947, p 330
45. F. Hellman, E.M. Gyorgy, D.W. Johnson, Jr., H.M. O'Bryan, and R.C. Sherwood, *J. Appl. Phys.*, Vol 63 (No. 2), 1988, p 447
46. V.W. Edwards and M.N. Wilson, *Cryogenics*, July 1978, p 423
47. T. Akachi, T. Ogasawara, and K. Yasukochi, *Jpn. J. Appl. Phys.*, Vol 20 (No. 8), 1981, p 1559
48. Z.J.J. Stekly and J.L. Zar, *IEEE Trans. Nuc. Sci.*, Vol 12, 1965, p 367
49. M.J. Leupold, R.J. Weggel, and Y. Iwasa, *Proceedings of the 6th International Conference on Magnet Technology*, 1977, p 400
50. J.M. Pfotenhauer, *IEEE Trans. Mag.*, Vol 23 (No. 2), 1987, p 926
51. P.N. Haubenrich, *IEEE Trans. Mag.*, Vol 23 (No. 2), 1987, p 800
52. E.W. Collings, *Adv. Cryog. Eng.*, Vol 34, 1988, p 639
53. J.L. de Reuver, J.M. Mulders, and L.J.M. van de Klundert, *IEEE Trans. Mag.*, Vol 19 (No. 3), 1983, p 252
54. W.H. Hartwig, *Proc. IEEE*, Vol 61 (No. 1), 1973, p 58
55. H. London, *Nature* Vol 133, 1934, p 497
56. J.P. Turneaure, *Proceedings of the 1972 Applied Superconductivity Conference*, 1972, p 621
57. B.S. Deaver and W.M. Fairbank, *Phys. Rev. Lett.*, Vol 7, 1961, p 43; R. Doll and M. Habauer, *Phys. Rev. Lett.*, Vol 7, 1961, p 51
58. B.D. Josephson, *Adv. Phys.*, Vol 14, 1965, p 419
59. M. Tinkham, *Introduction to Superconductivity*, McGraw-Hill, 1975
60. J. Clarke, *AIP Conference Proceedings*, Vol 29, American Institute of Physics, 1976, p 17
61. S.K. Lahiri, A.K. Gupta, and V.S. Tomar, *Indian J. Phys.*, Vol 59A (No. 4), 1985, p 247

Niobium-Titanium Superconductors

T. Scott Kreilick, Hudson International Conductors

NIOBIUM-TITANIUM ALLOYS became the superconductors of choice in the early 1960s, providing a viable alternative to the A-15 compounds and less ductile alloys of niobium-zirconium. The first NbTi alloys were titanium-rich with a composition of Nb-65Ti (Ref 1). The relative ease with which wire could be fabricated, better electrical properties, and greater compatibility with copper stabilizing materials were the primary forces behind this shift.

The monofilamentary wire utilized in the first magnets was subject to flux jumps or magnetic instabilities (Ref 2). By subdividing the superconducting core into individual fine filaments, adiabatic flux-jump stability was achieved. In addition, the incorporation of copper, or another conductive metal, as the interfilamentary matrix material provided not only dynamic flux-jump stability but also cryostability (Ref 3). Finer filaments also helped minimize hysteresis loss and inherent magnetization.

Multifilamentary composite wires, when subjected to a time-varying external magnetic field, exhibit coupling of the filaments by circulating currents. As a result, the effective filament diameter is larger than the actual filament diameter, thus negating the benefits of individual fine filaments. To counter the problem, multifilamentary wire must be twisted. Twisting of the strand reduces eddy-current losses in time-varying applied fields as does the incorporation of a resistive matrix.

In this article, the ramifications of the design requirements delineated above will be discussed in the context of multifilamentary NbTi superconducting composite fabrication.

Alloy Selection

As conductor instabilities became better understood, the titanium-rich NbTi alloy composition evolved into a more ductile, niobium-rich material. Historically, high-energy physics (HEP) pulsed accelerator-magnet applications have provided the impetus for alloy development. The alloy most widely used today is Nb-46.5Ti, although in some areas Nb-50Ti is preferred (Ref 4, 5). Binary NbTi compositions in the range of 45 to 50% Ti exhibit upper critical field ($\mu_0 H_{c2}$) values of 11.5 to 12.2 T at the boiling point of helium (4.2 K), combined with critical temperature (T_c) values of 9.0 to 9.3 K. For HEP applications this combination of operating parameters has provided sufficient operating margin, but for other applications (notably plasma fusion confinement), higher fields and larger temperature fluctuations are anticipated. For lower field applications, such as superconducting magnetic energy storage (SMES) and magnetic resonance imaging (MRI), there is a renewed interest in the higher-titanium-containing alloys (52 to 65% Ti) due to enhanced flux pinning characteristics (Ref 6). The tradeoff, however, for higher current-carrying capacity and lower alloy cost (by replacing niobium with titanium) is decreased ductility.

Theory predicts an upper critical field (H_{c2}) of 17 to 18 T for binary NbTi. As indicated above, experimental values are somewhat lower. The origin of the diminished H_{c2} lies in the appreciable orbital paramagnetism of NbTi, which makes a significant contribution to the free energy of the normal state. Spin-orbit scattering can relax this paramagnetic limitation, and it has been demonstrated that heavy element additions of tantalum (Z = 73) and hafnium (Z = 72) are very effective in this regard. The relative effects of adding tantalum or hafnium to the NbTi system have intrigued researchers for many years (Ref 7-9), and tantalum additions seem to be the concensus choice (Ref 10-13). While T_c is reduced for both additions, there is a small enhancement (0.3 T) of the upper critical field for the NbTiTa system at 4.2 K. At this temperature, no enhancement is observed for NbTiHf, whereas at 2 K a small increase of 0.3 T is noted. This rise in $\mu_0 H_{c2}$ is small compared to the 1.3 T enhancement at 2 K in the NbTiTa system. Experiments have shown that some quaternary alloys (for example, NbTiTaZr) can also extend the upper critical field (Ref 10).

Once the values for $\mu_0 H_{c2}$ and T_c are fixed during the alloy melting process, there is little the wire manufacturer can do to alter them. It is the critical current density (J_c) that is most affected by the processing from cast ingot to final wire and cable. The superconducting properties, therefore, are highly dependent on the cooling rate following melting or heat treatment and the degree of cold work. At high temperatures, the NbTi system forms a β solid solution.

The fine-scale dislocation cell structure achieved by cold working combined with the α-titanium precipitates produced in heat treatment yields an inhomogeneous microstructure. These defects are the flux pinning centers (see the article "Principles of Superconductivity" in this Volume) responsible for large critical current. The diameter of the fluxline (~11 nm in NbTi) and the distance between fluxlines dictate the size and spacing of the defect required to achieve the best current-carrying capacities at a given magnetic field. The process of heat treatment and cold working needed to develop these microstructures will be discussed in greater detail below.

Alloy Preparation

Niobium. High-purity niobium for superconducting applications comes from two primary sources: columbite-tantalite and pyrochlore (Ref 14). Columbite-tantalite contains Nb_2O_5 and Ta_2O_5 in ratios of 10:1 to 5:1 (depending on the specific geology) together with oxides of iron and manganese. After the recovery of the niobium pentoxide from the columbite concentrate by digestion in HF followed by solvent extraction, it is then processed in a manner identical to that used for the pyrochlore ores. Pyrochlore ore is mined at three principle locations:

- Araxa, in Minas Gerais, Brazil (3.0% Nb_2O_5)
- Catalao I, in Goias, Brazil (1 to 2% Nb_2O_5)
- Niobec, in Quebec, Canada (0.25 to 0.66% Nb_2O_5)

The niobium pentoxide is mixed with iron powder and reduced to ferrocolumbium, by the aluminothermic reaction. The specialty steel industry is the largest consumer of the material in this form (see the section "Nio-

Fig. 1 Consumable NbTi electrode being prepared for its final vacuum-arc remelt. Courtesy of Teledyne Wah Chang Albany

Fig. 2 NbTi ingot ready for sidewall machining prior to forging. Courtesy of Teledyne Wah Chang Albany

bium'' in the article ''Properties of Pure Metals'' in this Volume). $NbCl_5$ is then produced by chlorination and converted by hydrolysis to the oxide. The aluminothermic reaction is again used to reduce the oxide, now mixed with aluminum powder, to niobium. The metal is electron-beam (EB) melted under high vacuum for added purification.

Titanium. Most producers of commercially pure titanium utilize the Kroll process, which involves the chlorination of a mixture of carbon and the minerals rutile and ilmenite (found in beach sand). Liquid $TiCl_4$ is then mixed with liquid magnesium under an inert atmosphere in a closed heated reactor vessel. $MgCl_2$ is drained off and recycled electrolytically, leaving titanium sponge. The sponge is then consolidated by arc melting in a water-cooled copper crucible. Titanium cannot be effectively purified by melting because its vapor pressure is too high and its melting point too low for selective evaporation of most impurities. It is the purity of the niobium, therefore, that dictates the overall purity of the NbTi ingot. The allowable limits for metallic impurities and interstitials in NbTi alloy produced for the superconducting supercollider (SSC) program are listed below:

Impurity element	Allowable limit
Ti	45, 46.5, 48, 55 wt%, ± 1.5 wt%(a)
O	1000 ppm max
H	35 ppm max
C	200 ppm max
Fe	200 ppm max
Ta	1000 or 2500 ppm max(a)
N	150 ppm max
Ni	100 ppm max
Si	100 ppm max
Cu	100 ppm max
Al	100 ppm max
Cr	60 ppm max
Nb	Bal

(a) As specified by purchase order

Tantalum and oxygen are the principal impurities, and as was pointed out earlier, small additions of tantalum serve to enhance the high field properties. Furthermore, some impurities seem to be necessary for nucleation of pinning sites.

Melting of Composite. Consumable electrode vacuum arc remelting (VAR) or plasma arc melting techniques are employed to combine niobium and titanium ingots into the alloy composition of choice. Typically, two or three melts are required to yield a product with the desired homogeneity. Figure 1 shows an electrode ready for its final melt. The diameter of the cast product ranges from 203 to 584 mm (8 to 23 in.) in diameter, with weights exceeding 2000 kg (4400 lb) for the larger diameters. After sidewall machining (Fig. 2), the ingot undergoes bulk-chemical and electron-beam analysis to determine the exact alloy composition, in addition to metallographic examination.

Forging of NbTi Ingot. The NbTi ingot is then hot forged to a diameter of 152 mm (6 in.) followed by an anneal in air. The temperature of the anneal is approximately 870 °C (1600 °F). At this temperature the ingot is fully β-recrystallized. The time at temperature is typically 2 h, followed by a water quench. The ingot is then machined to a diameter of 146 mm (5.75 in.) and inspected. Further processing techniques will be described in the section ''Assembly Techniques'' in this article.

Ingot Inspection. Homogeneity of the ingot is determined by means of a high-resolution radiograph taken from a cross section of the material, which is compared to a series of standard radiographs. The importance of alloy homogeneity has been identified and discussed (Ref 15-17), and it is generally believed that the availability of more homogeneous alloys has led to the significant improvements in conductor performance observed in recent years. Mechanical properties testing includes tensile and hardness (less than 170 DPH is typically required). The internal structure is determined nondestructively by liquid penetrant examination and 100% ultrasonic inspection (see the articles ''Liquid Penetrant Inspection'' and ''Ultrasonic Inspection'' in Volume 17 of the 9th Edition of *Metals Handbook*) performed in three stages. The ultrasonic inspection techniques utilized are:

- Comparison with a flat bottom hole standard
- Longitudinal wave
- Circumferential angle-beam examination

Average grain size is determined using the applicable provisions of ASTM Standard E 112. Grain size 6 or smaller is typically specified. Large grain sizes lead to better workability but also to lower current densities.

Matrix Materials

Adequate stability has been identified as being of first-order importance in the operation of electromagnetic systems incorporating NbTi superconducting materials. The properties required in a successful stabilizing material are:

- High electrical and thermal conductivity
- High heat capacity
- Good mechanical strength at cryogenic temperature

- Good adherence to the superconductor
- Good ductility for forming and winding

Copper

High-purity copper satisfies *all* of these requirements to a high degree and is, therefore, the most frequently utilized stabilizing material.

Resistivity. The conductive metal surrounding a resistive spot in a superconductor stabilizes the coil by providing an alternative path to shunt the current around the spot until it has cooled down. The ability of a metal to conduct electricity at liquid helium temperatures is dependent on three factors. The resistivity of a metal will be the rough sum of:

- Resistivity of the pure metal
- Resistivity caused by impurities in solution in the metal and dislocations in the metal's crystal structure caused by stress
- Magnetoresistance of the metal

The residual resistance ratio (RRR) is often used as a figure of merit when comparing the relative purity of potential stabilizers. The residual resistance ratio is the ratio of the electrical resistivity at room temperature (293 K) divided by the resistivity at liquid helium temperature (4.2 K). In copper of nominal purity or better, the numerator depends essentially on the thermal vibrations of the copper lattice and not the impurities; the denominator depends only on the impurities. Resistance-at-field dictates the volume of stabilizer required to achieve a given stability criterion.

Magnetoresistance is the relative increase in resistivity of a metal in the presence of an external magnetic field. Copper tends to obey Kohler's rule, unlike most high-purity metals including aluminum. Kohler's rule is a mathematical statement of the fact that the magnetoresistance is observed to be an increasing function of the applied magnetic field per unit change in the electrical resistivity. With this information, it has been shown for different copper samples with RRR values of 150 or more and an applied field strength in excess of 4 MA · m^{-1} (50 kOe) that the at-field residual resistivities are nearly equal (Ref 18). In other words, higher-purity coppers, yielding higher RRR values have a diminishing return on investment in terms of magnetoresistance. Copper Development Association (CDA) alloy 10100 (ASTM designation B 170), oxygen-free copper is available from several commercial sources with RRR values ranging from 180 to over 500, depending on the quality of the cathode used in the continuous casting process.

Aluminum

High-purity aluminum, as mentioned above, does not obey Kohler's rule. Saturation occurs in the 800 to 1600 kA · m^{-1} (10 to 20 kOe) range after a very fast rise in the lower field region, the result being that in strong magnetic fields the resistivity of aluminum will be much less than that of copper. Thermal conductivity mimics electrical conductivity in the presence of a magnetic field. When RRR is used as the basis for comparison, any aluminum with an RRR greater than 1.57 times that of copper will have a higher thermal conductivity (at 4.2 K) and a higher electrical conductivity than the copper.

Aluminum offers improved flux-jump stability over copper due to its lower heat capacity. This, combined with a high thermal conductivity, reduces the time required to dissipate the heat generated during a flux jump. Aluminum offers several other advantages as a stabilizer of superconducting composites. Aluminum is one-third as dense as copper enabling the construction of lighter weight machines. In addition, aluminum has a high degree of transparency to many types of particles, thus radiation damage can be greatly reduced in environments where this is a factor.

One of the few areas that aluminum is not superior to copper as a stabilizer material is that of mechanical compatibility with the NbTi alloy during the reduction process. Attempts to coprocess aluminum and NbTi alloys have taken several directions. The three most common methods are:

- Strengthening of the aluminum by alloying or other metallurgical techniques
- Low-temperature hydrostatic extrusion that minimizes the mechanical dissimilarities
- Addition of the aluminum stabilizer, by extrusion cladding or soldering, after the NbTi alloy has been coprocessed in a copper matrix

Extrusion and cladding techniques will be discussed in the section "Processing of Superconductor Composites" in this article.

Higher-strength aluminum-base materials have been investigated with varied results. NbTi filaments embedded in commercial-purity aluminum (alloy 1100, 99.0% Al min) show signs of filament nonuniformities during extrusion or after relatively small strains (Ref 19, 20). Dispersion-hardened and heat-treatable aluminum alloys, such as those mentioned in Ref 20, offer higher strengths, but at the expense of ductility and conductivity. Some directionally solidified eutectic alloys containing randomly dispersed intermetallic whiskers reinforce the α-Al matrix without significantly reducing conductivity (Ref 21), but these materials have not been routinely used.

A great deal of effort has gone into producing aluminum alloys with good strength retention at high operating temperatures, such as those experienced by rotating engine components (Ref 22). Powder metallurgy, rapidly solidified, dynamically recrystallized aluminum alloys show promise in this area, in addition to use as a high-strength aluminum matrix material for cryoconducting and superconducting applications (Ref 23). Al-8.4Fe-3.6Ce (Alcoa's alloy CU78) has demonstrated the best combination of properties and is the most completely characterized composition of this type (Ref 24-27).

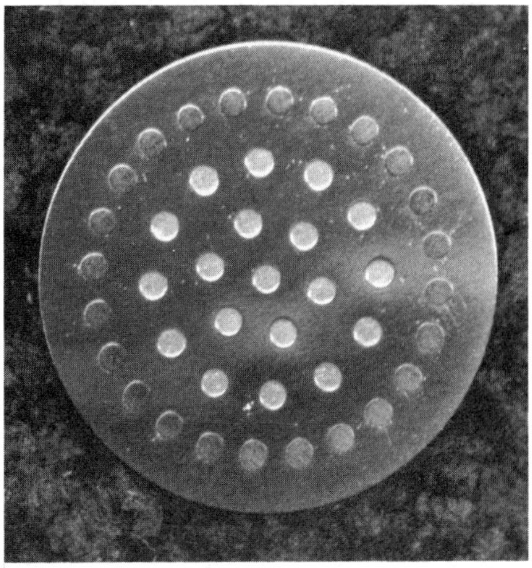

Fig. 3 Cross section of a prototype aluminum-stabilized composite for MRI applications fabricated using drilled-billet techniques. Outer ring of filaments are NbTi, inner filaments are high-purity aluminum, matrix is high-strength AlFeCe. Courtesy of Supercon, Inc.

The diffusionless alloying elements, iron and cerium, enable the retention of high conductivity in the aluminum of the final composite. Figure 3 shows a prototype conductor incorporating AlFeCe as the matrix, designed for MRI applications.

Copper-Nickel

Most applications of NbTi superconducting materials have been dc systems with relatively large filaments, and for these, copper and aluminum have been suitable matrix materials. For ac and pulsed-field environments, however, finer filaments are desired. The minimization of eddy current loss in the normal-metal matrix and hysteretic loss in the filaments must be considered. For these applications, the introduction of nickel into copper in the form of a high-resistivity solid-solution alloy matrix has found wide acceptance (Ref 28-32).

The copper-nickel alloy, typically in concentrations of 90:10 or 70:30, is used to isolate individual filaments or groups of filaments as a means of reducing intrafilament coupling leading to losses. Figure 4 shows a partial cross section of a composite in which a portion of the copper matrix (which remains in close contact with the NbTi core) is replaced by a web of copper-nickel. Figure 5 shows the cross section of a conductor designed for ac applications in which the stabilizing core area is subdivided with copper-nickel, as are the NbTi fila-

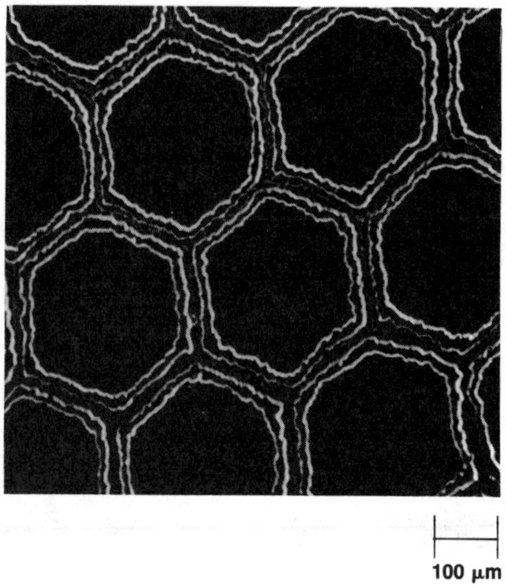

Fig. 4 Partial cross section of a mixed-matrix composite for ac applications. High-purity copper surrounds each NbTi filament for stability, and a web of resistive CuNi maintains the electrical integrity of each filament. Courtesy of Supercon, Inc.

ments themselves. Two limitations arise in the use of copper-nickel matrix materials. The first is a decreased fabricability and the second is the increase of resistivity and loss of thermal diffusivity resulting in a loss of stability.

Copper-Manganese

When the interfilamentary matrix material is pure copper, close spacing of the NbTi filaments will exhibit the proximity-effect coupling at lower fields, for as soon as the filament spacing and the superconducting coherence length become comparable, the wire will tend to behave as a monofilament (Ref 33). Several methods of reducing this coupling have been suggested. The first, mentioned above, operates by suppression of the mean free path of the electrons (that is, CuNi), but as also mentioned, this has an adverse effect on the stability of the composite. The second method employs electron-spin-flip scattering by solute ions carrying localized magnetic moments (for example, manganese, iron, or chromium) (Ref 34).

Several large billets have been fabricated to investigate the benefits (and limitations) of manganese additions to the copper matrix (Ref 35, 36). The evaluation has included mechanical, electrical, thermal, and magnetic properties, and to date no significant limitations appear to exist for composites containing 0.5% Mn additions to the copper (Ref 32, 35-41). This percentage of manganese is as effective as 30% Ni additions to the matrix in reducing proximity-effect coupling of the filaments, but without the added transverse resistivity. Susceptibility and magnetization measurements show that filaments as fine as 1.5 μm (60 μin.) in diameter with interfilament

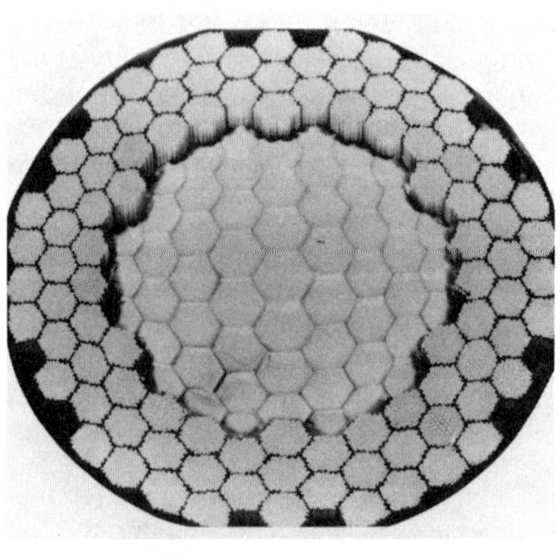

Fig. 5 Cross section of a 14 496-filament conductor designed for ac applications. Courtesy of Alsthom Atlantic

spacings of ≧280 nm are effectively decoupled.

Figure 6 shows the cross section of a prototype billet fabricated for the SSC strand-development program. The core and the outer can are high-purity copper. The interfilamentary region is Cu-0.5% Mn. This composite (Ref 36), containing 22 900 NbTi filaments, was designed to yield 2.5 μm (100 μin.) diam filaments when the overall conductor diameter was 0.65 mm (0.0255 in.).

Processing of Superconductor Composites

There are several methods used to fabricate NbTi superconducting composites depending primarily on the proposed application. The conditions of operation dictate the number and size of the superconducting filaments, as well as the volume of stabilizer necessary to adequately protect the magnet in the event of a flux jump. Minimum current density, piece length requirements, and cost considerations influence the size of the initial billet.

Assembly Techniques

The quality of the starting materials was emphasized in the section "Alloy Preparation" in this article. Well-characterized, homogeneous alloys and matrix materials are two of the prerequisites in the production of reliable wire products. NbTi alloy is usually procured in the β-recrystallized state, although the use of cold-worked alloy has been considered. When the stabilizing material is high-purity copper, special attention must be given to the size of the grains. Copper in the as-cast condition is not suitable for most applications of NbTi superconductors. The coarse grains of high-purity copper in the as-cast and/or insufficiently forged condition have been known to compromise the integrity of the superconducting composite during isostatic pressing and extrusion operations. The copper should

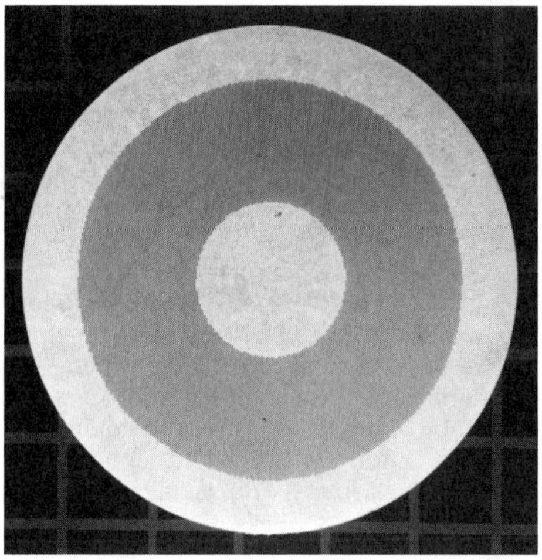

Fig. 6 Cross section of a 22 900-filament conductor designed to yield uncoupled 2.5 μm (100 μin.) diam filaments for the superconducting supercollider. Courtesy of Supercon, Inc.

have a fine-grain microstructure obtained by cross-grain forging or extrusion.

Billet Cleanliness. All billet components should be free of oxide, inclusions, or other surface defects prior to assembly. A nonductile inclusion at the time of assembly could become a significant fraction of a wire's cross section at final size resulting in breakage. Because O_2 has significant solubility in copper, diffusion from unclean surfaces can readily occur at temperatures encountered during processing. If the oxide is not completely removed by the cleaning procedures, such contamination can affect the entire copper cross section and even penetrate to the NbTi core. Some etching agents leave an oxide layer a few hundred angstroms in thickness, and calculations show that the total amount of oxygen that can be introduced into a fine filament material this way is enough to convert oxygen-free copper into a relatively high O_2 tough-pitch copper. The significance of this condition on the drawability of fine copper cross sections is well known to the copper industry (Ref 42, 43).

All cleaning procedures are not the same, but there are general similarities. All components are degreased, acid etched, rinsed in water, then in acetone and/or a drying agent, such as methanol. The parts are then blown dry and stored in a N_2 atmosphere until they are assembled. Assembly takes place in a low-level clean-room environment.

Monofilamentary conductors are used in small, low-field magnets (for example, nuclear magnetic resonance, NMR, applications) where persistent joints are required. Fewer filaments expedite the joining of superconducting wires with low resistance, enabling persistent mode operation. In other words, the power supply can be removed after current excitation and the current will

Fig. 7 Cross section of a 54-filament composite fabricated using gun-drilled billets. The Cu:Sc ratio is 1.35:1. Courtesy of Supercon, Inc.

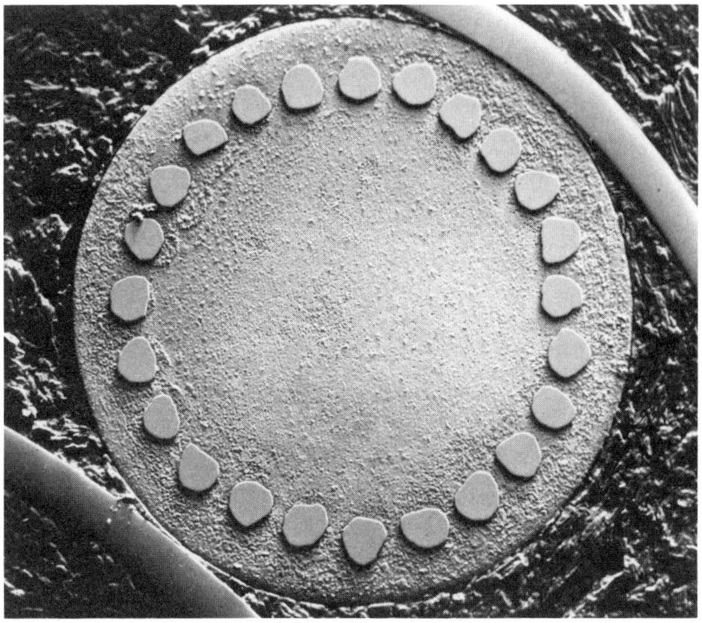

Fig. 8 Cross section of a 24-filament composite fabricated using gun-drilled billets. The Cu:Sc ratio is 7:1. Courtesy of Supercon, Inc.

continue to flow without significant degradation for long periods of time. Techniques employed to fabricate joints include resistance welding, spot welding, ultrasonic welding, pressure contacts, and soldered lap joints, among others. When a superconductor-superconductor joint becomes necessary, attempts are made to remove the joint from close proximity of the operating field of the magnet in order to maintain field uniformity. When these factors are considered it soon becomes clear why long lengths of conductor are requested from the wire manufacturer.

Monofilaments are made by coextruding a large ingot of NbTi alloy within a copper can. Billets of this type are frequently 200 mm (8 in.) in diameter, with a 120 mm (4¾ in.) diam core. The maximum billet length is dictated by the length of the extrusion press liner. Once extruded, the composite is reduced in diameter by rod reduction and wire-drawing techniques to final size, which is typically 0.5 to 1.0 mm (0.02 to 0.04 in.) in diameter. During the reduction process the wire is subjected to a series of intermediate heat treatments designed to precipitate the α-phase titanium of the alloy. The ratio of copper stabilizer to superconductor volume is approximately 1.8:1 but may vary depending on the application.

Multifilamentary Conductors. Conductors incorporating less than 200 NbTi filaments can utilize gun-drilling techniques. A solid copper billet is deep-hole drilled with either concentric circles or a hexagonal close-pack array to facilitate symmetry within the wire. Sufficient margin must be left between holes to allow for drift during the drilling operation. Longer lengths and smaller diameters make a difficult procedure self-limiting in terms of the number of holes that can be accommodated. NbTi rods are inserted and the billets processed to final wire. The copper-to-superconductor ratio (Cu:Sc) varies depending on the number and size of the filaments. Billet diameters are typically 250 to 280 mm (9.8 to 11 in.). Figure 7 shows the cross section of a 54-filament conductor with a Cu:Sc ratio of 1.35:1. Figure 8 shows the cross section of a 24-filament conductor with a Cu:Sc ratio of 7:1. It is this type of 24-filament conductor which is widely used in low field (0.6 T, or 6 kG) whole-body MRI magnets.

Composite conductors with more than 200 filaments are assembled using one of three basic techniques:

- Kit method
- Restacked monofilaments method
- Restacked drilled billets

Kit Method. The first approach is referred to as the CBA/Fermi kit method and is so named because it was used to manufacture the strand for the Fermilab Tevatron magnets. It was also considered for use in the Brookhaven National Laboratory colliding beam accelerator (CBA), a machine that was never built. Although it is a relatively simple and inexpensive method of assembly, it has significant drawbacks when the highest current-carrying capacity or long piece lengths are required.

Small-diameter NbTi rods (3.18 to 7.11 mm, or 0.125 to 0.280 in.) are inserted into copper hexagonal tubes with round inside diameters. Several hundred to 2000 of these tubes are stacked into a hexagonal close-pack array. A copper can, with an outside diameter of approximately 250 mm (10 in.), is lowered over the assembled filaments, the ends sealed, and the billet processed to final wire. Figure 9 is a schematic illustration of the kit approach to billet assembly. Figure 10 is a cross section of a 2070-filament conductor fabricated to the CBA specifications.

Restacked Monofilament Method. Inherent factors limit the quality of strand produced with the kit technique. The amount of exposed surface area that must be oxide-free prior to assembly is approximately twice that of a similar composite (with the same number of filaments) fabricated using the restacked monofilament method. Any remnant oxide, now in close proximity to the NbTi core, can lead to embrittlement problems, not only of the core but also of the copper matrix. Furthermore, the presence of oxide impurities in the copper reduces its effectiveness as a stabilizer, although this is probably a secondary effect.

Figure 11 is a micrograph of the NbTi filaments extracted from a composite, fabricated using a kit, at final size after undergoing several intermediate precipitation heat treatments. The nodules on the filament surfaces are a brittle intermetallic compound of titanium-copper. As the wire is drawn, the nonductile particles become a significant fraction of the filament cross section. Although the exact composition is not clear (Ti_2Cu, $TiCu_4$, Ti_2Cu_7 from Ref 15, 44, and 45, respectively), the effect on filament quality is to locally reduce the cross section and to eventually cause filament and wire failure.

Diffusion barrier technology has been employed to inhibit the growth of unwanted intermetallic compounds (Ref 46-48).

The most commonly used method of fabrication is to wrap one or more sheets of

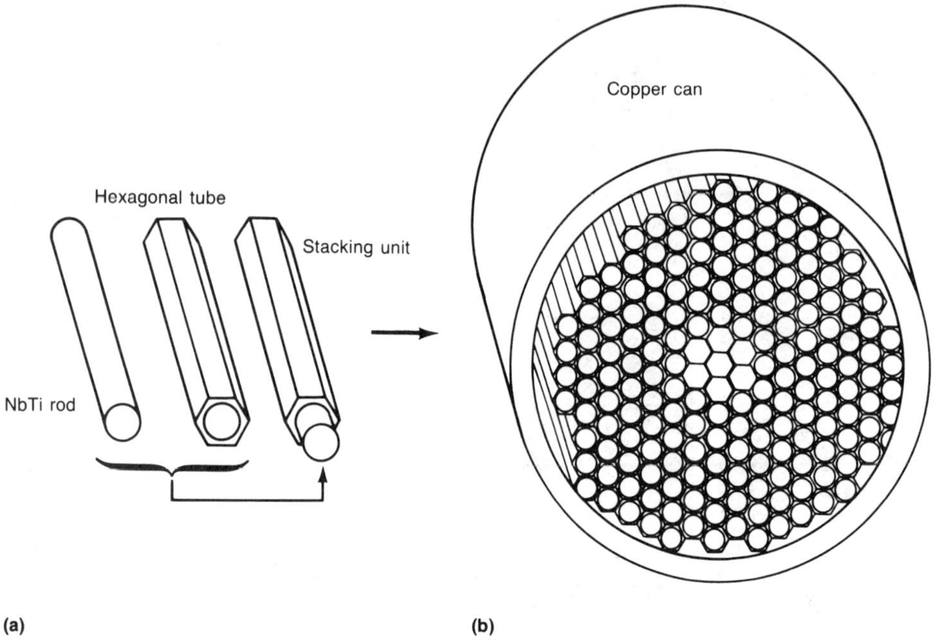

Fig. 9 Schematic illustration of the CBA/Fermi kit approach to billet fabrication. (a) NbTi rod is inserted into a copper tube with hexagonal OD and circular ID to yield an individual stacking unit. (b) Individual stacking units arranged in a hexagonal close-packed (hcp) array in a 250 mm (10 in.) diam copper can

Fig. 10 Cross section of a 2070-filament composite assembled using a kit. Courtesy of Supercon, Inc.

niobium foil around a large NbTi ingot and then to encase the whole assembly in a copper or copper-manganese can. Bonding is achieved during a high-temperature, large reduction extrusion process. Figure 12 shows niobium diffusion barriers in high relief around NbTi filaments within a deeply etched copper matrix. Figure 13 shows niobium barrier-protected NbTi filaments extracted from a composite in which the filaments were very closely spaced within a copper matrix.

The preparation of NbTi monofilament incorporating a diffusion barrier is shown schematically in Fig. 14. The NbTi ingot is procured at a diameter of 146 mm (5.75 in.) and a length of 610 mm (24 in.). Once extruded, the monofilament is drawn to the proper diameter, shaped by drawing through a hexagonal cross-sectional die, cut to the appropriate length, and restacked into the desired multifilamentary array. An alternative to this approach is to forego the hexagonal shaping step and to restack round monofilamentary subelements. Figure 15 shows a partial cross section of a conductor composed of hexagonal subelements, and Fig. 16 shows the result of restacked round subelements. The uniformity of the array obtained from restacking hexagonal subelements is evident. The voids and dislocations associated with the round subelement restack result in areas of the array in which filament spacing is not uniform.

Restacked Drilled Billets. The third method of billet assembly is to restack drilled billets. This approach is commonly used when many thousands of very fine filaments are required, as is the case for ac applications. A variation of this technique is to produce a monofilament and restack it twice, technically a triple extrusion. Figure 5 shows such a composite. The limitation to multiple restacks is, again, nonuniform fila-

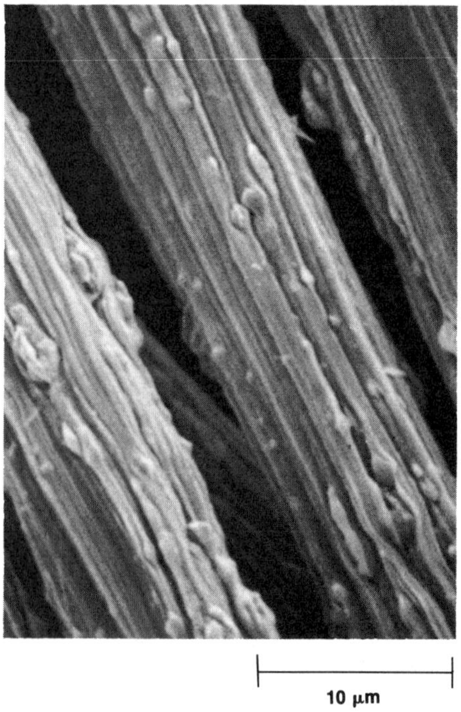

Fig. 11 NbTi filaments extracted from a composite at final size after several intermediate precipitation heat treatments. The nodules are an intermetallic compound of Cu-Ti. 2500×. Courtesy of Supercon, Inc.

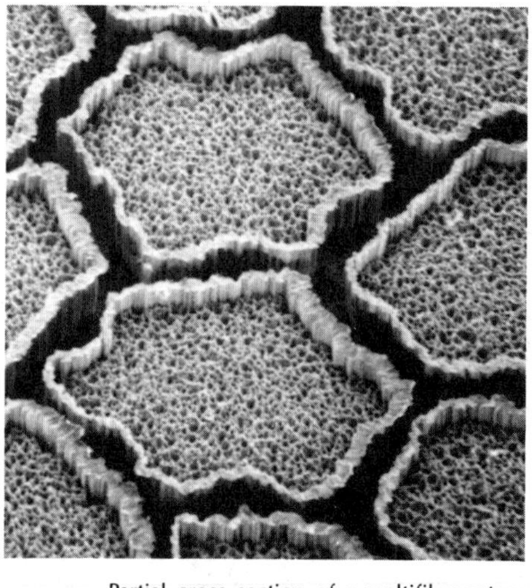

Fig. 12 Partial cross section of a multifilamentary NbTi composite with a niobium diffusion barrier (in high relief) around each filament. The copper interfilamentary matrix is deeply etched. Courtesy of Supercon, Inc.

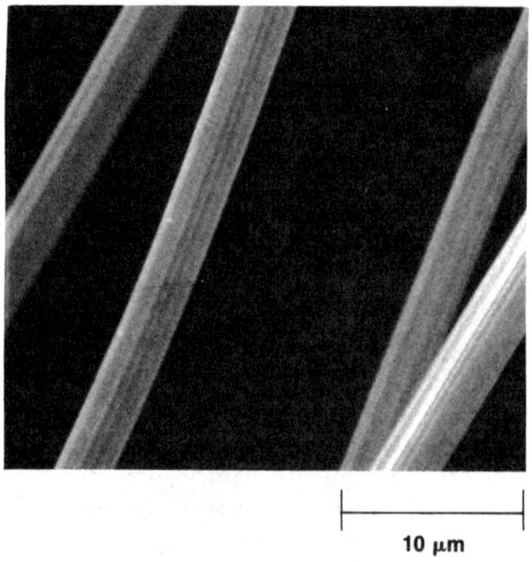

Fig. 13 Niobium diffusion barrier-clad NbTi filaments extracted from the copper matrix of a composite designed for the superconducting supercollider. Courtesy of Supercon, Inc.

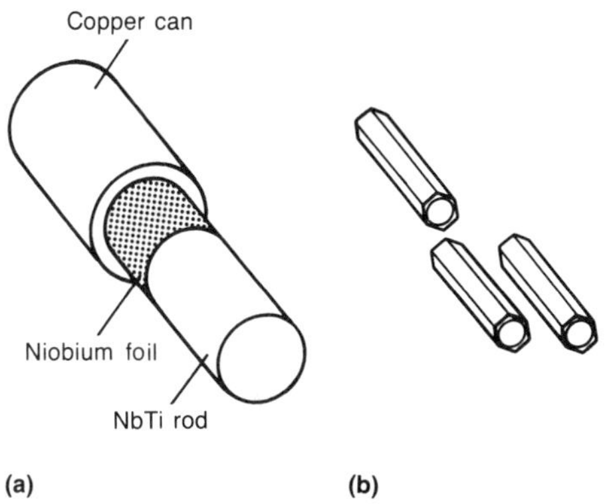

Fig. 14 Schematic illustration of diffusion barrier-clad monofilament assembly. (a) Diffusion barrier is obtained by wrapping an NbTi rod with niobium foil and placing foil-wrapped rod in a copper can. (b) Extruding and drawing of copper can assembly yields monofilamentary hexagonal rods with diffusion barrier.

ment spacing. The outermost filaments in each subbundle are mechanically unsupported. This instability is exemplified in Fig. 17.

It would seem, therefore, that the lowest percentage of unsupported filaments would be produced in a billet comprised of a single restack of hexagonal subelements. Unfortunately, there are limits to the number of hexed rods that can be practicably restacked. The cross section shown in Fig. 6 contains 22 900 NbTi filaments *plus* 11 200 copper hexes (which make up the core and part of the outer can). Each hexagonal subelement was 1.42 mm, or 0.056 in. (flat-to-flat). The can used to fabricate this composite had an outside diameter of 316 mm (12.4 in.). It is conceivable that with a larger can, and thus larger hexed rods, more filaments could be incorporated into a single restack, but this remains unexplored.

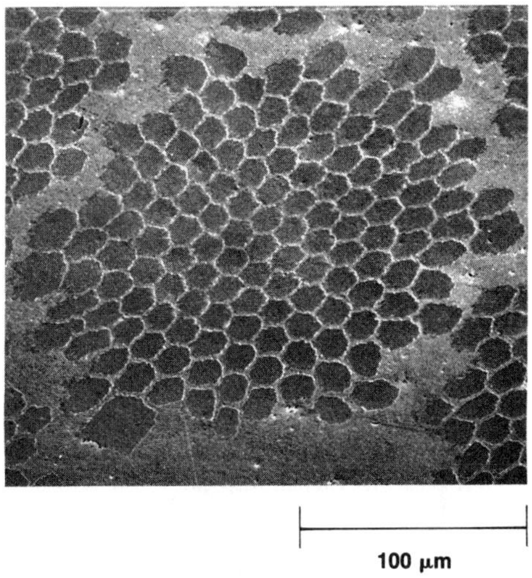

Fig. 17 Partial cross section of a multifilamentary NbTi composite assembled using a double restack of hexagonal monofilaments. Mechanically unsupported outermost filaments display erratic filament cross sections. Courtesy of Supercon, Inc.

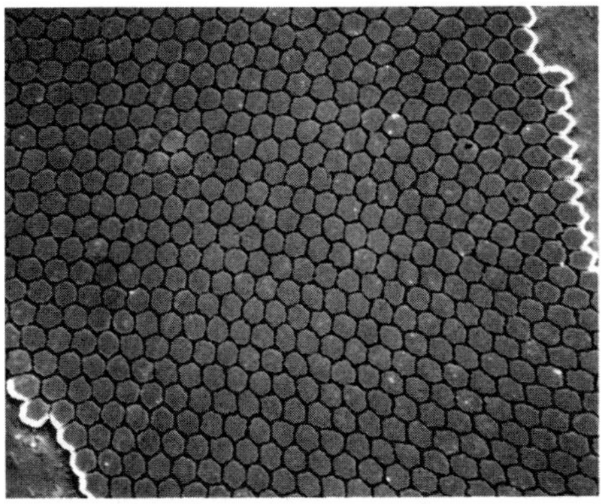

Fig. 15 Partial cross section of a multifilamentary NbTi composite assembled using hexagonal monofilaments. Courtesy of Supercon, Inc.

In addition, automation of the billet assembly process has yet to be undertaken.

Welding

Upon assembly, billets must be sealed to prevent oxidation of the freshly etched surfaces. The surface area increases as the inverse square of the filament size. Electron-beam (EB) welding and tungsten-inert-gas (TIG) welding are the two techniques most widely used. Trapped gases and contaminants will inhibit proper bonding of the components. It is, therefore, essential that a vacuum exist within the enclosed can.

Compared to TIG welding, EB welding provides a much deeper heat-affected zone (HAZ). This is a distinct advantage during the rigorous processing to follow. Furthermore, TIG welding requires an evacuation port, usually in the form of a copper tube protruding from the front or rear of the billet. After evacuation, this tube is sealed, but this site remains a point of weakness. Should a rupture occur during isostatic pressing or extrusion and go undetected, air can be forced into the composite during extrusion.

EB welding is not without its drawbacks. The lengthy residence time required in the EB vacuum chamber is significant. Good use of this time can be made by welding one end of the billet at the end of the work day, then allowing the heated billet to pump-down overnight and welding the other end in the morning. The slower cooling rate associated with EB welding yields larger grains in the HAZ. As mentioned in the section "Assembly Techniques" in this article, large copper grains are to be avoided.

Isostatic Compaction

The void space in a drilled billet as assembled is relatively small, and the upset experienced during extrusion is insignificant as long as the billet fits snugly in the liner. For composites containing larger numbers of filaments, the void space is much greater. For a billet containing several thousand hexagonal subelements the void space is on

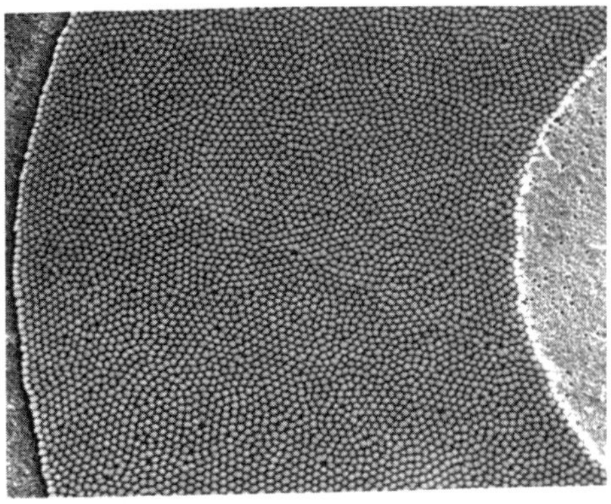

Fig. 16 Partial cross section of a multifilamentary NbTi composite assembled using round monofilaments. Courtesy of Supercon, Inc.

the order of 6%, and for the same number of round subelements the void space is approximately twice as much (Ref 49). To densify multifilamentary composites, isostatic compaction techniques have been employed prior to extrusion. The billets fabricated for the Fermilab Tevatron were isostatically pressed at ambient temperature (cold isostatic pressing, CIP), and sufficient bonding of the components was achieved during hot conventional extrusion.

Experience has shown that, by reducing the temperature of extrusion, higher current densities can be achieved. When reducing the extrusion temperature too much, however, complete bonding is not achieved (Ref 50). The use of hot isostatic pressing (HIP) as a means of densifying and prebonding extrusion components is not new (Ref 51), but from the perspective of achieving a complete bond followed by very low-temperature extrusion as a means of retaining cold work, it is relatively unexploited (Ref 52, 53). *In situ* billet consolidation is possible if hydrostatic extrusion techniques are used, due to the extremely high pressures involved (Ref 54). By compensating for void fraction in the billet design, a uniform circular cross section is obtained prior to extrusion.

HIP parameters, when applied to NbTi/Cu superconducting materials, include temperatures that mimic those of extrusion (500 to 650 °C, or 930 to 1200 °F), pressures in the range of 103 to 206 MPa (15 to 30 ksi), and residence times of several hours in an argon atmosphere.

Extrusion

Extrusion is utilized to transform the assembled array of components into a true composite structure. High area reduction in a single step and a homogeneous plastic flow are distinct advantages. The principles of conventional and hydrostatic extrusion have been discussed in the section "Processing of Superconductor Composites" in this article (Ref 15, 55, and 56). As mentioned earlier, NbTi alloys are sensitive to

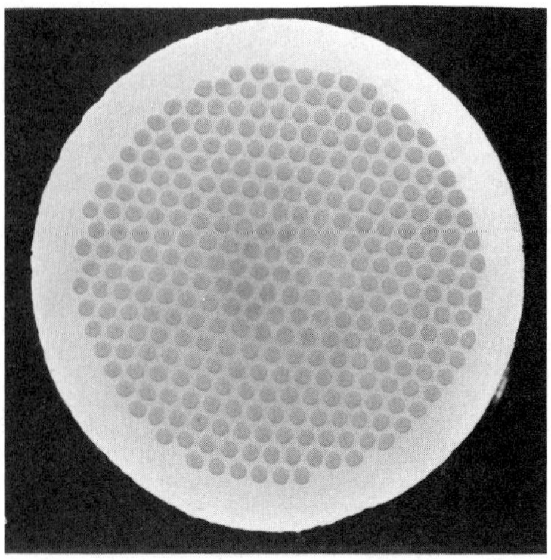

Fig. 18 Cross section of an aluminum-stabilized composite containing 294 NbTi filaments in a 6063-T6 aluminum alloy matrix. The 19 filaments in the center of the array are high-purity (99.999%) aluminum. This composite was assembled using kit technology and was hydrostatically extruded at ambient temperature in a conventional extrusion press converted for dual hydrostatic/conventional use. Courtesy of Supercon, Inc.

Fig. 19 122 m (400 ft) drawbench (useful draw of over 60 m, or 195 ft) with a pulling force of 700 kN (80 tonf) used in the processing of superconducting materials. Courtesy of IGC Advanced Superconductors Inc.

the temperatures encountered during processing. For this reason the temperature of extrusion is kept far below the recrystallization temperature of the superconductor. Those materials without diffusion barriers are also subject to intermetallic compound formation during preheating and extrusion.

Billet heating prior to extrusion can be accomplished several ways. Electric furnaces with forced convection, gas-fired furnaces, induction coils, and molten salt baths have all been used successfully. Manufacturers have their individual preferences. Conventional extrusion temperatures are usually in the range 500 to 650 °C (930 to 1200 °F). Hydrostatic extrusion temperatures range from ambient temperature for aluminum matrix composites to 200 to 400 °C (390 to 750 °F) for copper matrix composites (Ref 50, 52, and 53). The length of time required is determined by the heating method employed and is considered sufficient once the billet is uniformly heated. In addition to preheating the billet, the liner, cone, and die are often heated prior to extrusion.

The extrusion ratio is determined after considering a number of factors. Among these are: the available tonnage of the press, the equipment available to process the extrudate, and the deformation resistance of the billet. The deformation resistance is composition dependent and varies as a function of the temperature. Typical extrusion ratios range from 10:1 to 20:1. Too small a reduction can lead to a phenomenon referred to as center burst, in which the soft copper shell tends to flow faster than the harder NbTi core elements, causing an internal tensile tear. Billet diameters range from 178 to 203 mm (7 to 8 in.) for monofilaments, 203 to 280 mm (8 to 11 in.) for drilled billets, and 250 to 356 mm (9.8 to 14 in.) for restacked monofilaments.

The temperature dependence of the NbTi alloy again comes into play when one considers the speed of extrusion. As the extrusion proceeds, friction between the billet and the liner heats the billet. To facilitate a smooth extrusion, the billet and the extrusion tooling are coated with MoS_2 or graphite to reduce friction. Furthermore, if the deformation energy is transferred to the work piece and not the liner, the extrudate is subject to a significant heat differential from front to rear. These considerations necessitate relatively slow extrusion speeds. Stem speeds range from 3 to 7 mm/s (0.12 to 0.28 in./s). Upon complete extrusion the composite rods are water quenched to avoid coarsening of the grains.

Some benefits of low-temperature hydrostatic extrusion have been alluded to, that is, maximization of retained cold work, reduced intermetallic compound formation, and reduced friction during extrusion. Other advantages include:

- Ability to coextrude widely dissimilar metals such as aluminum-stabilized NbTi superconductors (Fig. 18)
- Larger length/diameter (*l*/*d*) ratios that translate into reduced extrusion end-effect losses and higher yields

The largest production hydrostatic extrusion presses have maximum capacities of 39.1 MN (4400 tonf) and are located in Europe and Japan. The feasibility of modifying an existing midsize (11.1 MN, or 1250 tonf) conventional extrusion press to permit, with a reversible tooling change, hydrostatic extrusion of superconducting materials, has been demonstrated (Ref 53).

Wire Drawing

Standard nonferrous rod and wire-drawing techniques are used to reduce the extruded rod to final wire diameter (see the article "Wire, Rod, and Tube Drawing" in Volume 14 of the 9th Edition of *Metals Handbook*). As-extruded superconducting composites are 50 to 90 mm (2 to 3½ in.) in diameter. The reduction in area per pass is 15 to 25%. Again, as with extrusion, centerbursting must be avoided. Rod is drawn straight, on a draw bench (Fig. 19), to 20 to 25 mm (25/32 to 1 in.) diameter. At this size the rod is coiled. The coil diameter should be large enough to prevent plastic straining of the filaments outside the neutral axis during bending. Coiling the material at this time serves a second purpose. It is at this size, 20 to 25 mm (25/32 to 1 in.), that the first of several intermediate precipitation heat treatments is administered, and it is much easier to uniformly heat treat a coil than it is to heat treat a long, straight rod. From 25 mm (1 in.) to approximately 8 mm (5/16 in.) the rod is drawn on single-capstan bull-blocks. Below this diameter, multiple die machines are used. Final wire diameters can range from 0.05 to 2.8 mm (0.002 to 0.11 in.) depending on the application.

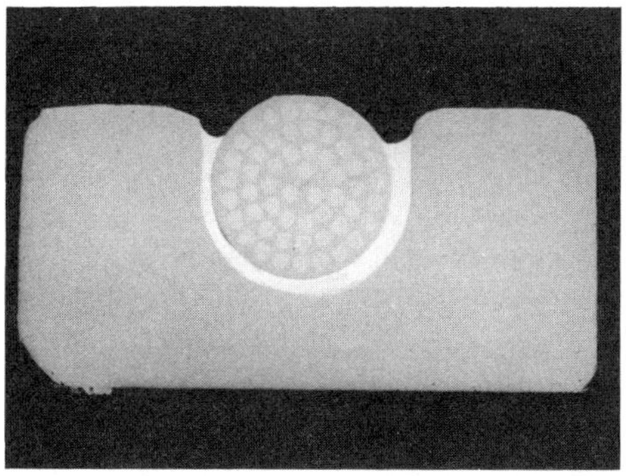

Fig. 20 Cross section of a 54-filament strand soldered in a copper channel (Cu:Sc is 15:1) for high-field (1 to 2 T, or 10 to 20 kG) MRI magnets. Courtesy of Supercon, Inc.

Fig. 21 Cross section of a Cu/NbTi strand continuously coextruded within high-purity aluminum with rectangular dimensions (2.1 × 3.2 mm, or 0.083 × 0.126 in.). Courtesy of IGC Advanced Superconductors Inc.

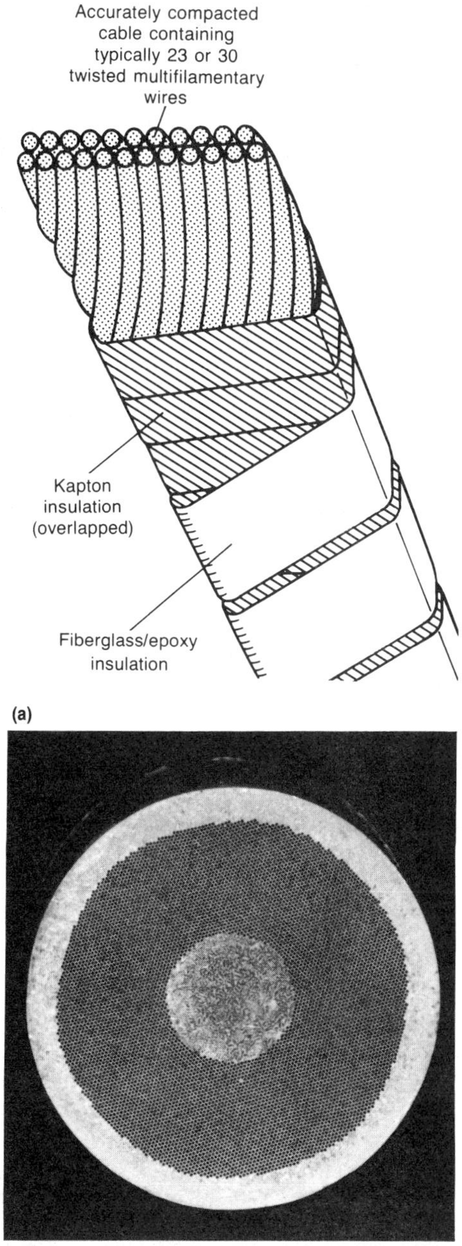

Fig. 22 Schematic representation and close-up photo of a 23-strand transposed Rutherford cable for the superconducting supercollider. Polyamide (Kapton) film wrap allows slippage with low friction as the coils are energized, reducing thermal transients in the conductor. Widely spaced fiberglass insulation promotes better cryogenic cooling. Courtesy of the SSC Laboratory. (b) Magnified conductor cross section contains 7248 NbTi filaments, each 6 μm (240 μin.) in diameter at a wire diameter of 0.81 mm (0.032 in.). This conductor was designed and fabricated by the author for the inner windings of an SSC dipole magnet. Courtesy of Supercon, Inc.

Twisting and Final Sizing

All multifilamentary superconducting composites must be twisted in order to:

- Minimize flux-jump instability in the presence of a time-varying external field
- Reduce eddy-current losses
- Eliminate self-field instabilities

The amount of twist required depends on the conditions of operation. Most applications envisioned for NbTi superconducting magnets operate in a steady-state mode, with minimal applied-field ramping, and therefore, require relatively little twist. For instance, the present SSC specification calls for 2 twists/inch, or 15.7 times the wire diameter (0.808 mm, or 0.0318 in.). When large applied-field ramp rates are required, as is the case for ac motors and generators, the twist pitch must be very small. Four to eight times the wire diameter is the norm.

Twisting is usually the second-to-last step in the mechanical processing of NbTi superconductors. The last step is final sizing or shaping. In this way, the twist is locked in. In some twisting machines where there is an external take-up the final sizing operation is performed in tandem with the twisting. If the take-up is an internal arrangement, the final drawing pass must be performed in a separate operation. Some magnet wire is shaped into a square or rectangular cross section. This can be accomplished with a Turk's head comprised of four independently adjusted rolls. The rolls can be passive or powered.

Additional Stabilizer

For those applications where greater stability is required, it is often easier and more economical to supplement the stabilizing matrix of a standard product than it is to design and fabricate a monolith with a large cross section with the necessary stabilizer-to-superconductor ratio. To reiterate a point made in the "Extrusion" section of the article, the length of a monolith is limited by the volume of the starting billet and the size of the extrusion press. Figure 20 shows the cross section of a 54-filament composite (see Fig. 7) that has been continuously soldered into the channel of a high-purity copper strip. The resulting Cu:Sc ratio is 15:1. Wire-in-channel conductors of this type are used in high field (1 to 2 T, or 10 to 20 kG) MRI magnets.

The advantages of aluminum stabilization have been outlined previously in the "Matrix Materials" section of this article. One of several techniques available to the wire manufacturer as a means of incorporating high-purity aluminum is to process a copper matrix composite to final wire and then to add the aluminum in a continuous coextrusion process (Ref 57). Figure 21 shows the cross section of a hybrid aluminum-clad Cu/NbTi conductor. Cables and aspected monoliths have also been clad with high-purity aluminum (Ref 57-59). Applications for aluminum-stabilized materials include mobile MRI units, particle detector magnets, isotope separation, magnetic energy storage, and fusion containment devices.

Cabling

For many large-scale applications, the required current-carrying capacity exceeds the capabilities of monolithic strand. For these situations, magnet designers have resorted to cabling or braiding many individual strands together to obtain a larger cross section. This assemblage must be fully transposed for the same reasons that individual strands must be twisted. Several methods have been employed including litz configurations and flat braid produced by rotary or maypole techniques. Both techniques have low filling factors, and flat braid has the added disadvantage that crossed wires can lead to filament degradation, especially when compacted.

The favored approach is the fabrication of a two-layer Rutherford cable. Strands run parallel over the entire length of the cable and suffer minimal damage during compaction. The early dipole magnets for the superconducting supercollider were designed to incorporate 23- and 30-strand keystoned cables of this type (Ref 60). Figure 22 shows a schematic illustration of an SSC cable. Figure 23 shows both inner and outer SSC cables in cross section. A unique cabling machine was designed and constructed for

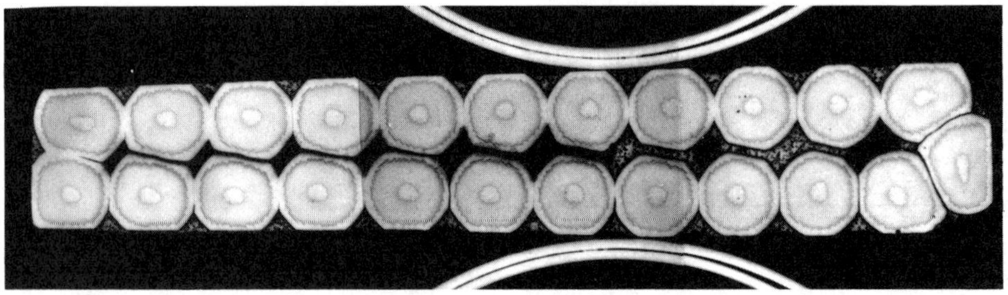

(a)

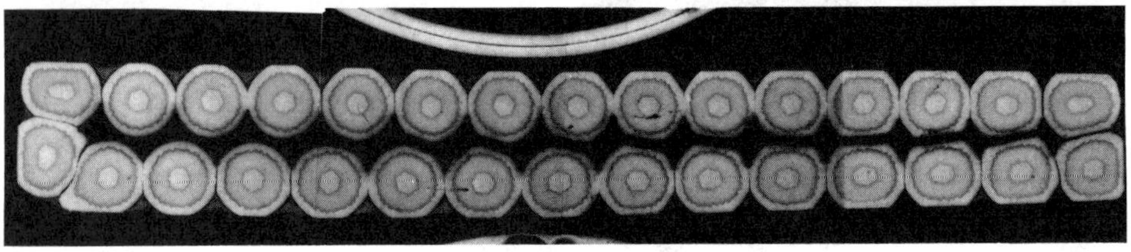

(b)

Fig. 23 Cross sections of cable used in windings of a 0.04 m SSC bore dipole magnet. (a) 23-strand cable for inner winding. (b) 30-strand cable for outer winding. Courtesy of Lawrence Berkeley Laboratory

the SSC project (Ref 61). Special features include:

- Planetary motion on the payoff spools to facilitate over- or under-twisting of the cable
- High rates of production (10.6 m/min, or 34.8 sfm)
- Power Turk's head for compaction/shaping and reduction of cable tension
- Broken-strand detectors
- In-line measurement capability and data acquisition

Figure 24 gives one perspective of the cabling-line arrangement. The cable produced on this machine shows little degradation of the strand due to cabling. Average values are 1 to 2% for 23-strand inner cable and 4 to 5% for 30-strand outer cable (Ref 62).

Superconductor Filament Properties

A number of factors are directly related to the ability to produce fine filaments of NbTi with very high current-carrying capacities. High-homogeneity NbTi alloy is frequently cited as a major factor. Of equal importance is the use of a diffusion barrier and its success in preventing the formation of brittle intermetallic compounds that degrade the filaments and impede the wire-drawing process. These considerations were discussed in the section "Assembly Techniques" in this article.

Other equally significant factors are:

- Uniformity of the array
- Filament spacing
- Employment of a carefully controlled strain-heat treatment cycle

It is insufficient to address one and not all of the design and process considerations mentioned when endeavoring to produce high-quality filaments.

Fig. 24 Primary components of the cabling production line for the superconducting supercollider. Shown (left-to-right) are the rotating drum with planetary payoffs, caterpuller supported by four air cushions for frictionless axial motion, and the in-line measuring machine. Courtesy of Lawrence Berkeley Laboratory

Geometric Uniformity

One of the primary causes of filament nonuniformity is the differential hardness of the filaments relative to that of the matrix. As nonheat-treated filaments are drawn down, both they and the matrix harden progressively and little sausaging (variation in cross-sectional area) occurs. When heat treatments are superimposed on the reduction process, hardening of the filaments and softening of the matrix takes place at each heat-treatment stage. During the reduction pass immediately following each heat treatment, the differential hardness is at its greatest, and this increases with each successive treatment until a point is reached where instability occurs and the filaments begin to reduce in cross section locally. This is known as sausaging. Less sausaging takes place if the average filament spacing (S) to filament diameter (D) ratio is small as more closely spaced filaments tend to mechanically support one another (Ref 63, 64).

This observation is a logical progression from the work described earlier for subbundles of filaments doubly stacked in which filaments on the outer edge of each subbundle tend to deform more readily than those in the center of the subbundle. By arranging hexagonally-shaped monofilaments in a single-stack array, the number of unsupported filaments is greatly reduced. Furthermore, the hexagonal subelements yield more uniform spacing between individual filaments. A qualitative examination of a wide range of superconducting materials produced by different manufacturers has shown that when the filaments are closer together they tend to sausage less than under otherwise similar conditions.

Sausaging results in low n values, where n is a measure of the broadening of the superconducting normal-state transition. This is frequently used as an indicator of the average filament quality (Ref 65) and can be represented by the equation:

$$\rho = VA/I = \text{Constant} \times I^n \quad \text{(Eq 1)}$$

where ρ is the resistivity, V is the voltage per unit length, A is the total wire cross section, and I is the measured current in the sample. The n value is inversely related to the width of the *V-I* transition. The influence of *n* on J_c was clearly shown for CBA wires in earlier work (Ref 66).

It is now generally agreed that more closely spaced filaments yield higher-quality filaments and are, therefore, able to carry higher currents. *S/D* ratios of 0.2 or less are now common in the fine-filament NbTi composites exhibiting the highest current-carrying capacities. Conversely, if the filaments are too closely spaced, proximity-effect coupling of the filaments can take place. The advent of copper-manganese alloys (see the section "Matrix Materials" in this article) offers a solution to proximity couplings. The present SSC specification calls for 6 μm (240 μin.) filaments with a minimum interfilamentary spacing of 1 μm (40 μin.) when the matrix is pure copper. This yields an *S/D* ratio of 0.167.

Heat-Treatment and Strain Cycles

The critical current density of Cu/NbTi composite conductors depends on two microstructure-related factors:

- Flux pinning by the cold-work-induced elongated grain boundaries
- Flux pinning by α-phase precipitates that are also the product of extreme cold work plus heat treatment

Manipulation of these precipitates by appropriate thermomechanical processing is responsible for J_c optimization.

The strand diameter required for the inner cable of an SSC dipole magnet is 0.81 mm (0.0319 in.). If one starts with a 305 mm (12 in.) diam billet, the total strain to which the material is subjected is approximately 12. Strain is defined as the true strain, ϵ:

$$\epsilon = \ln \left[\frac{A_0}{A} \right] \quad \text{(Eq 2)}$$

where A_0 and A are the initial and final cross-sectional areas, respectively. If the first precipitation heat treatment is delayed until a strain of 6, and the strain after the last heat treatment is 4, then there is room for only three heat treatments (Ref 67-68). The exact prestrain depends on heat-treatment temperature, alloy composition, and alloy homogeneity (Ref 69). It has been shown that the prestrain can be reduced to 4 if high-homogeneity Nb-46.5Ti alloy is used with a 420 °C (790 °F) heat treatment (Ref 70). Reduction of the prestrain to lower values can form Widmanstätten α-Ti and/or ω-phase precipitates that could lead to ductility problems (Ref 67).

An available strain of 12 assumes that hot extrusion has the same strain effect as cold drawing. If cold work is not retained throughout conventional extrusion, the total strain available is significantly reduced from the calculations referenced above. It has been demonstrated that employing HIP techniques (to densify and bond the composite) followed by hydrostatic extrusion allows one to retain significant amounts of cold work, as discussed previously (Ref 53).

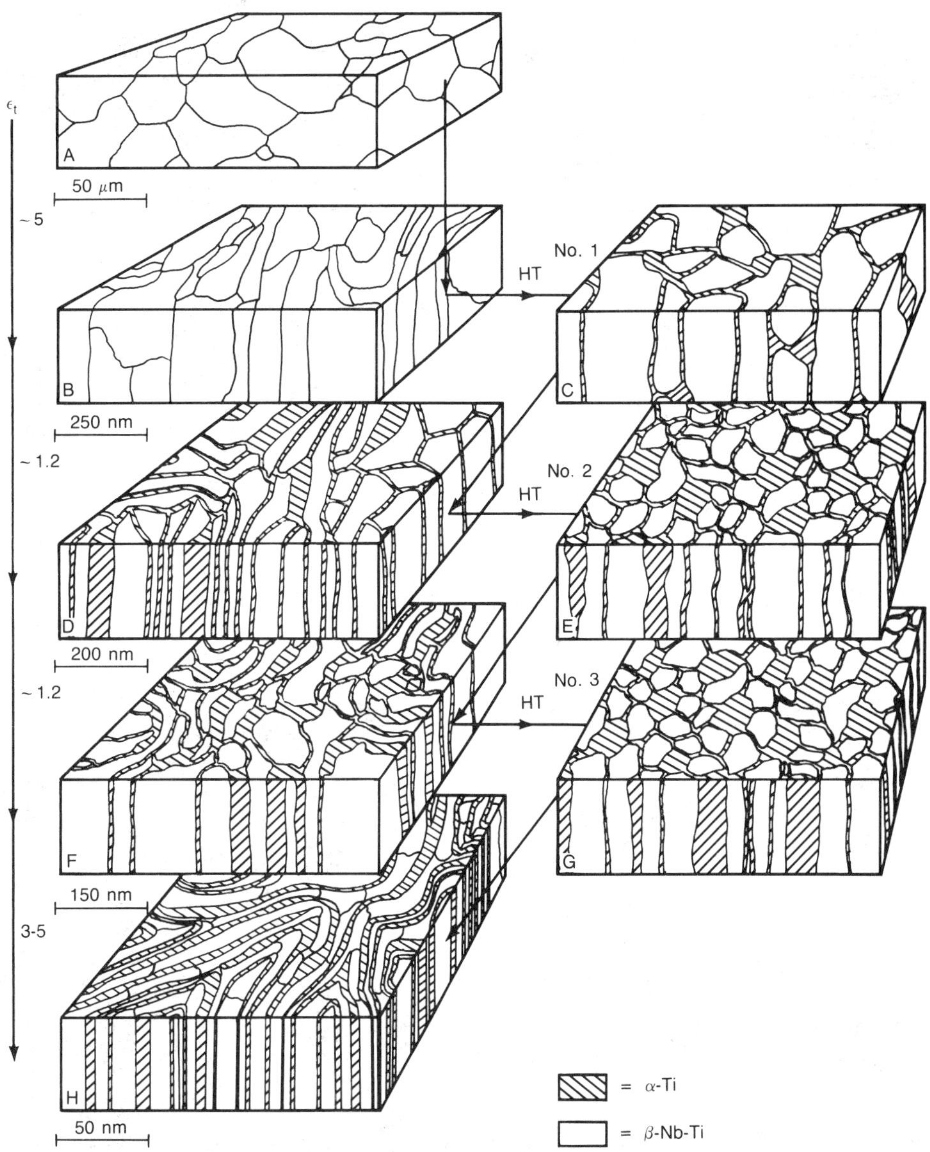

Fig. 25 Heat-treatment/strain sequence incorporating three intermediate precipitation heat treatments to illustrate the development of nanometer scale structures in composites of NbTi. Courtesy of the Applied Superconductivity Center, University of Wisconsin-Madison

The stored energy, achieved after sufficient prestrain, enables the precipitation of α-Ti in the β-NbTi structure after relatively low-temperature (375 to 420 °C, or 705 to 790 °F) heat treatments. Nucleation of the α-Ti precipitate takes place at the NbTi grain-boundary triple points, in addition to the formation of a thin titanium-rich film at the grain boundaries. The time-at-temperature ranges from 10 to 80 h, and varies from manufacturer to manufacturer. From an economic perspective, long residence times in a heat treat furnace increase cost and slow production. Although it is usually consistent within a single processing cycle (that is, three 40-h heat treatments at 375 °C, or 705 °F), recent work has shown that the first heat treatment or two need not be as long as subsequent heat treatments in order to maximize the volume percentage of α-Ti (Ref 71). Optimized composites contain approximately 20 vol% α-Ti.

A heat-treatment/strain sequence incorporating three intermediate precipitation heat treatments is represented in Fig. 25. As illustrated, the first heat treatment (section C) produces equiaxed precipitates approximately 50 to 100 nm in size. Figure 26(a) is a transmission electron microscopy (TEM) micrograph of a NbTi composite, in transverse cross section, after its first precipitation heat treatment (corresponding to section C of Fig. 25). Figure 26(b) is a longitudinal (parallel to the drawing axis) cross section of the same composite at the same stage of processing.

Further cold work and heat treatment deform the precipitates into ribbons with

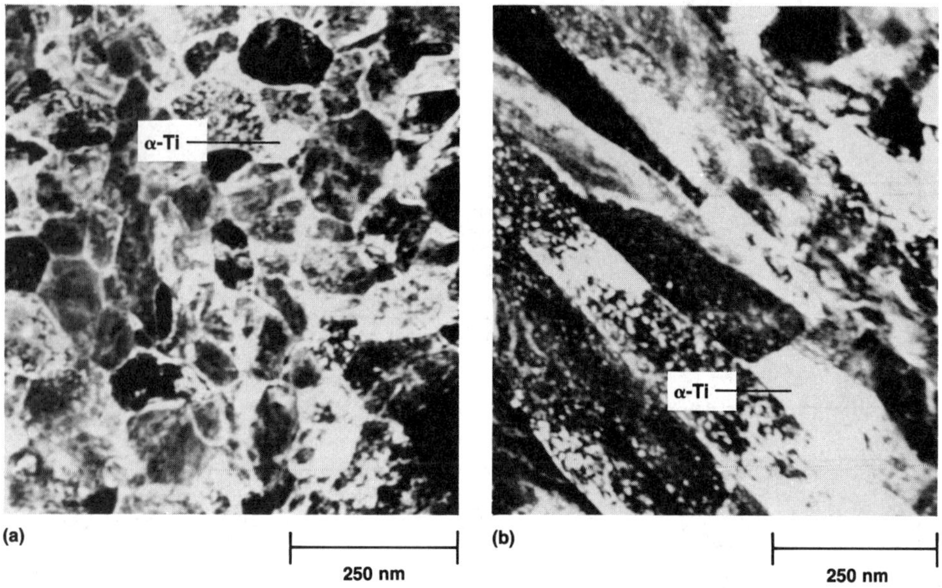

Fig. 26 TEM micrographs of NbTi composite in cross sections after first precipitation heat treatment. (a) Transverse (corresponding to section C of Fig. 25). (b) Longitudinal. Courtesy of the Applied Superconductivity Center, University of Wisconsin-Madison

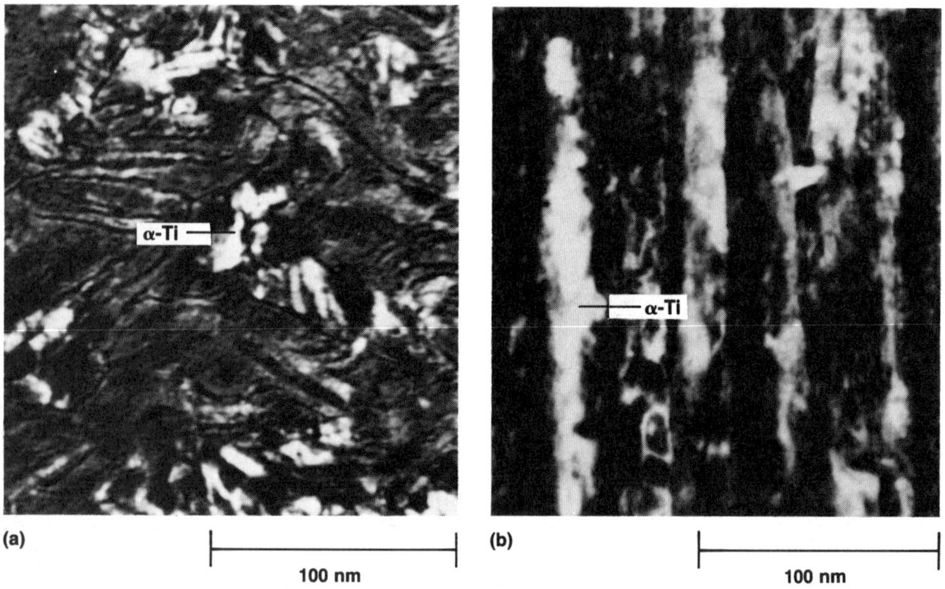

Fig. 27 TEM micrographs of NbTi composite cross sections, after third precipitation heat treatment and a final strain of 4.13. (a) Transverse (corresponding to section H of Fig. 25). (b) Longitudinal. Courtesy of the Applied Superconductivity Center, University of Wisconsin-Madison

large aspect ratios, as depicted in section H of Fig. 25. Typical ribbons are 1 to 2 nm thick, 10 to 300 nm in length, and spaced approximately 4 to 8 nm apart (Ref 72). It is this uniform ribbon morphology that acts to pin flux so efficiently. Figure 27(a) is a TEM micrograph showing a transverse cross section of the same composite that received two additional heat treatments followed by a drawing strain of 4.13 (corresponding to section H of Fig. 25). Figure 27(b) shows the same sample in longitudinal cross section.

Additional heat treatments (>3) yield higher current-carrying capacities, but overall ductility and piece length can be compromised (Ref 73). Furthermore, large final diameters do not have the requisite strain space to achieve the very high J_c values, unless hydrostatic extrusion is utilized. Figure 28 plots J_c versus strain (since the last precipitation heat treatment) for a multifilamentary composite that received 4, 5, 6, 7, and 8 heat treatments. Three observations can be made as follows:

- J_c increases with the number of heat treatments until saturation
- A saturation limit is reached for seven heat treatments

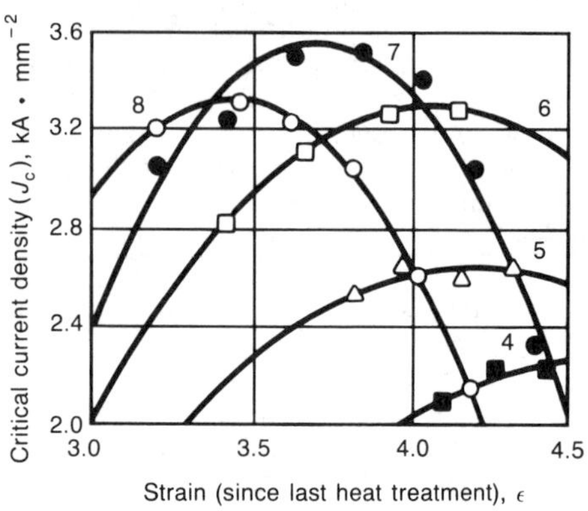

Fig. 28 Plot of critical current density versus strain (since the last precipitation heat treatment) for a multifilamentary composite that received 4, 5, 6, 7, and 8 heat treatments. Courtesy of Supercon, Inc.

- Strain required to obtain the peak J_c decreases with more heat treatments

This work did not benefit from microstructural analysis using TEM techniques; therefore, the amount of α-Ti present is unknown and the observations general in nature.

When standard three-heat-treatment processing is applied to NbTi alloys with higher percentages of Ti, Widmanstätten α-Ti and/or ω-phase precipitates can be produced (Ref 74). Higher-temperature heat treatments tend to suppress the ω-phase precipitation (Ref 69). It has been demonstrated that when titanium content increases by 1 wt%, the prestrain should be increased 0.74 to inhibit the precipitation of the hardening phases (Ref 69). Furthermore, the rate of precipitation increases markedly with increasing Ti content (Ref 69). It may, therefore, be possible to compensate for larger required prestrains with fewer intermediate heat treatments.

Superconductor Applications

Magnetic Resonance Imaging (MRI) is the first large-scale application of superconductivity and has now achieved the status of a mature industry with an annual turnover in excess of $150 000 000 (Ref 75). There are more than 1700 superconducting MRI units installed worldwide.

When the human body is exposed to a magnetic field, the protons in water and other molecules align themselves relative to this field. When a burst of radiofrequency energy having the correct resonant frequency is applied these protons are excited, and when the pulse decays they return to their former state with a release of energy. This energy is detected in several different ways and used to create an image, which in turn gives significant information about tissue that is frequently lacking in x-ray computerized axial tomography (CAT) scans. In addition, little patient preparation is re-

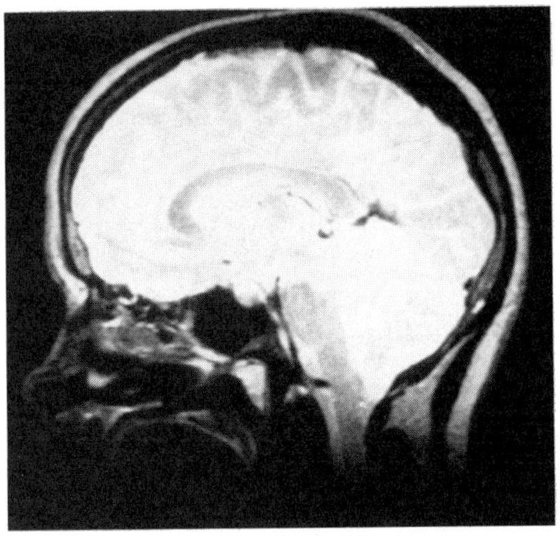

Fig. 29 MRI brain scan image. Courtesy of Oxford Superconducting Technology

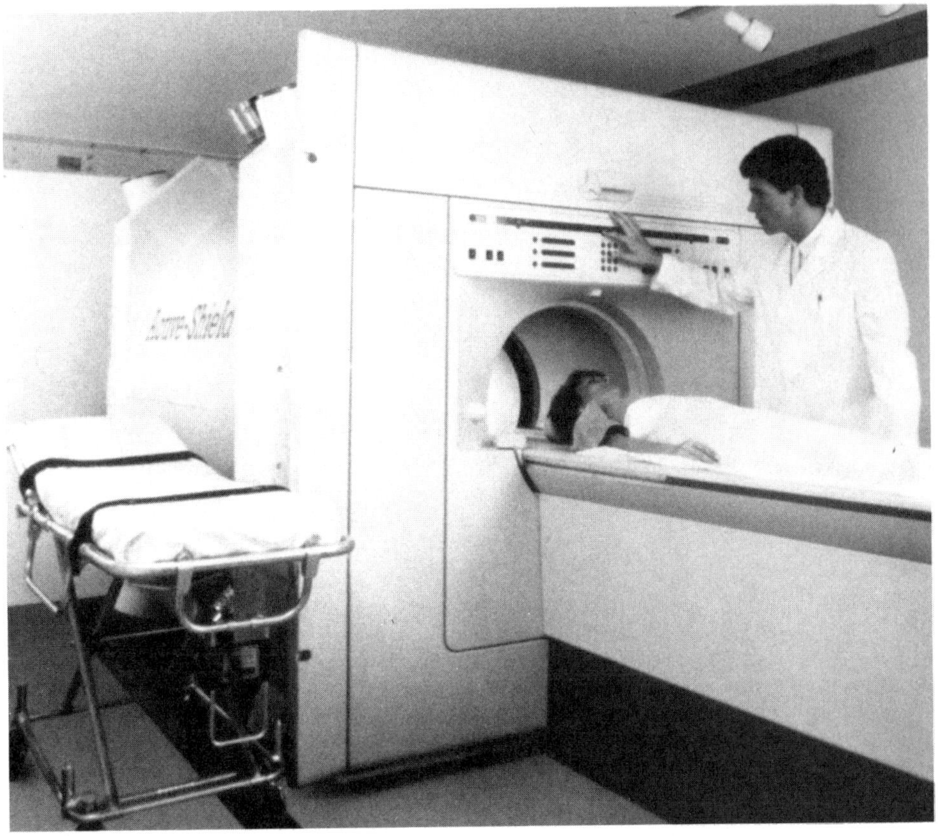

Fig. 30 Whole-body 2 T (20 kG) MRI unit with an active shield. Courtesy of Oxford Superconducting Technology

quired and there is no exposure to ionizing radiation.

Magnetic resonance imaging is particularly good for studies of the brain, liver, and kidneys. Figure 29 shows an MRI image of the brain. Higher fields for imaging produce better images faster. Higher fields are required for biochemistry where elements other than hydrogen (that is, ^{31}P, ^{23}Na, ^{39}K, ^{43}Ca, and ^{14}N) could be used to obtain additional information. Lower-field superconducting magnets with large homogeneous volumes and 1 m (3 ft) bores are available in a range of field strengths up to 2 T (20 kG) (Fig. 30).

High-energy physics (HEP) has for many years been concerned with obtaining an understanding of the nature of elementary particles of which all matter is composed and the forces through which matter interacts. The machines required to investigate these particles and forces have grown in size over the years and begun to consume more energy, primarily in the magnets required to confine the particles and control their behavior. The only way known to achieve the particle energies now required for further investigations at an acceptable cost is to employ superconducting magnets. High-energy physics has been the largest consumer of NbTi superconducting materials and has provided the impetus for much of the improved performance observed in recent years.

The world's first high-energy superconducting accelerator, Fermilab's Tevatron, was successfully tested in July 1983. The 6 km (3.7 mi) ring contains 774 superconducting dipoles and 216 quadrupoles, like those shown in Fig. 31. Other superconducting accelerators now in operation or under construction are the hadron-electron ring anordnung (HERA) at Deutsche Electronen Syncrotron (DESY) in Hamburg, FRG; the accelerating and storage complex (UNK) in Serpukhov, USSR; the TRISTAN facility at the Japanese atomic energy facility (KEK); the relativistic heavy ion collider (RHIC) at

Fig. 31 Portion of the Fermilab Tevatron main 6 km (3.7 mi) ring. Courtesy of Fermi National Accelerator Laboratory

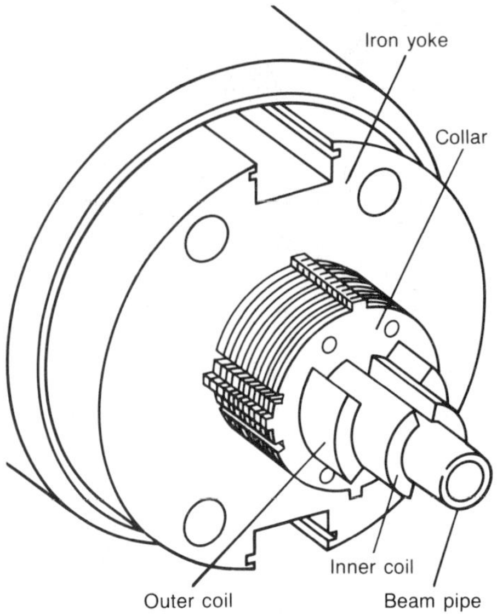

Fig. 32 Isometric drawing of the cold-mass assembly of a dipole magnet for the SSC project. Courtesy of Superconducting SuperCollider Laboratory

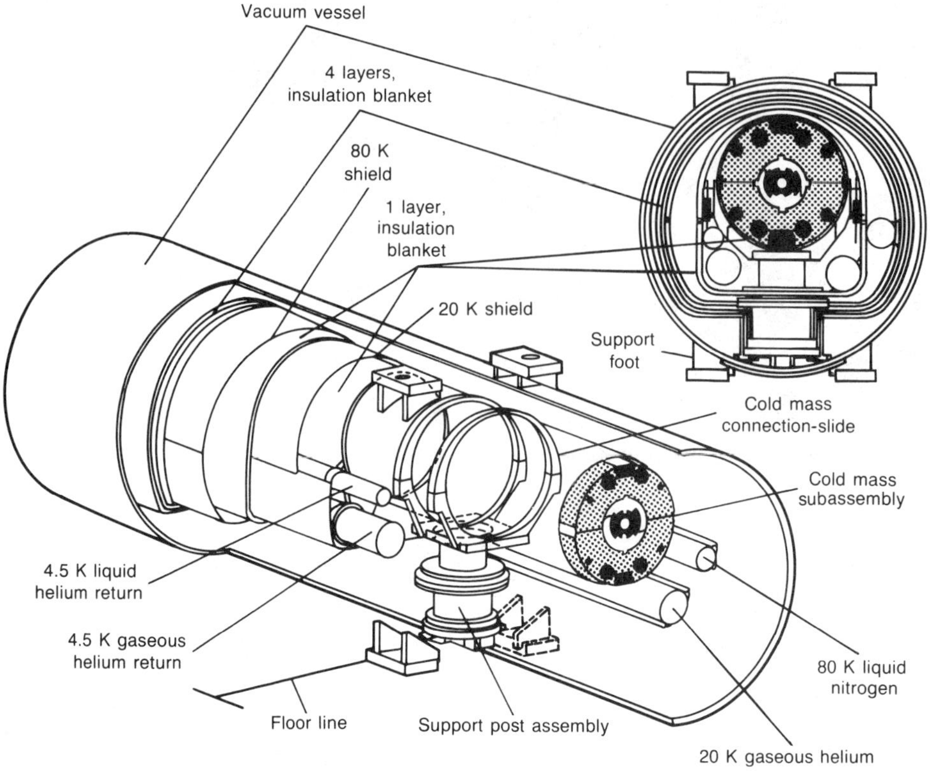

Fig. 33 Schematic drawing of a dipole magnet for the SSC project. Courtesy of Superconducting SuperCollider Laboratory

Brookhaven National Laboratory; the continuous electron beam accelerator facility (CEBAF); the large hadron collider (LHC) at Center for European Research (CERN); and the superconducting supercollider (SSC) in Texas.

The superconducting supercollider will be 83 km (51 mi) in circumference and generate 20 TeV energies in contrarotating beams of protons. The main rings will be composed of 7680 dipole magnets and 1776 quadrupole magnets. Figure 32 is an isometric drawing of the cold-mass assembly of an SSC dipole magnet. A schematic drawing of an SSC dipole magnet is shown in Fig. 33. It is estimated that more than 1.8 Gg of NbTi superconducting cable will be required. Construction is scheduled for completion at the end of this decade at a cost exceeding $6 000 000 000.

Magnetic Confinement for Thermonuclear Fusion. If thermonuclear fusion is to become a viable energy source, the plasma produced must be confined. The most highly developed fusion-containment machines are tokamaks, which operate on the principle of toroidal confinement by magnetic fields. Superconducting magnets are preferred over copper magnets, otherwise it is possible that more electrical energy will be used to confine the plasma than the machine produces.

The principal effort for the development of superconducting coils incorporating

Fig. 34 LCT facility during magnet installation with four of six coils shown. Courtesy of Oak Ridge National Laboratory

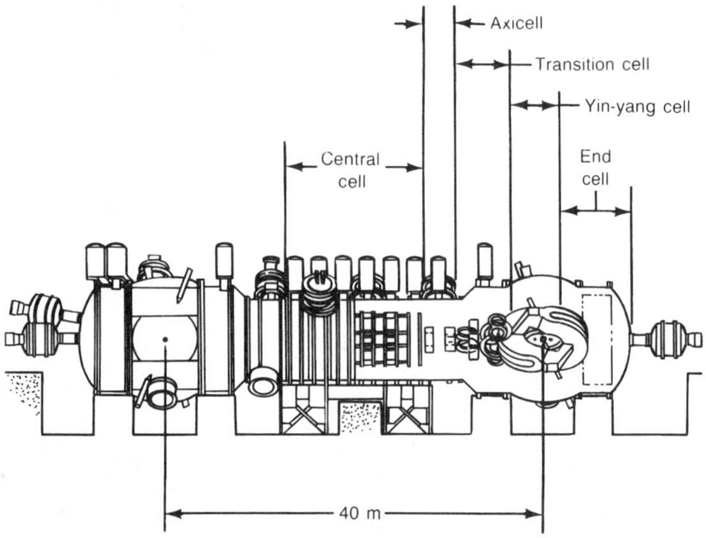

Fig. 35 Schematic configuration of magnets for Axicell MFTF-B. Courtesy of Lawrence Livermore National Laboratory

NbTi materials for tokamaks was the large coil test (LCT) at the Oak Ridge National Laboratory (ORNL) (Ref 76). Of the six coils in this array, five were fabricated using NbTi superconductors. All the coils have a peak field of 8 T (80 kG) with conductor currents in the range of 10 to 19 kA. Figure 34 shows the LCT test stand during magnet installation.

The alternative method of confinement is to restrict the plasma to a straight tube by means of plugs (or mirrors) at the end of the vessel. The Axicell Mirror Fusion Test Facility (Axicell MFTF-B) at Lawrence Livermore National Laboratory is an example of this approach. The configuration of the magnets is shown schematically in Fig. 35. The solenoids used in the central cell, the transition cell, and the Yin-yang cell use NbTi superconducting materials, and the Axicell magnets are made with Nb_3Sn.

Superconducting Magnetic Energy Storage. All electric utilities experience fluctuations in demand. Excess electrical energy, generated during off-peak periods, can be stored for use during the periods of highest demand. It is estimated that 5 to 15% of the total generating capacity could be in the form of stored energy. Superconducting magnetic energy storage (SMES) offers an efficient (~95%) method for storage and retrieval of the excess electric power. Presently, there is a joint program between the Electric Power Research Institute (EPRI) and the strategic defense initiative (SDI) to design and build an engineering test model (ETM), with an energy storage capacity of 20 MW · h (Ref 77).

Power Applications. Thus far, superconducting electric power generators have had only a superconducting rotor winding. The feasibility of fabricating low-loss 50 to 60 Hz conductors, as described above, leads to ac generators with superconducting armature windings. Making both the motor and the armature superconducting will allow a large reduction in the size of the device and simplify the cryogenic cooling requirement. Potentially 50% smaller and 1% more efficient than other types of generators with the same power output, superconducting motors and generators should find application not only in the utility industry but in aerospace and shipboard environments as well.

More efficient transformers with improved steady-state and transient stability, which when integrated over their life cycle result in appreciable cost savings, are possible with superconducting windings. The proposed application would eliminate the need for step-up transformers on a primary distribution system.

Additional Applications. Other NbTi applications of interest include nuclear magnetic resonance (NMR) for chemical spectroscopy, magnetic separation, magnetohydrodynamic (MHD) power generation, power transmission, magnetic levitation, and proton beam therapy.

REFERENCES

1. J.B. Vetrano and R.W. Boom, *J. Appl. Phys.*, Vol 16, 1965, p 1179
2. H. Riemersma, J.K. Hulm, and B.S. Chandasekhar, Flux Jumping and Degradation in Superconducting Solenoids, in *Proceedings of the Cryogenic Engineering Conference*, 1963
3. Z.J.J. Stekly and J.L. Zar, Stable Superconducting Coils, *IEEE Trans. Nucl. Sci.*, June 1965
4. L. Zhou, Recent Developments of Superconducting Wire in China, in *Advances in Cryogenic Engineering*, Vol 34, A.F. Clark and R.P. Reed, Ed., Plenum Press, 1988, p 983
5. V.Ya. Fil'kin, V.P. Kosenko, V.L. Mette, K.P. Myznikov, A.D. Nikulin, V.A. Vasiliev, G.K. Zelensky, and A.V. Zlobin, The Properties of Industrial Superconducting Composite Wires for the UNK Magnets, in *Advances in Cryogenic Engineering*, Vol 36, F.R. Fickett and R.P. Reed, Ed., Plenum Press, 1990, p 317
6. J. McKinnell, P.J. Lee, R. Remsbottom, D.C. Larbalestier, P.M. O'Larey, and W.K. McDonald, High Titanium NbTi Alloys—Initial High Critical Current Density Properties, in *Advances in Cryogenic Engineering*, Vol 34, A.F. Clark and R.P. Reed, Ed., Plenum Press, 1988, p 1001
7. T.G. Berlincourt and R.R. Hake, *Phys. Rev.*, Vol 131, 1963, p 140
8. L.J. Neuringer and V. Shapira, *Phys. Rev. Lett.*, Vol 17, 1966, p 81
9. M. Suenaga and K.M. Ralls, Some Superconducting Properties of Ti-Nb-Ta Ternary Alloys, *J. Appl. Phys.*, Vol 40, Oct 1969, p 4457
10. D.G. Hawksworth, "The Upper Critical Field and High Field Critical Current Density of Niobium-Titanium and Niobium-Titanium Based Alloys", Ph.D. thesis, University of Wisconsin, 1981; see also D.G. Hawksworth and D.C. Larbalestier, Further Investigations of the Upper Critical Field and the High Field Critical Current Density in Nb-Ti and its Alloys, *IEEE Trans. Magn.*, Vol 17 (No. 1), 1981, p 49
11. E. Gregory, T.S. Kreilick, F.S. von Goeler, and J. Wong, Preliminary Results on Properties of Ductile Superconducting Alloys for Operation to 10 Tesla and Above, in *Proceedings of the 12th International Cryogenic Engineering Conference*, R.G. Scurlock and C.A. Bailey, Ed., Butterworths, 1988, p 874
12. N. Zhou, X. Wu, Y. Li, Y. Yie, and L. Zhou, The Properties of Multifilamentary NbTi-25 Ta/Cu Superconductors in High Magnetic Fields, *Advances in Cryogenic Engineering*, Vol 34, A.F. Clark and R.P. Reed, Ed., Plenum Press, 1988, p 995
13. A.D. McInturff, J. Carson, D.C. Larbalestier, P.J. Lee, J. McKinnell, H. Kanithi, W.K. McDonald, and P.M. O'Larey, "Ternary Superconductor NbTiTa for High Field Superfluid Magnets," paper presented at INTERMAG (Brighton, England), April 1990
14. R.V. Gaines, Geology of Niobium and Tantalum Deposits, in *Proceedings of the International Symposium on Tantalum and Niobium*, Tantalum-Niobium International Study Center, Brussels, Nov 1988, p 99
15. H. Hillmann, Fabrication Technology of Superconducting Material, in *Superconducting Materials Science*, S. Foner and B.B. Schwartz, Ed., Plenum Publishing, 1981, p 295
16. D.C. Larbalestier, A.W. West, W.S. Starch, W.H. Warnes, P.J. Lee, W.K. McDonald, P.M. O'Larey, K. Hemachalam, B.A. Zeitlin, R.M. Scanlan, and C.E. Taylor, High Critical Current Densities in Industrial Scale Composites Made From High Homogeneity Nb 46.5 wt% Ti, *IEEE Trans. Magn.*, Vol 21, 1985, p 269
17. R.I. Asfahani, P. Kumar, and Y.V. Murty, Vacuum Arc Remelting (VAR) of Refractory Metals, in *Proceedings of the Vacuum Metallurgy Conference*, Iron & Steel Society, 1986
18. M.T. Taylor, A. Woolcock, and A.C. Barber, Strengthening Superconducting Composite Conductors for Large Magnet Construction, *Cryogenics*, Vol 8, 1968, p 317
19. M. Young, E. Gregory, E. Adam, and W. Marancick, Fabrication and Properties of an Aluminum Stabilized NbTi Multifilament Superconductor, in *Advances in Cryogenic Engineering*, Vol 24, K.D. Timmerhaus, R.P. Reed, and A.F. Clark, Ed., Plenum Press, 1978, p 383
20. J. Bishop, E. Gregory, and J. Wong, Aluminum Stabilized Multifilamentary Superconductors, in *Proceedings of ICEC 11*, IPC Science and Technology Press, 1986
21. K.T. Hartwig, D. Yu, and A. Khalil, Aluminum-Nickel and Aluminum-Calcium as Potential Stabilizer Alloys, in *Proceedings of ICMC*, Butterworths, 1982, p 489
22. W.M. Griffith, R.E. Sanders, Jr., and G.J. Hildeman, Elevated Temperature Aluminum Alloys for Aerospace Applications, in *High-Strength Powder Metallurgy Aluminum Alloys*, M.J. Koczak and G.J. Hildeman, Ed., The Metallurgical Society of AIME, 1982, p 209
23. C.E. Oberly and J.C. Ho, The Origin and Future of Composite Aluminum Conductors, in *Advances in Cryogenic*

Engineering, Vol 36, F.R. Fickett and R.P. Reed, Ed., Plenum Press, 1990, p 645

24. J.C. Ho, C.E. Oberly, H.L. Gegel, W.M. Griffith, J.T. Morgan, W.T. O'Hara, and Y.V.R.K. Prasad, A New Aluminum-Base Alloy With Potential Cryogenic Applications, in *Advances in Cryogenic Engineering*, Vol 32, A.F. Clark and R.P. Reed, Ed., Plenum Press, 1986, p 437
25. K.T. Hartwig and R.J. DeFrese, Mechanical and Electrical Testing of Composite Aluminum Cryoconductors, in *Advances in Cryogenic Engineering*, Vol 36, F.R. Fickett and R.P. Reed, Ed., Plenum Press, 1990, p 709
26. F.R. Fickett and C.A. Thompson, Anomalous Magnetoresistance in Al/Al Alloy Composite Conductors, in *Advances in Cryogenic Engineering*, Vol 36, F.R. Fickett and R.P. Reed, Ed., Plenum Press, 1990, p 671
27. T.S. Kreilick, E. Gregory, J. Bishop, F.S. von Goeler, and I. Levin, unpublished work
28. J.R. Cave, A. Fevrier, T. Verhaege, A. Lacaze, and Y. Laumond, Reduction of AC Loss in Ultra-Fine Multifilamentary NbTi Wires, *IEEE Trans. Magn.*, Vol 25 (No. 2), March 1989, p 1945
29. I. Hlasnik, S. Takacs, V.P. Burjak, M. Majoros, J. Krajcik, L. Krempasty, M. Polak, M. Jergei, T.A. Korneeva, O.N. Mironova, and I. Ivan, Properties of Superconducting NbTi Superfine Filament Composites With Diameter $\leqq$0.1 μm, *Cryogenics*, Vol 25, 1985, p 558
30. K. Ohmatsu, M. Nagata, M. Kawashima, H. Tateishi, and T. Onishi, AC Loss of NbTi Superconducting Wires With Fine Filament, *IEEE Trans. Magn.*, Vol 25 (No. 2), March 1989, p 2105
31. T.S. Kreilick, E. Gregory, and J. Wong, The Design and Fabrication of Multifilamentary NbTi Composites Utilizing Various Matrix Materials, *J. Less-Common Met.*, Vol 139 (No. 1), 1988, p 45
32. A.K. Ghosh, W.B. Sampson, E. Gregory, T.S. Kreilick, and J. Wong, The Effect of Magnetic Impurities and Barriers on the Magnetization and Critical Current of Fine Filament NbTi Composites, *IEEE Trans. Magn.*, Vol 24 (No. 2), March 1988, p 1145
33. A.K. Ghosh, W.B. Sampson, E. Gregory, and T.S. Kreilick, Anomalous Low Field Magnetization in Fine Filament NbTi Conductors, *IEEE Trans. Magn.*, Vol 23 (No. 2), March 1987, p 1724
34. E.W. Collings, Stabilizer Design Considerations in Fine-Filament Cu/NbTi Composites, in *Advances in Cryogenic Engineering*, Vol 34, A.F. Clark and R.P. Reed, Ed., Plenum Press, 1988, p 867
35. T.S. Kreilick, E. Gregory, R.M. Scanlan, A.K. Ghosh, W.B. Sampson, and E.W. Collings, Reduction of Coupling in Fine Filamentary Cu/NbTi Composites by the Addition of Manganese to the Matrix, in *Advances in Cryogenic Engineering*, Vol 34, A.F. Clark and R.P. Reed, Ed., Plenum Press, 1988, p 895
36. E. Gregory, T.S. Kreilick, J. Wong, E.W. Collings, K.R. Marken, Jr., R.M. Scanlan, and C.E. Taylor, A Conductor, With Uncoupled 2.5 μm Diameter Filaments, Designed for the Outer Cable of SSC Dipole Magnets, *IEEE Trans. Magn.*, Vol 25 (No. 2), March 1989, p 1926
37. T.S. Kreilick, E. Gregory, P. Valaris, and J. Wong, The Mechanical and Electrical Effects of Adding Manganese to the Copper Matrix of Multifilamentary NbTi Superconducting Composites, in *Proceedings of the 12th International Cryogenic Engineering Conference*, R.G. Scurlock and C.A. Bailey, Ed., Butterworths, 1988, p 857
38. R.B. Goldfarb, D.L. Ried, T.S. Kreilick, and E. Gregory, Magnetic Evaluation of Cu-Mn Matrix Material for Fine Filament Nb-Ti Superconductors, *IEEE Trans. Magn.*, Vol 25 (No. 2), March 1989, p 1953
39. S. Sakai, G. Iwaki, Y. Sawada, H. Moriai, and Y. Ishigami, Recent Development of the Cu/Nb-Ti Superconducting Cables for SSC in Hitachi Cable, Ltd., in *Proceedings of the 1st IISSC* (New Orleans), Feb 1989
40. E.W. Collings, K.R. Marken, Jr., A.J. Markworth, J.K. McCoy, M.D. Sumption, E. Gregory, and T.S. Kreilick, Critical Field Enhancement Due to Field Penetration in Fine-Filament Superconductors, in *Advances in Cryogenic Engineering*, Vol 36, F.R. Fickett and R.P. Reed, Ed., Plenum Press, 1990, p 255
41. E.W. Collings, K.R. Marken, Jr., M.D. Sumption, E. Gregory, and T.S. Kreilick, Magnetic Studies of Proximity-Effect Coupling in Very Closely Spaced Fine-Filament NbTi/CuMn Composites, in *Advances in Cryogenic Engineering*, Vol 36, F.R. Fickett and R.P. Reed, Ed., Plenum Press, 1990, p 231
42. W.R. Opie, P.W. Taubenblat, and Y.T. Hsu, A Fundamental Comparison of the Mechanical Behavior of Oxygen-Free and Tough-Pitch Coppers, *J. Inst. Met.*, Vol 98, 1970, p 245
43. J. Smets and R. Mortier, "The Influence of Oxygen During Hot Rolling and Drawing of Continuous Cast Rod," Metallurgie Hoboken Overpelt, Belgium, 1983
44. M. Garber, M. Suenaga, W.B. Sampson, and R.L. Sabatini, Effect of Cu_4Ti Compound Formation on the Characteristics of NbTi Accelerator Magnet Wire, *IEEE Trans. NS*, Vol 32, 1985, p 3681
45. D.C. Larbalestier, P.J. Lee, and R.W. Samuel, The Growth of Intermetallic Compounds at a Copper-Titanium Interface, in *Advances in Cryogenic Engineering*, Vol 32, A.F. Clark and R.P. Reed, Ed., Plenum Press, 1986, p 715
46. M.T. Taylor, C. Graeme-Barber, A.C. Barber, and R.B. Reed, Co-Processed Nb-25% Zr/Cu Composite, *Cryogenics*, June 1971, p 224
47. C.W. Curtis, "Production Development Program to Manufacture Cu/Nb 46.5 Ti Multi-filamentary Wire," Final report to Fermi National Accelerator Laboratory, Contract 50088, Feb 1976
48. T.S. Kreilick, E. Gregory, and J. Wong, Fine Filamentary NbTi Superconducting Wires, in *Advances in Cryogenic Engineering*, Vol 32, A.F. Clark and R.P. Reed, Ed., Plenum Press, 1986, p 739
49. P. Valaris, T.S. Kreilick, E. Gregory, and J. Wong, Refinements in the Billet Design for SSC Strand, *IEEE Trans. Magn.*, Vol 25 (No. 2), March 1989, p 1937
50. R.M. Scanlan, J. Royet, and R. Hannaford, Evaluation of Various Fabrication Techniques for Fabrication of Fine Filament NbTi Superconductors, *IEEE Trans. Magn.*, Vol 23 (No. 2), 1987, p 1719
51. W.A. Fietz, R.E. McDonald, and J.R. Miller, Preparation and Extrusion of Multifilamentary NbTi Conductor Billets, in *Proceedings of the 6th Symposium on Engineering Problems in Fusion Research*, Institute of Electrical and Electronics Engineers, 1976, p 256
52. S. Sakai, M. Seido, Y. Ishigami, K. Noguchi, H. Moriai, and A. Kobayashi, Production of Nb_3Sn and Nb-Ti Multifilamentary Superconducting Wires Using Warm Hydrostatic Extrusion, in *Proceedings of the International Cryogenic Materials Conference*, K. Tachikawa and A. Clark, Ed., Butterworths, 1982, p 301
53. T.S. Kreilick, R.J. Fiorentino, E.G. Smith, Jr., and W.W. Sunderland, Conversion of a 11MN Extrusion Press for Hydrostatic Extrusion of Superconducting Materials, in *Advances in Cryogenic Engineering*, Vol 36, F.R. Fickett and R.P. Reed, Ed., Plenum Press, 1990, p 51
54. G.E. Meyer, E.W. Collings, R.J. Fiorentino, F.J. Jelinek, and D.C. Carmichael, "Experimental Evaluation of Hydrostatic Extrusion for the Fabrication of Multifilament Superconducting Wire," Progress report to U.S. ERDA, Contract W-7405-eng-92, Task 93, Battelle-Columbus Laboratories, June 1975
55. *Forming and Forging*, Vol 14, 9th ed., *Metals Handbook*, ASM INTERNA-

TIONAL, 1988, p 315-326 and p 327-329
56. *Extrusion: Processes, Machinery, Tooling*, K. Laue and H. Stenger, Trans., American Society for Metals, 1981
57. H. Kanithi, D. Phillips, C. King, and B. Zeitlin, Development and Characterization of Aluminum Clad Superconductors, *IEEE Trans. Magn.*, Vol 24 (No. 2), March 1988, p 1029
58. H. Krauth, Recent Developments in NbTi Superconductors at Vacuumschmelze, *IEEE Trans. Magn.*, Vol 24 (No. 2), March 1988, p 1023
59. M. Ikeda, S. Meguro, I. Inoue, and A. Yamamoto, Aluminum-Stabilized Superconductor for the Topaz Thin Solenoid, in *Proceedings of ICEC 11*, G. Klipping and I. Klipping, Ed., Butterworths, 1986, p 675
60. R.M. Scanlan, J. Royet, and R. Hanaford, Fabrication of Rutherford-Type Superconducting Cables for Construction of Dipole Magnets, in *Proceedings of ICMC* (Shenyang, China), June 1988
61. J. Grisel, J.M. Royet, R.M. Scanlan, and R. Armer, A Unique Cabling Machine Designed to Produce Rutherford-Type Superconducting Cable for the SSC Project, *IEEE Trans. Magn.*, Vol 25 (No. 2), March 1989, p 1608
62. T.S. Kreilick, E. Gregory, D. Christopherson, G.P. Swenson, and J. Wong, Superconducting Wire and Cable for the Superconducting SuperCollider, in *Proceedings of IISSC* (New Orleans, LA), Feb 1989
63. E. Gregory, T.S. Kreilick, A.K. Ghosh, and W.B. Sampson, Importance of Spacing in the Development of High Current Densities in Multifilamentary Superconductors, *Cryogenics*, Vol 27, 1987, p 178
64. T.S. Kreilick and E. Gregory, Further Improvements in Current Density by the Reduction of Filament Spacing in Multifilamentary NbTi Superconductors, *Cryogenics*, Vol 27, 1987, p 401
65. W.H. Warnes, "The Resistive Critical Current Transition in Composite Superconductors," Ph.D. thesis, University of Wisconsin, May 1986; W.H. Warnes and D.C. Larbalestier, Determination of the Average Critical Current from Measurements of the Extended Resistive Transition, *IEEE Trans. Magn.*, Vol 23 (No. 2), March 1987, p 1183
66. M. Garber, M. Suenaga, W.B. Sampson, and R.L. Sabatini, Critical Current Studies on Fine Filamentary NbTi Accelerator Wires, in *Advances in Cryogenic Engineering*, Vol 32, R.P. Reed and A.F. Clark, Ed., Plenum Press, 1986, p 707
67. M.I. Buckett and D.C. Larbalestier, Precipitation at Low Strains in Nb 46.5 wt% Ti, *IEEE Trans. Magn.*, Vol 23 (No. 2), March 1987, p 1638
68. H. Kanithi, Expectations and Limitations of J_c in Practical NbTi Conductors, in *Advances in Cryogenic Engineering*, Vol 34, A.F. Clark and R.P. Reed, Ed., Plenum Press, 1988, p 951
69. P.J. Lee, J.C. McKinnell, and D.C. Larbalestier, Microstructure Control in High Ti NbTi Alloys, *IEEE Trans. Magn.*, Vol 25, 1989, p 1918
70. P.J. Lee, "Adventures in Heat Treatment," Paper presented to the 9th NbTi Workshop (Asilomar, CA), Jan 1989
71. P.J. Lee, J.C. McKinnell, and D.C. Larbalestier, Restricted, Novel Heat Treatments for Obtaining High J_c in Nb 46.5 wt% Ti, in *Advances in Cryogenic Engineering*, Vol 36, F.R. Fickett and R.P. Reed, Ed., Plenum Press, 1990, p 287
72. P.J. Lee and D.C. Larbalestier, Development of Nanometer Scale Structures in Composites of Nb-Ti and Their Effect on the Superconducting Critical Current Density, *Acta Metall.*, Vol 35 (No. 10), 1987, p 2523
73. E. Gregory, T.S. Kreilick, and J. Wong, Fine Filament Materials for Accelerator Dipoles and Quadrupoles, in *Proceedings of the ICFA Workshop*, P. Dahl, Ed., Brookhaven National Laboratory Report BNL 52006, 1986, p 85
74. P.J. Lee, D.C. Larbalestier, and J. McKinnell, High Titanium Nb-Ti Alloys—Initial Microstructural Studies, in *Advances in Cryogenic Engineering*, Vol 34, A.F. Clark and R.P. Reed, Ed., Plenum Press, 1988, p 967
75. D.G. Hawksworth, Development of Superconducting Magnet Systems for MRI, in *Advances in Cryogenic Engineering*, Vol 35, R. Fast, Ed., Plenum Press, 1990, p 529
76. D.S. Beard *et al.*, The IEA Large Coil Task: Development of Superconducting Toroidal Field Magnets for Fusion Power, *Fusion Eng. Des.*, Vol 7, 1988, p 1-232
77. W. Hassenzahl, Superconducting Magnetic Energy Storage, *IEEE Trans. Magn.*, Vol 25 (No. 2), March 1989, p 750

A15 Superconductors

David B. Smathers, Teledyne Wah Chang Albany

THE TERM A15 refers to a cubic crystal type in the Strukturbericht System represented by the example Cr_3Si. The intermetallic A_3B compound is formed by a body-centered cubic (bcc) arrangement of B atoms with two A atoms centered in every face yielding orthogonal chain structures running through the crystal (Fig. 1). Of 76 known A15 compounds, 46 are known to be superconducting (Fig. 2 and Table 1). The A atoms are from the groups IVA, VA, and VIA transition metals, whereas the B atoms are from groups IIIB, IVB, and VB and some transition metals including osmium, iridium, platinum, gold, and technetium.

V_3Si was the first A15 compound discovered to be superconducting. The compounds formed with niobium or vanadium have the best superconducting properties (Fig. 3). Nb_3Ge held the high critical temperature (T_c) record for 20 years at 23 K and has the highest critical temperature of the metal, or so-called low-temperature, superconductors. The high-T_c properties of this structure are related to its atomic arrangement. When niobium or vanadium (valence 5) are the A atoms, there are many choices for the B atom with valence 4 to give the compound an average electron-to-atom ratio of 4.75. The A atoms are spaced 10 to 15% closer together on the chains than in pure A metal, and this produces an enhanced *d*-band density of states near the Fermi level (also known as Fermi energy). This correlation between superconductivity and the *d*-band density of states was deduced from Knight Shift measurements on V_3X compounds (Ref 3).

The importance of the chain structure is also evidenced by the strong dependence on order in the compound. There can be compositional disorder or antisite disorder created by neutron irradiation or by the addition of third elements (Fig. 4). Beasley has pointed out that T_c in Nb_3Sn is dependent on disorder regardless of its origin (Ref 6). In general, A15 compounds exhibit a high degree of ordering (Ref 2).

Another intriguing feature of the A15 structure is its tendency to undergo a low-temperature martensitic transformation into a slightly tetragonal structure (the ratio of the lattice parameters, c/a, differs only by a few percent from unity). Why this transformation occurs only in V_3Si and Nb_3Sn is unknown (Ref 2). The cubic structure can be stabilized by the addition of small amounts of impurities.

The intermetallic compound is inherently brittle and presents unique handling problems when forming superconducting magnets. The compound fails brittlely at strains of about 0.5%.

The crystal structure is responsible for the high critical temperatures attained by the A15s but also leads to marked strain sensitivity of the superconducting properties T_c, B_{c2}, and I_c (Ref 7, 8, 9). Strain, ϵ, is the key parameter only because of the ex-

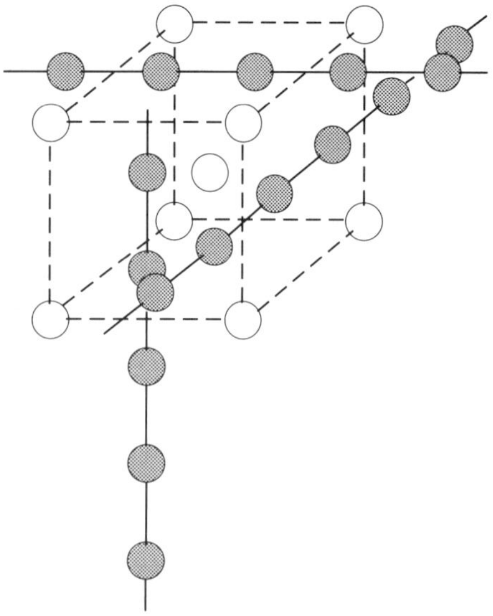

Fig. 1 Atomic arrangement for A_3B compounds of the A15 type structure. Shaded circles denote A-atom sites; open circles denote B-atom sites. For sake of clarity atoms on three of the six cube faces have been omitted. The extension of the A-chains is emphasized.

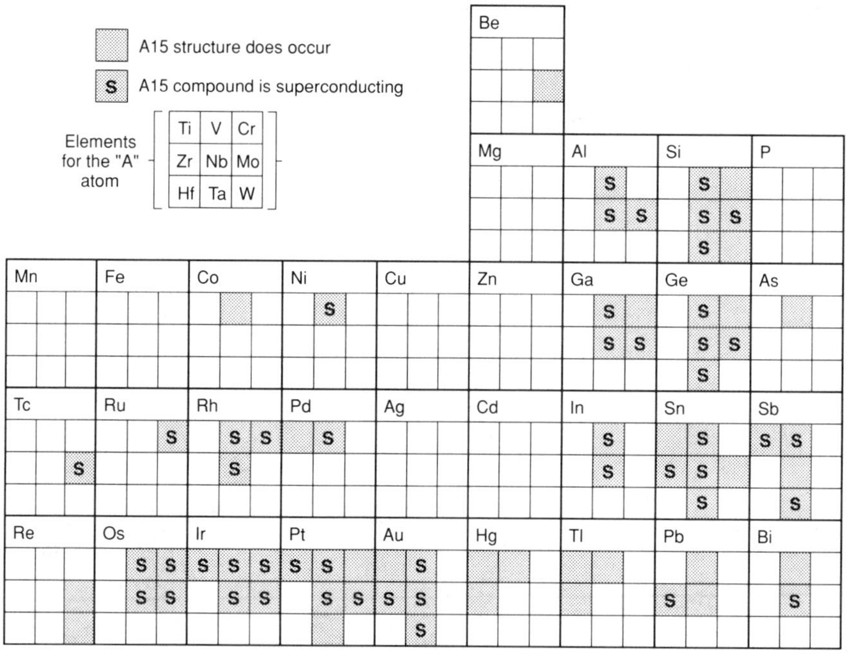

Fig. 2 Occurrence of the A15 (A_3B) crystal structure. Source: Ref 1

Table 1 Critical temperature of all A15 compounds known to be superconducting

Compound	T_c, K	Compound	T_c, K
Ti_3Ir	4.6	Nb_3Au	11
Ti_3Pt	0.49	Nb_3Al	18.9
Ti_3Sb	5.8	Nb_3Ga	20.3
Zr_3Au	0.92	Nb_3In	8
Zr_4Sn	0.92	Nb_3Ge	23
Zr_3Pb	0.76	Nb_3Sn	18.3
V_3Os	5.15	Nb_3Bi	2.25
V_3Rh	0.38	$Ta_{4.3}Au$	0.58
V_3Ir	1.39	Ta_3Ge	8
V_3Ni	0.57	Ta_3Sn	6.4
V_3Pd	0.08	Ta_3Sb	0.72
V_3Pb	3.7	Cr_3Ru	3.43
V_3Au	3.2	Cr_3Os	4.03
V_3Al	9.6	Cr_3Rh	0.07
V_3Ga	15.4	Cr_3Ir	0.17
V_3In	13.9	Mo_3Os	11.68
V_3Si	17.1	Mo_3Ir	8.1
V_3Ge	7	Mo_3Pt	4.56
V_3Sn	4.3	Mo_3Al	0.58
V_3Sb	0.8	Mo_3Ga	0.76
Nb_3Os	0.94	Mo_3Si	1.3
Nb_3Rh	2.5	Mo_3Ge	1.4
Nb_3Ir	1.76	Mo_2Tc_3	13.5
Nb_3Pt	10		

Source: Ref 1

perimental difficulties of measuring uniaxial stress on a superconducting sample at 4.2 K in a magnetic field. A more recent problem, transverse stress, keys on stress for the same reasons, chief of which is the experimental difficulty of measuring strain under the conditions of the experiment (Ref 10).

All the A15 compounds studied show the strain sensitivity to varying degrees (Fig. 5 and Fig. 6). The magnitude of the variation has been correlated with the order parameter, S (Ref 9). The properties degrade in either tensile or compressive conditions. Residual strains can be retained on the A15 compound from the manufacturing process and must be considered in both conductor design and utilization. Handling limits can be improved by processes that put a compressive strain on the compound formed at the expense of critical current and temperature. Because the upper critical field is affected, the loss of critical current with

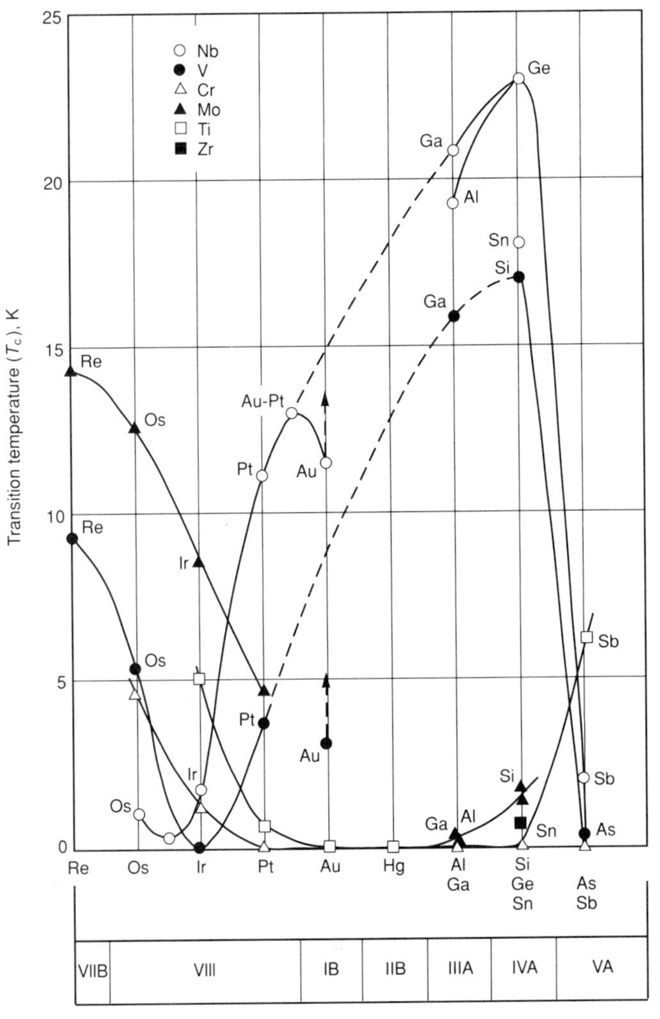

Fig. 3(a) Superconducting transition temperature, T_c, for different series of A15 compounds having the formula $A_{1-x}B_x$ as a function of the atomic number of the B atom. The expected T_c values for $Nb_{\sim3}Au$ and $V_{\sim3}Au$ would be higher if stoichoimetry could be reached. Source: Ref 2

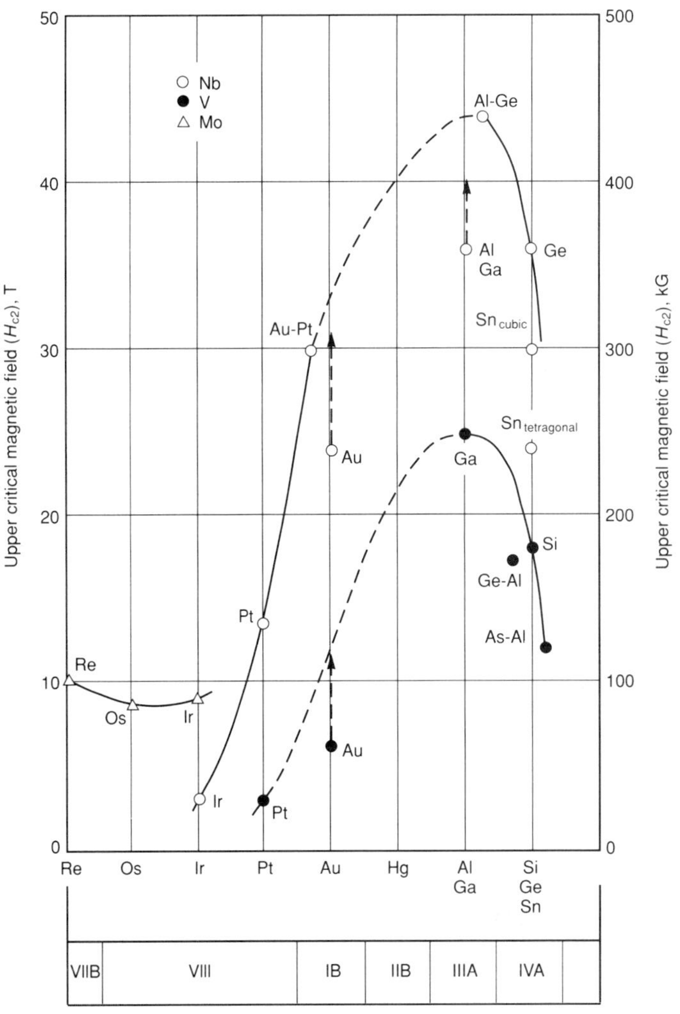

Fig. 3(b) Upper critical field, $H_{c2}(0)$, for different series of A15-type compounds having the formula A_3B as a function of the atomic number of the B atom. The $H_{c2}(0)$ values for stoichiometric V_3Au, Nb_3Au, and Nb_3Al are expected to be higher. Note the difference in $H_{c2}(0)$ between cubic and tetragonal Nb_3Sn. Source: Ref 2

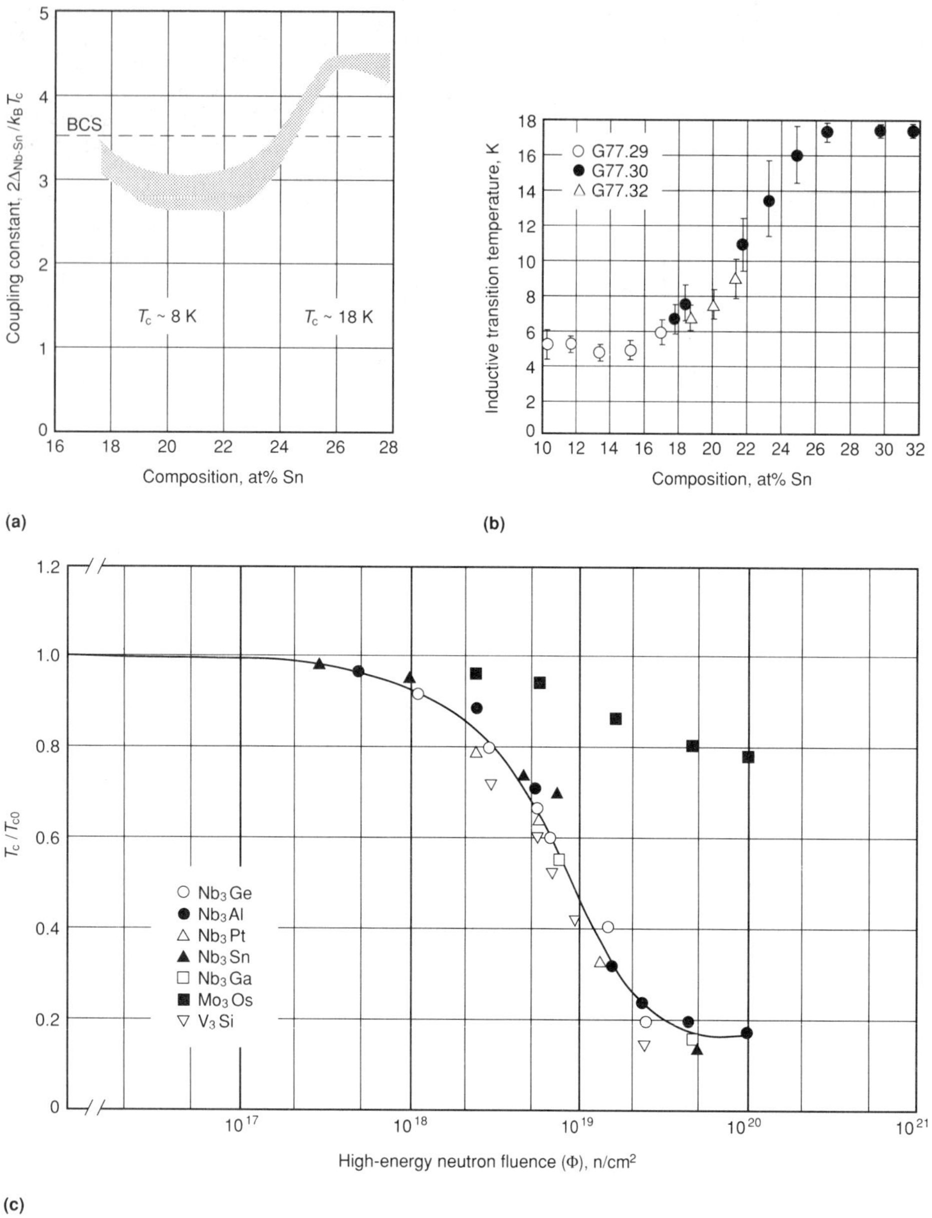

Fig. 4 Variation of superconducting properties with compositional or antisite disorder. (a) Variation of the coupling constant, $2\Delta/k_BT_c$ with composition of Nb_3Sn. The ratio is determined from tunneling data. Source: Ref 4. (b) Plot of superconducting transition temperature of optimum Nb-Sn films versus composition. The 5, 50, and 95% completion points are indicated on each transition. The films are 0.5 μm (20 μin.) thick and were deposited at ~1075 K at ~2.5 nm · s^{-1}. Source: Ref 4. (c) Reduced superconducting transition temperature normalized to the unirradiated value, T_{c0}, as a function of high-energy ($E > 1$ MeV) neutron fluence Φ for A15 compounds. Data are displayed on a semilogarithmic plot for easy comparison with other published data. Solid line is a visual aid. Source: Ref 5

strain is more severe at higher fields. Figure 7 shows the J_c-B-ϵ field for Nb_3Sn.

Phase Diagrams

Flükiger has made an extensive review of the A15 phase diagrams (Ref 2). As there are wide variations in the metallurgy, only those for the high-T_c A15 compounds will be discussed here. These, as mentioned earlier, are compounds of vanadium and niobium. The A15 phase field in a binary phase diagram varies from metastability to a line compound to a broad compositional range. Some compounds include the stoichiometric 3:1 ratio as stable at low temperatures, others are stable only near the melting temperature, and some are not stable at all.

The range of phase diagrams for vanadium is shown in Fig. 8. The compounds of commercial interest are V_3Si and V_3Ga, both because of their high T_c values and the stability of the compounds. The stability is important for fabrication purposes. The V_3Ga diagram is unique in that there is considerable composition range on either side of stoichiometry. Only V_3Pt and Nb_3Pt have similar behavior (Ref 2).

Niobium compounds of interest fall into two general types of phase diagrams. Compounds with gallium, aluminum, and germanium form stoichiometrically only near the melting temperature (Fig. 9). Phase segregation can be controlled by the annealing temperature, and the behavior is reversible (Fig. 10). This condition has a drastic influence on the fabrication of bulk conductors of these compounds.

Niobium compounds of tin and antimony form stoichiometric compositions at lower temperatures and are stable at very low temperatures (Fig. 11). Only Nb_3Sn receives much attention because of its very good superconducting properties. The entire Nb-Sn binary diagram is shown in Fig. 12. There has been a lot of work on this system because of its commercial interest and considerable disagreement on the shape of the phase field at lower temperature. The work of Charlesworth *et al.* (Ref 11) and Flükiger (Ref 2) are definitive and establish stability to low temperatures.

The stability of the stoichiometric compound at lower temperatures is important for forming this brittle structure into filamentary conductors. It is possible to form the compound by solid-state diffusion techniques at temperatures below 1000 °C (1830 °F) through direct couples between the A element and B element or a bronze containing copper and the B element. Copper appears to stabilize the A15 structure over other possible compounds, possibly by limiting the supply of B element (Ref 13). Though copper may be found in the A15 layers (Ref 14), it has been demonstrated to be in the grain boundaries and not in the bulk (Ref 15, 16).

Alloying With Third Element Additions

Historically, the route to improved superconductors is by elemental substitution designed to modify some desirable property. Third element additions to the A15s have been studied extensively with success. Nontransition metal additions generally substitute for the B element leaving the A chains unaffected. Such additions are often successful in raising the critical temperature, but only modestly (Table 2). Some additions serve to modify the growth kinetics without becoming incorporated into the A15 compound [magnesium (Ref 17, 18) and copper (Ref 15, 16)]. Transition metal additions prealloyed with the vanadium and niobium are thought to substitute on the A site, but this is still uncertain (Ref 13). The latter often increases the normal state resistivity faster than they cause a reduction in critical temperature. The result is an increased H_{c2} value for low alloy contents

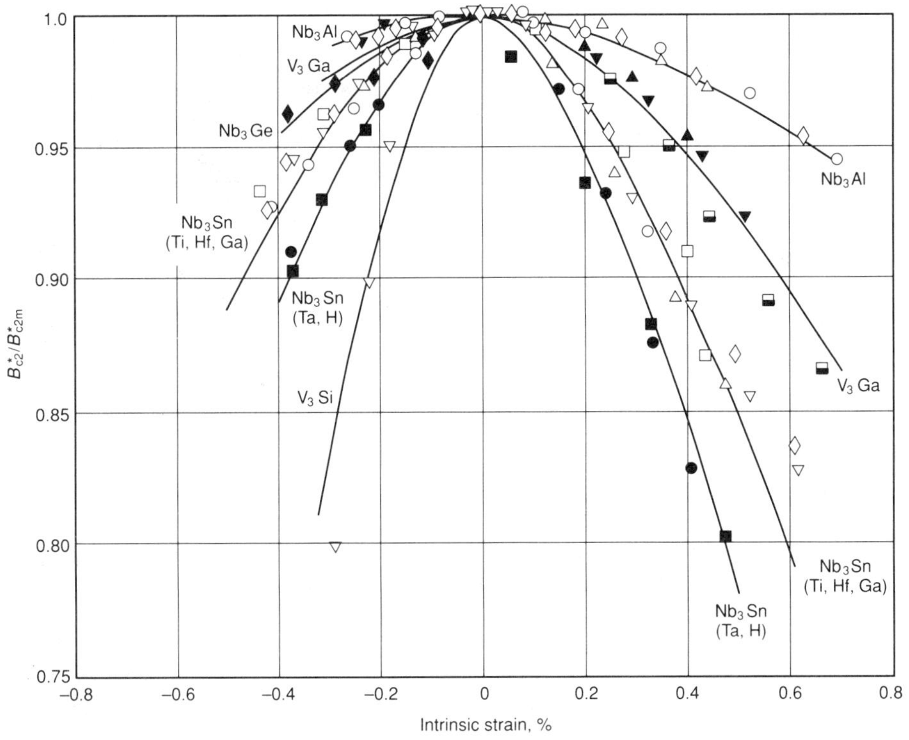

Fig. 5 Effect of uniaxial strain on the upper-critical field of practical A15 superconductors. B^*_{c2} has been normalized by its maximum (nearly strain-free) value B^*_{c2m}. Binary Nb_3Sn is represented by the same curve as for Nb_3Sn (Ti, Hf, Ga). Source: Ref 8

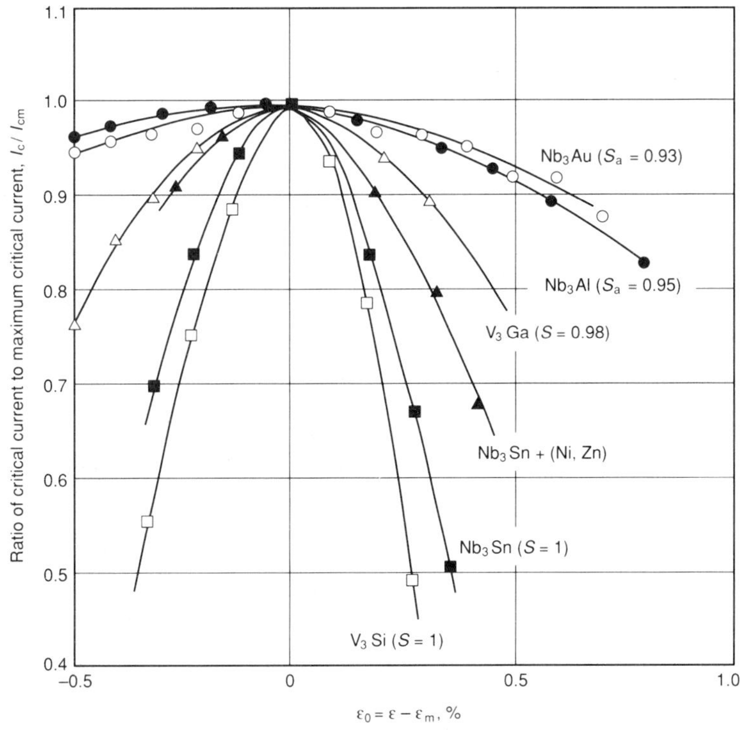

Fig. 6 I_c/I_{cm} versus ϵ for a series of mono- or multifilamentary wires based on A15-type compounds ($0.57 \leqq h \leqq 0.68$). S_a, order parameter for A-site atoms. Source: Ref 9

and a consequent increase in J_c at higher fields. These elements will affect grain morphology in compounds formed by solid-state diffusion. Elemental additions to Nb_3Sn that have been reported include hydrogen, beryllium, carbon, magnesium, aluminum, silicon, phosphorus, titanium, vanadium, manganese, iron, nickel, copper, zinc, gallium, germanium, arsenic, zirconium, indium, antimony, hafnium, tantalum, thallium, lead, molybdenum, tungsten, bismuth, and ZrO_2. Not all alloy additions are beneficial, however.

In the case of Nb_3Ge, additives are necessary to stabilize the A15 phase at stoichiometry. Aluminum and oxygen serve this purpose. $Nb_3(Al_{0.75}Ge_{0.25})$ has the highest B_{c2} of all the A15 compounds at 43.5 T (435 kG).

Layer Growth

A15 compounds are brittle and lack ductility. In all practical cases, the A15 compound is a reacted layer formation whether deposited on a substrate or grown by solid-state diffusion. This will be addressed more in the section "Processing and Properties of Superconductors" in this article, but the layer growth concept is important in understanding the generation of the superconducting properties and the conductor processing.

The importance of grain size will become clear in the section "Matrix Materials" in this article. Deposited layers will generally have small columnar grains but will become more equiaxed as the layer thickens. Processing parameters are extremely important. Diffusion grown layers are more complex. It has been demonstrated that solid-state diffusion occurs through grain boundaries and that grain growth occurs simultaneously with layer growth (Ref 13). In addition, alloying can greatly influence both grain and layer growth (Ref 19). Thus, the reaction conditions are very important in determining conductor performance.

Critical Current Density

Though the mechanism of flux pinning in the A15 compounds is not well understood (Ref 20), it is clear that grain boundaries are the important pinning centers (Fig. 13) (Ref 14). The critical current density generally scales well and the flux-pinning force has the field dependence,

$$F_p = |\mathbf{J}_c \times \mathbf{H}| = [H^{*5/2}_{c2}/A\kappa^2]h^{1/2}(1-h)^2 \qquad \text{(Eq 1)}$$

where H^*_{c2} is determined by the extrapolation of $J_c^{1/2}H^{1/4}$ versus H, A is a constant, and κ is the Ginsberg-Landau parameter. While grain boundaries provide the pinning, there is no quantitative theory that connects the maximum pinning force with grain size. The $H^{5/2}_{c2}/\kappa^2$ has limited theoretical justification but is reasonably well established experimentally (Ref 20, 21).

The field dependence of Equation 1 is always obeyed above $h = 0.6$ but may often deviate below $h = 0.6$. This deviation is correlated with grain morphology. The maximum pinning force that occurs at fields near $h \sim 0.25$ is diminished as the grains become less equiaxed (see Fig. 14). The general effect of alloy additions that increase H_{c2} is to reduce this $F_{p_{max}}$. Thus, there

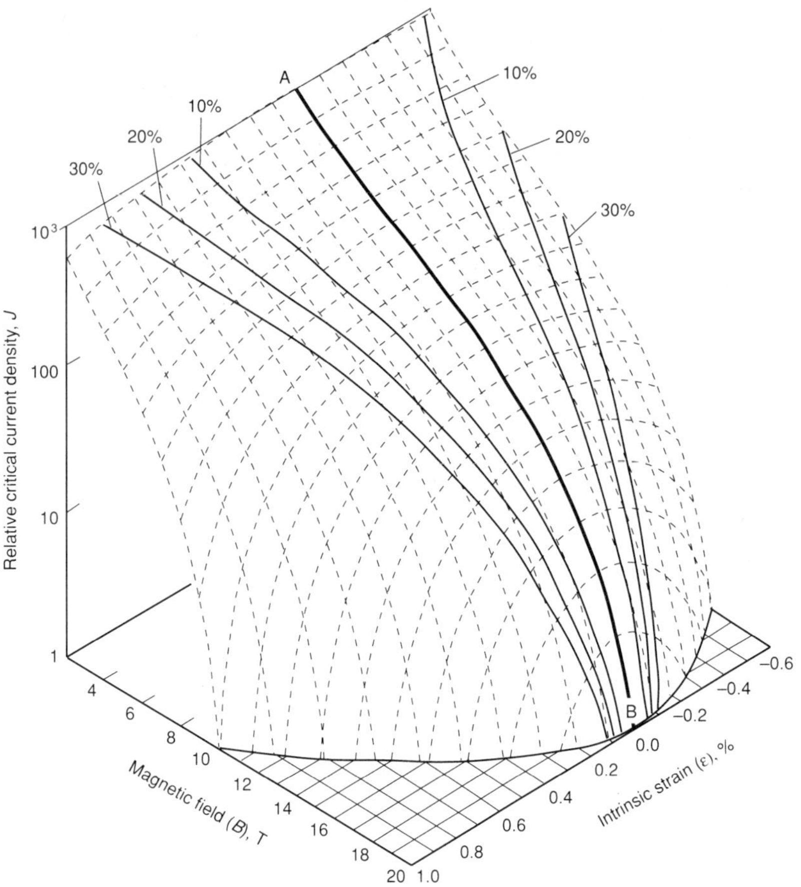

Fig. 7 *J-B-ε* critical surface for multifilamentary Nb_3Sn superconductors at T_c of 4.2 K. Line AB represents the maximum (nearly strain-free) value of the critical current as a function of magnetic field. Corresponding curves on each side of line AB represent the strain window for mechanical design that will result in a critical current within the indicated percentage of maximum. Source: Ref 8

is some crossover field where the critical current in alloyed material (relative to the binary compound) at high fields will be higher but lower at lower fields (Fig. 15).

There are two examples of precipitate pinning enhancements to the grain-boundary pinning. Nb-1%Zr can be internally oxidized to provide a fine ZrO_2 distribution that remains intact during the reaction heat treatment (Ref 22). Recent work in which niobium and tantalum are mechanically alloyed has been successful in increasing J_c by 30 to 50% at 10 T (100 kG) by reacting the wires such that the tantalum inclusions remain in the Nb_3Sn (Ref 23, 24).

Table 2 Summary of successful additions of third element to an A15 compound ($A_3B_{1-x}B'_x$) to raise T_c

Compound	Initial T_c, K	B′ substitution	Amount (x)	New T_c, K
V_3Ga	14.5	Al	0.1	15.0
V_3As	(a)	Al	0.7	10
V_3Sb	0.8	Al	0.4	7
Nb_3Al	...	Si	0.13	19.2
Nb_3Al	...	As	0.4	19.2
Nb_3Al	...	Ga	0.2	19.4
Nb_3Al	18.0	Be	0.05	19.6
Nb_3Al	18.0	Cu	0.1	19.0
Nb_3Al	18.0	B	0.1	19.1
Nb_3Al	18.0	Ga	0.2	19.5
Nb_3Sn	...	Al	0.1	18.6
Nb_3Sn	18.0	Ga	0.05	18.35
Nb_3Sn	18.0	In	0.15	18.3
Nb_3Sn	18.0	Tl	0.1	18.25
Nb_3Sn	18.0	Pb	0.15	18.25
Nb_3Sn	18.0	As	0.05	18.2
Nb_3Sn	18.0	Bi	0.15	18.25

(a) Not superconducting. Source: Ref 1

The intended application greatly influences the choices of (H_{c2} and J_c) conductor design. Alloying generally improves high field properties. The brittleness and strain sensitivity must also be considered at every stage of the device design process.

Matrix Materials

Stabilizers. Like the NbTi conductors, it is desirable to have a parallel conducting path of a good normal material (low electrical resistivity) to shunt currents and aid in heat diffusion to the coolant. The choices for materials are the same as for NbTi, but reaction temperatures must be considered. Copper is most often used because the melting temperature (1080 °C, or 1975 °F) is higher than for most required heat treatments. Aluminum, melting at 660 °C (1220 °F), is not satisfactory.

Aluminum may be used as the stabilizer if it is added to the conductor following the reaction heat treatment. This may be accomplished by soldering or coextruding. Such processing of the brittle conductor requires careful control over strain to avoid filament breakage.

Deposited thick-film conductors may have aluminum or copper added by soldering or electrodeposition. Copper or aluminum may also be added in a cabling step.

Noncopper. As described in the introduction, practical conductors are typically composite materials containing a reacted A15 compound layer plus either a substrate or matrix that provides one or more reaction components as well as structural strength. In filamentary composites, a matrix of low-solute-content bronze remains around the filaments after reaction. In addition, a material is generally added that keeps the reactive material from contaminating the high-purity stabilizer material. This material is called the diffusion barrier and is chosen from a group of refractory metals (niobium, tantalum, or vanadium) that react minimally or produce nonsuperconducting compounds at the barrier interface. These components must be present for reaction but serve little value after reaction. The matrix, contaminated with residual reactive B element, and the diffusion barrier are not useful stabilizing materials due to their high resistivities. This sum of the superconductor and reaction byproducts is generally called the noncopper. For conductors that have no stabilizer, the whole wire cross-section area is used as the overall area (Fig. 16).

In NbTi conductors, the specified critical current density is referred to the area of NbTi only. For A15 conductors, the specified critical current density is referred to the noncopper or nonstabilizer (conductors stabilized with other than copper) area for engineering reasons. In general, the A15 current density is diluted 2 to 5 times by the

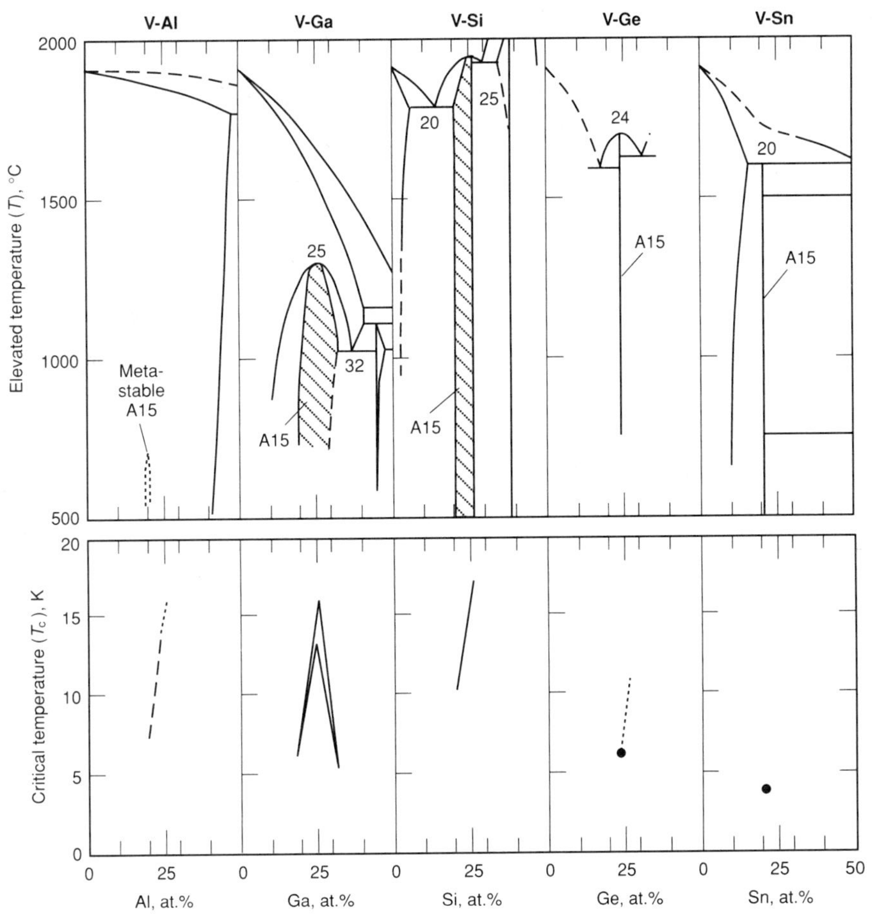

Fig. 8 A15 phase fields and superconductivity in V-Ga, V-Si, V-Ge, and V-Al. Source: Ref 2

nonsuperconducting components of the nonstabilizer. It should become clear that the A15s have extraordinarily high J_c values, though these cannot be fully utilized for engineering conductors.

Reactive Component. Diffusional reactions require the reactive B element be either carried in a matrix or present in its pure form in close proximity to the A element core. This is most often accomplished using a bronze. The conductor geometry can be multifilamentary, and the residual bronze serves to keep the filaments separated. The bronze can initially be pure copper and alloyed tin as separate elements in the cross section. Elemental additions such as titanium, magnesium, indium, antimony, gallium, and silicon may be added through the bronze or tin. In some cases the B element is carried next to the A element, but this is impractical for small filament conductors and does not then constitute a matrix material.

Processing and Properties of Superconductors

For the purposes of this chapter, the goal of the conductor design is presumed to be a stabilized magnet conductor or high-current conductor. For electromagnet applications a multifilament geometry with filaments 70 μm (0.0028 in.) or smaller is desired. For diffusion times that are reasonable, filaments less than 10 μm (400 μin.) are desired. With the exception of tape geometries, the fabrication processes for A15 conductors are very similar to those described earlier for NbTi. In all cases we are describing a thick-film approach, though geometries vary considerably.

Assembly Techniques for Tape Conductors

Superconductor windings can be produced from tape conductors. Niobium-tin and other more exotic A15 conductors are often produced in the form of thin tapes because this enables techniques such as chemical vapor deposition (CVD), surface diffusion, or vacuum sputtering to be used. Such tapes are always thin enough to satisfy the adiabatic flux-jumping criterion when the field is parallel to the broad face but not when the field is perpendicular to it.

Chloride Deposition. The first 10 T (100 kG) superconducting magnet was wound using a 7 μm (280 μin.) layer of Nb_3Sn deposited on a stainless steel substrate with a 7 μm (280 μin.) layer of copper to act as the stabilizer (Ref 25). The large aspect ratio (300:1) leads to magnetic instability at low fields. The chloride process is based on a hydrogen reduction of the gaseous chlorides of niobium and tin above 900 °C (1650 °F) (Ref 26), according to the reaction:

$$3NbCl_4 + SnCl_2 + 7H_2 \Leftrightarrow Nb_3Sn + 14HCl \quad \text{(Eq 2)}$$

This process is still exploited to make magnets aimed at 15 T (150 kG) and above.

Surface Diffusion. A thin niobium or niobium alloy strip (~0.025 mm, or 0.001 in.) may be coated with pure tin and reacted to form Nb_3Sn (Ref 27). The thin layer and ribbon allow the reacted tape to be bent without damaging the Nb_3Sn layer. The reaction occurs above 930 °C (1705 °F) to avoid the formation of Nb_6Sn_5 and $NbSn_2$ and occurs rapidly (recall the phase diagram in Fig. 12). If Nb-1%Zr foil is used, it may be anodized prior to tin coating. The zirconium will internally oxidize, and the fine ZrO_2 precipitates enhance the critical current density.

Bronze Tape. A bronze-niobium-bronze sandwich may be made either by corolling (Ref 28) or electron beam (EB) evaporation of bronze onto a niobium sheet (Ref 29). The Nb_3Sn is formed during a solid-state reaction and produces a material with low surface ac losses. The lower reaction temperature allows inclusion of a stabilizer prior to reaction.

Liquid Quenching. Liquid quenching on a hot substrate provides sufficient wetting, and thus, good cooling rates. Niobium-aluminum-germanium alloys have been prepared as supersaturated bcc-$Nb_3(AlGe)$ and subsequently transformed into fine-grained A15 phase with high J_c. Proper choices of nozzle material, orifice diameter, and degree of superheating above the liquidus temperature are important because of the very high melting temperature (see Fig. 9). The alloy is quenched directly onto a moving copper tape. The tape then acts as the stabilizer material (Ref 30).

Assembly Techniques for Multifilamentary Wires

Superconductor multifilamentary wires can be produced by a variety of processes:

- Rod process
- Modified jelly roll process
- Nb tube process
- *In-situ* process
- Powder metallurgy process
- Jelly roll method

Rod Process. The various rod processes for A15 conductors resemble very strongly those described for NbTi conductors (see the article "Niobium-Titanium Superconductors" in this Volume). The bronze process consists of stacking vanadium or niobium alloy rods in bronze tubes or drilled bronze billets (Fig. 17), processing these

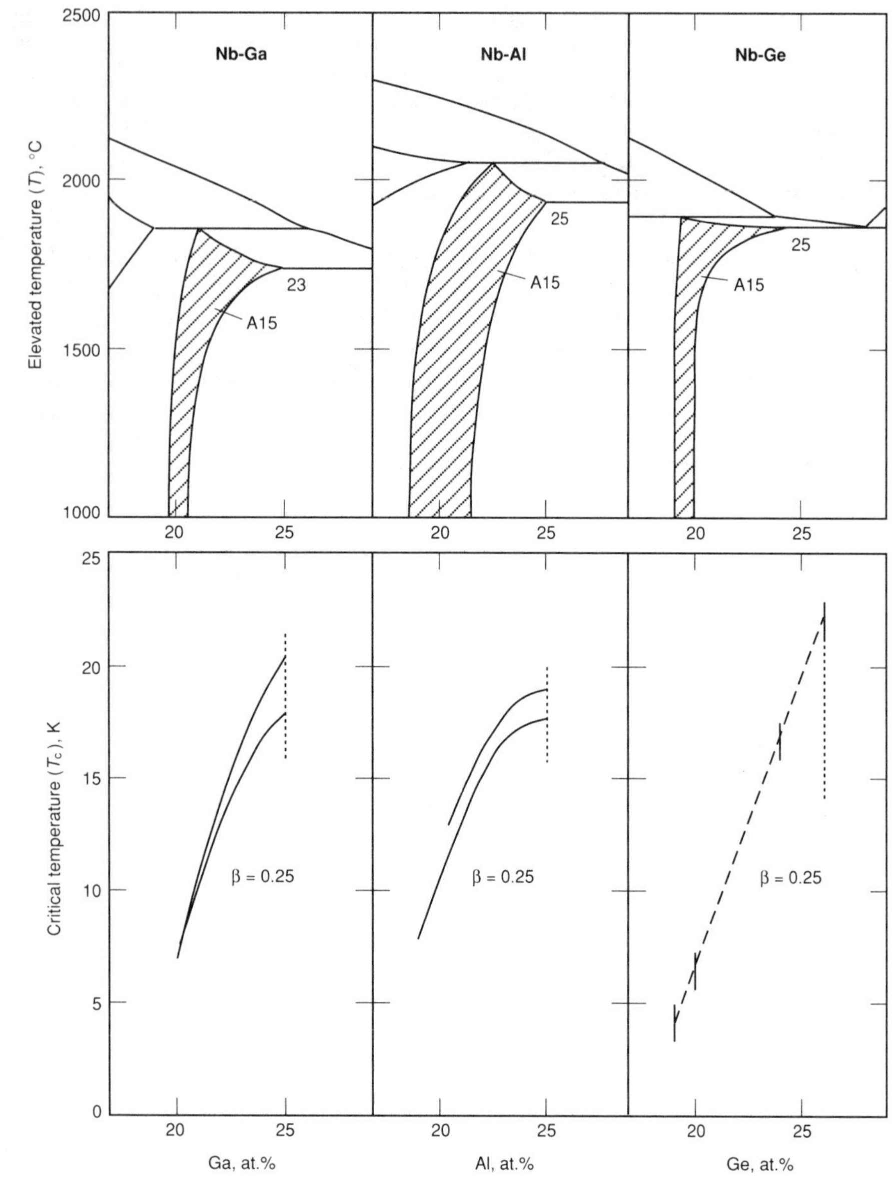

Fig. 9 A15 phase fields and superconductivity in Nb-Al, Nb-Ga, and Nb-Ge. Note the saturation of T_c, very marked for Nb_3Al but less pronounced than in Nb_3Ga. Source: Ref 2

rods to some intermediate-size wire, and restacking the wires. An excellent review is given by Hillman (Ref 31). The double extrusion route is required to reduce the filaments to micron dimensions. Cu-13%Sn or Cu-18%Ga both require frequent annealing during the drawing process to remain ductile. The unfortunate consequence of this is premature formation of the A15 phase at the interface, which can lead to severe filament distortion (Ref 32). This can be alleviated somewhat by including a copper layer between the bronze and the filament.

To keep the bronze ductile it must remain single phase, thus limiting the available tin content. One makes a trade-off in A15 current density versus overall current density by providing excess or insufficient tin or gallium supply. Commercial bronze supply is a problem since phosphorus is very detrimental to Nb_3Sn formation (Ref 33).

These shortcomings are eliminated in the internal tin process. The bronze consists of a copper matrix around the filaments surrounding a tin core (Fig. 18). Though a double stacking process is still required, there is no filament distortion because of in-process annealing. In fact, the processing can proceed with no anneals. The limitation of tin (or gallium) in copper is no longer a factor, and excess tin can be provided with minimal cost in overall J_c. The low melting temperature of gallium precludes an internal gallium process.

In the external diffusion process vanadium or niobium alloy filaments are drawn in a copper matrix, and gallium or tin is diffused in from the outside. For tin the wire dimensions must be less than 0.25 mm (0.010 in.) to avoid balling during the reaction process (Ref 34).

Modified Jelly Roll Process. The modified jelly roll (MJR) process is a unique method of producing micron-size filaments in a single extrusion, which limits the amount of in-process annealing required (Ref 35, 36). A bronze sheet (matrix) is cowrapped with niobium (Ref 37) or vanadium (Ref 38) expanded metal onto a bronze or copper (Ref 39) core rod (Fig. 19). Rolled sheet is expanded by conventional means. The strands of the expanded metal are crossdrawn in the composite to micron-size filaments. A 13 mm (½ in.) length of expanded metal is drawn out to roughly 800 m (0.5 mi), depending on composite design. As many as 4 or 5 different materials with as many as 3 separate crystal structures are being codrawn uniformly to wire. Bronze MJR wires have shown very good performance in short sample (Ref 39) and magnet tests (Ref 40).

The MJR method has also been successfully implemented in an internal diffusion version called the tin core MJR (Ref 41, 42). As noted earlier, this process is limited to Nb_3Sn. Versions of the tin-core MJR product are being used in the United States-demonstration poloidal coil (US-DPC) (Ref 43) and the Sultan III (Ref 44) project. There is also reasonable commercial interest for building compact high-field magnets. Figure 20 shows a cross section of multifilament wire manufactured by the tin-core MJR method.

There is a higher magnetization and ac loss with the internal diffusion conductors associated with filament bridging (Ref 45). The problem increases as the niobium filament content increases, and critical current density increases in a nonlinear fashion (Fig. 21) (Ref 46, 47).

The choices for filament alloy are less restricted in the internal tin MJR. Conductors with C103 (Nb-10Hf-1Ti-0.2W), Nb-1%Zr, and Nb-(1 to 3%)Ti filaments in a copper matrix have been successfully processed (Ref 41). The elimination of in-process annealing is greatly beneficial in this respect.

Nb Tube Process. The desirability of including a diffusion barrier has generated several variations in which a thick-wall niobium tube is partially consumed in the reaction to Nb_3Sn, and enough remains to serve as a barrier. The disadvantage is that the effective filament size will be limited by the niobium tube dimensions in the wire (Ref 48). Variations on the tube lining include bronze (Ref 48), copper around tin (Ref 49), and $NbSn_2$ powder (Ref 50). The latter process produces very fine-grain Nb_3Sn after reaction. Reduction of the effective filament size, however, requires reducing the niobium tube dimensions. The number of tubes in a wire cross section must be increased to accomplish this with

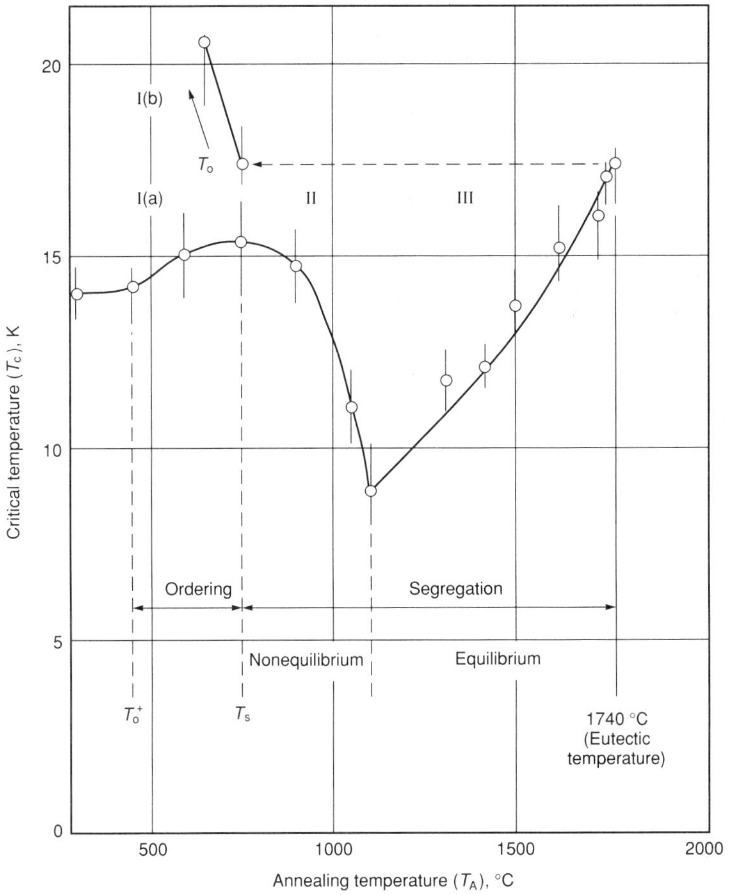

Fig. 10 Variation of T_c for an alloy of the total composition $Nb_{0.74}Ga_{0.26}$ as a function of the annealing temperature. Part I represents the region where only long-range order effects take place, whereas part II reflects the curved shape of the gallium-rich A15 phase boundary (see Fig. 9). T_o and T_s represent the minimum ordering and segregation temperature. Source: Ref 2

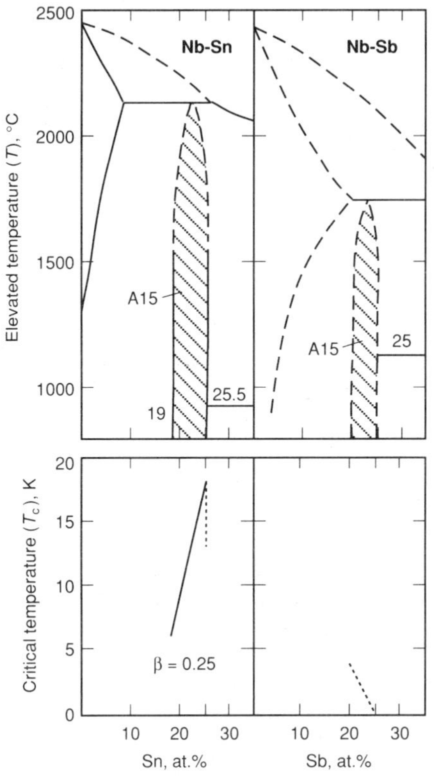

Fig. 11 A15 phase fields and superconductivity in the systems Nb-Sn and Nb-Sb. Source: Ref 2

very high area reductions. The quality of both the tubes and the internal material must be very good.

***In-Situ* Process.** The *in-situ* process makes use of the immiscibility of the niobium-copper and vanadium-copper systems. Alloys cast from the melt may segregate completely. The dendrites formed in the matrix draw out into highly aspected filaments. Tin or gallium may be added by internal or external methods, much as in other multifilament routes. The impurities present in the melt greatly affect the morphology of the precipitation, and thus, the filamentary nature of the product. Originally investigated by Tsuei, the technique has been refined and improved (Ref 34). A minimum of 15% Nb is necessary for the filaments to be interconnected. The composite is drawn to produce filaments of a few tenths of a micron thick. *In-situ* composites generally show excellent irreversible strain limits and high compressive strains because of fine filament reinforcement (Fig. 22) (Ref 34). The high intrinsic strain reduces the H_{c2} of the wire, and therefore, reduces the critical currents at all fields.

Powder Metallurgy. Powder metallurgy processes include the infiltrated tin and the mixed powder methods.

Infiltrated Tin. Niobium powder is sintered into a porous structure and then infiltrated with liquid tin (or aluminum) (Fig. 23). The result can be drawn to wire (or rolled to tape) and provide a very high critical current density (Ref 52).

Mixed Powders. Cold or hot powder processes (Fig. 24) can be used to produce structures very similar to the *in-situ* process. There is more flexibility in the elemental combinations since the natural immiscibility of two molten metals is not a requirement. The oxygen content of the powders can result in limited deformation, however (Ref 34, 51). This method has been pursued commercially for Nb_3Sn (Ref 53).

The mixed powder method has been a successful technique for Nb_3Al manufacture (Ref 51). To form metastable Nb_3Al at low temperatures (600 to 800 °C, or 1110 to 1470 °F), the niobium and aluminum elements must be extremely finely interdispersed with aluminum thicknesses much less than 1 μm (40 μin.). Critical current densities increase as the aluminum thickness becomes a few hundred angstroms. This requires very high area reductions or very fine starting elements. The oxygen content of the niobium powder is critical in processing. T_c values of about 16 K are achieved. Nb_3Al holds promise for higher field applications and a reduced strain sensitivity compared to Nb_3Sn (Fig. 5 and Fig. 6).

Jelly Roll Method. An alternative method for finely dividing niobium and aluminum has been pursued by Ceresara (Ref 54). Thin sheets of niobium and aluminum are cowrapped onto a core in a swiss roll style. The composite is then drawn to wire. Again, the oxygen content of the thin foils is critical. The optimum separation of the niobium and aluminum is the same in the jelly roll process as it is in the powder process.

Deformation

The processing of multifilamentary A15 wire geometries is very much the same as for NbTi. Bronze processed wires require frequent softening anneals in the 400 to 500 °C (750 to 930 °F) range. Internal or external tin processed wires do not require these anneals. Bronze and external tin billets may be hot extruded, whereas internal tin billets may not, though cold hydrostatic extrusion is an option. Note that the A15 phase is not intentionally formed during deformation processes.

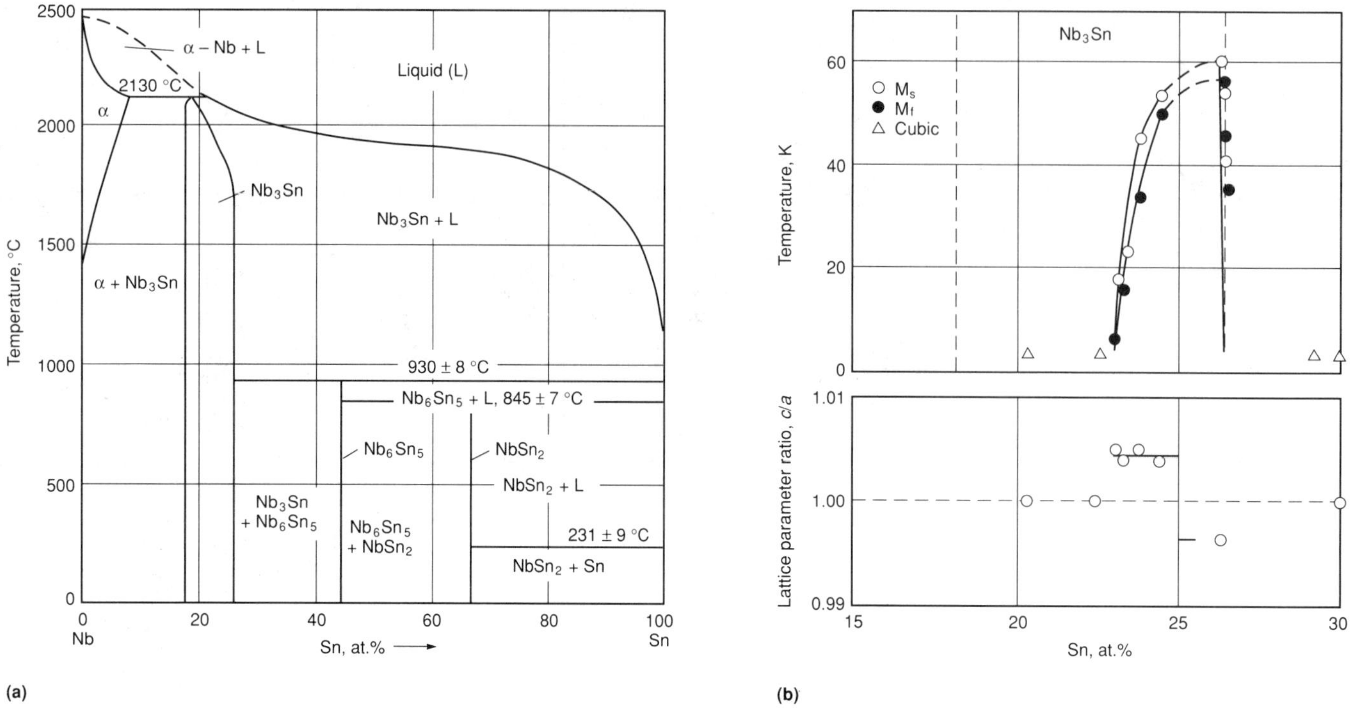

Fig. 12 Niobium-tin binary phase diagram. (a) Elevated temperatures. (b) Subzero temperatures. M_f, temperature at which martensite formation finishes during cooling; M_s, temperature at which martensite starts to form on cooling. Sources: Ref 11, 12

Reaction Heat Treatments

The reaction heat treatment is required at the end of wire processing to convert the ductile components to the desired, but brittle, superconductor. Subtle aspects of composite design are aimed at improving the critical current or strain tolerance as a result of this heat treatment. Bronze route conductors generally require a single-stage treatment, though two-stage treatments have often been recommended. The time and temperature employed must balance layer growth rate with A15 grain growth. The addition of third elements may increase both these rates, and lower reaction temperatures are then advised.

Internal and external tin composites require multiple-stage heat treatments. There must be several low-temperature steps (200 to 500 °C, or 390 to 930 °F) to alloy the copper and the tin before proceeding with Nb_3Sn formation. The copper-tin phase di-

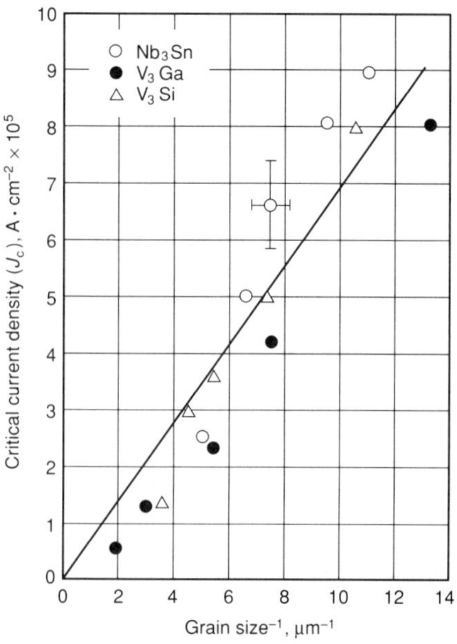

Fig. 13 Grain size dependence of critical current densities for bronze-processed Nb_3Ga, V_3Ga, and V_3Si at 4.2 K and H = 4 T (40 kG). Source: Ref 13

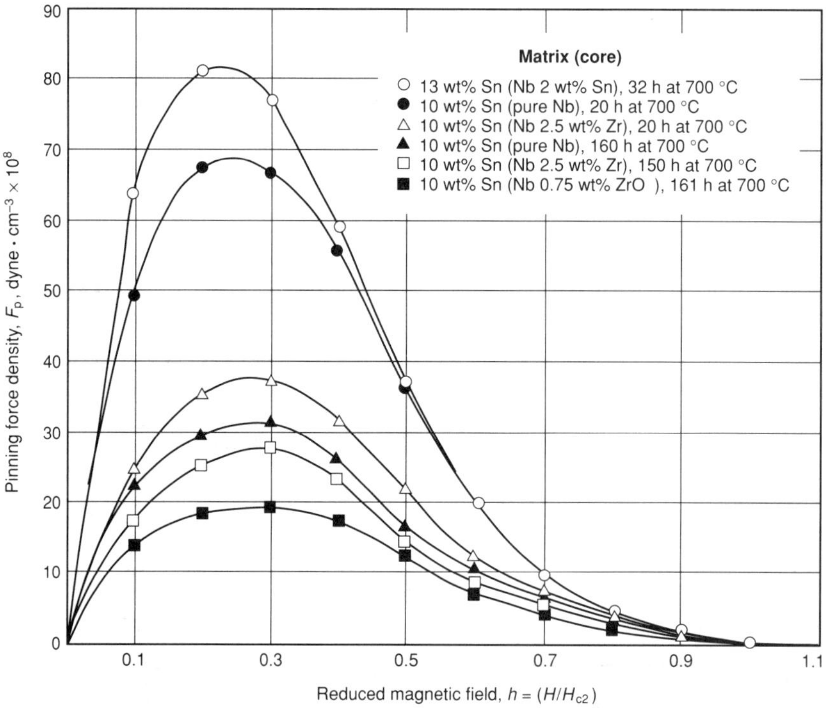

Fig. 14 Pinning force density as a function of reduced magnetic field for a series of bronze-processed Nb_3Sn wire conductors. Matrix (core). Source: Ref 22

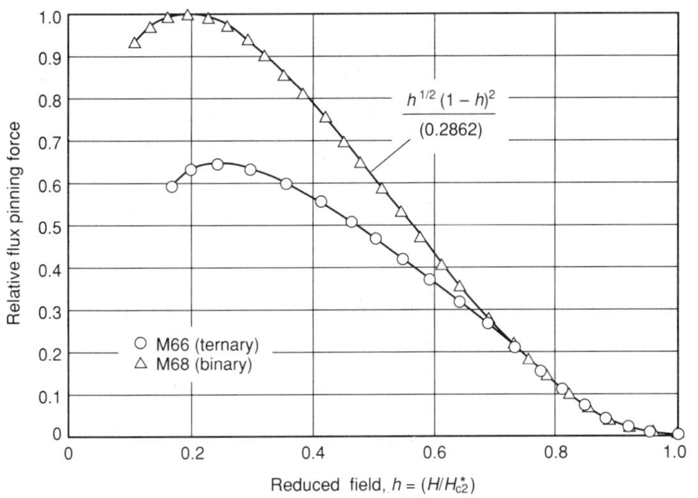

Fig. 15(a) Plot of the relative flux pinning force versus the reduced field. The relative flux pinning force is calculated from Equation 1 (see text). For binary Nb_3Sn, H^*_{c2} = 18.1 T and $[H^{*\,5/2}_{c2}/A\kappa^2]$ = 1.74 × 10^{10} dyne/cm^3. For ternary (Nb-1%Ti)$_3$Sn, H^*_{c2} = 21.4 T and $[H^{*\,5/2}_{c2}/A\kappa^2]$ = 1.54 × 10^{10} dyne/cm^3. Both conductors are bronze-processed modified jelly roll conductors.

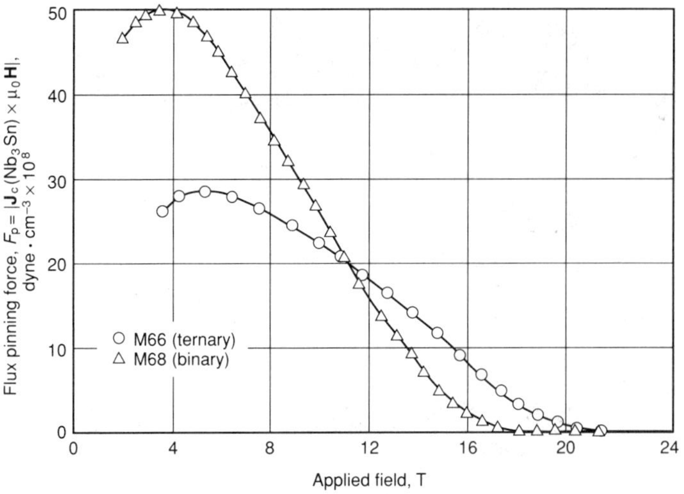

Fig. 15(b) Plot of actual flux pinning force versus applied field. Above 11 T, the ternary conductor will carry more supercurrent than the binary conductor. For scaling purposes, the relative flux pinning force is more instructive than plotting $F_p/F_{p_{max}}$. The term $[H^{*\,5/2}_{c2}/A\kappa^2]$ is determined by fitting F_p versus h between h = 0.6 and 1.0. H^*_{c2} is determined by the extrapolation of $J_c^{1/2} \cdot H^{1/4}$ versus H to the abscissa.

agram is complicated (Fig. 25). Generally a solid-state wetting treatment ($T < 227$ °C, or 441 °F) takes place followed by a higher-temperature treatment. Keeping the temperature less than 415 °C (780 °F) reduces the extent of molten phases present in the wire. Final treatment is identical to a bronze conductor with the same filament alloy and size. Generally, a three-stage treatment is encouraged. These conductor types require long furnace times because of the copper alloying steps. The elimination of wire processing anneals is at the expense of the reaction treatment complexity (Fig. 26).

React and Wind. The decision to react the wire before or after coil winding is governed by such factors as coil size, insulation, and complexity of winding. Though the wire is brittle, it may be bent so long as the maximum bending strain is kept below the irreversible strain limit, generally 0.7% or higher. Damage during handling is the most serious threat, but the heat treatment is more easily performed. Wire sticking is a concern, particularly if the wire has been cabled prior to reaction. A sintered cable has a greatly reduced flexibility and increased risk of wire damage. This technique has been successfully employed in several large coil constructions such as the Westinghouse Large Coil Project coil (Ref 56) and the Sultan facility (Ref 57).

Wind and React. It is by far more common to wind the coil first and then to heat treat the whole. This method greatly reduces the handling risks but requires the coil former and insulation to be dimensionally stable and temperature resistant. The insulation commonly used is a braided glass fiber around the bare wire. The US-DPC coil is wound with a 225-strand conductor, heat treated as a coil, and insulated afterward (Ref 58). Large coils require large furnaces with uniform heating.

Cable and Winding

Cabling A15 conductors is no different than for any other material. Cable design is influenced by reaction requirements, anticipated stresses, cooling method, ramp rate,

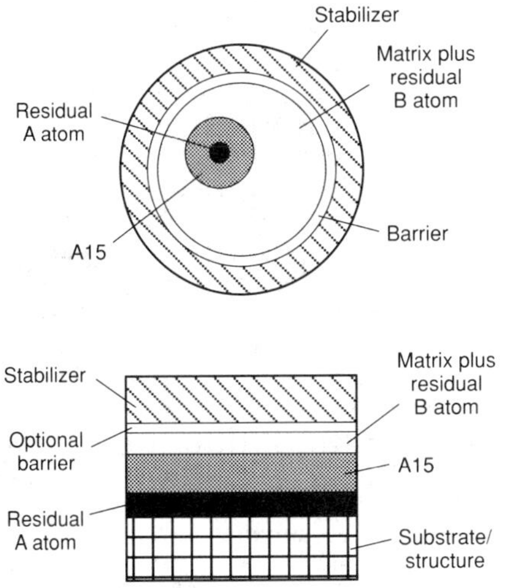

Fig. 16 Material components in a composite A15 conductor. Note all parts except stabilizer are included in calculation of noncopper current densities.

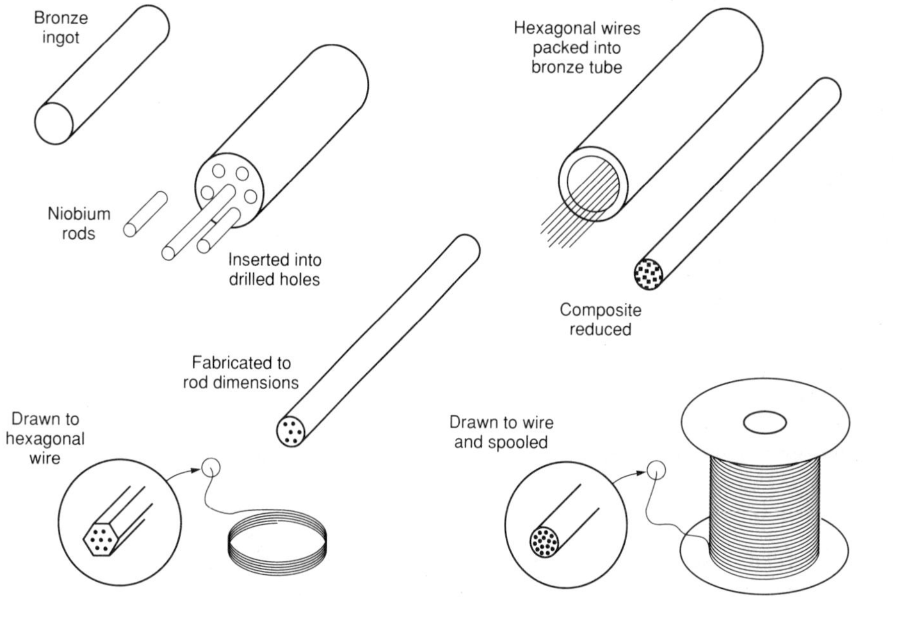

Fig. 17 Schematic illustration of the bronze-process multifilament conductor production routing. Source: Ref 22

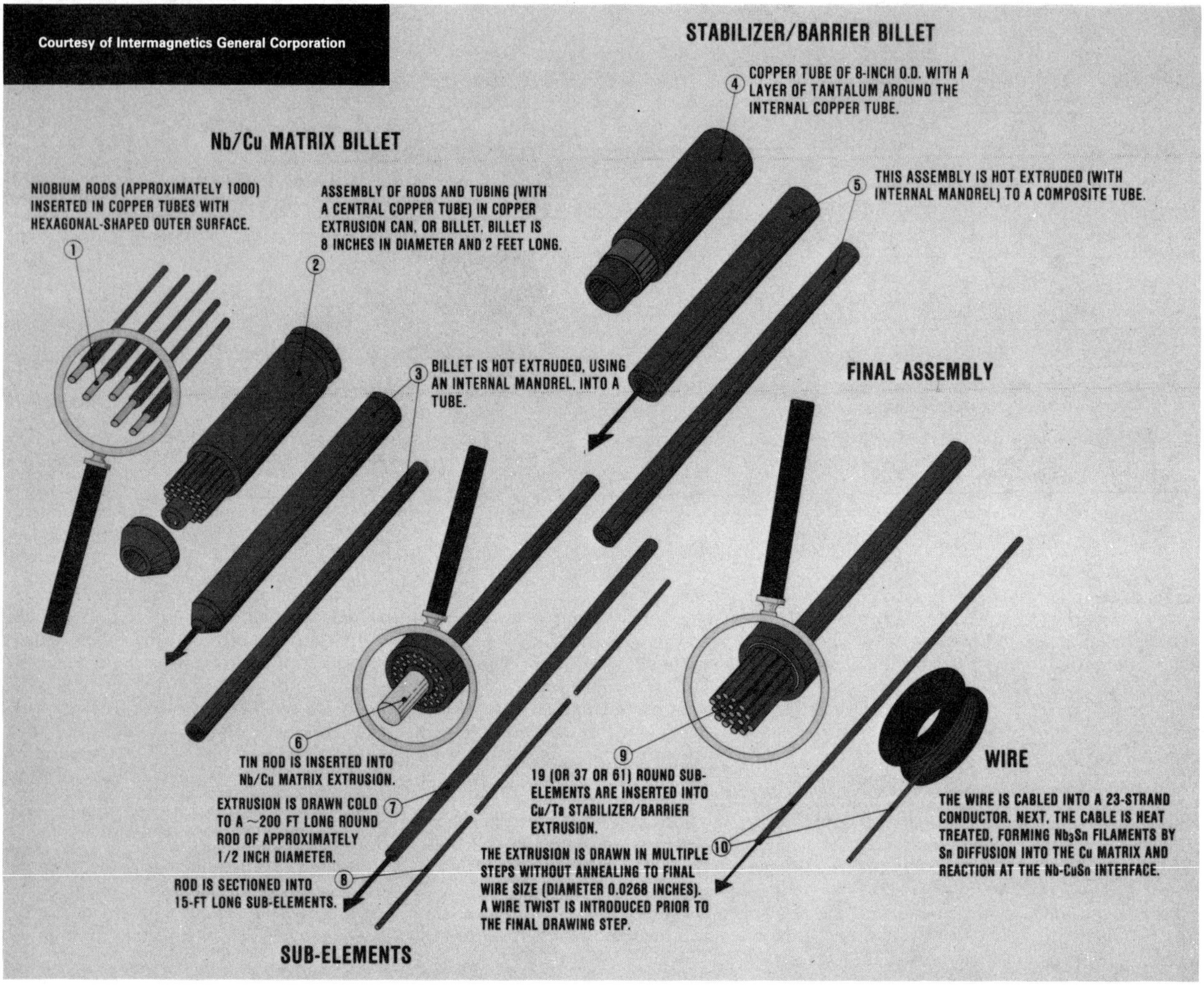

Fig. 18 Sequence of manufacturing operations involved in the formation of Nb_3Sn multifilamentary wire using the internal tin process.

and required current. As in NbTi cables, there is often some degradation caused by the cabling process.

Winding can be an anxious time if using prereacted cable (Ref 56). In the case of the US-DPC coil, however, a work-hardened superalloy jacket complicated the winding because of its high springback. During the reaction heat treatment, the superalloy precipitation hardened and retained the wound shape (Ref 58). To prevent strand sintering, the US-DPC conductor wire was chrome-plated prior to cabling (Fig. 27) (Ref 43, 46).

A15 Superconductor Applications

Commercial Magnets. It is efficient to wind solenoid magnets in concentric coils that make the most of the superconducting properties. To achieve fields above 9 T (90 kG) at 4.2 K, a high field material such as Nb_3Sn or V_3Ga is required, though a background field up to 9 T (90 kG) may be provided by NbTi. This is generally the case. Figure 28 shows the load lines for a hybrid 14.2 T (142 kG) coil system at the National Research Institute for Metals in Japan (N.R.I.M.) (Ref 59). V_3Ga inserts were used to boost the central field to over 18 T (180 kG) in an all-superconducting system (Ref 59).

Presently, 16 T (160 kG) hybrid (NbTi + Nb_3Sn) coils are commercially available with clear bores of greater than 50 mm (2 in.) and outer diameters less than 200 mm (8 in.) (Ref 60); 18 T (180 kG) magnets are advertised for sale as well (Ref 61). NMR systems with frequencies up to 600 MHz (14.1 T, or 141 kG) are available that utilize Nb_3Sn and NbTi in the persistent mode. The high field record for an all-superconducting system to date is 19.3 T (193 kG) using NbTi and alloyed Nb_3Sn (Ref 62).

Commercial solenoids are used for studying the magnetic behavior of materials and their chemistry by NMR techniques. A common application of superconducting solenoids is the study of superconducting materials. Magnetic refrigeration may use Nb_3Sn because of the desire to operate in the 10 to 15 K range.

Power Generation. Nb_3Sn has been pursued for ac motors and generators largely because of its 18 K critical temperature. The use of superconductors allows the current density of the armature to be raised because no iron is required and higher fields may be generated. The main advantages are the reduction in weight and volume with an increase in efficiency as well. Early proto-

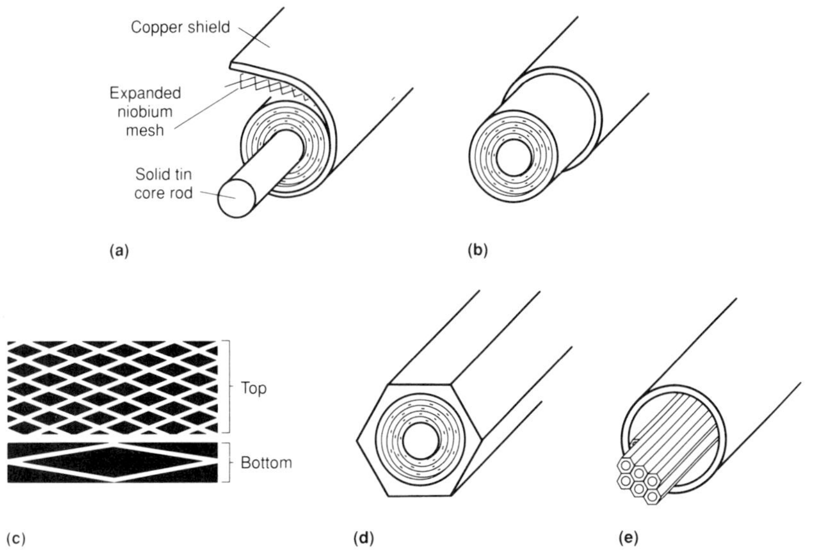

Fig. 19 Schematic of the modified jelly roll process. Because niobium-tin wire is fragile and brittle, multifilament superconducting wire cannot be made from these materials after the intermetallic niobium-tin compound has been formed. The two constituents, niobium and tin, as well as the copper for the matrix, are ductile materials. The multifilament wire, therefore, is made with the three metals in separate form. In this process, niobium, which has been formed into an expanded metal sheet, is interleaved with a sheet of copper and wound around a central mandrel made of tin. This initial preform is then drawn in successive stages to form the multifilament wire. The niobium-tin filaments are then formed in the wire by diffusing the tin into the niobium. This is accomplished by heating it to hundreds of degrees Celcius for ≧200 h. Because the resulting filaments are brittle, this is sometimes done after the wire is wound into its final coil form. The process yields superconducting wire in spools having 50 000 to 100 000 ft lengths. In (b), the jelly roll formed is inserted in outer copper tube. Billet is formed ready for drawing. In (c), the original niobium expanded mesh (top) becomes extremely long filaments (bottom) in the finished wire. A single diamond-shape segment will be reduced from ¼ in. to ⅓₀₀₀ in. in cross section and elongated from 1 in. to 1 mile. In (d), billet has been drawn down to <½ in. in diameter. In (e), rods ~½ in. in diameter are cut to shorter lengths and rebundled in another copper tube. Up to 50 additional drawing steps reduce material to wire ⅟₂₅ in. to ⅟₁₀₀ in. in diameter.

types were made using NbTi, but Nb_3Sn is preferred because of the higher temperature margin (Ref 63).

Power Transmission. Currently, overhead high-voltage lines remain the least expensive method for power transmission. However, concerns over electromagnetic fields (EMF) near these lines and public complaints may increase their cost of operation. There are also circumstances where underground transmission is desirable. Various high-power transmission cables have been researched, including versions employing Nb_3Sn. Nb_3Sn tapes, with smooth surfaces, have lower surface ac losses than NbTi or pure Nb. Brookhaven National Laboratory constructed and successfully operated a 100 m (330 ft) test facility using a Nb_3Sn-based cable (Fig. 29 and Fig. 30) (Ref 64).

High-Energy Physics. The desire to build particle accelerators with ever higher energies has spurred interest in using A15 conductors in dipole magnets. A twin aperture, 10 T (100 kG) dipole was successfully constructed and tested demonstrating the technical feasibility (Ref 65). Scale up from 1 to 17 m (0.3 to 55 ft) (superconducting supercollider size) has not yet been achieved. Other applications include high-gradient quadrupoles and focusing magnets near the interaction region where the temperature is likely to rise above 4.2 K.

Fusion. The high fields required for magnetic confinement of plasma energy and the

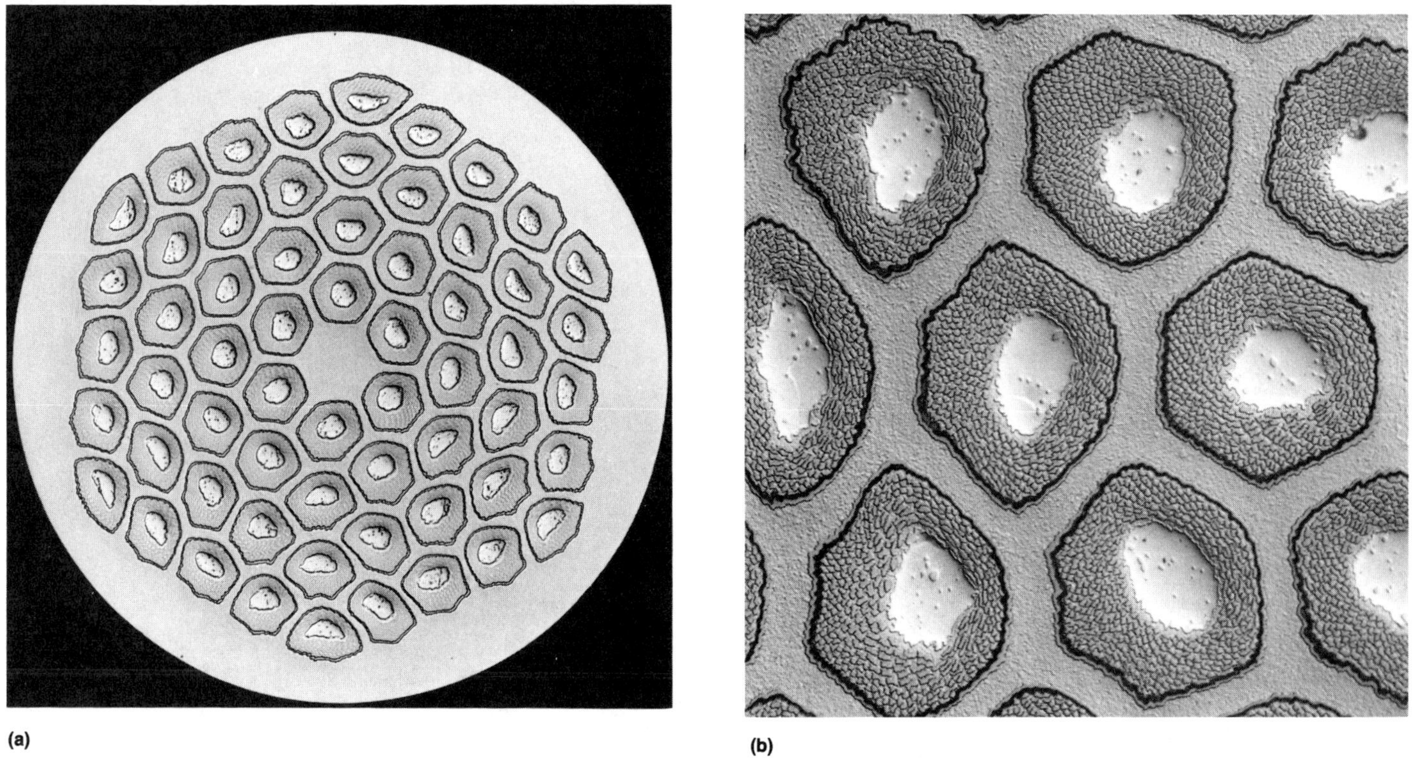

Fig. 20 Unreacted NbSn high-current density composite superconductor wire produced for high-field magnet application using tin-core MJR process. (a) 100× bright field illumination (B.F.). (b) 1000× differential interference contrast (D.I.C.). The 60 subelements in the 0.6 mm (0.024 in.) diam wire each have individual bimetal diffusion barriers composed of concentric rings of niobium around vanadium. The vanadium (inside layer) protects the niobium barrier (outside layer) from reacting with the tin while the niobium protects the copper from the vanadium at reaction temperatures above 700 °C (1290 °F). Courtesy of Paul E. Danielson, Teledyne Wah Chang Albany

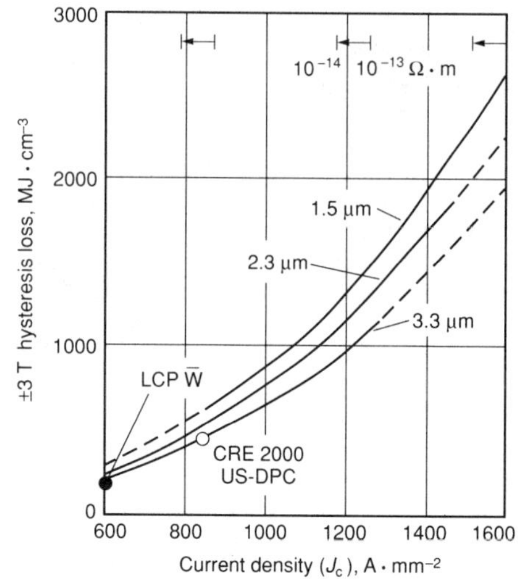

Fig. 21 Plot of noncopper hysteresis loss versus noncopper current density of the full range of current and filament size for tin-core $(Nb\text{-}1\%Ti)_3Sn$ superconductor. Modified jelly roll method was used. T_c was 4.2 K, $\mu_o H_{c2}$ of 10 T, and resistivity, ρ, of 10^{-13} Ω · m. US-DPC, United States-demonstration poloidal coil; LCP $\overline{W}$, Large coil project-Westinghouse

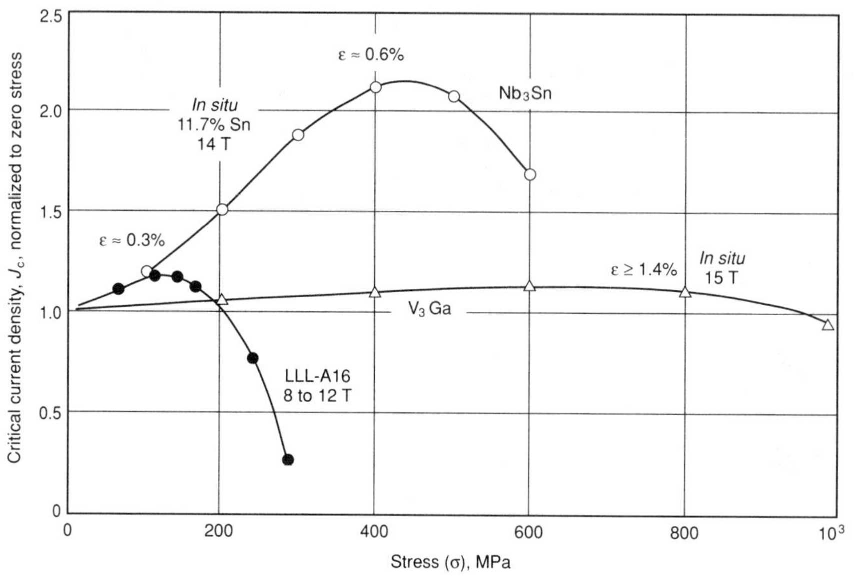

Fig. 22 J_c normalized to zero applied stress as a function of stress, σ, at indicated fields for Cu-Nb-Sn, Cu-V-Ga *in situ* composites, and a conventional commercial V_3Ga composite. Source: Ref 34

significant heat loads attained provide an exciting opportunity for the A15 materials, and fusion technology is the area most cited for their application. The US-DPC coil is a prototype cable for the central solenoid of a full-scale tokomak design (Ref 43, 46). The toroidal field coils (D shaped) have also been considered as Nb_3Sn coils. The next European torus (NET) design calls for Nb_3Sn in the poloidal field (PF) and toroidal field (TF) coils (Fig. 31). The Russians have built a tokomak, T15, using all Nb_3Sn coils and expect to test the system in 1990 (Ref 67). In competition with NET is the international thermonuclear experimental reactor (ITER), which is a joint European, American, Japanese, and Russian project in the design phase. This machine would also be a Nb_3Sn system (Ref 68).

The LCT project was an international demonstration project consisting of six D-shaped coils, one of which used Nb_3Sn conductors. The Westinghouse coil used forced flow technology with an internally cooled cable in conduit (ICCS) approach. This coil was a react and wind coil (Fig. 32) (Ref 56).

Additional Applications. Nb_3Sn, V_3Ga, $Nb_3(AlGe)$, and Nb_3Al conductors are use-

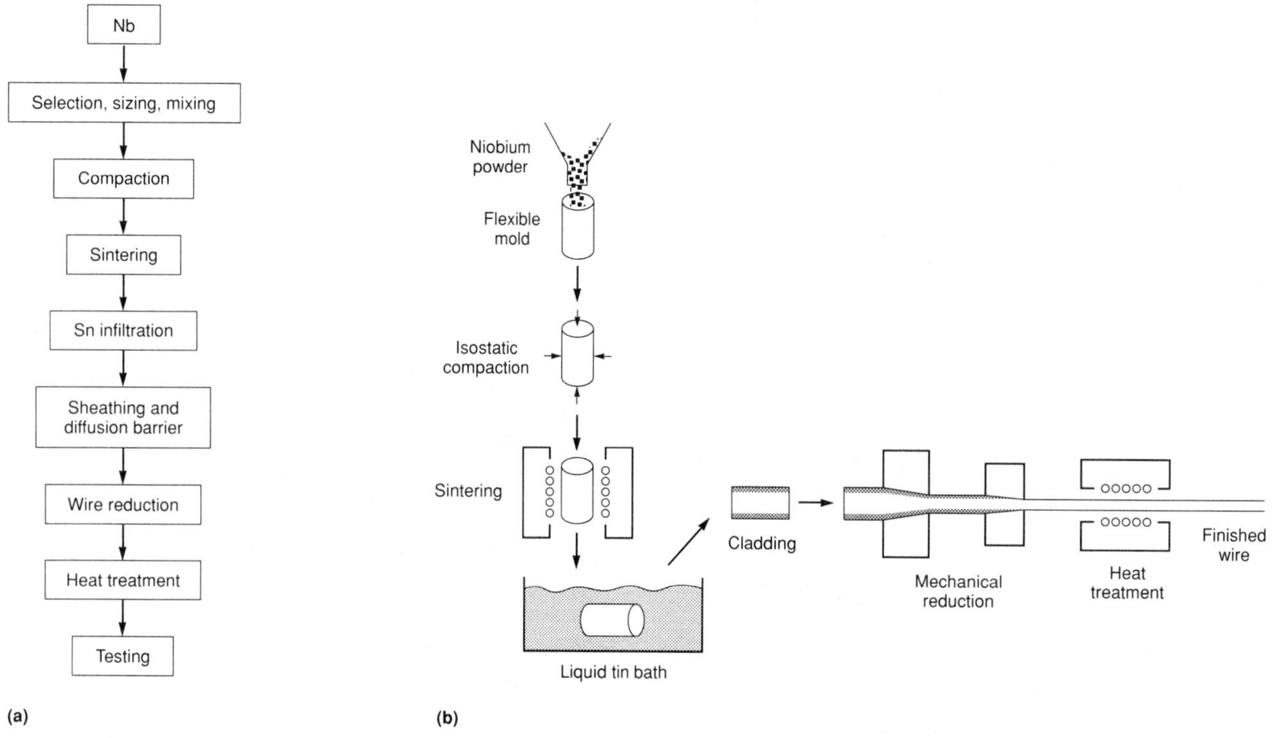

Fig. 23 Infiltrated tin P/M process for producing multifilamentary superconducting wire. (a) Flow diagram. (b) Schematic. Source: Ref 51

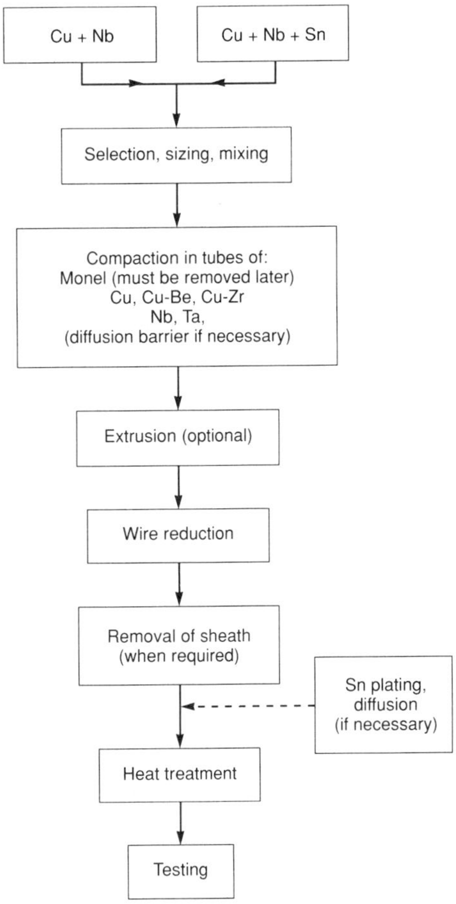

Fig. 24 Flow diagram for cold powder processes

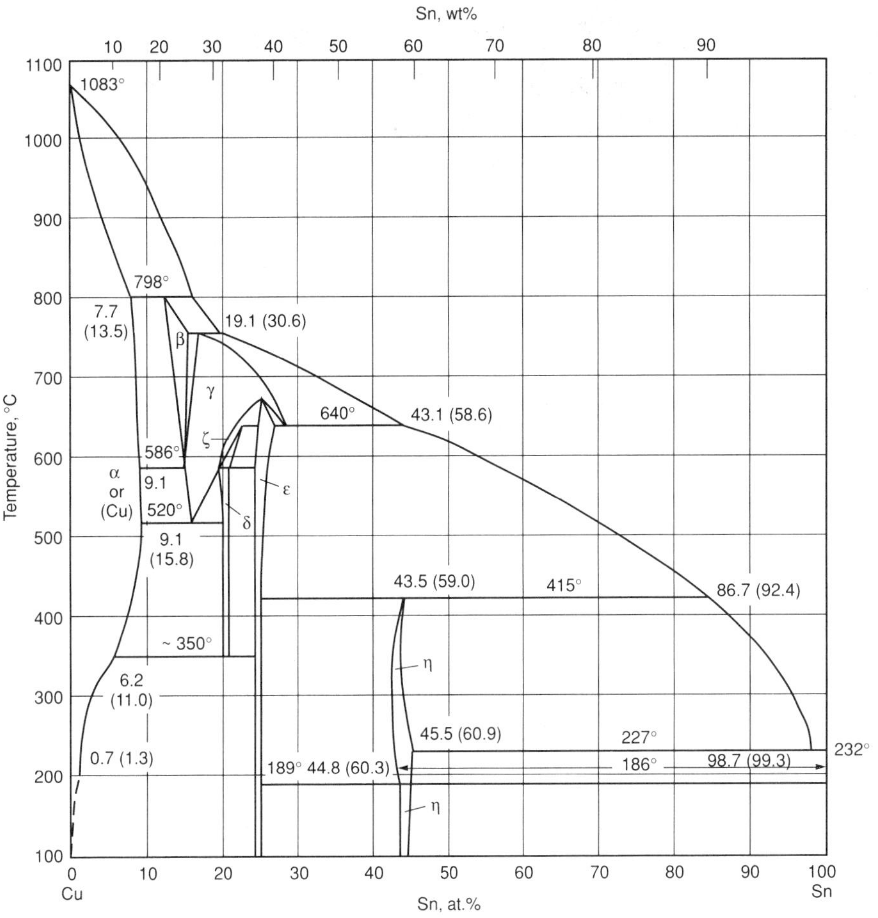

Fig. 25 Phase diagram of the Cu-Sn system. Source: Ref 55

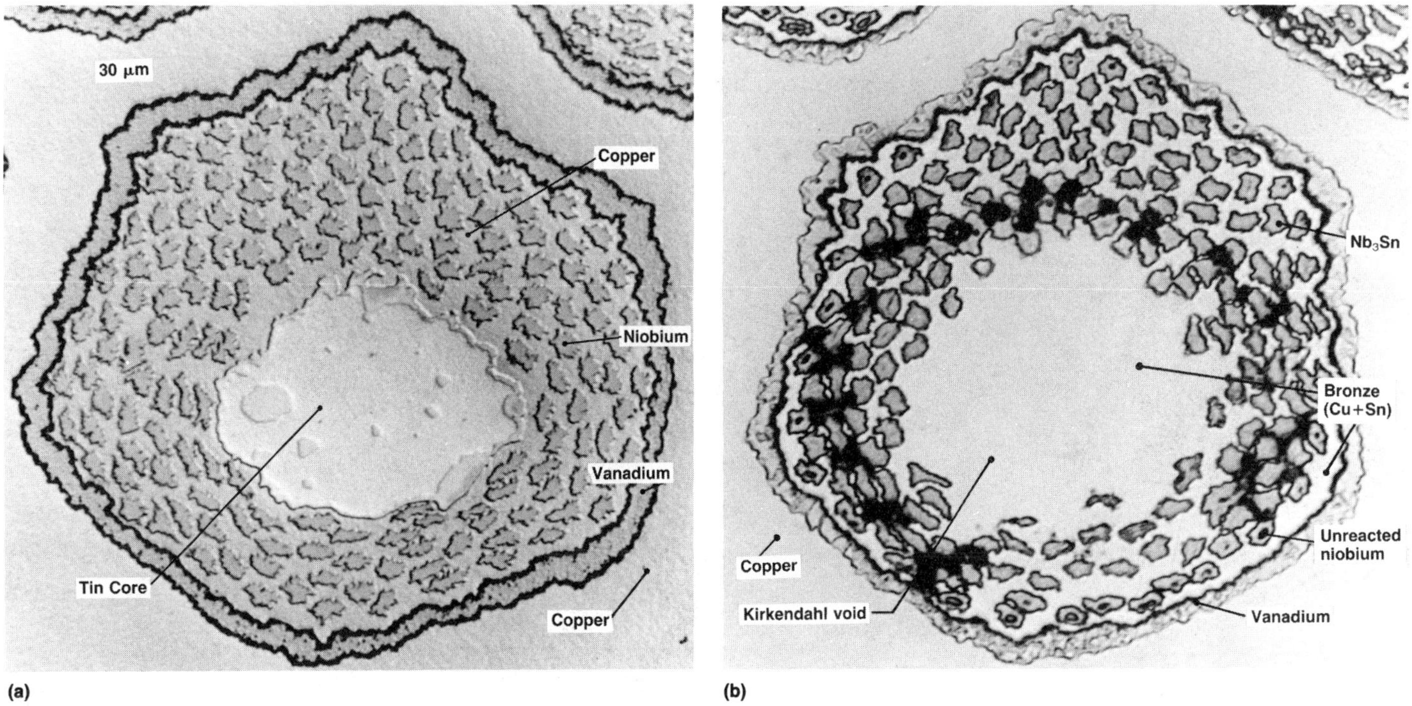

Fig. 26 Photomicrographs of an element in a tin-core MJR processed superconductor wire produced using internal tin process. (a) No reaction heat treatment. (b) Three-stage reaction heat treatment at 210 °C (410 °F), 340 °C (645 °F), and 650 °C (1200 °F). Courtesy of Paul E. Danielson, Teledyne Wah Chang Albany

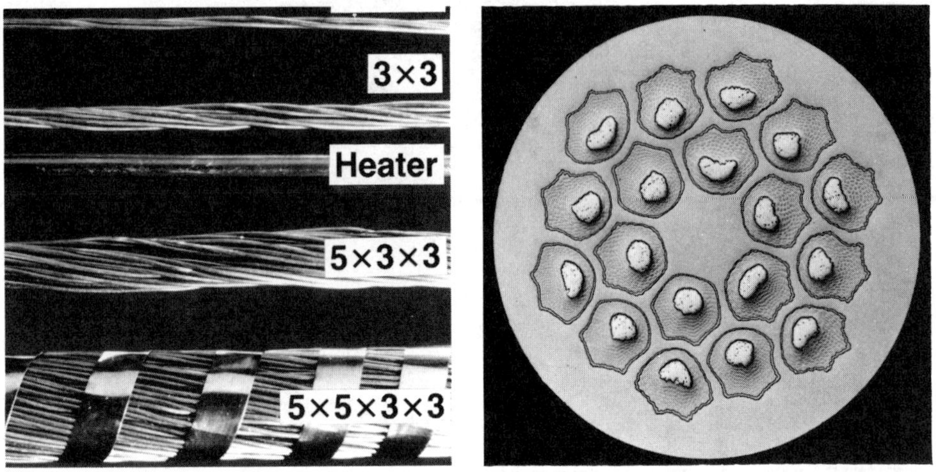

Fig. 27 Cabling elements for the 225-strand US-DPC conductor. The 3 mm (0.125 in.) heater shown in (a) was placed in the center of the cable in the final cabling operation. (b) Magnified view of a 20 mm (0.78 in.) diam conductor strand. ~75×. Courtesy of Paul E. Danielson, Teledyne Wah Chang Albany

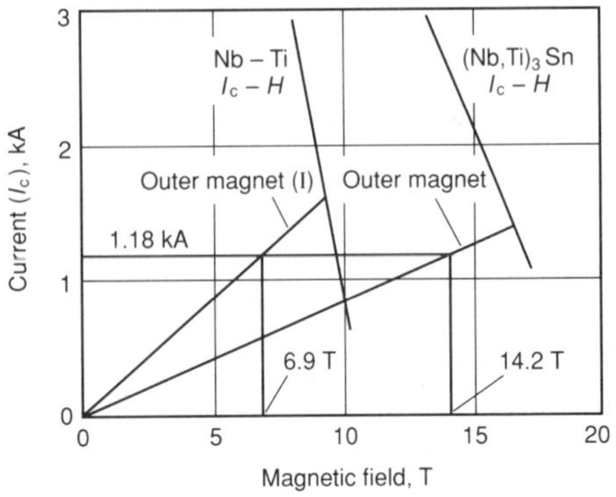

Fig. 28 Plot of current versus magnetic field at 4.2 K for the Nb-Ti and the $(Nb,Ti)_3Sn$ conductors in a hybrid coil system. Also shown are excitation load lines for the outer magnet (I) and the complete outer magnet. Source: Ref 59

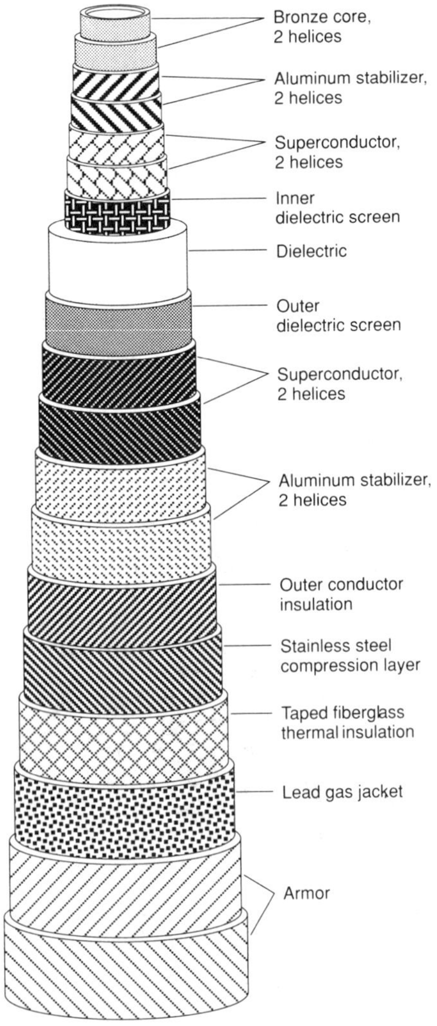

Fig. 29 Flexible Nb_3Sn ac coaxial power transmission cable being developed at Brookhaven National Laboratory. Source: Ref 63

ful whenever fields greater than 10 T (100 kG) or operating temperatures up to 15 K are required. The easiest to fabricate of this group is Nb_3Sn, and this fact alone accounts for the variety of fabrication routes and its preference for these applications. The lower strain sensitivity of V_3Ga and Nb_3Al is much more attractive, as is the higher critical temperature of Nb_3Ge and $Nb_3(AlGe)$. As fabrication routes become more economical, these materials will also see increasing use.

REFERENCES

1. D. Dew-Hughes, in *Treatise on Materials Science and Technology*, Vol 14, T. Luhman and D. Dew-Hughes, Ed., Academic Press, 1979, p 137
2. R. Flükiger, in *Superconductor Materials Science—Metallurgy, Fabrication and Applications*, S. Foner and B. Schwartz, Ed., Plenum Press, 1980, p 511
3. A.M. Clogston and V. Jaccarino, *Phys. Rev.*, Vol 121, 1961, p 1357
4. D. Moore, R. Zubeck, J. Powell, and M. Beasley, *Phys. Rev.*, Vol B20, 1979, p 2721
5. A. Sweedler, C. Snead, Jr., and D. Cox, in *Treatise on Materials Science and Technology*, Vol 14, T. Luhman and D. Dew-Hughes, Ed., Academic Press, 1979, p 349
6. M. Beasley, *Adv. Cryo. Eng.*, Vol 28, 1981, p 349
7. D. Welch, *Adv. Cryo. Eng.*, Vol 26, 1980, p 48
8. J. Ekin, *Adv. Cryo. Eng.*, Vol 30, 1984, p 823
9. R. Flükiger, R. Isernhagen, W. Gold-

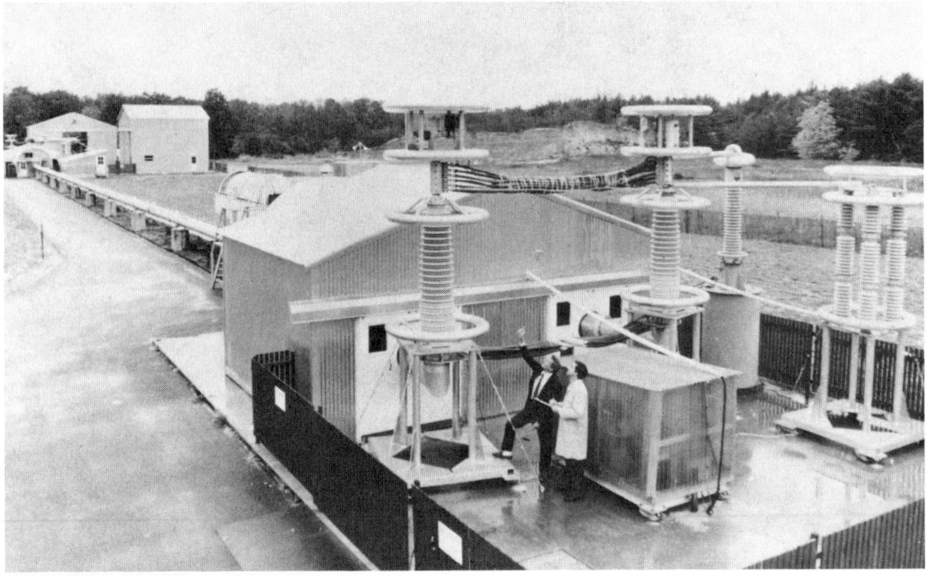

Fig. 30 The 100 m (330 ft) experimental test station at Brookhaven National Laboratory for power transmission

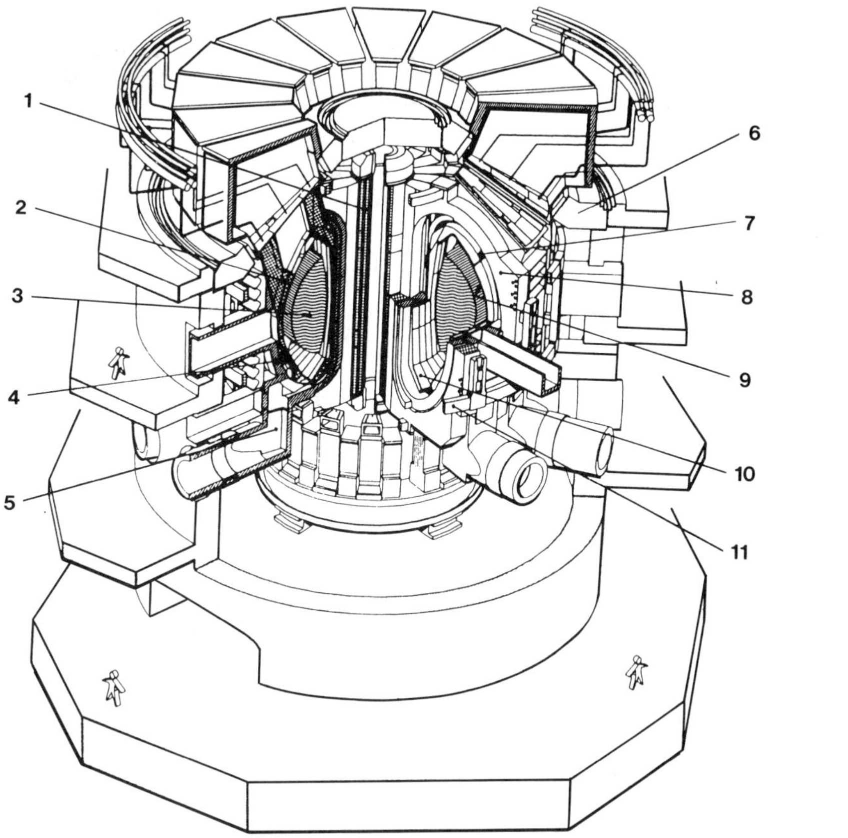

Fig. 31 An isometric view of the NET device showing the main components. (1) Inner PF coils (Nb_3Sn). (2) Blanket. (3) Plasma. (4) Vacuum vessel/shield. (5) Plasma exhaust. (6) Biological shield/cryostat. (7) Active control coils. (8) Toroidal (TF) coils (Nb_3Sn). (9) First wall. (10) Divertor plates. (11) Outer poloidal field (PF) coils. Source: Ref 66

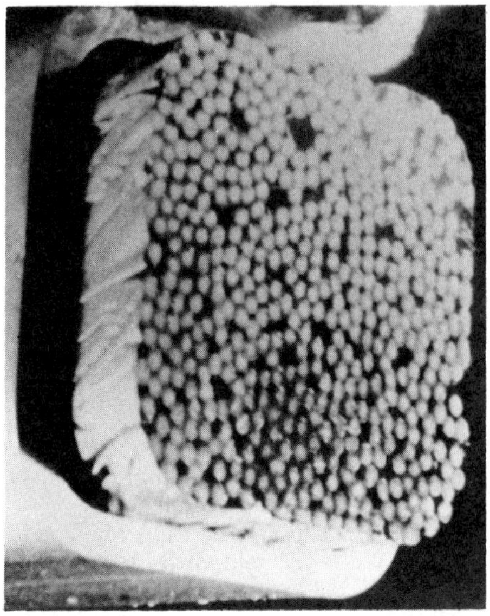

(a)

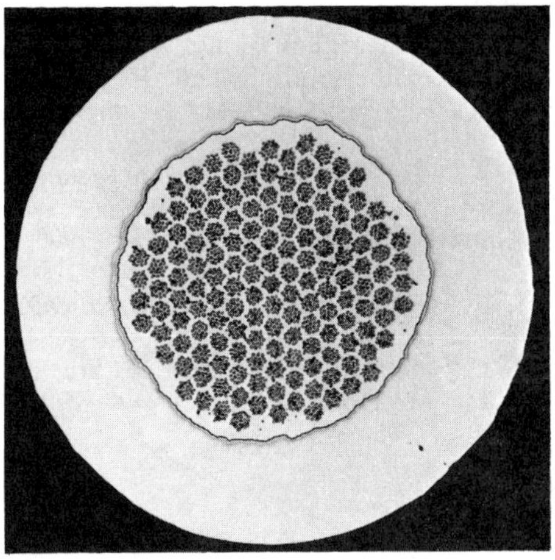

(b)

Fig. 32 Forced flow type conductor for 20.7 × 20.7 mm (0.815 × 0.815 in.) large coil program (LCP). (a) Conductor containing 486 (6 × 3^4) individual strands. (b) 0.7 mm (0.028 in.) diam individual strand containing 2869 (19 × 151) filaments with a tantalum barrier layer, 65% stabilizing copper, and 13% bronze matrix. Courtesy of Paul E. Danielson, Teledyne Wah Chang Albany

acker, and W. Specking, *Adv. Cryo. Eng.*, Vol 30, 1984, p 851

10. J. Ekin, *Adv. Cryo. Eng.*, Vol 34, 1988, p 547
11. J. Charlesworth, I. MacPhail, and P. Madsen, *J. Mater. Sci.*, Vol 5, 1970, p 580
12. H. King, in *Symposium on Phase Transformations*, Institute of Metals, 1968, p 196
13. M. Suenaga, in *Superconductor Materials Science—Metallurgy, Fabrication and Applications*, S. Foner and B. Schwartz, Ed., Plenum Press, 1980, p 201
14. J. Livingston, *Phys. Status Solidi*, Vol A44, 1977, p 295
15. D. Smathers and D. Larbalestier, *Adv. Cryo. Eng.*, Vol 28, 1982, p 415
16. M. Suenaga and W. Jansen, *Appl. Phys. Lett.*, Vol 43, 1983, p 791
17. K. Togano, T. Asano, and K. Tachikawa, *J. Less-Common Met.*, Vol 68, 1979, p 15
18. I. Wu, D. Dietderich, J. Holthuis, W. Hassenzahl, and J. Morris, Jr., *IEEE Trans. Magn.*, Vol 19, 1983, p 1437
19. D. Howe and L. Weinman, *IEEE Trans. Magn.*, Vol 11, 1975, p 251
20. D. Larbalestier, D. Smathers, M. Daeumling, C. Meingast, W. Warnes, and K. Marken, in *Proceedings of the International Symposium on Flux Pinning and Electromagnetic Properties of Superconductors* (Fukuoka, Japan), 1985
21. D. Hampshire, H. Jones, and E. Mitchell, *IEEE Trans. Magn.*, Vol 21, 1985, p 289
22. T. Luhman, in *Treatise on Materials Science and Technology*, Vol 14, T. Luhman and D. Dew-Hughes, Ed., Academic Press, 1979, p 221
23. S. Gauss and R. Flükiger, *IEEE Trans. Magn.*, Vol 23, 1987, p 657
24. M. Klemm, E. Seibt, W. Specking, J. Xu, and R. Flükiger, to be published in *Supercond. Sci. Technol.*
25. E. Schrader and F. Kolondra, *RCA Rev.*, Vol 25, 1964, p 582
26. J. Hanak, K. Strater, and G. Cullen, *RCA Rev.*, Vol 25, 1964, p 342
27. M. Benz, *IEEE Trans. Magn.*, Vol 2, 1966, p 760
28. M. Suenaga and M. Garber, *Science*, Vol 184, 1974, p 952
29. E. Adam, P. Beischer, W. Marancik, and M. Yound, *IEEE Trans. Magn.*, Vol 13, 1977, p 425
30. K. Togano, H. Kumakura, T. Takeuchi, and K. Tachikawa, *IEEE Trans. Magn.*, Vol 19, 1983, p 414
31. H. Hillman, in *Superconductor Materials Science—Metallurgy, Fabrication and Applications*, S. Foner and B. Schwartz, Ed., Plenum Press, 1980, p 275
32. D. Smathers, K. Marken, D. Larbalestier, and R. Scanlan, *IEEE Trans. Magn.*, Vol 19, 1983, p 1417
33. D. Larbalestier, V. Edwards, J. Lee, C. Scott, and M. Wilson, *IEEE Trans. Magn.*, Vol 11, 1975, p 555
34. R. Roberge, in *Superconductor Materials Science—Metallurgy, Fabrication and Applications*, S. Foner and B. Schwartz, Ed., Plenum Press, 1980, p 389

35. W. McDonald, U.S. Patent 4,262,412, 1981
36. W. McDonald, U.S. Patent 4,414,418, 1983
37. W. McDonald, C. Curtis, R. Scanlan, D. Larbalestier, K. Marken, and D. Smathers, *IEEE Trans. Magn.*, Vol 19, 1983, p 1124
38. D. Gubser, T. Francavilla, C. Pande, B. Rath, and W. McDonald, *J. Appl. Phys.*, Vol 56, 1984, p 1051
39. D. Smathers, K. Marken, P. Lee, D. Larbalestier, W. McDonald, and P. O'Larey, *IEEE Trans. Magn.*, Vol 21, 1985, p 1133
40. M. Siddall, W. McDonald, and K. Efferson, *IEEE Trans. Magn.*, Vol 19, 1983, p 907
41. D. Smathers, P. O'Larey, M. Siddall, and W. McDonald, *Adv. Cryo. Eng.*, Vol 34, 1988, p 515
42. D. Smathers, in *Proceedings of the International Symposium on Tantalum and Niobium* (Brussels), Tantalum-International Study Center, 1989, p 707
43. D. Smathers, M. Siddall, M. Steeves, M. Takayasu, and M. Hoenig, *Adv. Cryo. Eng.*, Vol 36, 1989
44. A. delle Corte *et al.*, *Fusion Technology 1988*, A. Van Ingen, A. Nijsen-Vis, and H. Klippel, Ed., Elsevier, 1989, p 1476
45. R. Goldfarb and J. Ekin, *Cryogenics*, Vol 26, 1986, p 478
46. D. Smathers, P. O'Larey, M. Steeves, and M. Hoenig, *IEEE Trans. Magn.*, Vol 24, 1988, p 1131
47. S. Shen, "Summary Report of the Expert Advisory Workshop on AC Losses in Superconducting Magnets for Fusion," Lawrence Livermore National Laboratory, SCMDG 89-54-80, Oct 1989
48. A.J. Zaleski, T.P. Orlando, A. Zieba, B.B. Schwartz, and S. Foner, *J. Appl. Phys.*, Vol 56, 1984, p 3278
49. Showa Electric Wire and Cable Company, Tokyo, Japan
50. H. Veringa, E. Hornsveld, and Hoogendam, *Adv. Cryo. Eng.*, Vol 30, 1988, p 813
51. S. Foner, C. Thieme, S. Pourrahimi, and B. Schwartz, *Adv. Cryo. Eng.*, Vol 32, 1986, p 1031
52. M. Pickus, J. Holthuis, and M. Rosen, in *Filamentary A15 Superconductors*, M. Suenaga and A. Clark, Ed., Plenum Press, 1980, p 331
53. A. Hecker, E. Gregory, J. Wong, C. Thieme, and S. Foner, *Adv. Cryo. Eng.*, Vol 34, 1988, p 485
54. S. Ceresara, M. Ricci, N. Sacchetti, and G. Sacerdoti, *IEEE Trans. Magn.*, Vol 11, 1975, p 263
55. M. Hansen, *Constitution of Binary Alloys*, McGraw-Hill, 1958, p 634
56. P.N. Haubenreich, *IEEE Trans. Magn.*, Vol 23, 1987, p 800
57. B. Jakob and G. Pasztor, *IEEE Trans. Magn.*, Vol 23, 1987, p 914
58. M. Steeves, M. Hoenig, M. Takayasu, R. Randall, J. Tracy, J. Hale, M. Morra, I. Hwang, and P. Marti, *IEEE Trans. Magn.*, Vol 25, 1989, p 1738
59. K. Tachikawa, K. Inoue, M. Sacki, K. Aihara, T. Fujinaga, H. Hashimoto, and R. Saito, *IEEE Trans. Magn.*, Vol 23, 1987, p 907
60. Cryogenic Consultants Limited, London, England
61. *Phys. Today*, Oxford Analytical Instruments, Oxford, England, Aug 1988, p 54
62. P. Turowski and T. Schneider, *IEEE Trans. Magn.*, Vol 24, 1988, p 1063
63. G. Bogner, in *Superconductor Materials Science—Metallurgy, Fabrication and Applications*, S. Foner and B. Schwartz, Ed., Plenum Press, 1980, p 757
64. E. Forsyth, A. McNerney, A. Muller, and S. Rigby, *IEEE Trans. Power Appar. Syst.*, PAS-97, 1978, p 737
65. A. Asner, R. Perin, S. Wengir, and F. Zerobin, Paper JA01 presented at MT-11 (Tscuba, Japan), Aug 1989, to be published by Elsevier, 1990
66. R. Toschi, M. Chazalow, F. Engelman, J. Nihoul, J. Raeder, and E. Scalpietro, *Fusion Technol.*, Vol 14, 1988, p 19
67. Ivanov, Klimenko, Lelekhov, Chernopleokov, Kiknadze, Kostenko, Malysheu, Monoszon, Trukhachev, and Churakov, Presented at MT-11 (Tscuba, Japan), Aug 1989, to be published by Elsevier, 1990
68. C. Henning and J. Miller, *IEEE Trans. Magn.*, Vol 25, 1989, p 1469

Ternary Molybdenum Chalcogenides (Chevrel Phases)

Luc Le Lay, Applied Superconductivity Center, University of Wisconsin—Madison

THE TERNARY MOLYBDENUM CHALCOGENIDES have generated many fundamental as well as applied research efforts since their discovery nearly two decades ago (Ref 1). This is essentially due to the originality of the materials and to the possible applications of some of them as high field superconductors (that is, >20 T, or 200 kG). For almost ten years, several research teams have been trying to fabricate high critical current density (J_c) superconducting filaments, solving first the wiring problem, then partially that of the transport of the current. Some research is still going on, due to the difficulties encountered for achieving good transport currents in the high-temperature oxides. Although the requirements for the construction of a coil are still not met, significant progress has been achieved year after year. After a very general presentation of the materials, an overview of the present status of the superconducting applications of the monofilaments at the dawn of this new decade will be discussed.

Ternary Molybdenum Chalcogenides

This name stands for a vast class of materials, whose general formula is $M_xMo_6X_8$, where M is a cation and X a chalcogen (sulfur, selenium, or tellurium). There are also some compounds in which Mo is partially substituted by Re, Ru, or Rh (Ref 2, 3). Figure 1(a) shows the fundamental structural unit: a Mo octahedron surrounded by an X cube; these cubes, set in a tridimensional network, are tilted, so as to create intercube Mo-X bonds (see Fig. 1b), that give the solid its cohesion (Ref 4). The M atoms are located in the channels generated by this network (Fig. 1c), and, mostly depending on their size, their location in the network, and their stoichiometry can vary (Ref 5, 6). It is even possible to insert them reversibly (Ref 7-9).

The great flexibility of this structure (where both M and X can vary) originated the diversity of both the chemical and physical properties observed: superconductivity (Ref 10), catalysis (Ref 11), ion conduction (Ref 7-9), and so on. In superconductivity alone, several compounds exhibit exciting characteristics such as reentrant superconductivity (Ref 12), interplay with magnetic order (Ref 13, 14), double domain superconductivity (Ref 15), Jaccarino-Peter effect (Ref 16), and high critical fields (Ref 17). The main applications of the ternary molybdenum chalcogenides appear to use the latter property. It is found mainly in $PbMo_6S_8$ (designated PMS), $SnMo_6S_8$ (designated SMS), and $LaMo_6Se_8$ whose respective upper critical fields at 0 K are 60, 34, and 45 T (600, 340, and 450 kG) (see Fig. 2). Because the latter is not considered as a possible application for practical reasons, the research has been focused on the wiring of PMS and SMS chalcogenides.

Fabrication Technology

The following describes only the PMS technology, but it is understood that the same technology applies to SMS. Two features determine the kind of wiring technology used:

- PMS is easily synthesized as a powder
- Its fusion (it is indeed a noncongruent fusion) occurs at high temperatures [>1500 °C, or 2730 °F (Ref 18)]

Therefore, the available processing technique is the powder metallurgy one. Its principle is to fit the powder into a drawable casing and then to draw the whole set. Whatever the drawing conditions, a final heat treatment is always necessary in order to generate or restore the superconducting properties of the wire.

The handling and the choice of the initial powder appear to be crucial for the superconducting properties expected. The first emphasis is that, as sulfides, PMS and its precursors (for example, MoS_2, PbS, Mo_2S_3, and so on) are particularly sensitive to moisture and oxygen (Ref 19, 20). Hence, their handling has to be carried out in a controlled atmosphere (either vacuum or inert gas). The powder that is effectively drawn can be either PMS or a mixture of its precursors, given that a final heat treatment of the wire is necessary anyway (otherwise, it would have been logical to deal with PMS alone). In both cases, a compaction step is used before insertion in the sheath. This is either a hot or a cold process. The first yields a 100% packing density (Ref 21), the second an 85% packing density (Ref 22). Both processes allow a good mechanical behavior during the wiring operation. Figure 3 shows a typical PMS monofilament.

In the past, gold and silver have been tried as matrices (Ref 23), but both yielded disappointing results. Today, molybdenum (Ref 24) or copper (Ref 25-28) is mostly used, with stainless steel sometimes (Ref 24, 25, 29) as an external prestress layer. In the case of copper, an antidiffusion barrier is compulsory, otherwise the powder eventually reacts with the copper in the final heat treatment, yielding $Cu_xMo_6S_8$, whose superconducting properties are not desired here. This compulsory antidiffusion barrier is either niobium or tantalum. These options categorize two types of PMS wire fabrication: hot processing and cold processing.

Hot processing is the fabrication method used with the molybdenum sheath. The powder is first hot or cold compacted in a controlled atmosphere (either vacuum or inert gas), then fitted into a molybdenum billet (typical size is 6 mm, or 1/4 in., ID, by 12 mm, or 15/32 in., OD, by 100 mm, or 4 in., length). Stainless steel surrounds it. This composite is then extruded (~1000 °C, or 1830 °F), hot swaged, and finally drawn down to the desired diameter (the final drawing temperature is around 600 °C, or 1110 °F). The wire lengths produced usually are on the order of 1 km (0.6 mile) (Ref 24). Despite an extensive technology and the limited powder choice (if it were used, any mixture of precursors would indeed be reacted during the drawing operation, making this kind of choice equivalent to the prereacted powder method), this kind of casing

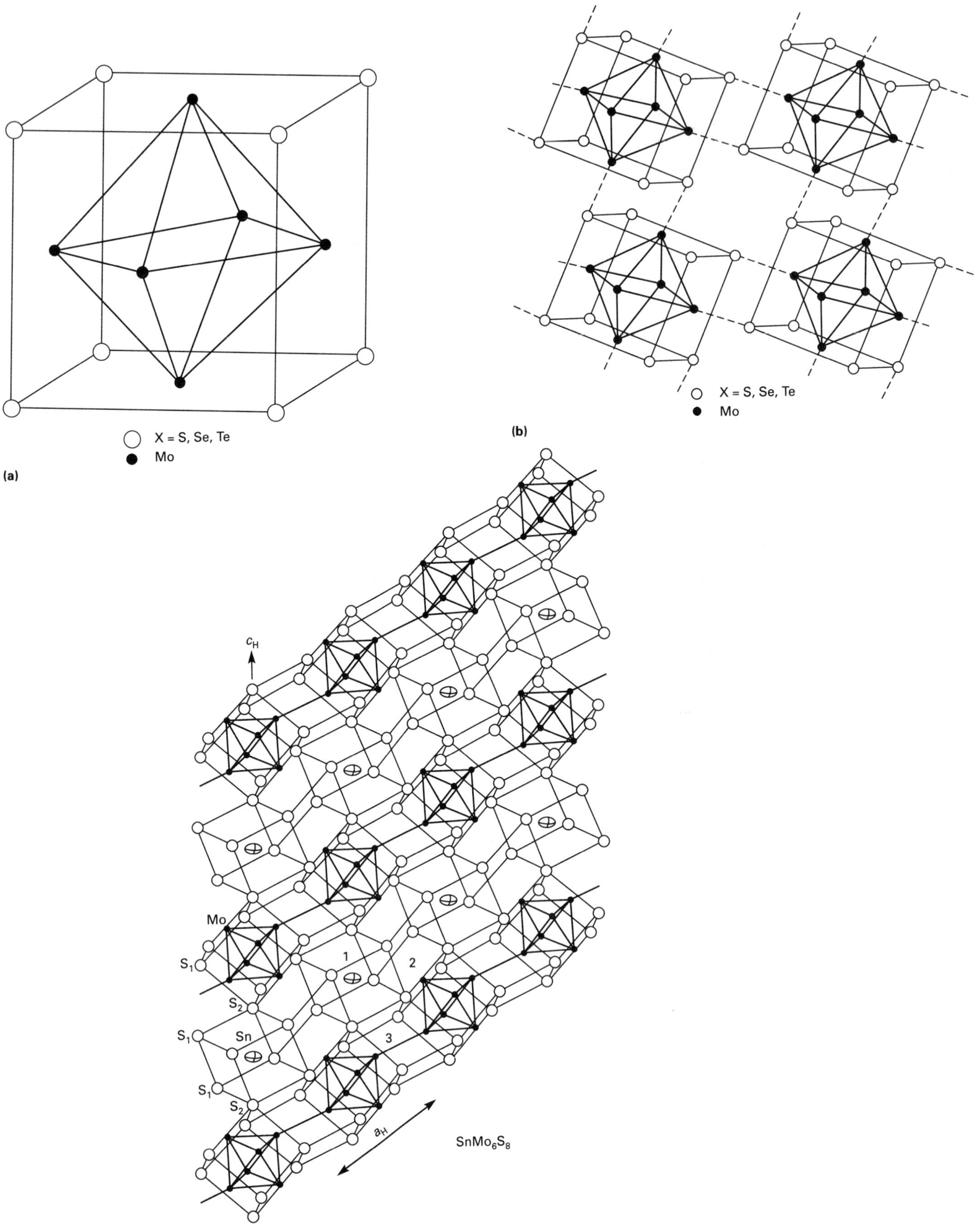

Fig. 1 Structure and bonding in ternary molybdenum chalcogenides. (a) Mo_6S_8 unit of the Chevrel phases. (b) Tilting of the fundamental structural units to form Mo-X bonds. (c) Generation of channel in $SnMo_6S_8$

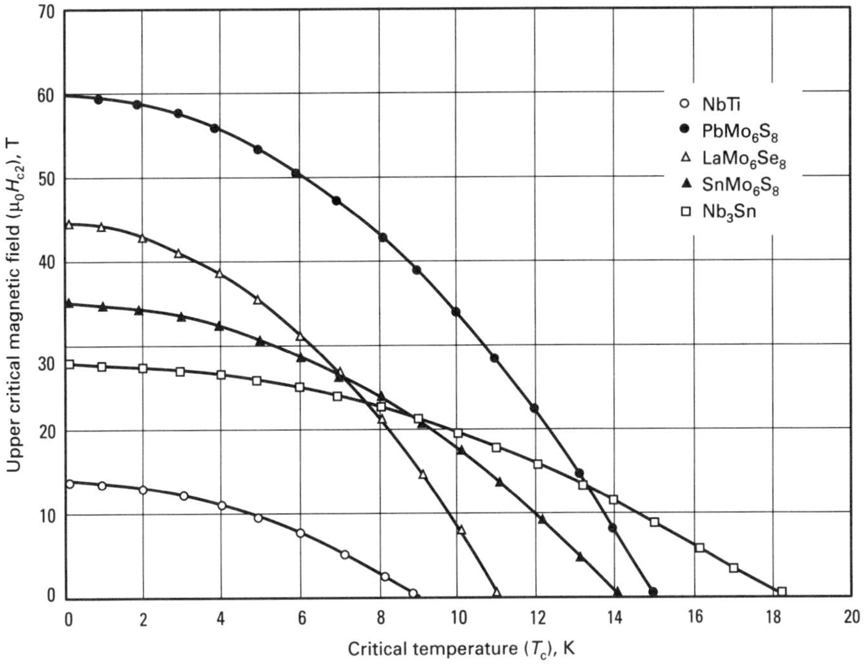

Fig. 2 Comparison of the critical fields of three ternary molybdenum chalcogenides ($PbMo_6S_8$, $LaMo_6Se_8$, $SnMo_6S_8$) and two commercially available superconductors (NbTi, Nb_3Sn)

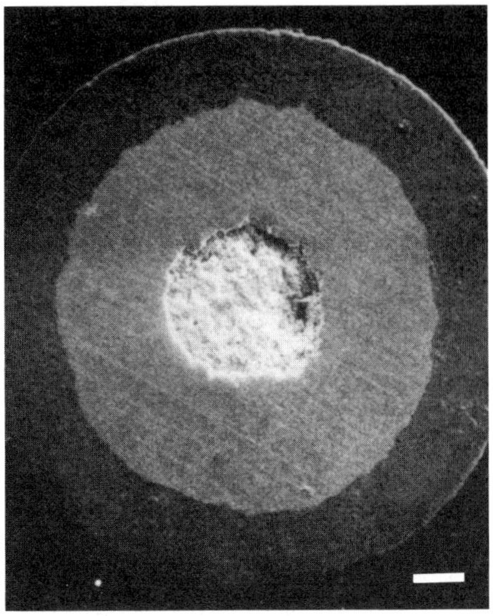

Fig. 3 Typical PMS monofilament with a niobium barrier (light gray) and a copper outer jacket (dark gray). The white line at the bottom is 100 μm (0.004 in.). Courtesy of P. Rabiller and M. Hirrien

possesses a fair advantage over niobium or tantalum. It is chemically nonreactive toward PMS, or one should rather say that molybdenum is in equilibrium with PMS (Ref 30). The usual temperature of the final heat treatment of the wire is on the order of 700 to 900 °C (1290 to 1650 °F) for 24 to 125 h (Ref 24, 29).

Cold processing refers to the fabrication method utilizing niobium and tantalum sheath technology. It is indeed similar to the hot processing fabrication method. The compacted powder is inserted in the anti-diffusion barrier (typical dimensions, 6 mm, or ¼ in., ID by 10 mm, or ⅜ in., OD by 250 mm, or 10 in., length) (Ref 25, 28), which is itself embedded in a copper matrix (10 mm, or ⅜ in., ID by 12 to 14 mm, or 15⁄32 to 9⁄16 in., OD); the whole set is then drawn at room temperature down to a diameter on the order of several tenths of a millimeter. Stainless steel is sometimes added during the process, when the wire diameter is close to 2 mm (0.08 in.) (Ref 25). The temperature of the final heat treatment is slightly higher (900 to 1000 °C, or 1650 to 1830 °F), and the time is much shorter (10 min to 2 h). This allows the formation of an intermediate phase at the sheath-powder interface (Ref 22, 31), resulting from several diffusion processes. Although we have little knowledge of this reaction, the primary effect is certainly to change the stoichiometry of the central powder and hence its properties. Nevertheless, no one has given up on this process, for this drawback is counterbalanced by both the ease of the cold drawing technique and the critical current densities obtained (see the section "Superconducting Properties of Wire Filaments"). The choice of the starting powder is also wider because either PMS or precursors may be used.

Superconducting Properties of Wire Filaments

Mechanical Behavior. It has been shown that the expansion coefficient of PMS is greater than that of molybdenum, niobium, or tantalum (Ref 32, 33). This leads to a tensile stress of PMS inside the wire when it is cooled down at helium temperature. In order to avoid this, an external stainless steel layer whose thickness is chosen so that it precompresses the wire can be used in order to make the thermal expansions of the whole casing and the PMS match. This results in improved critical current densities (Ref 21, 34). In general, the mechanical behavior of PMS is comparable to Nb_3Sn in the 10 to 15 T (100 to 150 kG) range (Ref 35), but because its upper critical field is about twice that of Nb_3Sn, PMS retains its mechanical characteristics (that is, the J_c dependence on the strain) in the 20 to 25 T (200 to 250 kG) range (Ref 25), where Nb_3Sn degrades dramatically.

Critical Current Densities. It has been ten years since PMS wires were first produced and their performance has steadily been improved. Several factors have contributed to that improvement:

- Advances in the handling and synthesis of the powders
- Use of an external stainless steel sheath
- Doping of PMS with Sn
- Better heat treatment conditions for the wires

Figure 4 summarizes the data of several researchers seeking to optimize chalcogenide properties. The transport J_c values of the best wires are slightly above 10^8 A · m^{-2} at 20 T (200 kG), which is calculated from the superconducting section only (as opposed to the usual literature that quotes the overall J_c values). It is important to point out that these data actually are for relatively short wires, wound in a coil shape (1 m, or 3.3 ft, long) or straight wires (several centimeters in length). It must be noted here that some empiricism prevails in obtaining optimum J_c values. It is still difficult to really pinpoint the factors that produce a given wire with better properties than another wire. One interesting hypothesis has recently been put forth (Ref 36) that relying on a phase transition of PMS would leave most of it nonsuperconducting at low temperatures. This would occur to a much lesser extent in SMS, thus explaining why the doping by tin improves the properties of SMS. Further investigation is required to confirm this hypothesis, which will be difficult because little is actually known about the microstructure of PMS and its influence on the superconducting properties of the wires. The objective is actually to raise the J_c values by a factor of 10 at 20 T (200 kG). While a tenfold increase in the present J_c values may be an overly ambitious goal, recent advances in SMS and PMS technology make it feasible.

Potential Applications

PMS wire applications include devices and processes that require high magnetic fields, such as high-energy physics, thermonuclear fusion, and nuclear magnetic reso-

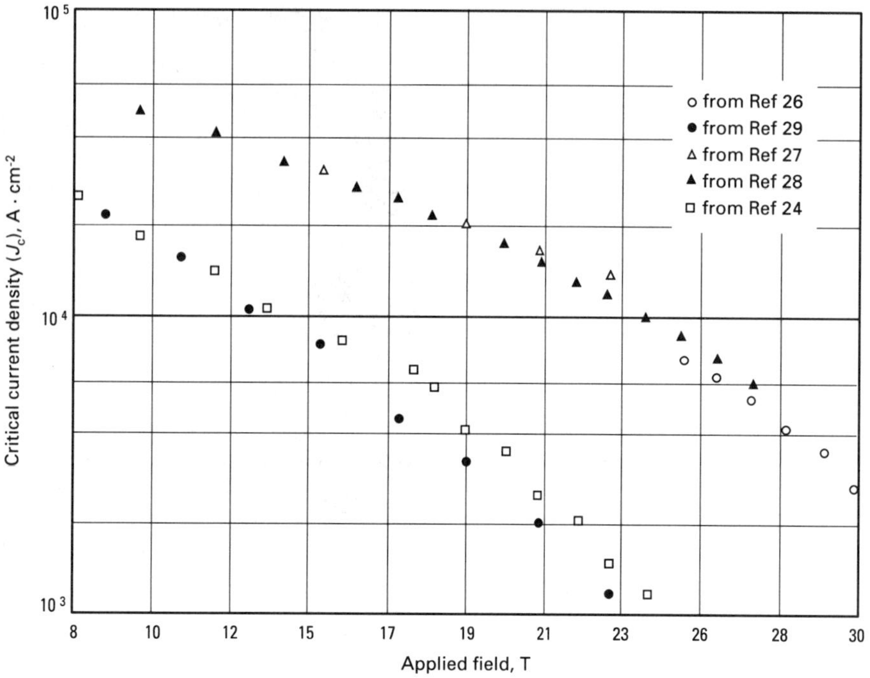

Fig. 4 Recent critical current density results in PMS wires

nance (NMR). However, research has not advanced to the stage where such devices and processes can utilize PMS technology efficiently. Current efforts are directed to improving the performance of the wires (which we know is possible because none of the published results showed that the wires have reached the expected upper critical field of about 50 T, or 500 kG, at 4.2 K) and eventually to construct small test coils. As of today, nobody has actually studied a coil longer than 2 m (6.7 ft) in length (Ref 37) and this will certainly be the focus of future SMS research and development, as well as seeking a better understanding of the factors that control the mechanism of the transport of the current.

REFERENCES

1. R. Chevrel, M. Sergent, and J. Prigent, *J. Solid State Chem.*, Vol 3, 1971, p 515
2. A. Perrin, M. Sergent, and O. Fischer, *Mater. Res. Bull.*, Vol 13, 1978, p 259
3. A. Perrin, R. Chevrel, M. Sergent, and O. Fischer, *J. Solid State Chem.*, Vol 33, 1980, p 43
4. T. Hughbanks and R. Hoffmann, *J. Am. Chem. Soc.*, Vol 105, 1983, p 1150
5. R. Chevrel and M. Sergent, in *Superconductivity in Ternary Compounds*, Vol I, O. Fischer and M.B. Maple, Ed., Springer-Verlag, 1982, p 25
6. K. Yvon, in *Superconductivity in Ternary Compounds*, Vol I, O. Fischer and M.B. Maple, Ed., Springer-Verlag, 1982, p 87
7. R. Schöllorn, *Agnew. Chem.*, Vol 92, 1980, p 1015
8. R. Schöllorn, *Agnew. Chem. Int. Ed. Engl.*, Vol 19, 1980, p 983
9. M. Wakihara, T. Uchida, K. Suzuki, and M. Taniguchi, *Electrochim. Acta*, Vol 34 (No. 6), 1989, p 867
10. O. Fischer, *J. Appl. Phys.*, Vol 16, 1978, p 1
11. M.E. Ekman, J.W. Anderegg, and G.L. Schrader, *J. Catal.*, Vol 117, 1989, p 246
12. M. Ishikawa, O. Fischer, and J. Muller, *J. Phys. (Orsay)*, Vol 39 (No. C6), 1978, p 1379
13. R.W. McCallum, D.C. Johnston, R.N. Shelton, and M.B. Maple, *Proceedings of the 2nd Rochester Conference on Superconductivity in d- and f-band Metals*, D.Y.H. Douglas, Ed., Plenum, 1976, p 265
14. M. Ishikawa, O. Fischer, and J. Muller, in *Superconductivity in Ternary Compounds*, Vol II, M.B. Maple and O. Fischer, Ed., Springer-Verlag, 1982, p 143
15. H.W. Meul, *Helv. Chim. Acta*, Vol 59, 1986, p 417
16. O. Fischer, M. Decroux, S. Roth, R. Chevrel, and M. Sergent, *J. Phys.*, Vol C8, 1975, p L474
17. R. Odermatt, O. Fischer, G. Bongi, and H. Jones, *J. Phys.*, Vol C7, 1974, p L13
18. J. Hauck, *Mater. Res. Bull.*, Vol 12 (No. 10), 1977, p 1015
19. S. Foner, E.J. McNiff, and D.G. Hinks, *Phys. Rev. B*, Vol 31 (No. 9), 1985, p 6108
20. D.W. Capone, R.P. Guertin, S. Foner, D.G. Hinks, and H.C. Li, *Phys. Rev. B*, Vol 29 (No. 11), 1984, p 6375
21. W. Goldacker, S. Miraglia, Y. Hariharan, T. Wolff, and R. Flükiger, *Adv. Cryog. Eng.*, Vol 34, 1988, p 655
22. L. Le Lay, thesis, University of Rennes, 1988
23. T. Luhman and D. Dew-Hughes, *J. Appl. Phys.*, Vol 49 (No. 2), 1978, p 936
24. B. Seeber, P. Herrmann, J. Zuccone, D. Cattani, J. Cors, M. Decroux, O. Fischer, E. Kny, and J.A.A.J. Perenboom, *MRS International Meeting on Advanced Materials*, Materials Research Society, 1988
25. W. Goldacker, W. Specking, F. Weiss, G. Rimikis, and R. Flükiger, *Cryogenics*, Vol 29, 1989, p 955
26. K. Hamasaki, Y. Shimizu, and K. Watanabe, to be published in *Adv. Cryog. Eng.*, 1990
27. Y. Kubo, K. Yoshizaki, F. Fujiwara, K. Noto, and K. Watanabe, *MRS International Meeting on Advanced Materials*, Materials Research Society, 1988
28. L. Le Lay, P. Rabiller, R. Chevrel, M. Sergent, T. Verhaege, J.-C. Vallier, and P. Genevey, to be published in *Adv. Cryog. Eng.*, 1990
29. H. Yamasaki and H. Kimura, *J. Appl. Phys.*, Vol 64 (No. 2), 1988, p 766
30. G. Krabbes and H. Oppermann, *Crys. Res. Technol.*, Vol 16, 1981, p 777
31. Y. Hamasaki, K. Noto, K. Watanabe, T. Yamashita, and T. Komata, *MRS International Meeting on Advanced Materials*, Materials Research Society, 1988
32. S. Miraglia, W. Goldacker, R. Flükiger, B. Seeber, and O. Fischer, *Mater. Res. Bull.*, Vol 1 (No. 22), 1987, p 795
33. N.E. Alekseevskii, V.I. Nizhankowskii, J. Beille, and G. du Trémolet, *J. Low Temp. Phys.*, Vol 72, (No. 3/4), 1988, p 241
34. B. Seeber, W. Glätzle, D. Cattani, R. Baillif, and O. Fischer, *IEEE Trans. on Magn.*, Vol 23, 1987, p 1740
35. J. Ekin, S. Yamashita, and K. Hamasaki, *IEEE Trans. on Magn.*, Vol 21, 1985, p 474
36. D.W. Capone, D.G. Hinks, and D.L. Brewe, submitted to *J. Appl. Phys.*
37. B. Seeber, M. Decroux, and O. Fischer, *Physica B*, Vol 155, 1989, p 129

Thin-Film Materials

Kenneth E. Kihlstrom, Department of Physics, Westmont College

THE DISCOVERY of 30 K superconductivity in the La-Ba-Cu-O system by Bednorz and Muller (Ref 1) and the subsequent dramatic increase in critical temperature (T_c) to 93 K in the Y-Ba-Cu-O (YBCO) system by Wu *et al.* (Ref 2) has led to an intense research effort to understand and expand on these results. High-quality thin films have been made by electron beam (EB) codeposition, sputtering, laser ablation, and most recently, chemical vapor deposition (CVD). Subsequently, high-temperature superconductivity was found in the Bi-Sr-Ca-Cu-O (BSCCO) system by Maeda *et al.* (Ref 3) with T_c values up to 110 K and in the Tl-Ba-Ca-Cu-O (TBCCO) system by Sheng and Hermann (Ref 4) with T_c values up to 125 K. Despite these advances in high-T_c compounds, low-T_c materials such as NbN and Nb continue to be important for many applications. This article will initially focus on the different thin-film deposition techniques used to make superconducting films and then will briefly discuss advantages of high-T_c versus low-T_c materials in a number of applications.

Superconducting Materials

The high-T_c systems have much in common. They each exhibit planes of Cu-O in the *a-b* plane, which seems critical to the superconductivity of the materials. As a result, they are all highly anisotropic with a substantial drop in superconducting properties (for example, critical current density, J_c), when the *c*-axis is parallel rather than perpendicular to the substrate. Thus, it is critical to be able to orient film growth. Films that are polycrystalline suffer further degradation of superconducting properties as it appears that the grains are weakly Josephson coupled. Correct oxygen stoichiometry is also important in all three systems. All three systems show high critical current densities (J_c values range from 10^6 to 10^7 A/cm^2) and enormous upper critical field, H_{c2} values ($dH_{c2}/dT > 4.5$ T/K at T_c), but adequate pinning seems difficult to obtain. Therefore, in both J_c and H_{c2} results there is not a clean break to true zero resistance.

The low-T_c materials are certainly much easier to work with. For both Nb ($T_c = 9$ K) and NbN ($T_c = 18$ K) there is no anisotropy, and thin films are obtainable by the different deposition techniques with little difficulty (that is, with low substrate temperature and no need for a post anneal). A15 materials such as Nb_3Sn ($T_c = 18$ K), V_3Si ($T_c = 17$ K), and Nb_3Ge ($T_c = 23$ K) can also be deposited in thin-film form but in general need elevated substrate temperatures that limit their appeal for layered structures. Critical current densities near 10^7 A/cm^2 can be obtained here as well, but at much lower temperatures (for Nb_3Ge, $J_c = 9 \times 10^6$ A/cm^2 at 14 K (Ref 5).

Substrates and Buffer Layers

Early attempts at film growth used the traditional substrates such as sapphire (single crystal Al_2O_3), but it was quickly determined (Ref 6) that at the annealing temperatures necessary to form the superconducting phase substantial diffusion occurred from the substrate into the material, with a substantial degradation of superconducting properties. This, combined with the desire for oriented growth (hopefully single crystal), led to other substrates such as $SrTiO_3$ (Ref 7, 8). But $SrTiO_3$ is an expensive substrate with a large dielectric constant and loss tangent, putting it at a disadvantage for high-frequency use. $LaAlO_3$ (Ref 9) and $LaGaO_3$ (Ref 10) both have significantly lower values for the dielectric constant and loss tangent. Both have produced films with sharp T_c values. A number of buffer layers (to prevent interdiffusion with the substrate) have been used to provide reasonably good films (Ref 11-13). Still, growing *in situ* films (without need of post anneal) is often preferable.

For the low-T_c materials, there is little problem with interdiffusion with the substrate, so convenient materials such as sapphire can be used. For the A15 materials where elevated temperatures are necessary, it can be important to clamp the sample to the backing plate securely to ensure a constant temperature during deposition (Ref 5).

Thin-Film Deposition Techniques

The main deposition techniques currently used are:

- Electron-beam coevaporation
- Sputtering from either a composite target or multiple sources
- Laser ablation (also called pulsed laser deposition)

Chemical vapor deposition has been the slowest of the methods to come on line, but recent results (Ref 14) suggest that in time it may become competitive.

Electron-Beam Coevaporation. The earliest successful high-T_c thin-film results came with EB-deposited material (Ref 6, 15). This procedure typically requires separate sources for each component of the superconductor. This multisource evaporation creates problems with accurate rate control of each of the sources (especially if a large oxygen pressure is present) as well as geometrically induced composition variation on the substrates (remember, because these are line compounds, stoichiometry is very important). Oxidation of the sources (affecting evaporation rates) is less of a problem here than in sputtering. This allows greater possibilities of *in situ* film growth. For single-layer superconductors (allowing for a post anneal) the use of BaF_2 (Ref 16) (which is much more stable than barium metal) allows patterning of deposited films by photolithography after which an anneal (with O_2 and H_2O) produces the superconducting properties. Even apart from patterning considerations, use of BaF_2 improves film quality but is not useful when making *in situ* films because of the necessity of removing the fluorine.

Sputtering. Film deposition by sputtering can be done with either a single composite target or with multiple sources.

Composite target sputtering avoids geometric compositional variation (inherent in a multiple-source configuration) as well as the need for good rate control. The main drawback involves the difficulty in obtaining the desired composition in the film (which does not necessarily match the target composition). This can be addressed either by adjusting target composition (Ref 17, 18) or by varying sputtering conditions such as location of the substrates (Ref 19). The latter results because of bombardment of the film surface by negative oxygen ions. By moving the substrates to the side, stoichiometric films were obtained. Recent results from the Karlsruhe (Ref 20) have superconducting films as thin as 3.6 nm (36 Å).

Multiple target sputtering (Ref 21-23) has greater flexibility to vary composition but has drawbacks due to oxidation of the barium target and rate control. In the end, composite sputtering seems to have the advantage.

Laser ablation (or pulsed laser deposition) uses a composite target of the desired composition that is exposed to a focused laser beam from a pulsed excimer laser (typically). The area vaporizes and is projected in a narrow forward plume to the substrate. Initial results (Ref 24) with laser ablation required a post anneal, but high-quality *in situ* films are now available from laser ablation (Ref 13, 25, 26) with $J_c = 4 \times 10^6$ to 5×10^6 A/cm^2. Substrate temperatures as low as 400 °C (750 °F) without post annealing resulted (Ref 27) in the $R = 0$ point, where R is the bulk resistance, at 85 K with $J_c = 10^5$ A/cm^2. The drawback of small sample size (due to the superconductor plume being very directional) can be overcome by rastering the substrate.

In Situ Film Growth

As mentioned earlier, films requiring a post anneal suffer substrate/film interdiffusion, causing degradation at high temperatures, limiting the choice of substrates. In addition, post-annealed samples tend to have very poor surfaces, limiting hope of layered devices (such as tunnel junctions) where a clean, abrupt interface is necessary. A number of *in situ* techniques have been used successfully. High-pressure (>1 mtorr, or 0.13 mPa) methods (Ref 28-30) cause problems with rate control, oxidation of sources, and shortened lifetimes of system components but have produced films with $T_c > 90$ K and J_c of 4×10^6 A/cm^2. To succeed with low pressures requires activating the oxygen in some way. Techniques include producing atomic oxygen by radio frequency (rf) excitation (Ref 31) or microwave discharge (Ref 32), or the use of ozone (Ref 33). Each of these techniques has produced excellent films.

Superconducting Materials Properties

$YBa_2Cu_3O_7$ (YBCO) Properties. Despite the discovery of yet higher-T_c compounds, YBCO continues to be the most studied material for several reasons. Because it came first, most groups have a working knowledge of it including both the pitfalls and established protocols for deposition conditions, oxidation, and annealing. Switching to the other materials may mean largely starting over and losing the progress already made. High-quality YBCO films have been made by several techniques. The $R = 0$ point is typically about 95 K. Critical current densities (J_c) for YBCO films at 77 K are typically in the mid 10^6 A/cm^2 range both for *in situ* [Venkatesan *et al.* (Ref 25) $J_c = 4 \times 10^6$ to 5×10^6 A/cm^2] and for post-annealed [Itozaki *et al.* (Ref 27) $J_c = 3.5 \times 10^6$ A/cm^2] films. There is a report out of Japan (Ref 34) that the Sumitomo group has obtained a J_c of 1.5×10^7 A/cm^2. It should be noted as well that in large fields (15 T, or 150 kG) the critical current is 10^5 A/cm^2 at 50 K (Ref 35). The upper critical field slope (dH_{c2}/dT at T_c) for parallel field was found by Chaudhari *et al.* (Ref 36) to be 4.5 T/K. Finally, much progress has been made on *in situ* growth of YBCO films (as already discussed).

There are some drawbacks of YBCO however:

- YBCO is a single line compound on the equilibrium phase diagram, which makes correct stoichiometry essential. Any deviation from stoichiometry results in a nonconducting second phase inclusion that can lead to losses. This is in contrast to the bismuth and thallium compounds that have a number of superconducting competing phases
- YBCO is especially sensitive to oxygen content. Homogeneous samples are thus difficult to obtain
- Barium is very reactive with water vapor and has a tendency to migrate to the surface, giving rise to a nonsuperconducting surface while leaving the bulk stoichiometry off the mark

Bi-Sr-Ca-Cu-O (BSCCO Properties). This system has not been as intensively studied as YBCO, in part because of the discovery of higher transition temperatures in the thallium compounds. There has been a strong effort in Japan where the system was first studied, and a number of interesting properties have been discovered. First, there are at least two superconducting phases (Ref 37): the $Bi_2Sr_2CaCu_2O_x$ (2212) phase with $T_c = 85$ K and the $Bi_2Sr_2Ca_2Cu_3O_x$ (2223) phase with $T_c = 110$ K. The lower-T_c phase material seems less sensitive to having exact stoichiometry, which is an advantage. The higher-T_c phase has been more difficult to synthesize in pure form, although the partial substitution of lead for bismuth is very helpful (Ref 38, 39). The absence of barium makes the BSCCO films less sensitive to atmospheric degradation than YBCO. The upper critical field slope (dH_{c2}/dT at T_c) for parallel field was found by Palstra *et al.* (Ref 40) to be 45 T/K. J_c at 77 K was found to be 1.9×10^6 A/cm^2 and at 40 K, $J_c = 2.1 \times 10^7$ A/cm^2 by the Sumitomo group (Ref 41).

Tl-Ba-Ca-Cu-O (TBCCO) Properties. The highest confirmed T_c values to date come in the thallium system (Ref 42) where $T_c = 125$ K including films with $R = 0$ at 120 K (Ref 43). Here, as with the bismuth superconductors, there are multiple superconducting phases, the highest of which ($T_c = 125$ K) is the $Tl_2Ba_2Ca_2Cu_3O_x$ (2223) phase. It is a distinct advantage that if a second phase is present it is also superconducting even if at a lower T_c. This is especially true for polycrystalline films. There are, however, disadvantages with the thallium compounds. As with BSCCO systems, there is an extra component versus YBCO systems, often requiring a four-source deposition system. The presence of several phases makes purifying a single phase difficult. Also, at the necessary annealing temperatures, thallium is very volatile, making it difficult to get the correct stoichiometry (although there are tricks such as annealing the film in a sealed quartz tube with bulk TBCCO material to set up an equilibrium vapor pressure). Of course the toxicity of thallium is also a major concern, and care must be taken. The thallium results certainly are not optimized, but some respectable values for J_c have been reported. Hong *et al.* (Ref 44) found $J_c = 10^4$ A/cm^2 at 110 K and $J_c = 10^5$ A/cm^2 at 100 K, suggesting a slope that would make the thallium compounds comparable to YBCO systems. The upper critical field slope (dH_{c2}/dT at T_c) for parallel field was found by Kang *et al.* (Ref 21) to be 70 T/K (for the 2212 phase).

Table 1 summarizes the results for high-T_c thin films.

Table 1 Properties of high-T_c thin-film systems

System	Critical temperature (T_c), K	Critical current density (J_c), A/cm^2	Upper critical magnetic field slope $\left(\frac{dH_{c2}}{dT}\bigg\vert_{T_c}\right)$, T/K
YBCO	95	1.5×10^7(a)	4.5
		10^5(b)	4.5
BSCCO	110	2×10^6(a)	45
TBCCO	125	10^5(c)	70

(a) Measurement temperature, 77 K. (b) Measurement temperature, 50 K; applied field (H_a), 15 T (150 kG). (c) Measurement temperature, 100 K. Source: Ref 35

Low-T_c Materials. Despite the advances in high-T_c materials, there is still a great deal of interest in the low-T_c superconductors such as Nb ($T_c = 9$ K) and NbN ($T_c = 18$ K). The A15 superconductors also offer possibilities, but the higher T_c (up to 23 K for Nb_3Ge) may not offset the difficulties in fabrication. For Nb and NbN, it is relatively easy to make high-quality thin films with low deposition temperatures (which allows much greater latitude in processing the films). In addition, there is no anisotropy to worry about and the surfaces tend to be of good quality. Critical current densities for the low-T_c materials can be substantial but only at liquid-helium temperatures ($T = 4.2$ K). For NbN, J_c values up to 1.5×10^6

A/cm^2 have been reported (Ref 45) when some tantalum was cosputtered with the NbN. The upper critical field (H_{c2}) for the same material at T = 4.2 K is as high as 24 T (240 kG). The obvious drawback for these materials is the need for liquid-helium cooling. Yet for some applications, especially detectors, the low temperature is necessary to reduce thermal noise. In this case, a high T_c is not a major advantage.

Applications of Thin-Film Superconductors

The promise of superconductive electronics is due to the inherent speed, low loss, low noise, and low power dissipation as compared with semiconductor technology. The advantages are already present in the low-T_c materials such as Nb (T_c = 9 K) and NbN (T_c = 18 K), which will continue to be important. There exists a well-established superconductive integrated circuit technology based on niobium with work being done on NbN. The Fujitsu group (Ref 46) in Japan has produced a four-bit chip that ran at 1.1 GHz dissipating only 6.1 mW of power, which surpasses gallium arsenide semiconductor circuits by factors of 15 and 150, respectively.

Introducing high-T_c materials to applications brings both advantages and disadvantages. The increased operating temperature also means increased thermal noise. The greater superconducting energy gap allows an increase in the potential frequency (and higher speed) but requires greater operating voltages (and greater power dissipation). Fabrication and reliability of the high-T_c superconductors is a major problem, but the option of liquid-nitrogen temperature operation would allow superconductor-semiconductor hybrid circuitry (where liquid-helium operation would freeze out semiconductor technology).

Thus both high- and low-T_c superconductors should have major roles in a number of areas. The high-speed, low-power dissipation has been discussed in digital electronics. Signal processing and analog electronic applications such as analog-to-digital (A/D) converters offer ultralinear high speed (Ref 47), high resolution (Ref 48) with low power dissipation. Josephson parametric amplifiers (Ref 49) and superconductor/insulating/superconductor (SIS) tunnel junction mixers (Ref 50) operate with very low noise levels. It is possible that even low-T_c superconductors could have high-frequency operation near 1 THz. The high-T_c superconductors hold the potential of 10 THz operation. Sensor applications such as infrared detectors and video detectors for millimeter radiometry also benefit from high-frequency broadband capabilities. Superconducting quantum interference device (SQUID) magnetometers represent another important application of superconductivity. These are the most sensitive detectors of small magnetic fields. High-T_c materials using liquid nitrogen rather than liquid helium could allow greater field use. Transmission lines operating at microwave frequencies make use of the orders of magnitude that lower-loss superconductors have versus gold or copper. Very high quality factor (Q) circuits and nearly ideal filters are possible.

Future Outlook

In the end, laser ablation and composite target sputtering seem to be the most promising deposition methods with EB codeposition also being competitive, especially for *in situ* films (a necessary process when multilayer structures are contemplated). In choosing the superconducting material, YBCO probably has the edge because of the greater wealth of knowledge available especially for *in situ* deposition. It should be noted, however, that the low-T_c materials such as Nb and NbN continue to be important for applications. Also, what is learned in developing Nb and NbN technology will carry over to the high-T_c compounds when the material's problems are fully mastered. Superconductivity does not promise inexpensive operation but does hold the potential of unrivaled performance.

ACKNOWLEDGMENT

Valuable input to this article was given by Stuart Wolf of the Naval Research Laboratory, Randy Simon of TRW Inc., and John Talvacchio of Westinghouse Corporation. The author is supported by National Science Foundation Grant DMR-8702994.

REFERENCES

1. J.G. Bednorz and K.A. Muller, *Z. Phys. B*, Vol 64, 1986, p 189
2. M.K. Wu *et al.*, *Phys Rev. Lett.*, Vol 58, 1987, p 908
3. H. Maeda, Y. Tanaka, M. Fukutomi, and T. Asano, *Jpn. J. Appl. Phys. Lett.*, Vol 27, 1988, p L209
4. Z.Z. Sheng and A. Herman, *Nature*, Vol 332, 1988, p 138
5. K.E. Kihlstrom *et al.*, *J. Appl. Phys.*, Vol 53, 1982, p 8907
6. M. Naito *et al.*, *J. Mater. Res.*, Vol 2, 1987, p 713
7. P. Chaudhari *et al.*, *Phys. Rev. Lett.*, Vol 58, 1987, p 2684
8. Y. Enomoto, *Jpn. J. Appl. Phys.*, Vol 26, 1987, p L1248
9. R.W. Simon *et al.*, *Appl. Phys. Lett.*, Vol 53, 1988, p 2677
10. R.L. Sandstrom *et al.*, *Appl. Phys. Lett.*, Vol 53, 1988, p 1874
11. R.W. Simon *et al.*, *IEEE Trans. Magn.*, Vol 25, 1989, p 2433
12. Myoren and Hiroaki *et al.*, *Jpn. J. Appl. Phys.*, Vol 27, 1988, p L1068
13. X.D. Wu *et al.*, *Appl. Phys. Lett.*, Vol 54, 1989, p 754
14. Y. Muto *et al.*, *Physica C*, Vol 162-164, 1989, p 105
15. R.B. Laibowitz *et al.*, *Phys. Rev. B*, Vol 35, 1987, p 8821
16. P.M. Mankiewich *et al.*, *Appl. Phys. Lett.*, Vol 51, 1987, p 1753
17. H. Itozaki *et al.*, in *Proceedings of the 5th International Workshop on Future Electron Devices*, Research and Development Association for Future Electron Devices, 1988, p 149
18. Adachi and Hideaki *et al.*, *Appl. Phys. Lett.*, Vol 51, 1987, p 2263
19. H.C. Li *et al.*, *Appl. Phys. Lett.*, Vol 52, 1988, p 1098
20. J. Geerk, private communication
21. J.H. Kang *et al.*, *Appl. Phys. Lett.*, Vol 53, 1988, p 2560
22. R.M. Silver, J. Talvacchio, and A.L. de Lozanne, *Appl. Phys. Lett.*, Vol 51, 1987, p 2149
23. K. Char *et al.*, *Appl. Phys. Lett.*, Vol 51, 1987, p 1370
24. C.C. Chang *et al.*, *Appl. Phys. Lett.*, Vol 53, 1988, p 517
25. T. Venkatesan *et al.*, *Appl. Phys. Lett.*, Vol 54, 1989, p 581
26. N. Klein *et al.*, *Appl. Phys. Lett.*, Vol 54, 1989, p 757
27. S. Witanachchi *et al.*, *Appl. Phys. Lett.*, Vol 53, 1988, p 234
28. D.K. Lathrop, S.E. Russek, and R.A. Buhrman, *Appl. Phys. Lett.*, Vol 51, 1987, p 1554
29. R.M. Silver *et al.*, *Appl. Phys. Lett.*, Vol 52, 1988, p 2174
30. Y. Bando *et al.*, in *Proceedings of the 5th International Workshop on Future Electron Devices*, Research and Development Association for Future Electron Devices, 1988, p 11
31. J. Kwo *et al.*, *Appl. Phys. Lett.*, Vol 53, 1988, p 2683
32. N. Missert *et al.*, *IEEE Trans. Magn.*, Vol 25, 1989, p 2418
33. D.D. Berkeley *et al.*, *Appl. Phys. Lett.*, Vol 53, 1988, p 1973
34. Kitozawa, private communication
35. Hettinger *et al.*, *Phys. Rev. Lett.*, Vol 62, 1989, p 2044
36. P. Chaudhari *et al.*, *Phys. Rev. B*, Vol 36, 1987, p 8903
37. R.M. Haven *et al.*, *Phys. Rev. Lett.*, Vol 60, 1988, p 1174
38. K. Doggone *et al.*, *Appl. Phys. Lett.*, Vol 53, 1988, p 1329
39. Takano *et al.*, *Jpn. J. Appl. Phys.*, Vol 27, 1988, p L1041
40. T.T.M. Palstra *et al.*, *Phys. Rev. B*, Vol 38, 1988, p 5102
41. S. Yazu, *Proceedings of the 1st International Symposium on Superconductivity*, in *Advances in Superconductivity*, K. Kitazawa and T. Ishiguro, Ed., International Superconductivity Technology Center, 1989

42. S.S.P. Parkin *et al.*, *Phys. Rev. Lett.*, Vol 60, 1988, p 2539
43. W.Y. Lee *et al.*, *Appl. Phys. Lett.*, Vol 53, 1988, p 329
44. M. Hong *et al.*, *Appl. Phys. Lett.*, Vol 53, 1988, p 2102
45. J.Y. Juang *et al.*, *J. Appl. Phys.*, Vol 66, 1989, p 3136
46. S. Kotani *et al.*, in *Digest of Technical Papers for the 31st International Solid State Circuit Conference*, L. Winner, Ed., Institute of Electrical and Electronics Engineers and the University of Pennsylvania, 1988, p 150-151
47. C.A. Hamilton and F.L. Lloyd, *IEEE-Electron Dev. Lett.*, Vol EDL-1, 1986, p 92
48. J.P. Hurrell, D.C. Pridemore-Brown, and A.H. Silver, *IEEE Trans. Electron. Dev.*, Vol ED-27, 1980, p 1887
49. A.D. Smith *et al.*, *IEEE Trans. Magn.*, Vol 21, 1985, p 1022
50. S.-K. Pan *et al.*, *IEEE Trans. Microwave Theory Tech.*, Vol 37, 1989, p 580

High-Temperature Superconductors for Wires and Tapes*

R.D. Blaugher, Intermagnetics General Corporation

THE INTEREST in applying superconductivity to power devices, transportation, electronics, and so on is directly related to predicted performance advantages and improved operating efficiency over conventional room-temperature (RT) approaches. The incorporation of superconducting wire or tape into large magnets and power generators, for example, provides the ability to transport large dc currents with no measurable resistive losses. High magnetic fields can thus be produced at a significantly reduced cost for the energy required for operation. Similar examples can be given for electronic applications with superconductivity offering lower losses, higher speed, and reduced signal dispersion at very high frequency (Ref 1, 2).

To demonstrate these predicted benefits, superconductivity, in fact, has been applied to many power-related and electronics applications with great success. Superconducting prototypes have been constructed, for power generators and motors, ac and dc transmission, energy storage, high-speed signal processing and computing, and high-sensitivity magnetic detectors, to name but a few (Ref 1, 2). It is also possible to operate power devices in an ac mode with acceptable losses, providing the superconductor is properly designed with respect to filament size, twist, and stabilizer. These past demonstrations, almost without exception, were tested and operated in liquid helium at 4.2 K. This requirement for liquid-helium cooling has, without a doubt, limited serious consideration for insertion of superconducting devices into existing power-generation equipment and electronic systems up to the present.

The discovery of the high-critical-temperature (high-T_c) oxide superconductors (Ref 3) in 1986 has accelerated the interest for superconducting applications because it offers the prospect for higher-temperature operation at liquid nitrogen (77 K) or above and thus reduces the refrigeration and/or liquid helium requirement.

The primary technical challenge that must be satisfied to permit usage of the high-T_c oxides in magnets or power applications is the successful demonstration of a high-current-carrying wire or tape with acceptable mechanical capability. The current-carrying performance of the oxide-base wire or tape must be functionally equivalent to the present liquid-helium-cooled conventional superconductors such as Nb-Ti or Nb_3Sn, which typically show a current density at 4.2 K of approximately 10^5 A/cm^2 at 5 T (50 kG). In addition, the high-T_c oxide conductor must have the mechanical ability to withstand the stresses produced during fabrication and winding, thermal contraction during low-temperature operation, and the Lorentz forces (F_L) due to the high magnetic fields (Ref 4).

Over the past three years, a large research effort has been directed at the understanding and processing of high-T_c oxide conductor materials. Much progress has been made, but to date there has been no actual demonstration of a technologically useful high-T_c wire or tape. The processing approaches have pursued many directions, but for the most part, follow either a powder precursor approach or a vapor deposition method. The powder techniques are mainly based on the production of an oxide powder precursor, which is then subjected to various processing and heat treatment schedules.

Processing of Primary Oxide Compounds

Y-Ba-Cu-O (YBCO) Systems. The worldwide efforts on producing wire and tape have concentrated for the most part on the $YBa_2Cu_3O_7$(123) or YBCO orthorhombic compounds or variants with other rare earths (REs) substituted for the yttrium. The 123 compound presents major processing difficulties:

- High reactivity with most metallic and ceramic interfaces
- Sensitivity to cation and anion stoichiometry, which degrades the superconducting properties
- Sensitivity to copper substitution, which degrades the superconducting properties

Once formed, the compound is highly brittle with strong crystalline anisotropy, which shows marked thermal expansion coefficient differences for its major axes with a resultant high tendency for microcracking. These processing problems, however, are balanced to some degree by the ability to produce a high percentage of single-phase material if the processing is properly followed. A fairly high 92 K superconducting transition and production of satisfactory critical current density in idealized thin-film samples also add to the interest for using this compound. Furthermore, an enormous amount of research has been conducted on the 123 compound, which provides a wealth of information with an almost unparalleled reference base for the materials scientist.

Bi-Sr-Ca-Cu-O (BSCCO) Systems. The $Bi_2Sr_2Ca_{n-1}Cu_nO_x$ system provides the other major oxide compounds under investigation for wire and tape development. The bismuth compound shows similar processing problems as the Y-123, but in contrast to Y-123, is more difficult to synthesize as a single phase. A high-T_c (110 K) phase is found at the composition 2223 with a lower-transition 85 K phase forming at 2212. Partial substitution of lead for the bismuth appears to promote the development of the 2223, 110 K phase. The bismuth compound's major advantage over the Y-123 is its relative insensitivity to oxygen loss during processing, and it does not require special low-temperature oxygenation to achieve optimum superconducting properties.

Tl-Ba-Ca-Cu-O (TBCCO) Systems. The thallium-base 125 K oxide superconductor

*This paper was presented as an invited talk by the author at The Metallurgical Society of AIME annual meeting in Anaheim, CA, on 19 February 1990.

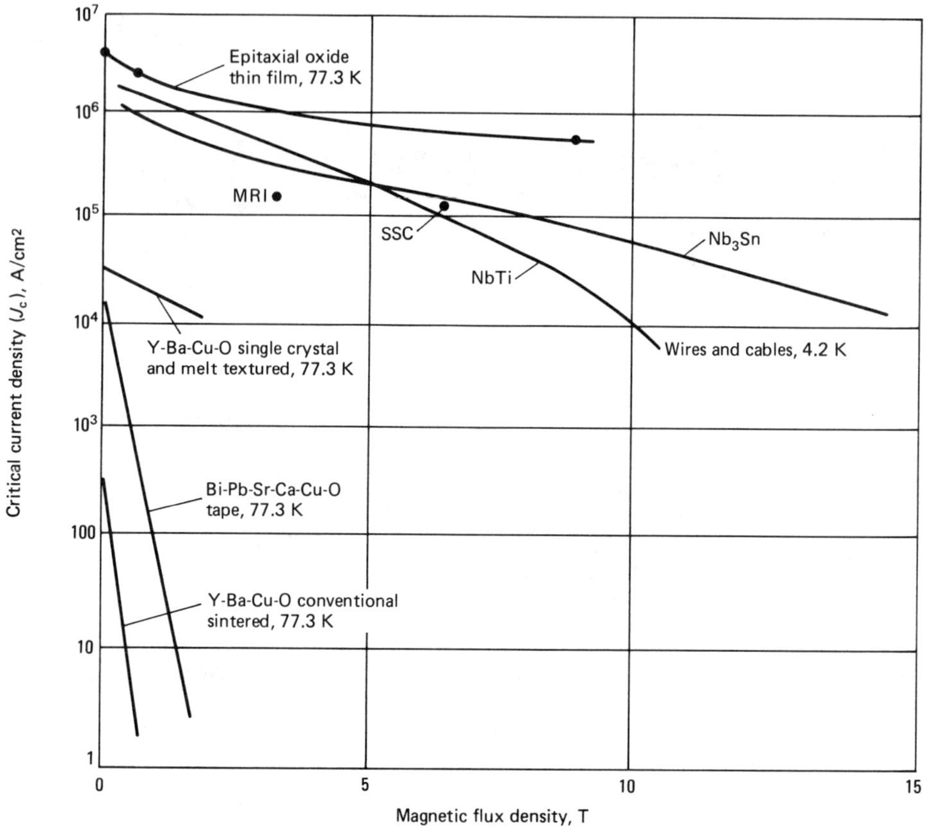

Fig. 1 Plot of critical current density versus magnetic flux density to compare properties of powder-in-tube process oxide-base superconductors with that of conventional superconductors. MRI, magnetic resonance imaging; SSC, superconducting supercollider

with barium, calcium, and copper provides the third major compound of interest for high-T_c wire and tape development. Processing for the thallium system has not been as active, primarily due to the high volatility of the thallium oxides and their high toxicity. Processing for the thallium-base superconductors must be conducted in a confined facility and much care followed to prevent toxic exposure.

Powder Precursor Preparation

Shake-and-Bake Method. The simplest method employed for producing oxide powder precursors is to use the so-called shake-and-bake method. The constituent powders, usually the metallic oxides (BaO, Y_2O_3, and CuO for example), are physically mixed and ground followed by calcining at 800 to 950 °C (1470 to 1740 °F) in air or flowing oxygen. This process is repeated a number of times with a final oxygenation (for the RE-Ba-Cu-O compounds) performed at 400 to 500 °C (750 to 930 °F) in flowing oxygen. The resulting powder obtained by this method is fairly close to stoichiometry for both the anion and cation composition and meets most processing requirements.

Additional Methods. More exotic techniques (sol-gel, coprecipitation, aerosol, and so on) have been developed that provide more precise control on phase purity and stoichiometry, minimize preparative contamination, and offer some control on the shape and size of the powder particulates.

Aerosol Pyrolysis Technique. One method that has achieved a high degree of success for the Y-Ba-Cu-O system is the aerosol pyrolysis technique. An aqueous metal nitrate solution is prepared from the respective oxides, which is then dispersed into a carrier flow stream (typically air or oxygen) to form an aerosol. The aerosol is passed through a high-temperature furnace that flash evaporates the solvent followed by nitrate decomposition and formation of the metal oxide. The oxide powder is either filtered or gravity collected and subsequently heat treated to complete the process. The superconducting properties of the powders produced by this method have been quite good, showing a very low percentage of impurity phases and fairly sharp superconducting transitions (Ref 5, 6). More importantly the powders are submicron in size, which is attractive for producing wire using the powder-in-tube method.

Powder-in-Tube Processing

The powder-in-tube approach is the most common method used to date for producing an oxide-base wire or tape. The oxide precursor powder is packed into a hollow metallic tube. Usually silver or gold are preferred due to their relative inertness to the oxide and ability to permit oxygen diffusion. The composite tube is then swaged or drawn into a wire, and if desired, rolled into a tape.

The most common heat treatment schedule for the powder-in-tube wire or tape provides a high-temperature reaction heat treatment near the melting point, which either sinters or partially melts the superconductor followed by a slow cool down, which may include a final low-temperature anneal required to completely oxygenate and equilibrate the RE-$Ba_2Cu_3O_7$ superconductor. The heating and cooling schedules must be carefully configured to minimize separation of the superconductor core due to the thermal expansion mismatch (Ref 7). It is possible with a silver sheath to diffuse sufficient oxygen to restore stoichiometry that may be lost during the high-temperature reaction. The best critical current densities (J_c) observed to date (at 77 K) for the various powder-in-tube processing approaches are shown in Fig. 1. The best critical current density data for an oxide wire or tape is $J_c = 1.7 \times 10^4$ A/cm^2 at 77 K, with the applied field (H_a) of $H_a = 0$ obtained by Sumitomo on silver-sheathed Bi-Pb-Sr-Ca-Cu-O material, which was drawn into wire and then cold rolled to a tape configuration (Ref 8). The current density for this tape, however, rapidly degraded in a magnetic field showing only 1.7×10^3 A/cm^2 at an applied field of 0.1 T (1.0 kG). The highest current density for a Y-Ba-Cu-O compound similarly processed in a silver sheath is 4×10^3 A/cm^2 at 77 K, for zero field, which was also reported by Sumitomo (Ref 9). Examples of $YBa_2Cu_3O_7$ rolled tape and multifilament wire are shown in Fig. 2.

The powder-in-tube current density and behavior in an applied magnetic field is significantly degraded compared to the conventional superconductors Nb-Ti and Nb_3Sn shown in Fig. 1 at 4.2 K. The critical current density requirements for two of the most prominent large-scale applications are also noted: magnetic resonance imaging (MRI) magnets and the superconducting supercollider (SSC). Both of these devices require current density performance near 10^5 A/cm^2 at a 5 T (50 kG) magnetic field.

It is significant to note two recent achievements for powder-in-tube superconducting oxide wire and tape measured at 4.2 K that indicate promise for oxide conductors at high magnetic fields. A silver sheathed Y-Ba-Cu-O wire showed 10^3 A/cm^2 at 4.2 K in a 10 T (100 kG) field (Ref 10). An even higher critical density of 10^4 A/cm^2 was observed for Bi-Sr-Ca-Cu-O wire measured at 4.2 K in magnetic fields up to 26 T (260 kG) (Ref 11). This latter result (see Fig. 3) shows higher critical current density at 4.2 K and 26 T (260 kG) than conventional Nb_3Sn wire or tape. This

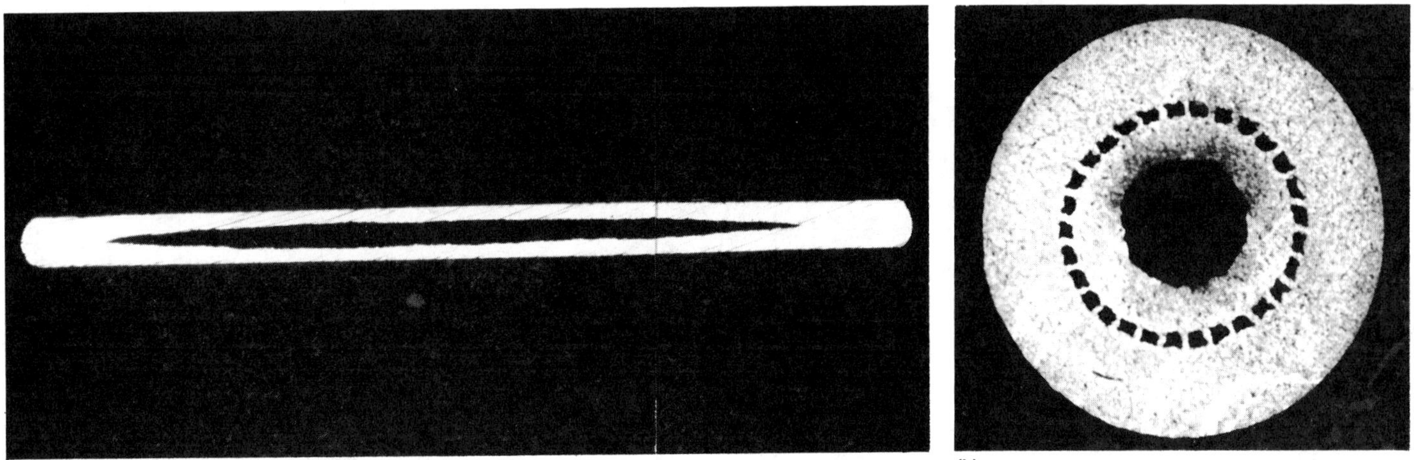

(a) (b)

Fig. 2 Cross sections of two YBCO powder-in-tube processed superconductors. (a) Silver-sheathed tape conductor with $YBa_2Cu_3O_7$ core. (b) 0.38 mm (0.015 in.) diam multifilament $YBa_2Cu_3O_7$ wire consisting of 29 filaments of 15 μm (600 μin.) diameter. Courtesy of Intermagnetics General Corporation

result thus presents a new opportunity for the oxide superconductors in providing a conductor for use in constructing very high field magnets, that is, $H > 25$ T (250 kG) operating at a 4.2 K temperature. Consideration of a superconducting design for a high field magnet is currently limited to 20 to 22 T (200 to 220 kG) using conventional, that is, nonoxide, superconductors.

Vapor Deposition Processing

The vapor deposition methods for producing a tape or wire have generally employed conventional physical deposition approaches such as radio frequency (rf) magnetron sputtering, laser ablation, and evaporation and chemical techniques such as metallo-organic chemical vapor deposition (MOCVD).

The vapor deposition approach offers some advantages in that the deposition can be performed at temperatures well below the oxide superconductor decomposition temperature. This minimizes substrate contamination, reduces postreaction heat treatment and oxygen equilibration, and minimizes the thermal expansion problem that occurs on heating. Thin films of approximately 0.5 μm (20 μin.) to a few μm are typical for the vapor deposition methods.

The critical current densities observed for thin films of the oxide superconductor have shown the highest values reported to date for high-T_c materials. These thin films, however, are highly idealized in that the film is of epitaxial grade grown on a specially prepared highly expensive substrate such as $SrTiO_3$. The critical current densities for these epitaxial films are shown in Fig. 1. Values in excess of 10^6 A/cm^2 have been observed at 77 K for RE-Ba-Cu-O oxide compounds with outstanding magnetic field properties comparable to the conventional superconductors (Ref 8). The thin-film results thus present some optimism that technologically useful current densities may be eventually produced in bulk wires and tapes. The mechanism leading to the high critical current densities in thin films is not presently understood but is related to the ability of the thin films to achieve higher pinning and reduced flux flow compared to bulk materials.

The critical current density for vapor deposited (VD) tapes or wires on a metallic substrate is considerably degraded compared to the epitaxial films. Prototype VD tapes produced to date typically show 10^3 to 10^4 A/cm^2 at 77 K and zero field, which is a direct result of their polycrystalline nature (Ref 12). The metallic substrate does not provide growth conditions comparable to the microelectronics grade substrates, which results in polycrystalline development and lower critical current density. Promising results for the VD approach have been obtained for MOCVD films on MgO (Ref 13) and for laser-ablated films also on MgO (Ref 14) with critical current densities near 10^5 A/cm^2 observed at 77 K. The microstructural development for these latter examples is apparently improved over the polycrystalline condition with a higher degree of texturing.

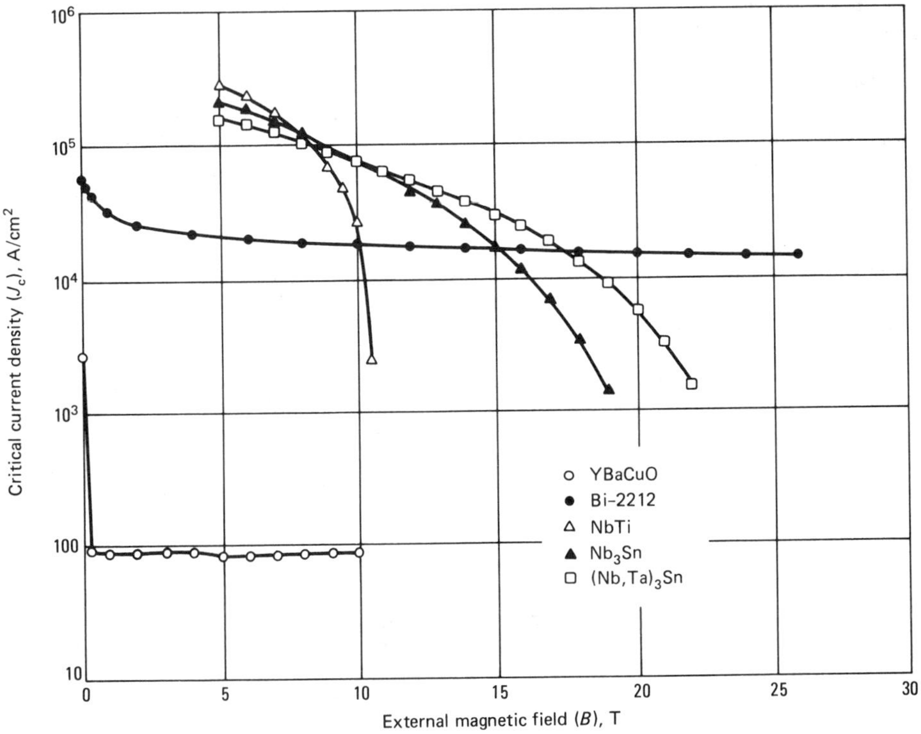

Fig. 3 Plot of critical current density versus external magnetic field at 4.2 K to compare two silver-sheathed powder-in-tube superconducting oxide wires (Bi-2212/Ag and $YBa_2Cu_3O_7$) with three conventional multifilamentary wires. J_c data is for superconductor cross section, also referred to as noncopper J_c. Source: Ref 11

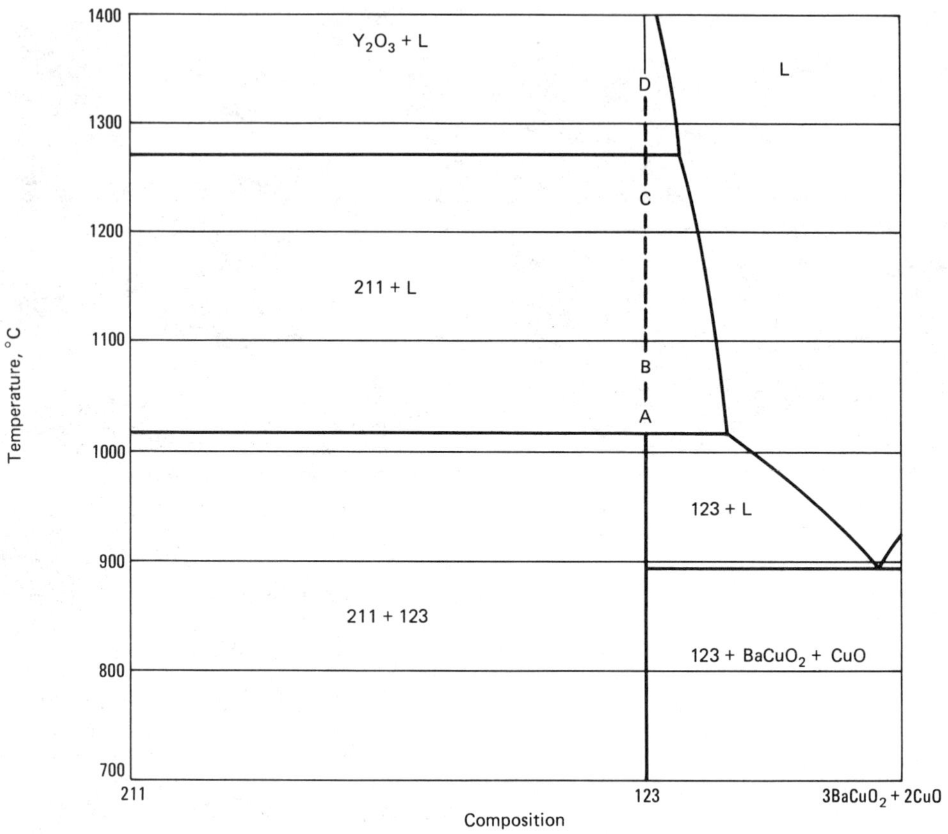

Fig. 4 Pseudobinary Y-Ba-Cu-O phase diagram along the tie line 211-123-(035)

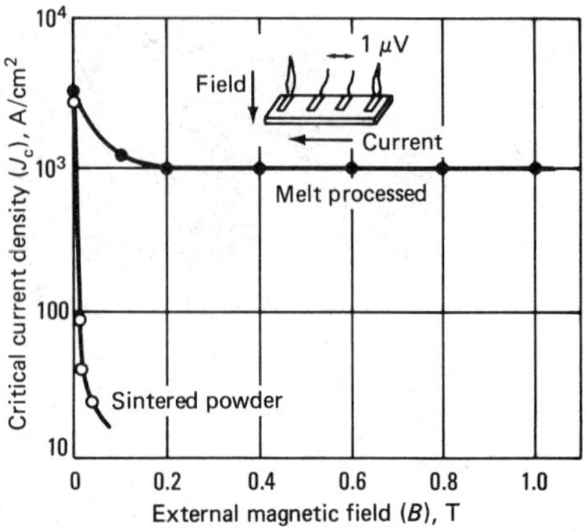

Fig. 5 Plot of critical current density versus external magnetic field at measurement temperature of 77 K to compare sintered powder YBCO tape-shaped wire with melt-processed YBCO tape-shaped wire. Source: Ref 21

Microstructural, Anisotropy, and Weak Link Influences

The critical current density for the best wire and tape shows over two orders of magnitude lower critical current and severe magnetic field degradation in contrast to the critical current density observed for high-quality thin films. The inability to realize bulk critical current densities comparable to thin films is presently attributed to microstructural and mechanical causes. The brittle nature of the ceramic material presents great difficulty in preserving the physical continuity necessary for optimum current transport. Physical separation or microcracking can easily occur, which severely degrades the critical current density. Even with perfect material, that is, with no microcracks evident, microstructural weak link problems are evident, which severely limits critical current density.

It is fairly well accepted that this degradation is mainly attributed to weak link Josephson-type coupling between grains. The oxides, which are either orthorhombic or tetragonal, exhibit crystalline anisotropy that results in strong anisotropy in the current density. Current flow in the *a-b* planes is orders of magnitude higher than current flow in the *c*-axis direction.

The poor coupling between grains or the weak link mechanism is dominated by this current density anisotropy and grain-boundary-related problems such as precipitate or impurities along the grain boundaries, compositional inhomogeneities, microcracking, and misalignment of the *a-b* planes across the grain boundary.

A superconducting weak link is normally associated with a weakly connected microscopic bridge or narrow constriction between two bulk superconductors. Weak link superconducting behavior was first predicted by Josephson and forms the basis for the Josephson junction. A weak link behaves much like an ordinary bulk superconductor with respect to a critical current and sensitivity to magnetic field, but because of its size it can only support a fraction of the current that can be carried by a bulk superconductor (Ref 15).

Recent work has shown that, even with essentially clean grain boundaries with no evidence of impurities or second phase, current flow across the boundary is still compromised. It is suspected that dislocation networks adjacent to the boundary create strain fields, which in turn limits the ability to transport current (Ref 16).

The presence of high-angle boundaries in the oxide superconductor even under the best conditions thus appears to limit the critical current density. It is important to note that the epitaxial films that exhibit high critical current density have very few high-angle boundaries. The processing followed for bulk materials must have an inherent capability for limiting high-angle development and achieving a high degree of crystalline development in the high-current *a-b* planes.

Melt Processing

In an attempt to reduce the problem of weak links, melt processing of bulk materials has been pursued at numerous laboratories. Jin *et al.* (Ref 17) and Salama *et al.* (Ref 18) have demonstrated elongated, oriented grain development in bulk 123 materials by processing (see Fig. 4) above the peritectic at 1000 to 1250 °C (1830 to 2280 °F) in the (211 + L) region. High-J_c properties have been reported using melt processing that may have been optimized by careful sample selection and their methods used for critical current measurements.

Murakami *et al.* (Ref 19) has recently expanded on the original melt processing approach used by Jin with the melt-quench growth (MQG) technique. A high degree of bulk-oriented 123 material consisting of 123 with a dispersion of 211 is obtained by the MQG method. This was done by an initial melting into the (Y_2O_3 + L) region above 1270 °C, or 2320 °F (point D in Fig. 4) and fast quenching to form a fine dispersion of Y_2O_3. Reheating into the (211 + L) regime (point B in Fig. 4) and slow cooling through the peritectic maximizes the formation of 123 from the reaction of the finely dispersed Y_2O_3 with the liquid to nucleate 211, which then reacts with the remaining liquid to form 123. As can be seen in Fig. 1, the resultant J_c values are much improved over the sintered powder values and earlier melt processing and show a much improved J_c versus magnetic field behavior. Blaugher *et al.* (Ref 20) have taken this one step further by successfully performing the MQG process on a metallic substrate representative of a prototype tape with steady-state magnetization J_c approaching 10 kA/cm^2 at 50 K in fields up to 4 T (40 kG). Large polycrystalline melt-processed samples (25 × 10 × 2 mm, or 1.0 × 0.4 × 0.08 in.) have recently been measured by four-probe steady-state dc current transport and indicate critical current density >1 kA/cm^2 at 77 K in a 2 T (20 kG) magnetic field (Ref 21). In addition, Okada *et al.* (Ref 22) of Hitachi have recently reported the fabrication of gold- (palladium-) sheathed melt-processed 123 tape with promising J_c characteristics (see Fig. 5).

These recent melt-processing results present highly encouraging data, indicating that large transport currents can in fact be realized for oxide-base conductors. Despite experiments related to flux flow and predicted poor pinning at 77 K, the melt-processing approach produces elongated grains with minimum high-angle boundaries that provide the ability to support high critical current density at significant magnetic field level (Ref 23). Further work on the melt-processing approach is being pursued at various institutions with the prospect of producing long lengths of oxide super- conductor suitable for numerous applications.

REFERENCES

1. R.D. Blaugher, Superconductivity Technology: The Impact of Oxide Superconductors, in *Proceedings of the Tokai University Symposium on Superconductivity*, World Scientific, Nov 1988, p 183-197
2. A.P. Molozemoff, W.J. Gallagher, and R.E. Schwall, Applications of High-Temperature Superconductors, in *Chemistry of High-Temperature Superconductors*, American Chemical Society Symposium Series, Vol 351, 1987, p 280-306
3. J.G. Bednorz and K.A. Mueller, *Z. Phys. B*, Vol 64, 1986, p 189
4. J.W. Ekin, Mechanical Properties and Strain Effects in Superconductors, in *Superconductor Materials Science: Metallurgy, Fabrication, and Applications*, S. Foner and B.B. Schwartz, Ed., Plenum Publishing, 1981
5. A. Pebler and R.G. Charles, *Mater. Res. Bull.*, Vol 23, 1988, p 1337-1344
6. T.T. Kodas, E.M. Engler, V.Y. Lee, R. Jacowitz, T.H. Baum, K. Roche, and S.S.P. Parkin, *Appl. Phys. Lett.*, 7 Jan 1988
7. O. Kohno, Y. Ikeno, N. Sadakota, and K. Goto, *J. Appl. Phys.*, Vol 27, 1988, p L77
8. H. Hitosuyanagi, K. Sato, S. Tokano, and M. Nagata, in *Proceedings of Magnet Technology*, 1989
9. M. Nagato, K. Ohmata, H. Mukai, T. Hikata, Y. Hosoda, N. Shibuta, K. Sato, H. Hitosuyanagi, and M. Kawashima, Paper presented at the Materials for Cryogenic Technology Symposium (Japan), May 1989
10. K. Osamura, T. Takayama, and S. Ochial, *Supercond. Sci. Technol.*, Vol 2, 1989, p 107
11. K. Heine, J. Tenbrink, and M. Thoener, High Field Critical Current Densities in $Bi_2Sr_2Ca_1Cu_2O_{8+x}$/Ag Wires, *Appl. Phys. Lett.*, Vol 55, 1989, p 2441-2443
12. M. Fukutomi, N. Akutsu, Y. Tamaka, T. Asano, and H. Maeda, in *Cryogenic Technology*, Vol 24, National Research Institute for Metals, 1989, p 98
13. A. Kaloyerous, M. Holma, and W.S. Williams, *Proceedings of Conference on Superconducting Materials and Applications*, 1989
14. D.T. Shaw *et al.*, Plasma-Assisted Laser Deposition of Superconducting Films Without Post-Annealing, to be published in *Superconductivity: Theory and Applications*
15. M.R. Beasley and C.J. Kircher, "Josephson Junction Electronics," *Superconducting Materials Science*, Plenum Publishing, 1981, p 605
16. D.C. Larbalestier, S.E. Babcock, X. Cai, L. Cooley, M. Daeumling, D.P. Hampshire, J. McKinnell, and J. Seuntjens, Recent Results on the Weak Link Problem in Bulk Polycrystalline RE-$Ba_2Cu_3O_7$, in *Proceedings of the Tokai University Workshop*, World Scientific, 1988, p 128
17. S. Jin, R.C. Sherwood, T.H. Tiefel, R.B. VanDover, R.A. Fastnacht, and M.E. Davis, *Mater. Res. Soc. Proc.*, Vol 99, 1988
18. K. Salama, V. Selvamanickam, L. Gao, and K. Sun, *Appl. Phys. Lett.*, Vol 54, 1989, p 2352
19. M. Murakami, M. Morita, K. Miyamoto, and S. Matlsuda, *Proceedings of Osaka University International Symposium on New Developments in Applied Superconductivity*, 1988
20. R.D. Blaugher, D.W. Hazelton, J.A. Rice, and M.S. Walker, Development of a Composite Tapc Conductor of Y-Ba-Cu-O, *Mater. Res. Soc. Proc.*, 30 Nov 1989
21. R.D. Blaugher, P. Haldar, D.W. Hazelton, M.S. Walker, and J.A. Rice, *Proceedings of Applied Superconductivity Conference*, 1990, to be published
22. M. Okada, T. Yuasa, T. Matsumoto, K. Aihara, M. Seido, and S. Matsuda, Texture Formation and Improvement of Grain Boundary Weak Links in Tape, *Mater. Res. Soc. Proc.*, 30 Nov 1989
23. D. Larbalestier, Critical Currents Pinned Down, *Nature*, Vol 343, 1990, p 210

Pure Metals

Preparation and Characterization of Pure Metals

G.T. Murray, Materials Engineering Department, California Polytechnic State University
T.A. Lograsso, Ames Laboratory, Iowa State University

AS A RESULT of the constant quest for the true values of physical and chemical properties of metals, there has been continual improvement in the purity levels attainable and in the accuracy and capability of techniques for measuring these levels. Therefore, the property values reported for pure metals in this section of the Handbook, which were determined at different times and in different laboratories, vary considerably in meaningfulness from one metal to another and from one property measurement to another.

The rapidly growing electronic microcircuit industry also has placed severe demands on metal suppliers to provide metals of the highest reproducible purity attainable. Trace impurity elements in concentrations below 1 ppm can prevent proper functioning of certain electronic devices.

The need for ultrapure metals for both the measurement of physical and chemical properties and the electronic microcircuit industry poses two important problems: how to obtain such purity and how best to measure levels of trace impurity elements.

Preparation Methods

Metal of the type commonly referred to as commercial-purity is normally used as the starting material in ultrapurification operations. Depending on the metal in question, commercial purity usually means a purity between 99.0 and 99.95%. Commercial-purity metal can be prepared by a variety of processes, of which such electrolytic processes as electrowinning and electrorefining are among the most common. In both of these processes, metal is deposited by electroplating from a bath. In electrowinning, the starting material usually is in the form of a concentrated ore or compound; in electrorefining, it is in metallic form. Many different types of baths are employed. For titanium and vanadium, fused salt baths are used, whereas chromium sometimes is produced by electrolysis of an aqueous solution of chromium-alum or chromic acid. For applications such as semiconductors, material produced by electrolytic processes is of insufficient purity and must be subsequently ultrapurified by one of the methods described below.

Fractional crystallization is a liquid-phase method that relies on differences in solubility in a liquid solvent among the various solid phases present in the impure metal. In this process, the metal to be purified is dissolved in a hot, often organic, solvent. The solvent selected is such that the metal is much more soluble at higher temperatures but that impurities are fairly soluble even at lower temperatures. On subsequent cooling of the solution, then, the pure metal precipitates out of solution, whereas most of the impurities remain. This process can be repeated many times, using fresh solvent each time. Gallium has been purified to the 99.9999% level using this method. This purity is required for the manufacture of semiconducting gallium arsenide, which is used in light-emitting diodes and as substrates for high-speed digital and monolithic microwave integrated circuits. Additional information is available in the article "Gallium and Gallium Compounds" in this Volume.

Fractional crystallization can also be used to produce ultrapure silver, gold, palladium, and platinum. In some instances, the metal being refined is precipitated and impurities are left in the solvent (as described above); in others, the impurities are precipitated (as compounds). Maximum purity in these metals, however, is obtained by zone refining following fractional crystallization.

Zone refining, also a liquid-phase technique, is probably the most widely used of all preparation methods. The classic zone refining experiments by Pfann (Ref 1) led to the production of germanium sufficiently pure to be used in the development of the first transistor.

In zone refining, a molten zone is made to move slowly from one end of a bar of impure metal to the other. During this zone pass, impurities are redistributed because of differences between the solubility limits of impurity elements (limiting impurity concentrations) in the liquid phase of the metal and the corresponding limits in the solid phase. Under equilibrium conditions, the resulting distribution is measured by the coefficient K_0, which is defined as follows:

$$K_0 = \frac{C_s}{C_l} \quad \text{(Eq 1)}$$

where C_s is the impurity concentration in the just-freezing solid phase and C_l is the impurity concentration in the liquid phase. In practically all instances of freezing, equilibrium is not attained. Therefore, it is more appropriate to use an effective distribution coefficient, K_e, which is a function of freezing velocity, impurity diffusion, and thickness of the diffusion layer, as well as the ratio C_s/C_l. When, as in most instances, K_e is less than 1, and when there is slow movement of the zone (for example, 10 mm/h or 0.39 in./h), the impurity concentration in the solid phase, C_s, at distance x from the starting end after a single pass of a liquid zone of length l, is as follows:

$$\frac{C_s}{C_0} = 1 - (1 - K_e) \exp\left[-\frac{K_e x}{l}\right] \quad \text{(Eq 2)}$$

where C_0 is the initial concentration in the liquid phase. Additional passes of the zone in the same direction cause further concentration of impurities at one end of the bar. After many zone passes, this end is removed and discarded.

Metals and semiconductors were first zone refined by placing a bar of the material in a long boat-type crucible. Later to be introduced was the floating-zone technique (Ref 2), in which the metal is suspended in a vertical position and the molten zone is held in place by its own surface tension. Heat sources commonly used for this technique include an electron beam and an induction coil. Although the diameter of the bar is limited to approximately 15 mm (0.59 in.) in the floating zone technique, this method has a distinct advantage in that the material being refined is not contacted by a crucible and thus not contaminated by a crucible reaction. This is particularly advantageous for high-melting-temperature reactive metals such as titanium, zirconium, niobium, tantalum, vanadium, tungsten, and molybdenum. By contrast, metals such as gold, silver, copper, aluminum, zinc, lead, tin, and bismuth, which melt below about 1200 °C (2190 °F) and are less reactive, are usually zone refined in a boat. However, silicon crystals as large as 150 mm (6 in.) in diameter have been made by the floating zone method (Ref 3).

Vacuum Melting. Zone refining of materials often is conducted in a dynamic vacuum in order to enhance the degree of purification. However, many metals—particularly those with high melting points—can be purified to a significant degree by the vacuum melting process alone. Although vacuum melting may not produce the degree of purity attainable with zone refining, it is less expensive and yields material of sufficient purity for a wide variety of applications.

In vacuum melting, purification occurs by degassing—that is, removal of oxygen, nitrogen, and hydrogen, as well as CO or CO_2 formed by side reactions of oxygen with carbon—and by vacuum distillation of high-vapor-pressure impurity elements. Degassing takes place because the solubility of gaseous elements in the liquid decreases when the partial pressure of the same elements in the surrounding gaseous medium is decreased. This was experimentally verified for partial pressures of about 10 to 100 kPa (75 to 750 mm Hg) in the early experiments of Sieverts, which led to the well-known relationship:

$$S \propto \sqrt{P} \quad \text{(Eq 3)}$$

where S is the solubility of a gas in the liquid phase and P is the partial pressure of the same gas in the surrounding medium.

This purification process is dependent on:

- Ability of the vacuum system to maintain a sufficiently low gas partial pressure near the molten surface
- Diffusion of gas atoms through the liquid to the surface
- Presence or absence of any stirring action that might enhance transport of gas atoms in the liquid phase
- Composition of the starting material

Vacuum melting can result in a purification process based on preferential evaporation of solute. The degree of purification is dependent on the ratio of the vapor pressure of the solute to that of the solvent. For a high degree of purification, the solute vapor pressure must be high relative to solute partial pressure in the gaseous medium in the immediate vicinity of the molten surface. As the solute concentration at the liquid/vapor interface diminishes, a concentration gradient is set up within the liquid. At this time, which may be very early in the melting operation, material transport in the liquid phase becomes the rate-controlling process. Thus, provided that vapor pressures are favorable and the pumping speed of the vacuum system is sufficient to maintain a low partial pressure of the solute element, purification should proceed at a rate that depends on the diffusivity of the solute in the liquid.

Distillation. Like vacuum-melting distillation, straight distillation (in which heated material changes from solid to liquid to vapor) is an important vapor-phase purification process. If the distillation is conducted under conditions of near-equilibrium between the liquid and vapor phases, impurity elements will concentrate in either the liquid phase or the vapor phase. The vapor or the liquid will then be a material of higher purity than that of the starting material. The most common distillation method is fractional distillation, in which the metal is repeatedly vaporized and condensed to liquid on a series of plates placed in a vertical column. A high reflux ratio (the ratio of the amount of liquid returning to the column from the condenser to the amount of vapor removed to the condenser) is desirable in this method. Some metals, however, can be purified in a single stage by simply condensing all the vapor produced by the still; this process has been used for alkali metals such as barium, calcium, lithium, and sodium (Ref 4). Distilled magnesium is further purified by zone refinement.

A variation of straight distillation is sublimation, in which the metal passes directly from the solid phase to the vapor phase. Only metals that have high vapor pressure when in the solid state are suited to this process. Such a metal usually has a higher vapor pressure than most impurity metals, so that impurities are left to concentrate in the remaining solid while the vapor is condensed to form a higher-purity metal.

Chemical Vapor Deposition. In purification by chemical vapor deposition (CVD), the starting material is reacted to form a gaseous compound, and that compound is subsequently decomposed in the vapor state. The metal vapor then is condensed to form a solid higher in purity than the starting material.

One of the more popular of the chemical vapor deposition processes is the iodide process, which has been used extensively to purify titanium, zirconium, and chromium (Ref 5). For each of these metals, the starting charge of metal is reacted to form a volatile metal-iodide compound, which in turn is thermally decomposed to liberate iodine vapor. The pure metal is allowed to condense onto a suitably heated substrate (glass tubes and wires of the base metal have been used), while the iodine returns to the metal charge to form more iodide compound. Hence, the iodine acts as a carrier of the metal, from the charge to the substrate.

In this process, some impurities are almost always carried over to the vapor phase along with the metal being purified. However, if a proper temperature is maintained, oxygen, nitrogen, hydrogen, and carbon, as well as many metallic impurities, will not be carried over. Typical purities obtained are about 99.96% for titanium, 99.98% for zirconium (plus hafnium, which is present at about the 200-ppm level), and 99.995% for chromium. In all cases, the starting metal has a purity of about 99.9%. Chromium has been purified to its highest state to date by this method. Only iron is carried over with these metals to a significant extent. Thus, if a low-iron starting metal is used, the condensed vapor will approach a purity level of 99.999%.

Other metals that have been purified by chemical vapor deposition include hafnium, thorium, vanadium, niobium, tantalum, molybdenum, and many less commercially important metals (Ref 5).

Solid State Refining Techniques. Solid state refining techniques have been used to prepare some of the purest metals in the world (Ref 6) for applications that require extreme purity. These methods rely on diffusion of impurities in the solid state, require long times at high temperatures, and are usually limited to small quantities of material (<100 g, or 0.20 lb). Furthermore, purification is restricted to those impurities that have high mobilities in the host metal, most notably carbon, nitrogen, oxygen, and hydrogen. Purification therefore requires starting materials that are relatively pure in nondiffusing elements. The most widely used techniques are external gettering, solid-state vacuum degassing, and electrotransport purification.

External gettering is the removal of impurities by reaction with chemically active elements through surface contact. The decarburizing of steel is a common example in which carbon can be removed by a hydrogen atmosphere to levels as low as 0.02 ppm. Hydrogen, oxygen, and nitrogen levels can be reduced to the parts-per-million range through heat treatment with materials that have greater affinity for the impurities than does the base metal. These materials include titanium, zirconium, yttrium, and liquid calcium.

Solid-State Vacuum Degassing. Reduction of interstitial levels can also be accomplished through vacuum degassing, which is a process similar to vacuum melting. This process can result in lower interstitial content because of the lower solubility of impurities in the solid than in the liquid state in equilibrium with the surrounding environment.

Electrotransport purification uses electricity to move impurities out of the metal.

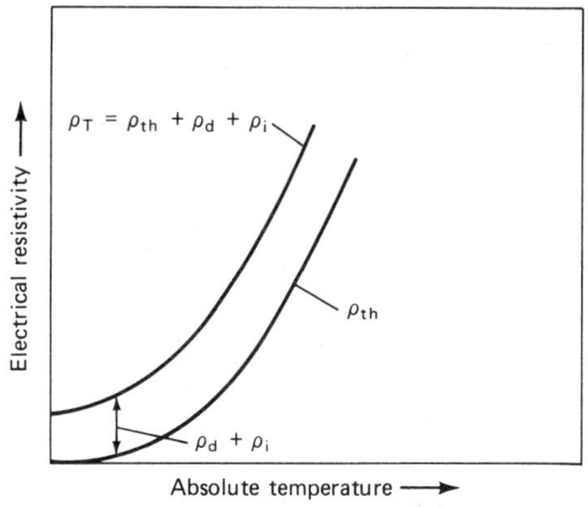

Fig. 1 Idealized graph of the components of total electrical resistivity of a metal as the temperature approaches absolute zero

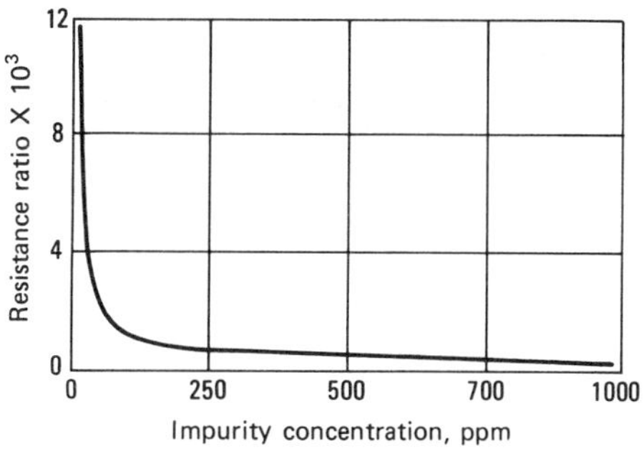

Fig. 2 Effect of interstitial impurity-atom concentration on resistance ratio of refractory metals

Table 1 Resistance ratios of samples of zone-refined metals

See Table 2 for impurity contents of these samples.

Metal	Resistance ratio × 10^3
Aluminum	40
Gold	2
Molybdenum	14
Nickel	3
Niobium	2
Niobium(a)	7.2
Rhenium	45
Tantalum	7
Tungsten	90
Vanadium	0.3
Vanadium(a)	1.88
Zirconium	0.2
Zirconium(a)	0.65

(a) Following electrotransport purification

A direct current of 10 to 20 MA/m^2 (6.5 to 13 kA/in.^{2}) is applied to the metal (heating it to 80% of its melting point) and, depending on the interaction with the direct current, an impurity will migrate toward one end or the other. The process was first used to purify zirconium metal and metals in the rare earth and actinide series.

Characterization of Purity

The traditional system of describing metal purity is based on measuring the total impurity-element content and subtracting this number from 100%. The result is reported in terms of number of nines—for example, five nines indicate a purity of 99.999% or total impurity content of 10 ppm. In order to characterize the purity to five nines, all impurity elements must be measurable to at least 100 to 200 ppb. Qualification of higher metal purity levels requires even greater measurement sensitivity. In many analyses, certain elements are not measured, and often the method employed is not sensitive enough to detect impurity levels near the low end of the parts-per-million range, let alone parts-per-billion. For many applications, this system is adequate. However, unless the method of measurement and its sensitivity are reported, the system is meaningless and unacceptable for some technical fields.

Trace Element Analysis. A number of analytical techniques are available for trace element detection. These techniques, which are described in detail in Volume 10, *Materials Characterization*, of the 9th Edition of *Metals Handbook*, generally employ one of the following methods for elemental identification.

- *Emission spectroscopy*, for simultaneous determination of metallic elements in the parts-per-million range and greater
- *Mass spectroscopy*, for determination of metallic elements in the parts-per-billion range. The analytical results are as accurate as the reference standard used
- *Neutron activation*, for determination of metallic elements, and particularly oxygen, in the parts-per-million range
- *Atomic absorption*, for sequential determination of metallic elements in the parts-per-billion range
- *Vacuum or inert gas fusion*, for determination of oxygen, hydrogen, and nitrogen in the parts-per-million range
- *Combustion technique for carbon*, sensitive in the parts-per-million range
- *Inductively coupled plasma* (ICP), for multielement qualitative and quantitative analysis of over 70 elements in the parts-per-billion to parts-per-million range

Emission spectroscopy is the most common analytical method and normally is used for detecting trace elements in concentrations of 10 to 1000 ppm. It is relatively inexpensive and yields results for most metallic elements in one analysis.

Mass spectroscopy is more sensitive (and more expensive) than emission spectroscopy and can easily detect impurity levels as low as 0.01 ppm. However, accuracy depends on the standards used, and the technique is not accurate above the 50- to 100-ppm level for some elements. Generally, emission spectroscopy is the best method for verifying purity at the five nines level and for obtaining information on all residual elements in the sample.

Neutron activation analysis can be more sensitive than mass spectroscopy but cannot detect many elements because of their inherent radioactive characteristics. It is an expensive method, but for some elements that are difficult to quantify, it can be extremely sensitive and accurate.

Atomic absorption analysis is excellent for concentrations of 0.1 to 10 ppm, when only a few elements are present. However, the specific elements being sought must be known, which generally requires emission spectroscopic analysis as a first step.

The combustion technique is commonly used for determining the carbon and sulfide content of metals. The sample is combusted with oxygen, and the resulting CO_2 or SO_2 are measured by infrared radiation (IR) absorption. The absorption of the IR signal is proportional to the CO_2 concentration.

In vacuum fusion, the sample is dissolved in a liquid platinum bath contained in a graphite crucible. Dissolved hydrogen and nitrogen are liberated as gases and the dissolved oxygen is reacted with carbon from the crucible to form CO gas. The partial pressure of each gas and the total pressure is measured and the level of each impurity in the original sample is determined.

In the ICP methods a metal sample is dissolved, for example, in an acid solution, and then injected into an argon plasma (at about 8000 K). Free atoms and ions are electronically excited to a higher energy state. When they return to a more stable state (within nanoseconds), the ultraviolet radiation emitted is measured in terms of both wavelength and intensity by a suitable spectrometer. An alternate method is to use the ionized sample as a source for a traditional quadrupole mass spectrometer.

All factors considered, mass spectroscopy is the preferred method for measuring trace elements in ultrahigh-purity metals. Several new mass spectroscopy techniques have recently been developed and represent the state of the art in analytical measurement techniques: glow discharge mass spectroscopy (GDMS), inductively coupled plasma mass spectroscopy (ICPMS), and laser ionization mass spectroscopy (LIMS). The GDMS and ICPMS techniques are bulk sample analyses and have detection limits in the range of 1 to 10 ppb, while LIMS provides surface analysis of sizes from 0.5 to 5 μm (20 to 200 μin.) with detection limits of 0.1 to 1 ppm. The main advantage of LIMS is that it places relatively few limitations on the material to be analyzed or sample shape. GDMS requires an electrically conductive or semiconducting material, and ICPMS samples must be dissolved. These techniques are capable of providing a quantitative survey of all elements, although, for the gaseous elements and car-

Table 2 Impurity concentrations of purified metals. Metallics were determined by glow discharge mass spectroscopy; carbon by combustion; and oxygen, nitrogen, and hydrogen by fusion method.

Impurity element	Impurity concentration of metals, ppm by weight(a)												
	Al	Au	Cu(b)	Cr(c)	Mo	Ni	Nb	Re	Ta	Ti	W	V	Zr
C	6	<1	...	...	10	40	8	5	10	40	5	57	20
H	0.09	<1	...	...	0.9	0.2	0.4	0.2	<0.1	1	0.1	3	3
O	14	2	<5	<10	4.3	25	23.4	0.5	3.5	570	0.8	250	200
N	<3	1	<2	<5	0.5	10	4	1	2.3	30	0.1	3	2
Ag	<0.08	4	<0.01	<0.01	<0.7	<0.01	<0.3	<0.001	<0.004	<0.01	<0.12	<0.002	<0.4
Al	...	<0.01	0.004	0.05	<0.03	0.3	0.15	0.05	0.05	6	0.07	0.1	3
As	<0.02	<0.01	0.34	<0.01	<0.01	<0.04	<0.01	<0.002	<0.002	<0.08	<0.005	<0.05	<0.01
Au	<0.02	...	<0.01	<0.01	<0.02	<0.15	<0.03	<0.15	<0.2	<0.01	<0.3	0.6	<0.2
Bi	<0.03	<0.01	<0.01	<0.01	<0.02	<0.01	<0.01	<0.101	<0.04	...	<0.12	<0.02	<0.007
Ca	<0.03	<0.01	0.028	<0.003	0.04	0.1	0.02	0.05	<0.008	<0.6	0.02	0.1	0.04
Cd	<0.02	<0.01	<0.008	<0.04	<1.0	<0.08	<0.5	<0.02	<0.007	0.14	<0.025	<0.03	0.5
Cl	3	<0.01	0.07	0.001	0.4	0.1	0.3	0.1	0.01	<1.8	0.2	0.1	2
Co	<0.02	<0.01	0.002	0.007	<0.06	<0.1	<0.01	0.06	0.3	<0.008	0.1	<0.15	<0.007
Cr	<0.01	<0.01	0.05	...	0.1	1.5	0.05	0.08	0.2	4.1	<0.001	<5	0.5
Cu	0.09	1	...	<0.02	<0.02	<0.04	<0.01	0.005	0.02	2.1	0.005	<0.3	0.01
Fe	0.05	2	1.7	5.4	12	12	0.12	3	0.3	1.5	0.01	<20	30
Ga	<0.01	<0.01	0.05	<0.08	<0.02	<0.4	<0.01	<0.004	<0.003	<0.003	<0.01	20	<0.02
Ge	<0.03	<0.01	<0.02	<0.09	<0.02	<0.7	<0.01	<0.02	<0.005	<0.005	<0.04	<0.6	<0.03
Hf	<0.01	<0.01	<0.002	<0.006	<0.03	<0.03	<0.02	<0.01	<0.4	0.25	<0.04	<0.03	40
In	0.09	<0.01	<0.012	<0.008	<1	<1	<0.07	<0.2	<0.02	0.05	<0.03	<0.03	<0.08
Ir	0.05	<0.01	<0.001	...	<0.03	<0.02	<0.01	<0.2	<0.2	...	<0.15	<0.06	<0.03
K	<0.01	<0.01	<0.001	<0.009	1	0.2	<0.04	0.01	0.02	<0.01	<0.02	0.4	0.004
Li	<0.01	<0.01	<0.001	<0.004	<0.02	<0.02	<0.01	0.004	0.001	<0.008	<0.001	<0.02	<0.001
Mg	0.18	<0.01	0.024	<0.004	<0.25	0.02	<0.05	0.02	0.006	<0.02	0.15	<0.25	<0.05
Mn	0.01	<0.01	0.31	<0.005	0.06	0.03	0.03	0.02	0.01	2	0.03	<0.15	<0.03
Mo	<0.02	<0.01	0.033	<0.02	...	0.5	<0.7	4	0.2	<0.05	<0.1	0.08	<0.6
Na	<0.01	<0.01	<0.002	<0.07	<1	<0.04	<0.03	<0.01	0.015	<0.01	<0.01	<0.05	<1
Nb	<0.02	<0.01	<0.001	<0.01	1	<0.02	...	1.2	25	<0.55	<1	0.8	<0.5
Ni	<0.01	<0.01	0.46	0.27	0.1	...	0.15	0.02	1.5	<0.02	<0.02	12	1.5
P	<0.01	<0.01	<0.9	<0.01	<0.03	<2	<30	0.02	<0.05	<0.07	<0.05	0.2	0.1
Pb	<0.01	0.5	0.26	<0.01	<0.03	<0.02	<0.02	<0.25	<0.08	<0.02	<0.25	<0.003	<0.015
Pd	<0.1	<0.01	<0.01	<0.01	<1	<0.03	<0.5	<0.02	<0.4	<0.01	<0.25	15	<0.8
Pt	<0.06	<0.01	<0.01	<0.01	<0.06	<0.03	0.02	<0.3	<0.2	<0.01	<0.5	<0.04	<0.2
Re	<0.04	<0.01	<0.01	<0.01	<0.4	<0.02	<0.02	...	<0.002	<0.01	<1	<0.5	<0.3
Rh	<0.04	<0.01	<0.01	<0.01	<0.1	<0.01	<0.06	<0.005	<0.001	<0.01	<0.06	<0.5	<0.2
Ru	<0.1	<0.01	<0.01	<0.01	<0.3	<0.03	<0.4	<0.04	0.02	<0.01	<0.2	0.1	<0.6
S	0.022	<0.01	<0.01	<0.01	1	<0.12	<0.07	1	<0.004	<0.01	0.07	<0.02	<1
Sb	<0.01	<0.01	0.34	4.3	<0.2	<2	<0.04	<0.004	<0.02	<0.05	<0.03	20	<0.15
Si	<0.02	0.5	0.04	5.6	0.08	<0.2	0.6	0.5	<0.002	2.3	0.3	<0.03	<0.25
Sn	<0.02	<0.01	0.06	<0.06	<0.4	<0.4	<0.3	<0.02	...	<0.04	<0.02	<0.3	<0.2
Ta	<0.02	<0.01	<0.01	<0.01	2	<0.5	50	3	0.01	<0.01	5	6	1
Ti	0.05	<0.01	<0.003	0.013	<1	<0.15	<0.02	<0.07	0.01	...	<0.01	...	0.05
V	<0.01	<0.01	<0.003	<0.01	<0.02	<0.01	<0.8	<0.001	1.2	2.1	<0.01	7	<0.7
W	<0.01	<0.01	<0.002	0.043	20	1.5	6.4	15	<0.004	0.09	...	<0.4	<0.5
Zn	0.1	<0.01	<0.002	<0.05	<0.02	<0.4	<0.02	0.005	<0.1	<0.03	<0.02	<0.12	...
Zr	<0.01	<0.01	<0.001	1.5	<0.04	<0.15	<0.03	<0.003	...	<0.2	0.2	<1	<0.5

(a) With exception of copper and chromium, all samples were zone refined. (b) Purified by electrolysis and vacuum melting. (c) Purified by chemical vapor deposition. Source: Materials Research Corporation

bon, residual gases and/or surface contamination may yield erroneous analyses.

In summary, the only way of describing purity that is both accurate and meaningful is to state the entire list of possible impurities, the amounts detected, and the limits of detection applicable to the specific analytical procedure used.

Resistance-Ratio Test. The resistance to passage of electrons through a sample of high-purity metal, particularly at low temperatures, is extremely sensitive to the amount of trace elements present in the sample. This fact gives rise to the resistance-ratio test, which is a very sensitive qualitative method of measuring purities of 99.999% and higher. This test is valuable not only because of its sensitivity but also because the measurement of electrical resistance is relatively simple.

Making a resistivity measurement at a single low temperature would require very accurate dimensional measurements. To avoid this requirement, resistance measurements are made both at the low temperature and at room temperature, and the ratio of the room-temperature value to the low-temperature value is reported. Unless otherwise stated, it can be assumed that the low-temperature measurement was made at liquid helium temperature (4.2 K).

The electrical resistivity of a metal can be conveniently divided into three parts:

$$\rho_T = \rho_{th} + \rho_d + \rho_i \qquad \text{(Eq 4)}$$

where ρ_T is total resistivity, ρ_{th} is resistivity due to thermal vibrations of the lattice, ρ_d is resistivity due to lattice imperfections (consisting primarily of vacancies, dislocations, and grain boundaries) and ρ_i is resistivity due to impurity atoms. Variation of ρ_T with temperature in terms of the components ρ_{th} and $\rho_d + \rho_i$ is depicted in Fig. 1. The sum $\rho_d + \rho_i$ is essentially temperature independent, whereas ρ_{th} is strongly temperature dependent ($\rho_{th} \propto T^5$), approaching zero at absolute zero temperature. Thus, resistivity near absolute zero affords a measure of $\rho_d + \rho_i$.

Point defects (vacancies) contribute to resistivity to about the same extent as impurity atoms. However, well-annealed metals contain far fewer point defects than do impurity atoms. Dislocations of a typical density of 10^{11} per m^2 contribute an insignificant amount to the resistance ratio. In the highest-purity metals obtained to date (impurity concentrations of 10^{-5} to 10^{-6} at.%), the contribution of ρ_d is still small compared with that of ρ_i. Total resistivity near 0 K, therefore, is a good measure of the impurity contribution.

For several reasons, caution should be exercised in using the resistance ratio as a characterization of purity. Most important is the fact that in resistance-ratio testing the impurity element in question is not determined (different impurity elements have vastly different effects on ρ_i). In addition,

Table 3 Impurity concentrations in two titanium samples and a chromium sample characterized using glow discharge mass spectroscopy (GDMS) method and Leco combustion methods

Impurity	Impurity concentration, ppm by weight — Titanium — Sample 1	Sample 2	Chromium
Metallic impurities, GDMS method			
Al	6	0.43	0.052
As	<0.8	0.25	. . .
B	<0.01	<0.008	<0.02
Ca	<0.6	<1	<0.003
Cd	0.14	<0.04	<0.04
Co	<0.008	0.081	0.007
Cr	4.1	<0.01	. . .
Cu	2.1	<0.04	<0.02
Fe	1.5	9.6	5.4
Ga	<0.003	<0.04	<0.08
Ge	<0.005	<0.06	<0.09
Hf	0.25	<0.01	<0.006
In	0.047	<0.02	<0.008
K	<0.01	<0.01	<0.009
Li	<0.008	<0.008	<0.004
Mg	<0.02	<0.01	<0.004
Mn	2	<0.007	<0.005
Mo	<0.05	0.36	<0.02
Na	<0.01	<0.01	<0.007
Nb	<0.55	<0.065	. . .
Ni	<0.02	0.23	0.27
Pb	<0.02	<0.02	<0.01
Sb	<0.05	5.3	4.3
Si	2.3	0.92	5.6
Sn	<0.04	2.3	<0.06
Ti	. . .	. . .	0.013
Th	<0.0009	<0.001	<0.0004
U	<0.001	<0.001	<0.0004
V	2.1	<0.004	<0.01
W	0.087	0.08	0.043
Zn	<0.03	0.03	<0.05
Zr	13	<0.01	1.5
Total detected metallic impurities	33.624	19.551	17.185
Nonmetallic impurities, Leco combustion method			
C(a)	40	39	. . .
H(b)	. . .	0.85	. . .
N(b)	30	<10	<5
O(b)	570	243	<10
S(a)	. . .	4	. . .
Nonmetallic impurities, GDMS method			
Cl	<1.8	<2.7	. . .
F	<0.4	<0.1	. . .
P	<0.07	<0.01	. . .
Overall purity	99.9965%	99.9979%	99.9982%

(a) Leco high-temperature combustion method. (b) Leco inert gas fusion method. Source: Materials Research Corporation

because only impurity atoms in solid solution are effective electron-scattering centers, nothing is learned about the impurity content in precipitate (compound) form. Finally, even for impurity atoms in solid solution, the resistance ratio is a sensitive measure of purity only when the impurity level is about 100 ppm or less. This is illustrated in Fig. 2, which is an estimate of variation in resistance ratio with concentration of interstitial atoms (O_2, N_2, and C) in refractory metals. This graph shows that the impurity level can be reduced from 500 to 250 ppm without appreciably affecting the ratio, whereas a reduction of 5 to 2.5 ppm has a marked effect.

Resistance ratios of zone-refined metals are listed in Table 1. Their corresponding chemical compositions, as measured by mass spectroscopy for metallic elements and Leco combustion method for carbon, oxygen, nitrogen, and hydrogen are listed in Table 2. Ratios higher than those shown in Table 1, and ratios for other metals, have been reported (Ref 7); however, they were not accompanied by chemical analyses. In fact, some of the ratios were so large that the impurity concentrations they indicated were too low to be detected by methods currently available.

Six Nines Characterization of Purity. Table 3 is an example of the detection of impurities to concentrations in the parts per billion range, utilizing a combination of the GDMS method and Leco combustion methods. Additional information on the Leco combustion methods is available in the articles "High-Temperature Combustion" and "Inert Gas Fusion" in Volume 10, *Materials Characterization* of the 9th Edition of *Metals Handbook.*

REFERENCES

1. W.G. Pfann, *Transactions of the American Institute of Mining, Metallurgical and Petroleum Engineers*, Vol 194, 1952, p 861
2. H.C. Theuerer, *Transactions of the American Institute of Mining, Metallurgical and Petroleum Engineers*, Vol 206, 1956, p 1316
3. R.N. Thomas, H.M. Hobgood, P.S. Ravishankar, and T.T. Braggins, *Solid State Technology*, Vol 33 (No. 3), 1990
4. P.A. Schmidt, *Journal of the Electrochemical Society*, Vol 113, 1966, p 201
5. R.F. Rolsten, *Iodide Metals*, Wiley, 1961
6. O.N. Carlson, High Temperature Materials and Processes, submitted for publication
7. W.G. Pfann, *Zone Melting*, 2nd ed., John Wiley & Sons, Inc., 1966

Periodic Table of the Elements

Metals / Nonmetals

Key: Symbol → **Al 13** ← Atomic number (Aluminum)

Rare earth metals

Actinide metals

IA	IIA	IIIB	IVB	VB	VIB	VIIB	VIII	VIII	VIII	IB	IIB	IIIA	IVA	VA	VIA	VIIA	0
H 1 Hydrogen																	**He** 2 Helium
Li 3 Lithium	**Be** 4 Beryllium											**B** 5 Boron	**C** 6 Carbon	**N** 7 Nitrogen	**O** 8 Oxygen	**F** 9 Fluorine	**Ne** 10 Neon
Na 11 Sodium	**Mg** 12 Magnesium											**Al** 13 Aluminum	**Si** 14 Silicon	**P** 15 Phosphorus	**S** 16 Sulfur	**Cl** 17 Chlorine	**Ar** 18 Argon
K 19 Potassium	**Ca** 20 Calcium	**Sc** 21 Scandium	**Ti** 22 Titanium	**V** 23 Vanadium	**Cr** 24 Chromium	**Mn** 25 Manganese	**Fe** 26 Iron	**Co** 27 Cobalt	**Ni** 28 Nickel	**Cu** 29 Copper	**Zn** 30 Zinc	**Ga** 31 Gallium	**Ge** 32 Germanium	**As** 33 Arsenic	**Se** 34 Selenium	**Br** 35 Bromine	**Kr** 36 Krypton
Rb 37 Rubidium	**Sr** 38 Strontium	**Y** 39 Yttrium	**Zr** 40 Zirconium	**Nb** 41 Niobium (Columbium)	**Mo** 42 Molybdenum	**Tc** 43 Technetium	**Ru** 44 Ruthenium	**Rh** 45 Rhodium	**Pd** 46 Palladium	**Ag** 47 Silver	**Cd** 48 Cadmium	**In** 49 Indium	**Sn** 50 Tin	**Sb** 51 Antimony	**Te** 52 Tellurium	**I** 53 Iodine	**Xe** 54 Xenon
Cs 55 Cesium	**Ba** 56 Barium	**La*** 57 Lanthanum	**Hf** 72 Hafnium	**Ta** 73 Tantalum	**W** 74 Tungsten	**Re** 75 Rhenium	**Os** 76 Osmium	**Ir** 77 Iridium	**Pt** 78 Platinum	**Au** 79 Gold	**Hg** 80 Mercury	**Tl** 81 Thallium	**Pb** 82 Lead	**Bi** 83 Bismuth	**Po** 84 Polonium	**At** 85 Astatine	**Rn** 86 Radon
Fr 87 Francium	**Ra** 88 Radium	**Ac†** 89 Actinium	**Rf** 104 Rutherfordium	**Ha** 105 Hahnium	106												

***Lanthanide series**	**Ce** 58 Cerium	**Pr** 59 Praseodymium	**Nd** 60 Neodymium	**Pm** 61 Promethium	**Sm** 62 Samarium	**Eu** 63 Europium	**Gd** 64 Gadolinium	**Tb** 65 Terbium	**Dy** 66 Dysprosium	**Ho** 67 Holmium	**Er** 68 Erbium	**Tm** 69 Thulium	**Yb** 70 Ytterbium	**Lu** 71 Lutetium
†Actinide series	**Th** 90 Thorium	**Pa** 91 Protactinium	**U** 92 Uranium	**Np** 93 Neptunium	**Pu** 94 Plutonium	**Am** 95 Americium	**Cm** 96 Curium	**Bk** 97 Berkelium	**Cf** 98 Californium	**Es** 99 Einsteinium	**Fm** 100 Fermium	**Md** 101 Mendelevium	**No** 102 Nobelium	**Lr** 103 Lawrencium

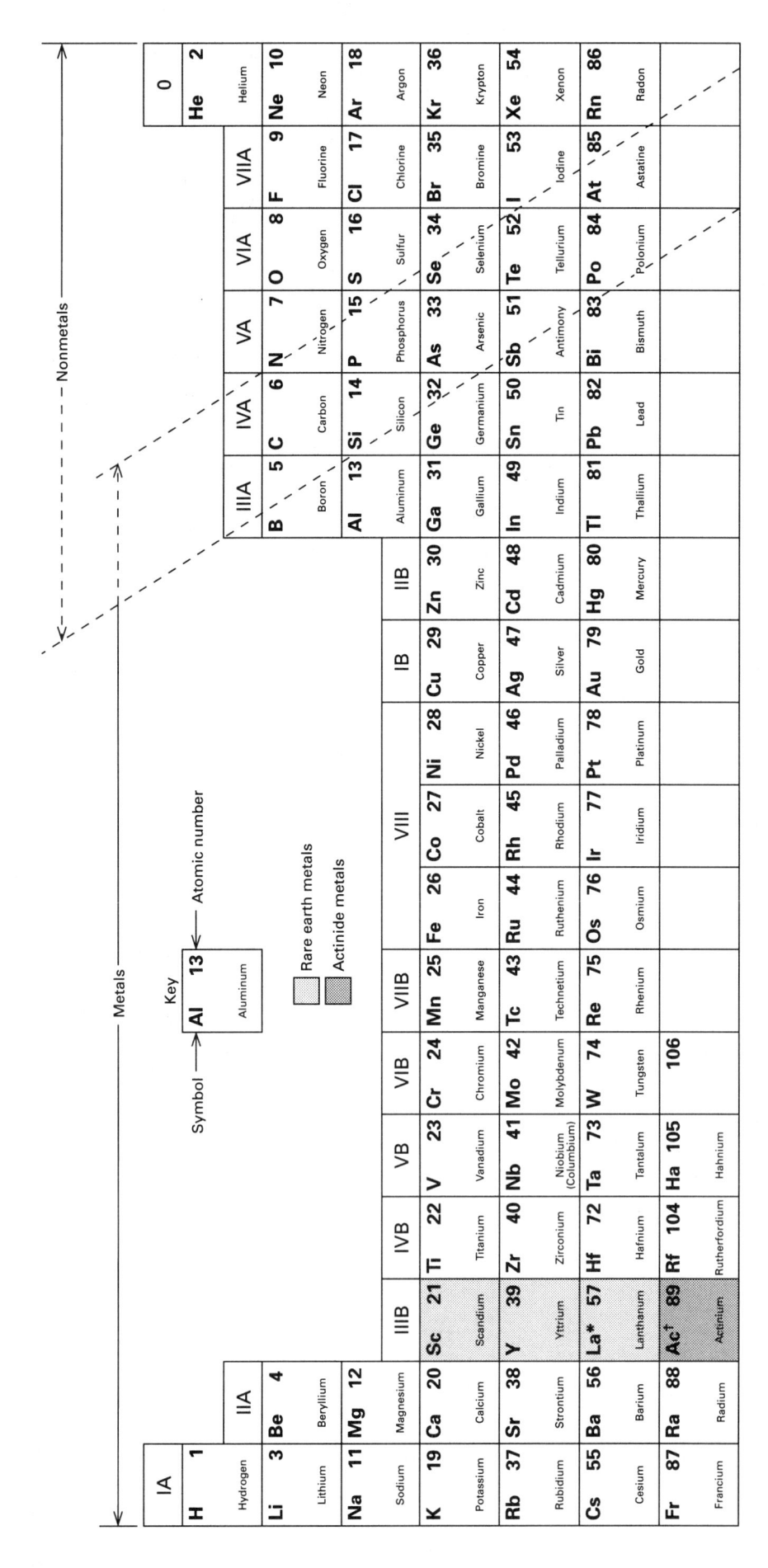

Properties of Pure Metals

Actinium (Ac)

See the section "Properties of the Actinide Metals (Ac-Pu)" in this article.

Aluminum (Al)

Compiled by H.Y. Hunsicker (retired), Aluminum Company of America;
L.F. Mondolfo, Consultant; and
P.A. Tomblin, The De Havilland Aircraft Company of Canada, Ltd.

Commercial Names

Common names. Unalloyed aluminum designated on the basis of purity:

Aluminum, %	Designation
99.50-99.79	Commercial purity
99.80-99.949	High purity
99.950-99.9959	Super purity
99.9960-99.9990	Extreme purity
>99.9990	Ultra purity

Structure

Crystal structure. Face-centered cubic (fcc); a = 0.404958 nm at 25 °C
Slip plane. (111)
Slip direction. [110]
Twinning plane. (111)

Mass Characteristics

Atomic weight. 26.98154
Density. 2.6989 g/cm^3 at 20 °C. Effect of temperature, see Fig. 1. Effect of deformation, 0.1 to 0.3% decrease at 90 to 99% plastic deformation
Volume change on freezing. 6.5% contraction

Thermal Properties

Melting point. 660.4 °C
Boiling point. 2494 °C
Thermal expansion:

Temperature range, °C	Average coefficient, µm/m · K
−200 to 20	18.0
−150 to 20	19.9
−100 to 20	21.0
−50 to 20	21.8
20 to 100	23.6
20 to 200	24.5
20 to 300	25.5
20 to 400	26.4
20 to 500	27.4

Specific heat. At 25 °C, 0.900 kJ/kg · K; at 660.4 °C (liquid), 1.18 kJ/kg · K
Latent heat of fusion. 397 kJ/kg
Latent heat of vaporization. 10.78 MJ/kg
Heat of combustion. 31.05 MJ/kg Al
Thermal conductivity. At 25 °C, 247 W/m ·

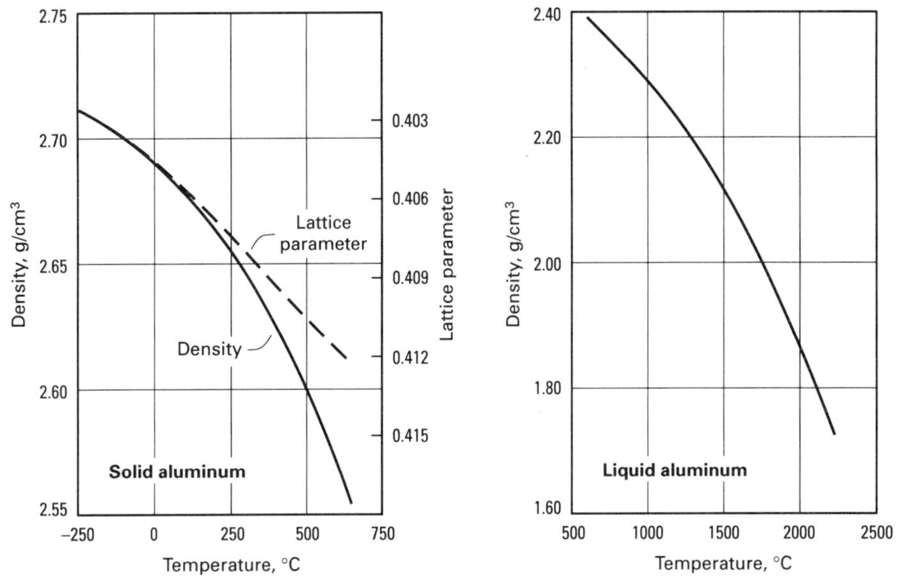

Fig. 1 Variation of density of pure aluminum with temperature. Lattice parameter data given for solid aluminum

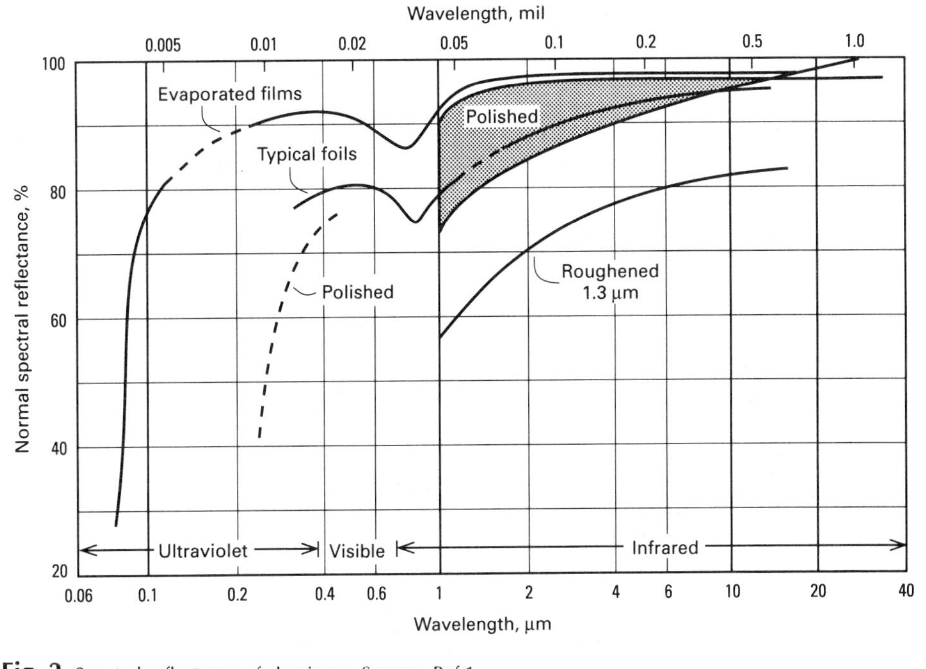

Fig. 2 Spectral reflectance of aluminum. Source: Ref 1

Table 1 Tensile properties of 99.999+% Al

Amount of cold work	Tensile strength, MPa	Yield strength, MPa	Elongation, %
Annealed	40–50	15–20	50–70
40%	80–90	50–60	15–20
70%	90–100	65–75	10–15
90%	120–140	100–120	8–12

Table 2 Mechanical properties of pure aluminum at room temperature

Purity, %	Tensile yield strength at 0.2% offset		Tensile strength		Elongation in 50 mm (2 in.), %	
	MPa	ksi	MPa	ksi		
99.99	10	1.4	45	6.5	50	65(a)
99.8	20	2.9	60	8.7	45	55(a)
99.6	30	4.4	70	10.2	43	. . .

(a) From Ref 2

K; at 660.4 °C (liquid), 90 W/m · K

Electrical Properties

Electrical conductivity. See table below.
Electrical resistivity. Temperature coefficient, 114.5 pΩ · m per K at 20 °C:

Aluminum purity, %	Electrical conductivity, volumetric %IACS(a)	Electrical resistivity, nΩ · m(b)
99.999+	65 to 66	26.2
99.8	62	26.55

(a) IACS, International Annealed Copper Standard. (b) At 20 °C (70 °F)

Electrolytic solution potential. −0.84 V versus 0.1 *N* calomel electrode in NaCl-H_2O_2 solution
Temperature of superconductivity. 1.2 K

Magnetic Properties

Magnetic susceptibility. Mass: 7.88×10^{-9} mks

Optical Properties

Spectral reflectance. 90% for white light from a tungsten filament, 86 to 87% for λ = 220 to 250 nm, 96% for λ = 1.0 μm, 97% for λ = 1.1 to 10 μm. See Fig. 2.
Spectral hemispherical emittance. At 9.3 μm: 3% at 25 °C, 15 to 20% at 660 °C (liquid)

Nuclear Properties

Neutron cross section. For neutron energy of 0.02 V, 0.2 b/atom; for 100 MV, 0.6 to 0.7 b/atom

Mechanical Properties

Tensile properties. See Table 1, Table 2, and Fig. 3.
Hardness. See Fig. 3.
Modulus of elasticity. 62 GPa
Elastic modulus. Shear, 25 GPa at 25 °C
Velocity of sound. At 25 °C, 6200 m/s; at 660 °C (liquid), 4650 m/s

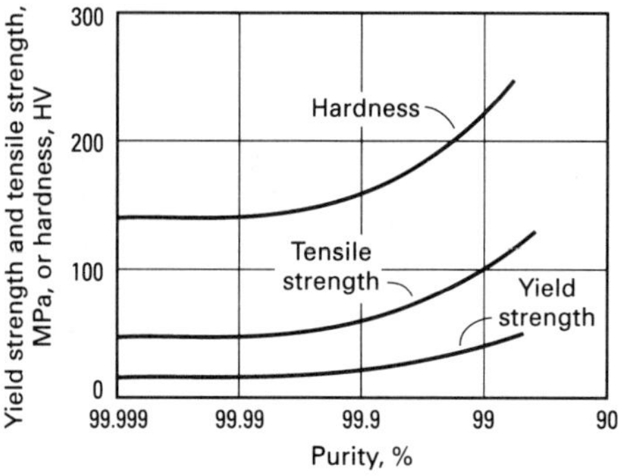

Fig. 3 Hardness and strength of aluminum as functions of purity

REFERENCES

1. Y.S. Touloukian and D.P. DeWitt, Ed., Thermophysical Properties of Matter, in *Thermal Radiative Properties—Metallic Elements and Alloys*, Vol 7, IFI/Plenum, 1970
2. T.G. Pearson and H.W.L. Phillips, The Production and Properties of Superpurity Aluminum, *Metall. Rev.*, Vol 2, 1957, p 305-360

Americium (Am)

See the section "Properties of the Transplutonium Actinide Metals (Am-Fm)" in this article.

Antimony (Sb)

Compiled by S.C. Carapella, Jr., ASARCO Inc.

Reviewed for this Volume by Douglas Hayduk, ASARCO Inc.

Antimony is used as an alloying element in lead alloys (for battery grids, printers' type, solder, bearings, cable sheathing, and ammunition) and in tin alloys (such as pewter and costume jewelry alloys). Trace quantities are added to copper-base alloys, to prevent dezincification; to ductile iron (in the amount of 50 ppm, preferably with some cerium added), to assist in forming nodular graphite; and to gray iron (in the amount of 0.05%), the antimony acts as a powerful pearlite former. In the form of Sb_2O_3, antimony is used in enamels, glass, pigments, catalysts, and flame retardants. Antimony is used as a component of III-V semiconductors such as InSb, AlSb, and GaSb, and as an alloying ingredient in thermoelectric alloys.

Chemical Composition

Composition limits. See Table 3.

Structure

Crystal structure. Hexagonal (rhombohedral equivalent); *a* = 0.4307 nm, *c* = 1.1273 nm

Mass Characteristics

Atomic weight. 121.75
Density. 6.697 g/cm³ at 26 °C

Thermal Properties

Melting point. 630.7 °C
Boiling point. 1587 °C
Coefficient of linear thermal expansion. 8 to 11 μm/m · K at 20 °C
Specific heat. 0.207 kJ/kg · K at 25 °C
Latent heat of fusion. 163.17 kJ/kg
Latent heat of vaporization. 1602 kJ/kg
Thermal conductivity. 25.9 W/m · K

Electrical Properties

Electrical resistivity. 370 nΩ · m at 0 °C

Magnetic Properties

Magnetic susceptibility. Volume: -10.2×10^{-6} mks

Optical Properties

Spectral reflectance. 70% for λ = 58.9 μm

Chemical Properties

Resistance to specific corroding agents:

Corroding agent	Resistance
Air	Moderate general attack when air is moist and light is present
Alkalis and alkali salts	General attack
Ammonia	Resistant
Aqua regia	Severe general attack
Carbon dioxide	Resistant
Chlorine	Severe general attack
Hydrochloric acid	Moderate attack in presence of air
Hydrofluoric acid	Resistant
Nitric acid	Severe general attack
Sulfuric acid	Severe general attack by warm concentrated acid; resistant to cold or dilute acid

Table 3 Nominal compositions and uses of antimony alloys

Uses	Composition, %			
	Sb	Sn	Pb	Other
Type metal	4–23	17–3	bal	0–2 Cu
Battery grids	2.5–5	0.25–0.50	bal	0.25–0.5 Sn
Bearing metal	4–15	89–0.9	bal	1–3 As, 0.5–8 Cu
Cable covering	1–6	. . .	bal	. . .
Sheet and pipe	2–6	. . .	bal	. . .
Collapsible tubes	2–3	. . .	bal	. . .
Plumber's solder	0–2	42–38	bal	. . .
Pewter	0–8	bal	20–2	0–7 Cu
Britannia metal	2–10	bal	0–9	0.2–5 Cu, 0–5 Zn
Bullets, shrapnel	0.5–12	0.25–1.0	bal	

Mechanical Properties

Tensile strength. 11.40 MPa
Hardness. 30 to 58 HB
Elastic modulus. Tension, 77.759 GPa; shear, 19 GPa

Arsenic (As)

Compiled by S.C. Carapella, Jr., ASARCO Inc.

Reviewed for this Volume by Douglas Hayduk, ASARCO Inc.

In quantities of 0.5 to 2%, arsenic improves the sphericity of lead shot. Small percentages are added to lead alloys for battery grids and cable sheathing to improve their hardness. In amounts up to 3%, arsenic improves the properties of lead-base bearing alloys. Minor additions improve the corrosion resistance and increase the recrystallization temperature of copper and stabilize pearlite in ductile and gray cast irons.

Phosphorized deoxidized arsenical copper (alloy 142) is used for locomotive fireboxes, staybolts, straps, and plates, and for heat exchangers and condenser tubes. Copper-arsenical leaded Muntz metal (alloy 366), Admiralty brass (alloy 443), naval brass (alloy 465), and aluminum brass (alloy 687) all find use in condensers, evaporators, ferrules, and heat exchanger and distillation tubes. The compositions of these alloys are listed in Table 4. Arsenic is a component of III-V semiconductors such as GaAs, GaAsP, and InAs.

Arsenic is normally available in the α-metallic and β-amorphous form. Other allotropes of arsenic have been reported, but supportive evidence is meager.

The toxicity of arsenic is related to its chemical state and can be extremely high. The degree of toxicity of arsenic in the elemental state is relatively low. Metallic arsenic on exposure to the atmosphere will develop an oxide coating, and, for this reason, care must be taken not to ingest or handle it. Inhalation of dust and fumes is to be avoided. Controlled exhaust ventilation of work areas is required to comply with the OSHA standard of 10 μg of arsenic per cubic meter of air.

Structure

Crystal structure. Hexagonal (rhombohedral equivalent); at 26 °C, a = 0.3760 nm and c = 1.0548 nm

Mass Characteristics

Atomic weight. 74.9216
Density. 5.778 g/cm^3 at 26 °C

Thermal Properties

Melting point. 816 °C at 3.91 MPa
Boiling point. Sublimes at 615 °C
Coefficient of linear thermal expansion. 5.6 μm/m · K at 20 °C
Specific heat. 0.328 kJ/kg · K

Table 4 Nominal composition of selected arsenical copper alloys

Alloy number	Composition, %						
	Cu	As	Pb	Fe	Sn	Zn	Other
142	99.4	0.015–0.50	...	...	...	...	0.015–0.040 P
366	58–61	0.02–0.10	0.40–0.9	0.15	0.25	bal	...
443	70–73	0.02–0.10	0.07	0.06	0.9–1.2	bal	...
465	59–62	0.02–0.10	0.20	0.10	0.5–1.0	bal	...
687	76–79	0.02–0.10	0.07	0.06	...	bal	1.8–2.5 Al

Latent heat of fusion. 370.3 kJ/kg
Latent heat of sublimation. 426.77 kJ/kg

Electrical Properties

Electrical resistivity. α-metallic form, 260 nΩ · m at 0 °C
Electrochemical equivalent. Valence 3, 0.15254 mg/C; valence 5, 0.25876 mg/C

Magnetic Properties

Magnetic susceptibility. Volume: −3.9 × 10^{-6} mks at 18 °C

Mechanical Properties

Hardness. 3.5 Mohs

Barium (Ba)

Revised by J.H. Westbrook, Sci-Tech Knowledge Systems

Structure

Crystal structure. Body-centered cubic, $cI2$ ($Im3m$); a = 0.5013 nm at 26 °C. High-pressure phases: Ba II, hexagonal, $hP2$ ($P6_3/mmc$), formed at 5.9 GPa, a = 0.3901 nm, c = 0.6155 nm; Ba III, structure not determined (Ref 1)

Mass Characteristics

Atomic weight. 137.3
Density. 3.5 g/cm^3 at 20 °C (Ref 2)

Thermal Properties

Melting point. 729 ± 2 °C (Ref 3)
Boiling point. 1637 °C (calculated from vapor pressure data) (Ref 2, 4)
Specific heat.

K	°C	kJ/kg · K
2	−271	1.232 × 10^{-4}
4	−269	7.805 × 1^{-4}
10	−263	0.01406
20	−253	0.06747
30	−243	0.1095
40	−233	0.1347
50	−223	0.1511
100	−173	0.1771
200	−73	0.1921
248	−25	0.2047
273–373	0–100	0.285

Source: Ref 5

Electronic coefficient (γ). 19 ± 3 mJ/kg · K^2 (Ref 3)
Latent heat of fusion. 56.4 ± 7.6 kJ/kg (Ref 3)
Latent heat of vaporization. 1290 kJ/kg (Ref 6)
Thermal conductivity. 18.4 W/m · K (Ref 7)
Temperature of superconductivity. Ba II, ~1.3 K at 5.5 GPa; Ba III, 3.05 K from 8.5 to 8.8 GPa; Ba III, ~5.2 K for pressures >14.0 GPa (Ref 8)
Coefficient of expansion. (0-100 °C) 18 × 10^{-6}/K

Electrical Properties

Resistivity. 60 μΩ · cm at 0 °C. 133 μΩ · cm for liquid barium (Ref 5)
Work function. 1.7 to 2.55 eV (0.27 to 0.41 aJ), depending on conditions and techniques of the experimental determination (Ref 9-11)

Magnetic Properties

Magnetic susceptibility. Molar: 0.254 mks at 20 °C (Ref 12)

Nuclear Properties

Stable isotopes. ^{130}Ba, isotope mass 129.90628, 0.10% abundant; ^{132}Ba, isotope mass 131.90505, 0.095% abundant; ^{134}Ba, isotope mass 133.90449, 2.4% abundant; ^{135}Ba, isotope mass 134.90567, 6.5% abundant; ^{136}Ba, isotope mass 135.90456, 7.8% abundant; ^{137}Ba, isotope mass 136.90582, 11.2% abundant; ^{138}Ba, isotope mass 137.90524, 71.9% abundant (Ref 13)
Unstable isotopes. ^{122}Ba, 12 m; ^{123}Ba, 12 m; ^{124}Ba, ~24 m, 11 m; ^{125}Ba, 8 m, ~6.5 m; ^{126}Ba, 97 m; ^{127}Ba, 10 m, 18m; ^{128}Ba, 2.42 d; ^{129}Ba, 2.5 h, 2.1 h; ^{130}Ba, 9 ms; ^{131}Ba, 14.3 m, 11.7 d; ^{136}Ba, 38.9 h, 10.4 y; ^{135}Ba, 28.7 h; ^{136}Ba, 0.308 s; ^{137}Ba, 2.55 m; ^{139}Ba, 83.3 m; ^{140}Ba, 12.79 d; ^{141}Ba, 18.3 m; ^{142}Ba, 10.7 m; ^{143}Ba, 13.6 s; ^{144}Ba, 11 s; ^{145}Ba, 6.2 s; ^{146}Ba, 2.2 s (Ref 13)
Thermal neutron cross section (0.025 eV). Absorption, 1.3 b; scattering, 8 b (Ref 14)

Mechanical Properties

Modulus. Bulk, 10.30 GPa; rigidity, 4.86 GPa; Young's, 12.8 GPa (Ref 5)
Poisson's ratio. 0.28 (Ref 5)

REFERENCES

1. P. Villars and L.D. Calvert, Ed., *Pearson's Handbook of Crystallographic Data for Intermetallic Phases*, American Society for Metals, 1985
2. R.J. Elliott, *Constitution of Binary Alloys*, First Supplement, McGraw-Hill, 1965
3. R. Hultgren, P.D. Desai, D.T. Hawkins, M. Gleiser, K.K. Kelley, and D.D.

Wagman, *Selected Values of the Thermodynamic Properties of the Elements*, American Society for Metals, 1973
4. K.A. Gschneider, Jr., Physical Properties and Interrelationships of Metallic and Semimetallic Elements, in *Solid State Physics*, Vol 16, F. Seitz and D.T. Turnbull, Ed., Academic Press, 1964, p 275
5. Eric A. Brandes, Ed., *Smithells Metals Reference Book*, 6th ed., Butterworths, 1983
6. R.B. Ross, *Metallic Materials Specification Handbook*, 3rd ed., E. & F.H. Spon, 1980
7. C.Y. Ho, R.W. Powell, and P.E. Liloy, *J. Phys. Chem. Ref. Data*, Vol 1, 1972, p 279-421
8. B.W. Roberts, "Superconductive Materials and Some of Their Properties," NBS Technical Note 724, National Bureau of Standards
9. V.S. Fomenko, *Handbook of Thermionic Properties–Electronic Work Functions and Richardson Constants of Elements and Compounds*, G.V. Samsonov, Ed., Plenum Press Data Division, 1966 (translation from the Russian)
10. G.A. Haas and R.E. Thomas, Thermionic Emission and Work Function, Chapter 2 in *Measurements of Physical Properties*, E. Passaglia, Ed., Vol 6, part 1, *Techniques of Metals Research*, R.F. Bunshah, Ed., Interscience, 1972, p 91
11. H.B. Michaelson, *Handbook of Chemistry and Physics*, 69th ed., R.C. Weast, Ed., CRC Press, 1988
12. Landolt-Börnstein Tables, II Band, 9. Teil, in *Magnetische Eigenschaften 1*, K.-H. Hellwege and A.M. Hellwege, Ed., Springer-Verlag, 1962
13. J.F. Parrington and F. Feiner, "Chart of the Nuclides," Knolls Atomic Power Laboratory, United States Atomic Energy Commission, Nov 1989
14. C.A. Hampel, Ed., *Rare Metals Handbook*, 2nd ed., Krieger, 1971

Berkelium (Bk)

See the section "Properties of the Transplutonium Actinide Metals (Am-Fm)" in this article.

Beryllium (Be)

Revised by J.M. Marder and A.J. Stonehouse, Brush Wellman Inc.

Unalloyed beryllium is used in weapons, spacecraft, nuclear reactor reflector segments, neutron sources, windows for x-ray tubes and radiation detection devices, rocket nozzles, aircraft brake discs, precision instruments, and mirrors.

Beryllium is used as an alloying addition to copper and nickel to produce an age-hardening alloy used for springs, electrical contacts, spot welding electrodes, and nonsparking tools. Beryllium is also added to aluminum and magnesium to achieve grain refinement and oxidation resistance.

Inhalation of respirable beryllium and its compounds should be avoided. Users should comply with the occupational safety and health standards applicable to beryllium in Title 29, Part 1910 of the Code of Federal Regulations.

Structure

Crystal structure. α phase, close-packed hexagonal; a = 0.22858 nm, c = 0.35842 nm at room temperature. β phase, body-centered cubic; a = 0.255 nm at 1270 °C
Slip planes. Primary: (0002), $(10\bar{1}0)$. Secondary: $(10\bar{1}1)$
Twinning planes. Primary: $(10\bar{1}2)$. Secondary: $(10\bar{1}\bar{1})$, $(10\bar{1}\bar{2})$, $(10\bar{1}\bar{3})$

Mass Characteristics

Atomic weight. 9.0122
Density. 1.848 g/cm^3 at 20 °C

Thermal Properties

Melting point. 1283 °C
Boiling point. 2770 °C
Phase transformation temperature. 1270 °C
Coefficient of linear thermal expansion:

Temperature range, °C	Average coefficient, µm/m · K
25–100	11.6
25–300	14.5
25–600	16.5
25–1000	18.4

Specific heat. 1.886 kJ/kg · K at 20 °C
Latent heat of fusion. 1.30 MJ/kg
Thermal conductivity. 210 W/m · K

Electrical Properties

Electrical conductivity. 38 to 43% IACS
Electrical resistivity. 40 nΩ · m at 20 °C
Temperature coefficient of electrical resistivity. 0.025 $\Delta\rho/\rho_0$/K

Magnetic Properties

Magnetic susceptibility. Mass: -0.79×10^{-9} mks at 93 K; -1.0×10^{-9} mks at 293 K; -1.2×10^{-9} mks at 573 K

Optical Properties

Color. Steel gray
Spectral hemispherical emittance. 61% for λ = 650 nm

Nuclear Properties

Thermal neutron cross section. 0.01 b

Chemical Properties

General corrosion behavior. Beryllium, a highly reactive metal, forms stable compounds with most other elements. It has excellent corrosion resistance at room temperature (except to certain acids and alkalies) because of a thin protective oxide coating.
Resistance to specific corroding agents. Beryllium reacts appreciably with oxygen and nitrogen above 760 °C. Impure beryllium containing carbide or chloride reacts with moist air at room temperature. Beryllium does not react with hydrogen at any temperature; it reacts with fluorine at room temperature and with Cl, B, I, HCl, HF, and CO_2 at elevated temperatures. Beryllium has excellent resistance to pure water at temperatures up to 300 °C if carbide and chloride are absent and if grain size is fine. Beryllium is attacked by dilute HF, HCl, H_2SO_4, and HNO_3 at room temperature.

Fabrication Characteristics

Machinability. Powder metallurgy material is readily machined to close tolerances. In machining operations, carbide tools are used. Chip formation is similar to that of cast iron. All machining, especially careless grinding, produces a damaged surface layer that must be removed by etching a minimum of 0.05 mm per surface for critical highly stressed applications.
Recrystallization temperature. 725 to 900 °C, depending on amount of cold work and annealing time
Hot-working temperature. 800 to 1100 °C

Mechanical Properties

Tensile properties. Tensile properties, espe-

Table 5 Tensile properties of beryllium

Temperature, °C	Tensile strength, MPa	Yield strength at 0.2% offset, MPa	Elongation, %
Vacuum hot-pressed block, grade S-200 F			
Room temperature	380–413(a)	262–269	2–5
400	317–331	255–262	11–26
800	234–241	186–200	45–55
1000	186–200	138–145	6–10
Hot-extruded billet			
Room temperature	483–690	310	5–20
Cross-rolled sheet			
Room temperature	483–621	345–414	10–40
Hot isostatically pressed billet (high purity), grade S-200 FH			
Room temperature	380–413	262–269	2–6

(a) Microyield stress (10^{-6} offset) at room temperature: grade I-220, 5–6 MPa; grade I-400, 9–10 MPa; grade I-250, 14 MPa

cially elongation, depend strongly on preferred orientation and grain size. Beryllium mill products are produced chiefly by powder metallurgy using different consolidation procedures. Wrought products are produced from powder metallurgy input materials. Extreme anisotropy in elongation occurs in wrought material; therefore, uniaxial tensile results are valueless as indications of behavior under conditions of complex stress. Tensile properties of different wrought forms can vary extremely, with the greatest uniformity occurring for vacuum hot-pressed and hot isostatically pressed powder. See Table 5 for some representative values.
Hardness. 75 to 85 HRB
Poisson's ratio. 0.07 to 0.075 (vacuum hot pressed)
Elastic modulus. 303 GPa
Impact strength. 1.4 to 5.4 J
Plane-strain fracture toughness. 9 to 13 $MPa\sqrt{m}$

SELECTED REFERENCES

- G.E. Darwin and J.H. Buddery, *Beryllium*, Academic Press, 1960
- D. Floyd and J. Lowe, Ed., *Beryllium Science and Technology*, Vol 2, Plenum Publishing, 1979
- H.H. Hausner, *Beryllium: Its Metallurgy and Properties*, University of California Press, 1965
- A.R. Kaufmann, P. Gordon, and D.W. Lillie, The Metallurgy of Beryllium, *Trans. ASM*, Vol 42, 1950, p 785
- D. Webster and G.J. London, Ed., *Beryllium Science and Technology*, Vol 1, Plenum Publishing, 1979

Bismuth (Bi)

Compiled by S.C. Carapella, Jr.,
ASARCO Inc.

Reviewed for this Volume by
Douglas Hayduk, ASARCO Inc.

Bismuth is used extensively in the production of fusible alloys (low-melting-point alloys) as a carbide stabilizer in the manufacture of malleable iron, and as an additive to low-carbon steel and aluminum to improve machinability. Compounds of bismuth are used for catalysts, in pharmaceutical and for semiconductor applications. Bismuth has shown potential to be an effective ingredient in the 123 copper oxide superconductive materials. However, it is not yet commercially available in this form.

Structure

Crystal structure. Hexagonal (rhombohedral equivalent); at 25 °C, a = 0.4546 nm and c = 1.1860 nm

Mass Characteristics

Atomic weight. 208.980
Density:

°C	g/cm³
25	9.808
271	1.0067
300	1.003
400	0.991
600	0.966
802	0.940
962	0.920

Thermal Properties

Melting point. 271.4 °C
Boiling point. 1564 °C
Coefficient of linear thermal expansion. 13.2 μm/m · K at 20 °C
Specific heat:

°C	kJ/kg · K
25	0.122
217.4	0.146
327	0.141
427	0.137
527	0.134

Latent heat of fusion. 53.976 kJ/kg
Latent heat of vaporization. 854.780 kJ/kg
Thermal conductivity:

°C	W/m · K
0	8.2
300	11.3
400	12.3
500	13.3
600	14.5

Vapor pressure:

°C	kPa
893	0.1013
1053	1.013
1266	10.13
1564	101.3

Electrical Properties

Electrical resistivity:

°C	nΩ · m
0	1050
300	1289
700	1535

Mechanical Properties

Hardness. 7.0 HB; 2.5 Mohs
Elastic modulus. Tension, 32 GPa
Liquid surface tension:

°C	mN/m
300	376
400	370
500	363

Magnetic Properties

Magnetic susceptibility. Volume: -1.68×10^{-5} mks

Nuclear Properties

Stable isotopes. ^{209}Bi
Thermal neutron cross section. For 2.2 km · s neutrons: absorption, 0.034 ± 0.002 b; scattering, 9 ± 1 b

Boron (B)

Compiled by James C. Schaefer,
JCS Consulting

Elemental boron can be prepared by several methods. The purest forms are black and are prepared by the reduction of boron halides on a hot tungsten wire. Electrolytic boron, prepared by molten salt electrolysis, is a black powder of 40 to 325 mesh. Electrolytic boron has a purity of 99% and above. Magnesium-reduced boron, which is prepared by the reduction of boric oxide (B_2O_3) with magnesium, is a light brown powder in the purity range of 95 to 97%.

Boron formed on hot tungsten wire is used for weight reduction and reinforcement of metals and plastics. Electrolytic boron powder is used for the preparation of borides for the deoxidation of alloys.

Boron compounds are used for medicinal and cleaning purposes. Traces of boron are beneficial to plant growth, but large concentrations are toxic. Boron compounds are added to aluminum alloys to improve electrical and thermal conductivity and for microstructure grain refining. Boron compounds have been used in rocket fuels and as diamond substitutes. Isotropic B-10 and its compounds are used for neutron absorption.

Elemental boron is nontoxic. Boron in the form of fine dust will slowly oxidize and should be kept under an inert gas. The dust is abrasive, and personnel should be protected from the dust. Normal air filtration and facial mask procedures should be followed. Metallic borides require no handling or storage precautions. Special precautions must be taken with the boron halides. Boron halides are very sensitive to shock and can detonate easily. The halides are toxic and corrosive.

Structure

Crystal structure. Material prepared at about 800 °C and below: amorphous. Prepared between about 800 and 1100 °C: α phase, rhombohedral, $R\bar{3}m$; a = 0.506 nm, α = 58°4′; unit cell contains a single B_{12} icosahedron. Prepared between about 1100 and 1300 °C: γ phase, tetragonal, $P4_2/nnm$; a = 0.875 nm, c = 0.506 nm; the 50 atoms per unit cell are distributed among four equivalent B_{12} icosahedrons of required symmetry $2/m$ and two tetrahedral positions $42m$. Prepared above about 1300 °C: β phase, rhombohedral, $R\bar{3}m$; a = 1.012 nm, α = 65°28′; unit cell contains approximately 108 atoms. Values differ slightly according to investigator.

Mass Characteristics

Atomic weight. 10.81
Density. Amorphous, 2.3 g/cm³ at <800 °C;

α phase, 2.46 g/cm^3 at 800 to 1100 °C; γ phase, 2.37 g/cm^3 at 1100 to 1300 °C; β phase, 2.35 g/cm^3 at >1300 °C

Thermal Properties

Melting point. Approximately 2300 °C
Boiling point. Approximately 2550 °C
Phase transformation temperature. Unknown
Coefficient of linear thermal expansion. 1.1 to 8.3 μm/m · K in temperature range from 20 to 750 °C
Specific heat:

K	°C	kJ/kg · K
82 to 195	−191 to −78	0.0297
197 to 273	−76 to 0	0.754
273 to 373	0 to 100	1.285
373	100	1.620
773	500	1.976
1173	900	2.135

Enthalpy:

K	°C	kJ/kg
400	127	120
600	327	416
800	527	786
1000	727	1200

Entropy. 604 J/kg · K
Latent heat of fusion. 22 000 kJ/kg
Latent heat of vaporization. 34 900 kJ/kg
Heat of combustion. 5.4 J/kg

Electrical Properties

Electrical resistivity:

K	°C	Resistivity
123	−150	4×10^5
263	−10	4×10^4
273	0	3×10^4
300	27	6.5×10^3
373	100	4×10^2
443	170	30
593	320	0.4×10^{-1}
793	520	1.2×10^{-2}
873	600	2×10^{-3}

Electrochemical equivalent. 37 μg/C
Standard electrode potential. At 25 °C, 0.87 V versus standard hydrogen electrode
Ionization potentials:

Degree of ionization	Potential, eV
I	8.296
II	23.98
III	37.75
IV	258.1
V	338

Semiconductor properties. *p*-type dopant for silicon and germanium. Intrinsic current carrier concentration, 5×10^{20} per m^3 at 160 °C to 9×10^{25} per m^3 at 850 °C
Dielectric constant. Approximately 12

Optical Properties

Color. Crystalline is black; amorphous is brown
Refractive index. 2.5 using mercury line (579 nm)

Nuclear Properties

Stable isotopes. B-10, atomic weight 10.01294, 19.9% abundant; B-11, atomic weight 11.00931, 80.1% abundant
Neutron absorption. B-10, 3850 b; B-11, 0.05 b; natural boron, 755 b
Unstable isotopes:

Isotope	Atomic weight	Half life, s	Particles emitted
^{8}B	...	0.78	β^+
^{9}B	9.01333	3×10^{-19}	P, (2α)
^{12}B	...	0.019	β^-
^{13}B	...	0.035	β^-

Chemical Properties

Effects of specific corroding agents. Reactivities and conditions for reaction of boron with several materials are:

- Fluorine, instantaneous at room temperature
- Chlorine, above 500 °C
- Bromine, above 600 °C
- Iodine, about 900 °C
- Hydrochloric acid, none
- Hydrofluoric acid, none
- Nitric acid (hot, concentrated), slow
- Oxygen, slight at room temperature, rapid above 1000 °C
- Hydrogen iodide, explosive
- Hydrogen, above 840 °C
- Nitrogen, bright red heat
- Sodium hydroxide, no reaction at room temperature, slow at 500 °C
- Boron nitride, none
- Metals, *caution*: above 900 °C, many metals react rapidly with boron, and the reactions are exothermic

Mechanical Properties

Tensile properties. Tensile strength: 98.8% pure, amorphous, 1.6 to 2.4 GPa; fibers, 2.6 to 3.1 GPa
Compressive properties. Compressive strength: with B_2O_3 present, up to 0.5 GPa
Hardness. 99.9% crystalline: 3300 HK (with 100 g load), 9.3 moh
Elastic modulus. Tension: amorphous, 440 GPa

Cadmium (Cd)

Compiled by Hugh Morrow,
Cadmium Council, Inc.

Cadmium is a relatively rare metal that is present in the crust of the earth at an average level of only 0.00005%, or 0.5 ppm. It is a soft and malleable silvery-white metal, and it is most often found in nature as the sulfide associated with zinc sulfide ores. Cadmium is, in fact, produced as a by-product of zinc-refining operations.

Metallic cadmium is used for corrosion-resistant coatings, mainly on iron and steel, but also on aluminum and magnesium. It is a minor but important alloying element in certain brazing and soldering alloys and in some copper-base alloys. Cadmium compounds (mainly cadmium oxide) are used in the manufacture of nickel-cadmium batteries; cadmium sulfides and organic cadmium salts are used as pigments or stabilizer additives to certain types of engineering plastics. At present, cadmium consumption in the western world can be roughly divided into these areas of application:

Use	%
Batteries	45
Coatings	16
Pigments	20
Stabilizers	14
Alloys and other	5

The appropriate ASTM specifications for cadmium products and processing are:

ASTM standard	Description
B 440	Standard specification for cadmium
B 201	Testing chromate coatings on zinc and cadmium surfaces
B 766	Electrodeposited coatings of cadmium
B 696	Coatings of cadmium mechanically deposited
B 635	Coatings of cadmium-tin mechanically deposited
B 699	Coatings of cadmium vacuum-deposited on iron and steel products

Cadmium fumes and dust can be toxic if inhaled in sufficient quantity. In September 1989, the Occupational Safety and Health Administration (OSHA) issued an interim standard, designated PUB 8-1.4A, that recommends a cadmium exposure level of no more than 50 μg/m^3, measured as an 8-h time-weighted average. Therefore, any occupational situation that generates cadmium fumes and/or dust should be well ventilated to remove these products. Cadmium compounds such as cadmium sulfide are equally regulated, but the solubility of these compounds is much less than that of other compounds such as cadmium chloride; therefore, they should pose much less of a health risk.

Structure

Crystal structure. Close-packed hexagonal, D^4_6h ($P6_3/mmc$); a = 0.29793 nm and c = 0.56181 nm at 26 °C
Slip plane and direction. (0001), [1120]
Twinning plane. (1012)
Distance of closest approach. 0.2973 nm

Mass Characteristics

Atomic weight. 112.40

Density:

°C	g/cm³
26	8.642
330 (liquid)	8.020
400	7.930
600	7.720

Volume contraction on freezing. 4.74%

Thermal Properties

Melting point. 321.1 °C
Boiling point. 767 °C
Coefficient of linear thermal expansion. 31.3 μm/m · K at 20 °C
Specific heat:

Temperature, °C	State	Specific heat, kJ/kg · K
−272	Solid	8×10^{-6}
−263	Solid	0.008
−253	Solid	0.046
−243	Solid	0.086
−233	Solid	0.117
−223	Solid	0.141
−213	Solid	0.159
−203	Solid	0.172
−193	Solid	0.182
−183	Solid	0.190
−173	Solid	0.196
20	Solid	0.230
100	Solid	0.239
300	Solid	0.260
321	Liquid	0.264
400	Liquid	0.264
500	Liquid	0.264
600	Liquid	0.264

Latent heat of fusion. 55 kJ/kg
Latent heat of vaporization. 887 kJ/kg
Thermal conductivity:

Temperature		Conductivity,
K	°C	W/m · K
100	−173	103 (solid)
200	−73	99 (solid)
273	0	97.5 (solid)
300	27	96.8 (solid)
400	127	94.7 (solid)
500	227	92.0 (solid)
600	327	42.0 (liquid)
700	427	49.0 (liquid)
800	527	55.9 (liquid)

Vapor pressure:

	Vapor pressure	
Temperature, C	mm Hg	kPa
148	10^{-5}	1.33×10^{-6}
180	10^{-4}	1.33×10^{-5}
220	10^{-3}	1.33×10^{-4}
264	10^{-2}	1.33×10^{-3}
321	10^{-1}	1.33×10^{-2}
394	1	0.133
484	10	1.33
563	50	6.66
610	100	13.33
711	400	53.32
767	760	101.3
830	1 520	202.6
930	3 800	506.5
1030	7 600	1013.0
1120	15 200	2026.0
1240	30 400	4052.0

Electrical Properties

Electrical conductivity. Volumetric, 25% IACS at 20 °C
Electrical resistivity:

State	Temperature, °C	Resistivity, nΩ · m
Solid	−252.9	1.7
Solid	−200	16.6
Solid	−100	48.0
Solid	0	68.3
Solid	18	75.0–75.4
Solid	20	73
Solid	22	72.7
Solid	100	96.0–98.2
Solid	300	165–180
Liquid	321	337
Liquid	400	337–343
Liquid	500	351.0–351.2
Liquid	600	348–360.7
Liquid	700	357.8–358.0

Electrochemical equivalent. Valence +2, 582.4 Mg/C
Electrode reduction potential. 0.40 V for the reaction $Cd = Cd^{2+} + 2e^-$, where potential for $H_2 = 0.0$ V
Temperature coefficient of electrical resistivity. 4.2×10^{-3} per °C at 0 °C; 4.24 to 4.3×10^{-3} per °C

Magnetic Properties

Magnetic susceptibility. Volume: -2.2×10^{-6} mks
Superconductivity. Critical temperature (T_c) of 0.54 K; critical magnetic field (H_c) of 300 kT (3000 MG)

Optical Properties

Color. Silver-gray
Spectral reflectance:

λ, nm	%
410	78
474	74
518	72.5
554	73

Refractive index. 1.8 at λ = 578 nm
Absorptive index. 1.17 at λ = 578 nm

Nuclear Properties

Stable isotopes:

Atomic weight	Relative abundance, %
106	1.22
108	0.88
110	12.39
111	12.75
112	24.07
113	12.26
114	28.86
116	7.58

Thermal neutron cross section. At 2200 m/s: absorption 2450 b; scattering 7 ± 1 b

Chemical Properties

General corrosion behavior. The chemical properties of cadmium resemble those of zinc, especially under reducing conditions and in covalent compounds. In oxides, fluorides, and carbonates and under oxidizing conditions, cadmium may behave similarly to calcium. It also forms a relatively large number of complex ions with other ions such as ammonia, cyanide, and chloride. Cadmium is a fairly reactive metal. It dissolves slowly in dilute hydrochloric or sulfuric acids, but dissolves rapidly in hot dilute nitric acid. Other elements that react readily with cadmium metal when heated include the halogens, phosphorus, sulfur, selenium, and tellurium.

Unlike zinc, cadmium is not markedly amphoteric, and cadmium hydroxide is virtually insoluble in alkaline media. Like zinc, however, cadmium forms a protective oxide that reduces its corrosion rate in atmospheric service. Both metals exhibit low corrosion rates over the range from approximately pH 5 to 10. In more acid or more alkaline environments, their corrosion rates increase dramatically. Cadmium is preferred for marine or alkaline service, whereas zinc is often as good or better in heavy industrial exposures.

As a protective coating on steel and cast iron parts, cadmium offers corrosion protection in marine atmospheres, under alkaline conditions, and in damp indoor applications. Cadmium-plated steel fasteners resist galvanic attack when used with aluminum parts.

Mechanical Properties

Tensile strength. 69 to 83 MPa
Elongation. 50% in 25 mm
Hardness. 16 to 24 HB
Poisson's ratio. 0.33 at room temperature
Elastic modulus. Tension, 55 GPa; shear, 19.2 GPa
Liquid surface tension. 0.564 N/m at 330 °C; 0.611 N/m at 450 °C

Calcium (Ca)

Compiled by J.F. Smith, Ames Laboratory, U.S. Department of Energy, Iowa State University

Metallic calcium is used as a reducing agent in the preparation of thorium, zirconium, uranium, chromium, vanadium, and the rare earths. It is also used as a deoxider, decarburizer, or desulfurizer for various ferrous and nonferrous alloys. Calcium is used as an alloying or modifying agent for aluminum, beryllium, copper, lead, tin, and magnesium alloys. Other uses for calcium include getters for residual gases in high vacuums and vacuum tube applications, and reagents for purification and scavenging of inert gases. Calcium reacts readily with atmospheric components, particularly water vapor, and is not inert to nitrogen. To avoid contamination, it must be handled in a dry inert-gas atmosphere or in a vacuum.

Structure

Crystal structure. α-phase, face-centered cubic, *cF*4 (*Fm*3*m*); *a* = 0.5588 nm at 26.6

°C. β-phase, body-centered cubic, *cI2* (*Im3m*); a = 0.4480 nm at 467 °C. Minor amounts of hydrogen stabilize a hexagonal form, and a low-symmetry form of undetermined structure results from contamination by nitrogen and/or carbon.
Minimum interatomic distance. 0.3952 nm at 25 °C

Mass Characteristics

Atomic weight. 40.08
Density. Solid, 1.55 g/cm^3 at 25 °C; liquid, 1.37 g/cm^3 at 842 °C
Density versus temperature. Solid, $\Delta d/d_0 \cdot K = -66.9 \times 10^{-6}$ at 0 to 400 °C; liquid, $\Delta d/d_0 \cdot K = -221 \times 10^{-6}$ at 842 to 1382 °C
Volume change on freezing. 4.7% contraction
Volume change on phase transformation. β to α phase, 0.04% contraction

Thermal Properties

Melting point. 842 °C at 1 atm; $\Delta T_m/\Delta P$ = 170 μK/Pa
Boiling point. 1495 °C
Phase transformation temperature. 443 °C; $\Delta T_{trans}/\Delta P$ = 33 μK/Pa
Coefficient of linear thermal expansion. α phase: up to −267 °C, $\Delta l/l_0 \cdot K = 5 \times 10^{-11} T + 81 \times 10^{-12} T^3$, where T is in K; 1 μm/m · K at −253 °C; 3.3 μm/m · K at −243 °C; 14.16 μm/m · K at −198 °C; 22.15 μm/m · K at 10 °C; 22.3 μm/m · K (average) for 0 to 400 °C. β phase: 33.6 μm/m · K for 467 to 603 °C. Liquid: 73.3 μm/m · K for 839 to 1382 °C
Specific heat:

°C	kJ/kg · K
25	0.6315
127	0.6549
327	0.7375
448 (fcc)	0.8079
448 (bcc)	0.7320
627	0.9174
839 (bcc)	1.136
839 (liquid)	0.7308
1027	0.7308

Enthalpy. $H_{298} - H_0$ = 142.4 kJ/kg; entropy, S_{298} = 1.03 kJ/kg · K
Heat of fusion. 213.1 kJ/kg
Latent heat of transformation. α to β phase, 23.20 kJ/kg
Latent heat of sublimation. 4.447 MJ/kg at 25 °C
Heat and free energy of formation of oxide. CaO: ΔH^0_{298} = −11.32 MJ/kg Ca; ΔG^0_{298} = −10.77 MJ/kg Ca
Thermal conductivity. W/m · K: $\lambda = 190 - 0.22\,T$ from 150 to 360 °C, $\lambda = 0.31\,T - 1.5$, from 360 to 600 °C, where T is in K
Vapor pressure. α phase, $\log P = 10.77 \times 9260/T$; β phase, $\log P = 10.38 \times 8980/T$; liquid, $\log P = 9.67 - 8190/T$, where T is in K and P is in Pa
Diffusion coefficients. At 500 to 800 °C (930 to 1470 °F):

Element	D_0, m^2/s	H, kJ/mol
Ca	8.3×10^{-4}	161
C	2.7×10^{-7}	97.5
Fe	3.2×10^{-9}	125
Ni	1.0×10^{-9}	121
U	1.1×10^{-9}	146

Electrical Properties

Electrical conductivity. Volumetric, 49.6% IACS
Electrical resistivity. α phase, 31.6 nΩ · m at 0 °C; liquid, 330 nΩ · m at 839 °C
Temperature coefficient of electrical resistivity. 4.02×10^{-3} per K at 0 °C
Pressure dependence of electrical resistivity. Unusual and currently believed to result from a manifestation of the electronic band structure, which has a degree of overlap that has been calculated to be highly sensitive to interatomic spacing. See Fig. 4.
Thermoelectric potential. Versus Cu: β phase, 9.6 μV/K at the melting point; liquid, 9.9 μV/K at the melting point
Electrochemical equivalent. 0.20762 mg/C
Electrolytic solution potential. Versus H_2, −2.87 V at 20 °C
First ionization potential. 6.11 eV
Hall coefficient. −0.228 nV · m/A · T (independent of temperature from −193 to 27 °C)
Work function. Thermionic, 0.359 aJ; photoelectric, 0.46 aJ

Magnetic Properties

Magnetic susceptibility. Volume: 2.71×10^{-5} mks
Magnetic permeability. 1.0000271

Optical Properties

Color. A fresh clean surface is a lustrous silvery white but darkens upon exposure to the atmosphere.
Dielectric constant. Ellipsometry has been used to determine the real (ϵ_1) and imaginary (ϵ_2) parts of the complex dielectric constant as a function of vacuum wavelength (λ):

Wavelength (λ), nm	$-\epsilon_1$	ϵ_2
1771.2	65.2	...
1549.8	51.8	11.7
1377.6	41.8	8.58
1239.8	34.2	7.08
1033.2	24.2	4.71
885.6	17.5	3.40
774.9	12.8	2.48
688.8	9.6	1.95
619.9	7.5	1.62
563.6	5.8	1.39
516.6	4.5	1.22
476.9	3.6	1.09
442.8	2.8	1.01
413.3	2.2	1.01
364.7	1.4	1.02
326.3	...	0.95

Reflectivity. At normal incidence (r), the refractive index (n) and adsorptive index (A) may be generated from these quantities

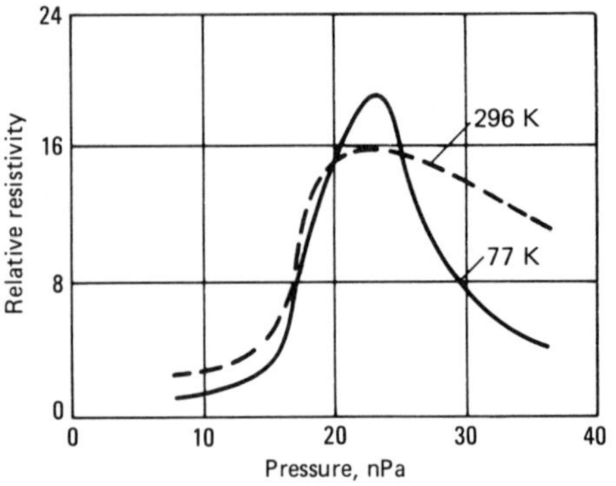

Fig. 4 Pressure dependence of the electrical resistivity of calcium. Relative resistivity is the ratio of the resistivity at high pressure to the resistivity at the same temperature, but at 1 atm pressure.

through the following relations: $\epsilon_2 = 2nK$; $\epsilon_1 = n^2 + K^2$; $A = 4\pi K/\lambda$; $r = [(n-1)^2 + K^2]/[(n+1)^2 + K^2]$

Nuclear Properties

Stable isotopes:

Mass number	Abundance, %
40	96.97
42	0.64
43	0.145
44	2.06
46	0.0033
48	0.185

Unstable isotopes. Isotopes with mass numbers 37, 38, 39, 41, 45, 47, 49, and 50 have been produced. Number 45 has the longest half-life (180 days); it is used in tracer experiments.

Chemical Properties

General corrosion behavior. Extremely poor corrosion resistance

Fabrication Characteristics

Recrystallization temperature. Below 300 °C; even at room temperature, x-ray diffraction patterns show no broadening or distortion after extensive deformation.

Mechanical Properties

Tensile properties. Annealed: tensile strength, 48.0 MPa; yield strength, 13.7 MPa; elongation, 51 to 53%; reduction in area, 58 to 62%. As rolled: tensile strength, 115 MPa; yield strength, 84.8 MPa; elongation, 7%; reduction in area, 35%
Hardness. Annealed: 16 to 18 HB
Poisson's ratio. 0.31
Elastic modulus. Tension, 19.6 GPa; shear, 7.38 GPa; bulk, 15.2 GPa
Compressibility. For 0 to 3900 MPa, $\Delta V/V_0$ Pa = $-6.578 \times 10^{-5} + 7.732 \times 10^{-11} P - 4.9 \times 10^{-13} P^2$, where P is in MPa
Liquid surface tension. For 839 to 1000 °C, $\gamma = 0.472 - 10^{-4}\,T$, where T is in K and γ is in N/m

Table 6 Selected physical and thermal properties of cesium

Temperature, °C	Coefficient of linear thermal expansion (α), 10^{-6}/°C	$1/\alpha^2[\partial\alpha/\partial T]_{p=0}$	Specific heat, kJ/kg · K	Bulk modulus (K), GPa
49.7	284	1.53	0.267	1.57
71.8	287	1.46	0.263	1.53
89.4	289	1.41	0.261	1.50
108.3	291	1.36	0.258	1.47
130.9	...	...	...	1.42
150.1	...	...	...	1.39

Sources: Ref 7, 8

Californium (Cf)

See the section "Properties of the Transplutonium Actinide Metals (Am-Fm)" in this article.

Cerium (Ce)

See the section "Properties of the Rare Earth Metals" in this article.

Cesium (Cs)

Compiled by John H. Madaus, Callery Chemical Company

Revised by Michael Stevens, Los Alamos National Laboratory

Cesium ignites immediately on contact with air if poured or sprayed, and it reacts explosively with water. Cesium may form peroxide compounds if allowed to oxidize in the absence of water normally present in atmosphere. The resulting peroxides may be shock sensitive with easily reduced compounds in the same manner that potassium superoxide is shock sensitive with mineral oil. Cesium metal must be contained under vacuum, inert gas, or anhydrous liquid hydrocarbons and protected from oxygen or air exposure. Safety and handling information is available from suppliers.

Because it produces electromagnetic energy that is accurate and stable in frequency, cesium is used in electronic devices and atomic clocks, providing accuracy of 5 s in 10 generations.

Structure

Crystal structure. Body-centered cubic; *Im*3*m*, *cI*2; a = 0.613 nm at −10 °C

Mass Characteristics

Atomic weight. 132.9054
Density. 1.903 g/cm³ at 0 °C, 1.892 g/cm³ at 18 °C (Ref 1), 1.827 g/cm³ at 40 °C (liquid) (Ref 2)

Thermal Properties

Melting point. 28.64 ± 0.17 °C (Ref 3)
Boiling point. 670 °C (Ref 4, 5)
Specific heat. 0.2016 kJ/kg · K at 20 °C (solid), 0.2395 kJ/kg · K at 670 °C (liquid) (Ref 6), 0.1557 kJ/kg · K at 670 °C (vapor) (Ref 6). See also Table 6.
Heat of fusion. 16.38 kJ/kg (Ref 3)
Heat of vaporization. 611.3 kJ/kg (Ref 6)
Thermal conductivity. 18.42 W/m · K at 28.64 °C (liquid) (Ref 9), 4.6 × 10^{-3} W/m · K at 670 °C (vapor) (Ref 6)
Coefficient of linear thermal expansion. See Table 6.
Vapor pressure. From −23 to 28.64 °C (solid):

$$\log P = \frac{-4120}{T} - 1.0 \log T + 8.32$$

From 28.64 to 377 °C (liquid):

$$\log P = \frac{-4042}{T} - 1.4 \log T + 9.05$$

From 1500 to 1780 °C:

$$\ln P = -61.09 + 4025 \left(\frac{1}{T}\right) + 10.07 \ln T - (3.51 \times 10^{-12})T^{3.5}$$

where P is in Pa and T is in K (Ref 10, 11)
Critical pressure. 11.75 MPa (Ref 11)
Critical temperature. 1779 °C (Ref 11)
Critical compressibility. 0.206 (Ref 11)

Electrical Properties

Electrical conductivity. 4.5 MS/m at 28.64 °C (solid)
Electrical resistivity. 200 nΩ · m at 20 °C
First ionization potential. 3.893 eV (Ref 6)

Magnetic Properties

Magnetic susceptibility. Volume: 167 × 10^{-6} mks (Ref 9)

Optical Properties

Color. 99.99% pure material is bronze colored.

Mechanical Properties

Viscosity. 0.686 mPa · s at 28.64 °C (Ref 6)
Surface tension. 0.0394 N/m at 28.64 °C (Ref 6)

REFERENCES

1. L. Losana, *Gazz. Chim. Ital.*, Vol 65, 1935, p 855
2. M. Eckardt and E. Graefe, *Zeitschrift Anorg. Allg. Chem.*, Vol 23, 1900, p 385
3. K. Clusius and H. Stern, *Zeitschrift Angew. Phys.*, Vol 6, 1954, p 194; *Chem. Abstr.*, Vol 48, 1954, p 6869a
4. O. Ruff and and O. Johannsen, *Chem. Ber.*, Vol 38, 1905, p 3608
5. Cesium, in *Gmelins Handbuch der Anorganischen Chemie*, Vol 25, 8th ed., Verlag Chemie, 1955
6. W.D. Weatherford, Jr., J.C. Tyler, and P.M. Ku, "Properties of Inorganic Fluids and Coolants for Space Applications," WADC Tech Report 59-598, Southwest Research Institute, 1959
7. M.G. Kim, K.A. Kemp, S.V. Letcher, *J. Acoust. Soc. Am.*, Vol 49, 1971, p 706
8. G.H. Shaw, D.A. Caldwell, *Phys. Rev. B, Condens. Matter*, Vol 32 (No. 12), 1985, p 7937
9. C.A. Hampel, Rubidium and Cesium, in *Rare Metals Handbook*, 2nd ed., Reinhold, 1961, p 434-440
10. J.W. Mellor, *Comprehensive Treatise on Inorganic and Theoretical Chemistry*, Vol 2, Supplement 3, John Wiley & Sons, 1963
11. I.L. Silver, Ph.D. thesis, Columbia University, 1968

Chromium (Cr)

Compiled by H.C. Aufderhaar, Union Carbide Corporation

Revised by F.H. Perfect, Reading Alloys, Inc.

Elemental chromium (also known as chrome metal or electrolytic chromium) is finding use in commercial quantities as an alloying element in titanium alloys. Earlier chromium alloys contained aluminum and were produced directly from chromium-bearing master alloys (Ref 1 to 3). Nitrogen is believed to be the element most detrimental to ductility in chromium metal (Ref 3 to 7).

About 98% of the total U.S. consumption of chromium is used in the form of ferrochromium for the production of stainless steels and heat-resistant alloys. Total chromium consumption in the first five months of 1989 was 96 281 Mg (metric tons). This total was consumed in the following forms (Ref 8):

Raw material	Consumption, %
High-carbon ferrochrome	88
Low-carbon ferrochrome	7
Ferrochrome-silicon	3
Chromium metal	2

The 2% (1849 Mg) consumed in the form of chromium metal was used for the following products:

Product	Consumption, %
Superalloys	55
Welding rods	22
Aluminum alloys	19
Miscellaneous	4

Chromium controls the microstructure in aluminum and copper. Chromizing is an application that takes advantage of the excellent corrosion resistance of chromium; chrome plating is the most widely known chromizing application. Table 7 lists typical compositions obtained by two processes used to produce commercially pure chromium metal.

Table 7 Nominal compositions of commercially pure chromium metal produced by two processes

Process	Composition, %								
	Al	C	Cr	Fe	N	O	P	Si	S
Electrolytic	0.015	0.05	99.50	0.35	0.02	0.05	0.01	0.04	0.01
Aluminothermic	0.10	0.05	99.40	0.20	0.02	0.10	0.01	0.10	0.01

Structure

Crystal structure. Body-centered cubic, *cI2* (*Im*3*m*); a = 0.28844 to 0.28848 nm at 20 °C (Ref 9-11). Above 1840 °C: face-centered cubic; a = 0.38 nm (Ref 9)

Mass Characteristics

Atomic weight. 51.996
Density. 7.19 g/cm^3 at 20 °C (Ref 10, 12)

Thermal Properties

Melting point. 1875 °C (Ref 10-12)
Boiling point. 2680 °C (Ref 10)
Phase transformation temperature. 1840 °C (Ref 9)
Coefficient of linear thermal expansion. 6.2 μm/m · K (Ref 10)
Specific heat. 0.4598 kJ/kg · K at 20 °C (Ref 10, 11)
Entropy:

Temperature, °C	Entropy, kJ/kg · K	
	Solid	Gas
25	0.46	3.35
227	0.71	3.56
727	1.09	3.83
1227	1.36	4.00
1727	1.59	4.13
2227	...	4.24
2727	...	4.31

Source: Ref 9

Latent heat of fusion. 258 to 283 kJ/kg (Ref 10, 11)
Latent heat of vaporization. 6168 kJ/kg (Ref 10, 11)
Thermal conductivity. 67 W/m · K at 20 °C, 76 W/m · K at 426 °C, 67 W/m · K at 760 °C (Ref 9)
Vapor pressure. From Ref 9:

°C	Pa
965	3.2×10^{-4}
1093	2.8×10^{-3}
1197	2.7×10^{-2}
1288	2.4×10^{-1}
1875	9.9×10^{2}

Electrical Properties

Electrical conductivity. 13% IACS at 20 °C (Ref 10-12)
Electrical resistivity. From Ref 9:

°C	nΩ · m
−260	5
20	130
152	180
200	200
407	310
600	400
652	470
1000	660

Table 8 Tensile properties of recrystallized, swaged, arc cast electrolytic chromium

Recrystallized at 1200 °C in hydrogen. Strain rate of testing, 0.017 m/m per min

Temperature, °C	Tensile strength, MPa	Proportional limit, MPa	Elongation in 25 mm, %	Reduction in area, %	Modulus of elasticity, GPa
20	83	...	0	0	248
200	234	...	0	0	...
300	154(a)	11.7	3	4	290
350	197	105	6	8	168
400	225(b)	132	51	89	227
500	...	...	30	75	...
600	242	69	42	81	200
700	203	...	33	85	...
800	180	97	47	92	255

(a) Yield strength, 131 MPa. (b) Yield strength, 140 MPa. Source: Ref 11

Temperature coefficient of electrical resistivity. At 0 °C, 0.03 nΩ · m per K (Ref 9)
Electrochemical equivalent. Valence 3, 0.17965 mg/C; valence 6, 0.08983 mg/C (Ref 11)
Electrolytic solution potential. For valence 3, 0.5 V versus hydrogen electrode (Ref 13)
Hydrogen overvoltage. 0.38 V (Ref 13)
Temperature of superconductivity. 0.08 K (Ref 11)
Work function. 0.7337 aJ (Ref 9)

Magnetic Properties

Magnetic susceptibility. Volume: 4.5×10^{-5} mks (Ref 10)

Optical Properties

Color. Steel gray
Reflectance. 67% at λ = 300 nm; 63% at λ = 1000 nm; 70% at λ = 500 nm; 88% at λ = 4000 nm (Ref 10)
Refractive index. 1.64 to 3.28 for λ from 257 to 608 nm (Ref 10)
Absorptive index. 3.69 to 4.30 for λ from 257 to 608 nm (Ref 14)

Nuclear Properties

Stable isotopes. ^{50}Cr, 4.31% abundant; ^{52}Cr, 83.76% abundant; ^{53}Cr, 9.55% abundant; ^{54}Cr, 2.38% abundant (Ref 9)
Unstable isotopes. From Ref 9:

Isotope number	Half-life
48	23-24 h
49	41.7-41.9 min
51	27.5-27.9 days
55	3.52-3.6 min
56	5.9 min

Chemical Properties

Resistance to specific corroding agents. (A 10% solution at 12 °C was used unless otherwise noted.) Chromium is resistant to the following acids: acetic, aqua regia, benzoic (saturated), butyric, carbonic, citric, fatty, formic, hydrobromic, hydroiodic, lactic, nitric, oleic, oxalic, palmitic, phosphoric, picric, salicylic, stearic, and tartaric. Chromium is not resistant to hydrochloric acid or other halogen acids.

Chromium is resistant to the following agents: acetone, air, ethyl and methyl alcohol, higher alcohols, aluminum chloride, aluminum sulfate, ammonia, ammonium chloride, barium chloride, beer, benzyl chloride (saturated and 100%), calcium chloride, carbon dioxide, carbon disulfide, carbon tetrachloride (saturated and 100%), dry chlorine, chlorobenzene (saturated and 100%), chloroform, copper sulfate, ferric chloride, ferrous chloride, foodstuffs, formaldehyde, fruit products, glue, hydrogen sulfide (100%), magnesium chloride, milk, mineral oils, motor fuels, crude petroleum products, phenols, photographic solutions, printing ink, sodium carbonate, sodium chloride, sodium hydroxide, sodium sulfate, sugar, sulfur (100%), sulfur dioxide (100%), chlorinated water, distilled water, rainwater, zinc chloride, and zinc sulfate (Ref 9, 11, 15, 16)

Mechanical Properties

Tensile properties. Iodide chromium at room temperature, as-swaged: tensile strength, 413 MPa; 0.2% yield strength, 362 MPa; elongation, 44%; reduction in area, 78% (Ref 10, 17). Iodide chromium at room temperature, swaged and recrystallized: tensile strength, 282 MPa; elongation, 0%; reduction in area, 0% (Ref 10, 17). Electrolytic chromium, see Table 8.
Hardness. As-cast, forged: room temperature, 125 HB; 700 °C, 70 HB. Electrodeposited, annealed: 500 to 1250 HB, depending on the amount of hydrogen in the deposit. Electrodeposited and annealed: 70 to 90 HB. Extruded, annealed at 1100 °C: 110

HV. Extruded, annealed, rolled at 400 °C: 160 HV (Ref 9)
Elastic modulus. Tension, see Table 8.
Impact strength. Unnotched Charpy, as-arc-cast electrolytic chromium: room temperature, 2 J; 400 °C, 160 J (Ref 11)

REFERENCES

1. E.F. Erbin *et al.*, "-Ti-13V-11Cr-3Al All Beta Alloy," Paper presented at the 6th Annual Titanium Metallurgical Conference (New York, NY), Sept 1960
2. P.J. Bania *et al.*, *Beta Titanium Alloys in the 1980's*, The Metallurgical Society, 1984, p 209
3. A.H. Sully *et al.*, *Chromium*, 2nd ed., Plenum Publishing, 1967
4. H.L. Wain *et al.*, *J. Inst. Met.*, Vol 83, 1952-53, p 585-598
5. W.H. Smith *et al.*, Ductile Chromium, *J. Electrochem. Soc.*, Vol 103, 1956, p 347
6. H.L. Wain *et al.*, *J. Inst. Met.*, Vol 86, 1957-58, p 281-288
7. P.M. Gruzensky *et al.*, Report 5305, U.S. Bureau of Mines, 1957
8. J.F. Papp, *Miner. Ind. Surv.*, *Chromium*, May 1989
9. J.C. Bailar *et al.*, Ed., *Comprehensive Inorganic Chemistry*, Vol 3, Pergamon Press, 1973, p 624
10. H F. Mark *et al.*, Ed., *Encyclopedia of Chemical Technology*, Vol 6, 3rd ed., John Wiley & Sons, 1979, p 54
11. C.A. Hampel, Ed., *The Encyclopedia of Chemical Elements*, Reinhold, 1968, p 145
12. C.J. Smithells, Ed., *Metals Reference Book*, Vol III, Plenum Publishing, 1967, p 685
13. H.S. Taylor, *Treatise on Physical Chemistry*, Vol 1, 1931, p 354
14. Freederickaz, *Ann. Phys.*, Vol 34, 1911, p 780
15. J.E. Hosdowich, *Mater. Methods*, Vol 24, 1946, p 896
16. McKay and Worthington, *Corrosion Resistance of Metals and Alloys*, Reinhold, 1936
17. Sully, Brandeis, and Mitchell, *J. Inst. Met.*, Vol 81, 1952-53, p 585

Cobalt (Co)

Compiled by D.J. Maykuth, Metals and Ceramics Information Center, Battelle Memorial Institute

Revised by M.J.H. Ruscoe, Sherritt Gordon Ltd.

Cobalt is used as an alloying element in:

- Permanent and soft magnetic materials
- High-temperature creep-resistant superalloys
- Hardfacing and wear-resistant alloys
- High-speed steels, tool steels, and other steels

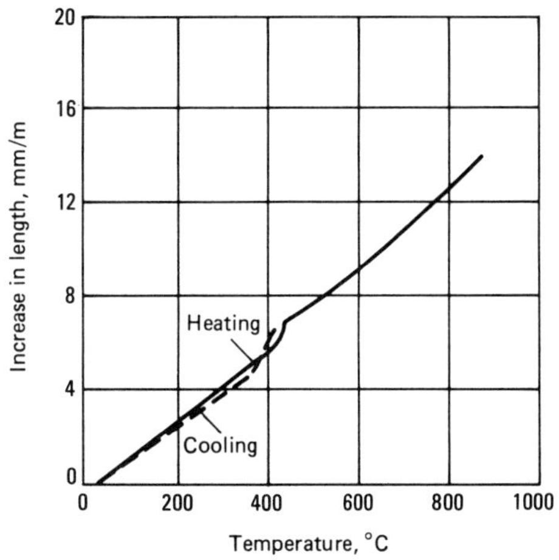

Fig. 5 Linear thermal expansion of cobalt (relative to 30 °C)

- Cobalt-base tool materials
- Electrical-resistant alloys
- High-temperature spring and bearing alloys
- Magnetostrictive alloys
- Special expansion and constant-modulus alloys

Structure

Crystal structure. α phase, close-packed hexagonal, *hP*2 ($P6_3/mmc$); a = 0.25071 nm, c = 0.40686 nm. β phase, face-centered cubic, *cF*4 (*Fm*3*m*); a = 0.35441 nm
Minimum interatomic distance. β phase, 0.25061 nm

Mass Characteristics

Atomic weight. 58.9332
Density. At 20 °C: 8.832 g/cm^3 for α phase; 8.80 g/m^3 for β phase
Volume change on phase transformation. β-1 to α phase (cooling), −0.3% (approximate)

Thermal Properties

Melting point. 1495 °C
Boiling point. 2900 °C (approximate)
Phase transformation temperature. β to α phase (cooling), 417 °C; a transformation near 1120 °C has not been confirmed.
Coefficient of linear thermal expansion. 13.8 μm/m · K near room temperature; 14.2 μm/m · K at 200 °C; see also Fig. 5.
Specific heat. 0.414 kJ/kg · K
Latent heat of fusion. 292 kJ/kg
Latent heat of vaporization. 7.209 MJ/kg
Thermal conductivity. 69.04 W/m · K at 20 °C

Electrical Properties

Electrical conductivity. 27.6% IACS at 20 °C
Electrical resistivity. 52.5 nΩ · m at 20 °C; temperature coefficient, 5.31 nΩ · m per K
Thermoelectric force. See Fig. 6.
Electrochemical equivalent. Valence +2, 0.03050 mg/C

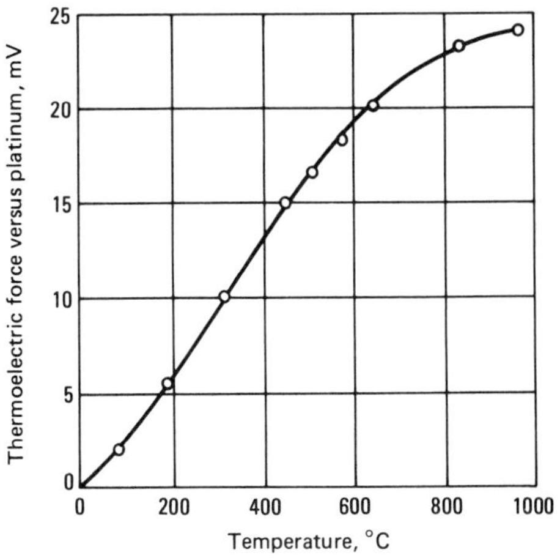

Fig. 6 Thermoelectric force of cobalt

Magnetic Properties

Magnetic permeability. Initial, 68; maximum, 245
Coercive force. 708.3 A · m^{-1} for H_{max} = 0.1 T
Saturation magnetization. 1.87 T ($4\pi I_s$)
Residual induction. 0.49 T for H_{max} = 0.1 T
Hysteresis loss. 690 J/m^3 · cycle for B_{max} = 0.5 T
Curie temperature. 1121 °C
Direct current magnetization and permeability. See Fig. 7.
Magnetostriction. See Fig. 8.

Optical Properties

Spectral reflectance:

λ, nm	%
200	37
1 060	67.5
6 750	92.7
12 030	96.6

Fabrication Characteristics

Workability. Annealed cobalt strip can be cold rolled to about 25% reduction in area between intermediate anneals (Ref 1).

Mechanical Properties

Tensile properties. See Table 9.
Hardness. Annealed strip: 65 HR45T (Ref 1)
Poisson's ratio. 0.32
Elastic modulus. Tension, 211 GPa; shear, 826 GPa; compression, 183 GPa
Velocity of sound. 5880 m/s for longitudinal bulk waves; 3100 m/s for shear waves

REFERENCE

1. R.W. Fraser, D.J.I. Evans, and V.N. Mackiw, The Production and Properties of Ductile Cobalt Strip, *Cobalt*, No. 23, June 1964

Columbium (Cb)

See the discussion of "Niobium (Nb)" in this article.

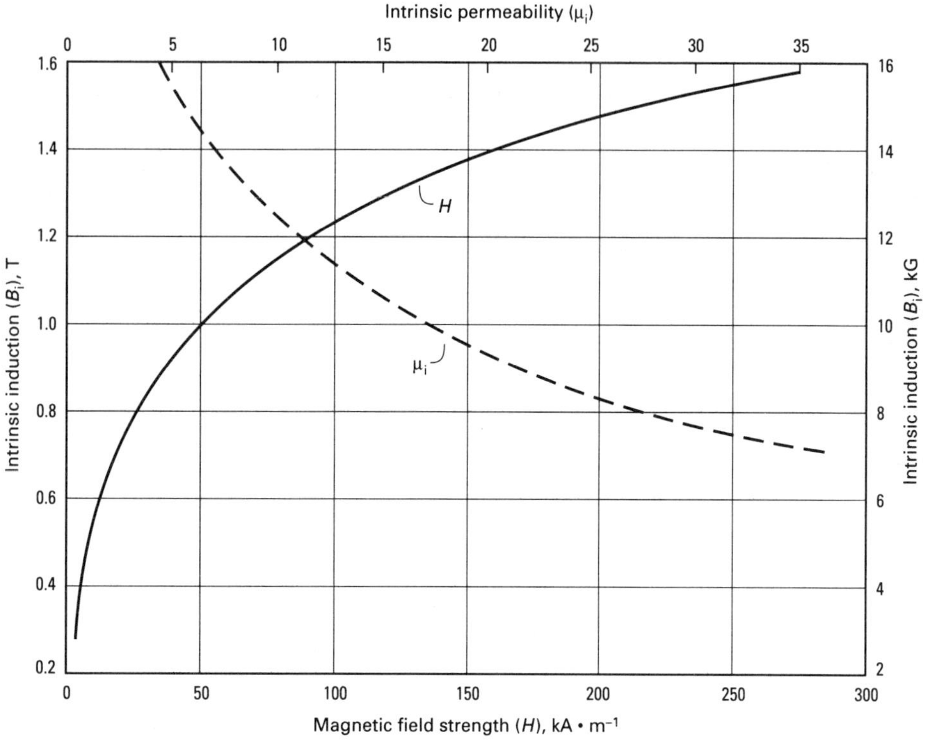

Fig. 7 Direct current magnetization and intrinsic permeability curves for annealed cobalt strip. Intrinsic permeability (μ_i) is the ratio of B to H. Source: Ref 1

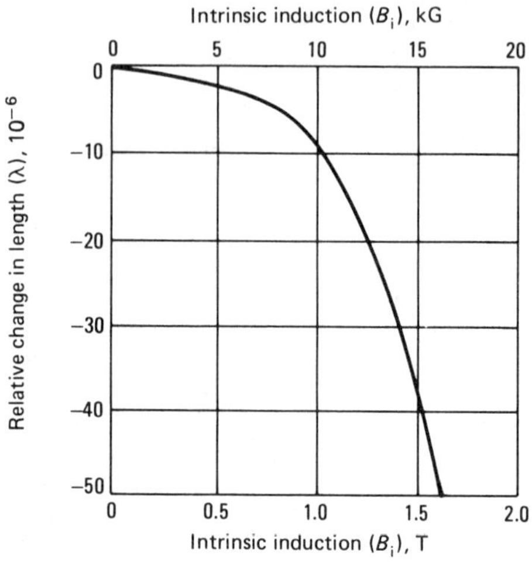

Fig. 8 Magnetostriction properties of annealed cobalt strip. Source: Ref 1

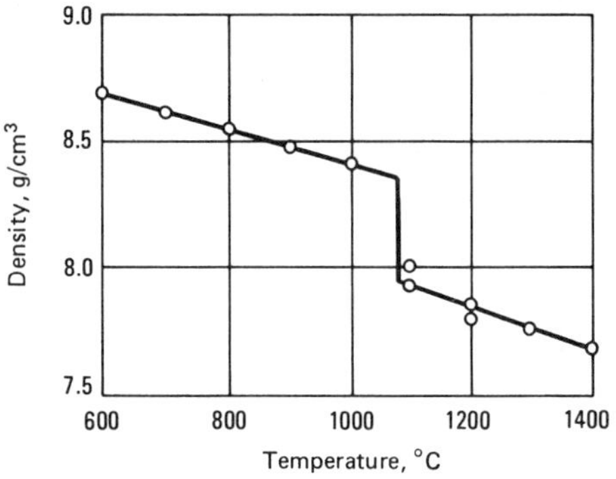

Fig. 9 Variation of density with temperature for pure copper

Table 9 Mechanical properties of cobalt

Form and purity	Tensile strength, MPa	0.2% yield strength, MPa	0.2% yield stress, MPa	Compressive yield strength, MPa	Elongation in 50 mm, %
As-cast (99.9)	234.4	. . .	. . .	291.0	. . .
Annealed (99.9)	255.1	. . .	. . .	386.8	. . .
Annealed strip(a)	760–860	. . .	310–345	. . .	15–22
Swaged (99.9)	689.5	. . .	. . .	. . .	. . .
Zone refined (99.8)	944.6	758.5	. . .	. . .	. . .

(a) Source: Ref 1

Table 10 Relative volume versus pressure for pure copper at 25 °C

Pressure, GPa	Relative volume, V/V_0	Pressure, GPa	Relative volume, V/V_0	Pressure, GPa	Relative volume, V/V_0
0.0	1.000	18.0	0.904	65.0	0.777
0.5	0.996	20.0	0.896	70.0	0.768
1.0	0.993	22.0	0.889	75.0	0.759
1.5	0.990	24.0	0.881	80.0	0.751
2.0	0.986	26.0	0.874	85.0	0.743
2.5	0.983	28.0	0.868	90.0	0.736
3.0	0.980	30.0	0.861	95.0	0.729
3.5	0.977	32.0	0.855	100	0.722
4.0	0.974	34.0	0.849	120	0.697
4.5	0.971	36.0	0.843	140	0.677
5.0	0.968	38.0	0.838	160	0.658
6.0	0.962	40.0	0.832	180	0.642
7.0	0.956	42.0	0.827	200	0.627
8.0	0.951	44.0	0.822	250	0.596
9.0	0.945	46.0	0.817	300	0.571
10.0	0.940	48.0	0.812	350	0.550
12.0	0.930	50.0	0.808	400	0.532
14.0	0.921	55.0	0.797	450	0.516
16.0	0.912	60.0	0.786		

Source: Ref 1

Copper (Cu)

Compiled by A.W. Blackwood, ASARCO Inc. and J.E. Casteras, Alpha Metals, Inc.

Structure

Crystal structure. Face-centered cubic, structure symbol, A1; Fm3m; cF4. Lattice parameter, 0.361509 ± 0.000004 nm at 25 °C (Ref 1, 2)

Twinning planes. (111) twin plane, [1$\bar{1}$2] twin direction; (111) twin plane, [112] twin direction (Ref 3)

Cleavage planes. None (Ref 4)

Minimum interatomic distance. 0.2551 nm (Ref 2)

Mass Characteristics

Atomic weight. 63.54

Density. (Ref 1):

°C	g/cm^3
20	8.93
Melting point	7.940
1100	7.924
1200	7.846
1300	7.764

See also Fig. 9. Density decreases 0.028% with a reduction of 50% by drawing.

Table 11 Mean coefficient of linear thermal expansion for pure copper

Temperature, K	Mean coefficient, μm/m · K
2	0.0006
4	0.0025
6	0.0074
8	0.016
10	0.030
12	0.052
14	0.083
16	0.128
18	0.186
20	0.26
25	0.6
50	3.8
75	7.6
100	10.5
150	13.6
200	15.2
250	16.1
293	16.7
350	17.3
400	17.6
500	18.3
600	18.9
700	19.6
800	20.4
1000	22.4
1200	24.8

Source: Ref 1

Specific volume. See Table 10.
Volume change on freezing. 4.92% contraction

Thermal Properties

Melting point. 1084.88 °C (Ref 5)
Boiling point. 2595 °C; 2567 °C (Ref 1)
Coefficient of thermal expansion. Linear, 16.5 μm/m · K at 20 °C (Ref 1). See also Table 11. Volumetric, 49.5 × 10^{-6}/K (Ref 2)
Specific heat. 0.494 kJ/kg · K at 2000 K; 0.386 kJ/kg · K at 293 K; 0.255 kJ/kg · K at 100 K (Ref 1). See also Fig. 10 and Table 12.
Enthalpy, entropy. See Table 12.
Latent heat of fusion. 205 kJ/kg (Ref 1); 204.9 kJ/kg (Ref 5); 206.8 kJ/kg (Ref 2)
Latent heat of vaporization. 4729 kJ/kg (Ref 1); 4726 kJ/kg (Ref 2); 4793 kJ/kg (Ref 6)
Thermal conductivity. 398 W/m · K at 27 °C (Ref 1). See also Fig. 10 and Table 13.
Recrystallization temperature. See Fig. 11.
Vapor pressure. From Ref 1:

°C	Pa
946	1.3 × 10^{-3}
1035	1.3 × 10^{-2}
1141	1.3 × 10^{-1}
1273	1.3
1432	13
1628	130
1879	1330
2067	5.33 × 10^{3}
2207	1.33 × 10^{4}
2465	5.33 × 10^{4}
2595	1.01 × 10^{5}
2760	2.02 × 10^{5}
3010	5.07 × 10^{5}
3500	1.01 × 10^{6}
3640	2.02 × 10^{6}
3740	4.05 × 10^{6}

Diffusion coefficient. See Table 14.

Table 12 Thermodynamic properties of copper

Temperature, K	C_p°, J/kg · K	$H_T^\circ - H_0^\circ$, J/kg(a)	$(H_T^\circ - H_0^\circ)/T$, J/kg · K	S_T°, J/kg · K	$-(G_T^\circ - H_0^\circ)$, J/kg	$(G_T^\circ - H_0^\circ)/T$, J/kg · K
1	0.0117	0.00565	0.00565	0.0112	0.00552	0.00552
2	0.0278	0.0249	0.0124	0.0239	0.0228	0.0114
3	0.0530	0.0644	0.0214	0.0395	0.0543	0.0181
4	0.0916	0.0135	0.0338	0.0596	0.103	0.0258
5	0.148	0.253	0.0507	0.0859	0.176	0.0351
6	0.228	0.439	0.0733	0.120	0.277	0.0463
7	0.335	0.717	0.103	0.162	0.417	0.0596
8	0.474	1.120	0.140	0.216	0.606	0.0757
9	0.651	1.684	0.187	0.282	0.853	0.0947
10	0.873	2.439	0.244	0.360	1.174	0.117
11	1.14	3.446	0.313	0.456	1.57	0.144
12	1.47	4.752	0.395	0.570	2.09	0.175
13	1.87	6.405	0.493	0.703	3.51	0.209
14	2.34	8.513	0.607	0.858	2.72	0.250
15	2.89	11.11	0.741	1.039	4.45	0.297
16	3.54	14.32	0.895	1.245	5.59	0.349
17	4.30	18.22	1.072	1.481	6.96	0.409
18	5.16	22.94	1.275	1.747	8.56	0.475
19	6.14	28.58	1.504	2.061	10.46	0.551
20	7.27	35.28	1.763	2.392	12.68	0.634
25	15.15	89.75	3.59	4.80	30.17	1.21
30	26.64	192.8	6.42	8.51	62.87	2.09
35	41.51	361.8	10.34	13.71	117.8	3.37
40	58.86	612.0	15.30	20.36	202.4	5.07
45	77.55	952.7	21.17	28.36	323.7	7.19
50	96.84	1388	27.78	37.53	488.0	9.757
60	135.3	2549	42.50	58.60	965.9	16.10
70	170.9	4084	58.35	82.18	1668	23.83
80	202.2	5955	74.42	107.1	2614	32.67
90	229.1	8115	90.17	132.5	3811	42.35
100	251.9	10520	105.2	157.8	5264	52.64
110	271.0	13140	119.5	182.9	6968	63.34
120	287.2	15940	132.8	207.1	8918	74.32
130	300.9	18870	145.2	230.7	11110	85.43
140	312.7	21950	156.7	253.4	13530	96.62
150	322.7	25130	167.4	275.2	16180	107.8
160	331.3	28390	177.5	296.5	19030	118.9
180	345.3	35170	195.5	336.3	25370	140.9
200	356.1	41290	210.9	373.3	32460	162.2
220	364.6	49400	224.6	407.6	40270	183.1
240	371.4	56760	236.5	439.7	48750	203.2
260	376.7	64240	247.1	469.6	57850	222.5
273.15	379.7	69210	253.4	488.2	64140	234.8
280	381.1	71820	256.5	497.6	67530	241.1
298.15	384.6	78760	264.2	521.7	76780	257.5
300	384.9	79490	265.0	524.0	77740	259.2

(a) H_0° is enthalpy at 0 K and 1 atm. Source: Ref 1

Table 13 Thermal conductivity of pure copper

Temperature, K	Conductivity, W/m · K	Temperature, K	Conductivity, W/m · K	Temperature, K	Conductivity, W/m · K
0	0	18	12400	300	398
1	2870	20	10500	350	394
2	5730	25	6800	400	392
3	8550	30	4300	500	388
4	11300	35	2900	600	383
5	13800	40	2050	700	377
6	15400	45	1530	800	371
7	17700	50	1220	900	364
8	18900	60	850	1000	357
9	19500	70	670	1100	350
10	19600	80	570	1200	342
11	19300	90	514	1300	334(a)
12	18500	100	483	1373	160
13	17600	150	428	1773	172
14	16600	200	413	1973	176
15	15600	250	404	2273	177
16	14500	273	401		

(a) Extrapolated value. Sources: Ref 1, 2

Electrical Properties

Electrical conductivity. Volumetric, 103.06% IACS. See also Fig. 11 and 12.
Electrical resistivity. 16.730 nΩ · m at 20 °C (Ref 1); temperature coefficient, 0.068 nΩ · m at 20 °C; pressure coefficient, −0.228 aΩ · m/Pa for pressure range 100 kPa to 9.8 GPa. See also Fig. 12. Effects of impurities are dealt with in Ref 7 to 10. Electrical resistivity for temperatures measured in Kelvin:

K	nΩ · m
250	14.0
220	12.0
200	10.6
180	9.2
160	7.75
140	6.35
120	4.90
100	3.50
90	2.80
80	2.15
70	1.53
60	0.95
50	0.50
40	0.22
30	0.063
25	0.025
20	0.008
15	0.001

Resistivity ratio. From Ref 2:

°C	R_T/R_0
−200	0.151
−100	0.557
0	1.000
100	1.431
200	1.862
300	2.299
400	2.747
500	3.210
600	3.695
800	4.750
1000	5.959

Thermoelectric potential versus platinum. From Ref 1, 2:

°C	mV
−200	−0.19
−100	0.37
0	0
100	0.76
200	1.83
300	3.15
400	4.68
500	6.41
600	8.34
700	10.47
800	12.81
900	15.37
1000	18.16

Electrochemical equivalent. 0.3294 mg/C for Cu^{2+}; 0.6588 mg/C for Cu^+
Electrolytic solution potential. All versus standard hydrogen electrode (Ref 1): $Cu^{2+} + e^- \rightleftarrows Cu^+$, 0.158 V; $Cu^{2+} + 2e^- \rightleftarrows Cu$, 0.3402 V; $Cu^+ + e^- \rightleftarrows Cu^+$, 0.522 V
Ionization potential. Cu(I), 7.724 eV; Cu(II), 20.29 eV; Cu(III), 36.83 eV (Ref 1)
Hydrogen overvoltage. In 1 *N* H_2SO_4. $\eta = a + b (\log i)$, where η is overvoltage in V, i is current density in A/cm^2, constant a is 0.80 V, and constant b is 0.115 V (Ref 6)
Hall effect. Hall voltage, -5.24×10^{-4} V at 0.30 to 0.8116 T; Hall coefficient, −5.5 mV · m/A · T (Ref 2)
Electron emission. Secondary electron emission: 1.3 max secondary electron yield; 600 eV primary electron energy for max yield; 200 eV for *E*(I) crossover; 1500 eV for *E*(II) crossover

Table 14 Radioactive tracer diffusion data for copper

Solute (tracer)	Crystalline form(a)	Purity, %	Temperature range, °C	Form of analysis(b)	Activation energy (Q), kJ/mole	Frequency factor (D_0), mm²/s(c)
^{110}Ag	S, P	. . .	580–980	RA	195	61
^{76}As	P	. . .	810–1075	RA	176.3	20
^{198}Au	S, P	. . .	400–1050	SS	178	3
^{115}Cd	S	99.98	725–950	SS	191	93.5
^{141}Ce	P	99.999	766–947	RA	115.5	21.7×10^{-7}
^{51}Cr	S, P	. . .	800–1070	RA	224	102
^{60}Co	S	99.998	701–1077	SS	226	193
^{67}Cu	S	99.999	698–1061	SS	211	78
^{152}Eu	P	99.999	750–970	SS, RA	112.4	11.7×10^{-6}
^{59}Fe	S, P	. . .	460–1070	RA	218	136
^{72}Ga	. . .	. . .	. . .	. . .	192.1	55
^{68}Ge	S	99.998	653–1015	SS	187.4	39.7
^{203}Hg	P	. . .	. . .	. . .	184	35
^{177}Lu	P	99.999	857–1010	RA	109.5	43×10^{-8}
^{54}Mn	S	99.99	754–950	SS	383	10^9
^{95}Nb	P	99.999	807–906	RA	251.4	204
^{63}Ni	P	. . .	620–1080	RA	225	110
^{102}Pd	S	99.999	807–1056	SS	227.6	171
^{147}Pm	P	99.999	720–955	RA	115	36.2×10^{-7}
^{195}Pt	P	. . .	843–997	SS	157	48×10^{-3}
^{35}S	S	99.999	800–1000	RA	206	23×10^{2}
^{124}Sb	S	99.999	600–1000	SS	176	34
^{113}Sn	P	. . .	680–910	. . .	188	11
^{160}Tb	P	99.999	770–980	RA	114.9	89.6×10^{-8}
^{204}Ti	S	99.999	785–996	SS	181	71
^{170}Tm	P	99.999	705–950	RA	101.1	72.8×10^{-8}
^{65}Zn	P	99.999	890–1000	SS	198.8	73

(a) P, polycrystalline; S, single crystal. (b) RA, residual activity; SS, serial section. (c) $D_T = D_0 \exp (Q/RT)$, where T is in K

Work function:

Work function, eV	Conditions	Method of determination
4.5	1160–1200 K	Thermionic
4.6	1350 K	Thermionic
4.4	1100–1300 K	Thermionic
4.76	⟨111⟩	Photoelectric
4.86	⟨111⟩	Photoelectric
5.61	⟨110⟩	Photoelectric
4.60	. . .	Contact potential
4.51	. . .	Contact potential

Source: Ref 2

Magnetic Properties

Magnetic susceptibility. Determined largely by the quantity of iron present as an impurity. If the copper is free from oxygen, the iron is present in solid solution and has a small effect. The presence of oxygen results in the precipitation of Fe_3O_4; in this form, iron has a greater effect on magnetic properties. The measurements below were probably made on oxygen-bearing coppers:

Temperature, °C	Volumetric, mks
18	-1.08×10^{-6}
1080	-0.97×10^{-6}
1090	-0.68×10^{-6}
−259 to −253	-1.22×10^{-6}

Optical Properties

Color. Reddish metallic (Ref 1)
Spectral reflection coefficient. For incandescent light, 0.63 (Ref 1)
Reflectance. Mirror coatings, see Table 15; calculated, see Table 16. Polished or electroplated surfaces (data are for polished surfaces at close-to-normal incidence) (Ref 2):

Wavelength, μm	Reflectance, %
0.25	25.9
0.30	25.3
0.35	27.5
0.40	30.0
0.50	43.7
0.60	71.8
0.70	83.1
0.80	88.6
1.0	90.1
2.0	95.5
4.0	97.3
6.0	98.0
8.0	98.3
10.0	98.4
12.0	98.4

Nominal spectral emittance. 0.15 for polished Cu at λ = 655 nm and 1080 K
Refractive index. See Table 16.
Absorptive index. Coefficient of absorption of solar radiation, 0.25 (Ref 1)

Nuclear Properties

Stable isotopes:

Isotope	Atomic weight	Natural abundance, %
^{63}Cu	62.9298	69.09
^{65}Cu	64.9278	30.91

Unstable isotopes:

Isotope	Half-life	Modes of decay(a)	Mean decay energy, MeV
^{58}Cu	3.20 s	β^+	8.569
^{59}Cu	82.0 ± 0.4 s	β^+, EC	4.8
^{60}Cu	23.0 ± 0.3 min	β^+, EC	6.12
^{61}Cu	3.41 h	β^+, EC	2.242
^{62}Cu	9.8 min	β^+, EC	3.939
^{64}Cu	12.9 h	β^-	0.573
		β^+, EC	1.677
^{66}Cu	5.10 ± 0.02 min	β^-	2.633
^{67}Cu	61.88 ± 0.11 h	β^-	0.576
^{68}Cu	30 s	β^-	4.6

(a) EC, electron capture

Chemical Properties

General corrosion behavior. Insoluble in hot and cold water (Ref 1)
Resistance to specific corroding agents. Soluble in HNO_3 and in hot H_2SO_4. Slightly soluble in HCl and NH_4OH (Ref 1)

Mechanical Properties

Tensile properties. Tensile strength: annealed, 209 MPa; cold drawn, 344 MPa (Ref 2); See also Fig. 11 and 13. Yield strength at 0.5% extension, under load; annealed, 33.3 MPa; cold drawn, 333.4 MPa (Ref 2). Elongation: annealed, 60%; cold drawn, 14% (Ref 2); see also Fig. 11. Reduction in area: annealed, 92%; cold drawn, 88% (Ref 2)
Hardness. Cold drawn, 37 HRB
Poisson's ratio. 0.308 calculated from elastic modulus; annealed, 0.343 (Ref 5); cold drawn, 0.364 (Ref 2)
Strain-hardening exponent. Annealed, 0.54 (Ref 3)
Elastic modulus:

Tension, GPa	Ref
128	2
112 (cold drawn)	2
125 (annealed)	3
129.8	5
Shear, GPa	**Ref**
46.8	2
46.4 (annealed)	3
48.3	5
Bulk, GPa	**Ref**
140	2
137.8	5

Elastic modulus along crystal axes. Tension: ⟨100⟩, 68 GPa; ⟨111⟩, 21 GPa. Shear: ⟨100⟩, 77 GPa (Ref 3)
Specific damping capacity. Log decrement: 3.2×10^{-3} (Ref 2)
Dynamic liquid viscosity. From Ref 1:

°C	mPa · s
1085	3.36
1100	3.33
1150	3.22
1200	3.12

Liquid surface tension. 99.99999% purity, in vacuum: 1.300 N/m at the melting point. 99.999% purity, in N_2: 1.341 N/m at 1100 °C; 1.338 N/m at 1150 °C; 1.335 N/m at 1200 °C. 99.997% purity, at the melting point: 1.355 N/m in He or H_2; 1.358 N/m in Ar; 1.352 N/m in vacuum
Coefficient of friction. Static. Cu on Cu: 4.0 in H_2 or N_2; 1.6 in air or O_2; 1.4 (clean); 0.8 in paraffin oil (Ref 1); 0.7 in paraffin oil plus 1% lauric acid (Ref 2)
Velocity of sound. 4759 m/s for longitudinal bulk waves; 3813 m/s for irrotational rod waves; 2325 m/s for shear waves; 2171 ms for Rayleigh waves

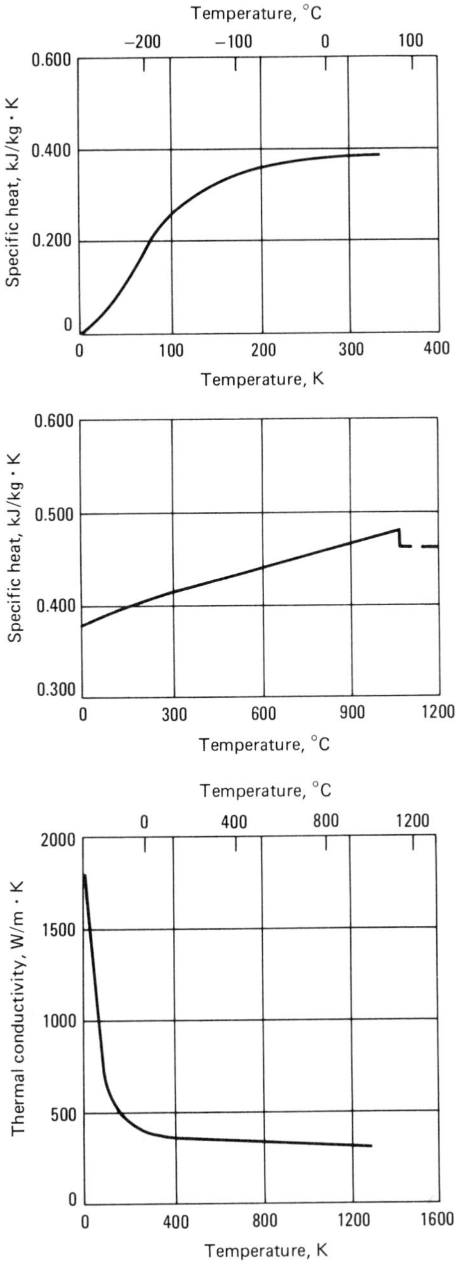

Fig. 10 Thermal properties of pure copper

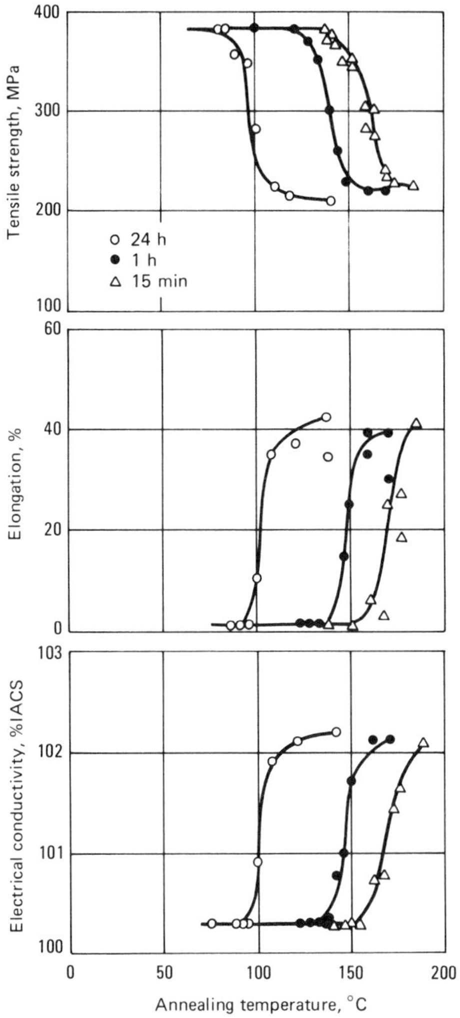

Fig. 11 Typical annealing curves for pure copper

Table 15 Normal-incidence reflectance of freshly evaporated mirror-coating copper

Wavelength, µm	Reflectance, %	Wavelength, µm	Reflectance, %
0.220	40.1	0.750	97.9
0.240	39.0	0.800	98.1
0.260	35.5	0.850	98.3
0.280	33.0	0.900	98.4
0.300	33.6	0.950	98.4
0.315	35.5	1.0	98.5
0.320	36.3	1.5	98.5
0.340	38.5	2.0	98.6
0.360	41.5	3.0	98.6
0.380	41.5	4.0	98.7
0.400	47.5	5.0	98.7
0.450	55.2	6.0	98.7
0.500	60.0	7.0	98.7
0.550	66.9	8.0	98.8
0.600	93.3	9.0	98.8
0.650	96.6	10.0	98.9
0.700	97.5	15.0	99.0

Source: Ref 2

REFERENCES

1. R.C. Weast, Ed., *CRC Handbook of Chemistry and Physics*, 55th ed., CRC Press, 1974
2. *American Institute of Physics Handbook*, 3rd ed., McGraw-Hill, 1972
3. W.J.M. Tegart, *Elements of Mechanical Metallurgy*, MacMillan, 1966
4. A.S. Tetalman and A.J. McEvily, *Fracture of Structural Materials*, John Wiley & Sons, 1967

Table 16 Optical properties of copper

Wavelength, μm	Index of refraction	Extinction coefficient	Reflectance (calculated)
Bulk copper			
0.3650	1.0719	2.0710	0.5004
0.4050	1.0769	2.2890	0.5491
0.4360	1.0707	2.4610	0.5860
0.5000	1.0308	2.7843	0.6528
0.5500	0.7911	2.7177	0.7013
0.5780	0.3250	2.8923	0.8716
0.6000	0.1491	3.2867	0.9508
0.6500	0.1074	3.9104	0.9740
0.7500	0.1034	4.8847	0.9835
1.0000	0.1471	6.9334	0.9881
Single-crystal copper			
0.4400	1.1070	2.5565	0.5965
0.4600	1.0942	2.6320	0.6131
0.4800	1.0618	2.7124	0.6341
0.5000	1.0836	2.7684	0.6390
0.5200	1.0438	2.7784	0.6490
0.5400	0.9324	2.7348	0.6674
0.5600	0.6470	2.7200	0.7440
0.5800	0.2805	2.9764	0.8931
0.6000	0.1360	3.3464	0.9565
0.6200	0.1040	3.6525	0.9714
0.6400	0.0972	4.0692	0.9798
0.6600	0.0897	4.0692	0.9798
Evaporated copper			
0.1025	1.05	0.70	0.098
0.1113	0.95	0.73	0.115
0.1215	0.95	0.78	0.137
0.1306	0.96	0.83	0.148
0.1392	1.00	0.91	0.165
0.1500	1.02	1.02	0.192
0.1603	0.98	1.04	0.219
0.1700	0.94	1.12	0.254
0.1800	0.90	1.21	0.296
0.1900	0.88	1.36	0.335
0.2000	0.94	1.51	0.378
0.500	0.88	2.42	0.625
0.600	0.186	2.980	0.928
0.700	0.150	4.049	0.966
0.800	0.170	4.840	0.973
0.900	0.190	5.569	0.977
1.000	0.197	6.272	0.981
1.35	0.45	7.81	0.971
1.69	0.58	9.96	0.977
2.28	0.82	13.0	0.981
3.00	1.22	17.1	0.984
3.4	1.53	20.3	0.985
3.97	1.94	23.1	0.986
4.87	2.86	28.9	0.987
5.0	2.92	27.45	0.985
5.8	3.71	34.6	0.988
7.00	5.25	40.7	0.988
7.3	5.79	43.2	0.988
8.35	7.28	49.2	0.988
9.6	9.76	57.2	0.988
10.25	11.0	60.6	0.988
10.8	12.6	64.3	0.988
12.25	15.5	71.9	0.989

Source: Ref 2

5. P.B. Coates and J.W. Andrews, A Precise Determination of the Freezing Point of Copper, *J. Phys. F, Met. Phys.*, Vol 8 (No. 2), 1978
6. G.W.C. Kaye and T.H. Laby, *Table of Physical and Chemistry Constants*, 14th ed., Longman Group, 1973
7. J.S. Smart, A.A. Smith, and A.J. Phillips, Preparation and Some Properties of High Purity Copper, *Trans. AIME*, Vol 143, 1941
8. J.S. Smart and A.A. Smith, Effect of Iron, Cobalt, and Nickel on Some Properties of High Purity Copper, *Trans. AIME*, Vol 147, 1942
9. J.S. Smart and A.A. Smith, Effect of Certain Fifth-Period Elements on Some Properties of High Purity Copper, *Trans. AIME*, Vol 152, 1943
10. J.S. Smart and A.A. Smith, Effect of Phosphorus, Arsenic, Sulfur, and Selenium on Some Properties of High Purity Copper, *Trans. AIME*, Vol 166, 1946

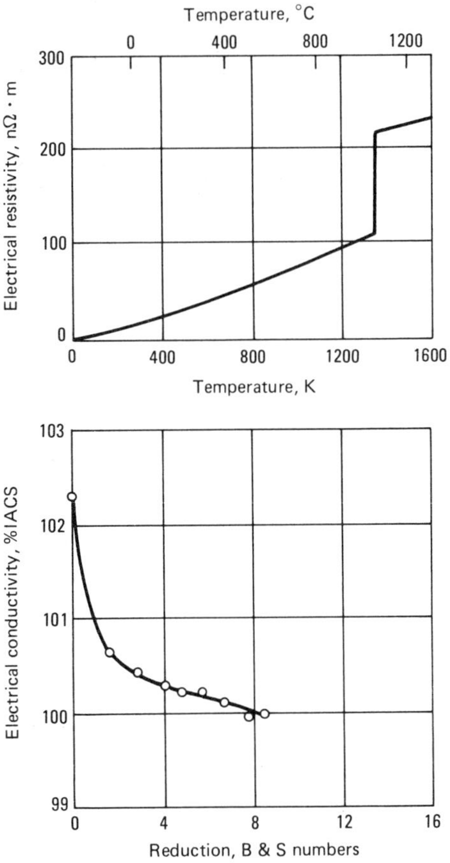

Fig. 12 Electrical properties of pure copper

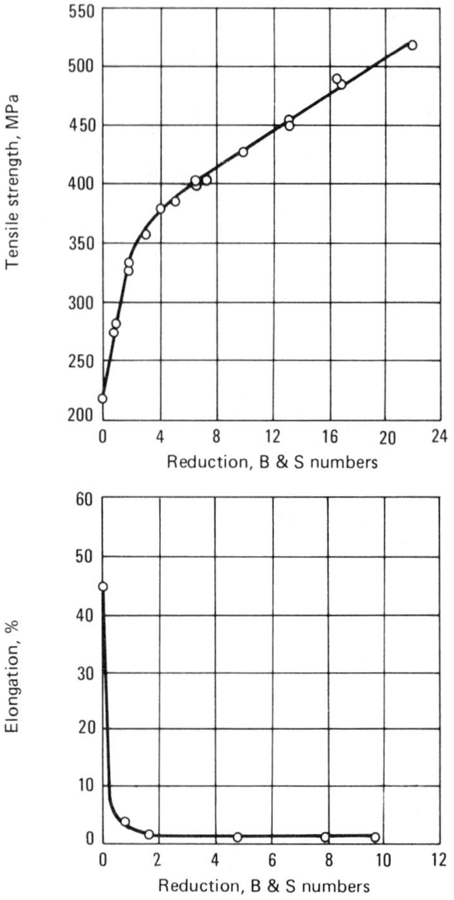

Fig. 13 Variation of tensile properties with amount of cold reduction for pure copper wire

Curium (Cm)

See the section "Properties of the Transplutonium Actinide Metals (Am-Fm)" in this article.

Dysprosium (Dy)

See the section "Properties of the Rare Earth Metals" in this article.

Einsteinium (Es)

See the section "Properties of the Transplutonium Actinide Metals (Am-Fm)" in this article.

Erbium (Er)

See the section "Properties of the Rare Earth Metals" in this article.

Europium (Eu)

See the section "Properties of the Rare Earth Metals" in this article.

Fermium (Fm)

See the section "Properties of the Transplutonium Actinide Metals (Am-Fm)" in this article.

Gadolinium (Gd)

See the section "Properties of the Rare Earth Metals" in this article.

Gallium (Ga)

Compiled by H. Clinton Snyder, Aluminum Company of America and R. Frankena, Ingal International Gallium, GmbH

Revised by M.W. Chase, National Institute of Standards and Technology

Gallium is used predominantly in the electronics industry, where it is combined with elements of group III, IV, or V of the periodic table to form semiconducting materials; most often, it is combined with arsenic and/or phosphorus for uses in light-emitting diodes, laser diodes, solar cells, transistors, and so on. In the oxide form, it is combined with other oxides in garnets for magnetic bubble domain devices; in metallic form, it is used for heat transfer media, eutectic alloys, liquid seals, and high-temperature lubricants; it is used in superconducting compounds such as GaV_3 and in compounds in organic reactions.

Commercially available gallium metal ranges in purity from 99.5% to 99.9999+%. The most common impurities are mercury, lead, tin, zinc, and copper. If certain impurity limits of high-purity gallium are exceeded, the optoelectric properties of electronic materials are degraded or destroyed. Gallium is tested for purity using emission spectrography and mass spectrography, and by residual resistivity measurement.

Gallium ordinarily is not considered to be hazardous, but it can be toxic in compounds or alloys, depending upon the nature of the other components or ions. Gallium in aluminum causes severe intergranular corrosion of the aluminum.

Structure

Crystal structure. Orthorhombic, *Cmca*: $a = 0.45258$ nm; $b = 0.45186$ nm; $c = 0.76570$ nm at 24 °C. Metastable high-pressure phases also exist.
Minimum interatomic distance. 0.2437 nm

Mass Characteristics

Atomic weight. 69.723 ± 0.001
Density:

°C	Phase	g/cm³
20	Solid	5.907
29.65	Solid	5.9037
29.8	Liquid	6.0947
32.4	Liquid	6.093
200	Liquid	5.972
500	Liquid	5.779
600	Liquid	5.720
1010	Liquid	5.492
1100	Liquid	5.445

Volume change on freezing. 3.2% expansion

Thermal Properties

Melting point. 29.78 °C
Triple point. 302.9169 ± 0.0005 K (29.7669 °C)
Boiling point. 2477 K (per International Practical Temperature Scale 48), or 2204 °C; some sources list 2237 °C as the boiling point, but this is reported to be an error caused by gallium suboxide pressure.
Coefficient of thermal expansion. Linear, along crystal axes, from 0 to 20 °C: 11.5 μm/m · K along *a* axis, 31.5 μm/m · K along *b* axis, 16.5 μm/m · K along *c* axis (Ref 1). Volumetric: solid from 0 to 29.7 °C, 58 000 mm^3/m^3 · K; liquid at 100 °C, 120 000 mm^3/m^3 · K; liquid at 900 °C, 97 000 mm^3/m^3 · K
Specific heat.

K	°C	Phase	kJ/kg · K
4.3	−268.9	Solid	1.22×10^{-4}
16.1	−257.1	Solid	0.01925
60.1	−213.1	Solid	0.1757
100	−173.2	Solid	0.2651
200	−73.2	Solid	0.3416
273–297	0–24	Solid	0.3723
298.15	25	Solid	0.3738
500	227	Liquid	0.3847
1000	727	Liquid	0.3811

Latent heat of fusion. 80.16 kJ/kg
Enthalpy of fusion. 79.82 kJ/kg at the temperature of fusion
Enthalpy of sublimation. 3887 kJ/kg at 298.15K
Enthalpy of combustion. To form Ga_2O_3: −15 648 kJ/kg
Thermal conductivity. Polycrystalline, at 29.8 °C: 33.49 W/m · K. Along crystal axes at 20 °C: 40.82 W/m · K along *a* axis; 88.47 W/m · K along *b* axis; 15.99 W/m · K along *c* axis. Liquid at 77 °C: 28.68 W/m · K along *a* axis; 34.04 W/m · K along *b* axis; 38.31 W/m · K along *c* axis
Vapor pressure:

K	Pa
1000	6.281×10^{-4}
1200	1.279×10^{-1}
1400	5.636
1600	9.572×10^{1}
1800	8.590×10^{2}
2000	4.932×10^{3}
2200	2.05×10^{4}
2477	1.0×10^{5}

Electrical Properties

Electrical resistivity. Polycrystalline, at 20 °C: 150.5 nΩ · m. Along crystal axes at 20 °C: 174 nΩ · m along *a* axis, 81 nΩ · m along *b* axis, 543 nΩ · m along *c* axis (Ref 2). Supercooled liquid: at 0 °C, 252 nΩ · m; at 20 °C, 256.1 nΩ · m. Liquid: at 40 °C, 260 nΩ · m; at 600 °C, 378 nΩ · m
Electrochemical equivalent. Valence +3: 0.241 mg/C
Electrolytic solution potential. Versus H_2: −0.56 V at 25 °C
Hydrogen overvoltage. Near melting point: solid, −0.31 V; liquid, −0.44 V
Temperature of superconductivity. 1.078 K

Magnetic Properties

Magnetic susceptibility. Volume (mks units): solid at 80 K (−193 °C), 3.07×10^{-4}; solid at 17 °C, 2.71×10^{-4}; liquid at 40 °C, 0.31×10^{-4}

Optical Properties

Color. Liquid metal is silvery white; solid metal is silvery with a bluish cast
Reflectance. Solid: 75.6% for λ = 436 nm; 71.3% for λ = 589 nm. Liquid: 88.8% for λ = 435 nm; 88.4% for λ = 546 nm; 88.6% for λ = 691 nm

Nuclear Properties

Stable isotopes. ^{69}Ga, isotope mass 68.9255809, 60.1% abundant; ^{71}Ga, isotope mass 70.9247006, 39.9% abundant

Chemical Properties

General corrosion behavior. Liquid gallium oxidizes rapidly to form a protective layer of oxide. The reaction of gallium with mineral acids depends on concentration and temperature. Gallium reacts with caustic—especially in the presence of iron metal. The rate of gallium corrosion is inversely related to the purity as is the tendency to super cool. At elevated temperatures, gallium is a corroding agent for many metals (Ref 3). At room temperature, diffusion of gallium into many metals takes place, and this results in the formation of an often low-melting compound in grain boundaries and grains of the corroded metal.

Mechanical Properties

Hardness. 1.5 to 2.5, Mohs scale
Elastic modulus. Compressibility, at 20 °C: 0.021 nm^3/m^3 · Pa between 15 and 50 MPa
Fracture behavior. Polycrystalline masses shatter easily.
Kinematic liquid viscosity. 287 m^2/s at 30 °C; 183 m^2/s at 500 °C
Liquid surface tension. In vacuum: 0.709 N/m at 30 °C; 0.712 N/m at 100 °C; 0.718 N/m at 200 °C; 0.743 N/m at 500 °C

REFERENCES

1. R.W. Powell, Electrical Resistivity of Gallium and Some Anisotropic Properties of the Metal, *Proc. R. Soc. (London) A*, Vol 209, 1951, p 525
2. R.W. Powell, M.J. Woodman, and R.P. Tye, Further Measurements Relating to the Anisotropic Thermal Conductivity of Gallium, *Br. J. Appl. Phys.*, Vol 14, 1963, p 432-435
3. L.R. Kelman, W.D. Wilkinson, and F.L. Yaggee, Resistance of Materials by the Attack of Liquid Metals, USAEC Report ANL-4417, Argonne National Laboratory, 1950

Germanium (Ge)

Compiled by C.D. Thurmond,
Bell Laboratories

Revised by J.H. Adams,
Eagle-Picher Industries, Inc.

Structure

Crystal structure. Face-centered cubic (diamond); $a = 0.565754$ nm at 25 °C (Ref 1)

Mass Characteristics

Atomic weight. 72.59
Density. 5.323 g/cm³ at 25 °C (Ref 2)

Thermal Properties

Melting point. 937.4 °C (Ref 3)
Boiling point. 2830 °C (Ref 2)

Coefficient of linear thermal expansion. At 25 °C (Ref 4):

K	µm/m · K
100	2.3
150	4.1
200	5.0
250	5.5
300	6.0

Specific heat. 0.3217 kJ/kg · K at 25 °C (Ref 3)
Entropy. 428.3 J/kg · K at 25 °C (Ref 3)
Latent heat of fusion. 466.5 J/kg (Ref 2)
Latent heat of vaporization. 4602 J/kg (Ref 2)
Vapor pressure. 0.140 MPa at 937.4 °C (Ref 3)
Thermal conductivity. From Ref 5:

K	W/m · K
100	232
200	96.8
300	59.9
400	43.2

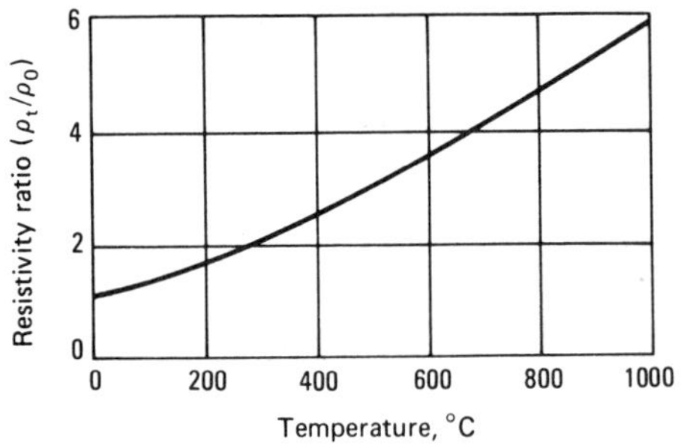

Fig. 14 Temperature dependence of the electrical resistivity of gold

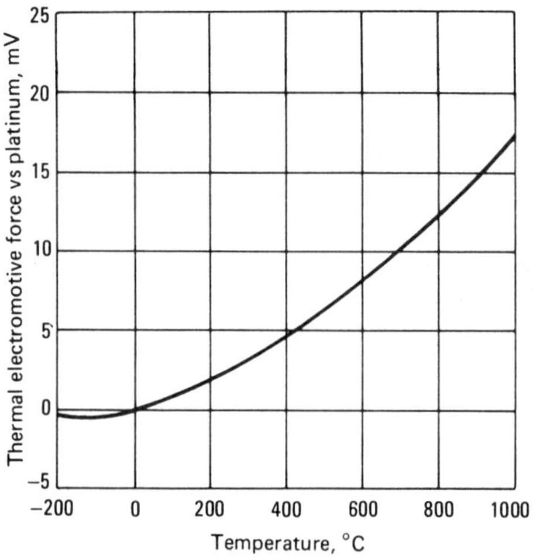

Fig. 15 Temperature dependence of the thermal electromotive force of gold versus platinum. Positive values indicate gold is positive to platinum.

Electrical Properties

Electrical resistivity. Intrinsic, 0.53 Ω · m at 25 °C (Ref 2)
Carrier density. Intrinsic at 25 °C, 2.12 × 10^{13} electrons/cm^3 (Ref 2)
Forbidden energy gap. 0.7437 eV at 0 K; 0.6642 eV at 25 °C (Ref 6)
Drift mobilities. At 25 °C. Electrons, 0.3800 m^2/V · s; holes, 0.1850 m^2/V · s (Ref 2)

Mechanical Properties

Hardness. 6.3 (Mohs scale)
Modulus of rupture. 110 MPa
Elastic coefficients. At 25 °C: C_{11}, 131.6 GPa; C_{22}, 50.9 GPa; C_{44}, 66.9 GPa
Elastic constants. At 25 °C: S_{11}, 9.685 × 10^{-12} m · N^{-1}; S_{22}, −2.70 × 10^{-12} m · N^{-1}; S_{44}, 14.94 × 10^{-12} m · N^{-1}
Young's moduli. At 25 °C: Y_{100}, 103.3 GPa; Y_{110}, 138.0 GPa; Y_{111}, 155.5 GPa
Shear moduli. At 25 °C: M_{100}, 66.9 GPa; M_{110}, 41.0 GPa; M_{111}, 49.0 GPa

REFERENCES

1. A.S. Cooper, *Acta Crystallogr.*, Vol 15, 1962, p 578
2. J.H. Adams, *Kirk-Othmer Encyclopedia of Chemical Technology*, Vol 11, 3rd ed., John Wiley & Sons, 1980, p 791
3. R. Hultgren, P.D. Desai, D.T. Hawkins, M. Gleiser, K.K. Kelley, and D.D. Wagman, *Selected Values of the Thermodynamic Properties of the Elements*, American Society for Metals, 1973, p 204
4. J.S. Browder and S.S. Ballard, *Appl. Opt.*, Vol 16, 1977, p 3214
5. C.Y. Ho *et al.*, *J. Phys. Chem. Ref. Data*, Vol 1, 1972, p 339
6. C.D. Thurmond, *J. Electrochem. Soc.*, Vol 122, 1975, p 1133

Gold (Au)

Compiled by S.C. Carapella, Jr., ASARCO Inc.

Reviewed for this Volume by Douglas Hayduk, ASARCO Inc.

Structure

Crystal structure. Face-centered cubic: *a* = 0.40786 nm
Minimum interatomic distance. 28.78 nm

Mass Characteristics

Atomic weight. 196.9665
Density. 19.302 g/cm^3 at 25 °C

Thermal Properties

Melting point. 1064.43 °C
Boiling point. 2857 °C
Coefficient of linear thermal expansion. At 20 °C, 14.2 µm/m · K; from 0 to 950 °C, $L_t = L_0\ [1 + (14.103t + 0.001628t^2 + 0.000001145t^3) \times 10^{-6}]$, where *t* is in °C
Specific heat:

°C	kJ/kg · K
25	0.128
227	0.133
627	0.142
1027	0.163
1063	0.170
1127	0.166
1227	0.159

Latent heat of fusion. 62.762 kJ/kg
Latent heat of vaporization. 1.6987 kJ/kg
Thermal conductivity. 317.9 W/m · K at 0 °C; 1749 W/m · K at 4.2 K
Vapor pressure:

°C	kPa
1770	0.1013
2036	1.013
2383	10.13
2857	101.3

Diffusion coefficients. At 20 °C:

Element	Matrix	Diffusion, m^2/s
Fe	Au	3×10^{-26}
Ni	Au	1×10^{-30}
Cu	Au	6×10^{-34}
Pd	Au	2×10^{-35}
Au	Au	2×10^{-40}
Pt	Au	1×10^{-45}
Au	Pb	1×10^{-15}
Au	Cu	5×10^{-24}
Au	Ag	5×10^{-30}
Au	Pd	5×10^{-35}
Au	Pt	1×10^{-35}

Electrical Properties

Electrical conductivity. 73.4% IACS at 20 °C
Electrical resistivity. 20.1 nΩ · m at 0 °C; 23.5 nΩ · m at 20 °C. Temperature coefficient: from 0 to 100 °C, 0.004 per K. See also Fig. 14.
Thermal electromotive force. See Fig. 15.
Effect of alloying elements on resistivity:

Element	Resistivity, nΩ · m	Increase in resistivity, %
1% Ag	28.2	28.2
1% Pd	29	31.8
1% Cd	30.7	39.5
1% Pt	33	50.0
1% Cu	35.9	63.3
1% In	45	104.0
1% Zn	49.3	124.0
1% Ni	51	132.0
1% Sn	76	245.0
1% Co	178	710.0
1% Fe	269	1220.0

Magnetic Properties

Magnetic susceptibility. Volume: −1.79 × 10^{-6} mks

Optical Properties

Reflectance. See Fig. 16.
Emittance. See Fig. 17.

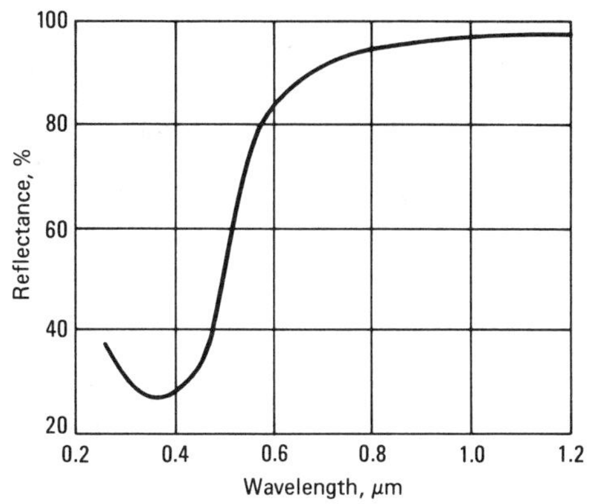

Fig. 16 Reflectance of gold as a function of wavelength

Mechanical Properties

Tensile properties. Tensile strength, 103 MPa (annealed wire). Elongation, 30%
Elastic modulus. Tension, 78 GPa
Liquid surface tension. At 1200 °C, 1070 mN/m; at 1300 °C, 1020 mN/m

Holmium (Ho)

See the section "Properties of the Rare Earth Metals" in this article.

Indium (In)

Compiled by C.E.T. White and L.G. Stevens, The Indium Corporation of America

The major uses of indium include solders and fusible alloys (low-melting-point alloys), production of compound semiconductors, nuclear reactor control rods, and bearings. Indium compounds, in particular the oxides and tin-doped oxides, find major applications in transparent electrodes for flat panel displays, conductive coatings for demisting windshields, and fluorescent panels. Minor applications include low-resistance contacts to aluminum wire, gaskets, and battery additives.

Structure

Crystal structure. Tetragonal; at 26 °C, a = 0.32512 nm and c = 0.49467 nm

Mass Characteristics

Atomic weight. 114.82
Density:

°C	g/cm^3
20	7.30
164	7.026
194	7.001
228	6.974
271	6.939
300	6.916

Volume change on freezing. 2.5% contraction

Thermal Properties

Melting point. 156.61 °C

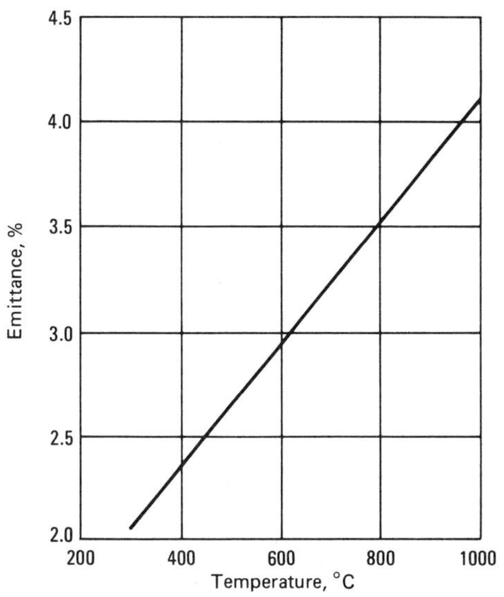

Fig. 17 Total hemispherical emittance of gold as a function of temperature

Boiling point. 2080 °C
Coefficient of linear thermal expansion. 24.8 μm/m · K at 20 °C
Specific heat:

°C	kJ/kg · K
25	0.233
127	0.252
156.63 (solid)	0.264
156.63 (liquid)	0.257
227	0.256
327	0.255
427	0.254

Latent heat of fusion. 28.47 kJ/kg
Latent heat of vaporization. 1959.42 kJ/kg
Thermal conductivity. 83.7 W/m · K at 0 °C
Vapor pressure:

°C	kPa
1215	0.1013
1421	1.013
1693	10.13
2080	101.3

Electrical Properties

Electrical resistivity:

°C	nΩ · m
−269.77 (3.38 K)	Superconducting
20	84
154	291
181	301
222	319
280	348

Electrochemical equivalent. Valence 3, 396.4 μg/C
Electrode potential. $In^0 \rightarrow In^{3+} + 3e$, 0.38 V

Magnetic Properties

Magnetic susceptibility. Volumetric: 7.0 × 10^{-6} mks

Nuclear Properties

Stable isotopes. 113, 115
Thermal neutron cross section. For 2.2 km · s neutrons: absorption, 190 ± 10 b; scattering, 2.2 ± 0.5 b

Mechanical Properties

Tensile strength:

K	MPa
295	1.6
76	15.0
4	31.9

Compressive strength. 2.14 MPa
Hardness. 0.9 HB
Elastic modulus. At 20 °C: 12.74 GPa in tension
Poisson's ratio. At 20 °C: 0.4498

Iridium (Ir)

Compiled by Leonard Bozza, Engelhard Corporation

Reviewed for this Volume by Louis Toth, Engelhard Corporation

Small crucibles made of iridium have been used for studying high-temperature reactions. A major use for iridium is crucibles for producing large, pure, defect-free man-made crystals for electronic and industrial applications. Single crystals so formed are used as substrates in magnetic bubble memory devices, solid-state lasers, insulating substrates for semiconductors, monoclinic filters, and substitutes for natural gemstones in jewelry. Iridium also is used as an alloying element to harden platinum; as electrodes in spark plugs for severe operating conditions, such as those experienced by jet engine igniters; as thermocouple elements; and as radioactive isotopes for industrial applications and cancer therapy.

Mass Characteristics

Atomic weight. 192.9
Density. 22.65 g/cm^3 at 20 °C

Thermal Properties

Melting point. 2447 °C (Ref 1)
Boiling point. 4500 °C (Ref 2)
Coefficient of linear thermal expansion. 6.8 μm/m · K at 20 °C (Ref 3)
Specific heat. 0.130 kJ/kg · K (Ref 4)
Thermal conductivity. 147 W/m · K at 0 to 100 °C (Ref 5)

Electrical Properties

Electrical resistivity. 47.1 nΩ · m at 0 °C, 53 nΩ · m at 20 °C (Ref 5). Temperature coefficient, 0.00427 nΩ · m per °C at 0 to 100 °C (Ref 6)
Thermal electromotive force. Pt 67 (reference junction at 0 °C): +3.626 mV at 400 °C; +6.271 mV at 600 °C; +12.741 mV at 1000 °C (Ref 7)

Table 17 Tensile properties of iridium annealed at 1500 °C

Temperature, °C	Tensile strength, MPa	0.2% yield strength, MPa	Reduction in area, %
24	623	234	6.8
500	530	234	12.7
750	450	142	51.0
1000	331	43.4	80.6

Magnetic Properties

Magnetic susceptibility. Mass: 0.19×10^{-8} mks at 18 °C (Ref 8)

Optical Properties

Reflectivity. 64% at λ = 0.45 μm; 70% at λ = 0.55 μm; 78% at λ = 0.75 μm (Ref 9)
Emissivity. 0.30 at 0.65 μm for solid unoxidized metal (Ref 6, 10)

Chemical Properties

General corrosion behavior. Iridium is the most corrosion-resistant element. It is not affected by common acids, including hot sulfuric acid. It is slightly attacked by sodium hypochlorite solutions but not by aqua regia at ordinary temperatures. However, at elevated temperatures and pressures, aqua regia does attack iridium, and it may be used under these conditions for dissolving iridium and its refractory alloys for analysis. Iridium is virtually insoluble in lead even at high temperatures, and use is often made of this fact in preliminary steps in chemical analysis.

Fabrication Characteristics

Working data. Iridium can be arc melted (inert-gas cover), electron beam melted, or consolidated by powder metallurgy techniques. It is hot worked using procedures similar to those used for tungsten. Final working is done at warm temperatures, which produce a fibrous structure. Iridium has limited malleability at room temperature.

Mechanical Properties

Tensile properties. Properties of 0.5 mm wire. Tensile strength: annealed at 1000 °C, 1100 to 1240 MPa; hot drawn, 2070 to 2480 MPa. Elongation: annealed, 20 to 22%; hot drawn, 13 to 18% (Ref 11). See also Table 17.
Hardness. Annealed at 1000 °C, 200 to 240 HV; as-cast, 210 to 240 HV; hot drawn, 600 to 700 HV (Ref 11)
Modulus of elasticity. Tension: static, 517 GPa; dynamic, 527 GPa. Compression: 210 GPa (Ref 12)
Poisson's ratio. 0.26

REFERENCES

1. International Practical Temperature Scale of 1968, Amended Edition of 1975, *Metrologia*, Vol 12, 1976, p 7-17
2. R.F. Hampson, Jr. and R.F. Walker, *J. Res. Natl. Bur. Stand.*, Vol 65A, 1961, p 289
3. P. Hidnert and W. Souder, NBS Circular 486, U.S. Dept of Commerce, 1950
4. F.M. Jaeger and E. Rosenbohn, *Proc. Acad. Sci. (Amsterdam)*, Vol 34, 1931, p 808
5. R.W. Powell *et al.*, *Platinum Met. Rev.*, Vol 6, 1962, p 138
6. D.L. Goldwater and W. Danforth, *Phys. Rev.*, Vol 103, 1956, p 871
7. G.F. Blackburn and F.R. Caldwell, *J. Res. Natl. Bur. Stand.*, Vol 66C, 1962, p 1
8. K. Honda, *Ann. Phys.*, Vol 32, 1910, p 1027
9. M. Auswarter, *Z. Tech. Phys.*, Vol 18, 1927, p 457
10. R.C. Weast, Ed., *CRC Handbook of Chemistry and Physics*, 58th ed., CRC Press, 1977, p E-230
11. *Engelhard Ind. Tech. Bull.*, Vol VI (No. 3), Dec 1965
12. R.I. Jaffee *et al.*, "High Temperature Properties and Alloying Behavior of the Platinum Group Metals," Contract 2547(00), NRO 39-067, Office of Naval Research

Iron (Fe)

Compiled by L.R. Smith, Ford Motor Company (formerly with University of Michigan) and W.C. Leslie (retired), University of Michigan

Iron of sufficient purity so that its properties are essentially those of the element is commonly called high-purity iron. Such iron is not an article of commerce; instead, it is employed almost exclusively in research. Iron of very high purity can be prepared by a variety of methods, but the last stage of the process is purification by floating-zone refining, often combined with treatment in oxidizing and reducing atmospheres. In order to maintain this purity, it is essential to avoid contamination of the iron, which can occur by reactions with the atmosphere or with containers.

Structure

Crystal structure. The various phases of iron are shown as a function of temperature and pressure in Fig. 18. The crystal symmetry and the space group for each phase are:

- α iron and δ iron: bcc, *Im*3*m*
- γ iron: fcc, *Fm*3*m*
- ε iron: hcp, $P6_3/mmc$

Lattice parameter. As a function of temperature, see Table 18. The pressure dependence of the lattice parameters of the bcc and hcp phases of iron at 23 ± 3 °C are $a(\text{bcc}) = 0.2866\ (1 + P/27.5)^{-0.056}$ and $a(\text{hcp}) = 0.2523\ (1 + P/32.5)^{-0.033}$, where c/a (hcp) = 1.603 ± 0.001 and is independent of pressure; P is the pressure in GPa; and a(bcc), a(hcp), and c are lattice parameters in nanometers (Ref 3).

Table 18 Lattice parameters for iron

Temperature, °C	Lattice parameter, nm
20	0.28665 α-Fe
53	0.28676
154	0.28708
248	0.28750
315	0.28775
378	0.28806
451	0.28840
523	0.28879
563	0.28882
588	0.28890
642	0.28922
660	0.28920
706	0.28923
730	0.28935
754	0.28940
764	0.28940
772	0.28943
799	0.28946
862	0.28988
898	0.29012
907	0.29005
950	0.36508 γ-Fe
1003	0.36535
1076	0.36599
1167	0.36660
1249	0.36720
1361	0.36810
1390	0.29315 δ-Fe
1439	0.29346
1480	0.29378
1508	0.29396

Source: Ref 2

Slip plane and direction.

Phase	Slip direction	Slip plane
α iron	⟨111⟩	{110}, {112}, {123}(a)
γ iron	⟨101⟩	{111}
ε iron	⟨11$\bar{2}$0⟩	{10$\bar{1}$0}

(a) It is generally considered that slip in body-centered cubic (bcc) iron at room temperature and above can occur on any plane containing ⟨111⟩.

Twinning plane. {112}, direction ⟨111⟩
Cleavage plane. {100}
Minimum interatomic distances: α iron: 20 °C, 0.24825 nm; 907 °C, 0.25119 nm. γ iron: 950 °C, 0.25815 nm; 1361 °C, 0.26029 nm. δ iron: 1390 °C, 0.25388 nm; 1508 °C, 0.25458 nm
Microstructure. Zone-refined iron shows an essentially featureless equiaxed grain structure.
Fracture behavior. Zone-refined iron can exhibit considerable ductility at −269 °C (4.2 K) (reduction in area of almost 100%), whereas less-pure irons become brittle at temperatures below −153 °C (120 K) (Ref 4, 5). Impure irons show either cube-face cleavage or conchoidal grain-boundary fracture, depending on grain size and impurities present (Ref 6).

Mass Characteristics

Atomic weight. 55.847 (based on ^{12}C = 12, International Union of Pure and Applied Chemistry, 1961)
Density. α-Fe at 20 °C, ρ = 7.870 g/cm^3;

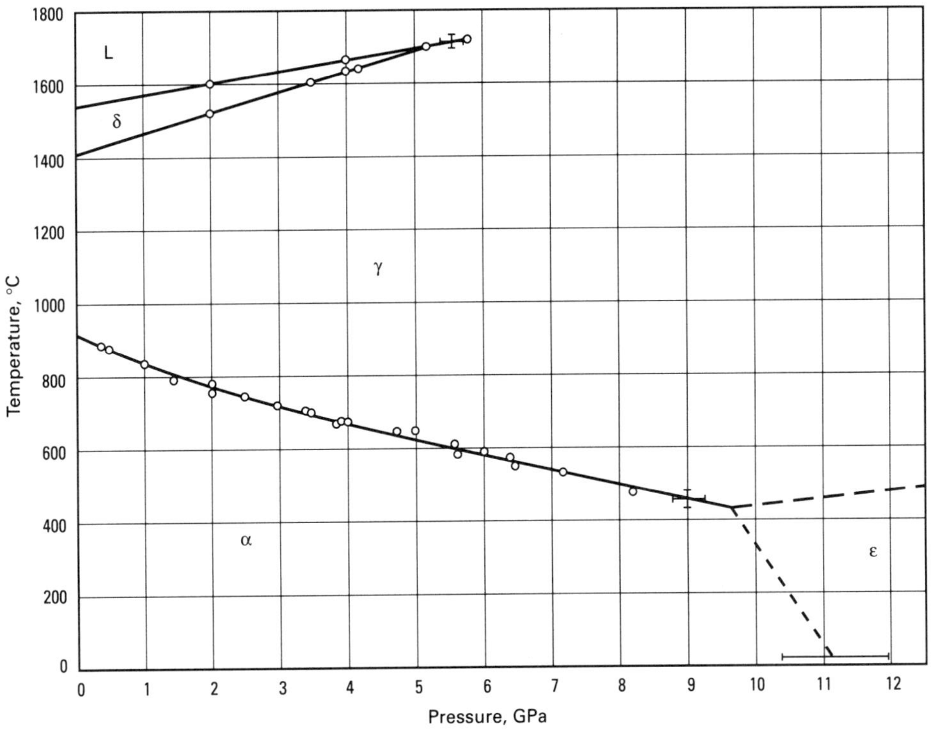

Fig. 18 Phase diagram for iron. Source: Ref 1

γ-Fe at 912 °C, 7.694 g/cm³; δ-Fe at 1394 °C, 7.406 g/cm³ (Ref 7). Liquid Fe at melting point (1538 °C), ρ = 7.035 g/cm³; at 1550 °C, 7.01 ± 0.03 g/cm³; at 1564 °C, 7.00 g/cm³ (Ref 7)

Density versus temperature. α-Fe (to 912 °C), $10^5 \cdot \Delta\rho/\rho_0 \cdot K = 4.3$; γ-Fe (912 to 1394 °C), 6.7; δ-Fe (1394 to 1539 °C), 4.8 (Ref 8)

Density versus deformation. The density of high-purity α iron will decrease very slightly with increasing cold work. The decrease in density can be estimated from the Stehle-Seeger relation (Ref 9): $N = (\Delta\rho/\rho_0)/2\mathbf{b}^2$, where N is the dislocation density and **b** is the Burgers vector. The appropriate value of N at a strain of 0.20, taken from Ref 10, yields an estimate of $\Delta\rho/\rho_0 = 2.5 \times 10^{-5}$.

Volume change on freezing. Liquid to δ, −3.4%

Volume change on phase transformation. δ to γ (1394 °C), −0.52%; γ to α (912 °C), +1.0% (computed from changes in lattice parameter)

Thermal Properties

Melting point. 1538 °C (Ref 11)

Boiling point. 2870 °C (Ref 11)

Phase transformation temperatures. α to γ, 912 °C; γ to δ, 1394 °C

Coefficient of linear thermal expansion. Values recommended in Ref 12 are shown in Fig. 19 and Table 19. These values are considered accurate to within ±3% at temperatures below 627 °C, ±5% below 912 °C, and ±20% above 912 °C. The volume per atom as a function of temperature is shown in Fig. 20.

Coefficient of volumetric thermal expansion (liquid). $10^6 \cdot \Delta v/v_0 \cdot K = 140$ (somewhat greater than three times the linear coefficient in the δ range) (Ref 14)

Specific heat. The specific heats at constant pressure of α, γ, δ, and liquid iron, as a function of temperature, are shown in Fig. 21 and Table 20. Data for temperatures below 27 °C are primarily taken from the compilation reported in Ref 6. Data for temperatures above 27 °C are taken from Ref 15.

Enthalpy, entropy, and free-energy function. See Table 20.

Latent heat of fusion. Adopted average value, 247 ± 7 kJ/kg for α iron at the melting point (1538 °C) (Ref 15)

Latent heat of phase transformation. Selected experimental averages are:

Transformation	Temperature, K	Latent heat (ΔH_t°), kJ/kg
α to γ	1185	16
γ to δ	1667	15

Source: Ref 15

Values at other temperatures can be derived from Table 20.

Latent heat of vaporization. 7018.3 kJ/kg at the boiling point of iron (2870 °C) (Ref 16)

Thermal conductivity. Figure 22 shows thermal conductivity as a function of temperature. The recommended values (Table 21) are for well-annealed high-purity iron, and those below −73 °C apply only to iron having a residual electrical resistivity of 0.143 nΩ · m. The values are accurate to within ±5% below −173 °C, ±3% from −173 °C to room temperature, ±2% from room temperature to +727 °C, and ±3 to 8% at higher temperatures.

Heat of combustion and free energy of formation. The following values are for 25 °C:

Oxide	Molecular weight	$-\Delta H_f^\circ$, kJ/kg Fe	$-\Delta F_f^\circ$, kJ/kg Fe
$Fe_{0.947}O$	68.89	4 758±14	4 357±18
Fe_2O_3	159.70	14 700±75	13 245±113
Fe_3O_4	231.55	19 990±150	18 146±165

Source: Ref 18

Vapor pressure. See Fig. 23. For the temperature range 1178 to 1394 °C: $\log p = m/T + b$, where m is −20 908 ± 109, b is 12.161 ± 0.070, T is in K, and p is in Pa (Ref 20)

Diffusion coefficient:

	Diffusion constant (D_0), mm²/s	Activation energy (Q), MJ/kg
In γ-Fe	70	5.12
In paramagnetic α-Fe	16	4.30
In ferromagnetic α-Fe	50	4.30

Source: Ref 21

The diffusion constant (D) can be calculated using the Arrhenius equation:

$$\ln D = \ln D_0 - \frac{Q}{RT}$$

In ferromagnetic iron, the Arrhenius plot of D cannot be approximated by a straight line; instead, it shows strong curvatures (see Fig. 24). For practical purposes, the diffusion coefficients for alloying elements differ from the self-diffusion of iron by factors that are independent of temperature. These factors (Fig. 25) can be used to determine the diffusion constants for other elements in γ iron or α iron. Grain-boundary diffusion, in γ iron and α iron, of iron and some alloying elements can be described by:

$$D_0 = \frac{0.054\ \mu m^3/s}{\delta}$$

where δ represents the thickness of the grain boundary in μm. For grain-boundary diffusion in both α iron and γ iron, Q is 2.78 MJ/kg.

Vacancy formation and migration. Reference 23 gives the following enthalpies for formation and migration of monovacancies in iron:

	Enthalpy (H), aJ	
	Formation	Migration(a)
In γ-Fe	0.245±0.024	0.147
In paramagnetic α-Fe	0.257±0.024	0.202
In ferromagnetic α-Fe	0.256±0.024	0.202±0.040

(a) The significance of these values is that monovacancy migration occurs only at temperatures well above ambient (>450 K, or >180 °C). Source: Ref 23

Table 19 Linear thermal expansion of iron

Temperature, °C	Temperature, K	Change in length, %(a)	Coefficient (α), $10^6 \cdot \Delta l/l_0 \cdot K$
−273	0	...	0
−268	5	−0.204	0.01
−248	25	−0.203	0.20
−223	50	−0.203	1.3
−173	100	−0.184	5.6
−73	200	−0.102	10.1
+20	293	0.000	11.8
127	400	0.134	13.4
227	500	0.274	14.4
327	600	0.421	15.1
427	700	0.575	15.7
527	800	0.735	16.2
627	900	0.899	16.4
727	1000	1.065	16.6
827	1100	1.230	16.7
912(b)	1185(b)	1.370	16.8
912(b)	1185(b)	0.993(c)	23.3
927	1200	1.028(c)	23.3
1127	1400	1.494(c)	23.3
1327	1600	1.960	23.3
1377	1650	2.077(c)	23.3
1394–1502(d)	1667–1775(d)	...	23.6

(a) Change from length at 20 °C. (b) α to γ phase transition. (c) Typical values. (d) γ to δ phase transition. Source: Ref 12

Electrical Properties

Volumetric electrical conductivity. 17.59% IACS at 25 °C

Electrical resistivity. The variation with temperature for solid iron is shown in Fig. 26(a) and 26(b). For liquid iron at the melting point (1536 °C), 1.39 μΩ · m (Ref 6)

Temperature coefficient of electrical resistivity. The fundamental coefficient is $(\rho_{100\ °C} - \rho_{0\ °C})/\rho_{0\ °C} = 0.00616$, or 0.616% increase per K (Ref 27). The temperature coefficient in the neighborhood of the Curie temperature is shown in Fig. 27. The resistivity ratio ($\rho_{300\ K}/\rho_{4\ K}$) of several high-purity irons is tabulated in Table 22. This ratio shows a fair correlation with purity, provided measurements are made in the proper longitudinal magnetic fields (Ref 29).

Pressure coefficient of electrical resistivity. The variation of resistivity with pressure at 25 °C is shown in Fig. 28. According to Fig. 28, the pressure coefficient is −0.18 per Pa. The resistance increases at the α to ε transition (at 13 GPa) by 366%.

Electrical properties versus deformation. The change in resistivity with true strain for iron deformed at −196 °C (ρ = 5.8 nΩ · m) is shown in Fig. 29. At 25 °C, the conductivity changed from 17.593% IACS (no reduction in area) to 17.577% IACS (9.5% reduction in area).

Thermoelectric power. The variation with temperature is shown in Fig. 30.

Electrochemical equivalent. Based on an atomic weight of 55.85 and a value of 96 495 coulombs (C) per gram equivalent weight for the Faraday, the electrochemical equivalents are 0.1929 and 0.2893 mg/C for Fe^{3+} and Fe^{2+}, respectively.

Standard electrode potential. −0.4402 V for the reaction $Fe = Fe^{2+} + 2e^-$ at 25 °C, where iron would be the negative terminal

Table 20 Thermal properties of iron

Temperature, °C	Temperature, K	Specific heat (c_p), J/mol · K	Enthalpy ($H_T^\circ - H_{298}^\circ$), kJ/mol	Entropy ($S_T - S_{298}^\circ$), J/mol · K	Free-energy function [$-(G_T^\circ - H_{298}^\circ)/T$], J/mol · K
α and δ phases					
−273.15	0	0	−4.498	−27.280	Infinite
−271.7	1.5	0.00749	...	...	...
−271.2	2.0	0.01017	...	...	...
−269.2	4.0	0.02134	...	...	...
−267.2	6.0	0.03452	...	...	...
−265.2	8.0	0.0502	...	...	...
−263.2	10.0	0.0682	...	...	...
−261.2	12.0	0.0916	...	...	...
−259.2	14.0	0.1188	...	...	...
−257.2	16.0	0.1556	...	...	...
−255.2	18.0	0.1987	...	...	...
−253	20	0.2573	...	...	...
−243	30	0.753	...	...	...
−233	40	1.57	...	...	...
−223	50	3.01	...	...	...
−213	60	4.81	...	...	...
−203	70	6.74	...	...	...
−193	80	8.62	...	...	...
−183	90	10.42	...	...	...
−173	100	12.05	−4.067	−21.150	46.800
−163	110	13.56	...	...	...
−153	120	14.90	...	...	...
−143	130	16.11	...	...	...
−133	140	17.20	...	...	...
−123	150	18.12	...	...	...
−113	160	18.91	...	...	...
−93	180	20.33	...	...	...
−73	200	21.46	−2.301	−9.280	29.505
−53	220	22.38	...	...	...
−33	240	23.18	...	...	...
−13	260	23.89	...	...	...
25.00	298.15	24.98	0	0	27.280
27	300	25.02	0.046	0.1544	27.281
127	400	27.36	2.665	7.673	28.290
227	500	29.71	5.518	14.031	30.273
327	600	32.05	8.606	19.654	32.590
427	700	34.60	11.934	24.778	35.009
527	800	37.90	15.548	29.560	37.405
577	850	40.15	17.498	31.923	38.167
627	900	43.03	19.573	34.291	39.823
677	950	47.15	21.819	36.717	41.030
727	1000	54.27	24.331	39.261	42.210
747	1020	59.77	25.466	40.385	42.698
757	1030	64.76	26.087	40.989	42.492
769	1042 (T_C)	83.51	26.978	41.847	43.236
777	1050	54.70	27.430	42.278	43.434
787	1060	51.55	27.964	42.795	43.694
807	1080	48.45	28.957	43.709	44.177
827	1100	46.37	29.904	44.578	44.673
912	1185 ($T_{\alpha-\gamma}$)	41.34	33.597	47.810	46.738
1394	1667 ($T_{\gamma-\delta}$)	41.03	52.601	61.244	56.970
1427	1700	41.36	53.963	62.052	57.589
1527	1800	42.36	58.149	64.443	59.418
1538	1811 (T_m)	42.47	58.626	64.706	59.614
γ phase					
912	1185 ($T_{\alpha-\gamma}$)	33.81	34.497	48.546	46.715
927	1200	33.93	35.010	48.976	47.081
1027	1300	34.77	38.442	51.717	49.426
1127	1400	35.60	41.957	54.319	51.630
1227	1500	36.44	45.559	56.800	53.707
1327	1600	37.27	49.247	59.176	55.677
1394	1667 ($T_{\gamma-\delta}$)	37.81	51.762	60.715	56.944
Liquid phase					
1538	1811 (T_m)	45.91	72.433	72.290	59.574
1627	1900	45.91	76.511	74.493	61.504
1727	2000	45.91	81.104	76.847	63.575
1827	2100	45.91	85.698	79.087	65.558
1927	2200	45.91	90.291	81.223	67.462
2027	2300	45.91	94.884	83.264	69.290
2127	2400	45.91	99.477	85.218	71.049
2227	2500	45.91	104.066	87.092	72.476
2327	2600	45.91	108.659	88.893	74.381
2427	2700	45.91	113.247	90.625	75.962
2527	2800	45.91	117.841	92.295	77.489
2627	2900	45.91	122.429	93.906	78.969
2727	3000	45.91	127.018	95.462	80.403
2827	3100	45.91	131.606	96.968	81.794
2927	3200	45.91	136.195	98.425	83.144

Sources: Ref 6, 15, 16

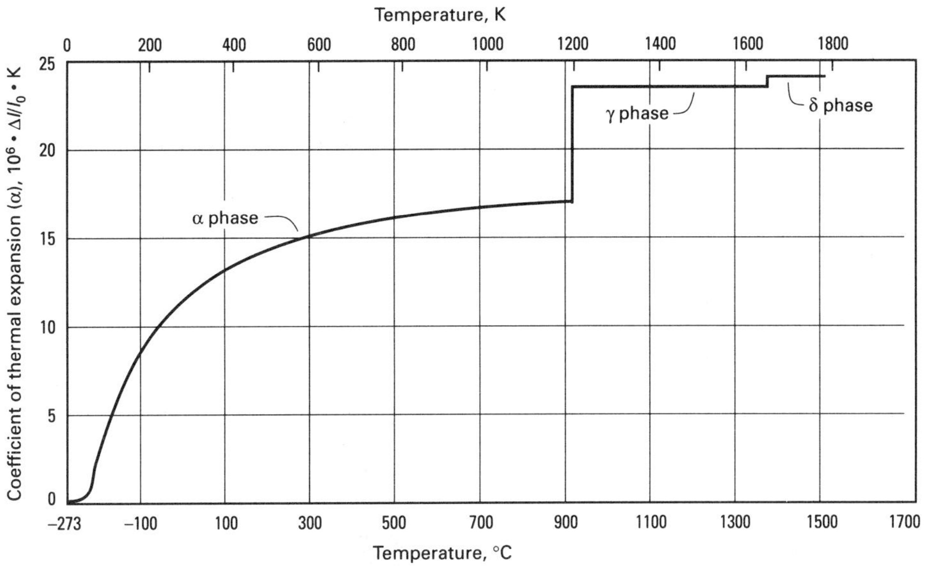

Fig. 19 Coefficient of linear thermal expansion for iron. Source: Ref 12

Table 21 Thermal conductivity of iron

Temperature, °C	Temperature, K	Conductivity, W/m · K
Solid		
−273	0	0
−272	1	171(a)
−271	2	342
−270	3	511
−269	4	677
−268	5	839
−267	6	993
−266	7	1140
−265	8	1270
−264	9	1390
−263	10	1480
−262	11	1560
−261	12	1630
−260	13	1670
−259	14	1690
−258	15	1700
−257	16	1690
−255	18	1630
−253	20	1540
−248	25	1270
−243	30	1000
−238	35	788
−233	40	623
−228	45	499
−223	50	405
−213	60	285
−203	70	216
−193	80	175
−183	90	150
−173	100	134
−150	123.2	115
−123	150	104
−100	173.2	99.1
−73	200	94.0
−50	223.2	90.4
Solid (continued)		
−23	250	86.5
0	273	83.5
25	298.2	80.4
27	300	80.2
50	323.2	77.4
77	350	74.4
100	373.2	72.0
127	400	69.5
200	473.2	63.4
227	500	61.3
300	573.2	56.4
327	600	54.7
400	673.2	50.4
427	700	48.8
500	773.2	44.8
527	800	43.3
600	873.2	39.4
627	900	38.0
700	973.2	34.2
727	1000	32.8
786	1059	29.7
800	1073.2	29.8
827	1100	29.8
900	1173.2	30.0
910	1183 (α)	30.0
912	1185 (γ)	28.0
927	1200	28.3
1000	1273.2	29.6
1027	1300	30.0
1100	1373.2	30.9
1127	1400	31.2
1200	1473.2	31.9
1227	1500	32.1
1300	1573.2	32.7
1327	1600 (γ)	33.0
1400	1673.2 (δ)	33.5(a)
1427	1700	33.8(a)
1500	1773.2	34.3(a)
1527	1800	34.5(a)
1537	1810	34.6(a)
Liquid(b)		
1537	1810	40.3(a)
1600	1873.2	41.3(a)
1627	1900	41.5(a)
1700	1973.2	42.3(a)
1727	2000	42.6(a)
1800	2073.2	43.2(a)
1900	2173.2	43.9(a)
1927	2200	44.1(a)
2000	2273.2	44.6(a)
2127	2400	45.0
2200	2473.2	45.2(a)
2327	2600	45.5(a)
2400	2673.2	45.6(a)
2527	2800	45.8(a)
2600	2873.2	45.9(a)
2727	3000	45.8(a)
2800	3073	45.8(a)
2927	3200	45.6(a)
3000	3273	45.4(a)
3127	3400	45.1(a)
3327	3600	44.2(a)
3527	3800	43.0(a)
3727	4000	41.5(a)
4227	4500	36.8(a)
4727	5000	30.8(a)
5227	5500	23.3(a)
5727	6000	14.7(a)
6227	6500	5.1(a)

(a) Extrapolated or estimated values. (b) Values for liquid iron are provisional. Source: Ref 17

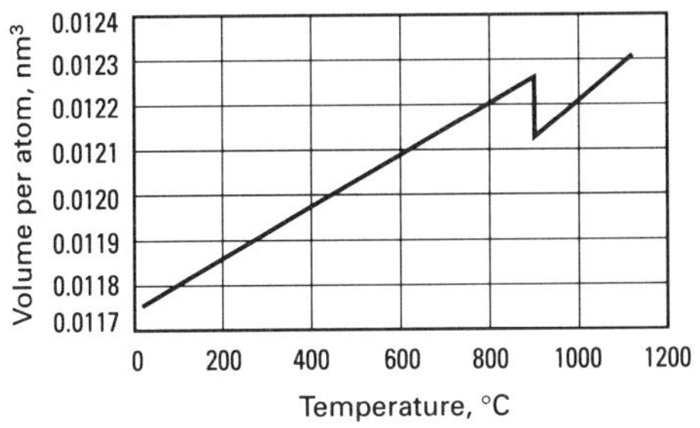

Fig. 20 Volume per atom for iron. Source: Ref 13

in a cell whose second electrode is a standard hydrogen electrode (SHE) and where the Fe^{2+} activity is unity (Ref 33)

Temperature coefficient of standard electrode potential. The thermal temperature coefficient, $(dV°/dT)_{th}$, at 25 °C is given in Ref 33 as +0.923 mV/K, and the isothermal temperature coefficient, $(dV°/dT)_{iso}$, at 25 °C is given as +0.052 mV/K. The thermal temperature coefficient is given a positive value because the hot electrode is the (+) terminal in a thermal cell. The isothermal temperature coefficient is given a positive value because the electromotive force of the isothermal cell, SHE/iron, increases with temperature.

First ionization potential. 7.87 eV (1.27 aJ) (Ref 34)

Hydrogen overvoltage. As given in Ref 35, the relationship between hydrogen overvoltage (activation) (η_a) and current density (i) is $\eta_a = -\beta \log (i/i_0)$, where β (the slope of the Tafel region) and i_0 (the exchange current density) are constants that depend on the environment. The current density consists of contributions from the external applied current density and the local-action current density. When the external applied current is zero, the overvoltage equals the corrosion potential (E_{corr}) and the local-action current density equals the corrosion current (i_{corr}). The variation in hydrogen overvoltage of pure iron in 4% NaCl (2.0 pH) is shown in Fig. 31. In 4% NaCl, the overvoltage is essentially constant for pH 1 to 4. Hydrogen overvoltage of pure iron in 0.1 *M* citric acid and 0.1 *M* malic acid is shown in Fig. 32. Overvoltage constants are given with the figures. Because there is a significant variation in overvoltage constants for different crystal orientations of pure iron, the data in the figures should be considered as average values for random orientation.

Hall effect. The Hall resistivity of polycrystalline ferromagnetic metals can be empirically expressed $\rho_H = R_0H + R_1M$, where R_0 and R_1 are the respective coefficients of ordinary and extraordinary Hall effect, H is the applied magnetic field, and M is the magnetization. The temperature dependence of ρ_H is shown in Fig. 33. Figures 34 and 35 show the temperature dependence of ordinary and extraordinary Hall coefficients. For single crystals, the extraordi-

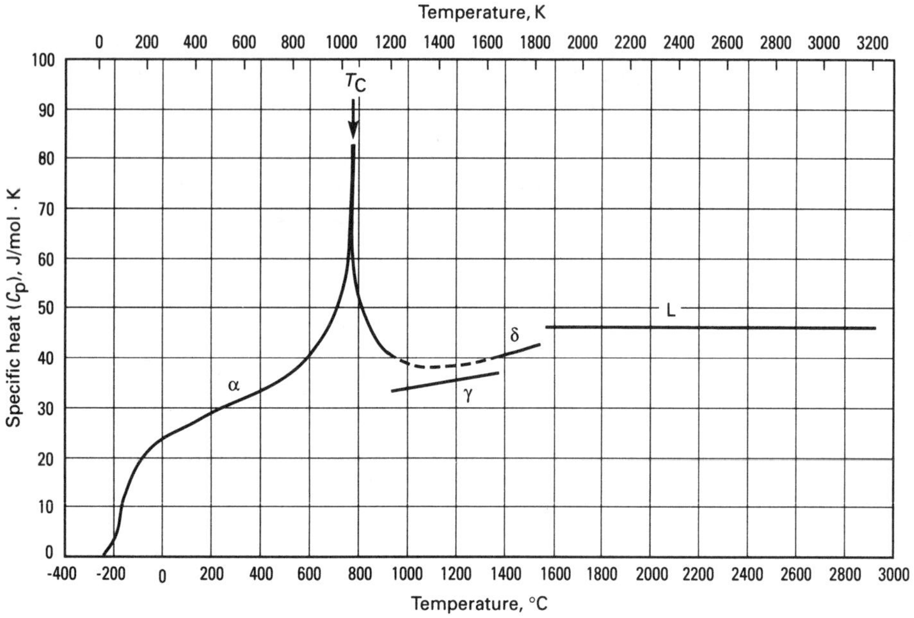

Fig. 21 Specific heat of iron from 0 to 3200 K. Sources: Ref 6, 15

nary Hall coefficient is dependent on the orientation of the crystal (Ref 38) (see Fig. 36). A detailed discussion of galvanomagnetic effects in iron whiskers (at −273 to 27 °C) is given in Ref 39 and 40.

Electrical resistivity versus alloying. The electrical resistivity of some binary iron alloys at −269 °C (4.2 K) versus solute concentration in atom percent is shown in Fig. 37. Results from the iron-nickel system overlap those of the iron-palladium system; the iron-tungsten system data are identical with the results for the iron-manganese system. For clarity, low-concentration points for some alloys have been omitted from the figure.

Electron emission. 260 kA/m^2 · K^2 for α iron and 15 kA/m^2 · K^2 for γ iron (Ref 42)

Work function. In high-purity iron, the work function is sensitive to impurities and surface condition. Using iron that was electropolished and then cleaned through repeated cycles of ion bombardment (with argon) and annealing, the value of the work function for the (100) plane of α iron was found to be 4.67 ± 0.03 eV (0.748 ± 0.005 aJ) (Ref 43). Using positive-ion-emission and electron-emission data, the change in work function at the α to γ transformation (ϕ_γ/ϕ_α) was found to be between +0.06 and +0.09 eV (+0.010 to +0.014 aJ) (Ref 44).

Magnetoresistance. The variation at −269 °C (4.2 K) of residual resistance in longitudinal and transverse magnetic fields is shown in Fig. 38. This effect can also be seen as a dependence of the residual electrical resistance upon measuring current density (see Fig. 39).

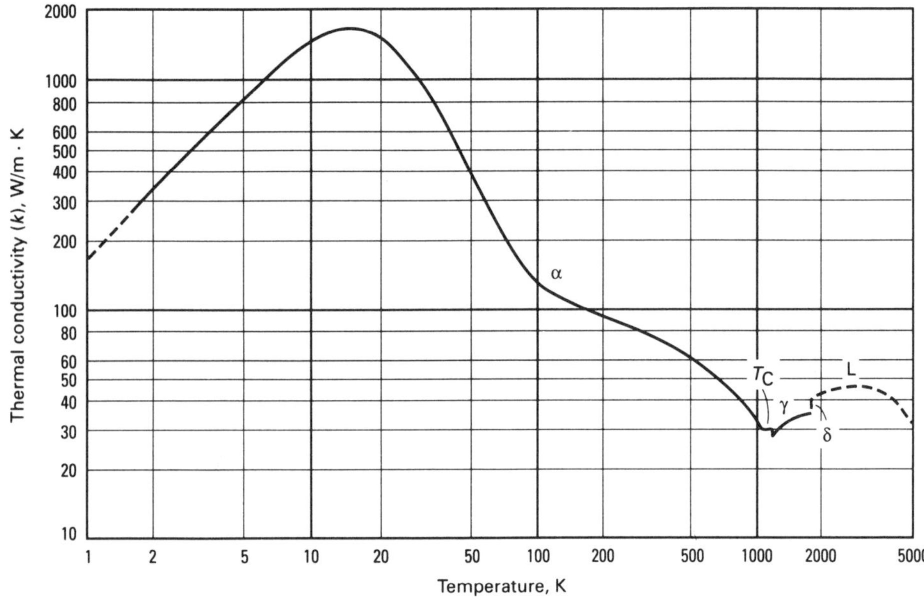

Fig. 22 Thermal conductivity of iron. Source: Ref 17

Table 22 Resistivity ratios of iron

Sample	Impurity, ppm	Ratio ($\rho_{300\,K}/\rho_{4\,K}$)
Arajs B (1965)	200–500	27
Rosenberg (1965)	~100	50, 83
Kempt *et al.* (1959)	~100	110
Arajs (1964)	~80	250
Arajs A (1965)	~25	300
Badiali *et al.* (1963)	~25	380
Takaki and Kimura (1973)	···	>5000

Sources: Ref 25, 28

Magnetic Properties

Magnetic properties versus treatment and composition. Magnetic properties can vary over a wide range, depending on such factors as impurity content (particularly of carbon, sulfur, nitrogen, and oxygen), impurity distribution (high-temperature solution anneal or low-temperature precipitation or aging), grain size, grain orientation, and strain or cold work.

Magnetic susceptibility. Temperature variations of the reciprocal of the mass paramagnetic susceptibility of iron are shown in Fig. 40.

Magnetic permeability. For polycrystal H_2-treated iron, 0.0176 H/m for the initial permeability; 0.314 to 0.352 H/m for the maximum permeability at room temperature (Ref 6). For an iron single crystal: maximum permeability in the [100] direction, 1.80 to 1.82 H/m

Coercive force. For zone-refined iron, the coercive force (H_c) follows the relation $H_c = 1.83 + 4.14/Q^{1/2}$, where Q is the grain size in mm^2. After heat treatment for 10 h at 880 °C in H_2 with a furnace cool, the coercive force at room temperature was reported to be 10.74 A/m. A different treatment (60 h at 1300 °C in H_2, followed by 20 h at 870 °C in H_2, then furnace cooling) gave a coercive force of 1.35 A/m (Ref 47).

Saturation magnetization. 2.158 T at room temperature (Ref 48). Magnetization per atom at 0 K (M_0): 2.216 μ_B (Bohr magnetons), or 2.055 × 10^{-23} J/T (Ref 49)

Residual induction. 1.183 T (Ref 48)

Hysteresis loss. 15 to 19 J/m^3 per cycle (Ref 6)

Magnetostriction. Data for single-crystal and polycrystalline annealed electrolytic iron are shown in Fig. 41. If the directions of magnetization and strain measurement relative to the [100] direction for a cubic crystal are ϕ and ψ, respectively, then the variation of the magnetostriction of a self-saturated domain (which depends on the angular position of the vector **M** at saturation) can be expressed by five constants, as

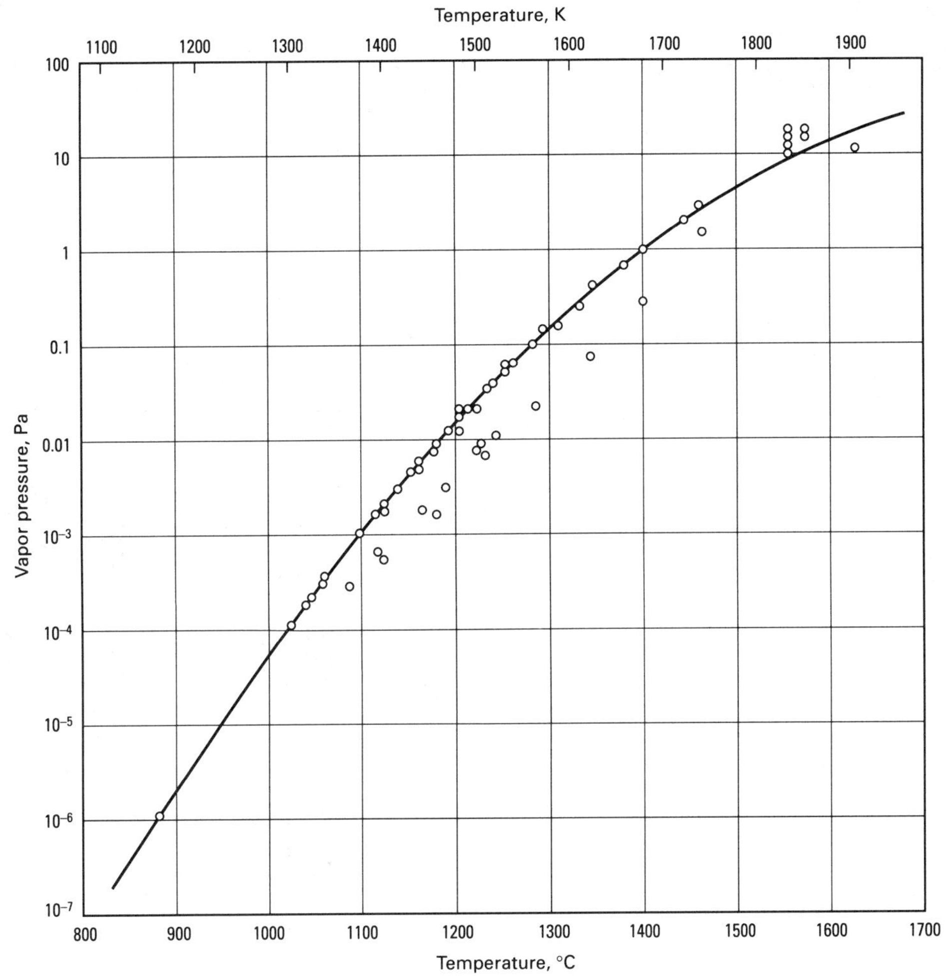

Fig. 23 Vapor pressure of iron. Source: Ref 19

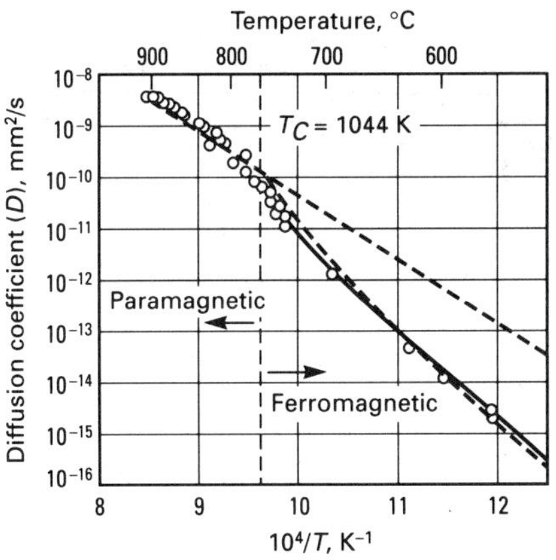

Fig. 24 Arrhenius plot of self-diffusion in para- and ferromagnetic α-iron. Source: Ref 22

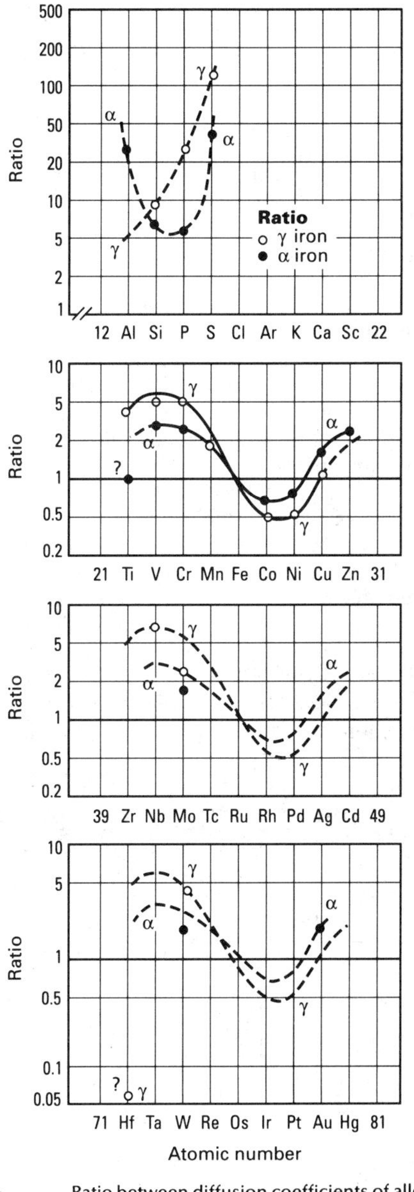

Fig. 25 Ratio between diffusion coefficients of alloy elements and self-diffusion of iron. Source: Ref 21

shown in the formula:

$$\Delta l/l_0 = \{1/8\,(h_1 - h_2 - h_3) + 5/48\,h_4\}\cos 2\phi$$
$$+ \{1/64\,(-6h_3 + h_4 - 2h_5\}\cos 4\phi$$
$$+ \{3/8\,h_1 + 1/8\,h_2 + 5/16\,h_4\}\cos 2\phi\cos 2\psi$$
$$+ \{1/2\,h_2 + 1/8\,h_5\}\sin 2\phi\sin 2\psi$$
$$+ \{3/64\,h_4 + 1/32\,h_5\}\cos 4\phi\cos 2\psi$$
$$+ \{-1/16\,h_5\}\sin 4\phi\sin 2\psi$$

where the room-temperature values for the constants are $h_1 = 36.1 \pm 2.1$, $h_2 = -34.5 \pm 1.4$, $h_3 = -1.2 \pm 0.5$, $h_4 = 3.3 \pm 0.7$, and $h_5 = 0.8 \pm 0.3$ (Ref 51). The variation of these constants with temperature is treated in Ref 52.

Magnetic transformation (Curie) temperature. 1044 ± 2 K (Ref 53).

Optical Properties

Color. Silvery white, resembling platinum more than ingot iron or steel

Reflectance. The normal spectral reflectance of polished iron varies from 65% at a wavelength of 1.5 μm to 97% at 15 μm (Ref 19, 54).

Absorption. The normal spectral absorptance of polished iron varies from about 0.33 at a wavelength of 1.5 μm to 0.03 at 15 μm (Ref 54).

Emittance. Normal spectral emittance of polished iron at about 927 °C: 35% at a wavelength of 0.65 μm, 26% at 1.5 μm, 11% at 15 μm (Ref 54)

Nuclear Properties

Stable isotopes:

Isotope	Atomic weight(a)	Percent of total
^{54}Fe	53.9396	5.82
^{56}Fe	55.9349	91.66
^{57}Fe	56.9354	2.19
^{58}Fe	57.9333	0.33

(a) Relative to ^{12}C. Source: Ref 55

Unstable isotopes:

Isotope	Half-life	Decay mode(a)	Particle energy, fJ
^{52}Fe	8.2 h	β^+, EC	130
^{53}Fe	8.5 min	β^+, EC	450, 380, 260
^{55}Fe	2.6 years	EC	...
^{59}Fe	45.1 ± 0.5 days	β^-	252.0, 76.1, 43.7
^{60}Fe	3 × 10^5 years	β^-	≤22
^{61}Fe	6.0 min	β^-	450

(a) EC, electron capture. Source: Ref 55

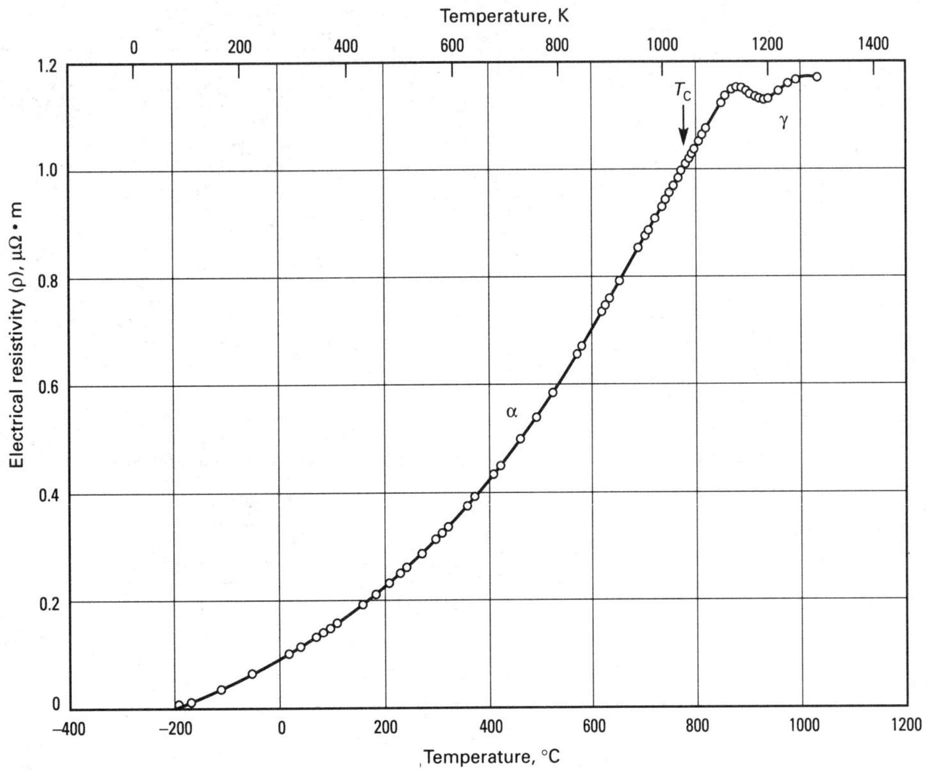

Fig. 26(a) Electrical resistivity of iron from −273 to 1027 °C. Sources: Ref 24, 25

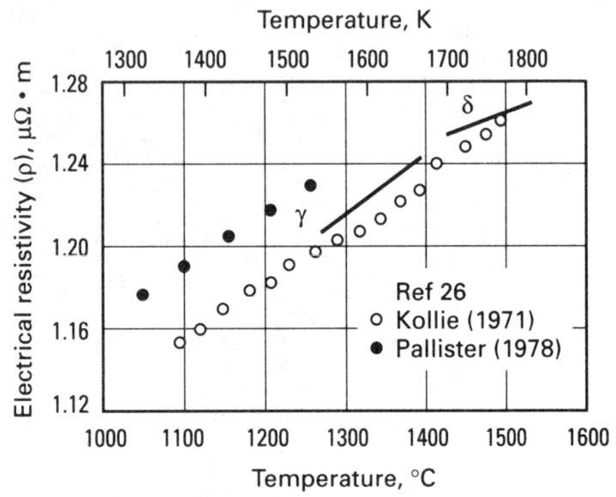

Fig. 26(b) Electrical resistivity of iron from 1027 to 1527 °C. Source: Ref 26

Effects of neutron irradiation. Neutron irradiation affects the properties of materials principally through the production of lattice defects. These defects influence structural, mechanical, electronic (associated with trapping of charge), and diffusion-controlled properties (see Ref 56). The effect of irradiation on yield strength and reduction in area of vacuum-melted iron (0.003% C, 0.0055% O_2, 0.0005% N) is shown in Fig. 42. For a review of mechanical properties of irradiated iron and iron alloys, see Ref 58. Corrosion effects are discussed in Ref 59.

Chemical Properties

General corrosion behavior. The corrosion behavior of iron in aqueous solutions is shown schematically in Fig. 43. Irons of high purity show a remarkably high resistance to corrosion, sometimes remaining untarnished in laboratory atmospheres for months or years (Ref 6). Zone-refined iron corrodes at the same rate in hydrochloric acid whether cold worked or annealed and is not affected by any heat treatment schedule (Ref 61). Results reported in Ref 35 indicate an orientation effect; that is, certain crystal faces are attacked more than others.

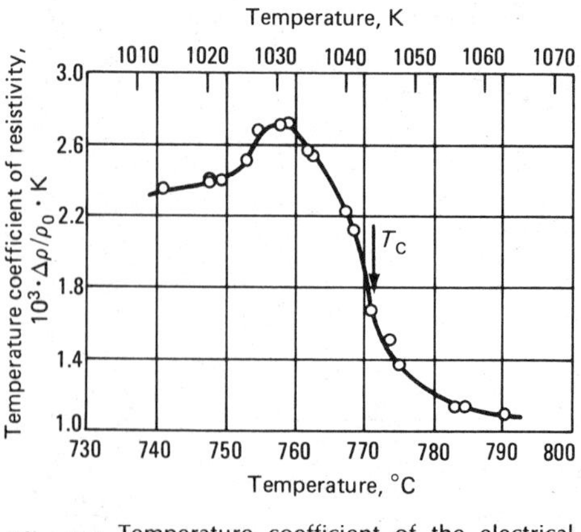

Fig. 27 Temperature coefficient of the electrical resistivity of iron in the neighborhood of the Curie temperature (T_C). Source: Ref 25

Effects of specific corroding agents. Effects of acids on zone-refined iron at 25 °C:

Acid (oxygen-free)	pH	Corrosion rate, g/m^2/day
0.1 *M* citric	2.06	2.9
0.1 *M* malic	2.24	0.3
4% NaCl	1–4	3.0
0.12 *N* HCl	1.01	2.0

Sources: Ref 35, 61, 62

Stability (Pourbaix) diagrams. The regions of stability of various species of iron in water at 25 °C are shown as a function of potential (relative to a standard hydrogen electrode) and pH in Fig. 44.

Mechanical Properties

Tensile properties. The data for zone-refined iron have been summarized in Ref 5. Plots of the temperature dependence of yield stress are shown in Fig. 45. The very strong dependence shown is an inherent characteristic of α iron. The yield stress, however, is not a smooth function of temperature; instead, there is a concave-downward region in the plot, centered at about −30 °C (about 240 K), that has been observed by several researchers (see Ref 5, 64, 65). The scatter in these data for zone-refined iron is due principally to the single-crystal or bamboo-structure specimens that

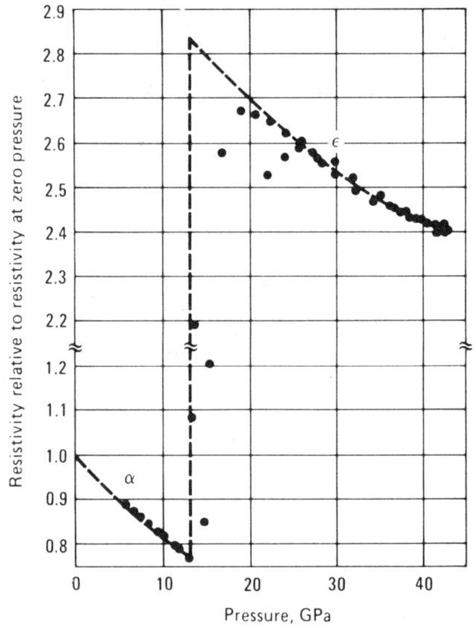

Fig. 28 Pressure dependence of the electrical resistivity of iron. Source: Ref 30

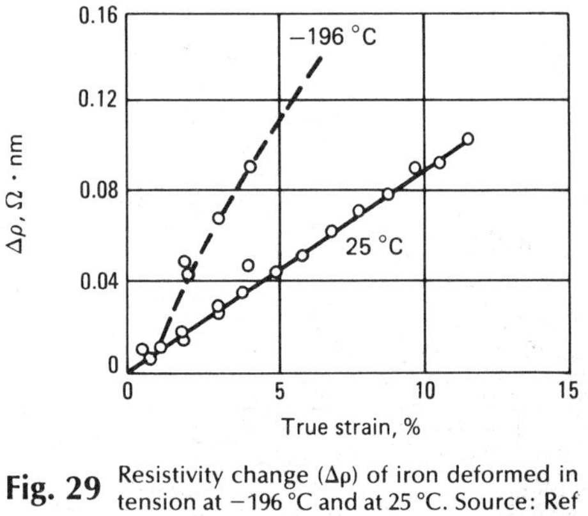

Fig. 29 Resistivity change (Δρ) of iron deformed in tension at −196 °C and at 25 °C. Source: Ref 31

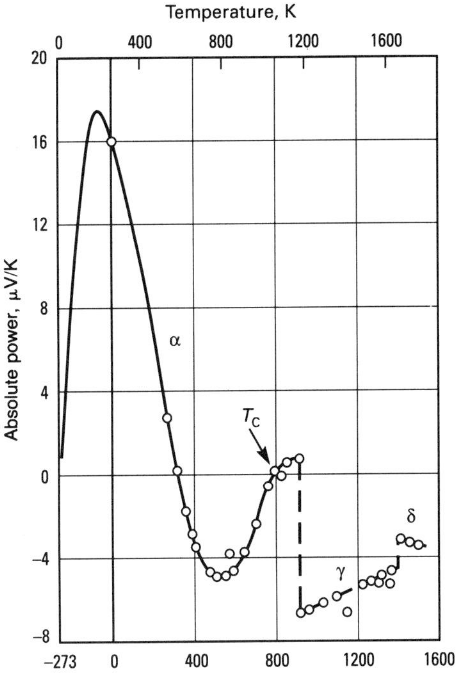

Fig. 30 Thermoelectric power of iron. Source: Ref 32

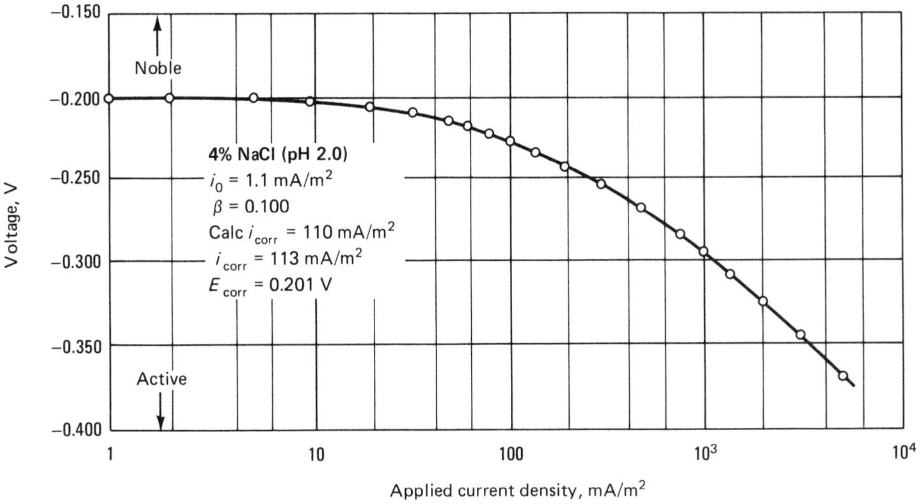

Fig. 31 Hydrogen overvoltage (activation) of iron in aqueous solution of NaCl. Source: Ref 35

were employed. The dependence of the yield stress and flow stress of interstitial-free iron of grain size ASTM No. 5 to 6 on strain rate and temperature is shown in Fig. 46 and 47. The grain size dependence of the yield stress is discussed in Ref 68 and 69.

Compressive properties. The yield strength in compression is the same as in tension. For a further discussion, see Ref 70.

Hardness. Vickers hardness as a function of temperature for pure iron single crystals and crystals containing 300 ppm carbon is shown in Fig. 48(a). The anisotropy of the hardness of pure iron single crystals is shown in Fig. 48(b).

Poisson's ratio. 0.291 at room temperature (Ref 69)

Strain-hardening exponent. About 0.3

Elastic moduli. At room temperature (Ref 72):

- Young's modulus: 208.2 GPa
- Bulk modulus: 166.0 GPa
- Shear modulus: 80.65 GPa
- Compressibility: 6.024 TPa^{-1}

First-, second-, and third-order elastic stiffness values are given in Ref 73.

Elastic moduli along crystal axes. The directional dependence of Young's modulus and shear modulus is shown in Fig. 49(a). The directional dependence of Poisson's ratio is shown in Fig. 49(b).

Properties of single crystals. Plastic-flow characteristics of single crystals are described in Ref 75. Three-stage hardening was observed at room temperature for all crystal orientations. At −130 and −196 °C (143 and 77 K), the critical resolved shear stress law was not obeyed. The critical resolved shear stress was greater in an antitwinning direction than in a twinning direction.

Ductile-to-brittle transition temperature (DBTT). For interstitial-free iron (Charpy V-notch, ½ size): −34 °C for ASTM grain

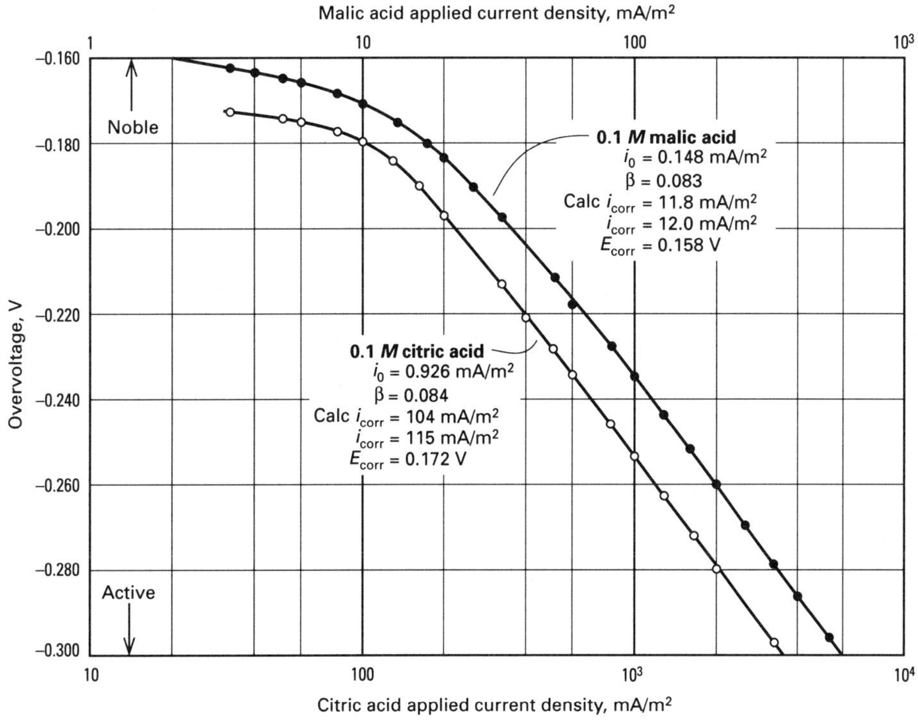

Fig. 32 Hydrogen overvoltage (activation) on iron in citric acid and malic acid. Source: Ref 35

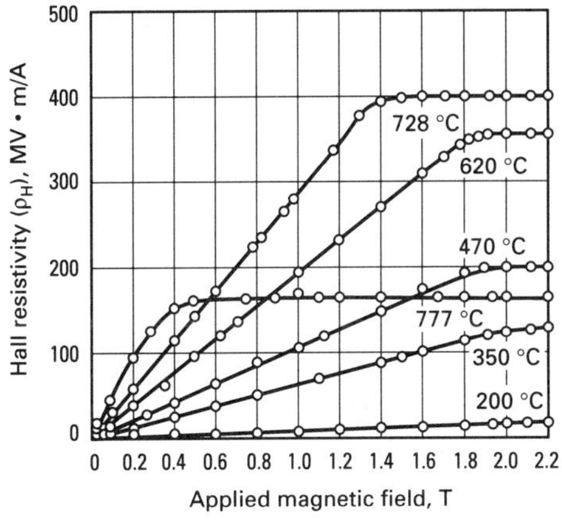

Fig. 33 Temperature dependence of the Hall resistivity of iron. Source: Ref 36

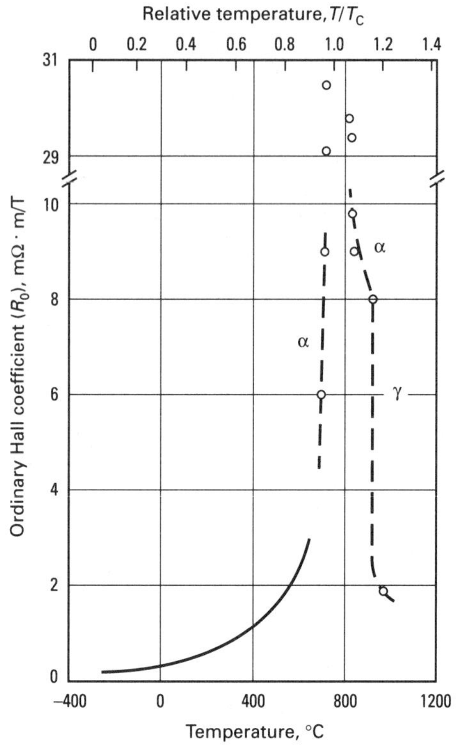

Fig. 34 Temperature dependence of the ordinary Hall coefficient of iron. Source: Ref 37

size No. 4 to 5; −29 °C for ASTM grain size No. 0 to 2 (Ref 67)
Viscosity of liquid. See Fig. 50.
Liquid surface tension. See Fig. 51.
Coefficient of friction. See Fig. 52.
Velocity of sound. At room temperature. Longitudinal, 5952 m/s; transverse, 3222 m/s (Ref 79)

REFERENCES

1. J.F. Cannon, *J. Phys. Chem. Ref. Data*, Vol 3 (No. 3), 1974, p 781
2. U.A. Kohlaas, P. Dünner, and N. Schmitz-Pranghe, *Z. Angew. Phys.*, Vol 23 (No. 4), 1967, p 245
3. H. Mao, W.A. Bassett, and T. Takahashi, *Jpn. J. Appl. Phys.*, Vol 38 (No. 1), 1967, p 272
4. J.R. Low, Jr., The Deformation and Fracture of Iron, in *Iron and Its Dilute Solid Solutions*, C.W. Spenser and F.E. Werner, Ed., John Wiley & Sons, 1963
5. H. Matsui, S. Moriya, S. Takaki, and H. Kimura, *Trans. Jpn. Inst. Met.*, Vol 19, 1978, p 163
6. G.A. Moore and T.R. Shives, Comp., Iron (99.9+%), in *Properties and Selection*, Vol 1, 8th ed., *Metals Handbook*, American Society for Metals, 1961, p 1206
7. L. Zwell, G.R. Speich, and W.C. Leslie, *Metall. Trans.*, Vol 4, 1973, p 1990
8. W. Hume-Rothery, Z.S. Basinski, and A.L. Sutton, *Proc. R. Soc. (London) A*, Vol A229, 1955, p 459
9. H. Stehle and A. Seeger, *Z. Phys.*, Vol 146, 1956, p 217
10. A.S. Keh and S. Weissmann, *Electron Microscopy and the Strength of Crystals*, Interscience, 1963, p 231
11. J. Chipman, *Metall. Trans.*, Vol 3, 1972, p 55
12. Y.S. Touloukian, R.K. Kirby, R.E. Taylor, and P.D. Desai, *Thermal Expansion*, Vol 12, *Thermophysical Properties of Matter*, Plenum Publishing, 1970
13. H. Stuart and N. Ridley, *J. Iron Steel Inst.*, Vol 204, 1966, p 711
14. S. Watanabe and T. Saito, *Trans. Jpn. Inst. Met.*, Vol 14, 1973, p 120
15. R.L. Orr and J. Chipman, *Trans. AIME*, Vol 239, 1967, p 630
16. *JANAF Thermochemical Tables*, 2nd ed., U.S. Government Printing Office, June 1971
17. C.Y. Ho, R.W. Powell, and P.E. Liley, *J. Phys. Chem. Ref. Data*, Vol 3, 1974, p 1
18. J.F. Elliott and M. Gleiser, *Thermochemistry for Steelmaking*, Addison-Wesley, 1960
19. Y.S. Touloukian, *Thermodynamic Properties of High Temperature Solid Materials*, Vol 1, MacMillan, 1967, p 604

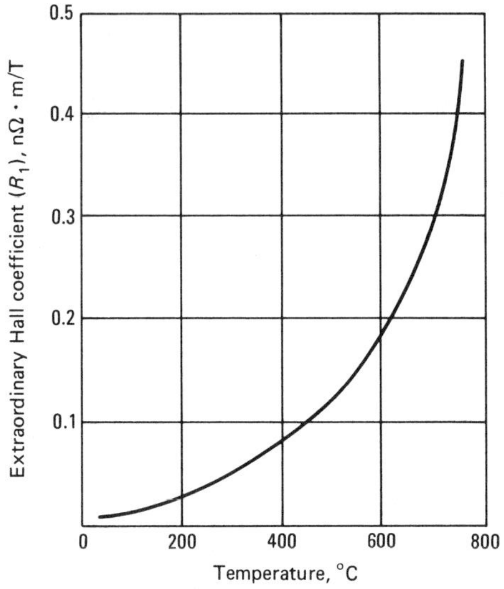

Fig. 35 Temperature dependence of the extraordinary Hall coefficient of iron. Source: Ref 36

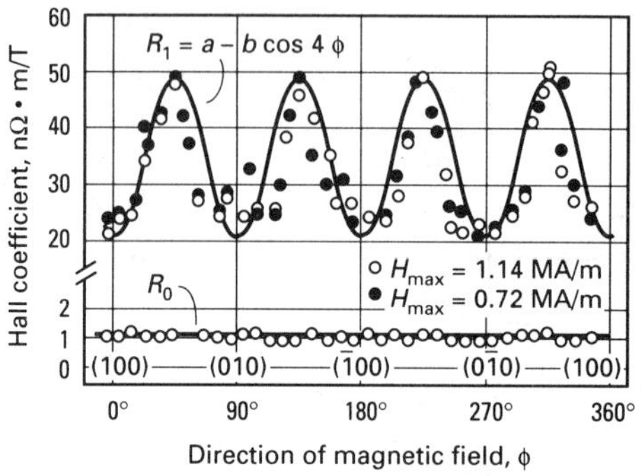

Fig. 36 Effect of magnetic field direction on Hall coefficients of iron at 27 °C. ϕ is the angle between the magnetic field and the [100] axis when current is passed along the [001] axis. Source: Ref 38

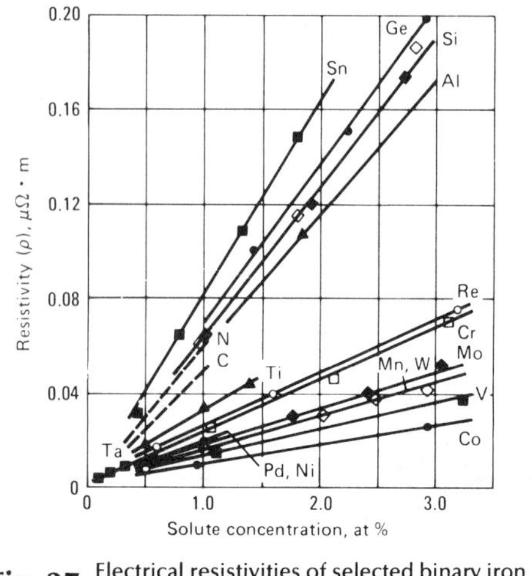

Fig. 37 Electrical resistivities of selected binary iron alloys at −269 °C (4.2 K). Source: Ref 41

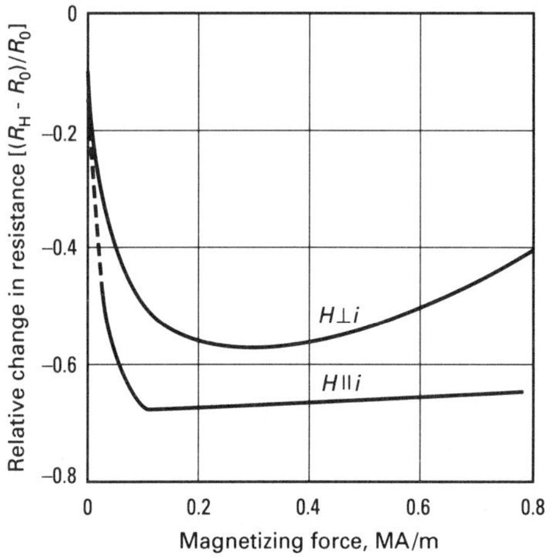

Fig. 38 Magnetoresistance of iron at −269 °C (4.2 K). Curves represent data for the application of the magnetic field strength (H) parallel and perpendicular to the electrical current (i). R_H, resistance in field H; R_0, resistance in no field. R_0 at 293 K is 258 times greater than R_0 at 4.2 K. Source: Ref 45

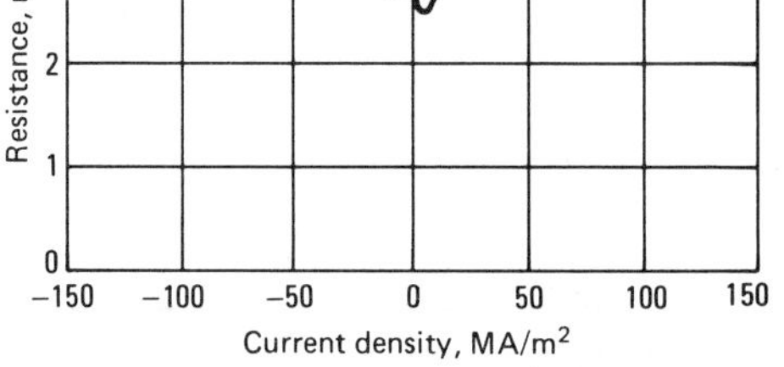

Fig. 39 Residual resistance of iron at −269 °C (4.2 K) as a function of measuring current density. Source: Ref 45

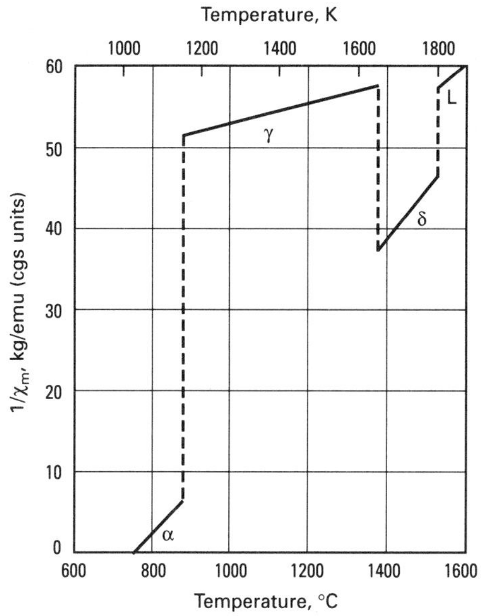

Fig. 40 Temperature dependence of the reciprocal of the mass paramagnetic susceptibility (χ_m) of iron. To change susceptibility values from cgs units to mks units, multiply by 4π. Source: Ref 46

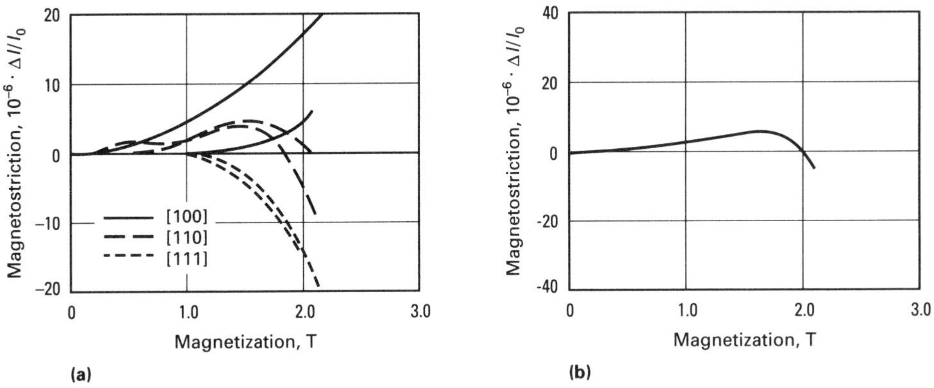

Fig. 41 Longitudinal magnetostriction in (a) single crystals of iron and (b) polycrystalline iron. Source: Ref 50

20. K.M. Myles and A.T. Aldred, *J. Phys. Chem.*, Vol 68 (No. 1), 1964, p 65
21. J. Fridberg, L. Törndahl, and M. Hillert, *Jernkontorets Ann.*, Vol 153, 1969, p 273
22. G. Hettich, H. Mehrer, and K. Maier, *Scr. Metall.*, Vol 11, 1977, p 795
23. H.-E. Shaefer, K. Maier, M. Weller, D. Herlach, A. Seeger, and J. Diehl, *Scr. Metall.*, Vol 11, 1977, p 803
24. S. Soffer, J.A. Dreesen, and E.M. Pugh, *Phys. Rev.*, Vol 140 (No. 2A), 1965, p A668
25. D.S. Miller and S. Arajs, *Mem. Soc. Rev. Met.*, Vol 65, 1968, p 103
26. A. Cezairliyan and J.L. McClure, *J. Res. Natl. Bur. Stand.*, Vol 78A (No. 1), 1974, p 1
27. J.G. Hust and P.J. Giarratano, Special Publication 260-50, National Bureau of Standards, 1975, p 32
28. S. Takaki and H. Kimura, *Scr. Metall.*, Vol 10, 1976, p 701
29. S. Arajs, B.F. Oliver, and J.T. Michalak, *J. Appl. Phys.*, Vol 38, 1967, p 1676
30. A.S. Balchan and H.G. Drickamer, *Rev. Sci. Instrum.*, Vol 32 (No. 3), 1961, p 308
31. K. Tanaka and T. Watanabe, *Jpn. J. Appl. Phys.*, Vol 11 (No. 10), 1972, p 1429
32. M. Shimizu and M. Sakoh, *J. Phys. Soc. Jpn.*, Vol 36 (No. 4), 1974, p 565
33. A.J. deBethune, T.S. Licht, and N. Swendeman, *J. Electrochem. Soc.*, Vol 106 (No. 7), 1959, p 616
34. G.V. Samsonov, *Handbook of the Physiochemical Properties of the Elements*, Plenum Publishing, 1968
35. M. Stern, *J. Electrochem. Soc.*, Vol 102 (No. 12), 1955, p 609
36. I.A. Tsoukalas, *Phys. Status Solidi (a)*, Vol 22, 1974, p K59
37. T. Okamoto, H. Tange, A. Nishimura, and E. Tatsumoto, *J. Phys. Soc. Jpn.*, Vol 17, 1962, p 717
38. A.A. Hirsch and Y. Weissman, *Phys. Lett. A*, Vol 44A (No. 4), 1973, p 239

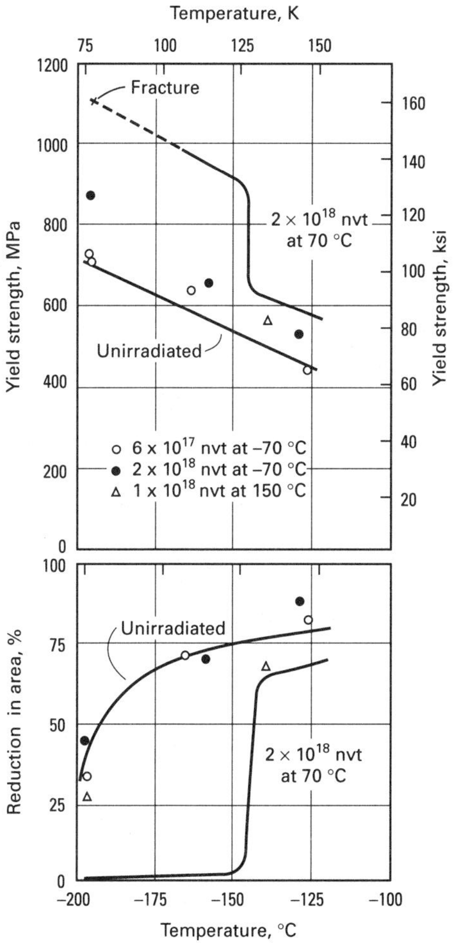

Fig. 42 Effect of temperature on yield strength and reduction in area of iron before and after irradiation. nvt, neutron dose (equivalent to the number of neutrons per square centimeter). Source: Ref 57

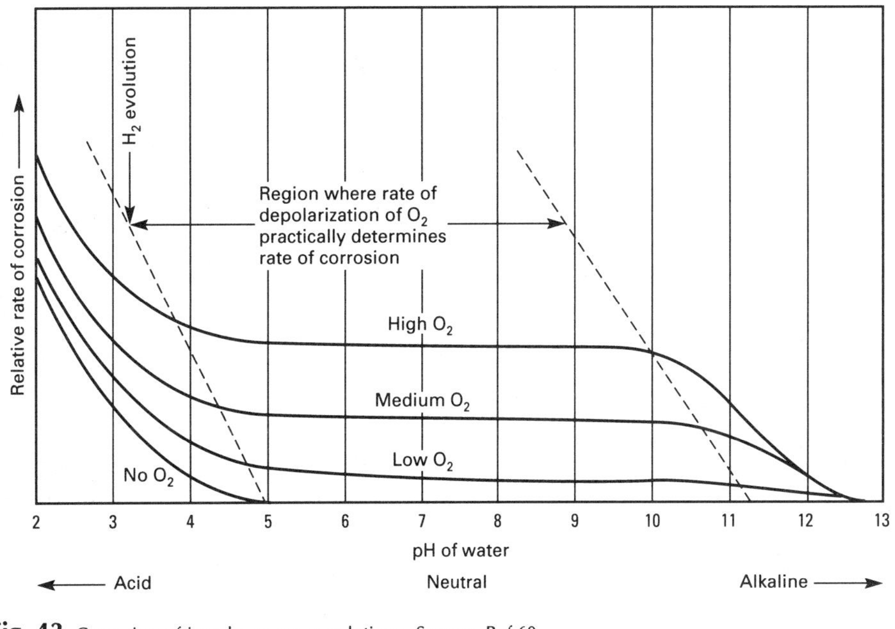

Fig. 43 Corrosion of iron by aqueous solutions. Source: Ref 60

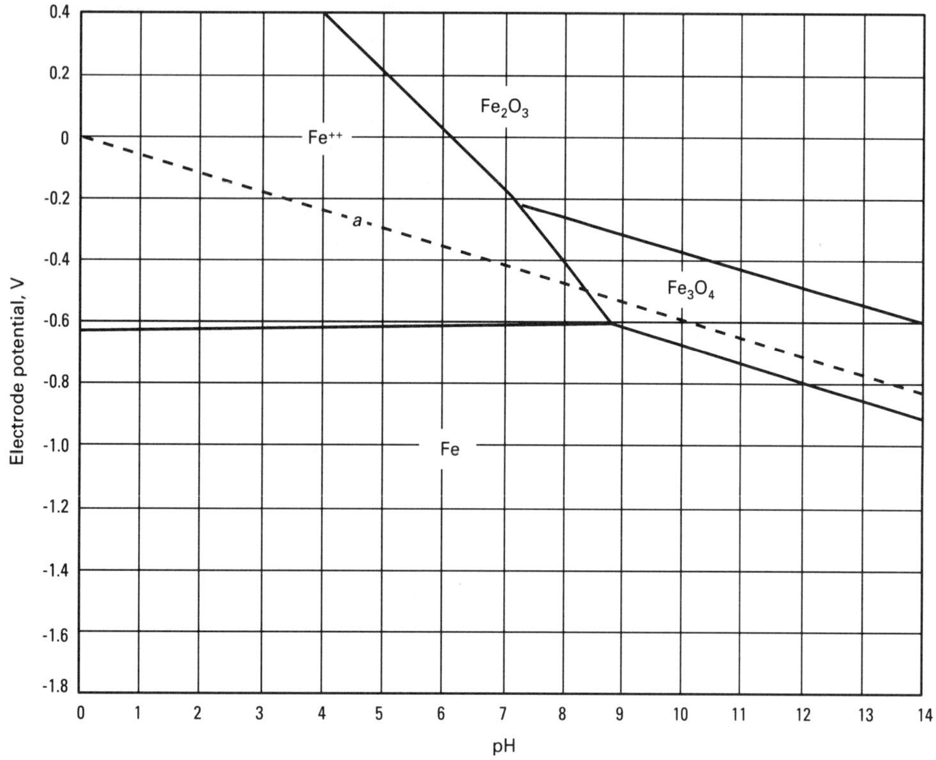

Fig. 44 Pourbaix diagram for the iron-water system at 25 °C. Fe, Fe_3O_4, and Fe_2O_3 are solid substances; water is stable above line *a*, H_2 gas is stable below line *a*. Source: Ref 63

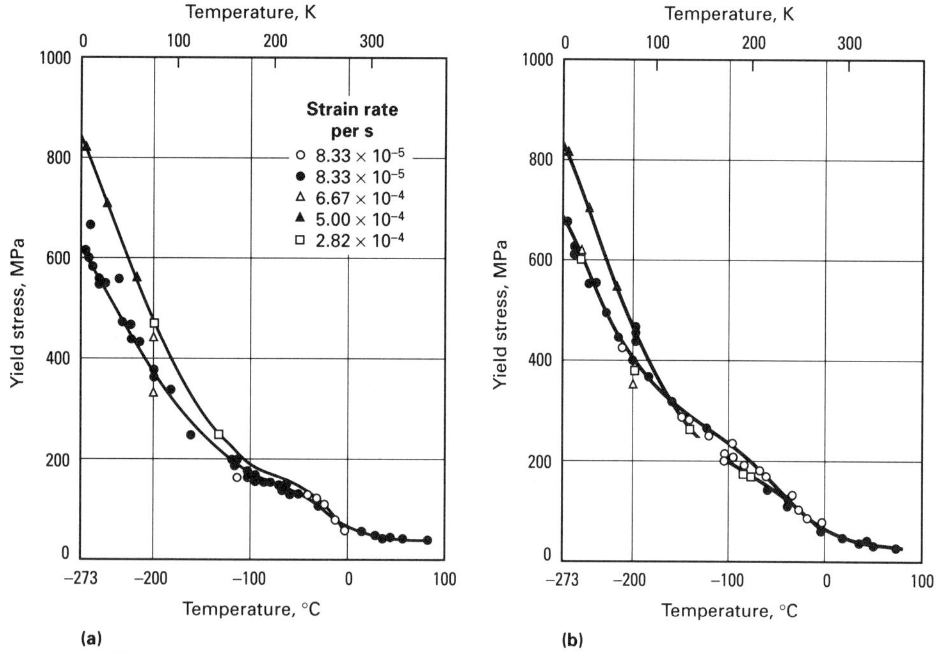

Fig. 45 Temperature dependence of the tensile yield stress of iron. (a) Stress at 0.1% strain. (b) Stress at 0.5% strain. The solid and open circles represent samples zone refined to residual resistivity ratio values of 3600 and >5000, respectively. Data from three other investigations are also shown. Source: Ref 5

39. P.N. Dheer, *Phys. Rev.*, Vol 156 (No. 2), 1967, p 637
40. R.W. Klaffky and R.V. Coleman, *Phys. Rev.*, Vol 10, 1974, p 2915
41. S. Arajs, F.C. Schwerer, and R.M. Fisher, *Phys. Status Solidi*, Vol 33, 1969, p 731
42. V.S. Fomenko, *Handbook of Thermoionic Properties*, Plenum Publishing, 1966
43. K. Ueda and R. Shimizu, *Jpn. J. Appl. Phys.*, Vol 11, 1972, p 916
44. R.V. Hill, E.K. Stefanakos, and R.F. Tinder, *J. Appl. Phys.*, Vol 42, 1971, p 4296
45. J. Frühauf and F. Günther, *Phys. Status Solidi*, Vol 23, 1974, p 399
46. Y. Nakagawa, *J. Phys. Soc. Jpn.*, Vol 11, 1956, p 855
47. A. Hoffman, *Arch. Eisenhüttenwes.*, Vol 40 (No. 12), 1969, p 999
48. H.E. Cleaves and J.M. Heigel, *J. Res. Natl. Bur. Stand.*, Vol 28 (No. 643), 1942, RP1472; J.G. Thompson and H.E. Cleaves, *J. Res. Natl. Bur. Stand.*, Vol 16 (No. 105), 1936, RP860
49. H. Danan, A. Herr, and A.J.P. Meyer, *J. Appl. Phys.*, Vol 39 (No. 2), 1968, p 669
50. F. Brailsford, *Physical Principles of Magnetism*, D. Van Nostrand, 1966, p 147
51. R.D. Greenough, C. Underhill, and P. Underhill, *Physica*, Vol 81B, 1976, p 24
52. G.M. Williams and A.S. Pavlovic, *J. Appl. Phys.*, Vol 39 (No. 2), 1968, p 571
53. S. Arajs and R.V. Colvin, *J. Appl. Phys.*, Vol 35, 1964, p 2424
54. Y.S. Touloukian and D.P. Dewitt, *Thermal Radiative Properties*, Vol 7, *Thermophysical Properties of Matter*, Plenum Publishing, 1970
55. R. Weast, Ed., *Handbook of Chemistry and Physics*, 55th ed., CRC Press, 1974
56. C.O. Smith, *Nav. Eng. J.*, Vol 78 (No. 5), 1966, p 789
57. S.B. McRickard and J.G.Y. Chow, *Acta Metall.*, Vol 14, 1966, p 1195
58. J.G.Y. Chow, S.B. McRickard, and D.H. Gurinsky, *Symposium on Radiation Effects on Metals and Neutron Dosimetry*, STP 341, American Society for Testing and Materials, 1963, p 46
59. V.I. Spitsyn, *Rec. Chem. Prog.*, Vol 31 (No. 1), 1970, p 27
60. F.L. LaQue and N.R. Copson, *Corrosion Resistance of Metals and Alloys*, Reinhold, 1963
61. Z.A. Foroulis and H.H. Uhlig, *J. Electrochem. Soc.*, Vol 111 (No. 5), 1964, p 522
62. M. Stern, *J. Electrochem. Soc.*, Vol 102 (No. 12), 1955, p 663
63. M. Pourbaix, *Atlas of Electrochemical Equilibria in Aqueous Solutions*, Pergamon Press, 1966
64. D. Tseng and K. Tangri, *Scr. Metall.*, Vol 11, 1977, p 719
65. I.J. Diehl, M. Schreiner, S. Staiger, and S. Zwiesele, *Scr. Metall.*, Vol 10, 1976, p 949
66. W.C. Leslie, R.J. Sober, S.G. Babcock, and S.J. Green, *Trans. ASM*, Vol 62, 1969, p 690
67. W.C. Leslie, *Metall. Trans.*, Vol 3, 1972, p 5
68. W.B. Morrison and W.C. Leslie, *Metall. Trans.*, Vol 4, 1973, p 379
69. N. Nagata, S. Yoshida, and Y. Sekino, *Trans. Iron Steel Inst. Jpn.*, Vol 10, 1970, p 173
70. T.L. Altshuler and J.W. Christian, *Philos. Trans. R. Soc. (London) A*, Vol A261, 1967, p 1121
71. T. Takeda, *Jpn. J. Appl. Phys.*, Vol 12

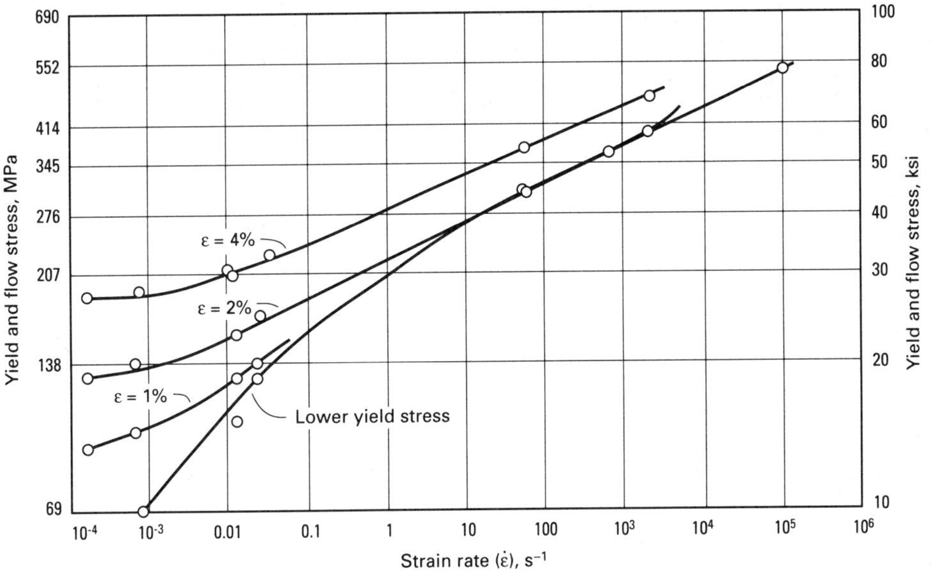

Fig. 46 The effect of strain rate ($\dot{\epsilon}$) on the strength of polycrystalline iron at room temperature. Source: Ref 66

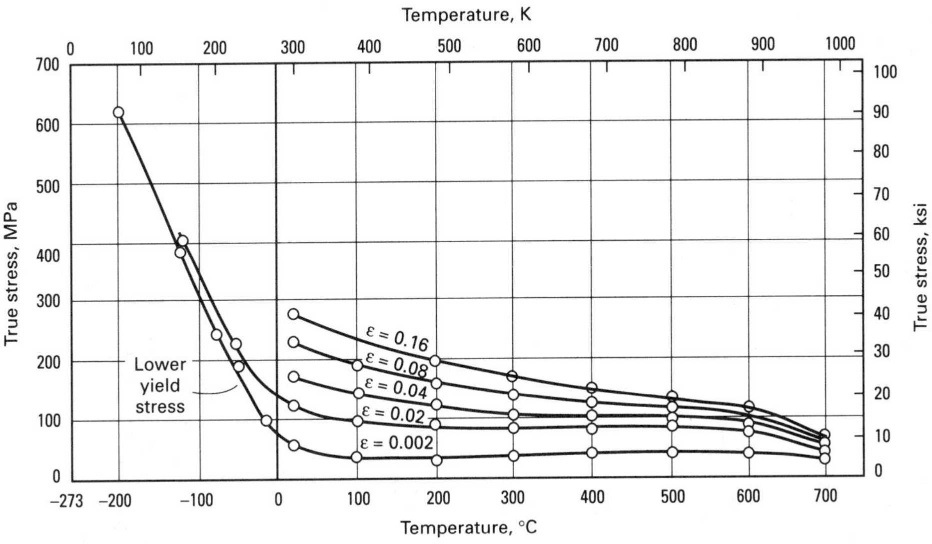

Fig. 47 Temperature dependence of yield and flow stresses in titanium-gettered iron. $\dot{\epsilon} \cong 2.5 \times 10^{-4}$ s^{-1}. Grain size, ASTM No. 5 to 6. Source: Ref 67

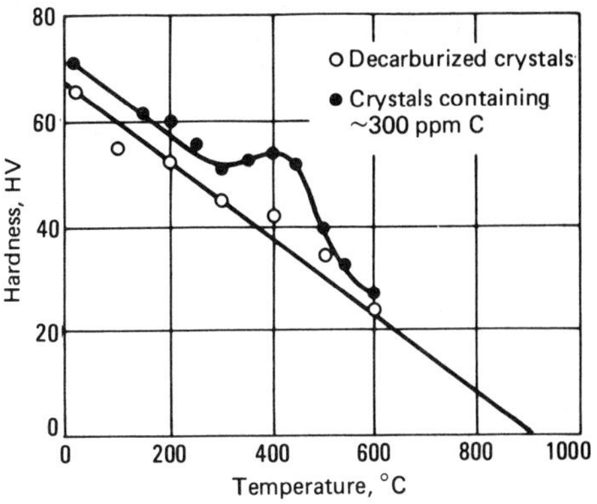

Fig. 48(a) Hardness of iron single crystals as a function of temperature. Source: Ref 71

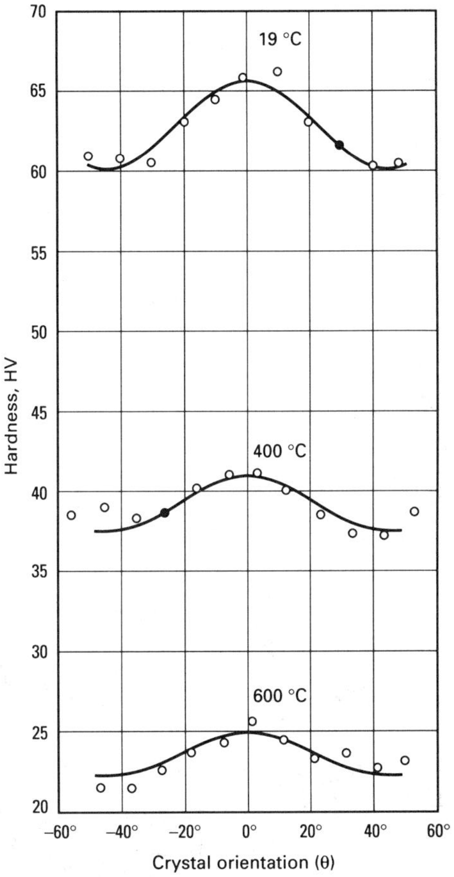

Fig. 48(b) Hardness of decarburized crystals of iron as a function of crystal orientation (θ). At θ = 0, the diagonal of the hardness impression is the projection of the ⟨100⟩ direction on the crystal surface. Source: Ref 71

(No. 7), 1973, p 974

72. G.R. Speich, A.J. Schwoeble, and W.C. Leslie, *Metall. Trans.*, Vol 3, 1972, p 2031
73. H.M. Ledbetter and R.P. Reed, *J. Phys. Chem. Ref. Data*, Vol 2 (No. 3), 1974, p 531
74. H.H. Wawra, *Arch. Eisenhüttenwes.*, Vol 45 (No. 5), 1974, p 317
75. W.A. Spitzig and A.S. Keh, *Acta Metall.*, Vol 18, 1970, p 611
76. Y. Ogino, F.O. Borgmann, and M.G. Frohberg, *Trans. Iron Steel Inst. Jpn.*, Vol 14, 1974, p 84
77. R. Murarka, W.-K. Lu, and A.E. Hamielec, *Metall. Trans.*, Vol 2, 1971, p 2949
78. M.B. Peterson, J.J. Florek, and R.E. Lee, *ASLE Trans.*, Vol 3, 1960, p 101
79. K.H. Schramm, *Z. Metallkd.*, Vol 53 (No. 11), 1962, p 729

Lanthanum (La)

See the section "Properties of the Rare Earth Metals" in this article.

Lead (Pb)

Compiled by J.F. Smith, Lead Industries Association, Inc. and A.T. Balcerzak, St. Joe Lead Company

Lead is used in lead acid storage batteries, ammunition, cable sheathing, pipe, sheet, counterweights, bearings, ballast, gaskets, type metal, low-melting alloys, steel coatings, and foil. Applications include sound and vibration control and x-ray shielding. Lead is used as an alloying ingredient in steel and copper alloys to improve machinability; it is also used in many chemicals.

Caution: Lead presents a health hazard, and should not be used to conduct very soft water for drinking, nor should it come in contact with foods. Inhalation of lead dust and fumes should be avoided.

Structure

Crystal structure. Face-centered cubic, a = 0.49489 nm

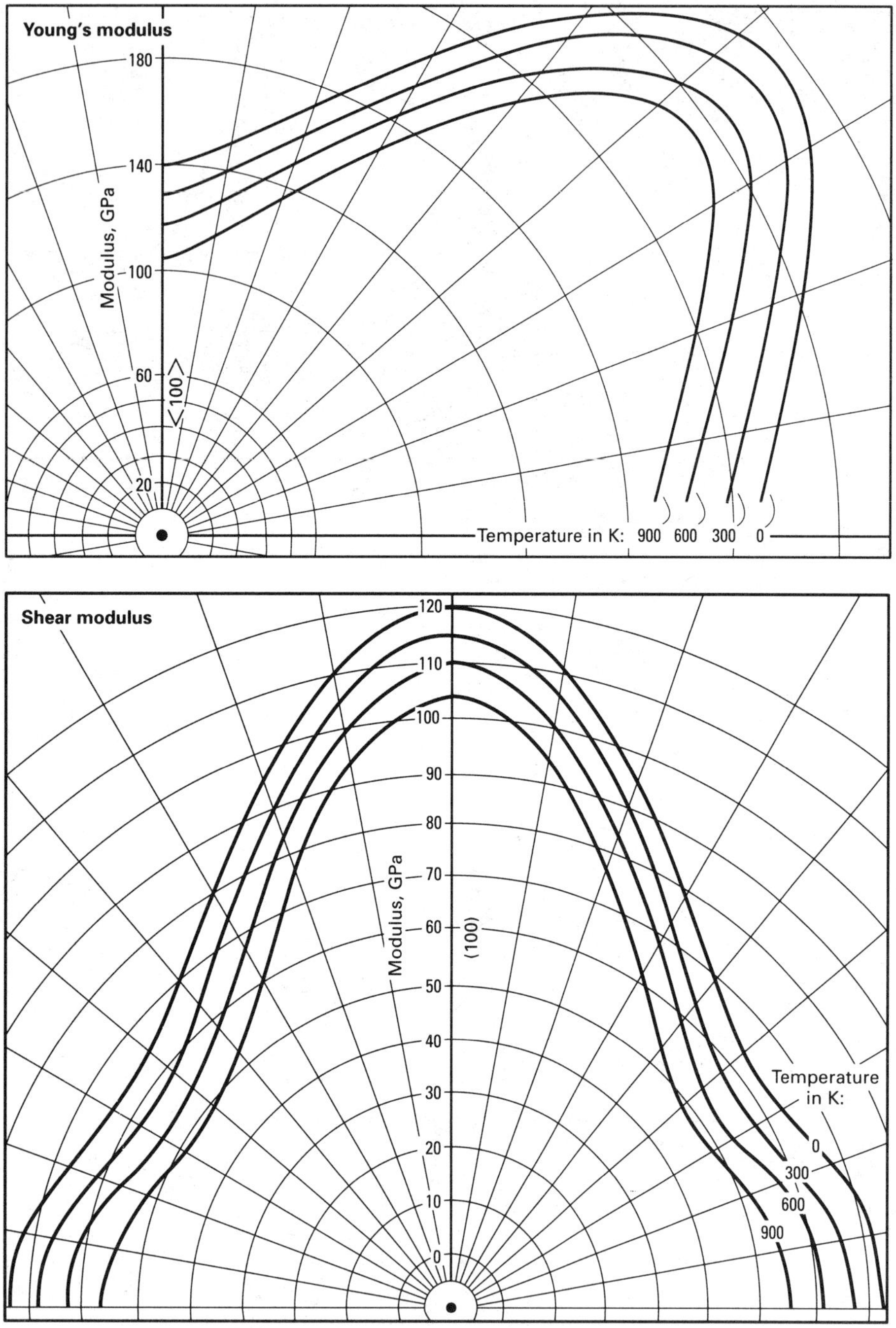

Fig. 49(a) Directional dependence of the Young's modulus and shear modulus values for iron in the (100) plane. Source: Ref 74

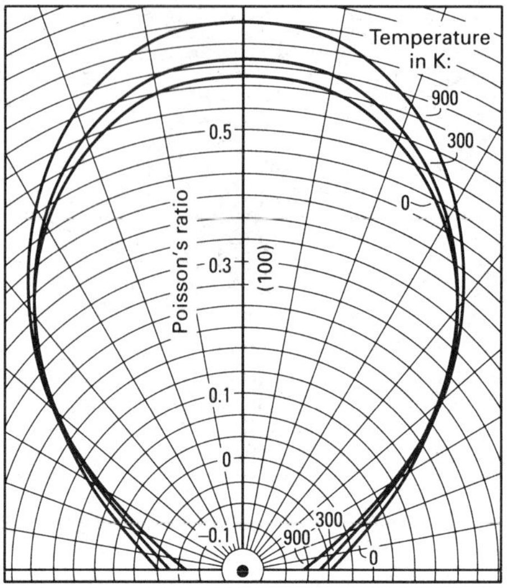

Fig. 49(b) Directional dependence of Poisson's ratio for iron in the (100) plane. Source: Ref 74

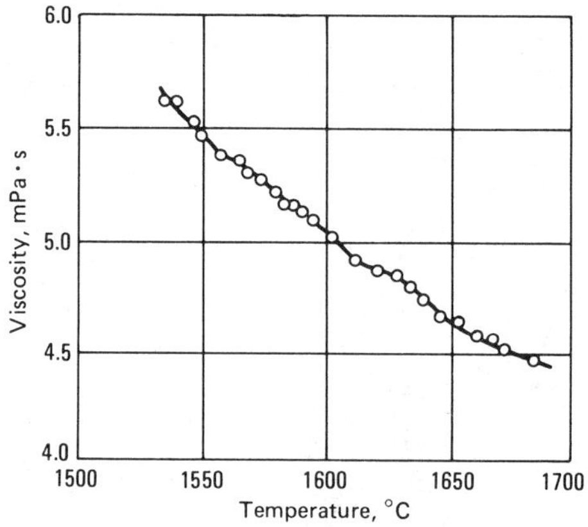

Fig. 50 Temperature dependency of viscosity for liquid iron. Source: Ref 76

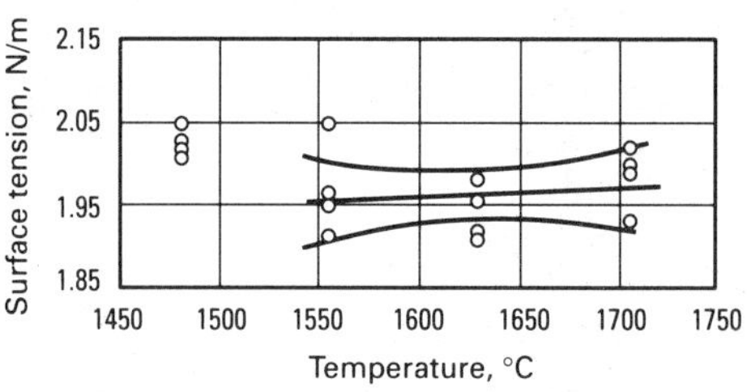

Fig. 51 Surface tension of pure iron. Source: Ref 77

Minimum interatomic distance. 0.3499 nm

Mass Characteristics

Atomic weight. 207.19
Density. See Fig. 53.
X-ray absorption characteristics. See Fig. 1 in the article "Lead and Lead Alloys" in this Volume.

Thermal Properties

Melting point. 327.4 °C
Boiling point. 1750 °C
Coefficient of linear thermal expansion. 26.5 μm/m · K at −190 to 19 °C; 29.3 μm/m · K at 17 to 100 °C
Specific heat and enthalpy:

°C	Specific heat (C_p), kJ/kg · K	Enthalpy (ΔH), kJ/kg
25	0.1287	0
127	0.1320	13.24
227	0.1368	26.67
327.4 (solid)	0.1421	40.76
327.4 (liquid)	0.1479	63.94
427	0.1465	78.53
527	0.1449	93.16
627	0.1433	107.54
727	0.1404	121.81
827	0.1390	135.94

Latent heat of fusion. 22.98 to 23.38 kJ/kg
Latent heat of vaporization. 945.34 kJ/kg
Thermal conductivity. See Fig. 54.
Vapor pressure:

°C	kPa
957	0.1013
1140	1.013
1389	10.13
1750	101.3

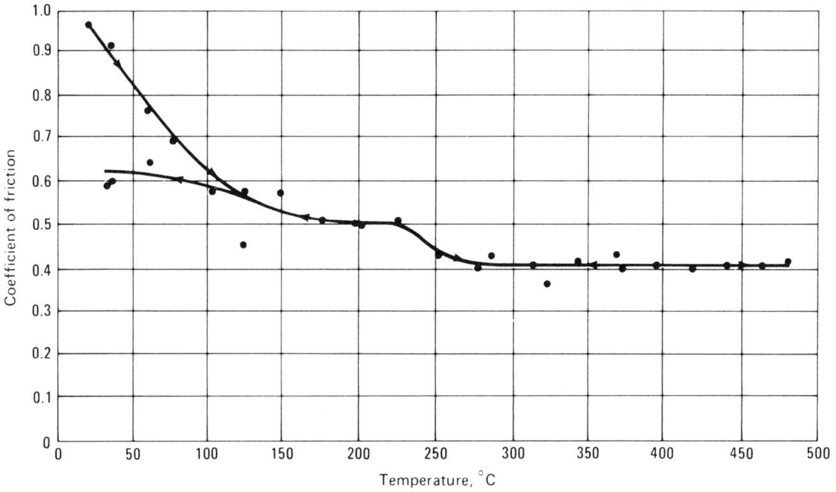

Fig. 52 Effect of temperature on the coefficient of friction for iron against iron during temperature cycling. Load, 1.88 kg; velocity, 7.6 mm/s. Source: Ref 78

Electrical Properties

Electrical resistivity:

°C	nΩ · m
20	206.43
100	270.21
200	363.78
300 (solid)	479.38
340 (liquid)	978.67
400	1014.18

Electrochemical equivalent. Valence +2, 1.0736 mg/C; valence +4, 0.5368 mg/C
Standard electrode potential. 0.122 V versus standard hydrogen electrode
Temperature of superconductivity. 4 K (−269 °C)

Magnetic Properties

Magnetic susceptibility. Volume: -1.5×10^{-6} mks

Optical Properties

Spectral reflectance. 62% at λ = 589 nm

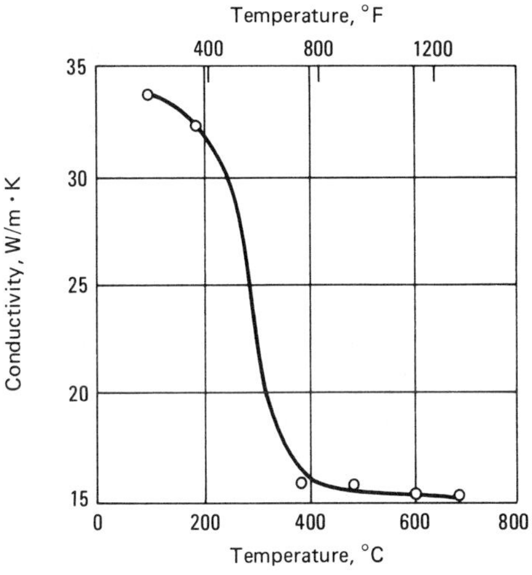

Fig. 54 Temperature dependence of the thermal conductivity of lead

Refractive index. Solid, 2.01 in yellow light; molten, 0.415 for λ = 602 nm
Absorptive index. Solid, 3.48 in yellow light

Chemical Properties

Resistance to specific corroding agents. See the article "Corrosion of Lead and Lead Alloys" in *Corrosion*, Volume 13 of the 9th Edition of *Metals Handbook*.

Mechanical Properties

Damping capacity. See Fig. 2 in the article "Lead and Lead Alloys" in this Volume.
Dynamic liquid viscosity. 1.67 mPa · s
Liquid surface tension. 438 kN/m at 400 °C
Velocity of sound. 1227 m/s at 18 °C

Lithium (Li)

Compiled by J.E. Selle, Oak Ridge National Laboratory

Revised by R.K. Williams, Oak Ridge National Laboratory

Lithium is used as a scavenging agent for inert gases; as an alloying element with aluminum, magnesium, zinc, and lead; in heat transfer applications; in tritium breeding; in the synthesis of organic compounds; and in battery anode material. Compounds containing lithium are used as refrigerant dryers and catalysts, high-temperature lubricants, and reagents in the ceramic and chemical industries.

Lithium is very reactive, and care must be taken to avoid reaction with air, water vapor, or other reactive gases. Airtight containers should be used for containment. Niobium, tantalum, and molybdenum containers are preferred for temperatures above 600 °C. Ferrous alloys are not recommended for long-term use above 550 °C but perform satisfactorily below this temperature.

Structure

Crystal structure. α phase, close-packed hexagonal: $a = 0.3111$ nm; $c = 0.5093$ nm. β phase, body-centered cubic: $a = 0.35089$ nm
Minimum interatomic distance. Distance of closest approach, β phase: 0.3039 nm. Goldschmidt atomic radii, 12-fold coordination: 0.157 nm

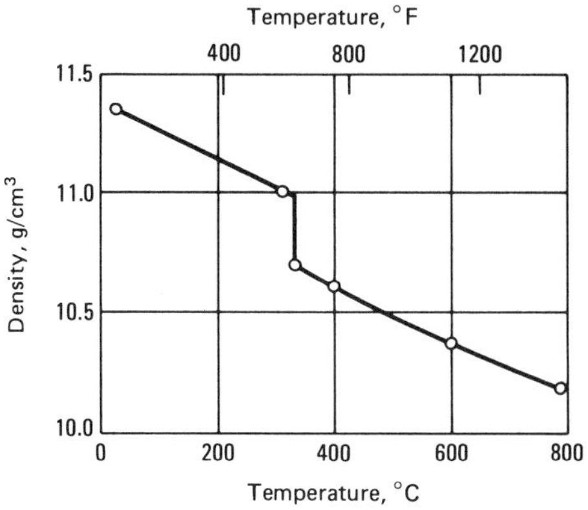

Fig. 53 Temperature dependence of the density of lead

Mass Characteristics

Atomic weight. 6.939
Density. 0.5334 g/cm³ at 20 °C
Density versus temperature. From 200 to 453.7 K:

$$\rho_S = 0.5633 - (8.898 \times 10^{-5})\,T - \frac{1.16}{T}$$

From 453.7 to 1700 K:

$$\rho_L = 0.5584 - (1.01 \times 10^{-4})T$$

where T is in K and ρ is in g/cm³
Expansion on melting. 1.5% of solid volume
Expansion on phase transformation. 0.12% (calculated)

Thermal Properties

Melting point. 180.7 °C
Boiling point. 1336 °C
Phase transformation temperature. α to β phase (cooling), −193 °C
Coefficient of linear thermal expansion. 56 μm/m · K at 20 °C
Specific heat. 3.3054 kJ/kg · K at 20 °C; 3.5146 kJ/kg · K from 0 to 100 °C; 4.2258 kJ/kg · K from 180 to 500 °C
Enthalpy. $H_{st} - H_0 = 666.88$ kJ/kg at 25 °C. Liquid: $H_l - H_{273} = -7.517 \times 10^5 + 4.169 \times 10^3\,T$ J/kg (T in K). Solid: $H_s - H_{273} = -1.030 \times 10^6 + 3.7799 \times 10^3\,T$ J/kg (where T is in K)
Entropy. $S_{st} = 4.219$ kJ/kg · K
Latent heat of fusion. 433.9 kJ/kg
Latent heat of phase transformation. 6.452 kJ/kg
Latent heat of vaporization. 22.73 MJ/kg
Thermal conductivity. 44.0 W/m · K at 180.7 °C
Thermal conductivity versus temperature. From 200 to 453.7 K:

Table 23 Typical chemical compositions of pure magnesium available commercially

Designations	Composition, %											
										Other metallics		
	Al	Ca	Cu	Fe	Mn	Ni	Pb	Si	Sn	Each	Total	Mg(a)
Primary electrolytic	0.005	0.0014	0.0014	0.029	0.06	<0.0005	0.0007	0.0015	<0.0001	<0.05	<0.13(b)	99.87
Magnesium 2	···	···	<0.02	<0.05	<0.01	<0.001	<0.01	···	<0.01	<0.05	<0.10	99.90
Magnesium 3	<0.004	<0.003	<0.005	<0.03	<0.01	<0.001	<0.01	<0.005	<0.005	<0.01	<0.08	99.92
Magnesium 4	<0.002	<0.003	<0.004	<0.03	<0.004	<0.001	<0.005	<0.005	<0.005	<0.01	<0.07	99.93
Magnesium 5	···	···	<0.003	<0.003	<0.004	<0.001	<0.005	<0.005	<0.005	<0.01	<0.05	99.95
Silicothermic	0.007	0.004	<0.001	0.001	0.002	<0.0005	0.001	0.006	0.001	<0.01	<0.04	99.96
High-purity sublimed(c)	0.0004	0.001	0.0002	0.0007	<0.001	<0.0005	<0.0005	<0.001	<0.001	<0.01	<0.02	99.98

(a) Magnesium by difference. (b) 0.006 H_2, 0.0025 N_2, 0.0022 O_2; hydrogen, nitrogen, and oxygen not reported for other grades. (c) Not available commercially. Sources: Ref 1 to 3

$$k_S = 44.00 + 0.02019\,T + \frac{8037}{T}$$

From 453.7 to 1700 K:

$$k_L = 21.42 + 0.05230T - (1.371 \times 10^{-5})T^2$$

where k is in W/m · K and T is in K

Vapor pressure. From 200 to 453.7 K:

$$\log_{10} P = \frac{-8310}{T} + 10.673$$

From 453.7 to 1700 K:

$$\log_{10} P = \frac{-7975.6}{T} + 9.9624$$

where P is in Pa and T is in K

Self-diffusion coefficient. For 195 to 450 °C:

$$D = 1.41\,(\pm 0.12) \times 10^{-7} \left(\exp \frac{2825 \pm 90}{RT}\right) \text{m}^2/\text{s}$$

Electrical Properties

Electrical resistivity. Solid: 93.5 nΩ · m at 20 °C. Liquid: 250 nΩ · m at 180.7 °C

Temperature dependence of electrical resistivity. From 200 to 453.7 K:

$$R_S = \frac{-2.508 \times 10^9}{T^4} + \frac{1.225 \times 10^5}{T^2} - 4.330 + 0.04271\,T$$

From 453.7 to 1700 K:

$$R_L = 5.819 + 0.05282\,T - (2.843 \times 10^{-5})T^2 + 9.474 \times 10^{-8}\,T^3$$

where R is in 10^{-8} Ω · m and T is in K

Thermoelectric potential. Versus platinum (reference junction, 0 °C):

Hot junction temperature, °C	Thermal emf, mV
−200	−1.12
−100	−1.00
100	1.82

Ionization potential. Li(I), 5.39 eV; Li(II), 75.619 eV; Li(III), 122.419 eV

Magnetic Properties

Magnetic susceptibility. Volume: 2.242 × 10^{-3} mks

Nuclear Properties

Stable isotopes. ^{6}Li, isotope mass 6.01512, 7.42% abundance; ^{7}Li, isotope mass 7.01600, 92.58% abundance

Unstable isotopes:

Isotope	Half-life, s	Decay	Energy, MeV
^{5}L(a)	10^{-21}	p	···
^{8}L	0.85	β^-	13
^{9}L	0.17	β^-	13.5 (75%); 11 (25%)

(a) Isotope mass, 5.0125

Thermal neutron cross section. ^{6}Li, 45 ± 10 mb; ^{7}Li, 37 ± 4 mb

Chemical Properties

General corrosion behavior. Lithium tarnishes quickly in oxygen, nitrogen, and moist air. Solubilities of various metallic elements are very sensitive to the lithium purity.

Mechanical Properties

Hardness. 0.6 (Mohs scale)

Dynamic viscosity. μ = 0.645 MPa · s at 180.7 °C; 0.140 MPa · s at 1335 °C. From 453.7 to 1700 K: $\mu = 0.1157 - 1.418 \times 10^{-4}\,T + 4.229 \times 10^{-8}$, where μ is in MPa · s and T is in K

Liquid surface tension. σ = 0.396 N/m at 180.7 °C; σ = 0.240 N/m at 1335 °C. From 453.7 to 1700 K: $\sigma = 0.4738 - (1.627 \times 10^{-4})T$, where σ is in N/m and T is in K

Velocity of sound. For 185 to 827 °C: $\nu = 4784.5 - 0.591\,T$, where ν is in m/s and T is in K

Lutetium (Lu)

See the section "Properties of the Rare Earth Metals" in this article.

Magnesium (Mg)

Compiled by S.C. Erickson, The Dow Chemical Company

Reviewed for this Volume by Carl Vass, Fansteel Wellman Dynamics

Primary magnesium has a minimum purity of 99.8% and must meet definite specifications limiting individual impurities. This purity is sufficient for most chemical and metallurgical uses. Most of the pure magnesium sold is produced electrolytically as primary magnesium. For applications requiring a minimum of specific impurities, special grades of electrolytic magnesium are available. Silicothermic magnesium is produced by thermal reduction of magnesium oxide. High-purity sublimed magnesium is produced by sublimation of primary electrolytic magnesium under vacuum. Typical analyses are shown in Table 23. Unless otherwise indicated, the properties listed for pure magnesium were determined on metal of 99.98+% purity.

Alloyed with small amounts of aluminum, manganese, rare earths, thorium, zinc, or zirconium, magnesium yields alloys with high ratios of strength to weight at both room and elevated temperatures. The alloys have unexcelled machinability, are workable by all common methods, are stable in many atmospheres, and have a high damping capacity.

Magnesium is an active chemical element and reacts with many common chemical oxidizing agents. A number of metals such as thorium, titanium, uranium, and zirconium are prepared by thermal reduction with magnesium. As a catalyst, magnesium is useful for promoting organic condensation, reduction, addition, and dehalogenation reactions. It is useful for the synthesis of complex and special organic compounds by the Grignard process. Its use in pyrotechnics is well established. Magnesium powder can be dispersed in hydrocarbons and mixed in solid propellants for high-energy fuels.

Magnesium alloyed with other metals, such as aluminum, copper, cast iron, lead, nickel, and zinc, improves their properties. It also deoxidizes copper and brass, desulfurizes iron and nickel, and debismuthizes lead.

As a galvanic anode, magnesium provides effective corrosion protection for water heaters, underground pipelines, ship hulls, ballast tanks, and other underground and underwater structures. Small lightweight high-current-output primary batteries use magnesium alloy as the anode. Magnesium has a low-capture cross section for thermal neutrons and a low-level retention of induced radioactivity; these properties make it suitable for varied uses in atomic energy applications.

Structure

Crystal structure. Close-packed hexagonal.

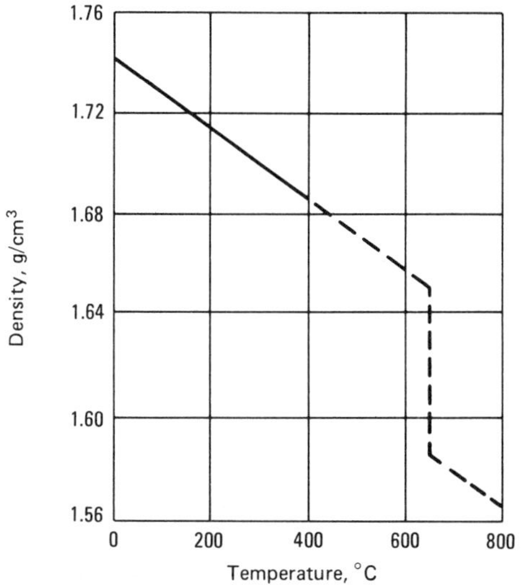

Fig. 55 Temperature dependence of the density of magnesium. Sources: Ref 13 to 15

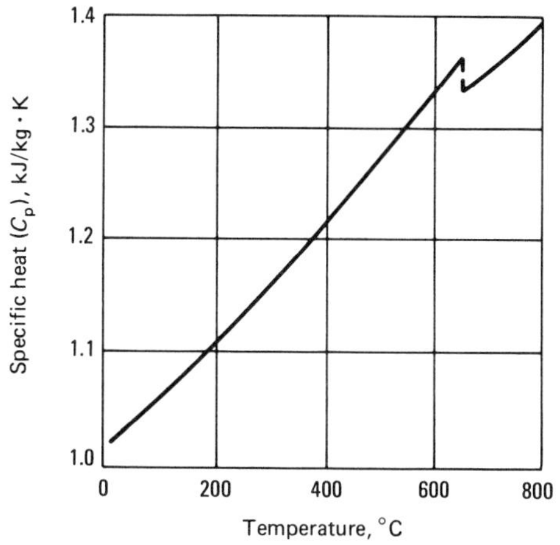

Fig. 56 Temperature dependence of the specific heat of magnesium. Source: Ref 25

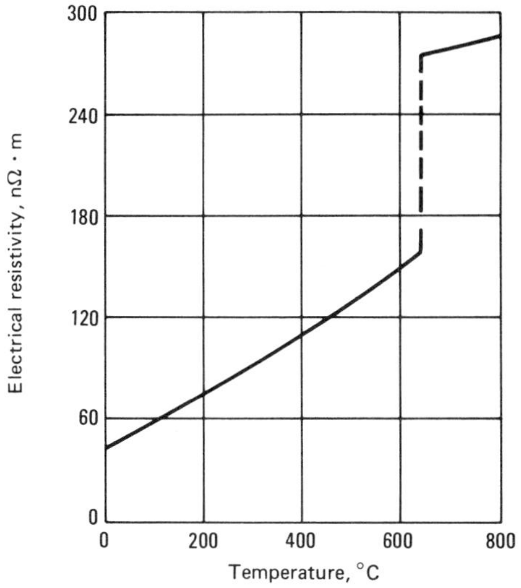

Fig. 57 Temperature dependence of the electrical resistivity of magnesium. Sources: Ref 28 to 30, 32 to 34

At 25 °C: a = 0.32087 ± 0.00009 nm; c = 0.5209 ± 0.00015 nm; c/a = 1.6236 (Ref 4, 5)
Slip planes. Primary (0001), ⟨11$\bar{2}$0⟩; secondary, {10$\bar{1}$0}, ⟨11$\bar{2}$0⟩; {10$\bar{1}$1}, ⟨11$\bar{2}$0⟩ at elevated temperatures (Ref 6 to 9)
Twinning planes. Primary {10$\bar{1}$2}; secondary, {30$\bar{3}$4}; {10$\bar{1}$3} at elevated temperatures (Ref 10, 11)
Cleavage plane. No definite cleavage plane (Ref 6, 12)
Minimum interatomic distance. 0.3196 nm
Fracture type. See Ref 6 and 12.

Mass Characteristics

Atomic weight. 24.312
Density. 1.738 g/cm^3 at 20 °C; solid, approximately 1.65 g/cm^3 at 650 °C; liquid, approximately 1.58 g/cm^3 (Ref 13, 14). See also Fig. 55.
Volume change on freezing. 4.2% shrinkage
Volume change during cooling. From 650 (solid) to 20 °C: 5% shrinkage

Thermal Properties

Melting point. 650 °C (Ref 16, 17)
Boiling point. 1107 ± 10 °C (Ref 3, 17, 18)
Thermal expansion. Polycrystalline at 20 °C: 25.2 μm/m · K (Ref 19 to 23). Values for all magnesium alloys are approximately the same: $L_t = L_0\,[1 + (24.8t + 0.0096t^2) \times 10^{-6}]$, where t is in °C. Along crystal axes, from 15 to 35 °C: 27.1 μm/m · K along a axis; 24.3 μm/m · K along c axis (Ref 24)
Specific heat. 1.025 kJ/kg · K at 20 °C (Ref 25). See also Fig. 56.
Latent heat of fusion. 360 to 377 kJ/kg (Ref 25)
Latent heat of sublimation. 6113 to 6238 kJ/kg at 25 °C (Ref 17)
Latent heat of vaporization. 5150 to 5400 kJ/kg (Ref 17)
Thermal conductivity. 418 W/m · K at 20 °C (Ref 26)
Temperature dependence: $K/T = 0.017 + 2.26 \times 10^{-5}/\rho$, where ρ is in nΩ · m, T is in K, and K is in W/m · K (Ref 2)
Heat of combustion. 24 900 to 25 200 kJ/kg Mg

Electrical Properties

Electrical conductivity. 38.6% IACS
Electrical resistivity. Polycrystalline at 20 °C: 44.5 nΩ · m (Ref 20, 27). Liquid at 650 °C: 247 nΩ · m (Ref 28 to 30). Along crystal axes at 20 °C: 44.8 nΩ · m along a axis; 37.4 nΩ · m along c axis (Ref 31). Temperature coefficient: polycrystalline at 20 °C, 0.165 nΩ · m per K (Ref 27). Along crystal axes at 20 °C: 0.165 nΩ · m per K along a axis; 0.143 nΩ · m per K along c axis (Ref 31). Effect of temperature, see Fig. 57. Effect of alloying, see Fig. 58
Contact potential. +0.44 mV versus platinum at 0 to 100 °C (Ref 35); −0.222 mV versus copper at 27 °C (Ref 27)
Electrochemical equivalent. 126 mg/C
Electrolytic solution potential. 1.63 mV versus saturated calomel electrode at 25 °C in aerated 3% NaCl solution (Ref 36)

Magnetic Properties

Magnetic susceptibility. Mass: 0.00627 to 0.00632 mks (Ref 37)
Magnetic permeability. 1.000012

Optical Properties

Reflectivity. 72% at λ = 0.500 μm; 74% at λ = 1.00 μm; 80% at λ = 3.0 μm; 93% at λ = 9.0 μm (Ref 38)
Refractive index. 0.37 at λ = 0.589 μm (Ref 39)
Absorption constant. 4.42 at λ = 0.589 μm

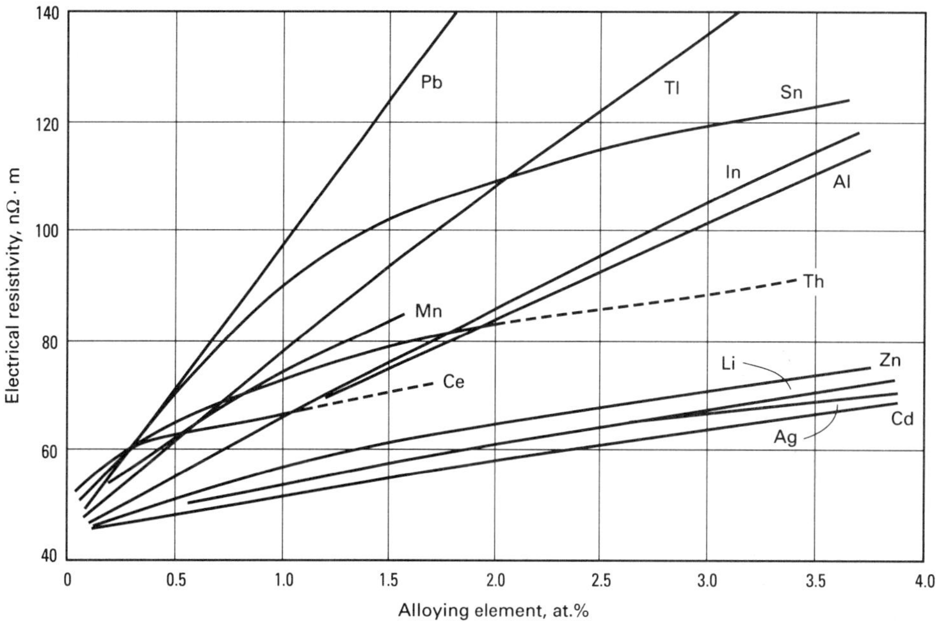

Fig. 58 Effect of alloying additions on the electrical resistivity of magnesium. Sources: Ref 14, 27

Table 24 Typical mechanical properties of magnesium at 20 °C

Form and section	Tensile strength, MPa	0.2% tensile yield strength, MPa	0.2% compressive yield strength, MPa	Elongation in 50 mm (2 in.), %	Hardness HRE	Hardness HB(a)
Sand cast, 13 mm (½ in.) diam	90	21	21	2–6	16	30
Extrusion, 13 mm (½ in.) diam	165–205	69–105	34–55	5–8	26	35
Hard rolled sheet	180–220	115–140	105–115	2–10	48–54	45–47
Annealed sheet	160–195	90–105	69–83	3–15	37–39	40–41

(a) 500 kg load; 10 mm diam ball

(Ref 39)
Color. Bright silvery white
Emissivity. 0.07 at 22 °C (Ref 40)

Nuclear Properties

Thermal neutron absorption cross section. 0.063 ± 0.004 b (Ref 41)

Chemical Properties

General resistance to corrosion. The corrosion resistance to magnesium is dependent on surface film formation; the rate of formation, solution, or chemical change of the film varies with the medium to which it is exposed and also with the alloying elements or impurities present in the metal. Magnesium has good resistance to both indoor and outdoor atmospheres and, in the absence of galvanic couples, even shows resistance to more aggressive environments such as seawater. Indoor tarnishing is controlled largely by the relative humidity. In mild marine and industrial inland atmospheres, the degree of corrosion resistance far exceeds that of mild steel. In stagnant distilled water at room temperature, magnesium forms a protective film that stops action (Ref 42).
Resistance to specific agents. The action of salt solutions on magnesium is dependent on both the anion and the cation of the dissolved salt. Neutral solutions of heavy metal salts will generally cause severe attack. Magnesium suffers little, if any, attack in alkalies, chromates, fluorides, nitrates, or phosphates; more vigorous corrosion occurs in solutions of chloride, bromides, iodides, and sulfates.

Mineral acids, except hydrofluoric and chromic acids, dissolve magnesium rapidly. Aqueous solutions of organic acids attack magnesium, whereas fatty acids (hot or cold, dry or containing water) do not.

Magnesium is not affected by aliphatic and aromatic hydrocarbons, ketones, ethers, glycols, and alcohols, with the exception of anhydrous methyl alcohol. The latter reaction is inhibited, but not completely suppressed, by the presence of water in the methyl alcohol.

Pure halogenated organic compounds do not attack magnesium at ordinary temperatures, but at elevated temperatures, or if water is present, corrosion can be severe. No marked reaction was found to occur between magnesium and methyl chloride, carbon tetrachloride, or chloroform, even after prolonged heating under increased pressures.

Lower alkyl halides, up to amyl derivatives, have been shown to react with magnesium only under pressure and at temperatures in excess of 270 °C, but higher alkyl halides are reported to react with magnesium at their boiling points. In general, the presence of water greatly stimulates the reaction between magnesium and halogenated compounds at elevated temperatures. Fluorinated hydrocarbons are generally without action on magnesium when dry (Ref 42). Additional information is available in the articles "Selection and Application of Magnesium and Magnesium Alloys" in this Volume and "Corrosion of Magnesium and Magnesium Alloys" in Volume 13 of the 9th Edition of *Metals Handbook*.

Fabrication Characteristics

Casting temperature. 705 to 760 °C
Type of flux. Open-pot melting, Dow No. 250; crucible melting, Dow No. 310
Precautions in melting. Molten metal must be protected from the atmosphere by the use of inert gas or protective fluxes. Molten magnesium does not react with carbon, silicon carbide, or combinations of these materials. There is little, if any, reaction with molybdenum, tungsten, or tantalum. Low-carbon (welded or cast) steel crucibles are used as containers for molten magnesium of commercial purity; nickel-bearing steels should not be used for this purpose. Use a protective agent (Dow No. 181) to prevent magnesium from burning when it is being poured in an open atmosphere. The usual safety precautions observed with any molten metal should be observed. Preheat all tools or metal introduced in molten magnesium. Keep pot settings free from iron scale.
Precautions in fabrication. A supply of an approved extinguishing agent should be readily accessible to any machining, grinding, or similar operations on magnesium (additional information is available in the article "Machining of Magnesium and Magnesium Alloys" in Volume 16 of the 9th Edition of *Metals Handbook*). Good housekeeping and sharp machine tools are the best deterrents to magnesium fires. Heat treating furnaces should have a protective atmosphere, such as SO_2 or BF_3, when operating at high temperatures. Magnesium powder must be kept dry (Ref 43).
Machinability index. For pure magnesium and all magnesium alloys, 500 (free-cutting brass = 100) (Ref 44)
Hot-working temperature range. 93 to above 482 °C for 99.98% Mg; 177 to above 482 °C for 99.80% Mg
Annealing temperature. 150 to 200 °C. Maximum reduction between anneals, 50 to 60% under suitable conditions
Forming temperature. 150 to 200 °C for best results
Joining. Rivet composition, aluminum alloy 5056. Oxyacetylene weld with pure magnesium welding rod, magnesium welding flux, and neutral flame. Resistance welding is satisfactory. Helium arc or argon arc welding is preferred. Use pure magnesium welding rod and no flux.
Recrystallization temperature. 93 °C for 1 h anneal after 30% cold reduction (99.98% Mg); 177 °C for a 1 h anneal after 30% cold reduction; 93 °C for a 1 h anneal after 60% cold reduction (99.80% Mg) (Ref 19)

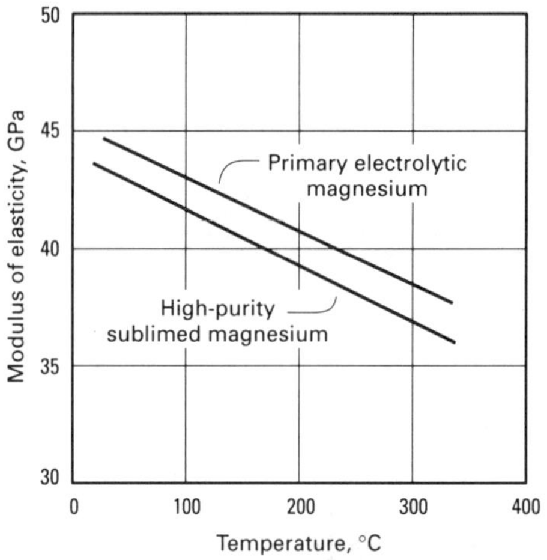

Fig. 59 Temperature dependence of the modulus of elasticity of magnesium. Sources: Ref 46 to 49

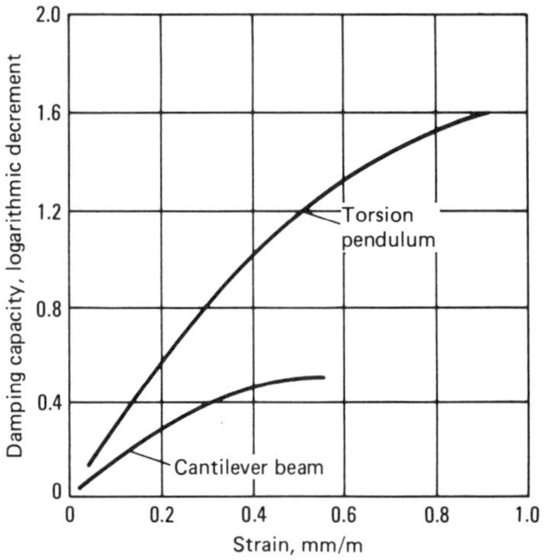

Fig. 60 Damping capacity of magnesium as a function of strain. Source: Ref 50, 51

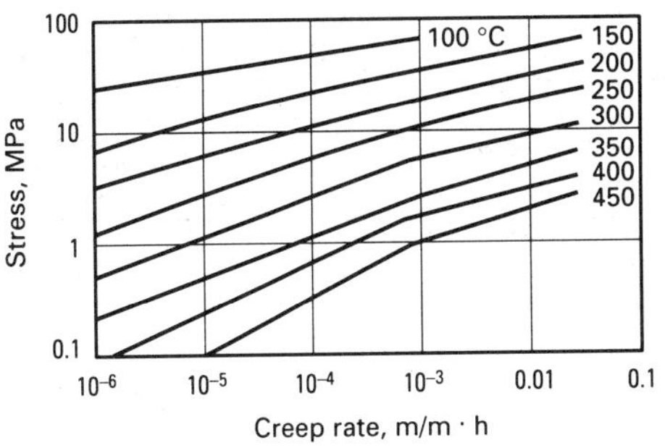

Fig. 61 Minimum creep rate of magnesium as a function of stress and temperature. Source: Ref 52

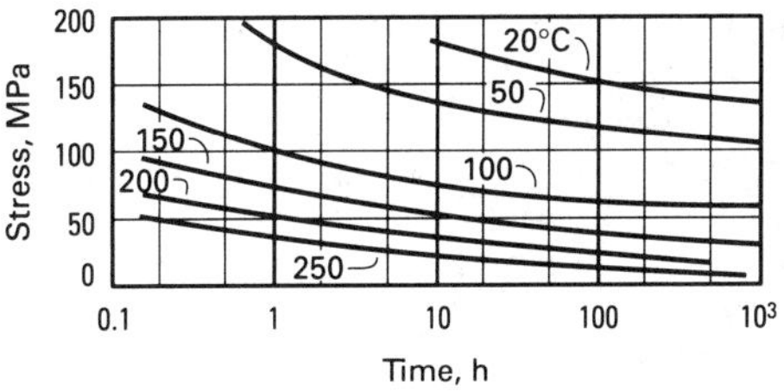

Fig. 62 Stress-rupture life of magnesium as a function of stress and temperature. Source: Ref 53

Mechanical Properties

Tensile properties. See Table 24; see also Ref 45.
Compressive properties. See Table 24.
Hardness. See Table 24.
Elastic modulus. Tension at 20 °C. 99.98% Mg: dynamic, 44 GPa; static, 40 GPa. 99.80% Mg: dynamic, 45 GPa; static, 43 GPa. See also Fig. 59.
Damping capacity. See Fig. 60.
Creep-rupture characteristics. See Fig. 61 and 62.
Dynamic liquid viscosity. At 650 °C, 1.23 mPa · s; at 700 °C, 1.13 mPa · s (approximate values)
Liquid surface tension. At 681 °C, 0.563 N/m; at 894 °C, 0.502 N/m (Ref 54)
Coefficient of friction. 0.36 at 20 °C, magnesium versus magnesium

REFERENCES

1. F.J. Krenske, J.W. Hays, and D.L. Spell, *J. Met.*, Jan 1958, p 28
2. "High-Purity Magnesium," Bulletin TIB 551, Dominion Magnesium, Ltd.
3. W. Leitgehel, *Z. Anorg. Allg. Chem.*, Vol 202, 1931, p 305
4. R.S. Busk, *Trans. AIME*, Vol 188, 1950, p 1460
5. F.W. Batchelder and R.F. Raeuckle, *Phys. Rev.*, Vol 105, 1957, p 59
6. F.E. Hauser, P.R. Landon, and J.E. Dorn, *Trans. ASM*, Vol 48, 1956, p 986; *Trans. ASM*, Vol 206, 1956, p 589
7. A.R. Chaduri, H.C. Chang, and N.J. Grant, *Trans. AIME*, Vol 203, 1955, p 682
8. Technical Report 55-241, Wright Air Development Center, Dow Chemical Company, Aug 1955
9. R.E. Reed-Hill and W.D. Robertson, *J. Met.*, Vol 209, April 1957, p 496
10. S.L. Couling and C.S. Roberts, *Acta. Crystallogr.*, Vol 9, 1956, p 972
11. R.E. Reed-Hill and W.D. Robertson, *Acta Metall.*, Vol 5, 1957, p 717
12. R.E. Reed-Hill and W.D. Robertson, *Acta Metall.*, Vol 5, 1957, p 728
13. R.S. Busk, *Trans. AIME*, Vol 194, 1952, p 207
14. Adolf Beck, *The Technology of Magnesium and Its Alloys*, F.A. Hughes and Company, Ltd., 1943
15. H. Grothe and C. Mangelsdorff, *Z. Metallkd.*, Vol 29, 1937, p 352
16. F.D. Rossini, D.D. Wagman, E.H. Evans, S. Levine, and I. Jaffe, Circular 500, National Bureau of Standards, 1952
17. D.R. Stull and G.C. Sinke, *Thermodynamic Properties of the Elements*, Advances in Chemistry Series, No. 18, American Chemical Society, 1956, p 124
18. H. Hartman and R. Schneider, *Z. Anorg. Allg. Chem.*, Vol 180, 1929, p 275
19. R.A. Townsend, in *Metals Handbook*, American Society for Metals, 1948, p 1013
20. P. Hidnert and W.T. Sweeney, *J. Res. Natl. Bur. Stand.*, Vol 1, 1928, p 771
21. K. Scheel, *Z. Phys.*, Vol 5, 1921, p 167
22. J.B. Austin, *Physics*, Vol 3, 1932, p 240
23. H. Esser and H. Eusterbrock, *Arch. Eisenhüttenwes.*, Vol 14, 1941, p 341
24. P.W. Bridgman, *Proc. Amer. Acad. Arts Sci.*, Vol 67, 1932, p 27
25. R.A. McDonald and D.R. Stull, *Am. Chem. Soc.*, Vol 77, 1955, p 5293
26. W. Bungardt and R. Kallenbach, *Metallwirtsch. Metallwiss. Tech.*, Vol 4, 1950, p 317
27. E.J. Salkovitz, A.J. Schindler, and F.W. Kammer, *Phys. Rev.*, Vol 105, 1957, p 887
28. F.H. Harn, *Phys. Rev.*, Vol 84 (No. 2), 1951, p 855
29. E. Scala and W.D. Robertson, *Trans. AIME*, Vol 197, 1953, p 1141
30. A. Roll and H. Motz, *Z. Metallkd.*, Vol 48 (No. 5), May 1957, p 272
31. J.L. Nichols, *J. Appl. Phys.*, Vol 26 (No. 4), 1955, p 470
32. R.W. Powell, *Philos. Mag.*, Series 7, Vol 27 (No. 185), 1939, p 677
33. G. Grube and E. Schiedt, *Z. Anorg. Allg. Chem.*, Vol 194, 1930, p 190
34. G. Grube, L. Mohr, and R. Bornhak, *Z. Elektrochem.*, Vol 40, 1934, p 160
35. *Temperature, Its Measurement and Control in Science and Industry*, American Institute of Physics, Reinhold, 1941, p 1308
36. R.E. McNulty and J.D. Hanawalt, *Trans. Electrochem. Soc.*, Vol 81, 1942, p 429
37. M. Gaber, Michigan State University, private communication, 1958
38. W.W. Coblenz, *J. Franklin Inst.*, Vol 170, p 169; *Bull. Natl. Bur. Stand.*, Vol 2, 1906, p 457 and Vol 7, 1911, p 197
39. P. Drude, *Ann. Phys.*, Vol 39, 1890, p 481
40. *Handbook, American Institute of Physics*, McGraw-Hill, 1957
41. "Chart of the Nuclides," Knolls Atomic Power Laboratory, General Electric Company, 1956
42. L. Whitby, Magnesium and Its Alloys, in *Corrosion Resistance of Metals and Alloys*, 2nd ed., Reinhold, 1963
43. "Standard for Magnesium," No. 48, National Fire Protection Association, 1957
44. Report of Independent Research Committee on Cutting Fluids, ASTE, *Automot. Ind.*, Vol 88 (No. 8), 1943, p 48
45. M.W. Toaz and E.J. Ripling, *Trans. AIME*, Vol 206, 1956, p 936
46. J.R. Frederick, Ph.D. dissertation, University of Michigan, 1947
47. J.R. Frederick and C.H. Church, University of Michigan, private communication, 1957
48. D.W. Levinson and W. Graft, Armour Research Foundation, private communication, 1957
49. R.W. Fenn, Jr., *Proc. ASTM*, Vol 58, 1958, p 826
50. R.E. Maringer, Battelle Memorial Institute, private communication, 1956
51. W.A. Babington and G.F. Weissman, Bell Telephone Laboratories, private communication, 1957
52. C.S. Roberts, *Trans. AIME*, Vol 197, 1953, p 1121
53. J.L. Bernard, R. Caillat, and R. Darras, *Progress in Nuclear Energy, Metallurgy and Fuels*, Vol 2, Pergamon Press, 1957
54. V.G. Givov, *Alum. Magnesium Inst.*, Vol 14, 1937, p 99

Manganese (Mn)

Compiled by Howard S. Avery, Consultant

Manganese is a silvery-gray or gray-white metal like iron, but with a faint pinkish tinge. It oxidizes in moist air and in powder form, and thus may appear black. The powder, especially if it contains impurities or iron, may be pyrophoric; therefore, manganese dust suspended in air may be an explosion hazard along with being toxic at levels above 5 mg/m^3. Manganese is attacked by weak acids.

Manganese is most widely used in alloys of iron; lesser amounts are used in other alloys, in batteries (where the oxide MnO_2, or pyrolusite, serves as a depolarizer), and in the chemical, glass, and ceramic industries. Most steels contain manganese at levels ranging from <1% up to several percent. Manganese serves as a scavenger in steel because of its affinity for sulfur; it also moderately enhances hardenability. At con-

Table 25 Specific heat of all phases of manganese at subzero and elevated temperatures

Temperature, K	Temperature, °C	Specific heat, kJ/kg · °C: Solid α	β	γ	δ	Liquid
10	−263	0.00295	0.010	0.0026	...	...
20	−253	0.00985	0.0226	0.0121	...	...
100	−173	0.271	0.282	0.274	...	...
150	−123	0.363	0.378	0.380	...	...
273	0	0.466	0.476	0.493	...	...
293	20	0.475	0.483	0.508	...	...
500	227	0.551	...	...	...	...
980	707	0.674	0.684	...	...	...
1360	1087	...	0.714	0.775	...	...
1400	1127	...	...	0.790	...	...
1500	1227	...	...	...	0.837	...
2000	1727	...	...	...	...	0.838

Sources: Ref 8, 10

centrations of 10 to 14%, it is the basis of the very tough and highly work-hardenable austenitic manganese steel (Ref 1).

For ironmaking and steelmaking, manganese is usually added in the form of spiegeleisen or various grades of ferromanganese. However, some alloys require the high purity of electrolytic manganese because manganese ores and derived ferromanganese may contain phosphorus as a tramp element, and phosphorus degrades the weldability of steels and can cause hot cracking. Iron-chromium-manganese stainless steels and such specialty nonferrous alloys as copper-manganese, manganese-containing brasses and bronzes, some nickel silvers, and manganese-bearing aluminum alloys also use electrolytic manganese, which typically contains (Ref 2):

Impurity	Wt%
Fe	0.0015
Cu	0.001
As	0.0005
Co	0.0025
Ni	0.0025
Pb	0.0025
Mo	0.001
S^{2-}	0.01
S^{6+}	0.014
C	0.002
H_2	0.015

The gas content of electrolytic manganese has a significant effect on some properties. Some degassed highly purified grades of the metal can reach a purity of 99.999%.

Thin films of manganese have interesting properties for strain gages (Ref 3) and for developing spectrally selective surfaces (Ref 4). Thin-film strain gages require a large gage factor, a low-temperature coefficient of resistance, low thermoelectric power, and excellent thermal and temporal stability. Manganese films exhibit very high resistivity (375 μΩ · cm), a very low temperature coefficient of resistance (5×10^{-6} per °C), extremely low thermoelectric power, and excellent thermal and temporal stability.

Quality control is crucial for thin film applications. Optical property measurements must be done with care because film properties are greatly affected by preparation factors such as substrate temperature during deposition, environmental conditions, storage, heat treatment, aging, thickness, surface contamination, and original purity.

Structure

Crystal structure. α phase: $a = 0.89139$ nm at 20 °C. β phase: $a = 0.63145$ nm at 20 °C. γ phase: $a = 0.38624$ nm; $c = 0.940$ nm at 1095 °C. δ phase: $a = 0.30806$ nm at 1134 °C (Ref 5)

Mass Characteristics

Atomic weight. 54.938
Density. Solid at 20 °C and 1 atm: α phase, 7.43 g/cm³; β phase, 7.29 g/cm³; γ phase, 7.18 g/cm³ (Ref 6). Solid at 1246 °C: δ phase, 6.11 g/cm³. Liquid at 1246 °C: 6.01 g/cm³
Volume change on freezing. 1.7% (Ref 6)

Thermal Properties

Melting point. 1246 °C
Boiling point. 2065 °C (Ref 7)
Phase transformation temperatures. α to β, 707 °C; β to γ, 1087 °C; γ to δ, 1138 °C
Coefficient of linear thermal expansion:

Temperature, K	Temperature, °C	Coefficient, μm/m · K
10	−263	0.19
20	−253	0.005
100	−173	9.602
150	−123	14.9
273	0	21.3
293	20	21.7
500	227	26.9
980	707	49.0(a)
1000	727	46.5(b)
1360	1087	45.7(b)
1400	1127	47.6(c)
1500	1227	41.3(d)

(a) α phase. (b) β phase. (c) γ phase. (d) δ phase. Sources: Ref 8, 9

Specific heat. See Table 25 and Fig. 63.
Latent heat of fusion. At melting point, 219 kJ/kg (Ref 8)
Latent heat of transformation. α to β, 40.5 kJ/kg; β to γ, 38.6 kJ/kg; γ to δ, 34.2 kJ/kg (Ref 8)

Table 26 Thermal conductivity for nominally 99% pure Mn at room temperature and subzero temperatures

Temperature, K	Temperature, °C	Thermal conductivity(a), W/m · K
10	−263	1.64
20	−253	2.42
50	−223	4.07
100	−173	5.81
150	−123	6.65
200	−73	7.18
250	−23	7.54
273	0	7.68
293	20	7.79
300	27	7.82

(a) Corrected for thermal expansion; the correction ranged from 0 to 0.02 W/m · K. Sources: Ref 8, 11 to 15

Table 27 Thermal diffusivity of well-annealed and degassed α phase manganese with a residual resistivity ratio of 13

Temperature, K	Temperature, °C	Thermal diffusivity, 10^4 m²/s(a)
10	−263	0.73
20	−253	0.33
50	−223	0.059
100	−173	0.029
150	−123	0.025
200	−73	0.023
250	−23	0.022
273	0	0.022
293	20	0.022
300	27	0.022

(a) Derived from values for thermal conductivity, specific heat, and density. Data corrected for thermal expansion; maximum correction of 0.01×10^4 m²/s. Source: Ref 8

Table 28 Effect of temperature on the electrical resistivity of well-annealed ≥99.99% pure manganese

Temperature, K	Temperature, °C	Resistivity, nΩ · m
0	−273	68.8(a)
10	−263	188.4(a)
20	−253	536(a)
50	−223	1261(a)
100	−173	1321(a)
150	−123	1359(a)
200	−73	1391(a)
250	−23	1419(a)
273	0	1430(a)
293	20	1440
300	27	1442
350	77	1461
400	127	1477
500	227	1501
600	327	1521
700	427	1536

(a) Data valid only for manganese with a residual resistivity of 69.0 nΩ · m. Sources: Ref 8, 18, 19

Latent heat of vaporization. At boiling point, 4110 kJ/kg (Ref 7)
Thermal conductivity. See Table 26.
Thermal diffusivity. See Table 27.
Critical temperature and pressure. 4337 K (4064 °C) and 56.1 MPa (Ref 16)
Vapor pressure. 5.98×10^{-38} Pa at 20 °C; 137 Pa at melting point (1246 °C) (Ref 7)

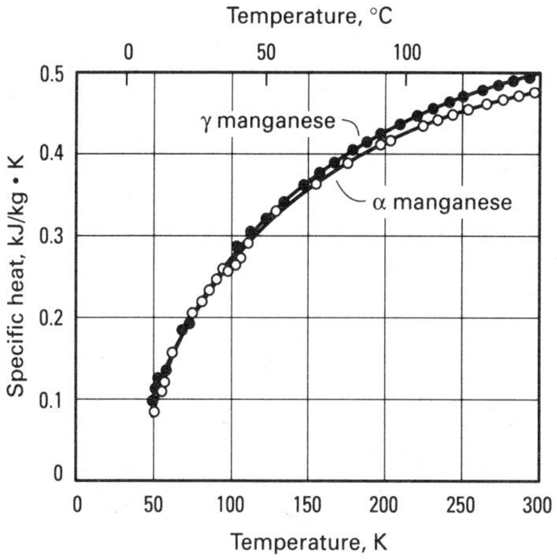

Fig. 63 Effect of temperature on the specific heat of manganese

Electrical Properties

Electrical conductivity. Volumetric, α phase: 0.9% IACS

Electrical resistivity. Largely dependent on gas absorbed. α phase, 1440 nΩ · m at 22 °C (Ref 17); β phase, 910 nΩ · m at 20 °C (Ref 18); γ phase, 400 nΩ · m (Ref 19). See Table 28 for resistivity at subzero and elevated temperatures.

Standard electrode potential. 1.134 V

Hall coefficient. See Table 29.

Thermoelectric power. See Fig. 64.

Work function. 3.83 eV (recommended value) (Ref 25); 4.24 ± 0.02 eV (photoelectric) (Ref 26)

Magnetic Properties

Magnetic susceptibility. See Table 30.

Influence of treatments. Manganese is ferromagnetic after certain treatments—for example, after absorption and expulsion of nitrogen, which leaves the lattice expanded.

Optical Properties

Spectral reflectance. See Fig. 65. See also Ref 27 and 28.

Spectral emittance. Based on emittance of 0.33% at λ = 650 nm for platinum as reference: 0.59 at 1473 K (1200 °C); 0.59 at 1723 K (1450 °C) (Ref 29, 30). Effect of wavelength on emittance (derived from absorption data):

Wavelength, λ μm	Emittance
2.0	0.509
3.0	0.427
4.0	0.362
5.0	0.315
6.0	0.218
7.0	0.256
8.0	0.237
9.0	0.222
10.0	0.210
15.0	0.172
20.0	0.152
23.0	0.143

Refractive index. See Fig. 66(a).

Table 29 Hall coefficients of manganese at subzero and room temperatures

Temperature K	°C	Hall coefficients (R_H), $m^3 \cdot A^{-1} \cdot s^{-1} \times 10^{-11}$ Recommended(a)	For 40-μm thick film
4	−269	−54	. . .
4.5	−268.5	. . .	4.3
5.6	−267.0	. . .	3.3
7.7	−265.3	. . .	1.3
10	−263	−8	. . .
14.0	−259.0	. . .	−1.8
17.1	−255.9	. . .	−2.9
20	−253	−42	. . .
21.3	−251.7	. . .	−6.9
30.6	−242.4	. . .	−9.9
36.8	−236.2	. . .	−11.6
50	−223	−65	. . .
80	−193	−57	. . .
88.6	−184.4	. . .	−10.6
100	−173	−48	. . .
108.8	−164.2	. . .	−7.9
150	−123	−25	. . .
200	−73	−7.0	. . .
250	−23	2.0	. . .
300	27	9.0	. . .

(a) Per Center for Information and Numerical Data Analysis and Synthesis (CINDAS), Purdue University. Sources: Ref 8, 22 to 24

Absorptive index. See Fig. 66(b).

Nuclear Properties

Stable isotopes. ^{55}Mn

Radioactive isotopes. ^{50m}Mn, ^{50}Mn, ^{51}Mn, ^{52m}Mn, ^{52}Mn, ^{53}Mn, ^{54}Mn, ^{56}Mn, ^{57}Mn, and ^{58}Mn, where superscript *m* indicates isomers of manganese (Ref 33)

Mechanical Properties

Tensile properties. α and β phases are quite brittle, but γ has useful mechanical properties (yield strength, 241 MPa; tensile strength, 496 MPa; 40% elongation) (Ref 6). The β phase can be preserved by quenching. Pure γ cannot be preserved at ambient temperature by quenching, but alloying with 2% Cu and 1% Ni (Ref 2) or 0.01 to 0.06% Cu (Ref 34), or electrolytic deposition can provide the γ toughness after quenching.

Hardness. γ phase, 35 HRC

Elastic modulus. Tension, 191 GPa (Ref 35); shear, 76.4 GPa (Ref 36); bulk, 92.6 GPa (Ref 35)

Viscosity:

Temperature K	°C	Viscosity, mPa · s(a)
1519	1246	5.28
1550	1277	5.02
1600	1327	4.65
1650	1377	4.32
1700	1427	4.04
1750	1477	3.79
1800	1527	3.57
1850	1577	3.27
1900	1627	3.19
1950	1677	3.02
2000	1727	2.88
2050	1777	2.75

(a) Per CINDAS data. Uncertainty of data, ±25%. Sources: Ref 8, 37

Velocity of sound. 4600 m/s (Ref 38)

REFERENCES

1. "Standard Specification for Austenitic Manganese Steel Castings," A 128, American Society for Testing and Materials
2. Manganese, in *Encyclopedia of Engineering Materials and Processes*, H.R. Clauser, Ed., Reinhold, 1963, p 414-415
3. M.A. Angadi, R. Whiting, and R. Angadi, The Electromechanical Properties of Thin Manganese Films, *J. Mater. Sci. Lett.*, Vol 8, 1989, p 555-558
4. M.A. Angadi and K. Nallamshetty, Optical Properties of Manganese Films, *J. Mater. Sci.*, Vol 22, 1987, p 1971-1974
5. W.B. Pearson, A Handbook of Lattice Spacings and Structures of Metals and Alloys, Vol 2, Pergamon Press, 1967, p 85
6. R.S. Dean, Manganese, in *Properties and Selection*, Vol 1, 8th ed., *Metals Handbook*, American Society for Metals, 1961, p 1215
7. R.H. Hultgren. P.D. Desai *et al.*, *Selected Values of Thermodynamic Properties of the Elements*, American Society for Metals, 1973
8. Y.S. Touloukian and C.Y. Ho, *Properties of Selected Ferrous Alloying Elements*, Vol III, McGraw-Hill/CINDAS Data Series on Material Properties, McGraw-Hill, 1981
9. G.K. White, *Proc. Phys. Soc. (London)*, Vol 86, 1965, p 159-169
10. P.D. Desai, Thermodynamic Properties of Manganese and Molybdenum, *J. Phys. Chem. Ref. Data*, Vol 16 (No. 1), 1987
11. K. Mendelssohn and H.M. Rosenberg, *Proc. R. Soc. (London) A*, Vol A65, 1952, p 385-394
12. K. Mendelssohn, *Bull. Int. Inst. Refrig.*, 1951-1952, p 69-79
13. G.K. White and S.B. Woods, *Can. J. Phys.*, Vol 35, 1957, p 346-348
14. H. Reddemann, *Ann. Phys.*, Vol 22 (No. 5), 1935, p 28-30
15. B.W. Jolliffe, R.P. Tye, and R.W. Powell, *J. Less-Common Met.*, Vol 11 (No. 6), 1966, p 388-394
16. R.W. Ohse and H. von Tippellskirch, *High-Temp.—High Press.*, Vol 9, 1977, p 367-385
17. G.T. Meaden and P. Pelloux-Gervais, *Cryogenics*, Vol 5, 1965, p 227
18. F. Bunke, *Ann. Phys.*, Vol 21, 1934, p 139
19. H.D. Efling, *Ann. Phys.*, Vol 37, 1940, p 162
20. G.T. Meaden and P. Pelloux-Gervais, *Cryogenics*, Vol 7 (No. 3), 1967, p 161-166
21. G.T. Meaden, *Cryogenics*, Vol 6 (No. 5), 1966, p 275-278
22. S. Foner, *Phys. Rev.*, Vol 107 (No. 6), 1957, p 1513-1516
23. G.T. Meaden and P. Pelloux-Gervais,

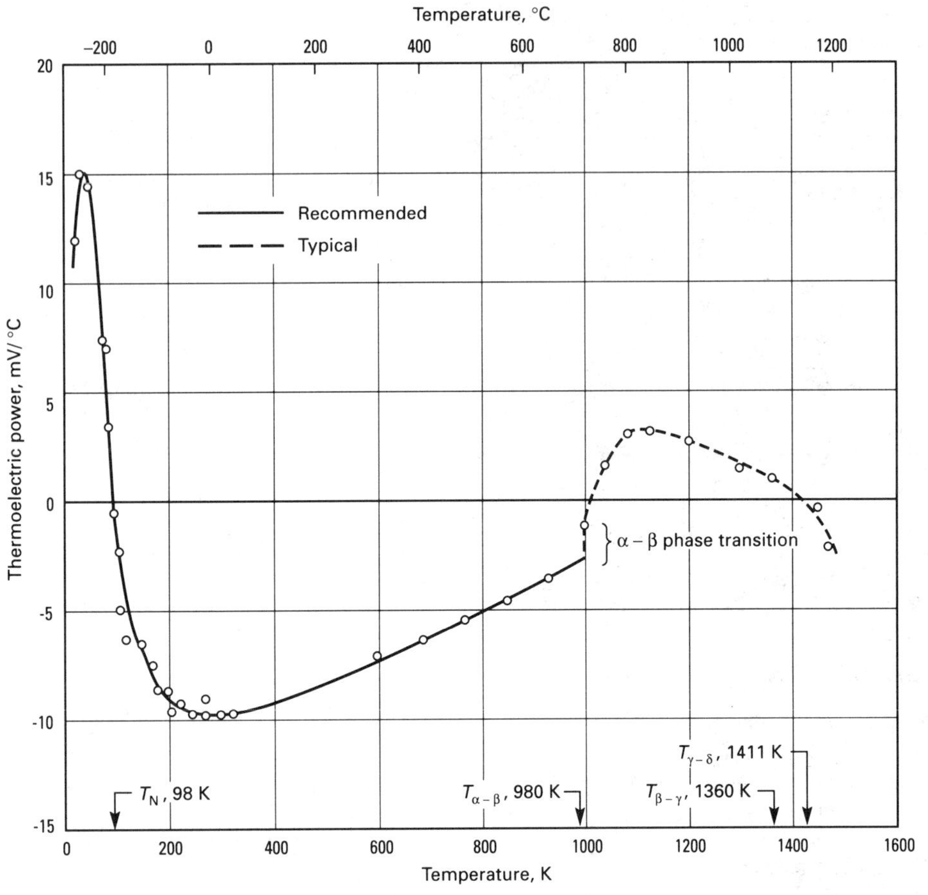

Fig. 64 Thermoelectric power of manganese. The Néel temperature (T_N) indicates the transition from ferromagnetic to paramagnetic behavior. Source: Ref 8

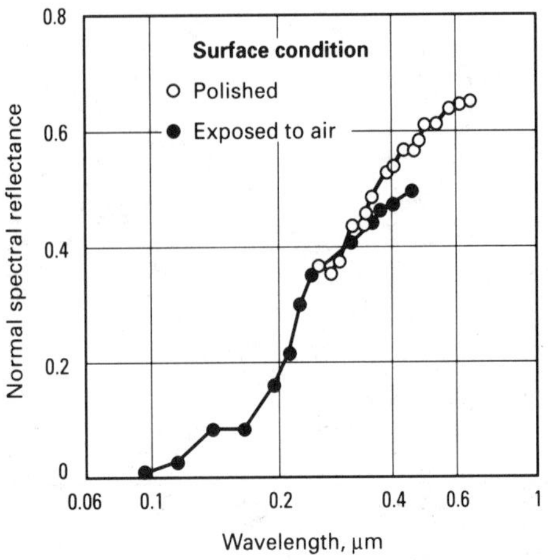

Fig. 65 Normal spectral reflectance of manganese at 20 °C. Source: Ref 8

Cryogenics, Vol 7 (No. 3), 1967, p 161-166
24. K.G. Adanu and A.D.C. Grassie, in *Transition Metals, 1977: Toronto*, Institute of Physics Conference Series, No. 39, M.J.G. Lee, Ed., Institute of Physics, 1978, p 200-204
25. V.S. Fomenko, *Handbook of Thermoionic Properties*, G.V. Samsonov, Ed., Plenum Publishing, 1966
26. G.K. Hall and C.H.B. Mee, *Phys. Status Solidi*, Vol 5 (No. 2), 1971, p 389-395
27. V. Freedericksz, *Ann. Phys. (Leipzig)*, Vol 34 (No. 4), 1911, p 780-796
28. G.B. Sabine, *Phys. Rev.*, Vol 55, 1939, p 1064-1069
29. G.K. Burgess and R.G. Waltenberg, *Bur. Stand. Bull.*, Vol 11, 1915, p 591-605
30. D.K. Edwards and N.B. deVolo, in *Advances in Thermophysical Properties at Extreme Temperatures and Pressures*, American Society of Mechanical Engineers, 1985, p 174-188
31. P.B. Johnson and R.W. Christy, *Phys. Rev. B.*, Vol 9 (No. 12), 1974, p 5056-5070
32. R.L. Aagard, *J. Opt. Soc. Am.*, Vol 64 (No. 11), 1974, p 1456-1458
33. R.C. Weast, Ed., *CRC Handbook of Chemistry and Physics*, 60th ed., CRC Press, 1979
34. J.C. Ho and N.E. Phillips, *Phys. Lett.*, Vol 10 (No. 1), 1964, p 34-35
35. M. Rosen, *Phys. Rev.*, Vol 165 (No. 2), 1971, p 357-359
36. K.A. Gachneider, Jr., in *Solid State Physics*, Vol 16, Seitz and Turnbull, Ed., Academic Press, 1964, p 275-426
37. E.S. Levin, V.N. Zamarayev, and P.V. Gel'd, *Russ. Metall.*, Vol 2, 1976, p 86-89
38. M.E. Delaney, in *Tables of Physical and Chemical Constants*, 13th ed., Kaye and Laby, Ed., John Wiley & Sons, 1966, p 60-72

Mercury (Hg)

Compiled by M. Nowak,
Troy Chemical Corporation

Mercury is the only common metal that is liquid at room temperature. It is rarely found in the free and uncombined state in

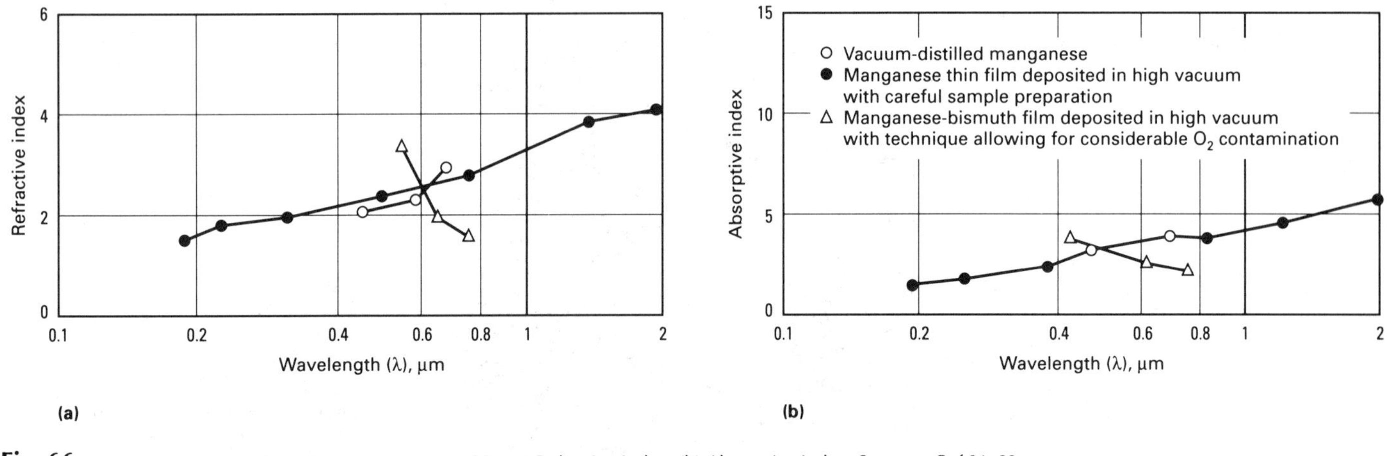

Fig. 66 Optical properties of manganese at 298 K (20 °C). (a) Refractive index. (b) Absorptive index. Sources: Ref 31, 32

Table 30 Effect of temperature on magnetic susceptibility and mass magnetic susceptibility of ≥99.99% pure manganese

Temperature					
K	°C	Phase(a)	Magnetic susceptibility (χ), 10^{-4}(b)	Mass magnetic susceptibility (χ_m), 10^{-8} m^3/kg	State
1	−272	α	9.61	12.8	solid
4	−269	α	9.58	12.8	solid
7	−266	α	9.55	12.7	solid
10	−263	α	9.52	12.7	solid
15	−258	α	9.48	12.6	solid
20	−253	α	9.43	12.6	solid
25	−248	α	9.39	12.5	solid
30	−243	α	9.35	12.5	solid
40	−233	α	9.27	12.4	solid
50	−223	α	9.21	12.3	solid
60	−213	α	9.21	12.3	solid
70	−203	α	9.26	12.3	solid
80	−193	α	9.34	12.4	solid
90	−183	α	9.43	12.6	solid
100	−173	α	9.49	12.6	solid
125	−148	α	9.60	12.8	solid
150	−123	α	9.60	12.8	solid
175	−98	α	9.56	12.8	solid
200	−73	α	9.47	12.7	solid
250	−23	α	9.32	12.5	solid
273	0	α	9.27	12.5	solid
293	20	α	9.22	12.4	solid
300	27	α	9.21	12.3	solid
350	77	α	9.08	12.2	solid
400	127	α	8.97	11.9	solid
500	227	α	8.71	11.7	solid
600	327	α	8.49	11.5	solid
700	427	α	8.24	11.3	solid
800	527	α	8.05	11.3	solid
900	627	α	7.91	11.3	solid
970	697	α	7.83	11.3	solid
980	707	α to β transition	...	...	solid
990	717	β	7.90	11.9	solid
1000	727	β	7.90	11.9	solid
1100	827	β	7.91	12.0	solid
1200	927	β	7.92	12.3	solid
1300	1027	β	8.04	12.6	solid
1340	1067	β	8.09	12.8	solid
1360	1087	β to γ transition			solid
1370	1097	γ	8.70	13.8	solid
1400	1127	γ	8.83	14.1	solid
1411	1138	γ to δ transition			solid
1420	1147	δ	9.28	15.0	solid
1500	1227	δ	9.40	15.3	solid
1519	1246	δ	9.42	15.4	solid
1519	1246	melt	9.30	15.4	liquid
1600	1327		9.32	15.7	liquid
1700	1427		9.34	16.0	liquid

(a) α, bcc (*A*12); β, complex cubic; γ, fct; δ, bcc (*A*2). (b) Uncertainty of values (supplied by CINDAS): $<$27 °C, ±15%; $>$27 °C, ±20%. Source: Ref 8

nature and most often occurs as the ore cinnabar (HgS). Mercury is widely used for thermometers, barometers, diffusion pumps, and other laboratory instruments. It is used commercially in mercury vapor lamps, in lamps and lamp tubes for advertising signs, in switches for instruments and control devices, in dental preparations, and in batteries. Mercury chemicals are widely used for making pesticides, antifouling paints, high-grade paint pigments, explosives, and medicines. Mercury cells are used in the production of caustic chlorine.

Prime virgin mercury as commonly obtained by refining directly from mercury has a purity of at least 99.9%, and in many instances 99.99%. Metal of lesser purity is generally obtained by reclaiming discarded mercury.

Precautions in Use. Mercury is toxic in both its organic and inorganic forms. Mercury vapor is readily absorbed through the respiratory tract, the gastrointestinal tract, or unbroken skin. Mercury acts as a heavy-metal poison, but its effect becomes known only after prolonged exposure. Acute poisoning from mercury vapor is extremely rare. Mercury absorbed from vapor is eliminated from the human body fairly quickly through the urinary and fecal tracts. Mercury levels resulting from exposure to mercury vapor or to inorganic mercury compounds (including aryl mercury compounds such as phenylmercuric acetate) are rapidly reduced and do not accumulate. On the other hand, mercury levels resulting from exposure to alkyl mercury compounds such as methyl mercury or ethyl mercury compounds cannot be eliminated quickly and tend to accumulate.

Mercury is a very volatile element, and dangerous levels of mercury vapor are readily attained at room temperature in enclosed spaces that are not adequately ventilated. The present toxicity limit for mercury vapor in air is 0.05 mg/m^3. At 20 °C, air saturated with mercury vapor contains a concentration more than 100 times this limit.

Because of the toxic nature of mercury and many mercury compounds, certain precautions are mandatory during handling and disposal. Containers should be securely covered. All operations involving mercury metal should be carried out in a well-ventilated area or in a closed system to prevent accumulation of mercury vapor in the workspace; this is of utmost importance if the operation involves heating mercury above room temperature. Workspaces should be continually monitored with special electronic instruments to detect any rise in mercury vapor concentration above the established safe working limit. Workers should be provided with masks or special breathing devices, and the level of mercury in the body of every worker should be periodically monitored by specially trained medical personnel. Any spills of liquid or escape of vapor from a closed heated system must be countered by immediate decontamination of the affected workspace.

Disposal is ordinarily accomplished by sending impure mercury or concentrated mercury compounds to reclamation centers, where purified metal is produced from the discards. Mercury compounds such as methyl mercury are dangerous pollutants and are required to be removed from effluents before they are discharged into natural waters. Sludges and other solid wastes that are contaminated with small concentrations of mercury are sometimes buried at approved sites.

Structure

Crystal system. Rhombohedral below −39 °C. Structure symbol, *A*10; space group, $R{:}\bar{3}m$; *hR*1

Mass Characteristics

Atomic weight. 200.59
Density. Solid: 14.193 g/cm^3 at −39 °C. Liquid: 14.43 g/cm^3 at melting point; 13.595 g/cm^3 at 0 °C; 13.546 g/cm^3 at 20 °C; 13.352 g/cm^3 at 100 °C; 13.115 g/cm^3 at 200 °C; 12.881 g/cm^3 at 300 °C
Compressibility. Volumetric, from 1 to 493 atm: 4×10^{-6} per atm at 20 °C

Thermal Properties

Boiling point. 356.58 °C at 1 atm
Freezing temperature. −38.87 °C
Vapor pressure:

Temperature, °C	Vapor pressure
−20	2.41 mPa
0	24.6 mPa
20	160.0 mPa
50	1.689 Pa
100	36.40 Pa
150	374.2 Pa
200	2.305 kPa
220	4.284 kPa
240	7.580 kPa
260	12.839 kPa
280	20.914 kPa
300	32.904 kPa
320	50.173 kPa
340	74.381 kPa
356.58	101.3 kPa
360	107.49 kPa
380	151.77 kPa
400	209.86 kPa

Table 31 Standard enthalpy, entropy, and Gibbs free energy of mercury

Temperature, K	Liquid			Gas(a)		
	Enthalpy, kJ/kg	Entropy, kJ/kg · K	Gibbs free energy, kJ/kg · K	Enthalpy, kJ/kg	Entropy, kJ/kg · K	Gibbs free energy, kJ/kg · K
298.15	0	0	−0.3786	0	0	−0.8723
350	7.180	0.0222	−0.3803	5.385	0.0166	−0.8736
400	14.05	0.0406	−0.3840	10.561	0.0305	−0.8764
450	20.87	0.0566	−0.3888	15.738	0.0427	−0.8800
500	27.66	0.0709	−0.3942	20.935	0.0536	−0.8841
550	34.42	0.0838	−0.3999	26.111	0.0635	−0.8883
600	41.20	0.0956	−0.4056	31.308	0.0725	−0.8927
630	45.27	0.1022	−0.4090	34.419	0.0776	−0.8952
650	47.99	0.1064	−0.4113	36.485	0.0808	−0.8970
700	54.77	0.1165	−0.4169	41.661	0.0885	−0.9013
750	61.57	0.1259	−0.4224	46.859	0.0957	−0.9055

(a) Ideal

Critical point. 1677 °C and 74.2 MPa (732 atm). Critical density: 3.56 g/cm^3
Coefficient of thermal expansion. Volumetric, liquid: 182 × 10^{-6} at 20 °C
Specific heat:

Temperature, K	Specific heat (C_p), kJ/kg · K Liquid	Gas(a)
234.28	0.1421(b)	...
234.28	0.1420	...
298.15	0.1396	0.1037
350	0.1377	0.1037
400	0.1368	0.1037
450	0.1360	0.1037
500	0.1356	0.1037
550	0.13536	0.1037
600	0.13538	0.1037
630	0.1355	0.1037
650	0.1356	0.1037
700	0.1361	0.1037
750	0.1367	0.1037

(a) Ideal. (b) Solid

Enthalpy. See Table 31.
Entropy. See Table 31.
Latent heat of fusion. 11.8 kJ/kg
Heat of vaporization. 61.42 kJ/kg at 25 °C
Latent heat of vaporization. 272 kJ/kg
Thermal conductivity. 8.21 W/m · K at 0 °C; 9.67 W/m · K at 60 °C; 10.9 W/m · K at 120 °C; 11.7 W/m · K at 160 °C; 12.7 W/m · K at 220 °C

Electrical Properties

Electrical resistivity:

Temperature, °C	Resistivity, nΩ · m
20	958
50	984
100	1032
200	1142
300	1275
350	1355

Temperature coefficient, 0.9 nΩ · m/K at 20 °C
Thermoelectric potential. Mercury versus platinum: −0.60 mV for 100 °C hot junction and 0 °C cold junction
Standard electrode potential. Versus hydrogen electrode at 20 °C: 0.851 V for $Hg^{2+} + 2e^- \rightleftarrows Hg$; 0.7961 V for $2Hg^{2+} + 2e^- \rightleftarrows 2Hg^+$; 0.905 V for $2Hg^{2+} + 2e^- \rightleftarrows (Hg_2)^{2+}$
Ionization potential. Hg(I), 10.43 eV; Hg(II), 18.75 eV; Hg(III), 34.30 eV; Hg(IV), 72 eV; Hg(V), 82 eV
Hydrogen overvoltage. 1.06 V
Contact potential. Mercury versus antimony, −0.26 V; mercury versus zinc, +0.17 V

Magnetic Properties

Magnetic susceptibility. Volume: −1.9 × 10^{-6} mks at 18 °C

Optical Properties

Color. Filtered mercury has a bright, clean, silvery appearance if the metal contains less than 1 ppm of impurities.
Spectral reflectance. 71.2% for λ = 550 μm
Refractive index. 1.6 to 1.9 at 20 °C

Nuclear Properties

Thermal neutron cross section. Capture, 420 b; scattering 5 to 15 b

Mechanical Properties

Dynamic viscosity. 1.55 mPa · s at 20 °C
Surface tension. 0.465 N/m at 20 °C; 0.454 N/m at 112 °C; 0.436 N/m at 200 °C; 0.405 N/m at 300 °C; 0.394 N/m at 354 °C. Angle of contact on glass: 128° at 18 °C

Molybdenum (Mo)

Compiled by J.Z. Briggs

Reviewed for this Volume by Joseph Linteau, Climax Specialty Metals

Molybdenum is used as an alloying addition in steels, and molybdenum and its alloys are used for electrical and electronic parts, missile and aircraft parts, high-temperature furnace parts, die casting cores, hot-working tools, boring bars, thermocouples, nuclear energy applications, corrosion-resistant equipment, equipment for glass-melting furnaces, and metallizing. Molybdenum also finds use as a catalyst in chemical reactions. Molybdenum is not suitable for continued service at temperatures above 500 °C in an oxidizing atmosphere unless protected by an adequate coating.

Structure

Crystal structure. Body-centered cubic, *a* = 0.31468 nm at 25 °C
Slip planes. {112} at 20 °C; {110} at 1000 °C
Slip direction. [111]
Interatomic distance. 0.27252 nm min
Metallography. Electrolytic polishing is preferred. Etching: (1) 10 g NaOH + 30 g $K_3Fe(CN)_6$ + 100 liters water; (2) 1 g NaOH + 35 g $K_3Fe(CH)_6$ + 600 liters water; (3) Murakami's reagent

Mass Characteristics

Atomic weight. 95.94
Density. At 20 °C: 10.22 g/cm^3; see also Fig. 67.
Compressibility. At 293 °C: 36 μm^2/N

Thermal Properties

Melting point. 2610 °C
Boiling point. 5560 °C
Thermal expansion. See Fig. 68.

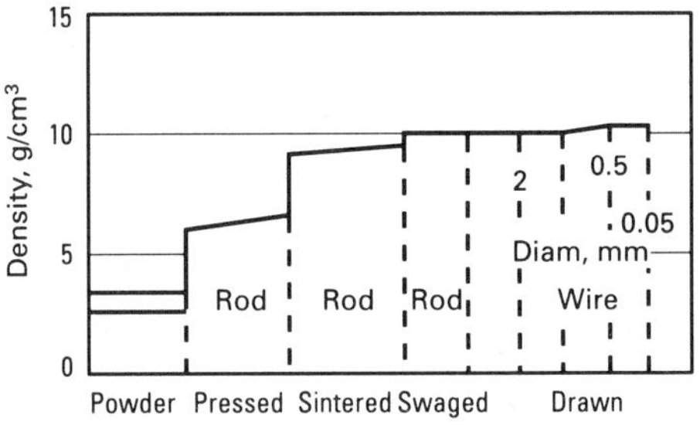

Fig. 67 Density of selected molybdenum products

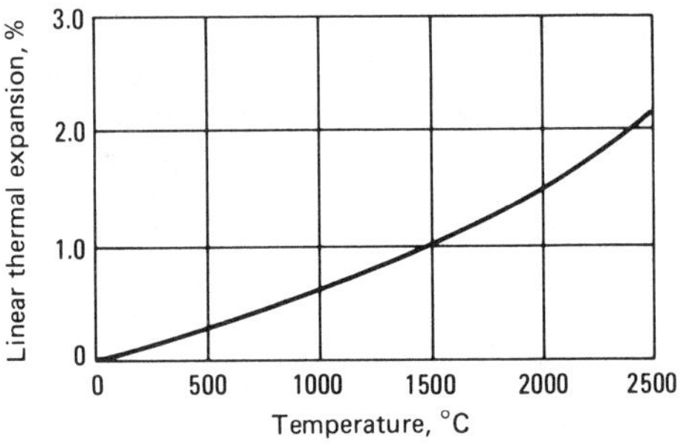

Fig. 68 Linear thermal expansion of molybdenum

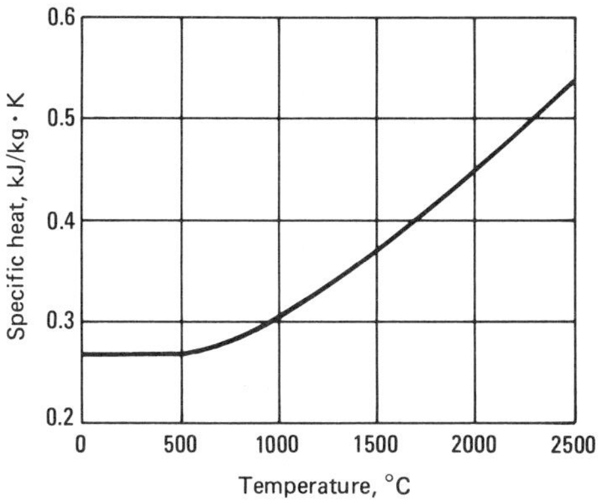

Fig. 69 Temperature dependence of the specific heat of molybdenum

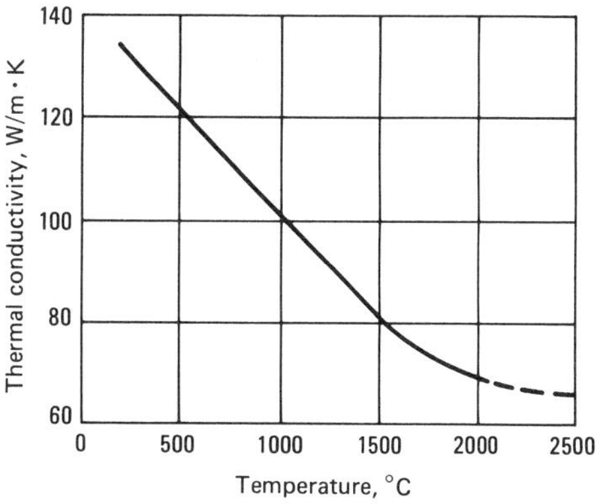

Fig. 70 Temperature dependence of the thermal conductivity of molybdenum

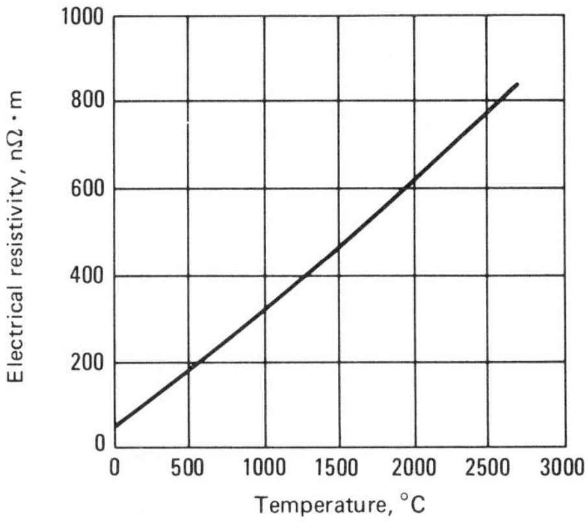

Fig. 71 Temperature dependence of the electrical resistivity of molybdenum

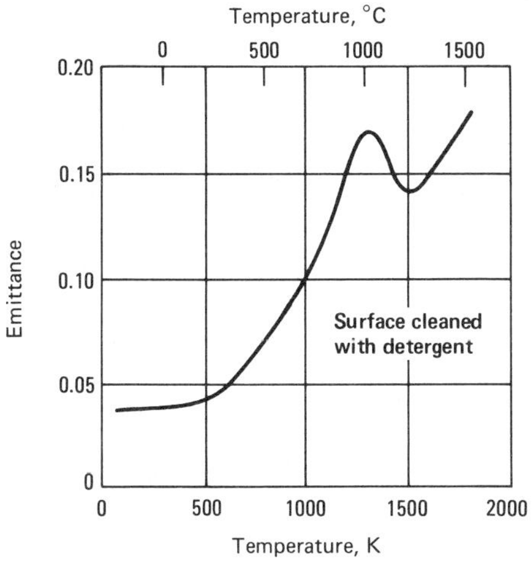

Fig. 72 Temperature dependence of the total normal emittance of molybdenum

Specific heat. At 20 °C: 0.276 kJ/kg · K. See also Fig. 69.
Latent heat of fusion. 270 kJ/kg (estimated)
Latent heat of vaporization. 5.123 MJ/kg
Thermal conductivity. At 20 °C: 142 W/m · K. See also Fig. 70.
Heat of combustion. 7.58 MJ/kg Mo
Recrystallization temperature. 900 °C min; commercial products normally require higher temperatures.

Electrical Properties

Electrical conductivity. At 0 °C: 34% IACS
Electrical resistivity. At 0 °C: 52 nΩ · m. See also Fig. 71.
Thermal electromotive force. Versus platinum, 0 to 100 °C: 1.45 mV
Electrochemical equivalent. Valence 6, 0.1658 mg/C
Hydrogen overpotential. At 100 A/m^2: 0.44 V

Magnetic Properties

Magnetic susceptibility. Mass: 1.17 × 10^{-8} mks at 25 °C; 1.39 × 10^{-8} mks at 1825 °C

Optical Properties

Reflectivity. 46% at 500 nm, 93% at 10 000 nm
Color. Silvery white
Total normal emittance. See Fig. 72.

Mechanical Properties

Tensile properties. See Fig. 73 to 75.
Shear properties. See Fig. 76.
Hardness. See Fig. 77.
Elastic modulus. See Fig. 78.
Impact strength. See Fig. 79.
Fatigue strength. See Fig. 80.
Creep-rupture characteristics. See Fig. 81 and 82.
Directional properties. If not cross rolled, the tensile strength of molybdenum sheet can be as much as 20% greater in the direction of rolling than when the inclination of the direction of tension to that of rolling is between 45 and 90°.
Properties of single crystals. Tensile strength, 350 MPa

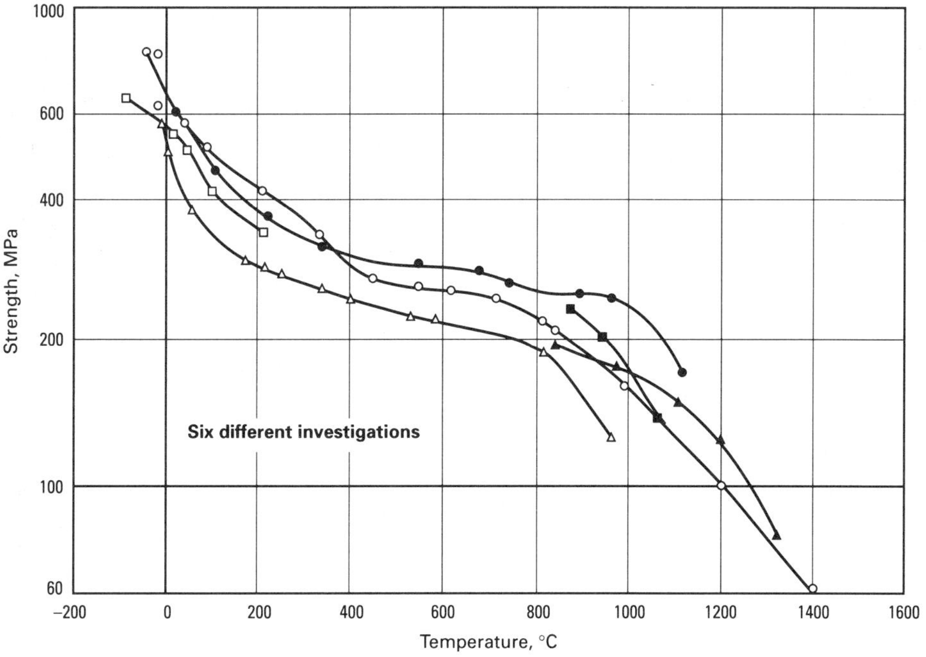

Fig. 73 Temperature dependence of the tensile strength of molybdenum

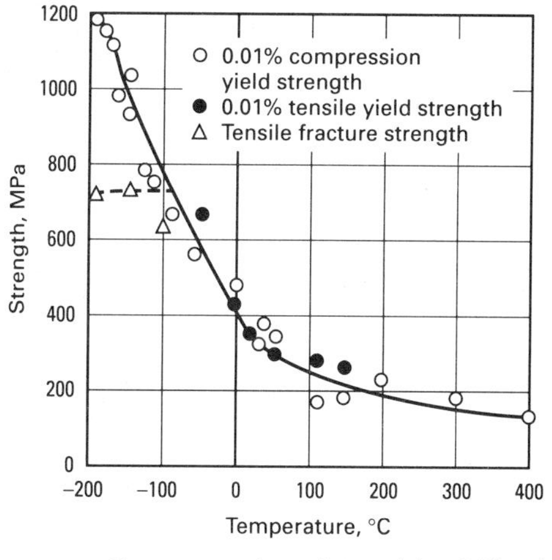

Fig. 74 Temperature dependence of the yield and fracture strengths of molybdenum

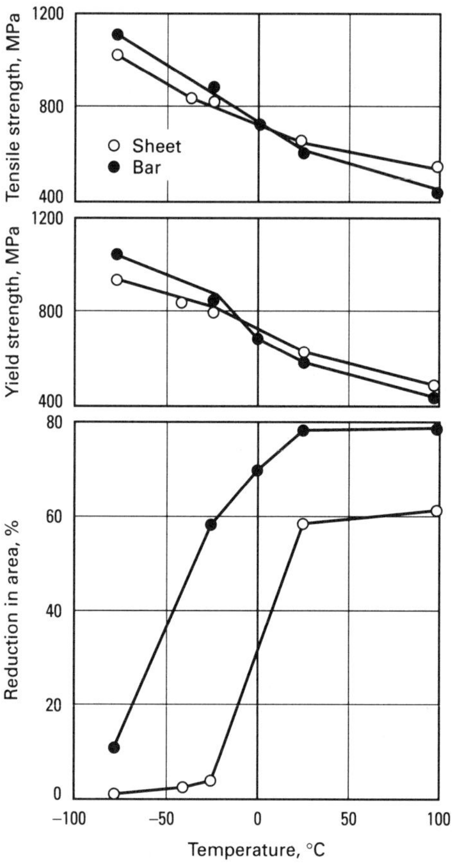

Fig. 75 Effect of product form on the tensile properties of molybdenum

Velocity of sound. Longitudinal wave, 6370 m/s; shear wave, 3410 m/s; thin rod, 5500 m/s

Chemical Properties

General corrosion behavior. Molybdenum has particularly good resistance to corrosion by mineral acids, provided oxidizing agents are not present. It is also resistant to many liquid metals and to most molten glasses. In inert atmospheres, it is unaffected up to 1760 °C by refractory oxides. Molybdenum is relatively inert in hydrogen, ammonia, and nitrogen up to about 1100 °C, but a superficial nitride case may be formed in ammonia or nitrogen.

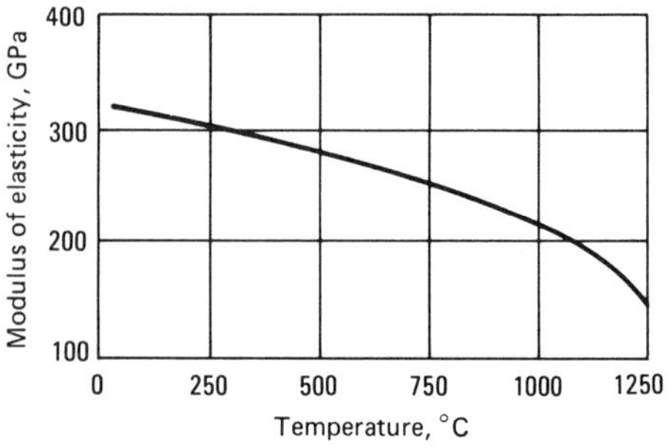

Fig. 78 Temperature dependence of the static modulus of elasticity of molybdenum

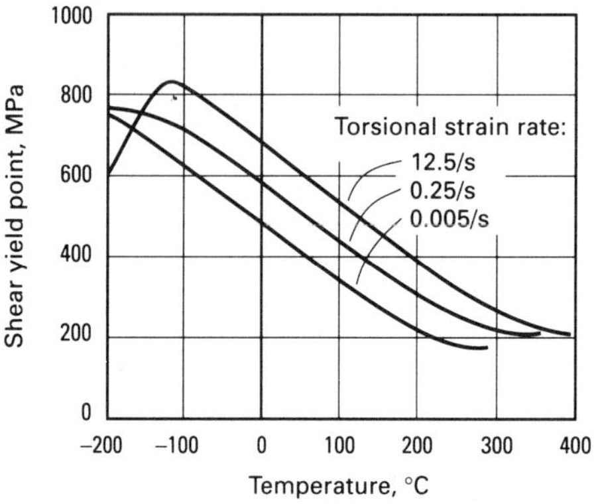

Fig. 76 Temperature dependence of the upper yield point in shear for molybdenum

Resistance to specific corroding agents:

Agent	Corrosion, mils/year
20% hydrochloric acid, boiling, unaerated	0.90
49% hydrochloric acid, room temperature, unaerated	0.14
85% phosphoric acid, 100 °C, unaerated	0.29
40% sulfuric acid, boiling, unaerated	1.5
Liquid lithium, sodium, and sodium-potassium, up to 900 °C	<1
Liquid bismuth, 980 °C	nil

Fabrication Characteristics

Consolidation. In most instances, molybdenum is consolidated from powder by compacting under pressure followed by sintering in the range from 1650 to 1900 °C. Some molybdenum is consolidated by a vacuum arc casting method in which a preformed electrode is melted by arc formation in a water-cooled mold.

Hot-working temperature. Generally forged between 1180 and 1290 °C down to 930 °C

Annealing temperature. Normal stress-relieving temperature is 870 to 980 °C.

Recrystallization temperature. Depends on prior working and condition; 1180 °C for full recrystallization in 1 h of a 16 mm (⅝ in.) bar reduced 97% by rolling

Suitable forming methods. Conventional methods

Formability. See Fig. 83 and 84.

Precautions in forming. Must be heated to the proper temperature relative to its thickness and forming speed

Heat treatment. Not hardenable by heat treatment but only by work hardening

Suitable joining methods. Can be brazed or joined mechanically, as well as welded by arc, resistance, percussion, flash, and electron beam methods. Arc cast molybdenum is preferred to a powder metallurgy product for welding. Absolute cleanliness of surface is essential. Fusion welding must be carried out in closely controlled inert atmosphere.

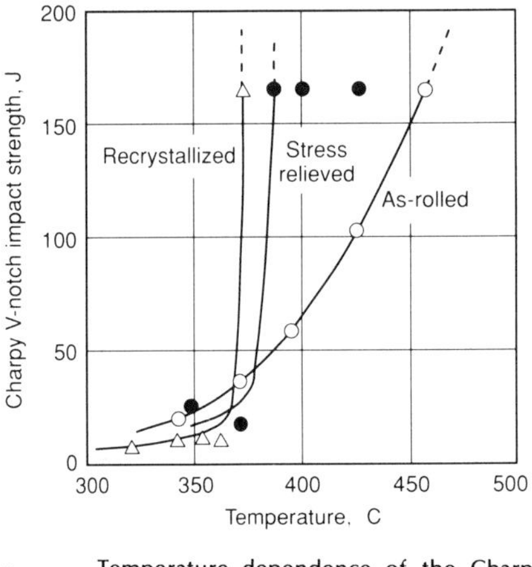

Fig. 79 Temperature dependence of the Charpy V-notch impact strength of molybdenum

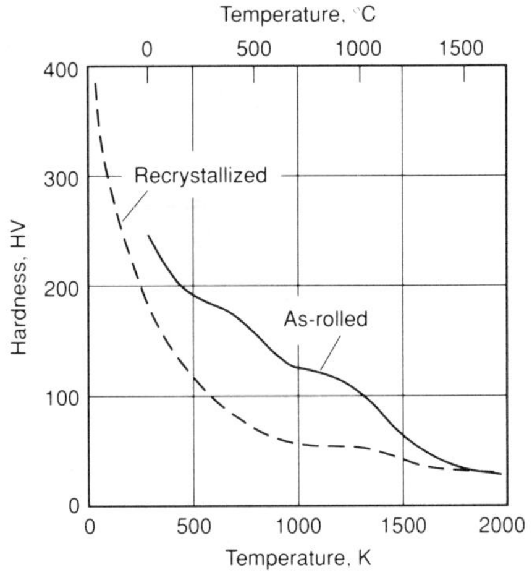

Fig. 77 Temperature dependence of the hardness of molybdenum

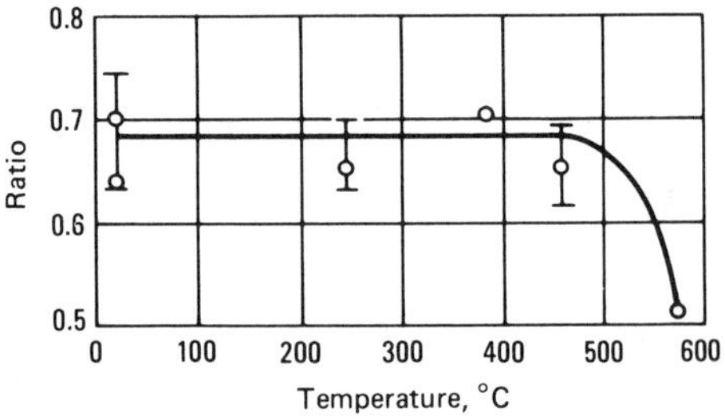

Fig. 80 Ratio at various temperatures of fatigue limit to tensile strength for molybdenum

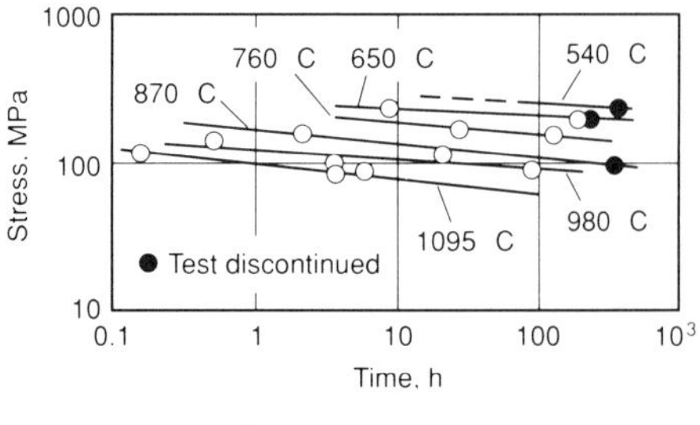

Fig. 81 Creep characteristics of recrystallized molybdenum

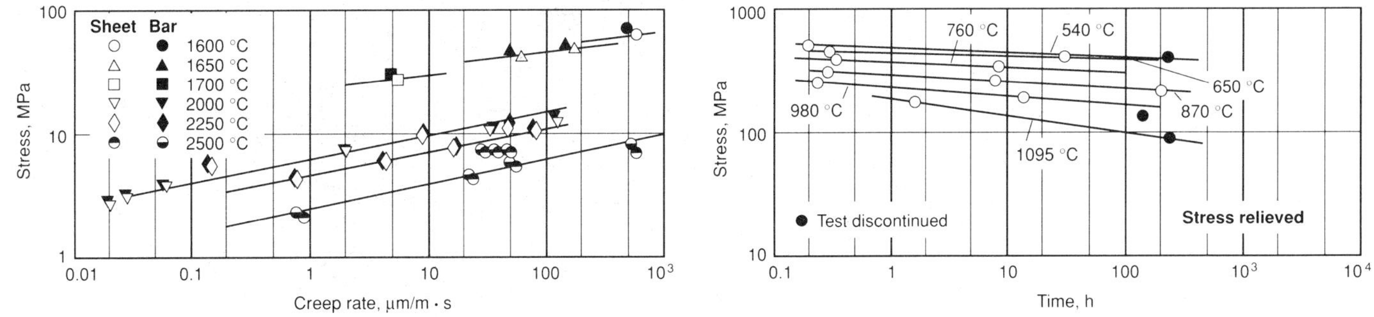

Fig. 82 Rupture strength of molybdenum

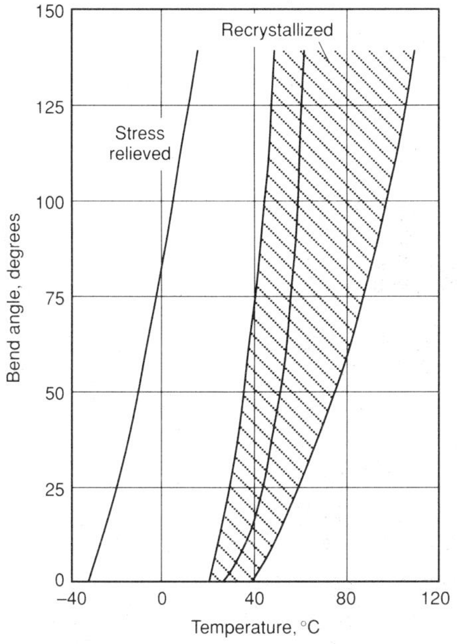

Fig. 83 Temperature dependence of the bend angle to fracture of molybdenum

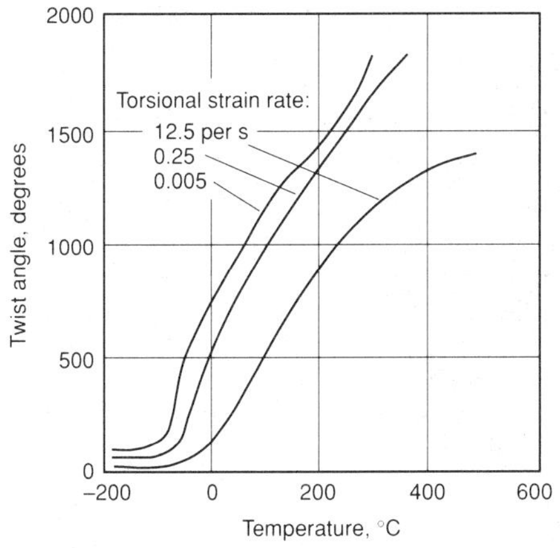

Fig. 84 Temperature dependence of the twist angle to fracture of molybdenum

SELECTED REFERENCES

- *Fabricating Molybdenum + TZM Alloys*, Climax Specialty Metals, 1988
- *Machining Molybdenum*, Climax Specialty Metals, 1986
- *Molybdenum Metal*, Climax Molybdenum Company, 1960

Neodymium (Nd)

See the section "Properties of the Rare Earth Metals" in this article.

Neptunium (Np)

See the section "Properties of the Actinide Metals (Ac-Pu)" in this article.

Nickel (Ni)

Compiled by INCO Alloys International, Inc.

Structure

Crystal structure. Face-centered cubic; a = 0.35167 nm at 20 °C

Mass Characteristics

Atomic weight. 58.71
Density. 8.902 g/cm^3 at 25 °C

Thermal Properties

Melting point. 1453 °C
Boiling point. Approximately 2730 °C
Coefficient of linear thermal expansion. 13.3 µm/m · K at 0 to 100 °C
Specific heat. 0.471 kJ/kg · K at 100 °C
Recrystallization temperature. 370 °C
Thermal conductivity. 82.9 W/m · K at 100 °C

Electrical Properties

Electrical conductivity. Volumetric, 25.2% IACS at 20 °C
Electrical resistivity. 68.44 nΩ · m at 20 °C; temperature coefficient, 69.2 nΩ · m per K at 0 to 100 °C

Magnetic Properties

Magnetic susceptibility. Ferromagnetic
Magnetic permeability. μ_{max} = 1240 at B = 1900 G
Coercive force. 167 A · m^{-1} from H = 4 kA · m^{-1}
Saturation magnetization. 0.616 T at 20 °C
Residual induction. 0.300 T
Hysteresis loss. 685 J/m^3 at B = 0.6 T
Curie temperature. 358 °C

Optical Properties

Color. Grayish white
Spectral reflectance. 41.3% for λ = 0.30 µm

Nuclear Properties

Effect of neutron irradiation. Results in small increase in tensile strength but large increase in yield strength

Chemical Properties

General corrosion behavior. Nickel is not an active element chemically and does not readily evolve hydrogen from acid solutions; the presence of an oxidizing agent is usually required for significant corrosion to occur. Generally, reducing conditions retard corrosion, whereas oxidizing conditions accelerate corrosion of nickel in chemical solutions. However, nickel may also form a protective corrosion-resistant, or passive, oxide film on exposure to some oxidizing conditions. Additional information is available in the article "Corrosion of Nickel-Base Alloys" in Volume 13 of the 9th Edition of *Metals Handbook*.

Mechanical Properties

Tensile properties. Typical: tensile strength, 317 MPa; 0.2% offset yield strength, 59 MPa; elongation, 30% in 50 mm
Hardness. 64 HV (annealed)
Poisson's ratio. 0.31 at 25 °C
Elastic modulus. Tension, 207 GPa; shear, 76 GPa; compression, 207 GPa
Velocity of sound. 4.7 km/s at 40 °C

Fabrication

Because of its excellent ductility, nickel can be readily hot and cold worked. After forming, the fabrication characteristics of a workpiece can be recovered through annealing. Additional information is available in the article "Machining of Nickel and Nickel Alloys" in Volume 16 of the 9th Edition of *Metals Handbook*.

SELECTED REFERENCES

- J. Crangle and G.M. Goodman, The Magnetization of Pure Iron and Nickel, *Proc. R. Soc. (London) A*, Vol 321, 1971
- R. Hultgren *et al.*, *Selected Values of the Thermodynamic Properties of the Ele-*

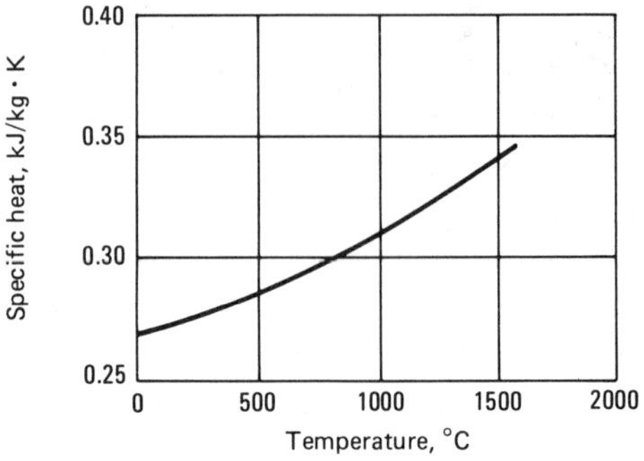

Fig. 85 Temperature dependence of the specific heat of niobium

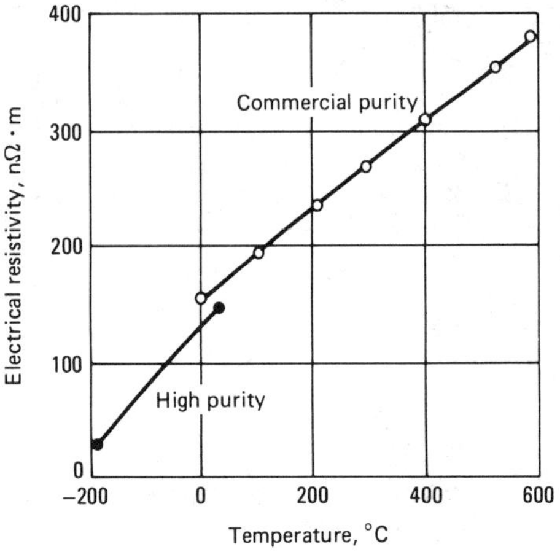

Fig. 86 Temperature dependence of the electrical resistivity of niobium

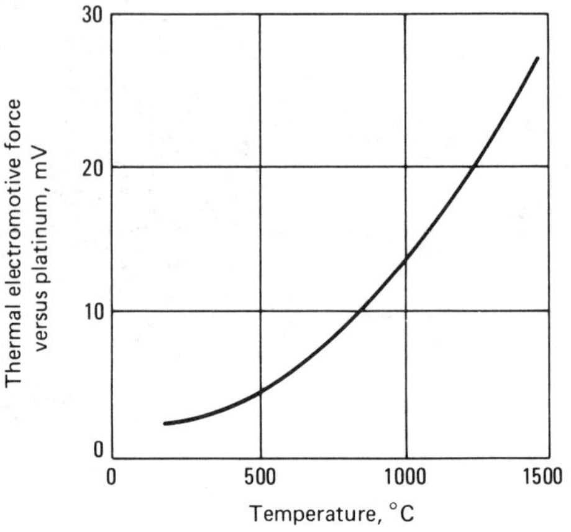

Fig. 87 Temperature dependence of the thermal electromotive force of niobium versus platinum. Cold junction at 0 °C

ments, American Society for Metals, 1973

- "Nickel and Its Alloys," Monograph 106, National Bureau of Standards, 1968
- G.W.P. Rengstorff, "High-Purity Metals," Report 222, Defense Metals Information Center, 1966
- W.A. Wesley, Preparation of Pure Nickel by Electrolysis of a Chloride Solution, *J. Electrochem. Soc.*, Vol 103 (No. 5), 1956
- E.M. Wise and R.H. Schaefer, The Properties of Pure Nickel, in *Metals and Alloys*, Vol 16, 1942

Niobium (Nb)

Compiled by E.S. Bartlett, Battelle Memorial Institute

Revised by R.E. Droegkamp, Fansteel Metals

Niobium is used as an alloying element in nickel- and cobalt-base superalloys as well as in some grades of stainless and low-alloy steels. It is also used as an alloy base for various combinations with zirconium, hafnium, tungsten, tantalum, and molybdenum to increase high-temperature mechanical properties. These alloys have found use in aerospace applications, but they invariably have to be coated, usually with a silicide, for elevated-temperature service. Corrosion/abrasion applications are being found for niobium alloyed with Group IV (titanium, zirconium, and hafnium) and Group VI (molybdenum and tungsten) elements wherein the subsequent reaction with carbon, oxygen, and nitrogen is such that a very hard surface keyed to the substrate is formed. Niobium, niobium-titanium alloys, and niobium-tin alloys are used as superconductors.

Niobium oxidizes and rapidly becomes contaminated with absorbed oxygen above about 400 °C in oxygen-containing atmospheres, including atmospheres normally considered neutral or reducing; it absorbs hydrogen at temperatures between about 250 and 950 °C from hydrogen-containing atmospheres. Contamination by interstitial elements results in a loss of ductility at ambient temperature. Consequences of high impurity levels include impaired fabricability, increased ductile-to-brittle transition temperature, considerable low-temperature strengthening with an attendant loss in ductility, intensified strain-aging effects at slightly elevated temperature, and slight strengthening at higher temperature.

Structure

Crystal structure. Body-centered cubic; a = 0.3294 nm; atomic diameter, 0.294 nm
Slip plane. 110
Metallography. (1) Grind through 000 emery; (2) rough polish with coarse diamond in kerosine; (3) standard finish polish with alumina; (4) etchant (all acids in parts by volume of laboratory reagent grades): 30 lactic—10 nitric—5 hydrofluoric acid solution (more HF for alloys); (5) chemical polish (for freedom) of distortion, if required): 30 lactic—30 nitric—1 to 2 hydrofluoric; (6) electrolytic etch (for particularly uniform grain-boundary definition): $90H_2SO_4$-10HF at 2 V

Mass Characteristics

Atomic weight. 92.9064
Density. At 20 °C: 8.57 g/cm^3

Thermal Properties

Melting point. 2468 °C
Boiling point. 4927 °C
Coefficient of linear thermal expansion:

°C	Coefficient, µm/m · K: Average(a)	Instantaneous
300	7.31	7.38
400	7.39	7.54
500	7.47	7.61
600	7.56	7.87
700	7.64	8.03
800	7.72	8.20
900	7.80	8.37
1000	7.88	8.52

(a) Mean value between 18 °C and indicated temperature

$\Delta l/l_0 = 6.892 \times 10^{-6}\ T + 8.17 \times 10^{-10}\ T^2$, where T is in K
Specific heat. See Fig. 85.
Latent heat of fusion. 290 kJ/kg
Latent heat of vaporization. 7490 kJ/kg
Thermal conductivity:

°C	W/m · K
0	52.3
100	54.4
200	56.5
300	58.6
400	60.7
500	63.2
600	65.3

Electrical Properties

Electrical conductivity. At 18 °C: 13.2% IACS
Electrical resistivity. See Fig. 86. Temperature coefficient, 0 to 600 °C: 0.395 nΩ · m per K
Thermal electromotive force. Versus platinum, see Fig. 87
Electrochemical equivalent. 0.1926 mg/C
Hall coefficient. 0 to 900 K, 0.09 nV · m/A · T

Magnetic Properties

Magnetic susceptibility. Volume: at 25 °C: 28 × 10^{-6} mks

Optical Properties

Refractive index. 1.80
Spectral emittance. λ = 650 nm. See also Fig. 88.

Nuclear Properties

Thermal neutron cross section. At 2200 m/s: 1.1 b

Chemical Properties

General corrosion behavior. Niobium is moderately to highly resistant to corrosion in most aqueous mediums that are usually considered highly corrosive, such as dilute mineral acids, organic acids, and organic liquids. Notable exceptions are dilute

Table 32 Creep and creep-rupture behavior of wrought niobium at various temperatures

Temperature, °C	Stress for rupture in 1 h, MPa	10-h stress, MPa, for				100-h stress, MPa, for				1000-h stress, MPa, for		
		0.05% creep	0.1% creep	0.2% creep	Rupture	0.05% creep	0.1% creep	0.2% creep	Rupture	0.05% creep	0.1% creep	0.2% creep
400	...	160	276	360	...	83	140	200	...	45	66	107
500	...	186	214	230	...	121	140	160	...	80	93	110
700	...	...	...	...	...	...	20	25	...	...	17	12
870	62	...	...	...	55	...	...	...	48	...	...	...
980	48	...	...	...	45	...	...	...	42	...	...	...
1200	35.8	...	...	...	32	...	...	...	28	...	...	...

strong alkalis, hot concentrated mineral acids, and hydrofluoric acid, all of which attack the metal rapidly. Gaseous atmospheres at high temperatures attack niobium rapidly, primarily by oxidation, although oxygen contents may be very low.

Niobium and its alloys are remarkably resistant to corrosion by certain liquid metals, notably lithium metal and sodium-potassium alloys, and to high temperatures (900 to 1010 °C). This resistance coupled with a low-capture cross section for thermal neutrons renders niobium materials most attractive for reactor applications. Additional information is available in the article "Corrosion of Niobium and Niobium Alloys" in Volume 13 of the 9th Edition of *Metals Handbook*.

Mechanical Properties

Tensile properties. Highly dependent on purity, particularly the content of interstitial elements. Values listed are for material of good commercial purity (only 100 to 200 ppm interstitial contaminants):

- Wrought: tensile strength, 585 MPa; elongation, 5%
- Annealed: tensile strength, 275 MPa; yield strength, 207 MPa; elongation, 30%; reduction in area, 80%

See also Fig. 89.
Hardness. Annealed: 80 HV. Wrought: 160 HV
Poisson's ratio. At 25 °C: 0.38

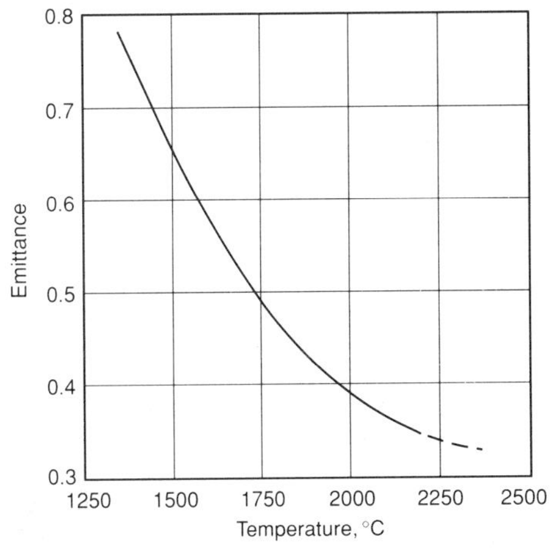

Fig. 88 Temperature dependence of emittance for niobium

Strain-hardening exponent. 0.24; similar to that of low-carbon steel
Elastic modulus. At 25 °C: tension, 103 GPa; shear, 37.5 GPa. At 870 °C: tension, 90 GPa
Ductile-to-brittle transition temperature. <147 K; increases sharply with lower purity
Creep-rupture characteristics. See Table 32.

Fabrication Characteristics

Alloying practice. High-vacuum powder metallurgy techniques can be utilized effectively. Consumable electrode vacuum arc melting and electron beam furnace melting can be used for alloying purposes. High-vacuum techniques purify niobium at temperatures above 1980 °C through volatilization of NbO_2.
Precautions in melting. Exclude atmospheric contaminants as completely as possible. Cold hearth techniques are required to prevent crucible reaction.
Recrystallization temperature. Material cold reduced 70 to 80% completely recrystallizes in 1 h at 1090 °C.
Hot-working temperature. 800 to 1100 °C may be necessary to break down the ingot structure of niobium. This process requires conditioning of the breakdown product to remove the contaminated surface layer. Subsequent working is done cold.
Maximum reduction between anneals. Virtually unlimited
Precautions in forming. Because of the high probability of seizure and galling, selection of lubricant and die material is important in extreme-pressure methods. Carbon tetrachloride (for machining) or sulfonated tallow or proprietary waxes (for spinning and drawing) are preferred lubricants. Polished aluminum bronze has been recommended as a die material for extreme-pressure processes.
Suitable joining methods. Welding processes capable of excluding interstitial contaminants from the hot zone are satisfactory.

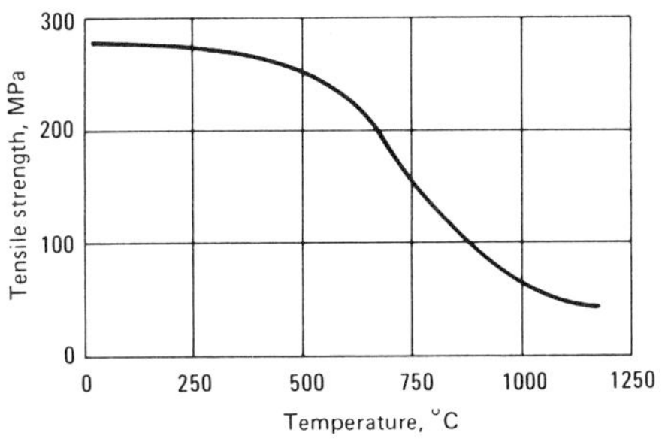

Fig. 89 Temperature dependence of the tensile strength of niobium

Osmium (Os)

Compiled by H.J. Albert, Engelhard Corporation

Reviewed for this Volume by Louis Toth, Engelhard Corporation

Osmium and its alloys are useful for their hardness and resistance to wear and corrosion. The resistance to rubbing wear is greater than would be expected on the basis of hardness; alloys of equal hardness have less resistance than osmium. Osmium is used in fountain nibs, phonograph needles, electrical contacts, and instrument pivots. Osmium should not be heated in the presence of oxygen because the toxic oxide OsO_4 boils off at 130 °C.

Structure

Crystal structure. Close-packed hexagonal, a = 0.27341 nm and c = 0.43197 nm at 26 °C (Ref 1). The space group is $C6/mmc\ D^4_{6h}$ (Ref 2)

Mass Characteristics

Atomic weight. 190.2
Density. At 26 °C, calculated from lattice constants: 22.583 g/cm^3 (Ref 1); 22.57 g/cm^3 obtained directly on an arc melted button of osmium

Thermal Properties

Melting point. Approximately 2700 °C
Boiling point. Approximately 5500 °C (Ref 3)
Coefficient of thermal expansion. 3.2 μm/m · K at 50 °C parallel to c axis; 2.2 μm/m · K at 50 °C parallel to a axis (Ref 4); mean value, 2.6 μm/m · K at 50 °C
Specific heat. At 0 °C, 0.12973 kJ/kg · K; at 100 °C, 0.131 kJ/kg · K; at 1600 °C, 0.161 kJ/kg · K. From 25 to 2727 °C: $C_p = 0.125 + 0.0190\ T$, where T is in K and C_p is in kJ/kg · K (Ref 5)

Electrical Properties

Electrical resistivity. Approximately 95 nΩ · m at 20 °C (Ref 6). Temperature coefficient, 0 to 100 °C: 0.0042 nΩ · m per K (Ref 7)

Magnetic Properties

Magnetic susceptibility. Approximately 0.93×10^{-6} mks (Ref 8)

Chemical Properties

Resistance to specific corroding agents. Easily oxidized and forms a tetroxide boiling at 130 °C. It is rapidly attacked by HNO_3 and aqua regia at room temperature. It is not attacked by H_2SO_4 at room temperature or 100 °C, nor by 36% HCl or 40% HF at 20 °C (Ref 9).

Fabrication Characteristics

Working data. Completely unworkable. Can be shaped by melting, powder metallurgy, and grinding

Mechanical Properties

Hardness. Approximately 800 HV, arc melted button
Elastic modulus. 560 GPa (estimated) (Ref 10)

REFERENCES

1. Swanson, Fuyat, and Ugrinic, *Standard X-Ray Diffraction Powder Patterns*, Circular 539, Vol 4, National Bureau of Standards, p 8-9
2. Barth and Lunde, *Z. Phys. Chem.*, Vol 121, 1926, p 78-102
3. D. Richardson, *Spectroscopy in Science and Industry*, John Wiley & Sons, 1938, p 64
4. Owen and Roberts, *Z. Kristallogr. Kristallgeom. Kristallphys. Kristallchem.*, Vol 69A, 1937, p 497-498
5. Jaeger and Rosenbohm, *Proc. Acad. Sci. (Amsterdam)*, Vol 34, 1931, p 85
6. Blau, *Elektrotech. Z.*, Vol 25, 1905, p 198
7. Lombardi, *Elektrotech. Z.*, Vol 25, 1902, p 42
8. Honda and Sone, *Sci. Rep. Tôhoku Imperial Univ.*, Vol 2, 1913, p 26
9. Wise, *Corrosion Handbook*, John Wiley & Sons, 1948, p 311-312
10. W. Koster, *Z. Electrochem.*, Vol 49, 1943, p 233

Palladium (Pd)

Compiled by E.M. Wise and R.F. Vines, INCO Alloys International, Inc.

Reviewed for this Volume by Louis Toth, Engelhard Corporation

Structure

Crystal structure. Face-centered cubic: a = 0.38902 nm at 20 °C (Ref 1)

Mass Characteristics

Atomic weight. 106.4
Density. 12.02 g/cm³ at 20 °C (Ref 2)

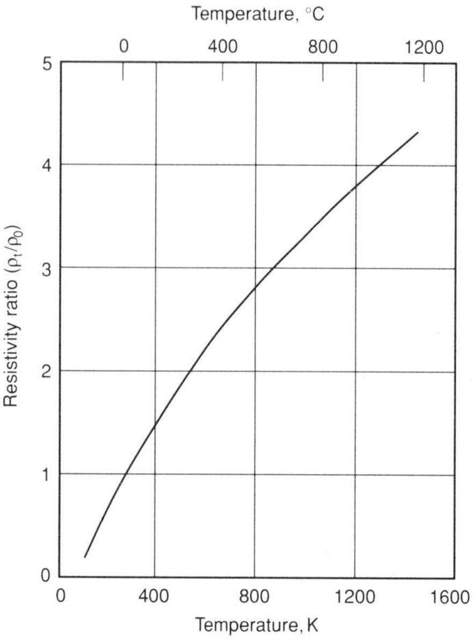

Fig. 90 Ratio of electrical resistivity at various temperatures (ρ_t) to resistivity at 0 °C (ρ_0) for palladium. t is in degrees Centigrade.

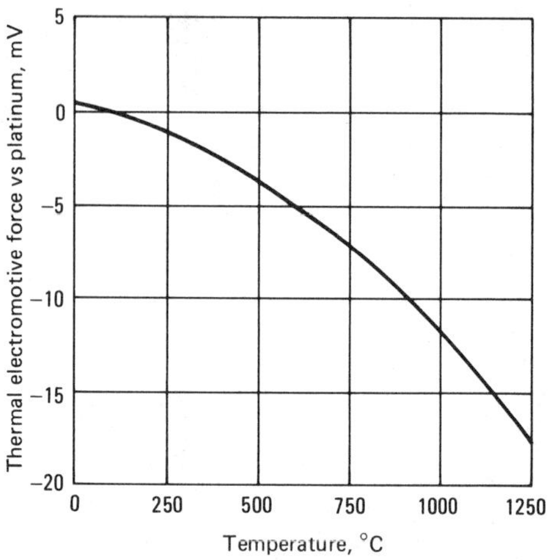

Fig. 91 Temperature dependence of the thermal electromotive force of palladium versus platinum. Cold junction at 0 °C. Source: Ref 11

Thermal Properties

Melting point. 1552 °C (Ref 3 to 5)
Boiling point. Approximately 3980 °C
Coefficient of thermal expansion. 11.76 µm/m · K at 20 °C; $L_t = L_0(1 + 1.167 \times 10^{-5}t + 2.187 \times 10^{-9}t^2)$, where t is in °C (Ref 6)
Specific heat. 0.245 kJ/kg · K at 0 °C; 0.296 and 0.311 kJ/kg · K at 1000 °C (Ref 7, 8)
Thermal conductivity. 70 W/m · K at 18 °C (Ref 9)
Vapor pressure. At 1000 °C, 1.53×10^{-3} Pa; at 1500 °C, 8.23 Pa; at 1554 °C, 15.7 Pa

Electrical Properties

Electrical conductivity. 16% IACS at 20 °C
Electrical resistivity. 108 nΩ · m at 20 °C; 100 nΩ · m at 0 °C. Temperature coefficient: 0.00377 per K (Ref 10). See also Fig. 90.
Thermal electromotive force. See Fig. 91.

Magnetic Properties

Magnetic susceptibility. Mass, at 18 °C: approximately 7.3×10^{-8} mks

Optical Properties

Reflectance. 62.8% in white light; increases slightly in going from blue to red
Emittance. At λ = 0.65 µm: solid, 0.33; liquid, 0.37 (Ref 12)

General Properties

General corrosion behavior. At room temperature, palladium is resistant to corrosion by hydrofluoric, perchloric, phosphoric, and acetic acids. It is attacked slightly by sulfuric, hydrochloric, and hydrobromic acids, especially in the presence of air; it is attacked readily by nitric acid, ferric chloride, hypochlorites, moist chlorine, bromine, and iodine. In ordinary atmospheres palladium is resistant to tarnish, but some discoloration may occur during exposure to moist industrial atmospheres that contain sulfur dioxide. Adding palladium to gold or silver alloys improves the tarnish resistance. Additional information is available in the article "Corrosion of the Noble Metals" in Volume 13 of the 9th Edition of *Metals Handbook.*

REFERENCES

1. C.S. Barrett, *Structure of Metals*, McGraw-Hill, 1952
2. E.A. Owen and E.L. Yates, *Philos. Mag.*, Vol 15, 1933, p 472
3. C.O. Fairchild, W.H. Hoover, and M.F. Peters, *J. Res. Natl. Bur. Stand.*, Vol 2, 1929, p 931
4. F.H. Schofield, *Proc. R. Soc. (London) A*, Vol 155, 1936, p 301
5. L.D. Morris and S.R. Scholes, *J. Am. Ceram. Soc.*, Vol 18, 1935, p 359
6. L. Holborn and A.L. Day, *Ann. Phys.*, Vol 4, 1901, p 104
7. F.M. Jaeger and W.E. Veenstra, *Proc. Acad. Sci. (Amsterdam)*, Vol 37, 1934, p 280
8. H. Holtzmann, "Festschrift 50 Jahriger," Siebert GmbH, 1931, p 147
9. W. Jaeger and H. Diesselhorst, *Wiss. Abh. Phys.-Tech. Reichsanstalt*, Vol 3 (No. 269), 1900, p 415
10. R.F. Vines and E.M. Wise, *Platinum Metals and Their Alloys*, International Nickel Company, 1941
11. L. Holborn and A.L. Day, *Sitzungsber. Akad. Wissen.*, 1899, p 694; *Ann. Phys.*, Vol 2, 1900, p 505
12. W.F. Roeser and H.T. Wensel, *Temperature—Its Measurement and Control in Science and Industry*, Reinhold, 1941, p 1293

Platinum (Pt)

Compiled by Edward D. Zysk (deceased), Engelhard Corporation

Reviewed for this Volume by
Lisa A. Dodson and James J. Klinzing, Johnson Matthey, Inc.

Structure

Crystal structure. Face-centered cubic, a = 0.39231 nm at 25 °C (Ref 1)

Mass Characteristics

Atomic weight. 195.09
Density. 21.45 g/cm^3 at 20 °C, calculated from lattice parameter (Ref 1). 21.46 g/cm^3 at 25 °C (measured)

Thermal Properties

Melting point. 1769 °C (Ref 2)
Boiling point. 3800 °C (Ref 3)
Coefficient of linear thermal expansion. 9.1 μm/m · K from 20 to 100 °C (Ref 4)
Specific heat. 0.132 kJ/kg · K at 0 °C (Ref 5)
Latent heat of fusion. 113 kJ/kg
Thermal conductivity. 71.1 W/m · K at 0 °C (Ref 6)
Vapor pressure. For 1377 to 1767 °C: p = 11.767 − 27 575/t, where t is in °C and p is in Pa (Ref 3, 7). See also Ref 5.
Oxide evaporation rate. Platinum has exceptional resistance to oxidation. Upon heating in air at all temperatures to the melting point, it remains untarnished. At temperatures above about 750 °C, slight weight loss occurs due to volatilization of the metal itself and the formation of a volatile oxide of platinum. This oxide is essentially PtO_2, although the presence of both PtO_2 and PtO has been noted (Ref 8); mass spectrometric techniques indicate that the main oxide molecules are PtO_2 and PtO_3 (Ref 9). As higher temperatures are reached, loss to volatilization of the oxide species becomes greater than that of the metal.

In determining weight loss of platinum at elevated temperatures, it is important that some consideration be given not only to the kinetic aspects of oxide formation, but also to the equilibrium between the metal and oxide species. The loss due to oxide formation is influenced by such factors as oxygen pressure, rate of gas flow over the metal surface, degree of saturation of the surrounding area with the oxide species and the geometry of the system. Should the evaporating species be removed, the evaporation rate is increased (Ref 10).

Equilibrium vapor pressures of the six platinum group metal oxides are dealt with in Ref 11; rate of evaporation of platinum in vacuum in low-pressure oxygen or air is covered in Ref 12 to 14; evaporation in air only is covered in Ref 15 to 17; evaporation in air, nitrogen, argon, hydrogen, and oxygen is covered in Ref 18. As indicated in the cited references, care should be taken in

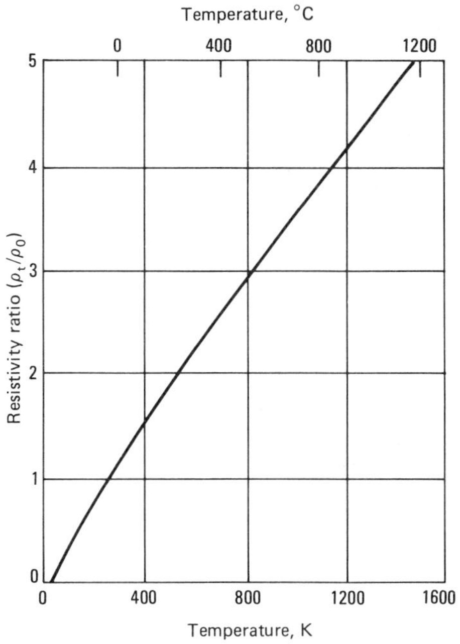

Fig. 92 Ratio of electrical resistivity at various temperatures to resistivity at 0 °C for platinum. t is in degrees Centigrade. Source: Ref 19

applying research data relating to oxide vaporization in actual commercial applications because operating conditions may not be similar to those in the experiments.

Electrical Properties

Electrical resistivity. 98.5 nΩ · m at 0 °C; 106 nΩ · m at 20 °C. Temperature coefficient: 0.003927 K from 0 to 100 °C; see also Fig. 92.

A laboratory standard platinum resistance thermometer is the standard instrument used on the International Practical Temperature Scale of 1968 from 13.81 K to 630.74 °C. The thermometer resistor must be strain-free annealed pure platinum (at least 99.998% Pt) to achieve a mandated temperature coefficient of at least 0.03925 nΩ · m per K. Below 0 °C, the resistance temperature relation of the thermometer is determined by a reference function and specified deviation equations. From 0 to 630.74 °C, two polynomial equations provide the resistance temperature relation (see Ref 2). For data on the effect of trace impurities on the temperature coefficient of electrical resistivity, see Ref 20.
Thermal electromotive force. Very pure platinum, U.S. Thermometric Standard Pt 67 (supplied as a standard reference material by the U.S. National Bureau of Standards), is used as the reference electrode in comparing the thermoelectric behavior of individual metals and alloys. Measurements of the thermal electromotive force of samples of platinum (with the joined hot end at 1200 °C and the two free ends at 0 °C) against Pt 67 are useful for estimating the purity of different lots of platinum. Small amounts of impurities (except gold) in solid solution in platinum make the slightly impure platinum thermoelectrically positive to the platinum without the additions. Iron is troublesome in this respect. For data on the effect of trace impurities on the thermal electromotive force of platinum, see Ref 20.
Work function. 0.852 to 0.876 aJ (5.32 to 5.47 eV) (Ref 21)

Magnetic Properties

Magnetic susceptibility. Mass: 0.012204 mks (Ref 22)

Optical Properties

Color. Silver-white
Reflectance. Bulk: 70% at 589 nm. Electrodeposited: 58.4% at 441 nm; 59.1% at 589 nm; 59.4% at 668 nm (Ref 23 to 25)
Spectral emittance. At 650 nm: solid, 0.30; liquid, 0.38 (Ref 26)
Total hemispherical emittance. From Ref 27:

°C	Emittance
25	0.037
100	0.047
500	0.096
1000	0.152
1500	0.191

Chemical Properties

General corrosion behavior. See the articles "Precious Metals" and "Properties of Precious Metals" in this Volume and "Corrosion of the Noble Metals" in Volume 13 of the 9th Edition of *Metals Handbook.*
Resistance to specific corroding agents. Platinum is resistant to ferric chloride at room temperature. It is attacked by hydrobromic acid plus bromine at room temperature. All of the free halogens attack platinum at elevated temperatures; however, hydrochloric acid in the absence of oxidizing agents does not attack it, and platinum is useful against this normally active gas up to 2000 °F. Sulfur dioxide does not attack platinum even at 2000 °F (Ref 28).

As an anode, platinum is outstanding and is used commercially in sulfuric and persulfuric acids, various sulfate chloride plating electrolytes and in chlorates with very little corrosion. If electrolyzed with alternating current, chlorides may attack it, and this is a characteristic exploited in etching platinum and platinum alloys.

Platinum is quite resistant to acid, to potassium sulfate and sodium carbonate, to potassium nitrate at moderate temperatures, and to sodium carbonate at 1475 to 1650 °F under nonoxidizing conditions. Although it is attacked vigorously by molten alkali cyanides and polysulfides, it is quite resistant to the normal sulfides plus alkali. Certain phosphates attack it at high temperatures, and care must be taken to avoid reducing conditions, particularly when compounds of arsenic, phosphorus, tin, lead, or iron are present. Platinum is resistant to molten glasses, especially to those low in lead and arsenic.

Platinum, even in the form of thin leaf, is resistant to corrosion and tarnishing on exposure to the atmosphere, including urban sulfur-bearing atmospheres.

REFERENCES

1. H.E. Swanson and E. Tatge, Circular 539, National Bureau of Standards, 1953
2. The International Practical Temperature Scale of 1968 Amended Edition of 1975, *Metrologia*, Vol 12, 1976, p 7-17
3. R.F. Hampson, Jr. and R.F. Walker, *J. Res. Natl. Bur. Stand.*, Vol 65A, 1961, p 289
4. P. Hidnert and W. Sander, NBS Circular 486, National Bureau of Standards, 1950
5. F.N. Jaeger and E. Rosenbohm, *Physics*, Vol 6, 1939, p 1123
6. R.W. Powell, R.P. Tye, and M.J. Woodman, *Platinum Met. Rev.*, Vol 6, 1962, p 138
7. L.H. Dreger and J.L. Margrave, *J. Phys. Chem.*, Vol 64, 1960, p 1323
8. J.H. Norman, H.G. Staley, and W.E. Bell, *J. Phys. Chem.*, Vol 71, 1967, p 3886
9. A. Olivei, *J. Less-Common Met.*, Vol 29, 1972, p 18
10. G.C. Fryburg and H.M. Petrus, *J. Electrochem. Soc.*, Vol 108, 1961, p 496
11. C.B. Alcock and G.W. Hooper, *Proc. R. Soc. (London) A*, Vol 254, 1960, p 557
12. G.K. Burgess and R.G. Waltenberg, Scientific Paper 254, National Bureau of Standards, 1916
13. T. Kubaschewski, *Z. Electrochem.*, Vol 49, 1943, p 446
14. E.K. Rideal and O.H. Wansborough-Jones, *Proc. R. Soc. (London) A*, Vol 123, 1933, p 202
15. W. Betteridge and D.W. Rhys, The High Temperature Oxidation of the Platinum Metals and Their Alloys, in *First International Congress on Metallic Corrosion 1961*, L. Kenworth, Ed., Butterworths, 1962, p 185
16. E. Raub and W. Plate, *Z. Metallkd.*, Vol 48, 1957, p 529
17. R.W. Douglass, C.A. Krier, and R.I. Jaffee, "Summary Report on High Temperature Properties and Alloying Behavior of the Refractory Platinum-Group Metals," NR 039-067, Battelle Memorial Institute, Aug 1961
18. J.S. Hill and H.J. Albert, *Engelhard Ind. Tech. Bull.*, Vol 4 (No. 2), 1963, p 59-63
19. W.F. Roeser and H.T. Wensel, in *Temperature, Its Measurement and Control in Science and Industry*, Reinhold, 1941, p 1312
20. J. Cochrane, *Engelhard Ind. Tech. Bull.*, Vol XI (No. 2), p 58-73
21. A. Ertel, *Phys. Rev.*, Vol 78, 1950, p 353
22. F.E. Hoare and J.C. Walling, *Proc. Phys. Soc.*, Section B, Vol 164, 1951, p 337
23. P. Drude, *Ann. Phys.*, Vol 39, 1890, p 481
24. W. Meier, *Ann. Phys.*, Vol 31, 1910, p 1017
25. G. Hass and L. Hadley, Optical Properties of Metals, in *American Institute of Physics Handbook*, 2nd ed., American Institute of Physics, 1965, p 6-107 to 6-118
26. W.F. Roeser and H.T. Wensel, in *Temperature, Its Measurement and Control in Science and Industry*, Reinhold, 1941, p 1313-1314
27. *Corrosion Handbook*, John Wiley & Sons, 1948
28. E.M. Wise and J.T. Eash, *Trans. AIME*, Vol 128, 1938, p 282

Plutonium (Pu)

See the section "Properties of the Actinide Metals (Ac-Pu)" in this article.

Potassium (K)

Compiled by J.R. Keiser,
Oak Ridge National Laboratory, and
Keith R. Willson, Geneva College

Few uses have been found for potassium metal, although an alloy of sodium and potassium is used as a heat transfer medium. Because potassium is an essential element in plant growth, potassium or K_2O is a main component of plant fertilizer. Potassium is also used as the super oxide KO_2 to produce oxygen in gas masks. Potassium is highly reactive and must be handled with great care. Use of a dry and oxygen-free inert-gas atmosphere is essential if reactions are to be avoided.

Structure

Crystal structure. Body-centered cubic, type *A*2; a = 0.5344 nm at 20 °C

Mass Characteristics

Atomic weight. 39.09
Density. (Ref 1):

°C	g/cm³
−273	0.909
−173	0.894
−73	0.873
20	0.855
100	0.820
200	0.797
300	0.774
400	0.751
600	0.702
800	0.653
1000	0.602

Volume change on freezing. 2.41% contraction (Ref 2)

Thermal Properties

Melting point. 63.2 °C (Ref 1)
Boiling point. 756.5 °C (Ref 1)
Coefficient of thermal expansion. Linear: 83 μm/m · K for 0 to 95 °C. Volumetric, (liquid): $V/V_0 = 1 + 2.58 \times 10^{-4} T + 13.08 \times 10^{-8} T^2 + 1.98 \times 10^{-12} T^3$ for 63.2 to 1250 °C, where T is in °C (Ref 1)
Specific heat. 0.770 kJ/kg · K at 20 °C
Specific heat versus temperature. Solid, from −173 to 63.2 °C: $C_p = 538.07 + 0.8004\,T$ J/kg · K, where T is in K. Liquid, from 63.2 to 1150 °C: $C_p = 839.14 - 0.3675\,T + 4.594 \times 10^{-4} T^2$ J/kg · K, where T is in °C (Ref 1)
Enthalpy. Where T is in °C and H_{0c} is the enthalpy of the solid at 0 °C: solid, $H_c - H_{0c} = 710.62\,T + 1.0388\,T^2$ J/kg; liquid, $H_l - H_{0c} = 56\,178 + 841.01\,T - 0.1585\,T^2 + 1.0502 \times 10^{-4} T^3$ J/kg (Ref 1)
Entropy. Where T is in K and S_{0c} is the entropy of the solid at 0 °C. Solid, from 0 to 63.2 °C: $S_c - S_{0c} = 329.47 \log T + 2.0776\,T - 13\,703$ J/kg · K. Liquid, from 63.2 to 800 °C: $S_l - S_{0c} = 2189.9 \log T + 0.4864\,T + 1.5723 \times 10^{-4} T^2 - 50\,481$ J/kg · K (Ref 1)
Latent heat of fusion. 59.45 kJ/kg (Ref 1)
Latent heat of vaporization. 1985 kJ/kg (Ref 1)
Thermal conductivity. 108.3 W/m · K at 293 K (Ref 1)
Thermal conductivity versus temperature. Solid: $k = 1.26 \times 10^2 - 6.03 \times 10^{-2}\,T$ W/m · K. Liquid:

$$k = 43.8 - 2.22 \times 10^{-2} T + \frac{3950}{T + 273.2}$$

where k is in W/m · K and T is in °C (Ref 1)
Vapor pressure:

$$\log \frac{p}{p_0} = \frac{-4625.3}{T} + 6.59817 - 0.700643 \log T$$

where T is in K and p_0 = 101.325 kPa = 1 atm; p = 3.95 kPa at 773 K; p = 6.24 kPa at 1273 K (Ref 1)
From Ref 3:

Temperature, °C	Pressure, kPa
590	15
710	50
770	100
850	200
950	500
1110	1000
1240	2000
1420	4000

Critical temperature and pressure. T_c = 2220 ± 25 K; P_c = 16.39 ± 0.02 MPa (Ref 3)
Diffusion characteristics:

Solute	Temperature range, °C	Activation energy, kJ/kg	Frequency factor, mm²/s
^{198}Au	5.6 to 52.5	345.89	0.129
^{42}K	−52.0 to 61.0	1002.3	16
^{22}Na	0 to 62.0	797.80	5.8
^{86}Rb	0.1 to 59.9	940.21	9.0

Source: Ref 4

Table 33 Electron binding energies of potassium

Shell	Electron configuration	Binding energy, eV
K	$1s$	3608.4
L(I)	$2s$	378.6
L(II)	$2p_{1/2}$	297.3
L(III)	$2p_{3/2}$	294.6
M(I)	$3s$	34.8
M(II)	$3p_{1/2}$	18.3
M(III)	$3p_{3/2}$	18.3

Source: Ref 3

Table 34 Electrical resistivity of potassium between 0 and 77 °C

Temperature K	Temperature °C	Resistivity, nΩ · m
273.15	0	64.9
293.15	20	72
336.35	63	92.2(a)
336.35	63	139.5(b)
350	77	146.4

(a) Solid. (b) Liquid. Source: Ref 3

Table 35 Unstable isotopes of potassium

Isotope	Half-life	Decay mode(a)	Particle energy, MeV	Particle intensity, %
^{35}K	0.19 s	β^+	...	...
		β^+, p	...	...
^{36}K	0.342 s	β^+	5.3	42
			9.9	44
^{37}K	1.23 s	β^+	5.13	...
^{38m}K	0.926 s	β^+	5.02	100
^{38}K	7.63 min	β^+	2.6	99.80
^{40}K	1.25×10^9 years	β^-	1.312	89
		β^+, EC	...	10.70
^{42}K	12.36 h	β^-	1.97	19
			3.523	81
^{43}K	22.3 h	β^-	0.465	8
			0.825	87
			1.24	3.50
			1.814	1.30
^{44}K	22.1 min	β^-	5.66	34
^{45}K	17.3 min	β^-	1.1	23
			2.1	69
			4	8
^{46}K	107 s	β^-	6.3	...
^{47}K	17.5 s	β^-	4.1	99
			6	1
^{48}K	69 s	β^-	5	...
^{49}K	1.3 s	β^-	...	...
^{50}K	~0.7 s	β^-	...	...
^{51}K	0.38 s	β^-	...	...

(a) p, proton; EC, electron capture. Source: Ref 3

Electrical Properties

Electron binding energy: See Table 33.
Electrical resistivity. From 0 to 77 °C, see Table 34. Liquid: 142.7 nΩ · m at 100 °C; 238.2 nΩ · m at 250 °C; 1096.7 nΩ · m at 1000 °C. For 93 to 1093 °C: $\rho = 79.898 + 0.6371\ T - 1.3959 \times 10^{-4}\ T^2 + 5.3020 \times 10^{-7}\ T^3$ nΩ · m, where T is in °C (Ref 1)
Thermoelectric potential. Versus platinum: 1.83×10^{-7} V/K at 25 °C; 1.988×10^{-6} V/K at 250 °C; 1.0168×10^{-5} V/K at 800 °C
Electrochemical equivalent. For K^+: 0.4052 mg/C (Ref 1)
Electrolytic solution potential. Versus H_2: −2.922 V (Ref 2)
Ionization potential: From Ref 3:

Degree of ionization	Potential, eV
I	4.341
II	31.625
III	45.72
IV	60.91
V	82.66
VI	100.0
VII	117.56
VIII	154.86
IX	175.814
X	503.44
XI	564.13
XII	629.09
XIII	714.02
XIV	787.13
XV	861.77
XVI	968
XVII	1034
XVIII	4610.955
XIX	4933.931

Hall coefficient. −4.9 aV · m/A · T (Ref 1)
Work function. 2.24 eV (0.359 aJ) (Ref 2)

Magnetic Properties

Magnetic susceptibility. Volume (mks units): at 30 °C, 4.94×10^{-6}; at 100 °C, 4.72×10^{-6}; at 250 °C, 4.61×10^{-6} (Ref 1)

Optical Properties

Color. Silver-white
Refractive index. 0.392 for λ = 313 nm; 0.924 for λ = 134 nm; 0.964 for λ = 128 nm (Ref 5)

Nuclear Properties

Stable isotopes. ^{39}K, isotope mass 38.96371, 93.10% abundant; ^{41}K, 6.88% abundant (Ref 6)
Unstable isotopes. See Table 35.

Chemical Properties

General corrosion behavior. Potassium is a highly reactive metal and consequently is found only in a combined state. It reacts vigorously with water to form the hydroxide and for this reason must be kept in a moisture-free environment. Potassium reacts with many other materials, including hydrogen, oxygen, sulfur, nitrogen, bromine, and graphite. It also forms alloys with many metals.

Mechanical Properties

Elastic constants. At 295 °C: C_{11}, 3.715 GPa; C_{12}, 3.153 GPa; C_{44}, 1.88 GPa (Ref 7)
Kinematic liquid viscosity. 0.00628 mm²/s at 69.6 °C; 0.00328 mm²/s at 250 °C; 0.00254 mm²/s at 400 °C (Ref 8)
Liquid surface tension. $\sigma = 0.1157 - 6.4 \times 10^{-5}\ T$, where T is in °C and σ is in N/m (Ref 1)
Speed of sound. In liquid: speed of sound at melting point, 1880 m/s. Speed of sound over liquid range: $v = 1880 - 0.53\ (T - T_m)$, where T_m is the melting point (Ref 9)

REFERENCES

1. O.J. Foust, Ed., *Sodium—NaK Engineering Handbook*, Vol 1, Gordon & Breach, 1972, p 10-89
2. T.P. Whaley, Sodium, Potassium, Rubidium, Cesium and Francium, in *Comprehensive Inorganic Chemistry*, Pergamon Press, 1973, p 369-381
3. R.C. Weast, Ed., *CRC Handbook of Chemistry and Physics*, 70th ed., CRC Press, 1989
4. R.C. Weast, Ed., *Handbook of Chemistry and Physics*, 55th ed., CRC Press, 1974, p F-65
5. J.C. Sutherland and E.T. Arakawa, *J. Opt. Soc. Am.*, Vol 58 (No. 8), 1968, p 1080-1083
6. R.C. Weast, Ed., *Handbook of Chemistry and Physics*, 55th ed., CRC Press, 1974, p B-253 to B-254
7. S.K. Sangal and P.K. Sharma, *Czech. J. Phys.*, Vol B19, 1969, p 1098
8. E.A. Schoeld, Potassium, in *The Encyclopedia of the Chemical Elements*, Reinhold, 1968, p 552-561
9. T. Iida and R.I.L. Guthrie, *The Physical Properties of Liquid Metals*, Clarendon Press, 1988

Praseodymium (Pr)

See the section "Properties of the Rare Earth Metals" in this article.

Table 36 Electrical properties of rhenium

Temperature, °C	Electrical resistivity, nΩ · m	Temperature coefficient of resistivity per K	Thermoelectric potential versus platinum, mV
20	193	...	0
100	254	0.00395	0
300	400	0.00383	0.61
500	526	0.00358	2.31
700	630	0.00333	4.9
900	725	0.00313	8.8
1100	805	0.00294	13.0
1300	870	0.00274	19.4
1500	930	0.00258	26.1
1700	985	0.00244	35.1
1900	1030	0.00231	...
2100	1065	0.00217	...
2300	1090	0.00204	...

Promethium (Pm)

See the section "Properties of the Rare Earth Metals" in this article.

Protactinium (Pa)

See the section "Properties of the Actinide Metals (Ac-Pu)" in this article.

Rhenium

Compiled by Toni Grobstein, NASA
Lewis Research Center

Rhenium is used for catalysts, thermocouples, x-ray tubes and targets, electrical contacts, filaments, and aerospace and nuclear applications. Rhenium is also used to increase ductility in tungsten and molybdenum alloys.

Structure

Crystal structure. Close-packed hexagonal; a = 0.2761 nm; c = 0.4458 nm; c/a = 1.615
Minimum interatomic distance. 0.2746 nm

Mass Characteristics

Atomic weight. 186.2
Density. 21.02 g/cm^3 at 20 °C

Thermal Properties

Liquidus temperature/melting point. 3180 °C
Boiling point. 5627 °C
Coefficient of linear thermal expansion. 6.6 μm/m · K from 20 to 100 °C; 6.8 μm/m · K from 20 to 1000 °C
Specific heat:

°C	kJ/kg · K
25	25.7
500	25
1000	30
1500	33
2000	37

Thermal conductivity. 71.2 W/m · K at 20 °C
Latent heat of fusion. 178 kJ/kg
Latent heat of vaporization. 3417 kJ/kg
Recrystallization temperature. 1200 to 1500 °C for 1 h, depending on the purity of the metal and the amount of cold work
Vapor pressure:

°C	mPa
1525	1 × 10^{-6}
2000	0.004
2200	0.11
2400	1.9
2600	16
2800	170
3000	1100

Electrical Properties

Electrical conductivity. 9.3% IACS
Electrical resistivity. See Table 36.
Temperature coefficient of electrical resistivity. See Table 36.
Thermoelectric force. See Table 36.
Work function. 4.8 eV (as high as 5.5 eV on the ⟨0001⟩ orientation)

Magnetic Properties

Magnetic susceptibility. Volume, 863 × 10^{-6} mks

Nuclear Properties

Thermal neutron absorption cross section. 85 b
Stable isotopes. ^{185}Re, atomic weight 184.953, 37.4% abundant; ^{187}Re, atomic weight 186.956, 62.6% abundant

Optical Properties

Spectral hemispherical emittance. 42% for λ = 655 nm from 0 to 2000 °C

Mechanical Properties

Tensile properties. At 20 °C: true stress at unit strain, 2.53 GPa; tensile strength, 1130 MPa; yield strength at 0.2% offset, 317 MPa; elongation, 24%
Hardness. Arc melted button, 135 HK; annealed rod, 270 HK; rod swaged 40% in cross-sectional area, 825 HK
Strain-hardening exponent. 0.353
Shear modulus. 155 GPa at 20 °C
Elastic modulus. Tension, 460 GPa at 20 °C
Proportional limit. 181 GPa
Poisson's ratio. 0.49
Creep strength. At 2200 °C: 10-h rupture stress, 20 MPa; 100-h rupture stress, 10 MPa

Chemical Properties

General resistance to corrosion. Oxidation in air is catastrophic above approximately 600 °C due to the formation of rhenium heptoxide (Re_2O_7), which has a melting point of 297 °C and a boiling point of 363 °C.

Rhenium is resistant to carburization (it does not form a carbide); it withstands arc corrosion well and has good wear resistance.
Resistance to specific agents. Rhenium is resistant to water cycle corrosion in high-temperature filaments in vacuum; to sulfuric acid and hydrochloric acid (but can be dissolved by nitric acid); to aqua regia at room temperature; to liquid alkali metal corrosion; and to attack by molten zinc, silver, copper, and aluminum.

Fabrication Characteristics

Consolidation. Rhenium can be consolidated by powder metallurgy techniques, inter-atmosphere arc melting, and thermal decomposition of volatile halides. The powder metallurgy product is usually made by pressing bars at 2000 MPa, followed by vacuum presintering at 1200 °C and hydrogen sintering at 2700 °C.
Hot fabrication. Rhenium cannot be hot worked in air due to penetration of Re_2O_7 into the grain boundaries, which causes hot shortness.
Cold fabrication. Rhenium has excellent room-temperature ductility; however, because of its high work-hardening coefficient, it must be annealed frequently in hydrogen for 1 to 2 h at 1700 °C between cold-working reduction steps. Primary working is by rolling, swaging, or forging. Wire drawing has been done. Strip and wire as thin as 2 mils are possible.
Welding. Rhenium can be welded, soldered, or brazed by conventional means. Welds made by inert-gas or electron beam methods are extremely ductile and can be formed further at room temperature.

SELECTED REFERENCES

- B.W. Gonser, Ed., *Rhenium*, Elsevier, 1962
- E.M. Savittskii and M.A. Tylkina, Ed., *The Study and Use of Rhenium Alloys*, Amerind Publishing, 1978 (translation from the Russian)
- C.T. Sims *et al.*, "Investigations of Rhenium," WADC TR 54-371, Battelle Memorial Institute, 1954
- T.E. Tietz and J.W. Wilson, *The Behavior and Properties of Refractory Metals*, Stanford University Press, 1965

Rhodium (Rh)

Compiled by Leonard Bozza,
Engelhard Corporation

Reviewed for this Volume by Louis Toth,
Engelhard Corporation

Pure rhodium is used principally where maintained high and uniform reflectivity is essential. These applications include mirrors, principally those made by electrodeposition of rhodium on metal, plus some made by subliming rhodium on glass. Various types of light filters can be made by applying very thin coatings of rhodium by subliming. A substantial amount of rhodium is electrodeposited as a nontarnishing finishing plate on jewelry articles, including those made of white gold, silver, and other metals. Some use is also made of rhodium for the plating of sliding electrical contact surfaces; for this purpose, rhodium is sometimes applied over a heavier plate of palladium. Some pure wrought rhodium is used, but rhodium is most important as an alloying element with platinum. These materials find much use at elevated temperatures, particularly for catalysts, but also for crucibles, furnace windings, glass-working equipment, and thermocouples.

Structure

Crystal structure. Face-centered cubic; lattice parameter a = 0.38044 nm at 20 °C (Ref 1)

Mass Characteristics

Atomic weight. 102.905
Density. 12.41 g/cm^3

Thermal Properties

Melting point. 1963 °C (Ref 2)
Boiling point. Approximately 3700 °C (Ref 3)
Coefficient of linear thermal expansion. 8.3 μm/m · K from 20 to 100 °C (Ref 4)
Specific heat. 0.247 kJ/kg · K (Ref 5)
Thermal conductivity. 150 W/m · K at 0 to 100 °C (Ref 6)

Electrical Properties

Electrical resistivity. 45.1 nΩ · m at 20 °C (Ref 6). Temperature coefficient, 0.043 per K from 20 to 100 °C (Ref 7)

Magnetic Properties

Magnetic susceptibility. Volume: 14.3 × 10^{-6} mks at 18 °C (Ref 8, 9)

Optical Properties

Reflectance. See Fig. 93.
Emittance. At λ = 0.65 μm: solid, 0.24; liquid, 0.30 (Ref 2)

Chemical Properties

General corrosion behavior. Rhodium remains bright and untarnished during atmospheric exposure, and its general resistance

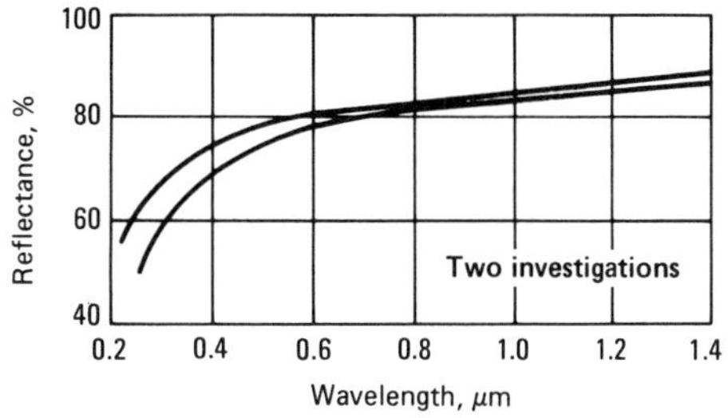

Fig. 93 Reflectance of rhodium as a function of wavelength. Sources: Ref 10, 11

to corrosion is exceptionally high. It is even resistant to boiling aqua regia. However, rhodium is attacked somewhat by hydrobromic acid, particularly when hot; by moist iodine; and by sodium hypochlorite. Hot sulfuric acid also attacks it, and sufficient corrosion occurs on electrolysis with alternating current in sulfuric acid to make this a useful metallographic etch. Additional information is available in the article "Corrosion of the Noble Metals" in Volume 13 of the 9th Edition of *Metals Handbook*.

Mechanical Properties

Tensile properties. Tensile strength: annealed, 951 MPa; hard, 2068 MPa
Hardness. 101 HB, 122 HV (Ref 12)
Elastic modulus. Hard wire, 293 GPa; 135 HB, 380 GPa (Ref 13)

REFERENCES

1. M.B. Bever, Ed., *Encyclopedia of Materials Science and Engineering*, Vol 5, MIT Press, 1986
2. International Practical Temperature Scale of 1968, Amended Edition of 1975, *Metrologia*, Vol 12, 1976, p 7-17
3. R.F. Hampson, Jr. and R.F. Walker, *J. Res. Natl. Bur. Stand.*, Vol 65A, 1961, p 289
4. W.H. Wanger, *J. Res. Natl. Bur. Stand.*, Vol 3, 1929, p 1029
5. F.M. Jaeger and E. Rosenbohm, *Proc. Acad. Sci. (Amsterdam)*, Vol 24, 1931, p 85
6. R.W. Powell *et al.*, *Platinum Met. Rev.*, Vol 6, 1962, p 138
7. E.G. Price and B. Taylor, *Nature*, Vol 195, 1962, p 272
8. K. Honda, *Ann. Phys.*, Vol 32, 1910, p 1027
9. K. Honda, *Sci. Rep. Tôhoku Imperial Univ.*, Vol 1, 1912, p 1
10. W.W. Coblentz and R. Stair, *J. Res. Natl. Bur. Stand.*, Vol 22, 1939, p 93
11. M. Auwarter, *J. Appl. Phys.*, Vol 10, 1939, p 705
12. W. Köster, *Z. Electrochem.*, Vol 49, 1943, p 233
13. J.S. Acken, *J. Res. Natl. Bur. Stand.*, Vol 12, 1934, p 249

Rubidium (Rb)

Compiled by J.R. Keiser, Oak Ridge National Laboratory, and
Keith R. Willson, Geneva College

Current uses of rubidium are limited but include such applications as vacuum tubes and photoelectric cells. Potential applications include use as a heat transfer medium and as a fuel for ion propulsion engines. The chemical behavior of rubidium is similar to that of potassium; it reacts vigorously with air or water and should be handled and stored in a dry inert environment.

Structure

Crystal structure. Body-centered cubic, *A*2; a = 0.562 ± 0.003 nm at −173 °C; a = 0.570 nm at 0 °C (Ref 1)

Mass Characteristics

Atomic weight. 85.467
Density. From Ref 1 and 2:

°C	g/cm^3
20	1.532
39	1.475
50	1.47
150	1.46
220	1.45

Density for liquid range. From Ref 3: $\rho = 1.48 - 4.5\ (T - T_m)$, where density is in g/cm^3 and T_m is the melting point
Volume change on melting. 2.5% increase (Ref 1)

Thermal Properties

Melting point. 38.89 °C (Ref 4)
Boiling point. 688 °C
Coefficient of thermal expansion. Linear: 90 μm/m · K at 20 °C; 340 μm/m · K at 40 to 100 °C (Ref 5). Volumetric: 3.38 × 10^{-4} m/m · K (Ref 1)
Specific heat. 0.33489 kJ/kg · K at 0 °C; 0.3797 kJ/kg · K at 50 °C (Ref 6)
Enthalpy. Solid: $H_T - H_{298} = 160.16\ T + 0.32080\ T^2 - 76\ 259$. Liquid: $H_T - H_{298} = 384.47\ T - 88\ 846$. T is in K and $H_T - H_{298}$ is in J/kg (Ref 7)
Entropy. Solid: $S = 368.84 \log T + 0.64162\ T - 288.96$. Liquid: $S = 885.45 \log T - 1297.9$. T is in K and S is in J/kg (Ref 7).
Latent heat of fusion. 25.535 kJ/kg (Ref 5)
Latent heat of vaporization. 887.46 kJ/kg (Ref 5)

Table 37 Electrical resistivity of rubidium

Temperature, °C	Temperature, K	Resistivity, nΩ · m
0	273	115.4
20	293	128.4
39.9	312.1	137(a)
39.9	312.1	220(b)(c)
100	373	274.7(c)

(a) Solid. (b) Liquid. (c) Equation for resistivity between 312.1 K and 373 K: $\rho = 0.086T - 4.8$, where T is in K. Sources: Ref 3, 10

Thermal conductivity:

°C	Phase	W/m · K
0	Solid	58.3
25	Solid	58.2
38.89	Solid	58.1
38.89	Liquid	33.3

Source: Ref 8

Vapor pressure: From Ref 6:

°C	kPa
294	0.1333
387	1.333
519	13.33
569	26.66
628	53.33

Saturated vapor pressure:

$$\log P = (-4688/T) - 1.76 \log T + 15.195$$

where P is in Pa and T is in K (Ref 3)
Critical temperature and pressure. T_c = 2083 ± 20 K and P_c = 14.54 ± 0.5 MPa (Ref 9)

Electrical Properties

Electrical resistivity. See Table 37. Temperature coefficient: 0.06 nΩ · m per K (Ref 5)
Electrochemical equivalent:

Valence	Equivalent, mg/C
1	0.8858
2	0.4429
3	0.2214
4	0.1107

Electron binding energy. See Table 38.
Electrolytic solution potential. Versus H_2: −2.924 V at 25 °C (Ref 10)
Ionization potential. From Ref 9:

Degree of ionization	Potential, eV
I	4.177
II	27.28
III	40
IV	52.6
V	71.0
VI	84.4
VII	99.2
VIII	136
IX	150
X	277.1

Table 38 Electron binding energy of rubidium

Shell	Electron configuration	Binding energy, eV
K	$1s$	15 200
L(I)	$2s$	2 065
L(II)	$2p_{1/2}$	1 864
L(III)	$2p_{3/2}$	1 804
M(I)	$3s$	326.7
M(II)	$3p_{1/2}$	246.7
M(III)	$3p_{3/2}$	239.1
M(IV)	$3d_{3/2}$	113
M(V)	$3d_{5/2}$	112
N(I)	$4s$	30.5
N(II)	$4p_{1/2}$	16.3
N(III)	$4p_{3/2}$	15.3

Source: Ref 9

Work function. 2.09 eV (0.335 aJ) (Ref 2)
Photoelectric threshold. 7.3 × 10^{-4} C (Ref 1)

Magnetic Properties

Magnetic susceptibility. Volume: 2.6 × 10^{-3} mks at 30 °C (Ref 11)

Optical Properties

Color: Silvery white

Nuclear Properties

Natural isotopes. ^{85}Rb: isotope mass, 84.9117; 72.15% abundant. ^{87}Rb: 27.85% abundant, decays with half-life of 5 × 10^{11} years (Ref 12)
Unstable isotopes. See Table 39.

Chemical Properties

General corrosion behavior. Rubidium is very active. It ignites in air at room temperature and will react explosively in oxygen. Rubidium can easily dissociate water to form the hydroxide and will react spontaneously with gaseous chlorine and fluorine and explosively with liquid bromine.

Mechanical Properties

Hardness. 0.3 Mohs scale (Ref 6)
Modulus of elasticity. 2.35 GPa (Ref 1)
Kinematic liquid viscosity. From Ref 1:

°C	mm²/s
38	0.4573
39	0.4561
40	0.4528
50	0.4267
99.7	0.3359
140.5	0.2904
179	0.2586
220.1	0.2332

Table 39 Unstable isotopes of rubidium

Isotope	Half-life	Decay mode(a)	Particle energy, MeV	Particle intensity, %
^{75}Rb	17 s	β^+	2.31	...
^{76}Rb	17 s	β^+	5.2	...
^{77}Rb	3.8 min	β^+	3.86	...
^{78m}Rb	5.7 min	β^+	...	...
		IT	...	...
		EC	...	...
^{78}Rb	17.66 min	β^+	...	...
		EC	...	...
^{79}Rb	23 min	β^+	...	...
		EC	...	...
^{80}Rb	34 s	β^+	3.86	22
			4.7	74
^{81m}Rb	32 min	IT	1.4	...
		β^+, EC	...	...
^{81}Rb	4.58 h	β^+	1.05	...
		EC	...	...
^{82m}Rb	6.47 h	β^+	0.8	...
		EC	...	...
^{82}Rb	1.273 min	β^+	3.3	...
		EC	...	...
^{83}Rb	86.2 days	EC	...	...
^{84m}Rb	20.3 min	IT	...	...
^{84}Rb	32.9 days	β^+	0.78	11
		EC	1.658	11
		β^-	0.893	...
^{86m}Rb	1.018 min	IT	...	...
^{86}Rb	18.63 days	β^-	1.774	8.80
^{87}Rb	4.9 × 10^{10} years	β^-	0.273	100
^{88}Rb	17.7 min	β^-	5.32	...
^{89}Rb	15.15 min	...	1.26	38
			1.9	5
			2.2	34
			4.49	18
^{90m}Rb	4.28 min	β^-	1.7	...
			6.5	...
^{90}Rb	2.6 min	β^-	...	...
^{91}Rb	58.4 s	β^-	...	...
^{92}Rb	...	β^-	8.1	94
^{93}Rb	5.85 s	β^-	7.4	...
		n	...	...
^{94}Rb	2.73 s	β^-	9.3	...
		n	...	...
^{95}Rb	0.38 s	β^-	8.6	...
		n	...	...
^{96}Rb	0.20 s	β^-	10.8	...
		n	...	...
^{97}Rb	0.17 s	β^-	...	...
		n	...	...
^{98}Rb	0.13 s	β^-	...	...
		n	...	...

(a) β^-, negative beta emission; β^+, positron emission; EC, orbital electron capture; IT, isomeric transition from upper to lower isomeric state; n, neutron emission. Source: Ref 9

Liquid surface tension. From Ref 2:

°C	N/m
39	0.0847
100	0.080
550	0.051
632	0.0468

Speed of sound. At melting point: v = 1260 m/s (Ref 3)

REFERENCES

1. M.A. Filyand and E.I. Semenova, *Handbook of the Rare Elements*, Vol 1, M.E. Alferieff, Trans. and Ed., Boston Technical Publishers, 1968, p 219-229
2. T.P. Whaley, Sodium, Potassium, Rubidium, Cesium, and Francium, in *Comprehensive Inorganic Chemistry*, Trotman-Dickenson *et al.*, Ed., Pergamon Press, 1973, p 369-381
3. T. Iida and R.I.L. Guthrie, *The Physical Properties of Liquid Metals*, Clarendon Press, 1988
4. R.C. Weast, Ed., *CRC Handbook of Chemistry and Physics*, 55th ed., CRC Press, 1974, p B-28
5. C.A. Hampel, Rubidium and Cesium, in *Rare Metals Handbook*, Reinhold, 1961, p 434-440
6. C.E. Mosheim, Rubidium, in *The Encyclopedia of the Chemical Elements*, C.A. Hampel, Ed., Reinhold, 1968, p 604-610
7. R.C. Weast, Ed., *CRC Handbook of Chemistry and Physics*, 55th ed., CRC Press, 1974, p D-57
8. R.C. Weast, Ed., *CRC Handbook of Chemistry and Physics*, 55th ed., CRC Press, 1974, p E-15
9. R.C. Weast, Ed., *CRC Handbook of Chemistry and Physics*, 70th ed., CRC Press, 1989
10. F.M. Perel'man, *Rubidium and Cesium*, R.G.P. Towndrow, Trans., R.W. Clarke, Ed., Pergamon Press, 1965, p 13
11. R.C. Weast, Ed., *CRC Handbook of Chemistry and Physics*, 55th ed., CRC Press, 1974, p E-124
12. R.C. Weast, Ed., *CRC Handbook of Chemistry and Physics*, 55th ed., CRC Press, 1974, p 265-266

Ruthenium (Ru)

Compiled by R.H. Atkinson (retired), INCO Alloys International, Inc.

Reviewed for this Volume by Michael F. Stevens, Los Alamos National Laboratory

Ruthenium is used as a hardener for platinum and palladium for jewelry and other applications; 10 to 11% Ru platinum is used to harden aircraft magneto contacts and similar contacts. Ruthenium can be used as an electric contact at temperatures up to around 500 °C because its oxide is conductive. Hard complex high-ruthenium alloys are also used for pen tipping and similar applications. Some of the alloys also contain osmium. A significant application for ruthenium is in the thick-film paste systems used for printed circuit resistance elements.

Structure

Crystal structure. Closed-packed hexagonal; a = 0.27058; c/a = 0.15824 (Ref 1)

Mass Characteristics

Atomic weight. 101.07
Density. At 20 °C, 12.45 g/cm^3; 12.37 g/cm^3 at 25 °C, computed from lattice constants

Thermal Properties

Melting point. 2310 ± 20 °C
Boiling point. Approximately 3900 °C
Coefficient of linear thermal expansion. 5.05 μm/m · K at 20 °C
Specific heat. 0.240 kJ/kg · K at 0 °C

Electrical Properties

Electrical resistivity. 76 nΩ · m at 0 °C
Electrochemical equivalent. Valence 2, 0.527 mg/C
Temperature of superconductivity. 2.04 K

Magnetic Properties

Magnetic susceptibility. Mass: 1.12×10^{-8} mks

Optical Properties

Reflectivity. 63% average in the visible range

Chemical Properties

General corrosion behavior. The corrosion resistance of ruthenium approaches that of iridium; it is unaffected by common acids, including aqua regia, at temperatures up to 100 °C, or by sulfuric acid up to 500 °C; it is moderately attacked by aqueous solutions of alkaline hypochlorites. Ruthenium exhibits good resistance to attack by certain molten metals. For example, at 200 °C above the melting points of the respective metals (under argon cover), it is not attacked by lithium, sodium, potassium, gold, silver, copper, lead, bismuth, tin, tellurium, indium, cadmium, calcium, or gallium. However, gold, silver, and copper wet the surface of ruthenium. Additional information is available in the article "Corrosion of the Noble Metals" in Volume 13 of the 9th Edition of *Metals Handbook.*

Fabrication Characteristics

Metallography. Etching: use ac electrolytic
Consolidation. Powder from refining process can be consolidated by powder metallurgy techniques or argon arc melting.
Hot-working temperature. 1500 to 2400 °C
Forming. Ruthenium is very difficult to work, but with care it can be forged at temperatures above 1500 °C. Ruthenium can be consolidated by powder metallurgy methods plus sintering above 1450 °C.
Compacting pressure. 275 MPa
Formability. Monocrystalline zone-refined rod can be bent over a small radius but, if worked, recrystallizes on annealing and becomes relatively brittle.

Mechanical Properties

Tensile properties. Tensile strength for compact powder bar, hot rolled 50%: 540 MPa at room temperature, 247 MPa at 1020 °C

REFERENCE

1. M.B. Bever, Ed., *Encyclopedia of Materials Science and Engineering*, Vol 5, MIT Press, 1986

Samarium (Sm)

See the section "Properties of the Rare Earth Metals" in this article.

Scandium (Sc)

See the section "Properties of the Rare Earth Metals" in this article.

Selenium (Se)

Compiled by S.C. Carapella, Jr., ASARCO Inc.

Reviewed for this Volume by Douglas Hayduk, ASARCO Inc.

Selenium is used in rectifiers, photovoltaic cells, and xerographic drums; as a colorizing and decolorizing agent in glass; as a color pigment used in paints, ceramics, and plastics; as an additive to improve machinability of low-carbon steels, stainless steels, copper alloys, and Invar; as an additive to lead-antimony battery grid metal to improve properties; and as a vulcanizing agent to improve the temperature and abrasion resistance of rubber.

Structure

Crystal structure. γ phase, hexagonal; at 20 °C, a = 0.43640 nm and c = 0.49594 nm. α and β phases, monoclinic

Mass Characteristics

Atomic weight. 78.96
Density:

Form	°C	g/cm^3
γ phase	25	4.809
α phase	25	4.389
β phase	25	4.470
Vitreous	20	4.280
Liquid	217	3.975
	267	4.060
	305	4.020
	406	3.910

Thermal Properties

Melting point. γ phase, 217 °C; vitreous softens at 40 °C
Boiling point. 684.9 °C

Phase transformation temperature. α to β unknown; β to γ, 209 °C (estimated)
Coefficient of thermal expansion. Linear at 20 °C: γ phase, 49 μm/m · K; vitreous, 37 μm/m · K
Specific heat. γ phase, 0.317 kJ/kg · K at 25 °C; vitreous, 0.462 kJ/kg · K at 22 °C
Latent heat of fusion. 84.93 kJ/kg
Latent heat of vaporization. 1213.3 kJ/kg
Thermal conductivity. At 25 °C: γ phase, 2.48 W/m · K; vitreous, 0.51 W/m · K
Vapor pressure:

°C	kPa
344	0.1013
431	1.013
540	10.13
684.9	101.3

Electrical Properties

Electrical resistivity. At 25 °C: γ phase, 100 MΩ · m; vitreous, 100 GΩ · m
Electrochemical equivalent. Valence +6, 136.4 μg/C

Magnetic Properties

Magnetic susceptibility. Volume: -3.9×10^{-6} mks

Optical Properties

Refractive index. Vitreous at wavelength of 1.152 μm, 2.4969. γ phase single crystals at 23 °C:

Wavelength, μm	Refractive index: Ordinary	Refractive index: Extraordinary
1.06	2.790	3.608
1.15	2.737	3.573
3.39	2.65	3.46
10.6	2.64	3.41

Nuclear Properties

Stable isotopes. ^{74}Se, ^{76}Se, ^{77}Se, ^{78}Se, ^{80}Se, and ^{82}Se
Thermal neutron cross section. For 2.2 km/s neutrons: absorption, 11.8 ± 0.4 b; scattering, 11 ± 2 b

Mechanical Properties

Hardness. 2.0 Mohs
Elastic modulus. Tension, 53.82 GPa; shear, 6.46 GPa
Surface tension:

°C	mN/m
220	105.5
250	100.5
280	98.0
310	98.2

Silicon (Si)

Compiled by H.M. Liaw, Motorola Inc. Semiconductor Product Sector

Silicon used in industry can be classified into metallurgical and semiconductor grades. Metallurgical-grade silicon is produced by the reduction of sand (SiO_2) in an electric arc furnace. It contains approximately 98% Si. Major impurities include aluminum, calcium, and iron (Ref 1). Metallurgical-grade silicon is used primarily in the aluminum, steel, and silicone industries. Metallurgical-grade silicon is also used for the production of chlorosilanes or fluorosilanes.

Polycrystalline semiconductor-grade silicon is produced from pure silane or chlorosilanes by the chemical vapor deposition (CVD) technique. Total impurity content of the semiconductor-grade silicon is generally less than 0.1 ppm. Single-crystal semiconductor-grade silicon ingots are grown by pulling them from the melt of polycrystalline semiconductor-grade silicon either by the floating-zone or Czochralski technique (Ref 2). The silicon ingots pulled from the melt in a quartz crucible (that is, in the Czochralski method) are unintentionally doped with oxygen and carbon at concentrations of approximately 10 ppm O_2 and 0.5 ppm C. Thin films of epitaxial silicon grown on silicon substrates have also been widely used for the fabrication of solid-state devices. They are primarily grown from vapor phase by the CVD technique (Ref 3).

Amorphous silicon can be produced from plasma-enhanced pyrolysis of silane at a low temperature (<550 °C) on the surface of a dielectric material such as silicon dioxide. Because the properties of amorphous silicon can be changed by a high-temperature treatment, the properties listed below are of crystalline silicon.

Structure

Crystal structure. At a pressure from 0 to 12.5 GPa: diamond cubic structure containing 8 atoms per unit cell. Lattice constant at 25 °C: $a = 0.54310626 \pm 0.00000008$ nm at 1 atm; $a = 0.54310644 \pm 0.00000008$ nm under vacuum (Ref 4 to 6)

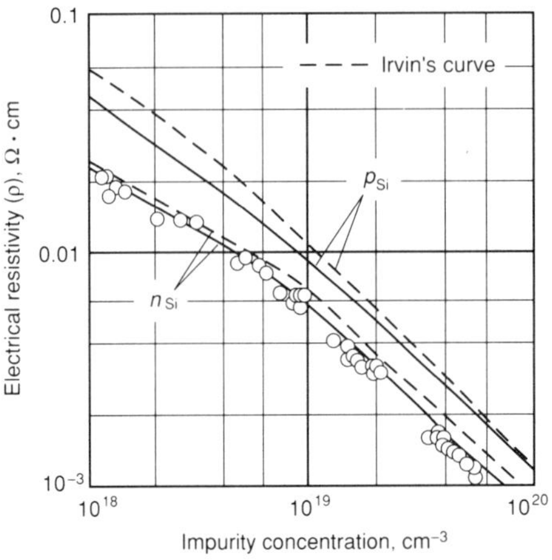

Fig. 94 Resistivity versus impurity concentration for *p*- and *n*-type silicon at 300 K

The lattice constant can be changed by the presence of impurities. For example, the presence of carbon in silicon causes the contraction of lattice constant Δa according to the empirical equation: $\Delta a/a = -6.5 \times 10^{-24} \times N_c$, where N_c is the carbon concentration in atoms per cm^3 (Ref 7).

The structure of silicon can be changed by applying pressure (Ref 8). The high-pressure phases observed include body-centered tetragonal, body-centered cubic, primitive hexagonal, hexagonal close-packed, and face-centered cubic.

Mass Characteristics

Atomic weight. 28.08
Density. Variation of density with temperature (Ref 9, 10):

Temperature, °C	Density, g/cm^3
25	2.3290
127	2.3269
427	2.3192
627	2.3136
827	2.3077
1027	2.3016

Thermal Properties

Melting point. 1414 °C (Ref 11)
Boiling point. 2355 °C at reduced pressure (10^{-3} atm); 3145 °C at 1 atm (Ref 12)
Thermal expansion coefficient. From Ref 13:

Temperature, K	Coefficient, μm/m · K
60	−0.400
100	−0.339
200	1.406
300	2.616
400	3.253
500	3.614
600	3.842
700	4.016
800	4.151
900	4.185
1000	4.258
1200	4.384
1300	4.442
1400	4.500
1500	4.556

Specific heat:

Temperature, K	Specific heat, kJ/kg · K
60	0.115
100	0.259
200	0.557
300	0.713
400	0.785
500	0.832
600	0.849
700	0.866
800	0.883
900	0.899
1000	0.916
1100	0.933
1200	0.950
1300	0.967
1400	0.983
1500	1.00

Latent heat of fusion. 1807.9 kJ/kg (Ref 14)
Latent heat of vaporization. 10 606 kJ/kg (Ref 15)

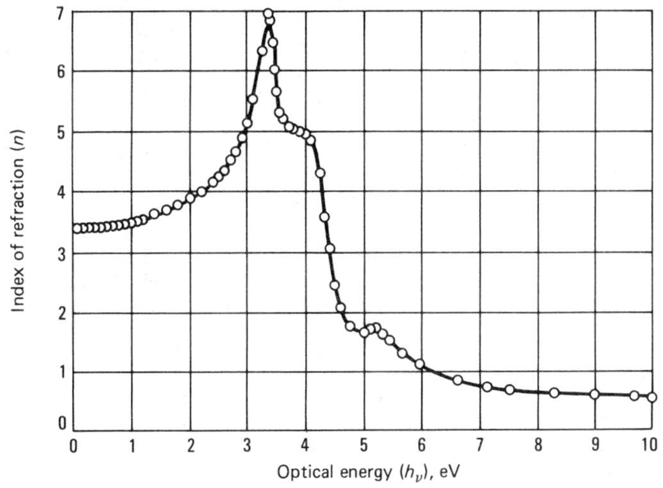

Fig. 95 Real part of the index of refraction versus energy for silicon. Source: Ref 27

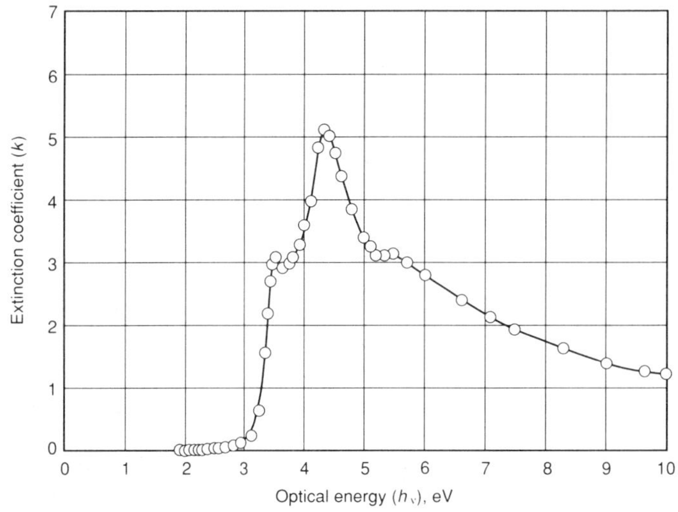

Fig. 96 Extinction coefficient (the imaginary part of the index of refraction) versus optical energy for silicon. Source: Ref 27

Heat of combustion. 31 350 kJ/kg Si (Ref 16)
Thermal conductivity. At temperatures below 150 K, silicon is very sensitive to variations in sample size, impurity level, orientation, and surface quality. The values listed below include typical data obtained in the 2 to 150 K range (Ref 17, 18):

Temperature, K	Conductivity, W/m · K
2	44
4	311
6	899
8	1640
10	2400
20	4770
30	4420
40	3660
50	2800
100	913
150	410
200	266
300	156
400	105
500	80
600	64
700	52
800	43
900	36
1000	31
1100	28
1200	26
1300	25
1400	24
1500	23

Electrical Properties (Ref 19)

Energy band gap. At 300 K: 1.12 eV
Intrinsic carrier concentration. 1.38×10^{10} cm^{-3}
Intrinsic Debye length. 28.7 μm
Intrinsic resistivity. At 300 K: 2.3×10^5 Ω · cm
Extrinsic resistivity. At 300 K, extrinsic resistivity is a function of impurity concentration. Figure 94 shows the plot of resistivity versus doping concentration for *p*- and *n*-doped silicon.
Electron mobility. The measured mean electron mobility in bulk silicon at 300 K is 1439 $cm^2/V \cdot s$.
Temperature dependence of electron mobility. Mobility (T) = 1439 $(T/300)^{-2.26}$ (Ref 20, 21)
Hole mobility. The measured mean hole mobility in bulk silicon at 300 K is 484 $cm^2/V \cdot s$.
Temperature dependence of hole mobility. Mobility (T) = 484 $(T/300)^{2.21}$ (Ref 20, 21)
Electrochemical equivalent. 0.07269 mg/°C (Ref 22)
Electrolytic solution potential. Versus H_2: −0.453 V (Ref 23)
Hydrogen overvoltage. Versus platinum: 0.192 ± 0.002 V (Ref 24)
Hall effect. 4100 V · m/A · T at 20 °C (Ref 25)
Dielectric constant. 11.695 ± 0.03 (Ref 26)

Magnetic Properties

Magnetic susceptibility. Volume: -1.63×10^{-6} mks (Ref 25)

Optical Properties (Ref 26)

Color. Dark steel gray
Refractive index. As a function of optical energy, see Fig. 95.
Extinction coefficient. As a function of optical energy, see Fig. 96.
Absorption coefficient. As a function of optical energy, see Fig. 97.
Reflectance. As a function of optical energy, see Fig. 98.
Effect of energy variations on optical parameters. A table of the refractive index extinction coefficient and the absorption coefficient versus energy (0 to 400 eV) has been compiled (Ref 28). Table 40 lists values from this table at selected energy levels.
Infrared refractive index. n as a function of wavelength (λ) (in μm) can be calculated with the empirical equation $n = A + BL + CL^2 + D\lambda^2 + E\lambda^4$, where $L = 1/(\lambda^2 - 0.028)$, A = 3.41983, B = 1.59906×10^{-1}, C = -1.23109×10^{-1}, D = 1.26878×10^{-6}, and E = -1.95104×10^{-9} (Ref 29).

Chemical Properties

Resistance to specific corroding agents. See Table 41.

Mechanical Properties*

Hardness. 10^9 kg/m^2 (Ref 30)
Bulk modulus. 98.74 GPa (Ref 31)
Modulus of rupture in bending. 7×10^6 to 35×10^6 kg/m^2 (Ref 32)
Breaking strength in compression. 4.9×10^7 to 5.6×10^7 kg/m^2 (Ref 32)

REFERENCES

1. H.M. Liaw, *Solar Cells*, Vol 10, 1983, p 119
2. H.M. Liaw, Crystal Growth of Silicon, in *Handbook of Semiconductors*, W.C. Omara and R. Herring, Ed., Noyes Publications, 1990
3. H.M. Liaw, Silicon Vapor-Phase Epitaxy, in *Epitaxial Silicon Technology*, B. Jayant Baliga, Ed., Academic Press, 1986
4. E.G. Kessler, Jr., R.D. Deslattes, and A. Henins, *Phys. Rev. A*, Vol 19, 1979, p 215
5. R.D. Deslattes, A. Henins, R.M. Schoonover, C.L. Carrol, and H.A. Bowman, *Phys. Rev. Lett.*, Vol 36, 1976, p 898
6. R.D. Deslattes and A. Henins, *Phys. Rev. Lett.*, Vol 31, 1973, p 927
7. M. Hart, *EMIS Datareview*, Series 4, 1987, p 7
8. J.Z. Hu, L.D. Merkle, C.S. Menoni, and I.L. Spain, *Phys. Rev. B*, Vol 34, 1986, p 4679

*The mechanical properties data were supplied by D.K. Schroder, Arizona State University.

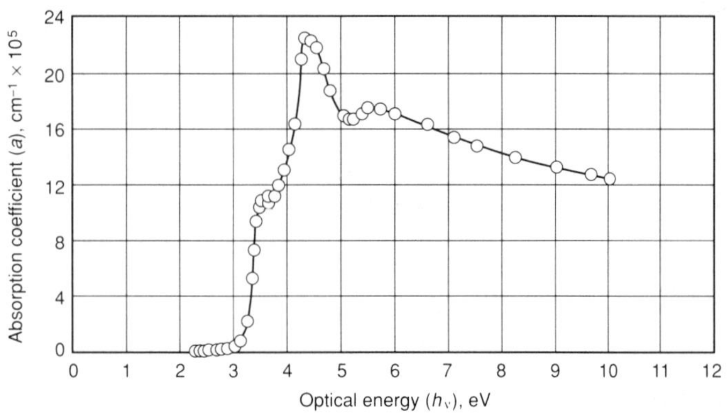

Fig. 97 Spectral dependence of the absorption coefficient of silicon as a function of optical energy. Source: Ref 27

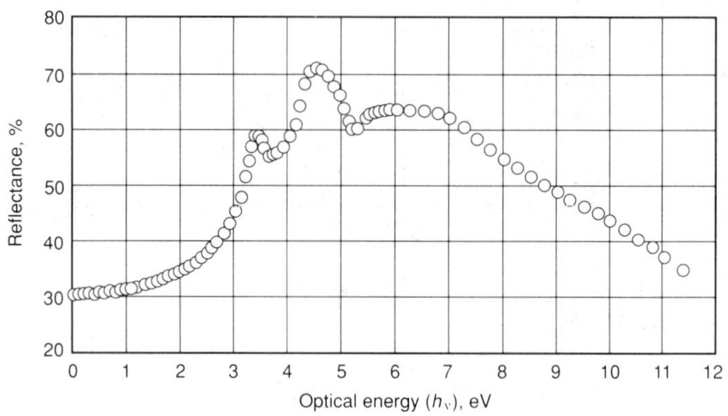

Fig. 98 Spectral dependence of the reflectance of silicon as a function of optical energy. Source: Ref 27

Table 40 Effect of 0 to 10 eV energy on selected optical index and coefficient parameters

Energy, eV	Refractive index (n)	Extinction coefficient (k)	Absorption coefficient (a)
0.20	3.4236	7.4×10^{-7}	0.015
1.15	3.550	8.1×10^{-5}	9.4
1.20	3.565	4.5×10^{-4}	5.4×10^{1}
2.00	3.906	0.022	4.5×10^{3}
3.00	5.222	0.269	8.2×10^{4}
4.00	5.010	3.587	1.45×10^{6}
5.00	1.57	3.565	1.81×10^{6}
10.0	0.306	1.38	1.4×10^{6}

Source: Ref 28

Table 41 Resistance of silicon to specific corroding agents

Corrosive agent	Resistance	Corrosive agent	Resistance
Air	Resistant	Hydrogen sulfide	Resistant
Ammonia	Resistant; reacts with vapors at bright red heat	Iodine	Resistant
Bromine	Resistant; burns at 500 °C	Nitric acid	Resistant (dilute or concentrate, cold or boiling)
Carbon dioxide	Resistant	Oxygen	Resistant
Chlorine	Resistant; burns at 340 °C	Potassium hydroxide	Attacked
Copper sulfate	Resistant (10% solution)	Sodium hydroxide	Attacked
Ferric chloride	Resistant (10% solution)	Sulfur	Resistant; reacts at elevated temperatures
Hydrochloric acid	Resistant (dilute or concentrate, cold or boiling)	Sulfur dioxide	Resistant
Hydrofluoric acid	Resistant (dilute or concentrate, cold or boiling)	Sulfuric acid	Resistant (dilute or concentrate, cold or boiling)
		Water, distilled	Resistant
		Water, rain	Resistant

9. K.G. Lyon, G.L. Salinger, C.A. Swenson, and G.K. White, *J. Appl. Phys.*, Vol 48, 1977, p 865
10. Y. Okada and Y. Tokumaru, *J. Appl. Phys.*, Vol 56, 1984, p 314
11. J.C. Brice, *EMIS Datareview*, Series 4, 1987, p 52
12. R.E. Honig, *RCA Rev.*, Vol 23, 1962, p 567
13. T. Soma and H.-M. Kagaya, *EMIS Datareview*, Series 4, 1987, p 33
14. Kubaschewski, Evans, and Alcock, *Metallurgical Thermochemistry*, 4th ed., Pergamon Press, 1967
15. L.L. Quill, *The Chemistry and Metallurgy of Miscellaneous Materials*, McGraw-Hill, 1950
16. J.F. Elliot and M. Gleiser, *Thermochemistry for Steelmaking*, Vol I, Addison-Wesley, 1960
17. M.G. Holland and L.G. Neuringer, in *Proceedings of an International Conference on the Physics of Semiconductors*, 1962, p 474
18. G.J. Glassbrenner and G.A. Slack, *Phys. Rev.*, Vol 134, 1964, p A1058
19. W.E. Beadle, J.C.C. Tsai, and P.D. Plummer, Ed., *Quick Reference Manual for Silicon Integrated Circuit Technology*, John Wiley & Sons, 1985, p 2-26
20. J.M. Dorkel and P. Letureq, *Solid State Electron.*, Vol 24, 1981, p 821
21. N.D. Arora, J.R. Hauser, and D.J. Rouston, *IEEE Trans. Electron. Devices*, Vol ED-29, 1982, p 292
22. G.A. Rousch, *Trans. Electrochem. Soc.*, Vol 70, 1938, p 293
23. G.W. Akimow and A.S. Oleschko, *Korros. Metallschutz*, Vol 10, 1934, p 134
24. Thiel and Hammerschmidt, *Z. Anorg. Allg. Chem.*, Vol 132, 1923, p 15
25. *International Critical Tables*, McGraw-Hill, 1926
26. D.E. Aspnes, *EMIS Datareview*, Series 4, 1987, p 63
27. H.R. Phillipp and E.A. Taft, *Phys. Rev.*, Vol 120, 1960, p 37
28. D.E. Aspnes, *EMIS Datareview*, Series 4, 1987, p 72
29. D.F. Edwards and E. Ochoa, *Appl. Opt.*, Vol 19, 1980, p 4130
30. P.J. Burnett, in *Properties of Silicon*, *EMIS Datareview*, Series 4, INSPEC, The Institution of Electrical Engineers, London, 1988
31. O.H. Nielsen, in *Properties of Silicon*, *EMIS Datareview*, Series 4, INSPEC, The Institution of Electrical Engineers, London, 1988
32. W.R. Runyan, *Silicon Semiconductor Technology*, McGraw-Hill, 1965

Silver (Ag)

Compiled by S.C. Carapella, Jr., ASARCO Inc., and D.A. Corrigan, Handy & Harman

Revised by G.M. Wityak, Handy & Harman

Silver is used in a very broad range of applications, including, but not limited to, jewelry, coinage, electrical and electronic devices, and photographic compounds. In some of these applications, silver is alloyed to improve its hardness. The primary alloying element is copper, which, in proportions of 7.5% and 10%, forms the recognized standards for sterling and coin silver, respectively.

The intrinsic qualities of silver (that is, excellent electrical and thermal conductivity, high reflectivity, good corrosion resistance, and ductility) make it a good material choice for a wide range of industrial applications. These applications include contact rivets, capacitor components, fuse links, thin-film coatings for optically and thermally efficient glass, conductive inks, plating anodes, and photographic emulsions.

Silver is also the primary ingredient in many brazing alloys, and many of these have historically been characterized as silver solders. The biological compatibility with human tissue of silver-base dental amalgams accounts, at least in part, for the long-term successful use of these compounds.

Structure

Crystal structure. Face-centered cubic at 25 °C, a = 0.408621 nm

Mass Characteristics

Atomic weight. 107.868
Density. At 20 °C: 10.49 g/cm^3. Liquid:

°C	g/cm^3
960.5	9.30
1000	9.26
1092	9.20
1195	9.10
1300	9.00

Volume change on freezing. 5% contraction

Thermal Properties

Melting point. 961.9 °C. Freezing point in approximate equilibrium with the oxygen in the atmosphere (partial pressure of oxygen 20 kPa) is about 950 °C (Ref 1, 2). Freezing point is not lowered by carbon.
Boiling point. 2163 °C (Ref 3)
Coefficient of linear thermal expansion. At 20 °C, 19.0 μm/m · K; −190 to 0 °C, 17.0 μm/m · K (Ref 4 to 6); 0 to 900 °C, $L_t = L_0$ $(1 + 19.494 \times 10^{-6}\, t + 1.0379 \times 10^{-6}\, t^2 + 2.375 \times 10^{-12}\, t^3$, where t is in °C (Ref 7); 0 to 100 °C, 19.68 μm/m · K (Ref 8); 0 to 500 °C, 20.61 μm/m · K
Specific heat. Solid: at 25 °C, 0.235 kJ/kg · K; 127 °C, 0.239 kJ/kg · K; 527 °C, 0.262 kJ/kg · K; 961 °C, 0.297 kJ/kg · K (Ref 3). Liquid: 961 to 2227 °C, 0.310 kJ/kg · K (Ref 3)
Latent heat of fusion. 104.2 kJ/kg (Ref 3)
Latent heat of vaporization. 2.63 MJ/kg (Ref 3)
Recrystallization temperature. 20 to 200 °C (68 to 392 °F), depending on purity and amount of cold work (Ref 9)
Thermal conductivity. 428 W/m · K at 20 °C; 356 W/m · K at 450 °C (Ref 10)

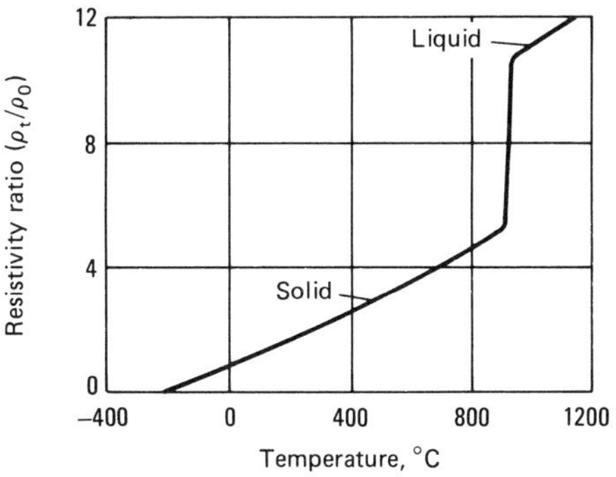

Fig. 99 Temperature dependence of the electrical resistivity ratio of silver. The ratio ρ_t/ρ_0 is about 0.10 at 20.4 K and 0.0068 at 1.3 K. Sources: Ref 14 to 16

Vapor pressure. Liquid (Ref 3):

°C	kPa
1304	1.013×10^2
1510	1.013×10^3
1783	1.013×10^4
2163	1.013×10^5

Liquid (Ref 11):

$$\log p = \frac{-13350}{T} + 10.486$$

Solid (Ref 12):

$$\log p = \frac{-14020}{T} + 11.012$$

where T is in K and p is in Pa. High rate of volatilization at high temperatures and with oxidizing gases rather than under reducing gases (Ref 13)

Electrical Properties

Electrical conductivity. 108.4% IACS for extremely pure silver (105% referred to very pure oxygen)
Electrical resistivity. 14.7 nΩ · m at 0 °C. Temperature coefficient: from 0 to 100 °C, 0.0041 per K. Temperature dependence: see Fig. 99.

Cold working of silver considerably increases resistivity: 5% for 90% reduction (Ref 17). Annealing commercially pure silver successively in air and hydrogen disrupts grain boundaries and increases resistance.

Tension reduces resistivity slightly as does hydrostatic pressure: 12 000 kg/cm^2 causes 4% reduction (Ref 18).
Thermal electromotive force. Versus platinum, +0.74 mV; cold junction at 0 °C, hot junction at 100 °C (Ref 19)

Magnetic Properties

Magnetic susceptibility. Volume: -2.27×10^{-6} mks

Optical Properties

Color. As a result of the high and fairly

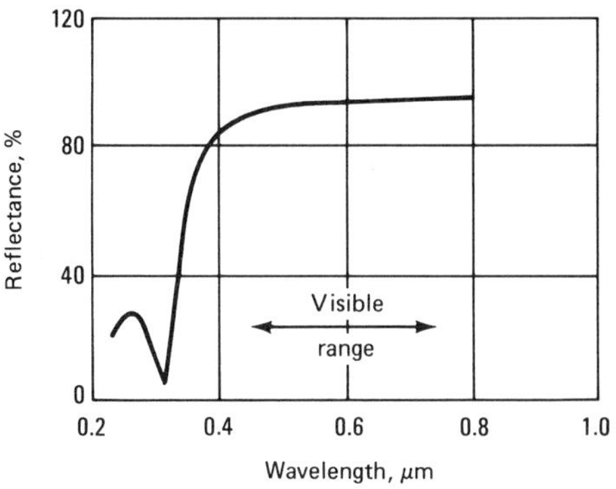

Fig. 100 Reflectance of silver as a function of wavelength. Sources: Ref 19, 20

uniform reflectance in the visible range, silver is considered white, but if human eyes were sensitive to a slightly shorter wavelength region, silver would appear to have color.
Reflectance. For clean silver, high in the visible and infrared but low in near-ultraviolet; see Fig. 100.
Emittance. Solid silver at 0.65 μm: extremely low and not known accurately; values of 0.044 μm at 940 °C and 0.072 μm at 980 °C have been observed for liquid silver (Ref 21). Other experiments showed no discontinuity at the melting point of silver, the emissivity being about 0.055 at about 700 °C (Ref 22).

Mechanical Properties

Tensile properties. Considerable spread in values for tensile strength and hardness of high-purity silver. Average tensile strength, 125 MPa for 5 mm wire annealed at 600 °C (Ref 23)
Hardness. High-purity silver; hydrogen anneal 650 °C, 25 HV; air anneal at 650 °C, 27 HV; electrodeposited silver (higher electrical resistivity than wrought silver), 100 HV
Elastic modulus. Strained 5%, then heated 0.5 h at 350 °C, 71.0 GPa (Ref 24)
Poisson's ratio. Annealed: 0.37; hard drawn, 0.39
Liquid surface tension. 0.923 N/m at 995 °C

REFERENCES

1. W.F. Roeser and A.I. Dahl, *J. Res. Natl. Bur. Stand.*, Vol 10, 1933, p 661
2. N.P. Allen, *J. Inst. Met.*, Vol 49, 1932, p 49
3. R. Hultgren, *et al.*, *Selected Values of the Thermodynamic Properties of the Elements*, American Society for Metals, 1973, p 17-21
4. R. Buffington and W.M. Latimer, *J. Am. Chem. Soc.*, Vol 48, 1926, p 2305
5. H. Ebert, *Z. Phys.*, Vol 47, 1928, p 712
6. F.C. Nix and D. MacNair, *Phys. Rev.*, Vol 61, 1942, p 74
7. H. Esser and H. Eusterbrock, *Arch. Eisenhüttenwes.*, Vol 14, 1941, p 341

8. B.A. Rodgers, I.C. Schoonover, and L. Jordan, "Silver: Its Properties and Industrial Uses," Circular 412, National Bureau of Standards, 1936; L. Addicks, Ed., *Silver in Industry*, Reinhold, 1940
9. A. Butts and C.D. Coxe, *Silver—Economics, Metallurgy and Use*, Van Nostrand, 1967, p 146-151
10. W. Hume-Rothery and P.W. Reynolds, *Proc. R. Soc. (London) A*, Vol A167, 1938, p 25
11. P.L. Woolf, G.R. Zellars, E. Foerster, and J.P. Morris, *US Bur. Mines Rep. Invest.*, No. 563, 1960
12. H.M. Schadel, Jr. and C.E. Birchenall, *Trans. Metall. Soc.*, Vol 188, 1950, p 1134-1138
13. I.N. Plaksin and A.Y. Brechsted, *Zh. Prikl. Khim.*, Vol 11 (No. 12), 1938, p 1055, 1158, 1262, 1556
14. A. Butts and C.D. Coxe, *Silver—Economics, Metallurgy and Use*, Van Nostrand, 1967, p 112
15. E.F. Northrup, *J. Franklin Inst.*, Vol 178, 1914, p 85
16. F. Pawlek and D. Rogalla, *Cryogenics*, Vol 6, 1966, p 14
17. G. Tammann and K.L. Dreyer, *Ann. Phys.*, Vol 5, 1933, p 16, 111
18. P.W. Bridgman, *Proc. Am. Acad. Arts Sci.*, Vol 52, 1917, p 573
19. W.W. Coblentz and R. Stair, *J. Res. Natl. Bur. Stand.*, Vol 2, 1929, p 343
20. M. Auwarter, *Z. Tech. Phys.*, Vol 18, 1937, p 457
21. G.K. Burgess and R.G. Waltenberg, *Bull. Natl. Bur. Stand.*, Vol 11, 1915, p 605
22. C.C. Bidwell, *Phys. Rev.*, Vol 3 (No. 2), 1914, p 439
23. F. Saeffel and G. Sachs, *Z. Metallkd.*, Vol 17, 1925, p 353
24. J. McKeown and O. Hudson, *J. Inst. Met.*, Vol 60, 1937, p 109

Sodium (Na)

Compiled by J.R. Keiser and J.H. DeVan, Oak Ridge National Laboratory, and Keith R. Willson, Geneva College

Sodium is used as a liquid metal heat transfer medium, a working fluid for evaporative heat pipes, and an electrical conductor in homopolar generators. It is also used in vapor lamps for highway lighting; as an alloying addition for lead, zinc, and aluminum; and as a reactant for deoxidation of metals and for reduction of metal fluorides.

Sodium is highly reactive with water; hydrogen released by the reaction is potentially explosive. Molten sodium will burn in ambient air. Iron-base alloys are usually selected as containers for the transport of liquid sodium. Argon, helium, and nitrogen are used as cover gases to minimize sodium oxidation.

Sodium fires are best extinguished by closing off air accesses or by blanketing with either nitrogen or inert solids such as carbon granules. Commercial extinguishing media include sodium chloride, sodium carbonate, and calcium phosphate. Carbon tetrachloride and solid carbon dioxide extinguishers should not be used on sodium fires.

Structure

Crystal structure. β phase, body-centered cubic, type *A*2; $a = 0.42906$ nm at 20 °C. On cooling below 36 K, sodium partially transforms to α phase, close-packed hexagonal, type *A*3; $a = 0.3767$ nm at 5 K (Ref 1)

Mass Characteristics

Atomic weight. 22.9898

Density. 0.9674 g/cm^3 at 25 °C; 0.9270 g/cm^3 at 100 °C; 0.7113 g/cm^3 at 1000 °C (Ref 2, 3)

Density versus temperature. Density at melting point: 0.927 g/cm^3. Density in the liquid range: $\rho = 0.927 - 2.35(T - T_m)$, where T_m is the melting point in K and density (ρ) is in g/cm^3 (Ref 4)

Volume change on melting. +2.71% (Ref 5)

Thermal Properties

Melting point. 97.82 °C (Ref 5)

Boiling point. 881.4 °C (Ref 5)

Phase transformation temperature. Incomplete transformation to α phase occurs on cooling (below 36 K) or on deforming (below 51 K) (Ref 6)

Coefficient of thermal expansion. Linear: 68.93 μm/m · K; at 0 to 96.6 °C, $l/l_0 = 1 + 6.893 \times 10^{-5}\,T + 0.63 \times 10^{-7}\,T^2$, where T is in 0 °C (Ref 4). Volumetric: 2.418 × 10^{-4}/°C; at 97.83 to 1350 °C, $V/V_0 = 1 + 2.4183 \times 10^{-4}\,T + 7.385 \times 10^{-8}\,T^2 + 15.64 \times 10^{-12}\,T^3$, where T is in °C (Ref 5)

Specific heat. $C_p = 1.2220$ kJ/kg · K at 25 °C (solid); $C_p = 1.3210$ kJ/kg · K at 250 °C (liquid); $C_p = 2.5100$ kJ/kg · K at 1000 °C (vapor) (Ref 7, 8)

Specific heat versus temperature. Solid: $C_p = 1198.72 + 0.64894\,T + 0.010527\,T^2$ J/kg · K for 0 to 97.8 °C (T is in °C); Liquid: $C_p = 1436.1 - 0.58026\,T + 4.6208 \times 10^{-4}\,T^2$ J/kg · K for 97.8 to 900 °C (T is in °C) (Ref 5)

Enthalpy. Where H_{0c} is enthalpy of the solid state at 0 °C: solid, $H_c - H_{0c} = 1199.26\,T + 0.3247\,T^2 + 3.510 \times 10^{-3}\,T^3$ J/kg for 0 to 97.8 °C (T is in °C); liquid, $H_l - H_{0c} = 98\,960 + 1436.7\,T - 0.29025\,T^2 + 1.5410 \times 10^{-4}\,T^3 + 2.400 \times 10^7 \times e^{-13600/(T+273)}$ J/kg (for 97.8 to 900 °C; T is in °C) (Ref 5)

Entropy. Solid at 0 to 97.8 °C: $S_c - S_{0c} = 4162.42 \log(T) - 5.1036\,T + 0.0052658\,T^2 - 9140.2$ J/kg · K, where S_{0c} is entropy at 0 °C and T is in K. Liquid at 97.8 to 900 °C: $S_l - S_{0c} = 3752.6 \log(T) - 0.8330\,T + 2.3112 \times 10^{-4}\,T^2 - 8673.9$ J/kg · K, where T is in K (Ref 5)

Latent heat of fusion. 113 kJ/kg (Ref 5)

Latent heat of vaporization. 3.874 MJ/kg (Ref 5)

Thermal conductivity. 131.4 W/m · K at 25 °C; 79.6 W/m · K at 250 °C (Ref 5)

Thermal conductivity versus temperature. For 0 to 95 °C: $k = 135.6 - 0.167\,T$, where k is in W/m · K and T is in °C. For 104 to 832 °C: $k = 91.8 - 0.049T$, where k is in W/m · K and T is in °C (Ref 5)

Table 42 Diffusion characteristics of sodium

Solute	Temperature range, °C	Activation energy, kJ/mol	Frequency factor, m^2/s × 10^{-8}
^{198}Au	1–77	9.25	3.34
^{42}K	0–91	35.3	0.08
^{22}Na	0–98	42.2	0.145
^{86}Rb	0–85	35.5	0.15

Source: Ref 10

Vapor pressure. From Ref 5 and 9:

°C	Pa
100	1.43 × 10^{-5}
200	1.81 × 10^{-2}
300	1.85
500	5.19 × 10^2
700	1.40 × 10^4
900	1.20 × 10^5
980	2.0 × 10^5
1120	5.1 × 10^5
1230	1.0 × 10^6
1370	2.0 × 10^6

Also, $\log P = -5780/T - 1.18 \log T + 13.625$, where P is in Pa and T is in K. Valid range: 298 K to boiling point

Diffusion characteristics. See Table 42.

Critical temperature and pressure. From Ref 9: $T_c = 2508.7 \pm 12.5$ K; $P_c = 25.64 \pm 0.02$ MPa

Electrical Properties

Electrical resistivity. From Ref 5, 9:

K	nΩ · m
273.15	43.3
293.15	47.7
350	62.3
371	68.6 (solid)
371	94.3 (liquid)
673	221.4
1073	463.5
1373	737.6

Temperature dependence of electrical resistivity. At −223 to 97.8 °C, $r_s = 42.9 + 0.1993\,T + 9.848 \times 10^{-5}\,T^2$; at 130 to 1090 °C, $r_s = 61.44 + 0.3504\,T + 5.695 \times 10^{-5}\,T^2 + 1.667 \times 10^{-7}\,T^3$, where r_s is in nΩ · m and T is in °C (Ref 5)

Thermoelectric potential. Versus platinum (Ref 5):

°C	mV
25	2.9 × 10^{-2}
50	5.3 × 10^{-2}
100	8.4 × 10^{-2}
200	1.36 × 10^{-1}
300	2.26 × 10^{-1}
400	3.90 × 10^{-1}
500	6.15 × 10^{-1}
600	9.96 × 10^{-1}
700	1.47
800	2.02
900	2.63

Table 43 Unstable isotopes of sodium

Isotope	Half-life	Decay mode(a)	Particle energy, MeV	Particle intensity, %
^{19}Na	0.03 s	β^+, p	...	...
^{20}Na	0.446 s	β^+	...	...
...		α	2.15	...
^{21}Na	22.5 s	β^+	2.5	95
^{22}Na	2.605 years	β^+	0.545	90
		EC	...	...
^{24m}Na	20.2 ms	IT, β^-	...	...
^{24}Na	14.97 h	β^-	1.389	>99
^{25}Na	59.3 s	β^-	2.6	7
			3.15	25
			4	65
^{26}Na	1.07 s	β^-	...	...
^{27}Na	0.29 s	β^-	7.95	...
		β^-, n	...	...
^{28}Na	30 ms	β^-	12.3	...
		β^-, n	...	...
^{29}Na	43 ms	β^-, n	11.5	...
^{30}Na	53 ms	β^-	...	...
^{31}Na	17 ms	β^-, n	...	...

(a) β^-, negative beta emission; β^+, positron emission; EC, orbital electron capture; IT, isomeric transition from upper to lower isomeric state; n, neutron emission; p, proton emission. Source: Ref 9

Electrochemical equivalent. For a valence of 1, 0.238 mg/C

Electrolytic solution potential. Versus H_2: −2.711 V at 25 °C (Ref 10)

Ionization potential. From Ref 9:

Ionization state	Potential, eV
I	5.139
II	47.286
III	71.64
IV	98.91
V	138.39
VI	172.15
VII	208.47
VIII	264.18
IX	299.87
X	1465.091
XI	1648.659

Electron binding energies:

Shell	Electron configuration	Binding energy, eV
K	1*s*	1070.8
L(I)	2*s*	63.5
L(II)	$2p_{1/2}$	30.4
L(III)	$2p_{3/2}$	30.5

Source: Ref 9

Hall coefficient. −2.5 aV · m/A · T (Ref 5)

Work function. 2.28 eV (0.365 aJ) (Ref 10)

Magnetic Properties

Magnetic susceptibility. From Ref 5:

°C	Volume susceptibility, mks units × 10^{-6}
30	7.29
95	7.24
150	7.04
250	6.95

Optical Properties

Color. Silver

Refractive index. Liquid, 0.0045 for λ = 589.3 μm; solid, 4.22 for λ = 589.3 μm (Ref 8)

Nuclear Properties

Stable isotopes. ^{23}Na, isotope mass 22.9898, 100% abundance (Ref 5)

Unstable isotopes. See Table 43.

Chemical Properties

Resistance to specific corroding agents. At 25 °C, sodium passivates in dry O_2 but oxidizes in moist air to form Na_2O, NaOH, and finally Na_2CO_3. Sodium is highly pyrophoric in air at or above 125 °C. Sodium reacts with CO_2 above 200 °C to form Na_2O, C, and possibly $Na_2C_2O_4$. Below 320 °C, water (gas or liquid) reacts with liquid sodium to produce NaOH (solid) and H_2. (H_2 is potentially explosive if O_2 is present.) Above 320 °C, products of the H_2O reaction include NaOH (liquid), NaH, Na_2O, and H_2. A useful technique for removing sodium residues is by reaction with water vapor in nitrogen or noble gas at 70 °C followed by water rinsing. Sodium undergoes metallic dissolution in anhydrous liquid ammonia, the solution ultimately converting to sodium amide. Sodium reacts with alcohols to form sodium alcoholates and hydrogen. *N*-butyl alcohol can be used to slowly dissolve sodium at 25 °C. Ethyl alcohol is much more reactive and can be ignited if sodium comes in contact with air. Solid sodium is relatively inert toward dry hydrocarbons that do not have an active hydrogen or acetylene hydrogen component. In contact with molten sodium, alkyne hydrogen atoms are liberated, and aryl hydrocarbons can be polymerized or decomposed.

Mechanical Properties

Kinematic liquid viscosity. From Ref 5:

°C	mm^2/s
100	0.7338
200	0.5001
300	0.3921
400	0.3323
500	0.2955
600	0.2568
700	0.2313
800	0.2134

Liquid surface tension. 0.192 N/m at 97.8 °C; 0.161 N/m at 400 °C; 0.146 N/m at 550 °C; 0.113 N/m at 881.4 °C; also, for 97.8 to 881.4 °C, $\sigma = 0.2067 - 1 \times 10^{-4}\ T$, where σ is in N/m and T is in °C (Ref 5)

Velocity of sound. $V = 2577.25 - 0.524\ T$, where V is in m/s and T is in °C (Ref 2)

REFERENCES

1. A. Taylor and Brenda J. Kagle, *Crystallographic Data on Metals and Alloy Structures*, Dover, 1963
2. M. Sittig, Physical and Thermodynamic Properties of Sodium, Chapter 9 in *Sodium: Its Manufacture, Properties and Uses*, G.W. Thomson and E. Garelis, Ed., American Chemical Society Monograph Series 133, Reinhold, 1956
3. J.P. Stone *et al.*, "High Temperature Properties of Sodium," NRL-6241, Naval Research Laboratory, 1965
4. T. Iida and R.I.L. Guthrie, *The Physical Properties of Liquid Metals*, Clarendon Press, Oxford, 1988
5. H.J. Bomelburg and C.R.F. Smith, *Physical Properties*, Vol 1, *Sodium-NaK Engineering Handbook*, O.J. Foust, Ed., Gordon & Breach, 1972, p 1-88
6. C.S. Barrett, X-Ray Study of the Alkali Metals at Low Temperature, *Acta Crystallogr.*, Vol 9, 1956, p 671
7. D.C. Ginnings *et al.*, Heat Capacity of Sodium Between 0° and 900 °C, The Triple Points and Heat of Fusion, *J. Res. Natl. Bur. Stand.*, Vol 45, 1950, p 23
8. G.H. Golden and J.G. Tokar, "Thermophysical Properties of Sodium," ANL-7323, Argonne National Laboratory, 1967
9. R.C. Weast, Ed., *CRC Handbook of Chemistry and Physics*, 70th ed., CRC Press, 1989
10. R.C. Weast, Ed., *Handbook of Chemistry and Physics*, 55th ed., CRC Press, 1974

Strontium (Sr)

Revised by J.H. Westbrook, Sci-Tech Knowledge Systems

Structure

Crystal structure. α phase, face-centered cubic, *cF*4 (*Fm*3*m*); a = 0.60849 nm at 25 °C. β phase (commonly referred to as γ phase in earlier literature), body-centered cubic, *cI*2 (*Im*3*m*); a = 0.4434 nm at 614 °C (Ref 1, 2). High-pressure phase: body-centered cubic, *cI*2 (*Im*3*m*) (Ref 2)

Mass Characteristics

Atomic weight. 87.62

Density. α phase, 2.6 g/cm^3 at 20 °C (Ref 3); β phase, 2.55 g/cm^3 at 614 °C (calculated from x-ray data) (Ref 1)

Thermal Properties

Melting point. 768 °C (Ref 4)
Boiling point. 1370 °C (calculated from vapor pressure data) (Ref 4, 5)
Phase transformation temperature. α to β, 557 °C (Ref 4)
Specific heat:

K	°C	kJ/kg · K
1	−272	4.77×10^{-5}
2	−271	1.39×10^{-4}
4	−269	6.12×10^{-4}
10	−263	8.46×10^{-3}
20	−253	5.44×10^{-2}

Electronic coefficient (γ). 41.5 ± 0.3 mJ/kg · K (Ref 4)
Latent heat of fusion. 104.7 kJ/kg (calculated from binary phase diagram data) (Ref 4)

Additional thermodynamic data are available in Ref 4.

Electrical Properties

Work function. 2.1 to 2.74 eV (0.34 to 0.44 aJ), depending on conditions and techniques of the experimental determination (Ref 6 to 8)

Magnetic Properties

Magnetic susceptibility. Molar: 1.16 mks at 22 °C (Ref 9)

Nuclear Properties

Stable isotopes. ^{84}Sr, isotope mass 83.913431, 0.56% abundant; ^{86}Sr, isotope mass 85.909276, 9.9% abundant; ^{87}Sr, isotope mass 86.908894, 7.0% abundant; ^{88}Sr, isotope mass 87.905628, 82.6% abundant (Ref 10)
Unstable isotopes. ^{78}Sr, 31 min; ^{79}Sr, 8.1 min; ^{80}Sr, 1.7 h; ^{81}Sr, 2.5 min; ^{82}Sr, 25.0 days; ^{83}Sr, 32.4 h; ^{85}Sr, 67.7 min, 65.2 days; ^{89}Sr, 50.5 days; ^{90}Sr, 29 years; ^{91}Sr, 9.48 h; ^{92}Sr, 2.71 h; ^{93}Sr, 7.5 min; ^{94}Sr, 1.29 min; ^{95}Sr, 26 s; ^{96}Sr, 4.0 s; ^{97}Sr, ≤0.2 s; ^{98}Sr, ~0.85 s; ^{89}Sr to ^{98}Sr, mode of decay by negative electron (Ref 10)

Additional nuclear data are available in Ref 10.

Mechanical Properties

Modulus. Bulk, 11.61 GPa; isothermal compressibility, 86.1 $\mu m^2/N$

REFERENCES

1. P. Eckerlin, H. Kandler, and A. Stegherr, Landolt-Börnstein Tables, New Series III/6, in *Structure Data of Elements and Intermetallic Phases*, K.-H. Hellwege and A.M. Hellwege, Ed., Springer-Verlag, 1971
2. P. Villars and L.D. Calvert, Ed., *Pearson's Handbook of Crystallographic Data for Intermetallic Phases*, American Society for Metals, 1985
3. R.J. Elliott, *Constitution of Binary Alloys, First Supplement*, McGraw-Hill, 1965
4. R. Hultgren, P.D. Desai, D.T. Hawkins, M. Gleiser, K.K. Kelley, and D.D. Wagman, *Selected Values of the Thermodynamic Properties of the Elements*, American Society for Metals, 1973
5. K.A. Gschneidner, Jr., Physical Properties and Interrelationships of Metallic and Semimetallic Elements, in *Solid State Physics*, Vol 16, F. Seitz and D.T. Turnbull, Ed., Academic Press, 1964, p 275
6. V.S. Fomenko, *Handbook of Thermionic Properties—Electronic Work Functions and Richardson Constants of Elements and Compounds*, G.V. Samsonov, Ed., Plenum Publishing, 1966 (translation from the Russian)
7. G.A. Haas and R.E. Thomas, Thermionic Emission and Work Function, chapter 2 in *Measurements of Physical Properties*, E. Passaglia, Ed., Vol 6, part 1, *Techniques of Metals Research*, R.F. Bunshah, Ed., Interscience, 1972, p 91
8. H.B. Michaelson, in *Handbook of Chemistry and Physics*, 69th ed., R.C. Weast, Ed., CRC Press, 1988
9. Landolt-Börnstein Tables, II Band, 9. Teil, in *Magnetische Eigenschaften 1*, K.-H. Hellwege and A.M. Hellwege, Ed., Springer-Verlag, 1962
10. F.W. Walker, J.R. Parrington, and F. Feiner, "Chart of the Nuclides," Knolls Atomic Power Laboratory, United States Atomic Energy Commission, distributed by Nuclear Energy Business Operation, General Electric Company, Nov 1989

Tantalum (Ta)

Compiled by Mortimer Schussler, Fansteel, Inc.

Revised by R.E. Droegkamp, Fansteel Metals

Tantalum provides a combination of properties not found in many refractory metals: excellent fabricability, a low ductile-to-brittle transition temperature, and a high melting point.

Figure 101 shows a breakdown of tantalum consumption by industry. The largest use of tantalum at this time is in electrolytic capacitors. Sizeable quantities of tantalum also are used in chemical process equipment (such as heat exchangers, condensers, thermowells, and lined vessels), notably for handling nitric, hydrochloric, bromic, and sulfuric acids, or combinations of these acids with many other chemicals. Spinnerettes for extruding man-made fibers constitute another important application of tantalum.

Because of its high melting point, tantalum is used for heating elements, heat shields, and other components of vacuum

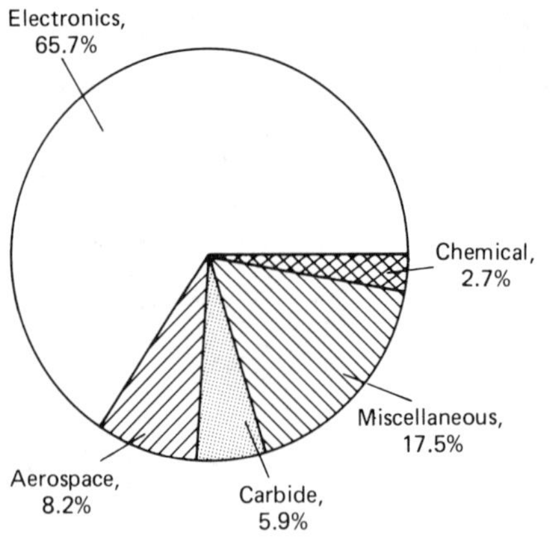

Fig. 101 1988 consumption of tantalum by specific industries. Source: Tantalum Producers Association

furnaces. Tantalum has been used in specialized aerospace and nuclear applications. Tantalum also is used in prosthetic devices in contact with body fluids and as an alloy component in superalloys. Tantalum carbide is an important constituent of cemented carbide cutting tools made from mixtures of titanium, tungsten, and tantalum carbides.

A new and important military application is evolving wherein tantalum is used as an armor penetrator. Also, emerging corrosion abrasion applications are being found for tantalum alloyed with Group IV elements (titanium, zirconium, and hafnium) and Group VI elements (molybdenum and tungsten); in these alloying combinations, the subsequent reaction with carbon, oxygen, and nitrogen is such that a very hard surface keyed to the substrate is formed.

Yield and ultimate strengths of tantalum are increased, and ductility is reduced, by increases in the amount of interstitial elements (oxygen, nitrogen, carbon, and hydrogen). Embrittlement of tantalum can occur if contamination by these elements is sufficiently severe. Maximum impurity limits (in ppm) for commercially available high-purity tantalum (99.90% min) are 500 Nb, 300 W, 100 to 200 O_2, 100 Fe, 100 Mo, 50 to 75 C, 50 to 75 N, 50 Ni, 50 Si, 50 Ti, and 10 H_2.

Structure

Crystal structure. Body-centered cubic:

$$I\frac{4}{m}\bar{3}\frac{2}{m}$$

a = 0.33026 nm at 20 °C
Slip planes. {110}
Cleavage planes. {110}
Minimum interatomic distance. 0.2854 nm

Mass Characteristics

Atomic weight. 180.948
Density. At 20 °C, 16.6 g/cm^3

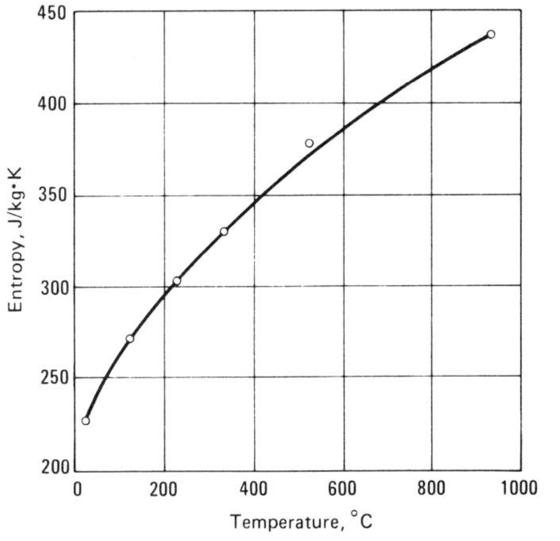

Fig. 102 Temperature dependence of the entropy of tantalum

Thermal Properties

Melting point. 2996 °C
Boiling point. 5427 °C
Coefficient of linear thermal expansion. 6.5 μm/m · K near 20 °C. Temperature dependence: $\alpha = 6.5 + 0.34 \times 10^{-3}\ T + 0.12 \times 10^{-6}\ T^2$, where T is in °C and α is in μm/m · K
Specific heat. At 0 °C, 0.1391 kJ/kg · K. Temperature dependence: $C_p = 139.04 + 1.757 \times 10^{-2}\ T + 1.375 \times 10^{-6}\ T^2$, where T is in K and C_p is in J/kg · K
Entropy. At 25 °C, 229 J/kg · K. See also Fig. 102.
Latent heat of fusion. 145 to 174 kJ/kg
Latent heat of vaporization. 4160 to 4270 kJ/kg
Heat of combustion. 5634 to 5772 kJ/kg Ta
Thermal conductivity:

°C	W/m · K
−73	56.1
20	54.4
127	59.9
527	66.6
927	72.9
1327	77.0
1727	80.8
2127	83.7
2527	85.8

Vapor pressure:

°C	mPa
2351	0.6298
2365	0.7488
2487	4.019
2566	9.820
2615	17.20
2652	24.40
2675	37.03

Electrical Properties

Electrical conductivity. 13% IACS
Electrical resistivity. At 20 °C: 135.0 nΩ · m. See also Fig. 103 and 104. Temperature coefficient: From 0 to 100 °C, 0.0038 per K. See also Fig. 104. Pressure coefficient (expressed as the ratio of resistivity at pressure to resistivity at zero pressure):

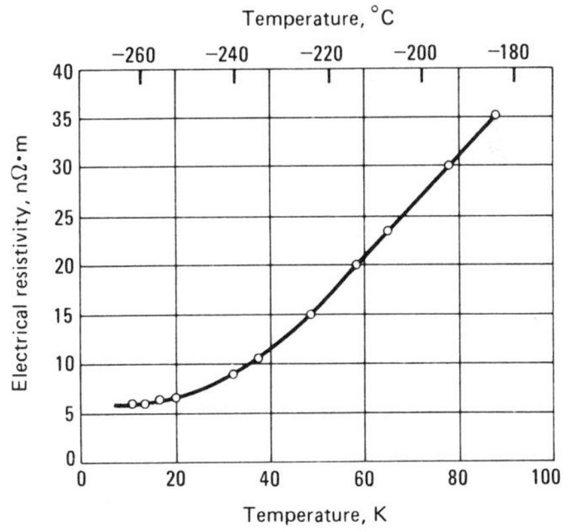

Fig. 103 Electrical resistivity of tantalum at low temperatures. Sample is unannealed 99.98% pure tantalum rod.

Gage pressure, kPa	Resistivity ratio
0	1.000
98	0.984
200	0.968
390	0.941
590	0.918
780	0.898
980	0.882

Thermal electromotive force. Versus platinum (cold junction at 0 °C):

Temperature at hot junction, °C	Thermal emf, mV
−200	+0.21
−100	−0.10
100	0.33
200	0.93
600	5.95
1000	15.20
1200	21.41

Electrochemical equivalent. Valence 5: 0.3749 mg/C
Standard electrode potential. Versus H_2: 1.12 V
Ionization potential. 7.89 eV
Hall coefficient. +0.095 nV · m/A · T (virtually independent of temperature from 100 to 900 K)
Temperature of superconductivity. 4.38 K
Electron emission. 600 kA/m² · T^2, where T is in K
Work function. 0.657 aJ (4.10 eV)
Positive ion emission. 1.60 aJ (10.0 eV)
Dielectric constant. For Ta_2O_5 layer: ~400 kV per mm (10 kV per mil)
Thickness of anodic oxide film. At 0 °C, 1.6 nm/V; at 100 °C, 2.5 nm/V; at 200 °C, 3.0 nm/V

Magnetic Properties

Magnetic susceptibility. At 25 °C: 10.4×10^{-6} mks

Optical Properties

Spectral emittance. 0.49 for λ = 650.0 nm
Total hemispherical emittance. At 1400 °C, 0.20; at 1500 °C, 0.21; at 2000 °C, 0.25. See also Fig. 105.
Total radiation. At 1300 °C, 73.0 kW/m²; at 1530 °C, 128.0 kW/m²; at 1730 °C, 212.0 kW/m²

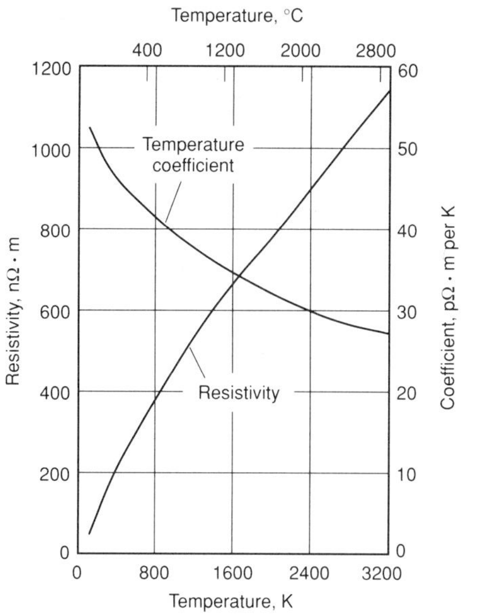

Fig. 104 Electrical resistivity and temperature coefficient of resistivity for tantalum

Nuclear Properties

Natural isotopes. ^{181}Ta, 99.988% natural abundance
Thermal neutron cross section. 21.3 b

Chemical Properties

General corrosion behavior. Tantalum oxidizes in air above 300 °C. It has excellent resistance to corrosion by a large number of acids, by most aqueous solutions of salts, by organic chemicals, and by various combinations and mixtures of these agents. Also, tantalum exhibits good resistance to many corrosive and common gases and to many liquid metals.

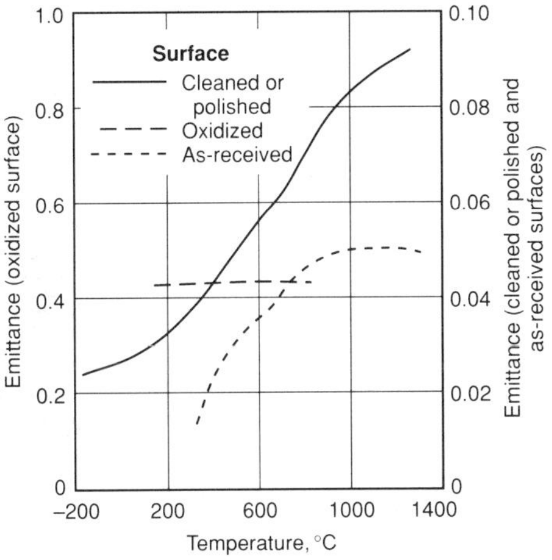

Fig. 105 Temperature dependence of the total emittance of commercially pure tantalum

Resistance to specific corroding agents. Tantalum is attacked by hydrofluoric acid, fuming sulfuric acid, and strong alkalis. The presence of salts that hydrolyze to form hydrofluoric acid or strong alkalis can also lead to attack of tantalum. Tantalum can become embrittled by hydrogen if it is the cathode in a galvanic couple exposed to an acid environment or to a hydrogen-containing atmosphere at elevated temperature. Halogen gases can attack tantalum: Fluorine causes attack at both room and elevated temperatures, chlorine at and above 250 °C, bromine at and above 300 °C, and iodine at somewhat higher temperatures. Bromine plus methanol also attacks tantalum.

Fabrication Characteristics

Precautions in melting. Exclude oxygen, hydrogen, nitrogen, and carbon. Melt in vacuum or inert atmosphere.
Hot-working temperature. None; tantalum is worked cold.
Annealing temperature. Above 1050 °C in high vacuum for complete recrystallization, with resulting grain size as given below. Material was cold rolled 75% after intermediate annealing, then annealed 1 h at the indicated temperature. The average ASTM grain size No. was determined by comparison with the ASTM grain size chart at a magnification of 100×:

Final annealing temperature, °C	Average ASTM grain size No.
1200	5-6
1300	4
1400	3-4
1425	3-4
1600	2
1700	1
1800	0-1

Maximum reduction between anneals. Greater than 95
Suitable forming methods. Tantalum can be formed by spinning, deep drawing, bulging, bending, blanking, punching, and stretch forming using conventional methods, and using equipment and tooling normally found in shops that fabricate heat-resistant alloys.
Compacting pressure. 10 to 85 MPa, depending on the physical properties of the powder
Sintering temperature. Sintering at 2300 to 2600 °C in high vacuum will essentially remove all detrimental impurities contained in the powder.
Machinability. Fully recrystallized unalloyed tantalum has a machinability similar to that of soft copper. Use chlorinated hydrocarbons, light oil, or water-soluble oil as a cutting fluid, and high-speed tool steel or cemented carbide tools. Tantalum can be successfully turned, bored, drilled, tapped, reamed, shaped, milled, sawed, and ground to desired tolerances and surface finishes.
Joining. Gas tungsten arc, gas metal arc, resistance, and electron beam welding techniques can be used for joining tantalum. High-purity inert gas (argon or helium) or vacuum must be used in fusion welding. Resistance spot and seam welding can be done in air or under water with proper precautions. Silver brazing alloys, copper, and several specially developed refractory metal brazing alloys can be used to braze tantalum to itself or to dissimilar metals such as stainless steels. Brazing is done in vacuum or under an inert atmosphere (high-purity argon or helium). Tantalum also can be bonded to dissimilar metals by explosive cladding and, in some instances, by roll bonding.
Cleaning. To avoid contamination of tantalum by interstitial elements and metallic impurities, it is mandatory that the material be chemically cleaned before any heating operation (such as annealing or welding). Such cleaning involves thorough degreasing with a detergent or solvent; chemical etching in 20 vol% HF, 20 vol% H_2SO_4, and 60 vol% HNO_3; rinsing with hot and cold water (deionized water recommended); and spot-free drying. The etching solution can be strengthened by adding HF or weakened by adding water to achieve the amount of stock removal necessary to ensure cleanness.

Mechanical Properties

Tensile properties. See Fig. 106 to 108.
Hardness. Electron beam melted, 110 HV; P/M compact, 120 HV
Poisson's ratio. 0.35 at 20 °C
Elastic modulus. Tension: 186 GPa at 20 °C; 159 GPa at 750 °C. Shear: 69 GPa at 20 °C.
Compressibility:

MPa	$\Delta V/V_0$ 99.9% Ta	$\Delta V/V_0$ 99.95 + % Ta
490	0.00244	0.00243
980	0.00488	0.00485
1470	0.00728	0.00726
1960	0.00969	0.00967
2450	0.01208	0.01208
2940	0.01447	0.01448

Impact strength. See Fig. 109.
Fatigue strength. See Fig. 110.
Creep-rupture characteristics. See Fig. 111.
Damping characteristics. In polycrystalline sheet vibrated at a frequency of 0.65 Hz, maximum damping occurs at 1100 °C.
Friction characteristics. Tantalum galls against itself and against type 18-8 stainless steel.

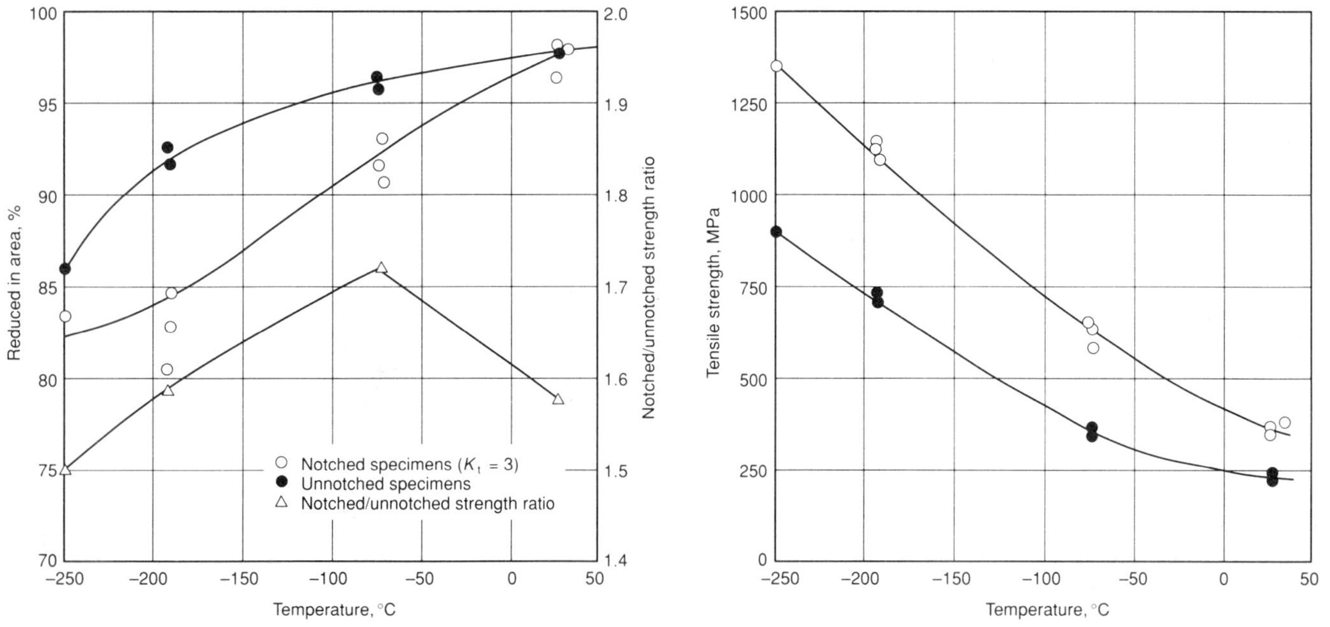

Fig. 106 Low-temperature tensile properties of electron-beam-melted tantalum bar. Sample impurities: <0.003% C, <0.003% O_2, 0.0008% N_2, <0.08% other. Bar was annealed for 3 h at 1200 °C: hardness, 83 HV; grain size, ASTM No. 5. Crosshead speed: unnotched specimens, 0.5 mm/min; notched specimens, 0.13 mm/min

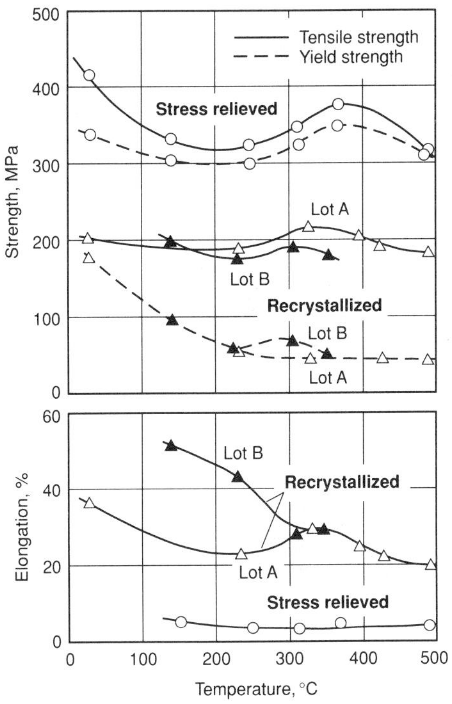

Fig. 107 Elevated-temperature tensile properties of 1 mm thick electron-beam-melted tantalum sheet. Sample impurities (both lots): 0.0030% C, 0.0016% O_2, 0.0010% N_2, <0.040% other. Stress-relieved sheet was cold rolled 95% and stress-relieved for ¼ h at 730 °C; recrystallized sheet was cold rolled 75% and recrystallized by heating for 1 h at 1200 °C. Crosshead speed, 1.3 mm/min

SELECTED REFERENCES

- D.R. Mash, D.W. Bauer, and M. Schussler, Fabricating the Refractory Metals, *Met. Prog.*, Feb-March-April 1971
- F.F. Schmidt and H.R. Ogden, "The Engineering Properties of Tantalum and Tantalum Alloys," DMIC Report 189, 13 Sept 1963
- J.G. Sessler and V. Weiss, *Nonferrous Heat Resistant Alloys*, Aerospace Structural Metals Handbook, Vol 11A, AFML-TR-68-115, Air Force Materials Laboratory, Jan 1968
- F.T. Sisco and E. Epremian, *Columbium and Tantalum*, Wiley, 1963
- R.C. Weast, Ed., *CRC Handbook of Chemistry and Physics*, 60th ed., CRC Press, 1979

Technetium (Tc)

Compiled by C.C. Koch, North Carolina State University

Technetium is used as a radioactive tracer in medicine and has potential uses arising from its favorable corrosion-inhibiting properties and catalytic behavior. and its high elemental superconducting transition temperature. There is contamination hazard from technetium due to its radioactivity, and the element is classed as moderately toxic. All processes such as sample preparation that could result in the dispersal of solid ^{99}Tc must be carried out in glove box facilities. The data that follow are for ^{99}Tc only.

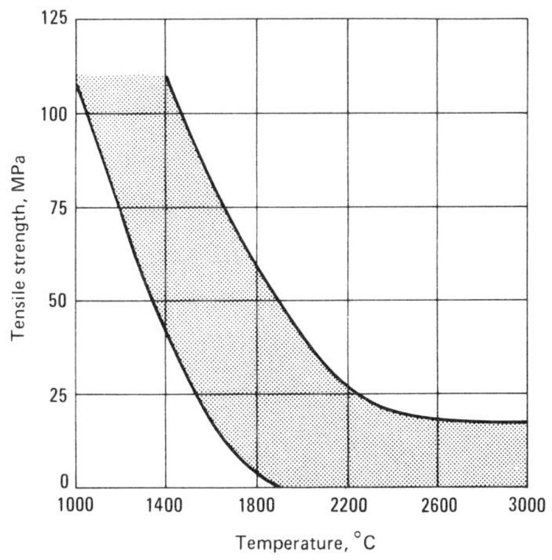

Fig. 108 High-temperature tensile strength of tantalum. The upper portion of the curve is characterized by high strain rates and high interstitial content, whereas the lower portion of the curve is characterized by low strain rates and low interstitial content.

Structure

Crystal structure. Close-packed hexagonal; A3; *a* = 0.27407 nm, *c* = 0.43980 nm, *c*/*a* = 1.6048 at 25 °C; *a* = 0.27364 nm, *c* = 0.43908 nm, *c*/*a* = 1.6046, at 4.2 K

Mass Characteristics

Atomic weight. 99.0000
Density. 11.5 g/cm^3 at 25 °C

Thermal Properties

Melting temperature. 2200 °C (Ref 1)
Coefficient of thermal expansion. Linear, from 150 K to 25 °C. Polycrystalline: 7.05 μm/m · K. Along crystal axes: 7.04 μm/m · K along *a* axis; 7.06 μm/m · K along *c* axis (Ref 2)
Thermal conductivity. 50.2 W/m · K at 25 °C (Ref 3)
Recrystallization temperature. 700 to 800 °C
Enthalpy of sublimation. 6.68 kJ/kg (Ref 4)
Vapor pressure. See Fig. 112.

Electrical Properties

Electrical resistivity. 185.0 nΩ · m at 25 °C (Ref 5). See also Fig. 113.
Temperature of superconductivity. 7.8 K (Ref 6)
Work function. 0.782 aJ (4.88 eV) (Ref 7)

Magnetic Properties

Magnetic susceptibility. Volume: 1.63 × 10^{-4} mks units (Ref 8)

Nuclear Properties

Isotopes. See Table 44.

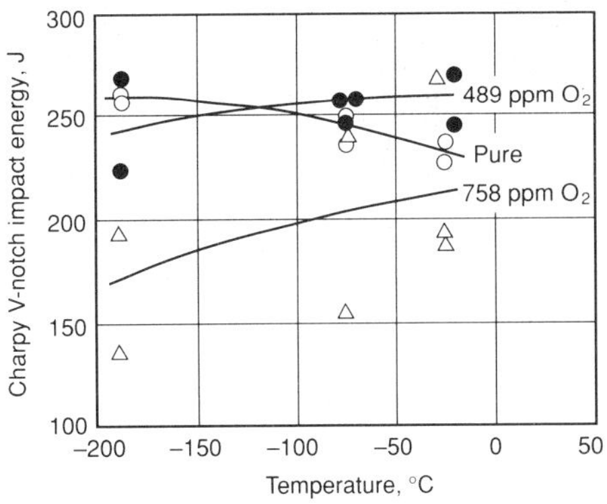

Fig. 109 Effects of temperature and oxygen content on the Charpy V-notch impact energy of wrought electron-beam-melted tantalum. Sample impurities: <44 ppm C + N_2. Temperature, °C

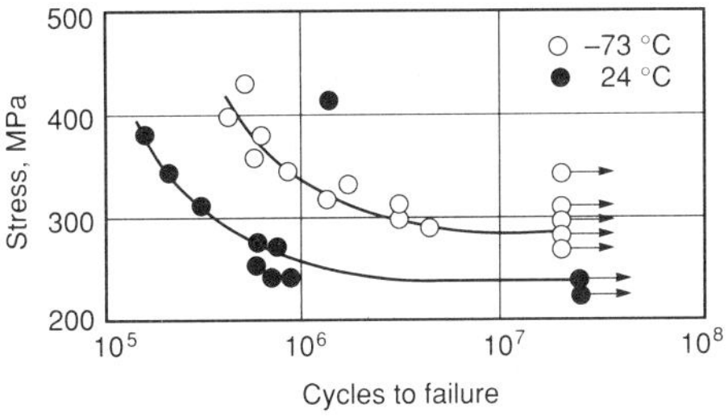

Fig. 110 Rotating-beam fatigue strength of wrought electron-beam-melted tantalum. Sample impurities: <44 ppm C + N_2

Thermal neutron cross section. 22 b at 2200 m/s

Chemical Properties

General corrosion behavior. Technetium metal tarnishes slowly in air and burns in oxygen. At 400 °C it reacts with fluorine to form the hexafluoride and with chlorine to form a mixture of the hexachloride and tetrachloride. Technetium dissolves in oxidizing acids such as nitric, sulfuric, and aqua regia but not in hydrochloric acid. The pertechnetate ion has been shown to be an efficient anticorrosion agent in solution.

Mechanical Properties

Tensile properties. Tensile strength: 1510 MPa as-rolled (46% reduction); 800 MPa after annealing 10 min at 950 °C. 0.2% offset yield strength: 1290 MPa as-rolled (46% reduction); 320 MPa after annealing 10 min at 950 °C (fully recrystallized). Elongation in 25 mm: 4% as-rolled (46% reduction); 30% after annealing 10 min at 950 °C (Ref 9)
Hardness. 46% cold worked: 394 HV, 442 HB. Annealed at 950 °C: 151 HV, 112 HB (Ref 9)
Poisson's ratio. 0.31 (Ref 10)
Elastic modulus. Tension, 322 GPa; shear, 123 GPa; bulk, 281 GPa (Ref 10)
Velocity of sound. At 25 °C: longitudinal velocity, 6220 m/s; shear velocity, 3270 m/s

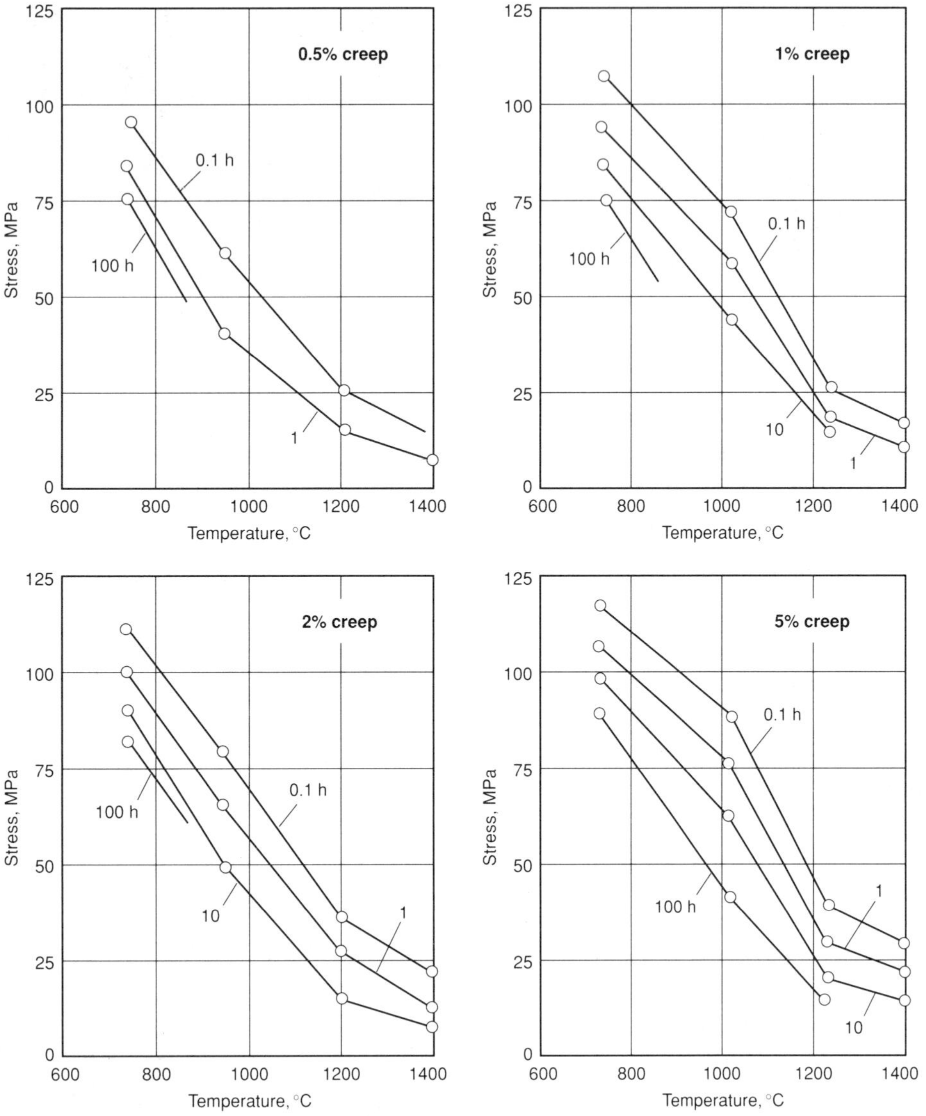

Fig. 111 Creep characteristics of 1 mm thick electron-beam-melted tantalum sheet. Sample impurities: 0.0030% C, 0.0016% O_2, 0.0010% N_2, <0.040% other. Sheet was cold rolled 75% and recrystallized by heating for 1 h at 1200 °C.

Table 44 Isotopes of technetium

Isotope	Half-life	Decay	Energy J × 10^{13}	Energy MeV
^{92}Tc	4.4 min	β^-, ϵ	3.96	2.47
^{93}Tc	2.7 h	ϵ, β^-	1.31	0.82
^{94}Tc	293 min	ϵ, β^+	1.31	0.816
^{95}Tc	20 h	ϵ	...	...
^{96}Tc	4.3 days	ϵ	...	...
^{97}Tc	2.6 × 10^6 years	ϵ	...	...
^{98}Tc	1.5 × 10^6 years	β^-	0.48	0.30
^{99}Tc	2.14 × 10^5 years	β^-	0.468	0.292
^{100}Tc	17 s	β^-	5.41	3.38
^{101}Tc	14.0 min	β^-	2.11	1.32
^{102}Tc	4.5 min	β^-	3.5	2.2
^{103}Tc	50 s	β^-	3.5	2.2
^{104}Tc	18 min	β^-	7.4	4.6
^{105}Tc	7.7 min	β^-	5.4	3.4
^{106}Tc	37 s	β^-	~10	~6.5

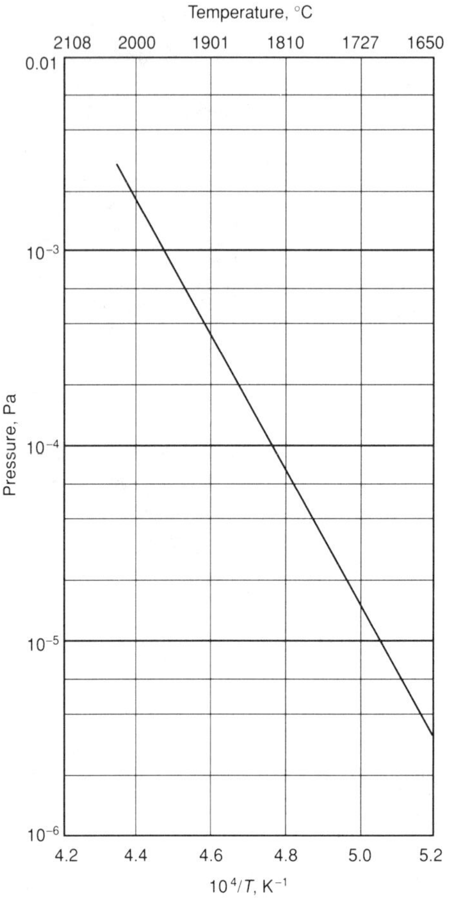

Fig. 112 Vapor pressure of ^{99}Tc. Source: Ref 4

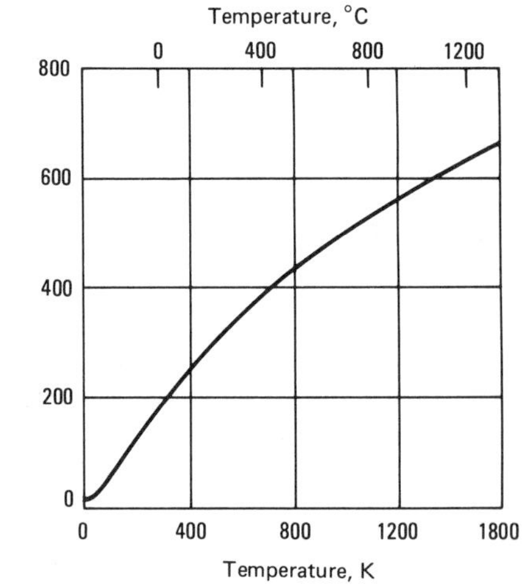

Fig. 113 Temperature dependence of the electrical resistivity of ^{99}Tc

REFERENCES

1. E. Anderson, R.A. Buckley, A. Hellawell, and W. Hume-Rothery, *Nature*, Vol 188, 1960, p 48-49
2. J.A.C. Marples and C.C. Koch, *Phys. Lett. A.*, Vol 41, 1972, p 307-308
3. D.E. Baker, *J. Less-Common Met.*, Vol 8, 1965, p 435
4. O.H. Krikarion, J.H. Carpenter, and R.S. Newbury, *High Temperature Science*, 1969, p 313-330
5. C.C. Koch and G.R. Love, *J. Less-Common Met.*, Vol 12, 1967, p 29-35
6. S.T. Sekula, R.H. Kernohan, and G.R. Love, *Phys. Rev.*, Vol 155, 1967, p 364-369

7. S. Trasatti, *Surf. Sci.*, Vol 32, 1972, p 735-738
8. C.C. Koch, W.E. Gardner, and M.J. Mortimer, *Low Temperature Physics-LT13*, Vol 2, K.D. Timmerhaus, W.J. O'Sullivan, and E.F. Hammel, Ed., Plenum Publishing, 1974, p 595-600
9. R.G. Nelson and D.P. O'Keefe, Concluding Progress Report, in *A Study of Tungsten-Technetium Alloys*, BNWL-865, Battelle Memorial Institute Pacific Northwest Laboratories, 1968
10. G.R. Love, C.C. Koch, H.L. Whaley, and Z.R. McNutt, *J. Less-Common Met.*, Vol 20, 1970, p 73-75

Tellurium (Te)

Compiled by S.C. Carapella, Jr., ASARCO Inc.

Reviewed for this Volume by Douglas Hayduk, ASARCO Inc.

Tellurium was not widely used until after World War II, when it found a modest usage in some nonferrous alloys and as a secondary vulcanizing agent in the natural rubber industry. In 1958, interest developed in the potentialities of tellurium as a thermoelectric material in the form of bismuth telluride and lead telluride; however, this interest has not continued to develop (Ref 1).

The metallurgical industry consumes about 80% of the tellurium produced. It is used as a carbide-stabilizing agent in the production of white cast iron and malleable iron castings, as an additive to low-carbon steels and stainless steels, and as an additive to copper to improve machinability. It is also used as an additive to improve the corrosion resistance of lead alloys (Ref 1).

Numerous other applications requiring small amounts of tellurium have been found. The principal uses are (Ref 1):

- *Chemicals*: for catalytic reactions; as an element of salts and other compounds
- *Electronics*: as a component in thermoelectric materials
- *Explosives*: as an ingredient in blasting caps
- *Metallurgy*: as an additive to copper to increase machinability without materially decreasing conductivity; to produce free-machining steels; added to lead to improve resistance to vibration and fatigue; as an additive to cast iron to help control the depth of chill; and as a carbide stabilizer for the production of malleable iron
- *Pigments*: in production of various colored glasses and ceramics
- *Rubber*: as a secondary vulcanizing agent

Table 45 lists impurity levels for various grades of elemental tellurium.

Structure

Crystal structure. Hexagonal; at 25 °C, a = 0.44565 nm and c = 0.59268 nm

Table 45 Nominal composition of selected grades of tellurium

Grade designation	Composition, wt%: Te	Se	Pb	Fe	Cu	As	Zn	Other impurities	Impurities max
Commercial or refined	99.0–99.4	0.5	0.3	0.1	0.02	...	...	...	...
Pure	99.95(a)	500(b)	25(b)	...	...	15(b)	10(b)	15(b)(c)	...
Highly pure	99.999	...	...	...	...	...	...	...	10(b)

(a) Some manufacturers claim 99.99%. (b) In ppm. (c) Total of iron, copper, nickel, chromium, antimony, bismuth, and mercury

Mass Characteristics

Atomic weight. 127.60
Density. 6.237 g/cm³ at 25 °C. Temperature dependence: solid (0 to 450 °C), $\rho = 6.250 - 0.000261\ T$; liquid (450 to 1000 °C), $\rho = 6.170 - 0.000777\ T$
Volume change on freezing. 5% contraction

Thermal Properties

Melting point. 449.5 °C
Boiling point. 988 °C
Coefficient of linear thermal expansion. 18.2 μm/m · K at 20 °C
Specific heat. 0.201 kJ/kg · K at 25 °C
Latent heat of fusion. 86.113 kJ/kg
Latent heat of vaporization. 446.43 kJ/kg
Thermal conductivity. Polycrystalline, between 5.98 and 6.02 W/m · K at about 20 to 28 °C. Single crystals: 3.3 W/m · K ∥ to c axis, 2.1 W/m · K ⊥ to c axis
Vapor pressure:

°C	kPa
505	0.1013
617	1.013
768	10.13
988	101.3

Electrical Properties

Electrical resistivity. Polycrystalline, between 1 and 50 mΩ · m at 25 °C. Single crystals, at 20 °C: 5 mΩ · m ∥ to c axis, 1.5 mΩ · m ⊥ to c axis

Magnetic Properties

Magnetic susceptibility. At 18 °C, -39.5×10^{-6} in cgs units (Ref 2)

Optical Properties

Index of refraction. Vapor, 1.002495 at λ = 589 nm; ∥ to plane of incidence, 1.9 to 2.4 at 350 to 400 nm; ⊥ to plane of incidence, 1.7 to 2.7

Nuclear Properties

Stable isotopes. ^{120}Te (0.089% natural abundance), ^{122}Te (2.46%), ^{123}Te (0.87%), ^{124}Te (4.61%), ^{125}Te (6.99%), ^{126}Te (18.71%), ^{128}Te (31.79%), and ^{130}Te (34.48%) (Ref 2, 3)
Unstable isotopes. ^{115}Te, ^{116}Te, ^{117}Te, ^{118}Te, ^{119m}Te, ^{119}Te, ^{121m}Te, ^{121}Te, ^{123}Te, ^{125m}Te, ^{127m}Te, ^{129m}Te, ^{129}Te, ^{131m}Te, ^{131}Te, ^{132}Te, ^{133m}Te, ^{133}Te, ^{134}Te, ^{135}Te (Ref 3)
Thermal neutron cross section. At 2200 m/s: absorption, 4.7 b; scattering, 5.0 b (Ref 2)

Mechanical Properties

Tensile strength. Approximately 11 MPa
Hardness. 25 HB, 2.3 Mohs
Elastic modulus. Tension (single crystals): 42.57 GPa ∥ to c axis, 20.45 GPa ⊥ to c axis. Shear (polycrystalline), 15.16 GPa
Liquid surface tension. 186 mN/m at 450 °C

REFERENCES

1. Selenium Tellurium Development Association, brochure, 1988
2. Bulletin L82 4-7-2M, ASARCO Inc., 1982
3. R.C. Weast, Ed., *CRC Handbook of Chemistry and Physics*, CRC Press, 1979

Terbium (Tb)

See the section "Properties of the Rare Earth Metals" in this article.

Thallium (Tl)

Compiled by S.C. Carapella, Jr., ASARCO, Inc.

Reviewed for this Volume by Douglas Hayduk, ASARCO, Inc.

Thallium is used in alloying to lower the freezing point of certain metals, such as mercury in arctic thermometers and low-temperature mercury switches. It is also a component of fusible alloys and glass, and can be used as an additive in the counter-electrode metal for selenium rectifiers. Compounds of thallium are used for catalysts and semiconductor applications.

Because of the high toxicity of thallium, skin contact and inhalation of dust and fumes are to be avoided. Impervious gloves and aprons should be worn and dust and fumes in work areas must be controlled by exhaust ventilation.

Structure

Crystal structure. Close-packed hexagonal below 230 °C; a = 0.34560 nm, c = 0.55248 nm. Body-centered cubic above 230 °C; a = 0.3874 nm

Mass Characteristics

Atomic weight. 204.37
Density. 11.872 g/cm³ at 20 °C
Volume change on freezing. 3.23% contraction

Thermal Properties

Melting point. 303 °C

Boiling point. 1473 °C
Coefficient of linear thermal expansion. 28 μm/m · K at 20 °C
Specific heat. 0.150 kJ/kg · K for liquid, 0.130 kJ/kg · K for solid
Latent heat of fusion. 20.27 kJ/kg
Latent heat of vaporization. 802.833 kJ · kg
Thermal conductivity. 47 W/m · K at 0 °C
Vapor pressure:

°C	kPa
818	0.1013
972	1.013
1179	10.13
1473	101.3

Electrical Properties

Electrical resistivity. 150 nΩ · m at 0 °C (solid), 740 nΩ · m at 303 °C (liquid)
Electrochemical equivalent. Valence +3, 706.01 μg/C
Standard electrode potential. 0.336 V versus standard hydrogen electrode

Magnetic Properties

Magnetic susceptibility. Volume: -3.1×10^{-6} mks

Optical Properties

Color. Dull gray

Nuclear Properties

Stable isotopes. ^{203}Tl (29.50% natural abundance), ^{205}Tl (70.50%)
Thermal neutron cross section. For 2.2 km/s neutrons: absorption, 3.3 ± 0.5 b; scattering, 14 ± 2 b

Mechanical Properties

Tensile strength. 8.9 MPa
Elongation. 40% in 125 mm
Hardness. 2 HB
Liquid surface tension. 467 mN/m at 303 °C, 450 mN/m at 450 °C

Thorium (Th)

See the section "Properties of the Actinide Metals (Ac-Pu)" in this article.

Thulium (Tm)

See the section "Properties of the Rare Earth Metals" in this article.

Tin (Sn)

Compiled by Joseph B. Long,
Tin Research Institute, Inc.

Reviewed for this Volume by William Hampshire, Tin Information Center

Under certain specific conditions at low temperature, tin can transform from the normal tetragonal metal (β tin or white tin) to a cubic form (α tin or gray tin) that has entirely different properties. Because this transformation is accompanied by an increase in volume, the resultant expansion causes disintegration of the metal to coarse powder or to local warts. The equilibrium temperature of transformation is 13.2 °C. In practice, it is extremely difficult to initiate the change, and the rate is slow even after transformation has started. Moreover, common impurities such as bismuth, antimony, and lead inhibit the change. The fear that fabricated products made of tin or high-tin alloys will fail at low temperatures or that tin in storage will disintegrate is largely unfounded.

The existence of γ tin, a brittle modification that is sometimes mentioned in technical literature, has been disproved (Ref 1). Abrupt changes in some properties at elevated temperatures, such as high-temperature ductility, are ascribed more accurately to impurities. Many of the properties listed here were determined on tin of 99.95% purity. Additional information is available in the article "Tin and Tin Alloys" in this Volume.

Structure

Crystal structure. α phase: face-centered cubic, *A*4, *cF*8 (*Fd*3*m*); *a* = 0.64912 nm (Ref 2, 3). β phase: body-centered tetragonal, *A*5, *tI*4 (*I*4*t*/*amd*); *a* = 0.58314 nm, *c* = 0.31815 nm (Ref 4 to 9)
Slip elements. β phase tin:

20 °C: Slip plane	20 °C: Slip direction	150 °C: Slip plane	150 °C: Slip direction
(110)	[001] [111]	(110)	[$\bar{1}$11]
(100)	[001] [010]		
(101)	[101]		
(121)			

Sources: Ref 10, 11

Twinning plane. (301) (Ref 3)
Interatomic distances:

Tin phase	Interatomic distance, nm	Number of neighbors
α tin	0.279	4
	0.456	8
β tin	0.302	4
	0.318	2
	0.376	4
Liquid tin (250 °C)	0.338	10

Source: Ref 12

Microstructure. Because tin is an extremely soft metal, it is often difficult to obtain an unworked surface free from scratches, and the low crystallization temperature of tin can result in false structure if distortion occurs during polishing. To overcome these difficulties, it is necessary to take special precautions during both mounting and polishing.
Fracture behavior. Ductile

Mass Characteristics

Atomic weight. 118.69
Density. α phase, 5.765 g/cm³ at 1 °C (Ref 13). β phase: 7.2984 g/cm³ at 15 °C (Ref 14), 7.168 g/cm³ at 20 °C (Ref 15, 16). Liquid:

°C	g/cm³	Ref
298	6.94	17
409	6.840	15, 16
474	6.789	15, 16
523	6.761	15, 16
538	6.77	17
574	6.729	15, 16
602	6.711	15, 16
648	6.671	15, 16
816	6.62	17
1093	6.45	17
1371	6.29	17
1573	6.16	17

Sources: Ref 15 to 17

Volume change on freezing. 2.8% contraction
Volume change on phase transformation. β phase to α phase, 27% expansion

Thermal Properties

Melting point. 231.9 °C (Ref 18)
Effect of pressure on melting point. From Ref 19:

Pressure, MPa	Temperature, °C
51	232.26
76	233.09
101	233.89
151	235.47
203	237.18

Boiling point. 2270 °C (Ref 20)
Phase transformation temperature. β phase to α phase, 13.2 °C (Ref 21)
Coefficient of thermal expansion. For α phase:

Temperature, °C	Linear, μm/m · K	Volumetric, mm³/m · K
−200	13.5	40 600
−150	16.6	49 900
−100	18.1	54 300
−50	19.2	57 500
0	19.9	59 800
50	23.1	69 200
100	23.8	71 400
150	26.7	80 200

Sources: Ref 22, 23

Volumetric for liquid phase (Ref 16, 17, 24): 106 000 mm³/m · K for 232 to 400 °C, 105 000 mm³/m · K for 400 to 700 °C, 100 000 mm³/m · K for 232 to 1600 °C
Coefficients of linear expansion along crystal axes:

Temperature, °C	Coefficient, μm/m · K: Parallel to *c* axis	Coefficient, μm/m · K: Perpendicular to *c* axis
−200	21.3	9.4
−150	24.1	12.8
−100	25.7	14.3
−50	27.0	15.2
0	28.4	15.8
50	32.9	16.6
100	35.7	17.9
150	38.4	19.2
200	40.4	19.9

Sources: Ref 22, 25

Specific heat. From Ref 20, 26: α phase at 10 °C, 0.205 kJ/kg · K; β phase at 25 °C, 0.222 kJ/kg · K; β phase from 25 to 231 °C, $C_p = 155 + 0.22 \times T$, where C_p is in J/kg · K and T is the temperature in K; liquid phase from 232 to 1000 °C, 0.257 kJ/kg · K
Latent heat of fusion. 59.5 kJ/kg (Ref 27)
Latent heat of phase transformation. 17.6 kJ/kg (Ref 27)
Latent heat of vaporization. 2.4 MJ/kg (Ref 20, 28 to 30)
Thermal conductivity:

Phase	°C	W/m · K
β	−170	80.8
β	0	62.8
β	100	60.7
β	200	56.5
Liquid	232–332	32.6

Sources: Ref 31 to 34

Vapor pressure. From Ref 30:

°C	Pa
727	9.9×10^4
927	0.16
1027	1.1
1127	5.9
1227	23
1327	89
1527	746
1727	4.08×10^3
1827	8.41×10^3
2027	2.9×10^4
2127	5.13×10^4
2227	8.51×10^4

Diffusion coefficient. Self-diffusion along crystal axes in β phase:

Temperature, °C	Coefficient, μm²/s: c axis	a axis
180.5	0.0111	0.00527
197.0	0.0148	0.00601
210.7	0.0213	0.00930
223.1	0.0265	0.00929

Source: Ref 35

Electrical Properties

Electrical conductivity. 15.6% IACS
Electrical resistivity.

°C	μΩ · m
0	0.110
100	0.155
200	0.200
231	0.220

For liquid phase (Ref 36):

°C	μΩ · m
232	0.450
300	0.468
400	0.490
500	0.515
600	0.540
700	0.563
800	0.587
900	0.612

Electrical resistivity along crystal axes. β phase (Ref 37, 38): 0.120 μΩ · m parallel to *c* axis, 0.092 μΩ · m perpendicular to *c* axis
Pressure coefficient of electrical resistance. -9.51×10^{-6} at 30 °C between 0 to 2.9 MPa (Ref 39)
Thermoelectric power. Between solid and liquid tin (liquid is at the higher potential) (Ref 40): 1.6 μV/K
Electrochemical equivalent. Valence +2, 615.03 g/C; valence +4, 307.51 g/C (Ref 41)
Standard electrode potential. Versus standard hydrogen electrode: $Sn^{2+} + 2e^- \rightleftarrows Sn$ − 0.14 V; $Sn^{4+} + 2e^- \rightleftarrows Sn^{2+}$ + 0.15 V (calculated) (Ref 42, 43)
Ionization potential. 7.297 eV, spectrographic (Ref 44)
Hydrogen overvoltage. See Table 46.
Hall effect. −0.02 nV · m/A · T at 0.4 T and room temperature (Ref 49 to 51)
Temperature of superconductivity. 3.73 K (Ref 51)
Photoelectric work function. 464 eV (Ref 52)

Table 46 Effect of current density on the hydrogen overvoltage of tin

Current density, A/m²	Overvoltage, V, in: Ref 1	Ref 45	Ref 46	Ref 47	Ref 48
1	...	...	...	...	0.50
5	...	...	...	...	0.65
10	0.66, 0.85	0.73	0.57, 0.59	...	0.73
20	...	...	...	0.71	0.89
50	...	...	...	0.87	1.09
100	0.85, 0.97	0.89	0.71, 0.72	0.98	1.29
200	...	...	...	1.04	...
500	...	...	...	1.13	...
1 000	0.89, 0.98	0.99	0.83, 0.86	1.19	...
2 000	...	...	...	1.37	...
5 000	...	1.00	...	...	...
10 000	0.98	...	...	0.88	...

Magnetic Properties

Magnetic susceptibility. Mass: α phase, -39×10^{-11} mks; β phase at 18 °C, $+34 \times 10^{-11}$ mks; liquid phase at 250 °C, -45×10^{-11} mks

Optical Properties

Color. White with bluish tinge
Reflectance. Refractive and absorption indices. See Table 47.
Emittance. 0.04 at 50 °C (Ref 14)

Nuclear Properties

Stable isotopes. Results from three measurements (Ref 57):

Mass number	Abundance, %
112	1.01, 0.90, 0.94
114	0.68, 0.61, 0.65
115	0.35, 0.33, ?
116	14.28, 14.07, 14.36
117	7.67, 7.54, 7.51
118	23.84, 23.98, 24.21
119	8.68, 8.62, 8.45
120	32.75, 33.03, 33.11
122	4.74, 4.78, 4.61
124	6.01, 6.11, 5.83

Unstable (radioactive) isotopes. See Table 48.

Mechanical Properties

Poisson's ratio. 0.33
Elastic modulus. At room temperature (Ref 58): as-cast (coarse grain), 41.6 GPa; self-annealed (fine grain), 44.3 GPa. For effect of temperature:

°C	Percentage of value at 16 °C: Polycrystalline	Single crystals
25	98	99
50	92	96
75	86	93
100	79	90
125	72	87

Source: Ref 59

Elastic modulus along crystal axes. Room temperature: 84.7 GPa along (001) plane (maximum value), 26.3 GPa along (110) plane (minimum value). Effect of temperature (Ref 60 to 62): 53.9 MPa/K at −180 to 0 °C, 75.5 MPa/K at 0 to 100 °C, 121.6 MPa/K at 100 to 200 °C
Dynamic liquid viscosity. From Ref 63 to 68:

°C	mPa · s
232	2.71
250	1.88
300	1.66
400	1.38
500	1.18
600	1.05
700	0.95
800	0.87

Liquid surface tension. From 400 to 800 °C: $\sigma = 700 - 0.17 \times T + (25 + 0.015 \times T)$, where σ is in mN/m and T is temperature in K (Ref 69)
Velocity of sound. In solid, 2.60 km/s at 18

Table 47 Optical properties of tin at λ = 546.1 nm

Surface	Reflectance	Refractive index	Absorptive index
Film, 42–200 nm thick(a)	0.70	2.4	1.9
Film, 2.5 nm thick(a)	...	3.0	0.17
Bulk solid	0.80	1.0	4.2
Liquid	0.80	1.7	3.1

(a) Vacuum-evaporated film. Sources: Ref 53 to 56

Table 48 Unstable (radioactive) isotopes of tin

Isotope	Half-life	Energy of radiation, MeV: Particles	Energy of radiation, MeV: γ rays
^{109}Sn	18 min	β^+: ~1.6	...
^{110}Sn	4 h	...	...
^{111}Sn	35 min	β^-: 1.45, 1.51	...
^{113}Sn	30–33, 25 min	β^-: 1.2	...
	112, ~100, 70 days		~0.09, 0.85, ?
^{117}Sn	14.5, 14, 13 days	...	0.175, 0.17, 0.159, 0.162, 0.152, 0.157
^{119}Sn	≥100, 279, ~250, 245 days	...	0.069, 0.064
^{121}Sn	27.0, 26.4, 27.5 h,	β^-: 0.383, 0.35, ~0.4	...
	>400 days	β^-: 0.41, 0.42	
^{123}Sn	130, 136 days	β^-: 1.42, 1.3, ~1.5	0.394, ?
	39.5, 41, 40 min	β^-: 1.26, 1.12, 1.32, ~1.7	0.153, ~0.17, ~0.4
^{124}Sn	0.4–0.9 × 10^{16} years?	β^-: 1.0–1.5, ?	...
^{125}Sn	10.0, 9.9, 11, 9.4 days	β^-: 2.38, 2.34, 2.1	?, ~1.9
		β^-: 2.6, 2.33, 2.06	
		β_1^-: 2.37	
		β_2^-: 0.40	
	9.5, 9.8, 10 min	β_1^-: 2.04, 2.05; β_2: 2.2, 2.06	γ_1: 0.326, 0.36
		2.06	
		1.17	γ_2: 1.86, >1
		β_3^-: 0.51, ?, 0.5, ~0.5	γ_3: 1.37
^{126}Sn	70, 80 min	β^-: 0.7, 2.7	~1.2

Sources: Ref 57 and *CRC Handbook of Chemistry and Physics* (1979–1980)

°C (Ref 3); in liquid, 2.27 km/s at 232 °C and 12 MHz (Ref 70)

REFERENCES

1. A. Hickling and F.W. Salt, *Trans. Faraday Soc.*, Vol 37, 1941, p 333
2. L.D. Brownlee, *Nature*, Vol 166, 1950, p 482
3. C.J. Smithells, *Metals Reference Book*, Butterworths, 1949
4. A.J.C. Wilson, Ed., *Structure Reports for 1949*, Vol 12, International Union of Crystallography, 1949
5. R. Clark, G.B. Craig, and B. Chalmers, *Acta Crystallogr.*, Vol 3, 1950, p 479
6. A. Ievins, M. Straumanis, and K. Karlsons, *Z. Phys. Chem. B.*, Vol 40, 1938, p 347
7. L.W. McKeehan and H.J. Hoge, *Z. Kristallogr.*, Vol 92, 1935, p 476
8. W. Stenzel and J. Weertz, *Z. Kristallogr.*, Vol 84, 1932, p 20
9. E.R. Jette and F. Foote, *J. Chem. Phys.*, Vol 3, 1935, p 605
10. E. Schmid and W. Boas, *Plasticity of Crystals*, E.A. Hughes, 1950
11. K. Brausch, *Z. Phys.*, Vol 93, 1935, p 479
12. C. Gamertsfelder, *J. Chem. Phys.*, Vol 9, 1941, p 450
13. H. Endo, *Bull. Chem. Soc. Jpn.*, Vol 2, 1927, p 131
14. C.L. Mantell, *Metals Handbook*, 1939, p 1714
15. Hess, *Ber. Dtsch. Phys. Ges.*, Vol 11 (No. 3), 1905, p 403
16. K. Bornemann and P. Siebe, *Z. Metallkd.*, Vol 14, 1922, p 329
17. A.L. Day, R.B. Sosman, and J.C. Hostetter, *Am. J. Sci.*, Vol 37 (No. IV), 1914, p 1
18. P.G.J. Gueterbock and G.N. Nicklin, *J. Soc. Chem. Ind. (London)*, Vol 44, 1925, p 370T
19. J. Johnston and L.H. Adams, *Am. J. Sci.*, Vol 31, 1911, p 501
20. K.K. Kelley, Bulletin 383, U.S. Bureau of Mines, 1935
21. C.E. Homer and H.C. Watkins, *Met. Ind. (London)*, Vol 60, 1942, p 364
22. H.D. Erfling, *Ann. Phys. V*, Vol 34, 1939, p 136
23. F.L. Uffelman, *Philos. Mag.*, Vol 10, 1930, p 633
24. T.R. Hogness, *J. Am. Chem. Soc.*, Vol 43, 1921, p 1621
25. B.G. Childs and S. Weintroub, *Proc. Phys. Soc. B*, Vol 63, 1950, p 267
26. E. Cohen and K.D. Dekker, *Z. Phys. Chem.*, Vol 127, 1927, p 214
27. O. Kubaschewski, *Z. Electrochem.*, Vol 54, 1950, p 275
28. A.W. Searoy and R.D. Freeman, *J. Am. Chem. Soc.*, Vol 76, 1954, p 5229
29. L. Brewer and R.F. Porter, *J. Chem. Phys.*, Vol 21, 1953, p 2012
30. E.C. Baughan, *Q. Rev. Chem. Soc.*, Vol 7, 1953, p 103
31. M. Jakob, *Z. Metallkd.*, Vol 16, 1924, p 353
32. W.B. Brown, *Phys. Rev.*, Vol 22, 1923, p 171
33. S. Konno, *Sci. Rep. Tôhoku Imperial Univ.*, Vol 8, 1919, p 169
34. C.H. Lees, *Philos. Trans. Soc.*, Vol 208, 1908, p 381
35. P.J. Fensham, *Aust. J. Sci. Res. A.*, Vol 4, 1951, p 229
36. "Properties of Tin," Tin Research Institute, 1965
37. B. Chalmers and R.H. Humphrey, *Philos. Mag.*, Vol 25, 1938, p 1108
38. P.W. Bridgeman, *Proc. Am. Acad. Arts Sci.*, Vol 68, 1933, p 95
39. P.W. Bridgeman, *Proc. Am. Acad. Arts Sci.*, Vol 72, 1938, p 157
40. F. Cirkler, *Z. Naturforsch. A*, Vol 8, 1953, p 646
41. G.A. Roush, *Trans. Electrochem. Soc.*, Vol 73, 1938, p 285
42. W.M. Latimer, *Oxidation Potentials*, Prentice-Hall, 1952
43. M.M. Haring and J.C. White, *Trans. Electrochem. Soc.*, Vol 73, 1938, p 211
44. S. Tolansky, *Proc. R. Soc. (London) A*, Vol 144, 1934, p 574
45. J.O. Bockris and S. Ignatowicz, *Trans. Faraday Soc.*, Vol 44, 1948, p 519
46. A.G. Pecherskaya and V.V. Stender, *Zh. Prikl. Khim.*, Vol 19, 1946, p 1303
47. H. Hunt, J.F. Chittum, and H.W. Ritchey, *Trans. Electrochem. Soc.*, Vol 73, 1938, p 299
48. G. Schmid and E.K. Stoll, *Z. Elektrochem.*, Vol 47, 1941, p 360
49. P. Räthjen, *Phys. Z.*, Vol 25, 1924, p 84
50. G. Busch, J. Wieland, and H. Zoller, *Helv. Phys. Acta*, Vol 24, 1951, p 49
51. D. Shoenberg, *Superconductivity*, Cambridge University Press, 1952
52. R. Hischberg and E. Lange, *Naturwissenschaften*, Vol 39, 1952, p 131
53. P.L. Clegg, *Proc. Phys. Soc. B*, Vol 65, 1952, p 774
54. D.G. Avery, *Philos. Mag.*, Vol 41, 1950, p 1018
55. C.V. Kent, *Phys. Rev.*, Vol 14, 1919, p 459
56. P. Erochin, *Ann. Phys. (IV)*, Vol 39, 1912, p 213
57. "Nuclear Data," Circular 499 and Supplements 1, 2, 3, U.S. National Bureau of Standards, 1951-1952
58. J.W. Cuthbertson, *J. Inst. Met.*, Vol 64, 1939, p 209
59. L. Rotherham, A.D.N. Smith, and G.B. Greenough, *J. Inst. Met.*, Vol 79, 1951, p 439
60. W. Koster, *Z. Metallkd.*, Vol 39, 1948, p 1
61. Y.L. Yousef, *Philos. Mag.*, Vol 37, 1946, p 490
62. S. Aoyama and T. Fukuroi, *Sci. Rep. Tôhoku Imperial Univ.*, Vol 28, 1940, p 423
63. T.P. Yao and V. Kondic, *J. Inst. Met.*, Vol 81, 1952, p 17
64. A.J. Lewis, *Proc. Phys. Soc.*, Vol 48, 1936, p 102
65. K. Gering and F. Sauerwald, *Z. Anorg. Allg. Chem.*, Vol 223, 1935, p 204
66. V.H. Stott, *Proc. Phys. Soc.*, Vol 45, 1933, p 530
67. F. Sauerwald and K. Topler, *Z. Anorg. Allg. Chem.*, Vol 157, 1926, p 117
68. M. Plüss, *Z. Anorg. Allg. Chem.*, Vol 93, 1915, p 1

Table 49 Impurity limits of electrolytic titanium and iodide titanium

	Typical concentration, %	
Impurity element	Electrolytic	Iodide crystal bar
Fe	0.009	0.002
Si	0.002	0.005
Ca	...	0.003
Cu	0.007	<0.001
Mg	<0.001	0.003
Mn	<0.001	0.003
Sn	<0.020	0.001
Zr	<0.001	0.050
C	0.008	0.001
O	0.037	0.03–0.06
N	0.004	0.002
Cl	0.073	0.002
Ti (by difference)	99.837	99.90–99.87

69. D.V. Atterton and T.P. Hoar, *J. Inst. Met.*, Vol 81, 1952-53, p 541
70. O.J. Kleppa, *J. Chem. Phys.*, Vol 18, 1950, p 1331

Titanium (Ti)

Compiled by W. Stuart Lyman, Copper Development Association, Inc. (formerly with Battelle Memorial Institute)

Reviewed for this Volume by Titanium Development Association

Designations

Common name. Iodide titanium, electrolytic titanium. See Table 49 for impurity limits.
Typical uses. Experimentation and research; commercial applications requiring freedom from interstitial alloying elements (oxygen, nitrogen, carbon, and hydrogen)

Structure

Crystal structure. α phase: close-packed hexagonal; a = 0.295030 nm, c = 0.468312, c/a = 1.5873. β phase: body-centered cubic, a = 0.332 nm at 900 °C

Mass Characteristics

Atomic weight. 47.9
Density. α phase: 4.507 g/cm³ at 20 °C. β phase: 4.35 g/cm³ at 885 °C (from indirect measurements)

Thermal Properties

Melting temperature. 1668 ± 10 °C
Boiling point. 3260 °C (estimated)
Vapor pressure. From 1587 to 1698 K:

$$\log P = 7.7960 - \frac{24\,644}{T} - 0.000227\,T$$

where P is in Pa and T is in K
Phase transformation temperature. α to β, 882.5 °C
Coefficient of thermal expansion. At 20 °C, 8.41 × 10⁻⁶/°C at 1000, 10.1 × 10⁻⁸/°C (estimated)
Thermal expansion in crystallographic direction. Calculated from lattice parameters:

Temperature, °C	Direction	Expansion, μm/m · K
20–400	Perpendicular to c axis	10.2
20–700	Perpendicular to c axis	11.0
20–700	Along c axis	12.8

Specific heat. Below 13 K, $C_p = 0.0706 + 5.43 \times 10^{-4}\,T^3$; above room temperature, $C_p = 669.0 - 0.037188\,T - 1.080 \times 10^7\,T^{-2}$, where C_p is in J/kg · K and T is in K:

Temperature, K	C_p, kJ/kg · K
50	0.0993
75	0.2100
100	0.3002
125	0.3632
150	0.4094
175	0.4408
200	0.4651
225	0.4841
250	0.4994
275	0.5126
298.15	0.5223
300	0.5378
350	0.5678
400	0.5866
450	0.5989
500	0.6072
550	0.6128
600	0.6167
650	0.6193
700	0.6209
750	0.6219
800	0.6224
850	0.6224
900	0.6223
950	0.6217
1000	0.6210
1050	0.6202
1100	0.6192
1150	0.6181

Latent heat of fusion. 440 kJ/kg (estimated)
Latent heat of transformation. 91.8 kJ/kg (estimated)
Latent heat of vaporization. 9.83 MJ/kg (estimated)
Thermal conductivity. 11.4 W/m · K at −240 °C

Electrical Properties

Electrical resistivity. 420 nΩ · m at 20 °C. See also Fig. 114.
Superconductivity. Critical temperature: 0.37 to 0.56 K

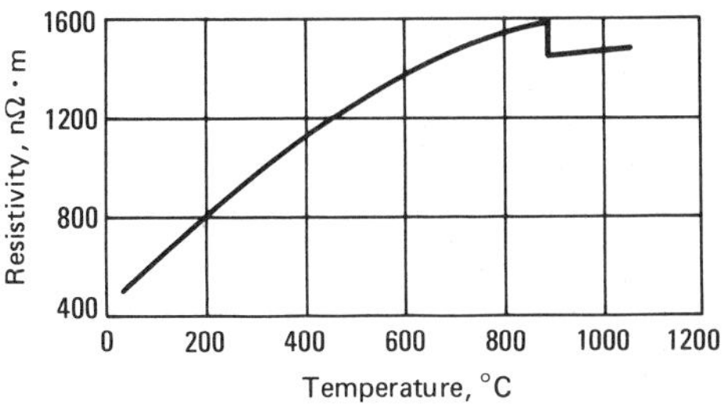

Fig. 114 Electrical resistivity of 99.9% pure titanium

Magnetic Properties

Magnetic susceptibility. Volume, at room temperature: 180 (±1.7) × 10⁻⁶ mks

Optical Properties

Total hemispherical emittance. 0.30 at 710 °C

Nuclear Properties

Stable isotopes:

Isotope	Natural abundance, %	Cross section, b(a)
^{46}Ti	7.95	0.6
^{47}Ti	7.75	1.6
^{48}Ti	73.43	8.0
^{49}Ti	5.51	1.8
^{50}Ti	5.34	0.2

(a) b, barns

Chemical Properties

General corrosion behavior. Greater resistance than that of commercial grades of unalloyed titanium. See the article "Corrosion of Titanium and Titanium Alloys" in Volume 13 of the 9th Edition of *Metals Handbook.*

Mechanical Properties

Tensile properties. Typical, at room temperature: tensile strength, 235 MPa; 0.2% yield strength, 140 MPa; elongation in 50 mm, 54%. See also Fig. 115.

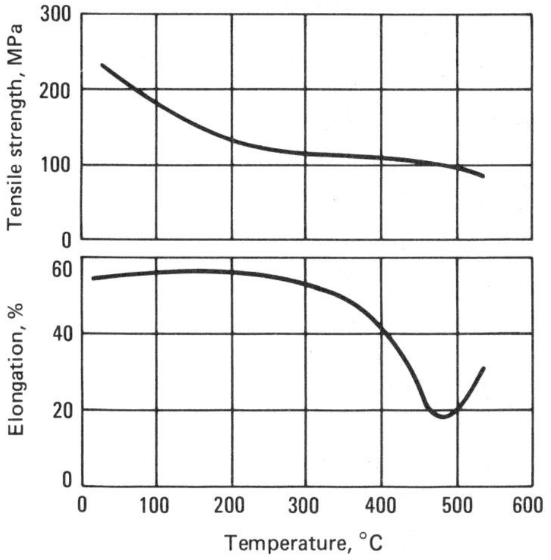

Fig. 115 Typical tensile properties of 99.9% pure titanium

Minimum bend radius. <1t
Hardness. Ingot melted from: electrolytic titanium, 70 to 74 HB; iodide titanium, 65 to 72 HB
Velocity of sound. 4970 m/s

Additional information is available in the articles "Introduction to Titanium and Titanium Alloys," "Wrought Titanium and Titanium Alloys," "Titanium and Titanium Alloy Castings," and "Titanium P/M Products" in this Volume.

Tungsten (W)

Compiled by Stephen W.H. Yih, Consultant

Reviewed for this Volume by Toni Grobstein, NASA Lewis Research Center

Structure

Crystal structure. α phase: body-centered cubic, *cI*2 (*Im*3*m*); a = 0.316522 ± 0.00009 nm at 25 °C. β phase occurs only in the presence of oxygen and is probably W_3O; it is stable below 630 °C and of type *A*15 or *cP*8 (*Pm*3*n*); a = 0.5046 nm at 25 °C
Minimum interatomic distance. α phase, 0.274116 nm; β phase, 0.252 nm

Mass Characteristics

Atomic weight. 183.85
Density. 19.254 g/cm^3

Thermal Properties

Melting point. 3410 ± 20 °C (Ref 1)
Boiling point. 5700 ± 200 °C (Ref 2)
Coefficient of thermal expansion. Linear at low temperatures:

Temperature		Coefficient, μm/m	
K	°C	X-ray spacing data	Extensometer data
180	−93	3.7	3.9
140	−133	3.2	3.4
100	−173	2.3	2.7
80	−193	1.8	2.2
60	−213	1.1	1.5
40	−233	0.4	0.6

Source: Ref 3 to 5

Thermal expansion. From 25 to 2500 °C:

$$\frac{L - L_{25\,°C}}{L_{25\,°C}} \times 100 = A_0 + A_1T + A_2T_2$$

where T is the Celsius temperature and the values of coefficients A_0, A_1, and A_2 are shown in Table 50

Table 50 Values of coefficients for calculation of the thermal expansion of tungsten from 25 to 2500 °C

Material	Coefficient A_0	A_1	A_2
Powder metallurgy rod	-8.69×10^{-3}	3.83×10^{-4}	7.92×10^{-8}
Powder metallurgy sheet	-4.58×10^{-3}	3.65×10^{-4}	9.81×10^{-8}
Arc cast sheet	-6.76×10^{-3}	3.91×10^{-4}	8.98×10^{-8}

Source: Ref 6

Specific heat. From 0 to 3000 °C (Ref 7):

$$C_p = 135.76\left(1 - \frac{4805}{T^2}\right) + (9.1159 \times 10^{-3})\,T + (2.3134 \times 10^{-9})\,T^3$$

where C_p is in J/kg · K and T is in K
Enthalpy. From 935 to 2975 °C (derived from specific heat equation):

$$H_T - H_{298} = 135.76\left(T + \frac{26.14}{T}\right) - 4.266 \times 10^3 + (4.5569 \times 10^{-3})\,T^2 + (5.78205 \times 10^{-10})\,T^4$$

where H is in J/kg and T is in K
Entropy. At 25 °C (Ref 8): 178.3 J/kg · K. Change with temperature (Ref 9):

°C	J/kg · K
25	0
127	37
727	163
1627	262

Latent heat of fusion. 220 ± 36 kJ/kg (Ref 2)
Latent heat of sublimation. 4680 ± 25 kJ/kg (Ref 10)
Thermal conductivity. See Fig. 116.
Vapor pressure. From 2327 to 2827 °C (Ref 10):

$$\log_{10} P = \frac{-45\,385}{T} + 2.865$$

where P is in Pa and T is in K

Electrical Properties

Electrical resistivity. 53 nΩ · m at 27 °C. From 27 to 967 °C (Ref 15): ρ = (4.33471 × 10^{-14}) T^2 + (2.19691 × 10^{-10}) T − (1.64011 × 10^{-8}, where ρ is in Ω · m and T is in K
Hall constant. 10.55 μV · m/A · T (Ref 2)
Temperature of superconductivity. 0.016 K (Ref 16)

Magnetic Properties

Magnetic susceptibility. See Fig. 117.

Optical Properties

Total emissivity. From Ref 18:

°C	Total emissivity
127	0.042
527	0.088
1327	0.207
1727	0.260
2127	0.296
2527	0.323
2927	0.341

Nuclear Properties

Stable isotopes. ^{180}W, isotope mass 179.9470, 0.14% abundant; ^{182}W, isotope mass 181.9483, 26.41% abundant; ^{183}W, isotope mass 182.9503, 14.40% abundant; ^{184}W, isotope mass 183.9510, 30.64% abundant; ^{186}W, isotope mass 185.9543, 28.41% abundant (Ref 19)

Mechanical Properties

Tensile properties. See Fig. 118.
Hardness. See Fig. 119.
Poisson's ratio. See Fig. 120.
Elastic modulus. See Fig. 120.
Ductile-to-brittle transition temperature. See Fig. 121.

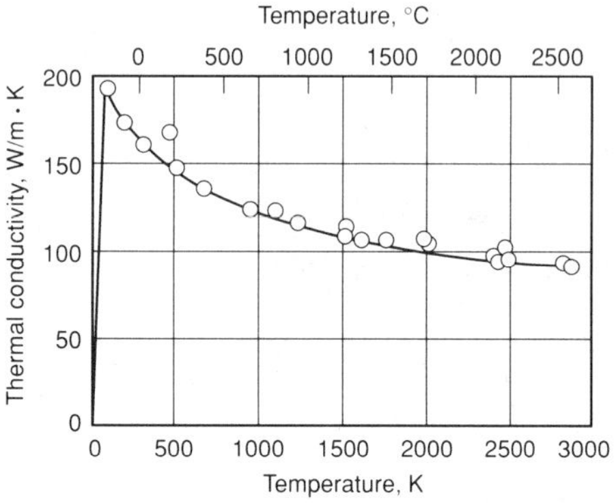

Fig. 116 Temperature dependence of the thermal conductivity of tungsten. Sources: Ref 11 to 15

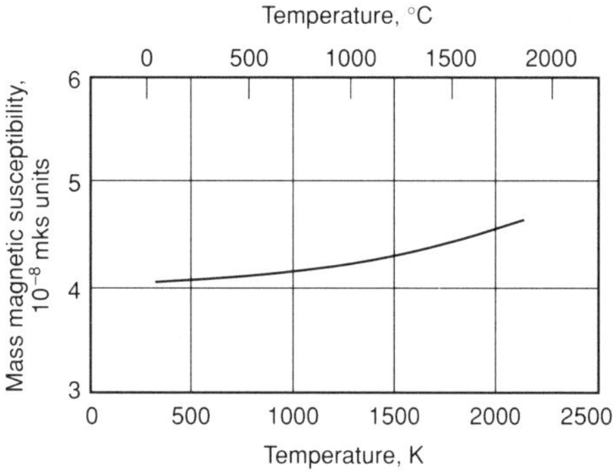

Fig. 117 Temperature dependence of the mass magnetic susceptibility (χ_m) of tungsten. Source: Ref 117

Creep-rupture characteristics. See Fig. 122 and 123.

REFERENCES

1. A. Cezairliyan, *High Temp. Sci.*, Vol 4 (No. 3) 1972, p 248-252
2. G.D. Rieck, *Tungsten and Its Compounds*, Pergamon Press, 1967
3. J.S. Shah and M.E. Straumanis, *J. Appl. Phys.*, Vol 42 (No. 9), 1971, p 3288
4. F.C. Nix and D. McNair, *Phys. Rev.*, Vol 61, 1942, p 74
5. R.J. Corruccini and J.J. Gniewek, *Natl. Bur. Stand. Monogr.*, No. 29, 1961
6. J.B. Conway and A.C. Losekamp, *Trans. TMS-AIME*, Vol 236, 1966, p 702-709
7. M. Hoch, *High Temp.—High Press.*, Vol 1, 1969, p 531-542
8. K. Clusius and P. Franzosini, *Z. Natur-*

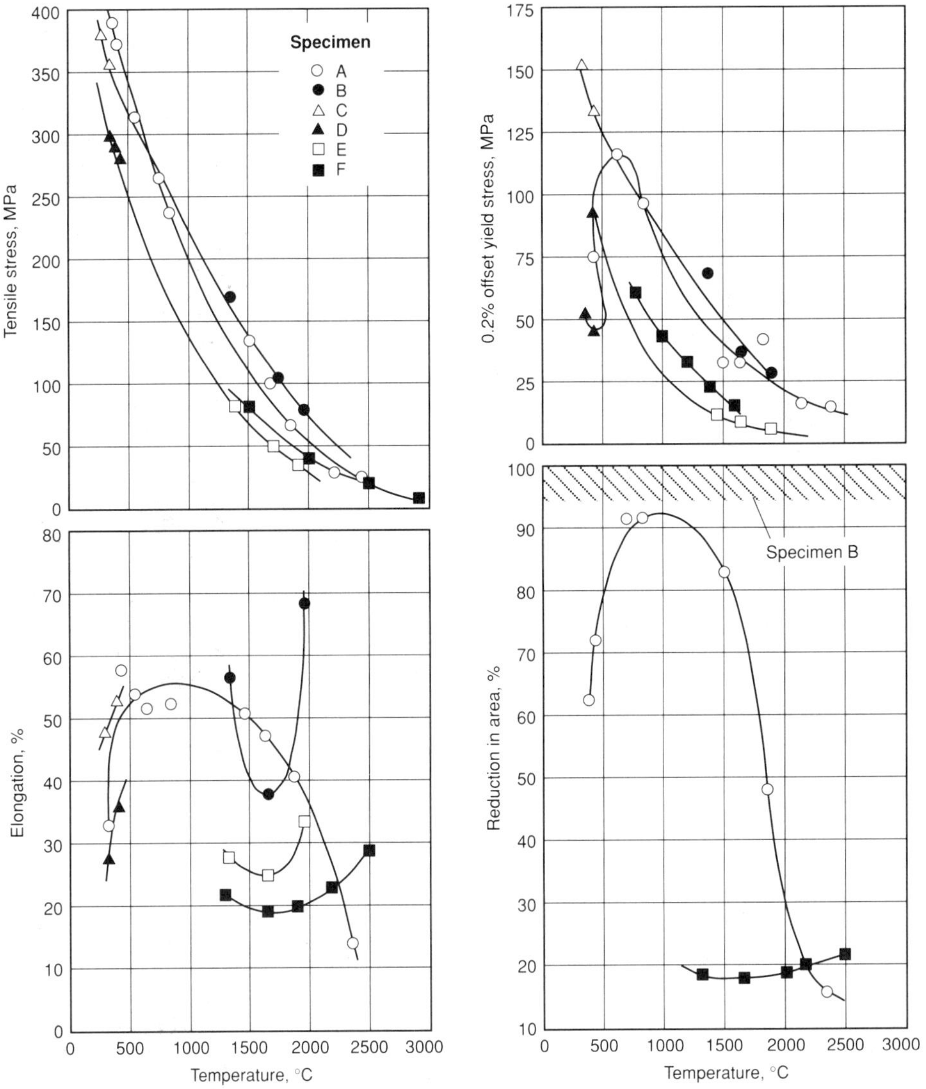

Fig. 118 Temperature dependence of the tensile strength of tungsten. Specimen A: P/M rod, 2.36 mm in diameter, annealed ½ h at 2400 °C. Specimen B: arc cast rod, 4.06 mm in diameter, annealed 1 h at 1982 °C. Specimen C: arc cast rod, 4.06 mm in diameter, annealed 1 h at 1648 °C. Specimen D: electron-beam-melted rod, 4.06 mm in diameter, annealed 1 h at 1371 °C. Specimen E: electron-beam-melted rod, 4.06 mm in diameter, annealed 1 h at 1982 °C. Specimen F: chemical-vapor-deposited rod, 4.06 mm in diameter, annealed 1 h at 2845 °C. Sources: Ref 11, 20 to 23

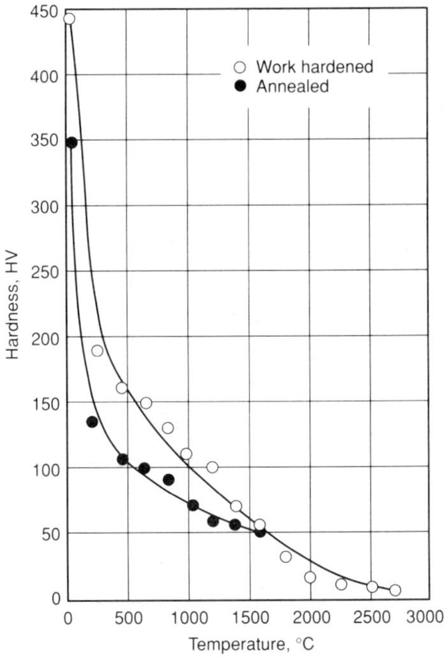

Fig. 119 Temperature dependence of the hardness of tungsten. Source: Ref 24

forsch., Vol 14, 1959, p 99
9. U. Schmidt, O. Volmer, and R. Kohlhass, *Z. Naturforsch.*, Vol 25, 1970, p 1258-1264
10. E.R. Plante and A.B. Sessoms, *J. Res. Natl. Bur. Stand.*, Vol 77A (No. 2), 1973, p 237-242
11. S.W.H. Yih and C.T. Wang, Chapter 6 in *Tungsten: Sources, Metallurgy, Properties and Applications*, Plenum Publishing, 1979
12. N.G. Backlund, *J. Phys. Chem. Solids*, Vol 28, 1967, p 2219-2223
13. B.E. Neimark and L.K. Voronin, *High Temperature*, Vol 6, 1968, p 999-1010
14. R.E. Taylor, F.E. Davis, and R.W. Powell, *High Temp.—High Press.*, Vol 1, 1969, p 663-673
15. V.A. Vertogradskii and V.Y. Chekhovskoi, *Teplofiz. Vys. Temp.*, Vol 8 (No. 4), 1970, p 784-788
16. B.B. Triplett *et al.*, *J. Low Temp. Phys.*, Vol 12 (No. 5/6), 1973, p 499-518
17. C. Kittel, *Introduction to Solid State Physics*, 3rd ed., John Wiley & Sons, 1971
18. D.E. Gray, *American Institute of Physics Handbook*, 3rd ed., McGraw-Hill, 1972, p 6-79
19. R.C. Weast, Ed., *Handbook of Chemistry and Physics*, CRC Press, 1977
20. H.G. Sell, W.R. Morcom, and G.W. King, "Development of Dispersion Strengthened Tungsten Base Alloys," AFML-TR-65-407, Part II, Westinghouse Lamp Division, 1966
21. W.D. Klopp and P.L. Raffo, "Effects of Purity and Structure on Recrystallization, Grain Growth, Ductility, Tensile and Creep Properties of Arc-Melted Tungsten," NASA-TND-2503, National Aeronautics and Space Administration Lewis Research Center, 1964
22. W.D. Klopp and W.R. Witzke, "Mechanical Properties and Recrystallization Behavior of Electron-Beam-Melted Tungsten Compared With Arc-Melted Tungsten," NASA-TND-3232, National Aeronautics and Space Administration Lewis Research Center, 1966
23. J.L. Taylor and D.H. Boone, *J. Less-Common Met.*, Vol 6, 1964, p 157-164
24. G.S. Pisarenki, V.A. Borisenko, and Y.A. Kashtalyan, *Sov. Powder Metall. Met. Ceram.*, Vol 5, 1962, p 371-374
25. R. Lowrie and A.M. Gonas, *J. Appl. Phys.*, Vol 38, 1967, p 4505-4509
26. R. Lowrie and A.M. Gonas, *J. Appl. Phys.*, Vol 36, 1965, p 2189-2192
27. H.R. Ogden, "Refractory Metals Sheet-Rolling Program," DMIC Report 176, Battelle Memorial Institute, 1962
28. A.C. Schaffhauser, "Low Temperature Ductility and Strength of Thermochemically Deposited Tungsten and Effects of Heat Treatment," AFML-TR-179, Oak Ridge National Laboratories, 1966
29. D.L. McDanels and R.A. Signorelli, "Stress-Rupture Properties of Tungsten Wire From 1200° to 2500 °F," NASA-TND-3467, National Aeronautics and Space Administration, 1966
30. J.K.Y. Hum and A. Donlevy, "Some Stress Rupture Properties of Columbium, Molybdenum, Tantalum and Tungsten Metals and Alloys Between 2400 °F and 5000 °F," Report 354D, Society of Automotive Engineers, 1961
31. C.A. Drury, R.C. Kay, A. Bennett, and

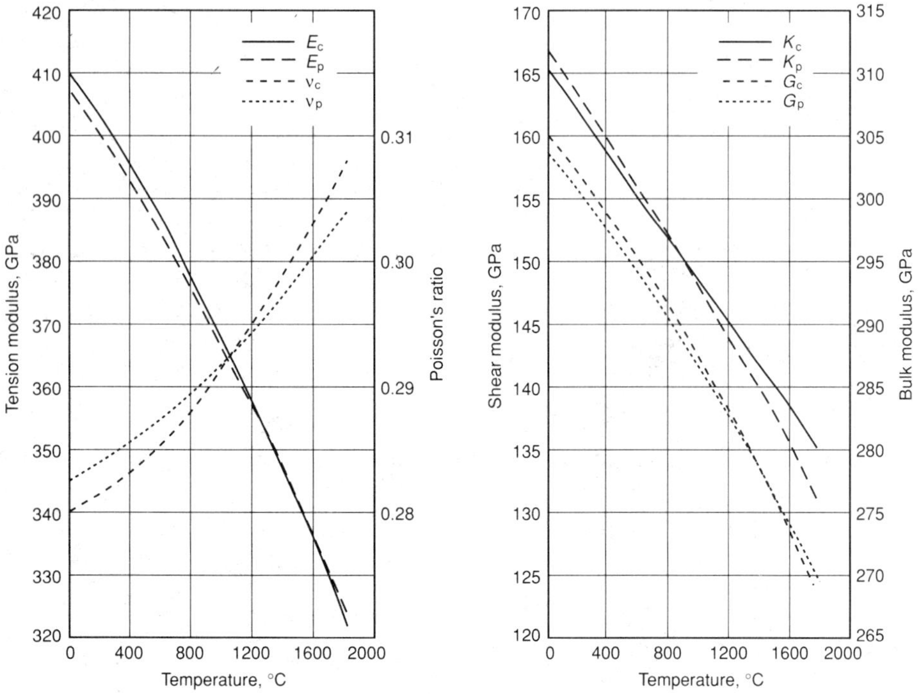

Fig. 120 Temperature dependence of the Poisson's ratio and the elastic moduli of tungsten. Poisson's ratio and elastic moduli calculated from single-crystal elastic constants (ν_c, E_c, G_c, and K_c) and from polycrystalline tungsten (ν_p, E_p, G_p, and K_p). Sources: Ref 11, 25, 26

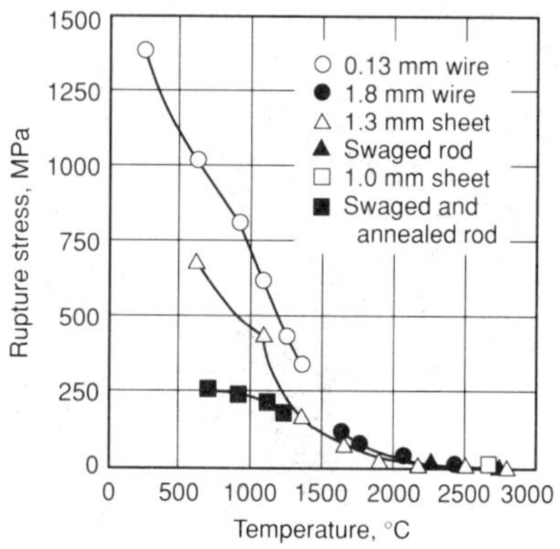

Fig. 122 Temperature dependence of the 1-h rupture strength of tungsten. Sources: Ref 11, 29 to 34

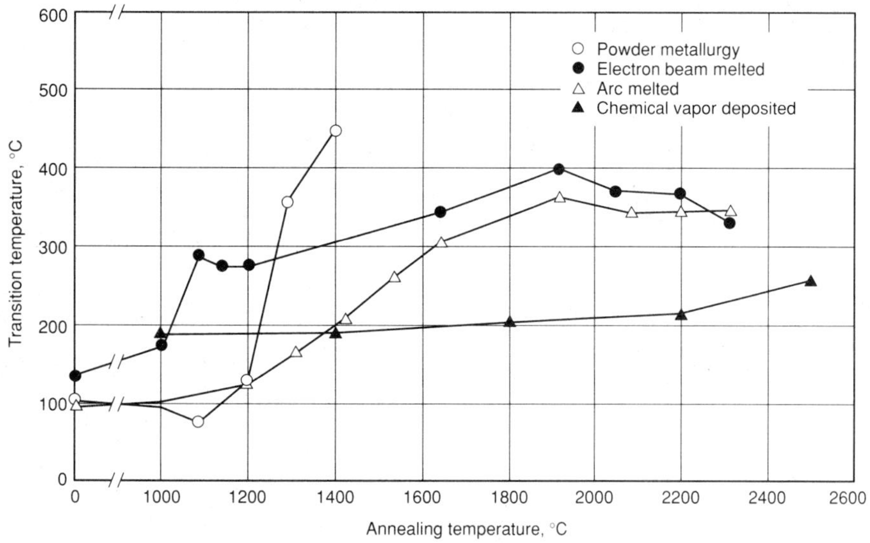

Fig. 121 Variation of ductile-to-brittle transition temperature of tungsten with annealing temperature. Ductile-to-brittle transition temperature determined by 4*t* bend for tungsten sheet. Sources: Ref 22, 27, 28

M.J. Albom, "Mechanical Properties of Wrought Tungsten," Report ASD-TDR-63-585, Vol 2, Marquardt Corporation, 1963
32. W.V. Green, *Trans. AIME*, Vol 215, 1959, p 1057-1060
33. E.C. Sutherland and W.D. Klopp, "Observations of Properties of Sintered Wrought Tungsten Sheet at Very High Temperatures," NASA-TND-1310, National Aeronautics and Space Administration, 1963
34. J.W. Pugh, *Proc. ASTM*, Vol 57, 1957, p 906-916

Uranium (U)

See the section "Properties of the Actinide Metals (Ac-Pu)" in this article.

Vanadium (V)

Revised by F.H. Perfect, Reading Alloys, Inc.

Elemental vanadium is finding use in commercial quantities in a titanium-base alloy. The nuclear industry is a major consumer of unalloyed vanadium; lesser amounts are used in the superconductor and electronics industries. U.S. consumption of vanadium in steel, usually in the form of ferrovanadium, is about 3605 tons annually (see Fig. 124). This constitutes 84% of the entire U.S. vanadium market. The titanium industry uses aluminum-vanadium, the most common nonferrous master alloy of vanadium; this use accounts for 15% of total U.S. consumption. The remaining 1% is used by the catalyst chemical industry (Ref 1).

At present, most vanadium metal is produced by electron beam remelting of aluminothermic aluminum-vanadium. The former process, calcium reduction of vanadium oxide, was used when ultrapurity was required (Ref 2). Currently, the less-pure metal is refined by a variation of the van Arkel-DeBore process (Ref 3).

The purity of the vanadium obtained by these processes depends on the initial purity of the oxide and reductant that are used. A typical nominal composition of vanadium metal produced by the electron beam remelting of aluminothermic aluminum-vanadium (a process that yields several tons of vanadium metal product) is listed below:

Element	Wt%
Aluminum	0.090
Carbon	0.017
Iron	0.025
Nitrogen	0.031
Oxygen	0.093
Silicon	0.190
Sulfur	0.006
Vanadium	99.52

Increased impurity levels, particularly of silicon, oxygen, nickel, carbon, and hydrogen, have an adverse effect on hardness and ductility.

Structure

Crystal structure. Body-centered cubic. Lattice parameters: calcium reduced, 0.30278 nm; iodide, 0.30258 nm

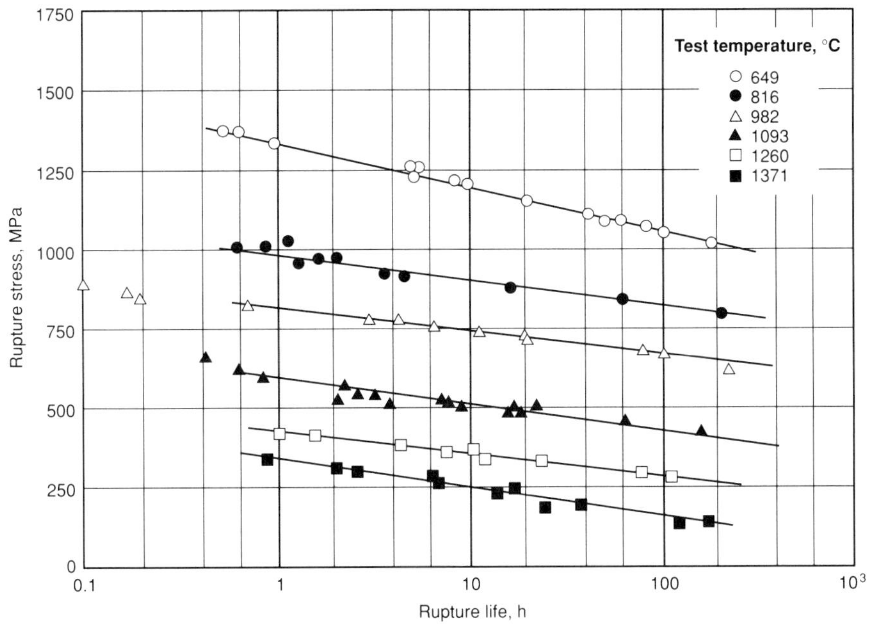

Fig. 123 Stress-rupture behavior of 0.127 mm diam as-drawn tungsten wire. Source: Ref 29

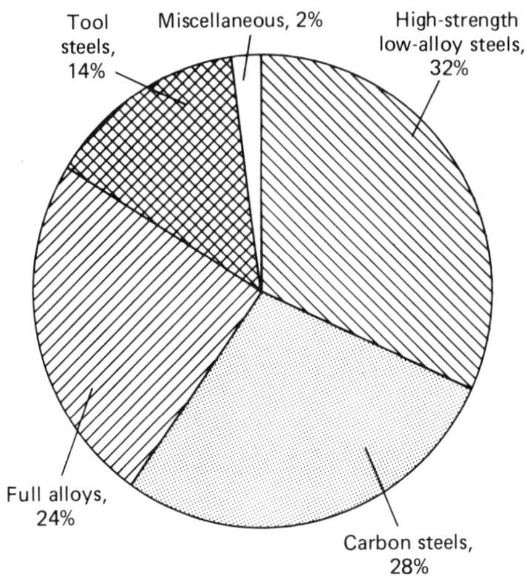

Fig. 124 Breakdown of unalloyed vanadium metal used in the ferrous industry in 1987. Miscellaneous includes vanadium used for stainless and heat-resistant steels, cast irons, superalloys, welding, and hardfacing. Source: Ref 1

Mass Characteristics

Atomic weight. 50.941
Density. 6.16 g/cm^3 at 20 °C (Ref 4); 5.55 g/cm^3 at melting point (Ref 5)

Thermal Properties

Melting point. 1910 °C (Ref 6, 7)
Boiling point. 3350 to 3400 °C (Ref 8)
Coefficient of linear thermal expansion:

Temperature range, °C	Average coefficient, μm/m · K
23–100	8.3
23–500	9.6
23–900	10.4
23–1100	10.9

Latent heat of fusion. 314 kJ/kg
Latent heat of vaporization. 9002 kJ/kg
Vapor pressure. From 1666 to 1882 K: $R \ln (P) = 1.21950 \times 10^5 T^{-1} - 5.123 \times 10^{-4} T + 38.3$, where R is gas constant, P is pressure in kPa, T is in K, and ln is logarithm to base e
Specific heat. 498 kJ/kg · K at 0 to 100 °C (Ref 9 to 11)
Thermal conductivity. 31.0 W/m · K at 100 °C (Ref 12 to 14)

Electrical Properties

Electrical resistivity. 248 to 260 nΩ · m at 20 °C (Ref 4, 9)

Nuclear Properties

Thermal neutron cross section. 4.7 ± 0.02 b; 5.06 ± 0.06 b (Ref 15); 5.1 ± 0.2 b (Ref 16)

Chemical Properties

Resistance to specific corroding agents. At room temperature, vanadium and its alloys have excellent resistance to corrosion in salt water and dilute hydrochloric acid; good corrosion resistance in sodium hydroxide solutions; and poor corrosion resistance in nitric acid solutions. Resistance to attack by liquid alkali metals is good.

Table 51 Typical mechanical properties for vanadium metal at room temperature

Condition	Tensile strength, MPa	Yield strength, MPa	Elongation in 50 mm, %	Reduction in area, %	Hardness HRA	Hardness HRB	Cold bend
Bar, 25.4 mm in diameter(a)							
Hot rolled	472	439	27.0	54.4	...	85	...
Wire, 3.9 mm in diameter(b)							
Vacuum annealed	538	463	25.0	87.5	48	...	180°
Cold drawn 80%	910.8	765	6.8	76.5	54	...	180°
Sheet, 1.9 mm thick(c)							
Vacuum annealed	536	454	20.0	53.0	...	83	180°
Cold rolled 84%	828	776.3	2.0	40.6	...	100	180°

(a) Specimen size: 12.8 mm diameter × 51 mm. (b) Specimen size: 3.9 mm diameter × 51 mm. (c) Specimen size: 1.9 mm × 12.7 mm × 51 mm

Mechanical Properties

Tensile properties. Typical at 1025 °C: tensile strength, 53 MPa; elongation, 37% in 25 mm
Hardness. 72 HB, electron beam ingot. See also Table 51.
Poisson's ratio. 0.36
Elastic modulus. Tension, 124 to 137 GPa; shear, 46.4 GPa
Creep-rupture characteristics. Limiting creep stress, 4.63 MPa for 1% deformation in 24 h at 1000 °C. Stress/density ratio at 1000 °C, 110

Fabrication Characteristics

Recrystallization temperature. 800 to 1010 °C
Standard finishes. The machining of vanadium metal is similar to that of stainless steel and presents no special problem except where the metal surface has been severely contaminated with oxygen and nitrogen.
Suitable joining methods. Satisfactory electric welding of vanadium requires adequate protection of the weld pool and heat-affected zone with a neutral gas, such as argon or helium, to prevent or minimize contamination with oxygen, hydrogen, and nitrogen. Flame welding is not practical because of the reactivity of any combustion gas mixture with molten vanadium. Vanadium can be joined by welding to ferritic and austenitic stainless steels, to titanium and titanium alloys, and to low-carbon steel.

REFERENCES

1. *Vanadium Minerals Yearbook*, U.S. Bureau of Mines, 1988, p 1005-1013
2. J.W. Marden and M.N. Rich, Vanadium, in *Ind. Eng. Chem.*, Vol 19 (No. 7),

1927, p 786-788
3. A.E. van Arkel, *Reine Metalle*, Edwards Brothers, 1943, p 222-223
4. A.B. Kinzel, *Met. Prog.*, Vol 58, 1950, p 344-B
5. C.E. Lacy *et al.*, *Trans. ASM*, Vol 48, 1956, p 579-593
6. H.K. Adenstedt *et al.*, *Trans. ASM*, Vol 44, 1952, p 990-1003
7. T.B. Massalski, Ed., *Binary Alloy Phase Diagrams*, American Society for Metals, 1986
8. C.J. Smithells, *Metals Reference Book*, 2nd ed., Interscience, 1955
9. K.K. Kelly, *Bull. US Bur. Mines*, No. 477, 1950
10. K. Clusius *et al.*, *Z. Naturforsch. A*, Vol 10A, 1955, p 930-934
11. *Bull. US Bur. Stand.*, Vol 7, 1911, p 197
12. C.A. Hampel, *Rare Metals Handbook*, Reinhold, 1961, p 634
13. H.M. Rosenburg, *Trans. R. Soc. (London) A*, Vol 247A, 1955, p 441-497
14. D.J. Hughes *et al.*, *Neutron Cross Sections*, McGraw-Hill, 1952
15. J. Emsley, *The Elements*, Clarendon Press, 1989, p 204-205
16. S.A. Bradford *et al.*, *ASM Trans. Q.*, Vol 55, 1962, p 493

SELECTED REFERENCES

- M.A. Guevich *et al.*, *Fiz. Met. Metalloved.*, Vol 4, 1957, p 112
- C.A. Hampel, Ed., *Encyclopedia of Chemical Elements*, Reinhold, 1968, p 790
- D.L. Harrod *et al.*, *Int. Met. Rev.*, 1980, p 163
- J.W. Marden *et al.*, *Ind. Eng. Chem.*, Vol 19, 1927, p 786-788
- W. Rostoker, *The Metallurgy of Vanadium*, John Wiley & Sons, 1958, p 33-45
- R.W. Thompson *et al.*, *J. Less-Common Met.*, Vol 9, 1965, p 354

Ytterbium (Yb)

See the section "Properties of the Rare Earth Metals" in this article.

Yttrium (Y)

See the section "Properties of the Rare Earth Metals" in this article.

Zinc (Zn)

Compiled by Ernest W. Horvick,
The Zinc Institute, Inc.

Revised by Dale H. Nevison,
Zinc Information Center, Ltd.

Structure

Crystal structure. Close-packed hexagonal; $a = 0.26648$ nm, $c = 0.49470$ nm; tests made with spectroscopically pure zinc (Ref 1)
Slip planes. Primary: (00.1) or (0001) at 25 °C

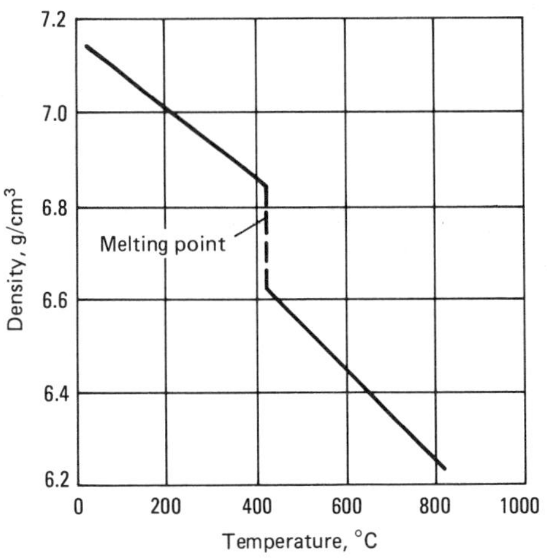

Fig. 125 Effect of temperature on the density of zinc. Source: Ref 2

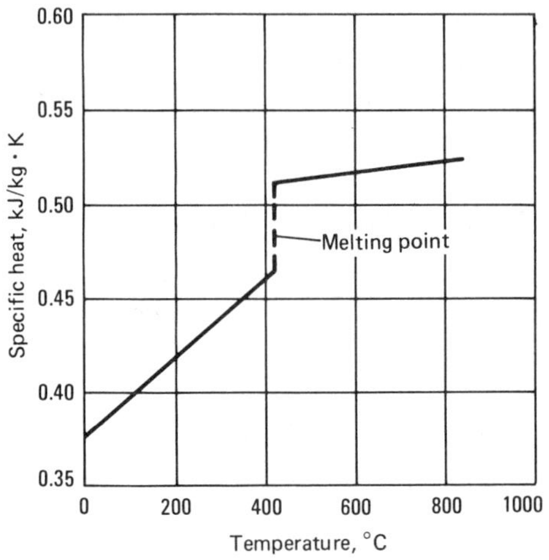

Fig. 126 Effect of temperature on the specific heat of zinc. Sources: Ref 5, 9

Twinning planes. $(10\bar{1}2)$
Cleavage planes. (00.1)
Minimum interatomic distance. 0.26594 nm (Ref 2)
Fracture type. Basal cleavage

Mass Characteristics

Atomic weight. 65.38
Density. 7.133 g/cm^3 at 25 °C; see also Fig. 125.
Volume change on freezing. 7.28% between 469 °C and 0 °C (Ref 2)

Thermal Properties

Melting point. 420 °C (Ref 3, 4)
Boiling point. 906 °C (Ref 5, 6, 7)
Coefficient of thermal expansion. Linear, single crystals at 0 to 100 °C: 15 μm/m · K along *a* axis, 61.5 μm/m · K along *c* axis. Polycrystalline solid at 20 to 250 °C: 39.7 μm/m · K; temperature effect, $L_T = L_0(1 + 35.4 \times 10^{-8} T + 1 \times 10^{-8} T^2)$ (Ref 8). Liquid at 500 to 600 °C, 60 μm/m · K (Ref 2)
Specific heat. 0.382 kJ/kg · K at 20 °C (Ref 5, 9); also see Fig. 126.
Latent heat of fusion. 100.9 kJ/kg (Ref 5)
Latent heat of vaporization. 1.782 MJ/kg (Ref 5)
Thermal conductivity. 113 W/m · K at 25 °C (Ref 3); see also Fig. 127.
Heat of combustion. −341 MJ/kg Zn (Ref 5)

Electrical Properties

Electrical conductivity. 28.27% IACS (Ref 12)
Electrical resistivity. Single crystals at 20 °C: 58.9 nΩ · m along *a* axis, 61.6 nΩ · m along *c* axis (Ref 13). Polycrystalline solid: 59.16 nΩ · m at 20 °C (Ref 3); temperature coefficient at 0 to 100 °C (Ref 3), 0.0419 nΩ · m per K; see also Fig. 128; pressure coefficient at room temperature (calculated), −25 nΩ · m per TPa at 100 kPa to 300 MPa; see also Fig. 129.
Electrochemical equivalent. 338.8 μg/C (Ref 15)
Electrolytic solution potential. −0.7618 V versus standard hydrogen electrode (Ref 3)
Hydrogen overvoltage. 0.75 V at 108 A/m^2 for metal rubbed with fine emery (Ref 16)

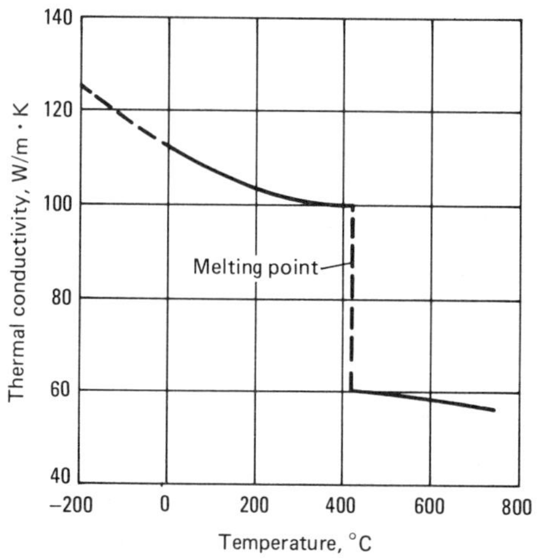

Fig. 127 Effect of temperature on the thermal conductivity of zinc. Sources: Ref 3, 10, 11

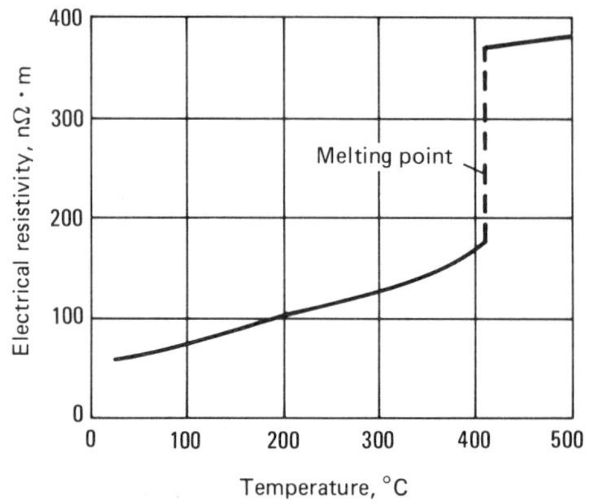

Fig. 128 Effect of temperature on the electrical resistivity of zinc (composition: 99.993% Zn, 0.005% Fe, 0.0004% Pb, 0.0018% Cd, traces of arsenic and sulfur)

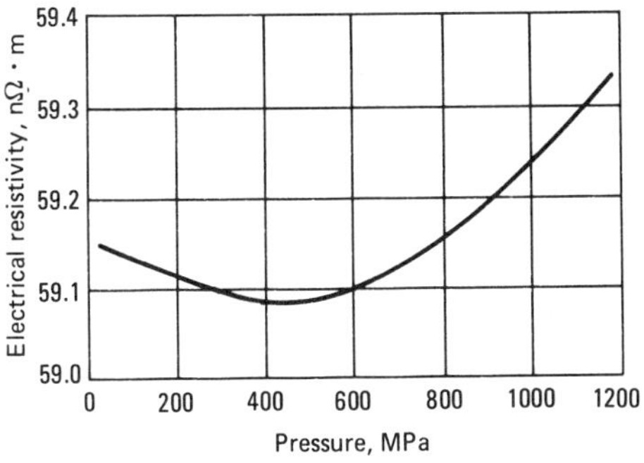

Fig. 129 Effect of pressure on the electrical resistivity of zinc at 21 °C. Source: Ref 14

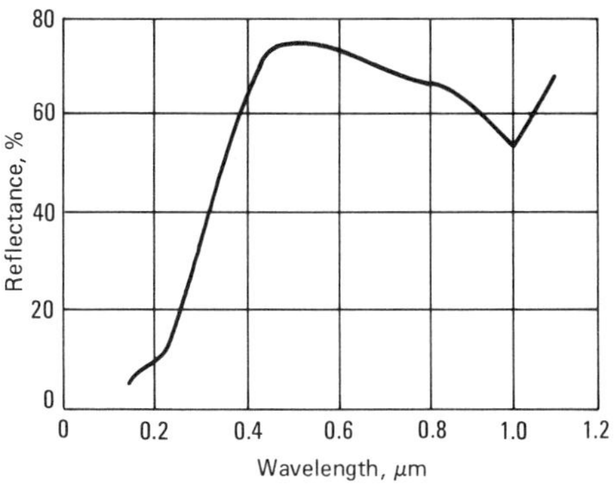

Fig. 130 Effect of wavelength on the spectral reflectance of zinc. Sources: Ref 18, 19

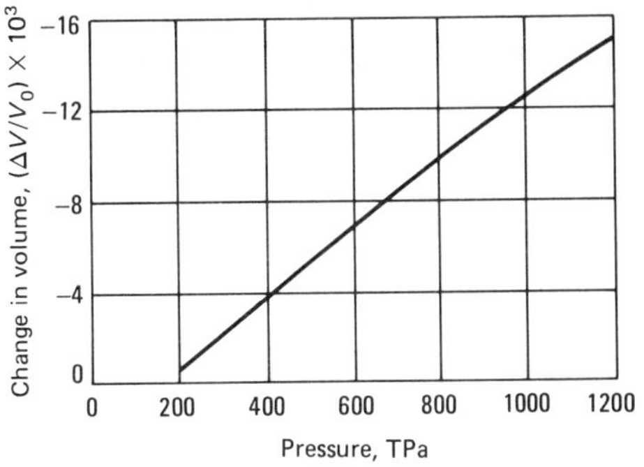

Fig. 131 Compressibility of zinc versus pressure

Temperature of superconductivity. 0.84 ± 0.05 K (Ref 17)

Magnetic Properties

Magnetic susceptibility. Volume: −123 × 10^{-6} mks (Ref 3)

Optical Properties

Color. Blue-white
Spectral reflectance. 74.7% at λ = 0.5000 μm; 69.9% at λ = 0.8000 μm; 53.3% at λ = 1.0100 μm; 70.0% at λ = 1.1300 μm (Ref 18); see also Fig. 130.
Refractive index. 1.19 in white light (λ = 0.5500 μm); ρ = 70°; 2ψ = 74°39′ (Ref 18)
Absorptive index. 3.71 in white light (λ = 0.5500 μm); ρ = 70°; 2ψ = 74°39′ (Ref 18)

Mechanical Properties

Elastic properties. Compressibility, see Fig. 131. Modulus of elasticity, 10 to 20 × 10^6 psi (actually no region of strict proportionality between stress and strain in polycrystalline zinc)
Coefficient of friction. 0.21, rolled zinc versus rolled zinc
Surface tension. Liquid, 0.755 N/m at 450 °C (Ref 3)
Velocity of sound. 3.67 km/s at room temperature (shape and size of specimen wire unknown) (Ref 20)

REFERENCES

1. E.R. Jette and F. Foote, *J. Chem. Phys.*, Vol 3, 1935, p 605
2. Erich Pelzel and Franz Sauerwald, *Z. Metallkd.*, Vol 33, 1941, p 229
3. "Zinc and Its Alloys," Circular 395, National Bureau of Standards, 6 Nov 1931
4. William Roeser and H.T. Wensel, *J. Res. Natl. Bur. Stand.*, Vol 14, 1935, p 247
5. C.G. Maier, *US Bur. Mines Bull.*, 1930, p 324
6. J. Fischer, *Z. Anorg. Allg. Chem.*, Vol 219, 1934, p 367
7. W. Leitgebel, *Z. Anorg. Allg. Chem.*, Vol 202, 1931, p 305
8. A. Schulze, *Phys. Z.*. Vol 22, 1921, p 403
9. K.K. Kelley, *US Bur. Mines Bull.*, 1934, p 371
10. L.C. Bailey, *Proc. R. Soc. (London) A*, Vol 134, 1931, p 51
11. C.C. Bidwell, *Phys. Rev.*, Series II, Vol 58, 1940, p 561
12. "Rolled Zinc," The New Jersey Zinc Company Bulletin, 1929
13. W.J. Poppe, *Phys. Rev.*, Vol 46, 1934, p 815
14. *International Critical Tables*, Vol 6, McGraw-Hill, 1933, p 136
15. H.J. Creighton and W.A. Koehler, *Electrochemistry*, John Wiley & Sons, 1944
16. C.L. Mantell, *Industrial Electrochemistry*, McGraw-Hill, 1931, p 52
17. D. Shoenberg, *Proc. Cambridge Philos. Soc.*, Vol 36 (No. 1), 1940, p 84
18. J. Bor, A. Hobson, and C. Wood, *Proc. Phys. Soc.*, Vol 51, 1939, p 932
19. G.B. Sabine, *Phys. Rev.*, Series II, Vol 55, 1939, p 1064
20. G. Gerosa, *Atti R. Accad. Naz. Lencei, Tendiconti*, Vol 4 (No. IV), 1888, p 127

Zirconium (Zr)

Compiled by R.T. Webster,
Teledyne Wah Chang Albany

Zirconium is principally used as a corrosion-resistant cladding for uranium in nuclear reactors. It is also used as a corrosion-resistant structural material in chemical processing equipment such as pressure vessels, heat exchangers, pumps, and valves.

Zirconium is nontoxic and, consequently, does not require serious limitations on its use because of health hazards (Ref 1).

Zirconium is pyrophoric because of its heat-producing reaction with oxidizing elements such as oxygen. Large pieces of sheet, plate, bar, tube, and ingot can be heated to high temperatures without excessive oxidation or burning, but small pieces with a high ratio of surface area to mass, such as machine chips and turnings, are easily ignited and burn at extremely high temperatures. It is recommended that large accumulations of chips and other finely divided material be avoided. Also, in storing the chips and turnings, care should be taken to place the material in nonflammable containers and isolated areas. One storage method that works quite well is to keep the material covered with water in the containers and use oil on the water to keep it from evaporating.

If a fire accidentally starts in zirconium, do not attempt to put it out with water or ordinary fire extinguishers, but use dry sand, powdered graphite, or commercially available Met-L-X powder. Large quantities of water can be used to control and extinguish fires in other flammables in the vicinity of a zirconium fire.

When zirconium is exposed to highly corrosive attack by a concentrated acid such as red fuming nitric acid, it is possible that in time the exposed surfaces of the zirconium will be converted to a finely divided powder. This powder can ignite, possibly with explosive force. If the proper balance between water vapor and nitrogen dioxide above the liquid is maintained, this hazard can be eliminated.

Structure

Crystal structure. α phase, close-packed hexagonal: at 20 °C, a = 0.323115 nm, c = 0.51477 nm, c/a = 1.5931. β phase, body-centered cubic: at 862 °C, a = 0.36090 nm (Ref 2)
Slip planes. $\{10\bar{1}0\}$ at 20 °C
Twinning planes. $\{10\bar{1}2\}$ $\{11\bar{2}1\}$ $\{11\bar{2}2\}$ $\{11\bar{2}3\}$ at 20 °C
Cleavage planes. α phase, {1000}. β phase, {100}
Minimum interatomic distance. α phase at 20 °C: d_1 = 0.316 nm, d_2 = 0.312 nm. β phase, d_1 = 0.322 nm
Microstructure. Polishing and etching zirconium to observe the microstructure is not difficult when the proper techniques are used. Due to the tendency of zirconium to smear during polishing, an attack-polish technique is used. A solution of alumina is used in conjunction with a dilute acid solution to attack the sample surface chemically at about the same rate that is being removed

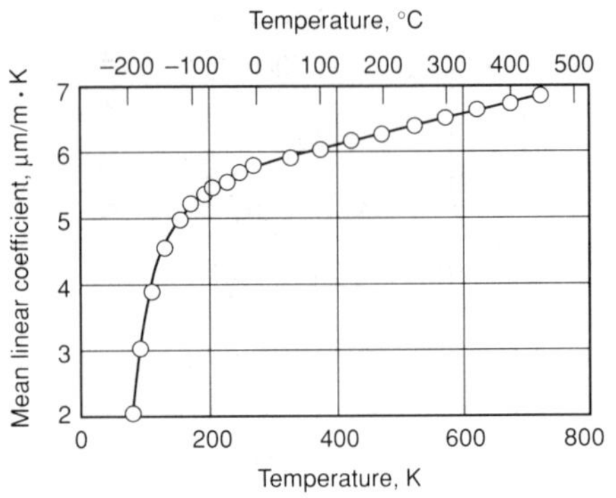

Fig. 132 Temperature dependence of the mean coefficient of linear thermal expansion for zirconium. Source: Ref 3

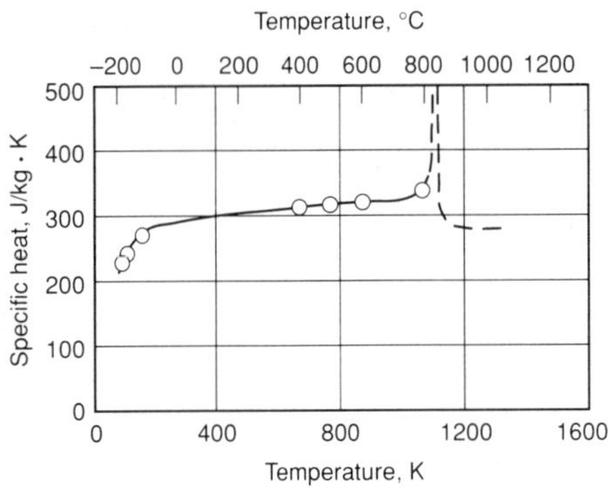

Fig. 133 Temperature dependence of the specific heat of zirconium. Source: Ref 3

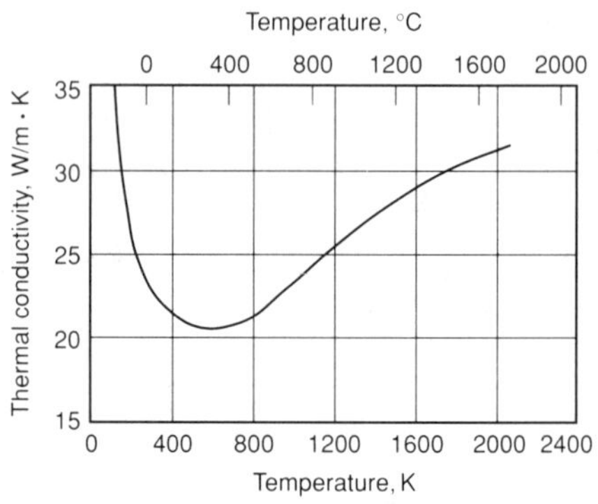

Fig. 134 Temperature dependence of the thermal conductivity of zirconium. Source: Ref 4

by abrasion. The right combination of wheel speed, hand pressure, acid solution, and abrasive will result in a true undisturbed microstructure.

The sample is first sanded on abrasive cloth down to a 3/0 grit. It is then polished on a wheel using nylon cloth over Metcloth (the nylon is fairly acid resistant), according to the following procedure:

1. Apply abrasive solution (5 g of 3 µm alumina/150 mL H_2O) to wheel
2. Apply acid solution (250 mL H_2O/22 mL HNO_3/3 mL HF) to wheel
3. Spin for several seconds (approximately 1000 rev/min) to allow an even film to form on the cloth
4. Reduce speed to approximately 550 rev/min and polish sample with light-to-moderate pressure between the wheel and the sample. When the sample is removed from the wheel, wash immediately with H_2O (squirt bottle works well) or overetching will occur. Repeat above polishing as necessary to obtain a surface with no disturbed metal

To maintain the proper acid concentration on the wheel, the polishing cloth should be thoroughly rinsed with water after using for 1 to 2 min. New alumina and acid should be applied to continue polishing.

For etching, the sample surface is swabbed with acid solution (22 mL H_2O/22 mL HNO_3/3 mL HF) for 5 to 10 s, then rinsed in water to prevent overetching.

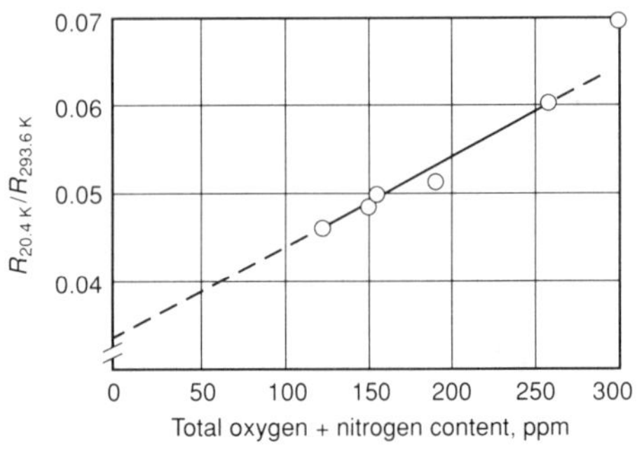

Fig. 135 Dependence of the electrical resistivity of zirconium on interstitial impurities. Source: Ref 5

Mass Characteristics

Atomic weight. 91.22
Density. α phase, 6.505 g/cm³ (low in hafnium) to 6.574 g/cm³ (high in hafnium); β phase at 979 °C, 6.046 g/cm³ (high-purity crystal bar zirconium)

Thermal Properties

Melting point. 1852 ± 2 °C
Boiling point. 4377 °C
Phase transformation temperature. 862 ± 5 °C
Coefficient of thermal expansion. Linear: 5.85 µm/m · K at 20 °C for heterogeneously oriented polycrystals; temperature dependence, see Fig. 132; along crystal axes, 5.65 µm/m · K perpendicular to *c* axis and 6.96 µm/m · K parallel to *c* axis. Volumetric: 17.68 µm/m · K for 0.005 at.% Hf; 17.47 µm/m · K for 1.2 at.% Hf
Specific heat. See Fig. 133.
Enthalpy. $H_T - H_{298}$:

°C	kJ/kg · K
100	23.1
200	52.2
300	80.3
400	112.6
500	146.4
600	182.0

Entropy. 426.1 J/kg · K at 25 °C

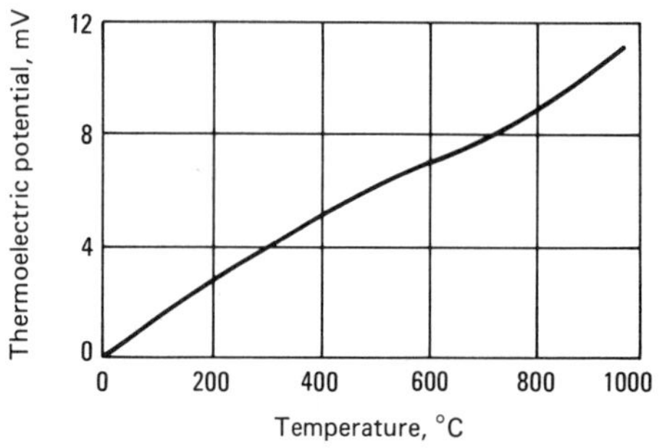

Fig. 136 Thermoelectric force of a zirconium-platinum thermocouple. Source: Ref 3

Latent heat of fusion. 25 kJ/kg
Latent heat of phase transformation. 42.2 kJ/kg
Latent heat of vaporization. 6520 kJ/kg
Heat of combustion. Heat of formation of ZrO_2, 5940 kJ/kg Zr
Thermal conductivity. From Ref 3: at 25 °C, 21.1 W/m · K; 100 °C, 20.4 W/m · K; 200 °C, 19.6 W/m · K. Temperature dependence: $k = 30.8\,(\sigma - 0.000327)\,T + 3.81$, where k is thermal conductivity in W/m · K, σ is electrical conductivity in reciprocal nΩ · m, and T is temperature in K (see also Fig. 134).
Vapor pressure:

°C	Pa
1574	1.013×10^{-5}
1690	1.013×10^{-4}
1822	1.013×10^{-3}
1976	1.013×10^{-2}
2156	1.013×10^{-1}
2367	1.013
2620	1.013×10
2926	1.013×10^{2}
3304	1.013×10^{3}
3783	1.013×10^{4}
4409	1.013×10^{5}

Diffusion coefficient. At 800 °C: 2×10^{-3} mm²/s for hydrogen; 2×10^{-7} mm²/s for oxygen; 1×10^{-7} mm²/s for nitrogen

Electrical Properties

Electrical conductivity. Volumetric, 4.1% IACS
Electrical resistivity. 450 nΩ · m; temperature coefficient, $44 \pm 1 \times 10^{-4}$ per K at 0 to 200 °C; pressure coefficient, average reduction of 0.2% in resistance for each 980 MPa; dependence on plastic deformation, 2 to 5% increase with cold reductions of 64 to 96%; dependence on impurities, see Fig. 135.
Thermoelectric potential. See Fig. 136.
Electrochemical equivalent. 0.2363 mg/C = coulomb
Ionization potential. 34.33 eV
Hydrogen overvoltage. 0.83 V at 10 A/m²
Hall coefficient. 0.118 nV · m/A · T at 77 K (−196 °C), 0.126 nV · m/A · T at 300 K (27 °C) (Ref 6)
Temperature of superconductivity. 0.63 K (−272.52 °C)

Table 52 Optical and electronic properties of zirconium

Temperature, K	Temperature, °C	Brightness temperature (λ = 652 nm), K	Brightness temperature (λ = 652 nm), °C	Total radiation, Kw/m²	Electron emission, A/m²
1000	727	967	694	16.8	...
1100	827	1059	786	22.7	...
1200	927	1151	878	30.3	...
1300	1027	1242	969	40.6	...
1400	1127	1332	1059	54.0	...
1500	1227	1423	1150	72.0	0.2
1600	1327	1513	1240	100	1.8
1700	1427	1602	1329	134	13
1800	1527	1691	1418	175	84
1900	1627	1779	1506	222	405
2000	1727	1866	1593	280	1600
2100	1827	1952	1679	345	5200
2130	1857	1980	1707	365	7200

Source: Ref 4

Electron emission. See Table 52.
Work function. 0.656 aJ

Magnetic Properties

Magnetic susceptibility. Volume: 16 × 10^{-6} mks units at 25 °C, 19.2 mks units at 700 °C, 24 mks units at 860 °C

Optical Properties

Brightness temperature. See Table 52.
Total radiation. See Table 52.

Nuclear Properties

Stable isotopes. See Table 53.
Unstable isotopes. See Table 54.
Effect of irradiation on properties. See Fig. 137 and 138.

Chemical Properties

Resistance to specific agents. Zirconium is able to maintain a bright surface permanently at room temperature. It is also able to form a stable oxide of high melting point and (possibly) to form a continuous oxide surface. However, its oxidation resistance in air or oxygen at moderately high temperatures is poor (Fig. 139).

Mechanical Properties

Tensile properties. See Fig. 137 and 138.
Poisson's ratio. 0.35 at room temperature
Elastic properties. Tension, 99.284 GPa. Pressure dependency of specific volume (compressibility) (Ref 8):

Pressure, MPa	Relative specific volume
0.10	1.000
2940	0.967
3920	0.956
5880	0.937
6860	0.929
7850	0.922
8830	0.916
9810	0.910

Specific damping capacity. 17 × 10^{-4} at 25 °C, decreasing to 6.2 × 10^{-4} at 260 °C with a sharp increase to 1 × 10^{-2} at 610 °C

Table 53 Stable isotopes of zirconium

Isotope	Abundance, %	Thermal neutron absorption cross section, b(a)
^{90}Zr	51.5	0.1
^{91}Zr	11.2	1.5
^{92}Zr	17.1	0.2
^{94}Zr	17.4	0.07
^{96}Zr	2.8	0.05

(a) b, barns. Source: Ref 3

Coefficient of friction. For zirconium sliding on zirconium, 0.42 at 20 °C (Ref 9)
Velocity of sound. 4.62 km/s

REFERENCES

1. H. Loevenstein and H.L. Gilbert, "Zirconium: A Review and Summary of Published Data," Technical Information Service Extension, Oak Ridge National Laboratory, Oct 1958
2. A. Taylor and Brenda J. Kagle, *Crystallographic Data on Metal and Alloy Structures*, Dover, 1963
3. G.L. Miller, *Zirconium*, Academic Press, 1957
4. C.Y. Ho, R.W. Powell, and P.E. Liley, "Thermal Conductivity of Selected Materials," NSRDS-NBS16, National Bureau of Standards, Feb 1968
5. D.L. Douglass, The Metallurgy of Zirconium, *At. Energy Rev. Suppl.*, 1971
6. Ted G. Berlincourt, Hall Effect, Resistivity, and Magnetoresistivity of Th, U, Zr, Ti, and Nb, *Phys. Rev.*, Vol 114 (No. 4), 15 May 1959
7. B. Lustman and K. Kerze, Jr., *The Metallurgy of Zirconium*, McGraw-Hill, 1955

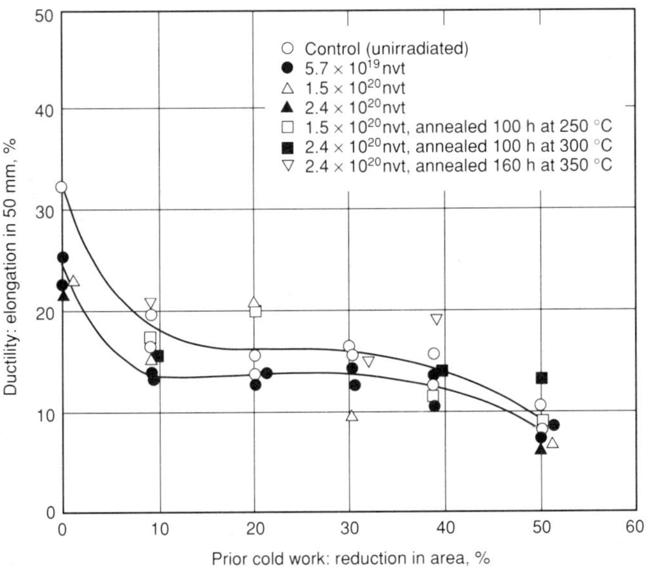

Fig. 137 Effect of irradiation and subsequent annealing on the ductility of sponge zirconium. Source: Ref 7

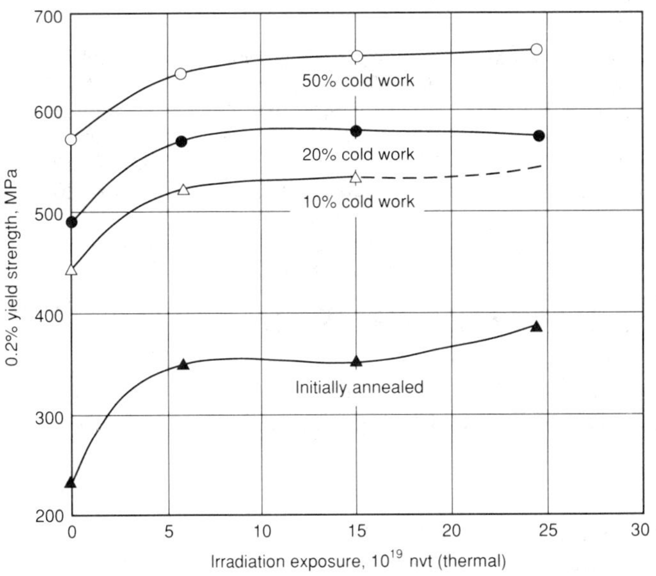

Fig. 138 Effect of irradiation on the yield strength of sponge zirconium. Exposure temperature was 50 to 60 °C. Source: Ref 7

Table 54 Unstable isotopes of zirconium

Isotope	Half-life	Mode of decay and radiation(a)	Energy of radiation, MeV	Correctness of mass number	Existence of element
^{86}Zr	17 h	K		Probable	Certain
^{87}Zr	1.6 h	β^-	2.10	Certain	Certain
		γ	0.6, 0.3		
^{88}Zr	85 days	K, γ	0.41	Probable	Certain
^{89}Zr(b)	4.4 min	IT	0.59	Certain	Certain
		β^-	0.9, 2.4		
		γ	1.5		
^{89}Zr(b)	78 h	K	0.91	Certain	Certain
		β^+	0.92	Radiation emitted by short-lived daughter	
		γ			
^{93}Zr(c)	~5×10^6 years	β^-	0.06	Probable	Certain
^{95}Zr(c)	65 days	β^-	0.39, 1.0; e^-(d)	Certain	Certain
		γ	0.73, 0.92		
^{97}Zr(c)	17 h	β^-	1.91	Certain	Certain
				Radiation emitted by	
		γ	0.75	short-lived daughter	

(a) K, K-electron capture; IT, isomeric transition. (b) Nuclide 89 exists in two isomeric states. (c) The nuclides 93, 95, and 97 have been identified as products of fusion of ^{235}U induced by slow neutrons. (d) e^-, internal conversion electron. Source: Ref 1

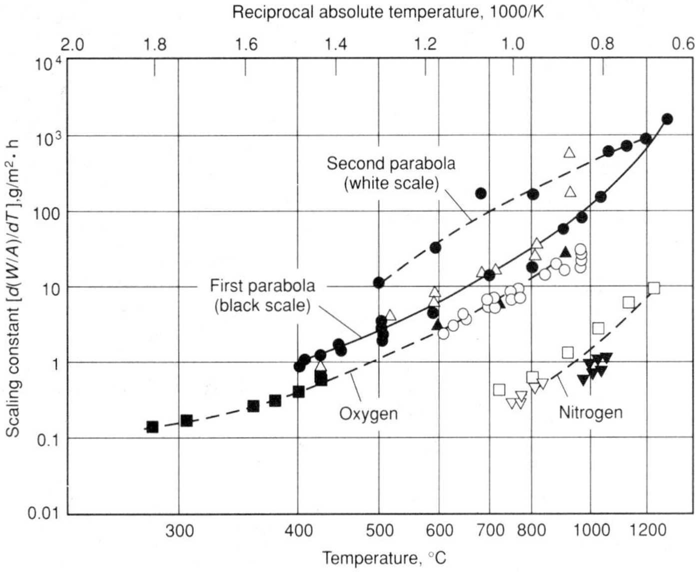

Fig. 139 Scaling rate of zirconium at elevated temperature. Source: Ref 7

8. J.M. Wash, M.H. Rice, R.G. McQueen, and F.L. Yarger, Shock-Wave Compressions of Twenty-Seven Metals, Equation of State of Metals, *Phys. Rev.*, Vol 108 (No. 2), 15 Oct 1957
9. D.H. Buckly and R.L. Johnson, "Relation of Lattice Parameters to Friction Characteristics of Beryllium, Hafnium, Zirconium, and Other Hexagonal Metals in Vacuum," NASA TND-2670, National Aeronautics and Space Administration, March 1965

Properties of the Rare Earth Metals

Compiled by K.A. Gschneidner, Jr. and B.J. Beaudry, Ames Laboratory, U.S. Department of Energy, Iowa State University

Cerium (Ce)

Compiled by K.A. Gschneidner, Jr. and B.J. Beaudry, Ames Laboratory, U.S. Department of Energy, Iowa State University

Cerium, as a component (~50%) of mischmetal, is used as an alloying additive to ferrous alloys to scavenge impurities such as sulfur and oxygen, to nodulize cast iron, and to strengthen magnesium. It improves the high-temperature oxidation resistance of superalloys. It is also used in glass-polishing compounds, petroleum cracking catalysts, catalytic converters, lighter flints, glass-decolorizing agents, carbon arc lights, ceramic capacitors, $CeCo_5$ permanent magnets, and pyrophoric ordnance devices.

Cerium readily oxidizes at room temperature in air. It should be stored in vacuum or an inert atmosphere; storage in oil is not recommended. Turnings can be ignited easily and burn white hot. Finely divided cerium should not be handled in air.

Structure

Crystal structure. α phase, face-centered cubic, $Fm3m$ O_5^h; a = 0.485 nm at 77 K. β phase, double close-packed hexagonal, $P6_3/mmc$ D_4^{6h}; a = 0.36810 nm, c = 1.1857 nm at 24 °C. γ phase, face-centered cubic, $Fm3m$ O_5^h; a = 0.51610 nm at 24 °C. δ phase, body-centered cubic, $Im3m$ O_h; a = 0.412 nm at 757 °C

Minimum interatomic distance. γ phase, 0.172 nm at 77 K; β phase, r_a = 0.18405 nm, r_c = 0.18237 nm, radius CN_{12} = 0.18321 nm at 24 °C, where CN_{12} indicates a coordination number of 12; γ phase, 0.18247 nm at 24 °C

Mass Characteristics

Atomic weight. 140.115

Density. α phase, 8.16 g/cm^3 at 77 K; β phase, 6.689 g/cm^3 at 24 °C; γ phase, 6.770 g/cm^3 at 24 °C; δ phase, 6.65 g/cm^3 at 768 °C; liquid, 6.68 g/cm^3 at 800 °C

Volume change on freezing. 1.1% expansion

Volume change on phase transformation. γ to α phase, 16.0% volume contraction on cooling at 110 K; γ to β phase, 1.2% volume expansion on cooling at 273 K; δ to γ phase, 0.3% volume expansion on cooling

Thermal Properties

Melting point. 798 °C

Boiling point. 3443 °C

Phase transformation temperature. γ to δ phase: 726 °C; γ to β phase: M_s = 237 to 278 K, M_f = ?. β to γ phase: A_s = 373 to 451 K, A_f = 420 to >451 K. γ to α phase: M_s = 89 to 116 K, $M_f \cong$ 4.2 K. α to γ + β phases: A_s = 158 to 180 K, A_f = 190 to 210 K. β to α phase: M_s = 45 K, M_f = 15K; α to β phase: A_s = 125 K, A_f = 200 K

Coefficient of thermal expansion. At 24 °C. Linear: γ phase, 6.3 μm/m · K. Linear along crystal axes: γ phase, 6.3 μm/m · K along a axis. Volumetric: γ phase, 18.9 × 10^{-6} per K

Specific heat. 0.192 kJ/kg · K at 25 °C

Entropy. At 25 °C, 513.9 J/kg · K

Latent heat of fusion. 38.97 kJ/kg

Latent heat of transformation. γ to δ phase, 21.34 kJ/kg

Latent heat of vaporization. 3.016 MJ/kg at 25 °C

Heat of combustion. For cubic CeO_2 at 25 °C: ΔH_c° = 777 MJ/kg Ce; ΔG_f° = −732

MJ/kg Ce
Recrystallization temperature. About 325 °C
Thermal conductivity. γ phase, 11.3 W/m · K at 25 °C
Vapor pressure. 0.001 Pa at 1290 °C; 0.101 Pa at 1554 °C; 10.1 Pa at 1926 °C; 1013 Pa at 2487 °C

Electrical Properties

Electrical resistivity. β phase, 828 nΩ · m at 25 °C; 41 nΩ · m at 2 K. γ phase, 744 nΩ · m at 25 °C. Liquid, 1300 nΩ · m at 800 °C
Ionization potential. Ce(I), 5.466 eV; Ce(II), 10.85 eV; Ce(III), 20.198 eV; Ce(IV), 36.758 eV
Hall coefficient. +0.181 nV · m/A · T at 25 °C
Temperature of superconductivity. Bulk cerium not superconducting down to 0.25 K at atmospheric pressure. α phase becomes superconducting at 0.022 K at 2.2 GPa.

Magnetic Properties

Magnetic susceptibility. Volume, mks units. β phase: 1.50×10^{-3} at 25 °C; obeys Curie-Weiss law from 50 to 320 K with an effective moment of 2.61 μ_B (Bohr magnetons) and θ = −41 K, where θ is the Curie temperature. γ phase: 1.38×10^{-3} at 25 °C; obeys Curie-Weiss law above 0 °C with an effective moment of 2.52 μ_B and θ = −50 K.
Magnetic transformation temperature. β phase: Néel temperatures at 12.5 K (cubic sites) and 13.7 K (hexagonal sites)

Optical Properties

Color. Metallic silver
Spectral hemispherical emittance. 32.2% for λ = 645 nm at 877 to 1547 °C

Nuclear Properties

Thermal neutron cross section. 0.7 b

Chemical Properties

General corrosion behavior. Cerium oxidizes readily in air at room temperature. Oxidation rates increase with temperature and humidity. Interstitial impurities increase the oxidation rate, whereas some solid-solution additives, such as scandium, decrease the oxidation rate. Hydrogen will react with cerium at room temperature.
Resistance to specific corroding agents. Cerium reacts vigorously with dilute acids. Cold water slowly attacks cerium; hot water reacts faster. The presence of the fluoride ion retards acid attack by the formation of CeF_3 on the surface of the metal.

Mechanical Properties

Tensile properties. β phase: tensile strength, 138 MPa; yield strength, 86 MPa; reduction in area, 24% at 24 °C. γ phase: tensile strength, 117 MPa; yield strength, 28 MPa; elongation, 22%; reduction in area, 30% at 24 °C
Hardness. 22 HV
Poisson's ratio. γ, 0.24
Strain-hardening exponent. 0.3
Elastic modulus. γ phase at 27 °C: tension, 33.6 GPa; shear, 13.5 GPa; bulk, 21.5 GPa
Kinematic liquid viscosity. 0.479 mm²/s at 804 °C
Liquid surface tension. 0.706 N/m at 804 °C

Dysprosium (Dy)

Compiled by K.A. Gschneidner, Jr. and B.J. Beaudry, Ames Laboratory, U.S. Department of Energy, Iowa State University

Dysprosium is used as a control rod in nuclear reactors; it is also used in magnetostrictive materials ($Tb_{0.3}Dy_{0.7}Fe_2$), Nd-Fe-B permanent magnets (to raise the magnetic ordering temperature), phosphors, catalysts, and garnet microwave devices. In addition, it is used to measure neutron fluxes. Dysprosium will remain shiny in air at room temperature. However, turnings can be ignited and will burn white hot. Finely divided metal should not be handled in air.

Structure

Crystal structure. α′ phase, orthorhombic, *Cmcm* D_{17}^{2h}; *a* = 0.3595 nm; *b* = 0.6184 nm; *c* = 0.5678 nm at 86 K. α phase, close-packed hexagonal, $P6_3/mmc$ D_4^{6h}; *a* = 0.35915 nm; *c* = 0.56501 nm at 24 °C. β phase, body-centered cubic, *Im*3*m* O_9^h; *a* = 0.403 nm at 1381 °C
Slip planes. At 24 °C: primary {$10\bar{1}0$}, secondary {0002}
Twinning planes. At 24 °C: primary {$11\bar{2}1$}, secondary {$10\bar{1}2$}
Minimum interatomic distance. r_a = 0.17958 nm; r_c = 0.17522 nm; radius CN_{12} = 0.17740 nm at 24 °C

Mass Characteristics

Atomic weight. 162.50
Density. α phase, 8.551 g/cm³ at 24 °C; β phase, 8.23 g/cm³ at 1381 °C; liquid, 8.2 g/cm³ at 1415 °C
Volume change on freezing. 4.5% contraction

Thermal Properties

Melting point. 1412 °C
Boiling point. 2567 °C
Phase transformation temperature. α to α′ phase, 86 K; α to β phase, 1381 °C
Coefficient of thermal expansion. At 24 °C. Linear: 9.9 μm/m · K. Linear along crystal axes: 7.1 μm/m · K along *a* axis, 15.6 μm/m · K along *c* axis. Volumetric: 29.8×10^{-6} per K
Specific heat. 0.1705 kJ/kg · K at 25 °C
Entropy. 465.2 J/kg · K at 25 °C
Latent heat of fusion. 68.1 kJ/kg
Latent heat of phase transformation. α to β phase, 25.62 kJ/kg
Latent heat of vaporization. 1.787 MJ/kg at 25 °C
Heat of combustion. For cubic Dy_2O_3 at 25 °C: ΔH_c° = −5.73 MJ/kg Dy; ΔG_f° = −5.46 MJ/kg Dy
Recrystallization temperature. About 550 °C
Thermal conductivity. 10.7 W/m · K at 25 °C
Vapor pressure. 0.001 Pa at 804 °C; 0.101 Pa at 988 °C; 10.1 Pa at 1252 °C; 1013 Pa at 1685 °C

Electrical Properties

Electrical resistivity. 926 nΩ · m at 25 °C; 24 nΩ · m at 4 K. Along crystal axes at 25 °C: 1110 nΩ · m along *a* axis, 766 nΩ · m along *c* axis. Liquid: 2100 nΩ · m at 1414 °C
Ionization potentials. Dy(I), 5.927 eV; Dy(II), 11.67 eV; Dy(III), 22.8 eV; Dy(IV), 41.47 eV
Hall coefficient. Along crystal axes at 20 °C: −0.03 nV · m/A · T along *b* axis; −0.37 nV · m/A · T along *c* axis
Temperature of superconductivity. Bulk dysprosium is not superconducting down to 0.45 K at atmospheric pressure.

Magnetic Properties

Magnetic susceptibility. Volume (mks units) at 27 °C: χ_a = 0.0717 and χ_c = 0.0511; obeys Curie-Weiss law above 250 K with an effective moment of 10.83 μ_B, θ_a = 169 K and θ_c = 121 K
Saturation magnetization. 3.71 T at 0 K along ⟨$11\bar{2}0$⟩
Magnetic transformation temperature. Curie temperature 89 K, Néel temperature 179 K

Optical Properties

Color. Metallic silver
Spectral hemispherical emittance. Liquid, 29.7% for λ = 645 nm at 1413 to 1437 °C

Nuclear Properties

Thermal neutron cross section. 1100 b

Chemical Properties

General corrosion behavior. Remains shiny in air at room temperature. The rate of oxidation is slow even at 1000 °C due to the formation of a dark, tightly adhering oxide on the surface. The presence of water vapor increases the rate of oxidation.
Resistance to specific corroding agents. Dysprosium does not react with cold or hot water, but it will react vigorously with dilute acids. It is attacked slowly by concentrated sulfuric acid. The presence of the fluoride ion retards acid attack due to the formation of DyF_3 on the surface of the metal.

Mechanical Properties

Tensile properties. Tensile strength, 139 MPa; yield strength, 43 MPa; elongation, 30%; reduction in area, 30%
Hardness. 44 HV
Poisson's ratio. 0.237
Strain-hardening exponent. 0.35

Elastic modulus. At 27 °C: tension, 61.4 GPa; shear, 24.7 GPa; bulk, 40.5 GPa
Elastic constants along crystal axes. At 27 °C: c_{11} = 73.0 GPa; c_{12} = 25.3 GPa; c_{13} = 22.1 GPa; c_{33} = 78.0 GPa; c_{44} = 24.0 GPa
Liquid surface tension. 0.648 N/m at 1415 °C

Europium (Eu)

Compiled by K.A. Gschneidner, Jr.
and B.J. Beaudry, Ames Laboratory,
U.S. Department of Energy,
Iowa State University

Europium is used as control rods in nuclear reactors and as phosphors, especially those that make up the red component in color television screens. Europium oxidizes rapidly in air at room temperature and should be handled and stored under an inert atmosphere; storage in oil is not recommended. Finely divided europium can ignite spontaneously in air.

Structure

Crystal structure. Body-centered cubic: *Im3m* O_9^h; a_0 = 0.45827 nm at 24 °C
Minimum interatomic distance. 0.19844 nm at 24 °C; radius CN_{12} = 0.20418 nm

Mass Characteristics

Atomic weight. 151.96
Density. 5.244 g/cm^3 at 24 °C; liquid, 4.87 g/cm^3 at 825 °C
Volume change on freezing. 4.8% contraction

Thermal Properties

Melting point. 822 °C
Boiling point. 1529 °C
Coefficient of thermal expansion. At 24 °C. Linear: 35.0 μm/m · K. Volumetric: 105 × 10^{-6} per K
Specific heat. 0.1823 kJ/kg · K
Entropy. At 25 °C: 512.0 J/kg · K
Latent heat of fusion. 60.6 kJ/kg
Latent heat of vaporization. 1.154 MJ/kg at 25 °C
Latent heat of combustion. For monoclinic Eu_2O_3 at 25 °C: ΔH_c° = −5.43 MJ/kg Eu; ΔG_f° = −5.12 MJ/kg Eu
Recrystallization temperature. 300 °C
Thermal conductivity. 13.9 W/m · K at 25 °C (estimated)
Vapor pressure. 0.001 Pa at 399 °C; 0.101 Pa at 515 °C; 10.1 Pa at 685 °C; 1013 Pa at 964 °C

Electrical Properties

Electrical resistivity. 900 nΩ · m at 25 °C; 6 nΩ · m at 4 K; liquid, 2420 nΩ · m at 822 °C
Ionization potential. Eu(I): 5.666 eV; Eu(II): 11.241 eV; Eu(III): 24.92 eV; Eu(IV): 42.6 eV
Hall coefficient. +2.44 nV · m/A · T
Temperature of superconductivity. Bulk europium is not superconducting down to 0.03 K at atmospheric pressure.

Magnetic Properties

Magnetic susceptibility. Volume: 0.0134 mks at 25 °C; obeys Curie-Weiss law above 100 K with an effective moment of 8.48 μ_B
Saturation magnetization. >6T at 4 K
Magnetic transformation temperature. Néel temperature, 90.4 K

Optical Properties

Color. Metallic silver when free from surface contamination

Nuclear Properties

Thermal neutron cross section. 4300 b

Chemical Properties

General corrosion behavior. Europium is the most air reactive of the rare earth metals, especially in moist air. In dry air, a dark coating is formed that retards oxidation. Hydrogen reacts with europium at about 250 °C.
Resistance to specific corroding agents. Europium reacts vigorously with cold water and dilute acids.

Mechanical Properties

Hardness. 17 HV
Poisson's ratio. 0.152
Elastic modulus. Tension, 18.2 GPa; shear, 7.9 GPa; bulk, 8.3 GPa at 27 °C
Liquid surface tension. 0.264 N/m at 825 °C

Erbium (Er)

Compiled by K.A. Gschneidner, Jr.
and B.J. Beaudry, Ames Laboratory,
U.S. Department of Energy,
Iowa State University

Erbium is used in lasers, phosphors, garnet microwave devices, ferrite bubble devices, and catalysts. Erbium will remain shiny in air at room temperature. However, turnings can be ignited and will burn white hot. Finely divided erbium metal should not be handled in air.

Structure

Crystal structure. Close-packed hexagonal: *P6₃/mmc* D_4^{6h}; *a* = 0.35592 nm; *c* = 0.55850 nm at 24 °C
Slip planes. Primary {10$\bar{1}$0}, secondary {0002} at 24 °C
Twinning planes. Primary {11$\bar{2}$1}, secondary {10$\bar{1}$2} at 24 °C
Minimum interatomic distance. r_a = 0.17796 nm; r_c = 0.17335 nm; radius CN_{12} = 0.17566 nm at 24 °C

Mass Characteristics

Atomic weight. 167.26
Density. 9.066 g/cm^3 at 24 °C; liquid, 8.6 g/cm^3 at 1530 °C
Volume change on freezing. 9.0% contraction

Thermal Properties

Melting point. 1529 °C
Boiling point. 2868 °C
Coefficient of thermal expansion. At 24 °C. Linear: 12.2 μm/m · K. Along crystal axes: 7.9 μm/m · K along *a* axis, 20.9 μm/m · K along *c* axis. Volumetric: 36.7 × 10^{-6} per K
Specific heat. 0.1680 kJ/kg · K at 25 °C
Entropy. 437.6 J/kg · K at 25 °C
Latent heat of fusion. 119.0 kJ/kg
Latent heat of vaporization. 1.896 MJ/kg
Heat of combustion. For cubic Er_2O_3 at 25 °C: ΔH_c° = −5.67 MJ/kg Er; ΔG_f° = −5.41 MJ/kg Er
Recrystallization temperature. About 520 °C
Thermal conductivity. 14.5 W/m · K at 25 °C
Vapor pressure. 0.001 Pa at 908 °C; 0.101 Pa at 1113 °C; 10.1 Pa at 1405 °C; 1013 Pa at 1896 °C

Electrical Properties

Electrical resistivity. 860 nΩ · m at 25 °C; 47 nΩ · m at 4 K. Along crystal axes at 25 °C: 945 nΩ · m along *a* axis, 603 nΩ · m along *c* axis. Liquid: 2260 nΩ · m at 1531 °C
Ionization potentials. Er(I), 6.101 eV; Er(II), 11.93 eV; Er(III), 22.74 eV; Er(IV), 42.65 eV
Hall coefficient. Along crystal axes at 20 °C: +0.03 nV · m/A · T along *b* axis; $R_{H,c}$ = −0.36 nV · m/A · T along *c* axis
Temperature of superconductivity. Bulk erbium is not superconducting down to 0.03 K at atmospheric pressure.

Magnetic Properties

Magnetic susceptibility. Volume (mks units): χ_a = 0.0314 and χ_c = 0.0353 at 25 °C; obeys Curie-Weiss law above 195 K with an effective moment of 9.9 μ_B, θ_a = 32.5 K and θ_c = 61.7 K
Saturization magnetization. >3.33 T at 4.2 K along ⟨10$\bar{1}$0⟩ and ⟨11$\bar{2}$0⟩; 3.33 T at 4.2 K along ⟨0001⟩
Magnetic transformation temperatures. Curie temperature, 20 K; a spin rearrangement at 53 K; Néel temperature, 85 K

Optical Properties

Color. Metallic silver
Spectral hemispherical emittance. Solid and liquid, 37.2% for λ = 645 nm from 1027 to 1587 °C

Nuclear Properties

Thermal neutron cross section. 170 b

Chemical Properties

General corrosion behavior. Erbium stays shiny in air at room temperature. The rate of oxidation is slow even at 1000 °C due to the formation of a dark, tightly adhering oxide on the surface of the metal.
Resistance to specific corroding agents. Erbium does not react with cold or hot water, but it will react vigorously with dilute acids. The attack by concentrated sulfuric acid is slow. The presence of the fluoride ion retards acid attack due to the

formation of ErF_3 on the surface of the metal.

Mechanical Properties

Tensile properties. Tensile strength, 136 MPa; yield strength, 60 MPa; elongation, 11.5%; reduction in area, 11.9%
Hardness. 42 HV
Poisson's ratio. 0.237
Strain-hardening exponent. 0.25
Elastic modulus. At 27 °C: tension, 69.9 GPa; shear, 28.3 GPa; bulk, 44.4 GPa
Elastic constants along crystal axes. At 27 °C: c_{11} = 83.67 GPa; c_{12} = 29.29 GPa; c_{13} = 22.22 GPa; c_{33} = 84.45 GPa; c_{44} = 27.53 GPa
Liquid surface tension. 0.637 N/m at 1530 °C

Gadolinium (Gd)

Compiled by K.A. Gschneidner, Jr.
and B.J. Beaudry, Ames Laboratory,
U.S. Department of Energy,
Iowa State University

Gadolinium is used as a burnable poison in shields and control rods in nuclear reactors; it is also used in host materials for rare earth phosphors, catalysts, and garnet microwave devices. Amorphous Gd-Co alloys are used as magnetooptic storage devices and gadolinium intermetallic compounds are used as magnetic refrigeration materials. Gadolinium will tarnish slightly in air. Turnings can be ignited and burn white hot. Finely divided gadolinium should not be handled in air.

Structure

Crystal structure. α phase, close-packed hexagonal, $P6_3/mmc\ D_4^{6h}$; a = 0.36336 nm, c = 0.57810 nm at 24 °C. β phase, body-centered cubic, $Im3m\ O_9^h$; a = 0.406 nm at 1265 °C
Slip planes. At 24 °C: primary $\{10\bar{1}0\}$, secondary {0002}
Twinning planes. At 24 °C: primary $\{11\bar{2}1\}$, secondary $\{10\bar{1}2\}$
Minimum interatomic distance. At 24 °C: r_a = 0.18168 nm; r_c = 0.17858 nm; radius CN_{12} = 0.18013 nm

Mass Characteristics

Atomic weight. 157.25
Density. α phase, 7.901 g/cm³ at 24 °C; β phase, 7.80 g/cm³ at 1265 °C; liquid, 7.4 g/cm³ at 1315 °C
Volume change on freezing. 2.0% contraction

Thermal Properties

Melting point. 1313 °C
Boiling point. 3273 °C
Phase transformation temperature. α to β phase, 1235 °C
Coefficient of thermal expansion. At 100 °C. Linear: 9.4 μm/m · K at 100 °C. Linear along crystal axes: 9.1 μm/m · K along a axis, 10.0 μm/m · K along c axis. Volumetric: 28.2 × 10^{-6} per K
Specific heat. 235.9 kJ/kg · K at 25 °C
Entropy. 431.8 J/kg · K at 25 °C
Latent heat of fusion. 63.6 kJ/kg
Latent heat of transformation. 24.9 kJ/kg
Latent heat of vaporization. 2.528 MJ/kg at 25 °C
Heat of combustion. For monoclinic Gd_2O_3 at 25 °C: ΔH_c° = −5.77 MJ/kg Gd; ΔG_f° = −5.50 MJ/kg Gd
Recrystallization temperature. About 500 °C
Thermal conductivity. 10.5 W/m · K at 25 °C
Vapor pressure. 0.001 Pa at 1167 °C; 0.101 Pa at 1408 °C; 10.1 Pa at 1760 °C; 1013 Pa at 2306 °C

Electrical Properties

Electrical resistivity. Solid: 1310 nΩ · m at 25 °C; 24 nΩ · m at 4 K. Liquid: 1950 nΩ · m at 1315 °C. Along crystal axes at 25 °C: 1351 nΩ · m along a axis, 1217 nΩ · m along c axis
Ionization potentials. Gd(I), 6.14 eV; Gd(II), 12.09 eV; Gd(III), 20.63 eV; Gd(IV), 44.0 eV
Hall coefficient. −0.448 nV · m/A · T at 77 °C. Along crystal axes at 20 °C: −1.0 nV · m/A · T along a axis; −5.4 nV · m/A · T along c axis
Temperature of superconductivity. Bulk gadolinium is not superconducting down to 0.37 K at atmospheric pressure.

Magnetic Properties

Magnetic susceptibility. Volume: 0.117 mks at 77 °C; obeys Curie-Weiss law above 77 °C with an effective moment of 7.98 μ_B and θ_a = θ_c = 317 K
Saturation magnetization. 2.63 T at 0 K
Magnetic transformation temperatures. Curie temperature, 293.4 K

Optical Properties

Color. Metallic silver
Spectral hemispherical emittance. Solid: 33.7% for λ = 645 nm at 1025 to 1313 °C. Liquid: 34.2% for λ = 645 nm at 1313 to 1600 °C

Nuclear Properties

Thermal neutron cross section. 40 000 b

Chemical Properties

General corrosion behavior. Gadolinium tarnishes slightly in air at room temperature. Even at 1000 °C the oxidation rate is slow because of the formation of a dark, tightly adhering oxide on the surface. The presence of water vapor increases the rate of oxidation. After heating to 550 °C in vacuum, hydrogen will react at 250 °C.
Resistance to specific corroding agents. Gadolinium does not react with cold or hot water, but it will react vigorously with dilute acids. It is attacked slowly by concentrated sulfuric acid. The presence of the fluoride ion retards acid attack due to the formation of GdF_3.

Mechanical Properties

Tensile properties. Tensile strength, 118 MPa; yield strength, 15 MPa; elongation, 37%; reduction in area, 56%
Hardness. 37 HV for polycrystalline; 23 HV for $\{10\bar{1}0\}$ prismatic face; 69 HV for {0001} basal plane
Poisson's ratio. 0.259
Strain-hardening exponent. 0.37
Elastic modulus. At 27 °C: tension, 54.8 GPa; shear, 21.8 GPa; bulk, 37.9 GPa
Elastic constants along crystal axes. At 27 °C: c_{11} = 67.83 GPa; c_{12} = 25.59 GPa; c_{13} = 20.73 GPa; c_{33} = 71.23 GPa; c_{44} = 20.77 GPa
Liquid surface tension. 0.664 N/m² at 1315 °C

Holmium (Ho)

Compiled by K.A. Gschneidner, Jr.
and B.J. Beaudry, Ames Laboratory,
U.S. Department of Energy,
Iowa State University

Holmium is used in phosphors and ferrite bubble devices. It will remain shiny in air at room temperature. Turnings can be ignited and will burn white hot. Finely divided holmium should not be handled in air.

Structure

Crystal structure. Close-packed hexagonal: $P6_3/mmc\ D_4^{6h}$; a = 0.35778 nm, c = 0.56178 nm at 24 °C
Slip planes. At 24 °C: primary $\{10\bar{1}0\}$, secondary {0002}
Twinning planes. At 24 °C: primary $\{11\bar{2}1\}$, secondary {1012}
Minimum interatomic distance. At 24 °C: r_a = 0.17889 nm; r_c = 17 433 nm; radius CN_{12} = 0.17661 nm

Mass Characteristics

Atomic weight. 164.93032
Density. 8.795 g/cm³ at 24 °C; liquid, 8.34 g/cm³ at 1480 °C
Volume change on freezing. 7.4% contraction

Thermal Properties

Melting point. 1474 °C
Boiling point. 2700 °C
Coefficient of thermal expansion. At 24 °C. Linear: 11.2 μm/m · K. Linear along crystal axes: 7.0 μm/m · K along a axis, 19.5 μm/m · K along c axis. Volumetric: 33.6 × 10^{-6} per K
Specific heat. 0.1649 kJ/kg · K at 25 °C
Entropy. At 25 °C: 454.7 J/kg · K
Latent heat of fusion. 103 kJ/kg (estimated)
Latent heat of vaporization. 1.824 MJ/kg at 25 °C
Latent heat of combustion. For cubic Ho_2O_3 at 25 °C: ΔH_c° = −5.70 MJ/kg Ho; ΔG_f° = −5.43 MJ/kg Ho

Thermal conductivity. 16.2 W/m · K at 25 °C
Recrystallization temperature. About 520 °C
Vapor pressure. 0.001 Pa at 845 °C; 0.101 Pa at 1036 °C; 10.1 Pa at 1313 °C; 1013 Pa at 1771 °C

Electrical Properties

Electrical resistivity. 814 nΩ · m at 25 °C; 70 nΩ · m at 4 K. Along crystal axes at 25 °C: 1015 nΩ · m along *a* axis, 605 nΩ · m along *c* axis. Liquid: 2210 nΩ · m at 1476 °C
Ionization potentials. Ho(I), 6.018 eV; Ho(II), 11.80 eV; Ho(III), 22.84 eV; Ho(IV), 42.5 eV
Hall coefficient. Along crystal axes at 20 °C: +0.02 nV · m/A · T along *b* axis; −0.32 nV · m/A · T along *c* axis
Temperature of superconductivity. Bulk holmium is not superconducting down to 0.38 K at atmospheric pressure.

Magnetic Properties

Magnetic susceptibility. Volume (mks units): $\chi_a = 0.0500$ and $\chi_c = 0.0466$ at 25 °C; obeys Curie-Weiss law above 140 K with an effective moment of 11.2 μ_B, θ_a = 88 K and θ_c = 73 K
Saturation magnetization. 3.87 T at 4.2 K along ⟨11$\bar{2}$0⟩ and 3.80 T at 4.2 K along ⟨10$\bar{1}$0⟩
Magnetic transformation temperatures. Curie temperature, 20 K; Néel temperature, 132 K

Optical Properties

Color. Metallic silver

Nuclear Properties

Thermal neutron cross section. 64 b

Chemical Properties

General corrosion behavior. Holmium stays shiny in air at room temperature. The rate of oxidation is slow even at 1000 °C due to the formation of a dark, tightly adhering oxide on the surface.
Resistance to specific corroding agents. Holmium does not react with cold or hot water, but it will react vigorously with dilute acids. It is attacked slowly by concentrated sulfuric acid. The presence of fluoride ion retards acid attack due to formation of HoF_3 on the surface of the metal.

Mechanical Properties

Tensile properties. About the same as those of dysprosium and erbium
Hardness. 46 HV
Poisson's ratio. 0.231
Elastic modulus. At 27 °C: tension, 64.8 GPa; shear, 26.3 GPa; bulk, 40.2 GPa
Elastic constants along crystal axes. At 27 °C: c_{11} = 76.2 GPa; c_{12} = 24.8 GPa; c_{13} = 20.6 GPa; c_{33} = 77.6 GPa; c_{44} = 25.7 GPa
Liquid surface tension. 0.650 N/m at 1475 °C

Lanthanum (La)

Compiled by K.A. Gschneidner, Jr. and B.J. Beaudry, Ames Laboratory, U.S. Department of Energy, Iowa State University

Lanthanum, as a component (~25%) of mischmetal, is used as an alloying additive to ferrous alloys to scavenge impurities such as sulfur and oxygen and to strengthen magnesium. It improves the high-temperature oxidation resistance of superalloys. Lanthanum is also used in optical lenses, petroleum cracking catalysts, carbon arc lights, lighter flints, hydrogen storage alloys, optical fibers, and ceramic capacitors.

Lanthanum readily oxidizes at room temperature in air. It should be stored in vacuum or an inert atmosphere; storage under oil is not recommended. Turnings can be ignited easily and burn white hot. Finely divided lanthanum should not be handled in air.

Structure

Crystal structure. α phase, double close-packed hexagonal, $P6_3/mmc$ D^4_{6h}. At 24 °C: *a* = 0.37740 nm; *c* = 1.2171 nm. β phase, face-centered cubic, $Fm3m$ O^5_h; *a* = 0.5303 nm at 325 °C. γ phase, body-centered cubic, $Im3m$ O^9_h; *a* = 0.426 nm at 887 °C
Minimum interatomic distance. At 24 °C: r_a = 0.18870 nm; r_c = 0.18711 nm; radius CN_{12} = 0.18791 nm

Mass Characteristics

Atomic weight. 138.9055
Density. α phase, 6.146 g/cm³ at 24 °C; β phase, 6.187 g/cm³ at 325 °C; γ phase, 5.97 g/cm³ at 865 °C; liquid, 5.96 g/cm³ at 920 °C
Volume change on freezing. 0.6% contraction
Volume change on phase transformation. α to β phase, 0.5% volume contraction on heating; β to γ phase, 1.3% volume expansion on heating

Thermal Properties

Melting point. 918 °C
Boiling point. 3464 °C
Phase transformation temperature. α to β phase: A_s = 330 °C, A_f = 336 °C. β to α phase: M_s = 251 °C, M_f = 247 °C. β to γ phase: 865 °C
Coefficient of thermal expansion. At 24 °C. Linear: 12.1 μm/m · K. Linear along crystal axes: 4.5 μm/m · K along *a* axis; 27.2 μm/m · K along *c* axis. Volumetric: 36.2×10^{-6} per K
Specific heat. 0.1951 kJ/kg · K at 25 °C
Entropy. At 298.15 K: 409.6 J/kg · K
Latent heat of fusion. 44.6 kJ/kg
Latent heat of transformation. α to β phase, 2.6 kJ/kg; β to γ phase, 22.5 kJ/kg
Latent heat of vaporization. 3.103 MJ/kg at 25 °C
Heat of combustion. For hexagonal La_2O_3 at 25 °C: ΔH°_c = −6.46 MJ/kg La; ΔG°_f = −6.14 MJ/kg La
Recrystallization temperature. About 300 °C
Thermal conductivity. α phase, 13.4 W/m · K at 25 °C
Vapor pressure. 0.001 Pa at 1301 °C; 0.101 Pa at 1566 °C; 10.1 Pa at 1938 °C; 1013 Pa at 2506 °C

Electrical Properties

Electrical resistivity. α phase: 615 nΩ · m at 25 °C, 3 nΩ · m at 7 K. Liquid: 1330 nΩ · m at 922 °C
Ionization potential. La(I), 5.5770 eV; La(II), 11.060 eV; La(III), 19.1774 eV; La(IV), 49.95 eV
Hall coefficient. −0.035 nV · m/A · T at 25 °C
Temperature of superconductivity. α phase, 5.10 K; β phase, 6.00 K

Magnetic Properties

Magnetic susceptibility. Volume (mks units) at 24 °C: α phase, 5.33×10^{-5}; β phase, 5.88×10^{-5}

Optical Properties

Color. Metallic silver
Spectral hemispherical emittance. Solid γ-La, 40.9% for λ = 645 nm from 867 to 918 °C; liquid, 25.4% for λ = 645 nm at 920 to 1287 °C

Nuclear Properties

Thermal neutron cross section. 8.9 b

Chemical Properties

General corrosion behavior. Lanthanum oxidizes readily in air at room temperature; oxidation rates increase with temperature and humidity. Hydrogen will react with lanthanum at room temperature.
Resistance to specific corroding agents. Lanthanum reacts vigorously with dilute acids. Cold water slowly attacks lanthanum; hot water reacts faster. The presence of the fluoride ion retards acid attack by the formation of LaF_3 on the surface of the metal.

Mechanical Properties

Tensile properties. Tensile strength, 130 MPa; yield strength, 126 MPa; elongation, 7.9%
Hardness. 28 HV
Poisson's ratio. 0.280
Elastic modulus. At 27 °C: tension, 36.6 GPa; shear, 14.3 GPa; bulk, 27.9 GPa
Kinematic liquid viscosity. 0.445 mm²/s at 922 °C
Liquid surface tension. 0.718 N/m at 922 °C

Lutetium (Lu)

Compiled by K.A. Gschneidner, Jr. and B.J. Beaudry, Ames Laboratory, U.S. Department of Energy, Iowa State University

Lutetium is used in ferrite and garnet bubble devices. It will remain shiny in air at room temperature. Turnings can be ignited and will burn white hot. Finely divided lutetium should not be handled in air.

Structure

Crystal structure. Close-packed hexagonal: $P6_3/mmc$ D^4_{6h}; a = 0.35052 nm, c = 0.55494 nm at 24 °C
Minimum interatomic distance. At 24 °C: r_a = 0.17526 nm; r_c = 0.17172 nm; radius CN_{12} = 0.17349 nm

Mass Characteristics

Atomic weight. 174.967
Density. 9.841 g/cm^3 at 24 °C; liquid, 9.3 g/cm^3 at 1670 °C
Volume change on freezing. 3.6% contraction

Thermal Properties

Melting point. 1663 °C
Boiling point. 3402 °C
Coefficient of thermal expansion. At 24 °C. Linear: 9.9 μm/m · K. Linear along crystal axes: 4.8 μm/m · K along a axis, 20.0 μm/m · K along c axis. Volumetric: 29.6 × 10^{-6} per K
Specific heat. 0.1503 kJ/kg · K at 25 °C
Entropy. 291.5 J/kg · K at 25 °C
Latent heat of fusion. 126 kJ/kg (estimated)
Latent heat of vaporization. 2.444 MJ/kg at 25 °C
Heat of combustion. For cubic Lu_2O_3 at 25 °C: ΔH°_c = −5.37 MJ/kg Lu; ΔG°_f = −5.11 MJ/kg Lu
Recrystallization temperature. About 600 °C
Thermal conductivity. 16.4 W/m · K at 25 °C
Vapor pressure. 0.001 Pa at 1241 °C; 0.101 Pa at 1483 °C; 10.1 Pa at 1832 °C; 1013 Pa at 2387 °C

Electrical Properties

Electrical resistivity. 582 nΩ · m at 25 °C; 45 nΩ · m at 4 K. Along crystal axes at 25 °C: ρ_b = 766 nΩ · m; ρ_c = 347 nΩ · m. Liquid: 2240 nΩ · m at 1665 °C
Ionization potentials. Lu(I), 5.42589 eV; Lu(II), 13.9 eV; Lu(III), 20.9596 eV; Lu(IV), 45.19 eV
Hall coefficient. At 20 °C. −0.0535 nV · m/A · T. Along crystal axes: 0.045 nV · m/A · T along a axis, −0.26 nV · m/A · T along c axis
Temperature of superconductivity. Bulk lutetium is not superconducting down to 0.03 K at atmospheric pressure; it becomes superconducting at 0.022 and 4.5 GPa.

Magnetic Properties

Magnetic susceptibility. Volume (mks units). At 27 °C: 1.293 × 10^{-4}. Along crystal axes: 1.353 × 10^{-4} along a axis, 1.173 × 10^{-4} along c axis

Optical Properties

Color. Metallic silver

Nuclear Properties

Thermal neutron cross section. 111 b

Chemical Properties

General corrosion behavior. Lutetium stays shiny in air at room temperature. Even at 1000 °C, the rate of oxidation is slow due to the formation of a pink, tightly adhering oxide on the surface of the metal.
Resistance to specific corroding agents. Lutetium does not react with cold or hot water, but it reacts vigorously with dilute acids. Concentrated sulfuric acid slowly attacks lutetium. The presence of the fluoride ion retards acid attack due to the formation of LuF_3 on the surface of the metal.

Mechanical Properties

Tensile properties. About the same as those of erbium
Hardness. 44 HV
Poisson's ratio. 0.261
Elastic modulus. At 27 °C: tension, 68.6 GPa; shear, 27.2 GPa; bulk, 47.6 GPa
Elastic constants along crystal axes. At 27 °C: c_{11} = 86.23 GPa; c_{12} = 32.0 GPa; c_{13} = 28.0 GPa; c_{33} = 80.86 GPa; c_{44} = 26.79 GPa
Liquid surface tension. 0.940 N/m at 1665 °C

Mischmetal (MM)

Compiled by K.A. Gschneidner, Jr.
and B.J. Beaudry, Ames Laboratory,
U.S. Department of Energy,
Iowa State University

Mischmetal is used as an alloying additive in ferrous alloys to scavenge sulfur, oxygen, and other substances. Mischmetal is also added to magnesium-base alloys to improve high-temperature strength and to ductile irons to nodularize graphite. It is used as a reductant to produce the volatile rare earth metals (samarium, europium, and ytterbium) from their oxides. Other uses of mischmetal include lighter flints, galvanizing alloys, and mischmetal cobalt (cerium-free mischmetal-nickel alloys are used as hydrogen storage alloys), and permanent magnets.

Mischmetal oxidizes at room temperature in air. Turnings can be ignited easily and burn white hot. Finely divided mischmetal should not be handled in air.

Because mischmetal is an indefinite mixture of rare earth metals, the properties of a particular mischmetal depend on its composition, which in turn depends on the mineral source for the mixture. Listed below are the properties of two of the most common high-purity mischmetal mixtures. Because most commercial mischmetals contain several atomic percent of iron and magnesium (about 1 wt%), their properties may differ somewhat from those listed below. Many values are estimated and are marked as such.

Bastnasite-Derived Mischmetal

Specifications

UNS number. E21000

Chemical Composition

Composition limits. Total mixed rare earths: 99.0 min; mixture consists of 39 La, 38 Ce, 16 Nd, 6 Pr, 1 other rare earth

Structure

Crystal structure. α phase, double close-packed hexagonal, $P6_3/mmc$ D^{6h}_4; a = 0.3718 nm; c = 1.1978 nm at 24 °C. β phase, body-centered cubic
Minimum interatomic distance. At 24 °C: r_a = 0.1859 nm; r_c = 0.1842 nm; radius CN_{12} = 0.1851

Mass Characteristics

Atomic weight. 140.1
Density. 6.490 g/cm^3 at 24 °C
Volume change on freezing. 0.1% expansion (estimated)

Thermal Properties

Melting point. Melts over a range of temperatures from 888 to 895 °C
Boiling point. Expected to evaporate incongruently with the initial loss of a major constituent (neodymium) by boiling at ~3100 °C
Phase transformation temperature. α to β phase, 801 °C
Coefficient of thermal expansion. At 24 °C. Linear: 8.7 μm/m · K. Volumetric: 26 × 10^{-6} per K (both estimated)
Specific heat. 0.193 kJ/kg · K at 25 °C
Entropy. 467 J/kg · K at 25 °C (estimated)
Latent heat of fusion. 47 kJ/kg
Latent heat of transformation. 22 kJ/mol (estimated)
Latent heat of combustion. For hexagonal R_2O_3 at 25 °C: ΔH°_c = −6.4 MJ/kg MM (estimated); ΔG°_f = −6.0 MJ/kg MM (estimated)
Recrystallization temperature. ~350 °C (estimated)
Thermal conductivity. 13 W/m · K at 25 °C (estimated)

Electrical Properties

Electrical resistivity. Solid: 800 nΩ · m at 25 °C (estimated). Liquid: 1300 nΩ · m at 900 °C (estimated)

Magnetic Properties

Magnetic susceptibility. Volume, 1.6 × 10^{-3} mks at 25 °C (estimated)

Optical Properties

Color. Metallic silver
Spectral hemispherical emittance. Liquid, 30% (estimated)

Chemical Properties

General corrosion behavior. Commercial mischmetal is stable in air at room tempera-

ture due to the presence of magnesium. Vacuum-melted mischmetal oxidizes in air at room temperature. Oxidation rates increase with increasing temperature and humidity.
Resistance to specific corroding agents. Mischmetal reacts vigorously with dilute acids. The presence of the fluoride ion retards acid attack by the formation of rare earth fluoride (RF_3) on the surface of the metal.

Mechanical Properties

Tensile properties. At 24 °C: tensile strength, 138 MPa; yield strength, 48 MPa; elongation, 25%; reduction in area, 50% (all estimated)
Hardness. 27 DPH
Poisson's ratio. 0.27 (estimated)
Elastic modulus. At 27 °C: tension, 35 GPa; shear, 14 GPa; bulk, 25 GPa (all estimated)
Kinematic liquid viscosity. 0.46 mm^2/s at 900 °C (estimated)
Liquid surface tension. 0.70 N/m at 900 °C (estimated)

Monazite-Derived Mischmetal

Specifications

UNS number. E31000

Chemical Composition

Composition limits. Total mixed rare earths, 99.0 min; mixture consists of 50 Ce, 20 La, 20 Nd, 6 Pr, 2 Gd, 1.6 Y, <1 other rare earths

Structure

Crystal structure. α phase, double close-packed hexagonal, $P6_3/mmc$ D^4_{6h}; a = 0.3695 nm; c = 1.1900 nm at 24 °C. β phase, probably body-centered cubic; a = 0.415 nm (estimated)
Minimum interatomic distance. At 24 °C: r_a = 0.1848 nm; r_c = 0.1830 nm; radius CN_{12} = 0.1839 nm

Mass Characteristics

Atomic weight. 140.1
Density. 6.612 g/cm^3 at 24 °C
Volume change on freezing. 0.1% expansion (estimated)

Thermal Properties

Melting point. Melts over a range of temperatures from 899 to 913 °C
Boiling point. Expected to evaporate incongruently with the initial loss of a major constituent (neodymium) by boiling at ~3100 °C
Phase transformation temperature. α to β phase, 782 °C
Coefficient of thermal expansion. At 24 °C. Linear: 8.6 μm/m · K. Volumetric: 26 × 10^{-6} per K (both estimated)
Specific heat. 0.195 kJ/kg · K at 25 °C (estimated)
Entropy. At 298.15 K, 477 J/kg · K (estimated)
Latent heat of fusion. 46 kJ/kg (estimated)
Latent heat of transformation. 22 kJ/kg (estimated)
Latent heat of combustion. For hexagonal R_2O_3 at 25 °C: ΔH°_c = −6.4 MJ/kg MM (estimated); ΔG°_f = −6.0 MJ/kg MM (estimated)
Recrystallization temperature. ~350 °C (estimated)
Thermal conductivity. 13 W/m · K at 25 °C (estimated)

Electrical Properties

Electrical resistivity. Solid, 800 nΩ · m at 25 °C (estimated); liquid, 1300 nΩ · m at 925 °C (estimated)

Magnetic Properties

Magnetic susceptibility. Volume: 5.2 × 10^{-3} mks at 25 °C (estimated)

Optical Properties

Color. Metallic silver
Spectral hemispherical emittance. Liquid, 30% (estimated)

Chemical Properties

General corrosion behavior. Commercial mischmetal is stable in air at room temperature due to the presence of magnesium. Vacuum-melted mischmetal oxidizes in air at room temperature. Oxidation rates increase with increasing temperature and humidity.
Resistance to specific corroding agents. Mischmetal reacts vigorously with dilute acids. The presence of the fluoride ion retards acid attack by the formation of rare earth fluoride (RF_3) on the surface of the metal.

Mechanical Properties

Tensile properties. At 24 °C: tensile strength, 138 MPa; yield strength, 48 MPa; elongation, 25%; reduction in area, 50% (all estimated)
Hardness. 28 HV
Poisson's ratio. 0.27 (estimated)
Elastic modulus. At 27 °C: tension, 37 GPa; shear, 15 GPa; bulk, 27 GPa (all estimated)
Kinematic liquid viscosity. 0.47 mm^2/s at 920 °C (estimated)
Liquid surface tension. 0.71 N/m at 920 °C (estimated)

Neodymium (Nd)

Compiled by K.A. Gschneidner, Jr.
and B.J. Beaudry, Ames Laboratory,
U.S. Department of Energy,
Iowa State University

The major use of neodymium is in high-strength Nd-Fe-B permanent magnets, which are the strongest magnets known. Neodymium, as a component (~20%) of mischmetal, is used as an alloying additive in ferrous alloys to scavenge sulfur, oxygen, and other elements, and to strengthen magnesium alloys. It is also used as a laser material and glass-coloring agent, and in petroleum cracking catalysts, carbon arc lights, lighter flints, and ceramic capacitors.

Neodymium oxidizes at room temperature in air. It should be stored in a vacuum or inert atmosphere; storage in oil is not recommended. Turnings can be ignited easily and will burn white hot. Finely divided neodymium should not be handled in air.

Structure

Crystal structure. α phase, double close-packed hexagonal, $P6_3/mmc$ D^{6h}_4; a = 0.36582 nm, c = 1.17966 nm at 24 °C. β phase, body-centered cubic, $Im3m$ O^h_9; a = 0.413 nm at 883 °C
Minimum interatomic distance. At 24 °C: r_a = 0.18291 nm; r_c = 0.18137 nm; radius CN_{12} = 0.18214 nm

Mass Characteristics

Atomic weight. 144.24
Density. α phase, 7.008 g/cm^3 at 24 °C; β phase, 6.80 g/cm^3 at 883 °C; liquid, 6.72 g/cm^3 at 1025 °C
Volume change on freezing. 0.9% contraction
Volume change on phase transformation. α to β phase, 0.1% volume expansion on heating

Thermal Properties

Melting point. 1021 °C
Boiling point. 3074 °C
Phase transformation temperature. α to β phase, 863 °C
Coefficient of thermal expansion. At 24 °C. Linear: 9.6 μm/m · K. Linear along crystal axes: 7.6 μm/m · K along a axis, 13.5 μm/m · K along c axis. Volumetric: 28.7 × 10^{-6} per K
Specific heat. 0.1900 kJ/kg · K at 25 °C
Entropy. At 25 °C: 492.9 J/kg · K
Latent heat of fusion. 49.5 kJ/kg
Latent heat of transformation. 21.0 kJ/kg
Latent heat of vaporization. 2.271 MJ/kg at 25 °C
Thermal conductivity. 16.5 W/m · K at 25 °C
Heat of combustion. Hexagonal Nd_2O_3 at 25 °C: ΔH°_c = −6.27 MJ/kg Nd; ΔG°_f = −5.96 MJ/kg Nd
Recrystallization temperature. 400 °C
Vapor pressure. 0.001 Pa at 955 °C; 0.101 Pa at 1175 °C; 10.1 Pa at 1500 °C; 1013 Pa at 2029 °C

Electrical Properties

Electrical resistivity. 643 nΩ · m at 25 °C; 68 nΩ · m at 4 K. Liquid: 1510 nΩ · m at 1022 °C
Ionization potential. Nd(I), 5.499 eV; Nd(II), 10.73 eV; Nd(III), 22.1 eV; Nd(IV), 40.41 eV
Hall coefficient. +0.0971 nV · m/A · T at 25 °C
Temperature of superconductivity. Bulk neodymium is not superconducting down to 0.25 K at atmospheric pressure.

Magnetic Properties

Magnetic susceptibility. Volume: 3.62 × 10^{-3} mks at 25 °C; obeys Curie-Weiss law above 35 K with an effective moment of 3.45 μ_B, θ_a = 5 K and θ_c = 0 K
Saturation magnetization. >35T at 2 K along ⟨11$\bar{2}$0⟩
Magnetic transformation temperature. Néel temperatures at 7.5 K (cubic sites) and 19.9 K (hexagonal sites)

Optical Properties

Color. Metallic silver
Spectral hemispherical emittance. 39.4% for λ = 645 nm from 1021 to 1567 °C

Nuclear Properties

Thermal neutron cross section. 48 b

Chemical Properties

General corrosion behavior. Neodymium oxidizes in air at room temperature, but at a slower rate than lanthanum or cerium. Oxidation rates increase with increasing temperature and humidity; interstitial impurities increase the rate of oxidation. Hydrogen will react with neodymium at room temperature.
Resistance to specific corroding agents. Neodymium reacts vigorously with dilute acids and slowly with concentrated sulfuric acid. The presence of the fluoride ion retards acid attack due to the formation of NdF_3 on the surface of the metal.

Mechanical Properties

Tensile properties. Tensile strength, 164 MPa; yield strength, 71 MPa; elongation, 25%; reduction in area, 72%
Hardness. 18 HV
Poisson's ratio. 0.281
Strain-hardening exponent. 0.28
Elastic modulus. At 27 °C: tension, 41.4 GPa; shear, 16.3 GPa; bulk, 31.8 GPa
Elastic constants along crystal axes. At 27 °C: c_{11} = 54.77 GPa; c_{12} = 24.60 GPa; c_{13} = 16.56 GPa; c_{33} = 60.80 GPa; c_{44} = 15.01 GPa
Liquid surface tension. 0.687 N/m at 1021 °C

Praseodymium (Pr)

Compiled by K.A. Gschneidner, Jr.
and B.J. Beaudry, Ames Laboratory,
U.S. Department of Energy,
Iowa State University

Praseodymium, as a component (~5%) of mischmetal, is used as an alloying additive to ferrous alloys to scavenge impurities such as sulfur and oxygen, and to strengthen magnesium. It is also used as a glass- and ceramic-coloring agent, and in petroleum cracking catalysts, carbon arc lights, and $PrCo_5$ permanent magnets. $PrNi_5$ is used in adiabatic demagnetization refrigerators to attain ultralow temperatures (<1 mK).

Praseodymium oxidizes at room temperature in air. It should be stored in vacuum or inert atmosphere; storage in oil is not recommended. Turnings can be ignited easily and will burn white hot. Finely divided praseodymium should not be handled in air.

Structure

Crystal structure. α phase, double close-packed hexagonal, $P6_3/mmc$ D_4^{6h}; a = 0.36721 nm, c = 1.18326 nm at 24 °C. β phase, body-centered cubic, $Im3m$ O_9^h; a = 0.413 nm at 821 °C
Minimum interatomic distance. At 24 °C: r_a = 0.18360 nm; r_c = 0.18197 nm; radius CN_{12} = 0.18279 nm

Mass Characteristics

Atomic weight. 140.90765
Density. α phase, 6.773 g/cm^3 at 24 °C; β phase, 6.64 g/cm^3 at 821 °C; liquid, 6.59 g/cm^3 at 935 °C
Volume change on freezing. 0.02% contraction
Volume change on phase transformation. α to β phase, 0.5% volume expansion on heating

Thermal Properties

Melting point. 931 °C
Boiling point. 3520 °C
Phase transformation temperature. α to β phase, 795 °C
Coefficient of thermal expansion. α phase at 24 °C. Linear: 6.7 μm/m · K. Linear along crystal axes: 4.5 μm/m · K along a axis, 11.2 μm/m · K along c axis. Volumetric: 20.2 × 10^{-6} per K
Specific heat. 0.1946 kJ/kg · K at 25 °C
Entropy. At 25 °C, 524.5 J/kg · K
Latent heat of fusion. 48.9 kJ/kg
Latent heat of transformation. 22.5 kJ/kg
Latent heat of vaporization. 2.524 MJ/kg at 25 °C
Heat of combustion. For cubic Pr_6O_{11} at 25 °C: ΔH_c° = −6.73 MJ/kg Pr; ΔG_f° = −6.34 MJ/kg Pr
Recrystallization temperature. About 400 °C
Thermal conductivity. 12.5 W/m · K at 25 °C
Vapor pressure. 0.001 Pa at 1083 °C; 0.101 Pa at 1333 °C; 10.1 Pa at 1701 °C; 1013 Pa at 2305 °C

Electrical Properties

Electrical resistivity. 700 nΩ · m at 25 °C; 22 nΩ · m at 4 K. Liquid: 1390 nΩ · m at 932 °C
Ionization potential. Pr(I), 5.422 eV; Pr(II), 10.55 eV; Pr(III), 21.624 eV; Pr(IV), 38.98 eV; Pr(V), 57.45 eV
Hall coefficient. +0.0709 nV · m/A · T at 25 °C
Temperature of superconductivity. Bulk praseodymium is not superconducting down to 0.25 K at atmospheric pressure.

Magnetic Properties

Magnetic susceptibility. Volume: 3.34 × 10^{-3} mks at 25 °C; obeys Curie-Weiss law above 100 K with an effective moment of 3.56 μ_B and θ ≅ 0 K
Saturation magnetization. >35 T at 4 K along ⟨11$\bar{2}$0⟩
Magnetic transformation temperature. Single-crystal strain-free praseodymium does not order magnetically; most polycrystalline samples order at various temperatures below 25 K.

Optical Properties

Color. Metallic silver
Spectral hemispherical emittance. 28.4% for λ = 645 nm from 931 to 1537 °C

Nuclear Properties

Thermal neutron cross section. 11 b

Chemical Properties

Corrosion behavior. Praseodymium oxidizes in air at room temperature, but at a lower rate than lanthanum or cerium. Oxidation rates increase with temperature and humidity. Interstitial impurities in the metal increase the rate of corrosion in air. Hydrogen will react with praseodymium at room temperature.
Resistance to specific corroding agents. Praseodymium reacts vigorously with dilute acids. It reacts slowly with concentrated sulfuric acid. The presence of the fluoride ion retards acid attack due to the formation of PrF_3 on the surface of the metal.

Mechanical Properties

Tensile properties. At 24 °C: tensile strength, 147 MPa; yield strength, 73 MPa; elongation, 15.4%; reduction in area, 67%
Hardness. 20 HV
Poisson's ratio. 0.281
Elastic modulus. At 27 °C: tension, 37.3 GPa; shear, 14.8 GPa; bulk, 28.8 GPa
Elastic constants along crystal axes. At 27 °C: c_{11} = 49.35 GPa; c_{12} = 22.95 GPa; c_{13} = 14.3 GPa; c_{33} = 57.40 GPa; c_{44} = 13.59 GPa
Kinematic liquid viscosity. 0.432 mm^2/s at 935 °C
Liquid surface tension. 0.707 N/m at 935 °C

Promethium (Pm)

Compiled by K.A. Gschneidner, Jr.
and B.J. Beaudry, Ames Laboratory,
U.S. Department of Energy,
Iowa State University

Promethium is used in luminous watch dials and as a lightly shielded radioisotope power source. It is also a highly radioactive β emitter (^{147}Pm).

Structure

Crystal structure. α phase, double close-packed hexagonal, $P6_3/mmc$ D_4^{6h}; a = 0.365 nm, c = 1.165 at 24 °C. β phase, probably body-centered cubic, a = 0.410 nm (estimated) at 890 °C
Minimum interatomic distance. At 24 °C: r_a

= 0.1825 nm; r_c = 0.1797 nm; radius CN_{12} = 0.1811

Mass Characteristics

Atomic weight. 145
Density. α phase, 7.264 g/cm³ at 24 °C; β phase, 6.99 g/cm³ (estimated) at 890 °C; liquid, 6.9 g/cm³ (estimated) at 1050 °C

Thermal Properties

Melting point. 1042 °C
Boiling point. 3000 °C (estimated)
Phase transformation temperature. 890 °C
Coefficient of thermal expansion. At 24 °C. Linear (estimated): 11 μm/m · K. Linear along crystal axes 9 μm/m · K along *a* axis, 16 μm/m · K along *c* axis. Volumetric: 34 × 10^{-6} per K
Specific heat. 0.188 kJ/kg · K at 25 °C (estimated)
Entropy. At 25 °C: 494 J/kg · K (estimated)
Latent heat of fusion. 53 kJ/kg (estimated)
Latent heat of transformation. 21 kJ/kg (estimated)
Latent heat of vaporization. 2.4 MJ/kg at 25 °C (estimated)
Thermal conductivity. 15 W/m · K at 27 °C (estimated)
Heat of combustion. For monoclinic Pm_2O_3 at 25 °C: ΔH°_c = −6.3 MJ/kg Pr; ΔG°_f = −6.0 MJ/kg Pm (estimated)
Recrystallization temperature. 400 °C (estimated)

Electrical Properties

Electrical resistivity. 750 nΩ · m at 25 °C (estimated)
Ionization potential. Pm(I), 5.554 eV; Pm(II), 10.90 eV; Pm(III), 22.3 eV; Pm(IV), 41.1 eV

Magnetic Properties

Magnetic susceptibility. Probably strongly paramagnetic with a susceptibility somewhat greater than that of cerium at 24 °C
Magnetic transformation temperature. Probably exhibits two Néel temperatures that fall between those observed in neodymium and samarium

Optical Properties

Color. Metallic silver

Chemical Properties

General corrosion behavior. About the same as that of neodymium
Resistance to specific corroding agents. About the same as that of neodymium

Mechanical Properties

Tensile properties. About the same as that of neodymium
Hardness. 63 HK
Poisson's ratio. 0.28 (estimated)
Elastic modulus. At 27 °C: tension, 46 GPa (estimated); shear, 18 GPa; bulk, 33 GPa
Liquid surface tension. 0.68 N/m (estimated) at 1045 °C

Samarium (Sm)

Compiled by K.A. Gschneidner, Jr.
and B.J. Beaudry, Ames Laboratory,
U.S. Department of Energy,
Iowa State University

Alloyed with cobalt, samarium is used as a permanent magnet, Sm_2Co_{17}-$SmCo_5$. Samarium is also used as a burnable poison in nuclear reactors, as a phosphor, and in catalysts and ceramic capacitors.

Samarium oxidizes slowly in air at room temperature. Storage in an inert atmosphere or vacuum is recommended. Turnings can be ignited easily. Finely divided samarium should not be handled in air.

Structure

Crystal structure. α phase, rhombohedral, $R3m$ D_5^{3d}; *a* = 0.89834 nm, α = 23.311° (hexagonal parameters, *a* = 0.36290 nm, *c* = 2.6207 nm at 24 °C). β phase, close-packed hexagonal, $P6_3/mmc$ D_4^{6h}; *a* = 0.36630 nm; *c* = 0.58448 nm at 450 °C. γ phase, body-centered cubic, $Im3m$ O_9^h; *a* = 0.410 nm (estimated) at 922 °C
Minimum interatomic distance. At 24 °C: r_a = 0.18145 nm; r_c = 0.17937 nm; radius CN_{12} = 0.18041 nm

Mass Characteristics

Atomic weight. 150.4
Density. α phase, 7.520 g/cm³ at 24 °C; β phase, 7.353 g/cm³ at 450 °C; γ phase, 7.25 g/cm³ (estimated) at 922 °C; liquid, 7.16 g/cm³ at 1075 °C
Volume change on freezing. 3.6% contraction

Thermal Properties

Melting point. 1074 °C
Boiling point. 1794 °C
Phase transformation temperature. α to β phase, 734 °C; β to α phase, 727 °C; β to γ phase, 922 °C
Coefficient of thermal expansion. At 24 °C. Linear: 12.7 μm/m · K. Linear along crystal axes: 9.6 μm/m · K along *a* axis; 19.0 μm/m · K along *c* axis. Volumetric, 38.1 × 10^{-6} per K
Specific heat. 0.1962 kJ/kg · K at 25 °C
Entropy. At 25 °C: 462.2 J/kg · K
Latent heat of fusion. 57.3 kJ/kg
Latent heat of transformation. β to γ phase, 20.7 kJ/kg
Latent heat of vaporization. 1.375 MJ/kg at 25 °C
Heat of combustion. For monoclinic Sm_2O_3 at 25 °C: ΔH°_c = −6.06 MJ/kg Sm; ΔG°_f = −5.77 MJ/kg Sm
Recrystallization temperature. About 440 °C
Thermal conductivity. 13.3 W/m · K at 25 °C
Vapor pressure. 0.001 Pa at 508 °C; 0.101 Pa at 642 °C; 10.1 Pa at 835 °C; 1013 Pa at 1150 °C

Electrical Properties

Electrical resistivity. 940 nΩ · m at 25 °C, 67 nΩ · m at 4 K. Liquid: 1820 nΩ · m at 1075 °C
Ionization potential. Sm(I), 5.631 eV; Sm(II), 11.07 eV; Sm(III), 23.4 eV; Sm(IV), 41.4 eV
Hall coefficient. −0.021 nV · m/A · T at 25 °C
Temperature of superconductivity. Bulk samarium is not superconducting down to 0.37 K at atmospheric pressure.

Magnetic Properties

Magnetic susceptibility. Volume: 8.03 × 10^{-4} mks at 17 °C; does not obey Curie-Weiss law
Saturation magnetization. >35 T at 4 K along ⟨0001⟩ and >30 T at 4 K along ⟨11$\bar{2}$0⟩
Magnetic transformation temperature. Ordering temperatures at 14 K (cubic sites) and 106 K (hexagonal sites)

Optical Properties

Color. Metallic silver
Spectral hemispherical emittance. Solid, 43.7% for λ = 645 nm from 852 to 1074 °C; liquid, 43.7% for λ = 645 nm at 1075 °C

Nuclear Properties

Thermal neutron cross section. 5600 b

Chemical Properties

General corrosion behavior. Samarium oxidizes slowly at room temperature in air. The rate of oxidation increases with temperature. Hydrogen will react at about 250 °C with samarium metal.
Resistance to specific corroding agents. Samarium reacts vigorously with dilute acids but only slowly with concentrated sulfuric acid. The presence of fluoride ion retards acid attack due to the formation of SmF_3 on the surface of the metal.

Mechanical Properties

Tensile properties. Tensile strength, 156 MPa; yield strength, 68 MPa; elongation, 17%; reduction in area, 29.5%
Hardness. 39 HV
Poisson's ratio. 0.274
Strain-hardening exponent. 0.23
Elastic modulus. At 27 °C: tension, 49.7 GPa; shear, 19.5 GPa, bulk, 37.8 GPa
Liquid surface tension. 0.431 N/m at 1075 °C

Scandium (Sc)

Compiled by K.A. Gschneidner, Jr.
and B.J. Beaudry, Ames Laboratory,
U.S. Department of Energy,
Iowa State University

Scandium is used as a neutron window or filter in reactors. It is also used in high-intensity lamps because of the multilined spectrum of incandescent scandium vapor. Turnings of scandium can be ignited and will burn white hot. Finely divided scandium should not be handled in air. Ingots of pure scandium can be stored in air.

Structure

Crystal structure. α phase, close-packed hexagonal, $P6_3/mmc$ D_4^{6h}; a = 0.33088 nm; c = 0.52680 nm at 24 °C. β phase, body-centered cubic, $Im3m$; a = 0.373 nm (estimated) at 1337 °C
Minimum interatomic distance. At 24 °C: r_a = 0.16544 nm; r_c = 0.16269 nm; radius CN_{12} = 0.16407 nm

Mass Characteristics

Atomic weight. 44.95591
Density. α phase, 2.989 g/cm^3 at 24 °C; β phase, 2.88 g/cm^3 at 1337 °C; liquid, 2.80 g/cm^3 at 1550 °C

Thermal Properties

Melting point. 1541 °C
Boiling point. 2836 °C
Phase transformation temperature. α to β phase, 1337 °C
Coefficient of thermal expansion. At 24 °C. Linear: 10.2 μm/m · K. Linear along crystal axis: 7.6 μm/m · K along a axis; 15.3 μm/m · K along c axis. Volumetric: 30.5 × 10^{-6} per K
Specific heat. 0.5674 kJ/kg · K at 25 °C
Entropy. 769.6 J/kg · K at 25 °C
Latent heat of fusion. 313.6 kJ/kg
Latent heat of transformation. 89.0 kJ/kg
Latent heat of vaporization. 8.404 MJ/kg at 25 °C
Heat of combustion. For cubic Sc_2O_3 at 25 °C: ΔH_c° = −21.23 MJ/kg Sc; ΔG_f° = −20.23 MJ/kg Sc
Recrystallization temperature. About 550 °C
Thermal conductivity. 15.8 W/m · K at 25 °C
Vapor pressure. 0.0010 Pa at 1036 °C; 0.101 Pa at 1243 °C; 10.1 Pa at 1533 °C; 1013 Pa at 1999 °C

Electrical Properties

Electrical resistivity. 562 nΩ · m at 25 °C; 1.6 nΩ · m at 4 K. Along crystal axes at 25 °C: 709 nΩ · m along a axis; 269 nΩ · m along c axis
Ionization potential. Sc(I), 6.54 eV; Sc(II), 12.80 eV; Sc(III), 24.76 eV; Sc(IV), 73.47 eV; Sc(V), 91.66 eV; Sc(VI), 111.1 eV
Hall coefficient. −0.013 nV · m/A · T at 25 °C
Temperature of superconductivity. Bulk scandium is not superconducting down to 0.032 K at atmospheric pressure; however, it is superconducting at 0.050 and 18.6 GPa.

Magnetic Properties

Magnetic susceptibility. Volume (mks units) at 24 °C: 2.466 × 10^{-4}. Along crystal axes: 2.490 × 10^{-4} along a axis; 2.419 × 10^{-4} along c axis

Optical Properties

Color. Metallic silver

Nuclear Properties

Thermal neutron cross section. 24 b

Chemical Properties

General corrosion behavior. Scandium remains shiny in air at room temperature; discoloration starts at about 300 °C. Oxidation proceeds slowly to completion at 1000 °C.
Resistance to specific corroding agents. Scandium reacts readily with most acids. The presence of fluoride ions causes the formation of ScF_3, which retards attack by nitric, hydrochloric, and other acids.

Mechanical Properties

Tensile properties. Tensile strength, 255 MPa; yield strength, 173 MPa; elongation, 5%; reduction in area, 8%
Hardness. Anisotropic: 132 HV (0001) and 36 HV (10$\bar{1}$0)
Poisson's ratio. 0.279
Elastic modulus. At 27 °C: Tension, 74.4 GPa; shear, 29.1 GPa; bulk, 56.6 GPa
Elastic constants along crystal axes. At 27 °C: c_{11} = 98.1 GPa; c_{12} = 45.7 GPa; c_{13} = 29.4 GPa; c_{33} = 105.1 GPa; c_{44} = 27.2 GPa
Liquid surface tension. 0.954 N/m at 1545 °C

Terbium (Tb)

Compiled by K.A. Gschneidner, Jr. and B.J. Beaudry, Ames Laboratory, U.S. Department of Energy, Iowa State University

Terbium is used as a phosphor and in magnetostrictive materials ($Tb_{0.3}Dy_{0.7}Fe_2$) and catalysts. Amorphous Tb-Co alloys are used as magnetooptic storage devices. Terbium will remain shiny in air at room temperature. Turnings can be ignited and will burn white hot. Finely divided terbium should not be handled in air.

Structure

Crystal structure. α′ phase, orthorhombic, $Cmcm$ D_{17}^{2h}; a = 0.3605 nm, b = 0.6244 nm, c = 0.5706 nm at −53 °C. α phase, close-packed hexagonal, $P6_3/mmc$ D_4^{6h}; a = 0.36055 nm, c = 0.56966 nm at 24 °C. β phase, body-centered cubic, $Im3m$ O_9^h; a = 0.407 nm at 1289 °C
Minimum interatomic distance. r_a = 0.18028; r_c = 0.17639; radius CN_{12} = 0.17833 nm

Mass Characteristics

Atomic weight. 158.92534
Density. α phase, 8.230 g/cm^3 at 24 °C; β phase, 7.82 g/cm^3 at 1289 °C; liquid, 765 g/cm^3 at 1360 °C
Volume change on freezing. 3.1% contraction

Thermal Properties

Melting point. 1356 °C
Boiling point. 3230 °C
Phase transformation temperature. α to α′ phase, −53 °C; α to β phase, 1289 °C
Coefficient of thermal expansion. At 24 °C. Linear: 10.3 μm/m · K. Linear along crystal axes: 9.3 μm/m · K along a axis, 12.4 μm/m · K along c axis. Volumetric: 31.0 × 10^{-6} per K
Specific heat. 0.1818 kJ/kg · K at 25 °C
Entropy. 461.2 J/kg · K at 25 °C
Latent heat of fusion. 67.9 kJ/kg
Latent heat of transformation. 31.6 kJ/kg
Latent heat of vaporization. 2.446 MJ/kg at 25 °C
Heat of combustion. For cubic Tb_2O_3 at 25 °C: ΔH_c° = −5.87 MJ/kg Tb; ΔG_f° = −5.59 MJ/kg Tb
Recrystallization temperature. 500 °C
Thermal conductivity. 11.1 W/m · K at 25 °C
Vapor pressure. 0.001 Pa at 1124 °C; 0.101 Pa at 1354 °C; 10.1 Pa at 1698 °C; 1013 Pa at 2237 °C

Electrical Properties

Electrical resistivity. 1150 nΩ · m at 25 °C; 35 nΩ · m at 4 K. Along crystal axes at 25 °C: 1235 nΩ · m along a axis, 1015 nΩ · m along c axis. Liquid: 1930 nΩ · m at 1358 °C
Ionization potentials. Tb(I), 5.842 eV; Tb(II), 11.52 eV; Tb(III), 21.91 eV; Tb(IV), 39.79 eV
Hall coefficient. Along crystal axes at 20 °C: −0.10 nV · m/A · T along b axis; −0.37 nV · m/A · T along c axis
Temperature of superconductivity. Bulk terbium is not superconducting down to 0.37 K at atmospheric pressure.

Magnetic Properties

Magnetic susceptibility. Volume (mks units): χ_a = 0.129 and χ_c = 0.0738 at 27 °C; obeys Curie-Weiss law above 240 K with an effective moment of 9.77 μ_B, θ_a = 239 K and θ_c = 195 K
Saturation magnetization. 13 T at 4.2 K along ⟨11$\bar{2}$0⟩
Magnetic transformation temperatures. Curie temperature, 219.5 K; Néel temperature, 230 K

Optical Properties

Color. Metallic silver

Nuclear Properties

Thermal neutron cross section. 45 b

Chemical Properties

General corrosion behavior. Terbium stays shiny in air at room temperature. The rate of oxidation is slow even at 1000 °C due to the formation of a dark, tightly adhering oxide on the surface. Water vapor increases the rate of oxidation. After heating to 550 °C in vacuum, hydrogen will react at 250 °C.
Resistance to specific corroding agents. Terbium does not react with cold or hot water, but it will react vigorously with dilute acids. It is slowly attacked by concentrated sulfuric acid. The presence of the fluoride ion retards acid attack due to the formation of TbF_3.

Mechanical Properties

Tensile properties. About the same as those of gadolinium
Hardness. 38 HV for polycrystalline; 30 HV for {1010} prismatic face; 80 HV for {0001} basal plane
Poisson's ratio. 0.261
Elastic modulus. At 27 °C: tension, 55.7 GPa; shear, 22.1 GPa; bulk, 38.7 GPa
Elastic constants along crystal axes. At 27 °C: c_{11} = 69.24 GPa; c_{12} = 24.98 GPa; c_{13} = 21.79 GPa; c_{33} = 74.39 GPa; c_{44} = 21.75 GPa
Liquid surface tension. 0.669 N/m at 1360 °C

Thulium (Tm)

Compiled by K.A. Gschneidner, Jr.
and B.J. Beaudry, Ames Laboratory,
U.S. Department of Energy,
Iowa State University

Thulium is used in phosphors, ferrite bubble devices, and catalysts. Irradiated thulium (^{169}Tm) is used as a portable radiographic source. Thulium will remain shiny in air at room temperature. Turnings can be ignited and will burn white hot. Finely divided thulium should not be handled in air. Because of its high vapor pressure at its melting point, thulium should not be arc melted.

Structure

Crystal structure. Close-packed hexagonal: $P6_3/mmc$ D_4^{6h}; a = 0.35375 nm, c = 0.55540 nm at 25 °C
Minimum interatomic distance. At 24 °C: r_a = 0.17688 nm; r_c = 0.17236 nm; radius CN_{12} = 0.17462 nm

Mass Characteristics

Atomic weight. 168.93421
Density. 9.321 g/cm^3 at 24 °C; liquid, 9.0 g/cm^3 (estimated) at 1550 °C
Volume change on freezing. 6.9% contraction

Thermal Properties

Melting point. 1545 °C
Boiling point. 1950 °C
Coefficient of thermal expansion. At 24 °C. Linear: 13.3 μm/m · K. Linear along crystal axes: 8.8 μm/m · K along a axis, 22.2 μm/m · K along c axis. Volumetric: 39.8 × 10^{-6} per K
Specific heat. 0.1598 kJ/kg · K at 25 °C
Entropy. At 25 °C: 438.0 J/kg · K
Latent heat of fusion. 99.4 kJ/kg
Latent heat of vaporization. 1.374 MJ/kg at 25 °C
Latent heat of combustion. For cubic Tm_2O_3 at 25 °C: ΔH_c° = −5.59 MJ/kg Tm; ΔG_c° = −5.32 MJ/kg Tm
Recrystallization temperature. About 600 °C
Thermal conductivity. 16.9 W/m · K at 25 °C
Vapor pressure. 0.001 Pa at 599 °C; 0.101 Pa at 748 °C; 10.1 Pa at 964 °C; 1013 Pa at 1300 °C

Electrical Properties

Electrical resistivity. 676 nΩ · m at 25 °C; 56 nΩ · m at 4 K. At 25 °C: 880 nΩ · m along c axis
Ionization potentials. Tm(I), 6.18436 eV; Tm(II), 12.05 eV; Tm(III), 23.68 eV; Tm(IV), 42.69 eV
Hall coefficient. −0.18 nV · m/A · T at 20 °C
Temperature of superconductivity. Bulk thulium is not superconducting down to 0.35 K at atmospheric pressure.

Magnetic Properties

Magnetic susceptibility. Volume (mks units): χ_a = 0.0160 and χ_c = 0.0195 at 25 °C; obeys Curie-Weiss law above 55 K with an effective moment of 7.61 μ_B, θ_a = −17 K and θ_c = 41 K
Saturation magnetization. 2.79 T at 4.2 K along ⟨0001⟩
Magnetic transformation temperatures. Curie temperature, 32 K; a spin rearrangement at 42 K, Néel temperature, 58 K

Optical Properties

Color. Metallic silver

Nuclear Properties

Thermal neutron cross section. 125 b

Chemical Properties

General corrosion behavior. Thulium stays shiny in air at room temperature. Even at 1000 °C, the rate of oxidation is slow due to the formation of a dark, tightly adhering oxide on the surface of the metal.
Resistance to specific corroding agents. Thulium does not react with cold or hot water, but it reacts vigorously with dilute acids. The attack by concentrated sulfuric acid is slow. The presence of the fluoride ion retards acid attack due to the formation of TmF_3 on the surface of the metal.

Mechanical Properties

Tensile properties. About the same as those of erbium
Hardness. 48 HV
Poisson's ratio. 0.213
Elastic modulus. At 27 °C: tension, 74.0 GPa; shear, 30.5 GPa; bulk, 44.5 GPa

Ytterbium (Yb)

Compiled by K.A. Gschneidner, Jr.
and B.J. Beaudry, Ames Laboratory,
U.S. Department of Energy,
Iowa State University

Ytterbium is used in phosphors, ceramic capacitors, ferrite devices, and catalysts. Ytterbium (^{170}Yb), which has been formed by neutron irradiation of thulium (^{169}Tm), is used as a portable radiograph source; ytterbium foils are used to measure pressure and as stress transducers. Ytterbium will tarnish slightly at room temperature in air. Massive ytterbium can be handled in air, but should be stored in an inert atmosphere or vacuum. Finely divided ytterbium should not be handled in air.

Structure

Crystal structure. α phase, close-packed hexagonal, $P6_3/mmc$ D_4^h; a = 0.38799 nm, c = 0.63859 nm at 24 °C. β phase, face-centered cubic, $Fm3m$ O_5^h; a = 0.54848 nm at 24 °C. γ phase, body-centered cubic, $Im3m$ O_9^h; a = 0.444 nm at 763 °C
Minimum interatomic distance. 0.19392 nm at 24 °C

Mass Characteristics

Atomic weight. 173.04
Density. α phase, 6.903 g/cm^3 at 23 °C; β phase, 6.966 g/cm^3 at 24 °C; γ phase, 6.57 g/cm^3 at 763 °C; liquid, 6.21 g/cm^3 at 820 °C
Volume change on freezing. 5.1% contraction
Volume change on phase transformation. β to γ phase, 0.1% volume contraction on heating

Thermal Properties

Melting point. 819 °C
Boiling point. 1196 °C
Phase transformation temperature. α to β phase: A_s = 280 K; β to α, M_s ≅ 260 K; β to γ, 795 °C
Coefficient of thermal expansion. At 24 °C. Linear: 26.3 μm/m · K. Linear along crystal axes: 26.3 μm/m · K along a axis. Volumetric: 78.9 × 10^{-6} per K
Specific heat. 0.1543 kJ/kg · K at 25 °C
Entropy. 345.6 J/kg · K at 25 °C
Latent heat of fusion. 44.3 kJ/kg
Latent heat of transformation. 10.1 J/kg
Latent heat of vaporization. 0.8790 kJ/kg at 25 °C
Heat of combustion. For cubic Yb_2O_3 at 25 °C: ΔH_c° = −5.24 MJ/kg Yb; ΔG_f° = −5.00 MJ/kg Yb
Recrystallization temperature. About 300 °C
Thermal conductivity. 38.5 W/m · K at 25 °C
Vapor pressure. 0.001 Pa at 301 °C; 0.101 Pa at 400 °C; 10.1 Pa at 541 °C; 1013 Pa at 776 °C

Electrical Properties

Electrical resistivity. 250 nΩ · m at 25 °C; 10 nΩ · m at 4 K. Liquid, 1130 nΩ · m at 821 °C
Ionization potentials. Yb(I), 6.25394 eV; Yb(II), 12.184 eV; Yb(III), 25.03 eV; Yb(IV), 43.74 eV
Hall coefficient. +0.377 nV · m/A · T at 20 °C
Temperature of superconductivity. Bulk ytterbium is not superconducting down to 0.015 K at atmospheric pressure.

Magnetic Properties

Magnetic susceptibility. Volume: 3.4 × 10^{-6} mks at 17 °C

Optical Properties

Color. Metallic silver

Nuclear Properties

Thermal neutron cross section. 37 b

Chemical Properties

General corrosion behavior. Ytterbium tarnishes slightly in moist air. It oxidizes slowly at elevated temperatures and reacts readily with hydrogen at 250 °C.
Resistance to specific corroding agents. Ytterbium does not react with cold water, but it will tarnish in hot water. It reacts vigorously with dilute acids.

Mechanical Properties

Tensile properties. Tensile strength, 58 MPa; yield strength, 7 MPa; elongation, 43%; reduction in area, 92%
Hardness. 17 HV
Poisson's ratio. 0.207
Strain-hardening exponent. 0.62
Elastic modulus. At 27 °C: tension, 23.9 GPa; shear, 9.9 GPa; bulk, 30.5 GPa
Kinematic liquid viscosity. 0.430 mm^2/s at 824 °C
Liquid surface tension. 0.320 N/m at 820 °C

Yttrium (Y)

Compiled by K.A. Gschneidner, Jr. and B.J. Beaudry, Ames Laboratory, U.S. Department of Energy, Iowa State University

Yttrium is used in magnesium alloys and oxidation-resistant alloys; it is also used in garnets and ferrites for electronic components. Yttrium is a host material for rare earth phosphors, including the red color (Eu) in color television screens. Yttrium oxide is used to stabilize cubic zirconia for structural and electronic ceramics, and as an oxide dispersant in superalloys; it is a major component in the high-temperature oxide superconductors ($YBa_2Cu_3O_{7-x}$).

Yttrium tarnishes slowly in air at room temperature. Turnings can be ignited quite easily and burn with great evolution of heat. Finely divided yttrium should be handled with great care and should be kept away from air and oxidizing agents.

Structure

Crystal structure. α phase, close-packed hexagonal, $P6_3/mmc$; a = 0.36482 nm, c = 0.57318 nm at 25 °C. β phase, body-centered cubic, $Im3m$; a = 0.410 nm above 1478 °C
Slip planes. $\{10\bar{1}0\}\langle\bar{1}2\bar{1}0\rangle$ from −196 to 224 °C; $\{0002\}\langle\bar{1}2\bar{1}0\rangle$ from −196 to 224 °C
Twinning planes. $\{11\bar{2}1\}\langle\bar{1}\bar{1}26\rangle$ at 25 °C
Minimum interatomic distance. At 24 °C: r_a = 1.824 nm; r_c = 1.7783 nm; radius CN_{12} = 1.8012 nm
Fracture behavior. Primarily ductile

Mass Characteristics

Atomic weight. 88.90585
Density. α phase, 4.469 g/cm^3 at 24 °C; β phase, 4.28 g/cm^3 at 1478 °C; liquid, 4.24 g/cm^3 at 1525 °C

Thermal Properties

Melting point. 1522 °C
Boiling point. 3345 °C
Phase transformation temperature. 1478 °C
Coefficient of thermal expansion. Linear: 10.6 μm/m · K. Linear along crystal axes: 6.0 μm/m · K along a axis; 19.7 μm/m · K along c axis. Volumetric: 31.7 × 10^{-6} per K
Specific heat. 0.2981 kJ/kg · K at 25 °C
Entropy. At 298.15 K: 499.4 J/kg · K
Latent heat of fusion. 128.2 kJ/kg
Latent heat of transformation. hcp → bcc, 56.1 kJ/kg
Latent heat of vaporization. 4.777 MJ/kg at 25 °C
Heat of combustion. For cubic Y_2O_3 at 25 °C: ΔH_f° = −10.72 MJ/kg Y; ΔG_f° = −10.22 MJ/kg Y
Recrystallization temperature. 550 °C
Thermal conductivity. 17.2 W/m · K at 25 °C
Vapor pressure. 0.001 Pa at 1222 °C; 0.101 Pa at 1460 °C; 10.1 Pa at 1812 °C; 1013 Pa at 2360 °C

Electrical Properties

Electrical resistivity. 596 nΩ · m at 25 °C; 32 nΩ · m at 4 K. Along crystal axes at 25 °C: 725 nΩ · m along a axis; 355 nΩ · m along c axis
Ionization potentials. Y(I), 6.38 eV; Y(II), 12.24 eV; Y(III), 20.52 eV; Y(IV), 61.8 eV; Y(V), 77.0 eV
Hall coefficient. $R_{H,b}$ = −0.027 nV · m/A · T and $R_{H,c}$ = −0.16 nV · m/A · T at 25 °C
Temperature of superconductivity. Bulk yttrium is not superconducting down to 0.006 K at atmospheric pressure; it becomes superconducting at 1.3 K and 11 GPa.

Magnetic Properties

Magnetic susceptibility. Volume at 25 °C (mks units): 1.186 × 10^{-4}. Along crystal axes: 1.233 × 10^{-4} along a axis; 1.109 × 10^{-4} along c axis

Nuclear Properties

Thermal neutron cross section. 1.3 b

Optical Properties

Color. Metallic silver
Spectral hemispherical emittance. Solid: 36.8% for λ = 645 nm at 1200 to 1522 °C. Liquid: 36.8% for λ = 645 nm at 1522 to 1647 °C

Chemical Properties

General corrosion behavior. Yttrium metal remains shiny in air at room temperature; discoloration starts at ~350 °C.
Resistance to specific chemical agents. Yttrium metal reacts vigorously with hydrochloric and nitric acids. It does not react with hydrofluoric acid or with HCl or HNO_3 in the presence of the fluoride ion.

Mechanical Properties

Tensile properties. At 25 °C, annealed rod: tensile strength, 129 MPa; yield strength, 42 MPa; elongation, 34% in 25 mm
Hardness. 40 HV; highly anisotropic
Poisson's ratio. 0.243
Strain-hardening exponent. n = 0.23
Elastic modulus. Tension, Young's, 63.5 GPa; shear, 25.6 GPa; bulk, 41.2 GPa
Elastic modulus along crystal axes. c_{11} = 77.9 GPa; c_{12} = 29.2 GPa; c_{13} = 21.0 GPa; c_{33} = 76.9 GPa; c_{44} = 24.7 GPa
Liquid surface tension. 0.871 N/m at 1525 °C

Properties of the Actinide Metals (Ac-Pu)

Actinium (Ac)

Compiled by Lester R. Morss, Chemistry Division, Argonne National Laboratory

Structure

Crystal structure. Face-centered cubic ($Fm3m$) a_0 = 0.5315 ± 0.0005 nm

Mass Characteristics

Atomic weight. 227.0277
Density. 10.1 g/cm^3 (calculated from x-ray lattice parameter). Variation in density with temperature: unknown
Volume change on freezing. Unknown

Thermal Properties

Melting point. 1430 °C (estimated)
Boiling point. 3200 ± 300 °C (estimated)
Phase transformation temperature. No known solid-solid phase transformation
Coefficient of thermal expansion. Unknown
Entropy. S°_{298K} = 61.9 ± 0.8 kJ/mol (predicted)
Specific heat. Unknown
Enthalpy. Unknown
Latent heat (enthalpy) of fusion. 10.9 kJ/mol (estimated)
Latent heat (enthalpy) of sublimation. 418 ± 13 kJ/mol at 298 K
Enthalpy of oxide. Ac_2O_3 formation: −1756 ± 80 kJ/mol (estimated)
Free energy of oxide formation. Unknown
Thermal conductivity of metal. Unknown
Thermal conductivity versus temperature.

Unknown
Vapor pressure. 0.9 Pa (0.007 torr) at 1600 °C (estimated)
Diffusion coefficient. Unknown

Electrical Properties

Unknown

Magnetic Properties

Magnetic susceptibility. Unknown
Magnetic permeability. Unknown

Optical Properties

Color. Silvery white, sometimes with golden cast
Emissivity. Unknown

Nuclear Properties

Radioactive isotopes. All isotopes (^{209}Ac through ^{232}Ac) are radioactive. The longest-lived isotope, ^{227}Ac, has a half-life of 21.773 years and decays by β^- emission (98.62%) and α emission (1.38%). It is the most abundant isotope, and has been recovered in milligram quantities from uranium ores and produced in gram quantities by thermal neutron irradiation of ^{226}Ra:

$$^{226}\text{Ra}\ (n, \gamma)\ ^{227}\text{Ra}\ (\beta^-, 41.2\ \text{min})\ ^{227}\text{Ac}$$

This reaction can also be expressed as the end product of two separate equations:

$$^{226}\text{Ra} + n \rightarrow ^{227}\text{Ra} + \gamma$$

and

$$^{227}\text{Ra} \xrightarrow{\beta^-,\ 41.2\ \text{min}} ^{227}\text{Ac}$$

Effect of neutron irradiation: Unknown
Thermal neutron cross sections. $\sigma_c = 900 \pm 150$ b; $\sigma_f = 3.5 \times 10^{-4}$ b

Chemical Properties

Oxidizes rapidly in moist air. Oxide coating somewhat inhibits further attack.

Fabrication Characteristics

Unknown

Mechanical Properties

Tensile properties. Unknown
Compressive properties. Unknown
Hardness. Unknown
Poisson's ratio. Unknown

ACKNOWLEDGMENT

This work was performed under the auspices of the Office of Basic Energy Sciences, Division of Chemical Sciences, U.S. Department of Energy, under Contract W-31-109-ENG-38.

SELECTED REFERENCES

- *Actinium*, Supplement Vol, *Gmelin Handbook of Inorganic Chemistry*, 8th ed., Springer-Verlag, 1981
- J.J. Katz, G.T. Seaborg, and L.R. Morss, *The Chemistry of the Actinide Elements*, 2nd ed., Chapman and Hall, 1986
- J.W. Ward *et al.*, Thermochemical Properties of the Actinide Elements and Selected Actinide-Noble Metal Intermetallics, Chapter 7 in *Handbook on the Physics and Chemistry of the Actinides*, Vol 4, A.J. Freeman and C. Keller, Ed., Elsevier, 1986

Neptunium (Np)

Compiled by J.A. Fahey,
Bronx Community College,
City University of New York

Neptunium was the first artificial element to be discovered. It was produced by the bombardment of uranium with slow neutrons. Many isotopes of neptunium are known, and all are radioactive. ^{237}Np is the most stable isotope, with an α decay half-life of 2.14×10^6 years. The 59.6 keV γ ray associated with the α decay of ^{237}Np to an excited state of ^{233}Pa makes it important in the investigation of the electronic, structural, and magnetic properties of the solid compounds and metallic phases of neptunium by Mössbauer spectroscopy. The isotope ^{237}Np is used for most studies because of its long half-life and its relative availability. ^{237}Np and ^{239}Np are produced, along with ^{239}Pu, in the operation of conventional nuclear reactors. ^{237}Np is also important because it is the source material for the production of ^{238}Pu for use in atomic-powered batteries. Most of the health concerns associated with neptunium are related to the possible presence of residual amounts (0.5%) of ^{239}Pu, a strong carcinogen, due to incomplete separation in the purification process.

Structure

Crystal structure. α phase: orthorhombic (*Pnma*); $a = 0.4723$ nm, $b = 0.4887$ nm, $c = 0.6663$ nm at 25 °C. β phase: tetragonal, *P4/nmm*; $a = 0.4897$ nm, $c = 0.3388$ nm at 313 °C. γ phase: bcc *Im3m*; $a = 0.3518$ nm at 600 °C

The unit cell of α neptunium contains eight atoms. Half the atoms are in a site with the seven nearest neighbors at an average distance of 0.2968 nm (2.968 Å), and the other half are in a site with the five nearest neighbors at an average distance of 0.2854 nm (2.854 Å).

The unit cell of β neptunium contains four atoms. Half the atoms are in a site with the four nearest neighbors at an average distance of 0.3206 nm (3.206 Å), and the other half are in a site with the four nearest neighbors at an average distance of 0.3232 nm (3.232 Å).

The unit cell of γ neptunium contains two atoms. Each atom has its eight nearest neighbors at a distance of 0.297 nm (2.97 Å) when extrapolated to 20 °C.

Mass Characteristics

Atomic weight. ^{237}Np, 237.0482
Density. α phase: 20.48 g/cm^3 at 25 °C (x-ray); 20.25 g/cm^3 at 25 °C (measured). β phase: 19.38 g/cm^3 at 313 °C (x-ray); 19.31 g/cm^3 at 25 °C (measured). γ phase: 18.08 g/cm^3 at 600 °C (x-ray)

Thermal Properties

Melting point. 637 °C
Boiling point. ~3902 °C
Phase transformation temperatures. At 1 atm pressure: α to β, 280 °C; β to γ, 577 °C; γ to liquid, 637 °C
Coefficient of thermal expansion. Volumetric, as determined by x-ray diffraction. α neptunium between 20 and 275 °C: $\alpha_{100} = 24 \times 10^{-6}$ per K; $\alpha_{010} = 25 \times 10^{-6}$ per K; $\alpha_{001} = 34 \times 10^{-6}$ per K. β neptunium between 278 and 530 °C: $\alpha_{100} = \alpha_{010} = 64 \times 10^{-6}$ per K; $\alpha_{001} \cong 0.0$.
The thermal expansion of neptunium metal has been measured by dilatometry. A typical dilatometric run is shown in Fig. 140. The values of the coefficient of linear expansion were found to be 27.5×10^{-6} per °C for the α phase (40 to 240 °C) and 41×10^{-6} °C for the β phase (300 to 540 °C). The dilatometric behavior also shows a discontinuous change of slope at the phase transitions.
Specific heat. The heat capacity of pure neptunium metal reaches the Dulong-Petit value of 3*R* (where *R* is the universal gas constant) at 140 °C and 29.68 J · K^{-1} at 300 K. Measuring the specific heat of neptunium metal between 7.4 and 300 K yields a γ (coefficient of the electronic term in the specific heat) of 14 mJ/mol · K^2. The specific heat of neptunium metal is a smooth function over this temperature range. The Debye temperature of neptunium metal is calculated to be 240 ± 4 K.
Latent heat of phase transformation. α to β, 5607 J/mol; β to γ, 5272 J/mol
Latent heat of fusion. 5230 J/mol
Enthalpy of solution. Neptunium metal in 1.5 *M* HCl: −165.7 ± 0.2 kcal/mol at 25 °C
Enthalpy of oxide formation. NpO_2, −1074 ± 3 kJ/mol
Free energy of oxide formation. NpO_2, −1022 ± 3 kJ/mol
Vapor pressure. Liquid: $\log P = -(20\,610 \pm 1280)/T + (5.10 \pm 0.70)$, where *P* is in atm and *T* is in K. See Fig. 141.
Heat of vaporization. At 1800 K: 94.3 ± 5.9 kcal/mol
Entropy of vaporization. At 1800 K: 23.3 ± 3.2 cal/mol · K
Phase diagram and compressibility. The compressibility of neptunium metal is similar to that of neighboring actinides such as uranium and plutonium. The phase diagram shown in Fig. 142 has one triple point (tetragonal body-centered cubic/liquid) at a temperature of 1000 K (725 °C) and a pressure of 3.2 GPa. A pressure-volume study at room temperature of neptunium metal es-

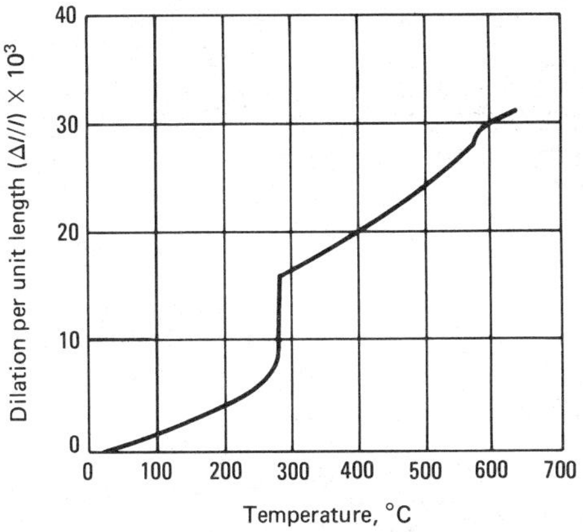

Fig. 140 Thermal expansion of neptunium metal. Source: Ref 1

tablished the bulk modulus to be 118 GPa and the pressure derivative to be 6.6 in the pressure range up to 52 GPa. At 1 atm pressure, the α phase is stable up to 280 °C, the β phase is stable up to 577 °C, and the γ phase is stable up to the melting point of 637 °C.

Electrical Properties

Temperature coefficients of resistivity. The resistivity of neptunium metal has been measured by standard potentiometric methods; the results are represented in Fig. 143 and Table 55. Table 55 also gives the variation of the temperature coefficient of resistivity with temperature in the three allotropes.

Thermoelectric power. The thermal electromotive force of a neptunium-platinum thermocouple has been measured and is represented in Fig. 144. The absolute thermoelectric power calculated from the slope of the electromotive force temperature curve and the absolute thermoelectric power versus platinum are shown in Fig. 145.

Magnetic Properties

The susceptibility of α-Np metal when measured on very pure samples has a room-temperature value of 560 × 10^{-6} emu/mol and is almost independent of temperature (Ref 5).

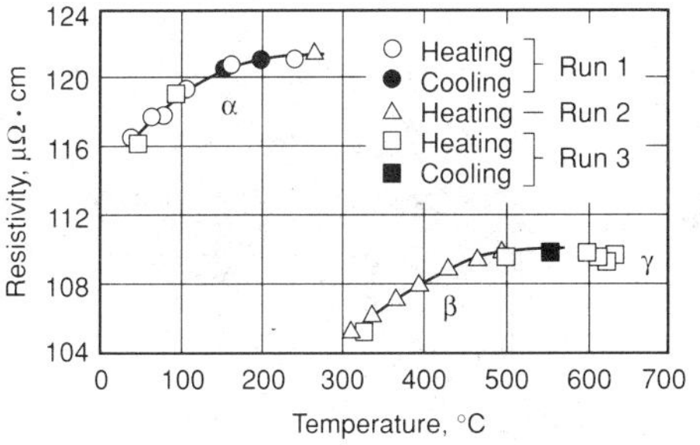

Fig. 143 Plot of resistivity versus temperature for neptunium metal. Source: Ref 4

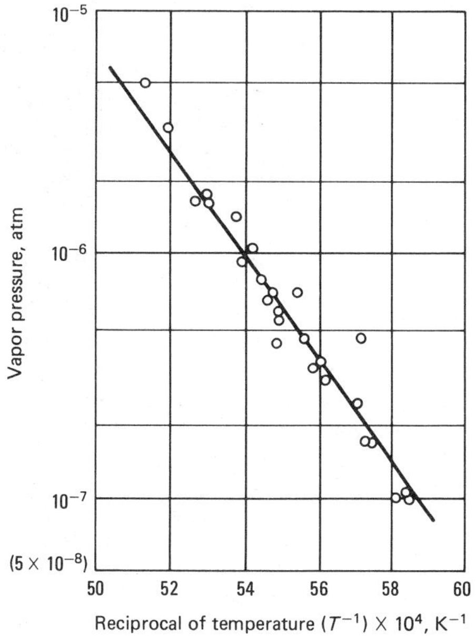

Fig. 141 Plot of vapor pressure versus temperature for neptunium metal. Source: Ref 2

Nuclear Properties

Isotopes from ^{227}Np to ^{242}Np are known; all are radioactive and have half-life periods ranging from ~60 s to 2.14 × 10^6 years. The latter half-life is associated with ^{237}Np, and accounts for the use of this isotope in most studies. The half-life of ^{237}Np is relatively short when compared with the estimated age of the earth; therefore, any primordial ^{237}Np would long since have decayed. However, small amounts of ^{239}Np are expected to occur in uranium minerals by continual formation from ^{238}U through the capture of neutrons from the spontaneous fission of ^{238}U.

Chemical Properties

Neptunium metal is silvery in appearance, about as malleable as uranium metal, and becomes covered with only a thin oxide layer when exposed to air for short periods. It reacts rapidly to form NpO_2 in air at high temperatures. NpO has been reported to form in vacuum at high temperature on the surface of the partially oxidized metal.

Direct reaction between hydrogen and neptunium metal results in the formation of at least two hydrides, NpH_2 and NpH_3. Neptunium hydride decomposes when heated in vacuum above 300 °C, yielding finely divided pyrophoric neptunium metal.

Phase diagrams for the Np-Pu and Np-U alloys all exhibit a common feature: the complete miscibility between γ neptunium and γ uranium and between γ neptunium and ε plutonium.

Several intermetallic compounds of neptunium with noble metals have been prepared by the reduction of NpO_2 with very pure hydrogen at 1300 °C in the presence of the noble metal. The intermetallics $NpAl_2$, $NpAl_3$, $NpBe_{13}$, NpB_6, $NpCd_6$, $NpCd_{11}$, and $NpPd_3$ have been prepared in this fashion (Ref 5).

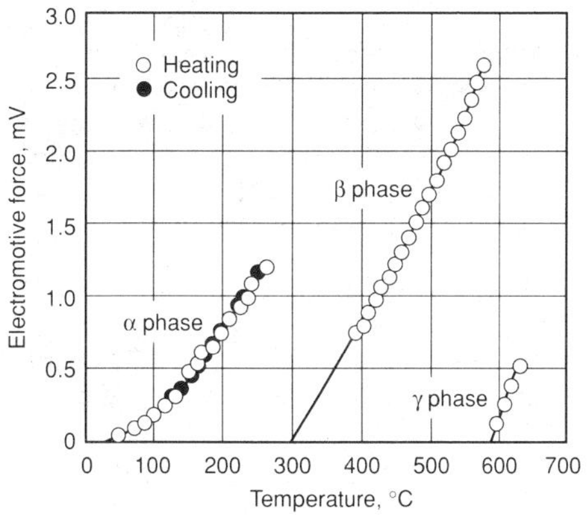

Fig. 144 Plot of thermal electromotive force versus temperature for neptunium metal. Source: Ref 4

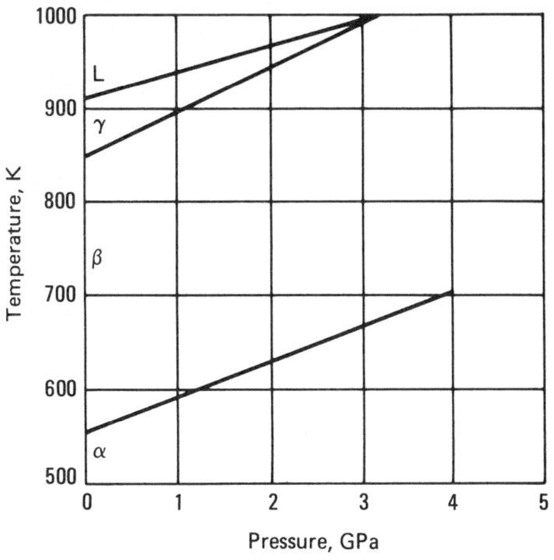

Fig. 142 Pressure versus temperature phase diagram for neptunium metal. Source: Ref 3

Mechanical Properties

Hardness. 346 HV

Elastic modulus. Shear modulus, 80 GPa; bulk modulus, 118 GPa

ACKNOWLEDGMENT

The author would like to thank the Chemists' Club Library of New York for the literature search they provided from the Chemical Abstracts data base.

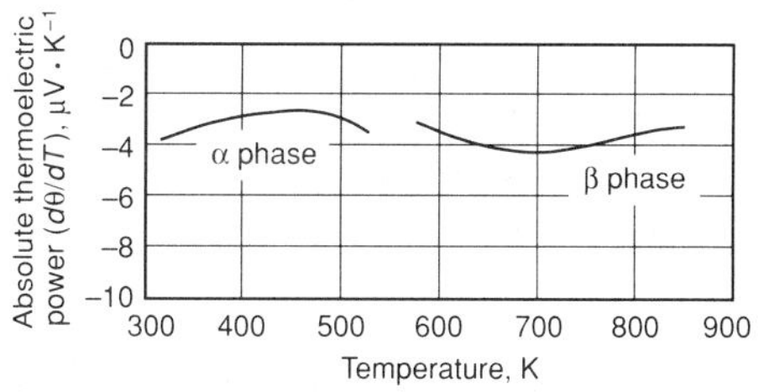

Fig. 145 Plot of absolute thermoelectric power versus temperature for neptunium metal. Source: Ref 4

Table 55 Effect of phase and temperature variation on the resistivity of neptunium

Temperature, K	Resistivity, μΩ · cm	Coefficient of resistivity × 10^5
α phase		
300	...	46
310	116.4	...
314	116.1	...
334	117.7	...
347	117.8	...
350	...	28
370	119.1	...
373	119.3	...
400	...	18
425	120.5	...
433	120.7	...
450	...	12
472	120.9	...
500	...	6
512	120.8	...
538	121.3	...
550	...	1
β phase		
586	105.3	...
600	105.3	31
612	106.2	...
641	107.1	...
669	107.9	...
705	108.9	...
740	109.5	...
771	109.9	...
773	109.7	...
828	109.8	...
840	...	0
γ phase		
873	109.8	...
875	...	−6
879	109.7	...
896	109.4	...
902	109.6	...

REFERENCES

1. J.A. Lee, P.G. Mardon, J.H. Pearce, and R.O.A. Hall, *J. Phys. Chem. Solids*, Vol 11, 1959, p 177-181
2. H.A. Eick and R.N.R. Mulford, *J. Chem. Phys.*, Vol 41, 1964, p 1475-1478
3. U. Benedict, *Handbook on the Physics and Diagram Chemistry of the Actinides*, A.J. Freeman and G.H. Lander, Ed., Elsevier, 1987
4. J.A. Lee, R.O. Evans, R.O.A. Hall, and E. King, *J. Phys. Chem. Solids*, Vol 11, 1959, p 278-283
5. J.-M. Fournier and R. Troc, Bulk Properties of the Actinides, Chapter 2 in *Handbook on the Physics and Chemistry of the Actinides*, A.J. Freeman and G.H. Lander, Ed., Elsevier, 1985

Plutonium (Pu)

Compiled by M.B. Brodsky,
Argonne National Laboratory

Revised for this Volume by
Dr. Brodsky and Michael Stevens,
Los Alamos National Laboratory

The term plutonium usually implies ^{239}Pu of at least 95% purity (generally 99.7 to 99.99 wt%). Small amounts of δ-phase stabilizers, such as 0.1 wt% Al, may cause retention of the δ phase at room temperature. The term plutonium, however, also implies ^{239}Pu sufficiently free of δ-phase stabilizers so that only the α phase is present at room temperature.

Typical uses of plutonium include nuclear weapons, nuclear fuel, neutron sources, heat sources for thermoelectric generators (especially ^{238}Pu), and production of higher isotopes and transplutonic elements. Plutonium is a highly radioactive α emitter, extremely poisonous, and is properly handled in glove boxes. It is about twice as poisonous as radium to the human system. The maximum permissible body burden is 0.6 μg. When handling quantities in excess of 300 g, the possibility of nuclear criticality must be considered.

Table 56 Crystal structure and density of various phases of plutonium

Phase	Lattice symmetry	Lattice constants, nm: a_0	b_0	c_0	Interaxial angle β	Density, g/cm³	Atoms per unit cell
α	Monoclinic	0.6183 (21 °C)	0.4822	1.0963	101.79°	19.86	16
β	Monoclinic	0.9284 (190 °C)	1.0463	0.7859	92.13°	17.70	34
γ	Orthorhombic	0.3159 (235 °C)	0.5768	1.0162	...	17.14	8
δ	Face-centered cubic	0.46371 (320 °C)	...	...	...	15.92	4
δ′	Body-centered tetragonal	0.334 (465 °C)	...	0.444	...	16.00	2
ε	Body-centered cubic	0.3636 (490 °C)	...	...	...	16.51	2

Sources: Ref 1 to 3

Structure

Crystal structure. See Table 56.
Slip planes. α phase (102), (112), (111), (101), (411), and (118) (Ref 4)
Twin planes. (102), (201) (Ref 5)
Minimum interatomic distance. From Ref 1 and 6:

Phase	Minimum distance, nm
α	0.25
β	0.297
γ	0.3026
δ	0.3279
δ′	0.3249
ε	0.3149

Metallography. Standard grinding procedures using kerosene or carbon tetrachloride as a lubricant are used. Polishing is done on microcloth charged with γ alumina or 1 μm diamond. Suitable etches contain tetraphosphoric acid, water, and 2-ethoxyethanol in the following proportions: 7:36:57 for low-temperature phases; 12:33:55 for δ phase; 2:3:5 for long etching. Tetraphosphoric acid can be replaced by orthophosphoric acid and less water. Additional etches include dimethylformamide and HNO_3 in an 80:20 mixture. This electrolyte can be used on α metal for electrolytic polishing and etching at room temperature and 10 to 12 V_{dc}. A 90:10 mixture of ethylene glycol and HNO_3 may be used on δ metal at 10 V_{dc}. Residual powders must be collected regularly to prevent the buildup of potentially critical amounts.

Fracture behavior. The metal exhibits little toughness ($K_{Ic} \cong 11.8$ MPa$\sqrt{m}$) (Ref 7); however, the fracture micromechanism is microvoid coalescence.

Mass Characteristics

Atomic weight. 239.052
Density. See Table 56.
Compressibility. α plutonium: approximately 0.2 per Pa at atmospheric pressure to 0.05 per Pa at 10 GPa; the volume at 10 GPa is about 90% of the volume at atmospheric pressure (Ref 1, 3). β plutonium at 200 °C: 0.23 ± 0.10/Pa in the range from 0 to 200 MPa
Volume change on transformation. See Table 57.

Thermal Properties

Melting point. 640 °C
Boiling point. 3235 °C (Ref 3)
Phase transformation temperature. See Table 57.
Coefficient of thermal expansion. See Table 58 and Fig. 146.
Specific heat. α phase, 33.9 kJ/kg · K at 25 °C; β phase, 41.4 kJ/kg · K at 160 °C; γ phase, 46.0 kJ/kg · K at 280 °C; δ phase, 45.6 kJ/kg · K at 350 °C; δ′ phase, 55.3 kJ/kg · K at 455 °C; ε phase, 43.5 kJ/kg · K at 500 °C (Ref 1)
Latent heat of fusion. See Table 57.
Latent heat of phase transformation. See Table 57.
Latent heat of vaporization. 336.9 kJ/mol
Latent heat of combustion. 1058.7 kJ/mol Pu
Recovery temperature. 109 °C

Table 57 Transformation properties of plutonium

Transformation	Temperature, °C	Volume change, %	Heat of transformation, kJ/kg
α to β	120	9	17.6
β to γ	210	2.5	2.6
γ to δ	315	6.9	2.5
δ to δ′	452	−0.4	0.4
δ′ to ε	480	−2	7.4
ε to liquid	640	−1 to −2	13.1

Sources: Ref 1, 3, 8

Table 58 Coefficient of linear thermal expansion for plutonium

Phase	From dilatometric data: Coefficient, μm/m · K	From dilatometric data: Temperature, °C	From x-ray data: Direction of expansion	From x-ray data: Coefficient, μm/m · K	From x-ray data: Temperature, °C
α	67	80–120	α_1 perpendicular to c axis	66	21–104
			α_2 parallel to b axis	73	
			α_3 parallel to c axis	29	
			Average	56	
β	41	160–200	α_1	94	93–190
			α_2 parallel to b axis	14	
			α_3 perpendicular to (101)	18	
			Average	42	
γ	35	220–280	α_1 parallel to a axis	−19.7 ± 1.0	210–310
			α_2 parallel to b axis	39.5 ± 0.6	
			α_3 parallel to c axis	84.3 ± 1.6	
			Average	34.7 ± 0.7	
δ	−8.6	340–440	α	−8.6 ± 0.3	320–420
δ′	−596(a)	470	α_1 parallel to a axis	444.8 ± 12.1	450–479
			α_2 parallel to c axis	−1063.5 ± 18.2	
			Average	−57.9 ± 10.1	
ϵ	15	490–550	α	36.5 ± 1.1	490–550
Liquid	50 ± 25(b)	665	. . .	. . .	. . .

(a) Value exaggerated by creep during residual transformation. (b) Volumetric. Source: Ref 6

Table 59 Chemical diffusion of plutonium

Composition, at.% Pu	Diffusion coefficient, cm²/s	Activation energy, kJ/mol
Magnesium-plutonium system		
0.045	1×10^{4}	150
0.562	2.45×10^{-2}	118.6
1.124	1.05×10^{-2}	118.5
1.686	3.6×10^{-4}	93.70
Plutonium-zinc system (δ-Pu)		
38.5	7.70×10^{-6}	98.39
46.2	3.56×10^{-7}	63.01
61.6	8.86×10^{-9}	49.40
69.3	9.60×10^{-9}	52.96
Uranium-plutonium system		
1.75	0.14×10^{-7}	56.1
3.50	0.15×10^{-7}	57.4
5.25	0.18×10^{-7}	59.0
7.00	0.28×10^{-7}	63.6
8.75	0.44×10^{-7}	68.2
10.50	0.88×10^{-7}	74.9
12.25	1.18×10^{-7}	78.7
14.00	2.00×10^{-7}	83.7
15.75	2.57×10^{-7}	86.2

Source: Ref 4

Thermal conductivity. 6.5 W/m · K at room temperature
Vapor pressure. $\log P = -17\,587/T + 10.02$, where P is in Pa and T is in K (Ref 3)
Chemical diffusion. See Table 59. Self-diffusion coefficient, ϵ phase: 1.2×10^{-7} cm²/s at 500 °C

Electrical Properties

Electrical resistivity. See Table 60.
Thermodynamic potential. See Table 60.
Electrochemical equivalent. Valence 3, 0.8256 mg/C; valence 4, 0.6142 mg/C
Hall coefficient. +35 pV · m/A · T at room temperature (Ref 4)
Superconductivity. None found at 1.3 K in metal that is 99.99% pure

Magnetic Properties

No magnetic ordering has been found in any of the phases of plutonium, whether pure or alloyed. However, compounds of plutonium are often magnetic, especially when the Pu-Pu distance increased beyond 0.34 nm.
Magnetic susceptibility. See Table 61.

Optical Properties

Color. White. When slightly oxidized, yellow tarnish; when heavily oxidized, green-black

Nuclear Properties

Unstable isotopes. See Table 62.
Thermal neutron cross section. See Table 62.

Chemical Properties

General corrosion behavior. Plutonium is a highly reactive metal, similar in reactivity to the rare earths.
Resistance to specific corroding agents. Relatively inert to dry air but corrodes rapidly if traces of moisture are present (Ref 1). Reacts slowly with water at room temperature

Table 60 Electrical properties of plutonium

Phase	Temperature, °C	Electrical resistivity, μΩ · cm	Temperature coefficient of resistivity, 10⁴/°C	Thermoelectric potential versus platinum(a), mV
α	−223	128.0	184.05	. . .
	107	141.4	−2.08	1.44
β	147	108.5	−0.62	2.23
γ	232	107.8	−0.50	3.81
δ	352	100.4	0.72	5.92
δ′	462	102.1	4.43	7.63
ϵ	501	110.6		8.31

(a) Reference junction at 20 °C. Sources: Ref 1, 3, 6, 9

Table 61 Magnetic susceptibility of plutonium

Allotrope	Temperature, °C	Mass susceptibility, emu/g	Temperature range, °C	Mean temperature coefficient × 10^{-5}
α	20	0.0280	20–118	−1.8
β	132	0.0290	132–198	−16.4
γ	224	0.0280	224–302	−11.5
δ	358	0.0268	358–446	−12.3
δ′	464	0.0266	464–477	36.3
ϵ	488	0.0270	488–570	−12.5

Sources: Ref 1, 3

Table 62 Nuclear properties of plutonium

Isotope	Half-life, years	Emitted particles	Cross sections, b: Capture	Cross sections, b: Fission
238	86.4	α (5.49, 4.45 MeV), γ	403	16.8
239	2.4×10^{4}	α (5.15 MeV), γ	315	746
240	6.6×10^{3}	α (5.16 MeV), γ	250	0.03
241	13.2	β^- (0.021 MeV), γ	390	1010
242	3.8×10^{5}	α (4.90 MeV), γ	19	<0.2

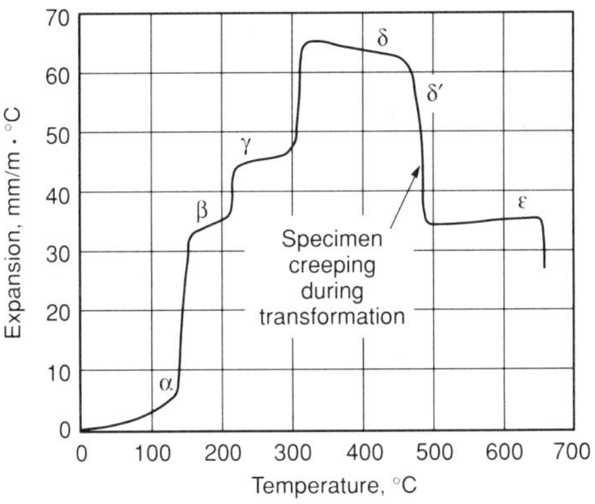

Fig. 146 Thermal expansion of plutonium. Source: Ref 6

Fabrication Characteristics

Machinability. Similar to that of 3003 Al
Recrystallization temperature. Approximately 120 °C
Casting temperature. In vacuum: 800 to 900 °C
Alloying practice. Low-melting elements are commonly added as pure metals to the molten bath. Alloys of refractory metals are added as master alloys.
Deoxidizers. Cerium and calcium have been used in deoxidizing plutonium. Plutonium melts have been made under potassium chloride-sodium chloride covers.
Melting practice. High-vacuum furnaces are commonly used to melt plutonium alloys. Magnesia and coated graphite crucibles are used to 1200 °C, and thoria crucibles are used to 1500 °C. Tantalum can be used to 1000 °C to contain molten plutonium. Magnesia, graphite, and copper are suitable mold materials.
Hot-working temperature. Can be worked readily in the δ (fcc) temperature range, 312 to 458 °C. β plutonium is ductile and can be worked (Ref 8).
Heat treatment. Plutonium is given a cold treatment at −23 °C to complete the β to α transformation (Ref 1).

Mechanical Properties

The mechanical properties of plutonium depend heavily on microstructure. They are especially sensitive to the presence of microcracks caused by the large volume change associated with the β-to-α phase transformation.
Tensile properties. Typical for cast α at 25 °C (Ref 1, 4, 9): tensile strength, 415 MPa; yield strength, 275 MPa; elongation, 0.2 to 0.5%; proportional limit, 160 MPa
Compressive properties. Typical for cast α at 25 °C (Ref 1, 4, 9): compressive strength, 830 MPa; compressive yield strength, 415 MPa
Hardness. 250 to 283 HV, 10 kg load (Ref 4)
Poisson's ratio. 0.15 to 0.21
Elastic modulus. Tension, 107 GPa; shear, 45 GPa
Fatigue strength. Typical, rotating beam: 90 MPa at 10^8 cycles
Liquid surface tension. 0.5 N/m
Viscosity. Dynamic, molten Pu: 7.4 mPa · s at 650 °C; 6.2 mPa · s at 750 °C

REFERENCES

1. W.N. Miner *et al.*, Plutonium, in *Rare Earth Metals Handbook*, 2nd ed., Reinhold, 1961
2. A.S. Coffinberry and W.N. Miner, Ed., *The Metal Plutonium*, American Society for Metals, 1961
3. A.S. Coffinberry and M.B. Waldron, Ed., The Physical Metallurgy of Plutonium, in *Progress in Nuclear Energy*, Vol I, Series V, Pergamon Press, 1956
4. J.H. Kittel *et al.*, Plutonium and Plutonium Alloys as Nuclear Fuel Materials, in *Nuclear Design and Engineering*, C.F. Bonilla and T.A. Jaegger, Ed., North-Holland, 1971
5. T.G. Zocco, R.I. Sheldon, and M.F. Stevens, *J. Nucl. Mater.*, Vol 165, 1989, p 238-246
6. E.L. Francis, "Plutonium Data Manual," I.G.R. 161 (RG/R), Industrial Group Headquarters, 1959
7. S. Beitcher and W.D. Ludemann, *Plutonium and Other Actinides*, H. Blank and R. Lindner, Ed., North-Holland, 1976, p 719-724
8. W.D. Wilkinson, Ed., *Extractive and Physical Metallurgy of Plutonium and Its Alloys*, Interscience, 1960
9. E. Grison and W.P.H. Lord, Ed., *Second International Conference on Plutonium Metallurgy*, Cleaver-Hume Press, 1960

SELECTED REFERENCE

• O.J. Wick, Ed., *Plutonium Handbook: A Guide to the Technology*, Gordon & Breach, 1967

Protactinium (Pa)

Compiled by Lester R. Morss, Chemistry Division, Argonne National Laboratory, and Barbara Cort, Los Alamos National Laboratory

Only in recent years has protactinium been available in sufficient amounts to characterize. It is radioactive and must be handled in a glove box. Although ^{231}Pa is an α emitter, daughters emit both β and γ radiation. Protactinium is also chemically reactive and oxidizes easily in air; it should be stored in an inert atmosphere.

Structure

Crystal structure. α phase: body-centered tetragonal (*I4/mmm*); a_0 = 0.3921 ± 0.0001 nm, c_0 = 0.3235 ± 0.0001 nm at 300 K. β phase: face-centered cubic (*Fm*3*m*), a_0 = 0.5018 ± 0.0001 nm at 1775 K (Ref 1)

Mass Characteristics

Atomic weight. ^{231}Pa, 231.0359
Density. 15.43 g/cm^3 at 300 K (Ref 2)
Volume change on freezing. Unknown

Thermal Properties

Melting point. 1845 ± 20 K (Ref 2)
Boiling point. 4300 K (Ref 3)
Phase transformation temperature. 1438 K
Coefficient of thermal expansion. Volume, 303 to 773 K: 18 × 10^{-6}/K. Linear, 303 to 973 K, 9.9 × 10^{-6}/K
Specific heat. 60 mJ/mol · K at 5 K, 2550 mJ/mol · K at 17 K; γ = 5 mJ/mol · K^2; Debye temperature, 185 ± 5 K; density of states = 1.54 states/eV/atom (solid) (Ref 4)
Entropy. S°_{298K} = 51.8 J/mol · K (estimated)
Enthalpy. Unknown
Heat of fusion. 12.3 kJ/mol (Ref 2)
Heat of vaporization. 569 kJ/mol (Ref 2)
Latent heat of phase transition. Unknown
Enthalpy of oxide formation. PaO_2, −1109 kJ/mol (solid, 298 K) (estimated); Pa_2O_5, unknown
Free energy of oxide formation. −1044 kJ/mol (estimated); Pa_2O_5, unknown. For PaO_2(s): $\Delta G^\circ = -1087 - 0.166\,T$, where ΔG° is in kJ/mol and T is in K
Thermal conductivity of metal. 0.47 W/cm · K (estimated)
Thermal conductivity versus temperature. Unknown
Vapor pressure. Estimated as 1 × 10^{-8} atm at 2300 K. Over protactinium (liquid) in the temperature region 2500 to 2900 K: $\log_{10}P = [(31328 \pm 375)/T] + (10.83 \pm 0.13)$, where P is in Pa and T is in K
Diffusion coefficient. Unknown

Electrical Properties

Electrical resistivity. 15 μΩ · cm at 300 K (Ref 5)
Superconducting transition temperature. 0.43 K (Ref 6)

Magnetic Properties

Magnetic susceptibility. 190 emu/mol from 20 to 298 K; property is independent of temperature.
Magnetic permeability. Unknown

Optical Properties

Color. Golden cast
Emissivity. Unknown

Nuclear Properties

Radioactive isotopes. All isotopes (^{215}Pa through ^{238}Pa) are radioactive. The longest-lived isotope, ^{231}Pa, has a half-life of 32 760 years and decays by α emission. It is the most abundant isotope, having been recovered in gram quantities from uranium ores.
Thermal neutron cross sections. σ_c = 210 ± 20 b; σ_f = 0.010 ± 0.005 b. Effect of thermal neutron irradiation is unknown.

Chemical Properties

Little or no tarnishing in air for several months; oxidation at 300 °C. Attacked by 8 *M* HCl, 2.5 *M* H_2SO_4, and 12 *M* HF, but reaction ceases. Best solvents are 8 *M* HCl-1 *M* HF and 12 *M* HCl-0.05 *M* HF.

Fabrication Characteristics

Malleable and ductile

Mechanical Properties

Tensile properties. Unknown
Compressive properties. Unknown
Hardness. Malleable and ductile
Poisson's ratio. Unknown
Elastic modulus. Bulk modulus, 157 ± 5 GPa (Ref 7)

ACKNOWLEDGMENT

Dr. Morss' work was performed under the auspices of the Office of Basic Energy Sciences, Division of Chemical Sciences, U.S. Department of Energy, under Contract W-31-109-ENG-38.

REFERENCES

1. J. Bohet and W. Muller, *J. Less-Common Met.*, Vol 57, 1978, p 185
2. J.W. Ward, P.D. Kleinschmidt, and D.E. Peterson, Thermochemical Properties of the Actinide Elements and Selected Actinide-Noble Metal Intermetallics, in *Handbook on the Physics and Chemistry of the Actinides*, Vol 4, A.J. Freeman and C. Keller, Ed., Elsevier, 1986, p 309-412
3. P.D. Kleinschmidt and J.W. Ward, *J. Less-Common Met.*, Vol 121, 1986, p 61
4. G.R. Stewart, J.L. Smith, J.C. Spirlet, and W. Muller, Low Temperature Specific Heat of Protactinium Metal, in *Superconductivity in d- and f-Band Metals*, H. Suhl and M.B. Maple, Ed., Academic Press, 1980, p 65
5. R.O.A. Hall and M.J. Mortimer, *J. Low Temp. Phys.*, Vol 27, 1977, p 313
6. J.L. Smith, J.C. Spirlet, and W. Muller, *Science*, Vol 205, 1979, p 188
7. U. Benedict, J.C. Spirlet, C. Dufour, I. Birkel, W.B. Holzapfel, and J.R. Peterson, *J. Magn. and Magn. Mater.*, Vol 19, 1982, p 287

SELECTED REFERENCES

- *Protactinium*, supplement volume 2, *Gmelin Handbook of Inorganic Chemistry*, 8th ed., Springer-Verlag, 1981
- J.J. Katz, G.T. Seaborg, and L.R. Morss, *The Chemistry of the Actinide Elements*, 2nd ed., Chapman and Hall, 1986
- J.W. Ward *et al.*, Thermochemical Properties of the Actinide Elements and Selected Actinide-Noble Metal Intermetallics, Chapter 7 in *Handbook on the Physics and Chemistry of the Actinides*, Vol 4, A.J. Freeman and C. Keller, Ed., Elsevier, 1986

Thorium (Th)

Compiled by J.F. Smith,
Ames Laboratory, U.S. Department of Energy, Iowa State University

Thorium, as a solid or fluid in elemental, intermetallic, or oxide form, is used as a fuel for nuclear reactors because it is a fertile material for the generation of fissionable ^{233}U. The oxide form of thorium is used for gas mantles. Thorium oxide additions control grain size in tungsten filaments. Thoria-dispersed nickel alloys (TD nickel) contain thorium oxide additions for increased strength. Thorium metal is used as an alloying addition in magnesium technology and as a deoxidant for molybdenum, iron, and other metals. Thorium has a variety of applications in electronic technology.

Thorium is radioactive. Pure, fresh thorium is a weak α emitter, but old thorium, with accumulated decay products, also emits β particles and penetrating γ rays. Thorium is chemically quite reactive. In finely divided form, thorium can be pyrophoric; in dust form, it may be explosive. Chemical toxicity of thorium and its compounds is generally low.

Structure

Crystal structure. α phase: face-centered cubic, *A*1, *cF*4 (*Fm*3*m*); $a = 0.5086$ nm at 25 °C. β phase: body-centered cubic, *A*2, *cI*2 (*Im*3*m*); $a = 0.411$ nm at 1450 °C
Slip planes. Deformation textures imply that {111} slip planes are active throughout the temperature range of study, −196 to 900 °C.
Minimum interatomic distance. 0.3596 nm at 25 °C
Microstructure. Common inclusions in thorium metal are gray ThO_2, a so-called white phase of debated identity that often surrounds cast grains of calcium-reduced thorium, gold-colored nitrides, and an occasional massive particle of tungsten in arc melted material. A fine Widmanstätten-like microstructure with cream-colored needles and angular inclusions, which turn deep blue after exposure to air for 1 day, can be produced by melting in graphite crucibles.

Mass Characteristics

Atomic weight. 232.038
Density. Solid: 11.8 g/cm^3 at −273 °C; 11.72 g/cm^3 at 25 °C; 10.89 g/cm^3 at 1755 °C. Liquid: 10.35 g/cm^3 at 1755 °C
Density versus temperature. $\Delta d/d_0 \cdot K = -34.2 \times 10^{-6}$ at 25 °C
Volume change on freezing. −5%

Thermal Properties

Melting point. 1755 °C
Boiling point. ~4800 °C
Phase transformation temperature. β to α phase (cooling), 1345 °C
Coefficient of thermal expansion. Linear: 10.9 μm/m · K at −193 °C; 11.4 μm/m · K at 25 °C; 12.6 μm/m · K at 600 °C; 13.3 μm/m · K at 750 °C; 14.0 μm/m · K at 850 °C; 14.9 μm/m · K at 950 °C
Specific heat. 0.11308 kJ/kg · K at 25 °C
Specific heat versus temperature:

°C	J/kg · K
−193	92.61
−173	97.79
−153	101.58
−113	106.1
−73	108.9
−23	110.8
27	113.17
127	116.7
227	120.2
327	123.9
427	127.8
527	132.0
627	136.4
727	141.0

Enthalpy. $H_{298} - H_0 = 25.16$ kJ/kg at 25 °C
Entropy. $S_{298} = 226.9$ J/kg · K at 25 °C
Latent heat of fusion. 59.50 kJ/kg
Latent heat of phase transformation. 15.6 kJ/kg
Latent heat of sublimation. 2.539 MJ/kg at 25 °C
Enthalpy of oxide formation. ThO_2: −4.6460 MJ/kg at 25 °C
Free energy of oxide formation. ThO_2: −4.4274 MJ/kg at 25 °C
Thermal conductivity. 77 W/m · K at 25 °C
Thermal conductivity versus temperature. See Fig. 147.
Vapor pressure. Solid, $\log P = -(28\,780 \pm 620)/T + 10.997 \pm 0.333$ at 1484 to 1683 °C; liquid, $\log P = -(29\,770 \pm 220)/T + 11.030 \pm 0.098$ at 1747 to 2187 °C, where P is in Pa and T is in K
Diffusion coefficient. Where D is in m^2/s, activation energy is in kJ/mol, T is in K. Self-diffusion: $D = 1.7 \times 10^{-4} \exp(-327/RT)$ for α phase; $D = 0.5 \times 10^{-4} \exp(-230/RT)$ for β phase. Hydrogen in α phase: $D = 2.92 \times 10^{-7} \exp(-40.8/RT)$ at infinite dilution. Carbon in β phase: $D = 2.2 \times 10^{-6} \exp(-113/RT)$. Nitrogen in β phase: $D = 3.2 \times 10^{-7} \exp(-71/RT)$. Oxygen in β phase: $D = 1.3 \times 10^{-7} \exp(-94.1/RT)$

Electrical Properties

Electrical conductivity. Volumetric, 11% IACS
Electrical resistivity. 157 nΩ · m at 25 °C
Temperature coefficient of electrical resistivity. 0.560 nΩ · m per K
Pressure coefficient of electrical resistivity. 5.7 aΩ · m per GPa
Thermoelectric potential. See Fig. 148.
First ionization potential. 6.08 eV
Hall coefficient. −0.088 to −0.13 nV · m/A · T at 25 °C
Temperature of superconductivity. 1.390 K at zero field
Increase of electrical resistivity with carbon addition. 55 pΩ · m/ppm C by weight
Work function. 0.559 aJ

Magnetic Properties

Magnetic susceptibility. Volume: 60.7×10^{-6} mks at 25 °C
Magnetic permeability. 1.0000607 at 25 °C

Optical Properties

Color. A fresh surface exhibits a bright silvery luster; the surface darkens after prolonged exposure to air.
Emissivity. Total: 30% at 1127 °C; 30.5% at 1227 °C; 31% at 1327 °C; 32% at 1427 °C; 34% at 1527 °C. Change in emissivity at the α-β transition, <0.5%

Nuclear Properties

Unstable isotopes. Isotopes from ^{223}Th through ^{235}Th are known; all are radioactive, with half-life periods ranging from 0.9 s to 1.4×10^{10} years. This latter half-life is

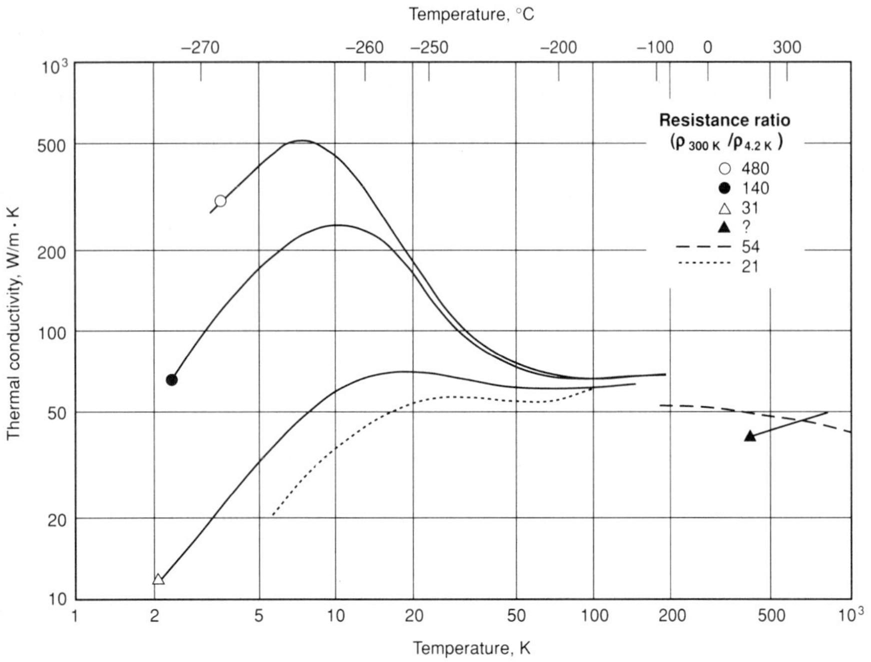

Fig. 147 Thermal conductivity of thorium

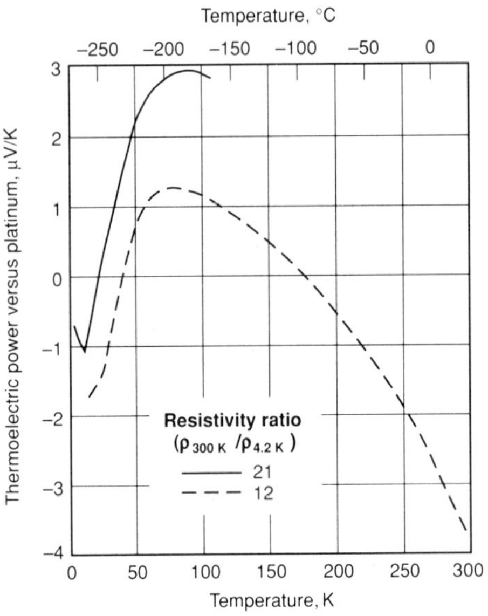

Fig. 148 Thermoelectric potential of thorium

associated with ^{232}Th, which is the isotope constituting essentially 100% of the natural abundance; ^{232}Th emits α particles with energies of 6.38×10^{-13} J.

Effect of neutron irradiation. Dimensional changes in irradiated thorium are essentially isotropic and relatively small. They are associated with the increase in volume from the accumulation of fission products in the material. Tensile strength has been found to increase 75% after a neutron exposure of 10^{19} nvt (where nvt is the neutron dose equivalent to the number of neutrons per square centimeter) and an additional 30% at double that exposure.

Chemical Properties

Resistance to specific corroding agents. In air between 100 and 900 °C, corrosion is principally oxidation and follows a linear reaction rate; at 800 °C, a weight gain of 2.88 kg/m^2 per day in air is typical. Above 850 °C, reaction with oxygen follows the parabolic law. Reaction with nitrogen also follows the parabolic law in the range of 671 to 1490 °C; in purified nitrogen, typical weight gain at 800 °C is 0.96 kg/m^2 per day. Thorium corrodes in water to form thorium oxide and hydrogen, and it loses weight by spallation; typical weight loss of unalloyed thorium in high-purity water at 178 °C is 0.109 kg/m^2 per day.

Fabrication Characteristics

Machinability. Thorium is readily machined. However, high-purity thorium tends to be gummy; its softness causes continuous turnings. Lower-purity material contains abrasive oxides, which cause rapid tool wear. It is advisable to use a water-soluble coolant when machining thorium.

Recrystallization temperature. ~650 °C (depends upon purity and amount of prior cold work)

Sintering temperature. 1100 to 1200 °C

Annealing temperature. Initial recovery: 525 °C. Recrystallization: 650 °C

Hot-working temperature. 750 to 900 °C

Mechanical Properties

Tensile properties. As-cast thorium (0.02 to 0.08 wt% C): tensile strength, 219 MPa; yield strength at 0.2% offset, 144 MPa;

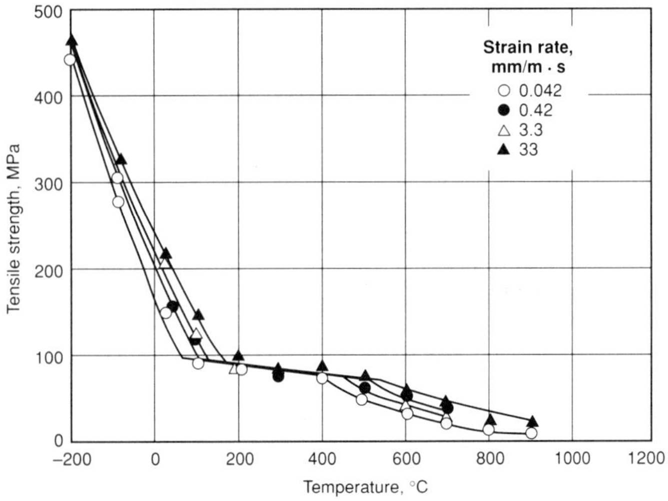

Fig. 149 Temperature dependence of the tensile strength of thorium

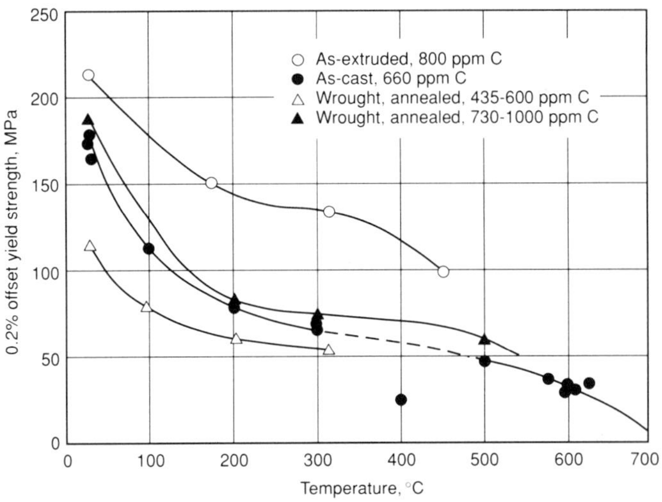

Fig. 150 Temperature dependence of the yield strength of thorium

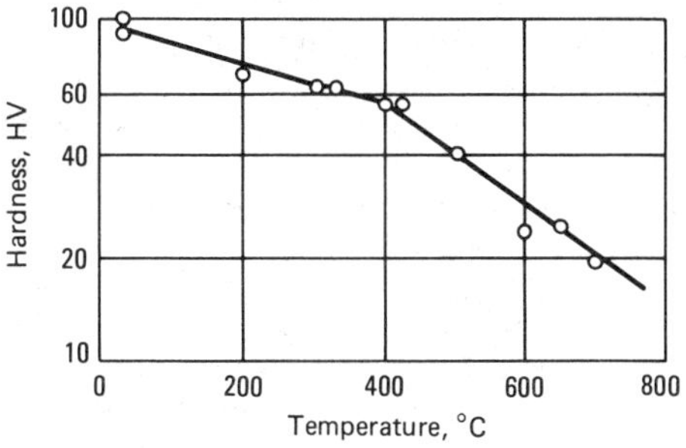

Fig. 151 Temperature dependence of the hardness of thorium

elongation, 34%; reduction in area, 35%. See also Fig. 149 and 150.
Compressive properties. Values closely comparable to tensile values
Hardness. 56 to 114 HV with 20 kg load; see also Fig. 151.
Poisson's ratio. 0.27
Strain-hardening exponent. 0.18 at 25 °C
Elastic modulus. Tension, 72.4 GPa; shear, 27.6 GPa; bulk, 57.7 GPa
Elastic properties along [100] crystal axis. Tension modulus, 60.1 GPa; shear modulus, 47.8 GPa; Poisson's ratio, 0.394
Impact strength. ~13 J for calcium-reduced thorium at 25 °C
Fatigue strength. Endurance limit: ~97 MPa for calcium-reduced thorium, reversed bending test
Specific damping capacity. 3×10^{-3} in 0 to 300 °C temperature range

SELECTED REFERENCE

- J.F. Smith *et al.*, *Thorium: Preparation and Properties*, Iowa State University Press, 1975

Uranium (U)

Compiled by Paul S. Dunn,
Los Alamos National Laboratory

Uranium can exist as a number of different isotopes. Natural uranium nominally contains 0.006% ^{234}U, 0.71% ^{235}U, and a balance of ^{238}U. The term enriched uranium designates uranium containing higher-than-natural ^{235}U; depleted uranium designates lower-than-natural ^{235}U; normal uranium contains the naturally occurring amount of ^{235}U. Other designations include α uranium, β uranium, and γ uranium for the three allotropic forms.

The most common use of uranium is in the nuclear and defense industries. Enriched uranium is used in nuclear reactor fuel elements and is usually a mixture of uranium oxide and plutonium dioxide. Enriched uranium is also used as a nuclear explosive. Depleted uranium is commonly used as a shielding material in particle accelerators and as a projectile material in conventional ordnance.

Massive depleted uranium offers no substantial problem in handling and storage. It oxidizes slowly in dry air and forms an adherent oxide. However, oxidation in moist air will result in a flaky oxide crust.

Finely divided uranium metal is pyrophoric, and care must be exercised to prevent fires. Uranium is a heavy-metal contaminant and must be processed under controlled conditions to avoid ingestion of fumes or dust.

Structure

Crystal structure. α: orthorhombic (*Cmcm*; *oC*4); a = 0.2854 nm, b = 0.5869 nm, c = 0.4856 nm at 298 K. β: complex tetragonal ($P4_2/mnm$; *tP*30); a = 1.0748 nm, c = 0.5652 nm at 950 K. γ: body-centered cubic (*Im*3*m*; *cI*2); a = 0.3535 nm at 1100 K
Slip planes. At 300 to 875 K, primary slip in α uranium is (010) [100], which cross slips onto (001).
Twinning planes. The most frequently observed type of twinning occurs on (130) and is operative as high as 873 K. The (172) twin is the second most frequently observed system in α uranium.

Mass Characteristics

Atomic weight. 238.029
Density. α: 19.05 g/cm³ at 298 K (from x-ray data); 18.7 to 19.0 g/cm³ for wrought metal. β: 18.3 g/cm³ at 973 K. γ: 17.91 g/cm³ at 1173 K. Liquid: 17.25 g/cm³ at 1410 K. Temperature coefficients: α, 0.001 g/cm³ · K; β, 0.0009 g/cm³ · K; γ, 0.0012 g/cm³ · K; liquid, 0.0016 g/cm³ · K
Volume change on melting. 2.2% expansion
Volume change on phase transformation. On cooling: β to α, 1.0% contraction; γ to β, 0.6% contraction

Thermal Properties

Melting point. 1406 K
Boiling point. 4091 K
Phase transformation temperature. α to β, 934 K; β to γ, 1042 K
Coefficient of thermal expansion. The thermal expansion of wrought α uranium is highly anisotropic and depends on fabrication history and the resultant preferred orientation. Linear: quenched α phase, 12 μm/m · K at 298 K and 28 μm/m · K at 900 K; β phase, 28 μm/m · K at 1000 K; γ phase, 20 μm/m · K between 1175 and 1400 K
Coefficients of linear thermal expansion (α) along crystal axes. α phase between 50 and 923 K: $\alpha_{[100]} = 2.422 \times 10^{-5} - 9.83 \times 10^{-9} T + 4.602 \times 10^{-11} T^2$; $\alpha_{[010]} = 3.07 \times 10^{-6} + 3.47 \times 10^{-9} T - 3.845 \times 10^{-11} T^2$; $\alpha_{[001]} = 8.72 \times 10^{-6} + 3.704 \times 10^{-8} T + 9.08 \times 10^{-12} T^2$, where T is in K. β phase: $\alpha_{[100]}$, 25 μm/m · K; $\alpha_{[001]}$, 5 μm/m · K
Specific heat versus temperature:

Temperature, K	Specific heat, J/kg · K
α phase	
300	117
600	145
800	172
900	190
β phase	
940	179
1040	179
γ phase	
1050	160
1300	160

The equation for specific heat of α uranium versus temperature is $C_p = 103.6 + 0.0180\,T + 8.49 \times 10^{-5} T^2$, where C_p is in J/kg · K and T is in K.
Latent heat of fusion. 38.72 kJ/kg at 1406 K
Latent heat of phase transformation. α to β, 12.3 kJ/kg at 943 K; β to γ, 20.1 kJ/kg at 1042 K
Latent heat of vaporization. 2.069 kJ/kg at 1406 K
Thermal conductivity:

Temperature, K	Thermal conductivity, W/m · K
α phase	
10	9.8
20	15.8
100	2.17
300	27.6
600	31.7
900	41.3
β phase	
1000	43.9
γ phase	
1100	46.3

Recrystallization temperature. Generally between 650 and 750 K, but highly dependent on purity and fabrication history
Vapor pressure. 1 μPa at 1500 K; 17.5 mPa at 2000 K
Enthalpy. At 298 K: 26.74 kJ/kg
Entropy. 211 J/kg · K at 298 K

Electrical Properties

Electrical resistivity. α phase, 300 nΩ · m at 300 K; β phase, 560 nΩ · m at 1000 K; γ phase, 540 nΩ · m at 1100 K; liquid, 66 nΩ · m at 1200 K. Temperature coefficient: α phase, 0.021 per K at 300 K, 0.039 per K at 900 K. Along crystal axes, α phase at 273 K: $\rho_{[100]}$, 390 nΩ · m; $\rho_{[010]}$, 240 nΩ · m; $\rho_{[001]}$, 262 nΩ · m
Hall coefficient. 380 nV · m/A · T at 300 K
Temperature of superconductivity. <0.5 K
Work function. 0.58 aJ

Magnetic Properties

Magnetic susceptibility. Volume: 390×10^{-6} mks at 300 K

Optical Properties

Spectral reflectance. 73.5% for $\lambda = 660$ nm
Spectral hemispherical emittance. 26.5% for $\lambda = 660$ nm

Nuclear Properties

Unstable isotopes. Table 63 lists data for α-particle emission.
Thermal neutron cross section. For 0.025 eV neutrons:

Isotope	Thermal neutron cross section, b Capture (σ_c)	Fission (σ_f)
Natural uranium	6.7	...
^{238}U	1.6	...
^{235}U	100	580

Mechanical Properties

Wide variations exist in all mechanical properties of α uranium, and these properties depend markedly on a large number of parameters, most notably preferred orientation, grain size, fabrication history, heat treatment, and type and distribution of impurities. For example, fracture stress decreases from approximately 600 MPa for a grain size of 1 μm to 130 MPa for a grain size of 10 μm.
Tensile properties. At ~293 K. As-cast: tensile strength, 400 MPa; 2% offset yield strength, 200 MPa; elongation, 4%; reduction in area, 10%. β annealed (grain size, 500 μm): tensile strength, 615 MPa. Wrought α uranium: tensile strength 1150 MPa; yield strength, 740 MPa; elongation, 7%; reduction in area, 14%
Hardness. Coarse-grain: α uranium, 185 HV at 300 K. Fine-grain: α uranium, 250 HV at 300 K; β uranium, 30 HV at 950 K; γ uranium, 1 HV at 1100 K
Poisson's ratio. See Table 64.
Elastic modulus. See Table 64.

Table 63 Properties of unstable uranium isotopes with α-particle emission

Isotope	Abundance, %	Half-life ($t_{1/2}$), years	Energy, MeV
^{234}U	0.0055	2.47×10^5	4.77, 4.72 4.58, 4.47,
^{235}U	0.720	7.1×10^6	4.40, 4.2
^{238}U	99.274	4.51×10^9	4.18

Table 64 Elastic properties of α uranium at room temperature

Condition	Modulus of elasticity(a) GPa	10^6 psi	Poisson's ratio
Swaged and annealed	201	29.1	0.22
Extruded	202	29.3	0.21
Cast	203	29.5	0.22
Hot rolled	203	29.5	0.20

(a) Young's modulus

Properties of the Transplutonium Actinide Metals (Am-Fm)

Compiled by R.G. Haire, Transuranium Research Laboratory, Oak Ridge National Laboratory

Availability and Nuclear Properties

The first six transplutonium metals, americium (Am), curium (Cm), berkelium (Bk), californium (Cf), einsteinium (Es), and fermium (Fm), are treated as a group rather than as individual elements because of the similarities in their physical properties and the limited amount of information available about them. All six are man-made radioactive elements with isotopes that decay mainly by α emission (5 to 7 MeV α particles); ^{249}Bk is an exception; it decays by β emission.

Beyond plutonium in the actinide series, the availability of each subsequent element diminishes rapidly. Einsteinium is the last element for which weighable (microgram) quantities are available. The low availability precludes obtaining solid-state properties for higher elements. The largest quantities of these six elements are produced using a neutron capture scheme in nuclear reactors; this method then governs the isotopes that are generated. Other isotopes of these elements can be produced in much smaller quantities using accelerators.

A limited summary of the commonly available isotopes, their availability, and other pertinent data are given in Table 65. Thermal neutron cross sections for selected transplutonium isotopes are listed in Table 66.

Applications and Basic Properties

Currently, the practical applications for these elements are limited, with most involving their radioactive nature. The ionizing radiation of ^{241}Am has been used in smoke detectors, and thermoelectric generators have employed the decay heat of curium isotopes. Also, the neutrons emitted (spontaneous fission decay branch) from ^{252}Cf have found use in cancer therapy, neutron radiography, and neutron activation analysis of remote areas (for example, the ocean floor).

In contrast to the earlier members of the actinide series, these six transplutonium elements tend to be more like the lanthanide elements in their properties and behavior. The tendency is for these six elements to be true *f* elements, formed by regularly adding a localized 5*f* electron in progression across the series. They differ from the lanthanide elements in their increased tendency toward being divalent metals when progressing to the higher members; divalency first occurs in the series at einsteinium. This latter tendency arises in the actinide series because of the increasing magnitude of the promotion energy ($f^n s^2$ to $f^{n-1} d s^2$) required to make a third electron available for bonding.

Some of the basic properties of these six metals are given in Table 67. Enthalpy of vaporization, which is a measure of cohesive energies, is plotted for each of the six transplutonium metals in Fig. 152; the enthalpy of each of their lanthanide homologs is plotted for comparison. The lower values for the actinides reflect this trend toward divalency. The lower enthalpy for the lanthanide europium is in accordance with its divalency. The vapor pressure over the solid phase of these transplutonium metals can be calculated from the relationships given in Table 68. The calculated boiling points of these metals are also listed in Table 68.

Crystal Structures

The first four transplutonium elements are trivalent metals; that is, they have three bonding or conduction electrons. They have a low-temperature phase with a double-hexagonal close-packed (dhcp) structure. These metals are isostructural with some of

Table 65 Nuclear properties for isotopes of the first six transplutonium elements

Isotope	Half-life	Available quantity	Specific heat, W/g
$^{241}_{95}$Am	433 years	kg	0.10
$^{243}_{95}$Am	7.38×10^3 years	g	6×10^{-3}
$^{244}_{96}$Cm	18.1 years	g	2.7
$^{248}_{96}$Cm	3.40×10^5 years	mg	5×10^{-4}
$^{249}_{97}$Bk	320 days	mg	1.1
$^{249}_{98}$Cf	351 years	mg	0.10
$^{252}_{98}$Cf	2.64 years	mg–g	49
$^{253}_{99}$Es	20.5 days	<1 mg	1.0×10^3
$^{254}_{99}$Es	276 days	μg	72
$^{255}_{100}$Fm	20.1 h	ng	2.4×10^4
$^{257}_{100}$Fm	100 days	10^{10} atoms	200

Table 66 Thermal neutron cross sections for selected transplutonium isotopes

Isotope	State	Thermal neutron cross section: Capture (σ_c), b	Thermal neutron cross section: Fission (σ_f), b
$^{241}_{95}$Am	Ground	54 (to $^{242m}_{95}$Am, metastable)	...
		533 (to $^{242}_{95}$Am, ground)	3.2(a)
$^{242m}_{95}$Am	Metastable	2000 (to $^{243}_{95}$Am, ground)	6950
$^{242}_{95}$Am	Ground	...	2100
$^{243}_{95}$Am	Ground	3.8 (to $^{244}_{95}$Am, ground)	0.198
		71.3 (to $^{244m}_{95}$Am, metastable)	...
$^{244}_{96}$Cm	Ground	15.2	1.04
$^{245}_{96}$Cm	Ground	369	2145
$^{246}_{96}$Cm	Ground	1.22	0.14
$^{247}_{96}$Cm	Ground	57	81.9
$^{248}_{96}$Cm	Ground	2.63	0.37
$^{249}_{97}$Bk	Ground	746	...
$^{249}_{98}$Cf	Ground	497	1642
$^{250}_{98}$Cf	Ground	2034	...
$^{251}_{98}$Cf	Ground	2850	4895
$^{252}_{98}$Cf	Ground	20.4	32
$^{253}_{99}$Es	Ground	178 (to $^{254m}_{99}$Es, metastable)	...
		5.8 (to $^{254}_{99}$Es, ground)	...
$^{254m}_{99}$Es	Metastable	...	1826
$^{254}_{99}$Es	Ground	28.3	1966
$^{255}_{100}$Fm	Ground	26	3360
$^{256}_{100}$Fm	Ground	45(b)	...
$^{257}_{100}$Fm	Ground	...	2950

(a) Total. (b) Estimated

Table 67 Physical properties of the first six transplutonium actinide metals

Element	Crystal structure at 298 K	Density, g/cm^3	Melting point, K	Enthalpy of vaporization at 298 K, kJ/mol	Entropy, ($S°$), J/mol · K(a)	Bulk modulus, GPa
Americium	dhcp	13.61	1446	284	55.4	45
Curium	dhcp	13.53	1620	387	72.0	37
Berkelium	dhcp	14.78	1323	310	77.0	52
Californium	dhcp	15.10	1173	196	80.3	50
Einsteinium	fcc	8.84	1133	134	89.4	15(b)
Fermium	fcc(b)	8.8(b)	1130(b)	142	87.2	15(b)

(a) Calculated crystal entropy at 298 K. (b) Estimated

Table 68 Vapor pressure (P) relationships for the solid phases and the calculated boiling points of the first six transplutonium actinide metals

P is obtained from the equation $\log_{10} P = a - b/T$, where P is in atm, T is in K, and a and b are coefficients listed in this table.

Element	a	b	Boiling point, K
Americium	6.578	14315	2340
Curium	6.082	19618	3383
Berkelium	5.78	15718	3173
Californium	5.675	9895	2018
Einsteinium(a)	5.642	7112	1269
Fermium(b)	5.474	7090	1350

(a) Calculated using data obtained from einsteinium-ytterbium alloys and assuming an ideal alloy behavior. (b) Calculated using data obtained from fermium-ytterbium alloys and assuming an ideal alloy behavior

the light lanthanide metals. Many of the lanthanide metals have a body-centered cubic (bcc) high-temperature phase; however, these transplutonium metals have a high-temperature fcc (β) phase. There is some evidence from dilatometry and differential thermal analysis to suggest that the fcc forms of americum and curium may transform to a bcc form just prior to melting, but x-ray confirmation of these bcc forms has not been obtained.

Einsteinium is the first divalent actinide metal, and it displays an fcc structure with a much larger lattice parameter than the fcc phases of the first four (Am, Cm, Bk, Cf) transplutonium metals. Structural data do not exist for fermium metal, although it is also a divalent metal (based on the magnitude of its enthalpy of vaporization), and it is likely to be isostructural with einsteinium metal. Both einsteinium and fermium metals are best compared to the divalent lanthanide metals europium and ytterbium. The lattice constants, atomic volumes, and metallic radii for these transplutonium metals are provided in Table 69. The phase and melting-point behaviors of the first six transplutonium actinide metals are illustrated in Fig. 153.

Under pressure, the first four transplutonium metals undergo two or three phase transitions and ultimately form the α-uranium-type, orthorhombic structure. The structural sequence is dhcp to fcc to distorted fcc (fcc′) to orthorhombic. Both americium and californium form this distorted fcc structure, whereas curium and berkelium metals do not. The formation of the α uranium structure has been interpreted to signify the partial delocalization of the $5f$ electrons and their participation in the metallic bonding. After release of the applied pressure, the sequence is reversed, but some hysteresis may occur between the fcc-to-dhcp transition.

The pressure behavior of the first four transplutonium metals is summarized in the block diagram in Fig. 154; the dashed line indicates the variation of the delocalization

Table 69 Structural properties of the first six transplutonium metals

Element	Structure	Lattice constants, nm: a_0	Lattice constants, nm: c_0	Atomic volume, nm^3 × 10^3	Metallic radius, nm(a)
Americium	dhcp	0.3468	1.1241	29.27	0.1725
	fcc	0.4894	...	29.30	0.1730
Curium	dhcp	0.3496	1.1331	29.98	0.1739
	fcc	0.493	...	30.0	0.174
Berkelium	dhcp	0.3416	1.1069	27.96	0.1699
	fcc	0.482	...	28.0	0.170
Californium	dhcp	0.3384	1.1040	27.37	0.1691
	fcc	0.478	...	27.3	0.169
Einsteinium(b)	fcc	0.575	...	47.5	0.203
Fermium(b)	fcc	0.57(c)	...	46.0(c)	0.20(c)

(a) Based on CN_{12}. (b) Divalent actinide metals; comparable to divalent europium and ytterbium metal in the lanthanide series. (c) Estimated

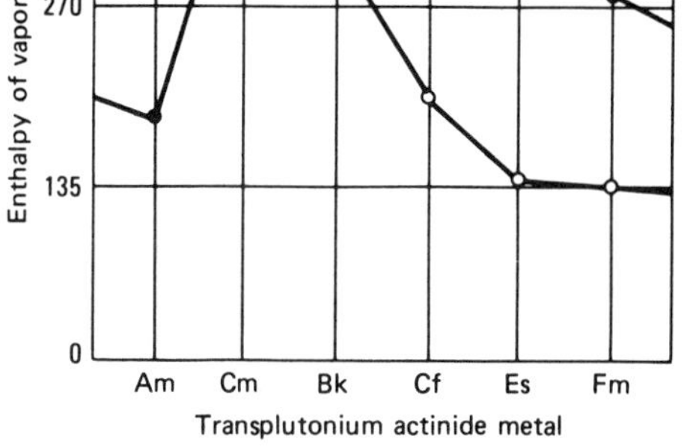

Fig. 152 Enthalpy of vaporization for the first six transplutonium actinide metals and their lanthanide homologs

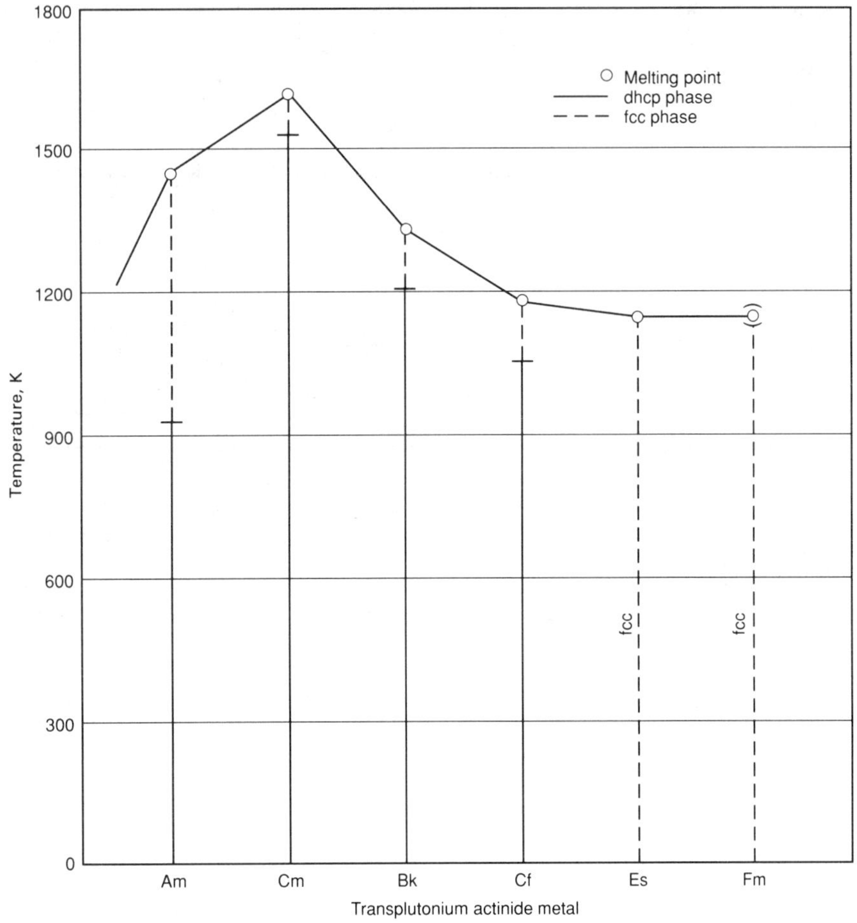

Fig. 153 Phase behavior and melting points of the first six transplutonium actinide metals

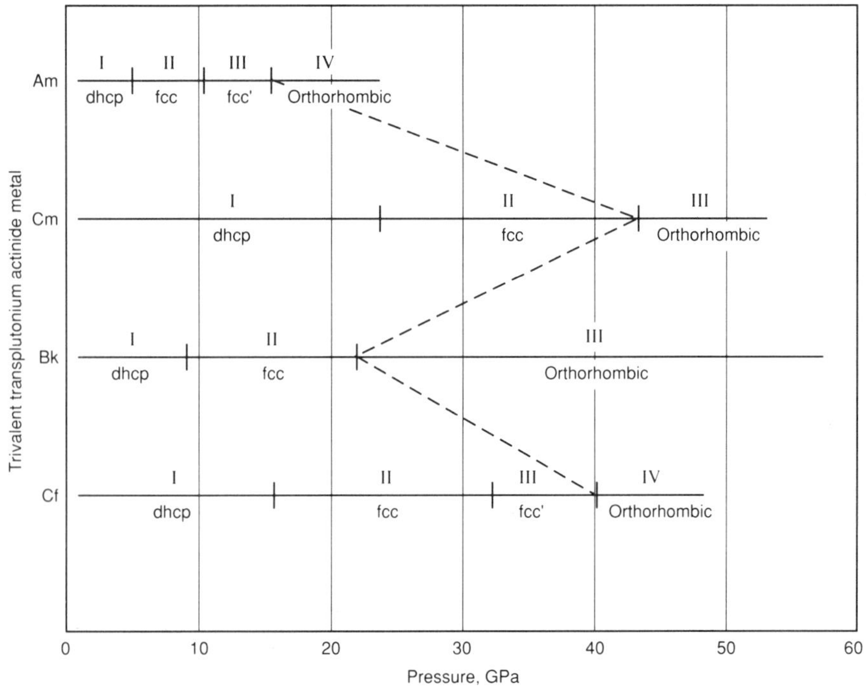

Fig. 154 Phase behavior of the first four transplutonium actinide metals under pressure. Dashed line indicates delocalization pressure variation of these trivalent transplutonium metals. Orthorhombic designation indicates an α-uranium structure (5*f* delocalization).

pressures for the metals. Data do not exist for einsteinium or fermium metals under pressure.

Chemical and Mechanical Properties

The first six transplutonium metals are electropositive metals that react with air and moisture to give oxides (sesquioxides or dioxides). They are usually prepared by the thermometallic reduction of their oxides with lanthanum or thorium metals, or by the reduction of their fluorides with lithium metal. At 25 °C, the rate of reaction for bulk forms of the first four metals (americium, curium, berkelium, and californium) with air is moderate and is comparable to that of the light lanthanide metals. The divalent metals, einsteinium and fermium, are more reactive, and their behavior is better compared to that of europium metal.

All of the first six transplutonium metals react rapidly with mineral acids to evolve hydrogen and yield trivalent ions (see Table 70 for thermodynamic values). They react with anhydrous hydrogen halides to yield the trihalides. The metals react slowly with dry nitrogen at room temperature to give mononitrides; at temperatures above 200 °C, they react with hydrogen to form di- to trihydrides. At elevated temperatures, the metals are more reactive; for example, californium metal will react with glass containers at temperatures as low as 200 °C. At elevated temperatures, the metals combine with the excess pnictogen elements (any member of the nitrogen family of elements, group V in the periodic family) and chalcogen elements (the elements that form group VI of the periodic table, including sulfur, selenium, tellurium, and polonium) to yield the monopnictides or chalcogenides of higher stoichiometries.

The pure metals are bright and metallic-looking, relatively ductile, and soft enough to be cut without much difficulty. Impurities tend to make the metals brittle. Coefficients of expansion have been determined for americium, curium, and berkelium metals:

Element	$\alpha_a \times 10^6$, K^{-1}	$\alpha_c \times 10^6$, K^{-1}
Americium	7.5	6.2
Curium	8.7	12.8
Berkelium	10.8	17.8

Note: Data for californium, einsteinium, and fermium are not available.

The molten metals are very reactive and difficult to contain; tungsten is the preferred container material, but tantalum is sometimes used. These transplutonium metals are known to form alloys with several other metals; their behavior in this respect is similar to that of the light trivalent lanthanide metals. The first four transplutonium metals appear to have only a limited solubility (estimated to be a few atomic percent) in the divalent lanthanide metals, europium and ytterbium.

Table 70 Thermodynamic values for An^{3+} ions at 298 K

Element	Enthalpy (ΔH_f°), kJ/mol	Entropy (S°), J/mol · K(a)	Gibbs free energy (ΔG_f°), kJ/mol
Americium	−617	−201	−599
Curium	−615	−194	−593
Berkelium	−601	−194	−578(a)
Californium	−577	−197	−553(a)
Einsteinium	−603(a)	−206	−573(a)
Fermium	−632(a)	−215	−599(a)

(a) Calculated or estimated values

Table 71 Magnetic properties of the first six transplutonium actinide metals

Element	Metallic valence	Localized 5f electrons	Magnetic moment (μ_{eff}), μ_B: Calculated(a)	Magnetic moment (μ_{eff}), μ_B: Experimental(b)	Transactions/ordering: Type(c)	Transactions/ordering: Temperature, K
Americium	3	6	0	0	P	...
Curium	3	7	7.94	7.6–8.1	AF	65
Berkelium	3	8	9.72	8.8–9.8	AF	34
Californium	3	9	10.63	9.7–10.2	FM	57
Einsteinium	2	11	10.60	11.3(d)	...	...
Fermium	2	12	7.57	...	...	...

(a) Moments calculated using the simple spin-orbit interaction ($L + S$, total angular momentum) model; μ_{eff} is expressed in Bohr magnetons (μ_B). (b) Experimental moments (in Bohr magnetons, μ_B) for low-temperature phases: Am-Cf, dhcp; ES, fcc. (c) P, paramagnetic; AF, antiferromagnetic; FM, ferromagnetic or ferrimagnetic. (d) Preliminary data

The transplutonium elements are classified as toxic because they are heavy metals and because of their radioactivity. Ingestion or inhalation of the metals or their compounds are to be avoided. In mammals, these materials tend to deposit in the bones or the liver.

Magnetic Properties

The elements beyond plutonium are believed to have localized $5f$ electrons, in contrast to the earlier actinides. Magnetic data that support this assumption have been acquired for the first five transplutonium metals. The magnetic moments (in Bohr magnetons, μ_B) of these metals are comparable to values found for their lanthanide homologs when the same electronic configurations are present (for example, the same number of localized f electrons). The magnetic moment for curium metal (8 μ_B; $5f^7 6d7s^2$ configuration) is thus essentially the same as that for gadolinium metal ($4f^7 5d6s^2$ configuration). In contrast, americium metal is a trivalent metal with a zero moment, whereas its homolog, europium, is divalent and has a moment of 7 μ_B. Magnetic data for these actinide metals is summarized in Table 71.

It has been determined that americium metal becomes superconducting at temperatures as high as 0.8 K. Although americium metal has localized $5f$ electrons, this superconductivity occurs because of its nonmagnetic ground state. The remaining transplutonium metals through einsteinium are magnetic, and they are not expected to be superconducting at low temperature. Measurements above 4.2 K on the metals from curium through einsteinium have shown no evidence for superconductivity.

Although the $5f$ electrons of these six transplutonium metals are reasonably localized, they still communicate by a slight overlapping of their wave functions and/or by interactions with the conduction electrons. This communication leads to the occurrence of magnetic transitions (ferro-, ferri-, or antiferromagnetic) and low-temperature saturated magnetic moments. Curium, berkelium, and californium exhibit transitions to ordered structures at low temperatures; americium is paramagnetic (see Table 70). Preliminary measurements on einsteinium metal have indicated it exhibits paramagnetism and does not exhibit low-temperature ordering, contrary to what would be expected for it. Data have not been obtained for fermium metal.

SELECTED REFERENCES

- U. Benedict, The Effect of High Pressures on Actinide Metals, in *Handbook of the Physics and Chemistry of the Actinides*, Vol 5, Freeman and Lander, Ed., Elsevier, 1987, p 227-269
- R.G. Haire, Preparation, Properties, and Some Recent Studies of the Actinide Metals, *J. Less-Common Met.*, Vol 121, 1986, p 379
- P.G. Huray and S.E. Nave, Magnetic Studies of Transplutonium Actinides, in *Handbook of the Physics and Chemistry of the Actinides*, Vol 5, Freeman and Lander, Ed., Elsevier, 1987, p 311-372
- Katz, Seaborg, and Morss, Ed., *The Chemistry of the Actinide Elements*, Vol I and II, Chapman and Hall, 1986
- S.F. Mughabghab, *Neutron Resonance Parameters and Thermal Cross Sections*, Vol I, Part B, Academic Press, 1984

Special Engineering Topics

Recycling of Nonferrous Alloys

Chairman: David V. Neff, Metaullics Systems

Recycling of Aluminum

Elwin L. Rooy, Aluminum Company of America
J.H.L. Van Linden, Alcoa Laboratories

ALUMINUM RECYCLING started less than 20 years after the commercialization of the Hall-Heroult process in 1888, driven by the high value and several unique properties of the metal. In the early days of the developing aluminum industry, the primary producers attempted to maximize new metal sales to reduce the unit price and make aluminum competitive with the traditional construction metals. They were not interested in scrap recycling, leaving that activity to others, who in time developed an independent secondary industry.

As both industries grew, their objectives changed. The primary producers started to use scrap of all forms and origins, and the secondary producers began producing more sophisticated end products, thus reducing the originally distinctive differences between the two industries. Similar developments occurred in the technology arena. The secondary producers were originally low-capital salvage operators. Some still operate in that mode, whereas others have grown into large corporations utilizing advanced technology. This latter group patterned its development after that of the primary industry, which has taken a more capital-intensive approach to solving recycling problems. This section will emphasize mainly these general technological trends and developments rather than the traditional methods and hardware.

Recyclability of Aluminum

Aluminum possesses many characteristics that make it highly compatible with recycling. It is resistant to corrosion, and a low ratio of energy is required to remelt aluminum compared with that required for its primary production. Also, the alloy versatility of aluminum has resulted in a large number of commercial compositions, many of which were designed to accommodate impurity contamination.

Aluminum is resistant to corrosion under most environmental conditions, and thus retains a high level of metal value after use, exposure, or storage. Once produced, aluminum can be considered a permanent resource for recycling, preferably into similar products.

The energy requirements for the conversion of aluminum scrap are low when compared with the energy consumed in primary aluminum production. For example, the actual ratio of primary total energy to recycled total energy for used beverage cans is 28.5:1 (Ref 1).

The objective of recycling is to produce a salable commercial aluminum alloy product. Currently, more than 300 compositions covering wrought and cast alloys are registered with the Aluminum Association. Many of these alloys are designed to tolerate the variations in composition and ranges in impurity content that may be experienced in the recovery of scrap. Even widely varying scraps can be melted to produce alloys that are commonly used in die and gravity casting, extrusion, and sheet rolling.

In the early decades of this century, the output of the secondary aluminum industry was largely tied to the consumption of castings by the automotive industry. As shown in Fig. 1, the recycling of aluminum has increased steadily from 1950 until the present despite recessions and energy crises. This increase is the result of growth in the automotive market as well as the development of new and significant recycling applications, such as the consumption of used beverage cans (UBCs) in the manufacture of sheet for new cans.

In recent years, environmental concerns have contributed to an increased awareness of the importance of scrap recycling. Today, the recyclability of aluminum is a major advantage to the aluminum industry in materials competitions for major product markets. The current driving forces for aluminum recycling are:

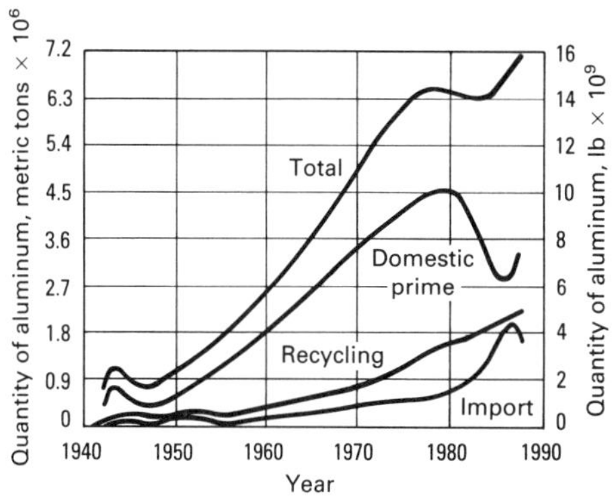

Fig. 1 U.S. aluminum supply distribution for the years 1940 to 1990. Source: Aluminum Association

- Regulatory actions taken by government agencies to encourage resource conservation, energy conservation, and waste reduction through mandatory segregation and deposit programs
- Consumer sensitivity to environmental issues and the solid-waste crisis
- Competitive pressures from other materials
- Economic advantages based on the relative value and availability of aluminum scrap

The Recycling Loop

The reclamation of aluminum scrap is a complex interactive process involving collection centers, primary producers, secondary smelters, metal processors, and consumers. Figure 2 depicts the flow of metal originating in primary smelting operations through various recycling activities. The initial reprocessing of scrap takes place in the facilities of primary producers. In-process scrap, generated both in casting and fabricating, is reprocessed by melting and recasting. Increasingly, primary producers are purchasing scrap to supplement primary

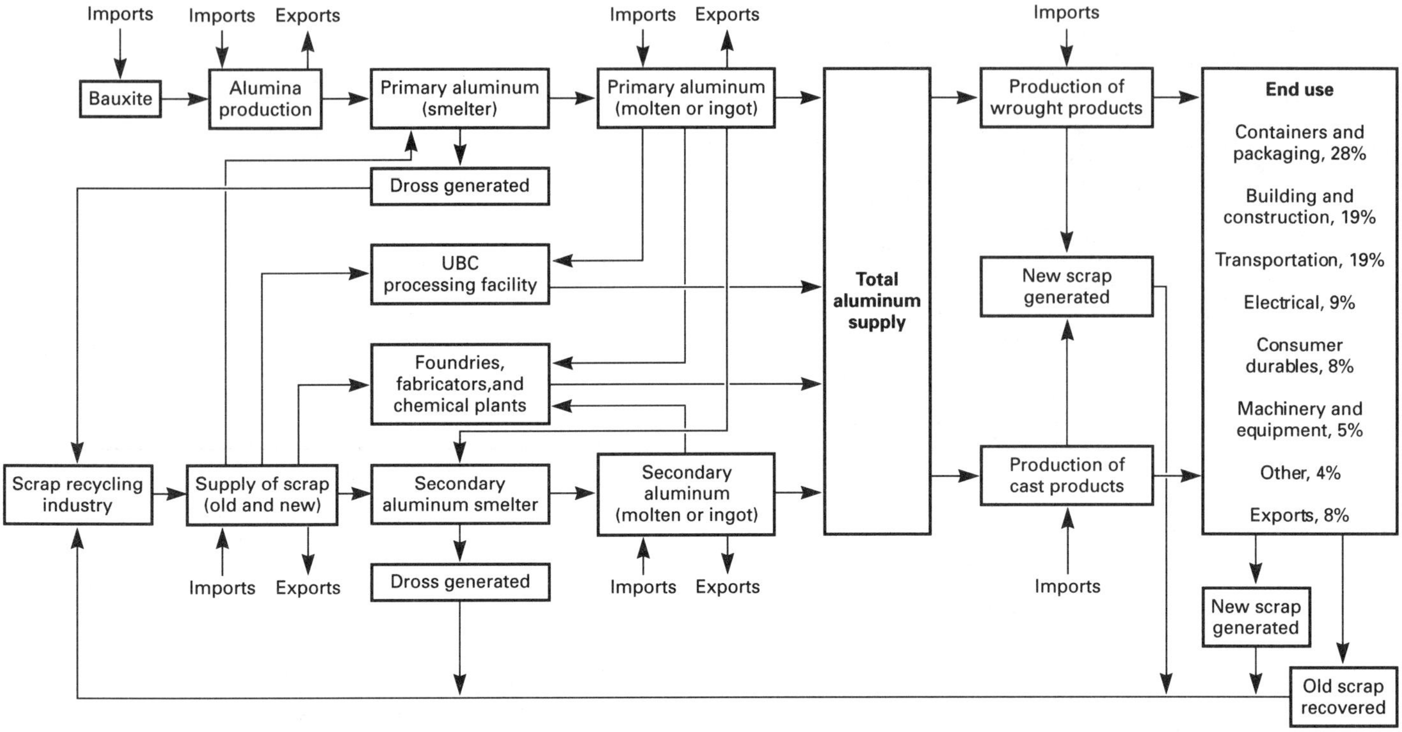

Fig. 2 Flow diagram for aluminum in the United States, showing the role of recycling in the industry. Scrap recycling (lower left) includes scrap collectors, processors, dealers and brokers, sweat furnace operators, and dross reclaimers. Source: U.S. Bureau of Mines

metal supply; an example of such activity is the purchase or toll conversion of UBCs by primary producers engaged in the production of rigid container stock.

Scrap incurred in the processing or fabrication of semifabricated aluminum products represents an additional source of recyclable aluminum. Traditionally, this form of new scrap has been returned to the supplier for recycling, or it has been disposed of through sale on the basis of competitive bidding by metal traders, primary producers, and secondary smelters.

Finished aluminum products, which include such items as consumer durable and nondurable goods; automotive, aerospace, and military products; machinery; miscellaneous transportation parts; and building and construction materials, have finite lives. In time, discarded aluminum becomes available for collection and recovery. So-called old scrap, metal product that has been discarded after use, can be segregated into classifications that facilitate recycling and recovery.

Scrap specifications have been developed that allow the convenient definition of scrap types for resale and subsequent reprocessing. Those developed by the Institute of Scrap Recycling Industries are in broad use (Ref 2). These specifications, however, are for the most part physical descriptions of scrap categories useful to dealers. Chemistry specifications and limits on harmful impurites are not defined, and treatment of extraneous contaminants is inconsistent. More comprehensive scrap specifications are being developed by the industry.

Recycling Trends

A number of industry segments are in competition for the available aluminum scrap, though not necessarily the same types of scrap. As shown in Fig. 3, the primary producers have experienced the largest increase in scrap consumption. The sporadic data available for the period prior to 1940 suggests that recycling of aluminum grew steadily to about 15% of total shipments. World War II disrupted the pattern drastically, but the stockpile reduction in the years following the war reduced the recycling rate to the prewar level (Fig. 4). Scrap consumption trends describing the relationship between the contributions of the primary and secondary industries to the recycling effort are illustrated in Fig. 5.

The objective of all collection activities is the conversion of scrap forms to products having the highest commercial value. A growing trend is the consumption of scrap in secondary fabricating facilities. A number of extruders, foundries, and minimill operations are producing billet, castings, and common alloy sheet products directly from scrap. The die casting industry relies heavily on secondary compositions for the production of automotive and other parts, and the larger die casters have, in many cases, expanded metal supply through direct scrap purchases.

Aluminum Industry Trends. Recent developments have strongly influenced the rate at which scrap is recycled and the nature of the markets served by recycling activity. Several generalizations concerning the U.S. and world aluminum industries can be made:

- *The aluminum industry today is truly international.* Primary aluminum is produced in virtually every global region, and the metal produced competes in global markets. Aluminum alloy scrap is now traded internationally as well
- *Domestic primary aluminum production will not be expanded beyond the capacity of existing smelting facilities.* Energy costs and labor rates in the United States suggest that there will be substantial decreases in primary output in the absence of any forseeable new, more economical smelting technologies
- *New risks and uncertainties.* The European and, to a lesser but significant extent, the U.S. aluminum industries face new risks and uncertainties caused by the growing geographic separation of primary production from major fabricating and consuming markets
- *Aluminum is a U.S.-dollar-based commodity.* Exchange rate fluctuations represent a major complication in stabilizing prices and regulating international competition and metal supply
- *The world aluminum production capacity will expand*, but such expansion will take

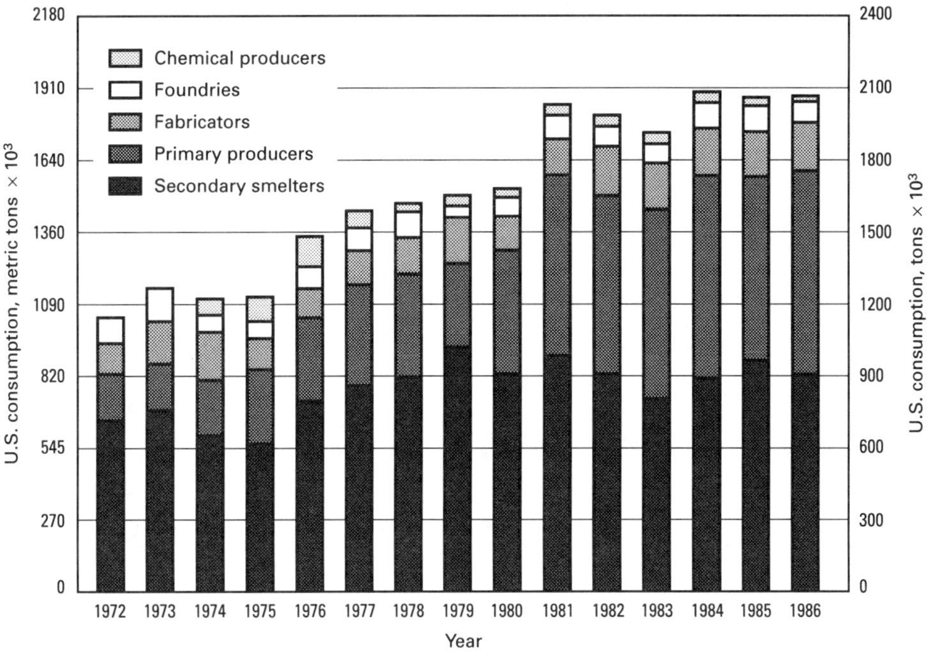

Fig. 3 U.S. aluminum scrap consumption by type of company for the years 1972 to 1986. Source: U.S. Bureau of Mines

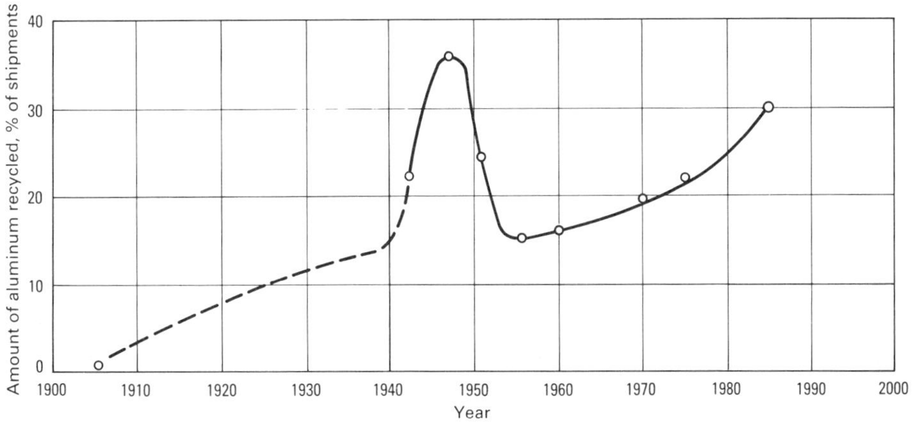

Fig. 4 Aluminum recycling trends in the United States. The percentage of shipments recycled is only now approaching the peak experienced during World War II.

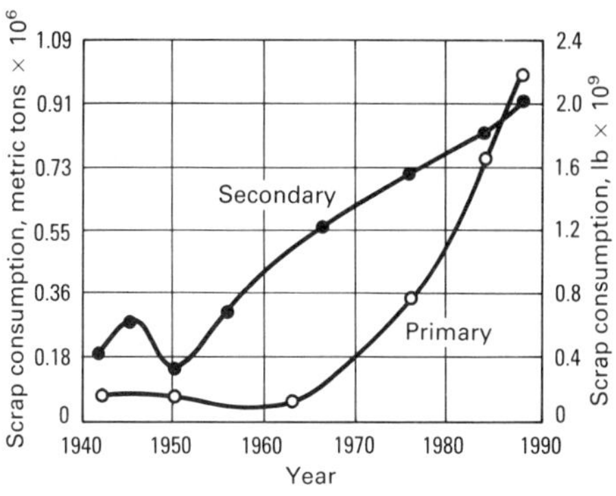

Fig. 5 U.S. aluminum recycling by industry type

place in countries with low energy costs, such as Canada, Venezuela, Brazil, and Australia

- *Recycling will increase in importance.* For the United States, and ultimately for the rest of the aluminum-consuming world, recycling and resource recovery will play an increasingly important strategic role in ensuring a reliable and economical metal supply
- *The U.S. will import aluminum.* On the basis of the best assumptions, the United States will become an importing nation as aluminum requirements exceed domestic smelting capacity. Most imports will consist of unalloyed smelter ingot for remelting, casting, and fabrication in North American facilities. Some products will be imported, but only for specialty applications with unique and/or cost advantages over domestic products
- *World competition for scrap units will intensify* based on the relative costs and availability of primary aluminum and scrap

Developing Scrap Streams

The traditional flow of scrap through the primary, consuming, and secondary industries is dominated by three major scrap streams: UBCs, automotive scrap, and municipal scrap.

Beverage Cans. The recycling of UBCs is a remarkable success story (Fig. 6). It is not an exaggeration to suggest that aluminum recycling has played a major role in the market growth of aluminum beverage cans and their penetration into a market previously dominated by competing materials. In 1976, the aluminum can accounted for 21% of the beverage container market; 46.4 billion containers were produced, of which 4.9 billion were recycled. In 1986, 72.9 billion cans were produced and 33.3 billion cans were recycled. It is projected that by 1991, the aluminum can market will have grown by 14% and the recycling rate may reach 60%. In 1988, 94% of all beverage can bodies were produced from aluminum and virtually 100% of all cans featured the aluminum easy-open end.

Aluminum producers, can makers, and the public have invested in the return system. Large-scale consumer advertising campaigns have emphasized energy savings and resource conservation. Under deposit legislation, cans have generally proven most convenient to handle by consumers, retailers, bottlers, and wholesalers. Collection activities include reverse vending machines, mobile return centers, and public information and educational campaigns. Recycling centers are active in the development of thematic programs to promote the concept of recycling; these programs often associate recycling with civic causes and medical programs, and encourage volunteer, service, and community groups and individuals to maximize the return of used beverage cans.

Recycling of UBCs is based on the inherently high scrap value of aluminum and its convertibility into new-can stock. For the most part, UBCs are consumed by primary producers of rigid-container sheet and are employed in the regeneration of can stock. The growth in markets such as Western Europe, coupled with high energy rates in some countries (Japan, for example), has also created an active market involving the export of UBCs and can scrap. The alloy compatibility of the components of the can makes the UBC uniquely suitable for the closed-loop recycling concept, and it is responsible for the consistent high value of UBCs as well as the ever-increasing volume of UBCs in new can sheet.

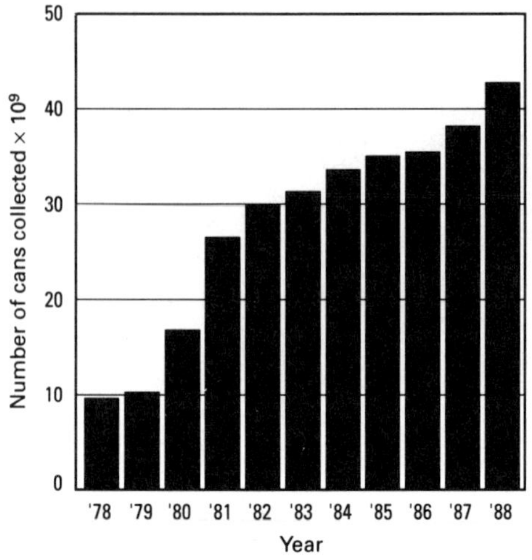

Fig. 6 Increase in recycling of aluminum UBCs from 1978 to 1988. The calculation for the number of cans collected is based on a can weight survey conducted by the Aluminum Association.

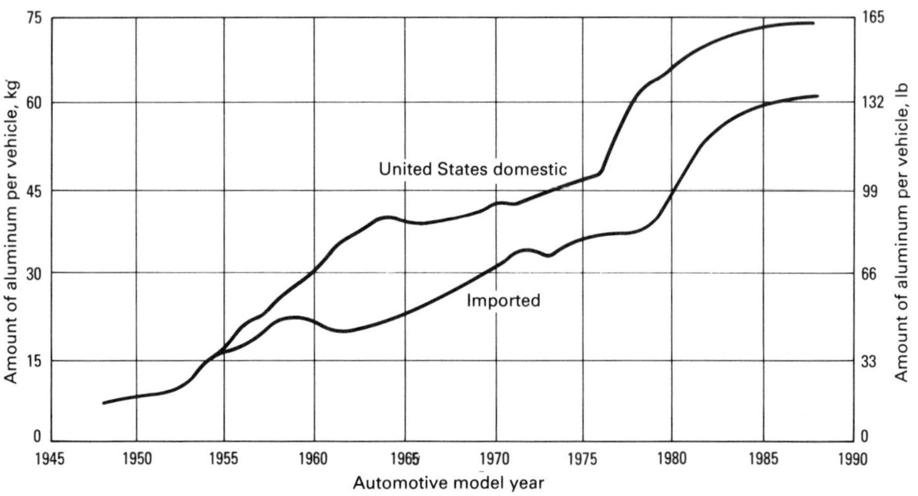

Fig. 7 Average use of aluminum in automobiles from 1946 to 1988. U.S. automakers have used more aluminum than world producers since the mid-1950s.

Automotive Scrap. Increased activity in the recycling of spent automobiles is based largely on new steel technologies that make scrap conversion economically attractive. However, the exceptional value of the nonferrous metals that are being separated through the efforts of metal recyclers and the secondary industry also contributes significantly to this recycling activity.

Because the use of aluminum in automobiles is increasing (Fig. 7), the aluminum value recovered from shredded automobiles is of special importance to the metal castings industry. The potential for aluminum use in automobiles is limited only by how aggressively material advantages are pursued by the manufacturers. Common informed predictions indicated that there could be an average use of 90 to 125 kg (200 to 275 lb) of aluminum per automobile in the near future (Ref 3). Growth to the present level of aluminum use in the automotive industry has been largely associated with construction concepts emphasizing castings. Aluminum alloy forgings, sheet, and extrusions are presently being developed by international automobile manufacturers in cooperation with primary aluminum producers to enable the increased use of these materials.

In 1990, 10 million retired automobiles may yield as much as 500 000 metric tons (550 000 tons) of aluminum; this figure is based on a recovery of 55 kg (120 lb) per vehicle, about the 1980 average. In 1995, because of the projected increase of the use of aluminum in automobiles, recoverable aluminum may increase to 820 000 metric tons (900 000 tons). Trucks and buses, in which aluminum is used more intensively, would add to these estimates.

Municipal Scrap. The separation of metals from municipal refuse has not grown as originally expected. The best available information indicates that more than 12 million metric tons (13.5 million tons) of metals are lost annually through refuse disposal in the United States. Metals make up only 9% of total refuse. Of the total metals available in metal municipal refuse, less than 1% is recovered. Because the cost of separating all components of the refuse stream must be based on the value of reusable materials and energy content, a logical conclusion is that municipal refuse processing has not yet become economically viable; future expansion of such processing will depend either on government subsidies or changes in energy costs and the availability of raw materials.

The solid waste crisis has exposed the need to increase all forms of recycling activity. In 1988, the United States generated 145 million metric tons (160 million tons) of waste. It is anticipated that it may produce more than 180 million metric tons (200 million tons) per year by the year 2000. In 1979, the nation had 18 500 landfills in operation. This number was reduced to 6500 by 1988, and the U.S. Environmental Protection Agency (EPA) estimates that 80% of existing landfills will close in the next twenty years. The EPA and other agencies have argued for responsible reductions in waste, increased emphasis on recycling, the use of incineration to generate energy and greatly reduce the reliance on landfill space. At present, only 10% of waste in the United States is recycled. A national goal established by EPA is to increase recycling to 25% by 1992. The value of aluminum scrap can be expected to significantly affect the success of plans to increase the percentage of waste that is recycled.

Clearly, the recyclability of any material enhances its attractiveness in commercial applications relative to materials that are not recyclable, or that can be recycled only at an excessive cost. The inherent recyclability of aluminum and its value after recycling enable it to be used in a manner that is supportive of national environmental and waste reduction goals.

Technological Aspects of Aluminum Recycling

Although the term recycling suggests a closed loop, or a set of endless material/product cycles, the widely accepted territory of aluminum recycling includes scrap collection and preparation, remelting, refining, and the upgrading of molten metal to a ready-to-cast condition. Molten metal treatment processes such as demagging, degassing, filtering, composition adjustment, and fluxing play an extremely important role in enabling the reuse of often heavily contaminated scrap in the production of new products; these processes are well described in the literature. A comprehensive overview of molten metal processes for nonferrous metals is available in the article "Nonferrous Molten Metal Processes" in Volume 15 of the 9th Edition of *Metals Handbook*. The following discussion will concentrate on issues involving scrap preparation and melting.

In the early days of the aluminum industry, recycling consisted entirely of hand charging unused parts of castings (for example, risers or cut-to-fit scrap pieces) into an oil-fired vessel that resembled an oil drum on its side and that could be tilted for tapping. After the invention of the resealable taphole, the reverberatory, or open hearth, furnace evolved as the predominant remelting facility.

There are two reasons for the long-term success of the direct-firing method in which the combustion chamber and the scrap charge room are combined into one open hearth. First, aluminum oxidation progresses slowly due to the protective, flexible thin oxide film that forms instantaneously when a fresh aluminum surface is exposed to any oxygen-containing atmosphere. Second, the surface-to-volume ratio of scrap particles was generally small because the scrap usually consisted of large sections or parts and heavy-gage sheet. Therefore, the penalty for direct exposure to the melting environment was acceptable when the scrap was generally coarse and the alloys were low in magnesium content.

The open hearth furnace was the mainstay of the scrap-remelting business for over 50 years, but as the application of aluminum became more widespread and diverse, the need for effective and efficient recycling technology grew rapidly. The corresponding growth and diversification of the scrap-recycling industry have been driven by several factors, including increased production requirements, alloy development, scarcity of energy and resources, and increased availability of mixed scrap.

Increased Production Requirements. Large modern casting stations may use up to 114 metric tons (125 tons) of metal per production cycle. This requires not only a high melt rate in large melt furnaces, but also a minimal charge time. The present rectangular open hearth furnace construction is practically limited in size to about a 30 m^2 (320 ft^2) bath area. Cylindrical furnace construction allows for hearth areas of up to 75 m^2 (800 ft^2) and features a removable lid. This configuration allows for very fast, well-distributed overhead charging with preloaded dump buckets. The old design relies on cumbersome door charging, which requires a special machine to push relatively small amounts of scrap as far back into the furnace as practically possible.

Alloy Developments. The use of aluminum-magnesium alloys in light-gage sheet for automotive and container products and aluminum-magnesium-zinc alloys for aerospace and extrusion applications results in large amounts of high-surface-area scrap that has a significantly higher oxidation rate at elevated temperatures than pure aluminum, aluminum-copper, or aluminum-silicon alloys. The prevalence of this type of scrap has stimulated the development and use of methods that minimize the exposure of the scrap to the hostile furnace atmosphere; these methods include continuous melting for large volumes and induction melting for more modest volumes.

Energy and Resource Scarcity. The energy crisis of 1973 highlighted the enormous energy advantage of using scrap rather than new metal and exposed the corresponding need to minimize melt losses. The primary producers started to keep more-difficult-to-melt scrap (using their newly developed continuous melters or induction furnaces), forcing the secondary producers to deal with even less-desirable scrap types. This scrap was processed mostly in rotary salt melters, which were already popular for skim and dross reclamation.

Increased Availability of Mixed Scrap. Since the early 1960s, the volume of scrap from discarded long-life-cycle products such as home construction materials, trailers, household goods, and so on—often inseparably mixed with other metals—increased steadily. Substantial amounts of otherwise lost aluminum are recovered from such scrap by sweat melting. This is a selective process that involves melting the scrap in a hearth with a sloped bottom; it works by exploiting the melt temperature differences of the metals present, melting, draining, and collecting at the base the metal with the lowest melt point first, followed by the metal with the next-lowest melt point, and so on.

Alloy Integrity. In addition to preservation and melt loss reduction, a key factor in successful recycling is the maintenance of alloy integrity. Obviously, the ideal way of reusing scrap is to recycle it into similar products, or in other words, process used cans into new-can sheet, auto scrap into new car parts, and so on. Melting UBCs, removing the magnesium by chlorination, and adding silicon to produce a casting alloy can be profitable at times and is technically less challenging than making new-can sheet, but the metal units involved will be permanently degraded by such an approach.

In primary aluminum production plants, a great effort is made to keep the scrap identified and segregated. This so-called turnaround scrap is pedigreed and therefore of the highest value. However, at the production stations of part fabrication (for example, a machining operation in an aircraft manufacturing plant), it is considered impractical and uneconomical to keep scrap segregated by alloy. Furthermore, in end-product fabrication, permanent attachment of some aluminum to other metals or nonmetals is inevitable. Therefore, no ideal process for recycling the latter two types of scrap material is readily available.

The greatest challenge for the recycling community is finding the most economical way to separate and prepare scrap for melting so that it can be used in the least-degraded form with the least number of postmelt treatments for alloy or quality adjustments. Each product has its own specific demands, which present obstacles and provide opportunities for meeting this challenge. A complete review of all possible paths would be impractical. Because the UBC and auto scrap loops are different in every aspect, this article will focus on several of the technological developments in preparation and melting in those loops. Municipal refuse and automotive recovery technologies are basically the same.

Can Recycling Technology

Considerable amounts of UBCs are either toll converted for primary producers or remelted by secondary operators, using open hearth or rotary salt furnaces for use in casting alloys. However, the majority (about 80%) of UBCs are returned directly to the primary industry, and UBC recycling is thus a prime example of closed-loop product recycling. The flow diagram in Fig. 8 shows the captive nature of the can manufacturing and recycling loop. The scrap preparation and melt technologies described in this section are considered the most advanced and should not be viewed as industry standards. However, many UBC converters are using similar, rather sophisticated technologies.

Collection. UBCs are received from collection centers as bales weighing 400 kg (880 lb) or as briquettes with a maximum density of 500 kg/m^3 (31 lb/ft^3). The briquettes can be stacked on skids and offer storage advantages to the supplier, but they can be hard on the equipment of the receiver. In the shredding operation, bales and briquettes are broken apart and the cans shredded to ensure that no trapped liquid or extraneous material will reach the melters and cause serious damage or injuries. From the shredder, the material passes, via a magnetic separator that removes ferrous contaminants, through an air knife. In the air knife, heavy nonferrous materials such as lead, zinc, and stainless steel scrap drop out. The shredded aluminum cans then pass on to the delacquering units.

Delacquering. There are two basic approaches to continuous thermal delacquering. One is based on a relatively long exposure time at a safe temperature, and the other is based on staged temperature increases to just below melting for as short an exposure time as possible. The first approach uses a pan conveyor on which a bed of precrushed and shredded UBCs approximately 200 mm (8 in.) deep moves through a chamber held at about 520 °C (970 °F). The chamber contains products of combustion (POC) gases that are diluted with air to provide the proper atmosphere and temperature for the delacquering process (part pyrolysis, part combustion). The second approach employs a rotary kiln with a sophisticated recirculating system for POC gases at various entry points. The temperature in the last stage is near 615 °C (1140 °F), which is very close to the temperature at which incipient melting occurs in the aluminum-magnesium (5*xxx* series) alloys typically used in can lids and tabs.

Both systems present inherent control problems that may result in nonuniform delacquering. A temperature that is too low or exposure times at the proper temperatures that are too short will leave a tar-like coating on the UBCs. This coating causes

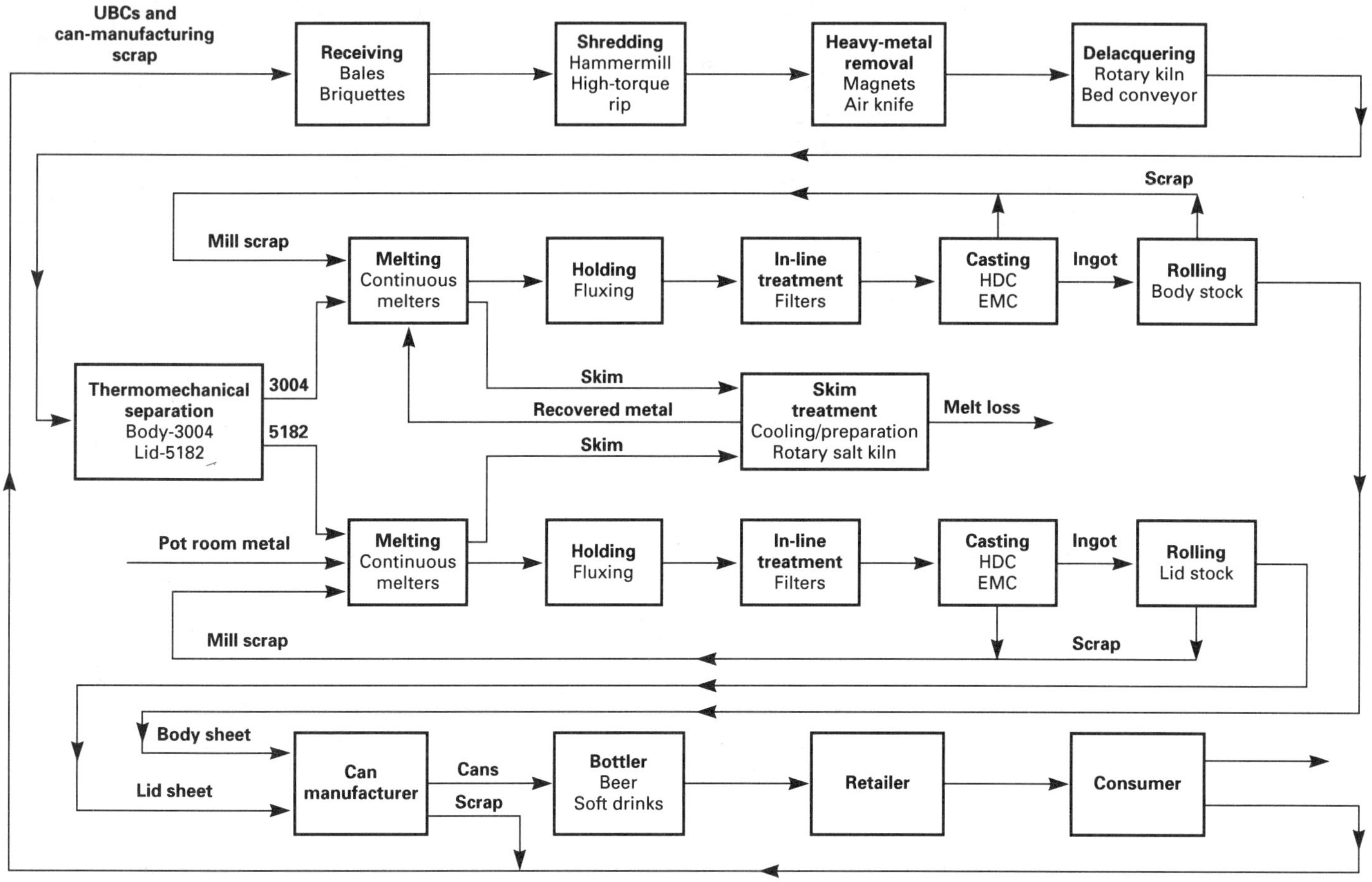

Fig. 8 Flow diagram of closed-loop UBC recycling and manufacturing. Source: Ref 4

premelt burning, which leads to increased metal losses upon melting. A temperature that is too high or exposure times at proper temperatures that are too long will cause considerable oxidation of the scrap, also resulting in increased melt losses.

In the pan delacquering system, large bed depths may result in temperature gradients that cause the above-mentioned problems. In the kiln method, gas flows are high and the UBC shreds are physically agitated, and these conditions lead to nonuniform residence times and the same problems. Proper operating controls for these delacquering units, which treat about 18 metric tons (20 tons) of scrap (~1.25 million UBCs) per hour, are vital for producing low-melt-loss feedstock.

Alloy Separation. The hot, delacquered UBCs then move into the thermomechanical separation chamber, which is held at a specific temperature and contains a nonoxidizing atmosphere. In this chamber, a gentle mechanical action breaks up the alloy 5182 lids into small fragments along grain boundaries, which have been weakened by the onset of incipient melting. An integrated screening action removes the fragments as soon as they can pass the screen to avoid overfragmentation. This process requires a very narrow operating control capability to avoid melting entire 5182 particles, which would then cluster with the still-solid alloy 3004 particles. The screened-out alloy 5182 particles are transported to the lid stock melters, and the large alloy 3004 particles continue directly into the body stock melters.

Melting, Preparation, and Casting. At present, most melting facilities for UBCs throughout the industry are dedicated units designed to handle the enormous volumes and to minimize the melt losses inherent in melting thin-walled material. Larger companies have developed their own processes, some of which are described later in this section. Significant amounts of skim—the mixture of metal, oxides, other contaminants, and trapped gas that floats on top of the melt—are removed and treated for metal recovery. A typical skim weight is 15% of the original charge. The recovered metal (6 to 8% of the original charge) from this skim will be used only in body stock manufacturing because of its high levels of manganese and contaminants.

The metal from these dedicated melters is often transferred to on-line melting furnaces, where additional bulky scrap is remelted and primary unalloyed metal is charged to create the desired volume of the proper alloy composition. From these melting furnaces, the metal is transferred to the holding furnaces, where minor composition adjustments are made and metal quality treatments are performed (for example, gas fluxing to remove hydrogen). Some metal treatment, for example, inclusion removal, can be done in so-called in-line treatment units; again, most major companies have developed their own preferred methods and technology. The clean and on-composition metal is cast into ingots weighing up to 13.5 metric tons (15 tons). During casting and rolling of the ingot to sheet, about 42% of the original melt weight may be shaved, cropped, or slit off in various stages. This metal is called the in-house, or turnaround, scrap, and it is directly returned to the remelters.

The body and lid sheet are shipped to a can manufacturer. As a result of can fabrication processes, about 20% of the sheet (or 13% of the original melt) is returned to the aluminum manufacturer as skeleton scrap. On a global basis, this means that 55% of a melt consists of new (production-related) scrap. If all cans were returned as UBCs and total melt losses were 7% of the melt, this 7% would be the only makeup metal

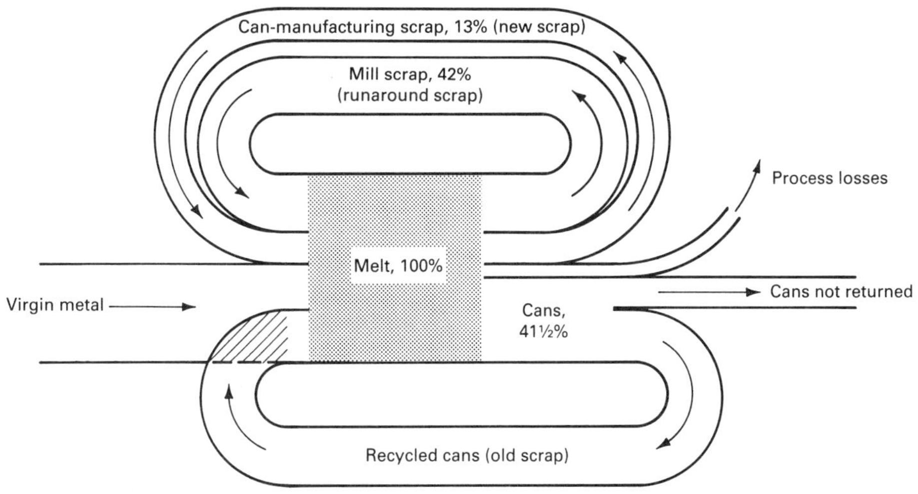

Fig. 9 Process by which recycled cans replace virgin metal in the beverage container market

required from primary smelters to close the loop (provided the market remained constant). Figure 9 illustrates this interrelationship.

As the recycling rate continues to increase, composition control and the corresponding contamination avoidance become technical challenges as important as melt loss reduction. These developments were predicted in a mathematical model of the recycling system (Ref 5). It appears that prudent use of salt fluxes may hold the key to improvements in these areas.

Process Developments

Although melt loss had become the major cost factor in ingot production, it was the soaring cost of energy during the 1973 energy crisis that triggered the search for more-efficient remelt processes. This effort also sought to develop processes that were less labor intensive and more productive, specifically with respect to handling UBCs. It was recognized that the open hearth furnace was designed for remelting bulky scrap, which not only requires extensive charging time but also creates large amounts of skim. This layer of skim acts as an insulating blanket between the burners and the melt, severely reducing the thermal efficiency of the process. It became clear that the commitment to beverage can recycling necessitated a fundamental review of processing methods.

Excessive Skim Formation. The open hearth furnace consists of a relatively shallow molten metal container with a large surface area; the combustion chamber is located directly above the container. The scrap is piled up in this single chamber through a door in the front wall. The burners are aimed at the periphery of the solid charge. Because aluminum is a good heat conductor, the temperature of the scrap increases fairly uniformly, provided that the volume-to-surface-area ratio of the scrap is sufficiently high for conductive heat transfer. The protective nature of aluminum oxide films will, under these circumstances, prevent accelerated high-temperature oxidation, and the entire charge will melt at about the same time. The bath can then be skimmed. The clean surface facilitates rapid heating of the melt to the desired operating temperature.

The situation is vastly different, however, when the scrap charge consists of shredded cans with a wall thickness of 0.13 mm (0.005 in.) and a heavy oxide layer acquired during thermal delacquering. The portion of the scrap contacted by the flame heats up and melts, but the unexposed metal remains cold, due to the poor heat transfer. The melting particles deform, breaking the oxide skin and exposing additional nascent metal to the harsh furnace environment. This metal immediately forms a new skin, while the particles solidify on the cold mass underneath. This melt/freeze cycle is repeated many times before the entire mass is melted. By then, a thick layer consisting of a mixture of oxide skins, trapped metal, and air covers the melt.

The Continuous Melting Concept. From the response of shredded can scrap melted in open hearth furnaces, it was reasoned that low-density scrap must not be exposed to the furnace atmosphere; this means that either the scrap must be submerged quickly in superheated molten metal inside the furnace, or molten metal has to be taken out of the furnace for external mixing with the scrap. The advantage of the latter procedure is that the inevitable skim (fortunately in much smaller amounts) can be captured outside the furnace as well. The resulting metal stream, cooled down but still molten, can be returned to the furnace for reheating by means of a molten metal pump. In this manner, a steady-state condition can be maintained if the heat required for melting a constant mass flow of scrap particles (plus makeup for heat losses) is equal to the net heat input into the furnace. The molten metal is mass balanced by equating metal overflow to the charge rate minus the skim generation rate times the oxidation constant (a correction for weight gain due to oxidation).

A few events can cause deviation from the steady-state conditions; for example, skim buildup on the metal in the furnace can change the heat transfer efficiency, and sludge buildup in the passageway can affect the circulation rate. Diagnostic sensors monitoring the process can detect the changes in the early stages and alert the operator.

Practical Applications. Several processes using continuous melting have been developed. For example, Alcoa developed three different methods (Ref 6-8) based on pump design variations; these designs create the scrap-ingesting vortex either in the pump bay itself or in an optimized adjacent charge bay (see Fig. 10). Although these methods differ in such areas as production capacity, scrap size tolerance, and hardware simplicity, the skim generation for a specific scrap type is equally low for all three methods. Several variations of this principle have been reported by other researchers (Ref 9-11), suggesting that the method is appealing to others in the industry who are also struggling to reduce melt losses.

Another advantage of continuous melting is that the constant supply of waste heat can be utilized very effectively to preheat the constant flow of scrap and/or combustion air. In the batch process of the open hearth furnace, the supply and demand of waste heat for scrap preheating are essentially out of phase. The new melters improve fuel efficiency 12% on the basis of constant-waste heat utilization alone. Also, the fact that the skim is confined to a relatively small chamber outside the furnace offers an opportunity to maximize metal recovery from skim, without interfering with production.

Automobile Scrap Recycling Technology

In contrast with the explosive recent development of closed-loop can recycling, automobile recycling is an established industry, and it has traditionally been a multifaceted scavenging activity carried out by many independent entrepreneurs without the focused objectives that prevail in can recycling. Material selection by car manufacturers is based on cost and performance; little consideration is given to what happens to the material at the end of the approximately 10-year life span of a car. Consumers are basically indifferent con-

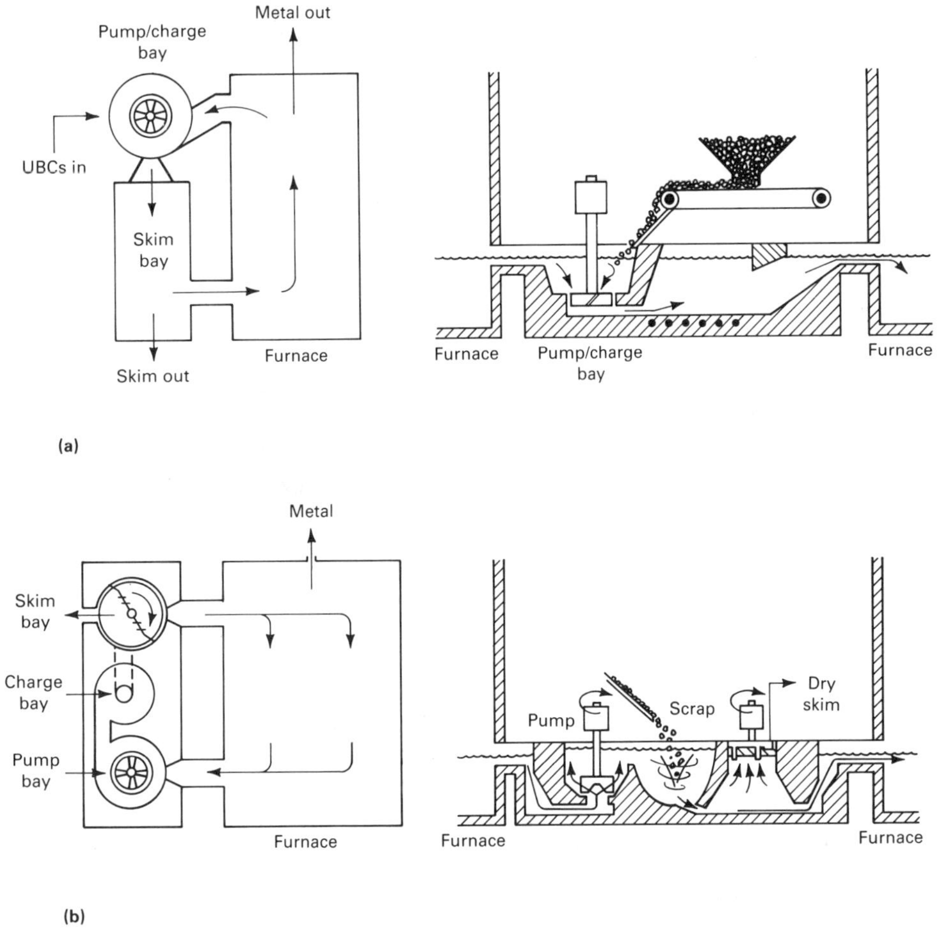

Fig. 10 Melting processes for UBC scrap. (a) Early can scrap melter. (b) More-advanced swirl scrap charge melter, which uses a continuous melting process

cerning material choice. Salvagers and dismantlers take what they can get. They may have preferences, but they have no say in material selection.

Substantial amounts of aluminum had been applied in early car manufacturing, but between 1925 and 1946, its use was minimal. As mentioned before, the amount of aluminum in cars has steadily increased since 1946 (Fig. 6). Until 1975, most growth was in castings, but as a consequence of the energy crisis, more wrought alloy hang-on parts began to be used for weight reduction, and sheet metal panels and space frame constructions are now under development.

An average 1980 model car contains approximately 70 kg (155 lb) of aluminum, which accounts for only 5% of the total weight of the average car. Yet it is economically an important fraction. Figure 11 is a schematic showing one of several possible paths for the recovery of the majority of the materials presently used in cars. In alternative schemes, a water-based classification step (elutriation) is added to remove plastic and a portion of the rubber and glass. The aluminum fraction is usually sold to an automotive cast shop, which uses open hearth as well as induction furnaces and occasionally rotary salt kilns. The fraction of inseparable multimetallic particles must be treated in a sweat furnace. In some schemes, without heavy-medium or eddy current separation, the entire nonferrous fraction goes to the sweat melter. The chemical composition of this sweat pig metal can vary substantially, and this variation greatly affects the value of the product.

The can recycling loop is totally dedicated to a single product of two compatible aluminum alloys. Automotive recyclers, on the other hand, must deal with a number of fractions with different destinations and relatively low values. These recyclers have not yet been driven to develop sophisticated technology to improve the quality and value of the fractions.

For automobile recycling to become as effective as can recycling, a cooperative effort is required by the scrap collectors, handlers, and manipulators to introduce advanced scrap separation and upgrading technology. A number of alternative scrap recovery methods have been developed for other mixed aluminum scrap sources, and these methods could be adapted for use in car recycling. They include improved preprocessing methods to concentrate aluminum fractions and mechanical, physical, and chemical separation processes for upgrading scrap mixtures and recovering aluminum from low-grade sources.

High-Temperature Processes. Other processes, besides the delacquering processes for can scrap described earlier, have been developed for separating aluminum scrap. A fluid-bed rotary furnace has been developed in England to remove paint, plastic, and other combustibles from aluminum in a heated bed of inert material, such as alumina. Flights move the scrap through the drum furnace, and the evolving gases and fumes are led to an afterburner. Rubber and wood, which do not evaporate in this furnace, need to be removed in advance (Ref 12).

In Sweden, the so-called Granges box is in use. This is an oven consisting of two chambers, one for containing scrap, the other for combustion of gases and fumes released from the scrap. The scrap is heated in part by external sources and in part by recirculation of the hot POC from the combustion chamber. This batch process may be suitable for certain auto scrap fractions.

The U.S. Bureau of Mines has developed a hot crushing process for separating wrought and cast alloys. It works on the same principle as the thermomechanical alloy separation process for UBCs. This process appears to be tailor-made for automotive scrap, which contains increasing amounts of sheet and die cast alloys.

Low-Temperature Separation. Cryogenic separation has been performed commercially in Belgium on shredded automotive scrap. The method is based on the difference in ductility of nonferrous and ferrous metals at extremely low temperatures. Below −65 °C (−85 °F), ferrous metals become very brittle and can be fragmentized easily, whereas nonferrous metals remain ductile. Simple screening achieves separation. This method is expensive and should be used only for separation of mixed shredder fragments (of steel attached to aluminum, for example).

Gravity separation methods include the common-heavy-media/sink-float process mentioned earlier and a process developed in the Netherlands (Ref 13). The latter uses the same heavy medium (ferrite/water suspension), but it is not a passive sink-float method. Instead, the scrap is charged in a cyclone through which the heavy medium is pumped. In this manner, the sensitivity of the process is increased, and inaccuracies due to shape differences of the particles are decreased.

Other Processes. Several separation processes used in the mining industry deserve consideration for adaptation by processors of automotive scrap and municipal refuse. The U.S. Bureau of Mines has experiment-

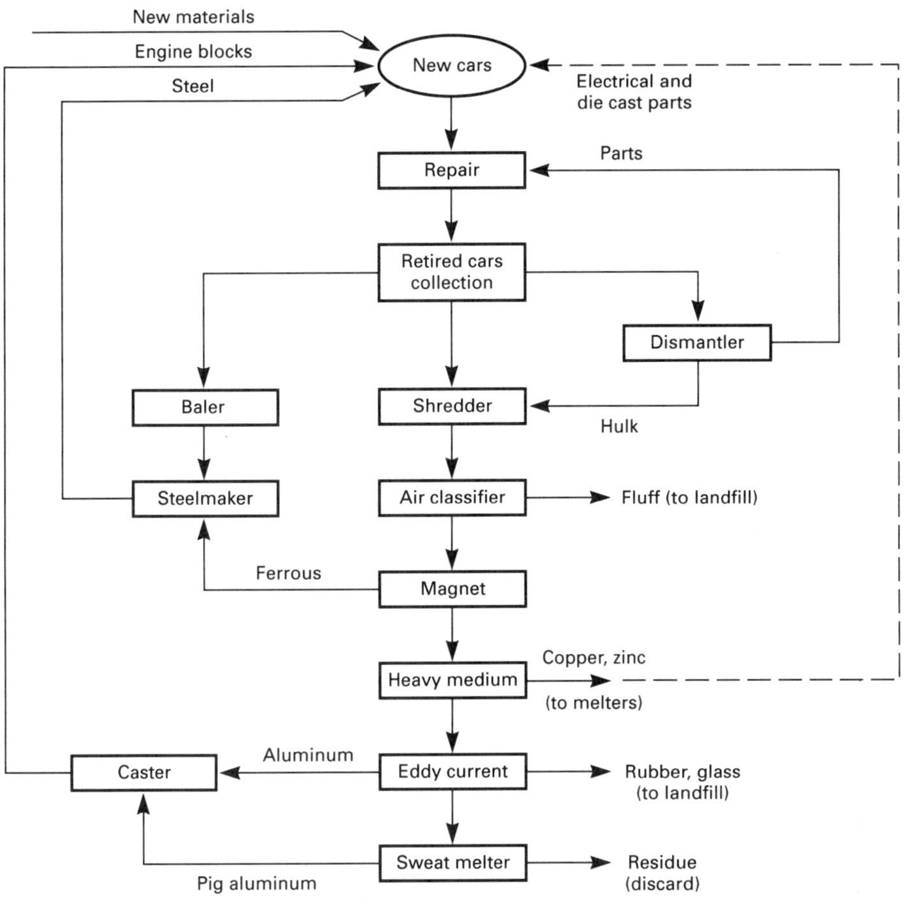

Fig. 11 Recycling loop for aluminum automotive components. Castings make up the bulk of aluminum automotive scrap.

ed with a jigging system in which gravity separation in a bed of shredded mixed scrap is obtained by pulsating liquid flows through the bed. A suspension is formed, and the less-dense material migrates to the top. A system of baffles, spigots, and screens is used to separate the fractions. The U.S. Bureau of Mines has also worked on a shaking-table process for recovering aluminum from municipal refuse incinerator residue. In this process, water flowing in a direction perpendicular to riffles on a slightly tilted vibrating table results in the less-dense material flowing over the riffles and washing downward, while the dense material travels upward between the riffles. This process could possibly be used for separating dirty, fine shredder fractions or dross.

A combined air classification/flotation process for separating wire and plastic insulation has been developed in Japan (Ref 12). After thorough shredding to detach the insulation from the wire and removing the larger metal pieces on a vibrating screen and air table, the light fraction (with up to 15% metal) is charged into a flotation separator to float off the plastic. If a method can be found to concentrate the substantial amount of wiring from the shredder output, this process could be useful for copper recovery, which, in turn would help prevent copper contamination of the aluminum fraction.

Physical separation methods already in use include electromagnetic, or eddy current, separators. These work by creating a force on a nonferrous particle that is traveling in a magnetic field; the created force is perpendicular to the direction of particle travel. The phenomenon is caused by induction of electrical currents when a conductor moves through a magnetic field, resulting in a Lorentz force. Equipment utilizing this phenomenon include the well-known scrap-carrying conveyor belt/electromagnet combination (Almag) and the lesser-known configuration of a sliding table with an array of permanent magnets mounted underneath to create an apparent alternating field for particles moving downward. These separators are presently used for separating nonferrous metal from nonmetallic particles. However, they can potentially be used for alloy separation, especially if shape differences are also characteristic, as in, for example, bulky cast alloy fragments mixed with light-gage wrought alloy sheet fragments (Ref 14).

Electrostatic separation methods, used in mineral separation, can be applied in metal/nonmetal particles separation. One possible application for these methods is the removal of glass, stones, and fiberglass from nonferrous and/or ferrous particles. Various other sorting systems are under study in various laboratories. One system is based on the color differences among copper, brass, and aluminum. Another uses x-ray fluorescence to detect specific elements. A third uses infrared thermal imaging. The success of these methods depends not only on the sensitivity and accuracy of the detection method, but also on coupling detection with a reliable removal method.

Recycling of Copper

David V. Neff, Metaullics Systems
Robert F. Schmidt, Colonial Metals Company

MANY DIFFERENT BRASSES, bronzes, and copper-base alloys exist in the world of scrap metals. Identification of these various grades can be made by description, chemistry, a Copper Development Association (CDA) number, or a universal product code. The universal product code, developed by the Institute of Scrap Recycling Industries (ISRI), serves as an internationally accepted language and greatly facilitates the trading and marketing of scrap metals.

Scrap Classification

The grades discussed in this section are among the most common grades of copper alloy scrap. In an ever-changing industry where new alloys are introduced yearly and old standard grades become obsolete, an all-inclusive listing is impractical, and such a listing might be misleading in some instances. Solid scrap is most prevalent, but almost all grades are available as turnings. In some cases, skims, spills, and drosses are also available and desirable.

Red brass, or composition, has a typical chemistry of Cu-5Sn-5Pb-5Zn. The ISRI code name is ebony. It exists in the form of valves, fittings, pump impellers, plumbing items, and so on. This grade has been progressively replaced over the years by a number of semired brass alloys, and thus current red brass scrap might have a typical

Table 1 U.S. consumption of copper scrap products from 1985 to 1988

Type of scrap	Consumption 1985 Metric tons	1985 Tons	1986 Metric tons	1986 Tons	1987 Metric tons	1987 Tons	1988 Metric tons	1988 Tons
No. 1 wire	348 087	383 694	389 198	429 010	410 636	452 641	416 655	459 275
No. 2 wire, mixed	278 047	306 489	338 031	409 870	383 862	423 128	409 332	451 204
Red brass	51 423	56 683	49 406	54 460	56 366	62 132	53 638	59 125
Cartridge brass	67 221	74 097	67 101	73 965	78 461	86 487	139 074	153 300
Yellow brass	332 143	366 119	299 766	330 430	323 969	357 109	332 212	365 433
Automobile radiators	77 230	85 130	67 101	73 965	62 260	68 629	104 364	366 195
Bronze	19 994	22 039	20 030	22 080	21 050	23 203	21 296	23 474
Nickel silver/cupronickel	15 819	17 437	13 229	14 582	9 617	10 600	14 968	16 499
Low brass	14 931	16 458	14 639	16 136	17 378	19 155	21 676	23 893
Aluminum bronze	969	1 068	970	1 069	965	1 064	1 010	1 113
Low-grade scrap, residues, and so on	201 142	221 717	241 492	266 195	209 216	230 617	101 223	111 577
Other	80 958	89 239	65 831	72 565	86 932	95 825	142 862	157 475

Source: U.S. Bureau of Mines

content of 80 to 83% Cu, 3 to 5% Sn, 3 to 6% Pb, and 5 to 8% Zn.

Yellow brass, or heavy brass, has a typical chemistry of 61 to 67% Cu, 1% Sn, 1 to 3% Pb, and 29 to 35% Zn. The ISRI code name for yellow brass scrap is honey. This scrap originates from lighting fixtures, valves, fittings, and other sources. Yellow brass should be free of automobile radiators and heater cores, as well as manganese and aluminum bronze.

Hard brass, or machinery brass, has a typical chemistry of Cu-10Sn-10Pb. Certain grades have higher lead contents; others contain a smaller percentage of zinc. The ISRI code name is engel. Hard brass scrap exists in the form of bearings, bushings, and other corrosion-resistant parts.

G metal, or high grade, has a typical chemistry of Cu-10Sn-2Zn. Some high grade contains a small percentage of lead or nickel. G metal scrap exists in the form of gears, pump bodies, fittings, and many naval applications.

M metal is a grade with a composition between those of red brass and G metal. Its typical chemistry is Cu-6Sn-1.5Pb-4Zn. M metal scrap exists in the form of fittings, pump bodies, high-pressure valves, and various naval applications.

Automobile radiators are a very common and popular grade of scrap, with a typical chemistry of Cu-3Sn-9Pb-11Zn. The ISRI code name is ocean. Automobile radiators should be unsweated and free of iron, heater cores, shredded radiators, and aluminum radiators. Locomotive and diesel radiators are included in this category and are usually accepted.

Other Brasses. Admiralty brass (ISRI code, pales) has a typical chemistry of Cu-1Sn-29Zn. Aluminum brass (ISRI code, pallu) has a typical chemistry of Cu-1.5Al-24Zn. Both are generally used as condenser tubing and in heat exchangers. They should be unplated and free of sediment. Because of their original application, they should also be checked for possible residues. Various grades of aluminum brass are used in electronic and industrial applications; these are slightly higher in aluminum content and are usually referred to by their CDA numbers.

Nickel silver, or German silver, has a nickel content that spans at broad range: 7.5 to 18%. The copper content is 55 to 65%, and zinc makes up the balance. It exists in the form of hardware fittings, valve trim, ornamental applications, eyeglass frames, and other items.

Phosphor bronze, a copper-tin alloy, is found mostly in electrical and industrial applications. Grade A (C51000) has a typical chemistry of Cu-5Sn. Grade C (C52100) has a typical chemistry of Cu-8Sn.

Aluminum bronze has a typical chemistry of 78 to 90% Cu, 8.5 to 11.5% Al, and 1 to 5% Fe. Certain grades also contain 4 to 5% Ni. It exists in the form of marine applications, bushings, pump impellers, and other applications.

Manganese Bronze. Regular, or low-tensile, manganese bronze has a typical chemistry of 55 to 60% Cu, 1% Sn max, 4 to 2% Fe, 0.5 to 1.5% Al, 1.5% Mn max, and a balance of zinc. It exists as gears, valve stems, marine applications, and propellers. High-tensile manganese bronze exists in similar forms including bushings and cams. Typical chemistry of the high-tensile varieties is 60 to 68% Cu, 3 to 4% Fe, 3 to 7.5% Al, 2.5 to 5% Mn, and a balance of zinc.

70/30 Brass. Material consisting of 70% Cu and 30% Zn is known simply as 70/30 brass (C26000). This grade is commonly found in the form of brass rifle shells (ISRI code, lake). They should be fired before shipment to a scrap consumer.

Free-Cutting Brass. Free-cutting rod brass solids (ISRI code, noble) and free-cutting rod brass turnings (ISRI code, night), like 70/30 brass, are popular brass mill items. The typical chemistry of rod brass is 60 to

Table 2 CDA designations and nominal compositions of copper-base alloys

CDA designation	Nominal composition, %(a) Ni	Sn	Pb	Zn	Mn	Al	Other
Wide freezing range alloys							
C83600	...	5	5	5	...	...	...
C84400	...	3	7	9	...	...	...
C92200	...	6	2	4	...	...	...
C93200	...	7	7	3	...	...	...
...	...	5	2	5	...	...	...
C83450	...	2	2	8	...	...	...
Narrow freezing range alloys							
C85700	...	1	1	35	...	...	...
C85800	...	1	1	40	...	...	...
C86500	...	...	...	40	...	1	1 Fe
C86200	...	...	...	26	3	4	3 Fe
C86300	...	...	...	25	3	6	3 Fe
C95200	...	...	...	...	9	...	3 Fe
C95300	...	...	...	...	...	10	1 Fe
C95400	...	...	...	...	...	10	4 Fe
C95500	5	...	...	...	...	10	4 Fe
C95800	5	...	...	...	1	9	4 Fe
C87300	...	...	...	...	1	...	4 Si
C87600	...	...	...	5	...	...	4 Si
C87500	...	...	...	14	...	...	4 Si
C96400	30	...	...	...	...	...	...
C99700	5	...	...	24	12	1	...

(a) All compositions contain a balance of copper.

63% Cu and 2.5 to 3.7% Pb, and a balance of zinc.

Other Scrap Grades. Other types of brass mill scrap include red brass pipe and yellow brass pipe (ISRI code, melon), with chemistries of Cu-15Zn and Cu-35Zn, respectively. Other nickel-bearing scrap items include Cu-10Ni, Cu-30Ni, and a variety of copper-nickel alloys. Also worth mentioning are lined railroad carbox journals (ISRI code, fence), silicon brass, low-brass house screen, paper mill screen, lined and unlined traction bearings, minnox metal, modine tubes, and a host of other grades that are usually identified by CDA numbers.

Pure Copper Scrap

There are three predominant grades of copper scrap: No. 1, No. 2, and light copper.

No. 1 copper (99+% Cu) consists of a number of categories including clean pipe and tubing, bare bright wire, burnt wire, green-line copper, busbar, and copper choppings. There are a number of ISRI codes (barley, berry, candy, and clove) that apply to this grade. All grades should be clean and free of any detrimental impurities; wire should have a minimum size of 16 gage. Certain grades of No. 1 copper, such as bare bright wire and heavy choppings, can command a premium price in the market place. Size and packaging are also factors in determining the price of the scrap.

No. 2 copper is usually sold with a minimum copper content of 96%. The ISRI codes for No. 2 copper are birch, cliff, and cobra. It should be free of excessively leaded, tinned, or soldered copper scrap; brass and bronze; excessive oil; iron; and ash and other nonmetallics.

Light copper is sold with a minimum content of 92% Cu. It is also referred to as sheet copper or No. 3 copper, or by the ISRI code name dream. It usually consists of sheet copper, gutters, downspouts, kettles, boilers, and similar items.

Other grades of copper scrap are referred to as refinery or copper-bearing scrap. Their value is determined by the percentage of copper recovered in a refinery process.

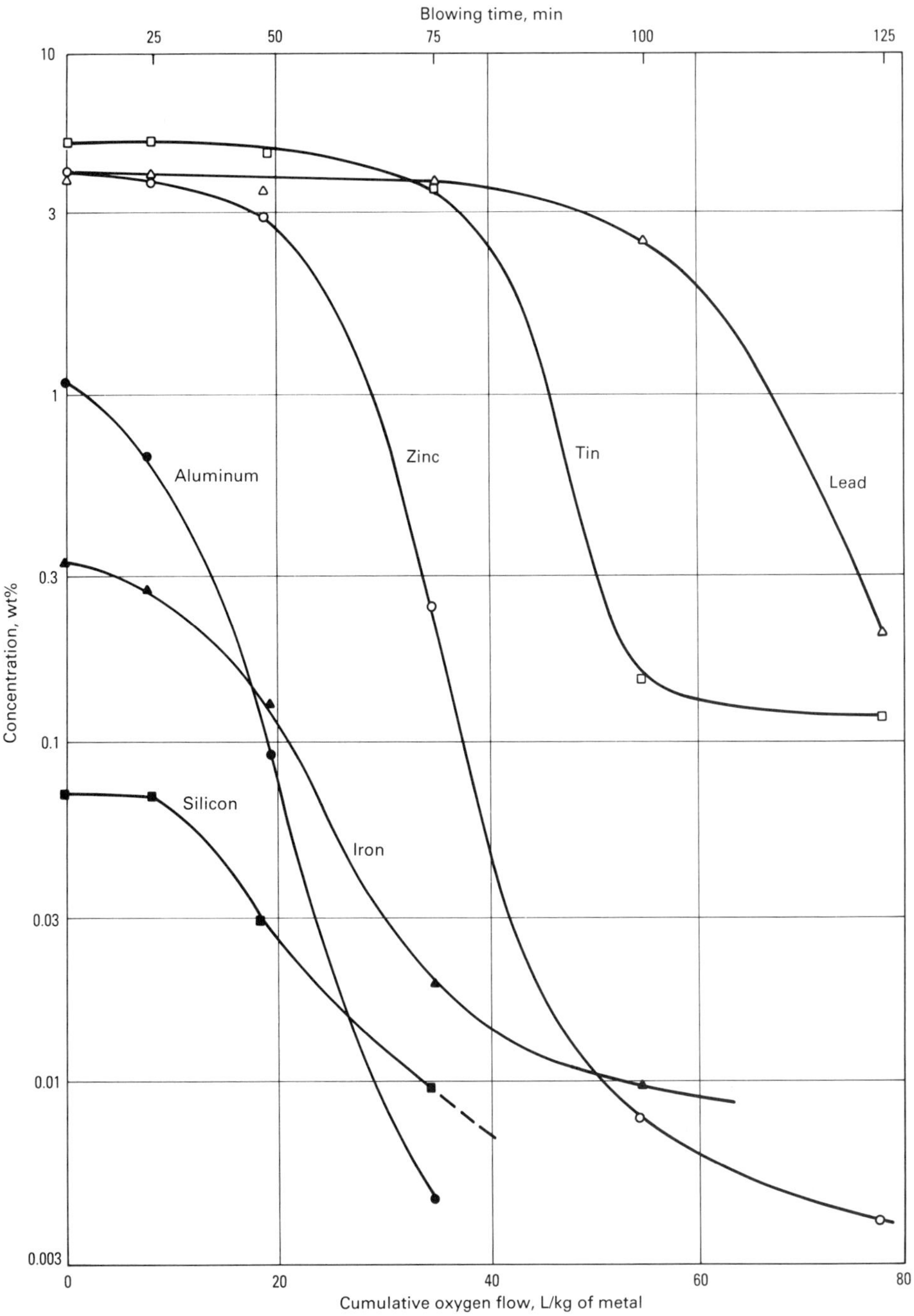

Fig. 12 Effect of fire refining on impurities in molten copper. Source: Ref 15

Recycling Technology

Collection Procedures. As shown in Table 1, huge amounts of copper-base scrap are recycled each year. Accumulating, processing, and preparing these amounts of scrap are not easy tasks. The peddlers and junkmen of yesterday have been replaced by the sophisticated and professional recyclers of today.

The material that finds its way to the scrap dealer comes from many sources. Obsolete scrap is perhaps the most common and abundant material. This category includes copper and brass from electrical and plumbing contractors, brass radiators from automobile wreckers and shredders, copper wire from telephone and power lines, railroad scrap, scrap from plant- and shop-dismantling operations, and generally any copper-base item from homes, farms, industries, and municipal solid-waste operations.

Industrial scrap is another category that consists of large volumes of scrap material. Industrial scrap usually consists of new production material, for example, the skeleton or strip from a stamping operation, the turnings from a machining process, rod or tube ends, wire, or any item that a manufacturer would consider a reject because of chemistry, size, shape, or other parameters.

The last major source of copper-base scrap is the military. Most of this scrap is in the form of fired brass shell cases. There is, however, a good amount of obsolete and dismantling scrap generated throughout the various military bases.

After material has been scrapped, the scrap dealer must process, identify, and sort it into saleable commodities. Uniform industrial scrap is probably the easiest to market. Most brass and copper mills, as well as secondary smelters, consider it a premium item. Therefore, it is imperative that uniform scrap be free of impurities, and extra care must be taken to ensure its purity. Once contaminated, it becomes an entirely different grade with a lower value.

A large percentage of the scrap entering the recycling chain is mixed. This material must be graded or sorted into the correct scrap category. This is done in a number of ways. The quickest method for a trained, seasoned sorter is by color. Spot testing with acid is another method used to identify alloys. Magnets, files, grinders, and a host of commercial metal identification products also assist in making proper identification. If these procedures prove to be inconclusive, laboratory work, either by wet process or by x-ray method, can result in positive identification.

Once identified, the various grades of scrap are packaged for shipment. Boxed and drummed material is usually accepted at most mills. Bulky items such as automobile radiators, copper wire, and numerous sheets and strips are baled. Rod turnings, as well as some larger solids, should be shipped loose; this facilitates unloading by dumping or by a front-end loader.

Drosses, skims, and spills that result from secondary smelting can be handled in a number of ways. If the dross is high in metallic yield and does not contain any detrimental impurities, it can be reused directly into the rotary furnaces. Low-yield material is less likely to be used directly. The majority of this material is best suited for the copper refineries; these refineries can recover the copper fraction of the material, and they are less affected by impurities and low metallic yield.

Melt refining practices depend on the materials being recycled, that is, whether they are high coppers or copper-base alloys. Secondary, or recycling, practice for the high coppers usually involves remelting only; it is performed in small induction furnaces. An inert flux cover such as graphite is used to prevent oxidation loss. A reactive flux cover containing fluoride salts is also often used to provide some fluidity. The reactive cover prevents oxygen transfer into the melt and also helps strip oxide films that can form from reactive elements such as chromium.

For refining purposes, copper-base alloys can be divided into two groups: wide melting range alloys and narrow melting range alloys. The first group melts over a range of about 165 °C (300 °F) and includes such alloys as the red brasses and the tin bronzes. The second group melts over a range of 70 °C (125 °F) or less; some alloys in this group melt over a range as narrow as 16 °C (30 °F). Table 2 lists the current CDA designations and nominal compositions of selected alloys from both groups.

Raw material for the first group usually consists of radiators, valves, fittings, bushings, bearings, machine turnings, grinding dust, and cupola slabs. Melting is done in reverberatory or rotary furnaces with capacities ranging from 13.5 to 63 metric tons (15 to 70 tons), depending on the monthly tonnage of the alloys sold. Refining is carried out by blowing air or oxygen into the molten charge and oxidizing all the lighter-element impurities. Impurities of sulfur, antimony, aluminum, iron, silicon, manganese, nickel, and phosphorus are usually present. Figure 12 shows the rate of reduction of the various impurities during the refining process.

Aluminum is usually the first element to be oxidized out, followed by manganese, silicon, phosphorus, iron, and then zinc. Although zinc is oxidized out, it is usually a desired element. If too much is lost during smelting, additional zinc and tin must be added after refining to bring the specific alloy up to specification.

For the most part, lead, nickel, tin, and antimony cannot be refined out; they must be diluted out if the alloy is not within specification limits. Care must be taken to use known raw material for the charge so that these elements are not over the allowable maximum. Some success in refining lead and antimony has been achieved with soda ash fluxes (see the article "Nonferrous Molten Metal Processes" in Volume 15 of the 9th Edition of *Metals Handbook*). Soda ash, borax, boron-containing mineral ores, silica sand, and coal screens are used as fluxes and covers to help with the refining processes and to control excess oxidation. During refining, continual chemical checks are used to monitor the removal of iron, aluminum, silicon, manganese, and phosphorus. These results can be obtained in less than 2 min with optical or x-ray spectrographs.

Alloys in the second group include the aluminum, manganese, silicon, and phosphor bronzes; copper-nickel; and nickel silver. Generally speaking, very close control is maintained over remelting these scrap materials, and contaminants and impurities are minimized in the charge materials. Consequently, alloys in this second group are generally not refined but are alloyed up from primary copper-base returns or virgin elements.

Because the total quantities of the second-group alloys are lower, they are usually made in smaller (4 to 22.5 metric ton, or 5 to 25 ton) rotary or high-frequency induction furnaces. Fluxes used include soda ash or borax, with fluoride salts to cut oxide skins. Virgin elements are usually used to produce aluminum bronze, high-strength manganese bronze, silicon bronze, brass, and copper-nickel alloys. Copper is melted first, and then manganese, iron, or silicon is added and worked in. Nickel, aluminum, or zinc is added next, depending on which alloy is being produced. All of these steps are checked as they occur by spectrographic analysis to ensure the right amount is added. The speed and accuracy of the chemical laboratory is an important factor in the production of these high-quality alloys.

Alloys sold on mechanical specifications are sand cast into test bars and checked to be sure they meet tensile property, yield strength, and elongation requirements. Many alloys in this second group have a wide range of acceptable compositions, but all of these may not meet certain restrictive mechanical property requirements. Consequently, mechanical testing is often mandatory.

Recycling of Magnesium

Michael Slovich, Garfield Alloys, Inc.

THE SECONDARY MAGNESIUM industry is quite small relative to other nonferrous recycling industries. There are two reasons for this: first, magnesium metal has not found the widespread acceptance of many other nonferrous metals, and second, approximately 80% of primary magnesium has historically been consumed by nonstructural uses for distributive and sacrificial purposes.

Scrap Sources

Magnesium scrap generally comes in forms similar to those of other nonferrous metals: new castings and the gates, runners, drippings, turnings, and drosses that result from such operations; and old scrap recovered at the end of the useful life of the magnesium-containing product. Since World War II, structural applications for magnesium have mainly been in the form of various wrought products and die and sand castings. Among the processes used to produce these products, die casting is the largest source of scrap magnesium.

Some forms of old scrap come from aircraft parts, such as wheels, fuselages, and control panels, and from military applications, such as incendiary bombs and tent support poles. Other significant sources of old scrap are discarded lawn mower decks, dock plates, chainsaws, and various hand tools.

By far, the single largest structural use of magnesium—and the largest source of mag-

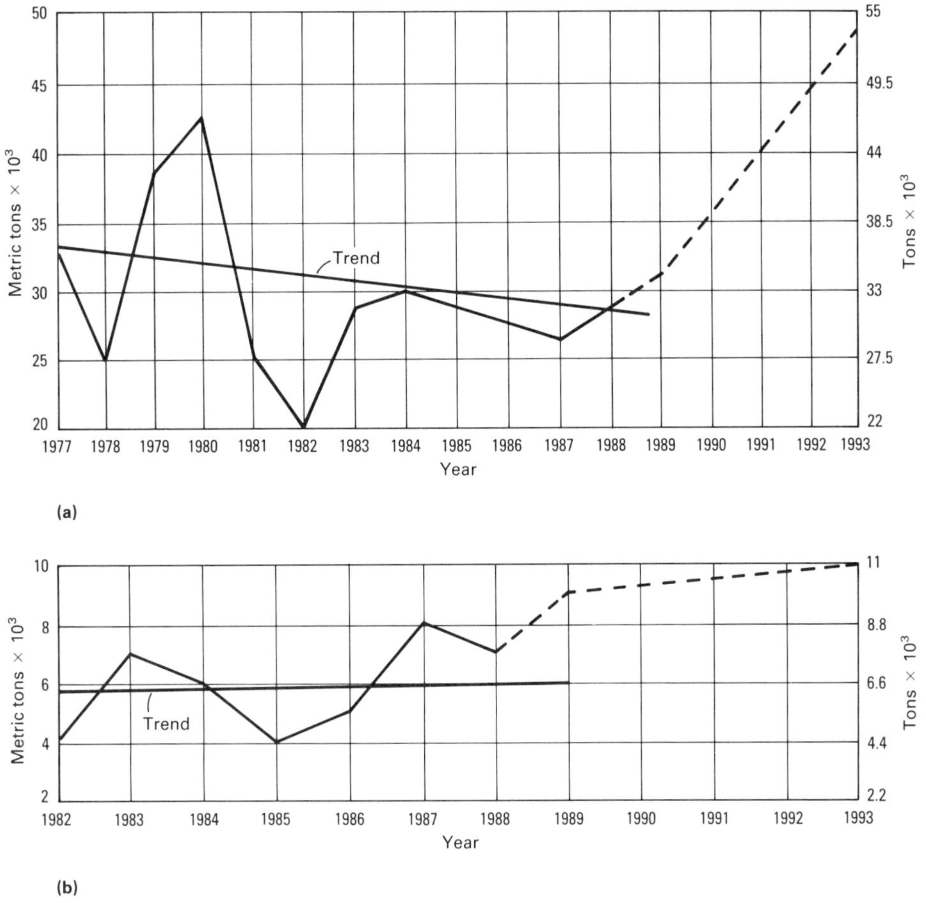

Fig. 13 Magnesium consumption in the United States. (a) Die castings. (b) Wrought products. Source: Ref 16

nesium scrap—has been Volkswagon Beetle die cast engine and transmission castings. When the use of magnesium in this application reached its peak in the mid-1970s, Volkswagon was consuming approximately 45 000 metric tons (50 000 tons) per year. Approximately 18 kg (40 lb) of magnesium was used in each vehicle.

Volkswagon continues to produce some magnesium engine blocks in Brazil and Mexico, but consumption in this application is significantly and permanently lowered. The main reason for the decrease in the use of magnesium in this application was the rapid price escalation of the metal that started in the mid-1970s; the price of magnesium quadrupled between 1973 and 1986. As a result, the secondary magnesium industry has seen a gradual reduction in the amount of old scrap available.

Because the supply of old magnesium scrap has come mainly from die cast and wrought products, it is useful to look at past consumption of primary magnesium by the producers of these items (Fig. 13). The data in Fig. 13 can be used to indicate the trend of the secondary magnesium industry because the amount of scrap generated is directly related to the amount of magnesium consumed in the production of these structural items. Although there are some rather large fluctuations in usage from year to year, Fig. 13(a) indicates an overall declining trend in die casting use. As shown in Fig. 13(b), wrought product consumption has virtually remained flat. All told, annual magnesium consumption has decreased by approximately 4500 metric tons (5000 tons) in recent years.

However, a recent resurgence in the use of die cast magnesium in the automotive industry has created additional sources of new magnesium scrap. As shown in Table 3, the amount of new magnesium-base scrap processed annually more than tripled from 1987 to 1988. On the other hand, the amount of old magnesium-base scrap processed has generally been declining since 1984, and this trend will continue in the foreseeable future. Eventually, the parts currently being placed in automobiles will return to the secondary market as the cars are scrapped, but this will not happen for many years.

At present, between 80 and 90% of recycled magnesium is used in recycled aluminum alloys, the most common of which are used to make aluminum beverage cans.

Recently, the price of magnesium die cast alloys has remained quite stable relative to the large fluctuations in the price of secondary aluminum alloys. The planned expansion of primary metal production will affect future pricing. The opening of two new facilities is expected to increase the free-world magnesium production capacity by approximately 20%. This projected new capacity should ensure stable or reduced prices for current and potential magnesium consumers.

On the technological front, the development of high-purity alloys that provide improved corrosion resistance has been a major factor in promoting the use of magnesium in applications where the metal is exposed to the elements. More information on these alloys is available in the articles "Selection and Application of Magnesium Alloys" and "Properties of Magnesium Alloys" in this Volume.

The two factors mentioned above—price stability and significant technological developments—coupled with the enlistment of new die casters to produce magnesium parts and continued market development by all participants, should provide the proper conditions for significant growth in the magnesium industry.

Table 3 Magnesium recovered from scrap processed in the United States

Kind of scrap	Amount recovered							
	1985		1986		1987		1988	
	Metric tons	Tons	Metric tons	Tons	Metric tons	Tons	Metric tons	Tons
New scrap								
Magnesium base	1 510	1 664	991	1 092	845	930	2 641	2 911
Aluminum base	16 252	17 914	17 822	19 645	20 868	23 000	19 926	21 964
Old scrap								
Magnesium base	4 630	5 104	3 958	4 363	3 857	4 251	3 882	4 279
Aluminum base	18 906	20 840	19 036	20 983	19 595	21 600	23 758	26 188
Total	**41 298**	**45 522**	**41 807**	**46 083**	**45 165**	**49 785**	**50 207**	**55 343**

Source: Ref 17

Magnesium Recycling Technology

Magnesium scrap is most often received loose on a dump trailer or in boxes on a van-type trailer. Because magnesium closely resembles aluminum, many suppliers have difficulty separating magnesium from aluminum. Consequently, a load of magnesium scrap will often contain a percentage of aluminum scrap. Therefore, it is imperative that the secondary magnesium facility have qualified personnel who can hand sort and identify foreign materials. The most-used method of separating aluminum from magnesium scrap involves scratching the metal with a sharp knife. Magnesium tends to flake, whereas aluminum will curl because of its relative softness. Once the extraneous materials have been removed, the remaining magnesium scrap must be sorted according to alloy. This process is absolutely critical in creating a product to specification.

Magnesium scrap storage piles, whether indoors or out, should have a width of at least 10 ft or one-half the pile height. For outdoor raw material storage, a pile should be no closer to any building than one-half the pile height. These practices help ensure safety as well as ease of access.

Melting. The sorted scrap is set up and charged into the furnace according to the type of alloy of the scrap relative to the desired composition for the finished metal. Once the scrap has been fully charged and melted, a sample is extracted for preliminary spectrographic analysis. If this analysis indicates a composition that does not meet desired specifications, adjustments are made by adding alloying elements such as aluminum, zinc, or manganese to the melt as needed. If the preliminary analysis indicates an excess of an alloying element or the presence of any tramp elements, the metal can be sweetened by adding pure magnesium to the melt to reduce the excess as a percentage of the whole.

Several proprietary techniques are involved in the economical and safe processing of fine magnesium turnings, borings, and powders. The same is true in processing magnesium dross for metal recovery. Drosses can be effectively recycled, assuming there is enough metal content in the dross itself. Generally, if the metal content is less than 15 to 20%, it is not economical to recover metal from magnesium dross. Anyone involved with such materials should be aware of the inherent dangers if they are mishandled. These materials pose the obvious danger of fire; in addition, they must be kept absolutely dry to avoid oxidation. Oxidation yields heat and hydrogen, which can result in spontaneous combustion. When handled properly, magnesium in both the solid or finely divided form is not dangerous. As with most other things, mishaps are usually the result of carelessness.

Magnesium scrap is melted in steel crucibles at a temperature of 675 °C (1250 °F). A fluxing agent is used to remove impurities such as oxides. The molten metal can then be handled in one of three ways: manual ladling, pumping, or tilt pouring. Because molten magnesium oxidizes rapidly, the metal should be cast immediately once the desired chemical specifications have been attained. Such a practice will minimize the danger of crucible failure and maximize metal recovery.

Overheating magnesium during melting causes several problems. First, even minor overheating will cause oxidation, thereby reducing metal recovery. Second, if the metal is at the appropriate temperature when alloying or sweetening materials are added, the alloying efficiency is increased and the danger of spit back is reduced. Third, significant overheating will cause the metal to burn, creating a potentially dangerous situation. Fourth, overheating, and the corresponding rise in the chamber temperature of the furnace, increases the oxidation of the crucible. As the crucible becomes heated to excess, pieces of the steel crucible begin to flake off in the form of iron oxide, increasing the possibility of leaking. If the crucible begins to leak, and if molten magnesium comes in contact with the iron oxide in the furnace chamber, a catastrophic thermic reaction could result.

Properties of Secondary Magnesium. The properties of secondary and primary magnesium die-cast alloy ingot are identical, assuming both are within chemical specification limits and all inclusions are removed from the metal. Any residual impurities left in the metal could cause a variety of problems for the die caster, depending on the type and amount of contamination. For example, flux inclusions in the ingot that are not removed in the die casting process will cause corrosion. Similarly, the presence of heavy metals (nickel, iron, copper, and silicon) in the alloy could cause premature oxidation.

The recent introduction of high-purity magnesium alloys has played a major role in gaining the attention of the worldwide automotive industry. However, these high-purity alloys are a double-edged sword for the die caster. The stringent specifications for these alloys have made it virtually impossible for the caster to recycle his own scrap in-house while maintaining high-purity specifications. Therefore, the die caster must sell the scrap to, or have it tolled by, a qualified secondary facility that can maintain quality by culling out any nonconforming materials. Indeed, such a facility is far more qualified to handle scrap and produce a quality metal alloy than a die caster; the die caster is geared toward producing castings of high integrity using source materials that meet chemical and physical specifications.

Recycling of Tin

William B. Hampshire, Tin Information Center

THE RECYCLING OF TIN is intimately interconnected with the recovery of metals other than tin. Even the briefest examination of the applications of metallic tin will reveal its ability to wet and coat a variety of base metals, including iron and copper. This property was the basis for the development of tinplate, which combines the strength and durability of steel with the nontoxic protection of a thin coating of tin to create the preeminent food container material. Soldering, the other substantial use for tin metal, also relies heavily on the ability of tin to wet a variety of metallic surfaces. In all cases, other metals must provide the required mechanical strength because tin itself is quite soft.

The cost of tin is relatively high compared with that of the more-common metals. Cost considerations keep the amount of tin used in its applications at a minimum. For example, bronze normally contains only a few percent of tin, perhaps up to 10%. Tinplate typically contains only 2 to 3 kg (4 to 6 lb) of tin per ton of steel (about 0.25% by weight). Because of this dilution of the tin content in so many of its major applications, it is very difficult to recover tin as tin metal, and it is often recycled as an alloy of the more-dominant metal.

Bronze and tin-containing brass are usually recycled by melting and realloying to produce alloys that are the same as, or similar to, the input metal. Tinplate is sometimes detinned, but the very low percentage of tin in the material makes the process less cost-effective than in earlier years, when tinplate was produced with a higher fraction of tin. Solders can be recycled; for example, solders from radiators (usually high-lead alloys), can be sweated out for recovery, but surface tension forces and alloying tend to severely limit the amount of solder that can be recovered in this way. In all these cases, tin is being recycled, but the larger proportion of recycled material consists of copper, steel, or lead.

Kinds of Scrap Tin Materials

Basically, two general types of tin-rich materials can serve as sources for secondary tin. The first type consists of mostly metallic tin that is usually fairly high in tin content. Prices for this type are quoted in *American Metal Market*. The specific scrap grades within this type—block tin, and high-tin babbitt and pewter—are consistent enough to identify and define.

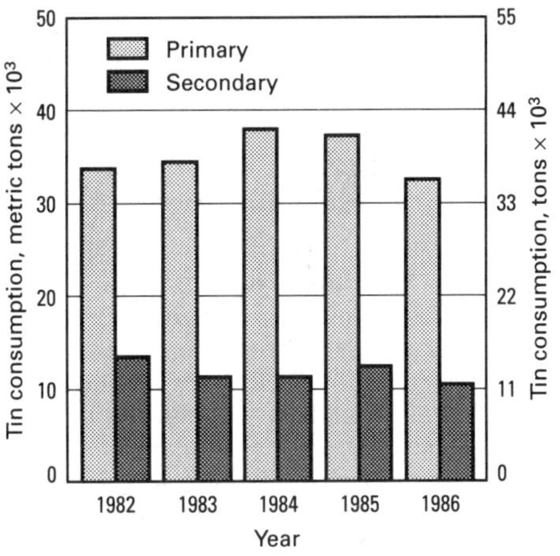

Fig. 14 Primary and secondary tin consumption in the United States from 1982 to 1986. Source: Ref 18

Block Tin. Pure tin is widely preferred as the containing material for high-quality distilled water. Less than 1 ppb Sn is expected to dissolve in freshly distilled water contained in tin. Large pipes and conduits of pure tin metal are found in some stills, and the general term for these pieces is block tin. Because block tin is of commercial purity and easily melted down, it commands a high premium as a secondary material. For cost considerations, block tin has often been replaced by heavily tinned copper tube; of course, this material has a much lower value for recycling.

High-Tin Babbitt and Pewter. These related alloys typically consist of about 90% Sn, with antimony and copper added as hardeners. The tin babbitts would likely come from worn-out or damaged bearings. It seems unlikely that much pewter would find its way into the secondary market; it is more likely to be repaired.

The U.S. Bureau of Mines makes the following stipulations concerning the content of these materials (Ref 18):

- *Block tin* must contain a minimum of 98% Sn, and it must be free of liquids, solder, brass connections, pewter, pumps, pot pieces, and dirt
- *High-tin-base babbitt* must contain a minimum of 78% Sn, and it must be free of brass-like metals and high-zinc metals
- *Pewter* can consist of tableware and soda fountain boxes, and it must contain a minimum of 84% Sn. Siphon tops should be accounted for separately. The material must be free of brass, zinc, and other foreign metals

Other Materials. The second general type of tin-rich secondary tin source material contains significant proportions of tin oxides. Drosses from hot-tinning pots or from soldering pots are examples of this type of material. The specific scrap grades lack any consistency and thus can be defined only by individual assay. The chemical stability of tin oxides makes these scrap forms more difficult to process for the recovery of the tin values. Therefore they are less popular as sources of secondary tin than materials from the first group, and they are usually resmelted with flue dusts and/or tin concentrates from ore.

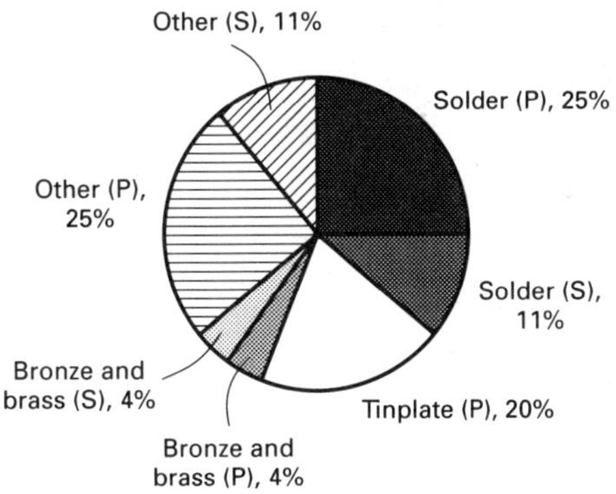

Fig. 15 U.S. primary and secondary tin consumption by sector for the year 1986. Consumption of secondary tin for tinplate withheld to avoid disclosing proprietary data; included in "Other (S)." P, primary; S, secondary. Source: Ref 18

Facts and Figures on Recycled Tin

Figure 14 shows data for annual tin consumption in the United States. The secondary tin market shows good stability, and this can be attributed to the relatively high cost of tin and the corresponding viability of tin recovery. The level of secondary tin consumption has remained fairly consistent at about 33% of primary tin consumption. In the 1970s, when tinplate accounted for a larger share of tin consumption, this percentage was about 25% because tinplate producers have always preferred to use primary tin for their products.

Figure 15 shows 1986 data for U.S. tin consumption by sector. These values are the latest complete set available at the time of publication. The solder market and the bronze and brass market consume substantial amounts of secondary tin. The "other" category includes tin chemicals, which are often made from secondary tin. In fact, in some detinning schemes tin is removed chemically from scrap tinplate and is converted into another chemical (often for use in tin-plating) without ever being recovered as tin metal.

Stott (Ref 19) divides the secondary tin industry into segments and provides estimates for the amount of tin recovered per year in each segment:

- Tinplate scrap processing, 1200 metric tons (1320 tons)
- Secondary copper industry, 6800 metric tons (7500 tons)
- Secondary lead industry, 6575 metric tons (7250 tons)
- Processing of clean tin residues and complex residues, 150 metric tons (165 tons)

The total is just under 15 000 metric tons (16 500 tons) per year, although, as Stott points out, statistics on this industry are subject to some variation.

The secondary copper and secondary lead segments are mentioned only briefly in the following section; they are covered in the sections "Recycling of Copper" and "Recycling of Lead" in this article. Processing of tinplate scrap is given relatively more attention. The recovery of tin-rich residues is discussed to the extent possible, given the complexity of such operations.

Recycling Technology

Collection of tin-containing secondary materials is often done industrially. Clean tinplate scrap generated by can manufacturers is the preferred input for a detinning operation because this feedstock is predictable and free of difficult contaminants. Recycled municipal waste is more heavily contaminated and may have been incinerated (which alloys the tin with the steel), making it more suitable for adding directly into the steelmaking process without detinning. Research is ongoing into scavenging additions that could remove tin from the basic structure of the steel. Such additions might greatly encourage the recycling of the ferrous fraction of municipal scrap.

Bronze and brass turnings, drosses, and so on are often collected by metal fabricators for return to the metal supplier. The supplier simply returns the scrap, after some cleaning, to the melting facility for input into similar alloys. Tinned copper wire scrap is also an obvious candidate for direct melting in copper alloys. In general, because tin is used in small proportions with other metals, successful recycling of the tin values depends on collection of the scrap metal at the earliest possible stage of fabrication, before additional dilution can take place.

The high-tin forms (block tin and high-tin babbitts) are collected after use during the dismantling of stills or bearings. The dismantlers will know the value of such parts (typically \$2 to 3 per pound, much higher than most scrap) and will remove them as completely and efficiently as possible to maintain the high value. Also, it is important to keep block tin separate from high-tin babbitt or pewter to maintain the scrap value. Block tin is made from Grade A tin; if it is collected without contamination, it can simply be substituted for primary Grade A tin, which is the normal commercial quality tin metal.

Used electronic equipment is sometimes dismantled to allow recovery of the solder (see the section "Recycling of Electronic Scrap" in this article). Miniaturization of electronic components is making this recovery more difficult. Stott (Ref 19) has estimated the tin content of electronic scrap to be about 1% by weight.

Processing. The high-tin scrap (metallic type) is most often recycled by pyrometallurgical techniques, usually by realloying. Block tin will probably need no processing, except to convert it in the same manner as purchased tin ingot. High-tin babbitt or pewter will ordinarily be recycled by remelting, then adding the recycled material, all or part, to new metal during the melting and casting process. Attempts to refine the metal by removing either the antimony or copper from the alloy would prove very difficult. It is the affinity of tin for wetting and alloying with other metals that makes recycling, in the sense of recovering the tin values as tin, difficult to the point of extreme inefficiency.

Tin removal from a copper surface, especially by sweating, might be contaminated with copper. This copper can be removed by treatment with sulfur, which then itself must be removed by oxidizing. When iron is the contaminant of concern, it is removed by boiling. This procedure is akin to poling as practiced in the copper industry, even to the point of using green wood poles to accomplish it. Antimony is removed by an aluminizing treatment. Finally, for removing zinc, careful additions of ammonium chloride or sulfate can be used. Nearly all of these procedures can generate noxious fumes or by-products, and they should be practiced only by those with some experience (see Ref 20, 21).

Tinning or solder drosses (oxide type) from hot tinning or soldering pots are usually recycled by smelting. These drosses will normally contain substantial proportions of metal entrained in the oxides, which again is due to the wetting ability of tin. First, a liquation process is used: The material is heated to above the metal (solder or tin) melting temperature and held at that temperature while the metal melts and is collected. Roasting or leaching may also be necessary as pretreatment steps. The resulting and remaining oxides can then be smelted by heating to a very high temperature with carbonaceous reducing material in a blast furnace or a reverberatory furnace (see Ref 19-21).

Detinning of tinplate scrap is usually accomplished by either an alkaline leaching method or chlorine detinning. The alkaline process has been the more popular in recent years. It involves dissolving the tin off the tinplate scrap by immersion into a caustic solution. This solution is then passed through electrolytic cells, where the tin is plated out of the solution. The tin thus recovered is easily refined to high purity, and the caustic solution is recycled. The steel that emerges from processing is a clean product that can be used directly by a steel mill or foundry.

Chlorine detinning uses dry chlorine gas at elevated temperature and pressure to react with the tin coating to produce stannic chloride in gaseous form. This reaction is exothermic; cooling is required to remove the heat and to condense the product. These complications, along with the relative difficulty of handling chlorine gas, have led to a decline in the use of this recovery method.

Tinplate is a significant portion of the ferrous fraction of municipal refuse. Therefore, the increasing interest in recycling municipal scrap is having a complex effect on the detinning industry (see the section "Current Trends and Future Expectations" in this article).

Recycled Metal Behavior

Recycled tin from a detinning operation, when it is electrolytically removed from the solution in a tankhouse, is comparable to or better than the fire-refined primary tin metal. In fact, some of the highest-purity tin has been produced by such a recovery process. Provided both the primary tin metal and the secondary tin metal meet Grade A requirements, there should be no noticeable performance differences. Composition requirements of Grade A tin are:

Element	Content, %
Tin, min	99.85
Antimony, max	0.04
Arsenic, max	0.05
Bismuth, max	0.030
Cadmium, max	0.001
Copper, max	0.04
Iron, max	0.010
Lead, max	0.05
Nickel plus cobalt, max	0.01
Sulfur, max	0.01
Zinc, max	0.005

There has emerged some concern in certain solder-using industries, notably electronics assembly, that the use of secondary solder may produce more soldering defects, although concrete data seem to be lacking. Nevertheless, to help reassure solder users, some recent specifications (especially U.S. Naval Weapons Specification WS 6536) now call for regulating the sulfur and phosphorus contents of the incoming solder alloy. Typical recovery of secondary solder might begin with a sulfur treatment to remove excess copper. An oxidation step comes next, followed by a phosphorus treatment to reduce the oxides; the presence of oxides could lead to higher sulfur and phosphorus contents than in virgin solder. No other specifications seem to try to distinguish secondary product in this way.

Current Trends and Future Expectations

The recycling of tin is well established. Processes developed over the years have gained a degree of stability and acceptance, and developments are more a matter of refinement than of truly new processing. The current interest in recycling of materials may make more tin available for processing, but because tin is so valuable, most of the easily recoverable tin is already being processed.

Tinplate can be easily separated magnetically with the ferrous fraction of municipal scrap, and the recently established Steel Can Recycling Institute is encouraging municipalities to do so. Much of this tinplate scrap will be fed into the steelmaking process without detinning because of its low tin content. Therefore, the tin will be recycled but not recovered as tin metal, unless it is separated from steelmaking residues.

Several recent developments in processing tinplate scrap are allowing increased rates of recycling for steel cans. Better pretreatment techniques allow detinners and recyclers to accept some food residues, paper labels, and aerosol and paint cans, all of which were previously unacceptable in the feed material (Ref 22). Some municipalities are beginning consumer curbside sorting. All these factors can lead to a cleaner, more dense steel input material for the steel mills. As scrap yards gain experience in the handling of steel can scrap, the recycling rate will no doubt continue to rise dramatically.

Research has shown other potential uses for tinplate scrap. Tin is well known as a useful addition to cast irons, functioning as a pearlite stabilizer at a level of about 0.1%. Adding tin can scrap to the melt can achieve this effect as efficiently as using tin metal additions. Tin is also added with copper to ferrous powder metallurgy parts in which it functions as a sintering aid. Research at the International Tin Research Institute has shown that tinplate scrap can be used for this purpose in place of tin metal with no detrimental effect on the product (Ref 23, 24).

It is much less clear how the recycling of solder can be increased. The recent trend in electronics assembly has been to use techniques, such as surface mounting, that use smaller and smaller joints of solder. Solder use has held steady due to the production of more joints, but these smaller joints make any attempts at recycling the metal very difficult.

Recycling of Lead

R. David Prengaman, RSR Corporation

LEAD HAS THE HIGHEST RATE of recycling of all metals. Because of its corrosion resistance, lead scrap is available for recycling decades or even centuries after it is produced. Environmental regulation in the United States has greatly reduced the dissipative uses for lead such as paint, leaded gasoline, pigments, stabilizers, solder, and ammunition. The remaining lead products are recyclable (Ref 25).

Production of Recycled Lead

Table 4 gives data for the world production of lead from recycled and mined (primary) sources for 1980 to 1988. At present, just under half of the total world lead production of 4.3 million metric tons (4.7 million tons) comes from recycling of scrap materials. As indicated in Table 4, there has been very little change in recent years in the total amount of lead production or in the percentage of recycled lead. Only in the past few years has the amount of recycled lead increased. This may be due to the increasing availability of batteries throughout the world.

Table 5 gives data for U.S. production of lead from recycled and mined sources for 1980 to 1988. Beginning in 1984, there was a dramatic increase in the percentage of lead produced from scrap in the U.S. despite a decrease in overall lead production. By 1984 the major dissipative use for lead, leaded gasoline, had decreased to a small percentage of the overall lead market. The percentage of recycled lead production has increased dramatically in recent years, from 50% of production in the late 1970s and early 1980s to about 65% in 1988. The rate of lead production from scrap materials is expected to increase dramatically in the future. The major market for lead in the United States is now lead acid batteries: Batteries constitute almost 80% of U.S. lead consumption.

Sources of Lead Scrap

The major source of scrap lead for recycling in the United States and throughout the world is lead acid batteries. In the United States, scrapped lead acid batteries and the associated manufacturing plant scrap represent over 90% of the contained lead available for recycling. Used automobile batteries represent about 85% of the lead acid battery scrap materials. Other lead recycled scrap materials are sheaths from telephone and power cable, lead pipe and sheet, weights (particularly automobile and truck wheel weights), anodes, printing metals, drosses, residues, sludges, and dusts.

In Europe and throughout most of the rest of the world, scrapped lead acid batteries represent only about half of the lead scrap input to recycling plants. Scrap cable covering, lead sheet and pipe, and miscellaneous metal scrap items represent a much higher percentage of input scrap to recyclers in these countries than those in the United States. As the number of vehicles increases, the percentage of scrap represented by lead acid batteries will increase.

Table 4 World lead production from 1980 to 1988

Type of lead	Production, metric tons × 10^3 (tons × 10^3)				
	1980	1982	1984	1986	1988
Primary	2300	2260	2230	2180	2270
	(2535)	(2490)	(2458)	(2403)	(2502)
Recycled	1810	1650	1850	1890	2120
	(1995)	(1820)	(2040)	(2083)	(2337)
Total	**4110**	**3910**	**4080**	**4070**	**4390**
	(4530)	**(4310)**	**(4498)**	**(4486)**	**(4839)**
Amount recycled, %	44	42	45	46	48

Source: International Lead-Zinc Study Group

Table 5 U.S. lead production from 1980 to 1988

Type of lead	Production, metric tons × 10^3 (tons × 10^3)				
	1980	1982	1984	1986	1988
Primary	551	513	396	370	392
	(607)	(565)	(437)	(408)	(432)
Recycled	602	515	590	562	698
	(664)	(567)	(650)	(619)	(769)
Total	**1153**	**1028**	**986**	**932**	**1090**
	(1271)	**(1133)**	**(1087)**	**(1027)**	**(1200)**
Amount recycled, %	52	50	60	61	64

Battery-Recycling Chain

The battery-recycling chain has changed dramatically over the past ten years. The changes have resulted from environmental regulation, changes in battery-processing technology, changes in battery distribution and sales techniques, changes in lead-smelting technology, and changes in the lead alloys used in the batteries.

Battery Scrap Collection and Processing. In the 1970s, batteries were distributed primarily through full-service gasoline stations. Smaller amounts were distributed through hardware stores, automobile supply stores, and mass merchandise outlets. The scrap batteries were recovered by the service stations and sold to scrap dealers, who also recovered batteries from wrecked or worn-out automobiles. The scrap dealers then sold the batteries to battery breakers and smelters. The battery breakers (numbering about 300 in 1975) decased the batteries, drained the acid, and recovered the plates for sale to the lead smelters. The higher lead content of the battery plates made it cost-effective to ship plates longer distances than whole batteries.

Environmental Regulations. In the 1980s, environmental legislation was passed regulating lead acid battery recycling. Rules were promulgated regarding the storage, processing, and transportation of batteries and battery scrap. Batteries and battery components are considered hazardous waste after arrival at a battery breaker or smelter if they are cracked or leaking acid, or if they are disposed of in landfills. Scrap batteries can be stored for only 90 days, after which they must be sent to a recycler or disposed of in a hazardous-waste landfill. Because only permitted processors can break batteries, the number of battery breakers has declined markedly. Only a few breakers still remain. Battery breaking is now performed mainly by lead smelters.

Battery-Breaking Processes. In the 1970s, most battery breakers used saws for decasing. In this process, the top is severed, the acid is drained, and the plates are dumped from the case. The lead posts are recovered from the tops by crushing and separation. This process is still utilized by many lead smelters in the United States and throughout the world.

In the late 1970s and early 1980s, several mechanical processes were developed to break the batteries. Technologies were developed to crush the whole batteries, separate the case from the lead-bearing materials, separate the hard rubber (ebonite) and separators from the plastic cases, and, in some cases, separate the paste portion of the battery from the metallics (Ref 26-28). The acid is neutralized in a separate procedure. A recent innovation desulfurizes the paste, produces lead carbonate, recovers sodium sulfate crystals, and recycles the

H_2O (Ref 29). Virtually all battery-wrecking processes now recycle the polypropylene battery cases. Battery breakers process from 5000 to more than 50 000 spent automobile batteries per day.

Lead-Smelting Processes

The major smelting processes to recycle lead scrap involve the use of blast furnaces, short rotary furnaces, long rotary kilns, reverberatory furnaces, electric furnaces, and top-blown rotary furnaces.

Blast Furnaces. For many years blast furnaces were the primary furnace for recycling lead. Today, about one-half of the U.S. recycling plants utilize blast furnaces as the primary smelting furnace (Ref 30). In other plants, blast furnaces are used to recycle slags, drosses, and residues from other processes. Blast furnaces require metallurgical coke, produce large volumes of gas that must be filtered, require a special charge, require afterburners to burn carbon monoxide contained in off-gases, and produce slag and matte that, in some cases, may be considered hazardous materials. Blast furnaces produce a bullion that is high in antimony; this bullion can be readily refined into lead-antimony alloys.

Rotary Furnaces. In most of the world other than the U.S., rotary furnaces (long, short, and top blown) have replaced blast furnaces as the major smelting vessels for lead recycling (Ref 31-33). Rotary furnaces are very versatile. They can accept virtually any type of lead-bearing feed material, including battery scrap, dust, drosses, scrap lead, and sludges. Rotary furnaces can use any carbon source such as coal, coke, or ebonite as reducing agent, and they can use a variety of fuels, such as oil, coal, or gas. Because they are batch furnaces, rotary furnaces can be operated in stages to produce low-impurity bullion for refining to pure lead, or they can completely reduce the charge to recover all metal values for production of lead-antimony alloys. Rotary furnaces generally use Na_2CO_3 and iron as fluxes, which produce a fluid, low-melting slag. Rotary furnace slags are generally considered hazardous wastes in the United States. Because of the hazardous nature of rotary furnace slag, these furnaces are not currently used by major lead-recycling companies in the United States. Most of the rest of the world does not classify rotary furnace slags as hazardous.

The reverberatory furnace is one of the major smelting furnaces used to recycle battery scrap in the United States. In the reverberatory furnace, battery scrap is melted and the alloying elements in the battery grids and scrap are oxidized to the slag. With the use of controlled reducing conditions, the reverberatory furnace can produce virtually all low-impurity bullion suitable for refining to pure lead. Slag from reverberatory furnaces is smelted in the second furnace, usually a blast furnace (Ref 34) or an electric furnace (Ref 35), to produce a high-antimony bullion for production of lead-antimony alloys. The electric furnace is capable of producing a nonhazardous slag.

Scrap as Charge for Primary-Lead Furnaces. Recycled battery scrap, particularly the paste portion, is often added in small amounts to the charge of sinter machines in primary-lead smelters. New lead-smelting processes can utilize lead battery paste as a substantial portion of the charge (Ref 36). Very little scrap is currently smelted by primary-lead companies in the United States.

Lead Sweat Furnaces. Small amounts of lead are recycled via lead sweat furnaces. The primary materials recycled in sweat furnaces are lead-coated power and communications cable, lead sheet and pipe, and other products that contain lead as a coating or as part of a complex part. The process is performed at relatively low temperatures and produces both metal for refining and dross; the dross is recycled to smelters. The total amount of lead recycled in sweat furnaces is estimated to be less than 18 000 metric tons (20 000 tons) per year.

Table 6 Composition specifications for pure lead in the United States

Element	Specified composition, max wt%: ASTM B 29	Common grade	Typical battery oxide grade	Typical highly refined recycled grade
Antimony	NS	NS	0.0010	0.0001
Arsenic	NS	NS	0.0010	0.0001
Tin	NS	NS	0.0010	0.0001
Antimony, arsenic, tin (combined)	0.002	0.002	NS	NS
Silver	0.0015	0.005	0.005	0.0017
Copper	0.0015	0.0015	0.0010	0.0005
Silver, copper (combined)	0.0025			
Bismuth	0.05	0.05	0.025	0.012
Zinc	0.001	0.001	0.0010	0.0001
Iron	0.002	0.002	0.0010	0.0001
Nickel	NS	NS	0.0002	0.0001
Tellurium	NS	NS	0.0002	0.00005
Lead, min	99.94	99.94	99.97	99.85

(a) NS, not specified

Refining of Recycled Lead

In most of the world, lead recyclers produce lead-antimony alloys for use as battery grids and straps. The recycled pure lead generally goes into nonbattery sources such as sheet, pipe, cable, and gasoline additives. The pure lead for battery oxide is generally supplied by primary-lead smelters. In the United States, maintenance-free batteries with lead-calcium alloy grids make up about 30% of the market, and hybrid batteries with lead-antimony positive grids and lead-calcium negative grids represent 60%; lead-antimony grid batteries constitute only 10% of the battery market. Lead-calcium alloys account for 60% of automotive battery grid production in the United States. Virtually all standby batteries use lead-calcium alloys.

Initially, the lead-calcium was supplied by the primary-lead companies. Lead recyclers, however, changed their smelting and refining techniques to produce pure lead and lead-calcium alloys. Currently, virtually all lead recyclers in the United States produce pure lead and/or lead-calcium alloys for batteries; however, very few recyclers throughout the rest of the world produce substantial amounts of refined pure lead for batteries.

Specifications for Recycled Lead

Throughout much of the world, two lead specifications prevail: one with a minimum of 99.99% Pb and the other with a minimum of 99.97% Pb. The major impurities in lead are antimony, arsenic, bismuth, copper, nickel, silver, tin, and zinc. Recently, selenium and tellurium have been added as important impurities in the United States. Primary-lead companies generally produce the 99.99% Pb grade, whereas recyclers produce the 99.97% Pb grade. The major difference in the lead grades is that recyclers generally do not remove the bismuth and silver in their refining process. Recycled lead generally contains sufficient bismuth to preclude reaching 99.99% purity.

U.S. Lead Specifications. The ASTM B 29 composition specification for refined pure lead is shown in Table 6. This specification permits a content of up to 0.05% Bi. Because of this bismuth content, the purity of the lead can be as low as 99.94%. In virtually every U.S. application, bismuth contents of 200 to 500 ppm are permitted. Thus, recycled refined pure lead can be utilized in almost every application if the other impurity limits are also met.

Gas-Producing Impurities. More important than restrictions of bismuth and silver in U.S. lead specifications has been the restriction of elements that increase gas generation in lead acid batteries. Many batteries in the United States are sealed or restrict access to the cells for water additions. Elements that promote decomposition of the electrolyte and production of gas upon charging are specified at very low levels regardless of the overall purity of the lead. The specification for pure lead for battery oxide given in Table 6 restricts antimony, arsenic, nickel, and tellurium to low levels, whereas nongassing impurities such as bismuth, silver, and copper are permitted at higher levels. In the most restrictive specifications, all the gas-producing impurities are restricted to a content of 1 ppm or less.

Recycled Pure Lead. Because the major concern of U.S. battery manufacturers is the gas-producing impurity elements in pure lead, lead-recycling companies have developed pyrometallurgical refining techniques to remove these elements to low levels (Ref 37). A typical analysis of highly refined recycled pure lead is given in Table 6. When refined pure lead is produced by recyclers, it is readily accepted by battery manufacturers. Only an occasional specification can be found that restricts bismuth or silver to levels not readily attained in recycled pure lead.

Government Regulations Regarding Recycling

The lead industry, and particularly the lead recycling industry, must conform to increasingly stringent environmental regulations. Lead acid batteries, the major raw material of recyclers, has been declared a hazardous waste. Because batteries are the largest source of lead, they constitute the major source of lead contamination in landfills and incinerators.

Proposed legislation would require recycling of all lead acid batteries, require manufacturers or distributors to accept spent lead acid batteries, and require return of the batteries to recyclers to prevent improper disposal. If such legislation is passed, even higher rates of lead recycling will result.

New Lead Recycling Processes

Several new processes have been developed to recycle lead acid battery scrap that use hydrometallurgical processes (Ref 38, 39), rather than conventional pyrometallurgical processes, to treat the paste portion. In these processes, $PbSO_4$ is converted to $PbCO_3$, PbO_2 is converted to soluble form, and the lead is leached into solution and electrowon to produce high-purity cathodes. These processes have not been developed beyond the pilot state; however, several recyclers have announced plans to construct large-scale plants for recovery of lead from battery paste by electrowinning.

Recycling of Zinc

Michael Bess, Certified Alloys, Inc.

ZINC, in its various forms, can be recycled like most other nonferrous metals. Recycling can take the form of recovery of the metallic content from scrap zinc and byproducts or by the conversion of various zinc-base residues into chemicals.

This article will deal with recovery of the metal content from zinc sources such as scrap and residues (such as galvanizing and die casting drosses).

The zinc recycling industry, worldwide, produces approximately 20% of total zinc production. Of this quantity, about 300 000 tons/year are in the form of secondary zinc metal. The scrap that feeds this industry is in the form of either process scrap (such as zinc sheet clips) or old scrap (discarded, used castings, and so forth), which becomes available after the products containing zinc are removed from service. Typical sources of scrap include: reject castings, flash, trim scrap, and dross from the die casting industry; clips, edge trim, and skeletons from zinc sheet processors; old zinc roofing; reject battery cans; and top and bottom dross from the galvanizing industry. Table 7 illustrates U.S. production of zinc from scrap for the years 1987 and 1988.

Sources of Zinc Scrap

The majority of zinc used is consumed by the galvanizing industry. Zinc, which is metallurgically combined with steel during galvanizing, cannot be readily separated from galvanized steel scrap. This scrap is often reprocessed in steel mills, leading to the generation of flue dust. This dust contains a relatively high quantity of zinc (typically 20%), which is generally recovered as crude zinc oxide for metal production (see the section "Recycling of Zinc From EAF Dust" in this article).

Die casting consumes approximately 25% of the zinc in North America. The die casting alloys typically contain about 4% Al and up to 1% Cu. Zinc die castings are generally small, from less than one ounce up to several pounds in weight. They are frequently found as components in complex assemblies such as in automobiles, appliances, and electronics. Separation of castings from these larger assemblies is difficult, as is the identification of zinc in the presence of other nonferrous metals.

All operations involving the melting of zinc or zinc-rich alloys will generate oxides. These oxides usually entrap additional metal and create products commonly referred to as dross, ash, or skim. These products are collected and are treated to release the metallic content. Often they also become the starting raw material for the production of zinc oxide and other chemicals.

During various manufacturing processes, internally generated metallic scrap can usually be recycled in-plant. Zinc enters the recycling stream as a function of product life cycle. Building products such as roofing and flashing are readily recoverable and are an excellent source of zinc (typically greater than 80% is recovered), but the product life is long, greater than 25 years. Old die cast components also provide high metal recovery (80%), but recovery of the products they are contained in is typically low, averaging from 10% in appliances and hardware to 50% in automotive and large machinery applications. Product life cycles are somewhat shorter than for rolled zinc, in the range of five to twenty years.

On a worldwide basis, zinc consumed for galvanizing is the highest end use for the metal. Therefore, residues from galvanizing

Table 7 Zinc recovered from scrap in the United States in 1987 and 1988

	Amount recovered			
	1987		1988	
Scrap	Metric tons	Tons	Metric tons	Tons
New scrap				
Zinc-base alloys	146 394	161 369	111 133	122 500
Copper-base alloys	123 969	136 650	133 881	147 576
Magnesium-base alloys	35	38.5	122	134
Old scrap				
Zinc-base alloys	59 964	66 098	74 632	82 266
Copper-base alloys	21 125	23 285	22 053	24 309
Aluminum-base alloys	262	289	349	385
Magnesium-base alloys	159	175	180	198
Total	**351 908**	**387 906**	**342 350**	**375 370**

Source: Ref 40

are the most important source of secondary zinc. Collection of drosses by steel mills is a good source, and the cycle time from initial zinc use to re-use is typically less than one year. Depending upon the steel and galvanizing process, drosses, ash, and skimmings range from 10 to 40% of the total zinc consumed in the galvanizing process.

Recycling Technology

As previously stated, the identification and separation of die castings from assemblies is difficult. Auto shredders and gravity separation are two of the most effective measures for recovering this type of recyclable zinc. For example, in the United States, die casting consumes about 25% of all zinc, but because of these difficulties, only 15% of recoverable zinc comes from this use sector. Where zinc scrap may be present with other, higher-melting metallics, the best processing equipment is a sweat furnace, which melts the zinc but not other materials such as copper, aluminum, and steel. Lower-melting nonmetallics (plastics and rubber) are simply burned off in this process. Before treatment in a sweat furnace, removal of free ferrous materials by magnetic separation is preferred, to minimize iron pickup and reduce energy consumption.

Steel mill flue dust is reduced in rotary kilns, with zinc-lead vapor being produced along with chromium and iron in a slag. The vapor is reoxidized to produce feedstock for subsequent furnace recovery to metal.

Galvanizer's dross is high in zinc (90 to 95%) but also contains lead, tin, iron, and aluminum. Being heavier than the molten galvanizing alloy, these elements sink to the bottom of the galvanizing vessel for subsequent removal. Recovery of zinc is usually accomplished by distillation from a retort. Either metallic zinc dust or zinc oxide can be produced by this method. The light, galvanizer's top dross contains primarily zinc with smaller quantities of aluminum, lead, and tin. Iron is usually present in much lower levels than in bottom dross. This dross can be reacted with various chemical fluxes to release the entrapped metallic zinc, which can be cast into blocks for subsequent use as secondary zinc or as starting material for products like galvanizing brightener (a zinc-aluminum alloy added to galvanizing baths).

Die casting drosses usually are of high purity, their composition being similar to the high-purity die casting alloys from which they are formed, except for generally higher aluminum and iron levels. Again, simple melting and treatment with chemical fluxes is practiced to release the metallic content. Recovery is normally in the range of 50 to 75%. Metal recovered in this process generally is used to produce galvanizing brightener or other zinc-aluminum master alloys.

Table 8 Typical chemical composition of EAF dust

Constituent	Amount, %
Valuable metals	
Zinc	5.0–37.8
Iron	13.0–44.0
Problem impurities	
Lead	0.7–7.0
Cadmium	0.01–0.16
Nickel	0.002–0.07
Chlorine	0.1–3.9
Fluorine	0.05–1.6
Sulfur	0.05–1.0
Sodium	0.2–2.0
Slag formers	
Manganese	1.5–7.0
Silica	0.15–2.8
Calcium	1.7–3.0
Aluminum	0.05–1.2
Chromium	0.04–3.1

Source: Ref 43

Table 9 EAF dust processing statistics

Country	Amount processed annually	
	Metric tons	Tons
West Germany	55 000	60 500
Italy	65 000	71 500
Spain	70 000	77 000
Japan	295 000	325 000
United States	350 000	385 000
Total	**835 000**	**920 000**

Source: Ref 44

Recycling of Zinc From EAF Dust

Charles O. Bounds, Rhône-Poulenc, Inc.

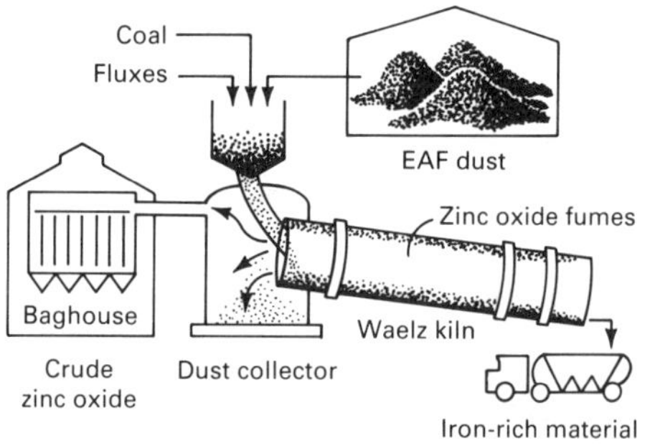

Fig. 16 Operation of a Waelz kiln for recycling zinc from EAF dust

ELECTRIC ARC FURNACE (EAF) dust, a baghouse product from environmental control activities, is generated by vaporization of metals such as zinc, lead, and cadmium during electric arc furnace steelmaking. Historically, the zinc content of the dust has been too low to permit economic recovery. As late as 1985, more than 70% of the dust was being discarded in landfills (Ref 41). However, the 1984 reauthorization of the U.S. Resource Conservation and Recovery Act (RCRA) recommended a moratorium on the landfilling of EAF dust after August 1988 (since delayed until August 1990). This legislation has substantially increased the cost of disposal, resulting in the emergence of a dust processing industry. Currently, more than 60% of the dust is being recycled in the United States (Ref 42); a similar pattern has emerged throughout the developed countries of the free world.

EAF Dust Characteristics

In the United States, carbon steel EAF dust, predominantly composed of submicron particles of magnetite, hematite, zinc oxide, and zinc ferrite (Ref 43), contains from about 5 to nearly 40% zinc (average ~20%) as well as the hazardous constituents lead, cadmium, and chromium (see Table 8). The lead, cadmium, and chromium contained in the dust are leachable in the Environmental Protection Agency (EPA) toxicity test, and thus, the dust is a listed hazardous waste.

Current estimates (see Table 9) of EAF dust generation and handling worldwide indicate that approximately 800 000 metric tons (880 000 tons) of dust are being processed annually in the free world, with more than 150 000 metric tons (165 000 tons) of zinc being recycled each year. The EAF dust generation rate is expected to grow substantially during the next decade. For example, in the United States the EAF market share is expected to grow from 30 to 50% by the year 2000, with an equivalent increase in EAF dust production (Ref 45). The quantity of zinc available for reclaiming could more than double if the percentage of galvanized steel production increases as expected by industry analysts.

Recycling Technology

The predominant technology used for the recovery of zinc from EAF dust is the Waelz kiln process, a technology employed since the early 1900s for the beneficiation of oxidic zinc ores. Currently 12 kilns are

Table 10 Compositions of typical Waelz kiln feed and products

Element	Composition, wt%: EAF dust	Oxide	Kiln slag
Zinc	17.5	52.4	1.9
Iron	23.1	4.2	33.8
Calcium	9.1	1.4	11.4
Silicon	1.7	...	6.6
Sodium	0.64	1.2	0.51
Potassium	0.47	0.48	0.14
Manganese	2.9	0.56	3.7
Magnesium	1.9	0.39	2.4
Aluminum	0.46	0.12	1.1
Lead	2.6	7.2	0.41
Cadmium	0.66	0.76	<0.001
Copper	0.23	0.087	0.47
Chromium	0.12	0.026	0.18
Nickel	0.03	0.01	0.07
Tin	0.04	0.09	0.02
Chlorine	1.03	3.58	<0.01

Source: Ref 42

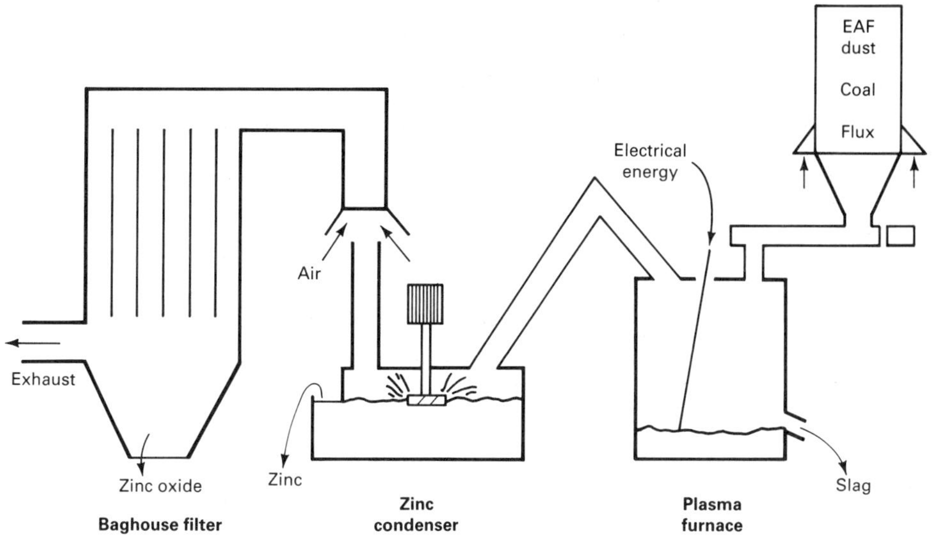

Fig. 17 Tetronics plasma treatment plant for recycling of zinc from EAF dust

reported to be operating in the free world (Ref 46). The kilns are operated similarly worldwide (Ref 47-51), with the EAF dust being mixed with coal and a flux and fed to the kiln as shown in Fig. 16. The materials in the kiln roll along the inside slope of the kiln; hence, the name Waelz from the German word *waelzen*, which means "to roll."

Several different reactions occur as the material flows through the kiln:

- Feeds are dried and preheated
- Halide and alkali compounds are volatilized
- Iron oxides are partially reduced
- Zinc, lead, and cadmium are reduced and fumed
- Metals are reoxidized above the bed and collected

Thus, both reduction and oxidation are simultaneously occurring in the kiln. Occasionally, the energy provided by coal combustion is supplemented by natural gas or oil combustion via a burner at the discharge end of the kiln.

Dust processing in a kiln generates two marketable products: a mixed zinc/lead oxide and an inert iron-rich slag. Typical analyses of Waelz kiln feed and products are presented in Table 10. The leachate from the EPA toxicity test on kiln slag is consistently below the toxicity limit; the material has been sold as a road base, antiskid agent, Portland cement additive, and an aggregate in highway blacktop.

The concentration of the alkali metals and halides, as well as other nonferrous metals, in the crude zinc oxide, limit its marketability as a zinc raw material. Generally, at least the alkalis and halides are removed by leaching, calcining, and/or hot briquetting. Even after preprocessing, the only significant outlet for the oxide is as a raw material for pyrometallurgical zinc plants, such as the Imperial Smelting Furnace or the Electrothermic Process. In fact, most Waelz kiln operations are integrated or closely aligned with a zinc smelter.

New Dust Processing Technologies

Alternative dust processing technologies have been under development throughout the 1980s. Hydrometallurgical techniques, both alkaline- and acid-base leaching processes, have been investigated but in general have rarely appeared economical (Ref 52). The new pyrometallurgical approaches have generally been targeted to directly produce metallic zinc and/or to be more economically scaled to small plants than the Waelz kiln. The most significant of these new developments are:

- *Tetronics Plasma Process*: This process (Fig. 17) unites a transferred arc plasma furnace to fume nonferrous metals with an Imperial Smelting Process (ISP) zinc condenser that recovers the metals in the molten state (Ref 53)
- *SKF Plasmadust Process*: Similarly, Plasmadust uses a coke bed to fume the nonferrous metals with the heat provided by the gas from a nontransferred arc plasma gun and the metal vapors being condensed in an ISP unit (Ref 54). One commercial plant was erected in Sweden for EAF carbon-steel dust processing, which has since been converted to process stainless dust because of repeated condenser failures
- *Elkem Electric Furnace Process*: In the Elkem Multi-Purpose Electric Furnace (Ref 55), agglomerated EAF dust is charged into the furnace slag from which the nonferrous metals are selected, reduced, fumed, and then condensed in an ISP condenser. To date, one steelmaker has committed to install the Elkem Process
- *HRD Flame Reactor Process*: HRD has developed a natural gas-fired version of its Flame Reactor technology (normally fired with coal or coke breeze) to provide a lower capital cost, on-site processing unit (Ref 56). Reactor temperatures in excess of 1450 °C (2640 °F) selectively reduce the nonferrous metals and produce a vitrified, nonfluxed slag and a marketable mixed metal oxide similar to that produced in a Waelz kiln. The process is shown in Fig. 18
- *ZIA IRRS Process*: The ZIA approach is to fully reduce and volatilize the ferrous and nonferrous metals from pelletized EAF dust, collecting the fumed fraction as an oxide and returning the iron (85% metallized) to the electric arc furnace. The mixed oxide is to be reduced to metal in a vertical retort (Ref 57)
- *BUS Circulating Fluid-Bed Process*: The CFB reactor is being developed for steel dusts with low zinc and lead contents (Ref 58). In the fluid bed, lead and zinc are volatilized, leaving behind a highly metallized iron product

Metallic zinc, condensed directly from reduced and volatilized EAF dust, will typically contain more lead and cadmium than even Prime Western grade zinc (GOB), for which markets are limited. Zinc purification via fractional distillation will likely be required for any direct condensation process to become widely adopted.

Summary

EAF dust, because of environmental legislation that has substantially raised the cost of its disposal, has become a significant source of reclaimed zinc in Europe, Japan, and the United States. To date, the dust generator has subsidized the reclamation activity by paying a processing fee that has varied from about $50 to more

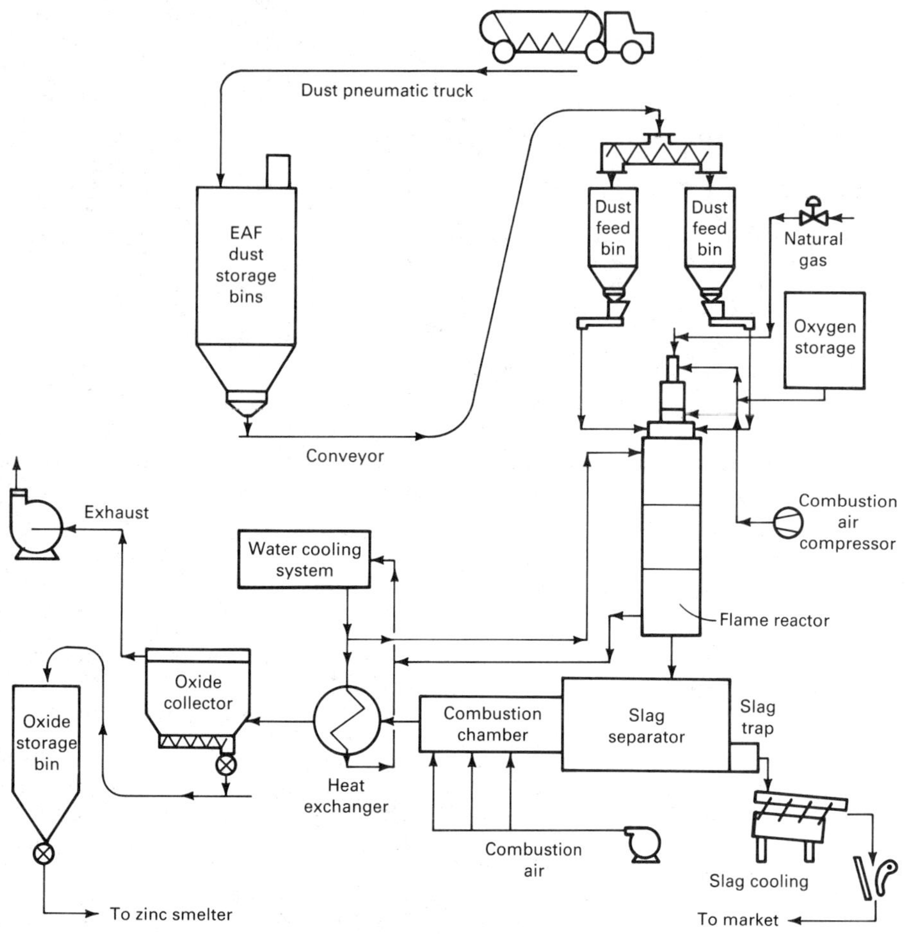

Fig. 18 The HRD flame reactor gas-fired process for zinc recycling from EAF dust. Source: Ref 56

than $100 per metric ton, depending on the zinc content of the dust and the local conditions.

The Waelz kiln process will most likely continue to dominate the dust processing industry until at least the mid-1990s. However, as EAF dust generation continues to grow, the high cost to expand kiln capacity and to ship dust to large central plants should permit the introduction of new technology. Successful processes most likely will be based on the production of only nonhazardous, beneficially used materials by recycling the lead and cadmium as well as the zinc contained in EAF dust.

Recycling of Titanium

Michael Suisman and Leonard Wasserman, Suisman Titanium Corporation

TITANIUM begins with its two most common ores, rutile and ilmenite. Leucoxene and anatase are minor titanium ores. All of these are utilized either to produce pigments or to produce metal. For pigments, the titanium dioxide (TiO_2) content of the ore is employed. In the case of metals, the ores are converted by complex chemical processes into purified titanium tetrachloride ($TiCl_4$), commonly referred to as "tickle," and then further reduced either by magnesium or sodium into sponge.

Titanium ores are predominantly used for pigments, with about 90% being used as basic raw materials for pigments, paints, paper, and plastic. The titanium ores are the materials of choice to produce white pigmentation in those materials. At this time only 10% of the ores result in metal. Recycling takes place in metal only.

The Kroll Process, the conversion of tickle by a magnesium reduction process, was the breakthrough that permitted the commercial and economic introduction of titanium metal in the late 1940s. Sodium reduction (the Hunter Process) is also a traditional method of reduction. Electrolytic processes have been pursued for over 30 years, and attempts to commercialize such processes continue to be investigated.

About 80% of titanium usage is destined for aerospace applications including civilian and military airframes, engines, helicopters, rockets, and missiles. Much of the balance is devoted to the fabrication of equipment for industrial and commercial applications such as chemical processing, pulp and paper, and marine and power generation. Recently titanium has found favor in competitive sports equipment such as racing cars, boats, sleds, golf clubs, tennis rackets, and bicycles. In the medical sector, due to biochemical compatibility, titanium has proven to be a valuable asset in prosthetic devices, pacemakers, artificial heart components, and implanted drug administration devices. Consumer products such as wristwatches, writing instruments, cameras, pocket knives, jewelry, and eyeglass frames have also found titanium applications. Significant quantities of titanium have been utilized in Japan for architectural purposes. More information on titanium applications is available in the article "Introduction to Titanium and Titanium Alloys" in this Volume.

Early Development of Titanium Recycling Technology

The first titanium scrap was generated in 1951. From the earliest years, recycling of titanium to the parent metal was hindered by a number of factors. Firstly, the chief end use of the metal was for aircraft, and consequently rigid controls or outright prohibitions were placed on scrap remelting. Then, as today, the ductility of titanium was critical, so that scrap that might contain high oxygen or other gases was suspect. The physical forms of titanium scrap, moreover, were often difficult to weld into the consumable electrodes of vacuum arc remelting (VAR) furnaces. Because titanium was melted under vacuum, there was little or no opportunity to adjust chemistry once the metal was molten.

Nonetheless, titanium recycling was begun in the 1950s. Timet attempted titanium turning recycling in 1952 but failed due to high-density inclusions (HDIs). Oremet began recycling bulk weldables in the late 1950s and in 1962 received approval to recycle rotor-quality turnings, chiefly for internal use. Timet also developed a system of scrap recycling known as hydrogenation. In the late 1960s, Viking Metallurgical-Quanex Corporation installed an electron beam furnace and became the first melter to recycle turnings for nonrotor purposes in a nonVAR furnace. In 1974, the U.S. Air Force sponsored a program to recycle aerospace metals, both titanium and superal-

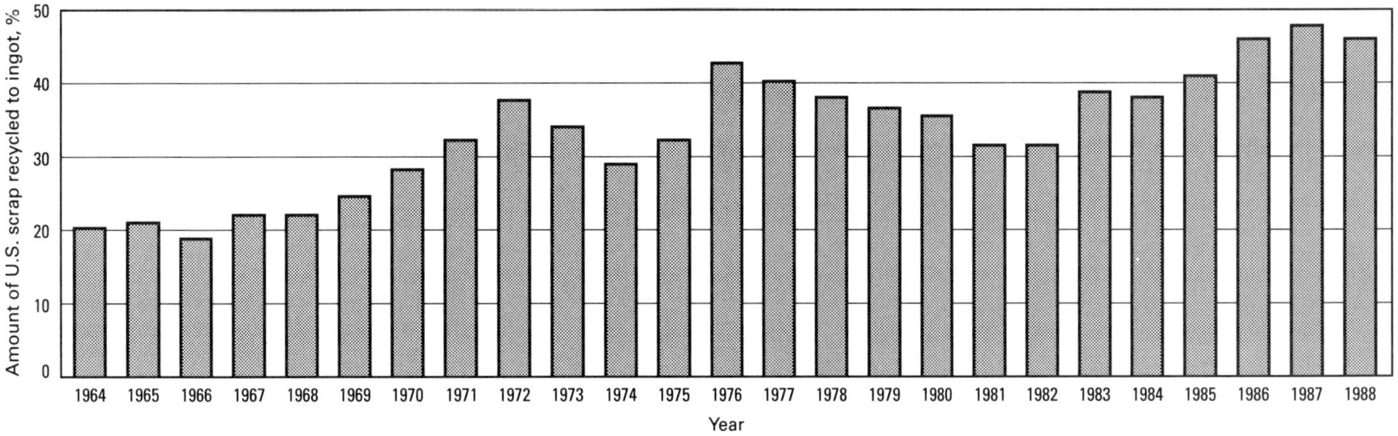

Fig. 19 Percentage of U.S. titanium scrap that was recycled to ingot from 1964 to 1988. Source: U.S. Bureau of Mines

loys, publishing an extensive report that urged melters, generators, and recyclers to develop systems to recycle titanium back into jet engines. Suisman Titanium Corporation received a patent in 1980 for the removal of HDIs from turnings, which allowed the first nonmelter processing of turnings for production of rotor-quality ingot.

Titanium Recycling

Sources. Titanium scrap is generated either during the melting process by ingot makers or companies producing castings, or during the fabrication process while converting mill products into semifinished or finished products. A small percentage, less than 2%, consists of obsolete ("old") scrap and is of little commercial significance; obsolete scrap results from aircraft overhaul, replacement of heat exchangers, and plating scrap.

U.S. melters have increased their use of scrap substantially over the last 25 years. In the 1960s, only 15 to 20% of U.S. ingot production was recycled metal; most of it was in-house scrap. Today 40 to 50% of ingot production is recycled metal (Fig. 19).

Scrap generated at the melter's plant consists of billet cutoffs, crops, bar ends, plate trimmings, skulls, and so forth, and is called in-house or turnaround scrap; melters attempt to recycle their own scrap to the greatest degree possible, but some forms have physical and chemical characteristics (high oxygen, enfoliation, chemistry intermixture, and so on) that make them impractical to remelt as titanium.

When mill products are shipped as billet, bar, sheet, and plate, they are processed by turning, shearing, welding, forming, and so on, to semifinished or finished parts. During this fabrication, continuous fractions of metal are generated as scrap and are known as new production or open market scrap. All of this scrap must in turn be processed before it is recycled. A large portion of scrap generated by fabrication is purchased by scrap dealers or scrap processors. Most of this metal must be classified, processed, and analyzed by a processor approved by aerospace end users and melters. Some titanium scrap is returned directly to melters, who process the scrap themselves. Table 11 illustrates the usage of scrap in titanium ingot production.

Forms of Scrap. Titanium scrap is generated in the form of solids (bars, sheets, forge flashings, weldments, and so forth) and turnings or chips. The turnings usually contain cutting lubricants; solids may also be contaminated with oil or grease. These lubricants must be completely removed prior to melting.

Recycling to Titanium Melters. Ingot melters and casting companies require a very tightly controlled recycled metal. They purchase recycled metal segregated by physical form and chemical purity. Physical forms of solid scrap are designated bulk weldables, feedstock, and cobbles, depending on density and size. Turnings have even more stringent specifications, requiring not only spectral and gaseous chemistry, cleanliness, and size control, but also freedom from HDIs, mostly tungsten carbide tool bits that become intermixed with the turnings during the machining process.

Newly developed cold hearth furnaces, whose power sources are electron beam or plasma, have been approved by major end users for their abilities to remove HDIs as well as low-density inclusions (oxides, and so on) pyrometallurgically.

Complexity of Titanium Scrap Recycling. Because titanium metal is heavily oriented toward aerospace, and for technical and metallurgical reasons, the quality requirements for recycled titanium are much stricter than for most other recycled metals. Processors must utilize high-technology sorting devices, sophisticated processing equipment, and full-scale laboratories with highly trained staffs. High motivation for quality must be imbued in employees. Titanium processors also must be approved (qualified) by titanium melters and often the end users.

Fire Hazard of Titanium Turnings. Titanium turnings and chips can ignite, and fires have taken place over the years. The finely divided form, fines, is the most hazardous. The reader is advised to use the utmost caution, and refer to Ref 59.

Differences Between Recycled Metal and Sponge. Sponge comes from a limited number of sources, with only seven producers in

Table 11 Use of scrap in production of titanium ingot from 1977 to 1988

	Consumption			
	Ingot		Scrap	
Year	kg × 10^6	lb × 10^6	kg × 10^6	lb × 10^6
1977	24 040	53 000	9 798	21 600
1978	29 030	64 000	11 068	24 400
1979	33 942	74 828	12 610	27 800
1980	38 682	85 278	13 938	30 728
1981	41 920	92 417	13 422	29 590
1982	24 073	53 072	7 736	17 056
1983	23 985	52 878	9 399	20 722
1984	36 256	79 929	14 106	31 098
1985	32 103	70 773	13 354	29 440
1986	31 836	70 186	14 957	32 974
1987	33 762	74 432	16 363	36 074
1988	38 857	85 663	18 058	39 810

Source: U.S. Department of Commerce

the Western world. It is seldom compromised by HDIs, although LDIs are a constant concern.

Recycled metal emanates from a wide variety of sources, rendering it more heterogeneous upon generation. After recycling, however, titanium solids can once again attain physical uniformity and in-spec chemistry. Turnings can also achieve physical and chemical homogeneity, and if processed for HDI removal can be melted for rotor-quality applications; they tend to run slightly high in oxygen, however, due to oxygen pickup during the earlier machining. Recycled metal has the advantage of providing the alloy components, such as aluminum and vanadium within the scrap itself, whereas sponge-produced titanium alloy must always be produced with master alloy additions to the heat.

Recycling for Sacrificial Uses

Ferrous Alloy Metallurgy. Titanium scrap is used in the steel and stainless steel industries either as a direct addition or after conversion into ferrotitanium. Titanium additions, either through alloying or as a gas scavenger, greatly improve the properties of ferrous products.

Aluminum Alloy Metallurgy. Titanium scrap additions to aluminum in the form of crushed, chemically cleaned turnings are used to make aluminum-titanium master alloys. Grain refinement of aluminum to which the master alloy has been introduced results in improvement in casting speed, eliminates cracking, and generally gives a more sound and uniform product.

Export of Titanium Scrap. Much of the titanium scrap utilized for sacrificial purposes is exported from the United States to the producers of ferrotitanium and master alloys. The volume of titanium scrap exports has risen by more than one-third in the last ten years.

New Trends in Titanium Recycling

Since the titanium industry is in its relative infancy, much can be expected technologically that will impact recycling. The aerospace industry's efforts to produce near net-shape components have fostered growth of titanium castings and to some degree have lessened scrap generation.

Cold hearth melting will also affect scrap recycling. The cold hearth furnaces are designed to melt high percentages of scrap. They also have the potential to cut recycling to traditional VAR furnaces.

The introduction of new titanium alloys, titanium aluminides, and metal matrix composites all carry with them new complexities to titanium recycling.

Recycling of Electronic Scrap

William D. Riley, Charles B. Daellenbach, and Robert C. Gabler, Jr.,
U.S. Bureau of Mines, Albany Research Center

PRECIOUS METALS (PM)—gold, silver, and the platinum-group metals (PGM) (platinum, palladium, rhodium, iridium, osmium, and ruthenium)—play a key role in the electronic and electrical industries. The recycling of these materials has a significant impact on the amounts that must be imported each year. For example, approximately 25% of the total demand for PM by all industrial uses is met by recycled material. The PM are essential in electrical and electronic applications because of their low contact resistance and resistance to corrosion and oxidation. A large amount of U.S. demand for PM is met by the recovery of metal values from electronic scrap. Typical processing steps for the recovery of PM from electronic scrap include hand disassembly or mechanical preparation followed by incineration, separation, and classification. Final recovery is usually performed by leaching and precipitation followed by flux melting, refining, and casting. The final purity and uses of the recycled materials are comparable to those produced from virgin ore.

Nature and Quantity of Scrap Material

Precious metals are contained in platings of various thicknesses, relay contact points, switch contacts, wires, solders, transistors, printed circuit boards, electron tubes, batteries, and thermocouples. Typically, this scrap is classified as precious metals-containing electronic scrap, telephone relays, coated plugs, computers, and so forth. The precious metals may be in an alloy form, or a near-pure form; however, they may only comprise a small percentage of the total amount of electronic scrap produced. The PM content of individual pieces of electronic scrap is thought to have dropped from the 1960s (Ref 60) to the present (Ref 61); however, the total number of parts has increased as has the total demand for precious metals in the electronics industry. Estimates place the current content of PM contained in electronic scrap in the range of 2 to 10 troy ounces/ton of gold, 40 to 100 troy ounces/ton of silver, and >1 troy ounce/ton PGM (Ref 62). The general composition for electronic scrap processed in the mid-1980s is as follows: plastics (30 wt%); refractory oxides (30 wt%); nonprecious metals including copper, iron, nickel, tin, lead, aluminum, and zinc (40 wt%); precious metals (0.305 wt%). Approximately 27.3 million troy ounces of PM (0.4 million oz PGM, 1.1 million oz Au, 25.8 million oz Ag) was consumed by the electronic manufacturing industry in 1987 (Ref 63). As shown in Table 12, this is expected to increase to at least 34.6 million troy ounces (1.1 million oz PGM, 2.5 million oz Au, 31 million oz Ag) by the year 2000. The palladium demand is expected to double as it gradually replaces gold in many electronics uses. The secondary refining industry supplies a significant portion of the total demand for PM each year. In 1987, 26 million troy ounces Ag, 1.5 million troy ounces PGM, and 2 million troy ounces Au were recovered from all sources of old scrap. The recovery of these materials from scrap resources is not expected to change much by the year 2000 (Ref 65). The amount of materials generated by secondary recovery efforts is expected to remain relatively flat between now and the turn of the century, with most PMs experiencing only 2 to 4% increases in amount recovered (Ref 66).

Table 12 Consumption of precious metals in electronic components

Metal	Consumption, troy ounce × 10^3 1986	2000 (estimated)
Gold	1 070	2 500
Silver	25 857	31 000
Platinum	75	150
Palladium	250	450
Rhodium	8	12
Iridium	1	3
Ruthenium	71	100
Osmium	1(a)	1

(a) All osmium is included in this number. Source: Ref 64

Scrap Electronics: A Complex Ore

Currently, other than the U.S. Department of Defense (DoD) Precious Metals Recovery Program, there is no well-organized and coordinated program for the collection of PM-containing electronic scrap, although a few commercial firms have established limited programs. The DoD is thought to be the major source of PM-containing scrap. The defense industry electronic scrap usually contains the highest amount of PM because of the need for maximum reliability. Prompt scrap, generated during the production of electronic components, is recycled in a reasonably effective manner, but obsolete scrap presents problems because of the expense associated with collecting, identifying, and separating the material. In fact, it has been

Table 13 Composition of hand-dismantled avionics units

Unit(a)	Composition, % Aluminum alloys	Copper alloys	Magnetic metals	Stainless steels	Nonmetals	Precious metals(b)
Receiver-transmitter	36.4	18.8	23.1	3.3	18.3	1.9
Receiver-transmitter	56.6	17.0	13.5	1.3	11.5	2.8
Receiver-transmitter	32.0	17.7	21.2	3.4	25.7	1.5
Receiver-transmitter	28.8	18.1	28.6	4.0	20.5	1.2
Receiver-transmitter	20.4	35.8	19.1	4.3	20.3	14.5
Tuner, radio	57.3	20.8	9.5	4.4	7.9	13.5
Tuner, radio	56.8	20.1	9.1	5.3	8.6	13.4
Tuner, radio	59.6	16.5	8.3	5.5	10.1	2.5
Tuner, radio	62.1	12.4	8.3	8.5	8.8	2.3
Tuner, radio	61.8	14.7	9.3	5.9	8.3	2.3
Tuner, radio	51.2	17.7	14.1	7.9	9.1	2.2
Tuner, radio	54.9	17.6	10.5	7.5	9.3	11.6
Radio receiver	28.6	25.9	23.5	1.5	20.4	3.5
Radio receiver	25.1	28.3	29.0	2.3	15.3	11.1
Converter	47.4	23.5	11.0	3.2	14.8	8.2
Keyer	28.5	23.6	21.0	5.7	21.2	5.2
Amplifier	32.9	22.2	23.8	2.3	18.7	5.2
Video amplifier	29.4	23.5	19.3	8.9	18.9	4.7
Video decoder	51.6	13.3	11.5	8.5	15.1	3.6
Control transmitter	32.1	22.4	12.9	13.7	18.9	2.8
Video coder	54.3	16.0	9.0	3.4	17.3	2.8
Indicator	43.9	17.0	19.6	3.0	16.5	1.4
Indicator	39.5	12.8	12.9	3.9	30.9	2.6
Receiver	25.9	21.7	30.1	3.4	18.9	1.8
Receiver	34.5	17.4	26.7	1.7	19.6	2.2
Coder transmitter set	45.3	15.3	22.5	1.4	15.5	2.2
Azimuth indicator	40.2	17.5	20.8	4.3	17.1	1.3
Azimuth indicator	36.2	19.8	21.1	4.4	18.5	1.8
Power supply	27.9	19.0	31.1	2.3	19.7	1.3
Power supply	6.3	24.4	47.6	2.5	19.2	1.1
Power supply	28.6	16.8	34.5	0.6	19.4	1.0
Power supply	25.1	15.9	33.3	3.7	21.9	0.8
Storage unit	58.4	14.1	15.7	0.5	11.3	0.3
N compass	16.1	5.1	43.8	33.3	1.7	0.2
Inverter	32.6	19.4	41.4	0.8	5.8	0.2
Electron tube	14.3	8.4	69.4	5.5	2.3	0

(a) Where units of the same kind are listed more than once, each listing represents an individual unit. (b) Percentage of the total black box weight that contained precious metals, based on hand segregation of silver- and gold-coated components through visual examination. Source: Ref 66

Table 14 Electronic scrap processing options

Process	Comments
Chemical stripping	Only of use on exposed gold surfaces. Recovery ~90%
Chemical treatment (acid dissolution)	Requires aggressive acids. Material encapsulated in ceramic not recovered. Produces large volumes of base metal effluent
Granulation and physical separation	PM from mixed scrap is dispersed in all the fractions, although most of it is concentrated in two fractions. Recoveries ~80%
Burn and smelt(a)	Suitable for all scrap. Recoveries approach 85%.

(a) Hand dismantling is always the first stage in the operation. Source: Ref 61

estimated that less than 10% of the PM was reclaimed from obsolete electrical contacts and electronic materials discarded in 1983 (Ref 67).

Precious-metals-containing scrap can be thought of as a low-grade complex ore. Both the complexity of the material and the low concentration of precious metals make getting a representative sample for analysis difficult. For example, Table 13 lists the results of an analysis for hand-dismantled avionic scrap. The scrap was thought to be pre-1957 because it included no printed circuit boards, but it is easy to see the wide variance in contained precious metals. The same sort of wide distribution of precious metals can be expected from consumer electronic scrap.

Preliminary Processing Options

Typically, various techniques are used to liberate the precious metals, or to concentrate them by removing as much of the base materials as possible. The general rule is the higher the precious metal concentration of the scrap, the greater the economic return. Technologies used to recover precious metals from electronic scrap include hand dismantling and segregation, mechanical processing to beneficiate the scrap, and various hydrometallurgical, pyrometallurgical, or electrometallurgical treatments to concentrate or recover the precious metals (Table 14). As a result of the complexity of recovery operations, there are numerous steps to the refining process, often carried out by different companies, with each step adding to the cost of the refined metal (Ref 68).

Possible recovery options for electronic scrap are shown in Fig. 20. Not all options are used for each type of scrap, and in some cases additional steps may be required. For example, precipitated PM is normally so finely divided that flux melting is required before refining and casting. In addition, the material may be sent to a copper refiner after assay and the sludge processed for PM recovery. In most cases the initial disassembly is done by either the manufacturer or a specialized company. Typically, this involves cutting of connectors, removing printed circuit boards, buses, and so forth. The other nonprecious metal-containing parts separated during the hand dismantling operation can often generate enough proceeds to pay for this step; this additional value is approximately $200 per gross ton of

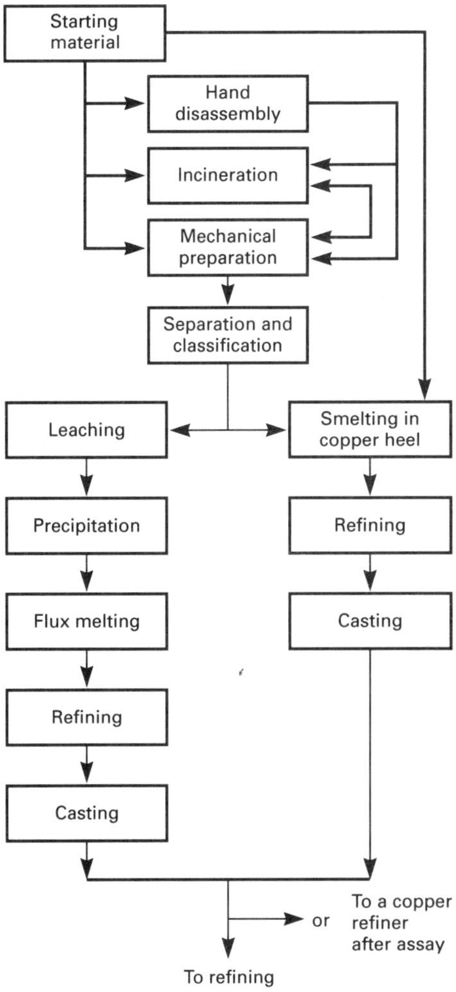

Fig. 20 Possible recovery options for the treatment of electronic scrap containing precious metals

Table 15 Material distribution and precious metal concentration in fractions of mechanically beneficiated printed-circuit cards

Fraction	Distribution, %	Contained precious metals, oz/ton Gold	Contained precious metals, oz/ton Silver
Baghouse lights	23.5	21.9	178
Wire bundles	1.2	19.6	286
Magnetic	30.7	8.7	11.3
Eddy-current aluminum	1.8	ND(a)	2.4
High-tension metal	35.2	43.8	451.8
High-tension rejects	7.8	6.6	108.3
Composite, total, or average	99.9	23.9	216.9

(a) ND, not detected

avionics material (Ref 69). Avionics contains a high percentage of aluminum in the cases; ground-based and Naval electronics would have the aluminum replaced by steel, a scrap value difference of $0.02/lb (steel) versus approximately $0.30/lb (aluminum). The fractions typically separated by disassembly are: iron, aluminum, and copper wire and solids, mixed breakage, precious metal-bearing material, and reusable parts. The removal of memory chips, tantalum capacitors, and so forth for sale as reusable merchandise can substantially improve the economics of this operation. It has been reported that a printed circuit board that contains several dollars worth of precious metals can have hundreds of dollars worth of memory chips on it (Ref 62). If no other processing is done, then the hand segregation is the most important step as far as the refiner is concerned (Ref 62).

Research has shown that electronic scrap can be mechanically upgraded (Ref 70-72). In this work, a series of unit operations was used to produce an iron-base fraction, an aluminum-base fraction, a wire fraction, and two precious metal fractions, one high-value and the other low. Table 15 lists the results from the application of a series of operations performed on printed circuit boards. The operations include hammer milling, air classification, magnetic separation, vibratory screen, wire separation screen, eddy current separation, and high-tension separation. The largest amounts of gold and silver and the PGM are always concentrated in the high-tension metal product. Most of the remaining precious metals can be found in the baghouse lights and wire bundles. This mechanical processing procedure or hand dismantling and segregation can greatly reduce shipping and toll refining costs by removing most of the extraneous materials.

The concentrated PM-containing material can then be processed by semirefiners who collect the material after initial disassembly and process it further. These companies may strip plastics from printed circuit boards chemically, or incinerate them at 400 to 500 °C (750 to 930 °F), followed by chopping or shredding the components. An additional step may include pulverizing capacitors to leave a pulp containing precious metals and smelting the material into a crude bullion for shipment to refiners. Another approach consists of incinerating plastics and other organics, followed by caustic leaching of the residue with NaOH to remove the aluminum, and finishing with smelting to produce a homogeneous product (Ref 73). A typical ingot analysis showed 85% Cu, 4% Fe, 0.2% Al, 333 oz/ton of silver, and 26 oz/ton of gold. These ingots can be electrorefined for copper recovery, and the precious metals, which concentrate in the anode sludge, can be processed for recovery by a precious metals refiner.

Another novel but costly metals separation approach proposes to avoid either mechanical processing or hand dismantling/segregation while still concentrating the precious metals. This process involves sweating the aluminum from the scrap. Both copper and the precious metals are dissolved in the ingot because of their high insolubility in the molten aluminum. The aluminum ingot can be refined in a three-layer electrorefining cell (Ref 73), where the precious metals are concentrated in the anode metal.

Still another processing approach involves some hand separation to remove bulky aluminum cases, shredding, and magnetic separation, followed by incineration and melting of the remaining material to form a brittle ingot. The procedure has been used on partial assemblies, and plugs and connectors. The ingot, consisting of aluminum, copper, iron, manganese, nickel, zinc, and precious metals, is crushed and the material is agitated with an acidified copper sulfate solution for copper cementation. The iron fraction of the shredded material is used as the cement agent in a tumble-type cementation system. The cement copper precipitate contains better than 90% of the precious metals (Ref 74). The analysis of a typical precipitate after cementation is 89% Cu, 1.1% Fe, 0.32% Pb, 0.36% Sn, 1.5% insolubles, 4.0 oz/ton of gold, and 11.5 oz/ton of silver. After melting and casting the cementation product into an anode, it can be electrorefined, and the anode sludge containing the precious metals sent to a toll refiner.

Research has also been conducted on recovering PGM from telephone relay scrap, reed switches (Ref 75), and gold from various contacts (Ref 76) by mechanical processing and hydrometallurgical treatments (Ref 77). Usually base metal substrates are first removed with a strong hydrochloric acid leach. Then the gold and silver are leached using either sulfuric/nitric or nitric acid solutions. The PGM require a pressure leach under air sparging in an autoclave. The leach solutions from each stage are concentrated by evaporation, and the concentrated residue from the hydrometallurgical treatments can be sent to a toll refiner for recovery of contained PGM.

A large portion of PM-containing scrap at one time consisted of tin-lead solders that were used to join electronic components to printed circuit boards. Soldering is usually accomplished by either dipping the board into a bath of solder or impinging a stationary wave of solder onto the circuit board. Some copper and precious metals are dissolved into the solder; as the content of the impurities increases, the solder loses its wetting property and is discarded. These solders on the average can contain 60 troy ounces/ton of PM. The increased use of integrated circuits has led to a diminishing of this technique, and it is reported that refiners receive less than 45 metric tons (50 tons) of the solder per year. Some of the contaminated solder is sold to automakers, and the precious metals are lost. It is possible to commercially recover the precious metals from the solder by fused-salt electrolysis in a molten chloride electrolyte (Ref 77).

Final Processing Options

Final recovery of the precious metals depends on the physical form, composition, grade, and associated metals in the concentrate. In general, most refiners specialize in the recovery of PM from certain kinds of scrap and semiprocess other types of scrap for recovery at other refinery operations. All of the scrap winds up as some kind of a bullion or sweeps (including dry precipitates). These materials are amenable to final recovery by a combination of smelting, fire, chemical, and electrolytic refining (Ref 78). Of all the recovery options for electronic scrap, the burn and smelt recovery systems seem to be the most effective. Table 14 summarizes the results of the research reported in the preceding sections.

Unlike most scrap metals, which are usually identified, segregated, and then remelted (oftentimes carrying along trace elements), the recycling of PM is unusual because they are reprocessed into extremely pure materials. For example, typical analysis for chemically refined gold produced from scrap is 99.95% while electrolytic gold ranges from 99.97 to 99.99% pure. Silver is typically 99.99% pure, and the PGM are 99.5% pure.

After refining, it is impossible to tell whether PM were recovered directly from ore or from printed circuit boards. There do not seem to be any reported problems with residual impurities in either pure materials or realloyed materials produced from PM-containing scrap. For example, because the precious metals are highly purified before reuse, palladium-nickel alloy for relay contacts can be produced from either recycled or virgin materials without any physical or compositional differences.

Future Use Trends

The rapid evolution of the electronics industry has lead to a continued improvement in the efficiency of the use of precious metals. For example, palladium was used for many years in telephone switching gear. This use is declining with the rapid introduction of solid-state switches. In addition, less expensive palladium-silver alloys are replacing the pure palladium that was previously used. However, the replacement of mechanical switches with solid-state switches has, in general, increased the demand for PGM because they are used as the contacts between the circuit and the outer package. In fact, 50% of the PGM and gold used in electronics industries goes into some kind of a contact. The introduction of ceramic capacitors containing gold-platinum-palladium or gold-palladium alloys also increased the demand for these metals. Inlay-clad gold and gold alloys are being increasingly used in electrical and electronic connections because they are an economic and reliable alternative to the use of electrodeposited gold (Ref 79). Other changes in the industry are the replacement of gold in printed circuit boards and edge connectors with palladium. The actual amount of precious metals contained in an individual item will decrease, for example, as the thickness of gold coatings decreases to a lower limit defined by reliability, and the ability increases to deposit any precious metal in a smaller, well-defined area. This will present new challenges to the recycling industry by requiring that new techniques for collecting PM-containing scrap be developed and better means of recovering the metal values from scrap be developed.

Currently large amounts of PM are recovered from obsolete computers, telecommunications, and DoD scrap. Once the existing supply of obsolete computers is scrapped, it is expected that recovery efforts will decline both because the precious metals content of the units is lower than in earlier-generation electronic scrap and because the smaller size of the units will cause this source to become dispersed. The same trends will be true in the telecommunications industry as existing circuitry is replaced with solid-state devices (Ref 80).

Electronics scrap is a complex material containing various recyclable fractions. The complexity of the material and the decrease in the amount of precious metals used in individual parts will make processing more difficult. Unless research is conducted on a continuing basis to develop technology to recover or concentrate PM from the new families of electronics entering the recycling stream, the increasing total consumption of precious metals for electronic equipment will have to be supplied by larger proportions of imports.

REFERENCES

1. P.R. Atkins, Recycling Can Cut Energy Demand Drastically, *Eng. Min. J.*, May 1973
2. Scrap Specifications Circular, Institute of Scrap Recycling Industries, Inc., 1988
3. A. Wrigley, *Ward's Auto World*, Sept 1981
4. J.H.L. van Linden, Aluminum Recycling—Everybody's Business, in *Light Metals 1990*, TMS, 1990
5. J.H.L. van Linden and R.E. Hannula, in *Light Metals 1981*, AIME, p 813-825
6. J.H.L. van Linden, J.R. Herrick, and M.J. Kinosz, Metal Scrap Melting System, U.S. Patent 3,997,366, 1976
7. J.H.L. van Linden, R.J. Claxton, J.R. Herrick, and R.J. Ormesher, Aluminum Scrap Reclamation, U.S. Patent 4,128,415, 1978
8. J.H.L. van Linden and J.B. Gorss, Vortex Melting System, U.S. Patent 4,286,985, 1981
9. A.G. Szekely, Vortex Reactor and Method for Adding Solids to Molten Metal Therewith, U.S. Patent 4,298,377, 1981
10. D.V. Neff, in *Proceedings of TMS Fall Meeting*, AIME, 1985, p 57-72
11. R.J. Claxton, Method for Submerging, Entrainment, Melting and Circulating Metal Charge in Molten Media, U.S. Patent 4,322,245, 1982
12. J. Butson, A Market Study for the Energy Efficiency Office, *Alum. Recycl.*, 1986
13. Observations at Stamicarbon Pilot Plant at Dalmeyer's Salvage Yard, Nieuwerkerk a/d Yssel, the Netherlands
14. B.C. Braam, W.L. Dalmyn, and W.P.C. Duyvenstein, "Recycle and Recovery of Secondary Metals," TMS, 1985, p 641
15. L.V. Whiting and D.A. Brown, "Air/Oxygen Injection Refining of Secondary Copper Alloys," Report MRP/PMRL 79-50(J), Physical Metallurgy Research Laboratories, CANMET, 1979
16. H.I. Kaplan, Magnesium Supply and Demand, in *Proceedings of the 46th Annual World Magnesium Conference*, May 1989, p 48
17. *Mineral Industry Survey*, U.S. Bureau of Mines, 15 Aug 1989
18. *Metal Statistics 1988*, Fairchild Publications, 1988, p 180
19. C.M. Stott, *The Secondary Tin Industry*, Paper A 85-3, presented at the 114th AIME Annual Meeting, Feb 1985
20. P.A. Wright, *The Extractive Metallurgy of Tin*, 2nd ed., Elsevier, 1982
21. C.L. Mantell, *Tin: Its Mining, Properties, Technology, and Applications*, Hafner Publishing, 1970
22. S. Apotheker, Recycle Steel Cans—Can Do, *Resour. Recycl.*, March 1990, p 22
23. S.K. Chatterjee and C.J. Thwaites, Sintered Iron Compacts Based on Atomized Powder From Tinplate Scrap, *Powder Metall. Int.*, Vol 13 (No. 3), 1981, p 118-120, 125; International Tin Research Institute Publication 610
24. M.E. Warwick, Developments in the Use of Tin in Ferrous Powder Metallurgy, *Met. Powder Rep.*, Vol 39 (No. 6), June 1984; International Tin Research Institute Publication 649
25. R.D. Prengaman, Secondary Lead in the United States, in *Pb 86 Ninth International Lead Conference—Goslar*, Lead Development Association, 1986, p 43
26. G. Tremolada, Automated Battery Wrecking Process, U.S. Patents 3,456,886, 1969, and 3,614,003, 1971
27. G. Schenker, in *Lead-Zinc '90*, T.S. Mackey and R.D. Prengaman, Ed., TMS, 1990, p 1001
28. R. Fischer, Treatment of Lead Battery Scrap at Stolberg Zinc, in *AIME World Symposium on Mining and Metallurgy of Lead and Zinc*, Vol 11, 1970, p 984
29. I.M. Olper and B. Asano, Improved Technology in Secondary Lead Processing—Engitec Lead Acid Battery Recycling System, in *Primary and Secondary Lead Processing*, M.L. Jaeck, Ed., Pergamon Press, 1989, p 119
30. K.N. Pike, Secondary Lead Blast Furnace Smelting at East Penn Manufacturing Co., Inc., in *Lead-Zinc '90*, T.S. Mackey and R.D. Prengaman, Ed., TMS, 1990, p 955
31. R. Egan, M. Rao, and K. Libsch, Rotary Kiln Smelting of Secondary Lead, in *Lead-Zinc-Tin '80*, J.M. Cigan, T.S. Mackey, and T.J. O'Keefe, Ed., TMS/AIME, 1980, p 953
32. J. Godfroi, Five Years Utilization of the Short Rotary Furnace in the Secondary Smelting of Lead, in *Lead-Zinc-Tin '80*, J.M. Cigan, T.S. Mackey, and T.J. O'Keefe, Ed., TMS, 1980, p 974
33. H. Forrest and J.D. Wilson, Lead Recycling Utilizing Short Rotary Furnaces, in *Lead-Zinc '90*, T.S. Mackey and R.D. Prengaman, Ed., TMS, 1990, p 971
34. R.D. Prengaman, Reverberatory Furnace-Blast Furnace Smelting of Battery Scrap at RSR, in *Lead-Zinc-Tin '80*, J.M. Cigan, T.S. Mackey, and T.J. O'Keefe, Ed., TMS/AIME, 1980, p 985
35. D.J. Eby, Electric Arc Smelting at RSR, in *Lead-Zinc '90*, T.S. Mackey and R.D. Prengaman, Ed., TMS, 1990, p 825
36. K. Moriya, Lead Smelting and Refining: Its Current Status and Future, in

Lead-Zinc '90, T.S. Mackey and R.D. Prengaman, Ed., TMS, 1990, p 23
37. T.R.A. Davey, The Physical Chemistry of Lead Refining, in *Lead-Zinc-Tin '80*, J.M. Cigan, T.S. Mackey, and T.J. O'Keefe, Ed., TMS/AIME, 1980, p 477
38. R.D. Prengaman and H.B. McDonald, RSR's Full Scale Plant to Electrowin Lead From Battery Scrap, *Lead-Zinc '90*, T.S. Mackey and R.D. Prengaman, Ed., TMS, 1990, p 1045
39. E.R. Cole, A.Y. Lee, and D.L. Paulson, Recovery of Lead From Battery Sludge by Electrowinning, *J. Met.*, Vol 8, 1983, p 42
40. J.H. Jolly, *Zinc*, in *Minerals Yearbook*, U.S. Bureau of Mines, 1988, p 12
41. D.R. MacRae, "Electric Arc Furnace Dust: Disposal, Recycle and Recovery," Center for Metals Production, May 1985
42. S.E. James and C.O. Bounds, Recycling Lead and Cadmium, as Well as Zinc, From EAF Dust, in *Lead-Zinc '90*, T.S. Mackey and R.D. Prengaman, Ed., TMS, 1990, p 447-496
43. "Characterization, Recovery and Recycling of Electric Arc Furnace Dust," Final Report for U.S. Department of Commerce, Lehigh University, Feb 1982
44. L.W. Lherbier, Jr., "Flame Reactor Process for EAF Dust," CMP Report 88-1, Center for Metals Production, Mellon Institute, Aug 1988
45. "Technological Assessment of Electric Steelmaking Through the Year 2000," Center for Metals Production, Oct 1987
46. "Processing of Steel Plant Flue Dusts," International Lead/Zinc Study Group, Oct 1989
47. R. Kola, The Processing of Steelworks Wastes, in *Lead-Zinc '90*, T.S. Mackey and R.D. Prengaman, Ed., TMS, 1990, p 453-464
48. N. Tsuneyama *et al.*, Production of Zinc Oxide for Zinc Smelting Process From EAF Dust at Shisaka Works, in *Lead-Zinc '90*, T.S. Mackey and R.D. Prengaman, Ed., TMS, 1990, p 465-476
49. K. Ikeda *et al.*, Production of Zinc Oxide for Zinc Smelting Process From EAF Dust at Shisaka Works, in *Proceedings of Zinc '85*, 1985, p 783-795
50. P.L. Kern and G.T. Mahler, Jr., "The Waelz Process for Recovering Zinc and Lead From Steelmaking Dusts," Paper presented at the 1988 TMS annual meeting (Phoenix, AZ), 1988
51. H. Maczek and R. Kola, Recovery of Zinc and Lead From Electric Furnace Steelmaking Dust at Berzelius, *J. Met.*, Vol 32 (No. 1), 1980, p 53-58
52. I.R. Geutskens, Pressure Leaching of Zinc-Bearing Blast Furnace Dust, in *Lead-Zinc '90*, T.S. Mackey and R.D. Prengaman, Ed., TMS, 1990, p 529-548
53. P. Cowx *et al.*, The Processing of Electric Arc Furnace Baghouse Dusts in the Tetronics Plasma Furnace, in *Lead-Zinc '90*, T.S. Mackey and R.D. Prengaman, Ed., TMS, 1990, p 497-510
54. S. Santen, Method of Recovering Volatile Metals From Material Containing Metal Oxides, U.S. Patent 4,488,905, 18 Dec 1984
55. T. Pedersen *et al.*, in *Lead-Zinc '90*, T.S. Mackey and R.D. Prengaman, Ed., TMS, 1990, p 857-879
56. C.O. Bounds and J.F. Pusateri, EAF Dust Processing in the Gas-Fired FLAME REACTOR Process, in *Lead-Zinc '90*, T.S. Mackey and R.D. Prengaman, Ed., TMS, 1990, p 511-528
57. N.L. Kotraba and N.G. Bishop, "Report on Pilot Work for a Continuous Feed Vertical Retort Smelting/Condensing Heavy Metals From Enriched Secondary Dust," Paper presented at the 1989 TMS annual meeting (Las Vegas, NV), Feb 1989
58. M. Hirsch *et al.*, "Recovery of Zinc and Lead From Steelmaking Dusts, in Particular by the Circulating Fluid Bed," Paper presented at the 28th Annual Conference of Metallurgists (Halifax, Nova Scotia), Canadian Institute of Mining, Aug 1989
59. "Production, Processing, Handling and Storage of Titanium," Document NFPA481, National Fire Protection Association
60. R.O. Dannenberg and G.M. Potter, Smelting of Military Electronic Scrap, in *Proceedings of the Second Mineral Waste Symposium*, ITT Research Institute, 18-19 March 1970, p 113-117
61. R.J. Garino, Making the Most of Electronic Scrap, *Scrap Process. Recycl.*, Jan/Feb 1989, p 65-71
62. J.H. Setchfield, Electronic Scrap Treatment at Englehard, in *Proceedings of the 11th IPMI Conference*, International Precious Metals Institute, June 1987, p 147-153
63. Chapters Gold, Silver, and Platinum-Group Metals, in *Minerals Yearbook 1987*, U.S. Bureau of Mines
64. Mineral Facts and Problems, *Bureau of Mines Bulletin*, U.S. Bureau of Mines, 1986
65. N.B. Coltoa, Prospects for Recycling Platinum, Paper presented at 2nd International Platinum Seminar, Metal Bulletin Journals Ltd.-Futures World, 14 Dec 1983
66. B.W. Dunning, Jr. and F. Ambrose, "Characterization of Pre-1957 Avionic Scrap for Resource Recovery," Report of Investigations 8499, Bureau of Mines, 1980
67. L. Johns, "Strategic Materials: Technologies to Reduce U.S. Import Vulnerability," Assessment OTA-1TE-248, U.S. Congress, Office of Technology, May 1985
68. B.W. Dunning, Jr., "Precious Metals Recovery From Electronic Scrap and Solder Used in Electronics Manufacture," Bureau of Mines Information, Circular 9059, 1986, p 44-56
69. Mechanical Processing of Electronic Scrap to Recover Precious-Metal-Bearing Concentrates, in *Precious Metals*, R.O. McGachie and A.G. Bradley, Ed., Pergamon, 1980, p 67-76
70. B.W. Dunning, Jr., F. Ambrose, and H.V. Makar, Distribution and Analyses of Gold and Silver in Mechanically Processed Mixed Electronic Scrap, Report of Investigations 8788, Bureau of Mines, 1983
71. F. Ambrose and B.W. Dunning, Jr., Precious Metals Recovery From Electronic Scrap, in *Proceedings of the Seventh Mineral Waste Utilization Symposium*, ITT Research Institute, 20-21 Oct 1980, p 184-197
72. R.O. Dannenberg, J.M. Maurice, and G.M. Potter, Recovery of Precious Metals From Electronic Scrap, Report of Investigations 7683, Bureau of Mines, 1972
73. T.A. Sullivan, R.L. deBeauchamp, and E.L. Singleton, Recovery of Aluminum, Base, and Precious Metals From Electronic Scrap, Report of Investigations 7617, Bureau of Mines, 1972
74. H.B. Salisbury, L.J. Duchene, and J.H. Bilbrey, Recovery of Copper and Associated Precious Metals From Electronic Scrap, Report of Investigations 8561, Bureau of Mines, 1981
75. J.S. Niederkorn and S. Huszar, Gold Recovery From Used Contactors, *Gold Bull.*, Vol 17 (No. 4), Oct 1984, p 128-130
76. E.K. Kleespies, J.P. Bennetts, and T.A. Henrie, Gold Recovery From Scrap Electronic Solders by Fused-Salt Electrolysis, in *Bureau of Mines Technical Progress*, TPR 9, U.S. Bureau of Mines, March 1969
77. H.E. Hilliard and B.W. Dunning, Jr., Recovery of Platinum-Group Metals and Gold From Electronic Scrap, in *The Platinum Group Metals—An In-Depth View of the Industry*, D.E. Lundy and E.D. Zysk, Ed., International Precious Metals Institute, April 1983, p 129-142
78. "Secondary Gold in the United States," Information Circular 8447, Bureau of Mines, 1970
79. R.J. Russel, Inlay-Clad Gold Alloys, *Gold Bull.*, Vol 9 (No. 1), Jan 1976, p 2-6
80. S.C. Malhotra, Future Opportunities in the Reclamation of Precious Metals From Major Sources of Obsolete Scrap, in *Proceedings of Precious Metals: Mining, Extraction, and Processing*, V. Kudyk, D.A. Corrigan, and W.W. Lang, Ed., International Precious Metals Institute, 27-29 Feb 1984, p 483-494

Toxicity of Metals

Robert A. Goyer, Department of Pathology, The University of Western Ontario

METALS DIFFER from other toxic substances in that they are neither created nor destroyed by humans. Nevertheless, utilization by humans influences the potential for health effects in at least two major ways: first, by environmental transport, that is, by human or anthropogenic contributions to air, water, soil, and food, and second, by altering the speciation or biochemical form of the element (Ref 1, 2).

Metals are redistributed naturally in the environment by both geologic and biologic cycles (Fig. 1). Rainwater dissolves rocks and ores and physically transports material to streams and rivers, adding and deleting from adjacent soil, and eventually to the ocean to be relocated elsewhere on earth. The biologic cycles include bioconcentration by plants and animals and incorporation into food cycles. Human industrial activity may greatly shorten the residence time of metals in ore, form new compounds, and greatly enhance worldwide distribution. These natural cycles may exceed the anthropogenic cycle, as is the case for mercury. However, the role of human activity in redistribution of metal is demonstrated by the 200-fold increase in lead content of Greenland ice, beginning with a "natural" low level (about 800 B.C.) and a gradual rise in lead content of ice through the evolution of the industrial age, followed by a nearly precipitous rise in lead corresponding to the period when lead was added to gasoline in the 1920s (Ref 3). Metal contamination of the environment, therefore, reflects both natural sources and contribution from industrial activity.

Metals emitted into the environment from combustion of fossil fuels in the United States are shown in Table 1. These include many of the metals most abundant in particulates in ambient air. The only metals or metal-like elements that may be emitted in gaseous discharges in measurable concentrations are mercury or selenium. Metals in raw surface water reflect erosion from natural sources, fallout from the atmosphere, and additions from industrial activities. Metals in soil and water may enter the food chain. For persons in the general population, food sources probably represent the largest source of exposure to metals, with an additional contribution from air. Further potential sources of human exposure include consumer products and industrial wastes as well as the working environment.

Occupational exposure to metals is restricted to "safe" levels defined as the threshold limit value for an eight-hour day, five-day work week. These levels are intended to provide a margin of safety between maximum exposure and minimum levels that will produce illness. Permissible levels vary widely, and the differences reflect, in a sense, the toxicologic potency of the metal. As a general rule, the metals that are most abundant in the environment have lesser potential for toxicity as evidenced by the prevailing standard for permissible occupation exposure.

Metals are probably the oldest toxins known to humans. Lead usage may have begun prior to 2000 B.C. when abundant supplies were obtained from ores as a byproduct of smelting silver. Hippocrates is credited in 370 B.C. with the first description of abdominal colic in a man who extracted metals. Arsenic and mercury are cited by Theophrastus of Erebus (387-372 B.C.) and Pliny the Elder (A.D. 23-79). Arsenic was obtained during the melting of copper and tin, and an early use was for decoration in Egyptian tombs. On the other hand, many of the metals of toxicologic concern today are only recently known to humans. Cadmium was first recognized in ores containing zinc carbonate in 1817. About 80 of the 105 elements in the periodic table are regarded as metals, but fewer than 30 have compounds that have been reported

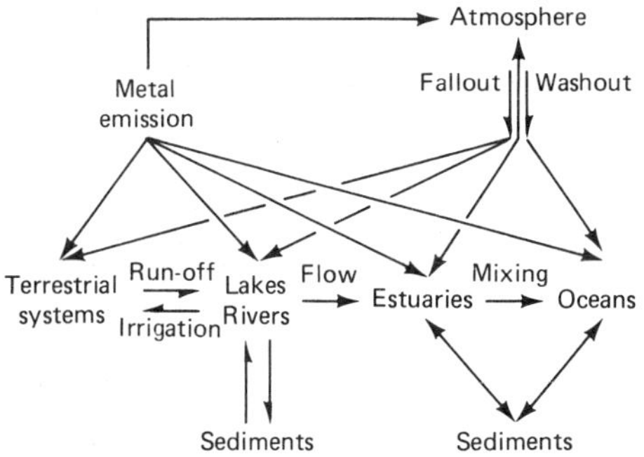

Fig. 1 Routes for transport of trace elements in the environment. Source: Ref 1

Table 1 Sources and standards of toxic metals in the United States

Element	Combustion of fossil fuels, 10^3 tons(a)	Particulates in air typical, ng/m^3(b)	Water, µg/L and frequency of detection(c): Maximum	Mean	Percent	Threshold limit values for 8-h occupational exposure, mg/m^3(d)
Al	6000	3080	2760	74	31	10
As	27	10	(e)			0.2
Ba	300	100	340	43	99	0.5
Be	15	0.2	1.22	0.19	5.4	0.002
Cd	1	1	120	10	2.5	0.02 dust
Co	15	5	48	17	2.8	0.05
Cr	0	20	112	10	25	0.05
Cu	9	500	280	15	75	0.1
Fe	6002	4000	4600	52	76	3.5
Li	39	4	...	...	...	0.025
Mg	1200	2000	...	...	...	5.0
Mn	33	100	3230	60	58	2.5
Mo	5	1	1500	68	38	10
Ni	11	20	30	19	16	0.1 soluble
Pb	126	2000	140	13	19	0.1
Sn	...	50	...	...	...	2.0 inorganic
Se	2	1	...	...	...	0.1
V	27	30	300	2	5	0.5
Zn	30	500	2010	79.2	80	1.0

(a) Data from 1977 fuel consumption. Source: Ref 4. (b) In 10-day period, six U.S. cities. Source: Ref 5. (c) Source: Ref 6. (d) sol; inorg., inorganic. Source: Ref 7. (e) Arsenic in water is extremely variable, 10 to 1100 µg/L 728 samples surface water 22% in 10 to 20 µg range.

to produce toxicity in humans. The importance of some of the rarer or lesser known metals such as indium or tantalum might increase with new applications in microelectronics or other new technologies.

The conceptual boundaries of what is regarded as toxicology of metals continues to broaden. Historically, metal toxicology has largely concerned acute or overt effects, such as abdominal colic from lead toxicity or the bloody diarrhea and suppression of urine formation from ingestion of corrosive (mercury) sublimate. There must still be knowledge and understanding of such effects, but with present-day occupational and environmental standards, such effects are uncommon. Beyond this, however, is growing inquiry regarding subtle, chronic, or long-term effects where cause-and-effect relationships are not obvious or may be subclinical. This might include a level of effect that causes a change that resides within the generally regarded norm of human performance, for example, lower I.Q. and childhood lead exposure. Assigning responsibility for such toxicologic effects is extremely difficult and not always possible, particularly when the end-point in question lacks specificity in that it may be caused by a number of agents or even combinations of substances.

The challenges, therefore, for the toxicologist are multiple. The major ones include the need for quantitative information regarding dose and tissue levels, greater understanding of the metabolism of metals particularly at the tissue and cellular level where effects that have specificity may occur, and finally, recognition of factors that influence toxicity of a particular level of exposure such as dietary factors or protein-complex formation that enhance or protect from toxicity. Treatment, particularly the administration of chelating agents, remains an important topic particularly for those metals that are cumulative and persistent, for example, Pb, Cd, Ni, and so forth. However, prevention of toxicity is the major objective of public health policies and occupational hygiene programs. There is increasing emphasis on the use of biologic indicators of toxicity such as heme enzymes in lead toxicity, renal tubular dysfunction in cadmium exposure, and neurologic effects in mercury toxicity, to serve as guidelines for preventive or therapeutic intervention.

Estimates of Dose-Effect Relationships

Estimates of the relationship of dose or level of exposure to a particular metal are, in many ways, a measure of dose-response relationships. Conceptual background for this topic is considered in Ref 8. Dose or estimate of exposure to a metal may be a multidimensional concept and is a function of time as well as concentration of metal. The most precise definition of dose is the amount of metal within cells of organs manifesting a toxicologic effect. Results from single measurements may reflect recent exposure or longer-term or past exposure, depending on retention time in the particular tissue. Blood, urine, and hair are the most accessible tissues in which to measure dose and are sometimes referred to as indicator tissues. *In vivo*, quantitation of metals within organs is not yet possible, although techniques such as neutron activation and fluorescence spectroscopy may hold promise for the future. Indirect estimates of quantities in specific organs may be calculated from metabolic models derived from autopsy data.

At the cellular level, toxicity is related to availability so that chemical form and ligand binding become critical factors. Alkyl compounds are lipid soluble and pass readily across biologic membranes unaltered by their surrounding medium. They are only slowly dealkylated or transformed to inorganic salts. Hence, their excretion tends to be slower than inorganic forms, and the pattern of organic toxicity differs. For example, alkyl mercury is primarily a neurotoxin versus the renal toxicity of mercuric chloride. Metals that have strong affinity for osseous tissue like lead and radium have a long retention time and tend to accumulate with age. Other metals are retained in soft tissues because of affinities for intracellular proteins, such as renal cadmium bound to metallothionein.

Blood and urine usually reflect recent exposure and correlate best with acute effects. An exception is urine cadmium where increased metal in urine reflects renal damage related to accumulation of cadmium in the kidney. Partitioning of metal between cells and plasma and between filterable and nonfilterable components of plasma should provide more precise information regarding the presence of biologically active forms of a particular metal. Such partitioning is now standard laboratory practice for blood calcium; ionic calcium is by far the most active form of the metal. Speciation of toxic metals in urine may also provide diagnostic insights. For example, cadmium metallothionein in urine may be of greater toxicologic significance than total cadmium.

Hair can be useful in assessing variations in exposure to metals over the long term. Analysis may be performed on segments so that metal content of the newest growth can be compared to past exposures. Correlation between blood levels of metal and concentration in hair is not expected because blood levels reflect only current exposures. Caution must be taken in washing hair prior to analysis to assure removal of metal deposits from external contamination (Ref 9).

Factors Influencing Toxicity of Metals

There are only a few general principles available that contribute to understanding the patho-physiology of metal toxicity. Most metals affect multiple organ systems, and the targets for toxicity are specific biochemical processes (enzymes) and/or membranes of cells and organelles. The toxic effect of the metal usually involves an interaction between the free metal ion and the toxicologic target. There may be multiple reasons for a particular toxic effect. For instance, the metabolism of the toxic metal may be similar to a metabolically related essential element. Such is the case for some of the effects of lead, for example, lead and calcium in the central nervous system, and lead, iron, and zinc in heme metabolism. Cells that are involved in the transport of metals, such as gastrointestinal, liver, or renal tubular cells, are particularly susceptible to toxicity. However, for many metals, these cells have protective mechanisms involving protein complex formation that permits intracellular accumulation of potentially toxic metals without causing cell injury.

Metal-protein complexes involved in detoxication or protection from toxicity have now been described for a few metals (Ref 10). Morphologically discernible cellular inclusion bodies are present with exposures to lead, bismuth, and a mercury-selenate mixture. Metallothioneins form complexes with cadmium, zinc, copper, and other metals, and ferritin and hemosiderin are intracellular iron-protein complexes. The protein complexes formed by lead, bismuth, mercury-selenate, and iron, at least for hemosiderin, have attracted interest because these complexes are insoluble in tissues and can be observed histologically. However, it is this lack of solubility that has made detailed biochemical study very difficult. On the other hand, for those metal-protein complexes that are stable and soluble in aqueous media, such as metallothionein and ferritin, there is considerable biochemical information. More is known about ferritin than perhaps any of these protein complexes because it is soluble and at the same time has a unique ultrastructural appearance so that it can be readily identified in cells and organelles. None of these proteins or metal-protein complexes have any known enzymatic activity. From these considerations it becomes clearer why speciation, that is, how much metal in a tissue that is in a particular biochemical form and what it is bound to, may be the ultimate determinant of toxicity.

Numerous exogenous factors influence the occurrence of toxicity in any particular subject (Ref 11). These include age, diet, and interactions and concurrent exposure with other toxic metals. Persons at either end of the life span, young children or elderly, are believed to be more susceptible to toxicity from exposure to a particular level of metal than adults. Rapid growth and cell division represent opportunities for genotoxic effects. Intrauterine toxicity to methyl mercury is well documented. Lead crosses the placenta, and it is recommended that maternal blood lead levels be lower than those of persons in the general population.

The major pathway of exposure to many toxic metals in children is with food, and children consume more calories per body weight than adults. Moreover, children have higher gastrointestinal absorption of metals, particularly lead. Experimental studies have extended these observations to other metals, and milk diet, probably because of lipid content, seems to increase metal absorption.

Effects of some dietary factors on metal toxicity are at the level of absorption from the gastrointestinal tract. There is an inverse relationship between protein content of diet and cadmium and lead toxicity. Vitamin C reduces lead and cadmium absorption, probably because of increased absorption of ferrous ion. On the other hand, metabolically related essential metals may alter toxicity by interaction at the cellular level. Lead, calcium, and vitamin D have a complex relationship affecting mineralization of bone and more directly through impairment of 1-25-dihydroxy vitamin D synthesis in the kidney. Metal-metal interaction may have considerable influence on dose-effect relationships and are commented on in discussions of specific metals.

Lifestyle factors such as smoking or alcohol ingestion may have indirect influences on toxicity. Cigarette smoke itself contains some toxic metals such as cadmium, and cigarette smoking may influence pulmonary effects. Alcohol ingestion may influence toxicity indirectly by altering diet and reducing essential mineral intake. For instance, a decrease in dietary calcium will influence toxicity of major toxic metals, including lead and cadmium.

Chemical form of the metal may be an important factor, not only for pulmonary and gastrointestinal absorption but in terms of body distribution and toxic effects. Dietary phosphate generally forms less soluble salts of metals than other anions. Alkyl compounds, such as tetraethyl lead and methyl mercury, are lipid soluble and more soluble in myelin than inorganic salts of these metals.

For metals that produce hypersensitivity reactions, the immune status of an individual becomes an additional toxicologic variable (Ref 12). Metals that provoke immune reactions include mercury, gold, platinum, beryllium, chromium, and nickel. Clinical effects are varied but usually involve any of four types of immune responses. In anaphylactic or immediate hypersensitivity reactions the antibody, IgE, reacts with the antigen on the surface of mast cells releasing vasoreactive amines. Clinical reactions include conjunctivitis, asthma, urticaria, or even systemic anaphylaxis. Cutaneous, mucosal, and bronchial reactions to platinum have been attributed to this type of hypersensitivity reaction. Cytotoxic hypersensitivity is the result of a complement-fixing reaction of IgG immunoglobulin with antigen or hapten bound to the cell surface. The thrombocytopenia sometimes occurring with exposure to organic gold salts may be brought about in this manner. Immune complex hypersensitivity occurs when soluble immune complex deposits (antigen, antibody, and complement) within tissues producing an acute inflammatory reaction. Immune complexes are typically deposited on the epithelial surface of glomerular basement membrane, resulting in proteinuria, and occur following exposure to mercury vapor or gold therapy. Cell-mediated hypersensitivity, also known as the delayed hypersensitivity reaction, is mediated by thymus-dependent lymphocytes and usually occurs 24 to 48 hours after exposure. The histologic reaction consists of mononuclear cells and is the typical reaction seen in the contact dermatitis following exposure to chromium or nickel. The granuloma formation occurring with beryllium and zirconium exposure may be a form of cell-mediated immune response.

Carcinogenesis

Given the long history of human exposure to metals, knowledge of the potential carcinogenicity of metal compounds has evolved slowly, and most of this information has only been obtained in recent years (Ref 13, 14).

Furthermore, predictive *in vitro* methods using nonmammalian systems, such as the Ames test, do not seem as responsive as for organic compounds (Ref 15). Evidence of carcinogenicity for metals relates more precisely with specific compounds of metals than with the metal itself. That is, some forms of the metal seem to be carcinogenic; for example, nickel subsulfide (Ni_3S_2) is more carcinogenic than amorphorus nickel monosulfide (NiS); but such differences may be explained on the basis of cell uptake rates or solubility. Similar debates concern various compounds of chromium. Nevertheless, if any form of a metal is carcinogenic, the metal itself must be regarded as a carcinogen.

Although only a few metals show any evidence of carcinogenicity, this is an exceedingly important topic because of the ubiquity of most metals, their wide industrial use, and their persistence in the environment. Identification of metal carcinogens in industry is made even more perplexing because seldom is exposure to a single metal, but it is usually to mixtures. Also, there is the added question of the role of metals as promoters or cocarcinogens with organic carcinogens because of their persistence in tissues, as may be the case for lead.

The chronology of observations on the carcinogenicity of metals is shown in Fig. 2. Specific details pertaining to the carcinogenicity of each metal are discussed later in the chapter along with other toxicologic effects. However, the figure does provide an overview. Human case reports of skin cancer due to arsenic exposure were recognized in the nineteenth century, but epidemiologic support from case study observations did not occur until over 50 years later, and there has not yet been confirmation in experimental animals. On the other hand, lead is the only metal shown to be carcinogenic in animal models by oral administration. Yet, evidence in humans is limited to a couple of recent case reports. How much of what kind of evidence, animal and/or human, is required to label a metal as a carcinogen must be decided for each metal. Animal studies that use routes of administration different from those by which humans may be exposed, such as by injection, have limitations for extrapolation to humans.

Chelation

Chelation is the formation of a metal ion complex in which the metal ion is associated with a charged or uncharged electron donor referred to as a ligand. The ligand may be monodentate, bidentate, or multidentate; that is, it may attach or coordinate using one or two or more donor atoms. Bidentate ligands form ring structures that include the metal ion and the two ligand atoms attached to the metal (Ref 16).

Chelating agents are generally nonspecific in regard to their affinity for metals. To varying degrees, they will mobilize and enhance the excretion of a rather wide range of metals, including essential metals such as calcium and zinc (Table 2). Their efficacy depends not solely on their affinity for the metal of interest, but also on their affinity for endogenous metals, mainly calcium, which compete in accordance with their own affinities for the chelator. Properties of a few of the commonly used chelators will be described.

BAL (British Anti Lewisite) or 2,3-dimercaptopropanol was the first clinically useful chelating agent. It was developed during World War II as a specific antagonist to vesicant arsenical war gases based on the observation that arsenic has an affinity for sulfhydryl-containing substances (Ref 17). BAL, a dithiol compound with two sulfur atoms on adjacent carbon atoms, competes with the critical binding sites responsible for the toxic effects. These observations led to

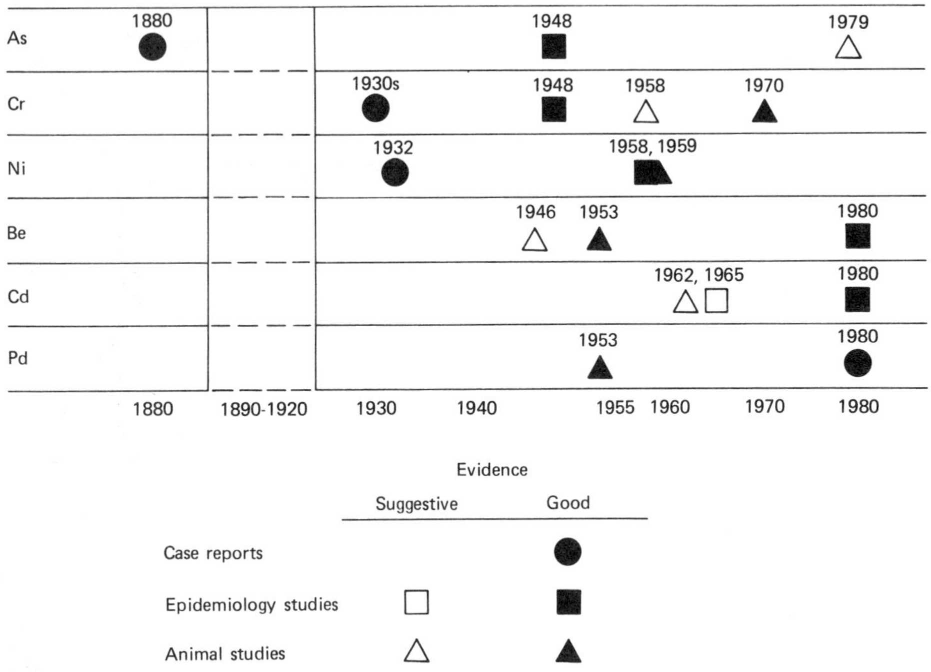

Fig. 2 Chronology of observations on the carcinogenicity of metals. Source: Ref 14

the prediction that the "biochemical lesion" of arsenic poisoning would prove to be a thiol with sulfhydryl groups separated by one or more intervening carbon atoms. This prediction was borne out a few years later with the discovery that arsenic interferes with the function of 6,8-dithiooctanoic acid in biologic oxidation (Ref 18).

BAL has been found to form stable chelates *in vivo* with many toxic metals including inorganic mercury, antimony, bismuth, cadmium, chromium, cobalt, gold, and nickel. However, it is not necessarily the treatment of choice for toxicity to these metals. BAL has been used as an adjunct in the treatment of the acute encephalopathy of lead toxicity. It is a potentially toxic drug, and its use may be accompanied by multiple side effects. Although BAL will increase the excretion of cadmium, there is a concomitant increase in renal cadmium concentration so that its use in cadmium toxicity is to be avoided. It does, however, remove inorganic mercury from kidneys but is not useful in treatment of alkyl or phenylmercury toxicity. BAL also enhances the toxicity of selenium and tellurium so it is not to be used to remove these metals.

DMPS (2,3-dimercapto-1-propanesulfonic acid) is a water-soluble derivative of BAL developed to reduce the toxicity and unpleasant side effects of BAL. A recent study has found that DMPS reduces blood lead levels in children (Ref 19). It has the advantage over EDTA in that it is administered orally and does not appear to have toxic side effects. It has been widely used in Russia to treat many different metal intoxications and even atherosclerosis by the adherents of the notion that this degenerative disorder of blood vessels is due to metal-ion accumulations in the blood vessel wall leading to inhibition of enzyme metabolism.

DMPS is effective in removal of both inorganic and methyl mercury, probably because it is not lipophilic like BAL and does not penetrate tissues but removes extracellular metal (Ref 20). The important point is that it does not increase the concentration of metal in the brain and reduces organ concentration of metal including the kidney. It may also be effective in removal of copper, nickel, and cadmium immediately after exposure but not from tissue stores.

Calcium EDTA is the calcium disodium salt of ethylene diamine tetraacetic acid. The calcium salt must be used clinically because the sodium salt has greater affinity for calcium and will produce hypocalcemic tetany. However, the calcium salt will bind lead with displacement of calcium from the chelate. It is poorly absorbed from the gastrointestinal tract so it must be given parenterally, and it becomes rapidly distributed in the body. It is the current method of choice for treatment of lead toxicity (Ref 21). The peak excretion is within the first 24 hours and represents excretion of lead from soft tissues. Removal from the skeletal system occurs more slowly with restoration of equilibrium with soft tissue compartments. Calcium EDTA does have the potential for nephrotoxicity, so it should be administered only when indicated clinically.

Table 2 Ligands (chelating agents) preferred for removal of toxic metals

Ligand	Metal
BAL(a)	Arsenic, lead (with Ca-EDTA), mercury, inorganic
DMPS(b)	Methyl mercury, inorganic mercury, cadmium, copper, and nickel
Calcium EDTA(c)	Lead
Penicillamine	Copper, lead
Calcium DTPA(d)	Cadmium (with BAL)
Desferrioxamine	Iron
Dithiocarb	Nickel carbonyl

(a) BAL, British Anti Lewisite. (b) DMPS, 2,3-dimercapto-1-propanesulfonic acid. (c) EDTA, ethylene diamine tetraacetic acid. (d) DTPA, diethylenetriamine-pentaacetic acid

Penicillamine (B,B[1]-dimethylcystein), a hydrolytic product of penicillin, is the choice for therapy of Wilson's disease (copper toxicity) and is effective in removal of lead, mercury, and iron (Ref 22). It is also important to note that penicillamine removes other physiologically essential metals including zinc, cobalt, and manganese. It also has the risk of inducing a hypersensitivity reaction with a wide spectrum of undesired immunologic effects including skin rash, blood dyscrasias, and possibly proteinuria and the nephrotic syndrome. It has cross-sensitivity to penicillin so it should be avoided by persons with penicillin hypersensitivity. Recent studies have shown the effectiveness of a new orally active chelating agent, triethylene tetramine 2HCl (Trien) in Wilson's disease, particularly in those persons who have developed sensitivity to penicillamine (Ref 23).

DTPA or diethylenetriamine-pentaacetic acid has chelating properties similar to those of EDTA. The calcium salt ($CaNa_2$ DTPA) must be used clinically because of its high affinity for calcium. It has been used for chelation of plutonium and other radioactive metals but with mixed success. More recently there has been considerable experimental study of BAL for removal of cadmium alone, or DTPA in combination with BAL, but with limited success (Ref 24).

Desferrioxamine is a hydroxylamine isolated as the iron chelate of *Streptomyces pilosus* and is used clinically in the metal-free form (Ref 25). It has a remarkable affinity for ferric iron and a low affinity for calcium and competes effectively for iron in ferritin and hemosiderin but not transferrin, or the iron in hemoglobin or heme-containing enzymes. It is poorly absorbed from the gastrointestinal tract so it must be given parenterally. Clinical usefulness is limited by a variety of toxic effects including hypotension, skin rashes, and possible cataract formation. It seems to be more effective in hemosiderosis due to blood transfusion but is less effective in treatment of hemochromatosis.

Dithiocarb (diethyldithiocarbamate) or DDC has been recommended as the drug of choice in the treatment of acute nickel carbonyl poisoning. The drug may be administered orally for mild toxicity but parenterally for acute or severe poisoning (Ref 26).

Major Toxic Metals With Multiple Effects

Overexposure to the metals described in this section can result in multiple toxicological effects. These metals include arsenic, beryllium, cadmium, chromium, lead, mercury, and nickel.

Arsenic

Arsenic is particularly difficult to characterize as a single element because its chemistry is so complex and there are many different compounds of arsenic. It may be trivalent or pentavalent and is widely distributed in nature. The most common inorganic trivalent arsenic compounds are arsenic trioxide, sodium arsenite, and arsenic trichloride. Pentavalent inorganic compounds are arsenic pentoxide, arsenic acid, and arsenates, such as lead arsenate and calcium arsenate. Organic compounds may also be trivalent or pentavalent such as arsanilic acid, or even in methylated forms as a consequence of bimethylation by organisms in soil and fresh and seawaters. A summary of environmental sources of arsenic as well as potential health effects is contained in a World Health Organization (WHO) criteria document (Ref 27). Mechanical and physical properties of arsenic are summarized in the article "Properties of Pure Metals" in this Volume.

Arsenic is transported in the environment mainly by water, and airborne arsenic is generally due to contributions from industrial contamination and may range from a few nanograms to a few tenths of a microgram per cubic meter. The 133 stations of the National Air Sampling Network reported in 1964 that the average annual concentration of arsenic in air ranges from 0.01 $\mu g/m^3$ to 0.75 $\mu g/m^3$ in smelters. Near point emissions, concentrations may exceed 1 $\mu g/m^3$. Drinking water usually contains a few micrograms per liter or less. More than 18 000 community water supplies in the United States have concentrations less than 0.01 mg/L, but levels exceeding 0.05 mg/L have been found in Nova Scotia where arsenic content of bed rock is high. Even higher concentrations have been reported from various mineral springs, for example, Japan—1.7 mg As/L; Cordoba, Argentina—3.4 mg/L; Taiwan (artesian well water)—1.8 mg/L. Most foods (meat and vegetables) contain some level of arsenic, but the daily diet in the United States contains below 0.04 mg, but may contain 0.2 mg per day if the diet contains seafood. The total daily intake of arsenic by humans without industrial exposure, however, is usually less than 0.3 mg/day.

The major source of occupational exposure to arsenic in the United States is in the manufacture of pesticides, herbicides, and other agricultural products (Ref 28). High exposure to arsenic fumes and dust may occur in the smelting industries; the highest concentrations most likely occur among roaster workers.

Disposition. Airborne arsenic is largely trivalent arsenic oxide, but deposition in airways and absorption from lungs is dependent on particle size and chemical form.

Studies show that 6 to 9% of orally administered ^{74}As-labeled trivalent or pentavalent arsenic is eliminated in feces in mice (Ref 29), indicating almost complete absorption from the gastrointestinal tract. Limited data also suggest nearly complete absorption of soluble forms of trivalent and pentavalent arsenic (Ref 30). Excretion of absorbed arsenic is mainly via urine. The biologic half-life of ingested inorganic arsenic is about 10 hours and 50 to 80% is excreted in about 3 days. The biologic half-life of methylated arsenic was found to be 30 hours in one study (Ref 31).

Arsenic has a predilection for skin and is excreted by desquamation of skin and in sweat, particularly during periods of profuse sweating. It also concentrates in nails and hair. Arsenic in nails produces Mee's lines (transverse white bands across fingernails) appearing about six weeks after onset of symptoms of toxicity. Time of exposure may be estimated from measuring the distance of the line from the base of the nail and the rate of nail growth, which is about 0.3 cm/month or 0.1 mm/day. Arsenic in hair may also reflect past exposure, but intrinsic or systematically absorbed arsenic in hair must be distinguished from arsenic that is deposited from external sources. Human milk contains about 3 $\mu g/L$ of arsenic.

Placental transfer of arsenic has been shown in hamsters injected intravenously with high doses (20 mg/kg body weight) of sodium arsenate (Ref 32) and studies of tissue levels of arsenic in fetuses and newborn babies in Japan show that the total amount of arsenic in the fetus tends to increase during gestation indicating placental transfer. A more recent study of women in the United States found cord blood levels of arsenic to be similar to maternal blood levels (Ref 33).

Biotransformation of arsenic has been difficult to study because of analytical problems. Pentavalent arsenic compounds are reduced *in vivo* to more toxic trivalent compounds (Ref 34). However, ingestion of trivalent arsenic by experimental animals and humans is followed by excretion of some percentage of administered dose as pentavalent arsenic (Ref 35). The major form of arsenic in urine is dimethylarsinic acid, indicating *in vivo* methylation in humans.

Ingestion of arsenic-containing seafood does not result in increased excretion of inorganic arsenic and methyl- and dimethylarsinic acid, suggesting that the unknown organic compounds of arsenic are not converted to methylarsinic acid *in vivo* (Ref 31).

Cellular Effects. It has been known for some years that trivalent compounds of arsenic are the principal toxic forms, and pentavalent arsenic compounds have little effect on enzyme activity (Ref 17). A number of sulfhydryl-containing proteins and enzyme systems have been found to be altered by exposure to arsenic. Some of these can be reversed by addition of an excess of a monothiol such as glutathione; those enzymes containing two thiol groups can be reversed by dithiols such as 2,3-dimercaptopropanol (BAL) but not by monothiols.

Arsenic affects mitochondrial enzymes and impairs tissue respiration (Ref 36), which seems to be related to the cellular toxicity of arsenic. Mitochondria accumulate arsenic, and respiration mediated by nicotinamide adenine dinucleotide (NAD)-linked substrates is particularly sensitive to arsenic and is thought to result from reaction between arsenite ion and dihydrolipoic acid cofactor, necessary for oxidation of the substrate (Ref 37). Arsenite also inhibits succinic dehydrogenase activity and uncouples oxidative phosphorylation, which results in stimulation of mitochondrial adenosinetriphosphatase (ATPase) activity. Mitchell *et al.* (Ref 38) proposed that arsenic inhibits energy-linked functions of mitochondria in two ways: competition with phosphate during oxidative phosphorylation and inhibition of energy-linked reduction of NAD.

Toxicology. Ingestion of large doses (70 to 180 mg) may be acutely fatal (Ref 39). Symptoms consist of fever, anorexia, hepatomegaly, melanosis, and cardiac arrhythmia with electrocardiograph changes that may be the prodroma of eventual cardiovascular failure. Other features include upper-respiratory-tract symptoms, peripheral neuropathy, and gastrointestinal, cardiovascular, and hematopoietic effects. Acute ingestion may be suspected from damage to mucous membranes such as irritation, vesicle formation, and even sloughing. Sensory loss in the peripheral nervous system is the most common neurologic effect, appearing one or two weeks after large exposures and consisting of Wallerian degeneration of axons, but is reversible if exposure is stopped. Anemia and leukopenia, particularly granulocytopenia, occur in a few days and are reversible.

Liver injury is characteristic of longer-term or chronic exposure, is initially reflected by jaundice, and may progress to cirrhosis and ascites. Toxicity to hepatic parenchymal cells results in elevations of liver enzymes in blood, and studies in experimental animals show granules and alterations in the ultrastructure of mitochondria, nonspecific manifestations of cell injury including loss of glycogen.

Peripheral vascular disease has been observed in persons with chronic exposure to arsenic in drinking water in Taiwan and Chile, is manifested by acrocyanosis and Raynaud's phenomenon, and may progress to endarteritis obliterans and gangrene of the lower extremities (blackfoot disease). This specific effect seems to be related to the cumulative dose of arsenic, but prevalence is uncertain because of difficulties in separating arsenic-induced peripheral vascular disease from other causes of gangrene (Ref 40).

Carcinogenicity. Arsenic has specific effects on the endothelial cells of the blood vessels in the liver, and hemangioendothelial tumors or angiosarcoma of the liver has been reported in vineyard workers following many years of exposure to arsenic-containing drinking water, Fowler's solution, wine, and arsenic-containing pesticides (Ref 41).

The skin is the critical organ of arsenic toxicity, and a variety of skin lesions have been associated with arsenic intoxication, particularly from chronic exposure in drinking water and from certain occupational exposures. A characteristic finding is symmetric verrucous hyperkeratosis of the palms and soles. Hyperpigmentation or melanosis is also common. Cancer of the skin related to arsenic exposure was first reported by an English physician, Sir Jonathan Hutchinson, in persons with long-continued ingestion of Fowler's solution. Skin cancers from occupational exposures have since been well documented, particularly in the last 20 years. Available data suggest a dose-response relationship (Ref 40).

Workers engaged in the production of arsenic-containing pesticides showed increased lung cancer mortality (Ref 42), and workers involved in copper smelting, where arsenic exposure may be very high, are also reported to have an increased risk of dying from lung cancer (Ref 43, 44). Nonworker populations living near point emission sources of arsenic to air may have increases in lung cancer as well, but the studies to date are not definitive (Ref 45). Nevertheless, the relationship of ingestion of arsenic with skin cancer and angiosarcoma and inhalation of arsenic containing particulates and lung cancer establishes arsenic as a human carcinogen. However, in contrast to most other human carcinogens, it has been difficult to confirm in experimental animals. In one study, rats given a mixture of calcium arsenate, copper sulfate, and calcium oxide by intratracheal instillation developed lung tumors (Ref 46), but other studies testing trivalent and pentavalent arsenic compounds by oral administration or skin application have not shown potential for either promotion or initiation of carcinogenicity. Similarly, experimental studies for carcinogenicity of organic arsenic compounds have been negative.

Table 3 Biologic indicators of arsenic exposure

Indicator	Normal	Excessive exposure
Whole blood	<10 μg/L	Up to 50 μg/L
Urine(a)	<50 μg/L	>100 μg/L
Hair	<1 μg/kg	

(a) Best indicator of current or recent exposure

Studies on mutagenic effects of arsenic have been generally negative. Inorganic arsenic compounds do interfere with deoxyribonucleic acid (DNA) repair mechanisms in bacteria and dermal cell cultures. An increased frequency of chromosomal aberrations has been found among workers exposed to inorganic arsenic compounds and patients taking drugs containing arsenic (Ref 47).

Reproductive Effects and Teratogenicity. High doses of inorganic arsenic compounds given to pregnant experimental animals produce various malformations somewhat dependent on time and route of administration. However, no such effects have been noted in people with excessive occupational exposures to arsenic compounds.

Arsine. Arsine gas is formed by the reaction of hydrogen with arsenic and is generated as a by-product in the refining of nonferrous metals. Arsine is a potent hemolytic agent, producing acute symptoms of nausea, vomiting, shortness of breath, and headache accompanying the hemolytic reaction. Exposure may be fatal and may be accompanied by hemoglobinuria and renal failure, and even jaundice and anemia in nonfatal cases where exposure persists (Ref 48).

Biologic indicators of arsenic exposure are blood, urine, and hair (Table 3). Because of the short half-life of arsenic, blood levels are only useful within a few days of acute exposure but are not useful to assess chronic exposure. Urine arsenic is the best indicator of current or recent exposure and has been noted to be several hundred micrograms per liter with occupational exposure. Hair or even fingernail concentration of arsenic may be helpful to evaluate past exposures, but interpretation is made difficult because of the problem of differentiating external contamination.

There are no specific biochemical parameters that reflect arsenic toxicity, but evaluation of clinical effects must be interpreted with knowledge of exposure history.

Treatment. BAL is used to treat acute dermatitis and pulmonary symptoms. BAL has also been used for the treatment of chronic arsenic poisoning, but there are no established biologic criteria or measures of effectiveness. BAL has been used most often in cases with dermatitis, but there is usually no change in the keratotic lesions or influence on progression to skin cancer.

Arsine toxicity is best treated symptomatically. BAL is not considered helpful (Ref 48).

Beryllium

The major toxicologic effects of beryllium are on the lung. It may produce an acute chemical pneumonitis, hypersensitivity, and chronic granulomatous pulmonary disease (berylliosis). A variety of beryllium compounds and some of its alloys have induced malignant tumors of the lung in rats and monkeys and osteogenic sarcoma in rabbits. Human epidemiologic studies are strongly suggestive of a carcinogenic effect in humans (Ref 49).

Beryllium in the environment largely results from coal combustion. Illinois and Appalachian coal contains an average of about 2.5 ppm; oil contains about 0.08 ppm. The combustion of coal and oil contributes about 1250 tons or more of beryllium to the environment each year (mostly from coal), which is about five times the annual production for industrial use. The major industrial processes that release beryllium into the environment are beryllium extraction plants, ceramic plants, and beryllium alloy manufacturers. These industries also provide the greatest potential for occupational exposure. A review published in 1959 states that inhalable beryllium in ore treatment rooms around baking furnaces or at the sites of fluorescent phosphor blending, milling, and salvaging must have been around 1 mg/m^3. The major current use is as an alloy (see the article "Beryllium-Copper and Other Beryllium-Containing Alloys" in this Volume), but about 20% of world production is for applications utilizing the free metal in nuclear reactions, x-ray windows, and other special applications related to space optics, missile fuel, and space vehicles (see the article "Beryllium" in this Volume).

Knowledge of the disposition of beryllium has largely been obtained from experimental animals, particularly the rat. Clearance of inhaled beryllium is multiphasic; half is cleared in about two weeks; the remainder is removed slowly, and a residuum becomes fixed in the tissues probably within fibrotic granulomata.

Absorption of ingested beryllium probably only occurs in the acidic milieu of the stomach, where it is in the ionized form, but passes through the intestinal tract as precipitated phosphate (Ref 50). Transport in plasma is in the form of a colloidal phosphate probably bound to an α-globulin. Removal of radiolabeled beryllium chloride from rat blood is rapid, having a half-life of about three hours. It is distributed to all tissues, but most goes to the skeleton. High doses go predominantly to liver, but it is gradually transferred to bone. A variable fraction of the administered dose is excreted in urine, probably by way of transtubular secretion, rather than glomerular filtration.

Skin Effects. Contact dermatitis is the commonest beryllium-related toxic effect. Exposure to soluble beryllium compounds may result in papulovesicular lesions on the skin. It is a delayed-type allergic reaction. If contact is made with an insoluble beryllium compound, a chronic granulomatous lesion develops, which may be necrotizing or ulcerative. If insoluble beryllium-containing material becomes embedded under the skin, the lesion will not heal and may progress in severity. Use of a beryllium patch test to identify beryllium-sensitive individuals may in itself be sensitizing, and use of this procedure as a diagnostic test is discouraged.

Beryllium combines with proteins in the skin to act as the antigen in the hypersensitivity reaction. The hypersensitivity is cell mediated, and passive transfer with lymphoid cells has been accomplished in guinea pigs.

Pulmonary Effects. Two types of pulmonary disease will be described in this section. These include acute chemical pneumonitis and chronic granulomatous pulmonary disease.

Acute pulmonary disease (Chemical Pneumonitis) from inhalation of beryllium is a fulminating inflammatory reaction of the entire respiratory tract, involving the nasal passages, pharynx, tracheobroncheal airways, and the alveoli, and in the most severe cases produces an acute fulminating pneumonitis. It occurs almost immediately following inhalation of aerosols of soluble beryllium compounds, particularly fluoride—an intermediate in the ore extraction process. Severity is dose related. Fatalities have occurred, although recovery is generally complete after a period of several weeks or even months.

Chronic Granulomatous Pulmonary Disease (Berylliosis). This syndrome was first described in 1946 by Hardy and Tabershaw (Ref 51) among fluorescent lamp workers exposed to insoluble beryllium compounds, particularly beryllium oxide. The major symptom is shortness of breath, but in severe cases may be accompanied by cyanosis and clubbing of fingers (hypertrophic osteoarthropathy—a characteristic manifestation of chronic pulmonary disease). Chest x-rays show miliary mottling. Histologically, the alveoli contain small interstitial granulomata, which resemble those seen in sarcoidosis. In the early stages, the lesions are composed of fluid, lymphocytes, and plasma cells. Multinucleated giant cells are common. Later, the granulomas become organized with proliferation of fibrosis tissue, eventually forming small, fibrous nodules. As the lesions progress, interstitial fibrosis increases with loss of functioning alveoli and effective air/capillary gas exchange and increasing respiratory dysfunction.

Beryllium is one metal in which evidence for carcinogenicity was observed in experimental studies, beginning in 1946, before the establishment of carcinogenicity in humans (Ref 49). Epidemiologic confirmation in humans has been evolving, so that there is increasing acceptance that beryllium is, in fact, a human carcinogen. Studies of humans with occupational exposure to beryllium prior to 1970 were negative. However, three recent reports of worker populations studied earlier show a small excess of lung cancer, but the total number of cases is small. It was the conclusion of a work group report in 1981 that beryllium is indeed "the cause of the excess mortality" in persons with excess occupational/environmental exposure to beryllium (Ref 52).

In vitro studies of genotoxicity have shown that beryllium will induce morphologic transformation in mammalian cells (Ref 53). Beryllium will also decrease fidelity of DNA synthesis, but is negative when tested as a mutagen in bacterial systems (Ref 54).

Cadmium

Cadmium is a modern toxic metal. It was only discovered as an element in 1817, and industrial use was minor until about 50 years ago. But now it is a very important metal with many applications. The main use is electroplating or galvanizing because of its noncorrosive properties. It is also used as a color pigment for paints and plastics, and cathode material for nickel-cadmium batteries. Cadmium is a by-product of zinc and lead mining and smelting, which are important sources of environmental pollution. The toxicology of cadmium is reviewed in detail in Ref 55. Mechanical and physical properties of cadmium are reviewed in the article "Properties of Pure Metals" in this Volume.

Air concentrations as high as 4 to 5 mg/m^3 have been detected in certain workplace environments such as battery factories (Ref 56), but airborne cadmium in the present-day workplace environment is generally less than 0.02 μg/m^3. Typical concentrations in ambient air in rural areas are 0.001 to 0.005 μg/m^3 and up to 0.050 or 0.060 μg/m^3 in urban areas (Ref 57).

Meat, fish, and fruit contain 1 to 50 μg/kg, grains contain 10 to 150 μg/kg, and the greatest concentrations are in liver and kidney of animals. Shellfish, such as mussels, scallops, and oysters, may be a major source of dietary cadmium and contain 100 to 1000 μg/kg. Shellfish accumulate cadmium from the water and then bind to cadmium-binding peptides (Ref 58). Total daily intake from food in North America and Europe varies considerably but is generally less than 100 μg/day, whereas in heavily polluted areas as in parts of Japan, cadmium intake from food and water may be up to 150 μg/day (Ref 59).

Rice grown in soil contaminated with cadmium and other grains contributes to dietary content. Cadmium is more readily taken up by plants than other metals such as lead (Ref 60). Factors contributing to soil content of cadmium are fallout from air, cadmium content of water irrigating fields, and cadmium added with fertilizers. Commercial phosphate fertilizers usually contain less than 20 mg/kg, but Anderson and Hahlin (Ref 61) found an annual increase in soil and barley grain from continued use of phosphate fertilizer over a 15-year period. Another concern is use of commercial sludge to fertilize agricultural fields (Ref 62). Commercial sludge may contain up to 1500 mg of cadmium per kilogram of dry material.

Respiratory absorption of cadmium is about 15 to 30%. Workplace exposure to cadmium is particularly hazardous where there are cadmium fumes or airborne cadmium. Most airborne cadmium is respirable (Ref 63). A major nonoccupational source of respirable cadmium is cigarettes. One cigarette contains 1 to 2 μg cadmium, and 10% of the cadmium in a cigarette is inhaled (0.1 to 0.2 μg) (Ref 64). Smoking one pack or more cigarettes a day may double the body burden of cadmium.

Disposition. Gastrointestinal absorption is less than respiratory absorption and is about 5 to 8%. It is enhanced by dietary deficiencies of calcium and iron, and diets low in protein. Low dietary calcium stimulates synthesis of calcium-binding protein, which enhances cadmium absorption. Women with low serum ferritin levels have been shown to have twice the normal absorption of cadmium (Ref 65). Zinc decreases cadmium absorption probably by stimulating production of metallothionein.

Cadmium is transported in blood bound to red blood cells and large-molecular-weight proteins in plasma, particularly albumin. A small fraction of blood cadmium may be transported by metallothionein. Blood cadmium levels in adults without excessive exposure is usually less than 1 μg/dL. Newborns have low body content of cadmium usually less than 1 mg total body burden. The placenta synthesizes metallothionein and may serve as a barrier to maternal cadmium, but the fetus may be exposed with increased maternal exposure (Ref 66). Cow's milk and human milk are low in cadmium content, less than 1 μg/kg of milk (Ref 67). About 50 to 75% of the body burden of cadmium is in liver and kidneys; half-life in the body is not exactly known but is many years and may be as long as 30 years. With continued retention, there is progressive accumulation in soft tissues, particularly kidney, through ages 50 to 60 years when it begins to decline slowly. Because of the potential for accumulation in kidney, there is considerable concern for levels of dietary intake of cadmium by persons in the general population. Studies from Sweden have shown a slow but steady increase in cadmium content of vegetables over the years (Ref 68). Increase in body burden has been determined from an historic autopsy study (Ref 69).

Toxicity. Acute toxicity may result from ingestion of relatively high concentrations of cadmium, as may occur in contaminated beverages or food. Nordberg (Ref 70) relates an instance in which nausea, vomiting, and abdominal pain occurred from consumption of drinks containing approximately 16 mg/L of cadmium. Recovery was rapid without apparent long-term effects. Inhalation of cadmium fumes or other heated cadmium-containing materials may produce an acute chemical pneumonitis and pulmonary edema.

The principal long-term effects of low-level exposure to cadmium are chronic obstructive pulmonary disease and emphysema and chronic renal tubular disease. There may also be effects on the cardiovascular and skeletal systems (Ref 71, 72).

Chronic Pulmonary Disease. Toxicity to the respiratory system is proportional to the time and level of exposure. Obstructive lung disease results from chronic bronchitis, progressive fibrosis of the lower airways, and accompanying alveolar damage leading to emphysema. The lung disease is manifested by dyspnea, reduced vital capacity, and increased residual volume. The pathogenesis of the lung lesion is turnover and necrosis of alveolar macrophages. Released enzymes produce irreversible damage to alveolar basement membranes including rupture of septa and interstitial fibrosis. It has been found that cadmium reduces α-1-antitrypsin activity, perhaps enhancing pulmonary toxicity (Ref 73). However, no difference in plasma α-1-antitrypsin activity could be found between cadmium-exposed workers with and without emphysema (Ref 74).

Kidney. The effects of cadmium on proximal renal tubular function are manifested by increased cadmium in the urine, proteinuria, aminoaciduria, glucosuria, and decreased renal tubular reabsorption of phosphate. Morphologic changes are nonspecific and consist of tubular cell degeneration in the initial stages, progressing to an interstitial inflammatory reaction and fibrosis. The nephropathy occurs when cadmium concentration reaches a level in the kidney (200 μg/g) that has been widely referred to as the critical concentration of cadmium.

The proteinuria is principally tubular, consisting of low-molecular-weight proteins whose tubular reabsorption has been impaired by cadmium injury to proximal tubular lining cells. The predominant protein is a β_2 microglobulin, but a number of other low-molecular-weight proteins have been identified in the urine of workers with excessive cadmium exposure, such as retinol-binding protein, lysozyme, ribonuclease, and immunoglobulin light chains (Ref 74). High-molecular-weight proteins in the urine, such as albumin and transferin, indicate that some workers may actually have a mixed proteinuria and suggesting a glomerular effect as well. The nature of the glomerular lesion in cadmium nephropathy has not been studied extensively, but circulating antiglomerular basement membrane antibodies have been identified in humans and rats chronically exposed to cadmium, suggesting the presence of immunologically induced glomerular disease in addition to the tubulonephropathy (Ref 75).

Aminoaciduria in cadmium toxicity is generalized, reflecting increased excretion of amino acids normally reabsorbed by proximal tubular lining cells. The severity of the aminoaciduria is increased in cadmium workers with increasing levels of cadmium exposure. In addition, particularly large increases in proline and hydroxyproline excretion have been noted in patients with chronic cadmium toxicity with bone disease or Itai-Itai disease, but this probably reflects the changes in bone metabolism found in these people. Glucosuria and decreased tubular reabsorption of phosphate parallel the occurrence of low-molecular-weight proteinuria and aminoaciduria, reflecting the proximal tubular cell effect. Proximal tubular dysfunction may be symptom-free for a number of years, but tubular dysfunction may progress resulting in hypercalcuria, renal calculi, and rarely osteomalacia and evidence of distal tubular dysfunction (Ref 76).

Although most of the data available to date related to cadmium exposure and cadmium nephropathy have been obtained from workers with occupational exposure, there is some evidence now that persons in the general population with nonoccupational exposure to cadmium may also have cadmium-related renal tubular dysfunction. Among inhabitants of cadmium-polluted areas of Japan where dietary content of cadmium is increased, the prevalence of proteinuria and glucosuria is higher than in control areas, and there is some association between increased excretion of low-molecular-weight proteins in urine and level of cadmium pollution (Ref 77). Also, in 1980 Lauwerys *et al.* (Ref 78) studied a group of Belgian women and found that a group of women living near a nonferrous metal smelter had a higher body burden as reflected by an increased excretion of cadmium in urine and a higher prevalence of signs of renal dysfunction than women from a control area.

Critical Concentration of Cadmium. With this awareness that cadmium-induced nephropathy may occur in persons in the general population, it becomes of major public health importance to know what is the maximum level of cadmium that a person can be exposed to without risk of renal tubular dysfunction and cadmium nephropathy. Also, the concept of a critical concentration of cadmium has very important implications with regard to establishing maximum levels of cadmium that human populations may be exposed to with some margin of safety.

Kjellstrom *et al.* (Ref 79), in 1977 established a metabolic model relating daily intake of cadmium and concentration of cadmium in renal cortex. The geometric average intake of cadmium was 14 μg cadmium per day, corresponding to a concentration of cadmium in the renal cortex at about age 50 of around 10 μg/g. The WHO Task Force estimated that daily ingestion of 200 to 300 μg cadmium per day would be required to reach the critical kidney cortex concentration of 200 μg cadmium per gram at age 50 for a 70 kg (150 lb) man. Rats given daily injections of cadmium also develop a nephropathy when renal cadmium concentration reaches about 200 μg/g kidney weight (Ref 80).

Role of Metallothionein in Cadmium Toxicity. Accumulation of cadmium in the kidney without apparent toxic effect is possible because of formation of cadmium-thionein or metallothionein, a metal protein complex with a low molecular weight (about 6500 atomic mass units, or Daltons) (Ref 81).

The amino acid composition of metallothionein is characterized by approximately 30% cysteine and the absence of aromatic amino acids. Specific optical absorption is due to location of metal thiolate complexes in the protein. Metallothionein contains 61 amino acids and 20 are cysteine. Structural studies using nuclear magnetic resonance spectroscopy and electron spin resonance spectroscopy have identified two distinct metal clusters in mammalian metallothionein. The clusters seem to have significant differences in their affinity for different metal ions; one of the clusters has a high level of specificity for zinc. Metal binding is by trimercaptide bridges (Ref 82). Metallothionein is primarily a tissue protein and is ubiquitous in most organs but is in highest concentration in liver, particularly following recent exposure, and in kidney where it accumulates with age in proportion to cadmium concentration.

A number of studies from experimental animals, as well as tissue culture models, confirm the protective role of metallothionein. It has been found that synthesis of metallothionein in tissues is directly related to exposure to metal, and toxicity to kidney probably only occurs when exposure exceeds the ability of that organ to either synthesize metallothionein or store additional cadmium. Toxic cell injury is thought to be caused by unbound cadmium or free cadmium ion. Administration of metallothionein prepared with different ratios of cadmium and zinc to rats has demonstrated that renal tubular necrosis is related to the cadmium content, not the amount of metallothionein (Ref 81, 83).

Pretreatment of experimental animals with small doses of cadmium has been shown to prevent acute toxic effects of a large dose of cadmium. This property is not restricted to protection from cadmium toxicity alone. Pretreatment of experimental animals with small doses of cadmium or mercury salts can prevent the nephrotoxic effects of high doses of mercury chloride. There is also some experimental evidence that suggests that the teratogenic effects of cadmium in Golden hamsters is prevented by pretreatment with zinc salts of small amounts of metallothionein. Other studies have shown that certain sulfhydryl-requiring enzymes are inhibited *in vitro* by small amounts of cadmium but not affected *in vivo* where intracellular cadmium is bound to metallothionein. Human cells in tissue culture, in which metallothionein has been induced by pretreatment with cadmium, become resistant to previously lethal exposure to cadmium (Ref 84).

Experimental studies have shown that cadmium administered parenterally as inorganic and cadmium bound to metallothionein has a different distribution in organs. Inorganic cadmium is largely recovered in liver whereas cadmium from cadmium metallothionein is preferentially taken up by kidney (Ref 85).

Skeletal System. Cadmium toxicity affects calcium metabolism, and individuals with severe cadmium nephropathy may have renal calculi and excess excretion of calcium, probably related to increased urinary loss, but with chronic exposure, urine calcium may be less than normal. Associated skeletal changes are probably related to calcium loss and include bone pain, osteomalacia, and/or osteoporosis. Bone changes are part of a syndrome recognized in postmenopausal multiparous women living in the Fuchu area of Japan prior to and during World War II. The syndrome consisted of severe bony deformities and chronic renal disease.

Excess cadmium exposure has been implicated in the pathogenesis of the syndrome, but vitamin D and perhaps other nutritional deficiencies are thought to be cofactors. "Itai-Itai" translates to "ouch-ouch," reflecting the accompanying bone pain (Ref 71).

Hypertension and Cardiovascular Disease. Schroeder and Balassa (Ref 67) first reported in 1961 that the chronic feeding of low levels (5 ppm) of cadmium in drinking water to rats could induce hypertension. These experimental results have been confirmed (Ref 86) and a number of mechanisms for the pathogenesis of the hypertension suggested, including increased sodium retention, direct vasoconstriction, hyperreninemia, and increased cardiac output. Recent studies from Japan found a twice-as-high cerebrovascular disease mortality rate among people who had cadmium-induced renal tubular proteinuria as among people in cadmium-polluted areas without proteinuria (Ref 87).

Carcinogenicity. An increase in carcinoma of the prostate was first noted in a mortality study of battery workers in England in 1965, but this was not found in a study of a large worker population (Ref 88). The problem of prostatic cancer is further complicated by the high incidence of latent (*in situ*) carcinoma of the prostate in elderly men in the general population and the implication of numerous other factors, such as marital status (singles), race (nonwhites), and even religion. There have been numerous experimental studies supporting the potential carcinogenicity of cadmium. Metallic cadmium or cadmium sulfide or sulfate given subcutaneously or intramuscularly will induce sarcomata at the site of injection in experimental animals. The tumors are truly malignant and have been found to metastasize to lymph nodes and lungs. Also, it was found many years ago that injection of several milligrams of cadmium per kilogram body weight to mice causes acute testicular necrosis followed by Leydig cell tumors. The pathogenesis of the Leydig cell tumors appears to be hormone dependent and is preceded by decreases in serum testosterone levels and stimulation of Leydig cell hyperplasia and tumors. Testicular necrosis as well as Leydig cell tumor formation is prevented by supplemental zinc (Ref 89). Carcinoma of lungs has recently been produced by exposing rats to cadmium aerosols (Ref 90).

Biologic Indicators. The most important measure of excessive cadmium exposure is increased cadmium excretion in urine. In persons in the general population, without excessive cadmium exposure, urine cadmium excretion is both small and constant. That is, it is usually of the order of only 1 or 2 μg/day, or less than 1 μg/g creatinine. With excessive exposure to cadmium as might occur in workers, increase in urine cadmium may not occur until all of the available cadmium binding sites are saturated. However, when binding sites (metallothionein) are saturated, increased urine cadmium reflects recent exposure and body burden and renal cadmium concentration so that urine cadmium measurement does provide a good index of excessive cadmium exposure. Nogawa *et al.* (Ref 87), determined the urinary concentration of cadmium corresponding to a 1% prevalence rate of a number of abnormal urinary findings (Table 4). Tubular proteinuria, as indicated by measurable excretion of β_2-microglobulin, occurred at the 1% prevalence rate with a urinary cadmium concentration of 3.2 μg/g of creatinine. This was at a slightly lower urine cadmium level than other signs of renal tubular dysfunction. Retinol binding protein may be a more practical and reliable test of proximal tubular function than β_2 microglobulin because sensitive immunologic analytic methods are now available, and it is more stable in urine (Ref 75). Changes in urinary excretion of low-molecular-weight proteins are mainly observed in workers excreting more than 10 μg cadmium per gram creatinine (Ref 91).

Table 4 Urinary cadmium concentration corresponding to 1% prevalence rate for parameters of renal dysfunction

Urinary finding	Urinary cadmium per μg/g creatinine Male	Female
Tubular proteinuria		
β_2-microglobulin	3.2	5.2
Retinal binding protein	4.4	7.4
Aminoaciduria (proline)	10.4	5.1
Proteinuria with glucosuria	7.4	7.4

Source: Ref 87

Most of the cadmium in urine is bound to metallothionein, and there is good correlation between metallothionein and cadmium in urine in cadmium workers with normal or abnormal renal function (Ref 92). Therefore, measurement of metallothionein in urine provides the same toxicologic information as measurement of cadmium and, in addition, does not have the problem of external contamination. Radioimmunoassay techniques for measurement of metallothionein are evolving rapidly (Ref 93, 94).

Recently, *in vivo* neutron activation analysis has been used to measure cadmium in liver and kidney in exposed workers. The detection limits are at least 15 mg/kg in kidney cortex and 1.5 mg/kg in liver, so that the method is not sufficiently sensitive to measure *in vivo* tissue levels in persons in the general population (Ref 95). Applying this technique to cadmium-exposed workers, Roels, Lauwerys, and co-workers found a wide range of variability and overlap in kidney cadmium concentration associated with and without renal disease. On the basis of their study of 309 workers, the critical concentration of cadmium in renal cortex may range from 215 to 390 ppm (Ref 96).

Treatment. Susceptibility to cadmium-induced toxicity is influenced by a number of factors, particularly ability of the body to provide binding sites on metallothionein. Protection is provided by dietary zinc, cobalt, or selenium. Treatment of the toxicity of cadmium on the kidney is to cease exposure to cadmium (Ref 11). What severity of cadmium-induced tubular dysfunction is reversible is still not certain.

Chelation therapy is not available for cadmium toxicity in humans. Experimental studies have shown that the action of chelating agents on the pharmacokinetics of cadmium depends on the time of administration of the chelators after cadmium exposure. When the chelators are given shortly after cadmium exposure, when no new metallothionein has been synthesized, the thiol-containing chelators such as BAL and penicillamine increase the biliary excretion of cadmium while EDTA, DTPA, and related chelators increase urinary excretion (Ref 97). For chronic cadmium exposure, when cadmium is bound to metallothionein, there is little effect from chelation therapy.

Chromium

Chromium is a generally abundant element in the earth's crust and occurs in oxidation states ranging from Cr^{2+} to Cr^{6+}, but only the trivalent and hexavalent forms are of biologic significance. The trivalent is the more common form. However, hexavalent forms of chromate compounds are of greater industrial importance. Sodium chromate and dichromate are the principal substances for the production of all chromium chemicals. Sodium dichromate is produced industrially by the reaction of sulfuric acid on sodium chromate. The major source of chromium is from chromite ore. Metallurgical-grade chromite is usually converted into one of several types of ferrochromium or processed for use in cobalt-base and nickel-base superalloys. Ferrochrome is used for the production of stainless steel (superalloys and stainless steels are described in Volume 1 of the 10th Edition of *Metals Handbook*). Chromates are produced by a smelting, roasting, and extraction process. The major uses of sodium dichromate are for the production of chrome pigments, for the production of chrome salts used for tanning leather, mordant dying, wood preservatives, and as an anticorrosive in cooking systems, boilers, and oil drilling muds (Ref 98, 99).

Chromium in ambient air originates from industrial sources, particularly ferrochrome production, ore refining, chemical and refractory processing, and combustion of fossil fuels. In rural areas, chromium in air is usually less than 0.1 $\mu g/m^3$ and from 0.01 to 0.03 $\mu g/m^3$ in industrial cities. Particulates from coal-fired power plants may contain from 2.3 to 31 ppm, but this is reduced to 0.19 to 6.6 ppm by fly-ash collection. Cement-producing plants are another important potential source of atmospheric chromium. Chromium precipitates and fallout are deposited on land and water; land fallout is eventually carried to water by runoff, where it is deposited in sediments. A controllable source of chromium is waste water from chrome-plating and metal-finishing industries, textile plants, and tanneries. Chromium in food is low, and estimates of daily intake by humans is under 100 μg, mostly from food, with trivial quantities from most water supplies and ambient air.

Disposition. Trivalent chromium is the most common form found in nature, and chromium in biologic materials is probably always trivalent. There is no evidence that trivalent chromium is converted to hexavalent forms in biologic systems. However, hexavalent chromium readily crosses cell membranes and is reduced intracellularly to trivalent chromium.

The known harmful effects of chromium in humans have been attributed to the hexavalent form, and it has been speculated that the biologic effects of hexavalent chromium may be related to the reduction to trivalent chromium and the formation of complexes with intracellular macromolecules. High concentrations of chromium are normally found in ribonucleic acid (RNA), but its role is unknown. Trace quantities of trivalent chromium are essential for carbohydrate metabolism in mammals. It is a cofactor for insulin action and has a role in the peripheral activities of this hormone by forming a ternary complex with insulin receptors, facilitating the attachment of insulin to these sites. The most biologically active form of insulin appears to be a naturally occurring complex containing niacin as well as glycine, glutamic acid, and cysteine (Ref 59).

Human chromium deficiency may be occurring in infants suffering from protein-caloric malnutrition and elderly people with impaired glucose tolerance, but this is not well documented. Prolonged use of a synthetic diet without chromium supplementation may lead to chromium deficiency, impaired glucose metabolism, and possibly effects on growth and on lipid and protein metabolism. Half-time for elimination of chromium from rats is 0.5, 5.9, and 83.4 days, according to a three-compartment model (Ref 100).

Human kinetic studies have identified an erythrocyte chromium compartment that corresponds to the survival time of the red blood cell and is almost exclusively excreted in urine.

Toxicology. Systemic toxicity to chromium compounds occurs largely from accidental exposures, occasional attempts to use chromium as a suicidal agent, and previous therapeutic uses. The major acute effect from ingested chromium is acute renal tubular necrosis (Ref 101).

Exposure to chromium, particularly in the chrome production and chrome pigment industries, is associated with cancer of the respiratory tract (Ref 102). As early as 1936, German health authorities recognized cancer of the lung among workers exposed to chromium dust. In a review paper from 1950, Baetjer described 109 cases of cancer in the chromate-producing industry, 11 cases in the chrome pigment industry, and 2 cases in other industries. In a 1966 review of the histologic classification of 123 cases of lung cancer in chromate workers, Hueper (Ref 103) found 46 squamous cell carcinomas, 66 anaplastic tumors, and 11 adenocarcinomas. The greatest risk to cancer is attributed to exposure to acid-soluble, water-insoluble hexavalent chromium as occurs in the roasting or refining processes. Other studies have supported the greater risk to cancer from exposure to slightly soluble, hexavalent compounds rather than trivalent chromium compounds. Hexavalent chromium is corrosive and causes chronic ulceration and perforation of the nasal septum. It also causes chronic ulceration of other skin surfaces, which is independent of hypersensitivity reactions on skin. Allergic chromium skin reactions readily occur with exposure and are independent of dose. Trivalent chromium compounds are considerably less toxic than the hexavalent compounds and are neither irritating nor corrosive. Nevertheless, nearly all workers in industries are exposed to both forms of chromium compounds, and at present, there is no information as to whether there is a gradient of risk from predominant exposure to hexavalent or insoluble forms of chromium to exposure to soluble trivalent forms. In a 1981 review, Norseth (Ref 102) suggests that if there are similar increased risks in both groups, as estimated from the death rates, trivalent chromium should be considered as an equally potent carcinogen as are the hexavalent compounds.

Whether chromium compounds cause cancer at sites other than the respiratory tract is not clear. A slight increase in cancer of the gastrointestinal tract has been reported in other studies, but each involved only small groups of workers.

Animal studies support the notion that the most potent carcinogenic chromium compounds are the slightly soluble hexavalent compounds. Studies on *in vitro* bacterial systems, however, show no difference between soluble and slightly soluble compounds. Trivalent chromium salts have little or no mutagenic activity in bacterial systems. Since there is preferred uptake of the hexavalent form by cells and it is the trivalent form that is metabolically active and binds with nucleic acids within the cell, it has been suggested that the causative agent in chromium mutagenesis is trivalent chromium bound to genetic material after reduction of the hexavalent form (Ref 102).

Human Body Burden. Tissue concentrations of chromium in the general population have considerable geographic variation, as high as 7 μg/kg in lungs of persons in New York or Chicago with lower concentrations in liver and kidney (Ref 104). In persons without excess exposure, blood chromium concentration is between 20 and 30 μg/L and is evenly distributed between erythrocytes and plasma. With occupational exposure, increase in blood chromium is related to increase in chromium in red blood cells. Urinary excretion is generally less than 10 μg/day in the absence of excess exposure (Ref 59).

Lead

Lead, the most ubiquitous toxic metal, is detectable in practically all phases of the inert environment and in all biologic systems. Because it is toxic to most living things at high exposures and there is no demonstrated biologic need, the major issue regarding lead is at what dose does it become toxic. Specific concerns vary with the age and circumstances of the host, and the major risk is toxicity to the nervous system. Several reviews and multiauthored books on the toxicology of lead are available (Ref 105-112). Applications and properties of lead are described in the articles "Lead and Lead Alloys" and "Properties of Pure Metals" in this Volume.

Sources. The principal route of exposure is food, but it is usually environmental and presumably controllable sources that produce excess exposure and toxic effects. These sources include lead-base indoor paint in old dwellings, lead in air from combustion of lead-containing auto exhausts or industrial emissions, lead-base paint, hand-to-mouth activities of young children living in polluted environments, and, less commonly, lead dust brought home by industrial workers on their clothes and shoes, and lead-glaze earthenware.

The total daily intake of lead for an adult in the United States varies from less than 0.1 mg/day to more than 2 mg/day (Ref 105, 113). The major source of daily intake of lead in adults and children (without excess exposure) is food and beverages. Lead content of food is extremely variable, but there are practically no lead-free food items. The average adult diet contains from 150 μg/day, 0.75 to 120 μg/day for infants and small children. Most municipal water supplies measured at the tap contain less than the WHO-recommended limit of 0.05 μg/ml, so that daily intake from water is usually about 10 μg, and unlikely to be more than 20 μg.

Air is a third source of lead exposure for persons in the general population. Concentrations of lead in air vary widely and may be lower than 1.0 μg/m^3 in rural areas to 10 μg/m^3 in certain urban environments. For the contemporary urbanite, the magnitude of respired lead is about one-half the intake from the diet.

Disposition. The gastrointestinal absorption of lead is influenced by a large number of factors of which age and nutritional factors are of particular importance. Adults absorb 5 to 15% of ingested lead and usually retain less than 5% of what is absorbed. Children are known to have a greater absorption of lead than adults; one study found an average net absorption of 41.5% and 31.8% net retention in infants on regular diets.

Lead in the atmosphere exists either in solid forms, dust or particulates of lead dioxide, or in the form of vapors, particularly alkyl lead that has escaped by evaporation from automobile fuel systems.

Lead absorption by the lungs also depends on a number of factors in addition to concentration. These include volume of air

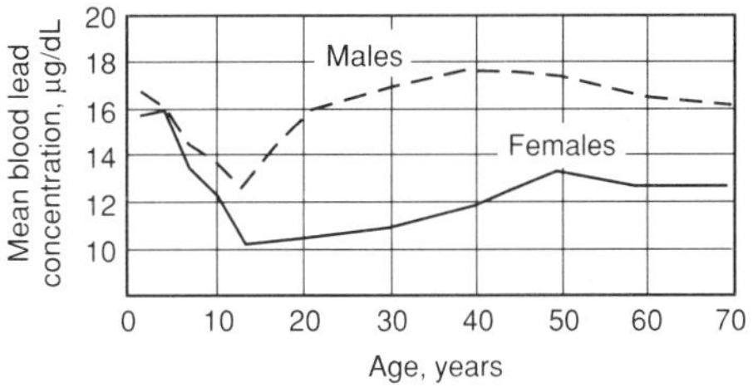

Fig. 3 National estimates of blood lead levels in the United States. Source: Ref 115

respired per day, whether the lead is in particle or vapor form, and size distribution of lead-containing particles. Only a very minor fraction of particles over 0.5 μm in mean maximal external diameter are retained in the lung but are cleared from the respiratory tract and swallowed. However, the percentage of particles less than 0.5 μm retained in the lung increases with reduction in particle size. About 90% of lead particles in ambient air that are deposited in the lungs are small enough to be retained. Absorption of retained lead through alveoli is relatively efficient and complete.

More than 90% of lead in blood is in the red blood cells. There seem to be at least two major compartments for lead in the red blood cell, one associated with the membrane and the other with hemoglobin (Ref 114). Small fractions may be related to other red blood cell components. Plasma ligands are not well defined, but it has been suggested that plasma and serum may contain diffusible fractions of lead in equilibrium with soft tissue or end-organ binding sites for lead. This fraction is difficult to measure accurately, but there is an equilibrium between red cell and plasma lead.

Blood lead levels are a good indicator of recent exposure to lead and are influenced by inhalation and ingestion. A number of recent studies suggest that inhalation of air containing 1 μg/m^3 in respirable particles will increase blood lead concentrations by about 1 μg/dL when air lead concentrations are in the range of 1 to 5 μg/m^3 (Ref 107).

Lead in blood varies with age (Fig. 3) (Ref 115). Children under seven years of age have significantly higher blood lead levels than older children, and there is no difference between boys and girls under age 12. Blood lead levels decline during adolescence probably related to bone growth and deposition of lead in bones with calcium. Blood lead levels are lower in adult females than adult males.

The total body burden of lead may be divided into at least two kinetic pools, which have different rates of turnover. The largest and kinetically slowest pool is the skeleton with a half-life of more than 20 years and a much more labile soft tissue pool. The total lifetime accumulation of lead may be about 200 mg and over 500 mg for an occupationally exposed worker. Kidney lead accumulates with age; lead in lung does not change. Lead in the central nervous system tends to concentrate in gray matter and certain nuclei. The highest concentrations are in the hippocampus, followed by cerebellum, cerebral cortex, and medulla. Cortical white matter seems to contain the least amount, but these comments are based on only a few reported human and animal studies.

Renal excretion of lead is usually with glomerular filtrate with some renal tubular resorption. With elevated blood lead levels, excretion may be augmented by transtubular transport.

Placental transfer of lead occurs. Cord blood generally correlates with maternal blood lead levels but is slightly lower. It is interesting that maternal blood lead decreases during pregnancy, suggesting that maternal lead is transferred to the fetus or excreted in some way.

Toxicity. The topic of greatest interest at the present time concerns the maximal level of lead exposure in the neonatal and young child that does not produce a cognitive or motor neurologic deficit. For the adult excess occupational exposure or even accidental exposure, the concerns are peripheral neuropathy and/or chronic nephropathy. Effects on the heme system provide biochemical indicators of lead exposure in the absence of chemically detectable effects, but anemia due to lead exposure is uncommon without other detectable effects or other synergistic factors. Other target organs are the gastrointestinal and reproductive systems.

Nearly all environmental exposure to lead is to inorganic compounds, even lead in food. Organolead exposures, including tetraethyl lead, have unique toxicologic patterns and will be discussed later.

Neurologic Effects. The central nervous system (CNS) effects of lead are the most significant in terms of human health and performance (Ref 110, 112). Manifestations of CNS effects are encephalopathy and/or peripheral neuropathy. Symptomatic encephalopathy is almost always a disease of childhood and varies from ataxia to stupor, coma, and convulsions. This form of lead intoxication has decreased appreciably in North America over the past 20 years with better understanding of factors that contribute to lead toxicity, particularly reduction of lead exposure in children. Morphologic effects of lead on the brain are nonspecific. Lead encephalopathy is accompanied by severe cerebral edema, increase in cerebral spinal fluid pressure, proliferation and swelling of endothelial cells in capillaries and arterioles, proliferation of glial cells, neuronal degeneration, and areas of focal cortical necrosis in fatal cases.

The pathogenesis of neuronal damage in lead encephalopathy is not well understood. In severe cases, there are obvious changes

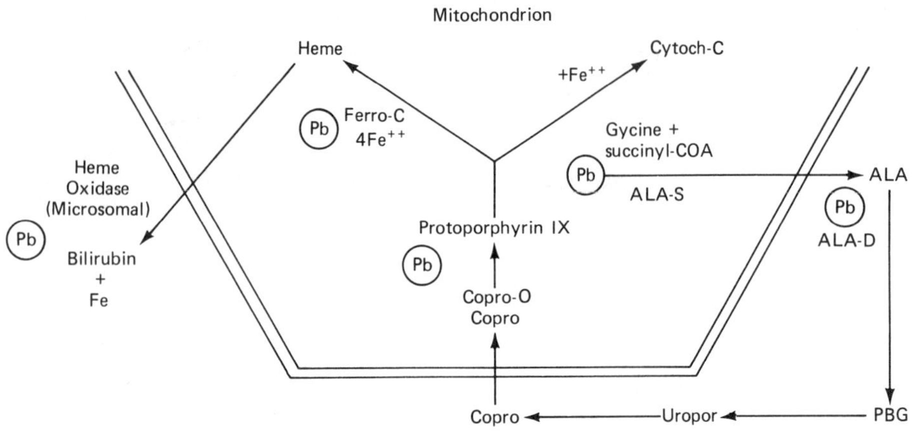

Fig. 4 Scheme of heme synthesis showing sites where lead has an effect. COA, coenzyme A; ALA-S, aminolevulinic acid synthetase; ALA, *d*-aminolevulinic acid; ALA-D, aminolevulinic acid dehydratase; PBG, porphobilinogen; Uropor, uroporphyrinogen; Copro, coproporphyrinogen; Copro-O, coproporphyrinogen oxidase; Ferro-C, Ferrochelatase; Cytoch-C, cytochrome c; Pb, site for lead effect

in hemodynamics (cerebral edema) and cellular hypoxia, but it is now apparent that there is a direct effect of lead on neuronal and possibly synaptic transmission at levels of lead exposure that do not produce apparent symptoms of intoxication. These effects are termed "low-level lead toxicity" because they are believed to be associated with blood lead levels of approximately 30 to 50 μg/dL or possibly even lower blood lead levels. These effects are also termed "subclinical lead toxicity" because they can only be detected by assessment of neuropsychologic behavior, such as hyperactivity, poor classroom behavior (decreased attention span), and even small decrements (point average in group studies) of four to five in I.Q. scores (Ref 116). Studies by others confirm the subclinical detrimental effects on I.Q., but possibly occurring at somewhat higher blood lead levels than found by Needleman (Ref 117). However, that low-level lead exposure (blood lead, 30 to 50 μg/dL) does, indeed, affect CNS function is further supported by the finding of changes in electroencephalogram (EEG) brain wave patterns and CNS evoked potential responses in children displaying neuropsychologic deficits (Ref 118). Neurochemical studies in experimental models have shown that lead in the absence of morphologic changes does produce deficits in neurotransmission through inhibition of cholinergic function, possibly by reduction of extracellular calcium. Other noted changes in neurotransmitter function include impairment of dopamine uptake by synaptosomes and impairment of the function of the inhibitory neurotransmitter γ-aminobutyric acid.

Peripheral neuropathy is a classic manifestation of lead toxicity, particularly the footdrop and wristdrop that characterized the house painter and other workers with excessive occupational exposure to lead more than a half-century ago (Ref 119). Segmental demyelination and possibly axonal degeneration follow lead-induced Schwann cell degeneration (Ref 120). Wallerian degeneration of posterior roots of sciatic and tibial nerves is possible, but sensory nerves are less sensitive to lead than motor nerve structure and function (Ref 121). Motor nerve dysfunction, assessed clinically by electrophysiologic measurement of nerve conduction velocities, has been shown to occur with blood lead levels in the 50 to 70 μg/dL range or lower (Ref 122).

Hematologic Effects. Lead has multiple hematologic effects. In lead-induced anemia, the red blood cells are microcytic and hypochromic, as in iron deficiency, and usually there are increased numbers of reticulocytes with basophilic stippling. This morphologic characteristic has long been recognized as a feature of lead-induced anemia and in the past (pre-World War II), it was employed as a method of monitoring workers in the lead industry (Ref 123).

The test is no longer useful because it is now known to be nonspecific and, most important, it is an uncommon occurrence with blood lead below 80 μg/dL, which is considerably above the present-day permissible industrial standard. Basophilic stippling results from inhibition of the enzyme pyrimidine-5-nucleotidase (Ref 124), which cleaves residual nucleotide chains remaining in erythrocytes after extrusion of the nucleus. The activity of this enzyme is decreased in persons with elevated blood lead levels even when stippling is not morphologically evident.

The anemia that occurs in lead poisoning results from two basic defects: shortened erythrocyte life span and impairment of heme synthesis. Shortened life span of the red blood cell is thought to be due to increased mechanical fragility of the cell membrane. The biochemical basis for this effect is not known but is accompanied by inhibition of sodium- and potassium-dependent ATPases (Ref 125).

A schematic presentation of effects of lead on heme synthesis is shown in Fig. 4. Probably the sensitive effect is inhibition of δ-aminolevulinic acid dehydratase (ALA-D), resulting in a negative exponential relationship between ALA-D and blood lead. There is also depression of coproporphyrinogen oxidase, resulting in increased coproporphyrin activity. Lead also decreases ferrochelatase activity. This enzyme catalyzes the incorporation of the ferrous ion into the porphyrin ring structure. Bessis and Jensen (Ref 126) have shown that iron in the form of apoferritin and ferruginous micelles may accumulate in mitochondria of bone marrow reticulocytes from lead-poisoned rats. Failure to insert iron into protoporphyrin results in depressed heme formation. The excess protoporphyrin takes the place of heme in the hemoglobin molecule and, as the red blood cells containing protoporphyrin circulate, zinc is chelated at the center of the molecule at the site usually occupied by iron. Red blood cells containing zinc-protoporphyrin are intensely fluorescent and may be used to diagnose lead toxicity. Depressed heme synthesis is thought to be the stimulus for increasing the rate of activity of the first step in the heme synthetic pathway, δ-aminolevulinic acid synthetase, by virtue of negative feedback control as proposed by Granick and Levere (Ref 127). As a consequence, the increased production of *d*-aminolevulinic acid and decreased activity of ALA-D result in a marked increase in circulating blood levels and urinary excretion of *d*-ALA. Prefeeding of lead to experimental animals also raises heme oxygenase activity, resulting in some increase in bilirubin formation. The change in rates of activity of these enzymes by lead produces a dose-related alteration in activity of affected enzymes, but anemia only occurs in very marked lead toxicity. The changes in enzyme activities, particularly ALA-D in peripheral blood and excretion of ALA in urine, correlate very closely with actual blood lead levels and serve as early biochemical indices of lead effect.

Renal Effects. Toxicologic effects of lead on the kidney divide into two major concerns: reversible renal tubular dysfunction that occurs mostly in children with acute exposure to lead, usually associated with overt central nervous system effects, and irreversible chronic interstitial nephropathy characterized by vascular sclerosis, tubular cell atrophy, interstitial fibrosis, and glomerular sclerosis (Ref 128). It is most often seen in workmen with years of exposure to lead. In the early stages of excess lead exposure, morphologic and functional changes in the kidney are confined to the renal tubules and are most pronounced in proximal tubular cells.

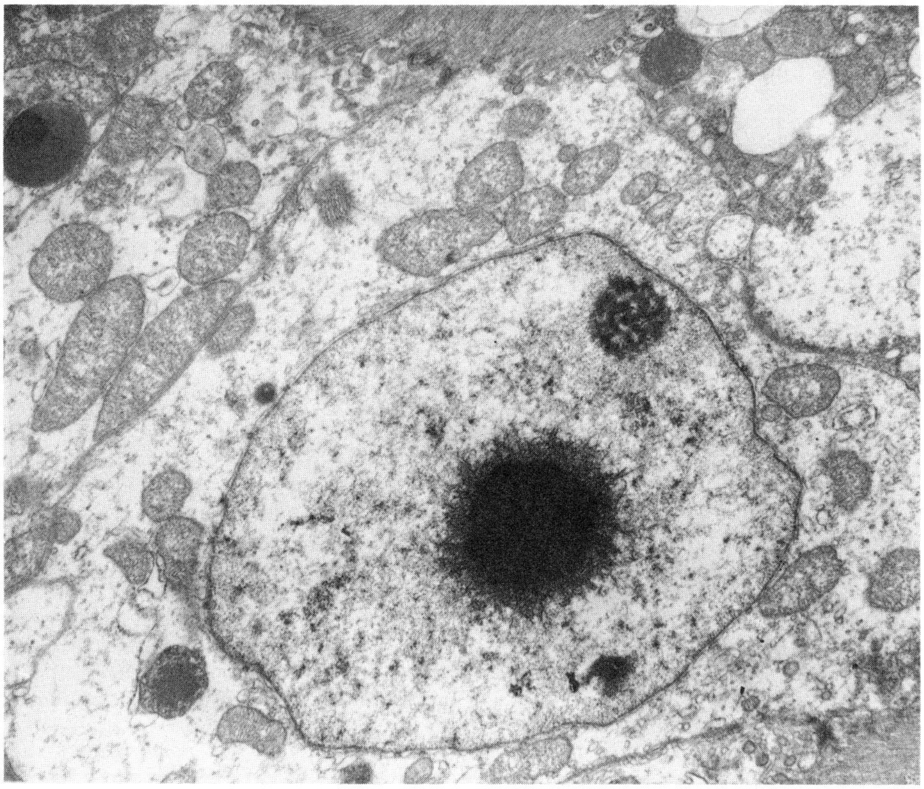

Fig. 5 Lead-induced inclusion bodies in nucleus of renal tubular lining cell

A pathognomonic feature of lead poisoning is the presence of characteristic nuclear inclusion bodies (Ref 129). By light microscopy the inclusions are dense, homogeneous eosinophilic bodies. They are acid-fast when stained with carbolfuchsin. Ultrastructurally the bodies have a dense central core and outer fibrillary region, as shown in Fig. 5. The bodies are composed of a lead-protein complex (Ref 130). The protein is acidic and contains large amounts of aspartic and glutamic acids and little cystine. It is suggested that lead binds loosely to the carboxyl groups of the acidic amino acids. Most of the lead in the tubular cell is bound to the inclusion body. The sequestering of lead in these complexes may protect more susceptible organelles like mitochondria and endoplasmic reticulum (Ref 129).

Experimental studies have shown that nuclear inclusion bodies are the earliest evidence of lead exposure and may be observed before any of the functional changes are detectable. Cells containing inclusion bodies are usually swollen and contain altered mitochondria. It has been shown that mitochondria isolated from kidneys of rats with lead toxicity have impaired oxidative and phosphorylative abilities. This may in part be responsible for the decrease in reabsorptive functions of proximal tubular cells. Experimental animals and people, that is, children and workmen with early exposure to lead, have a generalized amino aciduria, glycosuria, and hyperphosphaturia, and probably some impairment of sodium reabsorption. Some indirect evidence for an effect of lead on sodium transport is found in the clinical observation that the reninaldosterone response to sodium deprivation is altered in people with lead intoxication (Ref 131).

The pathogenesis of the inclusion bodies may be related to renal tubular cell transport and excretion of lead. Treatment of lead-exposed animals with chelating agents such as EDTA is accompanied by a sudden spike of urinary lead, which is maximum 12 to 24 hours after treatment (Ref 132). Also, no inclusion bodies can be found by morphologic study of renal tubular cells after EDTA therapy. The bodies may also be found intact in the urinary sediment of workmen with heavy exposure to lead (Ref 133). The bodies account for the major fraction of intracellular lead, and their loss in the urine may reflect a major pathway for lead excretion.

With continued exposure to lead there is a gradual change in morphology beginning with the appearance of peritubular and periglomerular fibrosis, particularly in the deep cortex of juxtamedullary zone (Ref 128). This is accompanied by atrophy of some tubules and hyperplasia of others. There are also fewer or no inclusion bodies present in the advanced stages of lead-induced nephrosclerosis. Recognition of interstitial fibrosis induced by lead, therefore, from any other forms of interstitial fibrosis is not possible morphologically, but must be made from history and knowledge of progression of the disease if this is available.

The most important feature of the changes associated with acute lead nephropathy is that they are reversible, either by reduction of lead exposure or by chelation therapy, but the progression into interstitial fibrosis is not reversible. It is very important, therefore, to diagnose this disorder as early as possible so that exposure to lead can be discontinued and decrease in function halted. At the present time there is no single definitive diagnostic test that will recognize lead-induced interstitial nephropathy except possibly renal biopsy, but this is not a practical measure. There have been several clinical studies in recent years of renal function in workmen with long-term occupational exposure to lead (Ref 134-136). If looked at together, the conclusion is reached that an early functional accompaniment of interstitial fibrosis is a reduction in glomerular filtration rate.

The relationship between chronic lead exposure and gouty nephropathy, suggested more than a hundred years ago by the English physician Garrod, has received recent support from studies showing that gout patients with renal disease have a greater chelate-provoked lead excretion than do renal patients without gout (Ref 137). Lead reduces uric acid excretion (Ref 138). Elevated blood uric acid has been demonstrated in rats with chronic lead nephropathy (Ref 128).

The relationship between lead and hypertension is uncertain; it has been associated with lead poisoning in a number of studies but not in others. The second National Health and Nutrition Examination Survey (1976-1980) found a statistically significant relationship between blood pressure and blood lead levels in white males age 40 to 59 years (Ref 139). Hypertension may follow the vascular changes associated with lead-induced chronic renal disease, or changes in renin-angiotensinaldosterone metabolism (Ref 131, 140).

Carcinogenesis. The possible carcinogenic effects of lead have been receiving increasing attention (Ref 13). It is clear that lead can induce cancer in kidneys of rodents fed high doses of lead (Ref 141). On the other hand, the evidence that lead is carcinogenic to humans is very limited. A study of workmen in England many years ago with occupational exposure to lead did not show an increased incidence of cancer (Ref 142). A more recent study of causes of mortality in 7000 lead workers in the United States showed a slight excess of deaths from cancer (Ref 143), but the statistical significance of these findings has been debated (Ref 144, 145). The most common tumors found were of the respiratory and digestive systems, not the kidney. However, case reports of renal adenocarcinoma in workmen with prolonged occupational exposure to lead have appeared (Ref 146, 147).

Other Effects. Severe lead toxicity has long been known to cause sterility, abortion, and neonatal mortality and morbidity. Studies have demonstrated gametotoxic effects in both male and female animals. However, the impact of levels of lead exposure occurring in today's society on reproductive effects is uncertain. The greatest concern is for intrauterine effects on the unborn fetus. Umbilical cord blood levels are the same as those of mother's blood, and because of the greater sensitivity of the fetus, pregnancy must be regarded as a period of increased susceptibility to lead.

A few clinical studies have found increased chromosomal defects in workers with blood lead levels above 60 μg/dL (Ref 148). Experimental studies suggest that lead alters the humoral immune system and lead-induced immunosuppression occurs at low dosages in experimental animals in which there is no apparent evidence of toxicity (Ref 149). Also, children with asymptomatic increase in blood lead levels appear to have more frequent febrile illness (Ref 150).

Lead lines (Burton's lines) or purple-blue discoloration of gingiva is a classical feature of severe lead toxicity in children with lead encephalopathy. However, this feature of lead toxicity as well as the presence of lead lines at the epiphyseal margins of long bones seen on x-rays of children with severe lead exposure are uncommon today.

Interaction With Other Minerals. Lead toxicity is enhanced by dietary deficiencies in calcium and iron and possibly zinc (Ref 151). The increased lead susceptibility found in animals on low calcium regimes is not a direct effect of calcium intake but of increased lead retention associated with decreased renal excretion of lead. In children, elevated blood lead is associated with decreased levels of 25-hydroxyvitamin D (synthesized in liver) and 1-to-25-dihydroxyvitamin D (synthesized in kidney). Iron deficiency is thought to increase gastrointestinal absorption of lead but does not otherwise affect lead metabolism. Relationships between lead and zinc have been studied in experimental animals and veterinary practice. Animals fed diets low in zinc accumulate greater amounts of lead in bones. Extra zinc in the diet has been found to protect horses grazing on lead-contaminated pastures from clinical manifestations of lead toxicity.

Organolead Compounds. Alkyl lead compounds used as gasoline additives, tetraethyl and tetramethyl lead, are rapidly absorbed into the nervous system and are much more severe neurotoxins on an equivalent dose basis than inorganic lead. Although these compounds may be emitted in small amounts from automobile exhaust, they degrade rapidly in the atmosphere. However, the practice of young people sniffing gasoline for psychodelic effects is particularly hazardous. Experimental studies have shown that tetraethyl lead is converted to triethyl lead and inorganic lead. Triethyl lead is relatively stable and becomes rapidly distributed between brain, liver, kidney, and blood (Ref 152).

Table 5 Blood lead levels below which the listed effects have not been detected(a)

Blood levels, μg Pb/dL	Effect	Population
>10	Erythrocyte ALA-D inhibition	Adults, children
20–25	FEP(b)	Children
20–30	FEP	Adult, female
25–35	FEP	Adult, male
30–40	Erythrocyte ATPase inhibition	General
40	ALA excretion in urine	Adults, children
40	CP(c) excretion in urine	Adults
40	Anemia	Children
40–50	Peripheral neuropathy	Adults
50	Anemia	Adults
30–40	Minimal brain dysfunction	Children
60–70	Minimal brain dysfunction	Adults
60–70	Encephalopathy	Children
>80	Encephalopathy	Adults

(a) Source: Ref 107. (b) FEP, free erythrocyte protoporphyrin (see Fig. 4). (c) CP, coproporphyrinogen (also Copro)

Biologic Indicators of Lead Toxicity. The most serious effects of lead are related to the central nervous system, although other effects such as chronic nephropathy may be important in some individuals with chronic exposure to high lead levels. Effects on heme synthesis are less evident from a clinical viewpoint, but these are the effects that in biochemical terms are best understood. With continual improvement in industrial hygiene and consciousness and control of potential environmental exposures to lead, concern for recognition of lead poisoning, in terms of traditional clinical signs and symptoms, is less relevant now than the matter of interpretation of more subtle neurologic and biochemical effects. Table 5 shows blood lead levels below which various parameters of lead toxicity are not detected. Blood lead level greater than 80 μg/dL is usually associated with clinical symptoms. Persons with blood lead levels above 60 μg/dL may also have clinical manifestations of lead poisoning, particularly in terms of minimal brain dysfunction. For these reasons children with blood levels in this range are sometimes treated with chelating agents. However, effects of lead on heme synthesis, along with increased urinary excretion of δ-ALA, are associated with blood lead levels of 40 μg/dL.

Heme Metabolism. The ALA-D activity of peripheral red blood cells may currently be the most sensitive biochemical parameter affected by lead. Samples collected from both children and adults have shown a negative correlation between the log of ALA-D activity and the blood lead concentration over a range to below 20 μg/dL. At present they are considered chemical effects of lead rather than adverse health effects. However, children with blood lead levels less than 40 μg/dL exhibit behavioral and fine-motor changes. Therefore, the maximum blood lead level that is not associated with harmful effects is not known. It may, indeed, differ for different segments of the population. Furthermore, blood lead levels and alterations in heme metabolism are indicative of recent exposure but do not reflect past exposure or body burden and may not correlate with the appearance and persistence of chronic effects of lead that are due to tissue levels. This is particularly true of irreversible effects on the central nervous system and kidney.

Chelatable Lead. Urinary excretion of lead after chelation with EDTA or penicillamine reflects the mobilizable pool of lead located in soft tissue. Urinary excretion of lead in the 24-hour period after administration of the chelating agent correlates with blood lead levels for persons in the steady state, without unusual past or recent exposure to lead. However, for persons exposed to excess lead for one year or more, urinary excretion of lead is considerably greater. It has been estimated that nearly 20% of the body burden of lead is mobilized into the urine during the 24-hour period after an injection of 40 mg of $CaNa_2$ EDTA per kilogram body weight (Ref 153).

Lead in Teeth. Lead concentration in circumpulpal dentine of deciduous teeth has been proposed as a method for estimating exposure to lead during childhood. It has been shown that dentine lead content is dose dependent and not reduced by chelation. The distribution of lead in various components of the tooth including enamel, root dentine, and coronal dentine is similar. However, lead content in secondary or circumpulpal dentine, the area of dentine adjacent to pulp and in immediate contact with blood, is higher than in other areas of the tooth and seems to reflect actual exposure to lead throughout the life of the tooth. Several studies now have found substantially higher dentine levels in teeth from urban children living in deteriorated housing or attending school in proximity to a major manufacturer of paint and lead products than in teeth from children considered to be a low risk for lead exposure (Ref 154).

Treatment. Treatment of lead toxicity must go beyond medical care for specific tissue and organ effects and chelation of lead. For both asymptomatic excess exposure to lead as well as the symptomatic child or worker with occupational exposure, the sources of lead must be identified and controlled. For the child, this might involve a review of lifestyle including diet, particularly iron deficiency, type of dwelling, play habits, and pica. Treatment might involve social services, modification of dwelling, and parent education. For the workman, industrial hygiene practices must be reviewed including appropriate environmental and biologic monitoring and deficiencies corrected. One must always be aware of the relationship between occupation and the home. Practices of changing or washing contaminated workclothes must be reviewed, and potential sources of transfer of metal and contamination need to be corrected.

Chelation usually has a role in the treatment of the symptomatic worker or child. Institution of chelation therapy is probably warranted in workmen with blood lead levels over 60 μg/100 mL, but this determination must be made after assessment of exposure factors including biologic estimates of clinical and biochemical parameters of toxicity.

For children, criteria have been established that may serve as guidelines to assist in evaluating the individual case (Ref 111). These include blood lead levels from 30 μg/dl up to 60 μg/dL depending on FEP levels, and results of a lead mobilization test.

Also, cautionary measures for the safe use of chelating agents have been expressed particularly for Ca EDTA (Ref 155). Serum blood urea nitrogen and creatinine are followed as indicators of renal function, and serum calcium is measured to monitor untoward effects of EDTA. In children with severe lead poisoning including encephalopathy, the mortality rate may be 25 to 38% when EDTA or BAL is used singly; combination therapy of EDTA and BAL has been shown to be effective in reducing mortality.

Mercury

No other metal better illustrates the diversity of effects caused by different biochemical forms than does mercury, properties and applications of which are described in the article "Properties of Pure Metals" in this Volume. On the basis of toxicologic characteristics, there are three forms of mercury: elemental, inorganic, and organic compounds. The major source of mercury is the natural degassing of the crust of the earth, including land areas, rivers, and the ocean, and is estimated to be of the order of 25 000 to 150 000 tons per year (Ref 156, 157, 158). Metallic mercury in the atmosphere represents the major pathway of global transport of mercury. Although anthropogenic sources of mercury have reached about 8000 to 10 000 tons per year since 1973, nonanthropogenic sources are the predominating factors. Nevertheless, mining, smelting, and industrial discharge have been factors in environmental contamination in the past. For instance, it is estimated that loss in water effluent from chloralkali plants, one of the largest users of mercury, has been reduced by 99% in recent years. Also, the use of mercury in the paper pulp industries has been reduced dramatically and has been banned in Sweden since 1966. Industrial activities not directly employing mercury or mercury products give rise to substantial quantities of this metal. Fossil fuel may contain as much as 1 ppm of mercury, and it is estimated that about 5000 tons of mercury per year may be emitted from burning coal, natural gas, and the refining of petroleum products. Calculations based on mercury content of the Greenland icecap show an increase from the year 1900 to the present day and suggest that the increment is related to increase in background levels in rainwater and is related to man-made release. As much as one-third of atmospheric mercury may be due to industrial release of organic or inorganic forms. Regardless of source, both organic and inorganic forms of mercury may undergo environmental transformation. Metallic mercury may be oxidized to inorganic divalent mercury, particularly in the presence of organic material such as in the aquatic environment. Divalent inorganic mercury may, in turn, be reduced to metallic mercury when conditions are appropriate for reducing reactions to occur. This is an important conversion in terms of the global cycle of mercury and a potential source of mercury vapor that may be released to the earth's atmosphere. A second potential conversion of divalent mercury is methylation to dimethyl mercury by anaerobic bacteria. This may diffuse into the atmosphere and return to earth crust or bodies of water as methyl mercury in rainfall. If taken up by fish in the food chain, it may eventually cycle through humans.

Disposition. Toxicity of various forms or salts of mercury is related to cationic mercury *per se* whereas solubility, biotransformation, and tissue distribution are influenced by valence state and anionic component (Ref 159, 160, 161). Metallic or elemental mercury volatilizes to mercury vapor at ambient air temperatures, and most human exposure is by inhalation. Mercury vapor readily diffuses across the alveolar membrane and is lipid soluble so that it has an affinity for red blood cells and central nervous system. Metallic mercury, such as may be swallowed from a broken thermometer, is only slowly absorbed by the gastrointestinal tract (0.01%) at a rate related to the vaporization of the elemental mercury and is generally thought to be of no toxicologic consequence.

Inorganic mercury salts may be divalent (mercuric) or monovalent (mercurous). Gastrointestinal absorption of inorganic salts of mercury from food is less than 15% in mice and about 7% in a study of human volunteers, whereas absorption of methyl mercury is of the order of 90 to 95%. Distribution between red blood cells and plasma also differs. For inorganic mercury salts cell-plasma ratio ranges from a high of 2 with high exposure to less than 1, but for methyl mercury it is about 10. The distribution ratio of the two forms of mercury between hair and blood also differs; for organic mercury it is about 250.

Kidneys contain the greatest concentrations of mercury following exposure to inorganic salts of mercury and mercury vapor, whereas organic mercury has a greater affinity for the brain, particularly the posterior cortex. However, mercury vapor has a greater predilection for the central nervous system than does inorganic mercury salts, but less than organic forms of mercury.

Excretion of mercury from the body is by way of urine and feces, again differing with the form of mercury, size of dose, and time after exposure. Exposure to mercury vapor is followed by exhalation of a small fraction, but fecal excretion releases more toxin and is predominant initially after exposure to inorganic mercury. Renal excretion increases with time. About 90% of methyl mercury is excreted in feces after acute or chronic exposure and does not change with time (Ref 162).

All forms of mercury cross the placenta to the fetus, but most of what is known has been learned from experimental animals. Fetal uptake of elemental mercury in rats probably because of lipid solubility has been shown to be 10 to 40 times higher than uptake after exposure to inorganic salts. Concentrations of mercury in the fetus after exposure to alkylmercuric compounds are twice those found in maternal tissues, and methyl mercury levels in fetal red blood cells are 30% higher than in maternal red cells. The positive fetal-maternal gradient and increased concentration of mercury in fetal red blood cells enhance fetal toxicity to mercury particularly following exposure to alkylmercury. Although maternal milk may contain only 5% of the mercury concentration of maternal blood, neonatal exposure to mercury may be greatly augmented by nursing.

Metabolic Transformation and Excretion. Elemental or metallic mercury is oxidized to divalent mercury after absorption to tissues in the body and is probably mediated by catalyses. Inhaled mercury vapor absorbed into red blood cells is transformed to divalent mercury, but a portion is also transported as metallic mercury to more distal tissues, particularly the brain where biotransformation may occur. Similarly, a fraction of absorbed metallic mercury may be carried across the placenta to the fetus. The oxidized divalent mercury is then accumulated by these tissues.

Alkylmercury also undergoes biotransformation to divalent mercuric compounds in tissues by cleavage of the carbon-mercury bond. There is no evidence of formation of any organic form of mercury by mammalian tissues. The aryl (phenyl) compounds are converted to inorganic mercury more rapidly than the shorter-chain alkyl (methyl) compounds. The relationship of these differences is rate of biotransformation versus rate of excretion and toxicity is not well understood. In those instances where the organomercurial is more rapidly excreted than inorganic mercury, increasing the rate of biotransformation will decrease the rate of excretion. Phenyl and methoxyethylmercury are excreted at about the same rate as inorganic mercury whereas methyl mercury excretion is slower.

Biologic half-times are available for a limited number of mercury compounds. Biologic half-time for methyl mercury is about 70 days and is virtually linear, whereas the half-time for retained salts of inorganic mercury is about 40 days. There are few studies on biologic half-times for elemental mercury or mercury vapor, but it also appears to be linear with a range of values from 35 to 90 days.

Cellular Metabolism. Within cells, mercury may bind to a variety of enzyme systems including those of microsomes and mitochondria, producing nonspecific cell injury or cell death. It has a particular affinity for ligands containing sulfhydryl groups. In liver cells, methyl mercury forms soluble complexes with cysteine and glutathione, which are secreted in bile and reabsorbed from the gastrointestinal tract. Organomercurial diuretics are thought to be absorbed in the proximal-tubule-binding specific receptor sites that inhibit sodium transport. In general, however, organomercury compounds undergo cleavage of the carbon-mercury bond, releasing ionic inorganic mercury.

Mercuric mercury, but not methyl mercury, induces synthesis of metallothionein probably only in kidney cells, but unlike cadmium-metallothionein it does not have a long biologic half-life. Mercury within renal cells becomes localized in lysosomes (Ref 163).

Toxicology. The toxic effects of mercury vapor, mercuric mercury, mercurous compounds, and organic mercury will be discussed in this section.

Mercury Vapor. Inhalation of mercury vapor may produce an acute, corrosive bronchitis and interstitial pneumonitis and, if not fatal, may be associated with symptoms of central nervous system effects such as tremor or increased excitability.

With chronic exposure to mercury vapor the major effects are on the central nervous system (Ref 164). Early signs are nonspecific and have been termed the "asthenic-vegetative syndrome" or "micromercurialism." Identification of the syndrome requires neuroasthenic symptoms and three or more of the following clinical findings: tremor, enlargement of the thyroid, increased uptake of radioiodine in the thyroid, labile pulse, tachycardia, dermographism, gingivitis, hematologic changes, or increased excretion of mercury in urine. With increasing exposure the symptoms become more characteristic beginning with intentional tremors of muscles that perform fine-motor functions (highly innervated), such as fingers, eyelids, and lips, and may progress to generalized trembling of the entire body and violent chronic spasms of the extremities. This is accompanied by changes in personality and behavior, with loss of memory, increased excitability (erethism), severe depression, and even delirium and hallucination. Another characteristic feature of mercury toxicity is severe salivation and gingivitis.

The triad of increased excitability, tremors, and gingivitis has been recognized historically as the major manifestation of mercury poisoning from inhalation of mercury vapor and exposure in the fur, felt, and hat industry to mercury nitrate (Ref 165).

Sporadic instances of proteinuria and even nephrotic syndrome may occur in persons with exposure to mercury vapor, particularly with chronic occupational exposure. The pathogenesis is probably immunologic similar to that which may occur following exposure to inorganic mercury (see below).

Mercuric Mercury. Bichloride of mercury (corrosive sublimate) is the best-known inorganic salt of mercury, and the trivial name suggests its most apparent toxicologic effect when ingested in concentrations greater than 10%. A reference from the Middle Ages in Goldwater's book on mercury describes oral ingestion of mercury as causing severe abdominal cramps, bloody diarrhea, and suppression of urine (Ref 165). This is an accurate report of effects following accidental or suicidal ingestion of mercuric chloride or other mercuric salts. Corrosive ulceration, bleeding, and necrosis of the gastrointestinal tract are usually accompanied by shock and circulatory collapse. If the patient survives the gastrointestinal damage, renal failure occurs within 24 hours owing to necrosis of the proximal tubular epithelium followed by oliguria, anuria, and uremia. If the patient can be maintained by dialysis, regeneration of tubular lining cells is possible. These may be followed by ultrastructural changes consistent with irreversible cell injury including actual disruption of mitochondria, release of lysosomal enzymes, and rupture of cell membranes.

Injection of mercuric chloride produces necrosis of the epithelium of the pars recta kidney (Ref 166). Cellular changes include fragmentation and disruption of the plasma membrane and its appendages, vesiculation and disruption of the endoplasmic reticulum and other cytoplasmic membranes, dissociation of polysomes and loss of ribosomes, mitochondrial swelling with appearance of amorphous intramatrical deposits, and condensation of nuclear chromatin. These changes are common to renal cell necrosis due to various causes.

Although exposure to a high dose of mercuric chloride is directly toxic to renal tubular lining cells, chronic low-dose exposure to mercuric salts or even elemental mercury vapor levels may induce an immunologic glomerular disease. This form of chronic mercury injury to the kidney is clinically the most common form of mercury-induced nephropathy. Exposed persons may develop a proteinuria that is reversible after workers are removed from exposure. It has been stated that chronic mercury-induced nephropathy seldom occurs without sufficient exposure to also produce detectable neuropathy.

Experimental studies have shown that the pathogenesis of chronic mercury nephropathy has two phases: an early phase characterized by an antibasement membrane glomerulonephritis followed by a superimposed immune-complex glomerulonephritis (Ref 167). The pathogenesis of the nephropathy in humans appears similar although antigens have not been characterized. Also, the early glomerular nephritis may progress in humans to an interstitial immune-complex nephritis (Ref 168).

Mercurous Compounds. Mercurous compounds of mercury are less corrosive and less toxic than mercuric salts, presumably because they are less soluble. Calomel, a powder containing mercurous chloride, has a long history of use in medicine. Perhaps the most notable modern usage has been as teething powder for children and is now known to be responsible for acrodynia or "pink disease." This is most likely a hypersensitivity response to the mercury salts in skin producing vasodilation, hyperkeratosis, and hypersecretion of sweat glands. Children develop fever, a pink-colored rash, swelling of the spleen and lymph nodes, and hyperkeratosis and swelling of fingers. The effects are independent of dose and are thought to be a hypersensitivity reaction (Ref 169).

Organic Mercury. Methyl mercury is the most important form of mercury in terms of toxicity, and health effects from environmental exposures and many of the effects produced by short-chain alkyls are unique in terms of mercury toxicity but are nonspecific in that they may be found in other disease states. Most of what is known about methyl mercury toxicity is from detailed epidemiologic studies of exposed populations (Ref 156).

Two major epidemics of methyl mercury poisoning have occurred in Japan in Minamata Bay and in Niigata. Both were caused by industrial release of methyl and other mercury compounds into Minamata Bay

Table 6 Time-weighted average air concentrations following long-term exposure to elemental mercury vapor(a)

Air, mg/m^3	Blood, µg/100 mL	Urine, µg/L	Earliest effects
0.05	3.5	150	Nonspecific symptoms
0.1–0.2.....	7–14	300–600	Tremor

(a) Blood and urine values may be used only on a group basis owing to gross individual variations. Furthermore, these average values reflect exposure only after exposure for a year or more. After shorter periods of exposure, air concentrations would be associated with lower concentrations in blood and urine. Source: Ref 156

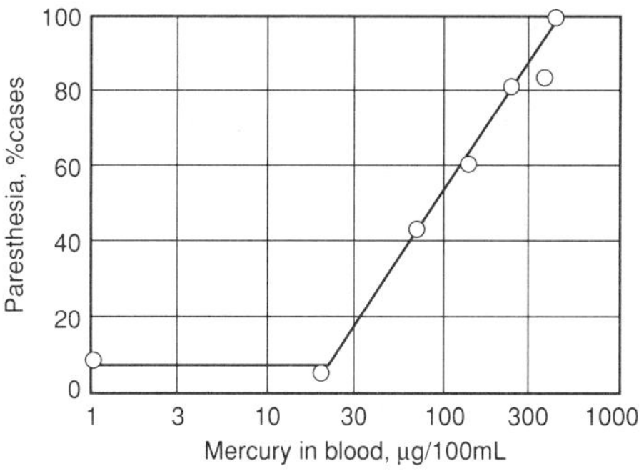

Fig. 6 Dose response relationship for methyl mercury using concentration of mercury in the blood as dose and paresthesia as response. Source: Ref 156, 170

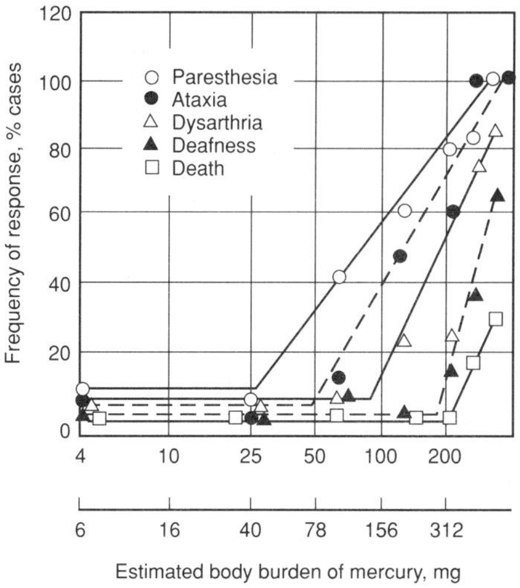

Fig. 7 Dose response relationships for methyl mercury. The upper scale of estimated body burden of mercury was based on the authors' actual estimate of intake. The lower scale, based on the body burden was calculated based on the concentration of mercury in the blood and its relationship to intake derived from radioisotopic studies of methyl mercury kinetics in human volunteers. Source: Ref 170

and into the Agano River, followed by accumulation of the mercury by edible fish. The median level of total mercury in fish caught in Minamata Bay during the epidemic was estimated to be about 11 mg/kg fresh weight and less than 10 mg/kg in fish from the Agano River. The largest recorded epidemic of methyl mercury poisoning took place in the winter of 1971-1972 in Iraq, resulting in admission of over 6000 patients to hospitals and over 500 deaths in hospitals (Ref 170). Methyl mercury exposure was from bread containing wheat imported as seed grain and dressed with methyl mercury fungicide. The mean methyl mercury content of wheat flour samples was 9.1 mg/kg (range 4.8 to 14.6 mg). In one village the average daily intake of contaminated loaves was 3.2 loaves per person, but it varied. On this basis, average daily intake of methyl mercury was calculated to be about 80 µg/kg ranging up to 250 µg/kg/day.

Several previous epidemics occurred in Iraq in 1961, in Pakistan in 1963, in Guatemala in 1966, and in other countries, but on a more limited scale. From studies of these episodes, data have been obtained to relate clinical manifestations, organ pathology, and level of exposure. The major clinical features are neurologic, consisting of paresthesia, ataxia, dysarthria, and deafness, appearing in that order. The main pathologic features of methyl mercury toxicity include degeneration and necrosis of neurons in focal areas of the cerebral cortex, particularly in the visual areas of the occipital cortex and in the granular layer of the cerebellum. Experimental studies of both organic and inorganic mercury-related peripheral neuropathy show degeneration of primary sensory ganglion cells. The particular distribution of lesions in the central nervous system is thought to reflect a propensity of mercury to damage small nerve cells in cerebellum and visual cortex (Ref 171).

Biologic Indicators. Both metallic mercury and alkyl mercury are described below.

Metallic Mercury. For persons in the general population, it is estimated that daily mercury exposure is about 1 µg/day from air, less than 2 µg/day from food, but may be up to 75 µg/day depending on the amount of fish in the diet. The recommended standard (time-weighted average, TWA) for permissible exposure limits for inorganic mercury in air in the workplace is 0.05 mg Hg/m^3 (Ref 172) and is equivalent to an ambient air level of 0.015 mg/m^3 for the general population (24-hour exposure).

The central nervous system is the major site of toxicity from exposure to elemental mercury. It is believed that a worker exposed to a constant average concentration of mercury vapor achieves a state of balance (steady state) after one year of exposure, and it can be expected that there is a consistent relationship between air levels and mercury content of blood or urine. Mercury content in blood and urine only reflects recent exposure to metallic mercury. Table 6 is an estimate of mercury concentration in blood and urine related to air concentration of elemental mercury vapor and earliest clinical effects (Ref 156). Urine mercury should be less than 100 µg/L content.

Alkyl Mercury. The federal standard for alkyl mercury exposure in the workplace is 0.01 mg/m^3 as an eight-hour TWA with an acceptable ceiling of 0.04 mg/m^3. Although a precise correlation has not been found between exposure levels and mercury content of blood and urine, study of the Iraq epidemic has provided estimates of the lowest mercury levels of mercury in blood due to alkyl mercury exposure associated with mild symptoms (Fig. 6). These studies do not indicate that percentage of people in the population who are sensitive at these levels, but other studies suggest that the frequency of paresthesia due to methyl mercury with these minimum mercury concentrations in blood and hair is 5% or less (Ref 156). However, studies reported from Minamata and elsewhere suggest that mothers with slight or no symptoms may have offspring who are retarded and have palsy. Mercury has been reported in breast milk of women exposed to methyl mercury from fish and from contaminated bread and averages about 5% of simultaneous concentrations in maternal blood. Figure 7 shows estimates of body burden of mercury onset and frequency of occurrence of these symptoms.

Mercury levels in hair may also be correlated with severity of clinical symptoms (Ref 170). Mild cases complained of numbness of extremities, slight tremors, and mild ataxia. Moderate cases had difficulty hearing, tunnel vision, and partial paralysis, and severe cases had some combination of complete paralysis, loss of vision, loss of hearing, loss of speech, and coma. Persons with no symptoms might have mercury hair levels up to 300 mg/kg, the mildly affected group in the range of 120 to 600 mg/kg, the moderate group in the range of 200 to 600 mg/kg, and the severely affected from 400 to 1600 mg/kg. Total dose of mercury for this group is not known.

It is not really understood why exposure to methyl mercury has such selective effect on cerebellum and visual cortex. Concentration of mercury in these areas of the brain is not very different from that in areas of the brain not affected by methyl mercury. Methyl mercury does inhibit protein synthesis in the brain before onset of signs of poisoning, and it has been found that recovery of protein synthesis does not occur in granular cells as it does in other types of neuronal cells (Ref 161).

Treatment. Therapy of mercury poisoning should be directed to lowering the concentration of mercury at the critical organ or site of injury. For the most severe cases, particularly with acute renal failure, hemodialysis may be the first measure along with infusion of chelating agents for mercury such as cysteine or penicillamine. For less severe cases of inorganic mercury poisoning, chelation with BAL may be effective.

However, chelation therapy is not very helpful for alkyl mercury exposure. Biliary excretion and reabsorption by the intestine and the enterohepatic cycling of mercury may be interrupted by surgically establishing gallbladder drainage or by the oral administration of a nonabsorbable thiol resin that binds mercury and enhances intestinal excretion (Ref 173).

Nickel

Nickel is a respiratory tract carcinogen in workmen in the nickel-refining industry. Other serious consequences of long-term exposure to nickel are not apparent, but severe acute and sometimes fatal toxicity may follow nickel carbonyl exposure. Allergic contact dermatitis is common among persons in the general population. Deficiency of nickel alters glucose metabolism and decreases tolerance to glucose. From studies on rats, there is growing evidence that nickel may be an essential trace metal for mammals (Ref 174, 175). For information on the uses and properties of nickel, see the article "Nickel and Nickel Alloys" in this Volume.

Disposition. Nickel is only sparsely absorbed from the gastrointestinal tract. It is transported in the plasma bound to serum albumin and multiple small organic ligands, amino acids, or polypeptides. Excretion in the urine is nearly complete in four or five days. Kinetics have been described in rodents as a two-compartment model.

Dietary nickel intake by adults in the United States was estimated by Schroeder *et al.* (Ref 176), to be in the range of 300 to 600 μg/day. In a more recent study of nickel content of diets prepared in university or hospital kitchens in the United States, Myron *et al.* (Ref 177) found standard diet nickel intake to average 165 ± 11 μg/day or 75 ± 10 μg/1000 calories.

In one study, serum nickel was found to be 2.6 ± 0.9 μg/L (range: 0.8 to 5.2) and mean excretion of nickel in urine of 2.6 ± 1.4 μg/day (range: 0.5 to 6.4) (Ref 178). Serum nickel is influenced by environmental nickel or nickel concentration in the ambient air. Serum nickel measured in persons living in Sudbury, Ontario, which is in the vicinity of a large nickel mine, showed concentrations of 4.6 ± 1.4 μg/liter (range: 2.0 to 7.3) and urinary concentrations were 7.9 ± 3.7 μg/day (range: 2.3 to 15.7). Generally, fecal nickel is about 100 times urine nickel concentration.

Nickel administered parenterally to animals is rapidly distributed to kidney, pituitary, lung, skin, adrenal, and ovary and testis (Ref 174).

The intracellular distribution and binding of nickel is not well understood. Ultrafiltrable ligands seem to be of major importance in transport in serum and bile and urinary excretion as well as intracellular binding. The ligands are not well characterized, but Sunderman (Ref 174) suggests that cysteine, histidine, and aspartic acid form nickel complexes either singly or as nickel-ligand species. *In vivo* binding with metallothionein has been demonstrated, but nickel at best induces metallothionein synthesis in liver or kidney only slightly.

A nickel-binding metalloprotein has also been identified in plasma with properties suggesting an α-2-glycoprotein with serum α-1-macroglobulin complex.

Evidence has accumulated over the past few years indicating that nickel is a nutritionally essential trace metal. Jackbean urease has been identified as a nickel metalloenzyme, and nickel is required for urea metabolism in cell cultures of soybean. However, a nickel-containing metalloenzyme has not yet been recovered from animal tissues. Nickel deficiency in rats is associated with retarded body growth and anemia, probably secondary to impaired absorption of iron from the gastrointestinal tract. In addition, there is significant reduction in serum glucose concentration. An interaction of nickel with copper and zinc is also suspected since anemia-induced nickel deficiency is only partially corrected with nickel supplementation in rats receiving low dietary copper and zinc (Ref 179).

Toxicology. This section will describe the carcinogenic effects of nickel, carbonyl poisoning, nickel dermatitis, and indicators of nickel toxicity.

Carcinogenesis. It has been known for 40 years that occupational exposure to nickel predisposes to lung and nasal cancer (Ref 180). Epidemiologic studies in 1958 showed that nickel refinery workers in Britain had a fivefold increase in risk to lung cancer and 150-times increase in risk to nasal cancers compared to people in the general population. More recently, increase in lung cancer among nickel workers has been reported from several different countries including suggestions of increased risks to laryngeal cancer in nickel refinery workers in Norway (Ref 181) and gastric carcinoma and soft tissue sarcomas from the Soviet Union. Six cases of renal cancer have been reported among Canadian and Norwegian workers employed in the electrolytic refining of nickel (Ref 182). McEwan (Ref 183) has been able to detect early cytologic changes in sputum of exposed workers prior to chest x-ray or clinical indicators of respiratory tract cancer.

Because the refining of nickel in the plants that were studied involved the Mond process with the formation of nickel carbonyl, it was believed for some time that nickel carbonyl was the principal carcinogen. However, additional epidemiologic studies of workers in refineries that do not use the Mond process also showed increased risk of respiratory cancer, suggesting that the source of the increased risk is the mixture of nickel sulfides present in molten ore. Indeed, studies with experimental animals have shown that the nickel subsulfide (Ni_3S_2) produces local tumors at injection sites and by inhalation in rats, and *in vitro* mammalian cell tests demonstrate that Ni_3S_2 and $NiSO_4$ compounds give rise to mammalian cell transformation (Ref 15).

Nickel Carbonyl Poisoning. Metallic nickel combines with carbon monoxide to form nickel carbonyl ($Ni[CO]_4$), which decomposes to pure nickel and carbon monoxide on heating to 200 °C (Mond process). This reaction provides a convenient and efficient method for the refinement of nickel. However, nickel carbonyl is extremely toxic, and many cases of acute toxicity have been reported. The illness begins with headache, nausea, vomiting, and epigastric or chest pain, followed by cough, hyperpnea, cyanosis, gastrointestinal symptoms, and weakness. The symptoms may be accompanied by fever and leukocytosis, and the more severe cases progress to pneumonia, respiratory failure, and eventually cerebral edema and death. Autopsy studies show the largest concentrations of nickel in lungs with lesser amounts in kidneys, liver, and brain (Ref 174).

Dermatitis. Nickel dermatitis is one of the most common forms of allergic contact dermatitis; 4 to 9% of persons with contact dermatitis react positively to nickel patch tests. Sensitization might occur from any of the numerous metal products in common use, such as coins and jewelry. The notion that increased ingestion of nickel-containing food increases the probability of external sensitization to nickel is supported by finding increased urinary nickel excretion in association with episodes of acute nickel dermatitis (Ref 184).

Indicators of Nickel Toxicity. Blood nickel levels immediately following exposure to nickel carbonyl provide a guideline as to severity of exposure and indication for chelation therapy (Ref 174). Sodium diethyldithiocarbamate is the preferred drug, but other chelating agents, such as *d*-penicillamine and triethylenetetraamine, provide some degree of protection from clinical effects.

Essential Metals With Potential For Toxicity

This group includes seven metals generally accepted as essential: cobalt, copper, iron, manganese, molybdenum, selenium, and zinc. Each of the seven essential metals has three levels of biologic activity, trace levels required for optimum growth and development, homeostatic levels (storage levels), and toxic levels. For these metals, environmental accumulations are generally less important routes of excess exposure than accidents or occupation.

Although chromium and arsenic are regarded as essential to humans and animals, respectively, the toxicologic significance of chromium and arsenic warrant their being discussed as major toxic metals in the context of this article. Tin and vanadium are also essential to animals but are of less importance toxicologically and are included in the group of minor toxic metals.

Cobalt

Cobalt is essential as a component of vitamin B_{12} required for the production of red blood cells and prevention of pernicious anemia. There is 0.434 μg of cobalt per microgram of vitamin B_{12}. If other requirements for cobalt exist, they are not well understood. Deficiency diseases of cattle and sheep, caused by insufficient natural levels of cobalt, are characterized by anemia and loss of weight or retarded growth.

Cobalt is a relatively rare metal produced primarily as a by-product of other metals, chiefly copper. It is used in high-temperature alloys, wear-resistant materials, and in permanent magnets (see the articles "Cobalt and Cobalt Alloys," "Permanent Magnet Materials" and "Cemented Carbides" in this Volume). Its salts are useful in paint driers, as catalysts, and in the production of numerous pigments.

Cobalt salts are generally well absorbed after oral ingestion, probably in the jejunum. Despite this fact, increased levels tend not to cause significant accumulation. About 80% of the ingested cobalt is excreted in the urine. Of the remaining, about 15% is excreted in the feces by an enterohepatic pathway, while the milk and sweat are other secondary routes of excretion. The total body burden has been estimated as 1.1 mg.

The muscle contains the largest total fraction, but the fat has the highest concentration. The liver, heart, and hair have significantly higher concentrations than other organs, but the concentration in these organs is relatively low. The normal levels in human urine and blood are about 98 and 0.18 μg/liter, respectively. The blood level is largely in association with the red cells.

Significant species differences have been observed in the excretion of radiocobalt. In rats and cattle, 80% is eliminated in the feces (Ref 185).

Polycythemia is the characteristic response of most mammals, including humans, to ingestion of excessive amounts of cobalt. Toxicity resulting from overzealous therapeutic administration has been reported to produce vomiting, diarrhea, and a sensation of warmth. Intravenous administration leads to flushing of the face, increased blood pressure, slowed respiration, giddiness, tinnitus, and deafness due to nerve damage (Ref 186).

High levels of chronic oral administration may result in the production of goiter. Epidemiologic studies suggest that the incidence of goiter is higher in regions containing increased levels of cobalt in the water and soil (Ref 187). The goitrogenic effect has been elicited by the oral administration of 3 to 4 mg/kg to children in the course of sickle cell anemia therapy (Ref 186).

Cardiomyopathy has been caused by excessive intake of cobalt, particularly from the drinking of beer to which 1 ppm cobalt was added to enhance its foaming qualities. Why such a low concentration should produce this effect in the absence of any similar change when cobalt is used therapeutically is unknown. The signs and symptoms were those of congestive heart failure. Autopsy findings revealed a tenfold increase in the cardiac levels of cobalt. Alcohol may have served to potentiate the effect of the cobalt (Ref 188).

Hyperglycemia due to β-cell pancreatic damage has been reported after injection into rats. Reduction of blood pressure has also been observed in rats after injection and has led to some experimental use in humans (Ref 185).

Occupational inhalation of cobalt salts in the cemented carbide industry may cause respiratory symptoms probably as a result of irritation of the pulmonary tract. Allergic dermatitis of an erythematous papular type may also occur, and affected persons may have positive skin tests.

Single and repeated subcutaneous or intramuscular injection of cobalt powder and salts to rats may cause sarcomas at the site of injection, but there is no evidence of carcinogenicity from any other route of exposure (Ref 189).

Copper

Copper is widely distributed in nature and is an essential element. Copper deficiency is characterized by hypochromic, microcytic anemia resulting from defective hemoglobin synthesis. Oxidative enzymes, such as catalase, peroxidase, cytochrome oxidase, and others, also require copper. Medicinally, copper sulfate is used as an emetic. It has also been used for its astringent and caustic action and as an anthelmintic. Copper sulfate mixed with lime has been used as a fungicide. As a structural material, copper has a wide variety of applications. A number of articles describing the properties and applications of wrought, cast, and powder metallurgy copper and copper alloys are provided in this Volume.

Gastrointestinal absorption of copper is normally regulated by body stores (Ref 190, 191). It is transported in serum bound initially to albumin and later more firmly bound to α-ceruloplasmin where it is exchanged in the cupric form. The normal serum level of copper is 120 to 145 μg/L. The bile is the normal excretory pathway and plays a primary role in copper homeostasis. Most copper is stored in liver and bone marrow where it may be bound to metallothionein. The amount of copper in milk is not enough to maintain adequate copper levels in the liver, lung, and spleen of the newborn. Tissue levels gradually decline up to about 10 years of age, remaining relatively constant thereafter. Brain levels, on the other hand, tend to almost double from infancy to adulthood. The ratios of newborn to adult liver copper levels show considerable species difference: human, 15:4; rat, 6:4, and rabbit, 16:6. Since urinary copper levels may be increased by soft water, under these conditions concentrations of approximately 60 μg/L are not uncommon.

Copper is an essential part of several enzymes, including tyrosinase, involved in the formation of melanin pigments, cytochrome oxidase, superoxide dismutase, amine oxidases, and uricase. It is essential for the utilization of iron. Iron deficiency anemia in infancy is sometimes accompanied by copper deficiency as well. Molybdenum also influences tissue levels of copper.

There are two genetically inherited inborn errors of copper metabolism that are in a sense a form of copper toxicity (Ref 191). Wilson's disease is characterized by excessive accumulation of copper in liver, brain, kidneys, and cornea. Serum ceruloplasmin is low, and serum copper, not bound to ceruloplasmin, is elevated. Urinary excretion of copper is high. The disorder is sometimes referred to as hepatolenticular degeneration in reference to the major symptoms. Clinical abnormalities of the nervous system, liver, kidneys, and cornea are related to copper accumulation. Although the etiology of this disorder is genetic, the basic defect at the biochemical level is not known. Increased binding of copper to an abnormal intracellular thionein or altered tissue excretion has been proposed. Cultured fibroblasts from persons with Wilson's disease have increased intracellular copper when cultured in Eagle's minimum essential medium with fetal bovine serum (Ref 192). Clinical improvement can be achieved by chelation of copper with penicillamine (Ref 22). Trien (triethylene tetramine, 2HCl) is also effective and has been used in patients with Wilson's disease who have toxic reactions to penicillamine (Ref 23).

Menkes' disease or Menkes' "kinky-hair syndrome" is a sex-linked trait characterized by peculiar hair, failure to thrive, severe mental retardation, neurologic impairment, and death before three years of age. There is extensive degeneration of the cerebral cortex and of white matter. Again, the basic defect is not known. There are low levels of copper in liver and brain but high concentrations in other tissues. Even in cells with increased copper concentration there is a relative deficiency in activities of some copper-dependent enzymes. Some laboratories have reported that larger-than-normal quantities of copper-thionein accumulated in fibroblasts so that the basic defect may be in regulation of metallothionein synthesis. The finding of increased amounts of other metallothionein binding metals (zinc, cadmium, mercury) in kidneys of patients with this disease supports this hypothesis (Ref 193).

Acute poisoning resulting from ingestion of excessive amounts of oral copper salts, most frequently copper sulfate, may produce death. The symptoms are vomiting, sometimes with a blue-green color observed in the vomitus, hematemesis, hypotension, melena, coma, and jaundice. Autopsy findings have revealed centrilobular hepatic necrosis (Ref 194). Few cases of copper intoxication as a result of burn treatment with copper compounds have resulted in hemolytic anemia. Copper poisoning producing hemolytic anemia has also been reported as the result of using copper-containing dialysis equipment (Ref 195).

Iron

The major interest in iron is as an essential metal, but toxicologic considerations are important in terms of accidental acute exposures and chronic iron overload due to idiopathic hemochromatosis or as a consequence of excess dietary iron or frequent blood transfusions. The complex metabolism of iron and mechanisms of toxicity are detailed by Jacobs and Worwood (Ref 196). Properties of pure iron are described in the article "Properties of Pure Metals" in this Volume. Its use in cast irons, steels, and superalloys is documented in Volume 1 of the 10th Edition of *Metals Handbook*.

Disposition. The disposition of iron is regulated by a complex mechanism to maintain homeostasis. Generally, about 2 to 15% is absorbed from the gastrointestinal tract, whereas elimination of absorbed iron is only about 0.01% per day (percent body burden or amount absorbed). During periods of increased iron need (childhood, pregnancy, blood loss) absorption of iron is greatly increased. Absorption occurs in two steps: absorption of ferrous ions from the intestinal lumen into the mucosal cells, and transfer from the mucosal cell to plasma where it is bound to transferrin for transfer to storage sites. Transferrin is a β_1-globulin with a molecular weight of 75 000 and is produced in the liver. As ferrous ion is released into plasma, it becomes oxidized by oxygen in the presence of ferroxidase I, which is identical to ceruloplasmin. There are 3 to 5 g of iron in the body. About two-thirds is bound to hemoglobin, 10% in myoglobin and iron-containing enzymes, and the remainder is bound to the iron storage proteins ferritin and hemosiderin. Exposure to iron induces synthesis of apoferritin, which then binds ferrous ions. The ferrous ion becomes oxidized, probably by histidine and cysteine residues and carbonyl groups. Iron may be released from ferritin by reducing agents; ascorbic acid, cysteine, and reduced glutathione release iron slowly. Normally, excess ingested iron is excreted, and some is contained within shed intestinal cells and in bile and urine and in even smaller amounts in sweat, nails, and hair. Total iron excretion is usually of the order of 0.5 mg/day.

With excess exposure to iron or iron overload, there may be a further increase in ferritin synthesis in hepatic parenchymal cells. In fact, the ability of the liver to synthesize ferritin exceeds the rate at which lysosomes can process iron for excretion. Lysosomes convert the protein from ferritin to hemosiderin, which then remains *in situ* (Ref 197). The formation of hemosiderin from ferritin is not well understood, but seems to involve denaturation of the apoferritin molecule. With increasing iron loading, ferritin concentration appears to reach a maximum and a greater portion of iron is found in hemosiderin. Both ferritin and hemosiderin are, in fact, storage sites for intracellular metal and are protective in that they maintain intracellular iron in bound form.

A portion of the iron taken up by cells of the reticuloendothelial system enters a labile iron pool available for erythropoiesis and part becomes stored as ferritin.

Toxicity. Acute iron toxicity is nearly always due to accidental ingestion of iron-containing medicines and most often occurs in children. As of 1970, there were about 2000 cases in the United States each year, generally among children aged one to five years, who eat ferrous sulfate tablets with candylike coatings. Decrease of this occurrence should follow use of "childproof" lids on prescription medicines. Severe toxicity occurs after ingestion of more than 0.5 g of iron or 2.5 g of ferrous sulfate. Toxicity becomes manifest with vomiting, one to six hours after ingestion. The vomitus may be bloody owing to ulceration of the gastrointestinal tract; stools may be black. This is followed by signs of shock and metabolic acidosis, liver damage, and coagulation defects within the next couple of days. Late effects may include renal failure and hepatic cirrhosis. The mechanism of the toxicity is thought to begin with acute mucosal cell damage, absorption of ferrous ions directly into the circulation, which then cause capillary endothelial cell damage in liver.

Chronic iron toxicity or iron overload in adults is a more common problem. There are three basic ways in which excessive amounts of iron can accumulate in the body. The first circumstance is idiopathic hemochromotosis due to abnormal absorption of iron from the intestinal tract. The condition may be genetic. A second possible cause of iron overload is excess dietary iron. The African Bantu who prepares his daily food and brews fermented beverages in iron pots is the classic example of this form of iron overload. Sporadic other cases occur owing to excessive ingestion of iron-containing tonics or medicines. The third circumstance in which iron overload may occur is from the regular requirement for blood transfusion for some form of refractory anemias and is sometimes referred to as transfusional siderosis (Ref 198).

The pathologic consequences of iron overload are similar regardless of basic cause. The body iron content is increased to between 20 and 40 g. Most of the extra iron is hemosiderin. Greatest concentrations are in parenchymal cells of liver and pancreas, as well as endocrine organs and heart. Iron in reticuloendothelial cells (spleen) is greatest in transfusional siderosis and in the Bantu. Further clinical effects may include disturbances in liver function, diabetes mellitus, and even endocrine disturbances and cardiovascular effects. At the cell level, increased lipid peroxidation occurs with consequent membrane damage to mitochondria, microsomes, and other cellular organelles (Ref 199).

Treatment of acute iron poisoning is directed toward removal of the ingested iron from the gastrointestinal tract by inducing vomiting or gastric lavage and providing corrective therapy for systemic effects such as acidosis and shock. Desferrioxamine is the chelating agent of choice for treatment of iron absorbed from acute exposure as well as for removal of tissue iron in hemosiderosis. Ascorbic acid will also increase iron excretion as much as twofold normal (Ref 200).

Inhalation of iron oxide fumes or dust by workers in metal industries may result in deposition of iron particles in lungs producing an x-ray appearance resembling silicosis. These effects are seen in hematite miners, iron and steel workers, and arc welders. Hematite is the most important iron ore (mainly Fe_2O_3). A report of autopsies of hematite miners noted an increase in lung cancer, as well as tuberculosis and interstitial fibrosis (Ref 201). The etiology of the lung cancer may be related to concomitant factors such as cigarettes or other workplace carcinogens. Hematite miners are also exposed to silica and other minerals, as well as radioactive materials; other iron workers have exposures to polycyclic hydrocarbons (Ref 202). Dose levels of iron among iron workers developing pneumoconiosis have been reported to exceed 10 mg Fe/m^3.

Manganese

Manganese is an essential element and is a cofactor for a number of enzymatic reactions, particularly those involved in phosphorylation, cholesterol, and fatty acids synthesis. Manganese is present in all living organisms. While it is present in urban air and in most water supplies, the principal portion of the intake is derived from food. Vegetables, the germinal portions of grains, fruits, nuts, tea, and some spices are rich in manganese (Ref 59, 203). The industrial use of manganese is primarily that of an alloying element in steel (see Volume 1 of the 10th Edition of *Metals Handbook*). Properties of manganese are described in the article "Properties of Pure Metals" in this Volume.

Daily manganese intake ranges from 2 to 9 mg. Gastrointestinal absorption is less than 5%. It is transported in plasma bound to a β_1-globulin, thought to be transferrin, and is widely distributed in the body. Manganese concentrates in mitochondria so that tissues rich in these organelles have the highest concentrations of manganese including pancreas, liver, kidney, and intestines. Biologic half-life in the body is 37 days. It readily crosses the blood-brain barrier and half-life in the brain is longer than in the whole body.

Manganese is eliminated in the bile and is reabsorbed in the intestine, but the principal route of excretion is with feces. This system apparently involves the liver, auxiliary gastrointestinal mechanisms for excreting excess manganese, and perhaps the adrenal cortex. This regulating mechanism, plus the tendency for extremely large doses of manganese salts to cause gastrointestinal irritation, accounts for the lack of systemic toxicity following oral administration or dermal application.

Manganese and its compounds are used in making steel alloys, dry-cell batteries, electrical coils, ceramics, matches, glass, dyes, in fertilizers, welding rods, as oxidizing agents, and as animal food additives.

Industrial toxicity from inhalation exposure, generally to manganese dioxide in mining or manufacturing, is of two types: The first, manganese pneumonitis, is the result of acute exposure. Men working in plants with high concentrations of manganese dust show an incidence of respiratory disease 30 times greater than normal. Pathologic changes include epithelial necrosis followed by mononuclear proliferation.

The second and more serious type of disease resulting from chronic inhalation exposure to manganese dioxide, generally over a period of more than two years, involves the central nervous system. In iron deficiency anemia, the oral absorption of manganese is increased, and it may be that variations in manganese transport related to iron deficiency account for individual susceptibility (Ref 204). Those who develop chronic manganese poisoning (manganism) exhibit a psychiatric disorder characterized by irritability, difficulty in walking, speech disturbances, and compulsive behavior that may include running, fighting, and singing. If the condition persists, a masklike face, retropulsion or propulsion, and a Parkinson-like syndrome develop (Ref 205). The outstanding feature of manganese encephalopathy has been classified as severe selective damage to the subthalamic nucleus and pallidum (Ref 206). These symptoms and the pathologic lesions, degenerative changes in the basal ganglia, make the analogy to Parkinson's disease feasible. In addition to the central nervous system changes, liver cirrhosis is frequently observed.

Victims of chronic manganese poisoning tend to recover slowly, even when removed from the excessive exposure. Metal-sequestering agents have not produced remarkable recovery; L-dopa, which is used in the treatment of Parkinson's disease, has been more consistently effective in the treatment of chronic manganese poisoning than in Parkinson's disease (Ref 207).

The syndrome of chronic nervous system effects has not been successfully duplicated in any experimental animals except monkeys and then only by inhalation or intraperitoneal injection. After intraperitoneal administration of manganese to squirrel monkeys, dopamine and serotonin levels markedly decreased in the caudate nucleus regardless of whether or not behavioral effects were present. Manganese levels were increased in the basal ganglia and cerebellum. Histopathologic examination of animals did not reveal any morphologic changes (Ref 208). Exposure of rats to manganese dioxide for 100 days does increase the brain manganese concentration but does not produce any hematologic, behavioral, or histologic effects.

Molybdenum

Molybdenum is an essential metal as a cofactor for the enzymes xanthine oxidase and aldehyde oxidase. In plants it is necessary for fixing of atmospheric nitrogen by bacteria at the start of protein synthesis. Because of these functions it is ubiquitous in food. Because plankton tend to concentrate molybdenum 25 times that of seawater, shellfish tend to have high concentrations of molybdenum. Molybdenum is added in trace amounts to fertilizers to stimulate plant growth. The average daily human intake in food is approximately 350 μg. The concentration of molybdenum in urban air is minimal, but it is present in more than one-third of fresh-water supplies (Table 1) and in certain areas the concentration may be near 1 μg/L. Excess exposure can result in toxicity to animals and humans (Ref 59, 209).

The most important mineral source of molybdenum is molybdenite (MoS_2). The United States is the major world producer of molybdenum. The industrial uses of this metal include the manufacture of high-strength steels, heat-resistant steels, and superalloys (see Volume 1 of the 10th Edition of *Metals Handbook*) as well as the production of catalysts, lubricants, and dyes.

Disposition. While molybdenum exists in various valence forms, biologic differences with respect to valence are not clear. The soluble hexavalent compounds are well absorbed from the gastrointestinal tract into the liver. It is a component of xanthine oxidase, which has a role in purine metabolism and has been shown to be a component of aldehyde oxidase and sulfite oxidase. Increased molybdenum intake in experimental animals has been shown to increase tissue levels of xanthine oxidase. In humans, molybdenum is contained principally in the liver, kidney, fat, and blood. Of the approximate total of 9 mg in the body, most is concentrated in the liver, kidney, adrenal, and omentum. More than 50% of molybdenum in the liver is contained in a nonprotein cofactor bound to the mitochondrial outer membrane and can be transferred to an apoenzyme, transforming it into an active enzyme molecule (Ref 210). The molybdenum level is relatively low in the newborn and increases until age 20, declining in concentration thereafter. More than half of the molybdenum excreted is in the urine. The blood level, at least in sheep, is in association with the red blood cells. However, molybdenum has been detected in only about 25% of the blood samples of the human urban population. The excretion of molybdenum is rapid, mainly as molybdate. Excesses may be excreted also by the bile, particularly the hexavalent forms.

Inhalation of molybdenum by guinea pigs has resulted in increased bone levels. Injected radiomolybdenum increased liver and kidney levels, but the endocrine glands were also exceptionally high in content.

Toxicity. Pastures containing 20 to 100 ppm molybdenum may produce a disease referred to as "teart" in cattle and sheep. It is characterized by anemia, poor growth rate, and diarrhea. Copper or sulfate in the diet prevents the disease, and removal of the animals from pastures containing high levels of molybdenum facilitates their rapid recovery. Prolonged exposure has led to deformities of the joints. Experimental studies have revealed differences in toxicity of molybdenum salts. Molybdenum sulfide was well tolerated in rats at 500 mg/kg/day and was not injurious to guinea pigs at 28 mg/m^3. Hexavalent compounds were more toxic. In rats molybdenum trioxide at a dose of 100 mg/kg/day, by inhalation, was irritating to the eyes and mucous membranes and subsequently lethal. After repeated oral administration at sufficient levels, fatty degeneration of the liver and kidney was induced. In comparison with chromium and tungsten salts, sodium molybdate by intraperitoneal injection was less toxic in mice.

Interesting relationships of molybdenum with other metals with respect to toxicity in cattle and sheep have been documented. For example, copper prevents the accumulation of molybdenum in the liver and may antagonize the absorption of molybdenum from food. It is reported that by alternating the intake of copper and molybdenum at weekly intervals, black sheep can be made to grow striped wool. White wool in black sheep is a sign of copper deficiency. The antagonism of copper is dependent on sulfate in the diet. It has been suggested that sulfate may displace molybdate in the body. It may be that the anemia caused by molybdenum is due to the reduction of sulfide oxidase in the liver, resulting in the formation of copper sulfide, thereby inducing a functional copper deficiency. Feeding of tungstate has also been shown to displace molybdate. In addition, it has been reported that molybdenum may promote fluoride retention and thereby decrease dental caries (Ref 59), but the incidences of caries in children living in high molybdenum areas compared to children living in normal or low molybdenum areas do not differ (Ref 211).

Selenium

The availability as well as the toxic potential for selenium and selenium compounds is related to chemical form and, most important, to solubility. Selenium occurs in nature and biologic systems as selenate (Se^{6+}), selenite (Se^{4+}), elemental selenium (Se^{0}), and selenide (Se^{2-}), and deficiency leads to a cardiomyopathy in mammals including humans (Ref 59, 212).

Selenium in foodstuffs provides a daily source of selenium (Ref 213). Seafoods, especially shrimp, meat, milk products, and grains provide the largest amounts in the diet. River water levels of selenium vary depending on environmental and geologic factors; 0.02 ppm has been reported as a representative estimate. Selenium has also been detected in urban air, presumably from sulfur-containing materials. Additional information on selenium can be found in the article "Properties of Pure Metals" in this Volume.

Disposition. Selenates are relatively soluble compounds, similar to sulfates, and are readily taken up by biologic systems, whereas selenites and elemental selenium are virtually insoluble. Because of their insolubility, these forms may be regarded as a form of inert selenium sink. Selenides of heavy metals are also very insoluble compounds, in fact, so insoluble that the *in vivo* formation of mercury selenide by dietary administration of selenite has been proposed as a method for detoxication of methyl mercury. Other metallic selenides such as arsenic, cadmium, and copper also have low solubility affecting absorption, retention, and distribution within the body of selenium and heavy metal. Elemental selenium is probably not absorbed from the gastrointestinal tract. Absorption of selenite is from the duodenum. Monogastric animals have a higher intestinal absorption than ruminants, probably because selenite is reduced to an insoluble form in rumen. Over 90% of milligram doses of sodium selenite may be absorbed by man and widely distributed in organs, with highest accumulation initially in liver and kidney, but appreciable levels remain in blood, brain, myocardium, and skeletal muscle and testis. Selenium is transferred through the placenta to the fetus, and it also appears in milk. Levels in milk are dependent on dietary intake. Selenium in red cells is associated with glutathione peroxidase and is about three times more concentrated than in plasma (Ref 214).

Selenium compounds may be biotransformed in the body by incorporation into amino acids or proteins or by methylation (Ref 215). Selenium amino acids, Se-cysteine, and Se-methionine are formed in plants and absorbed as free amino acid or from digested protein. Se-methionine can be directly incorporated into proteins in place of methionine (Ref 216). It is also suggested that selenite may be converted to Se-cysteine and incorporated into protein. Dimethyl selenium is an intermediate in the formation of a urinary metabolite, trimethyl selenium. It may be exhaled during acute selenium toxicity when its formation exceeds the rate of further methylation and urinary excretion (Ref 217).

The excretion pattern of a single exposure to selenite appears to have at least two phases: a rapid initial phase with as much as 15 to 40% of the absorbed dose excreted in the urine the first week. There is exponential excretion of the remainder of the dose with a half-life of 103 days. The half-life of Se-methionine is 234 days. In the steady state, urine contains about twice as much as feces and increased urinary levels provide a measure of exposure. Urinary selenium is usually less than 100 μg/liter.

Excretory products appear in sweat and expired air. The latter may have a garlicky odor due to dimethyl selenide. Within certain physiologic limits, the body appears to have a homeostatic mechanism for retaining trace amounts of selenium and excreting the excess material. Selenium toxicity occurs when the intake exceeds the excretory capacity (Ref 218, 219).

Essentiality. A biologic role for selenium is attributed to its incorporation in Se-cysteine at each of the four catalytic sites of the enzyme glutathione peroxidase. This enzyme uses glutathione to reduce organic hydroperoxides and protects membrane lipids and possibly proteins and nucleic acids against oxidant damage (Ref 220). Selenium is also a component of heme oxidase. The antioxidant activity of selenium-containing enzymes suggests a close relationship to vitamin E, but it may have a more subtle effect not yet defined in that selenium is beneficial to animals adequately supplied with vitamin E.

Selenium-deficient diets cause liver necrosis in rats and multiple organs (liver, heart, kidneys, skeletal muscle, and testes) in mice. In chicks, pancreatic fibrosis, exudative diathesis, and alopecia are responsive to selenium supplementation. Lambs and calves suffer from a muscle disease called stiff-lamb disease and white-muscle disease when raised in selenium ranges of selenium-deficient plants. Also, embryo mortality in ewes from selenium-deficient areas is reversed by supplementation. Liver necrosis and cardiac myopathy occur in young pigs on selenium-deficient diets and is prevented by the addition of selenium to the diet (Ref 59). While the role of selenium as an essential mineral seems certain in animals, the requirement for humans has been more difficult to establish. However, there are now reports of the efficacy of oral sodium selenite in the prophylaxis and treatment of an endemic cardiomyopathy in the People's Republic of China (Keshan disease) (Ref 221) and the alleviation by Se-methionine of muscle pain and tenderness in a New Zealand woman on intravenous feeding (Ref 222). Both reports are from regions where, for geochemical reasons, the indigenous population has a low intake of selenium. Selenium depletion has also been reported in association with cardiovascular disease and other cardiomyopathies. Although these case reports and the Chinese study are not the rigorous criteria required to establish the essentiality of a trace metal, for many it does seem that certain clinical situations may be improved with the administration of selenium (Ref 223).

Toxicity. Industrial exposure to hydrogen selenide, occurring as a result of a reaction to acid or water with metal selenides, produces "garlic" breath, nausea, dizziness, and lassitude. Eye and nasal irritation may occur. In experimental animals 10 ppm is fatal. Selenium oxychloride, a vesicant, presents an industrial hazard. In rabbits 0.01 ml applied dermally resulted in death. Percutaneous absorption increased blood and liver selenium concentrations.

Acute selenium poisoning produces central nervous system effects, which include nervousness, drowsiness, and sometimes convulsions. Symptoms of chronic inhalation exposure may include pallor, coated tongue, gastrointestinal disorders, nervousness, "garlic" breath, liver and spleen damage, anemia, mucosal irritation, and lumbar pain. It has been suggested that some of these symptoms are due to tellurium impurities (Ref 224).

"Blind stagger" caused by excess selenium in livestock consuming 100 to 1000 ppm is characterized by impairment of vision, weakness of limbs, and respiratory failure (Ref 225). Clear evidence of chronic selenium toxicity in humans occurs only in seleniferous areas when the local foods are processed. Signs of intoxication may include discolored or decayed teeth, skin eruptions, gastrointestinal distress, lassitude, and partial loss of hair and nails. Livestock foraging on plants containing about 25 ppm suffer from "alkali" disease, which is characterized by lack of vitality, loss of hair, sterility, atrophy of hooves, lameness, and anemia. Fatty necrosis of the liver is frequent. In rats given 3 ppm of the material in drinking water, selenite has been reported to be more toxic than selenate. Selenite produced increased numbers of aortic plaques and was found to be more toxic in female than male mice. Selenium has produced loss of fertility and congenital defects and is considered embryotoxic and teratogenic on the basis of animal experiments (Ref 219, 225). Selenium sulfide produced an increase in hepatocellular carcinomas and adenomas, but selenium sulfide suspension and Selsun, an antidandruff shampoo containing 2.5% selenium sulfide, applied to the skin of Swiss mice did not produce dermal tumors (Ref 226, 227).

Epidemiologic investigations have indicated a decrease in human cancer death rates (age and sex adjusted) correlated with increasing selenium content of forage crops (Ref 228). In addition, experimental evidence supports the antineoplastic effect of selenium with regard to benzo[a]pyrene- and benzanthracene-induced skin tumors in mice, N-2-fluorenylacetamide- and diethylaminoazobenzene-induced hepatic tumors in rats, and spontaneous mammary tumors in mice. A possible mechanism of the protective effects of selenium has been postulated to involve inhibition of the formation of malonaldehyde, a product of peroxidative tissue damage, which is carcinogenic.

In addition to the apparent protective effect against some carcinogenic agents, selenium is an antidote to the toxic effects of other metals, particularly arsenic, cadmium, mercury, copper, and thallium. The mechanism underlying these interactions is unknown (Ref 229).

Zinc

Zinc is a nutritionally essential metal, and deficiency results in severe health consequences. On the other hand, excessive exposure to zinc is relatively uncommon and requires heavy exposure. Zinc does not accumulate with continued exposure, but body content is modulated by homeostatic mechanisms that act principally on absorption and liver levels (Ref 59, 230, 231, 232).

Zinc is ubiquitous in the environment so that it is present in most foodstuffs, water, and air. Content may be increased in contact with galvanized copper or plastic pipes. Seafoods, meats, whole grains, dairy products, nuts, and legumes are high in zinc content. Vegetables are lower. Zinc applied to soil is taken up by growing vegetables. Zinc atmospheric levels are increased over industrial areas. The average American daily intake is approximately 12 to 15 mg, mostly from food. The industrial uses and properties of zinc are described in the article "Zinc and Zinc Alloys" in this Volume.

Disposition. About 20 to 30% of ingested zinc is absorbed. The mechanism is thought to be homeostatically controlled and is probably a carrier-mediated process (Ref 233). It is influenced by prostaglandins E_2 and F_2 and is chelated by picolinic acid—a tryptophan derivative. Deficiency of pyridoxine or tryptophan depresses zinc absorption. Within the mucosal cell, zinc induces metallothionein synthesis and, when saturated, may depress zinc absorption. In the blood, about two-thirds of the zinc is bound to albumin and most of the remainder is complexed with λ_2-macroglobulin. Zinc enters the gastrointestinal tract as a component of metallothionein secreted by the salivary glands, intestinal mucosa, pancreas, and liver. About 2 mg of zinc is filtered by the kidneys each day, and about 300 to 600 μg/day is actually excreted by normal adults. Renal tubular reabsorption is impaired by commonly prescribed drugs, such as thiazide diuretics, and is further influenced by dietary protein. There is good correlation between dietary zinc and urinary zinc excretion.

Zinc concentration in tissues varies widely. Liver receives up to about 40% of a tracer dose, declining to about 25% within five days. Liver concentration is influenced by humoral factors including adrenocorticotropic hormone, parathyroid hormone, and endotoxin. In the liver, as well as other tissues, zinc is bound to metallothionein. The greatest concentration of zinc in the body is in the prostate, probably related to the rich content of zinc-containing enzyme acid phosphatase.

Deficiency. More than 70 metalloenzymes require zinc as a cofactor, and deficiency results in a wide spectrum of clinical effects depending on age, stage of development, and deficiencies of related metals.

Zinc deficiency in humans was first characterized in 1963 by Prasad and co-workers (Ref 234) in adolescent Egyptian boys with growth failure and delayed sexual maturation and is accompanied by protein-caloric malnutrition, pellagra, iron, and folate deficiency. Zinc deficiency in the newborn may be manifested by dermatitis, loss of hair, impaired healing, susceptibility to infections, and neuropsychologic abnormalities. Dietary inadequacies coupled with liver disease from chronic alcoholism may be associated with dermatitis, night blindness, testicular atrophy, impotence, and poor wound healing. Other chronic clinical disorders, such as ulcerative colitis and the malabsorption syndromes, chronic renal disease, and the hemolytic anemias, are also prone to zinc deficiency. Many drugs affect zinc homeostasis particularly metal-chelating agents and some antibiotics, such as penicillin and isoniazid. Less common zinc deficiency may occur with myocardial infarction, arthritis, and even hypertension.

Biologic Indicators of Abnormal Zinc Homeostasis. The range of normal plasma zinc level is from 85 to 110 μg/dL. Severe deficiency may decrease plasma zinc to 40 to 60 μg/dl, accompanied by increased serum β_2-globulin and decreased α-globulin. Urine zinc excretion may decrease from over 300 μg/day to less than 100 μg/day. Zinc deficiency may exacerbate impaired copper nutrition and, of course, zinc interactions with cadmium and lead may modify the toxicity of these metals (Ref 231).

Toxicity. Zinc toxicity from excessive ingestion is uncommon, but gastrointestinal distress and diarrhea have been reported following ingestion of beverages standing in galvanized cans or from use of galvanized utensils. However, evidence of hematologic, hepatic, or renal toxicity has not been observed in individuals ingesting as much as 12 g of elemental zinc over a two-day period.

With regard to industrial exposure, metal fume fever resulting from inhalation of freshly formed fumes of zinc presents the most significant effect. The disorder has been most commonly associated with inhalation of zinc oxide fume, but it may be seen after inhalation of fumes of other metals, particularly magnesium, iron, and copper. Attacks usually begin after four to eight hours of exposure—chills and fever, profuse sweating, and weakness. Attacks usually last only 24 to 48 hours and are most common on Mondays or after holidays. The pathogenesis is not known, but is thought to be due to endogenous pyrogen released from cell lysis. Extracts prepared from tracheal mucosa and lungs of animals with experimentally induced metal fume fever produce similar symptoms when injected into other animals. Other aspects of zinc toxicity are not well established. Experimental animals have been given 100 times dietary requirements without discernible effects (Ref 235).

Exposure of guinea pigs three hours per day for six consecutive days to 5 mg/m^3 freshly formed ultrafine zinc oxide (the recommended threshold limit value, TLV) produced decrements in lung volumes and carbon monoxide diffusing capacity that persisted 72 hours after exposure. These functional changes were correlated with microscopic evidence of interstitial thickening and cellular infiltrate in alveolar ducts and alveoli (Ref 236).

Testicular tumors have been produced by direct intratesticular injection in rats and chickens. This effect is probably related to the concentration of zinc normally in the gonads and may be hormonally dependent. Zinc salts have not produced carcinogenic effects when administered to animals by other routes (Ref 237).

Metals With Toxicity Related To Medical Therapy

Metals considered in this group include aluminum, bismuth, gallium, gold, lithium, and platinum. Metals at one time were used to treat a number of human ills, particularly heavy metals like mercury and arsenic. Gold salts are still useful for the treatment of forms of rheumatism, and organic bismuth compounds are used to treat gastrointestinal disturbances. Lithium has become an important aid in the treatment of depression. The toxicologic hazards from aluminum are not from its use as an antacid but rather the accumulations that occur in bone and other tissues in patients with chronic renal failure receiving hemodialysis therapy. Platinum is receiving attention as an antitumor agent. Barium and gallium are used as a radiopaque and radiotracer material, respectively, so they do have importance in medical therapy. Toxicologic effects are unlikely and seldom occur.

Aluminum

Aluminum is one of the most abundant metals in the earth's crust, and it is ubiquitous in air and water, as well as in soil. The uses and properties associated with this metal are described extensively throughout this Volume (see in particular the section "Specific Metals and Alloys" in this Volume).

The toxicity of aluminum may be divided into three major categories: first, the effect of aluminum compounds on the gastrointestinal tract; second, the effect of inhalation of aluminum compounds; and third, systemic toxicity of aluminum (Ref 238).

Aluminum compounds can affect absorption of other elements in the gastrointestinal tract and alter intestinal function. Aluminum inhibits fluoride absorption and may decrease the absorption of calcium and iron compounds (Ref 239) and possibly the absorption of cholesterol by forming an aluminum-pectin complex that binds fats to nondigestible vegetable fibers (Ref 240). The binding of phosphorus in the intestinal tract can lead to phosphate depletion and osteomalacia (Ref 241). Aluminum may alter gastrointestinal tract motility by inhibition of acetylcholine-induced contractions and may be the explanation of why aluminum-containing antacids often produce constipation.

Pulmonary effects of aluminum occur following inhalation of bauxite (Al_2O_3-$3H_2O$) fumes. The resultant pulmonary fibrosis produces both restrictive and obstructive pulmonary disease (Ref 242). Interestingly, inhalation of aluminum mists was used in the 1930s to serve as prophylaxis of pulmonary fibrosis due to inhalation of silica particles. It is suggested that aluminum and silicic acid compete for a common reactive site in the oxidative phosphorylation pathway (Ref 243).

There has been increasing interest in the possible relationship of aluminum to dementia in humans (Ref 244). Intracerebral injection of aluminum phosphate or injection of aluminum powder in cerebrospinal fluid of animals has been noted to induce a progressive encephalopathy and neurofibrillary degeneration histologically comparable to the changes found in persons with senile and presenile dementia of the Alzheimer type (Ref 245). However, some morphologic differences have been noted at the ultrastructural level, and why specific individuals are affected by such a ubiquitous metal is an unresolved question.

A progressive fatal neurologic syndrome has also been reported in patients on long-term intermittent hemodialysis treatment for chronic renal failure (Ref 246). The first symptom in these patients is a speech disorder followed by dementia, convulsions, and myoclonus. The disorder, which typically arises after three to seven years of dialysis treatment, may be due to aluminum intoxication. Aluminum content of brain, muscle, and bone tissues is increased in these patients. Crapper (Ref 247) has shown that brain tissue of mammals normally contains 1 to 2 μg of aluminum per gram dry weight and that the toxic range is 4 to 8 μg/g dry weight of brain for the cat and rabbit.

Sources of the excess aluminum may be from oral aluminum hydroxide commonly given to these patients or from aluminum in dialysis fluid derived from tap water used to prepare the dialysate fluid. High serum and bone aluminum concentrations are generally present in these patients, and it is postulated that increased absorption may be related to increased parathyroid hormone due to low blood calcium and osteodystrophy common in patients with chronic renal disease. The syndrome may be prevented by avoidance of the use of aluminum-containing oral phosphate binders and monitoring of aluminum in the dialysate. Chelation of aluminum may be achieved with use of desferrioxamine, and progression of the dementia may be arrested or slowed (Ref 248).

Bismuth

Although bismuth is used for a variety of industrial uses (see the article "Indium and Bismuth" in this Volume), its primary use is in chemicals and pharmaceuticals. Both inorganic and organic salts have been used, depending on the specific application. There are three major categories of uses: antisyphylitic agents, topical creams, and antacids. Trivalent insoluble bismuth salts are used medicinally to control diarrhea and other types of gastrointestinal distress. Various bismuth salts have been used externally for their astringent and slight antiseptic property. Bismuth salts have also been used as radiocontrast agents. Further potential for exposure comes from the use of insoluble bismuth salts in cosmetics. Injections of soluble and insoluble salts, suspended in oil to maintain adequate blood levels, have been used to treat syphilis. Bismuth sodium thioglycollate, a water-soluble salt, was injected intramuscularly for malaria (*Plasmodium vivax*). Bismuth glycolyarsanilate is one of the few pentavalent salts that have been used medicinally. This material was formerly used for treatment of amebiasis (Ref 249). Exposure to various bismuth salts for medicinal use has decreased with the advent of newer therapeutic agents. However, in the 1970s reports appeared from France and Australia of unique encephalopathy occurring in colostomy and ileostomy patients using bismuth subgallate, bismuth subnitrate, and tripotassium-dicitrate-bismuthate for control of fecal odor and consistency. The symptoms included progressive mental confusion, irregular myoclonic jerks, a distinctive pattern of disordered gait, and a variable degree of dysarthria. The disorder was fatal to patients who continued use of the bismuth compounds, but full recovery was rapid in those in whom therapy was discontinued. The severity of the disorder seemed to be independent of dose and duration of therapy (Ref 250).

Most bismuth compounds are insoluble and poorly absorbed from the gastrointestinal tracts, or when applied to the skin, even if the skin is abraded or burned. Symptomatic patients taking bismuth subgallate had an elevated median blood bismuth level of 14.6 μg Bi/dL, patients without clinical symptoms had a median blood level of 3 μg/dL, and colostomy patients not on bismuth therapy had a median bismuth blood level of 0.8 μg/dL. Health laboratory workers had a median bismuth blood level of 1.0 μg/dL. Binding in blood is thought to be largely due to a plasma protein with a molecular weight greater than 50 000 daltons.

A diffusible equilibrium between tissues, blood, and urine is established. Tissue distribution, omitting injection depots, reveals the kidney as the site of the highest concentration. The liver concentration is considerably lower at therapeutic levels, but with massive doses in experimental animals (dogs), the kidney/liver ratio is decreased. Passage of bismuth into the amniotic fluid and into the fetus has been demonstrated. The urine is the major route of excretion. Traces of bismuth can be found in milk and saliva. The total elimination of bismuth after injection is slow and dependent on mobilization from the injection site.

Acute renal failure can occur following oral administration of such compounds as bismuth sodium triglycollamate or thioglycollate particularly in children (Ref 251). The tubular epithelium is the primary site of toxicity producing degeneration of renal tubular cells and nuclear inclusion bodies composed of a bismuth-protein complex analogous to those found in lead toxicity (Ref 252, 253).

The symptoms of chronic toxicity in humans consist of decreased appetite, weakness, rheumatic pain, diarrhea, fever, metal line on the gums, foul breath, gingivitis, and dermatitis. Jaundice and conjunctival hemorrhage are rare, but have been reported. Bismuth nephropathy with proteinuria may occur.

Chelation therapy using dimercaprol (BAL) is said to be helpful in removal of bismuth from children with acute toxicity (Ref 254).

Gallium

Gallium is of interest because of the use of radiogallium as a diagnostic tool for localization of bone lesions. It is obtained as a by-product of copper, zinc, lead, and aluminum refining and is used in high-temperature thermometers, as a substitute for mercury in arc lamps, as a component of metal alloys, as a seal for vacuum equipment, and in semiconductor applications (gallium arsenide) (see the article "Gallium and Gallium Compounds" in this Volume). It is only sparsely absorbed from the gastrointestinal tract, but concentrations of less than 1 ppm can be localized radiographically in bone lesions. Higher doses will visualize liver, spleen, and kidney as well.

Gallium is not readily absorbed by the oral route, but occurs in bone at concentrations less than 1 ppm. Increasing intake produces slight increases in gallium levels in the liver, spleen, kidney, and bone. The urine is the major route of excretion.

There are no reported adverse effects of gallium following industrial exposure. Therapeutic use of radiogallium produced some adverse effects, mild dermatitis, and gastrointestinal disturbances. Bone marrow depression has been reported and may be due largely to the radioactivity. In animals gallium acts as a neuromuscular poison and causes renal damage. Photophobia, blindness, and paralysis have been reported in rats. Renal damage ranging from cloudy swelling to tubular cell necrosis has been reported. Aplastic changes in the bone marrow have been observed in dogs (Ref 186).

Gold

Gold is widely distributed in small quantities but economically usable deposits occur as the free metal in quartz veins or alluvial gravel. Seawater contains 3 or 4 mg/ton and small amounts, 0.03 to 1 mg%, have been reported in many foods. Gold has a number of industrial uses because of its electrical and thermal conductivity and is described in the article "Precious Metals" as well as numerous other articles in this Volume.

While gold and its salts have been used for a wide variety of medicinal purposes, their present uses are limited to the treatment of rheumatoid arthritis and rare skin diseases such as discoid lupus. Gold salts are poorly absorbed from the gastrointestinal tract. Normal urine and fecal excretions of about 0.1 and 1 mg/day, respectively, have been reported. After injection of most of the soluble salts, gold is excreted via the urine, while the feces account for the major portion of insoluble compounds. Gold seems to have a long biologic half-life, and detectable blood levels can be demonstrated for ten months after cessation of treatment.

Dermatitis is the most frequently reported toxic reaction to gold and is sometimes accompanied by stomatitis. Use of gold in the form of organic salts to treat rheumatoid arthritis may be complicated by development of proteinuria and the nephrotic syndrome, which morphologically consists of an immune-complex glomerulonephritis with granular deposits along the glomerular basement membrane and in the mesangium. The pathogenesis of the immune-complex disease is not certain, but gold may behave as a hapten and generate the production of antibodies with subsequent disposition of gold protein-antibody complexes in the glomerular subepithelium. Another hypothesis is that antibodies are formed against damaged tubular structures, particularly mitochondria, providing immune complexes for the glomerular deposits (Ref 255).

The pathogenesis of the renal lesions induced by gold therapy is probably initiated by the direct toxicity of gold with tubular cell components. From experimental studies it appears that gold salts have an affinity for mitochondria of proximal tubular lining cells, which is followed by autophagocytosis and accumulation of gold in amorphous phagolysosomes (Ref 256), and gold particles can be identified in degenerating mitochondria in tubular lining cells and in glomerular epithelial cells by x-ray microanalysis (Ref 257).

Lithium

Lithium carbonate is an important aid in the treatment of depression. There must be careful monitoring of usage to provide optimal therapeutic value and not produce toxicity. Lithium is a common metal and present in many plant and animal tissues. Daily intake is about 2 mg. It is readily absorbed from the gastrointestinal tract. Distribution in the human organs is almost uniform. The normal plasma level is about 17 μg/L. The red cells contain less. Excretion is chiefly through the kidneys, but some is eliminated in the feces. The greater part of lithium is contained in the cells, perhaps at the expense of potassium. In general, the body distribution of lithium is quite similar to that of sodium, and it may be competing with sodium at certain sites, for example, in renal tubular reabsorption.

Lithium has some industrial uses, in alloys (see, for example, the article "Aluminum-Lithium Alloys" in this Volume), as a catalytic agent, and as a lubricant. Mechanical and physical characteristics of lithium are described in the article "Properties of Pure Metals" in this Volume. Lithium hydride produces hydrogen on contact with water and is used in manufacturing electronic tubes, in ceramics, and in chemical synthesis. From the industrial point of view, except for lithium hydride, none of the other salts or the metal itself is hazardous. Lithium hydride is intensely corrosive and may produce burns on the skin because of the formation of hydroxides (Ref 186, 258).

The therapeutic use of lithium carbonate may produce unusual toxic responses. These include neuromuscular changes (tremor, muscle hyperirritability, and ataxia), central nervous system changes (blackout spells, epileptic seizures, slurred speech, coma, psychosomatic retardation, and increased thirst), cardiovascular changes (cardiac arrhythmia, hypertension, and circulatory collapse), gastrointestinal changes (anorexia, nausea, and vomiting) and renal damage (albuminuria and glycosuria). The latter is believed to be due to temporary hypokalemic nephritis. These changes appear to be more frequent when the serum levels increase above 1.5 mEq/L, suggesting that careful monitoring of this parameter is needed rather than reliance on the amount given.

Chronic lithium nephrotoxicity can occur with long-term exposure even when lithium levels remain within the therapeutic range. Tubular defects, particularly nephrogenic diabetes insipidus, may occur. There is more recent awareness of the possible development of chronic interstitial nephritis (Ref 259).

The cardiovascular and nervous system changes may be due to the competitive relationship between lithium and potassium and may thus produce a disturbance in intracellular metabolism. Thyrotoxic reactions, including goiter formation, have also been suggested (Ref 260). While there has been some indication of adverse effects on fetuses following lithium treatment, none was observed in rats (4.05 mEq/kg), rabbits (1.08 mEq/kg), or primates (0.67 mEq/kg). This dose to rats was sufficient to produce maternal toxicity and effects on the pups of treated, lactating dams (Ref 261).

Lithium overdosage and toxicity may be treated by administration of diuretics and lowering of blood levels. Acetazolamide, a carbonic anhydrase inhibitor, has been used clinically. Animal studies have shown that urinary excretion of lithium can be further enhanced by the combined administration of acetazolamide and furosemide. Treatment with diuretics must be accompanied by replacement of water and electrolytes (Ref 262).

Platinum

Platinum-group metals include a relatively light triad of ruthenium, rhodium, and palladium, and the heavy metals osmium, iridium, and platinum (see the article "Precious Metals" in this Volume). They are found together in sparsely distributed mineral deposits or as a by-product of refining other metals, chiefly nickel and copper. Osmium and iridium are not important toxicologically. Osmium tetroxide, however, is a powerful eye irritant. The other metals are generally nontoxic in their metallic states but have been noted to have toxic effects in particular circumstances. Platinum is interesting because of its extensive industrial applications and use of certain complexes as antitumor agents.

Toxicological information for ruthenium is limited to references in the literature indicating that fumes may be injurious to eyes and lungs (Ref 186).

Rhodium trichloride produced death in rats and rabbits within 48 hours after intravenous administration at doses near the LD50 (approximately 200 mg/kg). It was suggested that death was attributable to central nervous system effects (Ref 263). In a single study, incorporation of rhodium (rhodium chloride) or palladium (palladous chloride) into the drinking water of mice at a concentration of 5 ppm over the lifetime of the animals produced a minimally significant increase in malignant tumors. Most of these tumors were classified as the lymphoma-leukemia type (Ref 264).

Palladium chloride is not readily absorbed from subcutaneous injection, and no adverse effects have been reported from industrial exposure. Colloid palladium ($Pd[OH]_2$) is reported to increase body temperature, produce discoloration and necrosis at the site of injection, decrease body weight, and cause slight hemolysis.

Platinum metal itself is generally harmless, but an allergic dermatitis can be produced in susceptible individuals. Skin changes are most common between the fingers and in the antecubital fossae. Symptoms of respiratory distress, ranging from irritation to an "asthmatic syndrome" with coughing, wheezing, and shortness of breath, have been reported following exposure to platinum dust. The skin and respiratory changes are termed platinosis. They are mainly confined to persons with a history of industrial exposure to soluble compounds such as sodium chloroplatinate, although cases resulting from wearing platinum jewelry have been reported.

The complex salts of platinum may act as powerful allergens, particularly ammonium hexachloroplatinate and hexachloroplatinic acid. The allergenicity appears to be related to the number of chlorine atoms present in the molecule, but other soluble nonchlorinated platinum compounds may also be allergenic. Biochemistry and antitumor activity of platinum complexes of major consideration for this group of metals are the potential antitumor and carcinogenic effects of certain neutral complexes of platinum such as *cis*-dichlorodiamine, platinum (II), and various analogs (Ref 265). They can inhibit cell division and have antibacterial properties as well. These compounds can react selectively with specific chemical sites in proteins such as disulfide bonds and terminal-NH_2 groups, with functional groups in amino acids, and in particular with receptor sites in nucleic acids. These compounds also exhibit neuromuscular toxicity and nephrotoxicity.

For antitumor activity, the complexes should be neutral and should have a pair of *cis*-leaving groups. Other metals in the group give complexes that are inactive or less active than the platinum analog. At dosages that are therapeutically effective (antitumor), these complexes produce severe and persistent inhibition of DNA synthesis and little inhibition of RNA and protein synthesis. DNA polymerase activity and transport of DNA precursors through plasma membranes are not inhibited. The complexes are thought to react directly with DNA in regions that are rich in guanosine and cytosine.

Mutagenic and Carcinogenic Effects of Platinum Complexes. *Cis*-platin platinum has been used clinically to treat some cancers of the head and neck, certain lymphomas, and testicular and ovarian tumors. *Cis*-DDP is a strong mutagen in bacterial systems and has been shown to form both intra- and interstrand cross-links probably involving the whole molecule with human DNA in HeLa cell cultures. There is also a correlation between antitumor activity of *cis*-DDP and its ability to bind DNA and induce phage from bacterial cells. It also causes chromosome aberration in cultured hamster cells and a dose-dependent increase in sister chromatid exchanges.

Although *cis*-DDP has antitumorigenic activity in experimental animals, it also seems to increase the frequency of lung adenomas and give rise to skin papillomas and carcinomas in mice. These observations are consistent with the activity of other alkylating agents used in cancer chemotherapy. There are no reports of increased risk to cancer from occupational exposure to platinum compounds.

Nephrotoxicity. *Cis*-DDP is a nephrotoxin. It produces compounds with antitumor activity and produces proximal and distal tubular cell injury mainly in the corticomedullary region where the concentration of platinum is highest (Ref 266). Although 90% of administered *cis*-platinum becomes tightly bound to plasma proteins, only unbound platinum is rapidly filtered by the glomerulus and has a half-life of only 48 minutes. Within tissues, platinum is protein bound with largest concentrations in kidney, liver, and spleen, and has a half-life of two or three days. Tubular cell toxicity seems to be directly related to dose, and prolonged weekly injection in rats causes atrophy of cortical portions of nephrons and cystic dilatation of inner cortical or medullary tubules and chronic renal failure due to tubulointerstitial nephritis (Ref 267).

Minor Toxic Metals

There are a number of metals that pose a relatively minor toxic threat. Those covered in this section include antimony, barium, indium, magnesium, silver, tellurium, thallium, tin, titanium, uranium, and vanadium.

Antimony

Antimony may have a tri- or pentavalence and it belongs to the same periodic group as arsenic. Its disposition metabolism is thought to resemble that of arsenic. It is absorbed slowly from the gastrointestinal tract, and many antimony compounds are gastrointestinal irritants. Antimony tartar has been used as an emetic. The disposition of the tri and penta forms differ. Trivalent antimony is concentrated in red blood cells and liver whereas the penta form is mostly in plasma. Both forms are excreted in feces and urine, but more trivalent antimony is excreted in urine whereas there is greater gastrointestinal excretion of pentavalent antimony. Antimony is a common air pollutant from industrial emissions, but exposure for the general population is largely from food.

Antimony is included (both ferrous and nonferrous) in alloys in the metals industry and is used for producing fireproofing chemicals, ceramics, glassware, and pigments (properties are described in the article "Properties of Pure Metals"). It has been used medicinally as an antiparasitic agent. Accidental poisonings can result in acute toxicity, which produces severe gastrointestinal symptoms including vomiting and diarrhea.

Most information about antimony toxicity has been obtained from industrial experiences. Occupational exposures are usually by inhalation of dust containing antimony compounds, antimony penta and trichloride, trioxide and trisulfide. Effects may be acute, particularly from the penta and trichloride exposures, producing a rhinitis and even acute pulmonary edema. Chronic exposures by inhalation of other antimony compounds result in rhinitis, pharyngitis, tracheitis, and, over the longer term, bronchitis and eventually pneumoconiosis with obstructive lung disease and emphysema. Antimony does accumulate in lung tissue (Ref 268).

Oral feeding of antimony to rats has not produced an excess of tumors. However, increased chromosome defects occur when human lymphocytes are incubated with a soluble antimony salt (Ref 269), and Syrian hamster embryo cells undergo neoplastic transformation when treated with antimony acetate (Ref 270). Transient skin eruptions, "antimony spots," may occur in workers with chronic exposure.

Antimony may also form an odorless toxic gas, stibine (H_3Sb), which, like arsine, causes hemolysis.

Barium

Barium is used in various alloys, in paints, soap, paper, and rubber, and in the manufacture of ceramics and glass. (Its properties are described in the article "Properties of Pure Metals" in this Volume.) Barium fluorosilicate and carbonate have been used as insecticides. Barium sulfate, an insoluble compound, is used as a radiopaque aid to x-ray diagnosis. Barium is relatively abundant in nature and is found in plants and animal tissue. Plants accumulate barium from the soil. Brazil nuts have very high concentrations (3000 to 4000 ppm). Some water contains barium from natural deposits.

The toxicity of barium compounds depends on their solubility. The soluble compounds of barium are absorbed, and small amounts are accumulated in the skeleton. The lung has an average concentration of 1 ppm (dry weight). The kidney, spleen, muscle, heart, brain, and liver concentrations are 0.10, 0.08, 0.08, 0.05, 0.05, and 0.03 ppm, respectively. Although some barium is excreted in urine, it is reabsorbed by the renal tubules. The major route of excretion is the feces. Occupational poisoning to barium is uncommon, but a benign pneumoconiosis (baritosis) may result from inhalation of barium sulfate (barite) dust and barium carbonate. It is not incapacitating and is usually reversible with cessation of exposure. Accidental poisoning from ingestion of soluble barium salts has resulted in gastroenteritis, muscular paralysis, decreased pulse rate, and ventricular fibrillation and extrasystoles. Potassium deficiency occurs in acute poisoning, and treatment with intravenous potassium appears beneficial. The digitalislike toxicity, muscle stimulation, and central nervous system effects have been confirmed by experimental investigation (Ref 271).

Indium

Indium is a rare metal whose toxiocologic importance was related to its use in alloys, solders, and as a hardening agent for bearings (see the article "Indium and Bismuth" in this Volume). Use in the electronic industry for production of semiconductors and photovoltaic cells may greatly expand worker exposure. It is currently being used in medicine for scanning of organs and treatment of tumors. Indium is poorly absorbed from the gastrointestinal tract. It is excreted in the urine and feces. Its tissue distribution is relatively uniform. The kidney, liver, bone, and spleen have relatively high concentrations. Intratracheal injections produce similar concentrations, but the concentration in the tracheobronchial lymph nodes is increased.

There are no meaningful reports of human toxicity to indium. From animal experiments it is apparent that toxicity is related to the chemical form. Indium chloride given intravenously to mice produces renal toxicity and liver necrosis. These effects are accompanied by induction of P-450-dependent microsomal enzyme activity and decreased activity of heme-synthesizing enzymes (Ref 272). Hydrated indium oxide produces damage to phagocytic cells in liver and the reticuloendothelial system (Ref 273).

Magnesium

Magnesium is used in lightweight alloys, as an electrical conductive material, and for incendiary devices such as flares. The properties and applications are described in several articles in this Volume (see in particular, "Selection and Application of Magnesium"). It is also an essential nutrient whose deficiency causes neuromuscular irritability, calcification, and cardiac and renal damage, which can be prevented by supplementation. The deficiency is called "grass staggers" in cattle and "magnesium tetany" in calves. Magnesium is a cofactor of many enzymes; it is apparently associated with phosphate in these functions.

Magnesium citrate, oxide, sulfate, hydroxide, and carbonate are widely taken as antacids or one of the constituents of the universal antidote for poisoning. Topically, the sulfate is also used widely to relieve inflammation. Magnesium sulfate may be used as a parenterally administered central depressant. Its most frequent use for this purpose is in the treatment of seizures associated with eclampsia of pregnancy and acute nephritis.

Nuts, cereals, seafoods, and meats are high dietary sources of magnesium. The average city water contains about 6.5 ppm, but varies considerably, increasing with the hardness of the water (Ref 274).

Disposition. Magnesium salts are poorly absorbed from the intestine. In cases of overload this may be due in part to their dehydrating action. Magnesium is absorbed mainly in the small intestine. The colon also absorbs some. Calcium and magnesium are competitive with respect to their absorptive sites, and excess calcium may partially inhibit the absorption of magnesium.

Magnesium is excreted into the digestive tract by the bile and pancreatic and intestinal juices. A small amount of radiomagnesium given intravenously appears in the gastrointestinal tract. The serum levels are remarkably constant. There is an apparent obligatory urinary loss of magnesium, which amounts to about 12 mg/day, and the urine is the major route of excretion under normal conditions. Magnesium found in the stool is probably not absorbed. Magnesium is filtered by the glomeruli and reabsorbed by the renal tubules. In the blood plasma about 65% is in the ionic form, while the remainder is bound to protein. The former is that which appears in the glomerular filtrate. Mercurial diuretics cause excretion of magnesium as well as potassium, sodium, and calcium. Excretion also occurs in the sweat and milk. Endocrine activity, particularly of the adrenocortical hormones, aldosterone, and parathyroid hormone, has an effect on magnesium levels, although these effects may be related to the interaction of calcium and magnesium.

Tissue distribution studies indicate that of the 20 g body burden, the majority is intracellular in the bone and muscle. Bone concentration of magnesium decreases as calcium increases. Most of the remaining tissues have higher concentrations than blood, except for fat and omentum. With age, the aorta tends to accumulate magnesium along with calcium, perhaps as a function of atherosclerotic disease.

Toxicity. Freshly generated magnesium oxide can cause metal fume fever if inhaled in sufficient amounts, analogous to the effect caused by zinc oxide. Both zinc and magnesium exposure of animals produced similar effects. It is reported that particles of magnesium in the subcutaneous tissue produce lesions that resist healing. In animals, magnesium subcutaneously or intramuscularly administered produces gas gangrene as a result of interaction with the body fluids and subsequent generation of hydrogen and magnesium hydroxide. The tissue lesion is reversible.

Conjunctivitis, nasal catarrh, and coughing up of discolored sputum results from industrial inhalation exposure. With industrial exposures, increases of serum magnesium up to twice the normal levels failed to produce ill effects but were accompanied by calcium increases. Intoxication occurring after oral administration of magnesium salts is rare, but may be present in the face of renal impairment. The symptoms include a sharp drop in blood pressure and respiratory paralysis due to central nervous system depression (Ref 186).

Silver

The principal industrial use of silver is as silver halide in the manufacture of photographic plates. Other uses are for jewelry, coins, electrical contact materials, and eating utensils (see the articles "Precious Metals" and "Electrical Contact Materials" in this Volume). Silver nitrate is used for making indelible inks and for medicinal purposes. The use of silver nitrate for prophylaxis of ophthalmia neonatorum is a legal requirement in some states. Other medicinal uses of silver salts are as a caustic, germicide, antiseptic, and astringent.

Silver does not occur regularly in animal or human tissue. The major effect of excessive absorption of silver is local or generalized impregnation of the tissues where it remains as silver sulfide, which forms an insoluble complex in elastic fibers resulting in argyria. Silver can be absorbed from the lungs and gastrointestinal tract. Complexes with serum albumin accumulate in the liver from which a fractional amount is excreted. Intravenous injection produces accumulation in the spleen, liver, bone marrow, lungs, muscle, and skin. The major route of excretion is via the gastrointestinal tract. Urinary excretion has not been reported to occur even after intravenous injection.

Industrial argyria, a chronic occupational disease, has two forms, local and generalized. The local form involves the formation of gray-blue patches on the skin or may manifest itself in the conjunctiva of the eye. In generalized argyria, the skin shows widespread pigmentation, often spreading from the face to most uncovered parts of the body. In some cases the skin may become black with a metallic luster. The eyes may be affected to such a point that the lens and vision are disturbed. The respiratory tract may also be affected in severe cases.

Large oral doses of silver nitrate cause severe gastrointestinal irritation due to its caustic action. Lesions of the kidneys and lungs and the possibility of arteriosclerosis have been attributed to both industrial and medicinal exposures. Large doses of colloidal silver administered intravenously to experimental animals produced death due to pulmonary edema and congestion. Hemolysis and resulting bone marrow hyperplasia have been reported. Chronic bronchitis has also been reported to result from medicinal use of colloidal silver (Ref 186, 275).

Tellurium

Tellurium is found in various sulfide ores along with selenium and is produced as a by-product of metal refineries. Its industrial uses include an additive in steels to improve machinability, an additive in cast irons, applications in the refining of copper and in the manufacture of rubber (see the article "Properties of Pure Metals" in this Volume for additional information). Tellurium vapor is used in "daylight" lamps. It is used in various alloys as a catalyst and as a semiconductor.

Condiments, dairy products, nuts, and fish have high concentrations of tellurium. Food packaging contains some tellurium; higher concentrations are found in aluminum cans than tin cans. Some plants, such as garlic, accumulate tellurium from the soil. Potassium tellurate has been used to reduce sweating.

The average body burden in humans is about 600 mg; the majority is in bone. The kidney is the highest in content among the soft tissues. Some data suggest that tellurites also accumulate in liver (Ref 276). Soluble tetravalent tellurites, absorbed into the body after oral administration, are reduced to tellurides, partly methylated, and then exhaled as dimethyl telluride. The latter is responsible for the garlic odor in persons exposed to tellurium compounds. Tellurium in the food is probably in the form of tellurates. The urine and bile are the principal routes of excretion. Sweat and milk are secondary routes of excretion.

Tellurates and tellurium are of low toxicity, but tellurites are generally more toxic. Acute inhalation exposure results in decreased sweating, nausea, a metallic taste, and sleeplessness. The typical garlic breath is a reasonable indicator of exposure to tellurium by the dermal, inhalation, or oral routes. Serious cases of tellurium intoxication from industrial exposure have not been reported. In rats, chronic exposure to high doses of tellurium dioxide has produced decreased growth and necrosis of the liver and kidney (Ref 186, 277).

Sodium tellurite at 2 ppm in drinking water or potassium tellurate at 2 ppm of tellurium plus 0.16 μg/g in the diet of mice for their lifetime produced no effects in the tellurate group. The females of the tellurite (tetravalent) group did not live as long. In rats, 500 ppm in the diet of pregnant females induced hydrocephalus in the offspring. Abnormalities of and reduction in numbers of mitochondria were thought to be possible cellular causes of the transplacental effect.

One of the few serious recorded cases of tellurium toxicity resulted from accidental poisoning by injection of tellurium into the ureters during retrograde pyelography. Two of the three victims died. Stupor, cyanosis, vomiting, garlic breath, and loss of consciousness were observed in this unlikely incident.

Dimercaprol treatment for tellurium increases the renal damage. While ascorbic acid decreases the characteristic garlic odor, it may also adversely affect the kidneys in the presence of increased amounts of tellurium (Ref 278).

Thallium

Thallium is one of the more toxic metals and can cause neural, hepatic, and renal injury. It may also cause deafness and loss of vision. It is obtained as a by-product of the refining of iron, cadmium, and zinc. It is used as a catalyst, in certain alloys, optical lenses, jewelry, low-temperature thermometers, semiconductors, dyes and pigments, and scintillation counters (properties of thallium are summarized in the article "Properties of Pure Metals" in this Volume). It has been used medicinally as a depilatory. Thallium compounds, chiefly thallous sulfate, have been used as rat poison and insecticides. This is one of the commonest sources of thallium poisoning.

Disposition. Thallium is not a normal constituent of animal tissues. It is absorbed through the skin and gastrointestinal tract. After parenteral administration a small amount can be identified in the urine within a few hours. The highest concentrations after poisoning are in the kidney and urine. The intestines, thyroids, testes, pancreas, skin, bone, and spleen have lesser amounts. The brain and liver concentrations are still lower. Following the initial exposure, large amounts are excreted in urine during the first 24 hours, but after that period excretion is slow and the feces may be an important route of excretion.

Toxicology. There are numerous clinical reports of acute thallium poisoning in humans characterized by gastrointestinal irritation, acute ascending paralysis, and psychic disturbances. Acute toxicity studies in rats have indicated that thallium is quite toxic. It has an oral LD_{50} of approximately 30 mg/kg. The estimated lethal dose in humans, however, is 8 to 12 mg/kg. Rat studies also indicate that thallium oxide, while relatively insoluble, is more toxic orally than by the intravenous or intraperitoneal route (Ref 279). The acute cardiovascular effects of thallium ions probably result from competition with potassium for membrane transport systems, inhibition of mitochondrial oxidative phosphorylation, and disruption of protein synthesis. It also alters heme metabolism.

The signs of subacute or chronic thallium poisoning in rats were hair loss, cataracts, and hindleg paralysis occurring with some delay after the initiation of dosing. Renal lesions were observed at gross necropsy. Histologic changes revealed damage of the proximal and distal renal tubules. The central nervous system changes were most severe in the mesencephalon where necrosis was observed. Perivascular cuffing was also reported in several other brain areas. Electron microscope examination indicated that the mitochondria in the kidney may have been the first organelles affected. Liver mitochondria also revealed degenerative changes. The livers of newborn rats whose dams had been treated throughout pregnancy showed these changes. Similar mitochondrial changes were observed in the intestine, brain, seminal vesicle, and pancreas. It has been suggested that thallium may combine with the sulfhydryl groups in the mitochondria and thereby interfere with oxidative phosphorylation (Ref 280). A teratogenic response to thallium salts characterized as achondroplasia (dwarfism) has been described in rats (Ref 281).

In humans, fatty infiltration and necrosis of the liver, nephritis, gastroenteritis, pulmonary edema, degenerative changes in the adrenals, degeneration of peripheral and central nervous system, alopecia, and in some cases death have been reported as a result of long-term systemic thallium intake. These cases usually are caused by the contamination of food or the use of thallium as a depilatory. Industrial poisoning is a special risk in the manufacture of fused halides for the production of lenses and windows. Loss of vision plus the other signs of thallium poisoning have been related to industrial exposures (Ref 180, 273).

Tin

Tin is used in the manufacture of tinplate, in food packaging, and in solder, bronze, and brass (see the article "Tin and Tin Alloys" in this Volume). Stannous and stannic chlorides are used in dyeing textiles. Organic tin compounds have been used in fungicides, bactericides, and slimicides, as well as in plastics as stabilizers. The disposition and possible health effects of inorganic and organic tin compounds have been summarized in a WHO report (Ref 282).

Disposition. There is only limited absorption of even soluble tin salts such as sodium stannous tartrate after oral administration. Ninety percent of the tin administered in this manner is recovered in feces. The small amounts absorbed are reflected by increases in the liver and kidneys. Injected tin is excreted by the kidneys, with smaller amounts in bile. A mean normal urine level of 16.6 μg/liter or 23.4 μg/day has been reported. The majority of inhaled tin or its salts remains in the lungs, most extracellularly, with some in the macrophages, in the form of SnO_2. The organic tins, particularly triethyltin, may be somewhat better absorbed. The tissue distribution of tin from this material shows highest concentrations in the blood and liver, with smaller amounts in the muscle, spleen, heart, or brain. Tetraethyltin is converted to triethyltin *in vivo*.

Chronic inhalation of tin in the form of dust or fumes leads to benign pneumoconiosis. Tin hydride (SnH_4) is more toxic to mice and guinea pigs than is arsine; however, its effects appear mainly in the central nervous system and no hemolysis is produced. Orally, tin or its inorganic compounds require relatively large doses (500 mg/kg for 14 months) to produce toxicity. The use of tin in food processing seems to demonstrate little hazard. The average United States daily intake, mostly from foods as a result of processing, is estimated at 17 mg. Inorganic tin salts given by injection produce diarrhea, muscle paralysis, and twitching.

Toxicology. Some organic tin compounds are highly toxic, particularly triethyltin. Trialkyl compounds including triethyltin cause an encephalopathy and cerebral edema. Toxicity declines as the number of carbon atoms in the chain increases. An outbreak of almost epidemic nature took place in France due to the oral ingestion of a preparation (Stalinon) containing diethyltin diodide for treatment of skin disorders (Ref 283).

Excessive industrial exposure to triethyltin has been reported to produce headaches, visual defects, and EEG changes that were very slowly reversed (Ref 284). Experimentally, triethyltin produces depression and cerebral edema. The resulting hyperglycemia may be related to the centrally mediated depletion of catecholamines from the adrenals. Acute burns or subacute dermal irritation has been reported among workers as a result of tributyltin. Triphenyltin has been shown to be a potent immunosuppressant (Ref 285). Inhibition in the hydrolysis of adenosine triphosphate and uncoupling of oxidative phosphorylation taking place in the mitochondria have been suggested as the cellular mechanisms of tin toxicity (Ref 282).

Titanium

Most titanium compounds are in the oxidation state +4 (titanic), but oxidation state +3 (titanous) and oxidation state +2 compounds as well as several organometallic compounds do occur. Titanium dioxide, the most widely used compound, is a white pigment used in paints and plastics, as a food additive to whiten flour, dairy products, and confections, and as a whitener in cosmetic products. Because of its high strength, low density, and resistance to corrosion, it has many metallurgical applications, including surgical implants and prostheses (see the articles on wrought, cast, and powder metallurgy titanium alloys in this Volume). It occurs widely in the environment; it is present in urban air, rivers, and drinking water and is detectable in many foods.

Disposition. Approximately 3% of an oral dose of titanium is absorbed. The majority of that absorbed is excreted in the urine. The normal urine concentration has been estimated at 10 μg/L (Ref 265, 286).

The estimated body burden of titanium is about 15 mg. Most of it is in the lungs, probably as a result of inhalation exposure. Inhaled titanium tends to remain in the lungs for long periods. It has been estimated that about one-third of the inhaled titanium is retained in the lungs. The geographic variation in lung burden is to some extent dependent on air concentration. For example, concentrations of 430, 1300, and 91 ppm in ashed lung tissue have been reported for the United States, Delhi, and Hong Kong, respectively. Mean concentrations of 8 and 6 ppm for the liver and kidney, respectively, were reported in the United States. Newborns have little titanium. Lung burdens tend to increase with age.

Toxicology. Occupational exposure to titanium may be heavy, and concentrations in air up to 50 mg/m^3 have been recorded. Titanium dioxide has been classified as a nuisance particulate with a TLV of 10 mg/m^3. Nevertheless, slight fibrosis of lung tissue has been reported following inhalation exposure to titanium dioxide pigment, but the injury was not disabling. Otherwise, titanium dioxide has been considered physiologically inert by all routes (ingestion, inhalation, dermal, and subcutaneous). The metal and other salts are also relatively nontoxic except for titanic acid, which, as might be expected, will produce irritation (Ref 287).

A titanium coordination complex, titanocene, suspended in trioctanoin, administered by intramuscular injection to rats and mice, produced fibrosarcomas at the site of injection and hepatomas and malignant lymphomas (Ref 288). A titanocene is a sandwich arrangement of titanium between two cyclopentadiene molecules. Titanium dioxide was found not to be carcinogenic in a bioassay study in rats and mice (Ref 289).

Uranium

The chief raw material of uranium is pitchblende or carnotite ore. This element is largely limited to use as a nuclear fuel although, as described in the article "Uranium and Uranium Alloys" in this Volume, it is also used in applications where its high density can be taken advantage of (for example, kinetic energy penetrators).

The uranyl ion is rapidly absorbed from the gastrointestinal tract. About 60% is carried as a soluble bicarbonate complex, while the remainder is bound to plasma protein. Sixty percent is excreted in the urine within 24 hours. About 25% may be fixed in the bone (Ref 290). Following inhalation of the insoluble salts, retention by the lungs is prolonged. Uranium tetrafluoride and uranyl fluoride can produce a typical toxicity because of hydrolysis to hydrogen fluoride (HF). Skin contact (burned skin) with uranyl nitrate has resulted in nephritis.

The soluble uranium compound (uranyl ion) and those that solubilize in the body by the formation of bicarbonate complex produce systemic toxicity in the form of acute renal damage and renal failure, which may be fatal. However, if exposure is not severe enough, the renal tubular epithelium is regenerated and recovery occurs. Renal toxicity with the classic signs of impairment, including albuminuria, elevated blood urea nitrogen, and loss of weight, is brought about by filtration of the bicarbonate complex through the glomerulus, reabsorption by the proximal tubule, liberation of uranyl ion, and subsequent damage to the proximal tubular cells. Uranyl ion is most likely concentrated intracellularly in lysosomes (Ref 291, 292, 293).

Inhalation of uranium dioxide dust by rats, dogs, and monkeys at a concentration of 5 mg U/m^3 for up to 5 years produced accumulation in the lungs and tracheobronchial lymph nodes that accounted for 90% of the body burden. No evidence of toxicity was observed despite the long duration of observation (Ref 294).

Vanadium

Vanadium is a ubiquitous element. It is a by-product of petroleum refining, and vanadium pentoxide is used as a catalyst in the various chemicals including sulfuric acid. It is used in the hardening of steel (see Volume 1, 10th Edition of *Metals Handbook*), in the manufacture of pigments, in photography, and in insecticides (properties are summarized in the article "Properties of Pure Metals" in this Volume). It is common in many foods; significant amounts are found in milk, seafoods, cereals, and vegetables. Vanadium has a natural affinity for fats and oils; food oils have high concentrations. Municipal water supplies may contain on the average about 1 to 6 ppb. Urban air contains some vanadium, perhaps due to the use of petroleum products or from refineries (Table 1), about 30 mg. The largest single compartment is the fat. Bone and teeth stores contribute to the body burden. It has been postulated that some homeostatic mechanism maintains the normal levels of vanadium in the face of excessive intake, since the element, in most forms, is moderately absorbed. The principal route of excretion of vanadium is the urine. The normal serum level is 35 to 48 μg/100 ml. When excess amounts of vanadium are in the diet, the concentration in the red cells tends to increase. Parenteral administration increases levels in the liver and kidney, but these increased amounts may only be transient. The lung tissue may contain some vanadium, depending on the exposure by that route, but normally the other organs contain negligible amounts.

The toxic action of vanadium is largely confined to the respiratory tract. Bronchitis and bronchopneumonia are more frequent in workers exposed to vanadium compounds. In industrial exposures to vanadium pentoxide dust a greenish-black discoloration of the tongue is characteristic. Irritant activity with respect to skin and eyes has also been ascribed to industrial exposure. Gastrointestinal distress, nausea, vomiting, abdominal pain, cardiac palpitation, tremor, nervous depression, and kidney damage, too, have been linked with industrial vanadium exposure.

Ingestion of vanadium compounds (V_2O_5) for medicinal purposes produced gastrointestinal disturbances, slight abnormalities of clinical chemistry related to renal function, and nervous system effects. Acute vanadium poisoning in animals is characterized by marked effects on the nervous system, hemorrhage, paralysis, convulsions, and respiratory depression. Short-term inhalation exposure of experimental animals tends to confirm the effects on the lungs as well as the effect on the kidney. In addition, experimental investigations have suggested that the liver, adrenals, and bone marrow may be adversely affected by subacute exposure at high levels (Ref 295).

ACKNOWLEDGMENT

This article was revised and reprinted with permission from "Toxic Effects of Metals" by Robert A. Goyer in *Casarett and Doull's Toxicology, The Basic Science of Poisons*, 3rd Ed., C.D. Klaassen, M.O. Amdur, and J. Doull, Ed., Copyright © 1986, Macmillan Publishing Company, a Division of Macmillan, Inc.

REFERENCES

1. K. Beijer and A. Jernelöv, Sources, Transport and Transformation of Metals in the Environment, in *Handbook on the Toxicology of Metals*, 2nd ed., Vol 1, *General Aspects*, L. Friberg, G.F. Nordberg, and V.B. Vouk, Ed., Elsevier, 1986, p 68-74
2. Y.-H. Li, Geochemical Cycles of Elements and Human Perturbation, *Geochim. Cosmochim. Acta*, Vol 45, 1981, p 2073-2084
3. A. Ng and C. Patterson, Natural Concentrations of Lead in Ancient Arctic and Antarctic Ice, *Geochim. Cosmochim. Acta*, Vol 45, 1981, p 2109-2121
4. V.B. Vouk and W.T. Piver, Metallic Elements in Fossil Fuel Combustion and Products: Amounts and Form of Emissions and Evaluation of Carcinogenicity and Mutagenicity, *Environ. Health Perspect.*, Vol 47, 1983, p 201-226
5. R.J. Thompson, Collection and Analysis of Airborne Metallic Elements, in *Ultratrace Metal Analysis in Biological Sciences and Environment*, T.H. Risby, Ed., American Chemical Society, 1979, p 54-72
6. Committee on Medical and Biological Effects of Atmospheric Pollutants, *Drinking Water and Health*, National Academy of Sciences, 1977
7. C. Levinson, *Threshold Limit Values, Best Prevailing Standards*, International Federation of Chemical, Energy and General Workers Unions, 1982
8. *Handbook on the Toxicology of Metals*, 2nd ed., Vol 1, *General Aspects*, L. Friberg, G.F. Nordberg, and V.B. Vouk, Ed., Elsevier, 1986
9. M. Laker, On Determining Trace Element Levels in Man: The Uses of Blood and Hair, *Lancet*, Vol 1, 31 July 1982, p 260-262
10. R.A. Goyer, Metal-Protein Complexes in Detoxification Processes, in *Clinical Chemistry and Clinical Toxicology*, Vol 2, S.S. Brown, Ed., Academic Press, 1984
11. G.F. Nordberg, B.A. Fowler, L. Friberg, A. Jernelov, N. Nelson, M. Piscator, H.H. Sandstead, J. Vostal, and V.B. Vouk, Factors Influencing Metabolism and Toxicity of Metals: A Consensus Report, *Environ. Health Perspect.*, Vol 25, 1978, p 3-42
12. G. Kazantzis, The Role of Hypersensitivity and the Immune Response in Influencing Susceptibility to Metal Toxicity, *Environ. Health Perspect.*, Vol 25, 1978, p 111-118
13. *IARC Monograph on the Evaluation of the Carcinogenic Risk of Chemicals to Humans. Some Metals and Metallic Compounds*, Vol 23, World Health Organization, International Agency for Research on Cancer, 1980
14. L. Friberg and N. Nelson, Introduction, General Findings and General Recommendations. Workshop/Conference on the Role of Metals in Carcinogenesis, *Environ. Health Perspect.*, Vol 40, 1981, p 5-10
15. M. Costa, *Metal Carcinogenesis Testing, Principles and In Vitro Methods*, The Humana Press, 1980, p 71
16. D.R. Williams and B.W. Halstead, Chelating Agents in Medicine, *Clin. Toxicol.*, Vol 19, 1982-83, p 1081-1115
17. R.A. Peters, *Biochemical Lesions and Lethal Synthesis*, Macmillan Publishing Co., New York, 1965, p 40-59
18. I.C. Gunsalus, The Chemistry and Function of the Pyruvate Oxidation Factor (Lipoic Acid), *J. Cell. Comp. Physiol.*, Vol 41 (Suppl. 1), 1953, p 113-136
19. J.J. Chisholm, Jr. and D. Thomas, Use of 2,3-Dimercaptopropane-1-Sulfonate in Treatment of Lead Poisoning in Children, *J. Pharmacol. Exp. Ther.*, Vol 235, 1985, p 665-669
20. B. Gabard, Treatment of Methyl Mercury Poisoning in the Rat With Sodium 2,3-Dimercaptopropane-i-Sulfonate: Influence of Dose and Mode of Administration, *Toxicol. Appl. Pharmacol.*, Vol 38, 1976, p 415-424
21. J.J. Chisolm, Jr., Chelation Therapy in Children With Subclinical Plumbism, *Pediatrics*, Vol 53, 1974, p 441-443
22. J.M. Walshe, Endogenous Copper Clearance in Wilson's Disease: A Study of the Mode of Action of Penicillamine, *Clin. Sci.*, Vol 26, 1964, p 461-469
23. J.M. Walshe, Assessment of Treatment of Wilson's Disease With Triethylene Tetramine 2HCl (Trien 2HCl), in *Biological Aspects of Metals and Metal-Related Diseases*, B. Sarkar, Ed., Raven Press, 1983, p 243-261
24. M.G. Cherian, Chelation of Cadmium With BAL and DTPA in Rats, *Nature*, Vol 287, 1980, p 871-872

25. H. Keberle, The Biochemistry of Desferrioxamine and its Relation to Iron Metabolism, *Ann. NY Acad. Sci.*, Vol 119, 1964, p 758-768
26. F.W. Sunderman, Sr., Efficacy of Sodium Diethyldithiocarbamate (Dithiocarb) in Acute Nickel Carbonyl Poisoning, *Ann. Clin. Lab. Sci.* Vol 9, 1979, p 1-10
27. *Environmental Health Criteria*, Vol 18, *Arsenic*, World Health Organization, 1981
28. P. Landrigan, Arsenic—State of the Art, *Am. J. Ind. Med.*, Vol 2, 1981, p 5-15
29. M. Vahter and H. Norin, Metabolism ^{74}As-Labeled Trivalent and Pentavalent Inorganic Arsenic in Mice, *Environ. Res.*, Vol 21, 1980, p 446-457
30. G.K.H. Tam, S.M. Charbonneau, F. Bryce, C. Pomroy, and E. Sandi, Metabolism of Inorganic Arsenic (^{74}As) in Humans Following Oral Ingestion, *Toxicol. Appl. Pharmacol.*, Vol 50, 1979, p 319-322
31. E.A. Crecelius, Changes in the Chemical Speciation of Arsenic Following Ingestion by Man, *Environ. Health Perspect.*, Vol 19, 1977, p 147-150
32. V.H. Ferm, Arsenic as a Teratogenic Agent, *Environ. Health Perspect.*, Vol 19, 1977, p 215-217
33. B.T. Kagey, J.E. Bumgarner, and J.P. Creason, Arsenic Levels in Maternal-Fetal Tissue Sets, in *Trace Substances in Environmental Health XI*, O.D. Hemphill, Ed., University of Missouri Press, 1977, p 252-256
34. R.M. Johnstone, Sulfhydryl Agent: Arsenicals, in *Metabolic Inhibitors: A Comprehensive Treatise*, Vol 2, R.M. Hochster and J.H. Quasital, Ed., Academic Press, 1963
35. V. Bencko, R. Benes, and M. Cikrt, Biotransformation of As(III) to As(V) and Arsenic Tolerance, *Arch. Toxicol.*, Vol 36, 1976, p 159-162
36. M.M. Brown, B.C. Rhyne, R.A. Goyer, and B.A. Fowler, Intracellular Effects of Chronic Arsenic Administration on Renal Proximal Tubule Cells, *J. Toxicol. Environ. Health*, Vol 1, 1976, p 505-514
37. A.L. Fluharty and D.R. Sanadi, On the Mechanism of Oxidative Phosphorylation. II. Effects of Arsenite Alone and in Combination With 2,3-Dimercaptopropanol., *J. Biol. Chem.*, Vol 236, 1961, p 2772-2778
38. R.A. Mitchell, B.F. Change, C.H. Huang, and E.G. DeMaster, Inhibition of Mitochondrial Energy-Linked Functions by Arsenate, *Biochemistry*, Vol 10, 1971, p 2049-2054
39. B.L. Vallee, D.D. Ulmer, and W.E.C. Wacker, Arsenic Toxicology and Biochemistry, *AMA Arch. Ind. Health*, Vol 21, 1960, p 132-151
40. W.-P. Tseng, Effects and Dose-Response Relationships of Skin Cancer and Blackfoot Disease With Arsenic, *Environ. Health Perspect.*, Vol 19, 1977, p 109-119
41. H. Popper, L.B. Thomas, N.C. Telles, H. Falk, and I.J. Selikoff, Development of Hepatic Angiosarcoma in Man Induced by Vinyl Chloride Thorotrast, and Arsenic, *Am. J. Pathol.*, Vol 92, 1978, p 349-369
42. M.G. Ott, B.B. Holder, and H.L. Gordon, Respiratory Cancer and Occupational Exposure to Arsenicals, *Arch. Environ. Health*, Vol 29, 1974, p 250-255
43. A.M. Lee and J.F. Fraumeni, Jr., Arsenic and Respiratory Cancer in Man, an Occupational Study, *JNCI*, Vol 42, 1969, p 1945-2052
44. S.S. Pinto, V. Henderson, and P.E. Enterline, Mortality Experience of Arsenic-Exposed Workers, *Arch. Environ. Health*, Vol 33, 1978, p 325-331
45. G. Pershagan, Carcinogenicity of Arsenic, *Environ. Health Perspect.*, Vol 40, 1981, p 93-100
46. S. Ivankovic, G. Eisenbrandt, and R. Preusmann, Lung Carcinoma Induction in BD Rats After Single Intratracheal Instillation of an Arsenic-Containing Pesticide Mixture Formerly Used in Vineyards, *Int. J. Cancer*, Vol 24, 1979, p 786-792
47. G. Lofroth and B.N. Ames, Mutagenicity of Inorganic Compounds in *Salmonella Typhimurium*: Arsenic, Chromium and Selenium, *Mutat. Res.*, Vol 53, 1978, p 65
48. B.A. Fowler and J.B. Weissberg, Arsine Poisoning, *New Engl. J. Med.*, Vol 291, 1974, p 1171-1174
49. M. Kuschner, The Carcinogenicity of Beryllium, *Environ. Health Perspect.*, Vol 40, 1981, p 101-106
50. A.L. Reeves, Absorption of Beryllium From the Gastrointestinal Tract, *Arch. Environ. Health*, Vol 11, 1965, p 209-214
51. H.L. Hardy and I.R. Tabershaw, Delayed Chemical Pneumonitis Occurring in Workers Exposed to Beryllium, *J. Ind. Hyg. Toxicol.*, Vol 28, 1946, p 197-216
52. R. Doll, L. Fishbein, P. Infante, P. Landrigan, J.W. Lloyd, T.J. Mason, E. Mastromalteo, T. Norseth, G. Pershagan, U. Saffiotti, and R. Saracci, Problems of Epidemiological Evidence, *Environ. Health Perspect.*, Vol 40, 1981, p 11-20
53. J.A. DiPaolo and B.C. Casto, Quantitative Studies of *In Vitro* Morphologic Transformation of Syrian Hamster Cells by Inorganic Metal Salts, *Cancer Res.*, Vol 39, 1976, p 1008-1019
54. H.S. Rosenkrantz and L.A. Poirier, Evaluation of the Mutagenicity and DNA-Modifying Activity of Carcinogens and Non-Carcinogens in Microbial Systems, *JNCI*, Vol 62, 1979, p 873-882
55. L. Friberg, C.G. Elinder, T. Kjellstrom, and G. Nordberg, *Cadmium and Health. A Toxicological and Epidemiological Appraisal*, Vol 1, *General Aspects*, and Vol 2, *Effects and Responses*, CRC Press, 1986
56. R.G. Adams, J.G. Harrison, and P. Scott, The Development of Cadmium-Induced Proteinuria, Impaired Renal Function and Osteomalacia in Alkaline Battery Workers, *J. Med.*, Vol 38, 1969, p 425-443
57. T.J. Kneip, M. Eisenbud, C.D. Strehlow, and P.C. Freudenthal, Airborne Particulates in New York City, *J. Air Pollut. Control Assoc.*, Vol 20, 1970, p 144-149
58. J.M. Frazier, Bioaccumulation of Cadmium in Marine Organisms, *Environ. Health Perspect.*, Vol 28, 1979, p 75-79
59. E.J. Underwood, *Trace Elements in Human and Animal Nutrition*, 4th ed., Academic Press, 1977
60. *Environmental Health Criteria for Cadmium*, Vol 6, *Ambio*, World Health Organization, 1977, p 287-290
61. A. Anderson and M. Hahlin, Cadmium Effects From Phosphorus Fertilization in Field Experiments, *Swed. J. Agric. Res.*, Vol 11, 1981, p 2
62. H.R. Pahren, J.B. Lucas, J.A. Ryan, and K.K. Dotson, Health Risks Associated With Land Application of Municipal Sludge, *J. Water Pollut. Control Fed.*, Vol 51, 1979, p 1588-1598
63. C.R. Dorn, J.O. Pierce, P.E. Phillips, and C.R. Chases, Airborne Pb, Cd, Zn, Cu Concentration by Particle Size Near a Pb Smelter, *Atmos. Environ.*, Vol 10, 1976, p 443-446
64. C.-G. Elinder, T. Kjellstrom, B. Lind, L. Linnman, M. Piscator, and K. Sundstedt, Cadmium Exposure From Smoking Cigarettes. Variations With Time and Country Where Purchased, *Environ. Res.*, Vol 32, 1983, p 220-227
65. P.R. Flanagan, J. McLellan, J. Haist, M.G. Cherian, M.J. Chamberlain, and L.S. Valberg, Increased Dietary Cadmium Absorption in Mice and Human Subjects With Iron Deficiency, *Gastroenterology*, Vol 74, 1978, p 841-846
66. N.E. Kowal, D.E. Johnson, D.F. Kaemer, and H.R. Pahren, Normal Levels of Cadmium in Diet, Urine, Blood, and Tissues of Inhabitants of the United States, *J. Toxicol. Environ. Health*, Vol 5, 1979, p 995-1012
67. H.A. Schroeder and J.J. Balassa, Hypertension Induced in Rats by Small Doses of Cadmium, *Am. J. Physiol.*, Vol 202, 1961, p 515-518
68. T. Kjellstrom, B. Lind, L. Linnman, and C.-G. Elinder, Variation of Cad-

mium Concentration in Swedish Wheat and Barley, an Indicator of Changes in Daily Cadmium Intake During the 20th Century, *Arch. Environ. Health*, Vol 30, 1975, p 321-328

69. C.-G. Elinder and T. Kjellstrom, Cadmium Concentration in Samples of Human Kidney Cortex From the 19th Century, *Ambio*, Vol 6, 1977, p 270
70. G.F. Nordberg, Cadmium Metabolism and Toxicity, *Environ. Physiol. Biochem.*, Vol 2, 1972, p 7-36
71. K. Nomiyama, Recent Progress and Perspectives in Cadmium Health Effects Studies, *Sci. Total Environ.*, Vol 14, 1980, p 199-232
72. L. Friberg and T. Kjellstrom, Cadmium, in *Disorders of Mineral Metabolism*, Vol 1, *Trace Minerals*, F. Bronner and J.W. Coburn, Ed., Academic Press, 1981, p 318-334
73. P. Chowdbury and D.B. Louria, Influence of Cadmium and Other Trace Elements on Human α_1-Antitrypsins: An *In Vitro* Study, *Science*, Vol 191, 1976, p 480-481
74. R.R. Lauwerys, H.A. Roels, J.-P. Buchet, A. Bernard, and D. Stanescu, Investigations on the Lung and Kidney Function in Workers Exposed to Cadmium, *Environ. Health Perspect.*, Vol 28, 1979, p 137-146
75. R.R. Lauwerys, A. Bernard, H.A. Roels, J.-P. Buchet, and C. Viau, Characterization of Cadmium Proteinuria in Man and Rat, *Environ. Health Perspect.*, Vol 54, 1984, p 147-152
76. G. Kazantzis, Renal Tubular Dysfunction and Abnormalities of Calcium Metabolism in Cadmium Workers, *Environ. Health Perspect.*, Vol 28, 1979, p 155-160
77. I. Shigematsu, Epidemiological Studies on Cadmium Pollution in Japan, in *Proceedings of First International Cadmium Conference*, San Francisco Metal Bulletin, 1978
78. R.R. Lauwerys, H.A. Roels, A. Bernard, and J.-P. Buchet, Renal Response to Cadmium in a Population Living in a Nonferrous Smelter Area in Belgium, *Int. Arch. Occup. Environ. Health*, Vol 45, 1980, p 271-274
79. T. Kjellstrom, P.-E. Ervin, and B. Rahnster, Dose-Response Relationship of Cadmium-Induced Tubular Proteinuria, *Environ. Res.*, Vol 13, 1977, p 303-317
80. R.A. Goyer, Cadmium Nephropathy, in *Nephrotoxic Mechanisms of Drugs and Environmental Toxins*, G.A. Porter, Ed., Plenum Medical Books, 1982, p 305-313
81. K.T. Suzuki, Induction and Degradation of Metallothioneins and Their Relation to the Toxicity of Cadmium, in *Biological Roles of Metallothionein*, E.C. Foulkes, Ed., Elsevier, 1982, p 215-235
82. Y. Boulanger, C.M. Goodman, C.P. Forte, S.W. Fesik, and I.M. Armitage, Model for Mammalian Metallothionein Structure, *Proc. Natl. Acad. Sci. U.S.A.*, Vol 80, 1983, p 1501-1505
83. K.T. Suzuki, S. Takenaka, and K. Kubota, Fate and Comparative Toxicity of Metallothionein With Differing Cadmium-Zinc Ratios in Rat Kidney, *Arch. Contemp. Toxicol.*, Vol 8, 1979, p 85-90
84. M.G. Cherian and M. Nordberg, Cellular Adaptation in Metal Toxicology and Metallothionein, *Toxicology*, Vol 28, 1983, p 1-15
85. M.G. Cherian, R.A. Goyer, and L. Delaquerriere-Richardson, Cadmium-Metallothionein Induced Nephropathy, *Toxicol. Appl. Pharmacol.*, Vol 38, 1976, p 399-408
86. H.M. Perry and M.W. Erlanger, Metal-Induced Hypertension Following Chronic Feeding of Low Doses of Cadmium and Mercury, *J. Lab. Clin. Med.*, Vol 83, 1974, p 541-547
87. K. Nogawa, E. Kobayashi, and R. Honda, A Study of the Relationship Between Cadmium Concentrations in Urine and Renal Effects of Cadmium, *Environ. Health Perspect.*, Vol 28, 1979, p 161-168
88. B.G. Armstrong and G. Kazantzis, The Mortality of Cadmium Workers, *Lancet*, Vol 1, 1983, p 1425-1427
89. M. Piscator, Role of Cadmium in Carcinogenesis With Special Reference to Cancer of the Prostate, *Environ. Health Perspect.*, Vol 40, 1981, p 107-120
90. S. Takenaka, H. Oldiges, H. Konig, D. Hochrainer, and G. Oberdorster, Carcinogenicity of Cadmium Chloride Aerosols in W Rats, *JNCI*, Vol 70, 1983, p 367-373
91. J.-P. Buchet, H. Roels, A. Bernard, and R. Lauwerys, Assessment of Renal Function of Workers Exposed to Inorganic Lead, Cadmium, or Mercury Vapor, *J. Occup. Med.*, Vol 22, 1980, p 741-750
92. Z.A. Shaikh and K. Hirayama, Metallothionein in the Extracellular Fluids as an Index of Cadmium Toxicity, *Environ. Health Perspect.*, Vol 28, 1979, p 267-371
93. C.C. Chang, R.J. Vander Mallie, and J.S. Garvey, A Radioimmunoassay for Human Metallothionein, *Toxicol. Appl. Pharmacol.*, Vol 55, 1980, p 94-102
94. C. Tohyama and Z.A. Shaikh, Metallothionein in Plasma and Urine of Cadmium-Exposed Rats Determined by a Single-Antibody Radioimmunoassay, *Fundam. Appl. Toxicol.*, Vol 1, 1981, p 1-7
95. K.J. Ellis, W.D. Morgan, I. Zanzi, S. Yasumura, D.D. Vartsky, and S.H. Cohn, Critical Concentrations of Cadmium in Human Renal Cortex: Dose-Effect Studies in Cadmium Smelter Workers, *J. Toxicol. Environ. Health*, Vol 7, 1981, p 691-698
96. N.J. Roels, R. Lauwerys, and A.N. Dardenne, The Critical Concentration of Cadmium in Human Renal Cortex: A Reevaluation, *Toxicol. Lett.*, Vol 15, 1983, p 357-360
97. M.G. Cherian and K. Rodgers, Chelation of Cadmium From Metallothionein *In Vivo* and its Excretion in Rats Repeatedly Injected With Cadmium Chloride, *J. Pharmacol. Exp. Ther.*, Vol 222, 1982, p 699-704
98. Committee on Medical and Biological Effects of Atmospheric Pollutants, *Chromium*, National Academy of Sciences, 1974
99. L. Fishbein, Sources, Transport, and Alteration of Metal Compounds: An Overview. I. Arsenic, Beryllium, Cadmium, Chromium, and Nickel, *Environ. Health Perspect.*, Vol 40, 1981, p 43-64
100. W. Mertz, Chromium Occurrence and Function in Biological Systems, *Physiol. Rev.*, Vol 49, 1969, p 163-239
101. S. Langard and T. Norseth, Chromium, in *Handbook on the Toxicology of Metals*, 2nd ed., Vol 2, *Specific Metals*, L. Friberg, G.F. Nordberg, and V.B. Vouk, Ed., 1986, p 185-210
102. T. Norseth, The Carcinogenicity of Chromium, *Environ. Health Perspect.*, Vol 40, 1981, p 121-130
103. W.C. Hueper, *Occupational and Environmental Cancers of the Respiratory System*, Springer-Verlag, 1966
104. H.A. Schroeder, J.J. Balassa, and I.H. Tipton, Abnormal Trace Metals in Man: Chromium, *J. Chronic Dis.*, Vol 15, 1962, p 941-964
105. Committee on Medical and Biological Effects of Atmospheric Pollutants, *Lead: Airborne Lead in Perspective*, National Academy of Sciences, 1972
106. R.A. Goyer and B. Rhyne, Pathological Effects of Lead, *Int. Rev. Exp. Pathol.*, Vol 12, 1973, p 1-77
107. *Environmental Health Criteria*, Vol 3, *Lead*, World Health Organization, 1977
108. J. Nriagu, *The Biogeochemistry of Lead*, Elsevier, 1978
109. *Lead Toxicity*, R.L. Singhal and J.A. Thomas, Ed., Urban & Schwarzenberg, 1980
110. H. Needleman, *Low Level Lead Exposure, The Clinical Implications of Current Research*, Raven Press, 1980
111. *Lead Absorption in Children. Management, Clinical and Environmental Aspects*, J.J. Chisolm, Jr. and D.M. O'Hara, Ed., Urban & Schwarzenberg, 1982

112. *Lead Versus Health Sources and Effects of Low Level Lead Exposure*, M. Rutter and R.R. Jones, Ed., John Wiley & Sons, 1983
113. R.A. Kehoe, The Metabolism of Lead in Health and Disease, The Harben Lectures, *J.R. Inst. Public Health Hyg.*, Vol 24, 1961, p 1-81
114. D. Barltrop and A. Smith, Interaction of Lead With Erythrocytes, *Experientia*, Vol 27, 1971, p 92-93
115. K.R. Mahaffey, J.L. Annest, J. Roberts, and R.S. Murphy, Estimates of Blood Lead Levels, United States 1976-1980. Association With Selected Demographic and Socioeconomic Factors, *New Engl. J. Med.*, Vol 307, 1982, p 573-579
116. H. Needleman, E.E. Gunnoe, A. Leviton, R. Reed, H. Peresie, C. Maher, and P. Barrett, Deficits in Psychologic and Classroom Performance of Children with Elevated Blood Lead Levels, *New Eng. J. Med.*, Vol 300, 1979, p 689-695
117. C.B. Ernhardt, B. Landa, and N.B. Schnell, Subclinical Levels of Lead and Development Deficit—A Multivariate Follow-Up Reassessment, *Pediatrics*, Vol 67, 1981, p 911-919
118. J.L. Burchfiel, F.H. Duffy, P.H. Bartels, and H.L. Needleman, The Combined Discriminating Power of Quantitative Electroencephalography and Neuropsychologic Measures in Evaluating Central Nervous System Effects of Lead at Low Levels, in *Low Level Lead Exposures*, H. Needleman, Ed., Raven Press, 1980, p 75-90
119. H.M. Thomas, A Case of Generalized Lead Paralysis, A Review of the Cases of Lead Palsy Seen in the Hospital, *Bull. Johns Hopkins Hosp.*, Vol 15, 1904, p 209-212
120. P.W. Lampert and S.S. Schochet, Demyelination and Remyelination in Lead Neuropathy, *J. Neuropathol. Exp. Neurol.*, Vol 27, 1968, p 527-545
121. W.W. Schlaepfer, Experimental Lead Neuropathy. A Disease of the Supporting Cells in the Peripheral System, *J. Neuropathol. Exp. Neurol.*, Vol 28, 1968, p 401-418
122. A.M. Seppalainen, S. Tola, S. Hernberg, and B. Kock, Subclinical Neuropathy at "Safe" Levels of Lead Exposure, *Arch. Environ. Health*, Vol 30, 1975, p 180-183
123. C.P. McCord, F.R. Holden, and J. Johnston, Basophilic Aggregation Test in the Lead Poisoning Epidemic of 1934-35, *Am. J. Public Health*, Vol 25, 1935, p 1089-1096
124. D.E. Paglia, W.N. Valentine, and J.G. Dahlgner, Effects of Low Level Lead Exposure on Pyrimidine-5′-Nucleotidase and Other Erythrocyte Enzymes, *J. Clin. Invest.*, Vol 56, 1975, p 1164-1169
125. S. Hernberg, M. Nurminen, and H. Hasan, Nonrandom Shortening of Red Cell Survival Times in Men Exposed to Lead, *Environ. Res.*, Vol 1, 1967, p 247-261
126. M.D. Bessis and W.N. Jensen, Sideroblastic Anemia, Mitochondria and Erythroblastic Iron, *Br. J. Haematol.*, Vol 11, 1965, p 49-51
127. J.L. Granick and R.D. Levere, Hemesynthesis in Erythroid Cells, *Prog. Hematol.*, Vol 4, 1964, p 1-47
128. R.A. Goyer, Lead and the Kidney, *Curr. Top. Pathol.*, Vol 55, 1971, p 147-176
129. R.A. Goyer, Lead Toxicity: A Problem in Environmental Pathology, *Am. J. Pathol.*, Vol 64, 1971, p 167-182
130. J.F. Moore, R.A. Goyer, and M.H. Wilson, Lead-Induced Inclusion Bodies, Solubility Amino Acid Content and Relationship to Residual Acidic Nuclear Proteins, *Lab. Invest.*, Vol 29, 1973, p 488-494
131. H.H. Sandstead, A.M. Michelakis, and T.E. Temple, Lead Intoxication. Its Effect on the Renin-Aldosterone Response to Sodium Deprivation, *Arch. Environ. Health*, Vol 20, 1970, p 356-363
132. R.A. Goyer and M.H. Wilson, Lead-Induced Inclusion Bodies: Results of EDTA Treatment, *Lab. Invest.*, Vol 32, 1975, p 149-156
133. G.B. Schumann, S.I. Lerner, M.A. Weiss, L. Gawronski, and G.K. Lohiya, Inclusion Bearing Cells in Industrial Workers Exposed to Lead, *Am. J. Clin. Pathol.*, Vol 74, 1980, p 192-196
134. R.P. Wedeen, J.K. Maesaka, B. Weiner, E.A. Lipat, M.M. Lyons, L.F. Vitale, and M.M. Joselow, Occupational Lead Nephropathy, *Am. J. Med.*, Vol 49, 1975, p 630-641
135. R. Lilis, J. Valciukas, A. Fischbein, G. Andrews, and I.J. Selikoff, Renal Function Impairment in Secondary Lead Smelter Workers: Correlations With Zinc Protoporphyrin and Blood Lead Levels, *J. Environ. Pathol. Toxicol.*, Vol 2, 1979, p 1447-1474
136. C.D. Hong, I.B. Hanenson, S. Lerner, P.B. Hammond, A.J. Pesce, and V.E. Pollack, Occupational Exposure to Lead: Effects on Renal Function, *Kidney Int.*, Vol 18, 1980, p 489-494
137. V. Batuman, J.K. Maesalsa, B. Haddad, E. Tepper, E. Landry, and R. Wedeen, The Role of Lead in Gout Nephropathy, *New Engl. J. Med.*, Vol 304, 1981, p 520-523
138. B.T. Emmerson, The Clinical Differentiation of Lead Gout From Primary Gout, *Arthritis Rheum.*, Vol 11, 1968, p 623-624
139. J.L. Pirkle, J. Schwartz, J.R. Landis, and W.R. Harlan, The Relationship Between Blood Lead Levels and Blood Pressure and Its Cardiovascular Risk Implications, *Am. J. Epidemiol.*, Vol 121, 1985, p 246-258
140. W. Victery, A.J. Vander, J.M. Shulak, P. Schoeps, and S. Julius, Lead, Hypertension and the Renin-Angiotensin System in Rats, *J. Lab. Clin. Med.*, Vol 99, 1982, p 354-362
141. M.R. Moore and P.A. Meredith, The Carcinogenicity of Lead, *Arch. Toxicol.*, Vol 42, 1979, p 87-94
142. I. Dingwall-Fordyce and R.E. Lane, A Follow-Up Study of Lead Workers, *Br. J. Ind. Med.*, Vol 20, 1963, p 313-315
143. W.C. Cooper and W.R. Gaffey, Mortality of Lead Workers, *J. Occup. Med.*, Vol 17, 1975, p 100-107
144. H.K. Kang, P.F. Infante, and J.S. Carra, Occupational Lead Exposure and Cancer, *Science*, Vol 207, 1980, p 935-936
145. W.C. Cooper, Occupational Lead Exposure. What Are the Risks?, *Science*, Vol 180, 1980, p 129
146. E.L. Baker, R.A. Goyer, B.A. Fowler, U. Khettry, O.B. Bernard, S. Adler, R. White, R. Babayan, and R.G. Feldman, Occupational Lead Exposure, Nephropathy and Renal Cancer, *Am. J. Ind. Med.*, Vol 1, 1980, p 139-148
147. R. Lilis, Long-Term Occupational Lead Exposure: Chronic Nephropathy and Renal Cancer, A Case Report, *Am. J. Ind. Med.*, Vol 2, 1981, p 293-297
148. G. Deknudt, Y. Manuel, and G.B. Gerber, Chromosomal Aberration in Workers Professionally Exposed to Lead, *J. Toxicol. Environ. Health*, Vol 3, 1977, p 885-891
149. L.D. Koller, J.H. Exon, S.A. Moore, and P.G. Watanabe, Evaluation of ELISA for Detecting *In Vivo* Chemical Immunomodulation, *J. Toxicol. Environ. Health*, Vol 11, 1983, p 15-22
150. M.A. Perlstein and R. Attala, Neurologic Sequelae of Plumbism in Children, *Clin. Pediatr.*, Vol 5, 1966, p 292-298
151. K.R. Mahaffey and J.A. Michaelson, The Interaction Between Lead and Nutrition, in *Low Level Lead Exposure: The Clinical Implications of Current Research*, H. Needleman, Ed., Raven Press, 1980, p 159-200
152. J.E. Cremer, Biochemical Studies on the Toxicity of Triethyl Lead and Other Organo-Lead Compounds, *Br. J. Ind. Med.*, Vol 16, 1959, p 191-199
153. S. Araki and K. Ushio, Mechanism of Increased Osmotic Resistance of Red Cells in Workers Exposed to Lead, *Br. J. Ind. Med.*, Vol 39, 1982, p 157-160
154. I.M. Shapiro, G. Mitchell, I. Davidson, and S.H. Katz, Lead Content of

Teeth, *Arch. Environ. Health*, Vol 30, 1975, p 483-486
155. R. Lilis and A. Fischbein, Chelation Therapy in Workers Exposed to Lead—A Critical Review, *JAMA*, Vol 235, 1976, p 2823-2824
156. *Environmental Health Criteria*, Vol 1, *Mercury*, World Health Organization, 1976
157. L.J. Goldwater and W. Stopford, Mercury, in *The Chemical Environment*, J. Lenihan and W.W. Fletcher, Ed., Blackie & Son, 1977, p 38-63
158. "Effects of Mercury in the Canadian Environment," National Research Council of Canada, 16739, 1979
159. T. Suzuki, Metabolism of Mercurial Compounds, in *Toxicology of Trace Elements*, R.A. Goyer and M.A. Mehlman, Ed., Hemisphere Publishing, 1977, p 1-39
160. M. Berlin, The Toxicokinetics of Mercury, in *Infant Formula and Junior Food*, E.H.F. Schmidt and A.G. Hildebrandt, Ed., Springer-Verlag, 1983, p 147-160
161. T.W. Clarkson, Methylmercury Toxicity to the Mature and Developing Nervous System: Possible Mechanisms, in *Biological Aspects of Metals and Metal-Related Diseases,* D. Sarkar, Ed., Raven Press, 1983, p 183-197
162. J.K. Miettinen, Absorption and Elimination of Dietary Mercury (Hg^{++}) and Methyl Mercury in Man, in *Mercury Mercurials and Mercaptans*, M.W. Miller and T.W. Clarkson, Ed., Charles C. Thomas, 1973, p 233
163. K.M. Madsen and E.F. Christensen, Effects of Mercury on Lysosomal Protein Digestion in the Kidney Proximal Tubule, *Lab. Invest.*, Vol 38, 1978, p 165-171
164. *Mercury in the Environment—Toxicological and Epidemiological Appraisal*, L. Friberg and J. Vostal, Ed., Chemical Rubber, 1972
165. L.J. Goldwater, *Mercury: A History of Quicksilver*, York Press, 1972, p 270-277
166. T.L. Gritzka and B.F. Trump, Renal Tubular Lesions Caused by Mercuric Chloride, *Am. J. Pathol.*, Vol 52, 1968, p 1225-1277
167. A.A. Roman-Franco, M. Twirello, B. Abini, and E. Ossi, Anti-Basement Membrane Antibodies With Antigenantibody Complexes in Rabbits Injected With Mercuric Chloride, *Clin. Immunol. Immunopathol.*, Vol 9, 1978, p 404-411
168. R.R. Tubbs, G.N. Gephardt, J.T. McMahon, M.C. Phol, D.G. Vidt, S.A. Barenberg, and R. Valenzuela, Membranous Glomerulonephritis Associated With Industrial Mercury Exposure, *Am. J. Clin. Pathol.*, Vol 77, 1982, p 409-413
169. D.S. Matheson, T.W. Clarkson, and E.W. Gelfand, Mercury Toxicity (Acrodynia) Induced by Long-Term Injection of Gamma Globulin, *J. Pediatr.*, Vol 97, 1980, p 153-155
170. F. Bakir, S.F. Damluji, L. Amin-Zaki, M. Murtadha, A. Khalidi, N.Y. Al-Rawi, S. Tikriti, H.I. Dhahir, T.W. Clarkson, J.C. Smith, and R.A. Doherty, Methyl Mercury Poisoning in Iraq, *Science*, Vol 181, 1973, p 230-241
171. L. Roizin, H. Shiraki, and N. Grceric, *Neurotoxicology*, Vol 1, Raven Press, 1977, p 658
172. "Occupational Diseases: A Guide to Their Recognition," U.S. Department of Health, Education and Welfare, 77-1811, 1977, p 305
173. M. Berlin, Mercury, in *Handbook on the Toxicology of Metals*, 2nd ed., Vol 2, *Specific Metals*, L. Friberg, G.F. Nordberg, and C. Nordman, Ed., Elsevier, 1986, p 386-445
174. F.W. Sunderman, Jr., Nickel, in *Disorders of Mineral Metabolism*, Vol 1, F. Bronner and J.W. Coburn, Ed., Academic Press, 1981, p 201-232
175. M. Anke, M. Grun, B. Gropped, and H. Kronemann, Nutritional Requirements of Nickel, in *Biologic Aspect of Metals and Metal-Related Diseases*, B. Sarkar, Ed., Raven Press, 1983, p 89-105
176. H.A. Schroeder, J.J. Balassa, and I.H. Tipton, Abnormal Trace Elements in Man: Nickel, *J. Chronic Dis.*, Vol 15, 1962, p 51-65
177. D.R. Myron, T.J. Zimmerman, T.R. Schuler, L.M. Klevay, D.E. Lee, and F.H. Nielsen, Intake of Nickel and Vanadium by Humans. A Survey of Selected Diet, *Am. J. Clin. Nutr.*, Vol 31, 1978, p 527-531
178. M.D. McNeely, M.W. Nechay, and F.W. Sunderman, Jr., Measurements of Nickel in Serum and Urine as Indices of Environmental Exposure to Nickel, *Clin. Chem.*, Vol 18, 1972, p 992-995
179. J.W. Spears, E.E. Hatfield, R.M. Forbes, and S.E. Koenig, Studies on the Role of Nickel in the Ruminant, *J. Nutr.*, Vol 108, 1978, p 313-320
180. R. Doll, J.D. Mathews, and L.G. Morgan, Cancers of the Lung and Nasal Sinuses in Nickel Workers: Reassessment of the Period of Risk, *Br. J. Ind. Med.*, Vol 34, 1977, p 102-106
181. E. Pedersen, A. Anderson, and A. Hogetveit, A Second Study of the Incidence and Mortality of Cancer of Respiratory Organs Among Workers at a Nickel Refinery, *Ann. Clin. Lab. Sci.*, Vol 8, 1978, p 503-510
182. F.W. Sunderman, Jr., Recent Research on Nickel Carcinogenesis, *Environ. Health Perspect.*, Vol 40, 1981, p 131-141
183. J.C. McEwan, Five-Year Review of Sputum Cytology in Workers at a Nickel Sinter Plant, *Ann. Clin. Lab. Sci.*, Vol 8, 1978, p 503-509
184. T. Menne and A. Thorboe, Nickel Dermatitis—Nickel Excretion, *Contact Dermatitis*, Vol 2, 1976, p 353-354
185. H.A. Schroeder, A.P. Nason, and I.H. Tipton, Essential Trace Metals in Man: Cobalt, *J. Chronic Dis.*, Vol 20, 1967, p 869-890
186. E. Browning, *Toxicity of Industrial Metals*, 2nd ed., Butterworths, 1969
187. J.H. Wills, Jr., Goitrogens in Foods, in *Toxicants Occurring Naturally in Foods*, Food Protection Committee, National Academy of Sciences, 1354, 1966, p 3-17
188. Y. Morin and P. Daniel, Quebec Beer-Drinkers Cardiomyopathy: Etiological Consideration, *J. Can. Med. Assoc.*, Vol 97, 1967, p 926-931
189. W. Gilman, Metal Carcinogenesis. II. Study on the Carcinogenicity of Cobalt, Copper, Iron, and Nickel Compounds, *Cancer Res.*, Vol 22, 1962, p 158-170
190. N. Aspin and A. Sass-Kortsak, Copper, in *Disorders of Mineral Metabolism*, Vol 1, *Trace Minerals*, F. Bronner and J.W. Coburn, Ed., Academic Press, 1981, p 60-86
191. B. Sarkar, J.-P. Laussac, and S. Lau, Transport Forms of Copper in Human Serum, in *Biological Aspects of Metals and Metal-Related Diseases*, B. Sarkar, Ed., Raven Press, 1983, p 23-40
192. W.Y. Chan, L.A. Tease, H.C. Liu, and O.M. Rennert, Cell Culture Studies in Wilson's Disease, in *Biological Aspects of Metals and Metal-Related Diseases*, B. Sarkar, Ed., Raven Press, 1983, p 147-158
193. J.R. Riordan, Handling of Heavy Metals by Cultured Cells From Patients With Menke's Disease, in *Biological Aspects of Metals and Metal-Related Diseases*, D. Sarkar, Ed., Raven Press, 1983, p 159-170
194. H.K. Chuttani, P.S. Gupti, and S. Gultati, Acute Copper Sulfate Poisoning, *Am. J. Med.*, Vol 39, 1965, p 849-854
195. A.D. Manzler and A.W. Schreiner, Copper-Induced Acute Hemolytic Anemia, A New Complication of Hemodialysis, *Ann. Intern. Med.*, Vol 73, 1970, p 409-412
196. A. Jacobs and M. Worwood, Iron, in *Disorders of Mineral Metabolism*, Vol 1, *Trace Minerals*, F. Bronner and J.W. Coburn, Ed., Academic Press, 1981, p 2-59
197. B.F. Trump, J.N. Valigersky, A.U. Arstila, W.J. Mergner, and T.D. Kinney, The Relationship of Intracellular

Pathways of Iron Metabolism to Cellular Iron Overload and the Iron Storage Diseases, *Am. J. Pathol.*, Vol 72, 1973, p 295-324
198. U. Muller-Eberhard, P.A. Miescher, and E.R. Jaffe, *Iron Excess. Aberrations of Iron and Porphyrin Metabolism*, Grune & Stratton, 1977
199. A. Jacobs, Iron Overload—Clinical and Pathological Aspects, *Semin. Hematol.*, Vol 14, 1977, p 89-113
200. E.B. Brown, Therapy for Disorders of Iron Excess, in *Biological Aspects of Metal-Related Diseases*, B. Sarkar, Ed., Raven Press, 1983, p 263-278
201. J.T. Boyd, R. Doll, J.S. Foulds, and J. Leiper, Cancer of the Lung in Iron Ore (Haematite) Miners, *Br. J. Ind. Med.*, Vol 27, 1970, p 97-103
202. A.I.G. McLaughlin and H.E. Harding, Pneumoconiosis and Other Causes of Death in Iron and Steel Foundry Workers, *Arch. Ind. Health*, Vol 14, 1956, p 350-362
203. Committee on Medical and Biological Effects of Atmospheric Pollutants, *Manganese*, National Academy of Sciences, 1973
204. I. Mena, H. Kazuko, K. Burke, and G.C. Cotzias, Chronic Manganese Poisoning. Individual Susceptibility and Absorption of Iron, *Neurology*, Vol 19, 1969, p 1000-1006
205. I. Mena, O. Meurin, S. Feunzobda, and G.C. Cotzias, Chronic Manganese Poisoning. Clinical Picture and Manganese Turnover, *Neurology*, Vol 17, 1967, p 128-136
206. W. Pentschew, F.F. Ebner, and R.M. Kovatch, Experimental Manganese Encephalopathy in Monkeys, *J. Neuropathol. Exp. Neurol.*, Vol 22, 1963, p 488-499
207. G.C. Cotzias, P.S. Papavasiliou, J. Ginos, A. Stechk, and S. Duby, Metabolic Modification of Parkinson's Disease and of Chronic Manganese Poisoning, *Annu. Rev. Med.*, Vol 22, 1971, p 305-326
208. N.H. Neff, R.E. Barrett, and E. Costa, Selective Depletion of Caudate Nucleus Dopamine and Serotonin During Chronic Manganese Dioxide Administration, *Experientia*, Vol 25, 1969, p 1140-1141
209. P.W. Winston, Molybdenum, in *Disorders of Mineral Metabolism*, Vol 1, *Trace Minerals*, F. Bonner and J.W. Coburn, Ed., Academic Press, 1981, p 295-315
210. J.L. Johnson, H.P. Jones, and K.V. Rajagopalan, *In Vitro* Reconstitution of Demolybdosulfite Oxidase by a Molybdenum Cofactor From Rat Liver and Other Sources, *J. Biol. Chem.*, Vol 252, 1977, p 4994-5003
211. M.E. Curzan, B.L. Adkins, B.G. Bibby, and F.L. Losee, Combined Effect of Trace Elements and Fluorine on Caries, *J. Dent. Res.*, Vol 49, 1970, p 526-528
212. C.G. Wilber, *Selenium: A Potential Environmental Poison and a Necessary Food Constituent*, Charles C. Thomas, 1983
213. Committee on Medical and Biological Effects of Atmospheric Pollutants, *Selenium*, National Academy of Sciences, 1975
214. R.F. Burk, Selenium in Man, in *Trace Elements in Human Health and Disease*, A.S. Prasad and D. Oberleas, Ed., Vol II, Academic Press, 1976, p 105-134
215. A.T. Diplock, Metabolic Aspects of Selenium Action and Toxicity, *Crit. Rev. Toxicol.*, Vol 4, 1976, p 271-329
216. K.P. McConnell and J.G. Hoffman, Methionine Selenomethionine Parallels in *E. Coli* Polypeptide Chain Initiation and Synthesis, *Proc. Soc. Exp. Biol. Med.*, Vol 140, 1972, p 638-641
217. I.S. Palmer, D.D. Fischer, A.W. Halverson, and O.E. Olson, Identification of a Major Selenium Excretory Product in Rat Urine, *Biochim. Biophys. Acta*, Vol 177, 1969, p 336-342
218. K.P. McConnell and O.W. Portman, Toxicity of Dimethyl Selenide in the Rat and Mouse, *Proc. Soc. Exp. Biol. Med.*, Vol 79, 1952, p 230-231
219. H.A. Schroeder and M. Mitchener, Selenium and Tellurium in Mice, *Arch. Environ. Health*, Vol 24, 1972, p 66-71
220. R.A. Sunde and W.G. Hoekstra, Structure, Synthesis, and Function of Glutathione Peroxidase, *Nutr. Rev.*, Vol 38, 1980, p 265-273
221. X. Chen, G. Yang, J. Chen, X. Chen, Z. Wen, and K. Ge, Studies on the Relations of Selenium and Keshan Disease, *Biol. Trace Elem. Res.*, Vol 2, 1980, p 91-107
222. A.M. Van Rij, C.R. Thomson, J.M. McKenzie, and M.F. Robinson, Selenium Deficiency in Total Parenteral Nutrition, *Am. J. Clin. Nutr.*, Vol 32, 1979, p 2076-2085
223. Selenium Perspective, *Lancet*, Vol 1, 1983, p 685
224. F.A. Patty, Arsenic, Phosphorous, Selenium, Sulfur, and Tellurium, in *Industrial Hygiene and Toxicology*, 2nd ed., D.W. Fassett and D.D. Irish, Ed., Interscience, 1963, p 871-910
225. A.L. Moxan and M. Rhian, Selenium Poisoning, *Physiol. Rev.*, Vol 203, 1943, p 305-337
226. "Bioassay of Selenium Sulfide (Dermal Study) for Possible Carcinogenicity," National Cancer Institute Technical Report Series 197, NTP 80-18, 1980
227. "Bioassay of Selsun[R] for Possible Carcinogenicity," National Cancer Institute Technical Report Series 199, NTP 80-19, 1980
228. R.J. Shamberger, S.A. Tytko, and C.E. Willis, Antioxidants and Cancer. Part VI. Selenium and Age-Adjusted Human Cancer Mortality, *Arch. Environ. Health*, Vol 31, 1976, p 231-235
229. G.O. Howell and C.H. Hill, Biological Interactions of Selenium With Other Trace Elements in Chicks, *Environ. Health Perspect.*, Vol 25, 1978, p 147-150
230. "Zinc in the Aquatic Environment," Chemistry, Distribution and Toxicology, National Research Council of Canada, 17589, 1981
231. H.H. Sandstead, Zinc in Human Nutrition, in *Disorders of Mineral Metabolism*, F. Bronner and J.W. Coburn, Ed., Academic Press, 1981, p 94-159
232. A.S. Prasad, Human Zinc Deficiency, in *Biological Aspects of Metals and Metal-Related Diseases*, B. Sarkar, Ed., Raven Press, 1983, p 107-119
233. N.T. Davies, Studies on the Absorption of Zinc by Rat Intestine, *Br. J. Nutr.*, Vol 43, 1980, p 189-203
234. A.S. Prasad, A. Miale, Jr., Z. Farid, H.H. Sandstead, A.R. Schulert, and W.J. Darby, Biochemical Studies on Dwarfism, Hypogonadism and Anemia, *Arch. Intern. Med.*, Vol 111, 1963, p 407-428
235. R.A. Goyer, J. Apgar, and M. Piscator, Toxicity of Zinc, in *Zinc*, R.I. Henkin and Committee, Ed., University Park Press, 1979, p 249-268
236. H.F. Lam, M.W. Conner, A.E. Rogers, S. Fitzgerald, and M.O. Amdur, Functional and Morphological Changes in the Lungs of Guinea Pigs Exposed to Freshly Generated Ultrafine Zinc Oxide, *Toxicol. Appl. Pharmacol.*, Vol 78, 1985, p 29-38
237. A. Furst, Bioassay of Metals for Carcinogenesis: Whole Animals, *Environ. Health Perspect.*, Vol 40, 1981, p 83-91
238. A.C. Alfrey, Aluminum and Tin, in *Disorders of Mineral Metabolism*, F. Bronner and J.W. Coburn, Ed., Academic Press, 1981, p 353-369
239. H. Spencer, I. Lewin, M.J. Belcher, and J. Samachson, Inhibition of Radiostrontium Absorption by Aluminum Phosphate Gel in Man and its Comparative Effect on Radiocalcium Absorption, *Int. J. Appl. Radiat. Isot.*, Vol 20, 1969, p 507-516
240. J. Nagyvary and E.L. Bradbury, Hypocholesterolemic Effects of Al^{3+} Complexes, *Biochem. Res. Commun.*, Vol 2, 1977, p 592-598
241. M. Lotz, E. Zisman, and F.C. Bartter, Evidence for Phosphorus-Depletion Syndrome in Man, *New Engl. J. Med.*, Vol 278, 1968, p 409-415
242. C.G. Schaver, Pulmonary Changes Encountered in Employees Engaged in the Manufacture of Aluminum Abra-

sives: Clinical and Roetgenologic Aspects, *Occup. Med.*, Vol 5, 1948, p 718-728
243. F.M. Engelbrecht and M.E. Jordaan, The Influence of Silica and Aluminum on the Cytochrome C Oxidase Activity of Rat Lung Homogenate, *S. Afr. Med. J.*, Vol 46, 1972, p 769-771
244. M.R. Wills and J. Savory, Aluminum Poisoning: Dialysis Encephalopathy, Osteomalacia, and Anemia, *Lancet*, Vol 2, 1983, p 29-33
245. U. De Boni, A. Otvos, J.W. Scott, and D.R. Crapper, Neurofibrillary Degeneration Induced by System Aluminum, *Acta Neuropathol.*, Vol 35, 1976, p 285-294
246. A.C. Alfrey, J.M. Mishell, J. Burks, S.R. Contiguglia, H. Rudolph, E. Lewin, and J.H. Holmes, Syndrome of Dyspraxia and Multifocal Seizures Associated With Chronic Hemodialysis, *Trans. Am. Soc. Artif. Intern. Organs*, Vol 18, 1972, p 257-261
247. D.R. Crapper, S.S. Krishnan, and S. Quittkat, Aluminum, Neurofibrillary Degeneration and Alzheimer's Disease, *Brain*, Vol 99, 1976, p 67-79
248. D.R. Crapper-McLachlan, B. Farnell, H. Galin, S. Kalik, G. Eichhorn, and U. DeBoni, Aluminum in Human Brain Disease, in *Biological Aspects of Metals and Metal-Related Diseases*, B. Sarkar, Ed., Raven Press, 1983, p 209-218
249. B.W. Fowler and V. Vouk, Bismuth, in *Handbook on the Toxicology of Metals*, 2nd ed., Vol 2, *Specific Metals*, L. Friberg, G.F. Nordberg, and V.B. Vouk, Ed., Elsevier, 1986, p 117-129
250. D.W. Thomas, T.F. Hartley, and S. Sobecki, Clinical and Laboratory Investigations of the Metabolism of Bismuth Containing Pharmaceuticals by Man and Dogs, in *Clinical Chemistry and Clinical Toxicology of Metals*, S.S. Brown, Ed., Elsevier, 1977, p 293-296
251. R. Urizar and R.L. Vernier, Bismuth Nephropathy, *JAMA*, Vol 198, 1966, p 187-189
252. D.L. Beaver and R.E. Burr, Electron Microscopy of Bismuth Inclusions, *Am. J. Pathol.*, Vol 42, 1975, p 609-614
253. B.A. Fowler and R.A. Goyer, Bismuth Localization Within Nuclear Inclusions by X-Ray Microanalysis, *J. Histochem. Cytochem.*, Vol 23, 1975, p 722-726
254. J.M. Arena, *Poisoning*, 3rd ed., Charles C. Thomas, 1974, p 81-82
255. G.W. Voil, J.A. Minielly, and T. Bistricki, Gold Nephropathy Tissue Analysis by X-Ray Fluorescent Spectroscopy, *Arch. Pathol. Lab. Med.*, Vol 101, 1977, p 635-640
256. J. Stuve and P. Galle, Role of Mitochondria in the Renal Handling of Gold by the Kidney, *J. Cell Biol.*, Vol 44, 1970, p 667-676
257. S.K. Ainsworth, R.P. Swain, N. Watabe, N.C. Brackett, P. Pilia, and G.R. Hennigar, Gold Nephropathy, Ultrastructural Fluorescent, and Energy-Dispersive X-Ray Microanalysis Study, *Arch. Pathol. Lab. Med.*, Vol 105, 1981, p 373-378
258. M. Cox and I. Singer, Lithium, in *Disorders of Mineral Metabolism*, F. Bronner and J.W. Coburn, Ed., Academic Press, 1981, p 369-438
259. I. Singer, Lithium and the Kidney, *Kidney Int.*, Vol 19, 1981, p 374-387
260. J.W. Davis and W.E. Fann, Lithium, *Annu. Rev. Pharmacol.*, Vol 11, 1971, p 285-298
261. E.J. Gralla and H.M. McIlhenny, Studies in Pregnant Rats, Rabbits, and Monkeys With Lithium Carbonate, *Toxicol. Appl. Pharmacol.*, Vol 21, 1972, p 428-433
262. T.N. Steele, Treatment of Lithium Intoxication With Diuretics, in *Clinical Chemistry and Chemical Toxicology of Metals*, S.S. Brown, Ed., Elsevier, 1977, p 289-292
263. R.R. Landoldt, H.W. Berk, and H.T. Russell, Studies on the Toxicity of Rhodium Trichloride in Rats and Rabbits, *Appl. Pharmacol.*, Vol 21, 1972, p 589-590
264. H.A. Schroeder and M. Mitchener, Scandium, Chromium (VI), Gallium, Yttrium, Rhodium, Palladium, Indium in Mice. Effects on Growth and Life Span, *J. Nutr.*, Vol 101, 1971, p 1431-1438
265. G. Kazantzis, Role of Cobalt, Iron, Lead, Manganese, Mercury, Platinum, Selenium and Titanium in Carcinogenesis, *Environ. Health Perspect.*, Vol 40, 1981, p 143-161
266. N.E. Madias and J.T. Harrington, Platinum Nephrotoxicity, *Am. J. Med.*, Vol 65, 1978, p 307-314
267. D.D. Choie, D.S. Longenecker, and A.A. Del Campo, Acute and Chronic Cisplatin Nephropathy in Rats, *Lab. Invest.*, Vol 44, 1981, p 397-402
268. C.-G. Elinder and L. Friberg, Antimony, in *Handbook on the Toxicology of Metals*, 2nd ed., Vol 2, *Specific Metals*, L. Friberg, G.F. Nordberg, and V.B. Vouk, Ed., Elsevier, 1986, p 211-232
269. F.R. Paton and A.C. Allison, Chromosome Damage in Human Cell Cultures Induced by Metal Salts, *Mutat. Res.*, Vol 16, 1972, p 332-336
270. B.C. Casto, J. Meyers, and J.A. DiPaolo, Enhancement of Viral Transformation for Evaluation of the Carcinogenic or Mutagenic Potential of Inorganic Metal Salts, *Cancer Res.*, Vol 39, 1979, p 193-198
271. A.L. Reeves, Barium, in *Handbook on the Toxicology of Metals*, 2nd ed., Vol 2, *Specific Metals*, L. Friberg, G.F. Nordberg, and V.B. Vouk, Ed., Elsevier, 1986, p 84-94
272. J.S. Woods, G.T. Carver, and B.A. Fowler, Altered Regulation of Hepatic Heme Metabolism by Indium Chloride, *Toxicol. Appl. Pharmacol.*, Vol 49, 1979, p 455-461
273. B.A. Fowler, Indium and Thallium in Health, in *Trace Metals in Human Health*, J. Rose, Ed., Butterworths, 1982
274. H.A. Schroeder, A.P. Nason, and I.H. Tipton, Essential Trace Metals in Man: Magnesium, *J. Chronic. Dis.*, Vol 21, 1969, p 815-841
275. T.D. Luckey, B. Venugopal, and D. Hutcheson, *Heavy Metal Toxicity Safety and Hormonology*, Academic Press, 1975
276. H.A. Schroeder, J. Buckman, and J.J. Balassa, Abnormal Trace Elements in Man: Tellurium, *J. Chronic Dis.*, Vol 20, 1967, p 147-161
277. E.A. Cerwenka and W.C. Cooper, Toxicology of Selenium and Tellurium and Their Compounds, *Arch. Environ. Health*, Vol 3, 1961, p 189-200
278. L. Fishbein, Toxicology of Selenium and Tellurium, in *Toxicology of Trace Metals*, R.A. Goyer and M.A. Mehlman, Ed., John Wiley & Sons, 1977, p 191-240
279. W.L. Downs, J.K. Scott, L.T. Steadman, and E.A. Maynard, Acute and Subacute Toxicity Studies of Thallium Compounds, *Am. Ind. Hyg. Assoc. J.*, Vol 21, 1960, p 399-406
280. M.M. Herman and K.G. Bensch, Light and Electron Microscopic Studies of Acute and Chronic Thallium Intoxication in Rats, *Toxicol. Appl. Pharmacol.*, Vol 10, 1967, p 199-222
281. H. Nogami and Y. Terashima, Thallium-Induced Achondroplasia in the Rat, *Teratology*, Vol 8, 1973, p 101-102
282. *Environmental Health Criteria*, Vol 15, *Tin and Organotin Compounds: A Preliminary Review*, World Health Organization, 1980
283. J.M. Barnes and H.B. Stoner, Toxicology of Tin Compounds, *Pharmacol. Rev.*, Vol 11, 1959, p 211-231
284. G. Prull and K. Rompel, EEG Changes in Acute Poisoning With Organic Tin Compounds, *Electroenceph. Clin. Neurophysiol.*, Vol 29, 1970, p 215-222
285. H.G. Verschuuren, E.J. Ruitenberg, F. Peetoom, P.W. Helleman, and G.J. Van Esch, Influence of Triphenyltin Acetate on Lymphatic Tissue and Immune Response in Guinea Pigs, *Toxicol. Appl. Pharmacol.*, Vol 16, 1970, p 400-410
286. H.A. Schroeder, J.J. Balassa, and

I.H. Tipton, Abnormal Trace Metals in Man: Titanium, *J. Chronic Dis.*, Vol 16, 1963, p 55-69

287. M. Berlin and C. Nordman, Titanium, in *Handbook on the Toxicology of Metals*, 2nd ed., Vol 2, *Specific Metals*, L. Friberg, G.F. Nordberg, and V.B. Vouk, Ed., Elsevier, 1986, p 594-609
288. A. Furst and R.T. Haro, A Survey of Metal Carcinogenesis, *Prog. Exp. Tumour Res.*, Vol 12, 1969, p 102-133
289. "Bioassay of Titanium Dioxide for Possible Carcinogenicity," National Cancer Institute Carcinogenesis Technical Report 97, Department of Health, Education and Welfare, NIH 79-1347, 1979
290. P.S. Chen, R. Terepka, and H.C. Hodge, The Pharmacology and Toxicology of the Bone Seekers, *Annu. Rev. Pharmacol.*, Vol 1, 1961, p 369-393
291. *The Pharmacology and Toxicology of Uranium Compounds*, Vol 1-4, C. Voegtlin and H.C. Hodge, Ed., McGraw-Hill, 1949-1951
292. H.A. Passaw, A. Rothstein, and T.W. Clarkson, The General Pharmacology of the Heavy Metals, *Pharmacol. Rev.*, Vol 13, 1961, p 185-224
293. F.N. Ghadially, J.A. Lalonde, and S. Yang-Steppuhn, Uraniosomes Produced in Cultured Rabbit Kidney Cells by Uranyl Acetate, *Virchows Arch. [Cell. Pathol.]*, Vol 39, 1982, p 21-30
294. L.J. Leach, E.A. Maynard, H.C. Hodge, J.K. Scott, C.L. Yuile, G.E. Sylvester, and H.B. Wilson, A Five Year Inhalation Study With Uranium Dioxide (UO_2) Dust. I. Retention and Biologic Effect in the Monkey, Dog and Rat, *Health Phys.*, Vol 18, 1970, p 599-612
295. M.D. Waters, Toxicology of Vanadium, in *Toxicology of Trace Metals*, R.A. Goyer and M.A. Mehlman, Ed., John Wiley & Sons, 1977, p 147-189

SELECTED REFERENCES

- A.M. Baetjer, Pulmonary Carcinoma in Chromatic Workers, *Arch. Ind. Hyg.*, Vol 2, 1950, p 487-493
- "Bioassay of Selenium Sulfide (Gavage) for Possible Carcinogenicity," National Cancer Institute Technical Report Series 194, NTP 80-17, 1980
- D. Hunter, The Diseases of Occupations, 5th ed., Little, Brown, 1975
- R.R. Lauwerys, *In Vivo* Tests to Monitor Body Burdens of Toxic Metals in Man, in *Clinical Toxicology and Clinical Chemistry of Metals*, S. Brown and J. Savory, Ed., Academic Press, 1983, p 113-122
- M.P. Orfilia, *A General System of Toxicology*, M. Carey & Sons, 1817, p 184
- A.J. Vander, D.L. Taylor, K. Kalitis, D.R. Mouw, and W. Victery, Renal Handling of Lead in Dogs: Clearance Studies, *Am. J. Physiol.*, Vol 233, 1977, p F532-F538
- W. Victery, A.J. Vander, and D.R. Mouw, Renal Handling of Lead in Dogs: Stop-Flow Analysis, *Am. J. Physiol.*, Vol 237, 1979, p F408-F414
- W. Victery, A.J. Vander, and D.R. Mouw, Effect of Acid-Base Status on Renal Excretion and Accumulation of Lead in Dogs and Rats, *Am. J. Physiol.*, Vol 237, 1979, p F398-F407
- J. Vostal and J. Heller, Renal Excretory Mechanisms of Heavy Metals. I. Transtubular Transport of Heavy Metal Ions in the Avian Kidney, *Environ. Res.*, Vol 2, 1968, p 1-10

Metric Conversion Guide

This Section is intended as a guide for expressing weights and measures in the Système International d'Unités (SI). The purpose of SI units, developed and maintained by the General Conference of Weights and Measures, is to provide a basis for worldwide standardization of units and measure. For more information on metric conversions, the reader should consult the following references:

- "Standard for Metric Practice," E 380, *Annual Book of ASTM Standards*, American Society for Testing and Materials, 1916 Race Street, Philadelphia, PA 19103
- "Metric Practice," ANSI/IEEE 268–1982, American National Standards Institute, 1430 Broadway, New York, NY 10018
- *The International System of Units*, SP 330, 1986, National Institute of Standards and Technology. Order from Superintendent of Documents, U.S. Government Printing Office, Washington, DC 20402-9325
- *Metric Editorial Guide*, 4th ed. (revised), 1985, American National Metric Council, 1010 Vermont Avenue NW, Suite 1000, Washington, DC 20005–4960
- *ASME Orientation and Guide for Use of SI (Metric) Units*, ASME Guide SI 1, 9th ed., 1982, The American Society of Mechanical Engineers, 345 East 47th Street, New York, NY 10017

Base, supplementary, and derived SI units

Measure	Unit	Symbol
Base units		
Amount of substance	mole	mol
Electric current	ampere	A
Length	meter	m
Luminous intensity	candela	cd
Mass	kilogram	kg
Thermodynamic temperature	kelvin	K
Time	second	s
Supplementary units		
Plane angle	radian	rad
Solid angle	steradian	sr
Derived units		
Absorbed dose	gray	Gy
Acceleration	meter per second squared	m/s^2
Activity (of radionuclides)	becquerel	Bq
Angular acceleration	radian per second squared	rad/s^2
Angular velocity	radian per second	rad/s
Area	square meter	m^2
Capacitance	farad	F
Concentration (of amount of substance)	mole per cubic meter	mol/m^3
Conductance	siemens	S
Current density	ampere per square meter	A/m^2
Density, mass	kilogram per cubic meter	kg/m^3
Electric charge density	coulomb per cubic meter	C/m^3
Electric field strength	volt per meter	V/m
Electric flux density	coulomb per square meter	C/m^2
Electric potential, potential difference, electromotive force	volt	V
Electric resistance	ohm	Ω
Energy, work, quantity of heat	joule	J
Energy density	joule per cubic meter	J/m^3
Entropy	joule per kelvin	J/K
Force	newton	N
Frequency	hertz	Hz
Heat capacity	joule per kelvin	J/K
Heat flux density	watt per square meter	W/m^2
Illuminance	lux	lx
Inductance	henry	H
Irradiance	watt per square meter	W/m^2
Luminance	candela per square meter	cd/m^2
Luminous flux	lumen	lm
Magnetic field strength	ampere per meter	A/m
Magnetic flux	weber	Wb
Magnetic flux density	tesla	T
Molar energy	joule per mole	J/mol
Molar entropy	joule per mole kelvin	J/mol · K
Molar heat capacity	joule per mole kelvin	J/mol · K
Moment of force	newton meter	N · m
Permeability	henry per meter	H/m
Permittivity	farad per meter	F/m
Power, radiant flux	watt	W
Pressure, stress	pascal	Pa
Quantity of electricity, electric charge	coulomb	C
Radiance	watt per square meter steradian	$W/m^2 \cdot sr$
Radiant intensity	watt per steradian	W/sr
Specific heat capacity	joule per kilogram kelvin	J/kg · K
Specific energy	joule per kilogram	J/kg
Specific entropy	joule per kilogram kelvin	J/kg · K
Specific volume	cubic meter per kilogram	m^3/kg
Surface tension	newton per meter	N/m
Thermal conductivity	watt per meter kelvin	W/m · K
Velocity	meter per second	m/s
Viscosity, dynamic	pascal second	Pa · s
Viscosity, kinematic	square meter per second	m^2/s
Volume	cubic meter	m^3
Wavenumber	1 per meter	1/m

Conversion factors

To convert from	to	multiply by
Angle		
degree	rad	1.745 329 E − 02
Area		
in.^{2}	mm^2	6.451 600 E + 02
in.^{2}	cm^2	6.451 600 E + 00
in.^{2}	m^2	6.451 600 E − 04
ft^2	m^2	9.290 304 E − 02
Bending moment or torque		
lbf · in.	N · m	1.129 848 E − 01
lbf · ft	N · m	1.355 818 E + 00
kgf · m	N · m	9.806 650 E + 00
ozf · in.	N · m	7.061 552 E − 03
Bending moment or torque per unit length		
lbf · in./in.	N · m/m	4.448 222 E + 00
lbf · ft/in.	N · m/m	5.337 866 E + 01
Current density		
A/in.^{2}	A/cm^2	1.550 003 E − 01
A/in.^{2}	A/mm^2	1.550 003 E − 03
A/ft^2	A/m^2	1.076 400 E + 01
Electricity and magnetism		
gauss	T	1.000 000 E − 04
maxwell	μWb	1.000 000 E − 02
mho	S	1.000 000 E + 00
Oersted	A/m	7.957 700 E + 01
Ω · cm	Ω · m	1.000 000 E − 02
Ω circular-mil/ft	μΩ · m	1.662 426 E − 03
Energy (impact, other)		
ft · lbf	J	1.355 818 E + 00
Btu (thermochemical)	J	1.054 350 E + 03
cal (thermochemical)	J	4.184 000 E + 00
kW · h	J	3.600 000 E + 06
W · h	J	3.600 000 E + 03
Flow rate		
ft^3/h	L/min	4.719 475 E − 01
ft^3/min	L/min	2.831 000 E + 01
gal/h	L/min	6.309 020 E − 02
gal/min	L/min	3.785 412 E + 00
Force		
lbf	N	4.448 222 E + 00
kip (1000 lbf)	N	4.448 222 E + 03
tonf	kN	8.896 443 E + 00
kgf	N	9.806 650 E + 00
Force per unit length		
lbf/ft	N/m	1.459 390 E + 01
lbf/in.	N/m	1.751 268 E + 02
Fracture toughness		
ksi $\sqrt{\text{in.}}$	MPA $\sqrt{\text{m}}$	1.098 800 E + 00
Heat content		
Btu/lb	kJ/kg	2.326 000 E + 00
cal/g	kJ/kg	4.186 800 E + 00
Heat input		
J/in.	J/m	3.937 008 E + 01
kJ/in.	kJ/m	3.937 008 E + 01
Length		
Å	nm	1.000 000 E − 01
μin.	μm	2.540 000 E − 02
mil	μm	2.540 000 E + 01
in.	mm	2.540 000 E + 01
in.	cm	2.540 000 E + 00
ft	m	3.048 000 E − 01
yd	m	9.144 000 E − 01
mile	km	1.609 300 E + 00
Mass		
oz	kg	2.834 952 E − 02
lb	kg	4.535 924 E − 01
ton (short, 2000 lb)	kg	9.071 847 E + 02
ton (short, 2000 lb)	kg × 10^3(a)	9.071 847 E − 01
ton (long, 2240 lb)	kg	1.016 047 E + 03
Mass per unit area		
oz/in.^{2}	kg/m^2	4.395 000 E + 01
oz/ft^2	kg/m^2	3.051 517 E − 01
oz/yd^2	kg/m^2	3.390 575 E − 02
lb/ft^2	kg/m^2	4.882 428 E + 00
Mass per unit length		
lb/ft	kg/m	1.488 164 E + 00
lb/in.	kg/m	1.785 797 E + 01
Mass per unit time		
lb/h	kg/s	1.259 979 E − 04
lb/min	kg/s	7.559 873 E − 03
lb/s	kg/s	4.535 924 E − 01
Mass per unit volume (includes density)		
g/cm^3	kg/m^3	1.000 000 E + 03
lb/ft^3	g/cm^3	1.601 846 E − 02
lb/ft^3	kg/m^3	1.601 846 E + 01
lb/in.^{3}	g/cm^3	2.767 990 E + 01
lb/in.^{3}	kg/m^3	2.767 990 E + 04
Power		
Btu/s	kW	1.055 056 E + 00
Btu/min	kW	1.758 426 E − 02
Btu/h	W	2.928 751 E − 01
erg/s	W	1.000 000 E − 07
ft · lbf/s	W	1.355 818 E + 00
ft · lbf/min	W	2.259 697 E − 02
ft · lbf/h	W	3.766 161 E − 04
hp (550 ft · lbf/s)	kW	7.456 999 E − 01
hp (electric)	kW	7.460 000 E − 01
Power density		
W/in.^{2}	W/m^2	1.550 003 E + 03
Press capacity		
See **Force**		
Pressure (fluid)		
atm (standard)	Pa	1.013 250 E + 05
bar	Pa	1.000 000 E + 05
in. Hg (32 °F)	Pa	3.386 380 E + 03
in. Hg (60 °F)	Pa	3.376 850 E + 03
lbf/in.^{2} (psi)	Pa	6.894 757 E + 03
torr (mm Hg, 0 °C)	Pa	1.333 220 E + 02
Specific heat		
Btu/lb · °F	J/kg · K	4.186 800 E + 03
cal/g · °C	J/kg · K	4.186 800 E + 03
Stress (force per unit area)		
tonf/in.^{2} (tsi)	MPa	1.378 951 E + 01
kgf/mm^2	MPa	9.806 650 E + 00
ksi	MPa	6.894 757 E + 00
lbf/in.^{2} (psi)	MPa	6.894 757 E − 03
MN/m^2	MPa	1.000 000 E + 00
Temperature		
°F	°C	5/9 · (°F − 32)
°R	°K	5/9
Temperature interval		
°F	°C	5/9
Thermal conductivity		
Btu · in./s · ft^2 · °F	W/m · K	5.192 204 E + 02
Btu/ft · h · °F	W/m · K	1.730 735 E + 00
Btu · in./h · ft^2 · °F	W/m · K	1.442 279 E − 01
cal/cm · s · °C	W/m · K	4.184 000 E + 02
Thermal expansion		
in./in. · °C	m/m · K	1.000 000 E + 00
in./in. · °F	m/m · K	1.800 000 E + 00
Velocity		
ft/h	m/s	8.466 667 E − 05
ft/min	m/s	5.080 000 E − 03
ft/s	m/s	3.048 000 E − 01
in./s	m/s	2.540 000 E − 02
km/h	m/s	2.777 778 E − 01
mph	km/h	1.609 344 E + 00
Velocity of rotation		
rev/min (rpm)	rad/s	1.047 164 E − 01
rev/s	rad/s	6.283 185 E + 00
Viscosity		
poise	Pa · s	1.000 000 E − 01
stokes	m^2/s	1.000 000 E − 04
ft^2/s	m^2/s	9.290 304 E − 02
in.^{2}/s	mm^2/s	6.451 600 E + 02
Volume		
in.^{3}	m^3	1.638 706 E − 05
ft^3	m^3	2.831 685 E − 02
fluid oz	m^3	2.957 353 E − 05
gal (U.S. liquid)	m^3	3.785 412 E − 03
Volume per unit time		
ft^3/min	m^3/s	4.719 474 E − 04
ft^3/s	m^3/s	2.831 685 E − 02
in.^{3}/min	m^3/s	2.731 177 E − 07
Wavelength		
Å	nm	1.000 000 E − 01

(a) kg × 10^3 = 1 metric ton or 1 megagram (Mg)

SI prefixes—names and symbols

Exponential expression	Multiplication factor	Prefix	Symbol
10^{18}	1 000 000 000 000 000 000	exa	E
10^{15}	1 000 000 000 000 000	peta	P
10^{12}	1 000 000 000 000	tera	T
10^{9}	1 000 000 000	giga	G
10^{6}	1 000 000	mega	M
10^{3}	1 000	kilo	k
10^{2}	100	hecto(a)	h
10^{1}	10	deka(a)	da
10^{0}	1	BASE UNIT	
10^{-1}	0.1	deci(a)	d
10^{-2}	0.01	centi(a)	c
10^{-3}	0.001	milli	m
10^{-6}	0.000 001	micro	μ
10^{-9}	0.000 000 001	nano	n
10^{-12}	0.000 000 000 001	pico	p
10^{-15}	0.000 000 000 000 001	femto	f
10^{-18}	0.000 000 000 000 000 001	atto	a

(a) Nonpreferred. Prefixes should be selected in steps of 10^3 so that the resultant number before the prefix is between 0.1 and 1000. These prefixes should not be used for units of linear measurement, but may be used for higher order units. For example, the linear measurement, decimeter, is nonpreferred, but square decimeter is acceptable.

Abbreviations, Symbols, and Tradenames

Abbreviations and Symbols

a edge length in crystal structure; crack length; absorption coefficient; half thickness of the conductor

a_0 flux line lattice spacing

A area

A ampere

Å angstrom

AA Aluminum Association

ABST α-β solution treatment

ac alternating current

Ac_{cm} in hypereutectoid steel, temperature at which cementite completes solution in austenite

Ac_1 temperature at which austenite begins to form on heating

Ac_3 temperature at which transformation of ferrite to austenite is completed on heating

Ae_{cm}, Ae_1, Ae_3 equilibrium transformation temperatures in steel

AECMA Association Européenne des Constructeurs de Matérial Aérospatial

AFS American Foundrymen's Society

AISI American Iron and Steel Institute

AKS aluminum-potassium-silicon

AMS Aerospace Material Specification

ANSI American National Standards Institute

Ar_{cm} temperature at which cementite begins to precipitate from austenite on cooling

Ar_1 temperature at which transformation to ferrite or to ferrite plus cementite is completed on cooling

Ar_3 temperature at which transformation of austenite to ferrite begins on cooling

ASME American Society of Mechanical Engineers

ASTM American Society for Testing and Materials

atm atmospheres (pressure)

at.% atomic percent

AWS American Welding Society

b barn; Burgers vector

B magnetic induction

B_d flux density

B_d total external permeance

$(B_dH_d)_{max}$ maximum magnetic energy product

B_i internal field; intrinsic induction

$(B_i)_P$ saturation induction

B_{is} intrinsic saturation induction

B_m amplitude of field change

B_p change in applied field

B_r remanent magnetization

B_s saturation induction

bal balance

bcc body-centered cubic

bct body-centered tetragonal

BDT brittle-ductile transition

BE blended elemental

B.F. bright field illumination

$(BH)_{max}$ maximum energy product

BSCCO Bi-Sr-Ca-Cu-O

BST β solution treatment

BUS broken-up structure

BWR boiling water reactors

c edge length in crystal structure; speed of light; specific heat

C coulomb; heat capacity

C capacitance of junction

C_l impurity concentration in the liquid phase

C_s impurity concentration in the just-freezing solid phase

C_0 initial concentration in the liquid phase

CANDU Canadian deuterium uranium (reactor)

CBA colliding beam accelerator

CCM Crucible ceramic mold

cd candela

CDA Copper Development Association

CEBAF continuous electron beam accelerator facility

CEN Comité Européen de Normalisation (European Committee for Standardization)

CERN Centre for European Research

C.G. centerless ground

CHIP cold and hot isostatic pressing

Ci curie

CINDAS Center for Information and Numerical Data Analysis and Synthesis

CIP cold isostatic pressing

cm centimeter

CN coordination number

COD crack opening displacement

CP commercially pure

cSt centiStokes

CST constitutional solution treatment

CTOD crack tip opening displacement

CVD chemical vapor deposition

CVN Charpy V-notch (impact test or specimen)

d day

d used in mathematical expressions involving a derivative (denotes rate of change); depth; diameter

D diameter; duration; dislocation density

da/dN fatigue crack growth rate

DARPA Defense Advanced Research Projects Agency

dB decibel

DB diffusion bonding

DBTT ductile-to-brittle transition temperature

dc direct current

DESY Deutsche Electronen Syncrotron

DFB diffusion brazing

DG directed grain

dhcp double hexagonal close-packed

Di didymium (a mixture of the rare earth elements praseodymium and neodymium)

diam diameter

D.I.C. differential interference contrast

DIN Deutsche Industrie-Normen (German Industrial Standards)

DoD Department of Defense

DPH diamond pyramid hardness

DS directionally solidified

DSA dispersion-strengthened alloy
DSC differential scanning calorimeter
DTA differential thermal analysis
e natural log base, 2.71828; charge of an electron
E Young's modulus; applied voltage; activation energy
E_{corr} corrosion potential
EAF electric arc furnace
EB electron beam
EBW electron beam welding
EC orbital electron capture
ECM electrochemical machining
EPC evaporative pattern casting
EDM electrical discharge machining
EEC European Economic Community
ELI extra-low interstitial
ELCl extra-low chlorine powder
EMF electromagnetic fields
EMI electromagnetic iron
EPA Environmental Protection Agency
EPR ethylene propylene rubber
EPRI Electric Power Research Institute
Eq equation
et al. and others
ETM engineering test model
eV electron volt
f frequency; transfer function
F_L Lorentz force
F_{ty} tensile yield strength
fcc face-centered cubic
FCGR fatigue crack growth rate
fct face-centered tetragonal
FEP fluorinated ethylene propylene
Fig. figure
ft foot
g gram
G gauss
G Gibbs free energy
$G_n(T)$ volumetric Gibbs free energy of the normal state
$G_s(T)$ volumetric Gibbs free energy of the superconducting state
GA gas atomization
gal gallon
GDMS glow discharge mass spectroscopy
GFC gas fan cooled
GFI ground fault interrupter
GGG gallium gadolinium garnet
GMAW gas metal arc welding
GP Guinier-Preston (precipitation zone)
GPa gigapascal
gr grain
GSGG gallium scandium gadolinium garnet
GTA gas tungsten arc
GTAW gas tungsten arc welding
Gy gray (unit of absorbed radiation)
h hour
h Planck's constant, 6.626×10^{-27} erg · s
H henry
H enthalpy; magnetic field
H_a applied magnetic field
H_c coercive force; thermodynamic critical field
H_{c1} lower critical magnetic field
H_{c2} upper critical magnetic field
H_{ci} intrinsic coercive force
H_d permeance of the space occupied by the magnet
HAZ heat-affected zone
HB Brinell hardness; horizontal Bridgeman
HBT heterojunction bipolar transistor
HCF high-cycle fatigue
hcp hexagonal close-packed
HDI high-density inclusion
HEMT high-electron-mobility transistor
HEP high-energy physics
HERA hadron-electron ring anordnung
HIP hot isostatic pressing
HK Knoop hardness
hp horsepower
HP Hunter (reduction) process
HPLT high-pressure low-temperature compaction
HR Rockwell hardness (requires scale designation, such as HRC for Rockwell C hardness)
HSLA high-strength low-alloy (steel)
HTH high-temperature hydrogenation
HV Vickers hardness
HVC hydrovac process
Hz hertz
i current density
i_0 exchange current density
i_{corr} corrosion current
I intensity; electrical current; bias current
I_{cj} critical current of the junction
I_r resistive current
I_s supercurrent through junction
IACS International Annealed Copper Standard
IC integrated circuit
ICCS internally cooled cable in conduit
ICP inductively coupled plasma
ICPMS inductively coupled plasma mass spectrometer
ID inside diameter
ILZRO International Lead Zinc Research Organization
I/M ingot metallurgy
in. inch
IPTS International Practical Temperature Scale
IR infrared
ISA Instrument Society of America
ISCC intergranular stress-corrosion cracking
ISO International Organization for Standardization
ISRI Institute of Scrap Recycling Industries
ITER international thermonuclear experimental reactor
ITS International Temperature Scale
J joule
J_c critical current density
k karat
k thermal conductivity; wave number; Boltzmann constant
K Kelvin
K stress intensity factor; thermal conductivity
K_f fatigue notch factor
K_{Ic} plane-strain fracture toughness
K_{ISCC} threshold stress intensity to produce stress-corrosion cracking
K_Q invalid fracture toughness values
K_t stress-concentration factor
K_v precracked Charpy
K_0 distribution coefficient at equilibrium
K_1 crystalline anisotropy constant
KEK Japanese atomic energy facility
kg kilogram
km kilometer
kN kilonewton
kPa kilopascal
ksi kips (1000 lbf) per square inch
kV kilovolt
kW kilowatt
l length
ℓ length
L longitudinal; liter
L twist pitch distance of the composite wire; length of straight bar magnet
lb pound
lbf pound force
LBW laser beam welding
LCD liquid crystal display
LCF low-cycle fatigue
LCP large coil program
LCT large coil test
LEC liquid-encapsulated Czochralski
LED light-emitting diode
LHC large hadron collider
LIMS laser ionization mass spectroscopy
LMP Larson-Miller parameter
ln natural logarithm (base *e*)
LNG liquefied natural gas
log common logarithm (base 10)
LPE liquid-phase epitaxy
LT long transverse (direction)

m meter
m strain rate sensitivity factor
M_f temperature at which martensite formation finishes during cooling
M_s temperature at which martensite starts to form from austenite on cooling
M magnetization
mA milliampere
MASTMAASIS modified ASTM acetic acid salt intermittent spray (exfoliation test)
MBE molecular beam epitaxy
MEK methyl ethyl ketone
MeV megaelectronvolt
MFTF mirror fusion test facility
mg milligram
Mg megagram (metric tonne, or kg $\times$ 10^3)
MHD magnetohydrodynamic
MIBK methyl isobutyl ketone
MIG metal inert gas
min minute; minimum
MJR modified jelly roll
mL milliliter
MM misch metal
mm millimeter
MMC metal-matrix composite
MMIC monolithic microwave integrated circuit
MOCVD metallo-organic chemical vapor deposition
mPa millipascal
MPa megapascal
mpg miles per gallon
mph miles per hour
MQG melt-quench technique
MRI magnetic resonance imaging
ms millisecond
MS megasiemens
mT millitesla
mV millivolt
MV megavolt
n neutrons
n strain-hardening exponent; index of refraction; number of flux quanta per unit area
n_s density of superconducting electron pairs
N newton
N number of cycles; normal solution
N_f number of cycles to failure
NASA National Aeronautics and Space Administration
NBS National Bureau of Standards
Nd:YAG neodymium: yttrium-aluminum-garnet (laser)
NET next European torus
NIST National Institute of Standards and Technology
nm nanometer
NMR nuclear magnetic resonance
No. number
N.R.I.M. National Research Institute for Metals in Japan
ns nanoseconds
OD outside diameter
ODS oxide dispersion strengthened
Oe oersted
ORNL Oak Ridge National Laboratory
oz ounce
p page
P pressure; thickness of superconductor outer layer; depth of penetration of applied field below superconductor surface
Pa pascal
PA prealloyed
PCA process control agent
PD preferred direction
PF poloidal field
PFC planar-flow casting
pH negative logarithm of hydrogen-ion activity
PH precipitation hardenable
PGM platinum-group metals
PM precious metal
P/M powder metallurgy
PMS $PbMo_6S_8$
POC products of combustion
ppb parts per billion
PPB prior particle boundary
ppba parts per billion atomic
ppm parts per million
ppt parts per trillion
PREP plasma rotating-electrode process
psi pounds per square inch
psig gage pressure (pressure relative to ambient pressure) in pounds per square inch
PTFE polytetrafluoroethylene
PVC polyvinyl chloride
PWR pressurized water reactors
r plastic strain ratio; radius
R roentgen
R stress (load) ratio; radius; gas constant; bulk resistance; reluctance (reciprocal of permeance)
R_H Hall coefficient
R_0 coefficient of ordinary Hall effect
R_1 coefficient of extraordinary Hall effect
RA reduction in area; recrystallization annealed
rad absorbed radiation dose
RE rare earth
Ref reference
rem roentgen equivalent man; remainder
REP rotating-electrode process
rf radio frequency
RHIC relativistic heavy ion collider
RHM roentgen per hour at one meter
rms root mean square
ROC rapid omnidirectional compaction
ROC rapid omnidirectional consolidation
RRR residual resistance ratio
RS rapid solidification
RSW resistance spot welding
RT room temperature
RWMA Resistance Welding Manufacturers Association
s second
S siemens
S filament spacing (in superconductor)
S_{max} maximum stress
S_{min} minimum stress
SAP sinter-aluminum-pulver
SAE Society of Automotive Engineers
SBC steel-bonded (titanium) carbide
Sc superconductor
SCC stress-corrosion cracking
SDC specific damping capacity
SDI strategic defense initiative
SEM scanning electron microscope
sfm surface feet per minute
SHE standard hydrogen electrode
SI Système International d'Unités
SIS superconductor/insulating/superconductor
SMA shape memory alloy
SMAW shielded metal-arc welding
SME shape memory effect
SMES superconducting magnetic energy storage
SMS $SnMo_6S_8$
SPC statistical process control
SPF superplastic forming
SPS standard pipe size
SQUID superconducting quantum interference device
SSC superconducting supercollider
ST short transverse (direction)
std standard
STA solution-treated and aged
Sv sievert
t thickness; time
T tesla
T temperature
T_A annealing temperature
T_b boiling temperature
T_c critical ordering temperature; Curie temperature; critical transition temperature
T_g glass transition temperature
T_m melting temperature
T_o minimum ordering temperature
T_s minimum segregation temperature

T_x crystallization temperature
TBCCO Tl-Ba-Ca-Cu-O
TCP thermochemical processing
TCT thermochemical treatment
TD thoria dispersed
TEM transmission electron microscopy
TF toroidal field
TIG tungsten inert gas (welding)
TIR total indicator reading
TMP thermomechanical processing
tsi tons per square inch
TTT time-temperature transformation
UBC used beverage can
UNF Unified Numbering System
US-DPC United States-demonstration poloidal coil
UTS ultimate tensile strength
v velocity
V volt
V volume; velocity
VAR vacuum arc remelting
VAW Vereinigte Aluminium Werke AG
VD vapor deposited
V-D vacuum degassing
VHP vacuum hot pressing
VIM vacuum induction melting
vol volume
vol% volume percent
VPE vapor-phase epitaxy
VPSD vacuum plasma structural deposition
W watt
W width; weight
W_h hysteresis losses
wt% weight percent
XLPE cross linked polyethylene
YBCO Y-Ba-Cu-O
Z impedance; atomic number
ZGS zirconia grain stabilized
ZTA zirconia-toughened alumina
° angular measure; degree
°C degree Celcius (centigrade)
°F degree Fahrenheit
$\leftrightarrows$ direction of reaction
÷ divided by
= equals
≅ approximately equals
≠ not equal to
≡ identical with
> greater than
≫ much greater than
≧ greater than or equal to
∞ infinity
∝ is proportional to; varies as
∫ integral of
< less than
≪ much less than
≦ less than or equal to
± maximum deviation
− minus; negative ion charge
× diameters (magnification); multiplied by
· multiplied by
/ per
% percent
+ plus; positive ion charge
√ square root of
~ approximately; similar to
∂ partial derivative
Δ change in quantity; an increment; a range
ε strain
$\dot{\epsilon}$ strain rate
η viscosity
θ Curie temperature
κ Ginzburg-Landau parameter; thermal diffusivity
$\lambda(T)$ magnetic penetration depth as a function of temperature
μ friction coefficient; magnetic permeability
μF microfarads
μin. microinch
μm micrometer (micron)
μs microsecond
ν Poisson's ratio
$\xi(T)$ temperature dependent coherence length
π pi (3.141592)
ρ electrical resistivity; density
σ liquid surface tension; stress
σ_c capture thermal neutron cross section
σ_f fission thermal neutron cross section
Σ summation of
τ time constant; applied stress
$\phi(x,t)$ phase of the wave function, dependent on position and time
Φ magnetic flux contained in the loop
Φ_0 magnetic flux quantum
χ magnetic susceptibility
Ψ macroscopic quantum mechanical wave function or order parameter
ω frequency of the ac magnetic field
Ω ohm

Greek Alphabet

Α, α alpha
Β, β beta
Γ, γ gamma
Δ, δ delta
Ε, ε epsilon
Ζ, ζ zeta
Η, η eta
Θ, θ theta
Ι, ι iota
Κ, κ kappa
Λ, λ lambda
Μ, μ mu
Ν, ν nu
Ξ, ξ xi
Ο, ο omicron
Π, π pi
Ρ, ρ rho
Σ, σ sigma
Τ, τ tau
Υ, υ upsilon
Φ, φ phi
Χ, χ chi
Ψ, ψ psi
Ω, ω omega

Tradenames

AF-56 is a registered tradename of Allison Gas Turbine, Division of General Motors Corporation.

AL-6X, AL-6XN, AL 29-4C, AL29-4-2, AL 904L, AL-36, AL-42, AL-52, AL 2205, AL-4750, ALFA IV, E-Brite 26-1, Sealmet, and **203 EZ** are registered tradenames of Allegheny Ludlum Steel, Division of Allegheny Ludlum Corporation.

AM1 is a registered tradename of SNECMA/ONERA.

CM 247 LC and **CMSX** are registered tradenames of Cannon-Muskegon Corporation.

Cronifer is a registered tradename of Vereingte Deutsche Metallwerks.

Cryogenic Tenelon and **Tenelon** are registered tradenames of USS, Division of USX Corporation.

Custom 450, Custom 455, Gall-Tough, Glass Sealing "49", Invar "36", Kovar, Low Expansion "42", Pyromet, TrimRite, 7-Mo PLUS, 18-18 PLUS, 20Cb-3, 20Mo-4, and **20Mo-6** are registered tradenames of Carpenter Technology Corporation.

CU78, CW67, and **CZ42** are registered tradenames of Aluminum Company of America.

Discaloy is a registered tradename of Westinghouse Electric Corporation.

DISPAL is a registered tradename of Sintermetallwerk Krebsöge GmbH.

DP3 is a registered tradename of Sumitomo Metal America, Inc.

Elinvar and **Invar** are registered tradenames of Imphy, S.A.

Esshete is a registered tradename of British Steel Corporation.

Ferralium is a registered tradename of Bonar Langley Alloy Ltd.

FVS-0611, FVS-0812, FVS-1212, and **Metglas** are registered tradenames of Allied-Signal Inc.

GlidCop is a registered tradename of SCM Metal Products, Inc.

Hastelloy and **Haynes** are registered tradenames of Haynes International, Inc.

Incoloy, Inconel, IncoMAP AL-905XL, IncoMAP AL-9052, Monel, Nimocast, Nimonic, NI-Rod, and **NI-Span-Cl** are registered tradenames of Inco Alloys International, Inc.

JS700 is a registered tradename of Jessop Steel Company.

Kapton, Teflon, and **Tefzell** are registered tradenames of E.I. Du Pont de Nemours & Company, Inc.

MAR-M is a registered tradename of Martin Marietta Corporation.

Monit is a registered tradename of Uddeholms Aktiebolag.

MP (Multiphase) is a registered tradename of Standard Pressed Steel Company.

Nextel is a registered tradename of 3M Company.

Nitronic and **PH 13-8 Mo** are registered tradenames of Baltimore Specialty Steels Corporation.

PH 15-7 MO, 12SR, 15-5 PH, 17-4 PH, 18 SR, and **21-6-9** are registered tradenames of Armco Advanced Materials Corporation.

PWA 1484 is a registered tradename of Pratt & Whitney Aircraft.

RA85H is a registered tradename of Rolled Alloys, Inc.

Refrasil is a registered tradename of Thompson Company.

René is a registered tradename of General Electric Company.

René 41 is a registered tradename of Allvac Metals Company, a Teledyne Company.

RR 2000 and **SRR 99** are registered tradenames of Rolls Royce, Inc.

Sanicro and **3RE60** are registered tradenames of Sandvik, Inc.

Sea-Cure is a registered tradename of Crucible, Inc.

Stellite is a registered tradename of Deloro Stellite, Inc.

Tantology and **Tribocor** are registered tradenames of Fansteel Inc.

Transage 134 and **Transage 175** are registered tradenames of Lockheed Missile and Space Company.

Udimet is a registered tradename of Special Metals Corporation.

Unitemp is a registered tradename of Universal Cyclops Steel Corporation.

Uranus is a registered tradename of Compagnie des Ateliers et Forges de la Loire.

Vitallium is a registered tradename of Pfizer Hospital Products Group, Inc.

Waspaloy is a registered tradename of United Technologies, Inc.

Weldalite is a registered tradename of Martin Marietta Corporation.

253MA and **254SMO** are registered tradenames of Avesta Stainless, Inc.

Index

A

E

F

G

J

K

L

M

O

P

Q

R

T

W

X

Y